MOLECULAR BIOLOGY OF THE CELL

fourth edition

MOLECULAR BIOLOGY OF

THE CELL

fourth edition

Bruce Alberts

Alexander Johnson

Julian Lewis

Martin Raff

Keith Roberts

Peter Walter

Garland
Vice President: Denise Schanck
Managing Editor: Sarah Gibbs
Senior Editorial Assistant: Kirsten Jenner
Managing Production Editor: Emma Hunt
Proofreader and Layout: Emma Hunt
Production Assistant: Angela Bennett
Text Editors: Marjorie Singer Anderson and Betsy Dilernia
Copy Editor: Bruce Goatly
Word Processors: Fran Dependahl, Misty Landers and Carol Winter
Designer: Blink Studio, London
Illustrator: Nigel Orme
Indexer: Janine Ross and Sherry Granum
Manufacturing: Nigel Eyre and Marion Morrow

Cell Biology Interactive
Artistic and Scientific Direction: Peter Walter
Narrated by: Julie Theriot
Production, Design, and Development: Mike Morales

Bruce Alberts received his Ph.D. from Harvard University and is President of the National Academy of Sciences and Professor of Biochemistry and Biophysics at the University of California, San Francisco. **Alexander Johnson** received his Ph.D. from Harvard University and is a Professor of Microbiology and Immunology at the University of California, San Francisco. **Julian Lewis** received his D.Phil. from the University of Oxford and is a Principal Scientist at the Imperial Cancer Research Fund, London. **Martin Raff** received his M.D. from McGill University and is at the Medical Research Council Laboratory for Molecular Cell Biology and Cell Biology Unit and in the Biology Department at University College London. **Keith Roberts** received his Ph.D. from the University of Cambridge and is Associate Research Director at the John Innes Centre, Norwich. **Peter Walter** received his Ph.D. from The Rockefeller University in New York and is Professor and Chairman of the Department of Biochemistry and Biophysics at the University of California, San Francisco, and an Investigator of the Howard Hughes Medical Institute.

Library of Congress Cataloging-in-Publicaton Data
Molecular biology of the cell / Bruce Alberts … [et al.].-- 4th ed.
p. cm
Includes bibliographical references and index.
ISBN 0-8153-3218-1 (hardbound) -- ISBN 0-8153-4072-9 (pbk.)
1. Cytology. 2. Molecular biology. I. Alberts, Bruce.
[DNLM: 1. Cells. 2. Molecular Biology.]
QH581.2 .M64 2002
571.6--dc21

2001054471 CIP

Published by Garland Science, a member of the Taylor & Francis Group, 29 West 35th Street, New York, NY 10001-2299

Printed in the United States of America

15 14 13 12 11 10 9 8 7 6 5 4 3 2 1

Front cover Human Genome: Reprinted by permission from *Nature,* International Human Genome Sequencing Consortium, 409:860–921, 2001 © Macmillan Magazines Ltd. Adapted from an image by Francis Collins, NHGRI; Jim Kent, UCSC; Ewan Birney, EBI; and Darryl Leja, NHGRI; showing a portion of Chromosome 1 from the initial sequencing of the human genome.

Back cover In 1967, the British artist Peter Blake created a design classic. Nearly 35 years later Nigel Orme (illustrator), Richard Denyer (photographer), and the authors have together produced an affectionate tribute to Mr Blake's image. With its gallery of icons and influences, its assembly created almost as much complexity, intrigue and mystery as the original. *Drosophila, Arabidopsis,* Dolly and the assembled company tempt you to dip inside where, as in the original, "a splendid time is guaranteed for all."
(Gunter Blobel, courtesy of The Rockefeller University; Marie Curie, Keystone Press Agency Inc; Darwin bust, by permission of the President and Council of the Royal Society; Rosalind Franklin, courtesy of Cold Spring Harbor Laboratory Archives; Dorothy Hodgkin, © The Nobel Foundation, 1964; James Joyce, etching by Peter Blake; Robert Johnson, photo booth self-portrait early 1930s, © 1986 Delta Haze Corporation all rights reserved, used by permission; Albert L. Lehninger, (unidentified photographer) courtesy of The Alan Mason Chesney Medical Archives of The Johns Hopkins Medical Institutions; Linus Pauling, from Ava Helen and Linus Pauling Papers, Special Collections, Oregon State University; Nicholas Poussin, courtesy of ArtToday.com; Barbara McClintock, © David Micklos, 1983; Andrei Sakharov, courtesy of Elena Bonner; Frederick Sanger, © The Nobel Foundation, 1958.)

Preface

"There is a paradox in the growth of scientific knowledge. As information accumulates in ever more intimidating quantities, disconnected facts and impenetrable mysteries give way to rational explanations, and simplicity emerges from chaos."

Thus began the preface of our first edition, written 18 years ago. Much of what we wrote in that preface holds for the present edition too. Our goals have not changed: we want to make cell biology comprehensible. We aim, as before, to give readers a perspective both on what is known, and on what is unknown. We have written the book for a wide range of students, but it should also prove useful for scientists wishing to follow progress outside their own specialized fields. As in past editions, each chapter is a joint composition of multiple authors and has been reviewed by a number of experts. This helps to explain why this edition, like the first, has been "a long time in gestation—three times longer than an elephant, five times longer than a whale."

Although information in the life sciences is expanding rapidly, human brain capacity is not. This discrepancy creates an increasing challenge for textbook writers, teachers, and especially students. We have been forced to think harder than ever in deciding what facts and concepts are essential. We have had to summarize—omitting much detail that, in overabundance, can prevent understanding. But specific examples are still needed to bring the subject to life. Thus, the most difficult part of the revision has been deciding what to leave out, while the most rewarding has been the opportunity that new discoveries provide to strengthen the conceptual framework.

What is new in the 4th edition?

Genomics has given us a new perspective that has demanded a complete recasting and expansion of the material on molecular genetics (Chapter 1 and Chapters 4 through 8). These six chapters can now be used as a stand-alone text in molecular biology. As before, the book ends with a major section on Cells in Their Social Context, but we have added a new chapter on Pathogens, Infection, and Innate Immunity. This reflects the remarkable advances in the understanding of the cell biology of infection, as well as the renewed awareness that infectious disease remains one of the greatest unconquered dangers in our world.

As our book makes clear, the complete sequencing of the genomes of hundreds of organisms, from bacteria to humans, has revolutionized our understanding of living things and the relationships between them. At last we can see what is there: the set of genes and proteins is finite, and we can list them. But we also recognize that these components are combined for use in marvelously subtle and complex ways, even in the simplest of organisms. Therefore, the traditional explanatory cartoons that we show on nearly every page of the book generally represent only the primitive first step toward an explanation. These drawings cannot capture the enormous complexity of the networks of protein–protein interactions that are responsible for most intracellular processes, whose understanding will require new and more quantitative forms of analysis. Thus, we are no longer as confident as we were 18 years ago that simplicity will eventually emerge from the complexity. The extreme sophistication of cellular mechanisms will challenge cell biologists throughout the new century, which is very good news for the many young scientists who will succeed us.

As never before, new imaging and computer technologies have changed the ways we can observe the inner workings of living cells. We have tried to capture some of the excitement of these observations in *Cell Biology Interactive*, a CD-ROM disk that is included with each book. It contains dozens of video clips, animations, molecular structures, and high-resolution micrographs, which complement the static material in individual book chapters. One cannot watch cells crawling, dividing, segregating their chromosomes, or rearranging their surface without feeling curious about the molecular mechanisms that underlie these processes. We hope that *Cell Biology Interactive* will stimulate this curiosity in students and thereby make the learning of cell biology easier and more rewarding. We also hope that instructors will use these visual resources in the classroom for the same purpose.

We are deeply indebted to the many people who have helped us with this revision. The experts who critically reviewed specific chapters are acknowledged separately on p. xxix. We are especially grateful to Julie Theriot, who largely wrote both Chapter 16 and Chapter 25; we have all benefited from her wisdom. We are also grateful to the other experts who made major contributions to individual chapters: Nancy Craig helped to revise part of Chapter 5; Maynard Olson drafted the section on genome evolution in Chapter 7; Peter Shaw helped to revise Chapter 9; David Morgan largely wrote Chapter 17; Lisa Satterwhite prepared initial drafts of Chapter 18; Robert Kypta largely revised Chapter 19; Cori Bargmann helped us to restructure Chapter 21; and Paul Edwards played a central part in the revision of Chapter 23. Karen Hopkin drafted large parts of Chapters 8 and 23. We are also very grateful to the many scientists who generously provided micrographs for the book and materials for the CD-ROM.

Finally we are indebted to the outstanding staff of Garland Publishing. Nigel Orme oversaw the production of the final artwork with remarkable skill and speed. Mike Morales put in endless hours tackling the diverse challenges of developing the CD-ROM. Emma Hunt skillfully and artfully set the entire text and figures into pages. Sarah Gibbs and Kirsten Jenner kept us organized. Adam Sendroff and Nasreen Arain connected us to our customers. Eleanor Lawrence prepared the Glossary, and Mary Purton collected many of the references. Through it all, Denise Schanck calmly directed the whole effort with great skill. Last, but not least, Tim Hunt and John Wilson once again created a masterful Problems Book that supplements the main text; they worked side-by-side with us and were a constant source of wise advice. Their clever problems illustrate how discoveries are made and provide a unique way of learning cell biology.

It goes without saying that our book could not have been written without the strong support of our families and the forbearance of our friends, colleagues, and students. For decades, they have had to put up with our absences at frequent and lengthy book meetings, where most of the writing was done. They have helped us in innumerable ways, and we are grateful to them all.

Contents

Special Features

List of Topics

Part I Introduction to the Cell

Part IV Internal Organization of the Cell

CHAPTER 10 MEMBRANE STRUCTURE 583

CHAPTER 11 MEMBRANE TRANSPORT OF SMALL MOLECULES AND THE ELECTRICAL PROPERTIES OF MEMBRANES 615

CHAPTER 12 INTRACELLULAR COMPARTMENTS AND PROTEIN SORTING 659

CHAPTER 13 INTRACELLULAR VESICULAR TRAFFIC 711

CAENORHABDITIS ELEGANS: DEVELOPMENT FROM THE PERSPECTIVE OF THE INDIVIDUAL CELL

DROSOPHILA AND THE MOLECULAR GENETICS OF PATTERN FORMATION: GENESIS OF THE BODY PLAN

HOMEOTIC SELECTOR GENES AND THE PATTERNING OF THE ANTEROPOSTERIOR AXIS

ORGANOGENESIS AND THE PATTERNING OF APPENDAGES

CELL MOVEMENTS AND THE SHAPING OF THE VERTEBRATE BODY

THE MOUSE

NEURAL DEVELOPMENT

Acknowledgments

In writing this book we have benefited greatly from the advice of many biologists and biochemists. In addition to those who advised us by telephone, we would like to thank the following for their written advice in preparing this edition, as well as those who helped in preparing the first, second, and third editions. (Those who helped on this edition are listed first, and those who helped with the first, second and third editions follow.)

Chapter 1 Eugene Koonin (National Institutes of Health, Bethesda), Maynard Olson (University of Washington, Seattle).

Chapter 3 Stephen Burley (The Rockefeller University), Christopher Dobson (University of Cambridge, UK), David Eisenberg (University of California, Los Angeles), Stephen Harrison (Harvard University), John Kuriyan (University of California, Berkeley), Greg Petsko (Brandeis University).

Chapter 4 Gary Felsenfeld (National Institutes of Health, Bethesda), Michael Grunstein (University of California, Los Angeles), Robert Kingston (Massachusetts General Hospital), Douglas Koshland (Carnegie Institution of Washington), Ulrich Laemmli (University of Geneva, Switzerland), Alan Wolffe (deceased), Keith Yamamoto (University of California, San Francisco).

Chapter 5 Nancy Craig (Johns Hopkins University), Joachim Li (University of California, San Francisco), Tomas Lindahl (Imperial Cancer Research Fund, Clare Hall Laboratories, UK), Maynard Olson (University of Washington, Seattle), Steven West (Imperial Cancer Research Fund, Clare Hall Laboratories, UK).

Chapter 6 Raul Andino (University of California, San Francisco), David Bartel (Massachusetts Institute of Technology), Jim Goodrich (University of Colorado, Boulder), Carol Gross (University of California, San Francisco), Christine Guthrie (University of California, San Francisco), Tom Maniatis (Harvard University), Harry Noller (University of California, Santa Cruz), Alan Sachs (University of California, Berkeley), Alexander Varshavsky (California Institute of Technology), Jonathan Weissman (University of California, San Francisco), Sandra Wolin (Yale University School of Medicine).

Chapter 7 Maynard Olson [substantial contribution] (University of Washington, Seattle), Michael Carey (University of California, Los Angeles), Jim Goodrich (University of Colorado, Boulder), Michael Green (University of Massachusetts at Amherst), Tom Maniatis (Harvard University), Barbara Panning (University of California, San Francisco), Roy Parker (University of Arizona, Tuscon), Alan Sachs (University of California, Berkeley), Keith Yamamoto (University of California, San Francisco).

Chapter 8 Karen Hopkin [major contribution] (science writer, PhD, Albert Einstein College of Medicine), Stephanie Mel (University of California, San Diego).

Chapter 9 Peter Shaw [major contribution] (John Innes Centre, Norwich, UK), Michael Sheetz (Columbia University, New York), Werner Kühlbrandt (Max Planck Institute for Biophysics, Frankfurt am Main, Germany).

Chapter 10 Richard Henderson (MRC Laboratory of Molecular Biology, Cambridge, UK), Robert Stroud (University of California, San Francisco).

Chapter 11 Clay Armstrong (University of Pennsylvania), Robert Edwards (University of California, San Francisco), Lily Jan (University of California, San Francisco), Ron Kaback (University of California, Los Angeles), Werner Kühlbrandt (Max Planck Institute for Biophysics, Frankfurt am Main, Germany), Chris Miller (Brandeis University), Joshua Sanes (Washington University in St. Louis), Nigel Unwin (MRC Laboratory of Molecular Biology, Cambridge, UK).

Chapter 12 Larry Gerace (The Scripps Research Institute), Reid Gilmore (University of Massachusetts at Amherst), Arthur Johnson (Texas A & M University), Paul Lazarow (Mount Sinai School of Medicine), Walter Neupert (University of Munich, Germany), Erin O'Shea (University of California, San Francisco), Karsten Weis (University of California, Berkeley).

Chapter 13 Scott Emr (University of California, San Diego), Regis Kelly (University of California, San Francisco), Hugh Pelham (MRC Laboratory of Molecular Biology, Cambridge, UK), Randy Schekman (University of California, Berkeley), Richard Scheller (Stanford University), Giampietro Schiavo (Imperial Cancer Research Fund, London), Jennifer Lippincott-Schwartz (National Institutes of Health, Bethesda), Kai Simons (Max Plank Institute of Molecular Cell Biology and Genetics, Dresden), Graham Warren (Yale University School of Medicine).

Chapter 14 Martin Brand (University of Cambridge, UK), Stuart Ferguson (University of Oxford, UK), Jodi Nunnari (University of California, Davis), Werner Kühlbrandt (Max Planck Institute for Biophysics, Frankfurt am Main, Germany).

Chapter 15 Nicholas Harberd [substantial contribution] (John Innes Centre, Norwich, UK), Spyros Artavanis-Tsakonas (Harvard Medical School), Michael J. Berridge (The Babraham Institute, Cambridge, UK), Henry Bourne (University of California, San Francisco), Lewis Cantley (Harvard Medical School), Julian Downward (Imperial Cancer Research Fund, London), Sankar Ghosh (Yale University School of Medicine), Alan Hall (MRC Laboratory for Molecular Biology and Cell Biology Unit, London), Adrian Harwood (MRC Laboratory for Molecular Cell Biology and Cell Biology Unit, London), John Hopfield (Princeton University),

Philip Ingham (University of Sheffield, UK), Hartmut Land (University of Rochester), Ottoline Leyser (University of York, UK), Joan Massagué (Memorial Sloan-Kettering Cancer Center), Elliot Meyerowitz [substantial contribution] (California Institute of Technology), Tony Pawson (Mount Sinai Hospital, Toronto), Julie Pitcher (University College London), Joseph Schlessinger (New York University Medical Center), Nick Tonks (Cold Spring Harbor Laboratory).

Chapter 16 Julie Theriot [major contribution] (Stanford University), Linda Amos (MRC Laboratory of Molecular Biology, Cambridge, UK), John Cooper (Washington University School of Medicine), Elaine Fuchs (University of Chicago), Frank Gertler (Massachusetts Institute of Technology), Larry Goldstein (University of California, San Diego), Jonathon Howard (University of Washington, Seattle), Laura Machesky (University of Birmingham, UK), Frank McNally (University of California, Davis), Tim Mitchison (Harvard Medical School), Frank Solomon (Massachusetts Institute of Technology), Clare Waterman-Storer (The Scripps Research Institute).

Chapter 17 David Morgan [major contribution] (University of California, San Francisco), Andrew Murray [substantial contribution] (Harvard University), Tim Hunt (Imperial Cancer Research Fund, Clare Hall Laboratories, UK), Paul Nurse (Imperial Cancer Research Fund, London).

Chapter 18 Lisa Satterwhite [substantial contribution] (Duke University Medical School), John Cooper (Washington University School of Medicine), Sharyn Endow (Duke University), Christine Field (Harvard Medical School), Michael Glotzer (University of Vienna, Austria), Tony Hyman (Max Planck Institute of Molecular Cell Biology & Genetics, Dresden), Tim Mitchison (Harvard Medical School), Andrew Murray (Harvard University), Jordan Raff (Wellcome/CRC Institute, Cambridge, UK), Conly Rieder (Wadsworth Center), Edward Salmon (University of North Carolina at Chapel Hill), William Sullivan (University of California, San Francisco), Yu-Lie Wang (Worcester Foundation for Biomedical Research).

Chapter 19 Robert Kypta [major contribution] (MRC Laboratory for Molecular Cell Biology, London), Merton Bernfield (Harvard Medical School), Keith Burridge (University of North Carolina at Chapel Hill), Benny Geiger (Weizmann Institute of Science, Rehovot, Israel), Daniel Goodenough (Harvard Medical School), Barry Gumbiner (Memorial Sloan-Kettering Cancer Center), Martin Humphries (University of Manchester, UK), Richard Hynes (Massachusetts Institute of Technology), Louis Reichhardt (University of California, San Francisco), Joel Rosenbloom (University of Pennsylvania), Erkki Ruoslahti (The Burnham Institute), Joshua Sanes (Washington University in St. Louis), Timothy Springer (Harvard Medical School), Peter Yurchenco (Robert Wood Johnson Medical School).

Chapter 20 Keith Dudley (King's College London), Harvey Florman (Tufts University), Nancy Hollingsworth (State University of New York, Stony Brook), Nancy Kleckner (Harvard University), Robin Lovell-Badge (National Institute for Medical Research, London), Anne McLaren (Wellcome/Cancer Research Campaign Institute, Cambridge, UK), Diana Myles (University of California, Davis), Terri Orr-Weaver (Massachusetts Institute of Technology), Paul Wassarman (Mount Sinai School of Medicine).

Chapter 21 Andre Brandli (Swiss Federal Institute of Technology, Zurich), Cornelia Bargmann [substantial contribution] (University of California, San Francisco), Enrico Coen (John Innes Centre, Norwich, UK), Stephen Cohen (EMBL Heidelberg, Germany), Leslie Dale (University College London), Richard Harland (University of California, Berkeley), David Ish-Horowicz (Imperial Cancer Research Fund, London), Jane Langdale (University of Oxford, UK), Michael Levine (University of California, Berkeley), William McGinnis (University of California, Davis), James Priess (University of Washington, Seattle), Janet Rossant (Mount Sinai Hospital, Toronto), Gary Ruvkun (Massachusetts General Hospital), Joshua Sanes (Washington University in St Louis), François Schweisguth (ENS, Paris), Clifford Tabin (Harvard Medical School), Diethard Tautz (University of Cologne, Germany).

Chapter 22 Paul Edwards (University of Cambridge, UK), Tariq Enver (Institute of Cancer Research, London), Simon Hughes (King's College London), Richard Gardner (University of Oxford, UK), Paul Martin (University College London), Peter Mombaerts (The Rockefeller University), William Otto (Imperial Cancer Research Fund, London), Terence Partridge (MRC Clinical Sciences Centre, London), Jeffrey Pollard (Albert Einstein College of Medicine), Elio Raviola (Harvard Medical School), Gregg Semenza (Johns Hopkins University), David Shima (Imperial Cancer Research Fund, London), Bruce Spiegelman (Harvard Medical School), Fiona Watt (Imperial Cancer Research Fund, London), Irving Weissman (Stanford University), Nick Wright (Imperial Cancer Research Fund, London).

Chapter 23 Karen Hopkin [substantial contribution] (science writer, Ph.D., Albert Einstein College of Medicine), Paul Edwards [substantial contribution] (University of Cambridge, UK), Adrian Harris (Imperial Cancer Research Fund, Oxford, UK), Richard Klausner (National Institutes of Health), Gordon Peters (Imperial Cancer Research Fund, London), Peter Selby (Imperial Cancer Research Fund, Leeds, UK), Margaret Stanley (University of Cambridge, UK).

Chapter 24 Anthony DeFranco (University of California, San Francisco), Jonathan Howard (University of Washington, Seattle), Charles Janeway (Yale University School of Medicine), Dan Littman (New York University School of Medicine), Philippa Marrack (National Jewish Medical and Research Center, Denver), N.A. Mitchison (University College London), Michael Neuberger (MRC Laboratory of Molecular Biology, Cambridge, UK), William E. Paul (National Institutes of Health, Bethesda), Klaus Rajewsky (University of Cologne, Germany), Ronald Schwartz (National Institutes of Health, Bethesda), Paul Travers (Anthony Nolan Research Institute, London).

Chapter 25 Julie Theriot [major contribution] (Stanford University), David Baldwin (Stanford University), Stanley Falkow (Stanford University), Warren Levinson (University of California, San Francisco), Shirley Lowe (University of California, San Francisco), Suzanne Noble (University of California, San Francisco), Dan Portnoy (University of California, Berkeley), Peter Sarnow (Stanford University).

References Mary Purton

Glossary Eleanor Lawrence

Student reviewers José Corona (University of California, Santa Cruz), David Kashatus (University of North Carolina at Chapel Hill), Carol Koyama (University of California, Santa Cruz), David States (University of California, Santa Cruz), Uyen Tram (University of California, Santa Cruz).

First, second, and third editions David Agard (University of California, San Francisco), Michael Akam (University of Cambridge, UK), Fred Alt (Columbia University), Martha Arnaud (University of California, San Francisco), Michael Ashburner (University of Cambridge, UK), Jonathan Ashmore (University College London), Tayna Awabdy (University of California, San Francisco), Peter Baker (deceased), Michael Banda (University of California, San Francisco), Ben Barres (Stanford University), Michael Bennett (Albert Einstein College of Medicine), Darwin Berg (University of California, San Diego), Michael J. Berridge (The Babraham Institute, Cambridge, UK), David Birk (UMNDJ—Robert Wood Johnson Medical School), Michael Bishop (University of California, San Francisco), Tim Bliss (National Institute for Medical Research, London), Hans Bode (University of California, Irvine), Piet Borst (Jan Swammerdam Institute, University of Amsterdam), Henry Bourne (University of California, San Francisco), Alan Boyde (University College London), Martin Brand (University of Cambridge, UK), Carl Branden (Karolinska Institute, Stockholm), Mark Bretscher (MRC Laboratory of Molecular Biology, Cambridge, UK), Marianne Bronner-Fraser (California Institute of Technology), Robert Brooks (King's College London), Barry Brown (King's College London), Michael Brown (University of Oxford, UK), Steve Burden (New York University of Medicine), Max Burger (University of Basel), Stephen Burley (The Rockefeller University), John Cairns (Radcliffe Infirmary, Oxford, UK), Zacheus Cande (University of California, Berkeley), Lewis Cantley (Harvard Medical School), Charles Cantor (Columbia University), Roderick Capaldi (University of Oregon), Mario Capecchi (University of Utah), Adelaide Carpenter (University of California, San Diego), Tom Cavalier-Smith (King's College London), Pierre Chambon (University of Strasbourg), Enrico Coen (John Innes Institute, Norwich, UK), Philip Cohen (University of Dundee, Scotland), Robert Cohen (University of California, San Francisco), Roger Cooke (University of California, San Francisco), Nancy Craig (Johns Hopkins University), James Crow (University of Wisconsin, Madison), Stuart Cull-Candy (University College London), Michael Dexter (The Wellcome Trust, UK), Russell Doolittle (University of California, San Diego), Julian Downward (Imperial Cancer Research Fund, London), Graham Dunn (MRC Cell Biophysics Unit, London), Jim Dunwell (John Innes Institute, Norwich, UK), Sarah Elgin (Washington University, St. Louis), Ruth Ellman (Institute of Cancer Research, Sutton, UK), Beverly Emerson (The Salk Institute), Charles Emerson (University of Virginia), David Epel (Stanford University), Gerard Evan (University of California, Comprehensive Cancer Center), Ray Evert (University of Wisconsin, Madison), Gary Felsenfeld (National Institutes of Health, Bethesda), Gary Firestone (University of California, Berkeley), Gerald Fischbach (Columbia University), Robert Fletterick (University of California, San Francisco), Judah Folkman (Harvard Medical School), Larry Fowke (University of Saskatchewan, Saskatoon), Daniel Friend (University of California, San Francisco), Joseph Gall (Yale University), Anthony Gardner-Medwin (University College London), Peter Garland (Institute of Cancer Research, London), Walter Gehring (Biozentrum, University of Basel), Benny Geiger (Weizmann Institute of Science, Rehovot, Israel), Larry Gerace (The Scripps Research Institute), John Gerhart (University of California, Berkeley), Günther Gerisch (Max Planck Institute of Biochemistry, Martinsried), Reid Gilmore (University of Massachusetts at Amherst), Bernie Gilula (deceased), Charles Gilvarg (Princeton University), Bastien Gomperts (University College Hospital Medical School, London), Daniel Goodenough (Harvard Medical School), Peter Gould (Middlesex Hospital Medical School, London), Alan Grafen (University of Oxford, UK), Walter Gratzer (King's College London), Howard Green (Harvard University), Leslie Grivell (University of Amsterdam), Frank Grosveld (Erasmus Universiteit, Rotterdam), Barry Gumbiner (Memorial Sloan-Kettering Cancer Center), Brian Gunning (Australian National University, Canberra), Christine Guthrie (University of California, San Francisco), Ernst Hafen (Universitat Zurich), David Haig (Harvard University), Jeffrey Hall (Brandeis University), John Hall (University of Southampton, UK), Zach Hall (University of California, San Francisco), David Hanke (University of Cambridge, UK), Graham Hardie (University of Dundee, Scotland), John Harris (University of Otago, Dunedin, New Zealand), Stephen Harrison (Harvard University), Leland Hartwell (University of Washington, Seattle), John Heath (University of Birmingham, UK), Ari Helenius (Yale University), Richard Henderson (MRC Laboratory of Molecular Biology, Cambridge, UK), Glenn Herrick (University of Utah), Ira Herskowitz (University of California, San Francisco), Bertil Hille (University of Washington, Seattle), Alan Hinnebusch (National Institutes of Health, Bethesda), Nancy Hollingsworth (State University of New York, Stoney Brook), Leroy Hood (University of Washington), Robert Horvitz (Massachusetts Institute of Technology), David Housman (Massachusetts Institute of Technology), James Hudspeth (The Rockefeller University), Tim Hunt (Imperial Cancer Research Fund, Clare Hall Laboratories, UK), Laurence Hurst (University of Bath, UK), Jeremy Hyams (University College London), Richard Hynes (Massachusetts Institute of Technology), Philip Ingham (University of Sheffield, UK), Norman Iscove (Ontario Cancer Institute, Toronto), David Ish-Horowicz (Imperial Cancer Research Fund, Oxford, UK), Tom Jessell (Columbia University), Sandy Johnson (University of California, San Francisco), Andy Johnston (John Innes Institute, Norwich, UK), E.G. Jordan (Queen Elizabeth College, London), Ron Kaback (University of California, Los Angeles), Ray Keller (University of California, Berkeley), Douglas Kellogg (University of California, Santa Cruz), Regis Kelly (University of California, San Francisco), John Kendrick-Jones (MRC Laboratory of Molecular Biology, Cambridge, UK), Cynthia Kenyon (University of California, San Francisco), Roger Keynes (University of Cambridge, UK), Judith Kimble (University of Wisconsin, Madison), Marc Kirschner (Harvard University), Nancy Kleckner (Harvard University), Mike Klymkowsky (University of Colorado, Boulder), Kelly Komachi (University of California, San Francisco), Juan Korenbrot (University of California, San Francisco), Tom Kornberg (University of California, San Francisco), Stuart Kornfeld (Washington University, St. Louis), Daniel Koshland (University of California, Berkeley), Marilyn Kozak (University of Pittsburgh), Mark Krasnow (Stanford University), Peter Lachmann (MRC Center, Cambridge, UK), Trevor Lamb (University of Cambridge, UK), Hartmut Land (Imperial Cancer Research Fund, London), David Lane (University of Dundee, Scotland), Jay Lash (University of Pennsylvania), Peter Lawrence (MRC Laboratory of Molecular Biology, Cambridge, UK), Robert J. Lefkowitz (Duke University), Mike Levine (University of California, Berkeley), Alex Levitzki (Hebrew University), Tomas Lindahl (Imperial Cancer Research Fund, Clare Hall Laboratories, UK), Vishu Lingappa (University of California, San Francisco), Clive Lloyd (John Innes Institute, Norwich, UK), Richard Losick (Harvard University), James Maller (University of Colorado Medical School), Colin Manoil (Harvard Medical School), Mark Marsh (Institute of Cancer Research, London), Gail Martin (University of California, San Francisco), Brian McCarthy (University of California, Irvine), Richard McCarty (Cornell University), Anne McLaren (Wellcome/Cancer Research Campaign Institute, Cambridge, UK), Freiderick Meins (Freiderich Miescher Institut, Basel), Ira Mellman (Yale University), Barbara Meyer (University of California, Berkeley), Robert Mishell (University of Birmingham, UK), Avrion Mitchison (University College London), N.A. Mitchison (Deutsches

Rheuma-Forschungszentrum Berlin), Tim Mitchison (Harvard Medical School), Mark Mooseker (Yale University), David Morgan (University of California, San Francisco), Michelle Moritz (University of California, San Francisco), Montrose Moses (Duke University), Keith Mostov (University of California, San Francisco), Anne Mudge (University College London), Hans Müller-Eberhard (Scripps Clinic and Research Institute), Alan Munro (University of Cambridge), J. Murdoch Mitchison (Harvard University), Andrew Murray (University of California, San Francisco), Richard Myers (Stanford University), Mark E. Nelson (University of Illinois, Urbana), Walter Neupert (University of Munich, Germany), David Nicholls (University of Dundee, Scotland), Harry Noller (University of California, Santa Cruz), Paul Nurse (Imperial Cancer Research Fund, London), Duncan O'Dell (deceased), Patrick O'Farrell (University of California, San Francisco), Stuart Orkin (Children's Hospital, Boston), John Owen (University of Birmingham, UK), Dale Oxender (University of Michigan), George Palade (University of California, San Diego), Roy Parker (University of Arizona, Tucson), William W. Parson (University of Washington, Seattle), William E. Paul (National Institutes of Health, Bethesda), Robert Perry (Institute of Cancer Research, Philadelphia), Greg Petsko (Brandeis University), David Phillips (The Rockefeller University), Jeremy Pickett-Heaps (University of Colorado), Tom Pollard (The Salk Institute), Bruce Ponder (University of Cambridge, UK), Darwin Prockop (Tulane University), Dale Purves (Duke University), Efraim Racker (Cornell University), Jordan Raff (Wellcome/CRC Institute, Cambridge, UK), Klaus Rajewsky (University of Cologne, Germany), George Ratcliffe (University of Oxford, UK), Martin Rechsteiner (University of Utah, Salt Lake City), David Rees (National Institute for Medical Research, Mill Hill, London), Louis Reichardt (University of California, San Francisco), Fred Richards (Yale University), Phillips Robbins (Massachusetts Institute of Technology), Elaine Robson (University of Reading, UK), Robert Roeder (The Rockefeller University), Joel Rosenbaum (Yale University), Jesse Roth (National Institutes of Health, Bethesda), Jim Rothman (Memorial Sloan-Kettering Cancer Center), Erkki Ruoslahti (La Jolla Cancer Research Foundation), David Sabatini (New York University), Alan Sachs (University of California, Berkeley), Howard Schachman (University of California, Berkeley), Gottfried Schatz (Biozentrum, University of Basel), Randy Schekman (University of California, Berkeley), Michael Schramm (Hebrew University), Robert Schreiber (Scripps Clinic and Research Institute), James Schwartz (Columbia University), Ronald Schwartz (National Institutes of Health, Bethesda), John Scott (University of Manchester, UK), John Sedat (University of California, San Francisco), Zvi Sellinger (Hebrew University), Philippe Sengel (University of Grenoble), Peter Shaw (John Innes Institute, Norwich, UK), Samuel Silverstein (Columbia University), Melvin I. Simon (California Institute of Technology, Pasadena), Kai Simons (Max-Plank Institute of Molecular Cell Biology and Genetics, Dresden), Jonathan Slack (Imperial Cancer Research Fund, Oxford, UK), Alison Smith (John Innes Institute, Norfolk, UK), John Maynard Smith (University of Sussex, UK), Michael Solursh (University of Iowa), Timothy Springer (Harvard Medical School), Mathias Sprinzl (University of Bayreuth), Scott Stachel (University of California, Berkeley), Andrew Staehelin (University of Colorado, Boulder), David Standring (University of California, San Francisco), Martha Stark (University of California, San Francisco), Wilfred Stein (Hebrew University), Malcolm Steinberg (Princeton University), Paul Sternberg (California Institute of Technology), Chuck Stevens (The Salk Institute, La Jolla, CA), Murray Stewart (MRC Laboratory of Molecular Biology, Cambridge, UK), Monroe Strickberger (University of Missouri, St. Louis), Michael Stryker (University of California, San Francisco), Daniel Szollosi (Institut National de la Recherche Agronomique, Jouy-en-Josas, France), Jack Szostak (Massachusetts General Hospital), Masatoshi Takeichi (Kyoto University), Roger Thomas (University of Bristol, Bristol, UK), Vernon Thornton (King's College London), Cheryll Tickle (University of Dundee, Scotland), Jim Till (Ontario Cancer Institute, Toronto), Lewis Tilney (University of Pennsylvania), Alain Townsend (Institute of Molecular Medicine, John Radcliffe Hospital, Oxford, UK), Robert Trelstad (UMDNJ—Robert Wood Johnson Medical School), Anthony Trewavas (Edinburgh University, Scotland), Victor Vacquier (University of California, San Diego), Harry van der Westen (Wageningen, The Netherlands), Tom Vanaman (University of Kentucky), Harold Varmus (Sloan-Kettering Institute, NY), Alexander Varshavsky (California Institute of Technology), Madhu Wahi (University of California, San Francisco), Virginia Walbot (Stanford University), Frank Walsh (Glaxo-Smithkline-Beecham, UK), Peter Walter (University of California, San Francisco), Trevor Wang (John Innes Institute, Norwich, UK), Anne Warner (University College London), Paul Wassarman (Mount Sinai School of Medicine), Fiona Watt (Imperial Cancer Research Fund, London), John Watts (John Innes Institute, Norwich, UK), Klaus Weber (Max Planck Institute for Biophysical Chemistry, Göttingen), Martin Weigert (Institute of Cancer Research, Philadelphia), Harold Weintraub (deceased), Norman Wessells (Stanford University), Judy White (University of Virginia), William Wickner (Dartmouth College), Michael Wilcox (deceased), Lewis T. Williams (Chiron Corporation), Keith Willison (Chester Beatty Laboratories, London), John Wilson (Baylor University), Richard Wolfenden (University of North Carolina), Sandra Wolin (Yale University School of Medicine), Lewis Wolpert (University College London), Rick Wood (Imperial Cancer Research Fund, Clare Hall Laboratories, UK), Abraham Worcel (University of Rochester), John Wyke (Beatson Institute, Glasgow), Keith Yamamoto (University of California, San Francisco), Charles Yocum (University of Michigan), Peter Yurchenco (UMDNJ—Robert Wood Johnson Medical School), Rosalind Zalin (University College London), Patricia Zambryski (University of California, Berkeley).

A Note to the Reader

Structure
Although the chapters of this book can be read independently of one another, they are arranged in a logical sequence of five parts. The first three chapters of **Part I** cover elementary principles and basic biochemistry. They can serve either as an introduction for those who have not studied biochemistry or as a refresher course for those who have.

Part II represents the central core of cell biology and is concerned mainly with those properties that are common to most eucaryotic cells. It deals with the expression and transmission of genetic information.

Part III deals with the principles of the main experimental methods for investigating cells. It is not necessary to read these two chapters in order to understand the later chapters, but a reader will find it a useful reference.

Part IV discusses the internal organization of the cell. **Part V** follows the behavior of cells in multicellular organisms, starting with cell–cell junctions and extracellular matrix and concluding with a new chapter on pathogens and infection.

References
A concise list of selected references is included at the end of each chapter. These are arranged in alphabetical order under the chapter section headings. These references frequently include the original papers in which important discoveries were first reported. Chapters 8 and 9 includes several tables giving the dates of crucial developments along with the names of the scientists involved. Elsewhere in the book the policy has been to avoid naming individual scientists.

Full references for each chapter arranged by concept headings can be found on the Garland Science website at: http://www.garlandscience.com

Glossary terms
Throughout the book, **boldface type** has been used to highlight key terms at the point in a chapter where the main discussion of them occurs. *Italic* is used to set off important terms with a lesser degree of emphasis. At the end of the book is the expanded **glossary**, covering technical terms that are part of the common currency of cell biology; it is intended as a first resort for a reader who encounters an unfamiliar term used without explanation.

Nomenclature
Each species has its own conventions for naming genes, proteins, and mutant phenotypes. In this book, for names of genes, we follow the established conventions when referring to a gene in a particular species (see the table below). When we refer to a gene or gene family generically, without intending restriction to a particular species, we use the same convention as for the mouse: italics, with first letter upper-case and subsequent letters lower-case (for example, *Engrailed, Wnt*); the corresponding protein, if it takes its name from the gene, is then given the same name, with the first letter upper-case, but not in italics (Engrailed, Wnt). For proteins not named after genes but given names in their own right (such as actin, tubulin), the initial letter is generally not capitalized.

ORGANISM	GENE	PROTEIN
Mouse	*Hoxa4*	Hoxa4
Human	*HOXA4*	HOXA4
Zebrafish	*cyclops, cyc*	Cyclops, Cyc
Caenorhabditis	*unc-6*	UNC-6
Drosophila	*sevenless, sev* (named after recessive mutant phenotype)	Sevenless, SEV
	Deformed, Dfd (named after dominant mutant phenotype, or named by homology with another species)	Deformed, DFD
Yeast		
Saccharomyces cerevisiae (budding yeast)	*CDC28*	Cdc28, Cdc28p
Schizosaccharomyces pombe (fission yeast)	*Cdc2*	Cdc2, Cdc2p
Arabidopsis	*GAI*	GAI
E. coli	*uvrA*	UvrA

Cell Biology Interactive

The *Cell Biology Interactive* CD-ROM is packaged with every copy of the book. Created by the *Molecular Biology of the Cell* author team, the CD-ROM contains over 90 video clips, animations, molecular structures and high-resolution micrographs. The authors have chosen to include material that not only reinforces basic concepts but also expands the content and scope of the book. A complete table of contents and overview of all electronic resources is contained in the *Viewing Guide,* located in the Appendix of the CD-ROM. For instructors, there is also a *Teaching Guide for Cell Biology Interactive,* which reviews the electronic resources from a pedagogical perspective, and provides practical suggestions for successfully incorporating multimedia in the classroom.

Molecular Biology of the Cell, Fourth Edition: A Problems Approach

Molecular Biology of the Cell, Fourth Edition: A Problems Approach is designed to help students appreciate the ways in which experiments and simple calculations can lead to an understanding of how cells work. It provides problems to accompany Chapters 1–8 and 10–18 of *Molecular Biology of the Cell.* Each chapter of problems is divided into sections that correspond to those of the main textbook and review key terms, test for understanding basic concepts, and pose research-based problems. *Molecular Biology of the Cell, Fourth Edition: A Problems Approach* should be useful for homework assignments and as a basis for class discussion. It could even provide ideas for exam questions. Answers for half of the problems are provided in the back of the book and the balance are available to instructors upon request.

The Art of MBoC4

A CD-ROM containing all the figures in the book for presentation purposes.

Teaching Supplements

Upon request, teaching supplements for *Molecular Biology of the Cell* are available to instructors.

MBoC4 Transparency Set

250 full-color overhead acetate transparencies of the most important figures from the book.

Garland Science Classwire™

Located at http://www.classwire.com/garlandscience offers instructional resources and course management tools (*Classwire* is a trademark of Chalkfree, Inc).

INTRODUCTION TO THE CELL

PART I

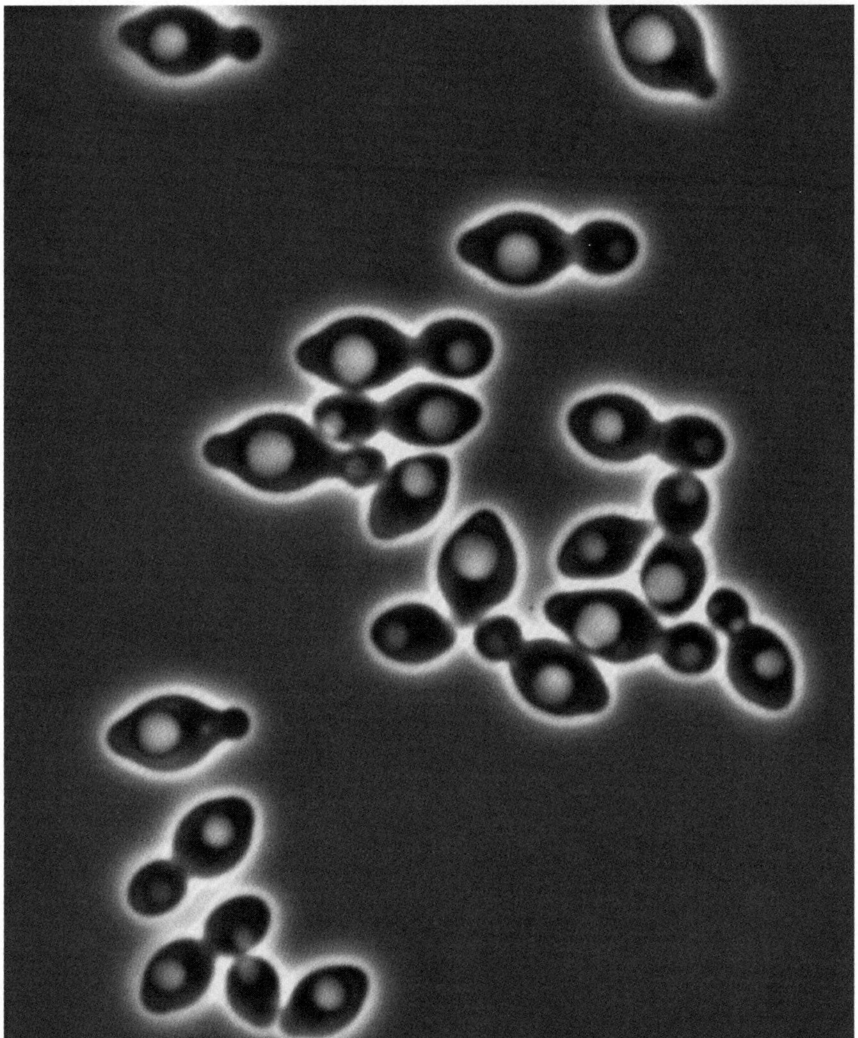

Dividing yeast cells. These single-celled organisms divide rapidly by budding off new daughter cells. Yeast has been widely exploited as a model organism, being especially valuable for studies of the cell cycle.

An egg cell. The DNA of this single cell contains the genetic information needed to specify construction of an entire multicellular animal—in this case, a frog, *Xenopus laevis*. (Courtesy of Tony Mills.)

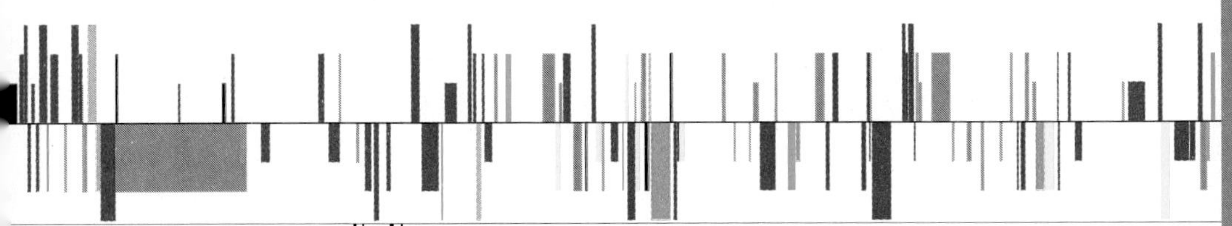

CELLS AND GENOMES

The surface of our planet is populated by living things—curious, intricately organized chemical factories that take in matter from their surroundings and use these raw materials to generate copies of themselves. The living organisms appear extraordinarily diverse in almost every way. What could be more different than a tiger and a piece of seaweed, or a bacterium and a tree? Yet our ancestors, knowing nothing of cells or DNA, saw that all these things had something in common. They called that something "life," marveled at it, struggled to define it, and despaired of explaining what it was or how it worked in terms that relate to nonliving matter.

The discoveries of the twentieth century have not diminished the marvel—quite the contrary. But they have lifted away the mystery surrounding the nature of life. We can now see that all living things are made of cells, and that these units of living matter all share the same machinery for their most basic functions. Living things, though infinitely varied when viewed from the outside, are fundamentally similar inside. The whole of biology is a counterpoint between the two themes: astonishing variety in individual particulars; astonishing constancy in fundamental mechanisms. In this first chapter we begin by outlining the features that are universal to all living things. We then survey, briefly, the diversity of cells. And we see how, thanks to the common code in which the specifications for all living organisms are written, it is possible to read, measure, and decipher these specifications to achieve a coherent understanding of all the forms of life, from the smallest to the greatest.

THE UNIVERSAL FEATURES OF CELLS ON EARTH

It is estimated that there are more than 10 million—perhaps 100 million—living species on Earth today. Each species is different, and each reproduces itself faithfully, yielding progeny that belong to the same species: the parent organism hands down information specifying, in extraordinary detail, the characteristics that the offspring shall have. This phenomenon of *heredity* is a central part of the definition of life: it distinguishes life from other processes, such as the growth of a crystal, or the burning of a candle, or the formation of waves on water, in which orderly structures are generated but without the same type of link between the peculiarities of parents and the peculiarities of offspring. Like the candle flame, the living organism must consume free energy to create and maintain

Figure 1–1 The hereditary information in the egg cell determines the nature of the whole multicellular organism. (A and B) A sea urchin egg gives rise to a sea urchin. (C and D) A mouse egg gives rise to a mouse. (E and F) An egg of the seaweed *Fucus* gives rise to a *Fucus* seaweed. (A, courtesy of David McClay; B, courtesy of M. Gibbs, Oxford Scientific Films; C, courtesy of Patricia Calarco, from G. Martin, *Science* 209:768–776, 1980. © AAAS; D, courtesy of O. Newman, Oxford Scientific Films; E and F, courtesy of Colin Brownlee.)

its organization; but the free energy drives a hugely complex system of chemical processes that is specified by the hereditary information.

Most living organisms are single cells; others, such as ourselves, are vast multicellular cities in which groups of cells perform specialized functions and are linked by intricate systems of communication. But in all cases, whether we discuss the solitary bacterium or the aggregate of more than 10^{13} cells that form a human body, the whole organism has been generated by cell divisions from a single cell. The single cell, therefore, is the vehicle for the hereditary information that defines the species (Figure 1–1). And specified by this information, the cell includes the machinery to gather raw materials from the environment, and to construct out of them a new cell in its own image, complete with a new copy of the hereditary information. Nothing less than a cell has this capability.

All Cells Store Their Hereditary Information in the Same Linear Chemical Code (DNA)

Computers have made us familiar with the concept of information as a measurable quantity—a million bytes (corresponding to about 200 pages of text) on a floppy disk, 600 million on a CD-ROM, and so on. They have also made us uncomfortably aware that the same information can be recorded in many different physical forms. A document that is written on one type of computer may be unreadable on another. As the computer world has evolved, the discs and tapes that we used 10 years ago for our electronic archives have become unreadable on present-day machines. Living cells, like computers, deal in information, and it is estimated that they have been evolving and diversifying for over 3.5 billion years. It is scarcely to be expected that they should all store their information in the same form, or that the archives of one type of cell should be readable by the information-handling machinery of another. And yet it is so. All living cells on Earth, without any known exception, store their hereditary information in the form of double-stranded molecules of DNA—long unbranched paired polymer chains, formed always of the same four types of monomers—A, T, C, G. These monomers are strung together in a long linear sequence that encodes the genetic information, just as the sequence of 1s and 0s encodes the information in a computer file. We can take a piece of DNA from a human cell and insert it into a bacterium, or a piece of bacterial DNA and insert it into a human cell, and the information will be successfully read, interpreted, and copied. Using chemical methods, scientists can read out the complete sequence of monomers in any DNA molecule—extending for millions of nucleotides—and thereby decipher the hereditary information that each organism contains.

All Cells Replicate Their Hereditary Information by Templated Polymerization

To understand the mechanisms that make life possible, one must understand the structure of the double-stranded DNA molecule. Each monomer in a single DNA strand—that is, each **nucleotide**—consists of two parts: a sugar (deoxyribose) with a phosphate group attached to it, and a *base*, which may be either adenine (A), guanine (G), cytosine (C) or thymine (T) (Figure 1–2). Each sugar is linked to the next via the phosphate group, creating a polymer chain composed of a repetitive sugar-phosphate backbone with a series of bases protruding from it. The DNA polymer is extended by adding monomers at one end. For a single isolated strand, these can, in principle, be added in any order, because each one links to the next in the same way, through the part of the molecule that is the same for all of them. In the living cell, however, there is a constraint: DNA is not

Figure 1–2 DNA and its building blocks. (A) DNA is made from simple subunits, called nucleotides, each consisting of a sugar-phosphate molecule with a nitrogen-containing sidegroup, or base, attached to it. The bases are of four types (adenine, guanine, cytosine, and thymine), corresponding to four distinct nucleotides, labeled A, G, C, and T. (B) A single strand of DNA consists of nucleotides joined together by sugar-phosphate linkages. Note that the individual sugar-phosphate units are asymmetric, giving the backbone of the strand a definite directionality, or polarity. This directionality guides the molecular processes by which the information in DNA is interpreted and copied in cells: the information is always "read" in a consistent order, just as written English text is read from left to right. (C) Through templated polymerization, the sequence of nucleotides in an existing DNA strand controls the sequence in which nucleotides are joined together in a new DNA strand; T in one strand pairs with A in the other, and G in one strand with C in the other. The new strand has a nucleotide sequence complementary to that of the old strand, and a backbone with opposite directionality: corresponding to the GTAA... of the original strand, it has ...TTAC. (D) A normal DNA molecule consists of two such complementary strands. The nucleotides within each strand are linked by strong (covalent) chemical bonds; the complementary nucleotides on opposite strands are held together more weakly, by hydrogen bonds. (E) The two strands twist around each other to form a double helix—a robust structure that can accommodate any sequence of nucleotides without altering its basic structure.

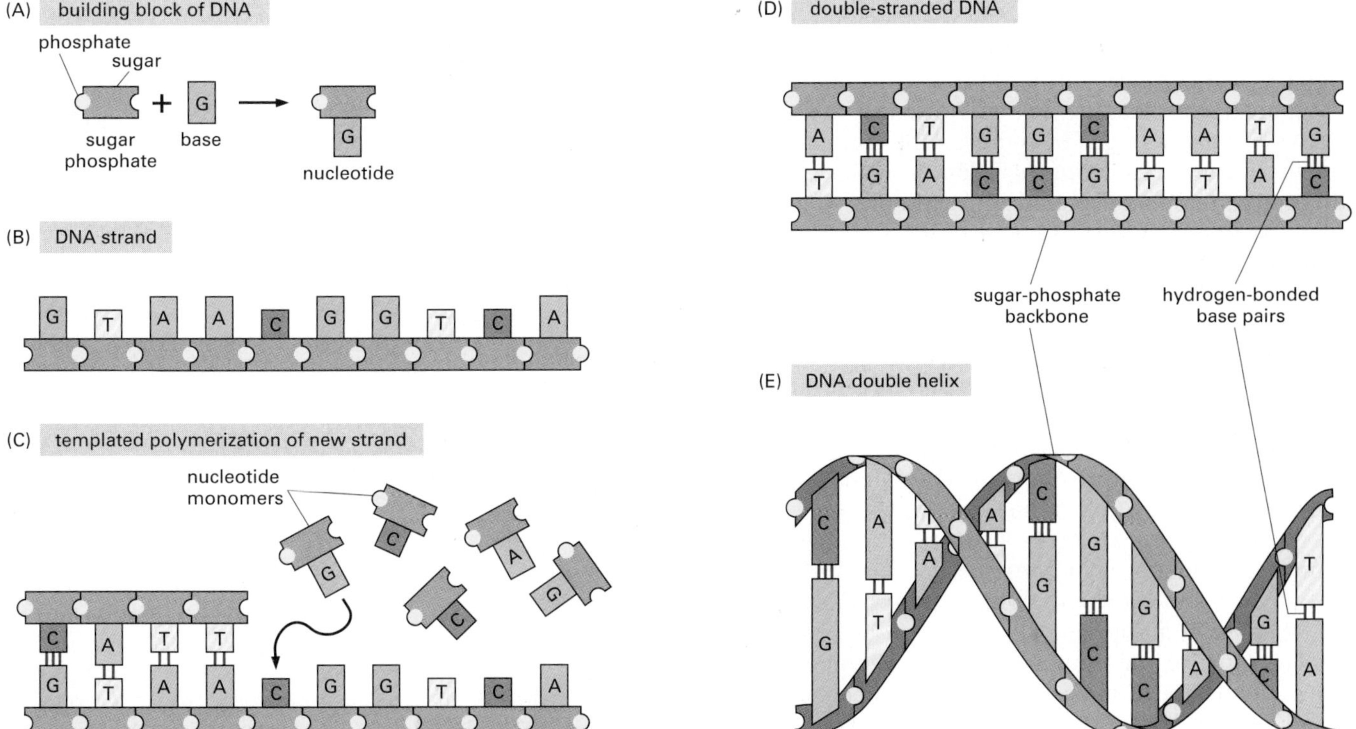

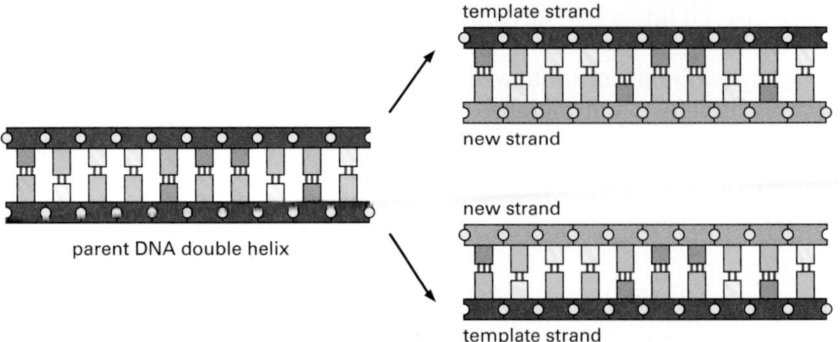

Figure 1–3 The duplication of genetic information by DNA replication. In this process, the two strands of a DNA double helix are pulled apart, and each serves as a template for synthesis of a new complementary strand.

synthesized as a free strand in isolation, but on a template formed by a preexisting DNA strand. The bases protruding from the existing strand bind to bases of the strand being synthesized, according to a strict rule defined by the complementary structures of the bases: A binds to T, and C binds to G. This base-pairing holds fresh monomers in place and thereby controls the selection of which one of the four monomers shall be added to the growing strand next. In this way, a double-stranded structure is created, consisting of two exactly complementary sequences of As, Cs, Ts, and Gs. The two strands twist around each other, forming a double helix (Figure 1–2E).

The bonds between the base pairs are weak compared with the sugar-phosphate links, and this allows the two DNA strands to be pulled apart without breakage of their backbones. Each strand then can serve as a template, in the way just described, for the synthesis of a fresh DNA strand complementary to itself—a fresh copy, that is, of the hereditary information (Figure 1–3). In different types of cells, this process of **DNA replication** occurs at different rates, with different controls to start it or stop it, and different auxiliary molecules to help it along. But the basics are universal: DNA is the information store, and *templated polymerization* is the way in which this information is copied throughout the living world.

All Cells Transcribe Portions of Their Hereditary Information into the Same Intermediary Form (RNA)

To carry out its information-storage function, DNA must do more than copy itself before each cell division by the mechanism just described. It must also express its information, putting it to use so as to guide the synthesis of other molecules in the cell. This also occurs by a mechanism that is the same in all living organisms, leading first and foremost to the production of two other key classes of polymers: RNAs and proteins. The process begins with a templated polymerization called **transcription**, in which segments of the DNA sequence are used as templates to guide the synthesis of shorter molecules of the closely related polymer **ribonucleic acid**, or **RNA**. Later, in the more complex process of **translation**, many of these RNA molecules serve to direct the synthesis of polymers of a radically different chemical class—the *proteins* (Figure 1–4).

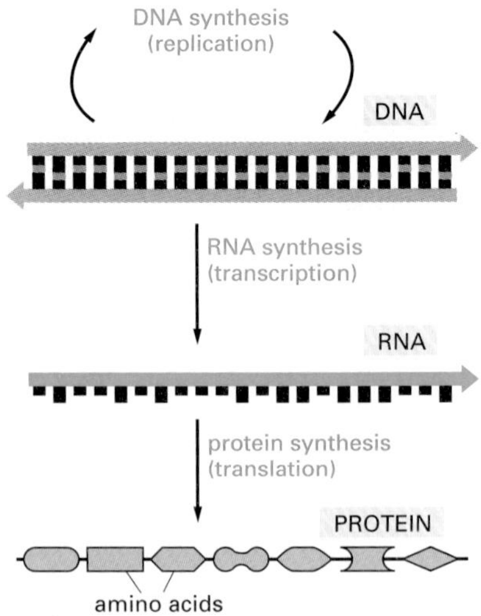

Figure 1–4 From DNA to protein. Genetic information is read out and put to use through a two-step process. First, in *transcription*, segments of the DNA sequence are used to guide the synthesis of molecules of RNA. Then, in *translation*, the RNA molecules are used to guide the synthesis of molecules of protein.

In RNA, the backbone is formed of a slightly different sugar from that of DNA—ribose instead of deoxyribose—and one of the four bases is slightly different—uracil (U) in place of thymine (T); but the other three bases—A, C, and G—are the same, and all four bases pair with their complementary counterparts in DNA—the A, U, C, and G of RNA with the T, A, G, and C of DNA. During transcription, RNA monomers are lined up and selected for polymerization on a template strand of DNA in the same way that DNA monomers are selected during replication. The outcome is therefore a polymer molecule whose sequence of nucleotides faithfully represents a part of the cell's genetic information, even though written in a slightly different alphabet, consisting of RNA monomers instead of DNA monomers.

The same segment of DNA can be used repeatedly to guide the synthesis of many identical RNA transcripts. Thus, whereas the cell's archive of genetic information in the form of DNA is fixed and sacrosanct, the RNA transcripts are

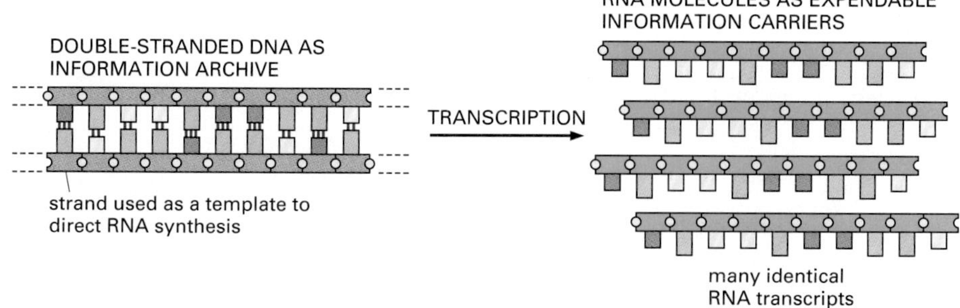

Figure 1–5 How genetic information is broadcast for use inside the cell. Each cell contains a fixed set of DNA molecules—its archive of genetic information. A given segment of this DNA serves to guide the synthesis of many identical RNA transcripts, which serve as working copies of the information stored in the archive. Many different sets of RNA molecules can be made by transcribing selected parts of a long DNA sequence, allowing each cell to use its information store differently.

mass-produced and disposable (Figure 1–5). As we shall see, the primary role of most of these transcripts is to serve as intermediates in the transfer of genetic information: they serve as **messenger RNA (mRNA)** to guide the synthesis of proteins according to the genetic instructions stored in the DNA.

RNA molecules have distinctive structures that can also give them other specialized chemical capabilities. Being single-stranded, their backbone is flexible, so that the polymer chain can bend back on itself to allow one part of the molecule to form weak bonds with another part of the same molecule. This occurs when segments of the sequence are locally complementary: a ...GGGG... segment, for example, will tend to associate with a ...CCCC... segment. These types of internal associations can cause an RNA chain to fold up into a specific shape that is dictated by its sequence (Figure 1–6). The shape of the RNA molecule, in turn, may enable it to recognize other molecules by binding to them selectively—and even, in certain cases, to catalyze chemical changes in the molecules that are bound. As we see later in this book, a few chemical reactions catalyzed by RNA molecules are crucial for several of the most ancient and fundamental processes in living cells, and it has been suggested that more extensive catalysis by RNA played a central part in the early evolution of life (discussed in Chapter 6).

All Cells Use Proteins as Catalysts

Protein molecules, like DNA and RNA molecules, are long unbranched polymer chains, formed by the stringing together of monomeric building blocks drawn from a standard repertoire that is the same for all living cells. Like DNA and RNA, they carry information in the form of a linear sequence of symbols, in the same way as a human message written in an alphabetic script. There are many different protein molecules in each cell, and—leaving out the water—they form most of the cell's mass.

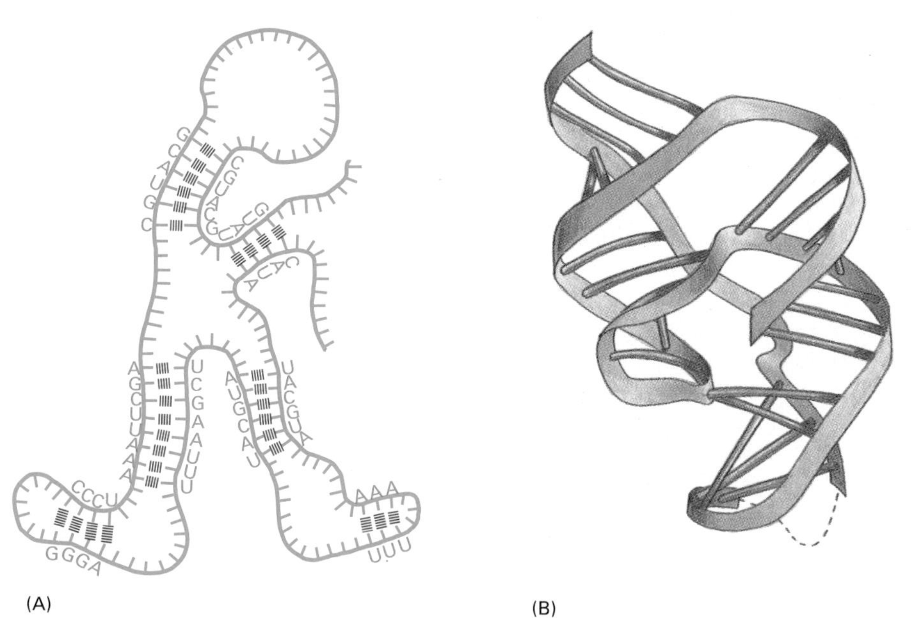

Figure 1–6 The conformation of an RNA molecule. (A) Nucleotide pairing between different regions of the same RNA polymer chain causes the molecule to adopt a distinctive shape. (B) The three-dimensional structure of an actual RNA molecule, from hepatitis delta virus, that catalyzes RNA strand cleavage. (B, based on A.R. Ferré D'Amaré, K. Zhou, and J.A. Doudna, *Nature* 395:567–574, 1998. © Macmillan Magazines Ltd.)

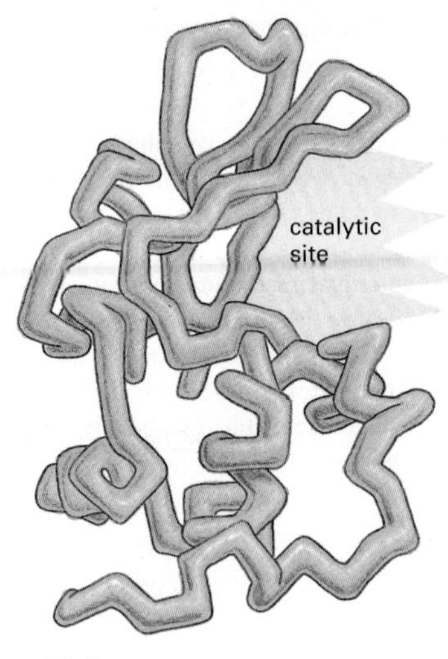

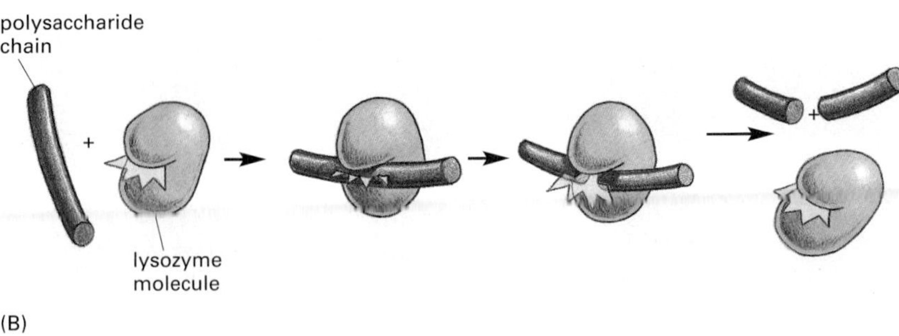

Figure 1–7 How a protein molecule acts as catalyst for a chemical reaction. (A) In a protein molecule the polymer chain folds up to into a specific shape defined by its amino acid sequence. A groove in the surface of this particular folded molecule, the enzyme lysozyme, forms a catalytic site. (B) A polysaccharide molecule *(red)*—a polymer chain of sugar monomers—binds to the catalytic site of lysozyme and is broken apart, as a result of a covalent bond-breaking reaction catalyzed by the amino acids lining the groove.

The monomers of protein, the **amino acids**, are quite different from those of DNA and RNA, and there are 20 types, instead of 4. Each amino acid is built around the same core structure through which it can be linked in a standard way to any other amino acid in the set; attached to this core is a side group that gives each amino acid a distinctive chemical character. Each of the protein molecules, or **polypeptides**, created by joining amino acids in a particular sequence folds into a precise three-dimensional form with reactive sites on its surface (Figure 1–7A). These amino acid polymers thereby bind with high specificity to other molecules and act as **enzymes** to catalyze reactions in which covalent bonds are made and broken. In this way they direct the vast majority of chemical processes in the cell (Figure 1–7B). Proteins have a host of other functions as well—maintaining structures, generating movements, sensing signals, and so on—each protein molecule performing a specific function according to its own genetically specified sequence of amino acids. Proteins, above all, are the molecules that put the cell's genetic information into action.

Thus, polynucleotides specify the amino acid sequences of proteins. Proteins, in turn, catalyze many chemical reactions, including those by which new DNA molecules are synthesized, and the genetic information in DNA is used to make both RNA and proteins. This feedback loop is the basis of the autocatalytic, self-reproducing behavior of living organisms (Figure 1–8).

All Cells Translate RNA into Protein in the Same Way

The translation of genetic information from the 4-letter alphabet of polynucleotides into the 20-letter alphabet of proteins is a complex process. The rules of this translation seem in some respects neat and rational, in other respects strangely arbitrary, given that they are (with minor exceptions) identical in all living things. These arbitrary features, it is thought, reflect frozen accidents in the early history of life—chance properties of the earliest organisms that were passed on by heredity and have become so deeply embedded in the constitution of all living cells that they cannot be changed without wrecking cell organization.

The information in the sequence of a messenger RNA molecule is read out in groups of three nucleotides at a time: each triplet of nucleotides, or *codon*, specifies (codes for) a single amino acid in a corresponding protein. Since there are 64 ($= 4 \times 4 \times 4$) possible codons, but only 20 amino acids, there are necessarily many cases in which several codons correspond to the same amino acid. The code is read out by a special class of small RNA molecules, the **transfer RNAs** (**tRNAs**). Each type of tRNA becomes attached at one end to a specific amino acid, and displays at its other end a specific sequence of three nucleotides—an *anticodon*—that enables it to recognize, through base-pairing, a particular codon or subset of codons in mRNA (Figure 1–9).

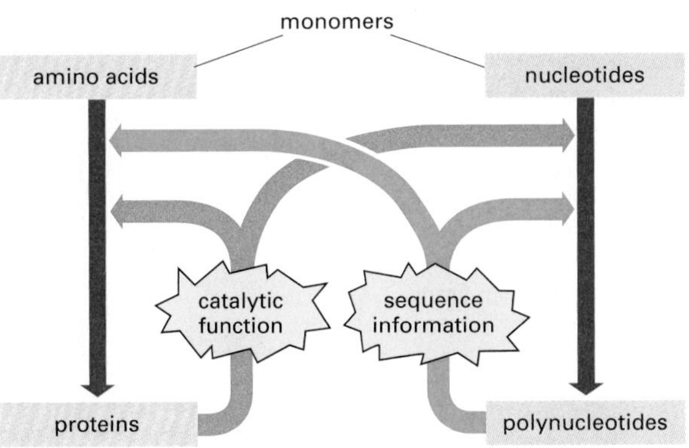

Figure 1–8 Life as an autocatalytic process. Polynucleotides (nucleotide polymers) and proteins (amino acid polymers) provide the sequence information and the catalytic functions that serve—through a complex set of chemical reactions—to bring about the synthesis of more polynucleotides and proteins of the same types.

For synthesis of protein, a succession of tRNA molecules charged with their appropriate amino acids have to be brought together with an mRNA molecule and matched up by base-pairing through their anticodons with each of its successive codons. The amino acids then have to be linked together to extend the growing protein chain, and the tRNAs, relieved of their burdens, have to be released. This whole complex of processes is carried out by a giant multimolecular machine, the ribosome, formed of two main chains of RNA, called **ribosomal RNAs (rRNAs)**, and more than 50 different proteins. This evolutionarily ancient molecular juggernaut latches onto the end of an mRNA molecule and then trundles along it, capturing loaded tRNA molecules and stitching together the amino acids they carry to form a new protein chain (Figure 1–10).

The Fragment of Genetic Information Corresponding to One Protein Is One Gene

DNA molecules as a rule are very large, containing the specifications for thousands of proteins. Segments of the entire DNA sequence are therefore transcribed into separate mRNA molecules, with each segment coding for a different protein. A **gene** is defined as the segment of DNA sequence corresponding to a single protein (or to a single catalytic or structural RNA molecule for those genes that produce RNA but not protein).

In all cells, the *expression* of individual genes is regulated: instead of manufacturing its full repertoire of possible proteins at full tilt all the time, the cell

Figure 1–9 Transfer RNA. (A) A tRNA molecule specific for the amino acid tryptophan. One end of the tRNA molecule has tryptophan attached to it, while the other end displays the triplet nucleotide sequence CCA (its anticodon), which recognizes the tryptophan codon in messenger RNA molecules. (B) The three-dimensional structure of the tryptophan tRNA molecule. Note that the codon and the anticodon in (A) are in antiparallel orientations, like the two strands in a DNA double helix (see Figure 1–2), so that the sequence of the anticodon in the tRNA is read from right to left, while that of the codon in the mRNA is read from left to right.

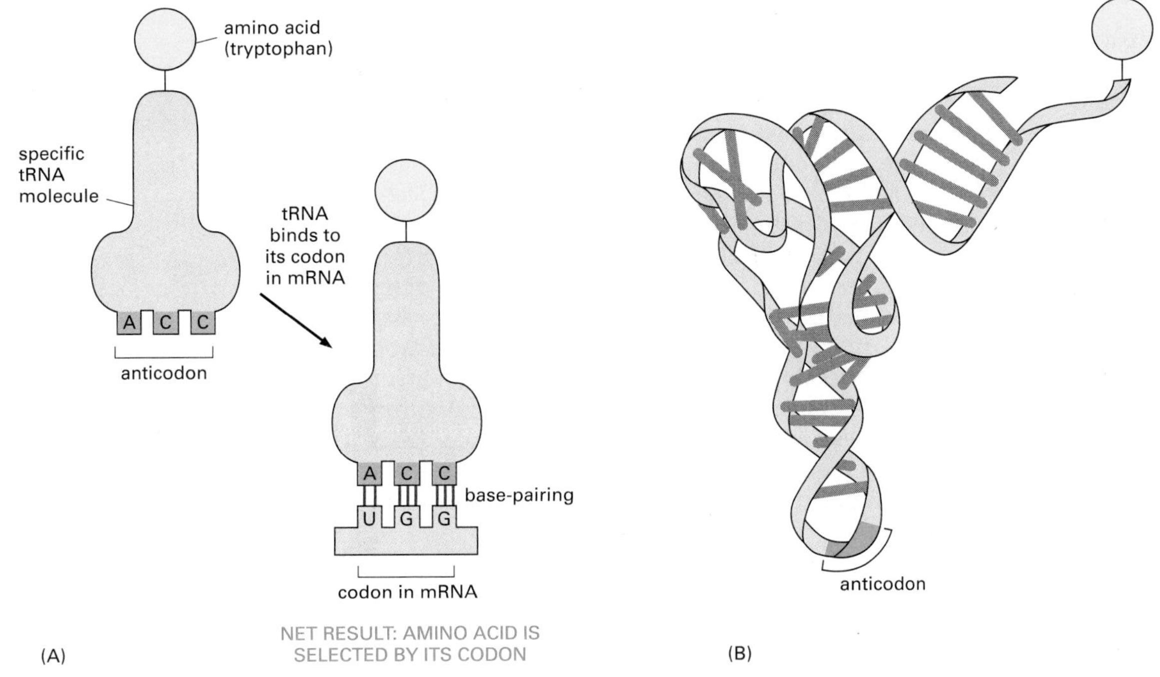

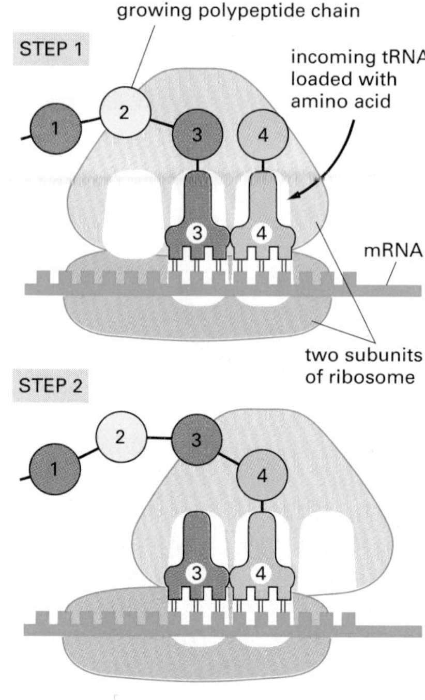

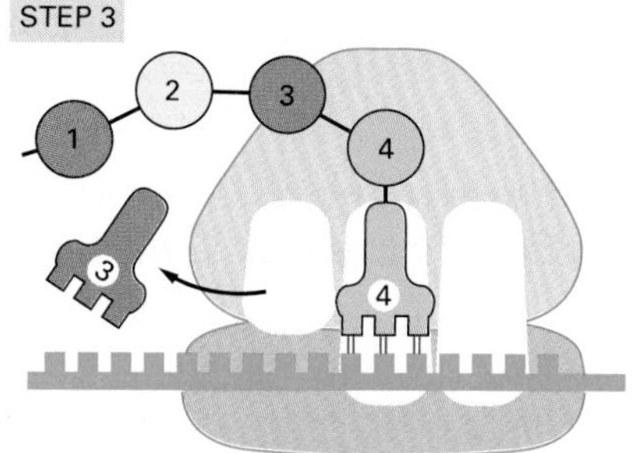

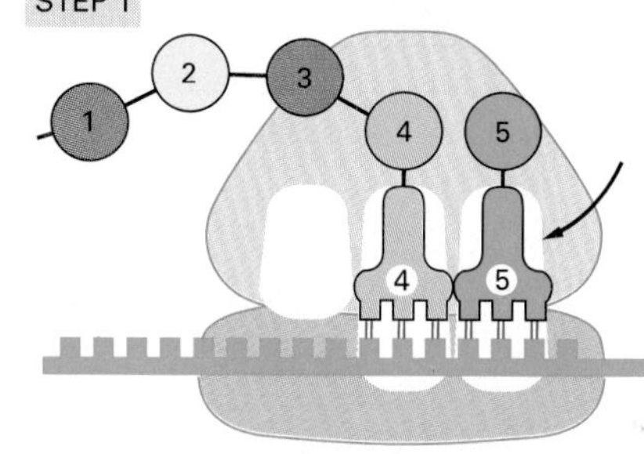

(A)

(B)

Figure 1–10 A ribosome at work. (A) The diagram shows how a ribosome moves along an mRNA molecule, capturing tRNA molecules that match the codons in the mRNA and using them to join amino acids into a protein chain. The mRNA specifies the sequence of amino acids. (B) The three-dimensional structure of a bacterial ribosome *(pale green* and *blue)*, moving along an mRNA molecule *(orange* beads*)*, with three tRNA molecules *(yellow, green*, and *pink)* at different stages in their process of capture and release. The ribosome is a giant assembly of more than 50 individual protein and RNA molecules. (B, courtesy of Joachim Frank, Yanhong Li, and Rajendra Agarwal.)

adjusts the rate of transcription and translation of different genes independently, according to need. Stretches of *regulatory DNA* are interspersed among the segments that code for protein, and these noncoding regions bind to special protein molecules that control the local rate of transcription (Figure 1–11). Other noncoding DNA is also present, some of it serving, for example, as punctuation, defining where the information for an individual protein begins and ends. The quantity and organization of the regulatory and other noncoding DNA vary widely from one class of organisms to another, but the basic strategy is universal. In this way, the **genome** of the cell—that is, the total of its genetic information as embodied in its complete DNA sequence—dictates not only the nature of the cell's proteins, but also when and where they are to be made.

Life Requires Free Energy

A living cell is a system far from chemical equilibrium: it has a large internal free energy, meaning that if it is allowed to die and decay towards chemical equilibrium, a great deal of energy is released to the environment as heat. For the cell to make a new cell in its own image, it must take in free energy from the environment, as well as raw materials, to drive the necessary synthetic reactions. This consumption of free energy is fundamental to life. When it stops, a cell dies. Genetic information is also fundamental to life. Is there any connection?

The answer is yes: free energy is required for the propagation of information, and there is, in fact, a precise quantitative relationship between the two entities. To specify one bit of information—that is, one yes/no choice between two equally probable alternatives—costs a defined amount of free energy (measured in joules), depending on the temperature. The proof of this abstract general principle of statistical thermodynamics is quite arduous, and depends on the precise definition of the term "free energy" (discussed in Chapter 2). The basic idea, however, is not difficult to understand intuitively in the context of DNA synthesis.

To create a new DNA molecule with the same sequence as an existing DNA molecule, nucleotide monomers must be lined up in the correct sequence on the DNA strand that is used as the template. At each point in the sequence, the selection of the appropriate nucleotide depends on the fact that the correctly matched nucleotide binds to the template more strongly than mismatched nucleotides. The greater the difference in binding energy, the rarer are the occasions on which a mismatched nucleotide is accidentally inserted in the sequence instead of the correct nucleotide. A high-fidelity match, whether it is achieved through the direct and simple mechanism just outlined, or in a more complex way, with the help of a set of auxiliary chemical reactions, requires that a lot of free energy be released and dissipated as heat as each correct nucleotide is slotted into its place in the structure. This cannot happen unless the system of molecules carries a large store of free energy at the outset. Eventually, after the newly recruited nucleotides have been joined together to form a new DNA strand, a fresh input of free energy is required to force the matched nucleotides apart again, since each new strand has to be separated from its old template strand to allow the next round of replication.

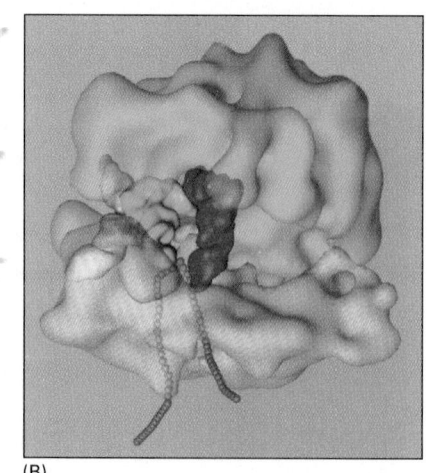

The cell therefore requires free energy, which has to be imported somehow from its surroundings, to replicate its genetic information faithfully. The same principle applies to the synthesis of most of the molecules in cells. For example, in the production of RNAs or proteins, the existing genetic information dictates the sequence of the new molecule through a process of molecular matching, and free energy is required to drive forward the many chemical reactions that construct the monomers from raw materials and link them together correctly.

All Cells Function as Biochemical Factories Dealing with the Same Basic Molecular Building Blocks

Because all cells make DNA, RNA, and protein, and these macromolecules are composed of the same set of subunits in every case, all cells have to contain and manipulate a similar collection of small molecules, including simple sugars, nucleotides, and amino acids, as well as other substances that are universally required for their synthesis. All cells, for example, require the phosphorylated nucleotide *ATP (adenosine triphosphate)* as a building block for the synthesis of DNA and RNA; and all cells also make and consume this molecule as a carrier of free energy and phosphate groups to drive many other chemical reactions.

Although all cells function as biochemical factories of a broadly similar type, many of the details of their small-molecule transactions differ, and it is not as easy as it is for the informational macromolecules to point out the features that are strictly universal. Some organisms, such as plants, require only the simplest of nutrients and harness the energy of sunlight to make from these almost all their own small organic molecules; other organisms, such as animals, feed on living things and obtain many of their organic molecules ready-made. We return to this point below.

All Cells Are Enclosed in a Plasma Membrane Across Which Nutrients and Waste Materials Must Pass

There is, however, at least one other feature of cells that is universal: each one is bounded by a membrane—the **plasma membrane**. This container acts as a selective barrier that enables the cell to concentrate nutrients gathered from its environment and retain the products it synthesizes for its own use, while excreting its waste products. Without a plasma membrane, the cell could not maintain its integrity as a coordinated chemical system.

This membrane is formed of a set of molecules that have the simple physicochemical property of being *amphipathic*—that is, consisting of one part that is hydrophobic (water-insoluble) and another part that is hydrophilic (water-soluble). When such molecules are placed in water, they aggregate spontaneously, arranging their hydrophobic portions to be as much in contact with one another as possible to hide them from the water, while keeping their hydrophilic portions exposed. Amphipathic molecules of appropriate shape, such as the phospholipid molecules that comprise most of the plasma membrane, spontaneously aggregate in water to form a *bilayer* that creates small closed vesicles (Figure 1–12). The phenomenon can be demonstrated in a test tube by simply mixing phospholipids and water together; under appropriate conditions, small vesicles form whose aqueous contents are isolated from the external medium.

Although the chemical details vary, the hydrophobic tails of the predominant membrane molecules in all cells are hydrocarbon polymers ($-CH_2-CH_2-CH_2-$), and their spontaneous assembly into a bilayered vesicle is but one of many examples of an important general principle: cells produce molecules whose chemical properties cause them to *self-assemble* into the structures that a cell needs.

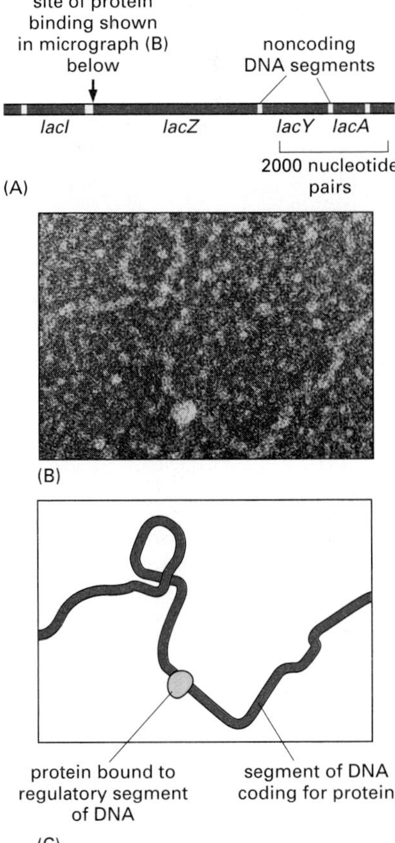

Figure 1–11 (A) A diagram of a small portion of the genome of the bacterium *Escherichia coli*, containing genes (called *lacI, lacZ, lacY,* and *lacA*) coding for four different proteins. The protein-coding DNA segments *(red)* have regulatory and other noncoding DNA segments *(yellow)* between them. (B) An electron micrograph of DNA from this region, with a protein molecule (encoded by the *lacI* gene) bound to the regulatory segment; this protein controls the rate of transcription of the *lacZ, lacY,* and *lacA* genes. (C) A drawing of the structures shown in (B). (B, courtesy of Jack Griffith.)

The boundary of the cell cannot be totally impermeable. If a cell is to grow and reproduce, it must be able to import raw materials and export waste across its plasma membrane. All cells therefore have specialized proteins embedded in their membrane that serve to transport specific molecules from one side to the other (Figure 1–13). Some of these *membrane transport proteins,* like some of the proteins that catalyze the fundamental small-molecule reactions inside the cell, have been so well preserved over the course of evolution that one can recognize the family resemblances between them in comparisons of even the most distantly related groups of living organisms.

The transport proteins in the membrane largely determine which molecules enter the cell, and the catalytic proteins inside the cell determine the reactions that those molecules undergo. Thus, by specifying the set of proteins that the cell is to manufacture, the genetic information recorded in the DNA sequence dictates the entire chemistry of the cell; and not only its chemistry, but also its form and its behavior, for these too are chiefly constructed and controlled by the cell's proteins.

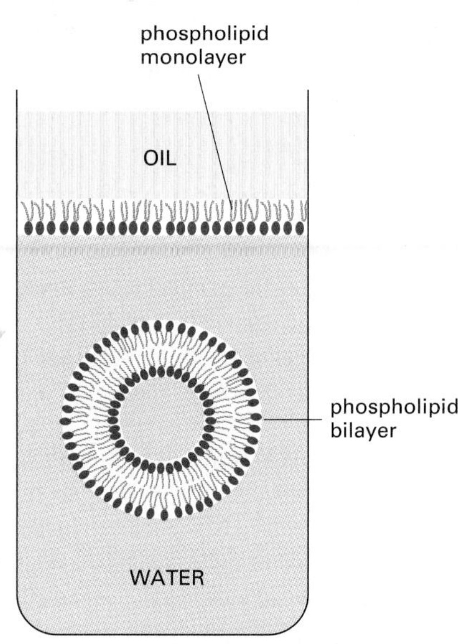

Figure 1–12 Formation of a membrane by amphipathic phospholipid molecules. These have a hydrophilic (water-loving, phosphate) head group and a hydrophobic (water-avoiding, hydrocarbon) tail. At an interface between oil and water, they arrange themselves as a single sheet with their head groups facing the water and their tail groups facing the oil. When immersed in water, they aggregate to form bilayers enclosing aqueous compartments.

A Living Cell Can Exist with Fewer Than 500 Genes

The basic principles of biological information transfer are simple enough, but how complex are real living cells? In particular, what are the minimum requirements? We can get a rough indication by considering the species that has the smallest known genome—the bacterium *Mycoplasma genitalium* (Figure 1–14). This organism lives as a parasite in mammals, and its environment provides it with many of its small molecules ready-made. Nevertheless, it still has to make all the large molecules—DNA, RNAs, and proteins—required for the basic processes of heredity. It has only 477 genes in its genome of 580,070 nucleotide pairs, representing 145,018 bytes of information—about as much as it takes to record the text of one chapter of this book. Cell biology may be complicated, but it is not impossibly so.

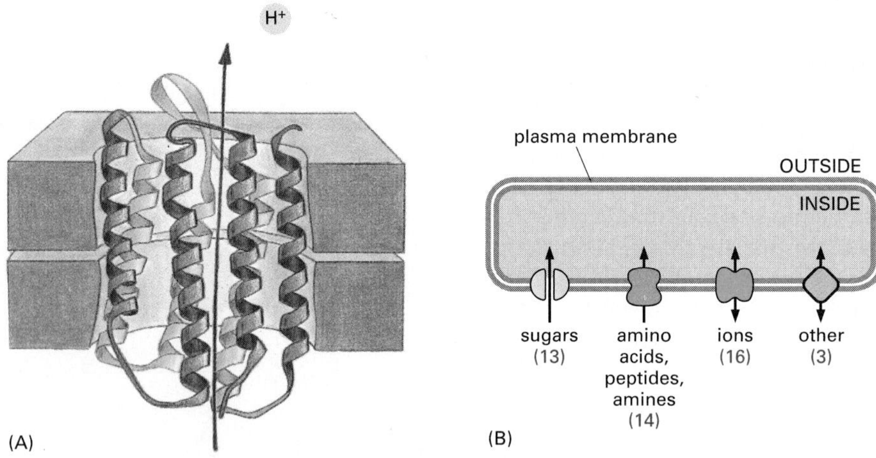

Figure 1–13 Membrane transport proteins. (A) Structure of a molecule of bacteriorhodopsin, from the archaean (archaebacterium) *Halobacterium halobium*. This transport protein uses the energy of absorbed light to pump protons (H^+ ions) out of the cell. The polypeptide chain threads to and fro across the membrane; in several regions it is twisted into a helical conformation, and the helical segments are arranged to form the walls of a channel through which ions are transported. (B) Diagram of the set of transport proteins found in the membrane of the bacterium *Thermotoga maritima*. The numbers in parentheses refer to the number of different membrane transport proteins of each type. Most of the proteins within each class are evolutionarily related to one another and to their counterparts in other species.

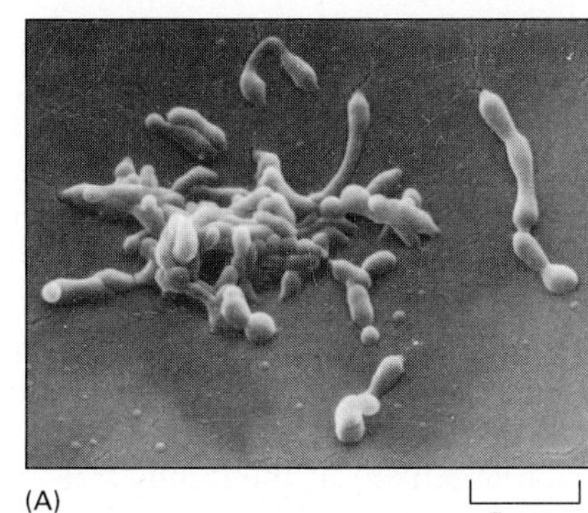

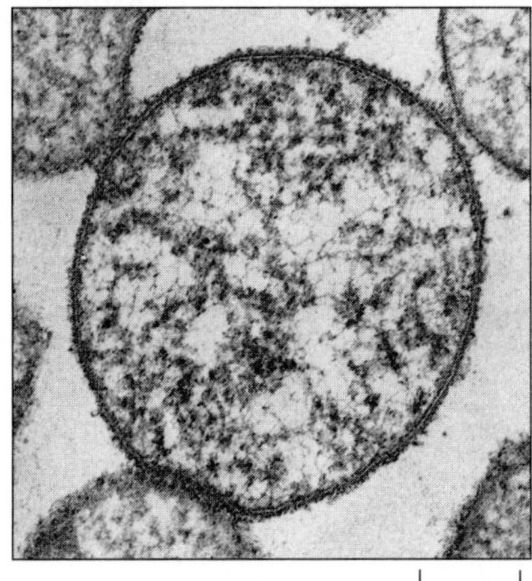

Figure 1–14 *Mycoplasma genitalium.* (A) Scanning electron micrograph showing the irregular shape of this small bacterium, reflecting the lack of any rigid wall. (B) Cross section (transmission electron micrograph) of a *Mycoplasma* cell. Of the 477 genes of *Mycoplasma genitalium*, 37 code for transfer, ribosomal, and other nonmessenger RNAs. Functions are known, or can be guessed, for 297 of the genes coding for protein: of these, 153 are involved in DNA replication, transcription, and translation and related processes involving DNA, RNA, and protein; 29 in the membrane and surface structures of the cell; 33 in the transport of nutrients and other molecules across the membrane; 71 in energy conversion and the synthesis and degradation of small molecules; and 11 in the regulation of cell division and other processes. (A, from S. Razin, M. Banai, H. Gamliel, A. Pollack, W. Bredt, and I. Kahane, *Infect. Immun.* 30:538–546, 1980; B, courtesy of Roger Cole, in Medical Microbiology, 4th edn., [S. Baron ed.]. Galveston: University of Texas Medical Branch, 1996.)

The minimum number of genes for a viable cell in today's environments is probably not less than 200–300. As we shall see in the next section, when we compare the most widely separated branches of the tree of life, we find that a core set of over 200 genes is common to them all.

Summary

Living organisms reproduce themselves by transmitting genetic information to their progeny. The individual cell is the minimal self-reproducing unit, and is the vehicle for transmission of the genetic information in all living species. Every cell on our planet stores its genetic information in the same chemical form—as double-stranded DNA. The cell replicates its information by separating the paired DNA strands and using each as a template for polymerization to make a new DNA strand with a complementary sequence of nucleotides. The same strategy of templated polymerization is used to transcribe portions of the information from DNA into molecules of the closely related polymer, RNA. These in turn guide the synthesis of protein molecules by the more complex machinery of translation, involving a large multimolecular machine, the ribosome, which is itself composed of RNA and protein. Proteins are the principal catalysts for almost all the chemical reactions in the cell; their other functions include the selective import and export of small molecules across the plasma membrane that forms the cell's boundary. The specific function of each protein depends on its amino acid sequence, which is specified by the nucleotide sequence of a corresponding segment of the DNA—the gene that codes for that protein. In this way, the genome of the cell determines its chemistry; and the chemistry of every living cell is fundamentally similar, because it must provide for the synthesis of DNA, RNA, and protein. The simplest known cells have just under 500 genes.

THE DIVERSITY OF GENOMES AND THE TREE OF LIFE

The success of living organisms based on DNA, RNA, and protein, out of the infinitude of other chemical forms that one might conceive of, has been spectacular. They have populated the oceans, covered the land, infiltrated the Earth's crust, and molded the surface of our planet. Our oxygen-rich atmosphere, the deposits of coal and oil, the layers of iron ores, the cliffs of chalk and limestone and marble—all these are products, directly or indirectly, of past biological activity on Earth.

Living things are not confined to the familiar temperate realm of land, water, and sunlight inhabited by plants and plant-eating animals. They can be found in the darkest depths of the ocean, in hot volcanic mud, in pools beneath the frozen surface of the Antarctic, and buried kilometers deep in the Earth's crust. The creatures that live in these extreme environments are unfamiliar, not only because they are inaccessible, but also because they are mostly microscopic. In more homely habitats, too, most organisms are too small for us to see without special equipment: they tend to go unnoticed, unless they cause a disease or rot the timbers of our houses. Yet microorganisms make up most of the total mass of living matter on our planet. Only recently, through new methods of molecular analysis and specifically through the analysis of DNA sequences, have we begun to get a picture of life on Earth that is not grossly distorted by our biased perspective as large animals living on dry land.

In this section we consider the diversity of organisms and the relationships among them. Because the genetic information for every organism is written in the universal language of DNA sequences, and the DNA sequence of any given organism can be obtained by standard biochemical techniques, it is now possible to characterize, catalogue, and compare any set of living organisms with reference to these sequences. From such comparisons we can estimate the place of each organism in the family tree of living species—the 'tree of life'. But before describing what this approach reveals, we need first to consider the routes by which cells in different environments obtain the matter and energy they require to survive and proliferate, and the ways in which some classes of organisms depend on others for their basic chemical needs.

Cells Can Be Powered by a Variety of Free Energy Sources

Living organisms obtain their free energy in different ways. Some, such as animals, fungi, and the bacteria that live in the human gut, get it by feeding on other living things or the organic chemicals they produce; such organisms are called *organotrophic* (from the Greek word *trophe*, meaning "food"). Others derive their energy directly from the nonliving world. These fall into two classes: those that harvest the energy of sunlight, and those that capture their energy from energy-rich systems of inorganic chemicals in the environment (chemical systems that are far from chemical equilibrium). Organisms of the former class are called *phototrophic* (feeding on sunlight); those of the latter are called *lithotrophic* (feeding on rock). Organotrophic organisms could not exist without these primary energy converters, which constitute the largest mass of living matter on Earth.

Phototrophic organisms include many types of bacteria, as well as algae and plants, on which we—and virtually all the living things that we ordinarily see around us—depend. Phototrophic organisms have changed the whole chemistry of our environment: the oxygen in the Earth's atmosphere is a by-product of their biosynthetic activities.

Lithotrophic organisms are not such an obvious feature of our world, because they are microscopic and mostly live in habitats that humans do not frequent—deep in the ocean, buried in the Earth's crust, or in various other inhospitable environments. But they are a major part of the living world, and are especially important in any consideration of the history of life on Earth.

Some lithotrophs get energy from *aerobic* reactions, which use molecular oxygen from the environment; since atmospheric O_2 is ultimately the product of living organisms, these aerobic lithotrophs are, in a sense, feeding on the products of past life. There are, however, other lithotrophs that live anaerobically, in places where little or no molecular oxygen is present, in circumstances similar to those that must have existed in the early days of life on Earth, before oxygen had accumulated.

The most dramatic of these sites are the hot *hydrothermal vents* found deep down on the floor of the Pacific and Atlantic Oceans, in regions where the ocean floor is spreading as new portions of the Earth's crust form by a gradual upwelling of material from the Earth's interior (Figure 1–15). Downward-percolating seawater is heated and driven back upward as a submarine geyser, carrying with it a current of chemicals from the hot rocks below. A typical cocktail might include H_2S, H_2, CO, Mn^{2+}, Fe^{2+}, Ni^{2+}, CH_2, NH_4^+, and phosphorus-containing compounds. A dense population of bacteria lives in the neighborhood of the vent, thriving on this austere diet and harvesting free energy from reactions between the available chemicals. Other organisms—clams, mussels, and giant marine worms—in turn live off the bacteria at the vent, forming an entire ecosystem analogous to the system of plants and animals that we belong to, but powered by geochemical energy instead of light (Figure 1–16).

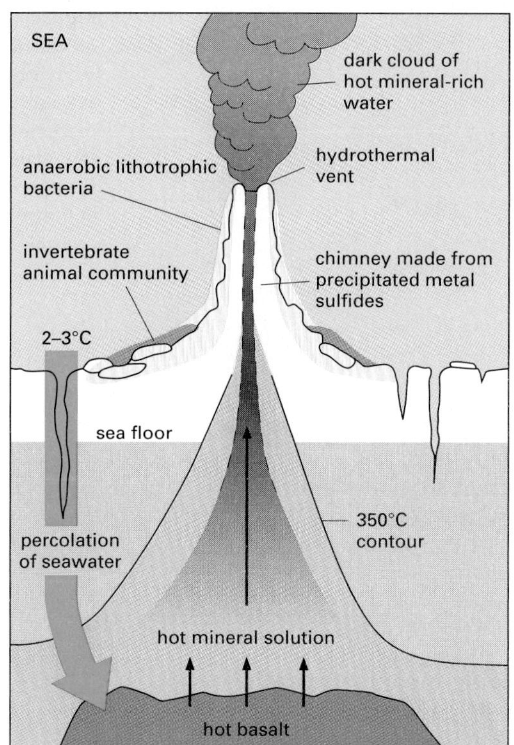

Figure 1–15 The geology of a hot hydrothermal vent in the ocean floor. Water percolates down toward the hot molten rock upwelling from the Earth's interior and is heated and driven back upward, carrying minerals leached from the hot rock. A temperature gradient is set up, from more than 350°C near the core of the vent, down to 2–3°C in the surrounding ocean. Minerals precipitate from the water as it cools, forming a chimney. Different classes of organisms, thriving at different temperatures, live in different neighborhoods of the chimney. A typical chimney might be a few meters tall, with a flow rate of 1–2 m/sec.

Some Cells Fix Nitrogen and Carbon Dioxide for Others

To make a living cell requires matter, as well as free energy. DNA, RNA, and protein are composed of just six elements: hydrogen, carbon, nitrogen, oxygen, sulfur, and phosphorus. These are all plentiful in the nonliving environment, in the Earth's rocks, water, and atmosphere, but not in chemical forms that allow easy incorporation into biological molecules. Atmospheric N_2 and CO_2, in particular, are extremely unreactive, and a large amount of free energy is required to drive the reactions that use these inorganic molecules to make the organic compounds needed for further biosynthesis—that is, to *fix* nitrogen and carbon dioxide, so as to make N and C available to living organisms. Many types of living cells lack the biochemical machinery to achieve this fixation, and rely on other classes of cells to do the job for them. We animals depend on plants for our supplies of organic carbon and nitrogen compounds. Plants in turn, although they can fix carbon dioxide from the atmosphere, lack the ability to fix atmospheric nitrogen, and they depend in part on nitrogen-fixing bacteria to supply their need for nitrogen compounds. Plants of the pea family, for example, harbor symbiotic nitrogen-fixing bacteria in nodules in their roots.

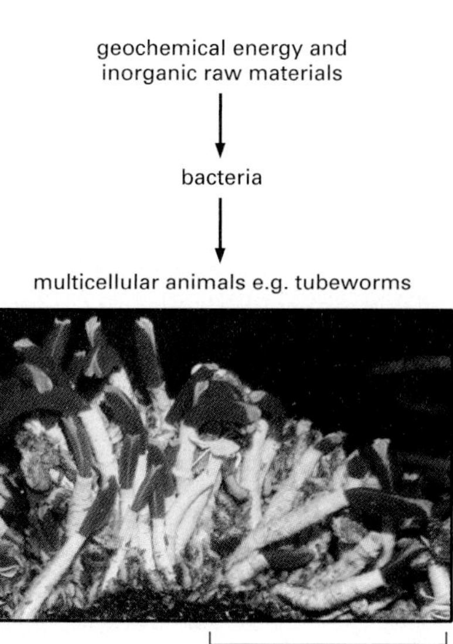

Figure 1–16 Living organisms at a hot hydrothermal vent. Close to the vent, at temperatures up to about 150°C, various lithotrophic species of bacteria and archaea (archaebacteria) live, directly fuelled by geochemical energy. A little further away, where the temperature is lower, various invertebrate animals live by feeding on these microorganisms. Most remarkable are the giant (2-meter) tube worms, which, rather than feed on the lithotrophic cells, live in symbiosis with them: specialized organs in the worms harbor huge numbers of symbiotic sulfur-oxidizing bacteria. These bacteria harness geochemical energy and supply nourishment to their hosts, which have no mouth, gut, or anus. The dependence of the tube worms on the bacteria for the harnessing of geothermal energy is analogous to the dependence of plants on chloroplasts for the harnessing of solar energy, discussed later in this chapter. The tube worms, however, are thought to have evolved from more conventional animals, and to have become secondarily adapted to life at hydrothermal vents. (Courtesy of Dudley Foster, Woods Hole Oceanographic Institution.)

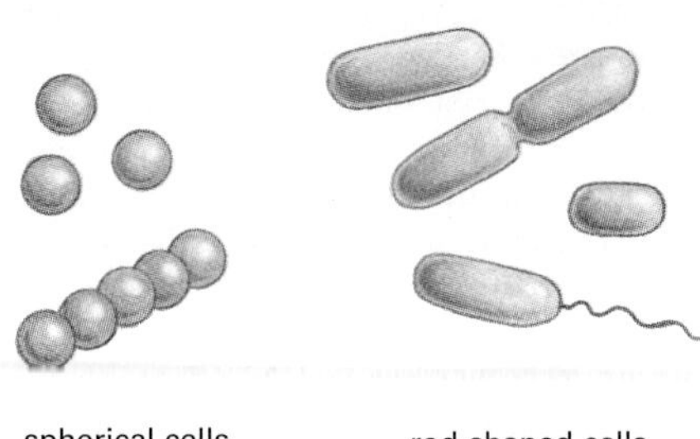
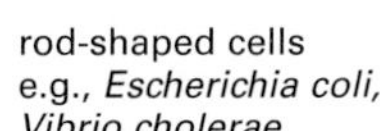
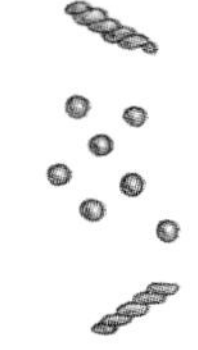
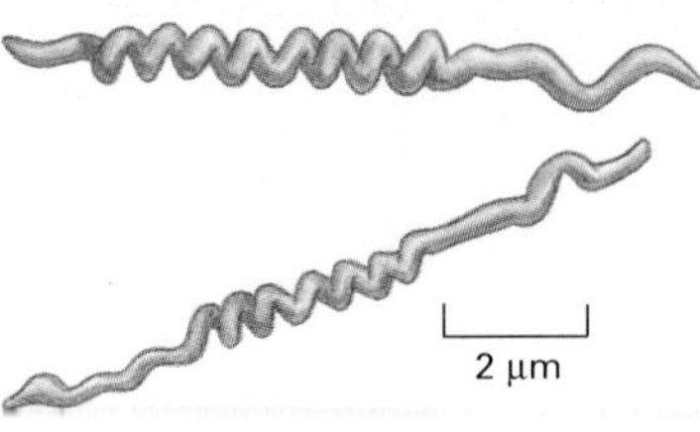

Figure 1–17 Shapes and sizes of some bacteria. Although most are small, as shown, there are also some giant species. An extreme example (not shown) is the cigar-shaped bacterium *Epulopiscium fishelsoni*, which lives in the gut of the surgeon fish and can be up to 600 μm long.

Living cells therefore differ widely in some of the most basic aspects of their biochemistry. Not surprisingly, cells with complementary needs and capabilities have developed close associations. Some of these associations, as we see below, have evolved to the point where the partners have lost their separate identities altogether: they have joined forces to form a single composite cell.

The Greatest Biochemical Diversity Is Seen Among Procaryotic Cells

From simple microscopy, it has long been clear that living organisms can be classified on the basis of cell structure into two groups: the **eucaryotes** and the **procaryotes**. Eucaryotes keep their DNA in a distinct membrane-bounded intracellular compartment called the nucleus. (The name is from the Greek, meaning "truly nucleated," from the words *eu*, "well" or "truly," and *karyon*, "kernel" or "nucleus".) Procaryotes have no distinct nuclear compartment to house their DNA. Plants, fungi, and animals are eucaryotes; bacteria are procaryotes.

Most procaryotic cells are small and simple in outward appearance, and they live mostly as independent individuals, rather than as multicellular organisms. They are typically spherical or rod-shaped and measure a few micrometers in linear dimension (Figure 1–17). They often have a tough protective coat, called a *cell wall,* beneath which a plasma membrane encloses a single cytoplasmic compartment containing DNA, RNA, proteins, and the many small molecules needed for life. In the electron microscope, this cell interior appears as a matrix of varying texture without any discernible organized internal structure (Figure 1–18).

Procaryotic cells live in an enormous variety of ecological niches, and they are astonishingly varied in their biochemical capabilities—far more so than eucaryotic cells. There are organotrophic species that can utilize virtually any type of organic molecule as food, from sugars and amino acids to hydrocarbons

Figure 1–18 The structure of a bacterium. (A) The bacterium *Vibrio cholerae*, showing its simple internal organization. Like many other species, *Vibrio* has a helical appendage at one end—a flagellum—that rotates as a propeller to drive the cell forward. (B) An electron micrograph of a longitudinal section through the widely studied bacterium *Escherichia coli (E. coli)*. This is related to *Vibrio* but lacks a flagellum. The cell's DNA is concentrated in the lightly stained region. (B, courtesy of E. Kellenberger.)

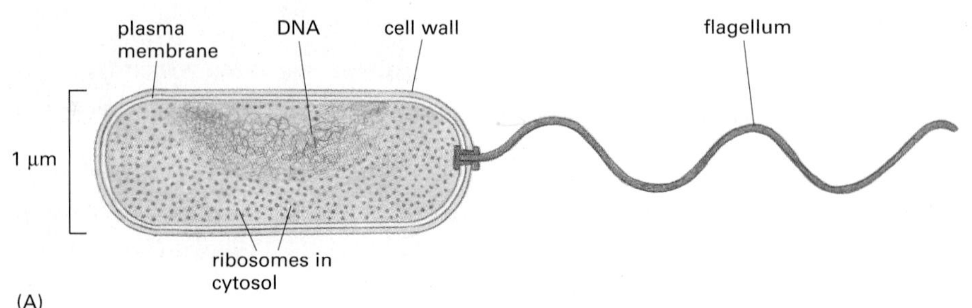

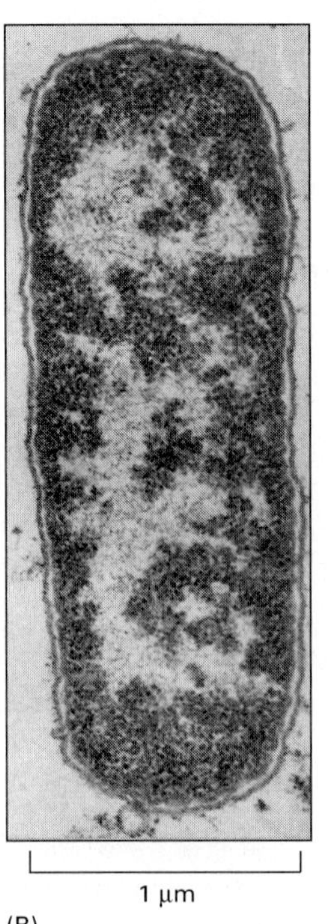

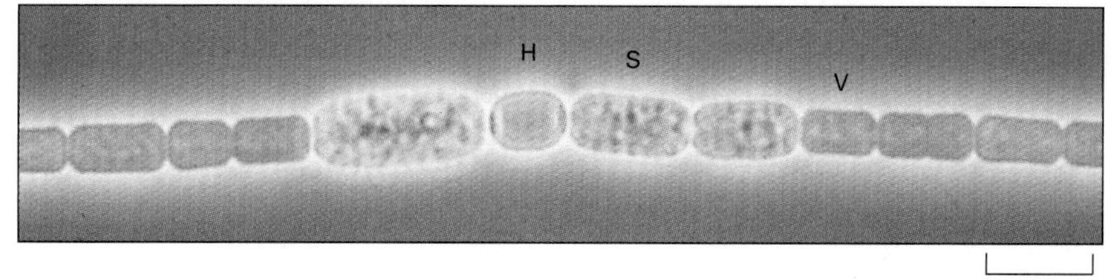

Figure 1–19 The phototrophic bacterium *Anabaena cylindrica* viewed in the light microscope. The cells of this species form long, multicellular filaments. Most of the cells (labeled *V*) perform photosynthesis, while others become specialized for nitrogen fixation (labeled *H*), or develop into resistant spores (labeled *S*). (Courtesy of Dave G. Adams.)

and methane gas. There are many phototrophic species (Figure 1–19), harvesting light energy in a variety of ways, some of them generating oxygen as a byproduct, others not. And there are lithotrophic species that can feed on a plain diet of inorganic nutrients, getting their carbon from CO_2, and relying on H_2S to fuel their energy needs (Figure 1–20)—or on H_2, or Fe^{2+}, or elemental sulfur, or any of a host of other chemicals that occur in the environment.

Many parts of this world of microscopic organisms are virtually unexplored. Traditional methods of bacteriology have given us a fair acquaintance with those species that can be isolated and cultured in the laboratory. But DNA sequence analysis of the populations of bacteria in fresh samples from natural habitats—such as soil or ocean water, or even the human mouth—has opened our eyes to the fact that most species cannot be cultured by standard laboratory techniques. According to one estimate, at least 99% of procaryotic species remain to be characterized.

The Tree of Life Has Three Primary Branches: Bacteria, Archaea, and Eucaryotes

The classification of living things has traditionally depended on comparisons of their outward appearances: we can see that a fish has eyes, jaws, backbone, brain, and so on, just as we do, and that a worm does not; that a rosebush is cousin to an apple tree, but less similar to a grass. We can readily interpret such close family resemblances in terms of evolution from common ancestors, and we can find the remains of many of these ancestors preserved in the fossil record. In this way, it has been possible to begin to draw a family tree of living organisms, showing the various lines of descent, as well as branch points in the history, where the ancestors of one group of species became different from those of another.

When the disparities between organisms become very great, however, these methods begin to fail. How are we to decide whether a fungus is closer kin to a plant or to an animal? When it comes to procaryotes, the task becomes harder still: one microscopic rod or sphere looks much like another. Microbiologists have therefore sought to classify procaryotes in terms of their biochemistry and nutritional requirements. But this approach also has its pitfalls. Amid the bewildering variety of biochemical behaviors, it is difficult to know which differences truly reflect differences of evolutionary history.

Genome analysis has transformed the problem, giving us a simpler, more direct, and more powerful way to determine evolutionary relationships. The complete DNA sequence of an organism defines the species with almost perfect precision and in exhaustive detail. Moreover, this specification, once we have determined it, is in a digital form—a string of letters—that can be fed directly into a computer and compared with the corresponding information for any other living thing. Because DNA is subject to random changes that accumulate over long periods of time (as we shall see shortly), the number of differences between the DNA sequences of two organisms can be used to provide a direct, objective, quantitative indication of the evolutionary distance between them.

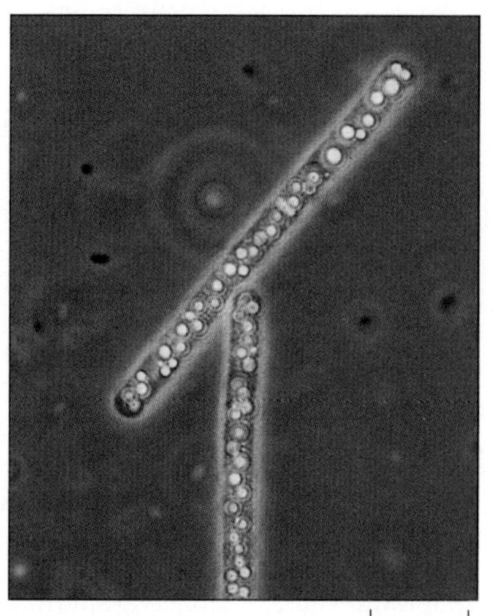

Figure 1–20 A lithotrophic bacterium. *Beggiatoa*, which lives in sulfurous environments, gets its energy by oxidizing H_2S and can fix carbon even in the dark. Note the yellow deposits of sulfur inside the cells. (Courtesy of Ralph W. Wolfe.)

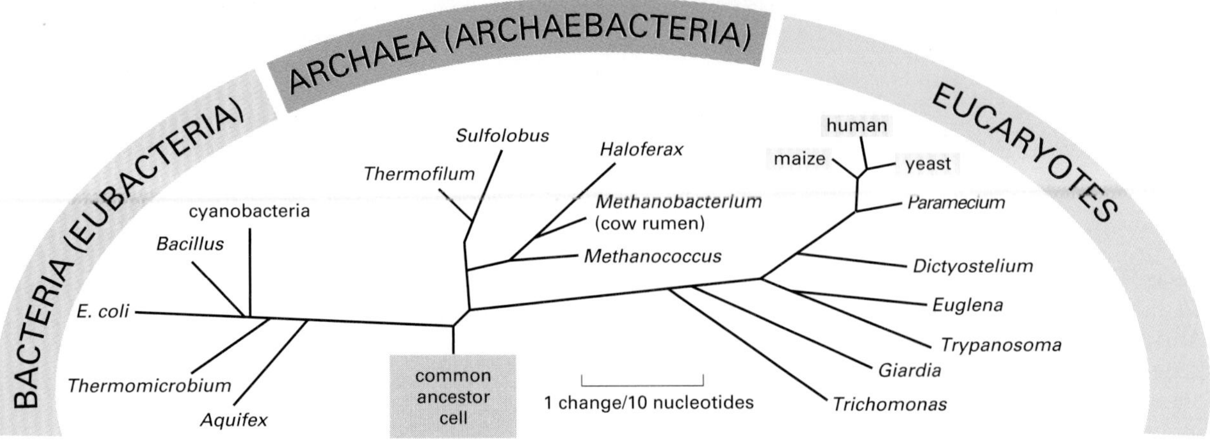

Figure 1–21 The three major divisions (domains) of the living world. Note that traditionally the word *bacteria* has been used to refer to procaryotes in general, but more recently has been redefined to refer to eubacteria specifically. Where there might be ambiguity, we use the term *eubacteria* when the narrow meaning is intended. The tree is based on comparisons of the nucleotide sequence of a ribosomal RNA subunit in the different species. The lengths of the lines represent the numbers of evolutionary changes that have occurred in this molecule in each lineage (see Figure 1–22).

This approach has shown that some of the organisms that were traditionally classed together as "bacteria" are as widely divergent in their evolutionary origins as is any procaryote from any eucaryote. It now appears that the procaryotes comprise two distinct groups that diverged early in the history of life on Earth, either before the ancestors of the eucaryotes diverged as a separate group or at about the same time. The two groups of procaryotes are called the **bacteria** (or eubacteria) and the **archaea** (or archaebacteria). The living world therefore has three major divisions or *domains*: bacteria, archaea, and eucaryotes (Figure 1–21).

Archaea were initially discovered as inhabitants of environments that we humans avoid, such as bogs, sewage farms, ocean depths, salt brines, and hot acid springs, although it is now known that they are also widespread in less extreme and more homely environments, from soils and lakes to the stomachs of cattle. In outward appearance they are not easily distinguished from the more familiar eubacteria. At a molecular level, archaea seem to resemble eucaryotes more closely in their machinery for handling genetic information (replication, transcription, and translation), but eubacteria more closely in their apparatus for metabolism and energy conversion. We discuss below how this might be explained.

Some Genes Evolve Rapidly; Others Are Highly Conserved

Both in the storage and in the copying of genetic information, random accidents and errors occur, altering the nucleotide sequence—that is, creating **mutations**. Therefore, when a cell divides, its two daughters are often not quite identical to one another or to their parent. On rare occasions, the error may represent a change for the better; more probably, it will cause no significant difference in the cell's prospects; and in many cases, the error will cause serious damage—for example, by disrupting the coding sequence for a key protein. Changes due to mistakes of the first type will tend to be perpetuated, because the altered cell has an increased likelihood of reproducing itself. Changes due to mistakes of the second type—*selectively neutral* changes—may be perpetuated or not: in the competition for limited resources, it is a matter of chance whether the altered cell or its cousins will succeed. But changes that cause serious damage lead nowhere: the cell that suffers them dies, leaving no progeny. Through endless repetition of this cycle of error and trial—of *mutation* and *natural selection*—

organisms evolve: their genetic specifications change, giving them new ways to exploit the environment more effectively, to survive in competition with others, and to reproduce successfully.

Clearly, some parts of the genome change more easily than others in the course of evolution. A segment of DNA that does not code for protein and has no significant regulatory role is free to change at a rate limited only by the frequency of random errors. In contrast, a gene that codes for a highly optimized essential protein or RNA molecule cannot alter so easily: when mistakes occur, the faulty cells are almost always eliminated. Genes of this latter sort are therefore *highly conserved*. Through 3.5 billion years or more of evolutionary history, many features of the genome have changed beyond all recognition; but the most highly conserved genes remain perfectly recognizable in all living species.

These latter genes are the ones that must be examined if we wish to trace family relationships between the most distantly related organisms in the tree of life. The studies that led to the classification of the living world into the three domains of bacteria, archaea, and eucaryotes were based chiefly on analysis of one of the ribosomal RNA subunits—the so-called 16S RNA, which is about 1500 nucleotides long. Because the process of translation is fundamental to all living cells, this component of the ribosome has been well conserved since early in the history of life on Earth (Figure 1–22).

Most Bacteria and Archaea Have 1000–4000 Genes

Natural selection has generally favored those procaryotic cells that can reproduce the fastest by taking up raw materials from their environment and replicating themselves most efficiently, at the maximal rate permitted by the available food supplies. Small size implies a large ratio of surface area to volume, thereby helping to maximize the uptake of nutrients across the plasma membrane and boosting a cell's reproductive rate.

Presumably for these reasons, most procaryotic cells carry very little superfluous baggage; their genomes are small and compact, with genes packed closely together and minimal quantities of regulatory DNA between them. The small genome size makes it relatively easy to determine the complete DNA sequence. We now have this information for many species of eubacteria and archaea, and a few species of eucaryotes. As shown in Table 1–1, most eubacterial and archaean genomes contain between 10^6 and 10^7 nucleotide pairs, encoding 1000–4000 genes.

A complete DNA sequence reveals both the genes an organism possesses and the genes it lacks. When we compare the three domains of the living world, we can begin to see which genes are common to all of them and must therefore have been present in the cell that was ancestral to all present-day living things, and which genes are peculiar to a single branch in the tree of life. To explain the findings, however, we need to consider a little more closely how new genes arise and genomes evolve.

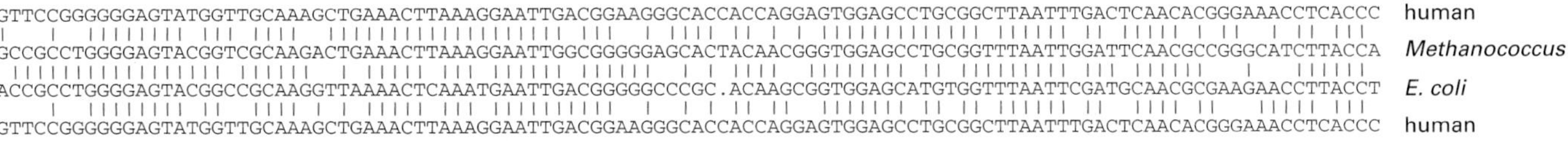

Figure 1–22 Genetic information conserved since the beginnings of life. A part of the gene for the smaller of the two main RNA components of the ribosome is shown. Corresponding segments of nucleotide sequence from an archaean *(Methanococcus jannaschii)*, a eubacterium *(Escherichia coli)* and a eucaryote *(Homo sapiens)* are aligned in parallel. Sites where the nucleotides are identical between species are indicated by a vertical line; the human sequence is repeated at the bottom of the alignment so that all three two-way comparisons can be seen. A dot halfway along the *E. coli* sequence denotes a site where a nucleotide has been either deleted from the eubacterial lineage in the course of evolution, or inserted in the other two lineages. Note that the sequences from these three organisms, representative of the three domains of the living world, all differ from one another to a roughly similar degree, while still retaining unmistakable similarities.

TABLE 1–1 Some Genomes That Have Been Completely Sequenced

SPECIES	SPECIAL FEATURES	HABITAT	GENOME SIZE (1000s OF NUCLEOTIDE PAIRS PER HAPLOID GENOME)	NUMBER OF GENES (PROTEINS)
EUBACTERIA				
Mycoplasma genitalium	smallest genome of any known cell	human genital tract	580	468
Synechocystis sp.	photosynthetic, oxygen-generating (cyanobacterium)	lakes and streams	3573	3168
Escherichia coli	laboratory favorite	human gut	4639	4289
Helicobacter pylori	causes stomach ulcers and predisposes to stomach cancer	human stomach	1667	1590
Bacillus subtilis	bacterium	soil	4214	4099
Aquifex aeolicus	lithotrophic; lives at high temperatures	hydrothermal vents	1551	1544
Mycobacterium tuberculosis	causes tuberculosis	human tissues	4447	4402
Treponema pallidum	spirochaete; causes syphilis	human tissues	1138	1041
Rickettsia prowazekii	bacterium most closely related to mitochondria; causes typhus	lice and humans (intracellular parasite)	1111	834
Thermotoga maritima	organotrophic; lives at high temperatures	hydrothermal vents	1860	1877
ARCHAEA				
Methanococcus jannaschii	lithotrophic, anaerobic, methane-producing	hydrothermal vents	1664	1750
Archaeoglobus fulgidus	lithotrophic or organotrophic, anaerobic, sulfate-reducing	hydrothermal vents	2178	2493
Aeropyrum pernix	aerobic, organotrophic hot-steam vents	coastal volcanic	669	2620
EUCARYOTES				
Saccharomyces cerevisiae (budding yeast)	minimal model eucaryote	grape skins, beer	12,069	~6300
Arabidopsis thaliana (wall cress)	model organism for flowering plants	soil and air	~142,000	~26,000
Caenorhabditis elegans (nematode worm)	simple animal with perfectly predictable development	soil	~97,000	~19,000
Drosophila melanogaster (fruit fly)	key to the genetics of animal development	rotting fruit	~137,000	~14,000
Homo sapiens (human)	most intensively studied mammal	houses	~3,200,000	~30,000

New Genes Are Generated from Preexisting Genes

The raw material of evolution is the DNA sequence that already exists: there is no natural mechanism for making long stretches of new random sequence. In this sense, no gene is ever entirely new. Innovation can, however, occur in several ways (Figure 1–23):

- *Intragenic mutation:* an existing gene can be modified by mutations in its DNA sequence.

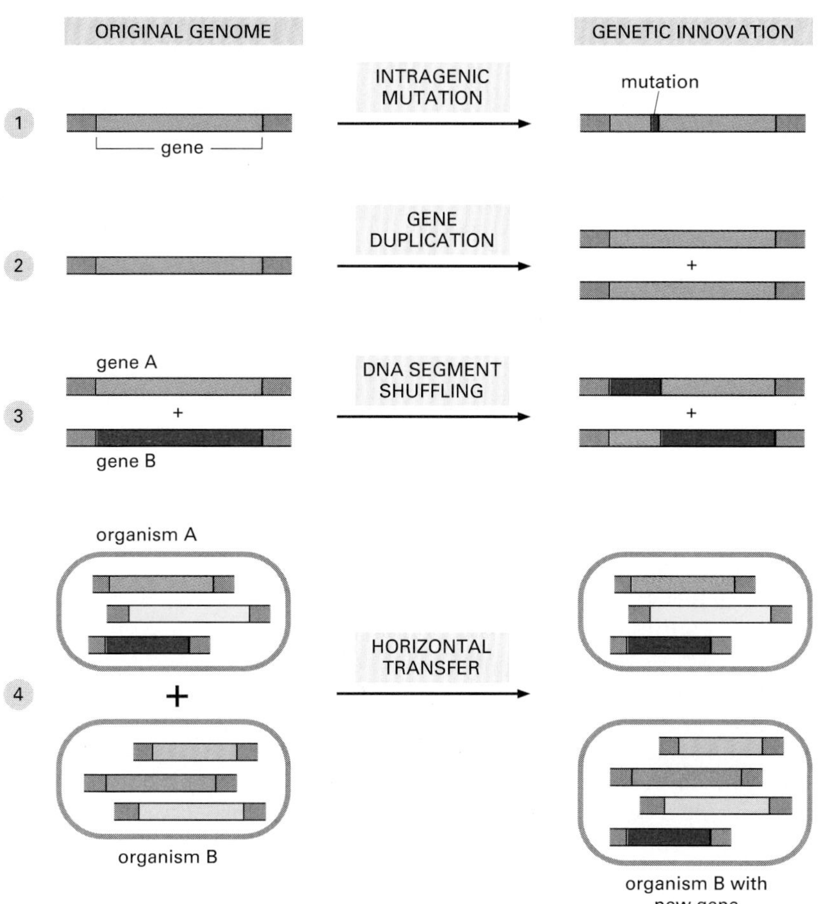

Figure 1–23 Four modes of genetic innovation and their effects on the DNA sequence of an organism.

- *Gene duplication:* an existing gene can be duplicated so as to create a pair of closely related genes within a single cell.
- *Segment shuffling*: two or more existing genes can be broken and rejoined to make a hybrid gene consisting of DNA segments that originally belonged to separate genes.
- *Horizontal (intercellular) transfer*: a piece of DNA can be transferred from the genome of one cell to that of another—even to that of another species. This process is in contrast with the usual *vertical transfer* of genetic information from parent to progeny.

Each of these types of change leaves a characteristic trace in the DNA sequence of the organism, providing clear evidence that all four processes have occurred. In later chapters we discuss the underlying mechanisms, but for the present we focus on the consequences.

Gene Duplications Give Rise to Families of Related Genes Within a Single Cell

A cell must duplicate its entire genome each time it divides into two daughter cells. However, accidents occasionally result in the duplication of just part of the genome, with retention of original and duplicate segments in a single cell. Once a gene has been duplicated in this way, one of the two gene copies is free to mutate and become specialized to perform a different function within the same cell. Repeated rounds of this process of duplication and divergence, over many millions of years, have enabled one gene to give rise to a whole family of genes within a single genome. Analysis of the DNA sequence of procaryotic genomes reveals many examples of such gene families: in *Bacillus subtilis*, for example, 47% of the genes have one or more obvious relatives (Figure 1–24).

When genes duplicate and diverge in this way, the individuals of one species become endowed with multiple variants of a primordial gene. This evolutionary

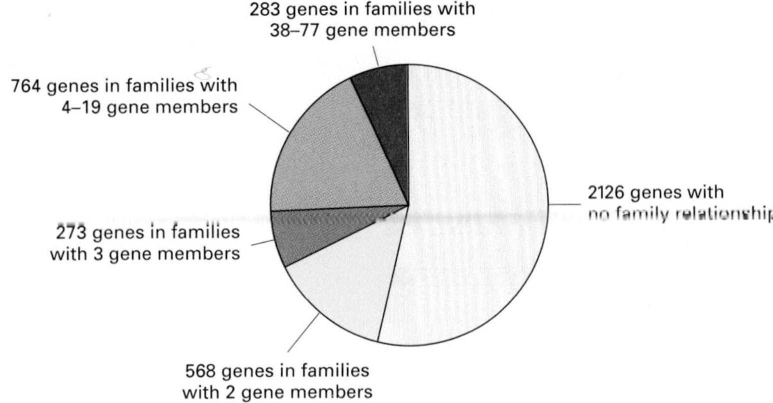

Figure 1–24 Families of evolutionarily related genes in the genome of *Bacillus subtilis*. The biggest family consists of 77 genes coding for varieties of ABC transporters—a class of membrane transport proteins found in all three domains of the living world. (After F. Kunst et al., *Nature* 390:249–256, 1997 © Macmillan Magazines Ltd.)

process has to be distinguished from the genetic divergence that occurs when one species of organism splits into two separate lines of descent at a branch point in the family tree—when the human line of descent became separate from that of chimpanzees, for example. There, the genes gradually become different in the course of evolution, but they are likely to continue to have corresponding functions in the two sister species. Genes that are related in this way—that is, genes in two separate species that derive from the same ancestral gene in the last common ancestor of those two species—are said to be **orthologs**. Related genes that have resulted from a gene duplication event within a single genome—and are likely to have diverged in their function—are said to be **paralogs**. Genes that are related by descent in either way are called **homologs**, a general term used to cover both types of relationship (Figure 1–25).

The family relationships between genes can become quite complex (Figure 1–26). For example, an organism that possesses a family of paralogous genes (for example, the seven hemoglobin genes α, β, γ, δ, ε, ζ, and θ) may evolve into two separate species (such as humans and chimpanzees) each possessing the entire set of paralogs. All 14 genes are homologs, with the human hemoglobin α orthologous to the chimpanzee hemoglobin α, but paralogous to the human or chimpanzee hemoglobin β, and so on. Moreover, the vertebrate hemoglobins (the oxygen-binding proteins of blood) are homologous to the vertebrate myoglobins (the oxygen-binding proteins of muscle), as well as to more distant

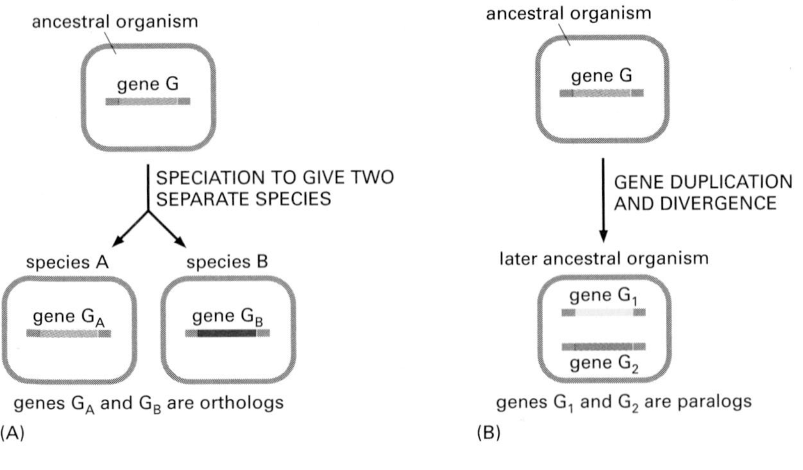

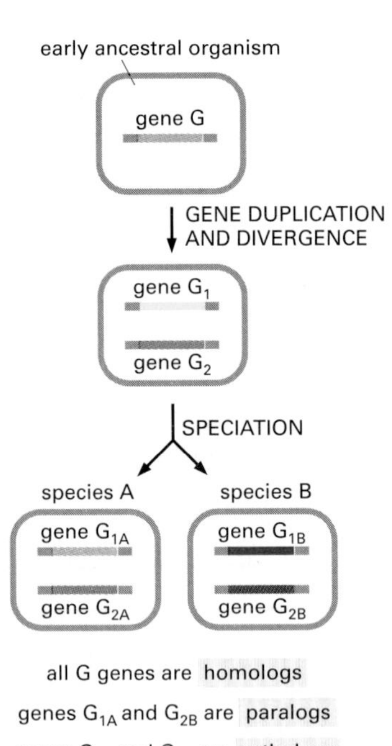

Figure 1–25 Paralogous genes and orthologous genes: two types of gene homology based on different evolutionary pathways. (A) and (B) The most basic possibilities. (C) A more complex pattern of events that can occur.

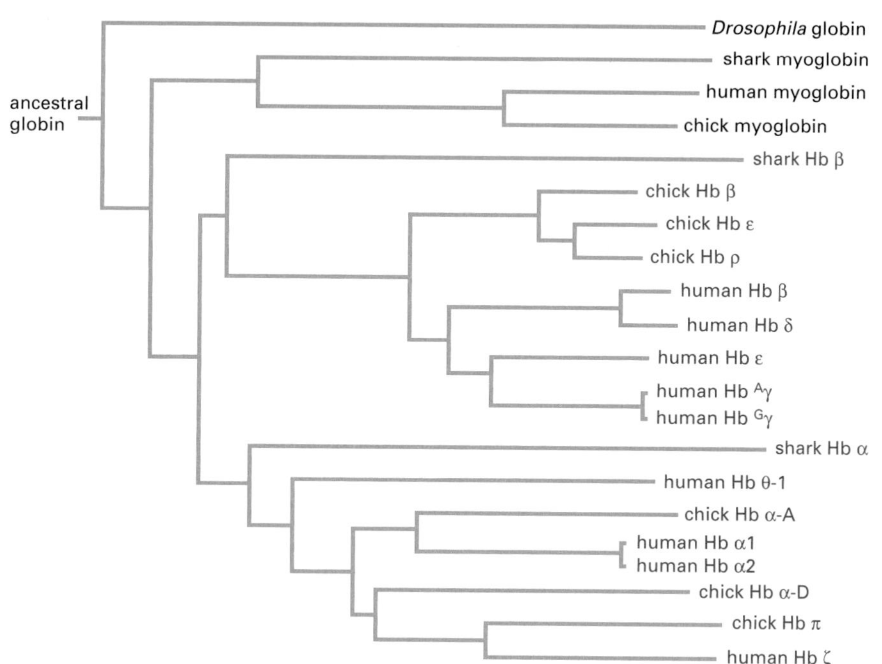

Figure 1–26 A complex family of homologous genes. This diagram shows the pedigree of the hemoglobin (Hb), myoglobin, and globin genes of human, chick, shark, and *Drosophila*. The lengths of the lines represent the amount of divergence in amino acid sequence.

genes that code for oxygen-binding proteins in invertebrates and plants. From the DNA sequences, it is usually easy to recognize that two genes in different species are homologous; it is much more difficult to decide, without other information, whether they are orthologs.

Genes Can Be Transferred Between Organisms, Both in the Laboratory and in Nature

Procaryotes also provide examples of the horizontal transfer of genes from one species of cell to another. The most obvious tell-tale signs are sequences recognizable as being derived from bacterial **viruses**, also called *bacteriophages* (Figure 1–27). These small packets of genetic material have evolved as parasites on the reproductive and biosynthetic machinery of host cells. They replicate in one cell, emerge from it with a protective wrapping, and then enter and infect another cell, which may be of the same or a different species. Inside a cell, they may either remain as separate fragments of DNA, known as *plasmids*, or insert themselves into the DNA of the host cell and become part of its regular genome. In their travels, viruses can accidentally pick up fragments of DNA from the genome of one host cell and ferry them into another cell. Such transfers of genetic material frequently occur in procaryotes, and they are common between eucaryotic cells of the same species.

Horizontal transfers of genes between eucaryotic cells of different species are very rare, and they do not seem to have played a significant part in eucaryote evolution. In contrast, horizontal gene transfers occur much more frequently between different species of procaryotes. Many procaryotes have a remarkable capacity to take up even nonviral DNA molecules from their surroundings and

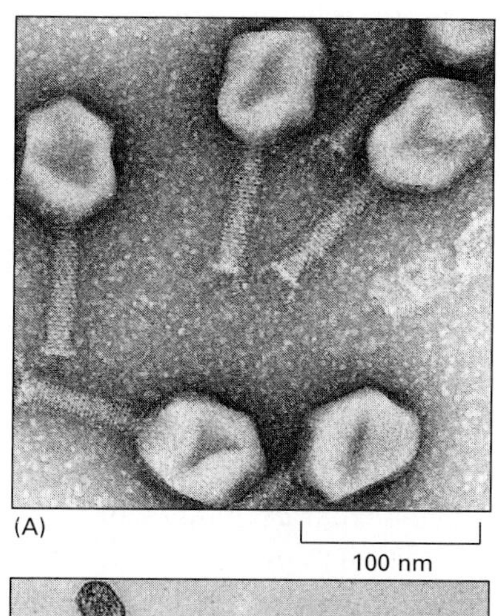

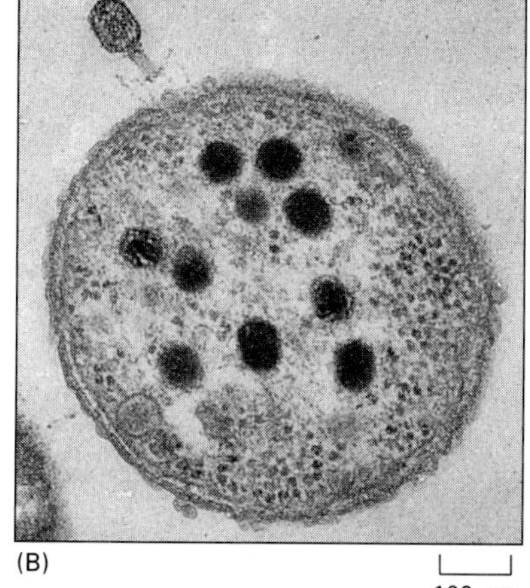

Figure 1–27 The viral transfer of DNA from one cell to another. (A) An electron micrograph of particles of a bacterial virus, the T4 bacteriophage. The head of this virus contains the viral DNA; the tail contains apparatus for injecting the DNA into a host bacterium. (B) A cross section of a bacterium with a T4 bacteriophage latched onto its surface. The large dark objects inside the bacterium are the heads of new T4 particles in course of assembly. When they are mature, the bacterium will burst open to release them. (A, courtesy of James Paulson; B, courtesy of Jonathan King and Erika Hartwig from G. Karp, Cell and Molecular Biology, 2nd edn. New York: John Wiley & Sons, 1999. © John Wiley.)

thereby capture the genetic information these molecules carry. This enables bacteria in the wild to acquire genes from neighboring cells relatively easily. Genes that confer resistance to an antibiotic or an ability to produce a toxin, for example, can be transferred from species to species and provide the recipient bacterium with a selective advantage. In this way, new and sometimes dangerous strains of bacteria have been observed to evolve in the bacterial ecosystems that inhabit hospitals or the various niches in the human body. For example, horizontal gene transfer is responsible for the spread over the past 40 years, of penicillin-resistant strains of *Neisseria gonorrheae,* the bacterium that causes gonorrhea. On a longer time scale, the results can be even more profound; it has been estimated that at least 18% of all of the genes in the present-day genome of *E. coli* have been acquired by horizontal transfer from another species within the past 100 million years.

Horizontal Exchanges of Genetic Information Within a Species Are Brought About by Sex

Horizontal exchanges of genetic information have an important role in bacterial evolution in today's world, and they may have occurred even more frequently and promiscuously in the early days of life on Earth. Indeed, it has been suggested that the genomes of present-day eubacteria, archaea, and eucaryotes originated not by divergent lines of descent from a single genome in a single ancestral type of cell, but rather as three independent anthologies of genes that have survived from the pool of genes in a primordial community of diverse cells in which genes were frequently exchanged (Figure 1–28). This could explain the otherwise puzzling observation that the eucaryotes seem more similar to archaea in their genes for the basic information-handling processes of DNA replication, transcription, and translation, but more similar to eubacteria in their genes for metabolic processes.

Horizontal gene transfer among bacteria may seem a surprising process, but it has a parallel in a phenomenon familiar to us all: sex. Sexual reproduction causes a large-scale horizontal transfer of genetic information between two initially separate cell lineages—those of the father and the mother. A key feature of sex, of course, is that the genetic exchange normally occurs only between individuals of the same species. But no matter whether they occur within a species or between species, horizontal gene transfers leave a characteristic imprint: they result in individuals who are related more closely to one set of relatives with respect to some genes, and more closely to another set of relatives with respect to others. By comparing the DNA sequences of individual human genomes, an intelligent visitor from outer space could deduce that humans reproduce sexually, even if it knew nothing about human behavior.

Sexual reproduction is a widespread (although not universal) phenomenon, especially among eucaryotes. Even bacteria indulge from time to time in controlled sexual exchanges of DNA with other members of their own species. Natural selection has clearly favored organisms that are capable of this behavior, although evolutionary theorists still dispute precisely what the selective advantage of sex is.

The Function of a Gene Can Often Be Deduced from Its Sequence

Family relationships among genes are important not just for their historical interest, but because they lead to a spectacular simplification in the task of

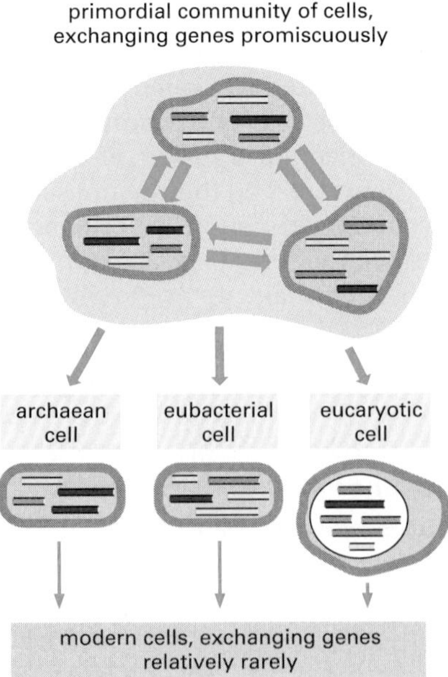

Figure 1–28 Horizontal gene transfers in early evolution. In the early days of life on Earth, cells may have been less capable of maintaining their separate identities and may have exchanged genes much more readily than now. In this way, the archaean, eubacterial, and eucaryotic lineages may have inherited different but overlapping subsets of genes from a primordial community of cells that were exchanging genes promiscuously.

deciphering gene functions. Once the sequence of a newly discovered gene has been determined, it is now possible, by tapping a few keys on a computer, to search the entire database of known gene sequences for genes related to it. In many cases, the function of one or more of these homologs will have been already determined experimentally, and thus, since gene sequence determines gene function, one can frequently make a good guess at the function of the new gene: it is likely to be similar to that of the already-known homologs.

In this way, it becomes possible to decipher a great deal of the biology of an organism simply by analyzing the DNA sequence of its genome and using the information we already have about the functions of genes in other organisms that have been more intensively studied. *Mycobacterium tuberculosis*, the eubacterium that causes tuberculosis, is extremely difficult to study experimentally in the laboratory and provides an example of the power of *comparative genomics*. DNA sequencing has revealed that this organism has a genome of 4,411,529 nucleotide pairs, containing approximately 4000 genes. Of these genes, 40% were immediately recognizable (when the genome was sequenced, in 1998) as homologs of known genes in other species, and could be tentatively assigned a function on that basis. Another 44% showed some informative similarity to other known genes—for example, containing a conserved protein domain within a longer amino acid sequence. Only 16% of the 4000 genes were totally unfamiliar. As we saw also for *Bacillus subtilis* (see Figure 1–24), about half the genes have sequences closely similar to those of other genes in the *M. tuberculosis* genome, showing that they must have arisen through relatively recent gene duplications. Compared with other bacteria, *M. tuberculosis* contains an exceptionally large number of genes coding for enzymes involved in the synthesis and degradation of lipid (fatty) molecules. This presumably reflects this bacterium's production of an unusual outer coat that is rich in these substances; the coat, and the enzymes that produce it, may explain how *M. tuberculosis* escapes destruction by the immune system of tuberculosis patients.

More Than 200 Gene Families Are Common to All Three Primary Branches of the Tree of Life

Given the complete genome sequences of representative organisms from all three domains—archaea, eubacteria, and eucaryotes—one can search systematically for homologies that span this enormous evolutionary divide. In this way we can begin to take stock of the common inheritance of all living things. There are considerable difficulties in this enterprise. For example, individual species have often lost some of the ancestral genes; other genes have probably been acquired by horizontal transfer from another species and therefore may not be truly ancestral, even though shared. Recent genome comparisons strongly suggest that both lineage-specific gene loss and horizontal gene transfer, in some cases between evolutionarily distant species, have been major factors of evolution, at least in the procaryotic world. Finally, in the course of 2 or 3 billion years, some genes that were initially shared will have changed beyond recognition by current methods.

Because of all these vagaries of the evolutionary process, it seems that only a small proportion of ancestral gene families have been universally retained in a recognizable form. Thus, out of 2264 protein-coding gene families recently defined by comparing the genomes of 18 bacteria, 6 archaeans and 1 eucaryote (yeast), only 76 are truly ubiquitous (that is, represented in all the genomes analyzed). The great majority of these universal families include components of the translation and transcription systems. This is not likely to be a realistic approximation of an ancestral gene set. A better—though still crude—idea of the latter can be obtained by tallying the gene families that have representatives in multiple, but not necessarily all, species from all three major kingdoms. Such an analysis reveals 239 ancient conserved families. With a single exception, these families can be assigned a function (at least in terms of general biochemical activity, but usually with more precision), with the largest number of shared gene families being involved in translation and ribosome production and in

TABLE 1–2 The Numbers of Gene Families, Classified by Function, That Are Common to All Three Domains of the Living World

GENE FAMILY FUNCTION	NUMBER OF "UNIVERSAL" FAMILIES
Translation, ribosomal structure and biogenesis	61
Transcription	5
Replication, repair, recombination	13
Cell division and chromosome partitioning	1
Molecule chaperones	9
Outer membrane, cell-wall biogenesis	3
Secretion	4
Inorganic ion transport	9
Signal transduction	1
Energy production and conversion	18
Carbohydrate metabolism and transport	14
Amino acid metabolism and transport	40
Nucleotide metabolism and transport	15
Coenzyme metabolism	23
Lipid metabolism	8
General biochemical function predicted; specific biological role unknown	33
Function unknown	1

For the purpose of this analysis, gene families are defined as "universal" if they are represented in the genomes of at least two diverse archaeans *(Archaeoglobus fulgidus* and *Aeropyrum pernix),* two evolutionarily distant bacteria *(Escherichia coli* and *Bacillus subtilis)* and one eucaryote (yeast, *Saccharomyces cerevisiae).* (Data from R.L. Tatusov, E.V. Koonin, and D.J. Lipman, *Science* 278:631–637, 1997; R.L. Tatusov, M.Y. Galperin, D.A. Natale, and E.V. Koonin, *Nucleic Acids Res.* 28:33–36, 2000; and E.V. Koonin, personal communication.)

amino acid metabolism and transport (Table 1–2). This set of highly conserved gene families represents only a very rough sketch of the common inheritance of all modern life; a more precise reconstruction of the gene complement of the last universal common ancestor might be feasible with further genome sequencing and more careful comparative analysis.

Mutations Reveal the Functions of Genes

Without additional information, no amount of gazing at genome sequences will reveal the functions of genes. We may recognize that gene B is like gene A, but how do we discover the function of gene A in the first place? And even if we know the function of gene A, how do we test whether the function of gene B is truly the same as the sequence similarity suggests? How do we make the connection between the world of abstract genetic information and the world of real living organisms?

The analysis of gene functions depends heavily on two complementary approaches: genetics and biochemistry. Genetics starts with the study of mutants: we either find or make an organism in which a gene is altered, and examine the effects on the organism's structure and performance (Figure 1–29). Biochemistry examines the functions of molecules: we extract molecules from an organism and then study their chemical activities. By putting genetics and biochemistry together and examining the chemical abnormalities in a mutant organism, it is possible to find those molecules whose production depends on a given gene. At the same time, studies of the performance of the mutant organism show us what role those molecules have in the operation of the organism as

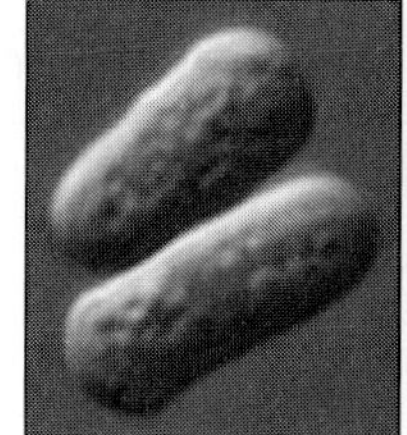
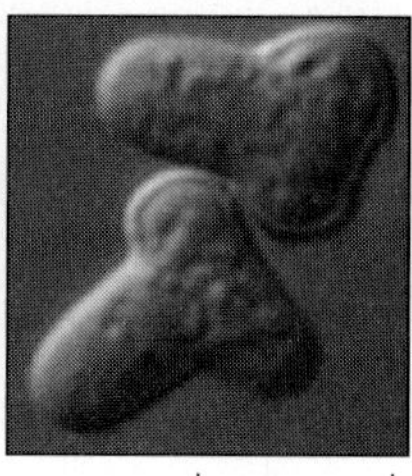

Figure 1–29 A mutant phenotype reflecting the function of a gene. A normal yeast (of the species *Schizosaccharomyces pombe*) is compared with a mutant in which a change in a single gene has converted the cell from a cigar shape *(left)* to a T shape *(right)*. The mutant gene therefore has a function in the control of cell shape. But how, in molecular terms, does the gene product perform that function? That is a harder question, and needs biochemical analysis to answer it. (Courtesy of Kenneth Sawin and Paul Nurse.)

a whole. Thus, genetics and biochemistry in combination provide a way to work out the connection between genes, molecules, and the structure and function of the organism.

In recent years, DNA sequence information and the powerful tools of molecular biology have allowed rapid progress. From sequence comparisons, one can often identify particular domains within a gene that have been preserved nearly unchanged over the course of evolution. These conserved domains are likely to be the most important parts of the gene in terms of function. We can test their individual contributions to the activity of the gene product by creating in the laboratory mutations of specific sites within the gene, or by constructing artificial hybrid genes that combine part of one gene with part of another. Organisms can be engineered to make either the RNA or the protein specified by the gene in large quantities to facilitate biochemical analysis. Specialists in molecular structure can determine the three-dimensional conformation of the gene product, revealing the exact position of every atom in it. Biochemists can determine how each of the parts of the genetically specified molecule contributes to its chemical behavior. Cell biologists can analyze the behavior of cells that are engineered to express a mutant version of the gene.

There is, however, no one simple recipe for discovering a gene's function, and no simple standard universal format for describing it. We may discover, for example, that the product of a given gene catalyzes a certain chemical reaction, and yet have no idea how or why that reaction is important to the organism. The functional characterization of each new family of gene products, unlike the description of the gene sequences, presents a fresh challenge to the biologist's ingenuity. Moreover, the function of a gene is never fully understood until we learn its role in the life of the organism as a whole. To make ultimate sense of gene functions, therefore, we have to study whole organisms, not just molecules or cells.

Molecular Biologists Have Focused a Spotlight on *E. coli*

Because living organisms are so complex, the more we learn about any particular species, the more attractive it becomes as an object for further study. Each discovery raises new questions and provides new tools with which to tackle questions in the context of the chosen organism. For this reason, large communities of biologists have become dedicated to studying different aspects of the same **model organism**.

In the enormously varied world of bacteria, the spotlight of molecular biology has for a long time focused intensely on just one species: *Escherichia coli*, or *E. coli* (see Figures 1–17 and 1–18). This small, rod-shaped eubacterial cell normally lives in the gut of humans and other vertebrates, but it can be grown easily in a simple nutrient broth in a culture bottle. Evolution has optimized it to cope with variable chemical conditions and to reproduce rapidly. Its genetic instructions are contained in a single, circular molecule of DNA that is 4,639,221 nucleotide-pairs long, and it makes approximately 4300 different kinds of proteins (Figure 1–30).

In molecular terms, we have a more thorough knowledge of the workings of *E. coli* than of any other living organism. Most of our understanding of the fundamental mechanisms of life—for example, how cells replicate their DNA to

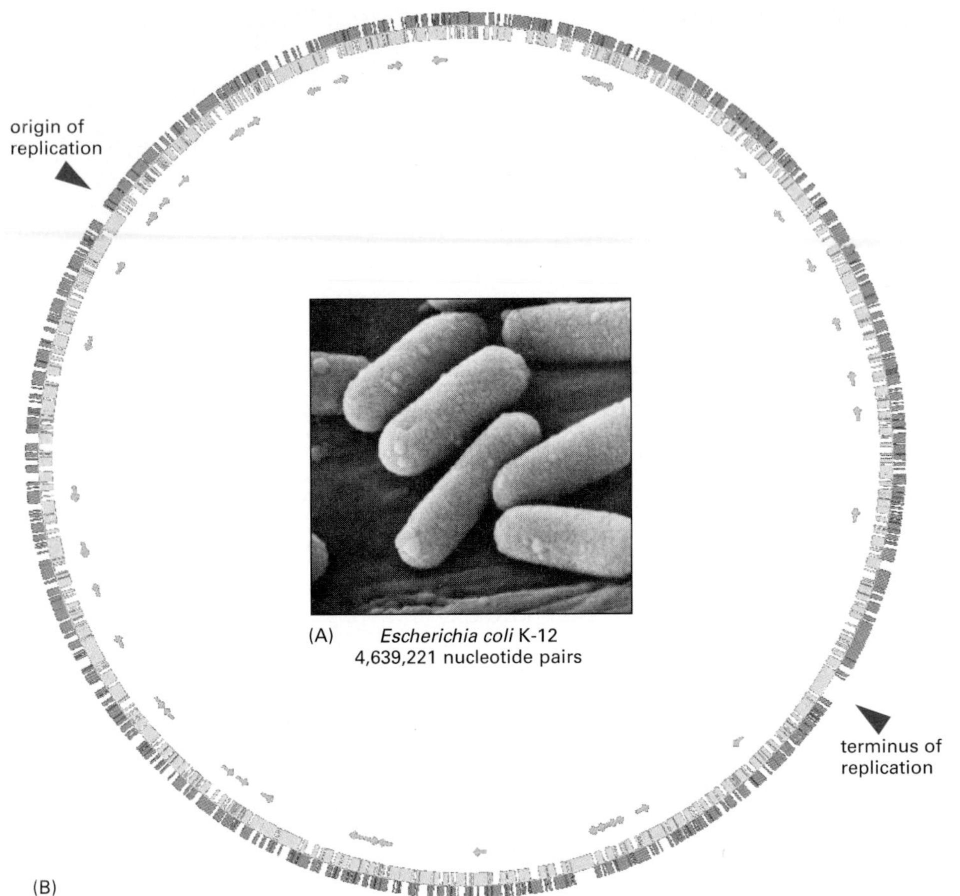

Figure 1–30 The genome of *E. coli*. (A) A cluster of *E. coli* cells. (B) A diagram of the *E. coli* genome of 4,639,221 nucleotide pairs (for *E. coli* strain K-12). The diagram is circular because the DNA of *E. coli*, like that of other procaryotes, forms a single, closed loop. Protein-coding genes are shown as *yellow* or *orange bars*, depending on the DNA strand from which they are transcribed; genes encoding only RNA molecules are indicated by *green arrows*. Some genes are transcribed from one strand of the DNA double helix (in a clockwise direction in this diagram), others from the other strand (counterclockwise). (A, courtesy of Tony Brain and the Science Photo Library; B, after F. R. Blattner et al., *Science* 277:1453–1462, 1997. © AAAS.)

pass on the genetic instructions to their progeny, or how they decode the instructions represented in the DNA to direct the synthesis of specific proteins—has come from studies of *E. coli*. The basic genetic mechanisms have turned out to be highly conserved throughout evolution: these mechanisms are therefore essentially the same in our own cells as in *E. coli*.

Summary

Procaryotes (cells without a distinct nucleus) are biochemically the most diverse organisms and include species that can obtain all their energy and nutrients from inorganic chemical sources, such as the reactive mixtures of minerals released at hydrothermal vents on the ocean floor—the sort of diet that may have nourished the first living cells 3.5 billion years ago. DNA sequence comparisons reveal the family relationships of living organisms and show that the procaryotes fall into two groups that diverged early in the course of evolution: the bacteria (or eubacteria) and the archaea. Together with the eucaryotes (cells with a membrane-bounded nucleus), these constitute the three primary branches of the tree of life. Most bacteria and archaea are small unicellular organisms with compact genomes comprising 1000–4000 genes. Many of the genes within a single organism show strong family resemblances in their DNA sequences, implying that they originated from the same ancestral gene through gene duplication and divergence. Family resemblances (homologies) are also clear when gene sequences are compared between different species, and more than 200 gene families have been so highly conserved that they can be recognized as common to all three domains of the living world. Thus, given the DNA sequence of a newly discovered gene, it is often possible to deduce the gene's function from the known function of a homologous gene in an intensively studied model organism, such as the bacterium **E. coli.**

GENETIC INFORMATION IN EUCARYOTES

Eucaryotic cells, in general, are bigger and more elaborate than procaryotic cells, and their genomes are bigger and more elaborate, too. The greater size is accompanied by radical differences in cell structure and function. Moreover, many classes of eucaryotic cells form multicellular organisms that attain a level of complexity unmatched by any procaryote.

Because they are so complex, eucaryotes confront molecular biologists with a special set of challenges, which will concern us in the rest of this book. Increasingly, biologists meet these challenges through the analysis and manipulation of the genetic information within cells and organisms. It is therefore important at the outset to know something of the special features of the eucaryotic genome. We begin by briefly reviewing how eucaryotic cells are organized, how this reflects their way of life, and how their genomes differ from those of procaryotes. This leads us to an outline of the strategy by which molecular biologists, by exploiting genetic information, are attempting to discover how eucaryotic organisms work.

Eucaryotic Cells May Have Originated as Predators

By definition, eucaryotic cells keep their DNA in a separate internal compartment, the nucleus. The DNA is separated from the cytoplasm by the *nuclear envelope,* which consists of a double layer of membrane. Eucaryotes also have other features that set them apart from procaryotes (Figure 1–31). Their cells are, typically, 10 times bigger in linear dimension, and 1000 times larger in volume. They have a *cytoskeleton*—a system of protein filaments crisscrossing the cytoplasm and forming, together with the many proteins that attach to them, a system of girders, ropes, and motors that gives the cell mechanical strength, controls its shape, and drives and guides its movements. The nuclear envelope is only one part of an elaborate set of *internal membranes,* each structurally similar to the plasma membrane and enclosing different types of spaces inside the cell, many of them involved in processes related to digestion and secretion. Lacking the tough cell wall of most bacteria, animal cells and the free-living eucaryotic cells called *protozoa* can change their shape rapidly and engulf other cells and small objects by *phagocytosis* (Figure 1–32).

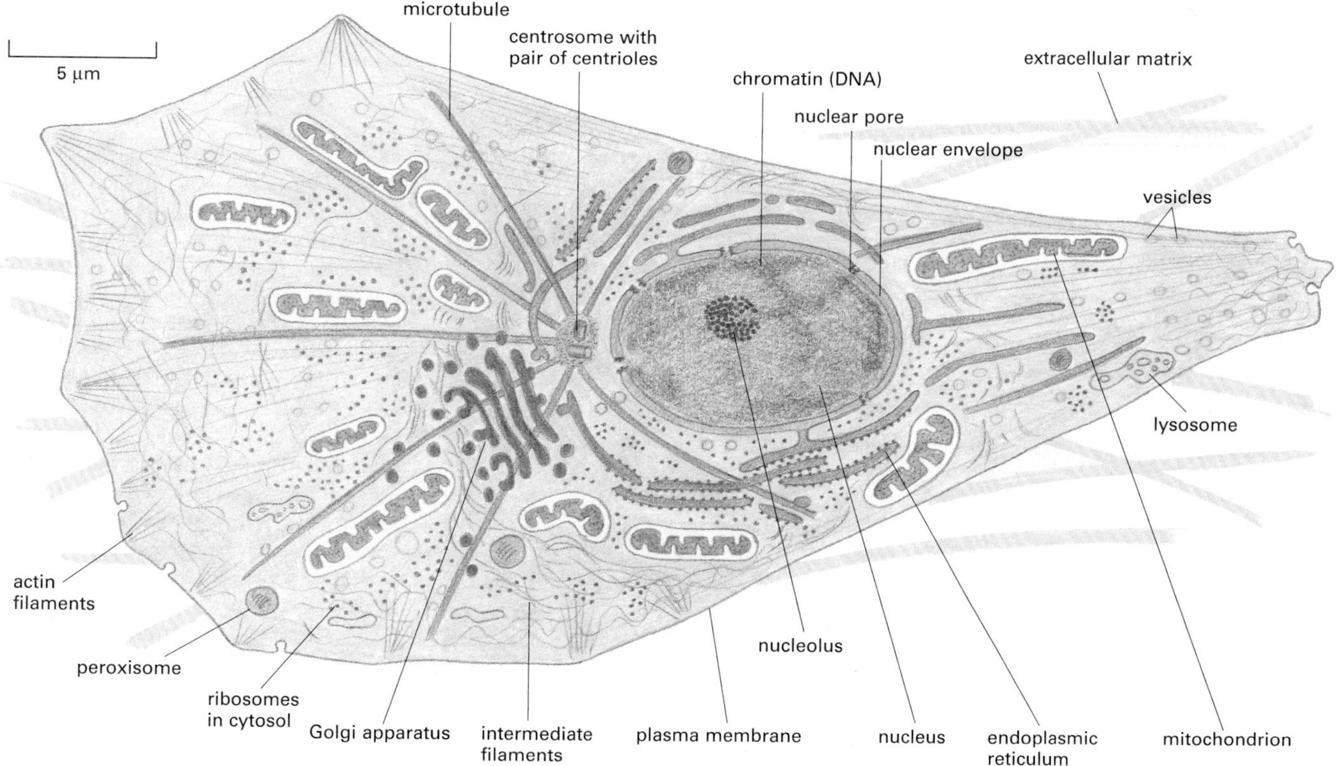

Figure 1–31 The major features of eucaryotic cells. The drawing depicts a typical animal cell, but almost all the same components are found in plants and fungi and in single-celled eucaryotes such as yeasts and protozoa. Plant cells contain chloroplasts in addition to the components shown here, and their plasma membrane is surrounded by a tough external wall formed of cellulose.

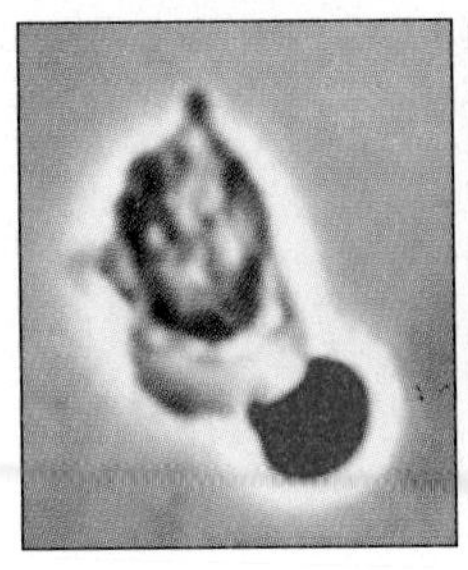
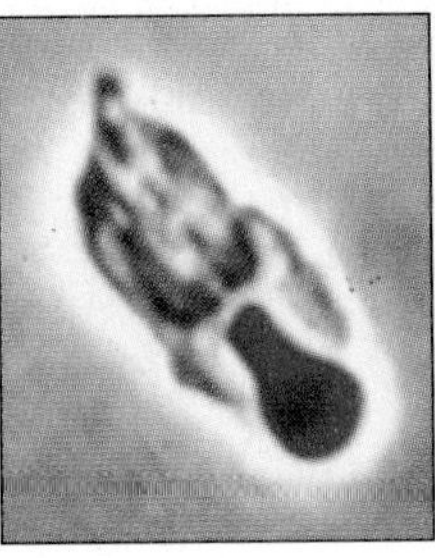
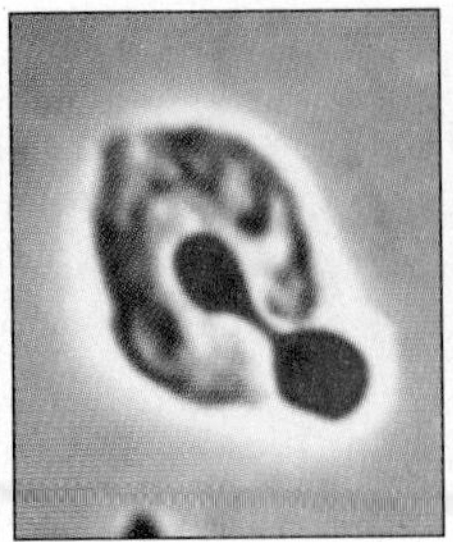
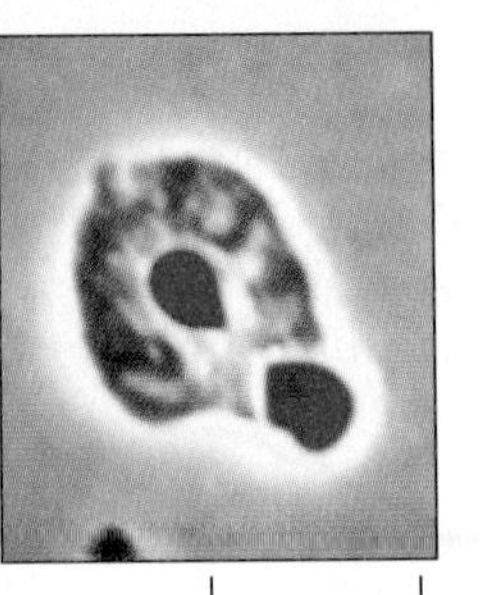

Figure 1–32 Phagocytosis. This series of stills from a movie shows a human white blood cell (a neutrophil) engulfing a red blood cell (artificially colored *red*) that has been treated with antibody. (Courtesy of Stephen E. Malawista and Anne de Boisfleury Chevance.)

It is still a mystery how all these properties evolved, and in what sequence. One plausible view, however, is that they are all reflections of the way of life of a primordial eucaryotic cell that was a predator, living by capturing other cells and eating them (Figure 1–33). Such a way of life requires a large cell with a flexible plasma membrane, as well as an elaborate cytoskeleton to support and move this membrane. It may also require that the cell's long, fragile DNA molecules be sequestered in a separate nuclear compartment, to protect the genome from damage by the movements of the cytoskeleton.

Eucaryotic Cells Evolved from a Symbiosis

A predatory way of life helps to explain another feature of eucaryotic cells. Almost all such cells also contain *mitochondria* (Figure 1–34). These are small bodies in the cytoplasm, enclosed by a double layer of membrane, that take up oxygen and harness energy from the oxidation of food molecules—such as sugars—to produce most of the ATP that powers the cell's activities. Mitochondria are similar in size to small bacteria, and, like bacteria, they have their own genome in the form of a circular DNA molecule, their own ribosomes that are different from those elsewhere in the eucaryotic cell, and their own transfer RNAs. It is virtually certain that mitochondria originated from free-living oxygen-metabolizing *(aerobic)* eubacteria that were engulfed by an ancestral eucaryotic cell that could otherwise make no such use of oxygen (that is, was *anaerobic)*. Escaping digestion, these bacteria evolved in symbiosis with the engulfing cell and its progeny, receiving shelter and nourishment in return for the power generation they performed for their hosts (Figure 1–35). This partnership between a primitive anaerobic eucaryotic predator cell and an aerobic bacterial cell is thought to have been established about 1.5 billion years ago, when the Earth's atmosphere first became rich in oxygen.

Many eucaryotic cells—specifically, those of plants and algae—also contain another class of small membrane-bounded organelles somewhat similar to

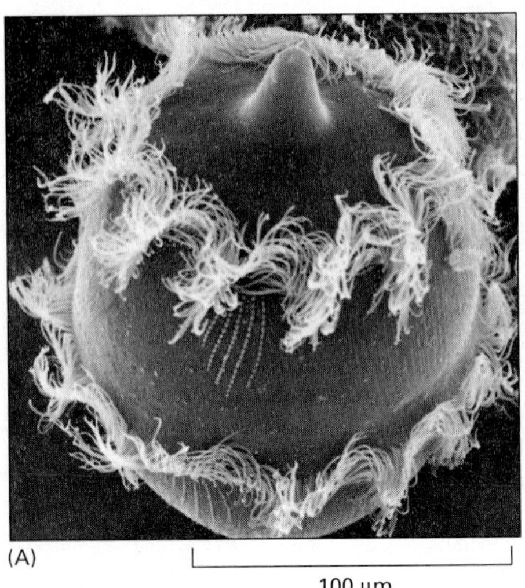

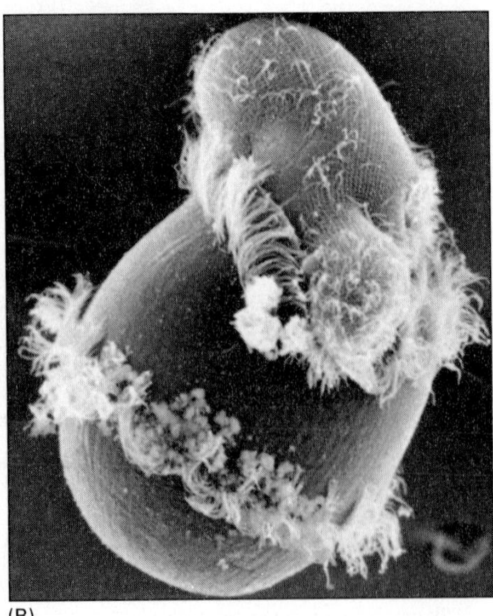

Figure 1–33 A single-celled eucaryote that eats other cells. (A) *Didinium* is a carnivorous protozoan, belonging to the group known as *ciliates*. It has a globular body, about 150 μm in diameter, encircled by two fringes of cilia—sinuous, whiplike appendages that beat continually; its front end is flattened except for a single protrusion, rather like a snout. (B) *Didinium* normally swims around in the water at high speed by means of the synchronous beating of its cilia. When it encounters a suitable prey, usually another type of protozoan, it releases numerous small paralyzing darts from its snout region. Then, the *Didinium* attaches to and devours the other cell by phagocytosis, inverting like a hollow ball to engulf its victim, which is almost as large as itself. (Courtesy of D. Barlow.)

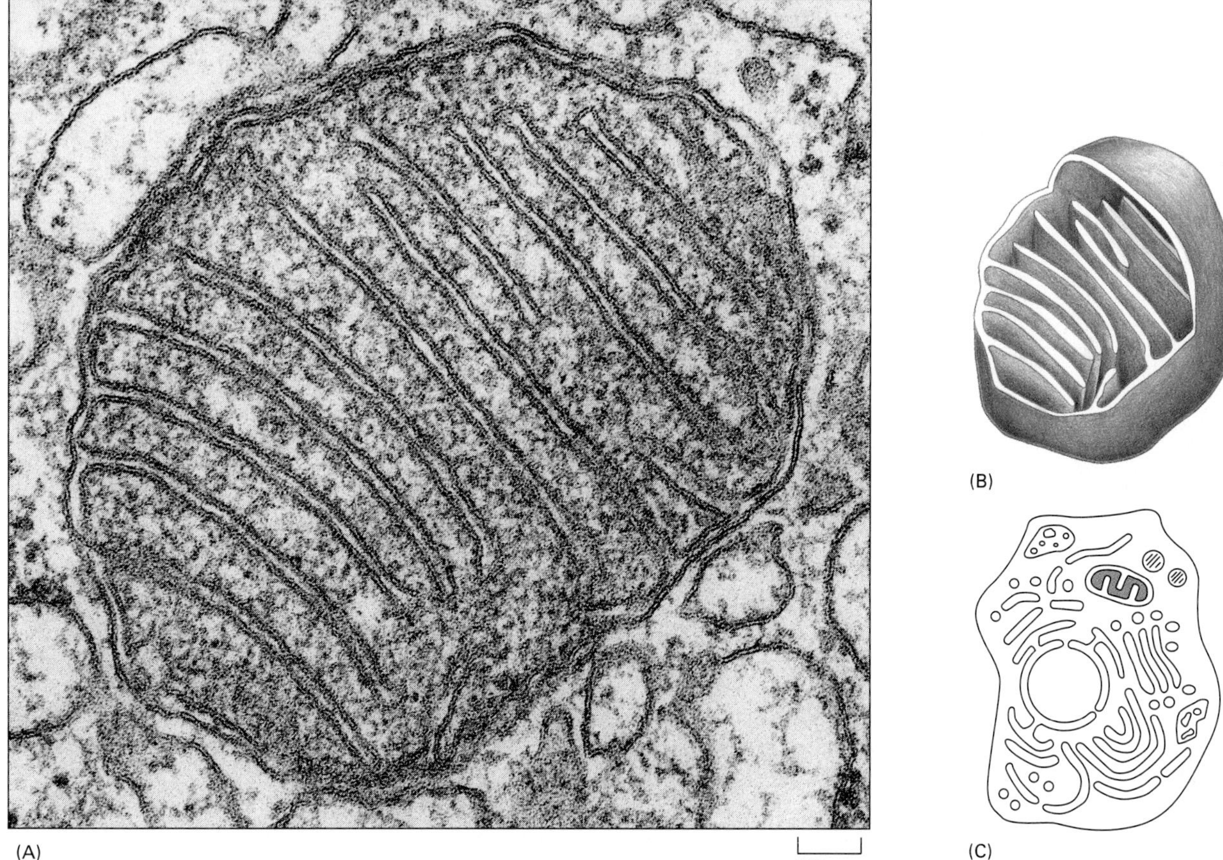

Figure 1–34 A mitochondrion. (A) A cross section, as seen in the electron microscope. (B) A drawing of a mitochondrion with part of it cut away to show the three-dimensional structure. (C) A schematic eucaryotic cell, with the interior space of a mitochondrion, containing the mitochondrial DNA and ribosomes, colored. Note the smooth outer membrane and the convoluted inner membrane, which houses the proteins that generate ATP from the oxidation of food molecules. (A, courtesy of Daniel S. Friend.)

mitochondria—the *chloroplasts* (Figure 1–36). Chloroplasts perform photosynthesis, using the energy of sunlight to synthesize carbohydrates from atmospheric carbon dioxide and water, and deliver the products to the host cell as food. Like mitochondria, chloroplasts have their own genome and almost certainly originated as symbiotic photosynthetic bacteria, acquired by cells that already possessed mitochondria (Figure 1–37).

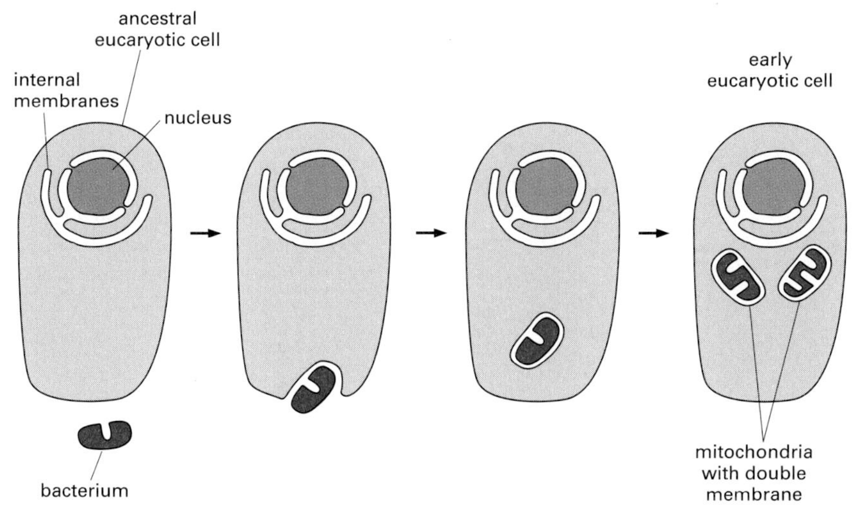

Figure 1–35 The origin of mitochondria. An ancestral eucaryotic cell is thought to have engulfed the bacterial ancestor of mitochondria, initiating a symbiotic relationship.

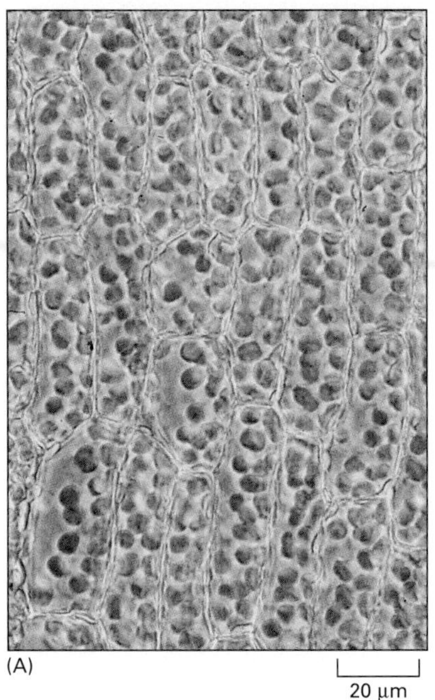

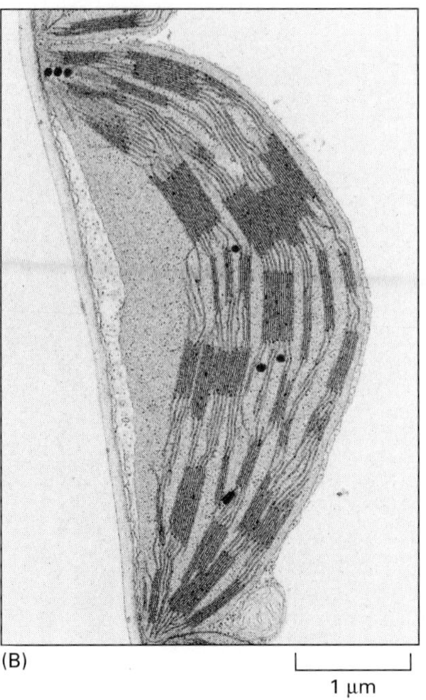

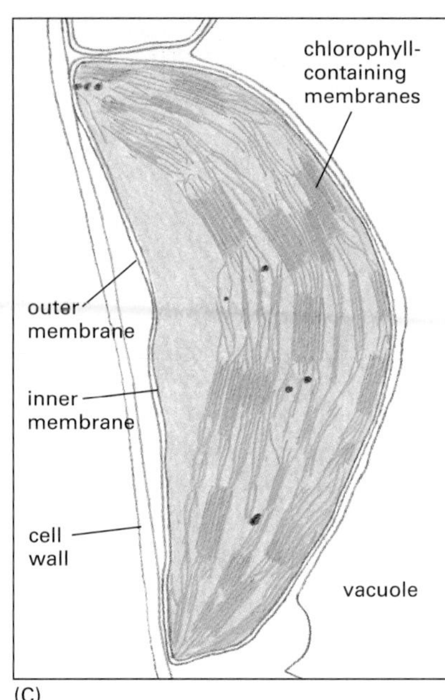

Figure 1–36 Chloroplasts. These organelles capture the energy of sunlight in plant cells and some single-celled eucaryotes. (A) Leaf cells in a moss seen in the light microscope, showing the green chloroplasts. (B) An electron micrograph of a chloroplast in a leaf of grass, showing the highly folded system of internal membranes containing the chlorophyll molecules by which light is absorbed. (C) An interpretative drawing of (B). (B, courtesy of Eldon Newcomb.)

A eucaryotic cell equipped with chloroplasts has no need to chase after other cells as prey; it is nourished by the captive chloroplasts it has inherited from its ancestors. Correspondingly, plant cells, although they possess the cytoskeletal equipment for movement, have lost the ability to change shape rapidly and to engulf other cells by phagocytosis. Instead, they create around themselves a tough, protective cell wall. If the ancestral eucaryote was indeed a predator on other organisms, we can view plant cells as eucaryotes that have made the transition from hunting to farming.

Fungi represent yet another eucaryotic way of life. Fungal cells, like animal cells, possess mitochondria but not chloroplasts; but in contrast with animal cells and protozoa, they have a tough outer wall that limits their ability to move rapidly or to swallow up other cells. Fungi, it seems, have turned from hunters into scavengers: other cells secrete nutrient molecules or release them upon death, and fungi feed on these leavings—performing whatever digestion is necessary extracellularly, by secreting digestive enzymes to the exterior.

Eucaryotes Have Hybrid Genomes

The genetic information of eucaryotic cells has a hybrid origin—from the ancestral anaerobic eucaryote, and from the bacteria that it adopted as symbionts. Most of this information is stored in the nucleus, but a small amount remains

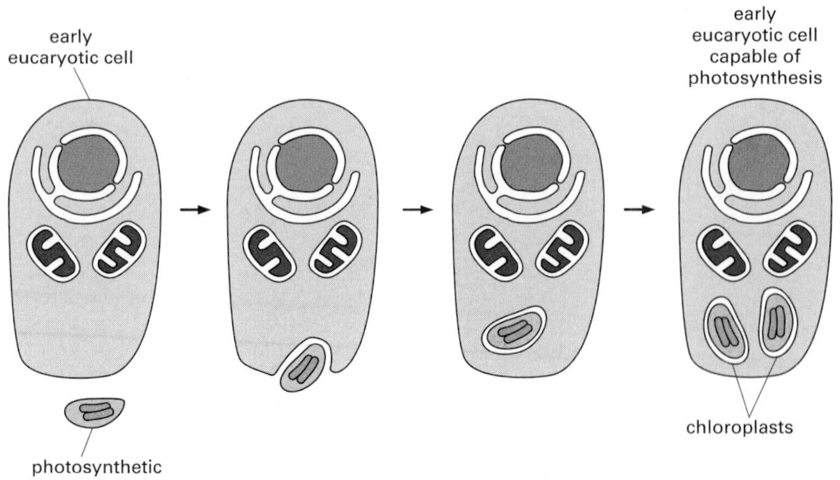

Figure 1–37 The origin of chloroplasts. An early eucaryotic cell, already possessing mitochondria, engulfed a photosynthetic bacterium (a cyanobacterium) and retained it in symbiosis. All present-day chloroplasts are thought to trace their ancestry back to a single species of cyanobacterium that was adopted as an internal symbiont (an endosymbiont) over a billion years ago.

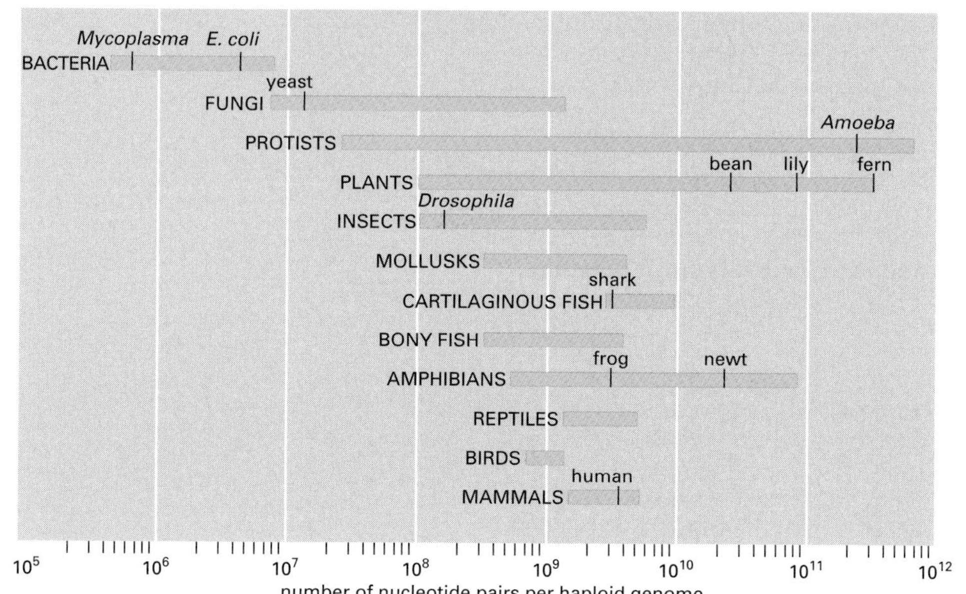

Figure 1–38 Genome sizes compared. Genome size is measured in nucleotide pairs of DNA per haploid genome, that is, per single copy of the genome. (The cells of sexually reproducing organisms such as ourselves are generally diploid: they contain two copies of the genome, one inherited from the mother, the other from the father.) Closely related organisms can vary widely in the quantity of DNA in their genomes, even though they contain similar numbers of functionally distinct genes. (Data from W.-H. Li, Molecular Evolution, pp. 380–383. Sunderland, MA: Sinauer, 1997.)

inside the mitochondria and, for plant and algal cells, in the chloroplasts. The mitochondrial DNA and the chloroplast DNA can be separated from the nuclear DNA and individually analyzed and sequenced. The mitochondrial and chloroplast genomes are found to be degenerate, cut-down versions of the corresponding bacterial genomes, lacking genes for many essential functions. In a human cell, for example, the mitochondrial genome consists of only 16,569 nucleotide pairs, and codes for only 13 proteins, two ribosomal RNA components, and 22 transfer RNAs.

The genes that are missing from the mitochondria and chloroplasts have not all been lost; instead, many of them have been somehow moved from the symbiont genome into the DNA of the host cell nucleus. The nuclear DNA of humans contains many genes coding for proteins that serve essential functions inside the mitochondria; in plants, the nuclear DNA also contains many genes specifying proteins required in chloroplasts.

Eucaryotic Genomes Are Big

Natural selection has evidently favored mitochondria with small genomes, just as it has favored bacteria with small genomes. By contrast, the nuclear genomes of most eucaryotes seem to have been free to enlarge. Perhaps the eucaryotic way of life has made large size an advantage: predators need to be bigger than their prey, and cell size generally increases in proportion to genome size. Whatever the explanation, the fact is that the genomes of most eucaryotes are orders of magnitude larger than those of bacteria and archaea (Figure 1–38). And the freedom to be extravagant with DNA has had profound implications.

Eucaryotes not only have more genes than procaryotes; they also have vastly more DNA that does not code for protein or for any other functional product molecule. The human genome contains a 1000 times as many nucleotide pairs as the genome of a typical bacterium, 20 times as many genes, and about 10,000 times as much noncoding DNA (~98.5% of the genome for a human is noncoding, as opposed to 11% of the genome for the bacterium *E. coli*).

Figure 1–39 The puffer fish *(Fugu rubripes)*. This organism has a genome size of 400 million nucleotide pairs—about one-quarter as much as a zebrafish, for example, even though the two species of fish have similar numbers of genes. (From a woodcut by Hiroshige, courtesy of Arts and Designs of Japan.)

Eucaryotic Genomes Are Rich in Regulatory DNA

Much of our noncoding DNA is almost certainly dispensable junk, retained like a mass of old papers because, when there is little pressure to keep an archive small, it is easier to retain everything than to sort out the valuable information and discard the rest. Certain exceptional eucaryotic species, such as the puffer fish (Figure 1–39), bear witness to the profligacy of their relatives; they have somehow managed either to rid themselves of large quantities of noncoding DNA, or to have avoided acquiring it in the first place. Yet they appear similar in structure, behavior, and fitness to related species that have vastly more such DNA.

Even in compact eucaryotic genomes such as that of puffer fish, there is more noncoding DNA than coding DNA, and at least some of the noncoding DNA certainly has important functions. In particular, it serves to regulate the expression of adjacent genes. With this regulatory DNA, eucaryotes have evolved distinctive ways of controlling when and where a gene is brought into play. This sophisticated gene regulation is crucial for the formation of complex multicellular organisms.

The Genome Defines the Program of Multicellular Development

The cells in an individual animal or plant are extraordinarily varied. Fat cells, skin cells, bone cells, nerve cells—they seem as dissimilar as any cells could be. Yet all these cell types are generated during embryonic development from a single fertilized egg cell, and all (with minor exceptions) contain identical copies of the genome of the species.

The explanation lies in the way in which these cells make selective use of their genetic instructions according to the cues they get from their surroundings. The DNA is not just a shopping list specifying the molecules that every cell must have, and the cell is not an assembly of all the items on the list. Rather, the cell behaves as a multipurpose machine, with sensors to receive environmental signals and highly developed abilities to call different sets of genes into action according to the sequences of signals to which the cell has been exposed. A large fraction of the genes in the eucaryotic genome code for proteins that serve to regulate the activities of other genes. Most of these *gene regulatory proteins* act by binding, directly or indirectly, to the regulatory DNA adjacent to the genes that are to be controlled (Figure 1–40), or by interfering with the abilities of other proteins to do so. The expanded genome of eucaryotes therefore serves not only to specify the hardware of the cell, but also to store the software that controls how that hardware is used.

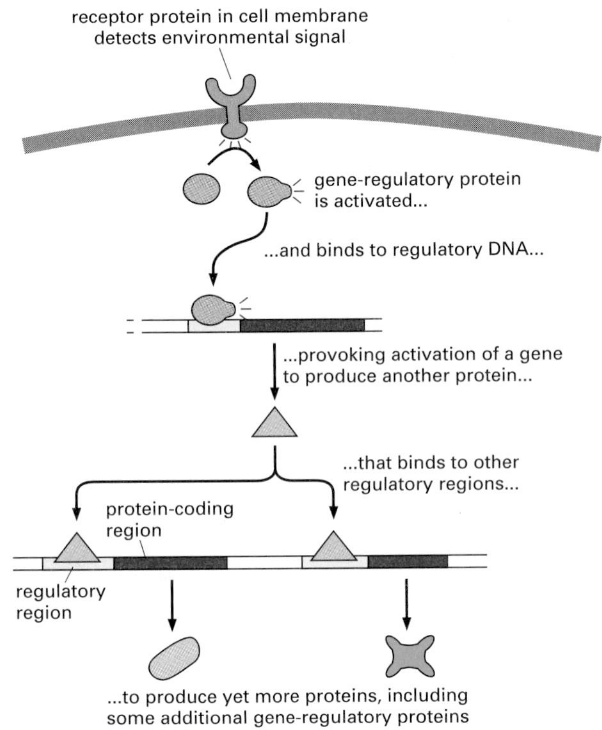
receptor protein in cell membrane detects environmental signal
gene-regulatory protein is activated...
...and binds to regulatory DNA...
...provoking activation of a gene to produce another protein...
...that binds to other regulatory regions...
protein-coding region
regulatory region
...to produce yet more proteins, including some additional gene-regulatory proteins

Figure 1–40 Controlling gene readout by environmental signals. Regulatory DNA allows gene expression to be controlled by regulatory proteins, which are in turn the products of other genes. This diagram shows how a cell's gene expression is adjusted according to a signal from the cell's environment. The initial effect of the signal is to activate a regulatory protein already present in the cell; the signal may, for example, trigger the attachment of a phosphate group to the regulatory protein, altering its chemical properties.

Figure 1–41 Genetic control of the program of multicellular development. The role of a regulatory gene is demonstrated in the snapdragon *Antirrhinum*. In this example, a mutation in a single gene coding for a regulatory protein causes leafy shoots to develop in place of flowers: because a regulatory protein has been changed, the cells adopt characters that would be appropriate to a different location in the normal plant. The mutant is on the left, the normal plant on the right. (Courtesy of Enrico Coen and Rosemary Carpenter.)

In a developing multicellular organism, each cell is governed by the same control system, but with different consequences depending on the signals the cell exchanges with its neighbors. The outcome, astonishingly, is a precisely patterned array of cells in different states, each displaying a character appropriate to its position in the multicellular structure. The genome in each cell is big enough to accommodate the information that specifies an entire multicellular organism, but in any individual cell only part of that information is used (Figure 1–41).

Many Eucaryotes Live as Solitary Cells: the Protists

Many species of eucaryotic cells lead a solitary life—some as hunters (the *protozoa*), some as photosynthesizers (the unicellular *algae*), some as scavengers (the unicellular fungi, or *yeasts*). Even though they do not form multicellular organisms, these single-celled eucaryotes, or *protists,* can be very complex. Figure 1–42 conveys something of the variety of forms they can take. The anatomy of protozoa, especially, is often elaborate and includes such structures as sensory bristles, photoreceptors, sinuously beating cilia, leglike appendages, mouth parts, stinging darts, and musclelike contractile bundles. Although they are single cells, protozoa can be as intricate, as versatile, and as complex in their behavior as many multicellular organisms (see Figure 1–33).

In terms of their ancestry and DNA sequences, protists are far more diverse than the multicellular animals, plants, and fungi, which arose as three comparatively late branches of the eucaryotic pedigree (see Figure 1–21). As with procaryotes, humans have tended to neglect the protists because they are microscopic. Only now, with the help of genome analysis, are we beginning to understand their positions in the tree of life, and to put into context the glimpses these strange creatures offer us of our distant evolutionary past.

A Yeast Serves as a Minimal Model Eucaryote

The molecular and genetic complexity of eucaryotes is daunting. Even more than for procaryotes, biologists need to concentrate their limited resources on a few selected model organisms to fathom this complexity.

To analyze the internal workings of the eucaryotic cell, without being distracted by the additional problems of multicellular development, it makes sense to use a species that is unicellular and as simple as possible. The popular choice for this role of minimal model eucaryote has been the yeast *Saccharomyces cerevisiae* (Figure 1–43)—the same species that is used by brewers of beer and bakers of bread.

S. cerevisiae is a small, single-celled member of the kingdom of fungi and thus, according to modern views, at least as closely related to animals as it is to plants. It is robust and easy to grow in a simple nutrient medium. Like other fungi, it has a tough cell wall, is relatively immobile, and possesses mitochondria but not chloroplasts. When nutrients are plentiful, it grows and divides almost as rapidly as a bacterium. It can reproduce either vegetatively (that is, by simple cell division), or sexually: two yeast cells that are *haploid* (possessing a single copy of the genome) can fuse to create a cell that is *diploid* (containing a double genome); and the diploid cell can undergo *meiosis* (a reduction division) to produce cells that are once again haploid (Figure 1–44). In contrast with higher plants and animals, the yeast can divide indefinitely in either the haploid or the diploid state, and the process leading from the one state to the other can be induced at will by changing the growth conditions.

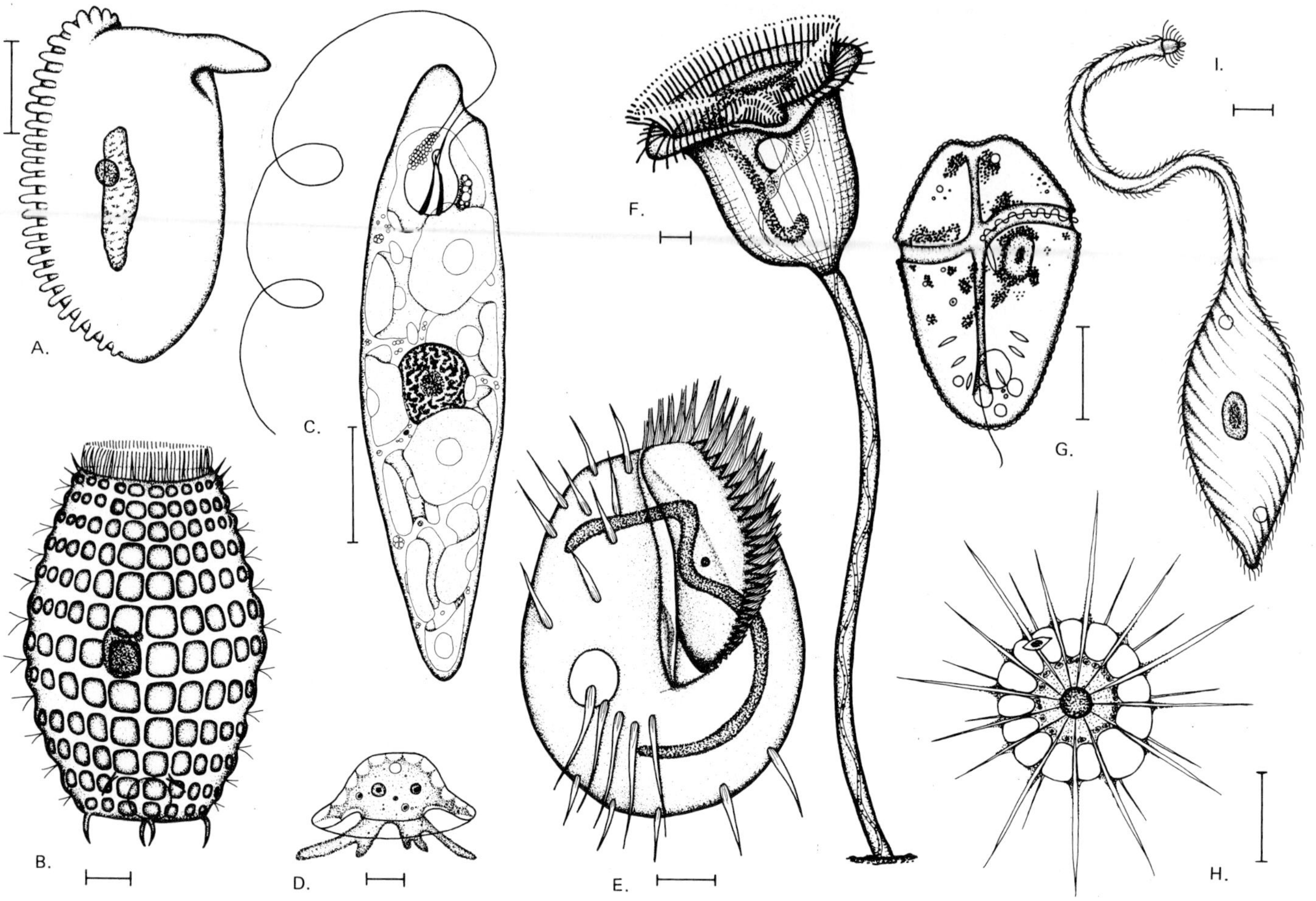

All these features make this yeast an extremely convenient organism for genetic studies, as does one further property: its genome, by eucaryotic standards,

Figure 1–42 An assortment of protists: a small sample of an extremely diverse class of organisms. The drawings are done to different scales, but in each case the scale bar represents 10 µm. The organisms in (A), (B), (E), (F), and (I) are ciliates; (C) is a euglenoid; (D) is an amoeba; (G) is a dinoflagellate; (H) is a heliozoan. (From M.A. Sleigh, Biology of Protozoa. Cambridge, UK: Cambridge University Press, 1973.)

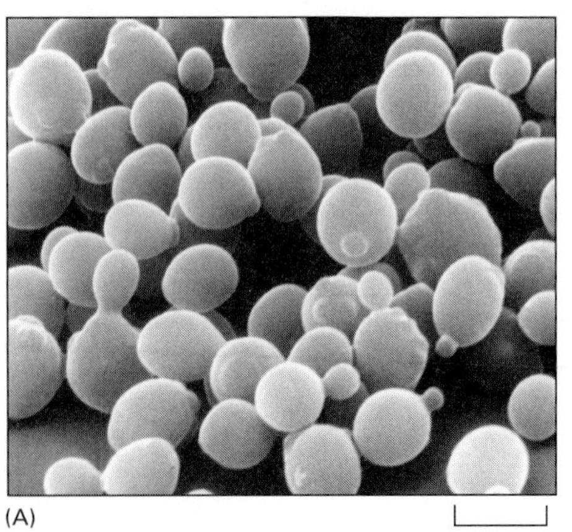

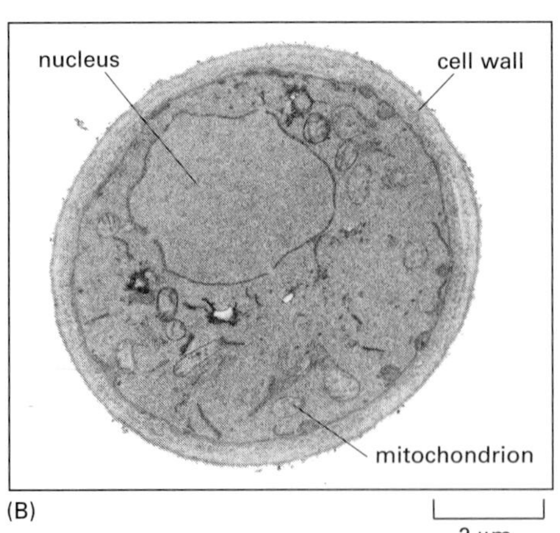

Figure 1–43 The yeast *Saccharomyces cerevisiae*. (A) A scanning electron micrograph of a cluster of the cells. This species is also known as budding yeast; it proliferates by forming a protrusion or bud that enlarges and then separates from the rest of the original cell. Many cells with buds are visible in this micrograph. (B) A transmission electron micrograph of a cross section of a yeast cell, showing its nucleus, mitochondrion, and thick cell wall. (A, courtesy of Ira Herskowitz and Eric Schabatach.)

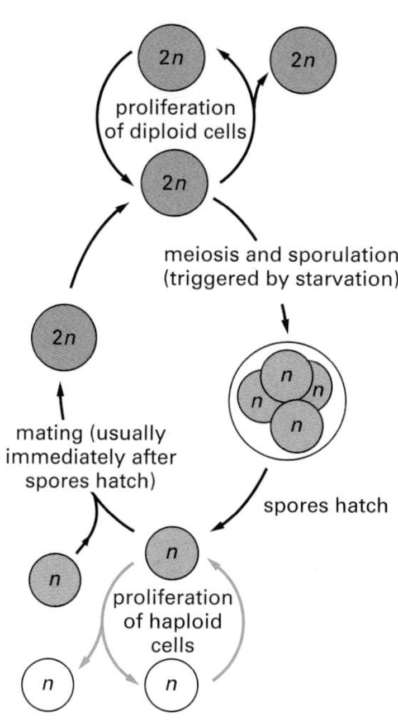

Figure 1–44 The reproductive cycles of the yeast *S. cerevisiae*. Depending on environmental conditions and on details of the genotype, cells of this species can exist in either a diploid state, with a double chromosome set, or a haploid state, with a single chromosome set. The diploid form can either proliferate by ordinary cell-division cycles or undergo meiosis to produce haploid cells. The haploid form can either proliferate by ordinary cell-division cycles or undergo sexual fusion with another haploid cell to become diploid. Meiosis is triggered by starvation and gives rise to spores—haploid cells in a dormant state, resistant to harsh environmental conditions.

is exceptionally small. Nevertheless, it suffices for all the basic tasks that every eucaryotic cell must perform. As we shall see later in this book, studies on yeasts (using both *S. cerevisiae* and other species) have provided a key to the understanding of many crucial processes. These include the eucaryotic cell-division cycle—the critical chain of events by which the nucleus and all the other components of a cell are duplicated and parceled out to create two daughter cells from one. The control system that governs this process has been so well conserved over the course of evolution that many of its components can function interchangeably in yeast and human cells: if a mutant yeast lacking an essential yeast cell-division-cycle gene is supplied with a copy of the homologous cell-division-cycle gene from a human, the yeast is cured of its defect and becomes able to divide normally.

The Expression Levels of All The Genes of An Organism Can Be Monitored Simultaneously

The complete genome sequence of *S. cerevisiae* was determined in 1997. It consists of approximately 13,117,000 nucleotide pairs, including the small contribution (78,520 nucleotide pairs) of the mitochondrial DNA. This total is only about 2.5 times as much DNA as there is in *E. coli*, and it codes for only 1.5 times as many distinct proteins (about 6300 in all). The way of life of *S. cerevisiae* is similar in many ways to that of a bacterium, and it seems that this yeast has likewise been subject to selection pressures that have kept its genome compact.

Knowledge of the complete genome sequence of any organism—be it a yeast or a human—opens up new perspectives on the workings of the cell: things that once seemed impossibly complex now seem within our grasp. One example will suffice. It is possible to array on a glass microscope slide a set of samples of DNA sequence corresponding to each of the ~6300 proteins the yeast has encoded in its genome. A DNA molecule will bind selectively, by base-pairing, with DNA or RNA that has a complementary sequence. Therefore, if a mixture of nucleic acid molecules is passed over such a *DNA array*, each type of molecule in the mixture will stick to its corresponding DNA spot. If the molecules in the mixture have a fluorescent chemical tag attached to them, the result will be a set of spots of fluorescence, each localized at a specific point in the array (Figure 1–45). In this way, it is possible to monitor, simultaneously, the amount of mRNA transcript that is produced from every gene in the genome by a yeast under any chosen conditions, and to see how this whole pattern of gene activity changes when conditions change. The analysis can be repeated with mRNA prepared from mutants lacking a chosen gene—any gene that we care to test. In principle, this approach provides a way to reveal the entire system of control relationships that govern gene expression—not only in these yeast cells, but in any organism whose genome sequence is known.

Arabidopsis Has Been Chosen Out of 300,000 Species As a Model Plant

The large multicellular organisms that we see around us—the flowers and trees and animals—seem fantastically varied, but they are much closer to one another in their evolutionary origins, and more similar in their basic cell biology, than the great host of microscopic single-celled organisms. Thus, while bacteria and

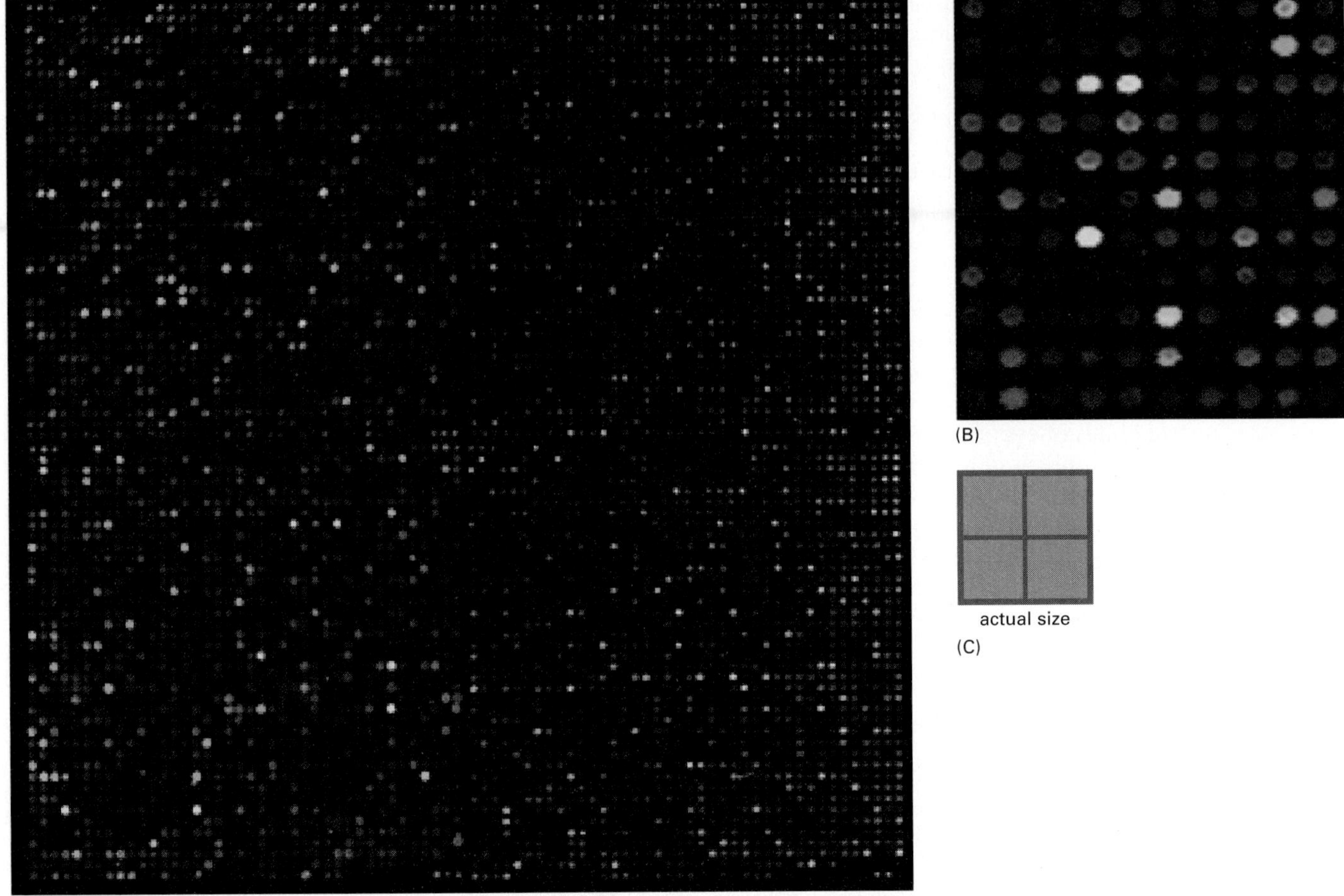

Figure 1–45 Monitoring changes in yeast gene expression using a DNA array. Samples of DNA sequence corresponding to each of the ~6300 protein-coding genes in the yeast genome are deposited in a square array on a glass microscope slide: (A) shows the entire array of 6400 spots; (B) shows a small region at higher magnification; (C) shows the actual size of array (A). The array has been used here to compare the sets of mRNA molecules produced by yeast cells with and without glucose in their culture medium. If the cells supplied with glucose transcribe a particular gene into mRNA, a preparation of mRNA from these cells will contain molecules that can bind selectively to the corresponding DNA spot in the array. In practice, the mRNA extracted from the cells is not applied directly to the array but is used (as we explain in Chapter 8) to generate another mixture of nucleic acid molecules, called cDNA molecules, that correspond in sequence to the mRNA molecules. In this experiment, they are tagged with a chemical label that fluoresces green. When the cDNA mixture is applied to the array, each cDNA in the mixture binds to its corresponding spot in the array: a bright *green* fluorescence at a given spot indicates strong expression of a specific gene. The whole procedure is now repeated with yeast cells deprived of glucose, tagging their cDNAs with a *red* fluorescent label and applying it to the same DNA array. Therefore, a *green* spot indicates a gene expressed more strongly in the presence of glucose than in its absence; a *red* spot indicates the opposite. Spots in the array corresponding to genes that are expressed equally in the two conditions bind both green and red cDNA molecules equally and therefore glow *yellow*. (From J.L. DeRisi et al., *Science* 278:680–686, 1997. © AAAS.)

eucaryotes are separated by more than 3000 million years of divergent evolution, vertebrates and insects are separated by about 700 million years, fish and mammals by about 450 million years, and the different species of flowering plants by only about 150 million years.

Because of the close evolutionary relationship between all flowering plants, we can, once again, get insight into the cell and molecular biology of this whole class of organisms by focusing on just one or a few species for detailed analysis. Out of the several hundred thousand species of flowering plants on Earth today, molecular biologists have chosen to concentrate their efforts on a small weed, the common wall cress *Arabidopsis thaliana* (Figure 1–46), which can be grown indoors in large numbers, and produces thousands of offspring per plant after

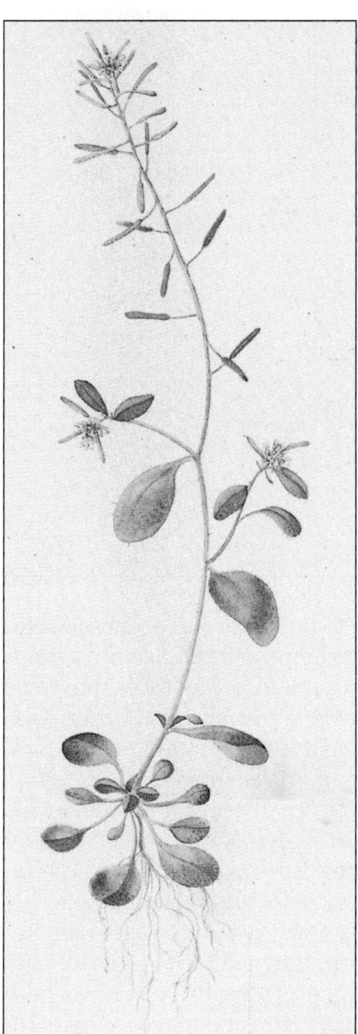

Figure 1–46 *Arabidopsis thaliana*, the plant chosen as the primary model for studying plant molecular genetics. (Courtesy of Toni Hayden and the John Innes Foundation.)

8–10 weeks. *Arabidopsis* has a genome of approximately 140 million nucleotide pairs, about 11 times as much as yeast, and its complete sequence is known.

The World of Animal Cells Is Represented By a Worm, a Fly, a Mouse, and a Human

Multicellular animals account for the majority of all named species of living organisms, and for the largest part of the biological research effort. Four species have emerged as the foremost model organisms for molecular genetic studies and have had their genomes sequenced. In order of increasing size, they are the nematode worm *Caenorhabditis elegans*, the fly *Drosophila melanogaster*, the mouse *Mus musculus*, and the human, *Homo sapiens*.

Caenorhabditis elegans (Figure 1–47) is a small, harmless relative of the eelworm that attacks crops. With a life cycle of only a few days, an ability to survive in a freezer indefinitely in a state of suspended animation, a simple body plan, and an unusual life cycle that is well suited for genetic studies (described in Chapter 21), it is an ideal model organism. *C. elegans* develops with clockwork precision from a fertilized egg cell into an adult worm with exactly 959 body cells (plus a variable number of egg and sperm cells)—an unusual degree of regularity for an animal. We now have a minutely detailed description of the sequence of events by which this occurs, as the cells divide, move, and change their characters according to strict and predictable rules. The genome of 97 million nucleotide pairs code for about 19,000 proteins and a wealth of mutants are available for the testing of gene functions. Although the worm has a body plan very different from our own, the conservation of biological mechanisms has been sufficient for the worm to provide a valuable model for many of the processes that occur in the human body. Studies of the worm help us to understand, for example, the programs of cell division and cell death that determine the numbers of cells in the body—a topic of great importance in developmental biology and cancer research.

Studies in *Drosophila* Provide a Key to Vertebrate Development

The fruitfly *Drosophila melanogaster* (Figure 1–48) has been in use as a model genetic organism for longer than any other; in fact, the foundations of classical genetics were built to a large extent on studies of this insect. Over 80 years ago, it provided, for example, definitive proof that genes—the abstract units of hereditary information—are carried on chromosomes, concrete physical objects

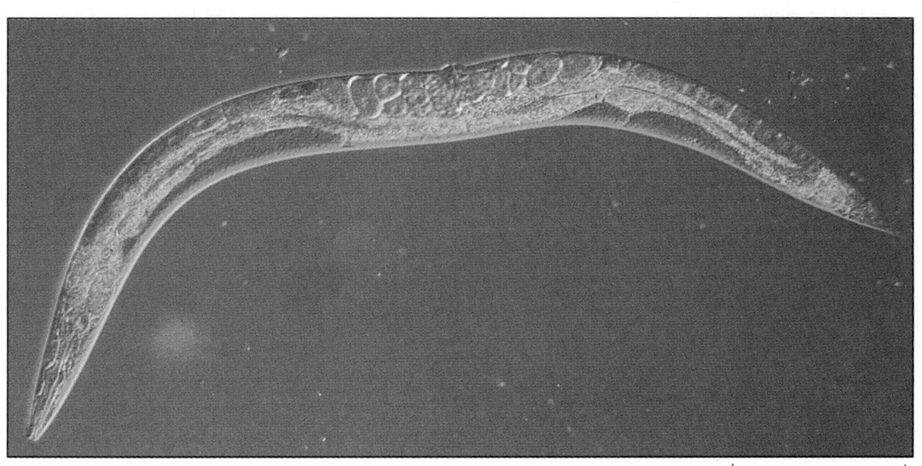

Figure 1–47 *Caenorhabditis elegans*, the first multicellular organism to have its complete genome sequence determined. This small nematode, about 1 mm long, lives in the soil. Most individuals are hermaphrodites, producing both eggs and sperm. (Courtesy of Ian Hope.)

Figure 1–48 ***Drosophila melanogaster.*** Molecular genetic studies on this fly have provided the main key to understanding how all animals develop from a fertilized egg into an adult. (From E.B. Lewis, *Science* 221:cover, 1983. © AAAS.)

whose behavior had been closely followed in the eucaryotic cell with the light microscope, but whose function was at first unknown. The proof depended on one of the many features that make *Drosophila* peculiarly convenient for genetics—the giant chromosomes, with characteristic banded appearance, that are visible in some of its cells (Figure 1–49). Specific changes in the hereditary information, manifest in families of mutant flies, were found to correlate exactly with the loss or alteration of specific giant-chromosome bands.

In more recent times, *Drosophila*, more than any other organism, has shown us how to trace the chain of cause and effect from the genetic instructions encoded in the chromosomal DNA to the structure of the adult multicellular body. *Drosophila* mutants with body parts strangely misplaced or mispatterned provided the key to the identification and characterization of the genes required to make a properly structured body, with gut, limbs, eyes, and all the other parts in their correct places. Once these *Drosophila* genes were sequenced, the genomes of vertebrates could be scanned for homologs. These were found, and their functions in vertebrates were then tested by analyzing mice in which the genes had been mutated. The results, as we see later in the book, reveal an astonishing degree of similarity in the molecular mechanisms of insect and vertebrate development.

The majority of all named species of living organisms are insects. Even if *Drosophila* had nothing in common with vertebrates, but only with insects, it would still be an important model organism. But if understanding the molecular genetics of vertebrates is the goal, why not simply tackle the problem head-on? Why sidle up to it obliquely, through studies in *Drosophila*?

Drosophila requires only 9 days to progress from a fertilized egg to an adult; it is vastly easier and cheaper to breed than any vertebrate, and its genome is much smaller—about 170 million nucleotide pairs, compared with 3200 million for a human. This codes for about 14,000 proteins, and mutants can now be obtained for essentially any gene. But there is also another, deeper reason why genetic mechanisms that are hard to discover in a vertebrate are often readily revealed in the fly. This relates, as we now explain, to the frequency of gene duplication, which is substantially greater in vertebrate genomes than in the fly genome and has probably been crucial in making vertebrates the complex and subtle creatures that they are.

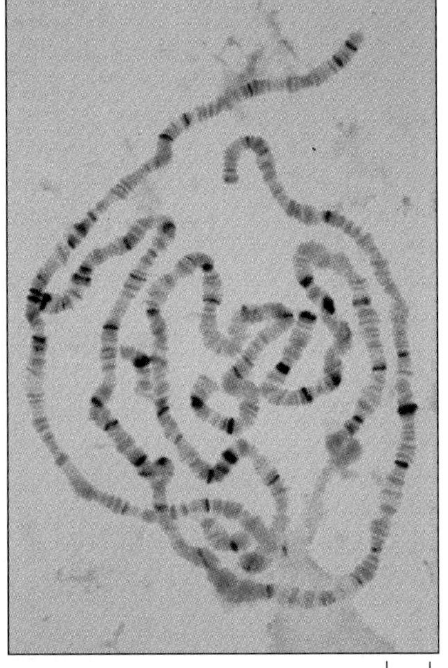

Figure 1–49 Giant chromosomes from salivary gland cells of *Drosophila.* Because many rounds of DNA replication have occurred without an intervening cell division, each of the chromosomes in these unusual cells contains over a 1000 identical DNA molecules, all aligned in register. This makes them easy to see in the light microscope, where they display a characteristic and reproducible banding pattern. Specific bands can be identified as the locations of specific genes: a mutant fly with a region of the banding pattern missing shows a phenotype reflecting loss of the genes in that region. Genes that are being transcribed at a high rate correspond to bands with a "puffed" appearance. The bands stained dark brown in the micrograph are sites where a particular regulatory protein is bound to the DNA. (Courtesy of B. Zink and R. Paro, from R. Paro, *Trends Genet.* 6:416–421, 1990. © Elsevier.)

The Vertebrate Genome Is a Product of Repeated Duplication

Almost every gene in the vertebrate genome is found to have paralogs—other genes in the same genome that are unmistakably related and must have arisen by gene duplication. In many cases, a whole cluster of genes is closely related to similar clusters present elsewhere in the genome, suggesting that genes have

Figure 1–50 Two species of the frog genus *Xenopus*. *X. tropicalis*, above, has an ordinary diploid genome; *X. laevis*, below, has twice as much DNA per cell. From the banding patterns of their chromosomes and the arrangement of genes along them, as well as from comparisons of gene sequences, it is clear that the high-ploidy species have evolved through duplications of the whole genome. These duplications are thought to have occurred in the aftermath of matings between frogs of slightly divergent *Xenopus* species. (Courtesy of E. Amaya, M. Offield, and R. Grainger, *Trends Gen.* 14:253–255, 1998. © Elsevier.)

been duplicated in linked groups rather than as isolated individuals. According to one hypothesis, at an early stage in the evolution of the vertebrates, the entire genome underwent duplication twice in succession, giving rise to four copies of every gene. In some groups of vertebrates, such as fish of the salmon and carp families (including the zebrafish, a popular research animal), it has been suggested that there was yet another duplication, creating an eightfold multiplicity of genes.

The precise course of vertebrate genome evolution remains uncertain, because many further evolutionary changes have occurred since these ancient events. Genes that were once identical have diverged; many of the gene copies have been lost through disruptive mutations; some have undergone further rounds of local duplication; and the genome, in each branch of the vertebrate family tree, has suffered repeated rearrangements, breaking up most of the original gene orderings. A careful comparison of the gene order in two related organisms, such as the human and the mouse, reveals that—on the time scale of vertebrate evolution—chromosomes frequently fuse and fragment to move large blocks of DNA sequence around. Indeed, it is entirely possible, as we shall discuss in Chapter 7, that the present state of affairs is the result of many separate duplications of fragments of the genome, rather than duplications of the genome as a whole.

There is, however, no doubt that such whole-genome duplications do occur from time to time in evolution, for we can see recent instances in which duplicated chromosome sets are still clearly identifiable as such. The frog genus *Xenopus*, for example, comprises a set of closely similar species related to one another by repeated duplications or triplications of the whole genome. Among these frogs are *X. tropicalis*, with an ordinary diploid genome; the common laboratory species *X. laevis*, with a duplicated genome and twice as much DNA per cell; and *X. ruwenzoriensis*, with a sixfold reduplicated genome and six times as much DNA per cell (108 chromosomes, compared with 36 in *X. laevis*, for example). These species are estimated to have diverged from one another within the past 120 million years (Figure 1–50).

Genetic Redundancy Is a Problem for Geneticists, But It Creates Opportunities for Evolving Organisms

Whatever the details of the evolutionary history, it is clear that most genes in the vertebrate genome exist in several versions that were once identical. The related genes often remain functionally interchangeable for many purposes. This phenomenon is called **genetic redundancy**. For the scientist struggling to discover all the genes involved in some particular process, it has a dire consequence. If gene A is mutated and no effect is seen, one can no longer conclude that gene A is functionally irrelevant—it may simply be that this gene normally works in parallel with its relatives, and these suffice for near-normal function even when gene A is defective (just as an airliner continues to fly when one of its engines fails). In the less repetitive genome of *Drosophila*, where gene duplication is less common, the analysis is more straightforward: single gene functions are revealed directly by the consequences of single-gene mutations (the single-engined plane stops flying when the engine fails).

Genome duplication has clearly allowed the development of more complex life forms; it provides an organism with a cornucopia of spare gene copies,

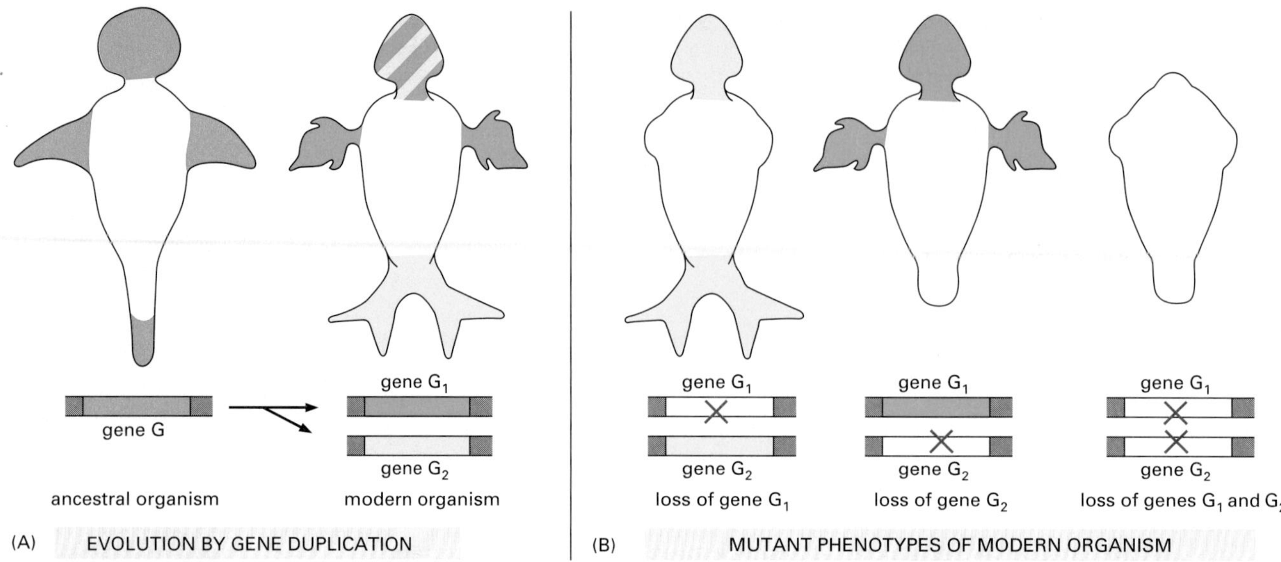

Figure 1–51 The consequences of gene duplication for mutational analyses of gene function. In this hypothetical example, an ancestral multicellular organism has a genome containing a single copy of gene G, which performs its function at several sites in the body, indicated in *green*. (A) Through gene duplication, a modern descendant of the ancestral organism has two copies of gene G, called G_1 and G_2. These have diverged somewhat in their patterns of expression and in their activities at the sites where they are expressed, but they still retain important similarities. At some sites, they are expressed together, and each independently performs the same old function as the ancestral gene G (alternating *green* and *yellow stripes); at* other sites, they are expressed alone and serve new purposes. (B) Because of a functional overlap, the loss of one of the two genes by mutation *(red cross)* reveals only a part of its role; only the loss of both genes in the double mutant reveals the full range of processes for which these genes are responsible. Analogous principles apply to duplicated genes that operate in the same place (for example, in a single-celled organism) but are called into action together or individually in response to varying circumstances. Thus, gene duplications complicate genetic analyses in all organisms.

which are free to mutate to serve divergent purposes. While one copy becomes optimized for use in the liver, say, another can become optimized for use in the brain or adapted for a novel purpose. In this way, the additional genes allow for increased complexity and sophistication. As the genes take on divergent functions, they cease to be redundant. Often, however, while the genes acquire individually specialized roles, they also continue to perform some aspects of their original core function in parallel, redundantly. Mutation of a single gene then causes a relatively minor abnormality that reveals only a part of the gene's function (Figure 1–51). Families of genes with divergent but partly overlapping functions are a pervasive feature of vertebrate molecular biology, and they are encountered repeatedly in this book.

The Mouse Serves as a Model for Mammals

Mammals have typically three or four times as many genes as *Drosophila,* a genome that is 20 times larger, and millions or billions of times as many cells in their adult bodies. In terms of genome size and function, cell biology, and molecular mechanisms, mammals are nevertheless a highly uniform group of organisms. Even anatomically, the differences among mammals are chiefly a matter of size and proportions; it is hard to think of a human body part that does not have a counterpart in elephants and mice, and vice versa. Evolution plays freely with quantitative features, but it does not readily change the logic of the structure.

To get a more exact measure of how closely mammalian species resemble one another genetically, we can compare the nucleotide sequences of corresponding (orthologous) genes, or the amino acid sequences of the proteins that these genes encode. The results for individual genes and proteins vary widely. But typically, if we line up the amino acid sequence of a human protein with that of the orthologous protein from, say, an elephant, about 85% of the amino acids are identical. A similar comparison between human and bird shows an amino acid identity of about 70%—twice as many differences, because the bird and the mammalian lineages have had twice as long to diverge as those of the elephant and the human (Figure 1–52).

The mouse, being small, hardy, and a rapid breeder, has become the foremost model organism for experimental studies of vertebrate molecular genetics. Many naturally occurring mutations are known, often mimicking the effects of corresponding mutations in humans (Figure 1–53). Methods have been developed, moreover, to test the function of any chosen mouse gene, or of any noncoding portion of the mouse genome, by artificially creating mutations in it, as we explain later in the book.

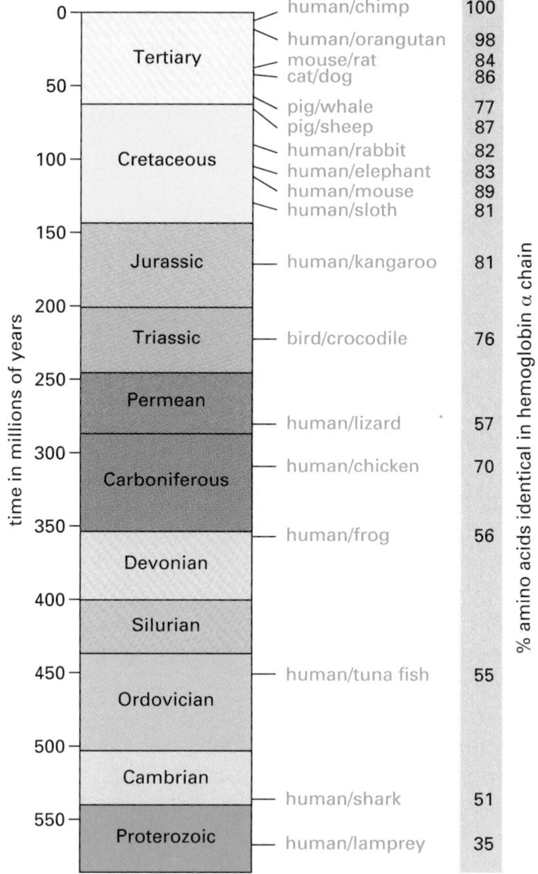

Figure 1–52 Times of divergence of different vertebrates. The scale on the left shows the estimated date and geological era of the last common ancestor of each specified pair of animals. Each time estimate is based on comparisons of the amino acid sequences of orthologous proteins; the longer a pair of animals have had to evolve independently, the smaller the percentage of amino acids that remain identical. Data from many different classes of proteins have been averaged to arrive at the final estimates, and the time scale has been calibrated to match the fossil evidence that the last common ancestor of mammals and birds lived 310 million years ago. The figures on the right give data on sequence divergence for one particular protein (chosen arbitrarily)—the α chain of hemoglobin. Note that although there is a clear general trend of increasing divergence with increasing time for this protein, there are also some irregularities. These reflect the randomness of the evolutionary process and, probably, the action of natural selection driving especially rapid changes of hemoglobin sequence in some organisms that experienced special physiological demands. On average, within any particular evolutionary lineage, hemoglobins accumulate changes at a rate of about 6 altered amino acids per 100 amino acids every 100 million years. Some proteins, subject to stricter functional constraints, evolve much more slowly than this, others as much as 5 times faster. All this gives rise to substantial uncertainties in estimates of divergence times, and some experts believe that the major groups of mammals diverged from one another as much as 60 million years more recently than shown here. (Adapted from S. Kumar and S.B. Hedges, *Nature* 392:917–920, 1998.)

One made-to-order mutant mouse can provide a wealth of information for the cell biologist. It reveals the effects of the chosen mutation in a host of different contexts, simultaneously testing the action of the gene in all the different kinds of cells in the body that could in principle be affected.

Humans Report on Their Own Peculiarities

As humans, we have, of course, a special interest in the human genome. We want to know the full set of parts from which we ourselves are made, and to discover how they work. But even if you were a mouse, preoccupied with the molecular biology of mice, humans would be attractive as model genetic organisms, because of one special property: through medical examinations and self-reporting, we catalog our own genetic (and other) disorders. The human population is enormous, consisting today of some 6 billion individuals, and this self-documenting property means that a huge database of information is available on human mutations. The complete human genome sequence of more than 3 billion

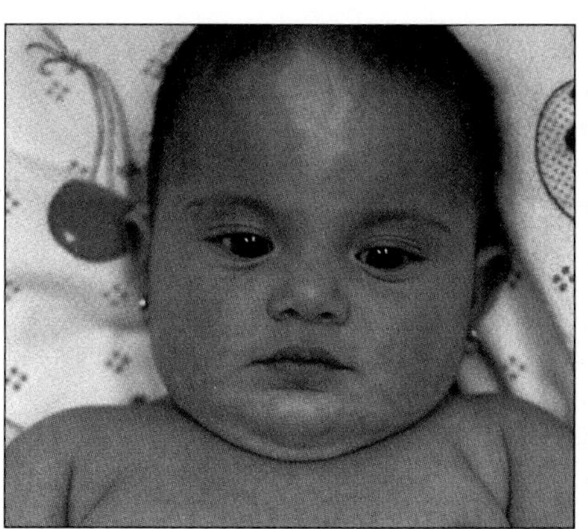

Figure 1–53 Human and mouse: similar genes and similar development. The human baby and the mouse shown here have similar white patches on their foreheads because both have mutations in the same gene (called *kit*), required for the development and maintenance of pigment cells. (From R.A. Fleischman, *Proc. Natl. Acad. Sci. USA* 88:10885–10889, 1991. © National Academy of Sciences.)

nucleotide pairs has recently been determined in a rough draft form, making it easier than ever before to identify at a molecular level the precise gene responsible for each human mutant characteristic.

By drawing together the insights from humans, mice, flies, worms, yeasts, plants, and bacteria—using gene sequence similarities to map out the correspondences between one model organism and another—we enrich our understanding of them all.

We Are All Different in Detail

What precisely do we mean when we speak of *the* human genome? Whose genome? On average, any two people taken at random differ in about one or two in every 1000 nucleotide pairs in their DNA sequence. The Human Genome Project has arbitrarily selected DNA from a small number of anonymous individuals for sequencing. The human genome—the genome of the human species—is, properly speaking, a more complex thing, embracing the entire pool of variant genes that are found in the human population and continually exchanged and reassorted in the course of sexual reproduction. Ultimately, we can hope to document this variation too. Knowledge of it will help us understand, for example, why some people are prone to one disease, others to another; why some respond well to a drug, others badly. It will also provide new clues to our history—the population movements and minglings of our ancestors, the infections they suffered, the diets they ate. All these things leave traces in the variant forms of genes that have survived in human communities.

Knowledge and understanding bring the power to intervene—with humans, to avoid or prevent disease; with plants, to create better crops; with bacteria, to turn them to our own uses. All these biological enterprises are linked, because the genetic information of all living organisms is written in the same language. The new-found ability of molecular biologists to read and decipher this language has already begun to transform our relationship to the living world. The cell biology to be presented in the subsequent chapters will, we hope, prepare you to understand, and possibly to contribute to, the great scientific adventure of the twenty-first century.

Summary

Eucaryotic cells, by definition, keep their DNA in a separate membrane-bounded compartment, the nucleus. They have, in addition, a cytoskeleton for movement, elaborate intracellular compartments for digestion and secretion, the capacity (in many species) to engulf other cells, and a metabolism that depends on the oxidation of organic molecules by mitochondria. These properties suggest that eucaryotes originated as predators on other cells. Mitochondria—and, in plants, chloroplasts—contain their own genetic material, and evidently evolved from bacteria that were taken up into the cytoplasm of the eucaryotic cell and survived as symbionts. Eucaryotic cells have typically 3–30 times as many genes as procaryotes, and often thousands of times more noncoding DNA. The noncoding DNA allows for complex regulation of gene expression, as required for the construction of complex multicellular organisms. Many eucaryotes are, however, unicellular, among them the yeast Saccharomyces cerevisiae, *which serves as a simple model organism for eucaryotic cell biology, revealing the molecular basis of conserved fundamental processes such as the eucaryotic cell division cycle. A small number of other organisms have been chosen as primary models for multicellular plants and animals, and the sequencing of their entire genomes has opened the way to systematic and comprehensive analysis of gene functions, gene regulation, and genetic diversity. As a result of gene duplications during vertebrate evolution, vertebrate genomes contain multiple closely related homologs of most genes. This genetic redundancy has allowed diversification and specialization of genes for new purposes, but it also makes gene functions harder to decipher. There is less genetic redundancy in the nematode* Caenorhabditis elegans *and the fly* Drosophila melanogaster, *which have thus played a key part in revealing universal genetic mechanisms of animal development.*

References

General

Alberts B, Bray D, Johnson A et al. (1997) Essential Cell Biology. London: Garland Publishing.

Darwin C (1859) On the Origin of Species. London: Murray.

Graur D & Li W-H (1999) Fundamentals of Molecular Evolution, 2nd edn. Sunderland, MA: Sinauer Associates.

Madigan MT, Martinko JM & Parker J (2000) Brock's Biology of Microorganisms, 9th edn. Englewood Cliffs, NJ: Prentice Hall.

Margulis L & Schwartz KV (1998) Five Kingdoms: An Illustrated Guide to the Phyla of Life on Earth, 3rd edn. New York: Freeman.

Watson JD, Hopkins NH, Roberts JW et al. (1987) Molecular Biology of the Gene, 4th edn. Menlo Park, CA: Benjamin-Cummings.

The Universal Features of Cells on Earth

Brenner S, Jacob F & Meselson M (1961) An unstable intermediate carrying information from genes to ribosomes for protein synthesis. *Nature* 190, 576–581.

Fraser CM, Gocayne JD, White O et al. (1995) The minimal gene complement of *Mycoplasma genitalium*. *Science* 270, 397–403.

Koonin EV (2000) How many genes can make a cell: the minimal-gene-set concept. *Annu. Rev. Genom. Hum. Genet.* 1, 99–116.

Watson JD & Crick FHC (1953) Molecular structure of nucleic acids. A structure for deoxyribose nucleic acid. *Nature* 171, 737–738.

Yusupov MM, Yusupova GZ, Baucom A et al. (2001) Crystal structure of the ribosome at 5.5Å resolution. *Science* 292, 883–896.

The Diversity of Genomes and the Tree of Life

Blattner FR, Plunkett G, Bloch CA et al. (1997) The complete genome sequence of *Escherichia coli* K-12. *Science* 277, 1453–1474.

Cole ST, Brosch R, Parkhill J et al. (1998) Deciphering the biology of Mycobacterium tuberculosis from the complete genome sequence. *Nature* 393, 537–544.

Dixon B (1994) Power Unseen: How Microbes Rule the World. Oxford: Freeman.

Doolittle RF (1998) Microbial genomes opened up. *Nature* 392, 339–342.

Doolittle WF (1999) Phylogenetic classification and the universal tree. *Science* 284, 2124–2129.

Henikoff S, Greene EA, Pietrokovski S et al. (1997) Gene families: the taxonomy of protein paralogs and chimeras. *Science* 278, 609–614.

Kerr RA (1997) Life goes to extremes in the deep earth—and elsewhere? *Science* 276, 703–704.

Koonin EV (2001) Computational genomics. *Curr. Biol.* 11, R155–R158.

Koonin EV, Aravind L & Kondrashov AS (2000) The impact of comparative genomics on our understanding of evolution. *Cell* 101, 573–576.

Mayr E (1998) Two empires or three? *Proc. Natl. Acad. Sci. USA* 95, 9720–9723.

Ochman H, Lawrence JG & Groisman EA (2000) Lateral gene transfer and the nature of bacterial innovation. *Nature* 405, 299–304.

Olsen GJ & Woese CR (1997) Archaeal genomics: an overview. *Cell* 89, 991–994.

Pace NR (1997) A molecular view of microbial diversity and the biosphere. *Science* 276, 734–740.

Tatusov RL, Koonin EV & Lipman DJ (1997) A genomic perspective on protein families. *Science* 278, 631–637.

The Institute of Genome Research (TIGR) Gene Indices. http://www.tigr.org.tdb/tgi.shtml

Woese C (1998) Default taxonomy: Ernst Mayr's view of the microbial world. *Proc. Natl. Acad. Sci. USA* 95, 11043–11046.

Woese C (1998) The universal ancestor. *Proc. Natl. Acad. Sci. USA* 95, 6854–6859.

Genetic Information in Eucaryotes

Adams MD, Celniker SE, Holt RA et al. (2000) The genome sequence of *Drosophila melanogaster*. *Science* 287, 2185–2195.

Andersson SG, Zomorodipour A, Andersson JO et al. (1998) The genome sequence of Rickettsia prowazekii and the origin of mitochondria. *Nature* 396, 133–140.

Arabidopsis Genome Sequencing Consortium (2000) Analysis of the genome sequence of the flowering plant *Arabidopsis thaliana*. *Nature* 408, 796–815.

Carroll SB, Grenier JK & Weatherbee SD (2001) From DNA to Diversity: Molecular Genetics and the Evolution of Animal Design. Maldon, MA: Blackwell Science.

Cavalier-Smith T (1987) The origin of eukaryote and archaebacterial cells. *Ann. NY Acad. Sci.* 503, 17–54.

DeRisi JL, Iyer VR & Brown PO (1997) Exploring the metabolic and genetic control of gene expression on a genomic scale. *Science* 278, 680–686.

Elgar G, Sandford R, Aparicio S et al. (1996) Small is beautiful: comparative genomics with the pufferfish (Fugu rubripes). *Trends Genet.* 12, 145–150.

Goffeau A, Barrell BG, Bussey H et al. (1996) Life with 6000 genes. *Science* 274, 546–567.

Gray MW, Burger G & Lang BF (1999) Mitochondrial evolution. *Science* 283, 1476–1481.

Holland PW (1999) Gene duplication: past, present and future. *Semin. Cell Dev. Biol.* 10, 541–547.

International Human Genome Sequencing Consortium (2001) Initial sequencing and analysis of the human genome. *Nature* 409, 860–921.

Lynch M & Conery JS (2000) The evolutionary fate and consequences of duplicate genes. *Science* 290, 1151–1155.

NCBI Genome Guide. http://www.ncbi.nlm.nih.gov/genome/guide/human/

Online Mendelian Inheritance in Man. http://www.ncbi.nlm.nih.gov/omim

Owens K & King M-C (1999) Genomic views of human history. *Science* 286, 451–453.

Palmer JD & Delwiche CF (1996) Second-hand chloroplasts and the case of the disappearing nucleus. *Proc. Natl. Acad. Sci. USA* 93, 7432–7435.

Philippe H, Germot A & Moreira D (2000) The new phylogeny of eukaryotes. *Curr. Opin. Genet. Dev.* 10, 596–601.

Plasterk RH (1999) The year of the worm. *Bioessays* 21, 105–109.

Rubin GM, Yandell MD, Wortman JR et al. (2000) Comparative genomics of the eukaryotes. *Science* 287, 2204–2215.

Sturtevant AH & Beadle GW (1939) An Introduction to Genetics. Philadelphia: Saunders. Reprinted 1988 New York: Garland.

The C. elegans Sequencing Consortium (1998) Genome sequence of the nematode *C. elegans:* a platform for investigating biology. *Science* 282, 2012–2018.

Tinsley RC & Kobel HR (eds) (1996) The Biology of *Xenopus*, p 16, pp 379, pp 391–401. Oxford: Clarendon Press.

Venter JC, Adams MD, Myers EW et al. (2001) The sequence of the human genome. *Science* 291, 1304–1351.

Walbot V (2000) *Arabidopsis thaliana* genome. A green chapter in the book of life. *Nature* 408, 794–795.

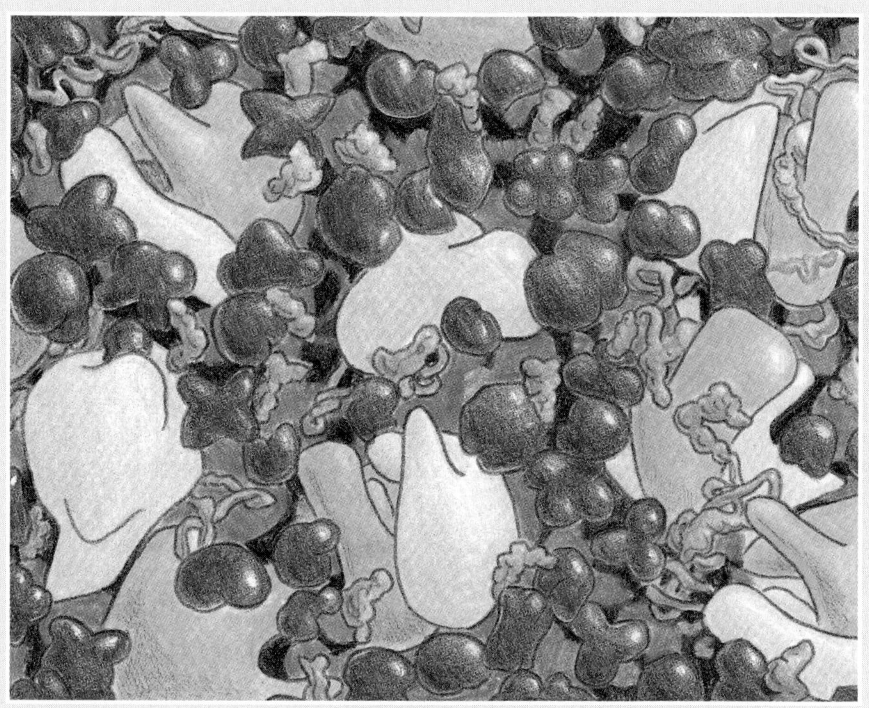

Crowded cytoplasm. This scale drawing, which shows only the macromolecules, gives a good impression of how crowded the cytoplasm is. RNAs are shown in *blue,* ribosomes in *green* and proteins in *red.* (Adapted from D.S. Goodsell, *Trends Biochem. Sci.* 16:203–206, 1991.)

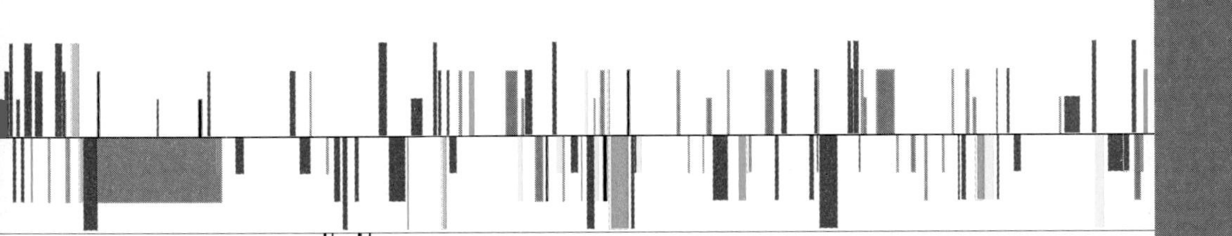

2

CELL CHEMISTRY AND BIOSYNTHESIS

It is at first sight difficult to accept the idea that each of the living creatures described in the previous chapter is merely a chemical system. The incredible diversity of living forms, their seemingly purposeful behavior, and their ability to grow and reproduce all seem to set them apart from the world of solids, liquids, and gases that chemistry normally describes. Indeed, until the nineteenth century it was widely accepted that animals contained a Vital Force—an "animus"—that was uniquely responsible for their distinctive properties.

We now know there is nothing in living organisms that disobeys chemical and physical laws. However, the chemistry of life is indeed of a special kind. First, it is based overwhelmingly on carbon compounds, whose study is therefore known as *organic chemistry*. Second, cells are 70 percent water, and life depends almost exclusively on chemical reactions that take place in aqueous solution. Third, and most importantly, cell chemistry is enormously complex: even the simplest cell is vastly more complicated in its chemistry than any other chemical system known. Although cells contain a variety of small carbon-containing molecules, most of the carbon atoms in cells are incorporated into enormous *polymeric molecules*—chains of chemical subunits linked end-to-end. It is the unique properties of these macromolecules that enable cells and organisms to grow and reproduce—as well as to do all the other things that are characteristic of life.

THE CHEMICAL COMPONENTS OF A CELL

Matter is made of combinations of *elements*—substances such as hydrogen or carbon that cannot be broken down or converted into other substances by chemical means. The smallest particle of an element that still retains its distinctive chemical properties is an *atom*. However, the characteristics of substances other than pure elements—including the materials from which living cells are made—depend on the way their atoms are linked together in groups to form *molecules*. In order to understand how living organisms are built from inanimate matter, therefore, it is crucial to know how all of the chemical bonds that hold atoms together in molecules are formed.

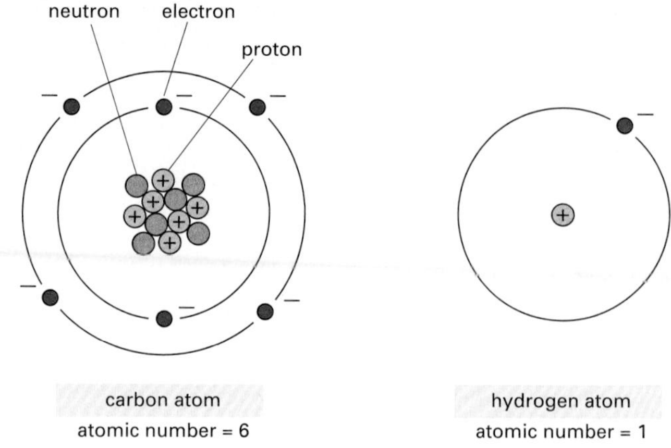

Figure 2–1 Highly schematic representations of an atom of carbon and an atom of hydrogen. Although the electrons are shown here as individual particles, in reality their behavior is governed by the laws of quantum mechanics, and there is no way of predicting exactly where an electron is at any given instant of time. The nucleus of every atom except hydrogen consists of both positively charged protons and electrically neutral neutrons. The number of electrons in an atom is equal to its number of protons (the atomic number), so that the atom has no net charge. The neutrons, protons, and electrons are in reality minute in relation to the atom as a whole; their size is greatly exaggerated here. In addition, the diameter of the nucleus is only about 10^{-4} that of the electron cloud.

Cells Are Made From a Few Types of Atoms

Each **atom** has at its center a positively charged nucleus, which is surrounded at some distance by a cloud of negatively charged **electrons**, held in a series of orbitals by electrostatic attraction to the nucleus. The nucleus in turn consists of two kinds of subatomic particles: **protons**, which are positively charged, and *neutrons*, which are electrically neutral. The number of protons in the atomic nucleus gives the *atomic number*. An atom of hydrogen has a nucleus composed of a single proton; so hydrogen, with an atomic number of 1, is the lightest element. An atom of carbon has six protons in its nucleus and an atomic number of 6 (Figure 2–1). The electric charge carried by each proton is exactly equal and opposite to the charge carried by a single electron. Since an atom as a whole is electrically neutral, the number of negatively charged electrons surrounding the nucleus is equal to the number of positively charged protons that the nucleus contains; thus the number of electrons in an atom also equals the atomic number. It is these electrons that determine the chemical behavior of an atom, and all of the atoms of a given element have the same atomic number.

Neutrons are uncharged subatomic particles of essentially the same mass as protons. They contribute to the structural stability of the nucleus—if there are too many or too few, the nucleus may disintegrate by radioactive decay—but they do not alter the chemical properties of the atom. Thus an element can exist in several physically distinguishable but chemically identical forms, called *isotopes*, each isotope having a different number of neutrons but the same number of protons. Multiple isotopes of almost all the elements occur naturally, including some that are unstable. For example, while most carbon on Earth exists as the stable isotope carbon 12, with six protons and six neutrons, there are also small amounts of an unstable isotope, the radioactive carbon 14, whose atoms have six protons and eight neutrons. Carbon 14 undergoes radioactive decay at a slow but steady rate. This forms the basis for a technique known as carbon 14 dating, which is used in archaeology to determine the time of origin of organic materials.

The **atomic weight** of an atom, or the **molecular weight** of a molecule, is its mass relative to that of a hydrogen atom. This is essentially equal to the number of protons plus neutrons that the atom or molecule contains, since the electrons are much lighter and contribute almost nothing to the total. Thus the major isotope of carbon has an atomic weight of 12 and is symbolized as ^{12}C, whereas the unstable isotope just discussed has an atomic weight of 14 and is written as ^{14}C. The mass of an atom or a molecule is often specified in *daltons*, one dalton being an atomic mass unit approximately equal to the mass of a hydrogen atom.

Atoms are so small that it is hard to imagine their size. An individual carbon atom is roughly 0.2 nm in diameter, so that it would take about 5 million of them, laid out in a straight line, to span a millimeter. One proton or neutron weighs approximately $1/(6 \times 10^{23})$ gram, so one gram of hydrogen contains 6×10^{23} atoms. This huge number (6×10^{23}, called **Avogadro's number**) is the key scale factor describing the relationship between everyday quantities and quantities measured in terms of individual atoms or molecules. If a substance has a molecular weight of X, 6×10^{23} molecules of it will have a mass of X grams. This quantity is called one mole of the substance (Figure 2–2).

There are 92 naturally occurring elements, each differing from the others in the number of protons and electrons in its atoms. Living organisms, however, are made of only a small selection of these elements, four of which—carbon (C), hydrogen (H), nitrogen (N), and oxygen (O)—make up 96.5% of an organism's weight. This composition differs markedly from that of the nonliving inorganic environment (Figure 2–3) and is evidence of a distinctive type of chemistry. The most common elements in living organisms are listed in Table 2–1 with some of their atomic characteristics.

A **mole** is X grams of a substance, where X is its relative molecular mass (molecular weight). A mole will contain 6×10^{23} molecules of the substance.

1 mole of carbon weighs 12 g
1 mole of glucose weighs 180 g
1 mole of sodium chloride weighs 58 g

Molar solutions have a concentration of 1 mole of the substance in 1 liter of solution. A molar solution (denoted as 1 M) of glucose, for example, has 180 g/l, while a millimolar solution (1 mM) has 180 mg/l.

The standard abbreviation for gram is g; the abbreviation for liter is l.

Figure 2–2 Moles and molar solutions.

The Outermost Electrons Determine How Atoms Interact

To understand how atoms bond together to form the molecules that make up living organisms, we have to pay special attention to their electrons. Protons and neutrons are welded tightly to one another in the nucleus and change partners only under extreme conditions—during radioactive decay, for example, or in the interior of the sun or of a nuclear reactor. In living tissues, it is only the electrons of an atom that undergo rearrangements. They form the exterior of an atom and specify the rules of chemistry by which atoms combine to form molecules.

Electrons are in continuous motion around the nucleus, but motions on this submicroscopic scale obey different laws from those we are familiar with in everyday life. These laws dictate that electrons in an atom can exist only in certain discrete states, called orbitals, and that there is a strict limit to the number of electrons that can be accommodated in an orbital of a given type—a so-called *electron shell*. The electrons closest on average to the positive nucleus are attracted most strongly to it and occupy the innermost, most tightly bound shell. This shell can hold a maximum of two electrons. The second shell is farther away from the nucleus, and its electrons are less tightly bound. This second shell can hold up to eight electrons. The third shell contains electrons that are even less tightly bound; it can also hold up to eight electrons. The fourth and fifth shells can hold 18 electrons each. Atoms with more than four shells are very rare in biological molecules.

The electron arrangement of an atom is most stable when all the electrons are in the most tightly bound states that are possible for them—that is, when they occupy the innermost shells. Therefore, with certain exceptions in the larger

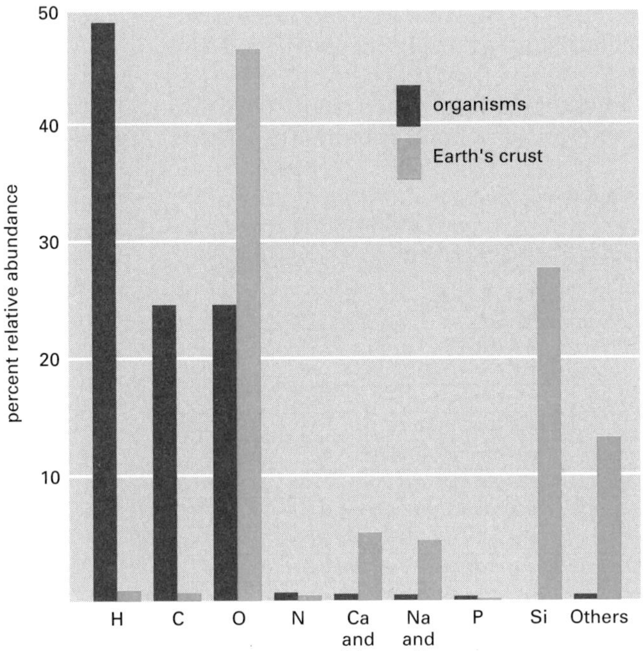

Figure 2–3 The abundances of some chemical elements in the nonliving world (the Earth's crust) compared with their abundances in the tissues of an animal. The abundance of each element is expressed as a percentage of the total number of atoms present in the sample. Thus, for example, nearly 50% of the atoms in a living organism are hydrogen atoms. The survey here excludes mineralized tissues such as bone and teeth, as they contain large amounts of inorganic salts of calcium and phosphorus. The relative abundance of elements is similar in all living organisms.

TABLE 2–1 Atomic Characteristics of the Most Abundant Elements in Living Tissues

COMMON ELEMENTS IN LIVING ORGANISMS

element		protons	neutrons	electrons	atomic number	atomic weight
Hydrogen	H	1	0	1	1	1
Carbon	C	6	6	6	6	12
Nitrogen	N	7	7	7	7	14
Oxygen	O	8	8	8	8	16

LESS COMMON ELEMENTS

element		protons	neutrons	electrons	atomic number	atomic weight
Sodium	Na	11	12	11	11	23
Magnesium	Mg	12	12	12	12	24
Phosphorus	P	15	16	15	15	31
Sulfur	S	16	16	16	16	32
Chlorine	Cl	17	18	17	17	35
Potassium	K	19	20	19	19	39
Calcium	Ca	20	20	20	20	40

atoms, the electrons of an atom fill the orbitals in order—the first shell before the second, the second before the third, and so on. An atom whose outermost shell is entirely filled with electrons is especially stable and therefore chemically unreactive. Examples are helium with 2 electrons, neon with 2 + 8, and argon with 2 + 8 + 8; these are all inert gases. Hydrogen, by contrast, with only one electron and therefore only a half-filled shell, is highly reactive. Likewise, the other atoms found in living tissues all have incomplete outer electron shells and are therefore able to donate, accept, or share electrons with each other to form both molecules and ions (Figure 2–4).

Because an unfilled electron shell is less stable than a filled one, atoms with incomplete outer shells have a strong tendency to interact with other atoms in a way that causes them to either gain or lose enough electrons to achieve a completed outermost shell. This electron exchange can be achieved either by transferring electrons from one atom to another or by sharing electrons between two atoms. These two strategies generate two types of **chemical bonds** between atoms: an *ionic bond* is formed when electrons are donated by one atom to another, whereas a *covalent bond* is formed when two atoms share a pair of electrons (Figure 2–5). Often, the pair of electrons is shared unequally, with a partial transfer between the atoms; this intermediate strategy results in a *polar covalent bond*, as we shall discuss later.

An H atom, which needs only one more electron to fill its shell, generally acquires it by electron sharing, forming one covalent bond with another atom;

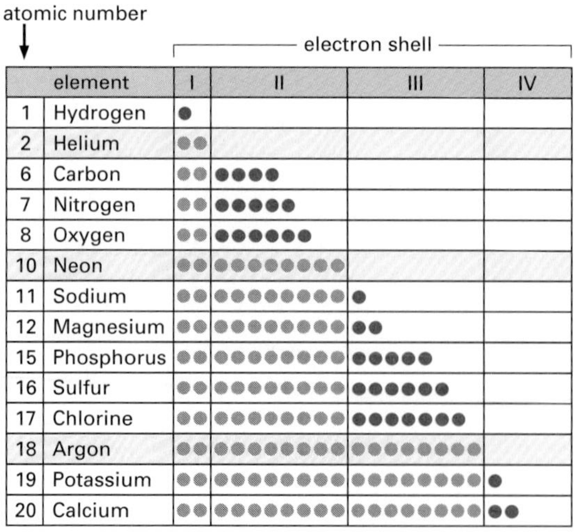

atomic number

electron shell

	element	I	II	III	IV
1	Hydrogen	●			
2	Helium	●●			
6	Carbon	●●	●●●●		
7	Nitrogen	●●	●●●●●		
8	Oxygen	●●	●●●●●●		
10	Neon	●●	●●●●●●●●		
11	Sodium	●●	●●●●●●●●	●	
12	Magnesium	●●	●●●●●●●●	●●	
15	Phosphorus	●●	●●●●●●●●	●●●●●	
16	Sulfur	●●	●●●●●●●●	●●●●●●	
17	Chlorine	●●	●●●●●●●●	●●●●●●●	
18	Argon	●●	●●●●●●●●	●●●●●●●●	
19	Potassium	●●	●●●●●●●●	●●●●●●●●	●
20	Calcium	●●	●●●●●●●●	●●●●●●●●	●●

Figure 2–4 Filled and unfilled electron shells in some common elements. All the elements commonly found in living organisms have unfilled outermost shells *(red)* and can thus participate in chemical reactions with other atoms. For comparison, some elements that have only filled shells *(yellow)* are shown; these are chemically unreactive.

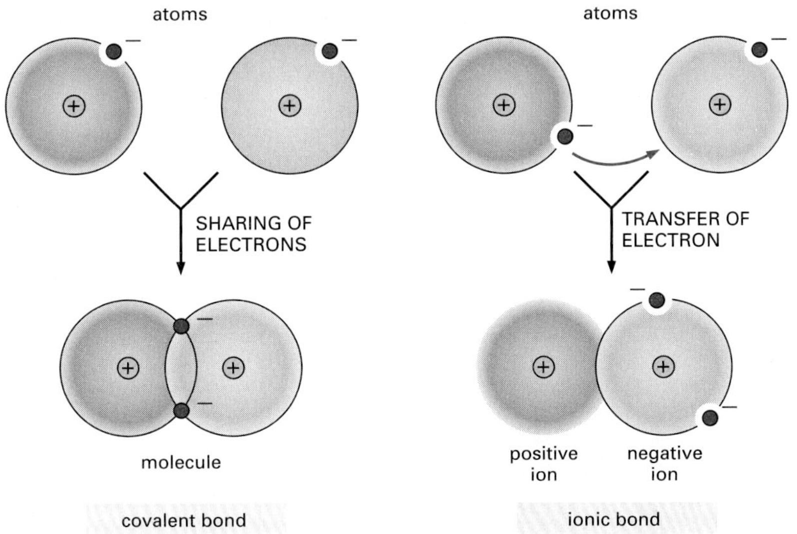

Figure 2–5 Comparison of covalent and ionic bonds. Atoms can attain a more stable arrangement of electrons in their outermost shell by interacting with one another. An ionic bond is formed when electrons are transferred from one atom to the other. A covalent bond is formed when electrons are shared between atoms. The two cases shown represent extremes; often, covalent bonds form with a partial transfer (unequal sharing of electrons), resulting in a polar covalent bond (see Figure 2–43).

in many cases this bond is polar. The other most common elements in living cells—C, N, and O, with an incomplete second shell, and P and S, with an incomplete third shell (see Figure 2–4)—generally share electrons and achieve a filled outer shell of eight electrons by forming several covalent bonds. The number of electrons that an atom must acquire or lose (either by sharing or by transfer) to attain a filled outer shell is known as its *valence*.

The crucial role of the outer electron shell in determining the chemical properties of an element means that, when the elements are listed in order of their atomic number, there is a periodic recurrence of elements with similar properties: an element with, say, an incomplete second shell containing one electron will behave in much the same way as an element that has filled its second shell and has an incomplete third shell containing one electron. The metals, for example, have incomplete outer shells with just one or a few electrons, whereas, as we have just seen, the inert gases have full outer shells.

Ionic Bonds Form by the Gain and Loss of Electrons

Ionic bonds are most likely to be formed by atoms that have just one or two electrons in addition to a filled outer shell or are just one or two electrons short of acquiring a filled outer shell. They can often attain a completely filled outer electron shell more easily by transferring electrons to or from another atom than by sharing electrons. For example, from Figure 2–4 we see that a sodium (Na) atom, with atomic number 11, can strip itself down to a filled shell by giving up the single electron external to its second shell. By contrast, a chlorine (Cl) atom, with atomic number 17, can complete its outer shell by gaining just one electron. Consequently, if a Na atom encounters a Cl atom, an electron can jump from the Na to the Cl, leaving both atoms with filled outer shells. The offspring of this marriage between sodium, a soft and intensely reactive metal, and chlorine, a toxic green gas, is table salt (NaCl).

When an electron jumps from Na to Cl, both atoms become electrically charged **ions**. The Na atom that lost an electron now has one less electron than it has protons in its nucleus; it therefore has a single positive charge (Na^+). The Cl atom that gained an electron now has one more electron than it has protons and has a single negative charge (Cl^-). Positive ions are called *cations*, and negative ions, *anions*. Ions can be further classified according to how many electrons are lost or gained. Thus sodium and potassium (K) have one electron to lose and form cations with a single positive charge (Na^+ and K^+), whereas magnesium and calcium have two electrons to lose and form cations with two positive charges (Mg^{2+} and Ca^{2+}).

Because of their opposite charges, Na^+ and Cl^- are attracted to each other and are thereby held together in an **ionic bond**. A salt crystal contains astronomical numbers of Na^+ and Cl^- (about 2×10^{19} ions of each type in a crystal

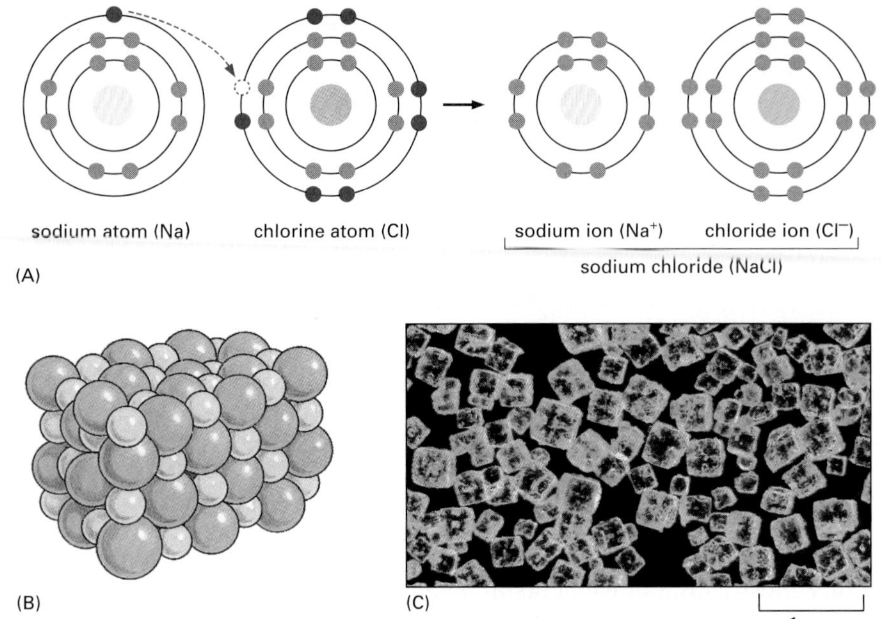

Figure 2–6 Sodium chloride: an example of ionic bond formation. (A) An atom of sodium (Na) reacts with an atom of chlorine (Cl). Electrons of each atom are shown schematically in their different energy levels; electrons in the chemically reactive (incompletely filled) shells are *red*. The reaction takes place with transfer of a single electron from sodium to chlorine, forming two electrically charged atoms, or ions, each with complete sets of electrons in their outermost levels. The two ions with opposite charge are held together by electrostatic attraction. (B) The product of the reaction between sodium and chlorine, crystalline sodium chloride, consists of sodium and chloride ions packed closely together in a regular array in which the charges are exactly balanced. (C) Color photograph of crystals of sodium chloride.

1 mm across) packed together in a precise three-dimensional array with their opposite charges exactly balanced (Figure 2–6). Substances such as NaCl, which are held together solely by ionic bonds, are generally called *salts* rather than molecules. Ionic bonds are just one of several types of *noncovalent bonds* that can exist between atoms, and we shall meet other examples.

Because of a favorable interaction between water molecules and ions, ionic bonds are greatly weakened by water; thus many salts (including NaCl) are highly soluble in water—dissociating into individual ions (such as Na^+ and Cl^-), each surrounded by a group of water molecules. In contrast, covalent bond strengths are not affected in this way.

Covalent Bonds Form by the Sharing of Electrons

All the characteristics of a cell depend on the molecules it contains. A **molecule** is defined as a cluster of atoms held together by **covalent bonds**; here electrons are shared between atoms to complete the outer shells, rather than being transferred between them. In the simplest possible molecule—a molecule of hydrogen (H_2)—two H atoms, each with a single electron, share two electrons, which is the number required to fill the first shell. These shared electrons form a cloud of negative charge that is densest between the two positively charged nuclei and helps to hold them together, in opposition to the mutual repulsion between like charges that would otherwise force them apart. The attractive and repulsive forces are in balance when the nuclei are separated by a characteristic distance, called the *bond length*.

A further crucial property of any bond—covalent or noncovalent—is its strength. *Bond strength* is measured by the amount of energy that must be supplied to break that bond. This is often expressed in units of kilocalories per mole (kcal/mole), where a kilocalorie is the amount of energy needed to raise the temperature of one liter of water by one degree centigrade. Thus if 1 kilocalorie must be supplied to break 6×10^{23} bonds of a specific type (that is, 1 mole of these bonds), then the strength of that bond is 1 kcal/mole. An equivalent, widely used measure of energy is the kilojoule, which is equal to 0.239 kilocalories.

To get an idea of what bond strengths mean, it is helpful to compare them with the average energies of the impacts that molecules are constantly undergoing from collisions with other molecules in their environment (their thermal, or heat, energy), as well as with other sources of biological energy such as light and glucose oxidation (Figure 2–7). Typical covalent bonds are stronger than the thermal energies by a factor of 100, so they are resistant to being pulled apart by

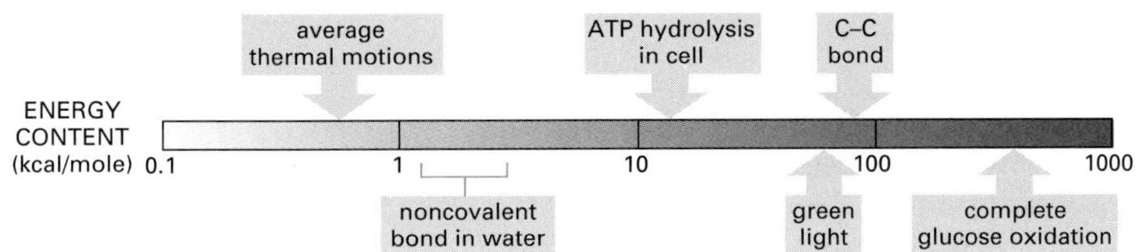

Figure 2–7 Some energies important for cells. Note that these energies are compared on a logarithmic scale.

thermal motions and are normally broken only during specific chemical reactions with other atoms and molecules. The making and breaking of covalent bonds are violent events, and in living cells they are carefully controlled by highly specific catalysts, called *enzymes*. Noncovalent bonds as a rule are much weaker; we shall see later that they are important in the cell in the many situations where molecules have to associate and dissociate readily to carry out their functions.

Whereas an H atom can form only a single covalent bond, the other common atoms that form covalent bonds in cells—O, N, S, and P, as well as the all-important C atom—can form more than one. The outermost shell of these atoms, as we have seen, can accommodate up to eight electrons, and they form covalent bonds with as many other atoms as necessary to reach this number. Oxygen, with six electrons in its outer shell, is most stable when it acquires an extra two electrons by sharing with other atoms and therefore forms up to two covalent bonds. Nitrogen, with five outer electrons, forms a maximum of three covalent bonds, while carbon, with four outer electrons, forms up to four covalent bonds—thus sharing four pairs of electrons (see Figure 2–4).

When one atom forms covalent bonds with several others, these multiple bonds have definite orientations in space relative to one another, reflecting the orientations of the orbits of the shared electrons. Covalent bonds between multiple atoms are therefore characterized by specific bond angles as well as bond lengths and bond energies (Figure 2–8). The four covalent bonds that can form around a carbon atom, for example, are arranged as if pointing to the four corners of a regular tetrahedron. The precise orientation of covalent bonds forms the basis for the three-dimensional geometry of organic molecules.

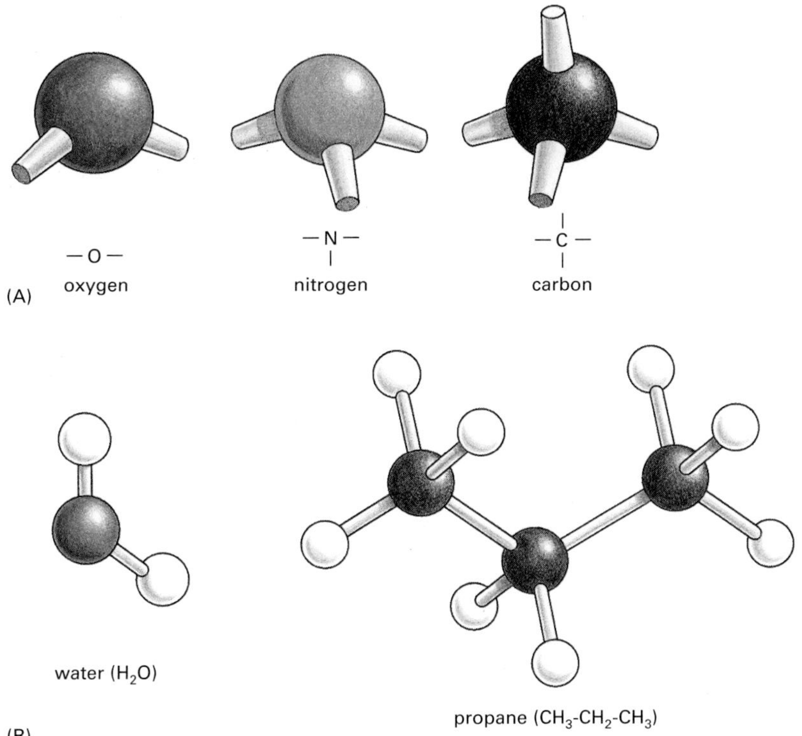

Figure 2–8 The geometry of covalent bonds. (A) The spatial arrangement of the covalent bonds that can be formed by oxygen, nitrogen, and carbon. (B) Molecules formed from these atoms have a precise three-dimensional structure, as shown here by ball and stick models for water and propane. A structure can be specified by the bond angles and bond lengths for each covalent linkage.

There Are Different Types of Covalent Bonds

Most covalent bonds involve the sharing of two electrons, one donated by each participating atom; these are called *single bonds*. Some covalent bonds, however, involve the sharing of more than one pair of electrons. Four electrons can be shared, for example, two coming from each participating atom; such a bond is called a *double bond*. Double bonds are shorter and stronger than single bonds and have a characteristic effect on the three-dimensional geometry of molecules containing them. A single covalent bond between two atoms generally allows the rotation of one part of a molecule relative to the other around the bond axis. A double bond prevents such rotation, producing a more rigid and less flexible arrangement of atoms (Figure 2–9 and Panel 2–1, pp. 111–112).

Some molecules share electrons between three or more atoms, producing bonds that have a hybrid character intermediate between single and double bonds. The highly stable benzene molecule, for example, comprises a ring of six carbon atoms in which the bonding electrons are evenly distributed (although usually depicted as an alternating sequence of single and double bonds, as shown in Panel 2–1).

When the atoms joined by a single covalent bond belong to different elements, the two atoms usually attract the shared electrons to different degrees. Compared with a C atom, for example, O and N atoms attract electrons relatively strongly, whereas an H atom attracts electrons more weakly. By definition, a **polar** structure (in the electrical sense) is one with positive charge concentrated toward one end (the positive pole) and negative charge concentrated toward the other (the negative pole). Covalent bonds in which the electrons are shared unequally in this way are therefore known as *polar covalent bonds* (Figure 2–10). For example, the covalent bond between oxygen and hydrogen, –O–H, or between nitrogen and hydrogen, –N–H, is polar, whereas that between carbon and hydrogen, –C–H, has the electrons attracted much more equally by both atoms and is relatively nonpolar.

Polar covalent bonds are extremely important in biology because they create *permanent dipoles* that allow molecules to interact through electrical forces. Any large molecule with many polar groups will have a pattern of partial positive and negative charges on its surface. When such a molecule encounters a second molecule with a complementary set of charges, the two molecules will be attracted to each other by permanent dipole interactions that resemble (but are weaker than) the ionic bonds discussed previously for NaCl.

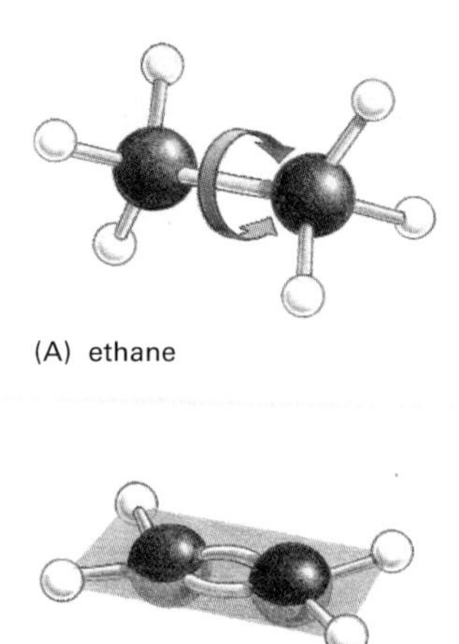

Figure 2–9 Carbon–carbon double bonds and single bonds compared. (A) The ethane molecule, with a single covalent bond between the two carbon atoms, illustrates the tetrahedral arrangement of single covalent bonds formed by carbon. One of the CH_3 groups joined by the covalent bond can rotate relative to the other around the bond axis. (B) The double bond between the two carbon atoms in a molecule of ethene (ethylene) alters the bond geometry of the carbon atoms and brings all the atoms into the same plane *(blue)*; the double bond prevents the rotation of one CH_2 group relative to the other.

An Atom Often Behaves as if It Has a Fixed Radius

When a covalent bond forms between two atoms, the sharing of electrons brings the nuclei of these atoms unusually close together. But most of the atoms that are rapidly jostling each other in cells are located in separate molecules. What happens when two such atoms touch?

For simplicity and clarity, atoms and molecules are usually represented in a highly schematic way—either as a line drawing of the structural formula or as a ball and stick model. However, a more accurate representation can be obtained through the use of so-called *space-filling models.* Here a solid envelope is used to represent the radius of the electron cloud at which strong repulsive forces prevent a closer approach of any second, non-bonded atom—the so-called *van der Waals radius* for an atom. This is possible because the amount of repulsion increases very steeply as two such atoms approach each other closely. At slightly greater distances, any two atoms will experience a weak attractive force, known as a *van der Waals attraction.* As a result, there is a distance at which repulsive and attractive forces precisely balance to produce an energy minimum in each atom's interaction with an atom of a second, non-bonded element (Figure 2–11).

Depending on the intended purpose, we shall represent small molecules either as line drawings, ball and stick models, or space filling models throughout this book. For comparison, the water molecule is represented in all three ways in Figure 2–12. When dealing with very large molecules, such as proteins, we shall

Figure 2–10 Polar and nonpolar covalent bonds. The electron distributions in the polar water molecule (H_2O) and the nonpolar oxygen molecule (O_2) are compared (δ^+, partial positive charge; δ^-, partial negative charge).

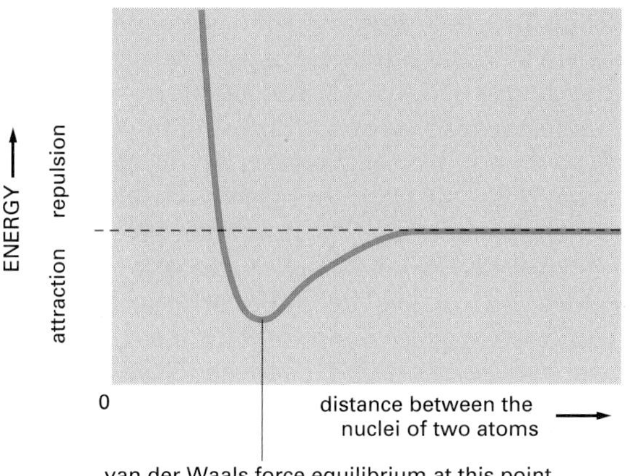

Figure 2–11 The balance of van der Waals forces between two atoms. As the nuclei of two atoms approach each other, they initially show a weak bonding interaction due to their fluctuating electric charges. However, the same atoms will strongly repel each other if they are brought too close together. The balance of these van der Waals attractive and repulsive forces occurs at the indicated energy minimum.

often need to further simplify the representation used (see, for example, Panel 3–2, pp. 138–139).

Water Is the Most Abundant Substance in Cells

Water accounts for about 70% of a cell's weight, and most intracellular reactions occur in an aqueous environment. Life on Earth began in the ocean, and the conditions in that primeval environment put a permanent stamp on the chemistry of living things. Life therefore hinges on the properties of water.

In each water molecule (H_2O) the two H atoms are linked to the O atom by covalent bonds (see Figure 2–12). The two bonds are highly polar because the O is strongly attractive for electrons, whereas the H is only weakly attractive. Consequently, there is an unequal distribution of electrons in a water molecule, with a preponderance of positive charge on the two H atoms and of negative charge on the O (see Figure 2–10). When a positively charged region of one water molecule (that is, one of its H atoms) comes close to a negatively charged region (that is, the O) of a second water molecule, the electrical attraction between them can result in a weak bond called a *hydrogen bond*. These bonds are much weaker than covalent bonds and are easily broken by the random thermal motions due to the heat energy of the molecules, so each bond lasts only an exceedingly short time. But the combined effect of many weak bonds is far from trivial. Each water molecule can form hydrogen bonds through its two H atoms to two other water molecules, producing a network in which hydrogen bonds are being continually broken and formed (Panel 2–2, pp. 112–113). It is only because of the hydrogen bonds that link water molecules together that water is a liquid at room temperature, with a high boiling point and high surface tension—rather than a gas.

Molecules, such as alcohols, that contain polar bonds and that can form hydrogen bonds with water dissolve readily in water. As mentioned previously, molecules carrying plus or minus charges (ions) likewise interact favorably with

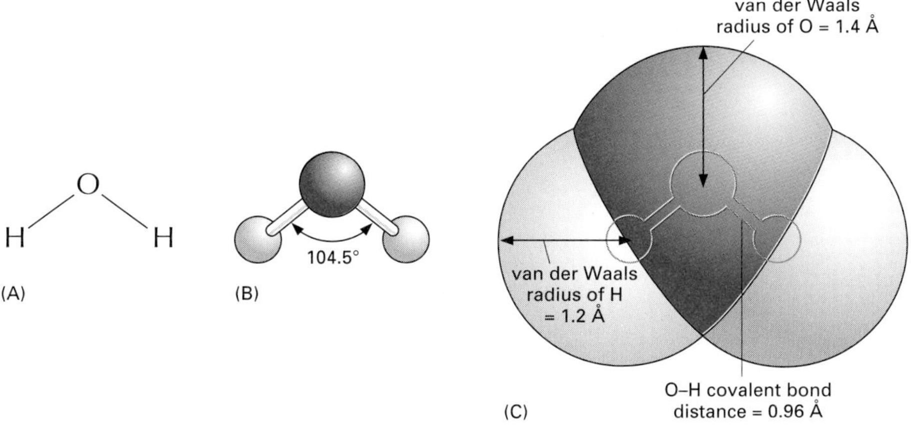

Figure 2–12 Three representations of a water molecule. (A) The usual line drawing of the structural formula, in which each atom is indicated by its standard symbol, and each line represents a covalent bond joining two atoms. (B) A ball and stick model, in which atoms are represented by spheres of arbitrary diameter, connected by sticks representing covalent bonds. Unlike (A), bond angles are accurately represented in this type of model (see also Figure 2–8). (C) A space-filling model, in which both bond geometry and van der Waals radii are accurately represented.

water. Such molecules are termed **hydrophilic**, meaning that they are water-loving. A large proportion of the molecules in the aqueous environment of a cell necessarily fall into this category, including sugars, DNA, RNA, and a majority of proteins. **Hydrophobic** (water-hating) molecules, by contrast, are uncharged and form few or no hydrogen bonds, and so do not dissolve in water. Hydrocarbons are an important example (see Panel 2–1, pp. 110–111). In these molecules the H atoms are covalently linked to C atoms by a largely nonpolar bond. Because the H atoms have almost no net positive charge, they cannot form effective hydrogen bonds to other molecules. This makes the hydrocarbon as a whole hydrophobic—a property that is exploited in cells, whose membranes are constructed from molecules that have long hydrocarbon tails, as we shall see in Chapter 10.

Some Polar Molecules Form Acids and Bases in Water

One of the simplest kinds of chemical reaction, and one that has profound significance in cells, takes place when a molecule possessing a highly polar covalent bond between a hydrogen and a second atom dissolves in water. The hydrogen atom in such a molecule has largely given up its electron to the companion atom and so exists as an almost naked positively charged hydrogen nucleus—in other words, a **proton (H^+)**. When the polar molecule becomes surrounded by water molecules, the proton is attracted to the partial negative charge on the O atom of an adjacent water molecule and can dissociate from its original partner to associate instead with the oxygen atoms of the water molecule to generate a **hydronium ion (H_3O^+)** (Figure 2–13A). The reverse reaction also takes place very readily, so one has to imagine an equilibrium state in which billions of protons are constantly flitting to and fro from one molecule in the solution to another.

Substances that release protons to form H_3O^+ when they dissolve in water are termed **acids**. The higher the concentration of H_3O^+, the more acidic the solution. H_3O^+ is present even in pure water, at a concentration of 10^{-7} M, as a result of the movement of protons from one water molecule to another (Figure 2–13B). By tradition, the H_3O^+ concentration is usually referred to as the H^+ concentration, even though most H^+ in an aqueous solution is present as H_3O^+. To avoid the use of unwieldy numbers, the concentration of H^+ is expressed using a logarithmic scale called the **pH scale**, as illustrated in Panel 2–2 (pp. 112–113). Pure water has a pH of 7.0.

Because the proton of a hydronium ion can be passed readily to many types of molecules in cells, altering their character, the concentration of H_3O^+ inside a cell (the acidity) must be closely regulated. Molecules that can give up protons will do so more readily if the concentration of H_3O^+ in solution is low and will tend to receive them back if the concentration in solution is high.

The opposite of an acid is a **base**. Just as the defining property of an acid is that it donates protons to a water molecule so as to raise the concentration of H_3O^+ ions, the defining property of a base is that it raises the concentration of hydroxyl (OH^-) ions—which are formed by removal of a proton from a water

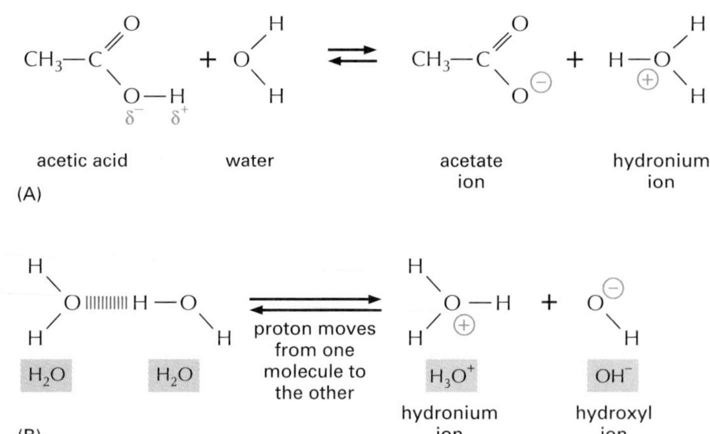

Figure 2–13 Acids in water. (A) The reaction that takes place when a molecule of acetic acid dissolves in water. (B) Water molecules are continuously exchanging protons with each other to form hydronium and hydroxyl ions. These ions in turn rapidly recombine to form water molecules.

TABLE 2–2 Covalent and Noncovalent Chemical Bonds

BOND TYPE	LENGTH (nm)	STRENGTH (kcal/mole) IN VACUUM	IN WATER
Covalent	0.15	90	90
Noncovalent: ionic	0.25	80	3
hydrogen	0.30	4	1
van der Waals attraction (per atom)	0.35	0.1	0.1

molecule. Thus sodium hydroxide (NaOH) is basic (the term *alkaline* is also used) because it dissociates in aqueous solution to form Na^+ ions and OH^- ions. Another class of bases, especially important in living cells, are those that contain NH_2 groups. These groups can generate OH^- by taking a proton from water: $-NH_2 + H_2O \rightarrow -NH_3^+ + OH^-$.

Because an OH^- ion combines with a H_3O^+ ion to form two water molecules, an increase in the OH^- concentration forces a decrease in the concentration of H_3O^+, and vice versa. A pure solution of water contains an equally low concentration (10^{-7} M) of both ions; it is neither acidic nor basic and is therefore said to be *neutral* with a pH of 7.0. The inside of cells is kept close to neutrality.

Four Types of Noncovalent Interactions Help Bring Molecules Together in Cells

In aqueous solutions, covalent bonds are 10 to 100 times stronger than the other attractive forces between atoms, allowing their connections to define the boundaries of one molecule from another. But much of biology depends on the specific binding of different molecules to each other. This binding is mediated by a group of noncovalent attractions that are individually quite weak, but whose bond energies can sum to create an effective force between two separate molecules. We have already introduced three of these noncovalent forces: ionic bonds, hydrogen bonds and van der Waals attractions. In Table 2–2, the strengths of these three types of bonds are compared to that of a typical covalent bond, both in the presence and the absence of water. Because of their fundamental importance in all biological systems, we shall summarize their properties here.

- **Ionic bonds**. These are purely electrostatic attractions between oppositely charged atoms. As we saw for NaCl, these forces are quite strong in the absence of water. However, the polar water molecules cluster around both fully charged ions and polar molecules that contain permanent dipoles (Figure 2–14). This greatly reduces the potential attractiveness of these charged species for each other (see Table 2–2).
- **Hydrogen bonds**. The structure of a typical hydrogen bond is illustrated in Figure 2–15. This bond represents a special form of polar interaction in which an electropositive hydrogen atom is partially shared by two electronegative atoms. Its hydrogen can be viewed as a proton that has partially dissociated from a donor atom, allowing it to be shared by a second acceptor atom. Unlike a typical electrostatic interaction, this bond is highly directional—being strongest when a straight line can be drawn between all three of the involved atoms. As already discussed, water weakens these bonds by forming competing hydrogen-bond interactions with the involved molecules (see Table 2–2).
- **van der Waals attractions**. The electron cloud around any nonpolar atom will fluctuate, producing a flickering dipole. Such dipoles will transiently induce an oppositely polarized flickering dipole in a nearby atom. This interaction generates an attraction between atoms that is very weak. But since many atoms can be simultaneously in contact when two surfaces fit closely, the net result is often significant. These so-called van der Waals attractions are not weakened by water (see Table 2–2).

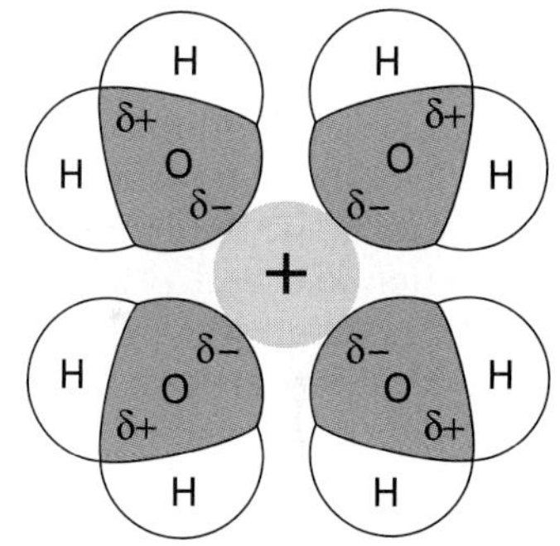

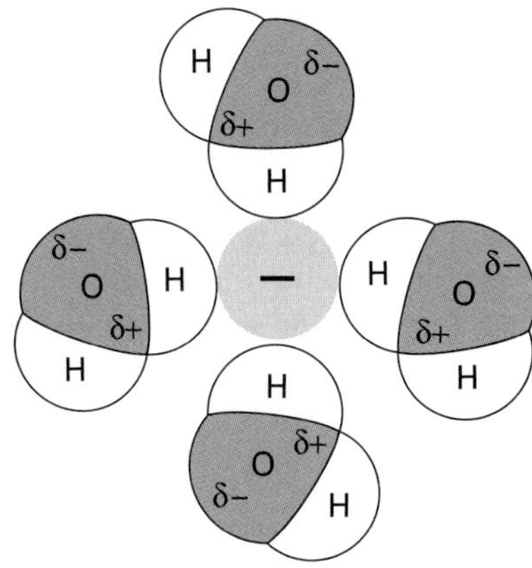

Figure 2–14 How the dipoles on water molecules orient to reduce the affinity of oppositely charged ions or polar groups for each other.

The fourth effect that can play an important part in bringing molecules together in water is a **hydrophobic force**. This force is caused by a pushing of nonpolar surfaces out of the hydrogen-bonded water network, where they would physically interfere with the highly favorable interactions between water molecules. Because bringing two nonpolar surfaces together reduces their contact with water, the force is a rather nonspecific one. Nevertheless, we shall see in Chapter 3 that hydrophobic forces are central to the proper folding of protein molecules.

Panel 2–3 provides an overview of the four types of interactions just described. And Figure 2–16 illustrates, in a schematic way, how many such interactions can sum to hold together the matching surfaces of two macromolecules, even though each interaction by itself would be much too weak to be effective.

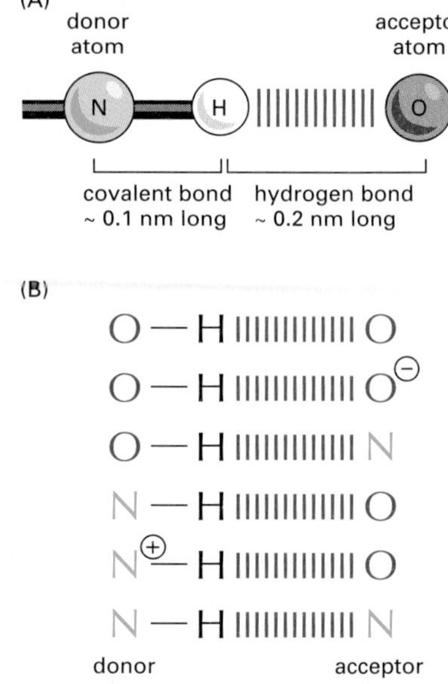

Figure 2–15 Hydrogen bonds. (A) Ball-and-stick model of a typical hydrogen bond. The distance between the hydrogen and the oxygen atom here is less than the sum of their van der Waals radii, indicating a partial sharing of electrons. (B) The most common hydrogen bonds in cells.

A Cell Is Formed from Carbon Compounds

Having looked at the ways atoms combine into small molecules and how these molecules behave in an aqueous environment, we now examine the main classes of small molecules found in cells and their biological roles. We shall see that a few basic categories of molecules, formed from a handful of different elements, give rise to all the extraordinary richness of form and behavior shown by living things.

If we disregard water, nearly all the molecules in a cell are based on carbon. Carbon is outstanding among all the elements in its ability to form large molecules; silicon is a poor second. Because it is small and has four electrons and four vacancies in its outermost shell, a carbon atom can form four covalent bonds with other atoms. Most important, one carbon atom can join to other carbon atoms through highly stable covalent C–C bonds to form chains and rings and hence generate large and complex molecules with no obvious upper limit to their size (see Panel 2–1, pp. 110–111). The small and large carbon compounds made by cells are called *organic molecules*.

Certain combinations of atoms, such as the methyl ($-CH_3$), hydroxyl (–OH), carboxyl (–COOH), carbonyl (–C=O), phosphate ($-PO_3^{2-}$), and amino ($-NH_2$) groups, occur repeatedly in organic molecules. Each such **chemical group** has distinct chemical and physical properties that influence the behavior of the molecule in which the group occurs. The most common chemical groups and some of their properties are summarized in Panel 2–1, pp. 110–111.

Cells Contain Four Major Families of Small Organic Molecules

The small organic molecules of the cell are carbon-based compounds that have molecular weights in the range 100 to 1000 and contain up to 30 or so carbon atoms. They are usually found free in solution and have many different fates. Some are used as *monomer* subunits to construct the giant polymeric *macromolecules*—the proteins, nucleic acids, and large polysaccharides—of the cell. Others act as energy sources and are broken down and transformed into other small molecules in a maze of intracellular metabolic pathways. Many small molecules have more than one role in the cell—for example, acting both as a potential subunit for a macromolecule and as an energy source. Small organic molecules are much less abundant than the organic macromolecules, accounting for only about one-tenth of the total mass of organic matter in a cell (Table 2–3). As a rough guess, there may be a thousand different kinds of these small molecules in a typical cell.

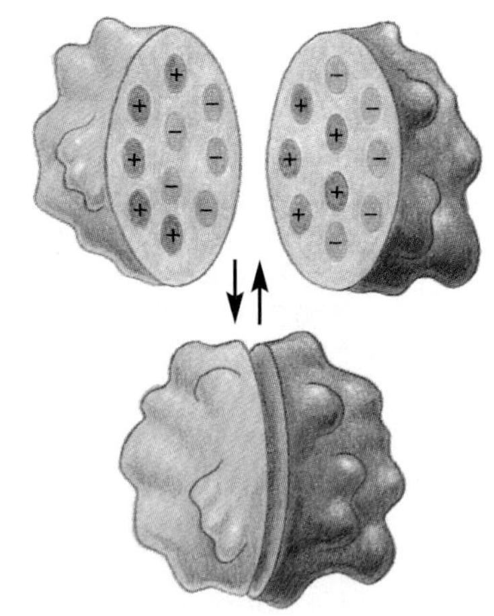

Figure 2–16 How two macromolecules with complementary surfaces can bind tightly to one another through noncovalent interactions. In this schematic illustration, plus and minus are used to mark chemical groups that can form attractive interactions when paired.

All organic molecules are synthesized from and are broken down into the same set of simple compounds. Both their synthesis and their breakdown occur through sequences of chemical changes that are limited in scope and follow definite rules. As a consequence, the compounds in a cell are chemically related and most can be classified into a small number of distinct families. Broadly speaking, cells contain four major families of small organic molecules: the *sugars*, the *fatty acids*, the *amino acids*, and the *nucleotides* (Figure 2–17). Although many compounds present in cells do not fit into these categories, these four

TABLE 2–3 The Approximate Chemical Composition of a Bacterial Cell

	PERCENT OF TOTAL CELL WEIGHT	NUMBER OF TYPES OF EACH MOLECULE
Water	70	1
Inorganic ions	1	20
Sugars and precursors	1	250
Amino acids and precursors	0.4	100
Nucleotides and precursors	0.4	100
Fatty acids and precursors	1	50
Other small molecules	0.2	~300
Macromolecules (proteins, nucleic acids, and polysaccharides)	26	~3000

families of small organic molecules, together with the macromolecules made by linking them into long chains, account for a large fraction of cell mass (see Table 2–3).

Sugars Provide an Energy Source for Cells and Are the Subunits of Polysaccharides

The simplest **sugars**—the *monosaccharides*—are compounds with the general formula $(CH_2O)_n$, where n is usually 3, 4, 5, 6, 7, or 8. Sugars, and the molecules made from them, are also called *carbohydrates* because of this simple formula. Glucose, for example, has the formula $C_6H_{12}O_6$ (Figure 2–18). The formula, however, does not fully define the molecule: the same set of carbons, hydrogens, and oxygens can be joined together by covalent bonds in a variety of ways, creating structures with different shapes. As shown in Panel 2–4 (pp. 116–117), for example, glucose can be converted into a different sugar—mannose or galactose—simply by switching the orientations of specific OH groups relative to the rest of the molecule. Each of these sugars, moreover, can exist in either of two forms, called the D-form and the L-form, which are mirror images of each other. Sets of molecules with the same chemical formula but different structures are called *isomers*, and the subset of such molecules that are mirror-image pairs are called *optical isomers*. Isomers are widespread among organic molecules in general, and they play a major part in generating the enormous variety of sugars.

An outline of sugar structures and chemistry is given in Panel 2–4. Sugars can exist in either a ring or an open-chain form. In their open-chain form, sugars contain a number of hydroxyl groups and either one aldehyde ($_{\text{H}}\!\!>$C=O) or one ketone (>C=O) group. The aldehyde or ketone group plays a special role. First, it can react with a hydroxyl group in the same molecule to convert the molecule into a ring; in the ring form the carbon of the original aldehyde or ketone group can be recognized as the only one that is bonded to two oxygens. Second, once the ring is formed, this same carbon can become further linked to one of the carbons bearing a hydroxyl group on another sugar molecule, creating a *disaccharide*; such as sucrose, which is composed of a glucose and a fructose unit. Larger sugar polymers range from the *oligosaccharides* (trisaccharides, tetrasaccharides, and so on) up to giant *polysaccharides*, which can contain thousands of monosaccharide units.

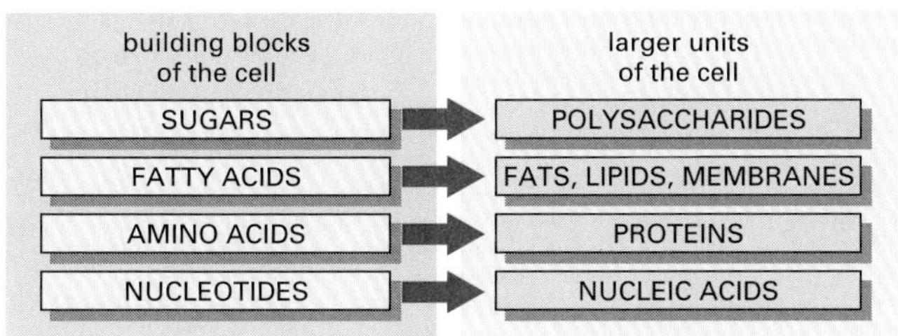

Figure 2–17 The four main families of small organic molecules in cells. These small molecules form the monomeric building blocks, or subunits, for most of the macromolecules and other assemblies of the cell. Some, like the sugars and the fatty acids, are also energy sources.

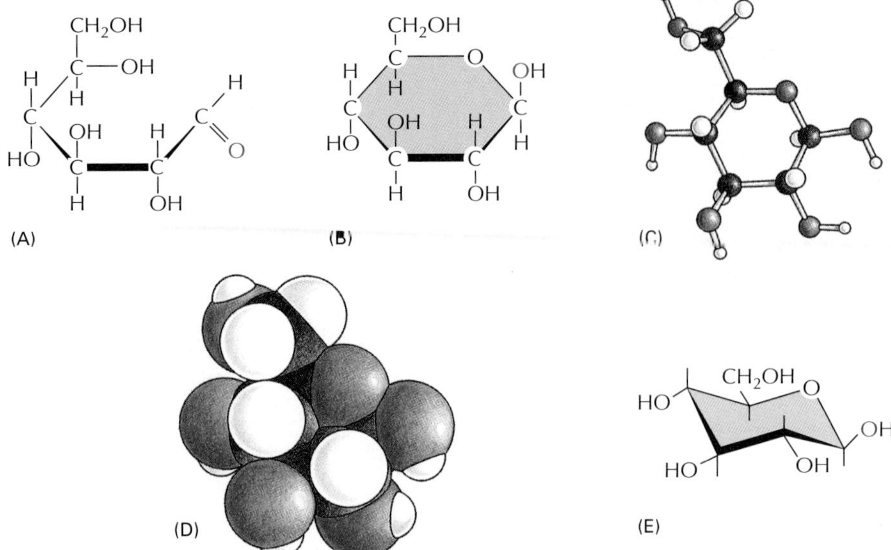

Figure 2–18 The structure of glucose, a simple sugar. As illustrated previously for water (see Figure 2–12), any molecule can be represented in several ways. In the structural formulas shown in (A), (B) and (E), the atoms are shown as chemical symbols linked together by lines representing the covalent bonds. The *thickened lines* here are used to indicate the plane of the sugar ring, in an attempt to emphasize that the –H and –OH groups are not in the same plane as the ring. (A) The open-chain form of this sugar, which is in equilibrium with the more stable cyclic or ring form in (B). (C) A ball-and-stick model in which the three-dimensional arrangement of the atoms in space is shown. (D) A space-filling model, which, as well as depicting the three-dimensional arrangement of the atoms, also uses the van der Waals radii to represent the surface contours of the molecule. (E) The chair form is an alternative way to draw the cyclic molecule that reflects the geometry more accurately than the structural formula in (B). The atoms in (C) and (D) are drawn according to the conventional color coding for atoms. For example, these colors are H, *white*; C, *black*; O, *red*; N, *blue* (see also Figure 2–8).

The way that sugars are linked together to form polymers illustrates some common features of biochemical bond formation. A bond is formed between an –OH group on one sugar and an –OH group on another by a **condensation reaction,** in which a molecule of water is expelled as the bond is formed (Figure 2–19). Subunits in other biological polymers, such as nucleic acids and proteins, are also linked by condensation reactions in which water is expelled. The bonds created by all of these condensation reactions can be broken by the reverse process of **hydrolysis**, in which a molecule of water is consumed (see Figure 2–19).

Because each monosaccharide has several free hydroxyl groups that can form a link to another monosaccharide (or to some other compound), sugar polymers can be branched, and the number of possible polysaccharide structures is extremely large. Even a simple disaccharide consisting of two glucose residues can exist in eleven different varieties (Figure 2–20), while three different hexoses ($C_6H_{12}O_6$) can join together to make several thousand trisaccharides. For this reason it is a much more complex task to determine the arrangement of sugars in a polysaccharide than to determine the nucleotide sequence of a DNA molecule, where each unit is joined to the next in exactly the same way.

The monosaccharide *glucose* has a central role as an energy source for cells. In a series of reactions, it is broken down to smaller molecules, releasing energy that the cell can harness to do useful work, as we shall explain later. Cells use

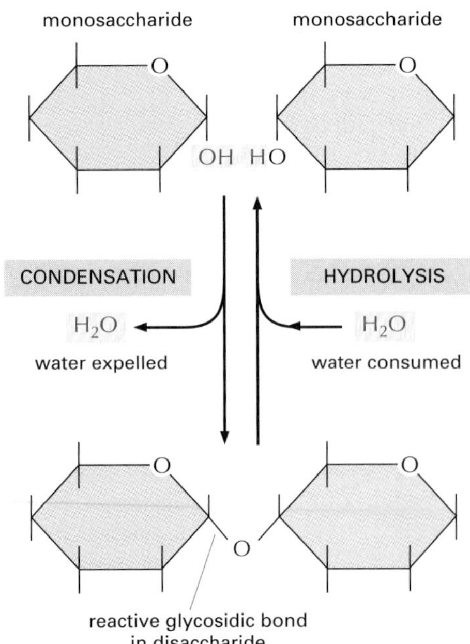

Figure 2–19 The reaction of two monosaccharides to form a disaccharide. This reaction belongs to a general category of reactions termed *condensation reactions,* in which two molecules join together as a result of the loss of a water molecule. The reverse reaction (in which water is added) is termed *hydrolysis.* Note that one of the two partners (the one on the *left* here) is the carbon joined to two oxygens through which the sugar ring forms (see Figure 2–18). As indicated, this common type of covalent bond between two sugar molecules is known as a *glycosidic bond* (see also Figure 2–20).

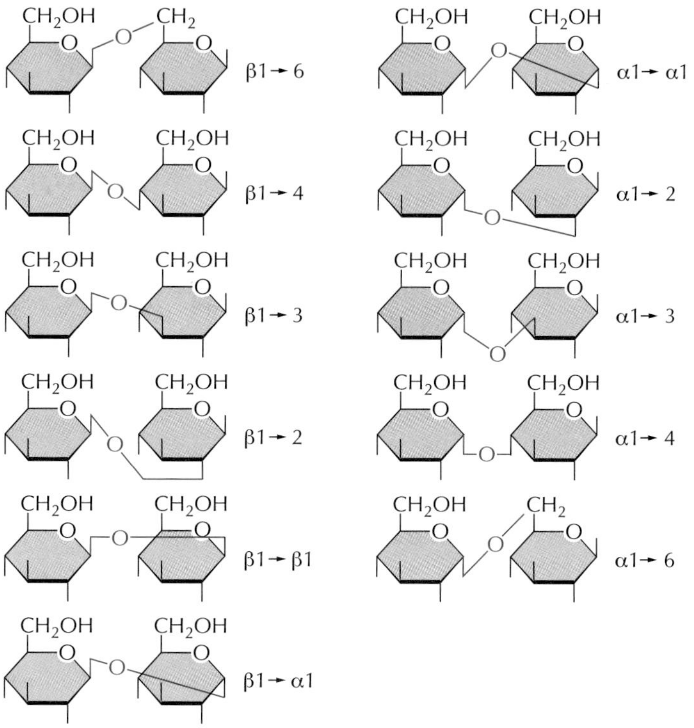

Figure 2–20 Eleven disaccharides consisting of two D-glucose units. Although these differ only in the type of linkage between the two glucose units, they are chemically distinct. Since the oligosaccharides associated with proteins and lipids may have six or more different kinds of sugar joined in both linear and branched arrangements through glycosidic bonds such as those illustrated here, the number of distinct types of oligosaccharides that can be used in cells is extremely large. For an explanation of α and β linkages, see Panel 2–4 (pp. 116–117).

simple polysaccharides composed only of glucose units—principally *glycogen* in animals and starch in plants—as long-term stores of energy.

Sugars do not function only in the production and storage of energy. They also can be used, for example, to make mechanical supports. Thus, the most abundant organic chemical on Earth—the *cellulose* of plant cell walls—is a polysaccharide of glucose. Another extraordinarily abundant organic substance, the *chitin* of insect exoskeletons and fungal cell walls, is also a polysaccharide—in this case a linear polymer of a sugar derivative called *N*-acetylglucosamine. Polysaccharides of various other sorts are the main components of slime, mucus, and gristle.

Smaller oligosaccharides can be covalently linked to proteins to form glycoproteins and to lipids to form *glycolipids*, which are found in cell membranes. As described in Chapter 10, the surfaces of most cells are clothed and decorated with sugar polymers belonging to glycoproteins and glycolipids in the cell membrane. These sugar side chains are often recognized selectively by other cells. And differences between people in the details of their cell-surface sugars are the molecular basis for the major different human blood groups.

Fatty Acids Are Components of Cell Membranes

A fatty acid molecule, such as *palmitic acid*, has two chemically distinct regions (Figure 2–21). One is a long hydrocarbon chain, which is hydrophobic and not very reactive chemically. The other is a carboxyl (–COOH) group, which behaves as an acid (carboxylic acid): it is ionized in solution ($-COO^-$), extremely hydrophilic, and chemically reactive. Almost all the fatty acid molecules in a cell are covalently linked to other molecules by their carboxylic acid group.

The hydrocarbon tail of *palmitic acid* is saturated: it has no double bonds between carbon atoms and contains the maximum possible number of hydrogens. Stearic acid, another one of the common fatty acids in animal fat, is also *saturated*. Some other fatty acids, such as oleic acid, have *unsaturated* tails, with one or more double bonds along their length. The double bonds create kinks in the molecules, interfering with their ability to pack together in a solid mass. It is this that accounts for the difference between hard (saturated) and soft (polyunsaturated) margarine. The many different fatty acids found in cells differ only in the length of their hydrocarbon chains and the number and position of the carbon–carbon double bonds (see Panel 2–5, pp. 118–119).

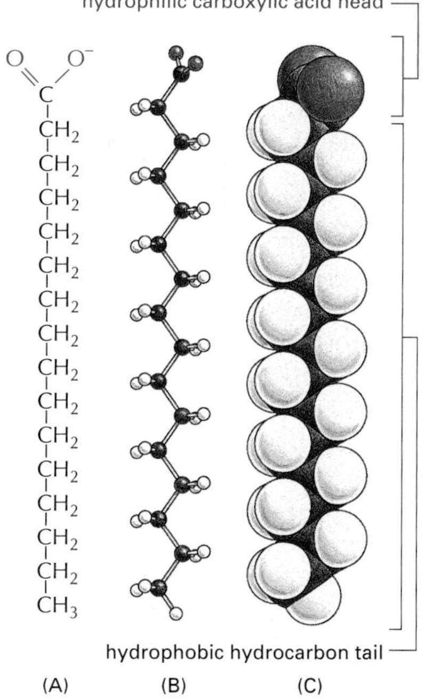

Figure 2–21 A fatty acid. A fatty acid is composed of a hydrophobic hydrocarbon chain to which is attached a hydrophilic carboxylic acid group. Palmitic acid is shown here. Different fatty acids have different hydrocarbon tails. (A) Structural formula. The carboxylic acid group is shown in its ionized form. (B) Ball-and-stick model. (C) Space-filling model.

Fatty acids serve as a concentrated food reserve in cells, because they can be broken down to produce about six times as much usable energy, weight for weight, as glucose. They are stored in the cytoplasm of many cells in the form of droplets of *triacylglycerol* molecules, which consist of three fatty acid chains joined to a glycerol molecule (see Panel 2–5); these molecules are the animal fats found in meat, butter, and cream, and the plant oils like corn oil and olive oil. When required to provide energy, the fatty acid chains are released from triacylglycerols and broken down into two-carbon units. These two-carbon units are identical to those derived from the breakdown of glucose and they enter the same energy-yielding reaction pathways, as will be described later in this chapter.

Fatty acids and their derivatives such as triacylglycerols are examples of *lipids*. Lipids comprise a loosely defined collection of biological molecules with the common feature that they are insoluble in water, while being soluble in fat and organic solvents such as benzene. They typically contain either long hydrocarbon chains, as in the fatty acids and isoprenes, or multiple linked aromatic rings, as in the *steroids*.

The most important function of fatty acids in cells is in the construction of cell membranes. These thin sheets enclose all cells and surround their internal organelles. They are composed largely of *phospholipids*, which are small molecules that, like triacylglycerols, are constructed mainly from fatty acids and glycerol. In phospholipids the glycerol is joined to two fatty acid chains, however, rather than to three as in triacylglycerols. The "third" site on the glycerol is linked to a hydrophilic phosphate group, which is in turn attached to a small hydrophilic compound such as choline (see Panel 2–5). Each phospholipid molecule, therefore, has a hydrophobic tail composed of the two fatty acid chains and a hydrophilic head, where the phosphate is located. This gives them different physical and chemical properties from triacylglycerols, which are predominantly hydrophobic. Molecules like phospholipids, with both hydrophobic and hydrophilic regions, are termed *amphipathic*.

The membrane-forming property of phospholipids results from their amphipathic nature. Phospholipids will spread over the surface of water to form a monolayer of phospholipid molecules, with the hydrophobic tails facing the air and the hydrophilic heads in contact with the water. Two such molecular layers can readily combine tail-to-tail in water to make a phospholipid sandwich, or *lipid bilayer*. This bilayer is the structural basis of all cell membranes (Figure 2–22).

Amino Acids Are the Subunits of Proteins

Amino acids are a varied class of molecules with one defining property: they all possess a carboxylic acid group and an amino group, both linked to a single carbon atom called the α-carbon (Figure 2–23). Their chemical variety comes from the side chain that is also attached to the α-carbon. The importance of amino

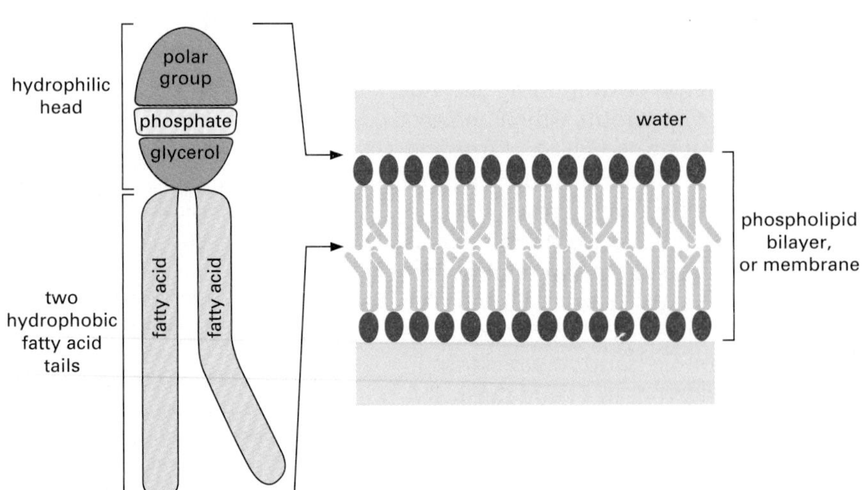

Figure 2–22 Phospholipid structure and the orientation of phospholipids in membranes. In an aqueous environment, the hydrophobic tails of phospholipids pack together to exclude water. Here they have formed a bilayer with the hydrophilic head of each phospholipid facing the water. Lipid bilayers are the basis for cell membranes, as discussed in detail in Chapter 10.

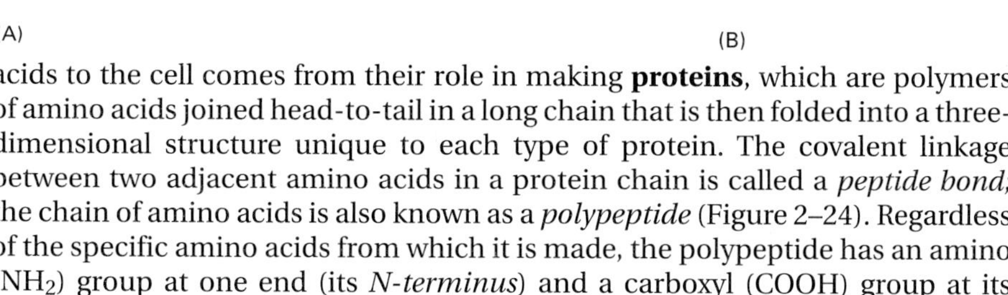

Figure 2–23 The amino acid alanine. (A) In the cell, where the pH is close to 7, the free amino acid exists in its ionized form; but when it is incorporated into a polypeptide chain, the charges on the amino and carboxyl groups disappear. (B) A ball-and-stick model and (C) a space-filling model of alanine (H, *white;* C, *black;* O, *red;* N, *blue*).

acids to the cell comes from their role in making **proteins**, which are polymers of amino acids joined head-to-tail in a long chain that is then folded into a three-dimensional structure unique to each type of protein. The covalent linkage between two adjacent amino acids in a protein chain is called a *peptide bond;* the chain of amino acids is also known as a *polypeptide* (Figure 2–24). Regardless of the specific amino acids from which it is made, the polypeptide has an amino (NH_2) group at one end (its *N-terminus*) and a carboxyl (COOH) group at its other end (its *C-terminus*). This gives it a definite directionality—a structural (as opposed to an electrical) polarity.

Twenty types of amino acids are found commonly in proteins, each with a different side chain attached to the α-carbon atom (see Panel 3–1, pp. 132–133). The same 20 amino acids occur over and over again in all proteins, whether from bacteria, plants, or animals. How this precise set of 20 amino acids came to be chosen is one of the mysteries surrounding the evolution of life; there is no obvious chemical reason why other amino acids could not have served just as well. But once the choice was established, it could not be changed; too much depended on it.

Like sugars, all amino acids, except glycine, exist as optical isomers in D- and L-forms (see Panel 3–1). But only L-forms are ever found in proteins (although D-amino acids occur as part of bacterial cell walls and in some antibiotics). The origin of this exclusive use of L-amino acids to make proteins is another evolutionary mystery.

The chemical versatility that the 20 standard amino acids provide is vitally important to the function of proteins. Five of the 20 amino acids have side chains that can form ions in solution and thereby can carry a charge (Figure 2–25). The others are uncharged; some are polar and hydrophilic, and some are nonpolar and hydrophobic. As we shall discuss in Chapter 3, the collective properties of the amino acid side chains underlie all the diverse and sophisticated functions of proteins.

N-terminal end of polypeptide chain

Phe, Ser, Glu, Lys

C-terminal end of polypeptide chain

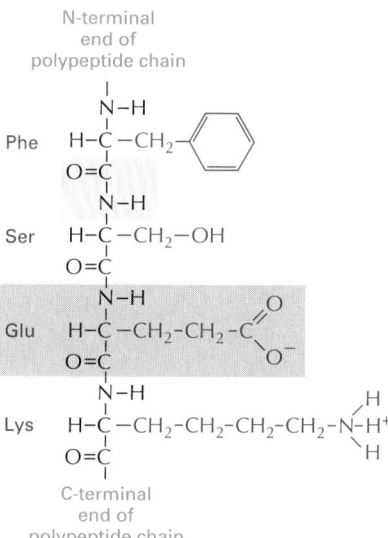

Figure 2–24 A small part of a protein molecule. The four amino acids shown are linked together by three peptide bonds, one of which is highlighted in *yellow.* One of the amino acids is shaded in *gray.* The amino acid side chains are shown in *red.* The two ends of a polypeptide chain are chemically distinct. One end, the N-terminus, terminates in an amino group, and the other, the C-terminus, in a carboxyl group. The sequence is always read from the N-terminal end; hence this sequence is Phe–Ser–Glu–Lys.

Nucleotides Are the Subunits of DNA and RNA

A **nucleotide** is a molecule made up of a nitrogen-containing ring compound linked to a five-carbon sugar. This sugar can be either ribose or deoxyribose, and it carries one or more phosphate groups (Panel 2–6, pp. 120–121). Nucleotides containing ribose are known as ribonucleotides, and those containing deoxyribose as deoxyribonucleotides. The nitrogen-containing rings are generally referred to as *bases* for historical reasons: under acidic conditions they can each bind an H^+ (proton) and thereby increase the concentration of OH^- ions in aqueous solution. There is a strong family resemblance between the different bases. *Cytosine (C), thymine (T),* and *uracil (U)* are called pyrimidines because they all derive from a six-membered pyrimidine ring; *guanine (G)* and *adenine (A)* are *purine* compounds, and they have a second, five-membered ring fused to the six-membered ring. Each nucleotide is named from the base it contains (see Panel 2–6).

Nucleotides can act as short-term carriers of chemical energy. Above all others, the ribonucleotide **adenosine triphosphate**, or **ATP** (Figure 2–26), is used to transfer energy in hundreds of different cellular reactions. ATP is formed through reactions that are driven by the energy released by the oxidative

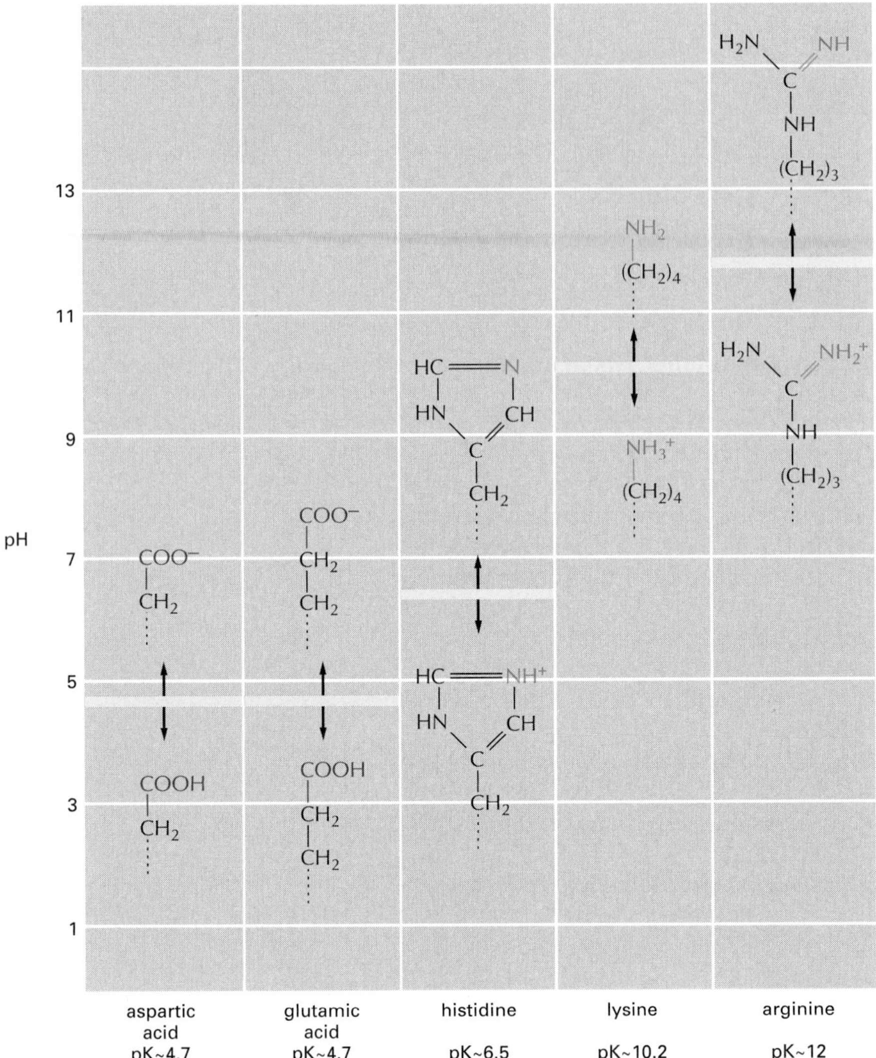

Figure 2–25 The charge on amino acid side chains depends on the pH. The five different side chains that can carry a charge are shown. Carboxylic acids can readily lose H^+ in aqueous solution to form a negatively charged ion, which is denoted by the suffix "-ate," as in aspart*ate* or glutam*ate*. A comparable situation exists for amines, which in aqueous solution can take up H^+ to form a positively charged ion (which does not have a special name). These reactions are rapidly reversible, and the amounts of the two forms, charged and uncharged, depend on the pH of the solution. At a high pH, carboxylic acids tend to be charged and amines uncharged. At a low pH, the opposite is true—the carboxylic acids are uncharged and amines are charged. The pH at which exactly *half* of the carboxylic acid or amine residues are charged is known as the pK of that amino acid side chain.

In the cell the pH is close to 7, and almost all carboxylic acids and amines are in their fully charged form.

breakdown of foodstuffs. Its three phosphates are linked in series by two *phosphoanhydride bonds,* whose rupture releases large amounts of useful energy. The terminal phosphate group in particular is frequently split off by hydrolysis, often transferring a phosphate to other molecules and releasing energy that drives energy-requiring biosynthetic reactions (Figure 2–27). Other nucleotide derivatives serve as carriers for the transfer of other chemical groups, as will be described later.

The most fundamental role of nucleotides in the cell, however, is in the storage and retrieval of biological information. Nucleotides serve as building blocks for the construction of *nucleic acids*—long polymers in which nucleotide subunits are covalently linked by the formation of a *phosphodiester bond* between the phosphate group attached to the sugar of one nucleotide and a hydroxyl

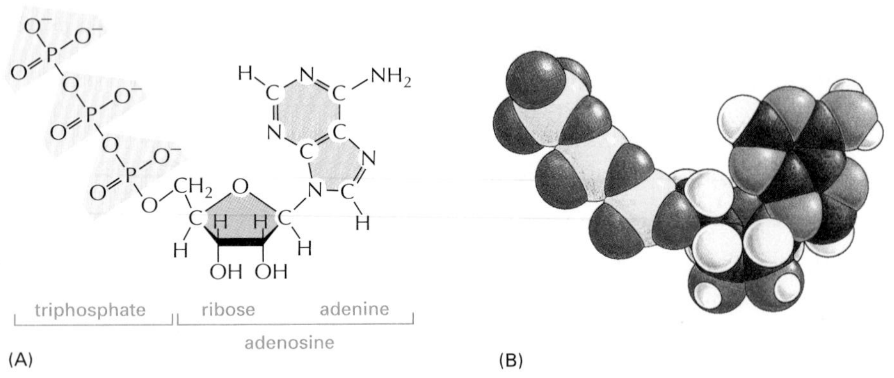

Figure 2–26 Chemical structure of adenosine triphosphate (ATP). (A) Structural formula. (B) Space-filling model. In (B) the colors of the atoms are C, *black;* N, *blue;* H, *white;* O, *red;* and P, *yellow.*

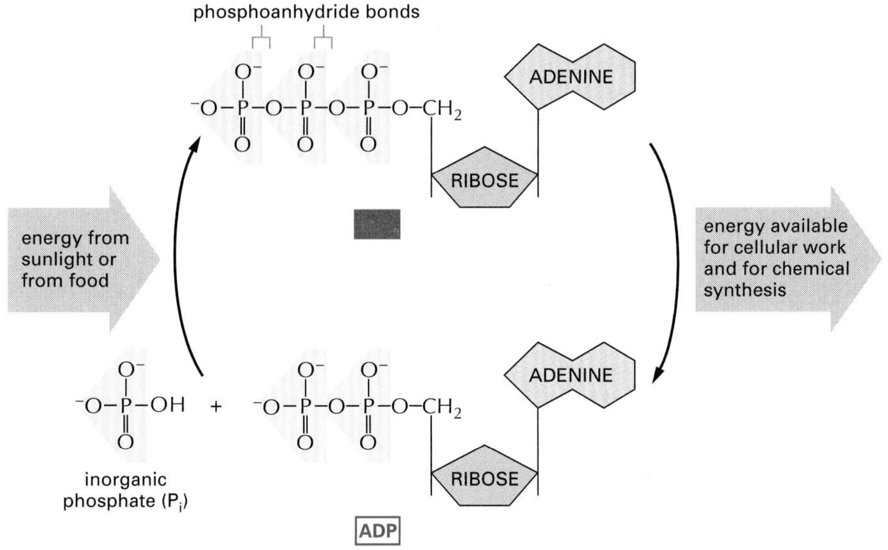

phosphoanhydride bonds

ATP

energy from sunlight or from food

energy available for cellular work and for chemical synthesis

inorganic phosphate (P_i)

ADP

Figure 2–27 The ATP molecule serves as an energy carrier in cells. The energy-requiring formation of ATP from ADP and inorganic phosphate is coupled to the energy-yielding oxidation of foodstuffs (in animal cells, fungi, and some bacteria) or to the capture of light energy (in plant cells and some bacteria). The hydrolysis of this ATP back to ADP and inorganic phosphate in turn provides the energy to drive many cellular reactions.

group on the sugar of the next nucleotide. Nucleic acid chains are synthesized from energy-rich nucleoside triphosphates by a condensation reaction that releases inorganic pyrophosphate during phosphodiester bond formation.

There are two main types of nucleic acids, differing in the type of sugar in their sugar-phosphate backbone. Those based on the sugar ribose are known as **ribonucleic acids**, or **RNA**, and contain the bases A, G, C, and U. Those based on *deoxyribose* (in which the hydroxyl at the 2′ position of the ribose carbon ring is replaced by a hydrogen (see Panel 2–6) are known as **deoxyribonucleic acids**, or **DNA**, and contain the bases A, G, C, and T (T is chemically similar to the U in RNA, merely adding the methyl group on the pyrimidine ring) (Figure 2–28). RNA usually occurs in cells in the form of a single polynucleotide chain, but DNA is virtually always in the form of a double-stranded molecule—a DNA double helix composed of two polynucleotide chains running antiparallel to each other and held together by hydrogen-bonding between the bases of the two chains.

The linear sequence of nucleotides in a DNA or an RNA encodes the genetic information of the cell. The ability of the bases in different nucleic acid molecules to recognize and pair with each other by hydrogen-bonding (called *base-pairing*)—G with C, and A with either T or U—underlies all of heredity and evolution, as explained in Chapter 4.

The Chemistry of Cells is Dominated by Macromolecules with Remarkable Properties

On a weight basis, macromolecules are by far the most abundant of the carbon-containing molecules in a living cell (Figure 2–29 and Table 2–4). They are the principal building blocks from which a cell is constructed and also the components that confer the most distinctive properties of living things. The macromolecules in cells are polymers that are constructed simply by covalently linking small organic molecules (called *monomers*, or *subunits*) into long chains (Figure 2–30). Yet they have many remarkable properties that could not have been predicted from their simple constituents.

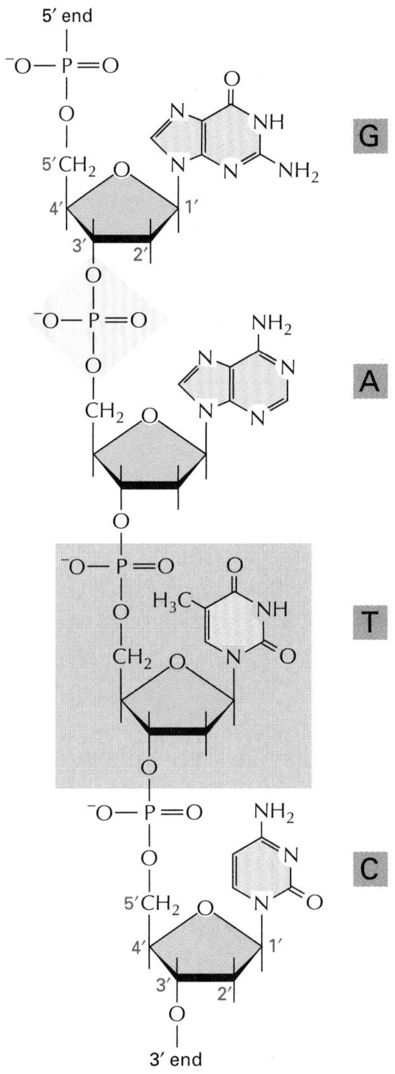

5′ end

G

A

T

C

3′ end

Figure 2–28 A small part of one chain of a deoxyribonucleic acid (DNA) molecule. Four nucleotides are shown. One of the phosphodiester bonds that links adjacent nucleotide residues is highlighted in *yellow*, and one of the nucleotides is shaded in *gray*. Nucleotides are linked together by a phosphodiester linkage between specific carbon atoms of the ribose, known as the 5′ and 3′ atoms. For this reason, one end of a polynucleotide chain, the 5′ end, will have a free phosphate group and the other, the 3′ end, a free hydroxyl group. The linear sequence of nucleotides in a polynucleotide chain is commonly abbreviated by a one-letter code, and the sequence is always read from the 5′ end. In the example illustrated the sequence is G–A–T–C.

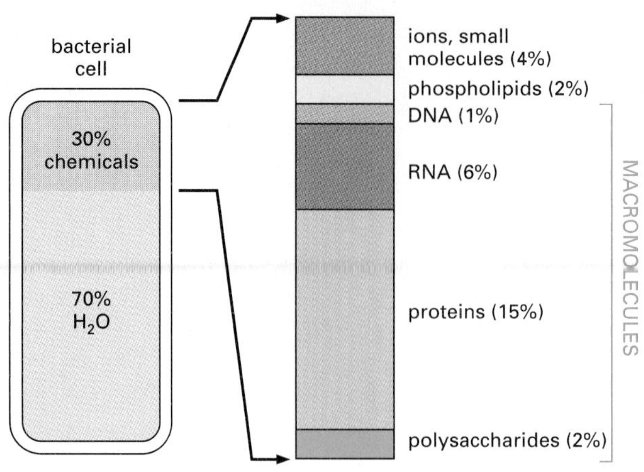

Figure 2–29 Macromolecules are abundant in cells. The approximate composition of a bacterial cell is shown by weight. The composition of an animal cell is similar (see Table 2–4).

Proteins are especially abundant and versatile. They perform thousands of distinct functions in cells. Many proteins serve as *enzymes,* the catalysts that direct the large number of covalent bond-making and bond-breaking reactions that the cell needs. All of the reactions whereby cells extract energy from food molecules are catalyzed by proteins serving as enzymes, for example, and an enzyme called ribulose bisphosphate carboxylase converts CO_2 to sugars in photosynthetic organisms, producing most of the organic matter needed for life on Earth. Other proteins are used to build structural components, such as tubulin, a protein that self-assembles to make the cell's long microtubules—or histones, proteins that compact the DNA in chromosomes. Yet other proteins act as molecular motors to produce force and movement, as in the case of myosin in muscle. Proteins can also have a wide variety of other functions, and we shall examine the molecular basis for many of them later in this book. Here we merely mention some general principles of macromolecular chemistry that make such functions possible.

Although the chemical reactions for adding subunits to each polymer are different in detail for proteins, nucleic acids, and polysaccharides, they share important features. Each polymer grows by the addition of a monomer onto the end of a growing polymer chain in a *condensation reaction,* in which a molecule of water is lost with each subunit added (see Figure 2–19). The stepwise

TABLE 2–4 Approximate Chemical Compositions of a Typical Bacterium and a Typical Mammalian Cell

	PERCENT OF TOTAL CELL WEIGHT	
COMPONENT	*E. COLI* BACTERIUM	MAMMALIAN CELL
H_2O	70	70
Inorganic ions (Na^+, K^+, Mg^{2+}, Ca^{2+}, Cl^-, etc.)	1	1
Miscellaneous small metabolites	3	3
Proteins	15	18
RNA	6	1.1
DNA	1	0.25
Phospholipids	2	3
Other lipids	–	2
Polysaccharides	2	2
Total cell volume	2×10^{-12} cm^3	4×10^{-9} cm^3
Relative cell volume	1	2000

Proteins, polysaccharides, DNA, and RNA are macromolecules. Lipids are not generally classed as macromolecules even though they share some of their features; for example, most are synthesized as linear polymers of a smaller molecule (the acetyl group on acetyl CoA), and they self-assemble into larger structures (membranes). Note that water and protein comprise most of the mass of both mammalian and bacterial cells.

polymerization of monomers into a long chain is a simple way to manufacture a large, complex molecule, since the subunits are added by the same reaction performed over and over again by the same set of enzymes. In a sense, the process resembles the repetitive operation of a machine in a factory—except in one crucial respect. Apart from some of the polysaccharides, most macromolecules are made from a set of monomers that are slightly different from one another—for example, the 20 different amino acids from which proteins are made. It is critical to life that the polymer chain is not assembled at random from these subunits; instead the subunits are added in a particular order, or *sequence*. The elaborate mechanisms that allow this to be accomplished by enzymes are described in detail in Chapters 5 and 6.

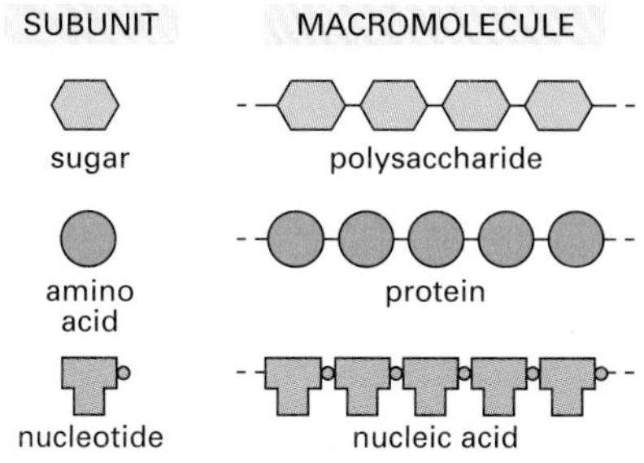

SUBUNIT | MACROMOLECULE
sugar | polysaccharide
amino acid | protein
nucleotide | nucleic acid

Figure 2–30 Three families of macromolecules. Each is a polymer formed from small molecules (called monomers, or subunits) linked together by covalent bonds.

Noncovalent Bonds Specify Both the Precise Shape of a Macromolecule and its Binding to Other Molecules

Most of the covalent bonds in a macromolecule allow rotation of the atoms they join, so that the polymer chain has great flexibility. In principle, this allows a macromolecule to adopt an almost unlimited number of shapes, or *conformations*, as the polymer chain writhes and rotates under the influence of random thermal energy. However, the shapes of most biological macromolecules are highly constrained because of the many weak *noncovalent bonds* that form between different parts of the molecule. If these noncovalent bonds are formed in sufficient numbers, the polymer chain can strongly prefer one particular conformation, determined by the linear sequence of monomers in its chain. Virtually all protein molecules and many of the small RNA molecules found in cells fold tightly into one highly preferred conformation in this way (Figure 2–31).

The four types of noncovalent interactions important in biological molecules have been previously described in this chapter, and they are reviewed in Panel 2–3 (pp. 114–115). Although individually very weak, these interactions not only cooperate to fold biological macromolecules into unique shapes: they can also add up to create a strong attraction between two different molecules when these molecules fit together very closely, like a hand in a glove. This form of molecular interaction provides for great specificity, inasmuch as the multipoint contacts required for strong binding make it possible for a macromolecule to select out—through binding—just one of the many thousands of other types of molecules present inside a cell. Moreover, because the strength of the binding depends on the number of noncovalent bonds that are formed, interactions of almost any affinity are possible—allowing rapid dissociation when necessary.

Binding of this type underlies all biological catalysis, making it possible for proteins to function as enzymes. Noncovalent interactions also allow macromolecules to be used as building blocks for the formation of larger structures. In cells, macromolecules often bind together into large complexes, thereby forming intricate machines with multiple moving parts that perform such complex tasks as DNA replication and protein synthesis (Figure 2–32).

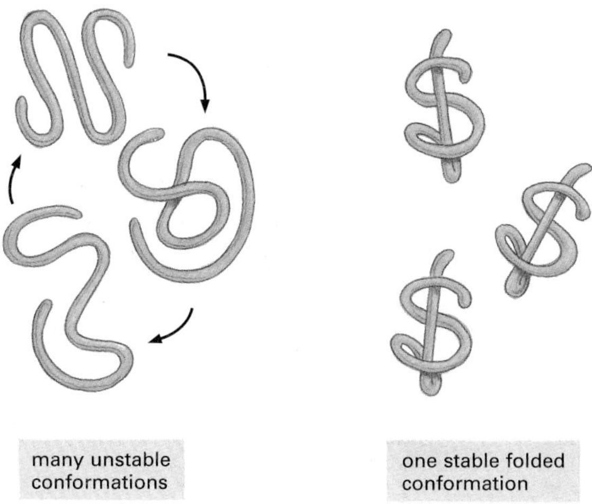

Figure 2–31 Most proteins and many RNA molecules fold into only one stable conformation. If the noncovalent bonds maintaining this stable conformation are disrupted, the molecule becomes a flexible chain that usually has no biological value.

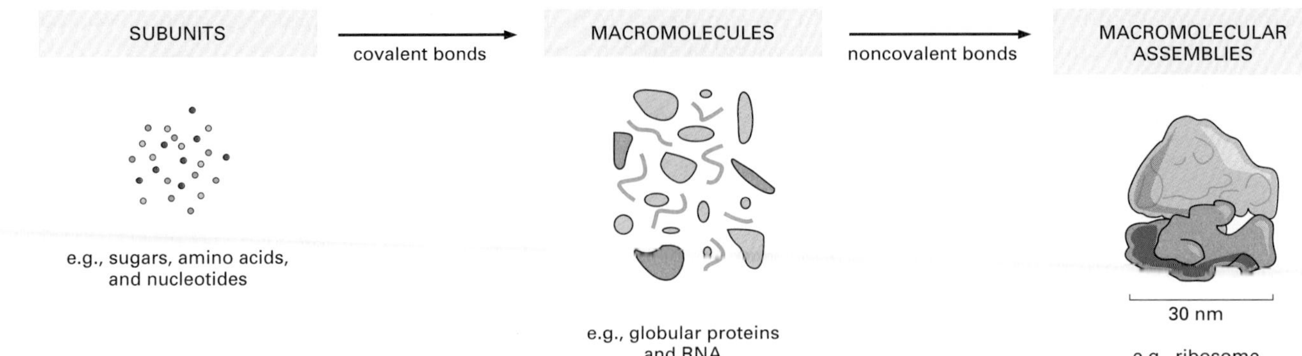

Figure 2–32 Small molecules, proteins, and a ribosome drawn approximately to scale. Ribosomes are a central part of the machinery that the cell uses to make proteins: each ribosome is formed as a complex of about 90 macromolecules (protein and RNA molecules).

Summary

Living organisms are autonomous, self-propagating chemical systems. They are made from a distinctive and restricted set of small carbon-based molecules that are essentially the same for every living species. Each of these molecules is composed of a small set of atoms linked to each other in a precise configuration through covalent bonds. The main categories are sugars, fatty acids, amino acids, and nucleotides. Sugars are a primary source of chemical energy for cells and can be incorporated into polysaccharides for energy storage. Fatty acids are also important for energy storage, but their most critical function is in the formation of cell membranes. Polymers consisting of amino acids constitute the remarkably diverse and versatile macromolecules known as proteins. Nucleotides play a central part in energy transfer. They are also the subunits from which the informational macromolecules, RNA and DNA, are made.

Most of the dry mass of a cell consists of macromolecules that have been produced as linear polymers of amino acids (proteins) or nucleotides (DNA and RNA), covalently linked to each other in an exact order. The protein molecules and many of the RNAs fold into a unique conformation that depends on their sequence of subunits. This folding process creates unique surfaces, and it depends on a large set of weak interactions produced by noncovalent forces between atoms. These forces are of four types: ionic bonds, hydrogen bonds, van der Waals attractions, and an interaction between nonpolar groups caused by their hydrophobic expulsion from water. The same set of weak forces governs the specific binding of other molecules to macromolecules, making possible the myriad associations between biological molecules that produce the structure and the chemistry of a cell.

CATALYSIS AND THE USE OF ENERGY BY CELLS

One property of living things above all makes them seem almost miraculously different from nonliving matter: they create and maintain order, in a universe that is tending always to greater disorder (Figure 2–33). To create this order, the cells in a living organism must perform a never-ending stream of chemical reactions. In some of these reactions, small organic molecules—amino acids, sugars, nucleotides, and lipids—are being taken apart or modified to supply the many other small molecules that the cell requires. In other reactions, these small

Figure 2–33 Order in biological structures. Well-defined, ornate, and beautiful spatial patterns can be found at every level of organization in living organisms. In order of increasing size: (A) protein molecules in the coat of a virus; (B) the regular array of microtubules seen in a cross section of a sperm tail; (C) surface contours of a pollen grain (a single cell); (D) close-up of the wing of a butterfly showing the pattern created by scales, each scale being the product of a single cell; (E) spiral array of seeds, made of millions of cells, in the head of a sunflower. (A, courtesy of R.A. Grant and J.M. Hogle; B, courtesy of Lewis Tilney; C, courtesy of Colin MacFarlane and Chris Jeffree; D and E, courtesy of Kjell B. Sandved.)

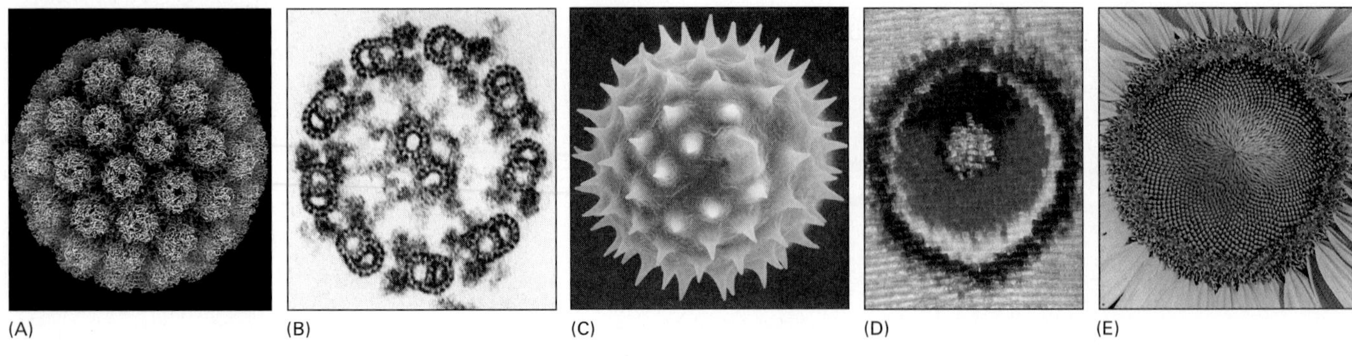

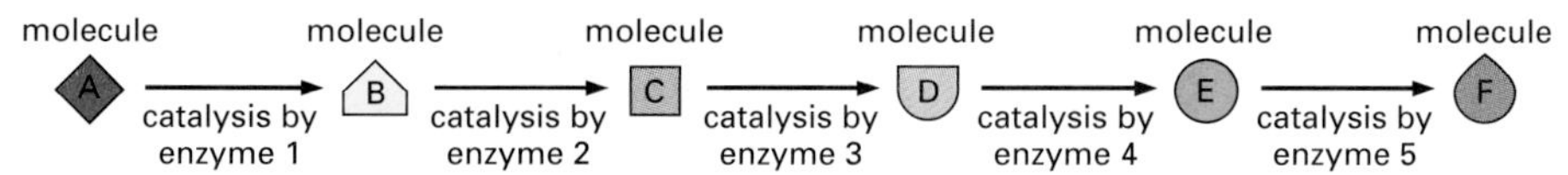

ABBREVIATED AS

Figure 2–34 How a set of enzyme-catalyzed reactions generates a metabolic pathway. Each enzyme catalyzes a particular chemical reaction, leaving the enzyme unchanged. In this example, a set of enzymes acting in series converts molecule A to molecule F, forming a metabolic pathway.

molecules are being used to construct an enormously diverse range of proteins, nucleic acids, and other macromolecules that endow living systems with all of their most distinctive properties. Each cell can be viewed as a tiny chemical factory, performing many millions of reactions every second.

Cell Metabolism Is Organized by Enzymes

The chemical reactions that a cell carries out would normally occur only at temperatures that are much higher than those existing inside cells. For this reason, each reaction requires a specific boost in chemical reactivity. This requirement is crucial, because it allows each reaction to be controlled by the cell. The control is exerted through the specialized proteins called *enzymes*, each of which accelerates, or *catalyzes*, just one of the many possible kinds of reactions that a particular molecule might undergo. Enzyme-catalyzed reactions are usually connected in series, so that the product of one reaction becomes the starting material, or *substrate*, for the next (Figure 2–34). These long linear reaction pathways are in turn linked to one another, forming a maze of interconnected reactions that enable the cell to survive, grow, and reproduce (Figure 2–35).

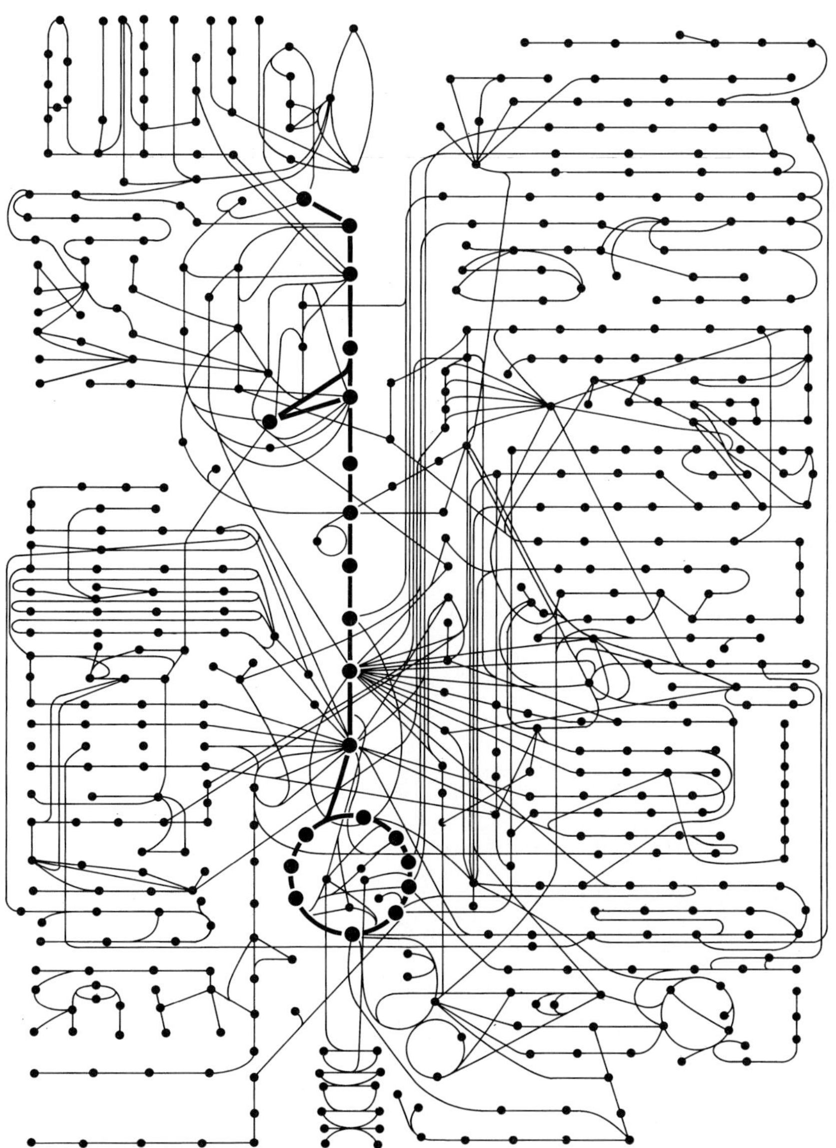

Figure 2–35 Some of the metabolic pathways and their interconnections in a typical cell. About 500 common metabolic reactions are shown diagrammatically, with each molecule in a metabolic pathway represented by a filled circle, as in the *yellow* box in Figure 2–34.

Two opposing streams of chemical reactions occur in cells: (1) the *catabolic* pathways break down foodstuffs into smaller molecules, thereby generating both a useful form of energy for the cell and some of the small molecules that the cell needs as building blocks, and (2) the *anabolic*, or *biosynthetic*, pathways use the energy harnessed by catabolism to drive the synthesis of the many other molecules that form the cell. Together these two sets of reactions constitute the **metabolism** of the cell (Figure 2–36).

Many of the details of cell metabolism form the traditional subject of *biochemistry* and need not concern us here. But the general principles by which cells obtain energy from their environment and use it to create order are central to cell biology. We begin with a discussion of why a constant input of energy is needed to sustain living organisms.

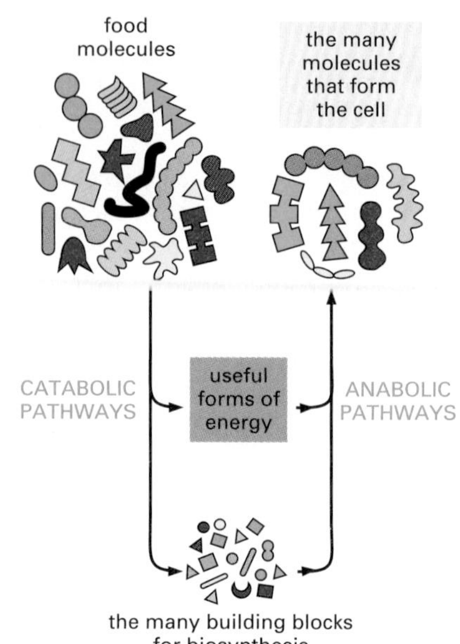

Figure 2–36 Schematic representation of the relationship between catabolic and anabolic pathways in metabolism. As suggested here, since a major portion of the energy stored in the chemical bonds of food molecules is dissipated as heat, the mass of food required by any organism that derives all of its energy from catabolism is much greater than the mass of the molecules that can be produced by anabolism.

Biological Order Is Made Possible by the Release of Heat Energy from Cells

The universal tendency of things to become disordered is expressed in a fundamental law of physics—the *second law of thermodynamics*—which states that in the universe, or in any isolated system (a collection of matter that is completely isolated from the rest of the universe), the degree of disorder can only increase. This law has such profound implications for all living things that it is worth restating in several ways.

For example, we can present the second law in terms of probability and state that systems will change spontaneously toward those arrangements that have the greatest probability. If we consider, for example, a box of 100 coins all lying heads up, a series of accidents that disturbs the box will tend to move the arrangement toward a mixture of 50 heads and 50 tails. The reason is simple: there is a huge number of possible arrangements of the individual coins in the mixture that can achieve the 50–50 result, but only one possible arrangement that keeps all of the coins oriented heads up. Because the 50–50 mixture is therefore the most probable, we say that it is more "disordered." For the same reason, it is a common experience that one's living space will become increasingly disordered without intentional effort: the movement toward disorder is a *spontaneous process*, requiring a periodic effort to reverse it (Figure 2–37).

The amount of disorder in a system can be quantified. The quantity that we use to measure this disorder is called the **entropy** of the system: the greater the

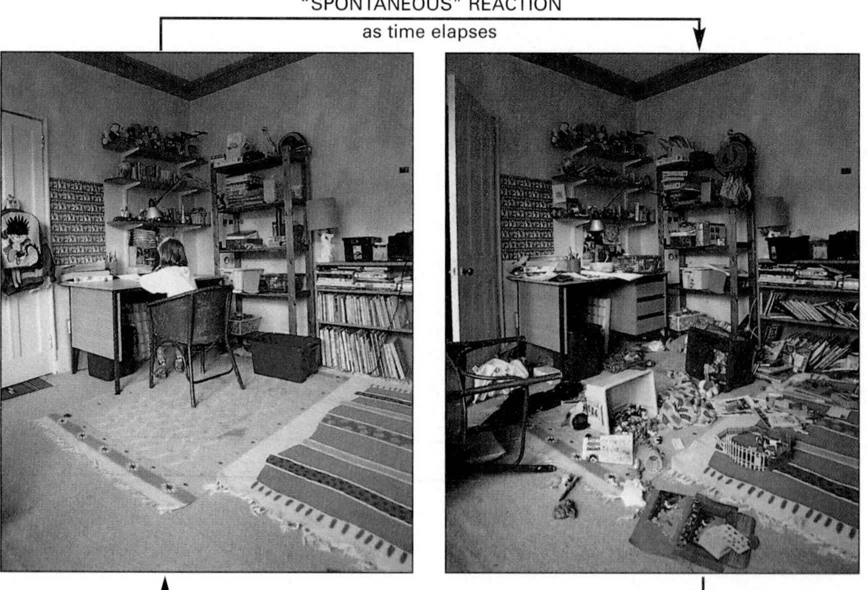

Figure 2–37 An everyday illustration of the spontaneous drive toward disorder. Reversing this tendency toward disorder requires an intentional effort and an input of energy: it is not spontaneous. In fact, from the second law of thermodynamics, we can be certain that the human intervention required will release enough heat to the environment to more than compensate for the reordering of the items in this room.

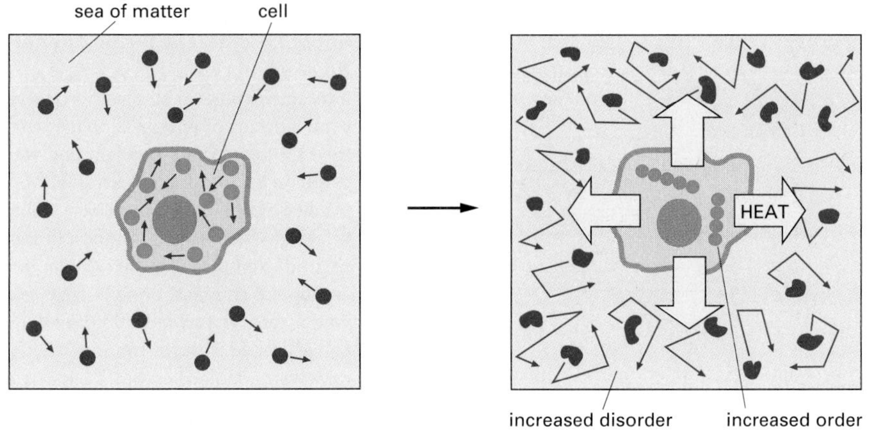

Figure 2–38 A simple thermodynamic analysis of a living cell. In the diagram on the left the molecules of both the cell and the rest of the universe (the sea of matter) are depicted in a relatively disordered state. In the diagram on the right the cell has taken in energy from food molecules and released heat by a reaction that orders the molecules the cell contains. Because the heat increases the disorder in the environment around the cell (depicted by the *jagged arrows* and distorted molecules, indicating the increased molecular motions caused by heat), the second law of thermodynamics—which states that the amount of disorder in the universe must always increase—is satisfied as the cell grows and divides. For a detailed discussion, see Panel 2–7 (pp. 122–123).

disorder, the greater the entropy. Thus, a third way to express the second law of thermodynamics is to say that systems will change spontaneously toward arrangements with greater entropy.

Living cells—by surviving, growing, and forming complex organisms—are generating order and thus might appear to defy the second law of thermodynamics. How is this possible? The answer is that a cell is not an isolated system: it takes in energy from its environment in the form of food, or as photons from the sun (or even, as in some chemosynthetic bacteria, from inorganic molecules alone), and it then uses this energy to generate order within itself. In the course of the chemical reactions that generate order, part of the energy that the cell uses is converted into heat. The heat is discharged into the cell's environment and disorders it, so that the total entropy—that of the cell plus its surroundings—increases, as demanded by the laws of physics.

To understand the principles governing these energy conversions, think of a cell as sitting in a sea of matter representing the rest of the universe. As the cell lives and grows, it creates internal order. But it releases heat energy as it synthesizes molecules and assembles them into cell structures. Heat is energy in its most disordered form—the random jostling of molecules. When the cell releases heat to the sea, it increases the intensity of molecular motions there (thermal motion)—thereby increasing the randomness, or disorder, of the sea. The second law of thermodynamics is satisfied because the increase in the amount of order inside the cell is more than compensated by a greater decrease in order (increase in entropy) in the surrounding sea of matter (Figure 2–38).

Where does the heat that the cell releases come from? Here we encounter another important law of thermodynamics. The *first law of thermodynamics* states that energy can be converted from one form to another, but that it cannot be created or destroyed. Some forms of energy are illustrated in Figure 2–39. The amount of energy in different forms will change as a result of the chemical reactions inside the cell, but the first law tells us that the total amount of energy must always be the same. For example, an animal cell takes in foodstuffs and converts some of the energy present in the chemical bonds between the atoms of these food molecules (chemical bond energy) into the random thermal motion of molecules (heat energy). This conversion of chemical energy into heat energy is essential if the reactions inside the cell are to cause the universe as a whole to become more disordered—as required by the second law.

The cell cannot derive any benefit from the heat energy it releases unless the heat-generating reactions inside the cell are directly linked to the processes that generate molecular order. It is the tight *coupling* of heat production to an increase in order that distinguishes the metabolism of a cell from the wasteful burning of fuel in a fire. Later in this chapter, we shall illustrate how this coupling occurs. For the moment, it is sufficient to recognize that a direct linkage of the "burning" of food molecules to the generation of biological order is required if cells are to be able to create and maintain an island of order in a universe tending toward chaos.

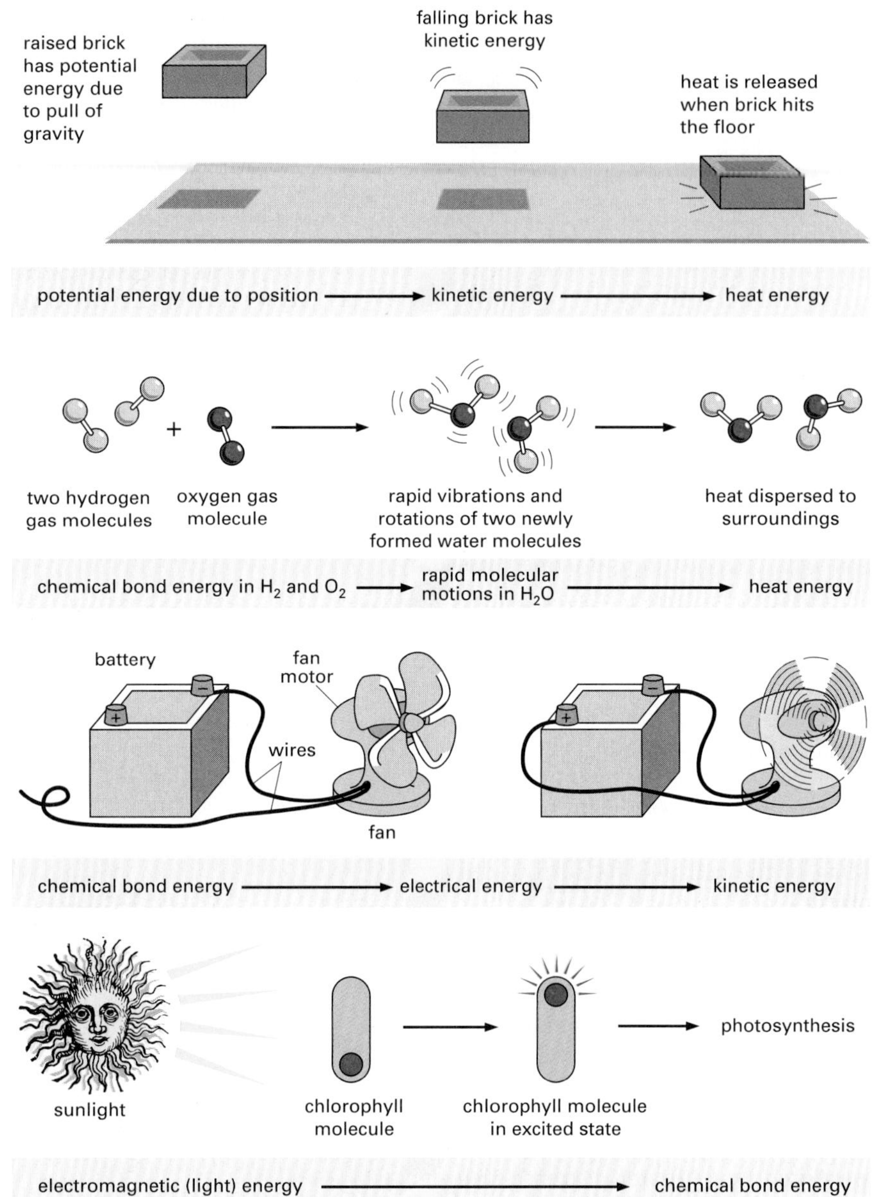

Figure 2–39 Some interconversions between different forms of energy. All energy forms are, in principle, interconvertible. In all these processes the total amount of energy is conserved; thus, for example, from the height and weight of the brick in the first example, we can predict exactly how much heat will be released when it hits the floor. In the second example, note that the large amount of chemical bond energy released when water is formed is initially converted to very rapid thermal motions in the two new water molecules; but collisions with other molecules almost instantaneously spread this kinetic energy evenly throughout the surroundings (heat transfer), making the new molecules indistinguishable from all the rest.

Photosynthetic Organisms Use Sunlight to Synthesize Organic Molecules

All animals live on energy stored in the chemical bonds of organic molecules made by other organisms, which they take in as food. The molecules in food also provide the atoms that animals need to construct new living matter. Some animals obtain their food by eating other animals. But at the bottom of the animal food chain are animals that eat plants. The plants, in turn, trap energy directly from sunlight. As a result, all of the energy used by animal cells is derived ultimately from the sun.

Solar energy enters the living world through **photosynthesis** in plants and photosynthetic bacteria. Photosynthesis allows the electromagnetic energy in sunlight to be converted into chemical bond energy in the cell. Plants are able to obtain all the atoms they need from inorganic sources: carbon from atmospheric carbon dioxide, hydrogen and oxygen from water, nitrogen from ammonia and nitrates in the soil, and other elements needed in smaller amounts from inorganic salts in the soil. They use the energy they derive from sunlight to build these atoms into sugars, amino acids, nucleotides, and fatty acids. These small molecules in turn are converted into the proteins, nucleic acids, polysaccharides, and lipids that form the plant. All of these substances serve as food molecules for animals, if the plants are later eaten.

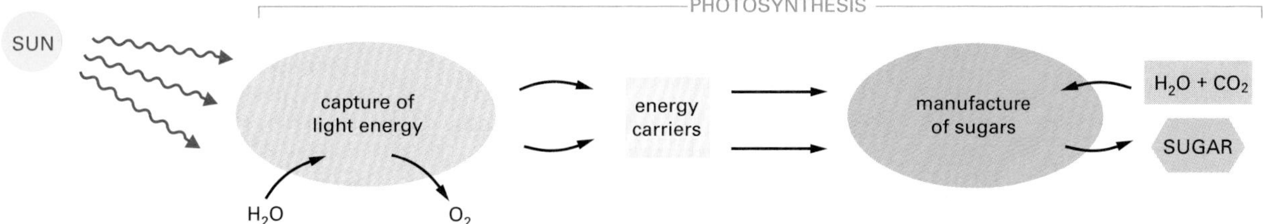

Figure 2–40 Photosynthesis. The two stages of photosynthesis. The energy carriers created in the first stage are two molecules that we discuss shortly—ATP and NADPH.

The reactions of photosynthesis take place in two stages (Figure 2–40). In the first stage, energy from sunlight is captured and transiently stored as chemical bond energy in specialized small molecules that act as carriers of energy and reactive chemical groups. (We discuss these activated carrier molecules later.) Molecular oxygen (O_2 gas) derived from the splitting of water by light is released as a waste product of this first stage.

In the second stage, the molecules that serve as energy carriers are used to help drive a *carbon fixation* process in which sugars are manufactured from carbon dioxide gas (CO_2) and water (H_2O), thereby providing a useful source of stored chemical bond energy and materials—both for the plant itself and for any animals that eat it. We describe the elegant mechanisms that underlie these two stages of photosynthesis in Chapter 14.

The net result of the entire process of photosynthesis, so far as the green plant is concerned, can be summarized simply in the equation

$$\text{light energy} + CO_2 + H_2O \rightarrow \text{sugars} + O_2 + \text{heat energy}$$

The sugars produced are then used both as a source of chemical bond energy and as a source of materials to make the many other small and large organic molecules that are essential to the plant cell.

Cells Obtain Energy by the Oxidation of Organic Molecules

All animal and plant cells are powered by energy stored in the chemical bonds of organic molecules, whether these be sugars that a plant has photosynthesized as food for itself or the mixture of large and small molecules that an animal has eaten. In order to use this energy to live, grow, and reproduce, organisms must extract it in a usable form. In both plants and animals, energy is extracted from food molecules by a process of gradual oxidation, or controlled burning.

The Earth's atmosphere contains a great deal of oxygen, and in the presence of oxygen the most energetically stable form of carbon is as CO_2 and that of hydrogen is as H_2O. A cell is therefore able to obtain energy from sugars or other organic molecules by allowing their carbon and hydrogen atoms to combine with oxygen to produce CO_2 and H_2O, respectively—a process called **respiration**.

Photosynthesis and respiration are complementary processes (Figure 2–41). This means that the transactions between plants and animals are not all one way. Plants, animals, and microorganisms have existed together on this planet for so long that many of them have become an essential part of the others' environments. The oxygen released by photosynthesis is consumed in the combustion of organic molecules by nearly all organisms. And some of the CO_2 molecules that are fixed today into organic molecules by photosynthesis in a green leaf were yesterday released into the atmosphere by the respiration of an animal—or by that of a fungus or bacterium decomposing dead organic matter. We therefore see that carbon utilization forms a huge cycle that involves the *biosphere* (all of the living organisms on Earth) as a whole, crossing boundaries between individual organisms (Figure 2–42). Similarly, atoms of nitrogen, phosphorus, and sulfur move between the living and nonliving worlds in cycles that involve plants, animals, fungi, and bacteria.

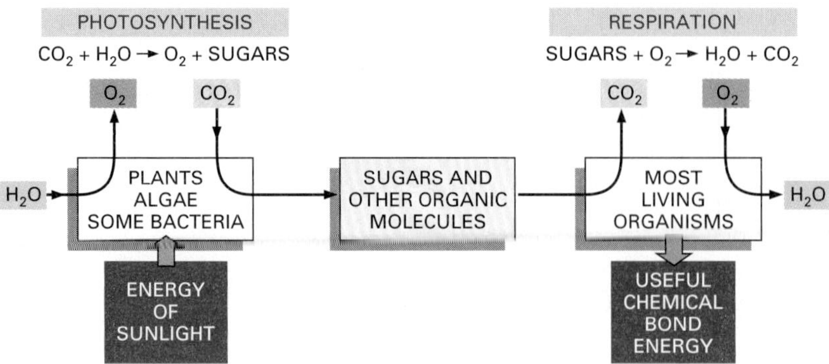

Figure 2–41 Photosynthesis and respiration as complementary processes in the living world. Photosynthesis uses the energy of sunlight to produce sugars and other organic molecules. These molecules in turn serve as food for other organisms. Many of these organisms carry out respiration, a process that uses O_2 to form CO_2 from the same carbon atoms that had been taken up as CO_2 and converted into sugars by photosynthesis. In the process, the organisms that respire obtain the chemical bond energy that they need to survive. The first cells on the Earth are thought to have been capable of neither photosynthesis nor respiration (discussed in Chapter 14). However, photosynthesis must have preceded respiration on the Earth, since there is strong evidence that billions of years of photosynthesis were required before O_2 had been released in sufficient quantity to create an atmosphere rich in this gas. (The Earth's atmosphere presently contains 20% O_2.)

Oxidation and Reduction Involve Electron Transfers

The cell does not oxidize organic molecules in one step, as occurs when organic material is burned in a fire. Through the use of enzyme catalysts, metabolism takes the molecules through a large number of reactions that only rarely involve the direct addition of oxygen. Before we consider some of these reactions and the purpose behind them, we need to discuss what is meant by the process of oxidation.

Oxidation, in the sense used above, does not mean only the addition of oxygen atoms; rather, it applies more generally to any reaction in which electrons are transferred from one atom to another. Oxidation in this sense refers to the removal of electrons, and **reduction**—the converse of oxidation—means the addition of electrons. Thus, Fe^{2+} is oxidized if it loses an electron to become Fe^{3+}, and a chlorine atom is reduced if it gains an electron to become Cl^-. Since the number of electrons is conserved (no loss or gain) in a chemical reaction, oxidation and reduction always occur simultaneously: that is, if one molecule gains an electron in a reaction (reduction), a second molecule loses the electron (oxidation). When a sugar molecule is oxidized to CO_2 and H_2O, for example, the O_2 molecules involved in forming H_2O gain electrons and thus are said to have been reduced.

The terms "oxidation" and "reduction" apply even when there is only a partial shift of electrons between atoms linked by a covalent bond (Figure 2–43). When a carbon atom becomes covalently bonded to an atom with a strong affinity for electrons, such as oxygen, chlorine, or sulfur, for example, it gives up more than its equal share of electrons and forms a *polar* covalent bond: the positive charge of the carbon nucleus is now somewhat greater than the negative charge of its electrons, and the atom therefore acquires a partial positive charge and is said to be oxidized. Conversely, a carbon atom in a C–H linkage has slightly more than its share of electrons, and so it is said to be reduced (see Figure 2–43).

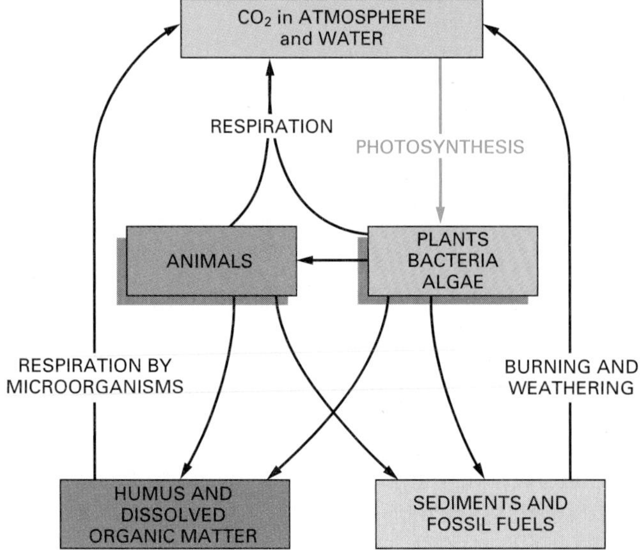

Figure 2–42 The carbon cycle. Individual carbon atoms are incorporated into organic molecules of the living world by the photosynthetic activity of plants, bacteria, and marine algae. They pass to animals, microorganisms, and organic material in soil and oceans in cyclic paths. CO_2 is restored to the atmosphere when organic molecules are oxidized by cells or burned by humans as fossil fuels.

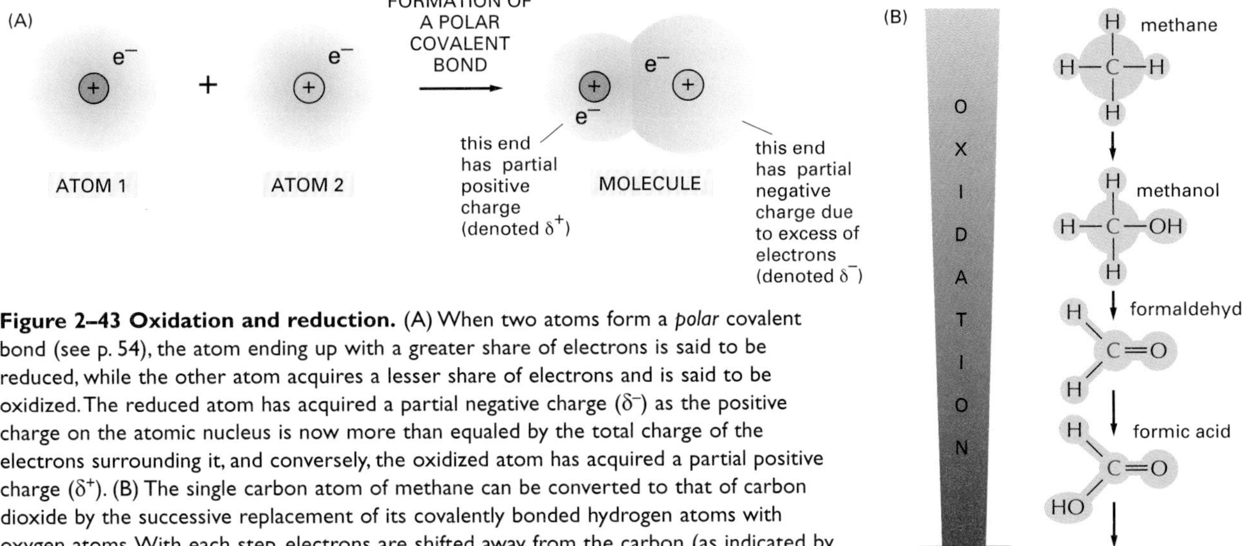

Figure 2–43 Oxidation and reduction. (A) When two atoms form a *polar* covalent bond (see p. 54), the atom ending up with a greater share of electrons is said to be reduced, while the other atom acquires a lesser share of electrons and is said to be oxidized. The reduced atom has acquired a partial negative charge (δ^-) as the positive charge on the atomic nucleus is now more than equaled by the total charge of the electrons surrounding it, and conversely, the oxidized atom has acquired a partial positive charge (δ^+). (B) The single carbon atom of methane can be converted to that of carbon dioxide by the successive replacement of its covalently bonded hydrogen atoms with oxygen atoms. With each step, electrons are shifted away from the carbon (as indicated by the *blue* shading), and the carbon atom becomes progressively more oxidized. Each of these steps is energetically favorable under the conditions present inside a cell.

When a molecule in a cell picks up an electron (e^-), it often picks up a proton (H^+) at the same time (protons being freely available in water). The net effect in this case is to add a hydrogen atom to the molecule

$$A + e^- + H^+ \rightarrow AH$$

Even though a proton plus an electron is involved (instead of just an electron), such *hydrogenation* reactions are reductions, and the reverse, *dehydrogenation* reactions, are oxidations. It is especially easy to tell whether an organic molecule is being oxidized or reduced: reduction is occurring if its number of C–H bonds increases, whereas oxidation is occurring if its number of C–H bonds decreases (see Figure 2–43B).

Cells use enzymes to catalyze the oxidation of organic molecules in small steps, through a sequence of reactions that allows useful energy to be harvested. We now need to explain how enzymes work and some of the constraints under which they operate.

Enzymes Lower the Barriers That Block Chemical Reactions

Consider the reaction

$$\text{paper} + O_2 \rightarrow \text{smoke} + \text{ashes} + \text{heat} + CO_2 + H_2O$$

The paper burns readily, releasing to the atmosphere both energy as heat and water and carbon dioxide as gases, but the smoke and ashes never spontaneously retrieve these entities from the heated atmosphere and reconstitute themselves into paper. When the paper burns, its chemical energy is dissipated as heat—not lost from the universe, since energy can never be created or destroyed, but irretrievably dispersed in the chaotic random thermal motions of molecules. At the same time, the atoms and molecules of the paper become dispersed and disordered. In the language of thermodynamics, there has been a loss of *free energy*, that is, of energy that can be harnessed to do work or drive chemical reactions. This loss reflects a loss of orderliness in the way the energy and molecules were stored in the paper. We shall discuss free energy in more detail shortly, but the general principle is clear enough intuitively: chemical reactions proceed only in the direction that leads to a loss of free energy; in other words, the spontaneous direction for any reaction is the direction that goes "downhill." A "downhill" reaction in this sense is often said to be *energetically favorable.*

Although the most energetically favorable form of carbon under ordinary conditions is as CO_2, and that of hydrogen is as H_2O, a living organism does not

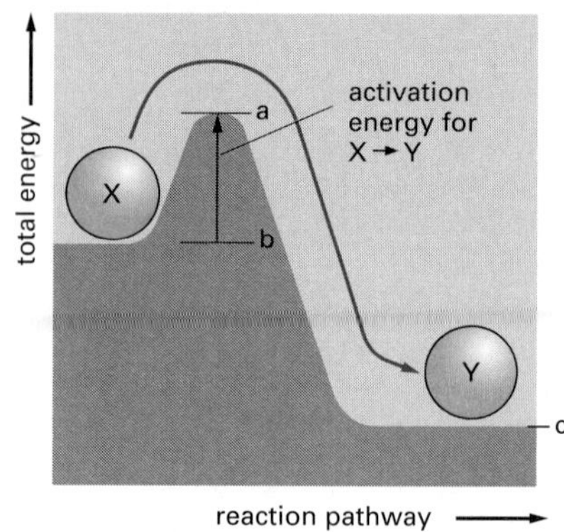

Figure 2–44 The important principle of activation energy. Compound X is in a stable state, and energy is required to convert it to compound Y, even though Y is at a lower overall energy level than X. This conversion will not take place, therefore, unless compound X can acquire enough activation energy (*energy a minus energy b*) from its surroundings to undergo the reaction that converts it into compound Y. This energy may be provided by means of an unusually energetic collision with other molecules. For the reverse reaction, Y → X, the activation energy will be much larger (*energy a minus energy c*); this reaction will therefore occur much more rarely. Activation energies are always positive; note, however, that the total energy change for the energetically favorable reaction X → Y is *energy c minus energy b*, a negative number.

disappear in a puff of smoke, and the book in your hands does not burst into flames. This is because the molecules both in the living organism and in the book are in a relatively stable state, and they cannot be changed to a state of lower energy without an input of energy: in other words, a molecule requires **activation energy**—a kick over an energy barrier—before it can undergo a chemical reaction that leaves it in a more stable state (Figure 2–44). In the case of a burning book, the activation energy is provided by the heat of a lighted match. For the molecules in the watery solution inside a cell, the kick is delivered by an unusually energetic random collision with surrounding molecules—collisions that become more violent as the temperature is raised.

In a living cell, the kick over the energy barrier is greatly aided by a specialized class of proteins—the **enzymes**. Each enzyme binds tightly to one or two molecules, called **substrates**, and holds them in a way that greatly reduces the activation energy of a particular chemical reaction that the bound substrates can undergo. A substance that can lower the activation energy of a reaction is termed a **catalyst**; catalysts increase the rate of chemical reactions because they allow a much larger proportion of the random collisions with surrounding molecules to kick the substrates over the energy barrier, as illustrated in Figure 2–45. Enzymes are among the most effective catalysts known, speeding up reactions by a factor of as much as 10^{14}, and they thereby allow reactions that would not otherwise occur to proceed rapidly at normal temperatures.

Enzymes are also highly selective. Each enzyme usually catalyzes only one particular reaction: in other words, it selectively lowers the activation energy of only one of the several possible chemical reactions that its bound substrate molecules could undergo. In this way, enzymes direct each of the many different molecules in a cell along specific reaction pathways (Figure 2–46).

The success of living organisms is attributable to a cell's ability to make enzymes of many types, each with precisely specified properties. Each enzyme has a unique shape containing an *active site*, a pocket or groove in the enzyme into which only particular substrates will fit (Figure 2–47). Like all other catalysts, enzyme molecules themselves remain unchanged after participating in a reaction and therefore can function over and over again. In Chapter 3, we discuss further how enzymes work, after we have looked in detail at the molecular structure of proteins.

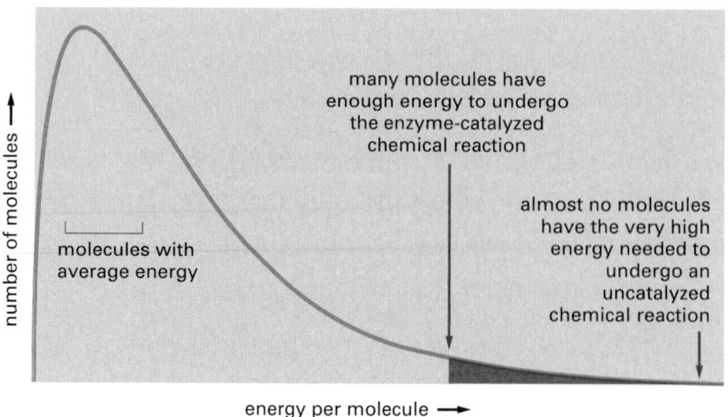

Figure 2–45 Lowering the activation energy greatly increases the probability of reaction. A population of identical substrate molecules will have a range of energies that is distributed as shown on the graph at any one instant. The varying energies come from collisions with surrounding molecules, which make the substrate molecules jiggle, vibrate, and spin. For a molecule to undergo a chemical reaction, the energy of the molecule must exceed the activation energy for that reaction; for most biological reactions, this almost never happens without enzyme catalysis. Even with enzyme catalysis, the substrate molecules must experience a particularly energetic collision to react, as indicated here.

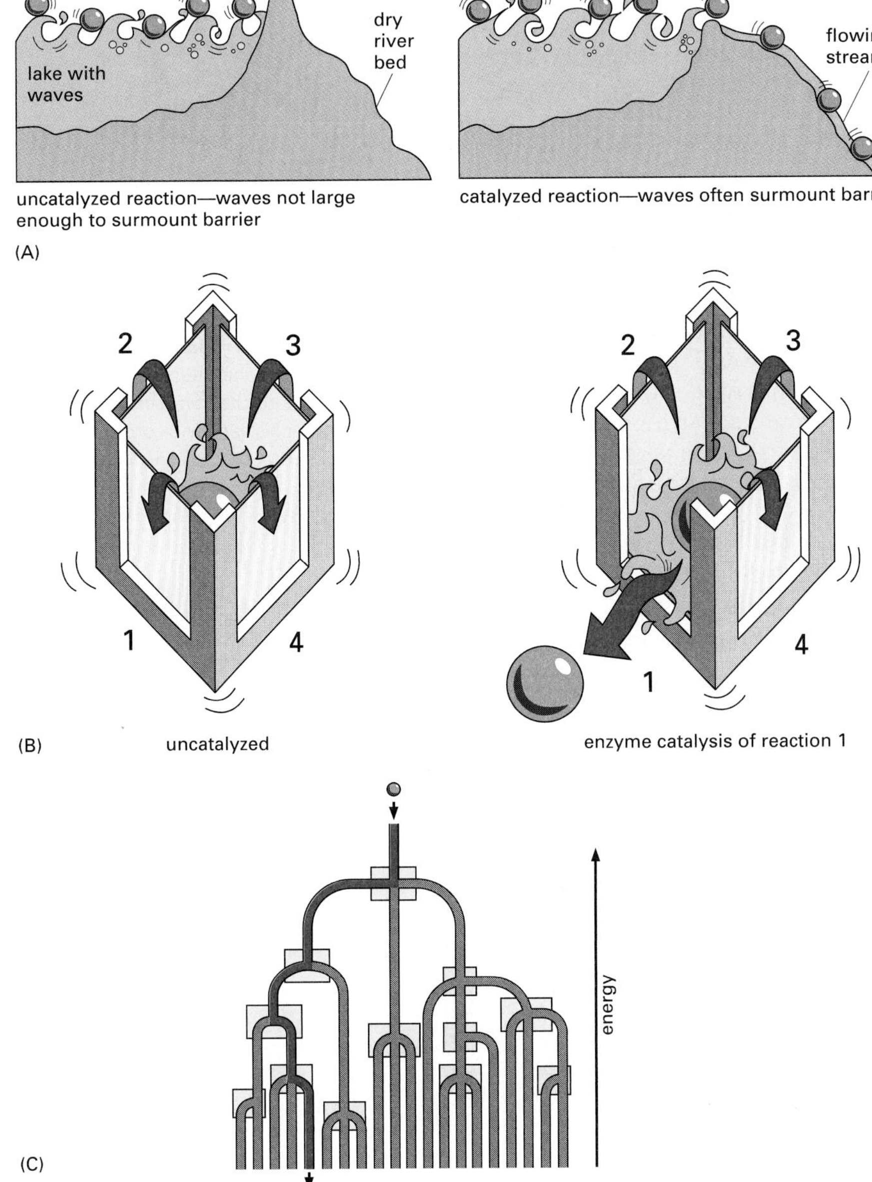

Figure 2–46 Floating ball analogies for enzyme catalysis. (A) A barrier dam is lowered to represent enzyme catalysis. The *green ball* represents a potential enzyme substrate (compound X) that is bouncing up and down in energy level due to constant encounters with waves (an analogy for the thermal bombardment of the substrate with the surrounding water molecules). When the barrier (activation energy) is lowered significantly, it allows the energetically favorable movement of the ball (the substrate) downhill. (B) The four walls of the box represent the activation energy barriers for four different chemical reactions that are all energetically favorable, in the sense that the products are at lower energy levels than the substrates. In the *left-hand* box, none of these reactions occurs because even the largest waves are not large enough to surmount any of the energy barriers. In the *right-hand* box, enzyme catalysis lowers the activation energy for reaction number 1 only; now the jostling of the waves allows passage of the molecule over this energy barrier only, inducing reaction 1. (C) A branching river with a set of barrier dams *(yellow boxes)* serves to illustrate how a series of enzyme-catalyzed reactions determines the exact reaction pathway followed by each molecule inside the cell.

How Enzymes Find Their Substrates: The Importance of Rapid Diffusion

A typical enzyme will catalyze the reaction of about a thousand substrate molecules every second. This means that it must be able to bind a new substrate molecule in a fraction of a millisecond. But both enzymes and their substrates are present in relatively small numbers in a cell. How do they find each other so fast? Rapid binding is possible because the motions caused by heat energy are enormously fast at the molecular level. These molecular motions can be classified

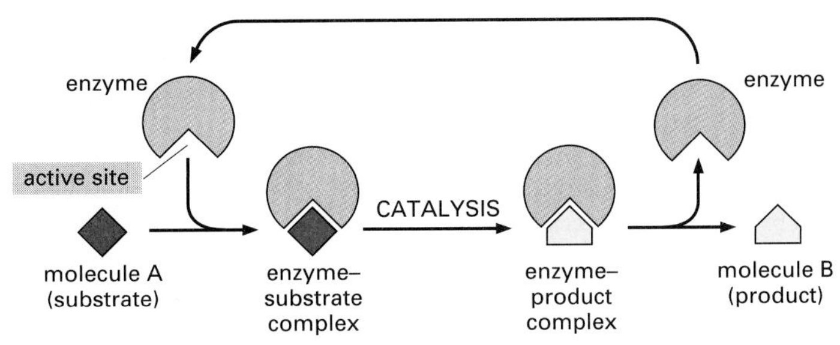

Figure 2–47 How enzymes work. Each enzyme has an active site to which one or two *substrate* molecules bind, forming an enzyme–substrate complex. A reaction occurs at the active site, producing an enzyme–product complex. The *product* is then released, allowing the enzyme to bind additional substrate molecules.

broadly into three kinds: (1) the movement of a molecule from one place to another (*translational motion*), (2) the rapid back-and-forth movement of covalently linked atoms with respect to one another (vibrations), and (3) rotations. All of these motions are important in bringing the surfaces of interacting molecules together.

These rates of molecular motions can be measured by a variety of spectroscopic techniques. These indicate that a large globular protein is constantly tumbling, rotating about its axis about a million times per second. Molecules are also in constant translational motion, which causes them to explore the space inside the cell very efficiently by wandering through it—a process called **diffusion**. In this way, every molecule in a cell collides with a huge number of other molecules each second. As the molecules in a liquid collide and bounce off one another, an individual molecule moves first one way and then another, its path constituting a *random walk* (Figure 2–48). In such a walk, the average distance that each molecule travels (as the crow flies) from its starting point is proportional to the square root of the time involved: that is, if it takes a molecule 1 second on average to travel 1 μm, it takes 4 seconds to travel 2 μm, 100 seconds to travel 10 μm, and so on.

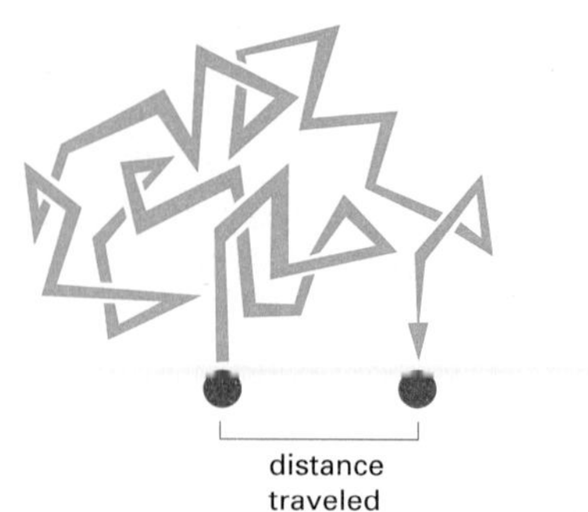

Figure 2–48 A random walk. Molecules in solution move in a random fashion due to the continual buffeting they receive in collisions with other molecules. This movement allows small molecules to diffuse rapidly from one part of the cell to another, as described in the text.

The inside of a cell is very crowded (Figure 2–49). Nevertheless, experiments in which fluorescent dyes and other labeled molecules are injected into cells show that small organic molecules diffuse through the watery gel of the cytosol nearly as rapidly as they do through water. A small organic molecule, for example, takes only about one-fifth of a second on average to diffuse a distance of 10 μm. Diffusion is therefore an efficient way for small molecules to move the limited distances in the cell (a typical animal cell is 15 μm in diameter).

Since enzymes move more slowly than substrates in cells, we can think of them as sitting still. The rate of encounter of each enzyme molecule with its substrate will depend on the concentration of the substrate molecule. For example, some abundant substrates are present at a concentration of 0.5 mM. Since pure water is 55 M, there is only about one such substrate molecule in the cell for every 10^5 water molecules. Nevertheless, the active site on an enzyme molecule that binds this substrate will be bombarded by about 500,000 random collisions with the substrate molecule per second. (For a substrate concentration tenfold lower, the number of collisions drops to 50,000 per second, and so on.) A random encounter between the surface of an enzyme and the matching surface of its substrate molecule often leads immediately to the formation of an enzyme–substrate complex that is ready to react. A reaction in which a covalent bond is broken or formed can now occur extremely rapidly. When one appreciates how quickly molecules move and react, the observed rates of enzymatic catalysis do not seem so amazing.

Once an enzyme and substrate have collided and snuggled together properly at the active site, they form multiple weak bonds with each other that persist until random thermal motion causes the molecules to dissociate again. In general, the stronger the binding of the enzyme and substrate, the slower their rate of dissociation. However, when two colliding molecules have poorly matching surfaces, few noncovalent bonds are formed and their total energy is negligible compared with that of thermal motion. In this case the two molecules dissociate as rapidly as they come together. This is what prevents incorrect and unwanted associations from forming between mismatched molecules, such as between an enzyme and the wrong substrate.

Figure 2–49 The structure of the cytoplasm. The drawing is approximately to scale and emphasizes the crowding in the cytoplasm. Only the macromolecules are shown: RNAs are shown in *blue*, ribosomes in *green*, and proteins in *red*. Enzymes and other macromolecules diffuse relatively slowly in the cytoplasm, in part because they interact with many other macromolecules; small molecules, by contrast, diffuse nearly as rapidly as they do in water. (Adapted from D.S. Goodsell, *Trends Biochem. Sci.* 16:203–206, 1991.)

The Free-Energy Change for a Reaction Determines Whether It Can Occur

We must now digress briefly to introduce some fundamental chemistry. Cells are chemical systems that must obey all chemical and physical laws. Although enzymes speed up reactions, they cannot by themselves force energetically unfavorable reactions to occur. In terms of a water analogy, enzymes by themselves cannot make water run uphill. Cells, however, must do just that in order to grow and divide: they must build highly ordered and energy-rich molecules from small and simple ones. We shall see that this is done through enzymes that directly *couple* energetically favorable reactions, which release energy and produce heat, to energetically unfavorable reactions, which produce biological order.

Before examining how such coupling is achieved, we must consider more carefully the term "energetically favorable." According to the second law of thermodynamics, a chemical reaction can proceed spontaneously only if it results in a net increase in the disorder of the universe (see Figure 2–38). The criterion for an increase in disorder of the universe can be expressed most conveniently in terms of a quantity called the **free energy, *G*,** of a system. The value of *G* is of interest only when a system undergoes a *change*, and the change in *G*, denoted ***ΔG*** (delta *G*), is critical. Suppose that the system being considered is a collection of molecules. As explained in Panel 2–7 (pp. 122–123), free energy has been defined such that *ΔG* directly measures the amount of disorder created in the universe when a reaction takes place that involves these molecules. *Energetically favorable reactions*, by definition, are those that decrease free energy, or, in other words, have a *negative ΔG* and disorder the universe (Figure 2–50).

A familiar example of an energetically favorable reaction on a macroscopic scale is the "reaction" by which a compressed spring relaxes to an expanded state, releasing its stored elastic energy as heat to its surroundings; an example on a microscopic scale is the dissolving of salt in water. Conversely, *energetically unfavorable reactions*, with a *positive ΔG*—such as those in which two amino acids are joined together to form a peptide bond—by themselves create order in the universe. Therefore, these reactions can take place only if they are coupled to a second reaction with a negative *ΔG* so large that the *ΔG* of the entire process is negative (Figure 2–51).

Y
ENERGETICALLY FAVORABLE REACTION
X
The free energy of Y is greater than the free energy of X. Therefore $\Delta G < 0$, and the disorder of the universe increases during the reaction.

this reaction can occur spontaneously

If the reaction X→Y occurred, ΔG would be > 0, and the universe would become more ordered.

this reaction can occur only if it is coupled to a second, energetically favorable reaction

Figure 2–50 The distinction between energetically favorable and energetically unfavorable reactions.

The Concentration of Reactants Influences *ΔG*

As we have just described, a reaction $A \rightleftharpoons B$ will go in the direction $A \rightarrow B$ when the associated free-energy change, *ΔG*, is negative, just as a tensed spring left to itself will relax and lose its stored energy to its surroundings as heat. For a chemical reaction, however, *ΔG* depends not only on the energy stored in each individual molecule, but also on the concentrations of the molecules in the reaction mixture. Remember that *ΔG* reflects the degree to which a reaction creates a more disordered—in other words, a more probable—state of the universe. Recalling our coin analogy, it is very likely that a coin will flip from a head to a tail orientation if a jiggling box contains 90 heads and 10 tails, but this is a less probable event if the box contains 10 heads and 90 tails. For exactly the same reason, for a reversible reaction $A \rightleftharpoons B$, a large excess of A over B will tend to drive the reaction in the direction $A \rightarrow B$; that is, there will be a tendency for there to be more molecules making the transition $A \rightarrow B$ than there are molecules making the transition $B \rightarrow A$. Therefore, the ΔG becomes more negative for the transition $A \rightarrow B$ (and more positive for the transition $B \rightarrow A$) as the ratio of A to B increases.

How much of a concentration difference is needed to compensate for a given decrease in chemical bond energy (and accompanying heat release)? The answer is not intuitively obvious, but it can be determined from a thermodynamic analysis that makes it possible to separate the concentration-dependent and the concentration-independent parts of the free-energy change. The *ΔG* for a given reaction can thereby be written as the sum of two parts: the first, called the *standard free-energy change, **ΔG°***, depends on the intrinsic characters of the

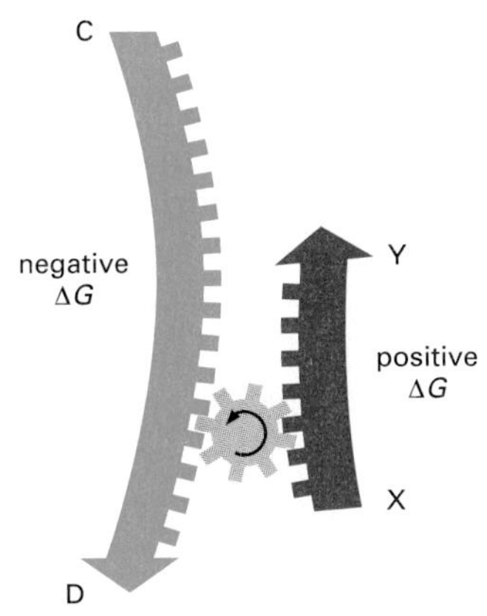

The energetically unfavorable reaction X→Y is driven by the energetically favorable reaction C→D, because the free-energy change for the pair of coupled reactions is less than zero.

Figure 2–51 How reaction coupling is used to drive energetically unfavorable reactions.

THE REACTION

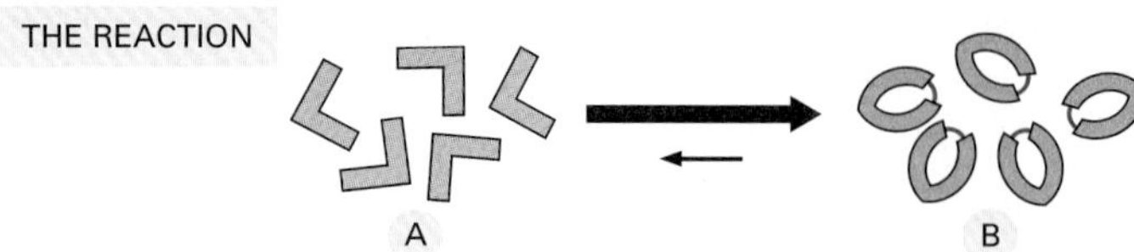

The formation of B is energetically favored in this example. But due to thermal bombardments, there will always be some B converting to A and vice versa.

SUPPOSE WE START WITH AN EQUAL NUMBER OF A AND B MOLECULES

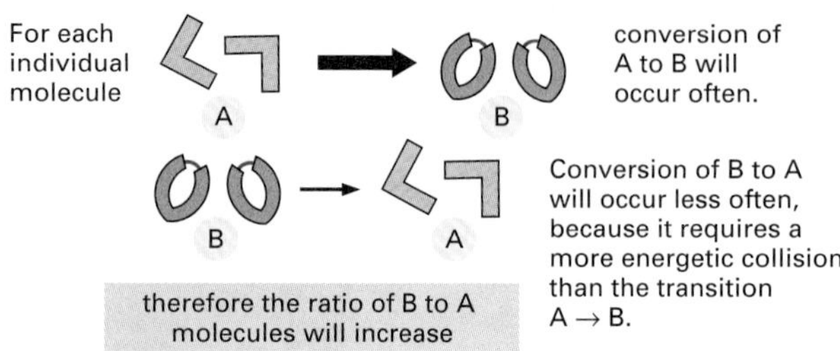

For each individual molecule
A
B
conversion of A to B will occur often.
B
A
Conversion of B to A will occur less often, because it requires a more energetic collision than the transition A → B.
therefore the ratio of B to A molecules will increase

EVENTUALLY there will be a large enough excess of B over A to just compensate for the slow rate of B → A. Equilibrium will then be attained.

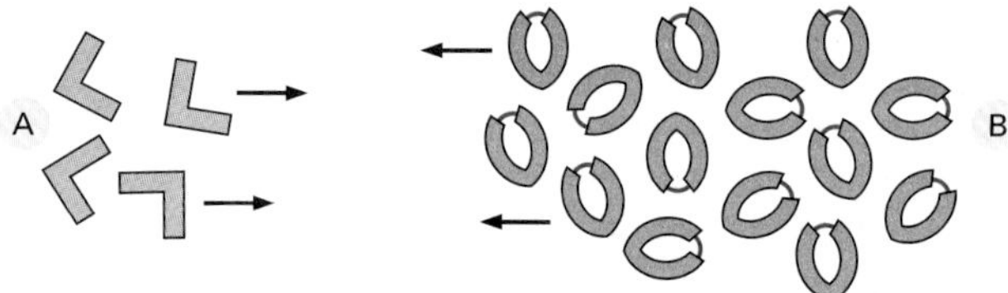

AT EQUILIBRIUM the number of A molecules being converted to B molecules each second is exactly equal to the number of B molecules being converted to A molecules each second, so that there is no net change in the ratio of A to B.

Figure 2–52 Chemical equilibrium. When a reaction reaches equilibrium, the forward and backward flux of reacting molecules are equal and opposite.

reacting molecules; the second depends on their concentrations. For the simple reaction A → B at 37°C,

$$\Delta G = \Delta G^\circ + 0.616 \ln \frac{[B]}{[A]}$$

where ΔG is in kilocalories per mole, [A] and [B] denote the concentrations of A and B, ln is the natural logarithm, and 0.616 is RT—the product of the gas constant, R, and the abolute temperature, T.

Note that ΔG equals the value of ΔG° when the molar concentrations of A and B are equal (ln 1 = 0). As expected, ΔG becomes more negative as the ratio of B to A decreases (the ln of a number < 1 is negative).

Chemical **equilibrium** is reached when the concentration effect just balances the push given to the reaction by ΔG°, so that there is no net change of free energy to drive the reaction in either direction (Figure 2–52). Here $\Delta G = 0$, and so the concentrations of A and B are such that

$$-0.616 \ln \frac{[B]}{[A]} = \Delta G^\circ = -1.42 \log \frac{[B]}{[A]}$$

which means that there is chemical equilibrium at 37°C when

$$\frac{[B]}{[A]} = e^{-\Delta G^\circ/0.616}$$

Table 2–5 shows how the equilibrium ratio of A to B (expressed as an **equilibrium constant, *K*)** depends on the value of ΔG°.

It is important to recognize that when an enzyme (or any catalyst) lowers the activation energy for the reaction A → B, it also lowers the activation energy for the reaction B → A by exactly the same amount (see Figure 2–44). The forward and backward reactions will therefore be accelerated by the same factor by an enzyme, and the equilibrium point for the reaction (and ΔG°) remains unchanged (Figure 2–53).

TABLE 2–5 Relationship Between the Standard Free-Energy Change, ΔG°, and Equilibrium Constant

EQUILIBRIUM CONSTANT $\frac{[B]}{[A]} = K$ (liters/mole)	FREE ENERGY OF B MINUS FREE ENERGY OF A (kcal/mole)
10^5	–7.1
10^4	–5.7
10^3	–4.3
10^2	–2.8
10	–1.4
1	0
10^{-1}	1.4
10^{-2}	2.8
10^{-3}	4.3
10^{-4}	5.7
10^{-5}	7.1

Values of the equilibrium constant were calculated for the simple chemical reaction A ⇌ B using the equation given in the text.

The ΔG° given here is in kilocalories per mole at 37°C (1 kilocalorie is equal to 4.184 kilojoules). As explained in the text, ΔG° represents the free-energy difference under standard conditions (where all components are present at a concentration of 1.0 mole/liter).

From this table, we see that if there is a favorable free-energy change of –4.3 kcal/mole for the transition A → B, there will be 1000 times more molecules in state B than in state A.

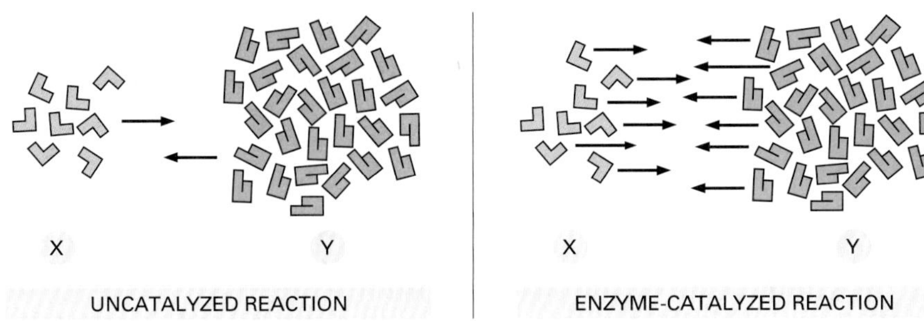

Figure 2–53 Enzymes cannot change the equilibrium point for reactions. Enzymes, like all catalysts, speed up the forward and backward rates of a reaction by the same factor. Therefore, for both the catalyzed and the uncatalyzed reactions shown here, the number of molecules undergoing the transition X → Y is equal to the number of molecules undergoing the transition Y → X when the ratio of Y molecules to X molecules is 3.5 to 1. In other words, the two reactions reach equilibrium at exactly the same point.

For Sequential Reactions, $\Delta G°$ Values Are Additive

The course of most reactions can be predicted quantitatively. A large body of thermodynamic data has been collected that makes it possible to calculate the standard change in free energy, $\Delta G°$, for most of the important metabolic reactions of the cell. The overall free-energy change for a metabolic pathway is then simply the sum of the free-energy changes in each of its component steps. Consider, for example, two sequential reactions

$$X \rightarrow Y \text{ and } Y \rightarrow Z$$

where the $\Delta G°$ values are +5 and –13 kcal/mole, respectively. (Recall that a mole is 6×10^{23} molecules of a substance.) If these two reactions occur sequentially, the $\Delta G°$ for the coupled reaction will be –8 kcal/mole. Thus, the unfavorable reaction X → Y, which will not occur spontaneously, can be driven by the favorable reaction Y → Z, provided that the second reaction follows the first.

Cells can therefore cause the energetically unfavorable transition, X → Y, to occur if an enzyme catalyzing the X → Y reaction is supplemented by a second enzyme that catalyzes the energetically *favorable* reaction, Y → Z. In effect, the reaction Y → Z will then act as a "siphon" to drive the conversion of all of molecule X to molecule Y, and thence to molecule Z (Figure 2–54). For example, several of the reactions in the long pathway that converts sugars into CO_2 and H_2O would be energetically unfavorable if considered on their own. But the pathway nevertheless proceeds rapidly to completion because the total $\Delta G°$ for the series of sequential reactions has a large negative value.

But forming a sequential pathway is not adequate for many purposes. Often the desired pathway is simply X → Y, without further conversion of Y to some other product. Fortunately, there are other more general ways of using enzymes to couple reactions together. How these work is the topic we discuss next.

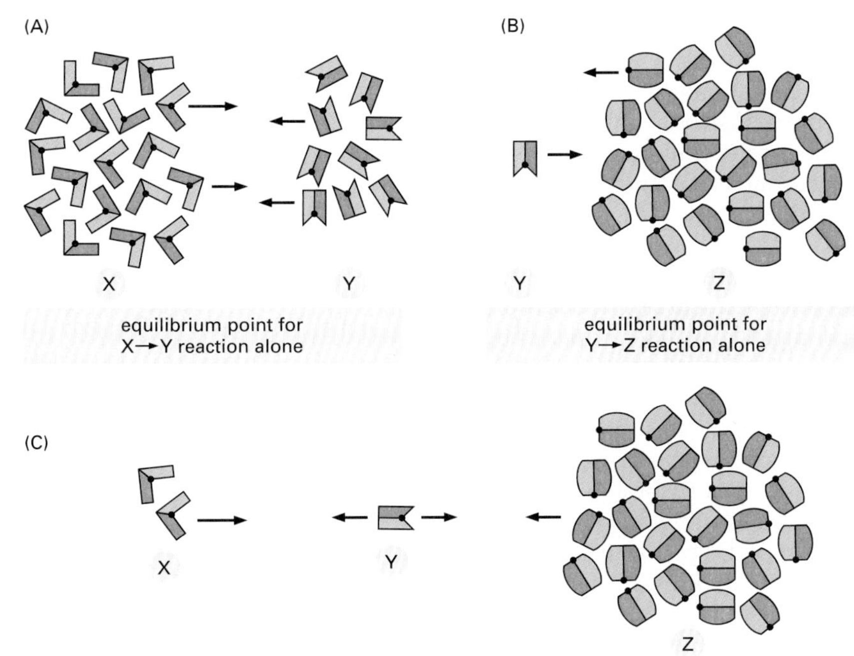

Figure 2–54 How an energetically unfavorable reaction can be driven by a second, following reaction. (A) At equilibrium, there are twice as many X molecules as Y molecules, because X is of lower energy than Y. (B) At equilibrium, there are 25 times more Z molecules than Y molecules, because Z is of much lower energy than Y. (C) If the reactions in (A) and (B) are coupled, nearly all of the X molecules will be converted to Z molecules, as shown.

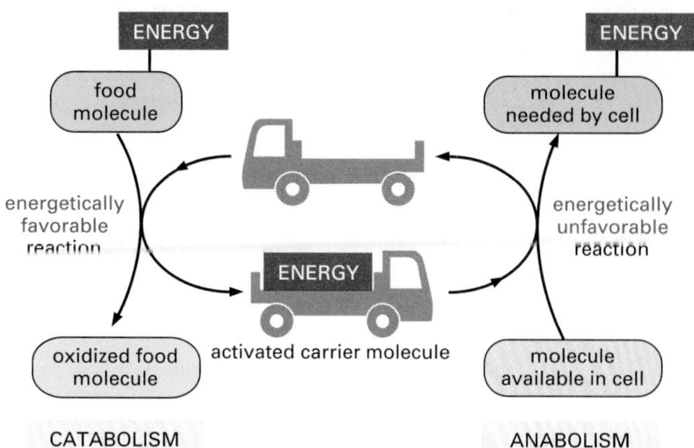

Figure 2–55 Energy transfer and the role of activated carriers in metabolism. By serving as energy shuttles, activated carrier molecules perform their function as go-betweens that link the breakdown of food molecules and the release of energy *(catabolism)* to the energy-requiring biosynthesis of small and large organic molecules *(anabolism)*.

Activated Carrier Molecules are Essential for Biosynthesis

The energy released by the oxidation of food molecules must be stored temporarily before it can be channeled into the construction of other small organic molecules and of the larger and more complex molecules needed by the cell. In most cases, the energy is stored as chemical bond energy in a small set of activated "carrier molecules," which contain one or more energy-rich covalent bonds. These molecules diffuse rapidly throughout the cell and thereby carry their bond energy from sites of energy generation to the sites where energy is used for biosynthesis and other needed cell activities (Figure 2–55).

The **activated carriers** store energy in an easily exchangeable form, either as a readily transferable chemical group or as high-energy electrons, and they can serve a dual role as a source of both energy and chemical groups in biosynthetic reactions. For historical reasons, these molecules are also sometimes referred to as *coenzymes*. The most important of the activated carrier molecules are ATP and two molecules that are closely related to each other, NADH and NADPH—as we discuss in detail shortly. We shall see that cells use activated carrier molecules like money to pay for reactions that otherwise could not take place.

The Formation of an Activated Carrier Is Coupled to an Energetically Favorable Reaction

When a fuel molecule such as glucose is oxidized in a cell, enzyme-catalyzed reactions ensure that a large part of the free energy that is released by oxidation is captured in a chemically useful form, rather than being released wastefully as heat. This is achieved by means of a **coupled reaction**, in which an energetically favorable reaction is used to drive an energetically unfavorable one that produces an activated carrier molecule or some other useful energy store. Coupling mechanisms require enzymes and are fundamental to all the energy transactions of the cell.

The nature of a coupled reaction is illustrated by a mechanical analogy in Figure 2–56, in which an energetically favorable chemical reaction is represented by rocks falling from a cliff. The energy of falling rocks would normally be entirely wasted in the form of heat generated by friction when the rocks hit the ground (see the falling brick diagram in Figure 2–39). By careful design, however, part of this energy could be used instead to drive a paddle wheel that lifts a bucket of water (Figure 2–56B). Because the rocks can now reach the ground only after moving the paddle wheel, we say that the energetically favorable reaction of rock falling has been directly *coupled* to the energetically unfavorable reaction of lifting the bucket of water. Note that because part of the energy is used to do work in (B), the rocks hit the ground with less velocity than in (A), and correspondingly less energy is wasted as heat.

Exactly analogous processes occur in cells, where enzymes play the role of the paddle wheel in our analogy. By mechanisms that will be discussed later in this chapter, they couple an energetically favorable reaction, such as the oxidation of

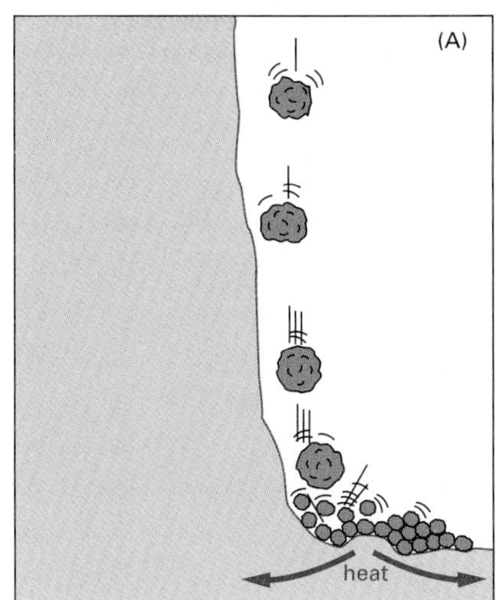

kinetic energy transformed into heat energy only

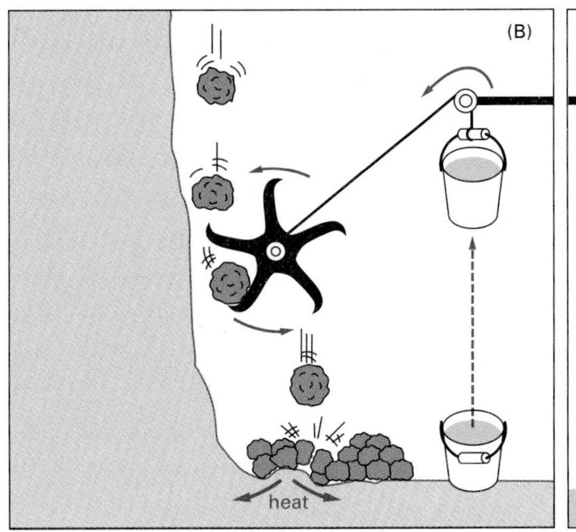

part of the kinetic energy is used to lift a bucket of water, and a correspondingly smaller amount is transformed into heat

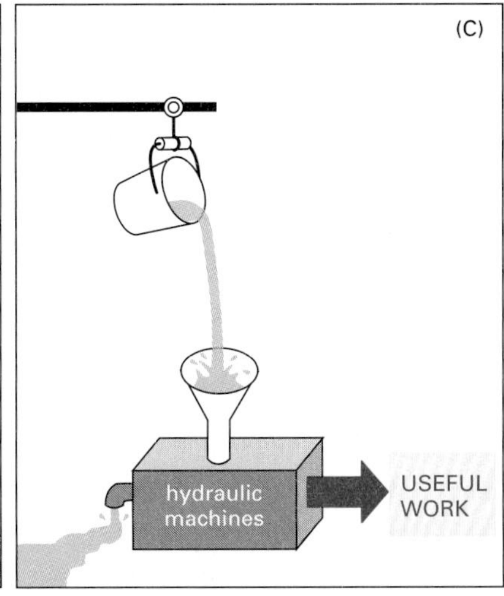

the potential kinetic energy stored in the raised bucket of water can be used to drive hydraulic machines that carry out a variety of useful tasks

Figure 2–56 A mechanical model illustrating the principle of coupled chemical reactions. The spontaneous reaction shown in (A) could serve as an analogy for the direct oxidation of glucose to CO_2 and H_2O, which produces heat only. In (B) the same reaction is coupled to a second reaction; this second reaction could serve as an analogy for the synthesis of activated carrier molecules. The energy produced in (B) is in a more useful form than in (A) and can be used to drive a variety of otherwise energetically unfavorable reactions (C).

foodstuffs, to an energetically unfavorable reaction, such as the generation of an activated carrier molecule. As a result, the amount of heat released by the oxidation reaction is reduced by exactly the amount of energy that is stored in the energy-rich covalent bonds of the activated carrier molecule. The activated carrier molecule in turn picks up a packet of energy of a size sufficient to power a chemical reaction elsewhere in the cell.

ATP Is the Most Widely Used Activated Carrier Molecule

The most important and versatile of the activated carriers in cells is **ATP** (adenosine triphosphate). Just as the energy stored in the raised bucket of water in Figure 2–56B can be used to drive a wide variety of hydraulic machines, ATP serves as a convenient and versatile store, or currency, of energy to drive a variety of chemical reactions in cells. ATP is synthesized in an energetically unfavorable phosphorylation reaction in which a phosphate group is added to **ADP** (adenosine diphosphate). When required, ATP gives up its energy packet through its energetically favorable hydrolysis to ADP and inorganic phosphate (Figure 2–57). The regenerated ADP is then available to be used for another round of the phosphorylation reaction that forms ATP.

Figure 2–57 The hydrolysis of ATP to ADP and inorganic phosphate. The two outermost phosphates in ATP are held to the rest of the molecule by high-energy phosphoanhydride bonds and are readily transferred. As indicated, water can be added to ATP to form ADP and inorganic phosphate (P_i). This hydrolysis of the terminal phosphate of ATP yields between 11 and 13 kcal/mole of usable energy, depending on the intracellular conditions. The large negative ΔG of this reaction arises from a number of factors. Release of the terminal phosphate group removes an unfavorable repulsion between adjacent negative charges; in addition, the inorganic phosphate ion (P_i) released is stabilized by resonance and by favorable hydrogen-bond formation with water.

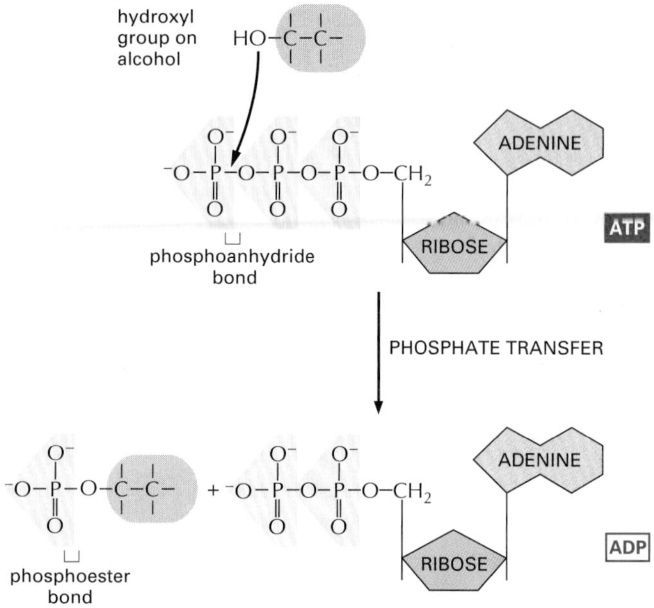

Figure 2–58 An example of a phosphate transfer reaction. Because an energy-rich phosphoanhydride bond in ATP is converted to a phosphoester bond, this reaction is energetically favorable, having a large negative ΔG. Reactions of this type are involved in the synthesis of phospholipids and in the initial steps of reactions that catabolize sugars.

The energetically favorable reaction of ATP hydrolysis is coupled to many otherwise unfavorable reactions through which other molecules are synthesized. We shall encounter several of these reactions later in this chapter. Many of them involve the transfer of the terminal phosphate in ATP to another molecule, as illustrated by the phosphorylation reaction in Figure 2–58.

ATP is the most abundant active carrier in cells. As one example, it is used to supply energy for many of the pumps that transport substances into and out of the cell (discussed in Chapter 11). It also powers the molecular motors that enable muscle cells to contract and nerve cells to transport materials from one end of their long axons to another (discussed in Chapter 16).

Energy Stored in ATP Is Often Harnessed to Join Two Molecules Together

We have previously discussed one way in which an energetically favorable reaction can be coupled to an energetically unfavorable reaction, $X \rightarrow Y$, so as to enable it to occur. In that scheme a second enzyme catalyzes the energetically favorable reaction $Y \rightarrow Z$, pulling all of the X to Y in the process (see Figure 2–54). But when the required product is Y and not Z, this mechanism is not useful.

A frequent type of reaction that is needed for biosynthesis is one in which two molecules, A and B, are joined together to produce A–B in the energetically unfavorable condensation reaction

$$\text{A–H} + \text{B–OH} \rightarrow \text{A–B} + H_2O$$

There is an indirect pathway that allows A–H and B–OH to form A–B, in which a coupling to ATP hydrolysis makes the reaction go. Here energy from ATP hydrolysis is first used to convert B–OH to a higher-energy intermediate compound, which then reacts directly with A–H to give A–B. The simplest possible mechanism involves the transfer of a phosphate from ATP to B–OH to make B–OPO_3, in which case the reaction pathway contains only two steps:

1. B–OH + ATP $\rightarrow$ B–O–PO_3 + ADP
2. A–H + B–O–PO_3 $\rightarrow$ A–B + P_i

Net result: B–OH + ATP + A–H $\rightarrow$ A–B + ADP + P_i

The condensation reaction, which by itself is energetically unfavorable, is forced to occur by being directly coupled to ATP hydrolysis in an enzyme-catalyzed reaction pathway (Figure 2–59A).

A biosynthetic reaction of exactly this type is employed to synthesize the amino acid glutamine, as illustrated in Figure 2–59B. We will see shortly that very

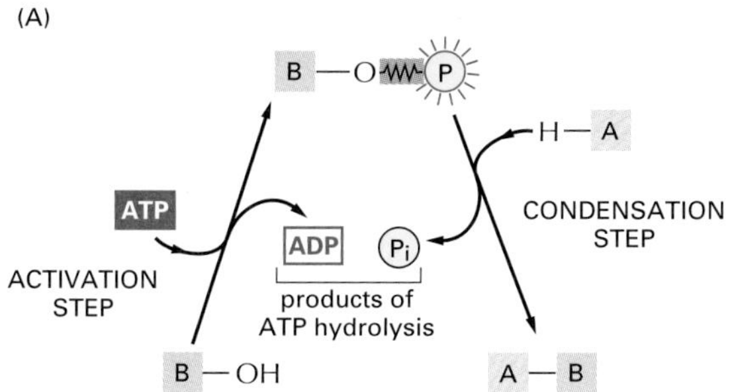

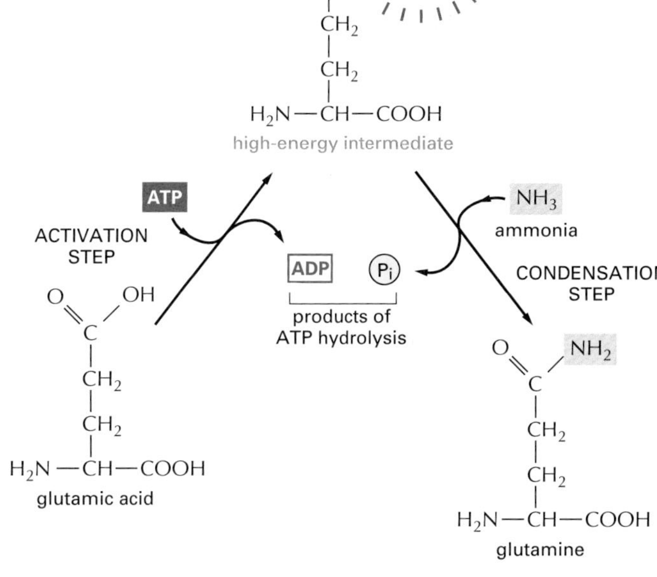

Figure 2–59 An example of an energetically unfavorable biosynthetic reaction driven by ATP hydrolysis. (A) Schematic illustration of the formation of A–B in the condensation reaction described in the text. (B) The biosynthesis of the common amino acid glutamine. Glutamic acid is first converted to a high-energy phosphorylated intermediate (corresponding to the compound B–O–PO_3 described in the text), which then reacts with ammonia (corresponding to A–H) to form glutamine. In this example both steps occur on the surface of the same enzyme, *glutamine synthase*. Note that, for clarity, the amino acids are shown in their uncharged form.

similar (but more complex) mechanisms are also used to produce nearly all of the large molecules of the cell.

NADH and NADPH Are Important Electron Carriers

Other important activated carrier molecules participate in oxidation–reduction reactions and are commonly part of coupled reactions in cells. These activated carriers are specialized to carry high-energy electrons and hydrogen atoms. The most important of these electron carriers are **NAD$^+$** (nicotinamide adenine dinucleotide) and the closely related molecule **NADP$^+$** (nicotinamide adenine dinucleotide phosphate). Later, we examine some of the reactions in which they participate. NAD$^+$ and NADP$^+$ each pick up a "packet of energy" corresponding to two high-energy electrons plus a proton (H^+)—being converted to **NADH** (*reduced* nicotinamide adenine dinucleotide) and **NADPH** (*reduced* nicotinamide adenine dinucleotide phosphate), respectively. These molecules can therefore also be regarded as carriers of hydride ions (the H^+ plus two electrons, or H^-).

Like ATP, NADPH is an activated carrier that participates in many important biosynthetic reactions that would otherwise be energetically unfavorable. The NADPH is produced according to the general scheme shown in Figure 2–60A. During a special set of energy-yielding catabolic reactions, a hydrogen atom plus two electrons are removed from the substrate molecule and added to the nicotinamide ring of NADP$^+$ to form NADPH. This is a typical oxidation–reduction reaction; the substrate is oxidized and NADP$^+$ is reduced. The structures of NADP$^+$ and NADPH are shown in Figure 2–60B.

The hydride ion carried by NADPH is given up readily in a subsequent oxidation–reduction reaction, because the ring can achieve a more stable arrangement of electrons without it. In this subsequent reaction, which regenerates NADP$^+$, it is the NADPH that becomes oxidized and the substrate that becomes reduced. The NADPH is an effective donor of its hydride ion to other molecules for the same reason that ATP readily transfers a phosphate: in both cases the transfer is accompanied by a large negative free-energy change. One example of the use of NADPH in biosynthesis is shown in Figure 2–61.

The difference of a single phosphate group has no effect on the electron-transfer properties of NADPH compared with NADH, but it is crucial for their distinctive roles. The extra phosphate group on NADPH is far from the region

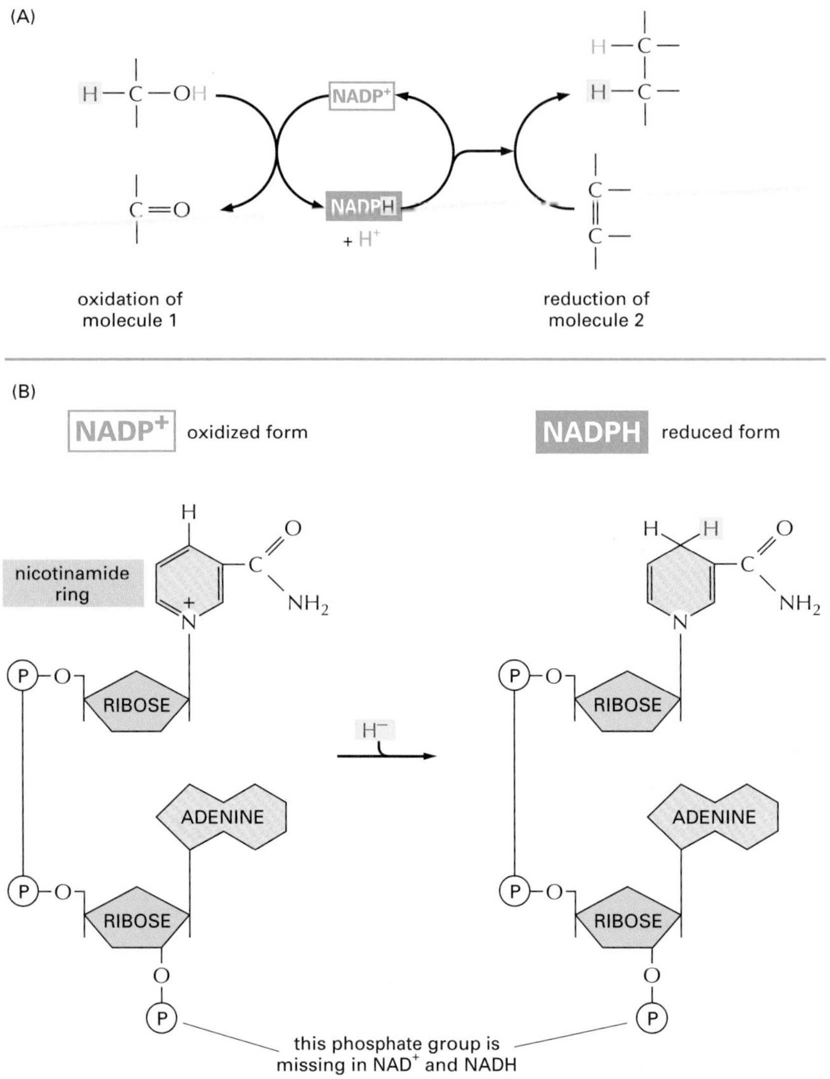

Figure 2–60 NADPH, an important carrier of electrons. (A) NADPH is produced in reactions of the general type shown on the left, in which two hydrogen atoms are removed from a substrate. The oxidized form of the carrier molecule, $NADP^+$, receives one hydrogen atom plus an electron (a hydride ion), and the proton (H^+) from the other H atom is released into solution. Because NADPH holds its hydride ion in a high-energy linkage, the added hydride ion can easily be transferred to other molecules, as shown on the right. (B) The structure of $NADP^+$ and NADPH. The part of the $NADP^+$ molecule known as the nicotinamide ring accepts two electrons together with a proton (the equivalent of a hydride ion, H^-), forming NADPH. The molecules NAD^+ and NADH are identical in structure to $NADP^+$ and NADPH, respectively, except that the indicated phosphate group is absent from both.

involved in electron transfer (see Figure 2–60B) and is of no importance to the transfer reaction. It does, however, give a molecule of NADPH a slightly different shape from that of NADH, and so NADPH and NADH bind as substrates to different sets of enzymes. Thus the two types of carriers are used to transfer electrons (or hydride ions) between different sets of molecules.

Why should there be this division of labor? The answer lies in the need to regulate two sets of electron-transfer reactions independently. NADPH operates chiefly with enzymes that catalyze anabolic reactions, supplying the high-energy electrons needed to synthesize energy-rich biological molecules. NADH, by contrast, has a special role as an intermediate in the catabolic system of reactions that generate ATP through the oxidation of food molecules, as we will discuss shortly. The genesis of NADH from NAD^+ and that of NADPH from $NADP^+$ occur by different pathways and are independently regulated, so that the cell can independently adjust the supply of electrons for these two contrasting purposes. Inside the cell the ratio of NAD^+ to NADH is kept high, whereas the ratio of $NADP^+$ to NADPH is kept low. This provides plenty of NAD^+ to act as an oxidizing agent and plenty of NADPH to act as a reducing agent—as required for their special roles in catabolism and anabolism, respectively.

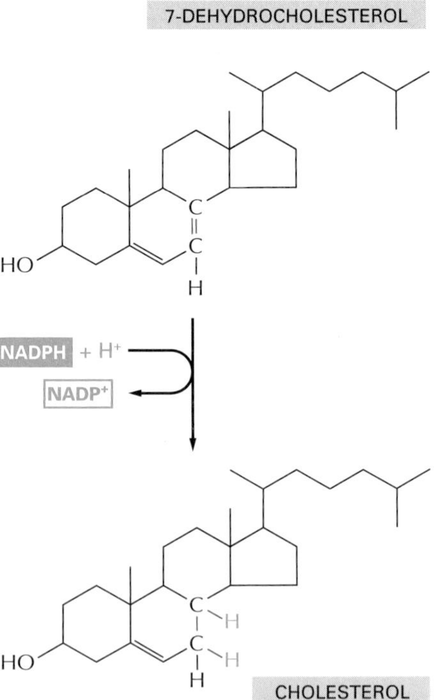

Figure 2–61 The final stage in one of the biosynthetic routes leading to cholesterol. As in many other biosynthetic reactions, the reduction of the C=C bond is achieved by the transfer of a hydride ion from the carrier molecule NADPH, plus a proton (H^+) from the solution.

There Are Many Other Activated Carrier Molecules in Cells

Other activated carriers also pick up and carry a chemical group in an easily transferred, high-energy linkage (Table 2–6). For example, coenzyme A carries an acetyl group in a readily transferable linkage, and in this activated form is

TABLE 2–6 Some Activated Carrier Molecules Widely Used in Metabolism

ACTIVATED CARRIER	GROUP CARRIED IN HIGH-ENERGY LINKAGE
ATP	phosphate
NADH, NADPH, $FADH_2$	electrons and hydrogens
Acetyl CoA	acetyl group
Carboxylated biotin	carboxyl group
S-Adenosylmethionine	methyl group
Uridine diphosphate glucose	glucose

known as **acetyl CoA** (acetyl coenzyme A). The structure of acetyl CoA is illustrated in Figure 2–62; it is used to add two carbon units in the biosynthesis of larger molecules.

In acetyl CoA and the other carrier molecules in Table 2–6, the transferable group makes up only a small part of the molecule. The rest consists of a large organic portion that serves as a convenient "handle," facilitating the recognition of the carrier molecule by specific enzymes. As with acetyl CoA, this handle portion very often contains a nucleotide, a curious fact that may be a relic from an early stage of evolution. It is currently thought that the main catalysts for early life-forms—before DNA or proteins—were RNA molecules (or their close relatives), as described in Chapter 6. It is tempting to speculate that many of the carrier molecules that we find today originated in this earlier RNA world, where their nucleotide portions could have been useful for binding them to RNA enzymes.

Examples of the type of transfer reactions catalyzed by the activated carrier molecules ATP (transfer of phosphate) and NADPH (transfer of electrons and hydrogen) have been presented in Figures 2–58 and 2–61, respectively. The reactions of other activated carrier molecules involve the transfers of methyl, carboxyl, or glucose group, for the purpose of biosynthesis. The activated carriers required are usually generated in reactions that are coupled to ATP hydrolysis, as in the example in Figure 2–63. Therefore, the energy that enables their groups to be used for biosynthesis ultimately comes from the catabolic reactions that

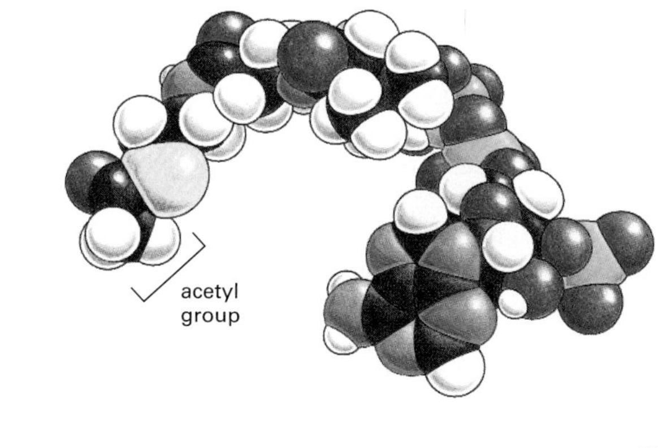

nucleotide

ADENINE

RIBOSE

high-energy bond

acetyl group

CoA

Figure 2–62 The structure of the important activated carrier molecule acetyl CoA. A space-filling model is shown above the structure. The sulfur atom *(yellow)* forms a thioester bond to acetate. Because this is a high-energy linkage, releasing a large amount of free energy when it is hydrolyzed, the acetate molecule can be readily transferred to other molecules.

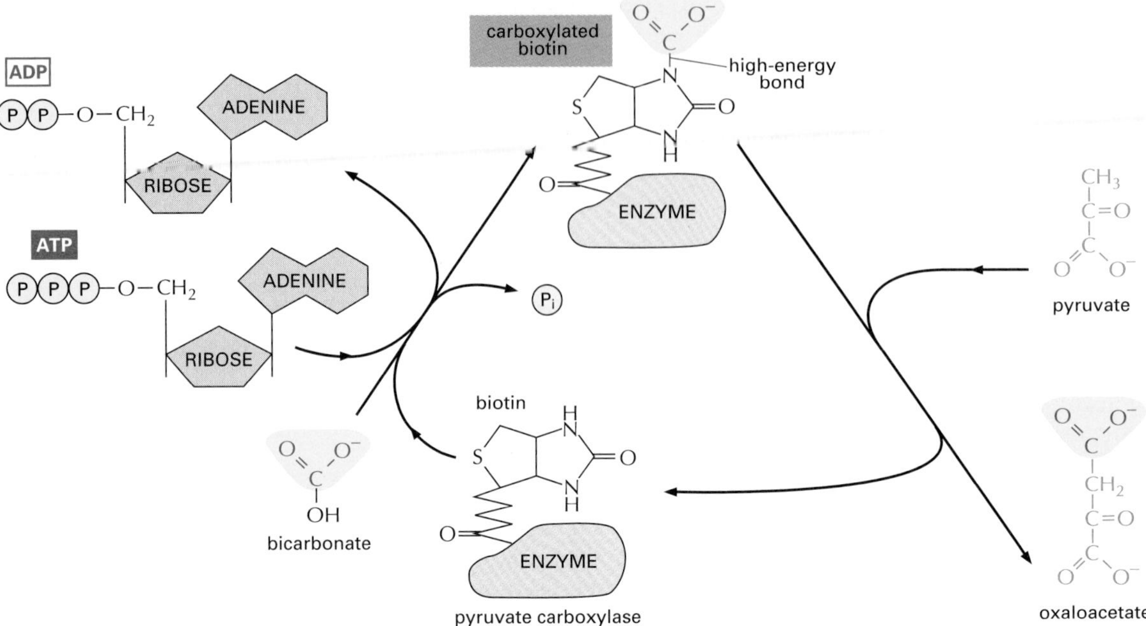

Figure 2–63 A carboxyl group transfer reaction using an activated carrier molecule. Carboxylated biotin is used by the enzyme *pyruvate carboxylase* to transfer a carboxyl group in the production of oxaloacetate, a molecule needed for the citric acid cycle. The acceptor molecule for this group transfer reaction is pyruvate. Other enzymes use biotin to transfer carboxyl groups to other acceptor molecules. Note that synthesis of carboxylated biotin requires energy that is derived from ATP—a general feature of many activated carriers.

generate ATP. Similar processes occur in the synthesis of the very large molecules of the cell—the nucleic acids, proteins, and polysaccharides—that we discuss next.

The Synthesis of Biological Polymers Requires an Energy Input

As discussed previously, the macromolecules of the cell constitute the vast majority of its dry mass—that is, of the mass not due to water (see Figure 2–29). These molecules are made from subunits (or monomers) that are linked together in a *condensation* reaction, in which the constituents of a water molecule (OH plus H) are removed from the two reactants. Consequently, the reverse reaction—the breakdown of all three types of polymers—occurs by the enzyme-catalyzed addition of water (*hydrolysis*). This hydrolysis reaction is energetically favorable, whereas the biosynthetic reactions require an energy input and are more complex (Figure 2–64).

The nucleic acids (DNA and RNA), proteins, and polysaccharides are all polymers that are produced by the repeated addition of a *subunit* (also called a monomer) onto one end of a growing chain. The synthesis reactions for these three types of macromolecules are outlined in Figure 2–65. As indicated, the condensation step in each case depends on energy from nucleoside triphosphate hydrolysis. And yet, except for the nucleic acids, there are no phosphate groups left in the final product molecules. How are the reactions that release the energy of ATP hydrolysis coupled to polymer synthesis?

For each type of macromolecule, an enzyme-catalyzed pathway exists which resembles that discussed previously for the synthesis of the amino acid glutamine (see Figure 2–59). The principle is exactly the same, in that the OH group that will be removed in the condensation reaction is first activated by becoming involved in a high-energy linkage to a second molecule. However, the actual

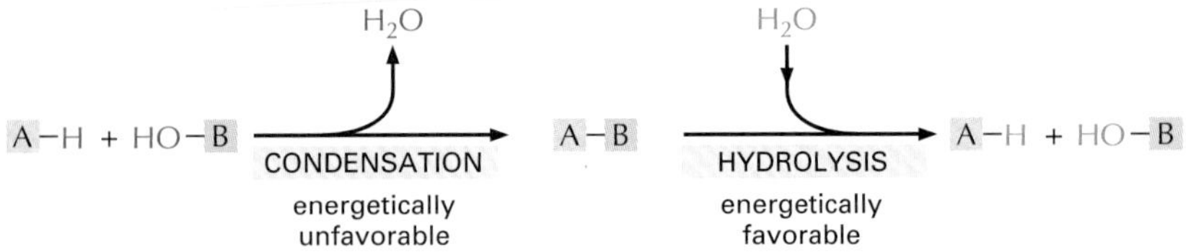

A–H + HO–B → (H_2O released) CONDENSATION, energetically unfavorable → A–B → (H_2O added) HYDROLYSIS, energetically favorable → A–H + HO–B

Figure 2–64 Condensation and hydrolysis as opposite reactions. The macromolecules of the cell are polymers that are formed from subunits (or monomers) by a condensation reaction and are broken down by hydrolysis. The condensation reactions are all energetically unfavorable.

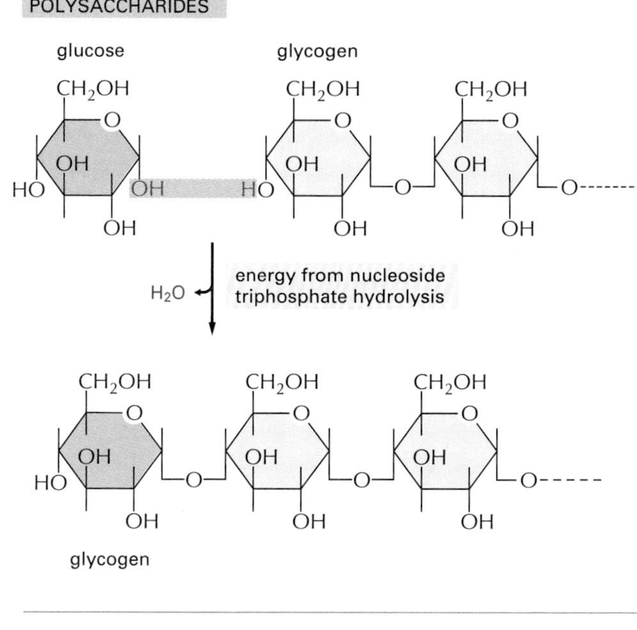

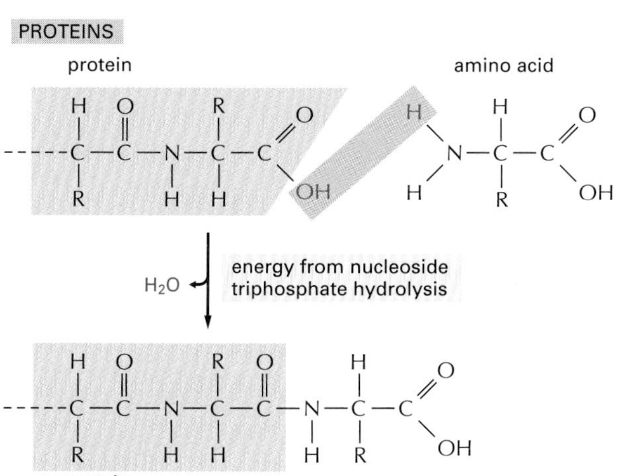

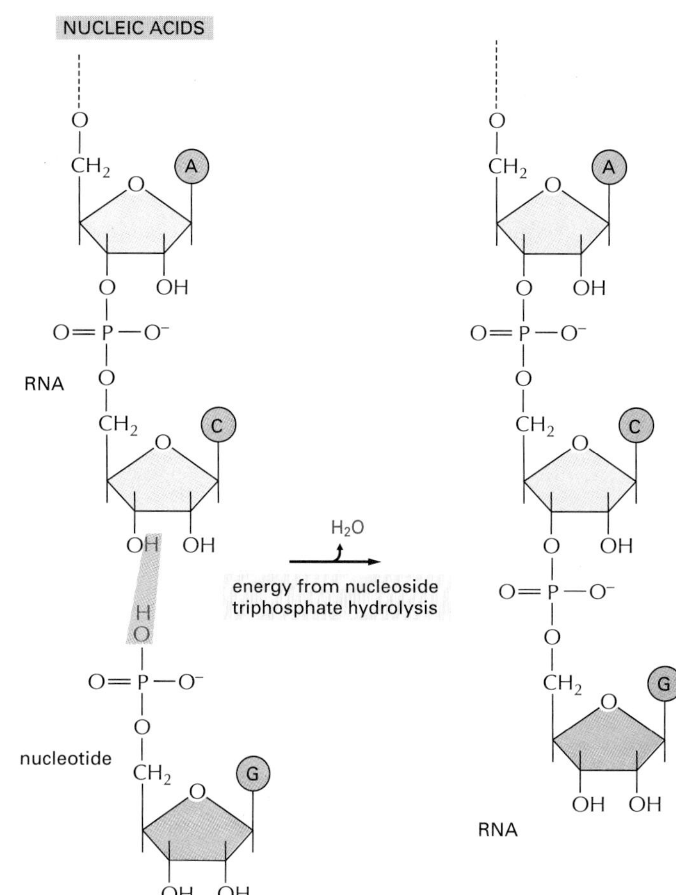

Figure 2–65 The synthesis of polysaccharides, proteins, and nucleic acids. Synthesis of each kind of biological polymer involves the loss of water in a condensation reaction. Not shown is the consumption of high-energy nucleoside triphosphates that is required to activate each monomer prior to its addition. In contrast, the reverse reaction—the breakdown of all three types of polymers—occurs by the simple addition of water (hydrolysis).

mechanisms used to link ATP hydrolysis to the synthesis of proteins and polysaccharides are more complex than that used for glutamine synthesis, since a series of high-energy intermediates is required to generate the final high-energy bond that is broken during the condensation step (discussed in Chapter 6 for protein synthesis).

There are limits to what each activated carrier can do in driving biosynthesis. The ΔG for the hydrolysis of ATP to ADP and inorganic phosphate (P_i) depends on the concentrations of all of the reactants, but under the usual conditions in a cell it is between –11 and –13 kcal/mole. In principle, this hydrolysis reaction can be used to drive an unfavorable reaction with a ΔG of, perhaps, +10 kcal/mole, provided that a suitable reaction path is available. For some biosynthetic reactions, however, even –13 kcal/mole may not be enough. In these cases the path of ATP hydrolysis can be altered so that it initially produces AMP and pyrophosphate (PP_i), which is itself then hydrolyzed in a subsequent step (Figure 2–66). The whole process makes available a total free-energy change of about –26 kcal/mole. An important biosynthetic reaction that is driven in this way is nucleic acid (polynucleotide) synthesis, as illustrated in Figure 2–67.

It is interesting to note that the polymerization reactions that produce macromolecules can be oriented in one of two ways, giving rise to either the head polymerization or the tail polymerization of monomers. In *head polymerization* the reactive bond required for the condensation reaction is carried on the end of the growing polymer, and it must therefore be regenerated each time

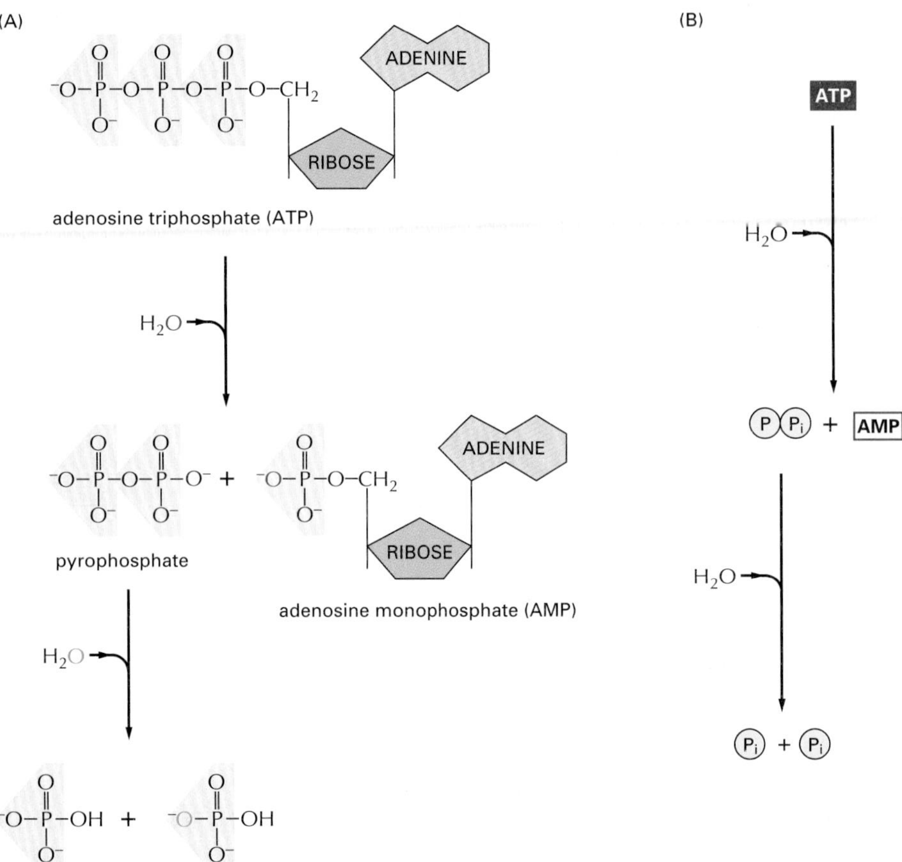

Figure 2–66 An alternative route for the hydrolysis of ATP, in which pyrophosphate is first formed and then hydrolyzed. This route releases about twice as much free energy as the reaction shown earlier in Figure 2–57. (A) In the two successive hydrolysis reactions, oxygen atoms from the participating water molecules are retained in the products, as indicated, whereas the hydrogen atoms dissociate to form free hydrogen ions, (H^+, not shown). (B) Diagram of overall reaction in summary form.

that a monomer is added. In this case, each monomer brings with it the reactive bond that will be used in adding the *next* monomer in the series. In *tail polymerization* the reactive bond carried by each monomer is instead used immediately for its own addition (Figure 2–68).

We shall see in later chapters that both these types of polymerization are used. The synthesis of polynucleotides and some simple polysaccharides occurs by tail polymerization, for example, whereas the synthesis of proteins occurs by a head polymerization process.

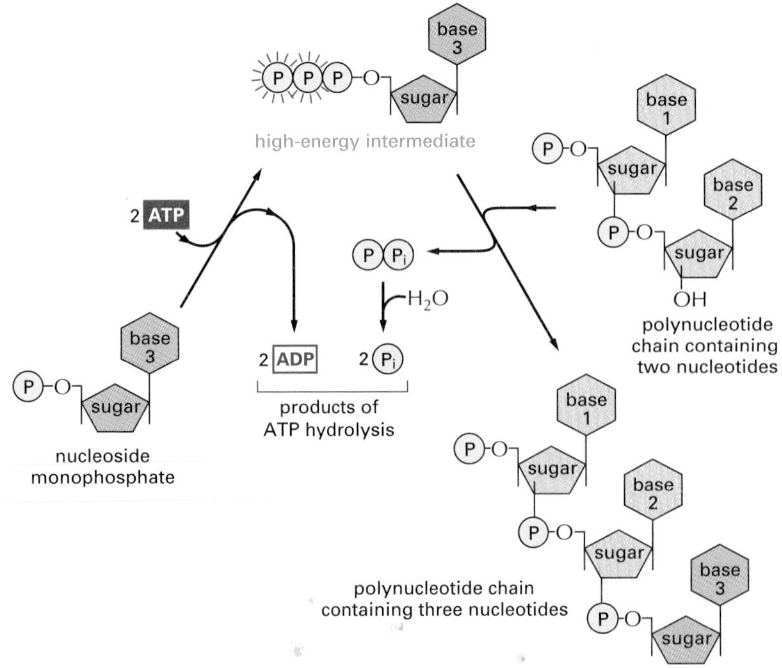

Figure 2–67 Synthesis of a polynucleotide, RNA or DNA, is a multistep process driven by ATP hydrolysis. In the first step, a nucleoside monophosphate is activated by the sequential transfer of the terminal phosphate groups from two ATP molecules. The high-energy intermediate formed—a nucleoside triphosphate—exists free in solution until it reacts with the growing end of an RNA or a DNA chain with release of pyrophosphate. Hydrolysis of the latter to inorganic phosphate is highly favorable and helps to drive the overall reaction in the direction of polynucleotide synthesis. For details, see Chapter 5.

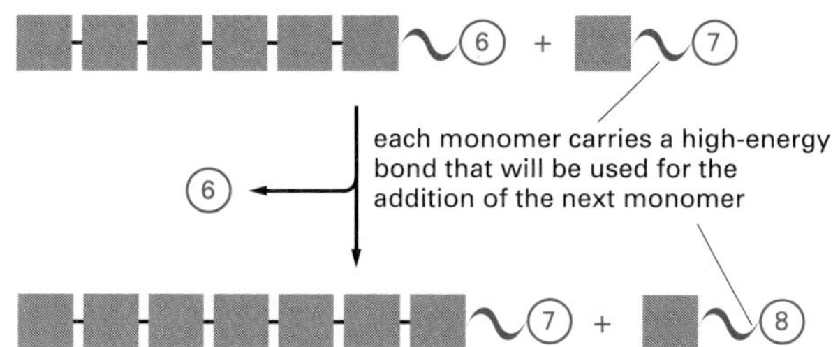

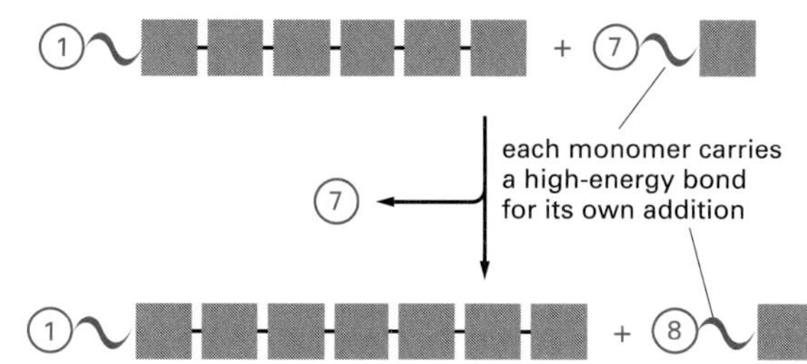

Figure 2–68 The orientation of the active intermediates in biological polymerization reactions. The head growth of polymers is compared with its alternative tail growth. As indicated, these two mechanisms are used to produce different biological macromolecules.

Summary

Living cells are highly ordered and need to create order within themselves in order to survive and grow. This is thermodynamically possible only because of a continual input of energy, part of which must be released from the cells to their environment as heat. The energy comes ultimately from the electromagnetic radiation of the sun, which drives the formation of organic molecules in photosynthetic organisms such as green plants. Animals obtain their energy by eating these organic molecules and oxidizing them in a series of enzyme-catalyzed reactions that are coupled to the formation of ATP—a common currency of energy in all cells.

To make possible the continual generation of order in cells, the energetically favorable hydrolysis of ATP is coupled to energetically unfavorable reactions. In the biosynthesis of macromolecules, this is accomplished by the transfer of phosphate groups to form reactive phosphorylated intermediates. Because the energetically unfavorable reaction now becomes energetically favorable, ATP hydrolysis is said to drive the reaction. Polymeric molecules such as proteins, nucleic acids, and polysaccharides are assembled from small activated precursor molecules by repetitive condensation reactions that are driven in this way. Other reactive molecules, called either active carriers or coenzymes, transfer other chemical groups in the course of biosynthesis: NADPH transfers hydrogen as a proton plus two electrons (a hydride ion), for example, whereas acetyl CoA transfers an acetyl group.

HOW CELLS OBTAIN ENERGY FROM FOOD

As we have just seen, cells require a constant supply of energy to generate and maintain the biological order that keeps them alive. This energy is derived from the chemical bond energy in food molecules, which thereby serve as fuel for cells.

Sugars are particularly important fuel molecules, and they are oxidized in small steps to carbon dioxide (CO_2) and water (Figure 2–69). In this section we trace the major steps in the breakdown, or catabolism, of sugars and show how they produce ATP, NADH, and other activated carrier molecules in animal cells. We concentrate on glucose breakdown, since it dominates energy production in most animal cells. A very similar pathway also operates in plants, fungi, and many bacteria. Other molecules, such as fatty acids and proteins, can also serve as energy sources when they are funneled through appropriate enzymatic pathways.

Food Molecules Are Broken Down in Three Stages to Produce ATP

The proteins, lipids, and polysaccharides that make up most of the food we eat must be broken down into smaller molecules before our cells can use them—either as a source of energy or as building blocks for other molecules. The breakdown processes must act on food taken in from outside, but not on the macromolecules inside our own cells. Stage 1 in the enzymatic breakdown of food molecules is therefore *digestion*, which occurs either in our intestine outside cells, or in a specialized organelle within cells, the lysosome. (A membrane that

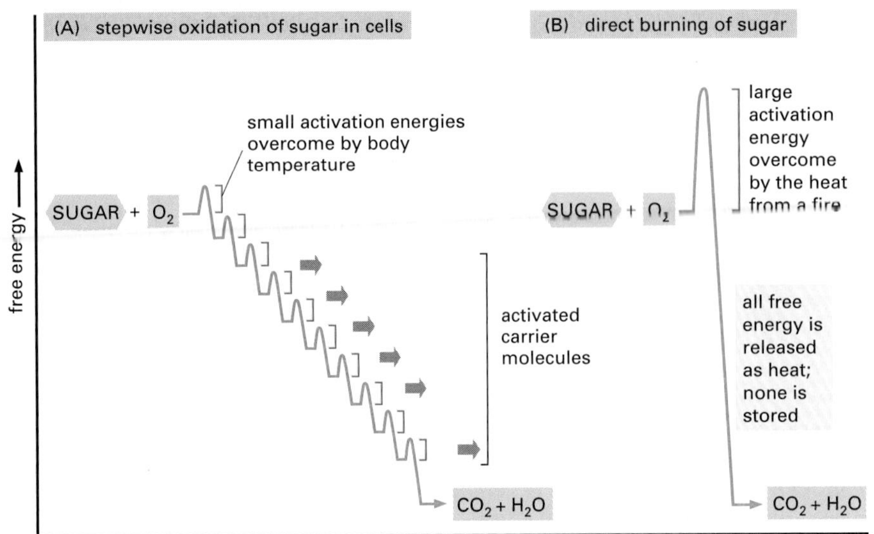

Figure 2–69 Schematic representation of the controlled stepwise oxidation of sugar in a cell, compared with ordinary burning. (A) In the cell, enzymes catalyze oxidation via a series of small steps in which free energy is transferred in conveniently sized packets to carrier molecules—most often ATP and NADH. At each step, an enzyme controls the reaction by reducing the activation energy barrier that has to be surmounted before the specific reaction can occur. The total free energy released is exactly the same in (A) and (B). But if the sugar was instead oxidized to CO_2 and H_2O in a single step, as in (B), it would release an amount of energy much larger than could be captured for useful purposes.

surrounds the lysosome keeps its digestive enzymes separated from the cytosol, as described in Chapter 13.) In either case, the large polymeric molecules in food are broken down during digestion into their monomer subunits—proteins into amino acids, polysaccharides into sugars, and fats into fatty acids and glycerol—through the action of enzymes. After digestion, the small organic molecules derived from food enter the cytosol of the cell, where their gradual oxidation begins. As illustrated in Figure 2–70, oxidation occurs in two further stages of cellular catabolism: stage 2 starts in the cytosol and ends in the major energy-converting organelle, the mitochondrion; stage 3 is entirely confined to the mitochondrion.

In stage 2 a chain of reactions called *glycolysis* converts each molecule of glucose into two smaller molecules of pyruvate. Sugars other than glucose are similarly converted to pyruvate after their conversion to one of the sugar intermediates in this glycolytic pathway. During pyruvate formation, two types of activated carrier molecules are produced—ATP and NADH. The pyruvate then passes from the cytosol into mitochondria. There, each pyruvate molecule is converted into CO_2 plus a two-carbon acetyl group—which becomes attached to coenzyme A (CoA), forming acetyl CoA, another activated carrier molecule (see Figure 2–62). Large amounts of acetyl CoA are also produced by the stepwise breakdown and oxidation of fatty acids derived from fats, which are carried in the bloodstream, imported into cells as fatty acids, and then moved into mitochondria for acetyl CoA production.

Stage 3 of the oxidative breakdown of food molecules takes place entirely in mitochondria. The acetyl group in acetyl CoA is linked to coenzyme A through a high-energy linkage, and it is therefore easily transferable to other molecules. After its transfer to the four-carbon molecule oxaloacetate, the acetyl group enters a series of reactions called the *citric acid cycle*. As we discuss shortly, the acetyl group is oxidized to CO_2 in these reactions, and large amounts of the electron carrier NADH are generated. Finally, the high-energy electrons from NADH are passed along an electron-transport chain within the mitochondrial inner membrane, where the energy released by their transfer is used to drive a process that produces ATP and consumes molecular oxygen (O_2). It is in these final steps that most of the energy released by oxidation is harnessed to produce most of the cell's ATP.

Because the energy to drive ATP synthesis in mitochondria ultimately derives from the oxidative breakdown of food molecules, the phosphorylation of ADP to form ATP that is driven by electron transport in the mitochondrion is known as *oxidative phosphorylation*. The fascinating events that occur within the mitochondrial inner membrane during oxidative phosphorylation are the major focus of Chapter 14.

Through the production of ATP, the energy derived from the breakdown of sugars and fats is redistributed as packets of chemical energy in a form convenient for use elsewhere in the cell. Roughly 10^9 molecules of ATP are in solution in a typical cell at any instant, and in many cells, all this ATP is turned over (that is, used up and replaced) every 1–2 minutes.

In all, nearly half of the energy that could in theory be derived from the oxidation of glucose or fatty acids to H_2O and CO_2 is captured and used to drive the energetically unfavorable reaction $P_i + ADP \rightarrow ATP$. (By contrast, a typical combustion engine, such as a car engine, can convert no more than 20% of the available energy in its fuel into useful work.) The rest of the energy is released by the cell as heat, making our bodies warm.

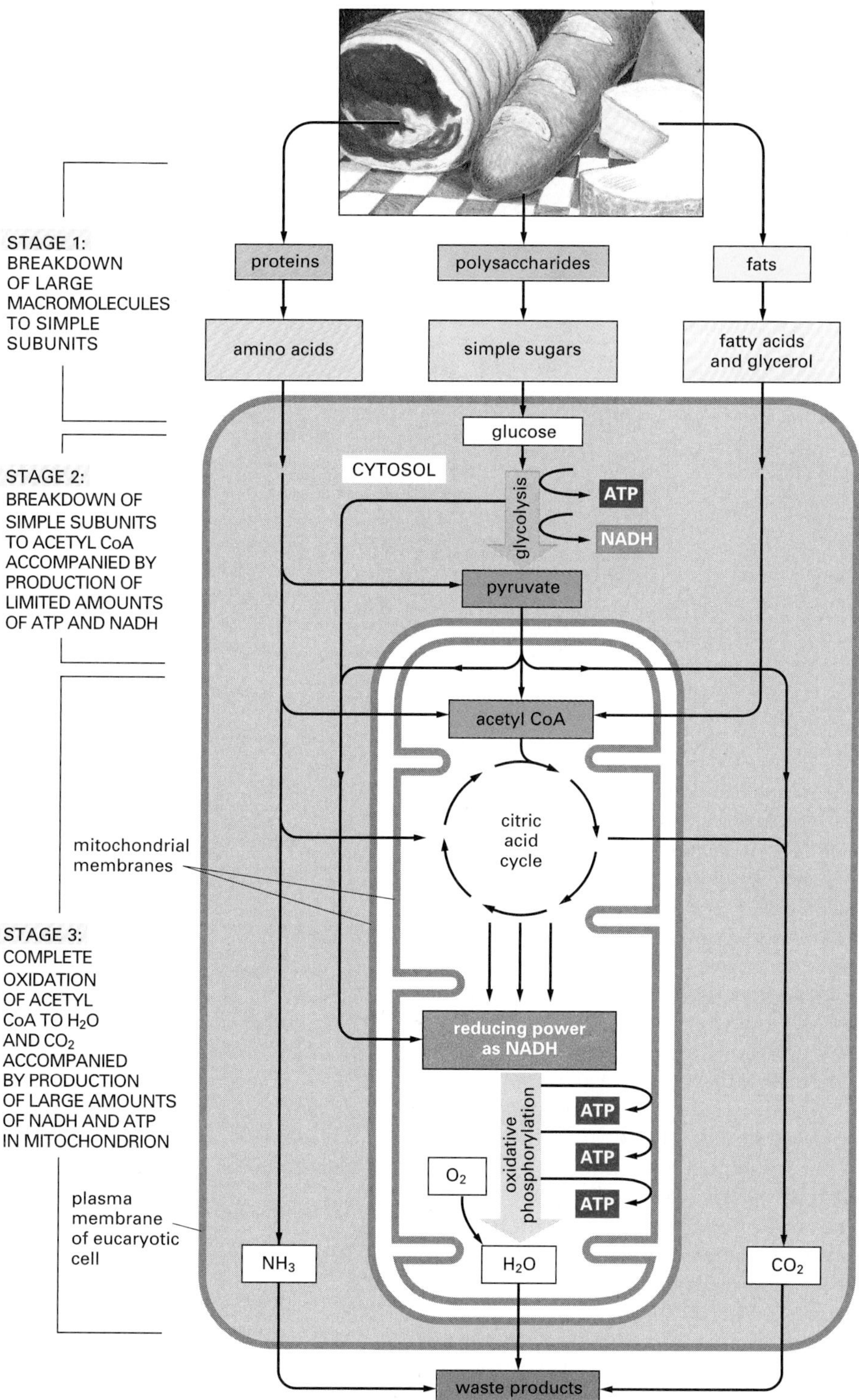

Figure 2–70 Simplified diagram of the three stages of cellular metabolism that lead from food to waste products in animal cells. This series of reactions produces ATP, which is then used to drive biosynthetic reactions and other energy-requiring processes in the cell. Stage I occurs outside cells. Stage 2 occurs mainly in the cytosol, except for the final step of conversion of pyruvate to acetyl groups on acetyl CoA, which occurs in mitochondria. Stage 3 occurs in mitochondria.

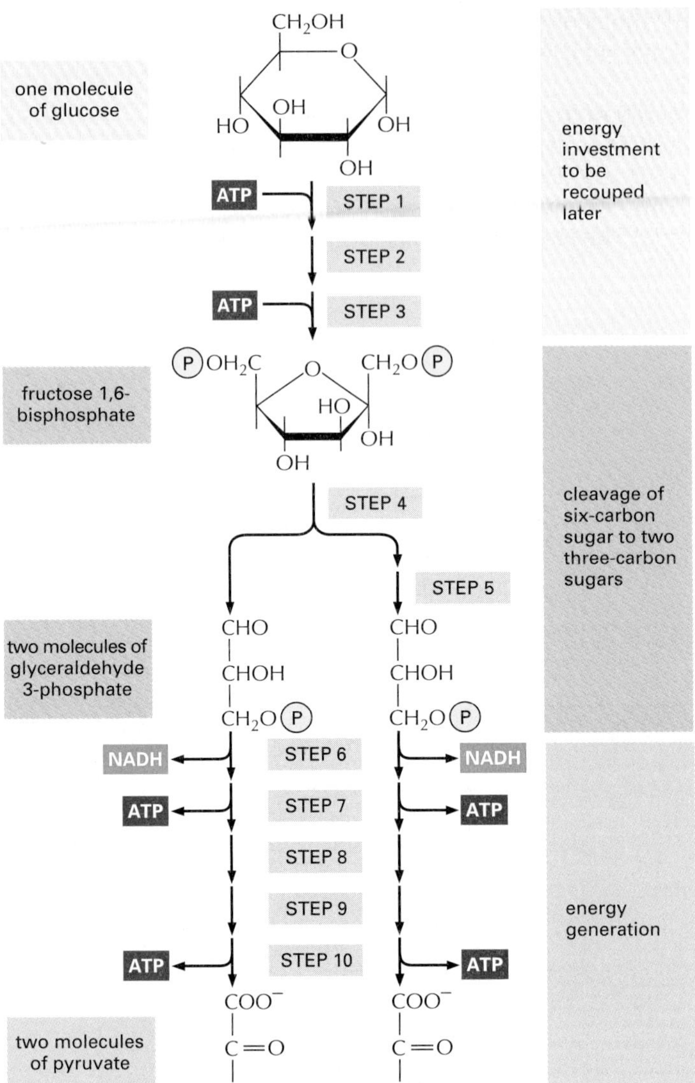

Figure 2–71 An outline of glycolysis. Each of the 10 steps shown is catalyzed by a different enzyme. Note that step 4 cleaves a six-carbon sugar into two three-carbon sugars, so that the number of molecules at every stage after this doubles. As indicated, step 6 begins the energy generation phase of glycolysis, which causes the net synthesis of ATP and NADH molecules (see also Panel 2–8).

Glycolysis Is a Central ATP-producing Pathway

The most important process in stage 2 of the breakdown of food molecules is the degradation of glucose in the sequence of reactions known as **glycolysis**—from the Greek *glykos*, "sugar," and *lysis*, "splitting." Glycolysis produces ATP without the involvement of molecular oxygen (O_2 gas). It occurs in the cytosol of most cells, including many anaerobic microorganisms (those that can live without utilizing molecular oxygen). Glycolysis probably evolved early in the history of life, before the activities of photosynthetic organisms introduced oxygen into the atmosphere. During glycolysis, a glucose molecule with six carbon atoms is converted into two molecules of *pyruvate*, each of which contains three carbon atoms. For each molecule of glucose, two molecules of ATP are hydrolyzed to provide energy to drive the early steps, but four molecules of ATP are produced in the later steps. At the end of glycolysis, there is consequently a net gain of two molecules of ATP for each glucose molecule broken down.

The glycolytic pathway is presented in outline in Figure 2–71, and in more detail in Panel 2–8 (pp. 124–125). Glycolysis involves a sequence of 10 separate reactions, each producing a different sugar intermediate and each catalyzed by a different enzyme. Like most enzymes, these enzymes all have names ending in *ase*—like isomer*ase* and dehydrogen*ase*—which indicate the type of reaction they catalyze.

Although no molecular oxygen is involved in glycolysis, oxidation occurs, in that electrons are removed by NAD^+ (producing NADH) from some of the carbons derived from the glucose molecule. The stepwise nature of the process allows the energy of oxidation to be released in small packets, so that much of it can be stored in activated carrier molecules rather than all of it being released as heat (see Figure 2–69). Thus, some of the energy released by oxidation drives the direct synthesis of ATP molecules from ADP and P_i, and some remains with the electrons in the high-energy electron carrier NADH.

Two molecules of NADH are formed per molecule of glucose in the course of glycolysis. In aerobic organisms (those that require molecular oxygen to live), these NADH molecules donate their electrons to the electron-transport chain described in Chapter 14, and the NAD^+ formed from the NADH is used again for glycolysis (see step 6 in Panel 2–8, pp. 124–125).

Fermentations Allow ATP to Be Produced in the Absence of Oxygen

For most animal and plant cells, glycolysis is only a prelude to the third and final stage of the breakdown of food molecules. In these cells, the pyruvate formed at the last step of stage 2 is rapidly transported into the mitochondria, where it is converted into CO_2 plus acetyl CoA, which is then completely oxidized to CO_2 and H_2O.

In contrast, for many anaerobic organisms—which do not utilize molecular oxygen and can grow and divide without it—glycolysis is the principal source of the cell's ATP. This is also true for certain animal tissues, such as skeletal muscle, that can continue to function when molecular oxygen is limiting. In these anaerobic conditions, the pyruvate and the NADH electrons stay in the cytosol. The pyruvate is converted into products excreted from the cell—for example, into ethanol and CO_2 in the yeasts used in brewing and breadmaking, or into lactate in muscle. In this process, the NADH gives up its electrons and is converted back into NAD^+. This regeneration of NAD^+ is required to maintain the reactions of glycolysis (Figure 2–72).

Anaerobic energy-yielding pathways like these are called **fermentations**. Studies of the commercially important fermentations carried out by yeasts inspired much of early biochemistry. Work in the nineteenth century led in 1896 to the then startling recognition that these processes could be studied outside living organisms, in cell extracts. This revolutionary discovery eventually made it possible to dissect out and study each of the individual reactions in the fermentation process. The piecing together of the complete glycolytic pathway in the 1930s was a major triumph of biochemistry, and it was quickly followed by the recognition of the central role of ATP in cellular processes. Thus, most of the fundamental concepts discussed in this chapter have been understood for more than 50 years.

Glycolysis Illustrates How Enzymes Couple Oxidation to Energy Storage

We have previously used a "paddle wheel" analogy to explain how cells harvest useful energy from the oxidation of organic molecules by using enzymes to couple an energetically unfavorable reaction to an energetically favorable one (see Figure 2–56). Enzymes play the part of the paddle wheel in our analogy, and we now return to a step in glycolysis that we have previously discussed, in order to illustrate exactly how coupled reactions occur.

Two central reactions in glycolysis (steps 6 and 7) convert the three-carbon sugar intermediate glyceraldehyde 3-phosphate (an aldehyde) into 3-phosphoglycerate (a carboxylic acid). This entails the oxidation of an aldehyde group to a carboxylic acid group, which occurs in two steps. The overall reaction releases enough free energy to convert a molecule of ADP to ATP and to transfer two electrons from the aldehyde to NAD^+ to form NADH, while still releasing enough heat to the environment to make the overall reaction energetically favorable ($\Delta G°$ for the overall reaction is –3.0 kcal/mole).

The pathway by which this remarkable feat is accomplished is outlined in Figure 2–73. The chemical reactions are guided by two enzymes to which the

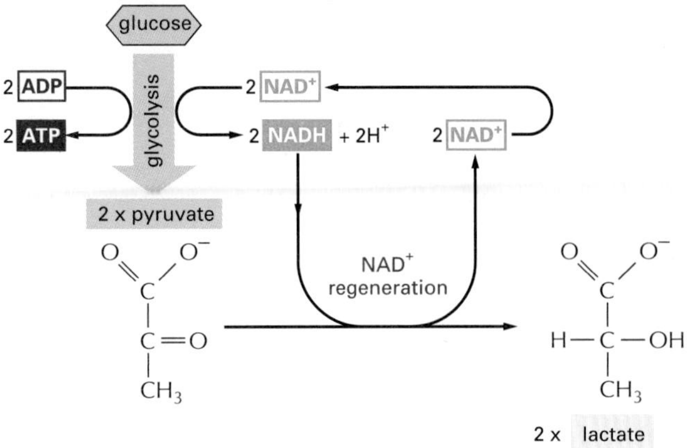

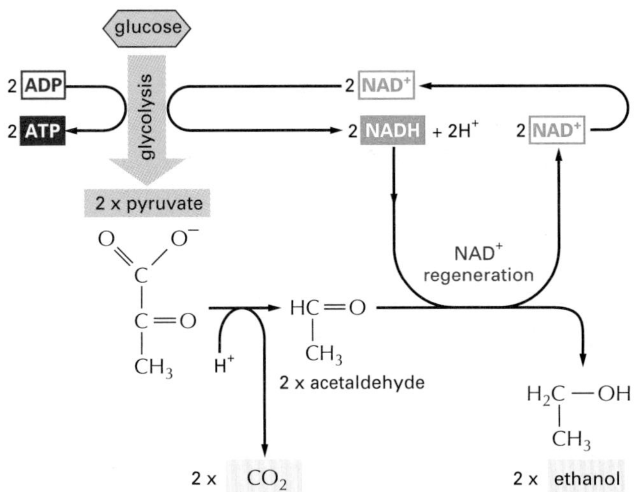

Figure 2–72 Two pathways for the anaerobic breakdown of pyruvate. (A) When inadequate oxygen is present, for example, in a muscle cell undergoing vigorous contraction, the pyruvate produced by glycolysis is converted to lactate as shown. This reaction regenerates the NAD^+ consumed in step 6 of glycolysis, but the whole pathway yields much less energy overall than complete oxidation. (B) In some organisms that can grow anaerobically, such as yeasts, pyruvate is converted via acetaldehyde into carbon dioxide and ethanol. Again, this pathway regenerates NAD^+ from NADH, as required to enable glycolysis to continue. Both (A) and (B) are examples of *fermentations*.

sugar intermediates are tightly bound. The first enzyme (*glyceraldehyde 3-phosphate dehydrogenase*) forms a short-lived covalent bond to the aldehyde through a reactive –SH group on the enzyme, and it catalyzes the oxidation of this aldehyde while still in the attached state. The high-energy enzyme–substrate bond created by the oxidation is then displaced by an inorganic phosphate ion to produce a high-energy sugar-phosphate intermediate, which is thereby released from the enzyme. This intermediate then binds to the second enzyme (*phosphoglycerate kinase*). This enzyme catalyzes the energetically favorable transfer of the high-energy phosphate just created to ADP, forming ATP and completing the process of oxidizing an aldehyde to a carboxylic acid (see Figure 2–73).

We have shown this particular oxidation process in some detail because it provides a clear example of enzyme-mediated energy storage through coupled reactions (Figure 2–74). These reactions (steps 6 and 7) are the only ones in glycolysis that create a high-energy phosphate linkage directly from inorganic phosphate. As such, they account for the net yield of two ATP molecules and two NADH molecules per molecule of glucose (see Panel 2–8, pp. 124–125).

As we have just seen, ATP can be formed readily from ADP when reaction intermediates are formed with higher-energy phosphate bonds than those in ATP. Phosphate bonds can be ordered in energy by comparing the standard free-energy change ($\Delta G°$) for the breakage of each bond by hydrolysis. Figure 2–75 compares the high-energy phosphoanhydride bonds in ATP with other phosphate bonds, several of which are generated during glycolysis.

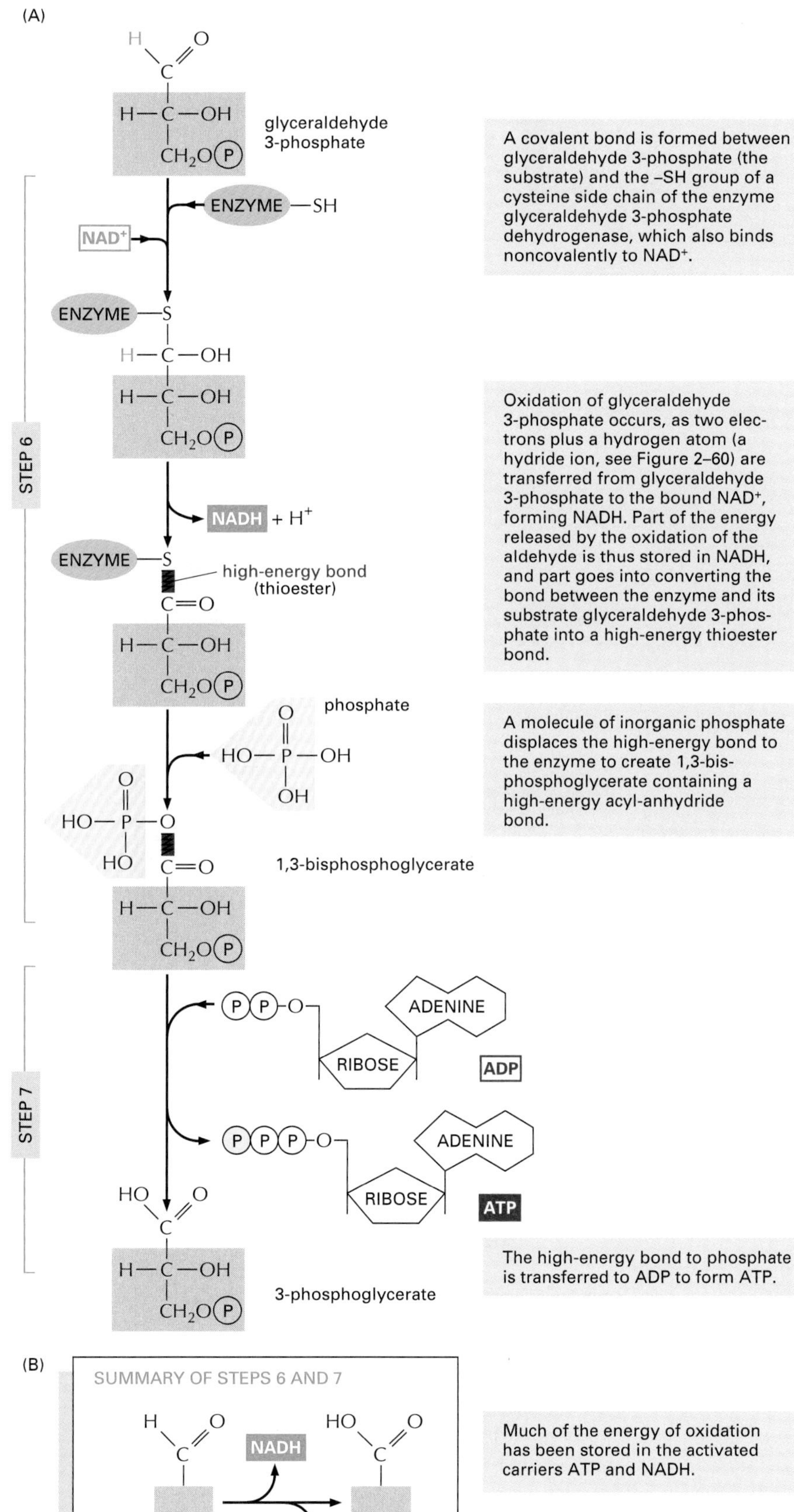
(A)
glyceraldehyde 3-phosphate
ENZYME —SH
NAD+
STEP 6
ENZYME —S
NADH + H+
high-energy bond (thioester)
phosphate
1,3-bisphosphoglycerate
STEP 7
ADENINE
RIBOSE
ADP
ATP
3-phosphoglycerate
A covalent bond is formed between glyceraldehyde 3-phosphate (the substrate) and the –SH group of a cysteine side chain of the enzyme glyceraldehyde 3-phosphate dehydrogenase, which also binds noncovalently to NAD+.
Oxidation of glyceraldehyde 3-phosphate occurs, as two electrons plus a hydrogen atom (a hydride ion, see Figure 2–60) are transferred from glyceraldehyde 3-phosphate to the bound NAD+, forming NADH. Part of the energy released by the oxidation of the aldehyde is thus stored in NADH, and part goes into converting the bond between the enzyme and its substrate glyceraldehyde 3-phosphate into a high-energy thioester bond.
A molecule of inorganic phosphate displaces the high-energy bond to the enzyme to create 1,3-bisphosphoglycerate containing a high-energy acyl-anhydride bond.
The high-energy bond to phosphate is transferred to ADP to form ATP.
(B)
SUMMARY OF STEPS 6 AND 7
NADH
aldehyde
ATP
carboxylic acid
Much of the energy of oxidation has been stored in the activated carriers ATP and NADH.

Figure 2–73 Energy storage in steps 6 and 7 of glycolysis. In these steps the oxidation of an aldehyde to a carboxylic acid is coupled to the formation of ATP and NADH. (A) Step 6 begins with the formation of a covalent bond between the substrate (glyceraldehyde 3-phosphate) and an –SH group exposed on the surface of the enzyme (glyceraldehyde 3-phosphate dehydrogenase). The enzyme then catalyzes transfer of hydrogen (as a hydride ion—a proton plus two electrons) from the bound glyceraldehyde 3-phosphate to a molecule of NAD^+. Part of the energy released in this oxidation is used to form a molecule of NADH and part is used to convert the original linkage between the enzyme and its substrate to a high-energy thioester bond (shown in *red*). A molecule of inorganic phosphate then displaces this high-energy bond on the enzyme, creating a high-energy sugar-phosphate bond instead *(red)*. At this point the enzyme has not only stored energy in NADH, but also coupled the energetically favorable oxidation of an aldehyde to the energetically unfavorable formation of a high-energy phosphate bond. The second reaction has been driven by the first, thereby acting like the "paddle wheel" coupler in Figure 2–56.

In reaction step 7, the high-energy sugar-phosphate intermediate just made, 1,3-bisphosphoglycerate, binds to a second enzyme, phosphoglycerate kinase. The reactive phosphate is transferred to ADP, forming a molecule of ATP and leaving a free carboxylic acid group on the oxidized sugar. (B) Summary of the overall chemical change produced by reactions 6 and 7.

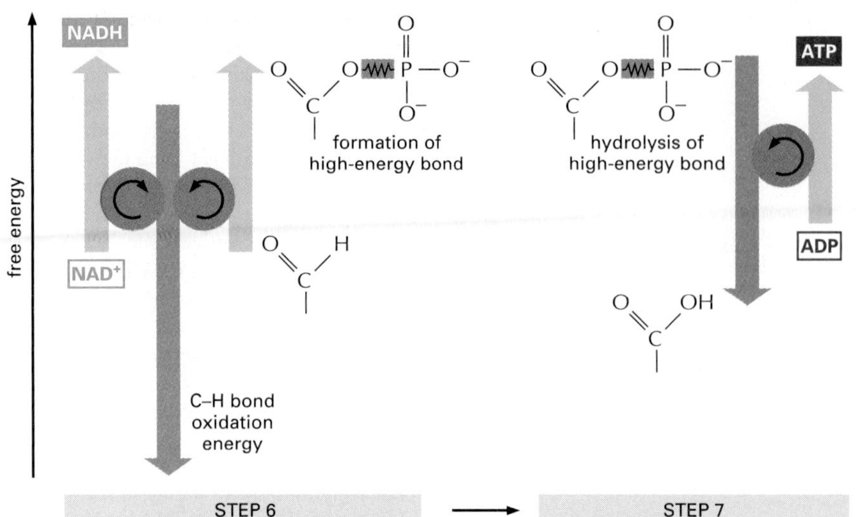

Figure 2–74 Schematic view of the coupled reactions that form NADH and ATP in steps 6 and 7 of glycolysis. The C–H bond oxidation energy drives the formation of both NADH and a high-energy phosphate bond. The breakage of the high-energy bond then drives ATP formation.

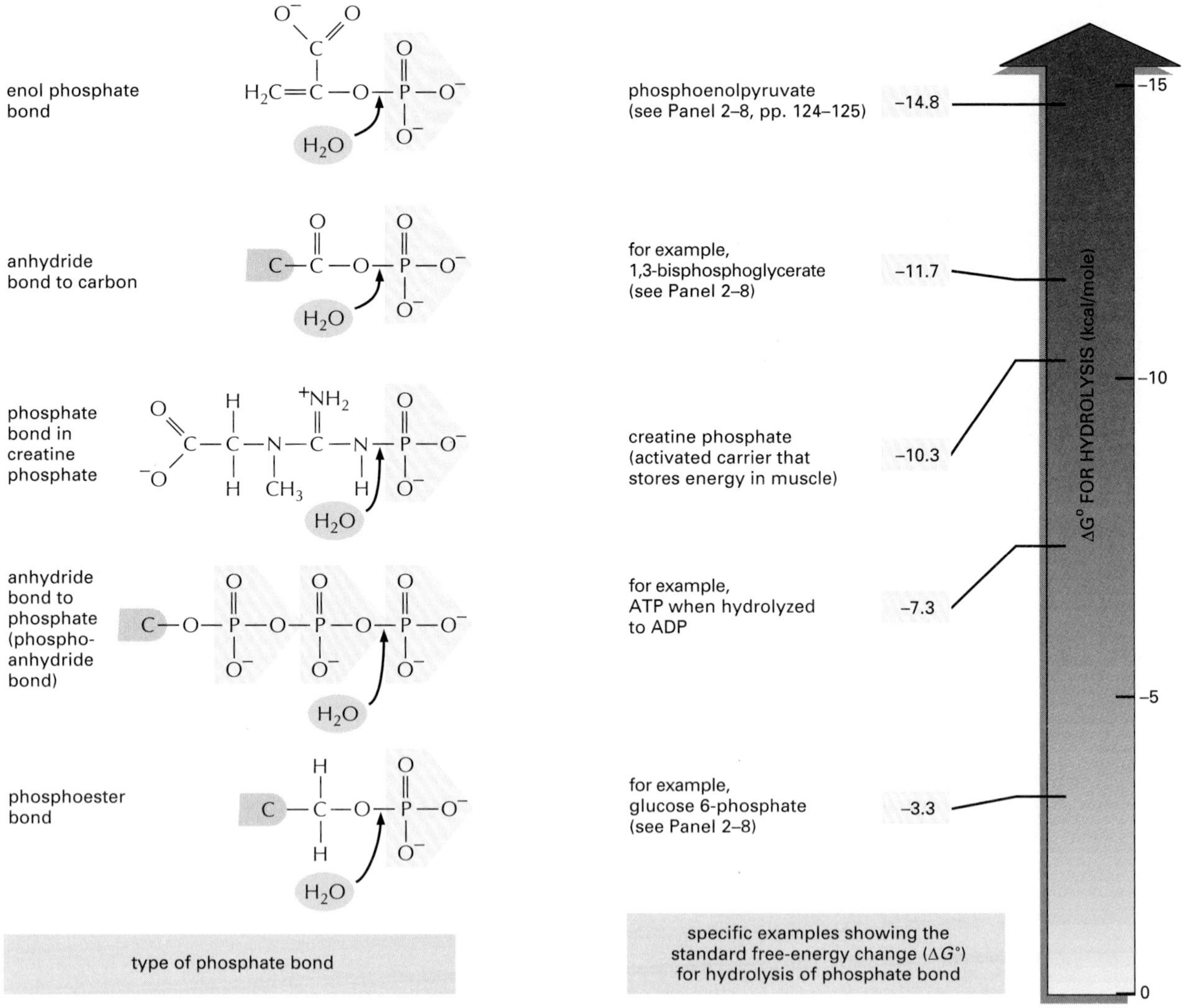

type of phosphate bond	specific examples showing the standard free-energy change ($\Delta G°$) for hydrolysis of phosphate bond	$\Delta G°$ FOR HYDROLYSIS (kcal/mole)
enol phosphate bond	phosphoenolpyruvate (see Panel 2–8, pp. 124–125)	–14.8
anhydride bond to carbon	for example, 1,3-bisphosphoglycerate (see Panel 2–8)	–11.7
phosphate bond in creatine phosphate	creatine phosphate (activated carrier that stores energy in muscle)	–10.3
anhydride bond to phosphate (phospho-anhydride bond)	for example, ATP when hydrolyzed to ADP	–7.3
phosphoester bond	for example, glucose 6-phosphate (see Panel 2–8)	–3.3

Figure 2–75 Some phosphate bond energies. The transfer of a phosphate group from any molecule 1 to any molecule 2 is energetically favorable if the standard free-energy change ($\Delta G°$) for the hydrolysis of the phosphate bond in molecule 1 is more negative than that for hydrolysis of the phosphate bond in molecule 2. Thus, for example, a phosphate group is readily transferred from 1,3-bisphosphoglycerate to ADP, forming ATP. Note that the hydrolysis reaction can be viewed as the transfer of the phosphate group to water.

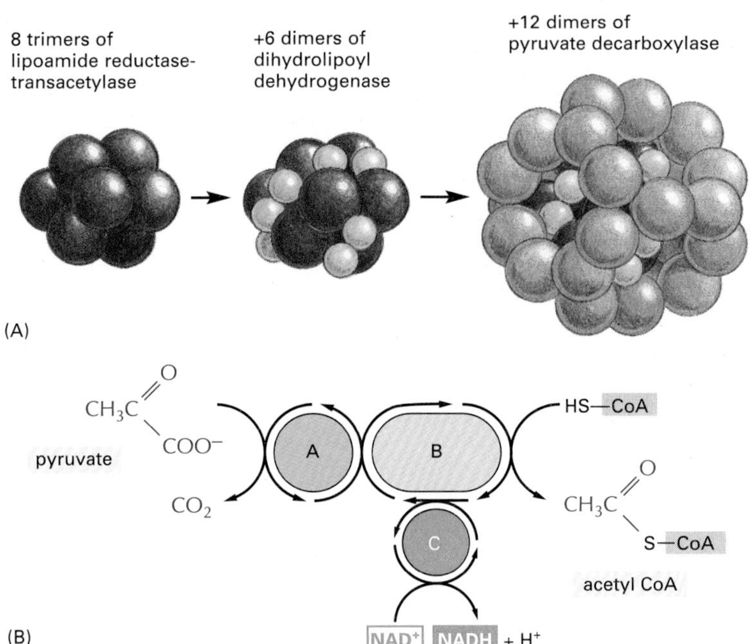

Figure 2–76 The oxidation of pyruvate to acetyl CoA and CO_2. (A) The structure of the pyruvate dehydrogenase complex, which contains 60 polypeptide chains. This is an example of a large multienzyme complex in which reaction intermediates are passed directly from one enzyme to another. In eucaryotic cells it is located in the mitochondrion. (B) The reactions carried out by the pyruvate dehydrogenase complex. The complex converts pyruvate to acetyl CoA in the mitochondrial matrix; NADH is also produced in this reaction. A, B, and C are the three enzymes *pyruvate decarboxylase, lipoamide reductase-transacetylase,* and *dihydrolipoyl dehydrogenase*, respectively. These enzymes are illustrated in (A); their activities are linked as shown.

Sugars and Fats Are Both Degraded to Acetyl CoA in Mitochondria

We now move on to consider stage 3 of catabolism, a process that requires abundant molecular oxygen (O_2 gas). Since the Earth is thought to have developed an atmosphere containing O_2 gas between one and two billion years ago, whereas abundant life-forms are known to have existed on the Earth for 3.5 billion years, the use of O_2 in the reactions that we discuss next is thought to be of relatively recent origin. In contrast, the mechanism used to produce ATP in Figure 2–73 does not require oxygen, and relatives of this elegant pair of coupled reactions could have arisen very early in the history of life on Earth.

In aerobic metabolism, the pyruvate produced by glycolysis is rapidly decarboxylated by a giant complex of three enzymes, called the *pyruvate dehydrogenase complex*. The products of pyruvate decarboxylation are a molecule of CO_2 (a waste product), a molecule of NADH, and acetyl CoA. The three-enzyme complex is located in the mitochondria of eucaryotic cells; its structure and mode of action are outlined in Figure 2–76.

The enzymes that degrade the fatty acids derived from fats likewise produce acetyl CoA in mitochondria. Each molecule of fatty acid (as the activated molecule *fatty acyl CoA*) is broken down completely by a cycle of reactions that trims two carbons at a time from its carboxyl end, generating one molecule of acetyl CoA for each turn of the cycle. A molecule of NADH and a molecule of $FADH_2$ are also produced in this process (Figure 2–77).

Sugars and fats provide the major energy sources for most non-photosynthetic organisms, including humans. However, the majority of the useful energy that can be extracted from the oxidation of both types of foodstuffs remains stored in the acetyl CoA molecules that are produced by the two types of reactions just described. The citric acid cycle of reactions, in which the acetyl group in acetyl CoA is oxidized to CO_2 and H_2O, is therefore central to the energy metabolism of aerobic organisms. In eucaryotes these reactions all take place in mitochondria, the organelle to which pyruvate and fatty acids are directed for acetyl CoA production (Figure 2–78). We should therefore not be surprised to discover that the mitochondrion is the place where most of the ATP is produced in animal cells. In contrast, aerobic bacteria carry out all of their reactions in a single compartment, the cytosol, and it is here that the citric acid cycle takes place in these cells.

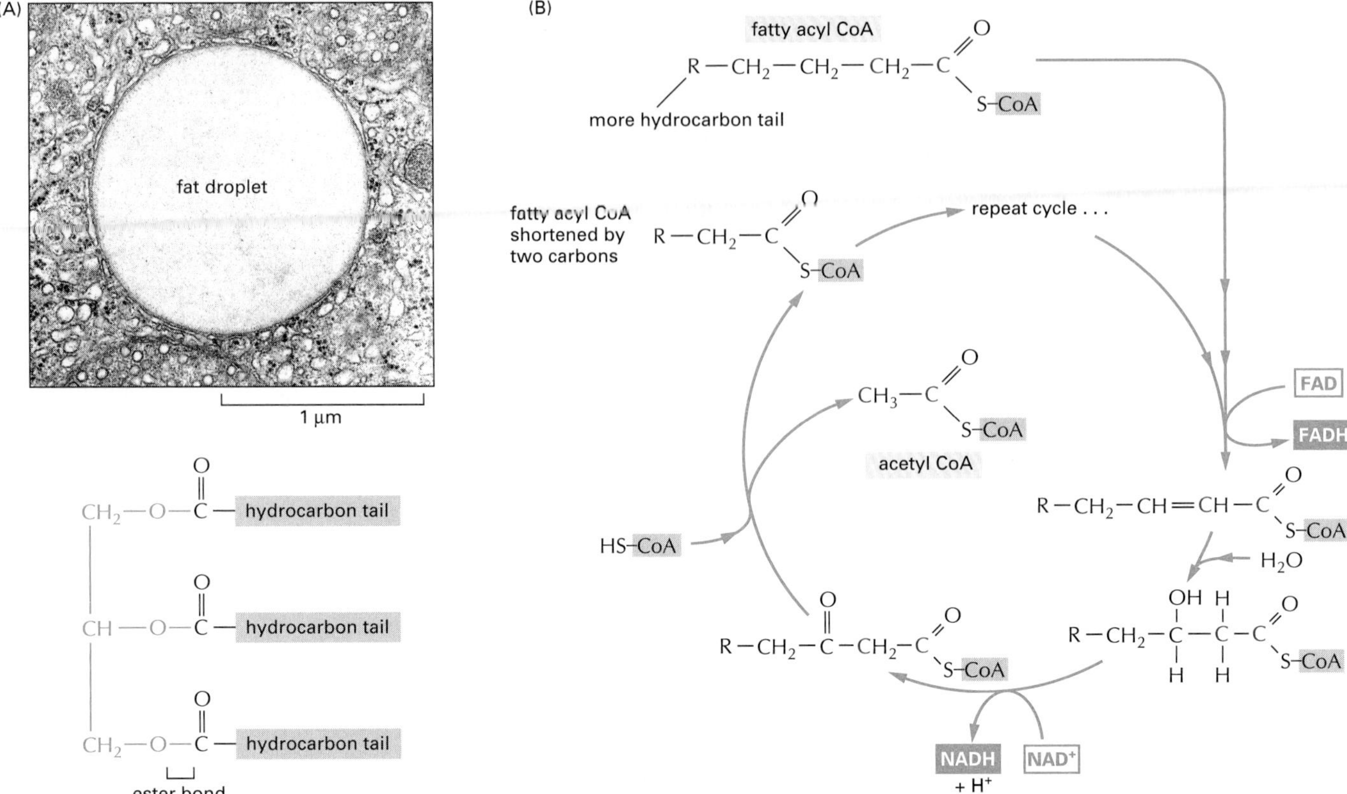

Figure 2–77 The oxidation of fatty acids to acetyl CoA. (A) Electron micrograph of a lipid droplet in the cytoplasm *(top)*, and the structure of fats *(bottom)*. Fats are triacylglycerols. The glycerol portion, to which three fatty acids are linked through ester bonds, is shown here in *green*. Fats are insoluble in water and form large lipid droplets in the specialized fat cells (called adipocytes) in which they are stored. (B) The fatty acid oxidation cycle. The cycle is catalyzed by a series of four enzymes in the mitochondrion. Each turn of the cycle shortens the fatty acid chain by two carbons (shown in *red*) and generates one molecule of acetyl CoA and one molecule each of NADH and $FADH_2$. The structure of $FADH_2$ will be presented in Figure 2–80B. (A, courtesy of Daniel S. Friend.)

The Citric Acid Cycle Generates NADH by Oxidizing Acetyl Groups to CO_2

In the nineteenth century, biologists noticed that in the absence of air (anaerobic conditions) cells produce lactic acid (for example, in muscle) or ethanol (for example, in yeast), while in its presence (aerobic conditions) they consume O_2 and produce CO_2 and H_2O. Intensive efforts to define the pathways of aerobic metabolism eventually focused on the oxidation of pyruvate and led in 1937 to the discovery of the **citric acid cycle**, also known as the *tricarboxylic acid cycle* or the *Krebs cycle*. The citric acid cycle accounts for about two-thirds of the total oxidation of carbon compounds in most cells, and its major end products are CO_2 and high-energy electrons in the form of NADH. The CO_2 is released as a waste product, while the high-energy electrons from NADH are passed to a membrane-bound electron-transport chain, eventually combining with O_2 to produce H_2O. Although the citric acid cycle itself does not use O_2, it requires O_2 in order to proceed because there is no other efficient way for the NADH to get rid of its electrons and thus regenerate the NAD^+ that is needed to keep the cycle going.

The citric acid cycle, which takes place inside mitochondria in eucaryotic cells, results in the complete oxidation of the carbon atoms of the acetyl groups

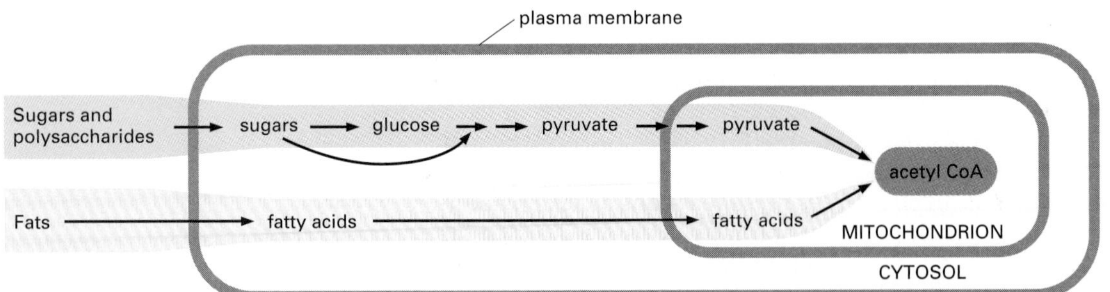

Figure 2–78 Pathways for the production of acetyl CoA from sugars and fats. The mitochondrion in eucaryotic cells is the place where acetyl CoA is produced from both types of major food molecules. It is therefore the place where most of the cell's oxidation reactions occur and where most of its ATP is made.

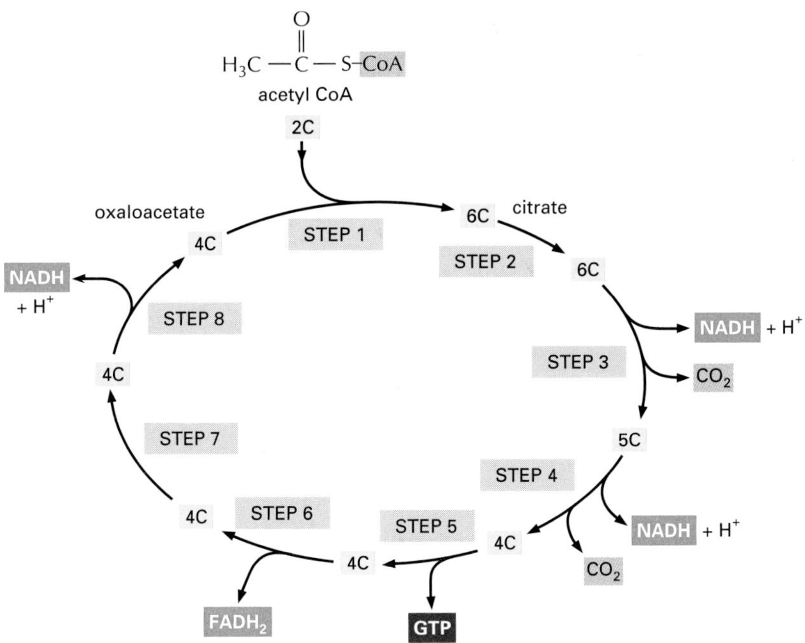

Figure 2–79 Simple overview of the citric acid cycle. The reaction of acetyl CoA with oxaloacetate starts the cycle by producing citrate (citric acid). In each turn of the cycle, two molecules of CO_2 are produced as waste products, plus three molecules of NADH, one molecule of GTP, and one molecule of $FADH_2$. The number of carbon atoms in each intermediate is shown in a *yellow box.* For details, see Panel 2–9 (pp. 126–127).

in acetyl CoA, converting them into CO_2. But the acetyl group is not oxidized directly. Instead, this group is transferred from acetyl CoA to a larger, four-carbon molecule, *oxaloacetate,* to form the six-carbon tricarboxylic acid, *citric acid,* for which the subsequent cycle of reactions is named. The citric acid molecule is then gradually oxidized, allowing the energy of this oxidation to be harnessed to produce energy-rich activated carrier molecules. The chain of eight reactions forms a cycle because at the end the oxaloacetate is regenerated and enters a new turn of the cycle, as shown in outline in Figure 2–79.

We have thus far discussed only one of the three types of activated carrier molecules that are produced by the citric acid cycle, the NAD^+–NADH pair (see Figure 2–60). In addition to three molecules of NADH, each turn of the cycle also produces one molecule of **$FADH_2$** (reduced flavin adenine dinucleotide) from FAD and one molecule of the ribonucleotide **GTP** (guanosine triphosphate) from GDP. The structures of these two activated carrier molecules are illustrated in Figure 2–80. GTP is a close relative of ATP, and the transfer of its terminal

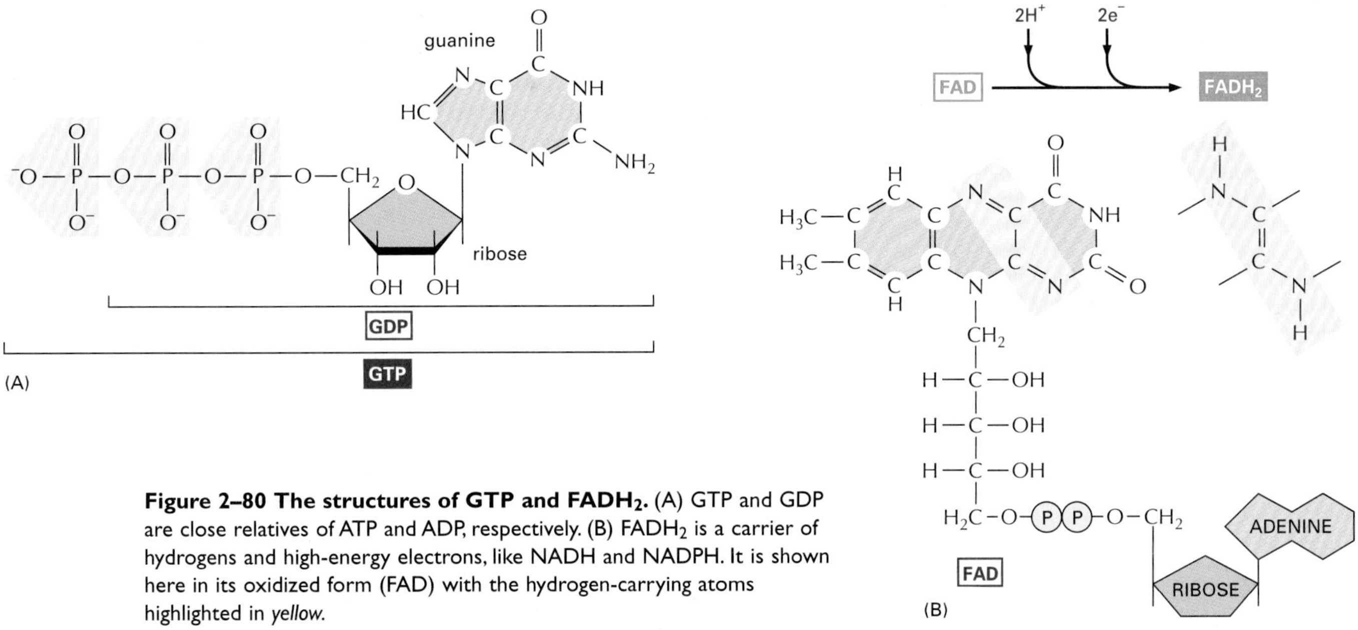

Figure 2–80 The structures of GTP and $FADH_2$. (A) GTP and GDP are close relatives of ATP and ADP, respectively. (B) $FADH_2$ is a carrier of hydrogens and high-energy electrons, like NADH and NADPH. It is shown here in its oxidized form (FAD) with the hydrogen-carrying atoms highlighted in *yellow.*

phosphate group to ADP produces one ATP molecule in each cycle. Like NADH, $FADH_2$ is a carrier of high-energy electrons and hydrogen. As we discuss shortly, the energy that is stored in the readily transferred high-energy electrons of NADH and $FADH_2$ will be utilized subsequently for ATP production through the process of *oxidative phosphorylation,* the only step in the oxidative catabolism of foodstuffs that directly requires gaseous oxygen (O_2) from the atmosphere.

The complete citric acid cycle is presented in Panel 2–9 (pp. 126–127). The extra oxygen atoms required to make CO_2 from the acetyl groups entering the citric acid cycle are supplied not by molecular oxygen, but by water. As illustrated in the panel, three molecules of water are split in each cycle, and the oxygen atoms of some of them are ultimately used to make CO_2.

In addition to pyruvate and fatty acids, some amino acids pass from the cytosol into mitochondria, where they are also converted into acetyl CoA or one of the other intermediates of the citric acid cycle. Thus, in the eucaryotic cell, the mitochondrion is the center toward which all energy-yielding processes lead, whether they begin with sugars, fats, or proteins.

The citric acid cycle also functions as a starting point for important biosynthetic reactions by producing vital carbon-containing intermediates, such as *oxaloacetate* and *α-ketoglutarate.* Some of these substances produced by catabolism are transferred back from the mitochondrion to the cytosol, where they serve in anabolic reactions as precursors for the synthesis of many essential molecules, such as amino acids.

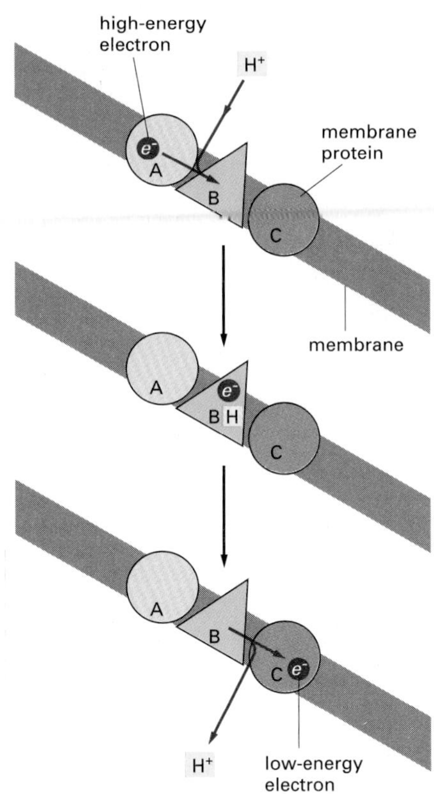

Figure 2–81 The generation of an H^+ gradient across a membrane by electron-transport reactions. A high-energy electron (derived, for example, from the oxidation of a metabolite) is passed sequentially by carriers A, B, and C to a lower energy state. In this diagram carrier B is arranged in the membrane in such a way that it takes up H^+ from one side and releases it to the other as the electron passes. The result is an H^+ gradient. This gradient represents a form of stored energy that is harnessed by other membrane proteins to drive the formation of ATP.

Electron Transport Drives the Synthesis of the Majority of the ATP in Most Cells

It is in the last step in the degradation of a food molecule that the major portion of its chemical energy is released. In this final process the electron carriers NADH and $FADH_2$ transfer the electrons that they have gained when oxidizing other molecules to the **electron-transport chain**, which is embedded in the inner membrane of the mitochondrion. As the electrons pass along this long chain of specialized electron acceptor and donor molecules, they fall to successively lower energy states. The energy that the electrons release in this process is used to pump H^+ ions (protons) across the membrane—from the inner mitochondrial compartment to the outside (Figure 2–81). A gradient of H^+ ions is thereby generated. This gradient serves as a source of energy, being tapped like a battery to drive a variety of energy-requiring reactions. The most prominent of these reactions is the generation of ATP by the phosphorylation of ADP.

At the end of this series of electron transfers, the electrons are passed to molecules of oxygen gas (O_2) that have diffused into the mitochondrion, which simultaneously combine with protons (H^+) from the surrounding solution to produce molecules of water. The electrons have now reached their lowest energy level, and therefore all the available energy has been extracted from the food molecule being oxidized. This process, termed **oxidative phosphorylation** (Figure 2–82), also occurs in the plasma membrane of bacteria. As one of the most remarkable achievements of cellular evolution, it will be a central topic of Chapter 14.

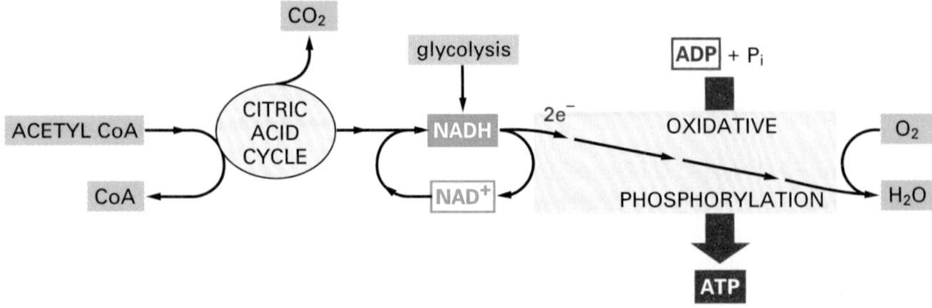

Figure 2–82 The final stages of oxidation of food molecules. Molecules of NADH and $FADH_2$ ($FADH_2$ is not shown) are produced by the citric acid cycle. These activated carriers donate high-energy electrons that are eventually used to reduce oxygen gas to water. A major portion of the energy released during the transfer of these electrons along an electron-transfer chain in the mitochondrial inner membrane (or in the plasma membrane of bacteria) is harnessed to drive the synthesis of ATP: hence the name oxidative phosphorylation.

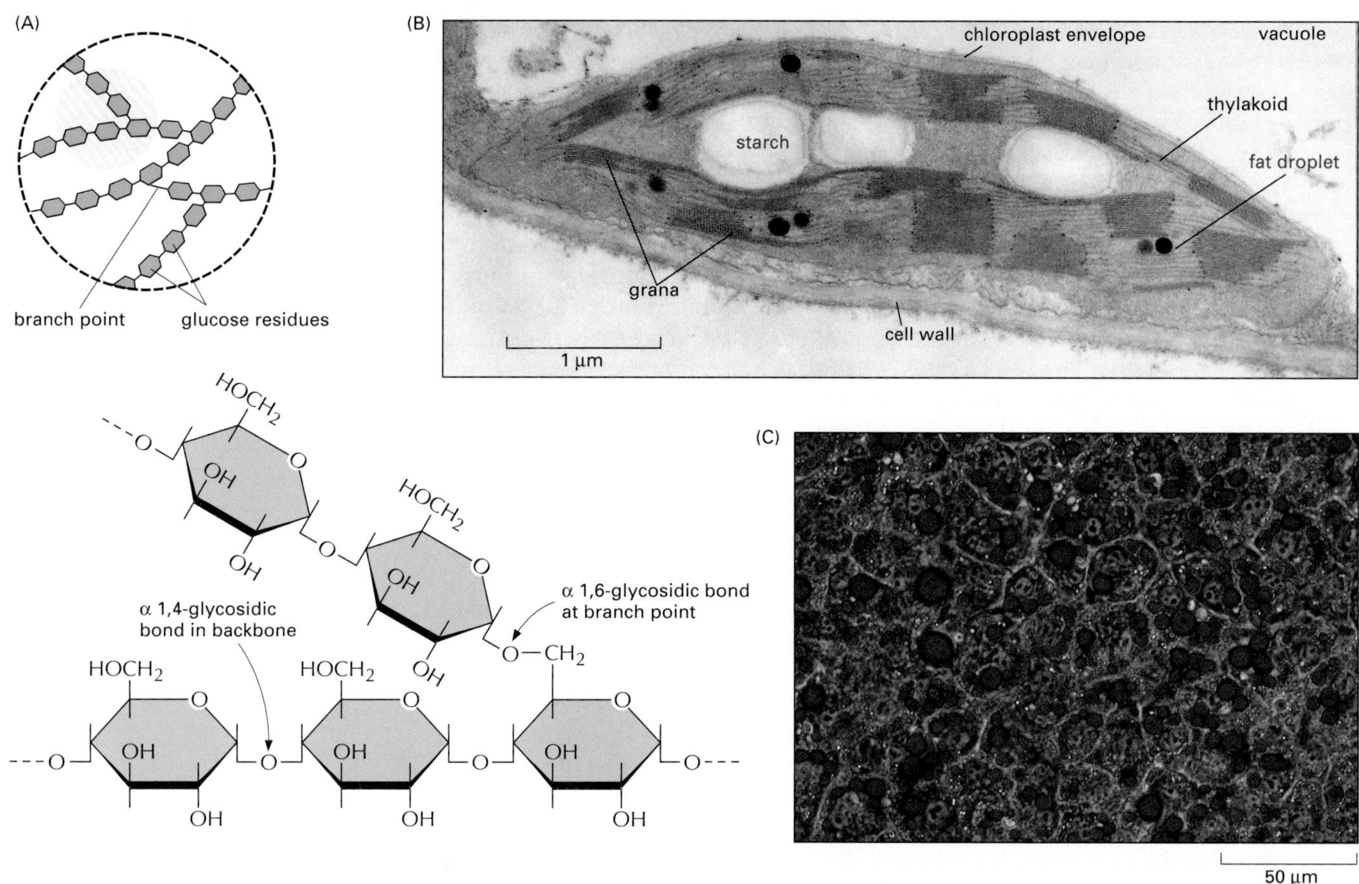

Figure 2–83 The storage of sugars and fats in animal and plant cells. (A) The structures of starch and glycogen, the storage form of sugars in plants and animals, respectively. Both are storage polymers of the sugar glucose and differ only in the frequency of branch points (the region in *yellow* is shown enlarged below). There are many more branches in glycogen than in starch. (B) A thin section of a single chloroplast from a plant cell, showing the starch granules and lipid droplets that have accumulated as a result of the biosyntheses occurring there. (C) Fat droplets (stained *red)* beginning to accumulate in developing fat cells of an animal. (B, courtesy of K. Plaskitt; C, courtesy of Ronald M. Evans and Peter Totonoz.)

In total, the complete oxidation of a molecule of glucose to H_2O and CO_2 is used by the cell to produce about 30 molecules of ATP. In contrast, only 2 molecules of ATP are produced per molecule of glucose by glycolysis alone.

Organisms Store Food Molecules in Special Reservoirs

All organisms need to maintain a high ATP/ADP ratio, if biological order is to be maintained in their cells. Yet animals have only periodic access to food, and plants need to survive overnight without sunlight, without the possibility of sugar production from photosynthesis. For this reason, both plants and animals convert sugars and fats to special forms for storage (Figure 2–83).

To compensate for long periods of fasting, animals store fatty acids as fat droplets composed of water-insoluble triacylglycerols, largely in specialized fat cells. And for shorter-term storage, sugar is stored as glucose subunits in the large branched polysaccharide **glycogen**, which is present as small granules in the cytoplasm of many cells, including liver and muscle. The synthesis and degradation of glycogen are rapidly regulated according to need. When more ATP is needed than can be generated from the food molecules taken in from the bloodstream, cells break down glycogen in a reaction that produces glucose 1-phosphate, which enters glycolysis.

Quantitatively, **fat** is a far more important storage form than glycogen, in part because the oxidation of a gram of fat releases about twice as much energy as the oxidation of a gram of glycogen. Moreover, glycogen differs from fat in binding a great deal of water, producing a sixfold difference in the actual mass of glycogen required to store the same amount of energy as fat. An average adult human stores enough glycogen for only about a day of normal activities but enough fat to last for nearly a month. If our main fuel reservoir had to be carried as glycogen instead of fat, body weight would need to be increased by an average of about 60 pounds.

Most of our fat is stored in adipose tissue, from which it is released into the bloodstream for other cells to utilize as needed. The need arises after a period of not eating; even a normal overnight fast results in the mobilization of fat, so that in the morning most of the acetyl CoA entering the citric acid cycle is derived from fatty acids rather than from glucose. After a meal, however, most of the acetyl CoA entering the citric acid cycle comes from glucose derived from food, and any excess glucose is used to replenish depleted glycogen stores or to synthesize fats. (While animal cells readily convert sugars to fats, they cannot convert fatty acids to sugars.)

Although plants produce NADPH and ATP by photosynthesis, this important process occurs in a specialized organelle, called a chloroplast, which is isolated from the rest of the plant cell by a membrane that is impermeable to both types of activated carrier molecules. Moreover, the plant contains many other cells—such as those in the roots—that lack chloroplasts and therefore cannot produce their own sugars or ATP. Therefore, for most of its ATP production, the plant relies on an export of sugars from its chloroplasts to the mitochondria that are located in all cells of the plant. Most of the ATP needed by the plant is synthesized in these mitochondria and exported from them to the rest of the plant cell, using exactly the same pathways for the oxidative breakdown of sugars that are utilized by nonphotosynthetic organisms (Figure 2–84).

During periods of excess photosynthetic capacity during the day, chloroplasts convert some of the sugars that they make into fats and into **starch**, a polymer of glucose analogous to the glycogen of animals. The fats in plants are triacylglycerols, just like the fats in animals, and differ only in the types of fatty acids that predominate. Fat and starch are both stored in the chloroplast as reservoirs to be mobilized as an energy source during periods of darkness (see Figure 2–83B).

The embryos inside plant seeds must live on stored sources of energy for a prolonged period, until they germinate to produce leaves that can harvest the energy in sunlight. For this reason plant seeds often contain especially large amounts of fats and starch—which makes them a major food source for animals, including ourselves (Figure 2–85).

Amino Acids and Nucleotides Are Part of the Nitrogen Cycle

In our discussion so far we have concentrated mainly on carbohydrate metabolism. We have not yet considered the metabolism of nitrogen or sulfur. These two elements are constituents of proteins and nucleic acids, which are the two most important classes of macromolecules in the cell and make up

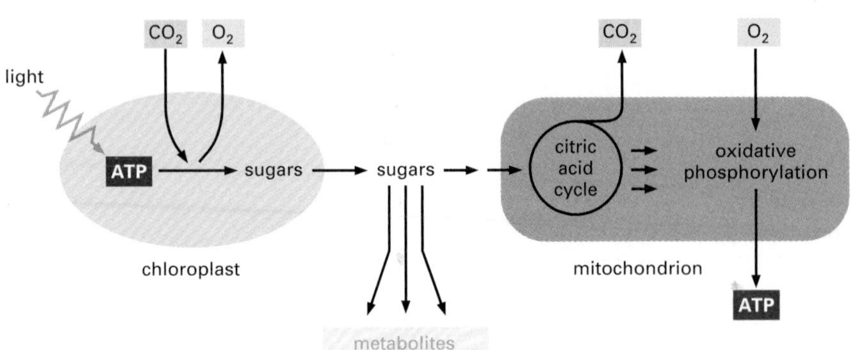

Figure 2–84 How the ATP needed for most plant cell metabolism is made. In plants, the chloroplasts and mitochondria collaborate to supply cells with metabolites and ATP.

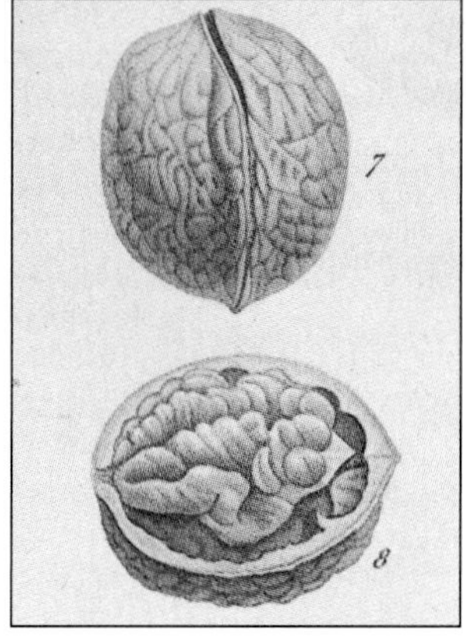

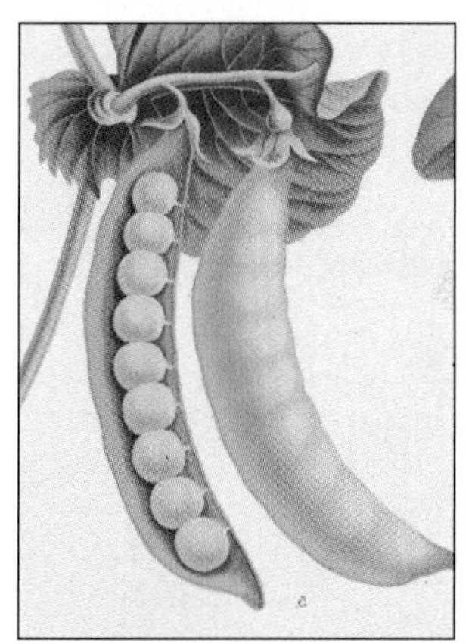

Figure 2–85 Some plant seeds that serve as important foods for humans. Corn, nuts, and peas all contain rich stores of starch and fat that provide the young plant embryo in the seed with energy and building blocks for biosynthesis. (Courtesy of the John Innes Foundation.)

approximately two-thirds of its dry weight. Atoms of nitrogen and sulfur pass from compound to compound and between organisms and their environment in a series of reversible cycles.

Although molecular nitrogen is abundant in the Earth's atmosphere, nitrogen is chemically unreactive as a gas. Only a few living species are able to incorporate it into organic molecules, a process called **nitrogen fixation**. Nitrogen fixation occurs in certain microorganisms and by some geophysical processes, such as lightning discharge. It is essential to the biosphere as a whole, for without it life would not exist on this planet. Only a small fraction of the nitrogenous compounds in today's organisms, however, is due to fresh products of nitrogen fixation from the atmosphere. Most organic nitrogen has been in circulation for some time, passing from one living organism to another. Thus present-day nitrogen-fixing reactions can be said to perform a "topping-up" function for the total nitrogen supply.

Vertebrates receive virtually all of their nitrogen in their dietary intake of proteins and nucleic acids. In the body these macromolecules are broken down to amino acids and the components of nucleotides, and the nitrogen they contain is used to produce new proteins and nucleic acids or utilized to make other molecules. About half of the 20 amino acids found in proteins are essential amino acids for vertebrates (Figure 2–86), which means that they cannot be synthesized from other ingredients of the diet. The others can be so synthesized, using a variety of raw materials, including intermediates of the citric acid cycle as described below. The essential amino acids are made by nonvertebrate organisms, usually by long and energetically expensive pathways that have been lost in the course of vertebrate evolution.

The nucleotides needed to make RNA and DNA can be synthesized using specialized biosynthetic pathways: there are no "essential nucleotides" that must be provided in the diet. All of the nitrogens in the purine and pyrimidine bases (as well as some of the carbons) are derived from the plentiful amino acids glutamine, aspartic acid, and glycine, whereas the ribose and deoxyribose sugars are derived from glucose.

Amino acids that are not utilized in biosynthesis can be oxidized to generate metabolic energy. Most of their carbon and hydrogen atoms eventually form CO_2 or H_2O, whereas their nitrogen atoms are shuttled through various forms and eventually appear as urea, which is excreted. Each amino acid is processed differently, and a whole constellation of enzymatic reactions exists for their catabolism.

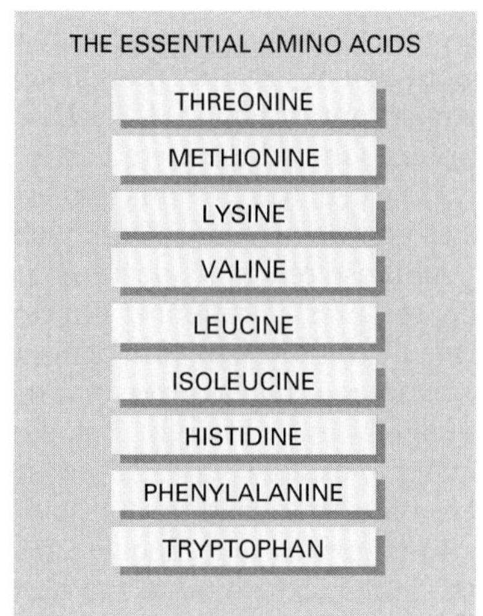

Figure 2–86 The nine essential amino acids. These cannot be synthesized by human cells and so must be supplied in the diet.

Many Biosynthetic Pathways Begin with Glycolysis or the Citric Acid Cycle

Catabolism produces both energy for the cell and the building blocks from which many other molecules of the cell are made (see Figure 2–36). Thus far, our discussions of glycolysis and the citric acid cycle have emphasized energy production, rather than the provision of the starting materials for biosynthesis. But

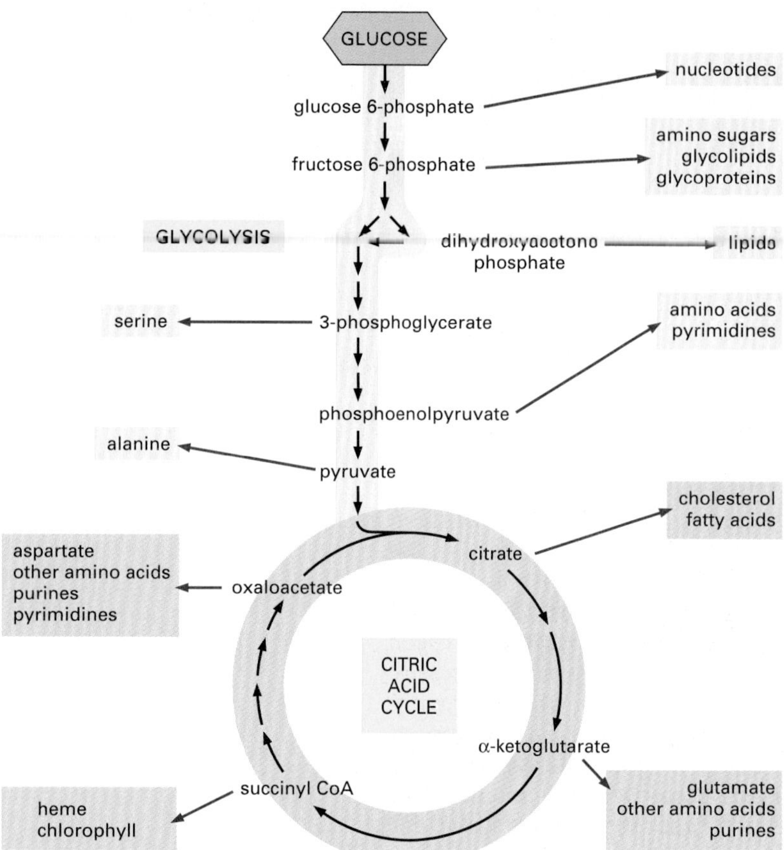

Figure 2–87 Glycolysis and the citric acid cycle provide the precursors needed to synthesize many important biological molecules. The amino acids, nucleotides, lipids, sugars, and other molecules—shown here as products—in turn serve as the precursors for the many macromolecules of the cell. Each *black arrow* in this diagram denotes a single enzyme-catalyzed reaction; the *red arrows* generally represent pathways with many steps that are required to produce the indicated products.

many of the intermediates formed in these reaction pathways are also siphoned off by other enzymes that use them to produce the amino acids, nucleotides, lipids, and other small organic molecules that the cell needs. Some idea of the complexity of this process can be gathered from Figure 2–87, which illustrates some of the branches from the central catabolic reactions that lead to biosyntheses.

The existence of so many branching pathways in the cell requires that the choices at each branch be carefully regulated, as we discuss next.

Metabolism Is Organized and Regulated

One gets a sense of the intricacy of a cell as a chemical machine from the relation of glycolysis and the citric acid cycle to the other metabolic pathways sketched out in Figure 2–88. This type of chart, which was used earlier in this chapter to introduce metabolism, represents only some of the enzymatic pathways in a cell. It is obvious that our discussion of cell metabolism has dealt with only a tiny fraction of cellular chemistry.

All these reactions occur in a cell that is less than 0.1 mm in diameter, and each requires a different enzyme. As is clear from Figure 2–88, the same molecule can often be part of many different pathways. Pyruvate, for example, is a substrate for half a dozen or more different enzymes, each of which modifies it chemically in a different way. One enzyme converts pyruvate to acetyl CoA, another to oxaloacetate; a third enzyme changes pyruvate to the amino acid alanine, a fourth to lactate, and so on. All of these different pathways compete for the same pyruvate molecule, and similar competitions for thousands of other small molecules go on at the same time. A better sense of this complexity can perhaps be attained from a three-dimensional metabolic map that allows the connections between pathways to be made more directly (Figure 2–89).

The situation is further complicated in a multicellular organism. Different cell types will in general require somewhat different sets of enzymes. And different tissues make distinct contributions to the chemistry of the organism as a

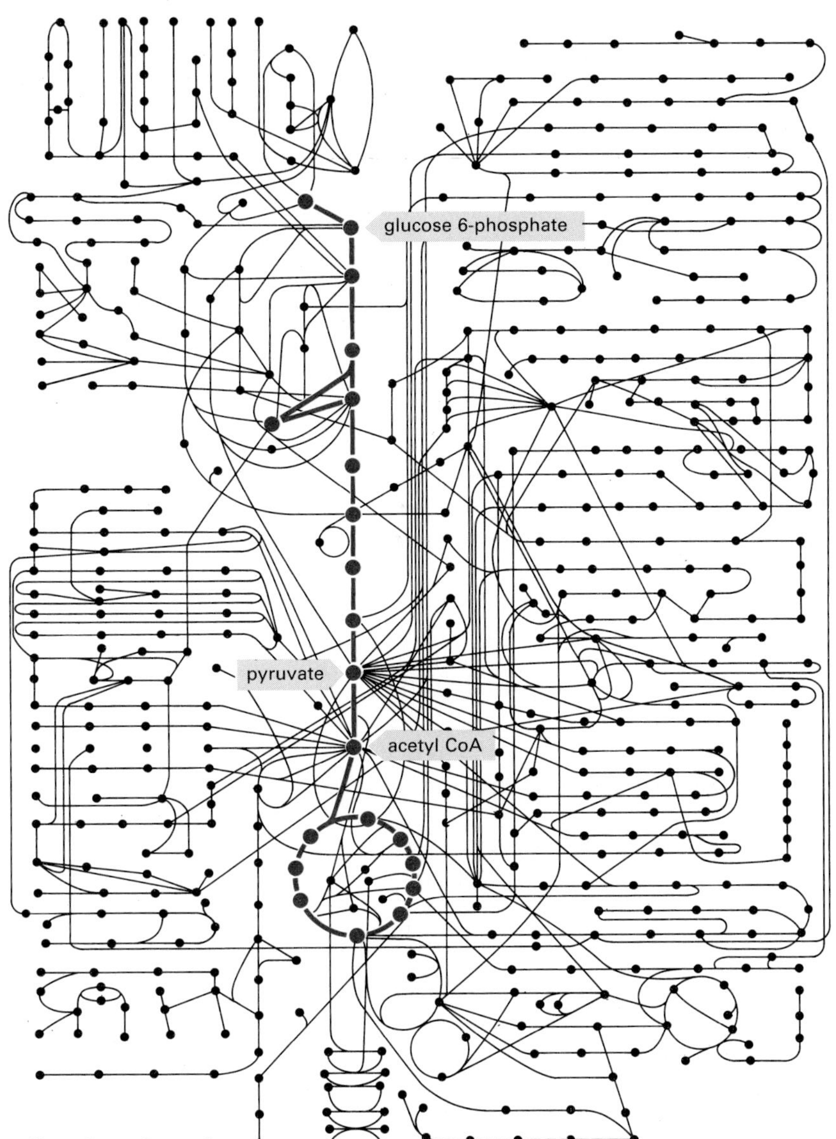

Figure 2–88 Glycolysis and the citric acid cycle are at the center of metabolism. Some 500 metabolic reactions of a typical cell are shown schematically with the reactions of glycolysis and the citric acid cycle in *red.* Other reactions either lead into these two central pathways—delivering small molecules to be catabolized with production of energy—or they lead outward and thereby supply carbon compounds for the purpose of biosynthesis.

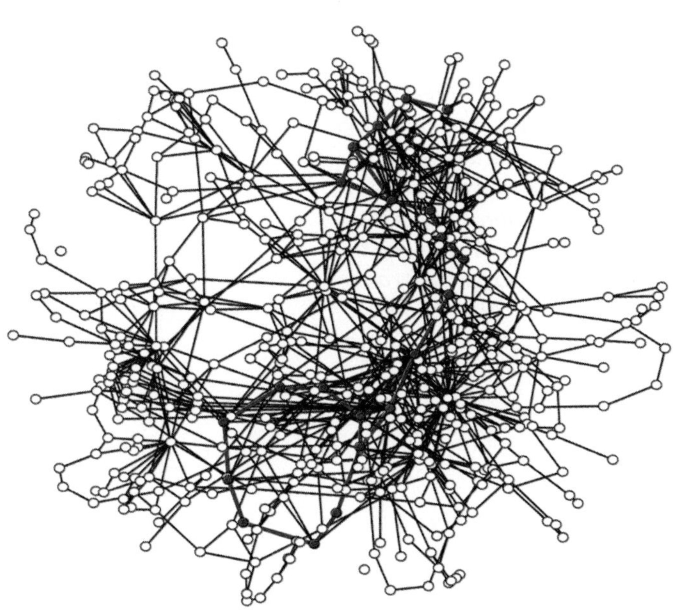

Figure 2–89 A representation of all of the known metabolic reactions involving small molecules in a yeast cell. As in Figure 2–88, the reactions of glycolysis and the citric acid cycle are highlighted in *red.* This metabolic map is unusual in making use of three-dimensions, so as to allow the many interactions between pathways to be emphasized. It is meant to be viewed on a computer screen, where it can be rotated and inspected from every angle. (From H. Jeong, S.P. Mason, A-L. Barabási and N. Oltava, *Nature* 411:41–42, 2001. © Macmillan Magazines Ltd.)

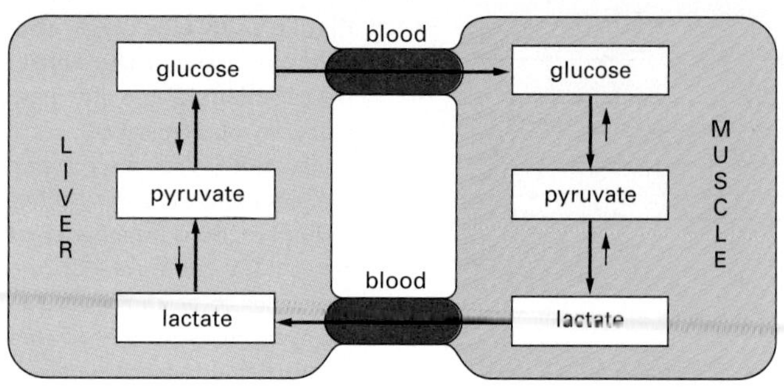

Figure 2–90 Schematic view of the metabolic cooperation between liver and muscle cells. The principal fuel of actively contracting muscle cells is glucose, much of which is supplied by liver cells. Lactic acid, the end product of anaerobic glucose breakdown by glycolysis in muscle, is converted back to glucose in the liver by the process of gluconeogenesis.

whole. In addition to differences in specialized products such as hormones or antibodies, there are significant differences in the "common" metabolic pathways among various types of cells in the same organism.

Although virtually all cells contain the enzymes of glycolysis, the citric acid cycle, lipid synthesis and breakdown, and amino acid metabolism, the levels of these processes required in different tissues are not the same. For example, nerve cells, which are probably the most fastidious cells in the body, maintain almost no reserves of glycogen or fatty acids and rely almost entirely on a constant supply of glucose from the bloodstream. In contrast, liver cells supply glucose to actively contracting muscle cells and recycle the lactic acid produced by muscle cells back into glucose (Figure 2–90). All types of cells have their distinctive metabolic traits, and they cooperate extensively in the normal state, as well as in response to stress and starvation. One might think that the whole system would need to be so finely balanced that any minor upset, such as a temporary change in dietary intake, would be disastrous.

In fact, the metabolic balance of a cell is amazingly stable. Whenever the balance is perturbed, the cell reacts so as to restore the initial state. The cell can adapt and continue to function during starvation or disease. Mutations of many kinds can damage or even eliminate particular reaction pathways, and yet—provided that certain minimum requirements are met—the cell survives. It does so because an elaborate network of *control mechanisms* regulates and coordinates the rates of all of its reactions. These controls rest, ultimately, on the remarkable abilities of proteins to change their shape and their chemistry in response to changes in their immediate environment. The principles that underlie how large molecules such as proteins are built and the chemistry behind their regulation will be our next concern.

Summary

Glucose and other food molecules are broken down by controlled stepwise oxidation to provide chemical energy in the form of ATP and NADH. These are three main sets of reactions that act in series—the products of each being the starting material for the next: glycolysis (which occurs in the cytosol), the citric acid cycle (in the mitochondrial matrix), and oxidative phosphorylation (on the inner mitochondrial membrane). The intermediate products of glycolysis and the citric acid cycle are used both as sources of metabolic energy and to produce many of the small molecules used as the raw materials for biosynthesis. Cells store sugar molecules as glycogen in animals and starch in plants; both plants and animals also use fats extensively as a food store. These storage materials in turn serve as a major source of food for humans, along with the proteins that comprise the majority of the dry mass of the cells we eat.

References

General

Garrett RH & Grisham CM (1998) Biochemistry, 2nd edn. Orlando: Saunders.

Horton HR, Moran LA, Ochs RS et al. (2001) Principles of Biochemistry, 3rd edn. Upper Saddle River, NJ: Prentice Hall.

Lehninger AL, Nelson DL & Cox MM (1993) Principles of Biochemistry, 2nd edn. New York: Worth.

Mathews CK, van Holde KE & Ahern K-G (2000) Biochemistry, 3rd edn. San Francisco: Benjamin Cummings.

Moore JA (1993) Science As a Way of Knowing. Cambridge, MA: Harvard University Press

Stryer L (1995) Biochemistry, 4th edn. New York: WH Freeman.

Voet D, Voet JG & Pratt CW (1999) Fundamentals of Biochemistry. New York: Wiley.

Zubay GL (1998) Biochemistry, 4th edn. Dubuque, IO: William C Brown.

The Chemical Components of a Cell

Abeles RH, Frey PA & Jencks WP (1992) Biochemistry. Boston: Jones & Bartlett.

Atkins PW (1996) Molecules. New York: WH Freeman.

Branden C & Tooze J (1999) Introduction to Protein Structure, 2nd edn. New York: Garland Publishing.

Bretscher MS (1985) The molecules of the cell membrane. *Sci. Am.* 253(4), 100–109.

Burley SK & Petsko GA (1988) Weakly polar interactions in proteins. *Adv. Protein Chem.* 39, 125–189.

Eisenberg D & Kauzman W (1969) The Structure and Properties of Water. Oxford: Oxford University Press.

Fersht AR (1987) The hydrogen bond in molecular recognition. *Trends Biochem. Sci.* 12, 301–304.

Franks F (1993) Water. Cambridge: Royal Society of Chemistry.

Henderson LJ (1927) The Fitness of the Environment, 1958 edn. Boston: Beacon.

Ingraham JL, Maaløe O & Neidhardt FC (1983) Growth of the Bacterial Cell. Sunderland, MA: Sinauer.

Pauling L (1960) The Nature of the Chemical Bond, 3rd edn. Ithaca, NY: Cornell University Press.

Saenger W (1984) Principles of Nucleic Acid Structure. New York: Springer.

Sharon N (1980) Carbohydrates. *Sci. Am.* 243(5), 90–116.

Stillinger FH (1980) Water revisited. *Science* 209, 451–457.

Tanford C (1978) The hydrophobic effect and the organization of living matter. *Science* 200, 1012–1018.

Tanford C (1980) The Hydrophobic Effect. Formation of Micelles and Biological Membranes, 2nd edn. New York: John Wiley.

Catalysis and the Use of Energy by Cells

Atkins PW (1994) The Second Law: Energy, Chaos and Form. New York: Scientific American Books.

Atkins PW (1998) Physical Chemistry, 6th edn. Oxford: Oxford University Press.

Berg HC (1983) Random Walks in Biology. Princeton, NJ: Princeton University Press.

Dickerson RE (1969) Molecular Thermodynamics. Menlo Park, CA: Benjamin Cummings.

Dressler D & Potter H (1991) Discovering Enzymes. New York: Scientific American Library.

Einstein A (1956) Investigations on the Theory of Brownian Movement. New York: Dover.

Eisenberg D & Crothers DM (1979) Physical Chemistry with Applications to the Life Sciences. Menlo Park, CA: Benjamin Cummings.

Fruton JS (1999) Proteins, Enzymes, Genes: The Interplay of Chemistry and Biology. New Haven: Yale University Press.

Goodsell DS (1991) Inside a living cell. *Trends Biochem. Sci.* 16, 203–206.

Karplus M & McCammon JA (1986) The dynamics of proteins. *Sci. Am.* 254(4), 42–51.

Karplus M & Petsko GA (1990) Molecular dynamics simulations in biology. *Nature* 347, 631–639.

Kauzmann W (1967) Thermodynamics and Statistics: with Applications to Gases. In Thermal Properties of Matter, vol 2, New York: Benjamin.

Klotz IM (1967) Energy Changes in Biochemical Reactions. New York: Academic Press.

Kornberg A (1989) For the Love of Enzymes. Cambridge, MA: Harvard University Press.

Lavenda BH (1985) Brownian Motion. *Sci. Am.* 252(2), 70–85.

Lawlor DW (2001) Photosynthesis, 3rd edn. Oxford: BIOS.

Lehninger AL (1971) The Molecular Basis of Biological Energy Transformations, 2nd edn. Menlo Park, CA: Benjamin Cummings.

Lipmann F (1941) Metabolic generation and utilization of phosphate bond energy. *Adv. Enzymol.* 1, 99–162.

Lipmann F (1971) Wanderings of a Biochemist. New York: Wiley.

Nisbet EE & Sleep NH (2001) The habitat and nature of early life. *Nature* 409, 1083–1091.

Racker E (1980) From Pasteur to Mitchell: a hundred years of bioenergetics. *Fed. Proc.* 39, 210–215.

Schrödinger E (1944 & 1958) What is Life?: The Physical Aspect of the Living Cell and Mind and Matter, 1992 combined edn. Cambridge: Cambridge University Press.

van Holde KE, Johnson WC & Ho PS (1998) Principles of Physical Biochemistry. Upper Saddle River, NJ: Prentice Hall.

Walsh C (2001) Enabling the chemistry of life. *Nature* 409, 226–231.

Westheimer FH (1987) Why nature chose phosphates. *Science* 235, 1173–1178.

Youvan DC & Marrs BL (1987) Molecular mechanisms of photosynthesis. *Sci. Am.* 256(6), 42–49.

How Cells Obtain Energy from Food

Cramer WA & Knaff DB (1990) Energy Transduction in Biological Membranes. New York: Springer-Verlag.

Fell D (1997) Understanding the Control of Metabolism. London: Portland Press.

Flatt JP (1995) Use and storage of carbohydrate and fat. *Am. J. Clin. Nutr.* 61, 952S–959S.

Fothergill-Gilmore LA (1986) The evolution of the glycolytic pathway. *Trends Biochem. Sci.* 11, 47–51.

Huynen MA, Dandekar T & Bork P (1999) Variation and evolution of the citric-acid cycle: a genomic perspective. *Trends Microbiol.* 7, 281–291.

Kornberg HL (1987) Tricarboxylic acid cycles. *Bioessays* 7, 236–238.

Krebs HA & Martin A (1981) Reminiscences and Reflections. Oxford/New York: Clarendon Press/Oxford University Press.

Krebs HA (1970) The history of the tricarboxylic acid cycle. *Perspect. Biol. Med.* 14, 154–170.

Martin BR (1987) Metabolic Regulation: A Molecular Approach. Oxford: Blackwell Scientific.

McGilvery RW (1983) Biochemistry: A Functional Approach, 3rd edn. Philadelphia: Saunders.

Newsholme EA & Stark C (1973) Regulation in Metabolism. New York: Wiley.

PANEL 2–1 Chemical Bonds and Groups Commonly Encountered in Biological Molecules

CARBON SKELETONS

Carbon has a unique role in the cell because of its ability to form strong covalent bonds with other carbon atoms. Thus carbon atoms can join to form chains.

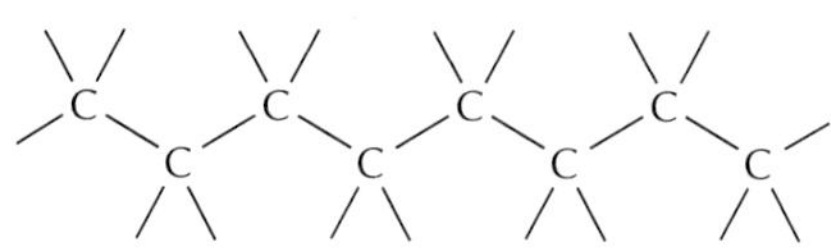

also written as

or branched trees

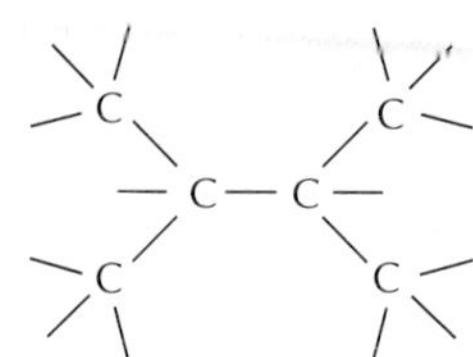

also written as

or rings

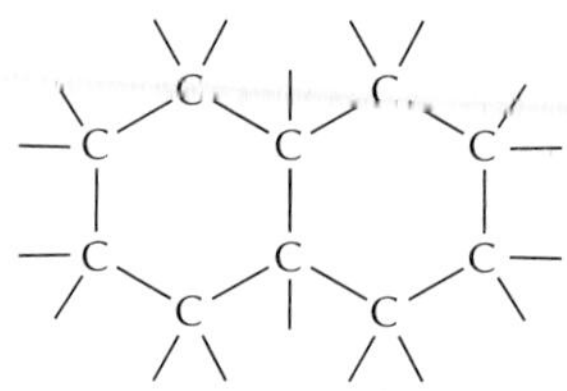

also written as

COVALENT BONDS

A covalent bond forms when two atoms come very close together and share one or more of their electrons. In a single bond one electron from each of the two atoms is shared; in a double bond a total of four electrons are shared.

Each atom forms a fixed number of covalent bonds in a defined spatial arrangement. For example, carbon forms four single bonds arranged tetrahedrally, whereas nitrogen forms three single bonds and oxygen forms two single bonds arranged as shown below.

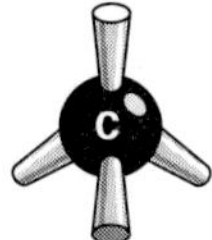

Double bonds exist and have a different spatial arrangement.

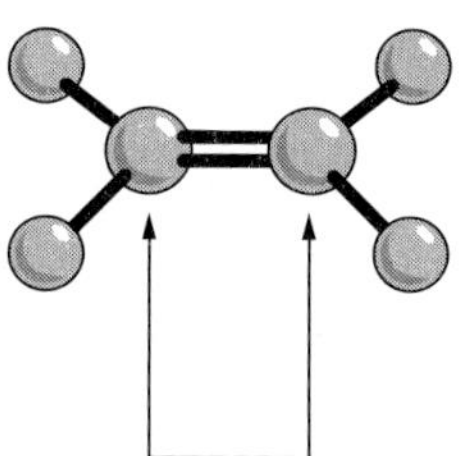

Atoms joined by two or more covalent bonds cannot rotate freely around the bond axis. This restriction is a major influence on the three-dimensional shape of many macromolecules.

HYDROCARBONS

Carbon and hydrogen combine together to make stable compounds (or chemical groups) called hydrocarbons. These are nonpolar, do not form hydrogen bonds, and are generally insoluble in water.

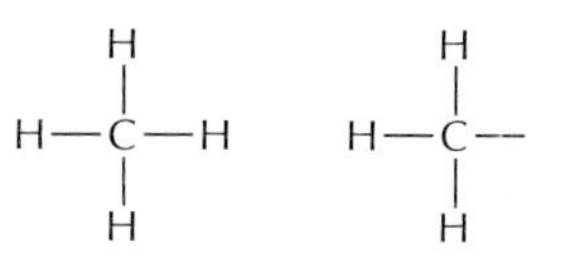

H_2C – CH_2 – H_2C – CH_2 – H_2C – CH_2 – H_2C – CH_2 – H_2C – CH_2 – H_2C – CH_2 – H_3C

part of the hydrocarbon "tail" of a fatty acid molecule

ALTERNATING DOUBLE BONDS

The carbon chain can include double bonds. If these are on alternate carbon atoms, the bonding electrons move within the molecule, stabilizing the structure by a phenomenon called resonance.

the truth is somewhere between these two structures

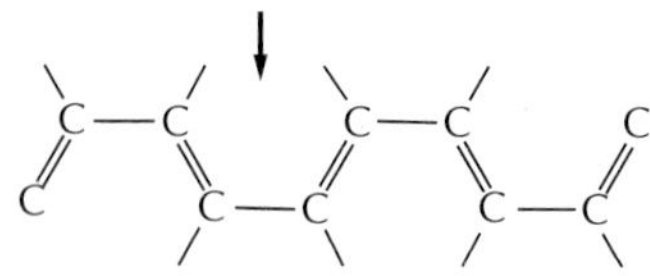

Alternating double bonds in a ring can generate a very stable structure.

benzene

often written as

C–O CHEMICAL GROUPS

Many biological compounds contain a carbon bonded to an oxygen. For example,

alcohol — The –OH is called a hydroxyl group.

aldehyde

ketone

The C=O is called a carbonyl group.

carboxylic acid — The –COOH is called a carboxyl group. In water this loses an H^+ ion to become –COO⁻.

esters — Esters are formed by combining an acid and an alcohol.

acid + alcohol → ester + H_2O

C–N CHEMICAL GROUPS

Amines and amides are two important examples of compounds containing a carbon linked to a nitrogen.

Amines in water combine with an H^+ ion to become positively charged.

Amides are formed by combining an acid and an amine. Unlike amines, amides are uncharged in water. An example is the peptide bond that joins amino acids in a protein.

acid + amine → amide + H_2O

Nitrogen also occurs in several ring compounds, including important constituents of nucleic acids: purines and pyrimidines.

cytosine (a pyrimidine)

PHOSPHATES

Inorganic phosphate is a stable ion formed from phosphoric acid, H_3PO_4. It is often written as P_i.

Phosphate esters can form between a phosphate and a free hydroxyl group. Phosphate groups are often attached to proteins in this way.

also written as

The combination of a phosphate and a carboxyl group, or two or more phosphate groups, gives an acid anhydride.

high-energy acyl phosphate bond (carboxylic–phosphoric acid anhydride) found in some metabolites

also written as

phosphoanhydride—a high-energy bond found in molecules such as ATP

also written as

PANEL 2–2 Water and Its Influence on the Behavior of Biological Molecules

WATER

Two atoms, connected by a covalent bond, may exert different attractions for the electrons of the bond. In such cases the bond is polar, with one end slightly negatively charged (δ−) and the other slightly positively charged (δ+).

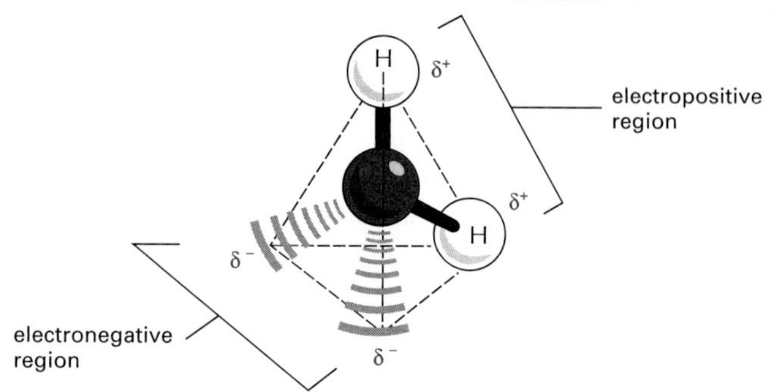

Although a water molecule has an overall neutral charge (having the same number of electrons and protons), the electrons are asymmetrically distributed, which makes the molecule polar. The oxygen nucleus draws electrons away from the hydrogen nuclei, leaving these nuclei with a small net positive charge. The excess of electron density on the oxygen atom creates weakly negative regions at the other two corners of an imaginary tetrahedron.

WATER STRUCTURE

Molecules of water join together transiently in a hydrogen-bonded lattice. Even at 37°C, 15% of the water molecules are joined to four others in a short-lived assembly known as a "flickering cluster."

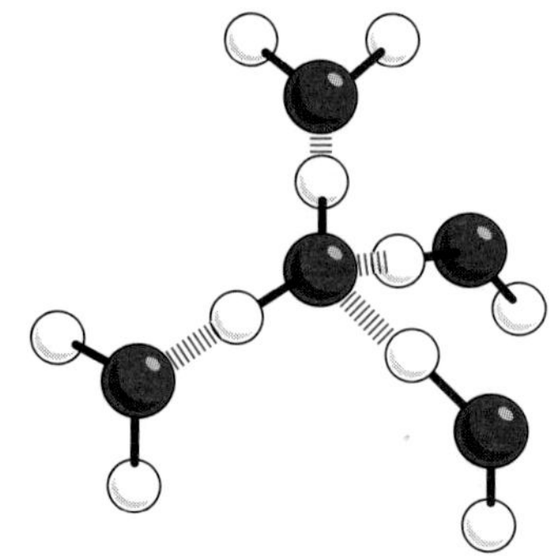

The cohesive nature of water is responsible for many of its unusual properties, such as high surface tension, specific heat, and heat of vaporization.

HYDROGEN BONDS

Because they are polarized, two adjacent H_2O molecules can form a linkage known as a hydrogen bond. Hydrogen bonds have only about 1/20 the strength of a covalent bond.

Hydrogen bonds are strongest when the three atoms lie in a straight line.

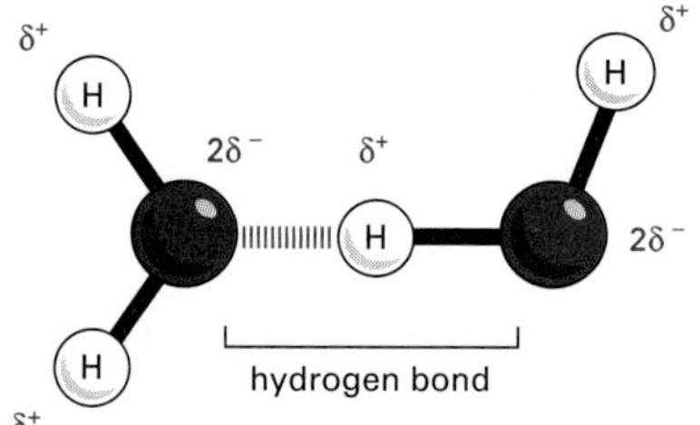

bond lengths

hydrogen bond 0.27 nm

0.10 nm covalent bond

HYDROPHILIC MOLECULES

Substances that dissolve readily in water are termed hydrophilic. They are composed of ions or polar molecules that attract water molecules through electrical charge effects. Water molecules surround each ion or polar molecule on the surface of a solid substance and carry it into solution.

Ionic substances such as sodium chloride dissolve because water molecules are attracted to the positive (Na^+) or negative (Cl^-) charge of each ion.

Polar substances such as urea dissolve because their molecules form hydrogen bonds with the surrounding water molecules.

HYDROPHOBIC MOLECULES

Molecules that contain a preponderance of non-polar bonds are usually insoluble in water and are termed hydrophobic. This is true, especially, of hydrocarbons, which contain many C–H bonds. Water molecules are not attracted to such molecules and so have little tendency to surround them and carry them into solution.

WATER AS A SOLVENT

Many substances, such as household sugar, dissolve in water. That is, their molecules separate from each other, each becoming surrounded by water molecules.

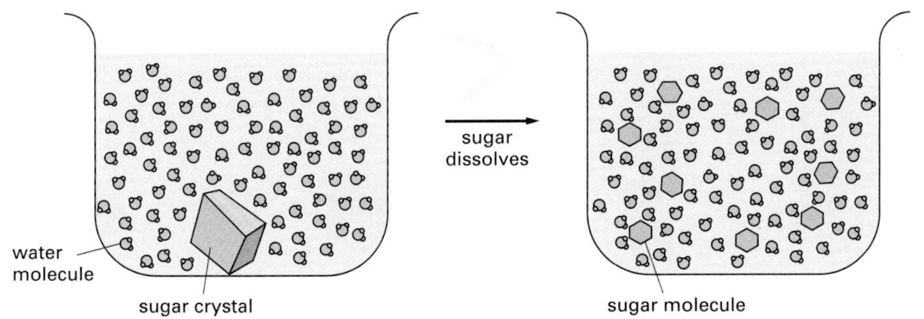

When a substance dissolves in a liquid, the mixture is termed a solution. The dissolved substance (in this case sugar) is the solute, and the liquid that does the dissolving (in this case water) is the solvent. Water is an excellent solvent for many substances because of its polar bonds.

ACIDS

Substances that release hydrogen ions into solution are called acids.

$$HCl \longrightarrow H^+ + Cl^-$$

HCl: hydrochloric acid (strong acid); H^+: hydrogen ion; Cl^-: chloride ion

Many of the acids important in the cell are only partially dissociated, and they are therefore weak acids—for example, the carboxyl group (–COOH), which dissociates to give a hydrogen ion in solution

$$-C(=O)OH \rightleftharpoons H^+ + -C(=O)O^-$$

(weak acid)

Note that this is a reversible reaction.

HYDROGEN ION EXCHANGE

Positively charged hydrogen ions (H^+) can spontaneously move from one water molecule to another, thereby creating two ionic species.

$$H_2O \cdots H{-}OH \rightleftharpoons H_3O^{+} + OH^{-}$$

hydronium ion (water acting as a weak base); hydroxyl ion (water acting as a weak acid)

often written as: $H_2O \rightleftharpoons H^+ + OH^-$

hydrogen ion; hydroxyl ion

Since the process is rapidly reversible, hydrogen ions are continually shuttling between water molecules. Pure water contains a steady-state concentration of hydrogen ions and hydroxyl ions (both 10^{-7} M).

pH

The acidity of a solution is defined by the concentration of H^+ ions it possesses. For convenience we use the pH scale, where

$$pH = -\log_{10}[H^+]$$

For pure water

$[H^+] = 10^{-7}$ moles/liter

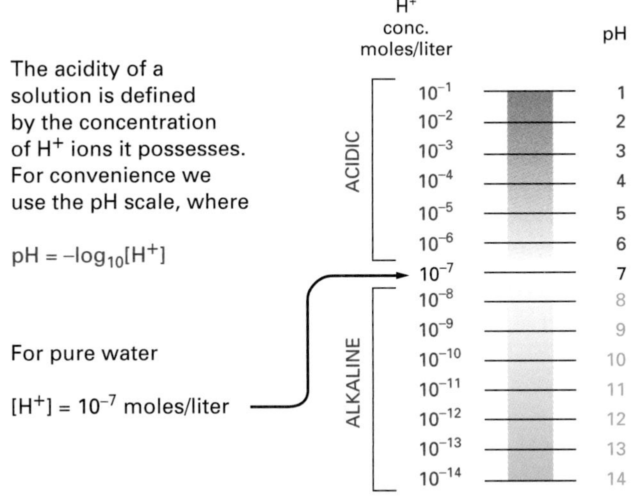

BASES

Substances that reduce the number of hydrogen ions in solution are called bases. Some bases, such as ammonia, combine directly with hydrogen ions.

$$NH_3 + H^+ \longrightarrow NH_4^+$$

ammonia; hydrogen ion; ammonium ion

Other bases, such as sodium hydroxide, reduce the number of H^+ ions indirectly, by making OH^- ions that then combine directly with H^+ ions to make H_2O.

$$NaOH \longrightarrow Na^+ + OH^-$$

sodium hydroxide (strong base); sodium ion; hydroxyl ion

Many bases found in cells are partially dissociated and are termed weak bases. This is true of compounds that contain an amino group ($-NH_2$), which has a weak tendency to reversibly accept an H^+ ion from water, increasing the quantity of free OH^- ions.

$$-NH_2 + H^+ \rightleftharpoons -NH_3^+$$

WEAK CHEMICAL BONDS

Organic molecules can interact with other molecules through short-range noncovalent forces.

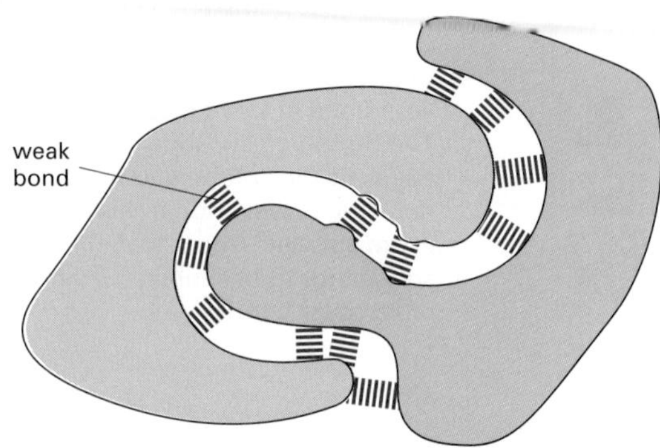

Weak chemical bonds have less than 1/20 the strength of a strong covalent bond. They are strong enough to provide tight binding only when many of them are formed simultaneously.

HYDROGEN BONDS

As already described for water (see Panel 2–2) hydrogen bonds form when a hydrogen atom is "sandwiched" between two electron-attracting atoms (usually oxygen or nitrogen).

Hydrogen bonds are strongest when the three atoms are in a straight line:

Examples in macromolecules:

Amino acids in polypeptide chains hydrogen-bonded together.

Two bases, G and C, hydrogen-bonded in DNA or RNA.

VAN DER WAALS ATTRACTIONS

If two atoms are too close together they repel each other very strongly. For this reason, an atom can often be treated as a sphere with a fixed radius. The characteristic "size" for each atom is specified by a unique van der Waals radius. The contact distance between any two non-covalently bonded atoms is the sum of their van der Waals radii.

0.12 nm radius

0.2 nm radius

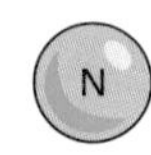

0.15 nm radius

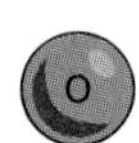

0.14 nm radius

At very short distances any two atoms show a weak bonding interaction due to their fluctuating electrical charges. The two atoms will be attracted to each other in this way until the distance between their nuclei is approximately equal to the sum of their van der Waals radii. Although they are individually very weak, van der Waals attractions can become important when two macromolecular surfaces fit very close together, because many atoms are involved.

Note that when two atoms form a covalent bond, the centers of the two atoms (the two atomic nuclei) are much closer together than the sum of the two van der Waals radii. Thus,

0.4 nm
two non-bonded carbon atoms

0.15 nm
single-bonded carbons

0.13 nm
double-bonded carbons

HYDROGEN BONDS IN WATER

Any molecules that can form hydrogen bonds to each other can alternatively form hydrogen bonds to water molecules. Because of this competition with water molecules, the hydrogen bonds formed between two molecules dissolved in water are relatively weak.

peptide bond

$2H_2O$

$2H_2O$

HYDROPHOBIC FORCES

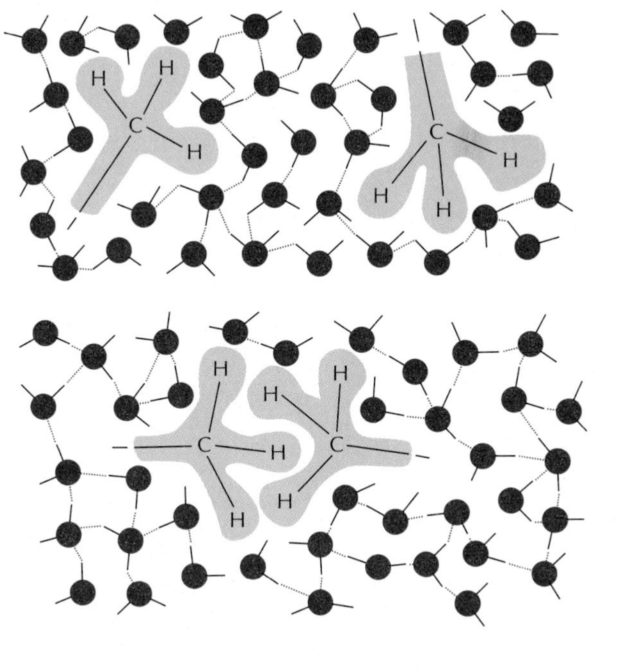

Water forces hydrophobic groups together, because doing so minimizes their disruptive effects on the hydrogen-bonded water network. Hydrophobic groups held together in this way are sometimes said to be held together by "hydrophobic bonds," even though the attraction is actually caused by a repulsion from the water.

IONIC BONDS IN AQUEOUS SOLUTIONS

Charged groups are shielded by their interactions with water molecules. Ionic bonds are therefore quite weak in water.

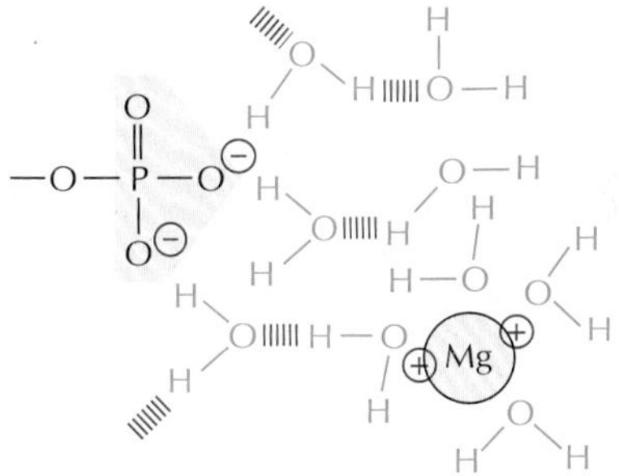

Similarly, other ions in solution can cluster around charged groups and further weaken ionic bonds.

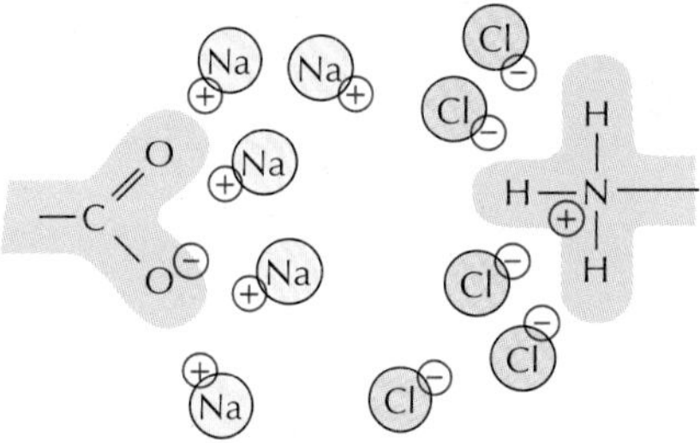

Despite being weakened by water and salt, ionic bonds are very important in biological systems; an enzyme that binds a positively charged substrate will often have a negatively charged amino acid side chain at the appropriate place.

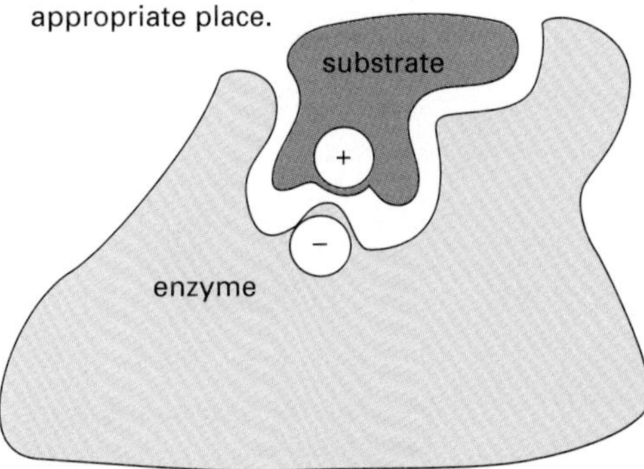

IONIC BONDS

Ionic interactions occur either between fully charged groups (ionic bond) or between partially charged groups.

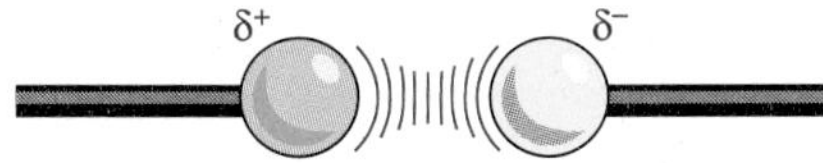

The force of attraction between the two charges, δ^+ and δ^-, falls off rapidly as the distance between the charges increases.

In the absence of water, ionic forces are very strong. They are responsible for the strength of such minerals as marble and agate.

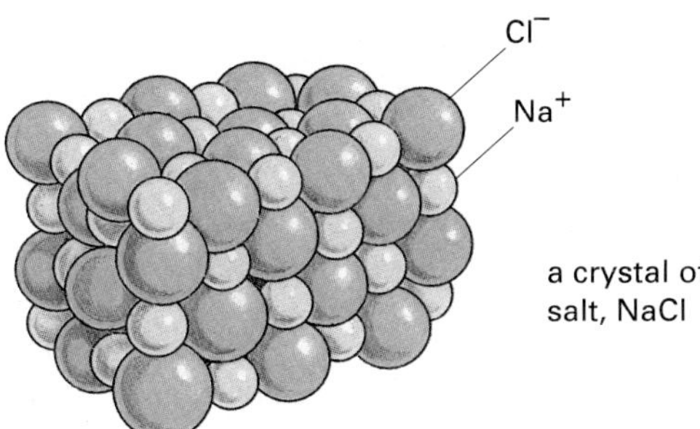

a crystal of salt, NaCl

MONOSACCHARIDES

Monosaccharides usually have the general formula $(CH_2O)_n$, where *n* can be 3, 4, 5, 6, 7, or 8, and have two or more hydroxyl groups. They either contain an aldehyde group ($-C\begin{smallmatrix}\nearrow O\\ \searrow H\end{smallmatrix}$) and are called aldoses or a ketone group ($\rangle C{=}O$) and are called ketoses.

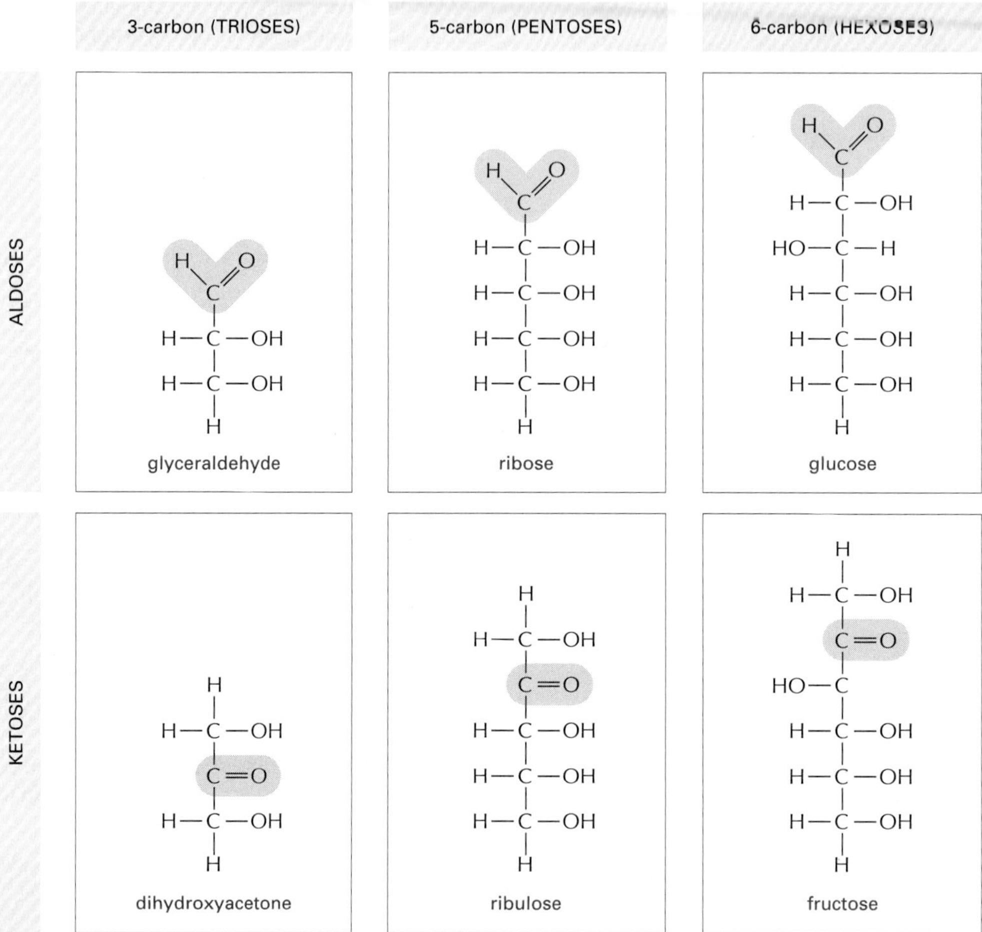

RING FORMATION

In aqueous solution, the aldehyde or ketone group of a sugar molecule tends to react with a hydroxyl group of the same molecule, thereby closing the molecule into a ring.

glucose

ribose

Note that each carbon atom has a number.

ISOMERS

Many monosaccharides differ only in the spatial arrangement of atoms—that is, they are isomers. For example, glucose, galactose, and mannose have the same formula ($C_6H_{12}O_6$) but differ in the arrangement of groups around one or two carbon atoms.

glucose

galactose

mannose

These small differences make only minor changes in the chemical properties of the sugars. But they are recognized by enzymes and other proteins and therefore can have important biological effects.

α AND β LINKS

The hydroxyl group on the carbon that carries the aldehyde or ketone can rapidly change from one position to the other. These two positions are called α and β.

β hydroxyl α hydroxyl

As soon as one sugar is linked to another, the α or β form is frozen.

SUGAR DERIVATIVES

The hydroxyl groups of a simple monosaccharide can be replaced by other groups. For example,

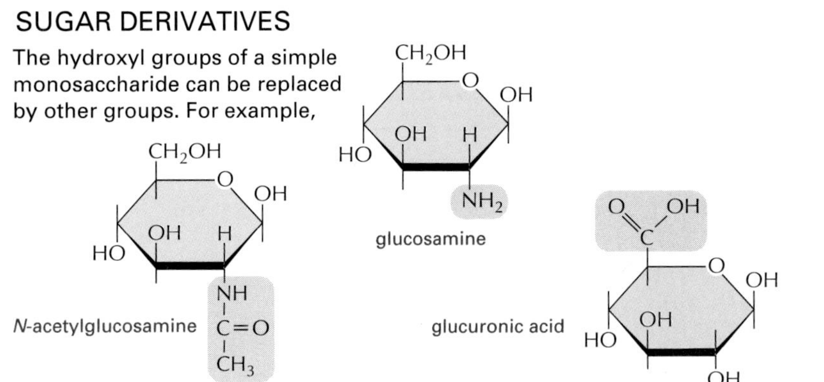

DISACCHARIDES

The carbon that carries the aldehyde or the ketone can react with any hydroxyl group on a second sugar molecule to form a disaccharide. The linkage is called a glycosidic bond.

Three common disaccharides are

maltose (glucose + glucose)
lactose (galactose + glucose)
sucrose (glucose + fructose)

The reaction forming sucrose is shown here.

α glucose β fructose

H_2O

sucrose

OLIGOSACCHARIDES AND POLYSACCHARIDES

Large linear and branched molecules can be made from simple repeating sugar subunits. Short chains are called oligosaccharides, while long chains are called polysaccharides. Glycogen, for example, is a polysaccharide made entirely of glucose units joined together.

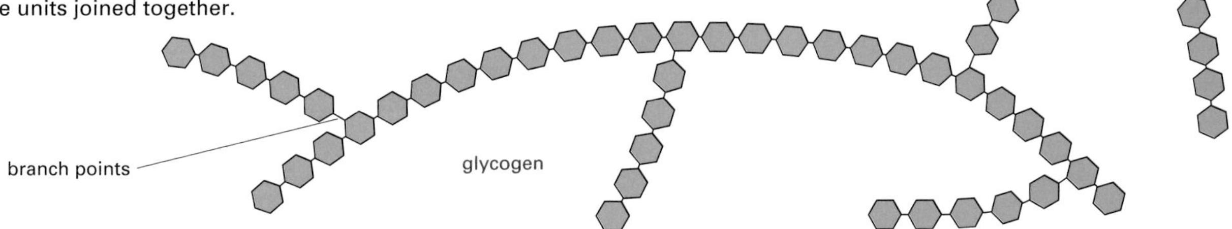

COMPLEX OLIGOSACCHARIDES

In many cases a sugar sequence is nonrepetitive. Many different molecules are possible. Such complex oligosaccharides are usually linked to proteins or to lipids, as is this oligosaccharide, which is part of a cell-surface molecule that defines a particular blood group.

COMMON FATTY ACIDS

These are carboxylic acids with long hydrocarbon tails.

stearic acid (C_{18})	palmitic acid (C_{16})	oleic acid (C_{18})
COOH	COOH	COOH
CH_2	CH_2	CH_2
CH_2	CH_2	CH_2
CH_2	CH_2	CH_2
CH_2	CH_2	CH_2
CH_2	CH_2	CH_2
CH_2	CH_2	CH_2
CH_2	CH_2	CH_2
CH_2	CH_2	CH
CH_2	CH_2	‖ CH
CH_2	CH_2	CH_2
CH_2	CH_2	CH_2
CH_2	CH_2	CH_2
CH_2	CH_2	CH_2
CH_2	CH_2	CH_2
CH_2	CH_3	CH_2
CH_2		CH_2
CH_3		CH_3

TRIACYLGLYCEROLS

Fatty acids are stored as an energy reserve (fats and oils) through an ester linkage to glycerol to form triacylglycerols, also known as triglycerides.

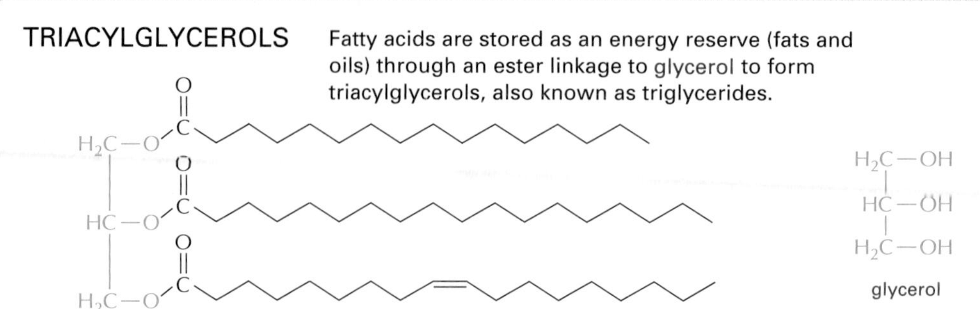

Hundreds of different kinds of fatty acids exist. Some have one or more double bonds in their hydrocarbon tail and are said to be unsaturated. Fatty acids with no double bonds are saturated.

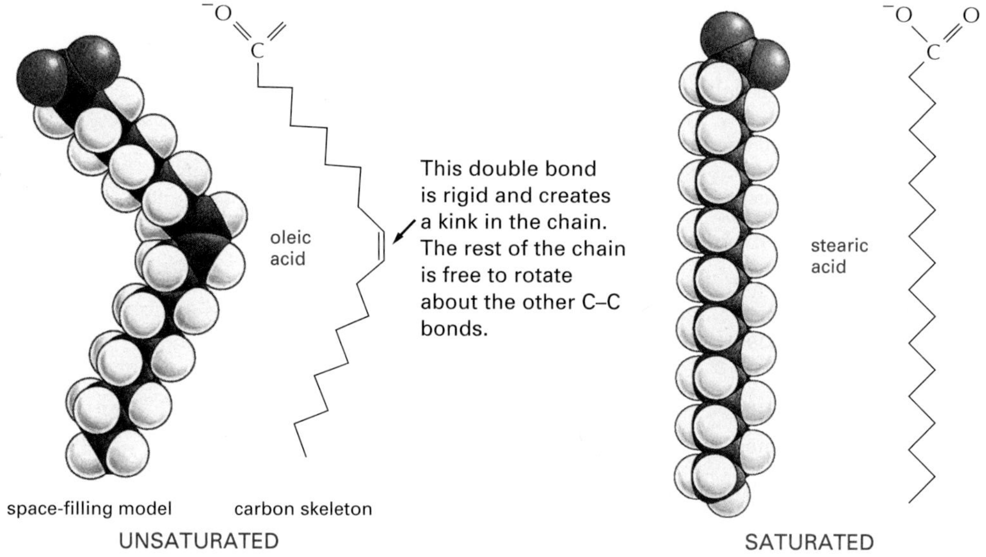

CARBOXYL GROUP

If free, the carboxyl group of a fatty acid will be ionized.

But more usually it is linked to other groups to form either esters

or amides.

PHOSPHOLIPIDS

Phospholipids are the major constituents of cell membranes.

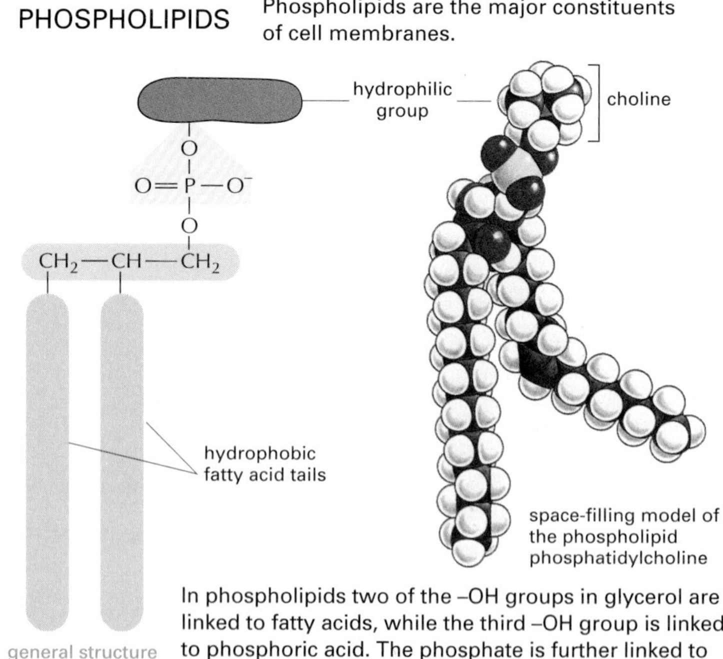

In phospholipids two of the –OH groups in glycerol are linked to fatty acids, while the third –OH group is linked to phosphoric acid. The phosphate is further linked to one of a variety of small polar groups (alcohols).

LIPID AGGREGATES

Fatty acids have a hydrophilic head and a hydrophobic tail.

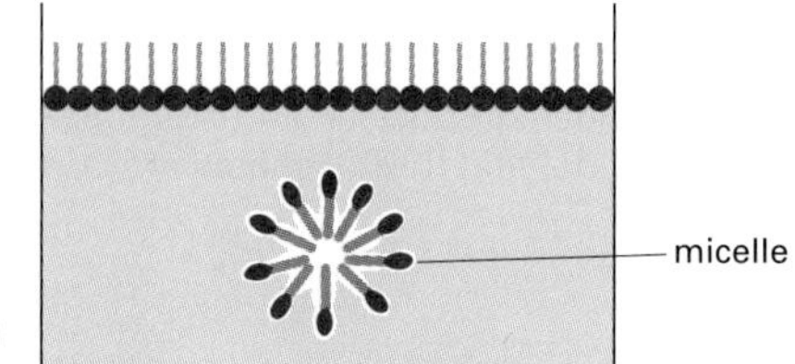

In water they can form a surface film or form small micelles.

Their derivatives can form larger aggregates held together by hydrophobic forces:

Triglycerides can form large spherical fat droplets in the cell cytoplasm.

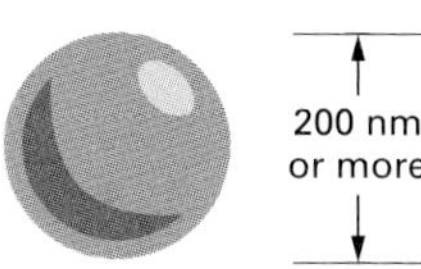

Phospholipids and glycolipids form self-sealing lipid bilayers that are the basis for all cell membranes.

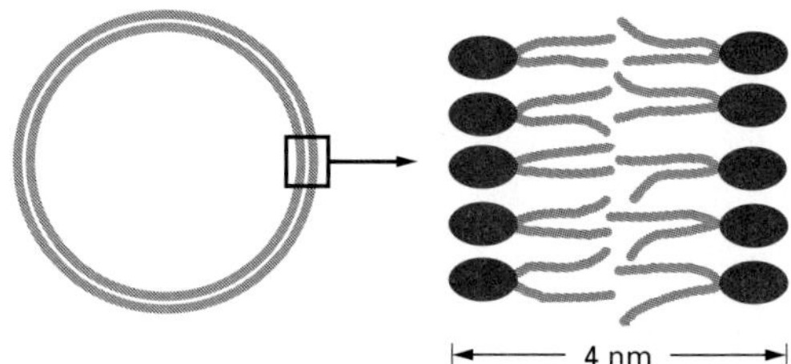

OTHER LIPIDS

Lipids are defined as the water-insoluble molecules in cells that are soluble in organic solvents. Two other common types of lipids are steroids and polyisoprenoids. Both are made from isoprene units.

CH_3
$C-CH=CH_2$
CH_2

isoprene

STEROIDS

Steroids have a common multiple-ring structure.

HO

cholesterol—found in many membranes

OH

O

testosterone—male steroid hormone

GLYCOLIPIDS

Like phospholipids, these compounds are composed of a hydrophobic region, containing two long hydrocarbon tails, and a polar region, which, however, contains one or more sugar residues and no phosphate.

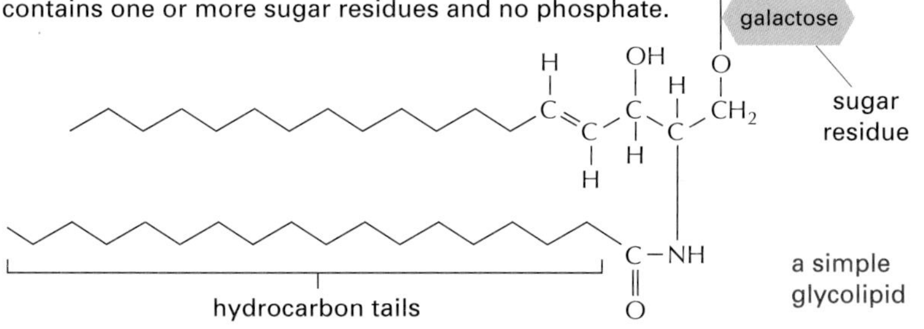

POLYISOPRENOIDS

long-chain polymers of isoprene

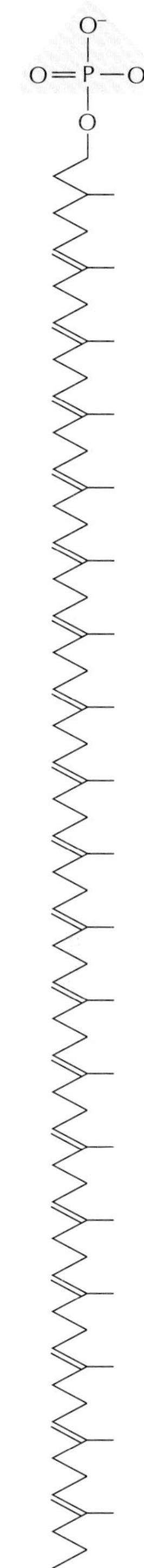

dolichol phosphate—used to carry activated sugars in the membrane-associated synthesis of glycoproteins and some polysaccharides

PANEL 2–6 A Survey of the Nucleotides

NUCLEOTIDES

A nucleotide consists of a nitrogen-containing base, a five-carbon sugar, and one or more phosphate groups.

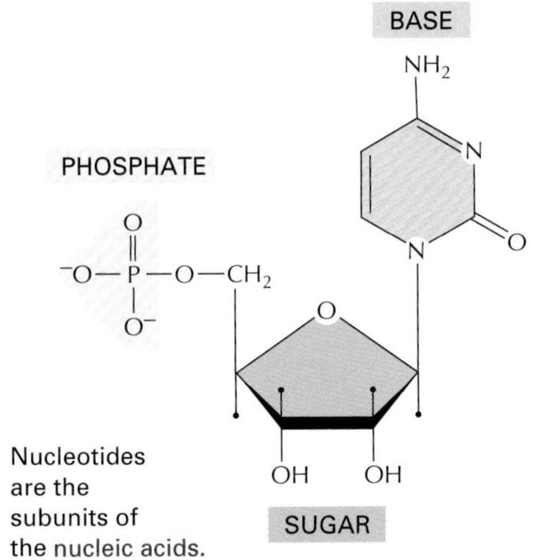

Nucleotides are the subunits of the nucleic acids.

PHOSPHATES

The phosphates are normally joined to the C5 hydroxyl of the ribose or deoxyribose sugar (designated 5'). Mono-, di-, and triphosphates are common.

$^-O-P(=O)(O^-)-O-CH_2$ as in AMP

$^-O-P(=O)(O^-)-O-P(=O)(O^-)-O-CH_2$ as in ADP

$^-O-P(=O)(O^-)-O-P(=O)(O^-)-O-P(=O)(O^-)-O-CH_2$ as in ATP

The phosphate makes a nucleotide negatively charged.

BASIC SUGAR LINKAGE

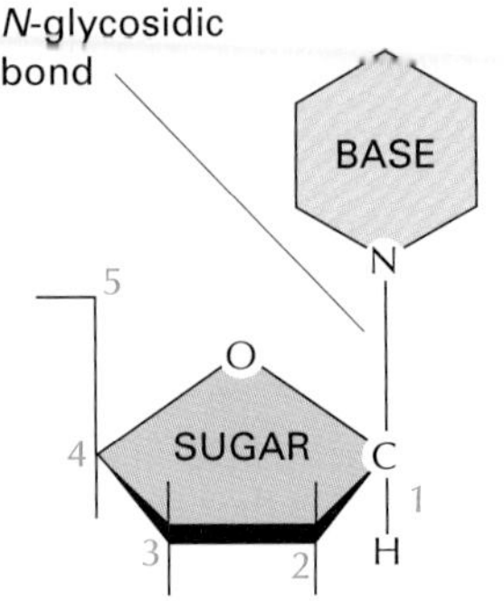

The base is linked to the same carbon (C1) used in sugar–sugar bonds.

BASES

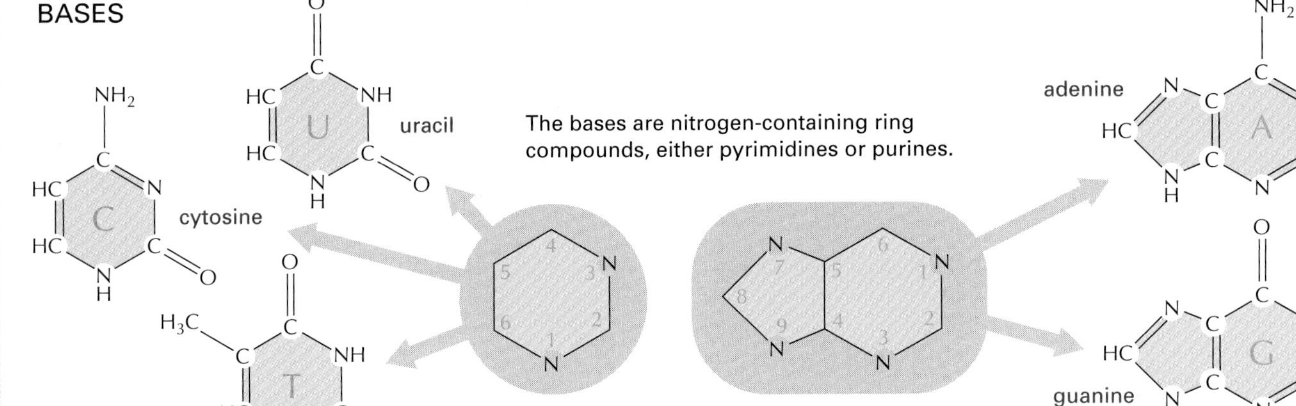

The bases are nitrogen-containing ring compounds, either pyrimidines or purines.

SUGARS

PENTOSE

a five-carbon sugar

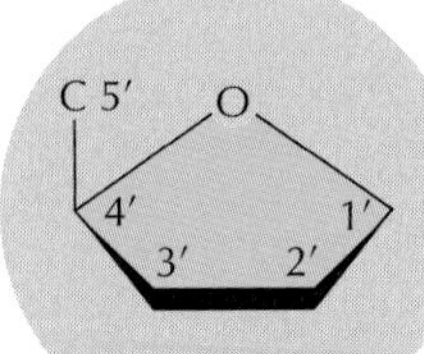

two kinds are used

$HOCH_2$ O OH; H H H H; OH OH

β-D-ribose used in ribonucleic acid

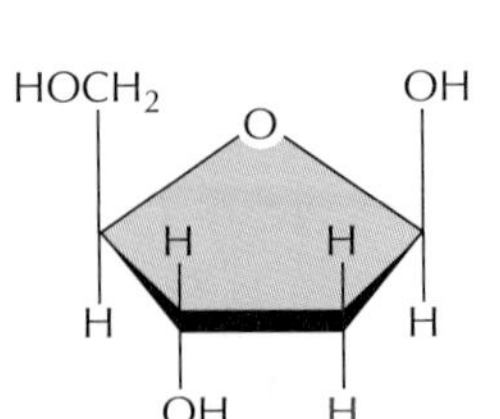

β-D-2-deoxyribose used in deoxyribonucleic acid

Each numbered carbon on the sugar of a nucleotide is followed by a prime mark; therefore, one speaks of the "5-prime carbon," etc.

NOMENCLATURE

The names can be confusing, but the abbreviations are clear.

BASE	NUCLEOSIDE	ABBR.
adenine	adenosine	A
guanine	guanosine	G
cytosine	cytidine	C
uracil	uridine	U
thymine	thymidine	T

Nucleotides are abbreviated by three capital letters. Some examples follow:

AMP = adenosine monophosphate
dAMP = deoxyadenosine monophosphate
UDP = uridine diphosphate
ATP = adenosine triphosphate

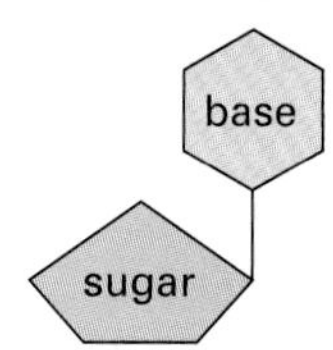

BASE + SUGAR = NUCLEOSIDE

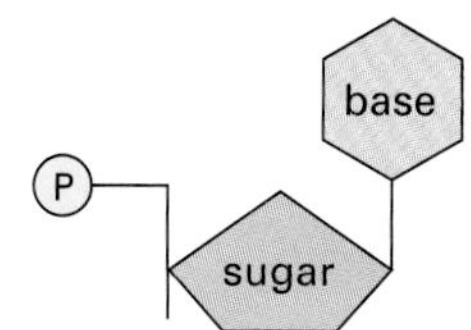

BASE + SUGAR + PHOSPHATE = NUCLEOTIDE

NUCLEIC ACIDS

Nucleotides are joined together by a phosphodiester linkage between 5′ and 3′ carbon atoms to form nucleic acids. The linear sequence of nucleotides in a nucleic acid chain is commonly abbreviated by a one-letter code, A—G—C—T—T—A—C—A, with the 5′ end of the chain at the left.

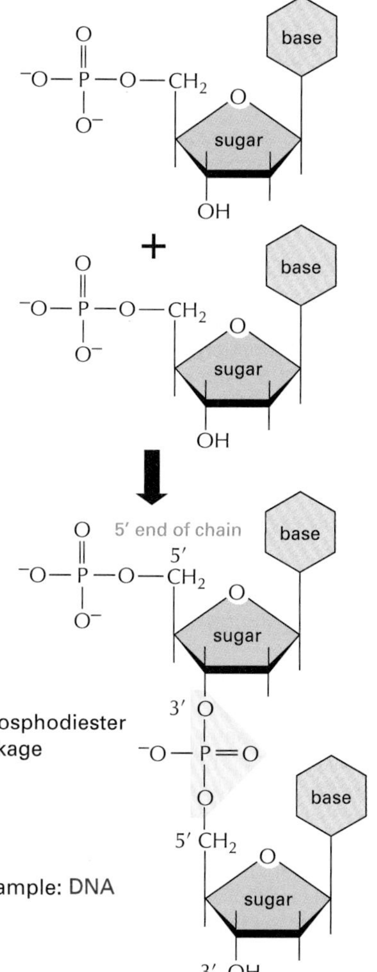

example: DNA

NUCLEOTIDES HAVE MANY OTHER FUNCTIONS

1 They carry chemical energy in their easily hydrolyzed phosphoanhydride bonds.

phosphoanhydride bonds

example: ATP (or ATP)

2 They combine with other groups to form coenzymes.

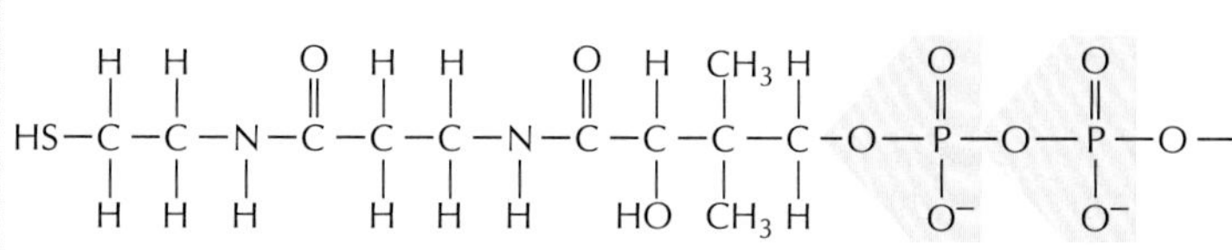

example: coenzyme A (CoA)

3 They are used as specific signaling molecules in the cell.

example: cyclic AMP (cAMP)

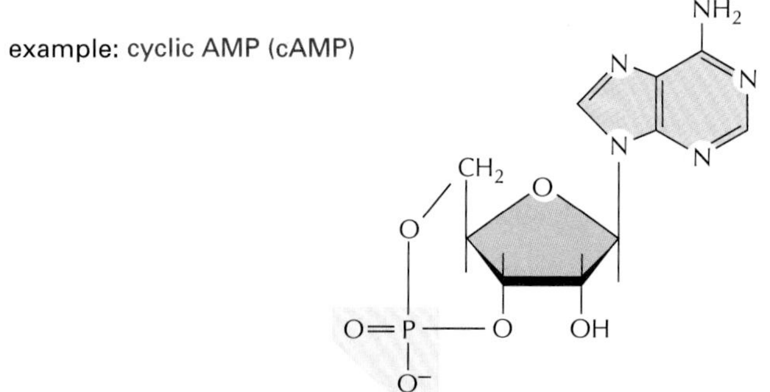

THE IMPORTANCE OF FREE ENERGY FOR CELLS

Life is possible because of the complex network of interacting chemical reactions occurring in every cell. In viewing the metabolic pathways that comprise this network, one might suspect that the cell has had the ability to evolve an enzyme to carry out any reaction that it needs. But this is not so. Although enzymes are powerful catalysts, they can speed up only those reactions that are thermodynamically possible; other reactions proceed in cells only because they are *coupled* to very favorable reactions that drive them. The question of whether a reaction can occur spontaneously, or instead needs to be coupled to another reaction, is central to cell biology. The answer is obtained by reference to a quantity called the *free energy:* the total change in free energy during a set of reactions determines whether or not the entire reaction sequence can occur. In this panel we shall explain some of the fundamental ideas—derived from a special branch of chemistry and physics called *thermodynamics*—that are required for understanding what free energy is and why it is so important to cells.

ENERGY RELEASED BY CHANGES IN CHEMICAL BONDING IS CONVERTED INTO HEAT

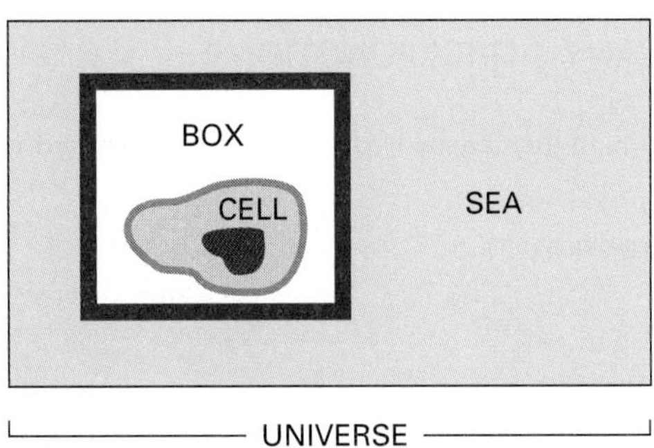

An *enclosed system* is defined as a collection of molecules that does not exchange matter with the rest of the universe (for example, the "cell in a box" shown above). Any such system will contain molecules with a total energy E. This energy will be distributed in a variety of ways: some as the translational energy of the molecules, some as their vibrational and rotational energies, but most as the bonding energies between the individual atoms that make up the molecules. Suppose that a reaction occurs in the system. The first law of thermodynamics places a constraint on what types of reactions are possible: it states that "in any process, the total energy of the universe remains constant." For example, suppose that reaction A→B occurs somewhere in the box and releases a great deal of chemical bond energy. This energy will initially increase the intensity of molecular motions (translational, vibrational, and rotational) in the system, which is equivalent to raising its temperature. However, these increased motions will soon be transferred out of the system by a series of molecular collisions that heat up first the walls of the box and then the outside world (represented by the sea in our example). In the end, the system returns to its initial temperature, by which time all the chemical bond energy released in the box has been converted into heat energy and transferred out of the box to the surroundings. According to the first law, the change in the energy in the box (ΔE_{box}, which we shall denote as ΔE) must be equal and opposite to the amount of heat energy transferred, which we shall designate as h: that is, $\Delta E = -h$. Thus, the energy in the box (E) decreases when heat leaves the system.

E also can change during a reaction due to work being done on the outside world. For example, suppose that there is a small increase in the volume (ΔV) of the box during a reaction. Since the walls of the box must push against the constant pressure (P) in the surroundings in order to expand, this does work on the outside world and requires energy. The energy used is $P(\Delta V)$, which according to the first law must decrease the energy in the box (E) by the same amount. In most reactions chemical bond energy is converted into both work and heat. *Enthalpy* (H) is a composite function that includes both of these ($H = E + PV$). To be rigorous, it is the change in enthalpy (ΔH) in an enclosed system and not the change in energy that is equal to the heat transferred to the outside world during a reaction. Reactions in which H decreases release heat to the surroundings and are said to be "exothermic," while reactions in which H increases absorb heat from the surroundings and are said to be "endothermic." Thus, $-h = \Delta H$. However, the volume change is negligible in most biological reactions, so to a good approximation

$$-h = \Delta H \cong \Delta E$$

THE SECOND LAW OF THERMODYNAMICS

Consider a container in which 1000 coins are all lying heads up. If the container is shaken vigorously, subjecting the coins to the types of random motions that all molecules experience due to their frequent collisions with other molecules, one will end up with about half the coins oriented heads down. The reason for this reorientation is that there is only a single way in which the original orderly state of the coins can be reinstated (every coin must lie heads up), whereas there are many different ways (about 10^{298}) to achieve a disorderly state in which there is an equal mixture of heads and tails; in fact, there are more ways to achieve a 50-50 state than to achieve any other state. Each state has a probability of occurrence that is proportional to the number of ways it can be realized. The second law of thermodynamics states that "systems will change spontaneously from states of lower probability to states of higher probability." Since states of lower probability are more "ordered" than states of high probability, the second law can be restated: "the universe constantly changes so as to become more disordered."

THE ENTROPY, S

The second law (but not the first law) allows one to predict the *direction* of a particular reaction. But to make it useful for this purpose, one needs a convenient measure of the probability or, equivalently, the degree of disorder of a state. The entropy (S) is such a measure. It is a logarithmic function of the probability such that the *change in entropy* (ΔS) that occurs when the reaction A → B converts one mole of A into one mole of B is

$$\Delta S = R \ln p_B / p_A$$

where p_A and p_B are the probabilities of the two states A and B, R is the gas constant (2 cal deg^{-1} $mole^{-1}$), and ΔS is measured in entropy units (eu). In our initial example of 1000 coins, the relative probability of all heads (state A) versus half heads and half tails (state B) is equal to the ratio of the number of different ways that the two results can be obtained. One can calculate that $p_A = 1$ and $p_B = 1000!(500! \times 500!) = 10^{298}$. Therefore, the entropy change for the reorientation of the coins when their container is vigorously shaken and an equal mixture of heads and tails is obtained is $R \ln (10^{298})$, or about 1370 eu per mole of such containers (6×10^{23} containers). We see that, because ΔS defined above is positive for the transition from state A to state B ($p_B / p_A > 1$), reactions with a large *increase* in S (that is, for which $\Delta S > 0$) are favored and will occur spontaneously.

As discussed in Chapter 2, heat energy causes the random commotion of molecules. Because the transfer of heat from an enclosed system to its surroundings increases the number of different arrangements that the molecules in the outside world can have, it increases their entropy. It can be shown that the release of a fixed quantity of heat energy has a greater disordering effect at low temperature than at high temperature and that the value of ΔS for the surroundings, as defined above (ΔS_{sea}), is precisely equal to the amount of heat transferred to the surroundings from the system (h) divided by the absolute temperature (T):

$$\Delta S_{sea} = h/T$$

THE GIBBS FREE ENERGY, G

When dealing with an enclosed biological system, one would like to have a simple way of predicting whether a given reaction will or will not occur spontaneously in the system. We have seen that the crucial question is whether the entropy change for the universe is positive or negative when that reaction occurs. In our idealized system, the cell in a box, there are two separate components to the entropy change of the universe—the entropy change for the system enclosed in the box and the entropy change for the surrounding "sea"—and both must be added together before any prediction can be made. For example, it is possible for a reaction to absorb heat and thereby decrease the entropy of the sea ($\Delta S_{sea} < 0$) and at the same time to cause such a large degree of disordering inside the box ($\Delta S_{box} > 0$) that the total $\Delta S_{universe} = \Delta S_{sea} + \Delta S_{box}$ is greater than 0. In this case the reaction will occur spontaneously, even though the sea gives up heat to the box during the reaction. An example of such a reaction is the dissolving of sodium chloride in a beaker containing water (the "box"), which is a spontaneous process even through the temperature of the water drops as the salt goes into solution.

Chemists have found it useful to define a number of new "composite functions" that describe *combinations* of physical properties of a system. The properties that can be combined include the temperature (T), pressure (P), volume (V), energy (E), and entropy (S). The enthalpy (H) is one such composite function. But by far the most useful composite function for biologists is the *Gibbs free energy, G*. It serves as an accounting device that allows one to deduce the entropy change of the universe resulting from a chemical reaction in the box, while avoiding any separate consideration of the entropy change in the sea. The definition of G is

$$G = H - TS$$

where, for a box of volume V, H is the enthalpy described above ($E + PV$), T is the absolute temperature, and S is the entropy. Each of these quantities applies to the inside of the box only. The change in free energy during a reaction in the box (the G of the products minus the G of the starting materials) is denoted as ΔG and, as we shall now demonstrate, it is a direct measure of the amount of disorder that is created in the universe when the reaction occurs.

At constant temperature the change in free energy (ΔG) during a reaction equals $\Delta H - T\Delta S$. Remembering that $\Delta H = -h$, the heat absorbed from the sea, we have

$$-\Delta G = -\Delta H + T\Delta S$$

$$-\Delta G = h + T\Delta S, \text{ so } -\Delta G/T = h/T + \Delta S$$

But h/T is equal to the entropy change of the sea (ΔS_{sea}), and the ΔS in the above equation is ΔS_{box}. Therefore

$$-\Delta G/T = \Delta S_{sea} + \Delta S_{box} = \Delta S_{universe}$$

We conclude that **the free-energy change is a direct measure of the entropy change of the universe**. A reaction will proceed in the direction that causes the change in the free energy (ΔG) to be less than zero, because in this case there will be a positive entropy change in the universe when the reaction occurs.

For a complex set of coupled reactions involving many different molecules, the total free-energy change can be computed simply by adding up the free energies of all the different molecular species after the reaction and comparing this value to the sum of free energies before the reaction; for common substances the required free-energy values can be found from published tables. In this way one can predict the direction of a reaction and thereby readily check the feasibility of any proposed mechanism. Thus, for example, from the observed values for the magnitude of the electrochemical proton gradient across the inner mitochondrial membrane and the ΔG for ATP hydrolysis inside the mitochondrion, one can be certain that ATP synthase requires the passage of more than one proton for each molecule of ATP that it synthesizes.

The value of ΔG for a reaction is a direct measure of how far the reaction is from equilibrium. The large negative value for ATP hydrolysis in a cell merely reflects the fact that cells keep the ATP hydrolysis reaction as much as 10 orders of magnitude away from equilibrium. If a reaction reaches equilibrium, $\Delta G = 0$, the reaction then proceeds at precisely equal rates in the forward and backward direction. For ATP hydrolysis, equilibrium is reached when the vast majority of the ATP has been hydrolyzed, as occurs in a dead cell.

PANEL 2–8 Details of the 10 Steps of Glycolysis

For each step, the part of the molecule that undergoes a change is shadowed in blue, and the name of the enzyme that catalyzes the reaction is in a yellow box.

STEP 1 Glucose is phosphorylated by ATP to form a sugar phosphate. The negative charge of the phosphate prevents passage of the sugar phosphate through the plasma membrane, trapping glucose inside the cell.

glucose + ATP —(hexokinase)→ glucose 6-phosphate + ADP + H^+

STEP 2 A readily reversible rearrangement of the chemical structure (isomerization) moves the carbonyl oxygen from carbon 1 to carbon 2, forming a ketose from an aldose sugar. (See Panel 2–4.)

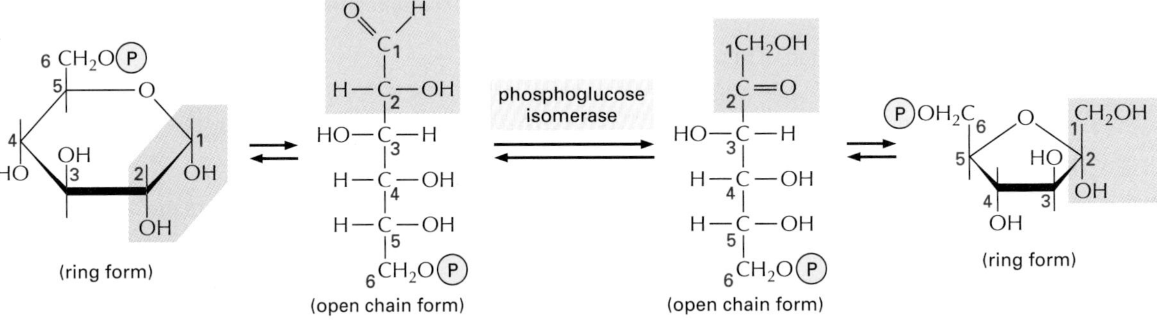

STEP 3 The new hydroxyl group on carbon 1 is phosphorylated by ATP, in preparation for the formation of two three-carbon sugar phosphates. The entry of sugars into glycolysis is controlled at this step, through regulation of the enzyme *phosphofructokinase*.

fructose 6-phosphate + ATP —(phosphofructokinase)→ fructose 1,6-bisphosphate + ADP + H^+

STEP 4 The six-carbon sugar is cleaved to produce two three-carbon molecules. Only the glyceraldehyde 3-phosphate can proceed immediately through glycolysis.

fructose 1,6-bisphosphate (ring form) ⇌ (open chain form) ⇌(aldolase) dihydroxyacetone phosphate + glyceraldehyde 3-phosphate

STEP 5 The other product of step 4, dihydroxyacetone phosphate, is isomerized to form glyceraldehyde 3-phosphate.

dihydroxyacetone phosphate ⇌(triose phosphate isomerase) glyceraldehyde 3-phosphate

STEP 6 The two molecules of glyceraldehyde 3-phosphate are oxidized. The energy generation phase of glycolysis begins, as NADH and a new high-energy anhydride linkage to phosphate are formed (see Figure 2–73).

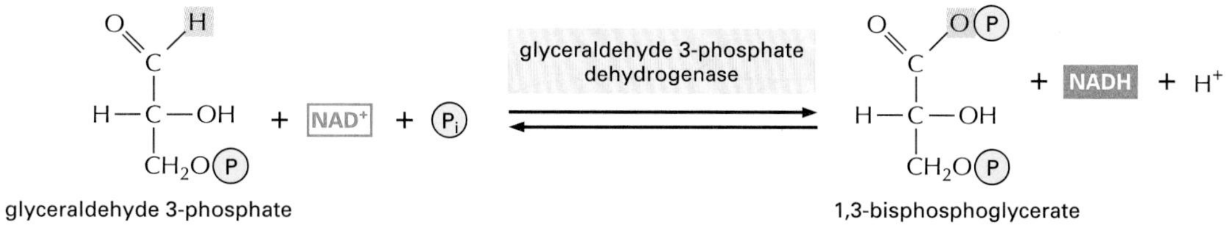

glyceraldehyde 3-phosphate + NAD^+ + P_i ⇌ (glyceraldehyde 3-phosphate dehydrogenase) 1,3-bisphosphoglycerate + NADH + H^+

STEP 7 The transfer to ADP of the high-energy phosphate group that was generated in step 6 forms ATP.

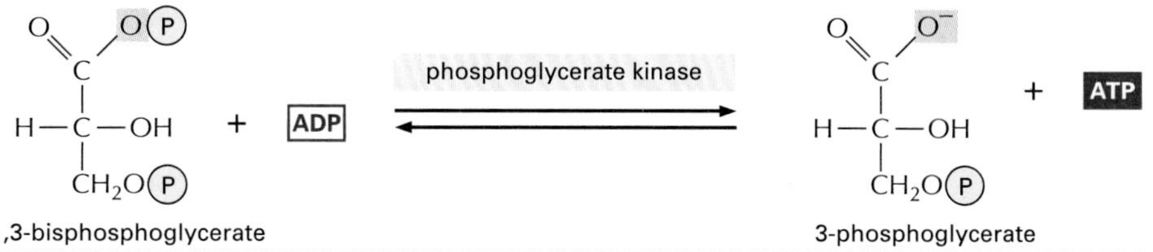

1,3-bisphosphoglycerate + ADP ⇌ (phosphoglycerate kinase) 3-phosphoglycerate + ATP

STEP 8 The remaining phosphate ester linkage in 3-phosphoglycerate, which has a relatively low free energy of hydrolysis, is moved from carbon 3 to carbon 2 to form 2-phosphoglycerate.

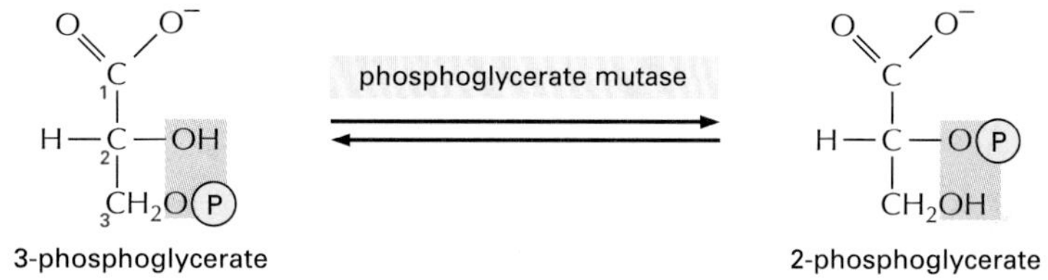

3-phosphoglycerate ⇌ (phosphoglycerate mutase) 2-phosphoglycerate

STEP 9 The removal of water from 2-phosphoglycerate creates a high-energy enol phosphate linkage.

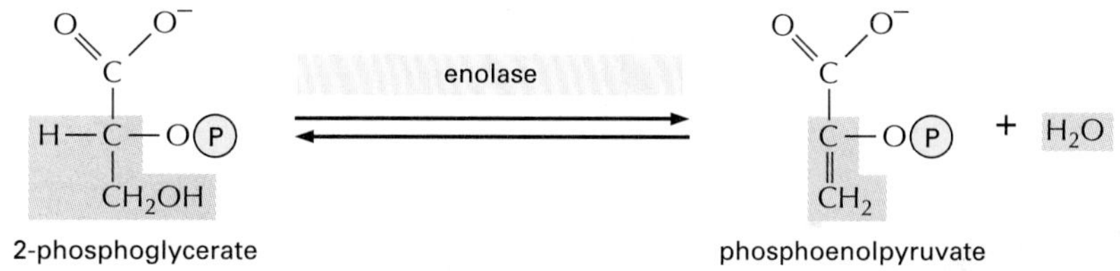

2-phosphoglycerate ⇌ (enolase) phosphoenolpyruvate + H_2O

STEP 10 The transfer to ADP of the high-energy phosphate group that was generated in step 9 forms ATP, completing glycolysis.

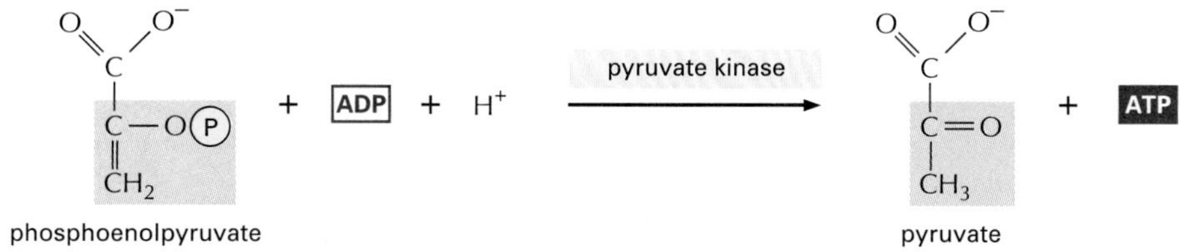

phosphoenolpyruvate + ADP + H^+ → (pyruvate kinase) pyruvate + ATP

NET RESULT OF GLYCOLYSIS

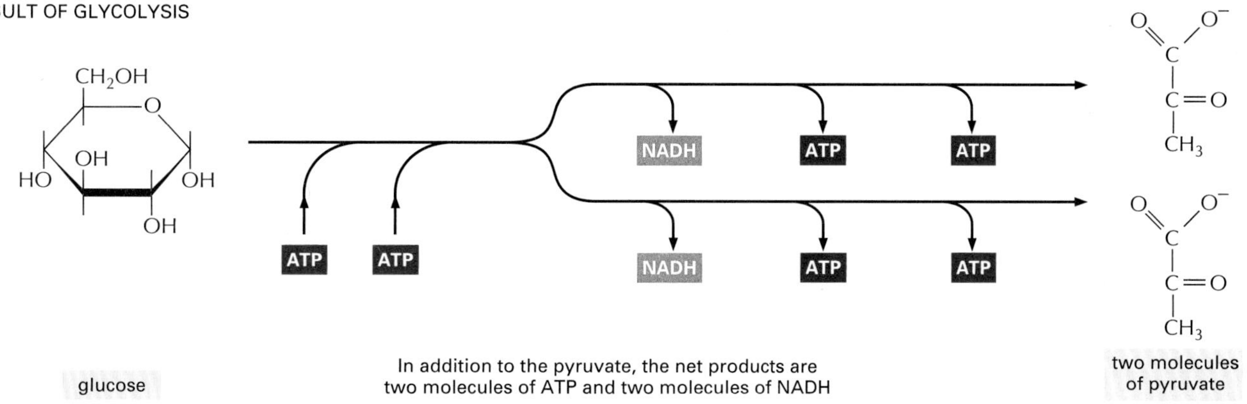

glucose

ATP ATP

NADH ATP ATP

NADH ATP ATP

two molecules of pyruvate

In addition to the pyruvate, the net products are two molecules of ATP and two molecules of NADH

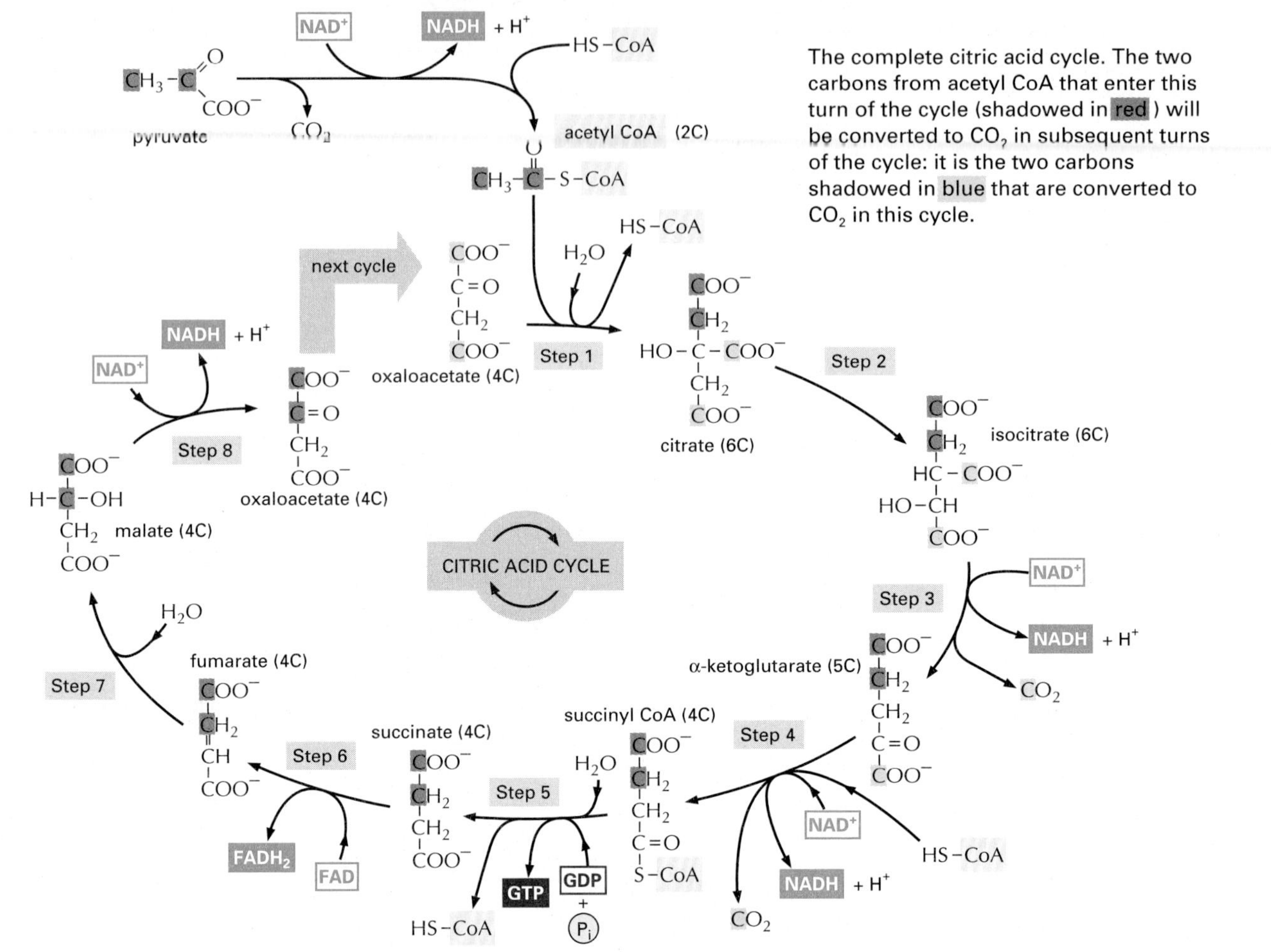

The complete citric acid cycle. The two carbons from acetyl CoA that enter this turn of the cycle (shadowed in red) will be converted to CO_2 in subsequent turns of the cycle: it is the two carbons shadowed in blue that are converted to CO_2 in this cycle.

Details of the eight steps are shown below. For each step, the part of the molecule that undergoes a change is shadowed in blue, and the name of the enzyme that catalyzes the reaction is in a yellow box.

STEP 1 After the enzyme removes a proton from the CH_3 group on acetyl CoA, the negatively charged CH_2^- forms a bond to a carbonyl carbon of oxaloacetate. The subsequent loss by hydrolysis of the coenzyme A (CoA) drives the reaction strongly forward.

citrate synthase

H_2O

+ HS–CoA + H^+

acetyl CoA

oxaloacetate

S-citryl-CoA intermediate

citrate

STEP 2 An isomerization reaction, in which water is first removed and then added back, moves the hydroxyl group from one carbon atom to its neighbor.

aconitase

H_2O

citrate

cis-aconitate intermediate

isocitrate

STEP 3 In the first of four oxidation steps in the cycle, the carbon carrying the hydroxyl group is converted to a carbonyl group. The immediate product is unstable, losing CO_2 while still bound to the enzyme.

isocitrate → (isocitrate dehydrogenase; NAD^+ → NADH + H^+) → oxalosuccinate intermediate → (H^+ in, CO_2 out) → α-ketoglutarate

STEP 4 The *α-ketoglutarate dehydrogenase complex* closely resembles the large enzyme complex that converts pyruvate to acetyl CoA (*pyruvate dehydrogenase*). It likewise catalyzes an oxidation that produces NADH, CO_2, and a high-energy thioester bond to coenzyme A (CoA).

α-ketoglutarate + HS–CoA → (α-ketoglutarate dehydrogenase complex; NAD^+ → NADH + H^+; CO_2) → succinyl-CoA

STEP 5 A phosphate molecule from solution displaces the CoA, forming a high-energy phosphate linkage to succinate. This phosphate is then passed to GDP to form GTP. (In bacteria and plants, ATP is formed instead.)

succinyl-CoA → (succinyl-CoA synthetase; H_2O, P_i, GDP → GTP) → succinate + HS–CoA

STEP 6 In the third oxidation step in the cycle, FAD removes two hydrogen atoms from succinate.

succinate → (succinate dehydrogenase; FAD → $FADH_2$) → fumarate

STEP 7 The addition of water to fumarate places a hydroxyl group next to a carbonyl carbon.

fumarate → (fumarase; H_2O) → malate

STEP 8 In the last of four oxidation steps in the cycle, the carbon carrying the hydroxyl group is converted to a carbonyl group, regenerating the oxaloacetate needed for step 1.

malate → (malate dehydrogenase; NAD^+ → NADH + H^+) → oxaloacetate

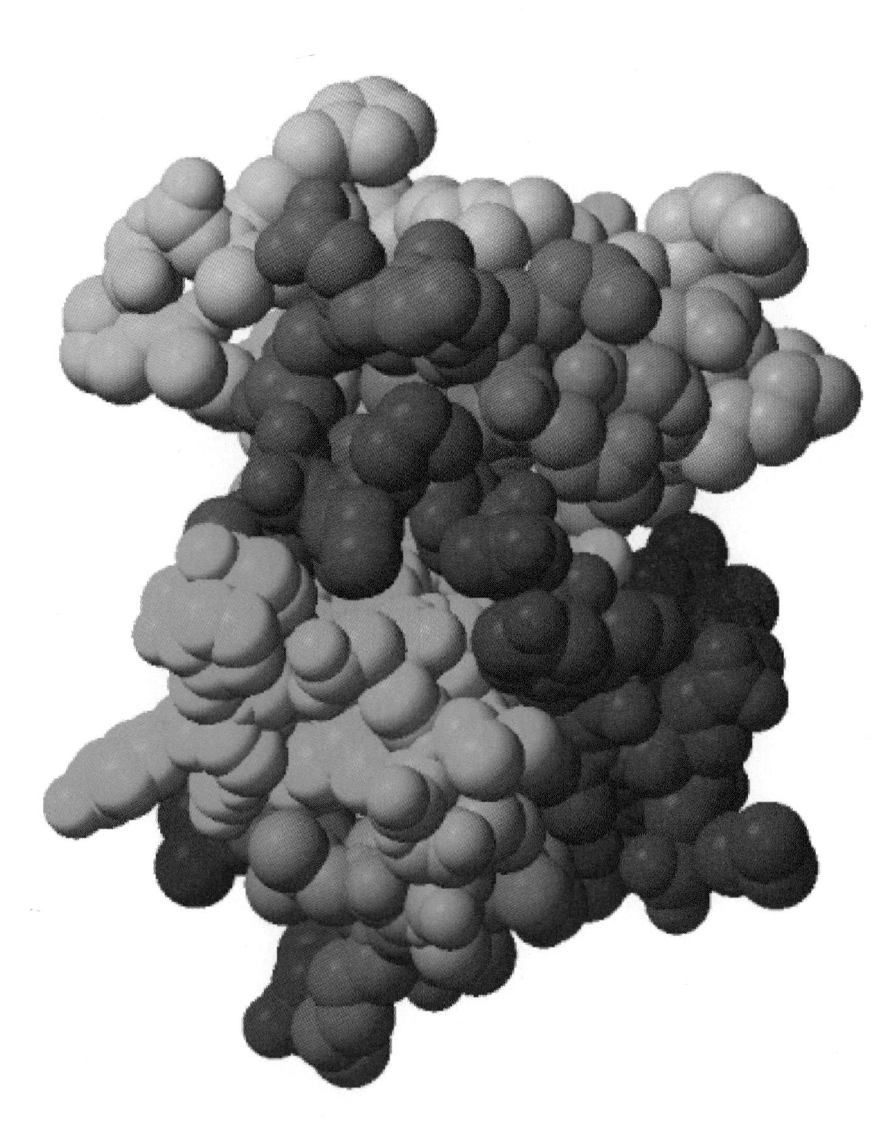

Protein structure. A small protein domain, the SH2 domain, is shown here in a space-filling representation. For further views, see Panel 3–2. (Courtesy of David Lawson.)

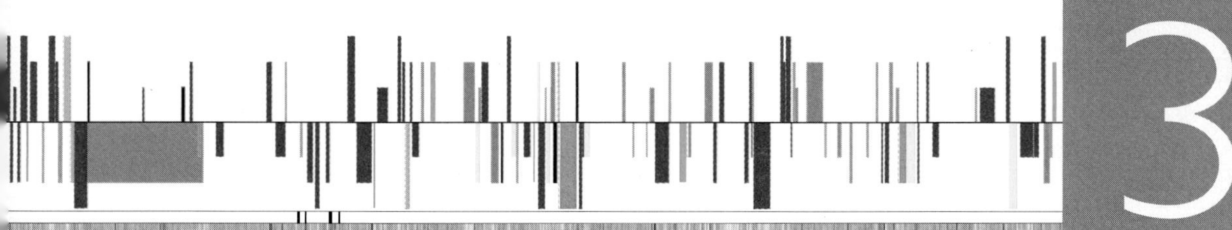

3

PROTEINS

When we look at a cell through a microscope or analyze its electrical or biochemical activity, we are, in essence, observing proteins. Proteins constitute most of a cell's dry mass. They are not only the building blocks from which cells are built; they also execute nearly all cell functions. Thus, enzymes provide the intricate molecular surfaces in a cell that promote its many chemical reactions. Proteins embedded in the plasma membrane form channels and pumps that control the passage of small molecules into and out of the cell. Other proteins carry messages from one cell to another, or act as signal integrators that relay sets of signals inward from the plasma membrane to the cell nucleus. Yet others serve as tiny molecular machines with moving parts: *kinesin*, for example, propels organelles through the cytoplasm; *topoisomerase* can untangle knotted DNA molecules. Other specialized proteins act as antibodies, toxins, hormones, antifreeze molecules, elastic fibers, ropes, or sources of luminescence. Before we can hope to understand how genes work, how muscles contract, how nerves conduct electricity, how embryos develop, or how our bodies function, we must attain a deep understanding of proteins.

THE SHAPE AND STRUCTURE OF PROTEINS

From a chemical point of view, proteins are by far the most structurally complex and functionally sophisticated molecules known. This is perhaps not surprising, once one realizes that the structure and chemistry of each protein has been developed and fine-tuned over billions of years of evolutionary history. We start this chapter by considering how the location of each amino acid in the long string of amino acids that forms a protein determines its three-dimensional shape. We will then use this understanding of protein structure at the atomic level to describe how the precise shape of each protein molecule determines its function in a cell.

The Shape of a Protein Is Specified by Its Amino Acid Sequence

Recall from Chapter 2 that there are 20 types of amino acids in proteins, each with different chemical properties. A **protein** molecule is made from a long chain of these amino acids, each linked to its neighbor through a covalent peptide bond (Figure 3–1). Proteins are therefore also known as *polypeptides*. Each type of protein has a unique sequence of amino acids, exactly the same from one molecule to the next. Many thousands of different proteins are known, each with its own particular amino acid sequence.

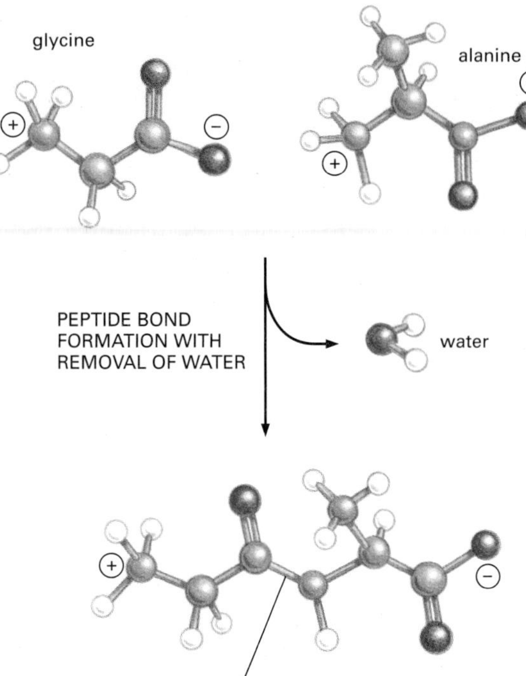

Figure 3–1 A peptide bond. This covalent bond forms when the carbon atom from the carboxyl group of one amino acid shares electrons with the nitrogen atom *(blue)* from the amino group of a second amino acid. As indicated, a molecule of water is lost in this condensation reaction.

The repeating sequence of atoms along the core of the polypeptide chain is referred to as the **polypeptide backbone**. Attached to this repetitive chain are those portions of the amino acids that are not involved in making a peptide bond and which give each amino acid its unique properties: the 20 different amino acid **side chains** (Figure 3–2). Some of these side chains are nonpolar and hydrophobic ("water-fearing"), others are negatively or positively charged, some are reactive, and so on. Their atomic structures are presented in Panel 3–1, and a brief list with abbreviations is provided in Figure 3–3.

As discussed in Chapter 2, atoms behave almost as if they were hard spheres with a definite radius (their *van der Waals radius*). The requirement that no two atoms overlap limits greatly the possible bond angles in a polypeptide chain (Figure 3–4). This constraint and other steric interactions severely restrict the variety of three-dimensional arrangements of atoms (or *conformations)* that are possible. Nevertheless, a long flexible chain, such as a protein, can still fold in an enormous number of ways.

The folding of a protein chain is, however, further constrained by many different sets of weak *noncovalent bonds* that form between one part of the chain and another. These involve atoms in the polypeptide backbone, as well as atoms in the amino acid side chains. The weak bonds are of three types: *hydrogen bonds, ionic bonds,* and *van der Waals attractions,* as explained in Chapter 2 (see p. 57). Individual noncovalent bonds are 30–300 times weaker than the typical covalent bonds that create biological molecules. But many weak bonds can act in parallel to hold two regions of a polypeptide chain tightly together. The stability of each folded shape is therefore determined by the combined strength of large numbers of such noncovalent bonds (Figure 3–5).

A fourth weak force also has a central role in determining the shape of a protein. As described in Chapter 2, hydrophobic molecules, including the nonpolar side chains of particular amino acids, tend to be forced together in an aqueous environment in order to minimize their disruptive effect on the hydrogen-bonded network of water molecules (see p. 58 and Panel 2–2, pp. 112–113). Therefore, an important factor governing the folding of any protein is the distribution of its polar and nonpolar amino acids. The nonpolar (hydrophobic) side chains in a protein—belonging to such amino acids as phenylalanine, leucine, valine, and tryptophan—tend to cluster in the interior of the molecule (just as hydrophobic oil droplets coalesce in water to form one large droplet). This enables them to

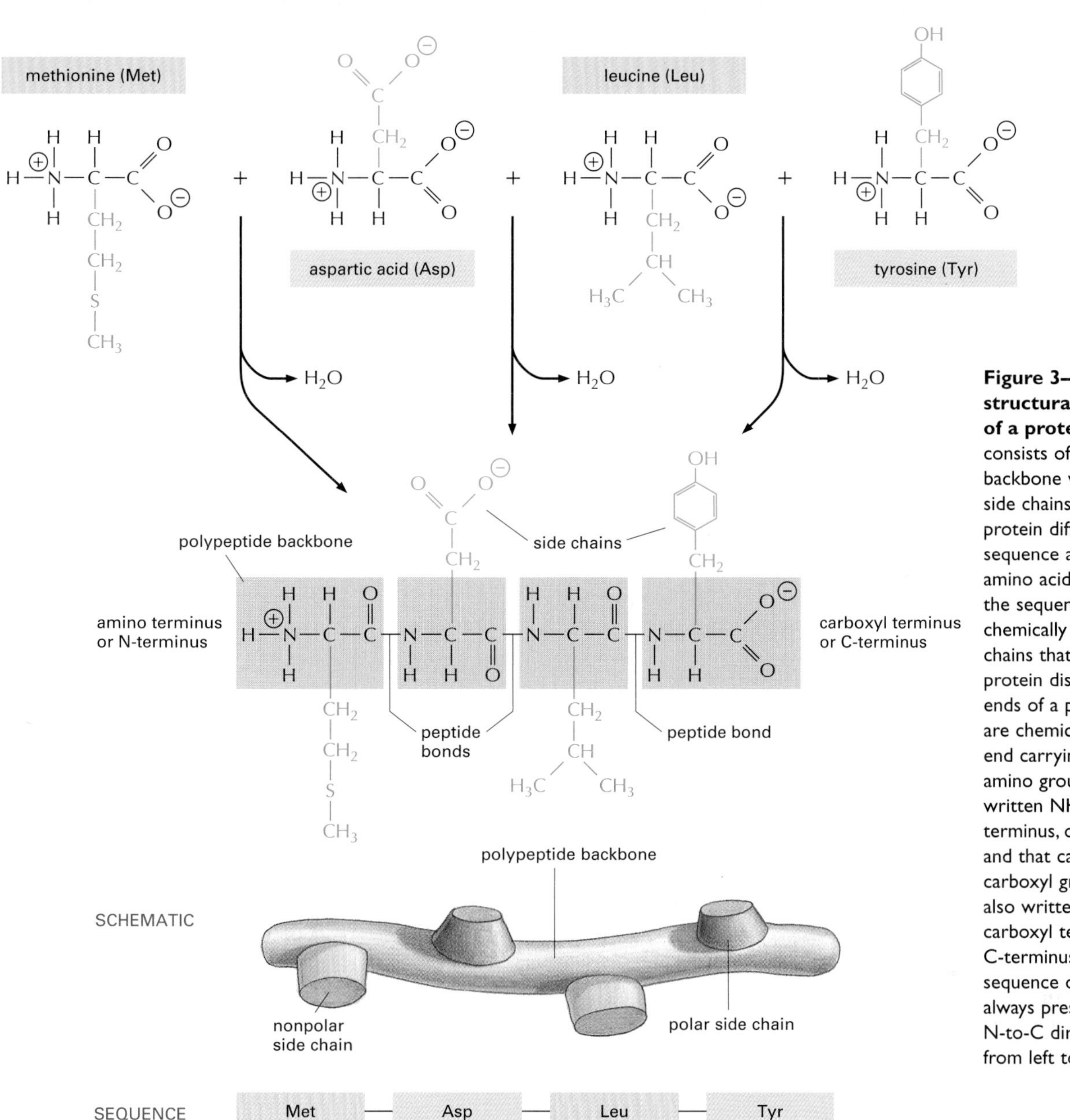

Figure 3–2 The structural components of a protein. A protein consists of a polypeptide backbone with attached side chains. Each type of protein differs in its sequence and number of amino acids; therefore, it is the sequence of the chemically different side chains that makes each protein distinct. The two ends of a polypeptide chain are chemically different: the end carrying the free amino group (NH_3^+, also written NH_2) is the amino terminus, or N-terminus, and that carrying the free carboxyl group (COO^-, also written COOH) is the carboxyl terminus or C-terminus. The amino acid sequence of a protein is always presented in the N-to-C direction, reading from left to right.

AMINO ACID			SIDE CHAIN	AMINO ACID			SIDE CHAIN
Aspartic acid	Asp	D	negative	Alanine	Ala	A	nonpolar
Glutamic acid	Glu	E	negative	Glycine	Gly	G	nonpolar
Arginine	Arg	R	positive	Valine	Val	V	nonpolar
Lysine	Lys	K	positive	Leucine	Leu	L	nonpolar
Histidine	His	H	positive	Isoleucine	Ile	I	nonpolar
Asparagine	Asn	N	uncharged polar	Proline	Pro	P	nonpolar
Glutamine	Gln	Q	uncharged polar	Phenylalanine	Phe	F	nonpolar
Serine	Ser	S	uncharged polar	Methionine	Met	M	nonpolar
Threonine	Thr	T	uncharged polar	Tryptophan	Trp	W	nonpolar
Tyrosine	Tyr	Y	uncharged polar	Cysteine	Cys	C	nonpolar
POLAR AMINO ACIDS				NONPOLAR AMINO ACIDS			

Figure 3–3 The 20 amino acids found in proteins. Both three-letter and one-letter abbreviations are listed. As shown, there are equal numbers of polar and nonpolar side chains. For their atomic structures, see Panel 3–1 (pp. 132–133).

PANEL 3–1 The 20 Amino Acids Found in Proteins

THE AMINO ACID

The general formula of an amino acid is

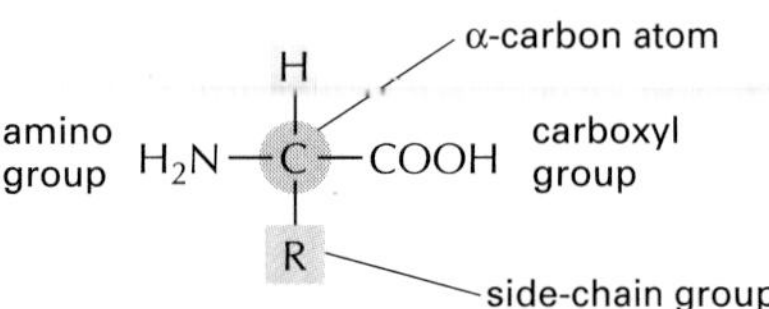

R is commonly one of 20 different side chains. At pH 7 both the amino and carboxyl groups are ionized.

$$H_3N^{+}—C(H)(R)—COO^{-}$$

OPTICAL ISOMERS

The α-carbon atom is asymmetric, which allows for two mirror image (or stereo-) isomers, L and D.

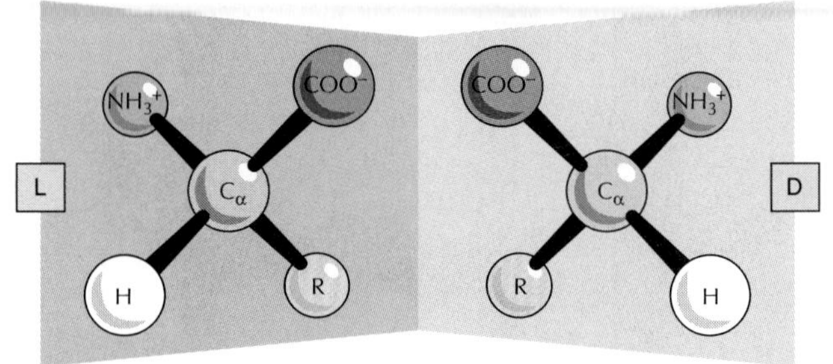

Proteins consist exclusively of L-amino acids.

FAMILIES OF AMINO ACIDS

The common amino acids are grouped according to whether their side chains are

- acidic
- basic
- uncharged polar
- nonpolar

These 20 amino acids are given both three-letter and one-letter abbreviations.

Thus: alanine = Ala = A

BASIC SIDE CHAINS

lysine (Lys, or K)

Side chain: CH_2—CH_2—CH_2—CH_2—NH_3^+

arginine (Arg, or R)

Side chain: CH_2—CH_2—CH_2—NH—C(=^+H_2N)—NH_2

This group is very basic because its positive charge is stabilized by resonance.

histidine (His, or H)

Side chain: CH_2—imidazole ring (C, HN, CH, HC=NH^+)

These nitrogens have a relatively weak affinity for an H^+ and are only partly positive at neutral pH.

PEPTIDE BONDS

Amino acids are commonly joined together by an amide linkage, called a peptide bond.

Peptide bond: The four atoms in each *gray box* form a rigid planar unit. There is no rotation around the C–N bond.

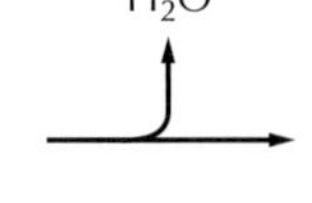

H_2N—CH(R)—COOH + H_2N—CH(R)—COOH → H_2N—CH(R)—C(=O)—NH—CH(R)—COOH + H_2O

Proteins are long polymers of amino acids linked by peptide bonds, and they are always written with the N-terminus toward the left. The sequence of this tripeptide is histidine-cysteine-valine.

amino- or N-terminus

^+H_3N—CH(CH_2-imidazole)—C(=O)—NH—CH(CH_2—SH)—C(=O)—NH—CH(CH(CH_3)$_2$)—COO^-

carboxyl- or C-terminus

These two single bonds allow rotation, so that long chains of amino acids are very flexible.

ACIDIC SIDE CHAINS

aspartic acid

(Asp, or D)

glutamic acid

(Glu, or E)

UNCHARGED POLAR SIDE CHAINS

asparagine

(Asn, or N)

glutamine

(Gln, or Q)

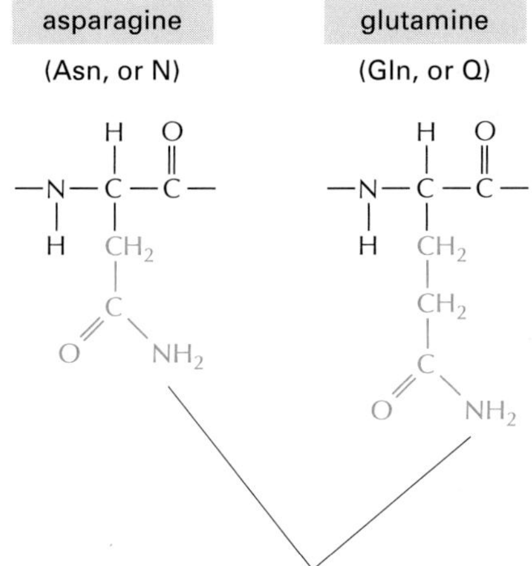

Although the amide N is not charged at neutral pH, it is polar.

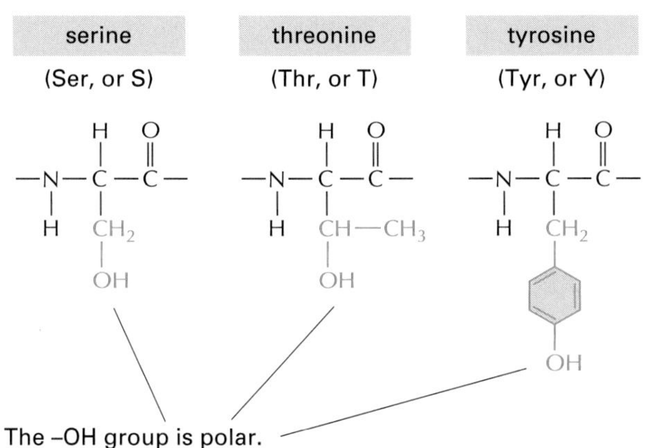

NONPOLAR SIDE CHAINS

alanine

(Ala, or A)

valine

(Val, or V)

leucine

(Leu, or L)

isoleucine

(Ile, or I)

proline

(Pro, or P)

(actually an imino acid)

phenylalanine

(Phe, or F)

methionine

(Met, or M)

tryptophan

(Trp, or W)

glycine

(Gly, or G)

cysteine

(Cys, or C)

Disulfide bonds can form between two cysteine side chains in proteins.

$$- -CH_2-S-S-CH_2- -$$

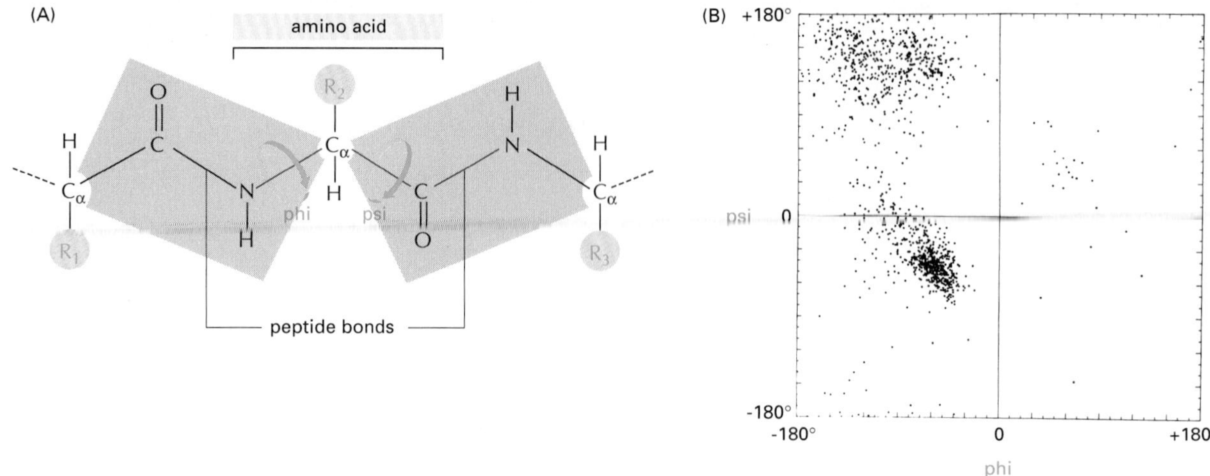

Figure 3–4 Steric limitations on the bond angles in a polypeptide chain. (A) Each amino acid contributes three bonds *(red)* to the backbone of the chain. The peptide bond is planar *(gray shading)* and does not permit rotation. By contrast, rotation can occur about the C_α–C bond, whose angle of rotation is called psi (ψ), and about the N–C_α bond, whose angle of rotation is called phi (ϕ). By convention, an R group is often used to denote an amino acid side chain *(green circles)*. (B) The conformation of the main-chain atoms in a protein is determined by one pair of ϕ and ψ angles for each amino acid; because of steric collisions between atoms within each amino acid, most pairs of ϕ and ψ angles do not occur. In this so-called Ramachandran plot, each dot represents an observed pair of angles in a protein. (B, from J. Richardson, *Adv. Prot. Chem.* 34:174–175, 1981. © Academic Press.)

avoid contact with the water that surrounds them inside a cell. In contrast, polar side chains—such as those belonging to arginine, glutamine, and histidine—tend to arrange themselves near the outside of the molecule, where they can form hydrogen bonds with water and with other polar molecules (Figure 3–6). When polar amino acids are buried within the protein, they are usually hydrogen-bonded to other polar amino acids or to the polypeptide backbone (Figure 3–7).

Proteins Fold into a Conformation of Lowest Energy

As a result of all of these interactions, each type of protein has a particular three-dimensional structure, which is determined by the order of the amino acids in its chain. The final folded structure, or **conformation**, adopted by any polypeptide chain is generally the one in which the free energy is minimized. Protein folding has been studied in a test tube by using highly purified proteins. A protein can be unfolded, or *denatured,* by treatment with certain solvents, which disrupt the noncovalent interactions holding the folded chain together. This treatment converts the protein into a flexible polypeptide chain that has lost its

Figure 3–5 Three types of noncovalent bonds that help proteins fold. Although a single one of these bonds is quite weak, many of them often form together to create a strong bonding arrangement, as in the example shown. As in the previous figure, R is used as a general designation for an amino acid side chain.

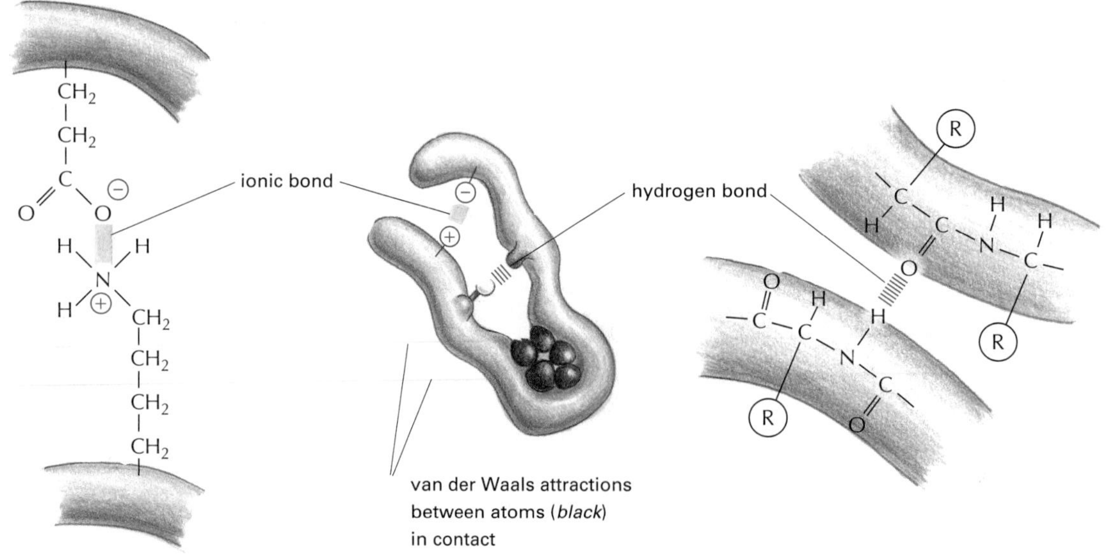

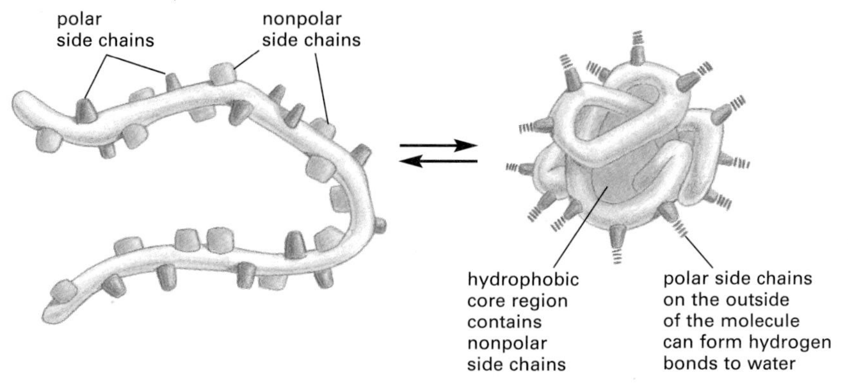

Figure 3–6 How a protein folds into a compact conformation. The polar amino acid side chains tend to gather on the outside of the protein, where they can interact with water; the nonpolar amino acid side chains are buried on the inside to form a tightly packed hydrophobic core of atoms that are hidden from water. In this schematic drawing, the protein contains only about 30 amino acids.

natural shape. When the denaturing solvent is removed, the protein often refolds spontaneously, or *renatures,* into its original conformation (Figure 3–8), indicating that all the information needed for specifying the three-dimensional shape of a protein is contained in its amino acid sequence.

Each protein normally folds up into a single stable conformation. However, the conformation often changes slightly when the protein interacts with other molecules in the cell. This change in shape is often crucial to the function of the protein, as we see later.

Although a protein chain can fold into its correct conformation without outside help, protein folding in a living cell is often assisted by special proteins called *molecular chaperones.* These proteins bind to partly folded polypeptide chains and help them progress along the most energetically favorable folding pathway. Chaperones are vital in the crowded conditions of the cytoplasm, since they prevent the temporarily exposed hydrophobic regions in newly synthesized protein chains from associating with each other to form protein aggregates (see p. 357). However, the final three-dimensional shape of the protein is still specified by its amino acid sequence: chaperones simply make the folding process more reliable.

Proteins come in a wide variety of shapes, and they are generally between 50 and 2000 amino acids long. Large proteins generally consist of several distinct *protein domains*—structural units that fold more or less independently of each other, as we discuss below. The detailed structure of any protein is complicated; for simplicity a protein's structure can be depicted in several different ways, each emphasizing different features of the protein.

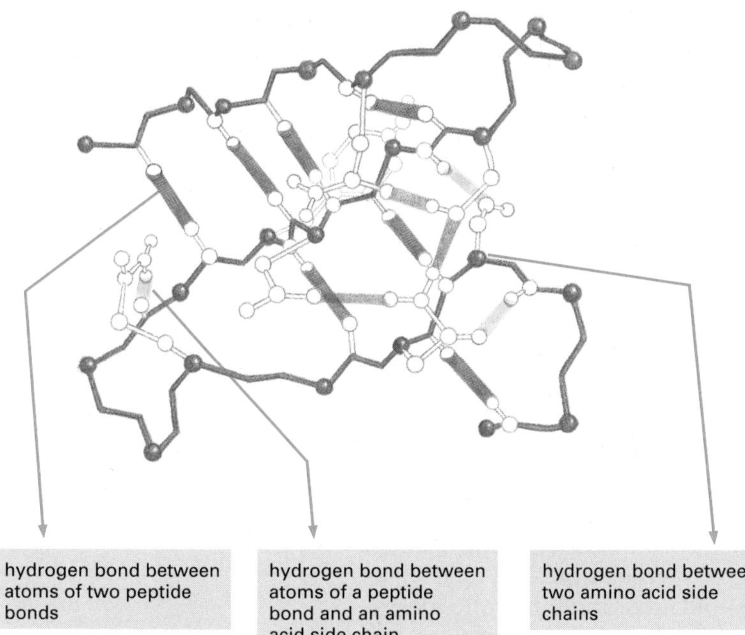

Figure 3–7 Hydrogen bonds in a protein molecule. Large numbers of hydrogen bonds form between adjacent regions of the folded polypeptide chain and help stabilize its three-dimensional shape. The protein depicted is a portion of the enzyme lysozyme, and the hydrogen bonds between the three possible pairs of partners have been differently colored, as indicated. (After C.K. Matthews and K.E. van Holde, Biochemistry. Redwood City, CA: Benjamin/Cummings, 1996.)

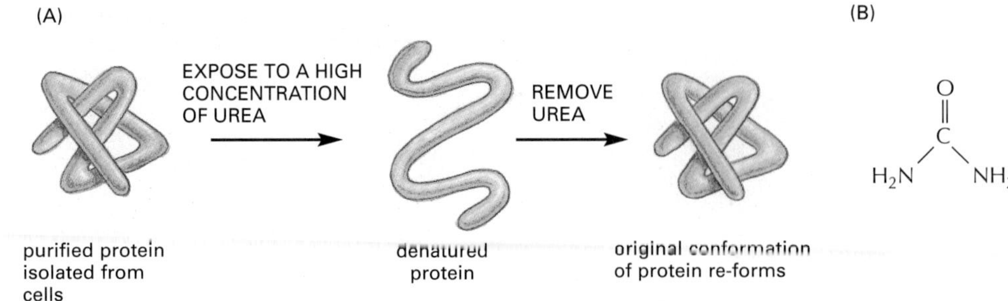

Figure 3–8 The refolding of a denatured protein. (A) This experiment demonstrates that the conformation of a protein is determined solely by its amino acid sequence. (B) The structure of urea. Urea is very soluble in water and unfolds proteins at high concentrations, where there is about one urea molecule for every six water molecules.

Panel 3–2 (pp. 138–139) presents four different depictions of a protein domain called SH2, which has important functions in eucaryotic cells. Constructed from a string of 100 amino acids, the structure is displayed as (A) a polypeptide backbone model, (B) a ribbon model, (C) a wire model that includes the amino acid side chains, and (D) a space-filling model. Each of the three horizontal rows shows the protein in a different orientation, and the image is colored in a way that allows the polypeptide chain to be followed from its N-terminus *(purple)* to its C-terminus *(red).*

Panel 3–2 shows that a protein's conformation is amazingly complex, even for a structure as small as the SH2 domain. But the description of protein structures can be simplified by the recognition that they are built up from several common structural motifs, as we discuss next.

The α Helix and the β Sheet Are Common Folding Patterns

When the three-dimensional structures of many different protein molecules are compared, it becomes clear that, although the overall conformation of each protein is unique, two regular folding patterns are often found in parts of them. Both patterns were discovered about 50 years ago from studies of hair and silk. The first folding pattern to be discovered, called the **α helix**, was found in the protein *α-keratin*, which is abundant in skin and its derivatives—such as hair, nails, and horns. Within a year of the discovery of the α helix, a second folded structure, called a **β sheet**, was found in the protein *fibroin*, the major constituent of silk. These two patterns are particularly common because they result from hydrogen-bonding between the N–H and C=O groups in the polypeptide backbone, without involving the side chains of the amino acids. Thus, they can be formed by many different amino acid sequences. In each case, the protein chain adopts a regular, repeating conformation. These two conformations, as well as the abbreviations that are used to denote them in ribbon models of proteins, are shown in Figure 3–9.

The core of many proteins contains extensive regions of β sheet. As shown in Figure 3–10, these β sheets can form either from neighboring polypeptide chains that run in the same orientation (parallel chains) or from a polypeptide chain that folds back and forth upon itself, with each section of the chain running in the direction opposite to that of its immediate neighbors (antiparallel chains). Both types of β sheet produce a very rigid structure, held together by hydrogen bonds that connect the peptide bonds in neighboring chains (see Figure 3–9D).

An α helix is generated when a single polypeptide chain twists around on itself to form a rigid cylinder. A hydrogen bond is made between every fourth peptide bond, linking the C=O of one peptide bond to the N–H of another (see Figure 3–9A). This gives rise to a regular helix with a complete turn every 3.6 amino acids. Note that the protein domain illustrated in Panel 3–2 contains two α helices, as well as β sheet structures.

Short regions of α helix are especially abundant in proteins located in cell membranes, such as transport proteins and receptors. As we discuss in Chapter 10, those portions of a transmembrane protein that cross the lipid bilayer usually cross as an α helix composed largely of amino acids with nonpolar side chains. The polypeptide backbone, which is hydrophilic, is hydrogen-bonded to itself in the α helix and shielded from the hydrophobic lipid environment of the membrane by its protruding nonpolar side chains (see also Figure 3–77).

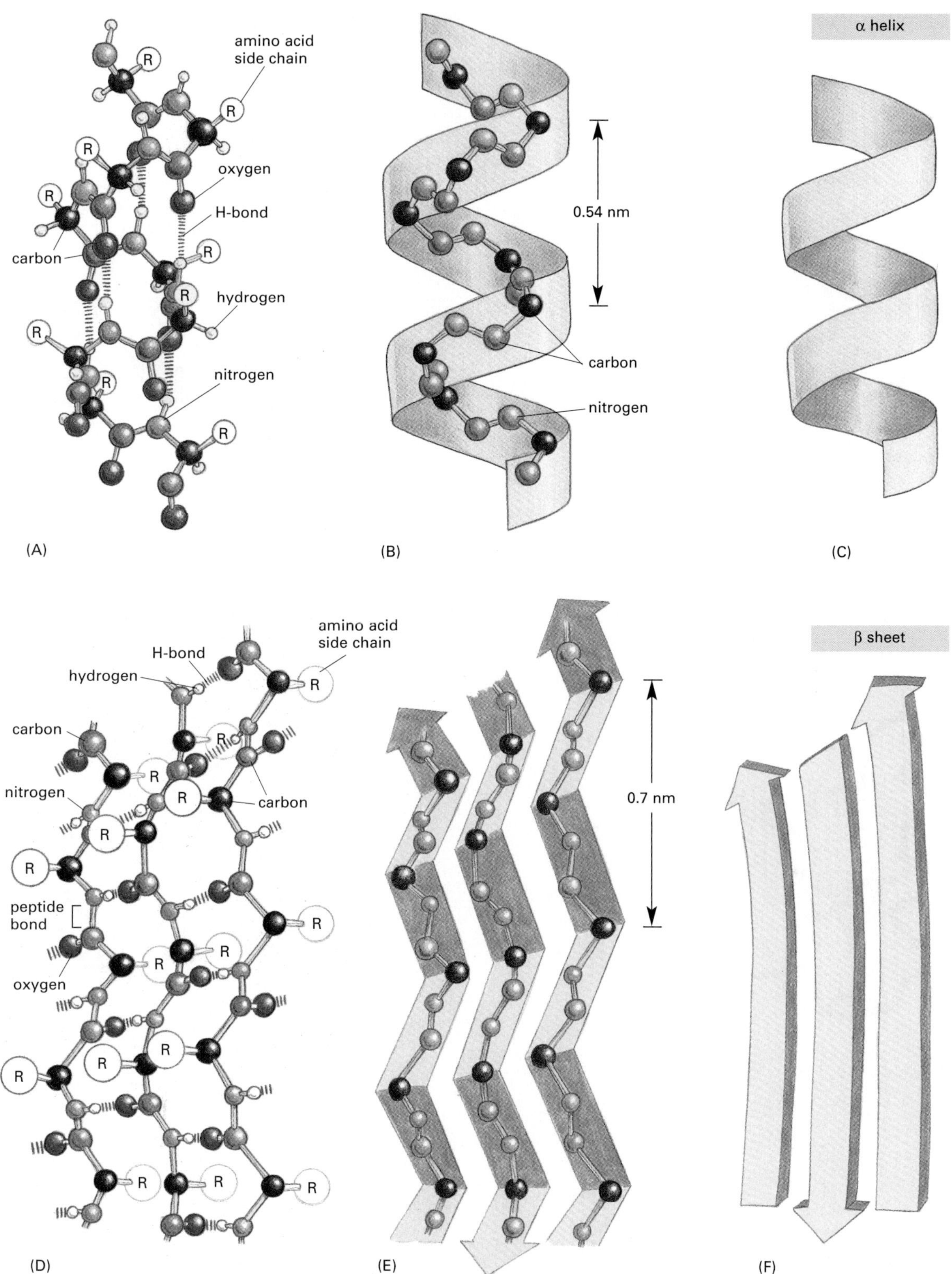

Figure 3–9 The regular conformation of the polypeptide backbone observed in the α helix and the β sheet. (A, B, and C) The α helix. The N–H of every peptide bond is hydrogen-bonded to the C=O of a neighboring peptide bond located four peptide bonds away in the same chain. (D, E, and F) The β sheet. In this example, adjacent peptide chains run in opposite (antiparallel) directions. The individual polypeptide chains (strands) in a β sheet are held together by hydrogen-bonding between peptide bonds in different strands, and the amino acid side chains in each strand alternately project above and below the plane of the sheet. (A) and (D) show all the atoms in the polypeptide backbone, but the amino acid side chains are truncated and denoted by R. In contrast, (B) and (E) show the backbone atoms only, while (C) and (F) display the shorthand symbols that are used to represent the α helix and the β sheet in ribbon drawings of proteins (see Panel 3–2B).

PANEL 3–2 Four Different Ways of Depicting a Small Protein Domain: the SH2 Domain. (Courtesy of David Lawson.)

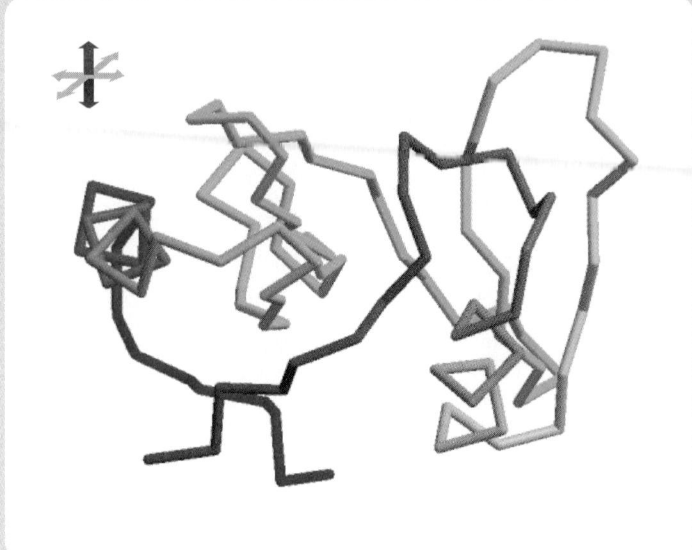

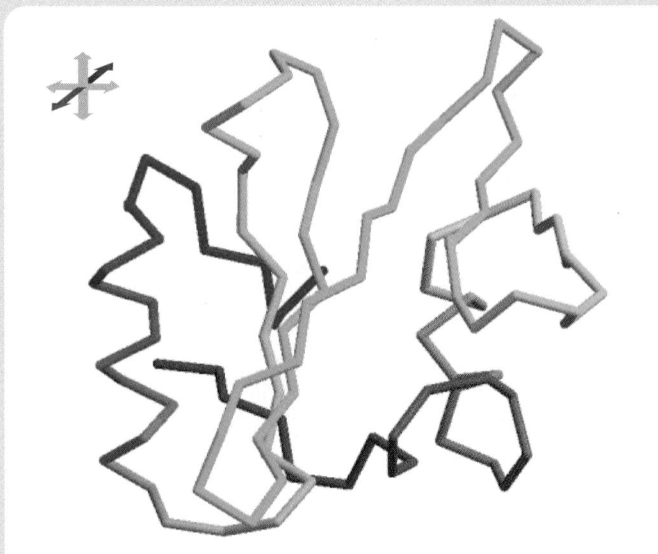

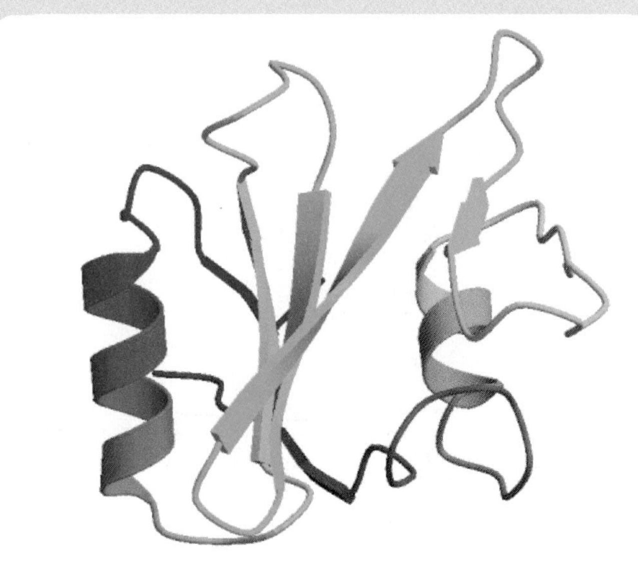

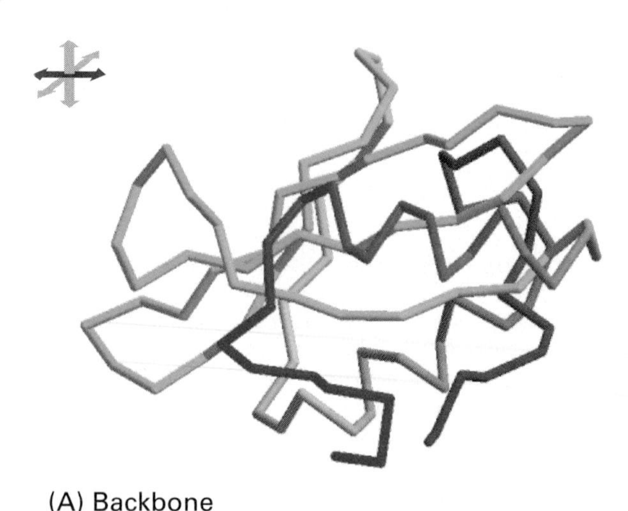

(A) Backbone

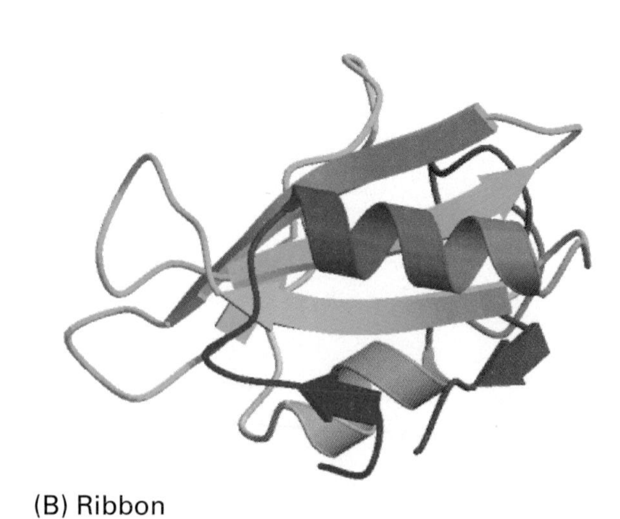

(B) Ribbon

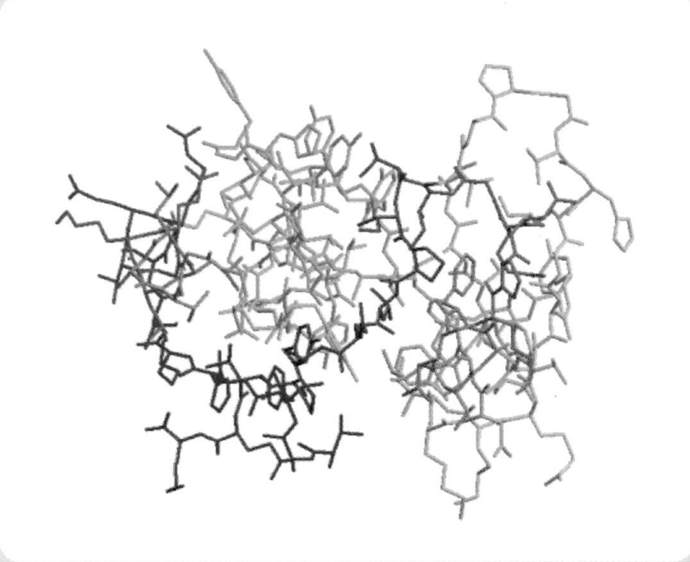

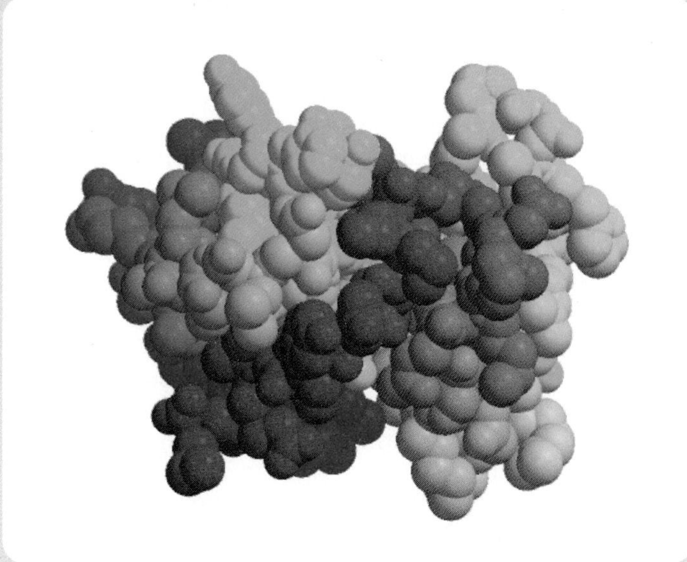

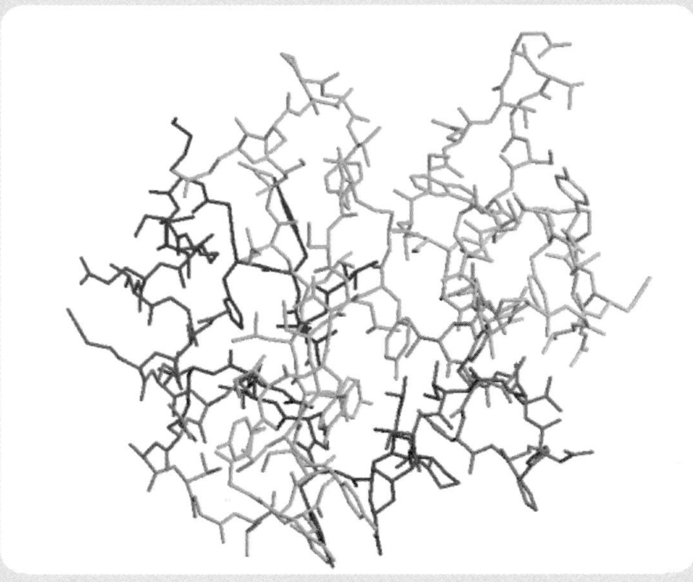

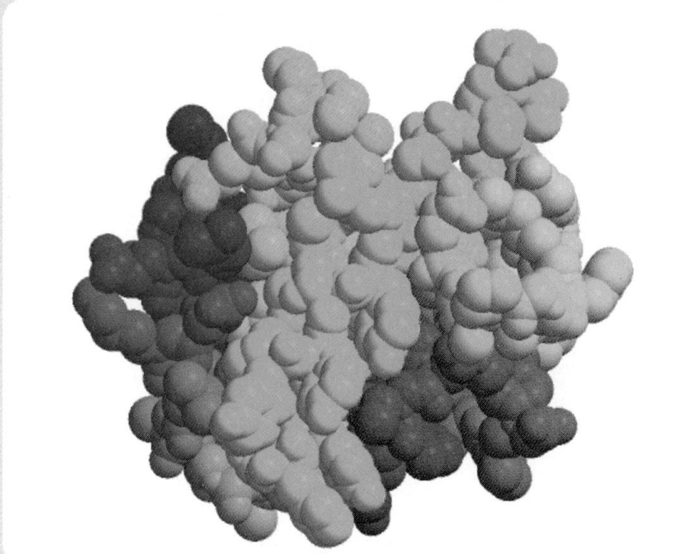

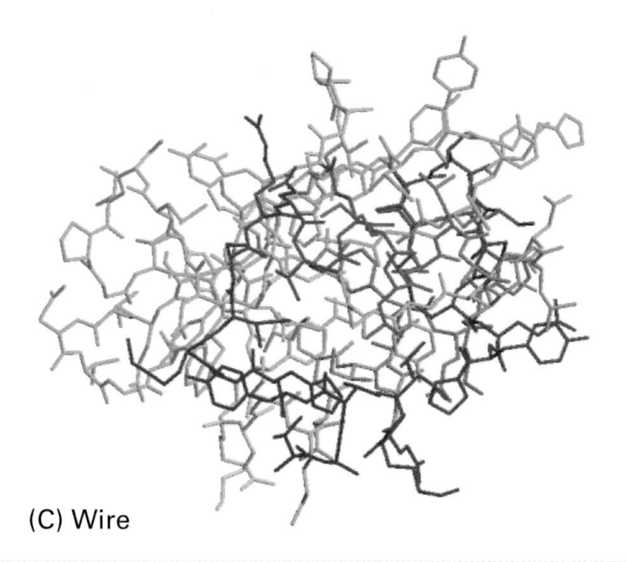

(C) Wire

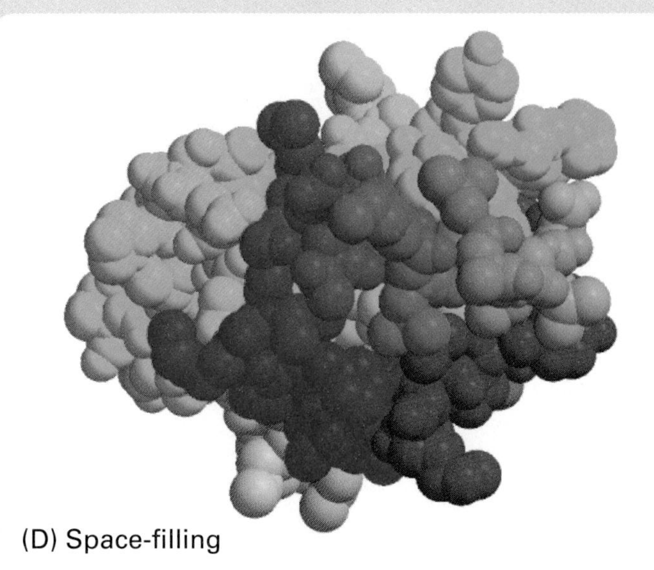

(D) Space-filling

In other proteins, α helices wrap around each other to form a particularly stable structure, known as a **coiled-coil**. This structure can form when the two (or in some cases three) α helices have most of their nonpolar (hydrophobic) side chains on one side, so that they can twist around each other with these side chains facing inward (Figure 3–11). Long rodlike coiled-coils provide the structural framework for many elongated proteins. Examples are α-keratin, which forms the intracellular fibers that reinforce the outer layer of the skin and its appendages, and the myosin molecules responsible for muscle contraction.

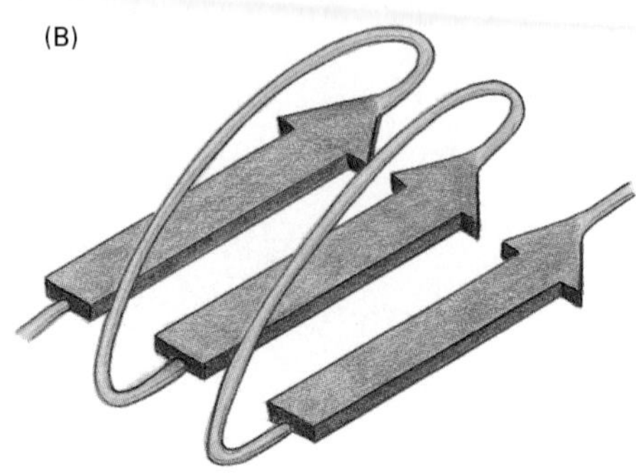

Figure 3–10 Two types of β sheet structures. (A) An antiparallel β sheet (see Figure 3–9D). (B) A parallel β sheet. Both of these structures are common in proteins.

The Protein Domain Is a Fundamental Unit of Organization

Even a small protein molecule is built from thousands of atoms linked together by precisely oriented covalent and noncovalent bonds, and it is extremely difficult to visualize such a complicated structure without a three-dimensional display. For this reason, various graphic and computer-based aids are used. A CD-ROM produced to accompany this book contains computer-generated images of selected proteins, designed to be displayed and rotated on the screen in a variety of formats.

Biologists distinguish four levels of organization in the structure of a protein. The amino acid sequence is known as the **primary structure** of the protein. Stretches of polypeptide chain that form α helices and β sheets constitute the protein's **secondary structure**. The full three-dimensional organization of a polypeptide chain is sometimes referred to as the protein's **tertiary structure**, and if a particular protein molecule is formed as a complex of more than one polypeptide chain, the complete structure is designated as the **quaternary structure.**

Studies of the conformation, function, and evolution of proteins have also revealed the central importance of a unit of organization distinct from the four just described. This is the **protein domain**, a substructure produced by any part of a polypeptide chain that can fold independently into a compact, stable structure. A domain usually contains between 40 and 350 amino acids, and it is the

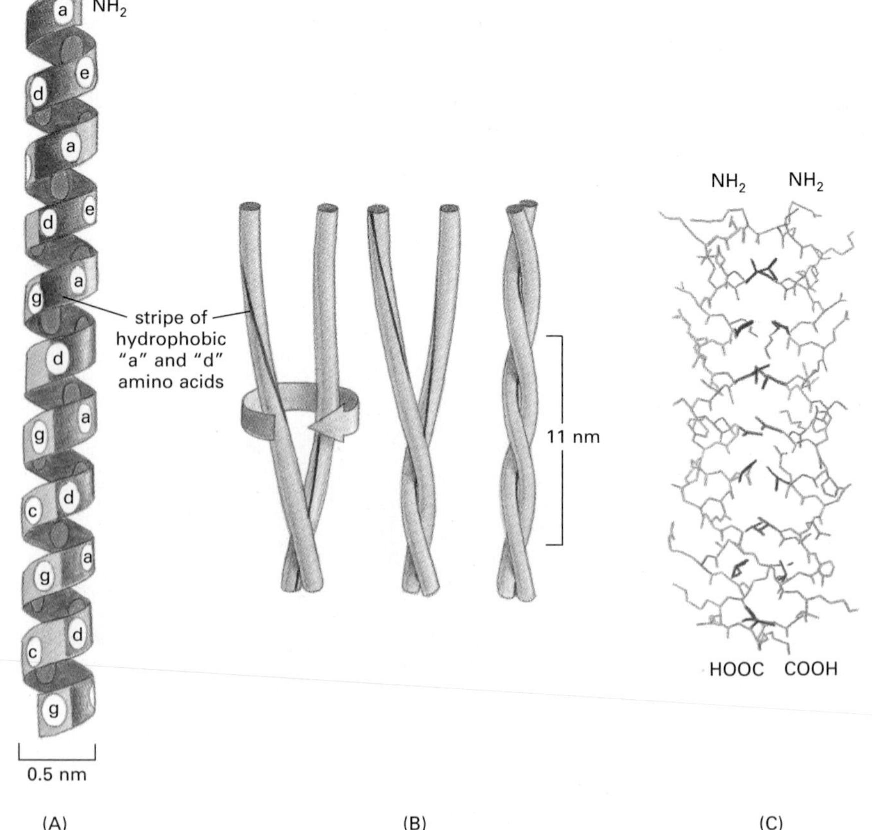

Figure 3–11 The structure of a coiled-coil. (A) A single α helix, with successive amino acid side chains labeled in a sevenfold sequence, "abcdefg" (from bottom to top). Amino acids "a" and "d" in such a sequence lie close together on the cylinder surface, forming a "stripe" (*red*) that winds slowly around the α helix. Proteins that form coiled-coils typically have nonpolar amino acids at positions "a" and "d." Consequently, as shown in (B), the two α helices can wrap around each other with the nonpolar side chains of one α helix interacting with the nonpolar side chains of the other, while the more hydrophilic amino acid side chains are left exposed to the aqueous environment. (C) The atomic structure of a coiled-coil determined by x-ray crystallography. The *red* side chains are nonpolar.

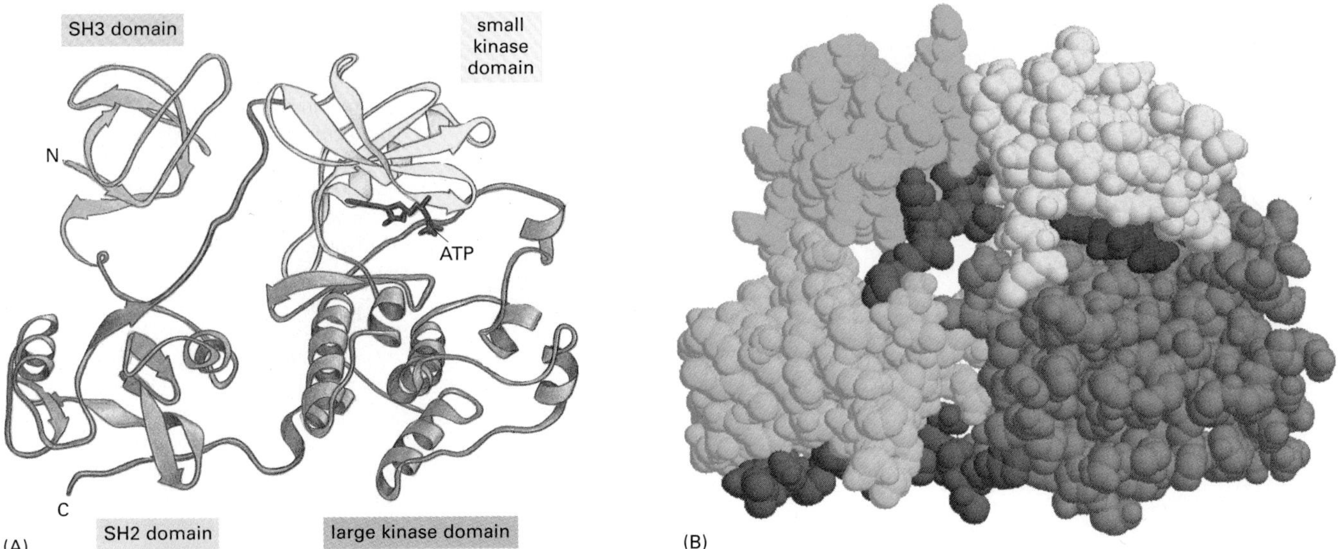

Figure 3–12 A protein formed from four domains. In the Src protein shown, two of the domains form a protein kinase enzyme, while the SH2 and SH3 domains perform regulatory functions. (A) A ribbon model, with ATP substrate in *red*. (B) A spacing-filling model, with ATP substrate in *red*. Note that the site that binds ATP is positioned at the interface of the two domains that form the kinase. The detailed structure of the SH2 domain is illustrated in Panel 3–2 (pp. 138–139).

modular unit from which many larger proteins are constructed. The different domains of a protein are often associated with different functions. Figure 3–12 shows an example—the Src protein kinase, which functions in signaling pathways inside vertebrate cells (Src is pronounced "sarc"). This protein has four domains: the SH2 and SH3 domains have regulatory roles, while the two remaining domains are responsible for the kinase catalytic activity. Later in the chapter, we shall return to this protein, in order to explain how proteins can form molecular switches that transmit information throughout cells.

The smallest protein molecules contain only a single domain, whereas larger proteins can contain as many as several dozen domains, usually connected to each other by short, relatively unstructured lengths of polypeptide chain. Figure 3–13 presents ribbon models of three differently organized protein domains. As these examples illustrate, the central core of a domain can be constructed from α helices, from β sheets, or from various combinations of these two fundamental folding elements. Each different combination is known as a *protein fold*. So far, about 1000 different protein folds have been identified among the ten thousand proteins whose detailed conformations are known.

Few of the Many Possible Polypeptide Chains Will Be Useful

Since each of the 20 amino acids is chemically distinct and each can, in principle, occur at any position in a protein chain, there are $20 \times 20 \times 20 \times 20 = 160{,}000$ different possible polypeptide chains four amino acids long, or 20^n different possible polypeptide chains n amino acids long. For a typical protein length of about 300 amino acids, more than 10^{390} (20^{300}) different polypeptide chains could theoretically be made. This is such an enormous number that to produce just one molecule of each kind would require many more atoms than exist in the universe.

Only a very small fraction of this vast set of conceivable polypeptide chains would adopt a single, stable three-dimensional conformation—by some estimates, less than one in a billion. The vast majority of possible protein molecules could adopt many conformations of roughly equal stability, each conformation having different chemical properties. And yet virtually all proteins present in cells adopt unique and stable conformations. How is this possible? The answer lies in natural selection. A protein with an unpredictably variable structure and biochemical activity is unlikely to help the survival of a cell that contains it. Such proteins would therefore have been eliminated by natural selection through the enormously long trial-and-error process that underlies biological evolution.

Because of natural selection, not only is the amino acid sequence of a present-day protein such that a single conformation is extremely stable, but this conformation has its chemical properties finely tuned to enable the protein to

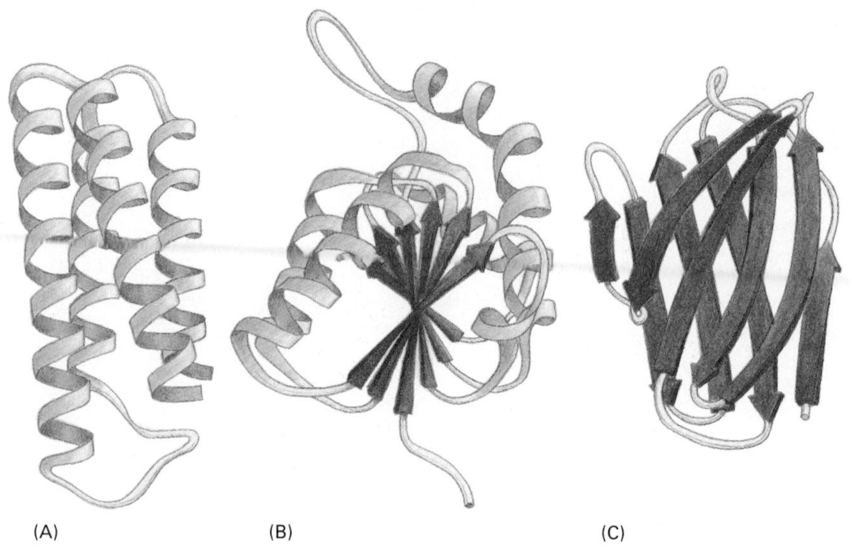

Figure 3–13 Ribbon models of three different protein domains. (A) Cytochrome b_{562}, a single-domain protein involved in electron transport in mitochondria. This protein is composed almost entirely of α helices. (B) The NAD-binding domain of the enzyme lactic dehydrogenase, which is composed of a mixture of α helices and β sheets. (C) The variable domain of an immunoglobulin (antibody) light chain, composed of a sandwich of two β sheets. In these examples, the α helices are shown in *green*, while strands organized as β sheets are denoted by *red arrows*.

Note that the polypeptide chain generally traverses back and forth across the entire domain, making sharp turns only at the protein surface. It is the protruding loop regions *(yellow)* that often form the binding sites for other molecules. (Adapted from drawings courtesy of Jane Richardson.)

perform a particular catalytic or structural function in the cell. Proteins are so precisely built that the change of even a few atoms in one amino acid can sometimes disrupt the structure of the whole molecule so severely that all function is lost.

Proteins Can Be Classified into Many Families

Once a protein had evolved that folded up into a stable conformation with useful properties, its structure could be modified during evolution to enable it to perform new functions. This process has been greatly accelerated by genetic mechanisms that occasionally produce duplicate copies of genes, allowing one gene copy to evolve independently to perform a new function (discussed in Chapter 7). This type of event has occurred quite often in the past; as a result, many present-day proteins can be grouped into protein families, each family member having an amino acid sequence and a three-dimensional conformation that resemble those of the other family members.

Consider, for example, the *serine proteases*, a large family of protein-cleaving (proteolytic) enzymes that includes the digestive enzymes chymotrypsin, trypsin, and elastase, and several proteases involved in blood clotting. When the protease portions of any two of these enzymes are compared, parts of their amino acid sequences are found to match. The similarity of their three-dimensional conformations is even more striking: most of the detailed twists and turns in their polypeptide chains, which are several hundred amino acids long, are virtually identical (Figure 3–14). The many different serine proteases nevertheless have distinct enzymatic activities, each cleaving different proteins or the peptide bonds between different types of amino acids. Each therefore performs a distinct function in an organism.

The story we have told for the serine proteases could be repeated for hundreds of other protein families. In many cases the amino acid sequences have diverged much further than for the serine proteases, so that one cannot be sure of a family relationship between two proteins without determining their three-dimensional structures. The yeast α2 protein and the *Drosophila* engrailed protein, for example, are both gene regulatory proteins in the homeodomain family. Because they are identical in only 17 of their 60 amino acid residues, their relationship became certain only when their three-dimensional structures were compared (Figure 3–15).

The various members of a large protein family often have distinct functions. Some of the amino acid changes that make family members different were no doubt selected in the course of evolution because they resulted in useful changes in biological activity, giving the individual family members the different functional properties they have today. But many other amino acid changes are effectively "neutral," having neither a beneficial nor a damaging effect on the

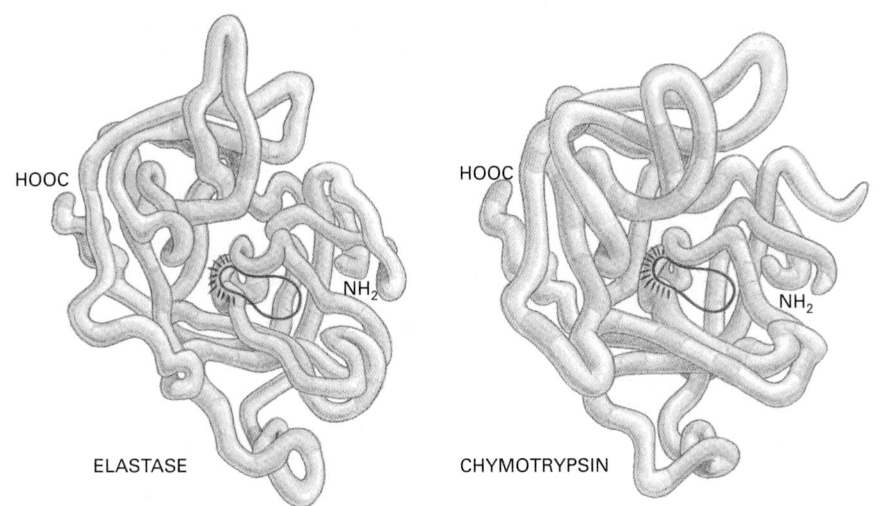

Figure 3–14 The conformations of two serine proteases compared. The backbone conformations of elastase and chymotrypsin. Although only those amino acids in the polypeptide chain shaded in *green* are the same in the two proteins, the two conformations are very similar nearly everywhere. The active site of each enzyme is circled in *red;* this is where the peptide bonds of the proteins that serve as substrates are bound and cleaved by hydrolysis. The serine proteases derive their name from the amino acid serine, whose side chain is part of the active site of each enzyme and directly participates in the cleavage reaction.

basic structure and function of the protein. In addition, since mutation is a random process, there must also have been many deleterious changes that altered the three-dimensional structure of these proteins sufficiently to harm them. Such faulty proteins would have been lost whenever the individual organisms making them were at enough of a disadvantage to be eliminated by natural selection.

Protein families are readily recognized when the genome of any organism is sequenced; for example, the determination of the DNA sequence for the entire genome of the nematode *Caenorhabditis elegans* has revealed that this tiny worm contains more than 18,000 genes. Through sequence comparisons, the products of a large fraction of these genes can be seen to contain domains from one or another protein family; for example, there appear to be 388 genes containing protein kinase domains, 66 genes containing DNA and RNA helicase domains, 43 genes containing SH2 domains, 70 genes containing immunoglobulin domains, and 88 genes containing DNA-binding homeodomains in this genome of 97 million base pairs (Figure 3–16).

Proteins Can Adopt a Limited Number of Different Protein Folds

It is astounding to consider the rapidity of the increase in our knowledge about cells. In 1950, we did not know the order of the amino acids in a single protein, and many even doubted that the amino acids in proteins are arranged in an exact sequence. In 1960, the first three-dimensional structure of a protein was determined by x-ray crystallography. Now that we have access to hundreds of

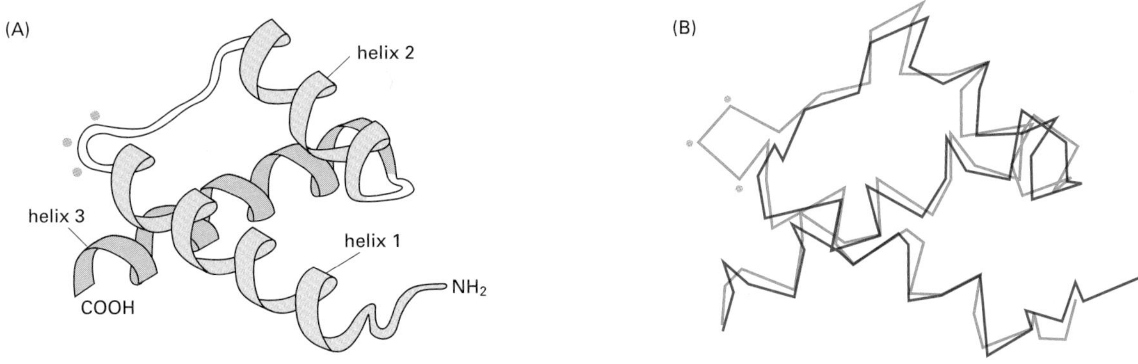

(C)

yeast
H_2N G H R F T K E N V R I L E S W F A K N I E N P Y L D T K G L E N L M K N T S L S R I Q I K N W V S N R R R K E K T I COOH
R T A F S S E O L A R L K R E F N E N - - - R Y L T E R R R Q Q L S S E L G L N E A Q I K I W F Q N K R A K I K K S
Drosophila

Figure 3–15 A comparison of a class of DNA-binding domains, called homeodomains, in a pair of proteins from two organisms separated by more than a billion years of evolution. (A) A ribbon model of the structure common to both proteins. (B) A trace of the α-carbon positions. The three-dimensional structures shown were determined by x-ray crystallography for the yeast α2 protein *(green)* and the *Drosophila* engrailed protein *(red).* (C) A comparison of amino acid sequences for the region of the proteins shown in (A) and (B). *Black dots* mark sites with identical amino acids. *Orange dots* indicate the position of a three amino acid insert in the α2 protein. (Adapted from C. Wolberger et al., *Cell* 67:517–528, 1991.)

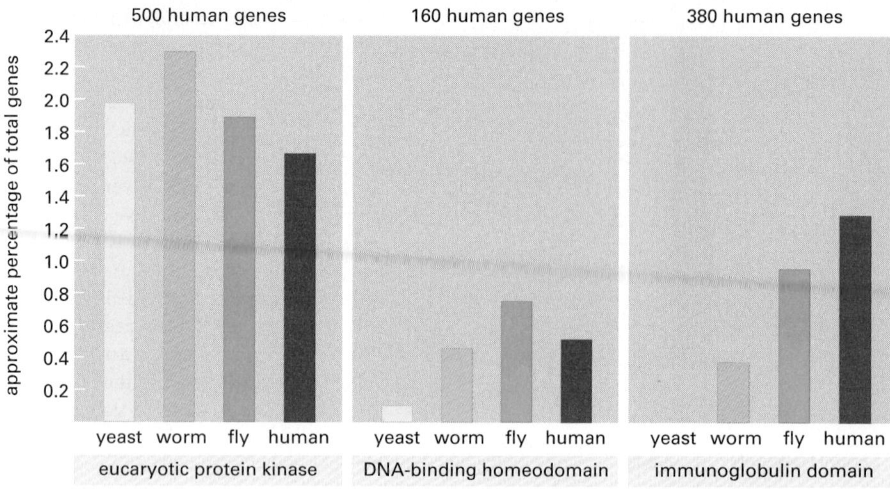

Figure 3–16 Percentage of total genes containing one or more copies of the indicated protein domain, as derived from complete genome sequences. Note that one of the three domains selected, the immunoglobulin domain, has been a relatively late addition, and its relative abundance has increased in the vertebrate lineage. The estimates of human gene numbers are approximate.

thousands of protein sequences from sequencing the genes that encode them, what technical developments can we look forward to next?

It is no longer a big step to progress from a gene sequence to the production of large amounts of the pure protein encoded by that gene. Thanks to DNA cloning and genetic engineering techniques (discussed in Chapter 8), this step is often routine. But there is still nothing routine about determining the complete three-dimensional structure of a protein. The standard technique based on x-ray diffraction requires that the protein be subjected to conditions that cause the molecules to aggregate into a large, perfectly ordered crystalline array—that is, a protein crystal. Each protein behaves quite differently in this respect, and protein crystals can be generated only through exhaustive trial-and-error methods that often take many years to succeed—if they succeed at all.

Membrane proteins and large protein complexes with many moving parts have generally been the most difficult to crystallize, which is why only a few such protein structures are displayed in this book. Increasingly, therefore, large proteins have been analyzed through determination of the structures of their individual domains: either by crystallizing isolated domains and then bombarding the crystals with x-rays, or by studying the conformations of isolated domains in concentrated aqueous solutions with powerful nuclear magnetic resonance (NMR) techniques (discussed in Chapter 8). From a combination of x-ray and NMR studies, we now know the three-dimensional shapes, or conformations, of thousands of different proteins.

By carefully comparing the conformations of known proteins, structural biologists (that is, experts on the structure of biological molecules) have concluded that there are a limited number of ways in which protein domains fold up—maybe as few as 2000. As we saw, the structures for about 1000 of these protein folds have thus far been determined; we may, therefore, already know half of the total number of possible structures for a protein domain. A complete catalog of all of the protein folds that exist in living organisms would therefore seem to be within our reach.

Sequence Homology Searches Can Identify Close Relatives

The present database of known protein sequences contains more than 500,000 entries, and it is growing very rapidly as more and more genomes are sequenced—revealing huge numbers of new genes that encode proteins. Powerful computer search programs are available that allow one to compare each newly discovered protein with this entire database, looking for possible relatives. Homologous proteins are defined as those whose genes have evolved from a common ancestral gene, and these are identified by the discovery of statistically significant similarities in amino acid sequences.

With such a large number of proteins in the database, the search programs find many nonsignificant matches, resulting in a background noise level that makes it very difficult to pick out all but the closest relatives. Generally speaking, a 30% identity in the sequence of two proteins is needed to be certain that a match has been found. However, many short signature sequences ("fingerprints") indicative of particular protein functions are known, and these are widely used to find more distant homologies (Figure 3–17).

These protein comparisons are important because related structures often imply related functions. Many years of experimentation can be saved by discovering that a new protein has an amino acid sequence homology with a protein of known function. Such sequence homologies, for example, first indicated that certain genes that cause mammalian cells to become cancerous are protein kinases. In the same way, many of the proteins that control pattern formation during the embryonic development of the fruit fly *Drosophila* were quickly recognized to be gene regulatory proteins.

Computational Methods Allow Amino Acid Sequences to Be Threaded into Known Protein Folds

We know that there are an enormous number of ways to make proteins with the same three-dimensional structure, and that—over evolutionary time—random mutations can cause amino acid sequences to change without a major change in the conformation of a protein. For this reason, one current goal of structural biologists is to determine all the different protein folds that proteins have in nature, and to devise computer-based methods to test the amino acid sequence of a domain to identify which one of these previously determined conformations the domain is likely to adopt.

A computational technique called threading can be used to fit an amino acid sequence to a particular protein fold. For each possible fold known, the computer searches for the best fit of the particular amino acid sequence to that structure. Are the hydrophobic residues on the inside? Are the sequences with a strong propensity to form an α helix in an α helix? And so on. The best fit gets a numerical score reflecting the estimated stability of the structure.

In many cases, one particular three-dimensional structure will stand out as a good fit for the amino acid sequence, suggesting an approximate conformation for the protein domain. In other cases, none of the known folds will seem possible. By applying x-ray and NMR studies to the latter class of proteins, structural biologists hope to able to expand the number of known folds rapidly, aiming for a database that contains the complete library of protein folds that exist in nature. With such a library, plus expected improvements in the computational methods used for threading, it may eventually become possible to obtain an approximate three-dimensional structure for a protein as soon as its amino acid sequence is known.

Some Protein Domains, Called Modules, Form Parts of Many Different Proteins

As previously stated, most proteins are composed of a series of protein domains, in which different regions of the polypeptide chain have folded independently to form compact structures. Such multidomain proteins are believed to have originated when the DNA sequences that encode each domain accidentally became joined, creating a new gene. Novel binding surfaces have often been created at the juxtaposition of domains, and many of the functional sites where

Figure 3–17 The use of short signature sequences to find homologous protein domains. The two short sequences of 15 and 9 amino acids shown *(green)* can be used to search large databases for a protein domain that is found in many proteins, the SH2 domain. Here, the first 50 amino acids of the SH2 domain of 100 amino acids is compared for the human and *Drosophila* Src protein (see Figure 3–12). In the computer-generated sequence comparison *(yellow row)*, exact matches between the human and *Drosophila* proteins are noted by the one-letter abbreviation for the amino acid; the positions with a similar but nonidentical amino acid are denoted by +, and nonmatches are blank. In this diagram, wherever one or both proteins contain an exact match to a position in the *green* sequences, both aligned sequences are colored *red*.

proteins bind to small molecules are found to be located there (for an example see Figure 3–12). Many large proteins show clear signs of having evolved by the joining of preexisting domains in new combinations, an evolutionary process called *domain shuffling* (Figure 3–18).

A subset of protein domains have been especially mobile during evolution; these so-called **protein modules** are generally somewhat smaller (40–200 amino acids) than an average domain, and they seem to have particularly versatile structures. The structure of one such module, the SH2 domain, was illustrated in Panel 3–2 (pp. 138–139). The structures of some additional protein modules are illustrated in Figure 3–19.

Each of the modules shown has a stable core structure formed from strands of β sheet, from which less-ordered loops of polypeptide chain protrude *(green)*. The loops are ideally situated to form binding sites for other molecules, as most flagrantly demonstrated for the immunoglobulin fold, which forms the basis for antibody molecules (see Figure 3–42). The evolutionary success of such β-sheet-based modules is likely to have been due to their providing a convenient framework for the generation of new binding sites for ligands through small changes to these protruding loops.

A second feature of protein modules that explains their utility is the ease with which they can be integrated into other proteins. Five of the six modules illustrated in Figure 3–19 have their N- and C-terminal ends at opposite poles of the module. This "in-line" arrangement means that when the DNA encoding such a module undergoes tandem duplication, which is not unusual in the evolution of genomes (discussed in Chapter 7), the duplicated modules can be readily linked in series to form extended structures—either with themselves or with other in-line modules (Figure 3–20). Stiff extended structures composed of a series of modules are especially common in extracellular matrix molecules and in the extracellular portions of cell-surface receptor proteins. Other modules, including the SH2 domain and the kringle module illustrated in Figure 3–19, are

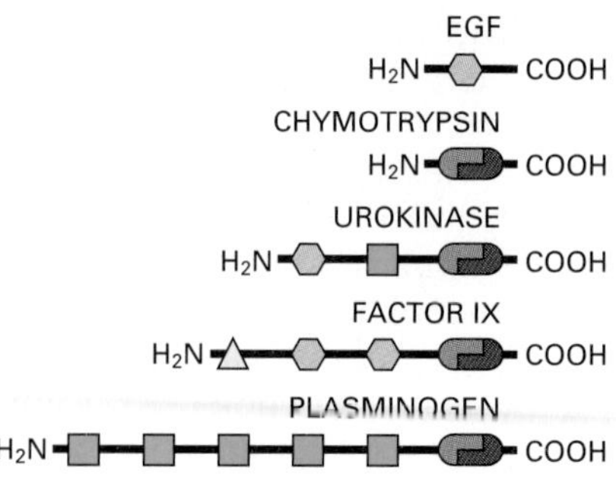

Figure 3–18 Domain shuffling. An extensive shuffling of blocks of protein sequence (protein domains) has occurred during protein evolution. Those portions of a protein denoted by the same shape and color in this diagram are evolutionarily related. Serine proteases like chymotrypsin are formed from two domains *(brown)*. In the three other proteases shown, which are highly regulated and more specialized, these two protease domains are connected to one or more domains homologous to domains found in epidermal growth factor (EGF; *green*), to a calcium-binding protein *(yellow)*, or to a "kringle" domain *(blue)* that contains three internal disulfide bridges. Chymotrypsin is illustrated in Figure 3–14.

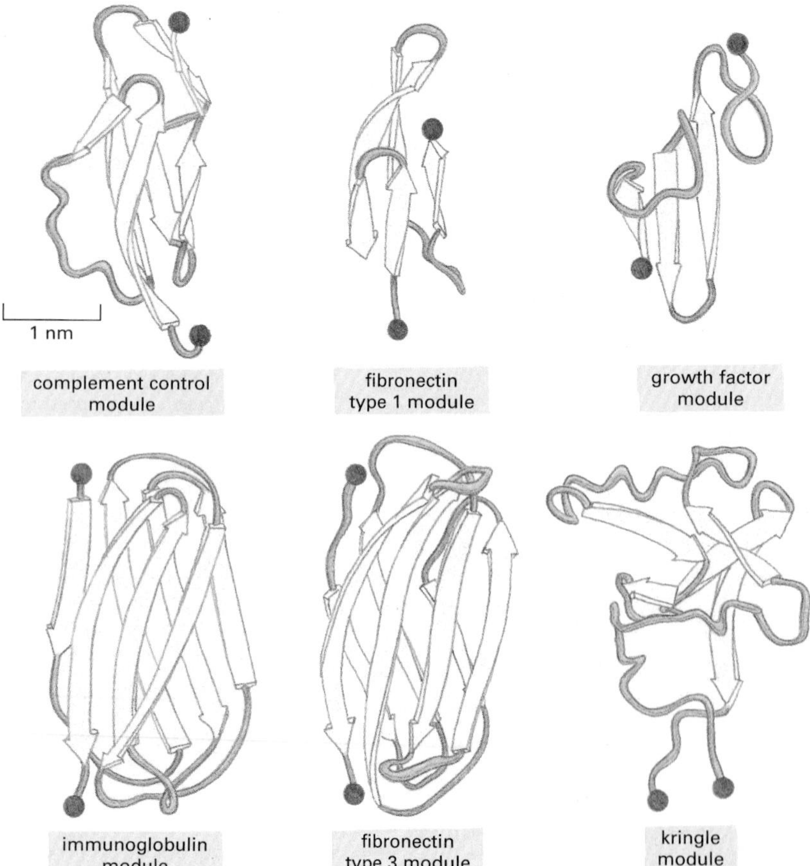

Figure 3–19 The three-dimensional structures of some protein modules. In these ribbon diagrams, β-sheet strands are shown as *arrows*, and the N- and C-termini are indicated by *red spheres*. (Adapted from M. Baron, D.G. Norman, and I.D. Campbell, *Trends Biochem. Sci.* 16:13–17, 1991, and D.J. Leahy et al., *Science* 258:987–991, 1992.)

of a "plug-in" type. After genomic rearrangements, such modules are usually accommodated as an insertion into a loop region of a second protein.

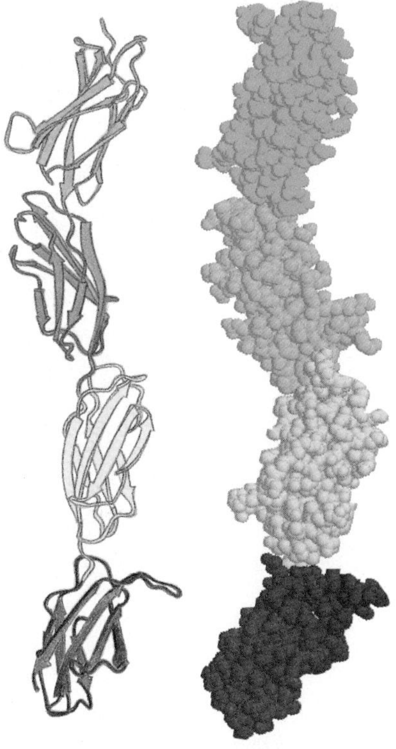

Figure 3–20 An extended structure formed from a series of in-line protein modules. Four fibronectin type 3 modules (see Figure 3–19) from the extracellular matrix molecule fibronectin are illustrated in (A) ribbon and (B) space-filling models. (Adapted from D.J. Leahy, I. Aukhil, and H.P. Erickson, *Cell* 84:155–164, 1996.)

The Human Genome Encodes a Complex Set of Proteins, Revealing Much That Remains Unknown

The result of sequencing the human genome has been surprising, because it reveals that our chromosomes contain only 30,000 to 35,000 genes. With regard to gene number, we would appear to be no more than 1.4-fold more complex than the tiny mustard weed, *Arabidopsis,* and less than 2-fold more complex than a nematode worm. The genome sequences also reveal that vertebrates have inherited nearly all of their protein domains from invertebrates—with only 7 percent of identified human domains being vertebrate-specific.

Each of our proteins is on average more complicated, however. A process of domain shuffling during vertebrate evolution has given rise to many novel combinations of protein domains, with the result that there are nearly twice as many combinations of domains found in human proteins as in a worm or a fly. Thus, for example, the trypsinlike serine protease domain is linked to at least 18 other types of protein domains in human proteins, whereas it is found covalently joined to only 5 different domains in the worm. This extra variety in our proteins greatly increases the range of protein–protein interactions possible (see Figure 3–78), but how it contributes to making us human is not known.

The complexity of living organisms is staggering, and it is quite sobering to note that we currently lack even the tiniest hint of what the function might be for more than 10,000 of the proteins that have thus far been identified in the human genome. There are certainly enormous challenges ahead for the next generation of cell biologists, with no shortage of fascinating mysteries to solve.

Larger Protein Molecules Often Contain More Than One Polypeptide Chain

The same weak noncovalent bonds that enable a protein chain to fold into a specific conformation also allow proteins to bind to each other to produce larger structures in the cell. Any region of a protein's surface that can interact with another molecule through sets of noncovalent bonds is called a **binding site**. A protein can contain binding sites for a variety of molecules, both large and small. If a binding site recognizes the surface of a second protein, the tight binding of two folded polypeptide chains at this site creates a larger protein molecule with a precisely defined geometry. Each polypeptide chain in such a protein is called a **protein subunit**.

In the simplest case, two identical folded polypeptide chains bind to each other in a "head-to-head" arrangement, forming a symmetric complex of two protein subunits (a *dimer*) held together by interactions between two identical binding sites. The *Cro repressor protein*—a gene regulatory protein that binds to DNA to turn genes off in a bacterial cell—provides an example (Figure 3–21). Many other types of symmetric protein complexes, formed from multiple copies of a single polypeptide chain, are commonly found in cells. The enzyme

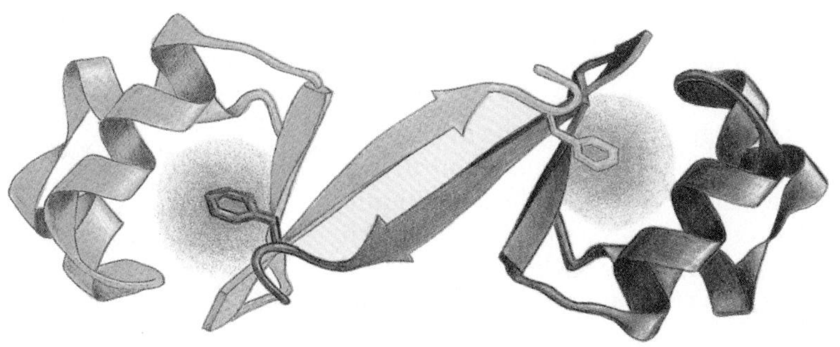

Figure 3–21 Two identical protein subunits binding together to form a symmetric protein dimer. The Cro repressor protein from bacteriophage lambda binds to DNA to turn off viral genes. Its two identical subunits bind head-to-head, held together by a combination of hydrophobic forces *(blue)* and a set of hydrogen bonds *(yellow* region). (Adapted from D.H. Ohlendorf, D.E. Tronrud, and B.W. Matthews, *J. Mol. Biol.* 280:129–136, 1998.)

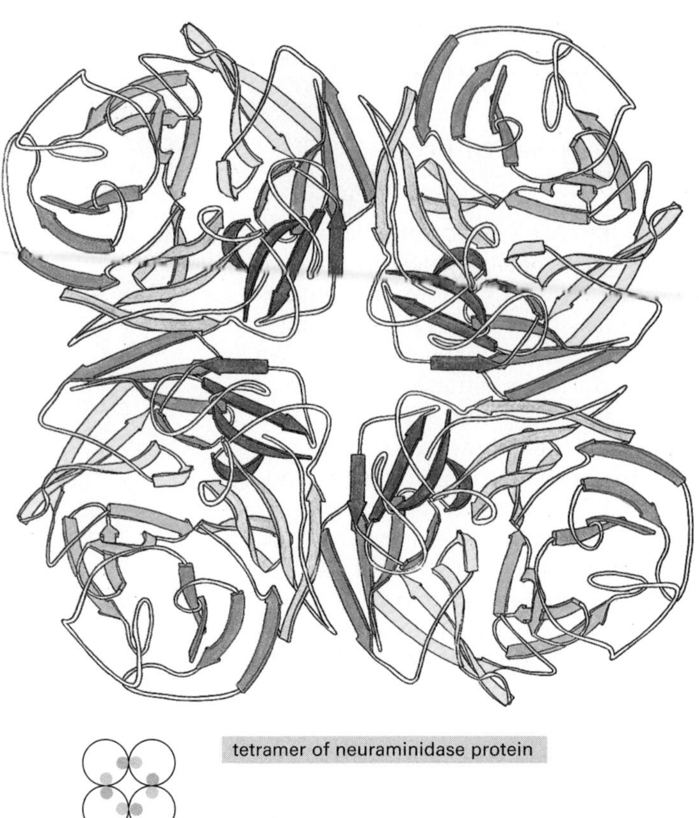

Figure 3–22 A protein molecule containing multiple copies of a single protein subunit. The enzyme neuraminidase exists as a ring of four identical polypeptide chains. The small diagram shows how the repeated use of the same binding interaction forms the structure.

neuraminidase, for example, consists of four identical protein subunits, each bound to the next in a "head-to-tail" arrangement that forms a closed ring (Figure 3–22).

Many of the proteins in cells contain two or more types of polypeptide chains. *Hemoglobin,* the protein that carries oxygen in red blood cells, is a particularly well-studied example (Figure 3–23). It contains two identical α-globin subunits and two identical β-globin subunits, symmetrically arranged. Such multisubunit proteins are very common in cells, and they can be very large. Figure 3–24 provides a sampling of proteins whose exact structures are known, allowing the sizes and shapes of a few larger proteins to be compared with the relatively small proteins that we have thus far presented as models.

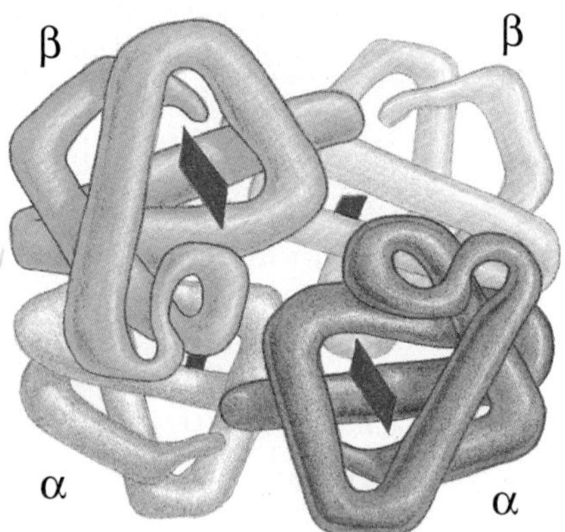

Figure 3–23 A protein formed as a symmetric assembly of two different subunits. Hemoglobin is an abundant protein in red blood cells that contains two copies of α globin and two copies of β globin. Each of these four polypeptide chains contains a heme molecule *(red),* which is the site where oxygen (O_2) is bound. Thus, each molecule of hemoglobin in the blood carries four molecules of oxygen.

Some Proteins Form Long Helical Filaments

Some protein molecules can assemble to form filaments that may span the entire length of a cell. Most simply, a long chain of identical protein molecules can be constructed if each molecule has a binding site complementary to another region of the surface of the same molecule (Figure 3–25). An actin filament, for example, is a long helical structure produced from many molecules of the protein *actin* (Figure 3–26). Actin is very abundant in eucaryotic cells, where it constitutes one of the major filament systems of the cytoskeleton (discussed in Chapter 16).

Why is a helix such a common structure in biology? As we have seen, biological structures are often formed by linking subunits that are very similar to each other—such as amino acids or protein molecules—into long, repetitive chains. If all the subunits are identical, the neighboring subunits in the chain can often fit together in only one way, adjusting their relative positions to minimize the free energy of the contact between them. As a result, each subunit is positioned in exactly the same way in relation to the next, so that subunit 3 fits onto subunit 2 in the same way that subunit 2 fits onto subunit 1, and so on. Because it is very rare for subunits to join up in a straight line, this arrangement generally results in a helix—a regular structure that resembles a spiral staircase,

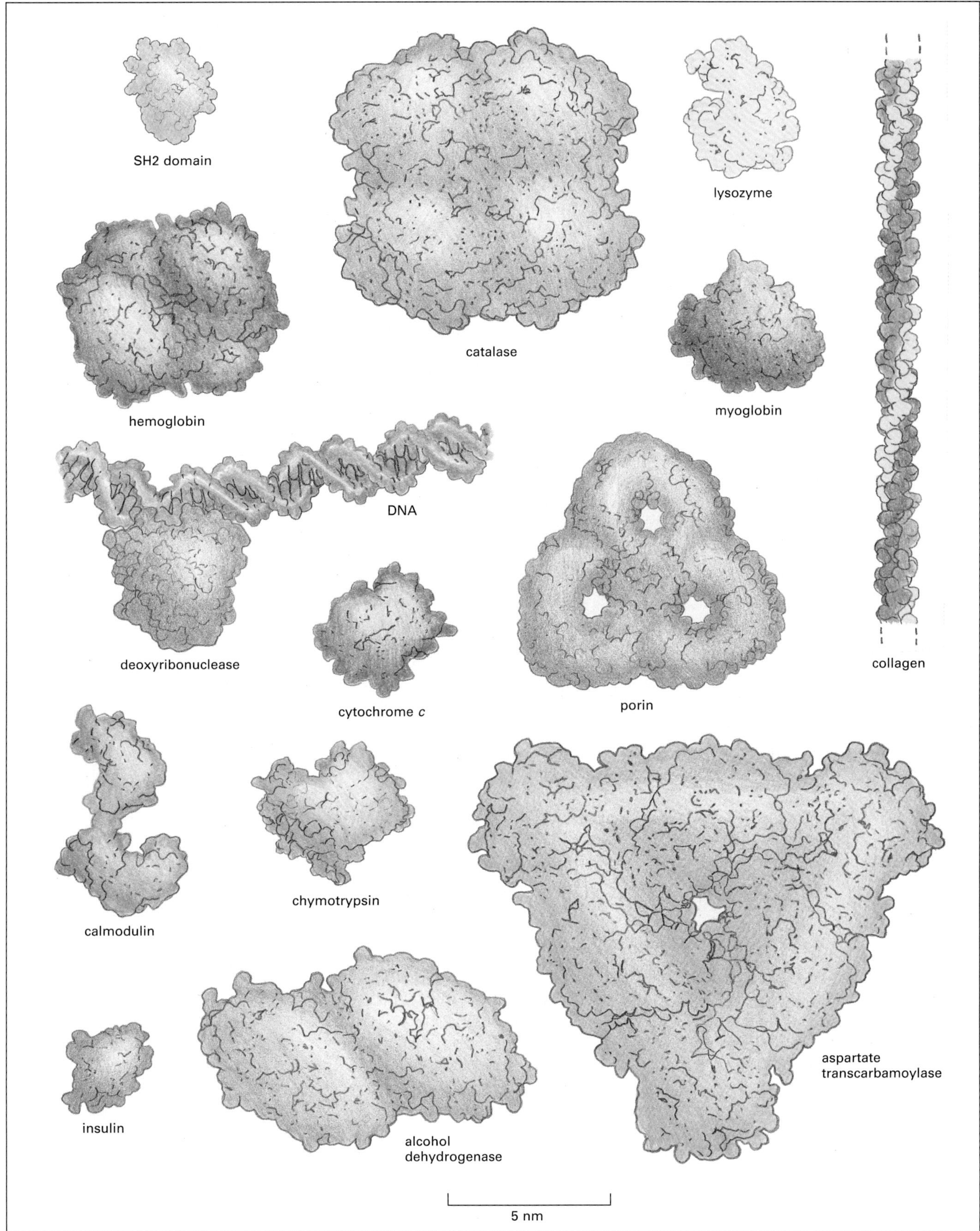

Figure 3–24 A collection of protein molecules, shown at the same scale. For comparison, a DNA molecule bound to a protein is also illustrated. These space-filling models represent a range of sizes and shapes. Hemoglobin, catalase, porin, alcohol dehydrogenase, and aspartate transcarbamoylase are formed from multiple copies of subunits. The SH2 domain *(top left)* is presented in detail in Panel 3–2 (pp. 138–139). (After David S. Goodsell, Our Molecular Nature. New York: Springer-Verlag, 1996.)

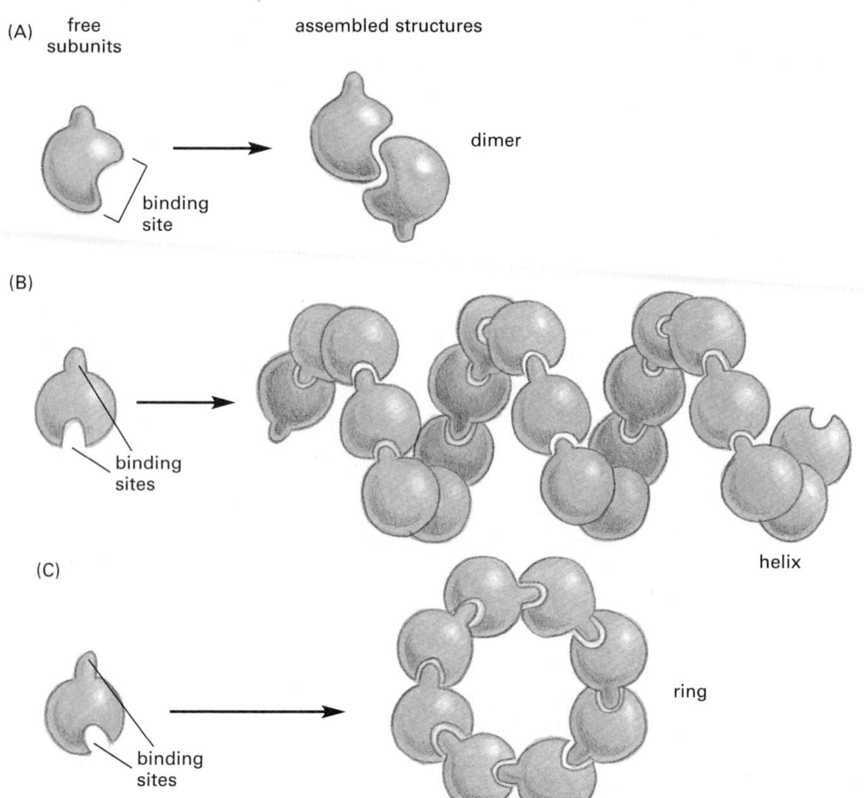

Figure 3–25 Protein assemblies. (A) A protein with just one binding site can form a dimer with another identical protein. (B) Identical proteins with two different binding sites often form a long helical filament. (C) If the two binding sites are disposed appropriately in relation to each other, the protein subunits may form a closed ring instead of a helix. (For an example of A, see Figure 3–21; for an example of C, see Figure 3–22.)

as illustrated in Figure 3–27. Depending on the twist of the staircase, a helix is said to be either right-handed or left-handed (Figure 3–27E). Handedness is not affected by turning the helix upside down, but it is reversed if the helix is reflected in the mirror.

Helices occur commonly in biological structures, whether the subunits are small molecules linked together by covalent bonds (for example, the amino acids in an α helix) or large protein molecules that are linked by noncovalent forces (for example, the actin molecules in actin filaments). This is not surprising. A helix is an unexceptional structure, and it is generated simply by placing many similar subunits next to each other, each in the same strictly repeated relationship to the one before.

A Protein Molecule Can Have an Elongated, Fibrous Shape

Most of the proteins we have discussed so far are *globular proteins,* in which the polypeptide chain folds up into a compact shape like a ball with an irregular surface. Enzymes tend to be globular proteins: even though many are large and complicated, with multiple subunits, most have an overall rounded shape (see Figure 3–24). In contrast, other proteins have roles in the cell requiring each individual protein molecule to span a large distance. These proteins generally have a relatively simple, elongated three-dimensional structure and are commonly referred to as *fibrous proteins.*

One large family of intracellular fibrous proteins consists of α-keratin, introduced earlier, and its relatives. Keratin filaments are extremely stable and are the main component in long-lived structures such as hair, horn, and nails. An α-keratin molecule is a dimer of two identical subunits, with the long α helices of each subunit forming a coiled-coil (see Figure 3–11). The coiled-coil regions are capped at each end by globular domains containing binding sites. This enables this class of protein to assemble into ropelike *intermediate filaments*—an important component of the cytoskeleton that creates the cell's internal structural scaffold (see Figure 16–16).

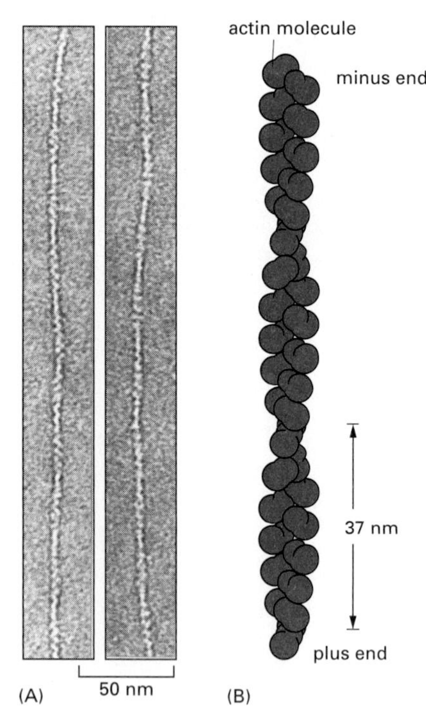

Figure 3–26 Actin filaments. (A) Transmission electron micrographs of negatively stained actin filaments. (B) The helical arrangement of actin molecules in an actin filament. (A, courtesy of Roger Craig.)

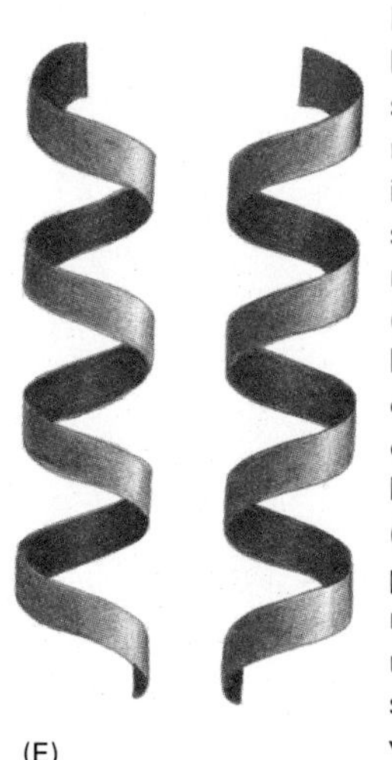

Figure 3–27 Some properties of a helix. (A–D) A helix forms when a series of subunits bind to each other in a regular way. At the bottom, the interaction between two subunits is shown; behind them are the helices that result. These helices have two (A), three (B), and six (C and D) subunits per helical turn. At the top, the arrangement of subunits has been photographed from directly above the helix. Note that the helix in (D) has a wider path than that in (C), but the same number of subunits per turn. (E) A helix can be either right-handed or left-handed. As a reference, it is useful to remember that standard metal screws, which insert when turned clockwise, are right-handed. Note that a helix retains the same handedness when it is turned upside down.

Fibrous proteins are especially abundant outside the cell, where they are a main component of the gel-like *extracellular matrix* that helps to bind collections of cells together to form tissues. Extracellular matrix proteins are secreted by the cells into their surroundings, where they often assemble into sheets or long fibrils. *Collagen* is the most abundant of these proteins in animal tissues. A collagen molecule consists of three long polypeptide chains, each containing the nonpolar amino acid glycine at every third position. This regular structure allows the chains to wind around one another to generate a long regular triple helix (Figure 3–28A). Many collagen molecules then bind to one another side-by-side and end-to-end to create long overlapping arrays—thereby generating the extremely tough collagen fibrils that give connective tissues their tensile strength, as described in Chapter 19.

In complete contrast to collagen is another protein in the extracellular matrix, *elastin*. Elastin molecules are formed from relatively loose and unstructured polypeptide chains that are covalently cross-linked into a rubberlike elastic meshwork: unlike most proteins, they do not have a uniquely defined stable structure, but can be reversibly pulled from one conformation to another, as illustrated in Figure 3–28B. The resulting elastic fibers enable skin and other tissues, such as arteries and lungs, to stretch and recoil without tearing.

Extracellular Proteins Are Often Stabilized by Covalent Cross-Linkages

Many protein molecules are either attached to the outside of a cell's plasma membrane or secreted as part of the extracellular matrix. All such proteins are directly exposed to extracellular conditions. To help maintain their structures, the polypeptide chains in such proteins are often stabilized by covalent cross-linkages. These linkages can either tie two amino acids in the same protein together, or connect different polypeptide chains in a multisubunit protein. The most common cross-linkages in proteins are covalent sulfur–sulfur bonds. These *disulfide bonds* (also called *S–S bonds*) form as proteins are being prepared for export from cells. As described in Chapter 12, their formation is catalyzed in the endoplasmic reticulum by an enzyme that links together two pairs of –SH groups of cysteine side chains that are adjacent in the folded protein (Figure 3–29). Disulfide bonds do not change the conformation of a protein but

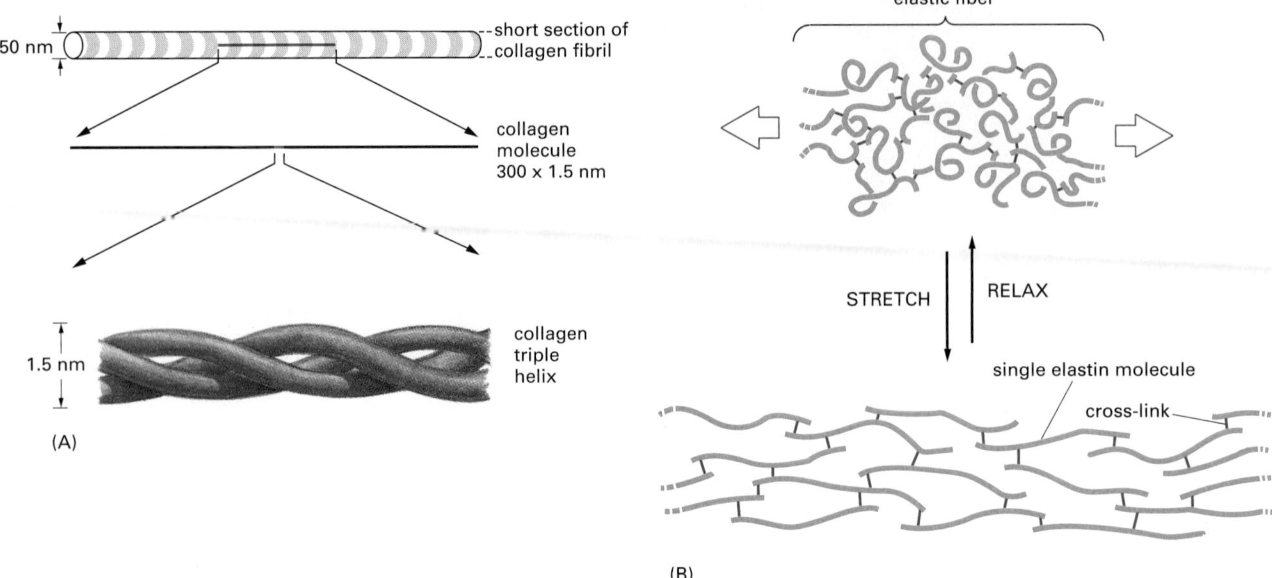

Figure 3–28 Collagen and elastin. (A) Collagen is a triple helix formed by three extended protein chains that wrap around one another *(bottom)*. Many rodlike collagen molecules are cross-linked together in the extracellular space to form unextendable collagen fibrils *(top)* that have the tensile strength of steel. The striping on the collagen fibril is caused by the regular repeating arrangement of the collagen molecules within the fibril. (B) Elastin polypeptide chains are cross-linked together to form rubberlike, elastic fibers. Each elastin molecule uncoils into a more extended conformation when the fiber is stretched and recoils spontaneously as soon as the stretching force is relaxed.

instead act as atomic staples to reinforce its most favored conformation. For example, lysozyme—an enzyme in tears that dissolves bacterial cell walls—retains its antibacterial activity for a long time because it is stabilized by such cross-linkages.

Disulfide bonds generally fail to form in the cell cytosol, where a high concentration of reducing agents converts S–S bonds back to cysteine –SH groups. Apparently, proteins do not require this type of reinforcement in the relatively mild environment inside the cell.

Protein Molecules Often Serve as Subunits for the Assembly of Large Structures

The same principles that enable a protein molecule to associate with itself to form rings or filaments operate to generate much larger structures in the cell—supramolecular structures such as enzyme complexes, ribosomes, protein filaments, viruses, and membranes. These large objects are not made as single, giant, covalently linked molecules. Instead they are formed by the noncovalent assembly of many separately manufactured molecules, which serve as the subunits of the final structure.

The use of smaller subunits to build larger structures has several advantages:

1. A large structure built from one or a few repeating smaller subunits requires only a small amount of genetic information.

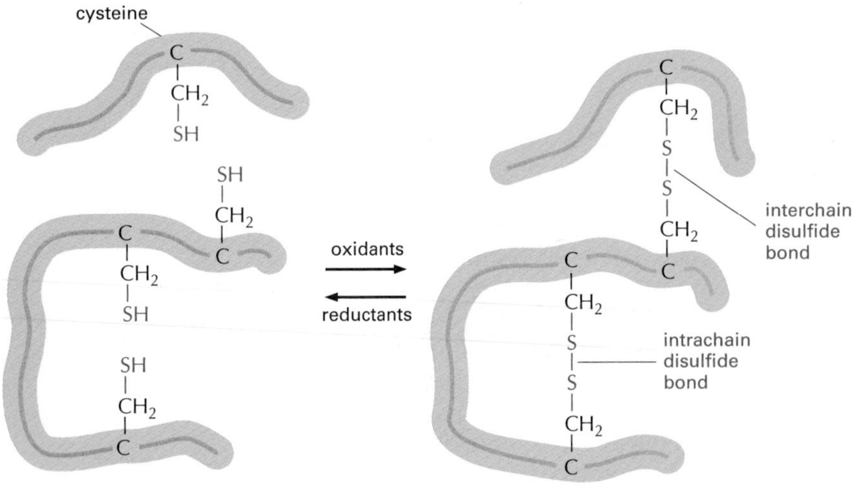

Figure 3–29 Disulfide bonds. This diagram illustrates how covalent disulfide bonds form between adjacent cysteine side chains. As indicated, these cross-linkages can join either two parts of the same polypeptide chain or two different polypeptide chains. Since the energy required to break one covalent bond is much larger than the energy required to break even a whole set of noncovalent bonds (see Table 2–2, p. 57), a disulfide bond can have a major stabilizing effect on a protein.

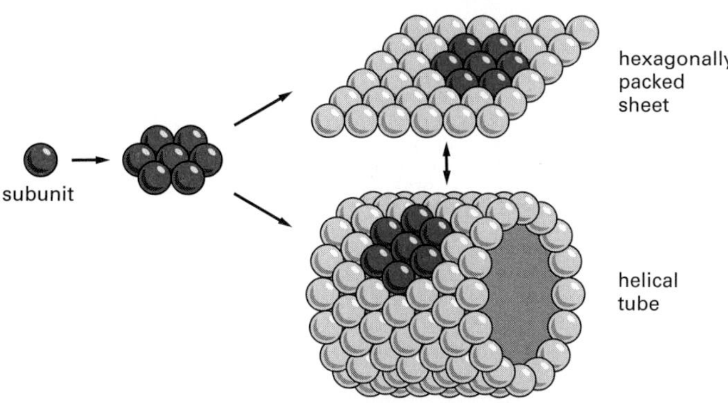

Figure 3–30 An example of single protein subunit assembly requiring multiple protein–protein contacts. Hexagonally packed globular protein subunits can form either a flat sheet or a tube.

2. Both assembly and disassembly can be readily controlled, reversible processes, since the subunits associate through multiple bonds of relatively low energy.
3. Errors in the synthesis of the structure can be more easily avoided, since correction mechanisms can operate during the course of assembly to exclude malformed subunits.

Some protein subunits assemble into flat sheets in which the subunits are arranged in hexagonal patterns. Specialized membrane proteins are sometimes arranged this way in lipid bilayers. With a slight change in the geometry of the individual subunits, a hexagonal sheet can be converted into a tube (Figure 3–30) or, with more changes, into a hollow sphere. Protein tubes and spheres that bind specific RNA and DNA molecules form the coats of viruses.

The formation of closed structures, such as rings, tubes, or spheres, provides additional stability because it increases the number of bonds between the protein subunits. Moreover, because such a structure is created by mutually dependent, cooperative interactions between subunits, it can be driven to assemble or disassemble by a relatively small change that affects each subunit individually. These principles are dramatically illustrated in the protein coat or *capsid* of many simple viruses, which takes the form of a hollow sphere (Figure 3–31).

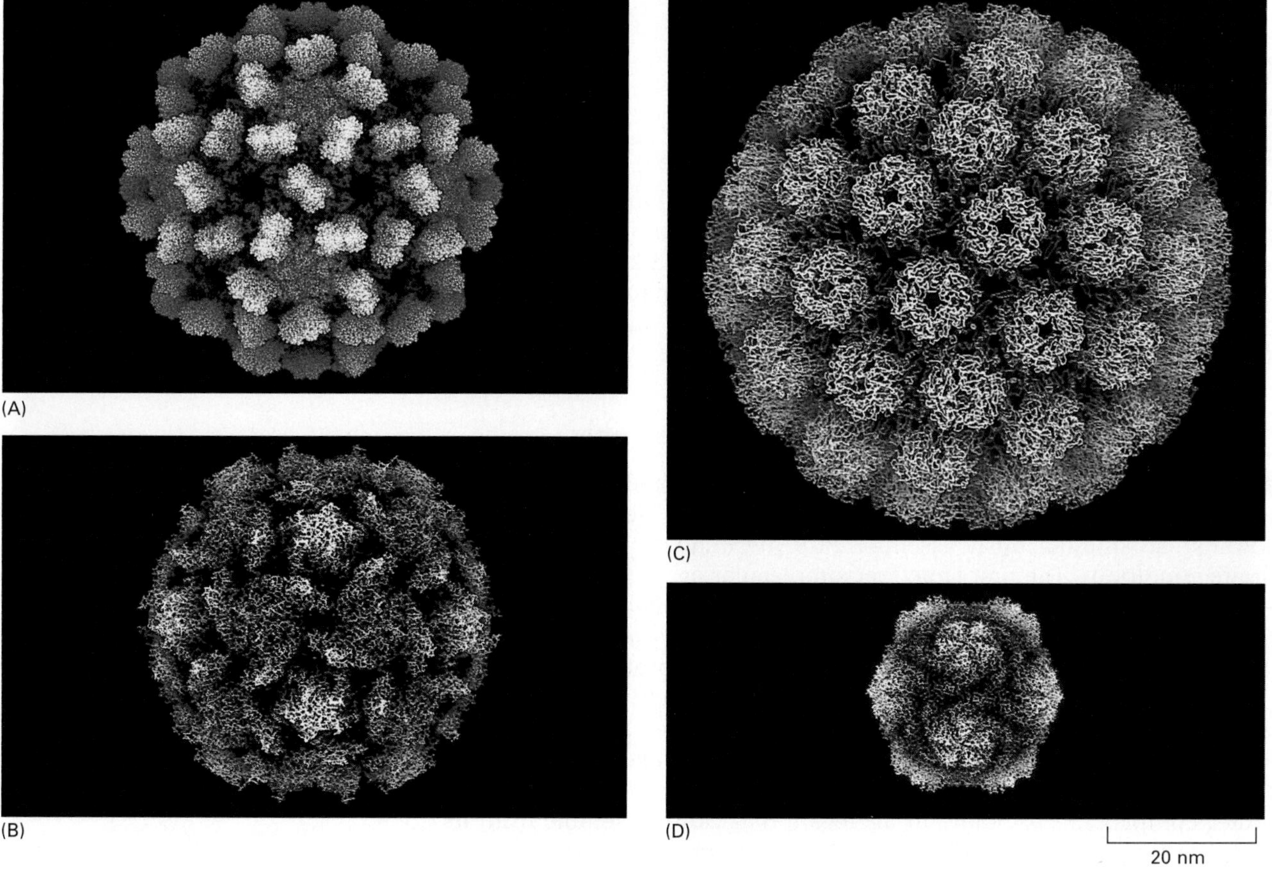

Figure 3–31 The capsids of some viruses, all shown at the same scale. (A) Tomato bushy stunt virus; (B) poliovirus; (C) simian virus 40 (SV40); (D) satellite tobacco necrosis virus. The structures of all of these capsids have been determined by x-ray crystallography and are known in atomic detail. (Courtesy of Robert Grant, Stephan Crainic, and James M. Hogle.)

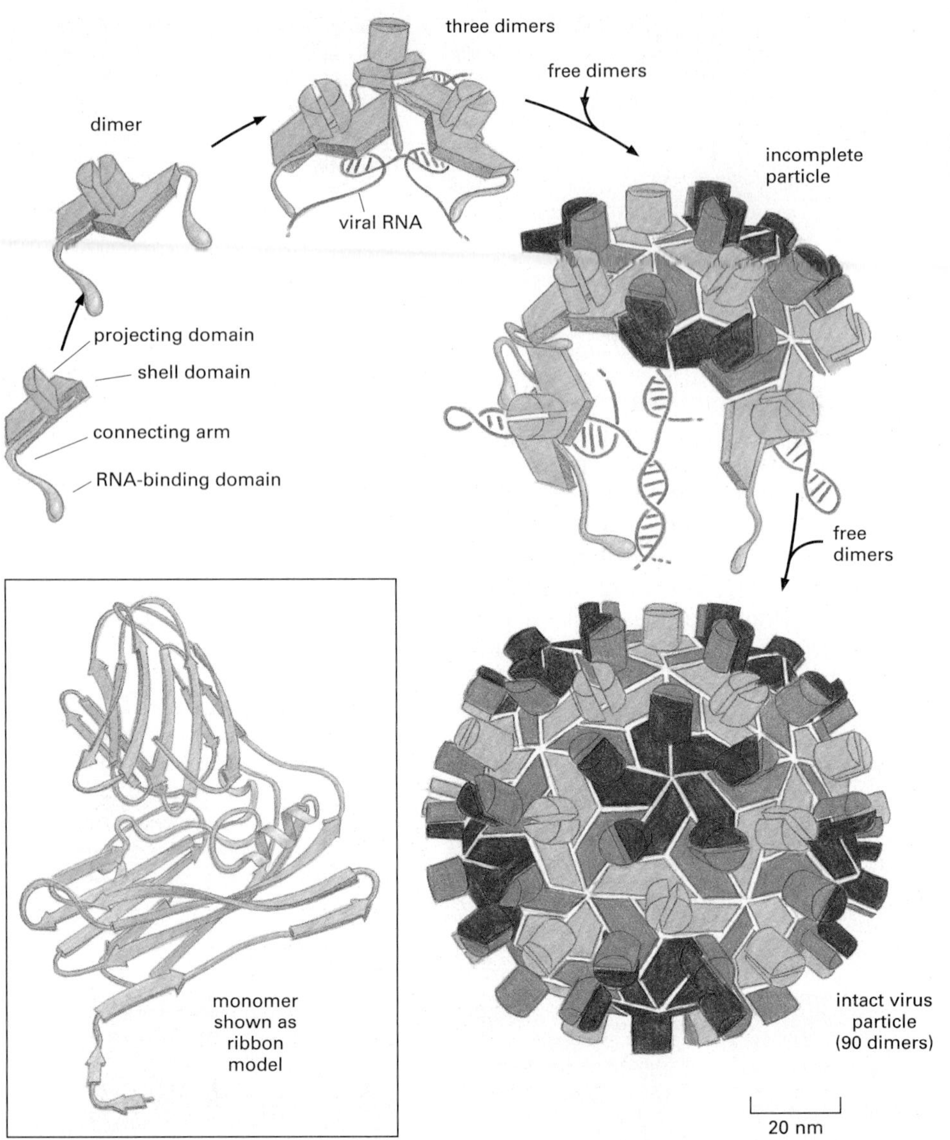

Figure 3–32 The structure of a spherical virus. In many viruses, identical protein subunits pack together to create a spherical shell (a capsid) that encloses the viral genome, composed of either RNA or DNA (see also Figure 3–31). For geometric reasons, no more than 60 identical subunits can pack together in a precisely symmetric way. If slight irregularities are allowed, however, more subunits can be used to produce a larger capsid. The tomato bushy stunt virus (TBSV) shown here, for example, is a spherical virus about 33 nm in diameter formed from 180 identical copies of a 386 amino acid capsid protein plus an RNA genome of 4500 nucleotides. To construct such a large capsid, the protein must be able to fit into three somewhat different environments, each of which is differently colored in the virus particle shown here. The postulated pathway of assembly is shown; the precise three-dimensional structure has been determined by x-ray diffraction. (Courtesy of Steve Harrison.)

Capsids are often made of hundreds of identical protein subunits that enclose and protect the viral nucleic acid (Figure 3–32). The protein in such a capsid must have a particularly adaptable structure: it must not only make several different kinds of contacts to create the sphere, it must also change this arrangement to let the nucleic acid out to initiate viral replication once the virus has entered a cell.

Many Structures in Cells Are Capable of Self-Assembly

The information for forming many of the complex assemblies of macromolecules in cells must be contained in the subunits themselves, because purified subunits can spontaneously assemble into the final structure under the appropriate conditions. The first large macromolecular aggregate shown to be capable of self-assembly from its component parts was *tobacco mosaic virus (TMV)*. This virus is a long rod in which a cylinder of protein is arranged around a helical RNA core (Figure 3–33). If the dissociated RNA and protein subunits are mixed together in solution, they recombine to form fully active viral particles. The assembly process is unexpectedly complex and includes the formation of double rings of protein, which serve as intermediates that add to the growing viral coat.

Another complex macromolecular aggregate that can reassemble from its component parts is the bacterial ribosome. This structure is composed of about

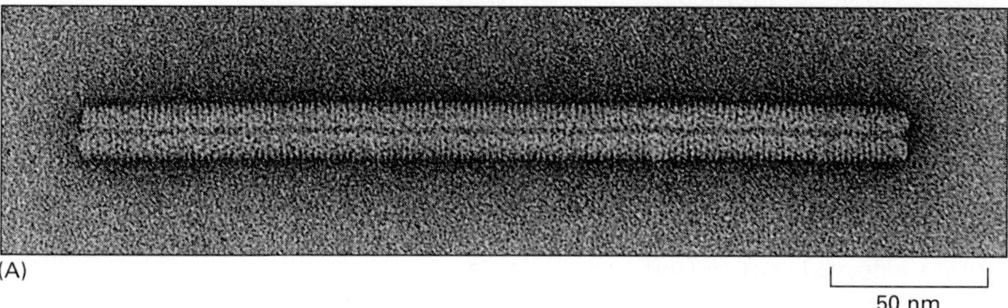

Figure 3–33 The structure of tobacco mosaic virus (TMV). (A) An electron micrograph of the viral particle, which consists of a single long RNA molecule enclosed in a cylindrical protein coat composed of identical protein subunits. (B) A model showing part of the structure of TMV. A single-stranded RNA molecule of 6000 nucleotides is packaged in a helical coat constructed from 2130 copies of a coat protein 158 amino acids long. Fully infective viral particles can self-assemble in a test tube from purified RNA and protein molecules. (A, courtesy of Robley Williams; B, courtesy of Richard J. Feldmann.)

55 different protein molecules and 3 different rRNA molecules. If the individual components are incubated under appropriate conditions in a test tube, they spontaneously re-form the original structure. Most importantly, such reconstituted ribosomes are able to perform protein synthesis. As might be expected, the reassembly of ribosomes follows a specific pathway: after certain proteins have bound to the RNA, this complex is then recognized by other proteins, and so on, until the structure is complete.

It is still not clear how some of the more elaborate self-assembly processes are regulated. Many structures in the cell, for example, seem to have a precisely defined length that is many times greater than that of their component macromolecules. How such length determination is achieved is in many cases a mystery. Three possible mechanisms are illustrated in Figure 3–34. In the simplest case, a long core protein or other macromolecule provides a scaffold that determines the extent of the final assembly. This is the mechanism that determines the length of the TMV particle, where the RNA chain provides the core. Similarly, a core protein is thought to determine the length of the thin filaments in muscle, as well as the length of the long tails of some bacterial viruses (Figure 3–35).

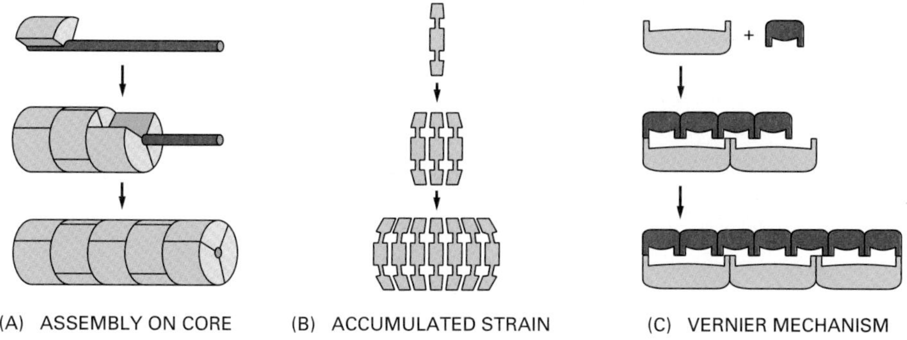

Figure 3–34 Three mechanisms of length determination for large protein assemblies. (A) Coassembly along an elongated core protein or other macromolecule that acts as a measuring device. (B) Termination of assembly because of strain that accumulates in the polymeric structure as additional subunits are added, so that beyond a certain length the energy required to fit another subunit onto the chain becomes excessively large. (C) A vernier type of assembly, in which two sets of rodlike molecules differing in length form a staggered complex that grows until their ends exactly match. The name derives from a measuring device based on the same principle, used in mechanical instruments.

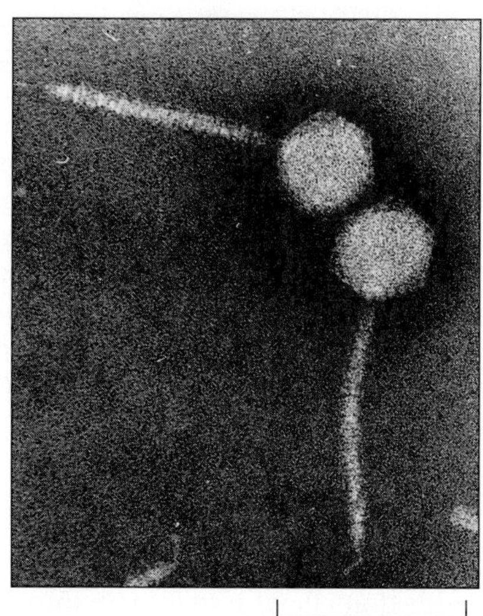

Figure 3–35 An electron micrograph of bacteriophage lambda. The tip of the virus tail attaches to a specific protein on the surface of a bacterial cell, after which the tightly packaged DNA in the head is injected through the tail into the cell. The tail has a precise length, determined by the mechanism shown in Figure 3–34A.

The Formation of Complex Biological Structures Is Often Aided by Assembly Factors

Not all cellular structures held together by noncovalent bonds are capable of self-assembly. A mitochondrion, a cilium, or a myofibril of a muscle cell, for example, cannot form spontaneously from a solution of its component macromolecules. In these cases, part of the assembly information is provided by special enzymes and other cellular proteins that perform the function of templates, guiding construction but taking no part in the final assembled structure.

Even relatively simple structures may lack some of the ingredients necessary for their own assembly. In the formation of certain bacterial viruses, for example, the head, which is composed of many copies of a single protein subunit, is assembled on a temporary scaffold composed of a second protein. Because the second protein is absent from the final viral particle, the head structure cannot spontaneously reassemble once it has been taken apart. Other examples are known in which proteolytic cleavage is an essential and irreversible step in the normal assembly process. This is even the case for some small protein assemblies, including the structural protein collagen and the hormone insulin (Figure 3–36). From these relatively simple examples, it seems very likely that the assembly of a structure as complex as a mitochondrion or a cilium will involve temporal and spatial ordering imparted by numerous other cell components.

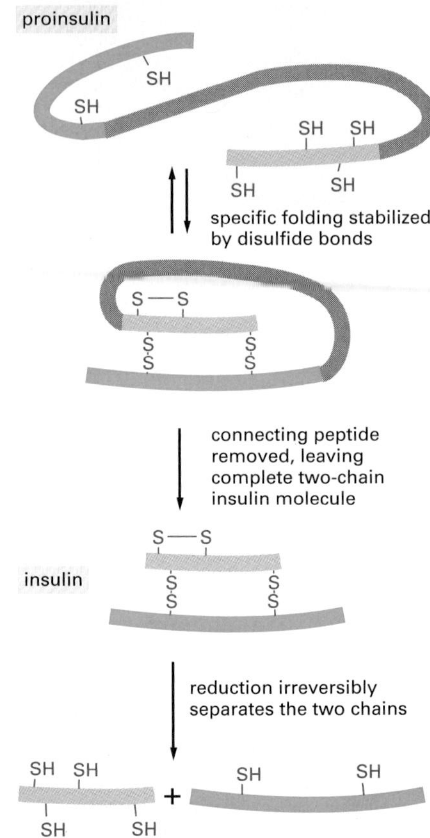

Figure 3–36 Proteolytic cleavage in insulin assembly. The polypeptide hormone insulin cannot spontaneously re-form efficiently if its disulfide bonds are disrupted. It is synthesized as a larger protein *(proinsulin)* that is cleaved by a proteolytic enzyme after the protein chain has folded into a specific shape. Excision of part of the proinsulin polypeptide chain removes some of the information needed for the protein to fold spontaneously into its normal conformation once it has been denatured and its two polypeptide chains separated.

Summary

The three-dimensional conformation of a protein molecule is determined by its amino acid sequence. The folded structure is stabilized by noncovalent interactions between different parts of the polypeptide chain. The amino acids with hydrophobic side chains tend to cluster in the interior of the molecule, and local hydrogen-bond interactions between neighboring peptide bonds give rise to α helices and β sheets.

Globular regions, known as domains, are the modular units from which many proteins are constructed; such domains generally contain 40–350 amino acids. Small proteins typically consist of only a single domain, while large proteins are formed from several domains linked together by short lengths of polypeptide chain. As proteins have evolved, domains have been modified and combined with other domains to construct new proteins. Domains that participate in the formation of large numbers of proteins are known as protein modules. Thus far, about 1000 different ways of folding up a domain have been observed, among more than about 10,000 known protein structures.

Proteins are brought together into larger structures by the same noncovalent forces that determine protein folding. Proteins with binding sites for their own surface can assemble into dimers, closed rings, spherical shells, or helical polymers. Although mixtures of proteins and nucleic acids can assemble spontaneously into complex structures in a test tube, many biological assembly processes involve irreversible steps. Consequently, not all structures in the cell are capable of spontaneous reassembly after they have been dissociated into their component parts.

PROTEIN FUNCTION

We have seen that each type of protein consists of a precise sequence of amino acids that allows it to fold up into a particular three-dimensional shape, or conformation. But proteins are not rigid lumps of material. They can have precisely engineered moving parts whose mechanical actions are coupled to chemical events. It is this coupling of chemistry and movement that gives proteins the extraordinary capabilities that underlie the dynamic processes in living cells.

In this section, we explain how proteins bind to other selected molecules and how their activity depends on such binding. We show that the ability to bind to other molecules enables proteins to act as catalysts, signal receptors, switches, motors, or tiny pumps. The examples we discuss in this chapter by no means exhaust the vast functional repertoire of proteins. However, the specialized

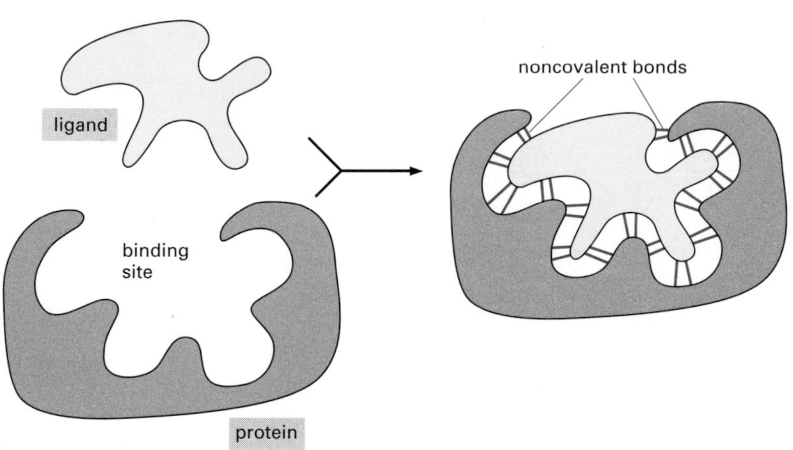

Figure 3–37 The selective binding of a protein to another molecule. Many weak bonds are needed to enable a protein to bind tightly to a second molecule, which is called a *ligand* for the protein. A ligand must therefore fit precisely into a protein's binding site, like a hand into a glove, so that a large number of noncovalent bonds can be formed between the protein and the ligand.

functions of many of the proteins you will encounter elsewhere in this book are based on similar principles.

All Proteins Bind to Other Molecules

The biological properties of a protein molecule depend on its physical interaction with other molecules. Thus, antibodies attach to viruses or bacteria to mark them for destruction, the enzyme hexokinase binds glucose and ATP so as to catalyze a reaction between them, actin molecules bind to each other to assemble into actin filaments, and so on. Indeed, all proteins stick, or *bind*, to other molecules. In some cases, this binding is very tight; in others, it is weak and short-lived. But the binding always shows great *specificity*, in the sense that each protein molecule can usually bind just one or a few molecules out of the many thousands of different types it encounters. The substance that is bound by the protein—no matter whether it is an ion, a small molecule, or a macromolecule— is referred to as a **ligand** for that protein (from the Latin word *ligare*, meaning "to bind").

The ability of a protein to bind selectively and with high affinity to a ligand depends on the formation of a set of weak, noncovalent bonds—hydrogen bonds, ionic bonds, and van der Waals attractions—plus favorable hydrophobic interactions (see Panel 2–3, pp. 114–115). Because each individual bond is weak, an effective binding interaction requires that many weak bonds be formed simultaneously. This is possible only if the surface contours of the ligand molecule fit very closely to the protein, matching it like a hand in a glove (Figure 3–37).

The region of a protein that associates with a ligand, known as the ligand's *binding site*, usually consists of a cavity in the protein surface formed by a particular arrangement of amino acids. These amino acids can belong to different portions of the polypeptide chain that are brought together when the protein folds (Figure 3–38). Separate regions of the protein surface generally provide binding sites for different ligands, allowing the protein's activity to be regulated, as we shall see later. And other parts of the protein can serve as a handle to place the protein in a particular location in the cell—an example is the SH2 domain discussed previously, which is often used to move a protein containing it to sites in the plasma membrane in response to particular signals.

Although the atoms buried in the interior of the protein have no direct contact with the ligand, they provide an essential scaffold that gives the surface its contours and chemical properties. Even small changes to the amino acids in the interior of a protein molecule can change its three-dimensional shape enough to destroy a binding site on the surface.

The Details of a Protein's Conformation Determine Its Chemistry

Proteins have impressive chemical capabilities because the neighboring chemical groups on their surface often interact in ways that enhance the chemical reactivity of amino acid side chains. These interactions fall into two main categories.

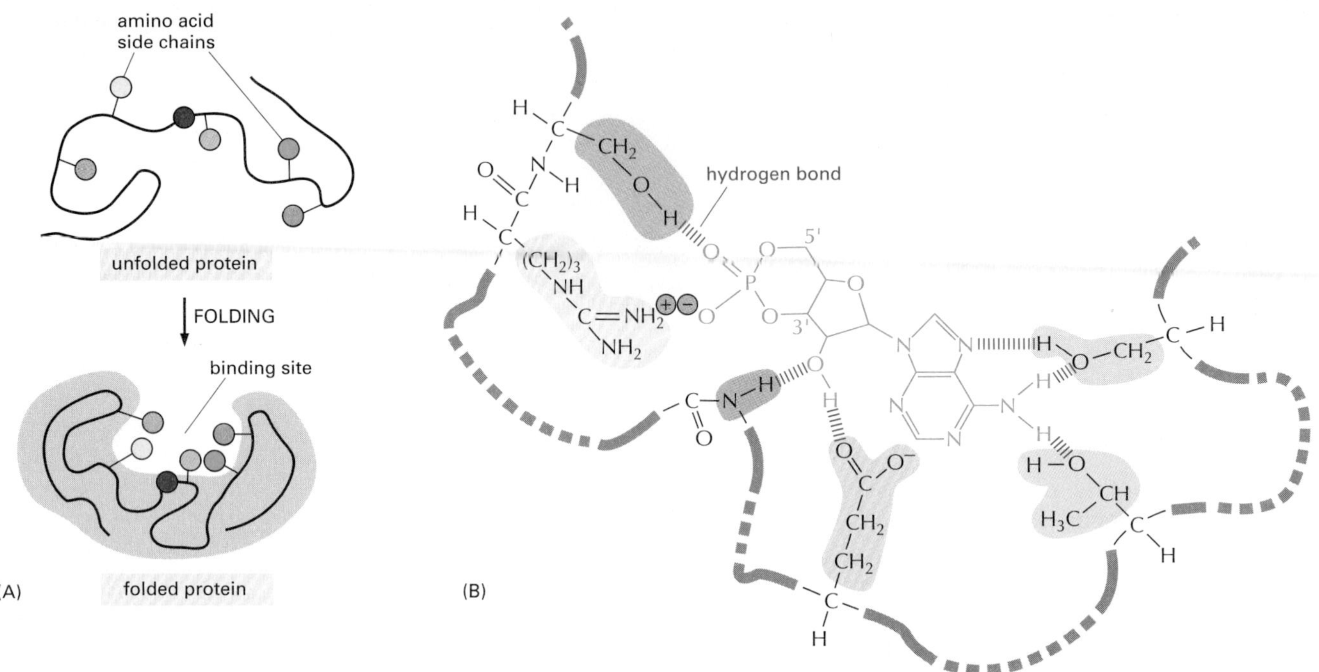

First, neighboring parts of the polypeptide chain may interact in a way that restricts the access of water molecules to a ligand binding site. Because water molecules tend to form hydrogen bonds, they can compete with ligands for sites on the protein surface. The tightness of hydrogen bonds (and ionic interactions) between proteins and their ligands is therefore greatly increased if water molecules are excluded. Initially, it is hard to imagine a mechanism that would exclude a molecule as small as water from a protein surface without affecting the access of the ligand itself. Because of the strong tendency of water molecules to form water–water hydrogen bonds, however, water molecules exist in a large hydrogen-bonded network (see Panel 2–2, pp. 112–113). In effect, a ligand binding site can be kept dry because it is energetically unfavorable for individual water molecules to break away from this network, as they must do to reach into a crevice on a protein's surface.

Second, the clustering of neighboring polar amino acid side chains can alter their reactivity. If a number of negatively charged side chains are forced together against their mutual repulsion by the way the protein folds, for example, the affinity of the site for a positively charged ion is greatly increased. In addition, when amino acid side chains interact with one another through hydrogen bonds, normally unreactive side groups (such as the $-CH_2OH$ on the serine shown in Figure 3–39) can become reactive, enabling them to enter into reactions that make or break selected covalent bonds.

The surface of each protein molecule therefore has a unique chemical reactivity that depends not only on which amino acid side chains are exposed, but also on their exact orientation relative to one another. For this reason, even two slightly different conformations of the same protein molecule may differ greatly in their chemistry.

Figure 3–38 The binding site of a protein. (A) The folding of the polypeptide chain typically creates a crevice or cavity on the protein surface. This crevice contains a set of amino acid side chains disposed in such a way that they can make noncovalent bonds only with certain ligands. (B) A close-up of an actual binding site showing the hydrogen bonds and ionic interactions formed between a protein and its ligand (in this example, cyclic AMP is the bound ligand).

Figure 3–39 An unusually reactive amino acid at the active site of an enzyme. This example is the "catalytic triad" found in chymotrypsin, elastase, and other serine proteases (see Figure 3–14). The aspartic acid side chain (Asp 102) induces the histidine (His 57) to remove the proton from serine 195. This activates the serine to form a covalent bond with the enzyme substrate, hydrolyzing a peptide bond.

Sequence Comparisons Between Protein Family Members Highlight Crucial Ligand Binding Sites

As we have described previously, many of the domains in proteins can be grouped into families that show clear evidence of their evolution from a common ancestor, and genome sequences reveal large numbers of proteins that contain one or more common domains. The three-dimensional structures of the members of the same domain family are remarkably similar. For example, even when the amino acid sequence identity falls to 25%, the backbone atoms in a domain have been found to follow a common protein fold within 0.2 nanometers (2 Å).

These facts allow a method called "evolutionary tracing" to be used to identify those sites in a protein domain that are the most crucial to the domain's function. For this purpose, those amino acids that are unchanged, or nearly unchanged, in all of the known protein family members are mapped onto a structural model of the three-dimensional structure of one family member. When this is done, the most invariant positions often form one or more clusters on the protein surface, as illustrated in Figure 3–40A for the SH2 domain described previously (see Panel 3–2, pp. 138–139). These clusters generally correspond to ligand binding sites.

The SH2 domain is a module that functions in protein–protein interactions. It binds the protein containing it to a second protein that contains a phosphorylated tyrosine side chain in a specific amino acid sequence context, as shown in Figure 3–40B. The amino acids located at the binding site for the phosphorylated polypeptide have been the slowest to change during the long evolutionary process that produced the large SH2 family of peptide recognition domains. Because mutation is a random process, this result is attributed to the preferential elimination during evolution of all organisms whose SH2 domains became altered in a way that inactivated the SH2-binding site, thereby destroying the function of the SH2 domain.

In this era of extensive genome sequencing, many new protein families have been discovered whose functions are unknown. By identifying the critical binding sites on a three-dimensional structure determined for one family member, the above method of evolutionary tracing is being used to help determine the functions of such proteins.

Proteins Bind to Other Proteins Through Several Types of Interfaces

Proteins can bind to other proteins in at least three ways. In many cases, a portion of the surface of one protein contacts an extended loop of polypeptide chain (a "string") on a second protein (Figure 3–41A). Such a surface–string interaction, for example, allows the SH2 domain to recognize a phosphorylated polypeptide as a loop on a second protein, as just described, and it also enables a protein kinase to recognize the proteins that it will phosphorylate (see below).

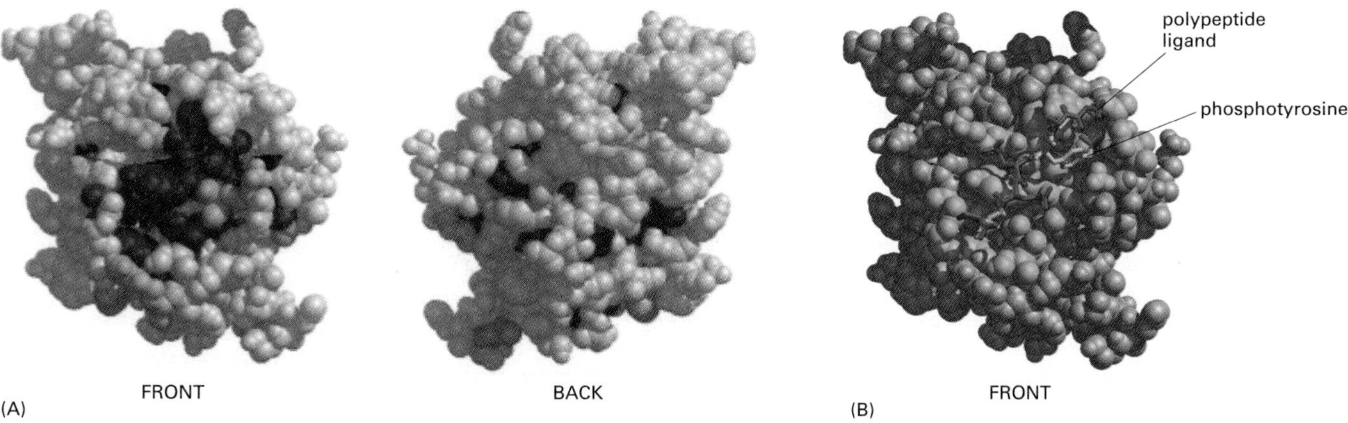

Figure 3–40 The evolutionary trace method applied to the SH2 domain. (A) Front and back views of a space-filling model of the SH2 domain, with evolutionarily conserved amino acids on the protein surface colored *yellow*, and those more toward the protein interior colored *red*. (B) The structure of the SH2 domain with its bound polypeptide. Here, those amino acids located within 0.4 nm of the bound ligand are colored *blue*. The two key amino acids of the ligand are *yellow*, and the others are *purple*. (Adapted from O. Lichtarge, H.R. Bourne, and F.E. Cohen, *J. Mol. Biol.* 257:342–358, 1996.)

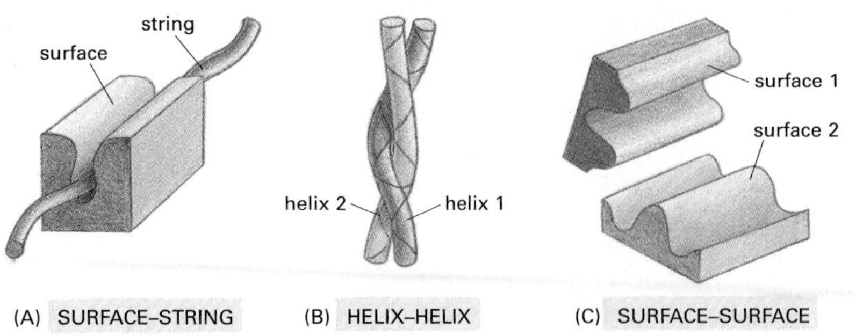

Figure 3–41 Three ways in which two proteins can bind to each other. Only the interacting parts of the two proteins are shown. (A) A rigid surface on one protein can bind to an extended loop of polypeptide chain (a "string") on a second protein. (B) Two α helices can bind together to form a coiled-coil. (C) Two complementary rigid surfaces often link two proteins together.

A second type of protein–protein interface is formed when two α helices, one from each protein, pair together to form a coiled-coil (Figure 3–41B). This type of protein interface is found in several families of gene regulatory proteins, as discussed in Chapter 7.

The most common way for proteins to interact, however, is by the precise matching of one rigid surface with that of another (Figure 3–41C). Such interactions can be very tight, since a large number of weak bonds can form between two surfaces that match well. For the same reason, such surface–surface interactions can be extremely specific, enabling a protein to select just one partner from the many thousands of different proteins found in a cell.

The Binding Sites of Antibodies Are Especially Versatile

All proteins must bind to particular ligands to carry out their various functions. This capacity for tight selective binding is displayed to an extraordinary degree by the antibody family, as discussed in detail in Chapter 24.

Antibodies, or immunoglobulins, are proteins produced by the immune system in response to foreign molecules, such as those on the surface of an invading microorganism. Each antibody binds to a particular target molecule extremely tightly, thereby either inactivating the target directly or marking it for destruction. An antibody recognizes its target (called an **antigen**) with remarkable specificity. Because there are potentially billions of different antigens we might encounter, we have to be able to produce billions of different antibodies.

Antibodies are Y-shaped molecules with two identical binding sites that are complementary to a small portion of the surface of the antigen molecule. A detailed examination of the antigen-binding sites of antibodies reveals that they are formed from several loops of polypeptide chain that protrude from the ends of a pair of closely juxtaposed protein domains (Figure 3–42). The enormous diversity of antigen-binding sites possessed by different antibodies is generated by changing only the length and amino acid sequence of these loops, without altering the basic protein structure.

Loops of this kind are ideal for grasping other molecules. They allow a large number of chemical groups to surround a ligand so that the protein can link to it with many weak bonds. For this reason, loops are often used to form the ligand-binding sites in proteins.

Binding Strength Is Measured by the Equilibrium Constant

Molecules in the cell encounter each other very frequently because of their continual random thermal movements. When colliding molecules have poorly matching surfaces, few noncovalent bonds form, and the two molecules dissociate as rapidly as they come together. At the other extreme, when many noncovalent bonds form, the association can persist for a very long time (Figure 3–43). Strong interactions occur in cells whenever a biological function requires that molecules remain associated for a long time—for example, when a group of RNA and protein molecules come together to make a subcellular structure such as a ribosome.

The strength with which any two molecules bind to each other can be measured directly. As an example, imagine a situation in which a population of

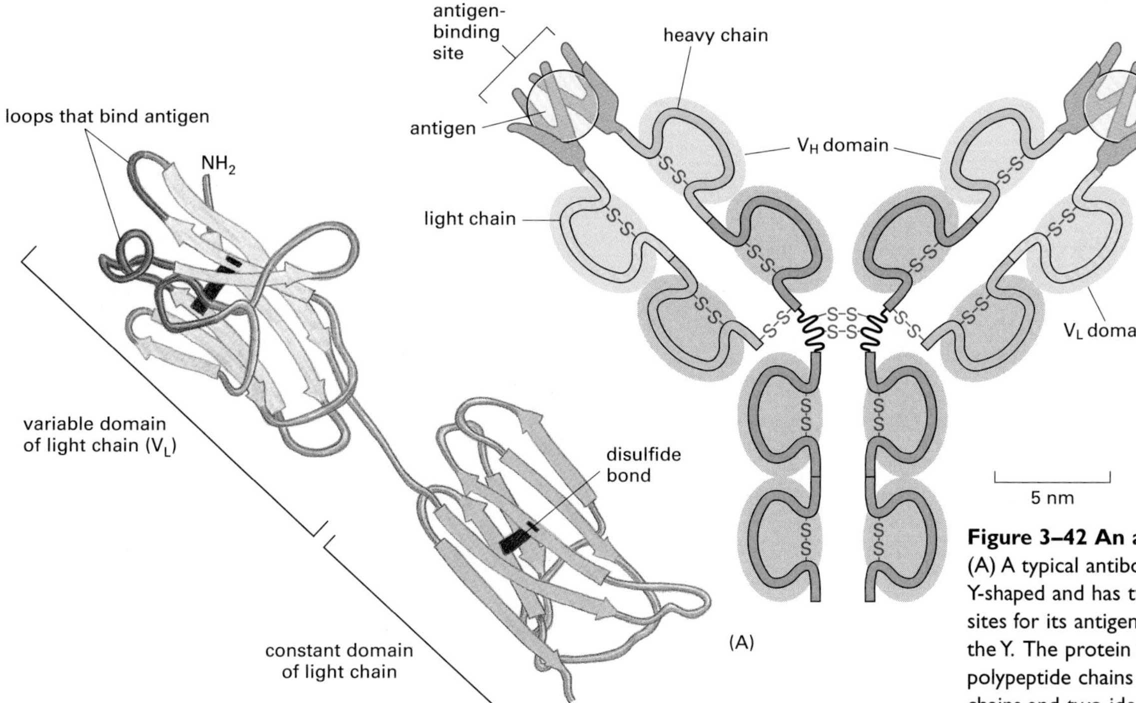

Figure 3–42 An antibody molecule. (A) A typical antibody molecule is Y-shaped and has two identical binding sites for its antigen, one on each arm of the Y. The protein is composed of four polypeptide chains (two identical heavy chains and two identical and smaller light chains) held together by disulfide bonds. Each chain is made up of several different immunoglobulin domains, here shaded either *blue* or *gray*. The antigen-binding site is formed where a heavy-chain variable domain (V_H) and a light-chain variable domain (V_L) come close together. These are the domains that differ most in their sequence and structure in different antibodies. (B) This ribbon model of a light chain shows the parts of the V_L domain most closely involved in binding to the antigen in *red*. They contribute half of the fingerlike loops that fold around each of the antigen molecules in (A).

identical antibody molecules suddenly encounters a population of ligands diffusing in the fluid surrounding them. At frequent intervals, one of the ligand molecules will bump into the binding site of an antibody and form an antibody–ligand complex. The population of antibody–ligand complexes will therefore increase, but not without limit: over time, a second process, in which individual complexes break apart because of thermally induced motion, will become increasingly important. Eventually, any population of antibody molecules and ligands will reach a steady state, or equilibrium, in which the number of binding (association) events per second is precisely equal to the number of "unbinding" (dissociation) events (see Figure 2–52).

From the concentrations of the ligand, antibody, and antibody–ligand complex at equilibrium, one can calculate a convenient measure—termed the **equilibrium constant *(K)***—of the strength of binding (Figure 3–44A). The equilibrium constant is greater the greater the binding strength, and it is a direct measure of the free-energy difference between the bound and free states (Figure 3–44B). Even a change of a few noncovalent bonds can have a striking effect on a binding interaction, as shown by the example in Figure 3–44C.

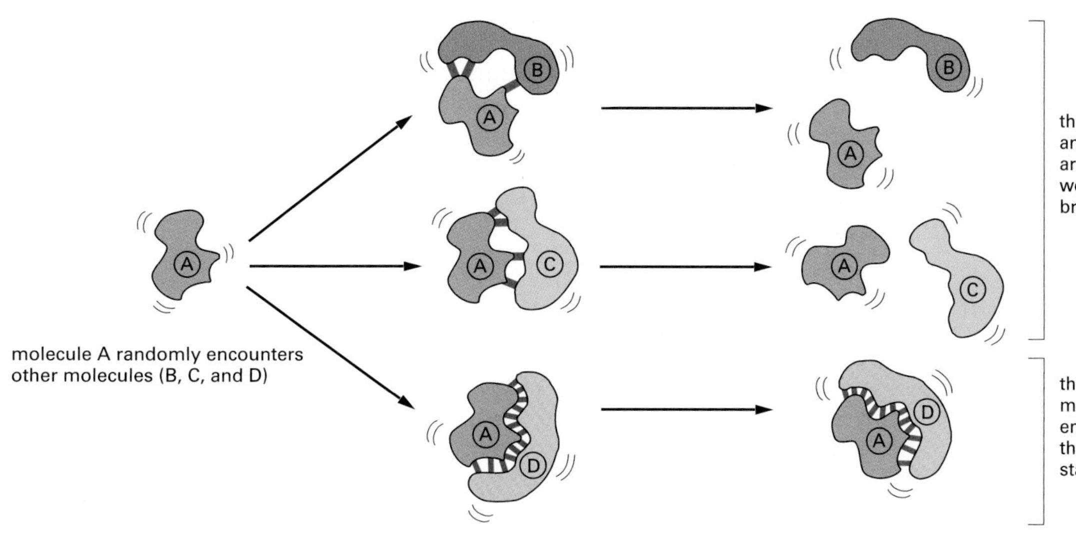

molecule A randomly encounters other molecules (B, C, and D)
the surfaces of molecules A and B, and A and C, are a poor match and are capable of forming only a few weak bonds; thermal motion rapidly breaks them apart
the surfaces of molecules A and D match well and therefore can form enough weak bonds to withstand thermal jolting; they therefore stay bound to each other

Figure 3–43 How noncovalent bonds mediate interactions between macromolecules.

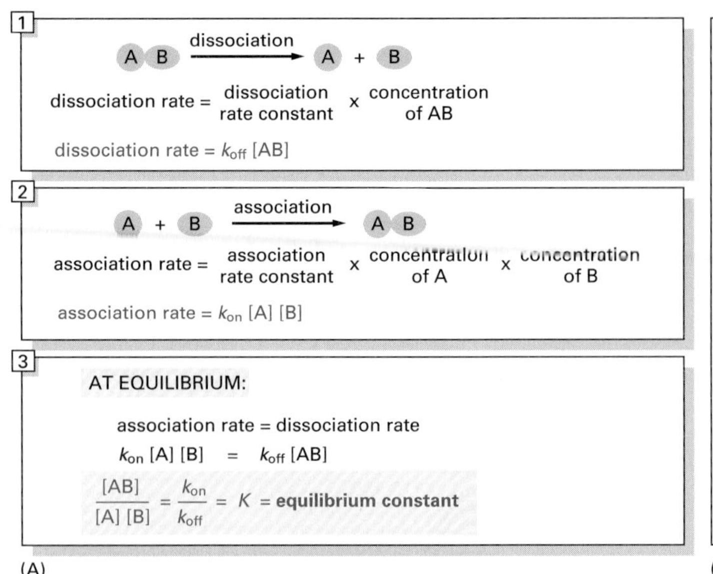

The relationship between free-energy differences and equilibrium constants

equilibrium constant $\frac{[AB]}{[A][B]} = K$ (liters/mole)	free energy of AB minus free energy of A + B (kcal/mole)
1	0
10	–1.4
10^2	–2.8
10^3	–4.3
10^4	–5.7
10^5	–7.1
10^6	–8.6
10^7	–10.0
10^8	–11.4
10^9	–12.9
10^{10}	–14.2
10^{11}	–15.6

(B)

Consider 1000 molecules of A and 1000 molecules of B in a eucaryotic cell. The concentration of both will be about 10^{-9} M.

If the equilibrium constant (K) for A + B ⇌ AB is 10^{10}, then at equilibrium there will be

270	270	730
A molecules	B molecules	AB molecules

If the equilibrium constant is a little weaker at 10^8, which represents a loss of 2.8 kcal/mole of binding energy from the example above, or 2–3 fewer hydrogen bonds, then there will be

915	915	85
A molecules	B molecules	AB molecules

(C)

Figure 3–44 Relating binding energies to the equilibrium constant. (A) The equilibrium between molecules A and B and the complex AB is maintained by a balance between the two opposing reactions shown in panels 1 and 2. Molecules A and B must collide if they are to react, and the association rate is therefore proportional to the product of their individual concentrations [A] × [B]. (Square brackets indicate concentration.) As shown in panel 3, the ratio of the rate constants for the association and the dissociation reactions is equal to the equilibrium constant (K) for the reaction. (B) The equilibrium constant in panel 3 is that for the association reaction A + B ↔ AB, and the larger its value, the stronger the binding between A and B. Note that, for every 1.4 kcal/mole of free-energy drop, the equilibrium constant increases by a factor of 10. (C) An example of the dramatic effect that the presence or absence of a few weak bonds can have in a biological context.

The equilibrium constant here has units of liters/mole: for simple binding interactions it is also called the *affinity constant* or *association constant*, denoted K_a. The reciprocal of K_a is called the *dissociation constant*, K_d (in units of moles/liter).

We have used the case of an antibody binding to its ligand to illustrate the effect of binding strength on the equilibrium state, but the same principles apply to any molecule and its ligand. Many proteins are enzymes, which, as we now discuss, first bind to their ligands and then catalyze the breakage or formation of covalent bonds in these molecules.

Enzymes Are Powerful and Highly Specific Catalysts

Many proteins can perform their function simply by binding to another molecule. An actin molecule, for example, need only associate with other actin molecules to form a filament. There are other proteins, however, for which ligand binding is only a necessary first step in their function. This is the case for the large and very important class of proteins called **enzymes**. As described in Chapter 2, enzymes are remarkable molecules that determine all the chemical transformations that make and break covalent bonds in cells. They bind to one or more ligands, called **substrates**, and convert them into one or more chemically modified *products*, doing this over and over again with amazing rapidity. Enzymes speed up reactions, often by a factor of a million or more, without themselves being changed—that is, they act as **catalysts** that permit cells to make or break covalent bonds in a controlled way. It is the catalysis of organized sets of chemical reactions by enzymes that creates and maintains the cell, making life possible.

Enzymes can be grouped into functional classes that perform similar chemical reactions (Table 3–1). Each type of enzyme within such a class is highly specific, catalyzing only a single type of reaction. Thus, *hexokinase* adds a phosphate group to D-glucose but ignores its optical isomer L-glucose; the blood-clotting enzyme *thrombin* cuts one type of blood protein between a particular arginine and its adjacent glycine and nowhere else, and so on. As discussed in detail in Chapter 2, enzymes work in teams, with the product of one enzyme becoming the substrate for the next. The result is an elaborate network of metabolic pathways that provides the cell with energy and generates the many large and small molecules that the cell needs (see Figure 2–35).

Substrate Binding Is the First Step in Enzyme Catalysis

For a protein that catalyzes a chemical reaction (an enzyme), the binding of each substrate molecule to the protein is an essential prelude. In the simplest case, if we denote the enzyme by E, the substrate by S, and the product by P, the basic reaction path is E + S → ES → EP → E + P. From this reaction path, we see that there is a limit to the amount of substrate that a single enzyme molecule can process in a given time. If the concentration of substrate is increased, the rate at

TABLE 3–1 Some Common Types of Enzymes

ENZYME	REACTION CATALYZED
Hydrolases	general term for enzymes that catalyze a hydrolytic cleavage reaction.
Nucleases	break down nucleic acids by hydrolyzing bonds between nucleotides.
Proteases	break down proteins by hydrolyzing bonds between amino acids.
Synthases	general name used for enzymes that synthesize molecules in anabolic reactions by condensing two smaller molecules together.
Isomerases	catalyze the rearrangement of bonds within a single molecule.
Polymerases	catalyze polymerization reactions such as the synthesis of DNA and RNA.
Kinases	catalyze the addition of phosphate groups to molecules. Protein kinases are an important group of kinases that attach phosphate groups to proteins.
Phosphatases	catalyze the hydrolytic removal of a phosphate group from a molecule.
Oxido-Reductases	general name for enzymes that catalyze reactions in which one molecule is oxidized while the other is reduced. Enzymes of this type are often called *oxidases, reductases,* and *dehydrogenases.*
ATPases	hydrolyze ATP. Many proteins with a wide range of roles have an energy-harnessing ATPase activity as part of their function, for example, motor proteins such as *myosin* and membrane transport proteins such as the *sodium–potassium pump.*

Enzyme names typically end in "-ase," with the exception of some enzymes, such as pepsin, trypsin, thrombin and lysozyme that were discovered and named before the convention became generally accepted at the end of the nineteenth century. The common name of an enzyme usually indicates the substrate and the nature of the reaction catalyzed. For example, citrate synthase catalyzes the synthesis of citrate by a reaction between acetyl CoA and oxaloacetate.

which product is formed also increases, up to a maximum value (Figure 3–45). At that point the enzyme molecule is saturated with substrate, and the rate of reaction (V_{max}) depends only on how rapidly the enzyme can process the substrate molecule. This maximum rate divided by the enzyme concentration is called the *turnover number*. The turnover number is often about 1000 substrate molecules processed per second per enzyme molecule, although turnover numbers between 1 and 10,000 are known.

The other kinetic parameter frequently used to characterize an enzyme is its K_m, the concentration of substrate that allows the reaction to proceed at one-half its maximum rate (0.5 V_{max}) (see Figure 3–45). A *low* K_m value means that the enzyme reaches its maximum catalytic rate at a *low concentration* of substrate and generally indicates that the enzyme binds to its substrate very tightly, whereas a *high* K_m value corresponds to weak binding. The methods used to characterize enzymes in this way are explained in Panel 3–3 (pp. 164–165).

Enzymes Speed Reactions by Selectively Stabilizing Transition States

Extremely high rates of chemical reaction are achieved by enzymes—far higher than for any synthetic catalysts. This efficiency is attributable to several factors. The enzyme serves, first, to increase the local concentration of substrate molecules at the catalytic site and to hold all the appropriate atoms in the correct orientation for the reaction that is to follow. More importantly, however, some of the binding energy contributes directly to the catalysis. Substrate molecules must pass through a series of intermediate states of altered geometry

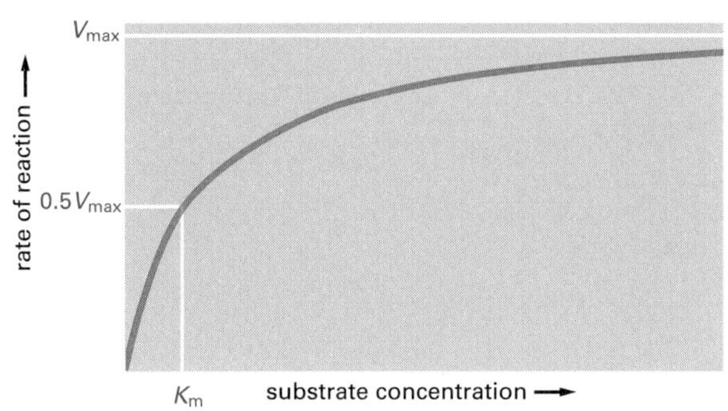

Figure 3–45 Enzyme kinetics. The rate of an enzyme reaction (*V*) increases as the substrate concentration increases until a maximum value (V_{max}) is reached. At this point all substrate-binding sites on the enzyme molecules are fully occupied, and the rate of reaction is limited by the rate of the catalytic process on the enzyme surface. For most enzymes, the concentration of substrate at which the reaction rate is half-maximal (K_m) is a measure of how tightly the substrate is bound, with a large value of K_m corresponding to weak binding.

WHY ANALYZE THE KINETICS OF ENZYMES?

Enzymes are the most selective and powerful catalysts known. An understanding of their detailed mechanisms provides a critical tool for the discovery of new drugs, for the large-scale industrial synthesis of useful chemicals, and for appreciating the chemistry of cells and organisms. A detailed study of the rates of the chemical reactions that are catalyzed by a purified enzyme—more specifically how these rates change with changes in conditions such as the concentrations of substrates, products, inhibitors, and regulatory ligands—allows biochemists to figure out exactly how each enzyme works. For example, this is the way that the ATP-producing reactions of glycolysis, shown previously in Figure 2–73, were deciphered—allowing us to appreciate the rationale for this critical enzymatic pathway.

In this Panel, we introduce the important field of enzyme kinetics, which has been indispensible for deriving much of the detailed knowledge that we now have about cellular chemistry.

STEADY STATE ENZYME KINETICS

Many enzymes have only one substrate, which they bind and then process to produce products according to the scheme outlined in Figure 3–50A. In this case, the reaction is written as

$$E + S \underset{k_{-1}}{\overset{k_1}{\rightleftharpoons}} ES \xrightarrow{k_{cat}} E + P$$

Here we have assumed that the reverse reaction, in which E + P recombine to form EP and then ES, occurs so rarely that we can ignore it. In this case, we can express the rate of the reaction—known as its velocity, V, as

$$V = k_{cat}\,[ES]$$

where [ES] is the concentration of the enzyme substrate complex, and k_{cat} is the turnover number: a rate constant that is equal to the number of substrate molecules processed per enzyme molecule each second.

But how does the value of [ES] relate to the concentrations that we know directly, which are the total concentration of the enzyme, $[E_o]$, and the concentration of the substrate, [S]? When enzyme and substrate are first mixed, the concentration [ES] will rise rapidly from zero to a so-called steady state level, as illustrated below

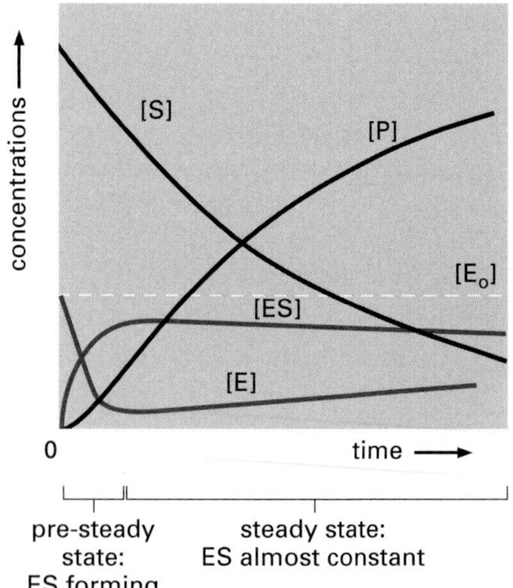

At this steady state, [ES] is nearly constant, so that

$$\underset{k_{-1}\,[ES] + k_{cat}\,[ES]}{\text{rate of ES breakdown}} = \underset{k_1\,[E][S]}{\text{rate of ES formation}}$$

or, since the concentration of the free enzyme, [E], is equal to $[E_o] - [ES]$

$$[ES] = \left(\frac{k_1}{k_{-1} + k_{cat}}\right)[E][S] = \left(\frac{k_1}{k_{-1} + k_{cat}}\right)\left([E_o] - [ES]\right)[S]$$

Rearranging, and defining the constant K_m as

$$\frac{k_{-1} + k_{cat}}{k_1}$$

we get

$$[ES] = \frac{[E_o][S]}{K_m + [S]}$$

or, remembering that $V = k_{cat}\,[ES]$, we obtain the famous Michaelis-Menton equation

$$V = \frac{k_{cat}\,[E_o][S]}{K_m + [S]}$$

As [S] is increased to higher and higher levels, essentially all of the enzyme will be bound to substrate at steady state; at this point, a maximum rate of reaction, V_{max}, will be reached where $V = V_{max} = k_{cat}\,[E_o]$. Thus, it is convenient to rewrite the Michaelis-Menton equation as

$$V = \frac{V_{max}\,[S]}{K_m + [S]}$$

THE DOUBLE RECIPROCAL PLOT

A typical plot of V versus [S] for an enzyme that follows Michaelis–Menton kinetics is shown below. From this plot, neither the value of V_{max} nor of K_m is immediately clear.

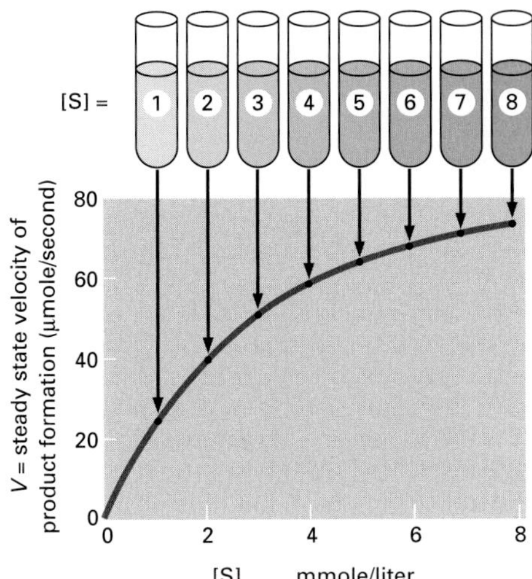

To obtain V_{max} and K_m from such data, a double-reciprocal plot is often used, in which the Michaelis–Menton equation has merely been rearranged, so that $1/V$ can be plotted versus $1/[S]$.

$$1/V = \left(\frac{K_m}{V_{max}}\right)\left(\frac{1}{[S]}\right) + 1/V_{max}$$

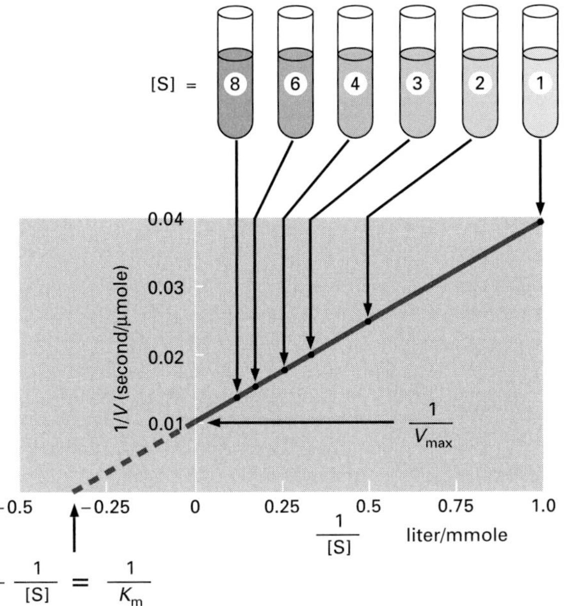

THE SIGNIFICANCE OF K_m, k_{cat}, and k_{cat}/K_m

As described in the text, K_m is an approximate measure of substrate affinity for the enzyme: it is numerically equal to the concentration of [S] at $V = 0.5\ V_{max}$. In general, a lower value of K_m means tighter substrate binding.

We have seen that k_{cat} is the turnover number for the enzyme. At very low substrate concentrations, where $[S] << K_m$, most of the enzyme is free. Thus we can think of $[E] = [E_o]$, so that the Michaelis-Menton equation becomes $V = k_{cat}/K_m\ [E][S]$. Thus, the ratio k_{cat}/K_m is equivalent to the rate constant for the reaction between free enzyme and free substrate.

A comparison of k_{cat}/K_m for the same enzyme with different substrates, or for two enzymes with their different substrates, is widely used as a measure of enzyme effectiveness.

For simplicity, in this Panel we have discussed enzymes that have only one substrate, such as the lysozyme enzyme described in the text (see p. 167). Most enzymes have two substrates, one of which is often an active carrier molecule—such as NADH or ATP.

A similar, but more complex analysis is used to determine the kinetics of such enzymes—allowing the order of substrate binding and the presence of covalent intermediates along the pathway to be revealed (see, for example, Figure 2–73).

SOME ENZYMES ARE DIFFUSION LIMITED

The values of k_{cat}, K_m, and k_{cat}/K_m for some selected enzymes are given below:

enzyme	substrate	k_{cat} (sec^{-1})	K_m (M)	k_{cat}/K_m ($sec^{-1}M^{-1}$)
acetylcholinesterase	acetylcholine	1.4×10^4	9×10^{-5}	1.6×10^8
catalase	H_2O_2	4×10^7	1	4×10^7
fumarase	fumarate	8×10^2	5×10^{-6}	1.6×10^8

Because an enzyme and its substrate must collide before they can react, k_{cat}/K_m has a maximum possible value that is limited by collision rates. If every collision forms an enzyme-substrate complex, one can calculate from diffusion theory that k_{cat}/K_m will be between 10^8 and 10^9 $sec^{-1}M^{-1}$, in the case where all subsequent steps proceed immediately. Thus, it is claimed that enzymes like acetylcholinesterase and fumarase are "perfect enzymes", each enzyme having evolved to the point where nearly every collision with its substrate converts the substrate to a product.

and electron distribution before they form the ultimate products of the reaction. The free energy required to attain the most unstable **transition state** is called the *activation energy* for the reaction, and it is the major determinant of the reaction rate. Enzymes have a much higher affinity for the transition state of the substrate than they have for the stable form. Because this tight binding greatly lowers the energies of the transition state, the enzyme greatly accelerates a particular reaction by lowering the activation energy that is required (Figure 3–46).

A dramatic proof that stabilizing a transition state can greatly increase a reaction rate is provided by the intentional production of antibodies that act like enzymes. Consider, for example, the hydrolysis of an amide bond, which is similar to the peptide bond that joins two adjacent amino acids in a protein. In an aqueous solution, an amide bond hydrolyzes very slowly by the mechanism shown in Figure 3–47A. In the central intermediate, or transition state, the carbonyl carbon is bonded to four atoms arranged at the corners of a tetrahedron. By generating monoclonal antibodies that bind tightly to a stable analog of this very unstable *tetrahedral intermediate*, an antibody that functions like an enzyme can be obtained (Figure 3–47B). Because this *catalytic antibody* binds to and stabilizes the tetrahedral intermediate, it increases the spontaneous rate of amide-bond hydrolysis by more than 10,000-fold.

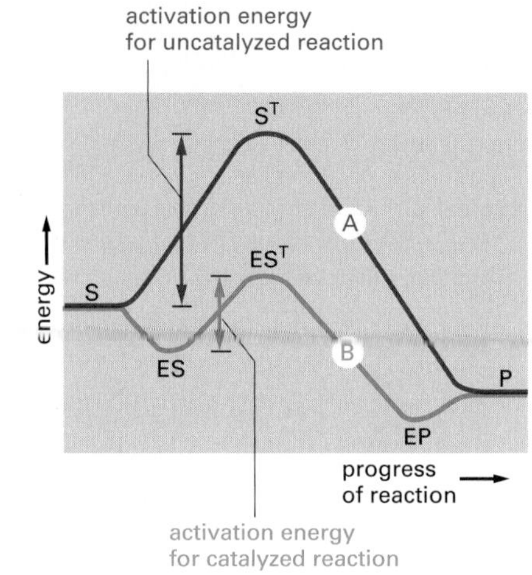

Figure 3–46 Enzymatic acceleration of chemical reactions by decreasing the activation energy. Often both the uncatalyzed reaction (A) and the enzyme-catalyzed reaction (B) can go through several transition states. It is the transition state with the highest energy (S^T and ES^T) that determines the activation energy and limits the rate of the reaction. (S = substrate; P = product of the reaction.)

Enzymes Can Use Simultaneous Acid and Base Catalysis

Figure 3–48 compares the spontaneous reaction rates and the corresponding enzyme-catalyzed rates for five enzymes. Rate accelerations of 10^9–10^{23} are observed. Clearly, enzymes are much better catalysts than catalytic antibodies. Enzymes not only bind tightly to a transition state, they also contain precisely positioned atoms that alter the electron distributions in those atoms that participate directly in the making and breaking of covalent bonds. Peptide bonds, for example, can be hydrolyzed in the absence of an enzyme by exposing a polypeptide to either a strong acid or a strong base, as explained in Figure 3–49. Enzymes are unique, however, in being able to use acid and base catalysis simultaneously, since the acidic and basic residues required are prevented from combining with each other (as they would do in solution) by being tied to the rigid framework of the protein itself (Figure 3–49D).

The fit between an enzyme and its substrate needs to be precise. A small change introduced by genetic engineering in the active site of an enzyme can have a profound effect. Replacing a glutamic acid with an aspartic acid in one enzyme, for example, shifts the position of the catalytic carboxylate ion by only 1 Å (about the radius of a hydrogen atom); yet this is enough to decrease the activity of the enzyme a thousandfold.

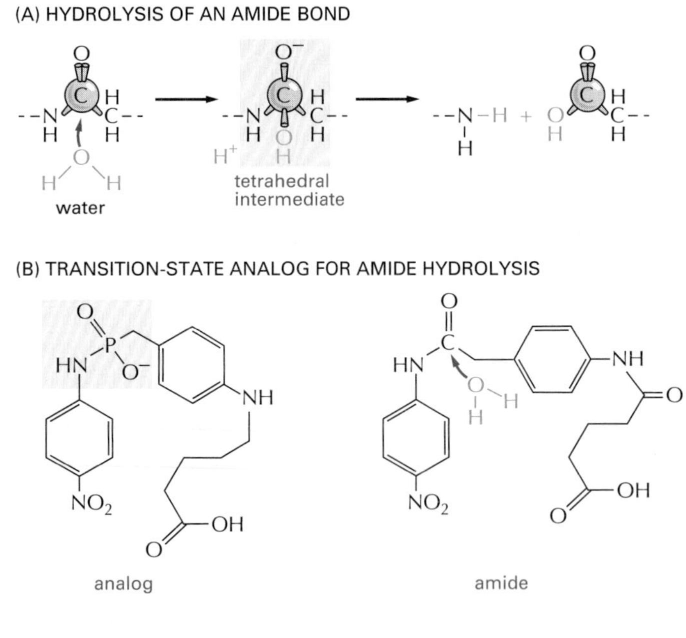

Figure 3–47 Catalytic antibodies. The stabilization of a transition state by an antibody creates an enzyme. (A) The reaction path for the hydrolysis of an amide bond goes through a tetrahedral intermediate, the high-energy transition state for the reaction. (B) The molecule on the left was covalently linked to a protein and used as an antigen to generate an antibody that binds tightly to the region of the molecule shown in *yellow*. Because this antibody also bound tightly to the transition state in (A), it was found to function as an enzyme that efficiently catalyzed the hydrolysis of the amide bond in the molecule on the right.

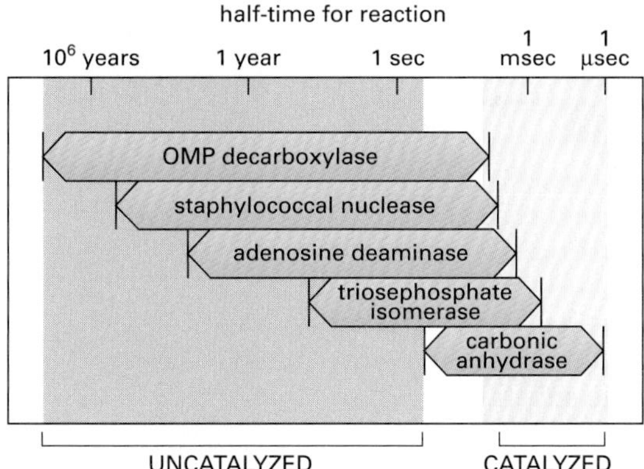

Figure 3–48 The rate accelerations caused by five different enzymes. (Modified from A. Radzicka and R. Wolfenden, *Science* 267:90–93, 1995.)

Lysozyme Illustrates How an Enzyme Works

To demonstrate how enzymes catalyze chemical reactions, we shall use the example of an enzyme that acts as a natural antibiotic in egg white, saliva, tears, and other secretions. **Lysozyme** is an enzyme that catalyzes the cutting of polysaccharide chains in the cell walls of bacteria. Because the bacterial cell is under pressure from osmotic forces, cutting even a small number of polysaccharide chains causes the cell wall to rupture and the cell to burst. Lysozyme is a relatively small and stable protein that can be easily isolated in large quantities. For these reasons, it has been intensively studied, and it was the first enzyme to have its structure worked out in atomic detail by x-ray crystallography.

The reaction catalyzed by lysozyme is a hydrolysis: a molecule of water is added to a single bond between two adjacent sugar groups in the polysaccharide chain, thereby causing the bond to break (see Figure 2–19). The reaction is energetically favorable because the free energy of the severed polysaccharide chain is lower than the free energy of the intact chain. However, the pure polysaccharide can sit for years in water without being hydrolyzed to any detectable degree. This is because there is an energy barrier to the reaction, as discussed in Chapter 2 (see Figure 2–46). For a colliding water molecule to break a bond linking two sugars, the polysaccharide molecule has to be distorted into a particular shape—the transition state—in which the atoms around the bond have an altered geometry and electron distribution. Because of this distortion, a large activation energy must be supplied through random collisions before the reaction can take place. In an aqueous solution at room temperature, the energy of collisions almost never exceeds the activation energy. Consequently, hydrolysis occurs extremely slowly, if at all.

This situation is drastically changed when the polysaccharide binds to lysozyme. The active site of lysozyme, because its substrate is a polymer, is a long groove that holds six linked sugars at the same time. As soon as the polysaccharide binds to form an enzyme–substrate complex, the enzyme cuts the polysaccharide by adding a water molecule across one of its sugar–sugar bonds. The product chains are then quickly released, freeing the enzyme for further cycles of reaction (Figure 3–50).

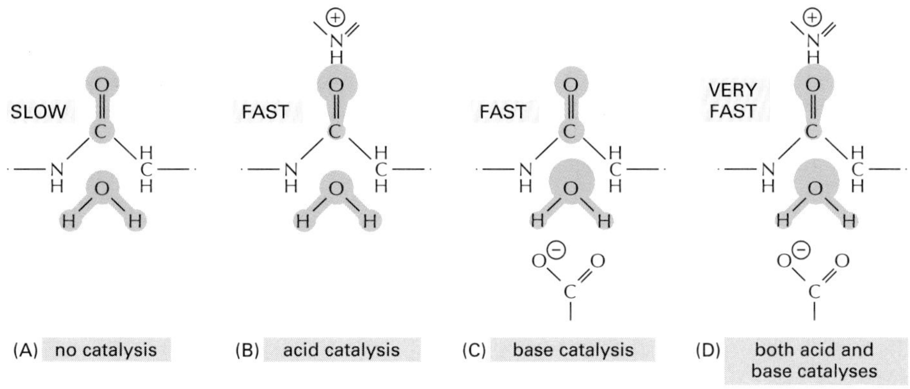

(A) no catalysis — SLOW
(B) acid catalysis — FAST
(C) base catalysis — FAST
(D) both acid and base catalyses — VERY FAST

Figure 3–49 Acid catalysis and base catalysis. (A) The start of the uncatalyzed reaction shown in Figure 3–47A is diagrammed, with *blue* indicating electron distribution in the water and carbonyl bonds. (B) An acid likes to donate a proton (H^+) to other atoms. By pairing with the carbonyl oxygen, an acid causes electrons to move away from the carbonyl carbon, making this atom much more attractive to the electronegative oxygen of an attacking water molecule. (C) A base likes to take up H^+. By pairing with a hydrogen of the attacking water molecule, a base causes electrons to move toward the water oxygen, making it a better attacking group for the carbonyl carbon. (D) By having appropriately positioned atoms on its surface, an enzyme can perform both acid catalysis and base catalysis at the same time.

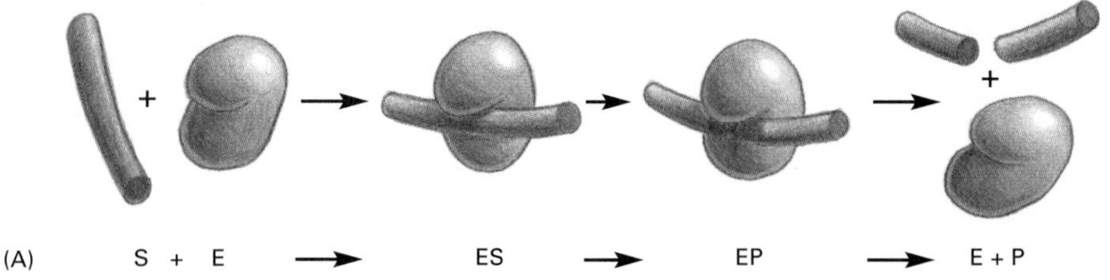

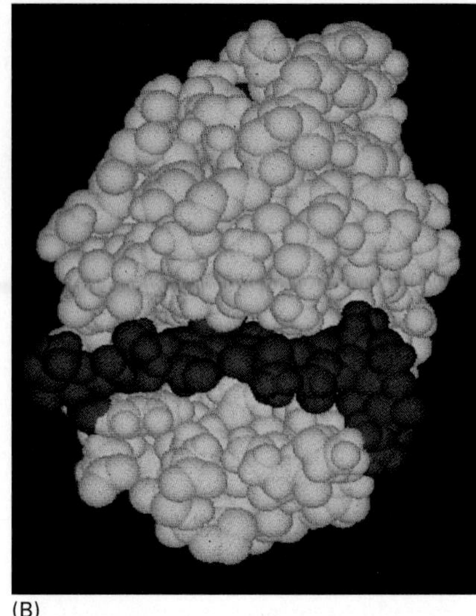

Figure 3–50 The reaction catalyzed by lysozyme. (A) The enzyme lysozyme (denoted as E) catalyzes the cutting of a polysaccharide chain, which is its substrate (S). The enzyme first binds to the chain to form an enzyme–substrate complex (ES) and then catalyzes the cleavage of a specific covalent bond in the backbone of the polysaccharide, forming an enzyme–product complex (EP) that rapidly dissociates. Release of the severed chain (the products P) leaves the enzyme free to act on another substrate molecule. (B) A space-filling model of the lysozyme molecule bound to a short length of polysaccharide chain before cleavage. (B, courtesy of Richard J. Feldmann.)

The chemistry that underlies the binding of lysozyme to its substrate is the same as that for antibody binding to its antigen—the formation of multiple noncovalent bonds. However, lysozyme holds its polysaccharide substrate in a particular way, so that one of the two sugars involved in the bond to be broken is distorted from its normal, most stable conformation. The bond to be broken is also held close to two amino acids with acidic side chains (a glutamic acid and an aspartic acid) within the active site.

Conditions are thereby created in the microenvironment of the lysozyme active site that greatly reduce the activation energy necessary for the hydrolysis to take place. Figure 3–51 shows the stages proposed in 1967 for this enzymatically catalyzed reaction.

1. The enzyme stresses its bound substrate by bending some critical chemical bonds that will participate in the chemical reaction, so that the shape of the substrate more closely resembles the shape of the high-energy transition state formed halfway through the reaction.
2. A precisely positioned acidic side chain of the glutamic acid within the active site speeds up the hydrolysis by providing a high concentration of acidifying H^+ ions (acid catalysis), even though the solution surrounding the enzyme is at neutral pH.
3. The negatively charged aspartic acid further stabilizes the positively charged transition state (base catalysis).

As indicated in the Figure legend, this classical view of the lysozyme reaction is in need of modification. In particular, a tetrahedral intermediate forms in the transition state—created by the transient covalent addition of the aspartic acid to the carbon atom at the point of cleavage.

Similar mechanisms are used by other enzymes to lower activation energies and speed up the reactions they catalyze. In reactions involving two or more reactants, the active site also acts like a template, or mold, that brings the substrates together in the proper orientation for a reaction to occur between them (Figure 3–52A). As we saw for lysozyme, the active site of an enzyme contains precisely positioned atoms that speed up a reaction by using charged groups to alter the distribution of electrons in the substrates (Figure 3–52B). The binding to the enzyme also changes substrate shapes, bending bonds so as to drive a substrate toward a particular transition state (Figure 3–52C). Finally, like lysozyme, many enzymes participate intimately in the reaction by briefly forming a covalent bond between the substrate and a side chain of the enzyme. Subsequent steps in the reaction restore the side chain to its original state, so that the enzyme remains unchanged after the reaction (see Figure 2–73).

Tightly Bound Small Molecules Add Extra Functions to Proteins

Although we have emphasized the versatility of proteins as chains of amino acids that perform different functions, there are many instances in which the amino acids by themselves are not enough. Just as humans employ tools to enhance and extend the capabilities of their hands, proteins often use small nonprotein molecules to perform functions that would be difficult or impossible to do with amino acids alone. Thus, the signal receptor protein *rhodopsin*, which is made by the photoreceptor cells in the retina, detects light by means of a small molecule, *retinal*, embedded in the protein (Figure 3–53A). Retinal changes its shape when it absorbs a photon of light, and this change causes the protein to trigger a cascade of enzymatic reactions that eventually leads to an electrical signal being carried to the brain.

substrate

This substrate is an oligosaccharide of six sugars, labeled A–F. Sugars D and E are shown.

products

The final products are an oligosaccharide of four sugars (*left*) and a disaccharide (*right*), produced by hydrolysis.

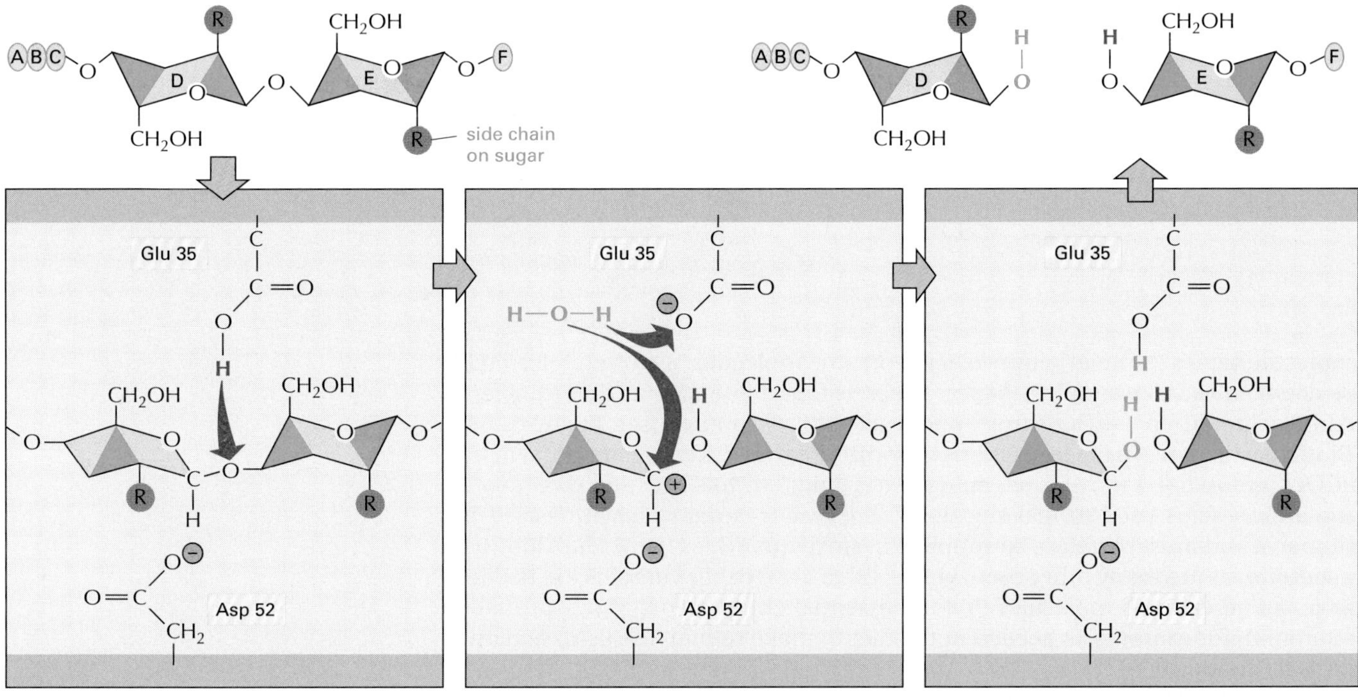

In the enzyme–substrate complex (ES), the enzyme forces sugar D into a strained conformation, with Glu 35 of lysozyme positioned to serve as an acid that attacks the adjacent sugar–sugar bond by donating a proton (H^+).

This is the unstable transition state, with a positive charge on sugar D. Both the strain on sugar D and the nearby negative charge on Asp 52 of lysozyme stabilize this intermediate, greatly lowering its energy on the enzyme surface.

The rapid addition of a water molecule (*green*) completes the hydrolysis and regenerates the protonated form of Glu 35, forming the enzyme–product complex (EP).

Figure 3–51 The classical model for events at the active site of lysozyme. The top left and top right drawings depict the free substrate and the free products, respectively, whereas the other three drawings depict the sequential events proposed at the enzyme active site. Note the change in the conformation of sugar D in the enzyme–substrate complex; this sugar has been postulated to acquire a positive charge in the unstable transition state, as shown in the central panel. However, very recent data has revealed that this standard textbook view of the reaction is incorrect. Instead, the Asp 52 forms a covalent bond to the carbon shown here with a positive charge. This covalent bond is rapidly broken in the next step, as water is added to complete the reaction.

Another example of a protein that contains a nonprotein portion is hemoglobin (see Figure 3–23). A molecule of hemoglobin carries four *heme* groups, ring-shaped molecules each with a single central iron atom (Figure 3–53B). Heme gives hemoglobin (and blood) its red color. By binding reversibly to oxygen gas through its iron atom, heme enables hemoglobin to pick up oxygen in the lungs and release it in the tissues.

Sometimes these small molecules are attached covalently and permanently to their protein, thereby becoming an integral part of the protein molecule itself. We see in Chapter 10 that proteins are often anchored to cell membranes through covalently attached lipid molecules. And membrane proteins exposed on the surface of the cell, as well as proteins secreted outside the cell, are often modified by the covalent addition of sugars and oligosaccharides.

Enzymes frequently have a small molecule or metal atom tightly associated with their active site that assists with their catalytic function. *Carboxypeptidase*, for example, an enzyme that cuts polypeptide chains, carries a tightly bound zinc ion in its active site. During the cleavage of a peptide bond by carboxypeptidase, the zinc ion forms a transient bond with one of the substrate atoms, thereby assisting the hydrolysis reaction. In other enzymes, a small organic

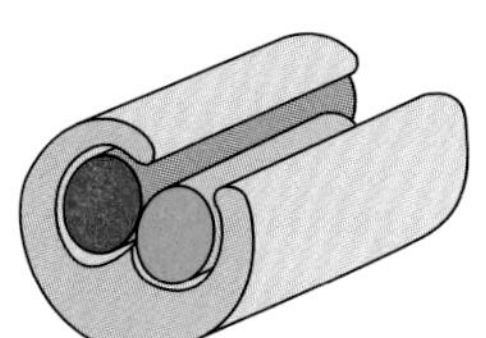

(A) enzyme binds to two substrate molecules and orients them precisely to encourage a reaction to occur between them

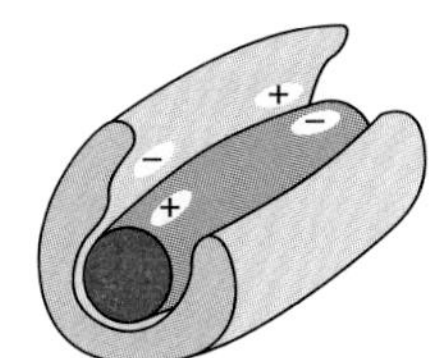

(B) binding of substrate to enzyme rearranges electrons in the substrate, creating partial negative and positive charges that favor a reaction

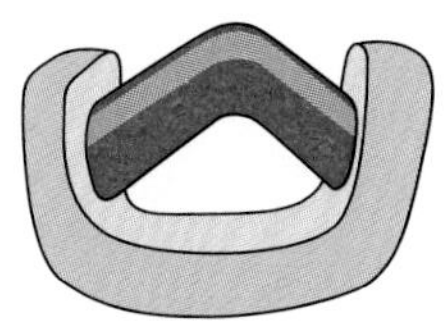

(C) enzyme strains the bound substrate molecule, forcing it toward a transition state to favor a reaction

Figure 3–52 Some general strategies of enzyme catalysis. (A) Holding substrates together in a precise alignment. (B) Charge stabilization of reaction intermediates. (C) Altering bond angles in the substrate to increase the rate of a particular reaction.

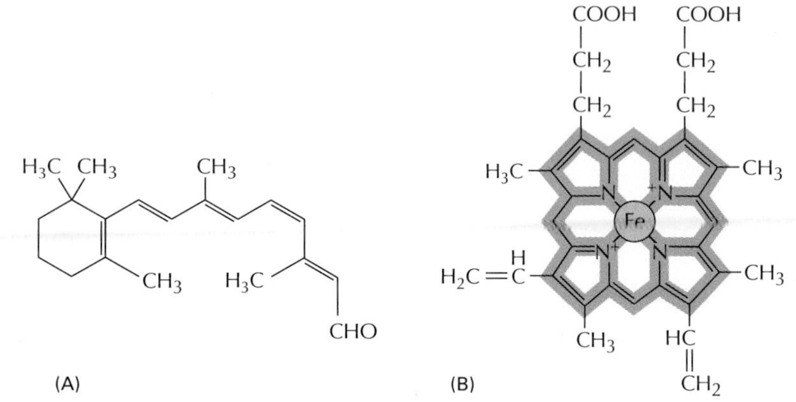

Figure 3–53 **Retinal and heme.** (A) The structure of retinal, the light-sensitive molecule attached to rhodopsin in the eye. (B) The structure of a heme group. The carbon-containing heme ring is *red* and the iron atom at its center is *orange*. A heme group is tightly bound to each of the four polypeptide chains in hemoglobin, the oxygen-carrying protein whose structure is shown in Figure 3–23.

molecule serves a similar purpose. Such organic molecules are often referred to as **coenzymes**. An example is *biotin*, which is found in enzymes that transfer a carboxylate group ($-COO^-$) from one molecule to another (see Figure 2–63). Biotin participates in these reactions by forming a transient covalent bond to the $-COO^-$ group to be transferred, being better suited to this function than any of the amino acids used to make proteins. Because it cannot be synthesized by humans, and must therefore be supplied in small quantities in our diet, biotin is a *vitamin*, as are many other coenzymes (Table 3–2). Other vitamins are needed to make other small molecules that are essential components of our proteins; vitamin A, for example, is needed in the diet to make retinal, the light-sensitive part of rhodopsin.

Multienzyme Complexes Help to Increase the Rate of Cell Metabolism

The efficiency of enzymes in accelerating chemical reactions is crucial to the maintenance of life. Cells, in effect, must race against the unavoidable processes of decay, which—if left unattended—cause macromolecules to run downhill toward greater and greater disorder. If the rates of desirable reactions were not greater than the rates of competing side reactions, a cell would soon die. Some idea of the rate at which cell metabolism proceeds can be obtained by measuring the rate of ATP utilization. A typical mammalian cell "turns over" (i.e., hydrolyzes and restores by phosphorylation) its entire ATP pool once every 1 or 2 minutes. For each cell this turnover represents the utilization of roughly 10^7 molecules of ATP per second (or, for the human body, about 1 gram of ATP every minute).

The rates of reactions in cells are rapid because of the effectiveness of enzyme catalysis. Many important enzymes have become so efficient that there

TABLE 3–2 Many Vitamins Provide Critical Coenzymes for Human Cells

VITAMIN	COENZYME	ENZYME-CATALYZED REACTIONS REQUIRING THESE COENZYMES
Thiamine (vitamin B_1)	thiamine pyrophosphate	activation and transfer of aldehydes
Riboflavin (vitamin B_2)	FADH	oxidation–reduction
Niacin	NADH, NADPH	oxidation–reduction
Pantothenic acid	coenzyme A	acyl group activation and transfer
Pyridoxine	pyridoxal phosphate	reactions involving amino acid activation
Biotin	biotin	CO_2 activation and transfer
Lipoic acid	lipoamide	acyl group activation; oxidation–reduction
Folic acid	tetrahydrofolate	activation and transfer of single carbon groups
Vitamin B_{12}	cobalamin coenzymes	isomerization and methyl group transfers

is no possibility of further useful improvement. The factor that limits the reaction rate is no longer the enzyme's intrinsic speed of action; rather, it is the frequency with which the enzyme collides with its substrate. Such a reaction is said to be *diffusion-limited.*

If an enzyme-catalyzed reaction is diffusion-limited, its rate depends on the concentration of both the enzyme and its substrate. If a sequence of reactions is to occur extremely rapidly, each metabolic intermediate and enzyme involved must be present in high concentration. However, given the enormous number of different reactions performed by a cell, there are limits to the concentrations of substrates that can be achieved. In fact, most metabolites are present in micromolar (10^{-6} M) concentrations, and most enzyme concentrations are much lower. How is it possible, therefore, to maintain very fast metabolic rates?

The answer lies in the spatial organization of cell components. Reaction rates can be increased without raising substrate concentrations by bringing the various enzymes involved in a reaction sequence together to form a large protein assembly known as a *multienzyme complex* (Figure 3–54). Because this allows the product of enzyme A to be passed directly to enzyme B, and so on, diffusion rates need not be limiting, even when the concentrations of the substrates in the cell as a whole are very low. It is perhaps not surprising, therefore, that such enzyme complexes are very common, and they are involved in nearly all aspects of metabolism—including the central genetic processes of DNA, RNA, and protein synthesis. In fact, few enzymes in eucaryotic cells may be left to diffuse freely in solution; instead, most seem to have evolved binding sites that concentrate them with other proteins of related function in particular regions of the cell, thereby increasing the rate and efficiency of the reactions that they catalyze.

Eucaryotic cells have yet another way of increasing the rate of metabolic reactions, using their intracellular membrane systems. These membranes can segregate particular substrates and the enzymes that act on them into the same membrane-enclosed compartment, such as the endoplasmic reticulum or the cell nucleus. If, for example, a compartment occupies a total of 10% of the volume of the cell, the concentration of reactants in the compartment may be increased as much as 10 times compared with the same cell with no compartmentalization. Reactions that would otherwise be limited by the speed of diffusion can thereby be speeded up by a factor of 10.

The Catalytic Activities of Enzymes Are Regulated

A living cell contains thousands of enzymes, many of which operate at the same time and in the same small volume of the cytosol. By their catalytic action, these enzymes generate a complex web of metabolic pathways, each composed of chains of chemical reactions in which the product of one enzyme becomes the substrate of the next. In this maze of pathways, there are many branch points where different enzymes compete for the same substrate. The system is so complex (see Figure 2–88) that elaborate controls are required to regulate when and how rapidly each reaction occurs.

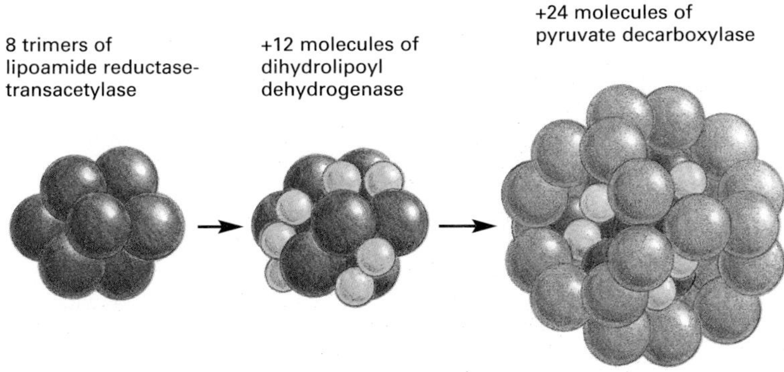

Figure 3–54 The structure of pyruvate dehydrogenase. This enzyme complex catalyzes the conversion of pyruvate to acetyl CoA, as part of the pathway that oxidizes sugars to CO_2 and H_2O. It is an example of a large multienzyme complex in which reaction intermediates are passed directly from one enzyme to another.

Regulation occurs at many levels. At one level, the cell controls how many molecules of each enzyme it makes by regulating the expression of the gene that encodes that enzyme (discussed in Chapter 7). The cell also controls enzymatic activities by confining sets of enzymes to particular subcellular compartments, enclosed by distinct membranes (discussed in Chapters 12 and 14). The rate of protein destruction by targeted proteolysis represents yet another important regulatory mechanism (see p. 361). But the most rapid and general process that adjusts reaction rates operates through a direct, reversible change in the activity of an enzyme in response to specific molecules that it encounters.

The most common type of control occurs when a molecule other than one of the substrates binds to an enzyme at a special regulatory site outside the active site, thereby altering the rate at which the enzyme converts its substrates to products. In **feedback inhibition**, an enzyme acting early in a reaction pathway is inhibited by a late product of that pathway. Thus, whenever large quantities of the final product begin to accumulate, this product binds to the first enzyme and slows down its catalytic action, thereby limiting the further entry of substrates into that reaction pathway (Figure 3–55). Where pathways branch or intersect, there are usually multiple points of control by different final products, each of which works to regulate its own synthesis (Figure 3–56). Feedback inhibition can work almost instantaneously and is rapidly reversed when the level of the product falls.

Feedback inhibition is *negative regulation:* it prevents an enzyme from acting. Enzymes can also be subject to *positive regulation,* in which the enzyme's activity is stimulated by a regulatory molecule rather than being shut down. Positive regulation occurs when a product in one branch of the metabolic maze stimulates the activity of an enzyme in another pathway. As one example, the accumulation of ADP activates several enzymes involved in the oxidation of sugar molecules, thereby stimulating the cell to convert more ADP to ATP.

Allosteric Enzymes Have Two or More Binding Sites That Interact

One feature of feedback inhibition was initially puzzling to those who discovered it: the regulatory molecule often has a shape totally different from the shape of the substrate of the enzyme. This is why this form of regulation is termed *allostery* (from the Greek words *allos,* meaning "other," and *stereos,* meaning "solid" or "three-dimensional"). As more was learned about feedback inhibition, it was recognized that many enzymes must have at least two different binding sites on their surface—an **active site** that recognizes the substrates, and a **regulatory site** that recognizes a regulatory molecule. These two sites must somehow communicate in a way that allows the catalytic events at the active site to be influenced by the binding of the regulatory molecule at its separate site on the protein's surface.

The interaction between separated sites on a protein molecule is now known to depend on a *conformational change* in the protein: binding at one of the sites causes a shift from one folded shape to a slightly different folded shape. During feedback inhibition, for example, the binding of an inhibitor at one site on the protein causes the protein to shift to a conformation in which its active site—located elsewhere in the protein—becomes incapacitated.

It is thought that most protein molecules are allosteric. They can adopt two or more slightly different conformations, and a shift from one to another caused by the binding of a ligand can alter their activity. This is true not only for enzymes but also for many other proteins—including receptors, structural proteins, and motor proteins. There is nothing mysterious about the allosteric regulation of these proteins: each conformation of the protein has somewhat different surface contours, and the protein's binding sites for ligands are altered when the protein changes shape. Moreover as we discuss next, each ligand stabilizes the conformation that it binds to most strongly, and thus—at high enough concentrations—tends to "switch" the protein toward the conformation it prefers.

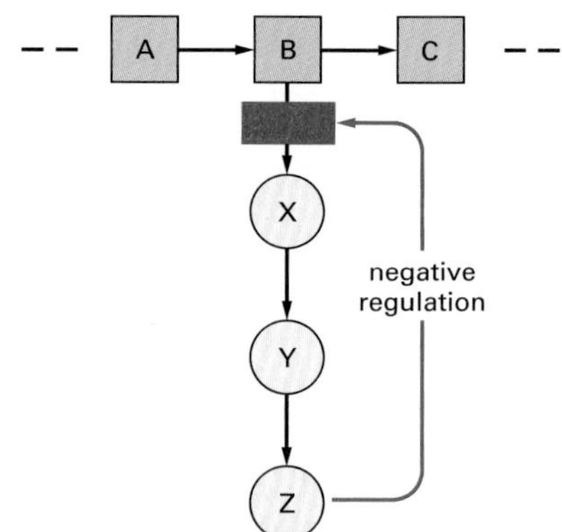

Figure 3–55 Feedback inhibition of a single biosynthetic pathway. The end-product Z inhibits the first enzyme that is unique to its synthesis and thereby controls its own level in the cell. This is an example of negative regulation.

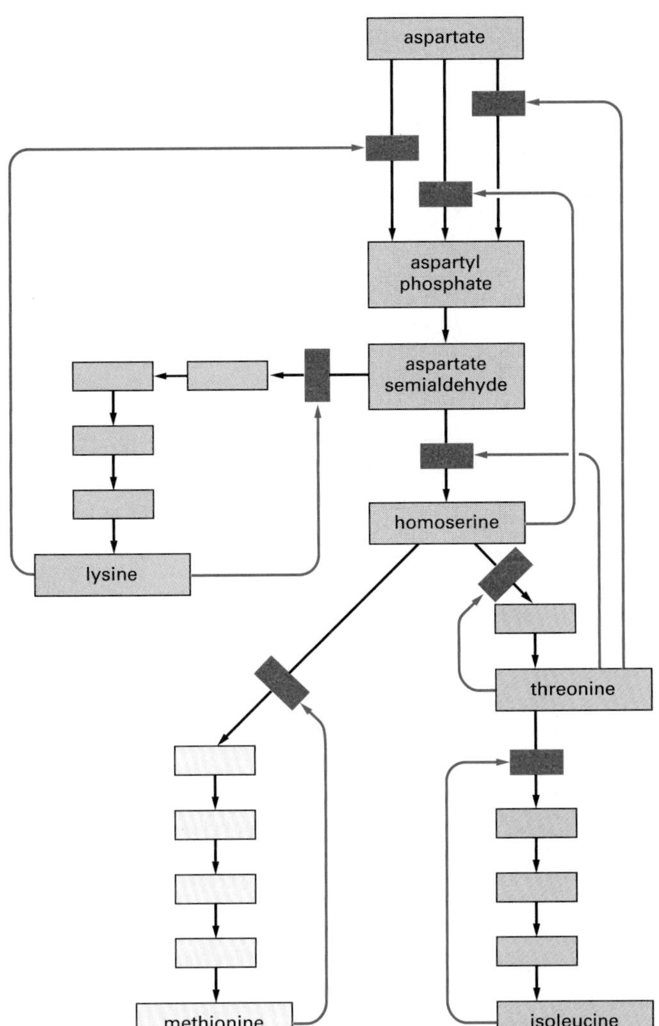

Figure 3–56 Multiple feedback inhibition. In this example, which shows the biosynthetic pathways for four different amino acids in bacteria, the *red arrows* indicate positions at which products feed back to inhibit enzymes. Each amino acid controls the first enzyme specific to its own synthesis, thereby controlling its own levels and avoiding a wasteful buildup of intermediates. The products can also separately inhibit the initial set of reactions common to all the syntheses; in this case, three different enzymes catalyze the initial reaction, each inhibited by a different product.

Two Ligands Whose Binding Sites Are Coupled Must Reciprocally Affect Each Other's Binding

The effects of ligand binding on a protein follow from a fundamental chemical principle known as **linkage**. Suppose, for example, that a protein that binds glucose also binds another molecule, X, at a distant site on the protein's surface. If the binding site for X changes shape as part of the conformational change induced by glucose binding, the binding sites for X and for glucose are said to be *coupled*. Whenever two ligands prefer to bind to the *same* conformation of an allosteric protein, it follows from basic thermodynamic principles that each ligand must increase the affinity of the protein for the other. Thus, if the shift of the protein in Figure 3–57 to the closed conformation that binds glucose best also causes the binding site for X to fit X better, then the protein will bind glucose more tightly when X is present than when X is absent.

Conversely, linkage operates in a negative way if two ligands prefer to bind to *different* conformations of the same protein. In this case, the binding of the first ligand discourages the binding of the second ligand. Thus, if a shape change caused by glucose binding decreases the affinity of a protein for molecule X, the binding of X must also decrease the protein's affinity for glucose (Figure 3–58). The linkage relationship is quantitatively reciprocal, so that, for example, if glucose has a very large effect on the binding of X, X has a very large effect on the binding of glucose.

The relationships shown in Figures 3–57 and 3–58 underlie all of cell biology. They seem so obvious in retrospect that we now take them for granted. But their discovery in the 1950s, followed by a general description of allostery in the early 1960s, was revolutionary at the time. Since molecule X in these examples

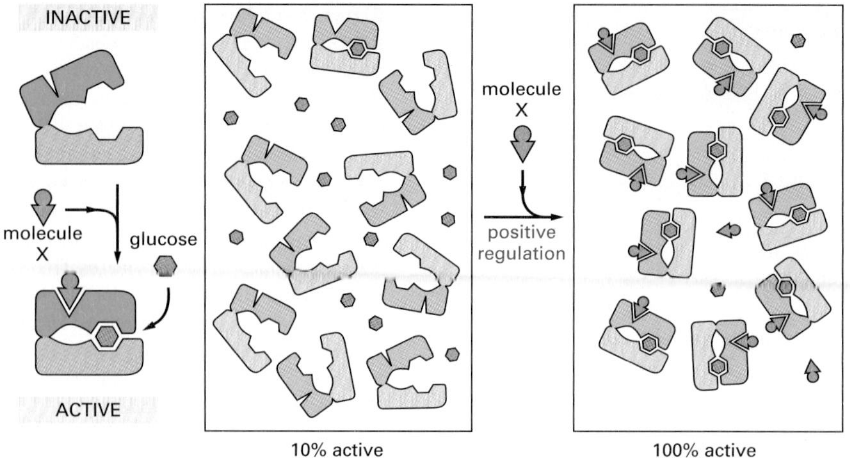

Figure 3–57 Positive regulation caused by conformational coupling between two distant binding sites. In this example, both glucose and molecule X bind best to the *closed* conformation of a protein with two domains. Because both glucose and molecule X drive the protein toward its closed conformation, each ligand helps the other to bind. Glucose and molecule X are therefore said to bind *cooperatively* to the protein.

binds at a site that is distinct from the site where catalysis occurs, it need have no chemical relationship to glucose or to any other ligand that binds at the active site. As we have just seen, for enzymes that are regulated in this way, molecule X can either turn the enzyme on (positive regulation) or turn it off (negative regulation). By such a mechanism, **allosteric proteins** serve as general switches that, in principle, allow one molecule in a cell to affect the fate of any other.

Symmetric Protein Assemblies Produce Cooperative Allosteric Transitions

A single subunit enzyme that is regulated by negative feedback can at most decrease from 90% to about 10% activity in response to a 100-fold increase in the concentration of an inhibitory ligand that it binds (Figure 3–59, *red line*). Responses of this type are apparently not sharp enough for optimal cell regulation, and most enzymes that are turned on or off by ligand binding consist of symmetric assemblies of identical subunits. With this arrangement, the binding of a molecule of ligand to a single site on one subunit can trigger an allosteric change in the subunit that can be transmitted to the neighboring subunits, helping them to bind the same ligand. As a result, a *cooperative allosteric transition* occurs (Figure 3–59, *blue line*), allowing a relatively small change in ligand concentration in the cell to switch the whole assembly from an almost fully active to an almost fully inactive conformation (or vice versa).

The principles involved in a cooperative "all-or-none" transition are easiest to visualize for an enzyme that forms a symmetric dimer. In the example shown in Figure 3–60, the first molecule of an inhibitory ligand binds with great difficulty since its binding destroys an energetically favorable interaction between the two identical monomers in the dimer. A second molecule of inhibitory ligand now binds more easily, however, because its binding restores the

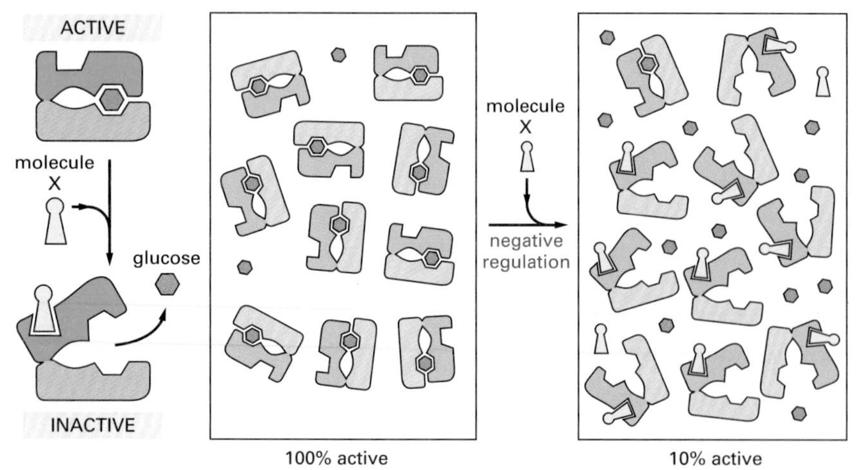

Figure 3–58 Negative regulation caused by conformational coupling between two distant binding sites. The scheme here resembles that in the previous figure, but here molecule X prefers the *open* conformation, while glucose prefers the *closed* conformation. Because glucose and molecule X drive the protein toward opposite conformations (closed and open, respectively), the presence of either ligand interferes with the binding of the other.

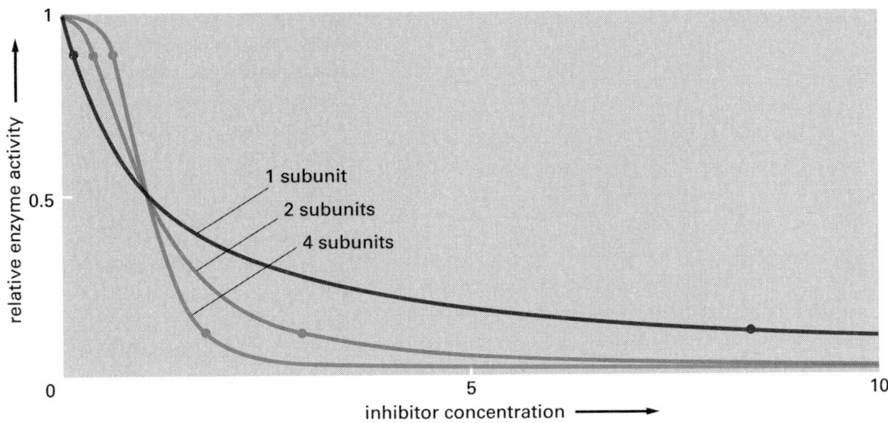

Figure 3–59 Enzyme activity versus the concentration of inhibitory ligand for single-subunit and multi-subunit allosteric enzymes. For an enzyme with a single subunit *(red line)*, a drop from 90% enzyme activity to 10% activity (indicated by the two dots on the curve) requires a 100-fold increase in the concentration of inhibitor. The enzyme activity is calculated from the simple equilibrium relationship $K = [I][P]/[IP]$, where P is active protein, I is inhibitor, and IP is the inactive protein bound to inhibitor. An identical curve applies to any simple binding interaction between two molecules, A and B. In contrast, a multisubunit allosteric enzyme can respond in a switchlike manner to a change in ligand concentration: the steep response is caused by a cooperative binding of the ligand molecules, as explained in Figure 3–60. Here, the *green line* represents the idealized result expected for the cooperative binding of two inhibitory ligand molecules to an allosteric enzyme with two subunits, and the *blue line* shows the idealized response of an enzyme with four subunits. As indicated by the two dots on each of these curves, the more complex enzymes drop from 90% to 10% activity over a much narrower range of inhibitor concentration than does the enzyme composed of a single subunit.

monomer–monomer contacts of a symmetric dimer (and also completely inactivates the enzyme).

An even sharper response to a ligand can be obtained with larger assemblies, such as the enzyme formed from 12 polypeptide chains that we discuss next.

The Allosteric Transition in Aspartate Transcarbamoylase Is Understood in Atomic Detail

One enzyme used in the early studies of allosteric regulation was aspartate transcarbamoylase from *E. coli*. It catalyzes the important reaction that begins the synthesis of the pyrimidine ring of C, U, and T nucleotides: carbamoylphosphate + aspartate → N-carbamoylaspartate. One of the final products of this pathway, cytosine triphosphate (CTP), binds to the enzyme to turn it off whenever CTP is plentiful.

Aspartate transcarbamoylase is a large complex of six regulatory and six catalytic subunits. The catalytic subunits are present as two trimers, each arranged like an equilateral triangle; the two trimers face each other and are held together by three regulatory dimers that form a bridge between them. The entire molecule is poised to undergo a concerted, all-or-none, allosteric transition between two conformations, designated as T (tense) and R (relaxed) states (Figure 3–61).

The binding of substrates (carbamoylphosphate and aspartate) to the catalytic trimers drives aspartate transcarbamoylase into its catalytically active R state, from which the regulatory CTP molecules dissociate. By contrast, the binding of CTP to the regulatory dimers converts the enzyme to the inactive T state, from which the substrates dissociate. This tug-of-war between CTP and substrates is identical in principle to that described previously in Figure 3–58 for a simpler allosteric protein. But because the tug-of-war occurs in a symmetric molecule with multiple binding sites, the enzyme undergoes a cooperative allosteric transition that will turn it on suddenly as substrates accumulate

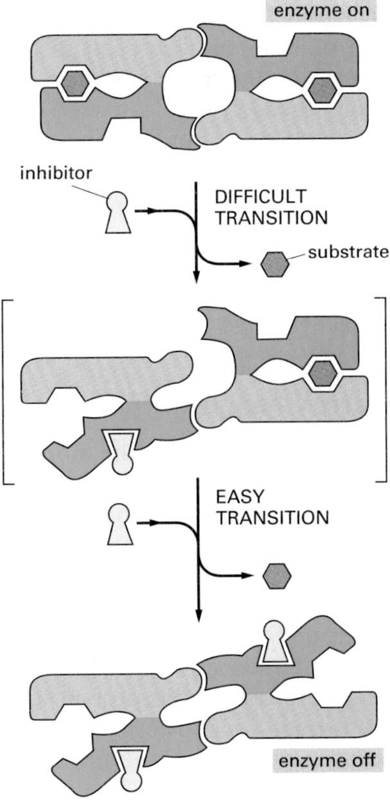

Figure 3–60 A cooperative allosteric transition in an enzyme composed of two identical subunits. This diagram illustrates how the conformation of one subunit can influence that of its neighbor. The binding of a single molecule of an inhibitory ligand *(yellow)* to one subunit of the enzyme occurs with difficulty because it changes the conformation of this subunit and thereby destroys the symmetry of the enzyme. Once this conformational change has occurred, however, the energy gained by restoring the symmetric pairing interaction between the two subunits makes it especially easy for the second subunit to bind the inhibitory ligand and undergo the same conformational change. Because the binding of the first molecule of ligand increases the affinity with which the other subunit binds the same ligand, the response of the enzyme to changes in the concentration of the ligand is much steeper than the response of an enzyme with only one subunit (see Figure 3–59).

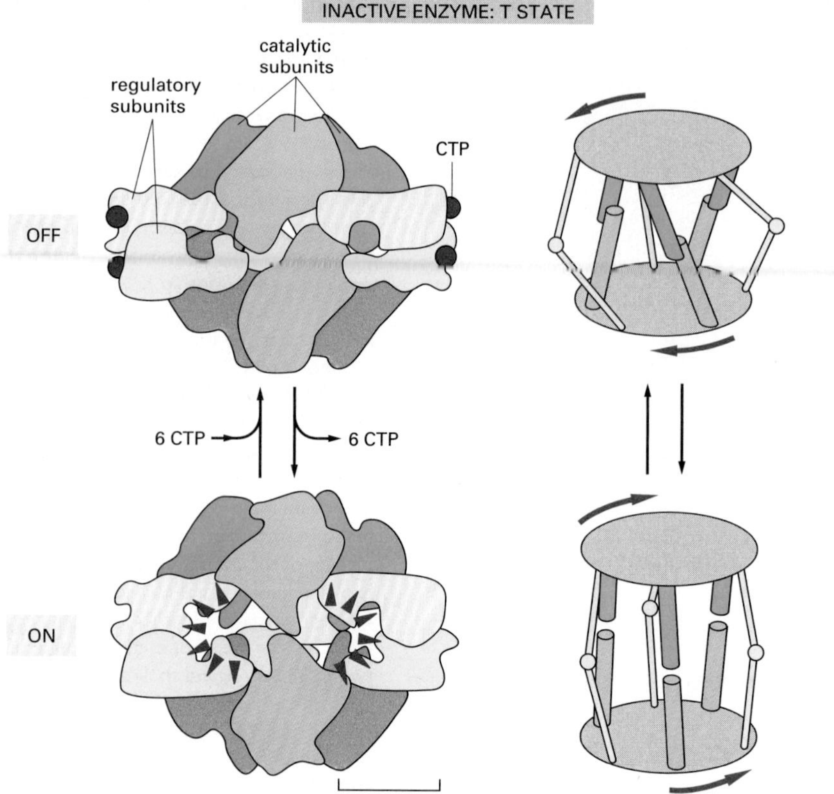

Figure 3–61 The transition between R and T states in the enzyme aspartate transcarbamoylase. The enzyme consists of a complex of six catalytic subunits and six regulatory subunits, and the structures of its inactive (T state) and active (R state) forms have been determined by x-ray crystallography. The enzyme is turned off by feedback inhibition when CTP concentrations rise. Each regulatory subunit can bind one molecule of CTP, which is one of the final products in the pathway. By means of this negative feedback regulation, the pathway is prevented from producing more CTP than the cell needs. (Based on K.L. Krause, K.W. Volz, and W.N. Lipscomb, *Proc. Natl. Acad. Sci. USA* 82:1643–1647, 1985.)

(forming the R state) or shut it off rapidly when CTP accumulates (forming the T state).

A combination of biochemistry and x-ray crystallography has revealed many fascinating details of this allosteric transition. Each regulatory subunit has two domains, and the binding of CTP causes the two domains to move relative to each other, so that they function like a lever that rotates the two catalytic trimers and pulls them closer together into the T state (see Figure 3–61). When this occurs, hydrogen bonds form between opposing catalytic subunits that help widen the cleft that forms the active site within each catalytic subunit, thereby destroying the binding sites for the substrates (Figure 3–62). Adding large amounts of substrate has the opposite effect, favoring the R state by binding in the cleft of each catalytic subunit and opposing the above conformational change. Conformations that are intermediate between R and T are unstable, so that the enzyme mostly clicks back and forth between its R and T forms, producing a mixture of these two species in proportions that depend on the relative concentrations of CTP and substrates.

Many Changes in Proteins Are Driven by Phosphorylation

Enzymes are regulated by more than the binding of small molecules. A second method that is commonly used by eucaryotic cells to regulate a protein's function is the covalent addition of a phosphate group to one of its amino acid side chains. Such *phosphorylation* events can affect the protein in two important ways.

First, because each phosphate group carries two negative charges, the enzyme-catalyzed addition of a phosphate group to a protein can cause a major conformational change in the protein by, for example, attracting a cluster of positively charged amino acid side chains. This can, in turn, affect the binding of ligands elsewhere on the protein surface, dramatically changing the protein's activity through an allosteric effect. Removal of the phosphate group by a second enzyme returns the protein to its original conformation and restores its initial activity.

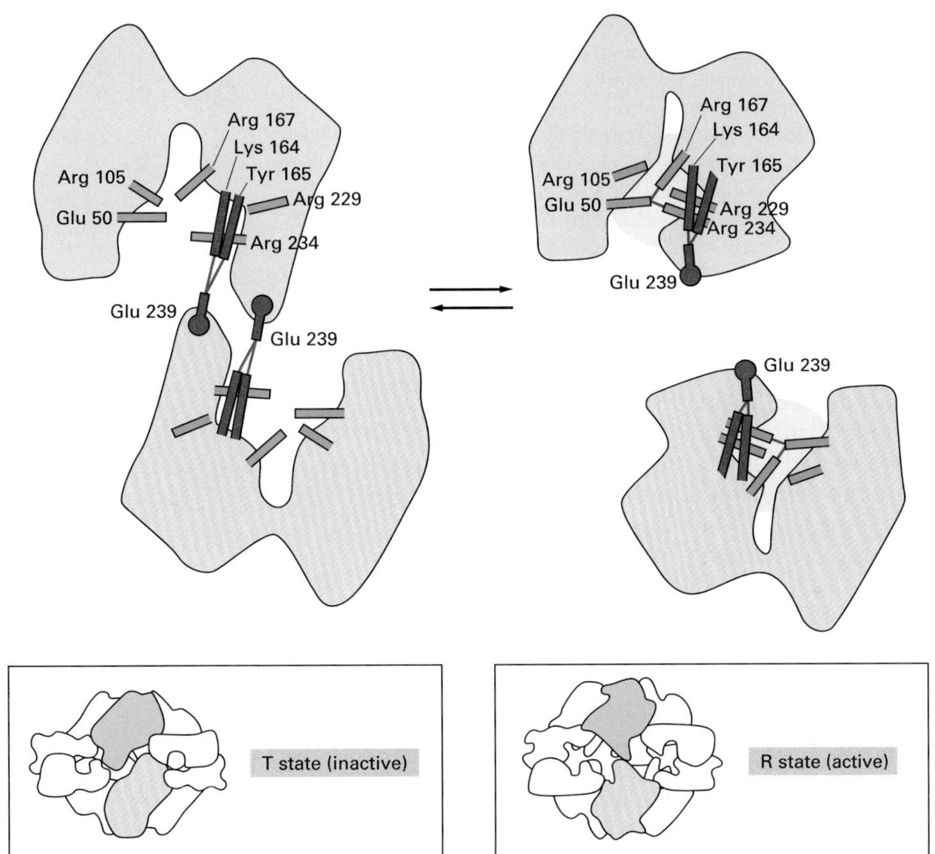

Figure 3–62 Part of the on–off switch in the catalytic subunits of aspartate transcarbamoylase. Changes in the indicated hydrogen-bonding interactions are partly responsible for switching this enzyme's active site between active *(yellow)* and inactive conformations. Hydrogen bonds are indicated by *thin red lines*. The amino acids involved in the subunit–subunit interaction are shown in *red*, while those that form the active site of the enzyme are shown in *blue*. The large drawings show the catalytic site in the interior of the enzyme; the boxed sketches show the same subunits viewed from the enzyme's external surface. (Adapted from E.R. Kantrowitz and W.N. Lipscomb, *Trends Biochem. Sci.* 15:53–59, 1990.)

Second, an attached phosphate group can form part of a structure that is directly recognized by binding sites of other proteins. As previously discussed, certain small protein domains, called modules, appear very frequently in larger proteins. A large number of these modules provide binding sites for attaching their protein to phosphorylated peptides in other protein molecules. One of these modules is the SH2 domain, featured previously in this chapter, which binds to a short peptide sequence containing a phosphorylated tyrosine side chain (see Figure 3–40B). Several other types of modules recognize phosphorylated serine or threonine side chains in a specific context. As a result, protein phosphorylation and dephosphorylation events have a major role in driving the regulated assembly and disassembly of protein complexes.

Through such effects, reversible protein phosphorylation controls the activity structure and cellular localization of many types of proteins in eucaryotic cells. In fact, this regulation is so extensive that more than one-third of the 10,000 or so proteins in a typical mammalian cell are thought to be phosphorylated at any given time—many with more than one phosphate. As might be expected, the addition and removal of phosphate groups from specific proteins often occur in response to signals that specify some change in a cell's state. For example, the complicated series of events that takes place as a eucaryotic cell divides is largely timed in this way (discussed in Chapter 17), and many of the signals mediating cell–cell interactions are relayed from the plasma membrane to the nucleus by a cascade of protein phosphorylation events (discussed in Chapter 15).

A Eucaryotic Cell Contains a Large Collection of Protein Kinases and Protein Phosphatases

Protein phosphorylation involves the enzyme-catalyzed transfer of the terminal phosphate group of an ATP molecule to the hydroxyl group on a serine, threonine, or tyrosine side chain of the protein (Figure 3–63). This reaction is catalyzed by a **protein kinase**, and the reaction is essentially unidirectional because

Figure 3–63 Protein phosphorylation. Many thousands of proteins in a typical eucaryotic cell are modified by the covalent addition of a phosphate group. (A) The general reaction, shown here, entails the transfer of a phosphate group from ATP to an amino acid side chain of the target protein by a protein kinase. Removal of the phosphate group is catalyzed by a second enzyme, a protein phosphatase. In this example, the phosphate is added to a serine side chain; in other cases, the phosphate is instead linked to the –OH group of a threonine or a tyrosine in the protein. (B) The phosphorylation of a protein by a protein kinase can either increase or decrease the protein's activity, depending on the site of phosphorylation and the structure of the protein.

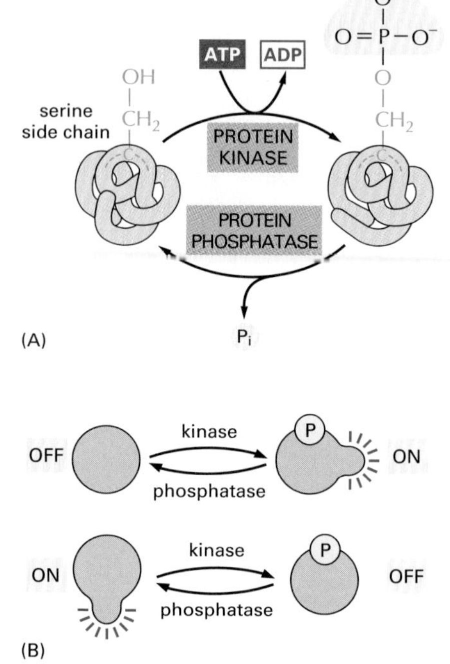

of the large amount of free energy released when the phosphate–phosphate bond in ATP is broken to produce ADP (discussed in Chapter 2). The reverse reaction of phosphate removal, or *dephosphorylation,* is instead catalyzed by a **protein phosphatase**. Cells contain hundreds of different protein kinases, each responsible for phosphorylating a different protein or set of proteins. There are also many different protein phosphatases; some of these are highly specific and remove phosphate groups from only one or a few proteins, whereas others act on a broad range of proteins and are targeted to specific substrates by regulatory subunits. The state of phosphorylation of a protein at any moment, and thus its activity, depends on the relative activities of the protein kinases and phosphatases that modify it.

The protein kinases that phosphorylate proteins in eucaryotic cells belong to a very large family of enzymes, which share a catalytic (kinase) sequence of 250 amino acids. The various family members contain different amino acid sequences on either side of the kinase sequence (see Figure 3–12), and often have short amino acid sequences inserted into loops within it *(red arrowheads* in Figure 3–64). Some of these additional amino acid sequences enable each kinase to recognize the specific set of proteins it phosphorylates, or to bind to structures that localize it in specific regions of the cell. Other parts of the protein allow the activity of each enzyme to be tightly regulated, so it can be turned on and off in response to different specific signals, as described below.

By comparing the number of amino acid sequence differences between the various members of a protein family, one can construct an "evolutionary tree" that is thought to reflect the pattern of gene duplication and divergence that gave rise to the family. An evolutionary tree of protein kinases is shown in Figure 3–65. Not surprisingly, kinases with related functions are often located on nearby branches of the tree: the protein kinases involved in cell signaling that phosphorylate tyrosine side chains, for example, are all clustered in the top left corner of the tree. The other kinases shown phosphorylate either a serine or a threonine side chain, and many are organized into clusters that seem to reflect their function—in transmembrane signal transduction, intracellular signal amplification, cell-cycle control, and so on.

As a result of the combined activities of protein kinases and protein phosphatases, the phosphate groups on proteins are continually turning over—being added and then rapidly removed. Such phosphorylation cycles may seem wasteful, but they are important in allowing the phosphorylated proteins to switch rapidly from one state to another: the more rapid the cycle, the faster the state of phosphorylation of a population of protein molecules can change in response to a sudden stimulus that changes the phosphorylation rate (see Figure 15–10). The

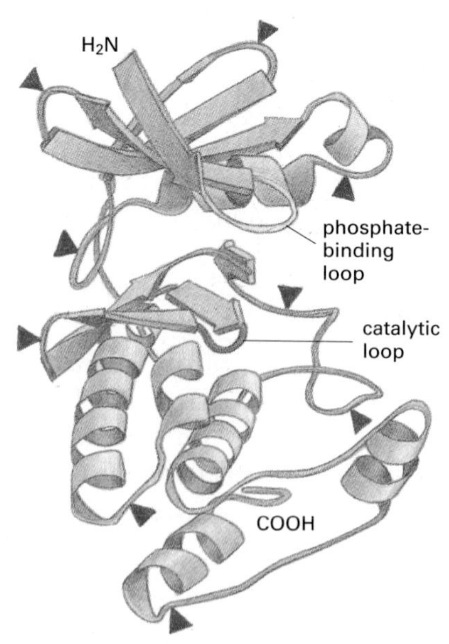

Figure 3–64 The three-dimensional structure of a protein kinase. Superimposed on this structure are *red arrowheads* to indicate sites where insertions of 5–100 amino acids are found in some members of the protein kinase family. These insertions are located in loops on the surface of the enzyme where other ligands interact with the protein. Thus, they distinguish different kinases and confer on them distinctive interactions with other proteins. The ATP (which donates a phosphate group) and the peptide to be phosphorylated are held in the active site, which extends between the phosphate-binding loop *(yellow)* and the catalytic loop *(orange).* See also Figure 3–12. (Adapted from D.R. Knighton et al., *Science* 253:407–414, 1991.)

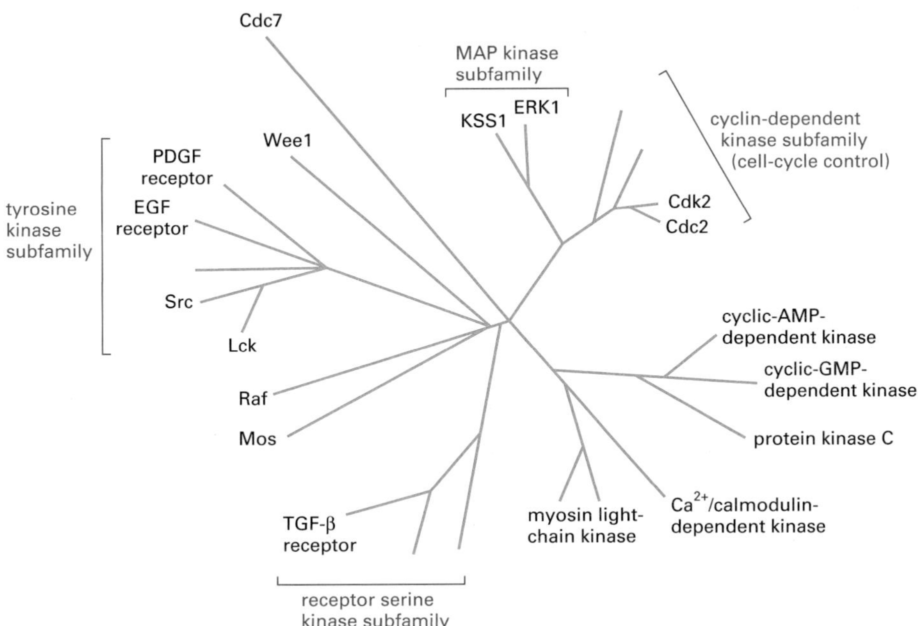

Figure 3–65 An evolutionary tree of selected protein kinases. Although a higher eucaryotic cell contains hundreds of such enzymes, and the human genome codes for more than 500, only some of those discussed in this book are shown.

energy required to drive the phosphorylation cycle is derived from the free energy of ATP hydrolysis, one molecule of which is consumed for each phosphorylation event.

The Regulation of Cdk and Src Protein Kinases Shows How a Protein Can Function as a Microchip

The hundreds of different protein kinases in a eucaryotic cell are organized into complex networks of signaling pathways that help to coordinate the cell's activities, drive the cell cycle, and relay signals into the cell from the cell's environment. Many of the extracellular signals involved need to be both integrated and amplified by the cell. Individual protein kinases (and other signaling proteins) serve as input–output devices, or "microchips," in the integration process. An important part of the input to these proteins comes from the control that is exerted by phosphates added and removed from them by protein kinases and protein phosphatases, respectively, in the signaling network.

For a protein that is phosphorylated at multiple sites, specific sets of phosphate groups serve to activate the protein, while other sets can inactivate it. A cyclin-dependent protein kinase (Cdk) represents a good example of such a signal processing device. Kinases in this class phosphorylate serines, and they are central components of the cell-cycle control system in eucaryotic cells, as discussed in detail in Chapter 17. In a vertebrate cell, individual Cdk enzymes turn on and off in succession, as a cell proceeds through the different phases of its division cycle. When a particular one of these kinases is on, it influences various aspects of cell behavior through effects on the proteins it phosphorylates.

A Cdk protein becomes active as a serine/threonine protein kinase only when it is bound to a second protein called a *cyclin*. But, as Figure 3–66 shows, the binding of cyclin is only one of three distinct "inputs" required to activate the Cdk. In addition to cyclin binding, a phosphate must be added to a specific threonine side chain, and a phosphate elsewhere in the protein (covalently bound to a specific tyrosine side chain) must be removed. Cdk thus monitors a specific set of cell components—a cyclin, a protein kinase, and a protein phosphatase—and it acts as an input–output device that turns on if, and only if, each of these components has attained its appropriate activity state. Some cyclins rise and fall in concentration in step with the cell cycle, increasing gradually in amount until they are suddenly destroyed at a particular point in the cycle. The sudden destruction of a cyclin (by targeted proteolysis) immediately shuts off its partner Cdk enzyme, and this triggers a specific step in the cell cycle.

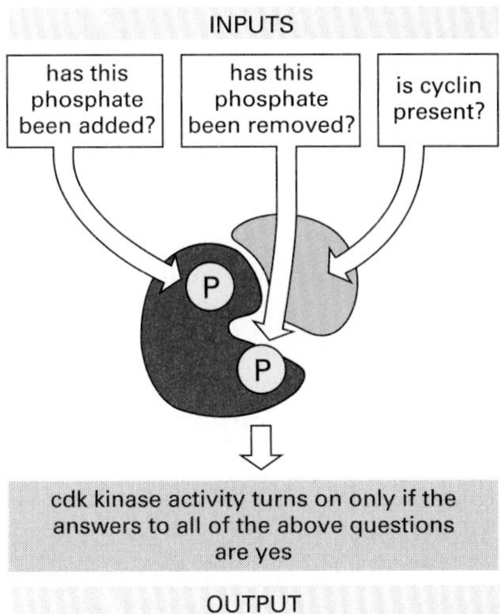

Figure 3–66 How a Cdk protein acts as an integrating device. The function of these central regulators of the cell cycle is discussed in Chapter 17.

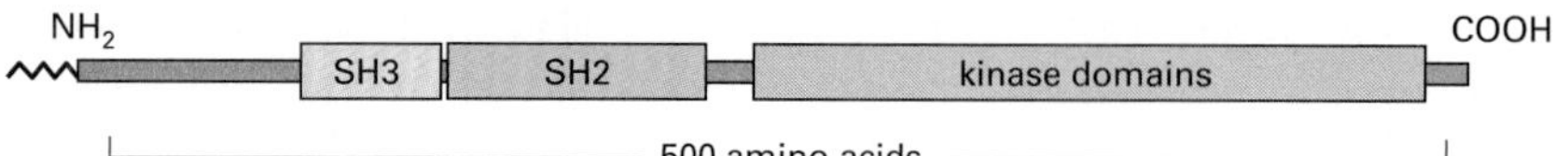

Figure 3–67 The domain structure of the Src family of protein kinases, mapped along the amino acid sequence.

A similar type of microchip behavior is exhibited by the Src family of protein kinases. The *Src protein* (pronounced "sarc") was the first tyrosine kinase to be discovered, and it is now known to be part of a subfamily of nine very similar protein kinases, which are found only in multicellular animals. As indicated by the evolutionary tree in Figure 3–65, sequence comparisons suggest that tyrosine kinases as a group were a relatively late innovation that branched off from the serine/threonine kinases, with the Src subfamily being only one subgroup of the tyrosine kinases created in this way.

The Src protein and its homologs contain a short N-terminal region that becomes covalently linked to a strongly hydrophobic fatty acid, which holds the kinase at the cytoplasmic face of the plasma membrane. Next come two peptide-binding modules, a Src homology 3 (SH3) domain and a SH2 domain, followed by the kinase catalytic domains (Figure 3–67). These kinases normally exist in an inactive conformation, in which a phosphorylated tyrosine near the C-terminus is bound to the SH2 domain, and the SH3 domain is bound to an internal peptide in a way that distorts the active site of the enzyme and helps to render it inactive (see Figure 3–12).

Turning the kinase on involves at least two specific inputs: removal of the C-terminal phosphate and the binding of the SH3 domain by a specific activating protein (Figure 3–68). As for the Cdk protein, the activation of the Src kinase signals that a particular set of separate upstream events has been completed (Figure 3–69). Thus, both the Cdk and Src families of proteins serve as specific signal integrators, helping to generate the complex web of information-processing events that enable the cell to compute logical responses to a complex set of conditions.

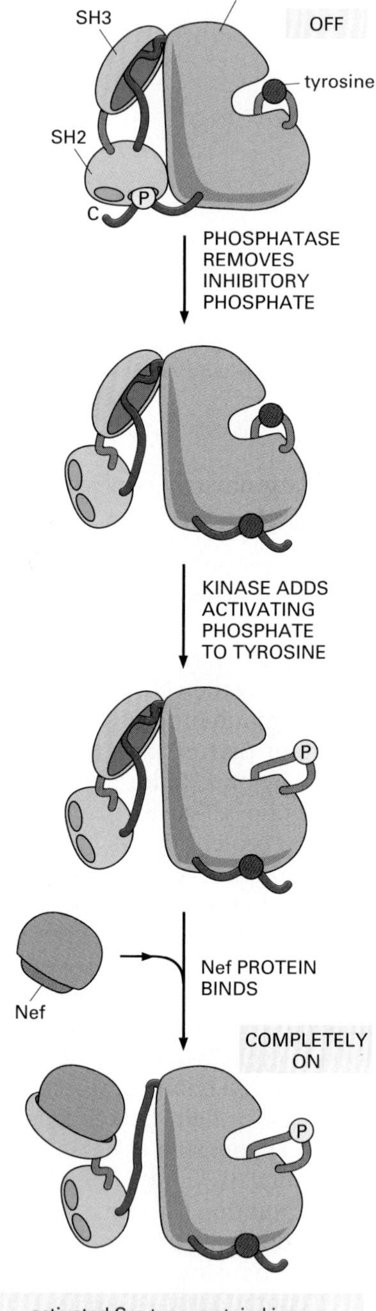

Figure 3–68 The activation of a Src-type protein kinase by two sequential events. (Adapted from I. Moareti, et al., *Nature* 385:650–653, 1997.)

Proteins That Bind and Hydrolyze GTP Are Ubiquitous Cellular Regulators

We have described how the addition or removal of phosphate groups on a protein can be used by a cell to control the protein's activity. In the examples discussed so far, the phosphate is transferred from an ATP molecule to an amino acid side chain of the protein in a reaction catalyzed by a specific protein kinase. Eucaryotic cells also have another way to control protein activity by phosphate addition and removal. In this case, the phosphate is not attached directly to the protein; instead, it is a part of the guanine nucleotide GTP, which binds very tightly to the protein. In general, proteins regulated in this way are in their active conformations with GTP bound. The loss of a phosphate group occurs when the bound GTP is hydrolyzed to GDP in a reaction catalyzed by the protein itself, and in its GDP bound state the protein is inactive. In this way, GTP-binding proteins act as on–off switches whose activity is determined by the presence or absence of an additional phosphate on a bound GDP molecule (Figure 3–70).

GTP-binding proteins (also called **GTPases** because of the GTP hydrolysis they catalyze) constitute a large family of proteins that all contain variations on the same GTP-binding globular domain. When the tightly bound GTP is hydrolyzed to GDP, this domain undergoes a conformational change that inactivates it. The three-dimensional structure of a prototypical member of this family, the monomeric GTPase called Ras, is shown in Figure 3–71.

The *Ras protein* has an important role in cell signaling (discussed in Chapter 15). In its GTP-bound form, it is active and stimulates a cascade of protein phosphorylations in the cell. Most of the time, however, the protein is in its inactive, GDP-bound form. It becomes active when it exchanges its GDP for a GTP molecule in response to extracellular signals, such as growth factors, that bind to receptors in the plasma membrane (see Figure 15–55).

Figure 3–69 How a Src-type protein kinase acts as an integrating device. The disruption of the SH3 domain interaction *(green)* can involve replacing its binding to the indicated *red* linker region with a tighter interaction with the Nef protein, as illustrated in Figure 3–68.

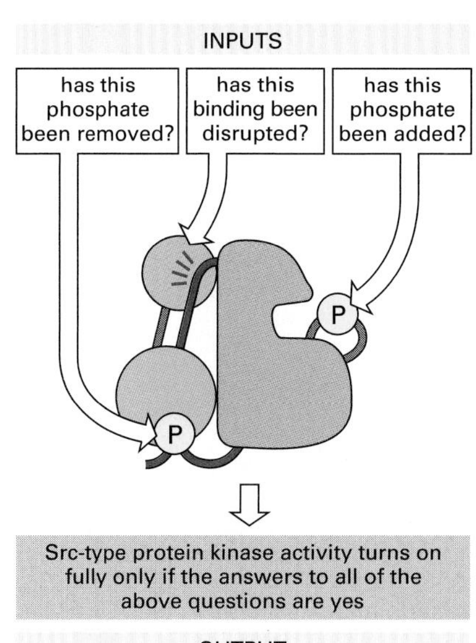

Regulatory Proteins Control the Activity of GTP-Binding Proteins by Determining Whether GTP or GDP Is Bound

GTP-binding proteins are controlled by regulatory proteins that determine whether GTP or GDP is bound, just as phosphorylated proteins are turned on and off by protein kinases and protein phosphatases. Thus, Ras is inactivated by a *GTPase-activating protein (GAP)*, which binds to the Ras protein and induces it to hydrolyze its bound GTP molecule to GDP—which remains tightly bound—and inorganic phosphate (P_i)—which is rapidly released. The Ras protein stays in its inactive, GDP-bound conformation until it encounters a *guanine nucleotide exchange factor (GEF)*, which binds to GDP-Ras and causes it to release its GDP. Because the empty nucleotide-binding site is immediately filled by a GTP molecule (GTP is present in large excess over GDP in cells), the GEF activates Ras by *indirectly* adding back the phosphate removed by GTP hydrolysis. Thus, in a sense, the roles of GAP and GEF are analogous to those of a protein phosphatase and a protein kinase, respectively (Figure 3–72).

Large Protein Movements Can Be Generated From Small Ones

The Ras protein belongs to a large superfamily of *monomeric GTPases*, each of which consists of a single GTP-binding domain of about 200 amino acids. Over the course of evolution, this domain has also become joined to larger proteins with additional domains, creating a large family of GTP-binding proteins. Family members include the receptor-associated trimeric G proteins involved in cell signaling (discussed in Chapter 15), proteins regulating the traffic of vesicles between intracellular compartments (discussed in Chapter 13), and proteins that bind to transfer RNA and are required as assembly factors for protein synthesis on the ribosome (discussed in Chapter 6). In each case, an important biological activity is controlled by a change in the protein's conformation that is caused by GTP hydrolysis in a Ras-like domain.

The *EF-Tu protein* provides a good example of how this family of proteins works. EF-Tu is an abundant molecule that serves as an elongation factor (hence the EF) in protein synthesis, loading each amino-acyl tRNA molecule onto the ribosome. The tRNA molecule forms a tight complex with the GTP-bound form of EF-Tu (Figure 3–73). In this complex, the amino acid attached to the tRNA is masked. Unmasking is required for the tRNA to transfer its amino acid in protein synthesis, and it occurs on the ribosome when the GTP bound to EF-Tu is

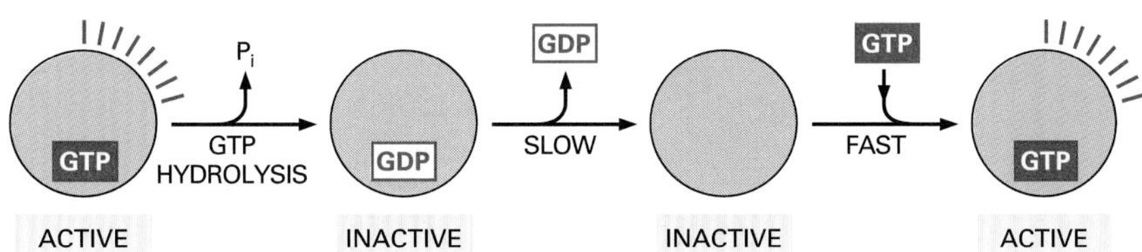

Figure 3–70 GTP-binding proteins as molecular switches. The activity of a GTP-binding protein (also called a GTPase) generally requires the presence of a tightly bound GTP molecule (switch "on"). Hydrolysis of this GTP molecule produces GDP and inorganic phosphate (P_i), and it causes the protein to convert to a different, usually inactive, conformation (switch "off"). As shown here, resetting the switch requires the tightly bound GDP to dissociate, a slow step that is greatly accelerated by specific signals; once the GDP has dissociated, a molecule of GTP is quickly rebound.

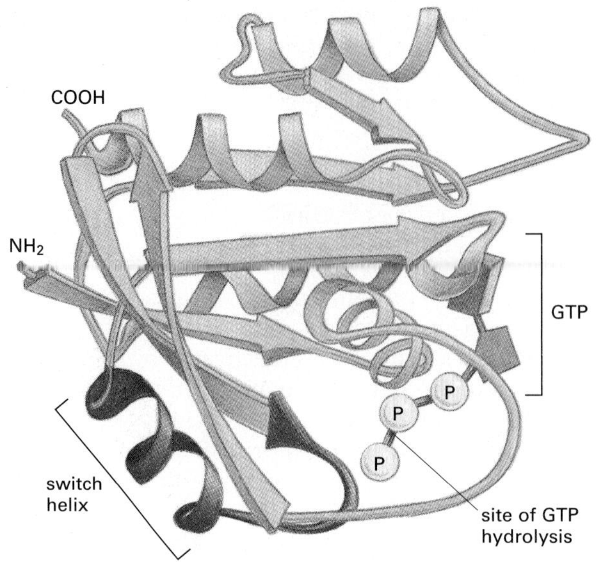

Figure 3–71 The structure of the Ras protein in its GTP-bound form. This monomeric GTPase illustrates the structure of a GTP-binding domain, which is present in a large family of GTP-binding proteins. The *red* regions change their conformation when the GTP molecule is hydrolyzed to GDP and inorganic phosphate by the protein; the GDP remains bound to the protein, while the inorganic phosphate is released. The special role of the "switch helix" in proteins related to Ras is explained next (see Figure 3–74).

hydrolyzed, allowing the tRNA to dissociate. Since the GTP hydrolysis is triggered by a proper fit of the tRNA to the mRNA molecule on the ribosome, the EF-Tu serves as an assembly factor that discriminates between correct and incorrect mRNA–tRNA pairings (see Figure 6–66 for a further discussion of this function of EF-Tu).

Comparison of the three-dimensional structure of EF-Tu in its GTP-bound and GDP-bound forms reveals how the unmasking of the tRNA occurs. The dissociation of the inorganic phosphate group (P_i), which follows the reaction GTP $\rightarrow$ GDP + P_i, causes a shift of a few tenths of a nanometer at the GTP-binding site, just as it does in the Ras protein. This tiny movement, equivalent to a few times the diameter of a hydrogen atom, causes a conformational change to propagate along a crucial piece of α helix, called the *switch helix*, in the Ras-like domain of the protein. The switch helix seems to serve as a latch that adheres to a specific site in another domain of the molecule, holding the protein in a "shut" conformation. The conformational change triggered by GTP hydrolysis causes the switch helix to detach, allowing separate domains of the protein to swing apart,

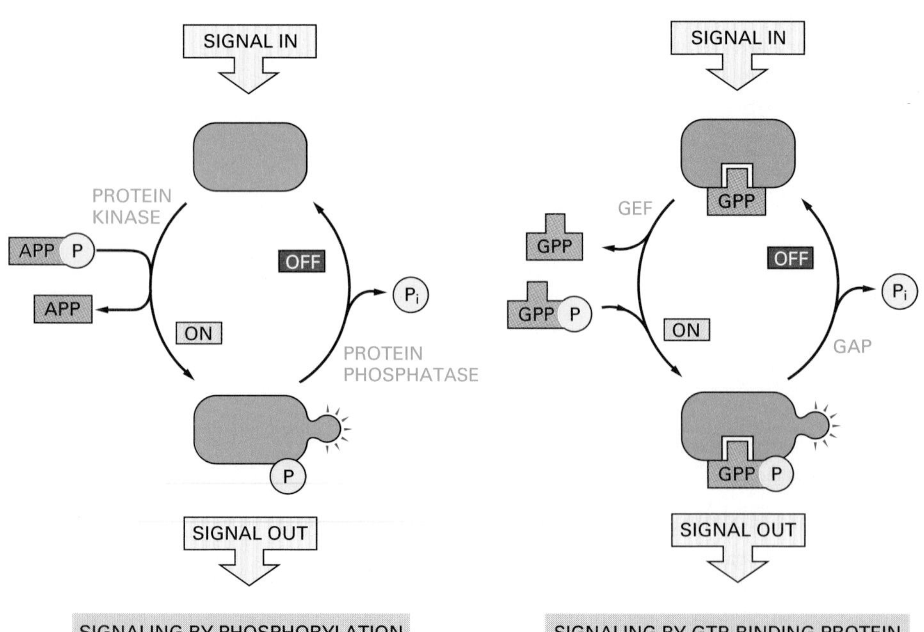

Figure 3–72 A comparison of the two major intracellular signaling mechanisms in eucaryotic cells. In both cases, a signaling protein is activated by the addition of a phosphate group and inactivated by the removal of this phosphate. To emphasize the similarities in the two pathways, ATP and GTP are drawn as APPP and GPPP, and ADP and GDP as APP and GPP, respectively. As shown in Figure 3–63, the addition of a phosphate to a protein can also be inhibitory.

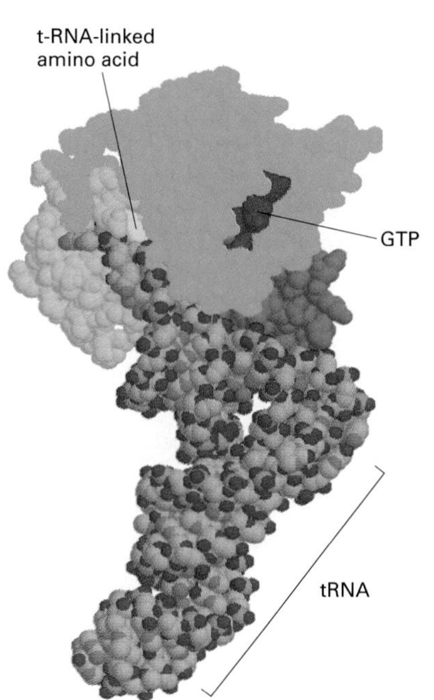

Figure 3–73 An aminoacyl tRNA molecule bound to EF-Tu. The three domains of the EF-Tu protein are colored differently, to match Figure 3–74. This is a bacterial protein; however, a very similar protein exists in eukaryotes, where it is called EF-1. (Coordinates determined by P. Nissen et al., *Science* 270:1464–1472, 1995.)

through a distance of about 4 nm. This releases the bound tRNA molecule, allowing its attached amino acid to be used (Figure 3–74).

One can see from this example how cells have exploited a simple chemical change that occurs on the surface of a small protein domain to create a movement 50 times larger. Dramatic shape changes of this type also underlie the very large movements that occur in motor proteins, as we discuss next.

Motor Proteins Produce Large Movements in Cells

We have seen how conformational changes in proteins have a central role in enzyme regulation and cell signaling. We now discuss proteins whose major function is to move other molecules. These **motor proteins** generate the forces responsible for muscle contraction and the crawling and swimming of cells. Motor proteins also power smaller-scale intracellular movements: they help to move chromosomes to opposite ends of the cell during mitosis (discussed in Chapter 18), to move organelles along molecular tracks within the cell (discussed in Chapter 16), and to move enzymes along a DNA strand during the synthesis of a new DNA molecule (discussed in Chapter 5). All these fundamental processes depend on proteins with moving parts that operate as force-generating machines.

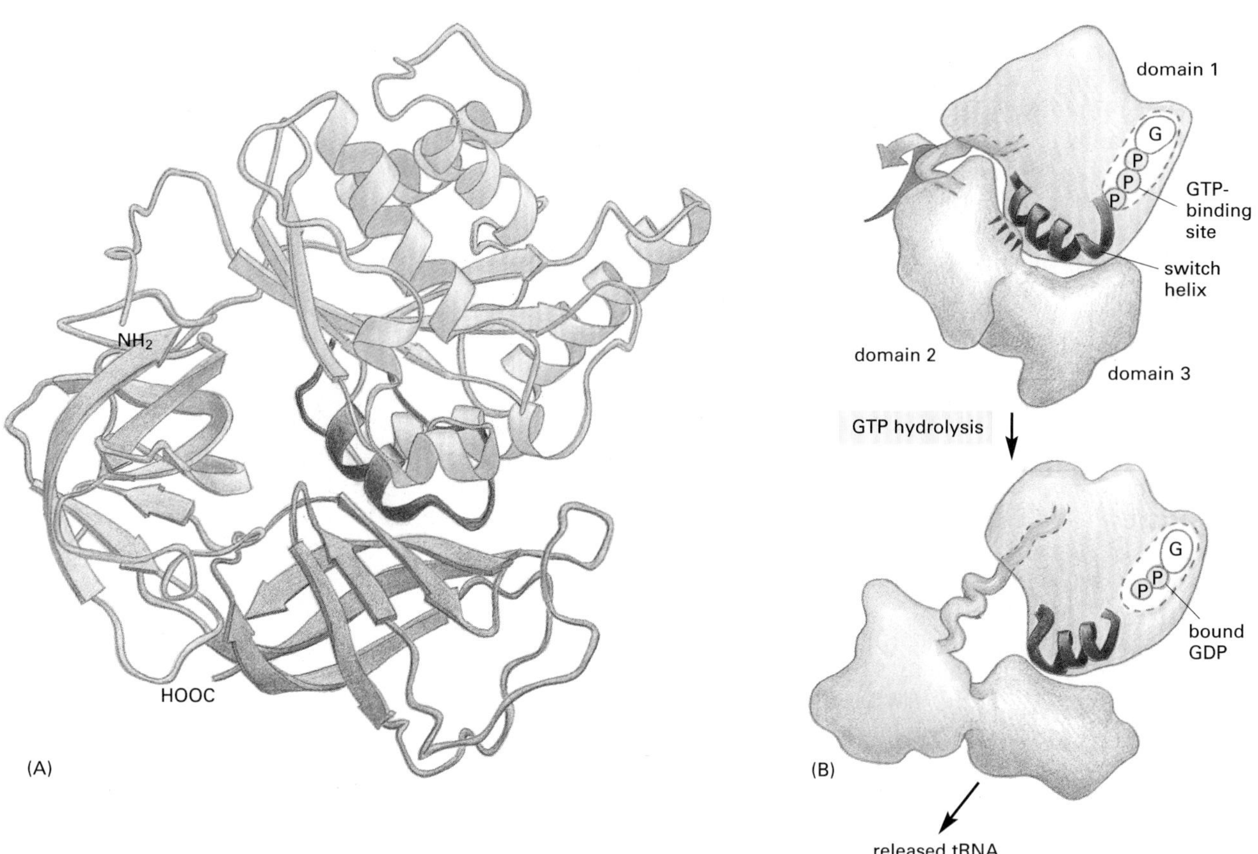

Figure 3–74 The large conformational change in EF-Tu caused by GTP hydrolysis. (A) The three-dimensional structure of EF-Tu with GTP bound. The domain at the top is homologous to the Ras protein, and its *red* α helix is the switch helix, which moves after GTP hydrolysis, as shown in Figure 3–71. (B) The change in the conformation of the switch helix in domain 1 causes domains 2 and 3 to rotate as a single unit by about 90° toward the viewer, which releases the tRNA that was shown bound to this structure in Figure 3–73. (A, adapted from H. Berchtold et al., *Nature* 365:126–132, 1993; B, courtesy of Mathias Sprinzl and Rolf Hilgenfeld.)

How do these machines work? In other words, how are shape changes in proteins used to generate directed movements in cells? If, for example, a protein is required to walk along a narrow thread such as a DNA molecule, it can do this by undergoing a series of conformational changes, such as those shown in Figure 3–75. With nothing to drive these changes in an orderly sequence, however, they are perfectly reversible, and the protein can wander randomly back and forth along the thread. We can look at this situation in another way. Since the directional movement of a protein does work, the laws of thermodynamics (discussed in Chapter 2) demand that such movement utilize free energy from some other source (otherwise the protein could be used to make a perpetual motion machine). Therefore, without an input of energy, the protein molecule can only wander aimlessly.

How, then, can one make the series of conformational changes unidirectional? To force the entire cycle to proceed in one direction, it is enough to make any one of the changes in shape irreversible. For most proteins that are able to walk in one direction for long distances, this is achieved by coupling one of the conformational changes to the hydrolysis of an ATP molecule bound to the protein. The mechanism is similar to the one just discussed that drives allosteric protein shape changes by GTP hydrolysis. Because a great deal of free energy is released when ATP (or GTP) is hydrolyzed, it is very unlikely that the nucleotide-binding protein will undergo the reverse shape change needed for moving backward—since this would require that it also reverse the ATP hydrolysis by adding a phosphate molecule to ADP to form ATP.

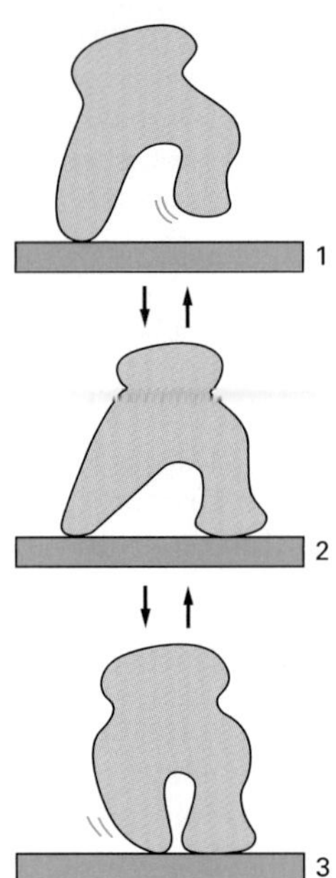

Figure 3–75 An allosteric "walking" protein. Although its three different conformations allow it to wander randomly back and forth while bound to a thread or a filament, the protein cannot move uniformly in a single direction.

In the model shown in Figure 3–76, ATP binding shifts a motor protein from conformation 1 to conformation 2. The bound ATP is then hydrolyzed to produce ADP and inorganic phosphate (P_i), causing a change from conformation 2 to conformation 3. Finally, the release of the bound ADP and P_i drives the protein back to conformation 1. Because the transition $2 \rightarrow 3$ is driven by the energy provided by ATP hydrolysis, this series of conformational changes is effectively irreversible. Thus, the entire cycle goes in only one direction, causing the protein molecule to walk continuously to the right in this example.

Many motor proteins generate directional movement in this general way, including the muscle motor protein *myosin*, which walks along actin filaments to generate muscle contraction, and the *kinesin* proteins that walk along microtubules (both discussed in Chapter 16). These movements can be rapid: some of the motor proteins involved in DNA replication (the DNA helicases) propel themselves along a DNA strand at rates as high as 1000 nucleotides per second.

Membrane-bound Transporters Harness Energy to Pump Molecules Through Membranes

We have thus far seen how allosteric proteins can act as microchips (Cdk and Src kinases), as assembly factors (EF-Tu), and as generators of mechanical force and motion (motor proteins). Allosteric proteins can also harness energy derived from ATP hydrolysis, ion gradients, or electron transport processes to pump specific ions or small molecules across a membrane. We consider one example here; others will be discussed in Chapter 11.

One of the best understood pump proteins is the calcium transport protein from muscle cells. This protein, called the **Ca^{2+} pump**, is embedded in the membrane of a specialized organelle in muscle cells called the *sarcoplasmic reticulum*. The Ca^{2+} pump (also known as the Ca^{2+} ATPase) maintains the low cytoplasmic calcium concentration of resting muscle by pumping calcium out of the cytosol into the membrane-enclosed sarcoplasmic reticulum; then, in response to a nerve impulse, Ca^{2+} is rapidly released (through other channels) back into the cytosol to trigger muscle contraction. The Ca^{2+} pump is homologous to the

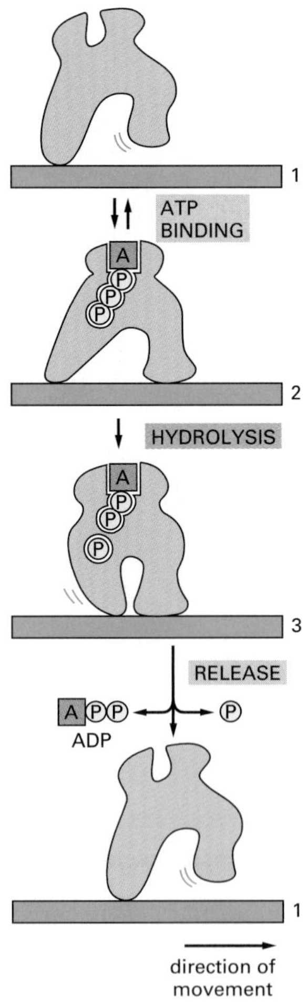

Figure 3–76 An allosteric motor protein. The transition between three different conformations includes a step driven by the hydrolysis of a bound ATP molecule, and this makes the entire cycle essentially irreversible. By repeated cycles, the protein therefore moves continuously to the right along the thread.

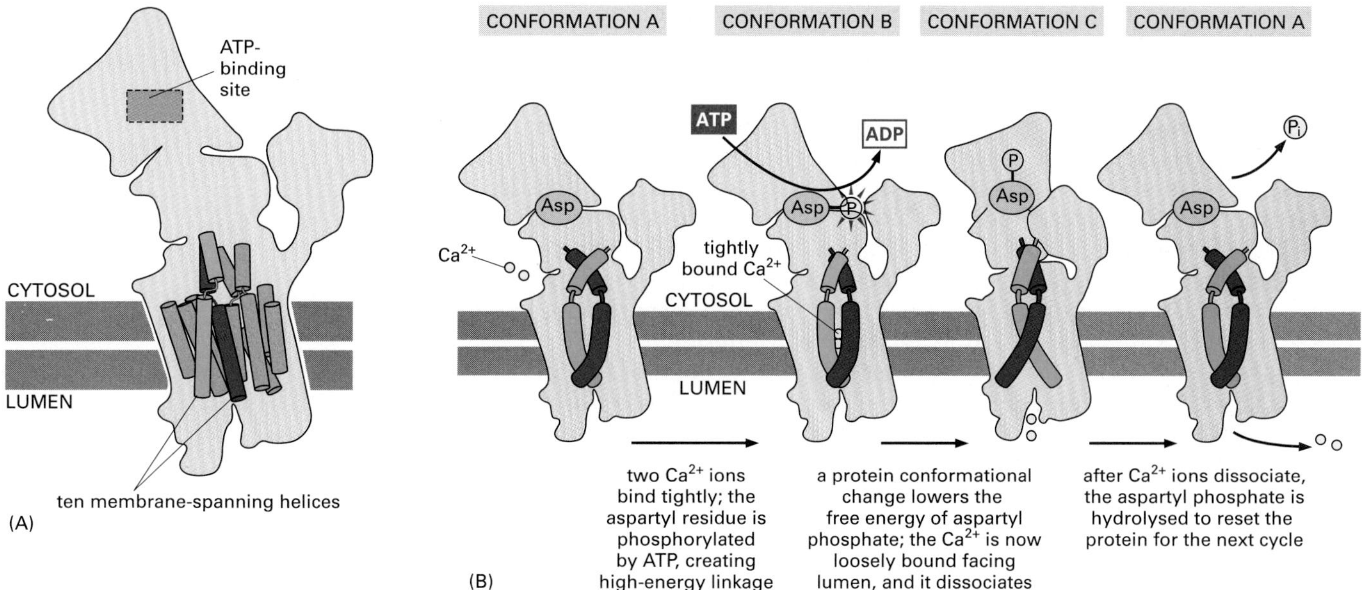
ATP-binding site
CYTOSOL
LUMEN
ten membrane-spanning helices
(A)
CONFORMATION A
CONFORMATION B
CONFORMATION C
CONFORMATION A
Ca2+
Asp
ATP
ADP
P
Pi
tightly bound Ca2+
CYTOSOL
LUMEN
two Ca2+ ions bind tightly; the aspartyl residue is phosphorylated by ATP, creating high-energy linkage
a protein conformational change lowers the free energy of aspartyl phosphate; the Ca2+ is now loosely bound facing lumen, and it dissociates
after Ca2+ ions dissociate, the aspartyl phosphate is hydrolysed to reset the protein for the next cycle
(B)

Figure 3–77 The transport of calcium ions by the Ca^{2+} pump. (A) The structure of the pump protein, formed from a single subunit of 994 amino acids (see Figure 11–15 for details). (B) A model for ion pumping. For simplicity, only two of the four transmembrane α helices that bind Ca^{2+} are shown.

Na^+-K^+ pump that maintains Na^+ and K^+ concentration differences across the plasma membrane, both being members of a family of P-type cation pumps—so named because these proteins are autophosphorylated during their reaction cycle. A large, cytoplasmic head region binds and hydrolyzes ATP, forming a covalent bond with the phosphate released. The protein thereby transiently shifts to a high-energy, phosphorylated state that is tightly bound to two Ca^{2+} ions that were picked up from the cytosol. This form of the protein then decays to a low-energy, phosphorylated state, which causes the Ca^{2+} ions to be released into the lumen. Dephosphorylation of the enzyme finally releases the phosphate and resets the protein for its next round of ion pumping.

The head region of all P-type cation pumps is linked to a series of ten transmembrane α helices, four of which form transmembrane Ca^{2+}-binding sites for the Ca^{2+} pump. The three-dimensional structure of this protein has been deciphered by high-resolution electron microscopy and x-ray diffraction (see Figure 11–15). This has enabled biologists to derive a molecular model for pump action based on extensive biochemical data on its normal and mutant forms.

As illustrated in Figure 3–77A, the transmembrane α helices that bind the Ca^{2+} wind around each other and create a cavity for Ca^{2+} between the helices. ATP hydrolysis generates a series of major conformational changes in the cytoplasmic head, which—through its stalk connection to the transmembrane helices—changes the structure and relative orientations of some of the helices in the membrane, thereby altering the cavity in a way that pushes the Ca^{2+} ions unidirectionally across the membrane (Figure 3–77B). As for a motor protein, unidirectional transport of an ion requires the cycle to use energy, so as to impart a preferred directionality to the protein's conformational changes.

Humans have invented many different types of mechanical pumps, and it should not be surprising that cells also contain membrane-bound pumps that function in other ways. Most notable are the rotary pumps that couple the hydrolysis of ATP to the transport of H^+ ions (protons). These pumps resemble miniature turbines, and they are used to acidify the interior of lysosomes and other eucaryotic organelles. Like other ion pumps that create ion gradients, they can function in reverse to catalyze the reaction ADP + $P_i \rightarrow$ ATP if there is a steep enough gradient across their membrane of the ion that they transport.

One such pump, the ATP synthase, harnesses a gradient of proton concentration produced by electron transport processes to produce most of the ATP used in the living world. Because this ubiquitous pump has such a central role in energy conversion, we postpone discussing its three-dimensional structure and mechanism until Chapter 14.

Proteins Often Form Large Complexes That Function as Protein Machines

As one progresses from small, single-domain proteins to large proteins formed from many domains, the functions the proteins can perform become more elaborate. The most impressive tasks, however, are carried out by large protein assemblies formed from many protein molecules. Now that it is possible to reconstruct most biological processes in cell-free systems in the laboratory, it is clear that each of the central processes in a cell—such as DNA replication, protein synthesis, vesicle budding, or transmembrane signaling—is catalyzed by a highly coordinated, linked set of ten or more proteins. In most such protein machines, the hydrolysis of bound nucleoside triphosphates (ATP or GTP) drives an ordered series of conformational changes in some of the individual protein subunits, enabling the ensemble of proteins to move coordinately. In this way, each of the appropriate enzymes can be placed directly into the positions where they are needed to perform successive reactions in a series. This is what occurs, for example, in protein synthesis on a ribosome (discussed in Chapter 6)—or in DNA replication, where a large multiprotein complex moves rapidly along the DNA (discussed in Chapter 5).

Cells have evolved protein machines for the same reason that humans have invented mechanical and electronic machines. For accomplishing almost any task, manipulations that are spatially and temporally coordinated through linked processes are much more efficient than the sequential use of individual tools.

A Complex Network of Protein Interactions Underlies Cell Function

There are many challenges facing cell biologists in this "post-genome" era when complete genome sequences are known. One is the need to dissect and reconstruct each one of the thousands of protein machines that exist in an organism such as ourselves. To understand these remarkable protein complexes, each must be reconstituted from its purified protein parts—so that its detailed mode of operation can be studied under controlled conditions in a test tube, free from all other cell components. This alone is a massive task. But we now know that each of these subcomponents of a cell also interacts with other sets of macromolecules, creating a large network of protein–protein and protein–nucleic acid interactions throughout the cell. To understand the cell, therefore, one will need to analyze most of these other interactions as well.

Some idea of the complexity of intracellular protein networks can be gained from a particularly well-studied example described in Chapter 16: the many dozens of proteins that interact with the actin cytoskeleton in the yeast *Saccharomyces cerevisiae* (see Figure 16–15).

The extent of such protein–protein interactions can also be estimated more generally. In particular, large-scale efforts have been undertaken to detect these interactions using the two-hybrid screening method described in Chapter 8 (see Figure 8–51). Thus, for example, this method is being applied to a large set of the 6000 gene products produced by *S. cerevisiae*. As expected, the majority of the interactions that have been observed are between proteins in the same functional group. That is, proteins involved in cell-cycle control tend to interact extensively with each other, as do proteins involved with DNA synthesis, or DNA repair, and so on. But there are also a surprisingly large number of interactions between the protein members of different functional groups (Figure 3–78). These interactions are presumably important for coordinating cell functions, but most of them are not understood.

An examination of the range of available data suggests that an average protein in a human cell may interact with somewhere between 5 and 15 different partners. Often, a different set of partners will be bound by each of the different domains in a multidomain protein; in fact, one can speculate that the unusually extensive multidomain structures observed for human proteins (see p. 462) may have evolved to generate these interactions. Given the enormous complexity of

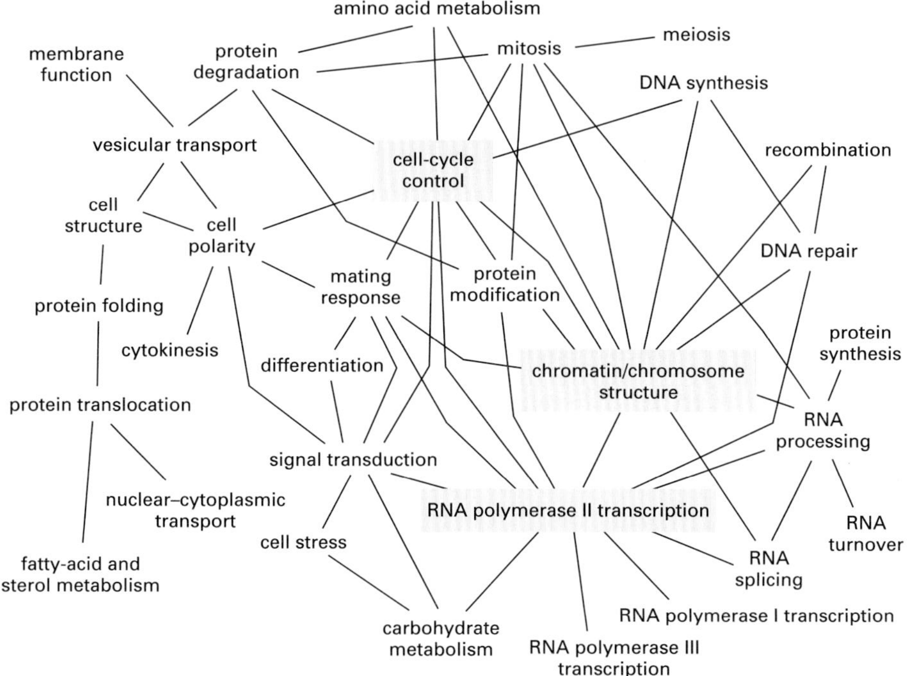

Figure 3–78 A map of the protein–protein interactions observed between different functional groups of proteins in the yeast *S. cerevisiae*. To produce this map, more than 1500 proteins were assigned to the indicated 31 functional groups. About 70 percent of the known protein–protein interactions were observed to occur between proteins assigned to the same functional group. However, as indicated, a surprisingly large number of interactions crossed between these groups. Each line in this diagram designates that more than 15 protein–protein interactions were observed between proteins in the two functional groups that are connected by that line. The three functional groups highlighted in *yellow* display at least 15 interactions with 10 or more of the 31 functional groups defined in the study. (From C.L. Tucker, J.F. Gera, and P. Vetz, *Trends Cell Biol.* 11:102–106, 2001.)

the interacting networks of macromolecules in cells, deciphering their full functional meaning may well keep scientists busy for centuries.

Summary

Proteins can form enormously sophisticated chemical devices, whose functions largely depend on the detailed chemical properties of their surfaces. Binding sites for ligands are formed as surface cavities in which precisely positioned amino acid side chains are brought together by protein folding. In the same way, normally unreactive amino acid side chains can be activated to make and break covalent bonds. Enzymes are catalytic proteins that greatly speed up reaction rates by binding the high-energy transition states for a specific reaction path; they also perform acid catalysis and base catalysis simultaneously. The rates of enzyme reactions are often so fast that they are limited only by diffusion; rates can be further increased if enzymes that act sequentially on a substrate are joined into a single multienzyme complex, or if the enzymes and their substrates are confined to the same compartment of the cell.

Proteins reversibly change their shape when ligands bind to their surface. The allosteric changes in protein conformation produced by one ligand affect the binding of a second ligand, and this linkage between two ligand-binding sites provides a crucial mechanism for regulating cell processes. Metabolic pathways, for example, are controlled by feedback regulation: some small molecules inhibit and other small molecules activate enzymes early in a pathway. Enzymes controlled in this way generally form symmetric assemblies, allowing cooperative conformational changes to create a steep response to changes in the concentrations of the ligands that regulate them.

Changes in protein shape can be driven in a unidirectional manner by the expenditure of chemical energy. By coupling allosteric shape changes to ATP hydrolysis, for example, proteins can do useful work, such as generating a mechanical force or moving for long distances in a single direction. The three-dimensional structures of proteins, determined by x-ray crystallography, have revealed how a small local change caused by nucleoside triphosphate hydrolysis is amplified to create major changes elsewhere in the protein. By such means, these proteins can serve as input–output devices that transmit information, as assembly factors, as motors, or as membrane-bound pumps. Highly efficient protein machines are formed by incorporating many different protein molecules into larger assemblies in which the allosteric movements of the individual components are coordinated. Such machines are now known to perform many of the most important reactions in cells.

References

General

Branden C & Tooze J (1999) Introduction to Protein Structure, 2nd edn. New York: Garland Publishing.

Creighton TE (1993) Proteins: Structures and Molecular Properties, 2nd edn. New York: WH Freeman.

Dickerson RE & Geis I (1969) The Structure and Action of Proteins. New York: Harper & Row.

Kyte J (1995) Structure in Protein Chemistry. New York: Garland Publishing.

Mathews CK, van Holde KE & Ahern K-G (2000) Biochemistry, 3rd edn. San Francisco: Benjamin Cummings.

Perutz M (1992) Protein Structure: New Approaches to Disease and Therapy. New York: WH Freeman.

Schulz GE & Schirmer RH (1990) Principles of Protein Structure, 2nd edn. New York: Springer.

Stryer L (1995) Biochemistry, 4th edn. New York: WH Freeman.

The Shape and Structure of Proteins

Anfinsen CB (1973) Principles that govern the folding of protein chains. *Science* 181, 223–230.

Bowie JU & Eisenberg D (1993) Inverted protein structure prediction. *Curr. Opin. Struct. Biol.* 3, 437–444.

Burkhard P, Stetefeld J & Strelkov SV (2001) Coiled coils: a highly versatile protein folding motif. *Trends Cell Biol.* 11, 82–88.

Caspar DLD & Klug A (1962) Physical principles in the construction of regular viruses. *Cold Spring Harb. Symp. Quant. Biol.* 27, 1–24.

Doolittle RF (1995) The multiplicity of domains in proteins. *Annu. Rev. Biochem.* 64, 287–314.

Fraenkel-Conrat H & Williams RC (1955) Reconstitution of active tobacco mosaic virus from its inactive protein and nucleic acid components. *Proc. Natl Acad. Sci. USA* 41, 690–698.

Fuchs E (1995) Keratins and the skin. *Annu. Rev. Cell Dev. Biol.* 11, 123–153.

Gehring WJ, Affolter M & Burglin T (1994) Homeodomain proteins. *Annu. Rev. Biochem.* 63, 487–526.

Harrison SC (1992) Viruses. *Curr. Opin. Struct. Biol.* 2, 293–299.

International Human Genome Sequencing Consortium (2001) Initial sequencing and analysis of the human genome. *Nature* 409, 860–921.

Marchler-Bauer A & Bryant SH (1997) A measure of success in fold recognition. *Trends Biochem. Sci.* 22, 236–240.

Nomura M (1973) Assembly of bacterial ribosomes. *Science* 179, 864–873.

Pauling L & Corey RB (1951) Configurations of polypeptide chains with favored orientations around single bonds: two new pleated sheets. *Proc. Natl. Acad. Sci. USA* 37, 729–740.

Pauling L, Corey RB & Branson HR (1951) The structure of proteins: two hydrogen-bonded helical configurations of the polypeptide chain. *Proc. Natl. Acad. Sci. USA* 27, 205–211.

Pawson T (1994) SH2 and SH3 domains in signal transduction. *Adv. Cancer Res.* 64, 87–110.

Ponting CP & Dickens NJ (2001) Genome cartography through domain annotation. *Genome Biol.* 2(7), comment2006.1–2006.7.

Ponting CP, Schultz J, Copley RR et al. (2000) Evolution of domain families. *Adv. Protein Chem.* 54, 185–244.

Richards FM (1991) The protein folding problem. *Sci. Am.* 264(1), 54–63.

Richardson JS (1981) The anatomy and taxonomy of protein structure. *Adv. Protein Chem.* 34, 167–339.

Sali A & Kuriyan J (1999) Challenges at the frontiers of structural biology. *Trends Cell Biol.* 9, M20–M24.

Steiner DF, Kemmler W, Tager HS & Peterson JD (1974) Proteolytic processing in the biosynthesis of insulin and other proteins. *Fed. Proc.* 33, 2105–2115.

Teichmann SA, Murzin AG & Chothia C (2001) Determination of protein function, evolution and interactions by structural genomics. *Curr. Opin. Struct. Biol.* 11, 354–363.

Trinick J (1992) Understanding the functions of titin and nebulin. *FEBS Lett.* 307, 44–48.

Protein Function

Alberts B (1998) The cell as a collection of protein machines: preparing the next generation of molecular biologists. *Cell* 92, 291–294.

Benkovic SJ (1992) Catalytic antibodies. *Annu. Rev. Biochem.* 61, 29–54.

Berg OG & von Hippel PH (1985) Diffusion-controlled macromolecular interactions. *Annu. Rev. Biophys. Biophys. Chem.* 14, 131–160.

Bourne HR (1995) GTPases: a family of molecular switches and clocks. *Philos. Trans. R. Soc. Lond. B. Biol. Sci.* 349, 283–289.

Braden BC & Poljak RJ (1995) Structural features of the reactions between antibodies and protein antigens. *FASEB J.* 9, 9–16.

Dickerson RE & Geis I (1983) Hemoglobin: Structure, Function, Evolution and Pathology. Menlo Park, CA: Benjamin Cummings.

Dressler D & Potter H (1991) Discovering Enzymes. New York: Scientific American Library.

Fersht AR (1999) Structure and Mechanisms in Protein Science: A Guide to Enzyme Catalysis. New York: WH Freeman.

Hazbun TR & Fields S (2001) Networking proteins in yeast. *Proc. Natl. Acad. Sci. USA* 98, 4277–4278.

Hunter T (1987) A thousand and one protein kinases. *Cell* 50, 823–829.

Jones S & Thornton JM (1996) Principles of protein–protein interactions. *Proc. Natl Acad. Sci. USA* 93, 13–20.

Kantrowitz ER & Lipscomb WN (1988) *Escherichia coli* aspartate transcarbamoylase: the relation between structure and function. *Science* 241, 669–674.

Khosla C & Harbury PB (2001) Modular enzymes. *Nature* 409, 247–252.

Koshland DE, Jr (1984) Control of enzyme activity and metabolic pathways. *Trends Biochem. Sci.* 9, 155–159.

Kraut J (1977) Serine proteases: structure and mechanism of catalysis. *Annu. Rev. Biochem.* 46, 331–358.

Lichtarge O, Bourne HR & Cohen FE (1996) An evolutionary trace method defines binding surfaces common to protein families. *J. Mol. Biol.* 257, 342–358.

Marcotte EM, Pellegrini M, Ng HL et al. (1999) Detecting protein function and protein–protein interactions from genome sequences. *Science* 285, 751–753.

Miles EW, Rhee S & Davies DR (1999) The molecular basis of substrate channeling. *J. Biol. Chem.* 274, 12193–12196.

Monod J, Changeux J-P & Jacob F (1963) Allosteric proteins and cellular control systems. *J. Mol. Biol.* 6, 306–329.

Pavletich NP (1999) Mechanisms of cyclin-dependent kinase regulation: structures of Cdks, their cyclin activators, and Cip and INK4 inhibitors. *J. Mol. Biol.* 287, 821–828.

Perutz M (1990) Mechanisms of Cooperativity and Allosteric Regulation in Proteins. Cambridge: Cambridge University Press.

Phillips DC (1967) The hen egg white lysozyme molecule. *Proc. Natl. Acad. Sci. USA* 57, 484–495.

Radzicka A & Wolfenden R (1995) A proficient enzyme. *Science* 267, 90–93.

Schramm VL (1998) Enzymatic transition states and transition state analog design. *Annu. Rev. Biochem.* 67, 693–720.

Schultz PG & Lerner RA (1995) From molecular diversity to catalysis: lessons from the immune system. *Science* 269, 1835–1842.

Sicheri F & Kuriyan J (1997) Structures of Src-family tyrosine kinases. *Curr. Opin. Struct. Biol.* 7, 777–785.

Todd AE, Orengo CA & Thornton JM (1999) Evolution of protein function, from a structural perspective. *Curr. Opin. Chem. Biol.* 3, 548–556.

Toyoshima C, Nakasako M, Nomura H & Ogawa H (2000) Crystal structure of the calcium pump of sarcoplasmic reticulum at 2.6Å resolution. *Nature* 405, 647–655.

Vale RD & Milligan RA (2000) The way things move: looking under the hood of molecular motor proteins. *Science* 288, 88–95.

Vocadlo DJ, Davies GJ, Laine R & Withers SG (2001) Catalysis by hen egg-white lysozyme proceeds via a covalent intermediate. *Nature* 412, 835–838.

Walsh C (2001) Enabling the chemistry of life. *Nature* 409, 226–231.

Wolfenden R (1972) Analog approaches to the structure of the transition state in enzyme reactions. *Acc. Chem. Res.* 5, 10–18.

BASIC GENETIC MECHANISMS

PART II

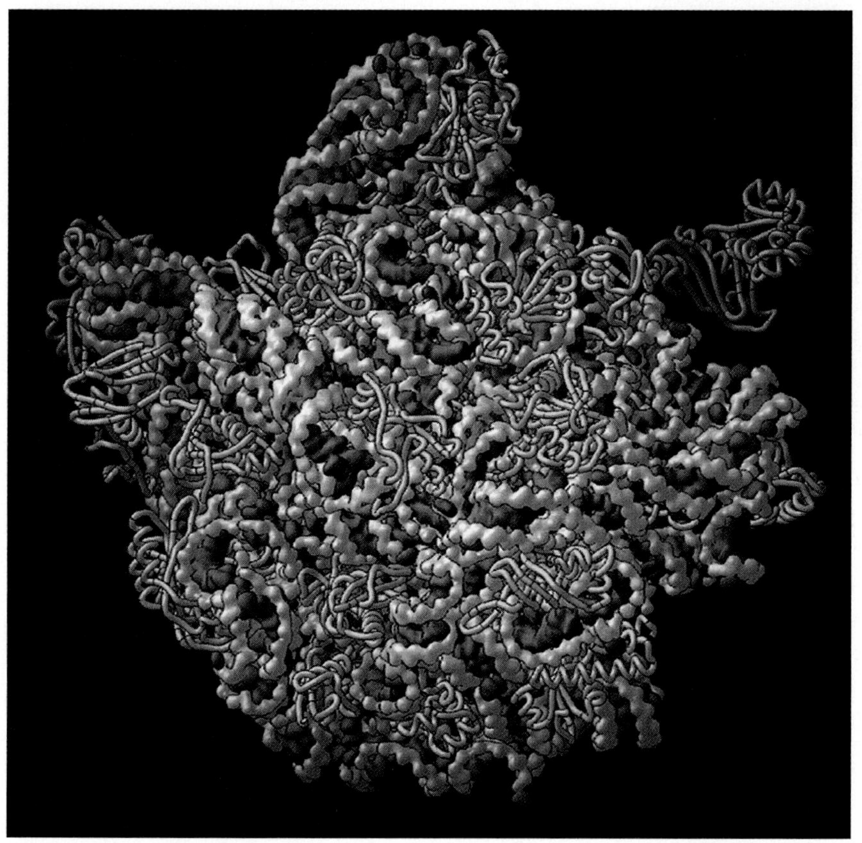

The ribosome. The large subunit of the bacterial ribosome is shown here with the ribosomal RNA shown in *gray* and the ribosomal proteins in *gold*. The ribosome is the complex catalytic machine at the heart of protein synthesis. (From N. Ban et al., *Science* 289:905–920, 2000. © AAAS.)

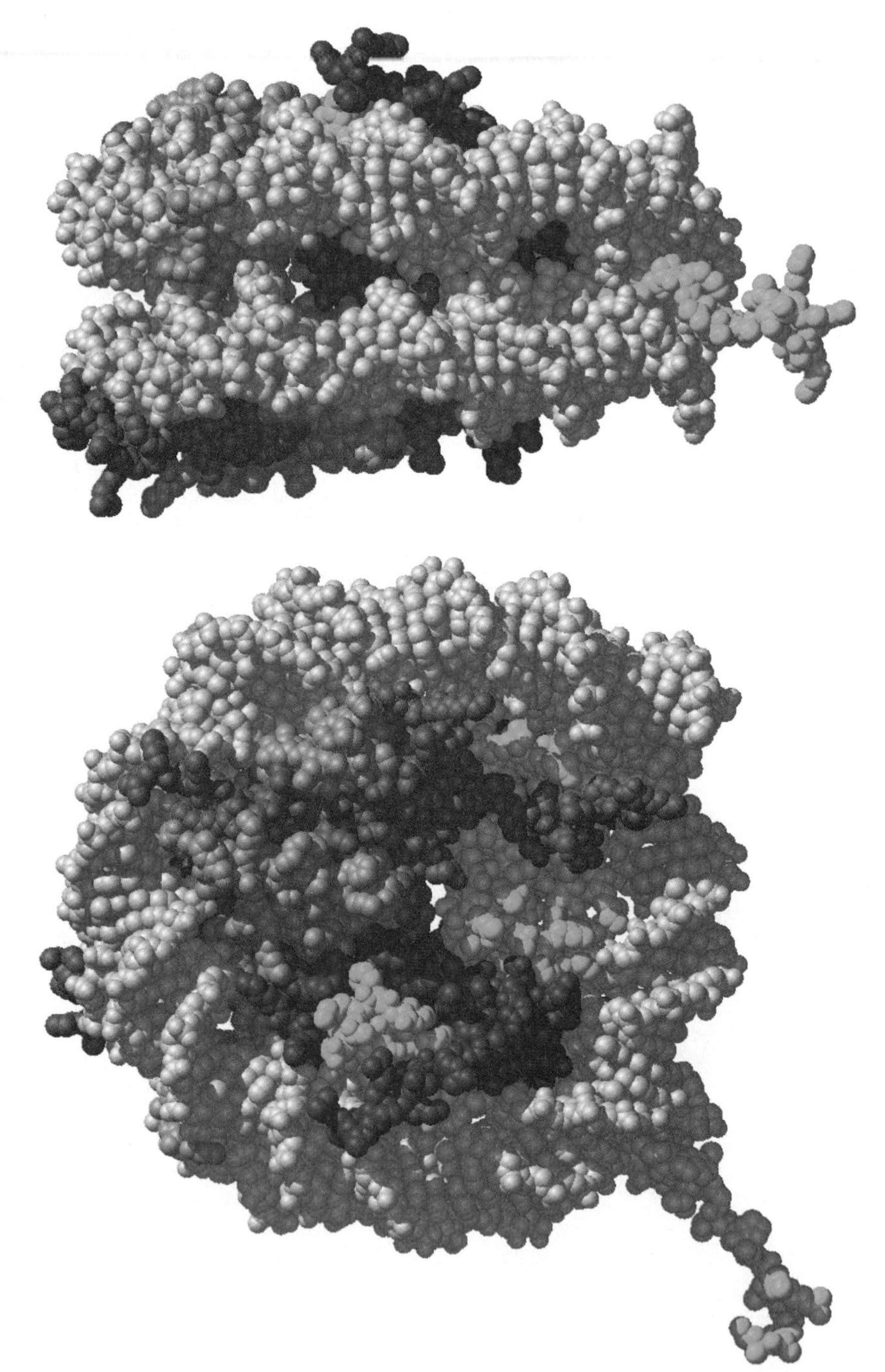

The nucleosome. The basic structural unit of all eucaryotic chromosomes is the nucleosome. The DNA double helix *(gray)* is wrapped around a core particle of histone proteins (colored) to create the nucleosome. Nucleosomes are spaced roughly 200 nucleotide pairs apart along the chromosomal DNA. (Reprinted by permission from K. Luger et al., *Nature* 389:251–260, 1997. © Macmillan Magazines Ltd.)

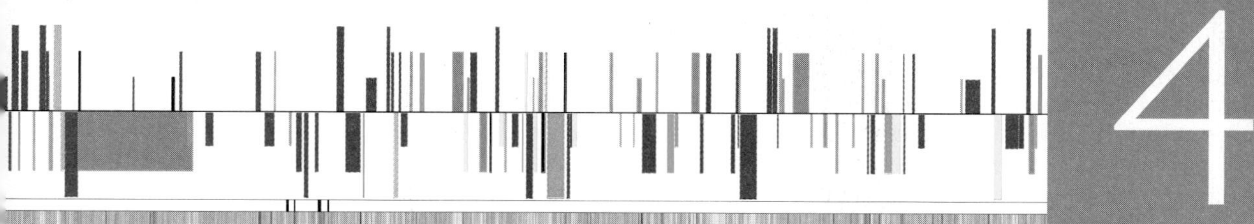

4

DNA AND CHROMOSOMES

Life depends on the ability of cells to store, retrieve, and translate the genetic instructions required to make and maintain a living organism. This *hereditary* information is passed on from a cell to its daughter cells at cell division, and from one generation of an organism to the next through the organism's reproductive cells. These instructions are stored within every living cell as its **genes**, the information-containing elements that determine the characteristics of a species as a whole and of the individuals within it.

As soon as genetics emerged as a science at the beginning of the twentieth century, scientists became intrigued by the chemical structure of genes. The information in genes is copied and transmitted from cell to daughter cell millions of times during the life of a multicellular organism, and it survives the process essentially unchanged. What form of molecule could be capable of such accurate and almost unlimited replication and also be able to direct the development of an organism and the daily life of a cell? What kind of instructions does the genetic information contain? How are these instructions physically organized so that the enormous amount of information required for the development and maintenance of even the simplest organism can be contained within the tiny space of a cell?

The answers to some of these questions began to emerge in the 1940s, when researchers discovered, from studies in simple fungi, that genetic information consists primarily of instructions for making proteins. Proteins are the macromolecules that perform most cellular functions: they serve as building blocks for cellular structures and form the enzymes that catalyze all of the cell's chemical reactions (Chapter 3), they regulate gene expression (Chapter 7), and they enable cells to move (Chapter 16) and to communicate with each other (Chapter 15). The properties and functions of a cell are determined almost entirely by the proteins it is able to make. With hindsight, it is hard to imagine what other type of instructions the genetic information could have contained.

The other crucial advance made in the 1940s was the identification of **deoxyribonucleic acid (DNA)** as the likely carrier of genetic information. But the mechanism whereby the hereditary information is copied for transmission from cell to cell, and how proteins are specified by the instructions in the DNA, remained completely mysterious. Suddenly, in 1953, the mystery was solved when the structure of DNA was determined by James Watson and Francis Crick. As mentioned in Chapter 1, the structure of DNA immediately solved the problem of how the information in this molecule might be copied, or *replicated*. It also provided the first clues as to how a molecule of DNA might encode the instructions for making proteins. Today, the fact that DNA is the genetic material is so fundamental to biological thought that it is difficult to realize what an enormous intellectual gap this discovery filled.

Well before biologists understood the structure of DNA, they had recognized that genes are carried on *chromosomes*, which were discovered in the nineteenth century as threadlike structures in the nucleus of a eucaryotic cell that become visible as the cell begins to divide (Figure 4–1). Later, as biochemical analysis became possible, chromosomes were found to consist of both DNA and protein. We now know that the DNA carries the hereditary information of the cell (Figure 4–2). In contrast, the protein components of chromosomes function largely to package and control the enormously long DNA molecules so that they fit inside cells and can easily be accessed by them.

In this chapter we begin by describing the structure of DNA. We see how, despite its chemical simplicity, the structure and chemical properties of DNA make it ideally suited as the raw material of genes. The genes of every cell on Earth are made of DNA, and insights into the relationship between DNA and genes have come from experiments in a wide variety of organisms. We then consider how genes and other important segments of DNA are arranged on the long molecules of DNA that are present in chromosomes. Finally, we discuss how eucaryotic cells fold these long DNA molecules into compact chromosomes. This packing has to be done in an orderly fashion so that the chromosomes can be replicated and apportioned correctly between the two daughter cells at each cell division. It must also allow access of chromosomal DNA to enzymes that repair it when it is damaged and to the specialized proteins that direct the expression of its many genes.

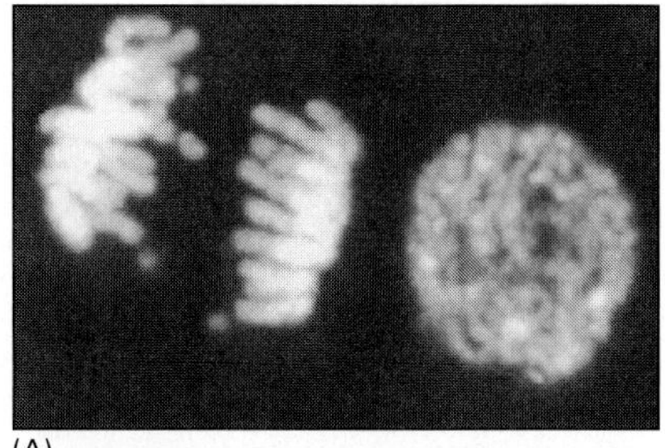

Figure 4–1 Chromosomes in cells. (A) Two adjacent plant cells photographed through a light microscope. The DNA has been stained with a fluorescent dye (DAPI) that binds to it. The DNA is present in chromosomes, which become visible as distinct structures in the light microscope only when they become compact structures in preparation for cell division, as shown on the left. The cell on the right, which is not dividing, contains identical chromosomes, but they cannot be clearly distinguished in the light microscope at this phase in the cell's life cycle, because they are in a more extended conformation. (B) Schematic diagram of the outlines of the two cells along with their chromosomes. (A, courtesy of Peter Shaw.)

This is the first of four chapters that deal with basic genetic mechanisms—the ways in which the cell maintains, replicates, expresses, and occasionally improves the genetic information carried in its DNA. In the following chapter (Chapter 5) we discuss the mechanisms by which the cell accurately replicates and repairs DNA; we also describe how DNA sequences can be rearranged through the process of genetic recombination. Gene expression—the process through which the information encoded in DNA is interpreted by the cell to guide the synthesis of proteins—is the main topic of Chapter 6. In Chapter 7, we describe how gene expression is controlled by the cell to ensure that each of the many thousands of proteins encrypted in its DNA is manufactured only at the proper time and place in the life of the cell. Following these four chapters on basic genetic mechanisms, we present an account of the experimental techniques used to study these and other processes that are fundamental to all cells (Chapter 8).

THE STRUCTURE AND FUNCTION OF DNA

Biologists in the 1940s had difficulty in accepting DNA as the genetic material because of the apparent simplicity of its chemistry. DNA was known to be a long polymer composed of only four types of subunits, which resemble one another chemically. Early in the 1950s, DNA was first examined by x-ray diffraction analysis, a technique for determining the three-dimensional atomic structure of a molecule (discussed in Chapter 8). The early x-ray diffraction results indicated that DNA was composed of two strands of the polymer wound into a helix. The observation that DNA was double-stranded was of crucial significance and

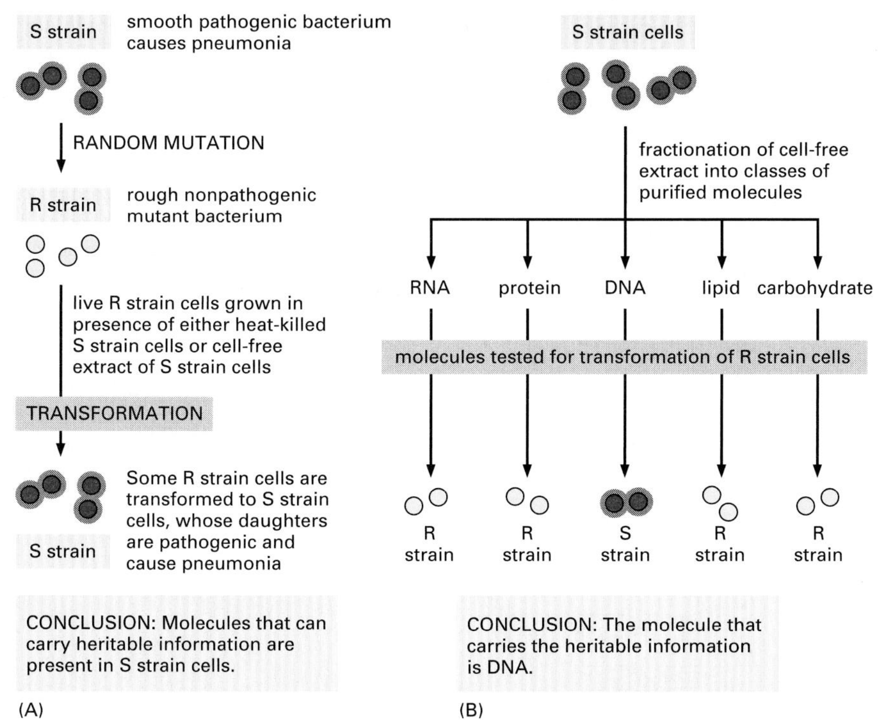
S strain
smooth pathogenic bacterium causes pneumonia
RANDOM MUTATION
R strain
rough nonpathogenic mutant bacterium
live R strain cells grown in presence of either heat-killed S strain cells or cell-free extract of S strain cells
TRANSFORMATION
Some R strain cells are transformed to S strain cells, whose daughters are pathogenic and cause pneumonia
S strain
CONCLUSION: Molecules that can carry heritable information are present in S strain cells.
(A)
S strain cells
fractionation of cell-free extract into classes of purified molecules
RNA
protein
DNA
lipid
carbohydrate
molecules tested for transformation of R strain cells
R strain
R strain
S strain
R strain
R strain
CONCLUSION: The molecule that carries the heritable information is DNA.
(B)

Figure 4–2 Experimental demonstration that DNA is the genetic material. These experiments, carried out in the 1940s, showed that adding purified DNA to a bacterium changed its properties and that this change was faithfully passed on to subsequent generations. Two closely related strains of the bacterium *Streptococcus pneumoniae* differ from each other in both their appearance under the microscope and their pathogenicity. One strain appears smooth (S) and causes death when injected into mice, and the other appears rough (R) and is nonlethal. (A) This experiment shows that a substance present in the S strain can change (or transform) the R strain into the S strain and that this change is inherited by subsequent generations of bacteria. (B) This experiment, in which the R strain has been incubated with various classes of biological molecules obtained from the S strain, identifies the substance as DNA.

provided one of the major clues that led to the Watson–Crick structure of DNA. Only when this model was proposed did DNA's potential for replication and information encoding become apparent. In this section we examine the structure of the DNA molecule and explain in general terms how it is able to store hereditary information.

A DNA Molecule Consists of Two Complementary Chains of Nucleotides

A DNA molecule consists of two long polynucleotide chains composed of four types of nucleotide subunits. Each of these chains is known as a *DNA chain*, or a *DNA strand*. *Hydrogen bonds* between the base portions of the nucleotides hold the two chains together (Figure 4–3). As we saw in Chapter 2 (Panel 2–6, pp. 120-121), nucleotides are composed of a five-carbon sugar to which are attached one or more phosphate groups and a nitrogen-containing base. In the case of the nucleotides in DNA, the sugar is deoxyribose attached to a single phosphate group (hence the name deoxyribonucleic acid), and the base may be either *adenine (A)*, *cytosine (C)*, *guanine (G)*, or *thymine (T)*. The nucleotides are covalently linked together in a chain through the sugars and phosphates, which thus form a "backbone" of alternating sugar–phosphate–sugar–phosphate (see Figure 4–3). Because only the base differs in each of the four types of subunits, each polynucleotide chain in DNA is analogous to a necklace (the backbone) strung with four types of beads (the four bases A, C, G, and T). These same symbols (A, C, G, and T) are also commonly used to denote the four different nucleotides—that is, the bases with their attached sugar and phosphate groups.

The way in which the nucleotide subunits are lined together gives a DNA strand a chemical polarity. If we think of each sugar as a block with a protruding knob (the 5′ phosphate) on one side and a hole (the 3′ hydroxyl) on the other (see Figure 4–3), each completed chain, formed by interlocking knobs with holes, will have all of its subunits lined up in the same orientation. Moreover, the two ends of the chain will be easily distinguishable, as one has a hole (the 3′ hydroxyl) and the other a knob (the 5′ phosphate) at its terminus. This polarity in a DNA chain is indicated by referring to one end as the *3′ end* and the other as the *5′ end*.

The three-dimensional structure of DNA—**the double helix**—arises from the chemical and structural features of its two polynucleotide chains. Because

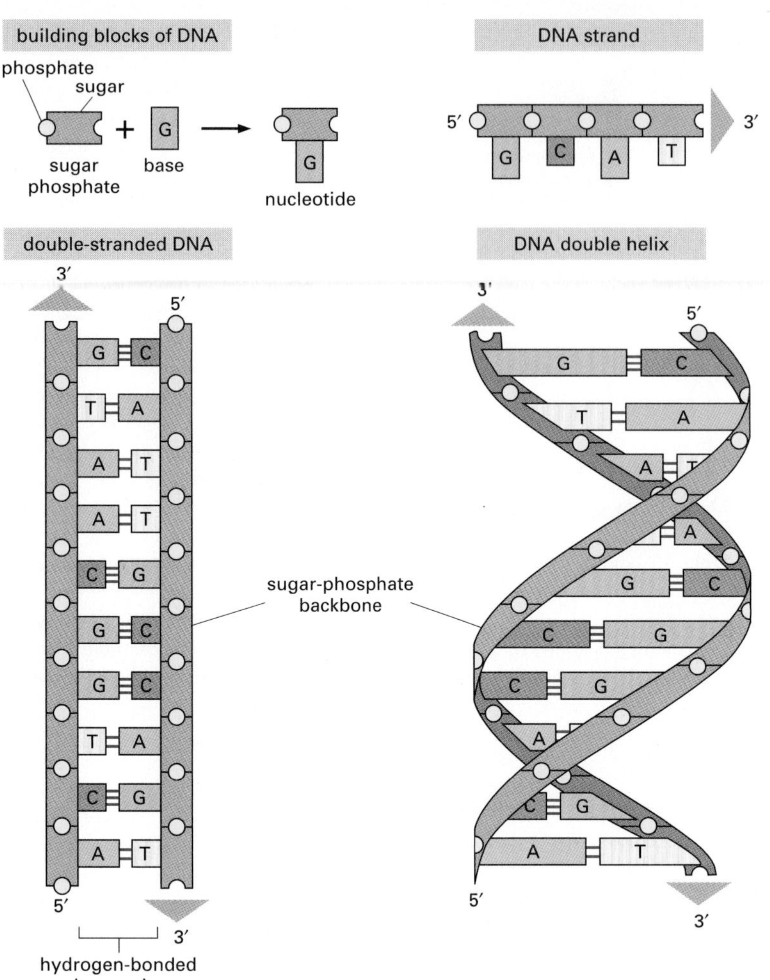

Figure 4–3 DNA and its building blocks. DNA is made of four types of nucleotides, which are linked covalently into a polynucleotide chain (a DNA strand) with a sugar-phosphate backbone from which the bases (A, C, G, and T) extend. A DNA molecule is composed of two DNA strands held together by hydrogen bonds between the paired bases. The *arrowheads* at the ends of the DNA strands indicate the polarities of the two strands, which run antiparallel to each other in the DNA molecule. In the diagram at the bottom left of the figure, the DNA molecule is shown straightened out; in reality, it is twisted into a double helix, as shown on the right. For details, see Figure 4–5.

these two chains are held together by hydrogen bonding between the bases on the different strands, all the bases are on the inside of the double helix, and the sugar-phosphate backbones are on the outside (see Figure 4–3). In each case, a bulkier two-ring base (a purine; see Panel 2–6, pp. 120–121) is paired with a single-ring base (a pyrimidine); A always pairs with T, and G with C (Figure 4–4). This *complementary base-pairing* enables the **base pairs** to be packed in the

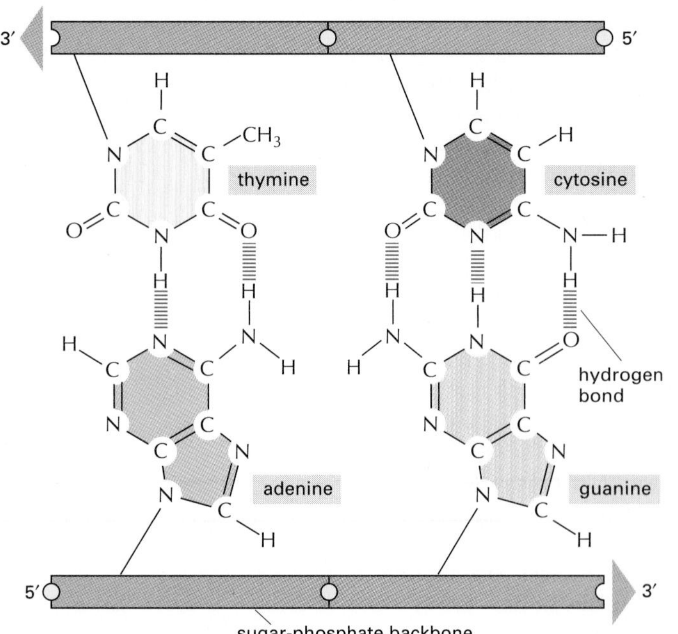

Figure 4–4 Complementary base pairs in the DNA double helix. The shapes and chemical structure of the bases allow hydrogen bonds to form efficiently only between A and T and between G and C, where atoms that are able to form hydrogen bonds (see Panel 2–3, pp. 114–115) can be brought close together without distorting the double helix. As indicated, two hydrogen bonds form between A and T, while three form between G and C. The bases can pair in this way only if the two polynucleotide chains that contain them are antiparallel to each other.

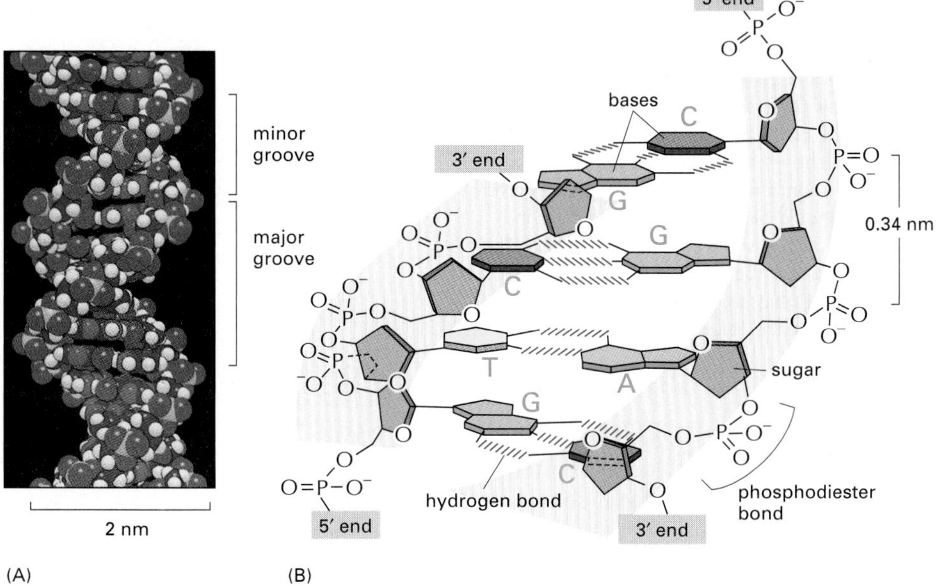

Figure 4–5 The DNA double helix. (A) A space-filling model of 1.5 turns of the DNA double helix. Each turn of DNA is made up of 10.4 nucleotide pairs and the center-to-center distance between adjacent nucleotide pairs is 3.4 nm. The coiling of the two strands around each other creates two grooves in the double helix. As indicated in the figure, the wider groove is called the major groove, and the smaller the minor groove. (B) A short section of the double helix viewed from its side, showing four base pairs. The nucleotides are linked together covalently by phosphodiester bonds through the 3′-hydroxyl (–OH) group of one sugar and the 5′-phosphate (P) of the next. Thus, each polynucleotide strand has a chemical polarity; that is, its two ends are chemically different. The 3′ end carries an unlinked –OH group attached to the 3′ position on the sugar ring; the 5′ end carries a free phosphate group attached to the 5′ position on the sugar ring.

energetically most favorable arrangement in the interior of the double helix. In this arrangement, each base pair is of similar width, thus holding the sugar-phosphate backbones an equal distance apart along the DNA molecule. To maximize the efficiency of base-pair packing, the two sugar-phosphate backbones wind around each other to form a double helix, with one complete turn every ten base pairs (Figure 4–5).

The members of each base pair can fit together within the double helix only if the two strands of the helix are **antiparallel**—that is, only if the polarity of one strand is oriented opposite to that of the other strand (see Figures 4–3 and 4–4). A consequence of these base-pairing requirements is that each strand of a DNA molecule contains a sequence of nucleotides that is exactly **complementary** to the nucleotide sequence of its partner strand.

The Structure of DNA Provides a Mechanism for Heredity

Genes carry biological information that must be copied accurately for transmission to the next generation each time a cell divides to form two daughter cells. Two central biological questions arise from these requirements: how can the information for specifying an organism be carried in chemical form, and how is it accurately copied? The discovery of the structure of the DNA double helix was a landmark in twentieth-century biology because it immediately suggested answers to both questions, thereby resolving at the molecular level the problem of heredity. We discuss briefly the answers to these questions in this section, and we shall examine them in more detail in subsequent chapters.

DNA encodes information through the order, or sequence, of the nucleotides along each strand. Each base—A, C, T, or G—can be considered as a letter in a four-letter alphabet that spells out biological messages in the chemical structure of the DNA. As we saw in Chapter 1, organisms differ from one another because their respective DNA molecules have different nucleotide sequences and, consequently, carry different biological messages. But how is the nucleotide alphabet used to make messages, and what do they spell out?

As discussed above, it was known well before the structure of DNA was determined that genes contain the instructions for producing proteins. The DNA messages must therefore somehow encode proteins (Figure 4–6). This relationship immediately makes the problem easier to understand, because of the chemical character of proteins. As discussed in Chapter 3, the properties of a protein, which are responsible for its biological function, are determined by its three-dimensional structure, and its structure is determined in turn by the linear

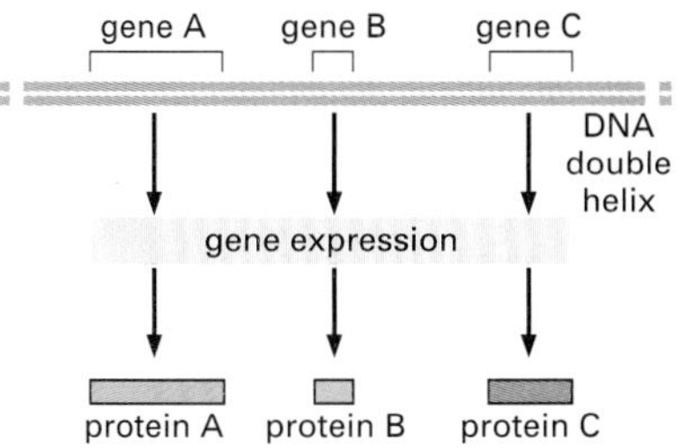

Figure 4–6 The relationship between genetic information carried in DNA and proteins.

Figure 4–7 The nucleotide sequence of the human β-globin gene. This gene carries the information for the amino acid sequence of one of the two types of subunits of the hemoglobin molecule, which carries oxygen in the blood. A different gene, the α-globin gene, carries the information for the other type of hemoglobin subunit (a hemoglobin molecule has four subunits, two of each type). Only one of the two strands of the DNA double helix containing the β-globin gene is shown; the other strand has the exact complementary sequence. By convention, a nucleotide sequence is written from its 5′ end to its 3′ end, and it should be read from left to right in successive lines down the page as though it were normal English text. The DNA sequences highlighted in *yellow* show the three regions of the gene that specify the amino sequence for the β-globin protein. We see in Chapter 6 how the cell connects these three sequences together to synthesize a full-length β-globin protein.

```
CCCTGTGGAGCCACACCCTAGGGTTGGCCA
ATCTACTCCCAGGAGCAGGGAGGGCAGGAG
CCAGGGCTGGGCATAAAAGTCAGGGCAGAG
CCATCTATTGCTTACATTTGCTTCTGACAC
AACTGTGTTCACTAGCAACTCAAACAGACA
CCATGGTGCACCTGACTCCTGAGGAGAAGT
CTGCCGTTACTGCCCTGTGGGGCAAGGTGA
ACGTGGATGAAGTTGGTGGTGAGGCCCTGG
GCAGGTTGGTATCAAGGTTACAAGACAGGT
TTAAGGAGACCAATAGAAACTGGGCATGTG
GAGACAGAGAAGACTCTTGGGTTTCTGATA
GGCACTGACTCTCTCTGCCTATTGGTCTAT
TTTCCCACCCTTAGGCTGCTGGTGGTCTAC
CCTTGGACCCAGAGGTTCTTTGAGTCCTTT
GGGGATCTGTCCACTCCTGATGCTGTTATG
GGCAACCCTAAGGTGAAGGCTCATGGCAAG
AAAGTGCTCGGTGCCTTTAGTGATGGCCTG
GCTCACCTGGACAACCTCAAGGGCACCTTT
GCCACACTGAGTGAGCTGCACTGTGACAAG
CTGCACGTGGATCCTGAGAACTTCAGGGTG
AGTCTATGGGACCCTTGATGTTTTCTTTCC
CCTTCTTTTCTATGGTTAAGTTCATGTCAT
AGGAAGGGGAGAAGTAACAGGGTACAGTTT
AGAATGGGAAACAGACGAATGATTGCATCA
GTGTGGAAGTCTCAGGATCGTTTTAGTTTC
TTTTATTTGCTGTTCATAACAATTGTTTTC
TTTTGTTTAATTCTTGCTTTCTTTTTTTTT
CTTCTCCGCAATTTTTACTATTATACTTAA
TGCCTTAACATTGTGTATAACAAAAGGAAA
TATCTCTGAGATACATTAAGTAACTTAAAA
AAAAACTTTACACAGTCTGCCTAGTACATT
ACTATTTGGAATATATGTGTGCTTATTTGC
ATATTCATAATCTCCCTACTTTATTTTCTT
TTATTTTTAATTGATACATAATCATTATAC
ATATTTATGGGTTAAAGTGTAATGTTTTAA
TATGTGTACACATATTGACCAAATCAGGGT
AATTTTGCATTTGTAATTTTAAAAAATGCT
TTCTTCTTTTAATATACTTTTTTGTTTATC
TTATTTCTAATACTTTCCCTAATCTCTTTC
TTTCAGGGCAATAATGATACAATGTATCAT
GCCTCTTTGCACCATTCTAAAGAATAACAG
TGATAATTTCTGGGTTAAGGCAATAGCAAT
ATTTCTGCATATAAATATTTCTGCATATAA
ATTGTAACTGATGTAAGAGGTTTCATATTG
CTAATAGCAGCTACAATCCAGCTACCATTC
TGCTTTTATTTTATGGTTGGGATAAGGCTG
GATTATTCTGAGTCCAAGCTAGGCCCTTTT
GCTAATCATGTTCATACCTCTTATCTTCCT
CCCACAGCTCCTGGGCAACGTGCTGGTCTG
TGTGCTGGCCCATCACTTTGGCAAAGAATT
CACCCCACCAGTGCAGGCTGCCTATCAGAA
AGTGGTGGCTGGTGTGGCTAATGCCCTGGC
CCACAAGTATCACTAAGCTCGCTTTCTTGC
TGTCCAATTTCTATTAAAGGTTCCTTTGTT
CCCTAAGTCCAACTACTAAACTGGGGGATA
TTATGAAGGGCCTTGAGCATCTGGATTCTG
CCTAATAAAAAACATTTATTTTCATTGCAA
TGATGTATTTAAATTATTTCTGAATATTTT
ACTAAAAAGGGAATGTGGGAGGTCAGTGCA
TTTAAAACATAAAGAAATGATGAGCTGTTC
AAACCTTGGGAAAATACACTATATCTTAAA
CTCCATGAAAGAAGGTGAGGCTGCAACCAG
CTAATGCACATTGGCAACAGCCCCTGATGC
CTATGCCTTATTCATCCCTCAGAAAAGGAT
TCTTGTAGAGGCTTGATTTGCAGGTTAAAG
TTTTGCTATGCTGTATTTTACATTACTTAT
TGTTTTAGCTGTCCTCATGAATGTCTTTTC
```

sequence of the amino acids of which it is composed. The linear sequence of nucleotides in a gene must therefore somehow spell out the linear sequence of amino acids in a protein. The exact correspondence between the four-letter nucleotide alphabet of DNA and the twenty-letter amino acid alphabet of proteins—the genetic code—is not obvious from the DNA structure, and it took over a decade after the discovery of the double helix before it was worked out. In Chapter 6 we describe this code in detail in the course of elaborating the process, known as *gene expression*, through which a cell translates the nucleotide sequence of a gene into the amino acid sequence of a protein.

The complete set of information in an organism's DNA is called its **genome**, and it carries the information for all the proteins the organism will ever synthesize. (The term genome is also used to describe the DNA that carries this information.) The amount of information contained in genomes is staggering: for example, a typical human cell contains 2 meters of DNA. Written out in the four-letter nucleotide alphabet, the nucleotide sequence of a very small human gene occupies a quarter of a page of text (Figure 4–7), while the complete sequence of nucleotides in the human genome would fill more than a thousand books the size of this one. In addition to other critical information, it carries the instructions for about 30,000 distinct proteins.

At each cell division, the cell must copy its genome to pass it to both daughter cells. The discovery of the structure of DNA also revealed the principle that makes this copying possible: because each strand of DNA contains a sequence of nucleotides that is exactly complementary to the nucleotide sequence of its partner strand, each strand can act as a **template**, or mold, for the synthesis of a new complementary strand. In other words, if we designate the two DNA strands as S and S′, strand S can serve as a template for making a new strand S′, while strand S′ can serve as a template for making a new strand S (Figure 4–8).

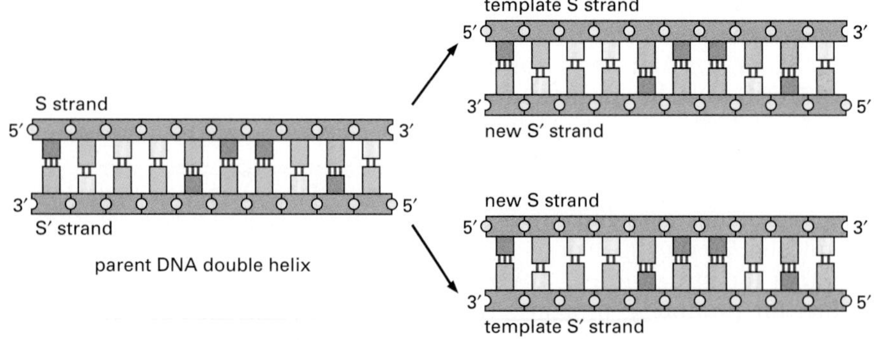

Figure 4–8 DNA as a template for its own duplication. As the nucleotide A successfully pairs only with T, and G with C, each strand of DNA can specify the sequence of nucleotides in its complementary strand. In this way, double-helical DNA can be copied precisely.

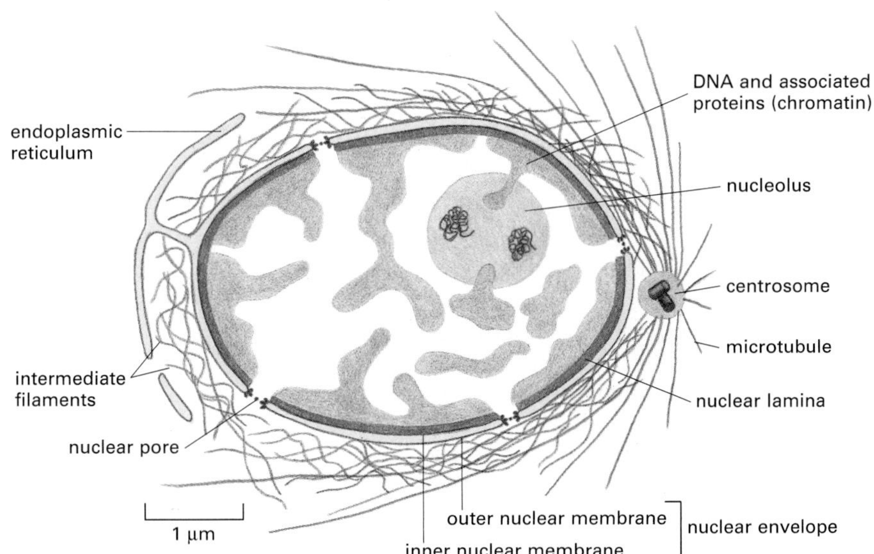

Figure 4–9 A cross-sectional view of a typical cell nucleus. The nuclear envelope consists of two membranes, the outer one being continuous with the endoplasmic reticulum membrane (see also Figure 12–9). The space inside the endoplasmic reticulum (the ER lumen) is colored *yellow;* it is continuous with the space between the two nuclear membranes. The lipid bilayers of the inner and outer nuclear membranes are connected at each nuclear pore. Two networks of intermediate filaments *(green)* provide mechanical support for the nuclear envelope; the intermediate filaments inside the nucleus form a special supporting structure called the nuclear lamina.

Thus, the genetic information in DNA can be accurately copied by the beautifully simple process in which strand S separates from strand S′, and each separated strand then serves as a template for the production of a new complementary partner strand that is identical to its former partner.

The ability of each strand of a DNA molecule to act as a template for producing a complementary strand enables a cell to copy, or *replicate,* its genes before passing them on to its descendants. In the next chapter we describe the elegant machinery the cell uses to perform this enormous task.

In Eucaryotes, DNA Is Enclosed in a Cell Nucleus

Nearly all the DNA in a eucaryotic cell is sequestered in a nucleus, which occupies about 10% of the total cell volume. This compartment is delimited by a *nuclear envelope* formed by two concentric lipid bilayer membranes that are punctured at intervals by large nuclear pores, which transport molecules between the nucleus and the cytosol. The nuclear envelope is directly connected to the extensive membranes of the endoplasmic reticulum. It is mechanically supported by two networks of intermediate filaments: one, called the *nuclear lamina,* forms a thin sheetlike meshwork inside the nucleus, just beneath the inner nuclear membrane; the other surrounds the outer nuclear membrane and is less regularly organized (Figure 4–9).

The nuclear envelope allows the many proteins that act on DNA to be concentrated where they are needed in the cell, and, as we see in subsequent chapters, it also keeps nuclear and cytosolic enzymes separate, a feature that is crucial for the proper functioning of eucaryotic cells. Compartmentalization, of which the nucleus is an example, is an important principle of biology; it serves to establish an environment in which biochemical reactions are facilitated by the high concentration of both substrates and the enzymes that act on them.

Summary

Genetic information is carried in the linear sequence of nucleotides in DNA. Each molecule of DNA is a double helix formed from two complementary strands of nucleotides held together by hydrogen bonds between G-C and A-T base pairs. Duplication of the genetic information occurs by the use of one DNA strand as a template for formation of a complementary strand. The genetic information stored in an organism's DNA contains the instructions for all the proteins the organism will ever synthesize. In eucaryotes, DNA is contained in the cell nucleus.

CHROMOSOMAL DNA AND ITS PACKAGING IN THE CHROMATIN FIBER

The most important function of DNA is to carry genes, the information that specifies all the proteins that make up an organism—including information about when, in what types of cells, and in what quantity each protein is to be made. The genomes of eucaryotes are divided up into chromosomes, and in this section we see how genes are typically arranged on each chromosome. In addition, we describe the specialized DNA sequences that allow a chromosome to be accurately duplicated and passed on from one generation to the next.

We also confront the serious challenge of DNA packaging. Each human cell contains approximately 2 meters of DNA if stretched end-to-end; yet the nucleus of a human cell, which contains the DNA, is only about 6 μm in diameter. This is geometrically equivalent to packing 40 km (24 miles) of extremely fine thread into a tennis ball! The complex task of packaging DNA is accomplished by specialized proteins that bind to and fold the DNA, generating a series of coils and loops that provide increasingly higher levels of organization, preventing the DNA from becoming an unmanageable tangle. Amazingly, although the DNA is very tightly folded, it is compacted in a way that allows it to easily become available to the many enzymes in the cell that replicate it, repair it, and use its genes to produce proteins.

Eucaryotic DNA Is Packaged into a Set of Chromosomes

In eucaryotes, the DNA in the nucleus is divided between a set of different **chromosomes**. For example, the human genome—approximately 3.2×10^9 nucleotides—is distributed over 24 different chromosomes. Each chromosome consists of a single, enormously long linear DNA molecule associated with proteins that fold and pack the fine DNA thread into a more compact structure. The complex of DNA and protein is called *chromatin* (from the Greek *chroma*, "colored," because of its staining properties). In addition to the proteins involved in packaging the DNA, chromosomes are also associated with many proteins required for the processes of gene expression, DNA replication, and DNA repair.

Bacteria carry their genes on a single DNA molecule, which is usually circular (see Figure 1–30). This DNA is associated with proteins that package and condense the DNA, but they are different from the proteins that perform these functions in eucaryotes. Although often called the bacterial "chromosome," it does not have the same structure as eucaryotic chromosomes, and less is known about how the bacterial DNA is packaged. Even less is known about how DNA is compacted in archaea. Therefore, our discussion of chromosome structure will focus almost entirely on eucaryotic chromosomes.

With the exception of the germ cells, and a few highly specialized cell types that cannot multiply and lack DNA altogether (for example, red blood cells), each human cell contains two copies of each chromosome, one inherited from the mother and one from the father. The maternal and paternal chromosomes of a pair are called **homologous chromosomes (homologs)**. The only nonhomologous chromosome pairs are the sex chromosomes in males, where a *Y chromosome* is inherited from the father and an *X chromosome* from the mother. Thus, each human cell contains a total of 46 chromosomes—22 pairs common to both males and females, plus two so-called sex chromosomes (X and Y in males, two Xs in females). *DNA hybridization* (described in detail in Chapter 8) can be used to distinguish these human chromosomes by "painting" each one a different color (Figure 4–10). Chromosome painting is typically done at the stage in the cell cycle when chromosomes are especially compacted and easy to visualize (mitosis, see below).

Another more traditional way to distinguish one chromosome from another is to stain them with dyes that produce a striking and reliable pattern of bands along each mitotic chromosome (Figure 4–11). The structural bases for these banding patterns are not well understood, and we return to this issue at the end of the chapter. Nevertheless, the pattern of bands on each type of chromosome is unique, allowing each chromosome to be identified and numbered.

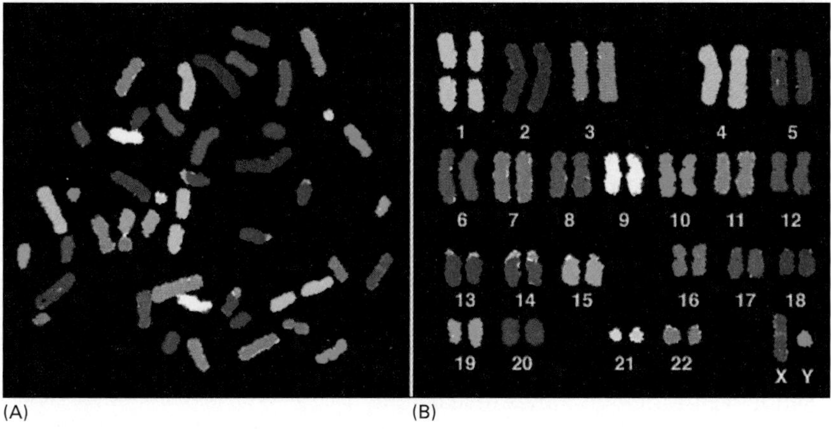

Figure 4–10 Human chromosomes. These chromosomes, from a male, were isolated from a cell undergoing nuclear division (mitosis) and are therefore highly compacted. Each chromosome has been "painted" a different color to permit its unambiguous identification under the light microscope. Chromosome painting is performed by exposing the chromosomes to a collection of human DNA molecules that have been coupled to a combination of fluorescent dyes. For example, DNA molecules derived from chromosome 1 are labeled with one specific dye combination, those from chromosome 2 with another, and so on. Because the labeled DNA can form base pairs, or hybridize, only to the chromosome from which it was derived (discussed in Chapter 8), each chromosome is differently labeled. For such experiments, the chromosomes are subjected to treatments that separate the double-helical DNA into individual strands, designed to permit base-pairing with the single-stranded labeled DNA while keeping the chromosome structure relatively intact. (A) The chromosomes visualized as they originally spilled from the lysed cell. (B) The same chromosomes artificially lined up in their numerical order. This arrangement of the full chromosome set is called a karyotype. (From E. Schröck et al., *Science* 273:494–497, 1996. © AAAS.)

The display of the 46 human chromosomes at mitosis is called the human **karyotype**. If parts of chromosomes are lost, or switched between chromosomes, these changes can be detected by changes in the banding patterns or by changes in the pattern of chromosome painting (Figure 4–12). Cytogeneticists use these alterations to detect chromosome abnormalities that are associated with inherited defects or with certain types of cancer that arise through the rearrangement of chromosomes in somatic cells.

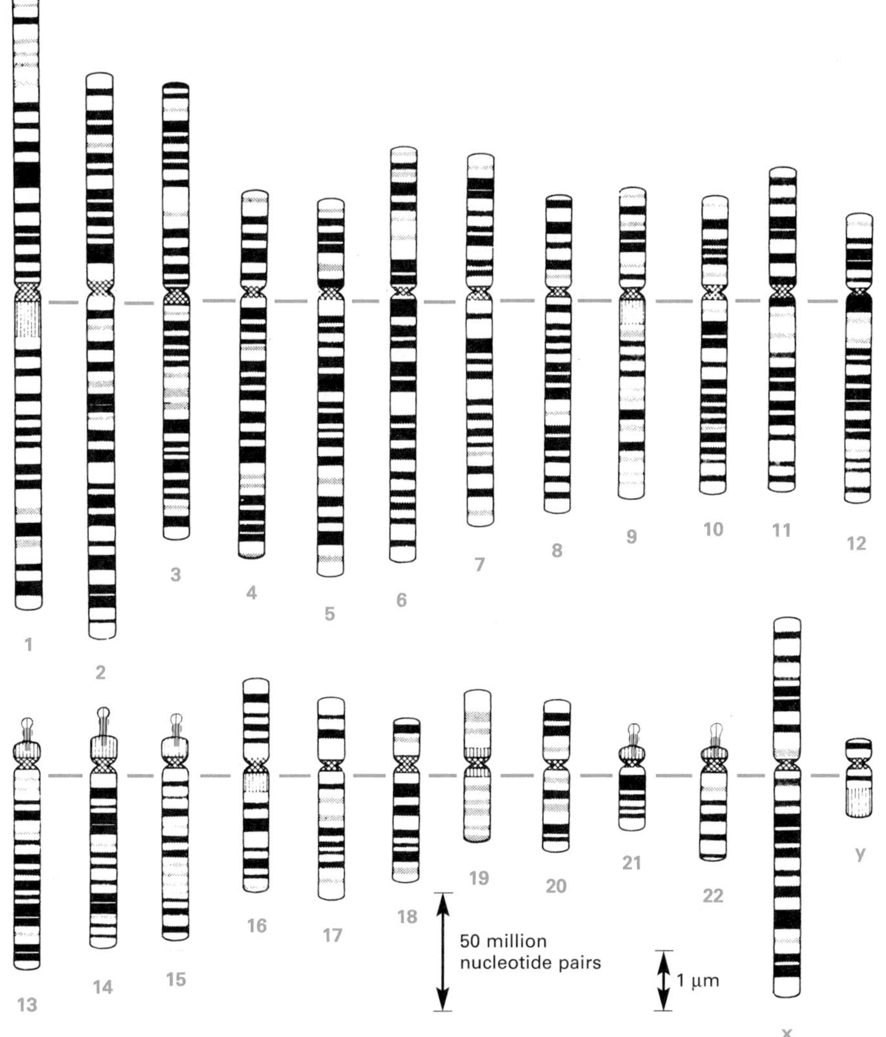

Figure 4–11 The banding patterns of human chromosomes. Chromosomes 1–22 are numbered in approximate order of size. A typical human somatic (non-germ line) cell contains two of each of these chromosomes, plus two sex chromosomes—two X chromosomes in a female, one X and one Y chromosome in a male. The chromosomes used to make these maps were stained at an early stage in mitosis, when the chromosomes are incompletely compacted. The *horizontal green line* represents the position of the centromere (see Figure 4–22), which appears as a constriction on mitotic chromosomes; the knobs on chromosomes 13, 14, 15, 21, and 22 indicate the positions of genes that code for the large ribosomal RNAs (discussed in Chapter 6). These patterns are obtained by staining chromosomes with Giemsa stain, and they can be observed under the light microscope. (Adapted from U. Franke, *Cytogenet. Cell Genet.* 31:24–32, 1981.)

Chromosomes Contain Long Strings of Genes

The most important function of chromosomes is to carry genes—the functional units of heredity. A gene is usually defined as a segment of DNA that contains the instructions for making a particular protein (or a set of closely related proteins). Although this definition holds for the majority of genes, several percent of genes produce an RNA molecule, instead of a protein, as their final product. Like proteins, these RNA molecules perform a diverse set of structural and catalytic functions in the cell, and we discuss them in detail in subsequent chapters.

As might be expected, a correlation exists between the complexity of an organism and the number of genes in its genome (see Table 1–1). For example, total gene numbers range from less than 500 for simple bacteria to about 30,000 for humans. Bacteria and some single-celled eucaryotes have especially compact genomes; the complete nucleotide sequence of their genomes reveals that the DNA molecules that make up their chromosomes are little more than strings of closely packed genes (Figure 4–13; see also Figure 1–30). However, chromosomes from many eucaryotes (including humans) contain, in addition to genes, a large excess of interspersed DNA that does not seem to carry critical information. Sometimes called junk DNA to signify that its usefulness to the cell has not been demonstrated, the particular nucleotide sequence of this DNA may not be important; but the DNA itself, by acting as spacer material, may be crucial for the long-term evolution of the species and for the proper expression of genes. These issues are taken up in detail in Chapter 7.

In general, the more complex the organism, the larger its genome, but because of differences in the amount of excess DNA, the relationship is not systematic (see Figure 1–38). For example, the human genome is 200 times larger than that of the yeast *S. cerevisiae*, but 30 times smaller than that of some plants and amphibians and 200 times smaller than a species of amoeba. Moreover, because of differences in the amount of excess DNA, the genomes of similar organisms (bony fish, for example) can vary several hundredfold in their DNA content, even though they contain roughly the same number of genes. Whatever the excess DNA may do, it seems clear that it is not a great handicap for a higher eucaryotic cell to carry a large amount of it.

The apportionment of the genome over chromosomes also differs from one eucaryotic species to the next. For example, compared with 46 for humans,

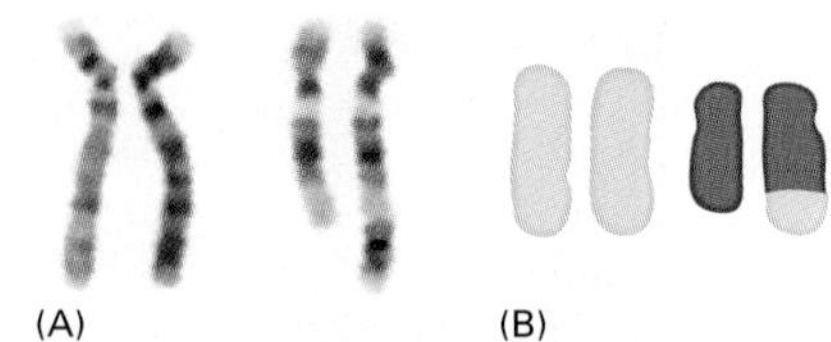

Figure 4–12 An aberrant human chromosome. (A) Two pairs of chromosomes, stained with Giemsa (see Figure 4–11), from a patient with ataxia, a disease characterized by progressive deterioration of motor skills. The patient has a normal pair of chromosome 4s *(left-hand pair)*, but one normal chromosome 12 and one aberrant chromosome 12, as seen by its greater length *(right-hand pair)*. The additional material contained on the aberrant chromosome 12 was deduced, from its pattern of bands, as a piece of chromosome 4 that had become attached to chromosome 12 through an abnormal recombination event, called a chromosomal translocation. (B) The same two chromosome pairs, "painted" *blue* for chromosome 4 DNA and *purple* for chromosome 12 DNA. The two techniques give rise to the same conclusion regarding the nature of the aberrant chromosome 12, but chromosome painting provides better resolution, and the clear identification of even short pieces of chromosomes that have become translocated. However, Giemsa staining is easier to perform. (From E. Schröck et al., *Science* 273:494–497, 1996. © AAAS.)

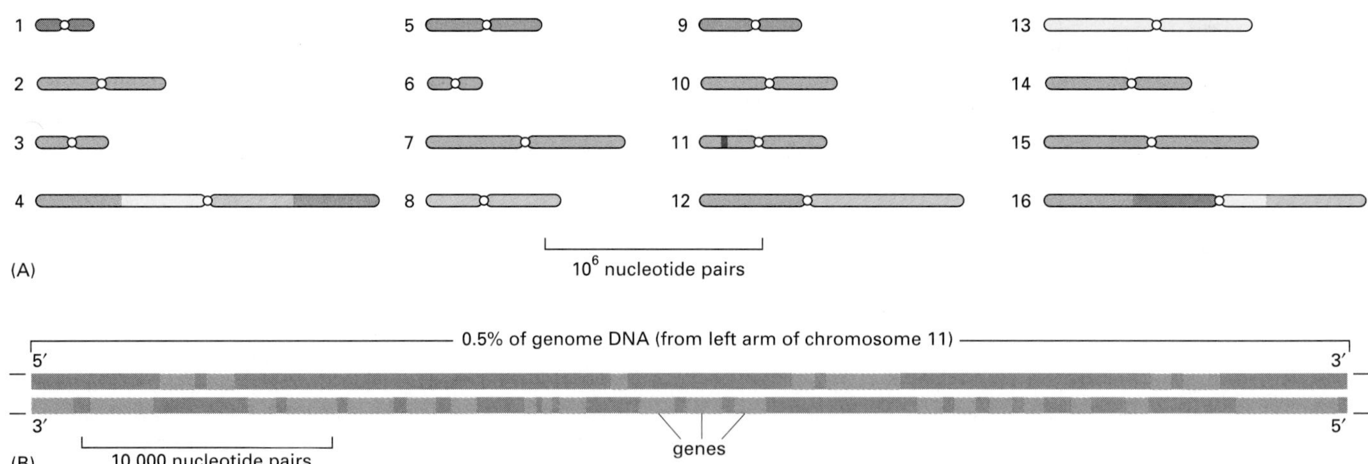

Figure 4–13 The genome of *S. cerevisiae* (budding yeast). (A) The genome is distributed over 16 chromosomes, and its complete nucleotide sequence was determined by a cooperative effort involving scientists working in many different locations, as indicated *(gray*, Canada; *orange*, European Union; *yellow*, United Kingdom; *blue*, Japan; *light green*, St Louis, Missouri; *dark green*, Stanford, California). The constriction present on each chromosome represents the position of its centromere (see Figure 4–22). (B) A small region of chromosome 11, highlighted in *red* in part A, is magnified to show the high density of genes characteristic of this species. As indicated by *orange*, some genes are transcribed from the lower strand (see Figure 1–5), while others are transcribed from the upper strand. There are about 6000 genes in the complete genome, which is 12,147,813 nucleotide pairs long.

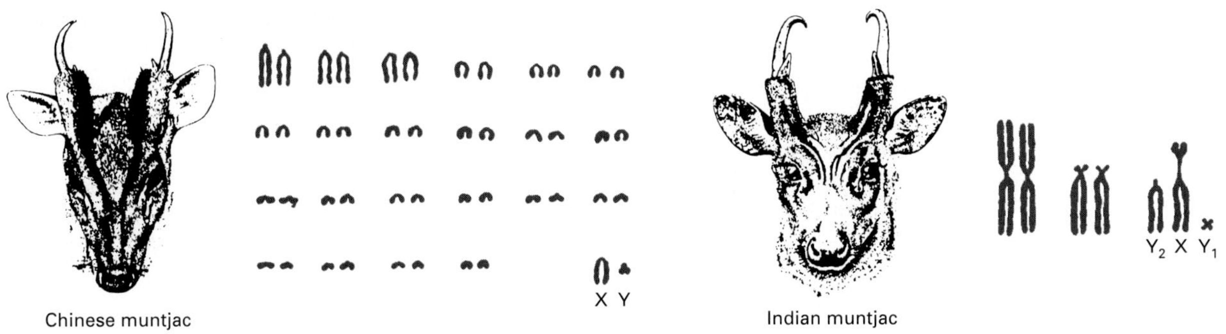

Figure 4–14 Two closely related species of deer with very different chromosome numbers. In the evolution of the Indian muntjac, initially separate chromosomes fused, without having a major effect on the animal. These two species have roughly the same number of genes. (Adapted from M.W. Strickberger, Evolution, 3rd edition, 2000, Sudbury, MA: Jones & Bartlett Publishers.)

somatic cells from a species of small deer contain only 6 chromosomes, while those from a species of carp contain over 100. Even closely related species with similar genome sizes can have very different numbers and sizes of chromosomes (Figure 4–14). Thus, there is no simple relationship between chromosome number, species complexity, and total genome size. Rather, the genomes and chromosomes of modern-day species have each been shaped by a unique history of seemingly random genetic events, acted on by selection pressures.

The Nucleotide Sequence of the Human Genome Shows How Genes Are Arranged in Humans

When the DNA sequence of human chromosome 22, one of the smallest human chromosomes (see Figure 4–11), was completed in 1999, it became possible for the first time to see exactly how genes are arranged along an entire vertebrate chromosome (Figure 4–15 and Table 4–1). With the publication of the "first draft" of the entire human genome in 2001, the genetic landscape of all human chromosomes suddenly came into sharp focus. The sheer quantity of information

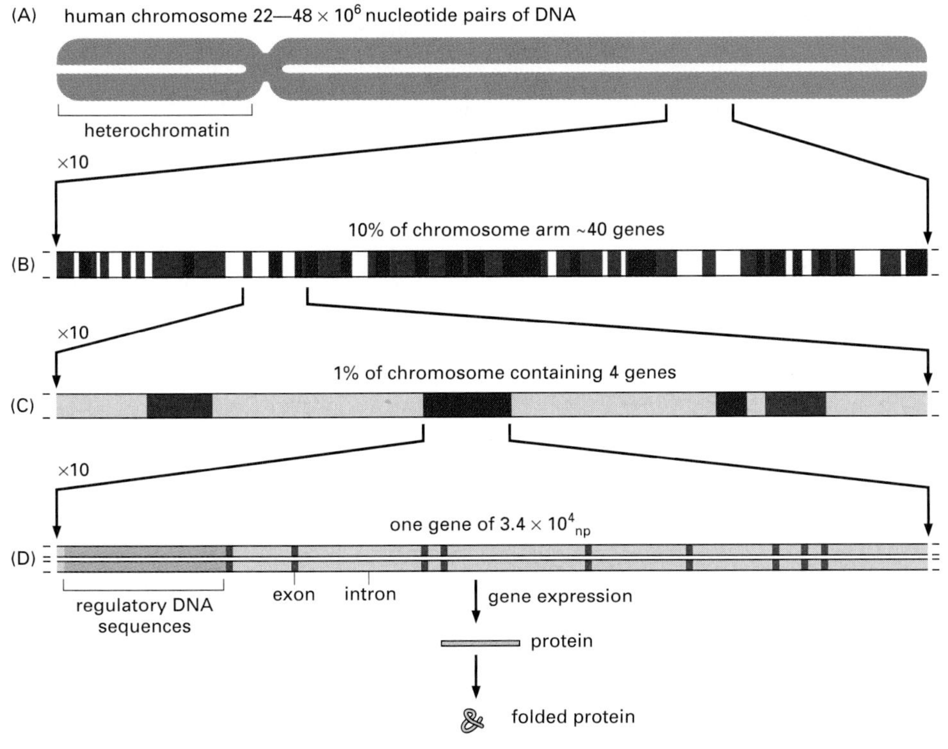

Figure 4–15 The organization of genes on a human chromosome. (A) Chromosome 22, one of the smallest human chromosomes, contains 48×10^6 nucleotide pairs and makes up approximately 1.5% of the entire human genome. Most of the left arm of chromosome 22 consists of short repeated sequences of DNA that are packaged in a particularly compact form of chromatin (heterochromatin), which is discussed later in this chapter. (B) A tenfold expansion of a portion of chromosome 22, with about 40 genes indicated. Those in *dark brown* are known genes and those in *light brown* are predicted genes. (C) An expanded portion of (B) shows the entire length of several genes. (D) The intron–exon arrangement of a typical gene is shown after a further tenfold expansion. Each exon *(red)* codes for a portion of the protein, while the DNA sequence of the introns *(gray)* is relatively unimportant. The entire human genome (3.2×10^9 nucleotide pairs) is distributed over 22 autosomes and 2 sex chromosomes (see Figures 4–10 and 4–11). The term *human genome sequence* refers to the complete nucleotide sequence of DNA in these 24 chromosomes. Being diploid, a human somatic cell therefore contains roughly twice this amount of DNA. Humans differ from one another by an average of one nucleotide in every thousand, and a wide variety of humans contributed DNA for the genome sequencing project. The published human genome sequence is therefore a composite of many individual sequences. (Adapted from International Human Genome Sequencing Consortium, *Nature* 409:860–921, 2001.)

provided by the Human Genome Project is unprecedented in biology (Figure 4–16 and Table 4–1); the human genome is 25 times larger than any other genome sequenced so far, and is 8 times as large as the sum of all previously sequenced genomes. At its peak, the Human Genome Project generated raw nucleotide sequences at a rate of 1000 nucleotides per second around the clock. It will be many decades before this information is fully analyzed, but it will continue to stimulate many new experiments and has already affected the content of all the chapters in this book.

Although there are many aspects to analyzing the human genome, here we simply make a few generalizations regarding the arrangement of genes in human chromosomes. The first striking feature of the human genome is how little of it (only a few percent) codes for proteins or structural and catalytic RNAs (Figure 4–17). Much of the remaining chromosomal DNA is made up of short, mobile pieces of DNA that have gradually inserted themselves in the chromosome over evolutionary time. We discuss these *transposable elements* in detail in later chapters.

A second notable feature of the human genome is the large average gene size of 27,000 nucleotide pairs. As discussed above, a typical gene carries in its linear sequence of nucleotides the information for the linear sequence of the amino acids of a protein. Only about 1300 nucleotide pairs are required to encode a protein of average size (about 430 amino acids in humans). Most of the remaining DNA in a gene consists of long stretches of noncoding DNA that interrupt the relatively short segments of DNA that code for protein. The coding sequences are called *exons*; the intervening (noncoding) sequences are called *introns* (see Figure 4–15 and Table 4–1).

The majority of human genes thus consist of a long string of alternating exons and introns, with most of the gene consisting of introns. In contrast, the majority of genes from organisms with compact genomes lack introns. This accounts for the much smaller size of their genes (about one-twentieth that of human genes), as well as for the much higher fraction of coding DNA in their chromosomes. In addition to introns and exons, each gene is associated with *regulatory DNA sequences,* which are responsible for ensuring that the gene is

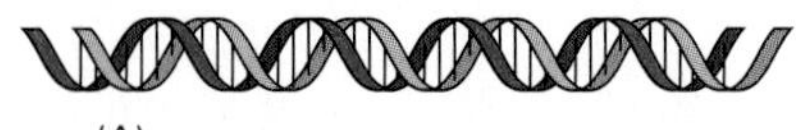
(A)

(B)

Figure 4–16 Scale of the human genome. If each nucleotide pair is drawn as 1 mm as in (A), then the human genome would extend 3200 km (approximately 2000 miles), far enough to stretch across the center of Africa, the site of our human origins (*red line* in B). At this scale, there would be, on average, a protein-coding gene every 300 m. An average gene would extend for 30 m, but the coding sequences in this gene would add up to only just over a meter.

TABLE 4–1 Vital Statistics of Human Chromosome 22 and the Entire Human Genome

	CHROMOSOME 22	HUMAN GENOME
DNA length	48×10^6 nucleotide pairs*	3.2×10^9
Number of genes	approximately 700	approximately 30,000
Smallest protein-coding gene	1000 nucleotide pairs	not analyzed
Largest gene	583,000 nucleotide pairs	2.4×10^6 nucleotide pairs
Mean gene size	19,000 nucleotide pairs	27,000 nucleotide pairs
Smallest number of exons per gene	1	1
Largest number of exons per gene	54	178
Mean number of exons per gene	5.4	8.8
Smallest exon size	8 nucleotide pairs	not analyzed
Largest exon size	7600 nucleotide pairs	17,106 nucleotide pairs
Mean exon size	266 nucleotide pairs	145 nucleotide pairs
Number of pseudogenes**	more than 134	not analyzed
Percentage of DNA sequence in exons (protein coding sequences)	3%	1.5%
Percentage of DNA in high-copy repetitive elements	42%	approximately 50%
Percentage of total human genome	1.5%	100%

* The nucleotide sequence of 33.8×10^6 nucleotides is known; the rest of the chromosome consists primarily of very short repeated sequences that do not code for proteins or RNA.
** A pseudogene is a nucleotide sequence of DNA closely resembling that of a functional gene, but containing numerous deletion mutations that prevent its proper expression. Most pseudogenes arise from the duplication of a functional gene followed by the accumulation of damaging mutations in one copy.

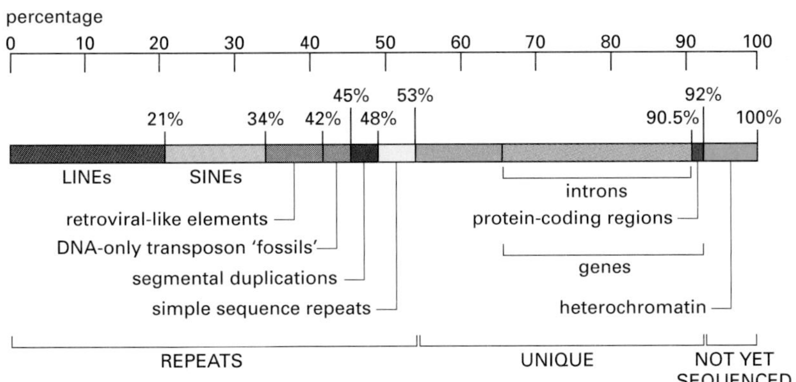

Figure 4–17 Representation of the nucleotide sequence content of the human genome. LINES, SINES, retroviral-like elements, and DNA-only transposons are all mobile genetic elements that have multiplied in our genome by replicating themselves and inserting the new copies in different positions. Mobile genetic elements are discussed in Chapter 5. Simple sequence repeats are short nucleotide sequences (less than 14 nucleotide pairs) that are repeated again and again for long stretches. Segmental duplications are large blocks of the genome (1000–200,000 nucleotide pairs) that are present at two or more locations in the genome. Over half of the unique sequence consists of genes and the remainder is probably regulatory DNA. Most of the DNA present in heterochromatin, a specialized type of chromatin (discussed later in this chapter) that contains relatively few genes, has not yet been sequenced. (Adapted from Unveiling the Human Genome, Supplement to the Wellcome Trust Newsletter. London: Wellcome Trust, February 2001.)

expressed at the proper level and time, and the proper type of cell. In humans, the regulatory sequences for a typical gene are spread out over tens of thousands of nucleotide pairs. As would be expected, these regulatory sequences are more compressed in organisms with compact genomes. We discuss in Chapter 7 how regulatory DNA sequences work.

Finally, the nucleotide sequence of the human genome has revealed that the critical information seems to be in an alarming state of disarray. As one commentator described our genome, "In some ways it may resemble your garage/bedroom/refrigerator/life: highly individualistic, but unkempt; little evidence of organization; much accumulated clutter (referred to by the uninitiated as 'junk'); virtually nothing ever discarded; and the few patently valuable items indiscriminately, apparently carelessly, scattered throughout."

Comparisons Between the DNAs of Related Organisms Distinguish Conserved and Nonconserved Regions of DNA Sequence

A major obstacle in interpreting the nucleotide sequences of human chromosomes is the fact that much of the sequence is probably unimportant. Moreover, the coding regions of the genome (the exons) are typically found in short segments (average size about 145 nucleotide pairs) floating in a sea of DNA whose exact nucleotide sequence is of little consequence. This arrangement makes it very difficult to identify all the exons in a stretch of DNA sequence; even harder is the determination of where a gene begins and ends and how many exons it spans. Accurate gene identification requires approaches that extract information from the inherently low signal-to-noise ratio of the human genome, and we describe some of them in Chapter 8. Here we discuss the most general approach, one that has the potential to identify not only coding sequences but also additional DNA sequences that are important. It is based on the observation that sequences that have a function are conserved during evolution, whereas those without a function are free to mutate randomly. The strategy is therefore to compare the human sequence with that of the corresponding regions of a related genome, such as that of the mouse. Humans and mice are thought to have diverged from a common mammalian ancestor about 100×10^6 years ago, which is long enough for the majority of nucleotides in their genomes to have been changed by random mutational events. Consequently, the only regions that will have remained closely similar in the two genomes are those in which mutations would have impaired function and put the animals carrying them at a disadvantage, resulting in their elimination from the population by natural selection. Such closely similar regions are known as *conserved regions*. In general, conserved regions represent functionally important exons and regulatory sequences. In contrast, *nonconserved regions* represent DNA whose sequence is generally not critical for function. By revealing in this way the results of a very long natural "experiment," comparative DNA sequencing studies highlight the most interesting regions in genomes.

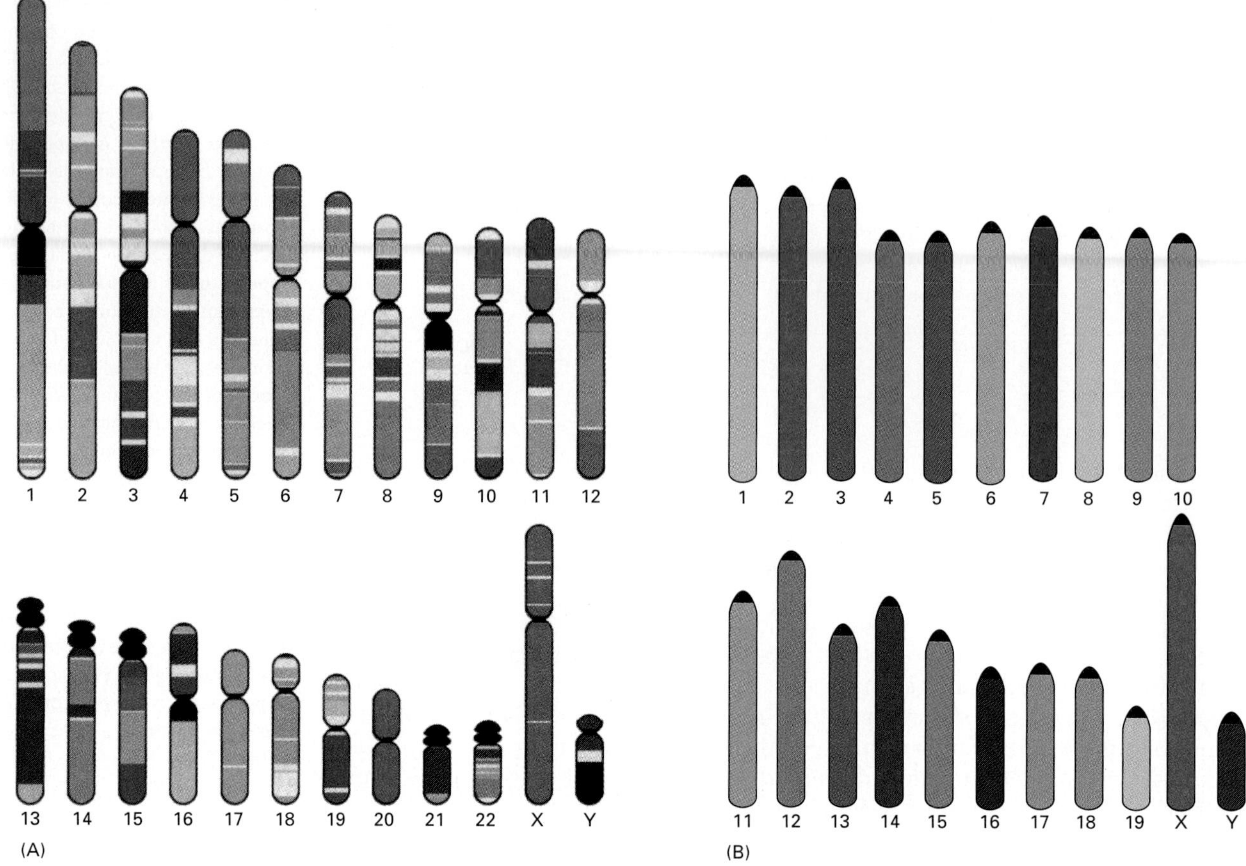

Figure 4–18 Conserved synteny between the human and mouse genomes. Regions from different mouse chromosomes (indicated by the colors of each mouse in B) show conserved synteny (gene order) with the indicated regions of the human genome (A). For example the genes present in the upper portion of human chromosome 1 *(orange)* are present in the same order in a portion of mouse chromosome 4. Regions of human chromosomes that are composed primarily of short, repeated sequences are shown in *black*. Mouse centromeres (indicated in *black* in B) are located at the ends of chromosomes; no known genes lie beyond the centromere on any mouse chromosome. For the most part, human centromeres, indicated by constrictions, occupy more internal positions on chromosomes (see Figure 4–11). (Adapted from International Human Genome Sequencing Consortium, *Nature* 409:860–921, 2001.)

Comparative studies of this kind have revealed not only that mice and humans share most of the same genes, but also that large blocks of the mouse and human genomes contain these genes in the same order, a feature called *conserved synteny* (Figure 4–18). Conserved synteny can also be revealed by chromosome painting, and this technique has been used to reconstruct the evolutionary history of our own chromosomes by comparing them with those from other mammals (Figure 4–19).

Chromosomes Exist in Different States Throughout the Life of a Cell

We have seen how genes are arranged in chromosomes, but to form a functional chromosome, a DNA molecule must be able to do more than simply carry genes: it must be able to replicate, and the replicated copies must be separated and reliably partitioned into daughter cells at each cell division. This process occurs through an ordered series of stages, collectively known as the **cell cycle**. The cell cycle is briefly summarized in Figure 4–20, and discussed in detail in Chapter 17. Only two of the stages of the cycle concern us in this chapter. During *interphase* chromosomes are replicated, and during *mitosis* they become highly condensed and then are separated and distributed to the two daughter nuclei. The highly condensed chromosomes in a dividing cell are known as *mitotic chromosomes*. This is the form in which chromosomes are most easily visualized; in fact, all the images of chromosomes shown so far in the chapter are of chromosomes in mitosis. This condensed state is important in allowing the duplicated chromosomes to be separated by the mitotic spindle during cell division, as discussed in Chapter 18.

During the portions of the cell cycle when the cell is not dividing, the chromosomes are extended and much of their chromatin exists as long, thin tangled

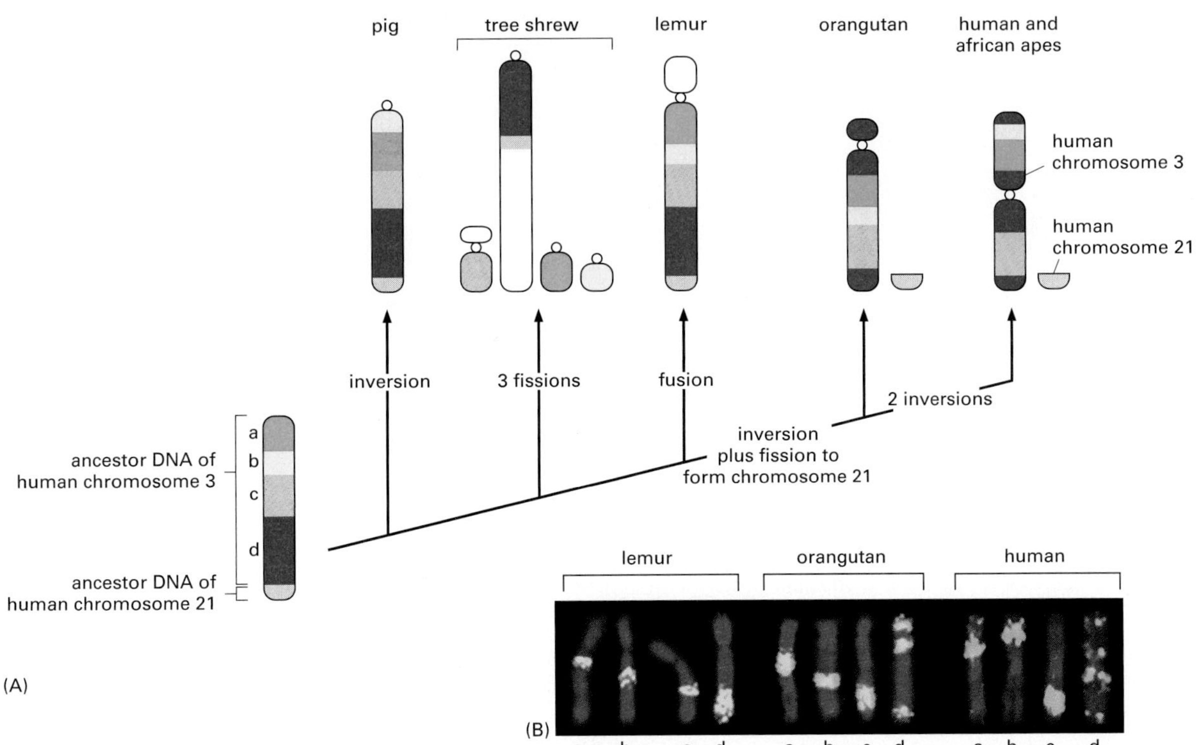

Figure 4–19 A proposed evolutionary history of human chromosome 3 and its relatives in other mammals. (A) At the *lower left* is the order of chromosome 3 segments hypothesized to be present on a chromosome of a mammalian ancestor. Along the *top* are the patterns of chromosome sequences found in the chromosomes of modern mammals. The minimum changes necessary to account for the appearance of the modern chromosomes from the hypothetical ancestor are marked along each branch. In mammals, these types of changes in chromosome organization are thought to occur once every $5–10 \times 10^6$ years. The *small circles* depicted in the modern chromosomes represent the positions of centromeres. (B) Some of the chromosome painting experiments that led to the diagram in (A). Each image shows the chromosome most closely related to human chromosome 3, painted *green* by hybridization with different segments of DNA, lettered a, b, c, and d along the *bottom* of the figure. These letters correspond to the colored segments of the diagram in (A). (From S. Müller et al., *Proc. Natl. Acad. Sci. USA* 97:206–211, 2000. © National Academy of Sciences.)

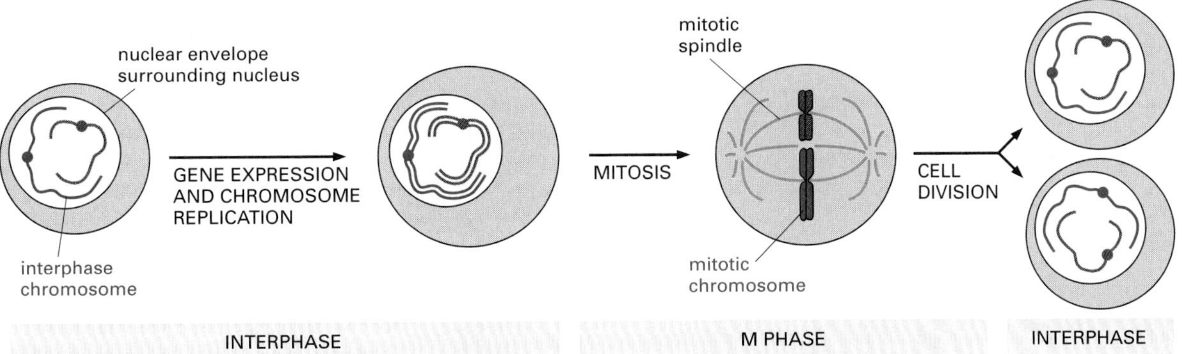

Figure 4–20 A simplified view of the eucaryotic cell cycle. During interphase, the cell is actively expressing its genes and is therefore synthesizing proteins. Also, during interphase and before cell division, the DNA is replicated and the chromosomes are duplicated. Once DNA replication is complete, the cell can enter *M phase,* when mitosis occurs and the nucleus is divided into two daughter nuclei. During this stage, the chromosomes condense, the nuclear envelope breaks down, and the mitotic spindle forms from microtubules and other proteins. The condensed mitotic chromosomes are captured by the mitotic spindle, and one complete set of chromosomes is then pulled to each end of the cell. A nuclear envelope re-forms around each chromosome set, and in the final step of M phase, the cell divides to produce two daughter cells. Most of the time in the cell cycle is spent in interphase; M phase is brief in comparison, occupying only about an hour in many mammalian cells.

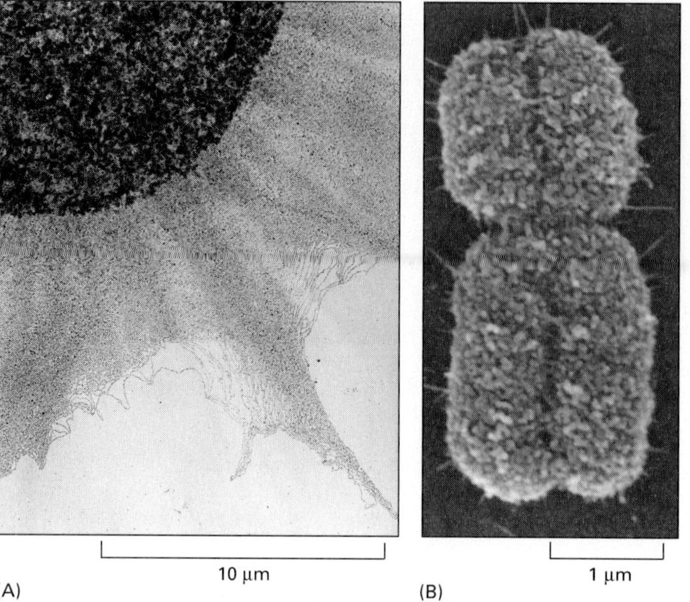

Figure 4–21 A comparison of extended interphase chromatin with the chromatin in a mitotic chromosome. (A) An electron micrograph showing an enormous tangle of chromatin spilling out of a lysed interphase nucleus. (B) A scanning electron micrograph of a mitotic chromosome: a condensed duplicated chromosome in which the two new chromosomes are still linked together (see Figure 4–22). The constricted region indicates the position of the centromere. Note the difference in scales. (A, courtesy of Victoria Foe; B, courtesy of Terry D. Allen.)

threads in the nucleus so that individual chromosomes cannot be easily distinguished (Figure 4–21). We refer to chromosomes in this extended state as *interphase chromosomes.*

Each DNA Molecule That Forms a Linear Chromosome Must Contain a Centromere, Two Telomeres, and Replication Origins

A chromosome operates as a distinct structural unit: for a copy to be passed on to each daughter cell at division, each chromosome must be able to replicate, and the newly replicated copies must subsequently be separated and partitioned correctly into the two daughter cells. These basic functions are controlled by three types of specialized nucleotide sequence in the DNA, each of which binds specific proteins that guide the machinery that replicates and segregates chromosomes (Figure 4–22).

Experiments in yeasts, whose chromosomes are relatively small and easy to manipulate, have identified the minimal DNA sequence elements responsible for each of these functions. One type of nucleotide sequence acts as a DNA **replication origin**, the location at which duplication of the DNA begins. Eucaryotic chromosomes contain many origins of replication to ensure that the entire chromosome can be replicated rapidly, as discussed in detail in Chapter 5.

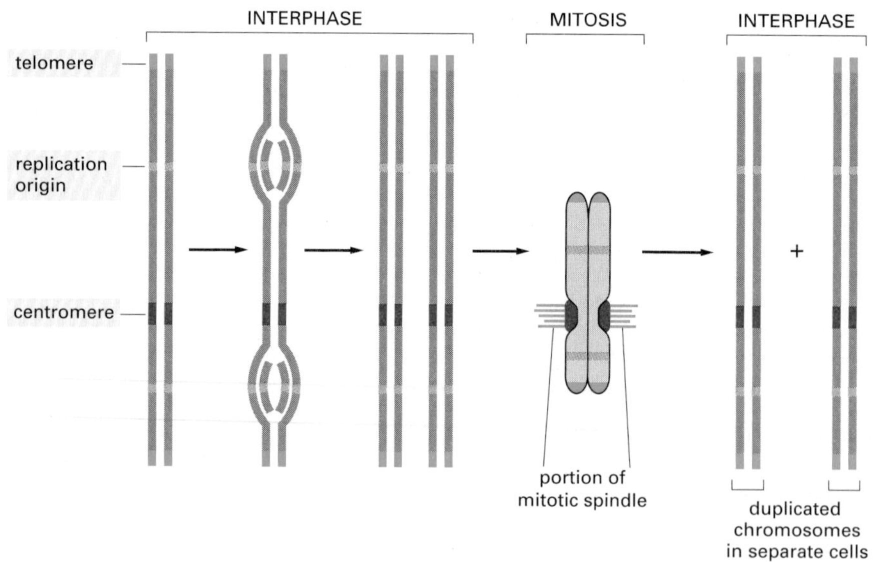

Figure 4–22 The three DNA sequences required to produce a eucaryotic chromosome that can be replicated and then segregated at mitosis. Each chromosome has multiple origins of replication, one centromere, and two telomeres. Shown here is the sequence of events a typical chromosome follows during the cell cycle. The DNA replicates in interphase beginning at the origins of replication and proceeding bidirectionally from the origins across the chromosome. In M phase, the centromere attaches the duplicated chromosomes to the mitotic spindle so that one copy is distributed to each daughter cell during mitosis. The centromere also helps to hold the duplicated chromosomes together until they are ready to be moved apart. The telomeres form special caps at each chromosome end.

After replication, the two daughter chromosomes remain attached to one another and, as the cell cycle proceeds, are condensed further to produce mitotic chromosomes. The presence of a second specialized DNA sequence, called a **centromere**, allows one copy of each duplicated and condensed chromosome to be pulled into each daughter cell when a cell divides. A protein complex called a *kinetochore* forms at the centromere and attaches the duplicated chromosomes to the mitotic spindle, allowing them to be pulled apart (discussed in Chapter 18).

The third specialized DNA sequence forms **telomeres**, the ends of a chromosome. Telomeres contain repeated nucleotide sequences that enable the ends of chromosomes to be efficiently replicated. Telomeres also perform another function: the repeated telomere DNA sequences, together with the regions adjoining them, form structures that protect the end of the chromosome from being recognized by the cell as a broken DNA molecule in need of repair. We discuss this type of repair and the other features of telomeres in Chapter 5.

In yeast cells, the three types of sequences required to propagate a chromosome are relatively short (typically less than 1000 base pairs each) and therefore use only a tiny fraction of the information-carrying capacity of a chromosome. Although telomere sequences are fairly simple and short in all eucaryotes, the DNA sequences that specify centromeres and replication origins in more complex organisms are much longer than their yeast counterparts. For example, experiments suggest that human centromeres may contain up to 100,000 nucleotide pairs. It has been proposed that human centromeres may not even require a stretch of DNA with a defined nucleotide sequence; instead, they may simply create a large, regularly repeating protein–nucleic acid structure. We return to this issue at the end of the chapter when we discuss in more general terms the proteins that, along with DNA, make up chromosomes.

DNA Molecules Are Highly Condensed in Chromosomes

All eucaryotic organisms have elaborate ways of packaging DNA into chromosomes. Recall from earlier in this chapter that human chromosome 22 contains about 48 million nucleotide pairs. Stretched out end to end, its DNA would extend about 1.5 cm. Yet, when it exists as a mitotic chromosome, chromosome 22 measures only about 2 μm in length (see Figures 4–10 and 4–11), giving an end-to-end compaction ratio of nearly 10,000-fold. This remarkable feat of compression is performed by proteins that successively coil and fold the DNA into higher and higher levels of organization. Although less condensed than mitotic chromosomes, the DNA of interphase chromosomes is still tightly packed, with an overall compaction ratio of approximately 1000-fold. In the next sections we discuss the specialized proteins that make the compression possible.

In reading these sections it is important to keep in mind that chromosome structure is dynamic. Not only do chromosomes globally condense in accord with the cell cycle, but different regions of the interphase chromosomes condense and decondense as the cells gain access to specific DNA sequences for gene expression, DNA repair, and replication. The packaging of chromosomes must therefore be accomplished in a way that allows rapid localized, on-demand access to the DNA.

Nucleosomes Are the Basic Unit of Eucaryotic Chromosome Structure

The proteins that bind to the DNA to form eucaryotic chromosomes are traditionally divided into two general classes: the **histones** and the *nonhistone chromosomal proteins*. The complex of both classes of protein with the nuclear DNA of eucaryotic cells is known as **chromatin**. Histones are present in such enormous quantities in the cell (about 60 million molecules of each type per human cell) that their total mass in chromatin is about equal to that of the DNA.

Histones are responsible for the first and most basic level of chromosome organization, the **nucleosome**, which was discovered in 1974. When interphase

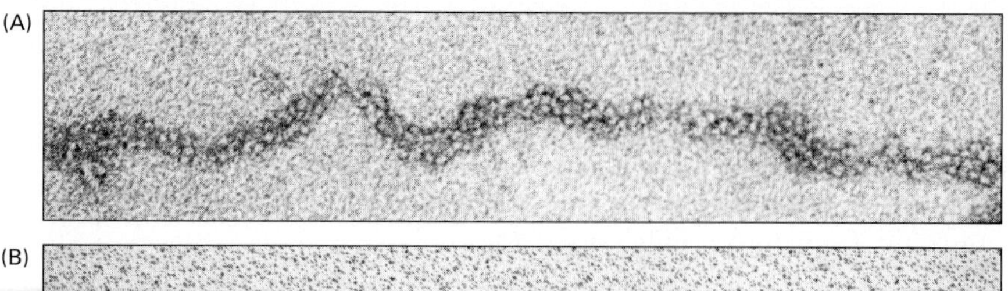

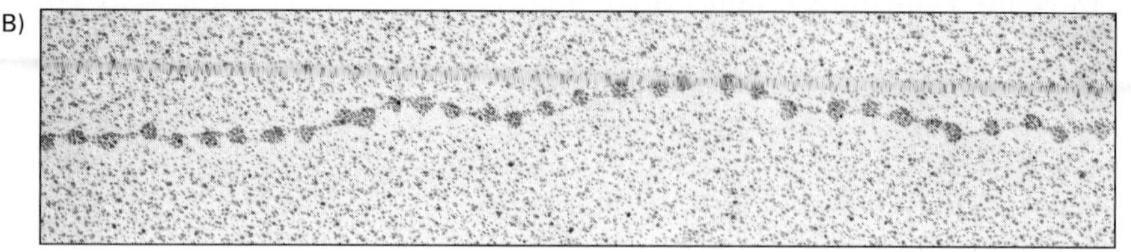

Figure 4–23 Nucleosomes as seen in the electron microscope. (A) Chromatin isolated directly from an interphase nucleus appears in the electron microscope as a thread 30 nm thick. (B) This electron micrograph shows a length of chromatin that has been experimentally unpacked, or decondensed, after isolation to show the nucleosomes. (A, courtesy of Barbara Hamkalo; B, courtesy of Victoria Foe.)

nuclei are broken open very gently and their contents examined under the electron microscope, most of the chromatin is in the form of a fiber with a diameter of about 30 nm (Figure 4–23A). If this chromatin is subjected to treatments that cause it to unfold partially, it can be seen under the electron microscope as a series of "beads on a string" (Figure 4–23B). The string is DNA, and each bead is a "nucleosome core particle" that consists of DNA wound around a protein core formed from histones. The beads on a string represent the first level of chromosomal DNA packing.

The structural organization of nucleosomes was determined after first isolating them from unfolded chromatin by digestion with particular enzymes (called nucleases) that break down DNA by cutting between the nucleosomes. After digestion for a short period, the exposed DNA between the nucleosome core particles, the *linker DNA*, is degraded. Each individual nucleosome core particle consists of a complex of eight histone proteins—two molecules each of histones H2A, H2B, H3, and H4—and double-stranded DNA that is 146 nucleotide pairs long. The *histone octamer* forms a protein core around which the double-stranded DNA is wound (Figure 4–24).

Each nucleosome core particle is separated from the next by a region of linker DNA, which can vary in length from a few nucleotide pairs up to about 80. (The term *nucleosome* technically refers to a nucleosome core particle plus one of its adjacent DNA linkers, but it is often used synonymously with nucleosome core particle.) On average, therefore, nucleosomes repeat at intervals of about 200 nucleotide pairs. For example, a diploid human cell with 6.4×10^9 nucleotide pairs contains approximately 30 million nucleosomes. The formation of nucleosomes converts a DNA molecule into a chromatin thread about one-third of its initial length, and this provides the first level of DNA packing.

The Structure of the Nucleosome Core Particle Reveals How DNA Is Packaged

The high-resolution structure of a nucleosome core particle, solved in 1997, revealed a disc-shaped histone core around which the DNA was tightly wrapped 1.65 turns in a left-handed coil (Figure 4–25). All four of the histones that make up the core of the nucleosome are relatively small proteins (102–135 amino

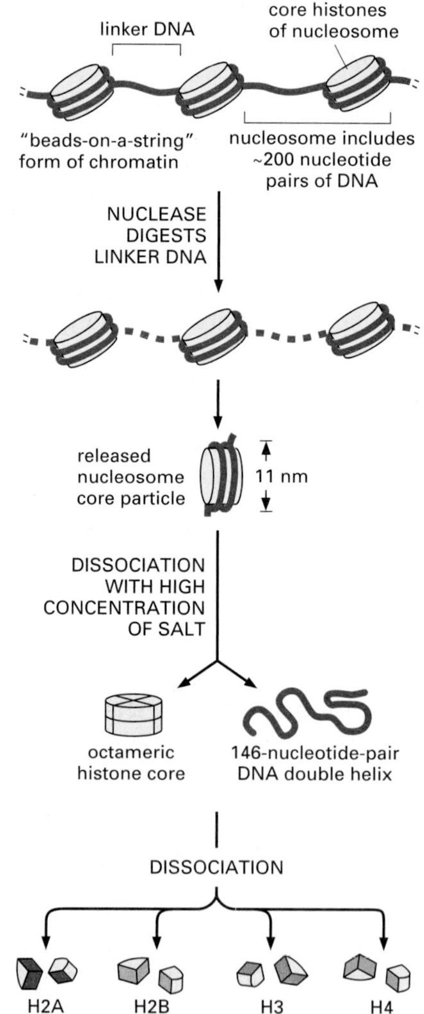

Figure 4–24 Structural organization of the nucleosome. A nucleosome contains a protein core made of eight histone molecules. As indicated, the nucleosome core particle is released from chromatin by digestion of the linker DNA with a nuclease, an enzyme that breaks down DNA. (The nuclease can degrade the exposed linker DNA but cannot attack the DNA wound tightly around the nucleosome core.) After dissociation of the isolated nucleosome into its protein core and DNA, the length of the DNA that was wound around the core can be determined. This length of 146 nucleotide pairs is sufficient to wrap 1.65 times around the histone core.

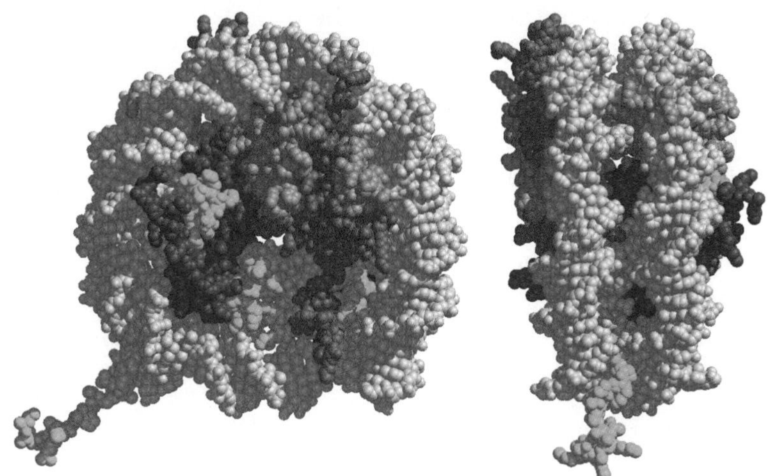

Figure 4–25 The structure of a nucleosome core particle, as determined by x-ray diffraction analyses of crystals. Each histone is colored according to the scheme of Figure 4–26, with the DNA double helix in *light gray.* (Reprinted by permission from K. Luger et al., *Nature* 389:251–260, 1997. © Macmillan Magazines Ltd.)

acids), and they share a structural motif, known as the *histone fold,* formed from three α helices connected by two loops (Figure 4–26). In assembling a nucleosome, the histone folds first bind to each other to form H3–H4 and H2A–H2B dimers, and the H3–H4 dimers combine to form tetramers. An H3–H4 tetramer then further combines with two H2A–H2B dimers to form the compact octamer core, around which the DNA is wound (Figure 4–27).

The interface between DNA and histone is extensive: 142 hydrogen bonds are formed between DNA and the histone core in each nucleosome. Nearly half of these bonds form between the amino acid backbone of the histones and the phosphodiester backbone of the DNA. Numerous hydrophobic interactions and salt linkages also hold DNA and protein together in the nucleosome. For example, all the core histones are rich in lysine and arginine (two amino acids with basic side chains), and their positive charges can effectively neutralize the negatively charged DNA backbone. These numerous interactions explain in part why DNA of virtually any sequence can be bound on a histone octamer core. The path of the DNA around the histone core is not smooth; rather, several kinks are seen in the DNA, as expected from the nonuniform surface of the core.

In addition to its histone fold, each of the core histones has a long N-terminal amino acid "tail", which extends out from the DNA-histone core (see Figure 4–27). These histone tails are subject to several different types of covalent modifications, which control many aspects of chromatin structure. We discuss these issues later in the chapter.

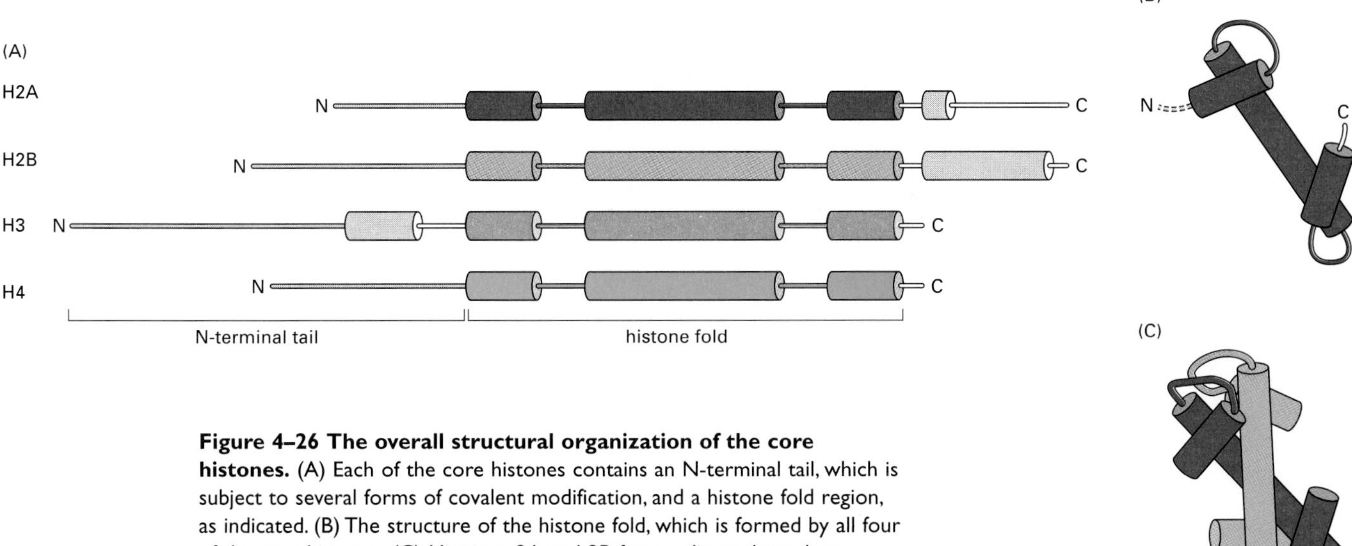

Figure 4–26 The overall structural organization of the core histones. (A) Each of the core histones contains an N-terminal tail, which is subject to several forms of covalent modification, and a histone fold region, as indicated. (B) The structure of the histone fold, which is formed by all four of the core histones. (C) Histones 2A and 2B form a dimer through an interaction known as the "handshake." Histones H3 and H4 form a dimer through the same type of interaction, as illustrated in Figure 4–27.

As might be expected from their fundamental role in DNA packaging, the histones are among the most highly conserved eucaryotic proteins. For example, the amino acid sequence of histone H4 from a pea and a cow differ at only at 2 of the 102 positions. This strong evolutionary conservation suggests that the functions of histones involve nearly all of their amino acids, so that a change in any position is deleterious to the cell. This suggestion has been tested directly in yeast cells, in which it is possible to mutate a given histone gene *in vitro* and introduce it into the yeast genome in place of the normal gene. As might be expected, most changes in histone sequences are lethal; the few that are not lethal cause changes in the normal pattern of gene expression, as well as other abnormalities.

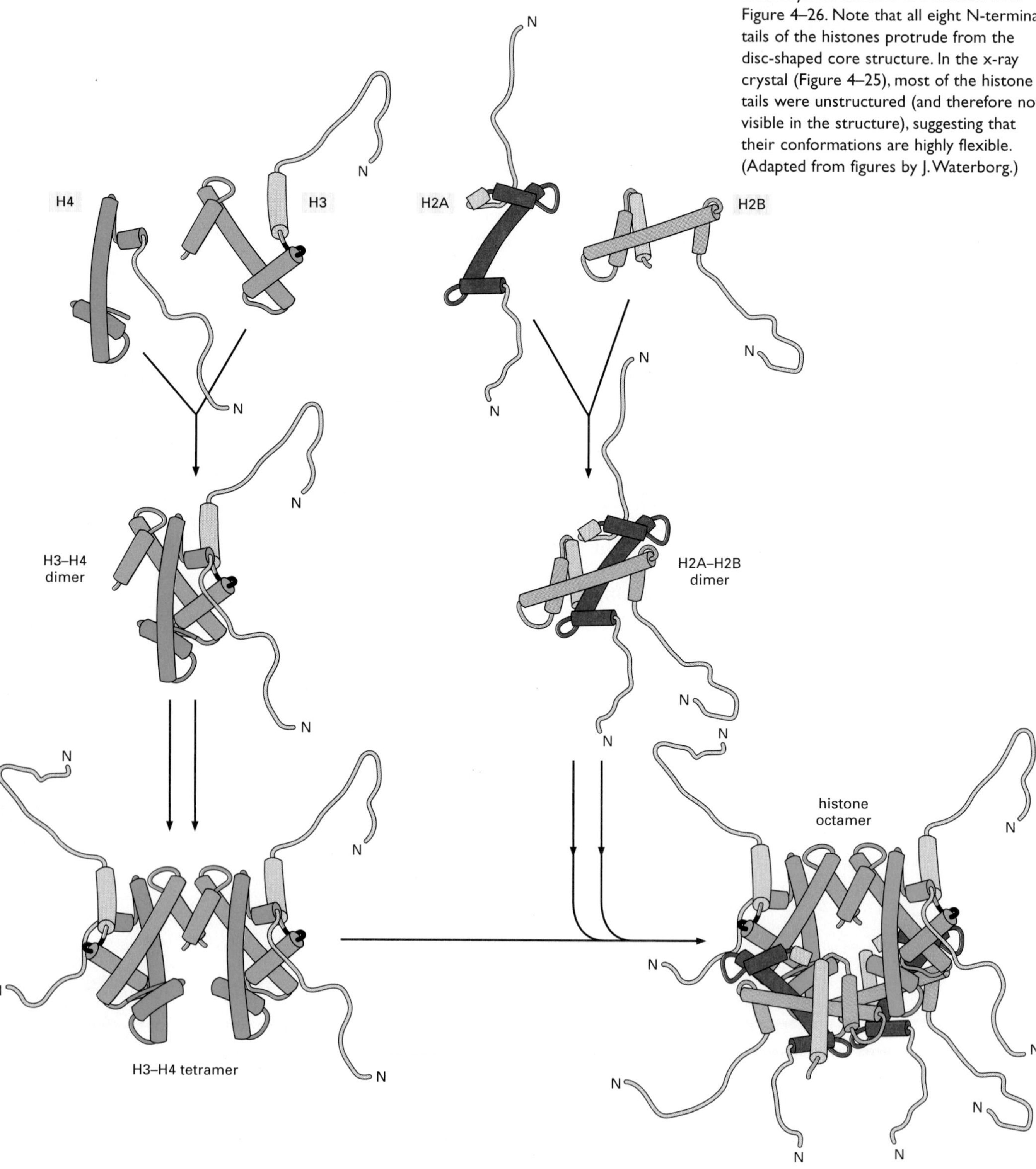

Figure 4–27 The assembly of a histone octamer. The histone H3–H4 dimer and the H2A–H2B dimer are formed from the handshake interaction. An H3–H4 tetramer forms the scaffold of the octamer onto which two H2A–H2B dimers are added, to complete the assembly. The histones are colored as in Figure 4–26. Note that all eight N-terminal tails of the histones protrude from the disc-shaped core structure. In the x-ray crystal (Figure 4–25), most of the histone tails were unstructured (and therefore not visible in the structure), suggesting that their conformations are highly flexible. (Adapted from figures by J. Waterborg.)

Despite the high conservation of the core histones, many eucaryotic organisms also produce specialized variant core histones that differ in amino acid sequence from the main ones. For example, the sea urchin has five histone H2A variants, each of which is expressed at a different time during development. It is thought that nucleosomes that have incorporated these variant histones differ in stability from regular nucleosomes, and they may be particularly well suited for the high rates of DNA transcription and DNA replication that occur during these early stages of development.

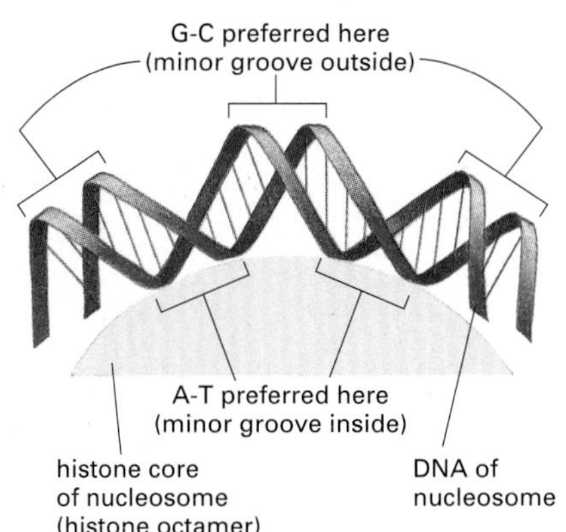

Figure 4–28 The bending of DNA in a nucleosome. The DNA helix makes 1.65 tight turns around the histone octamer. This diagram is drawn approximately to scale, illustrating how the minor groove is compressed on the inside of the turn. Owing to certain structural features of the DNA molecule, A-T base pairs are preferentially accommodated in such a narrow minor groove.

The Positioning of Nucleosomes on DNA Is Determined by Both DNA Flexibility and Other DNA-bound Proteins

Although nearly every DNA sequence can, in principle, be folded into a nucleosome, the spacing of nucleosomes in the cell can be irregular. Two main influences determine where nucleosomes form in the DNA. One is the difficulty of bending the DNA double helix into two tight turns around the outside of the histone octamer, a process that requires substantial compression of the minor groove of the DNA helix. Because A-T-rich sequences in the minor groove are easier to compress than G-C-rich sequences, each histone octamer tends to position itself on the DNA so as to maximize A-T-rich minor grooves on the inside of the DNA coil (Figure 4–28). Thus, a segment of DNA that contains short A-T-rich sequences spaced by an integral number of DNA turns is easier to bend around the nucleosome than a segment of DNA lacking this feature. In addition, because the DNA in a nucleosome is kinked in several places, the ability of a given nucleotide sequence to accommodate this deformation can also influence the position of DNA on the nucleosome.

These features of DNA probably explain some striking, but unusual, cases of very precise positioning of nucleosomes along a stretch of DNA. For most of the DNA sequences found in chromosomes, however, there is no strongly preferred nucleosome-binding site; a nucleosome can occupy any one of a number of positions relative to the DNA sequence.

The second, and probably most important, influence on nucleosome positioning is the presence of other tightly bound proteins on the DNA. Some bound proteins favor the formation of a nucleosome adjacent to them. Others create obstacles that force the nucleosomes to assemble at positions between them. Finally, some proteins can bind tightly to DNA even when their DNA-binding site is part of a nucleosome. The exact positions of nucleosomes along a stretch of DNA therefore depend on factors that include the DNA sequence and the presence and nature of other proteins bound to the DNA. Moreover, as we see below, the arrangement of nucleosomes on DNA is highly dynamic, changing rapidly according to the needs of the cell.

Nucleosomes Are Usually Packed Together into a Compact Chromatin Fiber

Although long strings of nucleosomes form on most chromosomal DNA, chromatin in a living cell probably rarely adopts the extended "beads on a string" form. Instead, the nucleosomes are packed on top of one another, generating regular arrays in which the DNA is even more highly condensed. Thus, when nuclei are very gently lysed onto an electron microscope grid, most of the chromatin is seen to be in the form of a fiber with a diameter of about 30 nm, which is considerably wider than chromatin in the "beads on a string" form (see Figure 4–23).

Several models have been proposed to explain how nucleosomes are packed in the 30-nm chromatin fiber; the one most consistent with the available data is a series of structural variations known collectively as the Zigzag model (Figure 4–29). In reality, the 30-nm structure found in chromosomes is probably a fluid mosaic of the different zigzag variations. We saw earlier that the linker DNA that connects adjacent nucleosomes can vary in length; these differences in linker length probably introduce further local perturbations into the zigzag structure. Finally, the presence of other DNA-binding proteins and DNA sequence that

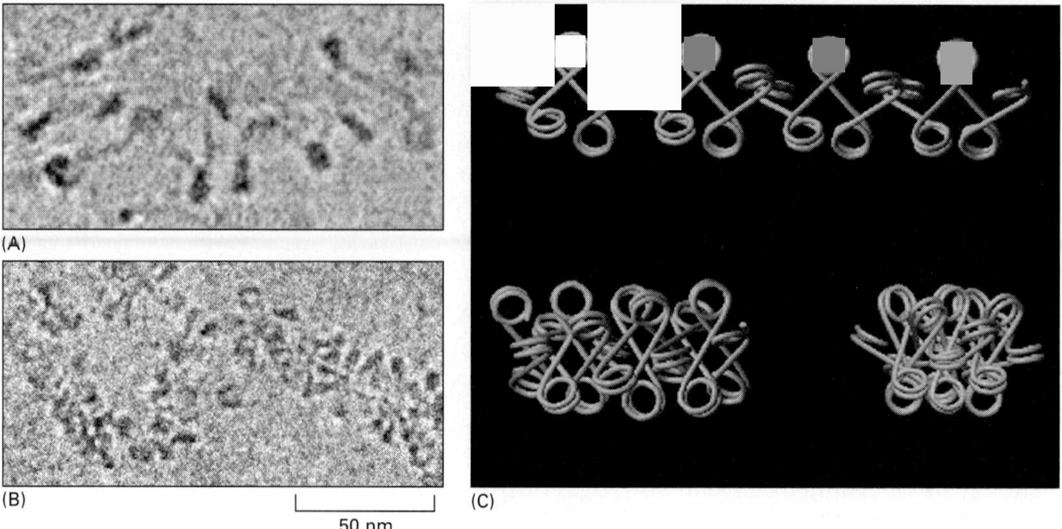

are difficult to fold into nucleosomes punctuate the 30-nm fiber with irregular features (Figure 4–30).

Several mechanisms probably act together to form the 30-nm fiber from a linear string of nucleosomes. First, an additional histone, called histone H1, is involved in this process. H1 is larger than the core histones and is considerably less well conserved. In fact, the cells of most eucaryotic organisms make several histone H1 proteins of related but quite distinct amino acid sequences. A single histone H1 molecule binds to each nucleosome, contacting both DNA and protein, and changing the path of the DNA as it exits from the nucleosome. Although it is not understood in detail how H1 pulls nucleosomes together into the 30-nm fiber, a change in the exit path in DNA seems crucial for compacting nucleosomal DNA so that it interlocks to form the 30-nm fiber (Figure 4–31).

A second mechanism for forming the 30-nm fiber probably involves the tails of the core histones, which, as we saw above, extend from the nucleosome. It is thought that these tails may help attach one nucleosome to another—thereby allowing a string of them, with the aid of histone H1, to condense into the 30-nm fiber (Figure 4–32).

Figure 4–29 Variations on the Zigzag model for the 30-nm chromatin fiber. (A and B) Electron microscopic evidence for the *top* and *bottom-left* model structures depicted in (C). (C) Zigzag variations. An interconversion between these three variations is proposed to occur by an accordion-like expansion and contraction of the fiber length. Differences in the length of the linker between adjacent nucleosome beads can be accommodated by snaking or coiling of the linker DNA, or by small local changes in the width of the fiber. Formation of the 30-nm fiber requires both histone H1 and the core histone tails; for simplicity, neither is shown here, but see Figures 4–30 and 4–32. (From J. Bednar et al., *Proc. Natl. Acad. Sci. USA* 95:14173–14178, 1998. © National Academy of Sciences.)

ATP-driven Chromatin Remodeling Machines Change Nucleosome Structure

For many years biologists thought that, once formed in a particular position on DNA, a nucleosome remained fixed in place because of the tight association

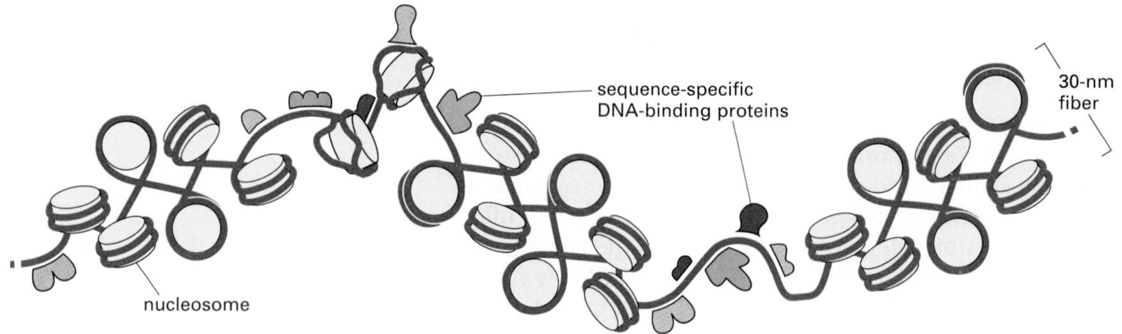

Figure 4–30 Irregularities in the 30-nm fiber. This schematic view of the 30-nm fiber illustrates its interruption by sequence-specific DNA-binding proteins. How these proteins bind tightly to DNA is explained in Chapter 7. The interruptions in the 30-nm fiber may be due to regions of DNA that lack nucleosomes altogether or, more probably, to regions that contain altered or remodeled nucleosomes. Regions of chromatin that are nucleosome free or contain remodeled nucleosome can often be detected experimentally by the unusually high susceptibility of their DNA to digestion by nucleases—as compared with the DNA in nucleosomes.

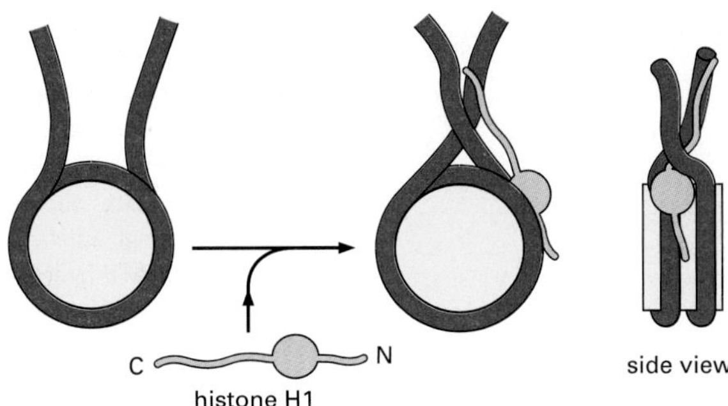

Figure 4–31 A speculative model for how histone H1 could change the path of DNA as it exits from the nucleosome. Histone H1 *(green)* consists of a globular core and two extended tails. Part of the effect of H1 on the compaction of nucleosome organization may result from charge neutralization: like the core histones, H1 is positively charged (especially its C-terminal tail), and this helps to compact the negatively charged DNA. Unlike the core histones, H1 does not seem to be essential for cell viability; in one ciliated protozoan the nucleus expands nearly twofold in the absence of H1, but the cells otherwise appear normal.

between the core histones and DNA. But it has recently been discovered that eucaryotic cells contain *chromatin remodeling complexes,* protein machines that use the energy of ATP hydrolysis to change the structure of nucleosomes temporarily so that DNA becomes less tightly bound to the histone core. The remodeled state may result from movement of the H2A–H2B dimers in the nucleosome core; the H3–H4 tetramer is particularly stable and would be difficult to rearrange (see Figure 4–27).

The remodeling of nucleosome structure has two important consequences. First, it permits ready access to nucleosomal DNA by other proteins in the cell, particularly those involved in gene expression, DNA replication, and repair. Even after the remodeling complex has dissociated, the nucleosome can remain in a "remodeled state" that contains DNA and the full complement of histones—but one in which the DNA–histone contacts have been loosened; only gradually does this remodeled state revert to that of a standard nucleosome. Second, remodeling complexes can catalyze changes in the positions of nucleosomes along DNA (Figure 4–33); some can even transfer a histone core from one DNA molecule to another.

Cells have several different chromatin remodeling complexes that differ subtly in their properties. Most are large protein complexes that can contain more than ten subunits. It is likely that they are used whenever a eucaryotic cell needs direct access to nucleosome DNA for gene expression, DNA replication, or DNA repair. Different remodeling complexes may have features specialized for each of these roles. It is thought that the primary role of some remodeling complexes is to allow access to nucleosomal DNA, whereas that of others is to re-form nucleosomes when access to DNA is no longer required (Figure 4–34).

Chromatin remodeling complexes are carefully controlled by the cell. We shall see in Chapter 7 that, when genes are turned on and off, these complexes can be brought to specific regions of DNA where they act locally to influence chromatin structure. During mitosis, at least some of the chromatin-remodeling complexes are inactivated by phosphorylation. This may help the tightly packaged mitotic chromosomes maintain their structure.

Covalent Modification of the Histone Tails Can Profoundly Affect Chromatin

The N-terminal tails of each of the four core histones are highly conserved in their sequence, and perform crucial functions in regulating chromatin structure.

Figure 4–32 A speculative model for histone tails in the formation of the 30-nm fiber. (A) The approximate exit points of the eight histone tails, four from each histone subunit, that extend from each nucleosome. In the high-resolution structure of the nucleosome (see Figure 4–25), the tails are largely unstructured, suggesting that they are highly flexible. (B) A speculative model showing how the histone tails may help to pack nucleosomes together into the 30-nm fiber. This model is based on (1) experimental evidence that histone tails aid in the formation of the 30-nm fiber, (2) the x-ray crystal structure of the nucleosome, which showed that the tails of one nucleosome contact the histone core of an adjacent nucleosome in the crystal lattice, and (3) evidence that the histone tails interact with DNA.

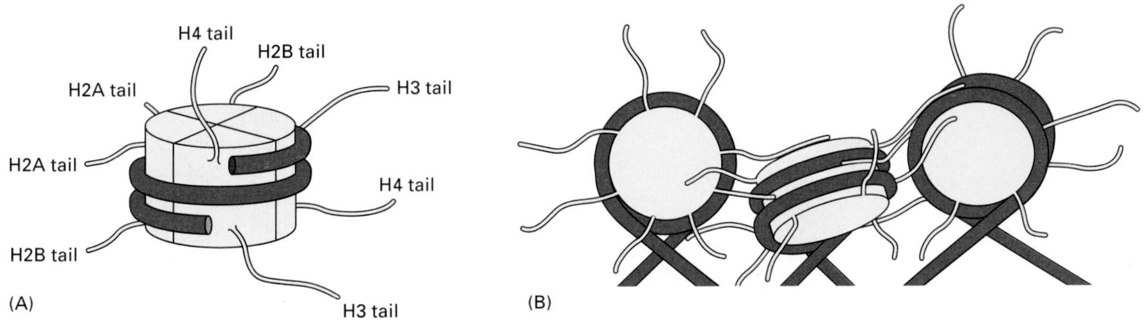

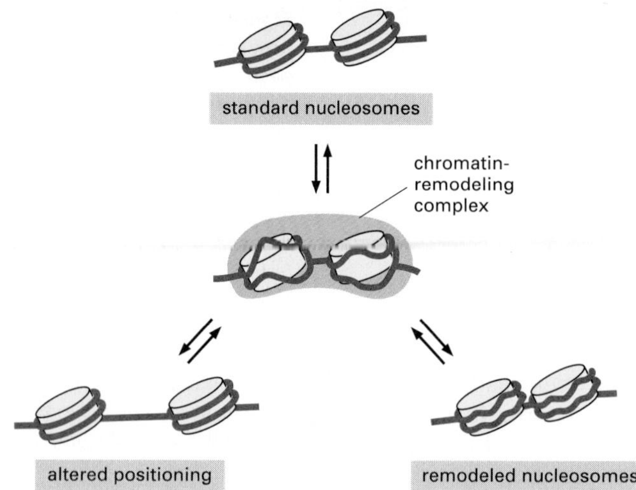

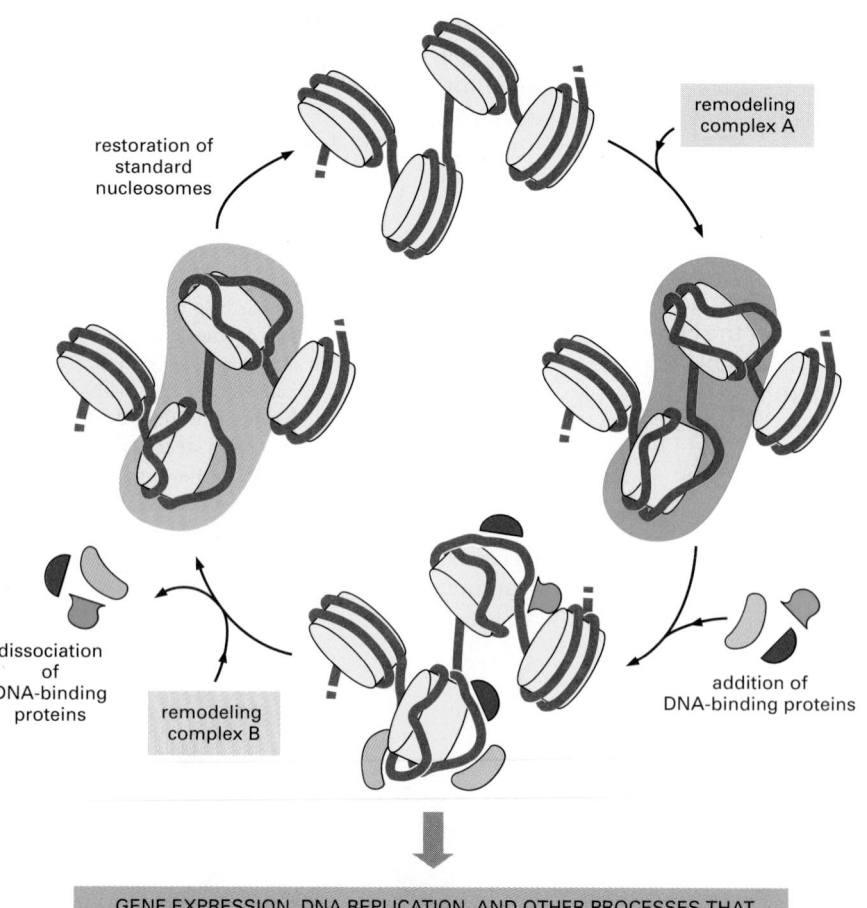

Figure 4–33 Model for the mechanism of some chromatin remodeling complexes. In the absence of remodeling complexes, the interconversion between the three nucleosomal states shown is very slow because of a high activation energy barrier. Using ATP hydrolysis, chromatin-remodeling complexes *(green)* create an activated intermediate (shown in the center of the figure) in which the histone–DNA contacts have been partly disrupted. This activated state can then decay to any one of the three nucleosomal configurations shown. In this way, the remodeling complexes greatly increase the rate of interconversion between different nucleosomal states. The remodeled state, in which the histone–DNA contacts have been loosened, has a higher free energy level than that of standard nucleosomes and will slowly revert to the standard nucleosome conformation, even in the absence of a remodeling complex. Cells have many different chromatin remodeling complexes, and they differ in their detailed biochemical properties; for example, not all can change the position of a nucleosome, but all use the energy of ATP hydrolysis to alter nucleosome structure. (Adapted from R.E. Kingston and G.J. Narlikar, *Genes Dev.* 13:2339–2352, 1999.)

Each tail is subject to several types of covalent modifications, including acetylation of lysines, methylation of lysines, and phosphorylation of serines (Figure 4–35A). Histones are synthesized in the cytosol and then assembled into nucleosomes. Some of the modifications of histone tails occur just after their synthesis, but before their assembly. The modifications that concern us, however, take place once the nucleosome has been assembled. These nucleosome modifications are added and removed by enzymes that reside in the nucleus; for example, acetyl groups are added to the histone tails by histone acetyl transferases (HATs) and taken off by histone deacetylases (HDACs).

The various modifications of the histone tails have several important consequences. Although modifications of the tails have little direct effect on the stability of an individual nucleosome, they seem to affect the stability of the

Figure 4–34 A cyclic mechanism for nucleosome disruption and re-formation. According to this model, different chromatin remodeling complexes disrupt and re-form nucleosomes, although, in principle, the same complex might catalyze both reactions. The DNA-binding proteins could function in gene expression, DNA replication, or DNA repair, and in some cases their binding could lead to the dissociation of the histone core to form nucleosome-free regions of DNA like those illustrated in Figure 4–30. (Adapted from A. Travers, *Cell* 96:311–314, 1999.)

30-nm chromatin fiber and of the higher-order structures discussed below. For example, histone acetylation tends to destabilize chromatin structure, perhaps in part because adding an acetyl group removes the positive charge from the lysine, thereby making it more difficult for histones to neutralize the charges on DNA as chromatin is compacted. However, the most profound effect of modified histone tails is their ability to attract specific proteins to a stretch of chromatin that has been appropriately modified. Depending on the precise tail modifications, these additional proteins can either cause further compaction of the chromatin or can facilitate access to the DNA. If combinations of modifications are taken into account, the number of possible distinct markings for each histone tail is very large. Thus, it has been proposed that, through covalent modification of the histone tails, a given stretch of chromatin can convey a particular meaning to

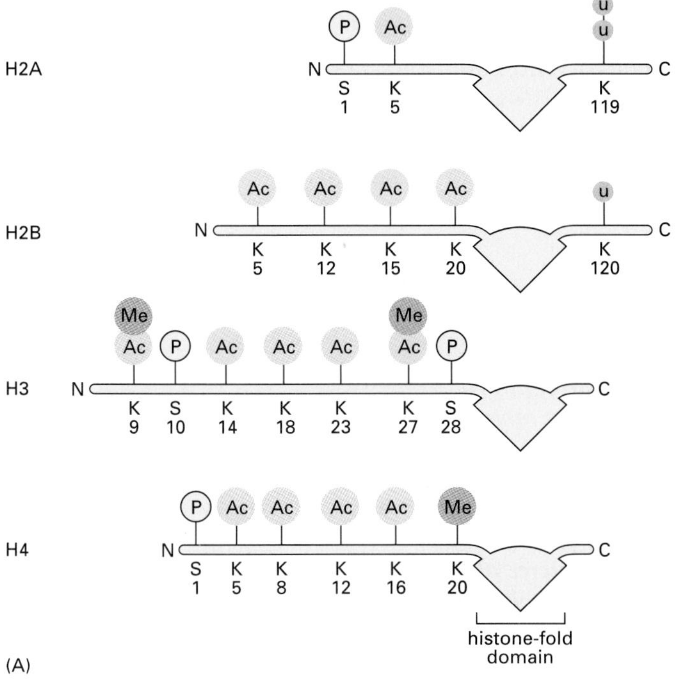

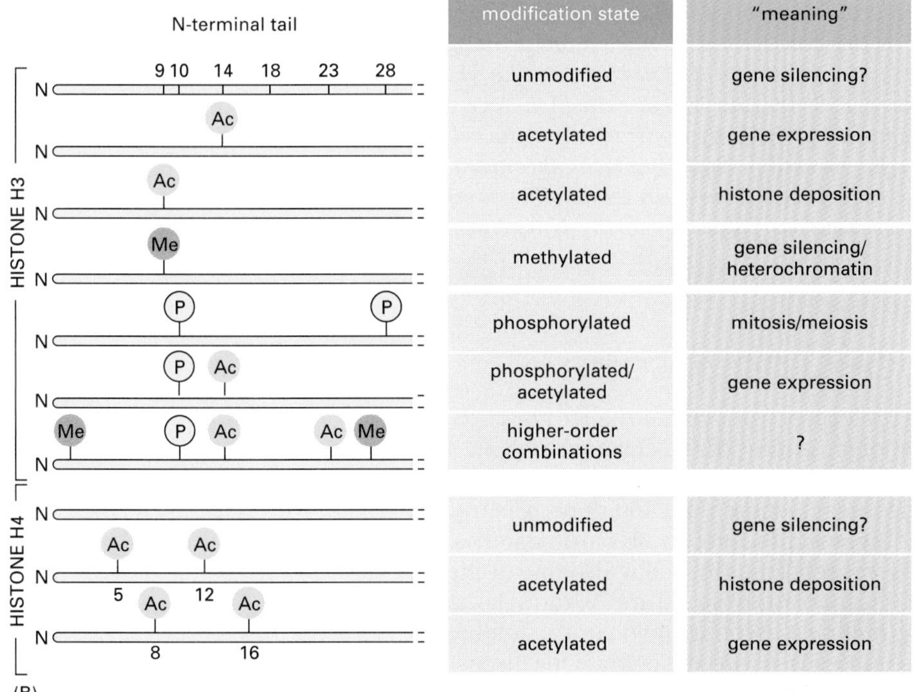

modification state	"meaning"
unmodified	gene silencing?
acetylated	gene expression
acetylated	histone deposition
methylated	gene silencing/ heterochromatin
phosphorylated	mitosis/meiosis
phosphorylated/ acetylated	gene expression
higher-order combinations	?
unmodified	gene silencing?
acetylated	histone deposition
acetylated	gene expression

Figure 4–35 Covalent modification of core histone tails. (A) Known modifications of the four histone core proteins are indicated: Me = methyl group, Ac = acetyl group, P = phosphate, u = ubiquitin. Note that some positions (e.g., lysine 9 of H3) can be modified in more than one way. Most of these modifications add a relatively small molecule onto the histone tails; the exception is ubiquitin, a 76 amino acid protein also used in other cellular processes (see Figure 6–87). The function of ubiquitin in chromatin is not well understood: histone H2B can be modified by a single ubiquitin molecule; H2A can be modified by the addition of several ubiquitins. (B) A histone code hypothesis. Histone tails can be marked by different combinations of modifications. According to this hypothesis, each marking conveys a specific meaning to the stretch of chromatin on which it occurs. Only a few of the meanings of the modifications are known. In Chapter 7, we discuss the way a doubly-acetylated H4 tail is "read" by a protein required for gene expression. In another well-studied case, an H3 tail methylated at lysine 9 is recognized by a set of proteins that create an especially compact form of chromatin, which silences gene expression.

The acetylation of lysine 14 of histone H3 and lysines 8 and 16 of histone H4—usually associated with gene expression—is performed by the type A histone acetylases (HATs) in the nucleus. In contrast, the acetylation of lysines 5 and 12 of histone H4 and a lysine of histone H3 takes place in the cytosol, after the histones have been synthesized but before they have been incorporated into nucleosomes; these modifications are catalyzed by type B HATs. These modified histones are deposited onto DNA after DNA replication (see Figure 5–41), and their acetyl groups are taken off shortly afterwards by histone deacetylases (HDACs). Thus, the acetylation at these positions signals newly replicated chromatin.

Modification of a particular position in a histone tail can take on different meanings depending on other features of the local chromatin structure. For example, the phosphorylation of position 10 of histone H3 is associated not only with the condensation of chromosomes that takes place in mitosis and meiosis but also with the expression of certain genes. Some histone tail modifications are interdependent. For example methylation of H3 position 9 blocks the phosphorylation of H3 position 10, and vice versa.

the cell (Figure 4–35B). For example, one type of marking could signal that the stretch of chromatin has been newly replicated, and another could signal that gene expression should not take place. According to this idea, each different marking would attract those proteins that would then execute the appropriate functions. Because the histone tails are extended, and are therefore probably accessible even when chromatin is condensed, they provide an especially apt format for such messages.

As with chromatin remodeling complexes, the enzymes that modify (and remove modifications from) histone tails are usually multisubunit proteins, and they are tightly regulated. They are brought to a particular region of chromatin by other cues, particularly by sequence-specific DNA-binding proteins. We can thus imagine how cycles of histone tail modification and demodification can allow chromatin structure to be dynamic—locally compacting and decompacting it, and, in addition, attracting other proteins specific for each modification state. It is likely that histone-modifying enzymes and chromatin remodeling complexes work in concert to condense and recondense stretches of chromatin; for example, evidence suggests that a particular modification of the histone tail attracts a particular type of remodeling complex. Moreover, some chromatin remodeling complexes contain histone modification enzymes as subunits, directly connecting the two processes.

Summary

A gene is a nucleotide sequence in a DNA molecule that acts as a functional unit for the production of a protein, a structural RNA, or a catalytic RNA molecule. In eucaryotes, protein-coding genes are usually composed of a string of alternating introns and exons. A chromosome is formed from a single, enormously long DNA molecule that contains a linear array of many genes. The human genome contains 3.2×10^9 DNA nucleotide pairs, divided between 22 different autosomes and 2 sex chromosomes. Only a small percentage of this DNA codes for proteins or structural and catalytic RNAs. A chromosomal DNA molecule also contains three other types of functionally important nucleotide sequences: replication origins and telomeres allow the DNA molecule to be completely replicated, while a centromere attaches the daughter DNA molecules to the mitotic spindle, ensuring their accurate segregation to daughter cells during the M phase of the cell cycle.

The DNA in eucaryotes is tightly bound to an equal mass of histones, which form a repeating array of DNA–protein particles called nucleosomes. The nucleosome is composed of an octameric core of histone proteins around which the DNA double helix is wrapped. Despite irregularities in the positioning of nucleosomes along DNA, nucleosomes are usually packed together (with the aid of histone H1 molecules) into quasi-regular arrays to form a 30-nm fiber. Despite the high degree of compaction in chromatin, its structure must be highly dynamic to allow the cell access to the DNA. Two general strategies for reversibly changing local chromatin structures are important for this purpose: ATP-driven chromatin remodeling complexes, and an enzymatically catalyzed covalent modification of the N-terminal tails of the four core histones.

THE GLOBAL STRUCTURE OF CHROMOSOMES

Having discussed the DNA and protein molecules from which the 30-nm chromatin fiber is made, we now turn to the organization of the chromosome on a more global scale. As a 30-nm fiber, the typical human chromosome would still be 0.1 cm in length and able to span the nucleus more than 100 times. Clearly, there must be a still higher level of folding, even in interphase chromosomes. This higher-order packaging is one of the most fascinating—but also one of the most poorly understood—aspects of chromosome structure. Although its molecular basis is still largely a mystery, it almost certainly involves the folding of the 30-nm fiber into a series of loops and coils, as we see below. Our discussion of this higher-order packing continues an important theme in chromosome architecture: interphase chromatin structure is fluid, exposing at any given moment the DNA sequences directly needed by the cell.

We first describe several rare cases in which the overall structure and organization of interphase chromosomes can be easily visualized, and we explain that certain features of these exceptional cases may be representative of the structures of all interphase chromosomes. Next we describe the different forms of chromatin that make up a typical interphase chromosome. Finally we discuss the additional compaction that interphase chromosomes undergo during the process of mitosis.

Lampbrush Chromosomes Contain Loops of Decondensed Chromatin

Most chromosomes in interphase cells are too fine and too tangled to be visualized clearly. In a few exceptional cases, however, *interphase chromosomes* can be seen to have a precisely defined higher-order structure, and it is thought that certain characteristics of these higher-order structures are representative of all interphase chromosomes. The meiotically paired chromosomes in growing amphibian oocytes (immature eggs), for example, are highly active in gene expression, and they form unusually stiff and extended chromatin loops. These so-called **lampbrush chromosomes** (the largest chromosomes known) are clearly visible even in the light microscope, where they are seen to be organized into a series of large chromatin loops emanating from a linear chromosomal axis (Figure 4–36).

The organization of a lampbrush chromosome is illustrated in Figure 4–37. A given loop always contains the same DNA sequence, and it remains extended in the same manner as the oocyte grows. Other experiments demonstrate that

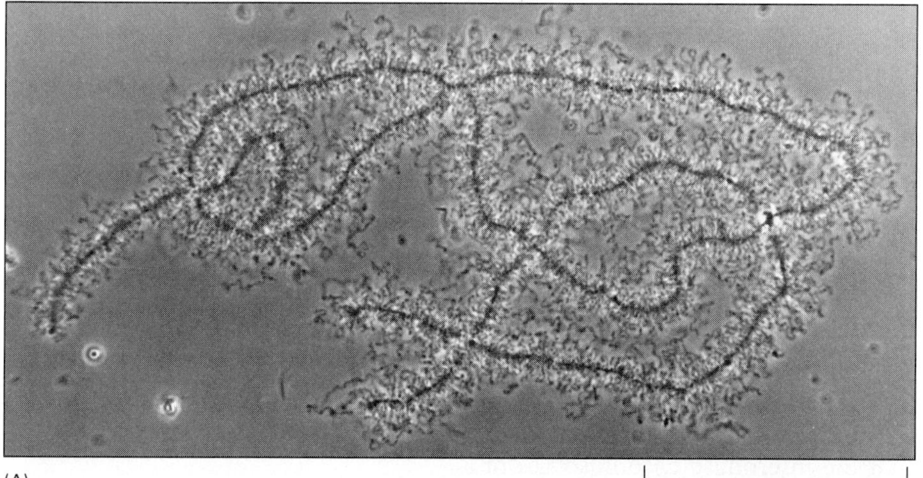

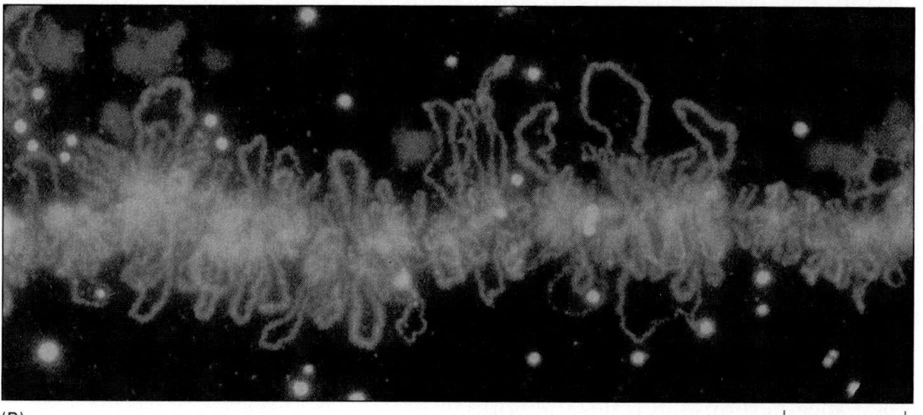

Figure 4–36 Lampbrush chromosomes. (A) A light micrograph of lampbrush chromosomes in an amphibian oocyte. Early in oocyte differentiation, each chromosome replicates to begin meiosis, and the homologous replicated chromosomes pair to form this highly extended structure containing a total of four replicated DNA molecules, or chromatids. The lampbrush chromosome stage persists for months or years, while the oocyte builds up a supply of materials required for its ultimate development into a new individual. (B) Fluorescence light micrograph showing a portion of an amphibian lampbrush chromosome. The regions of the chromosome that are being actively expressed are stained *green* by using antibodies against proteins that process RNA during one of the steps of gene expression (discussed in Chapter 6). The round granules are thought to correspond to large complexes of the RNA-splicing machinery that will also be discussed in Chapter 6. (A, courtesy of Joseph G. Gall; B, courtesy of Joseph G. Gall and Christine Murphy.)

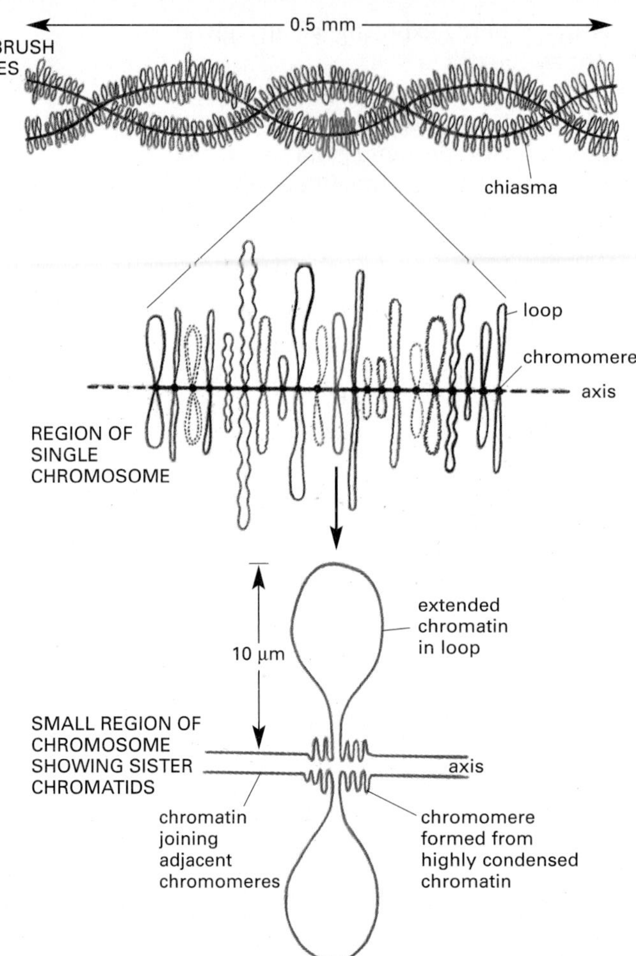

Figure 4–37 A model for the structure of a lampbrush chromosome. The set of lampbrush chromosomes in many amphibians contains a total of about 10,000 chromatin loops, although most of the DNA in each chromosome remains highly condensed in the chromomeres. Each loop corresponds to a particular DNA sequence. Four copies of each loop are present in each cell, since each of the two chromosomes shown at the top consists of two closely apposed, newly replicated chromosomes. This four-stranded structure is characteristic of this stage of development of the oocyte, the diplotene stage of meiosis; see Figure 20–12.

most of the genes present in the DNA loops are being actively expressed (see Figure 4–36B). Most of the DNA, however, is not in loops but remains highly condensed in the *chromomeres* on the axis, which are generally not expressed. Lampbrush chromosomes illustrate a recurrent theme of this chapter—when the DNA in a region of chromatin is in use (in this case, for gene expression), that part of the chromatin has an extended structure; otherwise, the chromatin is condensed. In lampbrush chromosomes, the structural units of this regulation are large, precisely defined loops.

Relatively few species undergo the specialization that produces lampbrush chromosomes. However, when injected into amphibian oocytes, the DNA from organisms that normally do not produce lampbrush chromosomes (e.g., DNA from a fish) is packaged into lampbrush chromosomes. On the basis of this type of experiment, it has been proposed that the interphase chromosomes of all eucaryotes are arranged in loops that are normally too small and fragile to be easily observed. It may be possible in the future to coax the DNA from a mammal such as a mouse to form lampbrush chromosomes by introducing it into amphibian oocytes. This could allow a detailed correlation of loop structure, gene arrangement, and DNA sequence, and we could begin to learn how the packaging into loops reflects the sequence content of our DNA.

Drosophila Polytene Chromosomes Are Arranged in Alternating Bands and Interbands

Certain insect cells also have specialized interphase chromosomes that are readily visible, although this type of specialization differs from that of lampbrush chromosomes. For example, many of the cells of certain fly larvae grow to an enormous size through multiple cycles of DNA synthesis without cell division. The resulting giant cells contain as much as several thousand times the normal DNA complement. Cells with more than the normal DNA complement are said to be *polyploid* when they contain increased numbers of standard chromosomes. In several types of secretory cells of fly larvae, however, all the homologous chromosome copies are held side by side, like drinking straws in a box, creating a single **polytene chromosome**. The fact that, in some large insect cells, polytene chromosomes can disperse to form a conventional polyploid cell demonstrates that these two chromosomal states are closely related, and that the basic structure of a polytene chromosome must be similar to that of a normal chromosome.

Polytene chromosomes are often easy to see in the light microscope because they are so large and because the precisely aligned side-by-side adherence of individual chromatin strands greatly elongates the chromosome axis and prevents tangling. Polyteny has been most studied in the salivary gland cells of *Drosophila* larvae, in which the DNA in each of the four *Drosophila* chromosomes has been replicated through 10 cycles without separation of the daughter chromosomes, so that 1024 (2^{10}) identical strands of chromatin are lined up side by side (Figure 4–38).

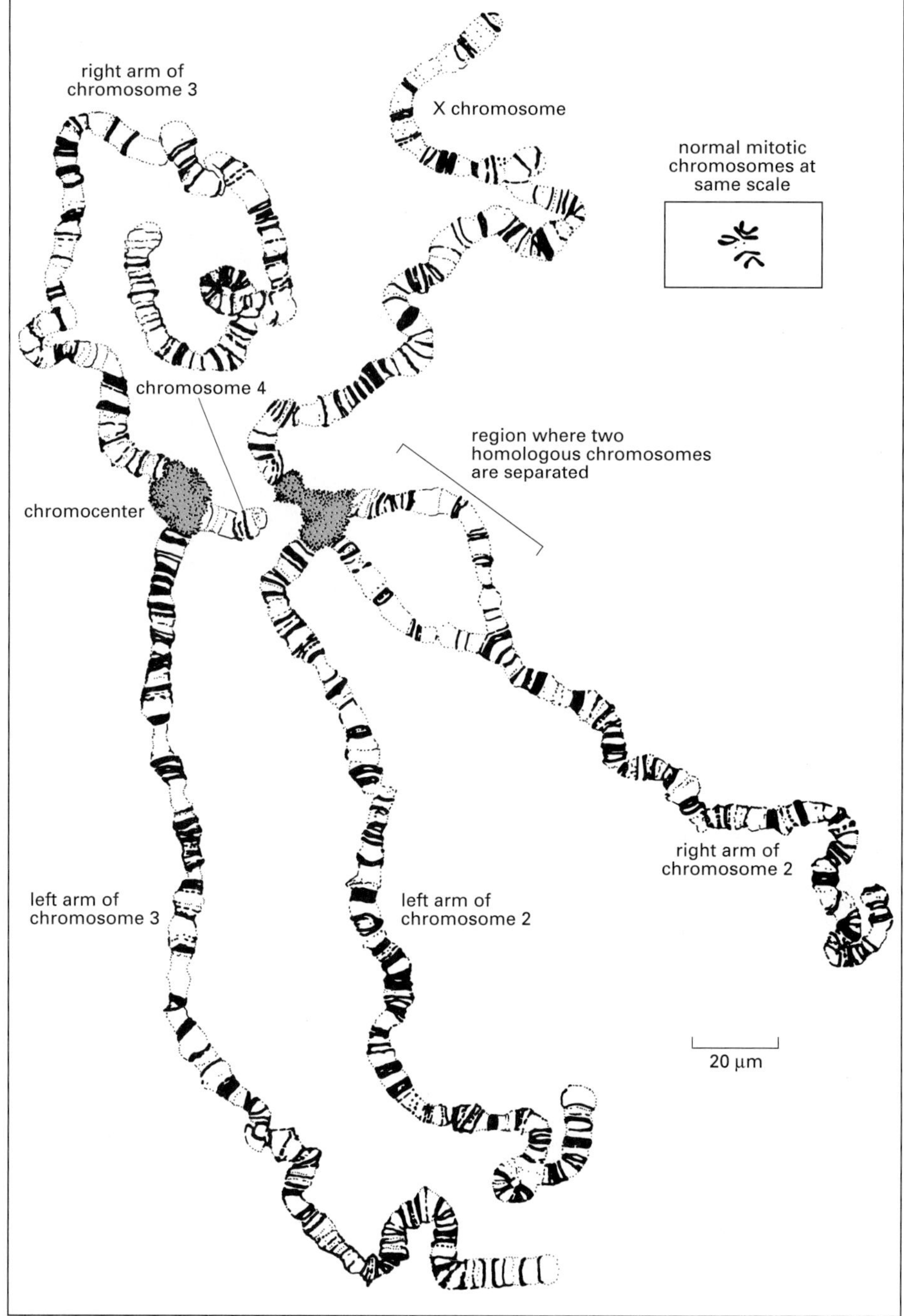

Figure 4–38 The entire set of polytene chromosomes in one *Drosophila* salivary cell. These chromosomes have been spread out for viewing by squashing them against a microscope slide. *Drosophila* has four chromosomes, and there are four different chromosome pairs present. But each chromosome is tightly paired with its homolog (so that each pair appears as a single structure), which is not true in most nuclei (except in meiosis). The four polytene chromosomes are normally linked together by regions near their centromeres that aggregate to create a single large chromocenter *(pink region)*. In this preparation, however, the chromocenter has been split into two halves by the squashing procedure used. (Modified from T.S. Painter, *J. Hered.* 25:465–476, 1934.)

When polytene chromosomes are viewed in the light microscope, distinct alternating dark *bands* and light *interbands* are visible (Figure 4–39). Each band and interband represents a set of 1024 identical DNA sequences arranged in register. About 95% of the DNA in polytene chromosomes is in bands, and 5% is in interbands. The chromatin in each band appears dark, either because it is much more condensed than the chromatin in the interbands, or because it contains a higher proportion of proteins, or both (Figure 4–40). Depending on their size, individual bands are estimated to contain 3000–300,000 nucleotide pairs in a chromatin strand. The bands of *Drosophila* polytene chromosomes can be recognized by their different thicknesses and spacings, and each one has been given a number to generate a chromosome "map." There are approximately 5000 bands and 5000 interbands in the complete set of *Drosophila* polytene chromosomes.

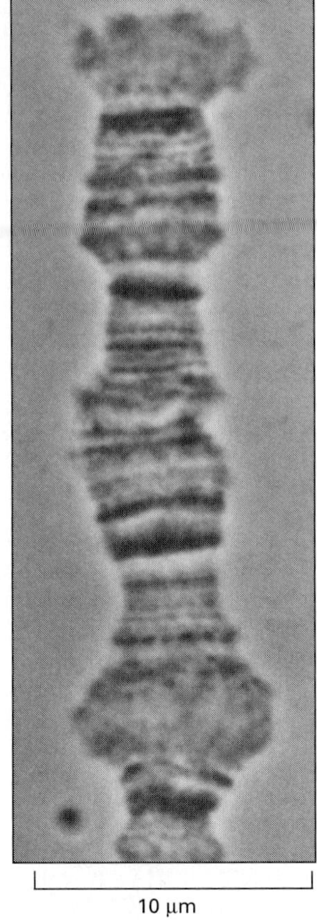

Figure 4–39 A light micrograph of a portion of a polytene chromosome from *Drosophila* salivary glands. The distinct pattern produced by bands and interbands is readily seen. The bands are regions of increased chromatin concentration that occur in interphase chromosomes. Although they are detectable only in polytene chromosomes, it is thought that they reflect a structure common to the chromosomes of most eucaryotes. (Courtesy of Joseph G. Gall.)

Both Bands and Interbands in Polytene Chromosomes Contain Genes

The reproducible pattern of bands and interbands seen in *Drosophila* polytene chromosomes means that these interphase chromosomes are highly organized. Since the 1930s, scientists have debated the nature of this organization, and we still do not have a clear answer. Because the number of bands in *Drosophila* chromosomes was once thought to be roughly equal to the number of genes in the genome, it was initially thought that each band might correspond to a single gene; however, we now know this simple idea is incorrect. There are nearly three times more genes in *Drosophila* than chromosome bands, and genes are found in both band and interband regions. Moreover, some bands contain multiple genes, and some bands seem to lack genes altogether.

It seems likely that the band–interband pattern reflects different levels of gene expression and chromatin structure along the chromosome, with genes in the less compact interbands being expressed more highly than those in the more compact bands. In any case, the remarkable appearance of fly polytene chromosomes is thought to reflect the heterogeneous nature of the chromatin compaction found along all interphase chromosomes. In the next section we see how the appearance of a band can change dramatically when the gene or genes within it become highly expressed.

Individual Polytene Chromosome Bands Can Unfold and Refold as a Unit

A major factor controlling gene expression in the polytene chromosomes of *Drosophila* is the insect steroid hormone *ecdysone*, the levels of which rise and fall periodically during larval development. When ecdysone concentrations rise,

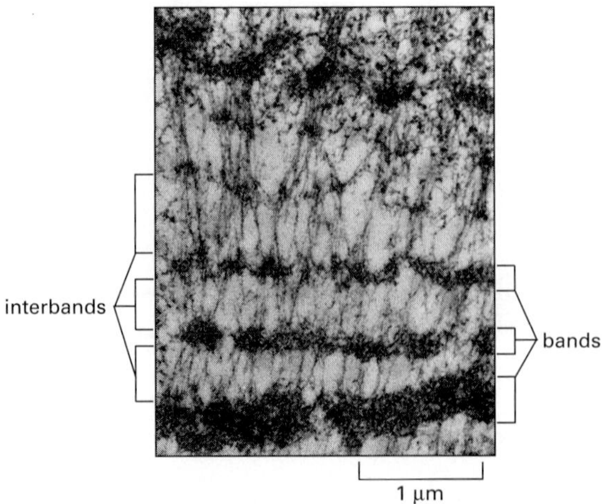

Figure 4–40 An electron micrograph of a small section of a *Drosophila* polytene chromosome seen in thin section. Bands of very different thickness can be readily distinguished, separated by interbands, which contain less condensed chromatin. (Courtesy of Veikko Sorsa.)

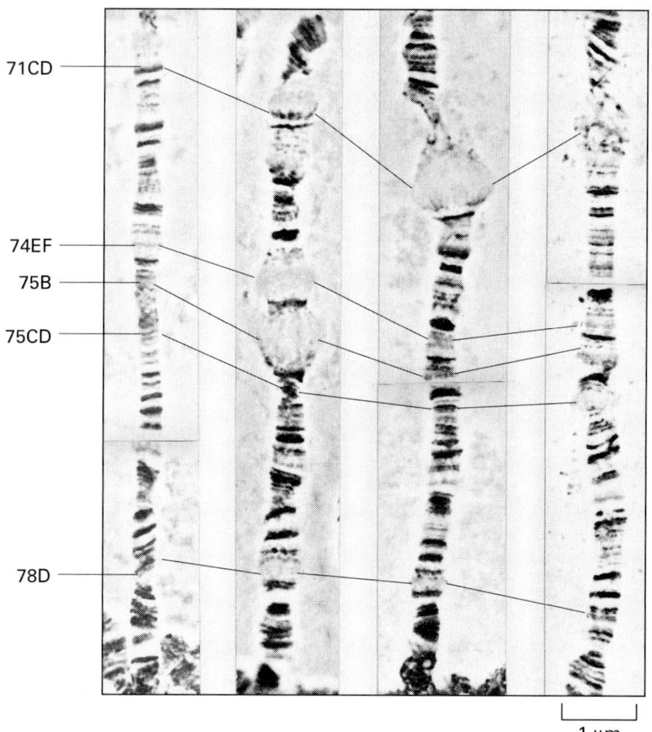

Figure 4–41 Chromosome puffs. This series of time-lapse photographs shows how puffs arise and recede in the polytene chromosomes of *Drosophila* during larval development. A region of the *left* arm of chromosome 3 is shown. It exhibits five very large puffs in salivary gland cells, each active for only a short developmental period. The series of changes shown occur over a period of 22 hours, appearing in a reproducible pattern as the organism develops. The designations of the indicated bands are given at the left of the photographs. (Courtesy of Michael Ashburner.)

they induce the expression of genes coding for proteins that the larva requires for each molt and for pupation. As the organism progresses from one developmental stage to another, distinctive *chromosome puffs* arise and old puffs recede as new genes become expressed and old ones are turned off (Figure 4–41). From inspection of each puff when it is relatively small and the banding pattern is still discernible, it seems that most puffs arise from the decondensation of a single chromosome band.

The individual chromatin fibers that make up a puff can be visualized with an electron microscope. For technical reasons, this is easier in the polytene chromosomes from a different insect, *Chironomus tentans,* a midge. Electron micrographs of certain puffs, called Balbiani rings, of *Chironomus* salivary gland polytene chromosomes show the chromatin arranged in loops (Figures 4–42 and 4–43), much like those observed in the amphibian lampbrush chromosomes discussed earlier. Additional experiments suggest that each loop contains a

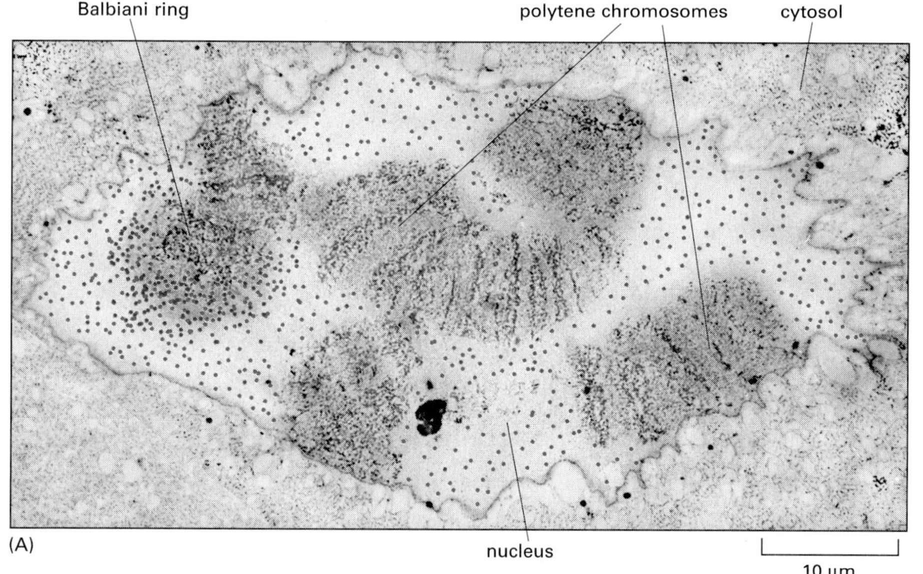

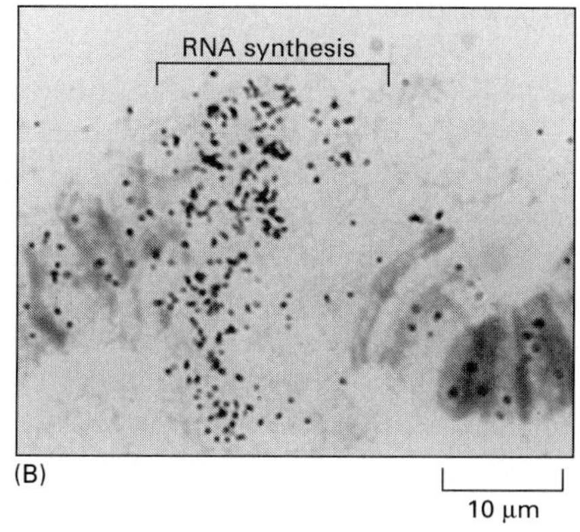

Figure 4–42 RNA synthesis in chromosome puffs. (A) Polytene chromosomes from the salivary glands of the insect *C. tentans.* As outlined in Chapter 1 and described in detail in Chapter 6, the first step in gene expression is the synthesis of an RNA molecule using the DNA as a template. In this electron micrograph, newly synthesized RNA from a Balbiani ring gene is indicated in *red.* Cells were exposed to a brief pulse of BrUTP (an analog of UTP), which was incorporated into RNA. Cells were then fixed and the newly synthesized RNAs were identified by using antibodies against BrU. Balbiani ring RNAs could be distinguished from other RNAs by their characteristic shape (see Figure 4–43). The *blue dots* in the figure represent positions of Balbiani ring RNAs that were synthesized before the addition of BrUTP. The experiment shows that Balbiani ring RNAs are synthesized in puffs and then diffuse through the nucleoplasm. (B) An autoradiogram of a single puff in a polytene chromosome. The portion of the chromosome indicated is undergoing RNA synthesis and has therefore become labeled with ^{3}H-uridine. (A, courtesy of B. Daneholt, from O.P Singh et al., *Exp. Cell Res.* 251:135–146, 1999. © Academic Press; B, courtesy of José Bonner.)

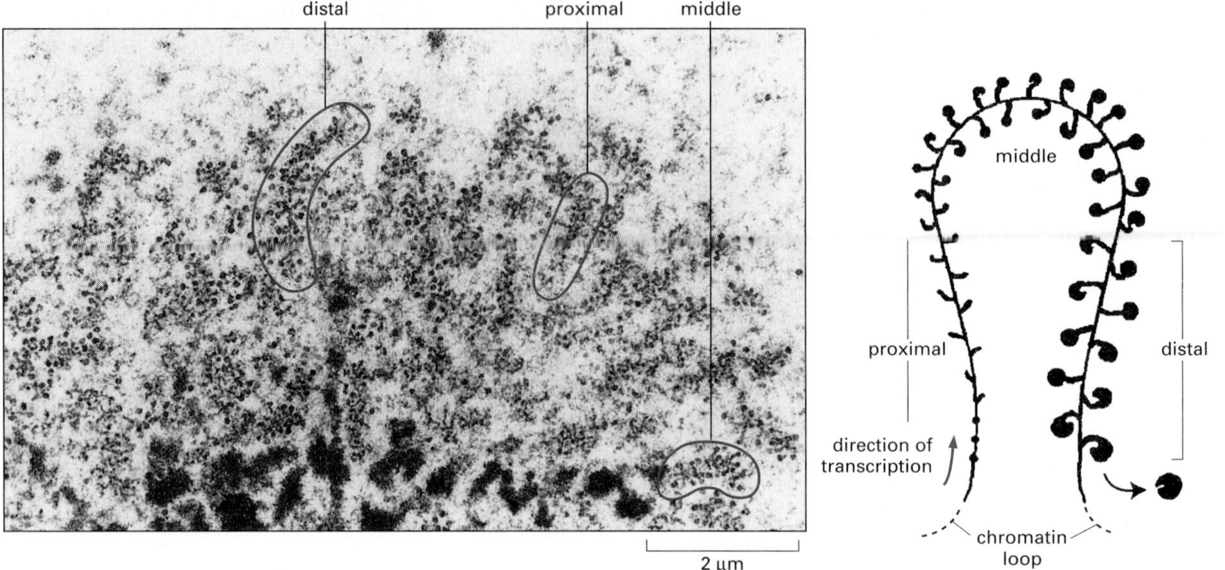

Figure 4–43 Polytene chromosomes from *C. tentans*. The electron micrograph shows a thin section of the chromatin in a Balbiani ring, a chromosome puff very active in gene expression. The Balbiani ring gene codes for secretory proteins the larvae uses to spin a protective tube. The chromatin is arranged in loops, but because the sample has been sectioned, only portions of the loops are visible. As they are synthesized on the chromatin, the RNA molecules are bound up by protein molecules, making them visible as knobs on stalks in the electron microscope. From the size of the RNA–protein complex, the extent of RNA synthesis (transcription) can be inferred; a whole chromatin loop (shown on the right) can then be reconstructed from a set of electron micrograph sections such as that shown here. (Courtesy of B. Daneholt, from U. Skoglund et al., *Cell* 34:847–855, 1983. © Elsevier.)

single gene. When not expressed, the loop of DNA assumes a thickened structure, possibly a folded 30-nm fiber, but when gene expression is occurring, the loop becomes more extended. Both types of loops contain the four core histones and histone H1.

It seems likely that the default loop structure is a folded 30-nm fiber and that the histone modifying enzymes, chromatin remodeling complexes, and other proteins required for gene expression all help to convert it to a more extended form whenever a gene is expressed. In electron micrographs, the chromatin located on either side of the decondensed loop appears considerably more compact, which is consistent with the idea that a loop constitutes an independent functional domain of chromatin structure.

Although controversial, it has been proposed that all of the DNA in polytene chromosomes is arranged in loops that condense and decondense according to when the genes within them are expressed. It may be that all interphase chromosomes from all eucaryotes are also packaged into an orderly series of looped domains, each containing a small number of genes whose expression is regulated in a coordinated way (Figure 4–44). We shall return to this issue in Chapter 7 when we discuss the ways in which gene expression is regulated by the cell.

Heterochromatin Is Highly Organized and Usually Resistant to Gene Expression

Having described some features of interphase chromosomes inferred from a few rare cases, we now turn to characteristics of interphase chromosomes that can be observed in a wide variety of organisms. Light-microscope studies in the 1930s distinguished between two types of chromatin in the interphase nuclei of many higher eucaryotic cells: a highly condensed form, called **heterochromatin**, and all the rest, which is less condensed, called **euchromatin**. Euchromatin is composed of the types of chromosomal structures—30-nm fibers and looped domains—that we have discussed so far. Heterochromatin, in contrast, includes additional proteins and probably represents more compact levels of organization that are just beginning to be understood. In a typical mammalian cell, approximately 10% of the genome is packaged into heterochromatin. Although present in many locations along chromosomes, it is concentrated in specific regions, including the centromeres and telomeres.

Most DNA that is folded into heterochromatin does not contain genes. However, genes that do become packaged into heterochromatin are usually resistant to being expressed, because heterochromatin is unusually compact. This does not mean that heterochromatin is useless or deleterious to the cell; as we see below, regions of heterochromatin are responsible for the proper functioning of telomeres and centromeres (which lack genes), and its formation may even help protect the genome from being overtaken by "parasitic" mobile elements of DNA. Moreover, a few genes require location in heterochromatin regions if they are to be expressed. In fact, the term *heterochromatin* (which was first defined cytologically) is likely to encompass several distinct types of chromatin structures whose common feature is an especially high degree of organization. Thus, heterochromatin should not be thought of as encapsulating "dead" DNA, but rather as creating different types of compact chromatin with distinct features and roles.

Heterochromatin's resistance to gene expression makes it amenable to study even in organisms in which it cannot be directly observed. When a gene that is normally expressed in euchromatin is experimentally relocated into a region of heterochromatin, it ceases to be expressed, and the gene is said to be *silenced*. These differences in gene expression are examples of **position effects**, in which the activity of a gene depends on its position along a chromosome. First recognized in *Drosophila*, position effects have now been observed in many organisms and they are thought to reflect an influence of the different states of chromatin structure along chromosomes on gene expression. Thus, chromosomes can be considered as mosaics of distinct forms of chromatin, each of which has a special effect on the ability of the DNA it contains to be addressed by the cell.

Many position effects exhibit an additional feature called *position effect variegation*, which is responsible for the mottled appearance of the fly eye and the sectoring of the yeast colony in the examples shown in Figure 4–45. These patterns can result from patches of cells in which a silenced gene has become reactivated; once reactivated, the gene is inherited stably in this form in daughter cells. Alternatively, a gene can start out in euchromatin early in development, and then be selected more or less randomly for packaging into heterochromatin, causing its inactivation in a cell and all of its daughters.

The study of position effect variegation has revealed two important characteristics of heterochromatin. First, heterochromatin is dynamic; it can "spread" into a region and later "retract" from it at low but observable frequencies. Second, the state of chromatin—whether heterochromatin or euchromatin—tends to be inherited from a cell to its progeny. These two features are responsible for position effect variegation, as explained in Figure 4–46. In the next section, we discuss several models to account for the self-sustaining nature of heterochromatin, once it has been formed.

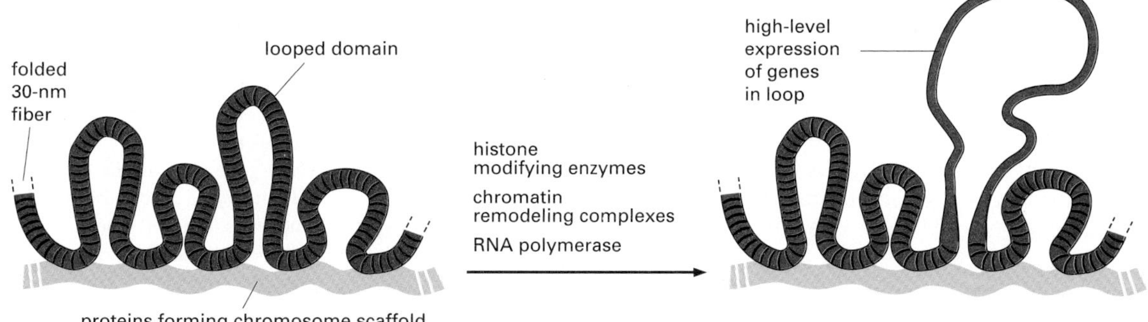

Figure 4–44 A model for the structure of an interphase chromosome. A section of an interphase chromosome is shown folded into a series of looped domains, each containing 20,000–100,000 nucleotide pairs of double-helical DNA condensed into a 30-nm fiber. Individual loops can decondense, perhaps in part through an accordionlike expansion of the 30-nm fiber (see Figure 4–29), when the cell requires direct access to the DNA packaged in these loops. This decondensation is brought about by enzymes that directly modify chromatin structure—as well as by proteins, such as RNA polymerase (discussed in Chapter 6), that act directly on the underlying DNA. It is not understood how the folded 30-nm fiber is anchored to the chromosome axis, but evidence suggests that the base of chromosomal loops is rich in DNA topoisomerases, which are enzymes that allow DNA to swivel when anchored (see pp. 251–253).

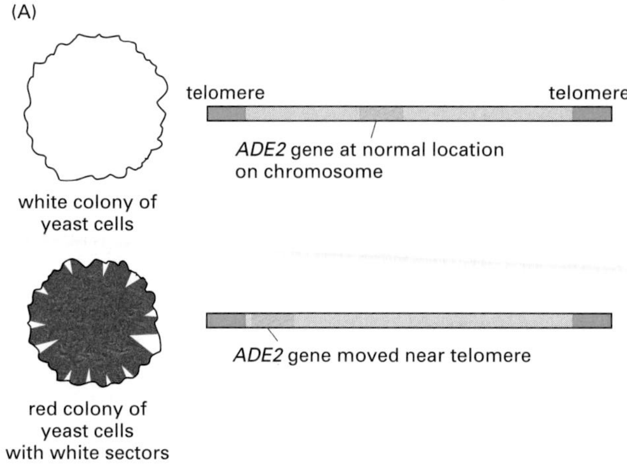

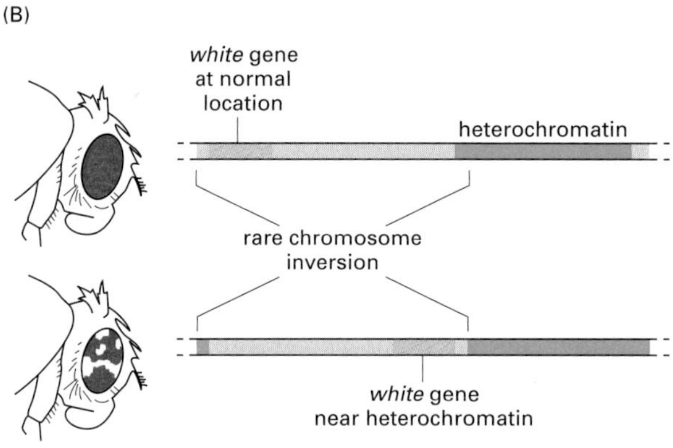

Figure 4–45 Position effects on gene expression in two different eucaryotic organisms. (A) The yeast *ADE2* gene at its normal chromosomal location is expressed in all cells. When moved near the end of a yeast chromosome, which is inferred to be folded into a form of heterochromatin, the gene is no longer expressed in most cells of the population. The *ADE2* gene codes for one of the enzymes of adenine biosynthesis, and the absence of the *ADE2* gene product leads to the accumulation of a red pigment. Therefore, a colony of cells that expresses *ADE2* is *white,* and one composed of cells where the *ADE2* gene is not expressed is *red.* The *white* sectors that fan out from the middle of the *red* colony grown on an agar surface represent the descendants of cells in which the *ADE2* gene has spontaneously become active. These white sectors are thought to result from a heritable change to a less tightly packed state of chromatin near the *ADE2* gene in these cells. Although yeast chromosomes are too small to be seen under the light microscope, the chromatin structure at the ends of yeast chromosomes is thought to have many of the same structural features as the heterochromatin in the chromosomes of larger organisms.

(B) Position effects can also be observed for the *white* gene in the fruit fly *Drosophila.* The *white* gene controls eye pigment production and is named after the mutation that first identified it. Wild-type flies with a normal *white* gene *(white$^-$)* have normal pigment production, which gives them red eyes, but if the *white* gene is mutated and inactivated, the mutant flies *(white$^-$)* make no pigment and have *white* eyes. In flies in which a normal *white$^+$* gene has been moved near a region of heterochromatin, the eyes are mottled, with both *red* and *white* patches. The *white* patches represent cells in which the *white$^+$* gene has been silenced by the effects of the heterochromatin. In contrast, the red patches represent cells that express the *white$^+$* gene because the heterochromatin did not spread across this gene at the time, early in development, when the heterochromatin first formed. As in the yeast, the presence of large patches of *red* and *white* cells indicates that the state of transcriptional activity of the gene is inherited, once determined by its chromatin packaging in the early embryo.

The Ends of Chromosomes Have a Special Form of Heterochromatin

Unlike the nucleosome and the 30-nm fiber, heterochromatin is not well understood structurally. It almost certainly involves an additional level of folding of 30-nm fiber and requires many proteins in addition to the histones. Although its chromosomes are too small to be seen under the light microscope, the molecular nature of heterochromatin is probably best understood in the simple yeast *S. cerevisiae.* Many experiments with yeast cells have shown that the chromatin extending inward roughly 5000 nucleotide pairs from each chromosome end is resistant to gene expression, and probably has a structure that corresponds to at least one type of heterochromatin in the chromosomes of more complex organisms. Extensive genetic analysis has led to the identification of many of the yeast proteins required for this type of gene silencing.

Mutations in any one of a set of yeast *S*ilent *i*nformation *r*egulator (Sir) proteins prevent the silencing of genes located near telomeres, thereby allowing these genes to be expressed. Analysis of these proteins has led to the discovery of a telomere-bound Sir protein complex that recognizes underacetylated N-terminal tails of selected histones (Figure 4–47A). One of the proteins in this complex is a highly conserved histone deacetylase known as Sir2, which has homologs in diverse organisms, including humans, and presumably has a major role in creating a pattern of histone underacetylation unique to heterochromatin. As discussed earlier in this chapter, deacetylation of the histone tails is thought to allow nucleosomes to pack together into tighter arrays and may also render them less susceptible to some chromatin remodeling complexes. In addition, heterochromatin-specific patterns of histone tail modification are likely to attract additional proteins involved in forming and maintaining heterochromatin (see Figure 4–35).

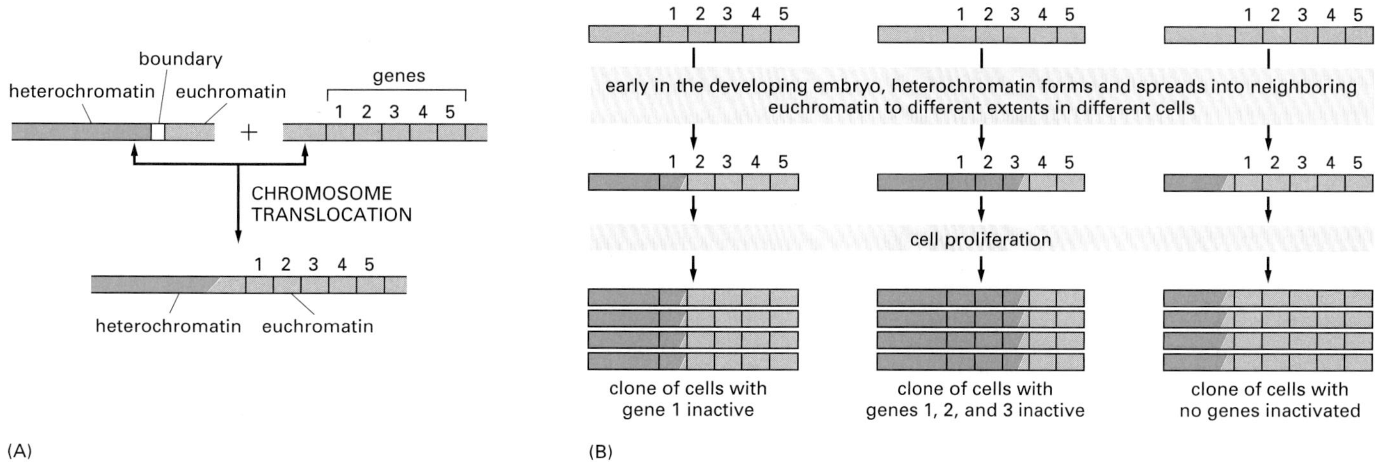

Figure 4–46 The cause of position effect variegation in *Drosophila*. (A) Heterochromatin *(blue)* is normally prevented from spreading into adjacent regions of euchromatin *(green)* by special boundary DNA sequences, which we discuss in Chapter 7. In flies that inherit certain chromosomal rearrangements, however, this barrier is no longer present. (B) During the early development of such flies, heterochromatin can spread into neighboring chromosomal DNA, proceeding for different distances in different cells. This spreading soon stops, but the established pattern of heterochromatin is inherited, so that large clones of progeny cells are produced that have the same neighboring genes condensed into heterochromatin and thereby inactivated (hence the "variegated" appearance of some of these flies; see Figure 4–45B). Although "spreading" is used to describe the formation of new heterochromatin near previously existing heterochromatin, the term may not be wholly accurate. There is evidence that during expansion, heterochromatin can "skip over" some regions of chromatin, sparing the genes that lie within them from repressive effects. One possibility is that heterochromatin can expand across the base of some DNA loops, thus bypassing the chromatin contained in the loop.

But how is the Sir2 protein delivered to the ends of chromosomes in the first place? Another series of experiments has suggested the model shown in Figure 4–47B. A DNA-binding protein that recognizes specific DNA sequences in yeast telomeres also binds to one of the Sir proteins, causing the entire Sir protein complex to assemble on the telomeric DNA. The Sir complex then spreads along the chromosome from this site, modifying the N-terminal tails of adjacent histones

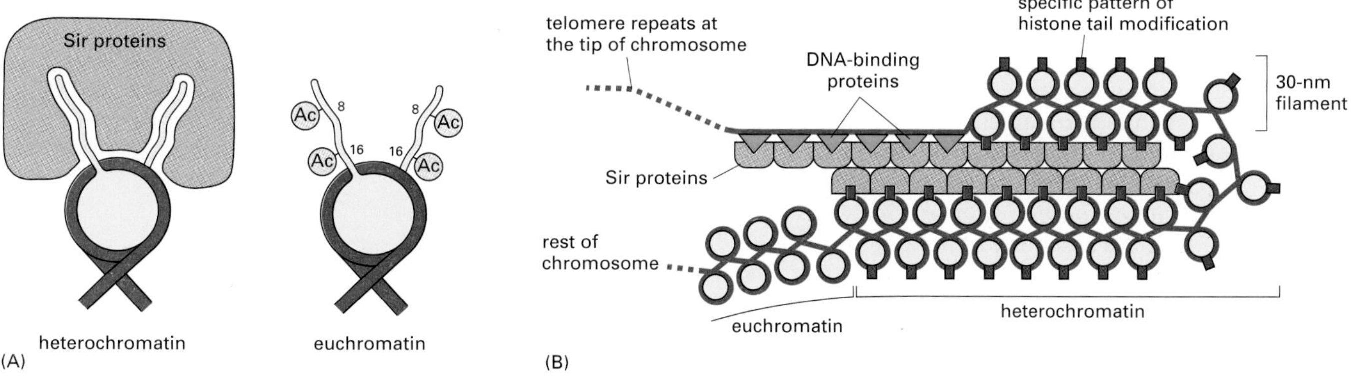

Figure 4–47 Speculative model for the heterochromatin at the ends of yeast chromosomes. (A) Heterochromatin is generally underacetylated, and underacetylated tails of histone H4 are proposed to interact with a complex of Sir proteins, thus stabilizing the association of these proteins with nucleosomes. Although shown as fully unacetylated, the exact pattern of histone H4 tail modification required to bind to the Sir complex is not known with certainty. In some organisms, the methylation of lysine 9 of histone H3 is also a critical signal for heterochromatin formation. In euchromatin, histone tails are typically highly acetylated. Those of H4 are shown as partially acetylated but, in reality, the acetylation state varies across euchromatin. (B) Specialized DNA-binding proteins *(blue triangles)* recognize DNA sequences near the ends of chromosomes and attract the Sir proteins, one of which (Sir2) is an NAD^+-dependent histone deacetylase. This then leads to the cooperative spreading of the Sir protein complex down the chromosome. As this complex spreads, the deacetylation catalyzed by Sir2 helps create new binding sites on nucleosomes for more Sir protein complexes. A "fold back" structure of the type shown may also form.

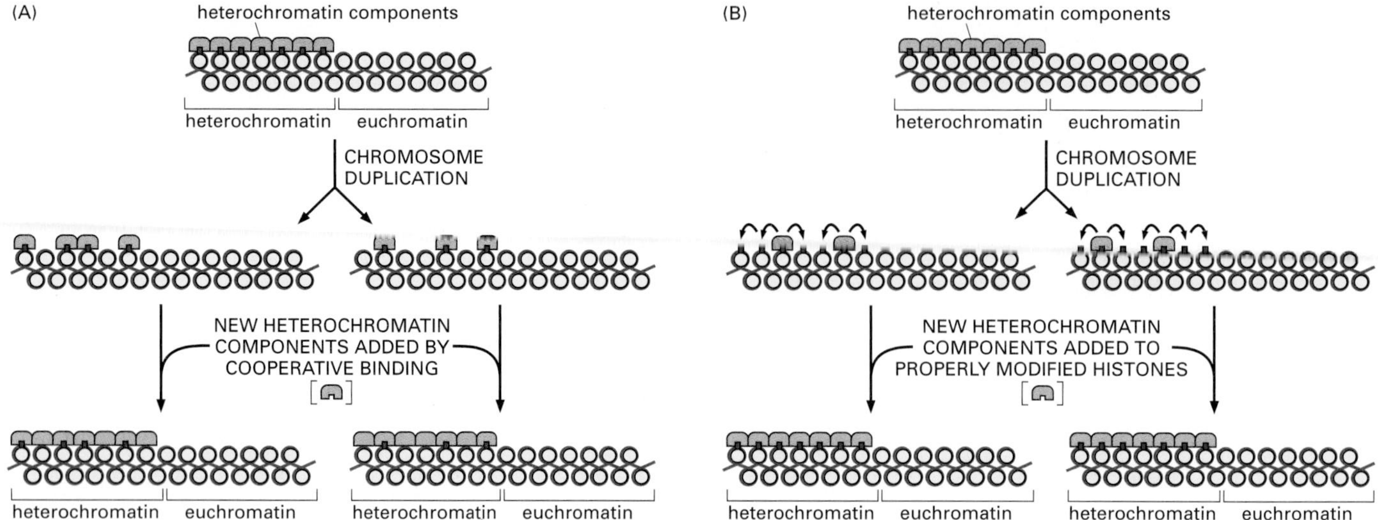

Figure 4–48 Two speculative models for how the tight packaging of DNA in heterochromatin can be inherited during chromosome replication. In both cases, half of the specialized heterochromatin components have been distributed to each daughter chromosome after DNA duplication. (A) In this model, new heterochromatin components bind cooperatively to the inherited components, thereby beginning the process of new heterochromatin formation. The process is completed with the assembly of additional proteins and the eventual covalent modification of the histones (not shown). (B) In this model, the inherited heterochromatin components change the pattern of histone modification on the newly formed daughter nucleosomes nearby, creating new binding sites for free heterochromatin components, which assemble and complete the structure. Both models can account for the spreading effects of heterochromatin, and indeed, both processes may occur simultaneously in cells.

to create the nucleosome-binding sites that the complex prefers. This "spreading effect" is thought to be driven by the cooperative binding of adjacent Sir protein complexes, as well as by the folding back of the chromosome on itself to promote Sir binding in nearby regions (see Figure 4–47B). In addition, the formation of heterochromatin probably requires the action of chromatin remodeling complexes to readjust the positions of nucleosomes as they are packed together.

Unlike most deacetylases, Sir2 requires NAD^+ as a cofactor (see Figure 2–60). The NAD^+ levels in the cell fluctuate with the nutritional health of the cell, increasing as cells become nutritionally deprived. This feature might cause the telomeric heterochromatin to spread in response to starvation (perhaps to silence the expression of genes that are not absolutely required for survival) and then to retract when conditions improve.

The properties of the yeast heterochromatin just described may resemble features of heterochromatin in more complex organisms. Certainly, the spreading of yeast heterochromatin from telomeres is similar in principle to the movement of heterochromatin that causes position effect variegation in animals (see Figure 4–46). Moreover, these properties can be used to explain the heritability of heterochromatin, as outlined in Figure 4–48. Whatever the precise mechanism of heterochromatin formation, it has become clear that covalent modifications of the nucleosome core histones have a critical role in this process. Of special importance in many organisms are the *histone methyl transferases,* enzymes that methylate specific lysines on histones including lysine 9 of histone H3 (see Figure 4–35). This modification is "read" by heterochromatin components (including HP1 in *Drosophila)* that specifically bind this modified form of histone H3 to induce the assembly of heterochromatin. It is likely that a spectrum of different histone modifications is used by the cell to distinguish heterochromatin from euchromatin (see Figure 4–35).

Having the ends of chromosomes packaged into heterochromatin provides several advantages to the cell: it helps to protect the ends of chromosomes from being recognized as broken chromosomes by the cellular repair machinery, it may help to regulate telomere length, and it may assist in the accurate pairing and segregation of chromosomes during mitosis. In Chapter 5 we see that telomeres have additional structural features that distinguish them from other parts of chromosomes.

Centromeres Are Also Packaged into Heterochromatin

Heterochromatin is also observed around centromeres, the DNA sequences that direct the movement of each chromosome into daughter cells every time a cell divides (see Figure 4–22). In many complex organisms, including humans, each centromere seems to be embedded in a very large stretch of heterochromatin that persists throughout interphase, even though the centromere-directed movement of DNA occurs only during mitosis. The structure and biochemical properties of this so-called *centric heterochromatin* are not well understood, but, like other forms of heterochromatin, it silences the expression of genes that are experimentally placed into it. It contains, in addition to histones (which are typically underacetylated and methylated in heterochromatin), several additional structural proteins that compact the nucleosomes into particularly dense arrangements.

As with telomeres, our best understanding of the chromatin structure of a centromere comes from studies of the much simpler centromeres of the yeast *S. cerevisiae*. Earlier in this chapter we saw that a simple DNA sequence of approximately 125 nucleotide pairs was sufficient to serve as a centromere in this organism. Despite its small size, more than a dozen different proteins assemble on this DNA sequence; the proteins include a histone H3 variant that, along with the other core histones, is believed to form a centromere-specific nucleosome (Figure 4–49A). We do not yet understand what properties this variant type of nucleosome provides to the cell, but similar specialized nucleosomes seem to be present in all eucaryotic centromeres (Figure 4–49B). The additional proteins at the yeast centromere attach it to the spindle microtubules and provide signals that ensure that this attachment is complete before the later stages of mitosis are allowed to proceed (discussed in Chapters 17 and 18).

The centromeres in more complex organisms are considerably larger than those in budding yeasts. For example, fly and human centromeres extend over hundreds of thousands of nucleotide pairs and do not seem to contain a centromere-specific DNA sequence. Rather, most consist largely of short, repeated DNA sequences, known as *alpha satellite DNA* in humans (Figure 4–50). But the same repeat sequences are also found at other (noncentromeric) positions on chromosomes, and how they specify a centromere is poorly understood. Somehow the formation of the inner plate of a kinetochore is "seeded," followed by the cooperative assembly of the entire group of special proteins that form the kinetochore (Figure 4–50B). It seems that centromeres in complex organisms are defined more by an assembly of proteins than by a specific DNA sequence.

There are some striking similarities between the formation and maintenance of centromeres and the formation and maintenance of other regions of heterochromatin. The entire centromere forms as an all-or-none entity, suggesting a highly cooperative addition of proteins after a seeding event. Moreover, once formed, the structure seems to be directly inherited on the DNA as part of each round of chromosome replication. Thus, for example, some regions of our chromosomes contain nonfunctional alpha satellite DNA sequences that seem to be identical to those at the centromere; these sequences are presumed to have arisen from a chromosome-joining event that initially created one chromosome with two centromeres (an unstable, dicentric

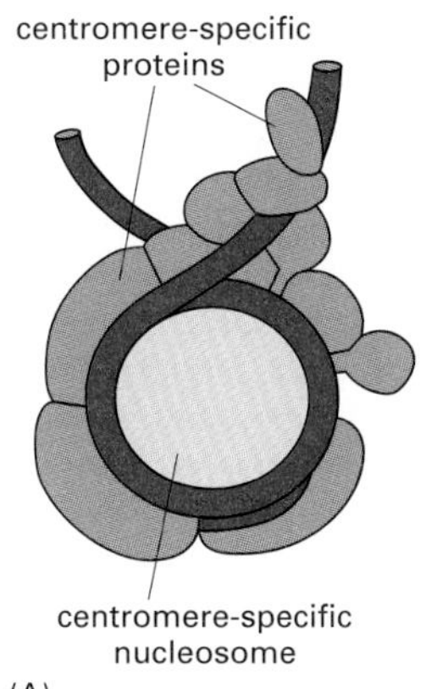

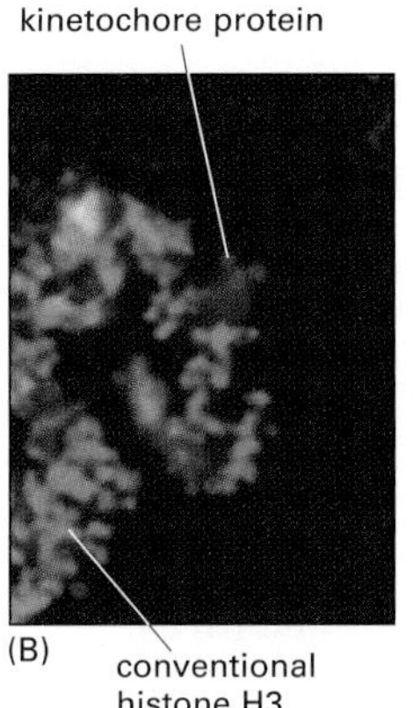

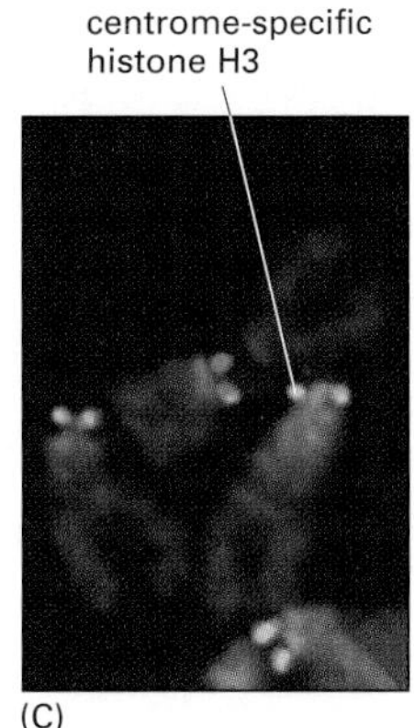

Figure 4–49 The specialized nucleosome formed on centromeres. (A) A model for the proteins that assemble on a yeast centromere. The specialized nucleosome contains an H3 variant (called CENP-A in most organisms), along with core histones H2A, H2B, and H4. The folding of DNA into this nucleosome facilitates the assembly of the other centromere-binding proteins, which form the kinetochore that attaches the centromere to the mitotic spindle (B) The localization of conventional histone H3 on *Drosophila* mitotic chromosomes. The conventional H3 has been fused to a fluorescent protein and appears *green*. A component of the kinetochore has been stained *red* with antibodies against a specific kinetochore protein. (C) The same experiment, but with the centromere-specific histone H3 (instead of the conventional H3) labeled *green*. When the *red* and *green* stains are coincident, the staining appears *yellow*. (A, adapted from P.B. Meluh et al., *Cell* 94:607–613, 1998; B and C, from S. Henikoff et al., *Proc. Natl. Acad. Sci. USA* 97:716–721, 2000. © National Academy of Sciences.)

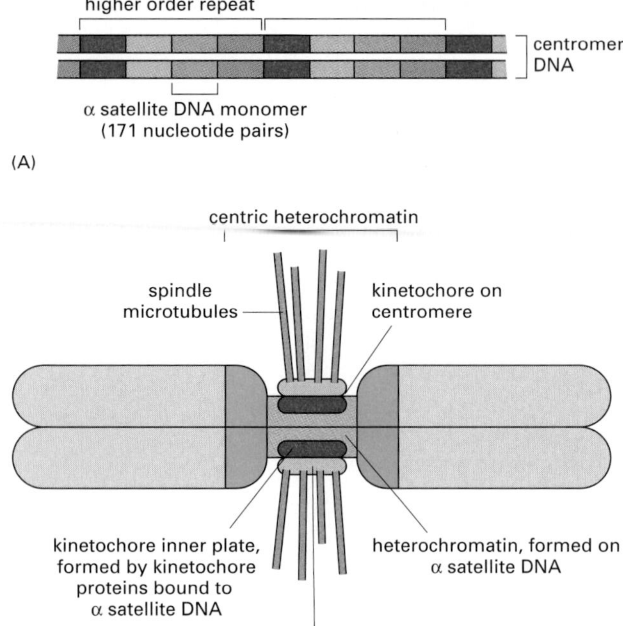

Figure 4–50 The structure of a human centromere. (A) The organization of the alpha satellite DNA sequences, which are repeated many thousands of times at a centromere. (B) An entire chromosome. The alpha satellite DNA sequences *(red)* are AT-rich and consist of a large number of repeats that vary slightly from one another in their DNA sequence. *Blue* represents the position of flanking centric heterochromatin, which contains DNA sequences composed of different types of repeats. As indicated, the kinetochore consists of an inner and an outer plate, formed by a set of kinetochore proteins. The spindle microtubules attach to the kinetochore in M phase of the cell cycle (see Figure 4–22). (B, adapted from T.D. Murphy and G.H. Karpen, *Cell* 93:317–320, 1998.)

chromosome; Figure 4–51A). Moreover, in some unusual cases, new human centromeres (called neocentromeres) have been observed to form spontaneously on fragmented chromosomes. Some of these new positions were originally euchromatic and lack alpha satellite DNA altogether (Figure 4–51B).

To explain these observations it has been proposed that *de novo* centromere formation requires an initial marking (or seeding) event, perhaps the formation of a specialized DNA–protein structure, which, in humans, happens more readily on arrays of alpha satellite DNA than on other DNA sequences. This mark would be duplicated when the chromosome divides, and the same centromere would then function in the next cell division. Very rarely, the mark would be lost after chromosome replication, in which case it would be very difficult to establish again (Figure 4–51C). Although the self-renewing nature of centromeres is not understood in detail, the type of models described for heterochromatin inheritance in Figure 4–48 could also be critical here.

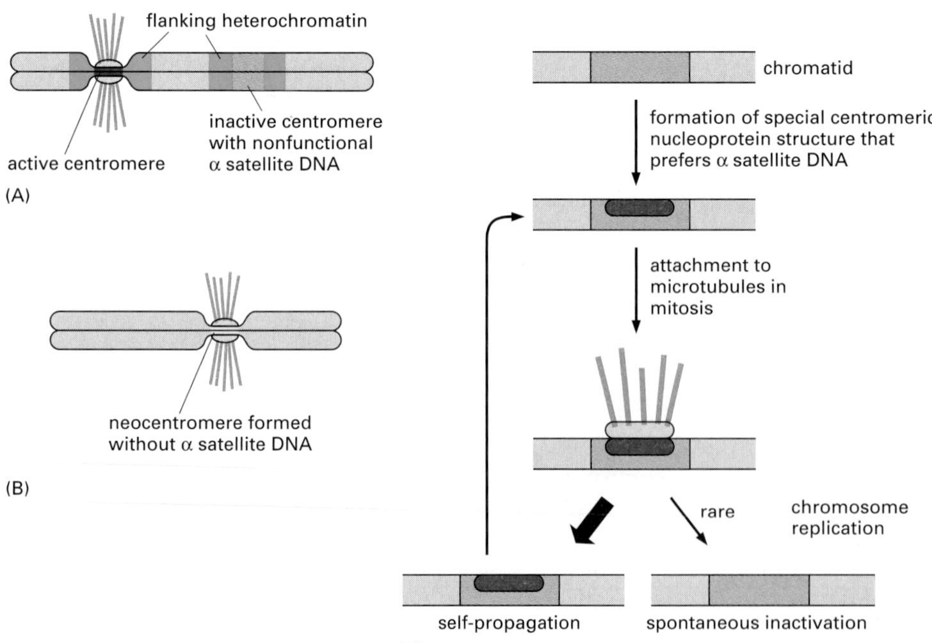

Figure 4–51 The plasticity of human centromere formation. (A) Owing to an ancient chromosome breakage and rejoining event, some human chromosomes contain two blocks of alpha satellite DNA *(red)*, each of which presumably functioned as a centromere in its original chromosome. Usually, these dicentric chromosomes are not stably propagated because they are attached improperly to the spindle and are broken apart during mitosis. In those chromosomes that survive, one of the centromeres has spontaneously inactivated, even though it contains all the necessary DNA sequences. This allows the chromosome to be stably propagated. (B) In a small fraction (1/2000) of human births, extra chromosomes are observed in cells of the offspring. Some of these extra chromosomes, which have formed from a breakage event, lack alpha satellite DNA altogether, yet new centromeres have arisen from what was originally euchromatic DNA. (C) A model to explain the plasticity and inheritance of centromeres.

The plasticity of centromeres may provide an important evolutionary advantage. We have seen that chromosomes evolve in part by breakage and rejoining events (see Figure 4–19). Many of these events produce chromosomes with two centromeres, or chromosome fragments with no centromeres at all. Although rare, the inactivation of centromeres and their ability to be activated *de novo* may occasionally allow newly formed chromosomes to be maintained stably and thereby facilitate the process of chromosome evolution.

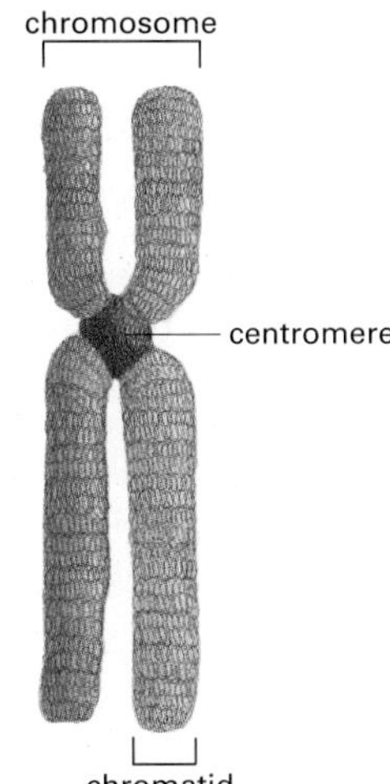

Figure 4–52 A typical mitotic chromosome at metaphase. Each sister chromatid contains one of two identical daughter DNA molecules generated earlier in the cell cycle by DNA replication.

Heterochromatin May Provide a Defense Mechanism Against Mobile DNA Elements

DNA packaged in heterochromatin often consists of large tandem arrays of short, repeated sequences that do not code for protein, as we saw above for the heterochromatin of mammalian centromeres. In contrast, euchromatic DNA is rich in genes and other single-copy DNA sequences. Although this correlation is not absolute (some arrays of repeated sequences exist in euchromatin and some genes are present in heterochromatin), this trend suggests that some types of repeated DNA may be a signal for heterochromatin formation. This idea is supported by experiments in which several hundred tandem copies of genes have been artificially introduced into the germ lines of flies and mice. In both organisms these gene arrays are often silenced, and in some cases, they can be observed under the microscope to have formed regions of heterochromatin. In contrast, when single copies of the same genes are introduced into the same position in the chromosome, they are actively expressed.

This feature, called *repeat-induced gene silencing*, may be a mechanism that cells have for protecting their genomes from being overtaken by mobile genetic elements. These elements, which are discussed in Chapter 5, can multiply and insert themselves throughout the genome. According to this idea, once a cluster of such mobile elements has formed, the DNA that contains them would be packaged into heterochromatin to prevent their further proliferation. The same mechanism could be responsible for forming the large regions of heterochromatin that contain large numbers of tandem repeats of a simple sequence, as occurs around centromeres.

Mitotic Chromosomes Are Formed from Chromatin in Its Most Condensed State

Having discussed the dynamic structure of interphase chromosomes, we now turn to the final level of DNA packaging, that observed for mitotic chromosomes. With the exception of a few specialized cases, such as the lampbrush and polytene chromosomes discussed above, most interphase chromosomes are too extended and entangled for their structures to be clearly seen. In contrast, the chromosomes from nearly all eucaryotic cells are readily visible during mitosis, when they coil up to form highly condensed structures. It is remarkable that this further condensation, which reduces the length of a typical interphase chromosome only about tenfold, produces such a dramatic change in the appearance of chromosomes.

Figure 4–52 depicts a typical **mitotic chromosome** at the metaphase stage of mitosis. The two daughter DNA molecules produced by DNA replication during interphase of the cell-division cycle are separately folded to produce two sister chromosomes, or *sister chromatids*, held together at their centromeres (see also Figure 4–21). These chromosomes are normally covered with a variety of molecules, including large amounts of RNA–protein complexes. Once this covering has been stripped away, each chromatid can be seen in electron micrographs to be organized into loops of chromatin emanating from a central scaffolding (Figures 4–53 and 4–54). Several types of experiment demonstrate that the order of visible features along a mitotic chromosome at least roughly reflects the order of the genes along the DNA molecule. Mitotic chromosome condensation can thus be thought of as the final level in the hierarchy of chromosome packaging (Figure 4–55).

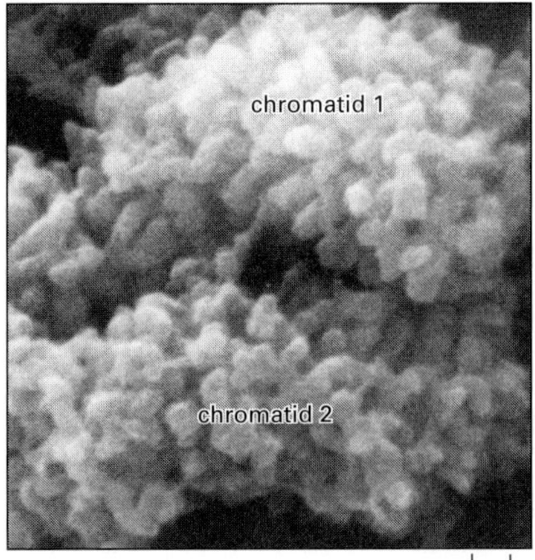

Figure 4–53 A scanning electron micrograph of a region near one end of a typical mitotic chromosome. Each knoblike projection is believed to represent the tip of a separate looped domain. Note that the two identical paired chromatids drawn in Figure 4–52 can be clearly distinguished. (From M.P. Marsden and U.K. Laemmli, *Cell* 17:849–858, 1979. © Elsevier.)

The compaction of chromosomes during mitosis is a highly organized and dynamic process that serves at least two important purposes. First, when condensation is complete (in metaphase), sister chromatids have been disentangled from each other and lie side by side. Thus, the sister chromatids can easily separate when the mitotic apparatus begins pulling them apart. Second, the compaction of chromosomes protects the relatively fragile DNA molecules from being broken as they are pulled to separate daughter cells.

The condensation of interphase chromosomes into mitotic chromosomes occurs in M phase, and it is intimately connected with the progression of the cell cycle, as discussed detail in Chapters 17 and 18. It requires a class of proteins called *condensins* which using the energy of ATP hydrolysis, drive the coiling of each interphase chromosome that produces a mitotic chromosome. Condensins are large protein complexes that contain SMC proteins: long, dimeric protein molecules hinged in the center, with globular domains at each end that bind DNA and hydrolyze ATP (Figure 4–56). When added to purified DNA, condensins use the energy of ATP hydrolysis to make large right-handed loops in the DNA. Although it is not yet known how they act on chromatin, the coiling model shown in Figure 4–56C is based on the fact that condensins are a major structural component of mitotic chromosomes, with one molecule of condensin being present for every 10,000 nucleotides of mitotic DNA.

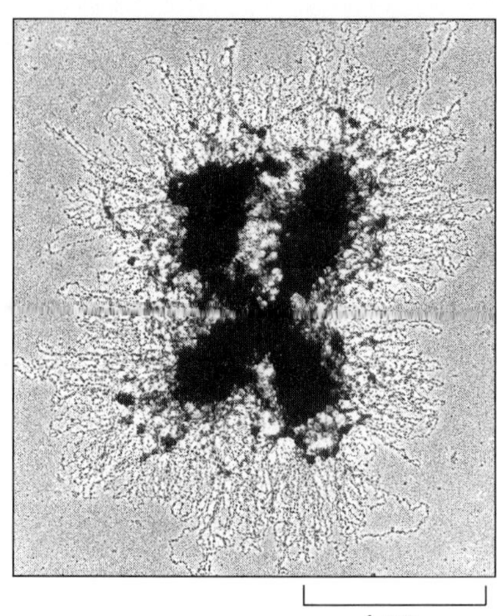

Figure 4–54 An electron micrograph of a mitotic chromosome. This chromosome (from an insect) was treated to reveal loops of chromatin fibers that emanate from a central scaffold of the chromatid. Such micrographs support the idea that the chromatin in all chromosomes is folded into a series of looped domains (see Figure 4–55). (Courtesy of Uli Laemmli.)

Each Mitotic Chromosome Contains a Characteristic Pattern of Very Large Domains

As mentioned earlier, the display of the 46 human chromosomes at mitosis is called the human karyotype. When stained with dyes such as Giemsa, mitotic

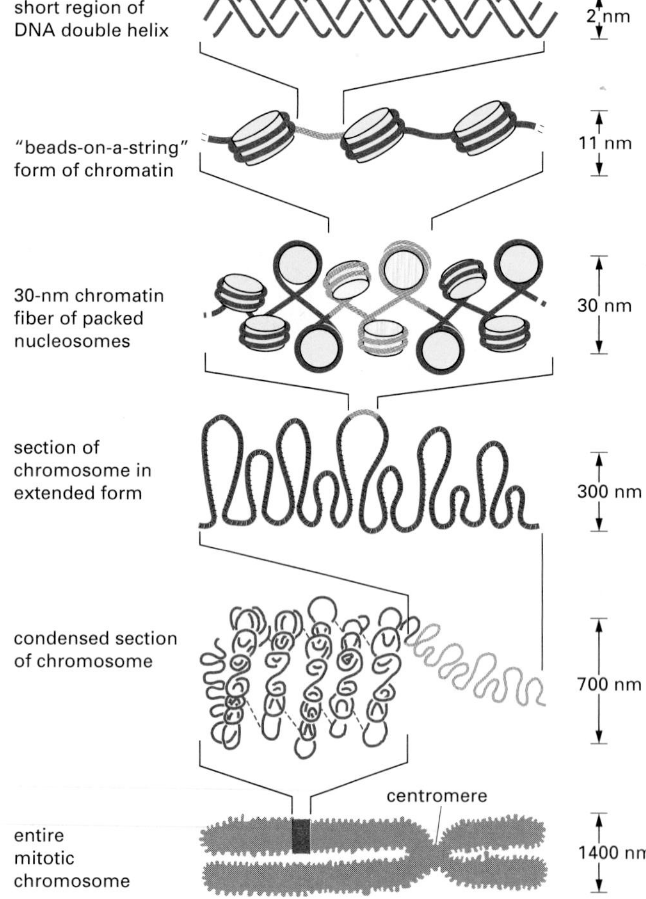

Figure 4–55 Chromatin packing. This model shows some of the many levels of chromatin packing postulated to give rise to the highly condensed mitotic chromosome.

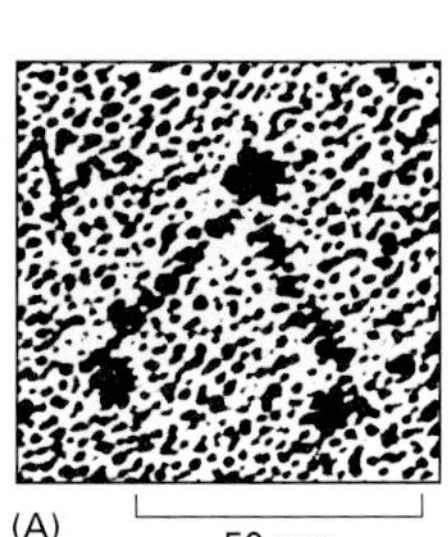

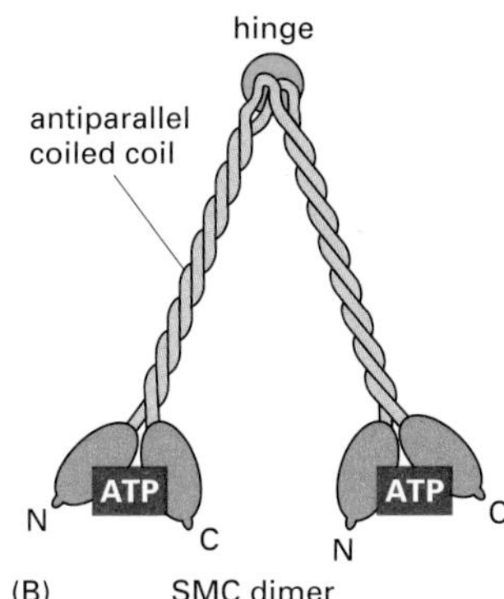

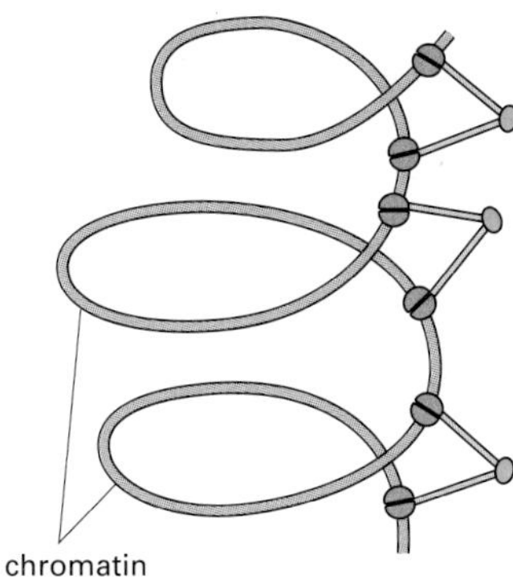

Figure 4–56 The SMC proteins in condensins. (A) Electron micrographs of a purified SMC dimer. (B) The structure of an SMC dimer. The long central region of this protein is an antiparallel coiled coil (see Figure 3–11) with a flexible hinge in its middle, as demonstrated by the electron micrograph in (A). (C) A model for the way in which the SMC proteins in condensins might compact chromatin. In reality, SMC proteins are components of a much larger condensin complex. It has been proposed that, in the cell, condensins coil long strings of looped chromatin domains (see Figure 4–55). In this way the condensins would form a structural framework that maintains the DNA in a highly organized state during M phase of the cell cycle. (A, courtesy of H.P. Erickson; B and C, adapted from T. Hirano, *Genes Dev.* 13:11–19, 1999.)

chromosomes show a striking and reproducible banding pattern along each chromosome, as shown in Figure 4–11. These bands are unrelated to those described earlier for the insect polytene chromosomes, which correspond to relatively small regions of interphase chromatin. In a human mitotic chromosome, all the chromatin is condensed and the bands represent a selective binding of the dyes.

By examining human chromosomes very early in mitosis, when they are less condensed than at metaphase, it has been possible to estimate that the total haploid genome contains about 2000 distinguishable bands. These coalesce progressively as condensation proceeds during mitosis, producing fewer and thicker bands. As we saw earlier, cytogeneticists routinely use the pattern of these chromosome bands to discover in patients genetic alterations such as chromosome inversions, translocations, and other types of chromosomal rearrangements (see Figure 4–12).

Mitotic chromosome bands are detected in chromosomes from species as diverse as humans and flies. Moreover, the pattern of bands in a chromosome has remained unchanged over long periods of evolutionary time. Each human chromosome, for example, has a clearly recognizable counterpart with a nearly identical banding pattern in the chromosomes of the chimpanzee, gorilla, and orangutan—although there are also clear differences, such as chromosome fusion, that give the human 46 chromosomes instead of the ape's 48 (Figure 4–57). This conservation suggests that chromosomes are organized into large domains that may be important for chromosomal function.

Even the thinnest of the bands in Figure 4–11 probably contains more than a million nucleotide pairs, which is nearly the size of a bacterial genome. These bands seem to reflect a rough division of chromosomes into regions of different GC content. The nucleotide sequence of the human genome has revealed large non-random blocks of sequence (some greater than 10^7 nucleotide pairs) that are significantly higher or lower in GC content than the genome-wide average of 41%. The blocks correlate roughly with the staining pattern of metaphase chromosomes. For example, bands that are darkly stained by Giemsa (the so-called G-bands) are correlated with DNA that is low in GC content, whereas lightly stained bands (the R-bands) correspond to DNA of higher than average GC content.

In general, GC-rich regions of the genome have a higher density of genes, especially of "house-keeping" genes, the genes that are expressed in virtually all cell types. On the basis of these observations, it has been proposed that the

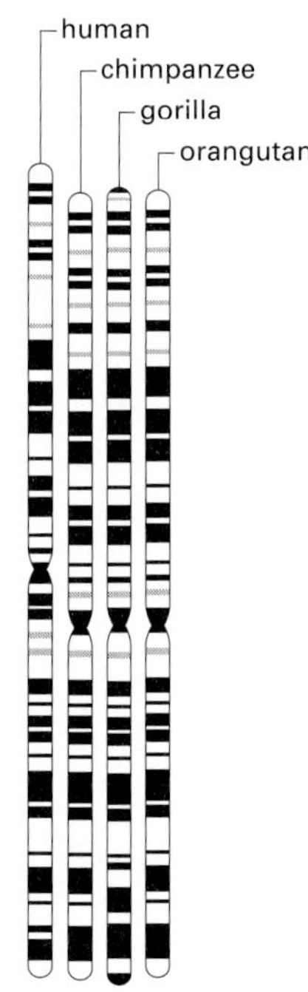

Figure 4–57 Comparison of the Giemsa pattern of the largest human chromosome (chromosome 1) with that of chimpanzee, gorilla, and orangutan. Comparisons among the staining patterns of all the chromosomes indicate that human chromosomes are more closely related to those of chimpanzee than to those of gorilla and that they are more distantly related to those of orangutan. (Adapted from M.W. Strickberger, Evolution, 3rd edn. Sudbury, MA: Jones & Bartlett Publishers, 2000.)

Figure 4–58 The polarized orientation of chromosomes found in certain interphase nuclei. (A) Fluorescent light micrograph of interphase nuclei from the rapidly growing root tip of a plant. Centromeres are stained *green* and teleomeres *red* by *in situ* hybridization of centromere- and telomere-specific DNA sequences coupled to the different fluorescent dyes. (B) Interpretation of (A) showing chromosomes in the Rabl orientation with all the centromeres facing one side of a nucleus and all the telomeres pointing toward the opposite side. (A, from R. Abranches et al., *J. Cell Biol.* 143:5–12, 1998. © The Rockefeller University Press.)

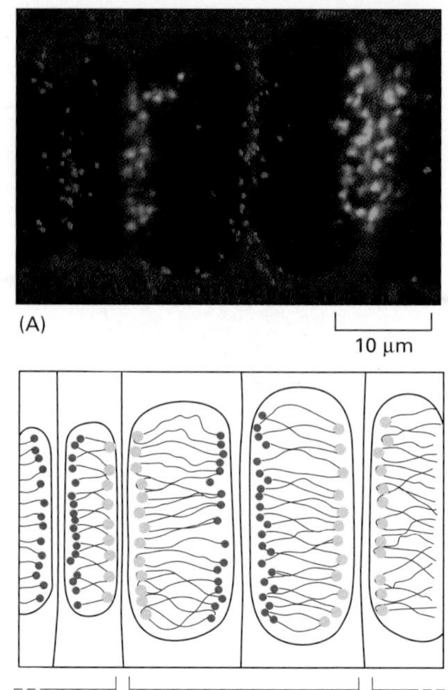

banding pattern may be related to gene expression. Perhaps the differentiation of chromosomes into G- and R-bands reflects subtle differences, determined by GC content, in the way in which chromatin loops are packaged in these areas. If this idea is correct, the rough division of chromosomes can be seen as a form of compartmentalization, in which the particular cellular components involved in gene expression are more concentrated in the R-bands where their activities are required. In any case, it should be obvious from this discussion that we are only beginning to glimpse the principles of large-scale chromosome organization.

Individual Chromosomes Occupy Discrete Territories in an Interphase Nucleus

We saw earlier in this chapter that chromosomes from eucaryotes are contained in the cell nucleus. However, the nucleus is not simply a bag of chromosomes; rather, the chromosomes—as well as the other components inside the nucleus which we shall encounter in subsequent chapters—are highly organized. The way in which chromosomes are organized in the nucleus during interphase, when they are active and difficult to see, has intrigued biologists since the nineteenth century. Although our understanding today is far from complete, we do know some interesting features of these chromosome arrangements.

A certain degree of chromosomal order results from the configuration that the chromosomes always have at the end of mitosis. Just before a cell divides, the condensed chromosomes are pulled to each spindle pole by microtubules attached to the centromeres; thus, as the chromosomes move, the centromeres lead the way and the distal arms (terminating in the telomeres) lag behind. The chromosomes in some nuclei tend to retain this so-called *Rabl orientation* throughout interphase, with their centromeres facing one pole of the nucleus and their telomeres pointing toward the opposite pole (Figures 4–58 and 4–59).

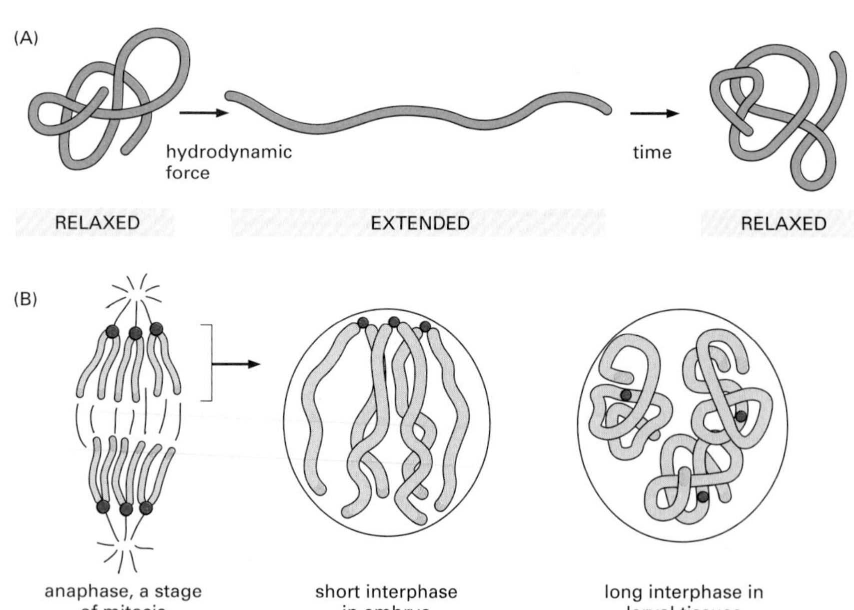

Figure 4–59 A polymer analogy for interphase chromosome organization. (A) The behavior of a polymer in solution. Entropy drives a long polymer into a compact conformation in the absence of an externally applied force. If the polymer is subjected to shear or hydrodynamic force, it becomes extended. But once the force is removed, the polymer chain returns to a more favorable, compact conformation. (B) The behavior of interphase chromosomes may reflect the same simple principles. In *Drosophila* embryos, for example, mitotic divisions occur at intervals of about 10 minutes; during the short intervening interphases, the chromosomes have little time to relax from the Rabl orientation induced by their movement during mitosis. However, in later stages of development, when interphase is much longer, the chromosomes have time to fold up. This folding may be strongly affected by specific associations between different regions of the same chromosome. (Adapted from A.F. Dernburg et al., *Cell* 85:745–759, 1996.)

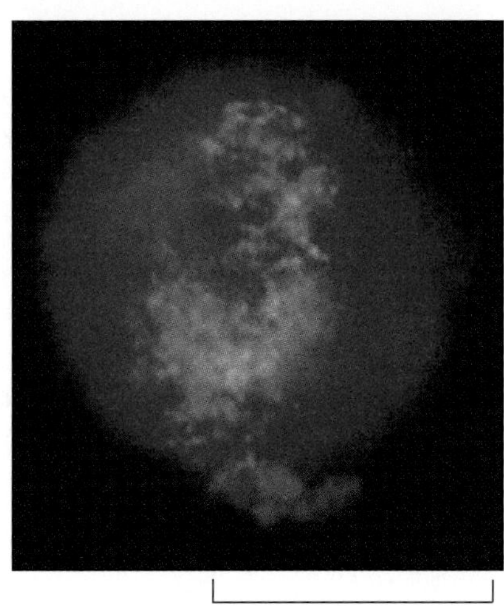

Figure 4–60 Selective "painting" of two interphase chromosomes in a human peripheral lymphocyte. The fluorescent light micrograph shows that the two copies of human chromosome 18 *(red)* and chromosome 19 *(turquoise)* occupy discrete territories of the nucleus. (From J.A. Croft et al., *J. Cell Biol.* 145:1119–1131, 1999. © The Rockefeller University Press.)

The chromosomes in most interphase cells are not found in the Rabl orientation; instead, the centromeres seem to be dispersed in the nucleus. Most cells have a longer interphase than the specialized cells illustrated above, and this presumably gives their chromosomes time to assume a different conformation (see Figure 4–59). Nevertheless, each interphase chromosome does tend to occupy a discrete and relatively small territory in the nucleus: that is, the different chromosomes are not extensively intertwined (Figure 4–60).

One device for organizing chromosomes in the nucleus may be the attachment of certain portions of each chromosome to the nuclear envelope (Figure 4–61). For example, in many cells, telomeres seem bound in this way. But the exact position of a chromosome in a nucleus is not fixed. In the same tissue, for example, two apparently identical cells can have different chromosomes as nearest neighbors.

Some cell biologists believe that there is an intranuclear framework, analogous to the cytoskeleton, on which chromosomes and other components of the nucleus are organized. The *nuclear matrix,* or *scaffold,* has been defined as the insoluble material left in the nucleus after a series of biochemical extraction steps. Some of the proteins that constitute it can be shown to bind specific DNA sequences called *SARs* or *MARs* (scaffold-associated or matrix-associated regions). These DNA sequences have been postulated to form the base of chromosomal loops (see Figure 4–44), or to attach chromosomes to the nuclear envelope and other structures in the nucleus. By means of such chromosomal attachment sites, the matrix might help to organize chromosomes, localize genes, and regulate gene expression and DNA replication. It still remains uncertain, however, whether the matrix isolated by cell biologists represents a structure that is present in intact cells.

Summary

Chromosomes are generally decondensed during interphase, so that their structure is difficult to visualize directly. Notable exceptions are the specialized lampbrush chromosomes of vertebrate oocytes and the polytene chromosomes in the giant secretory cells of insects. Studies of these two types of interphase chromosomes suggest that each long DNA molecule in a chromosome is divided into a large number of discrete domains organized as loops of chromatin, each loop probably consisting of a folded 30-nm chromatin fiber. When genes contained in a loop are expressed, the loop decondenses and allows the cell's machinery easy access to the DNA.

Euchromatin makes up most of interphase chromosomes and probably corresponds to looped domains of 30-nm fibers. However, euchromatin is interrupted by stretches of heterochromatin, in which 30-nm fibers are subjected to additional levels of packing that usually render it resistant to gene expression. Heterochromatin is commonly found around centromeres and near telomeres, but it is also present at other positions on chromosomes. Although considerably less condensed than mitotic chromosomes, interphase chromosomes occupy discrete territories in the cell nucleus; that is, they are not extensively intertwined.

All chromosomes adopt a highly condensed conformation during mitosis. When they are specially stained, these mitotic chromosomes have a banding structure that allows each individual chromosome to be recognized unambiguously. These bands contain millions of DNA nucleotide pairs, and they reflect a poorly-understood coarse heterogeneity of chromosome structure.

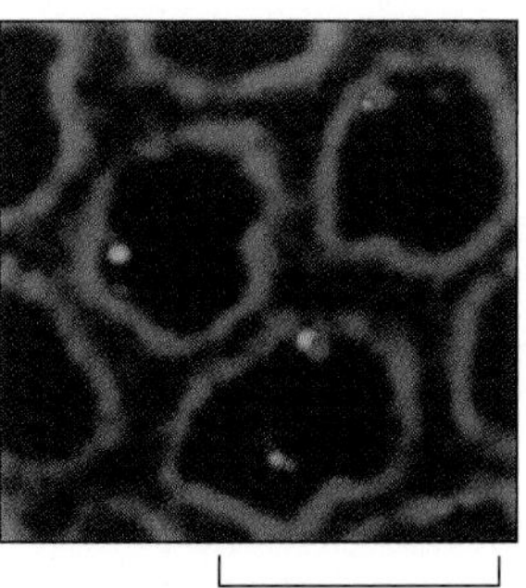

Figure 4–61 Specific regions of interphase chromosomes in close proximity to the nuclear envelope. This high-resolution microscopic view of nuclei from a *Drosophila* embryo shows the localization of two different regions of chromosome 2 *(yellow* and *magenta)* close to the nuclear envelope (stained green with antilamina antibodies). Other regions of the same chromosome are more distant from the envelope. (From W.F. Marshall et al., *Mol. Biol. Cell* 7:825–842, 1996.)

References

General

Hartwell L, Hood L, Goldberg ML et al. (2000) Genetics: from Genes to Genomes. Boston: McGraw Hill.

Lewin B (2000) Genes VII. Oxford: Oxford University Press.

Lodish H, Berk A, Zipursky SL et al. (2000) Molecular Cell Biology, 4th edn. New York: WH Freeman.

Wolfe A (1999) Chromatin: Structure and Function, 3rd edn. New York: Academic Press.

The Structure and Function of DNA

Avery OT, MacLeod CM & McCarty M (1944) Studies on the chemical nature of the substance inducing transformation of pneumococcal types. *J. Exp. Med.* 79, 137.

Meselson M & Stahl FW (1958) The replication of DNA in *E. coli*. *Proc. Natl Acad. Sci. USA* 44, 671–682.

Watson JD & Crick FHC (1953) Molecular structure of nucleic acids. A structure for deoxyribose nucleic acids. *Nature* 171, 737–738.

Chromosomal DNA and Its Packaging in the Chromatin Fiber

Aalfs JD & Kingston RE (2000) What does 'chromatin remodeling' mean? *Trends Biochem. Sci.* 25, 548–555.

Cairns BR (1998) Chromatin remodeling machines: similar motors, ulterior motives. *Trends Biochem. Sci.* 23, 20–25.

Carter NP (1994) Cytogenetic analysis by chromosome painting. *Cytometry* 18, 2–10.

Cheung P, Allis CD & Sassone-Corsi P (2000) Signaling to chromatin through histone modifications. *Cell* 103, 263–271.

Clark MS (1999) Comparative genomics: the key to understanding the Human Genome Project. *Bioessays* 21, 121–130.

DePamphilis ML (1999) Replication origins in metazoan chromosomes: fact or fiction? *Bioessays* 21, 5–16.

Dunham I, Shimizu N, Roe BA et al. (1999) The DNA sequence of human chromosome 22. *Nature* 402, 489–495.

Felsenfeld G (1985) DNA. *Sci. Am.* 253(4), 58–67.

Grunstein M (1992) Histones as regulators of genes. *Sci. Am.* 267(4), 68–74B.

International Human Genome Sequencing Consortium (2001) Initial sequencing and analysis of the human genome. *Nature* 409, 860–921.

Jenuwein T & Allis CD (2001) Translating the histone code. *Science* 293:1074–1080.

Kingston RE & Narlikar GJ (1999) ATP-dependent remodeling and acetylation as regulators of chromatin fluidity. *Genes Dev.* 13, 2339–2352.

Kornberg RD & Lorch Y (1999) Chromatin-modifying and -remodeling complexes. *Curr. Opin. Genet. Dev.* 9, 148–151.

Kornberg RD & Lorch Y (1999) Twenty-five years of the nucleosome, fundamental particle of the eukaryote chromosome. *Cell* 98, 285–294.

Luger K & Richmond TJ (1998) The histone tails of the nucleosome. *Curr. Opin. Genet. Dev.* 8, 140–146.

Luger K, Mader AW, Richmond RK et al. (1997) Crystal structure of the nucleosome core particle at 2.8 Å resolution. *Nature* 389, 251–260.

McEachern MJ, Krauskopf A & Blackburn EH (2000) Telomeres and their control. *Annu. Rev. Genet.* 34, 331–358.

Ng HH & Bird A (2000) Histone deacetylases: silencers for hire. *Trends Biochem. Sci.* 25, 121–126.

O'Brien S, Menotti-Raymond M, Murphy W et al. (1999) The promise of comparative genomics in mammals. *Science* 286, 458–480.

Pidoux AL & Allshire RC (2000) Centromeres: getting a grip of chromosomes. *Curr. Opin. Cell Biol.* 12, 308–319.

Rhodes D (1997) Chromatin structure. The nucleosome core all wrapped up. *Nature* 389, 231–233.

Rice JC & Allis CD (2001) Histone methylation versus histone acetylation: new insights into epigenetic regulation. *Curr. Opin. Cell Biol.* 13, 263–273.

Ried T, Schrock E, Ning Y & Wienberg J (1998) Chromosome painting: a useful art. *Hum. Mol. Genet.* 7, 1619–1626.

Rubin GM (2001) Comparing species. *Nature* 409, 820–821.

Stewart A (1990) The functional organization of chromosomes and the nucleus—a special issue. *Trends Genet.* 6, 377–379.

Strahl BD & Allis CD (2000) The language of covalent histone modifications. *Nature* 403, 41–45.

Travers AA (1987) DNA bending and nucleosome positioning. *Trends Biochem. Sci.* 12, 108–112.

Wu J & Grunstein M (2000) 25 years after the nucleosome model: chromatin modifications. *Trends Biochem. Sci.* 25, 619–623.

The Global Structure of Chromosomes

Agard DA & Sedat JW (1983) Three-dimensional architecture of a polytene nucleus. *Nature* 302, 676–681.

Ashburner M, Chihara C, Meltzer P & Richards G (1974) Temporal control of puffing activity in polytene chromosomes. *Cold Spring Harbor Symp. Quant. Biol.* 38, 655–662.

Bickmore WA & Sumner AT (1989) Mammalian chromosome banding—an expression genome organization. *Trends Genet.* 5, 144–148.

Birchler JA, Bhadra MP & Bhadra U (2000) Making noise about silence: repression of repeated genes in animals. *Curr. Opin. Genet. Dev.* 10, 211–216.

Callan HG (1982) Lampbrush chromosomes. *Proc. Roy. Soc. Lond. Ser. B. (Biol.)* 214, 417–448.

Croft JA, Bridger JM, Boyle S et al. (1999) Differences in the localization and morphology of chromosomes in the human nucleus. *J. Cell Biol.* 145, 1119–1131.

Griffith JD, Comeau L, Rosenfield S et al. (1999) Mammalian telomeres end in a large duplex loop. *Cell* 97, 503–514.

Grunstein M (1997) Molecular model for telomeric heterochromatin in yeast. *Curr. Opin. Cell Biol.* 9, 383–387.

Hart CM & Laemmli UK (1998) Facilitation of chromatin dynamics by SARs. *Curr. Opin. Genet. Dev.* 8, 519–525.

Henikoff S (1990) Position-effect variegation after 60 years. *Trends Genet.* 6, 422–426.

Henikoff S (1998) Conspiracy of silence among repeated transgenes. *Bioessays* 20, 532–535.

Hirano T (1999) SMC-mediated chromosome mechanics: a conserved scheme from bacteria to vertebrates. *Genes Dev.* 13, 11–19.

Hirano T (2000) Chromosome cohesion, condensation, and separation. *Annu. Rev. Biochem.* 69, 115–144.

Lamond AI & Earnshaw WC (1998) Structure and function in the nucleus. *Science* 280, 547–553.

Lyko F & Paro R (1999) Chromosomal elements conferring epigenetic inheritance. *Bioessays* 21, 824–832.

Marsden M & Laemmli UK (1979) Metaphase chromosome structure: evidence for a radial loop model. *Cell* 17, 849–858.

Pluta AF, Mackay AM, Ainsztein AM et al. (1995) The centromere: hub of chromosomal activities. *Science* 270, 1591–1594.

Saitoh N, Goldberg I & Earnshaw WC (1995) The SMC proteins and the coming of age of the chromosome scaffold hypothesis. *Bioessays* 17, 759–766.

Spector DL (1993) Macromolecular domains within the cell nucleus. *Annu. Rev. Cell Biol.* 9, 265–315.

Thummel CS (1990) Puffs and gene regulation—molecular insights into the Drosophila ecdysone regulatory hierarchy. *Bioessays* 12, 561–568.

Weiler KS & Wakimoto BT (1995) Heterochromatin and gene expression in Drosophila. *Annu. Rev. Genet.* 29, 577–605.

Zhimulev IF (1998) Morphology and structure of polytene chromosomes. *Adv. Genet.* 37, 1–566.

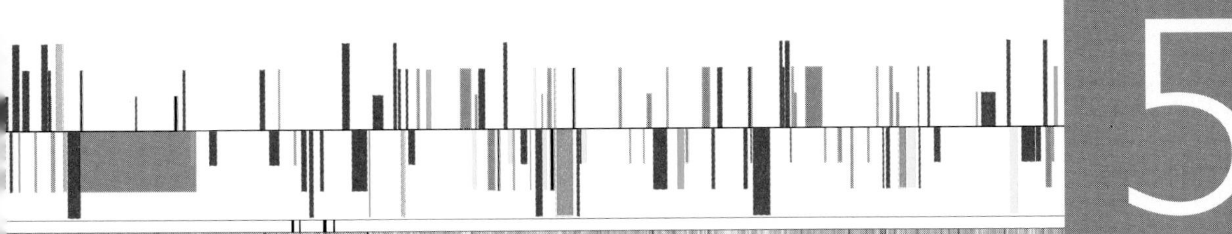

5

DNA REPLICATION, REPAIR, AND RECOMBINATION

The ability of cells to maintain a high degree of order in a chaotic universe depends upon the accurate duplication of vast quantities of genetic information carried in chemical form as DNA. This process, called *DNA replication*, must occur before a cell can produce two genetically identical daughter cells. Maintaining order also requires the continued surveillance and repair of this genetic information because DNA inside cells is repeatedly damaged by chemicals and radiation from the environment, as well as by thermal accidents and reactive molecules. In this chapter we describe the protein machines that replicate and repair the cell's DNA. These machines catalyze some of the most rapid and accurate processes that take place within cells, and their mechanisms clearly demonstrate the elegance and efficiency of cellular chemistry.

While the short-term survival of a cell can depend on preventing changes in its DNA, the long-term survival of a species requires that DNA sequences be changeable over many generations. Despite the great efforts that cells make to protect their DNA, occasional changes in DNA sequences do occur. Over time, these changes provide the genetic variation upon which selection pressures act during the evolution of organisms.

We begin this chapter with a brief discussion of the changes that occur in DNA as it is passed down from generation to generation. Next, we discuss the cellular mechanisms—DNA replication and DNA repair—that are responsible for keeping these changes to a minimum. Finally, we consider some of the most intriguing ways in which DNA sequences are altered by cells, with a focus on DNA recombination and the movement of special DNA sequences in our chromosomes called transposable elements.

THE MAINTENANCE OF DNA SEQUENCES

Although the long-term survival of a species is enhanced by occasional genetic changes, the survival of the individual demands genetic stability. Only rarely do the cell's DNA-maintenance processes fail, resulting in permanent change in the DNA. Such a change is called a **mutation**, and it can destroy an organism if it occurs in a vital position in the DNA sequence. Before examining the mechanisms responsible for genetic stability, we briefly discuss the accuracy with which DNA sequences are maintained from one generation to the next.

Mutation Rates Are Extremely Low

The **mutation rate**, the rate at which observable changes occur in DNA sequences, can be determined directly from experiments carried out with a bacterium such as *Escherichia coli*—a resident of our intestinal tract and a commonly used laboratory organism. Under laboratory conditions, *E. coli* divides about once every 30 minutes, and a very large population—several billion—can be obtained from a single cell in less than a day. In such a population, it is possible to detect the small fraction of bacteria that have suffered a damaging mutation in a particular gene, if that gene is not required for the survival of the bacterium. For example, the mutation rate of a gene specifically required for cells to utilize the sugar lactose as an energy source can be determined (using indicator dyes to identify the mutant cells), if the cells are grown in the presence of a different sugar, such as glucose. The fraction of damaged genes is an underestimate of the actual mutation rate because many mutations are *silent* (for example, those that change a codon but not the amino acid it specifies, or those that change an amino acid without affecting the activity of the protein coded for by the gene). After correcting for these silent mutations, a single gene that encodes an average-sized protein (~10^3 coding nucleotide pairs) is estimated to suffer a mutation (not necessarily one that would inactivate the protein) once in about 10^6 bacterial cell generations. Stated in a different way, bacteria display a mutation rate of 1 nucleotide change per 10^9 nucleotides per cell generation.

The germ-line mutation rate in mammals is more difficult to measure directly, but estimates can be obtained indirectly. One way is to compare the amino acid sequences of the same protein in several species. The fraction of the amino acids that are different between any two species can then be compared with the estimated number of years since that pair of species diverged from a common ancestor, as determined from the fossil record. Using this method, one can calculate the number of years that elapse, on average, before an inherited change in the amino acid sequence of a protein becomes fixed in an organism. Because each such change usually reflects the alteration of a single nucleotide in the DNA sequence of the gene encoding that protein, this value can be used to estimate the average number of years required to produce a single, stable mutation in the gene.

These calculations will nearly always substantially underestimate the actual mutation rate, because many mutations will spoil the function of the protein and vanish from the population because of natural selection—that is, by the preferential death of the organisms that contain them. But there is one family of protein fragments whose sequence does not seem to matter, allowing the genes that encode them to accumulate mutations without being selected against. These are the *fibrinopeptides*, 20 amino-acid fragments that are discarded from the protein *fibrinogen* when it is activated to form *fibrin* during blood clotting. Since the function of fibrinopeptides apparently does not depend on their amino acid sequence, they can tolerate almost any amino acid change. Sequence comparisons of the fibrinopeptides indicate that a typical protein 400 amino acids long would be randomly altered by an amino acid change in the germ line roughly once every 200,000 years.

Another way to estimate mutation rates is to use DNA sequencing to compare corresponding nucleotide sequences from different species in regions of the genome that do not carry critical information. Such comparisons produce estimates of the mutation rate that are in good agreement with those obtained from the fibrinopeptide studies.

E. coli and humans differ greatly in their modes of reproduction and in their generation times. Yet, when the mutation rates of each are normalized to a single round of DNA replication, they are found to be similar: roughly 1 nucleotide change per 10^9 nucleotides each time that DNA is replicated.

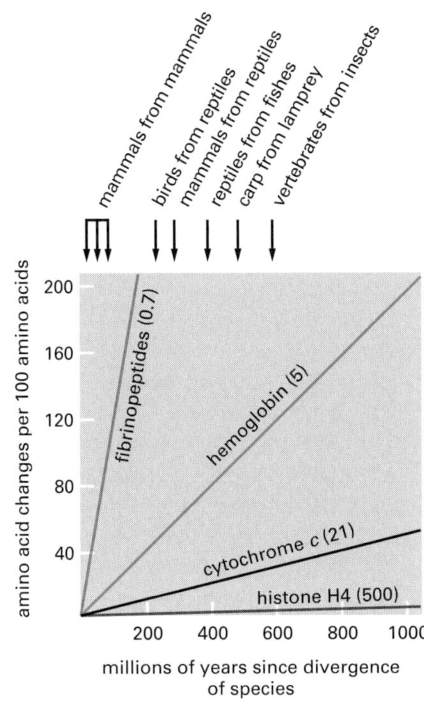

Figure 5–1 Different proteins evolve at very different rates. A comparison of the rates of amino acid change found in hemoglobin, histone H4, cytochrome *c*, and the fibrinopeptides. The first three proteins have changed much more slowly during evolution than the fibrinopeptides, the number in parentheses indicating how many million years it has taken, on average, for one *acceptable* amino acid change to appear for every 100 amino acids that the protein contains. In determining rates of change per year, it is important to realize that two species that diverged from a common ancestor 100 million years ago are separated from each other by 200 million years of evolutionary time.

Many Mutations in Proteins Are Deleterious and Are Eliminated by Natural Selection

When the number of amino acid differences in a particular protein is plotted for several pairs of species against the time that has elapsed since the pair of species diverged from a common ancestor, the result is a reasonably straight line: the longer the period since divergence, the larger the number of differences. For convenience, the slope of this line can be expressed in terms of a "unit evolutionary time" for that protein, which is the average time required for 1 amino acid change to appear in a sequence of 100 amino acid residues. When various proteins are compared, each shows a different but characteristic rate of evolution (Figure 5–1).

Since most DNA nucleotides are thought to be subject to roughly the same rate of random mutation, the different rates observed for different proteins must reflect differences in the probability that an amino acid change will be harmful for each protein. For example, from the data in Figure 5–1, we can estimate that about six of every seven random amino acid changes are harmful in cytochrome *c*, and that virtually all amino acid changes are harmful in histone H4. The individual animals that carried such harmful mutations were presumably eliminated from the population by natural selection.

Low Mutation Rates Are Necessary for Life as We Know It

Since so many mutations are deleterious, no species can afford to allow them to accumulate at a high rate in its germ cells. Although the observed mutation frequency is low, it is nevertheless thought to limit the number of essential proteins that any organism can encode to perhaps 60,000. By an extension of the same argument, a mutation frequency tenfold higher would limit an organism to about 6000 essential genes. In this case, evolution would probably have stopped at an organism less complex than a fruit fly.

Whereas *germ cells* must be protected against high rates of mutation to maintain the species, the *somatic cells* of multicellular organisms must be protected from genetic change to safeguard each individual. Nucleotide changes in somatic cells can give rise to variant cells, some of which, through natural selection, proliferate rapidly at the expense of the rest of the organism. In an extreme case, the result is an uncontrolled cell proliferation known as cancer, a disease that causes about 30% of the deaths each year in Europe and North America. These deaths are due largely to an accumulation of changes in the DNA sequences of somatic cells (discussed in Chapter 23). A significant increase in the mutation frequency would presumably cause a disastrous increase in the incidence of cancer by accelerating the rate at which somatic cell variants arise. Thus, both for the perpetuation of a species with a large number of genes (germ cell stability) and for the prevention of cancer resulting from mutations in somatic cells (somatic cell stability), multicellular organisms like ourselves depend on the remarkably high fidelity with which their DNA sequences are maintained.

As we see in subsequent sections, successful DNA maintenance depends both on the accuracy with which DNA sequences are duplicated and distributed to daughter cells, and on a set of enzymes that repair most of the changes in the DNA caused by radiation, chemicals, or other accidents.

Summary

In all cells, DNA sequences are maintained and replicated with high fidelity. The mutation rate, approximately 1 nucleotide change per 10^9 nucleotides each time the DNA is replicated, is roughly the same for organisms as different as bacteria and humans. Because of this remarkable accuracy, the sequence of the human genome (approximately 3×10^9 nucleotide pairs) is changed by only about 3 nucleotides each time a cell divides. This allows most humans to pass accurate genetic instructions from one generation to the next, and also to avoid the changes in somatic cells that lead to cancer.

DNA REPLICATION MECHANISMS

All organisms must duplicate their DNA with extraordinary accuracy before each cell division. In this section, we explore how an elaborate "replication machine" achieves this accuracy, while duplicating DNA at rates as high as 1000 nucleotides per second.

Base-Pairing Underlies DNA Replication and DNA Repair

As discussed briefly in Chapter 1, *DNA templating* is the process in which the nucleotide sequence of a DNA strand (or selected portions of a DNA strand) is copied by complementary base-pairing (A with T, and G with C) into a complementary DNA sequence (Figure 5–2). This process entails the recognition of each nucleotide in the DNA *template* strand by a free (unpolymerized) complementary nucleotide, and it requires that the two strands of the DNA helix be separated. This separation allows the hydrogen-bond donor and acceptor groups on each DNA base to become exposed for base-pairing with the appropriate incoming free nucleotide, aligning it for its enzyme-catalyzed polymerization into a new DNA chain.

The first nucleotide polymerizing enzyme, **DNA polymerase**, was discovered in 1957. The free nucleotides that serve as substrates for this enzyme were found to be deoxyribonucleoside triphosphates, and their polymerization into DNA required a single-stranded DNA template. The stepwise mechanism of this reaction is illustrated in Figures 5–3 and 5–4.

The DNA Replication Fork Is Asymmetrical

During DNA replication inside a cell, each of the two old DNA strands serves as a template for the formation of an entire new strand. Because each of the two daughters of a dividing cell inherits a new DNA double helix containing one old and one new strand (Figure 5–5), the DNA double helix is said to be replicated "semiconservatively" by DNA polymerase. How is this feat accomplished?

Analyses carried out in the early 1960s on whole replicating chromosomes revealed a localized region of replication that moves progressively along the parental DNA double helix. Because of its Y-shaped structure, this active region

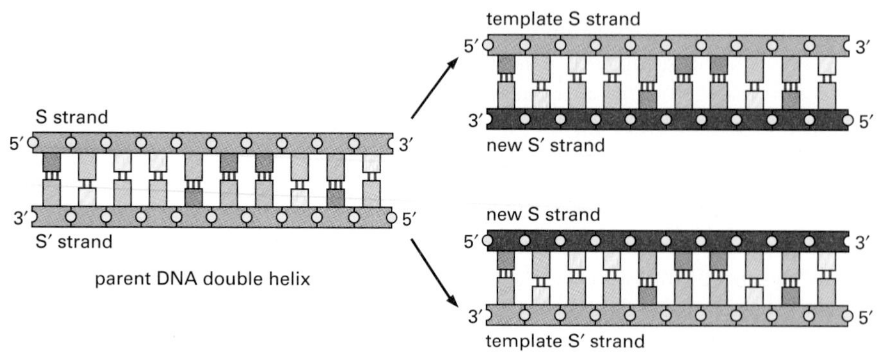

Figure 5–2 The DNA double helix acts as a template for its own duplication. Because the nucleotide A will successfully pair only with T, and G only with C, each strand of DNA can serve as a template to specify the sequence of nucleotides in its complementary strand by DNA base-pairing. In this way, a double-helical DNA molecule can be copied precisely.

Figure 5–3 The chemistry of DNA synthesis. The addition of a deoxyribonucleotide to the 3′ end of a polynucleotide chain (the *primer strand*) is the fundamental reaction by which DNA is synthesized. As shown, base-pairing between an incoming deoxyribonucleoside triphosphate and an existing strand of DNA (the *template strand*) guides the formation of the new strand of DNA and causes it to have a complementary nucleotide sequence.

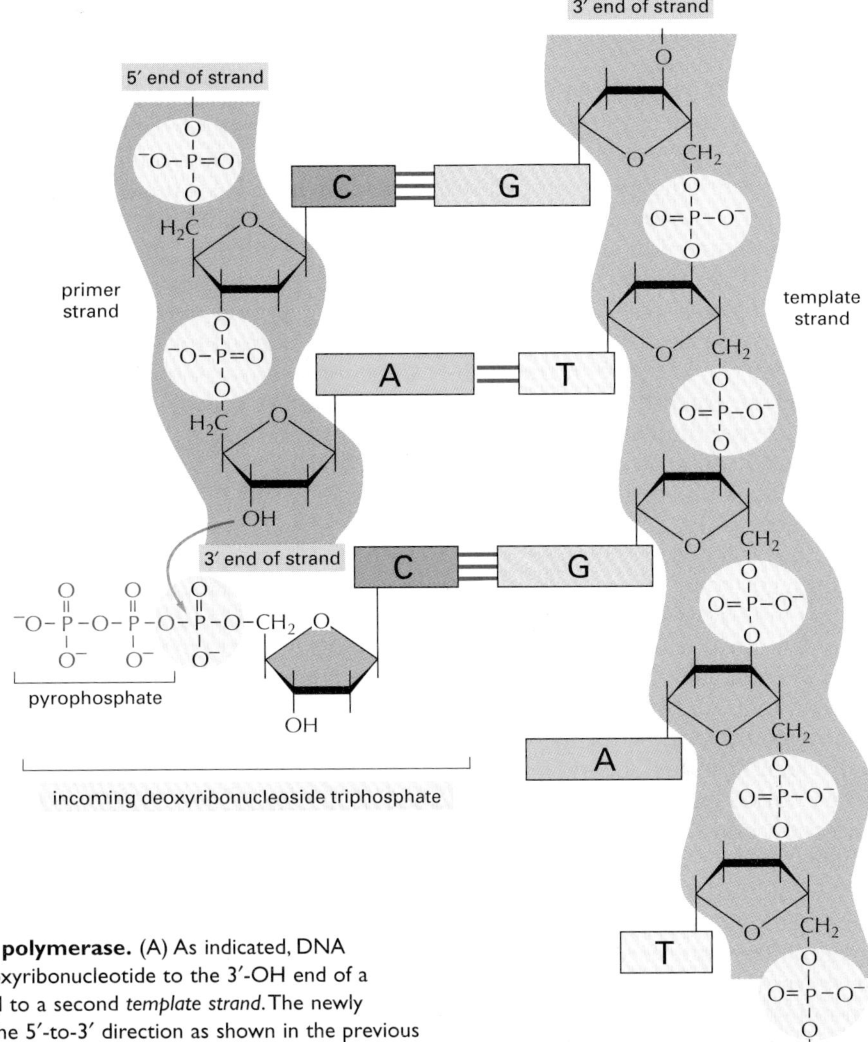

Figure 5–4 DNA synthesis catalyzed by DNA polymerase. (A) As indicated, DNA polymerase catalyzes the stepwise addition of a deoxyribonucleotide to the 3′-OH end of a polynucleotide chain, the *primer strand*, that is paired to a second *template strand*. The newly synthesized DNA strand therefore polymerizes in the 5′-to-3′ direction as shown in the previous figure. Because each incoming deoxyribonucleoside triphosphate must pair with the template strand to be recognized by the DNA polymerase, this strand determines which of the four possible deoxyribonucleotides (A, C, G, or T) will be added. The reaction is driven by a large, favorable free-energy change, caused by the release of pyrophosphate and its subsequent hydrolysis to two molecules of inorganic phosphate. (B) The structure of an *E. coli* DNA polymerase molecule, as determined by x-ray crystallography. Roughly speaking, it resembles a right hand in which the palm, fingers, and thumb grasp the DNA. This drawing illustrates a DNA polymerase that functions during DNA repair, but the enzymes that replicate DNA have similar features. (B, adapted from L.S. Beese, V. Derbyshire, and T.A. Steitz, *Science* 260:352–355, 1993.)

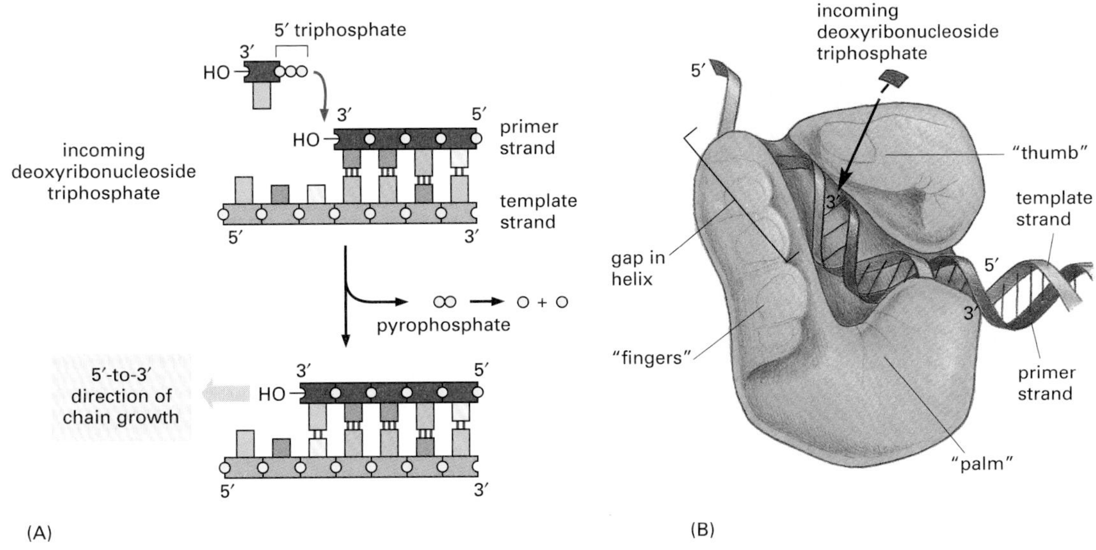

is called a **replication fork** (Figure 5–6). At a replication fork, the DNA of both new daughter strands is synthesized by a multienzyme complex that contains the DNA polymerase.

Initially, the simplest mechanism of DNA replication seemed to be the continuous growth of both new strands, nucleotide by nucleotide, at the replication fork as it moves from one end of a DNA molecule to the other. But because of the antiparallel orientation of the two DNA strands in the DNA double helix (see Figure 5–2), this mechanism would require one daughter strand to polymerize in the 5′-to-3′ direction and the other in the 3′-to-5′ direction. Such a replication fork would require two different DNA polymerase enzymes. One would polymerize in the 5′-to-3′ direction, where each incoming deoxyribonucleoside triphosphate carried the triphosphate activation needed for its own addition. The other would move in the 3′-to-5′ direction and work by so-called "head growth," in which the end of the growing DNA chain carried the triphosphate activation required for the addition of each subsequent nucleotide (Figure 5–7). Although head-growth polymerization occurs elsewhere in biochemistry (see pp. 89–90), it does not occur in DNA synthesis; no 3′-to-5′ DNA polymerase has ever been found.

How, then, is overall 3′-to-5′ DNA chain growth achieved? The answer was first suggested by the results of experiments in the late 1960s. Researchers added highly radioactive ^{3}H-thymidine to dividing bacteria for a few seconds, so that only the most recently replicated DNA—that just behind the replication fork—became radiolabeled. This experiment revealed the transient existence of pieces of DNA that were 1000–2000 nucleotides long, now commonly known as *Okazaki fragments,* at the growing replication fork. (Similar replication intermediates were later found in eucaryotes, where they are only 100–200 nucleotides long.) The Okazaki fragments were shown to be polymerized only in the 5′-to-3′chain direction and to be joined together after their synthesis to create long DNA chains.

A replication fork therefore has an asymmetric structure (Figure 5–8). The DNA daughter strand that is synthesized continuously is known as the **leading strand**. Its synthesis slightly precedes the synthesis of the daughter strand that is synthesized discontinuously, known as the **lagging strand**. For the lagging strand, the direction of nucleotide polymerization is opposite to the overall direction of DNA chain growth. Lagging-strand DNA synthesis is delayed because it must wait for the leading strand to expose the template strand on which each Okazaki fragment is synthesized. The synthesis of the lagging strand

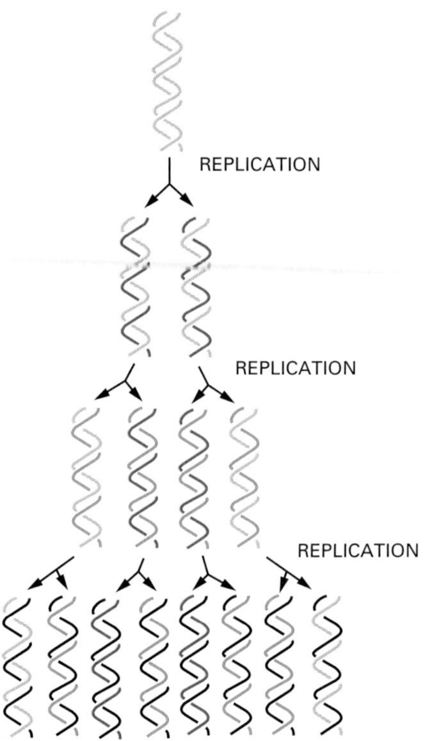

Figure 5–5 The semiconservative nature of DNA replication. In a round of replication, each of the two strands of DNA is used as a template for the formation of a complementary DNA strand. The original strands therefore remain intact through many cell generations.

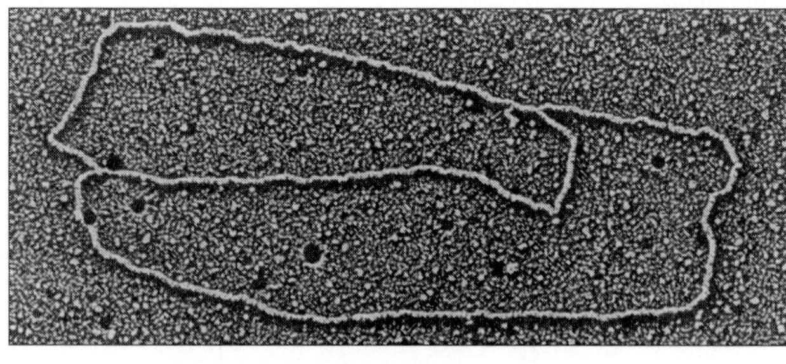

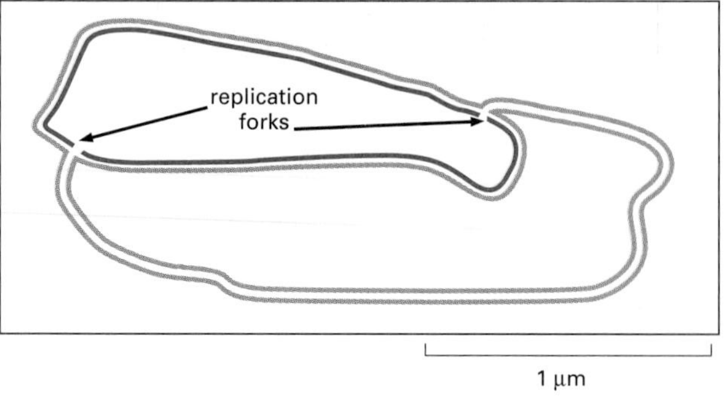

Figure 5–6 Two replication forks moving in opposite directions on a circular chromosome. An active zone of DNA replication moves progressively along a replicating DNA molecule, creating a Y-shaped DNA structure known as a replication fork: the two arms of each Y are the two daughter DNA molecules, and the stem of the Y is the parental DNA helix. In this diagram, parental strands are *orange;* newly synthesized strands are *red.* (Micrograph courtesy of Jerome Vinograd.)

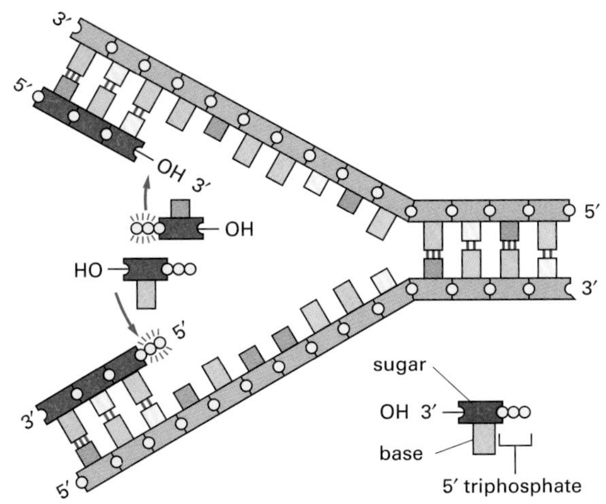

Figure 5–7 An incorrect model for DNA replication. Although it might seem to be the simplest possible model for DNA replication, the mechanism illustrated here is not the one that cells use. In this scheme, both daughter DNA strands would grow continuously, using the energy of hydrolysis of the two terminal phosphates *(yellow circles* highlighted by *red rays)* to add the next nucleotide on each strand. This would require chain growth in both the 5′-to-3′ direction *(top)* and the 3′-to-5′ direction *(bottom)*. No enzyme that catalyzes 3′-to-5′ nucleotide polymerization has ever been found.

by a discontinuous "backstitching" mechanism means that only the 5′-to-3′ type of DNA polymerase is needed for DNA replication.

The High Fidelity of DNA Replication Requires Several Proofreading Mechanisms

As discussed at the beginning of this chapter, the fidelity of copying DNA during replication is such that only about 1 mistake is made for every 10^9 nucleotides copied. This fidelity is much higher than one would expect, on the basis of the accuracy of complementary base-pairing. The standard complementary base pairs (see Figure 4–4) are not the only ones possible. For example, with small changes in helix geometry, two hydrogen bonds can form between G and T in DNA. In addition, rare tautomeric forms of the four DNA bases occur transiently in ratios of 1 part to 10^4 or 10^5. These forms mispair without a change in helix geometry: the rare tautomeric form of C pairs with A instead of G, for example.

If the DNA polymerase did nothing special when a mispairing occurred between an incoming deoxyribonucleoside triphosphate and the DNA template, the wrong nucleotide would often be incorporated into the new DNA chain, producing frequent mutations. The high fidelity of DNA replication, however, depends not only on complementary base-pairing but also on several "proofreading" mechanisms that act sequentially to correct any initial mispairing that might have occurred.

The first proofreading step is carried out by the DNA polymerase, and it occurs just before a new nucleotide is added to the growing chain. Our knowledge of this mechanism comes from studies of several different DNA polymerases, including one produced by a bacterial virus, T7, that replicates inside *E. coli*. The correct nucleotide has a higher affinity for the moving polymerase than does the incorrect nucleotide, because only the correct nucleotide can correctly base-pair with the template. Moreover, after nucleotide binding, but before the nucleotide is covalently added to the growing chain, the enzyme must undergo a conformational change. An incorrectly bound nucleotide is more likely to dissociate during this step than the correct one. This step therefore allows the polymerase to "double-check" the exact base-pair geometry before it catalyzes the addition of the nucleotide.

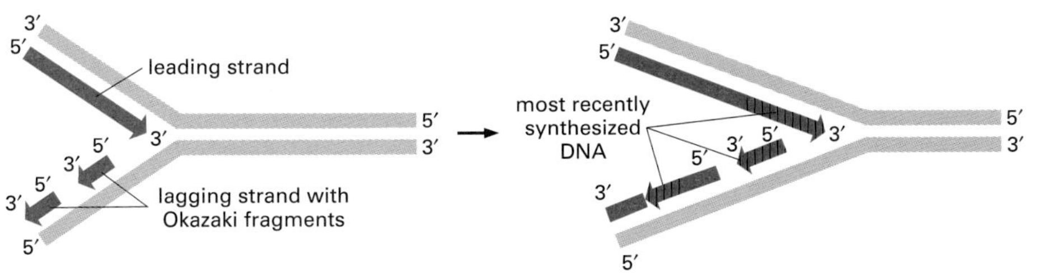

Figure 5–8 The structure of a DNA replication fork. Because both daughter DNA strands are polymerized in the 5′-to-3′ direction, the DNA synthesized on the lagging strand must be made initially as a series of short DNA molecules, called *Okazaki fragments*.

The next error-correcting reaction, known as *exonucleolytic proofreading*, takes place immediately after those rare instances in which an incorrect nucleotide is covalently added to the growing chain. DNA polymerase enzymes cannot begin a new polynucleotide chain by linking two nucleoside triphosphates together. Instead, they absolutely require a base-paired 3′-OH end of a *primer strand* on which to add further nucleotides (see Figure 5–4). Those DNA molecules with a mismatched (improperly base-paired) nucleotide at the 3′-OH end of the primer strand are not effective as templates because the polymerase cannot extend such a strand. DNA polymerase molecules deal with such a mismatched primer strand by means of a separate catalytic site (either in a separate subunit or in a separate domain of the polymerase molecule, depending on the polymerase). This *3′-to-5′ proofreading exonuclease* clips off any unpaired residues at the primer terminus, continuing until enough nucleotides have been removed to regenerate a base-paired 3′-OH terminus that can prime DNA synthesis. In this way, DNA polymerase functions as a "self-correcting" enzyme that removes its own polymerization errors as it moves along the DNA (Figures 5–9 and 5–10).

The requirement for a perfectly base-paired primer terminus is essential to the self-correcting properties of the DNA polymerase. It is apparently not possible for such an enzyme to start synthesis in the complete absence of a primer without losing any of its discrimination between base-paired and unpaired growing 3′-OH termini. By contrast, the RNA polymerase enzymes involved in gene transcription do not need efficient exonucleolytic proofreading: errors in making RNA are not passed on to the next generation, and the occasional defective RNA molecule that is produced has no long-term significance. RNA polymerases are thus able to start new polynucleotide chains without a primer.

An error frequency of about 1 in 10^4 is found both in RNA synthesis and in the separate process of translating mRNA sequences into protein sequences. This level of mistakes is 100,000 times greater than that in DNA replication, where a series of proofreading processes makes the process remarkably accurate (Table 5–1).

Only DNA Replication in the 5′-to-3′ Direction Allows Efficient Error Correction

The need for accuracy probably explains why DNA replication occurs only in the 5′-to-3′ direction. If there were a DNA polymerase that added deoxyribonucleoside triphosphates in the 3′-to-5′ direction, the growing 5′-chain end, rather than the incoming mononucleotide, would carry the activating triphosphate. In this

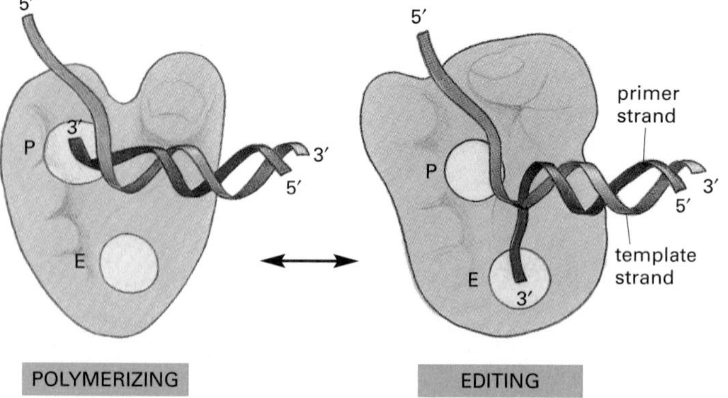

Figure 5–10 Editing by DNA polymerase. Outline of the structures of DNA polymerase complexed with the DNA template in the polymerizing mode *(left)* and the editing mode *(right)*. The catalytic site for the exonucleolytic (E) and the polymerization (P) reactions are indicated. To determine these structures by x-ray crystallography, researchers "froze" the polymerases in these two states, by using either a mutant polymerase defective in the exonucleolytic domain *(right)*, or by withholding the Mg^{2+} required for polymerization *(left)*.

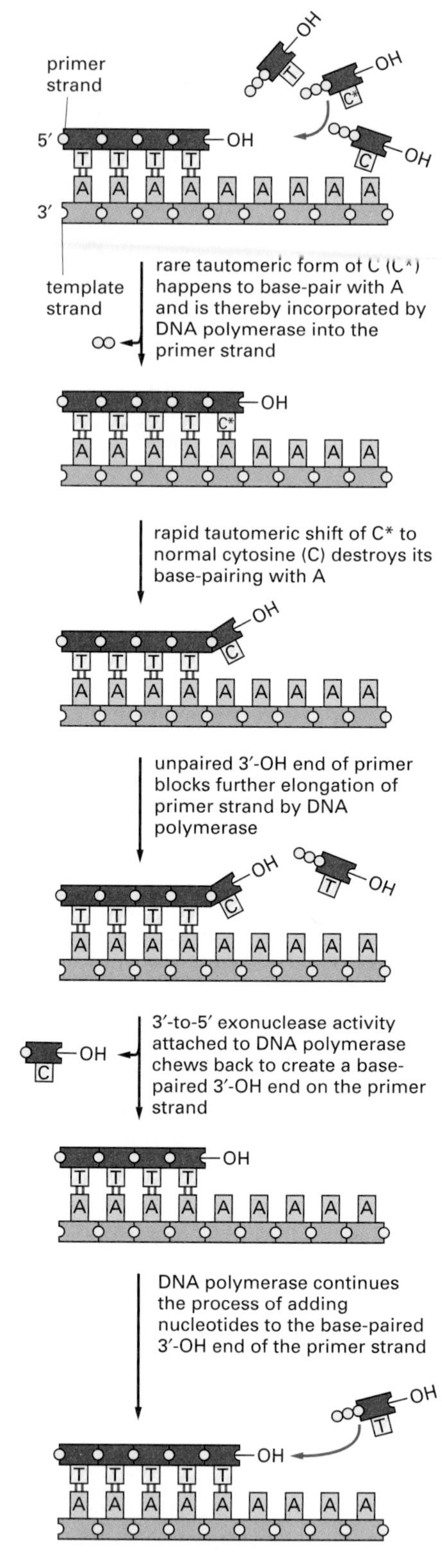

Figure 5–9 Exonucleolytic proofreading by DNA polymerase during DNA replication. In this example, the mismatch is due to the incorporation of a rare, transient tautomeric form of C, indicated by an asterisk. But the same proofreading mechanism applies to any misincorporation at the growing 3′-OH end.

TABLE 5–1 The Three Steps That Give Rise to High-Fidelity DNA Synthesis

REPLICATION STEP	ERRORS PER NUCLEOTIDE POLYMERIZED
5′→3′ polymerization	1×10^5
3′→5′ exonucleolytic proofreading	1×10^2
Strand-directed mismatch repair	1×10^2
Total	1×10^9

The third step, strand-directed mismatch repair, is described later in this chapter.

case, the mistakes in polymerization could not be simply hydrolyzed away, because the bare 5′-chain end thus created would immediately terminate DNA synthesis (Figure 5–11). It is therefore much easier to correct a mismatched base that has just been added to the 3′ end than one that has just been added to the 5′ end of a DNA chain. Although the mechanism for DNA replication (see Figure 5–8) seems at first sight much more complex than the incorrect mechanism depicted earlier in Figure 5–7, it is much more accurate because all DNA synthesis occurs in the 5′-to-3′ direction.

Despite these safeguards against DNA replication errors, DNA polymerases occasionally make mistakes. However, as we shall see later, cells have yet another chance to correct these errors by a process called *strand-directed mismatch repair.* Before discussing this mechanism, however, we describe the other types of proteins that function at the replication fork.

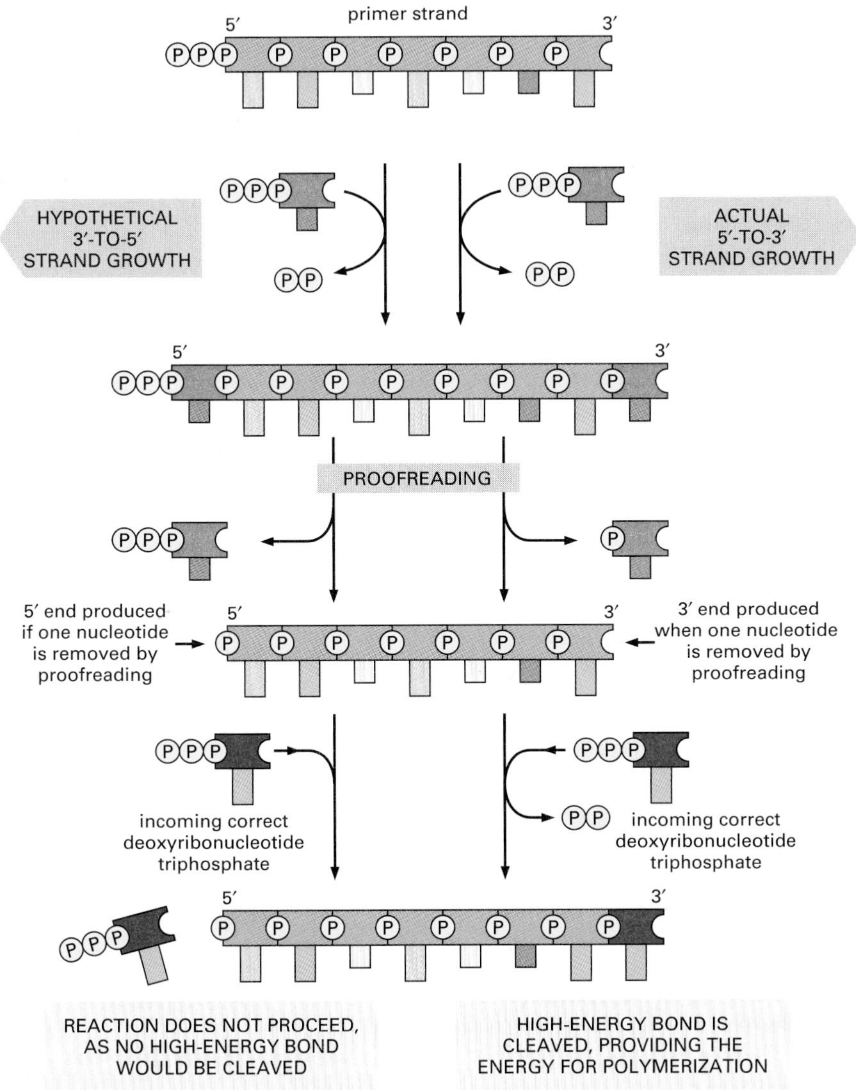

Figure 5–11 An explanation for the 5′-to-3′ direction of DNA chain growth. Growth in the 5′-to-3′ direction, shown on the right, allows the chain to continue to be elongated when a mistake in polymerization has been removed by exonucleolytic proofreading (see Figure 5–9). In contrast, exonucleolytic proofreading in the hypothetical 3′-to-5′ polymerization scheme, shown on the left, would block further chain elongation. For convenience, only the primer strand of the DNA double helix is shown.

A Special Nucleotide-Polymerizing Enzyme Synthesizes Short RNA Primer Molecules on the Lagging Strand

For the leading strand, a special primer is needed only at the start of replication: once a replication fork is established, the DNA polymerase is continuously presented with a base-paired chain end on which to add new nucleotides. On the lagging side of the fork, however, every time the DNA polymerase completes a short DNA Okazaki fragment (which takes a few seconds), it must start synthesizing a completely new fragment at a site further along the template strand (see Figure 5–8). A special mechanism is used to produce the base-paired primer strand required by this DNA polymerase molecule. The mechanism involves an enzyme called **DNA primase**, which uses ribonucleoside triphosphates to synthesize short **RNA primers** on the lagging strand (Figure 5–12). In eucaryotes, these primers are about 10 nucleotides long and are made at intervals of 100–200 nucleotides on the lagging strand.

The chemical structure of RNA was introduced in Chapter 1 and described in detail in Chapter 6. Here, we note only that RNA is very similar in structure to DNA. A strand of RNA can form base pairs with a strand of DNA, generating a DNA/RNA hybrid double helix if the two nucleotide sequences are complementary. The synthesis of RNA primers is thus guided by the same templating principle used for DNA synthesis (see Figures 1–5 and 5–2).

Because an RNA primer contains a properly base-paired nucleotide with a 3′-OH group at one end, it can be elongated by the DNA polymerase at this end to begin an Okazaki fragment. The synthesis of each Okazaki fragment ends when this DNA polymerase runs into the RNA primer attached to the 5′ end of the previous fragment. To produce a continuous DNA chain from the many DNA fragments made on the lagging strand, a special DNA repair system acts quickly to erase the old RNA primer and replace it with DNA. An enzyme called **DNA ligase** then joins the 3′ end of the new DNA fragment to the 5′ end of the previous one to complete the process (Figures 5–13 and 5–14).

Why might an erasable RNA primer be preferred to a DNA primer that would not need to be erased? The argument that a self-correcting polymerase cannot start chains *de novo* also implies its converse: an enzyme that starts chains anew cannot be efficient at self-correction. Thus, any enzyme that primes the synthesis of Okazaki fragments will of necessity make a relatively inaccurate copy (at least 1 error in 10^5). Even if the copies retained in the final product constituted as little as 5% of the total genome (for example, 10 nucleotides per 200-nucleotide DNA fragment), the resulting increase in the overall mutation rate would be enormous. It therefore seems likely that the evolution of RNA rather than DNA for priming brought a powerful advantage to the cell: the ribonucleotides in the primer automatically mark these sequences as "suspect copy" to be efficiently removed and replaced.

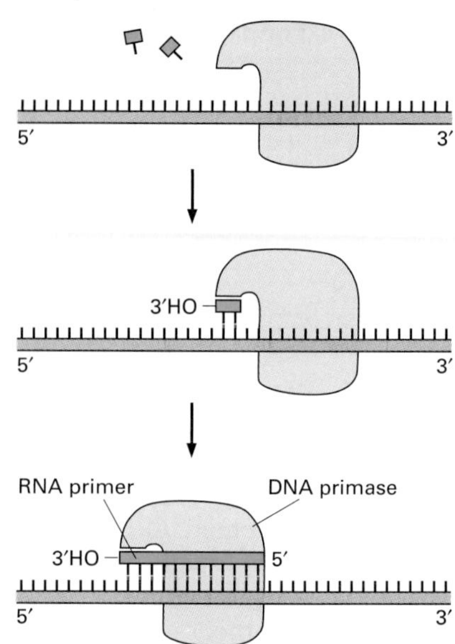

Figure 5–12 RNA primer synthesis. A schematic view of the reaction catalyzed by *DNA primase*, the enzyme that synthesizes the short RNA primers made on the lagging strand using DNA as a template. Unlike DNA polymerase, this enzyme can start a new polynucleotide chain by joining two nucleoside triphosphates together. The primase synthesizes a short polynucleotide in the 5′-to-3′ direction and then stops, making the 3′ end of this primer available for the DNA polymerase.

Special Proteins Help to Open Up the DNA Double Helix in Front of the Replication Fork

For DNA synthesis to proceed, the DNA double helix must be opened up ahead of the replication fork so that the incoming deoxyribonucleoside triphosphates can form base pairs with the template strand. However, the DNA double helix is very stable under normal conditions; the base pairs are locked in place so strongly that temperatures approaching that of boiling water are required to separate the two strands in a test tube. For this reason, DNA polymerases and

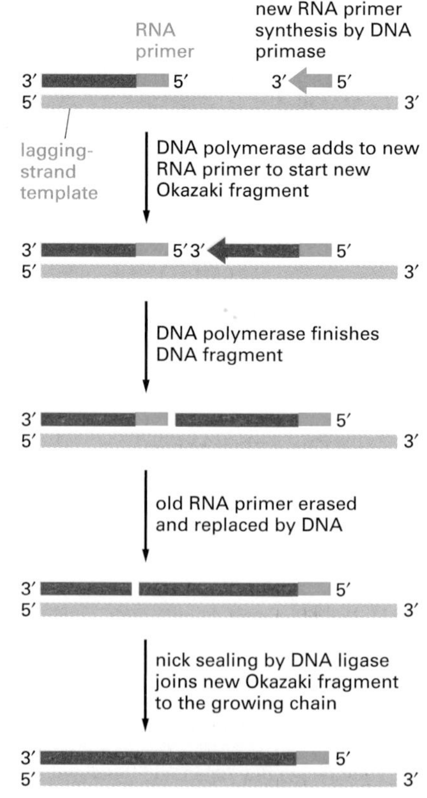

Figure 5–13 The synthesis of one of the many DNA fragments on the lagging strand. In eucaryotes, RNA primers are made at intervals spaced by about 200 nucleotides on the lagging strand, and each RNA primer is approximately 10 nucleotides long. This primer is erased by a special DNA repair enzyme (an RNAse H) that recognizes an RNA strand in an RNA/DNA helix and fragments it; this leaves gaps that are filled in by DNA polymerase and DNA ligase.

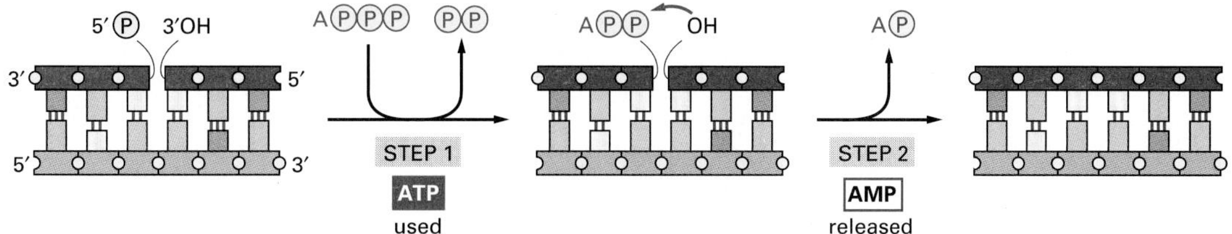

DNA primases can copy a DNA double helix only when the template strand has already been exposed by separating it from its complementary strand. Additional replication proteins are needed to help in opening the double helix and thus provide the appropriate single-stranded DNA template for the DNA polymerase to copy. Two types of protein contribute to this process—DNA helicases and single-strand DNA-binding proteins.

DNA helicases were first isolated as proteins that hydrolyze ATP when they are bound to single strands of DNA. As described in Chapter 3, the hydrolysis of ATP can change the shape of a protein molecule in a cyclical manner that allows the protein to perform mechanical work. DNA helicases use this principle to propel themselves rapidly along a DNA single strand. When they encounter a region of double helix, they continue to move along their strand, thereby prying apart the helix at rates of up to 1000 nucleotide pairs per second (Figures 5–15 and 5–16).

The unwinding of the template DNA helix at a replication fork could in principle be catalyzed by two DNA helicases acting in concert—one running along the leading strand template and one along the lagging strand template. Since the two strands have opposite polarities, these helicases would need to move in opposite directions along a DNA single strand and therefore would be different enzymes. Both types of DNA helicase exist. In the best understood replication systems, a helicase on the lagging-strand template appears to have the predominant role, for reasons that will become clear shortly.

Figure 5–14 The reaction catalyzed by DNA ligase. This enzyme seals a broken phosphodiester bond. As shown, DNA ligase uses a molecule of ATP to activate the 5′ end at the nick (step 1) before forming the new bond (step 2). In this way, the energetically unfavorable nick-sealing reaction is driven by being coupled to the energetically favorable process of ATP hydrolysis.

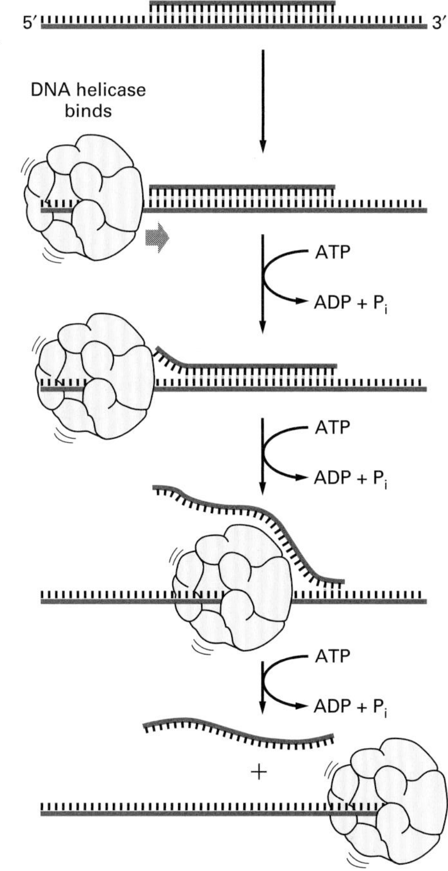

Figure 5–15 An assay used to test for DNA helicase enzymes. A short DNA fragment is annealed to a long DNA single strand to form a region of DNA double helix. The double helix is melted as the helicase runs along the DNA single strand, releasing the short DNA fragment in a reaction that requires the presence of both the helicase protein and ATP. The rapid step-wise movement of the helicase is powered by its ATP hydrolysis (see Figure 3–76).

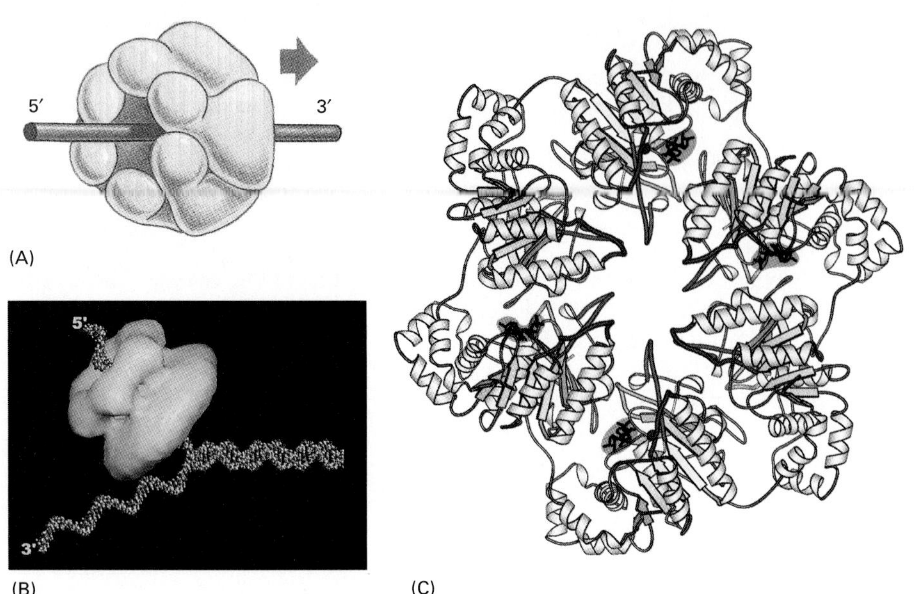

Figure 5–16 The structure of a DNA helicase. (A) A schematic diagram of the protein as a hexameric ring. (B) Schematic diagram showing a DNA replication fork and helicase to scale. (C) Detailed structure of the bacteriophage T7 replicative helicase, as determined by x-ray diffraction. Six identical subunits bind and hydrolyze ATP in an ordered fashion to propel this molecule along a DNA single strand that passes through the central hole. *Red* indicates bound ATP molecules in the structure. (B, courtesy of Edward H. Egelman; C, from M.R. Singleton et al., *Cell* 101:589–600, 2000. © Elsevier.)

Single-strand DNA-binding (SSB) proteins, also called *helix-destabilizing proteins,* bind tightly and cooperatively to exposed single-stranded DNA strands without covering the bases, which therefore remain available for templating. These proteins are unable to open a long DNA helix directly, but they aid helicases by stabilizing the unwound, single-stranded conformation. In addition, their cooperative binding coats and straightens out the regions of single-stranded DNA on the lagging-strand template, thereby preventing the formation of the short hairpin helices that readily form in single-strand DNA (Figures 5–17 and 5–18). These hairpin helices can impede the DNA synthesis catalyzed by DNA polymerase.

A Moving DNA Polymerase Molecule Stays Connected to the DNA by a Sliding Ring

On their own, most DNA polymerase molecules will synthesize only a short string of nucleotides before falling off the DNA template. The tendency to dissociate quickly from a DNA molecule allows a DNA polymerase molecule that has just finished synthesizing one Okazaki fragment on the lagging strand to be recycled quickly, so as to begin the synthesis of the next Okazaki fragment on the same strand. This rapid dissociation, however, would make it difficult for the polymerase to synthesize the long DNA strands produced at a replication fork were it not for an accessory protein that functions as a regulated clamp. This

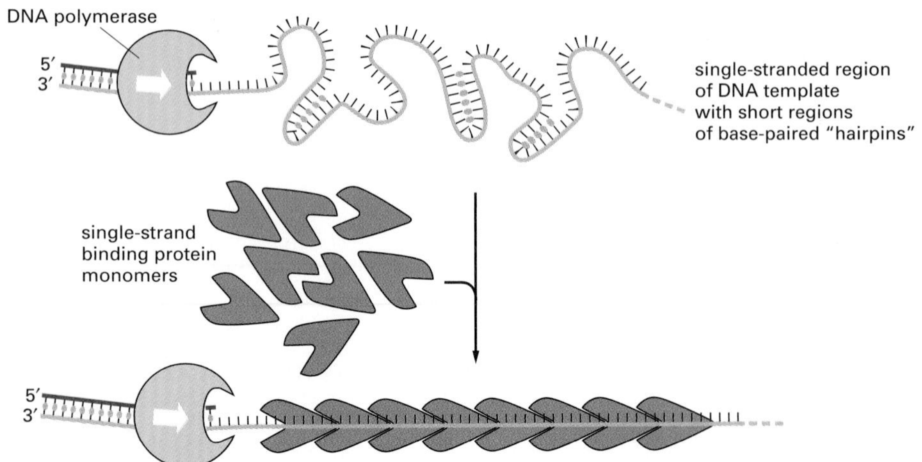

Figure 5–17 The effect of single-strand DNA-binding proteins (SSB proteins) on the structure of single-stranded DNA. Because each protein molecule prefers to bind next to a previously bound molecule, long rows of this protein form on a DNA single strand. This *cooperative binding* straightens out the DNA template and facilitates the DNA polymerization process. The "hairpin helices" shown in the bare, single-stranded DNA result from a chance matching of short regions of complementary nucleotide sequence; they are similar to the short helices that typically form in RNA molecules (see Figure 1–6).

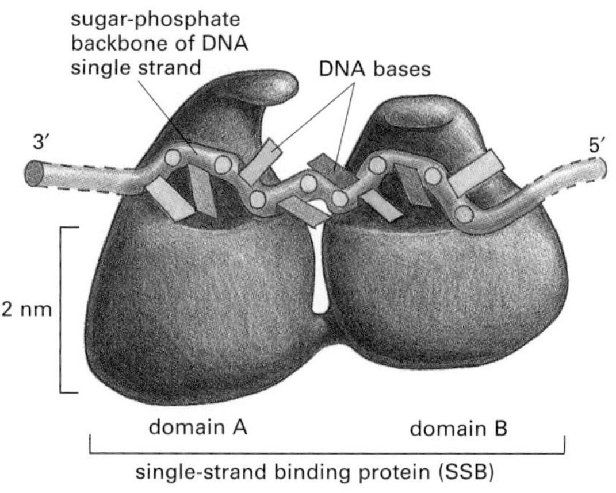

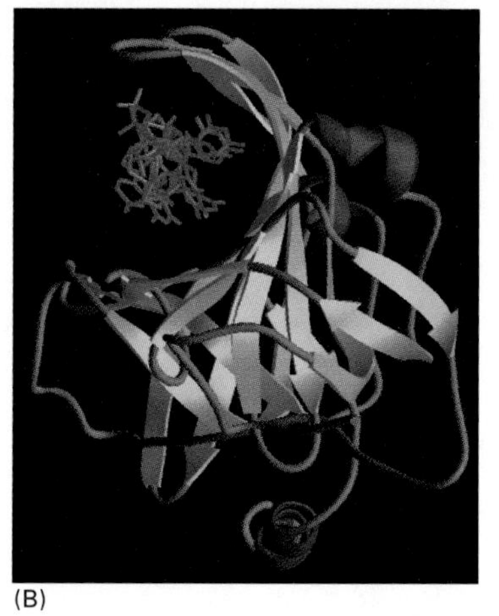

Figure 5–18 The structure of the single-strand binding protein from humans bound to DNA. (A) A front view of the two DNA binding domains of RPA protein, which cover a total of eight nucleotides. Note that the DNA bases remain exposed in this protein–DNA complex. (B) A diagram showing the three-dimensional structure, with the DNA strand *(red)* viewed end-on. (B, from A. Bochkarev et al., *Nature* 385:176–181, 1997. © Macmillan Magazines Ltd.)

clamp keeps the polymerase firmly on the DNA when it is moving, but releases it as soon as the polymerase runs into a double-stranded region of DNA ahead.

How can a clamp prevent the polymerase from dissociating without at the same time impeding the polymerase's rapid movement along the DNA molecule? The three-dimensional structure of the clamp protein, determined by x-ray diffraction, reveals that it forms a large ring around the DNA helix. One side of the ring binds to the back of the DNA polymerase, and the whole ring slides freely along the DNA as the polymerase moves. The assembly of the clamp around DNA requires ATP hydrolysis by a special protein complex, the clamp loader, which hydrolyzes ATP as it loads the clamp on to a primer-template junction (Figure 5–19).

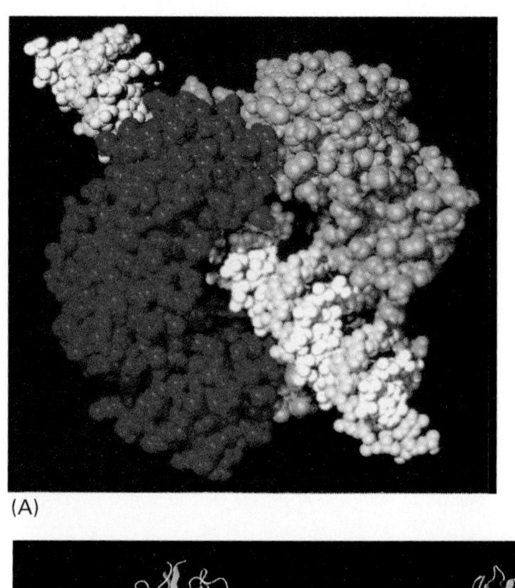

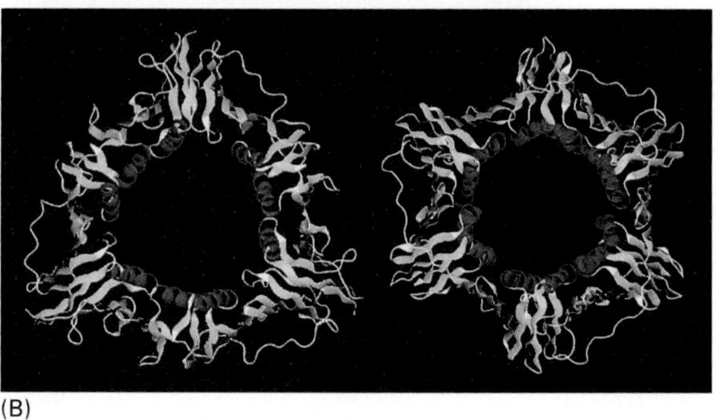

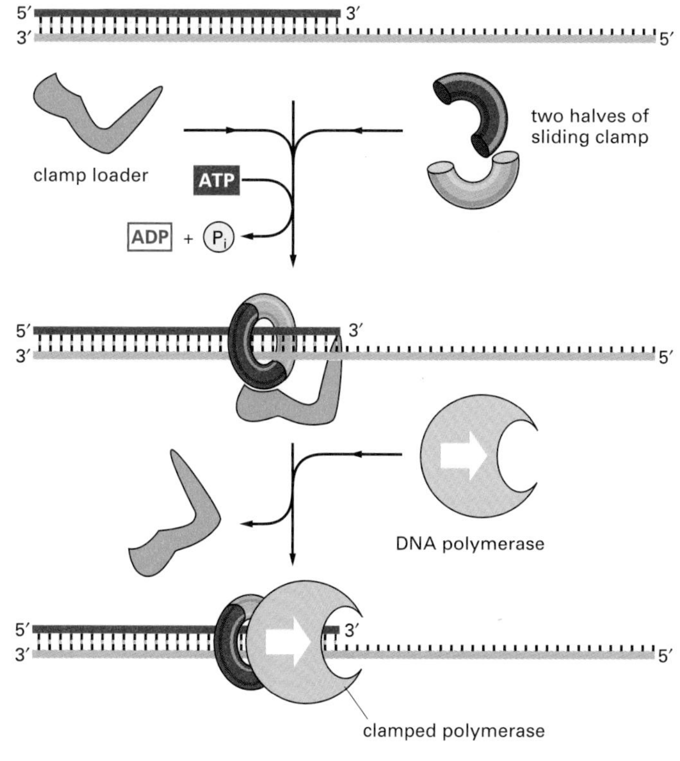

Figure 5–19 The regulated sliding clamp that holds DNA polymerase on the DNA. (A) The structure of the clamp protein from *E. coli*, as determined by x-ray crystallography, with a DNA helix added to indicate how the protein fits around DNA. (B) A similar protein is present in eucaryotes, as illustrated by this comparison of the *E. coli* sliding clamp *(left)* with the PCNA protein from humans *(right)*. (C) Schematic illustration showing how the clamp is assembled to hold a moving DNA polymerase molecule on the DNA. In the simplified reaction shown here, the clamp loader dissociates into solution once the clamp has been assembled. At a true replication fork, the clamp loader remains close to the lagging-strand polymerase, ready to assemble a new clamp at the start of each new Okazaki fragment (see Figure 5–22). (A and B, from X.-P. Kong et al., *Cell* 69:425–437, 1992. © Elsevier.)

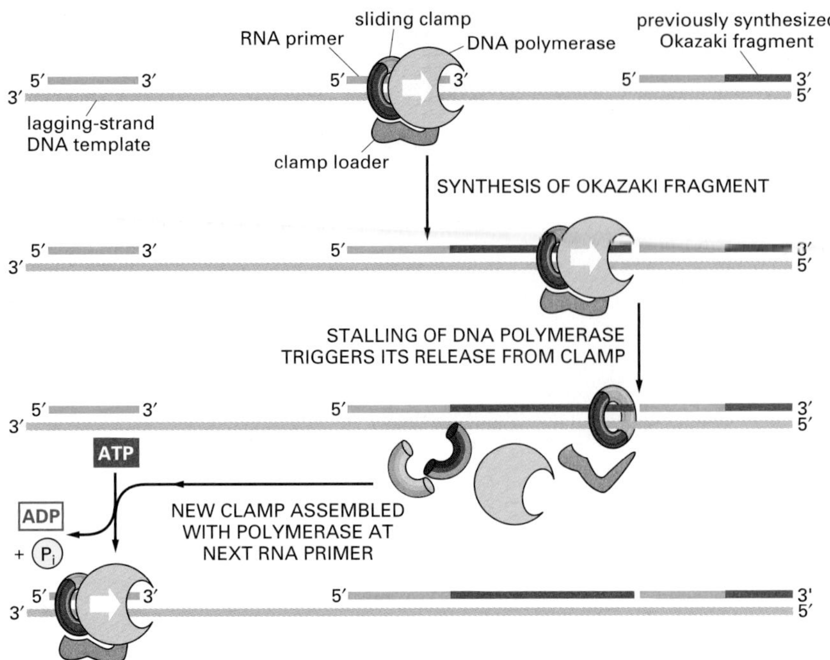

Figure 5–20 A cycle of loading and unloading of DNA polymerase and the clamp protein on the lagging strand. The association of the clamp loader with the lagging-strand polymerase shown here is for illustrative purposes only; in reality, the clamp loader is carried along with the replication fork in a complex that includes both the leading-strand and lagging-strand DNA polymerases (see Figure 5–22).

On the leading-strand template, the moving DNA polymerase is tightly bound to the clamp, and the two remain associated for a very long time. However, on the lagging-strand template, each time the polymerase reaches the 5′ end of the preceding Okazaki fragment, the polymerase is released; this polymerase molecule then associates with a new clamp that is assembled on the RNA primer of the next Okazaki fragment (Figure 5–20).

The Proteins at a Replication Fork Cooperate to Form a Replication Machine

Although we have discussed DNA replication as though it were performed by a mixture of proteins all acting independently, in reality, most of the proteins are held together in a large multienzyme complex that moves rapidly along the DNA. This complex can be likened to a tiny sewing machine composed of protein parts and powered by nucleoside triphosphate hydrolyses. Although the replication complex has been most intensively studied in *E. coli* and several of its viruses, a very similar complex also operates in eucaryotes, as we see below.

The functions of the subunits of the replication machine are summarized in Figure 5–21. Two DNA polymerase molecules work at the fork, one on the leading strand and one on the lagging strand. The DNA helix is opened by a DNA polymerase molecule clamped on the leading strand, acting in concert with one or more DNA helicase molecules running along the strands in front of it. Helix opening is aided by cooperatively bound molecules of single-strand DNA-binding protein. Whereas the DNA polymerase molecule on the leading strand can operate in a continuous fashion, the DNA polymerase molecule on the lagging strand must restart at short intervals, using a short RNA primer made by a DNA primase molecule.

The efficiency of replication is greatly increased by the close association of all these protein components. In procaryotes, the primase molecule is linked directly to a DNA helicase to form a unit on the lagging strand called a **primosome**. Powered by the DNA helicase, the primosome moves with the fork, synthesizing RNA primers as it goes. Similarly, the DNA polymerase molecule that synthesizes DNA on the lagging strand moves in concert with the rest of the proteins, synthesizing a succession of new Okazaki fragments. To accommodate this arrangement, the lagging strand seems to be folded back in the manner shown in Figure 5–22. This arrangement also facilitates the loading of the

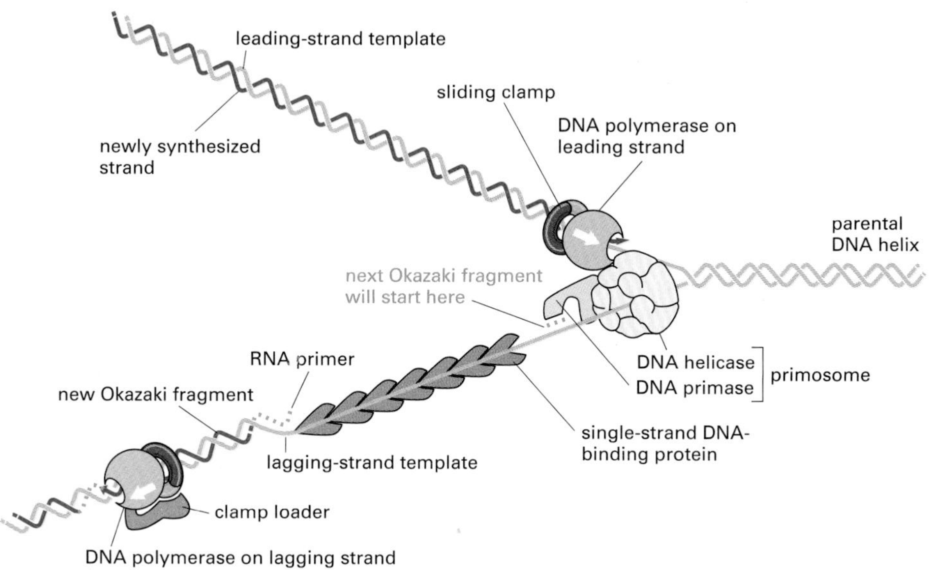

Figure 5–21 The proteins at a bacterial DNA replication fork. The major types of proteins that act at a DNA replication fork are illustrated, showing their approximate positions on the DNA.

polymerase clamp each time that an Okazaki fragment is synthesized: the clamp loader and the lagging-strand DNA polymerase molecule are kept in place as a part of the protein machine even when they detach from the DNA. The replication proteins are thus linked together into a single large unit (total molecular weight $>10^6$ daltons) that moves rapidly along the DNA, enabling DNA to be synthesized on both sides of the replication fork in a coordinated and efficient manner.

On the lagging strand, the DNA replication machine leaves behind a series of unsealed Okazaki fragments, which still contain the RNA that primed their synthesis at their 5′ ends. This RNA is removed and the resulting gap is filled in by DNA repair enzymes that operate behind the replication fork (see Figure 5–13).

A Strand-directed Mismatch Repair System Removes Replication Errors That Escape from the Replication Machine

As stated previously, bacteria such as *E. coli* are capable of dividing once every 30 minutes, making it relatively easy to screen large populations to find a rare mutant cell that is altered in a specific process. One interesting class of mutants contains alterations in so-called *mutator genes,* which greatly increase the rate of spontaneous mutation when they are inactivated. Not surprisingly, one such mutant makes a defective form of the 3′-to-5′ proofreading exonuclease that is a part of the DNA polymerase enzyme (see Figures 5–9 and 5–10). When this activity is defective, the DNA polymerase no longer proofreads effectively, and many replication errors that would otherwise have been removed accumulate in the DNA.

The study of other *E. coli* mutants exhibiting abnormally high mutation rates has uncovered another proofreading system that removes replication errors made by the polymerase that have been missed by the proofreading exonuclease. This **strand-directed mismatch repair** system detects the potential for distortion in the DNA helix that results from the misfit between noncomplementary base pairs. But if the proofreading system simply recognized a mismatch in newly replicated DNA and randomly corrected one of the two mismatched nucleotides, it would mistakingly "correct" the original template strand to match the error exactly half the time, thereby failing to lower the overall error rate. To be effective, such a proofreading system must be able to distinguish and remove the mismatched nucleotide only on the newly synthesized strand, where the replication error occurred.

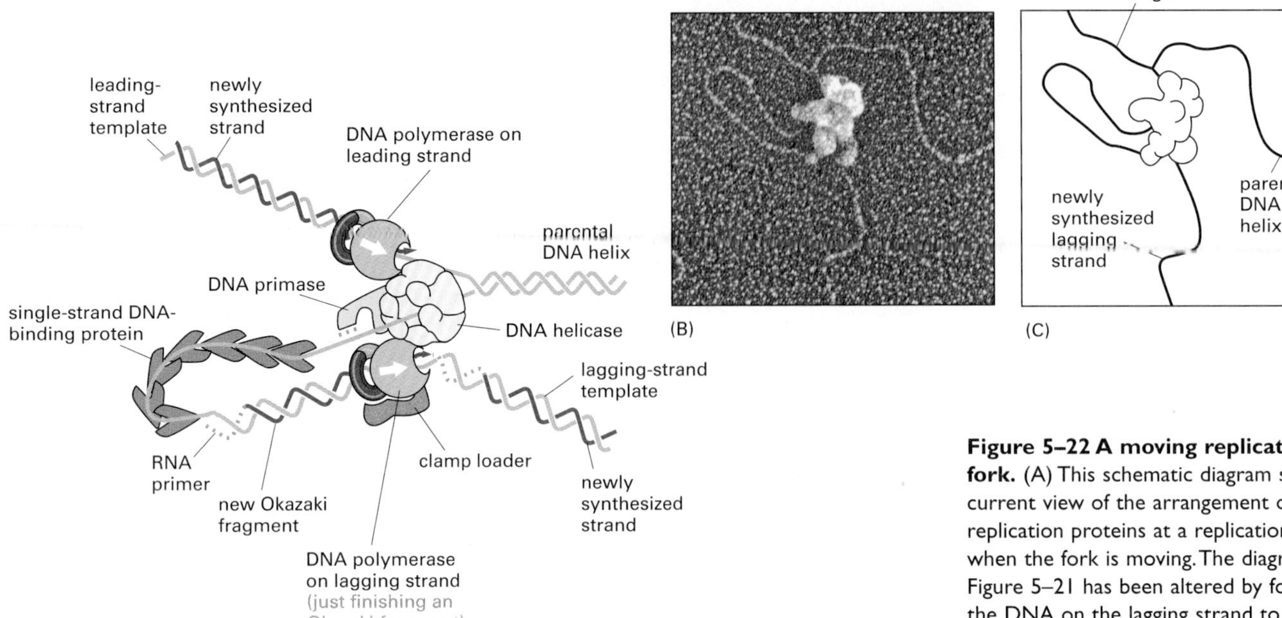

Figure 5–22 A moving replication fork. (A) This schematic diagram shows a current view of the arrangement of replication proteins at a replication fork when the fork is moving. The diagram in Figure 5–21 has been altered by folding the DNA on the lagging strand to bring the lagging-strand DNA polymerase molecule into a complex with the leading-strand DNA polymerase molecule. This folding process also brings the 3′ end of each completed Okazaki fragment close to the start site for the next Okazaki fragment (compare with Figure 5–21). Because the lagging-strand DNA polymerase molecule remains bound to the rest of the replication proteins, it can be reused to synthesize successive Okazaki fragments. In this diagram, it is about to let go of its completed DNA fragment and move to the RNA primer that will be synthesized nearby, as required to start the next DNA fragment. Note that one daughter DNA helix extends toward the bottom right and the other toward the top left in this diagram. Additional proteins help to hold the different protein components of the fork together, enabling them to function as a well-coordinated protein machine. The actual protein complex is more compact than indicated, and the clamp loader is held in place by interactions not shown here. (B) An electron micrograph showing the replication machine from the bacteriophage T4 as it moves along a template synthesizing DNA behind it. (C) An interpretation of the micrograph is given in the sketch: note especially the DNA loop on the lagging strand. Apparently, the replication proteins became partly detached from the very front of the replication fork during the preparation of this sample for electron microscopy. (B, courtesy of Jack Griffith.)

The strand-distinction mechanism used by the mismatch proofreading system in *E. coli* depends on the methylation of selected A residues in the DNA. Methyl groups are added to all A residues in the sequence GATC, but not until some time after the A has been incorporated into a newly synthesized DNA chain. As a result, the only GATC sequences that have not yet been methylated are in the new strands just behind a replication fork. The recognition of these unmethylated GATCs allows the new DNA strands to be transiently distinguished from old ones, as required if their mismatches are to be selectively removed. The three-step process involves recognition of a mismatch, excision of the segment of DNA containing the mismatch from the newly synthesized strand, and resynthesis of the excised segment using the old strand as a template—thereby removing the mismatch. This strand-directed mismatch repair system reduces the number of errors made during DNA replication by an additional factor of 10^2 (see Table 5–1, p. 243).

A similar mismatch proofreading system functions in human cells. The importance of this system is indicated by the fact that individuals who inherit one defective copy of a mismatch repair gene (along with a functional gene on the other copy of the chromosome) have a marked predisposition for certain types of cancers. In a type of colon cancer called *hereditary nonpolyposis colon cancer (HNPCC),* spontaneous mutation of the remaining functional gene produces a clone of somatic cells that, because they are deficient in mismatch proofreading, accumulate mutations unusually rapidly. Most cancers arise from cells that have accumulated multiple mutations (discussed in Chapter 23), and cells deficient in mismatch proofreading therefore have a greatly enhanced chance of becoming cancerous. Fortunately, most of us inherit two good copies of each gene that encodes a mismatch proofreading protein; this protects us, because it is highly unlikely that both copies would mutate in the same cell.

In eucaryotes, the mechanism for distinguishing the newly synthesized strand from the parental template strand at the site of a mismatch does not depend on DNA methylation. Indeed, some eucaryotes—including yeasts and *Drosophila*—do not methylate any of their DNA. Newly synthesized DNA strands are known to be preferentially *nicked,* and biochemical experiments reveal that such *nicks* (also called *single-strand breaks)* provide the signal that directs the mismatch proofreading system to the appropriate strand in a eucaryotic cell (Figure 5–23).

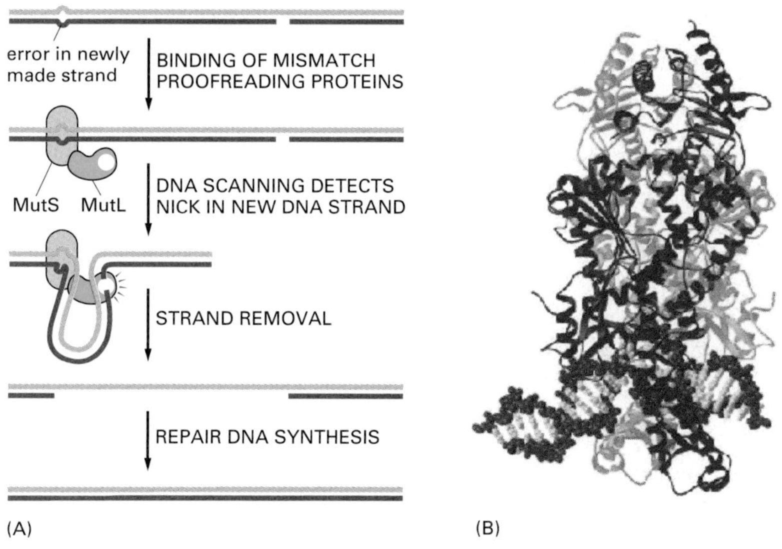

Figure 5–23 A model for strand-directed mismatch repair in eucaryotes. (A) The two proteins shown are present in both bacteria and eucaryotic cells: MutS binds specifically to a mismatched base pair, while MutL scans the nearby DNA for a nick. Once a nick is found, MutL triggers the degradation of the nicked strand all the way back through the mismatch. Because nicks are largely confined to newly replicated strands in eucaryotes, replication errors are selectively removed. In bacteria, the mechanism is the same, except that an additional protein in the complex (MutH) nicks unmethylated (and therefore newly replicated) GATC sequences, thereby beginning the process illustrated here. (B) The structure of the MutS protein bound to a DNA mismatch. This protein is a dimer, which grips the DNA double helix as shown, kinking the DNA at the mismatched base pair. It seems that the MutS protein scans the DNA for mismatches by testing for sites that can be readily kinked, which are those without a normal complementary base pair. (B, from G. Obmolova et al., *Nature* 407:703–710, 2000. © Macmillan Magazines Ltd.)

DNA Topoisomerases Prevent DNA Tangling During Replication

As a replication fork moves along double-stranded DNA, it creates what has been called the "winding problem." Every 10 base pairs replicated at the fork corresponds to one complete turn about the axis of the parental double helix. Therefore, for a replication fork to move, the entire chromosome ahead of the fork would normally have to rotate rapidly (Figure 5–24). This would require large amounts of energy for long chromosomes, and an alternative strategy is used instead: a swivel is formed in the DNA helix by proteins known as **DNA topoisomerases**.

A DNA topoisomerase can be viewed as a reversible nuclease that adds itself covalently to a DNA backbone phosphate, thereby breaking a phosphodiester bond in a DNA strand. This reaction is reversible, and the phosphodiester bond re-forms as the protein leaves.

One type of topoisomerase, called *topoisomerase I,* produces a transient single-strand break (or nick); this break in the phosphodiester backbone allows the two sections of DNA helix on either side of the nick to rotate freely relative to each other, using the phosphodiester bond in the strand opposite the nick as a swivel point (Figure 5–25). Any tension in the DNA helix will drive this rotation in the direction that relieves the tension. As a result, DNA replication can occur with the rotation of only a short length of helix—the part just ahead of the fork. The analogous winding problem that arises during DNA transcription (discussed in

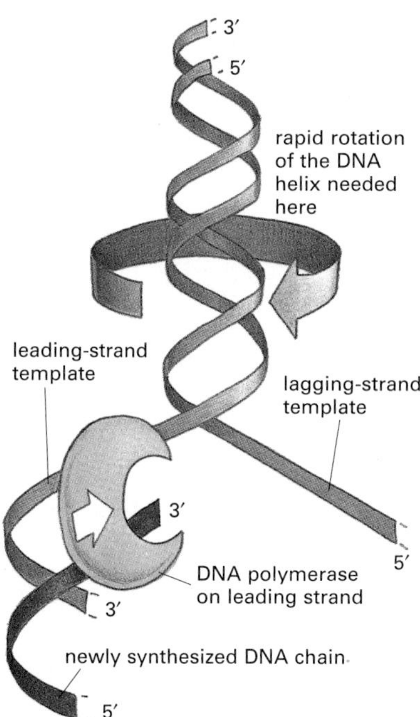

Figure 5–24 The "winding problem" that arises during DNA replication. For a bacterial replication fork moving at 500 nucleotides per second, the parental DNA helix ahead of the fork must rotate at 50 revolutions per second.

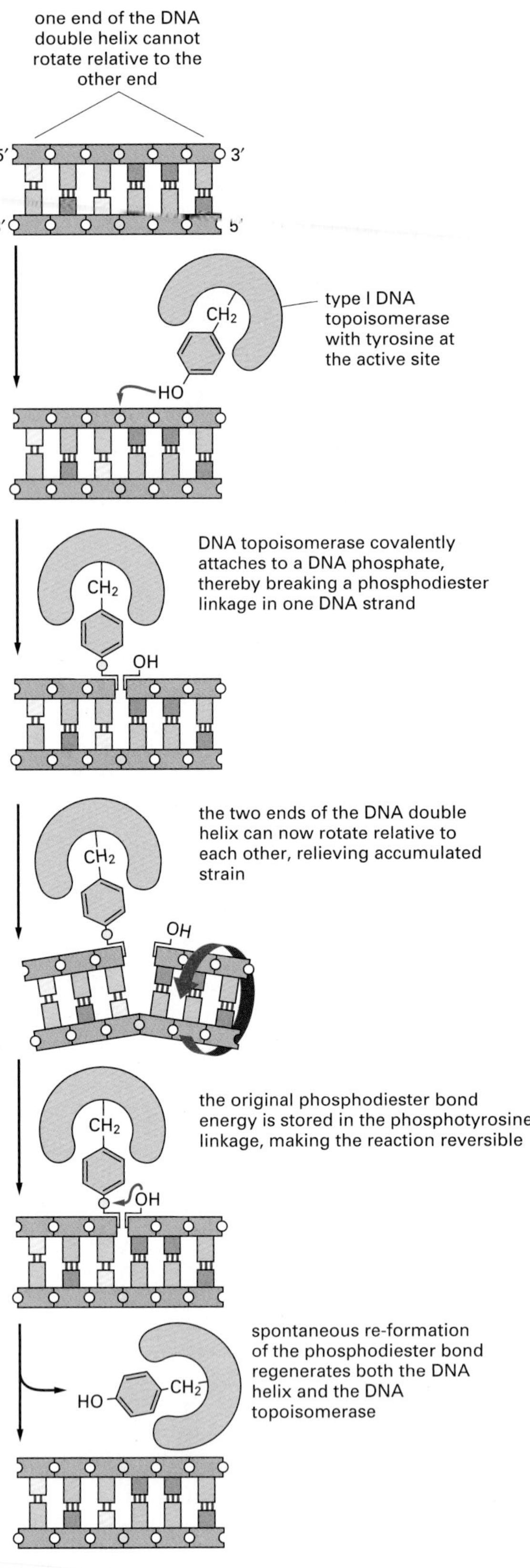
one end of the DNA double helix cannot rotate relative to the other end
5′
3′
3′
5′
type I DNA topoisomerase with tyrosine at the active site
CH2
HO
DNA topoisomerase covalently attaches to a DNA phosphate, thereby breaking a phosphodiester linkage in one DNA strand
CH2
OH
the two ends of the DNA double helix can now rotate relative to each other, relieving accumulated strain
CH2
OH
the original phosphodiester bond energy is stored in the phosphotyrosine linkage, making the reaction reversible
CH2
OH
spontaneous re-formation of the phosphodiester bond regenerates both the DNA helix and the DNA topoisomerase
HO
CH2

Figure 5–25 The reversible nicking reaction catalyzed by a eucaryotic DNA topoisomerase I enzyme. As indicated, these enzymes transiently form a single covalent bond with DNA; this allows free rotation of the DNA around the covalent backbone bonds linked to the *blue* phosphate.

Chapter 6) is solved in a similar way. Because the covalent linkage that joins the DNA topoisomerase protein to a DNA phosphate retains the energy of the cleaved phosphodiester bond, resealing is rapid and does not require additional energy input. In this respect, the rejoining mechanism is different from that catalyzed by the enzyme DNA ligase, discussed previously (see Figure 5–14).

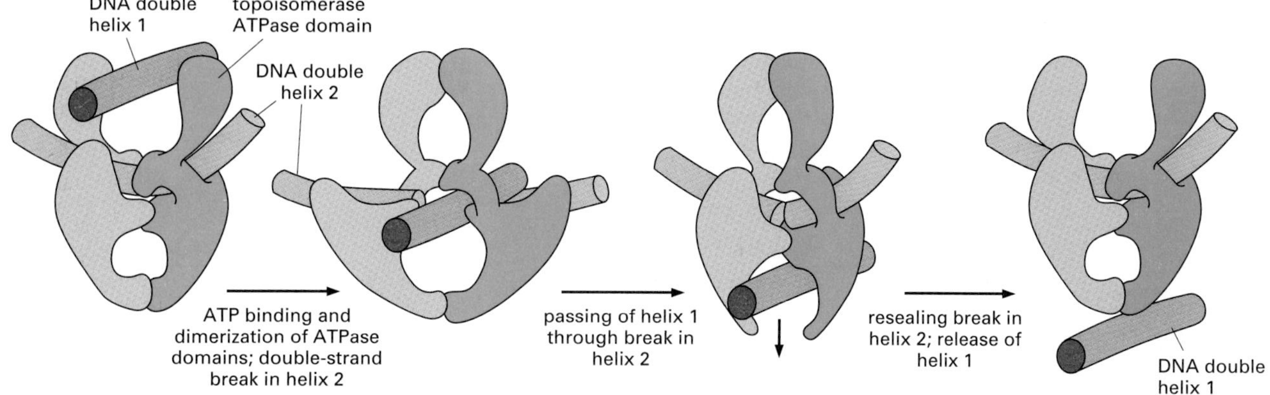

Figure 5–26 A model for topoisomerase II action. As indicated, ATP binding to the two ATPase domains causes them to dimerize and drives the reactions shown. Because a single cycle of this reaction can occur in the presence of a non-hydrolyzable ATP analog, ATP hydrolysis is thought to be needed only to reset the enzyme for each new reaction cycle. This model is based on structural and mechanistic studies of the enzyme. (Modified from J.M. Berger, *Curr. Opin. Struct. Biol.* 8:26–32, 1998.)

A second type of DNA topoisomerase, *topoisomerase II*, forms a covalent linkage to both strands of the DNA helix at the same time, making a transient *double-strand break* in the helix. These enzymes are activated by sites on chromosomes where two double helices cross over each other. Once a topoisomerase II molecule binds to such a crossing site, the protein uses ATP hydrolysis to perform the following set of reactions efficiently: (1) it breaks one double helix reversibly to create a DNA "gate;" (2) it causes the second, nearby double helix to pass through this break; and (3) it then reseals the break and dissociates from the DNA (Figure 5–26). In this way, type II DNA topoisomerases can efficiently separate two interlocked DNA circles (Figure 5–27).

The same reaction also prevents the severe DNA tangling problems that would otherwise arise during DNA replication. This role is nicely illustrated by mutant yeast cells that produce, in place of the normal topoisomerase II, a version that is inactive at 37°C. When the mutant cells are warmed to this temperature, their daughter chromosomes remain intertwined after DNA replication and are unable to separate. The enormous usefulness of topoisomerase II for untangling chromosomes can readily be appreciated by anyone who has struggled to remove a tangle from a fishing line without the aid of scissors.

DNA Replication Is Similar in Eucaryotes and Bacteria

Much of what we know about DNA replication was first derived from studies of purified bacterial and bacteriophage multienzyme systems capable of DNA replication *in vitro*. The development of these systems in the 1970s was greatly facilitated by the prior isolation of mutants in a variety of replication genes; these mutants were exploited to identify and purify the corresponding replication proteins. The first mammalian replication system that accurately replicated DNA *in vitro* was described in the mid-1980s, and mutations in genes encoding nearly all of the replication components have now been isolated and analyzed in the yeast *Saccharomyces cerevisiae*. As a result, a great deal is known about the detailed enzymology of DNA replication in eucaryotes, and it is clear that the fundamental features of DNA replication—including replication fork geometry and the use of a multiprotein replication machine—have been conserved during the long evolutionary process that separates bacteria and eucaryotes.

There are more protein components in eucaryotic replication machines than there are in the bacterial analogs, even though the basic functions are the same. Thus, for example, the eucaryotic single-strand binding (SSB) protein is formed from three subunits, whereas only a single subunit is found in bacteria. Similarly, the DNA primase is incorporated into a multisubunit enzyme called

Figure 5–27 The DNA-helix-passing reaction catalyzed by DNA topoisomerase II. Identical reactions are used to untangle DNA inside the cell. Unlike type I topoisomerases, type II enzymes use ATP hydrolysis and some of the bacterial versions can introduce superhelical tension into DNA. Type II topoisomerases are largely confined to proliferating cells in eucaryotes; partly for that reason, they have been popular targets for anticancer drugs.

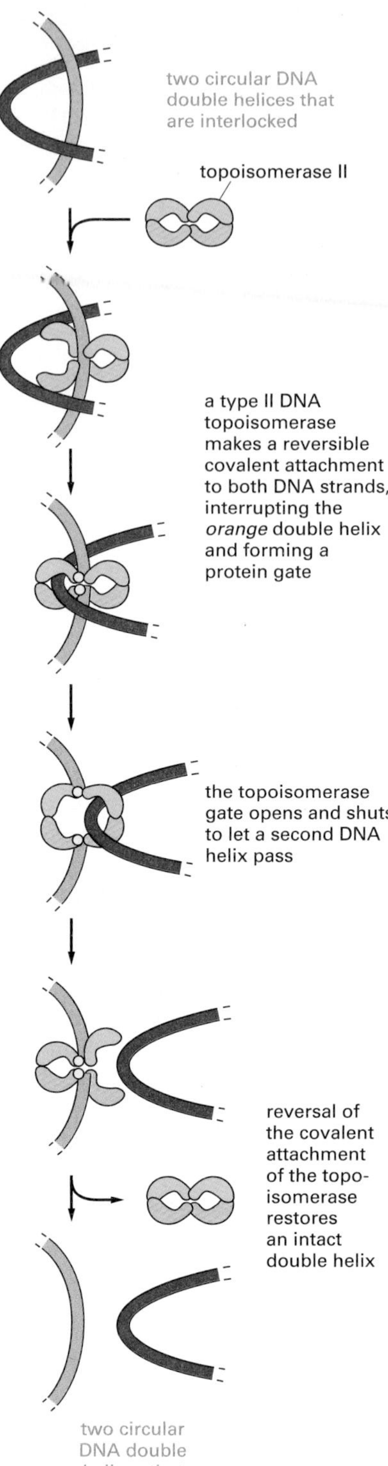
two circular DNA double helices that are interlocked
topoisomerase II
a type II DNA topoisomerase makes a reversible covalent attachment to both DNA strands, interrupting the *orange* double helix and forming a protein gate
the topoisomerase gate opens and shuts to let a second DNA helix pass
reversal of the covalent attachment of the topo-isomerase restores an intact double helix
two circular DNA double helices that are separated

DNA polymerase α. The polymerase α begins each Okazaki fragment on the lagging strand with RNA and then extends the RNA primer with a short length of DNA, before passing the 3′ end of this primer to a second enzyme, DNA polymerase δ. This second DNA polymerase then synthesizes the remainder of each Okazaki fragment with the help of a clamp protein (Figure 5–28).

As we see in the next section, the eucaryotic replication machinery has the added complication of having to replicate through nucleosomes, the repeating structural unit of chromosomes discussed in Chapter 4. Nucleosomes are spaced at intervals of about 200 nucleotide pairs along the DNA, which may explain why new Okazaki fragments are synthesized on the lagging strand at intervals of 100–200 nucleotides in eucaryotes, instead of 1000–2000 nucleotides as in bacteria. Nucleosomes may also act as barriers that slow down the movement of DNA polymerase molecules, which may be why eucaryotic replication forks move only one-tenth as fast as bacterial replication forks.

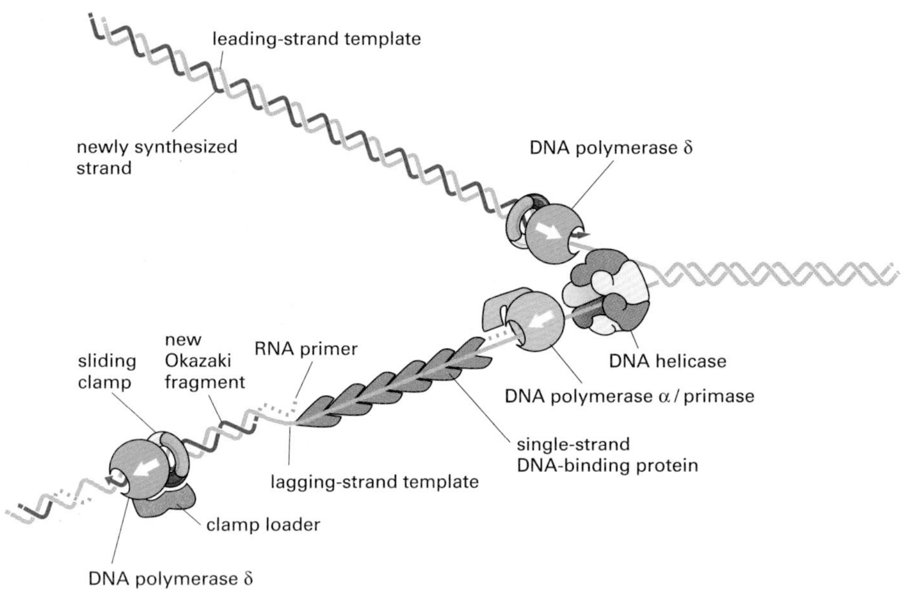
leading-strand template
newly synthesized strand
DNA polymerase δ
new Okazaki fragment
RNA primer
sliding clamp
DNA helicase
DNA polymerase α / primase
single-strand DNA-binding protein
lagging-strand template
clamp loader
DNA polymerase δ

Figure 5–28 A mammalian replication fork. The fork is drawn to emphasize its similarity to the bacterial replication fork depicted in Figure 5–21. Although both forks use the same basic components, the mammalian fork differs in at least two important respects. First, it uses two different DNA polymerases on the lagging strand. Second, the mammalian DNA primase is a subunit of one of the lagging-strand DNA polymerases, DNA polymerase α, while that of bacteria is associated with a DNA helicase in the primosome. The polymerase α (with its associated primase) begins chains with RNA, extends them with DNA, and then hands the chains over to the second polymerase (δ), which elongates them. It is not known why eucaryotic DNA replication requires two different polymerases on the lagging strand. The major mammalian DNA helicase seems to be based on a ring formed from six different Mcm proteins; this ring may move along the leading strand, rather than along the lagging-strand template shown here.

Summary

DNA replication takes place at a Y-shaped structure called a replication fork. A self-correcting DNA polymerase enzyme catalyzes nucleotide polymerization in a 5′-to-3′ direction, copying a DNA template strand with remarkable fidelity. Since the two strands of a DNA double helix are antiparallel, this 5′-to-3′ DNA synthesis can take place continuously on only one of the strands at a replication fork (the leading strand). On the lagging strand, short DNA fragments must be made by a "backstitching" process. Because the self-correcting DNA polymerase cannot start a new chain, these lagging-strand DNA fragments are primed by short RNA primer molecules that are subsequently erased and replaced with DNA.

DNA replication requires the cooperation of many proteins. These include (1) DNA polymerase and DNA primase to catalyze nucleoside triphosphate polymerization; (2) DNA helicases and single-strand DNA-binding (SSB) proteins to help in opening up the DNA helix so that it can be copied; (3) DNA ligase and an enzyme that degrades RNA primers to seal together the discontinuously synthesized lagging-strand DNA fragments; and (4) DNA topoisomerases to help to relieve helical winding and DNA tangling problems. Many of these proteins associate with each other at a replication fork to form a highly efficient "replication machine," through which the activities and spatial movements of the individual components are coordinated.

THE INITIATION AND COMPLETION OF DNA REPLICATION IN CHROMOSOMES

We have seen how a set of replication proteins rapidly and accurately generates two daughter DNA double helices behind a moving replication fork. But how is this replication machinery assembled in the first place, and how are replication forks created on a double-stranded DNA molecule? In this section, we discuss how DNA replication is initiated and how cells carefully regulate this process to ensure that it takes place at the proper positions on the chromosome and also at the appropriate time in the life of the cell. We also discuss a few of the special problems that the replication machinery in eucaryotic cells must overcome. These include the need to replicate the enormously long DNA molecules found in eucaryotic chromosomes, as well as the difficulty of copying DNA molecules that are tightly complexed with histones in nucleosomes.

DNA Synthesis Begins at Replication Origins

As discussed previously, the DNA double helix is normally very stable: the two DNA strands are locked together firmly by a large number of hydrogen bonds formed between the bases on each strand. To be used as a template, the double helix must first be opened up and the two strands separated to expose unpaired bases. As we shall see, the process of DNA replication is begun by special *initiator proteins* that bind to double-stranded DNA and pry the two strands apart, breaking the hydrogen bonds between the bases.

The positions at which the DNA helix is first opened are called **replication origins** (Figure 5–29). In simple cells like those of bacteria or yeast, origins are specified by DNA sequences several hundred nucleotide pairs in length. This DNA contains short sequences that attract initiator proteins, as well as stretches of DNA that are especially easy to open. We saw in Figure 4–4 that an A-T base pair is held together by fewer hydrogen bonds than a G-C base pair. Therefore, DNA rich in A-T base pairs is relatively easy to pull apart, and regions of DNA enriched in A-T pairs are typically found at replication origins.

Although the basic process of replication fork initiation, depicted in Figure 5–29 is the same for bacteria and eucaryotes, the detailed way in which this process is performed and regulated differs between these two groups of organisms. We first consider the simpler and better-understood case in bacteria and then turn to the more complex situation found in yeasts, mammals, and other eucaryotes.

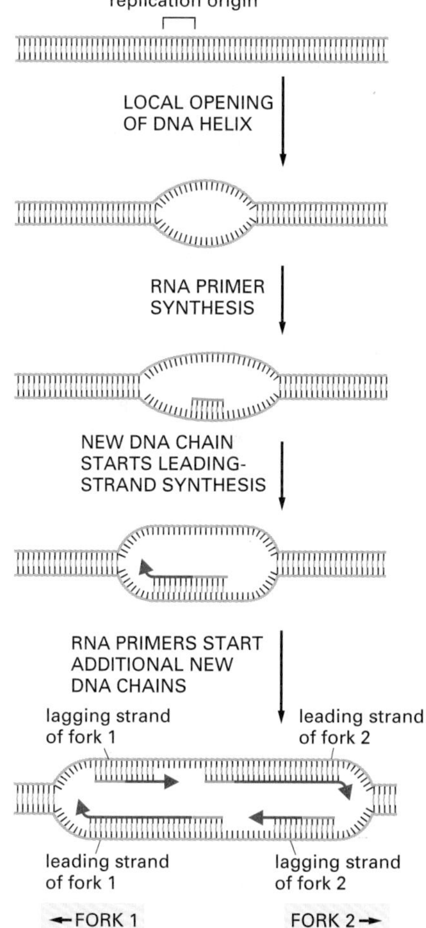

Figure 5–29 A replication bubble formed by replication fork initiation. This diagram outlines the major steps involved in the initiation of replication forks at replication origins. The structure formed at the last step, in which both strands of the parental DNA helix have been separated from each other and serve as templates for DNA synthesis, is called a *replication bubble*.

Figure 5–30 DNA replication of a bacterial genome. It takes *E. coli* about 40 minutes to duplicate its genome of 4.6×10^6 nucleotide pairs. For simplicity, no Okazaki fragments are shown on the lagging strand. What happens as the two replication forks approach each other and collide at the end of the replication cycle is not well understood, although the primosome is disassembled as part of the process.

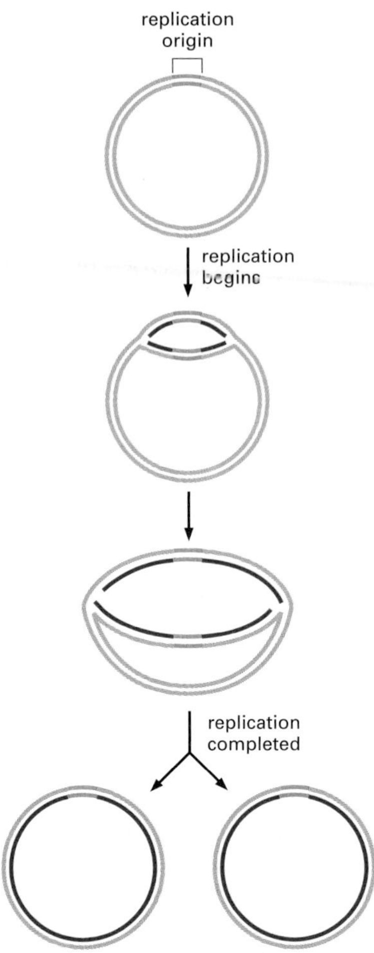

Bacterial Chromosomes Have a Single Origin of DNA Replication

The genome of *E. coli* is contained in a single circular DNA molecule of 4.6×10^6 nucleotide pairs. DNA replication begins at a single origin of replication, and the two replication forks assembled there proceed (at approximately 500–1000 nucleotides per second) in opposite directions until they meet up roughly halfway around the chromosome (Figure 5–30). The only point at which *E. coli* can control DNA replication is initiation: once the forks have been assembled at the origin, they move at a relatively constant speed until replication is finished. Therefore, it is not surprising that the initiation of DNA replication is a highly regulated process. It begins when initiator proteins bind in multiple copies to specific sites in the replication origin, wrapping the DNA around the proteins to form a large protein–DNA complex. This complex then binds a DNA helicase and loads it onto an adjacent DNA single strand whose bases have been exposed by the assembly of the initiator protein–DNA complex. The DNA primase joins the helicase, forming the primosome, which moves away from the origin and makes an RNA primer that starts the first DNA chain (Figure 5–31). This quickly leads to the assembly of the remaining proteins to create two replication forks, with protein complexes that move away from the origin in opposite directions. These protein machines continue to synthesize DNA until all of the DNA template downstream of each fork has been replicated.

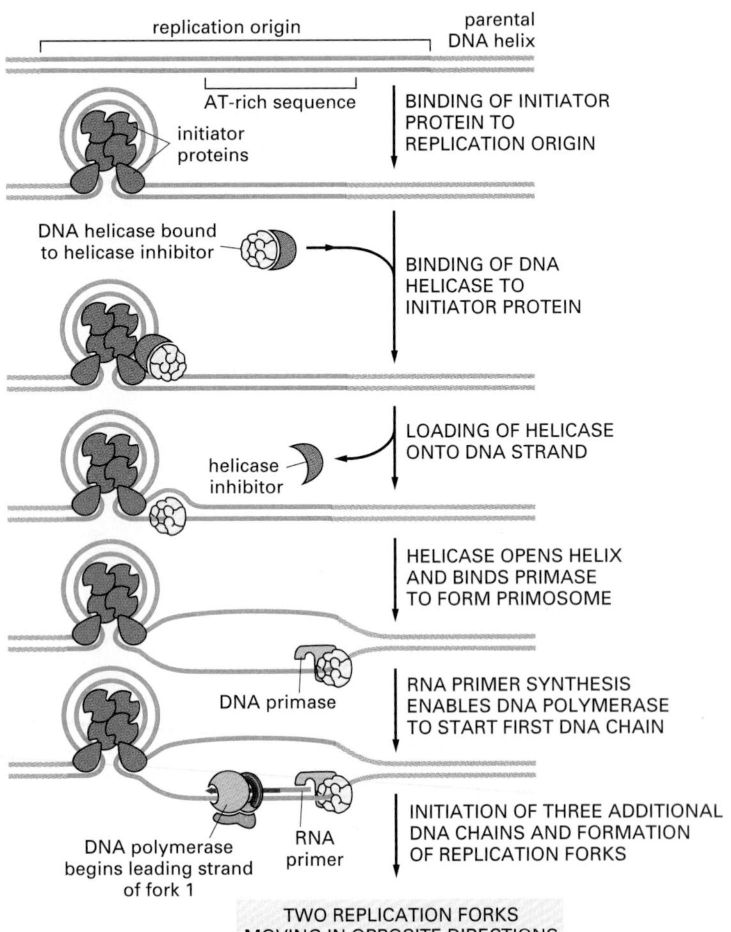

Figure 5–31 The proteins that initiate DNA replication in bacteria. The mechanism shown was established by studies *in vitro* with a mixture of highly purified proteins. For *E. coli* DNA replication, the major initiator protein is the dnaA protein; the primosome is composed of the dnaB (DNA helicase) and dnaG (DNA primase) proteins. In solution, the helicase is bound by an inhibitor protein (the dnaC protein), which is activated by the initiator proteins to load the helicase onto DNA at the replication origin and then released. This inhibitor prevents the helicase from inappropriately entering other single-stranded stretches of DNA in the bacterial genome. Subsequent steps result in the initiation of three more DNA chains (see Figure 5–29) by a pathway whose details are incompletely specified.

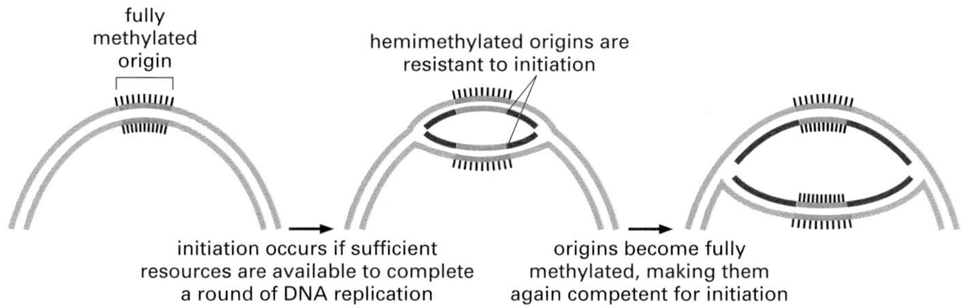

Figure 5–32 Methylation of the *E. coli* replication origin creates a refractory period for DNA initiation. DNA methylation occurs at GATC sequences, 11 of which are found in the origin of replication (spanning about 250 nucleotide pairs). About 10 minutes after replication is initiated, the hemimethylated origins become fully methylated by a DNA methylase enzyme. As discussed earlier, the lag in methylation after the replication of GATC sequences is also used by the *E. coli* mismatch proofreading system to distinguish the newly synthesized DNA strand from the parental DNA strand; in that case, the relevant GATC sequences are scattered throughout the chromosome. A single enzyme, the *dam* methylase, is responsible for methylating *E. coli* GATC sequences.

In *E. coli,* the interaction of the initiator protein with the replication origin is carefully regulated, with initiation occurring only when sufficient nutrients are available for the bacterium to complete an entire round of replication. Not only is the activity of the initiator protein controlled, but an origin of replication that has just been used experiences a "refractory period," caused by a delay in the methylation of newly synthesized A nucleotides. Further initiation of replication is blocked until these As are methylated (Figure 5–32).

Eucaryotic Chromosomes Contain Multiple Origins of Replication

We have seen how two replication forks begin at a single replication origin in bacteria and proceed in opposite directions, moving away from the origin until all of the DNA in the single circular chromosome is replicated. The bacterial genome is sufficiently small for these two replication forks to duplicate the genome in about 40 minutes. Because of the much greater size of most eucaryotic chromosomes, a different strategy is required to allow their replication in a timely manner.

A method for determining the general pattern of eucaryotic chromosome replication was developed in the early 1960s. Human cells growing in culture are labeled for a short time with ^{3}H-thymidine so that the DNA synthesized during this period becomes highly radioactive. The cells are then gently lysed, and the DNA is streaked on the surface of a glass slide coated with a photographic emulsion. Development of the emulsion reveals the pattern of labeled DNA through a technique known as *autoradiography*. The time allotted for radioactive labeling is chosen to allow each replication fork to move several micrometers along the DNA, so that the replicated DNA can be detected in the light microscope as lines of silver grains, even though the DNA molecule itself is too thin to be visible. In this way, both the rate and the direction of replication-fork movement can be determined (Figure 5–33). From the rate at which tracks of replicated DNA increase in length with increasing labeling time, the replication forks are

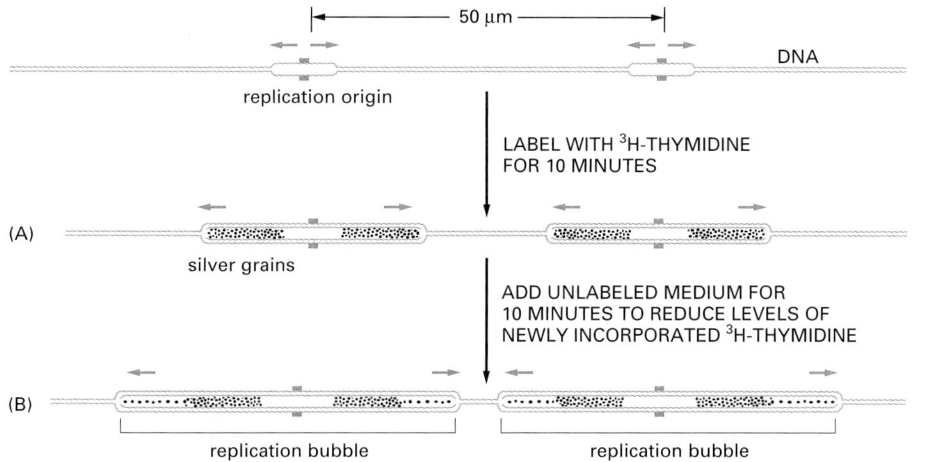

Figure 5–33 The experiments that demonstrated the pattern in which replication forks are formed and move on eucaryotic chromosomes. The new DNA made in human cells in culture was labeled briefly with a pulse of highly radioactive thymidine (^{3}H-thymidine). (A) In this experiment, the cells were lysed, and the DNA was stretched out on a glass slide that was subsequently covered with a photographic emulsion. After several months the emulsion was developed, revealing a line of silver grains over the radioactive DNA. The *brown* DNA in this figure is shown only to help with the interpretation of the autoradiograph; the unlabeled DNA is invisible in such experiments. (B) This experiment was the same except that a further incubation in unlabeled medium allowed additional DNA, with a lower level of radioactivity, to be replicated. The pairs of dark tracks in (B) were found to have silver grains tapering off in opposite directions, demonstrating bidirectional fork movement from a central replication origin where a replication bubble forms (see Figure 5–29). A replication fork is thought to stop only when it encounters a replication fork moving in the opposite direction or when it reaches the end of the chromosome; in this way, all the DNA is eventually replicated.

estimated to travel at about 50 nucleotides per second. This is approximately one-tenth of the rate at which bacterial replication forks move, possibly reflecting the increased difficulty of replicating DNA that is packaged tightly in chromatin.

An average-sized human chromosome contains a single linear DNA molecule of about 150 million nucleotide pairs. To replicate such a DNA molecule from end to end with a single replication fork moving at a rate of 50 nucleotides per second would require $0.02 \times 150 \times 10^6 = 3.0 \times 10^6$ seconds (about 800 hours). As expected, therefore, the autoradiographic experiments just described reveal that many forks are moving simultaneously on each eucaryotic chromosome. Moreover, many forks are found close together in the same DNA region, while other regions of the same chromosome have none.

Further experiments of this type have shown the following: (1) Replication origins tend to be activated in clusters, called *replication units,* of perhaps 20–80 origins. (2) New replication units seem to be activated at different times during the cell cycle until all of the DNA is replicated, a point that we return to below. (3) Within a replication unit, individual origins are spaced at intervals of 30,000–300,000 nucleotide pairs from one another. (4) As in bacteria, replication forks are formed in pairs and create a replication bubble as they move in opposite directions away from a common point of origin, stopping only when they collide head-on with a replication fork moving in the opposite direction (or when they reach a chromosome end). In this way, many replication forks can operate independently on each chromosome and yet form two complete daughter DNA helices.

In Eucaryotes DNA Replication Takes Place During Only One Part of the Cell Cycle

When growing rapidly, bacteria replicate their DNA continually, and they can begin a new round before the previous one is complete. In contrast, DNA replication in most eucaryotic cells occurs only during a specific part of the cell division cycle, called the *DNA synthesis phase* or **S phase** (Figure 5–34). In a mammalian cell, the S phase typically lasts for about 8 hours; in simpler eucaryotic cells such as yeasts, the S phase can be as short as 40 minutes. By its end, each chromosome has been replicated to produce two complete copies, which remain joined together at their centromeres until the *M phase* (M for *mitosis),* which soon follows. In Chapter 17, we describe the control system that runs the cell cycle and explain why entry into each phase of the cycle requires the cell to have successfully completed the previous phase.

In the following sections, we explore how chromosome replication is coordinated within the S phase of the cell cycle.

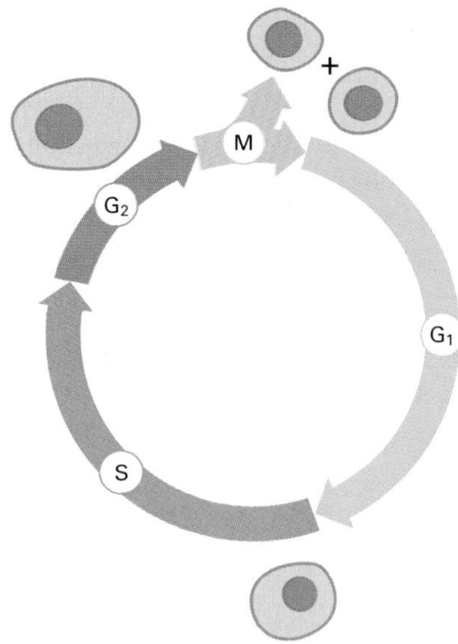

Figure 5–34 The four successive phases of a standard eucaryotic cell cycle. During the G_1, S, and G_2 phases, the cell grows continuously. During M phase growth stops, the nucleus divides, and the cell divides in two. DNA replication is confined to the part of interphase known as S phase. G_1 is the gap between M phase and S phase; G_2 is the gap between S phase and M phase.

Different Regions on the Same Chromosome Replicate at Distinct Times in S Phase

In mammalian cells, the replication of DNA in the region between one replication origin and the next should normally require only about an hour to complete, given the rate at which a replication fork moves and the largest distances measured between the replication origins in a replication unit. Yet S phase usually lasts for about 8 hours in a mammalian cell. This implies that the replication origins are not all activated simultaneously and that the DNA in each replication unit (which, as we noted above, contains a cluster of about 20–80 replication origins) is replicated during only a small part of the total S-phase interval.

Are different replication units activated at random, or are different regions of the genome replicated in a specified order? One way to answer this question is to use the thymidine analogue bromodeoxyuridine (BrdU) to label the newly synthesized DNA in synchronized cell populations, adding it for different short periods throughout S phase. Later, during M phase, those regions of the mitotic chromosomes that have incorporated BrdU into their DNA can be recognized by their altered staining properties or by means of anti-BrdU antibodies. The results show that different regions of each chromosome are replicated in a reproducible order during S phase (Figure 5–35). Moreover, as one would expect from the clusters of replication forks seen in DNA autoradiographs (see Figure 5–33), the timing of replication is coordinated over large regions of the chromosome.

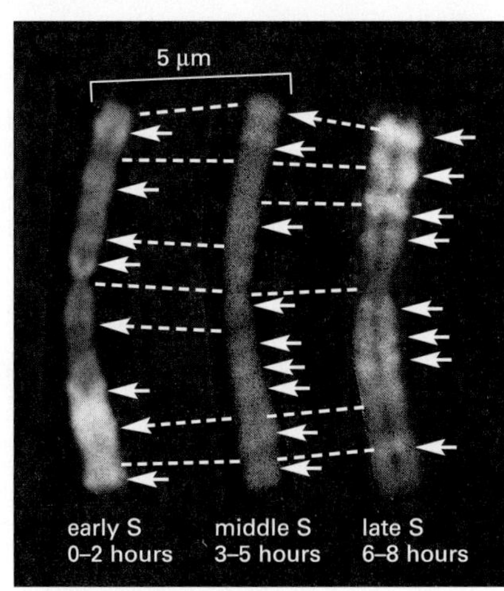

Figure 5–35 Different regions of a chromosome are replicated at different times in S phase. These light micrographs show stained mitotic chromosomes in which the replicating DNA has been differentially labeled during different defined intervals of the preceding S phase. In these experiments, cells were first grown in the presence of BrdU (a thymidine analog) and in the absence of thymidine to label the DNA uniformly. The cells were then briefly pulsed with thymidine in the absence of BrdU during early, middle, or late S phase. Because the DNA made during the thymidine pulse is a double helix with thymidine on one strand and BrdU on the other, it stains more darkly than the remaining DNA (which has BrdU on both strands) and shows up as a bright band *(arrows)* on these negatives. Broken lines connect corresponding positions on the three identical copies of the chromosome shown. (Courtesy of Elton Stubblefield.)

Highly Condensed Chromatin Replicates Late, While Genes in Less Condensed Chromatin Tend to Replicate Early

It seems that the order in which replication origins are activated depends, in part, on the chromatin structure in which the origins reside. We saw in Chapter 4 that heterochromatin is a particularly condensed state of chromatin, while transcriptionally active chromatin has a less condensed conformation that is apparently required to allow RNA synthesis. Heterochromatin tends to be replicated very late in S phase, suggesting that the timing of replication is related to the packing of the DNA in chromatin. This suggestion is supported by an examination of the two X chromosomes in a female mammalian cell. While these two chromosomes contain essentially the same DNA sequences, one is active for DNA transcription and the other is not (discussed in Chapter 7). Nearly all of the inactive X chromosome is condensed into heterochromatin, and its DNA replicates late in S phase. Its active homologue is less condensed and replicates throughout S phase.

These findings suggest that those regions of the genome whose chromatin is least condensed, and therefore most accessible to the replication machinery, are replicated first. Autoradiography shows that replication forks move at comparable rates throughout S phase, so that the extent of chromosome condensation seems to influence the time at which replication forks are initiated, rather than their speed once formed.

The above relationship between chromatin structure and the timing of DNA replication is also supported by studies in which the replication times of specific genes are measured. The results show that so-called "housekeeping" genes, which are those active in all cells, replicate very early in S phase in all cells tested. In contrast, genes that are active in only a few cell types generally replicate early in the cells in which the genes are active, and later in other types of cell.

The relationship between chromatin structure and the timing of replication has been tested directly in the yeast *S. cerevisiae*. In one case, an origin that functioned late in S phase, and was found in a transcriptionally silent region of a yeast chromosome, was experimentally relocated to a transcriptionally active region. After the relocation, the origin functioned early in the S phase, indicating that the time in S phase when this origin is used is determined by the origin's location in the chromosome. However, studies with additional yeast origins have revealed the existence of other origins that initiate replication late, even when present in normal chromatin. Thus, the time at which an origin is used can be determined both by its chromatin structure and by its DNA sequence.

Well-defined DNA Sequences Serve as Replication Origins in a Simple Eucaryote, the Budding Yeast

Having seen that a eucaryotic chromosome is replicated using many origins of replication, each of which "fires" at a characteristic time in S phase of the cell cycle, we turn to the nature of these origins of replication. We saw earlier in this chapter that replication origins have been precisely defined in bacteria as specific DNA sequences that allow the DNA replication machinery to assemble on the DNA double helix, form a replication bubble, and move in opposite directions to produce replication forks. By analogy, one would expect the replication origins in eucaryotic chromosomes to be specific DNA sequences too.

The search for replication origins in the chromosomes of eucaryotic cells has been most productive in the budding yeast *S. cerevisiae*. Powerful selection methods to find them have been devised that make use of mutant yeast cells defective for an essential gene. These cells can survive in a selective medium

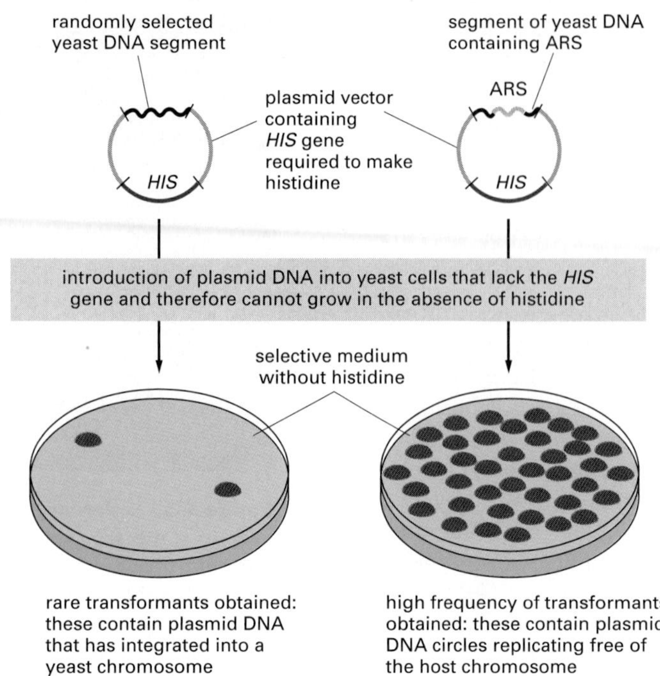
randomly selected yeast DNA segment
segment of yeast DNA containing ARS
ARS
plasmid vector containing HIS gene required to make histidine
HIS
HIS
introduction of plasmid DNA into yeast cells that lack the HIS gene and therefore cannot grow in the absence of histidine
selective medium without histidine
rare transformants obtained: these contain plasmid DNA that has integrated into a yeast chromosome
high frequency of transformants obtained: these contain plasmid DNA circles replicating free of the host chromosome

Figure 5–36 The strategy used to identify replication origins in yeast cells. Each of the yeast DNA sequences identified in this way was called an autonomously replicating sequence (ARS), since it enables a plasmid that contains it to replicate in the host cell without having to be incorporated into a host cell chromosome.

only if they are provided with DNA that carries a functional copy of the missing gene. If a circular bacterial plasmid with this gene is introduced into the mutant yeast cells directly, it will not be able to replicate because it lacks a functional origin. If random pieces of yeast DNA are inserted into this plasmid, however, only those few plasmid DNA molecules that contain a yeast replication origin can replicate. The yeast cells that carry such plasmids are able to proliferate because they have been provided with the essential gene in a form that can be replicated and passed on to progeny cells (Figure 5–36). A DNA sequence identified by its presence in a plasmid isolated from these surviving yeast cells is called an *autonomously replicating sequence (ARS)*. Most ARSs have been shown to be authentic chromosomal origins of replication, thereby validating the strategy used to obtain them.

For budding yeast, the location of every origin of replication on each chromosome can be determined (Figure 5–37). The particular chromosome shown—chromosome III from the yeast *S. cerevisiae*—is less than 1/100 the length of a typical human chromosome. Its origins are spaced an average of 30,000 nucleotides apart; this density of origins should permit a yeast chromosome to be replicated in about 8 minutes.

Genetic experiments in *S. cerevisiae* have tested the effect of deleting various sets of the replication origins on chromosome III. Removing a few origins has little effect, because replication forks that begin at neighboring origins of replication can continue into the regions that lack their own origins. However, as more replication origins are deleted from this chromosome, the chromosome is gradually lost as the cells divide, presumably because it is replicated too slowly.

A Large Multisubunit Complex Binds to Eucaryotic Origins of Replication

The minimal DNA sequence required for directing DNA replication initiation in the yeast *S. cerevisiae* has been determined by performing the experiment

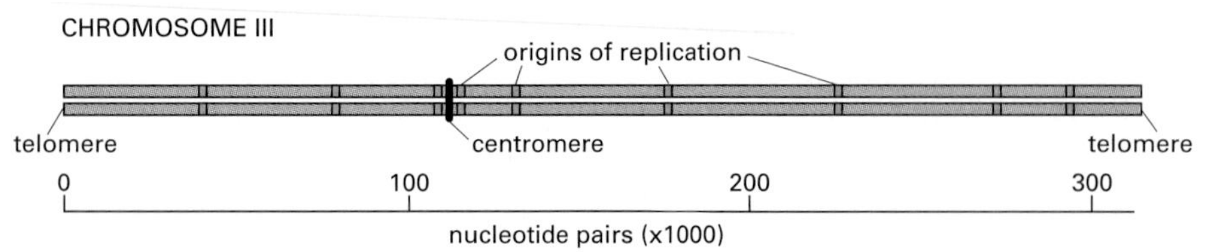
CHROMOSOME III
origins of replication
telomere
centromere
telomere
0
100
200
300
nucleotide pairs (x1000)

Figure 5–37 The origins of DNA replication on chromosome III of the yeast *S. cerevisiae*. This chromosome, one of the smallest eucaryotic chromosomes known, carries a total of 180 genes. As indicated, it contains nine replication origins.

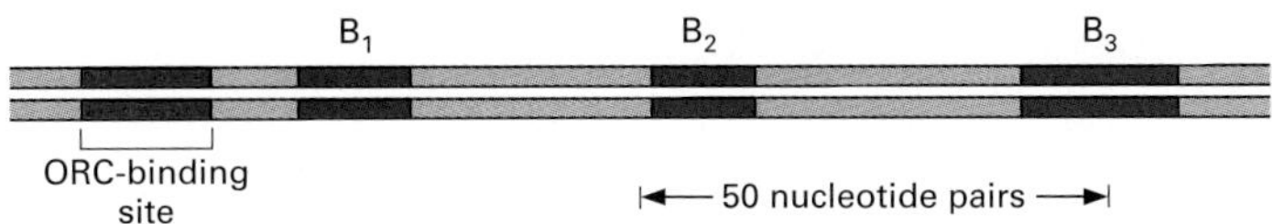

Figure 5–38 An origin of replication in yeast. Comprising about 150 nucleotide pairs, this yeast origin (identified by the procedure shown in Figure 5–36) has a binding site for ORC, a complex of proteins that binds to every origin of replication. The origin depicted also has binding sites (B1, B2, and B3) for other required proteins, which can differ between various origins. Although best characterized in yeast, a similar ORC is used to initiate DNA replication in more complex eucaryotes, including humans.

shown in Figure 5–36 with smaller and smaller DNA fragments. Each of the yeast replication origins contains a binding site for a large, multisubunit initiator protein called **ORC**, for **origin recognition complex**, and several auxiliary binding sites for proteins that help attract ORC to the origin DNA (Figure 5–38).

As we have seen, DNA replication in eucaryotes occurs only in the S phase. How is this DNA replication triggered, and how does the mechanism ensure that a replication origin is used only once during each cell cycle?

As we discuss in Chapter 17, the general answers to these two questions are now known. In brief, the ORC-origin interaction is a stable one that serves to mark a replication origin throughout the entire cell cycle. A prereplicative protein complex is assembled on each ORC during G_1 phase, containing both a hexameric DNA helicase and a helicase loading factor (the Mcm and Cdc6 proteins, respectively). S phase is triggered when a protein kinase is activated that assembles the rest of the replication machinery, allowing an Mcm helicase to start moving with each of the two replication forks that form at each origin. Simultaneously, the protein kinase that triggers S phase prevents all further assembly of the Mcm protein into prereplicative complexes, until this kinase is inactivated at the next M phase to reset the entire cycle (for details, see Figure 17–22).

The Mammalian DNA Sequences That Specify the Initiation of Replication Have Been Difficult to Identify

Compared with the situation in budding yeasts, DNA sequences that specify replication origins in other eucaryotes have been more difficult to define. In humans, for example, the DNA sequences that are required for proper origin function can extend over very large distances along the DNA.

Recently, however, it has been possible to identify specific human DNA sequences, each several thousand nucleotide pairs in length, that serve as replication origins. These origins continue to function when moved to a different chromosomal region by recombinant DNA methods, as long as they are placed in a region where the chromatin is relatively uncondensed. One of these origins is the sequence from the β-globin gene cluster. At its normal position in the genome, the function of this origin depends critically upon distant DNA sequences (Figure 5–39). As discussed in Chapter 7, this distant DNA is known to have a decondensing effect on the chromatin structure that surrounds the origin and includes the β-globin gene; the more open chromatin conformation that results is apparently required for this origin to function, as well as for the β-globin gene to be expressed.

We know now that a human ORC complex homologous to that in yeast cells is required for replication initiation and that the human Cdc6 and Mcm proteins likewise have central roles in the initiation process. It therefore seems likely that the yeast and human initiation mechanisms will turn out to be very similar. However, the binding sites for the ORC protein seem to be less specific in humans than they are in yeast, which may explain why the replication origins of humans are longer and less sharply defined than those of yeast.

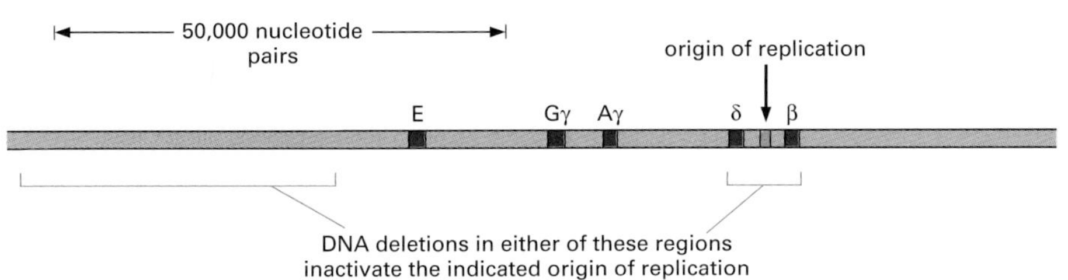

Figure 5–39 Deletions that inactivate an origin of replication in humans. These two deletions are found separately in two individuals who suffer from *thalassemia*, a disorder caused by the failure to express one or more of the genes in the β-globin gene cluster shown. In both of these deletion mutants, the DNA in this region is replicated by forks that begin at replication origins outside the β-globin gene cluster. As explained in the text, the deletion on the *left* removes DNA sequences that control the chromatin structure of the replication origin on the *right*.

New Nucleosomes Are Assembled Behind the Replication Fork

In this and the following section we consider several additional aspects of DNA replication that are specific to eucaryotes. As discussed in Chapter 4, eucaryotic chromosomes are composed of the mixture of DNA and protein known as chromatin. Chromosome duplication therefore requires not only that the DNA be replicated but also that new chromosomal proteins be assembled onto the DNA behind each replication fork. Although we are far from understanding this process in detail, we are beginning to learn how the *nucleosome,* the fundamental unit of chromatin packaging, is duplicated. A large amount of new histone protein, approximately equal in mass to the newly synthesized DNA, is required to make the new nucleosomes in each cell cycle. For this reason, most eucaryotic organisms possess multiple copies of the gene for each histone. Vertebrate cells, for example, have about 20 repeated gene sets, most sets containing the genes that encode all five histones (H1, H2A, H2B, H3, and H4).

Unlike most proteins, which are made continuously throughout interphase, histones are synthesized mainly in S phase, when the level of histone mRNA increases about fiftyfold as a result of both increased transcription and decreased mRNA degradation. By a mechanism that depends on special properties of their 3′ ends (discussed in Chapter 7), the major histone mRNAs become highly unstable and are degraded within minutes when DNA synthesis stops at the end of S phase (or when inhibitors are added to stop DNA synthesis prematurely). In contrast, the histone proteins themselves are remarkably stable and may survive for the entire life of a cell. The tight linkage between DNA synthesis and histone synthesis presumably depends on a feedback mechanism that monitors the level of free histone to ensure that the amount of histone made exactly matches the amount of new DNA synthesized.

As a replication fork advances, it must somehow pass through the parental nucleosomes. *In vitro* studies show that the replication apparatus has a poorly understood intrinsic ability to pass through parental nucleosomes without displacing them from the DNA. The chromatin-remodeling proteins discussed in Chapter 4, which destabilize the DNA–histone interface, likely facilitate this process in the cell.

Both of the newly synthesized DNA helices behind a replication fork inherit old histones (Figure 5–40). But since the amount of DNA has doubled, an equal amount of new histones is also needed to complete the packaging of DNA into chromatin. The addition of new histones to the newly synthesized DNA is aided by *chromatin assembly factors (CAFs),* which are proteins that associate with replication forks and package the newly synthesized DNA as soon as it emerges from the replication machinery. The newly synthesized H3 and H4 histones are rapidly acetylated on their N-terminal tails (discussed in Chapter 4); after they have been incorporated into chromatin, these acetyl groups are removed enzymatically from the histones (Figure 5–41).

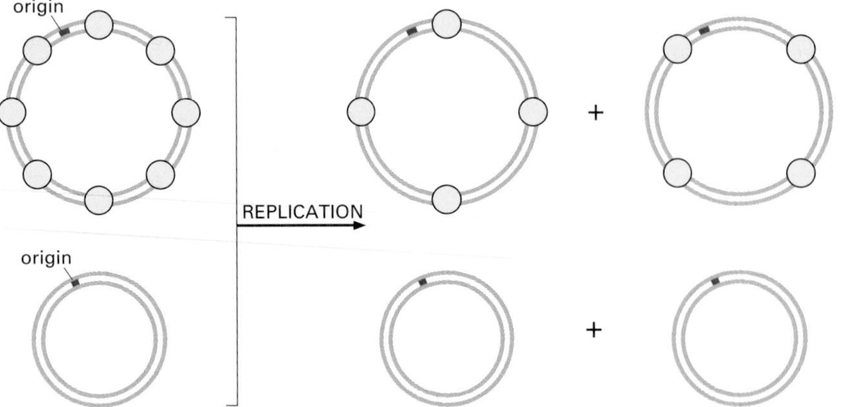

Figure 5–40 A demonstration that histones remain associated with DNA after the replication fork passes. In this experiment, performed *in vitro,* a mixture of two different-sized circular molecules of DNA (only one of which is assembled into nucleosomes) are replicated with purified proteins. After a round of DNA replication, only the daughter DNA molecules that derived from the nucleosomal parent have inherited nucleosomes. This experiment also demonstrates that both newly synthesized DNA helices inherit old histones.

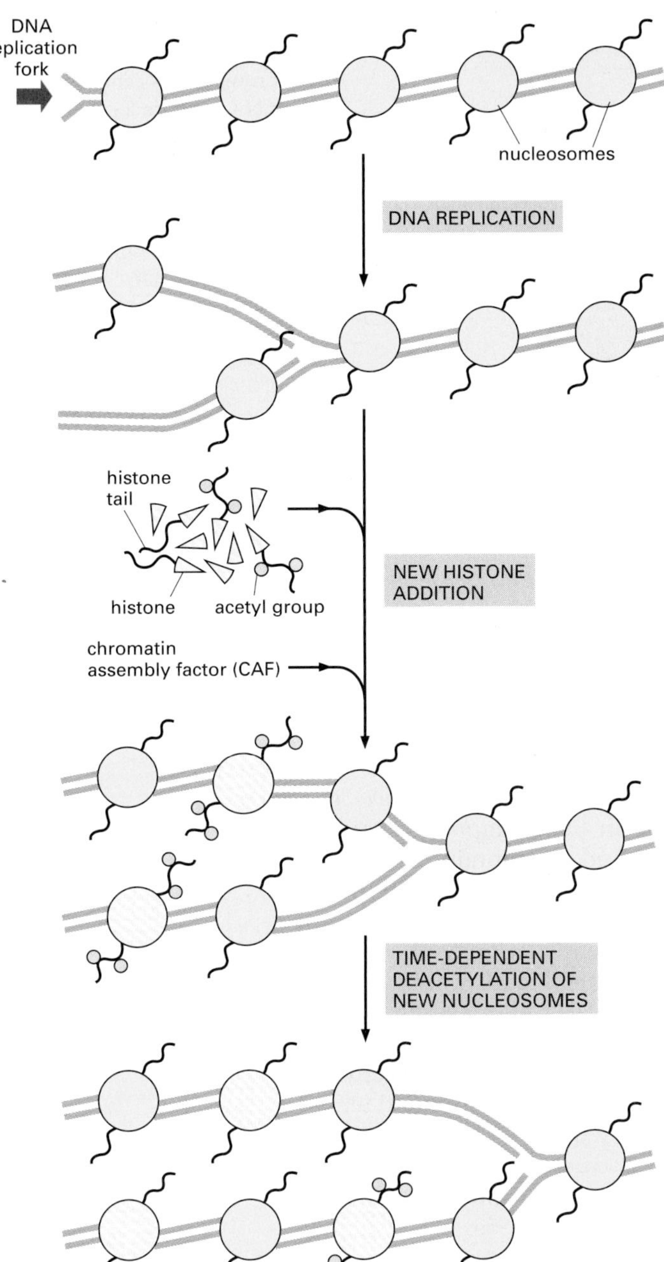

Figure 5–41 The addition of new histones to DNA behind a replication fork. The new nucleosomes are those colored *light yellow* in this diagram; as indicated, some of the histones that form them initially have specifically acetylated lysine side chains (see Figure 4–35), which are later removed.

Telomerase Replicates the Ends of Chromosomes

We saw earlier that, because DNA polymerases polymerize DNA only in the 5′-to-3′ direction, synthesis of the lagging strand at a replication fork must occur discontinuously through a backstitching mechanism that produces short DNA fragments. This mechanism encounters a special problem when the replication fork reaches an end of a linear chromosome: there is no place to produce the RNA primer needed to start the last Okazaki fragment at the very tip of a linear DNA molecule.

Bacteria solve this "end-replication" problem by having circular DNA molecules as chromosomes (see Figure 5–30). Eucaryotes solve it in an ingenious way: they have special nucleotide sequences at the ends of their chromosomes, which are incorporated into *telomeres* (discussed in Chapter 4), and attract an enzyme called **telomerase**. Telomere DNA sequences are similar in organisms as diverse as protozoa, fungi, plants, and mammals. They consist of many tandem repeats of a short sequence that contains a block of neighboring G nucleotides. In humans, this sequence is GGGTTA, extending for about 10,000 nucleotides.

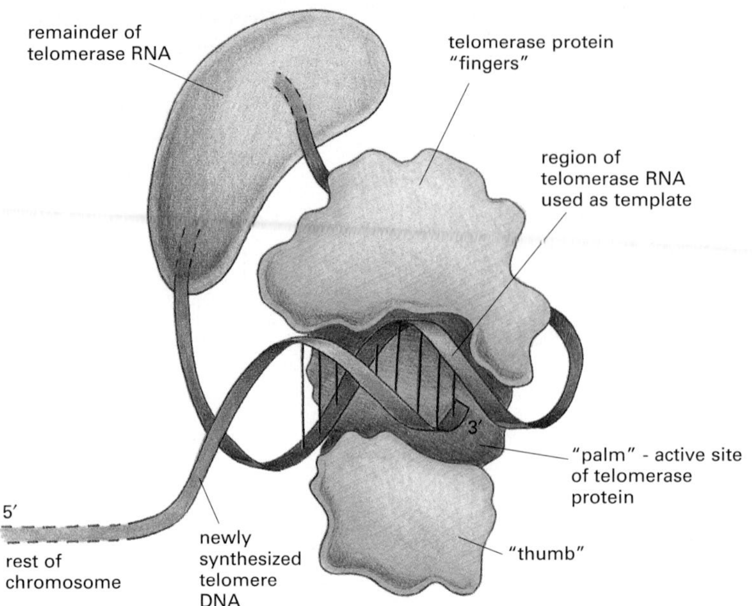

Figure 5–42 The structure of telomerase. The telomerase is a protein–RNA complex that carries an RNA template for synthesizing a repeating, G-rich telomere DNA sequence. Only the part of the telomerase protein homologous to reverse transcriptase is shown here *(green)*. A reverse transcriptase is a special form of polymerase enzyme that uses an RNA template to make a DNA strand; telomerase is unique in carrying its own RNA template with it at all times. (Modified from J. Lingner and T.R. Cech, *Curr. Opin. Genet. Dev.* 8:226–232, 1998.)

Telomerase recognizes the tip of a G-rich strand of an existing telomere DNA repeat sequence and elongates it in the 5′-to-3′ direction. The telomerase synthesizes a new copy of the repeat, using an RNA template that is a component of the enzyme itself. The telomerase enzyme otherwise resembles other *reverse transcriptases*, enzymes that synthesize DNA using an RNA template (Figure 5–42). The enzyme thus contains all the information used to maintain the characteristic telomere sequences. After several rounds of extension of the parental DNA strand by telomerase, replication of the lagging strand at the chromosome end can be completed by using these extensions as a template for synthesis of the complementary strand by a DNA polymerase molecule (Figure 5–43).

The mechanism just described ensures that the 3′ DNA end at each telomere is always slightly longer than the 5′ end with which it is paired, leaving a protruding single-stranded end (see Figure 5–43). Aided by specialized proteins, this protruding end has been shown to loop back to tuck its single-stranded terminus into the duplex DNA of the telomeric repeat sequence (Figure 5–44). Thus, the normal end of a chromosome has a unique structure, which protects it from degradative enzymes and clearly distinguishes it from the ends of the broken DNA molecules that the cell rapidly repairs (see Figure 5-53).

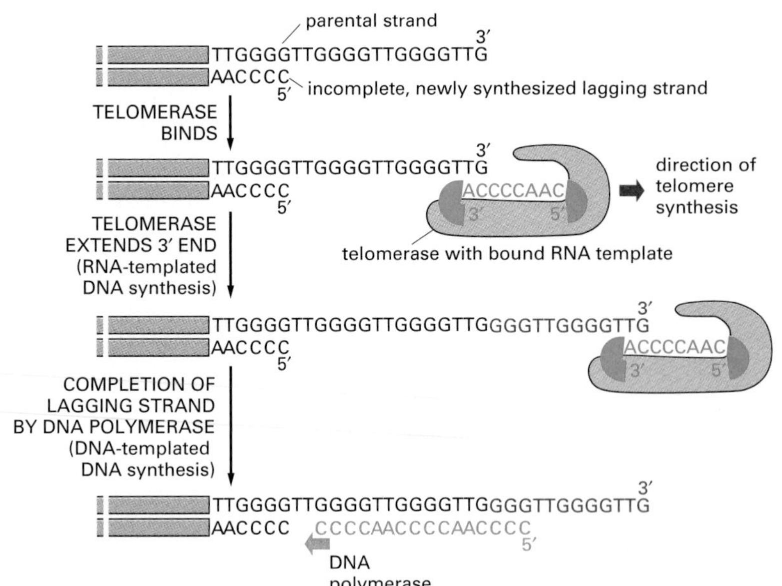

Figure 5–43 Telomere replication. Shown here are the reactions involved in synthesizing the repeating G-rich sequences that form the ends of the chromosomes (telomeres) of diverse eucaryotic organisms. The 3′ end of the parental DNA strand is extended by RNA-templated DNA synthesis; this allows the incomplete daughter DNA strand that is paired with it to be extended in its 5′ direction. This incomplete, lagging strand is presumed to be completed by DNA polymerase α, which carries a DNA primase as one of its subunits (see Figure 5–28). The telomere sequence illustrated is that of the ciliate *Tetrahymena*, in which these reactions were first discovered. The telomere repeats are GGGTTG in the ciliate *Tetrahymena*, GGGTTA in humans, and $G_{1-3}A$ in the yeast *S. cerevisiae*.

Figure 5–44 The t-loops at the end of mammalian chromosomes. (A) Electron micrograph of the DNA at the end of an interphase human chromosome. The chromosome was fixed, deproteinated, and artificially thickened before viewing. The loop seen here is approximately 15,000 nucleotide pairs in length. (B) Model for telomere structure. The insertion of the single-stranded end into the duplex repeats to form a t-loop is carried out and maintained by specialized proteins, schematized in *green.* In addition it is possible, as shown, that the chromosome end is looped once again on itself through the formation of heterochromatin adjacent to the t-loop (see Figure 4–47). (A, from J.D. Griffith et al., *Cell* 97:503–514, 1999. © Elsevier.)

(A)

1 μm

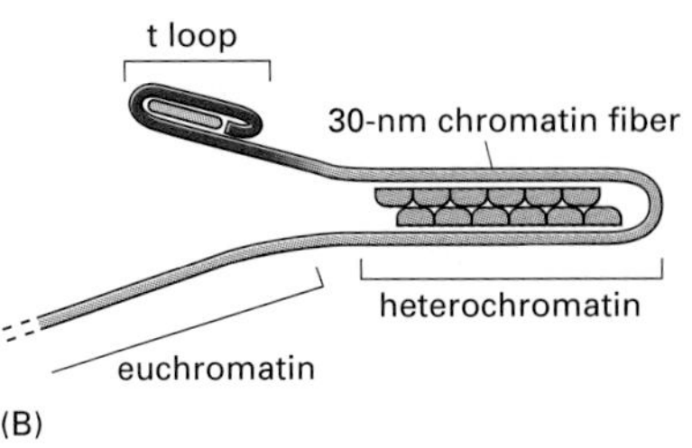

Telomere Length Is Regulated by Cells and Organisms

Because the processes that grow and shrink each telomere sequence are only approximately balanced, a chromosome end contains a variable number of telomeric repeats. Not surprisingly, experiments show that cells that proliferate indefinitely (such as yeast cells) have homeostatic mechanisms that maintain the number of these repeats within a limited range (Figure 5–45).

In the somatic cells of humans, the telomere repeats have been proposed to provide each cell with a counting mechanism that helps prevent the unlimited proliferation of wayward cells in adult tissues. According to this idea, our somatic cells are born with a full complement of telomeric repeats; however, the telomerase enzyme is turned off in a tissue like the skin, so that each time a cell divides, it loses 50–100 nucleotides from each of its telomeres. After many cell generations, the descendent cells will inherit defective chromosomes (because their tips cannot be replicated completely) and consequently will withdraw permanently from the cell cycle and cease dividing—a process called *replicative cell senescence* (discussed in Chapter 17). In theory, such a mechanism could provide a safeguard against the uncontrolled cell proliferation of abnormal cells in somatic tissues, thereby helping to protect us from cancer.

The idea that telomere length acts as a "measuring stick" to count cell divisions and thereby regulate the cell's lifetime has been tested in several ways. For certain types of human cells grown in tissue culture, the experimental results support such a theory. Human fibroblasts normally proliferate for about 60 cell divisions in culture before undergoing replicative senescence. Like most other somatic cells in humans, fibroblasts fail to produce telomerase, and their

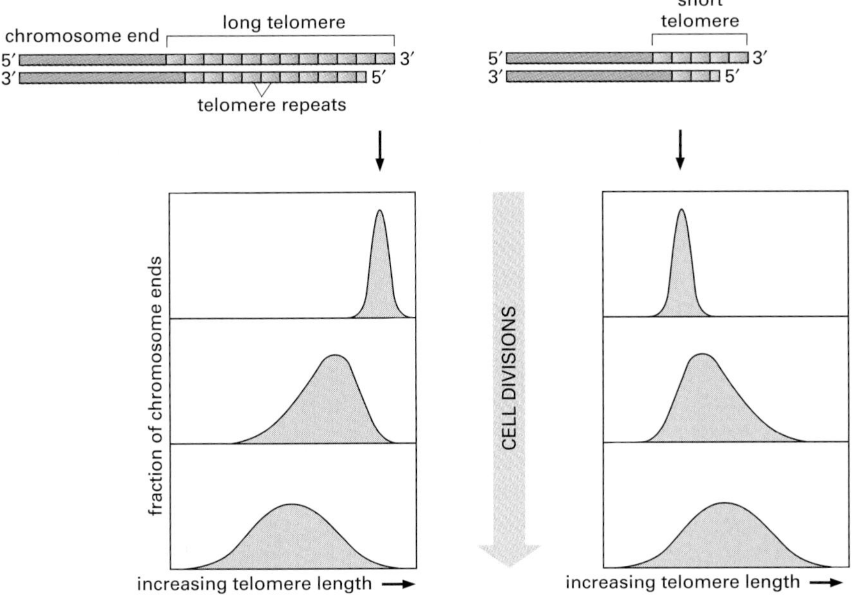

Figure 5–45 A demonstration that yeast cells control the length of their telomeres. In this experiment, the telomere at one end of a particular chromosome is artificially made either longer *(left)* or shorter *(right)* than average. After many cell divisions, the chromosome recovers, showing an average telomere length and a length distribution that is typical of the other chromosomes in the yeast cell. A similar feedback mechanism for controlling telomere length has been proposed to exist for the cells in the germ-line cells of animals.

telomeres gradually shorten each time they divide. When telomerase is provided to the fibroblasts by inserting an active telomerase gene, telomere length is maintained and many of the cells now continue to proliferate indefinitely. It therefore seems clear that telomere shortening can count cell divisions and trigger replicative senescence in human cells.

It has been proposed that this type of control on cell proliferation is important for the maintenance of tissue architecture and that it is also somehow responsible for the aging of animals like ourselves. These ideas have been tested by producing transgenic mice that lack telomerase. The telomeres in mouse chromosomes are about five times longer than human telomeres, and the mice must therefore be bred through three or more generations before their telomeres have shrunk to the normal human length. It is therefore perhaps not surprising that the mice initially develop normally. More importantly, the mice in later generations develop progressively more defects in some of their highly proliferative tissues. But these mice do not seem to age prematurely overall, and the older animals have a pronounced tendency to develop tumors. In these and other respects these mice resemble humans with the genetic disease *dyskeratosis congenita,* which has also been attributed to premature telomere shortening. Individuals afflicted with this disease show abnormalities in various epidermal structures (including skin, nails, and tear ducts) and in the production of red blood cells.

It is clear from the above observations that controlling cell proliferation by the removal of telomeres poses a risk to an organism, because not all of the cells that lack functional telomeres in a tissue will stop dividing. Others apparently become genetically unstable, but continue to divide giving rise to variant cells that can lead to cancer. Thus, one can question whether the observed absence of telomerase from most human somatic cells provides an evolutionary advantage, as suggested by those who postulate that telomere shortening tends to protect us from cancer and other proliferative diseases.

Summary

The proteins that initiate DNA replication bind to DNA sequences at a replication origin to catalyze the formation of a replication bubble with two outward-moving replication forks. The process begins when an initiator protein–DNA complex is formed that subsequently loads a DNA helicase onto the DNA template. Other proteins are then added to form the multienzyme "replication machine" that catalyzes DNA synthesis at each replication fork.

In bacteria and some simple eucaryotes, replication origins are specified by specific DNA sequences that are only several hundred nucleotide pairs long. In other eucaryotes, such as humans, the sequences needed to specify an origin of DNA replication seem to be less well defined, and the origin can span several thousand nucleotide pairs.

Bacteria typically have a single origin of replication in a circular chromosome. With fork speeds of up to 1000 nucleotides per second, they can replicate their genome in less than an hour. Eucaryotic DNA replication takes place in only one part of the cell cycle, the S phase. The replication fork in eucaryotes moves about 10 times more slowly than the bacterial replication fork, and the much longer eucaryotic chromosomes each require many replication origins to complete their replication in a typical 8-hour S phase. The different replication origins in these eucaryotic chromosomes are activated in a sequence, determined in part by the structure of the chromatin, with the most condensed regions of chromatin beginning their replication last. After the replication fork has passed, chromatin structure is re-formed by the addition of new histones to the old histones that are directly inherited as nucleosomes by each daughter DNA molecule.

Eucaryotes solve the problem of replicating the ends of their linear chromosomes by a specialized end structure, the telomere, which requires a special enzyme, telomerase. Telomerase extends the telomere DNA by using an RNA template that is an integral part of the enzyme itself, producing a highly repeated DNA sequence that typically extends for 10,000 nucleotide pairs or more at each chromosome end.

DNA REPAIR

Although genetic variation is important for evolution, the survival of the individual demands genetic stability. Maintaining genetic stability requires not only an extremely accurate mechanism for replicating DNA, but also mechanisms for repairing the many accidental lesions that occur continually in DNA. Most such spontaneous changes in DNA are temporary because they are immediately corrected by a set of processes that are collectively called **DNA repair**. Of the thousands of random changes created every day in the DNA of a human cell by heat, metabolic accidents, radiation of various sorts, and exposure to substances in the environment, only a few accumulate as mutations in the DNA sequence. We now know that fewer than one in 1000 accidental base changes in DNA results in a permanent mutation; the rest are eliminated with remarkable efficiency by DNA repair.

The importance of DNA repair is evident from the large investment that cells make in DNA repair enzymes. For example, analysis of the genomes of bacteria and yeasts has revealed that several percent of the coding capacity of these organisms is devoted solely to DNA repair functions. The importance of DNA repair is also demonstrated by the increased rate of mutation that follows the inactivation of a DNA repair gene. Many DNA repair pathways and the genes that encode them—which we now know operate in a wide variety of organisms, including humans—were originally identified in bacteria by the isolation and characterization of mutants that displayed an increased mutation rate or an increased sensitivity to DNA-damaging agents.

Recent studies of the consequences of a diminished capacity for DNA repair in humans have linked a variety of human diseases with decreased repair (Table 5–2). Thus, we saw previously that defects in a human gene that normally functions to repair the mismatched base pairs in DNA resulting from replication errors can lead to an inherited predisposition to certain cancers, reflecting an increased mutation rate. In another human disease, *xeroderma pigmentosum* (XP), the afflicted individuals have an extreme sensitivity to ultraviolet radiation because they are unable to repair certain DNA photoproducts. This repair defect results in an increased mutation rate that leads to serious skin lesions and an increased susceptibility to certain cancers.

Without DNA Repair, Spontaneous DNA Damage Would Rapidly Change DNA Sequences

Although DNA is a highly stable material, as required for the storage of genetic information, it is a complex organic molecule that is susceptible, even under

TABLE 5–2 Inherited Syndromes with Defects in DNA Repair

NAME	PHENOTYPE	ENZYME OR PROCESS AFFECTED
MSH2, 3, 6, MLH1, PMS2	colon cancer	mismatch repair
Xeroderma pigmentosum (XP) groups A–G	skin cancer, cellular UV sensitivity, neurological abnormalities	nucleotide excision-repair
XP variant	cellular UV sensitivity	translesion synthesis by DNA polymerase δ
Ataxia–telangiectasia (AT)	leukemia, lymphoma, cellular γ-ray sensitivity, genome instability	ATM protein, a protein kinase activated by double-strand breaks
BRCA-2	breast and ovarian cancer	repair by homologous recombination
Werner syndrome	premature aging, cancer at several sites, genome instability	accessory 3′-exonuclease and DNA helicase
Bloom syndrome	cancer at several sites, stunted growth, genome instability	accessory DNA helicase for replication
Fanconi anemia groups A–G	congenital abnormalities, leukemia, genome instability	DNA interstrand cross-link repair
46 BR patient	hypersensitivity to DNA-damaging agents, genome instability	DNA ligase I

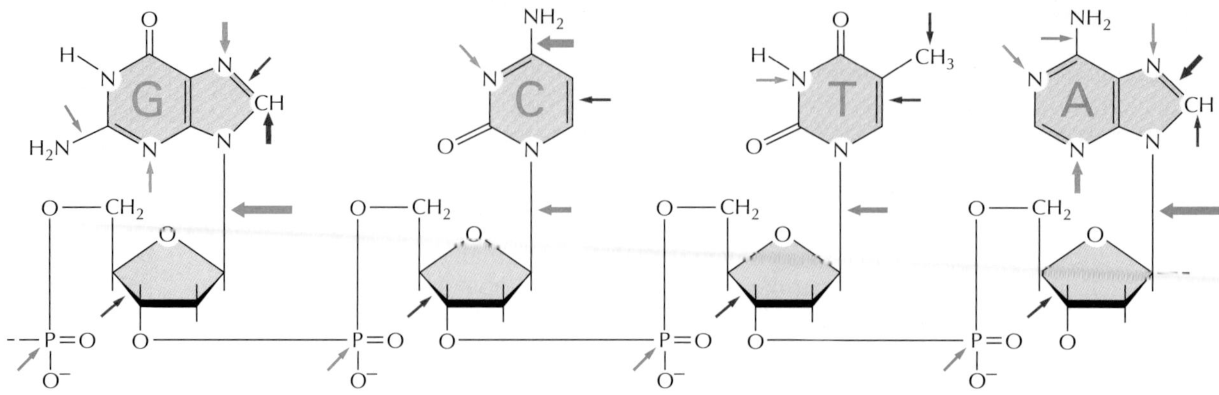

Figure 5–46 A summary of spontaneous alterations likely to require DNA repair. The sites on each nucleotide that are known to be modified by spontaneous oxidative damage *(red arrows)*, hydrolytic attack *(blue arrows)*, and uncontrolled methylation by the methyl group donor S-adenosylmethionine *(green arrows)* are shown, with the width of each arrow indicating the relative frequency of each event. (After T. Lindahl, *Nature* 362:709–715, 1993. © Macmillan Magazines Ltd.)

normal cellular conditions, to spontaneous changes that would lead to mutations if left unrepaired (Figure 5–46). DNA undergoes major changes as a result of thermal fluctuations: for example, about 5000 purine bases (adenine and guanine) are lost every day from the DNA of each human cell because their N-glycosyl linkages to deoxyribose hydrolyze, a spontaneous reaction called *depurination*. Similarly, a spontaneous *deamination* of cytosine to uracil in DNA occurs at a rate of about 100 bases per cell per day (Figure 5–47). DNA bases are also occasionally damaged by an encounter with reactive metabolites (including reactive forms of oxygen) or environmental chemicals. Likewise, ultraviolet radiation from the sun can produce a covalent linkage between two adjacent pyrimidine bases in DNA to form, for example, thymine dimers (Figure 5–48). If left uncorrected when the DNA is replicated, most of these changes would be expected to lead either to the deletion of one or more base pairs or to a base-pair substitution in the daughter DNA chain (Figure 5–49). The mutations would then be propagated throughout subsequent cell generations as the DNA is replicated. Such a high rate of random changes in the DNA sequence would have disastrous consequences for an organism.

Figure 5–47 Depurination and deamination. These two reactions are the most frequent spontaneous chemical reactions known to create serious DNA damage in cells. Depurination can release guanine (shown here), as well as adenine, from DNA. The major type of deamination reaction (shown here) converts cytosine to an altered DNA base, uracil, but deamination occurs on other bases as well. These reactions take place on double-helical DNA; for convenience, only one strand is shown.

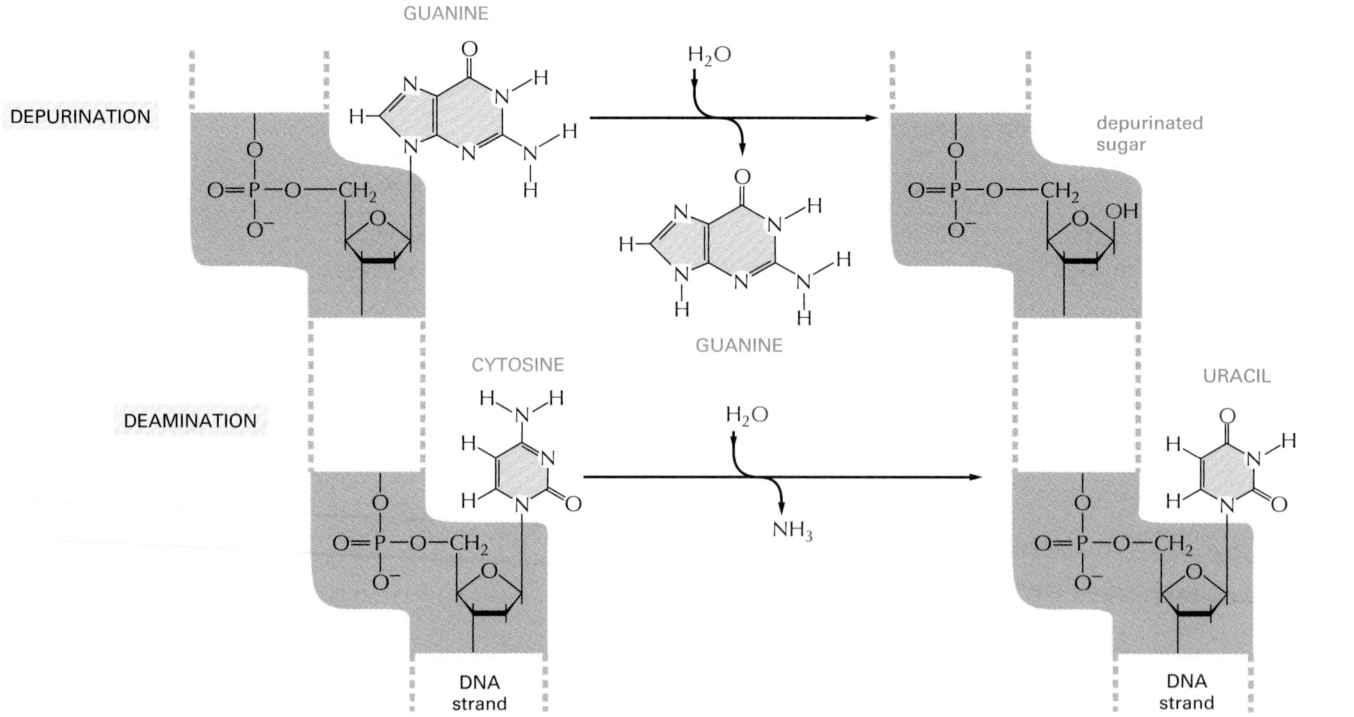

DEPURINATION
GUANINE
H_2O
depurinated sugar
GUANINE
DEAMINATION
CYTOSINE
H_2O
NH_3
URACIL
DNA strand
DNA strand

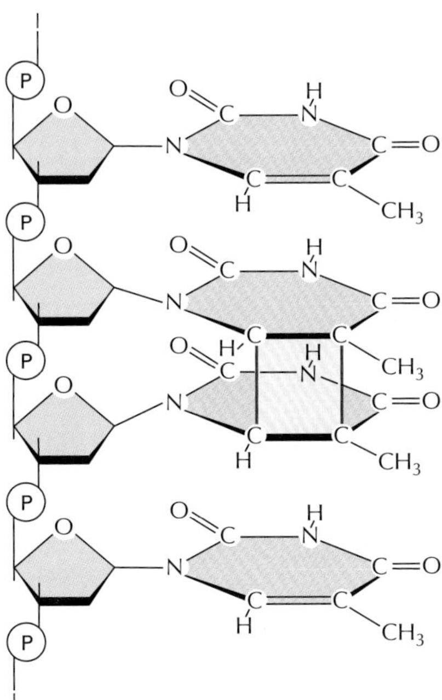

Figure 5–48 The thymine dimer. This type of damage is introduced into DNA in cells that are exposed to ultraviolet irradiation (as in sunlight). A similar dimer will form between any two neighboring pyrimidine bases (C or T residues) in DNA.

The DNA Double Helix Is Readily Repaired

The double-helical structure of DNA is ideally suited for repair because it carries two separate copies of all the genetic information—one in each of its two strands. Thus, when one strand is damaged, the complementary strand retains an intact copy of the same information, and this copy is generally used to restore the correct nucleotide sequences to the damaged strand.

An indication of the importance of a double-stranded helix to the safe storage of genetic information is that all cells use it; only a few small viruses use single-stranded DNA or RNA as their genetic material. The types of repair processes described in this section cannot operate on such nucleic acids, and the chance of a permanent nucleotide change occurring in these single-stranded genomes of viruses is thus very high. It seems that only organisms with tiny genomes can afford to encode their genetic information in any molecule other than a DNA double helix.

Each cell contains multiple DNA repair systems, each with its own enzymes and preferences for the type of damage recognized. As we see in the rest of this section, most of these systems use the undamaged strand of the double helix as a template to repair the damaged strand.

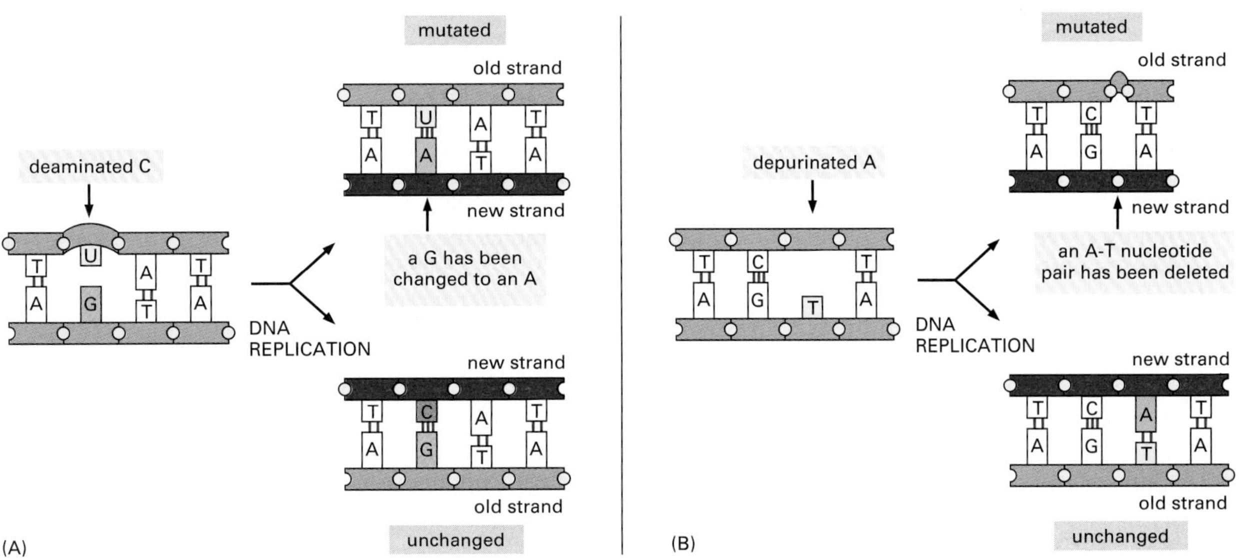

Figure 5–49 How chemical modifications of nucleotides produce mutations. (A) Deamination of cytosine, if uncorrected, results in the substitution of one base for another when the DNA is replicated. As shown in Figure 5–47, deamination of cytosine produces uracil. Uracil differs from cytosine in its base-pairing properties and preferentially base-pairs with adenine. The DNA replication machinery therefore adds an adenine when it encounters a uracil on the template strand. (B) Depurination, if uncorrected, can lead to either the substitution or the loss of a nucleotide pair. When the replication machinery encounters a missing purine on the template strand, it may skip to the next complete nucleotide as illustrated here, thus producing a nucleotide deletion in the newly synthesized strand. Many other types of DNA damage (see Figure 5–46) also produce mutations when the DNA is replicated if left uncorrected.

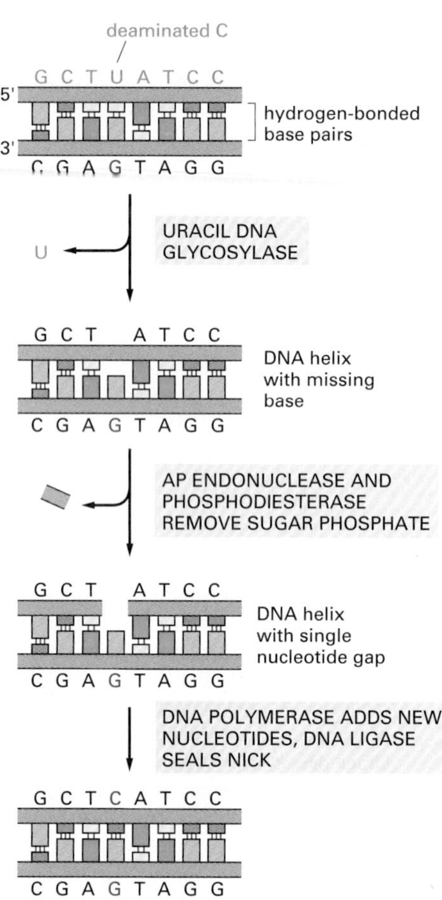

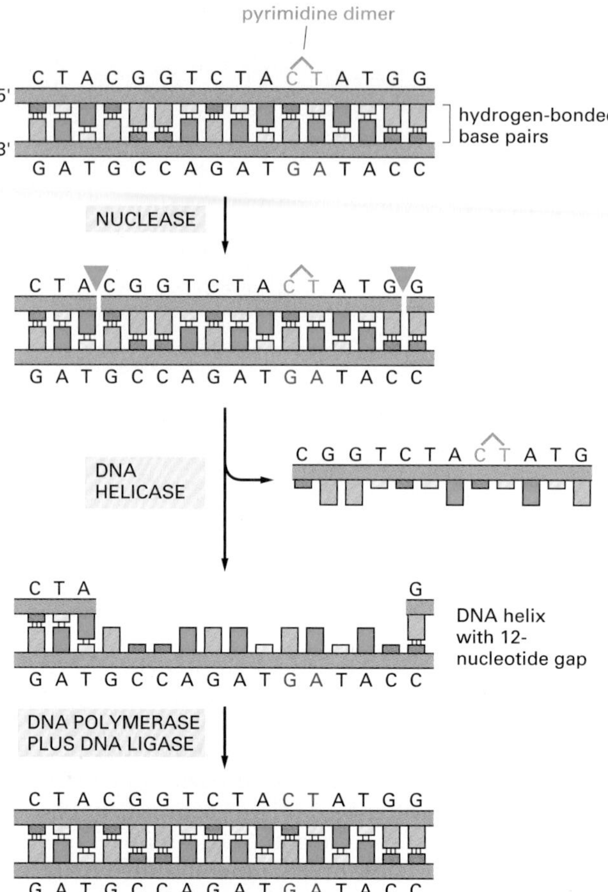

Figure 5–50 A comparison of two major DNA repair pathways. (A) *Base excision repair.* This pathway starts with a DNA glycosylase. Here the enzyme uracil DNA glycosylase removes an accidentally deaminated cytosine in DNA. After the action of this glycosylase (or another DNA glycosylase that recognizes a different kind of damage), the sugar phosphate with the missing base is cut out by the sequential action of AP endonuclease and a phosphodiesterase. (These same enzymes begin the repair of depurinated sites directly.) The gap of a single nucleotide is then filled by DNA polymerase and DNA ligase. The net result is that the U that was created by accidental deamination is restored to a C. The AP endonuclease derives its name from the fact that it recognizes any site in the DNA helix that contains a deoxyribose sugar with a missing base; such sites can arise either by the loss of a purine (*a*purinic sites) or by the loss of a pyrimidine (*a*pyrimidinic sites). (B) *Nucleotide excision repair.* After a multienzyme complex has recognized a bulky lesion such as a pyrimidine dimer (see Figure 5–48), one cut is made on each side of the lesion, and an associated DNA helicase then removes the entire portion of the damaged strand. The multienzyme complex in bacteria leaves the gap of 12 nucleotides shown; the gap produced in human DNA is more than twice this size. The nucleotide excision repair machinery can recognize and repair many different types of DNA damage.

DNA Damage Can Be Removed by More Than One Pathway

There are multiple pathways for DNA repair, using different enzymes that act upon different kinds of lesions. Two of the most common pathways are shown in Figure 5–50. In both, the damage is excised, the original DNA sequence is restored by a DNA polymerase that uses the undamaged strand as its template, and the remaining break in the double helix is sealed by DNA ligase (see Figure 5–14).

The two pathways differ in the way in which the damage is removed from DNA. The first pathway, called **base excision repair**, involves a battery of enzymes called *DNA glycosylases,* each of which can recognize a specific type of altered base in DNA and catalyze its hydrolytic removal. There are at least six types of these enzymes, including those that remove deaminated Cs, deaminated As, different types of alkylated or oxidized bases, bases with opened rings, and bases in which a carbon–carbon double bond has been accidentally converted to a carbon–carbon single bond.

As an example of the general mechanism of base excision repair, the removal of a deaminated C by uracil DNA glycosylase is shown in Figure 5–50A. How is the altered base detected within the context of the double helix? A key step is an enzyme-mediated "flipping-out" of the altered nucleotide from the helix, which allows the enzyme to probe all faces of the base for damage (Figure 5–51). It is thought that DNA glycosylases travel along DNA using base-flipping to evaluate the status of each base pair. Once a damaged base is recognized, the DNA glycosylase reaction creates a deoxyribose sugar that lacks its base. This "missing tooth" is recognized by an enzyme called *AP endonuclease,* which cuts the phosphodiester backbone, and the damage is then removed and repaired (see Figure 5–50A). Depurination, which is by far the most frequent type of damage suffered by DNA, also leaves a deoxyribose sugar with a missing base. Depurinations are directly repaired beginning with AP endonuclease, following the bottom half of the pathway in Figure 5–50A.

The second major repair pathway is called **nucleotide excision repair**. This mechanism can repair the damage caused by almost any large change in the structure of the DNA double helix. Such "bulky lesions" include those created by the covalent reaction of DNA bases with large hydrocarbons (such as the carcinogen benzopyrene), as well as the various pyrimidine dimers (T-T, T-C, and C-C) caused by sunlight. In this pathway, a large multienzyme complex scans the DNA for a distortion in the double helix, rather than for a specific base change. Once a bulky lesion has been found, the phosphodiester backbone of the abnormal strand is cleaved on both sides of the distortion, and an oligonucleotide containing the lesion is peeled away from the DNA double helix by a DNA helicase enzyme. The large gap produced in the DNA helix is then repaired by DNA polymerase and DNA ligase (Figure 5–50B).

The Chemistry of the DNA Bases Facilitates Damage Detection

The DNA double helix seems to be optimally constructed for repair. As noted above, it contains a backup copy of the genetic information, so that if one strand is damaged, the other undamaged strand can be used as a template for repair. The nature of the bases also facilitates the distinction between undamaged and damaged bases. Thus, every possible deamination event in DNA yields an unnatural base, which can therefore be directly recognized and removed by a specific DNA glycosylase. Hypoxanthine, for example, is the simplest purine base capable of pairing specifically with C, but hypoxanthine is the direct deamination product of A (Figure 5–52A). The addition of a second amino group to hypoxanthine produces G, which cannot be formed from A by spontaneous deamination, and whose deamination product is likewise unique.

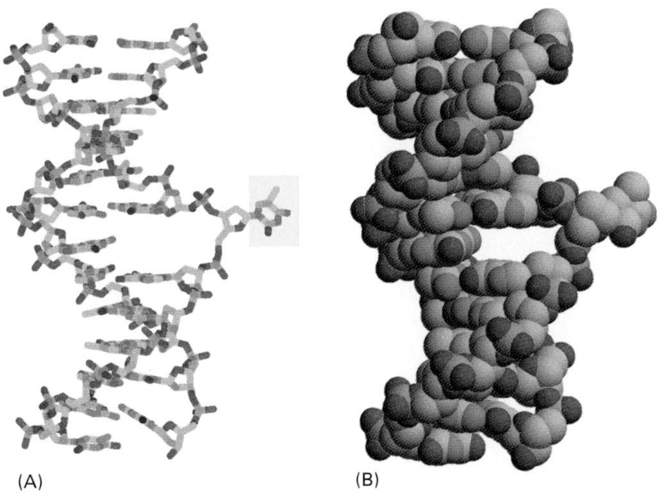

Figure 5–51 The recognition of an unusual nucleotide in DNA by base-flipping. The DNA glycosylase family of enzymes recognizes specific bases in the conformation shown. Each of these enzymes cleaves the glycosyl bond that connects a particular recognized base *(yellow)* to the backbone sugar, removing it from the DNA. (A) Stick model; (B) space-filling model.

As discussed in Chapter 6, RNA is thought, on an evolutionary time-scale, to have served as the genetic material before DNA, and it seems likely that the genetic code was initially carried in the four nucleotides A, C, G, and U. This raises the question of why the U in RNA was replaced in DNA by T (which is 5-methyl U). We have seen that the spontaneous deamination of C converts it to U, but that this event is rendered relatively harmless by uracil DNA glycosylase. However, if DNA contained U as a natural base, the repair system would be unable to distinguish a deaminated C from a naturally occuring U.

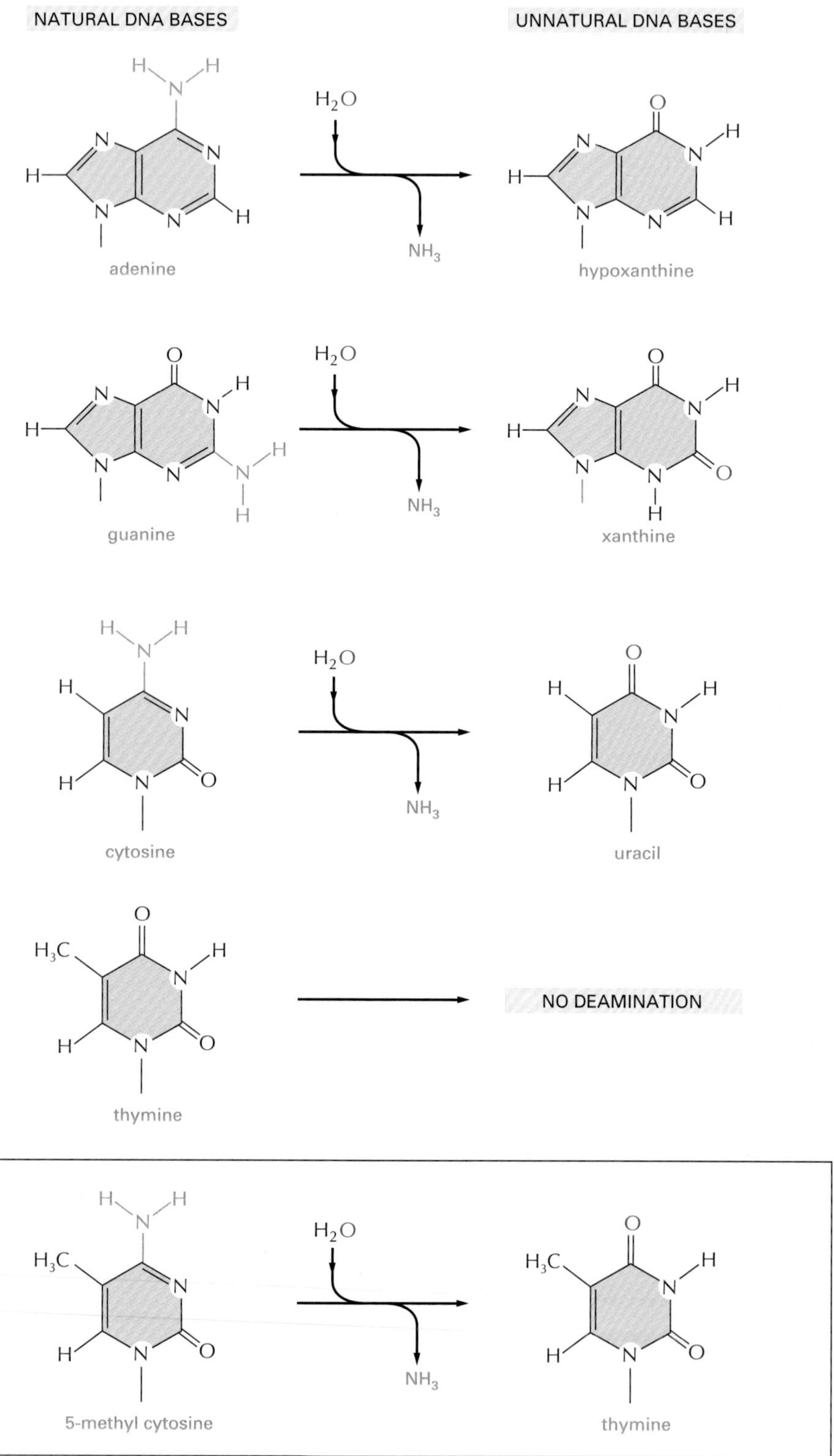

Figure 5–52 The deamination of DNA nucleotides. In each case the oxygen atom that is added in this reaction with water is colored *red*. (A) The spontaneous deamination products of A and G are recognizable as unnatural when they occur in DNA and thus are readily recognized and repaired. The deamination of C to U was previously illustrated in Figure 5–47; T has no amino group to deaminate. (B) About 3% of the C nucleotides in vertebrate DNAs are methylated to help in controlling gene expression (discussed in Chapter 7). When these 5-methyl C nucleotides are accidentally deaminated, they form the natural nucleotide T. This T would be paired with a G on the opposite strand, forming a mismatched base pair.

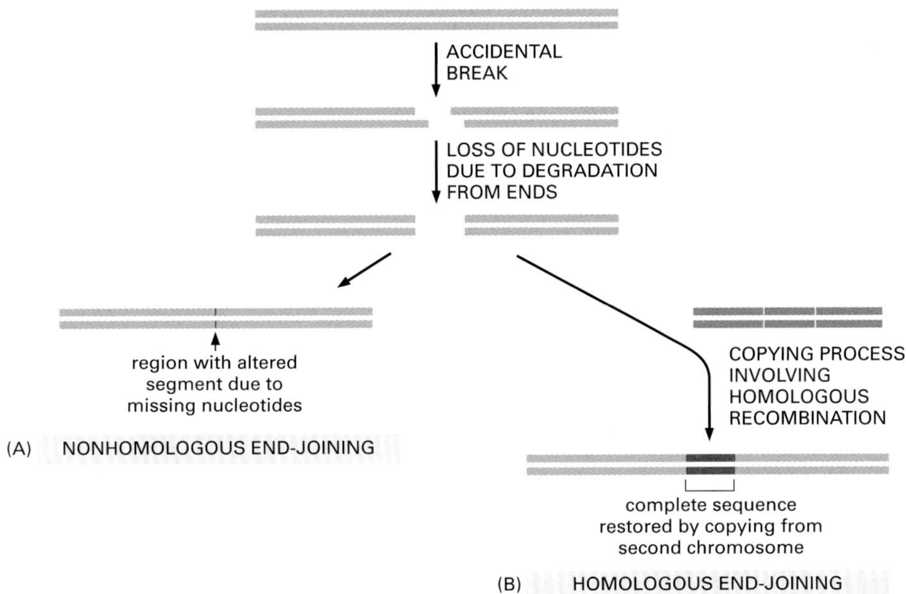

Figure 5–53 Two different types of end-joining for repairing double-strand breaks. (A) Nonhomologous end-joining alters the original DNA sequence when repairing broken chromosomes. These alterations can be either deletions (as shown) or short insertions. (B) Homologous end-joining is more difficult to accomplish, but is much more precise.

A special situation occurs in vertebrate DNA, in which selected C nucleotides are methylated at specific C-G sequences that are associated with inactive genes (discussed in Chapter 7). The accidental deamination of these methylated C nucleotides produces the natural nucleotide T (Figure 5–52B) in a mismatched base pair with a G on the opposite DNA strand. To help in repairing deaminated methylated C nucleotides, a special DNA glycosylase recognizes a mismatched base pair involving T in the sequence T-G and removes the T. This DNA repair mechanism must be relatively ineffective, however, because methylated C nucleotides are common sites for mutations in vertebrate DNA. It is striking that, even though only about 3% of the C nucleotides in human DNA are methylated, mutations in these methylated nucleotides account for about one-third of the single-base mutations that have been observed in inherited human diseases.

Double-Strand Breaks are Efficiently Repaired

A potentially dangerous type of DNA damage occurs when both strands of the double helix are broken, leaving no intact template strand for repair. Breaks of this type are caused by ionizing radiation, oxidizing agents, replication errors, and certain metabolic products in the cell. If these lesions were left unrepaired, they would quickly lead to the breakdown of chromosomes into smaller fragments. However, two distinct mechanisms have evolved to ameliorate the potential damage. The simplest to understand is *nonhomologous end-joining*, in which the broken ends are juxtaposed and rejoined by DNA ligation, generally with the loss of one or more nucleotides at the site of joining (Figure 5–53A). This end-joining mechanism, which can be viewed as an emergency solution to the repair of double-strand breaks, is a common outcome in mammalian cells. Although a change in the DNA sequence (a mutation) results at the site of breakage, so little of the mammalian genome codes for proteins that this mechanism is apparently an acceptable solution to the problem of keeping chromosomes intact. As previously discussed, the specialized structure of telomeres prevents the ends of chromosomes from being mistaken for broken DNA, thereby preserving natural DNA ends.

An even more effective type of double-strand break repair exploits the fact that cells that are diploid contain two copies of each double helix. In this second repair pathway, called *homologous end-joining*, general recombination mechanisms are called into play that transfer nucleotide sequence information from the intact DNA double helix to the site of the double-strand break in the broken helix. This type of reaction requires special recombination proteins that recognize areas of DNA sequence matching between the two chromosomes and bring them together. A DNA replication process then uses the undamaged chromosome as the template for transferring genetic information to the broken chromosome, repairing it with no change in the DNA sequence (Figure 5–53B). In cells that have replicated their DNA but not yet divided, this type of DNA repair can readily take place between the two sister DNA molecules in each chromosome; in this case, there is no need for the broken ends to find the matching DNA sequence in the homologous chromosome. The molecular details of the homologous end-joining reaction are considered later in this chapter because they require a general understanding of the way in which cells carry their genetic recombination events. Although present in humans, this type of DNA double-strand break repair predominates in bacteria, yeasts, and *Drosophila*—all organisms in which little nonhomologous DNA end-joining is observed.

Cells Can Produce DNA Repair Enzymes in Response to DNA Damage

Cells have evolved many mechanisms that help them survive in an unpredictably hazardous world. Often an extreme change in a cell's environment activates the expression of a set of genes whose protein products protect the cell from the deleterious effects of this change. One such mechanism shared by all cells is the *heat-shock response,* which is evoked by the exposure of cells to unusually high temperatures. The induced "heat-shock proteins" include some that help stabilize and repair partly denatured cell proteins, as discussed in Chapter 6.

Cells also have mechanisms that elevate the levels of DNA repair enzymes, as an emergency response to severe DNA damage. The best-studied example is the so-called **SOS response** in *E. coli*. In this bacterium, any block to DNA replication caused by DNA damage produces a signal that induces an increase in the transcription of more than 15 genes, many of which code for proteins that function in DNA repair. The signal (thought to be an excess of single-stranded DNA) first activates the *E. coli* RecA protein (see Figure 5–58), so that it destroys a gene regulatory protein that normally represses the transcription of a large set of SOS response genes.

Studies of mutant bacteria deficient in different parts of the SOS response demonstrate that the newly synthesized proteins have two effects. First, as would be expected, the induction of these additional DNA repair enzymes increases cell survival after DNA damage. Second, several of the induced proteins transiently increase the mutation rate by increasing the number of errors made in copying DNA sequences. The errors are caused by the production of low-fidelity DNA polymerases that can efficiently use damaged DNA as a template for DNA synthesis. While this "error-prone" DNA repair can be harmful to individual bacterial cells, it is presumed to be advantageous in the long term because it produces a burst of genetic variability in the bacterial population that increases the likelihood of a mutant cell arising that is better able to survive in the altered environment.

Human cells contain more than ten minor DNA polymerases, many of which are specifically called into play, as a last resort, to copy over unrepaired lesions in the DNA template. These enzymes can recognize a specific type of DNA damage and add the nucleotides that restore the initial sequence. Each such polymerase molecule is given a chance to add only one or a few nucleotides, because these enzymes are extremely error-prone when they copy a normal DNA sequence. Although the details of these fascinating reactions are still being worked out, they provide elegant testimony to the care with which organisms maintain their DNA sequences.

DNA Damage Delays Progression of the Cell Cycle

We have just seen that cells contain multiple enzyme systems that can recognize DNA damage and promote the repair of these lesions. Because of the importance of maintaining intact, undamaged DNA from generation to generation, cells have an additional mechanism that helps them respond to DNA damage: they delay progression of the cell cycle until DNA repair is complete. For example, one of the genes expressed in response to the *E. coli* SOS signal is *sulA*, which encodes an inhibitor of cell division. Thus, when the SOS functions are turned on in response to DNA damage, a block to cell division extends the time for repair. When DNA repair is complete, the expression of the SOS genes is repressed, the cell cyle resumes, and the undamaged DNA is segregated to the daughter cells.

Damaged DNA also generates signals that block cell-cycle progression in eucaryotes. As discussed in detail in Chapter 17, the orderly progression of the cell cycle is maintained through the use of *checkpoints* that ensure the completion of one step before the next step can begin. At several of these cell-cycle checkpoints, the cycle stops if damaged DNA is detected. Thus, in yeast, the presence of DNA damage can block entry into the G_1 phase; it can slow DNA replication once begun; and it can block the transition from S phase to M phase. The DNA damage results in an increased synthesis of some DNA repair enzymes, and the delays further facilitate repair by providing the time needed for repair to reach completion.

The importance of the special signaling mechanisms that respond to DNA damage is indicated by the phenotype of humans who are born with defects in the gene that encodes the *ATM protein,* a large protein kinase. These individuals have the disease *ataxia–telangiectasia (AT),* whose symptoms include neurodegeneration, a predisposition to cancer, and genome instability. In both humans and yeasts, the ATM protein is needed to generate the initial intracellular signals that produce a response to oxygen-inflicted DNA damage, and individual organisms with defects in this protein are hypersensitive to agents that cause such damage, such as ionizing radiation.

Summary

Genetic information can be stored stably in DNA sequences only because a large set of DNA repair enzymes continuously scan the DNA and replace any damaged nucleotides. Most types of DNA repair depend on the presence of a separate copy of the genetic information in each of the two strands of the DNA double helix. An accidental lesion on one strand can therefore be cut out by a repair enzyme and a corrected strand resynthesized by reference to the information in the undamaged strand.

Most of the damage to DNA bases is excised by one of two major DNA repair pathways. In base excision repair, the altered base is removed by a DNA glycosylase enzyme, followed by excision of the resulting sugar phosphate. In nucleotide excision repair, a small section of the DNA strand surrounding the damage is removed from the DNA double helix as an oligonucleotide. In both cases, the gap left in the DNA helix is filled in by the sequential action of DNA polymerase and DNA ligase, using the undamaged DNA strand as the template.

Other critical repair systems—based on either nonhomologous or homologous end-joining mechanisms—reseal the accidental double-strand breaks that occur in the DNA helix. In most cells, an elevated level of DNA damage causes both an increased synthesis of repair enzymes and a delay in the cell cycle. Both factors help to ensure that DNA damage is repaired before a cell divides.

GENERAL RECOMBINATION

In the two preceding sections, we discussed the mechanisms that allow the DNA sequences in cells to be maintained from generation to generation with very little change. However, it is also clear that these DNA sequences can occasionally be rearranged. The particular combination of genes present in any individual genome, as well as the timing and the level of expression of these genes, is often altered by such DNA rearrangements. In a population, this type of genetic

Figure 5–54 General recombination. The breaking and rejoining of two homologous DNA double helices creates two DNA molecules that have "crossed over." In meiosis, this process causes each chromosome in a germ cell to contain a mixture of maternally and paternally inherited genes.

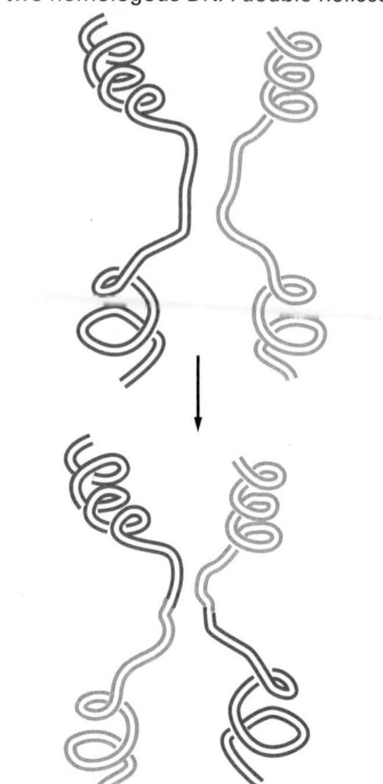

variation is crucial to allow organisms to evolve in response to a changing environment. The DNA rearrangements are caused by a set of mechanisms that are collectively called **genetic recombination**. Two broad classes are commonly recognized—*general recombination* and *site-specific recombination*. In this part of the chapter we discuss the first of these two mechanisms; in the next part, we consider the second mechanism.

In **general recombination** (also known as *homologous recombination*), genetic exchange takes place between a pair of homologous DNA sequences. These are usually located on two copies of the same chromosome, although other types of DNA molecules that share the same nucleotide sequence can also participate. The general recombination reaction is essential for every proliferating cell, because accidents occur during nearly every round of DNA replication that interrupt the replication fork and require general recombination mechanisms to repair. The details of the intimate interplay between replication and recombination are still incompletely understood, but they include using variations of the homologous end-joining reaction (see Figure 5–53) to restart replication forks that have run into a break in the parental DNA template.

General recombination is also essential for the accurate chromosome segregation that occurs during meiosis in fungi, plants, and animals (see Figure 20–11). The crossing-over of chromosomes that results causes bits of genetic information to be exchanged to create new combinations of DNA sequences in each chromosome. The evolutionary benefit of this type of gene mixing is apparently so great that the reassortment of genes by general recombination is not confined to multicellular organisms; it is also widespread in single-celled organisms.

The central features that lie at the heart of the general recombination mechanism seem to be the same in all organisms. Most of what we know about the biochemistry of genetic recombination was originally derived from studies of bacteria, especially of *E. coli* and its viruses, as well as from experiments with simple eucaryotes such as yeasts. For these organisms with short generation times and relatively small genomes, it was possible to isolate a large set of mutants with defects in their recombination processes. The identification of the protein altered in each mutant then allowed the collection of proteins that catalyze general recombination to be identified and characterized. More recently, close relatives of these proteins have been discovered and extensively characterized in *Drosophila*, mice, and humans as well.

General Recombination Is Guided by Base-pairing Interactions Between Two Homologous DNA Molecules

The abundant general recombination observed in meiosis has the following characteristics: (1) Two homologous DNA molecules that were originally part of different chromosomes "cross over;" that is, their double helices break and the two broken ends join to their opposite partners to re-form two intact double helices, each composed of parts of the two initial DNA molecules (Figure 5–54). (2) The site of exchange (that is, where a *red* double helix is joined to a *green* double helix in Figure 5–54) can occur anywhere in the homologous nucleotide sequences of the two participating DNA molecules. (3) At the site of exchange, a strand of one DNA molecule has become base-paired to a strand of the second DNA molecule to create a *heteroduplex joint* that links the two double helices (Figure 5–55). This heteroduplex region can be thousands of base pairs long; we explain later how it forms. (4) No nucleotide sequences are altered at the site of exchange; some DNA replication usually takes place, but the cleavage and rejoining events occur so precisely that not a single nucleotide is lost or gained. Despite its precision, general recombination creates DNA molecules of novel sequence: the heteroduplex joint can tolerate a small number of mismatched

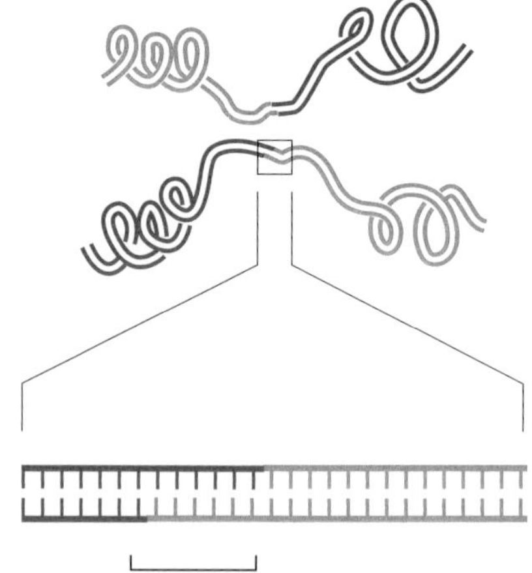

Figure 5–55 A heteroduplex joint. This structure unites two DNA molecules where they have crossed over. Such a joint is often thousands of nucleotides long.

base pairs, and, more importantly, the two DNA molecules that cross over are usually not exactly the same on either side of the joint. As a result, new recombinant DNA molecules (recombinant chromosomes) are generated.

The mechanism of general recombination ensures that two DNA double helices undergo an exchange reaction only if they contain an extensive region of sequence similarity (homology). The formation of a long heteroduplex joint requires such homology because it involves a long region of complementary base-pairing between a strand from one of the two original double helices and a complementary strand from the other double helix. But how does this heteroduplex joint arise, and how do the two homologous regions of DNA at the site of crossing-over recognize each other? As we shall see, recognition takes place during a process called **DNA synapsis**, in which base pairs form between complementary strands from the two DNA molecules. This base-pairing is then extended to guide the general recombination process, allowing it to occur only between DNA molecules that contain long regions of matching (or nearly matching) DNA sequence.

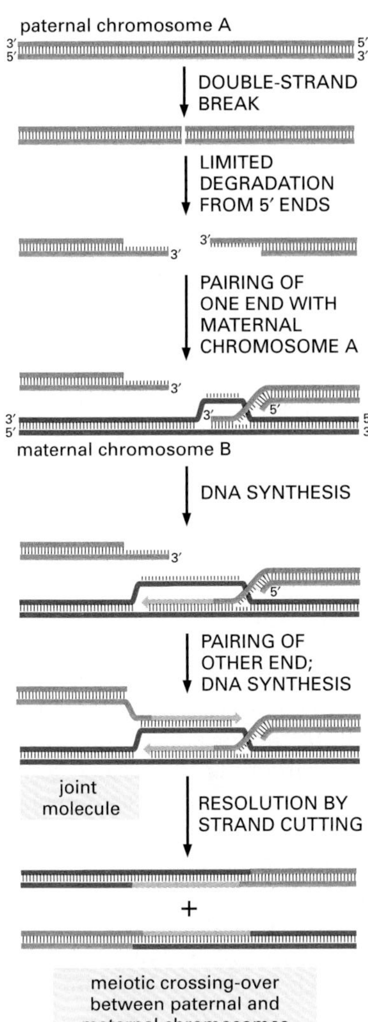

Figure 5–56 General recombination in meiosis. As indicated, the process begins when an endonuclease makes a double-strand break in a chromosome. An exonuclease then creates two protruding 3′ single-stranded ends, which find the homologous region of a second chromosome to begin DNA synapsis. The joint molecule formed can eventually be resolved by selective strand cuts to produce two chromosomes that have crossed over, as shown.

Meiotic Recombination Is Initiated by Double-strand DNA Breaks

Extensive base-pair interactions cannot occur between two intact DNA double helices. Thus, the DNA synapsis that is critical for general recombination in meiosis can begin only after a DNA strand from one DNA helix has been exposed and its nucleotides have been made available for pairing with another DNA helix. In the absence of direct experimental evidence, theoretical models were proposed based on the idea that a break needed to be made in just one of the two strands of a DNA helix to produce the exposed DNA strand required for DNA synapsis. This break in the phosphodiester backbone was thought to allow one of the nicked strand ends to separate from its base-paired partner strand, freeing it to form a short heteroduplex with a second intact DNA helix—thereby beginning synapsis. Models of this type are reasonable in theory, and they have been described in textbooks for nearly 30 years.

In the early 1990s, sensitive biochemical techniques became available for determining the actual structure of the recombination intermediates that form in yeast chromosomes at various stages of meiosis. These studies revealed that general recombination is initiated by a special endonuclease that simultaneously cuts *both* strands of the double helix, creating a complete break in the DNA molecule. The 5′ ends at the break are then chewed back by an exonuclease, creating protruding single-stranded 3′ ends. It is these single strands that search for a homologous DNA helix with which to pair—leading to the formation of a "joint molecule" between a maternal and a paternal chromosome (Figure 5–56).

In the next section, we begin to explain how a DNA single strand can "find" a homologous double-stranded DNA molecule to begin DNA synapsis.

DNA Hybridization Reactions Provide a Simple Model for the Base-pairing Step in General Recombination

In its simplest form, the type of base-pairing interaction central to the synapsis step of general recombination can be mimicked in a test tube by allowing a DNA double helix to re-form from its separated single strands. This process, called *DNA renaturation* or **hybridization**, occurs when a rare random collision juxtaposes complementary nucleotide sequences on two matching DNA single strands, allowing the formation of a short stretch of double helix between them. This relatively slow helix nucleation step is followed by a very rapid "zippering" step, as the region of double helix is extended to maximize the number of base-pairing interactions (Figure 5–57).

Formation of a new double helix in this way requires that the annealing strands be in an open, unfolded conformation. For this reason, *in vitro* hybridization reactions are performed at either high temperature or in the presence of an organic solvent such as formamide; these conditions "melt out" the

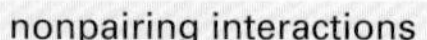

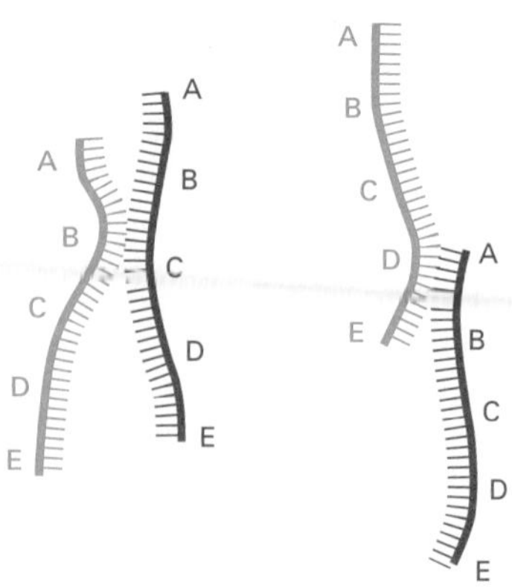

pairing interactions

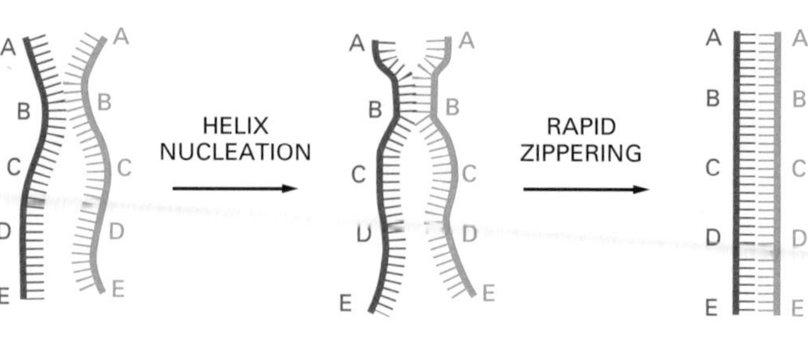

Figure 5–57 DNA hybridization. DNA double helices re-form from their separated strands in a reaction that depends on the random collision of two complementary DNA strands. The vast majority of such collisions are not productive, as shown on the left, but a few result in a short region where complementary base pairs have formed (helix nucleation). A rapid zippering then leads to the formation of a complete double helix. Through this trial-and-error process, a DNA strand will find its complementary partner even in the midst of millions of nonmatching DNA strands. A related, highly efficient trial-and-error recognition of a complementary partner DNA sequence seems to initiate all general recombination events.

short hairpin helices that result from the base-pairing interactions that occur within a single strand that folds back on itself. Most cells cannot survive such harsh conditions and instead use a single-strand DNA-binding (SSB) protein (see p. 246) to melt out the hairpin helices and help anneal their complementary single strands. This protein is essential for DNA replication (as described earlier) as well as for general recombination; it binds tightly and cooperatively to the sugar-phosphate backbone of all single-stranded DNA regions of DNA, holding them in an extended conformation with the bases exposed (see Figures 5–17 and 5–18). In this extended conformation, a DNA single strand can base-pair efficiently either with a nucleoside triphosphate molecule (in DNA replication) or with a complementary section of another DNA single strand (as part of a genetic recombination process).

The partner that a DNA single-strand needs to find in the synapsis step of general recombination is a DNA double helix, rather than a second single strand of DNA (see Figure 5–56). In the next section we see how the critical event that allows DNA hybridization to begin during recombination—the initial invasion of a single-stranded DNA into a DNA double helix—is achieved by the cell.

The RecA Protein and its Homologs Enable a DNA Single Strand to Pair with a Homologous Region of DNA Double Helix

General recombination is more complex than the simple hybridization reactions just described involving single-stranded DNA, and it requires several types of specialized proteins. In particular, the *E. coli* **RecA protein** has a central role in the recombination between chromosomes; it and its homologs in yeast, mice, and humans make synapsis possible (Figure 5–58).

Like a single-strand DNA-binding protein, the RecA type of protein binds tightly and in long cooperative clusters to single-stranded DNA to form a

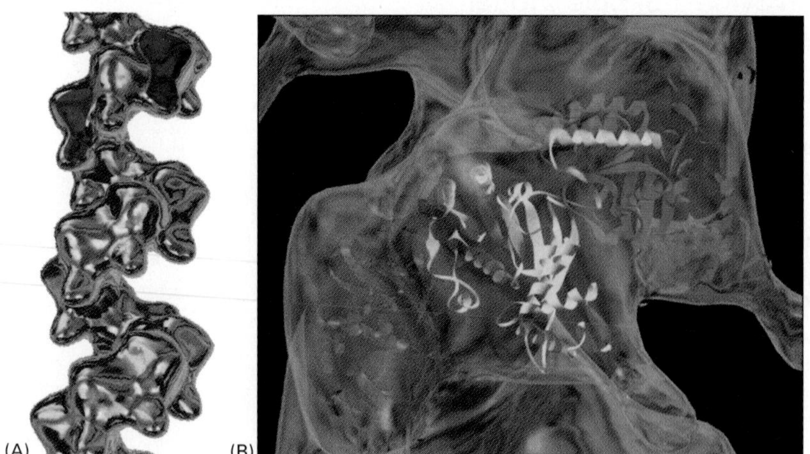

Figure 5–58 The structure of the RecA and Rad51 protein–DNA filaments. (A) The Rad51 protein bound to a DNA single strand. Rad51 is a human homolog of the bacterial RecA protein; three successive monomers in this helical filament are colored. (B) A short section of the RecA filament, with the three-dimensional structure of the protein fitted to the image of the filament determined by electron microscopy. The two DNA–protein filaments appear to be quite similar. There are about six RecA monomers per turn of the helix, holding 18 nucleotides of single-stranded DNA that is stretched out by the protein. The exact path of the DNA in this structure is not known. (A, courtesy of Edward Egelman; B, from X. Yu et al., *J. Mol. Biol.* 283:985–992, 1998.)

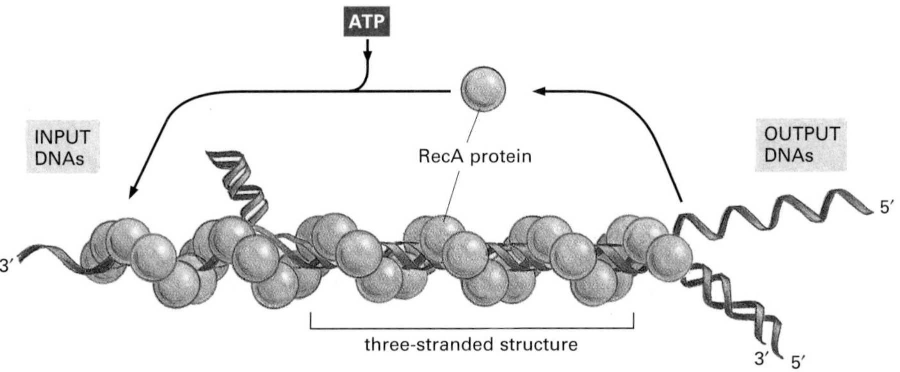

Figure 5–59 DNA synapsis catalyzed by the RecA protein. *In vitro* experiments show that several types of complex are formed between a DNA single strand covered with RecA protein *(red)* and a DNA double helix *(green)*. First a non-base-paired complex is formed, which is converted through transient base-flipping (see Figure 5–51) to a three-stranded structure as soon as a region of homologous sequence is found. This complex is unstable because it involves an unusual form of DNA, and it spins out a DNA heteroduplex (one strand *green* and the other strand *red*) plus a displaced single strand from the original helix *(green)*. Thus the structure shown in this diagram migrates to the left, reeling in the "input DNAs" while producing the "output DNAs." The net result is a DNA strand exchange identical to that diagrammed earlier in Figure 5–56. (Adapted from S.C. West, *Annu. Rev. Biochem.* 61:603–640, 1992.)

nucleoprotein filament. Because each RecA monomer has more than one DNA-binding site, a RecA filament can hold a single strand and a double helix together. This allows it to catalyze a multistep DNA synapsis reaction between a DNA double helix and a homologous region of single-stranded DNA. The region of homology is identified before the duplex DNA target has been opened up, through a three-stranded intermediate in which the DNA single strand forms transient base pairs with bases that flip out from the helix in the major groove of the double-stranded DNA molecule (Figure 5–59). This reaction begins the pairing shown previously in Figure 5–56, and it thereby initiates the exchange of strands between two recombining DNA double helices.

Once DNA synapsis has occurred, the short heteroduplex region where the strands from two different DNA molecules have begun to pair is enlarged through a process called *branch migration*. Branch migration can take place at any point where two single DNA strands with the same sequence are attempting to pair with the same complementary strand; in this reaction, an unpaired region of one of the single strands displaces a paired region of the other single strand, moving the branch point without changing the total number of DNA base pairs. Although spontaneous branch migration can occur, it proceeds equally in both directions, so it makes little progress and is unlikely to complete recombination efficiently (Figure 5–60A). The RecA protein catalyzes unidirectional branch migration, readily producing a region of heteroduplex DNA that is thousands of base pairs long (Figure 5–60B).

The catalysis of directional branch migration depends on a further property of the RecA protein. In addition to having two DNA-binding sites, the RecA protein is a DNA-dependent ATPase, with an additional site for binding and hydrolyzing ATP. The protein associates much more tightly with DNA when it has ATP bound than when it has ADP bound. Moreover, new RecA molecules

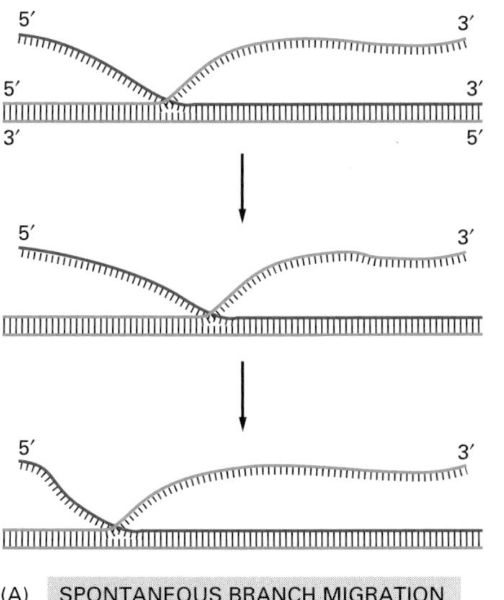

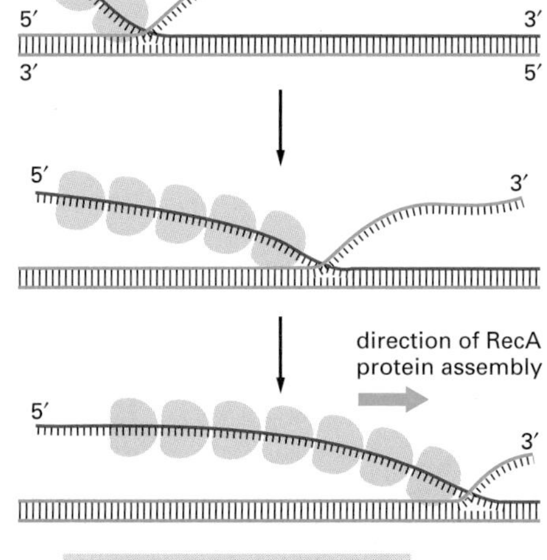

Figure 5–60 Two types of DNA branch migration observed in experiments *in vitro*. (A) Spontaneous branch migration is a back-and-forth, random-walk process, and it therefore makes little progress over long distances. (B) RecA-protein-directed branch migration proceeds at a uniform rate in one direction and may be driven by the polarized assembly of the RecA protein filament on a DNA single strand, which occurs in the direction indicated. Special DNA helicases that catalyze branch migration even more efficiently are also involved in recombination (for example see Figure 5–63).

with ATP bound are preferentially added at one end of the RecA protein filament, and the ATP is then hydrolyzed to ADP. The RecA protein filaments that form on DNA may therefore share some of the dynamic properties displayed by the cytoskeletal filaments formed from actin or tubulin (discussed in Chapter 16); an ability of the protein to "treadmill" unidirectionally along a DNA strand, for example, could drive the branch migration reaction shown in Figure 5–60B.

There Are Multiple Homologs of the RecA Protein in Eucaryotes, Each Specialized for a Specific Function

When one compares the proteins that catalyze the basic genetic functions in eucaryotes with those in bacteria such as *E. coli*, one generally finds that evolutionarily related proteins are present that catalyze similar reactions. In many cases, however, multiple eucaryotic homologs take the place of a particular bacterial protein, each specialized for a specific aspect of the bacterial protein's function.

This generalization applies to the *E. coli* RecA protein: humans and mice contain at least seven RecA homologs. Each homolog is presumed to have special catalytic activities and its own set of accessory proteins. The *Rad51 protein* is a particularly important RecA homolog in yeast, mice, and humans; it catalyzes a synaptic reaction between a DNA single strand and a DNA double helix in experiments *in vitro*. Genetic studies in which the Rad51 protein is mutated suggest that this protein is critical for the health of all three organisms, being required to repair replication forks that have been accidentally broken during the normal course of each S phase. Its proper function requires multiple accessory proteins. Two of these, the *Brca1* and *Brca2* proteins, were first discovered because mutations in their genes are inherited in a subset of human families with a greatly increased frequency of breast cancer. Whereas the removal of the Rad51 protein kills a cell, less drastic changes in its function caused by an alteration in such an accessory protein is thought to lead to an accumulation of DNA damage that often, in a small proportion of cells, gives rise to a cancer (see Figure 23–11).

Different RecA homologs in eucaryotes are specialized for meiosis, or for other unique types of DNA synaptic events that are less well understood. It is likely that each eucaryotic RecA homolog loads onto a DNA single strand to begin a general recombination event only when a particular DNA structure or cell condition allows the protein to bind there.

General Recombination Often Involves a Holliday Junction

The synapsis that exchanges the first single strand between two different DNA double helices is presumed to be the slow and difficult step in a general recombination event (see Figure 5–56). After this step, extending the region of pairing and establishing further strand exchanges between the two DNA helices is thought to proceed rapidly. In most cases, a key recombination intermediate, the *Holliday junction* (also called a *cross-strand exchange*) forms as a result.

In a **Holliday junction**, the two homologous DNA helices that have initially paired are held together by the reciprocal exchange of two of the four strands present, one originating from each of the helices. As shown in Figure 5–61A, a

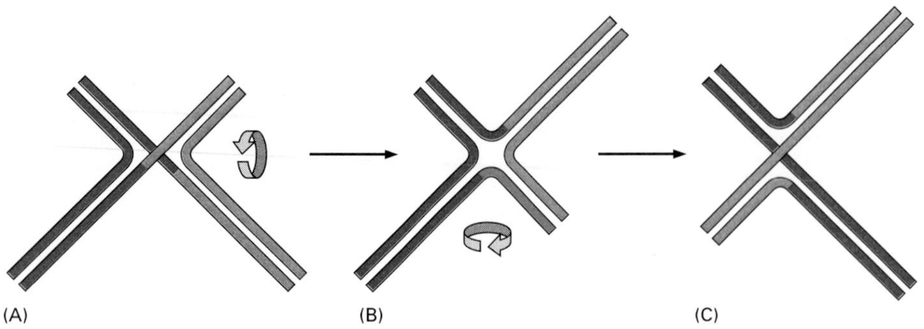

Figure 5–61 A Holliday junction and its isomerization. As described in the text, the synapsis step in general recombination is catalyzed by a RecA type of protein bound to a DNA single strand. This step is often followed by a reciprocal exchange of strands between two DNA double helices that have thereby paired with each other. This exchange produces a unique DNA structure known as a Holliday junction, named after the scientist who first proposed its formation. (A) The initially formed structure contains two crossing (inside) strands and two noncrossing (outside) strands. (B) An isomerization of the Holliday junction produces an open, symmetrical structure. (C) Further isomerization can interconvert the crossing and noncrossing strands, producing a structure that is otherwise the same as that in (A).

Holliday junction can be considered to contain two pairs of strands: one pair of crossing strands and one pair of noncrossing strands. The structure can *isomerize,* however, by undergoing a series of rotational movements, catalyzed by specialized proteins, to form a more open structure in which both pairs of strands occupy equivalent positions (Figures 5–61B and 5–62). This structure can, in turn, isomerize to a conformation that closely resembles the original junction, except that the crossing strands have been converted into noncrossing strands, and vice versa (Figure 5–61C).

Once the Holliday junction has formed an open structure, a special set of proteins can engage with the junction: one of these proteins uses the energy of ATP hydrolysis to move the crossover point (the point at which the two DNA helices are joined) rapidly along the two helices, extending the region of heteroduplex DNA (Figure 5–63).

To regenerate two separate DNA helices, and thus end the exchange process, the strands connecting the two helices in a Holliday junction must eventually be cut, a process referred to as *resolution*. There are two ways in which a Holliday junction can be resolved. In one, the original pair of crossing strands is cut (the invading, or inside, strands in Figure 5–61A). In this case, the two original DNA helices separate from each other nearly unaltered, exchanging the single-stranded DNA that formed the heteroduplex. In the other way, the original pair of noncrossing strands is cut (the inside strands in Figure 5–61C). Now the outcome is far more profound: two recombinant chromosomes are formed, having reciprocally exchanged major segments of double-stranded DNA with each other through a crossover event (Figure 5–64).

Genetic analyses reveal that heteroduplex regions of several thousand base pairs are readily formed during recombination. As described next, the processing of these heteroduplexes—which generally consist of nearly identical paired complementary strands—can further change the information in each resulting DNA helix.

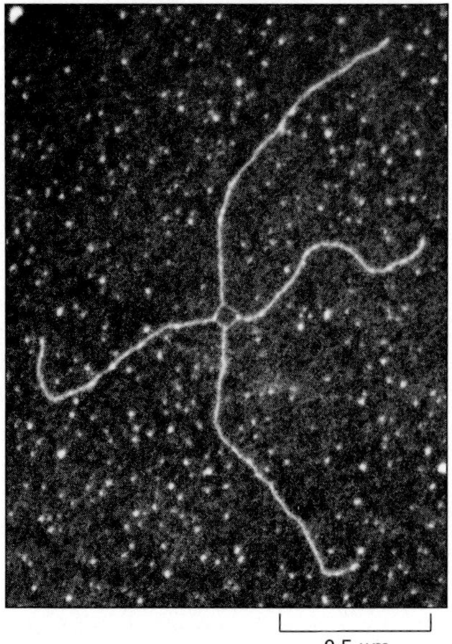

Figure 5–62 Electron micrograph of a Holliday junction. This view of the junction corresponds to the open structure illustrated in Figure 5–61B. (Courtesy of Huntington Potter and David Dressler.)

General Recombination Can Cause Gene Conversion

In sexually reproducing organisms, it is a fundamental law of genetics that each parent makes an equal genetic contribution to an offspring, which inherits one complete set of genes from the father and one complete set from the mother. Thus, when a diploid cell undergoes meiosis to produce four haploid cells (discussed in Chapter 20), exactly half of the genes in these cells should be maternal (genes that the diploid cell inherited from its father) and the other half paternal (genes that the diploid cell inherited from its father). In some organisms (fungi,

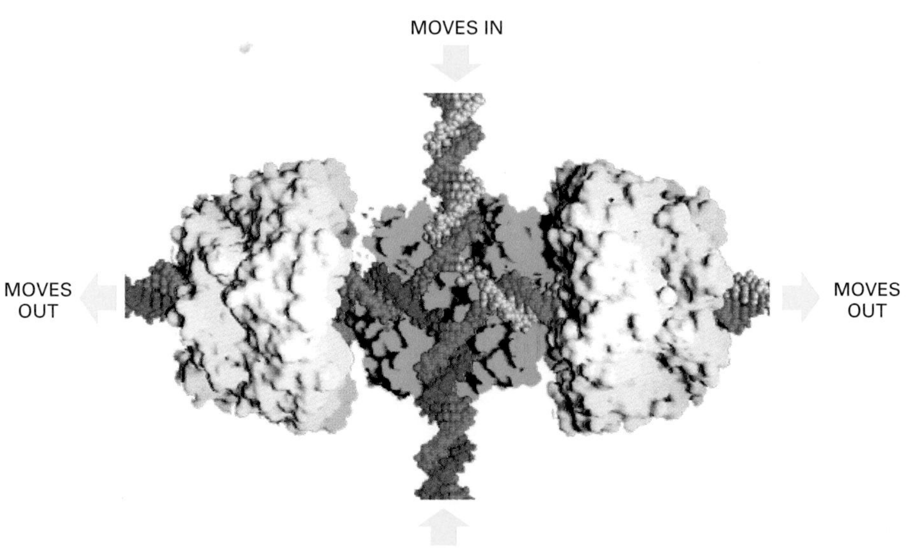

Figure 5–63 Enzyme-catalyzed double branch migration at a Holliday junction. In *E. coli,* a tetramer of the RuvA protein *(green)* and two hexamers of the RuvB protein *(pale gray)* bind to the open form of the junction. The RuvB protein uses the energy of ATP hydrolysis to move the crossover point rapidly along the paired DNA helices, extending the heteroduplex region as shown. There is evidence that similar proteins perform this function in vertebrate cells. (Image courtesy of P. Artymiuk; modified from S.C. West, *Cell* 94:699–701, 1998.)

Figure 5–64 The resolution of a Holliday junction to produce crossed-over chromosomes. In this example, homologous regions of a *red* and a *green* chromosome have formed a Holliday junction by exchanging two strands. Cutting these two strands would terminate the exchange without crossing-over. With isomerization of the Holliday junction (steps B and C), the original noncrossing strands become the two crossing strands; cutting them now creates two DNA molecules that have crossed over *(bottom)*. This type of isomerization may be involved in the breaking and rejoining of two homologous DNA double helices in meiotic general recombination. The grey bars in the central panels have been drawn to make it clear that the isomerization events shown can occur without disturbing the rest of the two chromosomes.

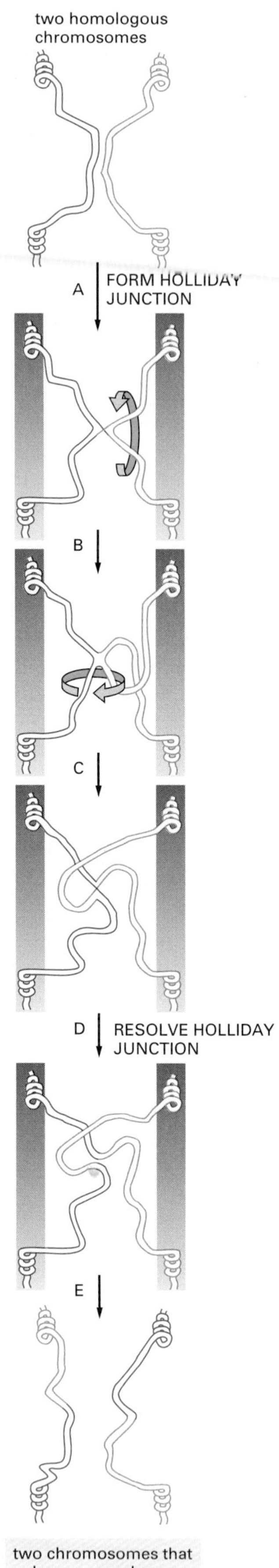

for example), it is possible to recover and analyze all four of the haploid gametes produced from a single cell by meiosis. Studies in such organisms have revealed rare cases in which the standard rules of genetics have been violated. Occasionally, for example, meiosis yields three copies of the maternal version of the gene and only one copy of the paternal allele (alternative versions of the same gene are called **alleles**). This phenomenon is known as **gene conversion** (Figure 5–65). Gene conversion often occurs in association with homologous genetic recombination events in meiosis (and more rarely in mitosis), and it is believed to be a straightforward consequence of the mechanisms of general recombination and DNA repair. Genetic studies show that only small sections of DNA typically undergo gene conversion, and in many cases only a part of a gene is changed.

In the process of gene conversion, DNA sequence information is transferred from one DNA helix that remains unchanged (a donor sequence) to another DNA helix whose sequence is altered (an acceptor sequence). There are several different ways this might happen, all of which involve the following two processes: (1) a homologous recombination event that juxtaposes two homologous DNA double helices, and (2) a limited amount of localized DNA synthesis, which is necessary to create an extra copy of one allele. In the simplest case, a general recombination process forms a heteroduplex joint (see Figure 5–55), in which the two paired DNA strands are not identical in sequence and therefore contain some mismatched base pairs. If the mispaired nucleotides in one of the two strands are recognized and removed by the DNA repair enzyme that catalyzes mismatch repair, an extra copy of the DNA sequence on the opposite strand is produced (Figure 5–66). The same gene conversion process can occur without crossover events, since it simply requires that a single DNA strand invade a

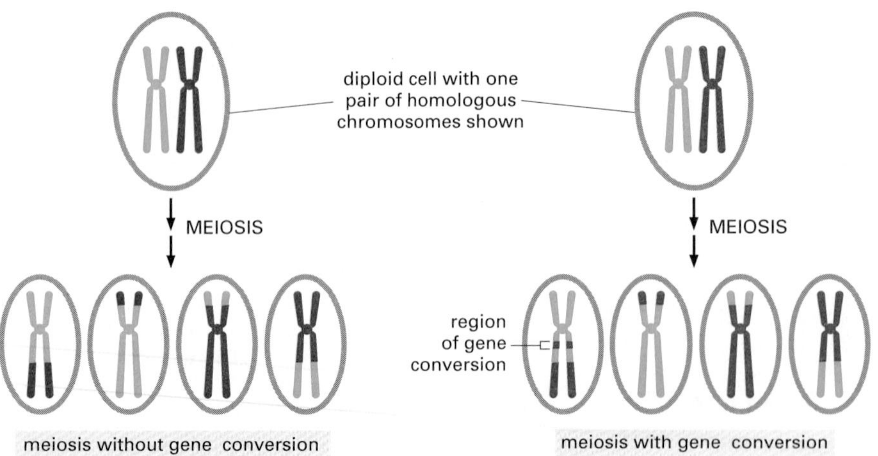

Figure 5–65 Gene conversion in meiosis. As described in Chapter 20, meiosis is the process through which a diploid cell gives rise to four haploid cells. Germ cells (eggs and sperm, for example) are produced by meiosis.

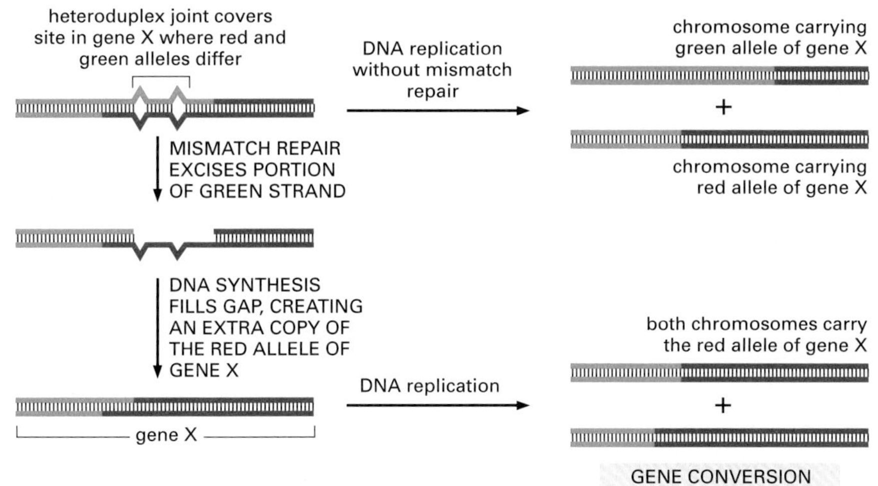

Figure 5–66 Gene conversion by mismatch correction. In this process, heteroduplex joints are formed at the sites of the crossing-over between homologous maternal and paternal chromosomes. If the maternal and paternal DNA sequences are slightly different, the heteroduplex joint will include some mismatched base pairs. The resulting mismatch in the double helix may then be corrected by the DNA mismatch repair machinery (see Figure 5–23), which can erase nucleotides on either the paternal or the maternal strand. The consequence of this mismatch repair is gene conversion, dectected as a deviation from the segregation of equal copies of maternal and paternal alleles that normally occurs in meiosis.

double helix to form a short heteroduplex region. The latter type of gene conversion is thought to be responsible for the unusually facile transfer of genetic information that is often observed between the different gene copies in a tandem array of repeated genes.

General Recombination Events Have Different Preferred Outcomes in Mitotic and Meiotic Cells

We have seen that meiotic recombination starts with a very bold stroke—the breakage of both strands of the double helix in one of the recombining chromosomes. How does the meiotic process that follows differ from the mechanism, also based on general recombination, that cells use for the precise repair of the accidental double-strand breaks that occur in chromosomes (the homologous end-joining reaction in Figure 5–53)? In both cases, the two new chromosome ends produced by a double-strand break are subjected to a degradative process, which exposes a single strand with an overhanging 3′ end. Moreover, in both cases, this strand seeks out a region of unbroken DNA double helix with the same nucleotide sequence and undergoes a synaptic reaction with it that is catalyzed by a RecA type of protein.

For double-strand break repair, DNA synthesis extends the invading 3′ end by thousands of nucleotides, using one of the strands of the recipient DNA helix as a template. If the second broken end becomes similarly engaged in the synaptic reaction, a joint molecule will be formed (see Figure 5–56). Depending on subsequent events, the final outcome can either be restoration of the two original DNA helices with repair of the double-strand break (the predominant reaction in mitotic cells), or a crossover event that leaves heteroduplex joints holding two different DNA helices together (the predominant reaction in meiotic cells). It is thought that the crossover events are created by a set of specific proteins that guide these reactions cells undergoing meiosis. These proteins not only ensure that a joint molecule with two Holliday junctions is formed but also cause a different pair of strands at each of the two junctions, thereby causing a crossover event (Figure 5–67).

With either outcome of general recombination, the DNA synthesis involved converts some of the genetic information at the site of the double-stranded break to that of the homologous chromosome. If these regions represent different alleles of the same gene, the nucleotide sequence in the broken helix is converted to that of the unbroken helix, causing a gene conversion. The yeast *Saccharomyces cerevisiae* exploits the gene conversion that accompanies double-strand break repair to switch from one mating type to another (discussed in Chapter 7). In this case, a double-strand break is intentionally induced by cleavage of a specific DNA sequence at the yeast *mating type locus* by an enzyme called HO endonuclease. After DNA degradation at the site of the break has removed the old sequence, the missing genetic information is restored by a synapsis of the broken ends with a "mating-type cassette" DNA sequence of the opposite mating type (a or α), followed by local DNA synthesis in the manner previously indicated to reseal the broken region of the chromosome. In fact, it is through a detailed study of this precisely positioned form of double-strand break repair that the general mechanism of homologous end-joining was revealed.

Mismatch Proofreading Prevents Promiscuous Recombination Between Two Poorly Matched DNA Sequences

As previously discussed, a critical step in recombination occurs when two DNA strands of complementary sequence pair to form a heteroduplex joint between two double helices. Experiments *in vitro* with purified RecA protein show that pairing can occur efficiently even when the sequences of the two DNA strands do not match well—when, for example, only four out of every five nucleotides on average can form base pairs. If recombination proceeded from these mismatched sequences, it would create havoc in cells, especially in those that contain a series of closely related DNA sequences in their genomes. How do cells prevent crossing over between these sequences?

Although the complete answer is not known, studies with bacteria and yeasts demonstrate that components of the same mismatch proofreading system that removes replication errors (see Figure 5–23) have the additional role of interrupting genetic recombination events between poorly matched DNA sequences. It has long been known, for example, that homologous genes in two closely related bacteria, *E. coli* and *Salmonella typhimurium*, generally will not recombine, even though their nucleotide sequences are 80% identical. However, when the mismatch proofreading system is inactivated by mutation, there is a 1000-fold increase in the frequency of such interspecies recombination events. It is thought that the mismatch proofreading system normally recognizes the mispaired bases in an initial strand exchange, and—if there are a significant number of mismatches—the subsequent steps required to break and rejoin the two paired DNA helices are prevented. This mechanism protects the bacterial genome from the sequence changes that would otherwise be caused by recombination with the foreign DNA molecules that occasionally enter the cell. In vertebrate cells, which contain many closely related DNA sequences, the same type of recombinational proofreading is thought to help prevent promiscuous recombination events that would otherwise scramble the genome (Figure 5–68).

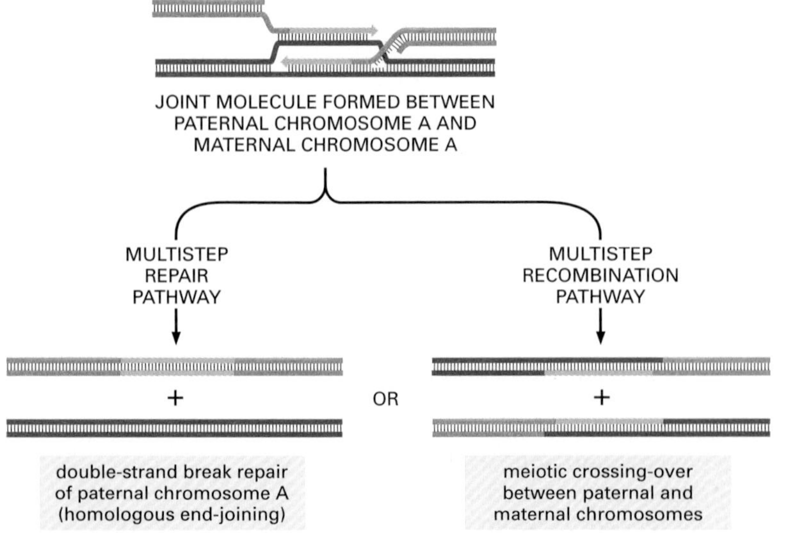

Figure 5–67 The different resolutions of a general recombination intermediate in mitotic and meiotic cells. As shown previously in Figure 5–56, general recombination begins when a double-strand break is generated in one double helix *(green)*, followed by DNA degradation and strand invasion into a homologous DNA duplex *(red)*. New DNA synthesis *(orange)* follows to generate the joint molecule shown. Depending on subsequent events, resolution of the joint molecule can lead either to a precise repair of the initial double-strand break *(left)* or to chromosome crossing-over *(right)*. The experimental induction of double-strand breaks at specific DNA sites has allowed the outcome of general recombination to be quantified in both mitotic and meiotic cells. More than 99% of these events fail to produce a crossover in mitotic cells, whereas crossovers are often the outcome in meiotic cells. In either case, if the maternal and paternal chromosomes differ in DNA sequence in the region of new DNA synthesis shown here, the sequence of the *green* DNA duplex in the region of new DNA synthesis is converted to that of the *red* duplex (a gene conversion).

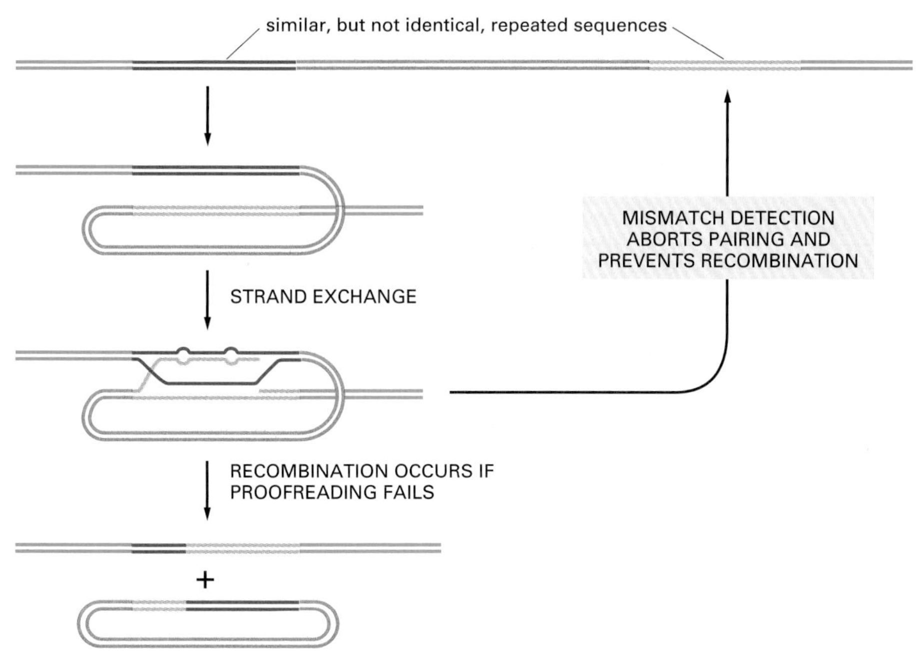
similar, but not identical, repeated sequences
STRAND EXCHANGE
MISMATCH DETECTION ABORTS PAIRING AND PREVENTS RECOMBINATION
RECOMBINATION OCCURS IF PROOFREADING FAILS
+

Figure 5–68 The mechanism that prevents general recombination from destabilizing a genome that contains repeated sequences. Studies with bacterial and yeast cells suggest that components of the mismatch proofreading system, diagrammed previously in Figure 5–23, have the additional function shown here.

Summary

General recombination (also called homologous recombination) allows large sections of the DNA double helix to move from one chromosome to another, and it is responsible for the crossing-over of chromosomes that occurs during meiosis in fungi, animals, and plants. General recombination is essential for the maintenance of chromosomes in all cells, and it usually begins with a double-strand break that is processed to expose a single-stranded DNA end. Synapsis between this single strand and a homologous region of DNA double helix is catalyzed by the bacterial RecA protein and its eucaryotic homologs, and it often leads to the formation of a four-stranded structure known as a Holliday junction. Depending on the pattern of strand cuts made to resolve this junction into two separate double helices, the products can be either a precisely repaired double-strand break or two chromosomes that have crossed over.

Because general recombination relies on extensive base-pairing interactions between the strands of the two DNA double helices that recombine, it occurs only between homologous DNA molecules. Gene conversion, the nonreciprocal transfer of genetic information from one chromosome to another, results from the mechanisms of general recombination, which involve a limited amount of associated DNA synthesis.

SITE-SPECIFIC RECOMBINATION

In general recombination, DNA rearrangements occur between DNA segments that are very similar in sequence. Although these rearrangements can result in the exchange of alleles between chromosomes, the order of the genes on the interacting chromosomes typically remains the same. A second type of recombination, called **site-specific recombination**, can alter gene order and also add new information to the genome. Site-specific recombination moves specialized nucleotide sequences, called *mobile genetic elements,* between nonhomologous sites within a genome. The movement can occur between two different positions in a single chromosome, as well as between two different chromosomes.

Mobile genetic elements range in size from a few hundred to tens of thousands of nucleotide pairs, and they have been identified in virtually all cells that have been examined. Some of these elements are viruses in which site-specific recombination is used to move their genomes into and out of the chromosomes

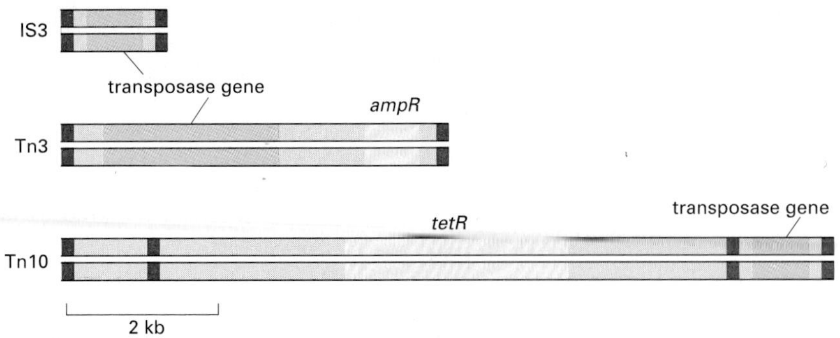

Figure 5–69 Three of the many types of mobile genetic elements found in bacteria. Each of these DNA elements contains a gene that encodes a *transposase*, an enzyme that conducts at least some of the DNA breakage and joining reactions needed for the element to move. Each mobile element also carries short DNA sequences (indicated in *red)* that are recognized only by the transposase encoded by that element and are necessary for movement of the element. In addition, two of the three mobile elements shown carry genes that encode enzymes that inactivate the antibiotics ampicillin *(ampR)* and tetracycline *(tetR)*. The transposable element Tn10, shown in the bottom diagram, is thought to have evolved from the chance landing of two short mobile elements on either side of a tetracyclin-resistance gene; the wide use of tetracycline as an antibiotic has aided the spread of this gene through bacterial populations. The three mobile elements shown are all examples of *DNA-only transposons* (see text).

of their host cell. A virus can package its nucleic acid into viral particles that can move from one cell to another through the extracellular environment. Many other mobile elements can move only within a single cell (and its descendents), lacking any intrinsic ability to leave the cell in which they reside.

The relics of site-specific recombination events can constitute a considerable fraction of a genome. The abundant repeated DNA sequences found in many vertebrate chromosomes are mostly derived from mobile genetic elements; in fact, these sequences account for more than 45% of the human genome (see Figure 4–17). Over time, the nucleotide sequences of these elements have been altered by random mutation. As a result, only a few of the many copies of these elements in our DNA are still active and capable of movement.

In addition to moving themselves, all types of mobile genetic elements occasionally move or rearrange neighboring DNA sequences of the host cell genome. These movements can cause deletions of adjacent nucleotide sequences, for example, or can carry these sequences to another site. In this way, site-specific recombination, like general recombination, produces many of the genetic variants upon which evolution depends. The translocation of mobile genetic elements gives rise to spontaneous mutations in a large range of organisms including humans; in some, such as the fruit fly *Drosophila*, these elements are known to produce most of the mutations observed. Over time, site-specific recombination has thereby been responsible for a large fraction of the important evolutionary changes in genomes.

Mobile Genetic Elements Can Move by Either Transpositional or Conservative Mechanisms

Unlike general recombination, site-specific recombination is guided by recombination enzymes that recognize short, specific nucleotide sequences present on one or both of the recombining DNA molecules. Extensive DNA homology is not required for a recombination event. Each type of mobile element generally encodes the enzyme that mediates its own movement and contains special sites upon which the enzyme acts. Many elements also carry other genes. For example, viruses encode coat proteins that enable them to exist outside cells, as well as essential viral enzymes. The spread of mobile elements that carry antibiotic resistance genes is a major factor underlying the widespread dissemination of antibiotic resistance in bacterial populations (Figure 5–69).

Site-specific recombination can proceed via either of two distinct mechanisms, each of which requires specialized recombination enzymes and specific DNA sites. (1) **Transpositional site-specific recombination** usually involves breakage reactions at the ends of the mobile DNA segments embedded in chromosomes and the attachment of those ends at one of many different nonhomologous target DNA sites. It does not involve the formation of heteroduplex DNA. (2) **Conservative site-specific recombination** involves the production of a very short heteroduplex joint, and it therefore requires a short DNA sequence that is the same on both donor and recipient DNA molecules. We first discuss transpositional site-specific recombination *(transposition* for short), returning to conservative site-specific recombination at the end of the chapter.

Transpositional Site-specific Recombination Can Insert Mobile Genetic Elements into Any DNA Sequence

Transposons, also called **transposable elements**, are mobile genetic elements that generally have only modest target site selectivity and can thus insert themselves into many different DNA sites. In transposition, a specific enzyme, usually encoded by the transposon and called a *transposase*, acts on a specific DNA sequence at each end of the transposon—first disconnecting it from the flanking DNA and then inserting it into a new target DNA site. There is no requirement for homology between the ends of the element and the insertion site.

Most transposons move only very rarely (once in 10^5 cell generations for many elements in bacteria), and for this reason it is often difficult to distinguish them from nonmobile parts of the chromosome. In most cases, it is not known what suddenly triggers their movement.

On the basis of their structure and transposition mechanisms, transposons can be grouped into three large classes (Table 5–3), each of which is discussed in detail in subsequent sections. Those in the first two of these classes use virtually identical DNA breakage and DNA joining reactions to translocate. However, for the *DNA-only transposons*, the mobile element exists as DNA throughout its life cycle: the translocating DNA segment is directly cut out of the donor DNA and joined to the target site by a transposase. In contrast, *retroviral-like retrotransposons* move by a less direct mechanism. An RNA polymerase first transcribes the DNA sequence of the mobile element into RNA. The enzyme reverse transcriptase then transcribes this RNA molecule back into DNA using the RNA as a template, and it is this DNA copy that is finally inserted into a new site in the genome. For historical reasons, the transposase-like enzyme that catalyzes this insertion reaction is called an *integrase* rather than a transposase. The third type of transposon in Table 5–3 also moves by making a DNA copy of an RNA

TABLE 5–3 Three Major Classes of Transposable Elements

CLASS DESCRIPTION AND STRUCTURE	GENES IN COMPLETE ELEMENT	MODE OF MOVEMENT	EXAMPLES
DNA-only transposons			
short inverted repeats at each end	encodes transposase	moves as DNA, either excising or following a replicative pathway	P element (*Drosophila*) Ac-Ds (maize) Tn3 and IS1 (*E.coli*) Tam3 (snapdragon)
Retroviral-like retrotransposons			
directly repeated long terminal repeats (LTRs) at ends	encodes reverse transcriptase and resembles retrovirus	moves via an RNA intermediate produced by promoter in LTR	Copia (*Drosophila*) Ty1 (yeast) THE-1 (human) Bs1 (maize)
Nonretroviral retrotransposons			
AAAA TTTT Poly A at 3′ end of RNA transcript; 5′ end is often truncated	encodes reverse transcriptase	moves via an RNA intermediate that is often produced from a neighboring promotor	F element (*Drosophila*) L1 (human) Cin4 (maize)

These elements range in length from 1000 to about 12,000 nucleotide pairs; each family contains many members, only a few of which are listed here. In addition to transposable elements, there are selected viruses that can move in and out of host cell chromosomes; these viruses are related to the first two classes of transposons.

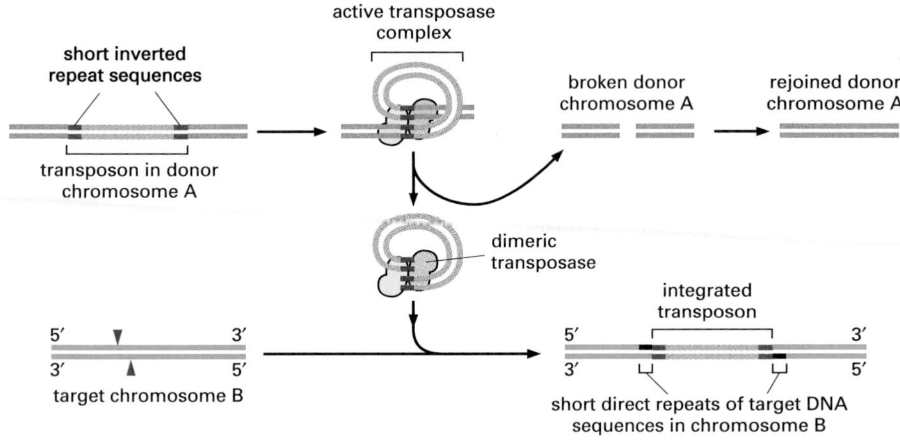

Figure 5–70 Cut-and-paste transposition. DNA-only transposons can be recognized in chromosomes by the "inverted repeat DNA sequences" *(red)* at their ends. Experiments show that these sequences, which can be as short as 20 nucleotides, are all that is necessary for the DNA between them to be transposed by the particular transposase enzyme associated with the element. The cut and-paste movement of a DNA-only transposable element from one chromosomal site to another begins when the transposase brings the two inverted DNA sequences together, forming a DNA loop. Insertion into the target chromosome, catalyzed by the transposase, occurs at a random site through the creation of staggered breaks in the target chromosome *(red arrowheads)*. As a result, the insertion site is marked by a short direct repeat of the target DNA sequence, as shown. Although the break in the donor chromosome *(green)* is resealed, the breakage-and-repair process often alters the DNA sequence, causing a mutation at the original site of the excised transposable element (not shown).

molecule that is transcribed from it. However, the mechanism involved for these *nonretroviral retrotransposons* is distinct from that just described in that the RNA molecule is directly involved in the transposition reaction.

DNA-only Transposons Move By DNA Breakage and Joining Mechanisms

Many **DNA-only transposons** move from a donor site to a target site by **cut-and-paste transposition**, using the mechanism outlined in Figure 5–70. Each subunit of a transposase recognizes the same specific DNA sequence at an end of the element; the joining together of these two subunits to form a dimeric transposase creates a DNA loop that brings the two ends of the element together. The transposase then introduces cuts at both ends of this DNA loop to expose the element termini and remove the element completely from its original chromosome (Figure 5–71). To complete the reaction, the transposase catalyses a direct attack of the element's two DNA termini on a target DNA molecule, breaking two phosphodiester bonds in the target molecule as it joins the element and target DNAs together.

Because the breaks made in the two target DNA strands are staggered *(red arrowheads* in Figure 5–70), two short, single-stranded gaps are initially formed in the product DNA molecule, one at each end of the inserted transposon. These gaps are filled-in by a host cell DNA polymerase and DNA ligase to complete the recombination process, producing a short duplication of the adjacent target DNA sequence. These flanking direct repeat sequences, whose length is different for different transposons, serve as convenient markers of a prior transpositional site-specific recombination event.

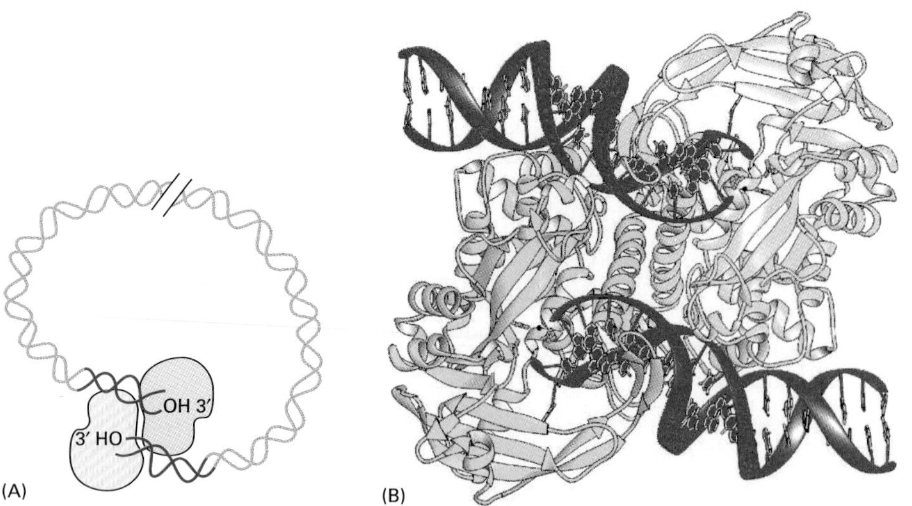

Figure 5–71 The structure of the central intermediate formed by a cut-and-paste transposase. (A) Schematic view of the overall structure. (B) The detailed structure of a transposase holding the two DNA ends, whose 3′-OH groups are poised to attack a target chromosome. (B, from D.R. Davies et al., *Science* 289:77–85, 2000. © AAAS.)

When a cut-and-paste DNA-only transposon is excised from the donor chromosome, a double-strand break is created in the vacated chromosome. This break can be perfectly "healed" by a homologous end-joining reaction. Alternatively, the break can be resealed by a nonhomologous end-joining reaction; in this case, the DNA sequence that flanked the transposon is often altered, producing a mutation at the chromosomal site from which the transposon was excised (see Figure 5–53).

Some DNA-only transposons move using a variation of the cut-and-paste mechanism called *replicative transposition*. In this case, the transposon DNA is replicated and a copy is inserted at a new chromosomal site, leaving the original chromosome intact (Figure 5–72). Although the mechanism used is more complex, it is closely related to the cut-and-paste mechanism just described; indeed, some transposons can move by either pathway.

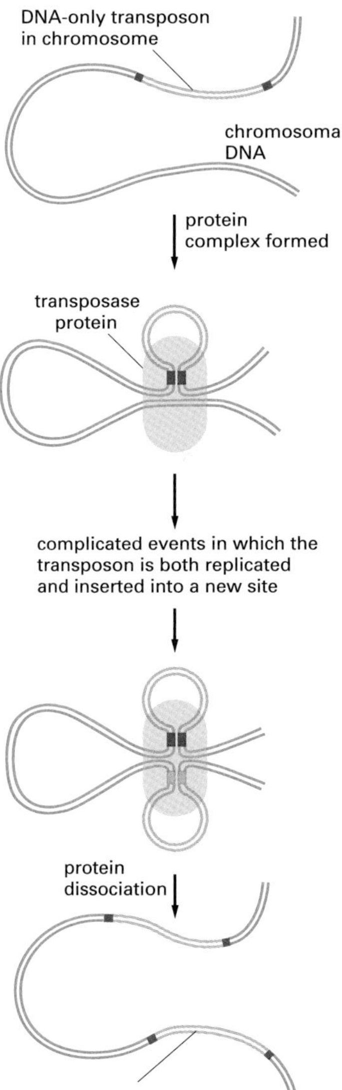

Figure 5–72 Replicative transposition. In the course of replicative transposition, the DNA sequence of the transposon is copied by DNA replication. The end products are a DNA molecule that is identical to the original donor and a target DNA molecule that has a transposon inserted into it. In general, a particular DNA-only transposon moves either by the cut-and-paste pathway shown in Figure 5–70 or by the replicative pathway outlined here. However, the two mechanisms have many enzymatic similarities, and a few transposons can move by either pathway.

Some Viruses Use Transpositional Site-specific Recombination to Move Themselves into Host Cell Chromosomes

Certain viruses are considered mobile genetic elements because they use transposition mechanisms to integrate their genomes into that of their host cell. However, these viruses also encode proteins that package their genetic information into virus particles that can infect other cells. Many of the viruses that insert themselves into a host chromosome do so by employing one of the first two mechanisms listed in Table 5–3. Indeed, much of our knowledge of these mechanisms has come from studies of particular viruses that employ them.

A virus that infects a bacterium is known as a **bacteriophage**. The *bacteriophage Mu* not only uses DNA-based transposition to integrate its genome into its host cell chromosome, it also uses the transposition process to initiate its viral DNA replication. The Mu transposase was the first to be purified in active form and characterized; it recognizes the sites of recombination at each end of the viral DNA by binding specifically to this DNA, and closely resembles the transposases just described.

Transposition also has a key role in the life cycle of many other viruses. Most notable are the **retroviruses**, which include the AIDS virus, called HIV, that infects human cells. Outside the cell, a retrovirus exists as a single-stranded RNA genome packed into a protein capsid along with a virus-encoded **reverse transcriptase** enzyme. During the infection process, the viral RNA enters a cell and is converted to a double-stranded DNA molecule by the action of this crucial enzyme, which is able to polymerize DNA on either an RNA or a DNA template (Figures 5–73 and 5–74). The term *retrovirus* refers to the fact that these viruses reverse the usual flow of genetic information, which is from DNA to RNA (see Figure 1–5).

Specific DNA sequences near the two ends of the double-stranded DNA product produced by reverse transcriptase are then held together by a virus-encoded integrase enzyme. This integrase creates activated 3′-OH viral DNA ends that can directly attack a target DNA molecule through a mechanism very similar to that used by the cut-and-paste DNA-only transposons (Figure 5–75). In fact, detailed analyses of the three-dimensional structures of bacterial transposases and HIV integrase have revealed remarkable similarities in these enzymes, even though their amino acid sequences have diverged considerably.

Retroviral-like Retrotransposons Resemble Retroviruses, but Lack a Protein Coat

Retroviruses move themselves in and out of chromosomes by a mechanism that is identical to that used by a large family of transposons called **retroviral-like retrotransposons** (see Table 5–3). These elements are present in organisms as diverse as yeasts, flies, and mammals. One of the best understood is the *Ty1 element* found in yeast. As with a retrovirus, the first step in its transposition is the transcription of the entire transposon, producing an RNA copy of the element that is more than 5000 nucleotides long. This transcript, which is translated as a

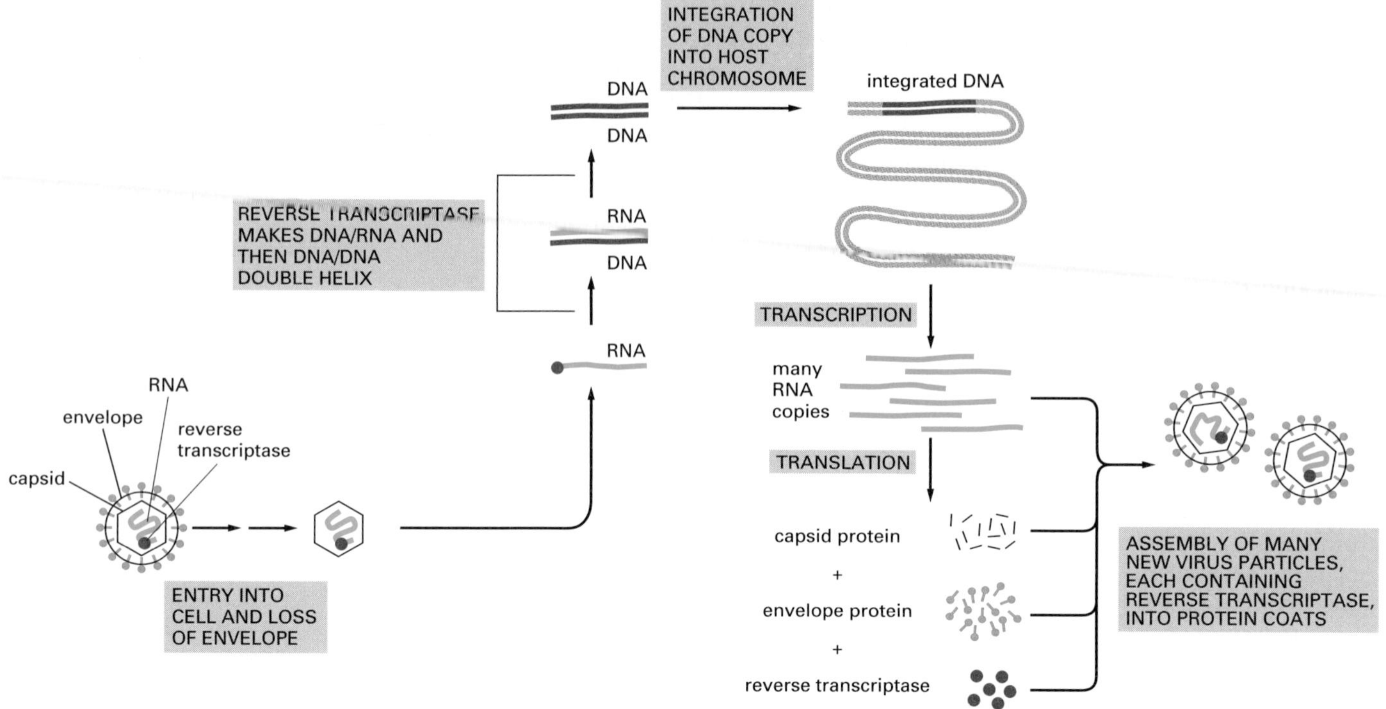

Figure 5–73 The life cycle of a retrovirus. The retrovirus genome consists of an RNA molecule of about 8500 nucleotides; two such molecules are packaged into each viral particle. The enzyme *reverse transcriptase* first makes a DNA copy of the viral RNA molecule and then a second DNA strand, generating a double-stranded DNA copy of the RNA genome. The integration of this DNA double helix into the host chromosome is then catalyzed by a virus-encoded *integrase* enzyme (see Figure 5–75). This integration is required for the synthesis of new viral RNA molecules by the host cell RNA polymerase, the enzyme that transcribes DNA into RNA (discussed in Chapter 6).

messenger RNA by the host cell, encodes a reverse transcriptase enzyme. This enzyme makes a double-stranded DNA copy of the RNA molecule via an RNA/DNA hybrid intermediate, precisely mimicking the early stages of infection by a retrovirus (see Figure 5–73). Like retroviruses, the linear double-stranded DNA molecule then integrates into a site on the chromosome by using an integrase enzyme that is also encoded by the *Ty1* DNA (see Figure 5–75). Although the resemblance to a retrovirus is striking, unlike a retrovirus, the *Ty1* element does not have a functional protein coat; it can therefore move only within a single cell and its descendants.

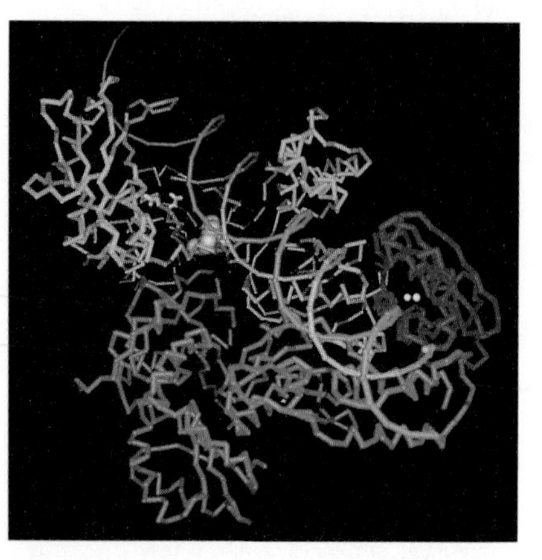
(A)

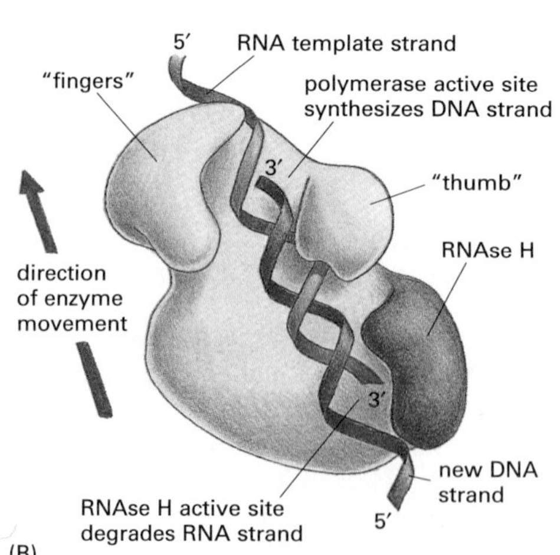

(B)

Figure 5–74 Reverse transcriptase. (A) The three-dimensional structure of the enzyme from HIV (the human AIDS virus) determined by x-ray crystallography. (B) A model showing the enzyme's activity on an RNA template. Note that the polymerase domain *(yellow* in B) has a covalently attached RNAse H (H for "Hybrid") domain *(red)* that degrades an RNA strand in an RNA/DNA helix. This activity helps the polymerase to convert the initial hybrid helix into a DNA double helix (A, courtesy of Tom Steitz; B, adapted from L.A. Kohlstaedt et al., *Science* 256:1783–1790, 1990.)

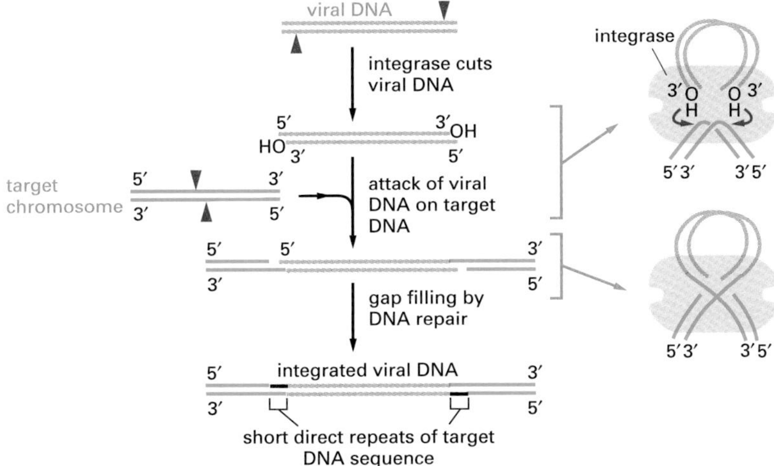

Figure 5–75 Transpositional site-specific recombination by a retrovirus or a retroviral-like retrotransposon. Outline of the strand-breaking and strand-rejoining events that lead to integration of the linear double-stranded DNA *(orange)* of a retrovirus (such as HIV) or a retroviral-like retrotransposon (such as *Ty1*) into the host cell chromosome *(blue)*. In an initial step, the integrase enzyme forms a DNA loop and cuts one strand at each end of the viral DNA sequence, exposing a protruding 3′-OH group. Each of these 3′-OH ends then directly attacks a phosphodiester bond on opposite strands of a randomly selected site on a target chromosome *(red arrowheads)*. This inserts the viral DNA sequence into the target chromosome, leaving short gaps on each side that are filled in by DNA repair processes. Because of the gap filling, this type of mechanism (like that of cut-and-paste transposons) leaves short repeats of target DNA sequence *(black)* on each side of the integrated DNA segment; these are 3–12 nucleotides long, depending on the integrase enzyme.

A Large Fraction of the Human Genome Is Composed of Nonretroviral Retrotransposons

A significant fraction of many vertebrate chromosomes is made up of repeated DNA sequences. In human chromosomes, these repeats are mostly mutated and truncated versions of a retrotransposon called an *L1 element* (sometimes referred to as a LINE or long interspersed nuclear element). Although most copies of the *L1* element are immobile, a few retain the ability to move. Translocations of the element have been identified, some of which result in human disease; for example, a particular type of hemophilia results from an *L1* insertion into the gene encoding a blood clotting factor, Factor VIII. Related mobile elements are found in other mammals and insects, as well as in yeast mitochondria. These **nonretroviral retrotransposons** (the third entry in Table 5–3) move via a distinct mechanism that requires a complex of an endonuclease and a reverse transcriptase. As illustrated in Figure 5–76, the RNA and reverse transcriptase have a much more direct role in the recombination event than for the mobile elements described above.

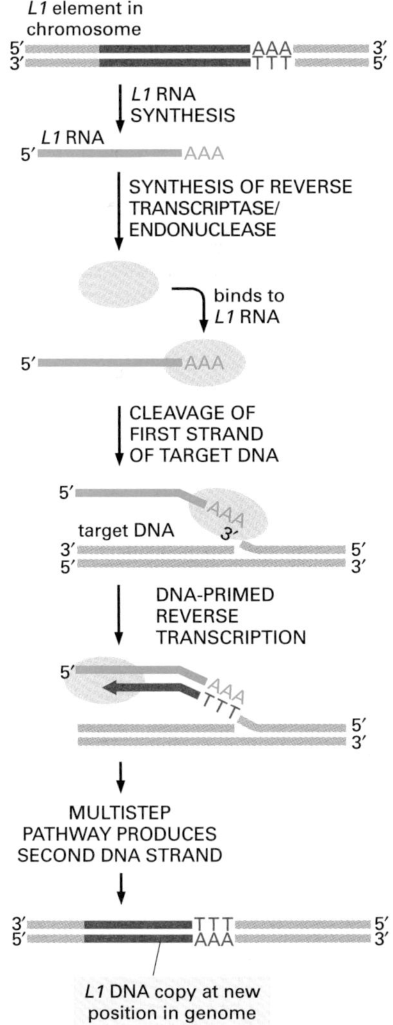

Figure 5–76 Transpositional site-specific recombination by a nonretroviral retrotransposon. Transposition by the *L1* element *(red)* begins when an endonuclease attached to the *L1* reverse transcriptase and the *L1* RNA *(blue)* makes a nick in the target DNA at the point at which insertion will occur. This cleavage releases a 3′-OH DNA end in the target DNA, which is then used as a primer for the reverse transcription step shown. This generates a single-stranded DNA copy of the element that is directly linked to the target DNA. In subsequent reactions, not yet understood in detail, further processing of the single-stranded DNA copy results in the generation of a new double-stranded DNA copy of the *L1* element that is inserted at the site where the initial nick was made.

Figure 5–77 The proposed pattern of expansion of the abundant *Alu* and *B1* sequences found in the human and mouse genomes, respectively. Both of these transposable DNA sequences are thought to have evolved from the essential 7SL RNA gene which encodes the SRP RNA (see Figure 12–41). On the basis of the species distribution and sequence similarity of these highly repeated elements, the major expansion in copy numbers seems to have occurred independently in mice and humans (see Figure 5–78). (Adapted from P.L. Deininger and G.R. Daniels, *Trends Genet.* 2:76–80, 1986 and International Human Genome Sequencing Consortium, *Nature* 409:860–921, 2001.)

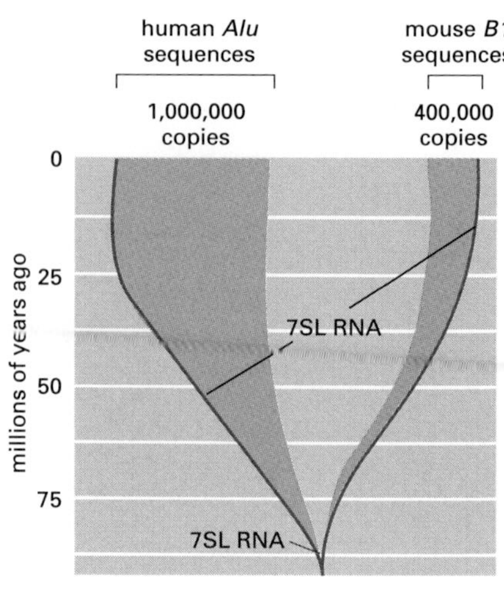

It is thought that other repeated DNAs that fail to encode an endonuclease or a reverse transcriptase in their own nucleotide sequence can multiply in chromosomes by a similar mechanism, using various endonucleases and reverse transcriptases present in the cell, including those encoded by *L1* elements. For example, the abundant *Alu* element lacks endonuclease or reverse transcriptase genes, yet it has amplified to become a major constituent of the human genome (Figure 5–77).

The *L1* and *Alu* elements seem to have multiplied in the human genome relatively recently. Thus, for example, the mouse contains sequences closely related to *L1* and *Alu*, but their placement in mouse chromosomes is very different from that in human chromosomes (Figure 5–78).

Different Transposable Elements Predominate in Different Organisms

We have described several types of transposable elements: (1) DNA-only transposons, the movement of which involves only DNA breakage and joining; (2) retroviral-like retrotransposons, which also move via DNA breakage and joining, but where RNA has a key role as a template to generate the DNA recombination substrate; and (3) nonretroviral retrotransposons, in which an RNA copy of the element is central to the incorporation of the element into the target DNA, acting as a direct template for a DNA target-primed reverse transcription event.

Interestingly, different types of transposons seem to predominate in different organisms. For example, the vast majority of bacterial transposons are DNA-only types, with a few related to the nonretroviral retrotransposons also present. In yeast, the main mobile elements that have been observed are retroviral-like retrotransposons. In *Drosophila*, DNA-based, retroviral, and nonretroviral transposons are all found. Finally, the human genome contains all three types of transposon, but as discussed below, their evolutionary histories are strikingly different.

Genome Sequences Reveal the Approximate Times when Transposable Elements Have Moved

The nucleotide sequence of the human genome provides a rich "fossil record" of the activity of transposons over evolutionary time spans. By carefully comparing the nucleotide sequences of the approximately 3 million transposable element

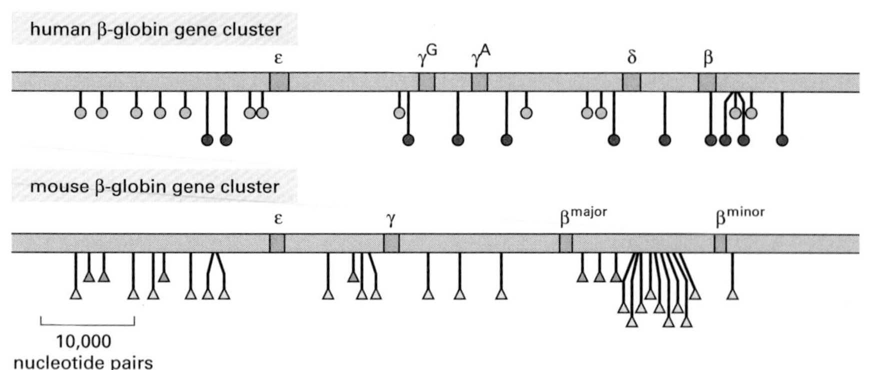

Figure 5–78 A comparison of the β-globin gene cluster in the human and mouse genomes, showing the location of transposable elements. This stretch of human genome contains five functional β-globin-like genes *(orange)*; the comparable region from the mouse genome has only four. The positions of the human *Alu* sequence are indicated by *green circles*, and the human *L1* sequences by *red circles*. The mouse genome contains different but related transposable elements: the positions of *B1* elements (which are related to the human Alu sequences) are indicated by *blue triangles*, and the positions of the mouse *L1* elements (which are related to the human *L1* sequences) are indicated by *yellow triangles*. Because the DNA sequences and positions of the transposable elements found in the mouse and human β-globin gene clusters are so different, it is believed that they accumulated in each of these genomes independently, relatively recently in evolutionary time. (Courtesy of Ross Hardison and Webb Miller.)

remnants in the human genome, it has been possible to broadly reconstruct the movements of transposons in our ancestor's genomes over the past several hundred million years. For example, the DNA-only transposons appear to have been very active well before the divergence of humans and old world monkeys (25–35 million years ago); but, because they gradually accumulated inactivating mutations, they have been inactive in the human lineage since that time. Likewise, although our genome is littered with relics of retroviral-like transposons, none appear to be active today. Only a single family of retroviral-like retrotransposons is believed to have transposed in the human genome since the divergence of human and chimpanzee approximately 7 million years ago. The nonretroviral retrotransposons are also very ancient, but in contrast to other types, some are still moving in our genome. As mentioned previously, they are responsible for a fraction of new human mutations—perhaps 2 mutations in every thousand.

The situation in mice is significantly different. Although the mouse and human genomes contain roughly the same density of the three types of transposon, both types of retrotransposon are still actively transposing in the mouse genome, being responsible for approximately ten per cent of new mutations. Clearly we are only beginning to understand how the movement of transposons have shaped the genomes of present-day mammals. It has been proposed that bursts in transposition activity could have been involved in critical speciation events during the radiation of mammalian lineages from a common ancestor, a process that began approximately 170 million years ago. At this point, we can only wonder how many of our uniquely human qualities are due to the past activity of the many mobile genetic elements whose remnants are found today in our chromosomes.

Conservative Site-specific Recombination Can Reversibly Rearrange DNA

A different kind of site-specific recombination known as *conservative site-specific recombination* mediates the rearrangements of other types of mobile DNA elements. In this pathway, breakage and joining occur at two special sites, one on each participating DNA molecule. Depending on the orientation of the two recombination sites, DNA integration, DNA excision, or DNA inversion can occur (Figure 5–79).

Site-specific recombination enzymes that break and rejoin two DNA double helices at specific sequences on each DNA molecule often do so in a reversible way: the same enzyme system that joins two DNA molecules can take them apart again, precisely restoring the sequence of the two original DNA molecules. This type of recombination is therefore called "conservative" site-specific recombination to distinguish it from the mechanistically distinct, transpositional site-specific recombination just discussed.

A bacterial virus, *bacteriophage lambda*, was the first mobile DNA element to be understood in biochemical detail. When this virus enters a cell, a virus-

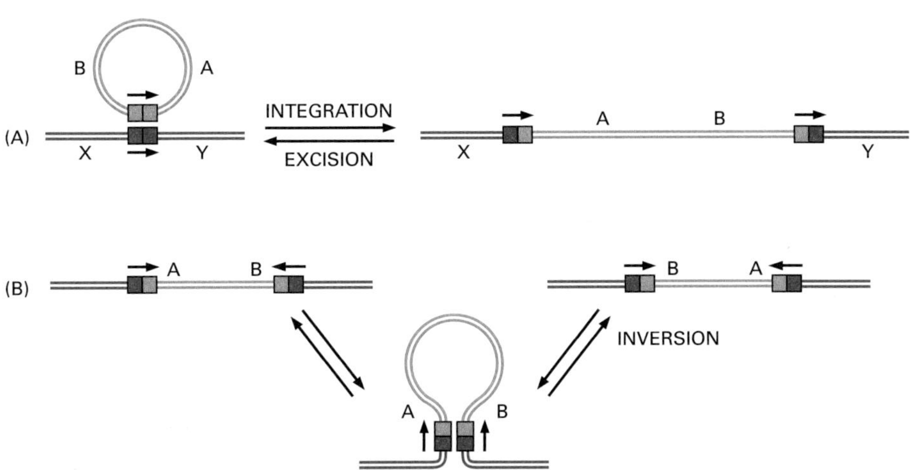

Figure 5–79 Two types of DNA rearrangement produced by conservative site-specific recombination. The only difference between the reactions in (A) and B) is the relative orientation of the two DNA sites (indicated by *arrows)* at which a site-specific recombination event occurs. (A) Through an integration reaction, a circular DNA molecule can become incorporated into a second DNA molecule; by the reverse reaction (excision), it can exit to reform the original DNA circle. Bacteriophage lambda and other bacterial viruses move in and out of their host chromosomes in precisely this way. (B) Conservative site-specific recombination can also invert a specific segment of DNA in a chromosome. A well-studied example of DNA inversion through site-specific recombination occurs in the bacterium *Salmonella typhimurium*, an organism that is a major cause of food poisoning in humans; the inversion of a DNA segment changes the type of flagellum that is produced by the bacterium (see Figure 7–64).

encoded enzyme called *lambda integrase* is synthesized. This enzyme mediates the covalent joining of the viral DNA to the bacterial chromosome, causing the virus to become part of this chromosome so that it is replicated automatically—as part of the host's DNA. A key feature of the lambda integrase reaction is that the site of recombination is determined by the recognition of two related but different DNA sequences—one on the bacteriophage chromosome and the other on the chromosome of the bacterial host. The recombination process begins when several molecules of the integrase protein bind tightly to a specific DNA sequence on the circular bacteriophage chromosome, along with several host proteins. This DNA–protein complex can now bind to an attachment site DNA sequence on the bacterial chromosome, bringing the bacterial and bacteriophage chromosomes together. The integrase then catalyzes the required cutting and resealing reactions that result in a site-specific strand exchange. Because of a short region of sequence homology in the two joined sequences, a tiny heteroduplex joint is formed at this point of exchange (Figure 5–80).

The lambda integrase resembles a DNA topoisomerase in forming a reversible covalent linkage to the DNA when it breaks a chain. Thus, this site-specific recombination event can occur in the absence of ATP and DNA ligase, which are normally required for phosphodiester bond formation.

The same type of site-specific recombination mechanism can also be used in reverse to promote the excision of a mobile DNA segment that is bounded by special recombination sites present as direct repeats. In bacteriophage lambda, excision enables it to exit from its integration site in the *E. coli* chromosome in

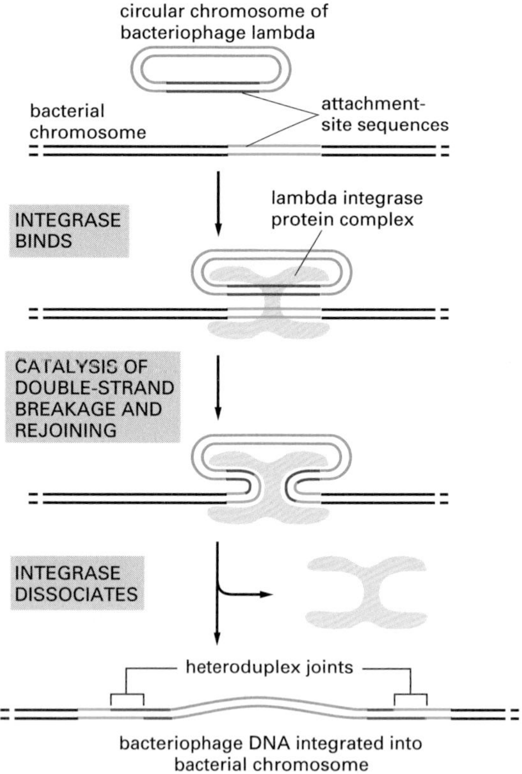

Figure 5–80 The insertion of a circular bacteriophage lambda DNA chromosome into the bacterial chromosome. In this example of site-specific recombination, the lambda integrase enzyme binds to a specific "attachment site" DNA sequence on each chromosome, where it makes cuts that bracket a short homologous DNA sequence. The integrase then switches the partner strands and reseals them to form a heteroduplex joint that is seven nucleotide pairs long. A total of four strand-breaking and strand-joining reactions is required; for each of them, the energy of the cleaved phosphodiester bond is stored in a transient covalent linkage between the DNA and the enzyme, so that DNA strand resealing occurs without a requirement for ATP or DNA ligase.

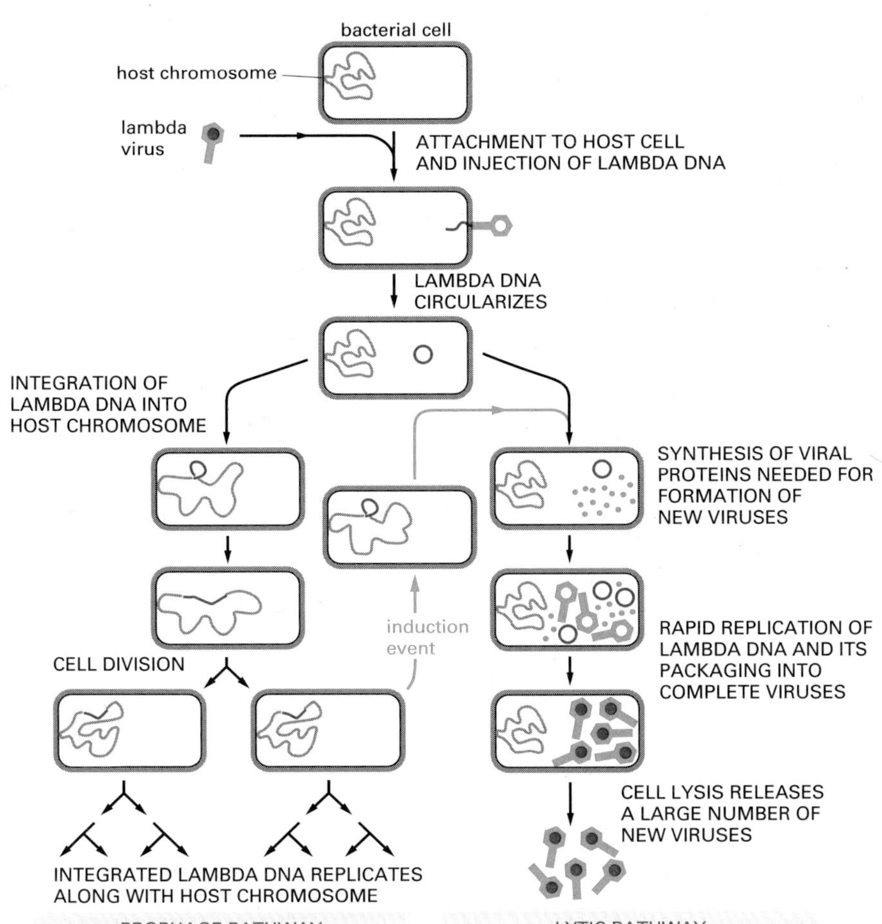

Figure 5–81 The life cycle of bacteriophage lambda. The double-stranded DNA lambda genome contains 50,000 nucleotide pairs and encodes 50–60 different proteins. When the lambda DNA enters the cell, the ends join to form a circular DNA molecule. This bacteriophage can multiply in *E. coli* by a lytic pathway, which destroys the cell, or it can enter a latent prophage state. Damage to a cell carrying a lambda prophage induces the prophage to exit from the host chromosome and shift to lytic growth *(green arrows)*. Both the entrance of the lambda DNA to, and its exit from, the bacterial chromosome are accomplished by a conservative site-specific recombination event, catalyzed by the lambda integrase enzyme (see Figure 5–80).

response to specific signals and multiply rapidly within the bacterial cell (Figure 5–81). Excision is catalyzed by a complex of integrase enzyme and host factors with a second bacteriophage protein, excisionase, which is produced by the virus only when its host cell is stressed—in which case, it is in the bacteriophage's interest to abandon the host cell and multiply again as a virus particle.

Conservative Site-Specific Recombination Can be Used to Turn Genes On or Off

When the special sites recognized by a conservative site-specific recombination enzyme are inverted in their orientation, the DNA sequence between them is inverted rather than excised (see Figure 5–79). Such inversion of a DNA sequence is used by many bacteria to control the gene expression of particular genes—for example, by assembling active genes from separated coding segments. This type of gene control has the advantage of being directly inheritable, since the new DNA arrangement is transferred to daughter chromosomes automatically when a cell divides.

These types of enzymes have also become powerful tools for cell and developmental biologists. To decipher the roles of specific genes and proteins in complex multicellular organisms, genetic engineering techniques can be used to introduce into a mouse a gene encoding a site-specific recombination enzyme and a carefully designed target DNA containing the DNA sites that are recognized by the enzyme. At an appropriate time, the gene encoding the enzyme can be activated to rearrange the target DNA sequence. This rearrangement is often used to cause the production of a specific protein in particular tissues of the mouse (Figure 5–82). By similar means, the technique can be used to turn off any specific gene in a tissue of interest. In this way, one can in principle determine the influence of any protein in any tissue of an intact animal.

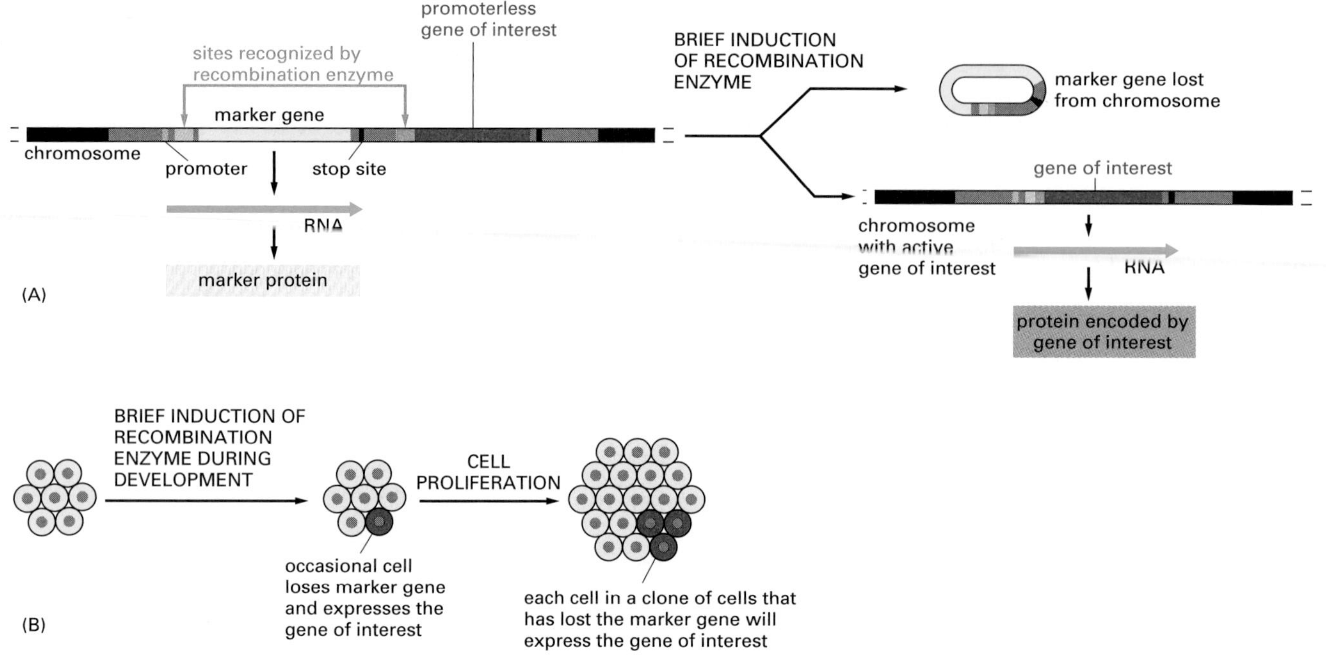

Figure 5–82 How a conservative site-specific recombination enzyme is used to turn on a specific gene in a group of cells in a transgenic animal. This technique requires the insertion of two specially engineered DNA molecules into the animal's germ line. (A) The DNA molecule shown has been engineered with specific recognition sites *(green)* so that the gene of interest *(red)* is transcribed only after a site-specific recombination enzyme that uses these sites is induced. As shown on the right, this induction removes a marker gene *(yellow)* and brings the promoter DNA *(orange)* adjacent to the gene of interest. The recombination enzyme is inducible, because it is encoded by a second DNA molecule (not shown) that has been engineered to ensure that the enzyme is made only when the animal is treated with a special small molecule or its temperature is raised. (B) Transient induction of the recombination enzyme causes a brief burst of synthesis of that enzyme, which in turn causes a DNA rearrangement in an occasional cell. For this cell and all its progeny, the marker gene is inactivated and the gene of interest is simultaneously activated (as shown in A). Those clones of cells in the developing animal that express the gene of interest can be identified by their loss of the marker protein. This technique is widely used in mice and *Drosophila*, because it allows one to study the effect of expressing any gene of interest in a group of cells in an intact animal. In one version of the technique, the Cre recombination enzyme of bacteriophage P1 is employed along with its loxP recognition sites (see pp. 542–543).

Summary

The genomes of nearly all organisms contain mobile genetic elements that can move from one position in the genome to another by either a transpositional or a conservative site-specific recombination process. In most cases this movement is random and happens at a very low frequency. Mobile genetic elements include transposons, which move only within a single cell (and its descendents), and those viruses whose genomes can integrate into the genome of their host cells.

There are three classes of transposons: the DNA-only transposons, the retroviral-like retrotransposons, and the nonretroviral retrotransposons. All but the last have close relatives among the viruses. Although viruses and transposable elements can be viewed as parasites, many of the new arrangements of DNA sequences that their site-specific recombination events produce have created the genetic variation crucial for the evolution of cells and organisms.

References

General

Friedberg EC, Walker GC & Siede W (1995) DNA Repair and Mutagenesis. Washington, DC: ASM Press.

Hartwell L, Hood L, Goldberg ML et al. (2000) Genetics: from Genes to Genomes. Boston: McGraw Hill.

Kornberg A & Baker TA (1992) DNA Replication, 2nd edn. New York: WH Freeman.

Lodish H, Berk A, Zipursky SL et al. (2000) Molecular Cell Biology, 4th edn. New York: WH Freeman.

Stent GS (1971) Molecular Genetics: An Introductory Narrative. San Francisco: WH Freeman.

The Maintenance of DNA Sequences

Crow JF (2000) The origins, patterns and implications of human spontaneous mutation. *Nat. Rev. Genet.* 1, 40–47.

Drake JW, Charlesworth B, Charlesworth D & Crow JF (1998) Rates of spontaneous mutation. *Genetics* 148, 1667–1686.

Ohta T & Kimura M (1971) Functional organization of genetic material as a product of molecular evolution. *Nature* 233, 118–119.

Wilson AC & Ochman H (1987) Molecular time scale for evolution. *Trends Genet.* 3, 241–247.

DNA Replication Mechanisms

Arezi B & Kuchta RD (2000) Eukaryotic DNA primase. *Trends Biochem. Sci.* 25, 572–576.

Baker TA & Bell SP (1998) Polymerases and the replisome: machines within machines. *Cell* 92, 295–305.

Buermeyer AB, Deschenes SM, Baker SM & Liskay RM (1999) Mammalian DNA mismatch repair. *Annu. Rev. Genet.* 33, 533–564.

Davey MJ & O'Donnell M (2000) Mechanisms of DNA replication. *Curr. Opin. Chem. Biol.* 4, 581–586.

Kelman Z & O'Donnell M (1995) DNA polymerase III holoenzyme: structure and function of a chromosomal replicating machine. *Annu. Rev. Biochem.* 64, 171–200.

Kolodner RD (2000) Guarding against mutation. *Nature* 407, 687, 689.

Kornberg A (1960) Biological synthesis of DNA. *Science* 131, 1503–1508.

Kunkel TA & Bebenek K (2000) DNA replication fidelity. *Annu. Rev. Biochem.* 69, 497–529.

Kuriyan J & O'Donnell M (1993) Sliding clamps of DNA polymerases. *J. Mol. Biol.* 234, 915–925.

Li JJ & Kelly TJ (1984) SV40 DNA replication *in vitro*. *Proc. Natl. Acad. Sci. USA* 81, 6973.

Marians KJ (2000) Crawling and wiggling on DNA: structural insights to the mechanism of DNA unwinding by helicases. *Structure Fold. Des.* 8, R227–R235.

Meselson M & Stahl FW (1958) The replication of DNA in *E. coli*. *Proc. Natl Acad. Sci. USA* 44, 671–682.

Modrich P & Lahue R (1996) Mismatch repair in replication fidelity, genetic recombination, and cancer biology. *Annu. Rev. Biochem.* 65, 101–133.

Ogawa T & Okazaki T (1980) Discontinuous DNA replication. *Annu. Rev. Biochem.* 49, 421–457.

Okazaki R, Okazaki T, Sakabe K et al. (1968) Mechanism of DNA chain growth. I. Possible discontinuity and unusual secondary structure of newly synthesized chains. *Proc. Natl. Acad. Sci. USA* 59, 598–605.

Postow L, Peter BJ & Cozzarelli NR (1999) Knot what we thought before: the twisted story of replication. *Bioessays* 21, 805–808.

von Hippel PH & Delagoutte E (2001) A general model for nucleic acid helicases and their "coupling" within macromolecular machines. *Cell* 104, 177–190.

Waga S & Stillman B (1998) The DNA replication fork in eukaryotic cells. *Annu. Rev. Biochem.* 67, 721–751.

Wang JC (1996) DNA topoisomerases. *Annu. Rev. Biochem.* 65, 635–692.

Young MC, Reddy MK & von Hippel PH (1992) Structure and function of the bacteriophage T4 DNA polymerase holoenzyme. *Biochemistry* 31, 8675–8690.

The Initiation and Completion of DNA Replication in Chromosomes

DePamphilis ML (1999) Replication origins in metazoan chromosomes: fact or fiction? *Bioessays* 21, 5–16.

Donaldson AD & Blow JJ (1999) The regulation of replication origin activation. *Curr. Opin. Genet. Dev.* 9, 62–68.

Dutta A & Bell SP (1997) Initiation of DNA replication in eukaryotic cells. *Annu. Rev. Cell Dev. Biol.* 13, 293–332.

Echols H (1990) Nucleoprotein structures initiation DNA replication, transcription and site-specific recombination. *J. Biol. Chem.* 265, 14697–14700.

Gasser SM (2000) A sense of the end. *Science* 288, 1377–1379.

Heun P, Laroche T, Raghuraman MK & Gasser SM (2001) The positioning and dynamics of origins of replication in the budding yeast nucleus. *J. Cell Biol.* 152, 385–400.

Huberman JA & Riggs AD (1968) On the mechanism of DNA replication in mammalian chromosomes. *J. Mol. Biol.* 32, 327–341.

Kass-Eisler A & Greider CW (2000) Recombination in telomere-length maintenance. *Trends Biochem. Sci.* 25, 200–204.

McEachern MJ, Krauskopf A & Blackburn EH (2000) Telomeres and their control. *Annu. Rev. Genet.* 34, 331–358.

Newlon CS & Theis JF (1993) The structure and function of yeast ARS elements. *Curr. Opin. Genet. Dev.* 3, 752–758.

Newlon CS (1997) Putting it all together: building a prereplicative complex. *Cell* 91, 717–720.

Randall SK & Kelly TJ (1992) The fate of parental nucleosomes during SV40 DNA replication. *J. Biol. Chem.* 267, 14259–14265.

Russev G & Hancock R (1982) Assembly of new histones into nucleosomes and their distribution in replicating chromatin. *Proc. Natl Acad. Sci. USA* 79, 3143–3147.

Wintersberger E (2000) Why is there late replication? *Chromosoma* 109, 300–307.

DNA Repair

Auerbach AD & Verlander PC (1997) Disorders of DNA replication and repair. *Curr. Opin. Pediatr.* 9, 600–616.

Critchlow SE & Jackson SP (1998) DNA end-joining: from yeast to man. *Trends Biochem. Sci.* 23, 394–398.

Lindahl T (1993) Instability and decay of the primary structure of DNA. *Nature* 362, 709–715.

Lindahl T, Karran P & Wood RD (1997) DNA excision repair pathways. *Curr. Opin. Genet. Dev.* 7, 158–169.

O'Connell MJ, Walworth NC & Carr AM (2000) The G2-phase DNA-damage checkpoint. *Trends Cell Biol.* 10, 296–303.

Parikh SS, Mol CD, Hosfield DJ et al. (1999) Envisioning the molecular choreography of DNA base excision repair. *Curr. Opin. Struct. Biol.* 9, 37–47.

Perutz MF (1990) Frequency of abnormal human haemoglobins caused by C→T transitions in CpGdinucleotides. *J. Mol. Biol.* 213, 203–206.

Sutton MD, Smith BT, Godoy VG et al. (2000) The SOS response: recent insights into umuDC-dependent mutagenesis and DNA damage tolerance. *Annu. Rev. Genet.* 34, 479–497.

Walker GC (1995) SOS-regulated proteins in translesion DNA synthesis and mutagenesis. *Trends Biochem. Sci.* 20, 416–420.

General Recombination

Baumann P & West SC (1998) Role of the human RAD51 protein in homologous recombination and double-stranded-break repair. *Trends Biochem. Sci.* 23, 247–251.

Flores-Rozas H & Kolodner RD (2000) Links between replication, recombination and genome instability in eukaryotes. *Trends Biochem. Sci.* 25, 196–200.

Haber JE (1998) Mating-type gene switching in Saccharomyces cerevisiae. *Annu. Rev. Genet.* 32, 561–599.

Haber JE (2000) Recombination: a frank view of exchanges and vice versa. *Curr. Opin. Cell Biol.* 12, 286–292.

Holliday R (1990) The history of the DNA heteroduplex. *Bioessays* 12, 133–142.

Kowalczykowski SC & Eggleston AK (1994) Homologous pairing and DNA strand-exchange proteins. *Annu. Rev. Biochem.* 63, 991–1043.

Kowalczykowski SC (2000) Initiation of genetic recombination and recombination-dependent replication. *Trends Biochem. Sci.* 25, 156–165.

Szostak JW, Orr-Weaver TK, Rothstein RJ et al. (1983) The double-strand break repair model for recombination. *Cell* 33, 25–35.

van Gent DC, Hoeijmakers JH & Kanaar R (2001) Chromosomal stability and the DNA double-stranded break connection. *Nat. Rev. Genet.* 2, 196–206.

Welcsh PL, Owens KN & King MC (2000) Insights into the functions of BRCA1 and BRCA2. *Trends Genet.* 16, 69–74.

West SC (1997) Processing of recombination intermediates by the RuvABC proteins. *Annu. Rev. Genet.* 31, 213–244.

Site-specific Recombination

Campbell AM (1993) Thirty years ago in genetics: prophage insertion into bacterial chromosomes. *Genetics* 133, 433–438.

Craig NL (1996) Transposition, in *Escherichia coli* and *Salmonella*, pp 2339–2362. Washington, DC: ASM Press.

Craig NL (1997) Target site selection in transposition. *Annu. Rev. Biochem.* 66, 437–474.

Gopaul DN & Duyne GD (1999) Structure and mechanism in site-specific recombination. *Curr. Opin. Struct. Biol.* 9, 14–20.

Gottesman M (1999) Bacteriophage lambda: the untold story. *J. Mol. Biol.* 293, 177–180.

Jiang R & Gridley T (1997) Gene targeting: things go better with Cre. *Curr. Biol.* 7, R321–323.

Mizuuchi K (1992) Transpositional recombination: mechanistic insights from studies of mu and other elements. *Annu. Rev. Biochem.* 61, 1011–1051.

Sandmeyer SB (1992) Yeast retrotransposons. *Curr. Opin. Genet. Dev.* 2, 705–711.

Smith AF (1999) Interspersed repeats and other mementos of transposable elements in mammalian genomes. *Curr. Opin. Genet. Dev.* 9, 657–663.

Stark WM, Boocock MR & Sherratt DJ (1992) Catalysis by site-specific recombinases. *Trends Genet.* 8, 432–439.

Varmus H (1988) Retroviruses. *Science* 240, 1427–1435.

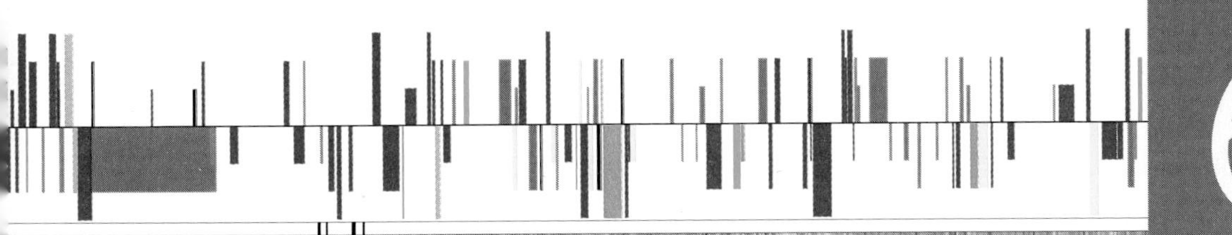

6

HOW CELLS READ THE GENOME: FROM DNA TO PROTEIN

Only when the structure of DNA was discovered in the early 1950s did it become clear how the hereditary information in cells is encoded in DNA's sequence of nucleotides. The progress since then has been astounding. Fifty years later, we have complete genome sequences for many organisms, including humans, and we therefore know the maximum amount of information that is required to produce a complex organism like ourselves. The limits on the hereditary information needed for life constrain the biochemical and structural features of cells and make it clear that biology is not infinitely complex.

In this chapter, we explain how cells decode and use the information in their genomes. We shall see that much has been learned about how the genetic instructions written in an alphabet of just four "letters"—the four different nucleotides in DNA—direct the formation of a bacterium, a fruitfly, or a human. Nevertheless, we still have a great deal to discover about how the information stored in an organism's genome produces even the simplest unicellular bacterium with 500 genes, let alone how it directs the development of a human with approximately 30,000 genes. An enormous amount of ignorance remains; many fascinating challenges therefore await the next generation of cell biologists.

The problems cells face in decoding genomes can be appreciated by considering a small portion of the genome of the fruit fly *Drosophila melanogaster* (Figure 6–1). Much of the DNA-encoded information present in this and other genomes is used to specify the linear order—the sequence—of amino acids for every protein the organism makes. As described in Chapter 3, the amino acid sequence in turn dictates how each protein folds to give a molecule with a distinctive shape and chemistry. When a particular protein is made by the cell, the corresponding region of the genome must therefore be accurately decoded. Additional information encoded in the DNA of the genome specifies exactly when in the life of an organism and in which cell types each gene is to be expressed into protein. Since proteins are the main constituents of cells, the decoding of the genome determines not only the size, shape, biochemical properties, and behavior of cells, but also the distinctive features of each species on Earth.

One might have predicted that the information present in genomes would be arranged in an orderly fashion, resembling a dictionary or a telephone directory.

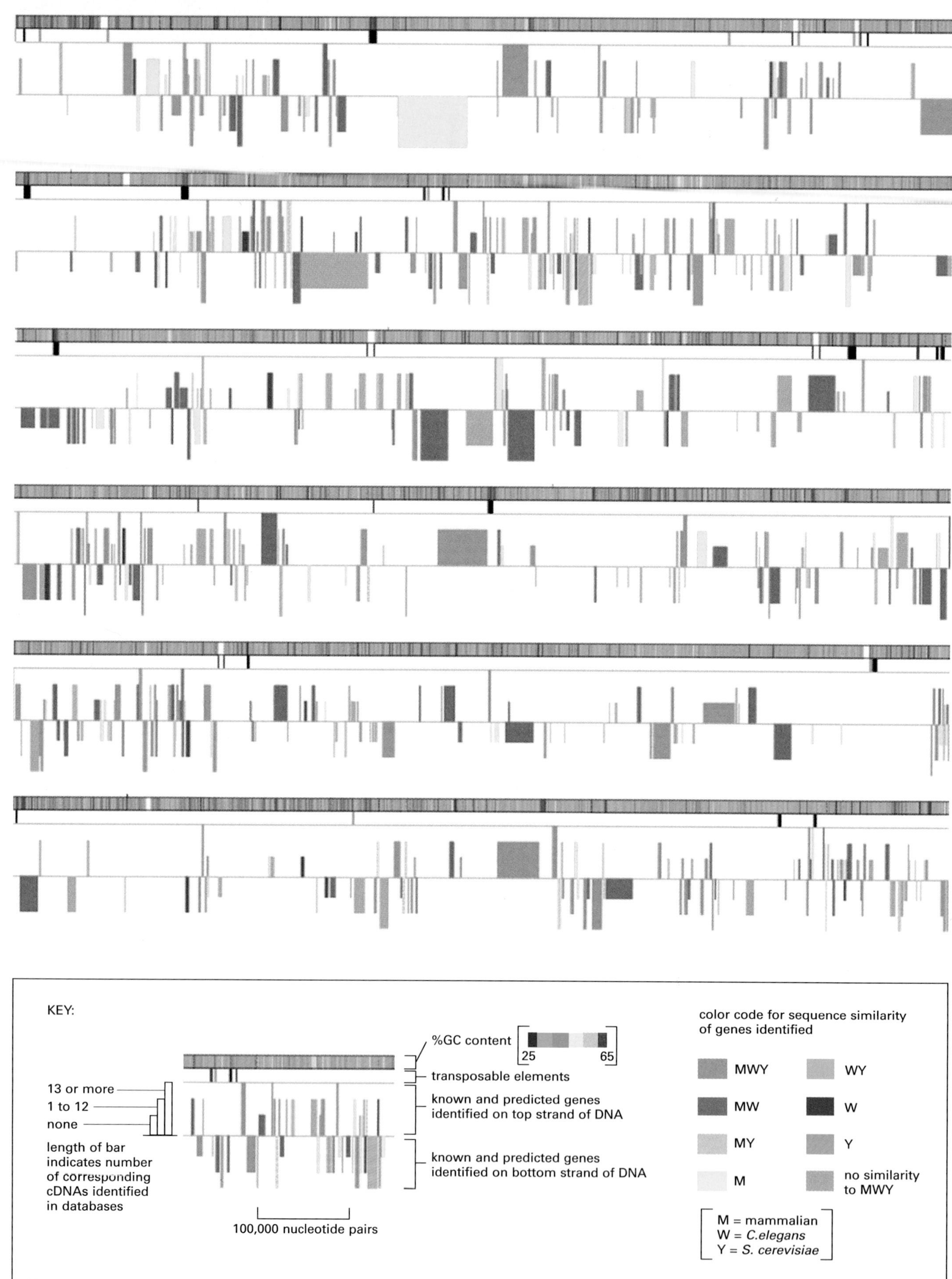
KEY:
%GC content
25
65
transposable elements
known and predicted genes identified on top strand of DNA
known and predicted genes identified on bottom strand of DNA
13 or more
1 to 12
none
length of bar indicates number of corresponding cDNAs identified in databases
100,000 nucleotide pairs
color code for sequence similarity of genes identified
MWY
WY
MW
W
MY
Y
M
no similarity to MWY
M = mammalian
W = C.elegans
Y = S. cerevisiae

Figure 6–1 *(opposite page)* Schematic depiction of a portion of chromosome 2 from the genome of the fruit fly *Drosophila melanogaster*. This figure represents approximately 3% of the total *Drosophila* genome, arranged as six contiguous segments. As summarized in the key, the symbolic representations are: *rainbow-colored bar*: G–C base-pair content; *black vertical lines* of various thicknesses: locations of transposable elements, with thicker bars indicating clusters of elements; *colored boxes*: genes (both known and predicted) coded on one strand of DNA (boxes *above* the midline) and genes coded on the other strand (boxes *below* the midline). The length of each predicted gene includes both its exons (protein-coding DNA) and its introns (non-coding DNA) (see Figure 4–25). As indicated in the key, the height of each gene box is proportional to the number of cDNAs in various databases that match the gene. As described in Chapter 8, cDNAs are DNA copies of mRNA molecules, and large collections of the nucleotide sequences of cDNAs have been deposited in a variety of databases. The higher the number of matches between the nucleotide sequences of cDNAs and that of a particular predicted gene, the higher the confidence that the predicted gene is transcribed into RNA and is thus a genuine gene. The color of each gene box (see *color code* in the key) indicates whether a closely related gene is known to occur in other organisms. For example, MWY means the gene has close relatives in mammals, in the nematode worm *Caenorhabditis elegans*, and in the yeast *Saccharomyces cerevisiae*. MW indicates the gene has close relatives in mammals and the worm but not in yeast. (From Mark D. Adams et al., *Science* 287:2185–2195, 2000.)

Although the genomes of some bacteria seem fairly well organized, the genomes of most multicellular organisms, such as our *Drosophila* example, are surprisingly disorderly. Small bits of coding DNA (that is, DNA that codes for protein) are interspersed with large blocks of seemingly meaningless DNA. Some sections of the genome contain many genes and others lack genes altogether. Proteins that work closely with one another in the cell often have their genes located on different chromosomes, and adjacent genes typically encode proteins that have little to do with each other in the cell. Decoding genomes is therefore no simple matter. Even with the aid of powerful computers, it is still difficult for researchers to locate definitively the beginning and end of genes in the DNA sequences of complex genomes, much less to predict when each gene is expressed in the life of the organism. Although the DNA sequence of the human genome is known, it will probably take at least a decade for humans to identify every gene and determine the precise amino acid sequence of the protein it produces. Yet the cells in our body do this thousands of times a second.

The DNA in genomes does not direct protein synthesis itself, but instead uses RNA as an intermediary molecule. When the cell needs a particular protein, the nucleotide sequence of the appropriate portion of the immensely long DNA molecule in a chromosome is first copied into RNA (a process called *transcription*). It is these RNA copies of segments of the DNA that are used directly as templates to direct the synthesis of the protein (a process called *translation*). The flow of genetic information in cells is therefore from DNA to RNA to protein (Figure 6–2). All cells, from bacteria to humans, express their genetic information in this way—a principle so fundamental that it is termed the *central dogma* of molecular biology.

Despite the universality of the central dogma, there are important variations in the way information flows from DNA to protein. Principal among these is that RNA transcripts in eucaryotic cells are subject to a series of processing steps in the nucleus, including *RNA splicing*, before they are permitted to exit from the nucleus and be translated into protein. These processing steps can critically change the "meaning" of an RNA molecule and are therefore crucial for understanding how eucaryotic cells read the genome. Finally, although we focus on the production of the proteins encoded by the genome in this chapter, we see that for some genes RNA is the final product. Like proteins, many of these RNAs fold into precise three-dimensional structures that have structural and catalytic roles in the cell.

We begin this chapter with the first step in decoding a genome: the process of transcription by which an RNA molecule is produced from the DNA of a gene. We then follow the fate of this RNA molecule through the cell, finishing when a correctly folded protein molecule has been formed. At the end of the chapter, we consider how the present, quite complex, scheme of information storage, transcription, and translation might have arisen from simpler systems in the earliest stages of cellular evolution.

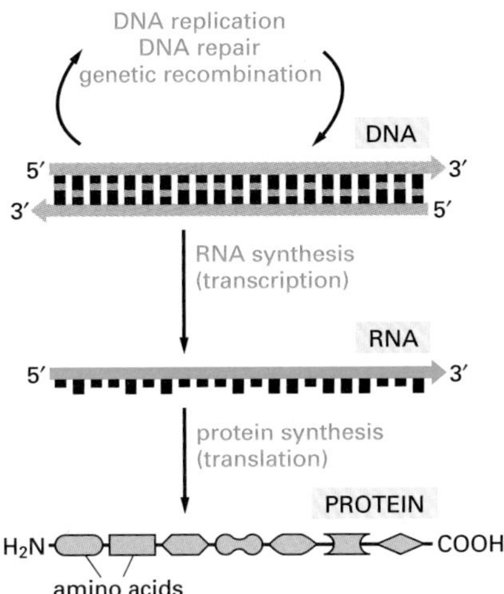

Figure 6–2 The pathway from DNA to protein. The flow of genetic information from DNA to RNA (transcription) and from RNA to protein (translation) occurs in all living cells.

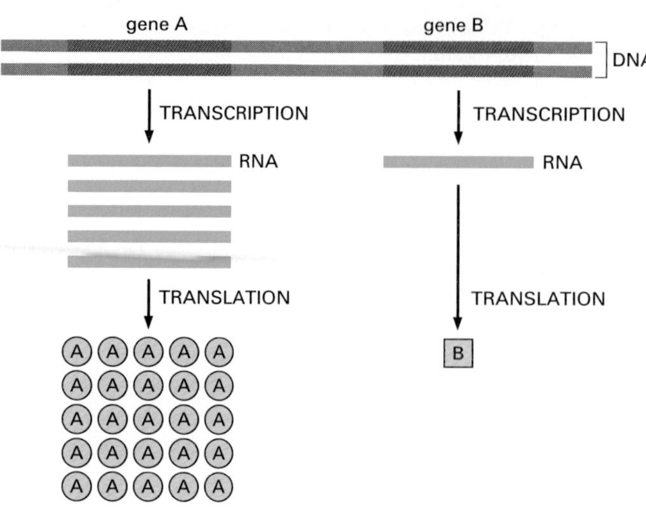

Figure 6–3 Genes can be expressed with different efficiencies. Gene A is transcribed and translated much more efficiently than gene B. This allows the amount of protein A in the cell to be much greater than that of protein B.

FROM DNA TO RNA

Transcription and translation are the means by which cells read out, or express, the genetic instructions in their genes. Because many identical RNA copies can be made from the same gene, and each RNA molecule can direct the synthesis of many identical protein molecules, cells can synthesize a large amount of protein rapidly when necessary. But each gene can also be transcribed and translated with a different efficiency, allowing the cell to make vast quantities of some proteins and tiny quantities of others (Figure 6–3). Moreover, as we see in the next chapter, a cell can change (or regulate) the expression of each of its genes according to the needs of the moment—most obviously by controlling the production of its RNA.

Portions of DNA Sequence Are Transcribed into RNA

The first step a cell takes in reading out a needed part of its genetic instructions is to copy a particular portion of its DNA nucleotide sequence—a gene—into an RNA nucleotide sequence. The information in RNA, although copied into another chemical form, is still written in essentially the same language as it is in DNA—the language of a nucleotide sequence. Hence the name **transcription**.

Like DNA, RNA is a linear polymer made of four different types of nucleotide subunits linked together by phosphodiester bonds (Figure 6–4). It differs from DNA chemically in two respects: (1) the nucleotides in RNA are *ribonucleotides*—that is, they contain the sugar ribose (hence the name *ribo*nucleic acid) rather than deoxyribose; (2) although, like DNA, RNA contains the bases adenine (A), guanine (G), and cytosine (C), it contains the base uracil (U) instead of the thymine (T) in DNA. Since U, like T, can base-pair by hydrogen-bonding with A (Figure 6–5), the complementary base-pairing properties described for DNA in Chapters 4 and 5 apply also to RNA (in RNA, G pairs with C, and A pairs with U). It is not uncommon, however, to find other types of base pairs in RNA: for example, G pairing with U occasionally.

Despite these small chemical differences, DNA and RNA differ quite dramatically in overall structure. Whereas DNA always occurs in cells as a double-stranded helix, RNA is single-stranded. RNA chains therefore fold up into a variety of shapes, just as a polypeptide chain folds up to form the final shape of a protein (Figure 6–6). As we see later in this chapter, the ability to fold into complex three-dimensional shapes allows some RNA molecules to have structural and catalytic functions.

Transcription Produces RNA Complementary to One Strand of DNA

All of the RNA in a cell is made by DNA transcription, a process that has certain similarities to the process of DNA replication discussed in Chapter 5.

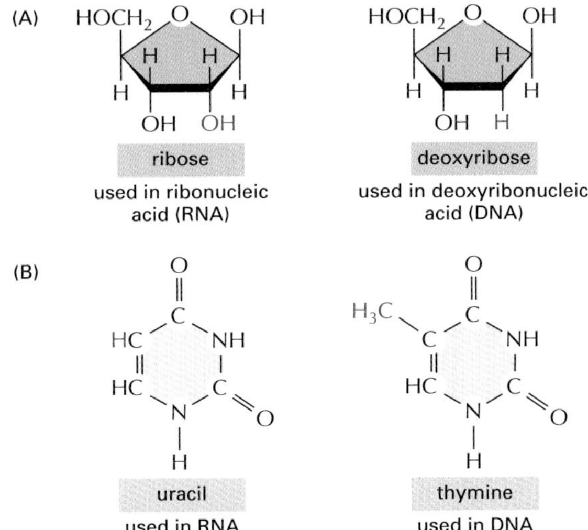

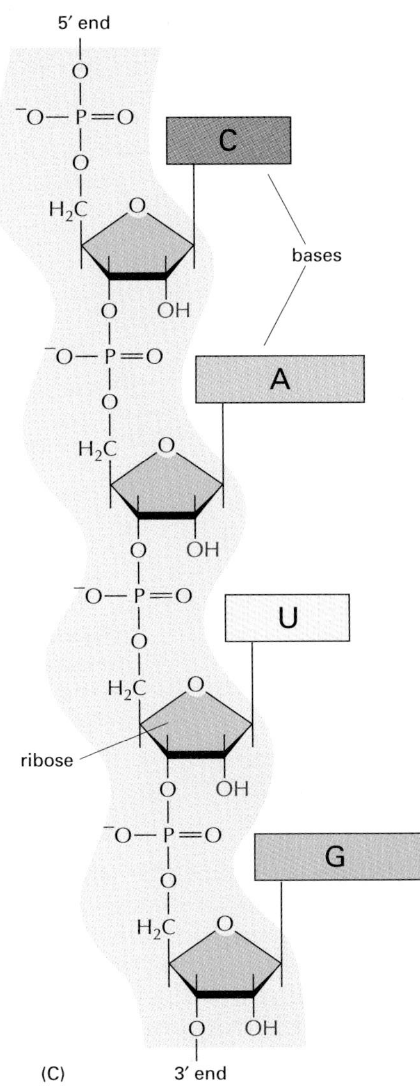

Figure 6–4 The chemical structure of RNA. (A) RNA contains the sugar ribose, which differs from deoxyribose, the sugar used in DNA, by the presence of an additional –OH group. (B) RNA contains the base uracil, which differs from thymine, the equivalent base in DNA, by the absence of a $-CH_3$ group. (C) A short length of RNA. The phosphodiester chemical linkage between nucleotides in RNA is the same as that in DNA.

Transcription begins with the opening and unwinding of a small portion of the DNA double helix to expose the bases on each DNA strand. One of the two strands of the DNA double helix then acts as a template for the synthesis of an RNA molecule. As in DNA replication, the nucleotide sequence of the RNA chain is determined by the complementary base-pairing between incoming nucleotides and the DNA template. When a good match is made, the incoming ribonucleotide is covalently linked to the growing RNA chain in an enzymatically catalyzed reaction. The RNA chain produced by transcription—the *transcript*—is therefore elongated one nucleotide at a time, and it has a nucleotide sequence that is exactly complementary to the strand of DNA used as the template (Figure 6–7).

Transcription, however, differs from DNA replication in several crucial ways. Unlike a newly formed DNA strand, the RNA strand does not remain hydrogen-bonded to the DNA template strand. Instead, just behind the region where the ribonucleotides are being added, the RNA chain is displaced and the DNA helix re-forms. Thus, the RNA molecules produced by transcription are released from the DNA template as single strands. In addition, because they are copied from only a limited region of the DNA, RNA molecules are much shorter than DNA molecules. A DNA molecule in a human chromosome can be up to 250 million nucleotide-pairs long; in contrast, most RNAs are no more than a few thousand nucleotides long, and many are considerably shorter.

The enzymes that perform transcription are called **RNA polymerases**. Like the DNA polymerase that catalyzes DNA replication (discussed in Chapter 5), RNA polymerases catalyze the formation of the phosphodiester bonds that link the nucleotides together to form a linear chain. The RNA polymerase moves stepwise along the DNA, unwinding the DNA helix just ahead of the active site for polymerization to expose a new region of the template strand for complementary base-pairing. In this way, the growing RNA chain is extended by one nucleotide at a time in the 5′-to-3′ direction (Figure 6–8). The substrates are nucleoside triphosphates (ATP, CTP, UTP, and GTP); as for DNA replication, a hydrolysis of high-energy bonds provides the energy needed to drive the reaction forward (see Figure 5–4).

The almost immediate release of the RNA strand from the DNA as it is synthesized means that many RNA copies can be made from the same gene in a

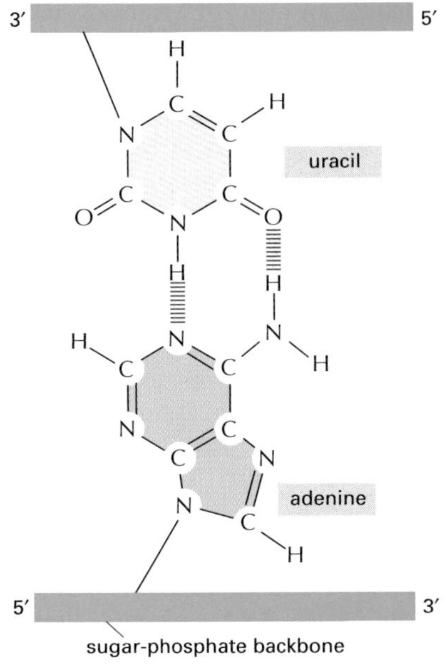

Figure 6–5 Uracil forms base pairs with adenine. The absence of a methyl group in U has no effect on base-pairing; thus, U–A base pairs closely resemble T–A base pairs (see Figure 4–4).

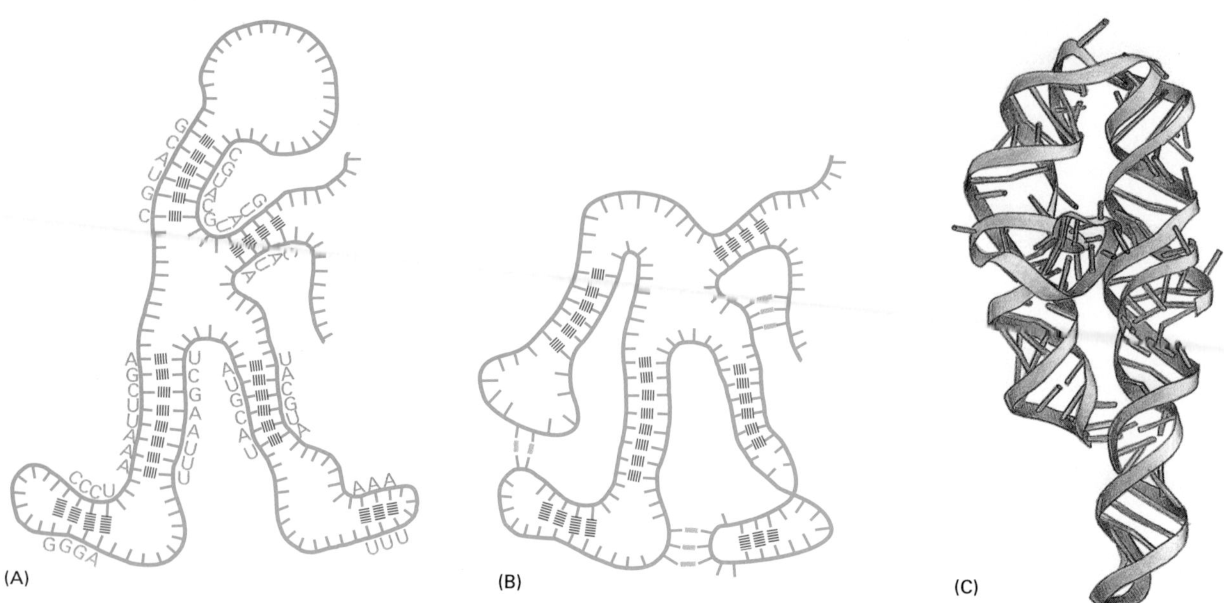

Figure 6–6 RNA can fold into specific structures. RNA is largely single-stranded, but it often contains short stretches of nucleotides that can form conventional base-pairs with complementary sequences found elsewhere on the same molecule. These interactions, along with additional "nonconventional" base-pair interactions, allow an RNA molecule to fold into a three-dimensional structure that is determined by its sequence of nucleotides. (A) Diagram of a folded RNA structure showing only conventional base-pair interactions; (B) structure with both conventional *(red)* and nonconventional *(green)* base-pair interactions; (C) structure of an actual RNA, a portion of a group I intron (see Figure 6–36). Each conventional base-pair interaction is indicated by a "rung" in the double helix. Bases in other configurations are indicated by broken rungs.

relatively short time, the synthesis of additional RNA molecules being started before the first RNA is completed (Figure 6–9). When RNA polymerase molecules follow hard on each other's heels in this way, each moving at about 20 nucleotides per second (the speed in eucaryotes), over a thousand transcripts can be synthesized in an hour from a single gene.

Although RNA polymerase catalyzes essentially the same chemical reaction as DNA polymerase, there are some important differences between the two enzymes. First, and most obvious, RNA polymerase catalyzes the linkage of ribonucleotides, not deoxyribonucleotides. Second, unlike the DNA polymerases involved in DNA replication, RNA polymerases can start an RNA chain without a primer. This difference may exist because transcription need not be as accurate as DNA replication (see Table 5–1, p. 243). Unlike DNA, RNA does not permanently store genetic information in cells. RNA polymerases make about one mistake for every 10^4 nucleotides copied into RNA (compared with an error rate for direct copying by DNA polymerase of about one in 10^7 nucleotides), and the consequences of an error in RNA transcription are much less significant than that in DNA replication.

Although RNA polymerases are not nearly as accurate as the DNA polymerases that replicate DNA, they nonetheless have a modest proofreading mechanism. If the incorrect ribonucleotide is added to the growing RNA chain, the polymerase can back up, and the active site of the enzyme can perform an excision reaction that mimics the reverse of the polymerization reaction, except that water instead of pyrophosphate is used (see Figure 5–4). RNA polymerase hovers around a misincorporated ribonucleotide longer than it does for a correct addition, causing excision to be favored for incorrect nucleotides. However, RNA polymerase also excises many correct bases as part of the cost for improved accuracy.

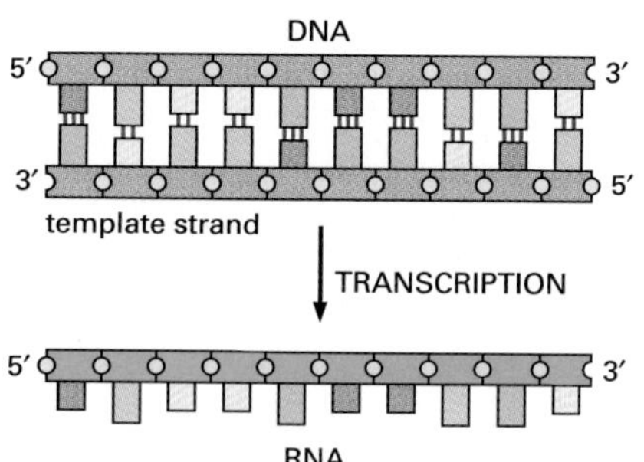

Figure 6–7 DNA transcription produces a single-stranded RNA molecule that is complementary to one strand of DNA.

Cells Produce Several Types of RNA

The majority of genes carried in a cell's DNA specify the amino acid sequence of proteins; the RNA molecules that are copied from these genes (which ultimately direct the synthesis of proteins) are called **messenger RNA (mRNA)** molecules.

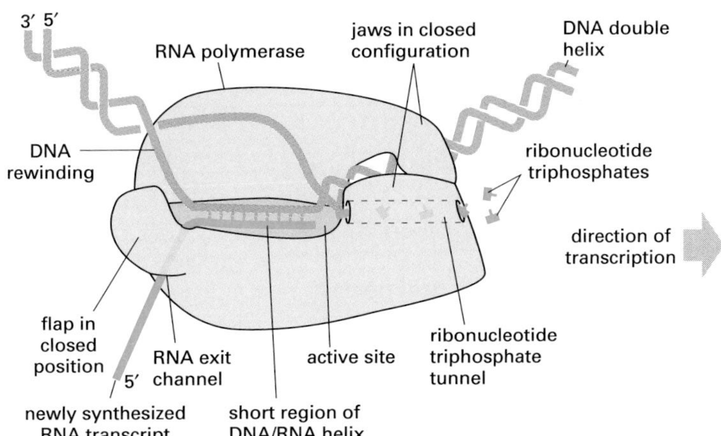

Figure 6–8 DNA is transcribed by the enzyme RNA polymerase. The RNA polymerase *(pale blue)* moves stepwise along the DNA, unwinding the DNA helix at its active site. As it progresses, the polymerase adds nucleotides (here, *small "T" shapes*) one by one to the RNA chain at the polymerization site using an exposed DNA strand as a template. The RNA transcript is thus a single-stranded complementary copy of one of the two DNA strands. The polymerase has a rudder (see Figure 6–11) that displaces the newly formed RNA, allowing the two strands of DNA behind the polymerase to rewind. A short region of DNA/RNA helix (approximately nine nucleotides in length) is therefore formed only transiently, and a "window" of DNA/RNA helix therefore moves along the DNA with the polymerase. The incoming nucleotides are in the form of ribonucleoside triphosphates (ATP, UTP, CTP, and GTP), and the energy stored in their phosphate–phosphate bonds provides the driving force for the polymerization reaction (see Figure 5–4). (Adapted from a figure kindly supplied by Robert Landick.)

The final product of a minority of genes, however, is the RNA itself. Careful analysis of the complete DNA sequence of the genome of the yeast *S. cerevisiae* has uncovered well over 750 genes (somewhat more than 10% of the total number of yeast genes) that produce RNA as their final product, although this number includes multiple copies of some highly repeated genes. These RNAs, like proteins, serve as enzymatic and structural components for a wide variety of processes in the cell. In Chapter 5 we encountered one of those RNAs, the template carried by the enzyme telomerase. Although not all of their functions are known, we see in this chapter that some *small nuclear RNA* (*snRNA*) molecules direct the splicing of pre-mRNA to form mRNA, that *ribosomal RNA (rRNA)* molecules form the core of ribosomes, and that *transfer RNA (tRNA)* molecules form the adaptors that select amino acids and hold them in place on a ribosome for incorporation into protein (Table 6–1).

Each transcribed segment of DNA is called a *transcription unit.* In eucaryotes, a transcription unit typically carries the information of just one gene, and therefore codes for either a single RNA molecule or a single protein (or group of related proteins if the initial RNA transcript is spliced in more than one way to produce different mRNAs). In bacteria, a set of adjacent genes is often transcribed as a unit; the resulting mRNA molecule therefore carries the information for several distinct proteins.

Overall, RNA makes up a few percent of a cell's dry weight. Most of the RNA in cells is rRNA; mRNA comprises only 3–5% of the total RNA in a typical mammalian cell. The mRNA population is made up of tens of thousands of different species, and there are on average only 10–15 molecules of each species of mRNA present in each cell.

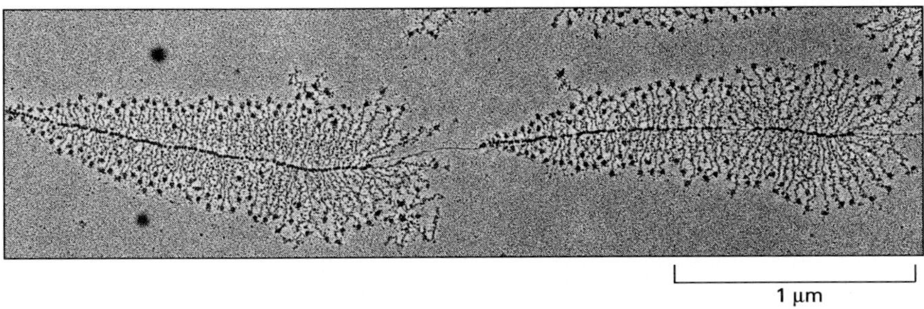

Figure 6–9 Transcription of two genes as observed under the electron microscope. The micrograph shows many molecules of RNA polymerase simultaneously transcribing each of two adjacent genes. Molecules of RNA polymerase are visible as a series of dots along the DNA with the newly synthesized transcripts (fine threads) attached to them. The RNA molecules (ribosomal RNAs) shown in this example are not translated into protein but are instead used directly as components of ribosomes, the machines on which translation takes place. The particles at the 5′ end (the free end) of each rRNA transcript are believed to reflect the beginnings of ribosome assembly. From the lengths of the newly synthesized transcripts, it can be deduced that the RNA polymerase molecules are transcribing from left to right. (Courtesy of Ulrich Scheer.)

TABLE 6–1 Principal Types of RNAs Produced in Cells

TYPE OF RNA	FUNCTION
mRNAs	messenger RNAs, code for proteins
rRNAs	ribosomal RNAs, form the basic structure of the ribosome and catalyze protein synthesis
tRNAs	transfer RNAs, central to protein synthesis as adaptors between mRNA and amino acids
snRNAs	small nuclear RNAs, function in a variety of nuclear processes, including the splicing of pre-mRNA
snoRNAs	small nucleolar RNAs, used to process and chemically modify rRNAs
Other noncoding RNAs	function in diverse cellular processes, including telomere synthesis, X-chromosome inactivation, and the transport of proteins into the ER

Signals Encoded in DNA Tell RNA Polymerase Where to Start and Stop

To transcribe a gene accurately, RNA polymerase must recognize where on the genome to start and where to finish. The way in which RNA polymerases perform these tasks differs somewhat between bacteria and eucaryotes. Because the process in bacteria is simpler, we look there first.

The initiation of transcription is an especially important step in gene expression because it is the main point at which the cell regulates which proteins are to be produced and at what rate. Bacterial RNA polymerase is a multisubunit complex. A detachable subunit, called *sigma (σ) factor*, is largely responsible for its ability to read the signals in the DNA that tell it where to begin transcribing (Figure 6–10). RNA polymerase molecules adhere only weakly to the bacterial DNA when they collide with it, and a polymerase molecule typically slides rapidly along the long DNA molecule until it dissociates again. However, when the polymerase slides into a region on the DNA double helix called a **promoter**, a special sequence of nucleotides indicating the starting point for RNA synthesis, it binds tightly to it. The polymerase, using its σ factor, recognizes this DNA sequence by making specific contacts with the portions of the bases that are exposed on the outside of the helix *(Step 1* in Figure 6–10).

After the RNA polymerase binds tightly to the promoter DNA in this way, it opens up the double helix to expose a short stretch of nucleotides on each strand *(Step 2* in Figure 6–10). Unlike a DNA helicase reaction (see Figure 5–15), this limited opening of the helix does not require the energy of ATP hydrolysis. Instead, the polymerase and DNA both undergo reversible structural changes that result in a more energetically favorable state. With the DNA unwound, one of the two exposed DNA strands acts as a template for complementary base-pairing with incoming ribonucleotides (see Figure 6–7), two of which are joined together by the polymerase to begin an RNA chain. After the first ten or so nucleotides of RNA have been synthesized (a relatively inefficient process during which polymerase synthesizes and discards short nucleotide oligomers), the σ factor relaxes its tight hold on the polymerase and evenutally dissociates from it. During this process, the polymerase undergoes additional structural changes that enable it to move forward rapidly, transcribing without the σ factor *(Step 4* in Figure 6–10). Chain elongation continues (at a speed of approximately 50 nucleotides/sec for bacterial RNA polymerases) until the enzyme encounters a second signal in the DNA, the **terminator** (described below), where the polymerase halts and releases both the DNA template and the newly made RNA chain *(Step 7* in Figure 6–10). After the polymerase has been released at a terminator, it reassociates with a free σ factor and searches for a new promoter, where it can begin the process of transcription again.

Several structural features of bacterial RNA polymerase make it particularly adept at performing the transcription cycle just described. Once the σ factor

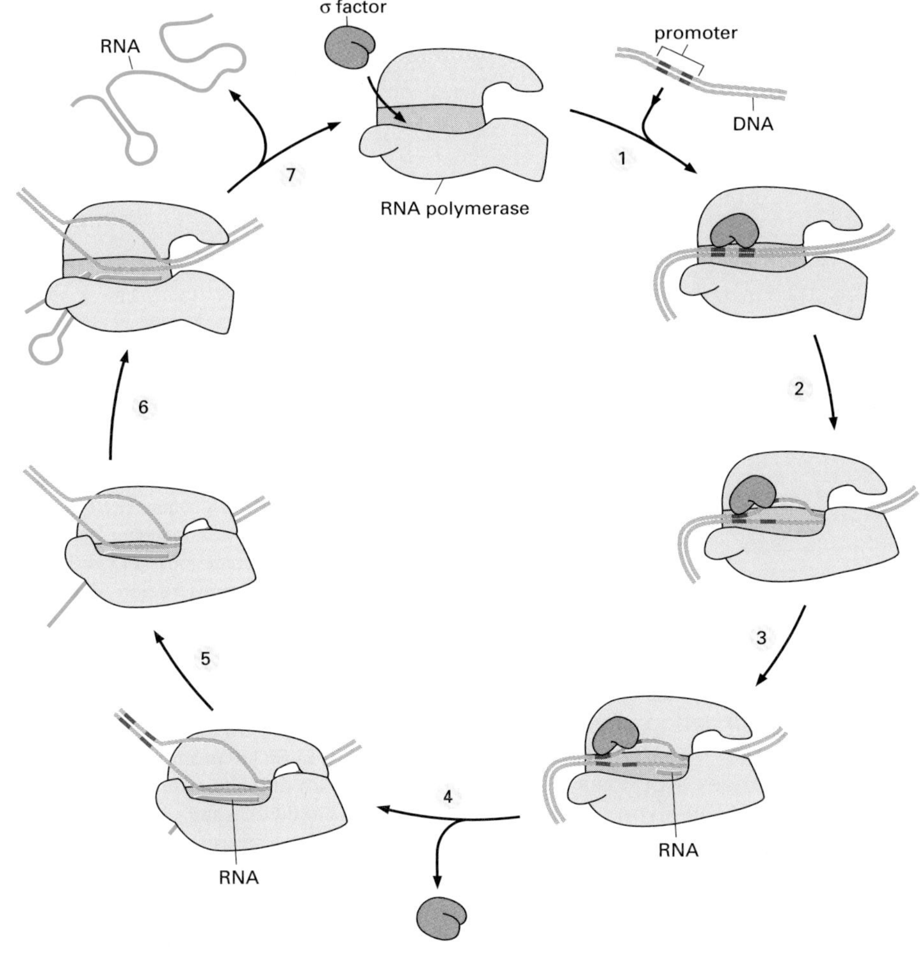

Figure 6–10 The transcription cycle of bacterial RNA polymerase. In step 1, the RNA polymerase holoenzyme (core polymerase plus σ factor) forms and then locates a promoter (see Figure 6–12). The polymerase unwinds the DNA at the position at which transcription is to begin (step 2) and begins transcribing (step 3). This initial RNA synthesis (sometimes called "abortive initiation") is relatively inefficient. However, once RNA polymerase has managed to synthesize about 10 nucleotides of RNA, σ relaxes its grip, and the polymerase undergoes a series of conformational changes (which probably includes a tightening of its jaws and the placement of RNA in the exit channel [see Figure 6–11]). The polymerase now shifts to the elongation mode of RNA synthesis (step 4), moving rightwards along the DNA in this diagram. During the elongation mode (step 5) transcription is highly processive, with the polymerase leaving the DNA template and releasing the newly transcribed RNA only when it encounters a termination signal (step 6). Termination signals are encoded in DNA and many function by forming an RNA structure that destabilizes the polymerase's hold on the RNA, as shown here. In bacteria, all RNA molecules are synthesized by a single type of RNA polymerase and the cycle depicted in the figure therefore applies to the production of mRNAs as well as structural and catalytic RNAs. (Adapted from a figure kindly supplied by Robert Landick.)

positions the polymerase on the promoter and the template DNA has been unwound and pushed to the active site, a pair of moveable jaws is thought to clamp onto the DNA (Figure 6–11). When the first 10 nucleotides have been transcribed, the dissociation of σ allows a flap at the back of the polymerase to

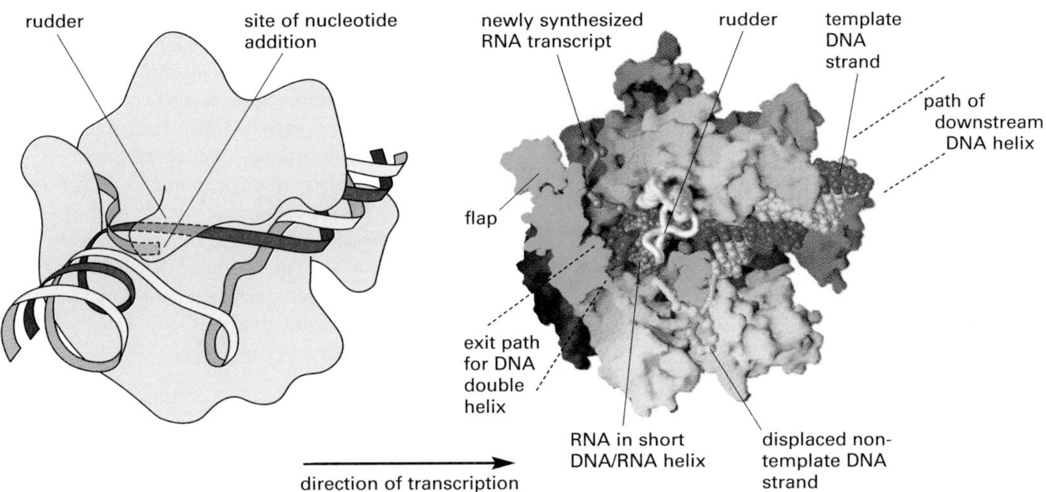

Figure 6–11 The structure of a bacterial RNA polymerase. Two depictions of the three-dimensional structure of a bacterial RNA polymerase, with the DNA and RNA modeled in. This RNA polymerase is formed from four different subunits, indicated by different colors *(right).* The DNA strand used as a template is *red*, and the non-template strand is *yellow*. The rudder wedges apart the DNA–RNA hybrid as the polymerase moves. For simplicity only the polypeptide backbone of the rudder is shown in the right-hand figure, and the DNA exiting from the polymerase has been omitted. Because the RNA polymerase is depicted in the elongation mode, the σ factor is absent. (Courtesy of Seth Darst.)

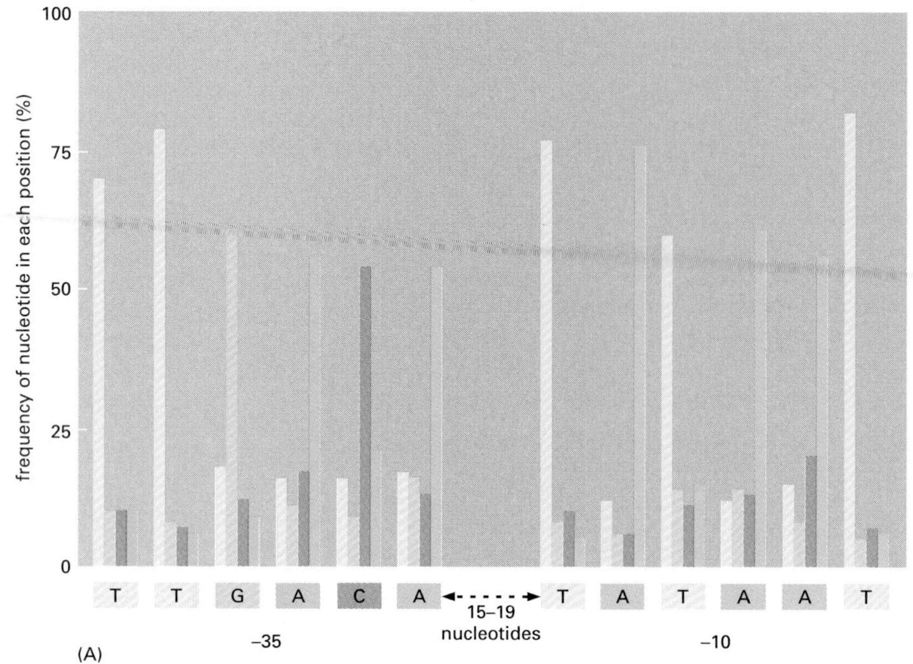

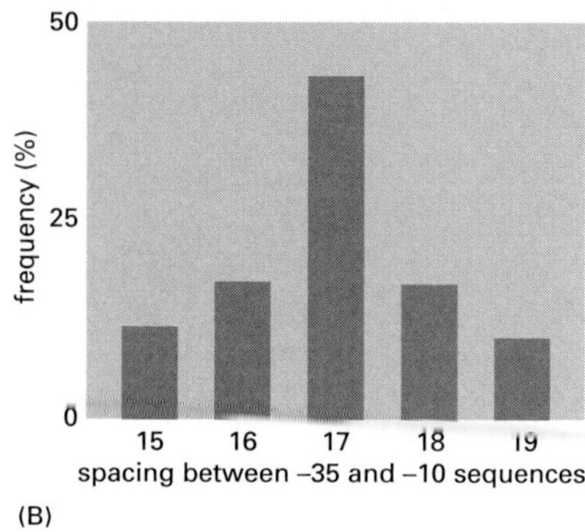

Figure 6–12 Consensus sequence for the major class of *E. coli* promoters. (A) The promoters are characterized by two hexameric DNA sequences, the –35 sequence and the –10 sequence named for their approximate location relative to the start point of transcription (designated +1). For convenience, the nucleotide sequence of a single strand of DNA is shown; in reality the RNA polymerase recognizes the promoter as double-stranded DNA. On the basis of a comparison of 300 promoters, the frequencies of the four nucleotides at each position in the –35 and –10 hexamers are given. The consensus sequence, shown *below* the graph, reflects the most common nucleotide found at each position in the collection of promoters. The sequence of nucleotides between the –35 and –10 hexamers shows no significant similarities among promoters. (B) The distribution of spacing between the –35 and –10 hexamers found in *E. coli* promoters. The information displayed in these two graphs applies to *E. coli* promoters that are recognized by RNA polymerase and the major σ factor (designated σ^{70}). As we shall see in the next chapter, bacteria also contain minor σ factors, each of which recognizes a different promoter sequence. Some particularly strong promoters recognized by RNA polymerase and σ^{70} have an additional sequence, located upstream (to the *left*, in the figure) of the –35 hexamer, which is recognized by another subunit of RNA polymerase.

close to form an exit tunnel through which the newly made RNA leaves the enzyme. With the polymerase now functioning in its elongation mode, a rudder-like structure in the enzyme continuously pries apart the DNA–RNA hybrid formed. We can view the series of conformational changes that takes place during transcription initiation as a successive tightening of the enzyme around the DNA and RNA to ensure that it does not dissociate before it has finished transcribing a gene. If an RNA polymerase does dissociate prematurely, it cannot resume synthesis but must start over again at the promoter.

How do the signals in the DNA (termination signals) stop the elongating polymerase? For most bacterial genes a termination signal consists of a string of A–T nucleotide pairs preceded by a two-fold symmetric DNA sequence, which, when transcribed into RNA, folds into a "hairpin" structure through Watson–Crick base-pairing (see Figure 6–10). As the polymerase transcribes across a terminator, the hairpin may help to wedge open the movable flap on the RNA polymerase and release the RNA transcript from the exit tunnel. At the same time, the DNA–RNA hybrid in the active site, which is held together predominantly by U–A base pairs (which are less stable than G–C base pairs because they form two rather than three hydrogen bonds per base pair), is not sufficiently strong enough to hold the RNA in place, and it dissociates causing the release of the polymerase from the DNA, perhaps by forcing open its jaws. Thus, in some respects, transcription termination seems to involve a reversal of the structural transitions that happen during initiation. The process of termination also is an example of a common theme in this chapter: the ability of RNA to fold into specific structures figures prominantly in many aspects of decoding the genome.

Transcription Start and Stop Signals Are Heterogeneous in Nucleotide Sequence

As we have just seen, the processes of transcription initiation and termination involve a complicated series of structural transitions in protein, DNA, and RNA molecules. It is perhaps not surprising that the signals encoded in DNA that specify these transitions are difficult for researchers to recognize. Indeed, a comparison of many different bacterial promoters reveals that they are heterogeneous in DNA sequence. Nevertheless, they all contain related sequences, reflecting in part aspects of the DNA that are recognized directly by the σ factor. These common features are often summarized in the form of a *consensus sequence* (Figure 6–12). In general, a consensus nucleotide sequence is derived

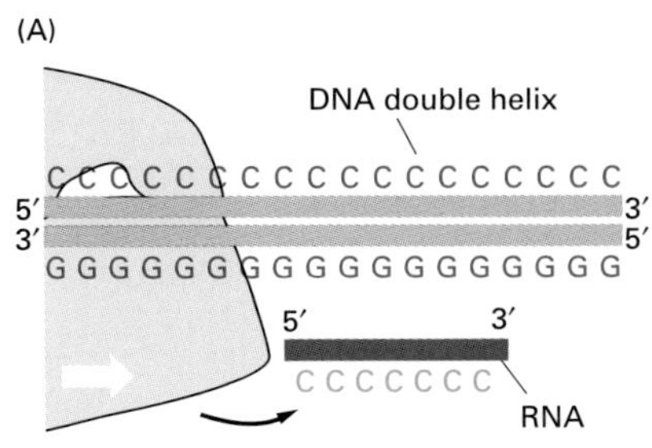

an RNA polymerase that moves from left to right makes RNA by using the bottom strand as a template

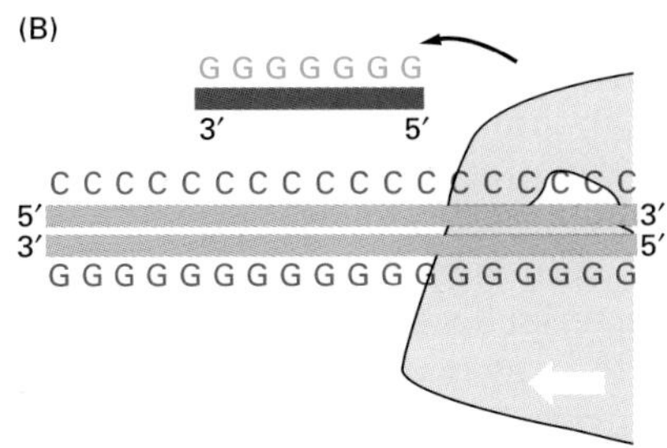

an RNA polymerase that moves from right to left makes RNA by using the top strand as a template

Figure 6–13 The importance of RNA polymerase orientation. The DNA strand serving as template must be traversed in a 3′ to 5′ direction, as illustrated in Figure 6–9. Thus, the direction of RNA polymerase movement determines which of the two DNA strands is to serve as a template for the synthesis of RNA, as shown in (A) and (B). Polymerase direction is, in turn, determined by the orientation of the promoter sequence, the site at which the RNA polymerase begins transcription.

by comparing many sequences with the same basic function and tallying up the most common nucleotide found at each position. It therefore serves as a summary or "average" of a large number of individual nucleotide sequences.

One reason that individual bacterial promoters differ in DNA sequence is that the precise sequence determines the strength (or number of initiation events per unit time) of the promoter. Evolutionary processes have thus fine-tuned each promoter to initiate as often as necessary and have created a wide spectrum of promoters. Promoters for genes that code for abundant proteins are much stronger than those associated with genes that encode rare proteins, and their nucleotide sequences are responsible for these differences.

Like bacterial promoters, transcription terminators also include a wide range of sequences, with the potential to form a simple RNA structure being the most important common feature. Since an almost unlimited number of nucleotide sequences have this potential, terminator sequences are much more heterogeneous than those of promoters.

We have discussed bacterial promoters and terminators in some detail to illustrate an important point regarding the analysis of genome sequences. Although we know a great deal about bacterial promoters and terminators and can develop consensus sequences that summarize their most salient features, their variation in nucleotide sequence makes it difficult for researchers (even when aided by powerful computers) to definitively locate them simply by inspection of the nucleotide sequence of a genome. When we encounter analogous types of sequences in eucaryotes, the problem of locating them is even more difficult. Often, additional information, some of it from direct experimentation, is needed to accurately locate the short DNA signals contained in genomes.

Promoter sequences are asymmetric (see Figure 6–12), and this feature has important consequences for their arrangement in genomes. Since DNA is double-stranded, two different RNA molecules could in principle be transcribed from any gene, using each of the two DNA strands as a template. However a gene typically has only a single promoter, and because the nucleotide sequences of bacterial (as well as eucaryotic) promoters are asymmetric the polymerase can bind in only one orientation. The polymerase thus has no option but to transcribe the one DNA strand, since it can synthesize RNA only in the 5′ to 3′ direction (Figure 6–13). The choice of template strand for each gene is therefore determined by the location and orientation of the promoter. Genome sequences reveal that the DNA strand used as the template for RNA synthesis varies from gene to gene (Figure 6–14; see also Figure 1–31).

Having considered transcription in bacteria, we now turn to the situation in eucaryotes, where the synthesis of RNA molecules is a much more elaborate affair.

Transcription Initiation in Eucaryotes Requires Many Proteins

In contrast to bacteria, which contain a single type of RNA polymerase, eucaryotic nuclei have three, called *RNA polymerase I, RNA polymerase II,* and *RNA*

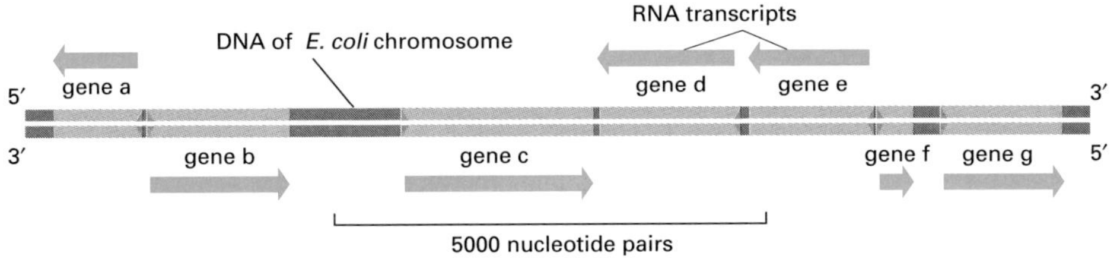

Figure 6–14 Directions of transcription along a short portion of a bacterial chromosome. Some genes are transcribed using one DNA strand as a template, while others are transcribed using the other DNA strand. The direction of transcription is determined by the promoter at the beginning of each gene *(green arrowheads)*. Approximately 0.2% (9000 base pairs) of the *E. coli* chromosome is depicted here. The genes transcribed from *left* to *right* use the bottom DNA strand as the template; those transcribed from *right* to *left* use the top strand as the template.

TABLE 6–2 The Three RNA Polymerases in Eucaryotic Cells

TYPE OF POLYMERASE	GENES TRANSCRIBED
RNA polymerase I	5.8S, 18S, and 28S rRNA genes
RNA polymerase II	all protein-coding genes, plus snoRNA genes and some snRNA genes
RNA polymerase III	tRNA genes, 5S rRNA genes, some snRNA genes and genes for other small RNAs

polymerase III. The three polymerases are structurally similar to one another (and to the bacterial enzyme). They share some common subunits and many structural features, but they transcribe different types of genes (Table 6–2). RNA polymerases I and III transcribe the genes encoding transfer RNA, ribosomal RNA, and various small RNAs. RNA polymerase II transcribes the vast majority of genes, including all those that encode proteins, and our subsequent discussion therefore focuses on this enzyme.

Although eucaryotic RNA polymerase II has many structural similarities to bacterial RNA polymerase (Figure 6–15), there are several important differences in the way in which the bacterial and eucaryotic enzymes function, two of which concern us immediately.

1. While bacterial RNA polymerase (with σ factor as one of its subunits) is able to initiate transcription on a DNA template *in vitro* without the help of additional proteins, eucaryotic RNA polymerases cannot. They require the help of a large set of proteins called *general transcription factors,* which must assemble at the promoter with the polymerase before the polymerase can begin transcription.
2. Eucaryotic transcription initiation must deal with the packing of DNA into nucleosomes and higher order forms of chromatin structure, features absent from bacterial chromosomes.

RNA Polymerase II Requires General Transcription Factors

The discovery that, unlike bacterial RNA polymerase, purified eucaryotic RNA polymerase II could not initiate transcription *in vitro* led to the discovery and purification of the additional factors required for this process. These **general transcription factors** help to position the RNA polymerase correctly at the promoter, aid in pulling apart the two strands of DNA to allow transcription to begin, and release RNA polymerase from the promoter into the elongation mode once transcription has begun. The proteins are "general" because they assemble on all promoters used by RNA polymerase II; consisting of a set of interacting proteins, they are designated as *TFII* (for transcription factor for polymerase II),

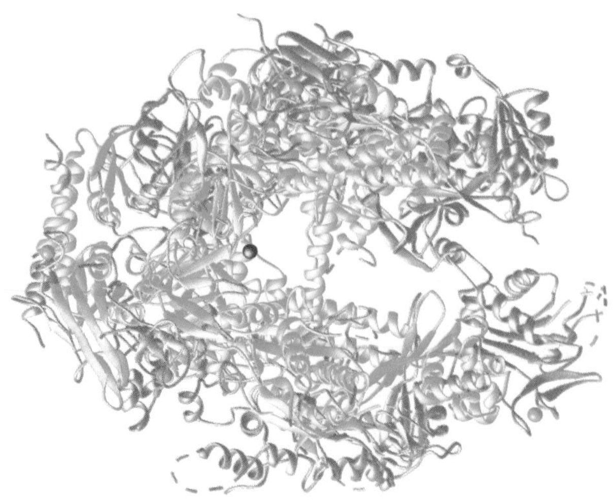

Figure 6–15 Structural similarity between a bacterial RNA polymerase and a eucaryotic RNA polymerase II. Regions of the two RNA polymerases that have similar structures are indicated in *green.* The eucaryotic polymerase is larger than the bacterial enzyme (12 subunits instead of 5), and some of the additional regions are shown in *gray.* The *blue* spheres represent Zn atoms that serve as structural components of the polymerases, and the *red* sphere represents the Mg atom present at the active site, where polymerization takes place. The RNA polymerases in all modern-day cells (bacteria, archaea, and eucaryotes) are closely related, indicating that the basic features of the enzyme were in place before the divergence of the three major branches of life. (Courtesy of P. Cramer and R. Kornberg.)

Figure 6–16 Initiation of transcription of a eucaryotic gene by RNA polymerase II. To begin transcription, RNA polymerase requires a number of general transcription factors (called TFIIA, TFIIB, and so on). (A) The promoter contains a DNA sequence called the TATA box, which is located 25 nucleotides away from the site at which transcription is initiated. (B) The TATA box is recognized and bound by transcription factor TFIID, which then enables the adjacent binding of TFIIB (C). For simplicity the DNA distortion produced by the binding of TFIID (see Figure 6–18) is not shown. (D) The rest of the general transcription factors, as well as the RNA polymerase itself, assemble at the promoter. (E) TFIIH then uses ATP to pry apart the DNA double helix at the transcription start point, allowing transcription to begin. TFIIH also phosphorylates RNA polymerase II, changing its conformation so that the polymerase is released from the general factors and can begin the elongation phase of transcription. As shown, the site of phosphorylation is a long C-terminal polypeptide tail that extends from the polymerase molecule. The assembly scheme shown in the figure was deduced from experiments performed *in vitro*, and the exact order in which the general transcription factors assemble on promoters in cells is not known with certainty. In some cases, the general factors are thought to first assemble with the polymerase, with the whole assembly subsequently binding to the DNA in a single step. The general transcription factors have been highly conserved in evolution; some of those from human cells can be replaced in biochemical experiments by the corresponding factors from simple yeasts.

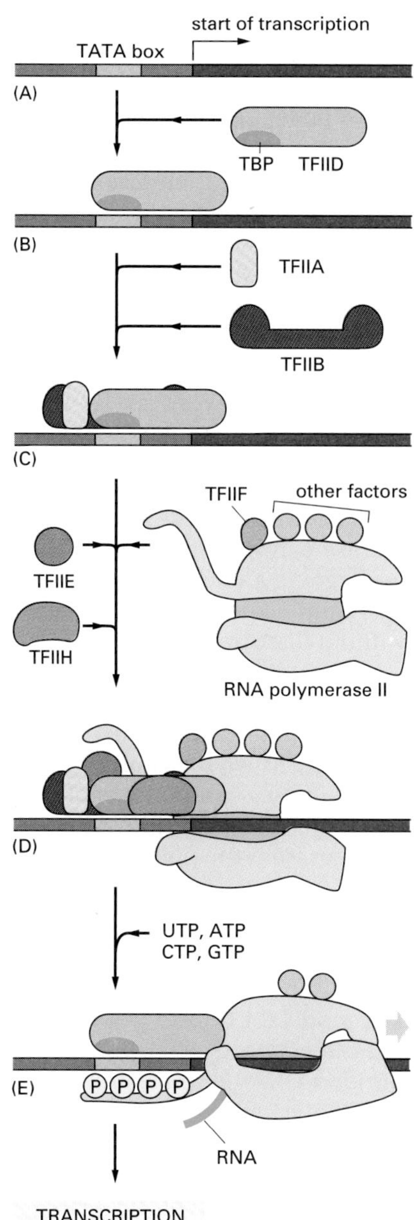

and listed as TFIIA, TFIIB, and so on. In a broad sense, the eucaryotic general transcription factors carry out functions equivalent to those of the σ factor in bacteria.

Figure 6–16 shows how the general transcription factors assemble *in vitro* at promoters used by RNA polymerase II. The assembly process starts with the binding of the general transcription factor TFIID to a short double-helical DNA sequence primarily composed of T and A nucleotides. For this reason, this sequence is known as the TATA sequence, or **TATA box**, and the subunit of TFIID that recognizes it is called TBP (for TATA-binding protein). The TATA box is typically located 25 nucleotides upstream from the transcription start site. It is not the only DNA sequence that signals the start of transcription (Figure 6–17), but

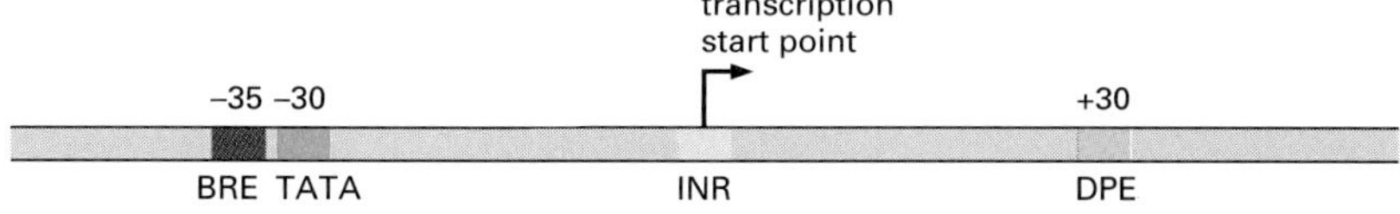

element	consensus sequence	general transcription factor
BRE	G/C G/C G/A C G C C	TFIIB
TATA	T A T A A/T A A/T	TBP
INR	C/T C/T A N T/A C/T C/T	TFIID
DPE	A/G G A/T C G T G	TFIID

Figure 6–17 Consensus sequences found in the vicinity of eucaryotic RNA polymerase II start points. The name given to each consensus sequence *(first column)* and the general transcription factor that recognizes it *(last column)* are indicated. N indicates any nucleotide, and two nucleotides separated by a slash indicate an equal probability of either nucleotide at the indicated position. In reality, each consensus sequence is a shorthand representation of a histogram similar to that of Figure 6–12. For most RNA polymerase II transcription start points, only two or three of the four sequences are present. For example, most polymerase II promoters have a TATA box sequence, and those that do not typically have a "strong" INR sequence. Although most of the DNA sequences that influence transcription initiation are located "upstream" of the transcription start point, a few, such as the DPE shown in the figure, are located in the transcribed region.

for most polymerase II promoters, it is the most important. The binding of TFIID causes a large distortion in the DNA of the TATA box (Figure 6–18). This distortion is thought to serve as a physical landmark for the location of an active promoter in the midst of a very large genome, and it brings DNA sequences on both sides of the distortion together to allow for subsequent protein assembly steps. Other factors are then assembled, along with RNA polymerase II, to form a complete *transcription initiation complex* (see Figure 6–16).

After RNA polymerase II has been guided onto the promoter DNA to form a transcription initiation complex, it must gain access to the template strand at the transcription start point. This step is aided by one of the general transcription factors, TFIIH, which contains a DNA helicase. Next, like the bacterial polymerase, polymerase II remains at the promoter, synthesizing short lengths of RNA until it undergoes a conformational change and is released to begin transcribing a gene. A key step in this release is the addition of phosphate groups to the "tail" of the RNA polymerase (known as the CTD or C-terminal domain). This phosphorylation is also catalyzed by TFIIH, which, in addition to a helicase, contains a protein kinase as one of its subunits (see Figure 6–16, D and E). The polymerase can then disengage from the cluster of general transcription factors, undergoing a series of conformational changes that tighten its interaction with DNA and acquiring new proteins that allow it to transcribe for long distances without dissociating.

Once the polymerase II has begun elongating the RNA transcript, most of the general transcription factors are released from the DNA so that they are available to initiate another round of transcription with a new RNA polymerase molecule. As we see shortly, the phosphorylation of the tail of RNA polymerase II also causes components of the RNA processing machinery to load onto the polymerase and thus be in position to modify the newly transcribed RNA as it emerges from the polymerase.

Polymerase II Also Requires Activator, Mediator, and Chromatin-modifying Proteins

The model for transcription initiation just described was established by studying the action of RNA polymerase II and its general transcription factors on purified DNA templates *in vitro*. However, as discussed in Chapter 4, DNA in eucaryotic cells is packaged into nucleosomes, which are further arranged in higher-order chromatin structures. As a result, transcription initiation in a eucaryotic cell is more complex and requires more proteins than it does on purified DNA. First, gene regulatory proteins known as *transcriptional activators*

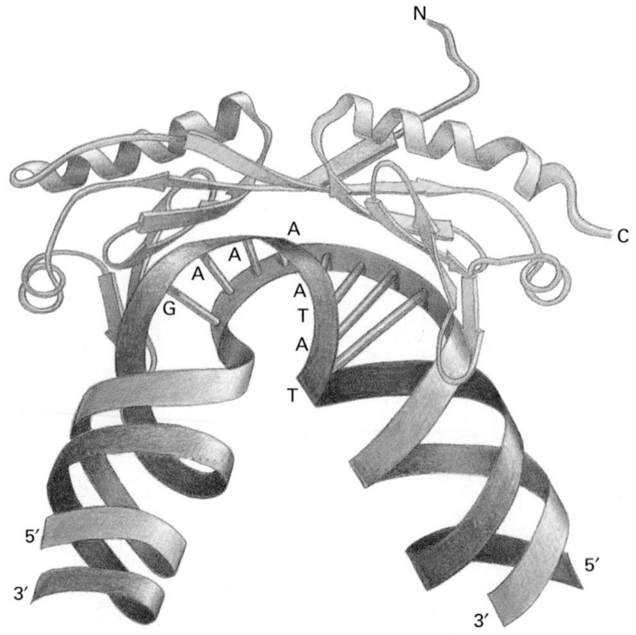

Figure 6–18 Three-dimensional structure of TBP (TATA-binding protein) bound to DNA. The TBP is the subunit of the general transcription factor TFIID that is responsible for recognizing and binding to the TATA box sequence in the DNA *(red)*. The unique DNA bending caused by TBP—two kinks in the double helix separated by partly unwound DNA—may serve as a landmark that helps to attract the other general transcription factors. TBP is a single polypeptide chain that is folded into two very similar domains *(blue* and *green)*. (Adapted from J.L. Kim et al., *Nature* 365:520–527, 1993.)

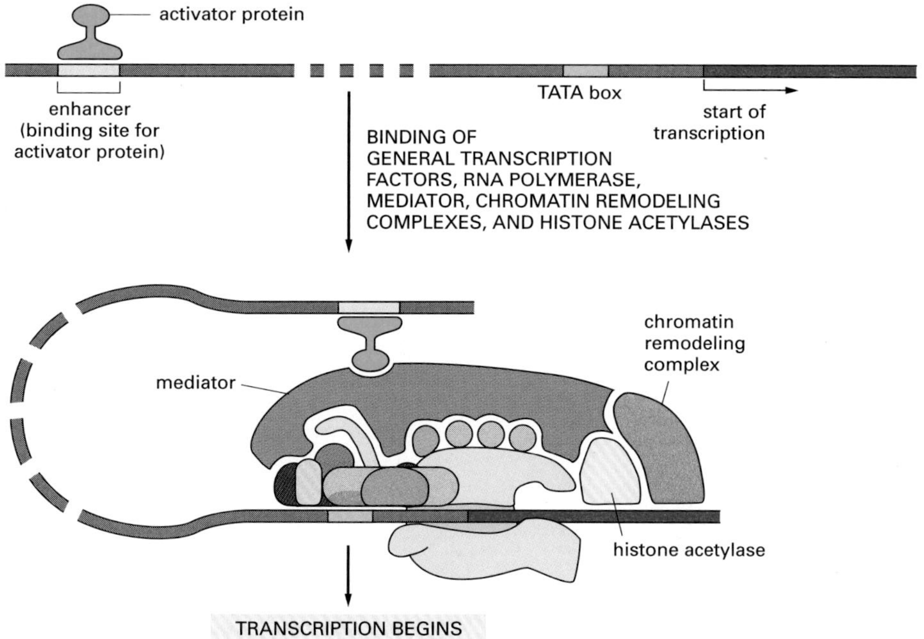

Figure 6–19 Transcription initiation by RNA polymerase II in a eucaryotic cell. Transcription initiation *in vivo* requires the presence of transcriptional activator proteins. As described in Chapter 7, these proteins bind to specific short sequences in DNA. Although only one is shown here, a typical eucaryotic gene has many activator proteins, which together determine its rate and pattern of transcription. Sometimes acting from a distance of several thousand nucleotide pairs (indicated by the dashed DNA molecule), these gene regulatory proteins help RNA polymerase, the general factors, and the mediator all to assemble at the promoter. In addition, activators attract ATP-dependent chromatin-remodeling complexes and histone acetylases.

As discussed in Chapter 4, the "default" state of chromatin is probably the 30-nm filament, and this is likely to be a form of DNA upon which transcription is initiated. For simplicity, it is not shown in the figure.

bind to specific sequences in DNA and help to attract RNA polymerase II to the start point of transcription (Figure 6–19). This attraction is needed to help the RNA polymerase and the general transcription factors in overcoming the difficulty of binding to DNA that is packaged in chromatin. We discuss the role of activators in Chapter 7, because they represent one of the main ways in which cells regulate expression of their genes. Here we simply note that their presence on DNA is required for transcription initiation in a eucaryotic cell. Second, eucaryotic transcription initiation *in vivo* requires the presence of a protein complex known as the *mediator*, which allows the activator proteins to communicate properly with the polymerase II and with the general transcription factors. Finally, transcription initiation in the cell often requires the local recruitment of chromatin-modifying enzymes, including chromatin remodeling complexes and histone acetylases (see Figure 6–19). As discussed in Chapter 4, both types of enzymes can allow greater accessibility to the DNA present in chromatin, and by doing so, they facilitate the assembly of the transcription initiation machinery onto DNA.

As illustrated in Figure 6–19, many proteins (well over one hundred individual subunits) must assemble at the start point of transcription to initiate transcription in a eucaryotic cell. The order of assembly of these proteins is probably different for different genes and therefore may not follow a prescribed pathway. In fact, some of these different protein assemblies may interact with each other away from the DNA and be brought to DNA as preformed subcomplexes. For example, the mediator, RNA polymerase II, and some of the general transcription factors can bind to each other in the nucleoplasm and be brought to the DNA as a unit. We return to this issue in Chapter 7, where we discuss the many ways eucaryotic cells can regulate the process of transcription initiation.

Transcription Elongation Produces Superhelical Tension in DNA

Once it has initiated transcription, RNA polymerase does not proceed smoothly along a DNA molecule; rather it moves jerkily, pausing at some sequences and rapidly transcribing through others. Elongating RNA polymerases, both bacterial and eucaryotic, are associated with a series of *elongation factors,* proteins that decrease the likelihood that RNA polymerase will dissociate before it reaches the end of a gene. These factors typically associate with RNA polymerase shortly after initiation has occurred and help polymerases to move through the wide

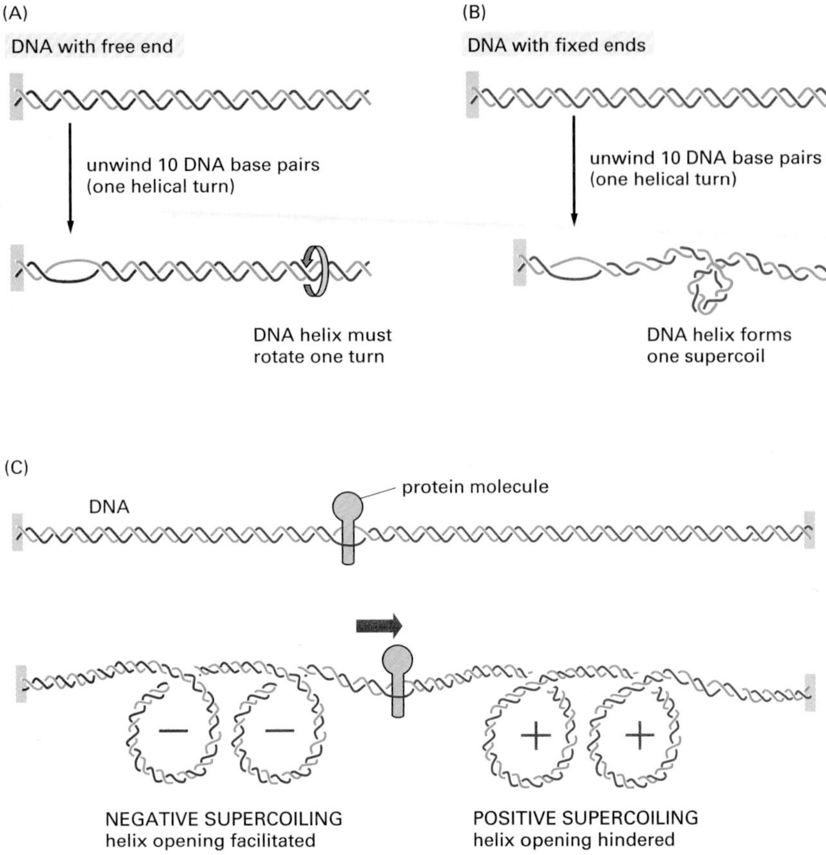

Figure 6–20 Superhelical tension in DNA causes DNA supercoiling. (A) For a DNA molecule with one free end (or a nick in one strand that serves as a swivel), the DNA double helix rotates by one turn for every 10 nucleotide pairs opened. (B) If rotation is prevented, superhelical tension is introduced into the DNA by helix opening. One way of accommodating this tension would be to increase the helical twist from 10 to 11 nucleotide pairs per turn in the double helix that remains in this example; the DNA helix, however, resists such a deformation in a springlike fashion, preferring to relieve the superhelical tension by bending into supercoiled loops. As a result, one DNA supercoil forms in the DNA double helix for every 10 nucleotide pairs opened. The supercoil formed in this case is a positive supercoil. (C) Supercoiling of DNA is induced by a protein tracking through the DNA double helix. The two ends of the DNA shown here are unable to rotate freely relative to each other, and the protein molecule is assumed also to be prevented from rotating freely as it moves. Under these conditions, the movement of the protein causes an excess of helical turns to accumulate in the DNA helix ahead of the protein and a deficit of helical turns to arise in the DNA behind the protein, as shown.

variety of different DNA sequences that are found in genes. Eucaryotic RNA polymerases must also contend with chromatin structure as they move along a DNA template. Experiments have shown that bacterial polymerases, which never encounter nucleosomes *in vivo*, can nonetheless transcribe through them *in vitro*, suggesting that a nucleosome is easily traversed. However, eucaryotic polymerases have to move through forms of chromatin that are more compact than a simple nucleosome. It therefore seems likely that they transcribe with the aid of chromatin remodeling complexes (see pp. 212–213). These complexes may move with the polymerase or may simply seek out and rescue the occasional stalled polymerase. In addition, some elongation factors associated with eucaryotic RNA polymerase facilitate transcription through nucleosomes without requiring additional energy. It is not yet understood how this is accomplished, but these proteins may help to dislodge parts of the nucleosome core as the polymerase transcribes the DNA of a nucleosome.

There is yet another barrier to elongating polymerases, both bacterial and eucaryotic. To discuss this issue, we need first to consider a subtle property inherent in the DNA double helix called **DNA supercoiling**. DNA supercoiling represents a conformation that DNA will adopt in response to superhelical tension; conversely, creating various loops or coils in the helix can create such tension. A simple way of visualizing the topological constraints that cause DNA supercoiling is illustrated in Figure 6–20A. There are approximately 10 nucleotide pairs for every helical turn in a DNA double helix. Imagine a helix whose two ends are fixed with respect to each other (as they are in a DNA circle, such as a bacterial chromosome, or in a tightly clamped loop, as is thought to exist in eucaryotic chromosomes). In this case, one large DNA supercoil will form to compensate for each 10 nucleotide pairs that are opened (unwound). The formation of this supercoil is energetically favorable because it restores a normal helical twist to the base-paired regions that remain, which would otherwise need to be overwound because of the fixed ends.

Superhelical tension is also created as RNA polymerase moves along a stretch of DNA that is anchored at its ends (Figure 6–20C). As long as the polymerase is not free to rotate rapidly (and such rotation is unlikely given the size

(A) EUCARYOTES

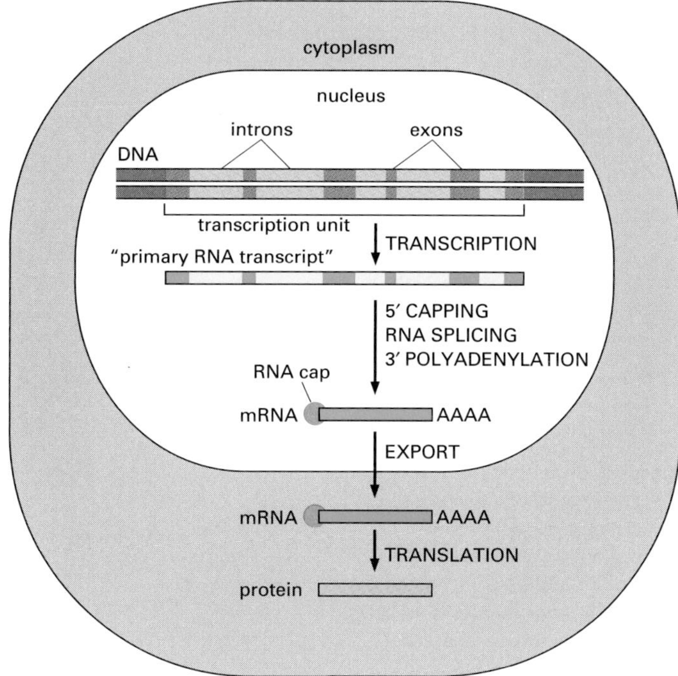

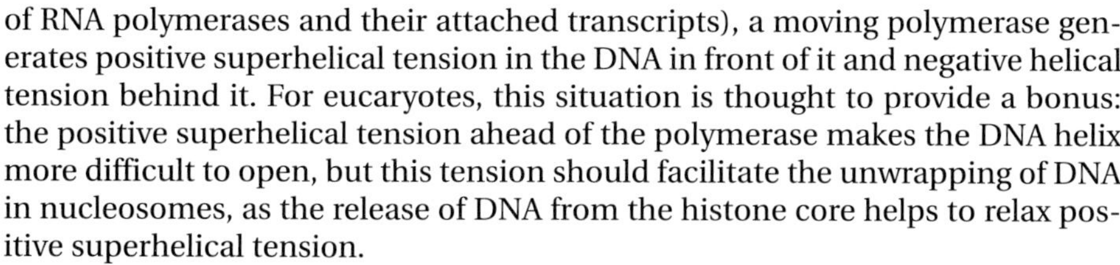

(B) PROCARYOTES

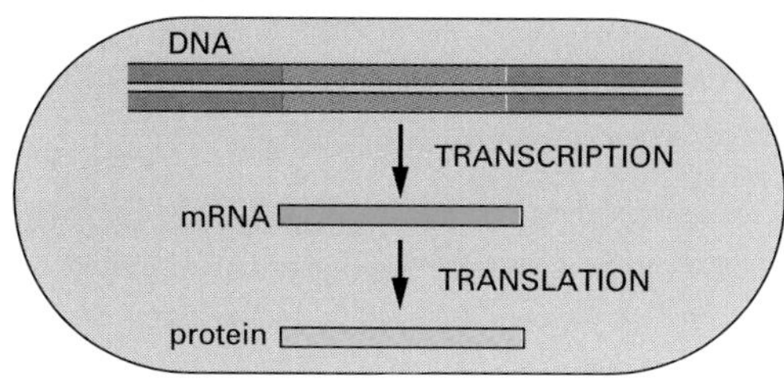

Figure 6–21 Summary of the steps leading from gene to protein in eucaryotes and bacteria. The final level of a protein in the cell depends on the efficiency of each step and on the rates of degradation of the RNA and protein molecules. (A) In eucaryotic cells the RNA molecule produced by transcription alone (sometimes referred to as the primary transcript) would contain both coding (exon) and noncoding (intron) sequences. Before it can be translated into protein, the two ends of the RNA are modified, the introns are removed by an enzymatically catalyzed RNA splicing reaction, and the resulting mRNA is transported from the nucleus to the cytoplasm. Although these steps are depicted as occurring one at a time, in a sequence, in reality they are coupled and different steps can occur simultaneously. For example, the RNA cap is added and splicing typically begins before transcription has been completed. Because of this coupling, complete primary RNA transcripts do not typically exist in the cell. (B) In procaryotes the production of mRNA molecules is much simpler. The 5′ end of an mRNA molecule is produced by the initiation of transcription by RNA polymerase, and the 3′ end is produced by the termination of transcription. Since procaryotic cells lack a nucleus, transcription and translation take place in a common compartment. In fact, translation of a bacterial mRNA often begins before its synthesis has been completed.

of RNA polymerases and their attached transcripts), a moving polymerase generates positive superhelical tension in the DNA in front of it and negative helical tension behind it. For eucaryotes, this situation is thought to provide a bonus: the positive superhelical tension ahead of the polymerase makes the DNA helix more difficult to open, but this tension should facilitate the unwrapping of DNA in nucleosomes, as the release of DNA from the histone core helps to relax positive superhelical tension.

Any protein that propels itself alone along a DNA strand of a double helix tends to generate superhelical tension. In eucaryotes, DNA topoisomerase enzymes rapidly remove this superhelical tension (see p. 251). But, in bacteria, a specialized topoisomerase called *DNA gyrase* uses the energy of ATP hydrolysis to pump supercoils continuously into the DNA, thereby maintaining the DNA under constant tension. These are *negative supercoils,* having the opposite handedness from the *positive supercoils* that form when a region of DNA helix opens (see Figure 6–20B). These negative supercoils are removed from bacterial DNA whenever a region of helix opens, reducing the superhelical tension. DNA gyrase therefore makes the opening of the DNA helix in bacteria energetically favorable compared with helix opening in DNA that is not supercoiled. For this reason, it usually facilitates those genetic processes in bacteria, including the initiation of transcription by bacterial RNA polymerase, that require helix opening (see Figure 6–10).

Transcription Elongation in Eucaryotes Is Tightly Coupled To RNA Processing

We have seen that bacterial mRNAs are synthesized solely by the RNA polymerase starting and stopping at specific spots on the genome. The situation in eucaryotes is substantially different. In particular, transcription is only the first step in a series of reactions that includes the covalent modification of both ends of the RNA and the removal of *intron sequences* that are discarded from the middle of the RNA transcript by the process of *RNA splicing* (Figure 6–21). The modifications of the ends of eucaryotic mRNA are *capping* on the 5′ end and *polyadenylation* of the 3′ end (Figure 6–22). These special ends allow the cell to assess whether both ends of an mRNA molecule are present (and the message is therefore intact) before it exports the RNA sequence from the nucleus for

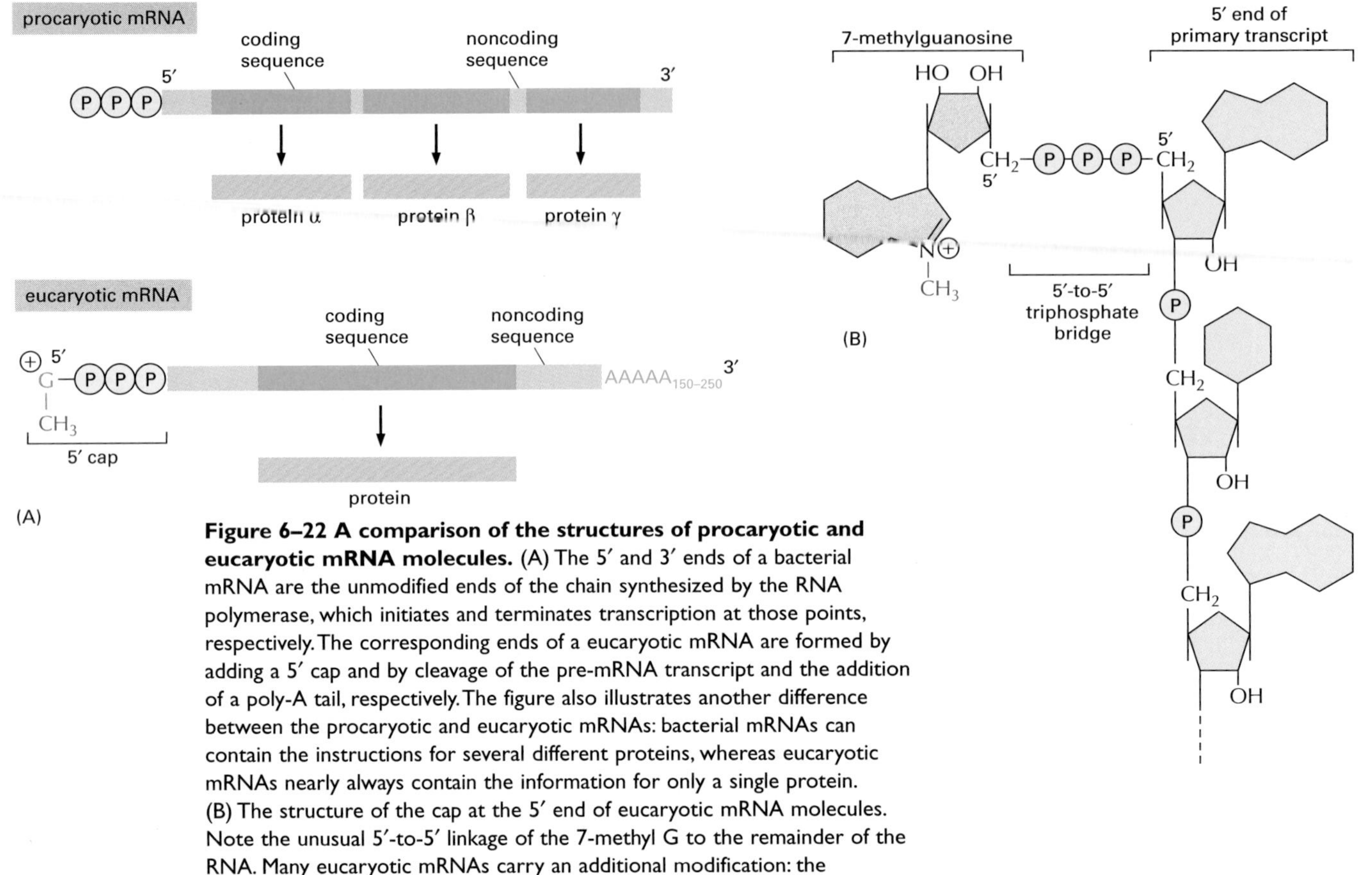

Figure 6–22 A comparison of the structures of procaryotic and eucaryotic mRNA molecules. (A) The 5′ and 3′ ends of a bacterial mRNA are the unmodified ends of the chain synthesized by the RNA polymerase, which initiates and terminates transcription at those points, respectively. The corresponding ends of a eucaryotic mRNA are formed by adding a 5′ cap and by cleavage of the pre-mRNA transcript and the addition of a poly-A tail, respectively. The figure also illustrates another difference between the procaryotic and eucaryotic mRNAs: bacterial mRNAs can contain the instructions for several different proteins, whereas eucaryotic mRNAs nearly always contain the information for only a single protein. (B) The structure of the cap at the 5′ end of eucaryotic mRNA molecules. Note the unusual 5′-to-5′ linkage of the 7-methyl G to the remainder of the RNA. Many eucaryotic mRNAs carry an additional modification: the 2′-hydroxyl group on the second ribose sugar in the mRNA is methylated (not shown).

translation into protein. In Chapter 4, we saw that a typical eucaryotic gene is present in the genome as short blocks of protein-coding sequence (exons) separated by long introns, and RNA splicing is the critically important step in which the different portions of a protein coding sequence are joined together. As we describe next, RNA splicing also provides higher eucaryotes with the ability to synthesize several different proteins from the same gene.

These RNA processing steps are tightly coupled to transcription elongation by an ingenious mechanism. As discussed previously, a key step of the transition of RNA polymerase II to the elongation mode of RNA synthesis is an extensive phosphorylation of the RNA polymerase II tail, called the CTD. This C-terminal domain of the largest subunit consists of a long tandem array of a repeated seven-amino-acid sequence, containing two serines per repeat that can be phosphorylated. Because there are 52 repeats in the CTD of human RNA polymerase II, its complete phosphorylation would add 104 negatively charged phosphate groups to the polymerase. This phosphorylation step not only dissociates the RNA polymerase II from other proteins present at the start point of transcription, it also allows a new set of proteins to associate with the RNA polymerase tail that function in transcription elongation and pre-mRNA processing. As discussed next, some of these processing proteins seem to "hop" from the polymerase tail onto the nascent RNA molecule to begin processing it as it emerges from the RNA polymerase. Thus, RNA polymerase II in its elongation mode can be viewed as an RNA factory that both transcribes DNA into RNA and processes the RNA it produces (Figure 6–23).

RNA Capping Is the First Modification of Eucaryotic Pre-mRNAs

As soon as RNA polymerase II has produced about 25 nucleotides of RNA, the 5′ end of the new RNA molecule is modified by addition of a "cap" that consists of

Figure 6–23 The "RNA factory" concept for eucaryotic RNA polymerase II. Not only does the polymerase transcribe DNA into RNA, but it also carries pre-mRNA-processing proteins on its tail, which are then transferred to the nascent RNA at the appropriate time. There are many RNA-processing enzymes, and not all travel with the polymerase. For RNA splicing, for example, only a few critical components are carried on the tail; once transferred to an RNA molecule, they serve as a nucleation site for the remaining components. The RNA-processing proteins first bind to the RNA polymerase tail when it is phosphorylated late in the process of transcription initiation (see Figure 6–16). Once RNA polymerase II finishes transcribing, it is released from DNA, the phosphates on its tail are removed by soluble phosphatases, and it can reinitiate transcription. Only this dephosphorylated form of RNA polymerase II is competent to start RNA synthesis at a promoter.

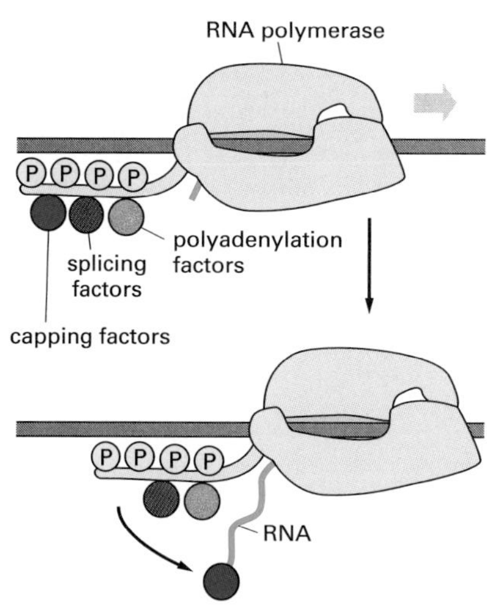

a modified guanine nucleotide (see Figure 6–22B). The capping reaction is performed by three enzymes acting in succession: one (a phosphatase) removes one phosphate from the 5′ end of the nascent RNA, another (a guanyl transferase) adds a GMP in a reverse linkage (5′ to 5′ instead of 5′ to 3′), and a third (a methyl transferase) adds a methyl group to the guanosine (Figure 6–24). Because all three enzymes bind to the phosphorylated RNA polymerase tail, they are poised to modify the 5′ end of the nascent transcript as soon as it emerges from the polymerase.

The 5′-methyl cap signals the 5′ end of eucaryotic mRNAs, and this landmark helps the cell to distinguish mRNAs from the other types of RNA molecules present in the cell. For example, RNA polymerases I and III produce uncapped RNAs during transcription, in part because these polymerases lack tails. In the nucleus, the cap binds a protein complex called CBC (cap-binding complex), which, as we discuss in subsequent sections, helps the RNA to be properly processed and exported. The 5′ methyl cap also has an important role in the translation of mRNAs in the cytosol as we discuss later in the chapter.

RNA Splicing Removes Intron Sequences from Newly Transcribed Pre-mRNAs

As discussed in Chapter 4, the protein coding sequences of eucaryotic genes are typically interrupted by noncoding intervening sequences (introns). Discovered in 1977, this feature of eucaryotic genes came as a surprise to scientists, who had been, until that time, familiar only with bacterial genes, which typically consist of a continuous stretch of coding DNA that is directly transcribed into mRNA. In marked contrast, eucaryotic genes were found to be broken up into small pieces of coding sequence (*expressed sequences* or **exons**) interspersed with much longer *intervening sequences* or **introns**; thus the coding portion of a eucaryotic gene is often only a small fraction of the length of the gene (Figure 6–25).

Both intron and exon sequences are transcribed into RNA. The intron sequences are removed from the newly synthesized RNA through the process of **RNA splicing**. The vast majority of RNA splicing that takes place in cells functions in the production of mRNA, and our discussion of splicing focuses on this type. It is termed precursor-mRNA (or pre-mRNA) splicing to denote that it occurs on RNA molecules destined to become mRNAs. Only after 5′ and 3′ end processing and splicing have taken place is such RNA termed mRNA.

Each splicing event removes one intron, proceeding through two sequential phosphoryl-transfer reactions known as transesterifications; these join two exons while removing the intron as a "lariat" (Figure 6–26). Since the number of phosphate bonds remains the same, these reactions could in principle take place without nucleoside triphosphate hydrolysis. However, the machinery that catalyzes pre-mRNA splicing is complex, consisting of 5 additional RNA molecules and over 50 proteins, and it hydrolyzes many ATP molecules per splicing event. This complexity is presumably needed to ensure that splicing is highly accurate, while also being sufficiently flexible to deal with the enormous variety of introns found in a typical eucaryotic cell. Frequent mistakes in RNA

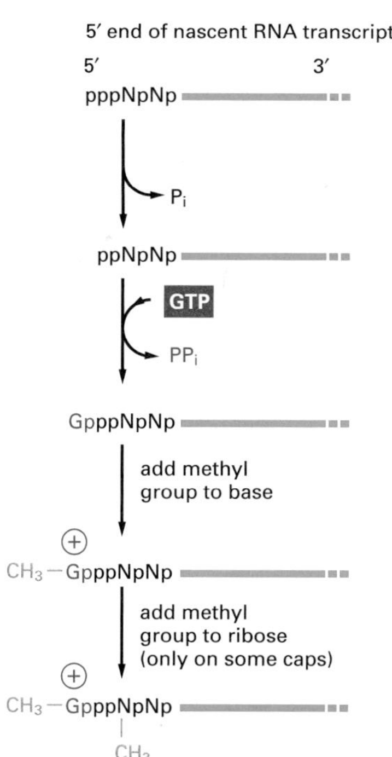

Figure 6–24 The reactions that cap the 5′ end of each RNA molecule synthesized by RNA polymerase II. The final cap contains a novel 5′-to-5′ linkage between the positively charged 7-methyl G residue and the 5′ end of the RNA transcript (see Figure 6–22B). The letter N represents any one of the four ribonucleotides, although the nucleotide that starts an RNA chain is usually a purine (an A or a G). (After A.J. Shatkin, *BioEssays* 7:275–277, 1987. © ICSU Press.)

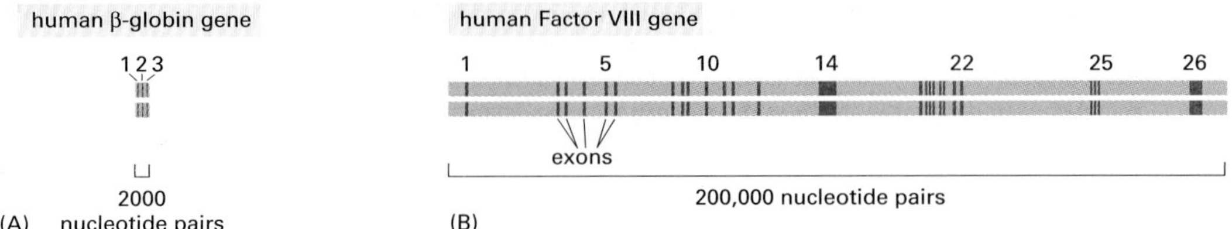

Figure 6–25 Structure of two human genes showing the arrangement of exons and introns. (A) The relatively small β-globin gene, which encodes one of the subunits of the oxygen-carrying protein hemoglobin, contains 3 exons (see also Figure 4–7). (B) The much larger Factor VIII gene contains 26 exons; it codes for a protein (Factor VIII) that functions in the blood-clotting pathway. Mutations in this gene are responsible for the most prevalent form of hemophilia.

splicing would severely harm the cell, as they would result in malfunctioning proteins. We see in Chapter 7 that when rare splicing mistakes do occur, the cell has a "fail-safe" device to eliminate the incorrectly spliced mRNAs.

It may seem wasteful to remove large numbers of introns by RNA splicing. In attempting to explain why it occurs, scientists have pointed out that the exon–intron arrangement would seem to facilitate the emergence of new and useful proteins. Thus, the presence of numerous introns in DNA allows genetic recombination to readily combine the exons of different genes (see p. 462), allowing genes for new proteins to evolve more easily by the combination of parts of preexisting genes. This idea is supported by the observation, described in Chapter 3, that many proteins in present-day cells resemble patchworks composed from a common set of protein pieces, called protein *domains*.

RNA splicing also has a present-day advantage. The transcripts of many eucaryotic genes (estimated at 60% of genes in humans) are spliced in a variety of different ways to produce a set of different mRNAs, thereby allowing a corresponding set of different proteins to be produced from the same gene (Figure 6–27). We discuss additional examples of alternative splicing in Chapter 7, as this is also one of the mechanisms that cells use to change expression of their genes. Rather than being the wasteful process it may have seemed at first sight, RNA splicing enables eucaryotes to increase the already enormous coding potential of their genomes. We shall return to this idea several times in this chapter and the next, but we first need to describe the cellular machinery that performs this remarkable task.

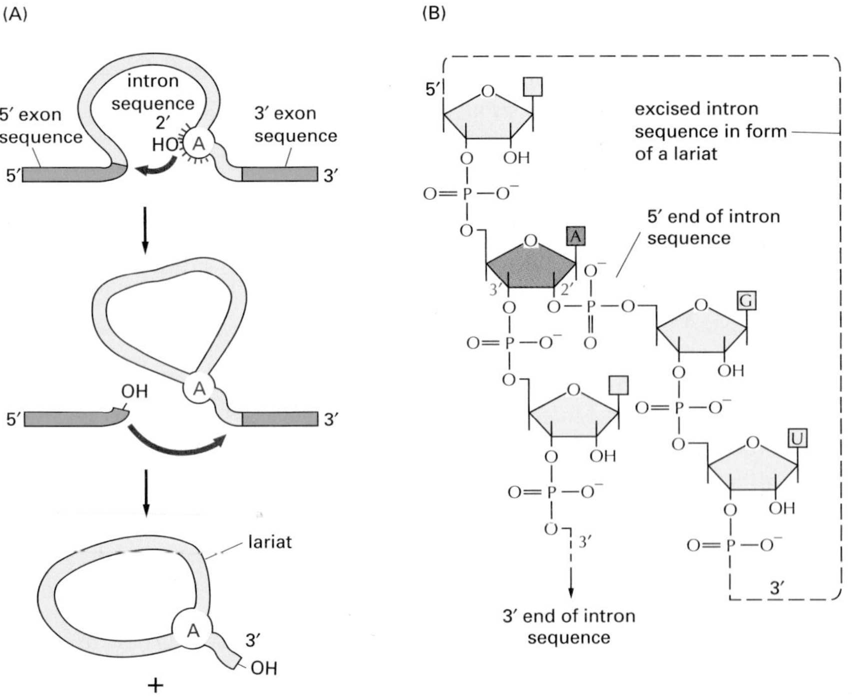

Figure 6–26 The RNA splicing reaction. (A) In the first step, a specific adenine nucleotide in the intron sequence (indicated in *red*) attacks the 5′ splice site and cuts the sugar-phosphate backbone of the RNA at this point. The cut 5′ end of the intron becomes covalently linked to the adenine nucleotide, as shown in detail in (B), thereby creating a loop in the RNA molecule. The released free 3′-OH end of the exon sequence then reacts with the start of the next exon sequence, joining the two exons together and releasing the intron sequence in the shape of a *lariat*. The two exon sequences thereby become joined into a continuous coding sequence; the released intron sequence is degraded in due course.

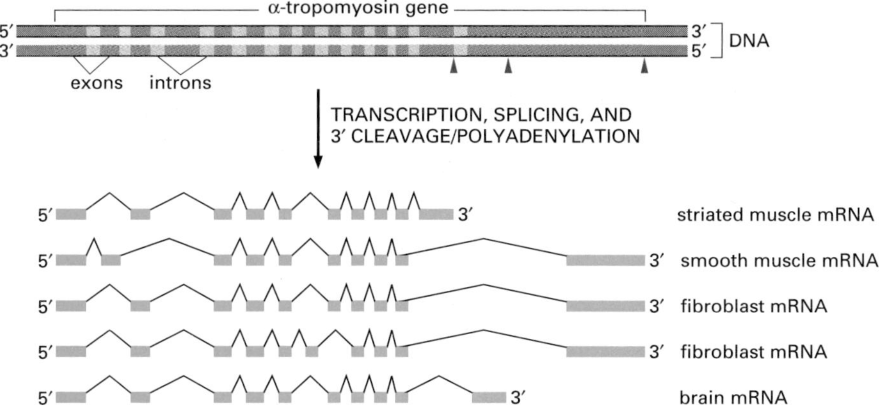

Figure 6–27 Alternative splicing of the α-tropomyosin gene from rat. α-tropomyosin is a coiled-coil protein (see Figure 3–11) that regulates contraction in muscle cells. The primary transcript can be spliced in different ways, as indicated in the figure, to produce distinct mRNAs, which then give rise to variant proteins. Some of the splicing patterns are specific for certain types of cells. For example, the α-tropomyosin made in striated muscle is different from that made from the same gene in smooth muscle. The arrowheads in the top part of the figure demark the sites where cleavage and poly-A addition can occur.

Nucleotide Sequences Signal Where Splicing Occurs

Introns range in size from about 10 nucleotides to over 100,000 nucleotides. Picking out the precise borders of an intron is very difficult for scientists to do (even with the aid of computers) when confronted by a complete genome sequence of a eucaryote. The possibility of alternative splicing compounds the problem of predicting protein sequences solely from a genome sequence. This difficulty constitutes one of the main barriers to identifying all of the genes in a complete genome sequence, and it is the primary reason that we know only the approximate number of genes in, for example, the human genome. Yet each cell in our body recognizes and rapidly excises the appropriate intron sequences with high fidelity. We have seen that intron sequence removal involves three positions on the RNA: the 5′ splice site, the 3′ splice site, and the branch point in the intron sequence that forms the base of the excised lariat. In pre-mRNA splicing, each of these three sites has a consensus nucleotide sequence that is similar from intron to intron, providing the cell with cues on where splicing is to take place (Figure 6–28). However, there is enough variation in each sequence to make it very difficult for scientists to pick out all of the many splicing signals in a genome sequence.

RNA Splicing Is Performed by the Spliceosome

Unlike the other steps of mRNA production we have discussed, RNA splicing is performed largely by RNA molecules instead of proteins. RNA molecules recognize intron–exon borders and participate in the chemistry of splicing. These RNA molecules are relatively short (less than 200 nucleotides each), and there are five of them (U1, U2, U4, U5, and U6) involved in the major form of pre-mRNA splicing. Known as **snRNAs (small nuclear RNAs)**, each is complexed with at least seven protein subunits to form a snRNP (small nuclear ribonucleoprotein). These snRNPs form the core of the **spliceosome**, the large assembly of RNA and protein molecules that performs pre-mRNA splicing in the cell.

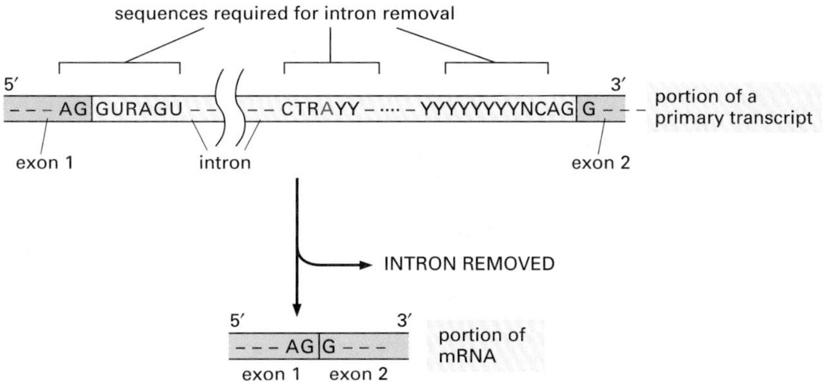

Figure 6–28 The consensus nucleotide sequences in an RNA molecule that signal the beginning and the end of most introns in humans. Only the three blocks of nucleotide sequences shown are required to remove an intron sequence; the rest of the intron can be occupied by any nucleotide. Here A, G, U, and C are the standard RNA nucleotides; R stands for either A or G; Y stands for either C or U. The A highlighted in *red* forms the branch point of the lariat produced by splicing. Only the GU at the start of the intron and the AG at its end are invariant nucleotides in the splicing consensus sequences. The remaining positions (even the branch point A) can be occupied by a variety of nucleotides, although the indicated nucleotides are preferred. The distances along the RNA between the three splicing consensus sequences are highly variable; however, the distance between the branch point and 3′ splice junction is typically much shorter than that between the 5′ splice junction and the branch point.

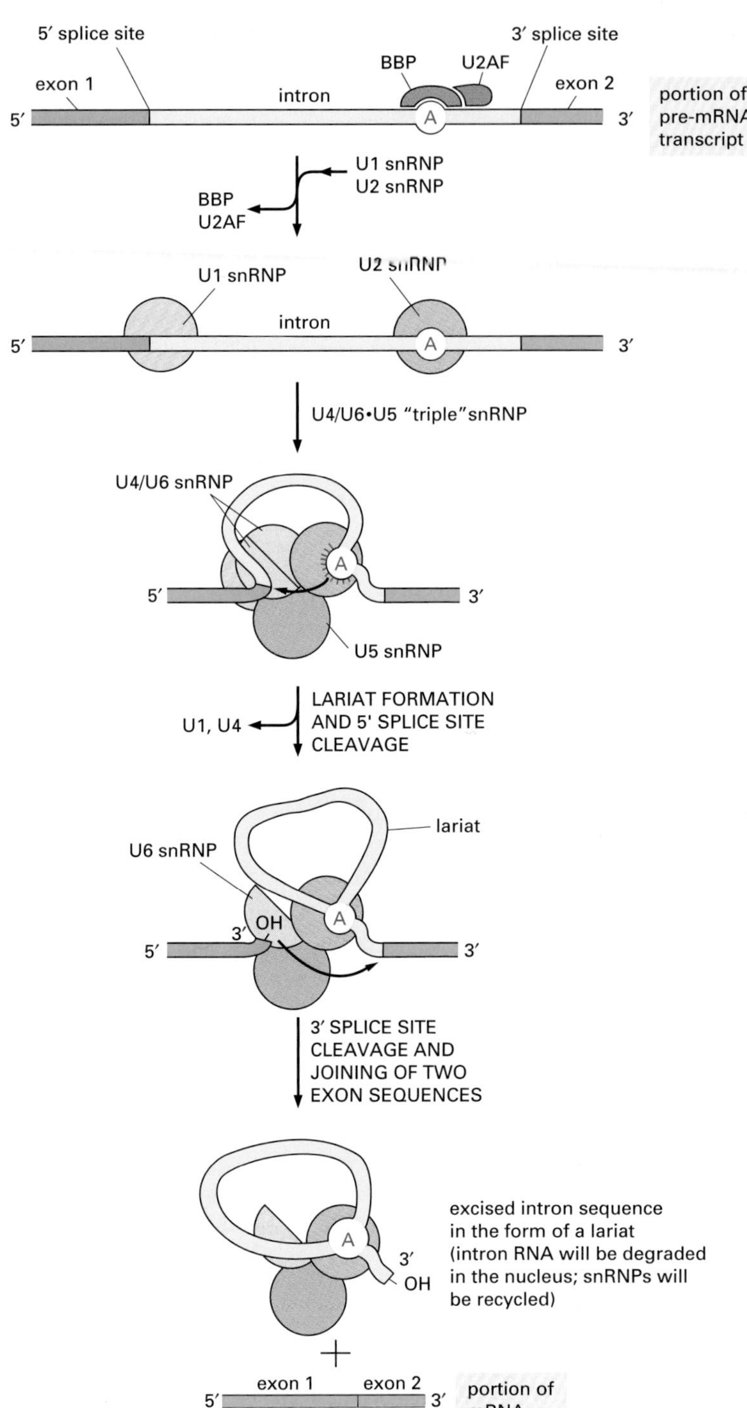

Figure 6–29 The RNA splicing mechanism. RNA splicing is catalyzed by an assembly of snRNPs (shown as *colored circles*) plus other proteins (most of which are not shown), which together constitute the spliceosome. The spliceosome recognizes the splicing signals on a pre-mRNA molecule, brings the two ends of the intron together, and provides the enzymatic activity for the two reaction steps (see Figure 6–26). The branch-point site is first recognized by the BBP (branch-point binding protein) and U2AF, a helper protein. In the next steps, the U2 snRNP displaces BBP and U2AF and forms base pairs with the branch-point site consensus sequence, and the U1 snRNP forms base-pairs with the 5′ splice junction (see Figure 6–30). At this point, the U4/U6•U5 "triple" snRNP enters the spliceosome. In this triple snRNP, the U4 and U6 snRNAs are held firmly together by base-pair interactions and the U5 snRNP is more loosely associated. Several RNA–RNA rearrangements then occur that break apart the U4/U6 base pairs (as shown, the U4 snRNP is ejected from the splicesome before splicing is complete) and allow the U6 snRNP to displace U1 at the 5′ splice junction (see Figure 6–30). Subsequent rearrangements create the active site of the spliceosome and position the appropriate portions of the pre-mRNA substrate for the splicing reaction to occur. Although not shown in the figure, each splicing event requires additional proteins, some of which hydrolyze ATP and promote the RNA–RNA rearrangements.

The spliceosome is a dynamic machine; as we see below, it is assembled on pre-mRNA from separate components, and parts enter and leave it as the splicing reaction proceeds (Figure 6–29). During the splicing reaction, recognition of the 5′ splice junction, the branch point site and the 3′ splice junction is performed largely through base-pairing between the snRNAs and the consensus RNA sequences in the pre-mRNA substrate (Figure 6–30). In the course of splicing, the spliceosome undergoes several shifts in which one set of base-pair interactions is broken and another is formed in its place. For example, U1 is replaced by U6 at the 5′ splice junction (see Figure 6–30A). As we shall see, this type of RNA–RNA rearrangement (in which the formation of one RNA–RNA interaction requires the disruption of another) occurs several times during the splicing reaction. It permits the checking and rechecking of RNA sequences before the chemical reaction is allowed to proceed, thereby increasing the accuracy of splicing.

The Spliceosome Uses ATP Hydrolysis to Produce a Complex Series of RNA–RNA Rearrangements

Although ATP hydrolysis is not required for the chemistry of RNA splicing *per se*, it is required for the stepwise assembly and rearrangements of the spliceosome. Some of the additional proteins that make up the spliceosome are RNA helicases, which use the energy of ATP hydrolysis to break existing RNA–RNA interactions so as to allow the formation of new ones. In fact, all the steps shown previously in Figure 6–29—except the association of BBP with the branch-point site and U1 snRNP with the 5′ splice site—require ATP hydrolysis and additional proteins. In all, more than 50 proteins, including those that form the snRNPs, are required for each splicing event.

The ATP-requiring RNA–RNA rearrangements that take place in the spliceosome occur within the snRNPs themselves and between the snRNPs and the pre-mRNA substrate. One of the most important roles of these rearrangements is the creation of the active catalytic site of the spliceosome. The strategy of creating an active site only after the assembly and rearrangement of splicing components on a pre-mRNA substrate is an important way of preventing wayward splicing.

Perhaps the most surprising feature of the spliceosome is the nature of the catalytic site itself: it is largely (if not exclusively) formed by RNA molecules instead of proteins. In the last section of this chapter we discuss in general terms the structural and chemical properties of RNA that allow it to perform catalysis; here we need only consider that the U2 and U6 snRNAs in the spliceosome form a precise three-dimensional RNA structure that juxtaposes the 5′ splice site of the pre-mRNA with the branch-point site and probably performs the first transesterification reaction (see Figure 6–30C). In a similar way, the 5′ and 3′ splice junctions are brought together (an event requiring the U5 snRNA) to facilitate the second transesterification.

Figure 6–30 Several of the rearrangements that take place in the spliceosome during pre-mRNA splicing. Shown here are the details for the yeast *Saccharomyces cerevisiae*, in which the nucleotide sequences involved are slightly different from those in human cells. (A) The exchange of U1 snRNP for U6 snRNP occurs before the first phosphoryl-transfer reaction (see Figure 6–29). This exchange allows the 5′ splice site to be read by two different snRNPs, thereby increasing the accuracy of 5′ splice site selection by the spliceosome. (B) The branch-point site is first recognized by BBP and subsequently by U2 snRNP; as in (A), this "check and recheck" strategy provides increased accuracy of site selection. The binding of U2 to the branch-point forces the appropriate adenine (in *red*) to be unpaired and thereby activates it for the attack on the 5′ splice site (see Figure 6–29). This, in combination with recognition by BBP, is the way in which the spliceosome accurately chooses the adenine that is ultimately to form the branch point. (C) After the first phosphoryl-transfer reaction *(left)* has occurred, U5 snRNP undergoes a rearrangement that brings the two exons into close proximity for the second phosphoryl-transfer reaction *(right)*. The snRNAs both position the reactants and provide (either all or in part) the catalytic site for the two reactions. The U5 snRNP is present in the spliceosome before this rearrangement occurs; for clarity it has been omitted from the left panel. As discussed in the text, all of the RNA–RNA rearrangements shown in this figure (as well as others that occur in the spliceosome but are not shown) require the participation of additional proteins and ATP hydrolysis.

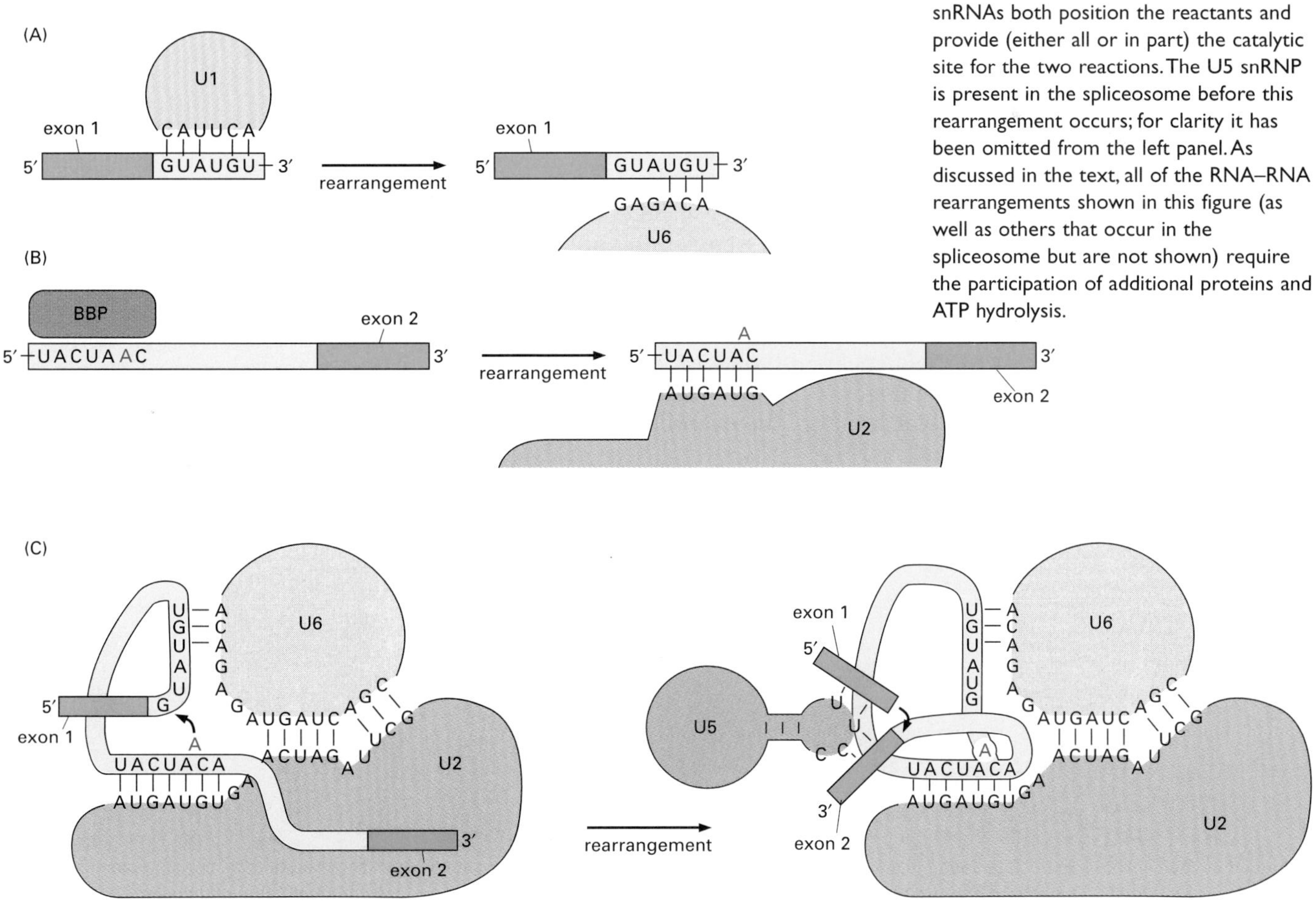

Figure 6–31 Two types of splicing errors. Both types might be expected to occur frequently if splice-site selection were performed by the spliceosome on a preformed, protein-free RNA molecule. "Cryptic" splicing signals are nucleotide sequences of RNA that closely resemble true splicing signals.

Once the splicing chemistry is completed, the snRNPs remain bound to the lariat and the spliced product is released. The disassembly of these snRNPs from the lariat (and from each other) requires another series of RNA–RNA rearrangements that require ATP hydrolysis, thereby returning the snRNAs to their original configuration so that they can be used again in a new reaction.

Ordering Influences in the Pre-mRNA Help to Explain How the Proper Splice Sites Are Chosen

As we have seen, intron sequences vary enormously in size, with some being in excess of 100,000 nucleotides. If splice-site selection were determined solely by the snRNPs acting on a preformed, protein-free RNA molecule, we would expect splicing mistakes—such as exon skipping and the use of cryptic splice sites—to be very common (Figure 6–31).

The fidelity mechanisms built into the spliceosome are supplemented by two additional factors that help ensure that splicing occurs accurately. These ordering influences in the pre-mRNA increase the probability that the appropriate pairs of 5′ and 3′ splice sites will be brought together in the spliceosome before the splicing chemistry begins. The first results from the assembly of the spliceosome occurring as the pre-mRNA emerges from a transcribing RNA polymerase II (see Figure 6–23). As for 5′ cap formation, several components of the spliceosome seem to be carried on the phosphorylated tail of RNA polymerase. Their transfer directly from the polymerase to the nascent pre-mRNA presumably helps the cell to keep track of introns and exons: the snRNPs at a 5′ splice site are initially presented with only a single 3′ splice site since the sites further downstream have not yet been synthesized. This feature helps to prevent inappropriate exon skipping.

The second factor that helps the cell to choose splice sites has been termed the "exon definition hypothesis," and it is understood only in outline. Exon size tends to be much more uniform than intron size, averaging about 150 nucleotide pairs across a wide variety of eucaryotic organisms (Figure 6–32). As RNA synthesis proceeds, a group of spliceosome components, called the SR proteins (so-named because they contain a domain rich in serines and arginines), are thought to assemble on exon sequences and mark off each 3′ and 5′ splice site starting at the 5′ end of the RNA (Figure 6–33). This assembly takes place in conjunction with the U1 snRNA, which marks one exon boundary, and U2AF,

Figure 6–32 Variation in intron and exon lengths in the human, worm, and fly genomes. (A) Size distribution of exons. (B) Size distribution of introns. Note that exon length is much more uniform than intron length. (Adapted from International Human Genome Sequencing Consortium, *Nature* 409:860–921, 2001.)

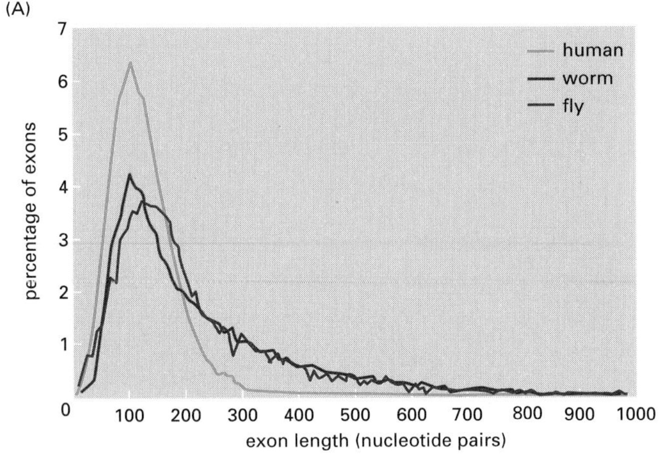

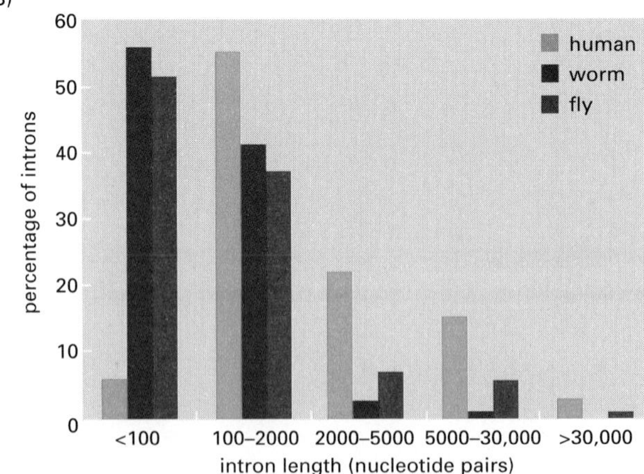

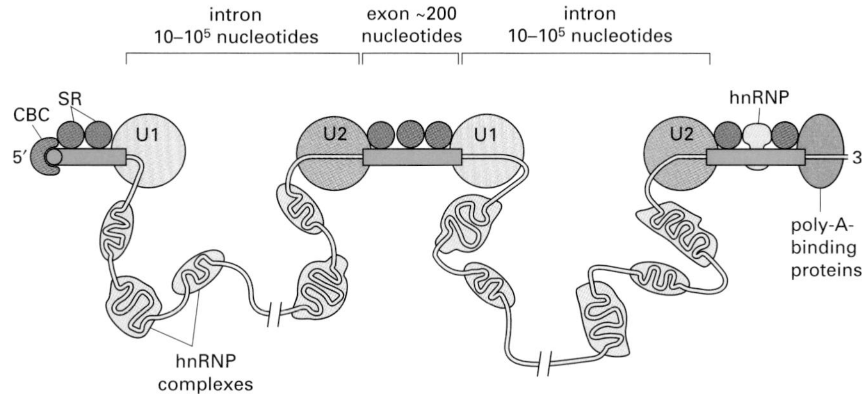

Figure 6–33 The exon definition hypothesis. According to one proposal, SR proteins bind to each exon sequence in the pre-mRNA and thereby help to guide the snRNPs to the proper intron/exon boundaries. This demarcation of exons by the SR proteins occurs co-transcriptionally, beginning at the CBC (cap-binding complex) at the 5′ end. As indicated, the intron sequences in the pre-mRNA, which can be extremely long, are packaged into hnRNP (heterogeneous nuclear ribonucleoprotein) complexes that compact them into more manageable structures and perhaps mask cryptic splice sites. Each hnRNP complex forms a particle approximately twice the diameter of a nucleosome, and the core is composed of a set of at least eight different proteins. It has been proposed that hnRNP proteins preferentially associate with intron sequences and that this preference also helps the spliceosome distinguish introns from exons. However, as shown, at least some hnRNP proteins may bind to exon sequences but their role, if any, in exon definition has yet to be established. (Adapted from R. Reed, *Curr. Opin. Cell Biol.* 12:340–345, 2000.)

which initially helps to specify the other. By specifically marking the exons in this way, the cell increases the accuracy with which the initial splicing components are deposited on the nascent RNA and thereby helps to avoid cryptic splice sites. How the SR proteins discriminate exon sequences from intron sequences is not understood; however, it is known that some of the SR proteins bind preferentially to RNA sequences in specific exons. In principle, the redundancy in the genetic code could have been exploited during evolution to select for binding sites for SR proteins in exons, allowing these sites to be created without constraining amino acid sequences.

Both the marking out of exon and intron boundaries and the assembly of the spliceosome begin on an RNA molecule while it is still being elongated by RNA polymerase at its 3′ end. However, the actual chemistry of splicing can take place much later. This delay means that intron sequences are not necessarily removed from a pre-mRNA molecule in the order in which they occur along the RNA chain. It also means that, although spliceosome assembly is co-transcriptional, the splicing reactions sometimes occur posttranscriptionally—that is, after a complete pre-mRNA molecule has been made.

A Second Set of snRNPs Splice a Small Fraction of Intron Sequences in Animals and Plants

Simple eucaryotes such as yeast have only one set of snRNPs that perform all pre-mRNA splicing. However, more complex eucaryotes such as flies, mammals, and plants have a second set of snRNPs that direct the splicing of a small fraction of their intron sequences. This minor form of spliceosome recognizes a different set of DNA sequences at the 5′ and 3′ splice junctions and at the branch point; it is called the *AT–AC spliceosome* because of the nucleotide sequence determinants at its intron–exon borders (Figure 6–34). Despite recognizing different nucleotide sequences, the snRNPs in this spliceosome make the same types of RNA–RNA interactions with the pre-mRNA and with each other as do the major snRNPs (Figure 6–34B). The recent discovery of this class of snRNPs gives us confidence in the base-pair interactions deduced for the major spliceosome, because it provides an independent set of molecules that undergo the same RNA–RNA interactions despite differences in the RNA sequences involved.

A particular variation on splicing, called **trans-splicing**, has been discovered in a few eucaryotic organisms. These include the single-celled trypanosomes—protozoans that cause African sleeping sickness in humans—and the model multicellular organism, the nematode worm. In trans-splicing, exons from two separate RNA transcripts are spliced together to form a mature mRNA molecule (see Figure 6–34). Trypanosomes produce all of their mRNAs in this way, whereas only about 1% of nematode mRNAs are produced by trans-splicing. In both cases, a single exon is spliced onto the 5′ end of many different RNA transcripts produced by the cell; in this way, all of the products of trans-splicing have the same 5′ exon and different 3′ exons. Many of the same snRNPs that function in conventional splicing are used in this reaction, although trans-splicing uses a unique snRNP (called the SL RNP) that brings in the common exon (see Figure 6–34).

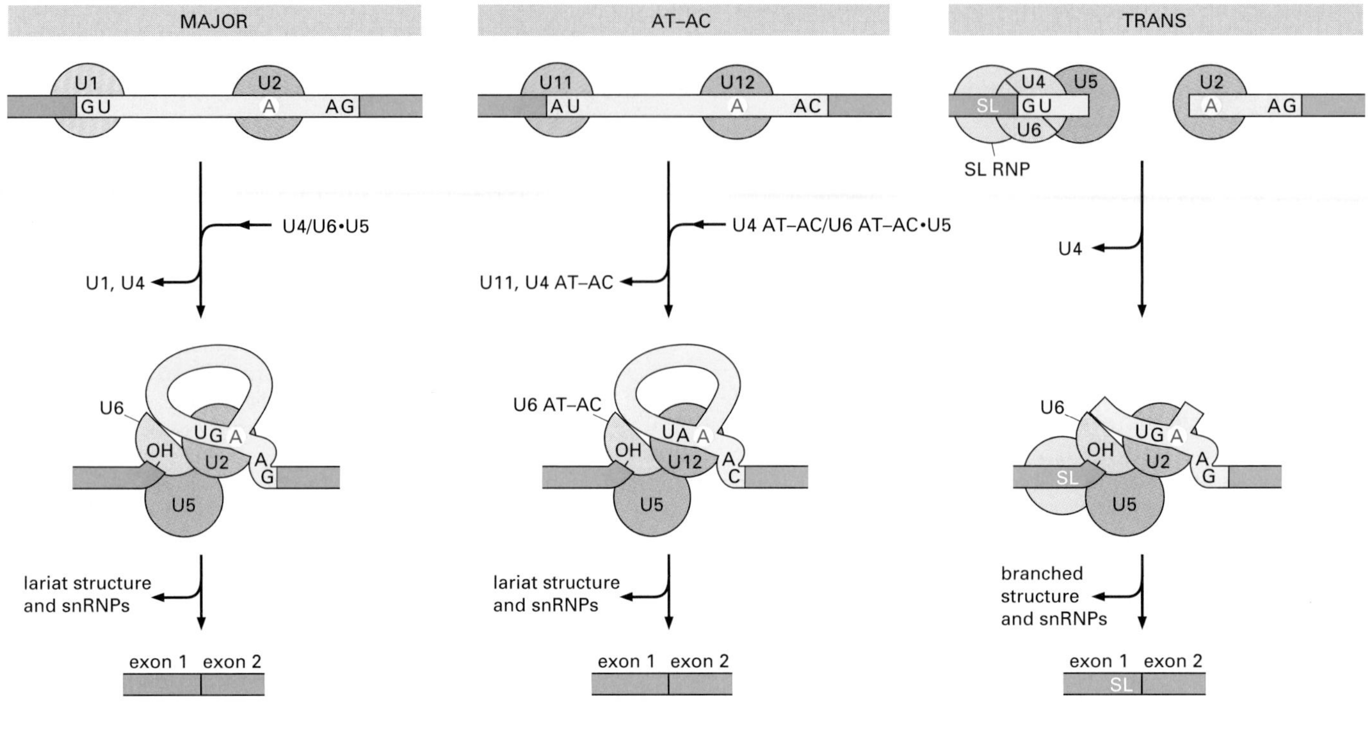

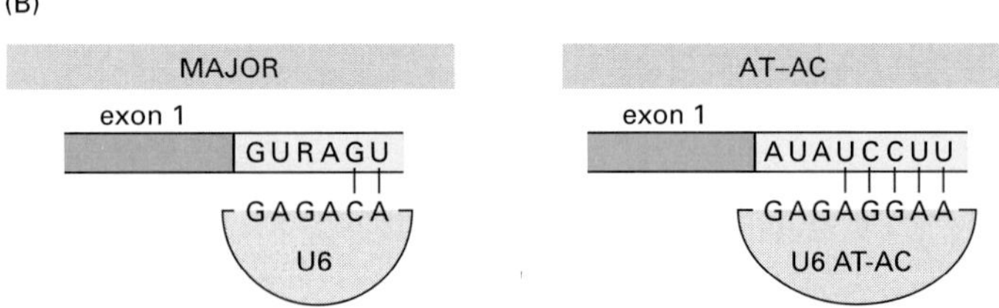

Figure 6–34 Outline of the mechanisms used for three types of RNA splicing. (A) Three types of spliceosomes. The major spliceosome *(left),* the AT–AC spliceosome (middle), and the trans-spliceosome *(right)* are each shown at two stages of assembly. The U5 snRNP is the only component that is common to all three spliceosomes. Introns removed by the AT–AC spliceosome have a different set of consensus nucleotide sequences from those removed by the major spliceosome. In humans, it is estimated that 0.1% of introns are removed by the AT–AC spliceosome. In trans-splicing, the SL snRNP is consumed in the reaction because a portion of the SL snRNA becomes the first exon of the mature mRNA. (B) The major U6 snRNP and the U6 AT–AC snRNP both recognize the 5′ splice junction, but they do so through a different set of base-pair interactions. The sequences shown are from humans. (Adapted from Y.-T. Yu et al., The RNA World, pp. 487–524. Cold Spring Harbor, New York: Cold Spring Harbor Laboratory Press, 1999.)

The reason that a few organisms use trans-splicing is not known; however, it is thought that the common 5′ exon may aid in the translation of the mRNA. Thus, the products of trans-splicing in nematodes seem to be translated with especially high efficiency.

RNA Splicing Shows Remarkable Plasticity

We have seen that the choice of splice sites depends on many features of the pre-mRNA transcript; these include the affinity of the three signals on the RNA (the 5′ and 3′ splice junctions and branch point) for the splicing machinery, the length and nucleotide sequence of the exon, the co-transcriptional assembly of the spliceosome, and the accuracy of the "bookkeeping" that underlies exon definition. So far we have emphasized the accuracy of the RNA splicing processes that occur in a cell. But it also seems that the mechanism has been selected for its flexibility, which allows the cell to try out new proteins on occasion. Thus, for example, when a mutation occurs in a nucleotide sequence critical for splicing of a particular intron, it does not necessarily prevent splicing of that intron altogether. Instead, the mutation typically creates a new pattern of splicing (Figure 6–35). Most commonly, an exon is simply skipped (Figure 6–35B). In other cases, the mutation causes a "cryptic" splice junction to be used (Figure 6–35C). Presumably, the splicing machinery has evolved to pick out the best possible pattern of splice junctions, and if the optimal one is damaged by mutation, it will seek out the next best pattern and so on. This flexibility in the process of RNA splicing suggests that changes in splicing patterns caused by random mutations have been an important pathway in the evolution of genes and organisms.

The plasticity of RNA splicing also means that the cell can easily regulate the pattern of RNA splicing. Earlier in this section we saw that alternative splicing

Figure 6–35 Abnormal processing of the β-globin primary RNA transcript in humans with the disease β thalassemia. In the examples shown, the disease is caused by splice-site mutations, denoted by *black arrowheads*. The *dark blue boxes* represent the three normal exon sequences; the *red lines* are used to indicate the 5′ and 3′ splice sites that are used in splicing the RNA transcript. The *light blue boxes* depict new nucleotide sequences included in the final mRNA molecule as a result of the mutation. Note that when a mutation leaves a normal splice site without a partner, an exon is skipped or one or more abnormal "cryptic" splice sites nearby is used as the partner site, as in (C) and (D). (Adapted in part from S.H. Orkin, in The Molecular Basis of Blood Diseases [G. Stamatoyannopoulos et al., eds.], pp. 106–126. Philadelphia: Saunders, 1987.)

(A) NORMAL ADULT β-GLOBIN PRIMARY RNA TRANSCRIPT

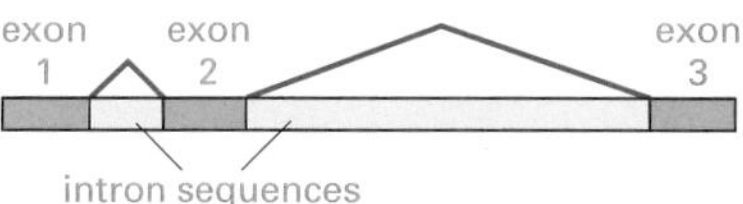

normal mRNA is formed from three exons

(B) SOME SINGLE-NUCLEOTIDE CHANGES THAT DESTROY A NORMAL SPLICE SITE CAUSE EXON SKIPPING

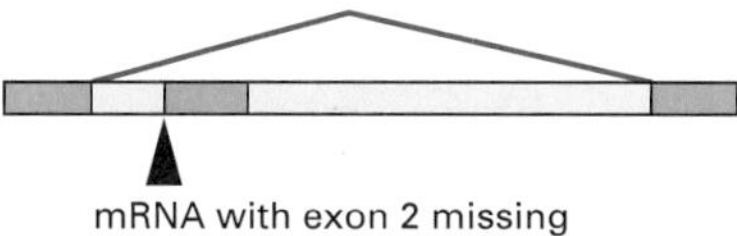

mRNA with exon 2 missing

(C) SOME SINGLE-NUCLEOTIDE CHANGES THAT DESTROY NORMAL SPLICE SITES ACTIVATE CRYPTIC SPLICE SITES

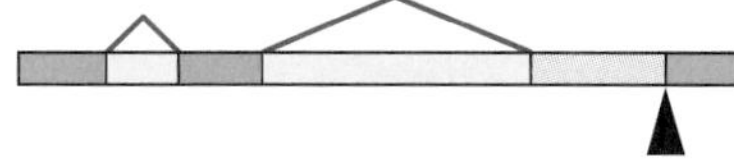

mRNA with extended exon 3

(D) SOME SINGLE-NUCLEOTIDE CHANGES THAT CREATE NEW SPLICE SITES CAUSE NEW EXONS TO BE INCORPORATED

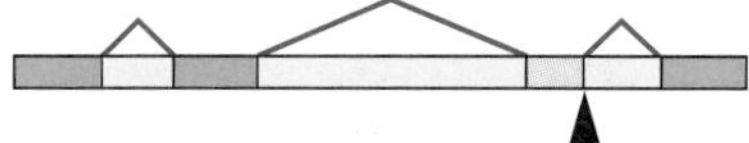

mRNA with extra exon inserted between exon 2 and exon 3

can give rise to different proteins from the same gene. Some examples of alternative splicing are constitutive; that is, the alternatively spliced mRNAs are produced continuously by cells of an organism. However, in most cases, the splicing patterns are regulated by the cell so that different forms of the protein are produced at different times and in different tissues (see Figure 6–27). In Chapter 7 we return to this issue to discuss some specific examples of regulated RNA splicing.

Spliceosome-catalyzed RNA Splicing Probably Evolved from Self-splicing Mechanisms

When the spliceosome was first discovered, it puzzled molecular biologists. Why do RNA molecules instead of proteins perform important roles in splice site recognition and in the chemistry of splicing? Why is a lariat intermediate used rather than the apparently simpler alternative of bringing the 5′ and 3′ splice sites together in a single step, followed by their direct cleavage and rejoining? The answers to these questions reflect the way in which the spliceosome is believed to have evolved.

As discussed briefly in Chapter 1 (and taken up again in more detail in the final section of this chapter), it is thought that early cells used RNA molecules rather than proteins as their major catalysts and that they stored their genetic information in RNA rather than in DNA sequences. RNA-catalyzed splicing reactions presumably had important roles in these early cells. As evidence, some *self-splicing RNA* introns (that is, intron sequences in RNA whose splicing out can occur in the absence of proteins or any other RNA molecules) remain today—for example, in the nuclear rRNA genes of the ciliate *Tetrahymena*, in a few bacteriophage T4 genes, and in some mitochondrial and chloroplast genes.

A self-splicing intron sequence can be identified in a test tube by incubating a pure RNA molecule that contains the intron sequence and observing the splicing reaction. Two major classes of self-splicing intron sequences can be distinguished in this way. *Group I intron sequences* begin the splicing reaction by binding a G nucleotide to the intron sequence; this G is thereby activated to form the attacking group that will break the first of the phosphodiester bonds cleaved during splicing (the bond at the 5′ splice site). In *group II intron sequences,* an especially reactive A residue in the intron sequence is the attacking group, and a lariat intermediate is generated. Otherwise the reaction pathways for the two types of self-splicing intron sequences are the same. Both are presumed to represent vestiges of very ancient mechanisms (Figure 6–36).

For both types of self-splicing reactions, the nucleotide sequence of the intron is critical; the intron RNA folds into a specific three-dimensional structure, which brings the 5′ and 3′ splice junctions together and provides precisely positioned reactive groups to perform the chemistry (see Figure 6–6C). Based on the fact that the chemistries of their splicing reactions are so similar, it has been proposed that the pre-mRNA splicing mechanism of the spliceosome evolved from group II splicing. According to this idea, when the spliceosomal snRNPs took over the structural and chemical roles of the group II introns, the strict sequence constraints on intron sequences would have disappeared, thereby permitting a vast expansion in the number of different RNAs that could be spliced.

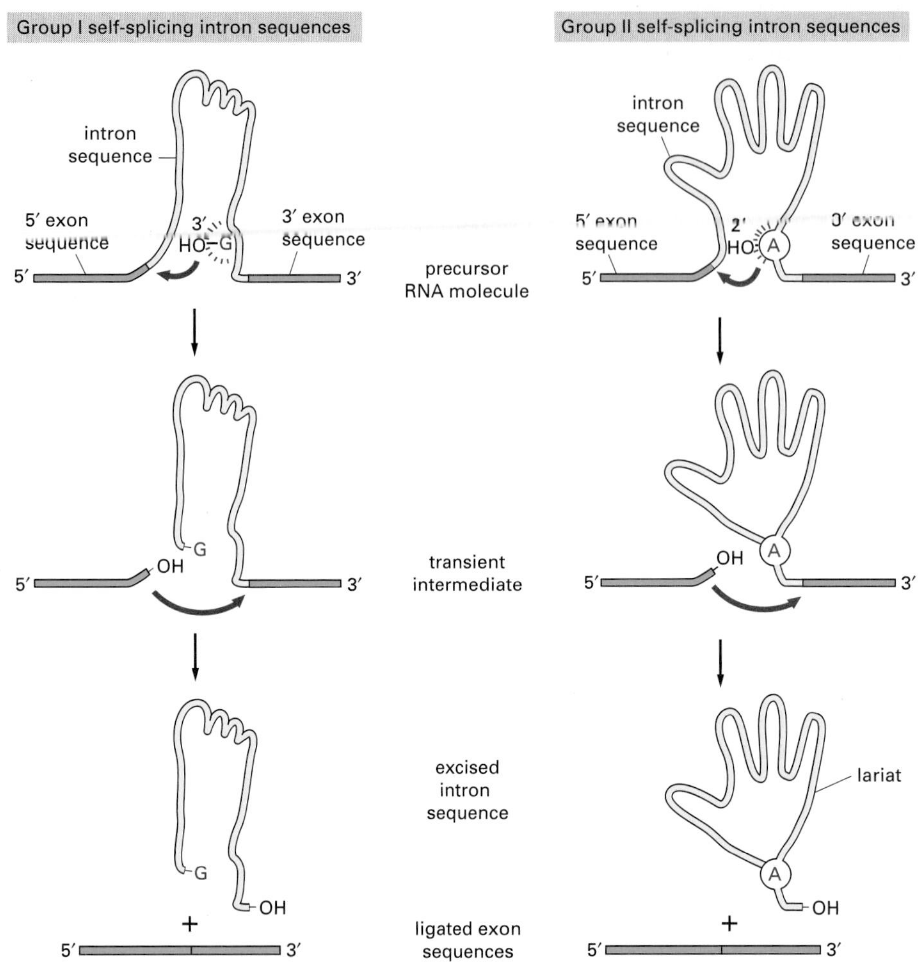

Figure 6–36 The two known classes of self-splicing intron sequences. The group I intron sequences bind a free G nucleotide to a specific site on the RNA to initiate splicing, while the group II intron sequences use an especially reactive A nucleotide in the intron sequence itself for the same purpose. The two mechanisms have been drawn to emphasize their similarities. Both are normally aided in the cell by proteins that speed up the reaction, but the catalysis is nevertheless mediated by the RNA in the intron sequence. Both types of self-splicing reactions require the intron to be folded into a highly specific three-dimensional structure that provides the catalytic activity for the reaction (see Figure 6–6). The mechanism used by group II intron sequences releases the intron as a lariat structure and closely resembles the pathway of pre-mRNA splicing catalyzed by the spliceosome (compare with Figure 6–29). The great majority of RNA splicing in eucaryotic cells is performed by the spliceosome, and self-splicing RNAs represent unusual cases. (Adapted from T.R. Cech, *Cell* 44:207–210, 1986.)

RNA-Processing Enzymes Generate the 3′ End of Eucaryotic mRNAs

As previously explained, the 5′ end of the pre-mRNA produced by RNA polymerase II is capped almost as soon as it emerges from the RNA polymerase. Then, as the polymerase continues its movement along a gene, the spliceosome components assemble on the RNA and delineate the intron and exon boundaries. The long C-terminal tail of the RNA polymerase coordinates these processes by transferring capping and splicing components directly to the RNA as the RNA emerges from the enzyme. As we see in this section, as RNA polymerase II terminates transcription at the end of a gene, it uses a similar mechanism to ensure that the 3′ end of the pre-mRNA becomes appropriately processed.

As might be expected, the 3′ ends of mRNAs are ultimately specified by DNA signals encoded in the genome (Figure 6–37). These DNA signals are transcribed

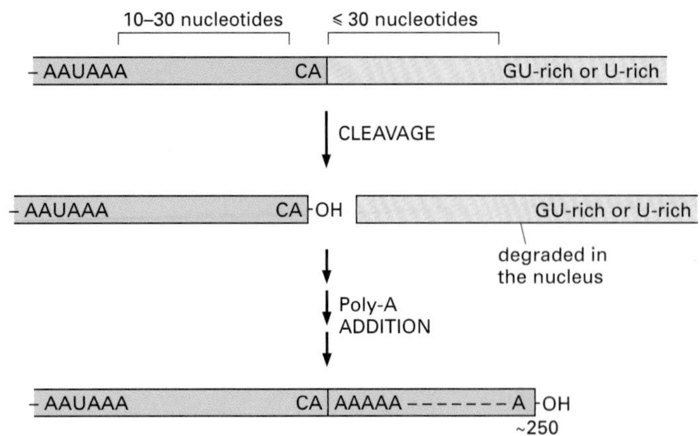

Figure 6–37 Consensus nucleotide sequences that direct cleavage and polyadenylation to form the 3′ end of a eucaryotic mRNA. These sequences are encoded in the genome and are recognized by specific proteins after they are transcribed into RNA. The hexamer AAUAAA is bound by CPSF, the GU-rich element beyond the cleavage site by CstF (see Figure 6–38), and the CA sequence by a third factor required for the cleavage step. Like other consensus nucleotide sequences discussed in this chapter (see Figure 6–12), the sequences shown in the figure represent a variety of individual cleavage and polyadenylation signals.

into RNA as the RNA polymerase II moves through them, and they are then recognized (as RNA) by a series of RNA-binding proteins and RNA-processing enzymes (Figure 6–38). Two multisubunit proteins, called CstF (cleavage stimulation factor F) and CPSF (cleavage and polyadenylation specificity factor), are of special importance. Both of these proteins travel with the RNA polymerase tail and are transferred to the 3′ end processing sequence on an RNA molecule as it emerges from the RNA polymerase. Some of the subunits of CPSF are associated with the general transcription factor TFIID, which, as we saw earlier in this chapter, is involved in transcription initiation. During transcription initiation, these subunits may be transferred from TFIID to the RNA polymerase tail, remaining associated there until the polymerase has transcribed through the end of a gene.

Once CstF and CPSF bind to specific nucleotide sequences on an emerging RNA molecule, additional proteins assemble with them to perform the processing that creates the 3′ end of the mRNA. First, the RNA is cleaved (see Figure 6–38). Next an enzyme called poly-A polymerase adds, one at a time, approximately 200 A nucleotides to the 3′ end produced by the cleavage. The nucleotide precursor for these additions is ATP, and the same type of 5′-to-3′ bonds are formed as in conventional RNA synthesis (see Figure 6–4). Unlike the usual RNA polymerases, poly-A polymerase does not require a template; hence the poly-A tail of eucaryotic mRNAs is not directly encoded in the genome. As the poly-A tail is synthesized, proteins called poly-A-binding proteins assemble onto it and, by a poorly understood mechanism, determine the final length of the tail. Poly-A-binding proteins remain bound to the poly-A tail as the mRNA makes its journey from the nucleus to the cytosol and they help to direct the synthesis of a protein on the ribosome, as we see later in this chapter.

After the 3′ end of a eucaryotic pre-mRNA molecule has been cleaved, the RNA polymerase II continues to transcribe, in some cases continuing as many as several hundred nucleotides beyond the DNA that contains the 3′ cleavage-site information. But the polymerase soon releases its grip on the template and transcription terminates; the piece of RNA downstream of the cleavage site is then degraded in the cell nucleus. It is not yet understood what triggers the loss in polymerase II processivity after the RNA is cleaved. One idea is that the transfer of the 3′ end processing factors from the RNA polymerase to the RNA causes a conformational change in the polymerase that loosens its hold on DNA; another is that the lack of a cap structure (and the CBC) on the 5′ end of the RNA that emerges from the polymerase somehow signals to the polymerase to terminate transcription.

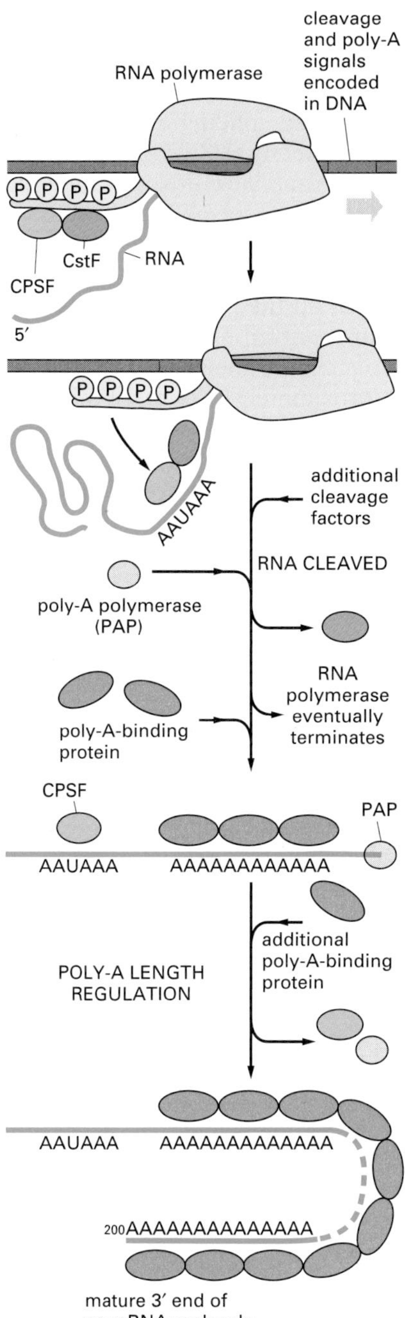

Figure 6–38 Some of the major steps in generating the 3′ end of a eucaryotic mRNA. This process is much more complicated than the analogous process in bacteria, where the RNA polymerase simply stops at a termination signal and releases both the 3′ end of its transcript and the DNA template (see Figure 6–10).

Mature Eucaryotic mRNAs Are Selectively Exported from the Nucleus

We have seen how eucaryotic pre-mRNA synthesis and processing takes place in an orderly fashion within the cell nucleus. However, these events create a special problem for eucaryotic cells, especially those of complex organisms where the introns are vastly longer than the exons. Of the pre-mRNA that is synthesized, only a small fraction—the mature mRNA—is of further use to the cell. The rest—excised introns, broken RNAs, and aberrantly spliced pre-mRNAs—is not only useless but could be dangerous if it was not destroyed. How then does the cell distinguish between the relatively rare mature mRNA molecules it wishes to keep and the overwhelming amount of debris from RNA processing? The answer is that transport of mRNA from the nucleus to the cytoplasm, where it is translated into protein, is highly selective—being closely coupled to correct RNA processing. This coupling is achieved by the *nuclear pore complex,* which recognizes and transports only completed mRNAs.

We have seen that as a pre-mRNA molecule is synthesized and processed, it is bound by a variety of proteins, including the cap-binding complex, the SR proteins, and the poly-A binding proteins. To be "export-ready," it seems than an mRNA must be bound by the appropriate set of proteins—with certain proteins such as the cap-binding complex being present, and others such as snRNP proteins absent. Additional proteins, placed on the RNA during splicing, seem to mark exon-exon boundaries and thereby signify completed splicing events. Only if the proper set of proteins is bound to an mRNA is it guided through the **nuclear pore complex** into the cytosol. As described in Chapter 12, nuclear pore complexes are aqueous channels in the nuclear membrane that directly connect the nucleoplasm and cytosol. Small molecules (less than 50,000 daltons) can diffuse freely through them. However, most of the macromolecules in cells, including mRNAs complexed with proteins, are far too large to pass through the pores without a special process to move them. An active transport of substances through the nuclear pore complexes occurs in both directions. As explained in Chapter 12, signals on the macromolecule determine whether it is exported from the nucleus (a mRNA, for example) or imported into it (an RNA polymerase, for example). For the case of mRNAs, the bound proteins that mark completed splicing events are of particular importance, as they are known to serve directly as RNA export factors (see Figure 12–16). mRNAs transcribed from genes that lack introns apparently contain nucleotide sequences that are directly recognized by other RNA export factors. Eucaryotic cells thus use their nuclear pore complexes as gates that allow only useful RNA molecules to enter the cytoplasm.

Of all the proteins that assemble on pre-mRNA molecules as they emerge from transcribing RNA polymerases, the most abundant are the hnRNPs (heterogeneous nuclear ribonuclear proteins). Some of these proteins (there are approximately 30 of them in humans) remove the hairpin helices from the RNA so that splicing and other signals on the RNA can be read more easily. Others package the RNA contained in the very long intron sequences typically found in genes of complex organisms (see Figure 6–33). Apart from histones, certain hnRNP proteins are the most abundant proteins in the cell nucleus, and they may play a particularly important role in distinguishing mature mRNA from processing debris. hnRNP particles (nucleosome-like complexes of hnRNP proteins and RNA—see Figure 6–33) are largely excluded from exon sequences, perhaps by prior binding of spliceosome components. They remain on excised introns and probably help mark them for nuclear retention and eventual destruction.

According to this view, introns—like mature mRNAs—carry signals in the form of bound proteins that specify their eventual fate. Not all hnRNP proteins, however remain in the nucleus; some are known to remain bound to fully processed mRNA molecules and to accompany them into the cytoplasm. It is presumably the entire set of bound proteins, rather than any single protein, that ultimately determines whether an RNA molecule will leave the nucleus.

The export of mRNA–protein complexes from the nucleus can be observed with an electron microscope for the unusually abundant mRNA of the insect Balbiani Ring genes. As these genes are transcribed, the newly formed RNA is seen to be packaged by proteins (including hnRNP and SR proteins). This protein–RNA complex undergoes a series of structural transitions, probably reflecting RNA processing events, culminating in a curved fiber (Figure 6–39). This curved fiber then moves through the nucleoplasm and enters the nuclear pore complex (with its 5′ cap proceeding first), and it undergoes another series of structural transitions as it moves through the NPC. These and other observations reveal that the pre-mRNA–protein and mRNA–protein complexes are

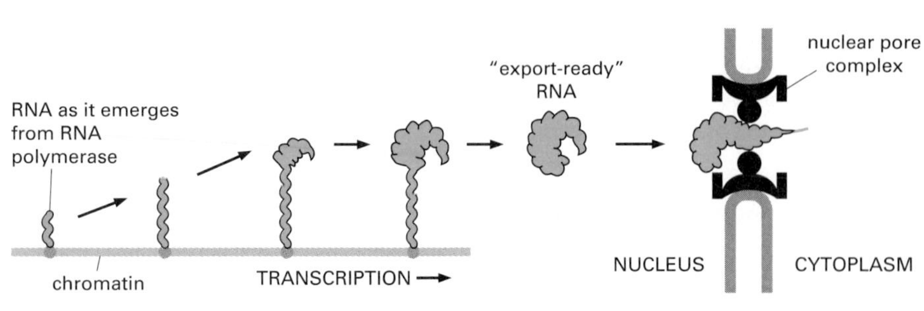

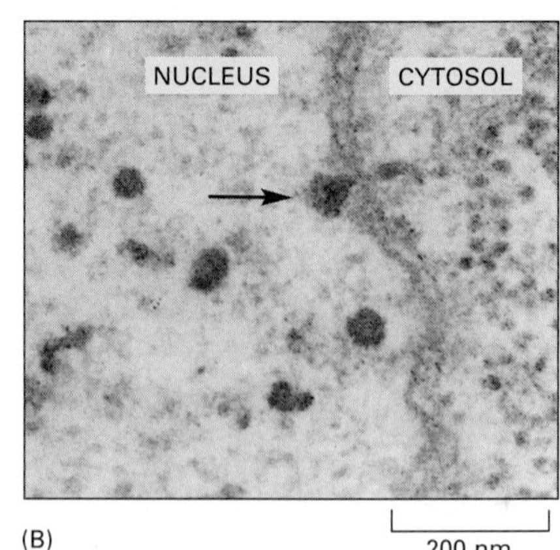

Figure 6–39 Transport of a large mRNA molecule through the nuclear pore complex. (A) The maturation of a Balbiani Ring mRNA molecule as it is synthesized by RNA polymerase and packaged by a variety of nuclear proteins. This drawing of unusually abundant RNA produced by an insect cell is based on EM micrographs such as that shown in (B). Balbiani Rings are described in Chapter 4. (A, adapted from B. Daneholt, *Cell* 88:585–588, 1997; B, from B.J. Stevens and H. Swift, *J. Cell Biol.* 31:55–77, 1966. © The Rockefeller University Press.)

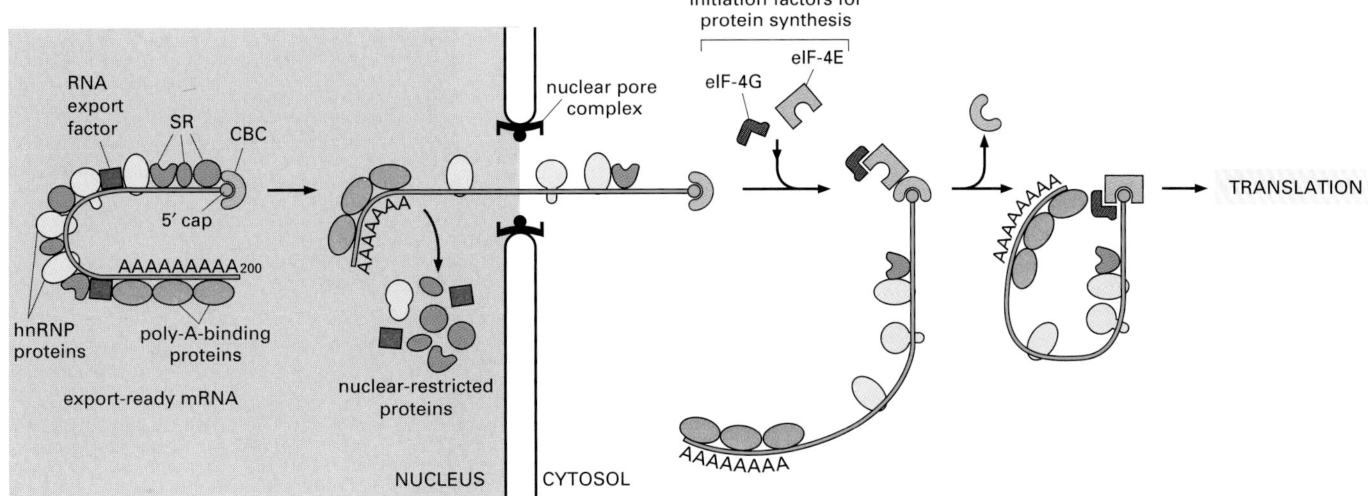

Figure 6–40 Schematic illustration of an "export-ready" mRNA molecule and its transport through the nuclear pore. As indicated, some proteins travel with the mRNA as it moves through the pore, whereas others remain in the nucleus. Once in the cytoplasm, the mRNA continues to shed previously bound proteins and acquire new ones; these substitutions affect the subsequent translation of the message. Because some are transported with the RNA, the proteins that become bound to an mRNA in the nucleus can influence its subsequent stability and translation in the cytosol. RNA export factors, shown in the nucleus, play an active role in transporting the mRNA to the cytosol (see Figure 12–16). Some are deposited at exon-exon boundaries as splicing is completed, thus signifying those regions of the RNA that have been properly spliced.

dynamic structures that gain and lose numerous specific proteins during RNA synthesis, processing, export, and translation (Figure 6–40).

Before discussing what happens to mRNAs after they leave the nucleus, we briefly consider how the synthesis and processing of noncoding RNA molecules occurs. Although there are many other examples, our discussion focuses on the rRNAs that are critically important for the translation of mRNAs into protein.

Many Noncoding RNAs Are Also Synthesized and Processed in the Nucleus

A few per cent of the dry weight of a mammalian cell is RNA; of that, only about 3–5% is mRNA. A fraction of the remainder represents intron sequences before they have been degraded, but most of the RNA in cells performs structural and catalytic functions (see Table 6–1, p. 306). The most abundant RNAs in cells are the ribosomal RNAs (rRNAs)—constituting approximately 80% of the RNA in rapidly dividing cells. As discussed later in this chapter, these RNAs form the core of the ribosome. Unlike bacteria—in which all RNAs in the cell are synthesized by a single RNA polymerase—eucaryotes have a separate, specialized polymerase, RNA polymerase I, that is dedicated to producing rRNAs. RNA polymerase I is similar structurally to the RNA polymerase II discussed previously; however, the absence of a C-terminal tail in polymerase I helps to explain why its transcripts are neither capped nor polyadenylated. As discussed earlier, this difference helps the cell distinguish between noncoding RNAs and mRNAs.

Because multiple rounds of translation of each mRNA molecule can provide an enormous amplification in the production of protein molecules, many of the proteins that are very abundant in a cell can be synthesized from genes that are present in a single copy per haploid genome. In contrast, the RNA components of the ribosome are final gene products, and a growing mammalian cell must synthesize approximately 10 million copies of each type of ribosomal RNA in each cell generation to construct its 10 million ribosomes. Adequate quantities of ribosomal RNAs can be produced only because the cell contains multiple copies of the **rRNA genes** that code for ribosomal RNAs **(rRNAs)**. Even *E. coli* needs seven copies of its rRNA genes to meet the cell's need for ribosomes. Human cells contain about 200 rRNA gene copies per haploid genome, spread

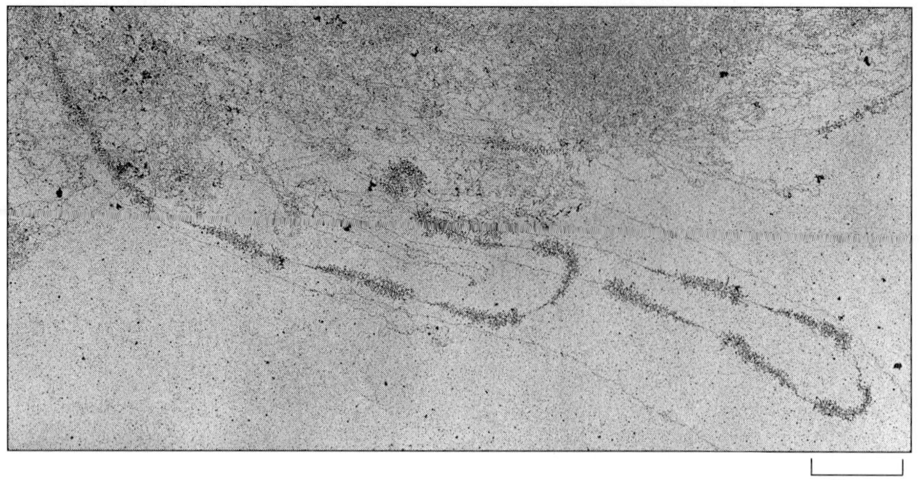

Figure 6–41 Transcription from tandemly arranged rRNA genes, as seen in the electron microscope. The pattern of alternating transcribed gene and nontranscribed spacer is readily seen. A higher-magnification view was shown in Figure 6–9. (From V.E. Foe, *Cold Spring Harbor Symp. Quant. Biol.* 42:723–740, 1978.)

out in small clusters on five different chromosomes (see Figure 4–11), while cells of the frog *Xenopus* contain about 600 rRNA gene copies per haploid genome in a single cluster on one chromosome (Figure 6–41).

There are four types of eucaryotic rRNAs, each present in one copy per ribosome. Three of the four rRNAs (18S, 5.8S, and 28S) are made by chemically modifying and cleaving a single large precursor rRNA (Figure 6–42); the fourth (5S RNA) is synthesized from a separate cluster of genes by a different polymerase, RNA polymerase III, and does not require chemical modification. It is not known why this one RNA is transcribed separately.

Extensive chemical modifications occur in the 13,000-nucleotide-long precursor rRNA before the rRNAs are cleaved out of it and assembled into ribosomes. These include about 100 methylations of the 2′-OH positions on nucleotide sugars and 100 isomerizations of uridine nucleotides to pseudouridine (Figure 6–43A). The functions of these modifications are not understood in detail, but they probably aid in the folding and assembly of the final rRNAs and may also subtly alter the function of ribosomes. Each modification is made at a specific position in the precursor rRNA. These positions are specified by several hundred "guide RNAs," which locate themselves through base-pairing to the precursor rRNA and thereby bring an RNA-modifying enzyme to the appropriate position (Figure 6–43B). Other guide RNAs promote, probably by causing conformational changes in the precursor rRNA, the cleavage of the precursor rRNA into the mature rRNAs. All of these guide RNAs are members of a large class of RNAs called **small nucleolar RNAs** (or **snoRNAs**), so named because these RNAs perform their functions in a subcompartment of the nucleus called the nucleolus. Many snoRNAs are encoded in the introns of other genes,

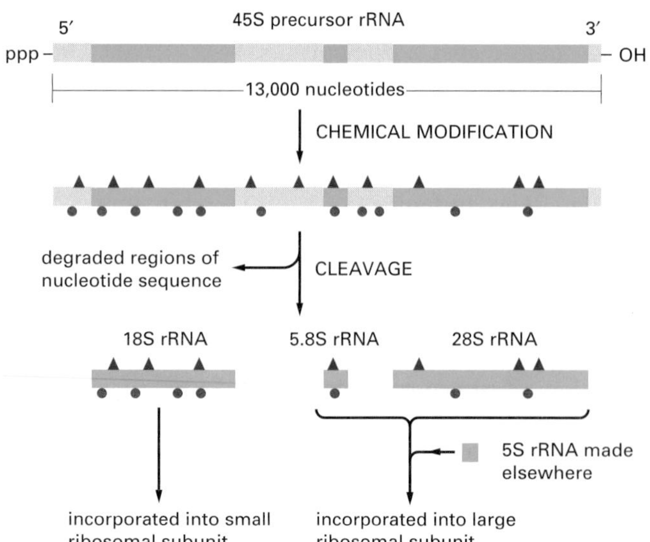

Figure 6–42 The chemical modification and nucleolytic processing of a eucaryotic 45S precursor rRNA molecule into three separate ribosomal RNAs. As indicated, two types of chemical modifications (shown in Figure 6–43) are made to the precursor rRNA before it is cleaved. Nearly half of the nucleotide sequences in this precursor rRNA are discarded and degraded in the nucleus. The rRNAs are named according to their "S" values, which refer to their rate of sedimentation in an ultra-centrifuge. The larger the S value, the larger the rRNA.

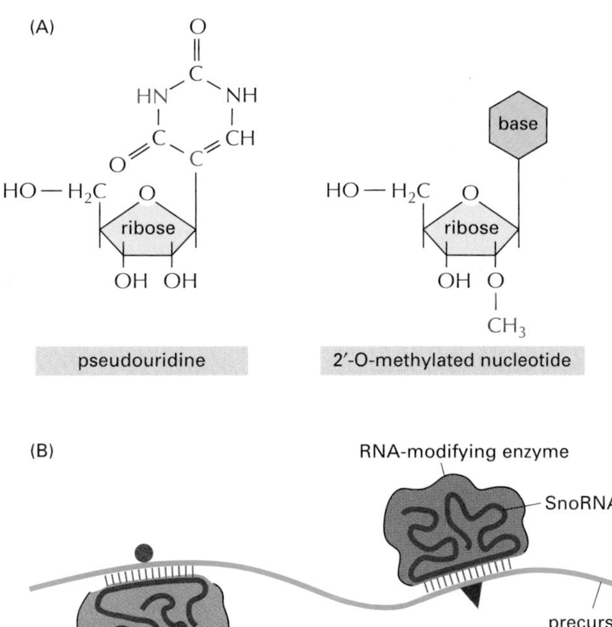

Figure 6–43 Modifications of the precursor rRNA by guide RNAs. (A) Two prominent covalent modifications occur after rRNA synthesis; the differences from the initially incorporated nucleotide are indicated by *red* atoms. (B) As indicated, snoRNAs locate the sites of modification by base-pairing to complementary sequences on the precursor rRNA. The snoRNAs are bound to proteins, and the complexes are called snoRNPs. snoRNPs contain the RNA modification activities, presumably contributed by the proteins but possibly by the snoRNAs themselves.

especially those encoding ribosomal proteins. They are therefore synthesized by RNA polymerase II and processed from excised intron sequences.

The Nucleolus Is a Ribosome-Producing Factory

The nucleolus is the most obvious structure seen in the nucleus of a eucaryotic cell when viewed in the light microscope. Consequently, it was so closely scrutinized by early cytologists that an 1898 review could list some 700 references. We now know that the nucleolus is the site for the processing of rRNAs and their assembly into ribosomes. Unlike other organelles in the cell, it is not bound by a membrane (Figure 6–44); instead, it is a large aggregate of macromolecules, including the rRNA genes themselves, precursor rRNAs, mature rRNAs, rRNA-processing enzymes, snoRNPs, ribosomal protein subunits and partly assembled

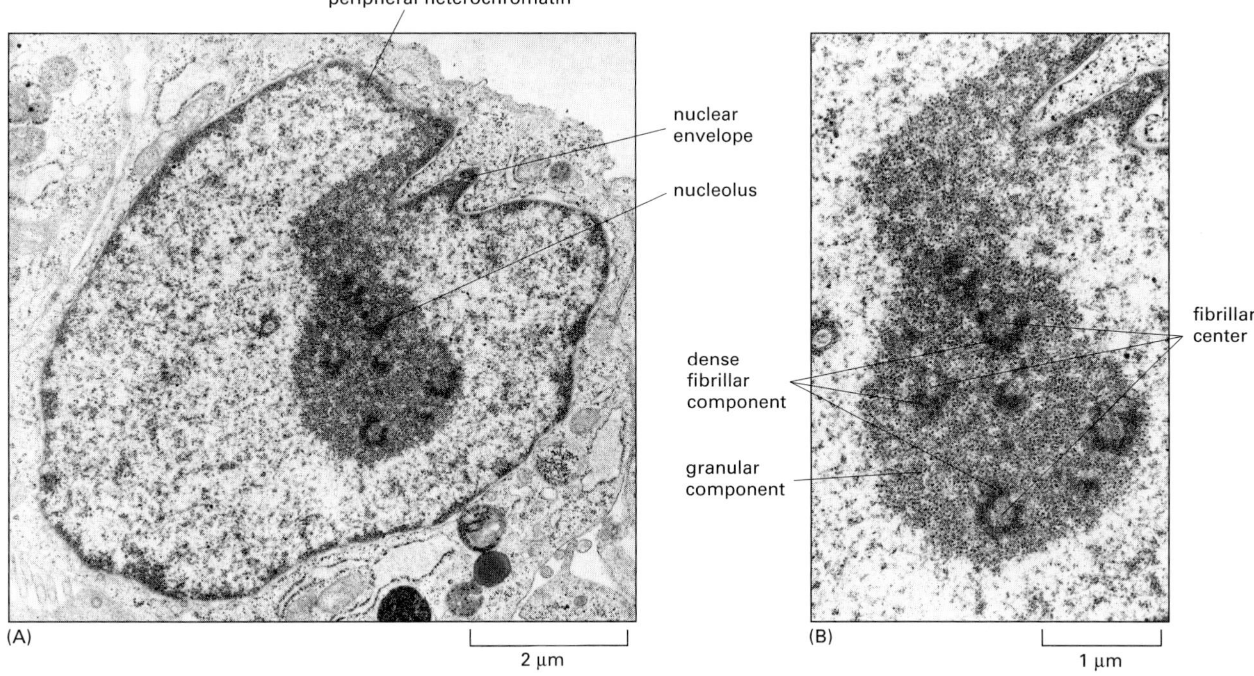

Figure 6–44 Electron micrograph of a thin section of a nucleolus in a human fibroblast, showing its three distinct zones. (A) View of entire nucleus. (B) High-power view of the nucleolus. It is believed that transcription of the rRNA genes takes place between the fibrillar center and the dense fibrillar component and that processing of the rRNAs and their assembly into ribosomes proceeds outward from the dense fibrillar component to the surrounding granular components. (Courtesy of E.G. Jordan and J. McGovern.)

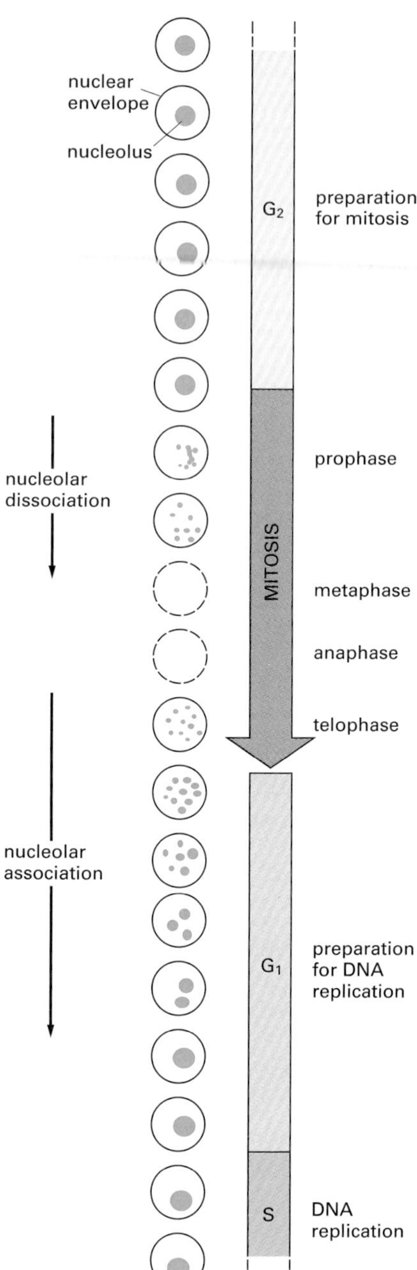

Figure 6–45 Changes in the appearance of the nucleolus in a human cell during the cell cycle. Only the cell nucleus is represented in this diagram. In most eucaryotic cells the nuclear membrane breaks down during mitosis, as indicated by the dashed circles.

ribosomes. The close association of all these components presumably allows the assembly of ribosomes to occur rapidly and smoothly.

It is not yet understood how the nucleolus is held together and organized, but various types of RNA molecules play a central part in its chemistry and structure, suggesting that the nucleolus may have evolved from an ancient structure present in cells dominated by RNA catalysis. In present-day cells, the rRNA genes also have an important role in forming the nucleolus. In a diploid human cell, the rRNA genes are distributed into 10 clusters, each of which is located near the tip of one of the two copies of five different chromosomes (see Figure 4–11). Each time a human cell undergoes mitosis, the chromosomes disperse and the nucleolus breaks up; after mitosis, the tips of the 10 chromosomes coalesce as the nucleolus reforms (Figures 6–45 and 6–46). The transcription of the rRNA genes by RNA polymerase I is necessary for this process.

As might be expected, the size of the nucleolus reflects the number of ribosomes that the cell is producing. Its size therefore varies greatly in different cells and can change in a single cell, occupying 25% of the total nuclear volume in cells that are making unusually large amounts of protein.

A schematic diagram of the assembly of ribosomes is shown in Figure 6–47. In addition to its important role in ribosome biogenesis, the nucleolus is also the site where other RNAs are produced and other RNA–protein complexes are assembled. For example, the U6 snRNP, which, as we have seen, functions in pre-mRNA splicing (see Figure 6–29), is composed of one RNA molecule and at least seven proteins. The U6 snRNA is chemically modified by snoRNAs in the nucleolus before its final assembly there into the U6 snRNP. Other important RNA protein complexes, including telomerase (encountered in Chapter 5) and the signal recognition particle (which we discuss in Chapter 12), are also believed to be assembled at the nucleolus. Finally, the tRNAs (transfer RNAs) that carry the amino acids for protein synthesis are processed there as well. Thus, the nucleolus can be thought of as a large factory at which many different noncoding RNAs are processed and assembled with proteins to form a large variety of ribonucleoprotein complexes.

The Nucleus Contains a Variety of Subnuclear Structures

Although the nucleolus is the most prominent structure in the nucleus, several other nuclear bodies have been visualized and studied (Figure 6–48). These include Cajal bodies (named for the scientist who first described them in 1906), GEMS (Gemini of coiled bodies), and interchromatin granule clusters (also called "speckles"). Like the nucleolus, these other nuclear structures lack membranes

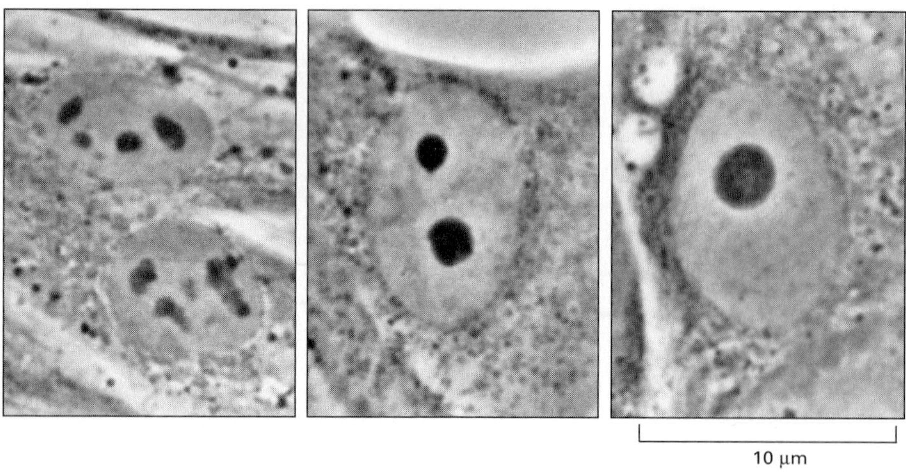

Figure 6–46 Nucleolar fusion. These light micrographs of human fibroblasts grown in culture show various stages of nucleolar fusion. After mitosis, each of the ten human chromosomes that carry a cluster of rRNA genes begins to form a tiny nucleolus, but these rapidly coalesce as they grow to form the single large nucleolus typical of many interphase cells. (Courtesy of E.G. Jordan and J. McGovern.)

and are highly dynamic; their appearance is probably the result of the tight association of protein and RNA (and perhaps DNA) components involved in the synthesis, assembly, and storage of macromolecules involved in gene expression. Cajal bodies and GEMS resemble one another and are frequently paired in the nucleus; it is not clear whether they are truly distinct structures. They may be sites where snRNAs and snoRNAs undergo their final modifications and assembly with protein. Both the RNAs and the proteins that make up the snRNPs are partly assembled in the cytoplasm, but they are transported into the nucleus for their final modifications. It has been proposed that Cajal bodies/GEMS are also sites where the snRNPs are recycled and their RNAs are "reset" after the rearrangements that occur during splicing (see p. 322). In contrast, the interchromatin granule clusters have been proposed to be stockpiles of fully mature snRNPs that are ready to be used in splicing of pre-mRNAs (Figure 6–49).

Scientists have had difficulties in working out the function of the small subnuclear structures just described. Much of the progress now being made depends on genetic tools—examination of the effects of designed mutations in mice or of spontaneous mutations in humans. As one example, GEMS contain the SMN (survival of motor neurons) protein. Certain mutations of the gene encoding this protein are the cause of inherited spinal muscular atrophy, a human disease characterized by a wasting away of the muscles. The disease seems

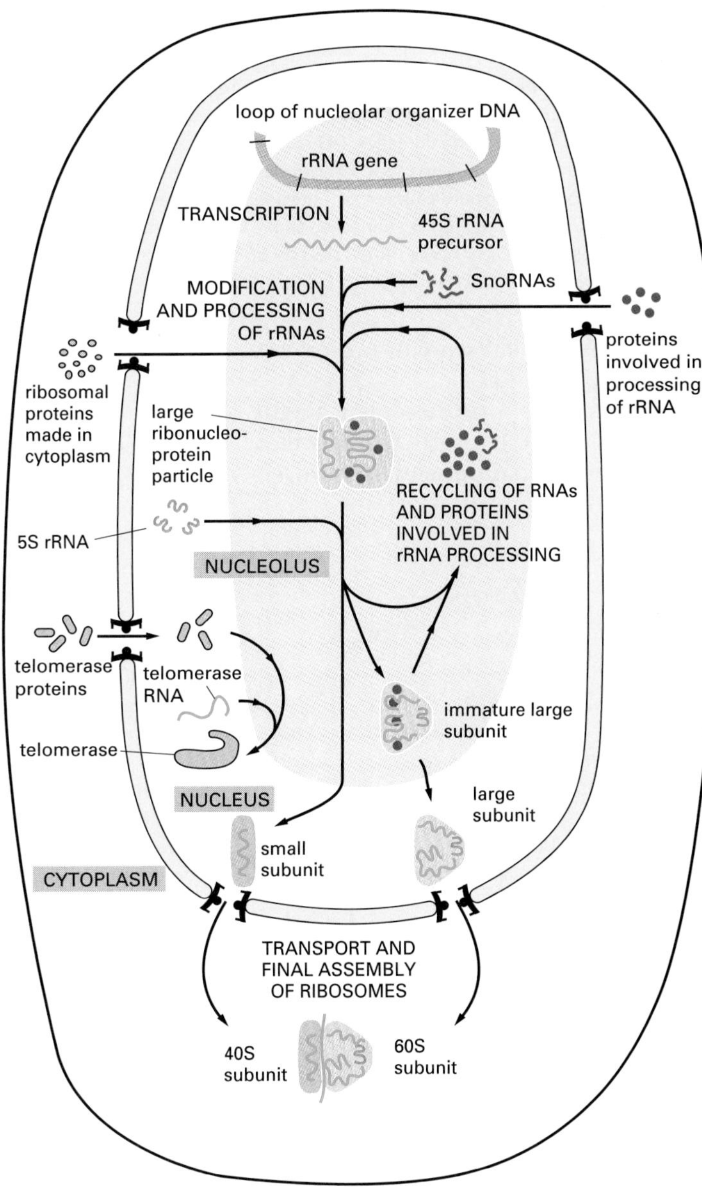

Figure 6–47 The function of the nucleolus in ribosome and other ribonucleoprotein synthesis. The 45S precursor rRNA is packaged in a large ribonucleoprotein particle containing many ribosomal proteins imported from the cytoplasm. While this particle remains in the nucleolus, selected pieces are added and others discarded as it is processed into immature large and small ribosomal subunits. The two ribosomal subunits are thought to attain their final functional form only as each is individually transported through the nuclear pores into the cytoplasm. Other ribonucleoprotein complexes, including telomerase shown here, are also assembled in the nucleolus.

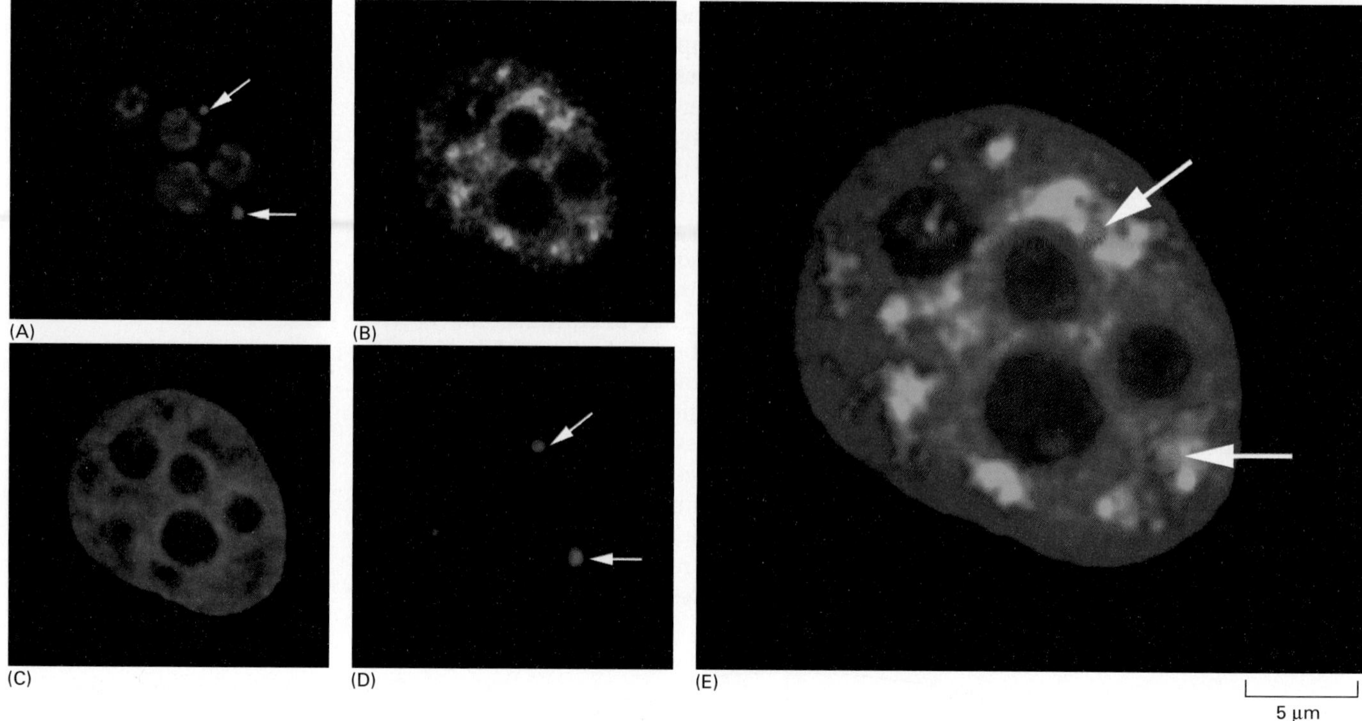

Figure 6–48 Visualization of chromatin and nuclear bodies. (A)–(D) show micrographs of the same human cell nucleus, each processed differently to show a particular set of nuclear structures. (E) shows an enlarged superposition of all four individual images. (A) shows the location of the protein fibrillarin (a component of several snoRNPs), which is present at both nucleoli and Cajal bodies, the latter indicated by arrows. (B) shows interchromatin granule clusters or "speckles" detected by using antibodies against a protein involved in pre-mRNA splicing. (C) is stained to show bulk chromatin. (D) shows the location of the protein coilin, which is present at Cajal bodies (indicated by arrows). (From J.R. Swedlow and A.I. Lamond, *Gen. Biol.* 2:1–7, 2001; micrographs courtesy of Judith Sleeman.)

to be caused by a subtle defect in snRNP assembly and subsequent pre-mRNA splicing. More severe defects would be expected to be lethal.

Given the importance of nuclear subdomains in RNA processing, it might have been expected that pre-mRNA splicing would occur in a particular location in the nucleus, as it requires numerous RNA and protein components. However,

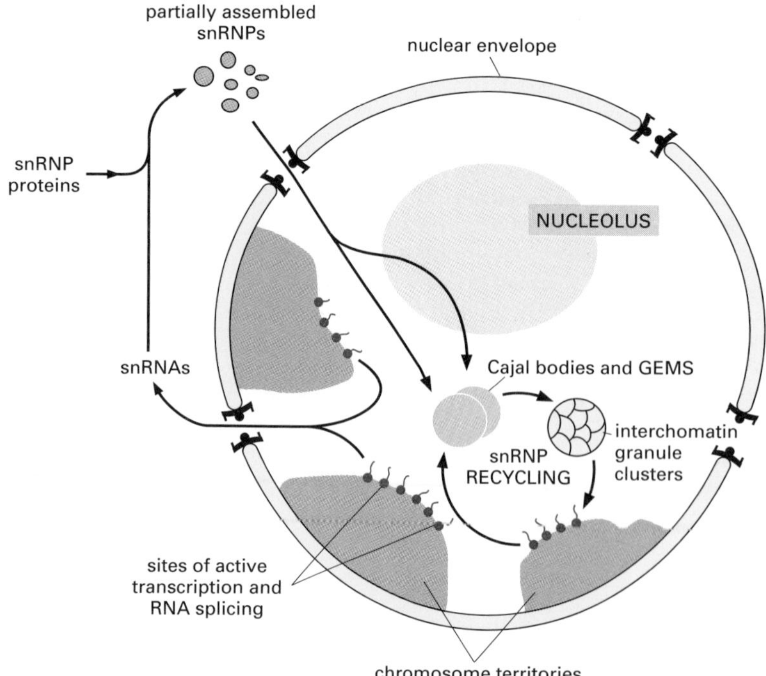

Figure 6–49 Schematic view of subnuclear structures. A typical vertebrate nucleus has several Cajal bodies, which are proposed to be the sites where snRNPs and snoRNPs undergo their final modifications. Interchromatin granule clusters are proposed to be storage sites for fully mature snRNPs. A typical vertebrate nucleus has 20–50 interchromatin granule clusters.

After their initial synthesis, snRNAs are exported from the nucleus, after which they undergo 5′ and 3′ end-processing and assemble with the seven common snRNP proteins (called Sm proteins). These complexes are reimported into the nucleus and the snRNPs undergo their final modification in Cajal bodies. In addition, the U6 snRNP requires chemical modification by snoRNAs in the nucleolus. The sites of active transcription and splicing (approximately 2000–3000 sites per vertebrate nucleus) correspond to the "perichromatin fibers" seen under the electron microscope. (Adapted from J.D. Lewis and D. Tollervey, *Science* 288:1385–1389, 2000.)

we have seen that the assembly of splicing components on pre-mRNA is co-transcriptional; thus splicing must occur at many locations along chromosomes. We saw in Chapter 4 that interphase chromosomes occupy discrete territories in the nucleus, and transcription and pre-mRNA splicing must take place within these territories. However, interphase chromosomes are themselves dynamic and their exact positioning in the nucleus correlates with gene expression. For example, transcriptionally silent regions of interphase chromosomes are often associated with the nuclear envelope where the concentration of heterochromatin components is believed to be especially high. When these same regions become transcriptionally active, they relocate towards the interior of the nucleus, which is richer in the components required for mRNA synthesis. It has been proposed that, although a typical mammalian cell may be expressing on the order of 15,000 genes, transcription and RNA splicing may be localized to only several thousand sites in the nucleus. These sites themselves are highly dynamic and probably result from the association of transcription and splicing components to create small "assembly lines" where the local concentration of these components is very high. As a result, the nucleus seems to be highly organized into subdomains, with snRNPs, snoRNPs, and other nuclear components moving between them in an orderly fashion according to the needs of the cell (Figure 6–49).

Summary

Before the synthesis of a particular protein can begin, the corresponding mRNA molecule must be produced by transcription. Bacteria contain a single type of RNA polymerase (the enzyme that carries out the transcription of DNA into RNA). An mRNA molecule is produced when this enzyme initiates transcription at a promoter, synthesizes the RNA by chain elongation, stops transcription at a terminator, and releases both the DNA template and the completed mRNA molecule. In eucaryotic cells, the process of transcription is much more complex, and there are three RNA polymerases—designated polymerase I, II, and III—that are related evolutionarily to one another and to the bacterial polymerase.

Eucaryotic mRNA is synthesized by RNA polymerase II. This enzyme requires a series of additional proteins, termed the general transcription factors, to initiate transcription on a purified DNA template and still more proteins (including chromatin-remodeling complexes and histone acetyltransferases) to initiate transcription on its chromatin template inside the cell. During the elongation phase of transcription, the nascent RNA undergoes three types of processing events: a special nucleotide is added to its 5′ end (capping), intron sequences are removed from the middle of the RNA molecule (splicing), and the 3′ end of the RNA is generated (cleavage and polyadenylation). Some of these RNA processing events that modify the initial RNA transcript (for example, those involved in RNA splicing) are carried out primarily by special small RNA molecules.

For some genes, RNA is the final product. In eucaryotes, these genes are usually transcribed by either RNA polymerase I or RNA polymerase III. RNA polymerase I makes the ribosomal RNAs. After their synthesis as a large precursor, the rRNAs are chemically modified, cleaved, and assembled into ribosomes in the nucleolus—a distinct subnuclear structure that also helps to process some smaller RNA–protein complexes in the cell. Additional subnuclear structures (including Cajal bodies and interchromatin granule clusters) are sites where components involved in RNA processing are assembled, stored, and recycled.

FROM RNA TO PROTEIN

In the preceding section we have seen that the final product of some genes is an RNA molecule itself, such as those present in the snRNPs and in ribosomes. However, most genes in a cell produce mRNA molecules that serve as intermediaries on the pathway to proteins. In this section we examine how the cell converts the information carried in an mRNA molecule into a protein molecule. This feat of translation first attracted the attention of biologists in the late 1950s,

GCA GCC GCG GCU	AGA AGG CGA CGC CGG CGU	GAC GAU	AAC AAU	UGC UGU	GAA GAG	CAA CAG	GGA GGC GGG GGU	CAC CAU	AUA AUC AUU	UUA UUG CUA CUC CUG CUU	AAA AAG	AUG	UUC UUU	CCA CCC CCG CCU	AGC AGU UCA UCC UCG UCU	ACA ACC ACG ACU	UGG	UAC UAU	GUA GUC GUG GUU	UAA UAG UGA
Ala	Arg	Asp	Asn	Cys	Glu	Gln	Gly	His	Ile	Leu	Lys	Met	Phe	Pro	Ser	Thr	Trp	Tyr	Val	stop
A	R	D	N	C	E	Q	G	H	I	L	K	M	F	P	S	T	W	Y	V	

Figure 6–50 The genetic code. The standard one-letter abbreviation for each amino acid is presented below its three-letter abbreviation (see Panel 3–1, pp. 132–133, for the full name of each amino acid and its structure). By convention, codons are always written with the 5′-terminal nucleotide to the left. Note that most amino acids are represented by more than one codon, and that there are some regularities in the set of codons that specifies each amino acid. Codons for the same amino acid tend to contain the same nucleotides at the first and second positions, and vary at the third position. Three codons do not specify any amino acid but act as termination sites (stop codons), signaling the end of the protein-coding sequence. One codon—AUG—acts both as an initiation codon, signaling the start of a protein-coding message, and also as the codon that specifies methionine.

when it was posed as the "coding problem": how is the information in a linear sequence of nucleotides in RNA translated into the linear sequence of a chemically quite different set of subunits—the amino acids in proteins? This fascinating question stimulated great excitement among scientists at the time. Here was a cryptogram set up by nature that, after more than 3 billion years of evolution, could finally be solved by one of the products of evolution—human beings. And indeed, not only has the code been cracked step by step, but in the year 2000 the elaborate machinery by which cells read this code—the ribosome—was finally revealed in atomic detail.

An mRNA Sequence Is Decoded in Sets of Three Nucleotides

Once an mRNA has been produced, by transcription and processing the information present in its nucleotide sequence is used to synthesize a protein. Transcription is simple to understand as a means of information transfer: since DNA and RNA are chemically and structurally similar, the DNA can act as a direct template for the synthesis of RNA by complementary base-pairing. As the term *transcription* signifies, it is as if a message written out by hand is being converted, say, into a typewritten text. The language itself and the form of the message do not change, and the symbols used are closely related.

In contrast, the conversion of the information in RNA into protein represents a **translation** of the information into another language that uses quite different symbols. Moreover, since there are only four different nucleotides in mRNA and twenty different types of amino acids in a protein, this translation cannot be accounted for by a direct one-to-one correspondence between a nucleotide in RNA and an amino acid in protein. The nucleotide sequence of a gene, through the medium of mRNA, is translated into the amino acid sequence of a protein by rules that are known as the **genetic code**. This code was deciphered in the early 1960s.

The sequence of nucleotides in the mRNA molecule is read consecutively in groups of three. RNA is a linear polymer of four different nucleotides, so there are $4 \times 4 \times 4 = 64$ possible combinations of three nucleotides: the triplets AAA, AUA, AUG, and so on. However, only 20 different amino acids are commonly found in proteins. Either some nucleotide triplets are never used, or the code is redundant and some amino acids are specified by more than one triplet. The second possibility is, in fact, the correct one, as shown by the completely deciphered genetic code in Figure 6–50. Each group of three consecutive nucleotides in RNA is called a **codon**, and each codon specifies either one amino acid or a stop to the translation process.

This genetic code is used universally in all present-day organisms. Although a few slight differences in the code have been found, these are chiefly in the DNA of mitochondria. Mitochondria have their own transcription and protein synthesis systems that operate quite independently from those of the rest of the cell, and it is understandable that their small genomes have been able to accommodate minor changes to the code (discussed in Chapter 14).

In principle, an RNA sequence can be translated in any one of three different **reading frames**, depending on where the decoding process begins (Figure 6–51). However, only one of the three possible reading frames in an mRNA encodes the required protein. We see later how a special punctuation signal at the beginning of each RNA message sets the correct reading frame at the start of protein synthesis.

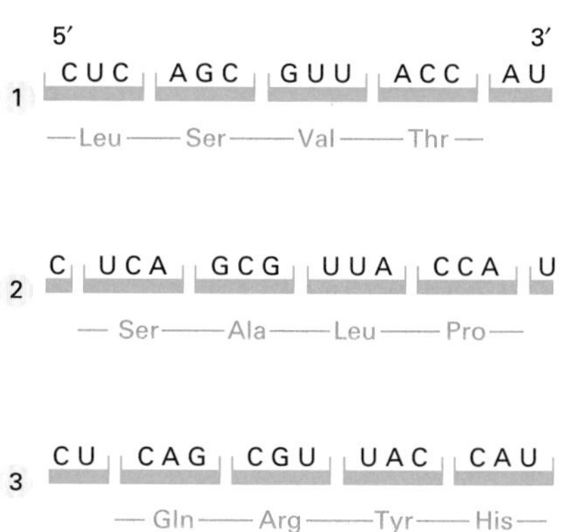

Figure 6–51 The three possible reading frames in protein synthesis. In the process of translating a nucleotide sequence *(blue)* into an amino acid sequence *(green)*, the sequence of nucleotides in an mRNA molecule is read from the 5′ to the 3′ end in sequential sets of three nucleotides. In principle, therefore, the same RNA sequence can specify three completely different amino acid sequences, depending on the reading frame. In reality, however, only one of these reading frames contains the actual message.

tRNA Molecules Match Amino Acids to Codons in mRNA

The codons in an mRNA molecule do not directly recognize the amino acids they specify: the group of three nucleotides does not, for example, bind directly to the amino acid. Rather, the translation of mRNA into protein depends on adaptor molecules that can recognize and bind both to the codon and, at another site on their surface, to the amino acid. These adaptors consist of a set of small RNA molecules known as **transfer RNAs (tRNAs)**, each about 80 nucleotides in length.

We saw earlier in this chapter that RNA molecules can fold up into precisely defined three-dimensional structures, and the tRNA molecules provide a striking example. Four short segments of the folded tRNA are double-helical, producing a molecule that looks like a cloverleaf when drawn schematically (Figure 6–52A). For example, a 5′-GCUC-3′ sequence in one part of a polynucleotide chain can form a relatively strong association with a 5′-GAGC-3′ sequence in another region of the same molecule. The cloverleaf undergoes further folding to form a compact L-shaped structure that is held together by additional hydrogen bonds between different regions of the molecule (Figure 6–52B,C).

Two regions of unpaired nucleotides situated at either end of the L-shaped molecule are crucial to the function of tRNA in protein synthesis. One of these regions forms the **anticodon**, a set of three consecutive nucleotides that pairs with the complementary codon in an mRNA molecule. The other is a short single-stranded region at the 3′ end of the molecule; this is the site where the amino acid that matches the codon is attached to the tRNA.

We have seen in the previous section that the genetic code is redundant; that is, several different codons can specify a single amino acid (see Figure 6–50). This redundancy implies either that there is more than one tRNA for many of the amino acids or that some tRNA molecules can base-pair with more than one

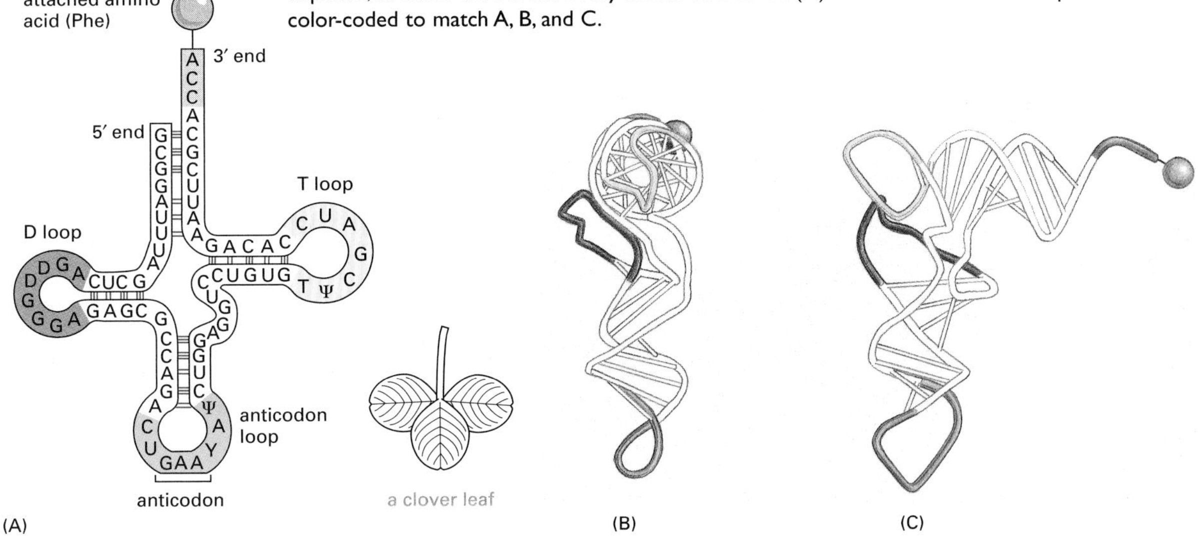

Figure 6–52 A tRNA molecule. In this series of diagrams, the same tRNA molecule—in this case a tRNA specific for the amino acid phenylalanine (Phe)—is depicted in various ways. (A) The cloverleaf structure, a convention used to show the complementary base-pairing *(red lines)* that creates the double-helical regions of the molecule. The anticodon is the sequence of three nucleotides that base-pairs with a codon in mRNA. The amino acid matching the codon/anticodon pair is attached at the 3′ end of the tRNA. tRNAs contain some unusual bases, which are produced by chemical modification after the tRNA has been synthesized. For example, the bases denoted ψ (for pseudouridine—see Figure 6–43) and D (for dihydrouridine—see Figure 6–55) are derived from uracil. (B and C) Views of the actual L-shaped molecule, based on x-ray diffraction analysis. Although a particular tRNA, that for the amino acid phenylalanine, is depicted, all other tRNAs have very similar structures. (D) The linear nucleotide sequence of the molecule, color-coded to match A, B, and C.

Figure 6–53 Wobble base-pairing between codons and anticodons. If the nucleotide listed in the first column is present at the third, or wobble, position of the codon, it can base-pair with any of the nucleotides listed in the second column. Thus, for example, when inosine (I) is present in the wobble position of the tRNA anticodon, the tRNA can recognize any one of three different codons in bacteria and either of two codons in eucaryotes. The inosine in tRNAs is formed from the deamination of guanine (see Figure 6–55), a chemical modification which takes place after the tRNA has been synthesized. The nonstandard base pairs, including those made with inosine, are generally weaker than conventional base pairs. Note that codon–anticodon base pairing is more stringent at positions 1 and 2 of the codon: here only conventional base pairs are permitted. The differences in wobble base-pairing interactions between bacteria and eucaryotes presumably result from subtle structural differences between bacterial and eucaryotic ribosomes, the molecular machines that perform protein synthesis. (Adapted from C. Guthrie and J. Abelson, in The Molecular Biology of the Yeast *Saccharomyces*: Metabolism and Gene Expression, pp. 487–528. Cold Spring Harbor, New York: Cold Spring Harbor Laboratory Press, 1982.)

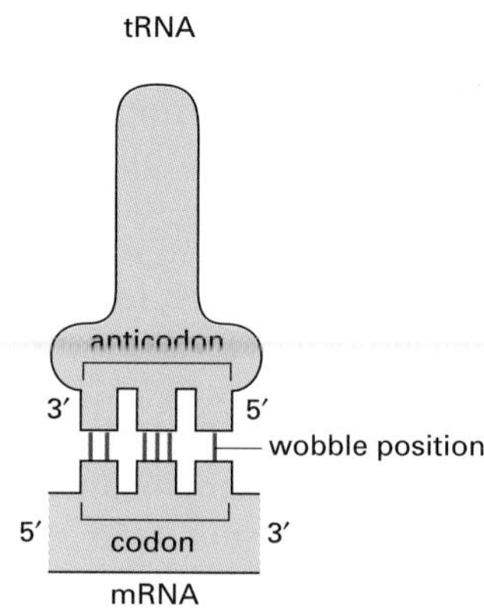

bacteria

wobble codon base	possible anticodon bases
U	A, G, or I
C	G or I
A	U or I
G	C or U

eucaryotes

wobble codon base	possible anticodon bases
U	G or I
C	G or I
A	U
G	C

codon. In fact, both situations occur. Some amino acids have more than one tRNA and some tRNAs are constructed so that they require accurate base-pairing only at the first two positions of the codon and can tolerate a mismatch (or *wobble*) at the third position (Figure 6–53). This wobble base-pairing explains why so many of the alternative codons for an amino acid differ only in their third nucleotide (see Figure 6–50). In bacteria, wobble base-pairings make it possible to fit the 20 amino acids to their 61 codons with as few as 31 kinds of tRNA molecules. The exact number of different kinds of tRNAs, however, differs from one species to the next. For example, humans have 497 tRNA genes but, among them, only 48 different anticodons are represented.

tRNAs Are Covalently Modified Before They Exit from the Nucleus

We have seen that most eucaryotic RNAs are covalently altered before they are allowed to exit from the nucleus, and tRNAs are no exception. Eucaryotic tRNAs are synthesized by RNA polymerase III. Both bacterial and eucaryotic tRNAs are typically synthesized as larger precursor tRNAs, and these are then trimmed to produce the mature tRNA. In addition, some tRNA precursors (from both bacteria and eucaryotes) contain introns that must be spliced out. This splicing reaction is chemically distinct from that of pre-mRNA splicing; rather than generating a lariat intermediate, tRNA splicing occurs through a cut-and-paste mechanism that is catalyzed by proteins (Figure 6–54). Trimming and splicing both require the precursor tRNA to be correctly folded in its cloverleaf configuration. Because misfolded tRNA precursors will not be processed properly, the

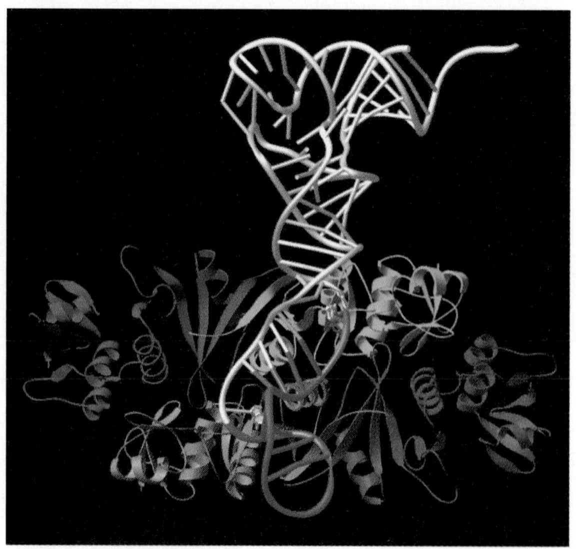

Figure 6–54 Structure of a tRNA-splicing endonuclease docked to a precursor tRNA. The endonuclease (a four-subunit enzyme) removes the tRNA intron *(blue)*. A second enzyme, a multifunctional tRNA ligase (not shown), then joins the two tRNA halves together. (Courtesy of Hong Li, Christopher Trotta, and John Abelson.)

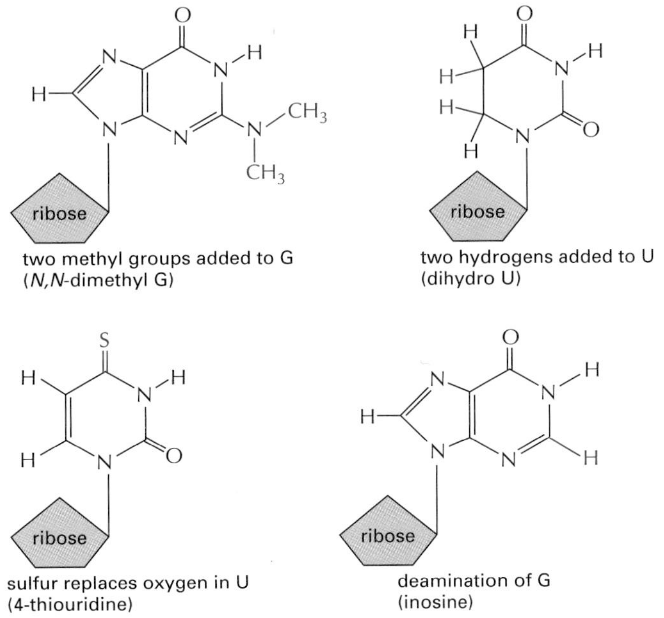

ribose
two methyl groups added to G
(*N,N*-dimethyl G)
ribose
two hydrogens added to U
(dihydro U)
ribose
sulfur replaces oxygen in U
(4-thiouridine)
ribose
deamination of G
(inosine)

Figure 6–55 A few of the unusual nucleotides found in tRNA molecules. These nucleotides are produced by covalent modification of a normal nucleotide after it has been incorporated into an RNA chain. In most tRNA molecules about 10% of the nucleotides are modified (see Figure 6–52).

trimming and splicing reactions are thought to act as quality-control steps in the generation of tRNAs.

All tRNAs are also subject to a variety of chemical modifications—nearly one in 10 nucleotides in each mature tRNA molecule is an altered version of a standard G, U, C, or A ribonucleotide. Over 50 different types of tRNA modifications are known; a few are shown in Figure 6–55. Some of the modified nucleotides—most notably inosine, produced by the deamination of guanosine—affect the conformation and base-pairing of the anticodon and thereby facilitate the recognition of the appropriate mRNA codon by the tRNA molecule (see Figure 6–53). Others affect the accuracy with which the tRNA is attached to the correct amino acid.

Specific Enzymes Couple Each Amino Acid to Its Appropriate tRNA Molecule

We have seen that, to read the genetic code in DNA, cells make a series of different tRNAs. We now consider how each tRNA molecule becomes linked to the one amino acid in 20 that is its appropriate partner. Recognition and attachment of the correct amino acid depends on enzymes called **aminoacyl-tRNA synthetases**, which covalently couple each amino acid to its appropriate set of tRNA molecules (Figures 6–56 and 6–57). For most cells there is a different synthetase enzyme for each amino acid (that is, 20 synthetases in all); one attaches glycine

ATP
2 P
amino acid
adenylated amino acid
P ribose - adenine
P ribose - adenine
AMP
tRNA
aminoacyl-
tRNA

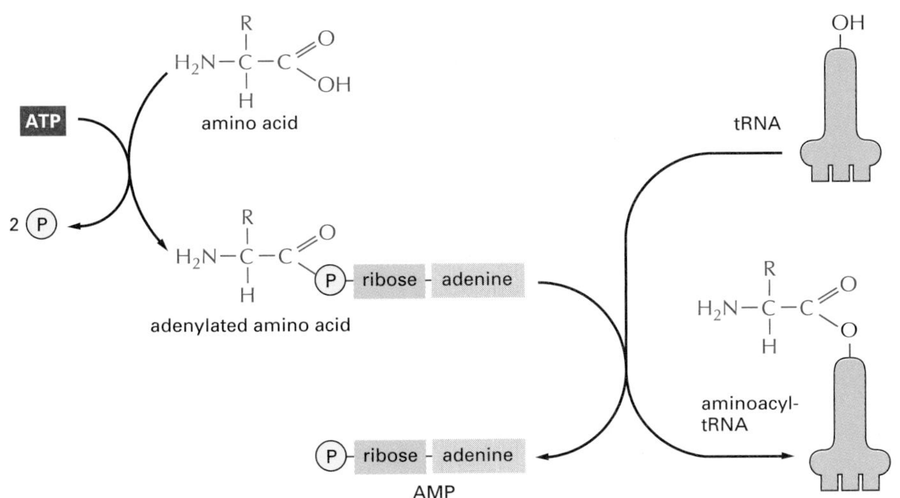

Figure 6–56 Amino acid activation. The two-step process in which an amino acid (with its side chain denoted by R) is activated for protein synthesis by an aminoacyl-tRNA synthetase enzyme is shown. As indicated, the energy of ATP hydrolysis is used to attach each amino acid to its tRNA molecule in a high-energy linkage. The amino acid is first activated through the linkage of its carboxyl group directly to an AMP moiety, forming an *adenylated amino acid*; the linkage of the AMP, normally an unfavorable reaction, is driven by the hydrolysis of the ATP molecule that donates the AMP. Without leaving the synthetase enzyme, the AMP-linked carboxyl group on the amino acid is then transferred to a hydroxyl group on the sugar at the 3′ end of the tRNA molecule. This transfer joins the amino acid by an activated ester linkage to the tRNA and forms the final aminoacyl-tRNA molecule. The synthetase enzyme is not shown in this diagram.

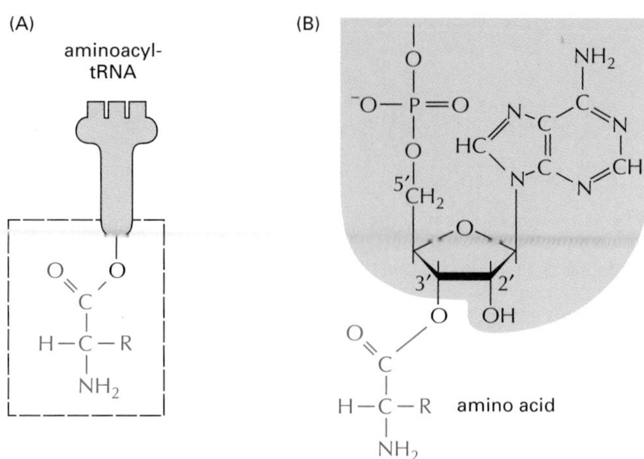

Figure 6–57 The structure of the aminoacyl-tRNA linkage. The carboxyl end of the amino acid forms an ester bond to ribose. Because the hydrolysis of this ester bond is associated with a large favorable change in free energy, an amino acid held in this way is said to be activated. (A) Schematic drawing of the structure. The amino acid is linked to the nucleotide at the 3′ end of the tRNA (see Figure 6–52). (B) Actual structure corresponding to boxed region in (A). These are two major classes of synthetase enzymes: one links the amino acid directly to the 3′-OH group of the ribose, and the other links it initially to the 2′-OH group. In the latter case, a subsequent transesterification reaction shifts the amino acid to the 3′ position. As in Figure 6–56, the "R-group" indicates the side chain of the amino acid.

to all tRNAs that recognize codons for glycine, another attaches alanine to all tRNAs that recognize codons for alanine, and so on. Many bacteria, however, have fewer than 20 synthetases, and the same synthetase enzyme is responsible for coupling more than one amino acid to the appropriate tRNAs. In these cases, a single synthetase places the identical amino acid on two different types of tRNAs, only one of which has an anticodon that matches the amino acid. A second enzyme then chemically modifies each "incorrectly" attached amino acid so that it now corresponds to the anticodon displayed by its covalently linked tRNA.

The synthetase-catalyzed reaction that attaches the amino acid to the 3′ end of the tRNA is one of many cellular reactions coupled to the energy-releasing hydrolysis of ATP (see pp. 83–84), and it produces a high-energy bond between the tRNA and the amino acid. The energy of this bond is used at a later stage in protein synthesis to link the amino acid covalently to the growing polypeptide chain.

Although the tRNA molecules serve as the final adaptors in converting nucleotide sequences into amino acid sequences, the aminoacyl-tRNA synthetase enzymes are adaptors of equal importance in the decoding process (Figure 6–58). This was established by an ingenious experiment in which an amino acid (cysteine) was chemically converted into a different amino acid (alanine) after it already had been attached to its specific tRNA. When such "hybrid" aminoacyl-tRNA molecules were used for protein synthesis in a cell-free system, the wrong amino acid was inserted at every point in the protein chain where that tRNA was used. Although cells have several quality control mechanisms to avoid this type of mishap, the experiment clearly establishes that the genetic code is translated by two sets of adaptors that act sequentially. Each matches one molecular surface to another with great specificity, and it is their combined

Figure 6–58 The genetic code is translated by means of two adaptors that act one after another. The first adaptor is the aminoacyl-tRNA synthetase, which couples a particular amino acid to its corresponding tRNA; the second adaptor is the tRNA molecule itself, whose *anticodon* forms base pairs with the appropriate *codon* on the mRNA. An error in either step would cause the wrong amino acid to be incorporated into a protein chain. In the sequence of events shown, the amino acid tryptophan (Trp) is selected by the codon UGG on the mRNA.

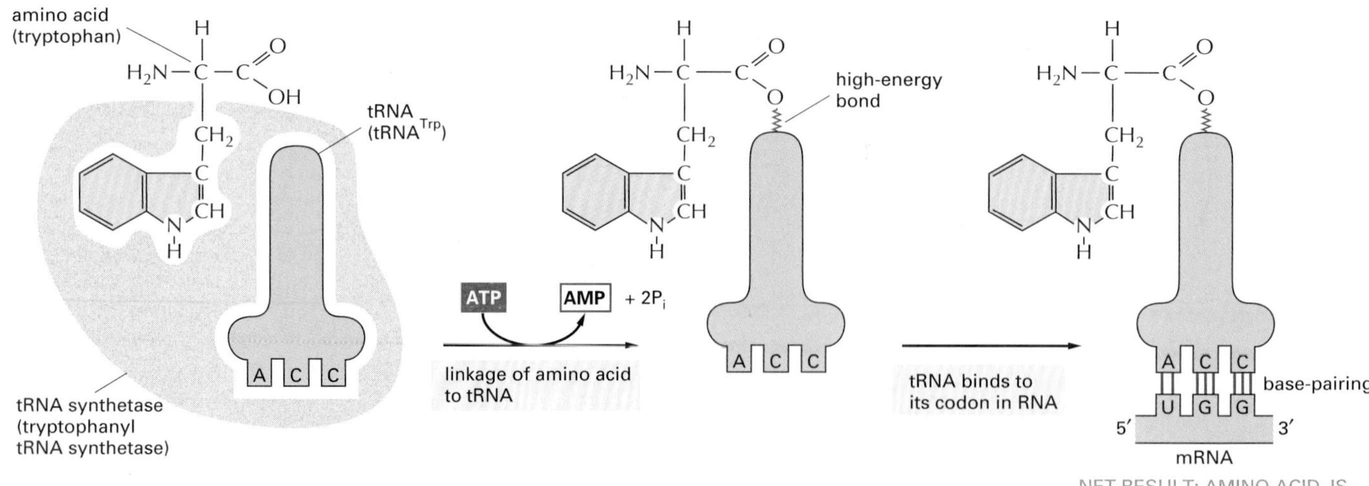

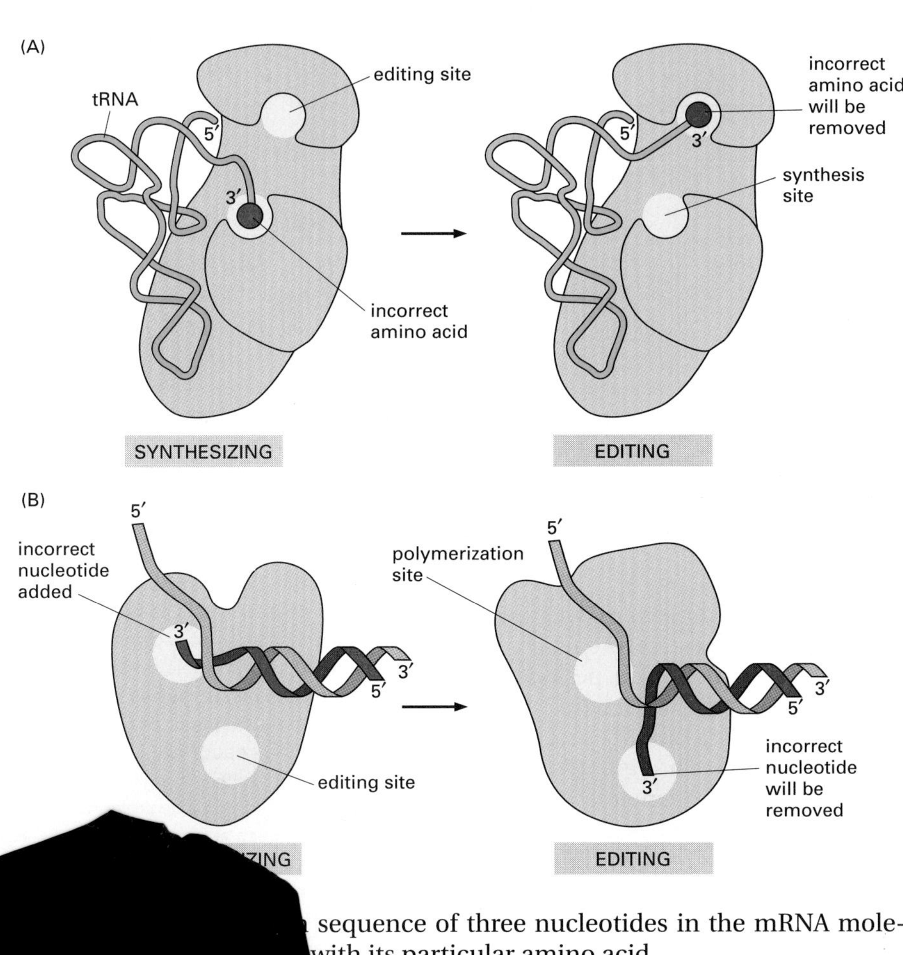

Figure 6–59 Hydrolytic editing. (A) tRNA synthetases remove their own coupling errors through hydrolytic editing of incorrectly attached amino acids. As described in the text, the correct amino acid is rejected by the editing site. (B) The error-correction process performed by DNA polymerase shows some similarities; however, it differs so far as the removal process depends strongly on a mispairing with the template (see Figure 5–9).

… sequence of three nucleotides in the mRNA mole- … with its particular amino acid.

…tases Ensures Accuracy

…g together ensure that the tRNA synthetase links the …ch tRNA. The synthetase must first select the correct …d, … do so by a two-step mechanism. First, the correct amino …he … affinity for the active-site pocket of its synthetase and is …e favor… over the other 19. In particular, amino acids larger than the c… one are effectively excluded from the active site. However, accurate discrimination between two similar amino acids, such as isoleucine and valine (which differ by only a methyl group), is very difficult to achieve by a one-step recognition mechanism. A second discrimination step occurs after the amino acid has been covalently linked to AMP (see Figure 6–56). When tRNA binds the synthetase, it forces the amino acid into a second pocket in the synthetase, the precise dimensions of which exclude the correct amino acid but allow access by closely related amino acids. Once an amino acid enters this editing pocket, it is hydrolyzed from the AMP (or from the tRNA itself if the aminoacyl-tRNA bond has already formed) and released from the enzyme. This hydrolytic editing, which is analogous to the editing by DNA polymerases (Figure 6–59), raises the overall accuracy of tRNA charging to approximately one mistake in 40,000 couplings.

The tRNA synthetase must also recognize the correct set of tRNAs, and extensive structural and chemical complementarity between the synthetase and the tRNA allows various features of the tRNA to be sensed (Figure 6–60). Most tRNA synthetases directly recognize the matching tRNA anticodon; these synthetases contain three adjacent nucleotide-binding pockets, each of which is complementary in shape and charge to the nucleotide in the anticodon. For other synthetases it is the nucleotide sequence of the acceptor stem that is the key recognition determinant. In most cases, however, nucleotides at several positions on the tRNA are "read" by the synthetase.

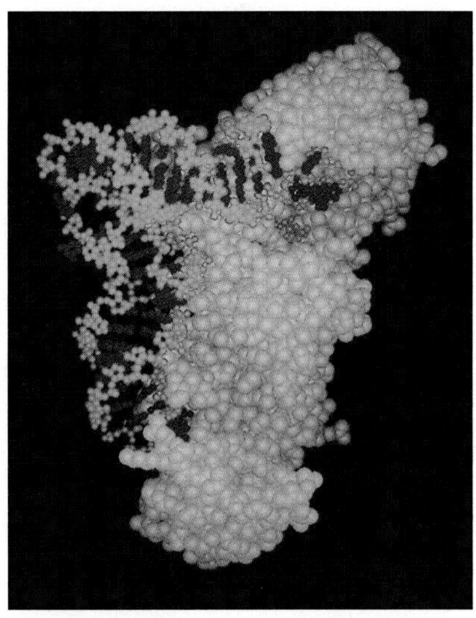

Figure 6–60 The recognition of a tRNA molecule by its aminoacyl-tRNA synthetase. For this tRNA ($tRNA^{Gln}$), specific nucleotides in both the anticodon *(bottom)* and the amino acid-accepting arm allow the correct tRNA to be recognized by the synthetase enzyme *(blue)*. (Courtesy of Tom Steitz.)

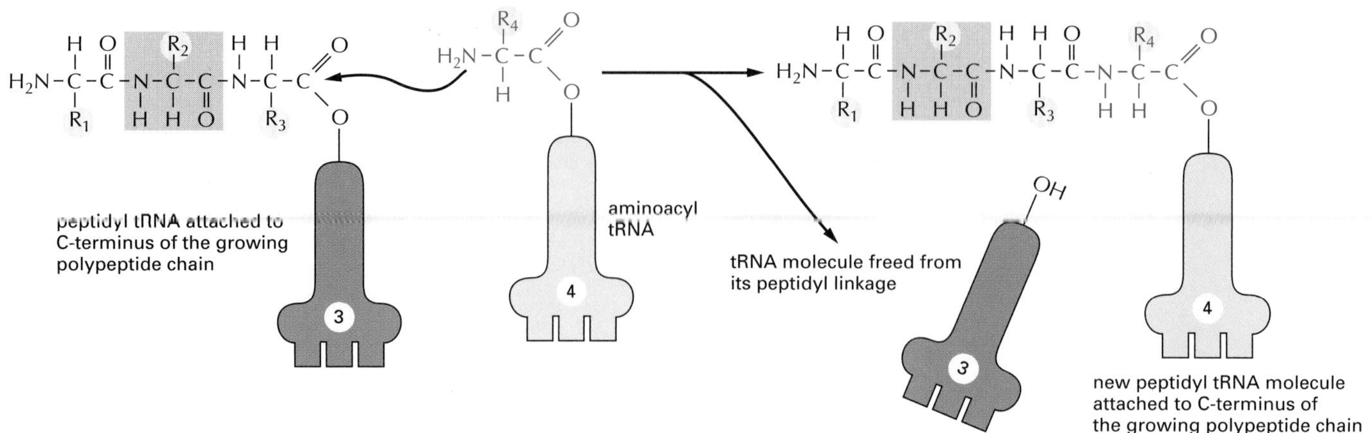

Figure 6–61 The incorporation of an amino acid into a protein. A polypeptide chain grows by the stepwise addition of amino acids to its C-terminal end. The formation of each peptide bond is energetically favorable because the growing C-terminus has been activated by the covalent attachment of a tRNA molecule. The peptidyl-tRNA linkage that activates the growing end is regenerated during each addition. The amino acid side chains have been abbreviated as R_1, R_2, R_3, and R_4; as a reference point, all of the atoms in the second amino acid in the polypeptide chain are shaded *gray*. The figure shows the addition of the fourth amino acid to the growing chain.

Amino Acids Are Added to the C-terminal End of a Growing Polypeptide Chain

Having seen that amino acids are first coupled to tRNA molecules, we now turn to the mechanism by which they are joined together to form proteins. The fundamental reaction of protein synthesis is the formation of a peptide bond between the carboxyl group at the end of a growing polypeptide chain and a free amino group on an incoming amino acid. Consequently, a protein is synthesized stepwise from its N-terminal end to its C-terminal end. Throughout the entire process the growing carboxyl end of the polypeptide chain remains activated by its covalent attachment to a tRNA molecule (a peptidyl-tRNA molecule). This high-energy covalent linkage is disrupted during each addition but is immediately replaced by the identical linkage on the most recently added amino acid (Figure 6–61). In this way, each amino acid added carries with it the activation energy for the addition of the next amino acid rather than the energy for its own addition—an example of the "head growth" type of polymerization described in Figure 2–68.

The RNA Message Is Decoded on Ribosomes

As we have seen, the synthesis of proteins is guided by information carried by mRNA molecules. To maintain the correct reading frame and to ensure accuracy (about 1 mistake every 10,000 amino acids), protein synthesis is performed in the **ribosome**, a complex catalytic machine made from more than 50 different proteins (the *ribosomal proteins*) and several RNA molecules, the **ribosomal RNAs (rRNAs)**. A typical eucaryotic cell contains millions of ribosomes in its cytoplasm (Figure 6–62). As we have seen, eucaryotic ribosomal subunits are

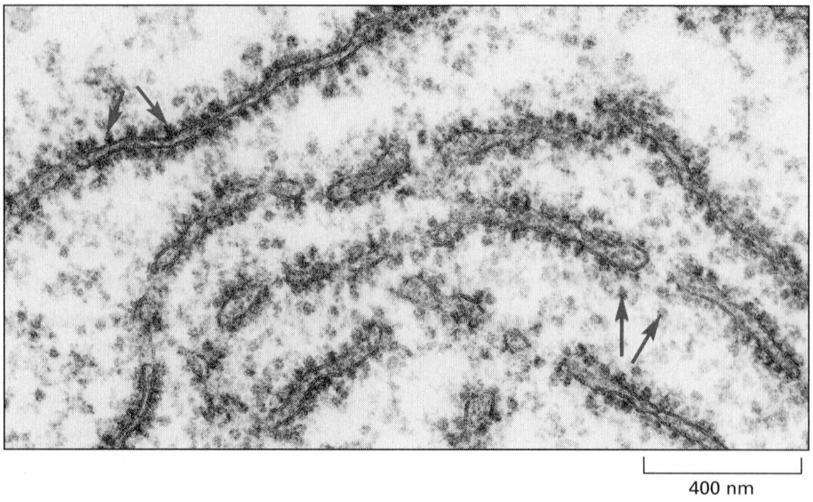

400 nm

Figure 6–62 Ribosomes in the cytoplasm of a eucaryotic cell. This electron micrograph shows a thin section of a small region of cytoplasm. The ribosomes appear as black dots *(red arrows)*. Some are free in the cytosol; others are attached to membranes of the endoplasmic reticulum. (Courtesy of Daniel S. Friend.)

assembled at the nucleolus, by the association of newly transcribed and modified rRNAs with ribosomal proteins, which have been transported into the nucleus after their synthesis in the cytoplasm. The two ribosomal subunits are then exported to the cytoplasm, where they perform protein synthesis.

Eucaryotic and procaryotic ribosomes are very similar in design and function. Both are composed of one large and one small subunit that fit together to form a complete ribosome with a mass of several million daltons (Figure 6–63). The small subunit provides a framework on which the tRNAs can be accurately matched to the codons of the mRNA (see Figure 6–58), while the large subunit catalyzes the formation of the peptide bonds that link the amino acids together into a polypeptide chain (see Figure 6–61).

When not actively synthesizing proteins, the two subunits of the ribosome are separate. They join together on an mRNA molecule, usually near its 5′ end, to initiate the synthesis of a protein. The mRNA is then pulled through the ribosome; as its codons encounter the ribosome's active site, the mRNA nucleotide sequence is translated into an amino acid sequence using the tRNAs as adaptors to add each amino acid in the correct sequence to the end of the growing polypeptide chain. When a stop codon is encountered, the ribosome releases the finished protein, its two subunits separate again. These subunits can then be used to start the synthesis of another protein on another mRNA molecule.

Ribosomes operate with remarkable efficiency: in one second, a single ribosome of a eucaryotic cell adds about 2 amino acids to a polypeptide chain; the ribosomes of bacterial cells operate even faster, at a rate of about 20 amino acids

Figure 6–63 A comparison of the structures of procaryotic and eucaryotic ribosomes. Ribosomal components are commonly designated by their "S values," which refer to their rate of sedimentation in an ultracentrifuge. Despite the differences in the number and size of their rRNA and protein components, both procaryotic and eucaryotic ribosomes have nearly the same structure and they function similarly. Although the 18S and 28S rRNAs of the eucaryotic ribosome contain many extra nucleotides not present in their bacterial counterparts, these nucleotides are present as multiple insertions that form extra domains and leave the basic structure of each rRNA largely unchanged.

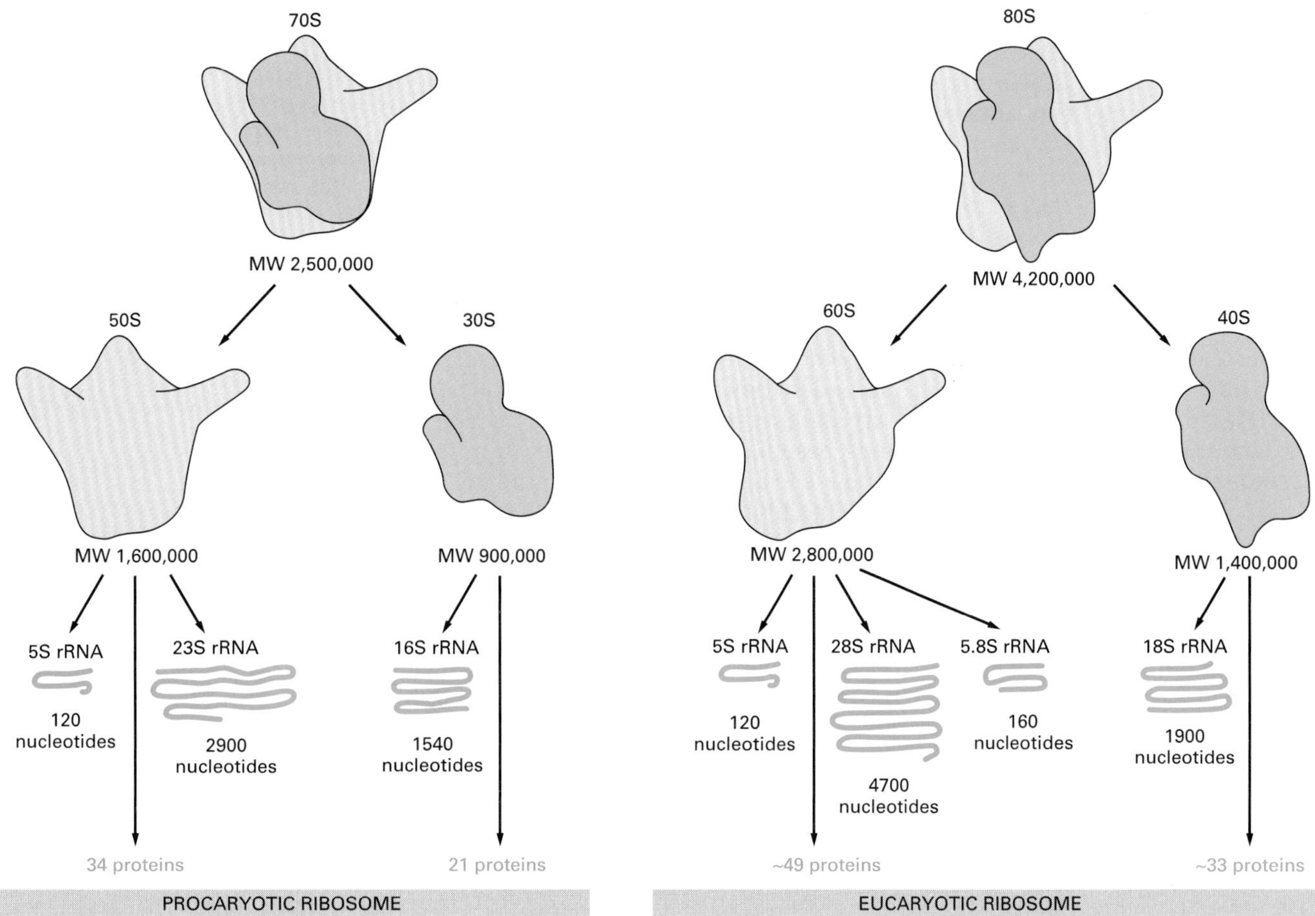

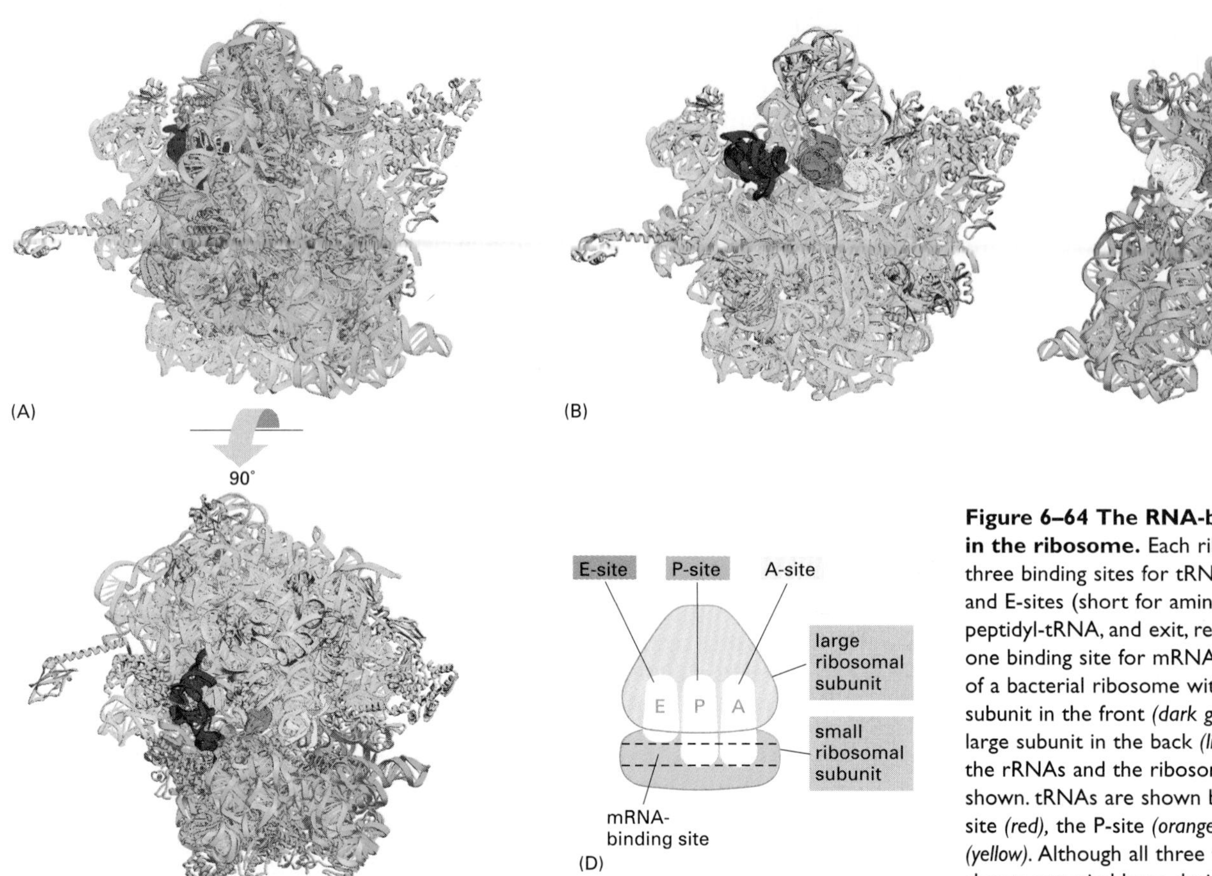

Figure 6–64 The RNA-binding sites in the ribosome. Each ribosome has three binding sites for tRNA: the A-, P-, and E-sites (short for aminoacyl-tRNA, peptidyl-tRNA, and exit, respectively) and one binding site for mRNA. (A) Structure of a bacterial ribosome with the small subunit in the front *(dark green)* and the large subunit in the back *(light green)*. Both the rRNAs and the ribosomal proteins are shown. tRNAs are shown bound in the E-site *(red)*, the P-site *(orange)* and the A-site *(yellow)*. Although all three tRNA sites are shown occupied here, during the process of protein synthesis not more than two of these sites are thought to contain tRNA molecules at any one time (see Figure 6–65). (B) Structure of the large *(left)* and small *(right)* ribosomal subunits arranged as though the ribosome in (A) were opened like a book. (C) Structure of the ribosome in (A) viewed from the top. (D) Highly schematic representation of a ribosome (in the same orientation as C), which will be used in subsequent figures. (A, B, and C, adapted from M.M. Yusupov et al., *Science* 292:883–896, 2001, courtesy of Albion Bausom and Harry Noller.)

per second. How does the ribosome choreograph the many coordinated movements required for efficient translation? A ribosome contains four binding sites for RNA molecules: one is for the mRNA and three (called the A-site, the P-site, and the E-site) are for tRNAs (Figure 6–64). A tRNA molecule is held tightly at the A- and P-sites only if its anticodon forms base pairs with a complementary codon (allowing for wobble) on the mRNA molecule that is bound to the ribosome. The A- and P-sites are close enough together for their two tRNA molecules to be forced to form base pairs with adjacent codons on the mRNA molecule. This feature of the ribosome maintains the correct reading frame on the mRNA.

Once protein synthesis has been initiated, each new amino acid is added to the elongating chain in a cycle of reactions containing three major steps. Our description of the chain elongation process begins at a point at which some amino acids have already been linked together and there is a tRNA molecule in the P-site on the ribosome, covalently joined to the end of the growing polypeptide (Figure 6–65). In step 1, a tRNA carrying the next amino acid in the chain binds to the ribosomal A-site by forming base pairs with the codon in mRNA positioned there, so that the P-site and the A-site contain adjacent bound tRNAs. In step 2, the carboxyl end of the polypeptide chain is released from the tRNA at the P-site (by breakage of the high-energy bond between the tRNA and its amino acid) and joined to the free amino group of the amino acid linked to the tRNA at the A-site, forming a new peptide bond. This central reaction of protein synthesis is catalyzed by a *peptidyl transferase* catalytic activity contained in the large ribosomal subunit. This reaction is accompanied by several conformational changes in the ribosome, which shift the two tRNAs into the E- and P-sites of the large subunit. In step 3, another series of conformational changes moves the mRNA exactly three nucleotides through the ribosome and resets the ribosome so it is ready to receive the next amino acyl tRNA. Step 1 is then repeated with a new incoming aminoacyl tRNA, and so on.

This three-step cycle is repeated each time an amino acid is added to the polypeptide chain, and the chain grows from its amino to its carboxyl end until a stop codon is encountered.

Elongation Factors Drive Translation Forward

The basic cycle of polypeptide elongation shown in outline in Figure 6–65 has an additional feature that makes translation especially efficient and accurate. Two *elongation factors* (EF-Tu and EF-G) enter and leave the ribosome during each cycle, each hydrolyzing GTP to GDP and undergoing conformational changes in the process. Under some conditions, ribosomes can be made to perform protein synthesis without the aid of the elongation factors and GTP hydrolysis, but this synthesis is very slow, inefficient, and inaccurate. The process is speeded up enormously by coupling conformational changes in the elongation factors to transitions between different conformational states of the ribosome. Although these conformational changes in the ribosome are not yet understood in detail, some may involve RNA rearrangements similar to those occurring in the RNAs of the spliceosome (see Figure 6–30). The cycles of elongation factor association, GTP hydrolysis, and dissociation ensures that the conformational changes occur in the "forward" direction and translation thereby proceeds efficiently (Figure 6–66).

In addition to helping move translation forward, EF-Tu is thought to increase the accuracy of translation by monitoring the initial interaction between a charged tRNA and a codon (see Figure 6–66). Charged tRNAs enter the ribosome bound to the GTP-form of EF-Tu. Although the bound elongation factor allows codon–anticodon pairing to occur, it prevents the amino acid from being incorporated into the growing polypeptide chain. The initial codon recognition, however, triggers the elongation factor to hydrolyze its bound GTP (to GDP and inorganic phosphate), whereupon the factor dissociates from the ribosome without its tRNA, allowing protein synthesis to proceed. The elongation factor introduces two short delays between codon–anticodon base pairing and polypeptide chain elongation; these delays selectively permit incorrectly bound tRNAs to exit from the ribosome before the irreversible step of chain elongation occurs. The first delay is the time required for GTP hydrolysis. The rate of GTP hydrolysis by EF-Tu is faster for a correct codon–anticodon pair than for an incorrect pair; hence an incorrectly bound tRNA molecule has a longer window of opportunity to dissociate from the ribosome. In other words, GTP hydrolysis selectively captures the correctly bound tRNAs. A second lag occurs between EF-Tu dissociation and the full accommodation of the tRNA in the A site of the ribosome. Although this lag is believed to be the same for correctly and incorrectly bound tRNAs, an incorrect tRNA molecule forms a smaller number of codon–anticodon hydrogen bonds than does a correctly matched pair and is therefore more likely to dissociate during this period. These two delays introduced by the elongation factor cause most incorrectly bound tRNA molecules (as well as a significant number of correctly bound molecules) to leave the

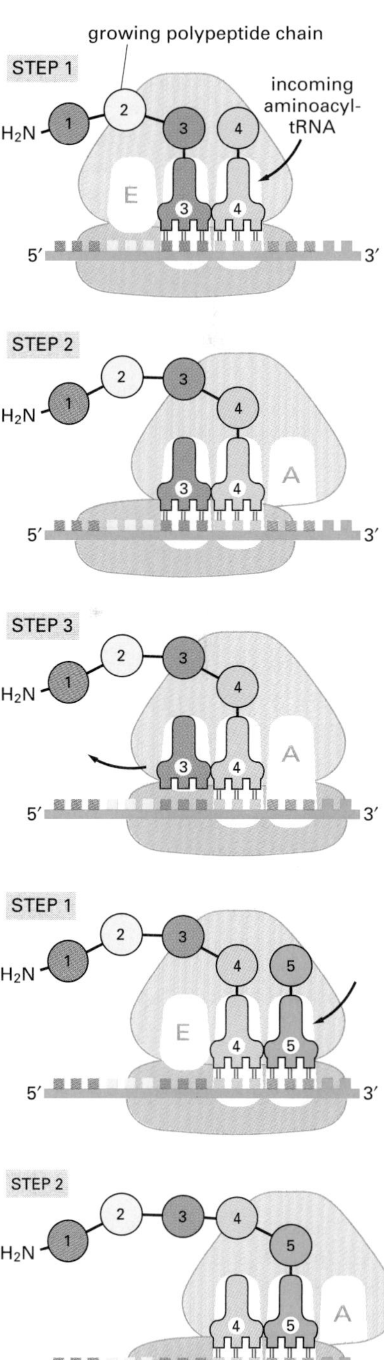

Figure 6–65 Translating an mRNA molecule. Each amino acid added to the growing end of a polypeptide chain is selected by complementary base-pairing between the anticodon on its attached tRNA molecule and the next codon on the mRNA chain. Because only one of the many types of tRNA molecules in a cell can base-pair with each codon, the codon determines the specific amino acid to be added to the growing polypeptide chain. The three-step cycle shown is repeated over and over during the synthesis of a protein. An aminoacyl-tRNA molecule binds to a vacant A-site on the ribosome in step 1, a new peptide bond is formed in step 2, and the mRNA moves a distance of three nucleotides through the small-subunit chain in step 3, ejecting the spent tRNA molecule and "resetting" the ribosome so that the next aminoacyl-tRNA molecule can bind. Although the figure shows a large movement of the small ribosome subunit relative to the large subunit, the conformational changes that actually take place in the ribosome during translation are more subtle. It is likely that they involve a series of small rearrangements within each subunit as well as several small shifts between the two subunits. As indicated, the mRNA is translated in the 5′-to-3′ direction, and the N-terminal end of a protein is made first, with each cycle adding one amino acid to the C-terminus of the polypeptide chain. The position at which the growing peptide chain is attached to a tRNA does not change during the elongation cycle: it is always linked to the tRNA present in the P site of the large subunit.

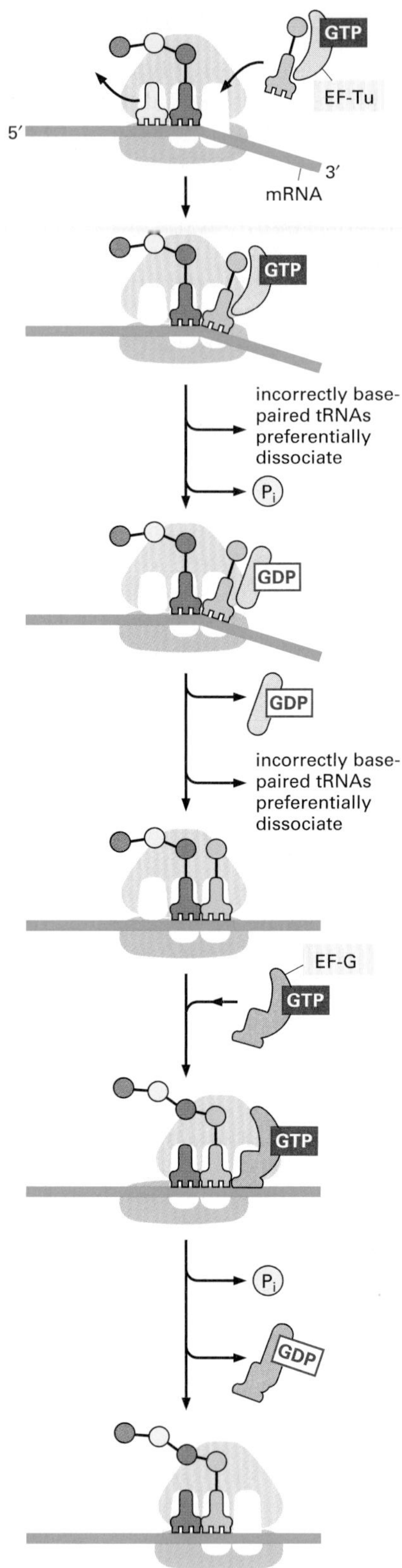

Figure 6–66 Detailed view of the translation cycle. The outline of translation presented in Figure 6–65 has been supplemented with additional features, including the participation of elongation factors and a mechanism by which translational accuracy is improved. In the initial binding event *(top panel)* an aminoacyl-tRNA molecule that is tightly bound to EF-Tu pairs transiently with the codon at the A-site in the small subunit. During this step *(second panel)*, the tRNA occupies a hybrid-binding site on the ribosome. The codon–anticodon pairing triggers GTP hydrolysis by EF-Tu causing it to dissociate from the aminoacyl-tRNA, which now enters the A-site *(fourth panel)* and can participate in chain elongation. A delay between aminoacyl-tRNA binding and its availability for protein synthesis is thereby inserted into the protein synthesis mechanism. As described in the text, this delay increases the accuracy of translation. In subsequent steps, elongation factor EF-G in the GTP-bound form enters the ribosome and binds in or near the A-site on the large ribosomal subunit, accelerating the movement of the two bound tRNAs into the A/P and P/E hybrid states. Contact with the ribosome stimulates the GTPase activity of EF-G, causing a dramatic conformational change in EF-G as it switches from the GTP to the GDP-bound form. This change moves the tRNA bound to the A/P hybrid state to the P-site and advances the cycle of translation forward by one codon.

During each cycle of translation elongation, the tRNAs molecules move through the ribosome in an elaborate series of gyrations during which they transiently occupy several "hybrid" binding states. In one, the tRNA is simultaneously bound to the A site of the small subunit and the P site of the large subunit; in another, the tRNA is bound to the P site of the small subunit and the E site of the large subunit. In a single cycle, a tRNA molecule is considered to occupy six different sites, the initial binding site (called the A/T hybrid state), the A/A site, the A/P hybrid state, the P/P site, the P/E hybrid state, and the E-site. Each tRNA is thought to ratchet through these positions, undergoing rotations along its long axis at each change in location.

EF-Tu and EF-G are the designations used for the bacterial elongation factors; in eucaryotes, they are called EF-1 and EF-2, respectively. The dramatic change in the three-dimensional structure of EF-Tu that is caused by GTP hydrolysis was illustrated in Figure 3–74. For each peptide bond formed, a molecule of EF-Tu and EF-G are each released in their inactive, GDP-bound forms. To be used again, these proteins must have their GDP exchanged for GTP. In the case of EF-Tu, this exchange is performed by a specific member of a large class of proteins known as *GTP exchange factors.*

ribosome without being used for protein synthesis, and this two-step mechanism is largely responsible for the 99.99% accuracy of the ribosome in translating proteins.

Recent discoveries indicate that EF-Tu may have an additional role in raising the overall accuracy of translation. Earlier in this chapter, we discussed the key role of aminoacyl synthetases in accurately matching amino acids to tRNAs. As the GTP-bound form of EF-Tu escorts aminoacyl-tRNAs to the ribosome (see Figure 6–66), it apparently double-checks for the proper correspondence between amino acid and tRNA and rejects those that are mismatched. Exactly how this is accomplished is not well-understood, but it may involve the overall binding energy between EF-Tu and the aminoacyl-tRNA. According to this idea, correct matches have a narrowly defined affinity for EF-Tu, and incorrect matches bind either too strongly or too weakly. EF-Tu thus appears to discriminate, albeit crudely, among many different amino acid-tRNA combinations, selectively allowing only the correct ones to enter the ribosome.

The Ribosome Is a Ribozyme

The ribosome is a very large and complex structure, composed of two-thirds RNA and one-third protein. The determination, in 2000, of the entire three-dimensional structure of its large and small subunits is a major triumph of modern structural biology. The structure strongly confirms the earlier evidence that rRNAs—and not proteins—are responsible for the ribosome's overall structure, its ability to position tRNAs on the mRNA, and its catalytic activity in forming covalent peptide bonds. Thus, for example, the ribosomal RNAs are folded into

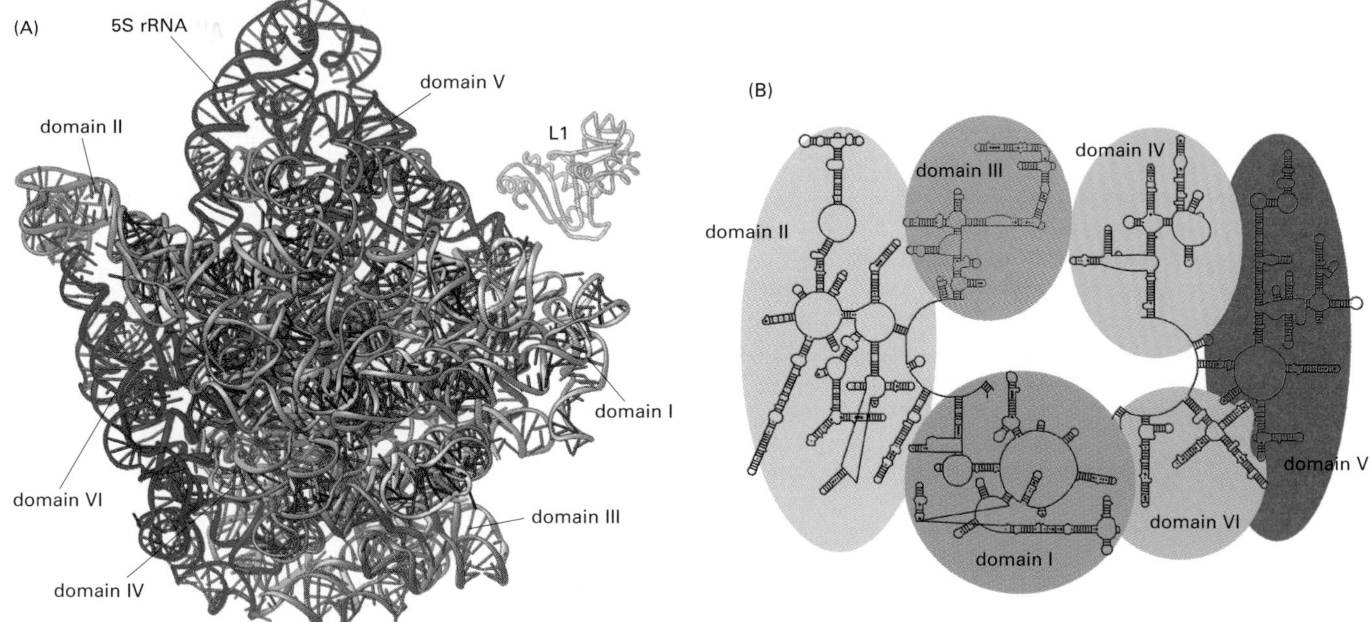

Figure 6–67 Structure of the rRNAs in the large subunit of a bacterial ribosome, as determined by x-ray crystallography. (A) Three-dimensional structures of the large-subunit rRNAs (5S and 23S) as they appear in the ribosome. One of the protein subunits of the ribosome (L1) is also shown as a reference point, since it forms a characteristic protrusion on the ribosome. (B) Schematic diagram of the secondary structure of the 23S rRNA showing the extensive network of base-pairing. The structure has been divided into six structural 'domains' whose colors correspond to those of the three-dimensional structure in (A). The secondary-structure diagram is highly schematized to represent as much of the structure as possible in two dimensions. To do this, several discontinuities in the RNA chain have been introduced, although in reality the 23S RNA is a single RNA molecule. For example, the base of Domain III is continuous with the base of Domain IV even though a gap appears in the diagram. (Adapted from N. Ban et al., *Science* 289:905–920, 2000.)

highly compact, precise three-dimensional structures that form the compact core of the ribosome and thereby determine its overall shape (Figure 6–67).

In marked contrast to the central positions of the rRNA, the ribosomal proteins are generally located on the surface and fill in the gaps and crevices of the folded RNA (Figure 6–68). Some of these proteins contain globular domains on the ribosome surface that send out extended regions of polypeptide chain that penetrate short distances into holes in the RNA core (Figure 6–69). The main role

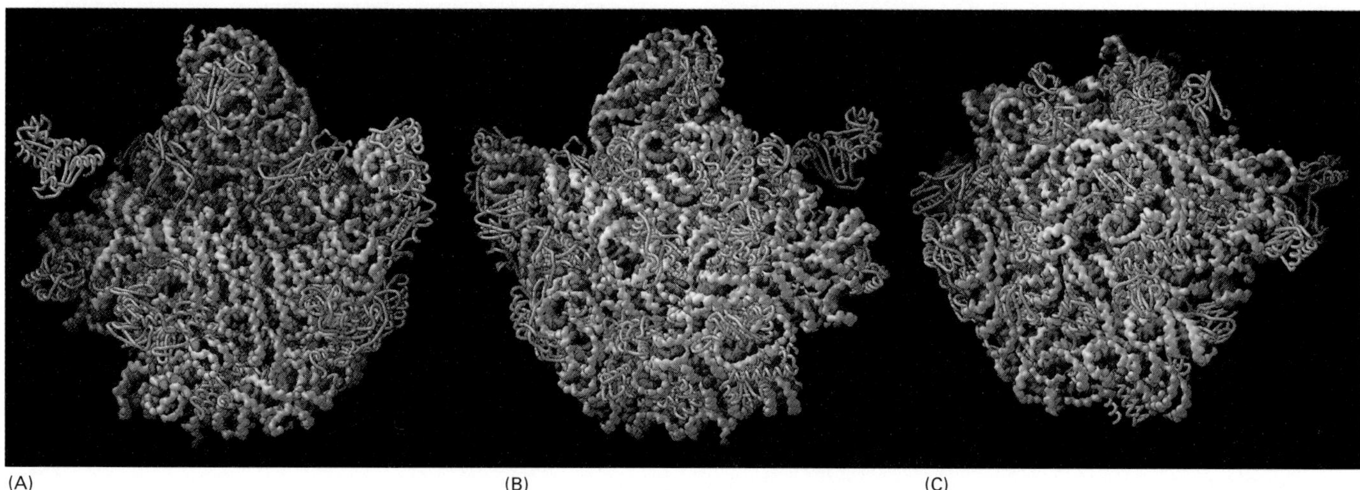

Figure 6–68 Location of the protein components of the bacterial large ribosomal subunit. The rRNAs (5S and 23S) are depicted in *gray* and the large-subunit proteins (27 of the 31 total) in *gold*. For convenience, the protein structures depict only the polypeptide backbones. (A) View of the interface with the small subunit, the same view shown in Figure 6–64B. (B) View of the back of the large subunit, obtained by rotating (A) by 180° around a vertical axis. (C) View of the bottom of the large subunit showing the peptide exit channel in the center of the structure. (From N. Ban et al., *Science* 289:905–920, 2000. © AAAS.)

of the ribosomal proteins seems to be to stabilize the RNA core, while permitting the changes in rRNA conformation that are necessary for this RNA to catalyze efficient protein synthesis.

Not only are the three binding sites for tRNAs (the A-, P-, and E-sites) on the ribosome formed primarily by the ribosomal RNAs, but the catalytic site for peptide bond formation is clearly formed by the 23S RNA, with the nearest amino acid located more than 1.8 nm away. This RNA-based catalytic site for peptidyl transferase is similar in many respects to those found in some proteins; it is a highly structured pocket that precisely orients the two reactants (the growing peptide chain and an aminoacyl-tRNA), and it provides a functional group to act as a general acid–base catalyst—in this case apparently, a ring nitrogen of adenine, instead of an amino acid side chain such as histidine (Figure 6–70). The ability of an RNA molecule to act as such a catalyst was initially surprising because RNA was thought to lack an appropriate chemical group that could both accept and donate a proton. Although the p*K* of adenine-ring nitrogens is usually around 3.5, the three-dimensional structure and charge distribution of the 23S rRNA active site force the p*K* of this apparently critical adenine into the neutral range and thereby create the enzymatic activity.

RNA molecules that possess catalytic activity are known as **ribozymes**. We saw earlier in this chapter how other ribozymes function in RNA-splicing reactions (for example, see Figure 6–36). In the final section of this chapter, we consider what the recently recognized ability of RNA molecules to function as catalysts for a wide variety of different reactions might mean for the early evolution of living cells. Here we need only note that there is good reason to suspect that RNA rather than protein molecules served as the first catalysts for living cells. If so, the ribosome, with its RNA core, might be viewed as a relic of an earlier time in life's history—when protein synthesis evolved in cells that were run almost entirely by ribozymes.

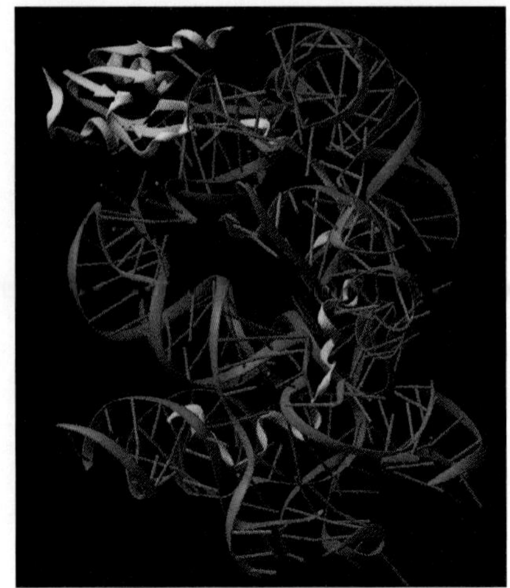

Figure 6–69 Structure of the L15 protein in the large subunit of the bacterial ribosome. The globular domain of the protein lies on the surface of the ribosome and an extended region penetrates deeply into the RNA core of the ribosome. The L15 protein is shown in *yellow* and a portion of the ribosomal RNA core is shown in *red*. (Courtesy of D. Klein, P.B. Moore and T.A. Steitz.)

Nucleotide Sequences in mRNA Signal Where to Start Protein Synthesis

The initiation and termination of translation occur through variations on the translation elongation cycle described above. The site at which protein synthesis begins on the mRNA is especially crucial, since it sets the reading frame for the whole length of the message. An error of one nucleotide either way at this stage would cause every subsequent codon in the message to be misread, so that a nonfunctional protein with a garbled sequence of amino acids would result. The initiation step is also of great importance in another respect, since for most genes it is the last point at which the cell can decide whether the mRNA is to be translated and the protein synthesized; the rate of initiation thus determines the rate at which the protein is synthesized. We shall see in Chapter 7 that cells use several mechanisms to regulate translation initiation.

The translation of an mRNA begins with the codon AUG, and a special tRNA is required to initiate translation. This **initiator tRNA** always carries the amino acid methionine (in bacteria, a modified form of methionine—formylmethionine—is used) so that all newly made proteins have methionine as the first amino acid at their N-terminal end, the end of a protein that is synthesized first. This

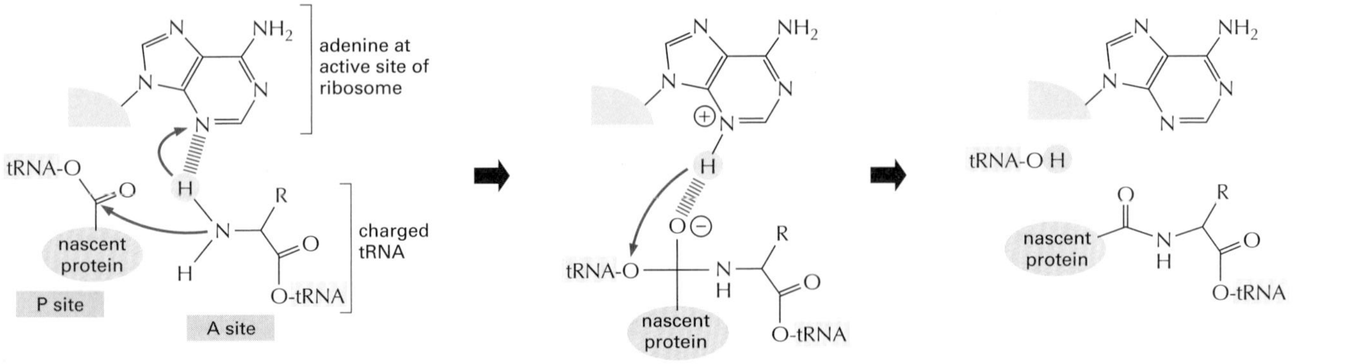

Figure 6–70 A possible reaction mechanism for the peptidyl transferase activity present in the large ribosomal subunit. The overall reaction is catalyzed by an active site in the 23S rRNA. In the first step of the proposed mechanism, the N3 of the active-site adenine abstracts a proton from the amino acid attached to the tRNA at the ribosome's A-site, allowing its amino nitrogen to attack the carboxyl group at the end of the growing peptide chain. In the next step this protonated adenine donates its hydrogen to the oxygen linked to the peptidyl-tRNA, causing this tRNA's release from the peptide chain. This leaves a polypeptide chain that is one amino acid longer than the starting reactants. The entire reaction cycle would then repeat with the next aminoacyl tRNA that enters the A-site. (Adapted from P. Nissen et al., *Science* 289:920–930, 2000.)

adenine at active site of ribosome

charged tRNA

tRNA-O

nascent protein

P site

A site

O-tRNA

tRNA-O H

Figure 6–71 The initiation phase of protein synthesis in eucaryotes. Only three of the many translation initiation factors required for this process are shown. Efficient translation initiation also requires the poly-A tail of the mRNA bound by poly-A-binding proteins which, in turn, interact with eIF4G. In this way, the translation apparatus ascertains that both ends of the mRNA are intact before initiating (see Figure 6–40). Although only one GTP hydrolysis event is shown in the figure, a second is known to occur just before the large and small ribosomal subunits join.

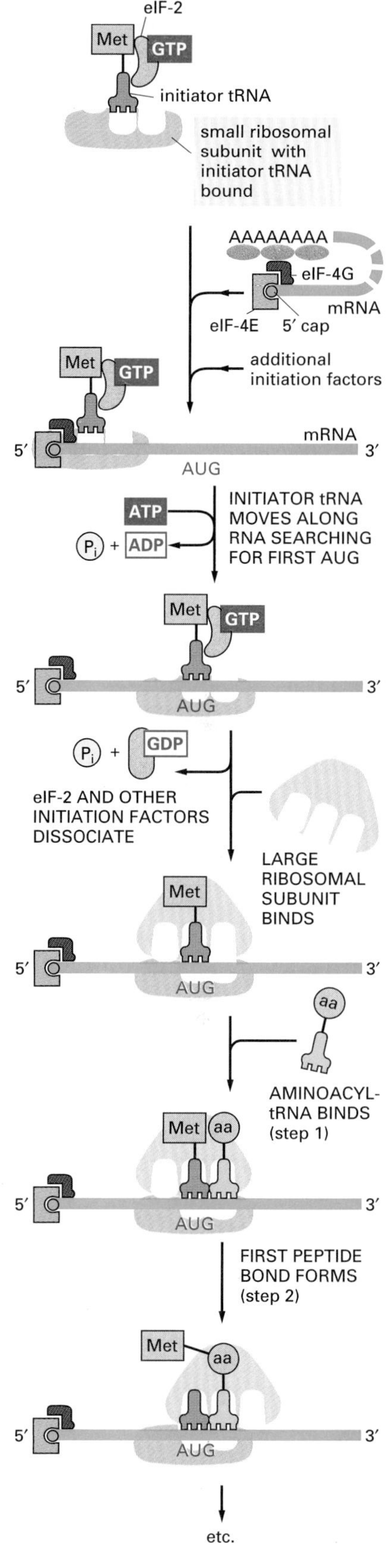

methionine is usually removed later by a specific protease. The initiator tRNA has a nucleotide sequence distinct from that of the tRNA that normally carries methionine.

In eucaryotes, the initiator tRNA (which is coupled to methionine) is first loaded into the small ribosomal subunit along with additional proteins called **eucaryotic initiation factors**, or **eIFs** (Figure 6–71). Of all the aminoacyl tRNAs in the cell, only the methionine-charged initiator tRNA is capable of tightly binding the small ribosome subunit without the complete ribosome present. Next, the small ribosomal subunit binds to the 5′ end of an mRNA molecule, which is recognized by virtue of its 5′ cap and its two bound initiation factors, eIF4E (which directly binds the cap) and eIF4G (see Figure 6–40). The small ribosomal subunit then moves forward (5′ to 3′) along the mRNA, searching for the first AUG. This movement is facilitated by additional initiation factors that act as ATP-powered helicases, allowing the small subunit to scan through RNA secondary structure. In 90% of mRNAs, translation begins at the first AUG encountered by the small subunit. At this point, the initiation factors dissociate from the small ribosomal subunit to make way for the large ribosomal subunit to assemble with it and complete the ribosome. The initiator tRNA is now bound to the P-site, leaving the A-site vacant. Protein synthesis is therefore ready to begin with the addition of the next aminoacyl tRNA molecule (see Figure 6–71).

The nucleotides immediately surrounding the start site in eucaryotic mRNAs influence the efficiency of AUG recognition during the above scanning process. If this recognition site is quite different from the consensus recognition sequence, scanning ribosomal subunits will sometimes ignore the first AUG codon in the mRNA and skip to the second or third AUG codon instead. Cells frequently use this phenomenon, known as “leaky scanning,” to produce two or more proteins, differing in their N-termini, from the same mRNA molecule. It allows some genes to produce the same protein with and without a signal sequence attached at its N-terminus, for example, so that the protein is directed to two different compartments in the cell.

The mechanism for selecting a start codon in bacteria is different. Bacterial mRNAs have no 5′ caps to tell the ribosome where to begin searching for the start of translation. Instead, each bacterial mRNA contains a specific ribosome-binding site (called the Shine–Dalgarno sequence, named after its discoverers) that is located a few nucleotides upstream of the AUG at which translation is to begin. This nucleotide sequence, with the consensus 5′-AGGAGGU-3′, forms base pairs with the 16S rRNA of the small ribosomal subunit to position the initiating AUG codon in the ribosome. A set of translation initiation factors orchestrates this interaction, as well as the subsequent assembly of the large ribosomal subunit to complete the ribosome.

Unlike a eucaryotic ribosome, a bacterial ribosome can therefore readily assemble directly on a start codon that lies in the interior of an mRNA molecule, so long as a ribosome-binding site precedes it by several nucleotides. As a result, bacterial mRNAs are often *polycistronic*—that is, they encode several different proteins, each of which is translated from the same mRNA molecule (Figure 6–72). In contrast, a eucaryotic mRNA generally encodes only a single protein.

Stop Codons Mark the End of Translation

The end of the protein-coding message is signaled by the presence of one of three codons (UAA, UAG, or UGA) called *stop codons* (see Figure 6–50). These are not recognized by a tRNA and do not specify an amino acid, but instead signal

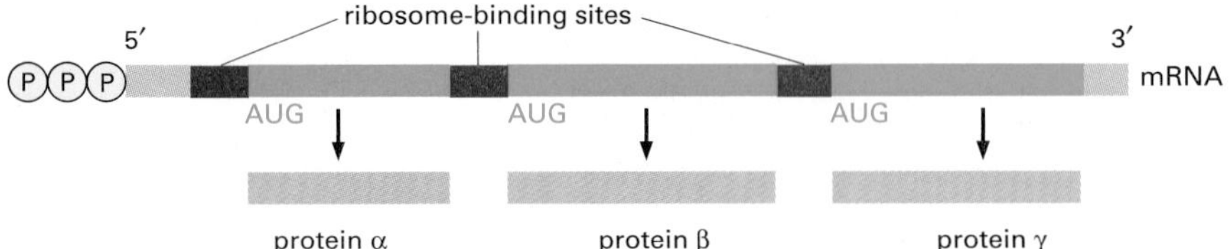

Figure 6–72 Structure of a typical bacterial mRNA molecule. Unlike eucaryotic ribosomes, which typically require a capped 5′ end, procaryotic ribosomes initiate transcription at ribosome-binding sites (Shine–Dalgarno sequences), which can be located anywhere along an mRNA molecule. This property of ribosomes permits bacteria to synthesize more than one type of protein from a single mRNA molecule.

to the ribosome to stop translation. Proteins known as *release factors* bind to any ribosome with a stop codon positioned in the A site, and this binding forces the peptidyl transferase in the ribosome to catalyze the addition of a water molecule instead of an amino acid to the peptidyl-tRNA (Figure 6–73). This reaction frees the carboxyl end of the growing polypeptide chain from its attachment to a tRNA molecule, and since only this attachment normally holds the growing polypeptide to the ribosome, the completed protein chain is immediately released into the cytoplasm. The ribosome then releases the mRNA and separates into the large and small subunits, which can assemble on another mRNA molecule to begin a new round of protein synthesis.

Release factors provide a dramatic example of *molecular mimicry*, whereby one type of macromolecule resembles the shape of a chemically unrelated molecule. In this case, the three-dimensional structure of release factors (made entirely of protein) bears an uncanny resemblance to the shape and charge distribution of a tRNA molecule (Figure 6–74). This shape and charge mimicry allows the release factor to enter the A-site on the ribosome and cause translation termination.

During translation, the nascent polypeptide moves through a large, water-filled tunnel (approximately 10 nm × 1.5 nm) in the large subunit of the ribosome (see Figure 6–68C). The walls of this tunnel, made primarily of 23S rRNA, are a patchwork of tiny hydrophobic surfaces embedded in a more extensive hydrophilic surface. This structure, because it is not complementary to any peptide structure, provides a "Teflon" coating through which a polypeptide chain can easily slide. The dimensions of the tunnel suggest that nascent proteins are largely unstructured as they pass through the ribosome, although some α-helical regions of the protein can form before leaving the ribosome tunnel. As it leaves the ribosome, a newly-synthesized protein must fold into its proper three-dimensional structure to be useful to the cell, and later in this chapter we discuss how this folding occurs. First, however, we review several additional aspects of the translation process itself.

Proteins Are Made on Polyribosomes

The synthesis of most protein molecules takes between 20 seconds and several minutes. But even during this very short period, multiple initiations usually take place on each mRNA molecule being translated. As soon as the preceding ribosome has translated enough of the nucleotide sequence to move out of the way, the 5′ end of the mRNA is threaded into a new ribosome. The mRNA molecules being translated are therefore usually found in the form of *polyribosomes* (also known as *polysomes*), large cytoplasmic assemblies made up of several ribosomes spaced as close as 80 nucleotides apart along a single mRNA molecule (Figure 6–75). These multiple initiations mean that many more protein

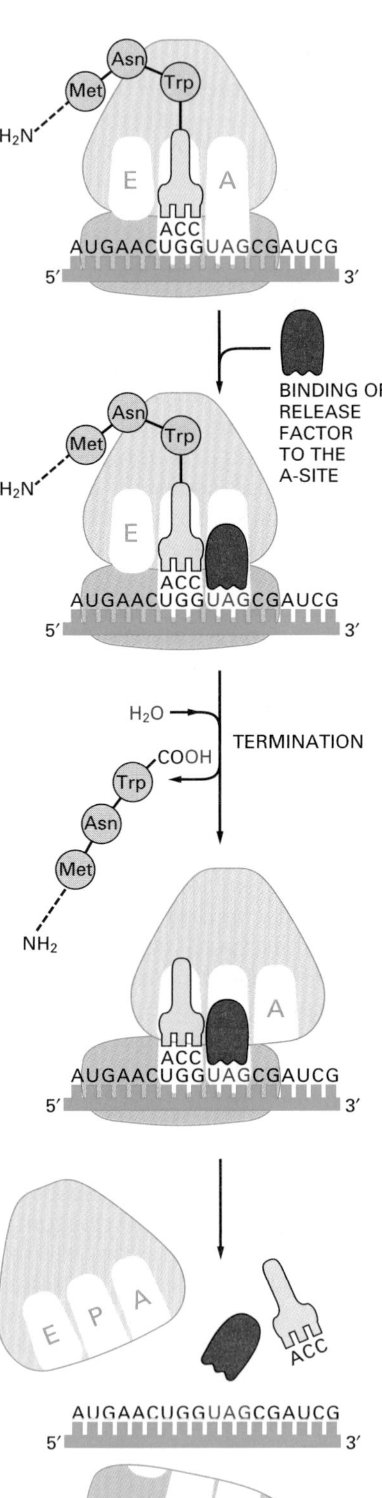

Figure 6–73 The final phase of protein synthesis. The binding of a release factor to an A-site bearing a stop codon terminates translation. The completed polypeptide is released and, after the action of a *ribosome recycling factor* (not shown), the ribosome dissociates into its two separate subunits.

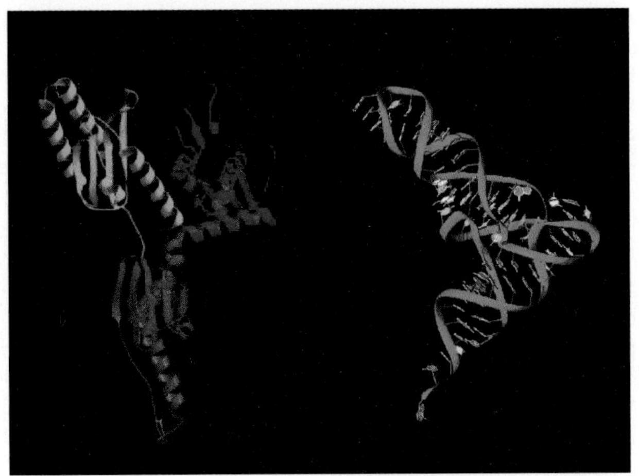

Figure 6–74 The structure of a human translation release factor (eRF1) and its resemblance to a tRNA molecule. The protein is on the *left* and the tRNA on the *right*. (From H. Song et al., *Cell* 100:311–321, 2000. © Elsevier.)

molecules can be made in a given time than would be possible if each had to be completed before the next could start.

Both bacteria and eucaryotes utilize polysomes, and both employ additional strategies to speed up the rate of protein synthesis even further. Because bacterial mRNA does not need to be processed and is accessible to ribosomes while it is being made, ribosomes can attach to the free end of a bacterial mRNA molecule and start translating it even before the transcription of that RNA is complete, following closely behind the RNA polymerase as it moves along DNA. In eucaryotes, as we have seen, the 5′ and 3′ ends of the mRNA interact (see Figures 6–40 and 6–75A); therefore, as soon as a ribosome dissociates, its two subunits are in an optimal position to reinitiate translation on the same mRNA molecule.

Quality-Control Mechanisms Operate at Many Stages of Translation

Translation by the ribosome is a compromise between the opposing constraints of accuracy and speed. We have seen, for example, that the accuracy of translation (1 mistake per 10^4 amino acids joined) requires a time delay each time a new amino acid is added to a growing polypeptide chain, producing an overall

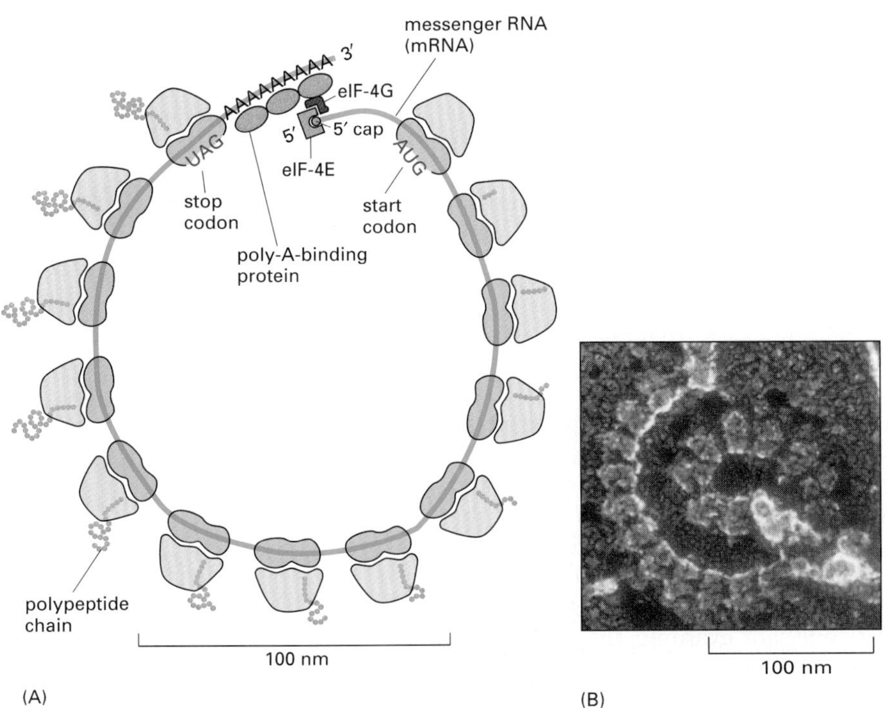

Figure 6–75 A polyribosome. (A) Schematic drawing showing how a series of ribosomes can simultaneously translate the same eucaryotic mRNA molecule. (B) Electron micrograph of a polyribosome from a eucaryotic cell. (B, courtesy of John Heuser.)

Figure 6–76 The rescue of a bacterial ribosome stalled on an incomplete mRNA molecule. The tmRNA shown is a 363-nucleotide RNA with both tRNA and mRNA functions, hence its name. It carries an alanine and can enter the vacant A-site of a stalled ribosome to add this alanine to a polypeptide chain, mimicking a tRNA except that no codon is present to guide it. The ribosome then translates ten codons from the tmRNA, completing an 11-amino acid tag on the protein. This tag is recognized by proteases that then degrade the entire protein.

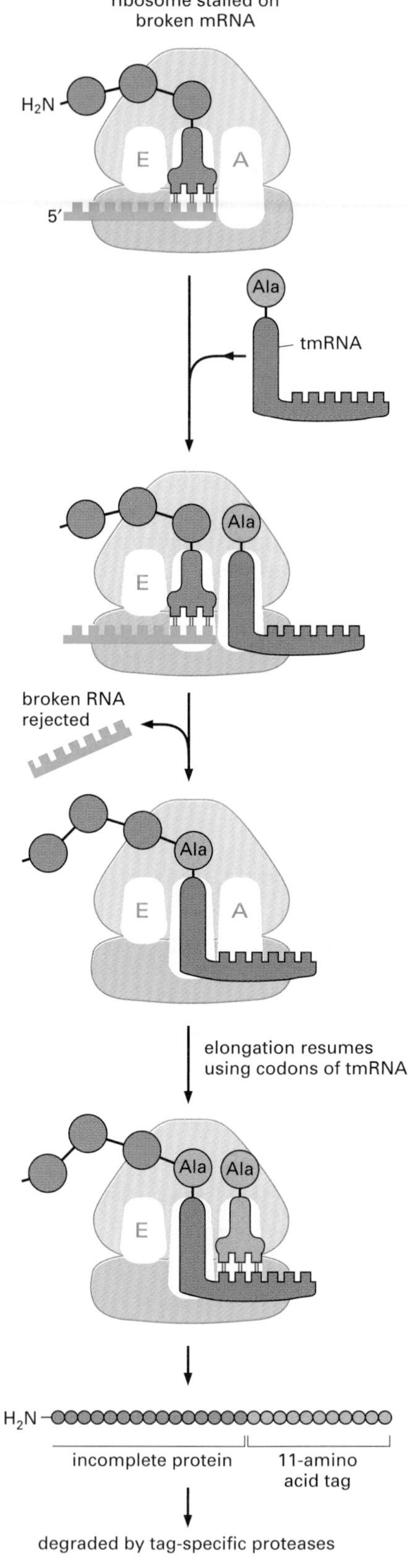

speed of translation of 20 amino acids incorporated per second in bacteria. Mutant bacteria with a specific alteration in their small ribosomal subunit translate mRNA into protein with an accuracy considerably higher than this; however, protein synthesis is so slow in these mutants that the bacteria are barely able to survive.

We have also seen that attaining the observed accuracy of protein synthesis requires the expenditure of a great deal of free energy; this is expected, since, as discussed in Chapter 2, a price must be paid for any increase in order in the cell. In most cells, protein synthesis consumes more energy than any other biosynthetic process. At least four high-energy phosphate bonds are split to make each new peptide bond: two are consumed in charging a tRNA molecule with an amino acid (see Figure 6–56), and two more drive steps in the cycle of reactions occurring on the ribosome during synthesis itself (see Figure 6–66). In addition, extra energy is consumed each time that an incorrect amino acid linkage is hydrolyzed by a tRNA synthetase (see Figure 6–59) and each time that an incorrect tRNA enters the ribosome, triggers GTP hydrolysis, and is rejected (Figure 6–66). To be effective, these proofreading mechanisms must also remove an appreciable fraction of correct interactions; for this reason they are even more costly in energy than they might seem.

Other quality control mechanisms ensure that a eucaryotic mRNA molecule is complete before ribosomes even begin to translate it. Translating broken or partly processed mRNAs would be harmful to the cell, because truncated or otherwise aberrant proteins would be produced. In eucaryotes, we have seen that mRNA production involves not only transcription but also a series of elaborate RNA-processing steps; these take place in the nucleus, segregated from ribosomes, and only when the processing is complete are the mRNAs transported to the cytoplasm to be translated (see Figure 6–40). An mRNA molecule that was intact when it left the nucleus can, however, become broken in the cytosol. To avoid translating such broken mRNA molecules, the 5′ cap and the poly-A tail are both recognized by the translation-initiation apparatus before translation begins (see Figures 6–71 and 6–75).

Bacteria solve the problem of incomplete mRNAs in an entirely different way. Not only are there no signals at the 3′ ends of bacterial mRNAs, but also, as we have seen, translation often begins before the synthesis of the transcript has been completed. When the bacterial ribosome translates to the end of an incomplete RNA, a special RNA (called *tmRNA*) enters the A-site of the ribosome and is itself translated; this adds a special 11 amino acid tag to the C terminus of the truncated protein that signals to proteases that the entire protein is to be degraded (Figure 6–76).

There Are Minor Variations in the Standard Genetic Code

As discussed in Chapter 1, the genetic code (shown in Figure 6–50) applies to all three major branches of life, providing important evidence for the common ancestry of all life on Earth. Although rare, there are exceptions to this code, and we discuss some of them in this section. For example, *Candida albicans*, the most prevalent human fungal pathogen, translates the codon CUG as serine, whereas nearly all other organisms translate it as leucine. Mitochondria (which have their own genomes and encode much of their translational apparatus) also show several deviations from the standard code. For example, in mammalian mitochondria AUA is translated as methionine, whereas in the cytosol of the cell it is translated as isoleucine (see Table 14–3, p. 352).

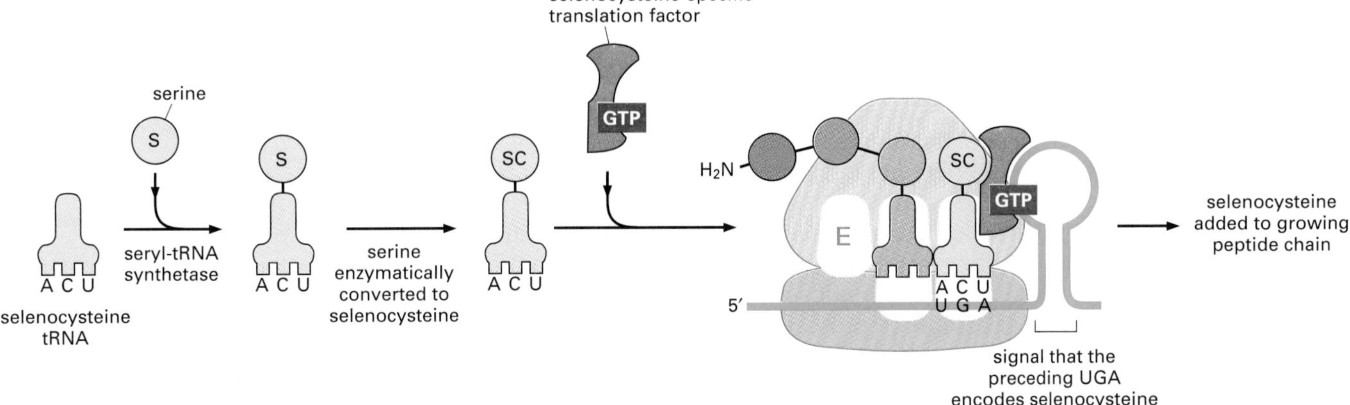

Figure 6–77 Incorporation of selenocysteine into a growing polypeptide chain. A specialized tRNA is charged with serine by the normal seryl-tRNA synthetase, and the serine is subsequently converted enzymatically to selenocysteine. A specific RNA structure in the mRNA (a stem and loop structure with a particular nucleotide sequence) signals that selenocysteine is to be inserted at the neighboring UGA codon. As indicated, this event requires the participation of a selenocysteine-specific translation factor.

The type of deviation in the genetic code discussed above is "hardwired" into the organisms or the organelles in which it occurs. A different type of variation, sometimes called *translational recoding,* occurs in many cells. In this case, other nucleotide sequence information present in an mRNA can change the meaning of the genetic code at a particular site in the mRNA molecule. The standard code allows cells to manufacture proteins using only 20 amino acids. However, bacteria, archaea, and eucaryotes have available to them a twenty-first amino acid that can be incorporated directly into a growing polypeptide chain through translational recoding. Selenocysteine, which is essential for the efficient function of a variety of enzymes, contains a selenium atom in place of the sulfur atom of cysteine. Selenocysteine is produced from a serine attached to a special tRNA molecule that base-pairs with the UGA codon, a codon normally used to signal a translation stop. The mRNAs for proteins in which selenocysteine is to be inserted at a UGA codon carry an additional nucleotide sequence in the mRNA nearby that causes this recoding event (Figure 6–77).

Another form of recoding is *translational frameshifting.* This type of recoding is commonly used by retroviruses, a large group of eucaryotic viruses, in which it allows more than one protein to be synthesized from a single mRNA. These viruses commonly make both the capsid proteins *(Gag proteins)* and the viral reverse transcriptase and integrase *(Pol proteins)* from the same RNA transcript (see Figure 5–73). Such a virus needs many more copies of the Gag proteins than it does of the Pol proteins, and they achieve this quantitative adjustment by encoding the *pol* genes just after the *gag* genes but in a different reading frame. A stop codon at the end of the *gag* coding sequence can be bypassed on occasion by an intentional translational frameshift that occurs upstream of it. This frameshift occurs at a particular codon in the mRNA and requires a specific *recoding signal,* which seems to be a structural feature of the RNA sequence downstream of this site (Figure 6–78).

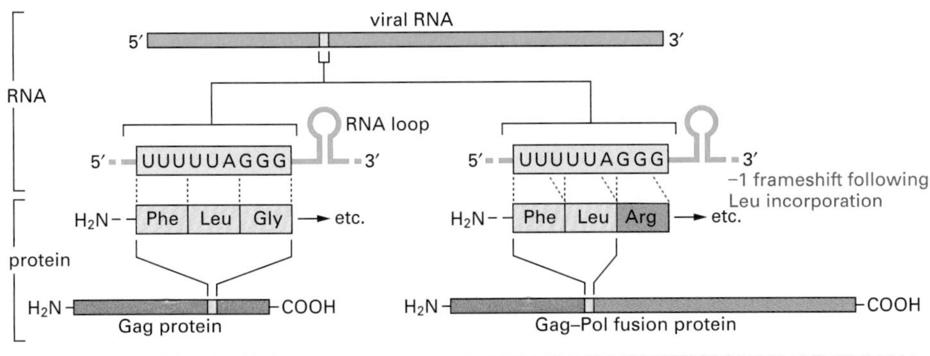

Figure 6–78 The translational frameshifting that produces the reverse transcriptase and integrase of a retrovirus. The viral reverse transcriptase and integrase are produced by proteolytic processing of a large protein (the Gag–Pol fusion protein) consisting of both the Gag and Pol amino acid sequences. The viral capsid proteins are produced by proteolytic processing of the more abundant Gag protein. Both the Gag and the Gag–Pol fusion proteins start identically, but the Gag protein terminates at an in-frame stop codon (not shown); the indicated frameshift bypasses this stop codon, allowing the synthesis of the longer Gag–Pol fusion protein. The frameshift occurs because features in the local RNA structure (including the RNA loop shown) cause the $tRNA^{Leu}$ attached to the C-terminus of the growing polypeptide chain occasionally to slip backward by one nucleotide on the ribosome, so that it pairs with a UUU codon instead of the UUA codon that had initially specified its incorporation; the next codon (AGG) in the new reading frame specifies an arginine rather than a glycine. This controlled slippage is due in part to a stem and loop structure that forms in the viral mRNA, as indicated in the figure. The sequence shown is from the human AIDS virus, HIV. (Adapted from T. Jacks et al., *Nature* 331:280–283, 1988.)

Many Inhibitors of Procaryotic Protein Synthesis Are Useful as Antibiotics

Many of the most effective antibiotics used in modern medicine are compounds made by fungi that act by inhibiting bacterial protein synthesis. Some of these drugs exploit the structural and functional differences between bacterial and eucaryotic ribosomes so as to interfere preferentially with the function of bacterial ribosomes. Thus some of these compounds can be taken in high doses without undue toxicity to humans. Because different antibiotics bind to different regions of bacterial ribosomes, they often inhibit different steps in the synthetic process. Some of the more common antibiotics of this kind are listed in Table 6–3 along with several other inhibitors of protein synthesis, some of which act on eucaryotic cells and therefore cannot be used as antibiotics.

Because they block specific steps in the processes that lead from DNA to protein, many of the compounds listed in Table 6–3 are useful for cell biological studies. Among the most commonly used drugs in such experimental studies are *chloramphenicol, cycloheximide,* and *puromycin,* all of which specifically inhibit protein synthesis. In a eucaryotic cell, for example, chloramphenicol inhibits protein synthesis on ribosomes only in mitochondria (and in chloroplasts in plants), presumably reflecting the procaryotic origins of these organelles (discussed in Chapter 14). Cycloheximide, in contrast, affects only ribosomes in the cytosol. Puromycin is especially interesting because it is a structural analog of a tRNA molecule linked to an amino acid and is therefore another example of molecular mimicry; the ribosome mistakes it for an authentic amino acid and covalently incorporates it at the C-terminus of the growing peptide chain, thereby causing the premature termination and release of the polypeptide. As might be expected, puromycin inhibits protein synthesis in both procaryotes and eucaryotes.

Having described the translation process itself, we now discuss how its products—the proteins of the cell—fold into their correct three-dimensional conformations.

A Protein Begins to Fold While It Is Still Being Synthesized

The process of gene expression is not over when the genetic code has been used to create the sequence of amino acids that constitutes a protein. To be useful to the cell, this new polypeptide chain must fold up into its unique three-dimensional

TABLE 6–3 Inhibitors of Protein or RNA Synthesis

INHIBITOR	SPECIFIC EFFECT
Acting only on bacteria	
Tetracycline	blocks binding of aminoacyl-tRNA to A-site of ribosome
Streptomycin	prevents the transition from initiation complex to chain-elongating ribosome and also causes miscoding
Chloramphenicol	blocks the peptidyl transferase reaction on ribosomes (step 2 in Figure 6–65)
Erythromycin	blocks the translocation reaction on ribosomes (step 3 in Figure 6–65)
Rifamycin	blocks initiation of RNA chains by binding to RNA polymerase (prevents RNA synthesis)
Acting on bacteria and eucaryotes	
Puromycin	causes the premature release of nascent polypeptide chains by its addition to growing chain end
Actinomycin D	binds to DNA and blocks the movement of RNA polymerase (prevents RNA synthesis)
Acting on eucaryotes but not bacteria	
Cycloheximide	blocks the translocation reaction on ribosomes (step 3 in Figure 6–65)
Anisomycin	blocks the peptidyl transferase reaction on ribosomes (step 2 in Figure 6–65)
α-Amanitin	blocks mRNA synthesis by binding preferentially to RNA polymerase II

The ribosomes of eucaryotic mitochondria (and chloroplasts) often resemble those of bacteria in their sensitivity to inhibitors. Therefore, some of these antibiotics can have a deleterious effect on human mitochondria.

conformation, bind any small-molecule cofactors required for its activity, be appropriately modified by protein kinases or other protein-modifying enzymes, and assemble correctly with the other protein subunits with which it functions (Figure 6–79).

The information needed for all of the protein maturation steps listed above is ultimately contained in the sequence of linked amino acids that the ribosome produces when it translates an mRNA molecule into a polypeptide chain. As discussed in Chapter 3, when a protein folds into a compact structure, it buries most of its hydrophobic residues in an interior core. In addition, large numbers of noncovalent interactions form between various parts of the molecule. It is the sum of all of these energetically favorable arrangements that determines the final folding pattern of the polypeptide chain—as the conformation of lowest free energy (see p. 134).

Through many millions of years of evolutionary time, the amino acid sequence of each protein has been selected not only for the conformation that it adopts but also for an ability to fold rapidly, as its polypeptide chain spins out of the ribosome starting from the N-terminal end. Experiments have demonstrated that once a protein domain in a multi-domain protein emerges from the ribosome, it forms a compact structure within a few seconds that contains most of the final secondary structure (α helices and β sheets) aligned in roughly the right way (Figure 6–80). For many protein domains, this unusually open and flexible structure, which is called a *molten globule,* is the starting point for a relatively slow process in which many side-chain adjustments occur that eventually form the correct tertiary structure. Nevertheless, because it takes several minutes to synthesize a protein of average size, a great deal of the folding process is complete by the time the ribosome releases the C-terminal end of a protein (Figure 6–81).

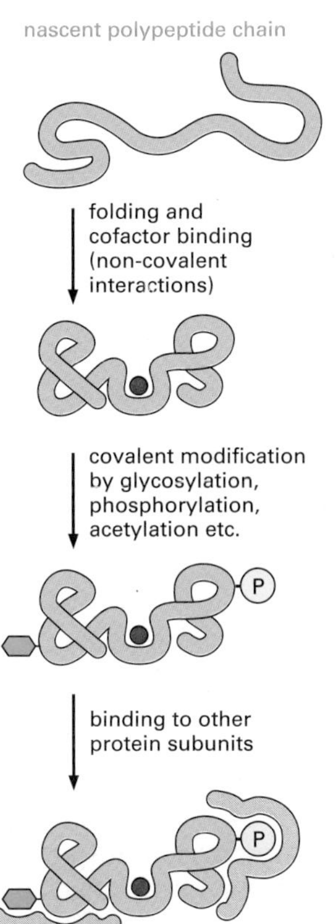

Figure 6–79 Steps in the creation of a functional protein. As indicated, translation of an mRNA sequence into an amino acid sequence on the ribosome is not the end of the process of forming a protein. To be useful to the cell, the completed polypeptide chain must fold correctly into its three-dimensional conformation, bind any cofactors required, and assemble with its partner protein chains (if any). These changes are driven by noncovalent bond formation. As indicated, many proteins also have covalent modifications made to selected amino acids. Although the most frequent of these are protein glycosylation and protein phosphorylation, more than 100 different types of covalent modifications are known (see, for example, Figure 4–35).

Molecular Chaperones Help Guide the Folding of Many Proteins

The folding of many proteins is made more efficient by a special class of proteins called **molecular chaperones**. The latter proteins are useful for cells because there are a variety of different paths that can be taken to convert the molten globule form of a protein to the protein's final compact conformation. For many proteins, some of the intermediates formed along the way would aggregate and be left as off-pathway dead ends without the intervention of a chaperone that resets the folding process (Figure 6–82).

Molecular chaperones were first identified in bacteria when *E. coli* mutants that failed to allow bacteriophage lambda to replicate in them were studied.

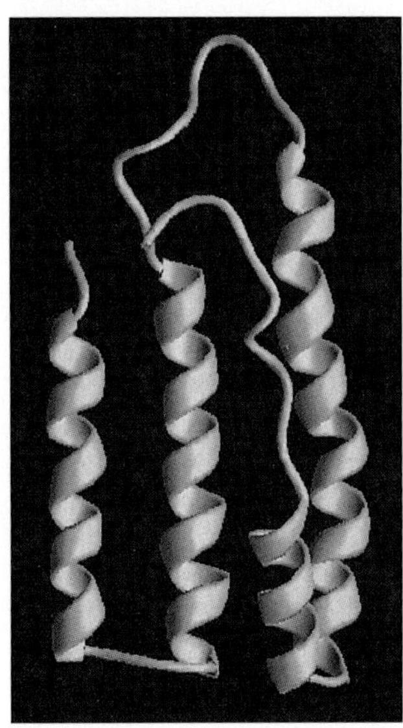

Figure 6–80 The structure of a molten globule. (A) A molten globule form of cytochrome b_{562} is more open and less highly ordered than the final folded form of the protein, shown in (B). Note that the molten globule contains most of the secondary structure of the final form, although the ends of the α helices are frayed and one of the helices is only partly formed. (Courtesy of Joshua Wand, from Y. Feng et al., *Nat. Struct. Biol.* 1:30–35, 1994.)

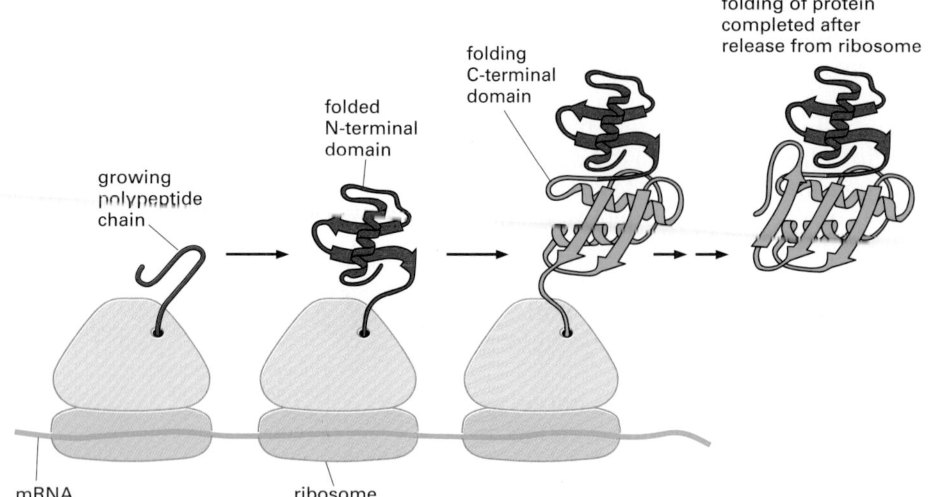

Figure 6–81 The co-translational folding of a protein. A growing polypeptide chain is shown acquiring its secondary and tertiary structure as it emerges from a ribosome. The N-terminal domain folds first, while the C-terminal domain is still being synthesized. In this case, the protein has not yet achieved its final conformation by the time it is released from the ribosome. (Modified from A.N. Federov and T.O. Baldwin, *J. Biol. Chem.* 272:32715–32718, 1997.)

These mutant cells produce slightly altered versions of the chaperone machinery, and as a result they are defective in specific steps in the assembly of the viral proteins. The molecular chaperones are included among the *heat-shock proteins* (hence their designation as *hsp),* because they are synthesized in dramatically increased amounts after a brief exposure of cells to an elevated temperature (for example, 42°C for cells that normally live at 37°C). This reflects the operation of a feedback system that responds to any increase in misfolded proteins (such as those produced by elevated temperatures) by boosting the synthesis of the chaperones that help these proteins refold.

Eucaryotic cells have at least two major families of molecular chaperones—known as the hsp60 and hsp70 proteins. Different family members function in different organelles. Thus, as discussed in Chapter 12, mitochondria contain their own hsp60 and hsp70 molecules that are distinct from those that function in the cytosol, and a special hsp70 (called *BIP*) helps to fold proteins in the endoplasmic reticulum.

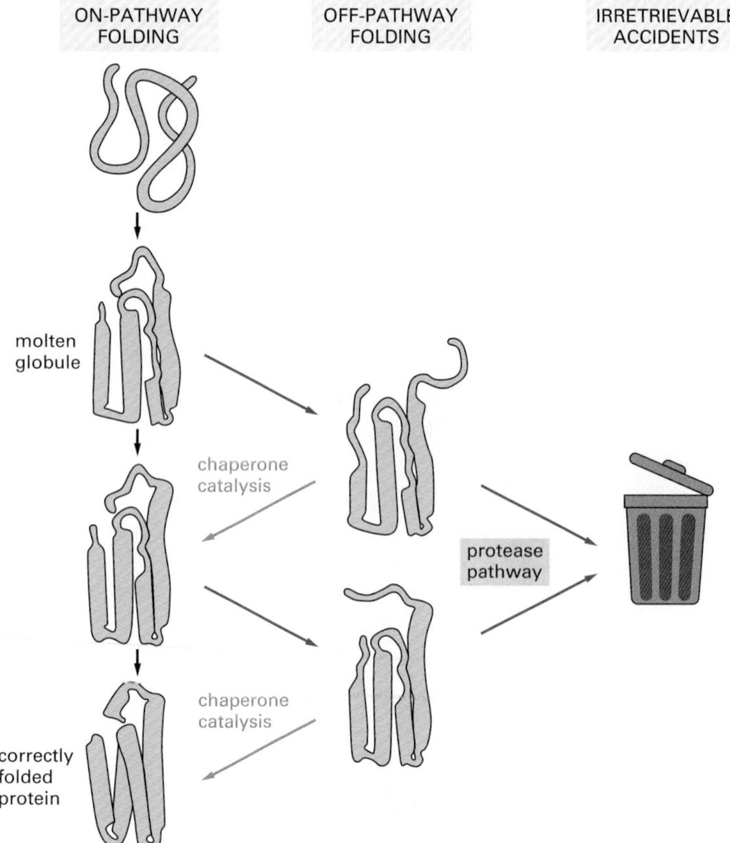

Figure 6–82 A current view of protein folding. Each domain of a newly synthesized protein rapidly attains a "molten globule" state. Subsequent folding occurs more slowly and by multiple pathways, often involving the help of a molecular chaperone. Some molecules may still fail to fold correctly; as explained shortly, these are recognized and degraded by specific proteases.

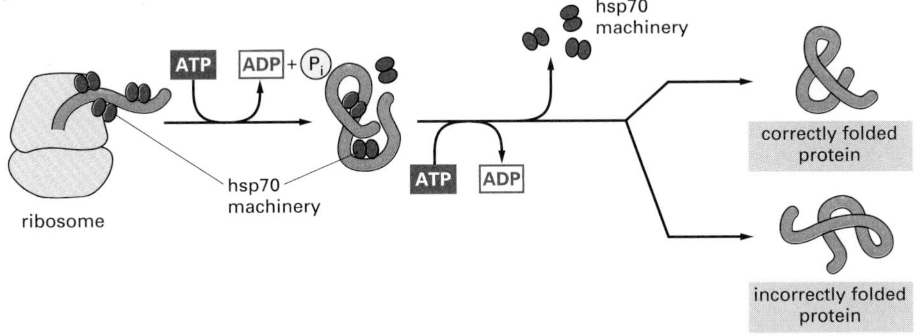

Figure 6–83 The hsp70 family of molecular chaperones. These proteins act early, recognizing a small stretch of hydrophobic amino acids on a protein's surface. Aided by a set of smaller hsp40 proteins, an hsp70 monomer binds to its target protein and then hydrolyzes a molecule of ATP to ADP, undergoing a conformational change that causes the hsp70 to clamp down very tightly on the target. After the hsp40 dissociates, the dissociation of the hsp70 protein is induced by the rapid re-binding of ATP after ADP release. Repeated cycles of hsp protein binding and release help the target protein to refold, as schematically illustrated in Figure 6–82.

The hsp60-like and hsp70 proteins each work with their own small set of associated proteins when they help other proteins to fold. They share an affinity for the exposed hydrophobic patches on incompletely folded proteins, and they hydrolyze ATP, often binding and releasing their protein with each cycle of ATP hydrolysis. In other respects, the two types of hsp proteins function differently. The hsp70 machinery acts early in the life of many proteins, binding to a string of about seven hydrophobic amino acids before the protein leaves the ribosome (Figure 6–83). In contrast, hsp60-like proteins form a large barrel-shaped structure that acts later in a protein's life, after it has been fully synthesized. This type of chaperone forms an "isolation chamber" into which misfolded proteins are fed, preventing their aggregation and providing them with a favorable environment in which to attempt to refold (Figure 6–84).

Exposed Hydrophobic Regions Provide Critical Signals for Protein Quality Control

If radioactive amino acids are added to cells for a brief period, the newly synthesized proteins can be followed as they mature into their final functional form. It is this type of experiment that shows that the hsp70 proteins act first, beginning when a protein is still being synthesized on a ribosome, and that the

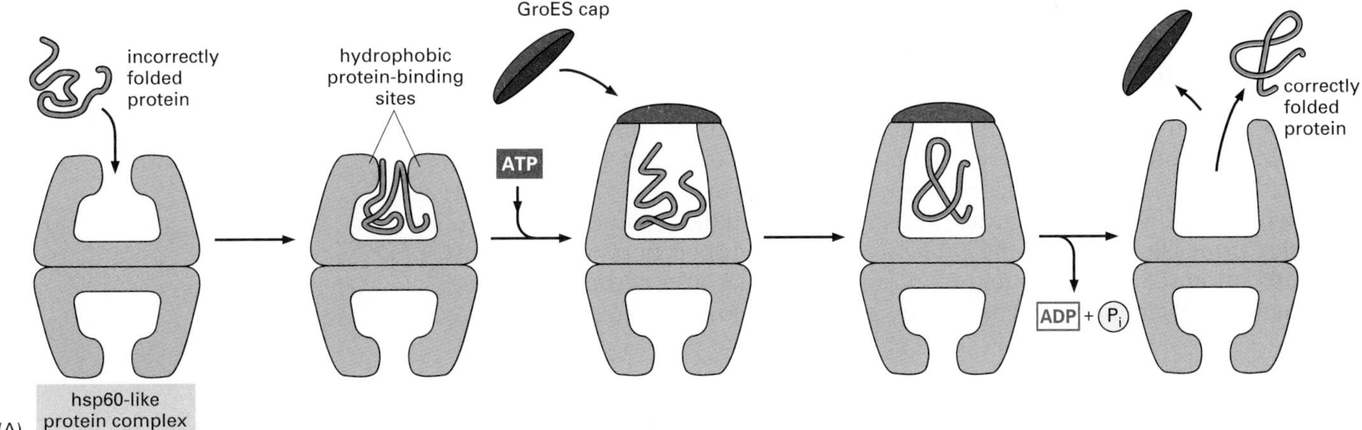

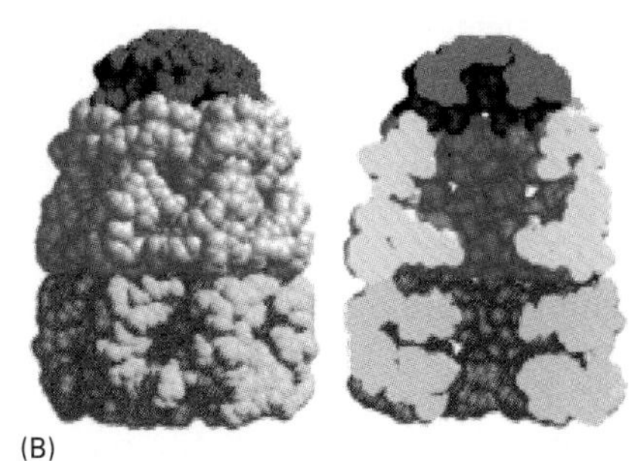

Figure 6–84 The structure and function of the hsp60 family of molecular chaperones. (A) The catalysis of protein refolding. As indicated, a misfolded protein is initially captured by hydrophobic interactions along one rim of the barrel. The subsequent binding of ATP plus a protein cap increases the diameter of the barrel rim, which may transiently stretch (partly unfold) the client protein. This also confines the protein in an enclosed space, where it has a new opportunity to fold. After about 15 seconds, ATP hydrolysis ejects the protein, whether folded or not, and the cycle repeats. This type of molecular chaperone is also known as a chaperonin; it is designated as hsp60 in mitochondria, TCP-1 in the cytosol of vertebrate cells, and GroEL in bacteria. As indicated, only half of the symmetrical barrel operates on a client protein at any one time. (B) The structure of GroEL bound to its GroES cap, as determined by x-ray crystallography. On the *left* is shown the outside of the barrel-like structure and on the *right* a cross section through its center. (B, adapted from B. Bukace and A.L. Horwich, *Cell* 92:351–366, 1998.)

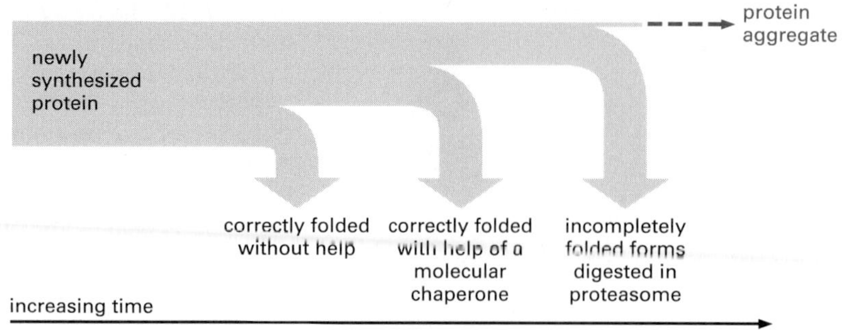

Figure 6–85 The cellular mechanisms that monitor protein quality after protein synthesis. As indicated, a newly synthesized protein sometimes folds correctly and assembles with its partners on its own, in which case it is left alone. Incompletely folded proteins are helped to refold by molecular chaparones: first by a family of hsp70 proteins, and if this fails, then by hsp60-like proteins. In both cases the client proteins are recognized by an abnormally exposed patch of hydrophobic amino acids on their surface. These processes compete with a different system that recognizes an abnormally exposed patch and transfers the protein that contains it to a proteasome for complete destruction. The combination of all of these processes is needed to prevent massive protein aggregation in a cell, which can occur when many hydrophobic regions on proteins clump together and precipitate the entire mass out of solution.

hsp60-like proteins are called into play only later to help in folding completed proteins. However, the same experiments reveal that only a subset of the newly synthesized proteins becomes involved: perhaps 20% of all proteins with the hsp70 and 10% with the hsp60-like molecular chaperones. How are these proteins selected for this ATP-catalyzed refolding?

Before answering, we need to pause to consider the post-translational fate of proteins more broadly. A protein that has a sizable exposed patch of hydrophobic amino acids on its surface is usually abnormal: it has either failed to fold correctly after leaving the ribosome, suffered an accident that partly unfolded it at a later time, or failed to find its normal partner subunit in a larger protein complex. Such a protein is not merely useless to the cell, it can be dangerous. Many proteins with an abnormally exposed hydrophobic region can form large aggregates, precipitating out of solution. We shall see that, in rare cases, such aggregates do form and cause severe human diseases. But in the vast majority of cells, powerful protein quality control mechanisms prevent such disasters.

Given this background, it is not surprising that cells have evolved elaborate mechanisms that recognize and remove the hydrophobic patches on proteins. Two of these mechanisms depend on the molecular chaperones just discussed, which bind to the patch and attempt to repair the defective protein by giving it another chance to fold. At the same time, by covering the hydrophobic patches, these chaperones transiently prevent protein aggregation. Proteins that very rapidly fold correctly on their own do not display such patches and are therefore bypassed by chaperones.

Figure 6–85 outlines all of the quality control choices that a cell makes for a difficult-to-fold, newly synthesized protein. As indicated, when attempts to refold a protein fail, a third mechanism is called into play that completely destroys the protein by proteolysis. The proteolytic pathway begins with the recognition of an abnormal hydrophobic patch on a protein's surface, and it ends with the delivery of the entire protein to a protein destruction machine, a complex protease known as the *proteasome*. As described next, this process depends on an elaborate protein-marking system that also carries out other central functions in the cell by destroying selected normal proteins.

The Proteasome Degrades a Substantial Fraction of the Newly Synthesized Proteins in Cells

Cells quickly remove the failures of their translation processes. Recent experiments suggest that as many as one-third of the newly made polypeptide chains are selected for rapid degradation as a result of the protein quality control mechanisms just described. The final disposal apparatus in eucaryotes is the **proteasome**, an abundant ATP-dependent protease that constitutes nearly 1% of cellular protein. Present in many copies dispersed throughout the cytosol and the nucleus, the proteasome also targets proteins of the endoplasmic reticulum (ER): those proteins that fail either to fold or to be assembled properly after they enter the ER are detected by an ER-based surveillance system that *retrotranslocate* them back to the cytosol for degradation (discussed in Chapter 12).

Each proteasome consists of a central hollow cylinder (the 20S core proteasome) formed from multiple protein subunits that assemble as a cylindrical stack of four heptameric rings. Some of these subunits are distinct proteases whose active sites face the cylinder's inner chamber (Figure 6–86A). Each end of the cylinder is normally associated with a large protein complex (the 19S cap) containing approximately 20 distinct polypeptides (Figure 6–86B). The cap subunits include at least six proteins that hydrolyze ATP; located near the edge of the cylinder, these ATPases are thought to unfold the proteins to be digested and move them into the interior chamber for proteolysis. A crucial property of the proteasome, and one reason for the complexity of its design, is the *processivity* of its mechanism: in contrast to a "simple" protease that cleaves a substrate's polypeptide chain just once before dissociating, the proteasome keeps the entire substrate bound until all of it is converted into short peptides.

The 19S caps act as regulated "gates" at the entrances to the inner proteolytic chamber, being also responsible for binding a targeted protein substrate to the proteasome. With a few exceptions, the proteasomes act on proteins that have been specifically marked for destruction by the covalent attachment of multiple copies of a small protein called *ubiquitin* (Figure 6–87A). Ubiquitin exists in cells either free or covalently linked to a huge variety of intracellular proteins. For most of these proteins, this tagging by ubiquitin results in their destruction by the proteasome.

An Elaborate Ubiquitin-conjugating System Marks Proteins for Destruction

Ubiquitin is prepared for conjugation to other proteins by the ATP-dependent *ubiquitin-activating* enzyme (E1), which creates an activated ubiquitin that is transferred to one of a set of ubiquitin-conjugating (E2) enzymes. The E2 enzymes act in conjunction with accessory (E3) proteins. In the E2–E3 complex, called *ubiquitin ligase*, the E3 component binds to specific degradation signals in protein substrates, helping E2 to form a *multiubiquitin* chain linked to a lysine of the substrate protein. In this chain, the C-terminal residue of each ubiquitin is linked to a specific lysine of the preceding ubiquitin molecule, producing a linear series of ubiquitin–ubiquitin conjugates (Figure 6–87B). It is this multiubiquitin chain on a target protein that is recognized by a specific receptor in the proteasome.

There are roughly 30 structurally similar but distinct E2 enzymes in mammals, and hundreds of different E3 proteins that form complexes with specific E2 enzymes. The ubiquitin–proteasome system thus consists of many distinct but similarly organized proteolytic pathways, which have in common both the El enzyme at the "top" and the proteasome at the "bottom," and differ by the compositions of their E2–E3 ubiquitin ligases and accessory factors. Distinct ubiquitin ligases recognize different degradation signals, and therefore target for degradation distinct subsets of intracellular proteins that bear these signals.

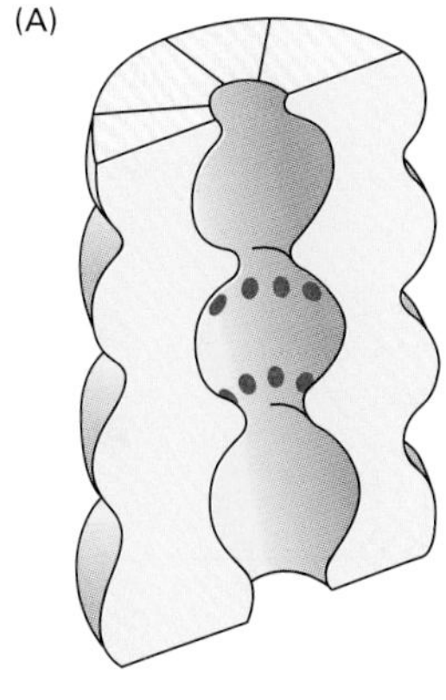

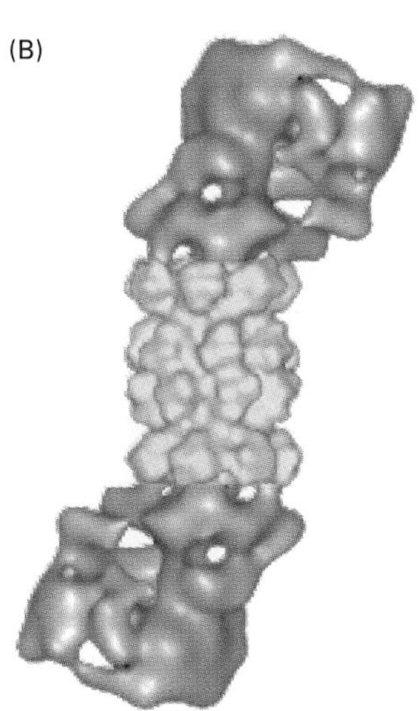

Figure 6–86 The proteasome. (A) A cut-away view of the structure of the central 20S cylinder, as determined by x-ray crystallography, with the active sites of the proteases indicated by *red dots*. (B) The structure of the entire proteasome, in which the central cylinder *(yellow)* is supplemented by a 19S cap *(blue)* at each end, whose structure has been determined by computer processing of electron microscope images. The complex cap structure selectively binds those proteins that have been marked for destruction; it then uses ATP hydrolysis to unfold their polypeptide chains and feed them into the inner chamber of the 20S cylinder for digestion to short peptides. (B, from W. Baumeister et al., *Cell* 92:367–380, 1998. © Elsevier.)

Denatured or otherwise misfolded proteins, as well as proteins containing oxidized or other abnormal amino acids, are recognized and destroyed because abnormal proteins tend to present on their surface amino acid sequences or conformational motifs that are recognized as degradation signals by a set of E3 molecules in the ubiquitin–proteasome system; these sequences must of course be buried and therefore inaccessible in the normal counterparts of these proteins. However, a proteolytic pathway that recognizes and destroys abnormal proteins must be able to distinguish between *completed* proteins that have "wrong" conformations and the many growing polypeptides on ribosomes (as well as polypeptides just released from ribosomes) that have not yet achieved their normal folded conformation. This is not a trivial problem; the ubiquitin–proteasome system is thought to destroy some of the nascent and newly formed protein molecules not because these proteins are abnormal as such but because they transiently expose degradation signals that are buried in their mature (folded) state.

Many Proteins Are Controlled by Regulated Destruction

One function of intracellular proteolytic mechanisms is to recognize and eliminate misfolded or otherwise abnormal proteins, as just described. Yet another function of these proteolytic pathways is to confer short half-lives on specific normal proteins whose concentrations must change promptly with alterations

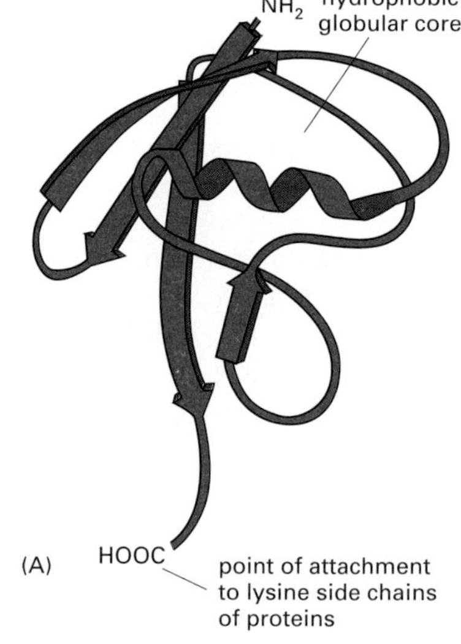

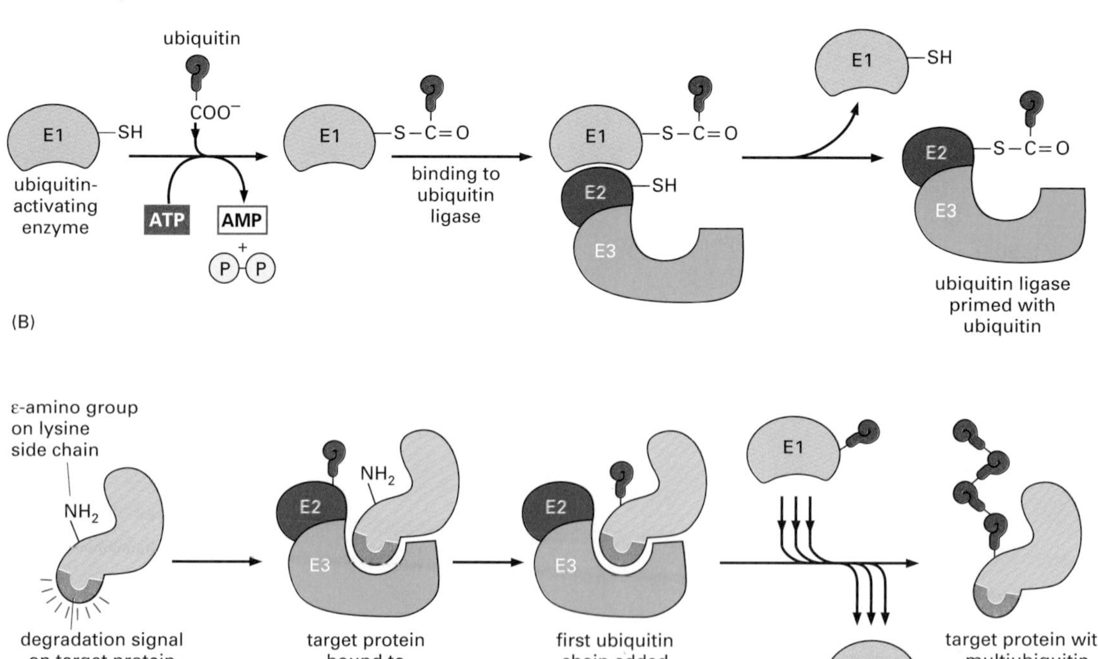

Figure 6–87 Ubiquitin and the marking of proteins with multiubiquitin chains. (A) The three-dimensional structure of ubiquitin; this relatively small protein contains 76 amino acids. (B) The C-terminus of ubiquitin is initially activated through its high-energy thioester linkage to a cysteine side chain on the E1 protein. This reaction requires ATP, and it proceeds via a covalent AMP-ubiquitin intermediate. The activated ubiquitin on E1, also known as the ubiquitin-activating enzyme, is then transferred to the cysteines on a set of E2 molecules. These E2s exist as complexes with an even larger family of E3 molecules. (C) The addition of a multiubiquitin chain to a target protein. In a mammalian cell there are roughly 300 distinct E2–E3 complexes, each of which recognizes a different degradation signal on a target protein by means of its E3 component. The E2s are called ubiquitin-conjugating enzymes. The E3s have been referred to traditionally as ubiquitin ligases, but it is more accurate to reserve this name for the functional E2–E3 complex.

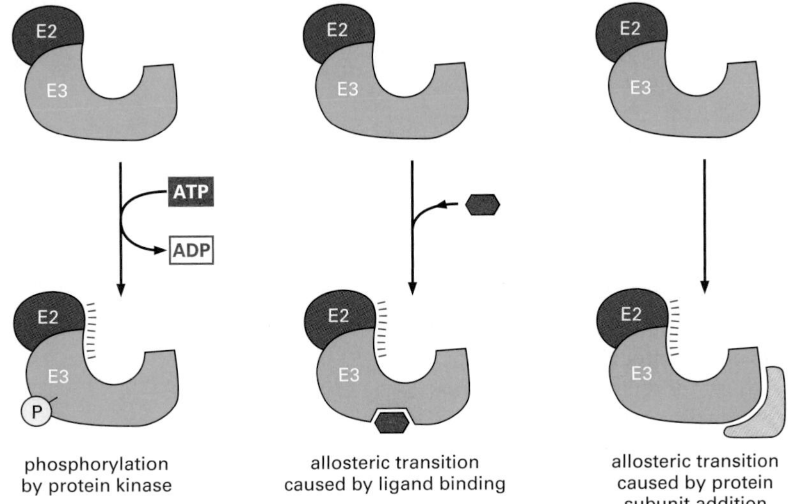

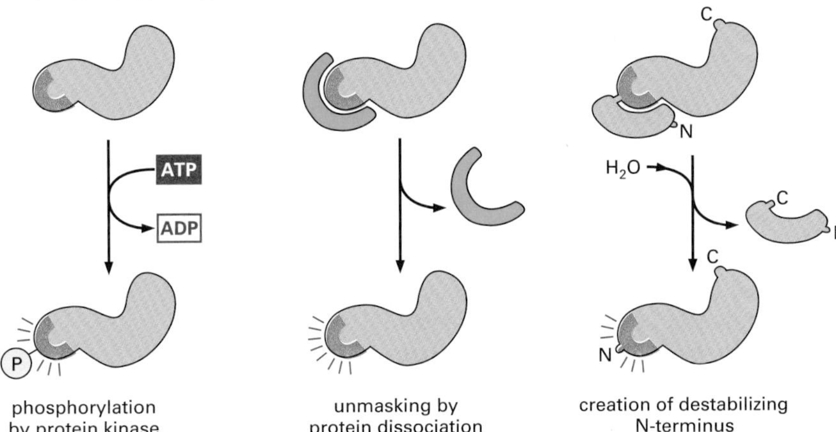

Figure 6–88 Two general ways of inducing the degradation of a specific protein. (A) Activation of a specific E3 molecule creates a new ubiquitin ligase. (B) Creation of an exposed degradation signal in the protein to be degraded. This signal binds a ubiquitin ligase, causing the addition of a multiubiquitin chain to a nearby lysine on the target protein. All six pathways shown are known to be used by cells to induce the movement of selected proteins into the proteasome.

in the state of a cell. Some of these short-lived proteins are degraded rapidly at all times, while many others are *conditionally* short-lived, that is, they are metabolically stable under some conditions, but become unstable upon a change in the cell's state. For example, mitotic cyclins are long-lived throughout the cell cycle until their sudden degradation at the end of mitosis, as explained in Chapter 17.

How is such a regulated destruction of a protein controlled? A variety of mechanisms are known, as illustrated through specific examples later in this book. In one general class of mechanism (Figure 6–88A), the activity of a ubiquitin ligase is turned on either by E3 phosphorylation or by an allosteric transition in an E3 protein caused by its binding to a specific small or large molecule. For example, the anaphase-promoting complex (APC) is a multisubunit ubiquitin ligase that is activated by a cell-cycle-timed subunit addition at mitosis. The activated APC then causes the degradation of mitotic cyclins and several other regulators of the metaphase–anaphase transition (see Figure 17–20).

Alternatively, in response either to intracellular signals or to signals from the environment, a degradation signal can be created in a protein, causing its rapid ubiquitylation and destruction by the proteasome. One common way to create such a signal is to phosphorylate a specific site on a protein that unmasks a normally hidden degradation signal. Another way to unmask such a signal is by the regulated dissociation of a protein subunit. Finally, powerful degradation signals can be created by a single cleavage of a peptide bond, provided that this cleavage creates a new N-terminus that is recognized by a specific E3 as a "destabilizing" N-terminal residue (Figure 6–88B).

The N-terminal type of degradation signal arises because of the "N-end rule," which relates the half-life of a protein *in vivo* to the identity of its N-terminal residue. There are 12 destabilizing residues in the N-end rule of the yeast *S. cerevisiae* (Arg, Lys, His, Phe, Leu, Tyr, Trp, Ile, Asp, Glu, Asn, and Gln), out of the 20 standard amino acids. The destabilizing N-terminal residues are recognized by a special ubiquitin ligase that is conserved from yeast to humans.

As we have seen, all proteins are initially synthesized bearing methionine (or formylmethionine in bacteria), as their N-terminal residue, which is a stabilizing residue in the N-end rule. Special proteases, called methionine aminopeptidases, will often remove the first methionine of a nascent protein, but they will do so only if the second residue is also stabilizing in the yeast-type N-end rule. Therefore, it was initially unclear how N-end rule substrates form *in vivo*. However, it has recently been discovered that a subunit of cohesin, a protein complex that holds sister chromatids together, is cleaved by a site-specific protease at the metaphase–anaphase transition. This cell-cycle-regulated cleavage allows separation of the sister chromatids and leads to the completion of mitosis (see Figure 17–26). The C-terminal fragment of the cleaved subunit bears an N-terminal arginine, a destabilizing residue in the N-end rule. Mutant cells lacking the N-end rule pathway exhibit a greatly increased frequency of chromosome loss, presumably because a failure to degrade this fragment of the cohesin subunit interferes with the formation of new chromatid-associated cohesin complexes in the next cell cycle.

Abnormally Folded Proteins Can Aggregate to Cause Destructive Human Diseases

When all of a cell's protein quality controls fail, large protein aggregates tend to accumulate in the affected cell (see Figure 6–85). Some of these aggregates, by adsorbing critical macromolecules to them, can severely damage cells and even cause cell death. The protein aggregates released from dead cells tend to accumulate in the extracellular matrix that surrounds the cells in a tissue, and in extreme cases they can also damage tissues. Because the brain is composed of a highly organized collection of nerve cells, it is especially vulnerable. Not surprisingly, therefore, protein aggregates primarily cause diseases of neurodegeneration. Prominent among these are Huntington's disease and Alzheimer's disease—the latter causing age-related dementia in more than 20 million people in today's world.

For a particular type of protein aggregate to survive, grow, and damage an organism, it must be highly resistant to proteolysis both inside and outside the cell. Many of the protein aggregates that cause problems form fibrils built from a series of polypeptide chains that are layered one over the other as a continuous stack of β sheets. This so-called *cross-beta filament* (Figure 6–89C) tends to be highly resistant to proteolysis. This resistance presumably explains why this structure is observed in so many of the neurological disorders caused by protein aggregates, where it produces abnormally staining deposits known as *amyloid*.

One particular variety of these diseases has attained special notoriety. These are the **prion diseases**. Unlike Huntington's or Alzheimer's disease, a prion disease can spread from one organism to another, providing that the second organism eats a tissue containing the protein aggregate. A set of diseases—called scrapie in sheep, Creutzfeldt–Jacob disease (CJD) in humans, and bovine spongiform encephalopathy (BSE) in cattle—are caused by a misfolded, aggregated form of a protein called PrP (for prion protein). The PrP is normally located on the outer surface of the plasma membrane, most prominently in neurons. Its normal function is not known. However, PrP has the unfortunate property of being convertible to a very special abnormal conformation (Figure 6–89A). This conformation not only forms protease-resistant, cross-beta filaments; it also is "infectious" because it converts normally folded molecules of PrP to the same form. This property creates a positive feedback loop that propagates the abnormal form of PrP, called PrP* (Figure 6–89B) and thereby allows PrP to spread rapidly from cell to cell in the brain, causing the death of both

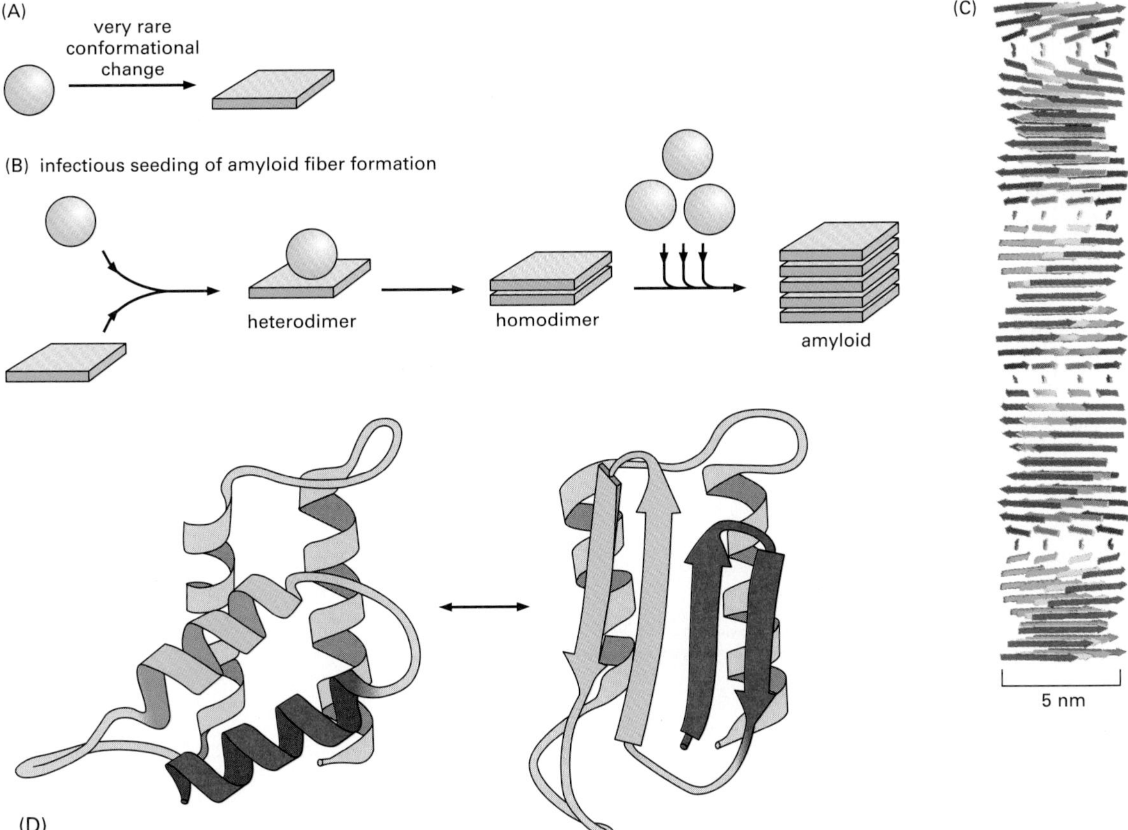

Figure 6–89 Protein aggregates that cause human disease. (A) Schematic illustration of the type of conformational change in a protein that produces material for a cross-beta filament. (B) Diagram illustrating the self-infectious nature of the protein aggregation that is central to prion diseases. PrP is highly unusual because the misfolded version of the protein, called PrP*, induces the normal PrP protein it contacts to change its conformation, as shown. Most of the human diseases caused by protein aggregation are caused by the overproduction of a variant protein that is especially prone to aggregation, but because this structure is not infectious in this way, it cannot spread from one animal to another. (C) Drawing of a cross-beta filament, a common type of protease-resistant protein aggregate found in a variety of human neurological diseases. Because the hydrogen-bond interactions in a β sheet form between polypeptide backbone atoms (see Figure 3–9), a number of different abnormally folded proteins can produce this structure. (D) One of several possible models for the conversion of PrP to PrP*, showing the likely change of two α-helices into four β-strands. Although the structure of the normal protein has been determined accurately, the structure of the infectious form is not yet known with certainty because the aggregation has prevented the use of standard structural techniques. (C, courtesy of Louise Serpell, adapted from M. Sunde et al., *J. Mol. Biol.* 273:729–739, 1997; D, adapted from S.B. Prusiner, *Trends Biochem. Sci.* 21:482–487, 1996.)

animals and humans. It can be dangerous to eat the tissues of animals that contain PrP*, as witnessed most recently by the spread of BSE (commonly referred to as the "mad cow disease") from cattle to humans in Great Britain.

Fortunately, in the absence of PrP*, PrP is extraordinarily difficult to convert to its abnormal form. Although very few proteins have the potential to misfold into an infectious conformation, a similar transformation has been discovered to be the cause of an otherwise mysterious "protein-only inheritance" observed in yeast cells.

There Are Many Steps From DNA to Protein

We have seen so far in this chapter that many different types of chemical reactions are required to produce a properly folded protein from the information contained in a gene (Figure 6–90). The final level of a properly folded protein in a cell therefore depends upon the efficiency with which each of the many steps is performed.

We discuss in Chapter 7 that cells have the ability to change the levels of their proteins according to their needs. In principle, any or all of the steps in Fig-

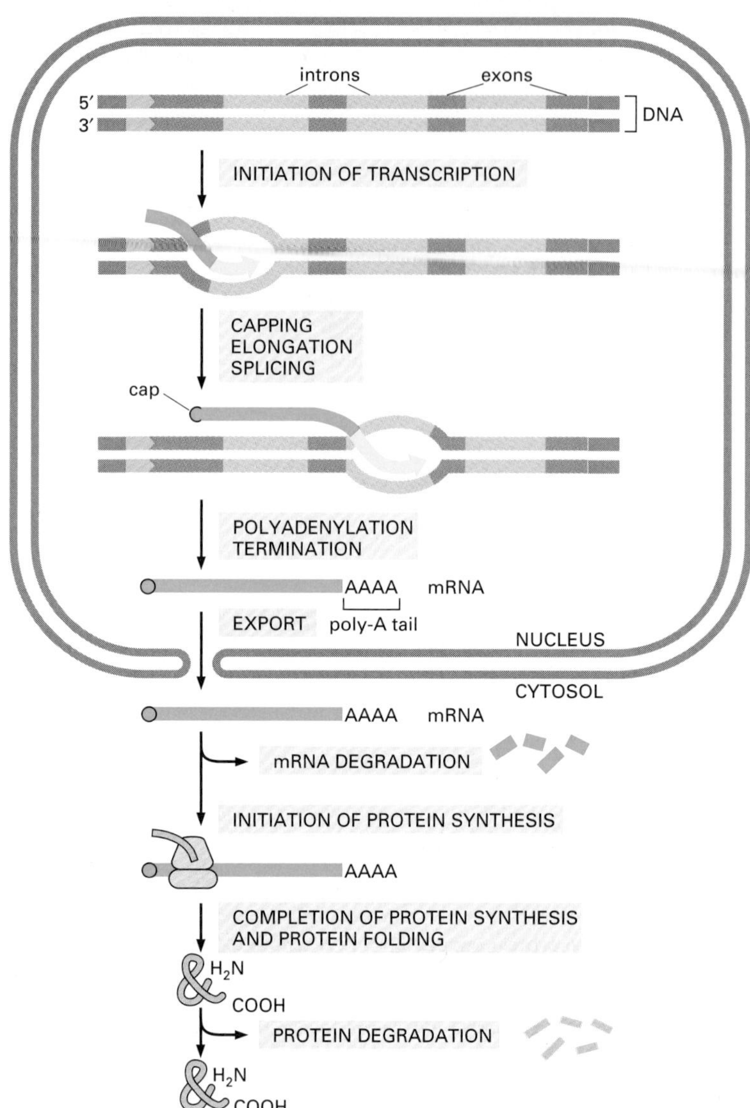

Figure 6–90 The production of a protein by a eucaryotic cell. The final level of each protein in a eucaryotic cell depends upon the efficiency of each step depicted.

ure 6–90) could be regulated by the cell for each individual protein. However, as we shall see in Chapter 7, the initiation of transcription is the most common point for a cell to regulate the expression of each of its genes. This makes sense, inasmuch as the most efficient way to keep a gene from being expressed is to block the very first step—the transcription of its DNA sequence into an RNA molecule.

Summary

The translation of the nucleotide sequence of an mRNA molecule into protein takes place in the cytoplasm on a large ribonucleoprotein assembly called a ribosome. The amino acids used for protein synthesis are first attached to a family of tRNA molecules, each of which recognizes, by complementary base-pair interactions, particular sets of three nucleotides in the mRNA (codons). The sequence of nucleotides in the mRNA is then read from one end to the other in sets of three according to the genetic code.

To initiate translation, a small ribosomal subunit binds to the mRNA molecule at a start codon (AUG) that is recognized by a unique initiator tRNA molecule. A large ribosomal subunit binds to complete the ribosome and begin the elongation phase of protein synthesis. During this phase, aminoacyl tRNAs—each bearing a specific amino acid bind sequentially to the appropriate codon in mRNA by forming complementary base pairs with the tRNA anticodon. Each amino acid is added to the C-terminal end of the growing polypeptide by means of a cycle of three sequential steps: aminoacyl-tRNA binding, followed by peptide bond formation, followed by ribosome translocation. The mRNA molecule progresses codon by codon through the ribosome in the 5′-to-3′ direction until one of three stop codons is reached. A release factor then binds to the ribosome, terminating translation and releasing the completed polypeptide.

Eucaryotic and bacterial ribosomes are closely related, despite differences in the number and size of their rRNA and protein components. The rRNA has the dominant role in translation, determining the overall structure of the ribosome, forming the binding sites for the tRNAs, matching the tRNAs to codons in the mRNA, and providing the peptidyl transferase enzyme activity that links amino acids together during translation.

In the final steps of protein synthesis, two distinct types of molecular chaperones guide the folding of polypeptide chains. These chaperones, known as hsp60 and hsp70, recognize exposed hydrophobic patches on proteins and serve to prevent the protein aggregation that would otherwise compete with the folding of newly synthesized proteins into their correct three-dimensional conformations. This protein folding process must also compete with a highly elaborate quality control mechanism that destroys proteins with abnormally exposed hydrophobic patches. In this case, ubiquitin is covalently added to a misfolded protein by a ubiquitin ligase, and the resulting multiubiquitin chain is recognized by the cap on a proteasome to move the entire protein to the interior of the proteasome for proteolytic degradation. A closely related proteolytic mechanism, based on special degradation signals recognized by ubiquitin ligases, is used to determine the lifetime of many normally folded proteins. By this method, selected normal proteins are removed from the cell in response to specific signals.

THE RNA WORLD AND THE ORIGINS OF LIFE

To fully understand the processes occurring in present-day living cells, we need to consider how they arose in evolution. The most fundamental of all such problems is the expression of hereditary information, which today requires extraordinarily complex machinery and proceeds from DNA to protein through an RNA intermediate. How did this machinery arise? One view is that an *RNA world* existed on Earth before modern cells arose (Figure 6–91). According to this hypothesis, RNA stored both genetic information and catalyzed the chemical reactions in primitive cells. Only later in evolutionary time did DNA take over as the genetic material and proteins become the major catalyst and structural component of cells. If this idea is correct, then the transition out of the RNA world was never complete; as we have seen in this chapter, RNA still catalyzes several fundamental reactions in modern-day cells, which can be viewed as molecular fossils of an earlier world.

In this section we outline some of the arguments in support of the RNA world hypothesis. We will see that several of the more surprising features of modern-day cells, such as the ribosome and the pre-mRNA splicing machinery, are most easily explained by viewing them as descendants of a complex network of RNA-mediated interactions that dominated cell metabolism in the RNA world. We also discuss how DNA may have taken over as the genetic material, how the genetic code may have arisen, and how proteins may have eclipsed RNA to perform the bulk of biochemical catalysis in modern-day cells.

Life Requires Autocatalysis

It has been proposed that the first "biological" molecules on Earth were formed by metal-based catalysis on the crystalline surfaces of minerals. In principle, an elaborate system of molecular synthesis and breakdown (metabolism) could have existed on these surfaces long before the first cells arose. But life requires molecules that possess a crucial property: the ability to catalyze reactions that lead, directly or indirectly, to the production of more molecules like themselves. Catalysts with this special self-promoting property can use raw materials to

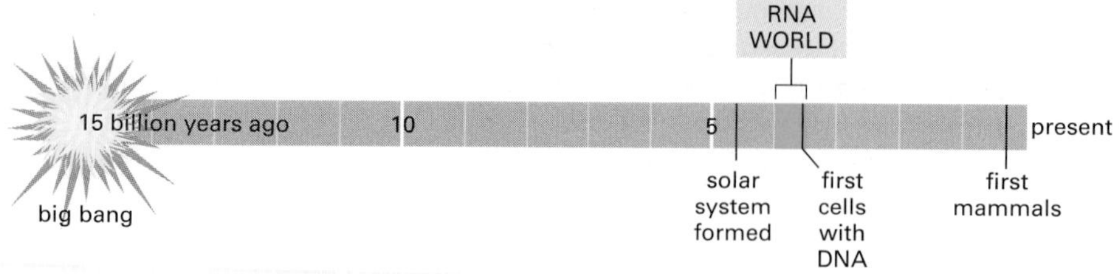

Figure 6–91 Time line for the universe, suggesting the early existence of an RNA world of living systems.

reproduce themselves and thereby divert these same materials from the production of other substances. But what molecules could have had such autocatalytic properties in early cells? In present-day cells the most versatile catalysts are polypeptides, composed of many different amino acids with chemically diverse side chains and, consequently, able to adopt diverse three-dimensional forms that bristle with reactive chemical groups. But, although polypeptides are versatile as catalysts, there is no known way in which one such molecule can reproduce itself by directly specifying the formation of another of precisely the same sequence.

Polynucleotides Can Both Store Information and Catalyze Chemical Reactions

Polynucleotides have one property that contrasts with those of polypeptides: they can directly guide the formation of exact copies of their own sequence. This capacity depends on complementary base pairing of nucleotide subunits, which enables one polynucleotide to act as a template for the formation of another. As we have seen in this and the preceding chapter, such complementary templating mechanisms lie at the heart of DNA replication and transcription in modern-day cells.

But the efficient synthesis of polynucleotides by such complementary templating mechanisms requires catalysts to promote the polymerization reaction: without catalysts, polymer formation is slow, error-prone, and inefficient. Today, template-based nucleotide polymerization is rapidly catalyzed by protein enzymes—such as the DNA and RNA polymerases. How could it be catalyzed before proteins with the appropriate enzymatic specificity existed? The beginnings of an answer to this question were obtained in 1982, when it was discovered that RNA molecules themselves can act as catalysts. We have seen in this chapter, for example, that a molecule of RNA is the catalyst for the peptidyl transferase reaction that takes place on the ribosome. The unique potential of RNA molecules to act both as information carrier and as catalyst forms the basis of the RNA world hypothesis.

RNA therefore has all the properties required of a molecule that could catalyze its own synthesis (Figure 6–92). Although self-replicating systems of RNA molecules have not been found in nature, scientists are hopeful that they can be constructed in the laboratory. While this demonstration would not prove that self-replicating RNA molecules were essential in the origin of life on Earth, it would certainly suggest that such a scenario is possible.

A Pre-RNA World Probably Predates the RNA World

Although RNA seems well suited to form the basis for a self-replicating set of biochemical catalysts, it is unlikely that RNA was the first kind of molecule to do so.

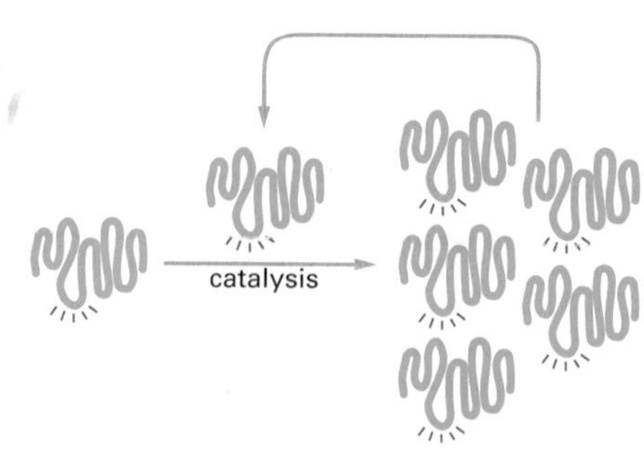

Figure 6–92 An RNA molecule that can catalyze its own synthesis. This hypothetical process would require catalysis of the production of both a second RNA strand of complementary nucleotide sequence and the use of this second RNA molecule as a template to form many molecules of RNA with the original sequence. The *red* rays represent the active site of this hypothetical RNA enzyme.

From a purely chemical standpoint, it is difficult to imagine how long RNA molecules could be formed initially by purely nonenzymatic means. For one thing, the precursors of RNA, the ribonucleotides, are difficult to form nonenzymatically. Moreover, the formation of RNA requires that a long series of 3′ to 5′ phosphodiester linkages form in the face of a set of competing reactions, including hydrolysis, 2′ to 5′ linkages, 5′ to 5′ linkages, and so on. Given these problems, it has been suggested that the first molecules to possess both catalytic activity and information storage capabilities may have been polymers that resemble RNA but are chemically simpler (Figure 6–93). We do not have any remnants of these compounds in present-day cells, nor do such compounds leave fossil records. Nonetheless, the relative simplicity of these "RNA-like polymers" make them better candidates than RNA itself for the first biopolymers on Earth that had both information storage capacity and catalytic activity.

The transition between the pre-RNA world and the RNA world would have occurred through the synthesis of RNA using one of these simpler compounds as both template and catalyst. The plausibility of this scheme is supported by laboratory experiments showing that one of these simpler forms (PNA—see Figure 6–93) can act as a template for the synthesis of complementary RNA molecules, because the overall geometry of the bases is similar in the two molecules. Presumably, pre-RNA polymers also catalyzed the formation of ribonucleotide precursors from simpler molecules. Once the first RNA molecules had been produced, they could have diversified gradually to take over the functions originally carried out by the pre-RNA polymers, leading eventually to the postulated RNA world.

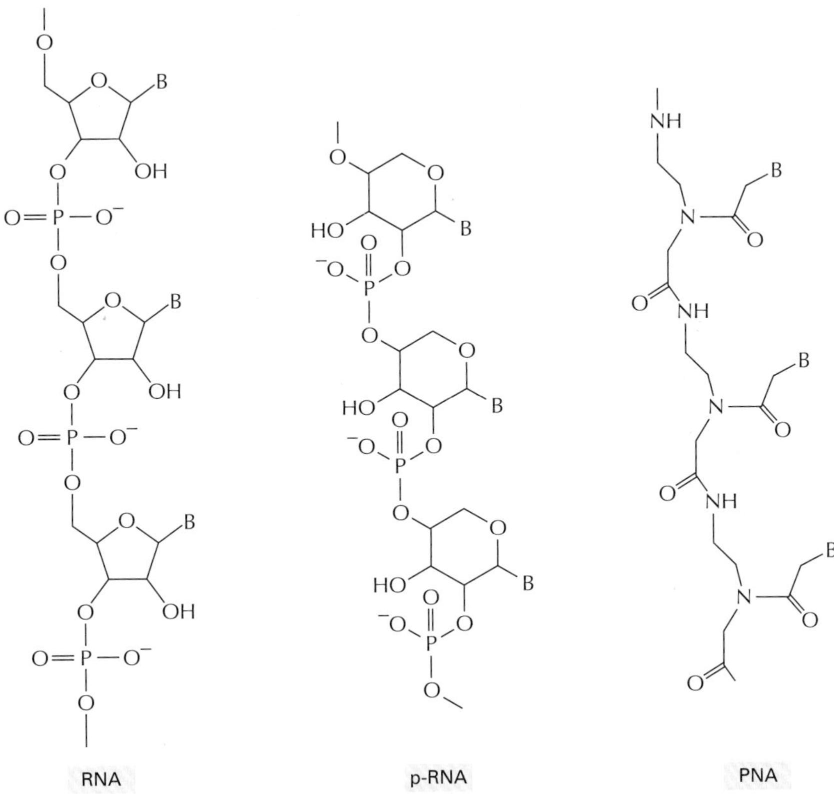

Figure 6–93 Structures of RNA and two related information-carrying polymers. In each case, B indicates the positions of purine and pyrimidine bases. The polymer p-RNA (pyranosyl-RNA) is RNA in which the furanose (five-membered ring) form of ribose has been replaced by the pyranose (six-membered ring) form. In PNA (peptide nucleic acid), the ribose phosphate backbone of RNA has been replaced by the peptide backbone found in proteins. Like RNA, both p-RNA and PNA can form double helices through complementary base-pairing, and each could therefore in principle serve as a template for its own synthesis (see Figure 6–92).

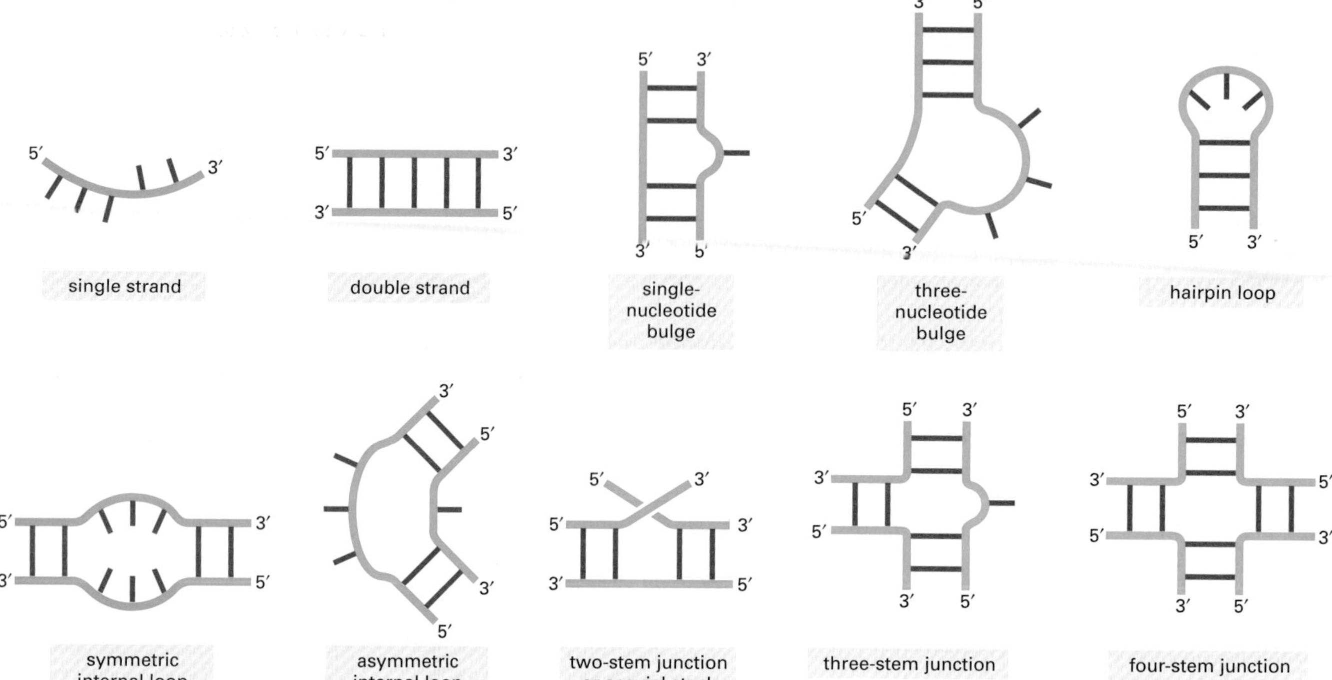

Single-stranded RNA Molecules Can Fold into Highly Elaborate Structures

Figure 6–94 Common elements of RNA secondary structure. Conventional, complementary base-pairing interactions are indicated by *red* "rungs" in double-helical portions of the RNA.

We have seen that complementary base-pairing and other types of hydrogen bonds can occur between nucleotides in the same chain, causing an RNA molecule to fold up in a unique way determined by its nucleotide sequence (see, for example, Figures 6–6, 6–52, and 6–67). Comparisons of many RNA structures have revealed conserved motifs, short structural elements that are used over and over again as parts of larger structures. Some of these RNA secondary structural motifs are illustrated in Figure 6–94. In addition, a few common examples of more complex and often longer-range interactions, known as RNA tertiary interactions, are shown in Figure 6–95.

Protein catalysts require a surface with unique contours and chemical properties on which a given set of substrates can react (discussed in Chapter 3). In exactly the same way, an RNA molecule with an appropriately folded shape can serve as an enzyme (Figure 6–96). Like some proteins, many of these ribozymes work by positioning metal ions at their active sites. This feature gives them a wider range of catalytic activities than can be accounted for solely by the limited chemical groups of the polynucleotide chain.

Relatively few catalytic RNAs exist in modern-day cells, however, and much of our inference about the RNA world has come from experiments in which large pools of RNA molecules of random nucleotide sequences are generated in the laboratory. Those rare RNA molecules with a property specified by the experimenter are then selected out and studied (Figure 6–97). Experiments of this type

Figure 6–95 Examples of RNA tertiary interactions. Some of these interactions can join distant parts of the same RNA molecule or bring two separate RNA molecules together.

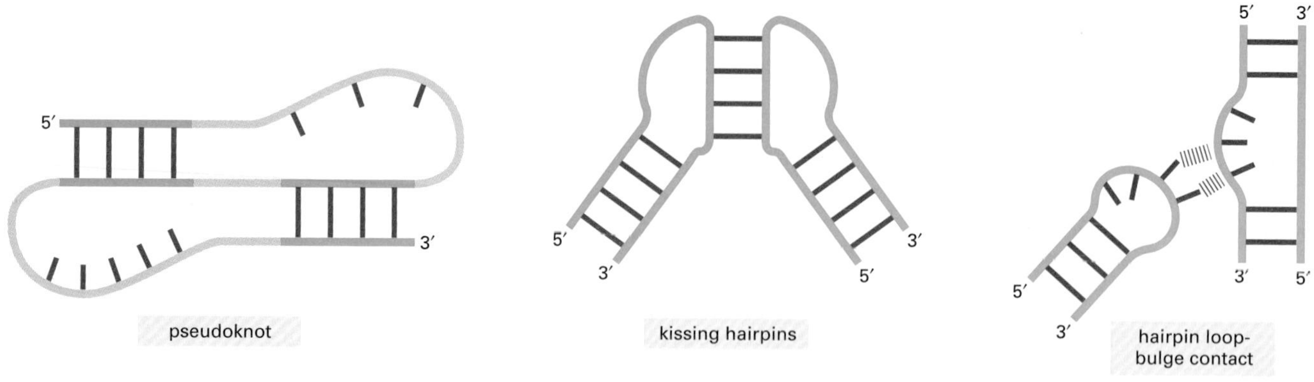

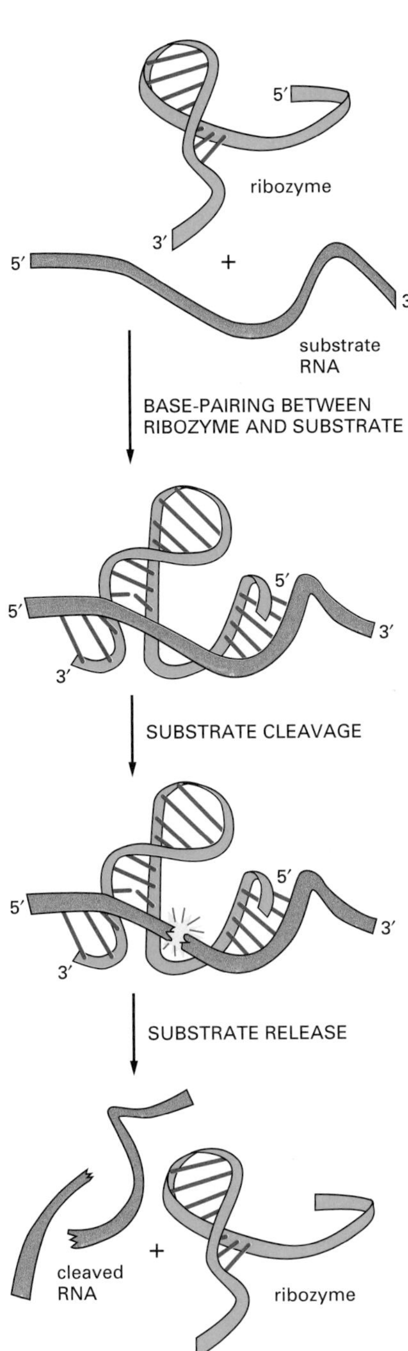

Figure 6–96 *(above)* A ribozyme. This simple RNA molecule catalyzes the cleavage of a second RNA at a specific site. This ribozyme is found embedded in larger RNA genomes—called viroids—which infect plants. The cleavage, which occurs in nature at a distant location on the same RNA molecule that contains the ribozyme, is a step in the replication of the viroid genome. Although not shown in the figure, the reaction requires a molecule of Mg positioned at the active site. (Adapted from T.R. Cech and O.C. Uhlenbeck, *Nature* 372:39–40, 1994.)

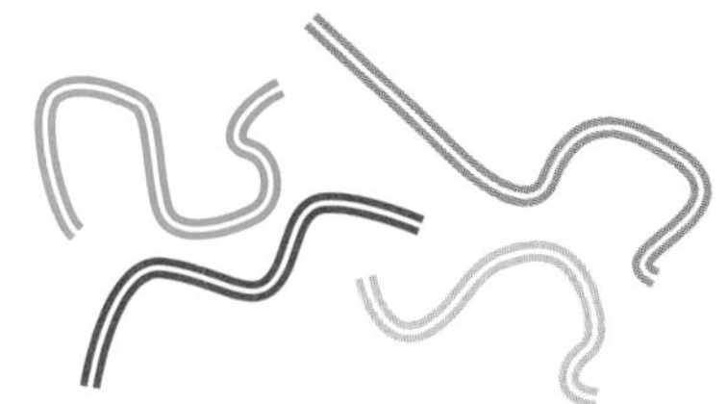

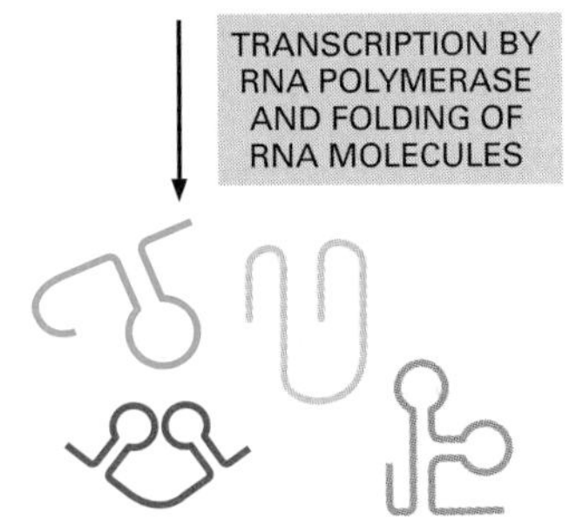

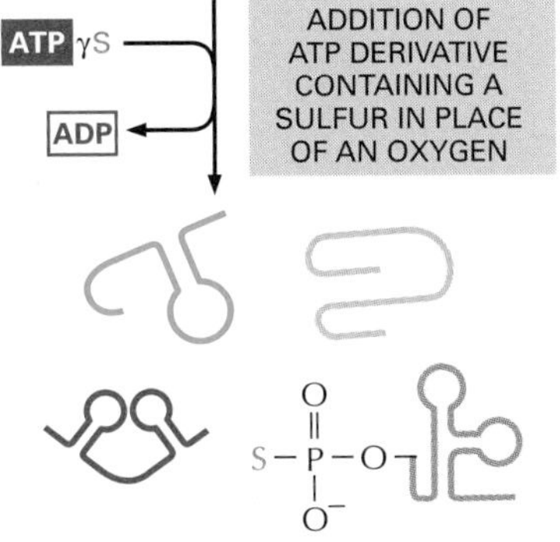

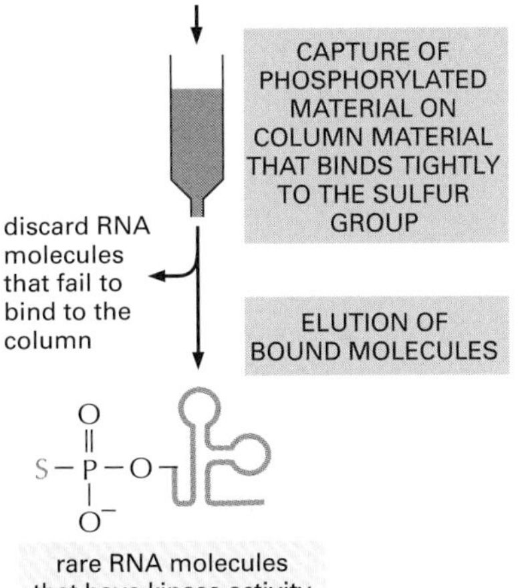

Figure 6–97 *(left)* *In vitro* selection of a synthetic ribozyme. Beginning with a large pool of nucleic acid molecules synthesized in the laboratory, those rare RNA molecules that possess a specified catalytic activity can be isolated and studied. Although a specific example (that of an autophosphorylating ribozyme) is shown, variations of this procedure have been used to generate many of the ribozymes listed in Table 6–4. During the autophosphorylation step, the RNA molecules are sufficiently dilute to prevent the "cross"-phosphorylation of additional RNA molecules. In reality, several repetitions of this procedure are necessary to select the very rare RNA molecules with catalytic activity. Thus the material initially eluted from the column is converted back into DNA, amplified many fold (using reverse transcriptase and PCR as explained in Chapter 8), transcribed back into RNA, and subjected to repeated rounds of selection. (Adapted from J.R. Lorsch and J.W. Szostak, *Nature* 371:31–36, 1994.)

TABLE 6–4 Some Biochemical Reactions That Can Be Catalyzed by Ribozymes = RNA enzymes

ACTIVITY	RIBOZYMES
Peptide bond formation in protein synthesis	ribosomal RNA
RNA cleavage, RNA ligation	self-splicing RNAs; also *in vitro* selected RNA
DNA cleavage	self-splicing RNAs
RNA splicing	self-splicing RNAs, perhaps RNAs of the spliceosome
RNA polymerizaton	*in vitro* selected RNA
RNA and DNA phosphorylation	*in vitro* selected RNA
RNA aminoacylation	*in vitro* selected RNA
RNA alkylation	*in vitro* selected RNA
Amide bond formation	*in vitro* selected RNA
Amide bond cleavage	*in vitro* selected RNA
Glycosidic bond formation	*in vitro* selected RNA
Porphyrin metalation	*in vitro* selected RNA

have created RNAs that can catalyze a wide variety of biochemical reactions (Table 6–4), and suggest that the main difference between protein enzymes and ribozymes lies in their maximum reaction speed, rather than in the diversity of the reactions that they can catalyze.

Like proteins, RNAs can undergo allosteric conformational changes, either in response to small molecules or to other RNAs. One artificially created ribozyme can exist in two entirely different conformations, each with a different catalytic activity (Figure 6–98). Moreover, the structure and function of the rRNAs in the ribosome alone have made it clear that RNA is an enormously versatile molecule. It is therefore easy to imagine that an RNA world could reach a high level of biochemical sophistication.

Self-Replicating Molecules Undergo Natural Selection

The three-dimensional folded structure of a polynucleotide affects its stability, its actions on other molecules, and its ability to replicate. Therefore, certain polynucleotides will be especially successful in any self-replicating mixture. Because errors inevitably occur in any copying process, new variant sequences of these polynucleotides will be generated over time.

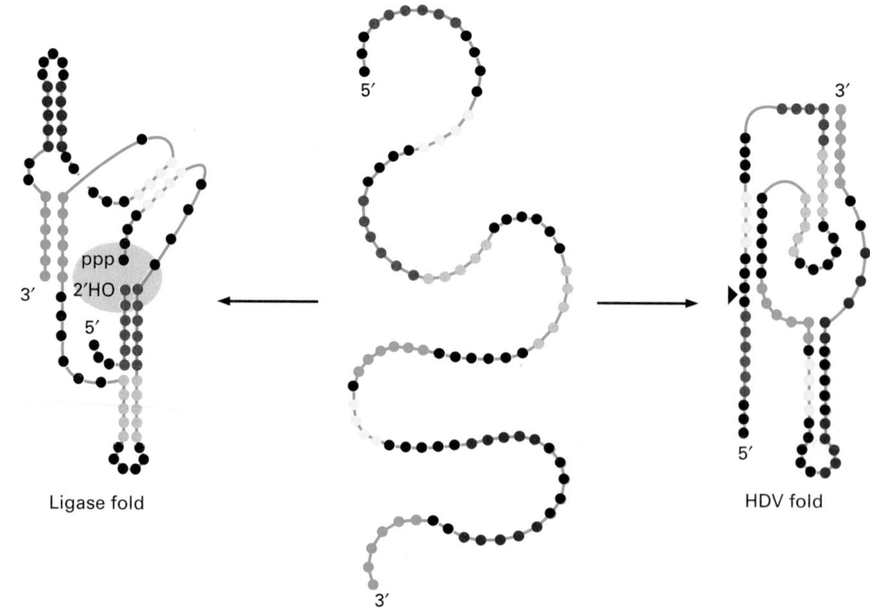

Figure 6–98 An RNA molecule that folds into two different ribozymes. This 88-nucleotide RNA, created in the laboratory, can fold into a ribozyme that carries out a self-ligation reaction *(left)* or a self-cleavage reaction *(right)*. The ligation reaction forms a 2′,5′ phosphodiester linkage with the release of pyrophosphate. This reaction seals the gap (*gray* shading), which was experimentally introduced into the RNA molecule. In the reaction carried out by the HDV fold, the RNA is cleaved at this same position, indicated by the *arrowhead*. This cleavage resembles that used in the life cycle of HDV, a hepatitis B satellite virus, hence the name of the fold. Each nucleotide is represented by a *colored dot*, with the colors used simply to clarify the two different folding patterns. The folded structures illustrate the secondary structures of the two ribozymes with regions of base-pairing indicated by close oppositions of the *colored dots*. Note that the two ribozyme folds have no secondary structure in common. (Adapted from E.A. Schultes and D.P. Bartel, *Science* 289:448–452, 2000.)

Certain catalytic activities would have had a cardinal importance in the early evolution of life. Consider in particular an RNA molecule that helps to catalyze the process of templated polymerization, taking any given RNA molecule as a template. (This ribozyme activity has been directly demonstrated *in vitro*, albeit in a rudimentary form that can only synthesize moderate lengths of RNA.) Such a molecule, by acting on copies of itself, can replicate. At the same time, it can promote the replication of other types of RNA molecules in its neighborhood (Figure 6–99). If some of these neighboring RNAs have catalytic actions that help the survival of RNA in other ways (catalyzing ribonucleotide production, for example), a set of different types of RNA molecules, each specialized for a different activity, may evolve into a cooperative system that replicates with unusually great efficiency.

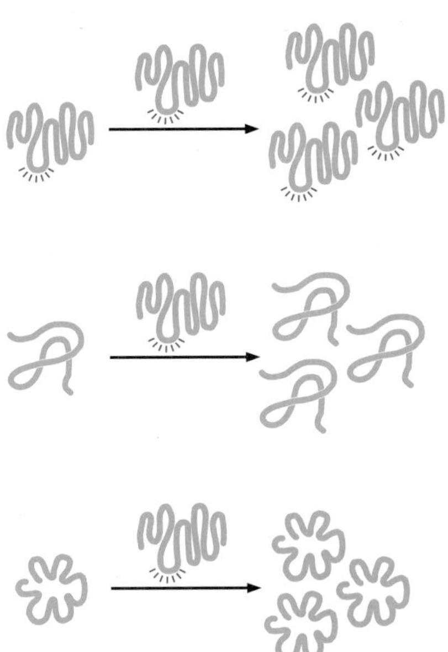

Figure 6–99 A family of mutually supportive RNA molecules, one catalyzing the reproduction of the others.

One of the crucial events leading to the formation of effective self-replicating systems must have been the development of individual compartments. For example, a set of mutually beneficial RNAs (such as those of Figure 6–99) could replicate themselves only if all the RNAs were to remain in the neighborhood of the RNA that is specialized for templated polymerization. Moreover, if these RNAs were free to diffuse among a large population of other RNA molecules, they could be co-opted by other replicating systems, which would then compete with the original RNA system for raw materials. Selection of a set of RNA molecules according to the quality of the self-replicating systems they generated could not occur efficiently until some form of compartment evolved to contain them and thereby make them available only to the RNA that had generated them. An early, crude form of compartmentalization may have been simple adsorption on surfaces or particles.

The need for more sophisticated types of containment is easily fulfilled by a class of small molecules that has the simple physicochemical property of being *amphipathic*, that is, consisting of one part that is hydrophobic (water insoluble) and another part that is hydrophilic (water soluble). When such molecules are placed in water they aggregate, arranging their hydrophobic portions as much in contact with one another as possible and their hydrophilic portions in contact with the water. Amphipathic molecules of appropriate shape spontaneously aggregate to form *bilayers*, creating small closed vesicles whose aqueous contents are isolated from the external medium (Figure 6–100). The phenomenon can be demonstrated in a test tube by simply mixing phospholipids and water together: under appropriate conditions, small vesicles will form. All present-day cells are surrounded by a *plasma membrane* consisting of amphipathic molecules—mainly phospholipids—in this configuration; we discuss these molecules in detail in Chapter 10.

Presumably, the first membrane-bounded cells were formed by the spontaneous assembly of a set of amphipathic molecules, enclosing a self-replicating mixture of RNA (or pre-RNA) and other molecules. It is not clear at what point in the evolution of biological catalysts this first occurred. In any case, once RNA molecules were sealed within a closed membrane, they could begin to evolve in earnest as carriers of genetic instructions: they could be selected not merely on the basis of their own structure, but also according to their effect on the other molecules in the same compartment. The nucleotide sequences of the RNA molecules could now be expressed in the character of a unitary living cell.

How Did Protein Synthesis Evolve?

The molecular processes underlying protein synthesis in present-day cells seem inextricably complex. Although we understand most of them, they do not make conceptual sense in the way that DNA transcription, DNA repair, and DNA

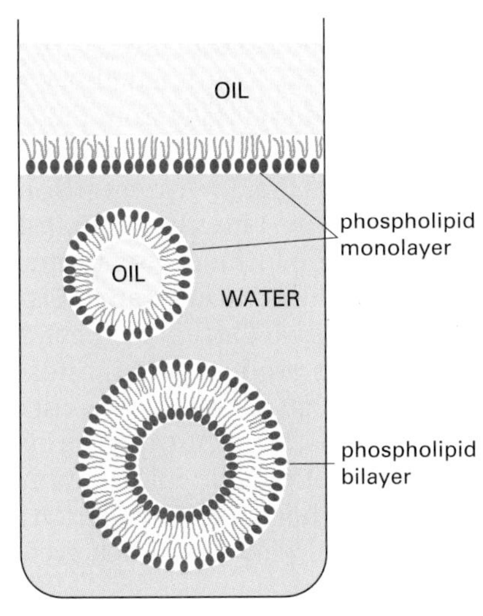

Figure 6–100 Formation of membrane by phospholipids. Because these molecules have hydrophilic heads and lipophilic tails, they align themselves at an oil/water interface with their heads in the water and their tails in the oil. In the water they associate to form closed bilayer vesicles in which the lipophilic tails are in contact with one another and the hydrophilic heads are exposed to the water.

replication do. It is especially difficult to imagine how protein synthesis evolved because it is now performed by a complex interlocking system of protein and RNA molecules; obviously the proteins could not have existed until an early version of the translation apparatus was already in place. Although we can only speculate on the origins of protein synthesis and the genetic code, several experimental approaches have provided possible scenarios.

In vitro RNA selection experiments of the type summarized previously in Figure 6–97 have produced RNA molecules that can bind tightly to amino acids. The nucleotide sequences of these RNAs often contain a disproportionately high frequency of codons for the amino acid that is recognized. For example, RNA molecules that bind selectively to arginine have a preponderance of Arg codons and those that bind tyrosine have a preponderance of Tyr codons. This correlation is not perfect for all the amino acids, and its interpretation is controversial, but it raises the possibility that a limited genetic code could have arisen from the direct association of amino acids with specific sequences of RNA, with RNAs serving as a crude template to direct the non-random polymerization of a few different amino acids. In the RNA world described previously, any RNA that helped guide the synthesis of a useful polypeptide would have a great advantage in the evolutionary struggle for survival.

In present-day cells, tRNA adaptors are used to match amino acids to codons, and proteins catalyze tRNA aminoacylation. However, ribozymes created in the laboratory can perform specific tRNA aminoacylation reactions, so it is plausible that tRNA-like adaptors could have arisen in an RNA world. This development would have made the matching of "mRNA" sequences to amino acids more efficient, and it perhaps allowed an increase in the number of amino acids that could be used in templated protein synthesis.

Finally, the efficiency of early forms of protein synthesis would be increased dramatically by the catalysis of peptide bond formation. This evolutionary development presents no conceptual problem since, as we have seen, this reaction is catalyzed by rRNA in present-day cells. One can envision a crude peptidyl transferase ribozyme, which, over time, grew larger and acquired the ability to position charged tRNAs accurately on RNA templates—leading eventually to the modern ribosome. Once protein synthesis evolved, the transition to a protein-dominated world could proceed, with proteins eventually taking over the majority of catalytic and structural tasks because of their greater versatility, with 20 rather than 4 different subunits.

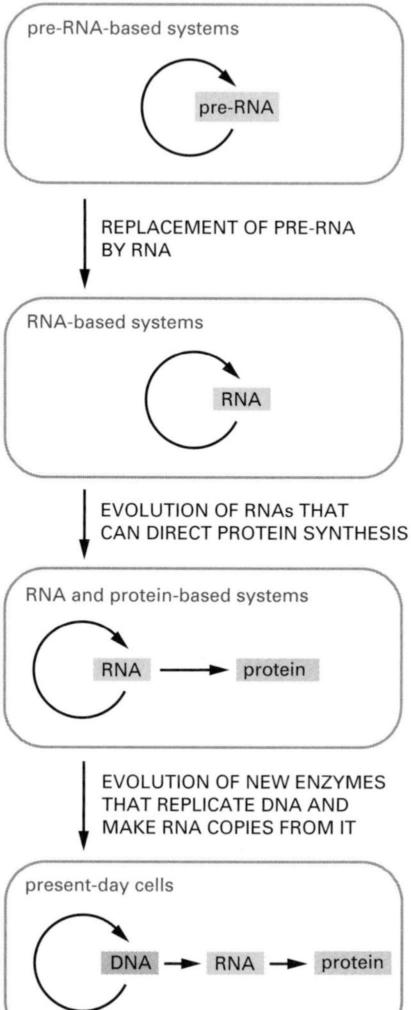

Figure 6–101 The hypothesis that RNA preceded DNA and proteins in evolution. In the earliest cells, pre-RNA molecules would have had combined genetic, structural, and catalytic functions and these functions would have gradually been replcaed by RNA. In present-day cells, DNA is the repository of genetic information, and proteins perform the vast majority of catalytic functions in cells. RNA primarily functions today as a go-between in protein synthesis, although it remains a catalyst for a number of crucial reactions.

All Present-day Cells Use DNA as Their Hereditary Material

The cells of the RNA world would presumably have been much less complex and less efficient in reproducing themselves than even the simplest present-day cells, since catalysis by RNA molecules is less efficient than that by proteins. They would have consisted of little more than a simple membrane enclosing a set of self-replicating molecules and a few other components required to provide the materials and energy for their replication. If the evolutionary speculations about RNA outlined above are correct, these early cells would also have differed fundamentally from the cells we know today in having their hereditary information stored in RNA rather than in DNA (Figure 6–101).

Evidence that RNA arose before DNA in evolution can be found in the chemical differences between them. Ribose, like glucose and other simple carbohydrates, can be formed from formaldehyde (HCHO), a simple chemical which is readily produced in laboratory experiments that attempt to simulate conditions on the primitive Earth. The sugar deoxyribose is harder to make, and in present-day cells it is produced from ribose in a reaction catalyzed by a protein enzyme, suggesting that ribose predates deoxyribose in cells. Presumably, DNA appeared on the scene later, but then proved more suitable than RNA as a permanent repository of genetic information. In particular, the deoxyribose in its sugar-phosphate backbone makes chains of DNA chemically more stable than chains of RNA, so that much greater lengths of DNA can be maintained without breakage.

The other differences between RNA and DNA—the double-helical structure of DNA and the use of thymine rather than uracil—further enhance DNA stability by making the many unavoidable accidents that occur to the molecule much easier to repair, as discussed in detail in Chapter 5 (see pp. 269–272).

Summary

From our knowledge of present-day organisms and the molecules they contain, it seems likely that the development of the directly autocatalytic mechanisms fundamental to living systems began with the evolution of families of molecules that could catalyze their own replication. With time, a family of cooperating RNA catalysts probably developed the ability to direct synthesis of polypeptides. DNA is likely to have been a late addition: as the accumulation of additional protein catalysts allowed more efficient and complex cells to evolve, the DNA double helix replaced RNA as a more stable molecule for storing the increased amounts of genetic information required by such cells.

References

General

Gesteland RF, Cech TR & Atkins JF (eds) (1999) The RNA World, 2nd edn. Cold Spring Harbor, NY: Cold Spring Harbor Laboratory Press.

Hartwell L, Hood L, Goldberg ML et al. (2000) Genetics: from Genes to Genomes. Boston: McGraw Hill.

Lewin B (2000) Genes VII. Oxford: Oxford University Press.

Lodish H, Berk A, Zipursky SL et al. (2000) Molecular Cell Biology, 4th edn. New York: WH Freeman.

Stent GS (1971) Molecular Genetics: An Introductory Narrative. San Francisco: WH Freeman.

Stryer L (1995) Biochemistry, 4th edn. New York: WH Freeman.

Watson JD, Hopkins NH, Roberts JW et al. (1987) Molecular Biology of the Gene, 4th edn. Menlo Park, CA: Benjamin/Cummings.

From DNA to RNA

Berget SM, Moore C & Sharp PA (1977) Spliced segments at the 5′ terminus of adenovirus 2 late mRNA. *Proc. Natl. Acad. Sci. USA* 74, 3171–3175.

Black DL (2000) Protein diversity from alternative splicing: a challenge for bioinformatics and post-genome biology. *Cell* 103, 367–370.

Brenner S, Jacob F & Meselson M (1961) An unstable intermediate carrying information from genes to ribosomes for protein synthesis. *Nature* 190, 576–581.

Cech TR (1990) Nobel lecture. Self-splicing and enzymatic activity of an intervening sequence RNA from Tetrahymena. *Biosci. Rep.* 10, 239–261.

Chow LT, Gelinas RE, Broker TR et al. (1977) An amazing sequence arrangement at the 5′ ends of adenovirus 2 messenger RNA. *Cell* 12, 1–8.

Conaway JW, Shilatifard A, Dvir A & Conaway RC (2000) Control of elongation by RNA polymerase II. *Trends Biochem. Sci.* 25, 375–380.

Cramer P, Bushnell DA, Fu J et al. (2000) Architecture of RNA polymerase II and implications for the transcription mechanism. *Science* 288, 640–649.

Crick F (1979) Split genes and RNA splicing. *Science* 204, 264–271.

Daneholt B (1997) A look at messenger RNP moving through the nuclear pore. *Cell* 88, 585–588.

Darnell JE, Jr. (1985) RNA. *Sci. Am.* 253(4), 68–78.

Dvir A, Conaway JW & Conaway RC (2001) Mechanism of transcription initiation and promoter escape by RNA polymerase II. *Curr. Opin. Genet. Dev.* 11, 209–214.

Ebright RH (2000) RNA polymerase: structural similarities between bacterial RNA polymerase and eukaryotic RNA polymerase II. *J. Mol. Biol.* 304, 687–698.

Eddy SR (1999) Noncoding RNA genes. *Curr. Opin. Genet. Dev.* 9, 695–699.

Green MR (2000) TBP-associated factors (TAFIIs): multiple, selective transcriptional mediators in common complexes. *Trends Biochem. Sci.* 25, 59–63.

Harley CB & Reynolds RP (1987) Analysis of E. coli promoter sequences. *Nucleic Acids Res.* 15, 2343–2361.

Hirose Y & Manley JL (2000) RNA polymerase II and the integration of nuclear events. *Genes Dev.* 14, 1415–1429.

Kadonaga JT (1998) Eukaryotic transcription: an interlaced network of transcription factors and chromatin-modifying machines. *Cell* 92, 307–313.

Lewis JD & Tollervey D (2000) Like attracts like: getting RNA processing together in the nucleus. *Science* 288, 1385–1389.

Lisser S & Margalit H (1993) Compilation of E. coli mRNA promoter sequences. *Nucleic Acids Res.* 21, 1507–1516.

Minvielle-Sebastia L & Keller W (1999) mRNA polyadenylation and its coupling to other RNA processing reactions and to transcription. *Curr. Opin. Cell Biol.* 11, 352–357.

Mooney RA & Landick R (1999) RNA polymerase unveiled. *Cell* 98, 687–690.

Olson MO, Dundr M & Szebeni A (2000) The nucleolus: an old factory with unexpected capabilities. *Trends Cell Biol.* 10, 189–196.

Proudfoot N (2000) Connecting transcription to messenger RNA processing. *Trends Biochem. Sci.* 25, 290–293.

Reed R (2000) Mechanisms of fidelity in pre-mRNA splicing. *Curr. Opin. Cell Biol.* 12, 340–345.

Roeder RG (1996) The role of general initiation factors in transcription by RNA polymerase II. *Trends Biochem. Sci.* 21, 327–335.

Shatkin AJ & Manley JL (2000) The ends of the affair: capping and polyadenylation. *Nat. Struct. Biol.* 7, 838–842.

Smith CW & Valcarcel J (2000) Alternative pre-mRNA splicing: the logic of combinatorial control. *Trends Biochem. Sci.* 25, 381–388.

Staley JP & Guthrie C (1998) Mechanical devices of the spliceosome: motors, clocks, springs, and things. *Cell* 92, 315–326.

Tarn WY & Steitz JA (1997) Pre-mRNA splicing: the discovery of a new spliceosome doubles the challenge. *Trends Biochem. Sci.* 22, 132–137.

von Hippel PH (1998) An integrated model of the transcription complex in elongation, termination, and editing. *Science* 281, 660–665.

From RNA to Protein

Abelson J, Trotta CR & Li H (1998) tRNA splicing. *J. Biol. Chem.* 273, 12685–12688.

Anfinsen CB (1973) Principles that govern the folding of protein chains. *Science* 181, 223–230.

Cohen FE (1999) Protein misfolding and prion diseases. *J. Mol. Biol.* 293, 313–320.

Crick FHC (1966) The genetic code: III. *Sci. Am.* 215(4), 55–62.

Fedorov AN & Baldwin TO (1997) Cotranslational protein folding. *J. Biol. Chem.* 272, 32715–32718.

Frank J (2000) The ribosome—a macromolecular machine par excellence. *Chem. Biol.* 7, R133–141.

Green R (2000) Ribosomal translocation: EF-G turns the crank. *Curr. Biol.* 10, R369–373.

Hartl FU (1996) Molecular chaperones in cellular protein folding. *Nature* 381, 571–580.

Hershko A, Ciechanover A & Varshavsky A (2000) The ubiquitin system. *Nat. Med.* 6, 1073–1081.

Ibba M & Soll D (2000) Aminoacyl-tRNA synthesis. *Annu. Rev. Biochem.* 69, 617–650.

Kozak M (1999) Initiation of translation in prokaryotes and eukaryotes. *Gene* 234, 187–208.

Nissen P, Hansen J, Ban N et al. (2000) The structural basis of ribosome activity in peptide bond synthesis. *Science* 289, 920–930.

Nureki O, Vassylyev DG, Tateno M et al. (1998) Enzyme structure with two catalytic sites for double-sieve selection of substrate. *Science* 280, 578–582.

Prusiner SB (1998) Nobel lecture. Prions. *Proc. Natl. Acad. Sci. USA* 95, 13363–13383.

Rich A & Kim SH (1978) The three-dimensional structure of transfer RNA. *Sci. Am.* 238(1), 52–62.

Sachs AB & Varani G (2000) Eukaryotic translation initiation: there are (at least) two side to every story. *Nat. Struct. Biol.* 7, 356–361.

The Genetic Code. (1966) Cold Spring Harbor Symposium on Quantitative Biology, vol XXXI. Cold Spring Harbor, NY: Cold Spring Harbor Laboratory Press.

Turner GC & Varshavsky A (2000) Detecting and measuring cotranslational protein degradation *in vivo*. *Science* 289, 2117–2120.

Varshavsky A, Turner G, Du F et al. (2000) The ubiquitin system and the N-end rule pathway. *Biol. Chem.* 381, 779–789.

Voges D, Zwickl P & Baumeister W (1999) The 26S proteasome: a molecular machine designed for controlled proteolysis. *Annu. Rev. Biochem.* 68, 1015–1068.

Wilson KS & Noller HF (1998) Molecular movement inside the translational engine. *Cell* 92, 337–349.

Wimberly BT, Brodersen DE, Clemons WM et al. (2000) Structure of the 30S ribosomal subunit. *Nature* 407, 327–339.

Yusupov MM, Yusupova GZ, Baucom A et al. (2001) Crystal structure of the ribosome at 5.5A resolution. *Science* 292, 883–896.

The RNA World and the Origins of Life

Bartel DP & Unrau PJ (1999) Constructing an RNA world. *Trends Cell Biol.* 9, M9–M13.

Joyce GF (1992) Directed molecular evolution. *Sci. Am.* 267(6), 90–97.

Knight RD & Landweber LF (2000) The early evolution of the genetic code. *Cell* 101, 569–572.

Orgel L (2000) Origin of life. A simpler nucleic acid. *Science* 290, 1306–1307.

Szathmary E (1999) The origin of the genetic code: amino acids as cofactors in an RNA world. *Trends Genet.* 15, 223–229.

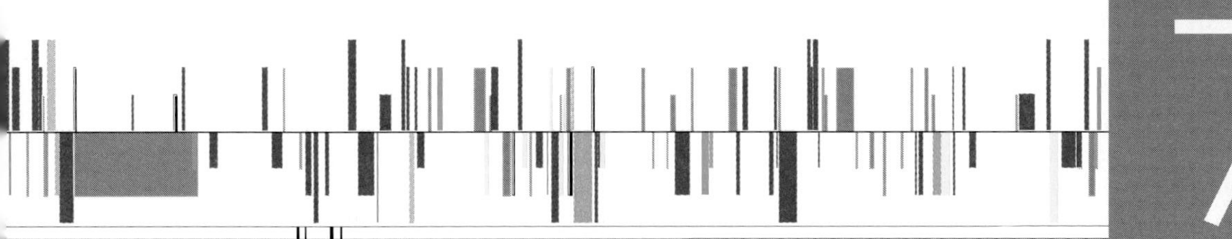

7

CONTROL OF GENE EXPRESSION

An organism's DNA encodes all of the RNA and protein molecules required to construct its cells. Yet a complete description of the DNA sequence of an organism—be it the few million nucleotides of a bacterium or the few billion nucleotides of a human—no more enables us to reconstruct the organism than a list of English words enables us to reconstruct a play by Shakespeare. In both cases the problem is to know how the elements in the DNA sequence or the words on the list are used. Under what conditions is each gene product made, and, once made, what does it do?

In this chapter we discuss the first half of this problem—the rules and mechanisms by which a subset of the genes is selectively expressed in each cell. The mechanisms that control the expression of genes operate at many levels, and we discuss the different levels in turn. At the end of the chapter, we examine how modern-day genomes and their systems of regulation have been shaped by evolutionary processes. We begin with an overview of some basic principles of gene control in multicellular organisms.

AN OVERVIEW OF GENE CONTROL

The different cell types in a multicellular organism differ dramatically in both structure and function. If we compare a mammalian neuron with a lymphocyte, for example, the differences are so extreme that it is difficult to imagine that the two cells contain the same genome (Figure 7–1). For this reason, and because cell differentiation is often irreversible, biologists originally suspected that genes might be selectively lost when a cell differentiates. We now know, however, that cell differentiation generally depends on changes in gene expression rather than on any changes in the nucleotide sequence of the cell's genome.

The Different Cell Types of a Multicellular Organism Contain the Same DNA

The cell types in a multicellular organism become different from one another because they synthesize and accumulate different sets of RNA and protein molecules. They generally do this without altering the sequence of their DNA. Evidence for the preservation of the genome during cell differentiation comes from a classic set of experiments in frogs. When the nucleus of a fully differentiated frog cell is injected into a frog egg whose nucleus has been removed, the injected donor nucleus is capable of directing the recipient egg to produce a normal tadpole (Figure 7–2A). Because the tadpole contains a full range of differentiated cells that derived their DNA sequences from the nucleus of the original donor cell, it follows that the differentiated donor cell cannot have lost any important DNA sequences. A similar conclusion has been reached in experiments performed with various plants. Here differentiated pieces of plant tissue are placed in culture and then dissociated into single cells. Often, one of these individual cells can regenerate an entire adult plant (Figure 7–2B). Finally, this same principle has been recently demonstrated in mammals, including sheep, cattle, pigs, goats, and mice by introducing nuclei from somatic cells into enucleated eggs; when placed into surrogate mothers, some of these eggs (called reconstructed zygotes) develop into healthy animals (Figure 7–2C).

Further evidence that large blocks of DNA are not lost or rearranged during vertebrate development comes from comparing the detailed banding patterns detectable in condensed chromosomes at mitosis (see Figure 4–11). By this criterion the chromosome sets of all differentiated cells in the human body appear to be identical. Moreover, comparisons of the genomes of different cells based on recombinant DNA technology have shown, as a general rule, that the changes in gene expression that underlie the development of multicellular organisms are not accompanied by changes in the DNA sequences of the corresponding genes. There are, however, a few cases where DNA rearrangements of the genome take place during the development of an organism—most notably, in generating the diversity of the immune system of mammals (discussed in Chapter 24).

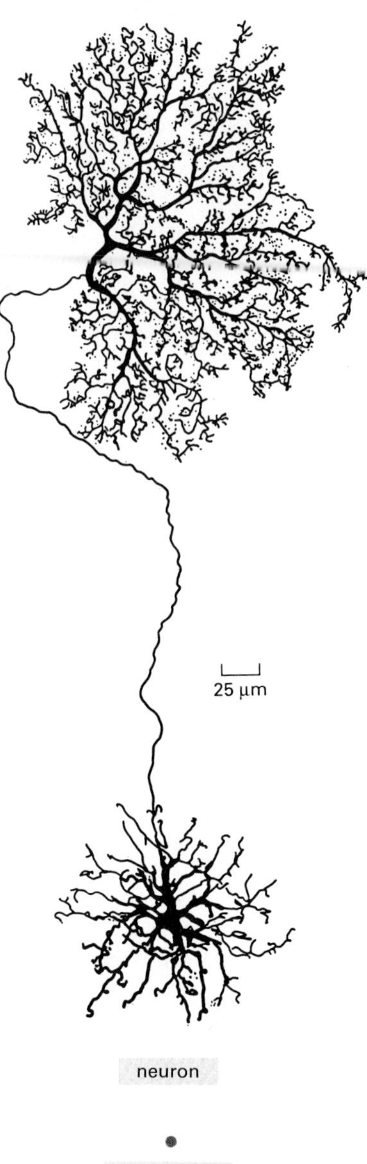

Figure 7–1 A mammalian neuron and a lymphocyte. The long branches of this neuron from the retina enable it to receive electrical signals from many cells and carry those signals to many neighboring cells. The lymphocyte is a white blood cell involved in the immune response to infection and moves freely through the body. Both of these cells contain the same genome, but they express different RNAs and proteins. (From B.B. Boycott, Essays on the Nervous System [R. Bellairs and E.G. Gray, eds.]. Oxford, UK: Clarendon Press, 1974.)

Different Cell Types Synthesize Different Sets of Proteins

As a first step in understanding cell differentiation, we would like to know how many differences there are between any one cell type and another. Although we still do not know the answer to this fundamental question, we can make certain general statements.

1. Many processes are common to all cells, and any two cells in a single organism therefore have many proteins in common. These include the structural proteins of chromosomes, RNA polymerases, DNA repair enzymes, ribosomal proteins, enzymes involved in the central reactions of metabolism, and many of the proteins that form the cytoskeleton.
2. Some proteins are abundant in the specialized cells in which they function and cannot be detected elsewhere, even by sensitive tests. Hemoglobin, for example, can be detected only in red blood cells.
3. Studies of the number of different mRNAs suggest that, at any one time, a typical human cell expresses approximately 10,000–20,000 of its approximately 30,000 genes. When the patterns of mRNAs in a series of different human cell lines are compared, it is found that the level of expression of almost every active gene varies from one cell type to another. A few of these differences are striking, like that of hemoglobin noted above but most are much more subtle. The patterns of mRNA abundance (determined using DNA microarrays, discussed in Chapter 8) are so characteristic of cell type that they can be used to type human cancer cells of uncertain tissue origin (Figure 7–3).
4. Although the differences in mRNAs among specialized cell types are striking, they nonetheless underestimate the full range of differences in the pattern of protein production. As we shall see in this chapter, there are many steps after transcription at which gene expression can be regulated. In addition, alternative splicing can produce a whole family of proteins from a single gene. Finally, proteins can be covalently modified after they are synthesized. Therefore a better way of appreciating the radical differences in gene expression between cell types is through the use of two-dimensional gel electrophoresis, where protein levels are directly measured and some of the most common posttranslational modifications are displayed (Figure 7–4).

A Cell Can Change the Expression of Its Genes in Response to External Signals

Most of the specialized cells in a multicellular organism are capable of altering their patterns of gene expression in response to extracellular cues. If a liver cell is exposed to a glucocorticoid hormone, for example, the production of several specific proteins is dramatically increased. Glucocorticoids are released in the body during periods of starvation or intense exercise and signal the liver to increase the production of glucose from amino acids and other small molecules;

Figure 7–2 Evidence that a differentiated cell contains all the genetic instructions necessary to direct the formation of a complete organism. (A) The nucleus of a skin cell from an adult frog transplanted into an enucleated egg can give rise to an entire tadpole. The *broken arrow* indicates that, to give the transplanted genome time to adjust to an embryonic environment, a further transfer step is required in which one of the nuclei is taken from the early embryo that begins to develop and is put back into a second enucleated egg. (B) In many types of plants, differentiated cells retain the ability to "dedifferentiate," so that a single cell can form a clone of progeny cells that later give rise to an entire plant. (C) A differentiated cell from an adult cow introduced into an enucleated egg from a different cow can give rise to a calf. Different calves produced from the same differentiated cell donor are genetically identical and are therefore clones of one another. (A, modified from J.B. Gurdon, *Sci. Am.* 219(6):24–35, 1968.)

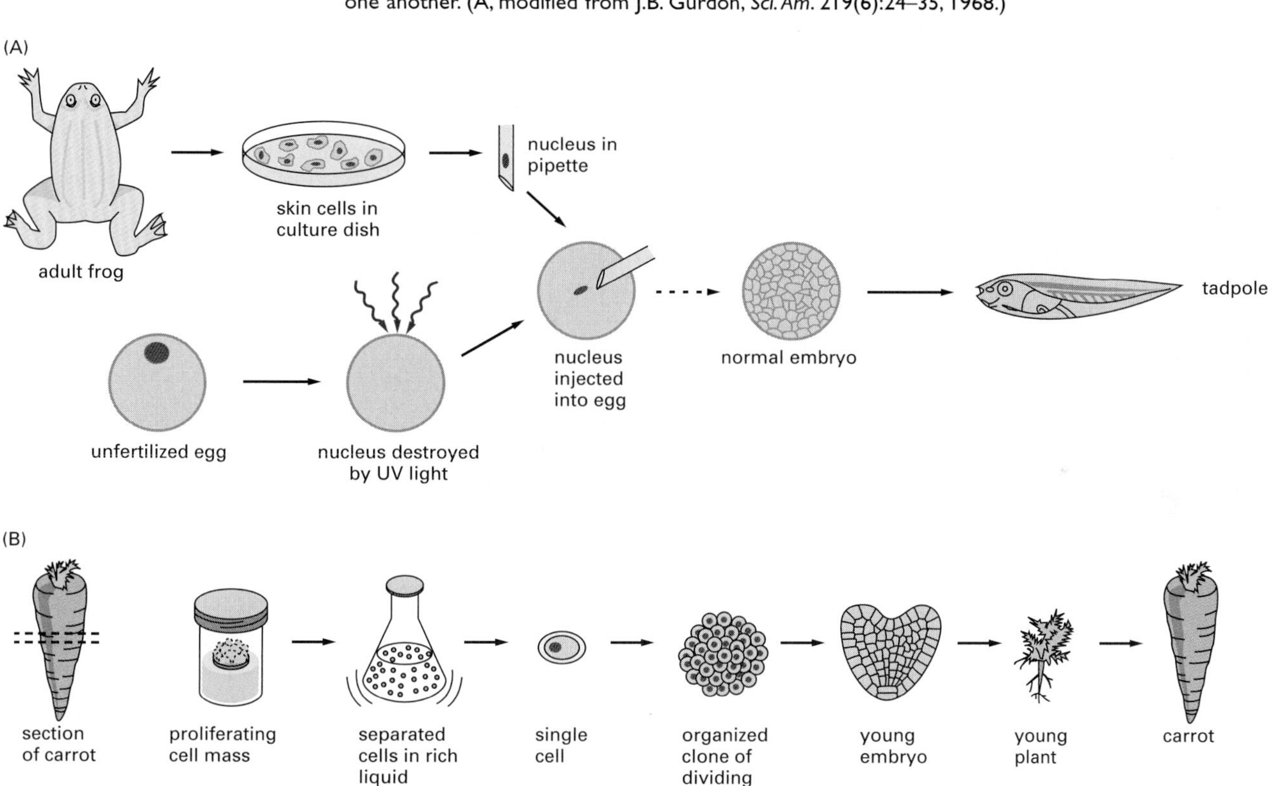

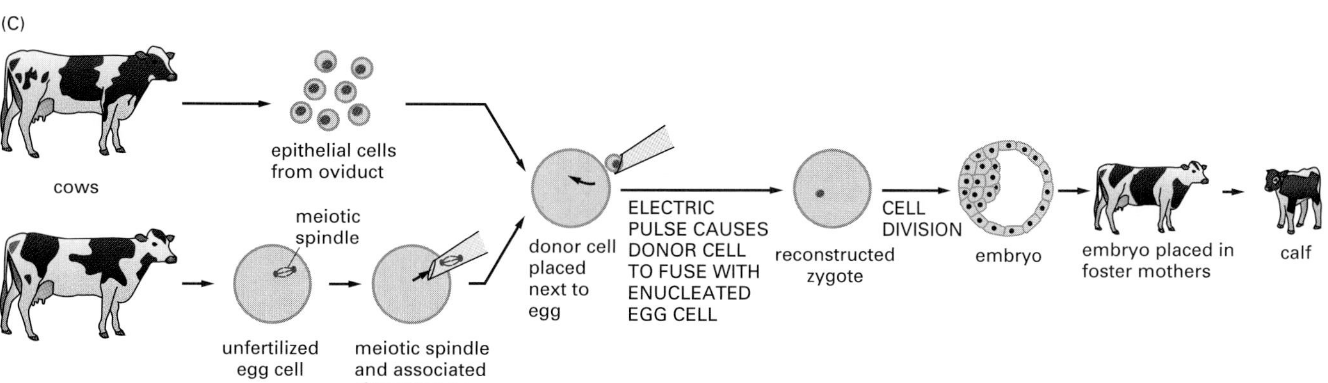

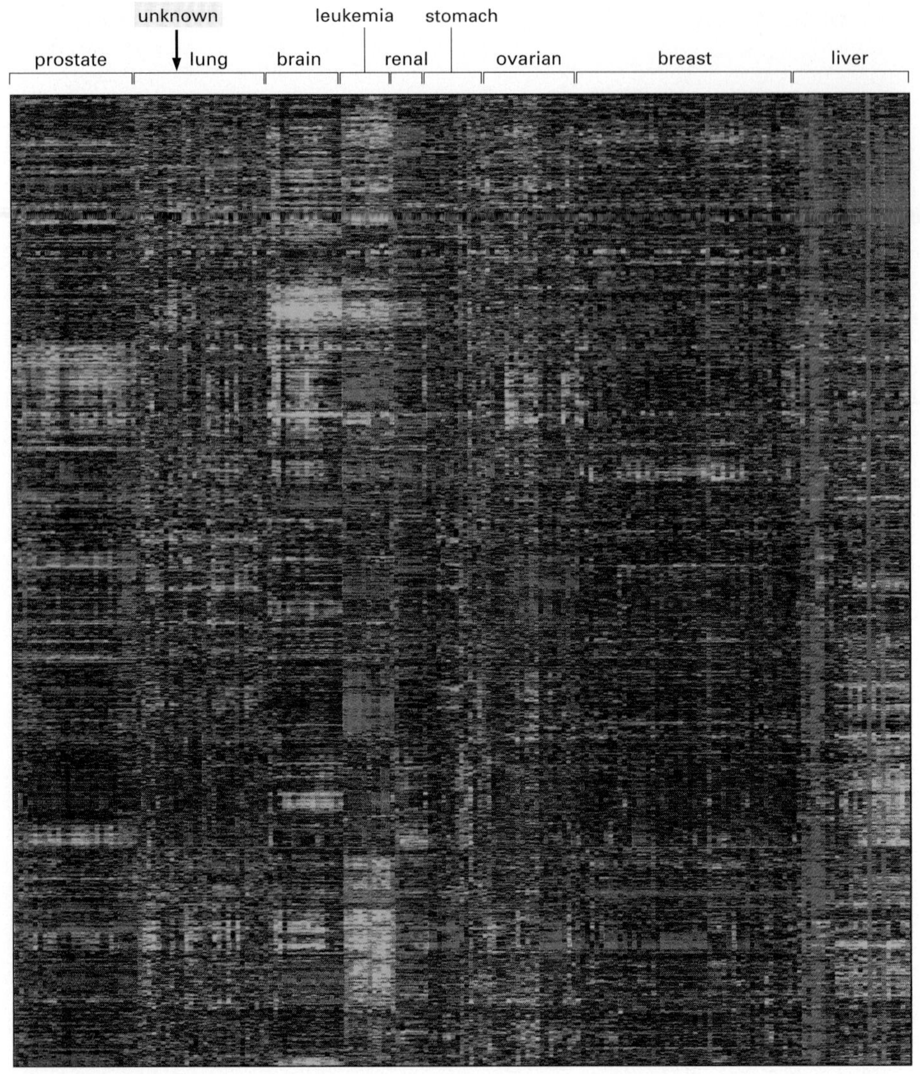

Figure 7–3 Differences in mRNA expression patterns among different types of human cancer cells. This figure summarizes a very large set of measurements in which the mRNA levels of 1800 selected genes (arranged *top* to *bottom)* were determined for 142 different human tumors (arranged *left* to *right*), each from a different patient. Each small *red* bar indicates that the given gene in the given tumor is transcribed at a level significantly higher than the average across all the cell lines. Each small *green* bar indicates a less-than-average expression level, and each *black* bar denotes an expression level that is close to average across the different tumors. The procedure used to generate these data—mRNA isolation followed by hybridization to DNA microarrays—is described in Chapter 8 (see pp. 533–535). The figure shows that the relative expression levels of each of the 1800 genes analyzed vary among the different tumors (seen by following a given gene *left* to *right* across the figure). This analysis also shows that each type of tumor has a characteristic gene expression pattern. This information can be used to "type" cancer cells of unknown tissue origin by matching the gene expression profiles to those of known tumors. For example, the unknown sample in the figure has been identified as a lung cancer. (Courtesy of Patrick O. Brown, David Botstein, and the Stanford Expression Collaboration.)

the set of proteins whose production is induced includes enzymes such as tyrosine aminotransferase, which helps to convert tyrosine to glucose. When the hormone is no longer present, the production of these proteins drops to its normal level.

Other cell types respond to glucocorticoids differently. In fat cells, for example, the production of tyrosine aminotransferase is reduced, while some other cell types do not respond to glucocorticoids at all. These examples illustrate a general feature of cell specialization: different cell types often respond in different ways to the same extracellular signal. Underlying such adjustments that occur in response to extracellular signals, there are features of the gene expression pattern that do not change and give each cell type its permanently distinctive character.

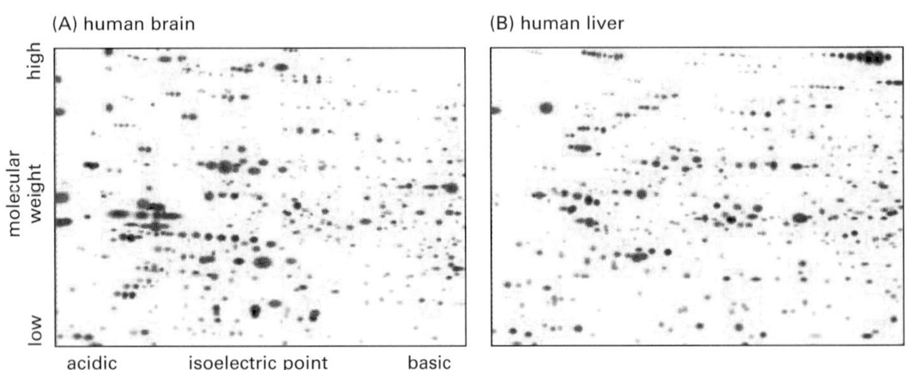

Figure 7–4 Differences in the proteins expressed by two human tissues. In each panel, the proteins have been displayed using two-dimensional polyacrylamide gel electrophoresis (see pp. 485–487). The proteins have been separated by molecular weight *(top* to *bottom)* and isoelectric point, the pH at which the protein has no net charge *(right* to *left*). The protein spots artificially colored *red* are common to both samples; those in *blue* are specific to one of the two tissues. The differences between the two tissue samples vastly outweigh their similarities: even for proteins that are shared between the two tissues, their relative abundance is usually different. Note that this technique separates proteins both by size and charge; therefore a protein that has, for example, several different phosphorylation states will appear as a series of *horizontal spots* (see *upper right-hand* portion of *right* panel). Only a small portion of the complete protein spectrum is shown for each sample. (Courtesy of Tim Myers and Leigh Anderson, Large Scale Biology Corporation.)

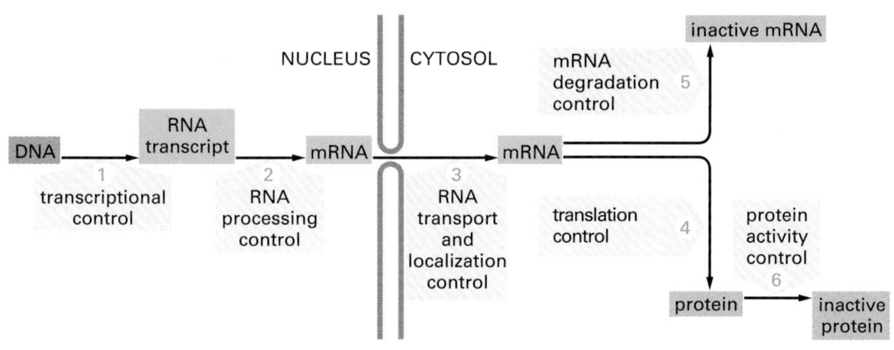

Figure 7–5 Six steps at which eucaryotic gene expression can be controlled. Controls that operate at steps 1 through 5 are discussed in this chapter. Step 6, the regulation of protein activity, includes reversible activation or inactivation by protein phosphorylation (discussed in Chapter 3) as well as irreversible inactivation by proteolytic degradation (discussed in Chapter 6).

Gene Expression Can Be Regulated at Many of the Steps in the Pathway from DNA to RNA to Protein

If differences among the various cell types of an organism depend on the particular genes that the cells express, at what level is the control of gene expression exercised? As we saw in the last chapter, there are many steps in the pathway leading from DNA to protein, and all of them can in principle be regulated. Thus a cell can control the proteins it makes by (1) controlling when and how often a given gene is transcribed (**transcriptional control**), (2) controlling how the RNA transcript is spliced or otherwise processed (**RNA processing control**), (3) selecting which completed mRNAs in the cell nucleus are exported to the cytosol and determining where in the cytosol they are localized (**RNA transport and localization control**), (4) selecting which mRNAs in the cytoplasm are translated by ribosomes (**translational control**), (5) selectively destabilizing certain mRNA molecules in the cytoplasm (**mRNA degradation control**), or (6) selectively activating, inactivating, degrading, or compartmentalizing specific protein molecules after they have been made (**protein activity control**) (Figure 7–5).

For most genes transcriptional controls are paramount. This makes sense because, of all the possible control points illustrated in Figure 7–5, only transcriptional control ensures that the cell will not synthesize superfluous intermediates. In the following sections we discuss the DNA and protein components that perform this function by regulating the initiation of gene transcription. We shall return at the end of the chapter to the additional ways of regulating gene expression.

Summary

The genome of a cell contains in its DNA sequence the information to make many thousands of different protein and RNA molecules. A cell typically expresses only a fraction of its genes, and the different types of cells in multicellular organisms arise because different sets of genes are expressed. Moreover, cells can change the pattern of genes they express in response to changes in their environment, such as signals from other cells. Although all of the steps involved in expressing a gene can in principle be regulated, for most genes the initiation of RNA transcription is the most important point of control.

DNA-BINDING MOTIFS IN GENE REGULATORY PROTEINS

How does a cell determine which of its thousands of genes to transcribe? As mentioned briefly in Chapters 4 and 6, the transcription of each gene is controlled by a regulatory region of DNA relatively near the site where transcription begins. Some regulatory regions are simple and act as switches that are thrown by a single signal. Many others are complex and act as tiny microprocessors, responding to a variety of signals that they interpret and integrate to switch the neighboring gene on or off. Whether complex or simple, these switching devices contain two types of fundamental components: (1) short stretches of DNA of defined sequence and (2) *gene regulatory proteins* that recognize and bind to them.

We begin our discussion of gene regulatory proteins by describing how these proteins were discovered.

Gene Regulatory Proteins Were Discovered Using Bacterial Genetics

Genetic analyses in bacteria carried out in the 1950s provided the first evidence for the existence of **gene regulatory proteins** that turn specific sets of genes on or off. One of these regulators, the *lambda repressor,* is encoded by a bacterial virus, *bacteriophage lambda.* The repressor shuts off the viral genes that code for the protein components of new virus particles and thereby enables the viral genome to remain a silent passenger in the bacterial chromosome, multiplying with the bacterium when conditions are favorable for bacterial growth (see Figure 5–81). The lambda repressor was among the first gene regulatory proteins to be characterized, and it remains one of the best understood, as we discuss later. Other bacterial regulators respond to nutritional conditions by shutting off genes encoding specific sets of metabolic enzymes when they are not needed. The *lac repressor,* the first of these bacterial proteins to be recognized, turns off the production of the proteins responsible for lactose metabolism when this sugar is absent from the medium.

The first step toward understanding gene regulation was the isolation of mutant strains of bacteria and bacteriophage lambda that were unable to shut off specific sets of genes. It was proposed at the time, and later proven, that most of these mutants were deficient in proteins acting as specific repressors for these sets of genes. Because these proteins, like most gene regulatory proteins, are present in small quantities, it was difficult and time-consuming to isolate them. They were eventually purified by fractionating cell extracts. Once isolated, the proteins were shown to bind to specific DNA sequences close to the genes that they regulate. The precise DNA sequences that they recognized were then determined by a combination of classical genetics, DNA sequencing, and DNA-footprinting experiments (discussed in Chapter 8).

The Outside of the DNA Helix Can Be Read by Proteins

As discussed in Chapter 4, the DNA in a chromosome consists of a very long double helix (Figure 7–6). Gene regulatory proteins must recognize specific nucleotide sequences embedded within this structure. It was originally thought that these proteins might require direct access to the hydrogen bonds between base pairs in the interior of the double helix to distinguish between one DNA sequence and another. It is now clear, however, that the outside of the double helix is studded with DNA sequence information that gene regulatory proteins can recognize without having to open the double helix. The edge of each base pair is exposed at the surface of the double helix, presenting a distinctive pattern of hydrogen bond donors, hydrogen bond acceptors, and hydrophobic patches for proteins to recognize in both the major and minor groove (Figure 7–7). But only in the major groove are the patterns markedly different for each of the four base-pair arrangements (Figure 7–8). For this reason, gene regulatory proteins generally bind to the major groove—as we shall see.

Although the patterns of hydrogen bond donor and acceptor groups are the most important features recognized by gene regulatory proteins, they are not the only ones: the nucleotide sequence also determines the overall geometry of the double helix, creating distortions of the "idealized" helix that can also be recognized.

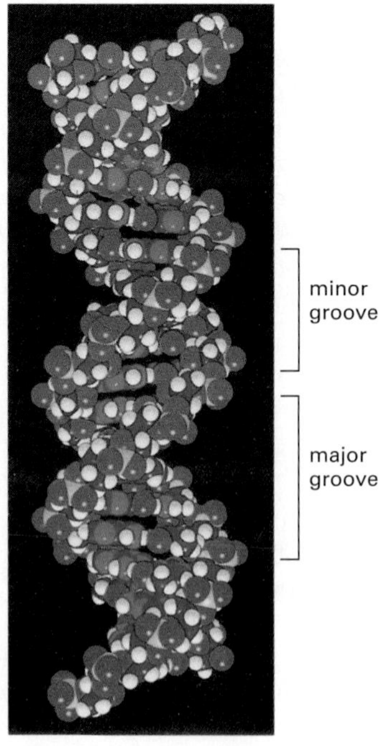

Figure 7–6 Double-helical structure of DNA. The major and minor grooves on the outside of the double helix are indicated. The atoms are colored as follows: carbon, *dark blue;* nitrogen, *light blue;* hydrogen, *white;* oxygen, *red;* phosphorus, *yellow.*

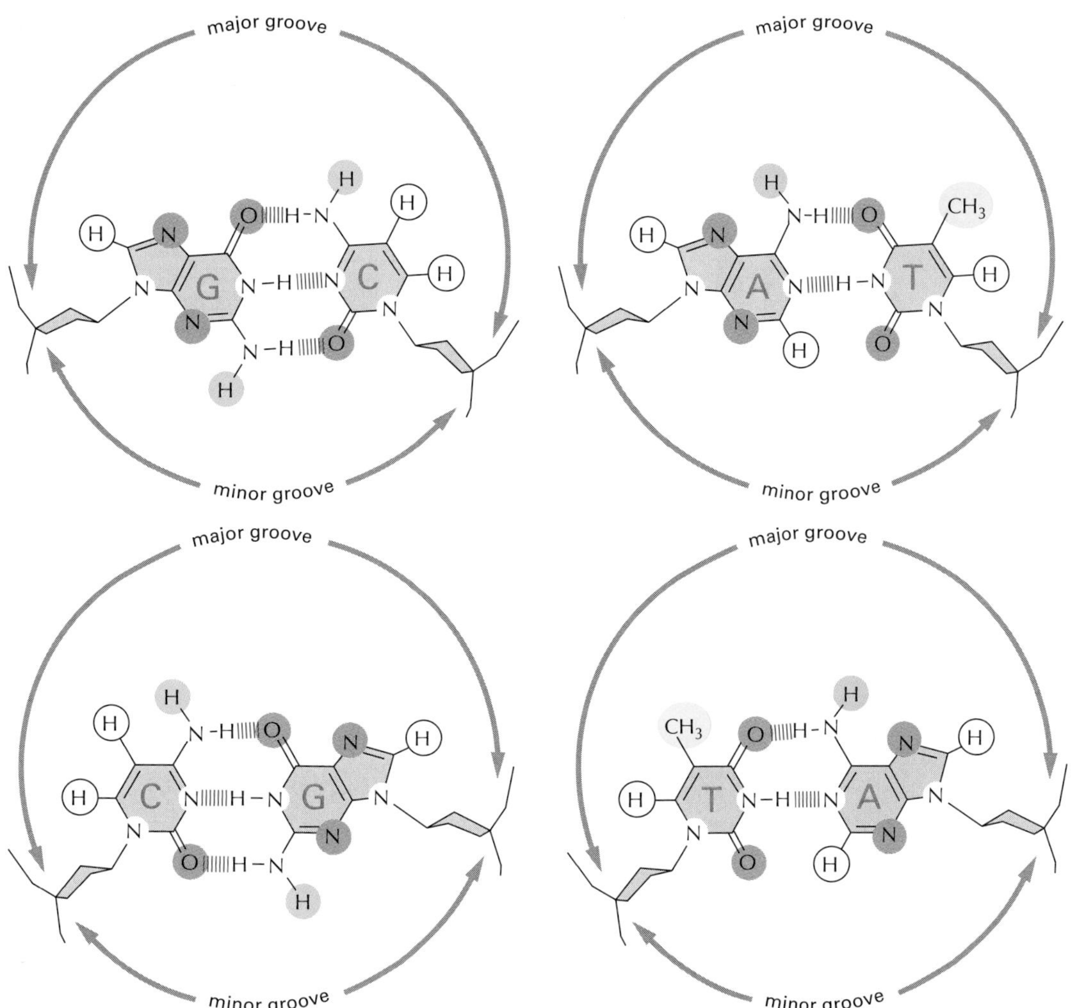

Figure 7–7 How the different base pairs in DNA can be recognized from their edges without the need to open the double helix. The four possible configurations of base pairs are shown, with potential hydrogen bond donors indicated in *blue,* potential hydrogen bond acceptors in *red,* and hydrogen bonds of the base pairs themselves as a series of short parallel *red* lines. Methyl groups, which form hydrophobic protuberances, are shown in *yellow,* and hydrogen atoms that are attached to carbons, and are therefore unavailable for hydrogen bonding, are *white.* (From C. Branden and J. Tooze, Introduction to Protein Structure, 2nd edn. New York: Garland Publishing, 1999.)

The Geometry of the DNA Double Helix Depends on the Nucleotide Sequence

For 20 years after the discovery of the DNA double helix in 1953, DNA was thought to have the same monotonous structure, with exactly 36° of helical twist between its adjacent nucleotide pairs (10 nucleotide pairs per helical turn) and a uniform helix geometry. This view was based on structural studies of heterogeneous mixtures of DNA molecules, however, and it changed once the three-dimensional structures of short DNA molecules of defined nucleotide sequence

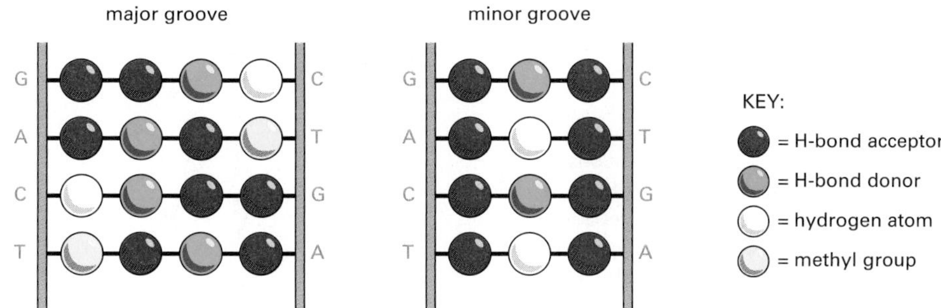

Figure 7–8 A DNA recognition code. The edge of each base pair, seen here looking directly at the major or minor groove, contains a distinctive pattern of hydrogen bond donors, hydrogen bond acceptors, and methyl groups. From the major groove, each of the four base-pair configurations projects a unique pattern of features. From the minor groove, however, the patterns are similar for G–C and C–G as well as for A–T and T–A. The color code is the same as that in Figure 7–7. (From C. Branden and J. Tooze, Introduction to Protein Structure, 2nd edn. New York: Garland Publishing, 1999.)

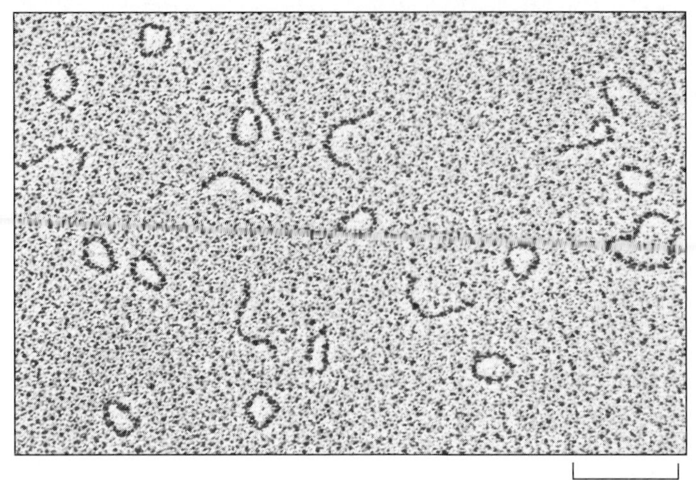

Figure 7–9 Electron micrograph of fragments of a highly bent segment of DNA double helix. The DNA fragments are derived from the small, circular mitochondrial DNA molecules of a trypanosome. Although the fragments are only about 200 nucleotide pairs long, many of them have bent to form a complete circle. On average, a normal DNA helix of this length would bend only enough to produce one-fourth of a circle (one smooth right-angle turn). (From J. Griffith, M. Bleyman, C.A. Raugh, P.A. Kitchin, and P.T. Englund, *Cell* 46:717–724, 1986. © Elsevier.)

were determined using x-ray crystallography and NMR spectroscopy. Whereas the earlier studies provided a picture of an average, idealized DNA molecule, the later studies showed that any given nucleotide sequence had local irregularities, such as tilted nucleotide pairs or a helical twist angle larger or smaller than 36°. These unique features can be recognized by specific DNA-binding proteins.

An especially striking departure from the average structure occurs in nucleotide sequences that cause the DNA double helix to bend. Some sequences (for example, AAAANNN, where N can be any base except A) form a double helix with a pronounced irregularity that causes a slight bend; if this sequence is repeated at 10-nucleotide-pair intervals in a long DNA molecule, the small bends add together so that the DNA molecule appears unusually curved when viewed in the electron microscope (Figure 7–9).

A related and equally important variable feature of DNA structure is the extent to which the double helix is deformable. For a protein to recognize and bind to a specific DNA sequence, there must be a tight fit between the DNA and the protein, and often the normal DNA conformation must be distorted to maximize this fit (Figure 7–10). The energetic cost of such distortion depends on the local nucleotide sequence. We encountered an example of this in the discussion of nucleosome assembly in Chapter 4: some DNA sequences can accommodate the tight DNA wrapping required for nucleosome formation better than others. Similarly, a few gene regulatory proteins induce a striking bend in the DNA when they bind to it (Figure 7–11). In general, these proteins recognize DNA sequences that are easily bent.

Short DNA Sequences Are Fundamental Components of Genetic Switches

We have seen how a specific nucleotide sequence can be detected as a pattern of structural features on the surface of the DNA double helix. Particular nucleotide sequences, each typically less than 20 nucleotide pairs in length, function as fundamental components of genetic switches by serving as recognition sites for the binding of specific gene regulatory proteins. Thousands of such DNA sequences have been identified, each recognized by a different gene regulatory protein (or by a set of related gene regulatory proteins). Some of the gene regulatory proteins that are discussed in the course of this chapter are listed in Table 7–1, along with the DNA sequences that they recognize.

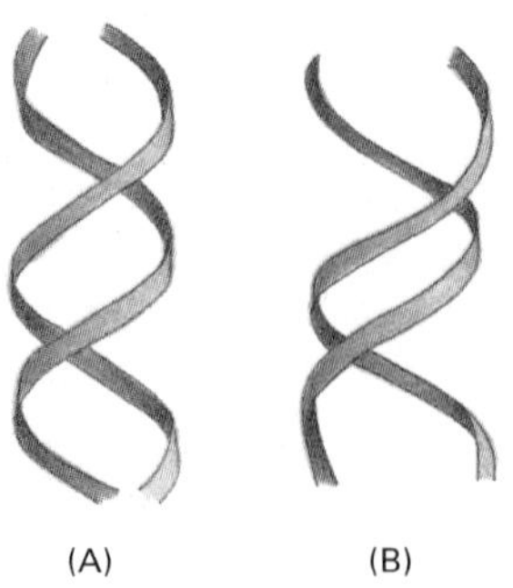

Figure 7–10 DNA deformation induced by protein binding. The figure shows the changes of DNA structure, from the conventional double-helix (A) to a distorted form (B) observed when a well-studied gene regulatory protein (the bacteriophage 434 repressor, a close relative of the lambda repressor) binds to specific sequences of DNA. The ease with which a DNA sequence can be deformed often affects the affinity of protein binding.

We now turn to the gene regulatory proteins themselves, the second fundamental component of genetic switches. We begin with the structural features that allows these proteins to recognize short, specific DNA sequences contained in a much longer double helix.

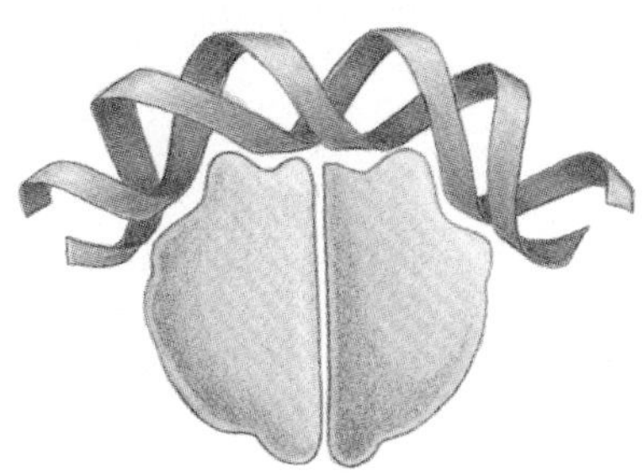

Figure 7–11 The bending of DNA induced by the binding of the catabolite activator protein (CAP). CAP is a gene regulatory protein from *E. coli*. In the absence of the bound protein, this DNA helix is straight.

Gene Regulatory Proteins Contain Structural Motifs That Can Read DNA Sequences

Molecular recognition in biology generally relies on an exact fit between the surfaces of two molecules, and the study of gene regulatory proteins has provided some of the clearest examples of this principle. A gene regulatory protein recognizes a specific DNA sequence because the surface of the protein is extensively complementary to the special surface features of the double helix in that region. In most cases the protein makes a large number of contacts with the DNA, involving hydrogen bonds, ionic bonds, and hydrophobic interactions. Although each individual contact is weak, the 20 or so contacts that are typically formed at the protein–DNA interface add together to ensure that the interaction is both highly specific and very strong (Figure 7–12). In fact, DNA–protein interactions include some of the tightest and most specific molecular interactions known in biology.

Although each example of protein–DNA recognition is unique in detail, x-ray crystallographic and NMR spectroscopic studies of several hundred gene regulatory proteins have revealed that many of the proteins contain one or another of a small set of DNA-binding structural motifs. These motifs generally use

TABLE 7–1 Some Gene Regulatory Proteins and the DNA Sequences That They Recognize

	NAME	DNA SEQUENCE RECOGNIZED*
Bacteria	lac repressor	5′ AATTGTGAGCGGATAACAATT 3′ TTAACACTCGCCTATTGTTAA
	CAP	TGTGAGTTAGCTCACT ACACTCAATCGAGTGA
	lambda repressor	TATCACCGCCAGAGGTA ATAGTGGCGGTCTCCAT
Yeast	Gal4	CGGAGGACTGTCCTCCG GCCTCCTGACAGGAGGC
	Matα2	CATGTAATT GTACATTAA
	Gcn4	ATGACTCAT TACTGAGTA
Drosophila	Kruppel	AACGGGTTAA TTGCCCAATT
	Bicoid	GGGATTAGA CCCTAATCT
Mammals	Sp1	GGGCGG CCCGCC
	Oct-1 Pou domain	ATGCAAAT TACGTTTA
	GATA-1	TGATAG ACTATC
	MyoD	CAAATG GTTTAC
	p53	GGGCAAGTCT CCCGTTCAGA

*Each protein in this table can recognize a set of closely related DNA sequences (see Figure 6–12); for convenience, only one recognition sequence, rather than a consensus sequence, is given for each protein.

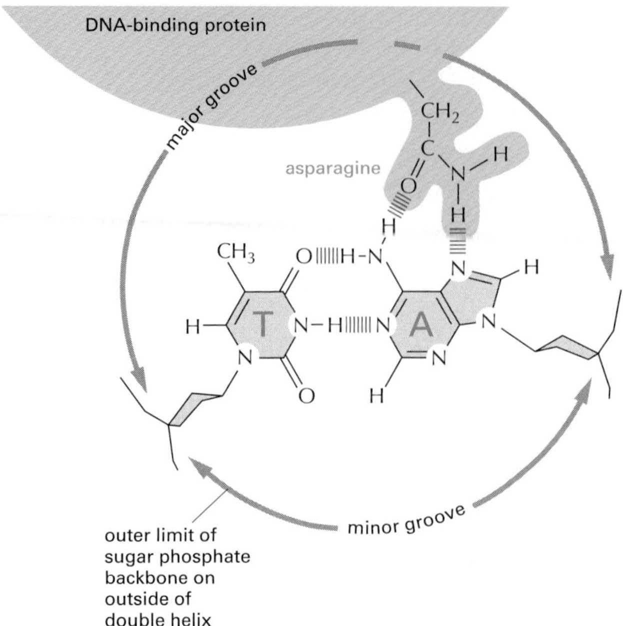

Figure 7–12 The binding of a gene regulatory protein to the major groove of DNA. Only a single contact is shown. Typically, the protein-DNA interface would consist of 10 to 20 such contacts, involving different amino acids, each contributing to the strength of the protein–DNA interaction.

either α helices or β sheets to bind to the major groove of DNA; this groove, as we have seen, contains sufficient information to distinguish one DNA sequence from any other. The fit is so good that it has been suggested that the dimensions of the basic structural units of nucleic acids and proteins evolved together to permit these molecules to interlock.

The Helix–Turn–Helix Motif Is One of the Simplest and Most Common DNA-binding Motifs

The first DNA-binding protein motif to be recognized was the **helix–turn–helix**. Originally identified in bacterial proteins, this motif has since been found in hundreds of DNA-binding proteins from both eucaryotes and procaryotes. It is constructed from two α helices connected by a short extended chain of amino acids, which constitutes the "turn" (Figure 7–13). The two helices are held at a fixed angle, primarily through interactions between the two helices. The more C-terminal helix is called the *recognition helix* because it fits into the major groove of DNA; its amino acid side chains, which differ from protein to protein, play an important part in recognizing the specific DNA sequence to which the protein binds.

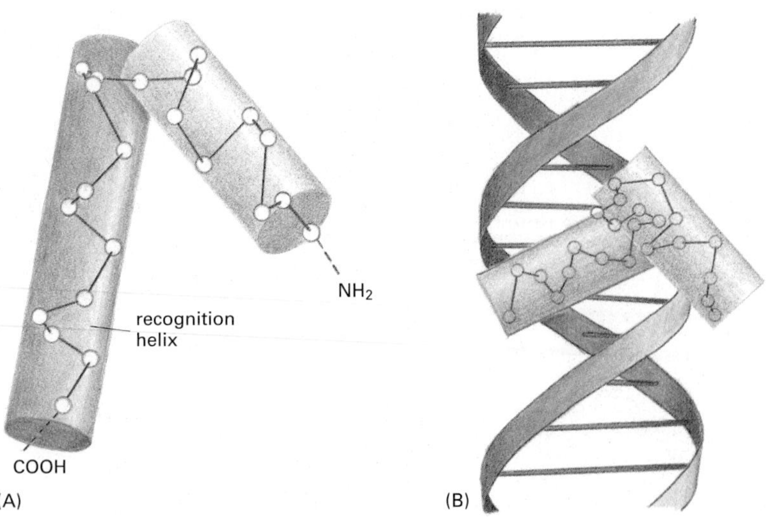

Figure 7–13 The DNA-binding helix–turn–helix motif. The motif is shown in (A), where each *white* circle denotes the central carbon of an amino acid. The C-terminal α helix *(red)* is called the recognition helix because it participates in sequence-specific recognition of DNA. As shown in (B), this helix fits into the major groove of DNA, where it contacts the edges of the base pairs (see also Figure 7–7). The N-terminal α-helix *(blue)* functions primarily as a structural component that helps to position the recognition helix.

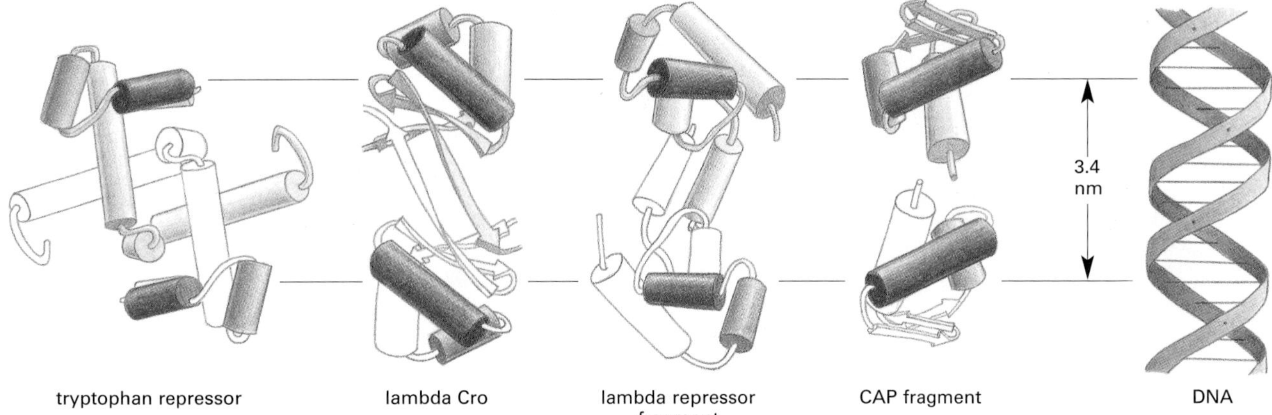

Figure 7–14 Some helix–turn–helix DNA-binding proteins. All of the proteins bind DNA as dimers in which the two copies of the recognition helix *(red cylinder)* are separated by exactly one turn of the DNA helix (3.4 nm). The other helix of the helix–turn–helix motif is colored *blue*, as in Figure 7–13. The lambda repressor and Cro proteins control bacteriophage lambda gene expression, and the tryptophan repressor and the catabolite activator protein (CAP) control the expression of sets of *E. coli* genes.

Outside the helix–turn–helix region, the structure of the various proteins that contain this motif can vary enormously (Figure 7–14). Thus each protein "presents" its helix–turn–helix motif to the DNA in a unique way, a feature thought to enhance the versatility of the helix–turn–helix motif by increasing the number of DNA sequences that the motif can be used to recognize. Moreover, in most of these proteins, parts of the polypeptide chain outside the helix–turn–helix domain also make important contacts with the DNA, helping to fine-tune the interaction.

The group of helix–turn–helix proteins shown in Figure 7–14 demonstrates a feature that is common to many sequence-specific DNA-binding proteins. They bind as symmetric dimers to DNA sequences that are composed of two very similar "half-sites," which are also arranged symmetrically (Figure 7–15). This arrangement allows each protein monomer to make a nearly identical set of contacts and enormously increases the binding affinity: as a first approximation, doubling the number of contacts doubles the free energy of the interaction and thereby *squares* the affinity constant.

Homeodomain Proteins Constitute a Special Class of Helix–Turn–Helix Proteins

Not long after the first gene regulatory proteins were discovered in bacteria, genetic analyses in the fruit fly *Drosophila* led to the characterization of an important class of genes, the *homeotic selector genes*, that play a critical part in orchestrating fly development. As discussed in Chapter 21, they have since proved to have a fundamental role in the development of higher animals as well. Mutations in these genes cause one body part in the fly to be converted into another, showing that the proteins they encode control critical developmental decisions.

When the nucleotide sequences of several homeotic selector genes were determined in the early 1980s, each proved to contain an almost identical stretch of 60 amino acids that defines this class of proteins and is termed the **homeodomain**. When the three-dimensional structure of the homeodomain was determined, it was seen to contain a helix–turn–helix motif related to that of the bacterial gene regulatory proteins, providing one of the first indications that the principles of gene regulation established in bacteria are relevant to higher organisms as well. More than 60 homeodomain proteins have now been discovered in *Drosophila* alone, and homeodomain proteins have been identified in virtually all eucaryotic organisms that have been studied, from yeasts to plants to humans.

5′ T A A C A C C G T G C G T G T T G 3′
3′ A T T G T G G C A C G C A C A A C 5′

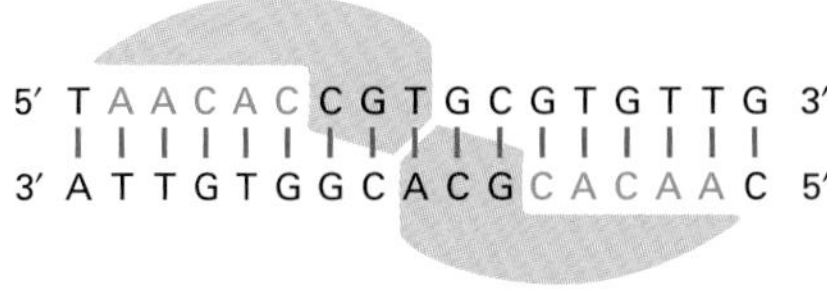

Figure 7–15 A specific DNA sequence recognized by the bacteriophage lambda Cro protein. The nucleotides labeled in *green* in this sequence are arranged symmetrically, allowing each half of the DNA site to be recognized in the same way by each protein monomer, also shown in *green*. See Figure 7–14 for the actual structure of the protein.

The structure of a homeodomain bound to its specific DNA sequence is shown in Figure 7–16. Whereas the helix–turn–helix motif of bacterial gene regulatory proteins is often embedded in different structural contexts, the helix–turn–helix motif of homeodomains is always surrounded by the same structure (which forms the rest of the homeodomain), suggesting that the motif is always presented to DNA in the same way. Indeed, structural studies have

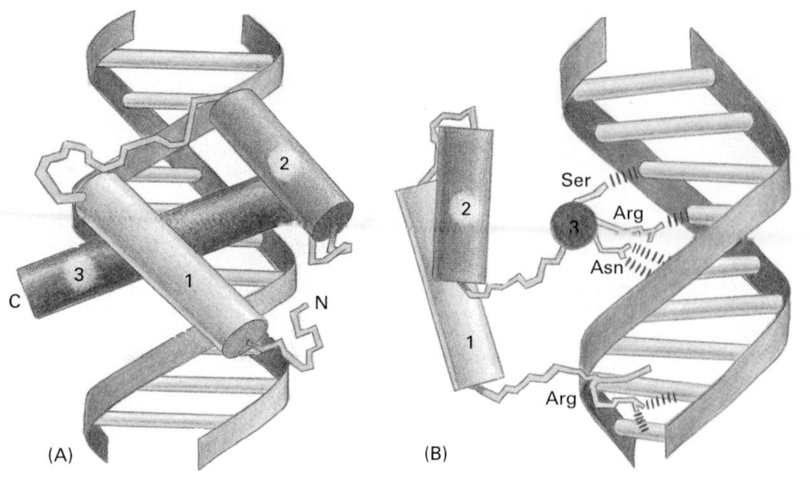

Figure 7–16 A homeodomain bound to its specific DNA sequence. Two different views of the same structure are shown. (A) The homeodomain is folded into three α helices, which are packed tightly together by hydrophobic interactions. The part containing helix 2 and 3 closely resembles the helix–turn–helix motif. (B) The recognition helix (helix 3, *red*) makes important contacts with the major groove of DNA. The asparagine (Asn) of helix 3, for example, contacts an adenine, as shown in Figure 7–12. Nucleotide pairs are also contacted in the minor groove by a flexible arm attached to helix 1. The homeodomain shown here is from a yeast gene regulatory protein, but it closely resembles homeodomains from many eucaryotic organisms. (Adapted from C. Wolberger et al., *Cell* 67:517–528, 1991.)

shown that a yeast homeodomain protein and a *Drosophila* homeodomain protein have very similar conformations and recognize DNA in almost exactly the same manner, although they are identical at only 17 of 60 amino acid positions (see Figure 3–15).

There Are Several Types of DNA-binding Zinc Finger Motifs

The helix–turn–helix motif is composed solely of amino acids. A second important group of DNA-binding motifs adds one or more zinc atoms as structural components. Although all such zinc-coordinated DNA-binding motifs are called **zinc fingers**, this description refers only to their appearance in schematic drawings dating from their initial discovery (Figure 7–17A). Subsequent structural studies have shown that they fall into several distinct structural groups, two of which are considered here. The first type was initially discovered in the protein that activates the transcription of a eucaryotic ribosomal RNA gene. It is a simple structure, consisting of an α helix and a β sheet held together by the zinc (Figure 7–17B). This type of zinc finger is often found in a cluster with additional zinc fingers, arranged one after the other so that the α helix of each can contact the major groove of the DNA, forming a nearly continuous stretch of α helices along the groove. In this way, a strong and specific DNA-protein interaction is built up through a repeating basic structural unit (Figure 7–18). A particular advantage of this motif is that the strength and specificity of the DNA-protein interaction can be adjusted during evolution by changes in the number of zinc finger repeats. By contrast, it is difficult to imagine how any of the other DNA-binding motifs discussed in this chapter could be formed into repeating chains.

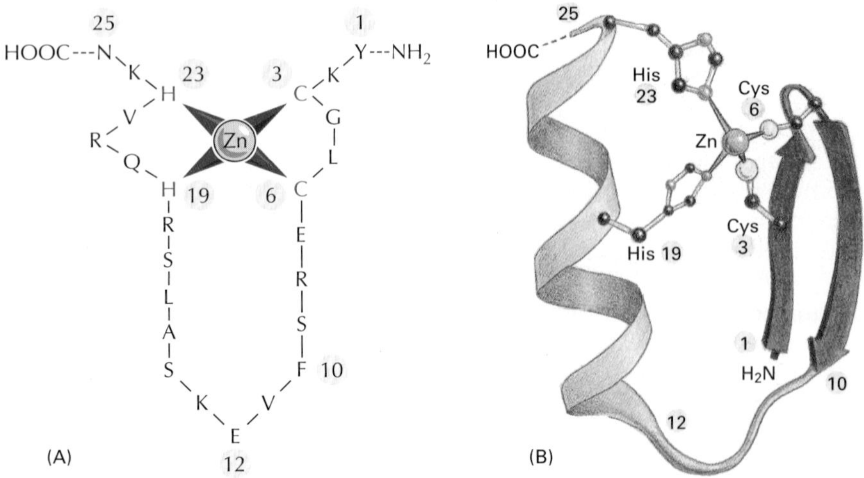

Figure 7–17 One type of zinc finger protein. This protein belongs to the Cys–Cys–His–His family of zinc finger proteins, named after the amino acids that grasp the zinc. (A) Schematic drawing of the amino acid sequence of a zinc finger from a frog protein of this class. (B) The three-dimensional structure of this type of zinc finger is constructed from an antiparallel β sheet (amino acids 1 to 10) followed by an α helix (amino acids 12 to 24). The four amino acids that bind the zinc (Cys 3, Cys 6, His 19, and His 23) hold one end of the α helix firmly to one end of the β sheet. (Adapted from M.S. Lee et al., *Science* 245:635–637, 1989.)

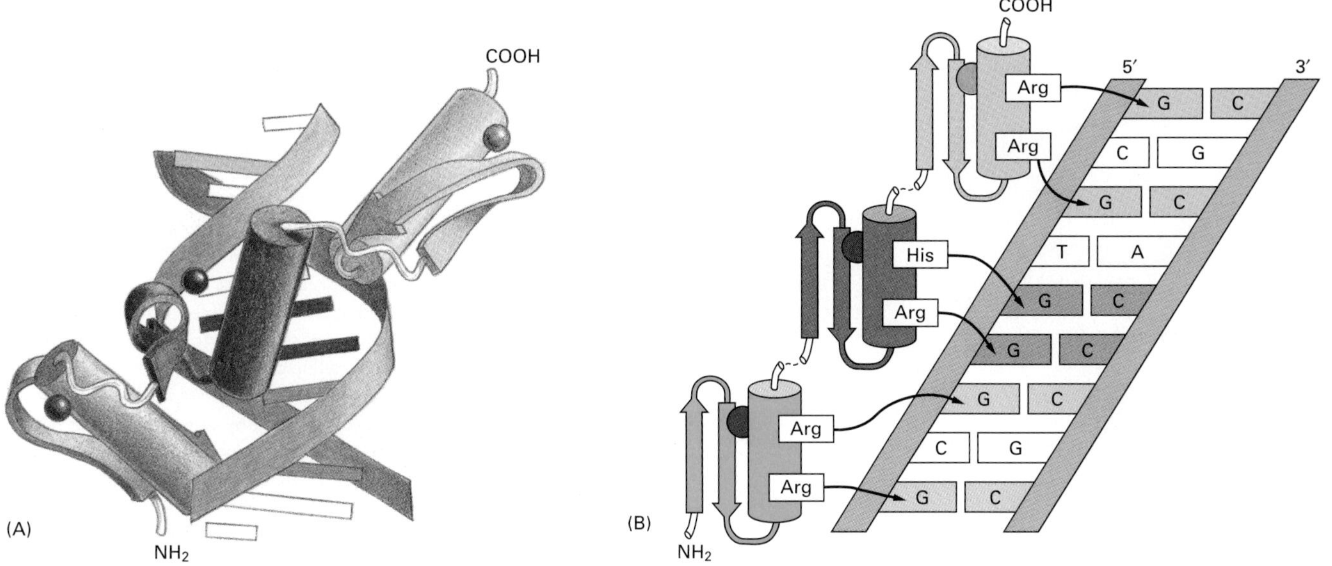

Figure 7–18 DNA binding by a zinc finger protein. (A) The structure of a fragment of a mouse gene regulatory protein bound to a specific DNA site. This protein recognizes DNA using three zinc fingers of the Cys–Cys–His–His type (see Figure 7–17) arranged as direct repeats. (B) The three fingers have similar amino acid sequences and contact the DNA in similar ways. In both (A) and (B) the zinc atom in each finger is represented by a small sphere. (Adapted from N. Pavletich and C. Pabo, *Science* 252:810–817, 1991.)

Another type of zinc finger is found in the large family of intracellular receptor proteins (discussed in detail in Chapter 15). It forms a different type of structure (similar in some respects to the helix–turn–helix motif) in which two α helices are packed together with zinc atoms (Figure 7–19). Like the helix–turn– helix proteins, these proteins usually form dimers that allow one of the two α helices of each subunit to interact with the major groove of the DNA (see Figure 7–14). Although the two types of zinc finger structures discussed in this section are structurally distinct, they share two important features: both use zinc as a structural element, and both use an α helix to recognize the major groove of the DNA.

β sheets Can Also Recognize DNA

In the DNA-binding motifs discussed so far, α helices are the primary mechanism used to recognize specific DNA sequences. One group of gene regulatory proteins, however, has evolved an entirely different and no less ingenious recognition strategy. In this case the information on the surface of the major groove is read by a two-stranded β sheet, with side chains of the amino acids extending from the sheet toward the DNA as shown in Figure 7–20. As in the case of a

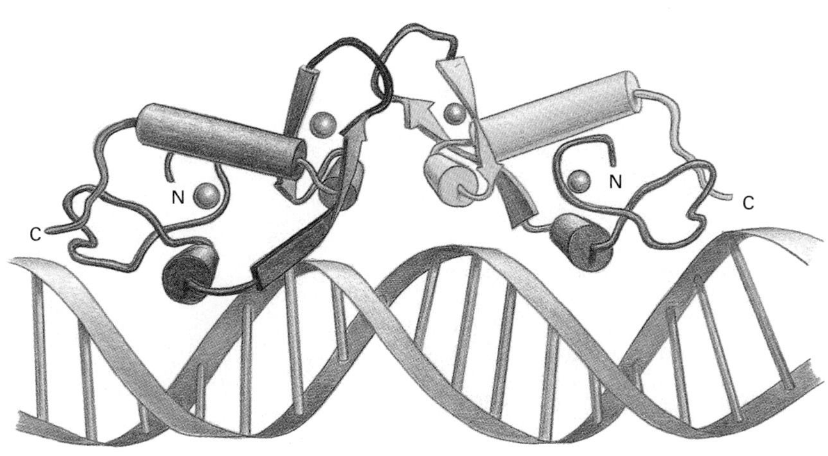

Figure 7–19 A dimer of the zinc finger domain of the intracellular receptor family bound to its specific DNA sequence. Each zinc finger domain contains two atoms of Zn (indicated by the small *gray spheres);* one stabilizes the DNA recognition helix (shown in *brown* in one subunit and *red* in the other), and one stabilizes a loop (shown in *purple)* involved in dimer formation. Each Zn atom is coordinated by four appropriately spaced cysteine residues. Like the helix–turn–helix proteins shown in Figure 7–14, the two recognition helices of the dimer are held apart by a distance corresponding to one turn of the DNA double helix. The specific example shown is a fragment of the glucocorticoid receptor. This is the protein through which cells detect and respond transcriptionally to the glucocorticoid hormones produced in the adrenal gland in response to stress. (Adapted from B.F. Luisi et al., *Nature* 352:497–505, 1991.)

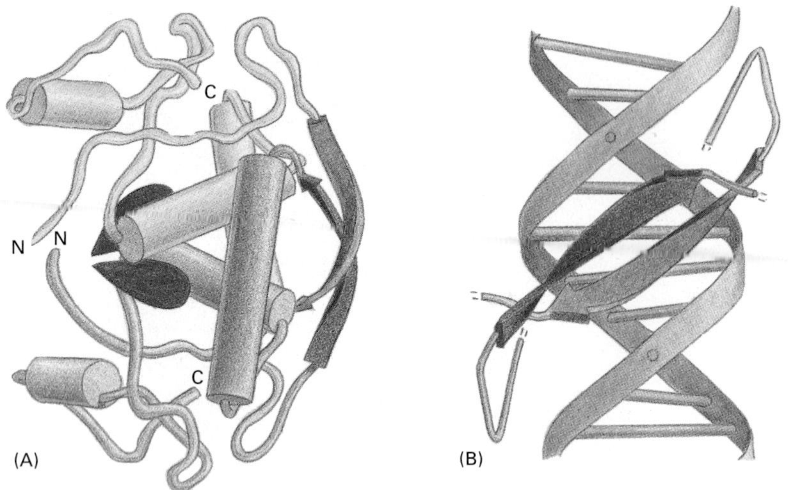

Figure 7–20 The bacterial *met* repressor protein. The bacterial *met* repressor regulates the genes encoding the enzymes that catalyze methionine synthesis. When this amino acid is abundant, it binds to the repressor, causing a change in the structure of the protein that enables it to bind to DNA tightly, shutting off the synthesis of the enzymes. (A) In order to bind to DNA tightly, the *met* repressor must be complexed with S-adenosyl methionine, shown in *red*. One subunit of the dimeric protein is shown in *green*, while the other is shown in *blue*. The two-stranded β sheet that binds to DNA is formed by one strand from each subunit and is shown in *dark green* and *dark blue*. (B) Simplified diagram of the *met* repressor bound to DNA, showing how the two-stranded β sheet of the repressor binds to the major groove of DNA. For clarity, the other regions of the repressor have been omitted. (A, adapted from S. Phillips, *Curr. Opin. Struct. Biol.* 1:89–98, 1991; B, adapted from W. Somers and S. Phillips, *Nature* 359:387–393, 1992.)

recognition α helix, this β-sheet motif can be used to recognize many different DNA sequences; the exact DNA sequence recognized depends on the sequence of amino acids that make up the β sheet.

The Leucine Zipper Motif Mediates Both DNA Binding and Protein Dimerization

Many gene regulatory proteins recognize DNA as homodimers, probably because, as we have seen, this is a simple way of achieving strong specific binding (see Figure 7–15). Usually, the portion of the protein responsible for dimerization is distinct from the portion that is responsible for DNA binding (see Figure 7–14). One motif, however, combines these two functions in an elegant and economical way. It is called the **leucine zipper motif**, so named because of the way the two α helices, one from each monomer, are joined together to form a short coiled-coil (see Figure 3–11). The helices are held together by interactions between hydrophobic amino acid side chains (often on leucines) that extend from one side of each helix. Just beyond the dimerization interface the two α helices separate from each other to form a Y-shaped structure, which allows their side chains to contact the major groove of DNA. The dimer thus grips the double helix like a clothespin on a clothesline (Figure 7–21).

Heterodimerization Expands the Repertoire of DNA Sequences Recognized by Gene Regulatory Proteins

Many of the gene regulatory proteins we have seen thus far bind DNA as homodimers, that is, dimers made up of two identical subunits. However, many gene regulatory proteins, including leucine zipper proteins, can also associate with nonidentical partners to form heterodimers composed of two different subunits. Because heterodimers typically form from two proteins with distinct DNA-binding specificities, the mixing and matching of gene regulatory proteins to form heterodimers greatly expands the repertoire of DNA-binding specificities that these proteins can display. As illustrated in Figure 7–22, three distinct DNA-binding specificities could, in principle, be generated from two types of

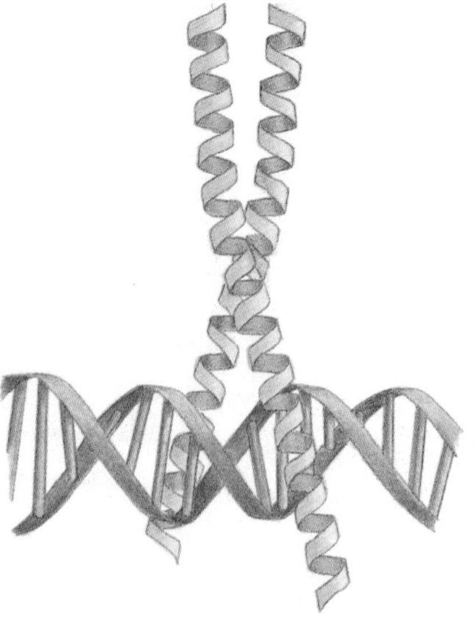

Figure 7–21 A leucine zipper dimer bound to DNA. Two α-helical DNA-binding domains (*bottom*) dimerize through their α-helical leucine zipper region (*top*) to form an inverted Y-shaped structure. Each arm of the Y is formed by a single α helix, one from each monomer, that mediates binding to a specific DNA sequence in the major groove of DNA. Each α helix binds to one-half of a symmetric DNA structure. The structure shown is of the yeast Gcn4 protein, which regulates transcription in response to the availability of amino acids in the environment. (Adapted from T.E. Ellenberger et al., *Cell* 71:1223–1237, 1992.)

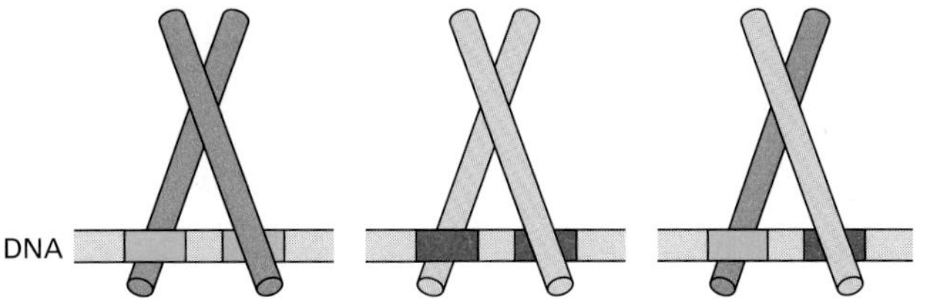

Figure 7–22 Heterodimerization of leucine zipper proteins can alter their DNA-binding specificity. Leucine zipper homodimers bind to symmetric DNA sequences, as shown in the left-hand and center drawings. These two proteins recognize different DNA sequences, as indicated by the *red* and *blue* regions in the DNA. The two different monomers can combine to form a heterodimer, which now recognizes a hybrid DNA sequence, composed from one *red* and one *blue* region.

leucine zipper monomer, while six could be created from three types of monomer, and so on.

There are, however, limits to this promiscuity: if all the many types of leucine zipper proteins in a typical eucaryotic cell formed heterodimers, the amount of "cross-talk" between the gene regulatory circuits of a cell would be so great as to cause chaos. Whether or not a particular heterodimer can form depends on how well the hydrophobic surfaces of the two leucine zipper α helices mesh with each other, which, in turn, depends on the exact amino acid sequences of the two zipper regions. Thus each leucine zipper protein in the cell can form dimers with only a small set of other leucine zipper proteins.

Heterodimerization is an example of **combinatorial control**, in which combinations of different proteins, rather than individual proteins, control a cellular process. Heterodimerization is one of the mechanisms used by eucaryotic cells to control gene expression in this way, and it occurs in a wide variety of different types of gene regulatory proteins (Figure 7–23). As we discuss later, however, the formation of heterodimeric gene regulatory complexes is only one of several combinatorial mechanisms for controlling gene expression.

During the evolution of gene regulatory proteins, similar combinatorial principles have produced new DNA-binding specificities by joining two distinct DNA-binding domains into a single polypeptide chain (Figure 7–24).

The Helix–Loop–Helix Motif Also Mediates Dimerization and DNA Binding

Another important DNA-binding motif, related to the leucine zipper, is the **helix–loop–helix (HLH) motif**, which should not be confused with the

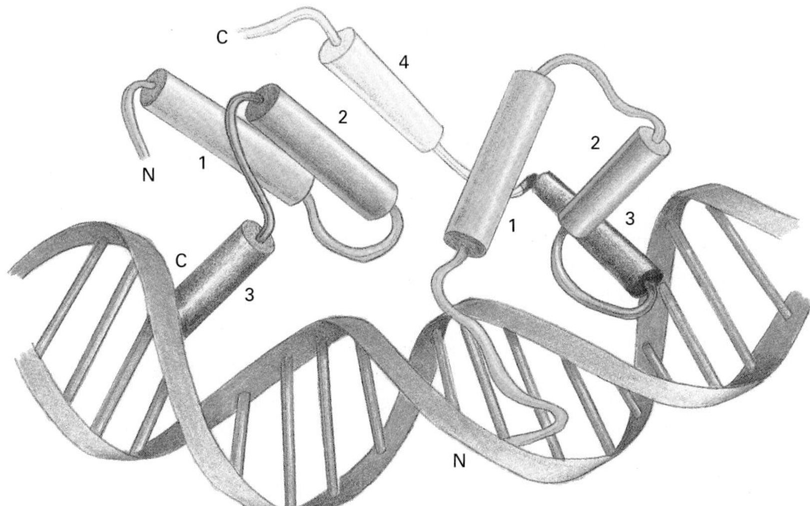

Figure 7–23 A heterodimer composed of two homeodomain proteins bound to its DNA recognition site. The *yellow* helix 4 of the protein on the right (Matα2) is unstructured in the absence of the protein on the left (Mata1), forming a helix only upon heterodimerization. The DNA sequence is thereby recognized jointly by both proteins; some of the protein–DNA contacts made by α2 were shown in Figure 7–16. These two proteins are from budding yeast, where the heterodimer specifies a particular cell type (see Figure 7–65). The helices are numbered in accordance with Figure 7–16. (Adapted from T. Li et al., *Science* 270:262–269, 1995.)

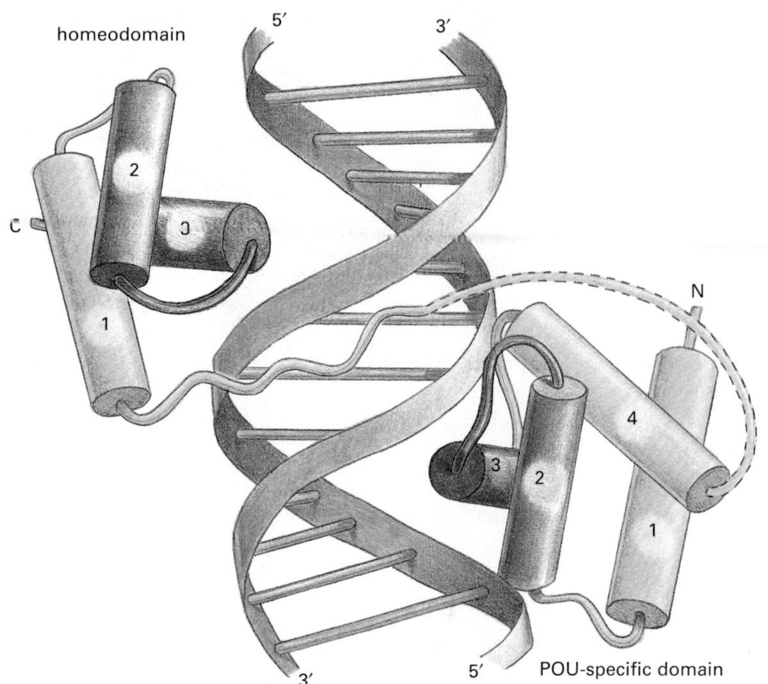

Figure 7–24 Two DNA-binding domains covalently joined by a flexible polypeptide. The structure shown (called a POU-domain) consists of both a homeodomain and a helix–turn–helix structure (closely related to the bacteriophage λ repressor—see Figure 7–14) joined by a flexible polypeptide "leash," indicated by the broken lines. The entire protein is encoded in a single gene and is synthesized as a continuous polypeptide chain. The covalent joining of two structures in this way results in a large increase in the affinity of the protein for its specific DNA sequence compared with the DNA affinity of either separate structure. The group of mammalian gene regulatory proteins exemplified by this structure regulate the production of growth factors, immunoglobulins, and other molecules involved in development. The particular example shown is from the Oct-1 protein. (Adapted from J.D. Klemm et al., *Cell* 77:21–32, 1994.)

helix–turn–helix motif discussed earlier. An HLH motif consists of a short α helix connected by a loop to a second, longer α helix. The flexibility of the loop allows one helix to fold back and pack against the other. As shown in Figure 7–25, this two-helix structure binds both to DNA and to the HLH motif of a second HLH protein. As with leucine zipper proteins, the second HLH protein can be the same (creating a homodimer) or different (creating a heterodimer). In either case, two α helices that extend from the dimerization interface make specific contacts with the DNA.

Several HLH proteins lack the α-helical extension responsible for binding to DNA. These truncated proteins can form heterodimers with full-length HLH proteins, but the heterodimers are unable to bind DNA tightly because they form only half of the necessary contacts. Thus, in addition to creating active dimers, heterodimerization provides a way to hold specific gene regulatory proteins in check (Figure 7–26).

It Is Not Yet Possible to Accurately Predict the DNA Sequences Recognized by All Gene Regulatory Proteins

The various DNA-binding motifs that we have discussed provide structural frameworks from which specific amino acid side chains extend to contact specific base pairs in the DNA. It is reasonable to ask, therefore, whether there is a

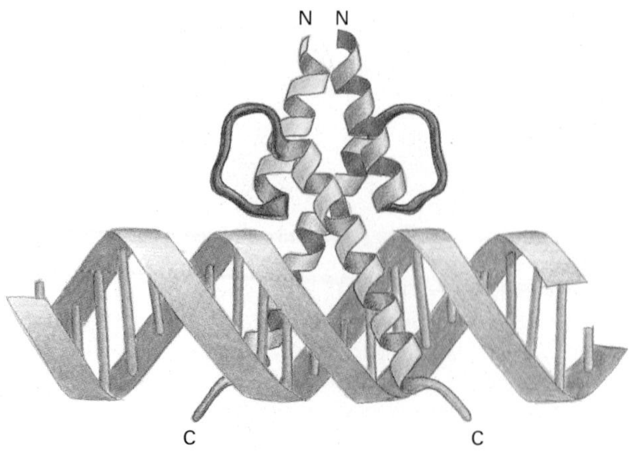

Figure 7–25 A helix–loop–helix dimer bound to DNA. The two monomers are held together in a four-helix bundle: each monomer contributes two α helices connected by a flexible loop of protein (*red*). A specific DNA sequence is bound by the two α helices that project from the four-helix bundle. (Adapted from A.R. Ferre-D'Amare et al., *Nature* 363:38–45, 1993.)

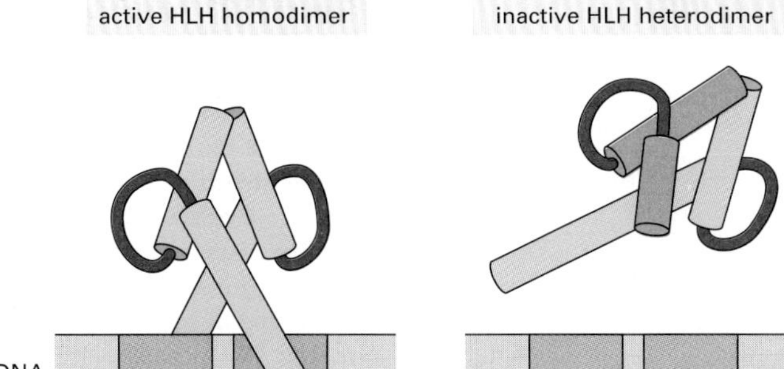

Figure 7–26 Inhibitory regulation by truncated HLH proteins. The HLH motif is responsible for both dimerization and DNA binding. On the *left*, an HLH homodimer recognizes a symmetric DNA sequence. On the *right*, the binding of a full-length HLH protein *(blue)* to a truncated HLH protein *(green)* that lacks the DNA-binding α helix generates a heterodimer that is unable to bind DNA tightly. If present in excess, the truncated protein molecule blocks the homodimerization of the full-length HLH protein and thereby prevents it from binding to DNA.

simple amino acid–base pair recognition code: is a G–C base pair, for example, always contacted by a particular amino acid side chain? The answer appears to be no, although certain types of amino acid-base interactions appear much more frequently than others (Figure 7–27). As we saw in Chapter 3, protein surfaces of virtually any shape and chemistry can be made from just 20 different amino acids, and a gene regulatory protein uses different combinations of these to create a surface that is precisely complementary to a particular DNA sequence. We know that the same base pair can thereby be recognized in many ways depending on its context (Figure 7–28). Nevertheless, molecular biologists are beginning to understand protein–DNA recognition well enough that we should soon be able to design proteins that will recognize any desired DNA sequence.

A Gel-Mobility Shift Assay Allows Sequence-specific DNA-binding Proteins to Be Detected Readily

Genetic analyses, which provided a route to the gene regulatory proteins of bacteria, yeast, and *Drosophila*, is much more difficult in vertebrates. Therefore, the isolation of vertebrate gene regulatory proteins had to await the development of different approaches. Many of these approaches rely on the detection in a cell extract of a DNA-binding protein that specifically recognizes a DNA sequence known to control the expression of a particular gene. The most common way to detect sequence-specific DNA-binding proteins is to use a technique that is based on the effect of a bound protein on the migration of DNA molecules in an electric field.

A DNA molecule is highly negatively charged and will therefore move rapidly toward a positive electrode when it is subjected to an electric field. When analyzed by polyacrylamide-gel electrophoresis, DNA molecules are separated according to their size because smaller molecules are able to penetrate the fine gel meshwork more easily than large ones. Protein molecules bound to a DNA molecule will cause it to move more slowly through the gel; in general, the larger the bound protein, the greater the retardation of the DNA molecule. This phenomenon provides the basis for the **gel-mobility shift assay**, which allows even trace amounts of a sequence-specific DNA-binding protein to be readily detected. In this assay, a short DNA fragment of specific length and sequence (produced either by DNA cloning or by chemical synthesis) is radioactively labeled and mixed with a cell extract; the mixture is then loaded onto a polyacrylamide gel and subjected to electrophoresis. If the DNA fragment corresponds to a chromosomal region where, for example, several sequence-specific

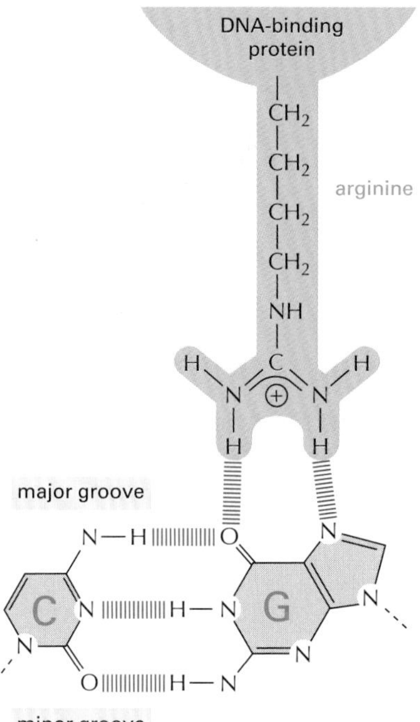

Figure 7–27 One of the most common protein–DNA interactions. Because of its specific geometry of hydrogen-bond acceptors (see Figure 7–7), guanine can be unambiguously recognized by the side chain of arginine. Another common protein–DNA interaction was shown in Figure 7–12.

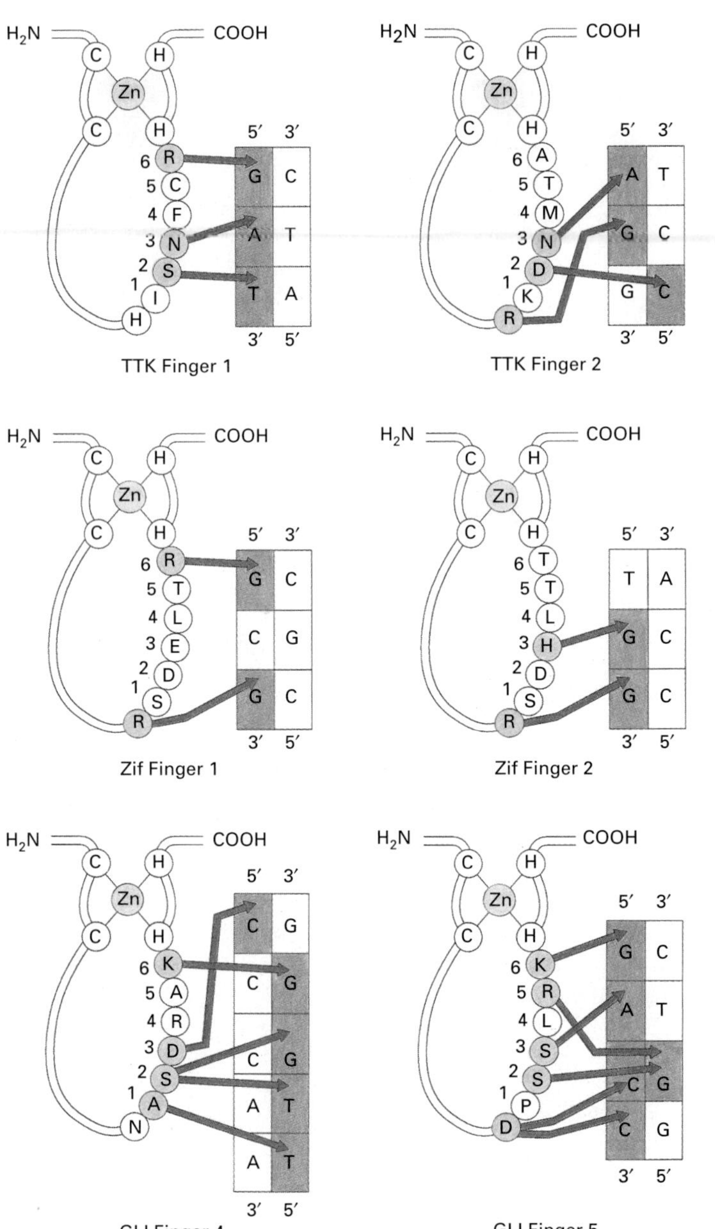

Figure 7–28 Summary of sequence-specific interactions between different six zinc fingers and their DNA recognition sequences. Even though all six Zn fingers have the same overall structure (see Figure 7–17), each binds to a different DNA sequence. The numbered amino acids form the α helix that recognizes DNA (Figures 7–17 and 7–18), and those that make sequence-specific DNA contacts are colored *green*. Bases contacted by protein are *orange*. Although arginine–guanine contacts are common (see Figure 7–27), guanine can also be recognized by serine, histidine, and lysine, as shown. Moreover, the same amino acid (serine, in this example) can recognize more than one base. Two of the Zn fingers depicted are from the TTK protein (a *Drosophila* protein that functions in development); two are from the mouse protein (Zif268) that was shown in Figure 7–18; and two are from a human protein (GLI), whose aberrant forms can cause certain types of cancers. (Adapted from C. Branden and J. Tooze, Introduction to Protein Structure, 2nd edn. New York: Garland Publishing, 1999.)

proteins bind, autoradiography will reveal a series of DNA bands, each retarded to a different extent and representing a distinct DNA–protein complex. The proteins responsible for each band on the gel can then be separated from one another by subsequent fractionations of the cell extract (Figure 7–29).

DNA Affinity Chromatography Facilitates the Purification of Sequence-specific DNA-binding Proteins

A particularly powerful purification method called **DNA affinity chromatography** can be used once the DNA sequence that a gene regulatory protein recognizes has been determined. A double-stranded oligonucleotide of the correct sequence is synthesized by chemical methods and linked to an insoluble porous matrix such as agarose; the matrix with the oligonucleotide attached is then used to construct a column that selectively binds proteins that recognize the particular DNA sequence (Figure 7–30). Purifications as great as 10,000-fold can be achieved by this means with relatively little effort.

Although most proteins that bind to a specific DNA sequence are present in a few thousand copies per higher eucaryotic cell (and generally represent only

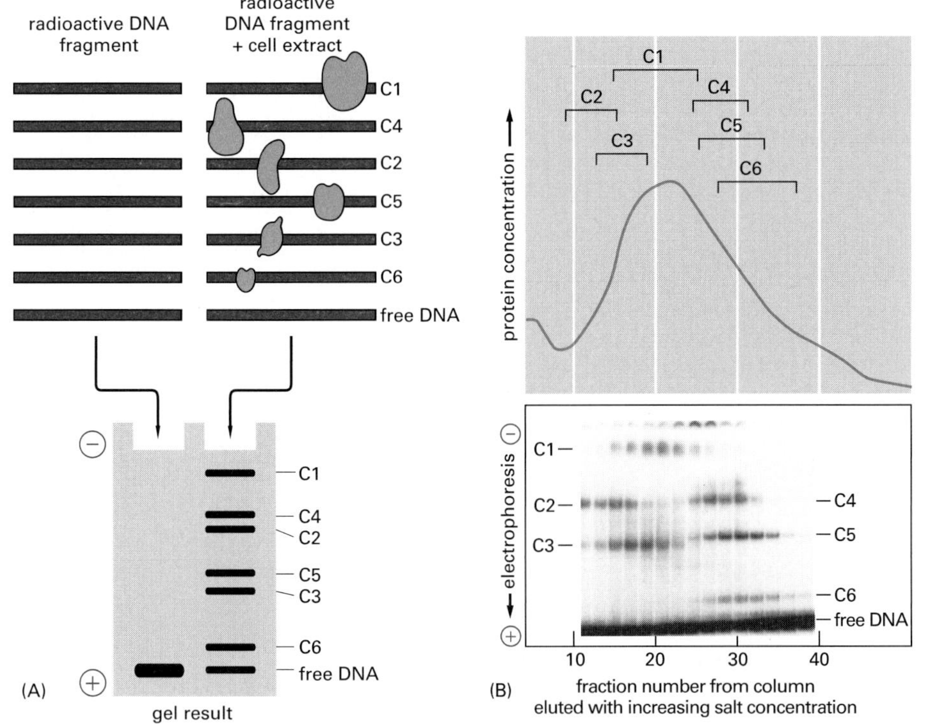

Figure 7–29 A gel-mobility shift assay. The principle of the assay is shown schematically in (A). In this example an extract of an antibody-producing cell line is mixed with a radioactive DNA fragment containing about 160 nucleotides of a regulatory DNA sequence from a gene encoding the light chain of the antibody made by the cell line. The effect of the proteins in the extract on the mobility of the DNA fragment is analyzed by polyacrylamide-gel electrophoresis followed by autoradiography. The free DNA fragments migrate rapidly to the bottom of the gel, while those fragments bound to proteins are retarded; the finding of six retarded bands suggests that the extract contains six different sequence-specific DNA-binding proteins (indicated as C1–C6) that bind to this DNA sequence. (For simplicity, any DNA fragments with more than one protein bound have been omitted from the figure.) In (B) the extract was fractionated by a standard chromatographic technique *(top)*, and each fraction was mixed with the radioactive DNA fragment, applied to one lane of a polyacrylamide gel, and analyzed as in (A). (B, modified from C. Scheidereit, A. Heguy, and R.G. Roeder, *Cell* 51:783–793, 1987.)

about one part in 50,000 of the total cell protein), enough pure protein can usually be isolated by affinity chromatography to obtain a partial amino acid sequence by mass spectrometry or other means (discussed in Chapter 8). If the complete genome sequence of the organism is known, the partial amino acid sequence can be used to identify the gene. The gene provides the complete amino acid sequence of the protein, and any uncertainties regarding exon and intron boundaries can be resolved by analyzing the mRNA produced by the gene, as described in Chapter 8. The gene also provides the means to produce the protein in unlimited amounts through genetic engineering techniques, as discussed in Chapter 8).

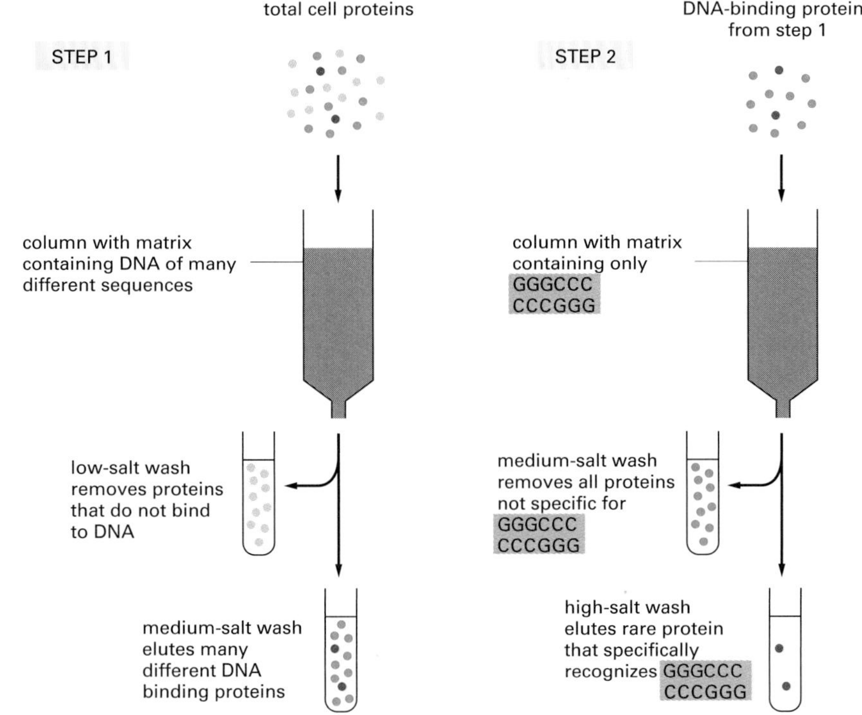

Figure 7–30 DNA affinity chromatography. In the first step, all the proteins that can bind DNA are separated from the remainder of the cellular proteins on a column containing a huge number of different DNA sequences. Most sequence-specific DNA-binding proteins have a weak (nonspecific) affinity for bulk DNA and are therefore retained on the column. This affinity is due largely to ionic attractions, and the proteins can be washed off the DNA by a solution that contains a moderate concentration of salt. In the second step, the mixture of DNA-binding proteins is passed through a column that contains only DNA of a particular sequence. Typically, all the DNA-binding proteins will stick to the column, the great majority by nonspecific interactions. These are again eluted by solutions of moderate salt concentration, leaving on the column only those proteins (typically one or only a few) that bind specifically and therefore very tightly to the particular DNA sequence. These remaining proteins can be eluted from the column by solutions containing a very high concentration of salt.

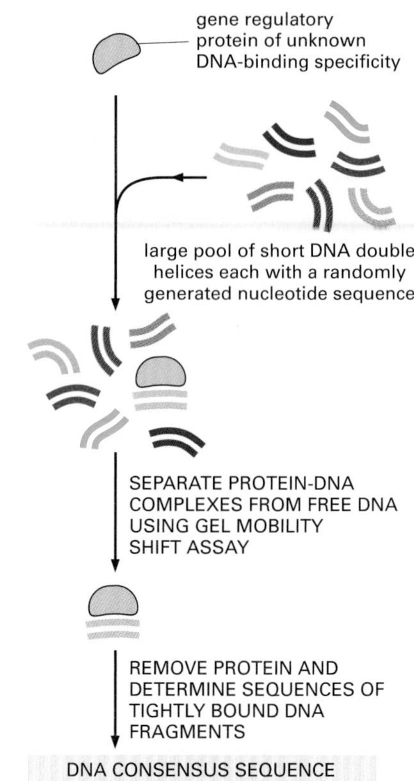

Figure 7–31 A method for determining the DNA sequence recognized by a gene regulatory protein. A purified gene regulatory protein is mixed with millions of different short DNA fragments, each with a different sequence of nucleotides. A collection of such DNA fragments can be produced by programming a DNA synthesizer, a machine that chemically synthesizes DNA of any desired sequence (discussed in Chapter 8). For example, there are 4^{11}, or approximately 4.2 million possible sequences for a DNA fragment of 11 nucleotides. The double-stranded DNA fragments that bind tightly to the gene regulatory protein are then separated from the DNA fragments that fail to bind. One method for accomplishing this separation is through gel-mobility shifts, as described in Figure 7–29. After separation of the DNA–protein complexes from the free DNA, the DNA fragments are removed from the protein, and several additional rounds of the same selection process are carried out. The nucleotide sequences of those DNA fragments that remain through multiple rounds of selection can be determined, and a consensus DNA recognition sequence can be generated.

The DNA Sequence Recognized by a Gene Regulatory Protein Can Be Determined

Some gene regulatory proteins were discovered before the DNA sequence to which they bound was known. For example, many of the *Drosophila* homeodomain proteins were discovered through the isolation of mutations that altered fly development. This allowed the genes encoding the proteins to be identified, and the proteins could then be over-expressed in cultured cells and easily purified. One method of determining the DNA sequences recognized by a gene regulatory protein is to use the purified protein to select out from a large pool of short nucleotides of differing sequence only those that bind tightly to it. After several rounds of selection, the nucleotide sequences of the tightly bound DNAs can be determined, and a consensus DNA recognition sequence for the gene regulatory protein can be formulated (Figure 7–31). The consensus sequence can be used to search genome sequences by computer and thereby identify candidate genes whose transcription might be regulated by the gene regulatory protein of interest. However, this strategy is not foolproof. For example, many organisms produce a set of closely related gene regulatory proteins that recognize very similar DNA sequences, and this approach cannot resolve them. In most cases, predictions of the sites of action of gene regulatory proteins obtained from searching genome sequences must be tested by more direct approaches, such as the one described in the next section.

A Chromatin Immunoprecipitation Technique Identifies DNA Sites Occupied by Gene Regulatory Proteins in Living Cells

In general, a given gene regulatory protein does not occupy all its potential DNA-binding sites in the genome all the time. Under some conditions, the protein may simply not be synthesized, and so be absent from the cell; or, for example, it may be present but may have to form a heterodimer with another protein to bind DNA efficiently in a living cell; or it may be excluded from the nucleus until an appropriate signal is received from the cell's environment. One method for empirically determining the sites on DNA occupied by a given gene regulatory protein under a particular set of conditions is called **chromatin immunoprecipitation** (Figure 7–32). Proteins are covalently cross-linked to DNA in living cells, the cells are lysed, and the DNA is mechanically broken into small fragments. Then, antibodies directed against a given gene regulatory protein are used to purify DNA that was covalently cross-linked to the gene regulatory protein due to the protein's close proximity to that DNA at the time of cross-linking. In this way, the DNA sites occupied by the gene regulatory protein in the original cells can be determined.

This method is also routinely used to identify the positions along a genome that are packaged by the various types of modified histones (see Figure 4–35). In this case, antibodies specific to a particular histone modification are employed.

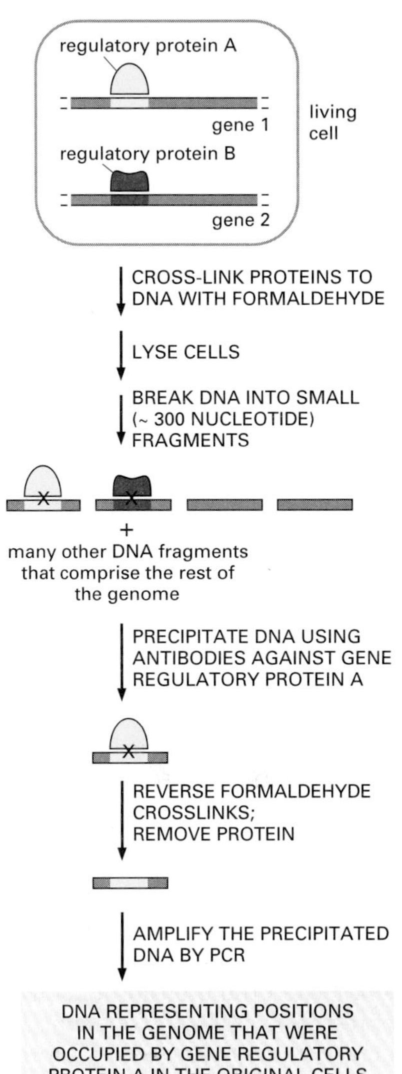

Figure 7–32 Chromatin immunoprecipitation. This methodology allows the identification of the sites in a genome that are occupied *in vivo* by a gene regulatory protein. The amplification of DNA by the polymerase chain reaction (PCR) is described in Chapter 8. The identities of the precipitated, amplified DNA fragments can be determined by hybridizing the mixture of fragments to DNA microarrays, described in Chapter 8.

Summary

Gene regulatory proteins recognize short stretches of double-helical DNA of defined sequence and thereby determine which of the thousands of genes in a cell will be transcribed. Thousands of gene regulatory proteins have been identified in a wide variety of organisms. Although each of these proteins has unique features, most bind to DNA as homodimers or heterodimers and recognize DNA through one of a small number of structural motifs. The common motifs include the helix–turn–helix, the homeodomain, the leucine zipper, the helix–loop–helix, and zinc fingers of several types. The precise amino acid sequence that is folded into a motif determines the particular DNA sequence that is recognized. Heterodimerization increases the range of DNA sequences that can be recognized by gene regulatory proteins. Powerful techniques are available that make use of the DNA-sequence specificity of gene regulatory proteins to identify and isolate these proteins, the genes that encode them, the DNA sequences they recognize, and the genes that they regulate.

HOW GENETIC SWITCHES WORK

In the previous section, we described the basic components of genetic switches—gene regulatory proteins and the specific DNA sequences that these proteins recognize. We shall now discuss how these components operate to turn genes on and off in response to a variety of signals.

Only 40 years ago the idea that genes could be switched on and off was revolutionary. This concept was a major advance, and it came originally from the study of how *E. coli* bacteria adapt to changes in the composition of their growth medium. Parallel studies on the lambda bacteriophage led to many of the same conclusions and helped to establish the underlying mechanism. Many of the same principles apply to eucaryotic cells. However, the enormous complexity of gene regulation in higher organisms, combined with the packaging of their DNA into chromatin, creates special challenges and some novel opportunities for control—as we shall see. We begin with the simplest example—an on-off switch in bacteria that responds to a single signal.

The Tryptophan Repressor Is a Simple Switch That Turns Genes On and Off in Bacteria

The chromosome of the bacterium *E. coli*, a single-celled organism, consists of a single circular DNA molecule of about 4.6×10^6 nucleotide pairs. This DNA encodes approximately 4300 proteins, although only a fraction of these are made at any one time. The expression of many of them is regulated according to the available food in the environment. This is illustrated by the five *E. coli* genes that code for enzymes that manufacture the amino acid tryptophan. These genes are arranged as a single **operon;** that is, they are adjacent to one another on the chromosome and are transcribed from a single *promoter* as one long mRNA molecule (Figure 7–33). But when tryptophan is present in the growth

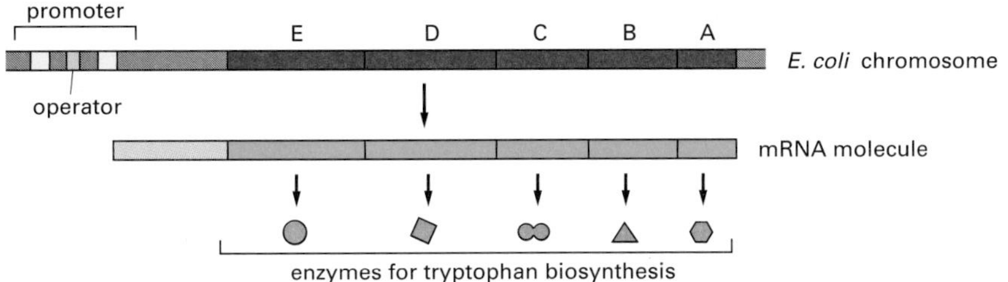

Figure 7–33 The clustered genes in *E. coli* that code for enzymes that manufacture the amino acid tryptophan. These five genes are transcribed as a single mRNA molecule, a feature that allows their expression to be controlled coordinately. Clusters of genes transcribed as a single mRNA molecule are common in bacteria. Each such cluster is called an operon.

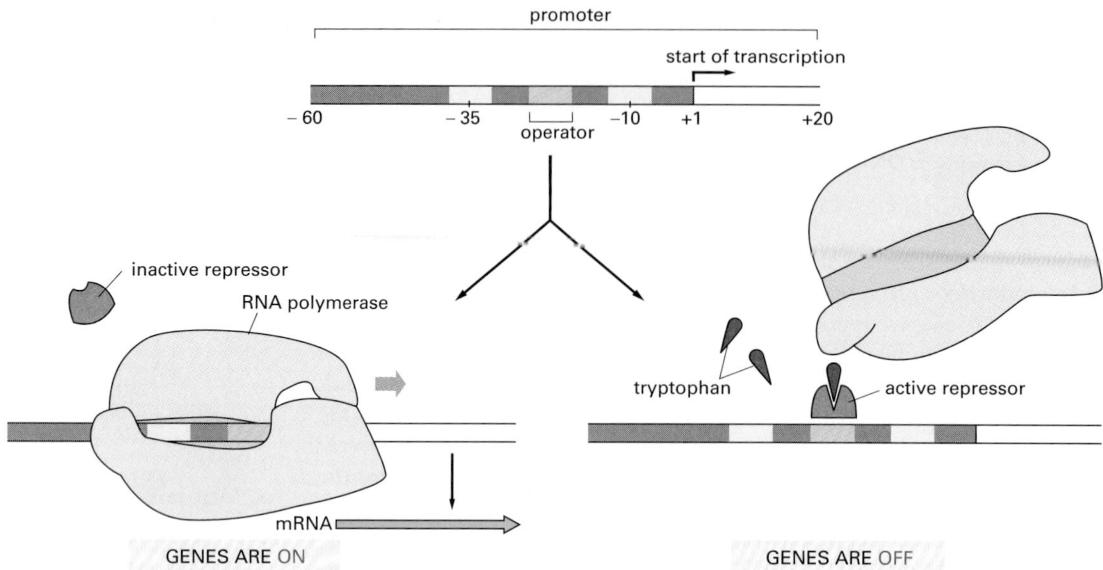

Figure 7–34 Switching the tryptophan genes on and off. If the level of tryptophan inside the cell is low, RNA polymerase binds to the promoter and transcribes the five genes of the tryptophan *(trp)* operon. If the level of tryptophan is high, however, the tryptophan repressor is activated to bind to the operator, where it blocks the binding of RNA polymerase to the promoter. Whenever the level of intracellular tryptophan drops, the repressor releases its tryptophan and becomes inactive, allowing the polymerase to begin transcribing these genes. The promoter includes two key blocks of DNA sequence information, the –35 and –10 regions highlighted in *yellow* (see Figure 6–12).

medium and enters the cell (when the bacterium is in the gut of a mammal that has just eaten a meal of protein, for example), the cell no longer needs these enzymes and shuts off their production.

The molecular basis for this switch is understood in considerable detail. As described in Chapter 6, a promoter is a specific DNA sequence that directs RNA polymerase to bind to DNA, to open the DNA double helix, and to begin synthesizing an RNA molecule. Within the promoter that directs transcription of the tryptophan biosynthetic genes lies a regulating element called an **operator** (see Figure 7–33). This is simply a short region of regulatory DNA of defined nucleotide sequence that is recognized by a repressor protein, in this case the **tryptophan repressor**, a member of the helix–turn–helix family (see Figure 7–14). The promoter and operator are arranged so that when the tryptophan repressor occupies the operator, it blocks access to the promoter by RNA polymerase, thereby preventing expression of the tryptophan-producing enzymes (Figure 7–34).

The block to gene expression is regulated in an ingenious way: to bind to its operator DNA, the repressor protein has to have two molecules of the amino acid tryptophan bound to it. As shown in Figure 7–35, tryptophan binding tilts the helix–turn–helix motif of the repressor so that it is presented properly to the DNA major groove; without tryptophan, the motif swings inward and the protein is unable to bind to the operator. Thus the tryptophan repressor and operator form a simple device that switches production of the tryptophan biosynthetic enzymes on and off according to the availability of free tryptophan. Because the active, DNA-binding form of the protein serves to turn genes off, this mode of gene regulation is called **negative control**, and the gene regulatory proteins that function in this way are called *transcriptional repressors* or *gene repressor proteins.*

Transcriptional Activators Turn Genes On

We saw in Chapter 6 that purified *E. coli* RNA polymerase (including the σ subunit) can bind to a promoter and initiate DNA transcription. Some bacterial promoters, however, are only marginally functional on their own, either because they are recognized poorly by RNA polymerase or because the polymerase has difficulty opening the DNA helix and beginning transcription. In either case these poorly functioning promoters can be rescued by gene regulatory proteins that bind to a nearby site on the DNA and contact the RNA polymerase in a way that dramatically increases the probability that a transcript will be initiated. Because the active, DNA-binding form of such a protein turns genes on, this mode of gene regulation is called **positive control**, and the gene regulatory proteins that function in this manner are known as *transcriptional activators* or *gene activator proteins*. In some cases, bacterial gene activator proteins aid RNA polymerase in binding to the promoter by providing an additional contact surface for the polymerase. In other cases, they facilitate the transition from the initial DNA-bound conformation of polymerase to the actively transcribing form, perhaps by stabilizing a transition state.

As in negative control by a transcriptional repressor, a transcriptional activator can operate as part of a simple on–off genetic switch. The bacterial activator protein *CAP (catabolite activator protein),* for example, activates genes that enable *E. coli* to use alternative carbon sources when glucose, its preferred carbon source, is not available. Falling levels of glucose induce an increase in the intracellular signaling molecule cyclic AMP, which binds to the CAP protein, enabling it to bind to its specific DNA sequence near target promoters and thereby turn on the appropriate genes. In this way the expression of a target gene is switched on or off, depending on whether cyclic AMP levels in the cell are high or low, respectively. Figure 7–36 summarizes the different ways that positive and negative control can be used to regulate genes.

In many respects transcriptional activators and transcriptional repressors are similar in design. The tryptophan repressor and the transcriptional activator CAP, for example, both use a helix–turn–helix motif (see Figure 7–14) and both require a small cofactor in order to bind DNA. In fact, some bacterial proteins (including CAP and the bacteriophage lambda repressor) can act as either activators or repressors, depending on the exact placement of the DNA sequence they recognize in relation to the promoter: if the binding site for the protein overlaps the promoter, the polymerase cannot bind and the protein acts as a repressor (Figure 7–37).

A Transcriptional Activator and a Transcriptional Repressor Control the *lac* Operon

More complicated types of genetic switches combine positive and negative controls. The *lac operon* in *E. coli*, for example, unlike the *trp operon,* is under both negative and positive transcriptional controls by the lac repressor protein and CAP, respectively. The *lac* operon codes for proteins required to transport the disaccharide lactose into the cell and to break it down. CAP, as we have seen, enables bacteria to use alternative carbon sources such as lactose in the absence

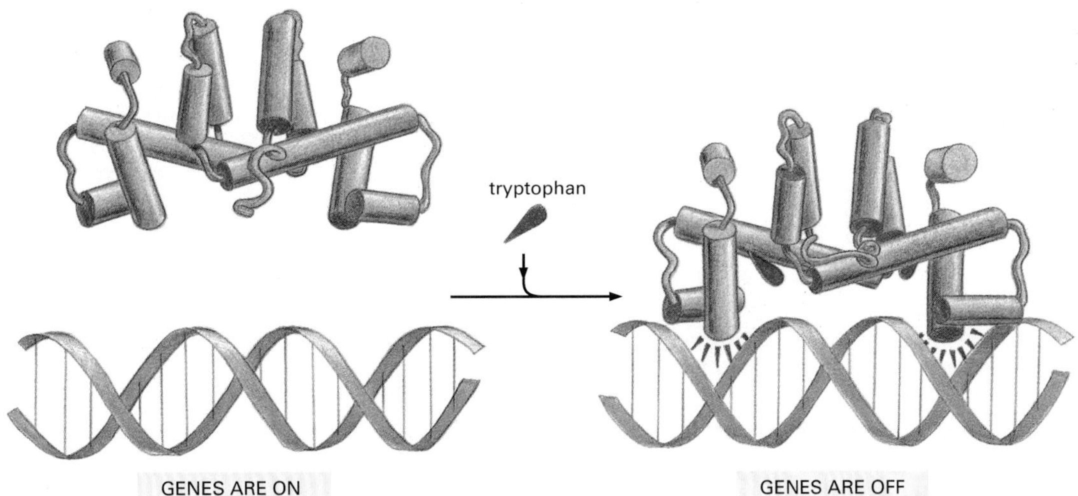

Figure 7–35 The binding of tryptophan to the tryptophan repressor protein changes the conformation of the repressor. The conformational change enables this gene regulatory protein to bind tightly to a specific DNA sequence (the operator), thereby blocking transcription of the genes encoding the enzymes required to produce tryptophan (the *trp* operon). The three-dimensional structure of this bacterial helix–turn–helix protein, as determined by x-ray diffraction with and without tryptophan bound, is illustrated. Tryptophan binding increases the distance between the two recognition helices in the homodimer, allowing the repressor to fit snugly on the operator. (Adapted from R. Zhang et al., *Nature* 327:591–597, 1987.)

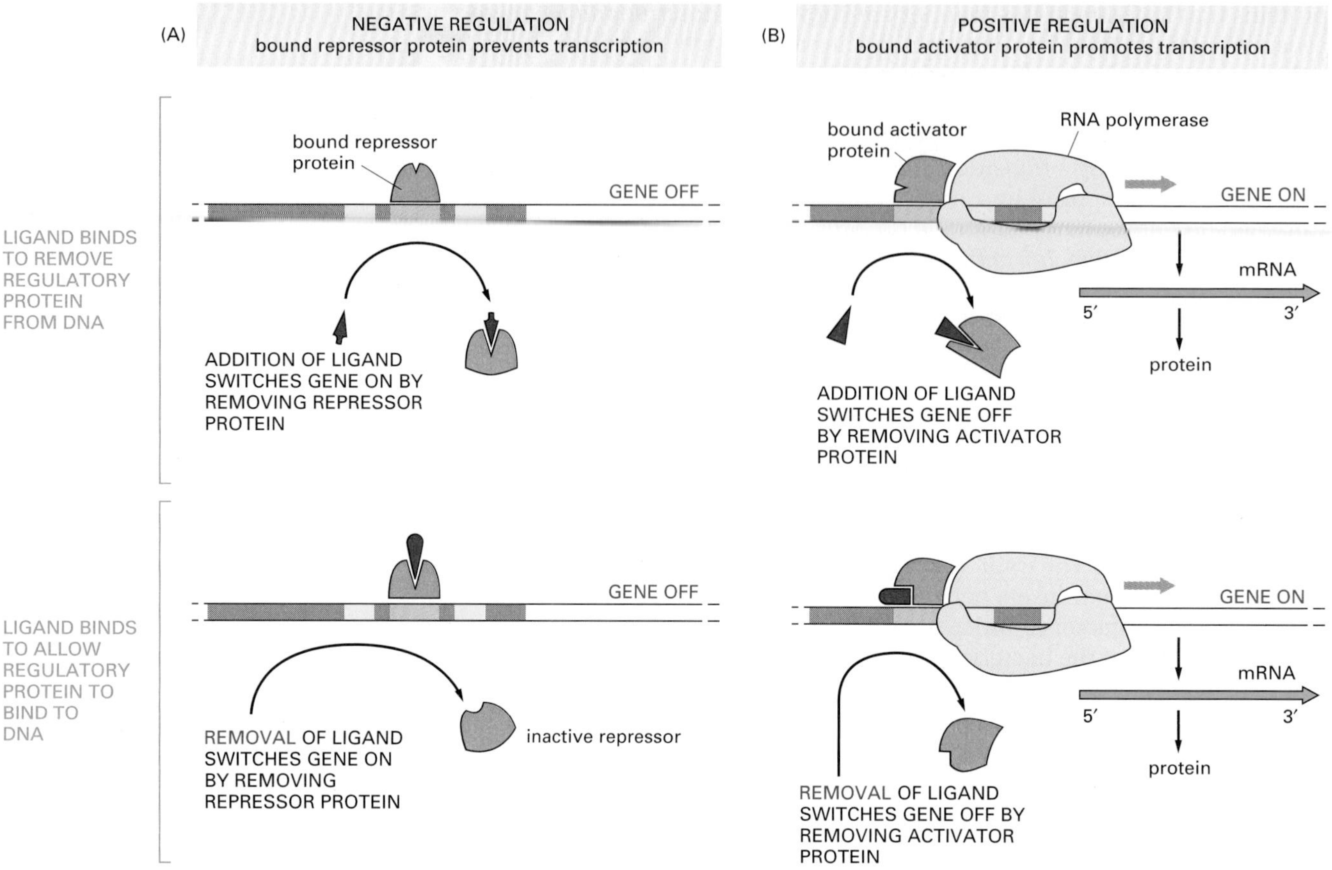

(A) NEGATIVE REGULATION
bound repressor protein prevents transcription

(B) POSITIVE REGULATION
bound activator protein promotes transcription

LIGAND BINDS TO REMOVE REGULATORY PROTEIN FROM DNA

bound repressor protein

GENE OFF

ADDITION OF LIGAND SWITCHES GENE ON BY REMOVING REPRESSOR PROTEIN

bound activator protein

RNA polymerase

GENE ON

mRNA

5′ 3′

protein

ADDITION OF LIGAND SWITCHES GENE OFF BY REMOVING ACTIVATOR PROTEIN

LIGAND BINDS TO ALLOW REGULATORY PROTEIN TO BIND TO DNA

GENE OFF

REMOVAL OF LIGAND SWITCHES GENE ON BY REMOVING REPRESSOR PROTEIN

inactive repressor

GENE ON

mRNA

5′ 3′

protein

REMOVAL OF LIGAND SWITCHES GENE OFF BY REMOVING ACTIVATOR PROTEIN

Figure 7–36 Summary of the mechanisms by which specific gene regulatory proteins control gene transcription in procaryotes. (A) Negative regulation; (B) positive regulation. Note that the addition of an "inducing" ligand can turn on a gene either by removing a gene repressor protein from the DNA *(upper left panel)* or by causing a gene activator protein to bind *(lower right panel).* Likewise, the addition of an "inhibitory" ligand can turn off a gene either by removing a gene activator protein from the DNA *(upper right panel)* or by causing a gene repressor protein to bind *(lower left panel).*

of glucose. It would be wasteful, however, for CAP to induce expression of the *lac* operon if lactose is not present, and the lac repressor ensures that the *lac* operon is shut off in the absence of lactose. This arrangement enables the control region of *lac* operon to respond to and integrate two different signals, so that the operon is highly expressed only when two conditions are met: lactose must be present and glucose must be absent. Any of the other three possible signal combinations maintain the cluster of genes in the off state (Figure 7–38).

The simple logic of this genetic switch first attracted the attention of biologists over 50 years ago. As explained above, the molecular basis of the switch was uncovered by a combination of genetics and biochemistry, providing the first insight into how gene expression is controlled. Although the same basic strategies are used to control gene expression in higher organisms, the genetic switches that are used are usually much more complex.

Regulation of Transcription in Eucaryotic Cells Is Complex

The two-signal switching mechanism that regulates the *lac* operon is elegant and simple. However, it is difficult to imagine how it could grow in complexity to allow dozens of signals to regulate transcription from the operon: there is not enough room in the neighborhood of the promoter to pack in a sufficient number of regulatory DNA sequences. How then have eucaryotes overcome such limitations to create their more complex genetic switches?

The regulation of transcription in eucaryotes differs in three important ways from that typically found in bacteria.

- First, eucaryotes make use of gene regulatory proteins that can act even when they are bound to DNA thousands of nucleotide pairs away from the promoter that they influence, which means that a single promoter can be controlled by an almost unlimited number of regulatory sequences scattered along the DNA.
- Second, as we saw in the last chapter, eucaryotic RNA polymerase II, which transcribes all protein-coding genes, cannot initiate transcription on its

Figure 7–37 Some bacterial gene regulatory proteins can act as both a transcriptional activator and a repressor, depending on the precise placement of its binding sites in DNA. An example is the bacteriophage lambda repressor. For some genes, the protein acts as a transcriptional activator by providing a favorable contact for RNA polymerase *(top)*. At other genes *(bottom)*, the operator is located one base pair closer to the promoter, and, instead of helping polymerase, the repressor now competes with it for binding to the DNA. The lambda repressor recognizes its operator by a helix–turn–helix motif, as shown in Figure 7–14.

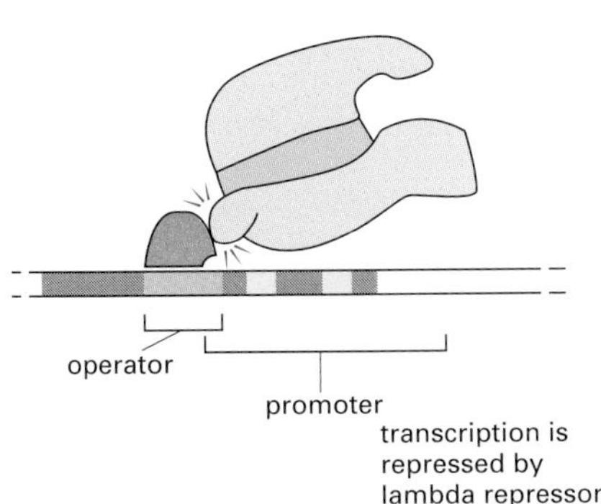

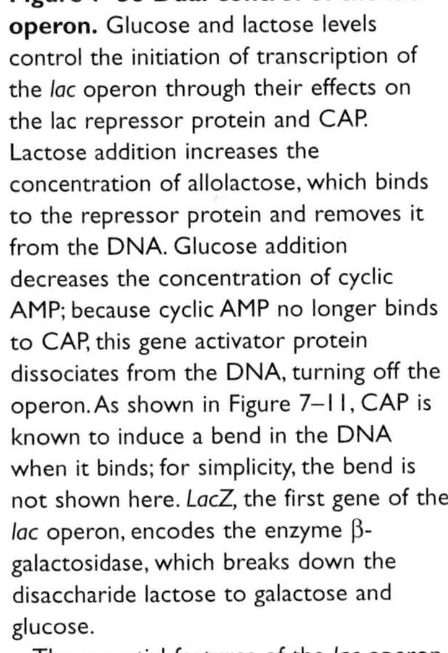

own. It requires a set of proteins called *general transcription factors*, which must be assembled at the promoter before transcription can begin. (The term "general" refers to the fact that these proteins assemble on all promoters transcribed by RNA polymerase II; in this they differ from gene regulatory proteins, which act only at particular genes.) This assembly process provides, in principle, multiple steps at which the rate of transcription initiation can be speeded up or slowed down in response to regulatory signals, and many eucaryotic gene regulatory proteins influence these steps.

- Third, the packaging of eucaryotic DNA into chromatin provides opportunities for regulation not available to bacteria.

Having discussed the general transcription factors for RNA polymerase II in Chapter 6 (see pp. 309–312), we focus here on the first and third of these features and how they are used to control eucaryotic gene expression selectively.

Eucaryotic Gene Regulatory Proteins Control Gene Expression from a Distance

Like bacteria, eucaryotes use gene regulatory proteins (activators and repressors) to regulate the expression of their genes but in a somewhat different way. The DNA sites to which the eucaryotic gene activators bound were originally termed **enhancers**, since their presence "enhanced," or increased, the rate of transcription dramatically. It came as a surprise when, in 1979, it was discovered that these activator proteins could be bound thousands of nucleotide pairs away from the promoter. Moreover, eucaryotic activators could influence transcription

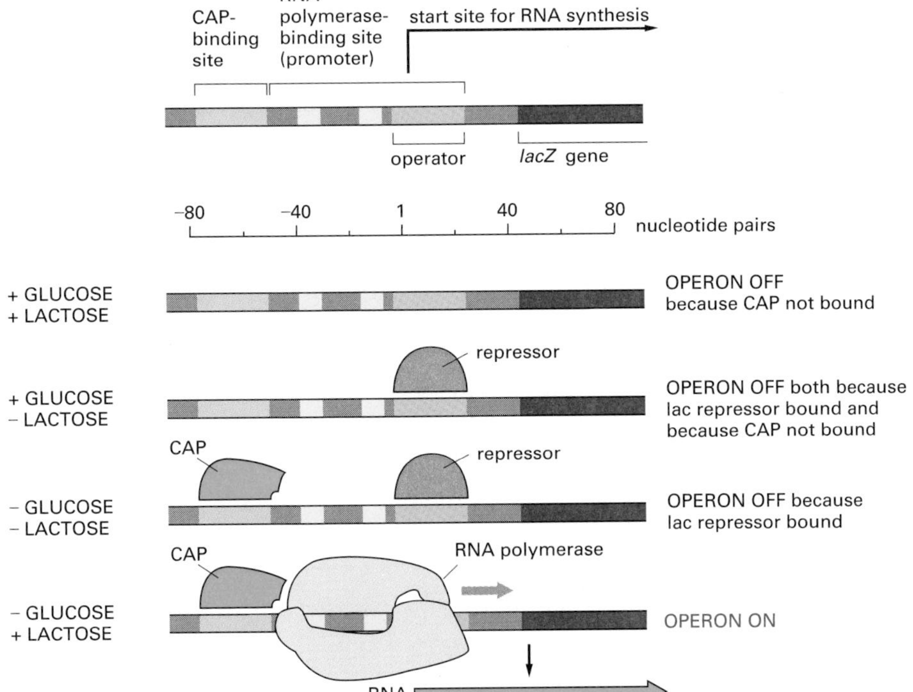

Figure 7–38 Dual control of the *lac* operon. Glucose and lactose levels control the initiation of transcription of the *lac* operon through their effects on the lac repressor protein and CAP. Lactose addition increases the concentration of allolactose, which binds to the repressor protein and removes it from the DNA. Glucose addition decreases the concentration of cyclic AMP; because cyclic AMP no longer binds to CAP, this gene activator protein dissociates from the DNA, turning off the operon. As shown in Figure 7–11, CAP is known to induce a bend in the DNA when it binds; for simplicity, the bend is not shown here. *LacZ*, the first gene of the *lac* operon, encodes the enzyme β-galactosidase, which breaks down the disaccharide lactose to galactose and glucose.

The essential features of the *lac* operon are summarized in the figure, but in reality the situation is more complex. For one thing, there are several *lac* repressor binding sites located at different positions along the DNA. Although the one illustrated exerts the greatest effect, the others are required for full repression. In addition, expression of the *lac* operon is never completely shut down. A small amount of the enzyme β-galactosidase is required to convert lactose to allolactose thereby permitting the *lac* repressor to be inactivated when lactose is added to the growth medium.

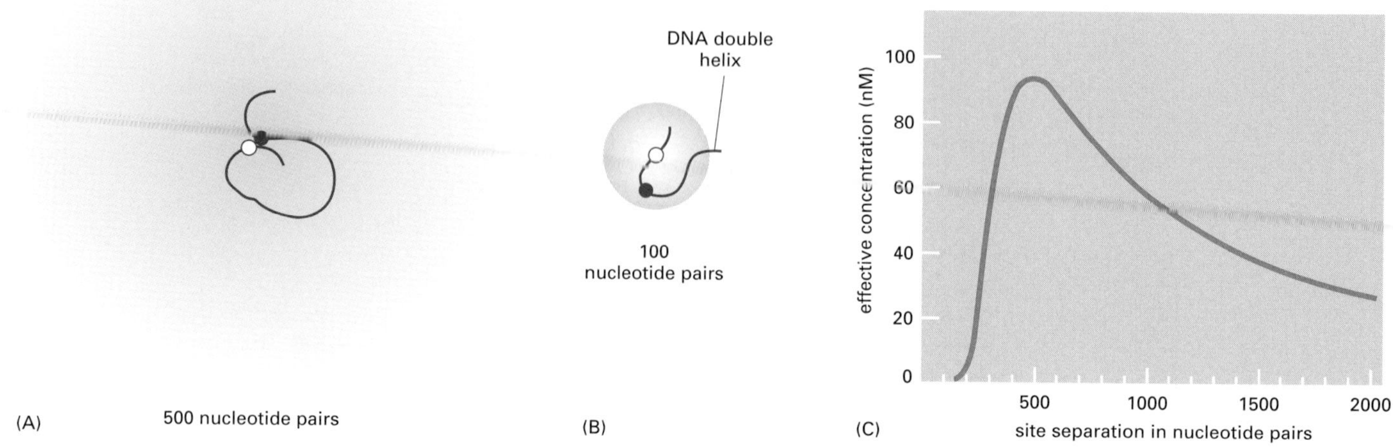

Figure 7–39 Binding of two proteins to separate sites on the DNA double helix can greatly increase their probability of interacting. (A) The tethering of one protein to the other via an intervening DNA loop of 500 nucleotide pairs increases their frequency of collision. The intensity of *blue* coloring reflects the probability that the *red* protein will be located at each position in space relative to the *white* protein. (B) The flexibility of DNA is such that an average sequence makes a smoothly graded 90° bend (a curved turn) about once every 200 nucleotide pairs. Thus, when two proteins are tethered by only 100 nucleotide pairs, their contact is relatively restricted. In such cases the protein interaction is facilitated when the two protein-binding sites are separated by a multiple of about 10 nucleotide pairs, which places both proteins on the same side of the DNA helix (which has about 10 nucleotides per turn) and thus on the inside of the DNA loop, where they can best reach each other. (C) The theoretical effective concentration of the *red* protein at the site where the *white* protein is bound, as a function of their separation. (C, courtesy of Gregory Bellomy, modified from M.C. Mossing and M.T. Record, *Science* 233:889–892, 1986. © AAAS.)

of a gene when bound either upstream or downstream from it. How do enhancer sequences and the proteins bound to them function over these long distances? How do they communicate with the promoter?

Many models for "action at a distance" have been proposed, but the simplest of these seems to apply in most cases. The DNA between the enhancer and the promoter loops out to allow the activator proteins bound to the enhancer to come into contact with proteins (RNA polymerase, one of the general transcription factors, or other proteins) bound to the promoter (see Figure 6–19). The DNA thus acts as a tether, helping a protein bound to an enhancer even thousands of nucleotide pairs away to interact with the complex of proteins bound to the promoter (Figure 7–39). This phenomenon also occurs in bacteria, although less commonly and over much shorter lengths of DNA (Figure 7–40).

A Eucaryotic Gene Control Region Consists of a Promoter Plus Regulatory DNA Sequences

Because eucaryotic gene regulatory proteins can control transcription when bound to DNA far away from the promoter, the DNA sequences that control the expression of a gene are often spread over long stretches of DNA. We shall use the term **gene control region** to refer to the whole expanse of DNA involved in regulating transcription of a gene, including the **promoter**, where the general transcription factors and the polymerase assemble, and all of the **regulatory sequences** to which gene regulatory proteins bind to control the rate of the assembly processes at the promoter (Figure 7–41). In higher eucaryotes it is not unusual to find the regulatory sequences of a gene dotted over distances as great as 50,000 nucleotide pairs. Although much of this DNA serves as "spacer" sequence and is not recognized by gene regulatory proteins, this spacer DNA may facilitate transcription by providing the flexibility needed for communication between DNA-bound proteins. It is also important to keep in mind that, like other regions of eucaryotic chromosomes, much of the DNA in gene control regions is packaged into nucleosomes and higher-order forms of chromatin, thereby compacting its length.

In this chapter we generally use the term **gene** to refer only to a segment of DNA that is transcribed into RNA (see Figure 7–41). However, the classical view of a gene would include the gene control region as well. The different definitions arise from the different ways in which genes were historically identified. The discovery of alternative RNA splicing has further complicated the definition of a gene—a point we discussed briefly in Chapter 6 and will return to later in this chapter.

Although many gene regulatory proteins bind to enhancer sequences and activate gene transcription, many others function as negative regulators, as we

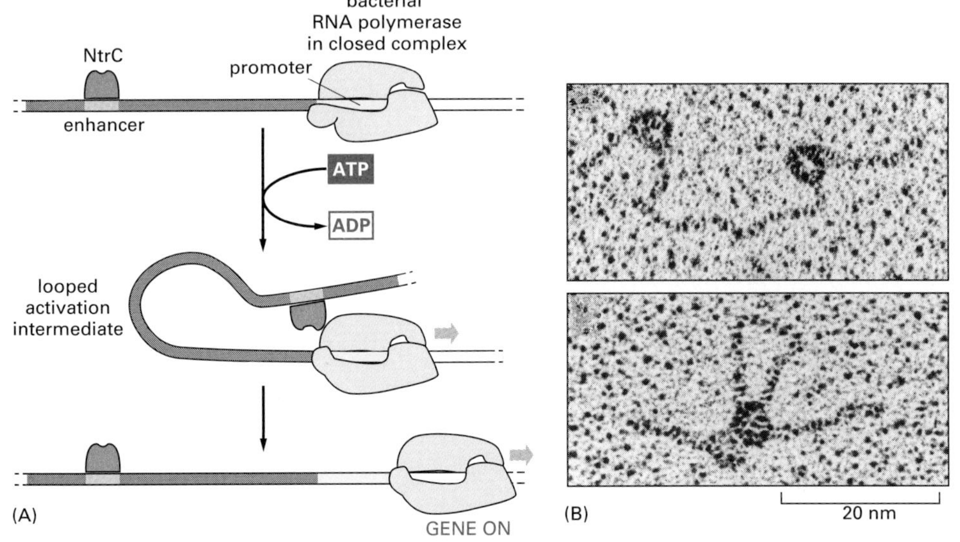

Figure 7–40 Gene activation at a distance. (A) NtrC is a bacterial gene regulatory protein that activates transcription by facilitating the transition between the initial binding of RNA polymerase to the promoter and the formation of an initiating complex (discussed in Chapter 6). As indicated, the transition stimulated by NtrC requires the energy produced by ATP hydrolysis, although this requirement is unusual for bacterial transcription initiation. (B) The interaction of NtrC and RNA polymerase, with the intervening DNA looped out, can be seen in the electron microscope. Although transcriptional activation by DNA looping is unusual in bacteria, it is typical of eucaryotic gene regulatory proteins. (B, courtesy of Harrison Echols and Sydney Kustu.)

see below. In contrast to the small number of general transcription factors, which are abundant proteins that assemble on the promoters of all genes transcribed by RNA polymerase II, there are thousands of different gene regulatory proteins. For example, of the roughly 30,000 human genes, an estimated 5–10% encode gene regulatory proteins. These regulatory proteins vary from one gene control region to the next, and each is usually present in very small amounts in a cell, often less than 0.01% of the total protein. Most of them recognize their specific DNA sequences using one of the DNA-binding motifs discussed previously, although as we discuss below, some do not recognize DNA directly but instead assemble on other DNA-bound proteins.

The gene regulatory proteins allow the individual genes of an organism to be turned on or off specifically. Different selections of gene regulatory proteins are present in different cell types and thereby direct the patterns of gene expression that give each cell type its unique characteristics. Each gene in a eucaryotic cell is regulated differently from nearly every other gene. Given the number of genes in eucaryotes and the complexity of their regulation, it has been difficult to formulate simple rules for gene regulation that apply in every case. We can, however, make some generalizations, about how gene regulatory proteins, once bound to a gene control region on DNA, influence the rate of transcription initiation, as we now explain.

Eucaryotic Gene Activator Proteins Promote the Assembly of RNA Polymerase and the General Transcription Factors at the Startpoint of Transcription

Most gene regulatory proteins that activate gene transcription—that is, most **gene activator proteins**—have a modular design consisting of at least two distinct domains. One domain usually contains one of the structural motifs discussed previously that recognizes a specific regulatory DNA sequence. In the simplest

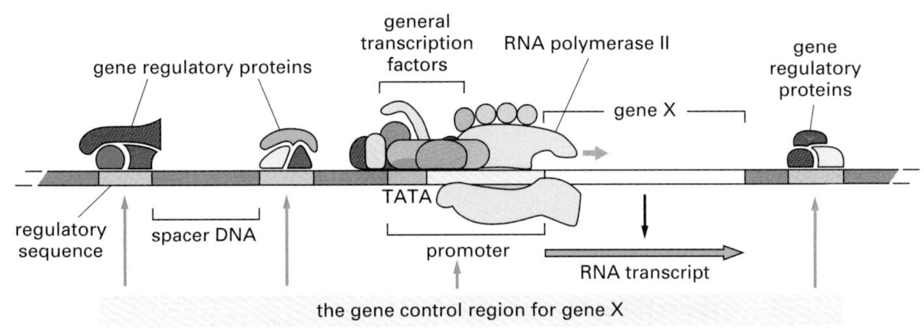

Figure 7–41 The gene control region of a typical eucaryotic gene. The *promoter* is the DNA sequence where the general transcription factors and the polymerase assemble (see Figure 6–16). The *regulatory sequences* serve as binding sites for gene regulatory proteins, whose presence on the DNA affects the rate of transcription initiation. These sequences can be located adjacent to the promoter, far upstream of it, or even within introns or downstream of the gene. DNA looping is thought to allow gene regulatory proteins bound at any of these positions to interact with the proteins that assemble at the promoter. Whereas the general transcription factors that assemble at the promoter are similar for all polymerase II transcribed genes, the gene regulatory proteins and the locations of their binding sites relative to the promoter are different for each gene.

cases, a second domain—sometimes called an *activation domain*—accelerates the rate of transcription initiation. This type of modular design was first revealed by experiments in which genetic engineering techniques were used to create a hybrid protein containing the activation domain of one protein fused to the DNA-binding domain of a different protein (Figure 7–42).

Once bound to DNA, how do eucaryotic gene activator proteins increase the rate of transcription initiation? As we will see shortly, there are several mechanisms by which this can occur, and, in many cases, these different mechanisms work in concert at a single promoter. But, regardless of the precise biochemical pathway, the main function of activators is to attract, position, and modify the general transcription factors and RNA polymerase II at the promoter so that transcription can begin. They do this both by acting directly on the transcription machinery itself and by changing the chromatin structure around the promoter.

We consider first the ways in which activators directly influence the positioning of the general transcription factors and RNA polymerase at promoters and help kick them into action. Although the general transcription factors and RNA polymerase II assemble in a stepwise, prescribed order *in vitro* (see Figure 6–16), there are cases in living cells where some of them are brought to the promoter as a large pre-assembled complex that is sometimes called the *RNA polymerase II holoenzyme.* In addition to some of the general transcription factors and RNA polymerase, the holoenzyme typically contains a 20-subunit protein complex called the *mediator,* which was first identified biochemically as being required for activators to stimulate transcription initiation.

Many activator proteins interact with the holoenzyme complex and thereby make it more energetically favorable for it to assemble on a promoter that is linked through DNA to the site where the activator protein is bound (Figure 7–43A). In this sense, eucaryotic activators resemble those of bacteria in helping to attract and position RNA polymerase on specific sites on DNA (see Figure 7–36). One type of experiment that supports the idea that activators attract the holoenzyme complex to promoters creates an "activator bypass" (Figure 7–43B). Here, a sequence-specific DNA-binding domain is experimentally fused directly to a component of the mediator; this hybrid protein, which lacks an activation domain, strongly stimulates transcription initiation when the DNA sequence to which it binds is placed in proximity to a promoter.

Although recruitment of the holoenzyme complex to promoters provides a conceptually simple mechanism for envisioning gene activation, the effect of activators on the holoenzyme complex is probably more complicated. For example, a stepwise assembly of the general transcription factors (see Figure 6–16)

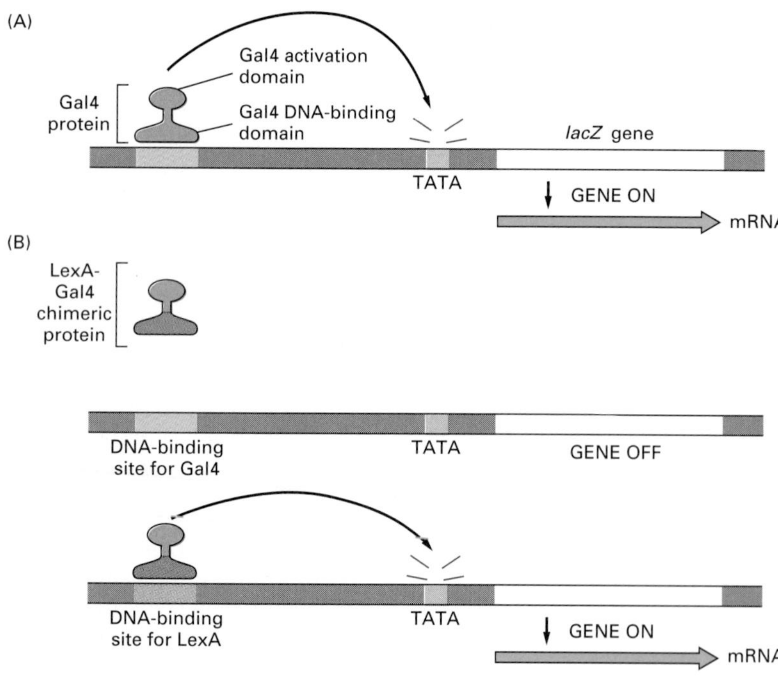

Figure 7–42 The modular structure of a gene activator protein. Outline of an experiment that reveals the presence of independent DNA-binding and transcription-activating domains in the yeast gene activator protein Gal4. A functional activator can be reconstituted from the C-terminal portion of the yeast Gal4 protein if it is attached to the DNA-binding domain of a bacterial gene regulatory protein (the LexA protein) by gene fusion techniques. When the resulting bacterial-yeast hybrid protein is produced in yeast cells, it will activate transcription from yeast genes provided that the specific DNA-binding site for the bacterial protein has been inserted next to them. (A) The normal activation of gene transcription produced by the Gal4 protein. (B) The chimeric gene regulatory protein requires the LexA protein DNA-binding site for its activity.

Gal4 is normally responsible for activating the transcription of yeast genes that code for the enzymes that convert galactose to glucose. In the experiments shown here, the control region for one of these genes was fused to the *E. coli lacZ* gene, which codes for the enzyme β-galactosidase (see Figure 7–38). β-galactosidase is very simple to detect biochemically and thus provides a convenient way to monitor the expression level specified by a gene control region; *lacZ* thus serves as a *reporter gene* since it "reports" the activity of a gene control region.

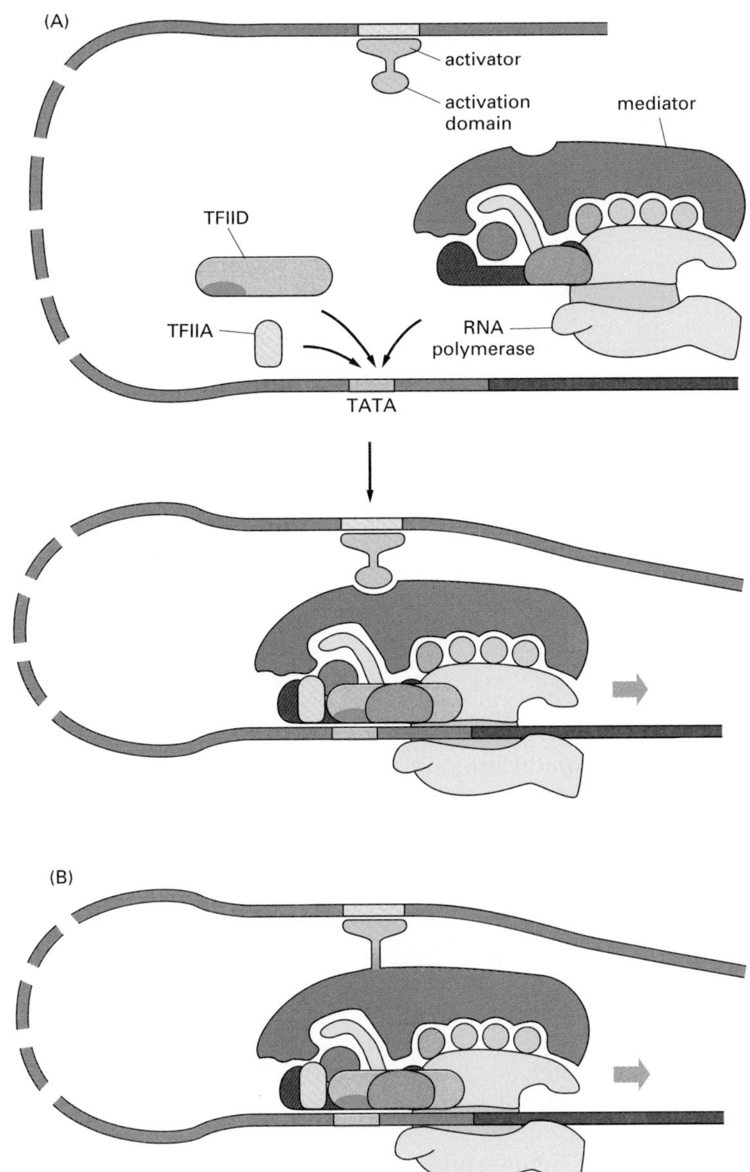

Figure 7–43 Activation of transcription initiation in eucaryotes by recruitment of the eucaryotic RNA polymerase II holoenzyme complex. (A) An activator protein bound in proximity to a promoter attracts the holoenzyme complex to the promoter. According to this model, the holoenzyme (which contains over 100 protein subunits) is brought to the promoter separately from the general transcription factors TFIID and TFIIA. The "broken" DNA in this and subsequent figures indicates that this portion of the DNA molecule can be very long and of variable length. (B) Diagram of an *in vivo* experiment whose outcome supports the holoenzyme recruitment model for gene activator proteins. The DNA-binding domain of a protein has been fused directly to a protein component of the mediator, a 20-subunit protein complex which is part of the holoenzyme complex, but which is easily dissociable from the remainder of the holoenzyme. When the binding site for the hybrid protein is experimentally inserted near a promoter, transcription initiation is strongly increased. In this experiment, the "activation domain" of the activator (see Figure 7–42) has been omitted, suggesting that an important function of the activation domain is simply to interact with the RNA polymerase holoenzyme complex and thereby aid in its assembly at the promoter. The ability of gene activator proteins to recruit the transcription machinery to promoters has also been demonstrated directly, using chromatin immunoprecipitation (see Figure 7–32).

DNA-bound activator proteins typically increase the rate of transcription by up to 1000-fold, which is consistent with a relatively weak and nonspecific interaction between the activator and the holoenzyme (a 1000-fold change in affinity corresponds to a change in ΔG of ~4 kcal/mole, which could be accounted for by just a few weak, noncovalent bonds).

may occur on some promoters. On others, their rearrangement, once brought to DNA as part of the holoenzyme, may be required. In addition, most forms of the holoenzyme complex lacks some of the general transcription factors (notably TFIID and TFIIA), and these must be assembled on the promoter separately (see Figure 7–43A). In principle, any of these assembly processes could be a slow step on the pathway to transcription initiation, and activators could facilitate their completion. In fact, many activators have been shown to interact with one or more of the general transcription factors, and several have been shown to directly accelerate their assembly at the promoter (Figure 7–44).

Eucaryotic Gene Activator Proteins Modify Local Chromatin Structure

In addition to their direct actions in assembling the RNA polymerase holoenzyme and the general transcription factors on DNA, gene activator proteins also

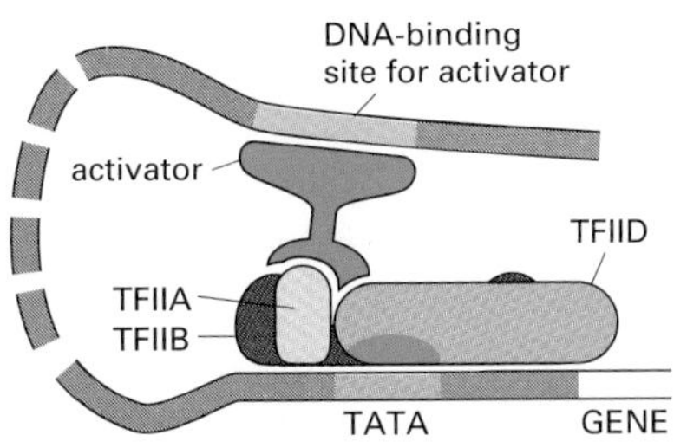

Figure 7–44 A model for the action of some eucaryotic transcriptional activators. The gene activator protein, bound to DNA in the rough vicinity of the promoter, facilitates the assembly of some of the general transcription factors. Although some activator proteins may be dedicated to particular steps in the pathway for transcription initiation, many seem to be capable of acting at several steps.

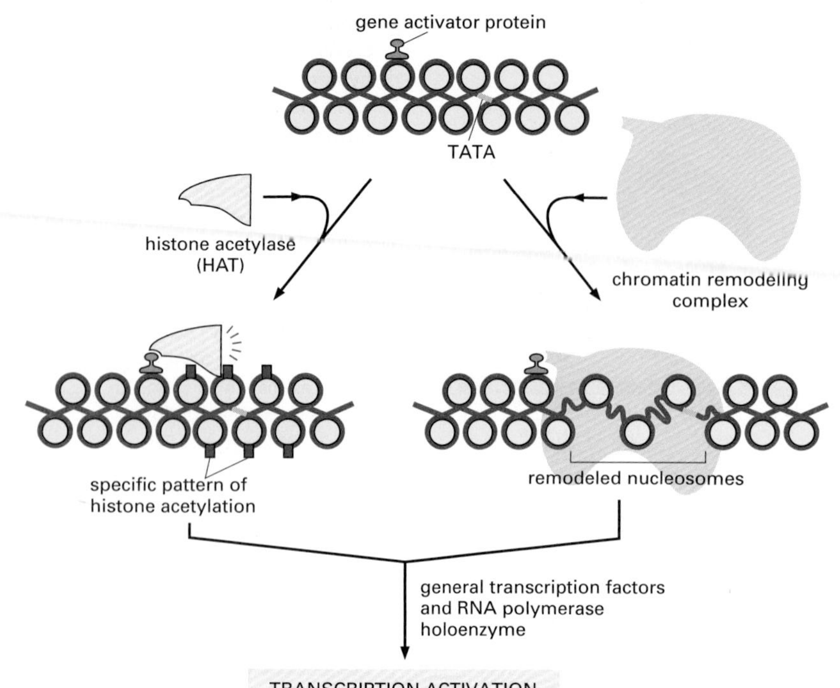

Figure 7–45 Local alterations in chromatin structure directed by eucaryotic gene activator proteins. Histone acetylation and nucleosome remodeling generally render the DNA packaged in chromatin more accessible to other proteins in the cell, including those required for transcription initiation. In addition, specific patterns of histone modification directly aid in the assembly of the general transcription factors at the promoter (see Figure 7–46).

Transcription initiation and the formation of a compact chromatin structure can be regarded as competing biochemical assembly reactions. Enzymes that increase, even transiently, the accessibility of DNA in chromatin will tend to favor transcription initiation (see Figure 4–34).

promote transcription initiation by changing the chromatin structure of the regulatory sequences and promoters of genes. As we saw in Chapter 4, the two most important ways of locally altering chromatin structure are through covalent histone modifications and nucleosome remodeling (see Figures 4–34 and 4–35). Many gene activator proteins make use of both these mechanisms by binding to and thereby recruiting histone acetyl transferases (HATs), commonly known as histone acetylases, and ATP-dependent chromatin remodeling complexes (Figure 7–45) to work on nearby chromatin. In general terms, the local alterations in chromatin structure that ensue allow greater accessibility to the underlying DNA. This accessibility facilitates the assembly of the general transcription factors and the RNA polymerase holoenzyme at the promoter, and it also allows the binding of additional gene regulatory proteins to the control region of the gene (Figure 7–46A).

The general transcription factors seem unable to assemble onto a promoter that is packaged in a conventional nucleosome. In fact, such packaging may have evolved in part to ensure that leaky, or basal, transcription initiation (initiation at a promoter in the absence of gene activator protein bound upstream of it) does

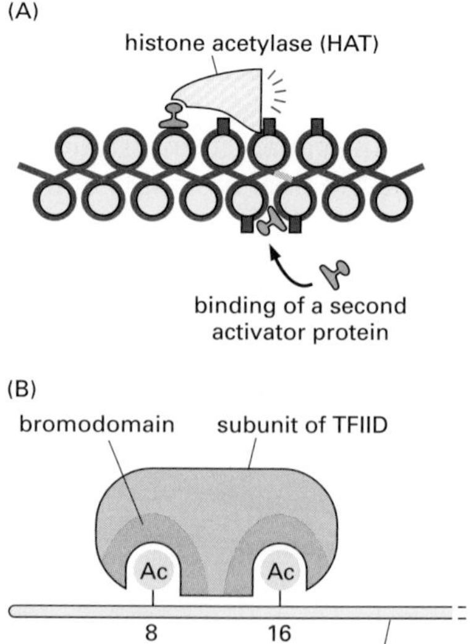

Figure 7–46 Two specific ways that local histone acetylation can stimulate transcription initiation. (A) Some gene activator proteins can bind directly to DNA that is packaged in unmodified chromatin. By attracting histone acetylases (and nucleosome remodeling complexes), these "pioneer" activators can facilitate the binding to DNA of additional activator proteins that cannot bind to unmodified chromatin. These additional proteins can in turn carry out additional modifications of chromatin or act directly on the transcription machinery as shown in Figures 7–43 and 7–44. (B) A subunit of the general transcription factor TFIID contains two 120-amino acid protein domains called *bromodomains*. Each bromodomain forms a binding pocket for an acetylated lysine side chain (designated Ac in the figure); in TFIID the two pockets are separated by 25 Å, which is the optimal spacing for recognizing a pair of acetylated lysines separated by six or seven amino acids on the N-terminal tail of histone H4. In addition to the pattern of acetylation shown, this subunit of TFIID also recognizes the histone H4 tail acetylated at positions 5 and 12 and the fully acetylated tail. It has no appreciable affinity for an unacetylated H4 tail and only low affinity for an H4 tail acetylated at a single lysine. As shown in Figure 4–35, certain patterns of histone H4 acetylation, including those recognized by TFIID, are associated with transcriptionally active regions of chromatin.

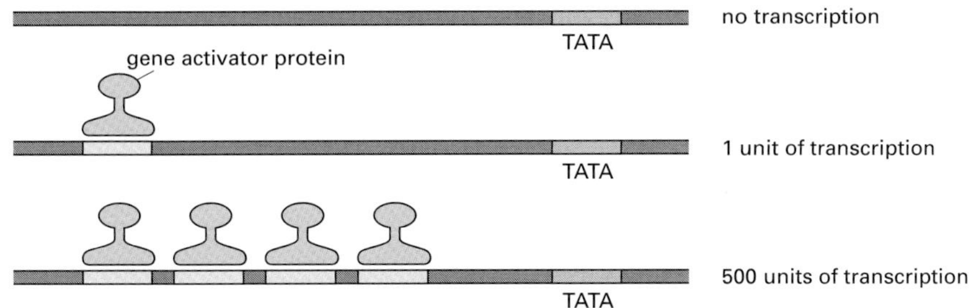

Figure 7–47 Transcriptional synergy. In this experiment, the rate of transcription produced by three experimentally constructed regulatory regions is compared in a eucaryotic cell. Transcriptional synergy, the greater than additive effect of the activators, is observed when several molecules of gene activator protein are bound upstream of the promoter. Synergy is also typically observed between different gene activator proteins from the same organism and even between activator proteins from widely different eucaryotic species when they are experimentally introduced into the same cell. This last observation reflects the high degree of conservation of the transcription machinery.

not occur. As well as making the DNA more generally accessible, local histone acetylation has a more specialized role in promoting transcription initiation. As discussed in Chapter 4 (see Figure 4–35), certain patterns of histone acetylation are associated with transcriptionally active chromatin, and gene activator proteins, by recruiting histone acetylases, produce these patterns. One such pattern (Figure 7–46B) is directly recognized by one of the subunits of the general transcription factor TFIID, and this recognition apparently helps the factor assemble DNA that is packaged in chromatin. Thus gene activator proteins, through the action of histone acetylases, can indirectly aid in the assembly of the general transcription factors at a promoter and thereby stimulate transcription initiation.

Gene Activator Proteins Work Synergistically

We have seen that eucaryotic gene activator proteins can influence several different steps in transcription initiation, and this property has important consequences when different activator proteins work together. In general, where several factors work together to enhance a reaction rate, the joint effect is generally not merely the sum of the enhancements caused by each factor alone, but the product. If, for example, factor A lowers the free-energy barrier for a reaction by a certain amount and thereby speeds up the reaction 100-fold, and factor B, by acting on another aspect of the reaction, does likewise, then A and B acting in parallel will lower the barrier by a double amount and speed up the reaction 10,000-fold. Similar multiplicative effects occur if A and B speed the reaction by each helping to recruit necessary proteins to the reaction site. Thus, gene activator proteins often exhibit what is called *transcriptional synergy*, where the transcription rate produced by several activator proteins working together is much higher than that produced by any of the activators working alone (Figure 7–47). Transcriptional synergy is observed both between different gene activator proteins bound upstream of a gene and between multiple DNA-bound molecules of the same activator. It is therefore not difficult to see how multiple gene regulatory proteins, each binding to a different regulatory DNA sequence, could control the final rate of transcription of a eucaryotic gene.

Since gene activator proteins can influence many different steps on the pathway to transcriptional activation, it is worth considering whether these steps always occur in a prescribed order. For example does chromatin remodeling necessarily precede histone acetylation or vice versa? When does recruitment of the holoenzyme complex occur relative to the chromatin modifying steps? The answers to these questions appears to be different for different genes—and even for the same gene under different conditions (Figure 7–48). Whatever the precise mechanisms and the order in which they are carried out, a gene regulatory protein must be bound to DNA either directly or indirectly to influence transcription of its target promoter, and the rate of transcription of a gene ultimately depends upon the spectrum of regulatory proteins bound upstream and downstream of its transcription start site.

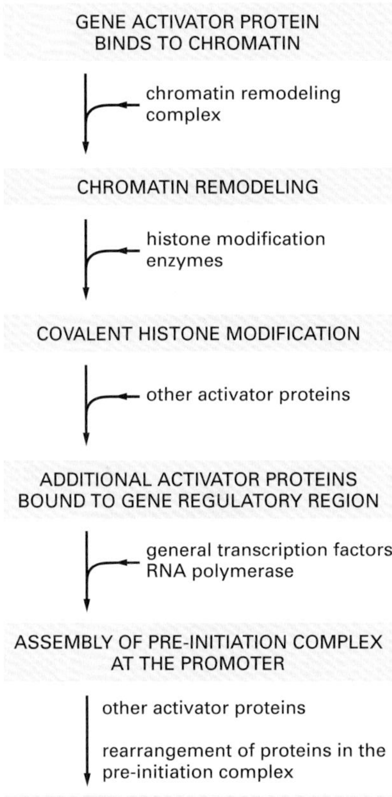

Figure 7–48 An order of events leading to transcription initiation at a specific promoter. The well-studied example shown is from a promoter in the budding yeast *S. cerevisiae*. The chromatin remodeling complex and histone acetylase apparently dissociate from the DNA after they sequentially act. The order of steps on the pathway to transcription initiation appears to be different for different promoters. For example, in a well-studied example from humans, histone acetylases function first, followed by RNA polymerase recruitment, followed by chromatin remodeling complex recruitment.

Eucaryotic Gene Repressor Proteins Can Inhibit Transcription in Various Ways

Like bacteria, eucaryotes use **gene repressor proteins** in addition to activator proteins to regulate transcription of their genes. However, because of differences in the way transcription is initiated in eucaryotes and bacteria, eucaryotic repressors have many more possible mechanisms of action. For example, we saw in Chapter 4 that whole regions of eucaryotic chromosomes can be packaged into *heterochromatin*, a form of chromatin that is normally resistant to transcription. We will return to this feature of eucaryotic chromosomes later in this chapter. In addition to molecules that shut down large regions of chromatin, eucaryotic cells also contain gene regulatory proteins that act only locally to repress transcription of nearby genes. Unlike bacterial repressors, most do not directly compete with the RNA polymerase for access to the DNA; rather they work by a variety of other mechanisms, some of which are illustrated in Figure 7–49. Like gene activator proteins, many eucaryotic repressor proteins act through more than one mechanism, thereby ensuring robust and efficient repression.

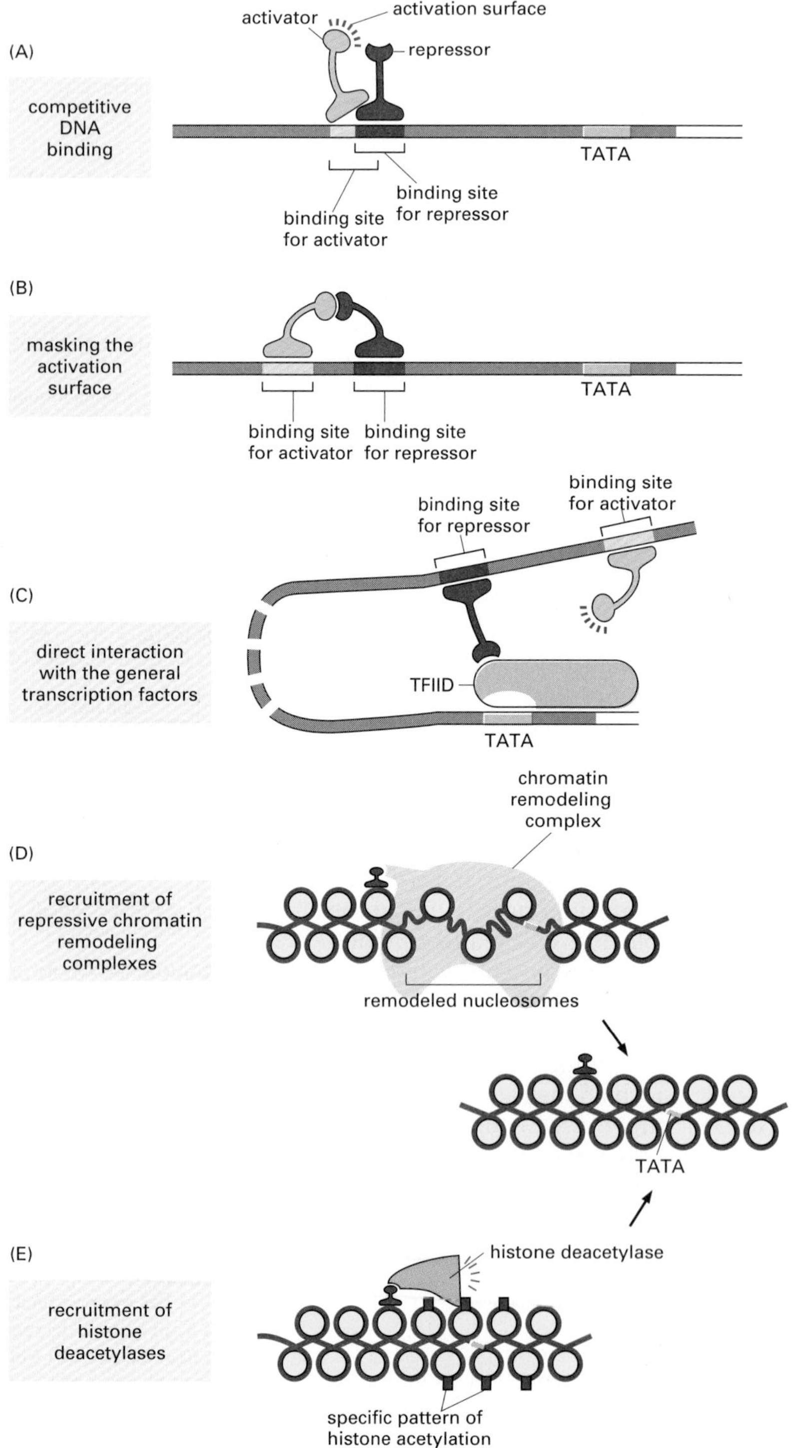

Figure 7–49 Five ways in which eucaryotic gene repressor proteins can operate. (A) Gene activator proteins and gene repressor proteins compete for binding to the same regulatory DNA sequence. (B) Both proteins can bind DNA, but the repressor binds to the activation domain of the activator protein thereby preventing it from carrying out its activation functions. In a variation of this strategy, the repressor binds tightly to the activator without having to be bound to DNA directly. (C) The repressor interacts with an early stage of the assembling complex of general transcription factors, blocking further assembly. Some repressors also act at late stages in transcription initiation, for example, by preventing the release of the RNA polymerase from the general transcription factors. (D) The repressor recruits a chromatin remodeling complex which returns the nucleosomal state of the promoter region to its pre-transcriptional form. Certain types of remodeling complexes appear dedicated to restoring the repressed nucleosomal state of a promoter, whereas others (for example, those recruited by activator proteins) render DNA packaged in nucleosomes more accessible (see Figure 4–34). However the same remodeling complex could in principle be used either to activate or repress transcription: depending on the concentration of other proteins in the nucleus, either the remodeled state or the repressed state could be stabilized. According to this view, the remodeling complex simply allows chromatin structure to change. (E) The repressor attracts a histone deacetylase to the promoter. Local histone deacetylation reduces the affinity of TFIID for the promoter (see Figure 7–46) and decreases the accessibility of DNA in the affected chromatin. A sixth mechanism of negative control—inactivation of a transcriptional activator by heterodimerization—was illustrated in Figure 7–26. For simplicity, nucleosomes have been omitted from (A)–(C), and the scale of (D) and (E) has been reduced relative to (A)–(C).

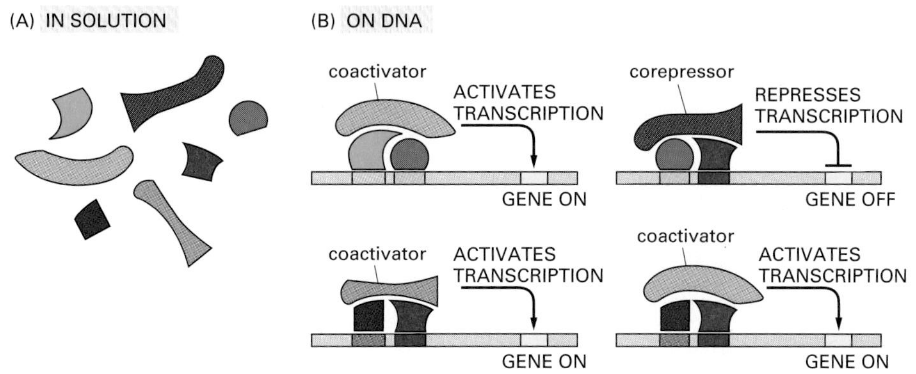

Figure 7–50 Eucaryotic gene regulatory proteins often assemble into complexes on DNA. Seven gene regulatory proteins are shown in (A). The nature and function of the complex they form depends on the specific DNA sequence that seeds their assembly. In (B), some assembled complexes activate gene transcription, while another represses transcription. Note that the *red* protein is shared by both activating and repressing complexes.

Eucaryotic Gene Regulatory Proteins Often Assemble into Complexes on DNA

So far we have been discussing eucaryotic gene regulatory proteins as though they work as individual polypeptides. In reality, most act as parts of complexes composed of several (and sometimes many) polypeptides, each with a distinct function. These complexes often assemble only in the presence of the appropriate DNA sequence. In some well-studied cases, for example, two gene regulatory proteins with a weak affinity for each other cooperate to bind to a DNA sequence, neither protein having a sufficient affinity for DNA to efficiently bind to the DNA site on its own. Once bound to DNA, the protein dimer creates a distinct surface that is recognized by a third protein that carries an activator domain that stimulates transcription (Figure 7–50). This example illustrates an important general point: protein–protein interactions that are too weak to cause proteins to assemble in solution can cause the proteins to assemble on DNA; in this way the DNA sequence acts as a "crystallization" site or seed for the assembly of a protein complex.

An individual gene regulatory protein can often participate in more than one type of regulatory complex. A protein might function, for example, in one case as part of a complex that activates transcription and in another case as part of a complex that represses transcription (see Figure 7–50). Thus individual eucaryotic gene regulatory proteins are not necessarily dedicated activators or repressors; instead, they function as regulatory units that are used to generate complexes whose function depends on the final assembly of all of the individual components. This final assembly, in turn, depends both on the arrangement of control region DNA sequences and on which gene regulatory proteins are present in the cell.

Gene regulatory proteins that do not themselves bind DNA but assemble on DNA-bound gene regulatory proteins are often termed **coactivators** or **corepressors**, depending on their effect on transcription initiation. As shown in Figure 7–50, the same coactivator or corepressor can assemble on different DNA binding proteins. Coactivators and corepressors typically carry out multiple functions: they can interact with chromatin remodeling complexes, histone modifying enzymes, the RNA polymerase holoenzyme, and several of the general transcription factors.

In some cases, the precise DNA sequence to which a regulatory protein directly binds can affect the conformation of this protein and thereby influence its subsequent transcriptional activity. When bound to one type of DNA sequence, for example, a steroid hormone receptor interacts with a corepressor and ultimately turns off transcription. When bound to a slightly different DNA sequence, it assumes a different conformation and interacts with a coactivator, thereby stimulating transcription.

Typically, the assembly of a group of regulatory proteins on DNA is guided by a few relatively short stretches of nucleotide sequence (see Figure 7–50). However, in some cases, a more elaborate protein–DNA structure, termed an *enhancesome*, is formed (Figure 7–51). A hallmark of enhancesomes is the participation of *architectural proteins* that bend the DNA by a defined angle and

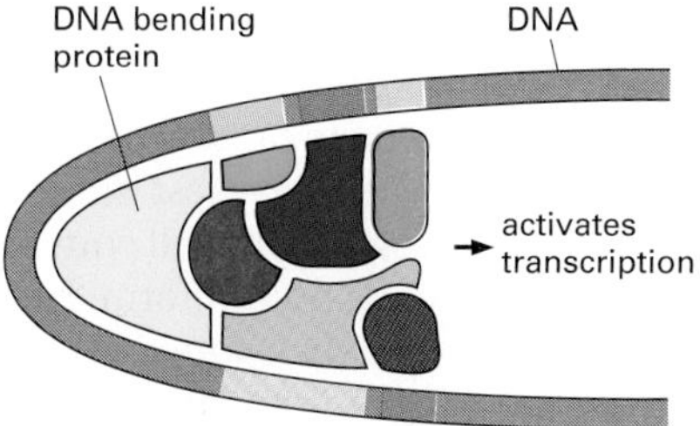

Figure 7–51 Schematic depiction of an enhancesome. The protein depicted in *yellow* is termed an architectural protein since its main role is to bend the DNA to allow the cooperative assembly of the other components. The protein surface of this enhancesome interacts with a coactivator which activates transcription at a nearby promoter. The enhancesome depicted here is based on that found in the control region of the gene that codes for a subunit of the T cell receptor (discussed in Chapter 24). The complete set of protein components for the enhancesome are present only in certain cells of the developing immune system, which eventually give rise to mature T cells.

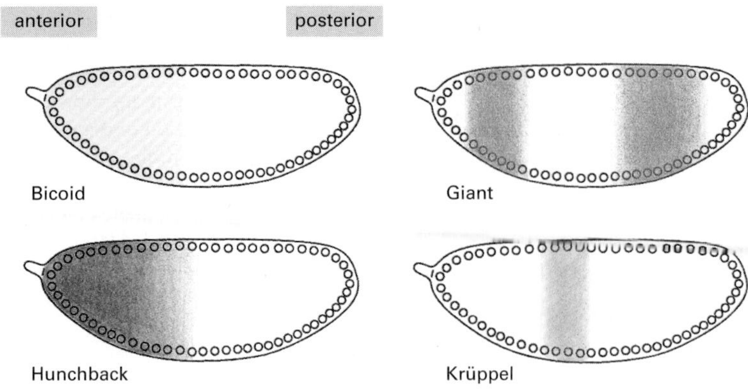

Figure 7–52 The nonuniform distribution of four gene regulatory proteins in an early *Drosophila* embryo. At this stage the embryo is a syncytium, with multiple nuclei in a common cytoplasm. Although not illustrated in these drawings, all of these proteins are concentrated in the nuclei.

thereby promote the assembly of the other enhancesome proteins. Since formation of the enhancesome requires the presence of many gene regulatory proteins, it provides a simple way to ensure that a gene is expressed only when the correct combination of these proteins is present in the cell. We saw earlier how the formation of gene regulatory heterodimers in solution provides a mechanism for the combinatorial control of gene expression. The assembly of larger complexes of gene regulatory proteins on DNA provides a second important mechanism for combinatorial control, offering far richer opportunities.

Complex Genetic Switches That Regulate *Drosophila* Development Are Built Up from Smaller Modules

Given that gene regulatory proteins can be positioned at multiple sites along long stretches of DNA, that these proteins can assemble into complexes at each site, and that the complexes can influence the chromatin structure and the recruitment and assembly of the general transcription machinery at the promoter, there would seem to be almost limitless possibilities for the elaboration of control devices to regulate eucaryotic gene transcription.

A particularly striking example of a complex, multicomponent genetic switch is that controlling the transcription of the *Drosophila even-skipped* (*eve*) gene, whose expression plays an important part in the development of the *Drosophila* embryo. If this gene is inactivated by mutation, many parts of the embryo fail to form, and the embryo dies early in development. As discussed in Chapter 21, at the earliest stage of development where *eve* is expressed, the embryo is a single giant cell containing multiple nuclei in a common cytoplasm. This cytoplasm is not uniform, however: it contains a mixture of gene regulatory proteins that are distributed unevenly along the length of the embryo, thus providing positional information that distinguishes one part of the embryo from another (Figure 7–52). (The way these differences are initially set up is discussed in Chapter 21.) Although the nuclei are initially identical, they rapidly begin to express different genes because they are exposed to different gene regulatory proteins. The nuclei near the anterior end of the developing embryo, for example, are exposed to a set of gene regulatory proteins that is distinct from the set that influences nuclei at the posterior end of the embryo.

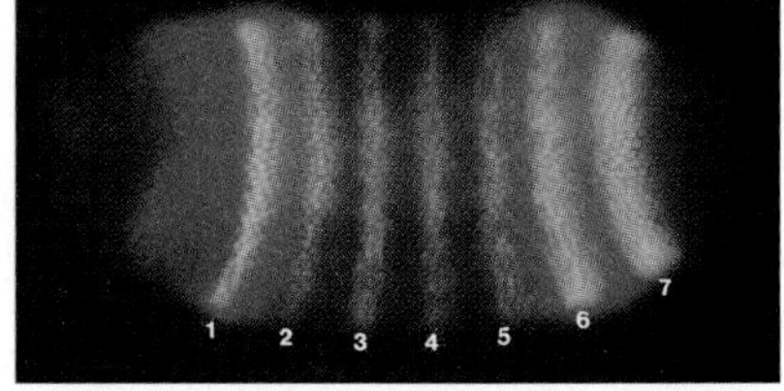

Figure 7–53 The seven stripes of the protein encoded by the *even-skipped* (*eve*) gene in a developing *Drosophila* embryo. Two and one-half hours after fertilization, the egg was fixed and stained with antibodies that recognize the Eve protein *(green)* and antibodies that recognize the Giant protein *(red)*. Where Eve and Giant proteins are both present, the staining appears *yellow*. At this stage in development, the egg contains approximately 4000 nuclei. The Eve and Giant proteins are both located in the nuclei, and the Eve stripes are about four nuclei wide. The staining pattern of the Giant protein is also shown in Figure 7–52. (Courtesy of Michael Levine.)

The regulatory DNA sequences of the *eve* gene are designed to read the concentrations of gene regulatory proteins at each position along the length of the embryo and to interpret this information in such a way that the *eve* gene is expressed in seven stripes, each initially five to six nuclei wide and positioned precisely along the anterior–posterior axis of the embryo (Figure 7–53). How is this remarkable feat of information processing carried out? Although the molecular details are not yet all understood, several general principles have emerged from studies of *eve* and other *Drosophila* genes that are similarly regulated.

The regulatory region of the *eve* gene is very large (approximately 20,000 nucleotide pairs). It is formed from a series of relatively simple regulatory modules, each of which contains multiple regulatory sequences and is responsible

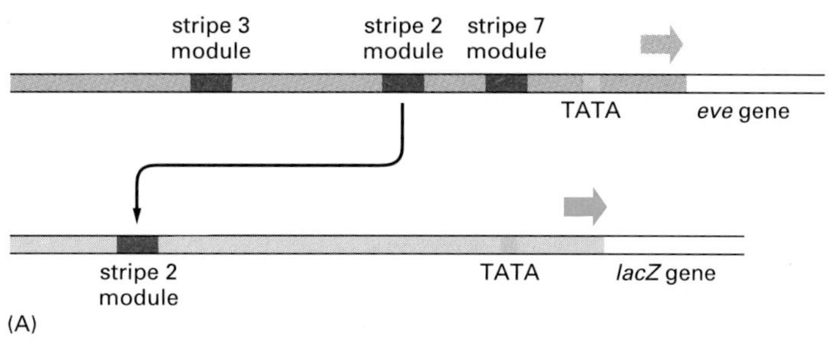

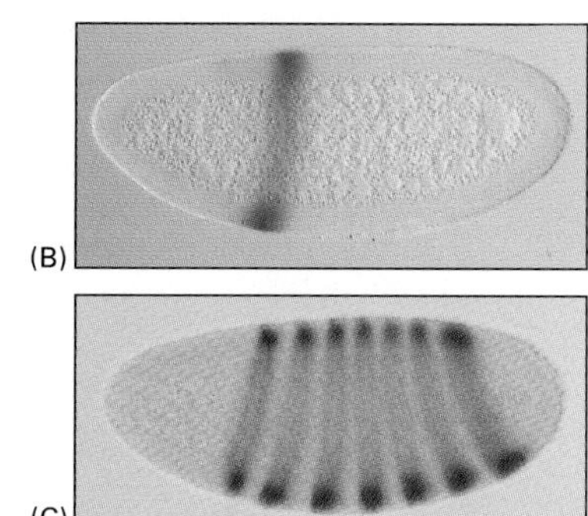

Figure 7–54 Experiment demonstrating the modular construction of the eve gene regulatory region. (A) A 480-nucleotide-pair piece of the *eve* regulatory region was removed and inserted upstream of a test promoter that directs the synthesis of the enzyme β-galactosidase (the product of the *E. coli lacZ* gene). (B) When this artificial construct was reintroduced into the genome of *Drosophila* embryos, the embryos expressed β-galactosidase (detectable by histochemical staining) precisely in the position of the second of the seven Eve stripes (C). (B and C, courtesy of Stephen Small and Michael Levine.)

for specifying a particular stripe of *eve* expression along the embryo. This modular organization of the *eve* gene control region is revealed by experiments in which a particular regulatory module (say, that specifying stripe 2) is removed from its normal setting upstream of the *eve* gene, placed in front of a reporter gene (see Figure 7–42), and reintroduced into the *Drosophila* genome (Figure 7–54A). When developing embryos derived from flies carrying this genetic construct are examined, the reporter gene is found to be expressed in precisely the position of stripe 2 (see Figure 7–54). Similar experiments reveal the existence of other regulatory modules, each of which specifies one of the other six stripes or some part of the expression pattern that the gene displays at later stages of development.

The *Drosophila eve* Gene Is Regulated by Combinatorial Controls

A detailed study of the stripe 2 regulatory module has provided insights into how it reads and interprets positional information. It contains recognition sequences for two gene regulatory proteins (Bicoid and Hunchback) that activate *eve* transcription and two (Krüppel and Giant) that repress it (Figure 7–55). (The gene regulatory proteins of *Drosophila* often have colorful names reflecting the phenotype that results if the gene encoding the protein is inactivated by mutation.) The relative concentrations of these four proteins determine whether protein complexes forming at the stripe 2 module turn on transcription of the *eve* gene. Figure 7–56 shows the distributions of the four gene regulatory proteins across the region of a *Drosophila* embryo where stripe 2 forms. Although the precise details are not known, it seems likely that either one of the two repressor proteins, when bound to the DNA, will turn off the stripe 2 module, whereas both Bicoid and Hunchback must bind for its maximal activation. This simple regulatory unit thereby combines these four positional signals so as to turn on the stripe 2 module (and therefore the expression of the *eve* gene) only in those nuclei that are located where the levels of both Bicoid and Hunchback are high and both Krüppel and Giant are absent. This combination of activators and repressors occurs only in one region of the early embryo; everywhere else, therefore, the stripe 2 module is silent.

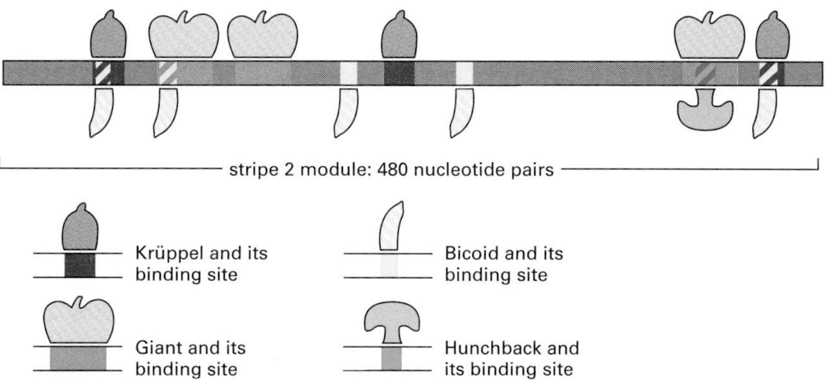

Figure 7–55 Close-up view of the eve stripe 2 unit. The segment of the *eve* gene control region identified in the previous figure contains regulatory sequences, each of which binds one or another of four gene regulatory proteins. It is known from genetic experiments that these four regulatory proteins are responsible for the proper expression of *eve* in stripe 2. Flies that are deficient in the two gene activators Bicoid and Hunchback, for example, fail to efficiently express *eve in* stripe 2. In flies deficient in either of the two gene repressors, Giant and Krüppel, stripe 2 expands and covers an abnormally broad region of the embryo. The DNA-binding sites for these gene regulatory proteins were determined by cloning the genes encoding the proteins, overexpressing the proteins in *E. coli,* purifying them, and performing DNA-footprinting experiments as described in Chapter 8. The *top* diagram indicates that, in some cases, the binding sites for the gene regulatory proteins overlap and the proteins can compete for binding to the DNA. For example, binding of Krüppel and binding of Bicoid to the site at the far right are thought to be mutually exclusive.

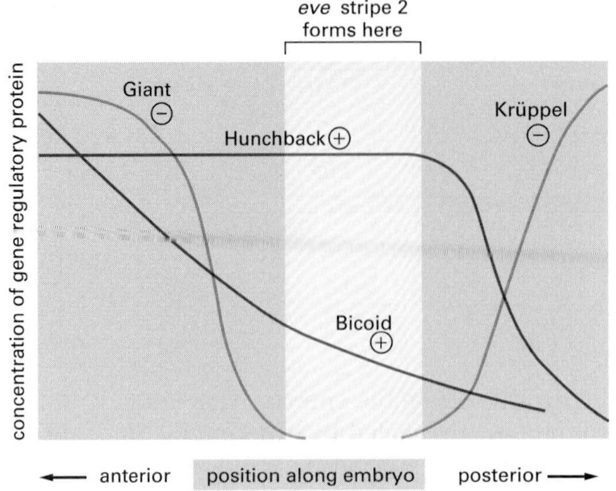

Figure 7–56 Distribution of the gene regulatory proteins responsible for ensuring that eve is expressed in stripe 2. The distributions of these proteins were visualized by staining a developing *Drosophila* embryo with antibodies directed against each of the four proteins (see Figures 7–52 and 7–53). The expression of *eve* in stripe 2 occurs only at the position where the two activators (Bicoid and Hunchback) are present and the two repressors (Giant and Krüppel) are absent. In fly embryos that lack Krüppel, for example, stripe 2 expands posteriorly. Likewise, stripe 2 expands posteriorly if the DNA-binding sites for Krüppel in the stripe 2 module (see Figure 7–55) are inactivated by mutation and this regulatory region is reintroduced into the genome.

The *eve* gene itself encodes a gene regulatory protein, which, after its pattern of expression is set up in seven stripes, regulates the expression of other *Drosophila* genes. As development proceeds, the embryo is thus subdivided into finer and finer regions that eventually give rise to the different body parts of the adult fly, as discussed in Chapter 21.

This example from *Drosophila* embryos is unusual in that the nuclei are exposed directly to positional cues in the form of concentrations of gene regulatory proteins. In embryos of most other organisms, individual nuclei are in separate cells, and extracellular positional information must either pass across the plasma membrane or, more usually, generate signals in the cytosol in order to influence the genome.

We have already discussed two mechanisms of combinatorial control of gene expression—heterodimerization of gene regulatory proteins in solution (see Figure 7–22) and the assembly of combinations of gene regulatory proteins into small complexes on DNA (see Figure 7–50). It is likely that both mechanisms participate in the complex regulation of *eve* expression. In addition, the regulation of stripe 2 just described illustrates a third type of combinatorial control. Because the individual regulatory sequences in the *eve* stripe 2 module are strung out along the DNA, many sets of gene regulatory proteins can be bound simultaneously and influence the promoter of a gene. The promoter integrates the transcriptional cues provided by all of the bound proteins (Figure 7–57).

The regulation of *eve* expression is an impressive example of combinatorial control. Seven combinations of gene regulatory proteins—one combination for each stripe—activate *eve* expression, while many other combinations (all those found in the interstripe regions of the embryo) keep the stripe elements silent. The other stripe regulatory modules are thought to be constructed along lines similar to those described for stripe 2, being designed to read positional information provided by other combinations of gene regulatory proteins. The entire gene control region, strung out over 20,000 nucleotide pairs of DNA, binds more than 20 different proteins. A large and complex control region is thereby built from a series of smaller modules, each of which consists of a unique arrangement of short DNA sequences recognized by specific gene regulatory proteins. Although the details are not yet understood, these gene regulatory proteins are thought to employ a number of the mechanisms previously described for activators and repressors. In this way, a single gene can respond to an enormous number of combinatorial inputs.

Complex Mammalian Gene Control Regions Are Also Constructed from Simple Regulatory Modules

It has been estimated that 5–10% of the coding capacity of a mammalian genome is devoted to the synthesis of proteins that serve as regulators of gene

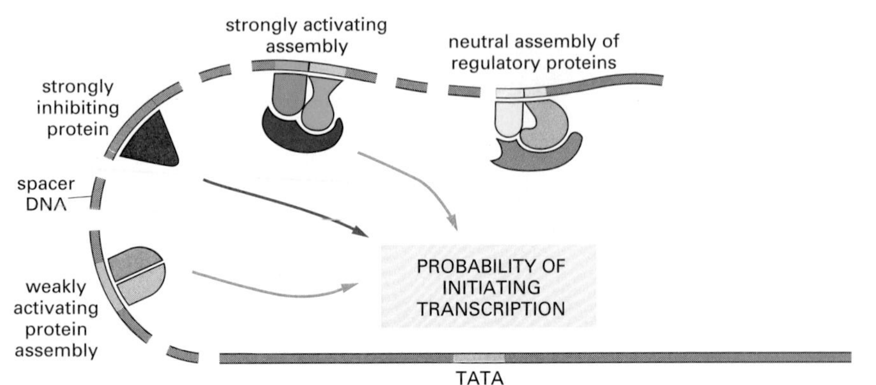

Figure 7–57 Integration at a promoter. Multiple sets of gene regulatory proteins can work together to influence transcription initiation at a promoter, as they do in the *eve* stripe 2 module illustrated previously in Figure 7–55. It is not yet understood in detail how the integration of multiple inputs is achieved, but it is likely that the final transcriptional activity of the gene results from a competition between activators and repressors that act by the mechanisms summarized in Figures 7–43, 7–44, 7–45, 7–46, and 7–49.

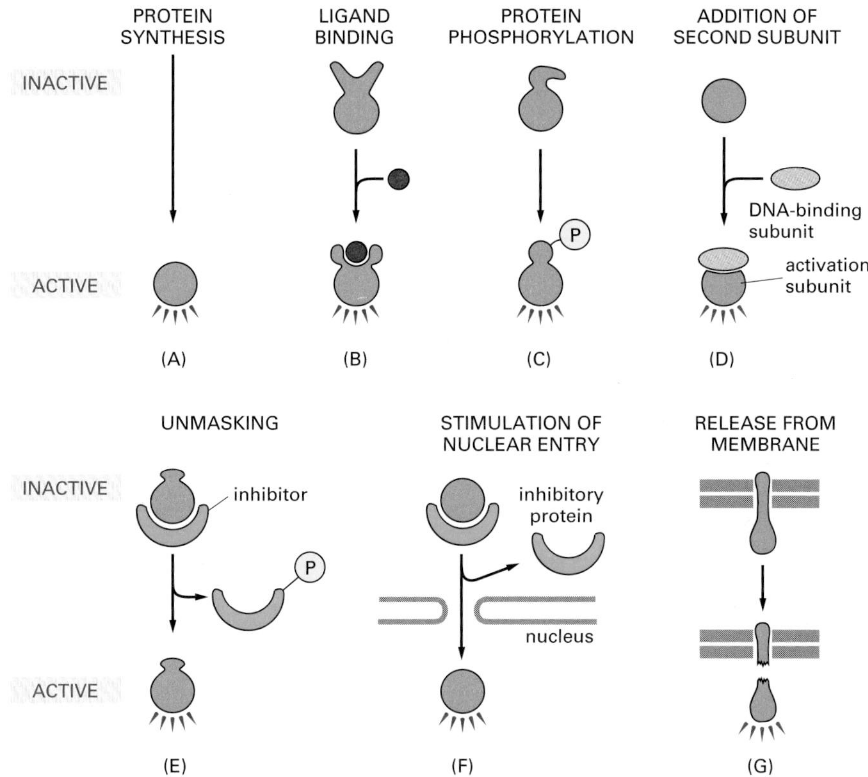

Figure 7–58 Some ways in which the activity of gene regulatory proteins is regulated in eucaryotic cells. (A) The protein is synthesized only when needed and is rapidly degraded by proteolysis so that it does not accumulate. (B) Activation by ligand binding. (C) Activation by phosphorylation. (D) Formation of a complex between a DNA-binding protein and a separate protein with a transcription-activating domain. (E) Unmasking of an activation domain by the phosphorylation of an inhibitor protein. (F) Stimulation of nuclear entry by removal of an inhibitory protein that otherwise keeps the regulatory protein from entering the nucleus. (G) Release of a gene regulatory protein from a membrane bilayer by regulated proteolysis.

Each of these mechanisms is typically controlled by extracellular signals which are communicated across the plasma membrane to the gene regulatory proteins in the cell. The ways in which this signaling occurs is discussed in Chapter 15. Mechanisms (A)–(F) are readily reversible and therefore also provide the means to selectively inactivate gene regulatory proteins.

transcription. This large number of genes reflects the exceedingly complex network of controls governing expression of mammalian genes. Each gene is regulated by a set of gene regulatory proteins; each of those proteins is the product of a gene that is in turn regulated by a whole set of other proteins, and so on. Moreover, the regulatory protein molecules are themselves influenced by signals from outside the cell, which can make them active or inactive in a whole variety of ways (Figure 7–58). Thus, pattern of gene expression in a cell can be viewed as the result of a complicated molecular computation that the intracellular gene control network performs in response to information from the cell's surroundings. We shall discuss this further in Chapter 21, dealing with multicellular development, but the complexity is remarkable even at the level of the individual genetic switch, regulating activity of a single gene. It is not unusual, for example, to find a mammalian gene with a control region that is 50,000 nucleotide pairs in length, in which many modules, each containing a number of regulatory sequences that bind gene regulatory proteins, are interspersed with long stretches of spacer DNA.

One of the best-understood examples of a complex mammalian regulatory region is found in the human β-globin gene, which is expressed exclusively in red blood cells and at a specific time in their development. A complex array of gene regulatory proteins controls the expression of the gene, some acting as activators and others as repressors (Figure 7–59). The concentrations (or activities) of many of these gene regulatory proteins are thought to change during development, and only a particular combination of all the proteins triggers transcription of the gene. The human β-globin gene is part of a cluster of globin genes (Figure 7–60A). The five genes of the cluster are transcribed exclusively in erythroid cells, that is, cells of the red blood cell lineage. Moreover, each gene is turned on at a different stage of development (see Figure 7–60B) and in different organs: the ε-globin gene is expressed in the embryonic yolk sac, γ in the yolk sac and the fetal liver, and δ and β primarily in the adult bone marrow. Each of the globin genes has its own set of regulatory proteins that are necessary to turn the gene on at the appropriate time and tissue. In addition to the individual regulation of each of the globin genes, the entire cluster appears to be subject to a shared control region called a *locus control region (LCR).* The LCR lies far

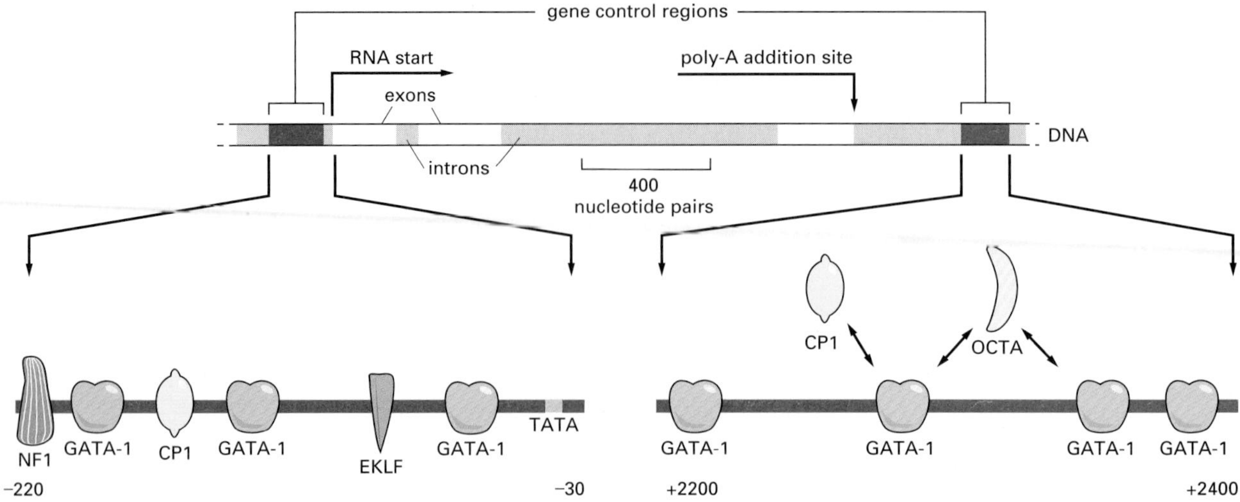

Figure 7–59 Model for the control of the human β-globin gene. The diagram shows some of the gene regulatory proteins thought to control expression of the gene during red blood cell development (see Figure 7–60). Some of the gene regulatory proteins shown, such as CP1, are found in many types of cells, while others, such as GATA-1, are present in only a few types of cells—including red blood cells—and therefore are thought to contribute to the cell-type specificity of β-globin gene expression. As indicated by the *double-headed arrows*, several of the binding sites for GATA-1 overlap those of other gene regulatory proteins; it is thought that occupancy of these sites by GATA-1 excludes binding of other proteins. Once bound to DNA, the gene regulatory proteins recruit chromatin remodeling complexes, histone modifying enzymes, the general transcription factors and RNA polymerase to the promoter. (Adapted from B. Emerson, in Gene Expression: General and Cell-Type Specific [M. Karin, ed.], pp. 116–161. Boston: Birkhauser, 1993.)

upstream from the gene cluster (see Figure 7–60A), and we shall discuss its function next.

In cells where the globin genes are not expressed (such as brain or skin cells), the whole gene cluster appears tightly packaged into chromatin. In erythroid cells, by contrast, the entire gene cluster is still folded into nucleosomes, but the higher-order packing of the chromatin has become decondensed This change occurs even before the individual globin genes are transcribed, suggesting that there are two steps of regulation. In the first, the chromatin of the entire globin locus becomes decondensed, which is presumed to allow additional gene regulatory proteins access to the DNA. In the second step, the remaining gene regulatory proteins assemble on the DNA and direct the expression of individual genes.

The LCR appears to act by controlling chromatin condensation, and its importance can be seen in patients with a certain type of thalassemia, a severe inherited form of anemia. In these patients, the β-globin locus is found to have undergone deletions that remove all or part of the LCR, and although the β-globin gene and its nearby regulatory regions are intact, the gene remains transcriptionally silent even in erythroid cells. Moreover, the β-globin gene in the erythroid cells fails to undergo the normal chromatin decondensation step that occurs during erythroid cell development.

Many LCRs (that is, DNA regulatory sequences that regulate the accessibility and expression of distant genes or gene clusters) are present in the human genome, and they regulate a wide variety of cell-type specific genes. The way in which they function is not understood in detail, but several models have been proposed. The simplest is based on principles we have already discussed in this chapter: the gene regulatory proteins that bind to the LCR interact through DNA

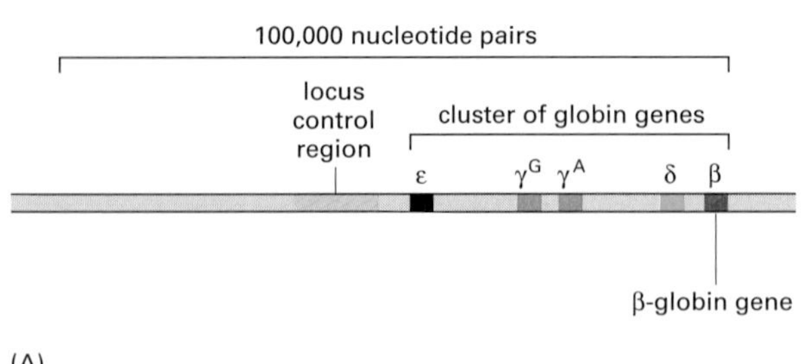

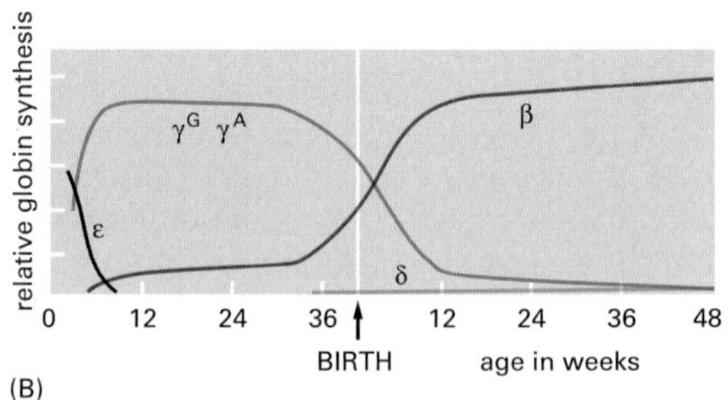

Figure 7–60 The cluster of β-like globin genes in humans. (A) The large chromosomal region shown spans 100,000 nucleotide pairs and contains the five globin genes and a locus control region (LCR). (B) Changes in the expression of the β-like globin genes at various stages of human development. Each of the globin chains encoded by these genes combines with an α-globin chain to form the hemoglobin in red blood cells (see Figure 7–115). (A, after F. Grosveld, G.B. van Assendelft, D.R. Greaves, and G. Kollias, *Cell* 51:975–985, 1987. © Elsevier.)

looping with proteins bound to the control regions of the genes they regulate. In this way, the proteins bound at the LCR could attract chromatin remodeling complexes and histone modifying enzymes that could alter the chromatin structure of the locus before transcription begins. Other models for LCRs propose a mechanism by which proteins initially bound at the LCR attract other proteins that assemble cooperatively and therefore spread along the DNA toward the genes they control, modifying the chromatin as they proceed.

Insulators Are DNA Sequences That Prevent Eucaryotic Gene Regulatory Proteins from Influencing Distant Genes

All genes have control regions, which dictate at which times, under what conditions, and in what tissues the gene will be expressed. We also have seen that eucaryotic gene regulatory proteins can act across very long stretches of DNA. How then are control regions of different genes kept from interfering with one another? In other words, what keeps a gene regulatory protein bound on the control region of one gene from inappropriately influencing transcription of adjacent genes?

Several mechanisms have been proposed to account for this regulatory compartmentalization, but the best understood rely on **insulator elements**, also called *boundary elements*. Insulator elements (insulators, for short) are DNA sequences that bind specialized proteins and have two specific properties (Figure 7–61). First, they buffer genes from the repressing effects of heterochromatin. When a gene (from a fly or a mouse, for example) and its normal control region is inserted into different positions in the genome, it is often expressed at levels that vary depending on its site of insertion in the genome and are especially low when it is inserted amid heterochromatin. We saw an example of this *position effect* in Chapter 4, where genes inserted into heterochromatin are transcriptionally silenced (see Figure 4–45). When insulator elements that flank the gene and its control region are included, however, the gene is usually expressed normally, irrespective of its new position in the genome. The second property of insulators is in some sense the converse of this: they can block the action of enhancers (see Figure 7–61). For this to occur, the insulator must be located between the enhancer and the promoter of the target gene.

Thus insulators can define domains of gene expression, both buffering the gene from outside effects and preventing the control region of the gene (or cluster of genes) from acting outside the domain. For example, the globin LCR (discussed above) is associated with a neighboring insulator which allows the LCR to influence only the cluster of globin genes. Presumably, another insulator is located on the distal side of the globin cluster, serving to define the other end of the domain.

The distribution of insulators in a genome is therefore thought to divide it into independent domains of gene regulation and chromatin structure. Consistent with this idea, the distribution of insulators across a genome is roughly correlated with variations in chromatin structure. For example, an insulator-binding protein from flies is localized preferentially to interbands (and also to the edges of puffs) in polytene chromosomes (Figure 7–62).

The mechanisms by which insulators work are not currently understood, and different insulators may function in different ways. At least some pairs of insulators may define the basis of a looped chromosomal domain (see Figure 4–44). It has been proposed that chromosomes of all eucaryotes are divided by insulators into independent looped domains, each regulated separately from all the others.

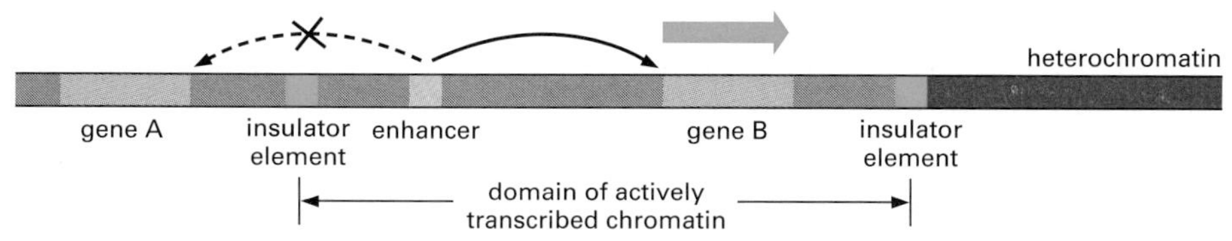

Figure 7–61 Schematic diagram summarizing the properties of insulators. Insulators both prevent the spread of heterochromatin *(right-hand side* of diagram) and directionally block the action of enhancers *(left-hand side).* Thus gene B is properly regulated and gene B's enhancer is prevented from influencing the transcription of gene A.

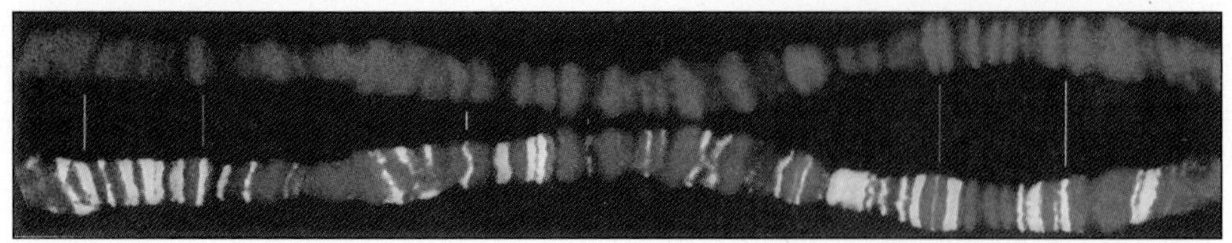

Figure 7–62 Localization of a *Drosophila* insulator-binding protein on polytene chromosomes. A polytene chromosome (see pp. 218–220) was stained with propidium iodide *(red)* to show its banding patterns—with bands appearing *bright red* and interbands as dark gaps in the pattern *(top)*. The positions on this polytene chromosome that are bound by a particular insulator protein (called BEAF) are stained *bright green* using antibodies directed against the protein *(bottom)*. BEAF is seen to be preferentially localized to interband regions suggesting a role for it in organizing domains of chromatin. For convenience, these two micrographs of the same polytene chromosome are arranged as mirror images. (Courtesy of Uli Laemmli, from K. Zhao et al., *Cell* 81:879–889, 1995. © Elsevier.)

Bacteria Use Interchangeable RNA Polymerase Subunits to Help Regulate Gene Transcription

We have seen the importance of gene regulatory proteins that bind to regulatory sequences in DNA and signal to the transcription apparatus whether or not to start the synthesis of an RNA chain. Although this is the main way of controlling transcriptional initiation in both eucaryotes and procaryotes, some bacteria and their viruses use an additional strategy based on interchangeable subunits of RNA polymerase. As described in Chapter 6, a sigma (σ) subunit is required for the bacterial RNA polymerase to recognize a promoter. Many bacteria make several different sigma subunits, each of which can interact with the RNA polymerase core and direct it to a different set of promoters (Table 7–2). This scheme permits one large set of genes to be turned off and a new set to be turned on simply by replacing one sigma subunit with another; the strategy is efficient because it bypasses the need to deal with the genes one by one. It is often used subversively by bacterial viruses to take over the host polymerase and activate several sets of viral genes rapidly and sequentially (Figure 7–63).

In a sense, eucaryotes employ an analogous strategy through the use of three distinct RNA polymerases (I, II, and III) that share some of their subunits. Procaryotes, in contrast, use only one type of core RNA polymerase molecule, but they modify it with different sigma subunits.

Gene Switches Have Gradually Evolved

We have seen that the control regions of eucaryotic genes are often spread out over long stretches of DNA, whereas those of procaryotic genes are typically closely packed around the start point of transcription. Several bacterial gene regulatory proteins, however, recognize DNA sequences that are located many nucleotide pairs away from the promoter, as we saw in Figure 7–40. This case provided one of the first examples of DNA looping in gene regulation and greatly influenced later studies of eucaryotic gene regulatory proteins.

It seems likely that the close-packed arrangement of bacterial genetic switches developed from more extended forms of switches in response to the evolutionary pressure on bacteria to maintain a small genome size. This compression comes at a price, however, as it restricts the complexity and adaptability of the control device. The extended form of eucaryotic control regions, in contrast, with discrete regulatory modules separated by long stretches of spacer DNA, would be expected to facilitate a reshuffling of the regulatory modules during evolution, both to create new regulatory circuits and to modify old ones. Unraveling the history of how gene control regions evolved presents a fascinating challenge, and many clues can be found in present-day DNA sequences. We shall take up this issue again at the end of this chapter.

TABLE 7–2 Sigma Factors of *E. coli*

SIGMA FACTOR	PROMOTERS RECOGNIZED
σ^{70}	most genes
σ^{32}	genes induced by heat shock
σ^{28}	genes for stationary phase and stress response
σ^{28}	genes involved in motility and chemotaxis
σ^{54}	genes for nitrogen metabolism

The sigma factor designations refer to their approximate molecular weights, in kilodaltons.

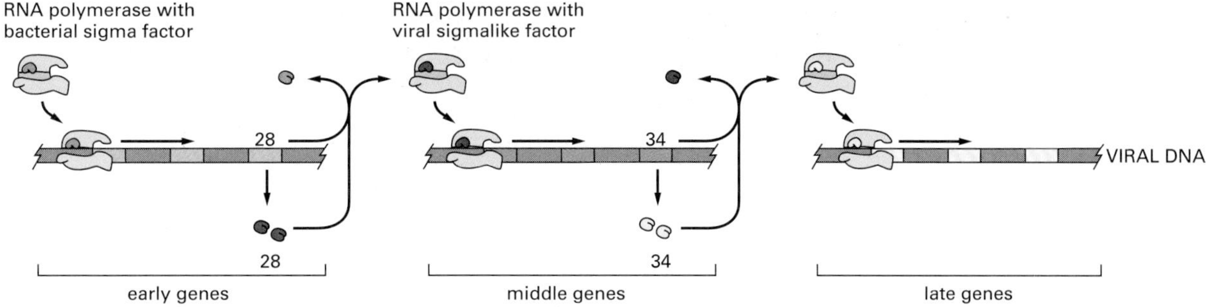

Figure 7–63 Interchangeable RNA polymerase subunits as a strategy to control gene expression in a bacterial virus. The bacterial virus SPO1, which infects the bacterium *B. subtilis,* uses the bacterial polymerase to transcribe its early genes immediately after the viral DNA enters the cell. One of the early genes, called 28, encodes a sigmalike factor that binds to RNA polymerase and displaces the bacterial sigma factor. This new form of polymerase specifically initiates transcription of the SPO1 "middle" genes. One of the middle genes encodes a second sigmalike factor, 34, that displaces the 28 product and directs RNA polymerase to transcribe the "late" genes. This last set of genes produces the proteins that package the virus chromosome into a virus coat and lyse the cell. By this strategy, sets of virus genes are expressed in the order in which they are needed; this ensures a rapid and efficient viral replication.

Summary

The transcription of individual genes is switched on and off in cells by gene regulatory proteins. In procaryotes these proteins usually bind to specific DNA sequences close to the RNA polymerase start site and, depending on the nature of the regulatory protein and the precise location of its binding site relative to the start site, either activate or repress transcription of the gene. The flexibility of the DNA helix, however, also allows proteins bound at distant sites to affect the RNA polymerase at the promoter by the looping out of the intervening DNA. Such action at a distance is extremely common in eucaryotic cells, where gene regulatory proteins bound to sequences thousands of nucleotide pairs from the promoter generally control gene expression. Eucaryotic activators and repressors act by a wide variety of mechanisms—generally causing the local modification of chromatin structure, the assembly of the general transcription factors at the promoter, and the recruitment of RNA polymerase.

Whereas the transcription of a typical procaryotic gene is controlled by only one or two gene regulatory proteins, the regulation of higher eucaryotic genes is much more complex, commensurate with the larger genome size and the large variety of cell types that are formed. The control region of the Drosophila eve gene, for example, encompasses 20,000 nucleotide pairs of DNA and has binding sites for over 20 gene regulatory proteins. Some of these proteins are transcriptional activators, whereas others are transcriptional repressors. These proteins bind to regulatory sequences organized in a series of regulatory modules strung together along the DNA, and together they cause the correct spatial and temporal pattern of gene expression. Eucaryotic genes and their control regions are often surrounded by insulators, DNA sequences recognized by proteins that prevent cross-talk between independently regulated genes.

THE MOLECULAR GENETIC MECHANISMS THAT CREATE SPECIALIZED CELL TYPES

Although all cells must be able to switch genes on and off in response to changes in their environments, the cells of multicellular organisms have evolved this capacity to an extreme degree and in highly specialized ways to form an organized array of differentiated cell types. In particular, once a cell in a multicellular organism becomes committed to differentiate into a specific cell type, the choice of fate is generally maintained through many subsequent cell generations, which means that the changes in gene expression involved in the choice must be remembered. This phenomenon of *cell memory* is a prerequisite for the creation of organized tissues and for the maintenance of stably differentiated cell types. In contrast, the simplest changes in gene expression in both eucaryotes and bacteria are only transient; the tryptophan repressor, for example, switches off the tryptophan genes in bacteria only in the presence of tryptophan; as soon as tryptophan is removed from the medium, the genes are switched back on, and the descendants of the cell will have no memory that their ancestors had been exposed to tryptophan. Even in bacteria, however, a few types of changes in gene expression can be inherited stably.

In this section we examine how gene regulatory devices can be combined to create "logic" circuits through which cells differentiate, keep time, remember events in their past, keep time, and adjust the levels of gene expression over whole chromosomes. We begin by considering some of the best-understood genetic mechanisms of cell differentiation, which operate in bacterial and yeast cells.

DNA Rearrangements Mediate Phase Variation in Bacteria

We have seen that cell differentiation in higher eucaryotes usually occurs without detectable changes in DNA sequence. In some procaryotes, in contrast, a stably inherited pattern of gene regulation is achieved by DNA rearrangements that activate or inactivate specific genes. Since changes in DNA sequence are copied faithfully during subsequent DNA replications, an altered state of gene activity will be inherited by all the progeny of the cell in which the rearrangement occurred. Some of these DNA rearrangements are, however, reversible so that occasional individuals switch back to original DNA configurations. The result is an alternating pattern of gene activity that can be detected by observations over long time periods and many generations.

A well-studied example of this differentiation mechanism occurs in *Salmonella* bacteria and is known as **phase variation**. Although this mode of differentiation has no known counterpart in higher eucaryotes, it can nevertheless have considerable impact on them because disease-causing bacteria use it to evade detection by the immune system. The switch in *Salmonella* gene expression is brought about by the occasional inversion of a specific 1000-nucleotide-pair piece of DNA. This change alters the expression of the cell-surface protein flagellin, for which the bacterium has two different genes (Figure 7–64). The inversion is catalyzed by a site-specific recombination enzyme and changes the orientation of a promoter that is within the 1000 nucleotide pairs. With the promoter in one orientation, the bacteria synthesize one type of flagellin; with the promoter in the other orientation, they synthesize the other type. Because inversions occur only rarely, whole clones of bacteria will grow up with one type of flagellin or the other.

Phase variation almost certainly evolved because it protects the bacterial population against the immune response of its vertebrate host. If the host makes antibodies against one type of flagellin, a few bacteria whose flagellin has been altered by gene inversion will still be able to survive and multiply.

Bacteria isolated from the wild very often exhibit phase variation for one or more phenotypic traits. These "instabilities" are usually lost with time from standard laboratory strains of bacteria, and underlying mechanisms have been studied in only a few cases. Not all involve DNA inversion. A bacterium that

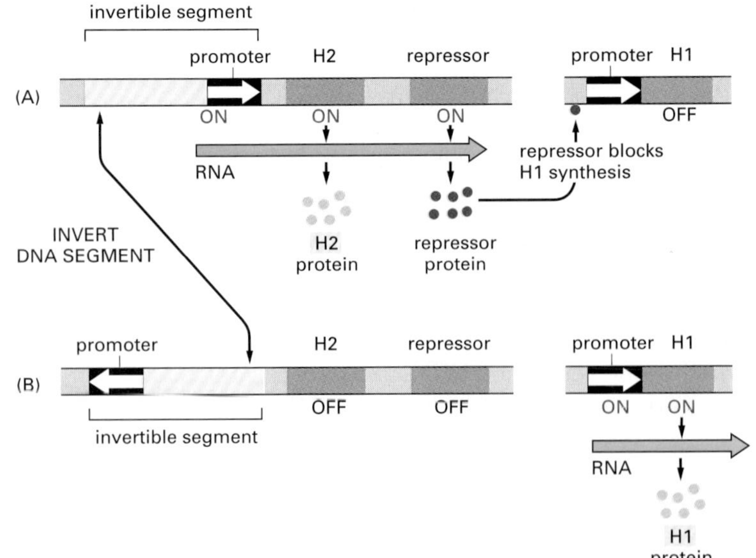

Figure 7–64 Switching gene expression by DNA inversion in bacteria. Alternating transcription of two flagellin genes in a *Salmonella* bacterium is caused by a simple site-specific recombination event that inverts a small DNA segment containing a promoter. (A) In one orientation, the promoter activates transcription of the H2 flagellin gene as well as that of a repressor protein that blocks the expression of the H1 flagellin gene. (B) When the promoter is inverted, it no longer turns on H2 or the repressor, and the H1 gene, which is thereby released from repression, is expressed instead. The recombination mechanism is activated only rarely (about once every 10^5 cell divisions). Therefore, the production of one or other flagellin tends to be faithfully inherited in each clone of cells.

causes a common sexually transmitted human disease *(Neisseria gonorrhoeae)*, for example, avoids immune attack by means of a heritable change in its surface properties that is generated by gene conversion (discussed in Chapter 5) rather than by inversion. This mechanism transfers DNA sequences from a library of silent "gene cassettes" to a site in the genome where the genes are expressed; it has the advantage of creating many variants of the major bacterial surface protein.

A Set of Gene Regulatory Proteins Determines Cell Type in a Budding Yeast

Because they are so easy to grow and to manipulate genetically, yeasts have served as model organisms for studying the mechanisms of gene control in eucaryotic cells. The common baker's yeast, *Saccharomyces cerevisiae*, has attracted special interest because of its ability to differentiate into three distinct cell types. *S. cerevisiae* is a single-celled eucaryote that exists in either a haploid or a diploid state. Diploid cells form by a process known as **mating**, in which two haploid cells fuse. In order for two haploid cells to mate, they must differ in *mating type* (sex). In *S. cerevisiae* there are two mating types, α and **a**, which are specialized for mating with each other. Each produces a specific diffusible signaling molecule (mating factor) and a specific cell-surface receptor protein. These jointly enable a cell to recognize and be recognized by its opposite cell type, with which it then fuses. The resulting diploid cells, called **a**/α, are distinct from either parent: they are unable to mate but can form spores (sporulate) when they run out of food, giving rise to haploid cells by the process of meiosis (discussed in Chapter 20).

The mechanisms by which these three cell types are established and maintained illustrate several of the strategies we have discussed for changing the pattern of gene expression. The mating type of the haploid cell is determined by a single locus, the **mating-type (MAT) locus**, which in an **a**-type cell encodes a single gene regulatory protein, a1, and in an α cell encodes two gene regulatory proteins, Matα1 and Matα2. The a1 protein has no effect in the **a**-type haploid cell that produces it but becomes important later in the diploid cell that results from mating; meanwhile, the α-type haploid cell produces the proteins specific to its mating type by default. In contrast, the α2 protein acts in the α cell as a transcriptional repressor that turns off the **a**-specific genes, while the α1 protein acts as a transcriptional activator that turns on the α-specific genes. Once cells of the two mating types have fused, the combination of the a1 and α2 regulatory proteins generates a completely new pattern of gene expression, unlike that of either parent cell. Figure 7–65 illustrates the mechanism by which the mating-type-specific genes are expressed in different patterns in the three cell types. This was among the first examples of combinatorial gene control to be identified, and it remains one of the best understood at the molecular level.

Although in most laboratory strains of *S. cerevisiae*, the **a** and α cell types are stably maintained through many cell divisions, some strains isolated from the wild can switch repeatedly between the **a** and α cell types by a mechanism of gene rearrangement whose effects are reminiscent of the DNA rearrangements in *N. gonorrhoeae*, although the exact mechanism seems to be peculiar to yeast. On either side of the *MAT* locus in the yeast chromosome, there is a silent locus encoding the mating-type gene regulatory proteins: the silent locus on one side encodes α1 and α2; the silent locus on the other side encodes a1. Approximately every other cell division, the active gene in the *MAT* locus is excised and replaced by a newly synthesized copy of the silent locus determining the opposite mating type. Because the change involves the removal of one gene from the active "slot" and its replacement by another, this mechanism is called the *cassette mechanism*. The change is reversible because, although the original gene at the *MAT* locus is discarded, a silent copy remains in the genome. New DNA copies made from the silent genes function as disposable cassettes that will be inserted in alternation into the *MAT* locus, which serves as the "playing head" (Figure 7–66).

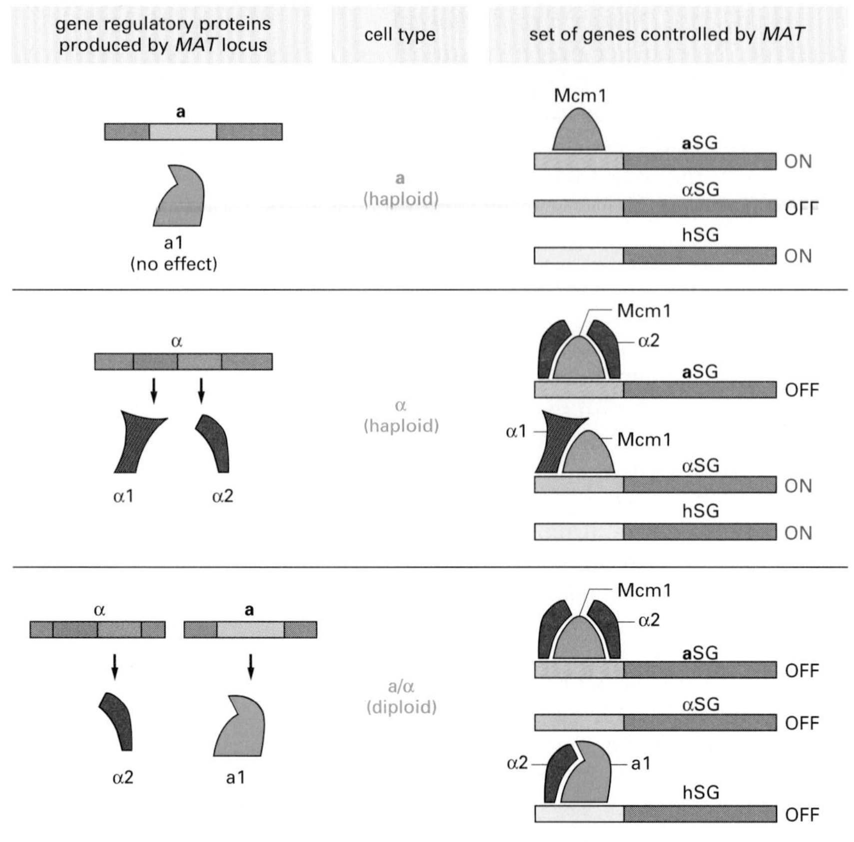

Figure 7–65 Control of cell type in yeasts. Yeast cell type is determined by three gene regulatory proteins (α1, α2, and a1) produced by the *MAT* locus. Different sets of genes are transcribed in haploid cells of type **a**, in haploid cells of type α, and in diploid cells (type a/α). The haploid cells express a set of haploid-specific genes (hSG) and either a set of α-specific genes (αSG) or a set of a-specific genes (aSG). The diploid cells express none of these genes. The α1, α2, and a1 proteins control many target genes in each type of cell by binding, in various combinations, to specific regulatory sequences upstream of these genes. Note that the α1 protein is a gene activator protein, whereas the α2 protein is a gene repressor protein. Both work in combination with a gene regulatory protein called Mcm1 that is present in all three cell types. In the diploid cell type, α2 and a1 form a heterodimer (shown in detail in Figure 7–23) that turns off a set of genes (including the gene encoding the α1 activator protein) different from that turned off by the α2 and Mcm1 proteins. This relatively simple system of gene regulatory proteins is an example of combinatorial control of gene expression (see Figure 7–50). The a1 and α2 proteins each recognize their DNA-binding sites using a homeodomain motif (see Figure 7–16).

The silent cassettes are maintained in a transcriptionally inactive form by the same mechanism that is responsible for silencing genes located at the ends of the yeast chromosomes (see Figure 4–47); that is, the DNA at a silent locus is packaged into a highly organized form of chromatin that is resistant to transcription.

Two Proteins That Repress Each Other's Synthesis Determine the Heritable State of Bacteriophage Lambda

The observation that a whole vertebrate or plant can be specified by the genetic information present in a single somatic cell nucleus (see Figure 7–2) eliminates the possibility that an irreversible change in DNA sequence is a major

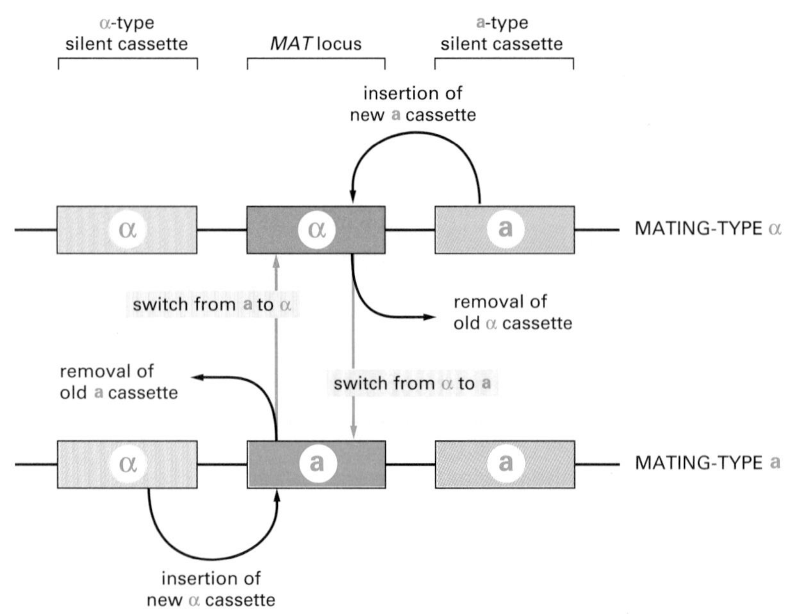

Figure 7–66 Cassette model of yeast mating-type switching. Cassette switching occurs by a gene-conversion process that involves a specialized enzyme (the HO endonuclease) that makes a double-stranded cut at a specific DNA sequence in the *MAT* locus. The DNA near the cut is then excised and replaced by a copy of the silent cassette of opposite mating type. The mechanism of this specialized form of gene conversion is similar to the general mechanism of homologous end joining discussed in Chapter 5 (see pp. 283–284).

mechanism in the differentiation of higher eucaryotic cells (although such changes are a crucial part of lymphocyte differentiation—discussed in Chapter 24). Reversible DNA sequence changes, resembling those just described for *Salmonella* and yeasts, in principle could still be responsible for some of the inherited changes in gene expression observed in higher organisms, but there is currently no evidence that such mechanisms are widely used.

Other mechanisms that we have touched upon in this chapter, however, are also capable of producing patterns of gene regulation that can be inherited by subsequent cell generations. Perhaps the simplest example is found in the bacterial virus (bacteriophage) lambda where a switch causes the virus to flip-flop between two stable self-maintaining states. This type of switch can be viewed as a prototype for similar, but more complex, switches that operate in the development of higher eucaryotes.

We mentioned earlier that bacteriophage lambda can in favorable conditions become integrated into the *E. coli* cell DNA, to be replicated automatically each time the bacterium divides. Alternatively, the virus can multiply in the cytoplasm, killing its host (see Figure 5–81). The switch between these two states is mediated by proteins encoded by the bacteriophage genome. The genome contains a total of about 50 genes, which are transcribed in very different patterns in the two states. A virus destined to integrate, for example, must produce the lambda *integrase* protein, which is needed to insert the lambda DNA into the bacterial chromosome, but must repress production of the viral proteins responsible for virus multiplication. Once one transcriptional pattern or the other has been established, it is stably maintained.

We cannot discuss the details of this complex gene regulatory system here and instead outline a few of its general features. At the heart of the switch are two gene regulatory proteins synthesized by the virus: the **lambda repressor protein** (cI protein), which we have already encountered, and the **Cro protein**. These proteins repress each other's synthesis, an arrangement giving rise to just two stable states (Figure 7–67). In state 1 (the *prophage state)* the lambda repressor occupies the operator, blocking the synthesis of Cro and also activating its own synthesis. In state 2 (the *lytic state)* the Cro protein occupies a different site in the operator, blocking the synthesis of repressor but allowing its own synthesis. In the prophage state most of the DNA of the stably integrated bacteriophage is not transcribed; in the lytic state, this DNA is extensively transcribed, replicated, packaged into new bacteriophage, and released by host cell lysis.

When the host bacteria are growing well, an infecting virus tends to adopt state 1, allowing the DNA of the virus to multiply along with the host chromosome. When the host cell is damaged, an integrated virus converts from state 1 to state 2 in order to multiply in the cell cytoplasm and make a quick exit. This conversion is triggered by the host response to DNA damage, which inactivates the repressor protein. In the absence of such interference, however, the lambda repressor both turns off production of the Cro protein and turns on its own synthesis, and this *positive feedback loop* helps to maintain the prophage state.

Gene Regulatory Circuits Can Be Used to Make Memory Devices As Well As Oscillators

Positive feedback loops provide a simple general strategy for cell memory—that is, for the establishment and maintenance of heritable patterns of gene

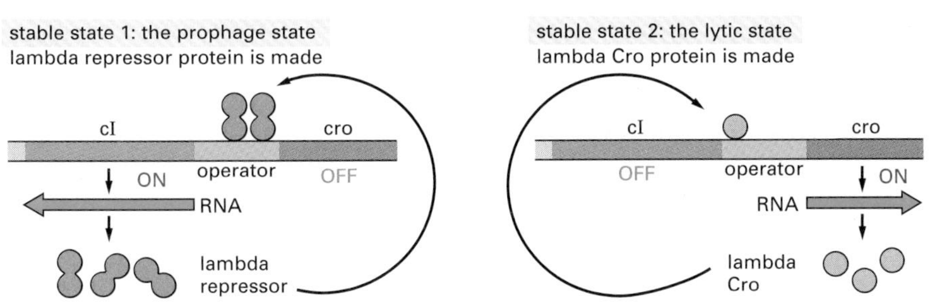

Figure 7–67 A simplified version of the regulatory system that determines the mode of growth of bacteriophage lambda in the *E. coli* host cell. In stable state 1 (the prophage state) the bacteriophage synthesizes a repressor protein, which activates its own synthesis and turns off the synthesis of several other bacteriophage proteins, including the Cro protein. In state 2 (the lytic state) the bacteriophage synthesizes the Cro protein, which turns off the synthesis of the repressor protein, so that many bacteriophage proteins are made and the viral DNA replicates freely in the *E. coli* cell, eventually producing many new bacteriophage particles and killing the cell. This example shows how two gene regulatory proteins can be combined in a circuit to produce two heritable states. Both the lambda repressor and the Cro protein recognize the operator through a helix–turn–helix motif (see Figure 7–14).

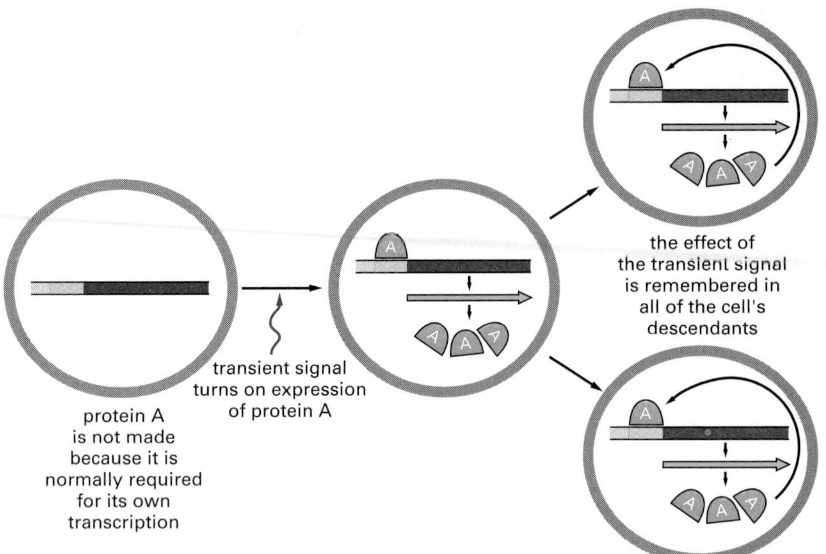

A
the effect of
the transient signal
is remembered in
all of the cell's
descendants
transient signal
turns on expression
of protein A
protein A
is not made
because it is
normally required
for its own
transcription

Figure 7–68 Schematic diagram showing how a positive feedback loop can create cell memory. Protein A is a gene regulatory protein that activates its own transcription. All of the descendants of the original cell will therefore "remember" that the progenitor cell had experienced a transient signal that initiated the production of the protein.

transcription. Figure 7–68 shows the basic principle, stripped to its barest essentials. Variations of this simple strategy are widely used by eucaryotic cells. Several gene regulatory proteins that are involved in establishing the *Drosophila* body plan (discussed in Chapter 21), for example, stimulate their own transcription, thereby creating a positive feedback loop that promotes their continued synthesis; at the same time many of these proteins repress the transcription of genes encoding other important gene regulatory proteins. In this way, a sophisticated pattern of inherited behavior can be achieved with only a few gene regulatory proteins that reciprocally affect one another's synthesis and activities.

Simple gene regulatory circuits can be combined to create all sorts of control devices, just as simple electronic switching elements in a computer are combined to perform all sorts of complex logical operations. Bacteriophage lambda, as we have seen, provides an example of a circuit that can flip-flop between two stable states. More complex types of regulatory networks are not only found in nature, but can also be designed and constructed in the laboratory. Figure 7–69 shows, for example, how an engineered bacterial cell can switch between three states in a prescribed order, thus functioning as an oscillator or "clock."

Circadian Clocks Are Based on Feedback Loops in Gene Regulation

Life on Earth evolved in the presence of a daily cycle of day and night, and many present-day organisms (ranging from archaea to plants to humans) have come to possess an internal rhythm that dictates different behaviors at different times of day. These behaviors range from the cyclical change in metabolic enzyme activities of a fungus to the elaborate sleep-wake cycles of humans. The internal oscillators that control such diurnal rhythms are called circadian clocks.

By carrying its own circadian clock, an organism can anticipate the regular daily changes in its environment and take appropriate action in advance. Of course, the internal clock cannot be perfectly accurate, and so it must be capable of being reset by external cues such as the light of day. Thus circadian clocks keep running even when the environmental cues (changes in light and dark) are removed, but the period of this free-running rhythm is generally a little less or a little more than 24 hours. External signals indicating the time of day cause small adjustments in the running of the clock, so as to keep the organism in synchrony with its environment. Following more drastic shifts, circadian cycles become gradually reset (entrained) by the new cycle of light and dark, as anyone who has experienced jet lag can attest.

One might expect that the circadian clock in a complex multicellular creature such as a human would itself be a complex multicellular device, with different groups of cells responsible for different parts of the oscillation mechanism. Remarkably, however, it turns out that in almost all organisms, including humans, the timekeepers are individual cells. Thus, our diurnal cycles of sleeping and waking, body temperature, and hormone release are controlled by a clock that operates in each member of a specialized group of cells (the SCN cells) in the hypothalamus (a part of the brain). Even if these cells are removed from the brain and dispersed in a culture dish, they will continue to oscillate individually, showing a cyclic pattern of gene expression with a period of approximately 24 hours. In the intact body, the SCN cells receive neural cues from the retina, entraining them to the daily cycle of light and dark, and they send information about the time of day to other tissues such as the pineal gland, which relays the time signal to the rest of the body by releasing the hormone melatonin in time with the clock.

Although the SCN cells have a central role as timekeepers in mammals, it has been shown that they are not the only cells in the mammalian body that have an internal circadian rhythm or an ability to reset it in response to light. Similarly, in *Drosophila,* many different types of cells, including those of the thorax, abdomen, antenna, leg, wing, and testis all continue a circadian cycle when they have been dissected away from the rest of the fly. The clocks in these isolated tissues, like those in the SCN cells, can be reset by externally imposed light and dark cycles.

The working of circadian clocks, therefore, is a fundamental problem in cell biology. Although we do not yet understand all the details, studies in a wide variety of organisms have revealed many of the basic principles and molecular components. For animals, much of what we know has come from searches in *Drosophila* for mutations that make the fly's circadian clock run fast, or slow, or not all; and this work has led to the discovery that many of the same components are involved in the circadian clock of mammals. The mechanism of the clock in *Drosophila* is outlined in Figure 7–70. At the heart of the oscillator is a transcriptional feedback loop that has a time delay built into it: accumulation of certain key gene products switches off their transcription, but with a delay, so that—crudely speaking—the cell oscillates between a state where the products are present and transcription is switched off, and one where the products are absent and transcription is switched on.

Despite the relative simplicity of the basic principle behind circadian clocks, the details are complex. One reason for this complexity is that clocks must be buffered against changes in temperature, which typically speed up or slow down macromolecular association. They must also run accurately but be capable of being reset. Although it is not yet understood how biological clocks run at a constant speed despite changes in temperature, the mechanism for resetting the *Drosophila* clock is the light-induced destruction of one of the key gene regulatory proteins, as indicated in Figure 7–70.

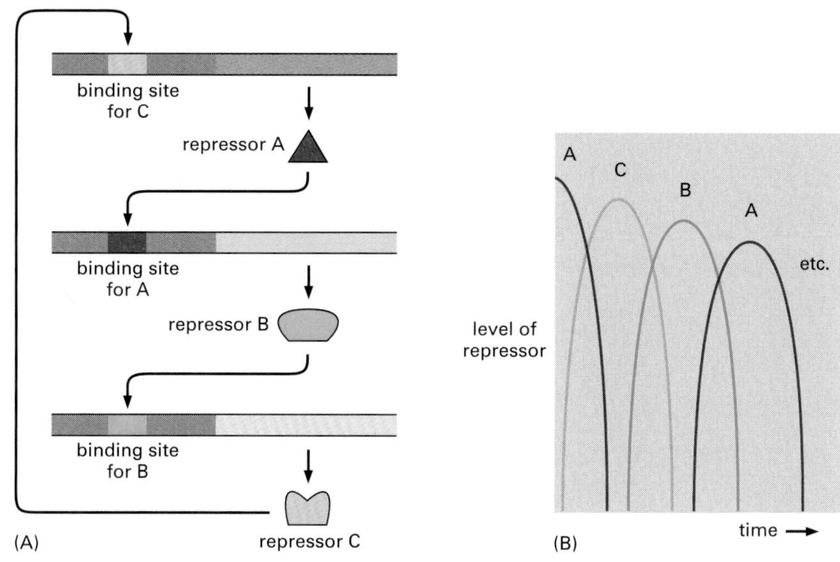

Figure 7–69 A simple gene clock designed in the laboratory. (A) Recombinant DNA techniques were used to place the genes for each of three different bacterial repressor proteins under the control of a different repressor. These repressors (denoted A, B, and C in the figure) are the *lac* repressor (see Figure 7–38), the lambda repressor (see Figure 7–67), and the *tet* repressor, which regulates genes in response to tetracycline. (B) A population of cells will cycle between the three different states shown in (A). For example, if the cells start in a state where only repressor A has accumulated to high levels, the gene for repressor B will be fully repressed. As repressor C is gradually synthesized, it begins to shut off production of repressor A, and repressor B begins to accumulate and eventually shuts off production of repressor C. As this cycling continues a synchronized population of cells oscillates among three states in a specified order. (Adapted from S. Leibler and N. Barkai, *Nature* 403:267–268, 2000.)

The Expression of a Set of Genes Can Be Coordinated by a Single Protein

Cells need to be able to switch genes on and off individually but they also need to coordinate the expression of large groups of different genes. For example, when a quiescent eucaryotic cell receives a signal to divide, many hitherto unexpressed genes are turned on together to set in motion the events that lead eventually to cell division (discussed in Chapter 18). One way bacteria coordinate the expression of a set of genes is by having them clustered together in an *operon* under control of a single promoter (see Figure 7–33). In eucaryotes, however, each gene is transcribed from a separate promoter.

How do eucaryotes coordinate gene expression? This is an especially important question because, as we have seen, most eucaryotic gene regulatory proteins act as part of a "committee" of regulatory proteins, all of which are necessary to express the gene in the right cell, at the right time, in response to the proper signals, and to the proper level. How then can a eucaryotic cell rapidly and decisively switch whole groups of genes on or off? The answer is that even though control of gene expression is combinatorial, the effect of a single gene regulatory protein can still be decisive in switching any particular gene on or off, simply by completing the combination needed to maximally activate or repress that gene. This situation is analogous to dialing in the final number of a combination lock: the lock will spring open if the other numbers have been previously entered. Just as the same number can complete the combination for different locks, the same protein can complete the combination for several different genes. If a number of different genes contain the regulatory site for the same gene regulatory protein, it can be used to regulate the expression of all of them.

An example of this in humans is the control of gene expression by the *glucocorticoid receptor protein*. To bind to regulatory sites in DNA, this gene regulatory protein must first form a complex with a molecule of a glucocorticoid steroid hormone, such as cortisol (see Figures 15–12 and 15–13). This hormone is released in the body during times of starvation and intense physical activity, and among its other activities, it stimulates cells in the liver to increase the production of glucose from amino acids and other small molecules. To make this response, liver cells increase the expression of many different genes, coding for metabolic enzymes and other products. Although these genes all have different and complex control regions, their maximal expression depends on the binding of the hormone-glucocorticoid receptor complex to a regulatory site in the DNA of each gene. When the body has recovered and the hormone is no longer present, the expression of each of these genes drops to its normal level in the liver. In this way a single gene regulatory protein can control the expression of many different genes (Figure 7–71).

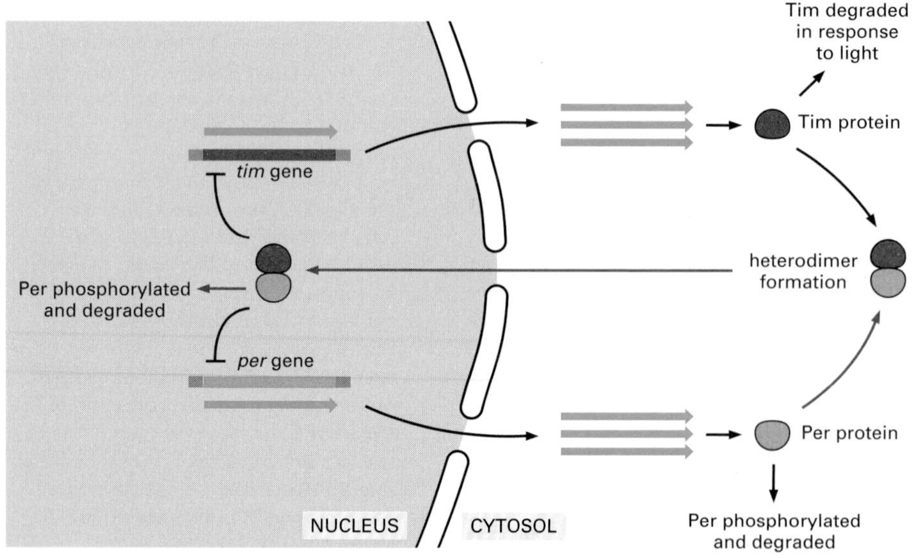

Figure 7–70 Outline of the mechanism of the circadian clock in *Drosophila* cells. The central feature of the clock is the periodic accumulation and decay of two gene regulatory proteins, Tim (short for timeless, based on the phenotype of a gene mutation) and Per (short for period). These proteins are translated in the cytosol, and, when they have accumulated to critical levels, they form a heterodimer. This heterodimer is transported into the nucleus where it regulates a number of genes in concert with the clock. The Tim–Per heterodimer also represses the *tim* and *per* genes, creating a feedback system that causes the levels of Tim and Per to rise and fall periodically. In addition to this transcriptional feedback, the clock depends on the phosphorylation and subsequent degradation of the Per protein, which occurs in both the nucleus and the cytoplasm and is regulated by an additional clock protein, Dbt (short for double-time). This degradation imposes delays in the periodic accumulation of Tim and Per, which are crucial to the functioning of the clock. For example, the accumulation of Per in the cytoplasm is delayed by the phosphorylation and degradation of free Per monomers. Steps at which specific delays are imposed are shown in *red*.

Entrainment (or resetting) of the clock occurs in response to new light-dark cycles. Although most *Drosophila* cells do not have true photoreceptors, light is sensed by intracellular flavoproteins, and it rapidly causes the destruction of the Tim protein, thus resetting the clock.

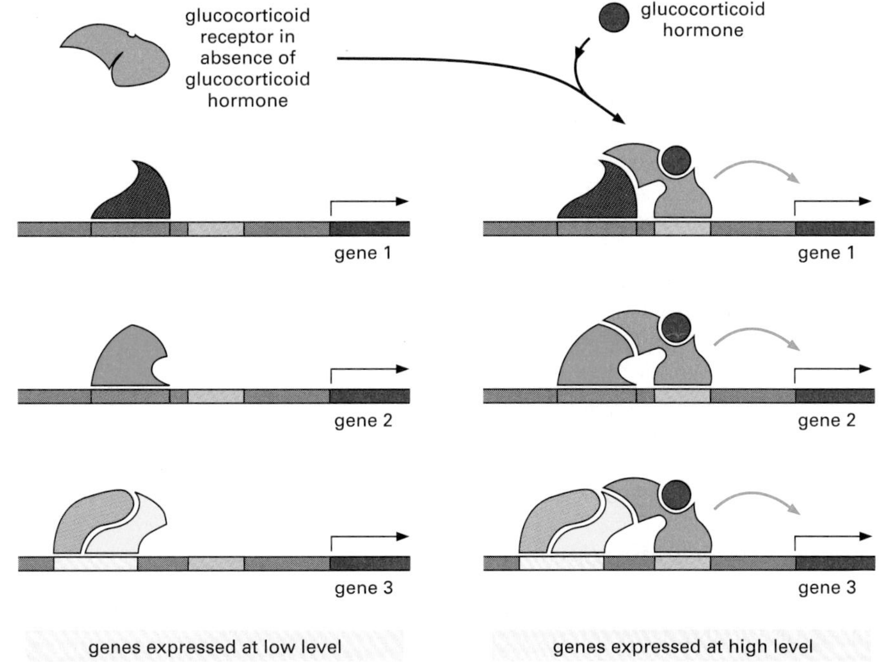

Figure 7–71 A single gene regulatory protein can coordinate the expression of several different genes. The action of the glucocorticoid receptor is illustrated schematically. On the *left* is a series of genes, each of which has various gene activator proteins bound to its regulatory region. However, these bound proteins are not sufficient on their own to fully activate transcription. On the *right* is shown the effect of adding an additional gene regulatory protein—the glucocorticoid receptor in a complex with glucocorticoid hormone—that can bind to the regulatory region of each gene. The glucocorticoid receptor completes the combination of gene regulatory proteins required for maximal initiation of transcription, and the genes are now switched on as a set. In the absence of the hormone, the glucocorticoid receptor is retained in the cytosol and is therefore unavailable to bind to DNA. In addition to activating gene expression, the hormone-bound form of the glucocorticoid receptor represses transcription of certain genes, depending on the gene regulatory proteins already present on their control regions. The effect of the glucocorticoid receptor on any given gene therefore depends upon the type of cell, the gene regulatory proteins contained within it, and the regulatory region of the gene. The structure of the DNA-binding portion of the glucocorticoid receptor is shown in Figure 7–19.

The effects of the glucocorticoid receptor are not confined to cells of the liver. In other cell types, activation of this gene regulatory protein by hormone also causes changes in the expression levels of many genes; the genes affected, however, are often different from those affected in liver cells. As we have seen, each cell type has an individualized set of gene regulatory proteins, and because of combinatorial control, these critically affect the action of the glucocorticoid receptor. Because the receptor is able to assemble with many different sets of cell-type specific gene regulatory proteins, it can produce a distinct spectrum of effects in different cell types.

Expression of a Critical Gene Regulatory Protein Can Trigger Expression of a Whole Battery of Downstream Genes

The ability to switch many genes on or off coordinately is important not only in the day-to-day regulation of cell function. It is also the means by which eucaryotic cells differentiate into specialized cell types during embryonic development. The development of muscle cells provides a striking example.

A mammalian skeletal muscle cell is a highly distinctive giant cell, formed by the fusion of many muscle precursor cells called *myoblasts,* and therefore containing many nuclei. The mature muscle cell is distinguished from other cells by a large number of characteristic proteins, including specific types of actin, myosin, tropomyosin, and troponin (all part of the contractile apparatus), creatine phosphokinase (for the specialized metabolism of muscle cells), and acetylcholine receptors (to make the membrane sensitive to nerve stimulation). In proliferating myoblasts these muscle-specific proteins and their mRNAs are absent or are present in very low concentrations. As myoblasts begin to fuse with one another, the corresponding genes are all switched on coordinately as part of a general transformation of the pattern of gene expression.

This entire program of muscle differentiation can be triggered in cultured skin fibroblasts and certain other cell types by introducing any one of a family of helix–loop–helix proteins—the so-called myogenic proteins (MyoD, Myf5, myogenin, and Mrf4)—normally expressed only in muscle cells (Figure 7–72A). Binding sites for these regulatory proteins are present in the regulatory DNA sequences adjacent to many muscle-specific genes, and the myogenic proteins thereby directly activate transcription of many muscle-specific structural genes. In addition, the myogenic proteins stimulate their own transcription as well as that of various other gene regulatory proteins involved in muscle development, creating an elaborate series of positive feedback loops that amplify and maintain the muscle developmental program, even after the initiating signal has dissipated (Figure 7–72B; see also Chapter 22).

It is probable that the fibroblasts and other cell types that are converted to muscle cells by the addition of myogenic proteins have already accumulated a number of gene regulatory proteins that can cooperate with the myogenic proteins to switch on muscle-specific genes. In this view it is a specific combination of gene regulatory proteins, rather than a single protein, that determines muscle differentiation. This idea is consistent with the finding that some cell types fail to be converted to muscle by myogenin or its relatives; these cells presumably have not accumulated the other gene regulatory proteins required.

The conversion of one cell type (fibroblast) to another (skeletal muscle) by a single gene regulatory protein reemphasizes one of the most important principles discussed in this chapter: dramatic differences between cell types—in size, shape, chemistry, and function—can be produced by differences in gene expression.

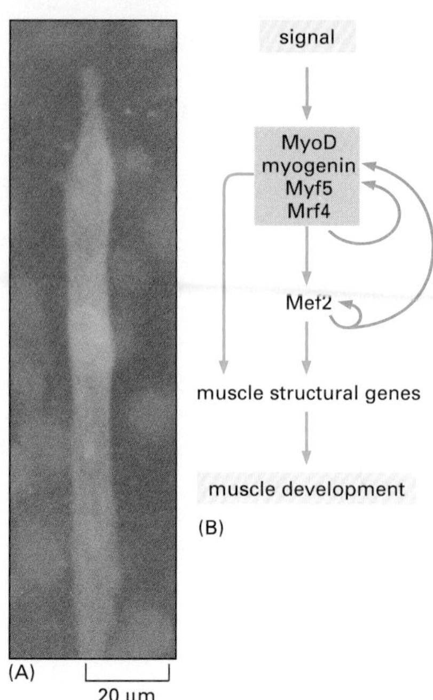

Figure 7–72 Role of the myogenic regulatory proteins in muscle development. (A) The effect of expressing the MyoD protein in fibroblasts. As shown in this immunofluorescence micrograph, fibroblasts from the skin of a chick embryo have been converted to muscle cells by the experimentally induced expression of the *myoD* gene. The fibroblasts that have been induced to express the *myoD* gene have fused to form elongated multinucleate muscle-like cells, which are stained *green* with an antibody that detects a muscle-specific protein. Fibroblasts that do not express the *myoD* gene are barely visible in the background. (B) Simplified scheme for some of the gene regulatory proteins involved in skeletal muscle development. The commitment of mesodermal progenitor cells to the muscle-specific pathway involves the synthesis of the four myogenic gene regulatory proteins, MyoD, Myf5, myogenin and Mrf4. These proteins directly activate transcription of muscle structural genes as well as the *MEF2* gene, which encodes an additional gene regulatory protein. Mef2 acts in combination with the myogenic proteins to further activate transcription of muscle structural genes and to create a positive feedback loop that acts to maintain transcription of the myogenic genes. (A, courtesy of Stephen Tapscott and Harold Weintraub; B, adapted from J.D. Molkentin and E.N. Olson, *Proc. Natl. Acad. Sci. USA* 93:9366–9373, 1996.)

Combinatorial Gene Control Creates Many Different Cell Types in Eucaryotes

We have already discussed how multiple gene regulatory proteins can act in combination to regulate the expression of an individual gene. But, as the example of the myogenic proteins shows, combinatorial gene control means more than this: not only does each gene have many gene regulatory proteins to control it, but each regulatory protein contributes to the control of many genes. Moreover, although some gene regulatory proteins are specific to a single cell type, most are switched on in a variety of cell types, at several sites in the body, and at several times in development. This point is illustrated schematically in Figure 7–73, which shows how combinatorial gene control makes it possible to generate a great deal of biological complexity with relatively few gene regulatory proteins.

With combinatorial control, a given gene regulatory protein does not necessarily have a single, simply definable function as commander of a particular battery of genes or specifier of a particular cell type. Rather, gene regulatory proteins can be likened to the words of a language: they are used with different meanings in a variety of contexts and rarely alone; it is the well-chosen combination that conveys the information that specifies a gene regulatory event. One requirement of combinatorial control is that many gene regulatory proteins must be able to work together to influence the final rate of transcription. To a remarkable extent, this principal is true: even unrelated gene regulatory proteins from widely different eucaryotic species can cooperate when experimentally introduced into the same cell. This situation reflects both the high degree of conservation of the transcription machinery and the nature of transcriptional activation itself. As we have seen, *transcriptional synergy,* in which multiple activator proteins can show more than additive effects on the final state of transcription, results from the ability of the transcription machinery to respond to multiple inputs (see Figure 7–47). It seems that the multifunctional, combinatorial mode of action of gene regulatory proteins has put a tight constraint on their evolution: they must interact with other gene regulatory proteins, the general transcription factors, the RNA polymerase holoenzyme, and the chromatin-modifying enzymes.

An important consequence of combinatorial gene control is that the effect of adding a new gene regulatory protein to a cell will depend on the cell's past history, since this history will determine which gene regulatory proteins are already present. Thus during development a cell can accumulate a series of gene regulatory proteins that need not initially alter gene expression. When the final member of the requisite combination of gene regulatory proteins is added, however, the regulatory message is completed, leading to large changes in gene expression. Such a scheme, as we have seen, helps to explain how the addition of a single regulatory protein to a fibroblast can produce the dramatic transformation of the fibroblast into a muscle cell. It also can account for the important difference, discussed in Chapter 21, between the process of cell determination—where a cell becomes committed to a particular developmental fate—and the process of cell differentiation, where a committed cell expresses its specialized character.

The Formation of an Entire Organ Can Be Triggered by a Single Gene Regulatory Protein

We have seen that even though combinatorial control is the norm for eucaryotic genes, a single gene regulatory protein, if it completes the appropriate combination, can be decisive in switching a whole set of genes on or off, and we have seen how this can convert one cell type into another. A dramatic extension of the principle comes from studies of eye development in *Drosophila*, mice, and humans. Here, a gene regulatory protein (called Ey in flies and Pax-6 in vertebrates) is crucial. When expressed in the proper context, Ey can trigger the formation of not just a single cell type but a whole organ (an eye), composed of different types of cells, all properly organized in three-dimensional space.

The most striking evidence for the role of Ey comes from experiments in fruit flies in which the *ey* gene is artificially expressed early in development in groups of cells that normally will go on to form leg parts. This abnormal gene expression causes eyes to develop in the middle of the legs (Figure 7–74). The *Drosophila* eye is composed of thousands of cells, and the question of how a regulatory protein coordinates the specification of a whole array in a tissue is a central topic in *developmental biology*. As discussed in Chapter 21, it involves cell–cell interactions as well as intracellular gene regulatory proteins. Here, we note that Ey directly controls the expression of many other genes by binding to

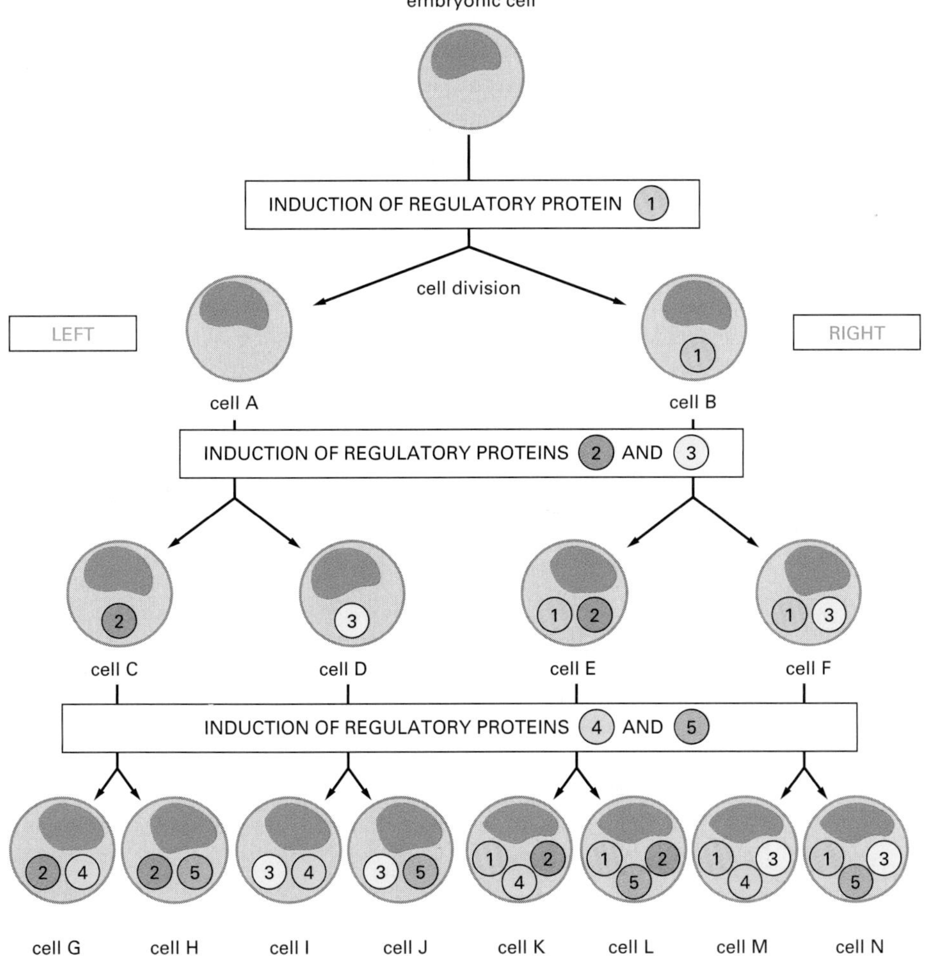

Figure 7–73 The importance of combinatorial gene control for development. Combinations of a few gene regulatory proteins can generate many cell types during development. In this simple, idealized scheme a "decision" to make one of a pair of different gene regulatory proteins (shown as numbered *circles)* is made after each cell division. Sensing its relative position in the embryo, the daughter cell toward the *left side* of the embryo is always induced to synthesize the even-numbered protein of each pair, while the daughter cell toward the *right side* of the embryo is induced to synthesize the odd-numbered protein. The production of each gene regulatory protein is assumed to be self-perpetuating once it has become initiated (see Figure 7–68). In this way, through cell memory, the final combinatorial specification is built up step by step. In this purely hypothetical example, eight final cell types (G–N) have been created using five different gene regulatory proteins.

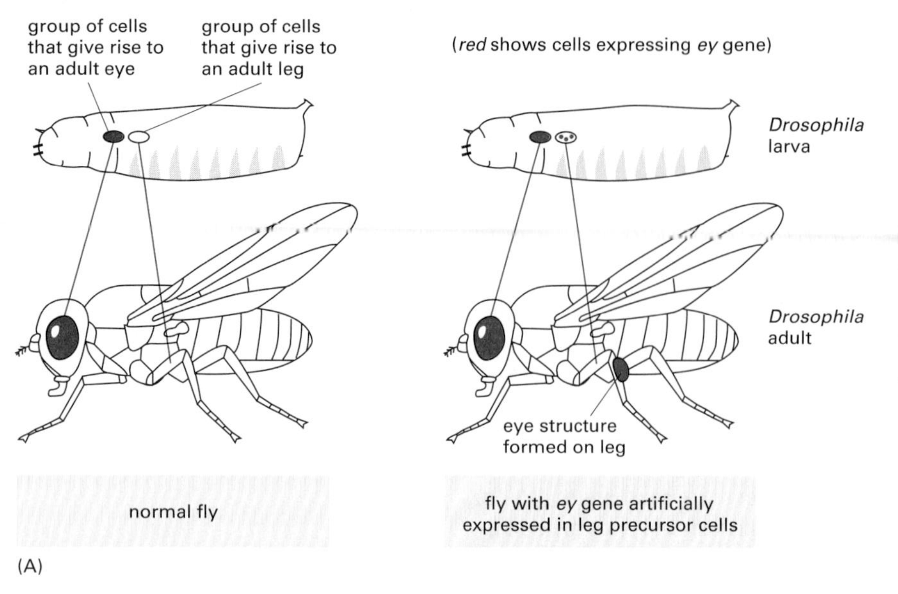

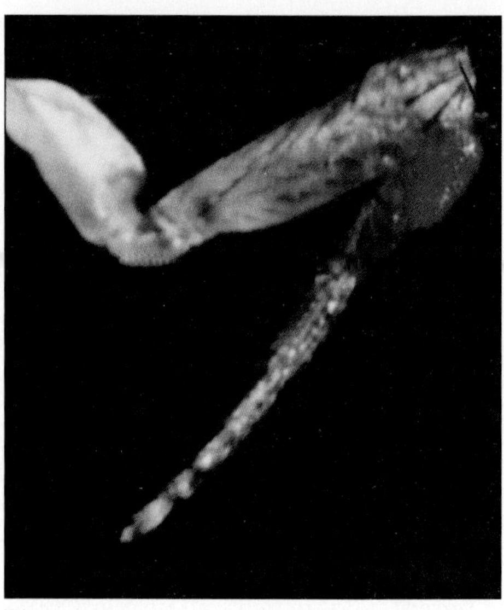

Figure 7–74 Expression of the *Drosophila* ey gene in precursor cells of the leg triggers the development of an eye on the leg. (A) Simplified diagrams showing the result when a fruit fly larva contains either the normally expressed *ey* gene (*left*) or an *ey* gene that is additionally expressed artificially in cells that normally give rise to leg tissue (*right*). (B) Photograph of an abnormal leg that contains a misplaced eye. (B, courtesy of Walter Gehring.)

their regulatory regions. Some of the genes controlled by Ey are themselves gene regulatory proteins that, in turn, control the expression of other genes. Moreover, some of these regulatory genes act back on *ey* itself to create a positive feedback loop that ensures the continued synthesis of the Ey protein (Figure 7–75). In this way, the action of just one regulatory protein can turn on a cascade of gene regulatory proteins and cell–cell interaction mechanims whose actions result in an organized group of many different types of cells. One can begin to imagine how, by repeated applications of this principle, a complex organism is built up piece by piece.

Stable Patterns of Gene Expression Can Be Transmitted to Daughter Cells

Once a cell in an organism has become differentiated into a particular cell type, it generally remains specialized in that way, and if it divides, its daughters inherit the same specialized character. For example, liver cells, pigment cells, and endothelial cells (discussed in Chapter 22) divide many times in the life of an individual. This means that the pattern of gene expression specific to a differentiated cell must be remembered and passed on to its progeny through all subsequent cell divisions.

We have already described several ways of ensuring that daughter cells can "remember" what kind of cells they are supposed to be. One of the simplest is through a positive feedback loop (see Figures 7–68, 7–72B and 7–75) where a key gene regulatory protein activates transcription of its own gene (either directly or indirectly) in addition to that of other cell-type specific genes. The simple flip-flop switch shown in Figure 7–67 is a variation on this theme: by inhibiting expression of its own inhibitor, a gene product indirectly activates and maintains its own expression. Another very different way of maintaining cell type in

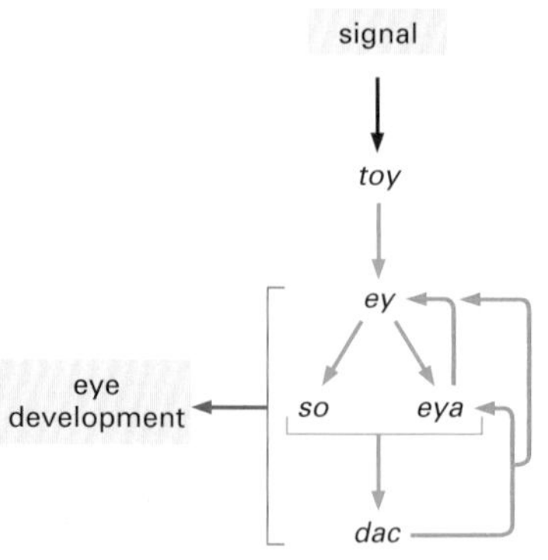

Figure 7–75 Gene regulatory proteins that specify eye development in *Drosophila*. *toy (twin of eyeless)* and *ey (eyeless)* encode similar gene regulatory proteins, Toy and Ey, either of which, when ectopically expressed, can trigger eye development. In normal eye development, expression of *ey* requires the *toy* gene. Once its transcription is activated by Toy, Ey activates transcription of *so (sine oculis)* and *eya (eyes absent)* which act together to express the *dac (dachshund)* gene. As indicated by the *green* arrows, some of the gene regulatory proteins form positive feedback loops which reinforce the initial commitment to eye development. The Ey protein is known to bind directly to numerous target genes for eye development, including those encoding lens crystallins (see Figure 7–119), rhodopsins, and other photoreceptor proteins. (Adapted from T. Czerny et al., *Mol. Cell* 3:297–307, 1999.)

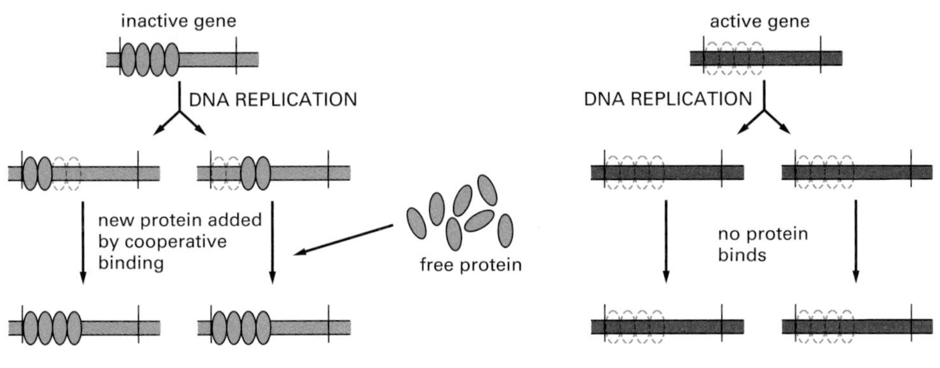

Figure 7–76 A general scheme that permits the direct inheritance of states of gene expression during DNA replication. In this hypothetical model, portions of a cooperatively bound cluster of chromosomal proteins are transferred directly from the parental DNA helix *(top left)* to both daughter helices. The inherited cluster then causes each of the daughter DNA helices to bind additional copies of the same proteins. Because the binding is cooperative, DNA synthesized from an identical parental DNA helix that lacks the bound proteins *(top right)* will remain free of them. If the bound proteins turn off gene transcription, then the inactive gene state will be directly inherited, as illustrated. If the bound proteins activate transcription, then the active gene state will be directly inherited (not shown).

When the cooperative protein binding requires specific DNA sequences, these events will be limited to specific gene control regions; if the binding can be propagated all along the chromosome, however, it could account for the spreading effect associated with the heritable chromatin states discussed in Chapter 4. Although the proteins are depicted as being identical, the same principle can explain how cooperatively assembling combinations of different proteins can be propagated stably.

eucaryotes is through the faithful propagation of chromatin structures from parent to daughter cells, as discussed in Chapter 4. Once a differentiated cell type has been specified by gene regulatory proteins, developmental decisions can be reinforced by packaging unexpressed genes into more compacted forms of chromatin and "marking" that chromatin as silent (see Figure 4–35). The chromatin of actively transcribed genes can also be marked and propagated by the same type of mechanism. The packing of selected regions of the genome into condensed chromatin is a genetic regulatory mechanism that is not available to bacteria, and it is thought to allow eucaryotes to maintain extraordinarily stable patterns of gene expression over many generations. This stability is particularly crucial in multicellular organisms, where abnormal gene expression in a single cell can have profound developmental consequences for the entire organism.

If maintenance of the pattern of gene expression depends on the pattern of chromatin packing, how is this chromatin configuration passed on faithfully from one cell to its daughters? Some possibilities have already been discussed in Chapter 4 (see Figure 4–48). One general mechanism depends on the cooperative binding of proteins to DNA (Figure 7–76). When the cell replicates its DNA, each DNA strand can inherit a share of the protein molecules bound to a given segment of the original double helix, and these inherited molecules can then recruit freshly made molecules to reconstruct a complete copy of the original chromatin complex in each daughter cell. This mechanism of cell memory can be based on cooperative binding of specific gene regulatory proteins, or of general chromatin structural components, or of both classes of molecules acting together. Thus, an initial pattern of binding of specific gene regulatory proteins can initiate a pattern of chromatin condensation that is subsequently maintained.

Yet another strategy of cell memory is based on self-propagating patterns of enzymatic modification of the chromatin proteins (as we saw in Chapter 4) or even of the DNA itself, as we explain later. But first we look more closely at a specific example in which cell memory clearly involves changes of chromatin structure.

Chromosome Wide Alterations in Chromatin Structure Can Be Inherited

We saw in Chapter 4 that chromatin states can be heritable, and that they can be used to establish and preserve patterns of gene expression over great distances along DNA and for many cell generations. A striking example of such long-range effects of chromatin organization occurs in mammals, where an alteration in the chromatin structure of an entire chromosome is used to modulate levels of expression of all genes on that chromosome.

Males and females differ in their *sex chromosomes*. Females have two X chromosomes, whereas males have one X and one Y chromosome. As a result, female cells contain twice as many copies of X-chromosome genes as do male cells. In mammals, the X and Y sex chromosomes differ radically in gene content: the X chromosome is large and contains more than a thousand genes, whereas the Y chromosome is smaller and contains less than 100 genes. Mammals have evolved a *dosage compensation* mechanism to equalize the dosage of X chromosome gene products between males and females. Mutations that interfere with dosage compensation are lethal, demonstrating the necessity of maintaining the correct ratio of X chromosome to *autosome* (non-sex chromosome) gene products.

In mammals dosage compensation is achieved by the transcriptional inactivation of one of the two X chromosomes in female somatic cells, a process known as **X-inactivation.** Early in the development of a female embryo, when it consists of a few thousand cells, one of the two X chromosomes in each cell becomes highly condensed into a type of heterochromatin. The condensed X chromosome can be easily seen under the light microscope in interphase cells; it was originally called a *Barr body* and is located near the nuclear membrane. As a result of X-inactivation, two X chromosomes can coexist within the same nucleus exposed to the same transcriptional regulatory proteins, yet differ entirely in their expression.

The initial choice of which X chromosome to inactivate, the maternally inherited one (X_m) or the paternally inherited one (X_p), is random. Once either X_p or X_m has been inactivated, it remains silent throughout all subsequent cell divisions of that cell and its progeny, indicating that the inactive state is faithfully maintained through many cycles of DNA replication and mitosis. Because X-inactivation is random and takes place after several thousand cells have already formed in the embryo, every female is a mosaic of clonal groups of cells in which either X_p or X_m is silenced (Figure 7–77). These clonal groups are distributed in small clusters in the adult animal because sister cells tend to remain close together during later stages of development. For example, X-chromosome inactivation causes the red and black "tortoise-shell" coat coloration of some female cats. In these cats, one X chromosome carries a gene that produces red hair

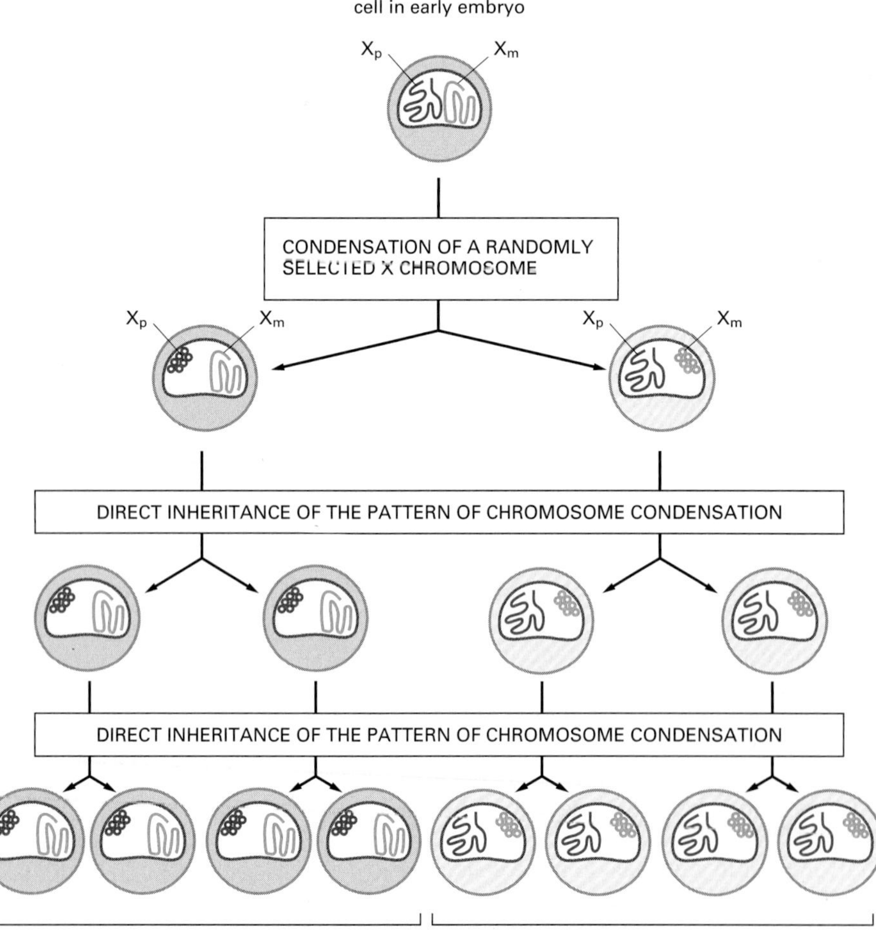

Figure 7–77 X-inactivation. The clonal inheritance of a condensed inactive X chromosome that occurs in female mammals.

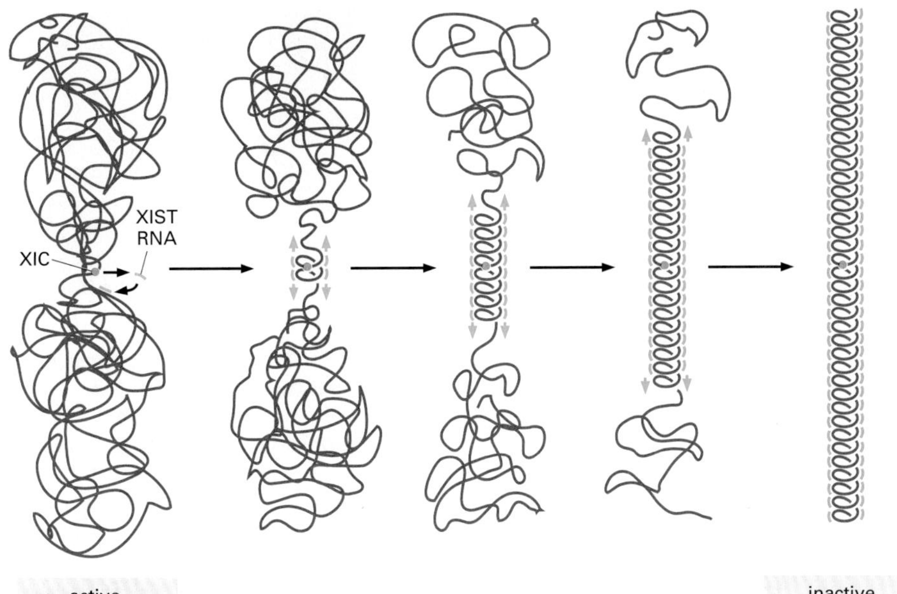

Figure 7–78 Mammalian X-chromosome inactivation. X-chromosome inactivation begins with the synthesis of XIST (X-inactivation specific transcript) RNA from the XIC (X-inactivation center) locus. The association of XIST RNA with the X chromosome is correlated with the condensation of the chromosome. Both XIST association and chromosome condensation gradually move from the XIC locus outward to the chromosome ends. The details of how this occurs remain to be deciphered.

color, and the other X chromosome carries an allele of the same gene that results in black hair color; it is the random X-inactivation that produces patches of cells of two distinctive colors. In contrast to the females, male cats of this genetic stock are either solid red or solid black, depending on which X chromosome they inherit from their mothers. Although X-chromosome inactivation is maintained over thousands of cell divisions, it is not always permanent. In particular, it is reversed during germ cell formation, so that all haploid oocytes contain an active X chromosome and can express X-linked gene products.

How is an entire chromosome transcriptionally inactivated? X-chromosome inactivation is initiated and spreads from a single site in the middle of the X chromosome, the **X-inactivation center** (**XIC**). Portions of the X chromosome that are removed from the XIC and fused to an autosome escape inactivation. In contrast, autosomes that are fused to the XIC of an inactive X chromosome are transcriptionally silenced. The XIC (a DNA sequence of approximately 10^6 nucleotide pairs) can therefore be considered as a large regulatory element that seeds the formation of heterochromatin and facilitates its bi-directional spread along the entire chromosome. Encoded within the XIC is an unusual RNA molecule, *XIST RNA*, which is expressed solely from the inactive X chromosome and whose expression is necessary for X-inactivation. It does not get translated into protein; rather the XIST RNA remains in the nucleus, where it eventually coats the inactive X chromosome. The spread of XIST RNA from the XIC over the entire chromosome correlates with the spread of gene silencing, indicating that XIST RNA participates in the formation and spread of heterochromatin (Figure 7–78). In addition to containing XIST RNA, the X-chromosome heterochromatin is characterized by a specific variant of histone 2A, by hypoacetylation of histones H3 and H4, by methylation of a specific position on histone H3 and by methylation of the underlying DNA, a topic we will discuss below. Presumably all these features make the inactive X chromosome unusually resistant to transcription.

Many features of mammalian X-chromosome inactivation remain to be discovered. How is the initial decision made as to which X chromosome to inactivate? What mechanism prevents the other X chromosome from also being inactivated? How does XIST RNA coordinate the formation of heterochromatin? How is the inactive chromosome maintained through many cell divisions? We are just beginning to understand this mechanism of gene regulation that is crucial for the survival of our own species.

X-chromosome inactivation in females is only one way that sexually reproducing organisms solve the problem of dosage compensation. In *Drosophila*, all the genes on the single X chromosome present in male cells are transcribed at two-fold higher levels than their counterparts in female cells. This male-specific "up-regulation" of transcription results from an alteration in chromatin structure over the entire male X chromosome. As in mammals, this alteration involves the association of a specific RNA molecule with the X chromosome; however, in *Drosophila,* the X-chromosome-associated RNA increases gene activity rather than blocking it. The male X chromosome also contains a specific pattern of histone acetylation which may help to attract the transcription machinery to this chromosome (see Figures 4–35 and 7–40).

Dosage compensation in the nematode worm occurs by a third strategy. Here, the two sexes are male (with one X chromosome) and hermaphrodite (with two X chromosomes), and dosage compensation occurs by a two-fold "down-regulation" of transcription from each of the two X chromosomes in cells of the hermaphrodite. This is brought about through chromosome-wide structural changes in the X chromosomes of hermaphrodites (Figure 7–79). These changes involve the X-specific assembly of proteins, some of which are shared with the *condensins* that helps condense chromosomes during mitosis (see Figures 4–56 and 18–3).

Although the strategies for dosage compensation differ between mammals, flies, and worms, they all involve structural alterations over the entire X chromosome. It is likely that features of chromosome structure that are quite general were adapted and harnessed during evolution to overcome a highly specific problem in gene regulation encountered by sexually reproducing animals.

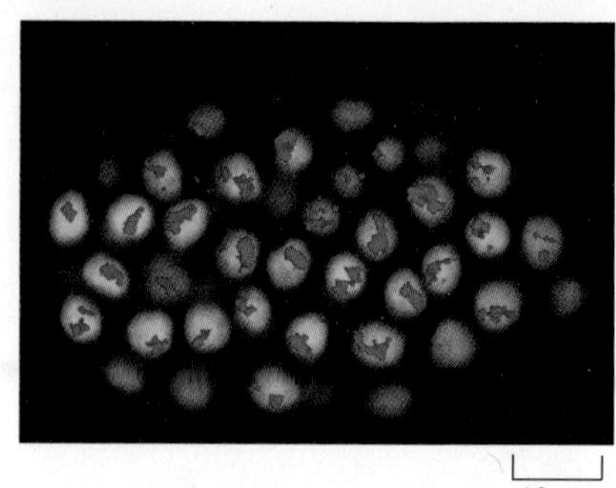

10 μm

Figure 7–79 Localization of dosage compensation proteins to the X chromosomes of *C. elegans* hermaphrodite (XX) nuclei. Many nuclei from a developing embryo are visible in this image. Total DNA is stained *blue* with the DNA-intercalating dye DAPI, and the Sdc-2 protein is stained *red* using anti-Sdc-2 antibodies coupled to a fluorescent dye. This experiment shows that the Sdc-2 protein associates with only a limited set of chromosomes, identified by other experiments to be the two X chromosomes. Sdc-2 is bound along the entire length of the X chromosome and attracts other proteins, including a condensin-like complex, that complete the specialized structure of these chromosomes. (From H.E. Dawes et al., *Science* 284:1800–1804, 1999. © AAAS.)

The Pattern of DNA Methylation Can Be Inherited When Vertebrate Cells Divide

Thus far, we have emphasized the regulation of gene transcription by proteins that associate with specific DNA sequences. However, DNA itself can be covalently modified, and in the following sections we shall see that this, too, provides opportunities for the regulation of gene expression. In vertebrate cells the methylation of cytosine seems to provide an important mechanism for distinguishing genes that are active from those that are not. The methylated form of cysteine, 5-methylcytosine (5-methyl C), has the same relation to cytosine that thymine has to uracil and the modification likewise has no effect on base-pairing (Figure 7–80). The methylation in vertebrate DNA is restricted to cytosine (C) nucleotides in the sequence CG, which is base-paired to exactly the same sequence (in opposite orientation) on the other strand of the DNA helix. Consequently, a simple mechanism permits the existing pattern of **DNA methylation** to be inherited directly by the daughter DNA strands. An enzyme called *maintenance methyltransferase* acts preferentially on those CG sequences that are base-paired with a CG sequence that is already methylated. As a result, the pattern of DNA methylation on the parental DNA strand serves as a template for the methylation of the daughter DNA strand, causing this pattern to be inherited directly following DNA replication (Figure 7–81).

The stable inheritance of DNA methylation patterns can be explained by maintenance DNA methyltransferases. DNA methylation patterns, however, are dynamic during vertebrate development. Shortly after fertilization there is a genome-wide wave of demethylation, when the vast majority of methyl groups are lost from the DNA. This demethylation may occur either by suppression of maintenance DNA methyltransferase activity, resulting in the passive loss of methyl groups during each round of DNA replication, or by a specific demethylating enzyme. Later in development, at the time that the embryo implants in the wall of the uterus, new methylation patterns are established by several *de novo DNA methyltransferases* that modify specific unmethylated CG dinucleotides. Once the new patterns of methylation are established, they can be propagated through rounds of DNA replication by the maintenance methyl transferases. Mutations in either the maintenance or the *de novo* methyltransferases

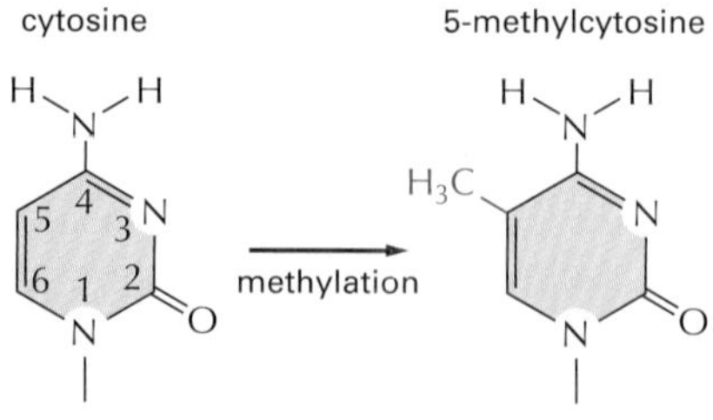

cytosine

methylation

5-methylcytosine

Figure 7–80 Formation of 5-methylcytosine occurs by methylation of a cytosine base in the DNA double helix. In vertebrates this event is confined to selected cytosine (C) nucleotides located in the sequence CG.

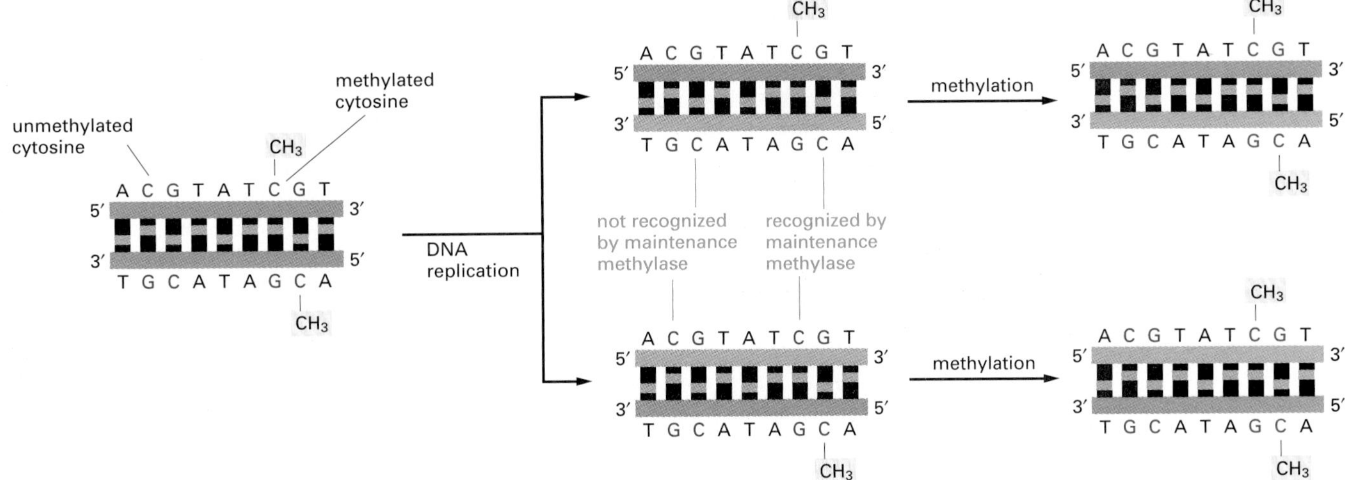

Figure 7–81 How DNA methylation patterns are faithfully inherited. In vertebrate DNAs a large fraction of the cytosine nucleotides in the sequence CG are methylated (see Figure 7–80). Because of the existence of a methyl-directed methylating enzyme (the maintenance methyl transferase), once a pattern of DNA methylation is established, each site of methylation is inherited in the progeny DNA, as shown.

result in early embryonic death in mice, indicating that establishing and maintaining correct methylation patterns is crucial for normal development.

Vertebrates Use DNA Methylation to Lock Genes in a Silent State

In vertebrates DNA methylation is found primarily on transcriptionally silent regions of the genome, such as the inactive X chromosome or genes that are inactivated in certain tissues, suggesting that it plays a role in gene silencing. Vertebrate cells contain a family of proteins that bind methylated DNA. These DNA-binding proteins, in turn, interact with chromatin remodeling complexes and histone deacetylases that condense chromatin so it becomes transcriptionally inactive. In spite of this, DNA methylation is not sufficient to signal the inactivation of a gene, as the following examples demonstrate. Plasmid DNA encoding a muscle-specific actin gene can be prepared *in vitro* in both fully methylated and fully unmethylated forms, using bacterial proteins that methylate or demethylate DNA. When these two versions of the plasmid are introduced into cultured muscle cells, the methylated plasmid is transcribed at the same high rate as the unmethylated copy. Moreover, when a silent, methylated gene is turned on during the normal course of development, methylation is lost only after the gene has been transcribed for some time. Finally, during X chromosome inactivation, condensation and silencing occur before an increase in levels of DNA methylation can be detected. These results all suggest that methylation reinforces transcriptional repression that is initially established by other mechanisms. DNA methylation seems to be used in vertebrates mainly to ensure that once a gene is turned off, it stays off completely (Figure 7–82).

Experiments designed to test whether a DNA sequence that is transcribed at high levels in one vertebrate cell type is transcribed at all in another have demonstrated that rates of gene transcription can differ between two cell types by a factor of more than 10^6. Thus unexpressed vertebrate genes are much less "leaky" in terms of transcription than are unexpressed genes in bacteria, in which the largest known differences in transcription rates between expressed and unexpressed gene states are about 1000-fold. DNA methylation of unexpressed vertebrate genes, with the consequent changes in their chromatin structures, accounts for at least part of this difference. Leaky transcription of the many thousands of genes that are normally turned off completely in each vertebrate cell may be the cause of early embryonic death in mice that lack the maintenance DNA methyltransferase.

Transcriptional silencing in vertebrate genomes is also particularly important to repress the proliferation of transposable elements (see Figure 4–17). While coding sequences make up only a few percent of a typical vertebrate genome, transposable elements can comprise nearly half of these genomes. As we saw in Chapter 5, transposable elements can make copies of themselves and

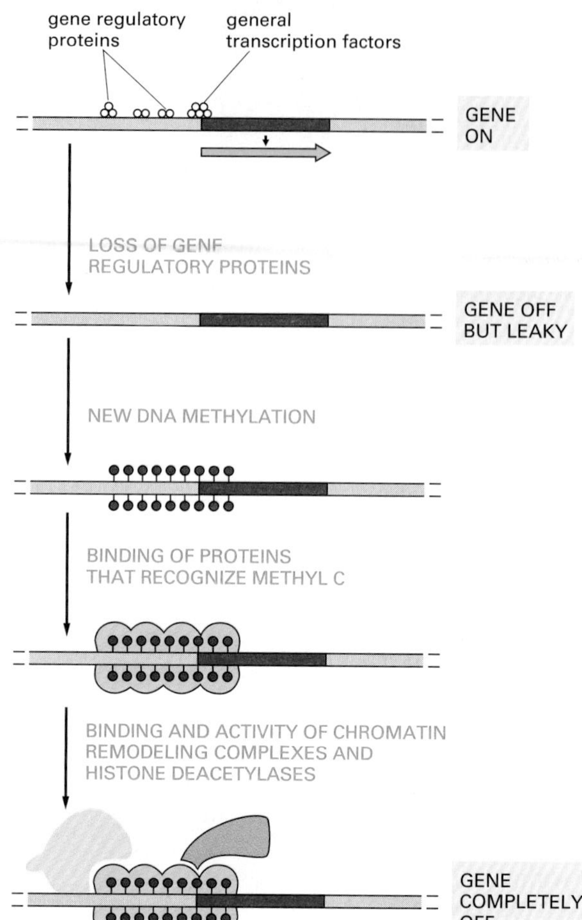

Figure 7–82 How DNA methylation may help turn off genes. The binding of gene regulatory proteins and the general transcription machinery near an active promoter may prevent DNA methylation by excluding *de novo* methylases. If most of these proteins dissociate from the DNA, however, as generally occurs when a cell no longer produces the required activator proteins, the DNA becomes methylated, which enables other proteins to bind, and these shut down the gene completely by further altering chromatin structure (see Figure 7–49).

insert these copies elsewhere in the genome, potentially disrupting genes or important regulatory sequences. By suppressing the transcription of transposable elements, DNA methylation limits their spread and thereby maintains the integrity of the genome. In addition to these varied uses, DNA methylation is also required for at least one special type of cellular memory, as we discuss next.

Genomic Imprinting Requires DNA Methylation

Mammalian cells are diploid, containing one set of genes inherited from the father and one set from the mother. In a few cases the expression of a gene has been found to depend on whether it is inherited from the mother or the father, a phenomenon called **genomic imprinting**. The gene for *insulin-like growth factor-2* (*Igf2*) is one example of an imprinted gene. *Igf2* is required for prenatal growth, and mice that do not express *Igf2* are born half the size of normal mice. Only the paternal copy of *Igf2* is transcribed. As a result, mice with a mutated paternally derived *Igf2* gene are stunted, while mice with a defective maternally derived *Igf2* gene are normal.

During the formation of germ cells, genes subject to imprinting are marked by methylation according to whether they are present in a sperm or an egg. In this way, the parental origin of the gene can be subsequently detected in the embryo; DNA methylation is thus used as a mark to distinguish two copies of a gene that may be otherwise identical (Figure 7–83). Because imprinted genes are not affected by the wave of demethylation that takes place shortly after fertilization (see p. 430), this mark enables somatic cells to "remember" the parental origin of each of the two copies of the gene and to regulate their expression accordingly. In most cases, the methyl imprint silences nearby gene expression using the mechanisms shown in Figure 7–82. In some cases, however, the methyl imprint can activate expression of a gene. In the case of *Igf2*, for example, methylation of an insulator element (see Figure 7–61) on the paternally derived

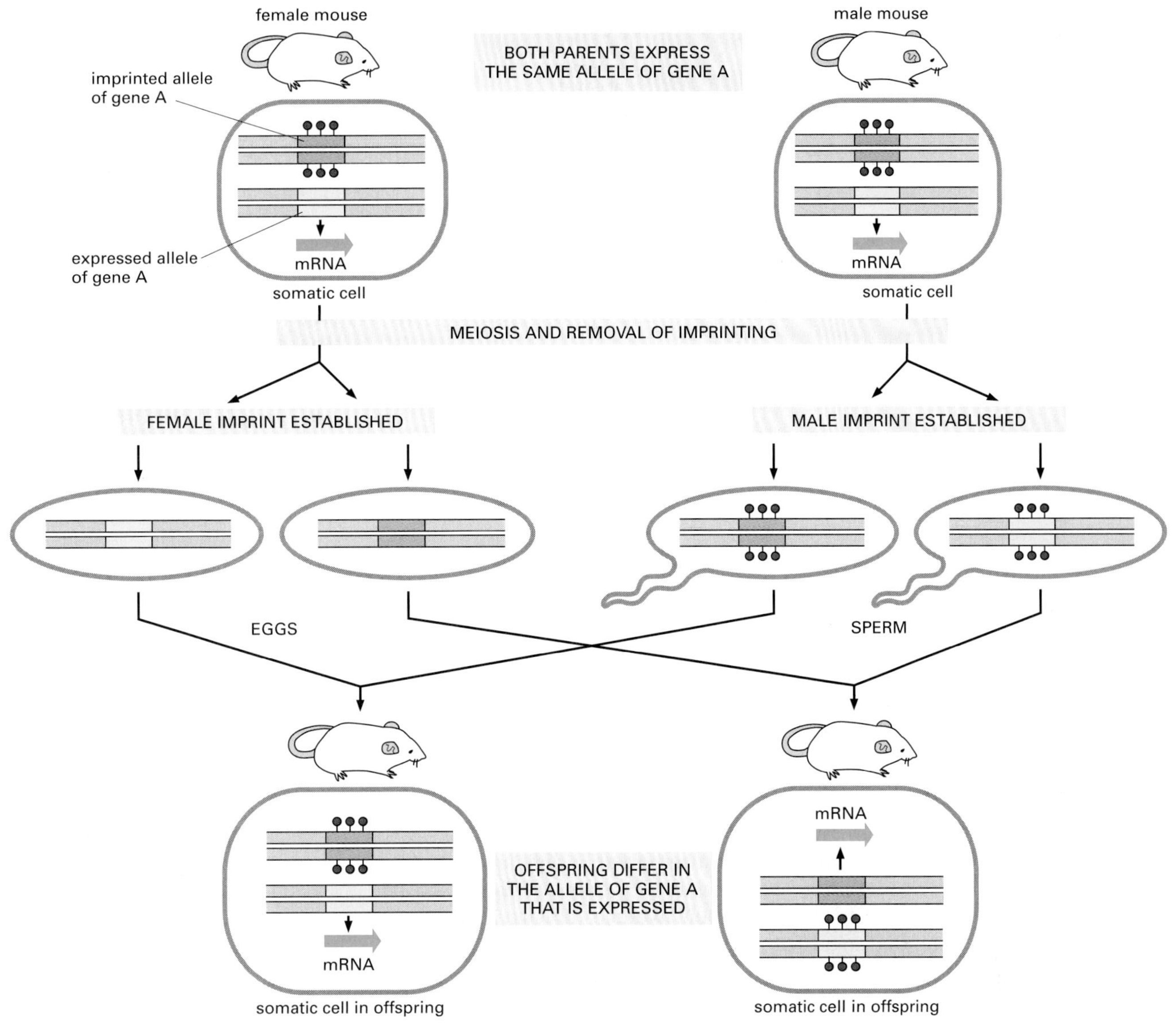

Figure 7–83 Imprinting in the mouse. The *top* portion of the figure shows a pair of homologous chromosomes in the somatic cells of two adult mice, one male and one female. In this example, both mice have inherited the top homolog from their father and the bottom homolog from their mother, and the paternal copy of a gene subject to imprinting (indicated in *orange*) is methylated, which prevents its expression. The maternally-derived copy of the same gene *(yellow)* is expressed. The remainder of the figure shows the outcome of a cross between these two mice. During meiosis and germ cell formation, the imprints are first erased and then reimposed *(middle* portion of figure). In eggs produced from the female, neither allele of the A gene is methylated. In sperm from the male, both alleles of gene A are methylated. Shown at the *bottom* of the figure are two of the possible imprinting patterns inherited by the progeny mice; the mouse on the *left* has the same imprinting pattern as each of the parents, whereas the mouse on the *right* has the opposite pattern. If the two alleles of A gene are distinct, these different imprinting patterns can cause phenotypic differences in the progeny mice, even though they carry exactly the same DNA sequences of the two A gene alleles. Imprinting provides an important exception to classical genetic behavior, and more than 100 mouse genes are thought to be affected in this way. However, the great majority of mouse genes are not imprinted, and therefore the rules of Mendelian inheritance apply to most of the mouse genome.

chromosome blocks its function and allows a distant enhancer to activate transcription of the *Igf2* gene. On the maternally derived chromosome, the insulator is not methylated and the *Igf2* gene is therefore not transcribed (Figure 7–84).

Imprinting is an example of an *epigenetic change,* that is, a heritable change in phenotype that does not result from a change in DNA nucleotide sequence. Why imprinting should exist at all is a mystery. In vertebrates, it is restricted to placental mammals, and all the imprinted genes are involved in fetal development. One idea is that imprinting reflects a middle ground in the evolutionary struggle between males to produce larger offspring and females to limit offspring

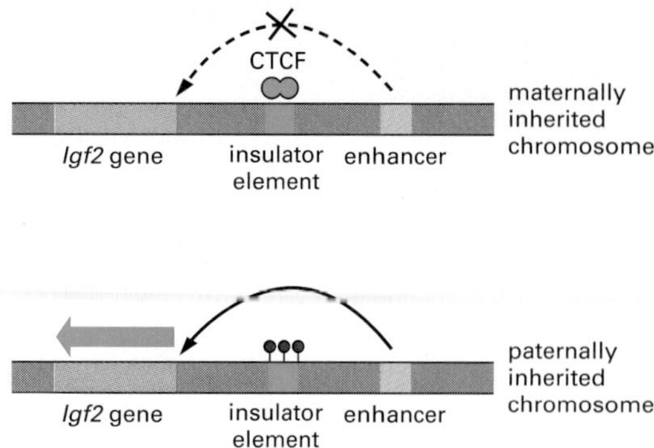

Figure 7–84 Mechanism of imprinting of the mouse *Igf2* gene. On chromosomes inherited from the female, a protein called CTCF binds to an insulator, (see Figure 7–61) blocking communication between the enhancer *(green)* and the *Igf2* gene *(orange)*. *Igf2* is therefore not expressed from the maternally inherited chromosome. Because of imprinting, the insulator on the male-derived chromosome is methylated; this inactivates the insulator, by blocking the binding of the CTCF protein, and allows the enhancer to activate transcription of the *Igf2* gene. In other examples of imprinting, methylation blocks gene expression by interfering with the binding of proteins required for the gene's transcription. The methylation patterns (imprints) on the chromosome, inherited by the zygote after fertilization, are maintained in subsequent generations by maintenance methyl transferases (see Figure 7–81).

size. Whatever its purpose might be, imprinting provides startling evidence that features of DNA other than its sequence of nucleotides can be inherited.

CG-rich Islands Are Associated with About 20,000 Genes in Mammals

Because of the way DNA repair enzymes work, methylated C nucleotides in the genome tend to be eliminated in the course of evolution. Accidental deamination of an unmethylated C gives rise to U, which is not normally present in DNA and thus is recognized easily by the DNA repair enzyme uracil DNA glycosylase, excised, and then replaced with a C (as discussed in Chapter 5). But accidental deamination of a 5-methyl C cannot be repaired in this way, for the deamination product is a T and so indistinguishable from the other, nonmutant T nucleotides in the DNA. Although a special repair system exists to remove these mutant T nucleotides, many of the deaminations escape detection, so that those C nucleotides in the genome that are methylated tend to mutate to T over evolutionary time.

During the course of evolution, more than three out of every four CGs have been lost in this way, leaving vertebrates with a remarkable deficiency of this dinucleotide. The CG sequences that remain are very unevenly distributed in the genome; they are present at 10 to 20 times their average density in selected regions, called **CG islands**, that are 1000 to 2000 nucleotide pairs long. These islands, with some important exceptions, seem to remain unmethylated in all cell types. They often surround the promoters of the so-called *housekeeping genes*—those genes that code for the many proteins that are essential for cell viability and are therefore expressed in most cells (Figure 7–85). In addition, some *tissue-specific genes*, which code for proteins needed only in selected types of cells, are also associated with CG islands.

The distribution of CG islands (also called CpG islands to distinguish the CG dinucleotides from CG nucleotide pairs) can be explained if we assume that CG methylation was adopted in vertebrates primarily as a way of maintaining DNA in a transcriptionally inactive state (Figures 7–82 and 7–86). In vertebrates, new methyl-C to T mutations can be transmitted to the next generation only if they

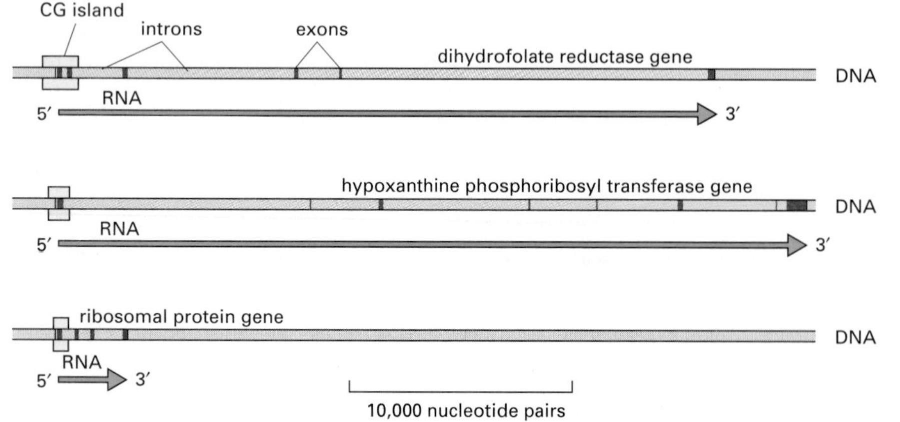

Figure 7–85 The CG islands surrounding the promoter in three mammalian housekeeping genes. The *yellow* boxes show the extent of each island. As for most genes in mammals (see Figure 6–25), the exons *(dark red)* are very short relative to the introns *(light red)*. (Adapted from A.P. Bird, *Trends Genet.* 3:342–347, 1987.)

occur in the germ line, the cell lineage that gives rise to sperm or eggs. Most of the DNA in vertebrate germ cells is inactive and highly methylated. Over long periods of evolutionary time, the methylated CG sequences in these inactive regions have presumably been lost through spontaneous deamination events that were not properly repaired. However promoters of genes that remain active in the germ cell lineages (including most housekeeping genes) are kept unmethylated, and therefore spontaneous deaminations of Cs that occur within them can be accurately repaired. Such regions are preserved in modern day vertebrate cells as CG islands. In addition, any mutation of a CG sequence in the genome that destroyed the function or regulation of a gene in the adult would be selected against, and some CG islands are simply the result of a higher than normal density of critical CG sequences.

The mammalian genome contains an estimated 20,000 CG islands. Most of the islands mark the 5′ ends of transcription units and thus, presumably, of genes. The presence of CG islands often provides a convenient way of identifying genes in the DNA sequences of vertebrate genomes.

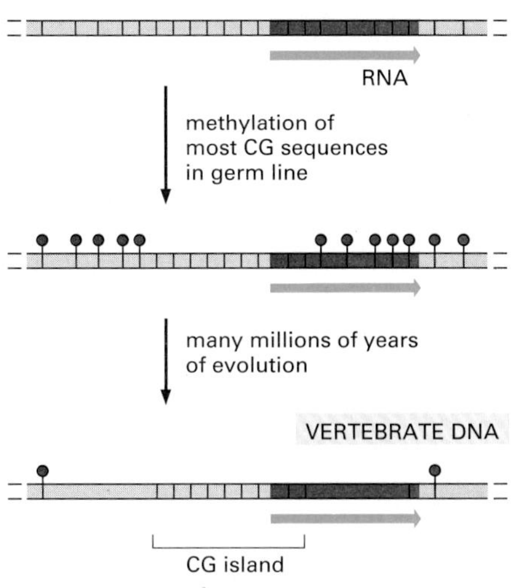

Figure 7–86 A mechanism to explain both the marked overall deficiency of CG sequences and their clustering into CG islands in vertebrate genomes. A *black line* marks the location of a CG dinucleotide in the DNA sequence, while a *red* "lollipop" indicates the presence of a methyl group on the CG dinucleotide. CG sequences that lie in regulatory sequences of genes that are transcribed in germ cells are unmethylated and therefore tend to be retained in evolution. Methylated CG sequences, on the other hand, tend to be lost through deamination of 5-methyl C to T, unless the CG sequence is critical for survival.

Summary

The many types of cells in animals and plants are created largely through mechanisms that cause different genes to be transcribed in different cells. Since many specialized animal cells can maintain their unique character through many cell division cycles and even when grown in culture, the gene regulatory mechanisms involved in creating them must be stable once established and heritable when the cell divides. These features endow the cell with a memory of its developmental history. Bacteria and yeasts provide unusually accessible model systems in which to study gene regulatory mechanisms. One such mechanism involves a competitive interaction between two gene regulatory proteins, each of which inhibits the synthesis of the other; this can create a flip-flop switch that switches a cell between two alternative patterns of gene expression. Direct or indirect positive feedback loops, which enable gene regulatory proteins to perpetuate their own synthesis, provide a general mechanism for cell memory. Negative feedback loops with programmed delays form the basis for cellular clocks.

In eucaryotes the transcription of a gene is generally controlled by combinations of gene regulatory proteins. It is thought that each type of cell in a higher eucaryotic organism contains a specific combination of gene regulatory proteins that ensures the expression of only those genes appropriate to that type of cell. A given gene regulatory protein may be active in a variety of circumstances and typically is involved in the regulation of many genes.

In addition to diffusible gene regulatory proteins, inherited states of chromatin condensation are also used by eucaryotic cells to regulate gene expression. An especially dramatic case is the inactivation of an entire X chromosome in female mammals. In vertebrates DNA methylation also functions in gene regulation, being used mainly as a device to reinforce decisions about gene expression that are made initially by other mechanisms. DNA methylation also underlies the phenomenon of genomic imprinting in mammals, in which the expression of a gene depends on whether it was inherited from the mother or the father.

POSTTRANSCRIPTIONAL CONTROLS

In principle, every step required for the process of gene expression could be controlled. Indeed, one can find examples of each type of regulation, although any one gene is likely to use only a few of them. Controls on the initiation of gene transcription are the predominant form of regulation for most genes. But other controls can act later in the pathway from DNA to protein to modulate the amount of gene product that is made. Although these **posttranscriptional controls**, which operate after RNA polymerase has bound to the gene's promoter and begun RNA synthesis, are less common than *transcriptional control*, for many genes they are crucial.

In the following sections, we consider the varieties of posttranscriptional regulation in temporal order, according to the sequence of events that might be experienced by an RNA molecule after its transcription has begun (Figure 7–87).

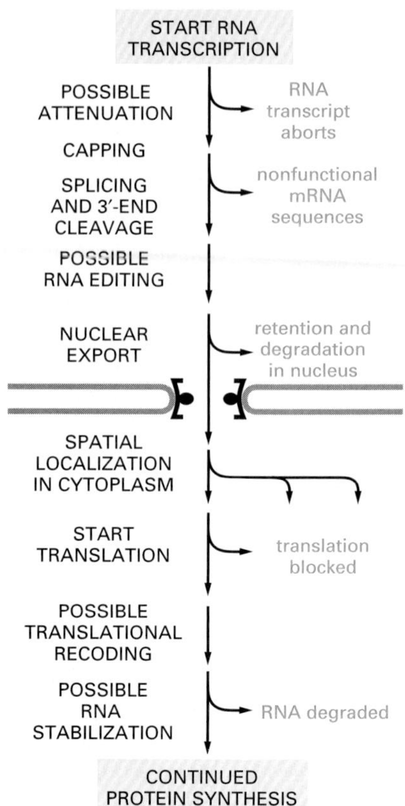

Figure 7–87 Possible post-transcriptional controls on gene expression. Only a few of these controls are likely to be important for any one gene.

Transcription Attenuation Causes the Premature Termination of Some RNA Molecules

In bacteria the expression of certain genes is inhibited by premature termination of transcription, a phenomenon called **transcription attenuation**. In some of these cases the nascent RNA chain adopts a structure that causes it to interact with the RNA polymerase in such a way as to abort its transcription. When the gene product is required, regulatory proteins bind to the nascent RNA chain and interfere with attenuation, allowing the transcription of a complete RNA molecule.

In eucaryotes transcription attenuation can occur by a number of distinct mechanisms. A well-studied example is found in HIV (the human AIDS virus). Once it has been integrated into the host genome, the viral DNA is transcribed by the cellular RNA polymerase II (see Figure 5–73). However, the host polymerase usually terminates transcription (for reasons that are not well-understood) after synthesizing transcripts of several hundred nucleotides and therefore does not efficiently transcribe the entire viral genome. When conditions for viral growth are optimal, this premature termination is prevented by a virus-encoded protein called Tat, which binds to a specific stem-loop structure in the nascent RNA that contains a "bulged base." Once bound to this specific RNA structure (called Tar), Tat assembles several cellular proteins which allow the RNA polymerase to continue transcribing. The normal role of at least some of these cellular proteins is to prevent pausing and premature termination by RNA polymerase when it transcribes normal cellular genes. Eucaryotic genes often contain long introns; to transcribe a gene efficiently, RNA polymerase II cannot afford to linger at nucleotide sequences that happen to promote pausing. Thus a normal cellular mechanism has apparently been adapted by HIV to permit transcription of its genome to be controlled by a single viral protein.

Alternative RNA Splicing Can Produce Different Forms of a Protein from the Same Gene

As discussed in Chapter 6, the transcripts of many eucaryotic genes are shortened by RNA splicing, in which the intron sequences are removed from the mRNA precursor. We saw that a cell can splice the "primary transcript" in different ways and thereby make different polypeptide chains from the same gene—a process called **alternative RNA splicing** (see Figures 6–27 and 7–88). A substantial proportion of higher eucaryotic genes (at least a third of human genes, it is estimated) produce multiple proteins in this way.

When different splicing possibilities exist at several positions in the transcript, a single gene can produce dozens of different proteins. In one extreme case, a *Drosophila* gene may produce as many as 38,000 different proteins from a single gene through alternative splicing (Figure 7–89), although only a small fraction of these forms have thus far been experimentally observed. Considering that the *Drosophila* genome has approximately 14,000 identified genes, it is clear that the protein complexity of an organism can greatly exceed the number of its genes. This example also illustrates the perils in equating gene number with organism complexity. For example, alternative splicing is relatively rare in single-celled budding yeasts but very common in flies. Budding yeast has ~6200

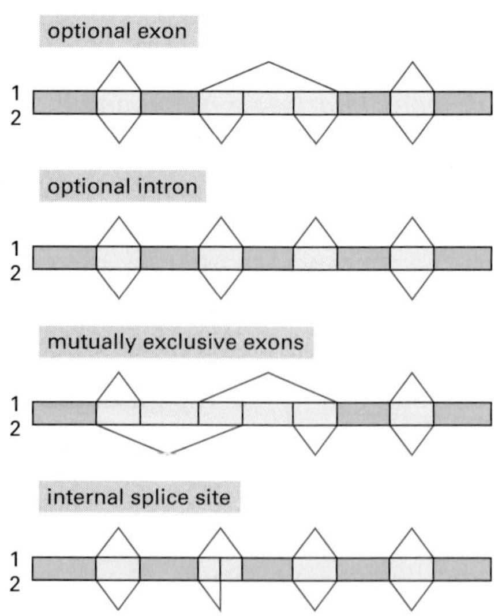

Figure 7–88 Four patterns of alternative RNA splicing. In each case a single type of RNA transcript is spliced in two alternative ways to produce two distinct mRNAs (1 and 2). The *dark blue boxes* mark exon sequences that are retained in both mRNAs. The *light blue boxes* mark possible exon sequences that are included in only one of the mRNAs. The boxes are joined by *red lines* to indicate where intron sequences *(yellow)* are removed. (Adapted with permission from A. Andreadis, M.E. Gallego, and B. Nadal-Ginard, *Annu. Rev. Cell Biol.* 3:207–242, 1987.)

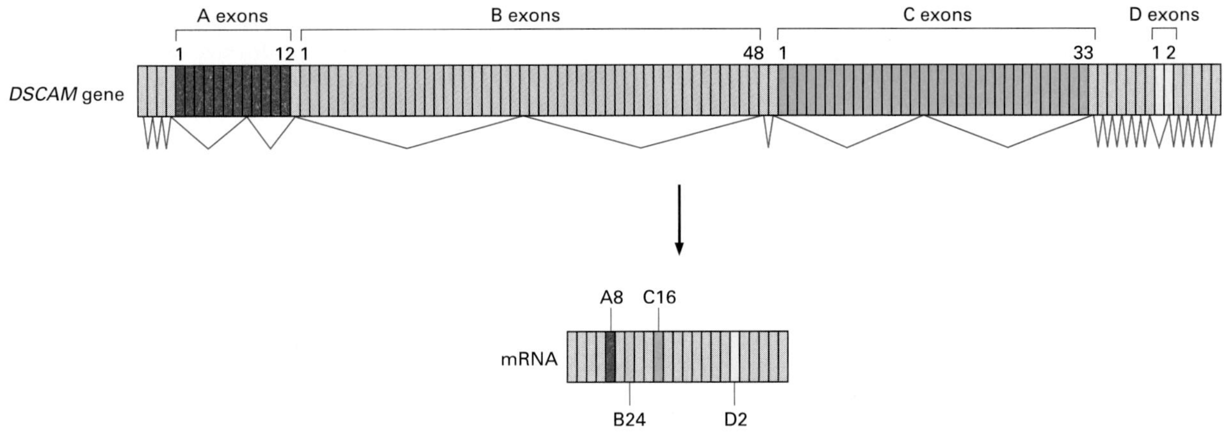

Figure 7–89 Alternative splicing of RNA transcripts of the *Drosophila DSCAM* gene. DSCAM proteins are axon guidance receptors that help to direct growth cones to their appropriate targets in the developing nervous system. The final mRNA contains 24 exons, four of which (denoted A, B, C, and D) are present in the *DSCAM* gene as arrays of alternative exons. Each RNA contains 1 of 12 alternatives for exon A *(red)*, 1 of 48 alternatives for exon B *(green)*, 1 of 33 alternatives for exon C *(blue)*, and 1 of 2 alternatives for exon D *(yellow)*. If all possible splicing combinations are used, 38,016 different proteins could in principle be produced from the *DSCAM* gene. Only one of the many possible splicing patterns (indicated by the *red line* and by the mature mRNA below it) is shown. Each variant DSCAM protein would fold into roughly the same structure [predominantly a series of extracellular immunoglobulin-like domains linked to a membrane-spanning region (see Figure 24–71)], but the amino acid sequence of the domains would vary according to the splicing pattern. It is possible that this receptor diversity contributes to the formation of complex neural circuits, but the precise properties and functions of the many DSCAM variants are not yet understood. (Adapted from D.L. Black, *Cell* 103:367–370, 2000.)

genes, only 327 of which are subject to splicing, and nearly all of these have only a single intron. To say that flies have only 2–3 times as many genes as yeasts is to greatly underestimate the difference in complexity of these two genomes.

In some cases alternative RNA splicing occurs because there is an *intron sequence ambiguity:* the standard spliceosome mechanism for removing intron sequences (discussed in Chapter 6) is unable to distinguish cleanly between two or more alternative pairings of 5′ and 3′ splice sites, so that different choices are made by chance on different transcripts. Where such constitutive alternative splicing occurs, several versions of the protein encoded by the gene are made in all cells in which the gene is expressed.

In many cases, however, alternative RNA splicing is regulated rather than constitutive. In the simplest examples, regulated splicing is used to switch from the production of a nonfunctional protein to the production of a functional one. The transposase that catalyzes the transposition of the *Drosophila* P element, for example, is produced in a functional form in germ cells and a nonfunctional form in somatic cells of the fly, allowing the P element to spread throughout the genome of the fly without causing damage in somatic cells (see Figure 5–70). The difference in transposon activity has been traced to the presence of an intron sequence in the transposase RNA that is removed only in germ cells.

In addition to switching from the production of a functional protein to the production of a nonfunctional one, the regulation of RNA splicing can generate different versions of a protein in different cell types, according to the needs of the cell. Tropomyosin, for example, is produced in specialized forms in different types of cells (see Figure 6–27). Cell-type-specific forms of many other proteins are produced in the same way.

RNA splicing can be regulated either negatively, by a regulatory molecule that prevents the splicing machinery from gaining access to a particular splice site on the RNA, or positively, by a regulatory molecule that helps direct the splicing machinery to an otherwise overlooked splice site (Figure 7–90). In the case of the *Drosophila* transposase, the key splicing event is blocked in somatic cells by negative regulation.

Because of the plasticity of RNA splicing (see pp. 324–325), the blocking of a "strong" splicing site will often expose a "weak" site and result in a different pattern of splicing. Likewise, activating a suboptimal splice site can result in alternative splicing by suppressing a competing splice site. Thus the splicing of a pre-mRNA molecule can be thought of as a delicate balance between competing splice sites—a balance that can easily be tipped by regulatory proteins.

The Definition of a Gene Has Had to Be Modified Since the Discovery of Alternative RNA Splicing

The discovery that eucaryotic genes usually contain introns and that their coding sequences can be assembled in more than one way raised new questions about the definition of a gene. A gene was first clearly defined in molecular

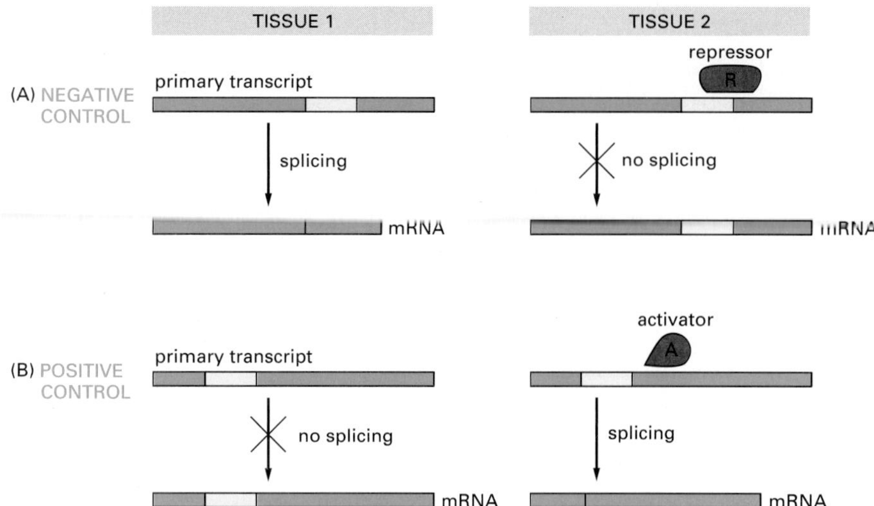

Figure 7–90 Negative and positive control of alternative RNA splicing. (A) Negative control, in which a repressor protein binds to the primary RNA transcript in tissue 2, thereby preventing the splicing machinery from removing an intron sequence. (B) Positive control, in which the splicing machinery is unable to efficiently remove a particular intron sequence without assistance from an activator protein.

terms in the early 1940s from work on the biochemical genetics of the fungus *Neurospora*. Until then, a gene had been defined operationally as a region of the genome that segregates as a single unit during meiosis and gives rise to a definable phenotypic trait, such as a red or a white eye in *Drosophila* or a round or wrinkled seed in peas. The work on *Neurospora* showed that most genes correspond to a region of the genome that directs the synthesis of a single enzyme. This led to the hypothesis that one gene encodes one polypeptide chain. The hypothesis proved fruitful for subsequent research; as more was learned about the mechanism of gene expression in the 1960s, a gene became identified as that stretch of DNA that was transcribed into the RNA coding for a single polypeptide chain (or a single structural RNA such as a tRNA or an rRNA molecule). The discovery of split genes and introns in the late 1970s could be readily accommodated by the original definition of a gene, provided that a single polypeptide chain was specified by the RNA transcribed from any one DNA sequence. But it is now clear that many DNA sequences in higher eucaryotic cells can produce a set of distinct (but related) proteins by means of alternative RNA splicing. How then is a gene to be defined?

In those relatively rare cases in which two very different eucaryotic proteins are produced from a single transcription unit, the two proteins are considered to be produced by distinct genes that overlap on the chromosome. It seems unnecessarily complex, however, to consider most of the protein variants produced by alternative RNA splicing as being derived from overlapping genes. A more sensible alternative is to modify the original definition to count as a gene any DNA sequence that is transcribed as a single unit and encodes one set of closely related polypeptide chains (protein isoforms). This definition of a gene also accommodates those DNA sequences that encode protein variants produced by posttranscriptional processes other than RNA splicing, such as translational frameshifting (see Figure 6–78), regulated poly-A addition, and RNA editing (to be discussed below).

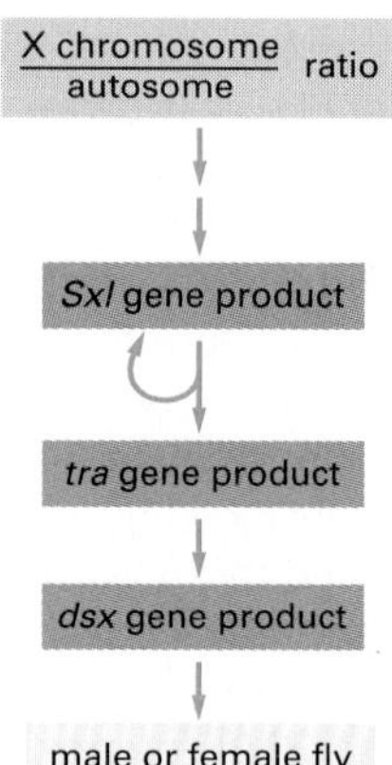

Figure 7–91 Sex determination in *Drosophila*. The gene products shown act in a sequential cascade to determine the sex of the fly according to the X chromosome/autosome ratio. The genes are called *sex-lethal (Sxl), transformer (tra),* and *doublesex (dsx)* because of the phenotypes that result when the gene is inactivated by mutation. The function of these gene products is to transmit the information about the X chromosome/autosome ratio to the many other genes that create the sex-related phenotypes. These other genes function as two alternative sets: those that specify female features and those that specify male features (see Figure 7–92).

Sex Determination in *Drosophila* Depends on a Regulated Series of RNA Splicing Events

We now turn to one of the best understood examples of regulated RNA splicing. In *Drosophila* the primary signal for determining whether the fly develops as a male or female is the X chromosome/autosome ratio. Individuals with an X chromosome/autosome ratio of 1 (normally two X chromosomes and two sets of autosomes) develop as females, whereas those with a ratio of 0.5 (normally one X chromosome and two sets of autosomes) develop as males. This ratio is assessed early in development and is remembered thereafter by each cell. Three crucial gene products transmit information about this ratio to the many other genes that specify male and female characteristics (Figure 7–91). As explained in

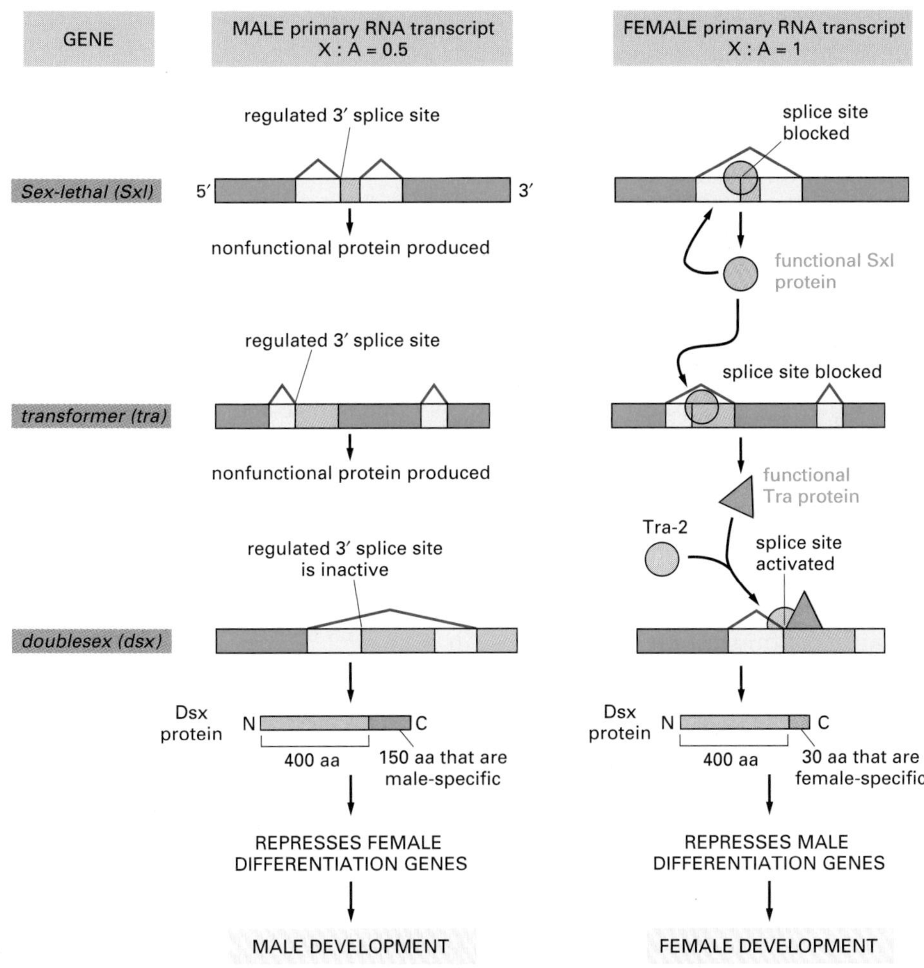

Figure 7–92 The cascade of changes in gene expression that determines the sex of a fly through alternative RNA splicing. An X chromosome/autosome ratio of 0.5 results in male development. Male is the "default" pathway in which the *Sxl* and *tra* genes are both transcribed, but the RNAs are spliced constitutively to produce only nonfunctional RNA molecules, and the *dsx* transcript is spliced to produce a protein that turns off the genes that specify female characteristics. An X chromosome/autosome ratio of 1 triggers the female differentiation pathway in the embryo by transiently activating a promoter within the *Sxl* gene that causes synthesis of a special class of *Sxl* transcripts that are constitutively spliced to give functional Sx1 protein. Sxl is a splicing regulatory protein with two sites of action: (1) it binds to the constitutively produced *Sxl* RNA transcript, causing a female-specific splice that continues the production of a functional Sxl protein, and (2) it binds to the constitutively produced *tra* RNA and causes an alternative splice of this transcript, which now produces an active Tra regulatory protein. The Tra protein acts with the constitutively produced Tra-2 protein to produce the female-specific spliced form of the *dsx* transcript; this encodes the female form of the Dsx protein, which turns off the genes that specify male features. The components in this pathway were all initially identified through the study of *Drosophila* mutants that are altered in their sexual development. The *dsx* gene, for example, derives its name *(doublesex)* from the observation that a fly lacking this gene product expresses both male- and female-specific features. Note that, although both the Sxl and the Tra proteins bind to specific RNA sites, Sxl is a repressor that acts negatively to block a splice site, whereas the Tra proteins are activators that act positively to induce a splice (see Figure 7–90). Sx1 binds to the pyrimidine-rich stretch of nucleotides that is part of the standard splicing consensus sequence (see Figure 6–28) and blocks access by the normal splicing factor, U2AF (see Figure 6–29). Tra binds to specific RNA sequences in an exon and activates a normally suboptimal splicing signal.

Figure 7–92, sex determination in *Drosophila* depends on a cascade of regulated RNA splicing events that involves these three gene products.

Although *Drosophila* sex determination provides one of the best-understood examples of a regulatory cascade based on RNA splicing, it is not clear why the fly should use this strategy. Other organisms (the nematode, for example) use an entirely different scheme for sex determination—one based on transcriptional and translational controls. Moreover, the *Drosophila* male-determination pathway requires that a number of nonfunctional RNA molecules be continually produced, which seems unnecessarily wasteful. One speculation is that this RNA-splicing cascade exploits an ancient control device, left over from the early stage of evolution where RNA was the predominant biological molecule, and controls of gene expression would have had to be based almost entirely on RNA–RNA interactions (discussed in Chapter 6).

A Change in the Site of RNA Transcript Cleavage and Poly-A Addition Can Change the C-terminus of a Protein

We saw in Chapter 6 that the 3′ end of a eucaryotic mRNA molecule is not formed by the termination of RNA synthesis by the RNA polymerase. Instead, it results from an RNA cleavage reaction that is catalyzed by additional factors while the transcript is elongating (see Figure 6–37). A cell can control the site of this cleavage so as to change the C-terminus of the resultant protein.

A well-studied example is the switch from the synthesis of membrane-bound to secreted antibody molecules that occurs during the development of B lymphocytes. Early in the life history of a B lymphocyte, the antibody it produces is anchored in the plasma membrane, where it serves as a receptor for antigen. Antigen stimulation causes B lymphocytes to multiply and to begin secreting their antibody. The secreted form of the antibody is identical to the membrane-bound form except at the extreme C-terminus. In this part of the protein, the membrane-bound form has a long string of hydrophobic amino acids that traverses the lipid bilayer of the membrane, whereas the secreted form has a much shorter string of hydrophilic amino acids. The switch from membrane-bound to secreted antibody therefore requires a different nucleotide sequence at the 3′ end of the mRNA; this difference is generated through a change in the length of the primary RNA transcript, caused by a change in the site of RNA cleavage, as shown in Figure 7–93. This change is caused by an increase of the concentration of a subunit of CStF, the protein that binds to the G/U rich sequences of RNA cleavage and poly-A addition sites and promotes RNA cleavage (see Figures 6–37 and 6–38). The first cleavage-poly-A addition site encountered by an RNA polymerase transcribing the antibody gene is suboptimal and is usually skipped in unstimulated B lymphocytes, leading to production of the longer RNA transcript. When antibody stimulation causes an increase in CSTF concentration, cleavage now occurs at the suboptimal site, and the shorter transcript is produced. In this way a change in concentration of a general RNA processing factor can produce specific effects on a relatively small number of genes.

RNA Editing Can Change the Meaning of the RNA Message

The molecular mechanisms used by cells are a continual source of surprises. An example is the process of **RNA editing**, which alters the nucleotide sequences of mRNA transcripts once they are transcribed. In Chapter 6, we saw that rRNAs and tRNAs are modified posttranscriptionally. In this section we see that some mRNAs are modified in ways that change the coded message they carry. The most dramatic form of RNA editing was discovered in RNA transcripts that code for proteins in the mitochondria of trypanosomes. Here, one or more U nucleotides are inserted (or, less frequently, removed) from selected regions of a

Figure 7–93 Regulation of the site of RNA cleavage and poly-A addition determines whether an antibody molecule is secreted or remains membrane-bound. In unstimulated B lymphocytes *(left)*, a long RNA transcript is produced, and the intron sequence near its 3′ end is removed by RNA splicing to give rise to an mRNA molecule that codes for a membrane-bound antibody molecule. In contrast, after antigen stimulation *(right)* the primary RNA transcript is cleaved upstream from the splice site in front of the last exon sequence. As a result, some of the intron sequence that is removed from the long transcript remains as coding sequence in the short transcript. These are the nucleotide sequences that encode the hydrophilic C-terminal portion of the secreted antibody molecule.

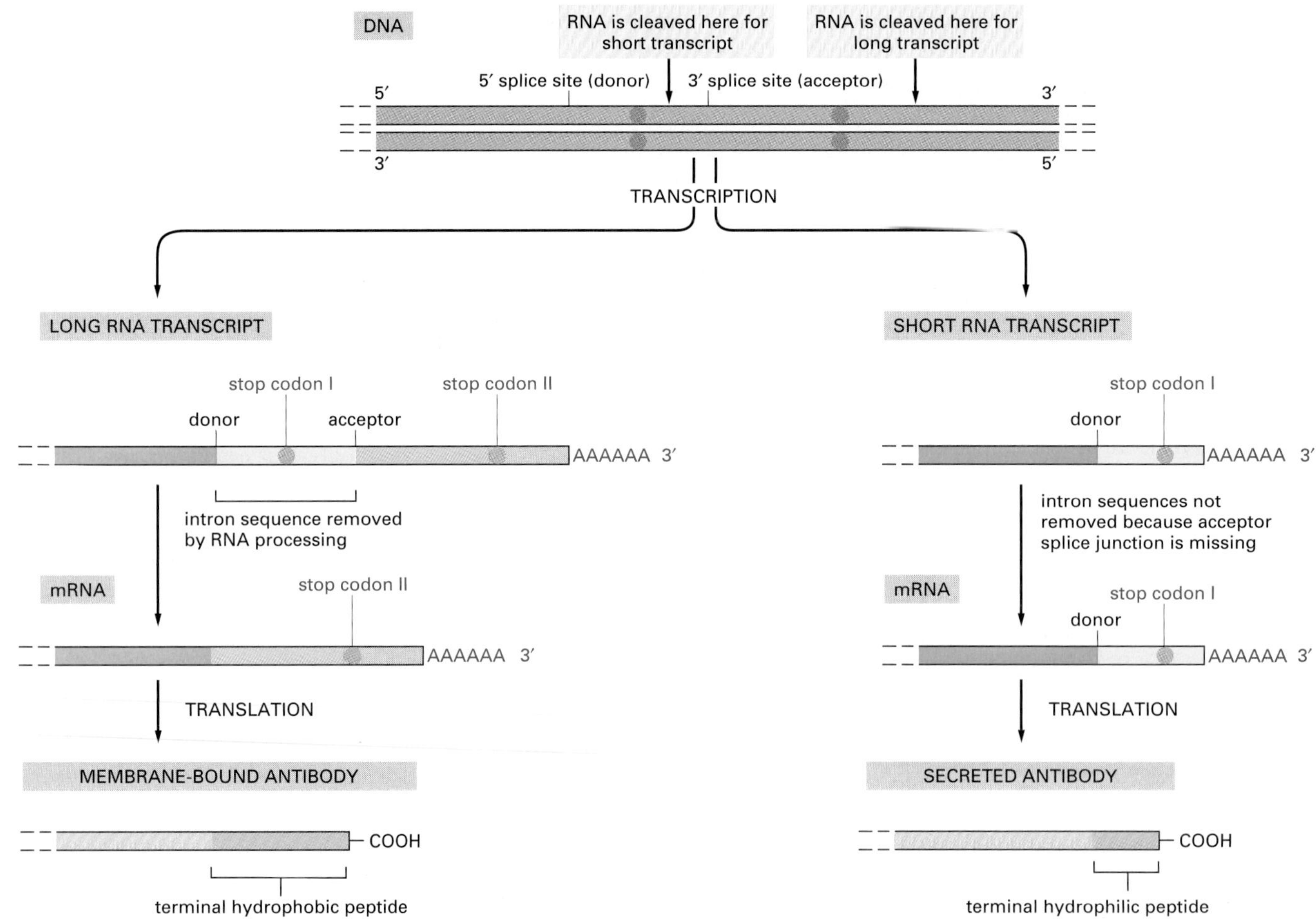

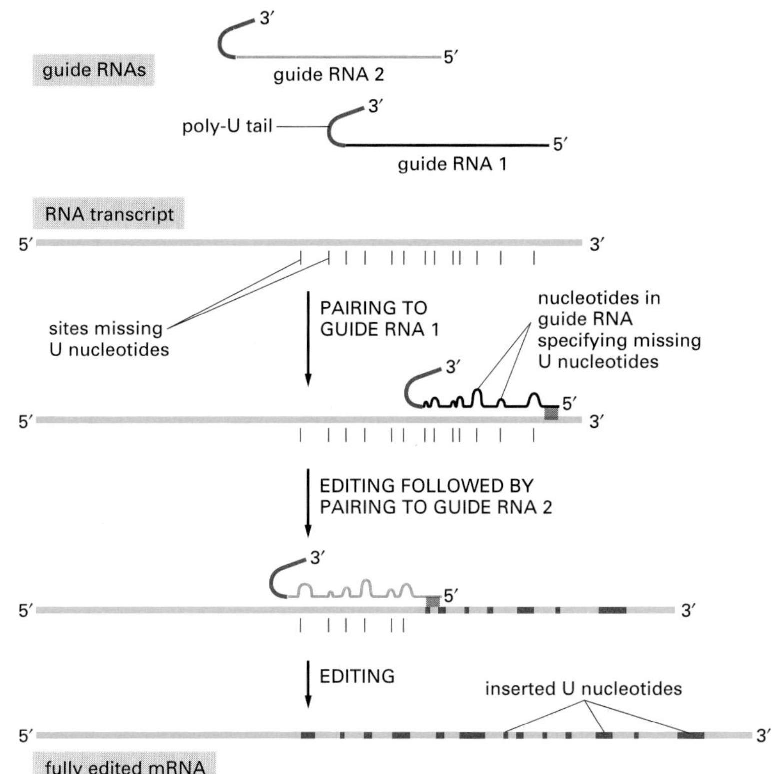

Figure 7–94 RNA editing in the mitochondria of trypanosomes. Guide RNAs contain at their 3′ end a stretch of poly U, which donates U nucleotides to sites on the RNA transcript that mispair with the guide RNA; thus the poly-U tail gets shorter as editing proceeds (not shown). Editing generally starts near the 3′ end and progresses toward the 5′ end of the RNA transcript, as shown, because the "anchor sequence" at the 5′ end of most guide RNAs can pair only with edited sequences.

transcript, causing major modifications in both the original reading frame and the sequence, thereby changing the meaning of the message. For some genes the editing is so extensive that over half of the nucleotides in the mature mRNA are U nucleotides that were inserted during the editing process. The information that specifies exactly how the initial RNA transcript is to be altered is contained in a set of 40– to 80–nucleotide-long RNA molecules that are transcribed separately. These so-called *guide RNAs* have a 5′ end that is complementary in sequence to one end of the region of the transcript to be edited; this is followed by a sequence that specifies the set of nucleotides to be inserted into the transcript, which is followed in turn by a continuous run of U nucleotides. The editing mechanism is remarkably complex, with the U nucleotides at the 3′ end of the guide RNA being transferred directly into the transcript, as illustrated in Figure 7–94.

Extensive editing of mRNA sequences has also been found in the mitochondria of many plants, with nearly every mRNA being edited to some extent. In this case, however, RNA bases are changed from C to U, without nucleotide insertions or deletions. Often many of the Cs in an mRNA are affected by editing, changing 10% or more of the amino acids that the mRNA encodes.

We can only speculate as to why the mitochondria of trypanosomes and plants make use of such extensive RNA editing. The suggestions that seem most reasonable are based on the premise that mitochondria contain a primitive genetic system that offers scanty opportunities for other forms of control. There is evidence that editing is regulated to produce different mRNAs under different conditions, so that RNA editing can be viewed as a primitive way to change the expression of genes, a relic, perhaps, of mechanisms that operated in very ancient cells, where most catalyses were probably carried out by RNA molecules rather than by proteins.

RNA editing of a much more limited kind occurs in mammals. One of the most important types is the enzymatic deamination of adenine to produce inosine (see Figure 6–55), which occurs at selected positions in some pre-mRNAs. In some cases, this modification changes the splicing pattern of the RNA; in others, it changes the meanings of codons. Because inosine base pairs with cytosine, A-to-I editing can result in a protein with an altered amino acid sequence.

This editing is carried out by protein enzymes called *ADARs* (adenosine deaminases acting on RNA); these enzymes recognize a double-stranded RNA structure that is formed by base pairing between the site to be edited and a complementary sequence located elsewhere on the same RNA molecule, typically in a 3′ intron (Figure 7–95). An especially important example of A-to-I editing takes place in the pre-mRNA that codes for a transmitter-gated ion channel in the brain. A single edit changes a glutamine to an arginine; the affected amino acid lies on the inner wall of the channel, and the editing change alters the Ca^{2+} permeability of the channel. The importance of this edit in mice has been demonstrated by deleting the relevant ADAR gene. The mutant mice are prone to epileptic seizures and die during or shortly after weaning. If the gene for the gated ion channel is mutated to produce the edited form of the protein directly, mice lacking the ADAR develop normally, showing that editing of the ion channel RNA is normally crucial for proper brain development. Mice and humans have several additional ADAR genes, and deletion of one of these in mice causes death of the mouse embryo before birth. Because these mice have severe defects in the production of red blood cell precursors, it is likely that RNA editing is also essential for the proper development of the hemopoietic system.

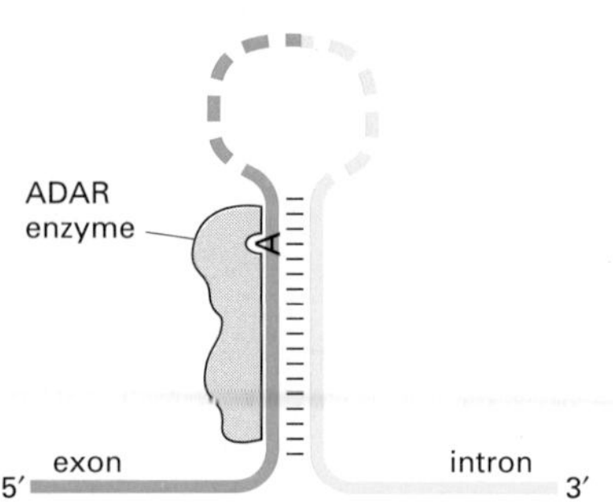

Figure 7–95 Mechanism of A-to-I RNA editing in mammals. The position of an edit is signaled by RNA sequences carried on the same RNA molecule. Typically, a sequence complementary to the position of the edit is present in an intron, and the resulting double-stranded RNA attracts the A-to-I editing enzyme ADAR. Mice and humans have three ADAR enzymes: ADR1 is required in the liver for proper red blood cell development, ADR2 is required for proper brain development (as described in the text), and the role of ADR3 is not yet known.

C-to-U editing has also been observed in mammals. In one example, that of the *apolipoprotein-B* mRNA, a C to U change creates a stop codon that causes a truncated version of this large protein to be made in a tissue-specific manner. Why editing in mammalian cells exists at all is a mystery. One idea is that it arose in evolution to correct "mistakes" in the genome. Another is that is provides yet another way for the cell to produce a variety of related proteins from a single gene. A third view is that editing is merely one of a large number of haphazard, makeshift devices that have originated through random mutation and have been perpetuated because they happen to contribute to a useful effect.

RNA Transport from the Nucleus Can Be Regulated

It has been estimated that in mammals only about one-twentieth of the total mass of RNA synthesized ever leaves the nucleus. We saw in Chapter 6 that most mammalian RNA molecules undergo extensive processing and the "left-over" RNA fragments (excised introns and RNA sequences 3′ to the cleavage/poly-A site) are degraded in the nucleus. Incompletely processed and otherwise damaged RNAs are also eventually degraded in the nucleus as part of the quality control system of RNA production. This degradation is carried out by the **exosome**, a large protein complex that contains, as subunits, several different RNA exonucleases.

As described in Chapter 6, the export of RNA molecules from the nucleus is delayed until processing has been completed. Therefore any mechanism that prevents the completion of RNA splicing on a particular RNA molecule could in principle block the exit of that RNA from the nucleus. This feature forms the basis for one of the best understood examples of **regulated nuclear transport** of mRNA, which occurs in HIV, the virus that causes AIDS.

HIV is a retrovirus—an RNA virus that, once inside a cell, directs the formation of a double-stranded DNA copy of its genome which is then inserted into the host genome (see Figure 5–73). Once inserted, the viral DNA is transcribed as one long RNA molecule by the host cell RNA polymerase II. This transcript is then spliced in many different ways to produce over 30 different species of mRNA, which in turn, are translated into a variety of different proteins (Figure 7–96). In order to make progeny virus, entire, unspliced viral transcripts must be exported from the nucleus to the cytosol where they are packaged into viral capsids (see Figure 5–73). Moreover, several of the HIV mRNAs are alternatively spliced in such a way that they still carry complete introns. The host cell's block to the nuclear export of unspliced RNA (and its subsequent degradation) therefore presents a special problem for HIV, and it is overcome in an ingenious way.

The virus encodes a protein (called Rev) that binds to a specific RNA sequence (called the Rev responsive element, RRE) located within a viral intron. The Rev protein interacts with a nuclear export receptor (exportin 1), which

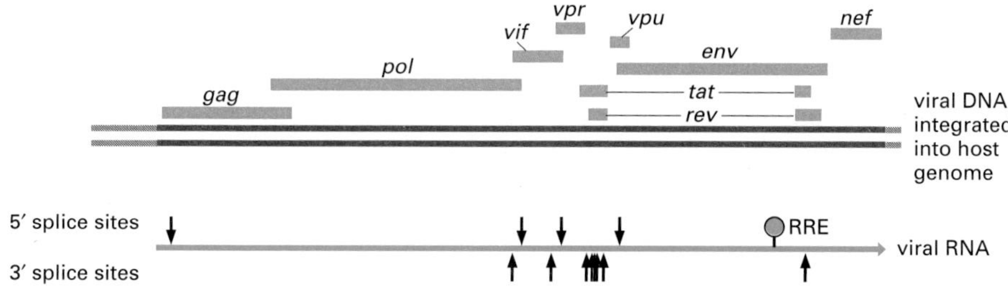

Figure 7–96 The compact genome of HIV, the human AIDS virus. The positions of the nine HIV genes are shown in *green*. The *red double line* indicates a DNA copy of the viral genome which has become integrated into the host DNA *(gray)*. Note that the coding regions of many genes overlap, and those of *tat* and *rev* are split by introns. The *blue line* at the *bottom* of the figure represents the pre-mRNA transcript of the viral DNA showing the locations of all the possible splice sites *(arrows)*. There are many alternative ways of splicing the viral transcript; for example the *env* mRNAs retain the intron that has been spliced out of the *tat* and *rev* mRNAs. The Rev response element (RRE) is indicated by a blue ball and stick. It is a 234-nucleotide long stretch of RNA that folds into a defined structure; Rev recognizes a particular hairpin (see Figure 6–94) within this larger structure.

The *gag* gene codes for a protein that is cleaved into several smaller proteins that form the viral capsid. The *pol* gene codes for a protein that is cleaved to produce reverse transcriptase (which transcribes RNA into DNA) as well as the integrase involved in integrating the viral genome (as double-stranded DNA) into the host genome. Pol is produced by ribosomal frameshifting of translation that begins at *gag* (see Figure 6–78). The *env* gene codes for the envelope proteins (see Figure 5–73). Tat, Rev, Vif, Vpr, Vpu, and Nef are small proteins with a variety of functions. For example, Rev regulates nuclear export (see Figure 7–97) and Tat regulates the elongation of transcription across the integrated viral genome (see p. 436).

directs the movement of viral RNAs through nuclear pores into the cytosol despite the presence of intron sequences. We discuss in detail the way that export receptors function in Chapter 12.

The regulation of nuclear export by Rev has several important consequences for HIV growth and pathogenesis. In addition to ensuring the nuclear export of specific unspliced RNAs, it divides the viral infection into an early phase (where Rev is translated from a fully spliced RNA and RNAs containing an intron are retained in the nucleus and degraded) and a late phase (where unspliced RNAs are exported due to Rev function). This timing helps the virus replicate by providing the gene products roughly in the order in which they are needed (Figure 7–97). It is also possible that regulation by Rev helps the HIV virus to achieve latency, a condition where the HIV genome has become integrated into the host cell genome but the production of viral proteins has temporarily ceased. If, after its initial entry into a host cell, conditions became unfavorable for viral transcription and replication, Rev is made at levels too low to promote export of unspliced RNA. This situation stalls the viral growth cycle. When conditions for viral replication improve, Rev levels increase, and the virus can enter the replication cycle.

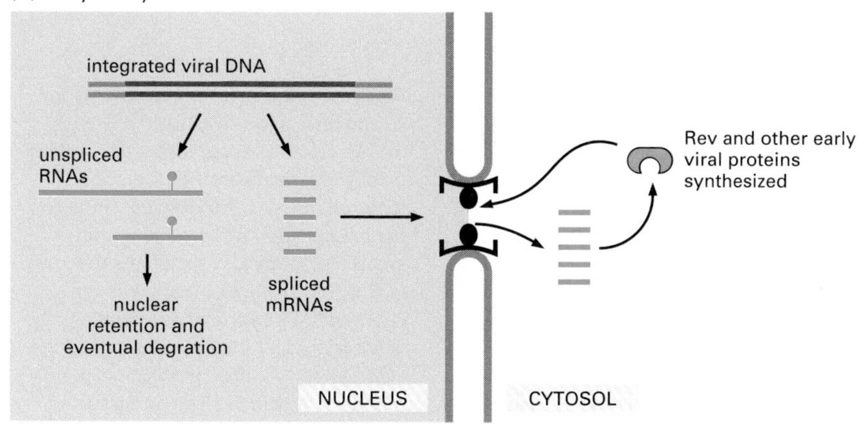

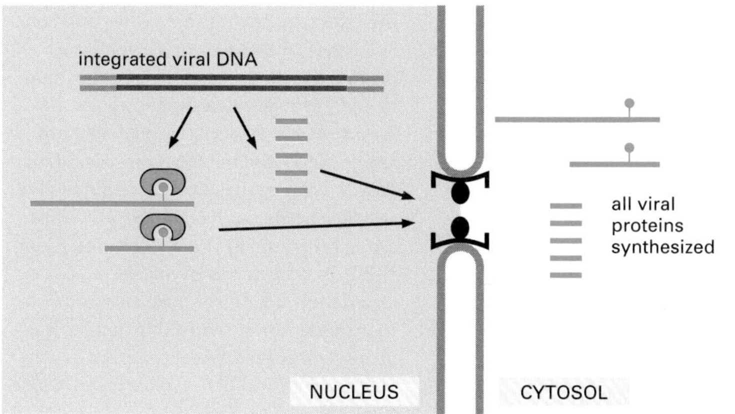

Figure 7–97 Regulation of nuclear export by the HIV Rev protein. Early in HIV infection (A), only the fully spliced RNAs (which contain the coding sequences for Rev, Tat, and Nef) are exported from the nucleus and translated. Once sufficient Rev protein has accumulated and been transported into the nucleus (B), unspliced viral RNAs can be exported from the nucleus. Many of these RNAs are translated into protein, and the full length transcripts are packaged into new viral particles.

Some mRNAs Are Localized to Specific Regions of the Cytoplasm

Once a newly made eucaryotic mRNA molecule has passed through a nuclear pore and entered the cytosol, it is typically met by ribosomes, which translate it into a polypeptide chain (see Figure 6–40). If the mRNA encodes a protein that is destined to be secreted or expressed on the cell surface, it will be directed to the endoplasmic reticulum (ER) by a signal sequence at the protein's amino terminus; components of the cell's protein-sorting apparatus recognize the signal sequence as soon as it emerges from the ribosome and direct the entire complex of ribosome, mRNA, and nascent protein to the membrane of the ER, where the remainder of the polypeptide chain is synthesized, as discussed in Chapter 12. In other cases the entire protein is synthesized by free ribosomes in the cytosol, and signals in the completed polypeptide chain may then direct the protein to other sites in the cell.

Some mRNAs are themselves directed to specific intracellular locations before translation begins. Presumably it is advantageous for the cell to position its mRNAs close to the sites where the protein produced from the mRNA is required. The signals that direct mRNA localization are typically located in the 3′ *untranslated region* (*UTR*) of the mRNA molecule—a region that extends from the stop codon, which terminates protein synthesis, to the start of the poly-A tail (see Figure 6–22A). A striking example of mRNA localization is seen in the *Drosophila* egg, where the mRNA encoding the bicoid gene regulatory protein is localized by attachment to the cytoskeleton at the anterior tip of the developing egg. When the translation of this mRNA is triggered by fertilization, a gradient of the bicoid protein is generated that plays a crucial part in directing the development of the anterior part of the embryo (shown in Figure 7–52 and discussed in more detail in Chapter 21). Many mRNAs in somatic cells are localized in a similar way. The mRNA that encodes actin, for example, is localized to the actin-filament-rich cell cortex in mammalian fibroblasts by means of a 3′ UTR signal.

RNA localization has been observed in many organisms, including unicellular fungi, plants, and animals, and it is likely to be a common mechanism used by cells to concentrate high-level production of proteins at specific sites. Several distinct mechanisms for mRNA localization have been discovered (Figure 7–98), but all of them require specific signals in the mRNA itself, usually concentrated in the 3′ UTR (Figure 7–99).

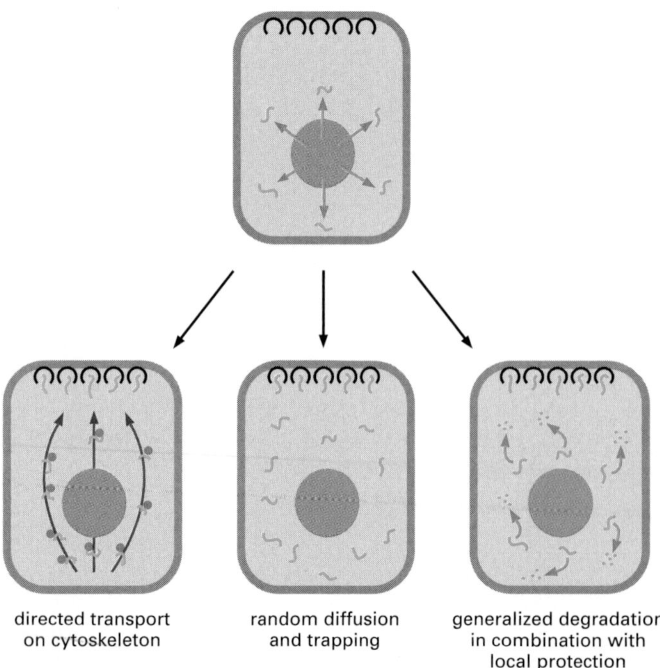

Figure 7–98 Three mechanisms for the localization of mRNAs. The mRNA to be localized leaves the nucleus through nuclear pores *(top)*. Some localized mRNAs *(left diagram)* travel to their destination by associating with cytoskeleton motors *(green)*. As described in Chapter 16, these motors use the energy of ATP hydrolysis to move unidirectionally along components of the cytoskeleton. At their destination, the mRNAs are held in place by anchor proteins *(black)*. Other mRNAs randomly diffuse through the cytosol and are simply trapped and therefore concentrated at their sites of localization *(center diagram)*. Still other mRNAs *(right diagram)* are degraded in the cytosol unless they have bound, through random diffusion, a localized protein complex that anchors and protects the mRNA from degradation *(black)*. Each of these mechanisms requires specific signals on the mRNA, which are typically located in the 3′ UTR (see Figure 7–99). In many cases of mRNA localization, additional mechanisms block the translation of the mRNA until it is properly localized. (Adapted from H.D. Lipshitz and C.A. Smibert, *Curr. Opin. Gen. Dev.* 10:476–488, 2000.)

We saw in Chapter 6 that the 5′ cap and the 3′ poly-A tail are necessary for efficient translation, and their presence on the same mRNA molecule thereby signals to the translation machinery that the mRNA molecule is intact. As just described, the 3′ UTR often contains a "zip code," which directs mRNAs to different places in the cell. In this chapter, we will also see that mRNAs also carry information specifying the average length of time each mRNA persists in the cytosol and the efficiency with which each mRNA is translated into protein. In a broad sense, the untranslated regions of eucaryotic mRNAs resemble the transcriptional control regions of genes: their nucleotide sequences contain information specifying the way the RNA is to be used, and proteins that interpret this information bind specifically to these sequences. Thus, over and above the specification of the amino acid sequences of proteins, mRNA molecules are rich with many additional types of information.

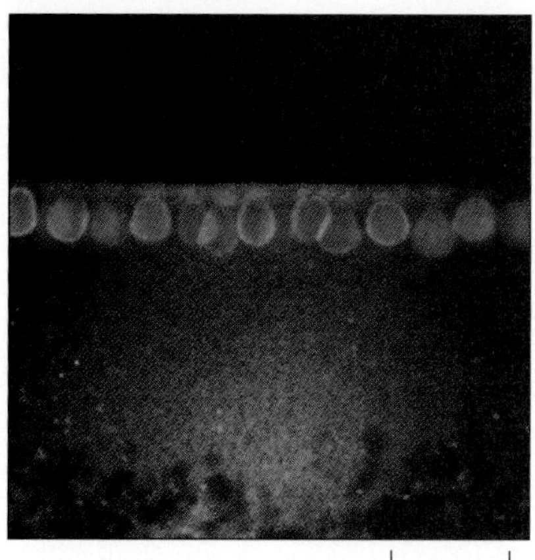

Figure 7–99 The importance of the 3′ UTR in localizing mRNAs to specific regions of the cytoplasm. For this experiment, two different fluorescently-labeled RNAs were prepared by transcribing DNA *in vitro* in the presence of fluorescently-labeled derivatives of UTP. One RNA (labeled with a *red* fluorochrome) contains the coding region for the *Drosophila* hairy protein and includes the adjacent 3′ UTR. The other RNA (labeled *green*) contains the hairy coding region but the 3′ UTR has been deleted. The two RNAs were mixed and injected into a *Drosophila* embryo at a stage of development when multiple nuclei reside in a common cytoplasm (see Figure 7–52). When the fluorescent RNAs were visualized 10 minutes later, the full-length hairy RNA *(red)* was localized to the apical side of nuclei *(blue)* but the transcript missing the 3′ UTR *(green)* failed to localize. Hairy is one of many gene regulatory proteins that specifies positional information in the developing *Drosophila* embryo discussed in Chapter 21). The localization of its mRNA (shown in this experiment to depend on its 3′ UTR) is thought to be critical for proper fly development. (Courtesy of Simon Bullock and David Ish-Horowicz.)

Proteins That Bind to the 5′ and 3′ Untranslated Regions of mRNAs Mediate Negative Translational Control

Once an mRNA has been synthesized, one of the most common ways of regulating the levels of its protein product is by controlling the step in which translation is initiated. Even though the mechanistic details of translation initiation differ between eucaryotes and bacteria (as we saw in Chapter 6), some of the same basic regulatory strategies are used.

In bacterial mRNAs a conserved stretch of six nucleotides, the *Shine-Dalgarno sequence*, is always found a few nucleotides upstream of the initiating AUG codon. This sequence forms base pairs with the 16S RNA in the small ribosomal subunit, correctly positioning the initiating AUG codon in the ribosome. Because this interaction makes a major contribution to the efficiency of initiation, it provides the bacterial cell with a simple way to regulate protein synthesis through **negative translational control** mechanisms. These mechanisms generally involve blocking the Shine-Dalgarno sequence, either by covering it with a bound protein or by incorporating it into a base-paired region in the mRNA molecule. Many bacterial mRNAs have specific *translational repressor proteins* that can bind in the vicinity of the Shine-Dalgarno sequence and thereby inhibit translation of only that species of mRNA. For example, some ribosomal proteins can repress translation of their own mRNAs by binding to the 5′ untranslated region. This mechanism comes into play only when the ribosomal proteins are produced in excess over ribosomal RNA and are therefore not incorporated into ribosomes. In this way, it allows the cell to maintain correctly balanced quantities of the various components needed to form ribosomes. It is not hard to guess how this mechanism might have evolved. Ribosomal proteins assemble into ribosomes by binding to specific sites in rRNA; ingeniously, some of them exploit this RNA-binding ability to regulate their own production by binding to similar sites present in their own mRNAs.

Eucaryotic mRNAs do not contain a Shine-Dalgarno sequence. Instead, as discussed in Chapter 6, the selection of an AUG codon as a translation start site is largely determined by its proximity to the cap at the 5′ end of the mRNA molecule, which is the site at which the small ribosomal subunit binds to the mRNA and begins scanning for an initiating AUG codon. Despite the differences in translation initiation, eucaryotes also utilize translational repressors. Some bind to the 5′ end of the mRNA and thereby inhibit translation initiation. Others recognize nucleotide sequences in the 3′ UTR of specific mRNAs and decrease translation initiation by interfering with the communication between the 5′ cap and 3′ poly-A tail, which is required for efficient translation (see Figure 6–72).

A well-studied form of negative translational control in eucaryotes allows the synthesis of the intracellular iron storage protein ferritin to be increased rapidly if the level of soluble iron atoms in the cytosol rises. The iron regulation depends on a sequence of about 30 nucleotides in the 5′ leader of the ferritin mRNA molecule. This iron-response element folds into a stem-loop structure that binds a translation repressor protein called aconitase, which blocks the translation of any RNA sequence downstream (Figure 7–100). Aconitase is an iron-binding protein, and exposure of the cell to iron causes it to dissociate from the ferritin mRNA, releasing the block to translation and increasing the production of ferritin by as much as hundredfold.

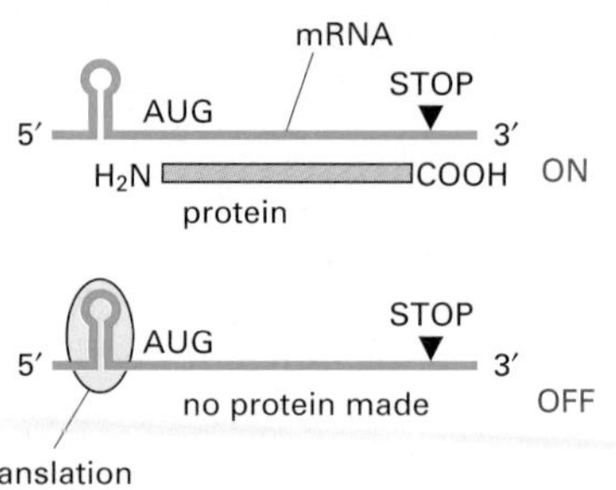

Figure 7–100 Negative translational control. This form of control is mediated by a sequence-specific RNA-binding protein that acts as a translation repressor. Binding of the protein to an mRNA molecule decreases the translation of the mRNA. Several cases of this type of translational control are known. The illustration is modeled on the mechanism that causes more ferritin (an iron storage protein) to be synthesized when the free iron concentration in the cytosol rises; the iron-sensitive translation repressor protein is called aconitase (see also Figure 7–105). In other examples, a complementary RNA molecule, rather than a protein, regulates translation initiation by blocking a critical region of the mRNA through the formation of a short region of double-helical RNA.

The Phosphorylation of an Initiation Factor Globally Regulates Protein Synthesis

Eucaryotic cells decrease their overall rate of protein synthesis in response to a variety of situations, including deprivation of growth factors or nutrients, infection by viruses, and sudden increases in temperature. Much of this decrease is caused by the phosphorylation of the translation initiation factor eIF-2 by specific protein kinases that respond to the changes in conditions.

The normal function of eIF-2 was outlined in Chapter 6. It forms a complex with GTP and mediates the binding of the methionyl initiator tRNA to the small ribosomal subunit, which then binds to the 5′ end of the mRNA and begins scanning along the mRNA. When an AUG codon is recognized, the bound GTP is hydrolyzed to GDP by the eIF-2 protein, causing a conformational change in the protein and releasing it from the small ribosomal subunit. The large ribosomal subunit then joins the small one to form a complete ribosome that begins protein synthesis (see Figure 6–71).

Because eIF-2 binds very tightly to GDP, a guanine nucleotide exchange factor (see Figure 15–54), designated eIF-2B, is required to cause GDP release so that a new GTP molecule can bind and eIF-2 can be reused (Figure 7–101A). The reuse of eIF-2 is inhibited when it is phosphorylated—the phosphorylated eIF-2 binds to eIF-2B unusually tightly, inactivating eIF-2B. There is more eIF-2 than eIF-2B in cells, and even a fraction of phosphorylated eIF-2 can trap nearly all of the eIF-2B. This prevents the reuse of the nonphosphorylated eIF-2 and greatly slows protein synthesis (Figure 7–101B).

Regulation of the level of active eIF-2 is especially important in mammalian cells, being part of the mechanism that allows them to enter a nonproliferating, resting state (called G_0)—in which the rate of total protein synthesis is reduced to about one-fifth the rate in proliferating cells (discussed in Chapter 17).

Initiation at AUG Codons Upstream of the Translation Start Can Regulate Eucaryotic Translation Initiation

We saw in Chapter 6 that eucaryotic translation typically begins at the first AUG downstream of the 5′ end of the mRNA, as it is the first AUG encountered by a

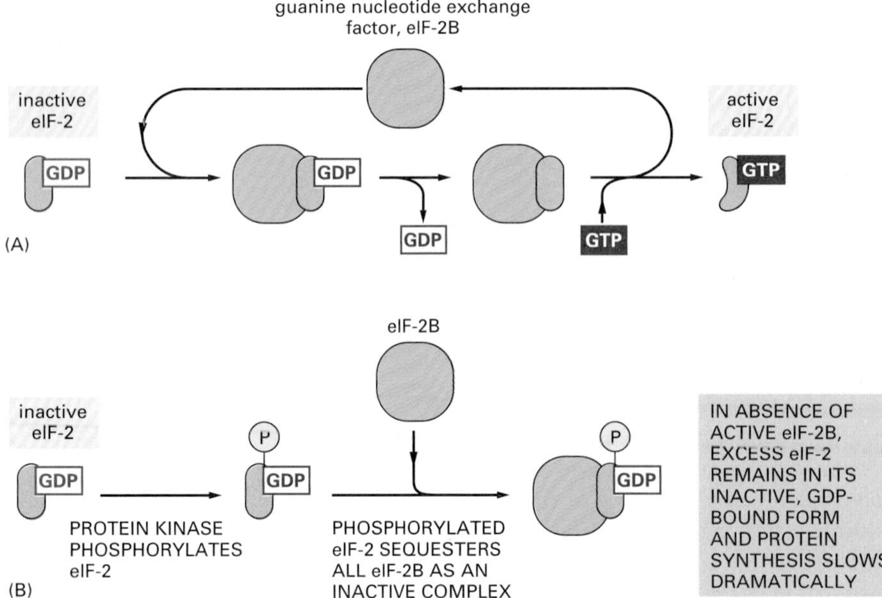

Figure 7–101 The eIF-2 cycle. (A) The recycling of used eIF-2 by a guanine nucleotide exchange factor (eIF-2B). (B) eIF-2 phosphorylation controls protein synthesis rates by tying up eIF-2B.

scanning small ribosomal subunit. But the nucleotides immediately surrounding the AUG also influence the efficiency of translation initiation. If the recognition site is poor enough, scanning ribosomal subunits will ignore the first AUG codon in the mRNA and skip to the second or third AUG codon instead. This phenomenon, known as "leaky scanning," is a strategy frequently used to produce two or more closely related proteins, differing only in their amino termini, from the same mRNA. For example, it allows some genes to produce the same protein with and without a signal sequence attached at its amino terminus so that the protein is directed to two different locations in the cell. In some cases, the cell can regulate the relative abundance of the protein isoforms produced by leaky scanning; for example, a cell-type specific increase in the abundance of the initiation factor eIF-4F favors usage of the AUG closest to the 5′ end of the mRNA.

Another type of control found in eucaryotes uses one or more short open reading frames that lie between the 5′ end of the mRNA and the beginning of the gene. Often, the amino acid sequences coded by these upstream open reading frames (uORFs) are not critical; rather the uORFs serve a purely regulatory function. An uORF present on an mRNA molecule will generally decrease translation of the downstream gene by trapping a scanning ribosome initiation complex and causing the ribosome to translate the uORF and dissociate from the mRNA before it reaches the protein coding sequences.

When the activity of a general translation factor, such as eIF-2 (discussed above), is reduced, one might expect that the translation of all mRNAs would be reduced equally. Contrary to this expectation, however, the phosphorylation of eIF-2 can have selective effects, even enhancing the translation of specific mRNAs that contain uORFs. This can enable yeast cells, for example, to adapt to starvation for specific nutrients by shutting down the synthesis of all proteins except those that are required for synthesis of the nutrients that are missing. The details of this mechanism have been worked out for a specific yeast mRNA that encodes a protein called Gcn4, a gene regulatory protein that is required for the activation of many genes encoding proteins that are important for amino acid synthesis.

The *GCN4* mRNA contains four short uORFs, and these are responsible for selectively increasing the translation of *GCN4* in response to eIF-2 phosphorylation provoked by amino acid starvation. The mechanism by which *GCN4* translation is increased is complex. In outline, ribosomal subunits move along the mRNA, encountering each of the uORFs but translating only a subset of them; if the fourth uORF is translated, as is the case in unstarved cells, the ribosomes dissociate at the end of the uORF, and translation of *GCN4* is inefficient. The decrease in eIF-2 activity makes it more likely that a scanning ribosome will move through the fourth uORF before it acquires the ability to initiate translation. Such a ribosome can then efficiently initiate translation on the *GCN4* sequences, leading to the production of proteins that promote amino acid synthesis inside the cell.

Internal Ribosome Entry Sites Provide Opportunities for Translation Control

Although approximately 90% of eucaryotic mRNAs are translated beginning with the first AUG downstream from the 5′ cap, certain AUGs, as we saw in the last section, can be skipped over during the scanning process. In this section, we discuss yet another way that cells can initiate translation at positions distant from the 5′ end of the mRNA. In these cases, translation is initiated directly at specialized RNA sequences, each of which is called an **internal ribosome entry site** (**IRES**). An IRES can occur in many different places in an mRNA. In some unusual cases, two distinct protein coding sequences are carried in tandem on the same eucaryotic mRNA; translation of the first occurs by the usual scanning mechanism and translation of the second through an IRES. IRESs are typically several hundred nucleotides in length and fold into specific structures that bind many, but not all, of the same proteins that are used to initiate normal cap-dependent translation (Figure 7–102). In fact, different IRESs require different subsets of initiation factors. However, all of them bypass the need for a 5′ cap structure and the translation initiation factor that recognizes it, eIF-4E.

IRESs were first discovered in certain mammalian viruses where they provide a clever way for the virus to take over its host cell's translation machinery. On infection, these viruses produce a protease (encoded in the viral genome) that cleaves the cellular translation factor eIF-4G and thereby renders it unable to bind to eIF-4E, the cap-binding complex. This shuts down the great majority of host cell translation and effectively diverts the translation machinery to the IRES sequences, which are present on many viral mRNAs. The truncated eIF-4G remains competent to initiate translation at these internal sites and may even stimulate the translation of certain IRES-containing viral mRNAs.

The selective activation of IRES-mediated translation also occurs on cellular mRNAs. For example, when eucaryotic cells enter M phase of the cell cycle, the overall rate of translation drops to approximately 25% that in interphase cells. This drop is largely caused by a cell-cycle dependent dephosphorylation of the cap binding complex , eIF-4E, which lowers its affinity for the 5′ cap. IRES-containing mRNAs, however, are immune to this effect, and their relative translation rates therefore increase as the cell enters M phase.

Finally, when mammalian cells enter the programmed cell death pathway (discussed in Chapter 17), eIF-4G is cleaved, and a general decrease in translation ensues. Some proteins critical for the control of cell death are translated from IRES-containing mRNAs, and they continue to be synthesized. It seems that one of the main advantages of the IRES mechanism for the cell is that it allows selected mRNAs to be translated at a high rate despite a general decrease in the cell's capacity to initiate protein synthesis.

Gene Expression Can Be Controlled By a Change In mRNA Stability

The vast majority of mRNAs in a bacterial cell are very unstable, having a half-life of about 3 minutes. Exonucleases, which degrade in the 3′ to 5′ direction, are usually responsible for the rapid destruction of these mRNAs. Because its mRNAs are both rapidly synthesized and rapidly degraded, a bacterium can adapt quickly to environmental changes.

The mRNAs in eucaryotic cells are more stable. Some, such as that encoding β-globin, have half-lives of more than 10 hours. Many, however, have half-lives of 30 minutes or less. These unstable mRNAs often code for regulatory proteins, such as growth factors and gene regulatory proteins, whose production rates need to change rapidly in cells.

Two major degradation pathways exist for eucaryotic mRNAs, and sequences in each mRNA molecule determine the pathway and kinetics of degradation. The most common pathway involves the gradual shortening of the poly-A tail. We saw in Chapter 6 that capping and polyadenylation of mRNA

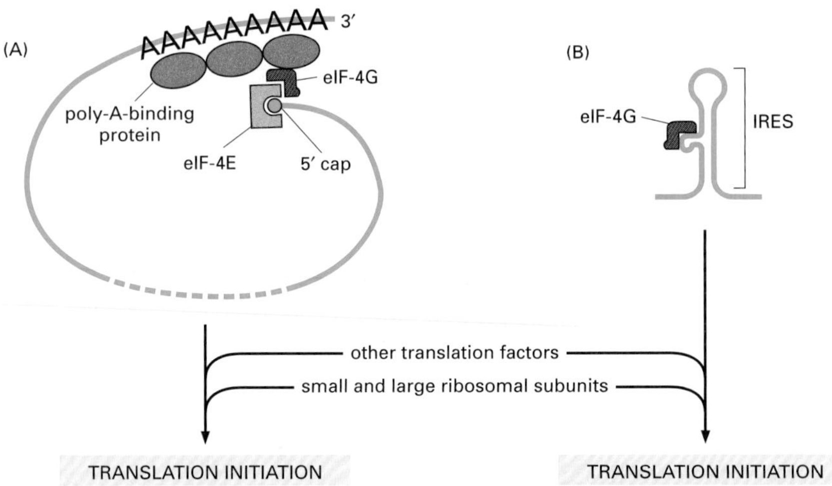

Figure 7–102 Two mechanisms of translation initiation. (A) The cap-dependent mechanism requires a set of initiation factors whose assembly on the mRNA is stimulated by the presence of a 5′ cap and a poly-A tail (see also Figure 6–71). (B) The IRES-dependent mechanism requires only a subset of the normal translation initiating factors, and these assemble directly on the folded IRES. (Adapted from A. Sachs, *Cell* 101:243–245, 2000.)

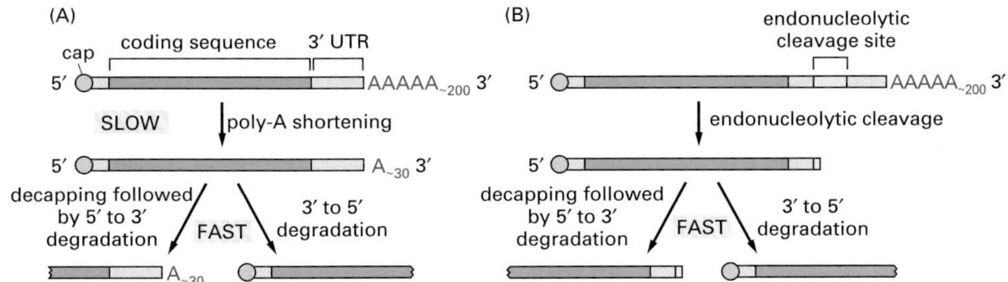

Figure 7–103 Two mechanisms of eucaryotic mRNA decay. (A) Deadenylation-dependent decay. Most eucaryotic mRNAs are degraded by this pathway. The critical threshold of poly-A tail length that induces decay may correspond to the loss of the poly-A binding proteins (see Figure 6–40). As shown in Figure 7–104, the deadenylation enzyme associates with both the 3′ poly-A tail and the 5′ cap, and this arrangement may coordinate decapping with poly-A shortening. Although 5′ to 3′ and 3′ to 5′ degradation are shown on separate RNA molecules, these two processes can occur together on the same molecule.
(B) Deadenylation-independent decay. It is not yet known with certainty whether decapping follows endonucleolytic cleavage of the mRNA. (Adapted from C.A. Beelman and R. Parker, *Cell* 81:179–183, 1995.)

molecules occurs in the nucleus. Once in the cytosol, the poly-A tails (which average about 200 As in length) are gradually shortened by an exonuclease that chews away the tail in the 3′ to 5′ direction. Once a critical threshold of tail shortening has been reached (approximately 30 A's remaining), the 5′ cap is removed (a process called "decapping"), and the RNA is rapidly degraded (Figure 7–103A).

Nearly all mRNAs are subject to poly-A tail shortening, decapping, and eventual degradation, but the rate at which this occurs differs from one species of mRNA to the next. The proteins that carry out tail-shortening compete directly with the machinery that catalyzes translation; therefore, any factors that affect the translation efficiency of an mRNA will tend to have the opposite effect on its degradation (Figure 7–104). In addition, many mRNAs carry in their 3′ UTR sequences binding sites for specific proteins that increase or decrease the rate of poly-A tail shortening. For example, many unstable mRNAs contain stretches of AU sequences, which greatly enhance the shortening rate.

A second pathway by which mRNA is degraded begins with the action of specific endonucleases, which simply cleave the poly-A tail from the rest of the mRNA in one step (see Figure 7–103). The mRNAs that are degraded in this way carry specific nucleotide sequences, typically in their 3′ UTR, that serve as recognition sequences for the endonucleases.

The stability of an mRNA can be changed in response to extracellular signals. For example, the addition of iron to cells decreases the stability of the mRNA that encodes the receptor protein that binds the iron-transporting protein transferrin, causing less of this receptor to be made. Interestingly, this effect is mediated by the iron-sensitive RNA-binding protein aconitase, which, as we discussed above, also controls ferritin mRNA translation. Aconitase can bind to the 3′ UTR of the transferrin receptor mRNA and cause an increase in receptor production by blocking endonucleolytic cleavage of the mRNA. On the addition of iron, aconitase is released from the mRNA, decreasing mRNA stability (Figure 7–105).

Cytoplasmic Poly-A Addition Can Regulate Translation

The initial polyadenylation of an RNA molecule (discussed in Chapter 6) occurs in the nucleus, apparently automatically for nearly all eucaryotic mRNA precursors. As we have just seen, the poly-A tails on most mRNAs gradually shorten in

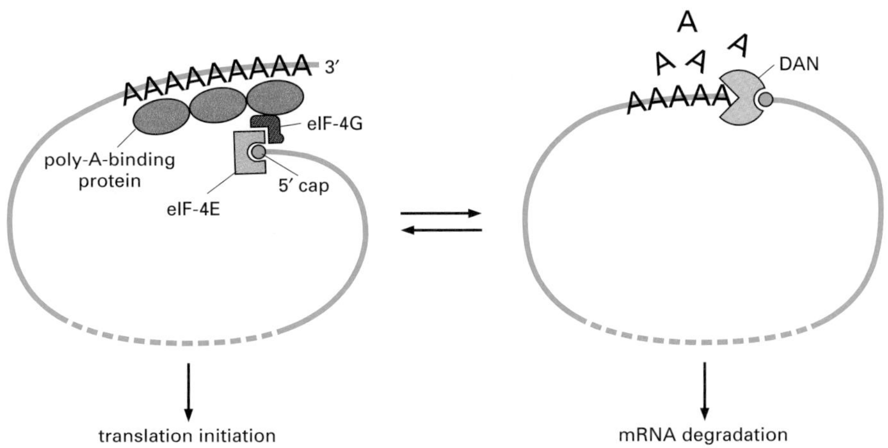

Figure 7–104 The competition between mRNA translation and mRNA decay. The same two features of mRNA—the 5′ cap and the 3′ poly-A site—are used in both translation initiation and deadenylation-dependent mRNA decay (see Figure 7–103). The enzyme (called DAN) that shortens the poly-A tail in the 3′ to 5′ direction associates with the 5′ cap. As described in Chapter 6 (see Figure 6–71), the translation initiation machinery also associates with both the 5′ cap and the poly-A tail. (Adapted from M. Gao et al., *Mol. Cell* 5:479–488, 2000.)

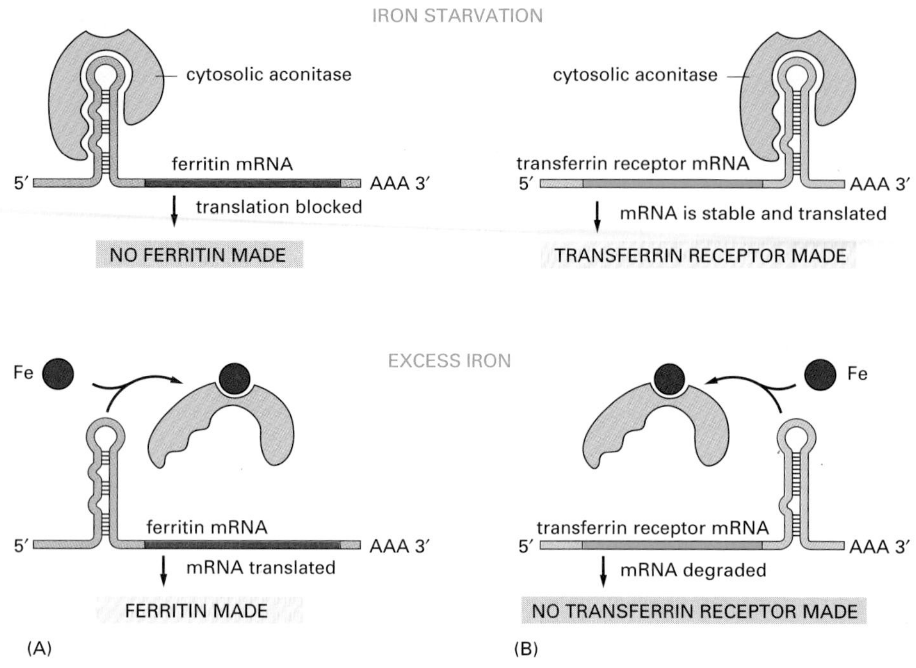

Figure 7–105 Two posttranslational controls mediated by iron. In response to an increase in iron concentration in the cytosol, a cell increases its synthesis of ferritin in order to bind the extra iron (A) and decreases its synthesis of transferrin receptors in order to import less iron across the plasma membrane (B). Both responses are mediated by the same iron-responsive regulatory protein, aconitase, which recognizes common features in a stem-and-loop structure in the mRNAs encoding ferritin and transferrin receptor. Aconitase dissociates from the mRNA when it binds iron. But because the transferrin receptor and ferritin are regulated by different types of mechanisms, their levels respond oppositely to iron concentrations even though they are regulated by the same iron-responsive regulatory protein. The binding of aconitase to the 5′ UTR of the ferritin receptor mRNA blocks translation initiation; its binding to the 3′ UTR of the ferritin receptor mRNA blocks an endonuclease cleavage site and thereby stabilizes the mRNA (see Figure 7–103). (Adapted from M.W. Hentze et al., *Science* 238:1570–1573, 1987 and J.L. Casey et al., *Science* 240:924–928, 1988.)

the cytosol, and the RNAs are eventually degraded. In some cases, however, the poly-A tails of specific mRNAs are lengthened in the cytosol, and this mechanism provides an additional form of translational regulation.

Maturing oocytes and eggs provide the most striking example. Many of the normal mRNA degradation pathways seem to be disabled in these giant cells, so that the cells can build up large stores of mRNAs in preparation for fertilization. Many mRNAs are stored in the cytoplasm with only 10 to 30 As at their 3′ end, and in this form they are not translated. At specific times during oocyte maturation and postfertilization, when the proteins encoded by these mRNAs are required, poly A is added to selected mRNAs, greatly stimulating the initiation of their translation.

Nonsense-mediated mRNA Decay Is Used as an mRNA Surveillance System in Eucaryotes

We saw in Chapter 6 that mRNA production in eucaryotes occurs by an elaborately choreographed series of synthesis and processing steps. Only when all of the steps of mRNA production have been completed are the mRNAs exported from the nucleus to the cytosol for translation into protein. If any of those steps go awry, the RNA is eventually degraded in the nucleus (along with excised introns) by the *exosome,* a large protein complex that contains at least ten 3′-to-5′ RNA exonucleases. The eucaryotic cell has an additional mechanism, called **nonsense-mediated mRNA decay,** that eliminates certain types of aberrant mRNAs before they can be efficiently translated into protein. This mechanism was discovered when mRNAs that contain misplaced in-frame translation stop codons (UAA, UAG, or UGA) were found to be rapidly degraded. These stop codons can arise either from mutation or from incomplete splicing: in both cases, the phenomenon was observed. This mRNA surveillance system therefore prevents the synthesis of abnormally truncated proteins which, as we have seen, can be especially dangerous to the cell. But how are these potentially harmful mRNAs recognized by the cell?

In vertebrates, the critical feature of mRNA that is sensed by the nonsense-mediated decay system is the spatial relationship between the first in-frame termination codon and the exon–exon boundaries formed by RNA splicing. If the stop codon lies downstream (3′) of all the exon–exon boundaries, the mRNA is spared from nonsense-mediated decay; if, on the other hand, a stop codon is located upstream (5′) to an exon–exon boundary, the mRNA is degraded. Translating ribosomes, in conjunction with other surveillance proteins, assess this relationship for each individual mRNA. Exactly how this is accomplished is not understood in detail, but it is easy to understand why ribosomes must play a part: only in-frame termination codons trigger nonsense-mediated decay, and it is the relationship between the ribosome and the mRNA that defines the reading frame. According to one model (Figure 7–106), proteins in the nucleus bind to and thereby mark the exon–exon junctions following RNA splicing. As the mRNA leaves the nucleus, it remains at the nuclear periphery and is joined by a set of additional surveillance proteins as translation begins. The first round of translation of an individual mRNA molecule would, in this view, be used simply to assess the fitness of the mRNA for further rounds of translation. If the mRNA passes this test, translation begins in earnest as the mRNA is released to diffuse through the cytosol.

Nonsense-mediated decay may have been especially important in evolution, allowing eucaryotic cells to more easily explore new genes formed by DNA rearrangements, mutations, or alternative patterns of splicing—by selecting only those mRNAs for translation that produce a full-length protein. Nonsense-mediated decay is also important in cells of the developing immune system, where the extensive DNA rearrangements that occur (see Figure 24–37) often generate premature termination codons. The mRNAs produced from such rearranged genes are degraded by this surveillance system, thereby avoiding the toxic effects of truncated proteins.

RNA Interference Is Used by Cells to Silence Gene Expression

Eucaryotic cells use a specialized type of RNA degradation as a defense mechanism to destroy foreign RNA molecules, specifically those that can be identified by virtue of their occurrence within the cell in double-stranded form. Termed **RNA interference** (**RNAi**), this mechanism is found in a wide variety of organisms, including single-celled fungi, plants, worms, mice, and probably humans—suggesting that it is an evolutionarily ancient defense mechanism. In plants, RNA interference protects cells against RNA viruses. In other types of organisms, it is thought to protect against the proliferation of transposable elements that replicate via RNA intermediates (see Figure 5–76). Many transposable elements and plant viruses produce double-stranded RNA, at least transiently, in their life cycles. RNAi not only helps to keep such infestations in check, but also provides scientists with a powerful experimental technique to turn off the expression of individual cellular genes (see Figure 8–65).

The presence of free, double-stranded RNA triggers RNAi by attracting a protein complex containing an RNA nuclease and an RNA helicase. This protein complex cleaves the double-stranded RNA into small (approximately 23 nucleotide pair) fragments which remain associated with the enzyme. The bound RNA fragments then direct the enzyme complex to other RNA molecules

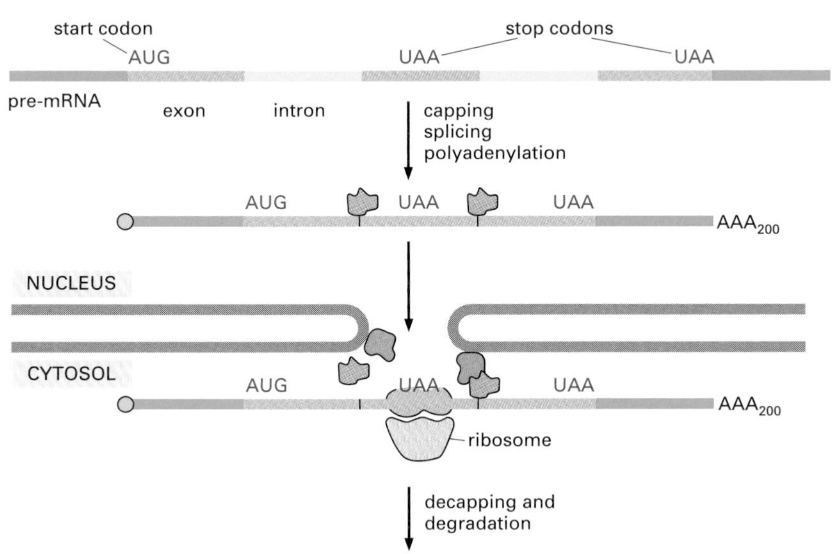

Figure 7–106 A model for nonsense-mediated mRNA decay. According to this model, nuclear proteins *(orange)* mark the exon–exon boundaries on a spliced mRNA molecule. These proteins are thought to assemble in concert with the splicing reaction and may also be involved in the transport of mature mRNAs from the nucleus (see Figure 6–40). A mature mRNA molecule is exported from the nucleus but remains in the vicinity of the nuclear envelope where a "test" round of translation is performed by the ribosome *(green)* aided by additional surveillance proteins *(dark green)*. If an in-frame stop codon is encountered before the final exon–exon boundary is reached, the mRNA is subject to nonsense-mediated decay. If not, the mRNA is released from the nuclear envelope (perhaps because of a displacement of the exon–exon marking proteins by the ribosome) and is free to undergo multiple rounds of translation in the cytosol. According to the model shown, the test round of translation occurs just outside the nucleus; however, it is also possible that it takes place within the nucleus, just before the mRNA is exported. (Adapted from J. Lykke-Anderson et al., *Cell* 103:1121–1131, 2000.)

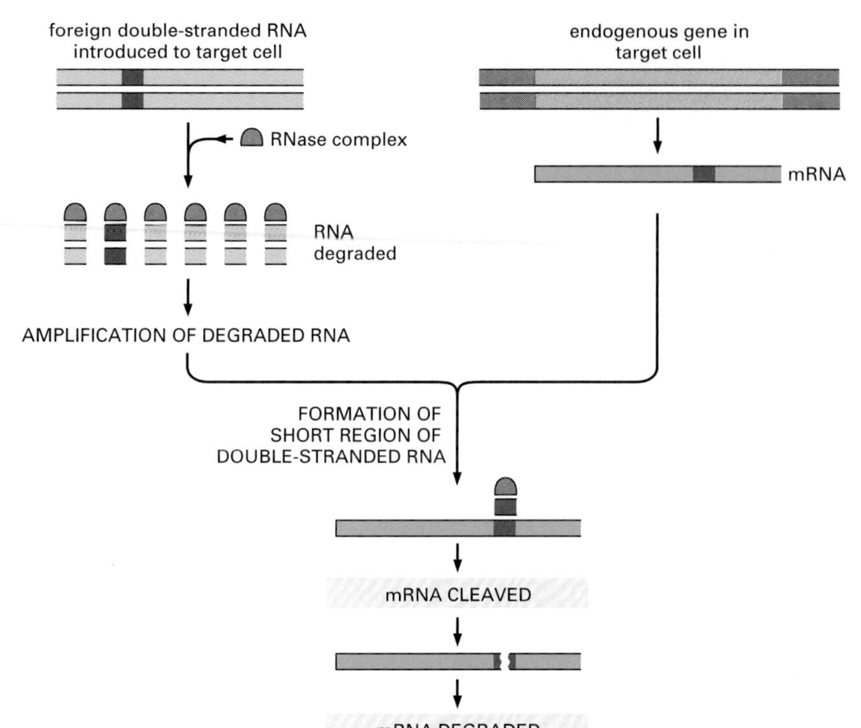

Figure 7–107 The mechanism of RNA interference. On the left is shown the fate of foreign double-stranded RNA molecules. They are recognized by an RNase, present in a large protein complex, and degraded into short fragments that are approximately 23 nucleotide pairs in length. These fragments are sometimes amplified by an RNA-dependent RNA polymerase and, in this case, can be efficiently transmitted to progeny cells. If the foreign RNA has a nucleotide sequence similar to that of a cellular gene (*right* side of figure), mRNA produced by this gene will also be degraded, by the pathway shown. In this way, the expression of a cellular gene can be experimentally shut off by introducing double-stranded RNA into the cell that matches the nucleotide sequence of the gene. RNA interference also requires ATP hydrolysis and RNA helicases, probably to produce single-stranded RNA molecules that can form base pairs with additional RNA molecules.

that have complementary nucleotide sequences, and the enzyme degrades these as well. These other molecules can be either single- or double-stranded (as long as they have a complementary strand). In this way, the experimental introduction of a double-stranded RNA molecule can be used by scientists to inactivate specific cellular mRNAs (Figure 7–107).

Each time it cleaves a new RNA, the enzyme complex is regenerated with a short RNA molecule, so that an original double-stranded RNA molecule can act catalytically to destroy many complementary RNAs. In addition, the short double-stranded RNA cleavage products themselves can be replicated by additional cellular enzymes, providing an even greater amplification of RNA interference activity (see Figure 7–107). This amplification ensures that once initiated, RNA interference can continue even after all the initiating double-stranded RNA has been degraded or diluted out. For example, it permits progeny cells to continue carrying out RNA interference that was provoked in the parent cells. In addition, the RNA interference activity can be spread by the transfer of RNA fragments from cell to cell. This is particularly important in plants (whose cells are linked by fine connecting channels, as discussed in Chapter 19), because it allows an entire plant to become resistant to an RNA virus after only a few of its cells have been infected.

Summary

Many steps in the pathway from RNA to protein are regulated by cells to control gene expression. Most genes are thought to be regulated at multiple levels, although control of the initiation of transcription (transcriptional control) usually predominates. Some genes, however, are transcribed at a constant level and turned on and off solely by posttranscriptional regulatory processes. These processes include (1) attenuation of the RNA transcript by its premature termination, (2) alternative RNA splice-site selection, (3) control of 3′-end formation by cleavage and poly-A addition, (4) RNA editing, (5) control of transport from the nucleus to the cytosol, (6) localization of mRNAs to particular parts of the cell, (7) control of translation initiation, and (8) regulated mRNA degradation. Most of these control processes require the recognition of specific sequences or structures in the RNA molecule being regulated. This recognition is accomplished by either a regulatory protein or a regulatory RNA molecule.

HOW GENOMES EVOLVE

In this and the preceding three chapters, we discussed the structure of genes, the way they are arranged in chromosomes, the intricate cellular machinery that converts genetic information into functional protein and RNA molecules, and the many ways in which gene expression is regulated by the cell. In this section, we discuss some of the ways that genes and genomes have evolved over time to produce the vast diversity of modern-day life forms on our planet. Genome sequencing has revolutionized our view of this process of *molecular evolution*, uncovering an astonishing wealth of information about the family relationships among organisms and evolutionary mechanisms.

It is perhaps not surprising that genes with similar functions can be found in a diverse range of living things. But the great revelation of the past 20 years has been the discovery that the actual nucleotide sequences of many genes are sufficiently well conserved that **homologous** genes—that is, genes that are similar in their nucleotide sequence because of a common ancestry—can often be recognized across vast phylogenetic distances. For example, unmistakable homologs of many human genes are easy to detect in such organisms as nematode worms, fruit flies, yeasts, and even bacteria.

As discussed in Chapter 3 and again in Chapter 8, the recognition of sequence homology has become a major tool for inferring gene and protein function. Although finding such a homology does not guarantee similarity in function, it has proven to be an excellent clue. Thus, it is often possible to predict the function of a gene in humans for which no biochemical or genetic information is available simply by comparing its sequence to that of an intensively studied gene in another organism.

Gene sequences are often far more tightly conserved than is overall genome structure. As discussed in Chapter 4, features of genome organization such as genome size, number of chromosomes, order of genes along chromosomes, abundance and size of introns, and amount of repetitive DNA are found to differ greatly among organisms, as does the actual number of genes.

The number of genes is only very roughly correlated with the phenotypic complexity of an organism. Thus, for example, current estimates of gene number are 6,000 for the yeast *Saccharomyces cerevisiae*, 18,000 for the nematode *Caenorhabditis elegans*, 13,000 for *Drosophila melanogaster*, and 30,000 for humans (see Table 1–1). As we shall soon see, much of the increase in gene number with increasing biological complexity involves the expansion of families of closely related genes, an observation that establishes gene duplication and divergence as major evolutionary processes. Indeed, it is likely that all present-day genes are descendants—via the processes of duplication, divergence, and reassortment of gene segments—of a few ancestral genes that existed in early life forms.

Genome Alterations are Caused by Failures of the Normal Mechanisms for Copying and Maintaining DNA

With a few exceptions, cells do not have specialized mechanisms for creating changes in the structures of their genomes: evolution depends instead on accidents and mistakes. Most of the genetic changes that occur result simply from failures in the normal mechanisms by which genomes are copied or repaired when damaged, although the movement of transposable DNA elements also plays an important role. As we discussed in Chapter 5, the mechanisms that maintain DNA sequences are remarkably precise—but they are not perfect. For example, because of the elaborate DNA-replication and DNA-repair mechanisms that enable DNA sequences to be inherited with extraordinary fidelity, only about one nucleotide pair in a thousand is randomly changed every 200,000 years. Even so, in a population of 10,000 individuals, every possible nucleotide substitution will have been "tried out" on about 50 occasions in the course of a million years—a short span of time in relation to the evolution of species.

Errors in DNA replication, DNA recombination, or DNA repair can lead either to simple changes in DNA sequence—such as the substitution of one base pair for another—or to large-scale genome rearrangements such as deletions, duplications, inversions, and translocations of DNA from one chromosome to another. It has been argued that the rates of occurrence of these mistakes have themselves been shaped by evolutionary processes to provide an acceptable balance between genome stability and change.

In addition to failures of the replication and repair machinery, the various mobile DNA elements described in Chapter 5 are an important source of genomic change. In particular, transposable DNA elements (transposons) play a major part as parasitic DNA sequences that colonize a genome and can spread within it. In the process, they often disrupt the function or alter the regulation of existing genes; and sometimes they even create altogether novel genes through fusions between transposon sequences and segments of existing genes. Examples of the three major classes of transposons were presented in Table 5–3, p. 287. Over long periods of evolutionary time, these transposons have profoundly affected the structure of genomes.

The Genome Sequences of Two Species Differ in Proportion to the Length of Time That They Have Separately Evolved

The differences between the genomes of species alive today have accumulated over more than 3 billion years. Lacking a direct record of changes over time, we can nevertheless reconstruct the process of genome evolution from detailed comparisons of the genomes of contemporary organisms.

The basic tool of comparative genomics is the phylogenetic tree. A simple example is the tree describing the divergence of humans from the great apes (Figure 7–108). The primary support for this tree comes from comparisons of gene and protein sequences. For example, comparisons between the sequences of human genes or proteins and those of the great apes typically reveal the fewest differences between human and chimpanzee and the most between human and orangutan.

For closely related organisms such as humans and chimpanzees, it is possible to reconstruct the gene sequences of the extinct, last common ancestor of the two species (Figure 7–109). The close similarity between human and chimpanzee genes is mainly due to the short time that has been available for the accumulation of mutations in the two diverging lineages, rather than to functional constraints that have kept the sequences the same. Evidence for this view comes from the observation that even DNA sequences whose nucleotide order is functionally unconstrained—such as the sequences that code for the fibrinopeptides (see p. 236) or the third position of "synonymous" codons (codons specifying the same amino acid—see Figure 7–109)—are nearly identical.

For less closely related organisms such as humans and mice, the sequence conservation found in genes is largely due to **purifying selection** (that is, selection that eliminates individuals carrying mutations that interfere with important genetic functions), rather than to an inadequate time for mutations to occur. As

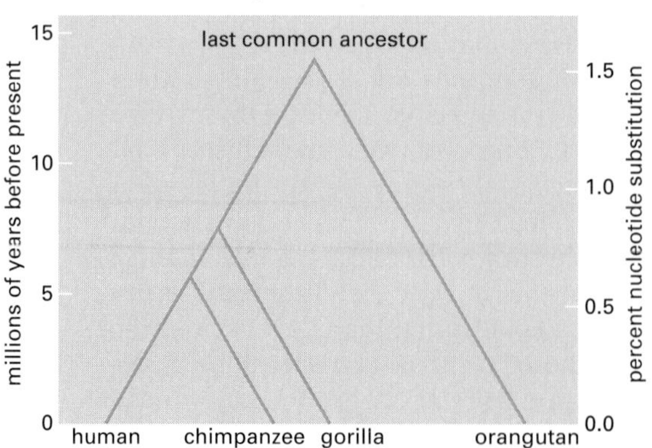

Figure 7–108 A phylogenetic tree showing the relationship between the human and the great apes based on nucleotide sequence data. As indicated, the sequences of the genomes of all four species are estimated to differ from the sequence of the genome of their last common ancestor by a little over 1.5%. Because changes occur independently on both diverging lineages, pairwise comparisons reveal twice the sequence divergence from the last common ancestor. For example, human–orangutan comparisons typically show sequence divergences of a little over 3%, while human–chimpanzee comparisons show divergences of approximately 1.2%. (Modified from F.-C. Chen and W.-H. Li, *Am. J. Hum. Genet.* 68:444–456, 2001.)

As indicated by the DNA sequence comparison in Figure 7–110, mutation has led to extensive sequence divergence between humans and mice at all sites that are not under selection—such as the nucleotide sequences of introns. Indeed, human–mouse-sequence comparisons are much more informative of the functional constraints on genes than are human–chimpanzee comparisons. In the latter case, nearly all sequence positions are the same simply because not enough time has elapsed since the last common ancestor for large numbers of changes to have occurred. In contrast, because of functional constraints in human–mouse comparisons the exons in genes stand out as small islands of conservation in a sea of introns.

As the number of sequenced genomes increases, comparative genome analysis is becoming an increasingly important method for identifying their functionally important sites. For example, conservation of open-reading frames between distantly related organisms provides much stronger evidence that these sequences are actually the exons of expressed genes than does a computational analysis of any one genome. In the future, detailed biological annotation of the sequences of complex genomes—such as those of the human and the mouse—will depend heavily on the identification of sequence features that are conserved across multiple, distantly related mammalian genomes.

In contrast to the situation for humans and chimpanzees, local gene order and overall chromosome organization have diverged greatly between humans and mice. According to rough estimates, a total of about 180 break-and-rejoin events have occurred in the human and mouse lineages since these two species last shared a common ancestor. In the process, although the number of chromosomes is similar in the two species (23 per haploid genome in the human versus 20 in the mouse), their overall structures differ greatly. For example, while the centromeres occupy relatively central positions on most human chromosomes, they lie next to an end of each chromosome in the mouse. Nonetheless, even after the extensive genomic shuffling, there are many large blocks of DNA in which the gene order is the same in the human and the mouse. These regions of conserved gene order in chromosomes are referred to as **synteny** blocks (see Figure 4–18).

Analysis of the transposon families in the human and the mouse provide additional evidence of the long divergence time separating the two species. Although the major retrotransposon families in the human have counterparts in the mouse—for example, human Alu repeats are similar in sequence and transposition mechanism to the mouse B1 family—the two families have undergone separate expansions in the two lineages. Even in regions where human and mouse sequences are sufficiently conserved to allow reliable alignment, there is no correlation between the positions of Alu elements in the human genome and the B1 elements in corresponding segments of the mouse genome (Figure 7–111).

It Is Difficult to Reconstruct the Structure of Ancient Genomes

The genomes of ancestral organisms can be inferred, but never directly observed: there are no ancient organisms alive today. Although a modern

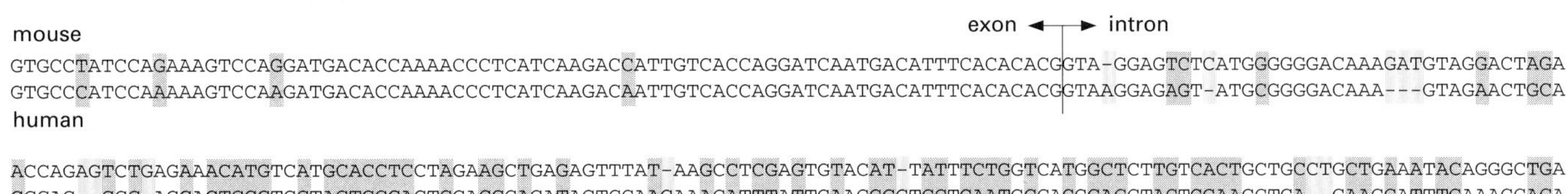

Figure 7–110 Comparison of a portion of the mouse and human leptin genes. Positions where the sequences differ by a single nucleotide substitution are boxed in *green*, and positions that differ by the addition or deletion of nucleotides are boxed in *yellow*. Note that the coding sequence of the exon is much more conserved than the adjacent intron sequence.

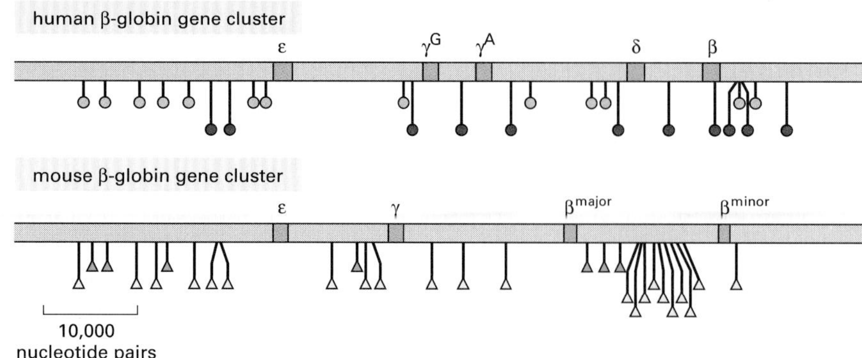

Figure 7–111 A comparison of the β-globin gene cluster in the human and mouse genomes, showing the location of transposable elements. This stretch of human genome contains five functional β-globin-like genes *(orange);* the comparable region from the mouse genome has only four. The positions of the human *Alu* sequence are indicated by *green circles*, and the human *L1* sequences by *red circles*. The mouse genome contains different but related transposable elements: the positions of *B1* elements (which are related to the human Alu sequences) are indicated by *blue triangles*, and the positions of the mouse *L1* elements (which are related to the human *L1* sequences) are indicated by *yellow triangles*. The absence of transposable elements from the globin structural genes can be attributed to purifying selection, which would have eliminated any insertion that compromised gene function. (Courtesy of Ross Hardison and Webb Miller.)

organism such as the horseshoe crab looks remarkably similar to fossil ancestors that lived 200 million years ago, there is every reason to believe that the horseshoe-crab genome has been changing during all that time at a rate similar to that occurring in other evolutionary lineages. Selective constraints must have maintained key functional properties of the horseshoe-crab genome to account for the morphological stability of the lineage. However, genome sequences reveal that the fraction of the genome subject to purifying selection is small; hence the genome of the modern horseshoe crab must differ greatly from that of its extinct ancestors, known to us only through the fossil record.

It is difficult to infer even gross features of the genomes of long-extinct organisms. An important example is the so-called introns-early versus introns-late controversy. Soon after the discovery in 1977 that the coding regions of most genes in metazoan organisms are interrupted by introns, a debate arose about whether introns reflect a late acquisition during the evolution of life on earth or whether they were instead present in the earliest genes. According to the introns-early model, fast-growing organisms such as bacteria lost the introns present in their ancestors because they were under selection for a compact genome adapted for rapid replication. This view is contested by an introns-late model, in which introns are viewed as having been inserted into intronless genes long after the evolution of single-cell organisms, perhaps through the agency of certain types of transposons.

There is presently no reliable way of resolving this controversy. Comparative studies of existing genomes provide estimates of rates of intron gain and loss in various evolutionary lineages. However, these estimates bear only indirectly on the question of how genomes were organized billions of years ago. Bacteria and humans are equally "modern" organisms, both of whose genomes differ so greatly from that of their last common ancestor that we can only speculate about the properties of this very ancient, ancestral genome.

When two modern organisms share nearly identical patterns of intron positions in their genes, we can be confident that the introns were present in the last common ancestor of the two species. An illuminating comparison involves humans and the puffer fish, *Fugu rubripes* (Figure 7–112). The *Fugu* genome is remarkable in having an unusually small size for a vertebrate (0.4 billion nucleotide pairs compared to 1 billion or more for many other fish and 3 billion for typical mammals). The small size of the *Fugu* genome is due almost entirely to the small size of its introns. Specifically, *Fugu* introns, as well as other noncoding segments of the *Fugu* genome, lack the repetitive DNA that makes up a large portion of the genomes of most well studied vertebrates. Nevertheless, the positions of Fugu introns are nearly perfectly conserved relative to their positions in mammalian genomes (Figure 7–113).

The question of why *Fugu* introns are so small is reminiscent of the introns-early versus introns-late debate. Obviously, either introns grew in many lineages while staying small in the *Fugu* lineage, or the *Fugu* lineage experienced massive loss of repetitive sequences from its introns. We have a clear understanding of how genomes can grow by active transposition since most transposition events are duplicative [*i.e.*, the original copy stays where it was while a copy inserts at

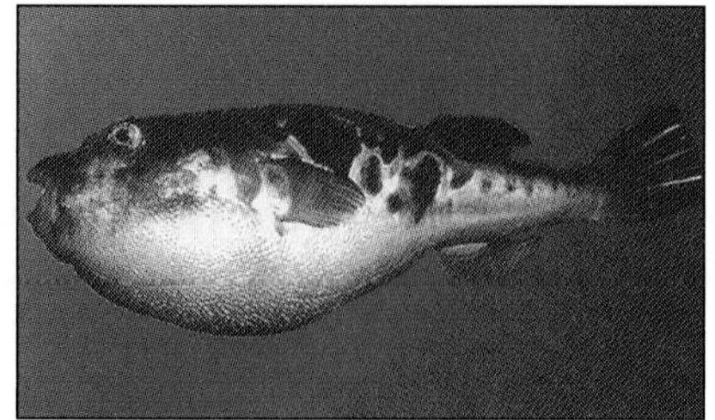

Figure 7–112 The puffer fish, *Fugu rubripes*. (Courtesy of Byrappa Venkatesh.)

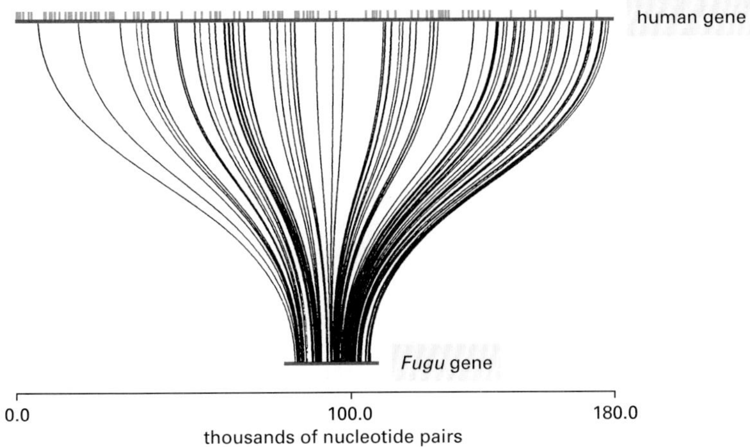

Figure 7–113 Comparison of the genomic sequences of the human and *Fugu* genes encoding the protein huntingtin. Both genes (indicated in *red*) contain 67 short exons that align in 1:1 correspondence to one another; these exons are connected by curved lines. The human gene is 7.5 times larger than the *Fugu* gene (180,000 versus 24,000 nucleotide pairs). The size difference is entirely due to larger introns in the human gene. The larger size of the human introns is due in part to the presence of retrotransposons, whose positons are represented by *green vertical lines;* the *Fugu* introns lack retrotransposons. In humans, mutation of the huntingtin gene causes Huntington's disease, an inherited neurodegenerative disorder (see p. 362). (Adapted from S. Baxendale et al., *Nat. Genet.* 10:67–76, 1995.)

the new site (see Figures 5–72 and 5–76)]. There is considerably less evidence in well-studied organisms for mutational processes that would efficiently delete transposons from immense numbers of sites without also deleting adjacent functionally critical sequences at rates that would threaten the survival of the lineage. Nonetheless, the origin of *Fugu*'s unusually small introns remains uncertain.

Gene Duplication and Divergence Provide a Critical Source of Genetic Novelty During Evolution

Much of our discussion of genome evolution so far has emphasized neutral change processes or the effects of purifying selection. However, the most important feature of genome evolution is the capacity for genomic change to create biological novelty that can be positively selected for during evolution, giving rise to new types of organisms.

Comparisons between organisms that seem very different illuminate some of the sources of genetic novelty. A striking feature of these comparisons is the relative scarcity of lineage-specific genes (for example, genes found in primates but not in rodents, or those found in mammals but not in other vertebrates). Much more prominent are selective expansions of preexisting gene families. The genes encoding nuclear hormone receptors in humans, a nematode worm, and a fruit fly, all of which have fully sequenced genomes, illustrate this point (Figure 7–114). Many of the subtypes of these nuclear receptors (also called intracellular receptors) have close homologs in all three organisms that are more similar to each other than they are to other family subtypes present in the same species. Therefore, much of the functional divergence of this large gene family must have preceded the divergence of these three evolutionary lineages. Subsequently, one major branch of the gene family underwent an enormous expansion only in the worm lineage. Similar, but smaller lineage-specific expansions of particular subtypes are evident throughout the gene family tree, but they are particularly evident in the human—suggesting that such expansions offer a path toward increased biological complexity.

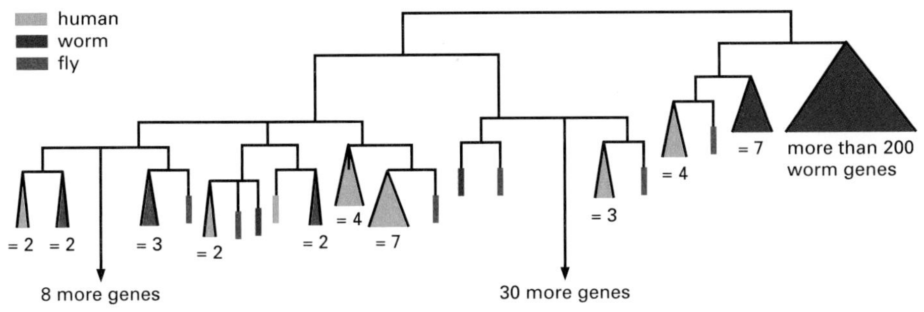

Figure 7–114 A phylogenetic tree based on the inferred protein sequences for all nuclear hormone receptors encoded in the genomes of human *(H. sapiens)*, a nematode worm *(C. elegans)*, and a fruit fly *(D. melanogaster)*. *Triangles* represent protein subfamilies that have expanded within individual evolutionary lineages; the width of these triangles indicates the number of genes encoding members of these subfamilies. *Colored vertical bars* represent a single gene. There is no simple pattern to the historical duplications and divergences that have created the gene families encoding nuclear receptors in the three contemporary organisms. The structure of a portion of a particular nuclear hormone receptor is shown in Figure 7–14, and a general description of their functions is discussed in Chapter 15. (Adapted from International Human Genome Sequencing Consortium, *Nature* 409:860–921, 2001.)

Gene duplication appears to occur at high rates in all evolutionary lineages. An examination of the abundance and rate of divergence of duplicated genes in many different eucaryotic genomes suggests that the probability that any particular gene will undergo a successful duplication event (*i.e.*, one that spreads to most or all individuals in a species) is approximately 1% every million years. Little is known about the precise mechanism of gene duplication. However, because the two copies of the gene are often adjacent to one another immediately following duplication, it is thought that the duplication frequently results from inexact repair of double-strand chromosome breaks (see Figure 5–53).

Duplicated Genes Diverge

A major question in genome evolution concerns the fate of newly duplicated genes. In most cases, there is presumed to be little or no selection—at least initially—to maintain the duplicated state since either copy can provide an equivalent function. Hence, many duplication events are likely to be followed by loss-of-function mutations in one or the other gene. This cycle would functionally restore the one-gene state that preceded the duplication. Indeed, there are many examples in contemporary genomes where one copy of a duplicated gene can be seen to have become irreversibly inactivated by multiple mutations. Over time, the sequence similarity between such a **pseudogene** and the functional gene whose duplication produced it would be expected to be eroded by the accumulation of many mutational changes in the pseudogene—eventually becoming undetectable.

An alternative fate for gene duplications is for both copies to remain functional, while diverging in their sequence and pattern of expression and taking on different roles. This process of "duplication and divergence" almost certainly explains the presence of large families of genes with related functions in biologically complex organisms, and it is thought to play a critical role in the evolution of increased biological complexity.

Whole-genome duplications offer particularly dramatic examples of the duplication-divergence cycle. A whole-genome duplication can occur quite simply: all that is required is one round of genome replication in a germline cell lineage without a corresponding cell division. Initially, the chromosome number simply doubles. Such abrupt increases in the **ploidy** of an organism are common, particularly in fungi and plants. After a whole-genome duplication, all genes exist as duplicate copies. However, unless the duplication event occurred so recently that there has been little time for subsequent alterations in genome structure, the results of a series of segmental duplications—occurring at different times—are very hard to distinguish from the end product of a whole-genome duplication. In the case of mammals, for example, the role of whole genome duplications versus a series of piecemeal duplications of DNA segments is quite uncertain. Nevertheless, it is clear that a great deal of gene duplication has ocurred in the distant past.

Analysis of the genome of the zebrafish, in which either a whole-genome duplication or a series of more local duplications occurred hundreds of millions of years ago, has cast some light on the process of gene duplication and divergence. Although many duplicates of zebrafish genes appear to have been lost by mutation, a significant fraction—perhaps as many as 30–50%—have diverged functionally while both copies have remained active. In many cases, the most obvious functional difference between the duplicated genes is that they are expressed in different tissues or at different stages of development (see Figure 21–45). One attractive theory to explain such an end result imagines that different, mildly deleterious mutations quickly occur in both copies of a duplicated gene set. For example, one copy might lose expression in a particular tissue due to a regulatory mutation, while the other copy loses expression in a second tissue. Following such an occurrence, both gene copies would be required to provide the full range of functions that were once supplied by a single gene; hence, both copies would now be protected from loss through inactivating mutations. Over a longer period of time, each copy could then undergo further changes through which it could acquire new, specialized features.

Figure 7–115 A comparison of the structure of one-chain and four-chain globins. The four-chain globin shown is hemoglobin, which is a complex of two α- and β-globin chains. The one-chain globin in some primitive vertebrates forms a dimer that dissociates when it binds oxygen, representing an intermediate in the evolution of the four-chain globin.

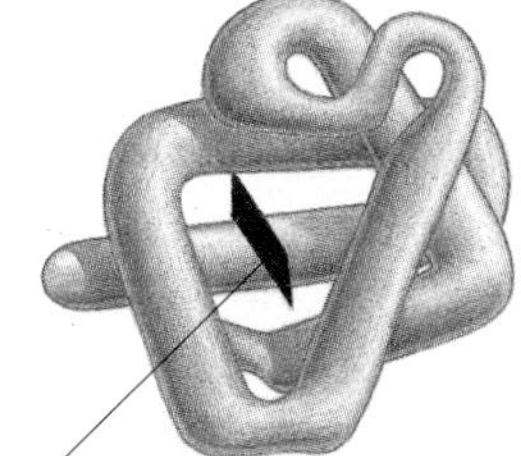

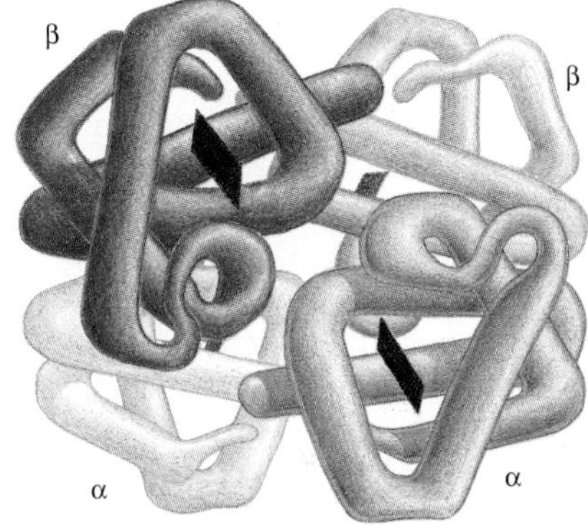

The Evolution of the Globin Gene Family Shows How DNA Duplications Contribute to the Evolution of Organisms

The globin gene family provides a particularly good example of how DNA duplication generates new proteins, because its evolutionary history has been worked out particularly well. The unmistakable homologies in amino acid sequence and structure among the present-day globins indicate that they all must derive from a common ancestral gene, even though some are now encoded by widely separated genes in the mammalian genome.

We can reconstruct some of the past events that produced the various types of oxygen-carrying hemoglobin molecules by considering the different forms of the protein in organisms at different positions on the phylogenetic tree of life. A molecule like hemoglobin was necessary to allow multicellular animals to grow to a large size, since large animals could no longer rely on the simple diffusion of oxygen through the body surface to oxygenate their tissues adequately. Consequently, hemoglobin-like molecules are found in all vertebrates and in many invertebrates. The most primitive oxygen-carrying molecule in animals is a globin polypeptide chain of about 150 amino acids, which is found in many marine worms, insects, and primitive fish. The hemoglobin molecule in higher vertebrates, however, is composed of two kinds of globin chains. It appears that about 500 million years ago, during the evolution of higher fish, a series of gene mutations and duplications occurred. These events established two slightly different globin genes, coding for the α- and β-globin chains in the genome of each individual. In modern higher vertebrates each hemoglobin molecule is a complex of two α chains and two β chains (Figure 7–115). The four oxygen-binding sites in the $\alpha_2\beta_2$ molecule interact, allowing a cooperative allosteric change in the molecule as it binds and releases oxygen, which enables hemoglobin to take up and to release oxygen more efficiently than the single-chain version.

Still later, during the evolution of mammals, the β-chain gene apparently underwent duplication and mutation to give rise to a second β-like chain that is synthesized specifically in the fetus. The resulting hemoglobin molecule has a higher affinity for oxygen than adult hemoglobin and thus helps in the transfer of oxygen from the mother to the fetus. The gene for the new β-like chain subsequently mutated and duplicated again to produce two new genes, ε and γ, the ε chain being produced earlier in development (to form $\alpha_2\varepsilon_2$) than the fetal γ chain, which forms $\alpha_2\gamma_2$. A duplication of the adult β-chain gene occurred still later, during primate evolution, to give rise to a δ-globin gene and thus to a minor form of hemoglobin ($\alpha_2\delta_2$) found only in adult primates (Figure 7–116).

Each of these duplicated genes has been modified by point mutations that affect the properties of the final hemoglobin molecule, as well as by changes in regulatory regions that determine the timing and level of expression of the gene. As a result, each globin is made in different amounts at different times of human development (see Figure 7–60B).

The end result of the gene duplication processes that have given rise to the diversity of globin chains is seen clearly in the human genes that arose from the original β gene, which are arranged as a series of homologous DNA sequences located within 50,000 nucleotide pairs of one another. A similar cluster of α-globin genes is located on a separate human chromosome. Because the α- and β-globin gene clusters are on separate chromosomes in birds and mammals but are together in the frog *Xenopus,* it is believed that a chromosome translocation event separated the two gene clusters about 300 million years ago (see Figure 7–116).

There are several duplicated globin DNA sequences in the α- and β-globin gene clusters that are not functional genes, but pseudogenes. These have a close homology to the functional genes but have been disabled by mutations that prevent their expression. The existence of such pseudogenes make it clear

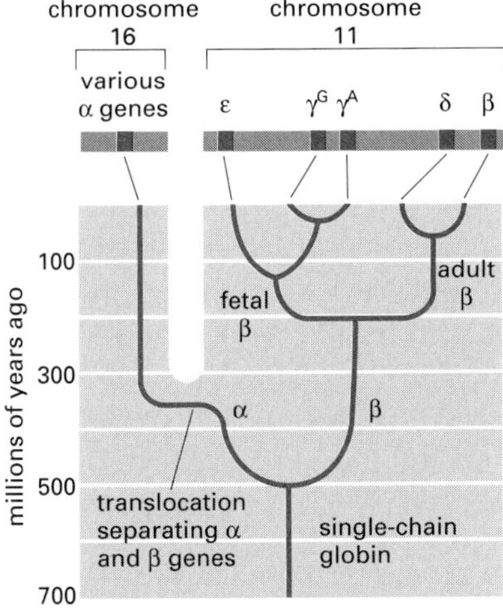

Figure 7–116 An evolutionary scheme for the globin chains that carry oxygen in the blood of animals. The scheme emphasizes the β-like globin gene family. A relatively recent gene duplication of the γ-chain gene produced γ^G and γ^A, which are fetal β-like chains of identical function. The location of the globin genes in the human genome is shown at the top of the figure (see also Figure 7–60).

that, as expected, not every DNA duplication leads to a new functional gene. We also know that nonfunctional DNA sequences are not rapidly discarded, as indicated by the large excess of noncoding DNA that is found in mammalian genomes.

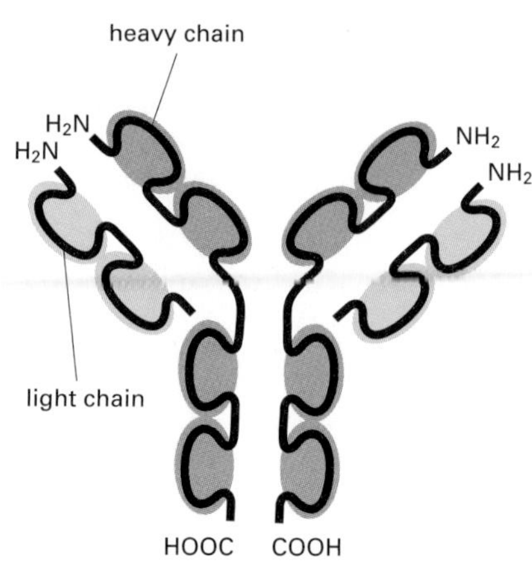

Figure 7–117 Schematic view of an antibody (immunoglobulin) molecule. This molecule is a complex of two identical heavy chains and two identical light chains. Each heavy chain contains four similar, covalently linked domains, while each light chain contains two such domains. Each domain is encoded by a separate exon, and all of the exons are thought to have evolved by the serial duplication of a single ancestral exon.

Genes Encoding New Proteins Can Be Created by the Recombination of Exons

The role of DNA duplication in evolution is not confined to the expansion of gene families. It can also act on a smaller scale to create single genes by stringing together short, duplicated segments of DNA. The proteins encoded by genes generated in this way can be recognized by the presence of repeating, similar protein domains, which are covalently linked to one another in series. The immunoglobulins (Figure 7–117) and albumins, for example, as well as most fibrous proteins (such as collagens) are encoded by genes that have evolved by repeated duplications of a primordial DNA sequence.

In genes that have evolved in this way, as well as in many other genes, each separate exon often encodes an individual protein folding unit, or domain. It is believed that the organization of DNA coding sequences as a series of such exons separated by long introns has greatly facilitated the evolution of new proteins. The duplications necessary to form a single gene coding for a protein with repeating domains, for example, can occur by breaking and rejoining the DNA anywhere in the long introns on either side of an exon encoding a useful protein domain; without introns there would be only a few sites in the original gene at which a recombinational exchange between DNA molecules could duplicate the domain. By enabling the duplication to occur by recombination at many potential sites rather than just a few, introns increase the probability of a favorable duplication event.

More generally, we know from genome sequences that component parts of genes—both their individual exons and their regulatory elements—have served as modular elements that have been duplicated and moved about the genome to create the present great diversity of living things. As a result, many present-day proteins are formed as a patchwork of domains from different domain families, reflecting their long evolutionary history (Figure 7–118).

Genome Sequences Have Left Scientists with Many Mysteries to Be Solved

Now that we know from genome sequences that a human and a mouse contain essentially the same genes, we are forced to confront one of the major problems that will challenge cell biologists throughout the next century. Given that a human and a mouse are formed from the same set of proteins, what has happened during the evolutionary process to make a mouse and a human so different? Although the answer is present somewhere among the three billion nucleotides in each sequenced genome, we do not yet know how to decipher this type of information—so that the answer to this critical, most fundamental question is not known.

Despite our ignorance, it is perhaps worth engaging in a bit of speculation, if only to help point the way forward to some of the hard problems ahead. In biology, timing is everything, as will become clear when we examine the elaborate mechanisms that allow a fertilized egg to develop into an embryo, and the embryo to develop into an adult (discussed in Chapter 21). The human body is formed as the result of many billions of decisions that are made during our development as to which RNA molecule and which protein are to be made where, as well as exactly when and in what amount each is to be produced.

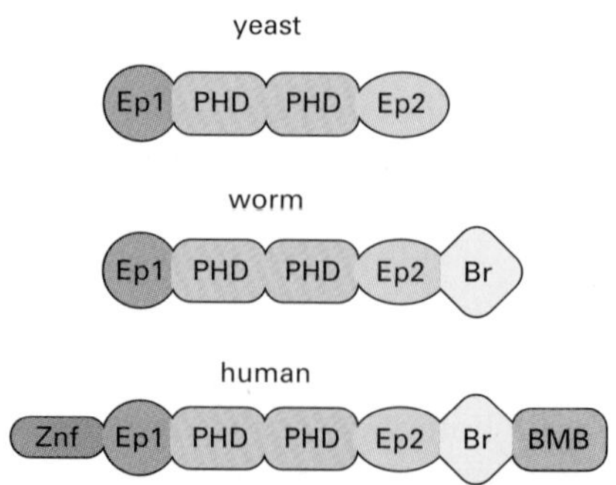

Figure 7–118 Domain structure of a group of evolutionary related proteins that are thought to have a similar function. In general, there is a tendency for the proteins in more complex organisms, such as ourselves, to contain additional domains—as is the case for the DNA-binding protein compared here.

These decisions are different for a human than for a chimpanzee or a mouse. The coding sequences of genomes represent a more or less standard set of the 30,000 or so basic parts from which all three organisms are made. It is therefore the many different types of controls on gene expression described in this Chapter that must largely create the difference between a human and other mammals.

Given these assumptions, it would be reasonable to expect genomes to have evolved in a way that allows organisms to experiment with altered gene timing and expression patterns in selected cells. We have already seen some evidence that this is so, when we discussed alternative RNA splicing and RNA editing mechanisms. There also appear to be mechanisms—some based on the movements of transposable DNA elements—that allow modules to be readily added to and subtracted from the regulatory regions of genes, so as to produce changes in the pattern of their transcription as organisms evolve. In fact, an analysis of these regulatory regions provides evidence to support the claim that most gene regulatory regions have been formed by the evolutionary mixing and matching of the DNA-binding sites that are recognized by gene regulatory proteins (Figure 7–119).

Genetic Variation within a Species Provides a Fine-Scale View of Genome Evolution

In comparisons between two species that have diverged from one another by millions of years, it makes little difference which individuals from each species are compared. For example, typical human and chimpanzee DNA sequences differ from one another by 1%. In contrast, when the same region of the genome is sampled from two different humans, the differences are typically less than 0.1%. For more distantly related organisms, the inter-species differences overshadow intra-species variation even more dramatically. However, each "fixed difference" between the human and the chimpanzee (*i.e.*, each difference that is now characteristic of all or nearly all individuals of each species) started out as a new mutation in a single individual. If the size of the interbreeding population in which the mutation occurred is N, the initial **allele frequency** of a new mutation would be ½N for a diploid organism. How does such a rare mutation become fixed in the population, and hence become a characteristic of the species rather than of a particular individual genome?

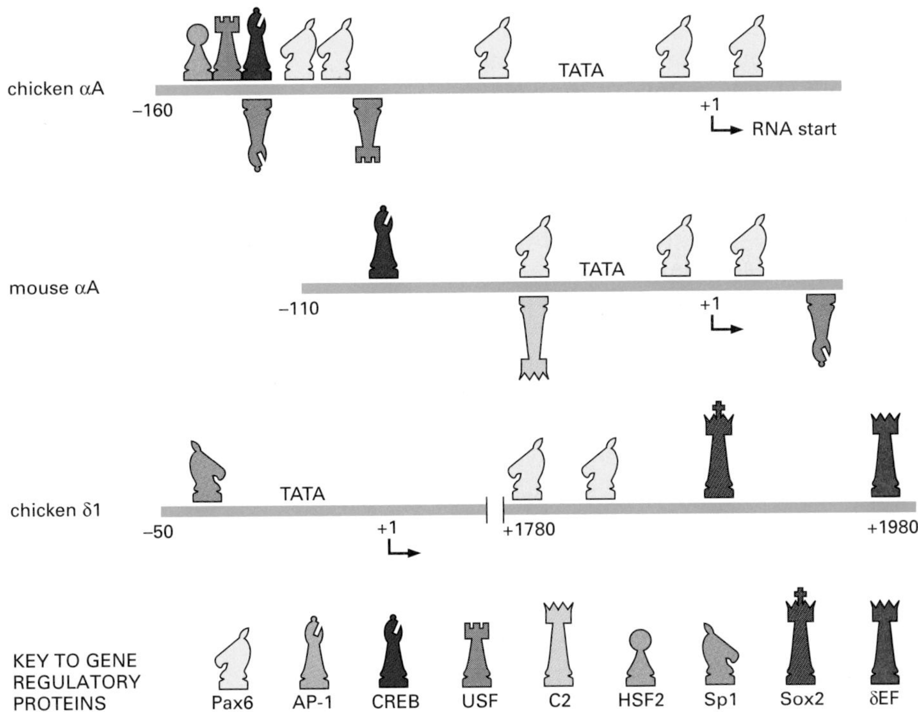

Figure 7–119 Gene control regions for mouse and chicken eye lens crystallins. Crystallins make up the bulk of the lens and are responsible for refracting and focusing light onto the retina. Many proteins in the cell have properties (high solubility, proper refractive index, etc.) suitable for lens function, and a wide variety of such proteins have been co-opted during evolution for use in the lens. For example, the α crystallins *(top two lines)* are closely related to heat shock proteins and are found in all vertebrate lenses. In contrast, δ crystallin *(third line)* is closely related to an enzyme involved in amino acid metabolism and is found only in birds and reptiles. The three crystallin gene control regions shown are a patchwork of different regulatory sequences that reflect the evolutionary history of each gene. The common feature of all three control regions is the presence of binding sites for the gene regulatory protein Pax6. Pax6 is the vertebrate homolog of the fly Toy and Eyeless proteins (see Figure 7–75) and is one of the key regulators that specifies eye development. Proteins above each gene control region are transcriptional activators and those below the line are repressors. (Adapted from E.H. Davidson, Genomic Regulatory Systems: Development and Evolution, pp. 191–201. San Diego: Academic Press, 2001 and A. Cvekl and J. Piatigarsky, *BioEssays* 18:621–630, 1996.)

The answer to this question depends on the functional consequences of the mutation. If the mutation has a significantly deleterious effect, it will simply be eliminated by purifying selection and will not become fixed. (In the most extreme case, the individual carrying the mutation will die without producing progeny.) Conversely, the rare mutations that confer a major reproductive advantage on individuals who inherit them will spread rapidly in the population. Because humans reproduce sexually and genetic recombination occurs each time a gamete is formed, the genome of each individual who has inherited the mutation will be a unique recombinational mosaic of segments inherited from a large number of ancestors. The selected mutation along with a modest amount of neighboring sequence—ultimately inherited from the individual in which the mutation occurred—will simply be one piece of this huge mosaic.

The great majority of mutations that are not harmful are not beneficial either. These *selectively neutral mutations* can also spread and become fixed in a population, and they make a large contribution to the evolutionary change in genomes. Their spread is not as rapid as the spread of the rare strongly advantageous mutations. The process by which such neutral genetic variation is passed down through an idealized interbreeding population can be described mathematically by equations that are surprisingly simple. The idealized model that has proven most useful for analyzing human genetic variation assumes a constant population size, and random mating, as well as selective neutrality for the mutations. While neither of these assumptions is a good description of human population history, they nonetheless provide a useful starting point for analyzing intra-species variation.

When a new neutral mutation occurs in a constant population of size N that is undergoing random mating, the probability that it will ultimately become fixed is approximately $½N$. For those mutations that do become fixed, the average time to fixation is approximately $4N$ generations. A detailed analysis of data on human genetic variation suggests an ancestral population size of approximately 10,000 during the period when the current pattern of genetic variation was largely established. Under these conditions, the probability that a new, selectively neutral mutation would become fixed was small (5×10^{-5}), while the average time to fixation was on the order of 800,000 years. Thus, while we know that the human population has grown enormously since the development of agriculture approximately 15,000 years ago, most human genetic variation arose and became established in the human population much earlier than this, when the human population was still small.

Even though most of the variation among modern humans originates from variation present in a comparatively tiny group of ancestors, the number of variations encountered is very large. Most of the variations take the form of **single-nucleotide polymorphisms (SNPs)**. These are simply points in the genome sequence where one large fraction of the human population has one nucleotide, while another large fraction has another. Two human genomes sampled from the modern world population at random will differ at approximately 2.5×10^6 sites (1 per 1300 nucleotide pairs). Mapped sites in the human genome that are polymorphic—meaning that there is a reasonable probability that the genomes of two individuals will differ at that site—are extremely useful for genetic analyses, in which one attempts to associate specific traits (phenotypes) with specific DNA sequences for medical or scientific purposes (see p. 531).

Against the background of ordinary SNPs inherited from our prehistoric ancestors, certain sequences with exceptionally high mutation rates stand out. A dramatic example is provided by *CA repeats,* which are ubiquitous in the human genome and in the genomes of other eucaryotes. Sequences with the motif $(CA)_n$ are replicated with relatively low fidelity because of a slippage that occurs between the template and the newly synthesized strands during DNA replication; hence, the precise value of n can vary over a considerable range from one genome to the next. These repeats make ideal DNA-based genetic markers, since most humans are heterozygous—carrying two values of n at any particular CA repeat, having inherited one repeat length *(n)* from their mother and a different repeat length from their father. While the value of n changes sufficiently rarely that most parent-child transmissions propagate CA repeats faithfully, the changes are sufficiently frequent to maintain high levels of heterozygosity in the human population. These and other simple repeats that display exceptionally high variability provide the basis for identifying individuals by DNA analysis in crime investigations, paternity suits, and other forensic applications (see Figure 8–41).

While most of the SNPs and other common variations in the human genome sequence are thought to have no effect on phenotype, a subset of them must be responsible for nearly all of the heritable aspects of human individuality. A major challenge in human genetics is to learn to recognize those relatively few variations that are functionally important—against the large background of neutral variation that distinguishes the genomes of any two human beings.

Summary

Comparisons of the nucleotide sequences of present-day genomes have revolutionized our understanding of gene and genome evolution. Due to the extremely high fidelity of DNA replication and DNA repair processes, random errors in maintaining the nucleotide sequences in genomes occur so rarely that only about 5 nucleotides in 1000 are altered every million years. Not surprisingly, therefore, a comparison of human and chimpanzee chromosomes—which are separated by about 5 million years of evolution—reveals very few changes. Not only are our genes essentially the same, but their order on each chromosome is almost identical. In addition, the positions of the transposable elements that make up a major portion of our noncoding DNA are mostly unchanged.

When one compares the genomes of two more distantly related organisms—such as a human and a mouse, separated by about 100 million years—one finds many more changes. Now the effects of natural selection can be clearly seen: through purifying selection, essential nucleotide sequences—both in regulatory regions and coding sequences (exon sequences)—have been highly conserved. In contrast, nonessential sequences (for example, intron sequences) have been altered to such an extent that an accurate alignment according to ancestry is often not possible.

Because of purifying selection, homologous genes can be recognized over large phylogenetic distances, and it is often possible to construct a detailed evolutionary history of a particular gene, tracing its history back to common ancestors of present-day species. We can thereby see that a great deal of the genetic complexity of present-day organisms is due to the expansion of ancient gene families. DNA duplication followed by sequence divergence has thus been a major source of genetic novelty during evolution.

References

General

Carey M & Smale ST (2000) Transcriptional Regulation in Eukaryotes: Concepts, Strategies and Techniques. Cold Spring Harbor, NY: Cold Spring Harbor Laboratory Press.

Hartwell L, Hood L, Goldberg ML et al. (2000) Genetics: from Genes to Genomes. Boston: McGraw Hill.

Lewin B (2000) Genes VII. Oxford: Oxford University Press.

Lodish H, Berk A, Zipursky SL et al. (2000) Molecular Cell Biology, 4th edn. New York: WH Freeman.

McKnight SL & Yamamoto KR (eds) (1992) Transcriptional Regulation. Cold Spring Harbor, NY: Cold Spring Harbor Laboratory Press.

Mechanisms of Transcription. (1998) *Cold Spring Harb. Symp. Quant. Biol.* 63.

An Overview of Gene Control

Campbell KH, McWhir J, Ritchie WA & Wilmut I (1996) Sheep cloned by nuclear transfer from a cultured cell line. *Nature* 380, 64–66.

Gurdon JB (1992) The generation of diversity and pattern in animal development. *Cell* 68, 185–199.

Ross DT, Scherf U, Eisen MB et al. (2000) Systematic variation in gene expression patterns in human cancer cell lines. *Nat. Genet.* 24, 227–235.

DNA-binding Motifs in Gene Regulatory Proteins

Bulger M & Groudine M (199) Looping versus linking: toward a model for long-distance gene activation. *Genes Dev.* 13, 2465–2477.

Choo Y & Klug A (1997) Physical basis of a protein–DNA recognition code. *Curr. Opin. Struct. Biol.* 7, 117–125.

Gehring WJ, Affolter M & Burglin T (1994) Homeodomain proteins. *Annu. Rev. Biochem.* 63, 487–526.

Harrison SC (1991) A structural taxonomy of DNA-binding domains. *Nature* 353, 715–719.

Jacob F & Monod J (1961) Genetic regulatory mechanisms in the synthesis of proteins. *J. Mol. Biol.* 3, 318–356.

Laity JH, Lee BM & Wright PE (2001) Zinc finger proteins: new insights into structural and functional diversity. *Curr. Opin. Struct. Biol.* 11, 39–46.

Lamb P & McKnight SL (1991) Diversity and specificity in transcriptional regulation: the benefits of heterotypic dimerization. *Trends Biochem. Sci.* 16, 417–422.

McKnight SL (1991) Molecular zippers in gene regulation. *Sci. Am.* 264, 54–64.

Muller CW (2001) Transcription factors: global and detailed views. *Curr. Opin. Struct. Biol.* 11, 26–32.

Orlando V (2000) Mapping chromosomal proteins *in vivo* by formaldehyde-crosslinked-chromatin. *Trends Biochem. Sci.* 25, 99–104.

Pabo CO & Sauer RT (1992) Transcription factors: structural families and principles of DNA recognition. *Annu. Rev. Biochem.* 61, 1053–1095.

Ptashne M (1992) A Genetic Switch, 2nd edn. Cambridge, MA: Cell Press and Blackwell Press.

Rhodes D, Schwabe JW, Chapman L et al. (1996) Towards an understanding of protein–DNA recognition. *Proc. R. Soc. Lond.* B 351, 501–509.

Seeman NC, Rosenberg JM & Rich A (1976) Sequence-specific recognition of double helical nucleic acids by proteins. *Proc. Natl Acad. Sci. USA* 73, 804–808.

Wolberger C (1996) Homeodomain interactions. *Curr. Opin. Struct. Biol.* 6, 62–68.

How Genetic Switches Work

Beckwith J (1987) The operon: an historical account. In *Escherichia coli* and *Salmonella typhimurium:* Cellular and Molecular Biology (Neidhart FC, Ingraham JL, Low KB et al. eds), vol 2, pp 1439–1443. Washington, DC: ASM Press.

Bell AC, West AG & Felsenfeld G (2001) Insulators and boundaries: versatile regulatory elements in the eukaryotic. *Science* 291, 447–450.

Buratowski S (2000) Snapshots of RNA polymerase II transcription initiation. *Curr. Opin. Cell Biol.* 12, 320–325.

Carey M (1998) The enhanceosome and transcriptional synergy. *Cell* 92, 5–8.

Fraser P & Grosveld F (1998) Locus control regions, chromatin activation and transcription. *Curr. Opin. Cell Biol.* 10, 361–365.

Kadonaga JT (1998) Eukaryotic transcription: an interlaced network of transcription factors and chromatin-modifying machines. *Cell* 92, 307–313.

Kercher MA, Lu P & Lewis M (1997) Lac repressor-operator complex. *Curr. Opin. Struct. Biol.* 7, 76–85.

Kornberg RD (1999) Eukaryotic transcriptional control. *Trends Cell Biol.* 9, M46–49.

Malik S & Roeder RG (2000) Transcriptional regulation through Mediator-like coactivators in yeast and metazoan cells. *Trends Biochem. Sci.* 25, 277–283.

Merika M & Thanos D (2001) Enhanceosomes. *Curr. Opin. Genet. Dev.* 11, 205–208.

Myers LC & Kornberg RD (2000) Mediator of transcriptional regulation. *Annu. Rev. Biochem.* 69, 729–749.

Ptashne M & Gann A (1998) Imposing specificity by localization: mechanism and evolvability. *Curr. Biol.* 8, R812–R822.

Schleif R (1992) DNA looping. *Annu. Rev. Biochem.* 61, 199–223.

St Johnston D & Nusslein-Volhard C (1992) The origin of pattern and polarity in the *Drosophila* embryo. *Cell* 68, 201–219.

Struhl K (1998) Histone acetylation and transcriptional regulatory mechanisms. *Genes Dev.* 12, 599–606.

The Molecular Genetic Mechanisms that Create Specialized Cell Types

Bird AP & Wolffe AP (1999) Methylation-induced repression—belts, braces, and chromatin. *Cell* 99, 451–454.

Cross SH & Bird AP (1995) CpG islands and genes. *Curr. Opin. Genet. Dev.* 5, 309–314.

Dasen J & Rosenfeld M (1999) Combinatorial codes in signaling and synergy: lessons from pituitary development. *Curr. Opin. Genet. Dev.* 9, 566–574.

Gehring WJ & Ikeo K (1999) Pax 6: mastering eye morphogenesis and eye evolution. *Trends Genet.* 15, 371–377.

Haber JE (1998) Mating-type gene switching in *Saccharomyces cerevisiae*. *Annu. Rev. Genet.* 32, 561–599.

Marin I, Siegal ML & Baker BS (2000) The evolution of dosage-compensation mechanisms. *Bioessays* 22, 1106–1114.

Meyer BJ (2000) Sex in the worm: counting and compensating X-chromosome dose. *Trends Genet.* 16, 247–253.

Robertson BD & Meyer TF (1992) Genetic variation in pathogenic bacteria. *Trends Genet.* 8, 422–427.

Surani MA (1998) Imprinting and the initiation of gene silencing in the germ line. *Cell* 93, 309–312.

Weintraub H (1993) The MyoD family and myogenesis: redundancy, networks, and thresholds. *Cell* 75, 1241–1244.

Wolberger C (1999) Multiprotein-DNA complexes in transcriptional regulation. *Annu. Rev. Biophys. Biomol. Struct.* 28, 29–56.

Young MW (1998) The molecular control of circadian behavioral rhythms and their entrainment in Drosophila. *Annu. Rev. Biochem.* 67, 135–152.

Posttranscriptional Controls

Baker BS (1989) Sex in flies: the splice of life. *Nature* 340, 521–524.

Benne R (1996) RNA editing: how a message is changed. *Curr. Opin. Genet. Dev.* 6, 221–231.

Cline TW & Meyer BJ (1996) Vive la difference: males vs females in flies vs. worms. *Annu. Rev. Genet.* 30, 637–702.

Dever TE (1999) Translation initiation: adept at adapting. *Trends Biochem. Sci.* 24, 398–403.

Frankel AD & Young JAT (1998) HIV-1: fifteen proteins and an RNA. *Annu. Rev. Biochem.* 67, 1–25.

Graveley BR (2001) Alternative splicing: increasing diversity in the proteomic world. *Trends Genet.* 17, 100–107.

Gray NK & Wickens M (1998) Control of translation initiation in animals. *Annu. Rev. Cell Dev. Biol.* 14, 399–458.

Hentze MW & Kulozik AE (1999) A perfect message: RNA surveillance and nonsense-mediated decay. *Cell* 96, 307–310.

Hinnebusch AG (1997) Translational regulation of yeast GCN4. A window on factors that control initiator-tRNA binding to the ribosome. *J. Biol. Chem.* 272, 21661–21664.

Holcik M, Sonenberg N & Korneluk RG (2000) Internal ribosome initiation of translation and the control of cell death. *Trends Genet.* 16, 469–473.

Jansen RP (2001) mRNA localization: message on the move. *Nat Rev Mol Cell Biol* 2, 247–256.

Pollard VW & Malim MH (1998) The HIV-1 Rev protein. *Annu. Rev. Microbiol.* 52, 491–532.

Sharp PA (2001) RNA interference – 2001. *Genes Dev.* 15, 485–490.

Wilusz CJ, Wormington M & Peltz SW (2001) The cap-to-tail guide to mRNA turnover. *Nat. Rev. Mol. Cell Biol.* 2, 237–246.

How Genomes Evolve

Dehal P, Predki P, Olsen AS et al. (2001) Human chromosome 19 and related regions in mouse: conservative and lineage-specific evolution. *Science* 293, 104–111.

Henikoff S, Greene EA, Pietrokovski S et al. (1997) Gene families: the taxonomy of protein paralogs and chimeras. *Science* 278, 609–614.

International Human Genome Sequencing Consortium (2001) Initial sequencing and analysis of the human genome. *Nature* 409, 860–921.

Kumar S & Hedges SB (1998) A molecular timescale for vertebrate evolution. *Nature* 392, 917–920.

Li WH, Gu Z, Wang H & Nekrutenko A (2001) Evolutionary analyses of the human genome. *Nature* 409, 847–849.

Long M, de Souza SJ & Gilbert W (1995) Evolution of the intron-exon structure of eukaryotic genes. *Curr. Opin. Genet. Dev.* 5, 774–778.

Rowold DJ & Herrera RJ (2000) Alu elements and the human genome. *Genetica* 108, 57–72.

Stoneking M (2001) Single nucleotide polymorphisms. From the evolutionary past. *Nature* 409, 821–822.

Wolfe KH (2001) Yesterday's polyploids and the mystery of diploidization. *Nat. Rev. Genet.* 2, 333–341.

METHODS

PART III

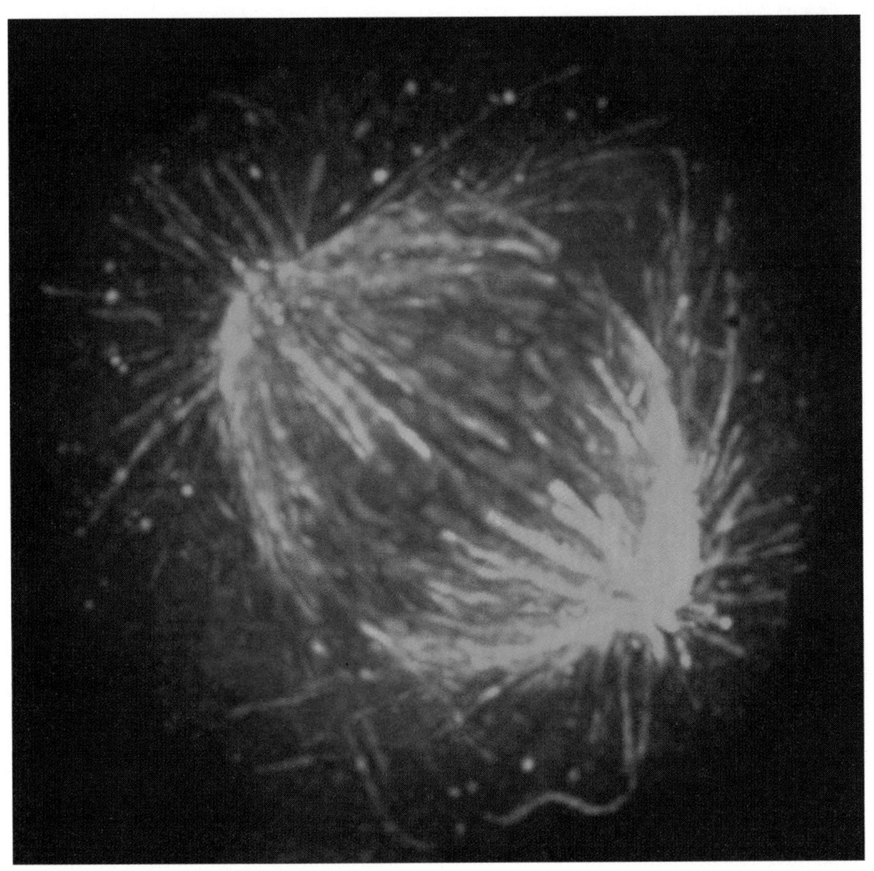

Fluorescence microscopy. New probes and improved fluorescence microscopes have revolutionized our ability to discriminate between different cellular components and to determine their subcellular location accurately. This mitotic spindle is stained to show the spindle microtubules *(green)*, the DNA in the chromosomes *(blue)*, and the centromeres *(red)*. (Courtesy of Kevin F. Sullivan.)

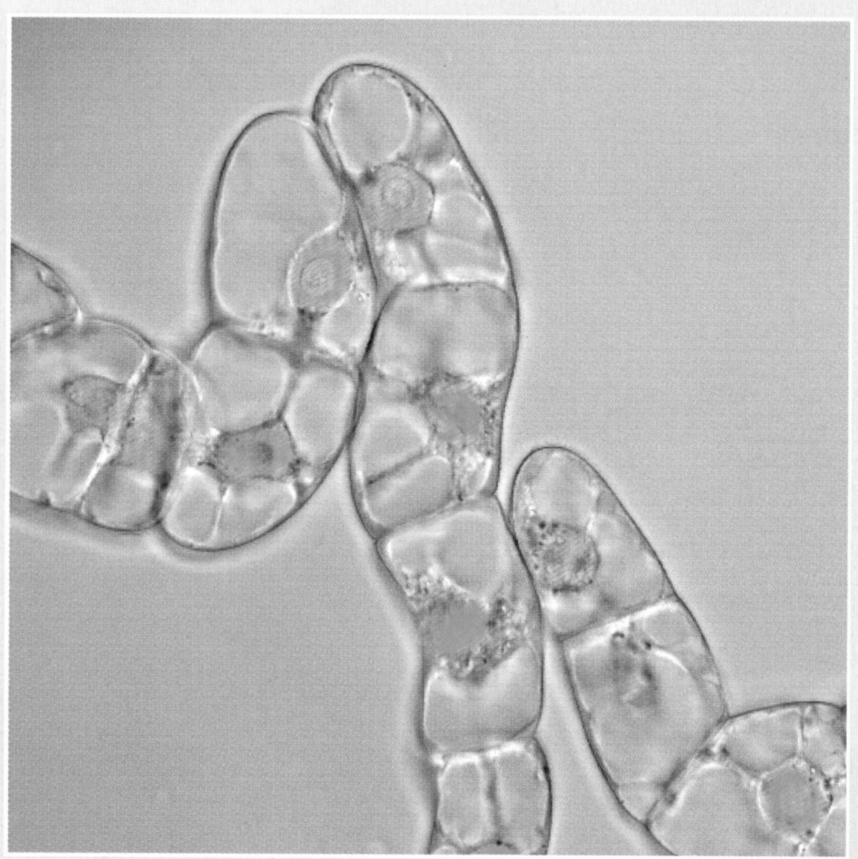

Plant cells in culture. These tobacco cells are growing in liquid culture. A histone protein that has been tagged with the green fluorescent protein (GFP) has been incorporated into chromatin. Some of the cells are dividing and are seen to have condensed chromosomes. Cultures of both animal and plant cells have been valuable research tools in all areas of cell biology. (Courtesy of Gethin Roberts.)

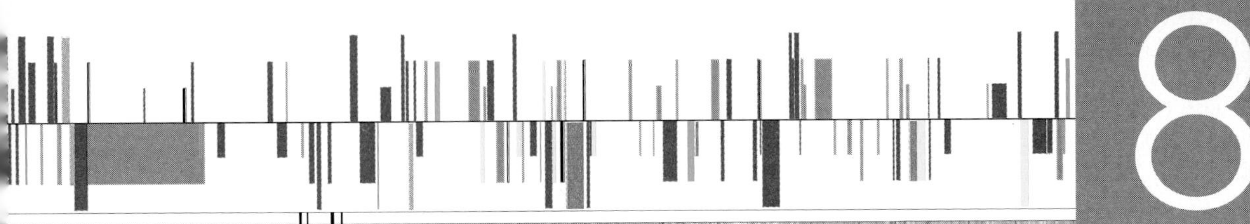

8

MANIPULATING PROTEINS, DNA, AND RNA

Progress in science is often driven by advances in technology. Biology, for example, entered a new era when Anton van Leeuwenhoek, a Dutch dry-goods dealer, ground the first microscope lens. Peering into his marvelous new looking glass, van Leeuwenhoek discovered a previously unseen cellular world, where tiny creatures tumble and twirl in a small droplet of water (Figure 8–1).

The 21st century promises to be a particularly exciting time for biology. New methods for analyzing proteins, DNA, and RNA are fueling an information explosion and allowing scientists to study cells and their macromolecules in previously unimagined ways. We now have access to the sequences of many billions of nucleotides, providing the complete molecular blueprints for dozens of organisms—from microbes and mustard weeds to worms, flies, and humans. And powerful new techniques are helping us to decipher that information, allowing us not only to compile huge, detailed catalogs of genes and proteins, but to begin to unravel how these components work together to form functional cells and organisms. The goal is nothing short of obtaining a complete understanding of what takes place inside a cell as it responds to its environment and interacts with its neighbors. We want to know which genes are switched on, which mRNA transcripts are present, and which proteins are active—where they are located, with whom they partner, and to which pathways or networks they belong. We also want to understand how the cell successfully manages this staggering number of variables and how it chooses among an almost unlimited number of possibilities in performing its diverse biological tasks. Possession of such information will permit us to begin to build a framework for delineating, and eventually predicting, how genes and proteins operate to lay the foundations for life.

In this chapter we briefly review some of the principal methods used to study cells and their components, particularly proteins, DNA, and RNA. We consider how cells of different types can be separated from tissues and grown outside the body and how cells can be disrupted and their organelles and constituent macromolecules isolated in pure form. We then review the breakthroughs in recombinant DNA technology that continue to revolutionize our understanding of cellular function. Finally we present the latest techniques used to determine the structures and functions of proteins and genes, as well as to dissect their complex interactions.

This chapter serves as a bridge from the basics of cell and molecular biology to the detailed discussion of how these macromolecules are organized and function together to coordinate the growth, development, and physiology of cells and organisms. The techniques and methods described here have made possible the discoveries that are presented throughout this book, and they are currently being used by tens of thousands of scientists each day.

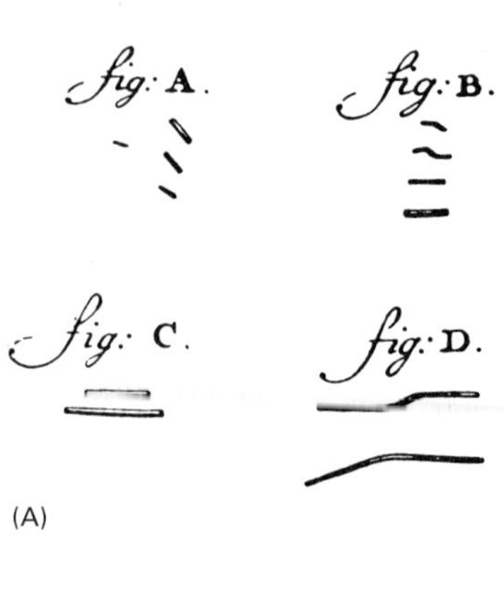

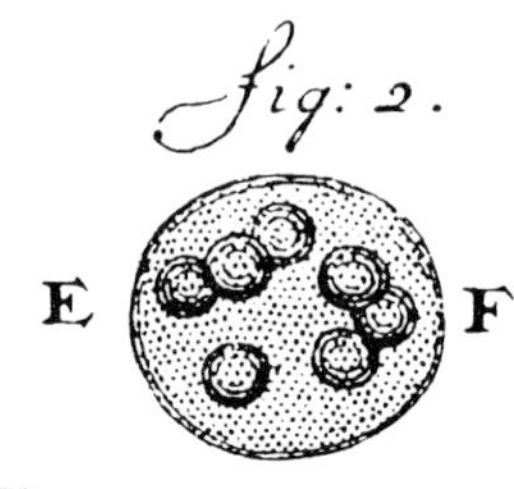

Figure 8–1 Microscopic life. A sample of "diverse animalcules" seen by van Leeuwenhoek using his simple microscope. (A) Bacteria seen in material he excavated from between his teeth. Those in fig. B he described as "swimming first forward and then backwards" (1692). (B) The eucaryotic green alga *Volvox* (1700). (Courtesy of the John Innes Foundation.)

ISOLATING CELLS AND GROWING THEM IN CULTURE

Although the organelles and large molecules in a cell can be visualized with microscopes, understanding how these components function requires a detailed biochemical analysis. Most biochemical procedures require obtaining large numbers of cells and then physically disrupting them to isolate their components. If the sample is a piece of tissue, composed of different types of cells, heterogeneous cell populations will be mixed together. To obtain as much information as possible about an individual cell type, biologists have developed ways of dissociating cells from tissues and separating the various types. These manipulations result in a relatively homogeneous population of cells that can then be analyzed—either directly or after their number has been greatly increased by allowing the cells to proliferate as a pure culture.

Cells Can Be Isolated from a Tissue Suspension and Separated into Different Types

The first step in isolating cells of a uniform type from a tissue that contains a mixture of cell types is to disrupt the extracellular matrix that holds the cells together. The best yields of viable dissociated cells are usually obtained from fetal or neonatal tissues. The tissue sample is typically treated with proteolytic enzymes (such as trypsin and collagenase) to digest proteins in the extracellular matrix and with agents (such as ethylenediaminetetraacetic acid, or EDTA) that bind, or chelate, the Ca^{2+} on which cell–cell adhesion depends. The tissue can then be teased apart into single living cells by gentle agitation.

Several approaches are used to separate the different cell types from a mixed cell suspension. One exploits differences in physical properties. Large cells can be separated from small cells and dense cells from light cells by centrifugation, for example. These techniques will be described below in connection with the separation of organelles and macromolecules, for which they were originally developed. Another approach is based on the tendency of some cell types to adhere strongly to glass or plastic, which allows them to be separated from cells that adhere less strongly.

An important refinement of this last technique depends on the specific binding properties of antibodies. Antibodies that bind specifically to the surface of only one cell type in a tissue can be coupled to various matrices—such as collagen, polysaccharide beads, or plastic—to form an affinity surface to which only cells recognized by the antibodies can adhere. The bound cells are then recovered by gentle shaking, by treatment with trypsin to digest the proteins that mediate the adhesion, or, in the case of a digestible matrix (such as collagen), by degrading the matrix itself with enzymes (such as collagenase).

One of the most sophisticated cell-separation technique uses an antibody coupled to a fluorescent dye to label specific cells. The labeled cells can then be separated from the unlabeled ones in an electronic *fluorescence-activated cell sorter.* In this remarkable machine, individual cells traveling single file in a fine stream pass through a laser beam and the fluorescence of each cell is rapidly measured. A vibrating nozzle generates tiny droplets, most containing either one cell or no cells. The droplets containing a single cell are automatically given

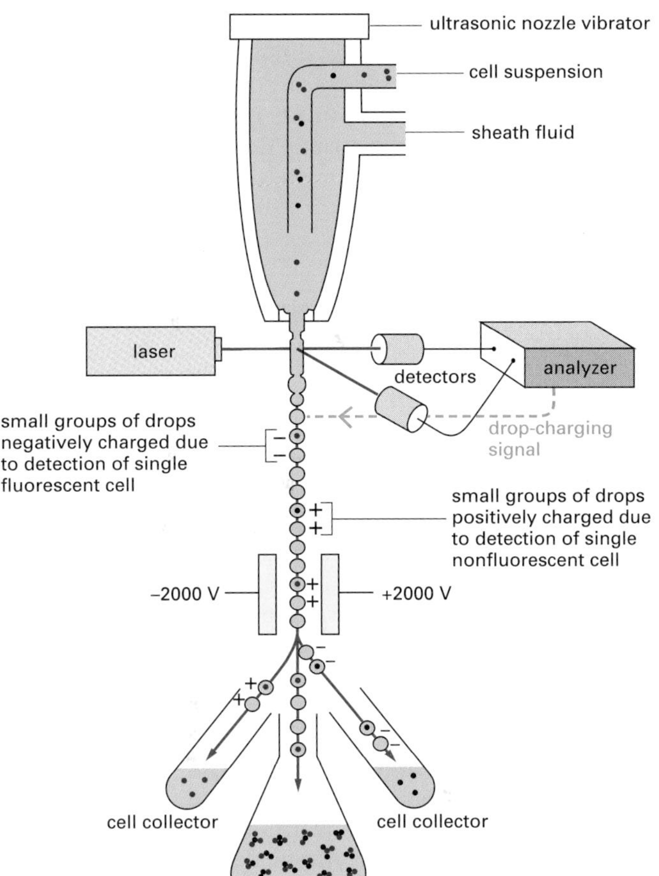

Figure 8–2 A fluorescence-activated cell sorter. A cell passing through the laser beam is monitored for fluorescence. Droplets containing single cells are given a negative or positive charge, depending on whether the cell is fluorescent or not. The droplets are then deflected by an electric field into collection tubes according to their charge. Note that the cell concentration must be adjusted so that most droplets contain no cells and flow to a waste container together with any cell clumps.

a positive or a negative charge at the moment of formation, depending on whether the cell they contain is fluorescent; they are then deflected by a strong electric field into an appropriate container. Occasional clumps of cells, detected by their increased light scattering, are left uncharged and are discarded into a waste container. Such machines can accurately select 1 fluorescent cell from a pool of 1000 unlabeled cells and sort several thousand cells each second (Figure 8–2).

Selected cells can also be obtained by carefully dissecting them from thin tissue slices that have been prepared for microscopic examination (discussed in Chapter 9). In one approach, a tissue section is coated with a thin plastic film and a region containing the cells of interest is irradiated with a focused pulse from an infrared laser. This light pulse melts a small circle of the film, binding the cells underneath. These captured cells are then removed for further analysis. The technique, called *laser capture microdissection*, can be used to separate and analyze cells from different areas of a tumor, allowing their properties to be compared. A related method uses a laser beam to directly cut out a group of cells and catapult them into an appropriate container for future analysis (Figure 8–3).

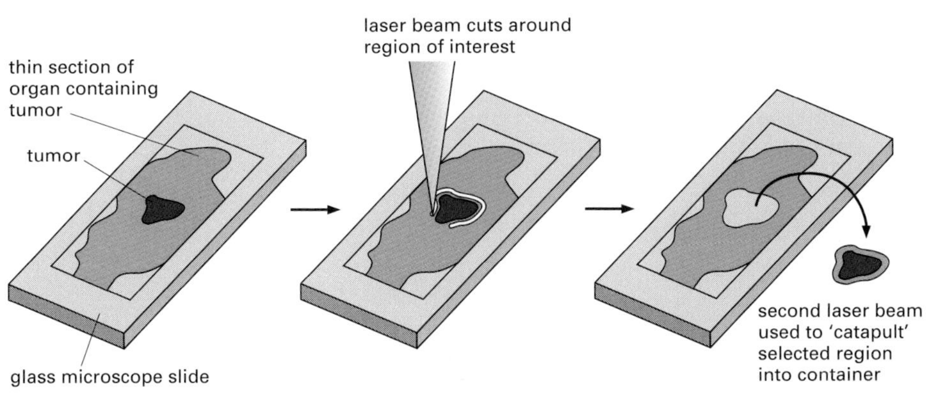

Figure 8–3 Microdissection techniques allow selected cells to be isolated from tissue slices. This method uses a laser beam to excise a region of interest and eject it into a container, and it permits the isolation of even a single cell from a tissue sample.

Once a uniform population of cells has been obtained—by microdissection or by any of the separation methods just described—it can be used directly for biochemical analysis. A homogeneous cell sample also provides a starting material for cell culture, thereby allowing the number of cells to be greatly increased and their complex behavior to be studied under the strictly defined conditions of a culture dish.

Cells Can Be Grown in a Culture Dish

Given appropriate surroundings, most plant and animal cells can live, multiply, and even express differentiated properties in a tissue-culture dish. The cells can be watched continuously under the microscope or analyzed biochemically, and the effects of adding or removing specific molecules, such as hormones or growth factors, can be explored. In addition, by mixing two cell types, the interactions between one cell type and another can be studied. Experiments performed on cultured cells are sometimes said to be carried out *in vitro* (literally, "in glass") to contrast them with experiments using intact organisms, which are said to be carried out *in vivo* (literally, "in the living organism"). These terms can be confusing, however, because they are often used in a very different sense by biochemists. In the biochemistry lab, *in vitro* refers to reactions carried out in a test tube in the absence of living cells, whereas *in vivo* refers to any reaction taking place inside a living cell (even cells that are growing in culture).

Tissue culture began in 1907 with an experiment designed to settle a controversy in neurobiology. The hypothesis under examination was known as the neuronal doctrine, which states that each nerve fiber is the outgrowth of a single nerve cell and not the product of the fusion of many cells. To test this contention, small pieces of spinal cord were placed on clotted tissue fluid in a warm, moist chamber and observed at regular intervals under the microscope. After a day or so, individual nerve cells could be seen extending long, thin filaments into the clot. Thus the neuronal doctrine received strong support, and the foundations for the cell-culture revolution were laid.

The original experiments on nerve fibers used cultures of small tissue fragments called explants. Today, cultures are more commonly made from suspensions of cells dissociated from tissues using the methods described earlier. Unlike bacteria, most tissue cells are not adapted to living in suspension and require a solid surface on which to grow and divide. For cell cultures, this support is usually provided by the surface of a plastic tissue-culture dish. Cells vary in their requirements, however, and many do not grow or differentiate unless the culture dish is coated with specific extracellular matrix components, such as collagen or laminin.

Cultures prepared directly from the tissues of an organism, that is, without cell proliferation *in vitro*, are called *primary cultures*. These can be made with or without an initial fractionation step to separate different cell types. In most cases, cells in primary cultures can be removed from the culture dish and made to proliferate to form a large number of so-called secondary cultures; in this way, they may be repeatedly subcultured for weeks or months. Such cells often display many of the differentiated properties appropriate to their origin: fibroblasts continue to secrete collagen; cells derived from embryonic skeletal muscle fuse to form muscle fibers that contract spontaneously in the culture dish; nerve cells extend axons that are electrically excitable and make synapses with other nerve cells; and epithelial cells form extensive sheets with many of the properties of an intact epithelium (Figure 8–4). Because these phenomena occur in culture, they are accessible to study in ways that are often not possible in intact tissues.

Serum-free, Chemically Defined Media Permit Identification of Specific Growth Factors

Until the early 1970s tissue culture seemed a blend of science and witchcraft. Although fluid clots were replaced by dishes of liquid media containing specified quantities of small molecules such as salts, glucose, amino acids, and vitamins,

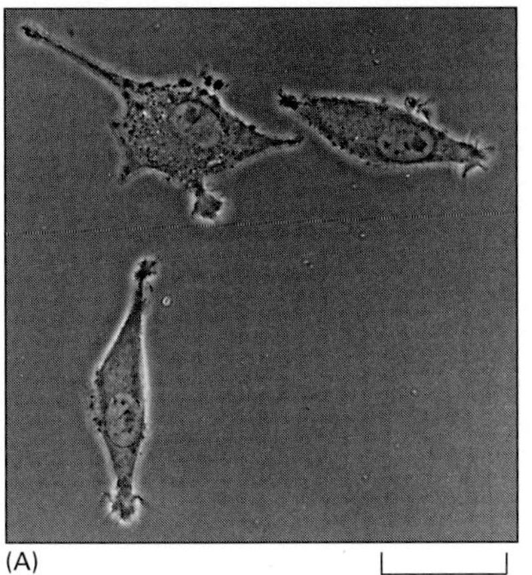

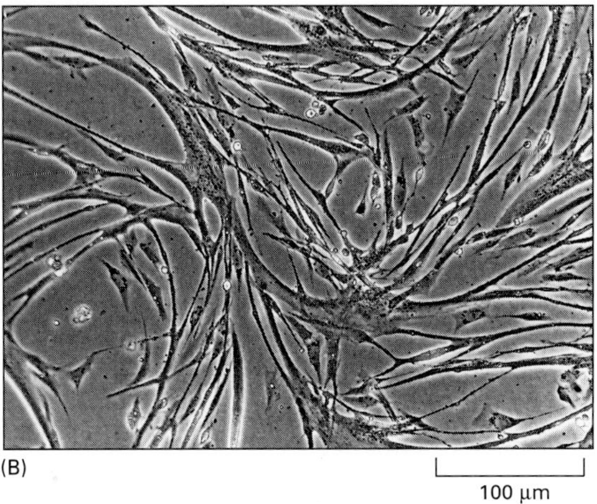

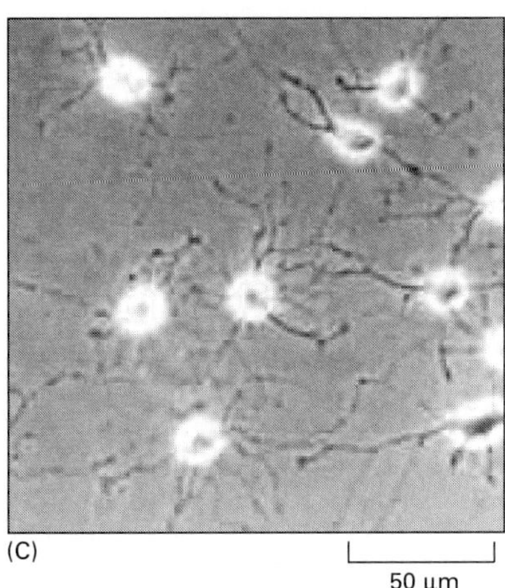

Figure 8–4 Cells in culture. (A) Phase-contrast micrograph of fibroblasts in culture. (B) Micrograph of myoblasts in culture shows cells fusing to form multinucleate muscle cells. (C) Oligodendrocyte precursor cells in culture. (D) Tobacco cells, from a fast-growing immortal cell line called BY2, in liquid culture. Nuclei and vacuoles can be seen in these cells. (A, courtesy of Daniel Zicha; B, courtesy of Rosalind Zalin; C, from D.G. Tang et al., *J. Cell Biol.* 148:971–984, 2000. © The Rockefeller University Press; D, courtesy of Gethin Roberts.)

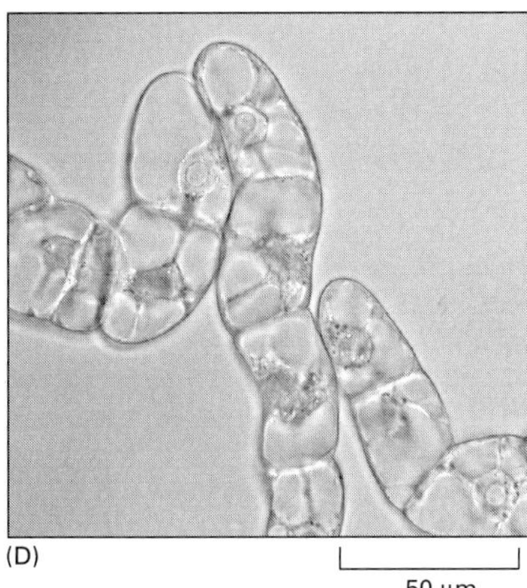

most media also included either a poorly defined mixture of macromolecules in the form of horse or fetal calf serum, or a crude extract made from chick embryos. Such media are still used today for most routine cell culture (Table 8–1), but they make it difficult for the investigator to know which specific macromolecules a particular type of cell requires to thrive and to function normally.

This difficulty led to the development of various serum-free, chemically defined media. In addition to the usual small molecules, such defined media contain one or more specific proteins that the cells require to survive and proliferate in culture. These added proteins include growth factors, which stimulate cell proliferation, and transferrin, which carries iron into cells. Many of the extracellular protein signaling molecules essential for the survival, development, and

TABLE 8–1 Composition of a Typical Medium Suitable for the Cultivation of Mammalian Cells

AMINO ACIDS	VITAMINS	SALTS	MISCELLANEOUS	PROTEINS (REQUIRED IN SERUM-FREE, CHEMICALLY DEFINED MEDIA)
Arginine	biotin	NaCl	glucose	insulin
Cystine	choline	KCl	penicillin	transferrin
Glutamine	folate	NaH_2PO_4	streptomycin	specific growth factors
Histidine	nicotinamide	$NaHCO_3$	phenol red	
Isoleucine	pantothenate	$CaCl_2$	whole serum	
Leucine	pyridoxal	$MgCl_2$		
Lysine	thiamine			
Methionine	riboflavin			
Phenylalanine				
Threonine				
Trytophan				
Tyrosine				
Valine				

Glucose is used at a concentration of 5–10 mM. The amino acids are all in the L form and, with one or two exceptions, are used at concentrations of 0.1 or 0.2 mM; vitamins are used at a 100-fold lower concentration, that is, about 1 μM. Serum, which is usually from horse or calf, is added to make up 10% of the total volume. Penicillin and streptomycin are antibiotics added to suppress the growth of bacteria. Phenol red is a pH indicator dye whose color is monitored to ensure a pH of about 7.4.

Cultures are usually grown in a plastic or glass container with a suitably prepared surface that allows the attachment of cells. The containers are kept in an incubator at 37°C in an atmosphere of 5% CO_2, 95% air.

proliferation of specific cell types were discovered by studies seeking minimal conditions under which the cell type behaved properly in culture. Thus, the search for new signaling molecules has been made much easier by the availability of chemically defined media.

Eucaryotic Cell Lines Are a Widely Used Source of Homogeneous Cells

Most vertebrate cells stop dividing after a finite number of cell divisions in culture, a process called *cell senescence* (discussed in Chapter 17). Normal human fibroblasts, for example, typically divide only 25–40 times in culture before they stop. In these cells, the limited proliferation capacity reflects a progressive shortening of the cell's telomeres, the repetitive DNA sequences and associated proteins that cap the ends of each chomosome (discussed in Chapter 5). Human somatic cells have turned off the enzyme, called *telomerase*, that normally maintains the telomeres, which is why their telomeres shorten with each cell division. Human fibroblasts can be coaxed to proliferate indefinitely by providing them with the gene that encodes the catalytic subunit of telomerase; they can then be propagated as an "immortalized" **cell line**.

Some human cells, however, are not immortalized by this trick. Although their telomeres remain long, they still stop dividing after a limited number of divisions because the culture conditions activate cell-cycle *checkpoint mechanisms* (discussed in Chapter 17) that arrest the cell cycle. In order to immortalize these cells, one has to do more than introduce telomerase. One must also inactivate the checkpoint mechanisms, which can be done by introducing certain cancer-promoting oncogenes derived from tumor viruses (discussed in Chapter 23). Unlike human cells, most rodent cells do not turn off telomerase and therefore their telomeres do not shorten with each cell division. In addition, rodent cells can undergo genetic changes in culture that inactivate their checkpoint mechanisms, thereby spontaneously producing immortalized cell lines.

Cell lines can often be most easily generated from cancer cells, but these cells differ from those prepared from normal cells in several ways. Cancer cell lines often grow without attaching to a surface, for example, and they can proliferate to a very much higher density in a culture dish. Similar properties can be induced experimentally in normal cells by transforming them with a tumor-inducing virus or chemical. The resulting transformed cell lines, in reciprocal fashion, can often cause tumors if injected into a susceptible animal. Both transformed and immortal cell lines are extremely useful in cell research as sources of very large numbers of cells of a uniform type, especially since they can be stored in liquid nitrogen at −196°C for an indefinite period and retain their viability when thawed. It is important to keep in mind, however, that the cells in both types of cell lines nearly always differ in important ways from their normal progenitors in the tissues from which they were derived. Some widely used cell lines are listed in Table 8–2.

Among the most promising cell cultures to be developed—from a medical point of view—are the human embryonic stem (ES) cell lines. These cells, harvested from the inner cell mass of the early embryo, can proliferate indefinitely while retaining the ability to give rise to any part of the body (discussed in Chapter 21). ES cells could potentially revolutionize medicine by providing a source of cells capable of replacing or repairing tissues that have been damaged by injury or disease.

Although all the cells in a cell line are very similar, they are often not identical. The genetic uniformity of a cell line can be improved by cell cloning, in which a single cell is isolated and allowed to proliferate to form a large colony. In such a colony, or clone, all the cells are descendants of a single ancestor cell. One of the most important uses of cell cloning has been the isolation of mutant cell lines with defects in specific genes. Studying cells that are defective in a specific protein often reveals valuable information about the function of that protein in normal cells.

Some important steps in the development of cell culture are listed in Table 8–3.

TABLE 8–2 Some Commonly Used Cell Lines

CELL LINE*	CELL TYPE AND ORIGIN
3T3	fibroblast (mouse)
BHK21	fibroblast (Syrian hamster)
MDCK	epithelial cell (dog)
HeLa	epithelial cell (human)
PtK1	epithelial cell (rat kangaroo)
L6	myoblast (rat)
PC12	chromaffin cell (rat)
SP2	plasma cell (mouse)
COS	kidney (monkey)
293	kidney (human); transformed with adenovirus
CHO	ovary (chinese hamster)
DT40	lymphoma cell for efficient targeted recombination (chick)
R1	embryonic stem cells (mouse)
E14.1	embryonic stem cells (mouse)
H1, H9	embryonic stem cells (human)
S2	macrophage-like cells *(Drosophila)*
BY2	undifferentiated meristematic cells (tobacco)

*Many of these cell lines were derived from tumors. All of them are capable of indefinite replication in culture and express at least some of the special characteristics of their cell of origin. BHK21 cells, HeLa cells, and SP2 cells are capable of efficient growth in suspension; most of the other cell lines require a solid culture substratum in order to multiply.

Cells Can Be Fused Together to Form Hybrid Cells

It is possible to fuse one cell with another to form a **heterocaryon**, a combined cell with two separate nuclei. Typically, a suspension of cells is treated with certain inactivated viruses or with polyethylene glycol, each of which alters the plasma membranes of cells in a way that induces them to fuse. Heterocaryons provide a way of mixing the components of two separate cells in order to study their interactions. The inert nucleus of a chicken red blood cell, for example, is reactivated to make RNA and eventually to replicate its DNA when it is exposed to the cytoplasm of a growing tissue-culture cell by fusion. The first direct evidence that membrane proteins are able to move in the plane of the plasma membrane (discussed in Chapter 10) came from an experiment in which mouse cells and human cells were fused: although the mouse and human cell-surface proteins were initially confined to their own halves of the heterocaryon plasma membrane, they quickly diffused and mixed over the entire surface of the cell.

Eventually, a heterocaryon proceeds to mitosis and produces a hybrid cell in which the two separate nuclear envelopes have been disassembled, allowing all the chromosomes to be brought together in a single large nucleus (Figure 8–5). Although such hybrid cells can be cloned to produce hybrid cell lines, the cells tend to lose chromosomes and are therefore genetically unstable. For unknown reasons, mouse–human hybrid cells predominantly lose human chromosomes. These chromosomes are lost at random, giving rise to a variety of mouse–human hybrid cell lines, each of which contains only one or a few human chromosomes. This phenomenon has been put to good use in mapping the locations of genes in the human genome: only hybrid cells containing human chromosome 11, for example, synthesize human insulin, indicating that the gene encoding insulin is located on chromosome 11. The same hybrid cells are also used as a source of human DNA for preparing chromosome-specific human DNA libraries.

TABLE 8–3 Some Landmarks in the Development of Tissue and Cell Culture

Year	Landmark
1885	**Roux** shows that embryonic chick cells can be maintained alive in a saline solution outside the animal body.
1907	**Harrison** cultivates amphibian spinal cord in a lymph clot, thereby demonstrating that axons are produced as extensions of single nerve cells.
1910	**Rous** induces a tumor by using a filtered extract of chicken tumor cells, later shown to contain an RNA virus (Rous sarcoma virus).
1913	**Carrel** shows that cells can grow for long periods in culture provided they are fed regularly under aseptic conditions.
1948	**Earle** and colleagues isolate single cells of the L cell line and show that they form clones of cells in tissue culture.
1952	**Gey** and colleagues establish a continuous line of cells derived from a human cervical carcinoma, which later become the well-known HeLa cell line.
1954	**Levi-Montalcini** and associates show that nerve growth factor (NGF) stimulates the growth of axons in tissue culture.
1955	**Eagle** makes the first systematic investigation of the essential nutritional requirements of cells in culture and finds that animal cells can propagate in a defined mixture of small molecules supplemented with a small proportion of serum proteins.
1956	**Puck** and associates select mutants with altered growth requirements from cultures of HeLa cells.
1958	**Temin and Rubin** develop a quantitative assay for the infection of chick cells in culture by purified Rous sarcoma virus. In the following decade the characteristics of this and other types of viral transformation are established by **Stoker**, **Dulbecco**, **Green**, and other virologists.
1961	**Hayflick and Moorhead** show that human fibroblasts die after a finite number of divisions in culture.
1964	**Littlefield** introduces HAT medium for the selective growth of somatic cell hybrids. Together with the technique of cell fusion, this makes somatic-cell genetics accessible.
	Kato and Takeuchi obtain a complete carrot plant from a single carrot root cell in tissue culture.
1965	**Ham** introduces a defined, serum-free medium able to support the clonal growth of certain mammalian cells.
	Harris and Watkins produce the first heterocaryons of mammalian cells by the virus-induced fusion of human and mouse cells.
1968	**Augusti-Tocco and Sato** adapt a mouse nerve cell tumor (neuroblastoma) to tissue culture and isolate clones that are electrically excitable and that extend nerve processes. A number of other differentiated cell lines are isolated at about this time, including skeletal muscle and liver cell lines.
1975	**Köhler and Milstein** produce the first monoclonal antibody-secreting hybridoma cell lines.
1976	**Sato** and associates publish the first of a series of papers showing that different cell lines require different mixtures of hormones and growth factors to grow in serum-free medium.
1977	**Wigler and Axel** and their associates develop an efficient method for introducing single-copy mammalian genes into cultured cells, adapting an earlier method developed by **Graham and van der Eb**.
1986	**Martin** and **Evans** and colleagues isolate and culture pluripotent embryonic stem cells from mouse.
1998	**Thomson** and **Gearhart** and their associates isolate human embryonic stem cells.

Hybridoma Cell Lines Provide a Permanent Source of Monoclonal Antibodies

In 1975 the development of a special type of hybrid cell line revolutionized the production of antibodies for use as tools in cell biology. The technique involves propagating a clone of cells from a single antibody-secreting B lymphocyte so that a homogeneous preparation of antibodies can be obtained in large quantities. The practical problem, however, is that B lymphocytes normally have a limited life-span in culture. To overcome this limitation, individual antibody-producing B lymphocytes from an immunized mouse or rat are fused with cells derived from an "immortal" B lymphocyte tumor. From the resulting heterogeneous mixture of hybrid cells, those hybrids that have both the ability to make a particular antibody and the ability to multiply indefinitely in culture are selected. These **hybridomas** are propagated as individual clones, each of which provides a permanent and stable source of a single type of **monoclonal antibody** (Figure 8–6). This antibody recognizes a single type of antigenic site—for example, a particular cluster of five or six amino acid side chains on the surface of a protein. Their uniform specificity makes monoclonal antibodies much more useful for most purposes than conventional antisera, which generally contain a mixture of

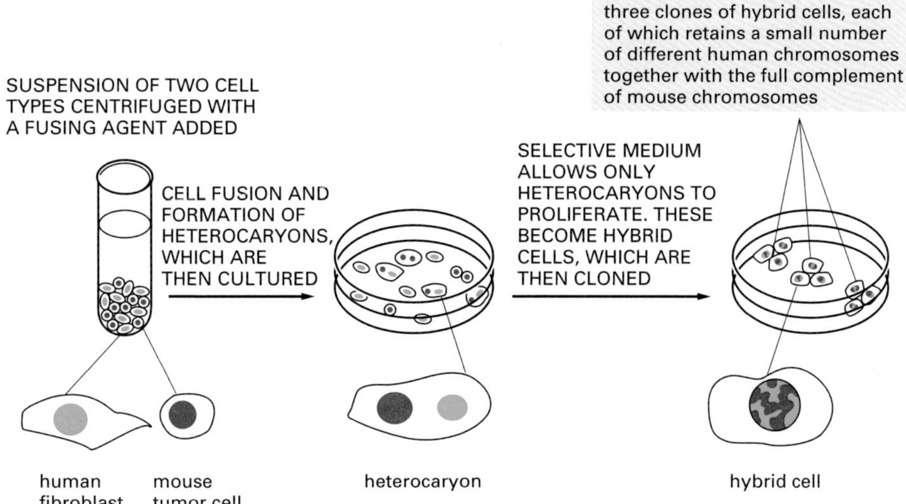
three clones of hybrid cells, each of which retains a small number of different human chromosomes together with the full complement of mouse chromosomes
SUSPENSION OF TWO CELL TYPES CENTRIFUGED WITH A FUSING AGENT ADDED
CELL FUSION AND FORMATION OF HETEROCARYONS, WHICH ARE THEN CULTURED
SELECTIVE MEDIUM ALLOWS ONLY HETEROCARYONS TO PROLIFERATE. THESE BECOME HYBRID CELLS, WHICH ARE THEN CLONED
human fibroblast
mouse tumor cell
heterocaryon
hybrid cell

Figure 8–5 The production of hybrid cells. Human cells and mouse cells are fused to produce heterocaryons (each with two or more nuclei), which eventually form hybrid cells (each with one fused nucleus). These particular hybrid cells are useful for mapping human genes on specific human chromosomes because most of the human chromosomes are quickly lost in a random manner, leaving clones that retain only one or a few. The hybrid cells produced by fusing other types of cells often retain most of their chromosomes.

antibodies that recognize a variety of different antigenic sites on a macromolecule.

The most important advantage of the hybridoma technique is that monoclonal antibodies can be made against molecules that constitute only a minor component of a complex mixture. In an ordinary antiserum made against such a mixture, the proportion of antibody molecules that recognize the minor

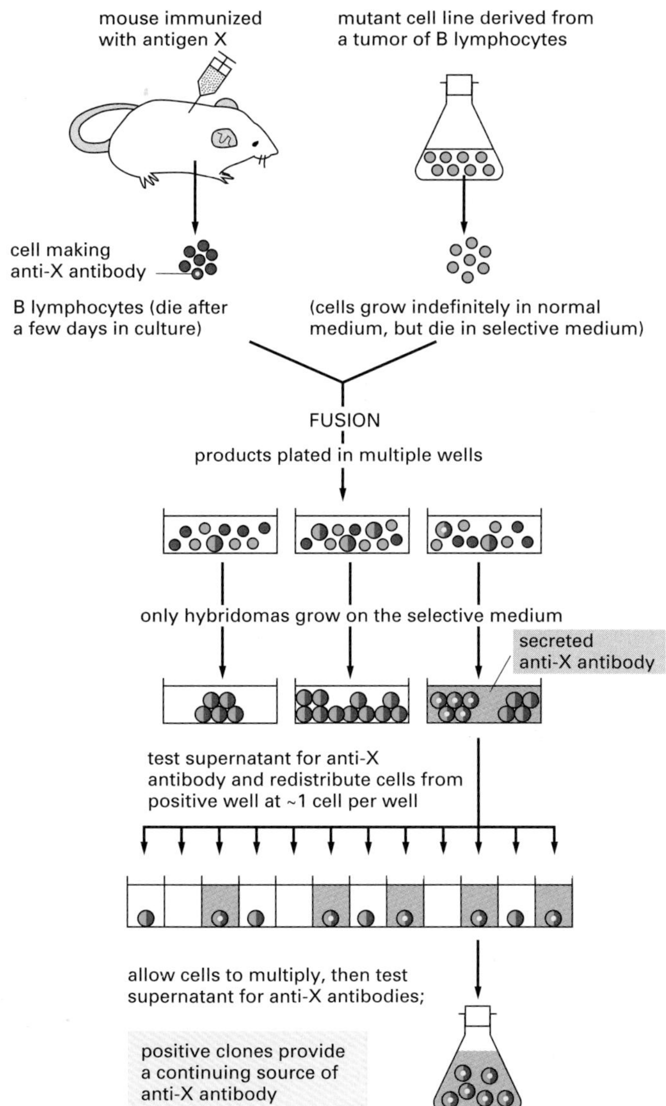
mouse immunized with antigen X
mutant cell line derived from a tumor of B lymphocytes
cell making anti-X antibody
B lymphocytes (die after a few days in culture)
(cells grow indefinitely in normal medium, but die in selective medium)
FUSION
products plated in multiple wells
only hybridomas grow on the selective medium
secreted anti-X antibody
test supernatant for anti-X antibody and redistribute cells from positive well at ~1 cell per well
allow cells to multiply, then test supernatant for anti-X antibodies;
positive clones provide a continuing source of anti-X antibody

Figure 8–6 Preparation of hybridomas that secrete monoclonal antibodies against a particular antigen. Here the antigen of interest is designated as "antigen X." The selective growth medium used after the cell fusion step contains an inhibitor (aminopterin) that blocks the normal biosynthetic pathways by which nucleotides are made. The cells must therefore use a bypass pathway to synthesize their nucleic acids. This pathway is defective in the mutant cell line derived from the tumor, but it is intact in the cells obtained from the immunized mouse. Because neither cell type used for the initial fusion can grow on its own, only the hybrid cells survive.

component would be too small to be useful. But if the B lymphocytes that produce the various components of this antiserum are made into hybridomas, it becomes possible to screen individual hybridoma clones from the large mixture to select one that produces the desired type of monoclonal antibody and to propagate the selected hybridoma indefinitely so as to produce that antibody in unlimited quantities. In principle, therefore, a monoclonal antibody can be made against any protein in a biological sample.

Once an antibody has been made, it can be used as a specific probe—both to track down and localize its protein antigen and to purify that protein in order to study its structure and function. Because only a small fraction of the estimated 10,000–20,000 proteins in a typical mammalian cell have thus far been isolated, many monoclonal antibodies made against impure protein mixtures in fractionated cell extracts identify new proteins. With the use of monoclonal antibodies and the rapid protein identification methods we shall describe shortly, it is no longer difficult to identify and characterize novel proteins and genes. The major problem is instead to determine their function, using a set of powerful tools that we discuss in the last sections of this chapter.

Summary

Tissues can be dissociated into their component cells, from which individual cell types can be purified and used for biochemical analysis or for the establishment of cell cultures. Many animal and plant cells survive and proliferate in a culture dish if they are provided with a suitable medium containing nutrients and specific protein growth factors. Although most animal cells die after a finite number of divisions, immortal cells that arise spontaneously in culture—or are generated by adding genes through genetic manipulation—can be maintained indefinitely as cell lines. Clones can be derived from a single ancestor cell, making it possible to isolate uniform populations of mutant cells with defects in a single protein. Two cells can be fused to produce heterocaryons with two nuclei, enabling interactions between the components of the original two cells to be examined. Heterocaryons eventually form hybrid cells with a single fused nucleus. Because such cells lose chromosomes, they can provide a convenient method for assigning genes to specific chromosomes. One type of hybrid cell, called a hybridoma, is widely employed to produce unlimited quantities of uniform monoclonal antibodies, which are widely used to detect and purify cellular proteins.

FRACTIONATION OF CELLS

Although biochemical analysis requires disruption of the anatomy of the cell, gentle fractionation techniques have been devised to separate the various cell components while preserving their individual functions. Just as a tissue can be separated into its living constituent cell types, so the cell can be separated into its functioning organelles and macromolecules. In this section we consider the methods that allow organelles and proteins to be purified and analyzed biochemically.

Organelles and Macromolecules Can Be Separated by Ultracentrifugation

Cells can be broken up in various ways: they can be subjected to osmotic shock or ultrasonic vibration, forced through a small orifice, or ground up in a blender. These procedures break many of the membranes of the cell (including the plasma membrane and membranes of the endoplasmic reticulum) into fragments that immediately reseal to form small closed vesicles. If carefully applied, however, the disruption procedures leave organelles such as nuclei, mitochondria, the Golgi apparatus, lysosomes, and peroxisomes largely intact. The suspension of cells is thereby reduced to a thick slurry (called a *homogenate* or *extract)* that contains a variety of membrane-enclosed organelles, each with a distinctive

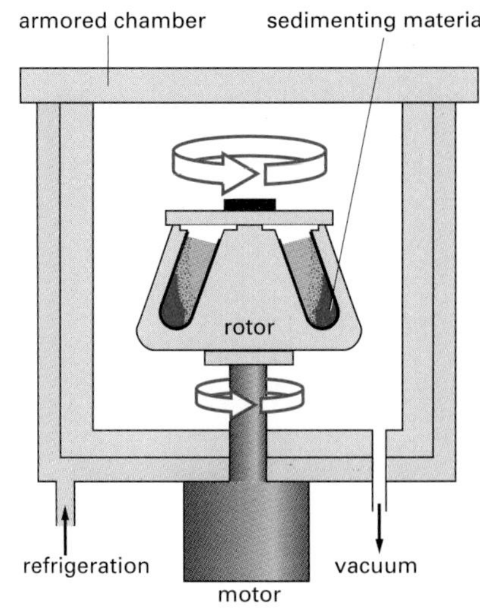

Figure 8–7 The preparative ultracentrifuge. The sample is contained in tubes that are inserted into a ring of cylindrical holes in a metal *rotor*. Rapid rotation of the rotor generates enormous centrifugal forces, which cause particles in the sample to sediment. The vacuum reduces friction, preventing heating of the rotor and allowing the refrigeration system to maintain the sample at 4°C.

size, charge, and density. Provided that the homogenization medium has been carefully chosen (by trial and error for each organelle), the various components—including the vesicles derived from the endoplasmic reticulum, called microsomes—retain most of their original biochemical properties.

The different components of the homogenate must then be separated. Such cell fractionations became possible only after the commercial development in the early 1940s of an instrument known as the *preparative ultracentrifuge*, in which extracts of broken cells are rotated at high speeds (Figure 8–7). This treatment separates cell components by size and density: in general, the largest units experience the largest centrifugal force and move the most rapidly. At relatively low speed, large components such as nuclei sediment to form a pellet at the bottom of the centrifuge tube; at slightly higher speed, a pellet of mitochondria is deposited; and at even higher speeds and with longer periods of centrifugation, first the small closed vesicles and then the ribosomes can be collected (Figure 8–8). All of these fractions are impure, but many of the contaminants can be removed by resuspending the pellet and repeating the centrifugation procedure several times.

Centrifugation is the first step in most fractionations, but it separates only components that differ greatly in size. A finer degree of separation can be achieved by layering the homogenate in a thin band on top of a dilute salt solution that fills a centrifuge tube. When centrifuged, the various components in the mixture move as a series of distinct bands through the salt solution, each at a different rate, in a process called *velocity sedimentation* (Figure 8–9A). For the procedure to work effectively, the bands must be protected from convective mixing, which would normally occur whenever a denser solution (for example, one containing organelles) finds itself on top of a lighter one (the salt solution). This is achieved by filling the centrifuge tube with a shallow gradient of sucrose prepared by a special mixing device. The resulting density gradient—with the dense end at the bottom of the tube—keeps each region of the salt solution denser than any solution above it, and it thereby prevents convective mixing from distorting the separation.

When sedimented through such dilute sucrose gradients, different cell components separate into distinct bands that can be collected individually. The relative rate at which each component sediments depends primarily on its size and shape—being normally described in terms of its sedimentation coefficient, or s value. Present-day ultracentrifuges rotate at speeds of up to 80,000 rpm and produce forces as high as 500,000 times gravity. With these enormous forces, even small macromolecules, such as tRNA molecules and simple enzymes, can be driven to sediment at an appreciable rate and so can be separated from one another by size. Measurements of sedimentation coefficients are routinely used to help in determining the size and subunit composition of the organized assemblies of macromolecules found in cells.

The ultracentrifuge is also used to separate cellular components on the basis of their buoyant density, independently of their size and shape. In this case the

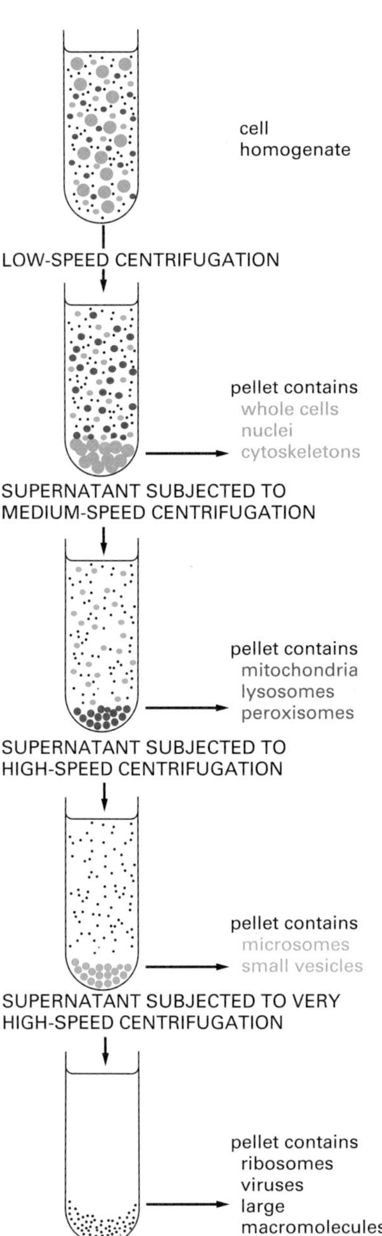

Figure 8–8 Cell fractionation by centrifugation. Repeated centrifugation at progressively higher speeds will fractionate homogenates of cells into their components. In general, the smaller the subcellular component, the greater is the centrifugal force required to sediment it. Typical values for the various centrifugation steps referred to in the figure are:

low speed	1000 times gravity for 10 minutes
medium speed	20,000 times gravity for 20 minutes
high speed	80,000 times gravity for 1 hour
very high speed	150,000 times gravity for 3 hours

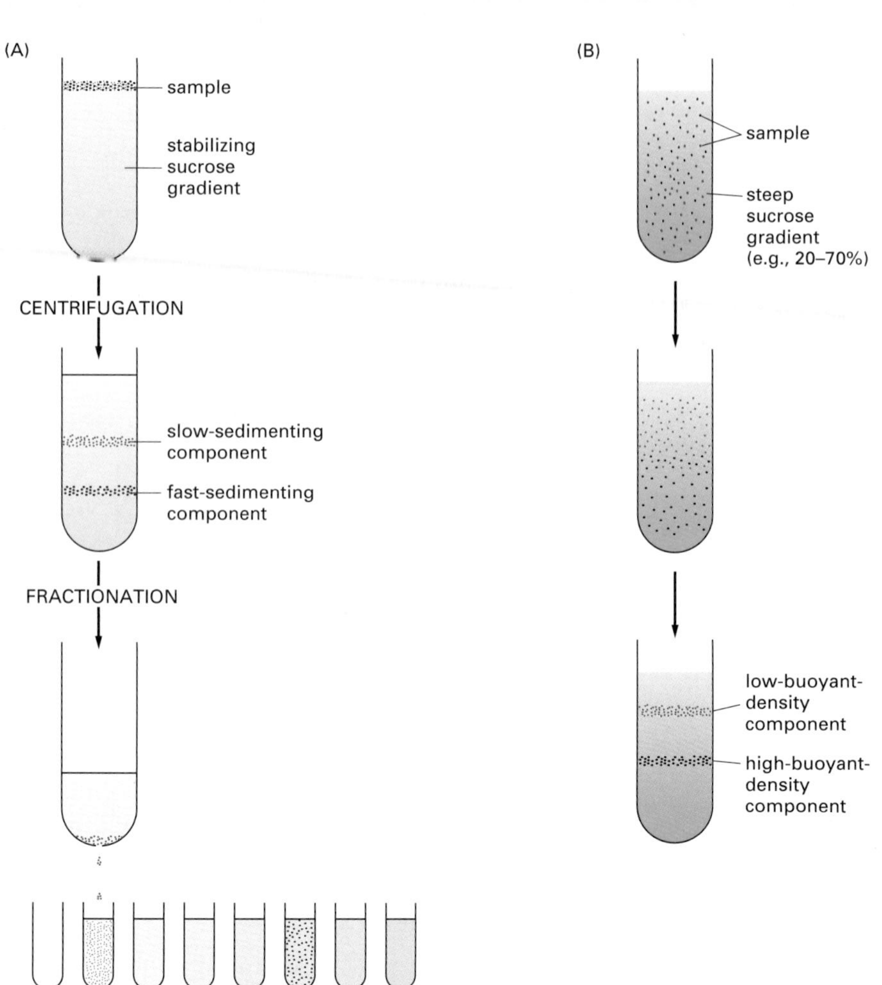

Figure 8–9 Comparison of velocity sedimentation and equilibrium sedimentation. In velocity sedimentation (A) subcellular components sediment at different speeds according to their size and shape when layered over a dilute solution containing sucrose. To stabilize the sedimenting bands against convective mixing caused by small differences in temperature or solute concentration, the tube contains a continuous shallow gradient of sucrose that increases in concentration toward the bottom of the tube (typically from 5% to 20% sucrose). Following centrifugation, the different components can be collected individually, most simply by puncturing the plastic centrifuge tube and collecting drops from the bottom, as illustrated here. In equilibrium sedimentation (B) subcellular components move up or down when centrifuged in a gradient until they reach a position where their density matches their surroundings. Although a sucrose gradient is shown here, denser gradients, which are especially useful for protein and nucleic acid separation, can be formed from cesium chloride. The final bands, at equilibrium, can be collected as in (A).

sample is usually sedimented through a steep density gradient that contains a very high concentration of sucrose or cesium chloride. Each cellular component begins to move down the gradient as in Figure 8–9A, but it eventually reaches a position where the density of the solution is equal to its own density. At this point the component floats and can move no farther. A series of distinct bands is thereby produced in the centrifuge tube, with the bands closest to the bottom of the tube containing the components of highest buoyant density (Figure 8–9B). This method, called *equilibrium sedimentation,* is so sensitive that it is capable of separating macromolecules that have incorporated heavy isotopes, such as ^{13}C or ^{15}N, from the same macromolecules that contain the lighter, common isotopes (^{12}C or ^{14}N). In fact, the cesium-chloride method was developed in 1957 to separate the labeled from the unlabeled DNA produced after exposure of a growing population of bacteria to nucleotide precursors containing ^{15}N; this classic experiment provided direct evidence for the semiconservative replication of DNA (see Figure 5–5).

The Molecular Details of Complex Cellular Processes Can Be Deciphered in Cell-Free Systems

Studies of organelles and other large subcellular components isolated in the ultracentrifuge have contributed enormously to our understanding of the functions of different cellular components. Experiments on mitochondria and chloroplasts purified by centrifugation, for example, demonstrated the central function of these organelles in converting energy into forms that the cell can use. Similarly, resealed vesicles formed from fragments of rough and smooth endoplasmic reticulum (microsomes) have been separated from each other and analyzed as functional models of these compartments of the intact cell.

An extension of this approach makes it possible to study many other biological processes free from all of the complex side reactions that occur in a living cell, by using purified **cell-free systems**. In this case, cell homogenates are fractionated with the aim of purifying each of the individual macromolecules that are needed to catalyze a biological process of interest. For example, the mechanisms of protein synthesis were deciphered in experiments that began with a cell homogenate that could translate RNA molecules to produce proteins. Fractionation of this homogenate, step by step, produced in turn the ribosomes, tRNAs, and various enzymes that together constitute the protein-synthetic machinery. Once individual pure components were available, each could be added or withheld separately to define its exact role in the overall process. A major goal today is the reconstitution of every biological process in a purified cell-free system, so as to be able to define all of its components and their mechanism of action. Some landmarks in the development of this critical approach for understanding the cell are listed in Table 8–4.

Much of what we know about the molecular biology of the cell has been discovered by studying cell-free systems. As a few of many examples, they have been used to decipher the molecular details of DNA replication and DNA transcription, RNA splicing, protein translation, muscle contraction, and particle transport along microtubules. Cell-free systems have even been used to study such complex and highly organized processes as the cell-division cycle, the separation of chromosomes on the mitotic spindle, and the vesicular-transport steps involved in the movement of proteins from the endoplasmic reticulum through the Golgi apparatus to the plasma membrane.

Cell homogenates also provide, in principle, the starting material for the complete separation of all of the individual macromolecular components from the cell. We now consider how this separation is achieved, focusing on proteins.

Proteins Can Be Separated by Chromatography

Proteins are most often fractionated by **column chromatography**, in which a mixture of proteins in solution is passed through a column containing a porous solid matrix. The different proteins are retarded to different extents by their

TABLE 8–4 Some Major Events in the Development of Cell-Free Systems

1897	**Buchner** shows that cell-free extracts of yeast can ferment sugars to form carbon dioxide and ethanol, laying the foundations of enzymology.
1926	**Svedberg** develops the first analytical ultracentrifuge and uses it to estimate the mass of hemoglobin as 68,000 daltons.
1935	**Pickels and Beams** introduce several new features of centrifuge design that lead to its use as a preparative instrument.
1938	**Behrens** employs differential centrifugation to separate nuclei and cytoplasm from liver cells, a technique further developed for the fractionation of cell organelles by **Claude**, **Brachet**, **Hogeboom**, and others in the 1940s and early 1950s.
1939	**Hill** shows that isolated chloroplasts, when illuminated, can perform the reactions of photosynthesis.
1949	**Szent-Györgyi** shows that isolated myofibrils from skeletal muscle cells contract upon the addition of ATP. In 1955 a similar cell-free system was developed for ciliary beating by **Hofmann-Berling**.
1951	**Brakke** uses density-gradient centrifugation in sucrose solutions to purify a plant virus.
1954	**de Duve** isolates lysosomes and, later, peroxisomes by centrifugation.
1954	**Zamecnik** and colleagues develop the first cell-free system to perform protein synthesis. A decade of intense research activity, during which the genetic code is elucidated, follows.
1957	**Meselson, Stahl, and Vinograd** develop equilibrium density-gradient centrifugation in cesium chloride solutions for separating nucleic acids.
1975	**Dobberstein and Blobel** demonstrate protein translocation across membranes in a cell-free system.
1976	**Neher and Sakmann** develop patch-clamp recording to measure the activity of single ion channels.
1983	**Lohka and Masui** makes concentrated extracts from frog eggs that performs the entire cell cycle *in vitro.*
1984	**Rothman** and colleagues reconstitute Golgi vesicle trafficking *in vitro* with a cell-free system.

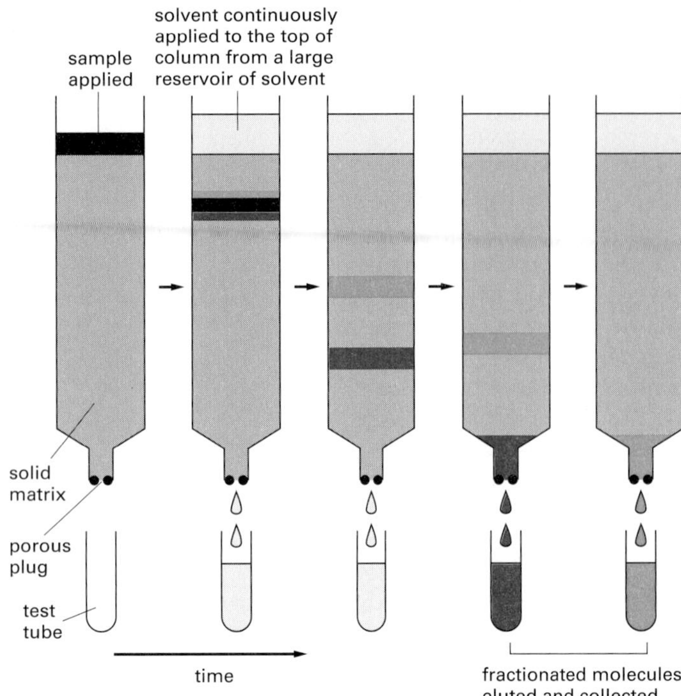

Figure 8–10 The separation of molecules by column chromatography. The sample, a mixture of different molecules, is applied to the top of a cylindrical glass or plastic column filled with a permeable solid matrix, such as cellulose, immersed in solvent. A large amount of solvent is then pumped slowly through the column and collected in separate tubes as it emerges from the bottom. Because various components of the sample travel at different rates through the column, they are fractionated into different tubes.

interaction with the matrix, and they can be collected separately as they flow out of the bottom of the column (Figure 8–10). Depending on the choice of matrix, proteins can be separated according to their charge *(ion-exchange chromatography),* their hydrophobicity *(hydrophobic chromatography),* their size *(gel-filtration chromatography),* or their ability to bind to particular small molecules or to other macromolecules *(affinity chromatography).*

Many types of matrices are commercially available (Figure 8–11). Ion-exchange columns are packed with small beads that carry either a positive or negative charge, so that proteins are fractionated according to the arrangement of charges on their surface. Hydrophobic columns are packed with beads from which hydrophobic side chains protrude, so that proteins with exposed hydrophobic regions are retarded. Gel-filtration columns, which separate proteins according to their size, are packed with tiny porous beads: molecules that are small enough to enter the pores linger inside successive beads as they pass down the column, while larger molecules remain in the solution flowing between the beads and therefore move more rapidly, emerging from the column first. Besides providing a means of separating molecules, gel-filtration chromatography is a convenient way to determine their size.

The resolution of conventional column chromatography is limited by inhomogeneities in the matrices (such as cellulose), which cause an uneven flow of solvent through the column. Newer chromatography resins (usually silica-based) have been developed in the form of tiny spheres (3 to 10 μm in diameter) that can be packed with a special apparatus to form a uniform column bed. A high degree of resolution is attainable on such **high-performance liquid chromatography (HPLC)** columns. Because they contain such tightly packed particles, HPLC columns have negligible flow rates unless high pressures are applied. For this reason these columns are typically packed in steel cylinders and require an elaborate system of pumps and valves to force the solvent through them at sufficient pressure to produce the desired rapid flow rates of about one column volume per minute. In conventional column chromatography, flow rates must be kept slow (often about one column volume per hour) to give the solutes being fractionated time to equilibrate with the interior of the large matrix particles. In HPLC the solutes equilibrate very rapidly with the interior of the tiny spheres, so solutes with different affinities for the matrix are efficiently separated from one another even at fast flow rates. This allows most fractionations to be carried out in minutes, whereas hours are required to obtain a poorer separation by conventional chromatography. HPLC has therefore become the method of choice for separating many proteins and small molecules.

Affinity Chromatography Exploits Specific Binding Sites on Proteins

If one starts with a complex mixture of proteins, these types of column chromatography do not produce very highly purified fractions: a single passage through the column generally increases the proportion of a given protein in the mixture no more than twentyfold. Because most individual proteins represent less than 1/1000 of the total cellular protein, it is usually necessary to use several different types of column in succession to attain sufficient purity (Figure 8–12). A more efficient procedure, known as **affinity chromatography**, takes advantage of the biologically important binding interactions that occur on protein surfaces. If a substrate molecule is covalently coupled to an inert matrix such as a polysaccharide bead, for example, the enzyme that operates on that substrate will often be specifically retained by the matrix and can then be eluted (washed out) in nearly pure form. Likewise, short DNA oligonucleotides of a specifically designed sequence can be immobilized in this way and used to purify DNA-binding proteins that normally recognize this sequence of nucleotides in chromosomes (see Figure 7–30). Alternatively, specific antibodies can be coupled to a matrix to purify protein molecules recognized by the antibodies. Because of the great specificity of all such affinity columns, 1000- to 10,000-fold purifications can sometimes be achieved in a single pass.

Any gene can be modified, using the recombinant DNA methods discussed in the next section, to produce its protein with a molecular tag attached to it, making subsequent purification of the protein by affinity chromatography simple and rapid (see Figure 8–48, below). For example, the amino acid histidine

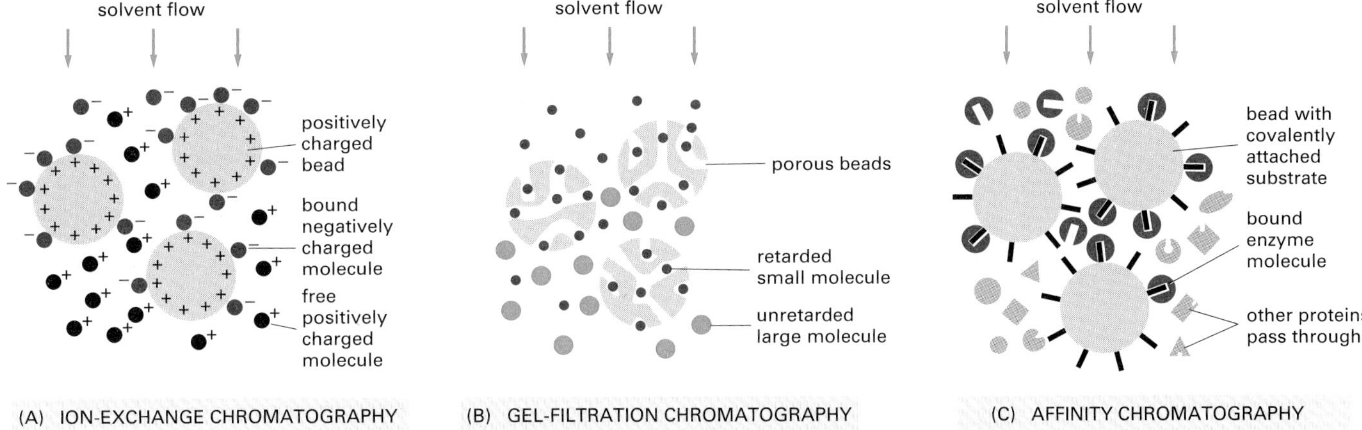

Figure 8–11 Three types of matrices used for chromatography. In ion-exchange chromatography (A) the insoluble matrix carries ionic charges that retard the movement of molecules of opposite charge. Matrices used for separating proteins include diethylaminoethylcellulose (DEAE-cellulose), which is positively charged, and carboxymethylcellulose (CM-cellulose) and phosphocellulose, which are negatively charged. Analogous matrices based on agarose or other polymers are also frequently used. The strength of the association between the dissolved molecules and the ion-exchange matrix depends on both the ionic strength and the pH of the solution that is passing down the column, which may therefore be varied systematically (as in Figure 8–12) to achieve an effective separation. In gel-filtration chromatography (B) the matrix is inert but porous. Molecules that are small enough to penetrate into the matrix are thereby delayed and travel more slowly through the column. Beads of cross-linked polysaccharide (dextran, agarose, or acrylamide) are available commercially in a wide range of pore sizes, making them suitable for the fractionation of molecules of various molecular weights, from less than 500 to more than 5×10^6. Affinity chromatography (C) uses an insoluble matrix that is covalently linked to a specific ligand, such as an antibody molecule or an enzyme substrate, that will bind a specific protein. Enzyme molecules that bind to immobilized substrates on such columns can be eluted with a concentrated solution of the free form of the substrate molecule, while molecules that bind to immobilized antibodies can be eluted by dissociating the antibody–antigen complex with concentrated salt solutions or solutions of high or low pH. High degrees of purification are often achieved in a single pass through an affinity column.

binds to certain metal ions, including nickel and copper. If genetic engineering techniques are used to attach a short string of histidine residues to either end of a protein, the slightly modified protein can be retained selectively on an affinity column containing immobilized nickel ions. Metal affinity chromatography can thereby be used to purify that modified protein from a complex molecular mixture. In other cases, an entire protein is used as the molecular tag. When the small enzyme glutathione S-transferase (GST) is attached to a target protein, the resulting fusion protein can be purified using an affinity column containing glutathione, a substrate molecule that binds specifically and tightly to GST (see Figure 8–50, below).

As a further refinement of this last technique, an amino acid sequence that forms a cleavage site for a highly specific protease can be engineered between the protein of choice and the histidine or GST tag. The cleavage sites for the proteases that are used, such as factor X that functions during blood clotting, are very rarely found by chance in proteins. Thus, the tag can later be specifically removed by cleavage at the cleavage site without destroying the purified protein.

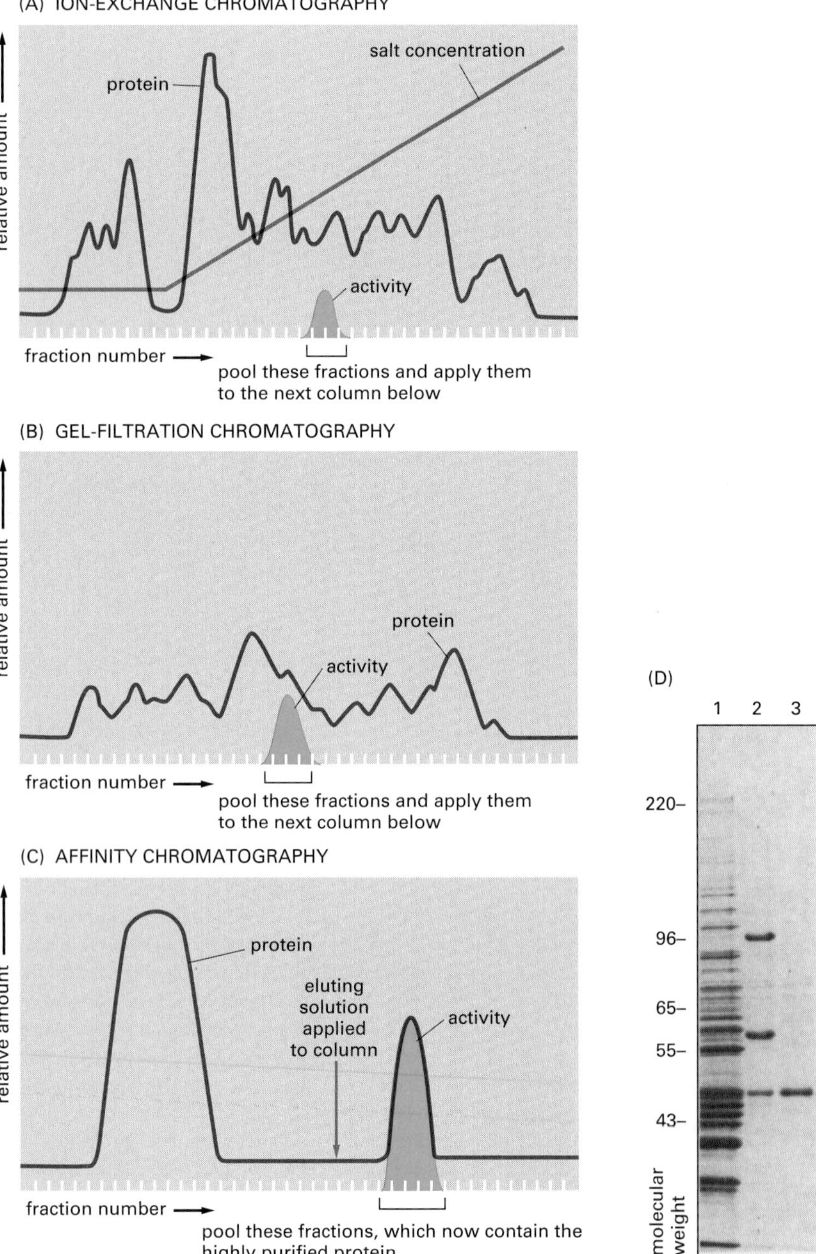

Figure 8–12 Protein purification by chromatography. Typical results obtained when three different chromatographic steps are used in succession to purify a protein. In this example a homogenate of cells was first fractionated by allowing it to percolate through an ion-exchange resin packed into a column (A). The column was washed, and the bound proteins were then eluted by passing a solution containing a gradually increasing concentration of salt onto the top of the column. Proteins with the lowest affinity for the ion-exchange resin passed directly through the column and were collected in the earliest fractions eluted from the bottom of the column. The remaining proteins were eluted in sequence according to their affinity for the resin—those proteins binding most tightly to the resin requiring the highest concentration of salt to remove them. The protein of interest was eluted in several fractions and was detected by its enzymatic activity. The fractions with activity were pooled and then applied to a second, gel-filtration column (B). The elution position of the still-impure protein was again determined by its enzymatic activity and the active fractions were pooled and purified to homogeneity on an affinity column (C) that contained an immobilized substrate of the enzyme. (D) Affinity purification of cyclin-binding proteins from *S. cerevisiae*, as analyzed by SDS polyacrylamide-gel electrophoresis (see Figure 8–14). Lane 1 is a total cell extract; lane 2 shows the proteins eluted from an affinity column containing cyclin B2; lane 3 shows one major protein eluted from a cyclin B3 affinity column. Proteins in lanes 2 and 3 were eluted with salt and the gels was stained with Coomassie blue. (D, from D. Kellogg et al., *J. Cell Biol.* 130:675–685, 1995. © The Rockefeller University Press.)

The Size and Subunit Composition of a Protein Can Be Determined by SDS Polyacrylamide-Gel Electrophoresis

Proteins usually possess a net positive or negative charge, depending on the mixture of charged amino acids they contain. When an electric field is applied to a solution containing a protein molecule, the protein migrates at a rate that depends on its net charge and on its size and shape. This technique, known as electrophoresis, was originally used to separate mixtures of proteins either in free aqueous solution or in solutions held in a solid porous matrix such as starch.

In the mid-1960s a modified version of this method—which is known as **SDS polyacrylamide-gel electrophoresis (SDS-PAGE)**—was developed that has revolutionized routine protein analysis. It uses a highly cross-linked gel of polyacrylamide as the inert matrix through which the proteins migrate. The gel is prepared by polymerization from monomers; the pore size of the gel can be adjusted so that it is small enough to retard the migration of the protein molecules of interest. The proteins themselves are not in a simple aqueous solution but in one that includes a powerful negatively charged detergent, sodium dodecyl sulfate, or SDS (Figure 8–13). Because this detergent binds to hydrophobic regions of the protein molecules, causing them to unfold into extended polypeptide chains, the individual protein molecules are released from their associations with other proteins or lipid molecules and rendered freely soluble in the detergent solution. In addition, a reducing agent such as β-mercaptoethanol (see Figure 8–13) is usually added to break any S–S linkages in the proteins, so that all of the constituent polypeptides in multisubunit molecules can be analyzed separately.

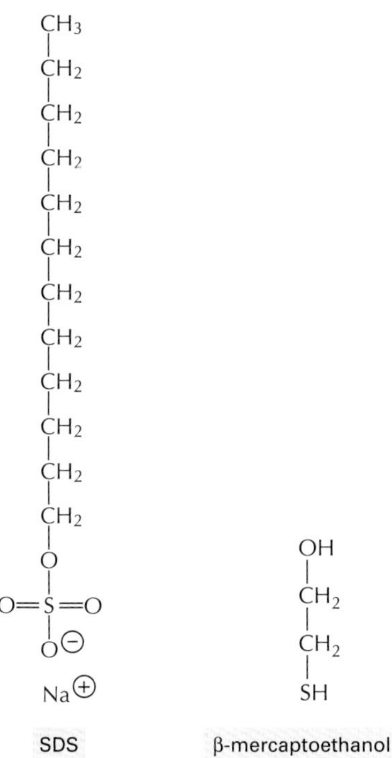

Figure 8–13 The detergent sodium dodecyl sulfate (SDS) and the reducing agent β-mercaptoethanol. These two chemicals are used to solubilize proteins for SDS polyacrylamide-gel electrophoresis. The SDS is shown here in its ionized form.

What happens when a mixture of SDS-solubilized proteins is run through a slab of polyacrylamide gel? Each protein molecule binds large numbers of the negatively charged detergent molecules, which mask the protein's intrinsic charge and cause it to migrate toward the positive electrode when a voltage is applied. Proteins of the same size tend to move through the gel with similar speeds because (1) their native structure is completely unfolded by the SDS, so that their shapes are the same, and (2) they bind the same amount of SDS and therefore have the same amount of negative charge. Larger proteins, with more charge, will be subjected to larger electrical forces and also to a larger drag. In free solution the two effects would cancel out, but in the mesh of the polyacrylamide gel, which acts as a molecular sieve, large proteins are retarded much more than small ones. As a result, a complex mixture of proteins is fractionated into a series of discrete protein bands arranged in order of molecular weight (Figure 8–14). The major proteins are readily detected by staining the proteins in the gel with a dye such as Coomassie blue, and even minor proteins are seen in gels treated with a silver or gold stain (with which as little as 10 ng of protein can be detected in a band).

SDS polyacrylamide-gel electrophoresis is a more powerful procedure than any previous method of protein analysis principally because it can be used to separate all types of proteins, including those that are insoluble in water. Membrane proteins, protein components of the cytoskeleton, and proteins that are part of large macromolecular aggregates can all be resolved. Because the method separates polypeptides by size, it also provides information about the molecular weight and the subunit composition of any protein complex. A photograph of a gel that has been used to analyze each of the successive stages in the purification of a protein is shown in Figure 8–15.

More Than 1000 Proteins Can Be Resolved on a Single Gel by Two-dimensional Polyacrylamide-Gel Electrophoresis

Because closely spaced protein bands or peaks tend to overlap, one-dimensional separation methods, such as SDS polyacrylamide-gel electrophoresis or chromatography, can resolve only a relatively small number of proteins (generally fewer than 50). In contrast, **two-dimensional gel electrophoresis**, which

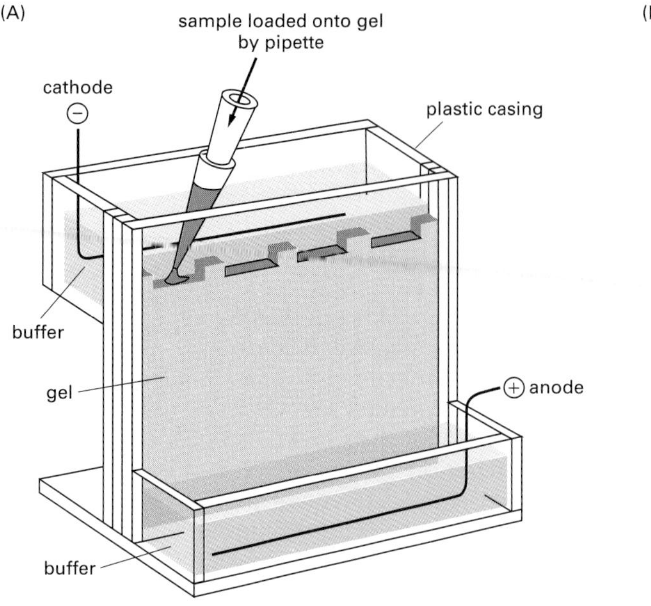

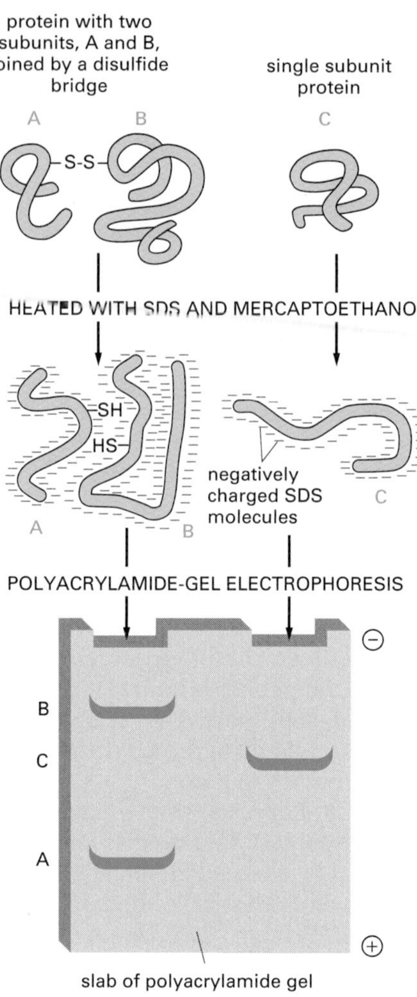

Figure 8–14 SDS polyacrylamide-gel electrophoresis (SDS-PAGE). (A) An electrophoresis apparatus. (B) Individual polypeptide chains form a complex with negatively charged molecules of sodium dodecyl sulfate (SDS) and therefore migrate as a negatively charged SDS-protein complex through a porous gel of polyacrylamide. Because the speed of migration under these conditions is greater the smaller the polypeptide, this technique can be used to determine the approximate molecular weight of a polypeptide chain as well as the subunit composition of a protein. If the protein contains a large amount of carbohydrate, however, it will move anomalously on the gel and its apparent molecular weight estimated by SDS-PAGE will be misleading.

combines two different separation procedures, can resolve up to 2000 proteins—the total number of different proteins in a simple bacterium—in the form of a two-dimensional protein map.

In the first step, the proteins are separated by their intrinsic charges. The sample is dissolved in a small volume of a solution containing a nonionic (uncharged) detergent, together with β-mercaptoethanol and the denaturing reagent urea. This solution solubilizes, denatures, and dissociates all the polypeptide chains but leaves their intrinsic charge unchanged. The polypeptide chains are then separated by a procedure called isoelectric focusing, which takes advantage of the fact that the net charge on a protein molecule varies with the pH of the surrounding solution. Every protein has a characteristic isoelectric point, the pH at which the protein has no net charge and therefore does not migrate in an electric field. In isoelectric focusing, proteins are separated electrophoretically in a narrow tube of polyacrylamide gel in which a gradient of pH is established by a mixture of special buffers. Each protein moves to a position in the gradient that corresponds to its isoelectric point and stays there (Figure 8–16). This is the first dimension of two-dimensional gel electrophoresis.

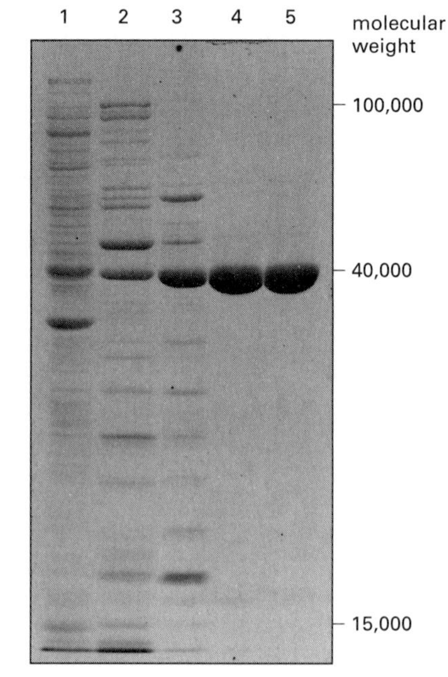

Figure 8–15 Analysis of protein samples by SDS polyacrylamide-gel electrophoresis. The photograph shows a Coomassie-stained gel that has been used to detect the proteins present at successive stages in the purification of an enzyme. The leftmost lane (lane 1) contains the complex mixture of proteins in the starting cell extract, and each succeeding lane analyzes the proteins obtained after a chromatographic fractionation of the protein sample analyzed in the previous lane (see Figure 8–12). The same total amount of protein (10 μg) was loaded onto the gel at the top of each lane. Individual proteins normally appear as sharp, dye-stained bands; a band broadens, however, when it contains too much protein. (From T. Formosa and B.M. Alberts, *J. Biol. Chem.* 261:6107–6118, 1986.)

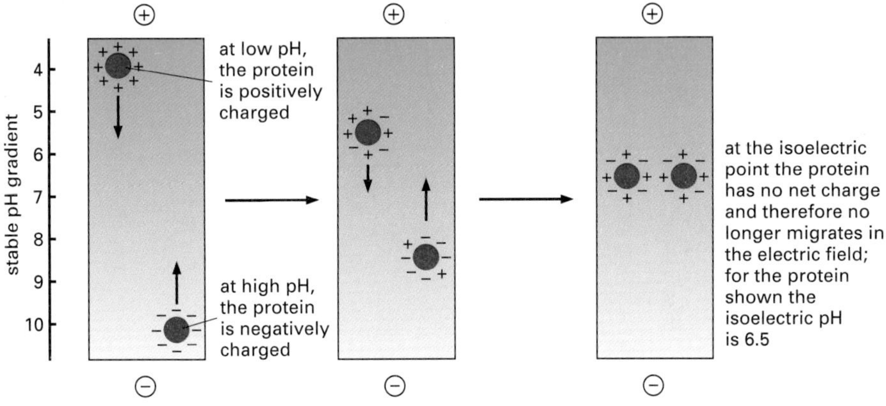

Figure 8–16 Separation of protein molecules by isoelectric focusing. At low pH (high H^+ concentration) the carboxylic acid groups of proteins tend to be uncharged (–COOH) and their nitrogen-containing basic groups fully charged (for example, $-NH_3^+$), giving most proteins a net positive charge. At high pH the carboxylic acid groups are negatively charged ($-COO^-$) and the basic groups tend to be uncharged (for example, $-NH_2$), giving most proteins a net negative charge. At its *isoelectric pH* a protein has no net charge since the positive and negative charges balance. Thus, when a tube containing a fixed pH gradient is subjected to a strong electric field in the appropriate direction, each protein species present migrates until it forms a sharp band at its isoelectric pH, as shown.

In the second step the narrow gel containing the separated proteins is again subjected to electrophoresis but in a direction that is at a right angle to the direction that used in the first step. This time SDS is added, and the proteins are separated according to their size, as in one-dimensional SDS-PAGE: the original narrow gel is soaked in SDS and then placed on one edge of an SDS polyacrylamide-gel slab, through which each polypeptide chain migrates to form a discrete spot. This is the second dimension of two-dimensional polyacrylamide-gel electrophoresis. The only proteins left unresolved are those that have both identical sizes and identical isoelectric points, a relatively rare situation. Even trace amounts of each polypeptide chain can be detected on the gel by various staining procedures—or by autoradiography if the protein sample was initially labeled with a radioisotope (Figure 8–17). The technique has such great resolving power that it can distinguish between two proteins that differ in only a single charged amino acid.

A specific protein can be identified after its fractionation on either one-dimensional or two-dimensional gels by exposing all the proteins present on the gel to a specific antibody that has been coupled to a radioactive isotope, to an easily detectable enzyme, or to a fluorescent dye. For convenience, this is normally done after all the separated proteins present in the gel have been transferred (by "blotting") onto a sheet of nitrocellulose paper, as described later for nucleic acids (see Figure 8–27). This protein-detection method is called **Western blotting** (Figure 8–18).

Some landmarks in the development of chromatography and electrophoresis are listed in Table 8–5.

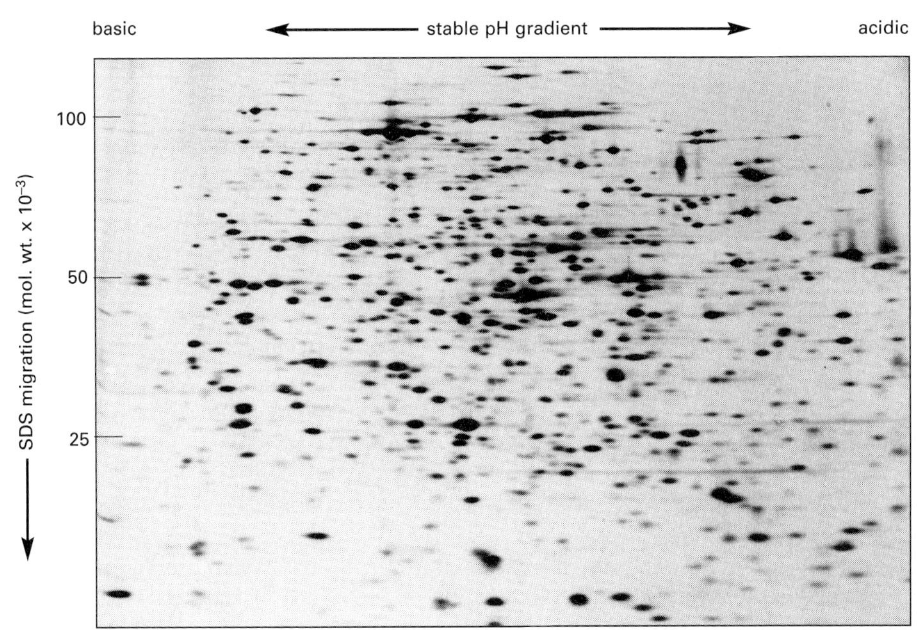

Figure 8–17 Two-dimensional polyacrylamide-gel electrophoresis. All the proteins in an *E. coli* bacterial cell are separated in this gel, in which each spot corresponds to a different polypeptide chain. The proteins were first separated on the basis of their isoelectric points by isoelectric focusing from left to right. They were then further fractionated according to their molecular weights by electrophoresis from top to bottom in the presence of SDS. Note that different proteins are present in very different amounts. The bacteria were fed with a mixture of radioisotope-labeled amino acids so that all of their proteins were radioactive and could be detected by autoradiography (see pp. 578–579). (Courtesy of Patrick O'Farrell.)

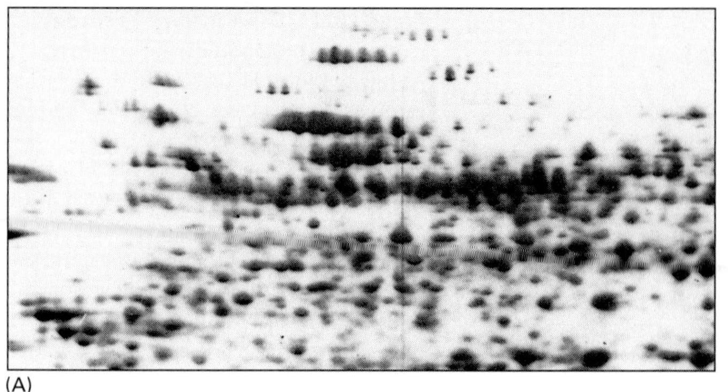
(A)

(B)

Figure 8–18 Western blotting. The total proteins from dividing tobacco cells in culture are first separated by two-dimensional polyacrylamide-gel electrophoresis and in (A) their positions are revealed by a sensitive protein stain. In (B) the separated proteins on an identical gel were then transferred to a sheet of nitrocellulose and exposed to an antibody that recognizes only those proteins that are phosphorylated on threonine residues during mitosis. The positions of the dozen or so proteins that are recognized by this antibody are revealed by an enzyme-linked second antibody. This technique is also known as immunoblotting. (From J.A. Traas, A.F. Bevan, J.H. Doonan, J. Cordewener, and P.J. Shaw, *Plant J.* 2:723–732, 1992. © Blackwell Scientific Publications.)

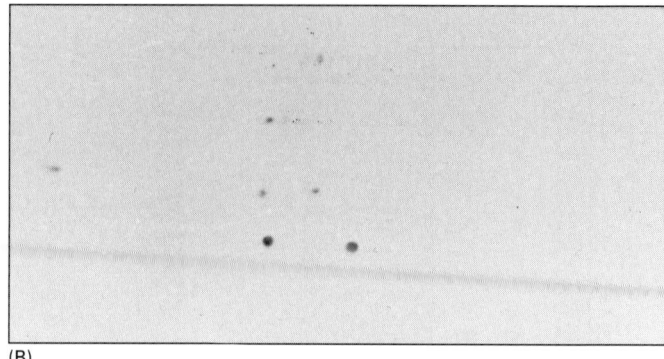

Selective Cleavage of a Protein Generates a Distinctive Set of Peptide Fragments

Although proteins have distinctive molecular weights and isoelectric points, unambiguous identification ultimately depends on determining their amino acid sequences. This can be most easily accomplished by determining the nucleotide sequence of the gene encoding the protein and using the genetic code to deduce the amino acid sequence of the protein, as discussed later in this chapter. It can also be done by directly analyzing the protein, although the complete amino acid sequences of proteins are rarely determined directly today.

There are several more rapid techniques that are used to reveal crucial information about the identity of purified proteins. For example, simply cleaving the protein into smaller fragments can provide information that helps to characterize the molecule. Proteolytic enzymes and chemical reagents are available that cleave proteins between specific amino acid residues (Table 8–6). The enzyme trypsin, for instance, cuts on the carboxyl side of lysine or arginine residues, whereas the chemical cyanogen bromide cuts peptide bonds next to methionine residues. Because these enzymes and chemicals cleave at relatively few sites, they tend to produce a few relatively large peptides when applied to a purified protein. If such a mixture of peptides is separated by chromatographic or electrophoretic procedures, the resulting pattern, or peptide map, is diagnostic of

TABLE 8–5 Landmarks in the Development of Chromatography and Electrophoresis and their Applications to Protein Molecules

1833	**Faraday** describes the fundamental laws concerning the passage of electricity through ionic solutions.
1850	**Runge** separates inorganic chemicals by their differential adsorption to paper, a forerunner of later chromatographic separations.
1906	**Tswett** invents column chromatography, passing petroleum extracts of plant leaves through columns of powdered chalk.
1933	**Tiselius** introduces electrophoresis for separating proteins in solution.
1942	**Martin and Synge** develop partition chromatography, leading to paper chromatography on ion-exchange resins.
1946	**Stein and Moore** determine for the first time the amino acid composition of a protein, initially using column chromatography on starch and later developing chromatography on ion-exchange resins.
1955	**Smithies** uses gels made of starch to separate proteins by electrophoresis.
	Sanger completes the analysis of the amino acid sequence of bovine insulin, the first protein to be sequenced.
1956	**Ingram** produces the first protein fingerprints, showing that the difference between sickle-cell and normal hemoglobin is due to a change in a single amino acid.
1959	**Raymond** introduces polyacrylamide gels, which are superior to starch gels for separating proteins by electrophoresis; improved buffer systems allowing high-resolution separations are developed in the next few years by **Ornstein and Davis**.
1966	**Maizel** introduces the use of sodium dodecyl sulfate (SDS) for improving the polyacrylamide-gel electrophoresis of proteins.
1975	**O'Farrell** devises a two-dimensional gel system for analyzing protein mixtures in which SDS polyacrylamide-gel electrophoresis is combined with separation according to isoelectric point.

TABLE 8–6 Some Reagents Commonly Used to Cleave Peptide Bonds in Proteins

	AMINO ACID 1	AMINO ACID 2
Enzyme		
Trypsin	Lys or Arg	any
Chymotrypsin	Phe, Trp, or Tyr	any
V8 protease	Glu	any
Chemical		
Cyanogen bromide	Met	any
2-Nitro-5-thiocyanobenzoate	any	Cys

The specificity for the amino acids on either side of the cleaved bond is indicated; amino acid 2 is linked to the C-terminus of amino acid 1.

the protein from which the peptides were generated and is sometimes referred to as the protein's "fingerprint" (Figure 8–19).

Protein fingerprinting was developed in 1956 to compare normal hemoglobin with the mutant form of the protein found in patients suffering from sickle-cell anemia. A single peptide difference was found and was eventually traced to a single amino acid change, providing the first demonstration that a mutation can change a single amino acid in a protein. Nowadays it is most often used to map the position of posttranslational modifications, such as phosphorylation sites.

Historically, cleaving a protein into a set of smaller peptides was an essential step in determining its amino acid sequence. This was ultimately accomplished through a series of repeated chemical reactions that removed one amino acid at a time from each peptide's N-terminus. After each cycle, the identity of the excised amino acid was determined by chromatographic methods. Now that the complete genome sequences for many organisms are available, mass spectrometry has become the method of choice for identifying proteins and matching each to its corresponding gene, thereby also determining its amino acid sequence as we discuss next.

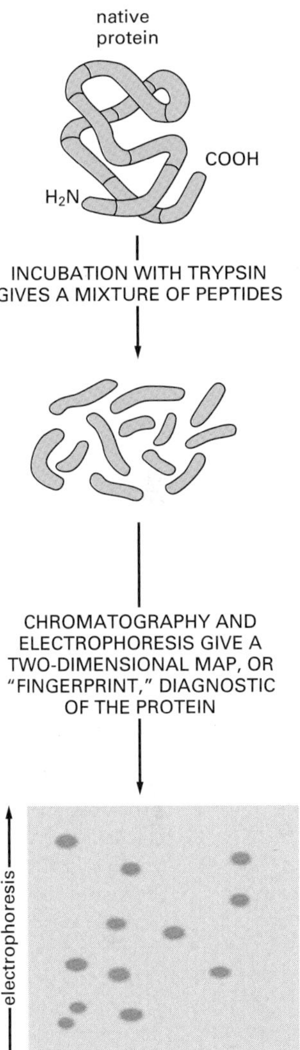

Figure 8–19 Production of a peptide map, or fingerprint, of a protein. Here, the protein was digested with trypsin to generate a mixture of polypeptide fragments, which was then fractionated in two dimensions by electrophoresis and partition chromatography. The latter technique separates peptides on the basis of their differential solubilities in water, which is preferentially bound to the solid matrix, as compared to the solvent in which they are applied. The resulting pattern of spots obtained from such a digest is diagnostic of the protein analyzed. It is also used to detect posttranslational modifications of proteins.

Mass Spectrometry Can Be Used to Sequence Peptide Fragments and Identify Proteins

Mass spectrometry allows one to determine the precise mass of intact proteins and of peptides derived from them by enzymatic or chemical cleavage. This information can then be used to search genomic databases, in which the masses of all proteins and of all their predicted peptide fragments have been tabulated (Figure 8–20A). An unambiguous match to a particular open reading frame can often be made knowing the mass of only a few peptides derived from a given protein. Mass spectrometric methods are therefore critically important for the field of *proteomics,* the large-scale effort to identify and characterize all of the proteins encoded in an organism's genome, including their posttranslational modifications.

Mass spectrometry is an enormously sensitive technique that requires very little material. Masses can be obtained with great accuracy, often with an error of less than one part in a million. The most commonly used mass spectrometric method is called *matrix-assisted laser desorption ionization–time-of-flight spectrometry (MALDI-TOF).* In this method, peptides are mixed with an organic acid and then dried onto a metal or ceramic slide. The sample is then blasted with a laser, causing the peptides to become ejected from the slide in the form of an ionized gas in which each molecule carries one or more positive charges. The ionized peptides are then accelerated in an electric field and fly toward a detector. The time it takes them to reach the detector is determined by their mass and their charge: large peptides move more slowly, and more highly charged molecules move more quickly. The precise mass is readily determined by analysis of those peptides with a single charge. MALDI–TOF can even be used

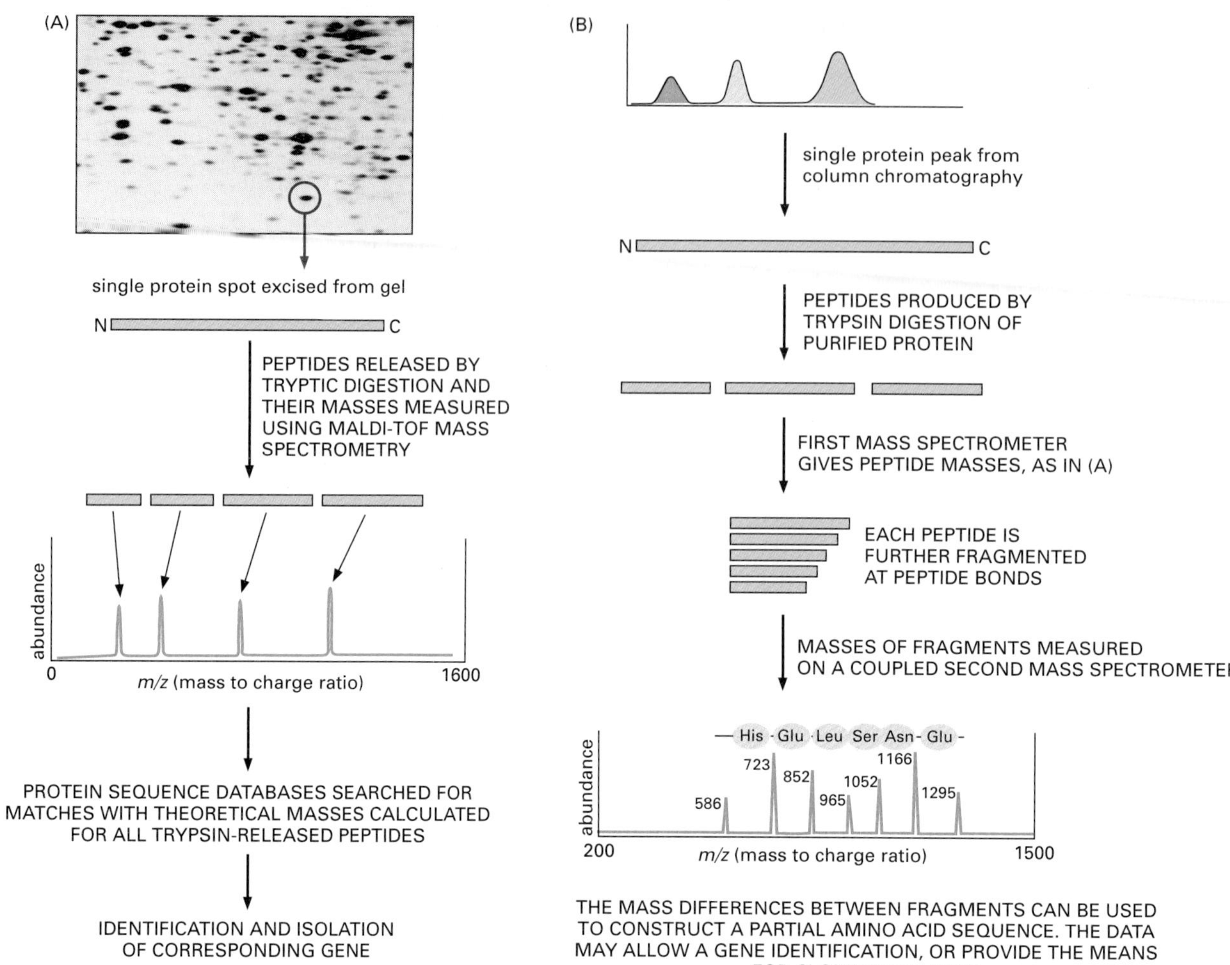

Figure 8–20 Mass-spectrometric approaches to identify proteins and sequence peptides. (A) Mass spectrometry can be used to identify proteins by determining their precise masses, and the masses of peptides derived from them, and using that information to search a genomic database for the corresponding gene. In this example, the protein of interest is excised from a two-dimensional polyacrylamide gel and then digested with trypsin. The peptide fragments are loaded into the mass spectrometer and their masses are measured. Sequence databases are then searched to find the gene that encodes a protein whose calculated tryptic digest profile matches these values. (B) Mass spectrometry can also be used to determine directly the amino acid sequence of peptide fragments. In this example, proteins that form a macromolecular complex have been separated by chromatography, and a single protein selected for digestion with trypsin. The masses of these tryptic fragments are then determined by mass spectrometry as in (A). To determine their exact amino acid sequence, each peptide is further fragmented, primarily by cleaving its peptide bonds. This treatment generates a nested set of peptides, each differing in size by one amino acid. These fragments are fed into a second coupled mass spectrometer and their masses are determined. The difference in masses between two closely related peptides can be used to deduce the "missing" amino acid. By repeated applications of this procedure, a partial amino acid sequence of the original protein can be determined. (Micrograph courtesy of Patrick O'Farrell.)

to measure the mass of intact proteins as large as 200,000 daltons, which corresponds to a polypeptide about 2000 amino acids in length.

Mass spectrometry is also used to determine the sequence of amino acids of individual peptide fragments. This method is particularly useful when the genome for the organism of interest has not yet been fully sequenced; the partial amino acid sequence obtained in this way can then be used to identify and clone the gene. Peptide sequencing is also important if proteins contain modifications, such as attached carbohydrates, phosphates, or methyl groups. In this case, the precise amino acids that are the sites of modifications can be determined.

To obtain such peptide sequence information, two mass spectrometers are required in tandem. The first separates peptides obtained after digestion of the protein of interest and allows one to zoom in on one peptide at a time. This peptide is then further fragmented by collision with high-energy gas atoms. This method of fragmentation preferentially cleaves the peptide bonds, generating a ladder of fragments, each differing by a single amino acid. The second mass spectrometer then separates these fragments and displays their masses. The amino acid sequence can be deduced from the differences in mass between the peptides (Figure 8–20B). Post-translational modifications are identified when the amino acid to which they are attached show a characteristically increased mass.

To learn more about the structure and function of a protein, one must obtain large amounts of the protein for analysis. This is most often accomplished by using the powerful recombinant DNA technologies discussed next.

Summary

Populations of cells can be analyzed biochemically by disrupting them and fractionating their contents by ultracentrifugation. Further fractionations allow functional cell-free systems to be developed; such systems are required to determine the molecular details of complex cellular processes. Protein synthesis, DNA replication, RNA splicing, the cell cycle, mitosis, and various types of intracellular transport can all be studied in this way. The molecular weight and subunit composition of even very small amounts of a protein can be determined by SDS polyacrylamide-gel electrophoresis. In two-dimensional gel electrophoresis, proteins are resolved as separate spots by isoelectric focusing in one dimension, followed by SDS polyacrylamide-gel electrophoresis in a second dimension. These electrophoretic separations can be applied even to proteins that are normally insoluble in water.

The major proteins in soluble cell extracts can be purified by column chromatography; depending on the type of column matrix, biologically active proteins can be separated on the basis of their molecular weight, hydrophobicity, charge characteristics, or affinity for other molecules. In a typical purification the sample is passed through several different columns in turn—the enriched fractions obtained from one column are applied to the next. Once a protein has been purified to homogeneity, its biological activities can be examined in detail. Using mass spectrometry, the masses of proteins and peptides derived from them can be rapidly determined. With this information one can refer to genome databases to deduce the remaining amino acid sequence of the protein from the nucleotide sequence of its gene.

ISOLATING, CLONING, AND SEQUENCING DNA

Until the early 1970s DNA was the most difficult cellular molecule for the biochemist to analyze. Enormously long and chemically monotonous, the string of nucleotides that forms the genetic material of an organism could be examined only indirectly, by protein or RNA sequencing or by genetic analysis. Today the situation has changed entirely. From being the most difficult macromolecule of the cell to analyze, DNA has become the easiest. It is now possible to isolate a specific region of a genome, to produce a virtually unlimited number of copies of it, and to determine the sequence of its nucleotides overnight. At the height of the Human Genome Project, large facilities with automated machines were generating DNA sequences at the rate of 1000 nucleotides per second, around the clock. By related techniques, an isolated gene can be altered (engineered) at will and transferred back into the germ line of an animal or plant, so as to become a functional and heritable part of the organism's genome.

These technical breakthroughs in genetic engineering—the ability to manipulate DNA with precision in a test tube or an organism—have had a dramatic impact on all aspects of cell biology by facilitating the study of cells and their macromolecules in previously unimagined ways. They have led to the discovery of whole new classes of genes and proteins, while revealing that many proteins have been much more highly conserved in evolution than had been suspected. They have provided new tools for determining the functions of proteins and of individual domains within proteins, revealing a host of unexpected relationships between them. By making available large amounts of any protein, they have shown the way to efficient mass production of protein hormones and vaccines. Finally, by allowing the regulatory regions of genes to be dissected, they provide biologists with an important tool for unraveling the complex regulatory networks by which eucaryotic gene expression is controlled.

Recombinant DNA technology comprises a mixture of techniques, some new and some borrowed from other fields such as microbial genetics (Table 8–7). Central to the technology are the following key techniques:

1. Cleavage of DNA at specific sites by restriction nucleases, which greatly facilitates the isolation and manipulation of individual genes.
2. DNA cloning either through the use of cloning vectors or the polymerase chain reaction, whereby a single DNA molecule can be copied to generate many billions of identical molecules.
3. Nucleic acid hybridization, which makes it possible to find a specific sequence of DNA or RNA with great accuracy and sensitivity on the basis of its ability to bind a complementary nucleic acid sequence.
4. Rapid sequencing of all the nucleotides in a purified DNA fragment, which makes it possible to identify genes and to deduce the amino acid sequence of the proteins they encode.

TABLE 8–7 Some Major Steps in the Development of Recombinant DNA and Transgenic Technology

1869	**Miescher** first isolates DNA from white blood cells harvested from pus-soaked bandages obtained from a nearby hospital.
1944	**Avery** provides evidence that DNA, rather than protein, carries the genetic information during bacterial transformation.
1953	**Watson and Crick** propose the double-helix model for DNA structure based on x-ray results of **Franklin and Wilkins**.
1955	**Kornberg** discovers DNA polymerase, the enzyme now used to produce labeled DNA probes.
1961	**Marmur and Doty** discover DNA renaturation, establishing the specificity and feasibility of nucleic acid hydridization reactions.
1962	**Arber** provides the first evidence for the existence of DNA restriction nucleases, leading to their purification and use in DNA sequence characterization by **Nathans and H. Smith**.
1966	**Nirenberg**, **Ochoa**, and **Khorana** elucidate the genetic code.
1967	**Gellert** discovers DNA ligase, the enzyme used to join DNA fragments together.
1972–1973	DNA cloning techniques are developed by the laboratories of **Boyer**, **Cohen**, **Berg**, and their colleagues at Stanford University and the University of California at San Francisco.
1975	**Southern** develops gel-transfer hybridization for the detection of specific DNA sequences.
1975–1977	**Sanger and Barrell** and **Maxam and Gilbert** develop rapid DNA-sequencing methods.
1981–1982	**Palmiter and Brinster** produce transgenic mice; **Spradling and Rubin** produce transgenic fruit flies.
1982	**GenBank**, NIH's public genetic sequence database, is established at Los Alamos National Laboratory.
1985	**Mullis** and co-workers invent the polymerase chain reaction (PCR).
1987	**Capecchi** and **Smithies** introduce methods for performing targeted gene replacement in mouse embryonic stem cells.
1989	**Fields and Song** develop the yeast two-hybrid system for identifying and studying protein interactions
1989	**Olson** and colleagues describe sequence-tagged sites, unique stretches of DNA that are used to make physical maps of human chromosomes.
1990	**Lipman** and colleagues release BLAST, an algorithm used to search for homology between DNA and protein sequences.
1990	**Simon** and colleagues study how to efficiently use bacterial artificial chromosomes, BACs, to carry large pieces of cloned human DNA for sequencing.
1991	**Hood and Hunkapillar** introduce new automated DNA sequence technology.
1995	**Venter** and colleagues sequence the first complete genome, that of the bacterium *Haemophilus influenzae.*
1996	**Goffeau** and an international consortium of researchers announce the completion of the first genome sequence of a eucaryote, the yeast *Saccharomyces cerevisiae.*
1996–1997	**Lockhart** and colleagues and **Brown and DeRisi** produce DNA microarrays, which allow the simultaneous monitoring of thousands of genes.
1998	**Sulston and Waterston** and colleagues produce the first complete sequence of a multicellular organism, the nematode worm *Caenorhabditis elegans.*
2001	Consortia of researchers announce the completion of the draft human genome sequence.

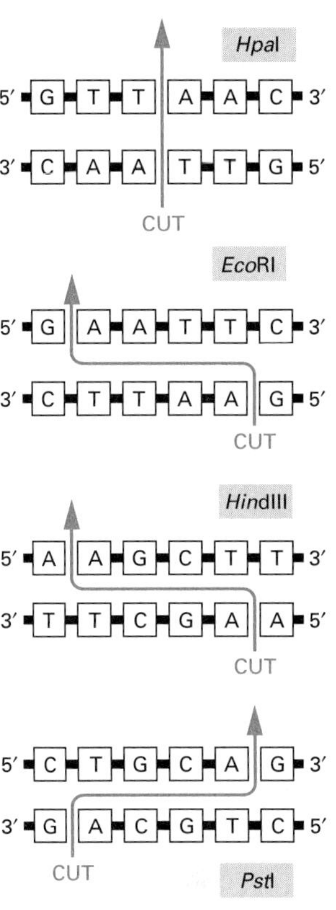

Figure 8–21 The DNA nucleotide sequences recognized by four widely used restriction nucleases. As in the examples shown, such sequences are often six base pairs long and "palindromic" (that is, the nucleotide sequence is the same if the helix is turned by 180 degrees around the center of the short region of helix that is recognized). The enzymes cut the two strands of DNA at or near the recognition sequence. For some enzymes, such as *Hpa*I, the cleavage leaves blunt ends; for others, such as *Eco*RI, *Hin*dIII, and *Pst*I, the cleavage is staggered and creates cohesive ends. Restriction nucleases are obtained from various species of bacteria: *Hpa*I is from *Hemophilus parainfluenzae*, *Eco*RI is from *Escherichia coli*, *Hin*dIII is from *Hemophilus influenzae*, and *Pst*I is from *Providencia stuartii*.

5. Simultaneous monitoring of the expression level of each gene in a cell, using nucleic acid microarrays that allow tens of thousands of hybridization reactions to be performed simultaneously.

In this chapter we describe each of these basic techniques, which together have revolutionized the study of cell biology.

Large DNA Molecules Are Cut into Fragments by Restriction Nucleases

Unlike a protein, a gene does not exist as a discrete entity in cells, but rather as a small region of a much longer DNA molecule. Although the DNA molecules in a cell can be randomly broken into small pieces by mechanical force, a fragment containing a single gene in a mammalian genome would still be only one among a hundred thousand or more DNA fragments, indistinguishable in their average size. How could such a gene be purified? Because all DNA molecules consist of an approximately equal mixture of the same four nucleotides, they cannot be readily separated, as proteins can, on the basis of their different charges and binding properties. Moreover, even if a purification scheme could be devised, vast amounts of DNA would be needed to yield enough of any particular gene to be useful for further experiments.

The solution to all of these problems began to emerge with the discovery of **restriction nucleases**. These enzymes, which can be purified from bacteria, cut the DNA double helix at specific sites defined by the local nucleotide sequence, thereby cleaving a long double-stranded DNA molecule into fragments of strictly defined sizes. Different restriction nucleases have different sequence specificities, and it is relatively simple to find an enzyme that can create a DNA fragment that includes a particular gene. The size of the DNA fragment can then be used as a basis for partial purification of the gene from a mixture.

Different species of bacteria make different restriction nucleases, which protect them from viruses by degrading incoming viral DNA. Each nuclease recognizes a specific sequence of four to eight nucleotides in DNA. These sequences, where they occur in the genome of the bacterium itself, are protected from cleavage by methylation at an A or a C residue; the sequences in foreign DNA are generally not methylated and so are cleaved by the restriction nucleases. Large numbers of restriction nucleases have been purified from various species of bacteria; several hundred, most of which recognize different nucleotide sequences, are now available commercially.

Some restriction nucleases produce staggered cuts, which leave short single-stranded tails at the two ends of each fragment (Figure 8–21). Ends of this type are known as *cohesive ends,* as each tail can form complementary base pairs with the tail at any other end produced by the same enzyme (Figure 8–22). The

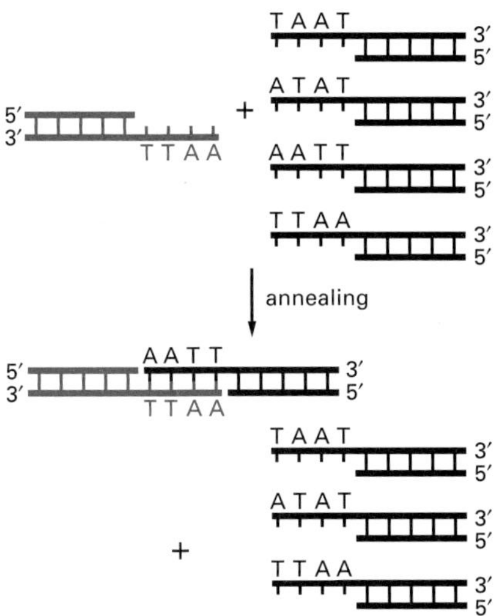

Figure 8–22 Restriction nucleases produce DNA fragments that can be easily joined together. Fragments with the same cohesive ends can readily join by complementary base-pairing between their cohesive ends, as illustrated. The two DNA fragments that join in this example were both produced by the *Eco*RI restriction nuclease, whereas the three other fragments were produced by different restriction nucleases that generated different cohesive ends (see Figure 8–21). Blunt-ended fragments, like those generated by *Hpa*I (see Figure 8–21), can be spliced together with more difficulty.

cohesive ends generated by restriction enzymes allow any two DNA fragments to be easily joined together, as long as the fragments were generated with the same restriction nuclease (or with another nuclease that produces the same cohesive ends). DNA molecules produced by splicing together two or more DNA fragments are called **recombinant DNA** molecules; they have made possible many new types of cell-biological studies.

Gel Electrophoresis Separates DNA Molecules of Different Sizes

The length and purity of DNA molecules can be accurately determined by the same types of gel electrophoresis methods that have proved so useful in the analysis of proteins. The procedure is actually simpler than for proteins: because each nucleotide in a nucleic acid molecule already carries a single negative charge, there is no need to add the negatively charged detergent SDS that is required to make protein molecules move uniformly toward the positive electrode. For DNA fragments less than 500 nucleotides long, specially designed polyacrylamide gels allow separation of molecules that differ in length by as little as a single nucleotide (Figure 8–23A). The pores in polyacrylamide gels, however, are too small to permit very large DNA molecules to pass; to separate these by size, the much more porous gels formed by dilute solutions of agarose (a polysaccharide isolated from seaweed) are used (Figure 8–23B). These DNA separation methods are widely used for both analytical and preparative purposes.

A variation of agarose gel electrophoresis, called *pulsed-field gel electrophoresis,* makes it possible to separate even extremely long DNA molecules. Ordinary gel electrophoresis fails to separate such molecules because the steady electric field stretches them out so that they travel end-first through the gel in snakelike configurations at a rate that is independent of their length. In pulsed-field gel electrophoresis, by contrast, the direction of the electric field is changed periodically, which forces the molecules to reorient before continuing to move snakelike through the gel. This reorientation takes much more time for larger molecules, so that longer molecules move more slowly than shorter ones. As a consequence, even entire bacterial or yeast chromosomes separate into discrete

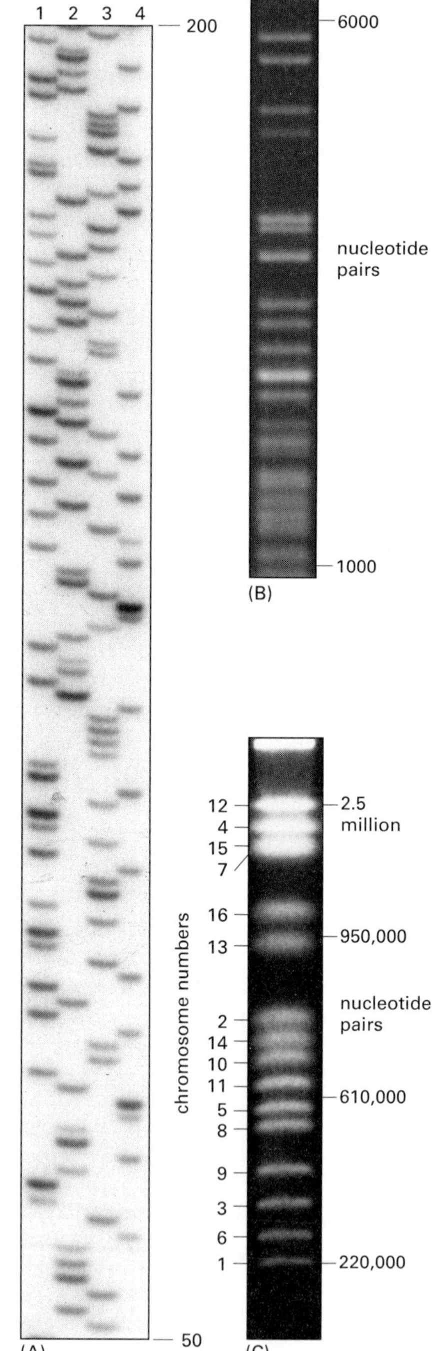

Figure 8–23 Gel electrophoresis techniques for separating DNA molecules by size. In the three examples shown, electrophoresis is from top to bottom, so that the largest—and thus slowest-moving—DNA molecules are near the top of the gel. In (A) a polyacrylamide gel with small pores is used to fractionate single-stranded DNA. In the size range 10 to 500 nucleotides, DNA molecules that differ in size by only a single nucleotide can be separated from each other. In the example, the four lanes represent sets of DNA molecules synthesized in the course of a DNA-sequencing procedure. The DNA to be sequenced has been artificially replicated from a fixed start site up to a variable stopping point, producing a set of partial replicas of differing lengths. (Figure 8–36 explains how such sets of partial replicas are synthesized.) Lane 1 shows all the partial replicas that terminate in a G, lane 2 all those that terminate in an A, lane 3 all those that terminate in a T, and lane 4 all those that terminate in a C. Since the DNA molecules used in these reactions are radiolabeled, their positions can be determined by autoradiography, as shown. In (B) an agarose gel with medium-sized pores is used to separate double-stranded DNA molecules. This method is most useful in the size range 300 to 10,000 nucleotide pairs. These DNA molecules are fragments produced by cleaving the genome of a bacterial virus with a restriction nuclease, and they have been detected by their fluorescence when stained with the dye ethidium bromide. In (C) the technique of pulsed-field agarose gel electrophoresis has been used to separate 16 different yeast *(Saccharomyces cerevisiae)* chromosomes, which range in size from 220,000 to 2.5 million nucleotide pairs (see Figure 4–13). The DNA was stained as in (B). DNA molecules as large as 10^7 nucleotide pairs can be separated in this way. (A, courtesy of Leander Lauffer and Peter Walter; B, courtesy of Ken Kreuzer; C, from D. Vollrath and R.W. Davis, *Nucleic Acids Res.* 15:7865–7876, 1987. © Oxford University Press.)

bands in pulsed-field gels and so can be sorted and identified on the basis of their size (Figure 8–23C). Although a typical mammalian chromosome of 10^8 base pairs is too large to be sorted even in this way, large segments of these chromosomes are readily separated and identified if the chromosomal DNA is first cut with a restriction nuclease selected to recognize sequences that occur only rarely (once every 10,000 or more nucleotide pairs).

The DNA bands on agarose or polyacrylamide gels are invisible unless the DNA is labeled or stained in some way. One sensitive method of staining DNA is to expose it to the dye *ethidium bromide,* which fluoresces under ultraviolet light when it is bound to DNA (see Figures 8–23B,C). An even more sensitive detection method incorporates a radioisotope into the DNA molecules before electrophoresis; ^{32}P is often used as it can be incorporated into DNA phosphates and emits an energetic β particle that is easily detected by autoradiography (as in Figure 8–23A).

Purified DNA Molecules Can Be Specifically Labeled with Radioisotopes or Chemical Markers *in vitro*

Two procedures are widely used to label isolated DNA molecules. In the first method a DNA polymerase copies the DNA in the presence of nucleotides that are either radioactive (usually labeled with ^{32}P) or chemically tagged (Figure 8–24A). In this way "DNA probes" containing many labeled nucleotides can be produced for nucleic acid hybridization reactions (discussed below). The second procedure uses the bacteriophage enzyme polynucleotide kinase to transfer a single ^{32}P-labeled phosphate from ATP to the 5′ end of each DNA chain (Figure 8–24B). Because only one ^{32}P atom is incorporated by the kinase into each DNA strand, the DNA molecules labeled in this way are often not radioactive enough to be used as DNA probes; because they are labeled at only one end, however, they have been invaluable for other applications including DNA footprinting, as we see shortly.

Today, radioactive labeling methods are being replaced by labeling with molecules that can be detected chemically or through fluorescence. To produce such nonradioactive DNA molecules, specially modified nucleotide precursors are used (Figure 8–24C). A DNA molecule made in this way is allowed to bind to its complementary DNA sequence by hybridization, as discussed in the next section, and is then detected with an antibody (or other ligand) that specifically recognizes its modified side chain (see Figure 8–28).

Nucleic Acid Hybridization Reactions Provide a Sensitive Way of Detecting Specific Nucleotide Sequences

When an aqueous solution of DNA is heated at 100°C or exposed to a very high pH (pH ≥ 13), the complementary base pairs that normally hold the two strands of the double helix together are disrupted and the double helix rapidly dissociates into two single strands. This process, called *DNA denaturation*, was for many years thought to be irreversible. In 1961, however, it was discovered that complementary single strands of DNA readily re-form double helices by a process called **hybridization** (also called *DNA renaturation*) if they are kept for a prolonged period at 65°C. Similar hybridization reactions can occur between any two single-stranded nucleic acid chains (DNA/DNA, RNA/RNA, or RNA/DNA), provided that they have complementary nucleotide sequences. These specific hybridization reactions are widely used to detect and characterize specific nucleotide sequences in both RNA and DNA molecules.

Single-stranded DNA molecules used to detect complementary sequences are known as **probes**; these molecules, which carry radioactive or chemical markers to facilitate their detection, can be anywhere from fifteen to thousands of nucleotides long. Hybridization reactions using DNA probes are so sensitive and selective that they can detect complementary sequences present at a concentration as low as one molecule per cell. It is thus possible to determine how many copies of any DNA sequence are present in a particular DNA sample. The

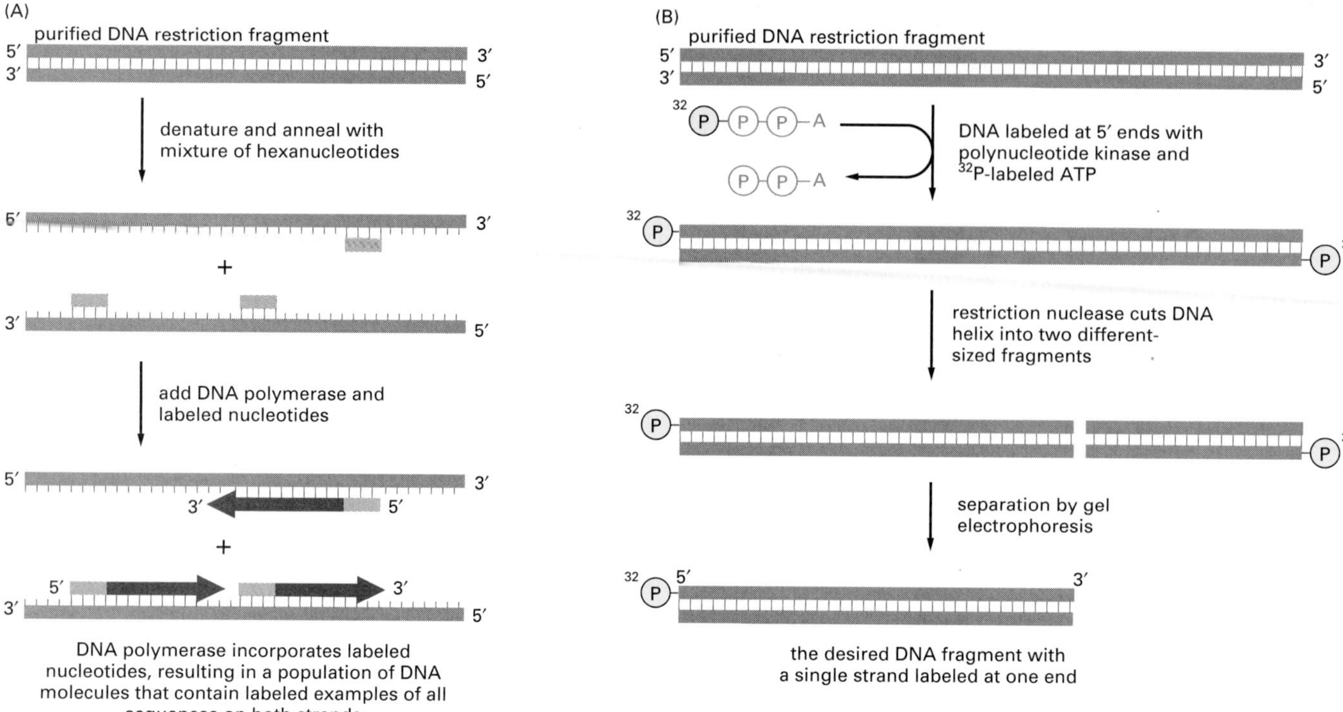

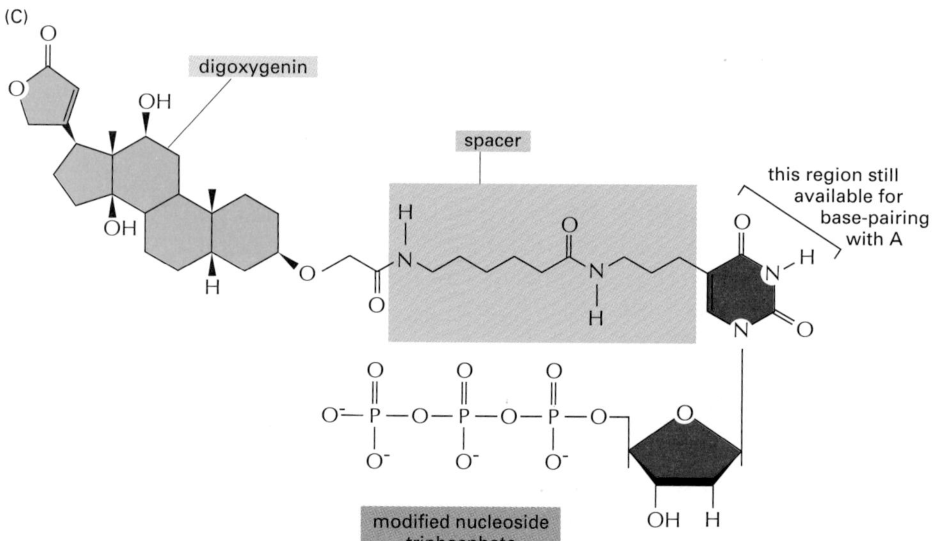

Figure 8–24 Methods for labeling DNA molecules *in vitro.* (A) A purified DNA polymerase enzyme labels all the nucleotides in a DNA molecule and can thereby produce highly radioactive DNA probes. (B) Polynucleotide kinase labels only the 5′ ends of DNA strands; therefore, when labeling is followed by restriction nuclease cleavage, as shown, DNA molecules containing a single 5′-end-labeled strand can be readily obtained. (C) The method in (A) is also used to produce nonradioactive DNA molecules that carry a specific chemical marker that can be detected with an appropriate antibody. The modified nucleotide shown can be incorporated into DNA by DNA polymerase so as to allow the DNA molecule to serve as a probe that can be readily detected. The base on the nucleoside triphosphate shown is an analog of thymine in which the methyl group on T has been replaced by a spacer arm linked to the plant steroid digoxigenin. To visualize the probe, the digoxigenin is detected by a specific antibody coupled to a visible marker such as a fluorescent dye. Other chemical labels such as biotin can be attached to nucleotides and used in essentially the same way.

same technique can be used to search for related but nonidentical genes. To find a gene of interest in an organism whose genome has not yet been sequenced, for example, a portion of a known gene can be used as a probe (Figure 8–25).

Alternatively, DNA probes can be used in hybridization reactions with RNA rather than DNA to find out whether a cell is expressing a given gene. In this case a DNA probe that contains part of the gene's sequence is hybridized with RNA purified from the cell in question to see whether the RNA includes molecules matching the probe DNA and, if so, in what quantities. In somewhat more elaborate procedures the DNA probe is treated with specific nucleases after the hybridization is complete, to determine the exact regions of the DNA probe that have paired with cellular RNA molecules. One can thereby determine the start and stop sites for RNA transcription, as well as the precise boundaries of the intron and exon sequences in a gene (Figure 8–26).

Today, the positions of intron/exon boundaries are usually determined by sequencing the cDNA sequences that represent the mRNAs expressed in a cell. Comparing this expressed sequence with the sequence of the whole gene reveals where the introns lie. We review later how cDNAs are prepared from mRNAs.

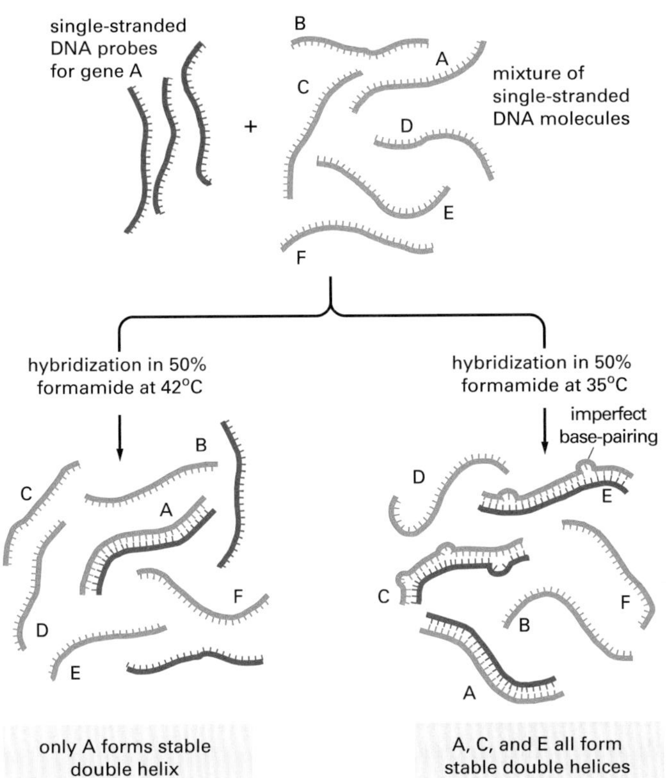

Figure 8–25 Different hybridization conditions allow less than perfect DNA matching. When only an identical match with a DNA probe is desired, the hybridization reaction is kept just a few degrees below the temperature at which a perfect DNA helix denatures in the solvent used (its *melting temperature*), so that all imperfect helices formed are unstable. When a DNA probe is being used to find DNAs that are related, but not identical, in sequence, hybridization is performed at a lower temperature. This allows even imperfectly paired double helices to form. Only the lower-temperature hybridization conditions can be used to search for genes (C and E in this example) that are nonidentical but related to gene A (see Figure 10–18).

We have seen that genes are switched on and off as a cell encounters new signals in its environment. The hybridization of DNA probes to cellular RNAs allows one to determine whether or not a particular gene is being transcribed; moreover, when the expression of a gene changes, one can determine whether the change is due to transcriptional or posttranscriptional controls (see Figure 7–87). These tests of gene expression were initially performed with one DNA probe at a time. DNA microarrays now allow the simultaneous monitoring of

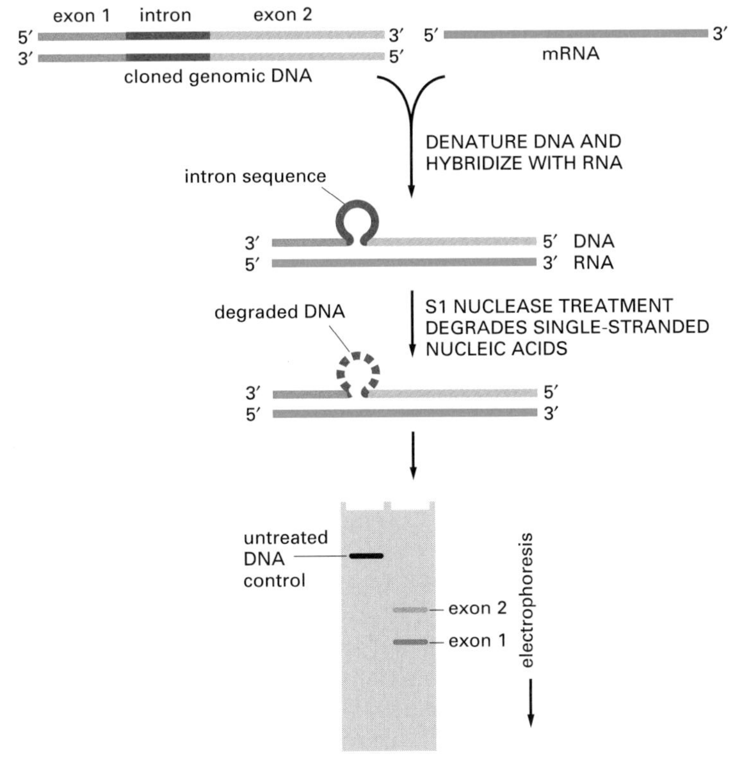

Figure 8–26 The use of nucleic acid hybridization to determine the region of a cloned DNA fragment that is present in an mRNA molecule. The method shown requires a nuclease that cuts the DNA chain only where it is not base-paired to a complementary RNA chain. The positions of the introns in eucaryotic genes are mapped by the method shown; the beginning and the end of an RNA molecule can be determined in the same way. For this type of analysis the DNA is electrophoresed through a denaturing agarose gel, which causes it to migrate as single-stranded molecules.

hundreds or thousands of genes at a time, as we discuss later. Hybridization methods are in such wide use in cell biology today that it is difficult to imagine how we could study gene structure and expression without them.

Northern and Southern Blotting Facilitate Hybridization with Electrophoretically Separated Nucleic Acid Molecules

DNA probes are often used to detect, in a complex mixture of nucleic acids, only those molecules with sequences that are complementary to all or part of the probe. Gel electrophoresis can be used to fractionate the many different RNA or DNA molecules in a crude mixture according to their size before the hybridization reaction is performed; if molecules of only one or a few sizes become labeled with the probe, one can be certain that the hybridization was indeed specific. Moreover, the size information obtained can be invaluable in itself. An example illustrates this point.

Suppose that one wishes to determine the nature of the defect in a mutant mouse that produces abnormally low amounts of albumin, a protein that liver cells normally secrete into the blood in large amounts. First, one collects identical samples of liver tissue from mutant and normal mice (the latter serving as controls) and disrupts the cells in a strong detergent to inactivate cellular nucleases that might otherwise degrade the nucleic acids. Next, one separates the RNA and DNA from all of the other cell components: the proteins present are completely denatured and removed by repeated extractions with phenol—a potent organic solvent that is partly miscible with water; the nucleic acids, which remain in the aqueous phase, are then precipitated with alcohol to separate them from the small molecules of the cell. Then one separates the DNA from the RNA by their different solubilities in alcohols and degrades any contaminating nucleic acid of the unwanted type by treatment with a highly specific enzyme—either an RNase or a DNase. The mRNAs are typically separated from bulk RNA by retention on a chromatography column that specifically binds the poly-A tails of mRNAs.

To analyze the albumin-encoding mRNAs with a DNA probe, a technique called **Northern blotting** is used. First, the intact mRNA molecules purified from mutant and control liver cells are fractionated on the basis of their sizes into a series of bands by gel electrophoresis. Then, to make the RNA molecules accessible to DNA probes, a replica of the pattern of RNA bands on the gel is made by transferring ("blotting") the fractionated RNA molecules onto a sheet of nitrocellulose or nylon paper. The paper is then incubated in a solution containing a labeled DNA probe whose sequence corresponds to part of the template strand that produces albumin mRNA. The RNA molecules that hybridize to the labeled DNA probe on the paper (because they are complementary to part of the normal albumin gene sequence) are then located by detecting the bound probe by autoradiography or by chemical means (Figure 8–27). The size of the RNA molecules in each band that binds the probe can be determined by reference to bands of RNA molecules of known sizes (RNA standards) that are electrophoresed side by side with the experimental sample. In this way one might discover that liver cells from the mutant mice make albumin RNA in normal amounts and of normal size; alternatively, albumin RNA of normal size might be detected in greatly reduced amounts. Another possibility is that the mutant albumin RNA molecules might be abnormally short and therefore move unusually quickly through the gel; in this case the gel blot could be retested with a series of shorter DNA probes, each corresponding to small portions of the gene, to reveal which part of the normal RNA is missing.

An analogous gel-transfer hybridization method, called **Southern blotting**, analyzes DNA rather than RNA. Isolated DNA is first cut into readily separable fragments with restriction nucleases. The double-stranded fragments are then separated on the basis of size by gel electrophoresis, and those complementary to a DNA probe are identified by blotting and hybridization, as just described for RNA (see Figure 8–27). To characterize the structure of the albumin gene in the mutant mice, an albumin-specific DNA probe would be used to construct a

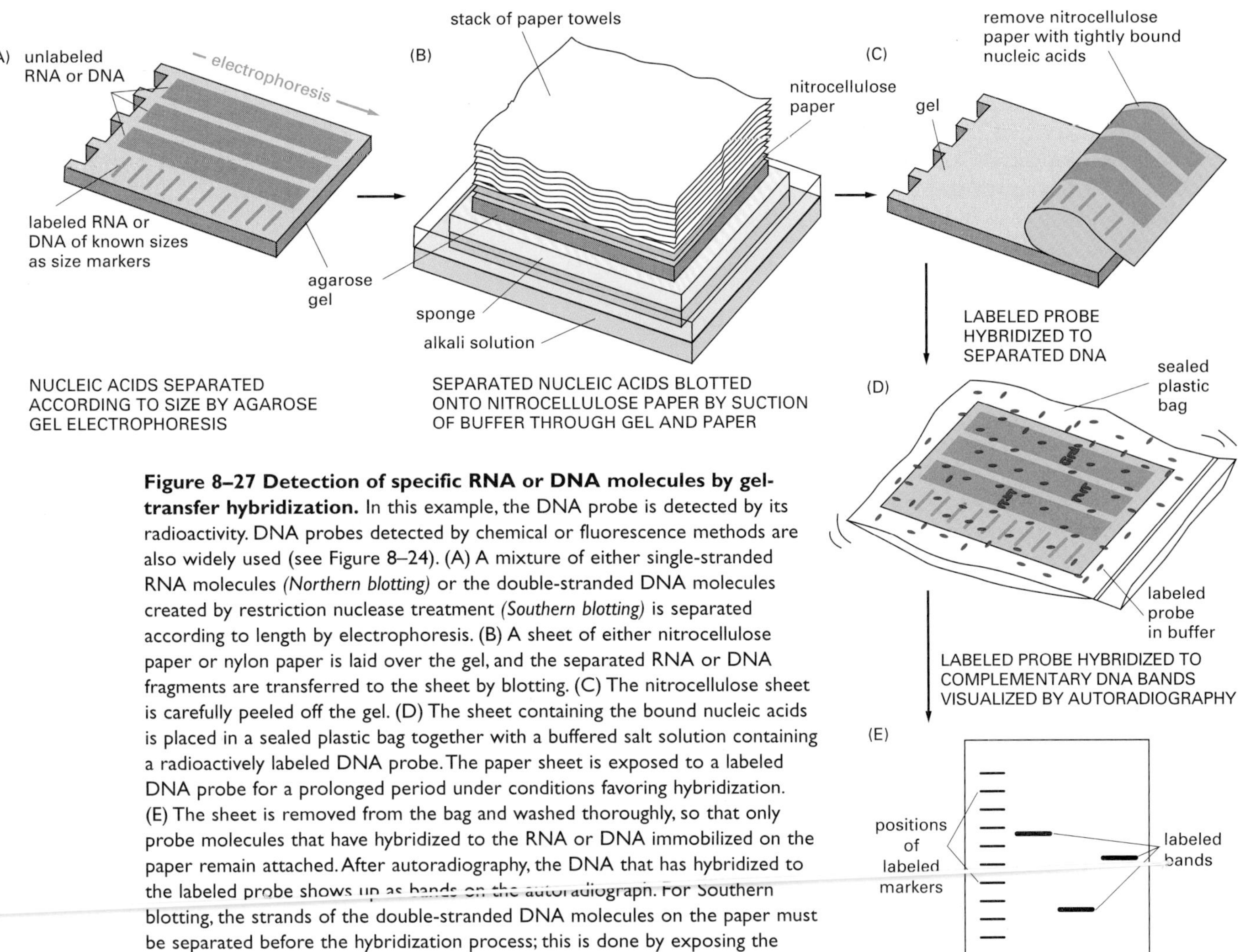

Figure 8–27 Detection of specific RNA or DNA molecules by gel-transfer hybridization. In this example, the DNA probe is detected by its radioactivity. DNA probes detected by chemical or fluorescence methods are also widely used (see Figure 8–24). (A) A mixture of either single-stranded RNA molecules *(Northern blotting)* or the double-stranded DNA molecules created by restriction nuclease treatment *(Southern blotting)* is separated according to length by electrophoresis. (B) A sheet of either nitrocellulose paper or nylon paper is laid over the gel, and the separated RNA or DNA fragments are transferred to the sheet by blotting. (C) The nitrocellulose sheet is carefully peeled off the gel. (D) The sheet containing the bound nucleic acids is placed in a sealed plastic bag together with a buffered salt solution containing a radioactively labeled DNA probe. The paper sheet is exposed to a labeled DNA probe for a prolonged period under conditions favoring hybridization. (E) The sheet is removed from the bag and washed thoroughly, so that only probe molecules that have hybridized to the RNA or DNA immobilized on the paper remain attached. After autoradiography, the DNA that has hybridized to the labeled probe shows up as bands on the autoradiograph. For Southern blotting, the strands of the double-stranded DNA molecules on the paper must be separated before the hybridization process; this is done by exposing the DNA to alkaline denaturing conditions after the gel has been run (not shown).

detailed *restriction map* of the genome in the region of the albumin gene. From this map one could determine if the albumin gene has been rearranged in the defective animals—for example, by the deletion or the insertion of a short DNA sequence; most single base changes, however, could not be detected in this way.

Hybridization Techniques Locate Specific Nucleic Acid Sequences in Cells or on Chromosomes

Nucleic acids, no less than other macromolecules, occupy precise positions in cells and tissues, and a great deal of potential information is lost when these molecules are extracted by homogenization. For this reason, techniques have been developed in which nucleic acid probes are used in much the same way as labeled antibodies to locate specific nucleic acid sequences *in situ*, a procedure called ***in situ* hybridization**. This procedure can now be done both for DNA in chromosomes and for RNA in cells. Labeled nucleic acid probes can be hybridized to chromosomes that have been exposed briefly to a very high pH to disrupt their DNA base pairs. The chromosomal regions that bind the probe during the hybridization step are then visualized. Originally, this technique was developed with highly radioactive DNA probes, which were detected by autoradiography. The spatial resolution of the technique, however, can be greatly improved by labeling the DNA probes chemically (Figure 8–28) instead of radioactively, as described earlier.

In situ hybridization methods have also been developed that reveal the distribution of specific RNA molecules in cells in tissues. In this case the tissues are not exposed to a high pH, so the chromosomal DNA remains double-stranded and cannot bind the probe. Instead the tissue is gently fixed so that its RNA is retained in an exposed form that can hybridize when the tissue is incubated with a complementary DNA or RNA probe. In this way the patterns of differential gene expression can be observed in tissues, and the location of specific RNAs can be determined in cells (Figure 8–29). In the *Drosophila* embryo, for example, such patterns have provided new insights into the mechanisms that create distinctions between cells in different positions during development (described in Chapter 21).

Figure 8–28 *In situ* hybridization to locate specific genes on chromosomes. Here, six different DNA probes have been used to mark the location of their respective nucleotide sequences on human chromosome 5 at metaphase. The probes have been chemically labeled and detected with fluorescent antibodies. Both copies of chromosome 5 are shown, aligned side by side. Each probe produces two dots on each chromosome, since a metaphase chromosome has replicated its DNA and therefore contains two identical DNA helices. (Courtesy of David C. Ward.)

Genes Can Be Cloned from a DNA Library

Any DNA fragment that contains a gene of interest can be cloned. In cell biology, the term **DNA cloning** is used in two senses. In one sense it literally refers to the act of making many identical copies of a DNA molecule—the amplification of a particular DNA sequence. However, the term is also used to describe the isolation of a particular stretch of DNA (often a particular gene) from the rest of a cell's DNA, because this isolation is greatly facilitated by making many identical copies of the DNA of interest.

DNA cloning in its most general sense can be accomplished in several ways. The simplest involves inserting a particular fragment of DNA into the purified DNA genome of a self-replicating genetic element—generally a virus or a plasmid. A DNA fragment containing a human gene, for example, can be joined in a test tube to the chromosome of a bacterial virus, and the new recombinant DNA molecule can then be introduced into a bacterial cell. Starting with only one such recombinant DNA molecule that infects a single cell, the normal replication mechanisms of the virus can produce more than 10^{12} identical virus DNA molecules in less than a day, thereby amplifying the amount of the inserted human DNA fragment by the same factor. A virus or plasmid used in this way is known as a *cloning vector*, and the DNA propagated by insertion into it is said to have been *cloned*.

To isolate a specific gene, one often begins by constructing a *DNA library*—a comprehensive collection of cloned DNA fragments from a cell, tissue, or organism. This library includes (one hopes) at least one fragment that contains the gene of interest. Libraries can be constructed with either a virus or a plasmid vector and are generally housed in a population of bacterial cells. The principles underlying the methods used for cloning genes are the same for either type of cloning vector, although the details may differ. Today most cloning is performed with plasmid vectors.

The **plasmid vectors** most widely used for gene cloning are small circular molecules of double-stranded DNA derived from larger plasmids that occur naturally in bacterial cells. They generally account for only a minor fraction of the total host bacterial cell DNA, but they can easily be separated owing to their

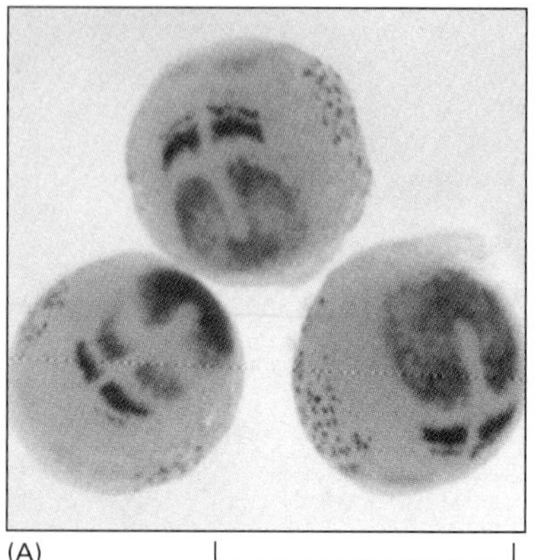

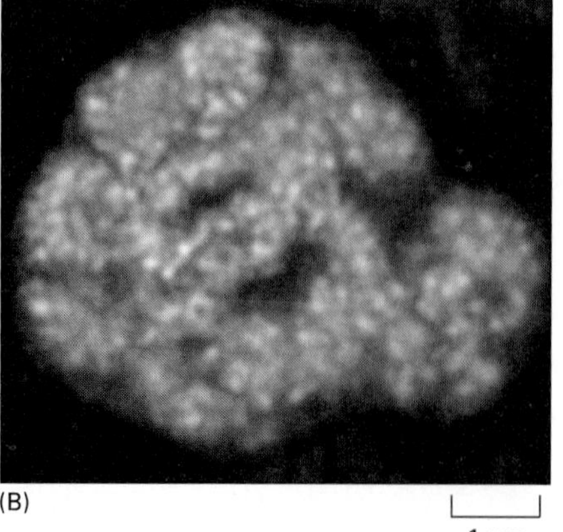

Figure 8–29 *In situ* hybridization for RNA localization. (A) Expression pattern of *deltaC* in the early zebrafish embryo. This gene codes for a ligand in the Notch signaling pathway (discussed in Chapter 15), and the pattern shown here reflects its role in the development of somites—the future segments of the vertebrate trunk and tail. (B) High-resolution RNA *in situ* localization reveals the sites within the nucleolus of a pea cell where ribosomal RNA is synthesized. The sausage-like structures, 0.5–1 µm in diameter, correspond to the loops of chromosomal DNA that contain the genes encoding rRNA. Each small, white spot represents transcription of a single rRNA gene. (A, courtesy of Yun-Jin Jiang; B, courtesy of Peter Shaw.)

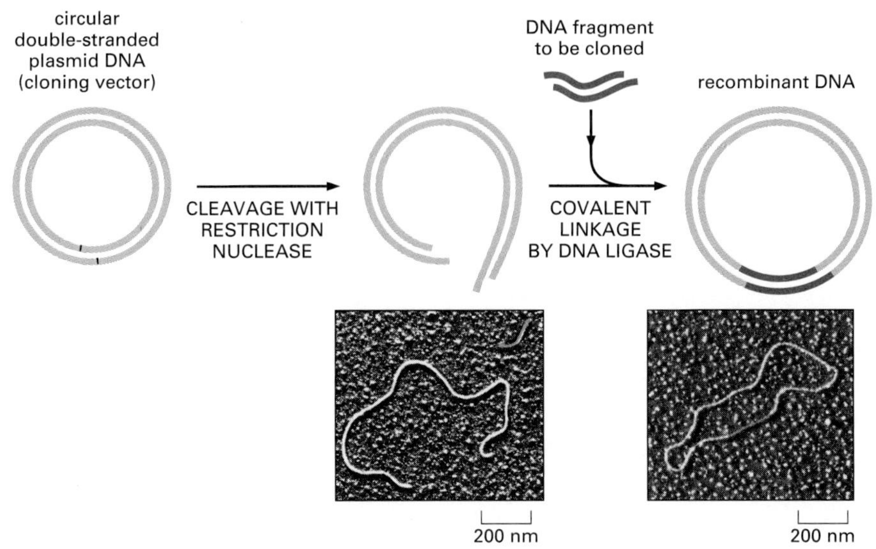

Figure 8–30 The insertion of a DNA fragment into a bacterial plasmid with the enzyme DNA ligase. The plasmid is cut open with a restriction nuclease (in this case one that produces cohesive ends) and is mixed with the DNA fragment to be cloned (which has been prepared with the same restriction nuclease), DNA ligase, and ATP. The cohesive ends base-pair, and DNA ligase seals the nicks in the DNA backbone, producing a complete recombinant DNA molecule. (Micrographs courtesy of Huntington Potter and David Dressler.)

small size from chromosomal DNA molecules, which are large and precipitate as a pellet upon centrifugation. For use as cloning vectors, the purified plasmid DNA circles are first cut with a restriction nuclease to create linear DNA molecules. The cellular DNA to be used in constructing the library is cut with the same restriction nuclease, and the resulting restriction fragments (including those containing the gene to be cloned) are then added to the cut plasmids and annealed via their cohesive ends to form recombinant DNA circles. These recombinant molecules containing foreign DNA inserts are then covalently sealed with the enzyme DNA ligase (Figure 8–30).

In the next step in preparing the library, the recombinant DNA circles are introduced into bacterial cells that have been made transiently permeable to DNA; such cells are said to be *transfected* with the plasmids. As these cells grow and divide, doubling in number every 30 minutes, the recombinant plasmids also replicate to produce an enormous number of copies of DNA circles containing the foreign DNA (Figure 8–31). Many bacterial plasmids carry genes for antibiotic resistance, a property that can be exploited to select those cells that have been successfully transfected; if the bacteria are grown in the presence of the antibiotic, only cells containing plasmids will survive. Each original bacterial cell that was initially transfected contains, in general, a different foreign DNA insert; this insert is inherited by all of the progeny cells of that bacterium, which together form a small colony in a culture dish.

For many years, plasmids were used to clone fragments of DNA of 1,000 to 30,000 nucleotide pairs. Larger DNA fragments are more difficult to handle and were harder to clone. Then researchers began to use yeast artificial chromosomes (YACs), which could handle very large pieces of DNA (Figure 8–32). Today, new plasmid vectors based on the naturally occurring F plasmid of *E. coli* are used to clone DNA fragments of 300,000 to 1 million nucleotide pairs. Unlike

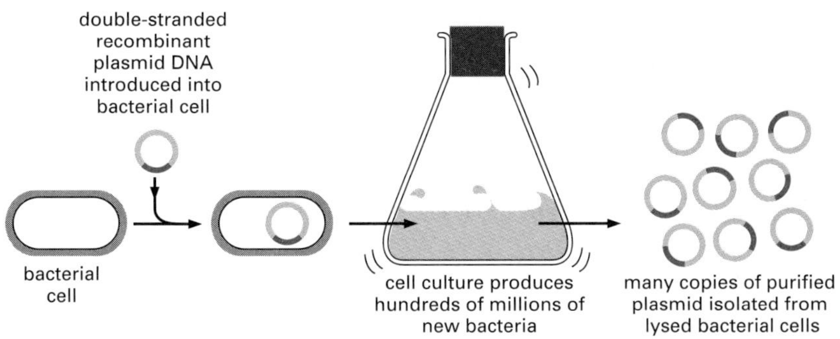

Figure 8–31 Purification and amplification of a specific DNA sequence by DNA cloning in a bacterium. To produce many copies of a particular DNA sequence, the fragment is first inserted into a plasmid vector, as shown in Figure 8–30. The resulting recombinant plasmid DNA is then introduced into a bacterium, where it can be replicated many millions of times as the bacterium multiplies.

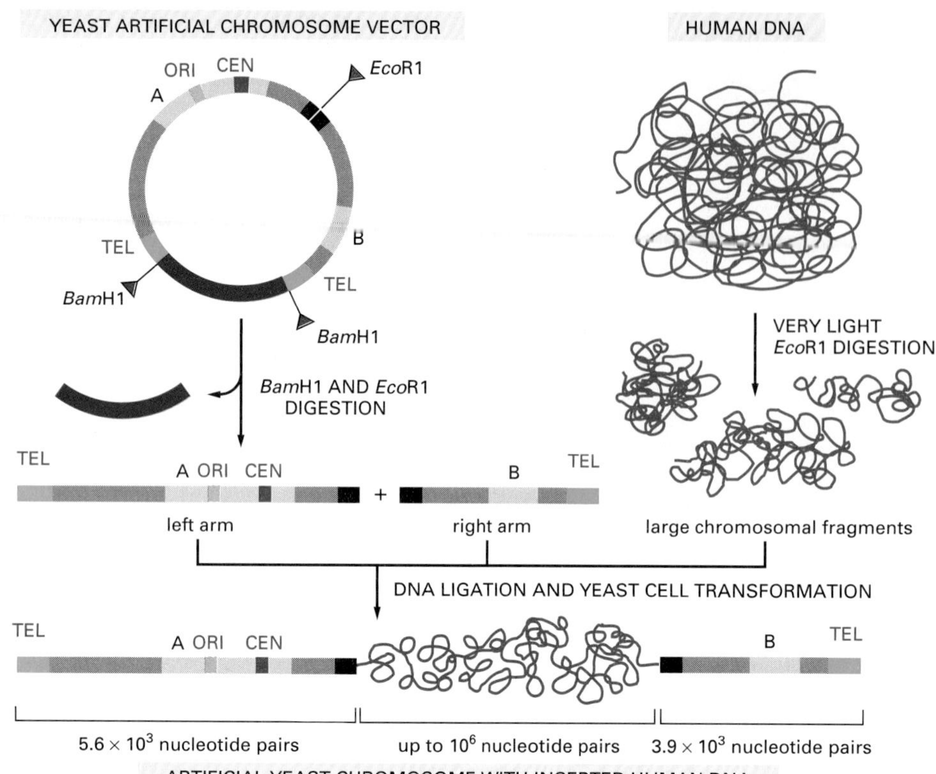

Figure 8–32 The making of a yeast artificial chromosome (YAC). A YAC vector allows the cloning of very large DNA molecules. TEL, CEN, and ORI are the telomere, centromere, and origin of replication sequences, respectively, for the yeast *Saccharomyces cerevisiae*. *Bam*H I and *Eco*R I are sites where the corresponding restriction nucleases cut the DNA double helix. The sequences denoted A and B encode enzymes that serve as selectable markers to allow the easy isolation of yeast cells that have taken up the artificial chromosome. (Adapted from D.T. Burke, G.F. Carle, and M.V. Olson, *Science* 236:806–812, 1987.)

smaller bacterial plasmids, the F plasmid—and its derivative, the **bacterial artificial chromosome (BAC)**—is present in only one or two copies per *E. coli* cell. The fact that BACs are kept in such low numbers in bacterial cells may contribute to their ability to maintain large cloned DNA sequences stably: with only a few BACs present, it is less likely that the cloned DNA fragments will become scrambled due to recombination with sequences carried on other copies of the plasmid. Because of their stability, ability to accept large DNA inserts, and ease of handling, BACs are now the preferred vector for building DNA libraries of complex organisms—including those representing the human and mouse genomes.

Two Types of DNA Libraries Serve Different Purposes

Cleaving the entire genome of a cell with a specific restriction nuclease and cloning each fragment as just described is sometimes called the "shotgun" approach to gene cloning. This technique can produce a very large number of DNA fragments—on the order of a million for a mammalian genome—which will generate millions of different colonies of transfected bacterial cells. (When working with BACs rather than typical plasmids, larger fragments can be inserted, so fewer transfected bacterial cells are required to cover the genome.) Each of these colonies is composed of a clone of cells derived from a single ancestor cell, and therefore harbors many copies of a particular stretch of the fragmented genome (Figure 8–33). Such a plasmid is said to contain a **genomic DNA clone**, and the entire collection of plasmids is called a **genomic DNA library**. But because the genomic DNA is cut into fragments at random, only some fragments contain genes. Many of the genomic DNA clones obtained from the DNA of a higher eucaryotic cell contain only noncoding DNA, which, as we discussed in Chapter 4, makes up most of the DNA in such genomes.

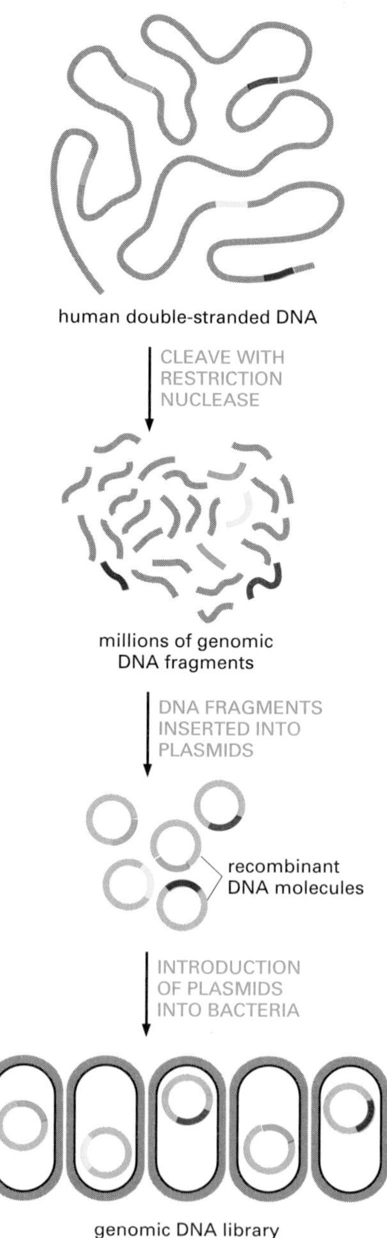

Figure 8–33 Construction of a human genomic DNA library. A genomic library is usually stored as a set of bacteria, each carrying a different fragment of human DNA. For simplicity, cloning of just a few representative fragments *(colored)* is shown. In reality, all of the *gray* DNA fragments would also be cloned.

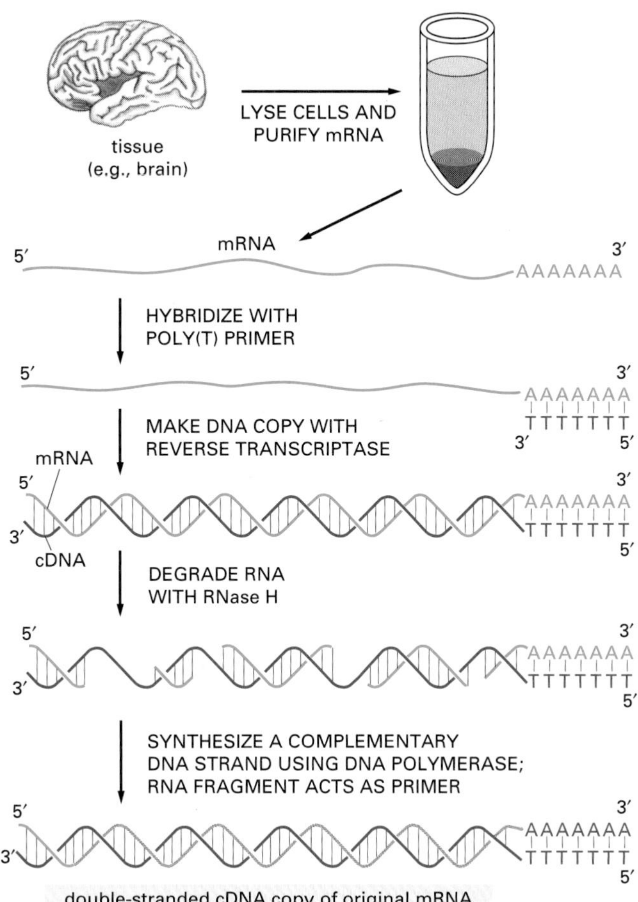

Figure 8–34 The synthesis of cDNA. Total mRNA is extracted from a particular tissue, and DNA copies (cDNA) of the mRNA molecules are produced by the enzyme reverse transcriptase (see p. 289). For simplicity, the copying of just one of these mRNAs into cDNA is illustrated. A short oligonucleotide complementary to the poly-A tail at the 3′ end of the mRNA (discussed in Chapter 6) is first hybridized to the RNA to act as a primer for the reverse transcriptase, which then copies the RNA into a complementary DNA chain, thereby forming a DNA/RNA hybrid helix. Treating the DNA/RNA hybrid with RNase H (see Figure 5–13) creates nicks and gaps in the RNA strand. The remaining single-stranded cDNA is then copied into double-stranded cDNA by the enzyme DNA polymerase. The primer for this synthesis reaction is provided by a fragment of the original mRNA, as shown. Because the DNA polymerase used to synthesize the second DNA strand can synthesize through the bound RNA molecules, the RNA fragment that is base-paired to the 3′ end of the first DNA strand usually acts as the primer for the final product of the second strand synthesis. This RNA is eventually degraded during subsequent cloning steps. As a result, the nucleotide sequences at the extreme 5′ ends of the original mRNA molecules are often absent from cDNA libraries.

An alternative strategy is to begin the cloning process by selecting only those DNA sequences that are transcribed into mRNA and thus are presumed to correspond to protein-encoding genes. This is done by extracting the mRNA (or a purified subfraction of the mRNA) from cells and then making a complementary DNA (cDNA) copy of each mRNA molecule present; this reaction is catalyzed by the reverse transcriptase enzyme of retroviruses, which synthesizes a DNA chain on an RNA template. The single-stranded DNA molecules synthesized by the reverse transcriptase are converted into double-stranded DNA molecules by DNA polymerase, and these molecules are inserted into a plasmid or virus vector and cloned (Figure 8–34). Each clone obtained in this way is called a **cDNA clone**, and the entire collection of clones derived from one mRNA preparation constitutes a **cDNA library**.

There are important differences between genomic DNA clones and cDNA clones, as illustrated in Figure 8–35. Genomic clones represent a random sample of all of the DNA sequences in an organism and, with very rare exceptions, are the same regardless of the cell type used to prepare them. By contrast, cDNA clones contain only those regions of the genome that have been transcribed into mRNA. Because the cells of different tissues produce distinct sets of mRNA molecules, a distinct cDNA library is obtained for each type of cell used to prepare the library.

cDNA Clones Contain Uninterrupted Coding Sequences

The use of a cDNA library for gene cloning has several advantages. First, some proteins are produced in very large quantities by specialized cells. In this case, the mRNA encoding the protein is likely to be produced in such large quantities that a cDNA library prepared from the cells is highly enriched for the cDNA molecules encoding the protein, greatly reducing the problem of identifying the desired clone in the library (see Figure 8–35). Hemoglobin, for example, is made

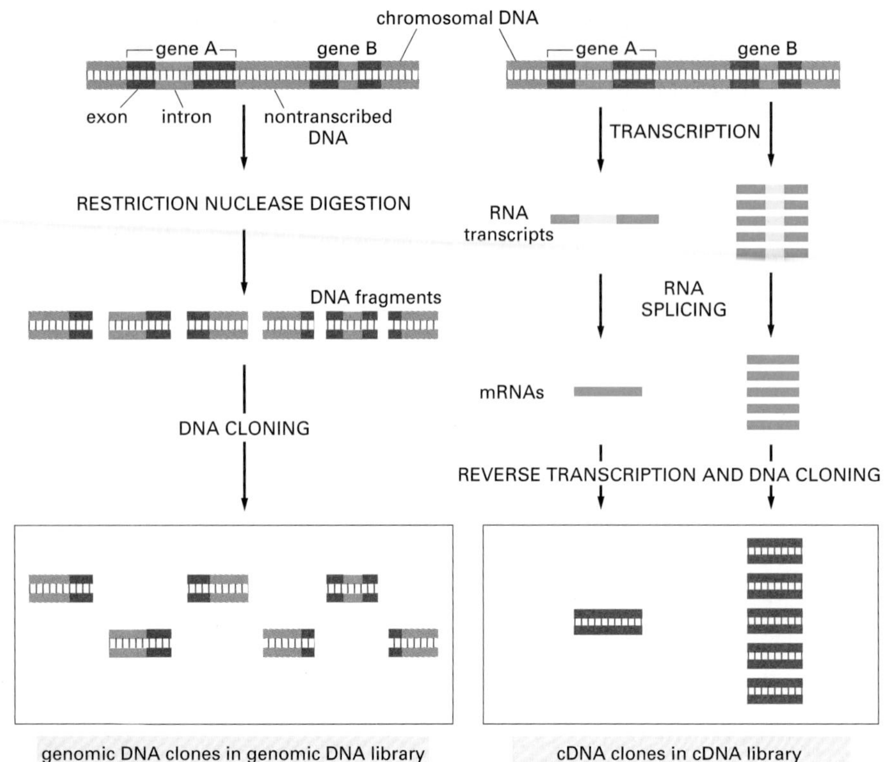

Figure 8–35 The differences between cDNA clones and genomic DNA clones derived from the same region of DNA. In this example gene A is infrequently transcribed, whereas gene B is frequently transcribed, and both genes contain introns *(green)*. In the genomic DNA library, both the introns and the nontranscribed DNA *(pink)* are included in the clones, and most clones contain, at most, only part of the coding sequence of a gene *(red)*. In the cDNA clones the intron sequences *(yellow)* have been removed by RNA splicing during the formation of the mRNA *(blue)*, and a continuous coding sequence is therefore present in each clone. Because gene B is transcribed more abundantly than in gene A in the cells from which the cDNA library was made, it is represented much more frequently than A in the cDNA library. In contrast, A and B are in principle represented equally in the genomic DNA library.

in large amounts by developing erythrocytes (red blood cells); for this reason the globin genes were among the first to be cloned.

By far the most important advantage of cDNA clones is that they contain the uninterrupted coding sequence of a gene. As we have seen, eucaryotic genes usually consist of short coding sequences of DNA (exons) separated by much longer noncoding sequences (introns); the production of mRNA entails the removal of the noncoding sequences from the initial RNA transcript and the splicing together of the coding sequences. Neither bacterial nor yeast cells will make these modifications to the RNA produced from a gene of a higher eucaryotic cell. Thus, when the aim of the cloning is either to deduce the amino acid sequence of the protein from the DNA sequence or to produce the protein in bulk by expressing the cloned gene in a bacterial or yeast cell, it is much preferable to start with cDNA.

Genomic and cDNA libraries are inexhaustible resources that are widely shared among investigators. Today, many such libraries are also available from commercial sources.

Isolated DNA Fragments Can Be Rapidly Sequenced

In the late 1970s methods were developed that allowed the nucleotide sequence of any purified DNA fragment to be determined simply and quickly. They have made it possible to determine the complete DNA sequences of tens of thousands of genes, and many organisms have had their DNA genomes fully sequenced (see Table 1–1, p. 20). The volume of DNA sequence information is now so large (many tens of billions of nucleotides) that powerful computers must be used to store and analyze it.

Large volume DNA sequencing was made possible through the development in the mid-1970s of the **dideoxy method** for sequencing DNA, which is based on *in vitro* DNA synthesis performed in the presence of chain-terminating dideoxyribonucleoside triphosphates (Figure 8–36).

Although the same basic method is still used today, many improvements have been made. DNA sequencing is now completely automated: robotic devices mix the reagents and then load, run, and read the order of the nucleotide bases from the gel. This is facilitated by using chain-terminating nucleotides that are each labeled with a different colored fluorescent dye; in this case, all four synthesis reactions can be performed in the same tube, and the products can be separated in a single lane of a gel. A detector positioned near the bottom of the

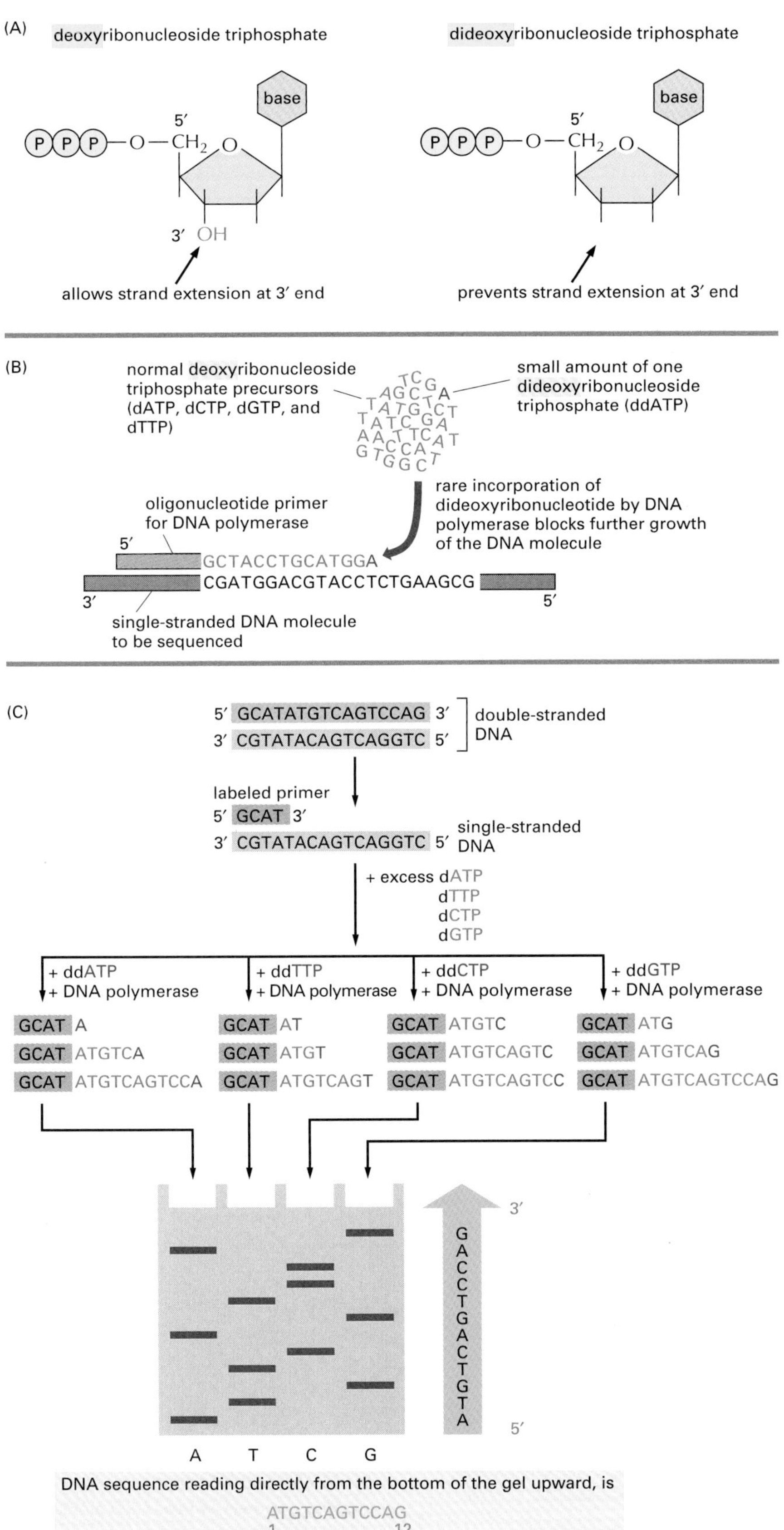

Figure 8–36 The enzymatic—or dideoxy—method of sequencing DNA. (A) This method relies on the use of dideoxyribonucleoside triphosphates, derivatives of the normal deoxyribonucleoside triphosphates that lack the 3′ hydroxyl group. (B) Purified DNA is synthesized *in vitro* in a mixture that contains single-stranded molecules of the DNA to be sequenced *(gray)*, the enzyme DNA polymerase, a short primer DNA *(orange)* to enable the polymerase to start DNA synthesis, and the four deoxyribonucleoside triphosphates (dATP, dCTP, dGTP, dTTP: *green* A, C, G, and T). If a dideoxyribonucleotide analog *(red)* of one of these nucleotides is also present in the nucleotide mixture, it can become incorporated into a growing DNA chain. Because this chain now lacks a 3′ OH group, the addition of the next nucleotide is blocked, and the DNA chain terminates at that point. In the example illustrated, a small amount of dideoxyATP (ddATP, symbolized here as a *red* A) has been included in the nucleotide mixture. It competes with an excess of the normal deoxyATP (dATP, *green* A), so that ddATP is occasionally incorporated, at random, into a growing DNA strand. This reaction mixture will eventually produce a set of DNAs of different lengths complementary to the template DNA that is being sequenced and terminating at each of the different A's. The exact lengths of the DNA synthesis products can then be used to determine the position of each A in the growing chain. (C) To determine the complete sequence of a DNA fragment, the double-stranded DNA is first separated into its single strands and one of the strands is used as the template for sequencing. Four different chain-terminating dideoxyribonucleoside triphosphates (ddATP, ddCTP, ddGTP, ddTTP, again shown in *red)* are used in four separate DNA synthesis reactions on copies of the same single-stranded DNA template *(gray)*. Each reaction produces a set of DNA copies that terminate at different points in the sequence. The products of these four reactions are separated by electrophoresis in four parallel lanes of a polyacrylamide gel (labeled here A, T, C, and G). The newly synthesized fragments are detected by a label (either radioactive or fluorescent) that has been incorporated either into the primer or into one of the deoxyribonucleoside triphosphates used to extend the DNA chain. In each lane, the bands represent fragments that have terminated at a given nucleotide (e.g., A in the leftmost lane) but at different positions in the DNA. By reading off the bands in order, starting at the bottom of the gel and working across all lanes, the DNA sequence of the newly synthesized strand can be determined. The sequence is given in the *green arrow* to the right of the gel. This sequence is identical to that of the 5′ → 3′ strand *(green)* of the original double-stranded DNA molecule.

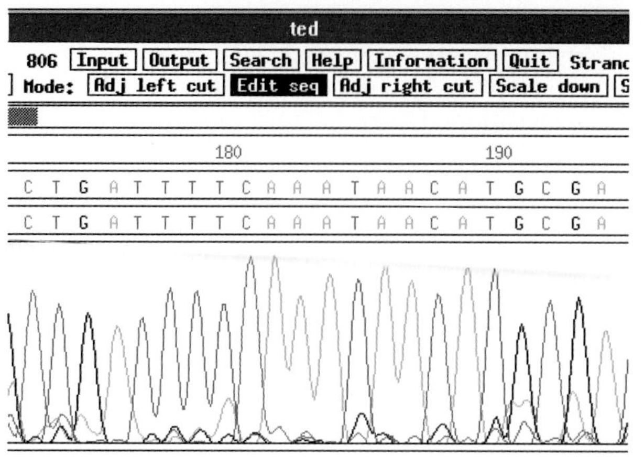

Figure 8–37 Automated DNA sequencing. Shown here is a tiny part of the data from an automated DNA-sequencing run as it appears on the computer screen. Each colored peak represents a nucleotide in the DNA sequence—a clear stretch of nucleotide sequence can be read here between positions 173 and 194 from the start of the sequence. This particular example is taken from the international project that determined the complete nucleotide sequence of the genome of the plant *Arabidopsis*. (Courtesy of George Murphy.)

gel reads and records the color of the fluorescent label on each band as it passes through a laser beam (Figure 8–37). A computer then reads and stores this nucleotide sequence.

Nucleotide Sequences Are Used to Predict the Amino Acid Sequences of Proteins

Now that DNA sequencing is so rapid and reliable, it has become the preferred method for determining, indirectly, the amino acid sequences of most proteins. Given a nucleotide sequence that encodes a protein, the procedure is quite straightforward. Although in principle there are six different reading frames in which a DNA sequence can be translated into protein (three on each strand), the correct one is generally recognizable as the only one lacking frequent stop codons (Figure 8–38). As we saw when we discussed the genetic code in Chapter 6, a random sequence of nucleotides, read in frame, will encode a stop signal for protein synthesis about once every 20 amino acids. Those nucleotide sequences that encode a stretch of amino acids much longer than this are candidates for presumptive exons, and they can be translated (by computer) into amino acid sequences and checked against databases for similarities to known proteins from other organisms. If necessary, a limited amount of amino acid sequence can then be determined from the purified protein to confirm the sequence predicted from the DNA.

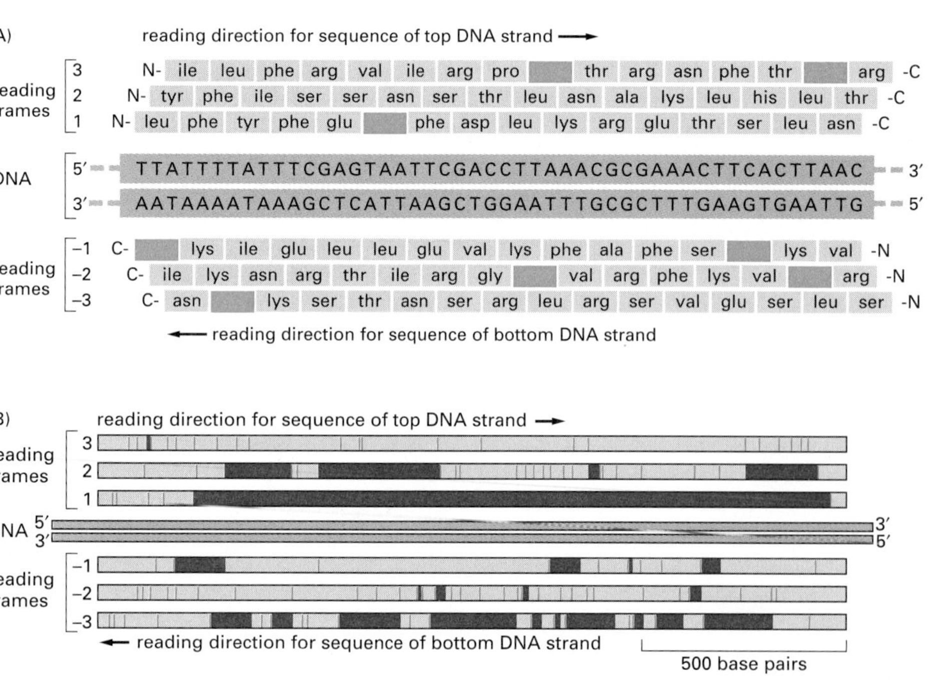

Figure 8–38 Finding the regions in a DNA sequence that encode a protein. (A) Any region of the DNA sequence can, in principle, code for six different amino acid sequences, because any one of three different reading frames can be used to interpret the nucleotide sequence on each strand. Note that a nucleotide sequence is always read in the 5′-to-3′ chain direction and encodes a polypeptide from the amino (N) to the carboxyl (C) terminus. For a random nucleotide sequence read in a particular frame, a stop signal for protein synthesis is encountered, on average, about once every 21 amino acids (once every 63 nucleotides). In this sample sequence of 48 base pairs, each such signal *(stop codon)* is colored *green*, and only reading frame 2 lacks a stop signal. (B) Search of a 1700 base-pair DNA sequence for a possible protein-encoding sequence. The information is displayed as in (A), with each stop signal for protein synthesis denoted by a *green line*. In addition, all of the regions between possible start and stop signals for protein synthesis (see pp. 348–350) are displayed as *red bars*. Only reading frame 1 actually encodes a protein, which is 475 amino acid residues long.

The problem comes, however, in determining which nucleotide sequences—within a whole genome sequence—represent genes that encode proteins. Identifying genes is easiest when the DNA sequence is from a bacterial or archeal chromosome, which lacks introns, or from a cDNA clone. The location of genes in these nucleotide sequences can be predicted by examining the DNA for certain distinctive features (discussed in Chapter 6). Briefly these genes that encode proteins are identified by searching the nucleotide sequence for open reading frames (ORFs) that begin with an initiation codon, usually ATG, and end with a termination codon, TAA, TAG, or TGA. To minimize errors, computers used to search for ORFs are often directed to count as genes only those sequences that are longer than, say, 100 codons in length.

For more complex genomes, such as those of eucaryotes, the process is complicated by the presence of large introns embedded within the coding portion of genes. In many multicellular organisms, including humans, the average exon is only 150 nucleotides long. Thus in eucaryotes, one must also search for other features that signal the presence of a gene, for example, sequences that signal an intron/exon boundary or distinctive upstream regulatory regions.

A second major approach to identifying the coding regions in chromosomes is through the characterization of the nucleotide sequences of the detectable mRNAs (in the form of cDNAs). The mRNAs (and the cDNAs produced from them) lack introns, regulatory DNA sequences, and the nonessential "spacer" DNA that lies between genes. It is therefore useful to sequence large numbers of cDNAs to produce a very large collection (called a database) of the coding sequences of an organism. These sequences are then readily used to distinguish the exons from the introns in the long chromosomal DNA sequences that correspond to genes.

Finally, nucleotide sequences that are conserved between closely related organisms usually encode proteins. Comparison of these conserved sequences in different species can also provide insight into the function of a particular protein or gene, as we see later in the chapter.

The Genomes of Many Organisms Have Been Fully Sequenced

Owing in large part to the automation of DNA sequencing, the genomes of many organisms have been fully sequenced; these include plant chloroplasts and animal mitochondria, large numbers of bacteria and archea, and many of the model organisms that are studied routinely in the laboratory, including several yeasts, a nematode worm, the fruit fly *Drosophila*, the model plant *Arabidopsis*, the mouse, and, last but not least, humans. Researchers have also deduced the complete DNA sequences for a wide variety of human pathogens. These include the bacteria that cause cholera, tuberculosis, syphilis, gonorrhea, Lyme disease, and stomach ulcers, as well as hundreds of viruses—including smallpox virus and Epstein–Barr virus (which causes infectious mononucleosis). Examination of the genomes of these pathogens should provide clues about what makes them virulent, and will also point the way to new and more effective treatments.

Haemophilus influenzae (a bacterium that can cause ear infections or meningitis in children) was the first organism to have its complete genome sequence—all 1.8 million nucleotides—determined by the shotgun sequencing method, the most common strategy used today. In the shotgun method, long sequences of DNA are broken apart randomly into many shorter fragments. Each fragment is then sequenced and a computer is used to order these pieces into a whole chromosome or genome, using sequence overlap to guide the assembly. The shotgun method is the technique of choice for sequencing small genomes. Although larger, more repetitive genome sequences are more tricky to assemble, the shotgun method has been useful for sequencing the genomes of *Drosophila melanogaster*, mouse, and human.

With new sequences appearing at a steadily accelerating pace in the scientific literature, comparison of the complete genome sequences of different organisms allows us to trace the evolutionary relationships among genes and organisms, and to discover genes and predict their functions. Assigning functions to genes often involves comparing their sequences with related sequences from model organisms that have been well characterized in the laboratory, such as the bacterium *E. coli*, the yeasts *S. cerevisiae* and *S. pombe*, the nematode worm *C. elegans*, and the fruit fly *Drosophila* (discussed in Chapter 1).

Although the organisms whose genomes have been sequenced share many cellular pathways and possess many proteins that are homologous in their amino acid sequences or structure, the functions of a very large number of newly identified proteins remain unknown. Some 15–40% of the proteins encoded by these sequenced genomes do not resemble any other protein that has been characterized functionally. This observation underscores one of the limitations of the emerging field of genomics: although comparative analysis of genomes reveals a great deal of information about the relationships between genes and organisms, it often does not provide immediate information about how these genes function, or what roles they have in the physiology of an organism. Comparison of the full gene complement of several thermophilic bacteria, for example, does not reveal why these bacteria thrive at temperatures exceeding 70°C. And examination of the genome of the incredibly radioresistant bacterium *Deinococcus radiodurans* does not explain how this organism can survive a blast of radiation that can shatter glass. Further biochemical and genetic studies, like those described in the final sections of this chapter, are required to determine how genes function in the context of living organisms.

Selected DNA Segments Can Be Cloned in a Test Tube by a Polymerase Chain Reaction

Now that so many genome sequences are available, genes can be cloned directly without the need to construct DNA libraries first. A technique called the **polymerase chain reaction (PCR)** makes this rapid cloning possible. PCR allows the DNA from a selected region of a genome to be amplified a billionfold, effectively "purifying" this DNA away from the remainder of the genome.

Two sets of DNA oligonucleotides, chosen to flank the desired nucleotide sequence of the gene, are synthesized by chemical methods. These oligonucleotides are then used to prime DNA synthesis on single strands generated by heating the DNA from the entire genome. The newly synthesized DNA is produced in a reaction catalyzed *in vitro* by a purified DNA polymerase, and the primers remain at the 5′ ends of the final DNA fragments that are made (Figure 8–39A).

Nothing special is produced in the first cycle of DNA synthesis; the power of the PCR method is revealed only after repeated rounds of DNA synthesis. Every cycle doubles the amount of DNA synthesized in the previous cycle. Because each cycle requires a brief heat treatment to separate the two strands of the template DNA double helix, the technique requires the use of a special DNA polymerase, isolated from a thermophilic bacterium, that is stable at much higher temperatures than normal, so that it is not denatured by the repeated heat treatments. With each round of DNA synthesis, the newly generated fragments serve as templates in their turn, and within a few cycles the predominant product is a single species of DNA fragment whose length corresponds to the distance between the two original primers (see Figure 8–39B).

In practice, 20–30 cycles of reaction are required for effective DNA amplification, with the products of each cycle serving as the DNA templates for the next—hence the term polymerase "chain reaction." A single cycle requires only about 5 minutes, and the entire procedure can be easily automated. PCR thereby makes possible the "cell-free molecular cloning" of a DNA fragment in a few hours, compared with the several days required for standard cloning procedures. This technique is now used routinely to clone DNA from genes of interest directly—starting either from genomic DNA or from mRNA isolated from cells (Figure 8–40).

The PCR method is extremely sensitive; it can detect a single DNA molecule in a sample. Trace amounts of RNA can be analyzed in the same way by first

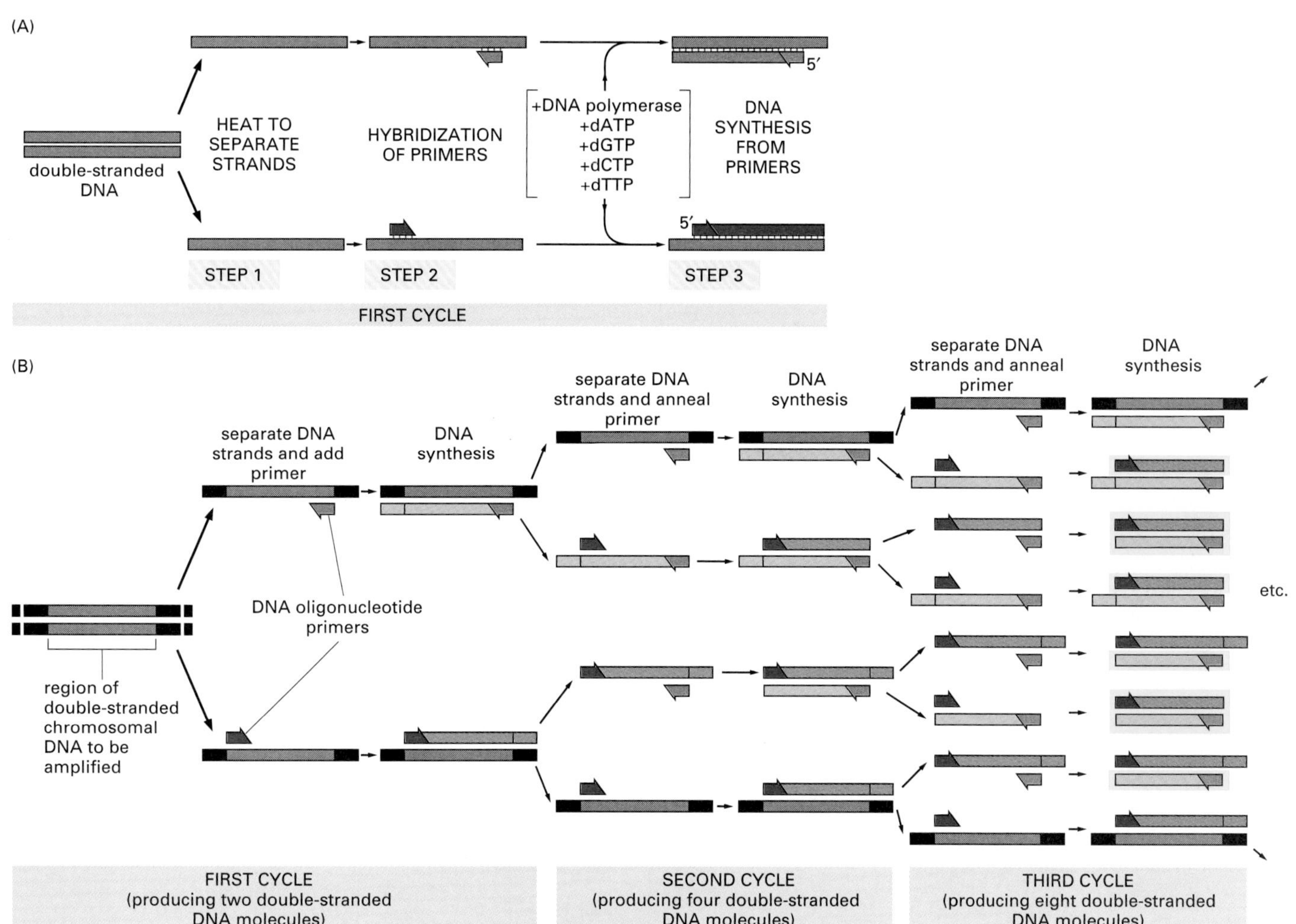

Figure 8–39 Amplification of DNA using the PCR technique. Knowledge of the DNA sequence to be amplified is used to design two synthetic DNA oligonucleotides, each complementary to the sequence on one strand of the DNA double helix at opposite ends of the region to be amplified. These oligonucleotides serve as primers for *in vitro* DNA synthesis, which is performed by a DNA polymerase, and they determine the segment of the DNA that is amplified. (A) PCR starts with a double-stranded DNA, and each cycle of the reaction begins with a brief heat treatment to separate the two strands (step 1). After strand separation, cooling of the DNA in the presence of a large excess of the two primer DNA oligonucleotides allows these primers to hybridize to complementary sequences in the two DNA strands (step 2). This mixture is then incubated with DNA polymerase and the four deoxyribonucleoside triphosphates so that DNA is synthesized, starting from the two primers (step 3). The entire cycle is then begun again by a heat treatment to separate the newly synthesized DNA strands. (B) As the procedure is performed over and over again, the newly synthesized fragments serve as templates in their turn, and within a few cycles the predominant DNA is identical to the sequence bracketed by and including the two primers in the original template. Of the DNA put into the original reaction, only the sequence bracketed by the two primers is amplified because there are no primers attached anywhere else. In the example illustrated in (B), three cycles of reaction produce 16 DNA chains, 8 of which *(boxed in yellow)* are the same length as and correspond exactly to one or the other strand of the original bracketed sequence shown at the far left; the other strands contain extra DNA downstream of the original sequence, which is replicated in the first few cycles. After three more cycles, 240 of the 256 DNA chains correspond exactly to the original bracketed sequence, and after several more cycles, essentially all of the DNA strands have this unique length.

transcribing them into DNA with reverse transcriptase. The PCR cloning technique has largely replaced Southern blotting for the diagnosis of genetic diseases and for the detection of low levels of viral infection. It also has great promise in forensic medicine as a means of analyzing minute traces of blood or other tissues—even as little as a single cell—and identifying the person from whom they came by his or her genetic "fingerprint" (Figure 8–41).

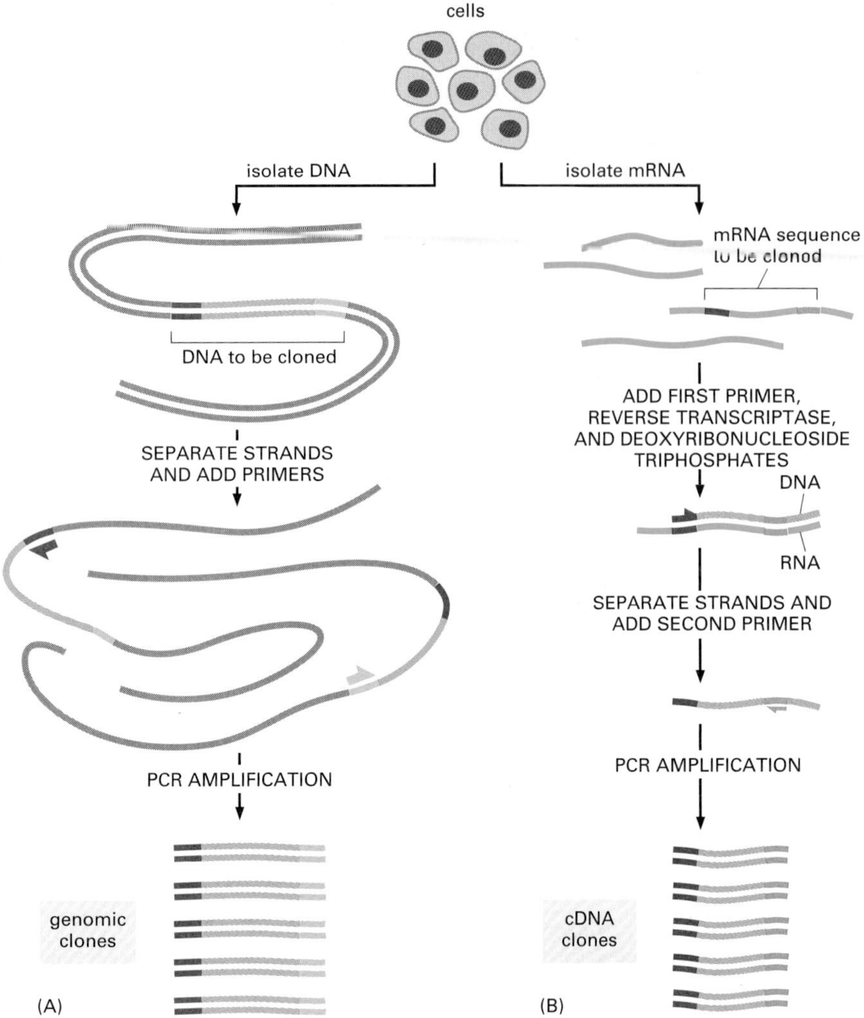

Figure 8–40 Use of PCR to obtain a genomic or cDNA clone. (A) To obtain a genomic clone by using PCR, chromosomal DNA is first purified from cells. PCR primers that flank the stretch of DNA to be cloned are added, and many cycles of the reaction are completed (see Figure 8–39). Since only the DNA between (and including) the primers is amplified, PCR provides a way to obtain a short stretch of chromosomal DNA selectively in a pure form. (B) To use PCR to obtain a cDNA clone of a gene, mRNA is first purified from cells. The first primer is then added to the population of mRNAs, and reverse transcriptase is used to make a complementary DNA strand. The second primer is then added, and the single-stranded DNA molecule is amplified through many cycles of PCR, as shown in Figure 8–39. For both types of cloning, the nucleotide sequence of at least part of the region to be cloned must be known beforehand.

Cellular Proteins Can Be Made in Large Amounts Through the Use of Expression Vectors

Fifteen years ago, the only proteins in a cell that could be studied easily were the relatively abundant ones. Starting with several hundred grams of cells, a major protein—one that constitutes 1% or more of the total cellular protein—can be purified by sequential chromatography steps to yield perhaps 0.1 g (100 mg) of pure protein. This amount was sufficient for conventional amino acid sequencing, for detailed analysis of biochemical activities, and for the production of antibodies, which could then be used to localize the protein in the cell. Moreover, if suitable crystals could be grown (often a difficult task), the three-dimensional structure of the protein could be determined by x-ray diffraction techniques, as we will discuss later. The structure and function of many abundant

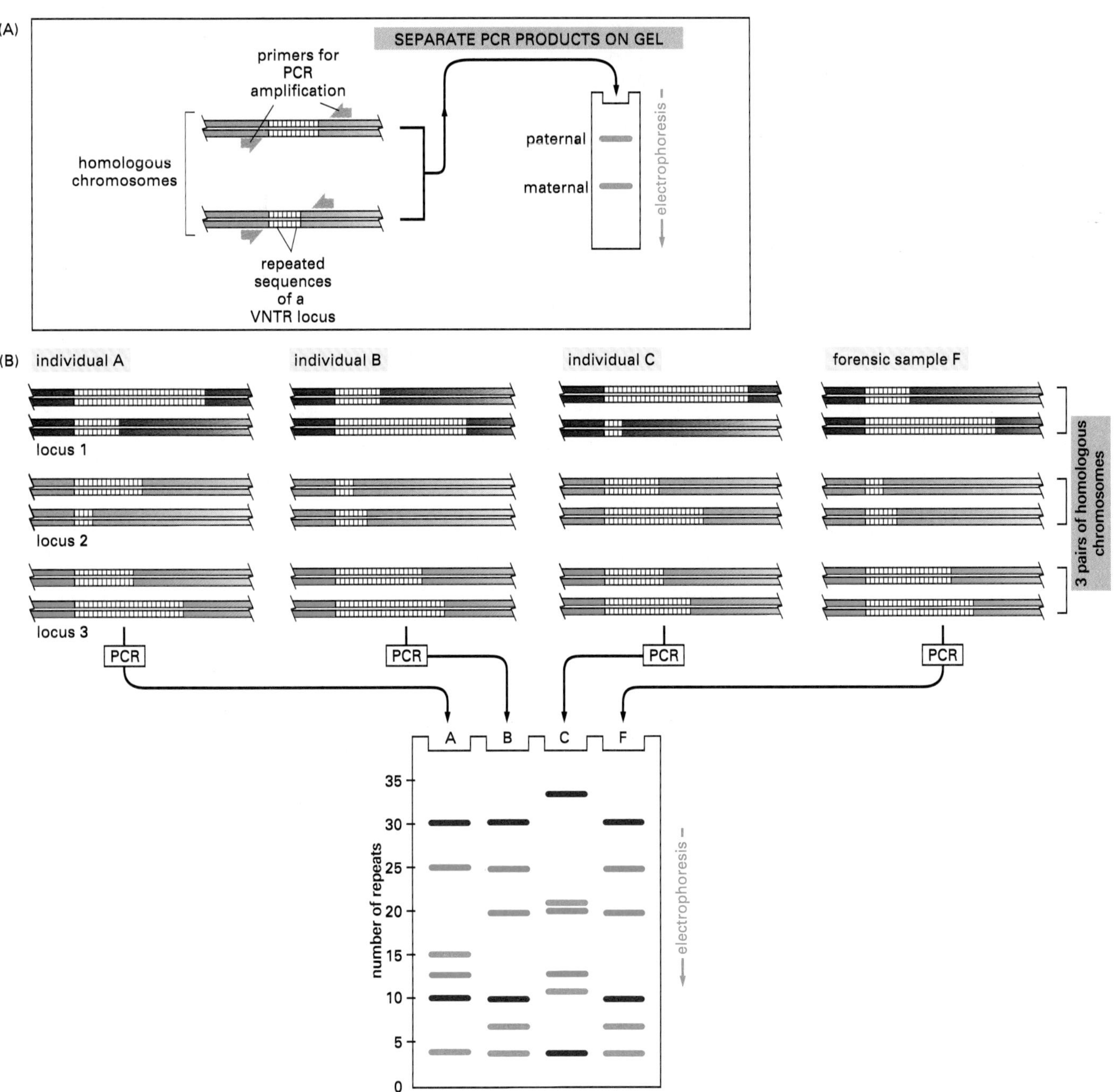

Figure 8–41 How PCR is used in forensic science. (A) The DNA sequences that create the variability used in this analysis contain runs of short, repeated sequences, such as CACACA . . . , which are found in various positions (loci) in the human genome. The number of repeats in each run can be highly variable in the population, ranging from 4 to 40 in different individuals. A run of repeated nucleotides of this type is commonly referred to as a *hypervariable microsatellite* sequence—also known as a VNTR *(variable number of tandem repeat)* sequence. Because of the variability in these sequences at each locus, individuals usually inherit a different variant from their mother and from their father; two unrelated individuals therefore do not usually contain the same pair of sequences. A PCR analysis using primers that bracket the locus produces a pair of bands of amplified DNA from each individual, one band representing the maternal variant and the other representing the paternal variant. The length of the amplified DNA, and thus the position of the band it produces after electrophoresis, depends on the exact number of repeats at the locus.
(B) In the schematic example shown here, the same three VNTR loci are analyzed (requiring three different pairs of specially selected oligonucleotide primers) from three suspects (individuals A, B, and C), producing six DNA bands for each person after polyacrylamide gel electrophoresis. Although some individuals have several bands in common, the overall pattern is quite distinctive for each. The band pattern can therefore serve as a "fingerprint" to identify an individual nearly uniquely. The fourth lane (F) contains the products of the same reactions carried out on a forensic sample. The starting material for such a PCR can be a single hair or a tiny sample of blood that was left at the crime scene. When examining the variability at 5 to 10 different VNTR loci, the odds that two random individuals would share the same genetic pattern by chance can be approximately one in 10 billion. In the case shown here, individuals A and C can be eliminated from further enquiries, whereas individual B remains a clear suspect for committing the crime. A similar approach is now routinely used for paternity testing.

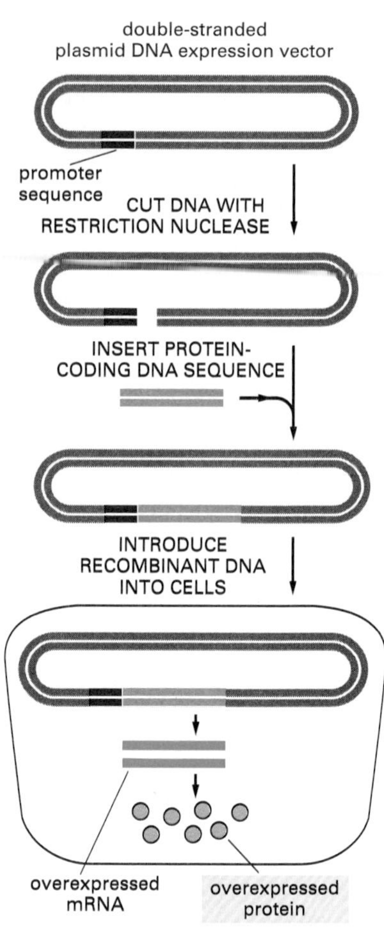

Figure 8–42 Production of large amounts of a protein from a protein-coding DNA sequence cloned into an expression vector and introduced into cells. A plasmid vector has been engineered to contain a highly active promoter, which causes unusually large amounts of mRNA to be produced from an adjacent protein-coding gene inserted into the plasmid vector. Depending on the characteristics of the cloning vector, the plasmid is introduced into bacterial, yeast, insect, or mammalian cells, where the inserted gene is efficiently transcribed and translated into protein.

proteins—including hemoglobin, trypsin, immunoglobulin, and lysozyme—were analyzed in this way.

The vast majority of the thousands of different proteins in a eucaryotic cell, however, including many with crucially important functions, are present in very small amounts. For most of them it is extremely difficult, if not impossible, to obtain more than a few micrograms of pure material. One of the most important contributions of DNA cloning and genetic engineering to cell biology is that they have made it possible to produce any of the cell's proteins in nearly unlimited amounts.

Large amounts of a desired protein are produced in living cells by using **expression vectors** (Figure 8–42). These are generally plasmids that have been designed to produce a large amount of a stable mRNA that can be efficiently translated into protein in the transfected bacterial, yeast, insect, or mammalian cell. To prevent the high level of the foreign protein from interfering with the transfected cell's growth, the expression vector is often designed so that the synthesis of the foreign mRNA and protein can be delayed until shortly before the cells are harvested (Figure 8–43).

Because the desired protein made from an expression vector is produced inside a cell, it must be purified away from the host cell proteins by chromatography following cell lysis; but because it is such a plentiful species in the cell lysate (often 1–10% of the total cell protein), the purification is usually easy to accomplish in only a few steps. Many expression vectors have been designed to add a molecular tag—a cluster of histidine residues or a small marker protein—to the expressed protein to make possible easy purification by affinity chromatography, as discussed previously (see pp. 483–484). A variety of expression vectors are available, each engineered to function in the type of cell in which the protein is to be made. In this way cells can be induced to make vast quantities of medically useful proteins—such as human insulin and growth hormone, interferon, and viral antigens for vaccines. More generally, these methods make it possible to produce every protein—even those that may be present in only a few copies per cell—in large enough amounts to be used in the kinds of detailed structural and functional studies that we discuss in the next section (Figure 8–44).

DNA technology can also be used to produce large amounts of any RNA molecule whose gene has been isolated. Studies of RNA splicing, protein synthesis, and RNA-based enzymes, for example, ar greatly facilitated by the availability of pure RNA molecules. Most RNAs are present in only tiny quantities in cells, and they are very difficult to purify away from other cellular components—especially from the many thousands of other RNAs present in the cell. But any RNA of interest can be synthesized efficiently *in vitro* by transcription of its DNA sequence with a highly efficient viral RNA polymerase. The single species of RNA produced is then easily purified away from the DNA template and the RNA polymerase.

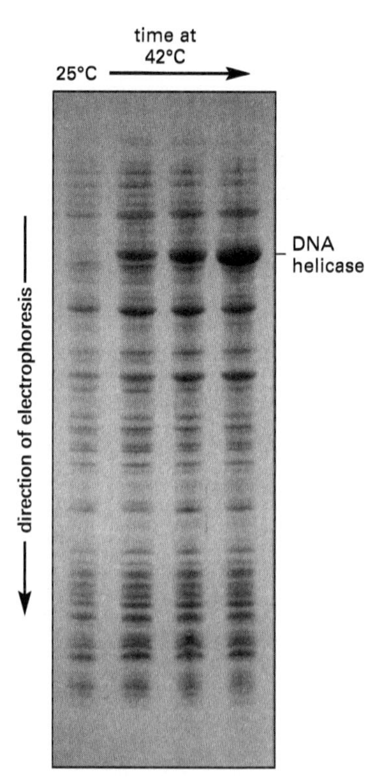

Figure 8–43 Production of large amounts of a protein by using a plasmid expression vector. In this example, bacterial cells have been transfected with the coding sequence for an enzyme, DNA helicase; transcription from this coding sequence is under the control of a viral promoter that becomes active only at temperatures of 37°C or higher. The total cell protein has been analyzed by SDS-polyacrylamide gel electrophoresis, either from bacteria grown at 25°C (no helicase protein made), or after a shift of the same bacteria to 42°C for up to 2 hours (helicase protein has become the most abundant protein species in the lysate). (Courtesy of Jack Barry.)

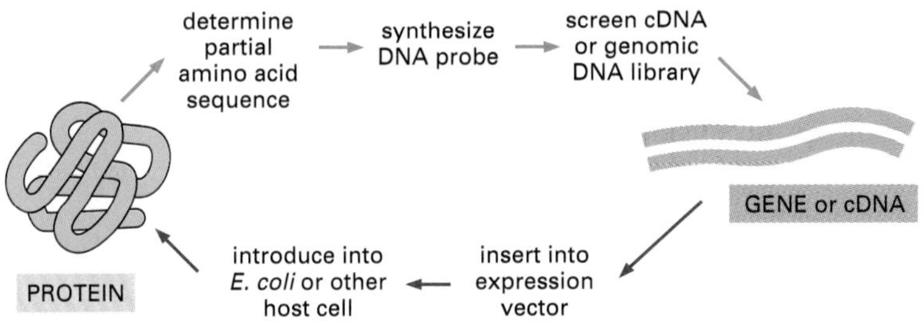

Figure 8–44 Knowledge of the molecular biology of cells makes it possible to experimentally move from gene to protein and from protein to gene. A small quantity of a purified protein is used to obtain a partial amino acid sequence. This provides sequence information that enables the corresponding gene to be cloned from a DNA library. Once the gene has been cloned, its protein-coding sequence can be inserted into an expression vector and used to produce large quantities of the protein from genetically engineered cells.

Summary

DNA cloning allows a copy of any specific part of a DNA or RNA sequence to be selected from the millions of other sequences in a cell and produced in unlimited amounts in pure form. DNA sequences can be amplified after cutting chromosomal DNA with a restriction nuclease and inserting the resulting DNA fragments into the chromosome of a self-replicating genetic element. Plasmid vectors are generally used and the resulting "genomic DNA library" is housed in millions of bacterial cells, each carrying a different cloned DNA fragment. Individual cells that are allowed to proliferate produce large amounts of a single cloned DNA fragment from this library. As an alternative, the polymerase chain reaction (PCR) allows DNA cloning to be performed directly with a purified, thermostable DNA polymerase—providing that the DNA sequence of interest is already known.

The procedures used to obtain DNA clones that correspond in sequence to mRNA molecules are the same except that a DNA copy of the mRNA sequence, called cDNA, is first made. Unlike genomic DNA clones, cDNA clones lack intron sequences, making them the clones of choice for analyzing the protein product of a gene.

Nucleic acid hybridization reactions provide a sensitive means of detecting a gene or any other nucleotide sequence of choice. Under stringent hybridization conditions (a combination of solvent and temperature where a perfect double helix is barely stable), two strands can pair to form a "hybrid" helix only if their nucleotide sequences are almost perfectly complementary. The enormous specificity of this hybridization reaction allows any single-stranded sequence of nucleotides to be labeled with a radioisotope or chemical and used as a probe to find a complementary partner strand, even in a cell or cell extract that contains millions of different DNA and RNA sequences. Probes of this type are widely used to detect the nucleic acids corresponding to specific genes, both to facilitate their purification and characterization and to localize them in cells, tissues, and organisms.

The nucleotide sequence of purified DNA fragments can be determined rapidly and simply by using highly automated techniques based on the dideoxy method for sequencing DNA. This technique has made it possible to determine the complete DNA sequences of tens of thousands of genes and to completely sequence the genomes of many organisms. Comparison of the genome sequences of different organisms allows us to trace the evolutionary relationships among genes and organisms, and it has proved valuable for discovering new genes and predicting their function.

Taken together, these techniques have made it possible to identify, isolate, and sequence genes from any organism of interest. Related technologies allow scientists to produce the protein products of these genes in the large quantities needed for detailed analyses of their structure and function, as well as for medical purposes.

ANALYZING PROTEIN STRUCTURE AND FUNCTION

Proteins perform most of the work of living cells. This versatile class of macromolecule is involved in virtually every cellular process: proteins replicate and transcribe DNA, and produce, process, and secrete other proteins. They control cell division, metabolism, and the flow of materials and information into and out of the cell. Understanding how cells work requires understanding how proteins function.

The question of what a protein does inside a living cell is not a simple one to answer. Imagine isolating an uncharacterized protein and discovering that its structure and amino acid sequence suggest that it acts as a protein kinase. Simply knowing that the protein can add a phosphate group to serine residues, for example, does not reveal how it functions in a living organism. Additional information is required to understand the context in which the biochemical activity is used. Where is this kinase located in the cell and what are its protein targets? In which tissues is it active? Which pathways does it influence? What role does it have in the growth or development of the organism?

In this section, we discuss the methods currently used to characterize protein structure and function. We begin with an examination of the techniques used to determine the three-dimensional structure of purified proteins. We then discuss methods that are used to predict how a protein functions, based on its homology to other known proteins and its location inside the cell. Finally, because most proteins act in concert with other proteins, we present techniques for detecting protein–protein interactions. But these approaches only begin to define how a protein might work inside a cell. In the last section of this chapter, we discuss how genetic approaches are used to dissect and analyze the biological processes in which a given protein functions.

The Diffraction of X-rays by Protein Crystals Can Reveal a Protein's Exact Structure

Starting with the amino acid sequence of a protein, one can often predict which secondary structural elements, such as membrane-spanning α helices, will be present in the protein. It is presently not possible, however, to deduce reliably the three-dimensional folded structure of a protein from its amino acid sequence unless its amino acid sequence is very similar to that of a protein whose three-dimensional structure is already known. The main technique that has been used to discover the three-dimensional structure of molecules, including proteins, at atomic resolution is **x-ray crystallography**.

X-rays, like light, are a form of electromagnetic radiation, but they have a much shorter wavelength, typically around 0.1 nm (the diameter of a hydrogen atom). If a narrow parallel beam of x-rays is directed at a sample of a pure protein, most of the x-rays pass straight through it. A small fraction, however, is scattered by the atoms in the sample. If the sample is a well-ordered crystal, the scattered waves reinforce one another at certain points and appear as diffraction spots when the x-rays are recorded by a suitable detector (Figure 8–45).

The position and intensity of each spot in the x-ray diffraction pattern contain information about the locations of the atoms in the crystal that gave rise to it. Deducing the three-dimensional structure of a large molecule from the diffraction pattern of its crystal is a complex task and was not achieved for a protein molecule until 1960. But in recent years x-ray diffraction analysis has become increasingly automated, and now the slowest step is likely to be the generation of suitable protein crystals. This requires large amounts of very pure protein and often involves years of trial and error, searching for the proper crystallization conditions. There are still many proteins, especially membrane proteins, that have so far resisted all attempts to crystallize them.

Analysis of the resulting diffraction pattern produces a complex three-dimensional electron-density map. Interpreting this map—translating its contours into a three-dimensional structure—is a complicated procedure that requires knowledge of the amino acid sequence of the protein. Largely by trial and error, the sequence and the electron-density map are correlated by computer to give the best possible fit. The reliability of the final atomic model depends on the resolution of the original crystallographic data: 0.5 nm resolution might produce a low-resolution map of the polypeptide backbone, whereas a resolution of 0.15 nm allows all of the non-hydrogen atoms in the molecule to be reliably positioned.

A complete atomic model is often too complex to appreciate directly, but simplified versions that show a protein's essential structural features can be

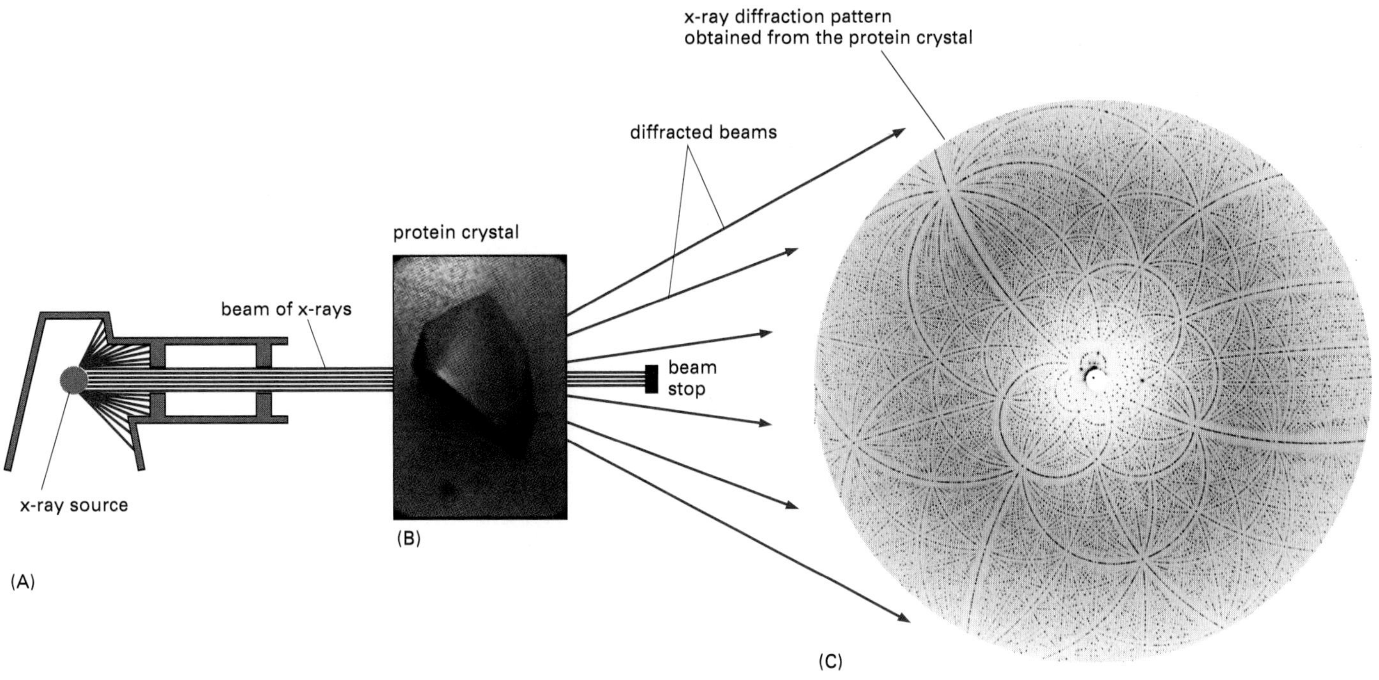

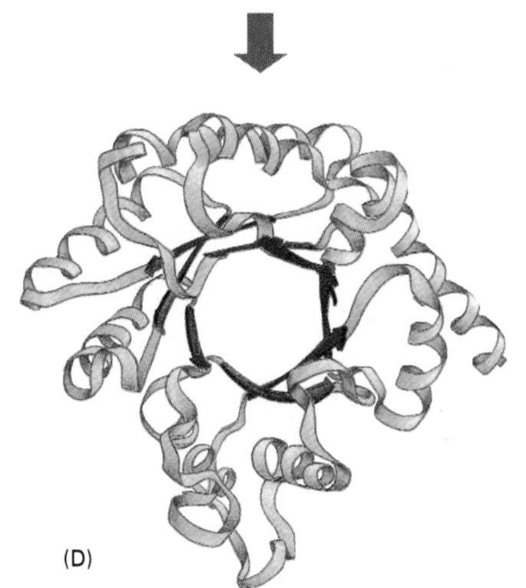

Figure 8–45 X-ray crystallography. (A) A narrow parallel beam of x-rays is directed at a well-ordered crystal (B). Shown here is a protein crystal of ribulose bisphosphate carboxylase, an enzyme with a central role in CO_2 fixation during photosynthesis. Some of the beam is scattered by the atoms in the crystal. The scattered waves reinforce one another at certain points and appear as a pattern of diffraction spots (C). This diffraction pattern, together with the amino acid sequence of the protein, can be used to produce an atomic model (D). The complete atomic model is hard to interpret, but this simplified version, derived from the x-ray diffraction data, shows the protein's structural features clearly (α helices, *green*; β strands, *red*). Note that the components pictured in A to D are not shown to scale. (B, courtesy of C. Branden; C, courtesy of J. Hajdu and I. Andersson; D, adapted from original provided by B. Furugren.)

readily derived from it (see Panel 3–2, pp. 138–139). The three-dimensional structures of about 10,000 different proteins have now been determined by x-ray crystallography or by NMR spectroscopy (see below)—enough to begin to see families of common structures emerging. These structures or protein folds often seem to be more conserved in evolution than are the amino acid sequences that form them (see Figure 3–15).

X-ray crystallographic techniques can also be applied to the study of macromolecular complexes. In a recent triumph, the method was used to solve the structure of the ribosome, a large and complex cellular machine made of several RNAs and more than 50 proteins (see Figure 6–64). The determination required the use of a synchrotron, a radiation source that generates x-rays with the intensity needed to analyze the crystals of such large macromolecular complexes.

Molecular Structure Can Also Be Determined Using Nuclear Magnetic Resonance (NMR) Spectroscopy

Nuclear magnetic resonance (NMR) spectroscopy has been widely used for many years to analyze the structure of small molecules. This technique is now also increasingly applied to the study of small proteins or protein domains. Unlike x-ray crystallography, NMR does not depend on having a crystalline sample; it simply requires a small volume of concentrated protein solution that is placed in a strong magnetic field.

Certain atomic nuclei, and in particular those of hydrogen, have a magnetic moment or spin: that is, they have an intrinsic magnetization, like a bar magnet. The spin aligns along the strong magnetic field, but it can be changed to a misaligned, excited state in response to applied radiofrequency (RF) pulses of electromagnetic radiation. When the excited hydrogen nuclei return to their aligned state, they emit RF radiation, which can be measured and displayed as a spectrum. The nature of the emitted radiation depends on the environment of each hydrogen nucleus, and if one nucleus is excited, it influences the absorption and emission of radiation by other nuclei that lie close to it. It is consequently possible, by an ingenious elaboration of the basic NMR technique known as two-dimensional NMR, to distinguish the signals from hydrogen nuclei in different amino acid residues and to identify and measure the small shifts in these signals that occur when these hydrogen nuclei lie close enough together to interact: the size of such a shift reveals the distance between the interacting pair of hydrogen atoms. In this way NMR can give information about the distances between the parts of the protein molecule. By combining this information with a knowledge of the amino acid sequence, it is possible in principle to compute the three-dimensional structure of the protein (Figure 8–46).

For technical reasons the structure of small proteins of about 20,000 daltons or less can readily be determined by NMR spectroscopy. Resolution is lost as the size of a macromolecule increases. But recent technical advances have now pushed the limit to about 100,000 daltons, thereby making the majority of proteins accessible for structural analysis by NMR.

The NMR method is especially useful when a protein of interest has resisted attempts at crystallization, a common problem for many membrane proteins. Because NMR studies are performed in solution, this method also offers a convenient means of monitoring changes in protein structure, for example during protein folding or when a substrate binds to the protein. NMR is also used widely to investigate molecules other than proteins and is valuable, for example, as a

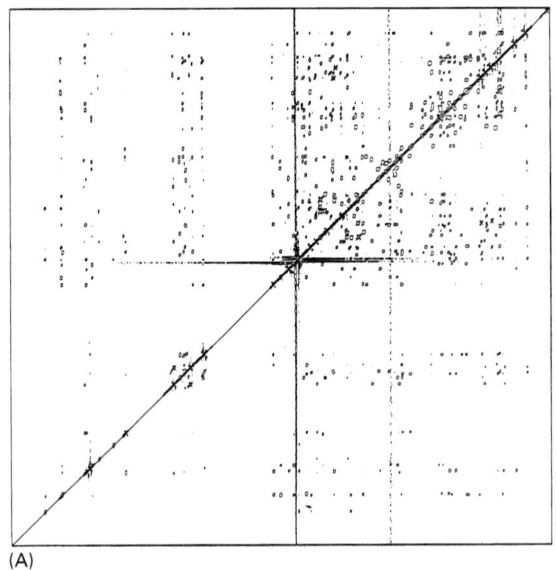

(A)

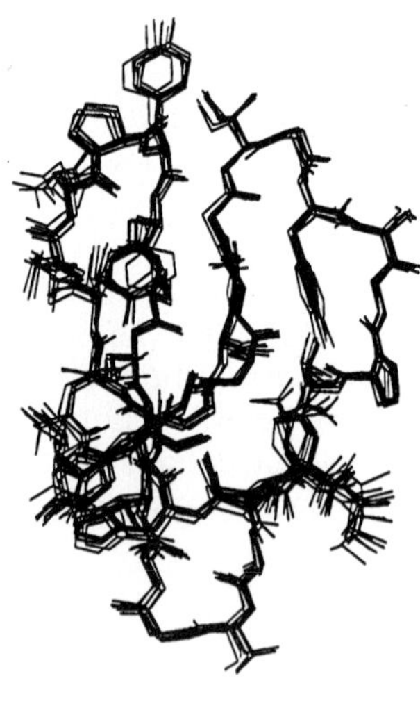

(B)

Figure 8–46 NMR spectroscopy. (A) An example of the data from an NMR machine. This two-dimensional NMR spectrum is derived from the C-terminal domain of the enzyme cellulase. The spots represent interactions between hydrogen atoms that are near neighbors in the protein and hence reflects the distance that separates them. Complex computing methods, in conjunction with the known amino acid sequence, enable possible compatible structures to be derived. In (B) 10 structures, which all satisfy the distance constraints equally well, are shown superimposed on one another, giving a good indication of the probable three-dimensional structure. (Courtesy of P. Kraulis.)

TABLE 8–8 Landmarks in the Development of X-ray Crystallography and NMR and Their Application to Biological Molecules

1864	**Hoppe-Seyler** crystallizes, and names, the protein hemoglobin.
1895	**Röntgen** observes that a new form of penetrating radiation, which he names x-rays, is produced when cathode rays (electrons) hit a metal target.
1912	**Von Laue** obtains the first x-ray diffraction patterns by passing x-rays through a crystal of zinc sulfide.
	W.L. Bragg proposes a simple relationship between an x-ray diffraction pattern and the arrangement of atoms in a crystal that produce the pattern.
1926	**Summer** obtains crystals of the enzyme urease from extracts of jack beans and demonstrates that proteins possess catalytic activity.
1931	**Pauling** publishes his first essays on 'The Nature of the Chemical Bond,' detailing the rules of covalent bonding.
1934	**Bernal and Crowfoot** present the first detailed x-ray diffraction patterns of a protein obtained from crystals of the enzyme pepsin.
1935	**Patterson** develops an analytical method for determining interatomic spacings from x-ray data.
1941	**Astbury** obtains the first x-ray diffraction pattern of DNA.
1951	**Pauling and Corey** propose the structure of a helical conformation of a chain of L-amino acids—the α helix—and the structure of the β sheet, both of which were later found in many proteins.
1953	**Watson and Crick** propose the double-helix model of DNA, based on x-ray diffraction patterns obtained by **Franklin and Wilkins.**
1954	**Perutz** and colleagues develop heavy-atom methods to solve the phase problem in protein crystallography.
1960	**Kendrew** describes the first detailed structure of a protein (sperm whale myoglobin) to a resolution of 0.2 nm, and **Perutz** presents a lower-resolution structure of the larger protein hemoglobin.
1966	**Phillips** describes the structure of lysozyme, the first enzyme to have its structure analyzed in detail.
1971	**Jeener** proposes the use of two-dimensional NMR, and **Wuthrich** and colleagues first use the method to solve a protein structure in the early 1980s.
1976	**Kim and Rich** and **Klug** and colleagues describe the detailed three-dimensional structure of tRNA determined by x-ray diffraction.
1977–1978	**Holmes and Klug** determine the structure of tobacco mosaic virus (TMV), and **Harrison** and **Rossman** determine the structure of two small spherical viruses.
1985	**Michel, Deisenhofer** and colleagues determine the first structure of a transmembrane protein (a bacterial reaction center) by x-ray crystallography. **Henderson** and colleagues obtain the structure of bacteriorhodopsin, a transmembrane protein, by high-resolution electron-microscopy methods between 1975 and 1990.

method to determine the three-dimensional structures of RNA molecules and the complex carbohydrate side chains of glycoproteins.

Some landmarks in the development of x-ray crystallography and NMR are listed in Table 8–8.

Sequence Similarity Can Provide Clues About Protein Function

Thanks to the proliferation of protein and nucleic acid sequences that are catalogued in genome databases, the function of a gene—and its encoded protein—can often be predicted by simply comparing its sequence with those of previously characterized genes. Because amino acid sequence determines protein structure and structure dictates biochemical function, proteins that share a similar amino acid sequence usually perform similar biochemical functions, even when they are found in distantly related organisms. At present, determining what a newly discovered protein does therefore usually begins with a search for previously identified proteins that are similar in their amino acid sequences.

Searching a collection of known sequences for homologous genes or proteins is typically done over the World-Wide Web, and it simply involves selecting a database and entering the desired sequence. A sequence alignment program—the most popular are BLAST and FASTA—scans the database for similar sequences by sliding the submitted sequence along the archived sequences until a cluster of residues falls into full or partial alignment (Figure 8–47). The results of even a complex search—which can be performed on either a

```
 Score =  399 bits (1025), Expect = e-111
 Identities = 198/290 (68%), Positives = 241/290 (82%), Gaps = 1/290

Query: 57   MENFQKVEKIGEGTYGVVYKARNKLTGEVVALKKIRLDTETEGVPSTAIREISLLKELNH 116
             ME ++KVEKIGEGTYGVVYKA +K T E +ALKKIRL+ E EGVPSTAIREISLLKE+NH
Sbjct: 1    MEQYEKVEKIGEGTYGVVYKALDKATNETIALKKIRLEQEDEGVPSTAIREISLLKEMNH 60

Query: 117  PNIVKLLDVIHTENKLYLVFEFLHQDLKKFMDASALTGIPLPLIKSYLFQLLQGLAFCHS 176
              NIV+L DV+H+E ++YLVFE+L  DLKKFMD+            LIKSYL+Q+L G+A+CHS
Sbjct: 61   GNIVRLHDVVHSEKRIYLVFEYLDLDLKKFMDSCPEFAKNPTLIKSYLYQILHGVAYCHS 120

                             ▼
Query: 177  HRVLHRDLKPQNLLINTE-GAIKLADFGLARAFGVPVRTYTHEVVTLWYRAPEILLGCKY 235
             HRVLHRDLKPQNLLI+    A+KLADFGLARAFG+PVRT+THEVVTLWYRAPEILLG +
Sbjct: 121  HRVLHRDLKPQNLLIDRRTNALKLADFGLARAFGIPVRTFTHEVVTLWYRAPEILLGARQ 180

Query: 236  YSTAVDIWSLGCIFAEMVTRRALFPGDSEIDQLFRIFRTLGTPDEVVWPGVTSMPDYKPS 295
             YST VD+WS+GCIFAEMV ++ LFPGDSEID+LF+IFR LGTP+E  WPGV+ +PD+K +
Sbjct: 181  YSTPVDVWSVGCIFAEMVNQKPLFPGDSEIDELFKIFRILGTPNEQSWPGVSCLPDFKTA 240

Query: 296  FPKWARQDFSKVVPPLDEDGRSLLSQMLHYDPNKRISAKAALAHPFFQDV 345
             FP+W  QD + VVP LD  G  LLS+ML Y+P+KRI+A+ AL H +F+D+
Sbjct: 241  FPRWQAQDLATVVPNLDPAGLDLLSKMLRYEPSKRITARQALEHEYFKDL 290
```

Figure 8–47 Results of a BLAST search. Sequence databases can be searched to find similar amino acid or nucleic acid sequences. Here a search for proteins similar to the human cell-cycle regulatory protein cdc2 *(Query)* locates maize cdc2 *(Subject)*, which is 68% identical (and 82% similar) to human cdc2 in its amino acid sequence. The alignment begins at residue 57 of the Query protein, suggesting that the human protein has an N-terminal region that is absent from the maize protein. The *green* blocks indicate differences in sequence, and the *yellow* bar summarizes the similarities: when the two amino acid sequences are identical, the residue is shown; conservative amino acid substitutions are indicated by a plus sign (+). Only one small gap has been introduced—indicated by the *red arrow* at position 194 in the Query sequence—to align the two sequences maximally. The alignment score *(Score)* takes into account penalties for substitution and gaps; the higher the alignment score, the better the match. The significance of the alignment is reflected in the *Expectation* (E) value, which represents the number of alignments with scores equal to or better than the given score that are expected to occur by chance. The lower the E value, the more significant the match; the extremely low value here indicates certain significance. E values much higher than 0.1 are unlikely to reflect true relatedness.

nucleotide or an amino acid sequence—are returned within minutes. Such comparisons can be used to predict the functions of individual proteins, families of proteins, or even the entire protein complement of a newly sequenced organism.

In the end, however, the predictions that emerge from sequence analysis are often only a tool to direct further experimental investigations.

Fusion Proteins Can Be Used to Analyze Protein Function and to Track Proteins in Living Cells

The location of a protein within the cell often suggests something about its function. Proteins that travel from the cytoplasm to the nucleus when a cell is exposed to a growth factor, for example, may have a role in regulating gene expression in response to that factor. A protein often contains short amino acid sequences that determine its location in a cell. Most nuclear proteins, for example, contain one or more specific short sequences of amino acids that serve as signals for their import into the nucleus after their synthesis in the cytosol (discussed in Chapter 12). These special regions of the protein can be identified by fusing them to an easily detectable protein that lacks such regions and then following the behavior of this surrogate protein in a cell. Such fusion proteins can be readily produced by the recombinant DNA techniques discussed previously.

Another common strategy used both to follow proteins in cells and to purify them rapidly is *epitope tagging*. In this case, a fusion protein is produced that contains the entire protein being analyzed plus a short peptide of 8 to 12 amino acids (an "epitope") that can be recognized by a commercially available antibody. The fusion protein can therefore be specifically detected, even in the presence of a large excess of the normal protein, using the anti-epitope antibody and a labeled secondary antibody that can be monitored by light or electron microscopy (Figure 8–48).

Today large numbers of proteins are being tracked in living cells by using a fluorescent marker called **green fluorescent protein** (**GFP**). Tagging proteins with GFP is as simple as attaching the gene for GFP to one end of the gene that encodes a protein of interest. In most cases, the resulting GFP fusion protein behaves in the same way as the original protein, and its movement can be monitored by following its fluorescence inside the cell by fluorescence microscopy.

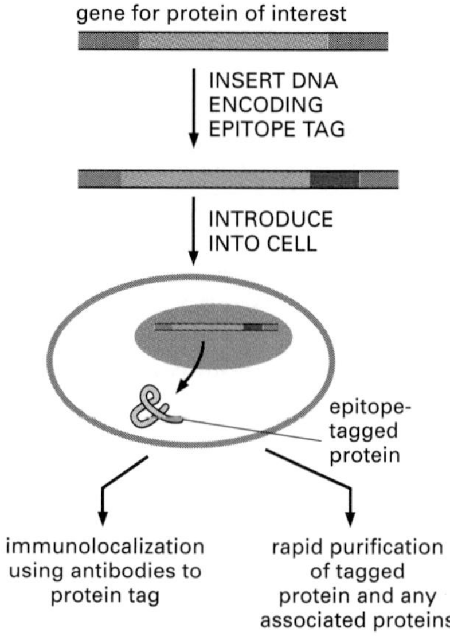

Figure 8–48 Epitope tagging allows the localization or purification of proteins. Using standard genetic engineering techniques, a short epitope tag can be added to a protein of interest. The resulting protein contains the protein being analyzed plus a short peptide that can be recognized by commercially available antibodies. The labeled antibody can be used to follow the cellular localization of the protein or to purify it by immunoprecipitation or affinity chromatography.

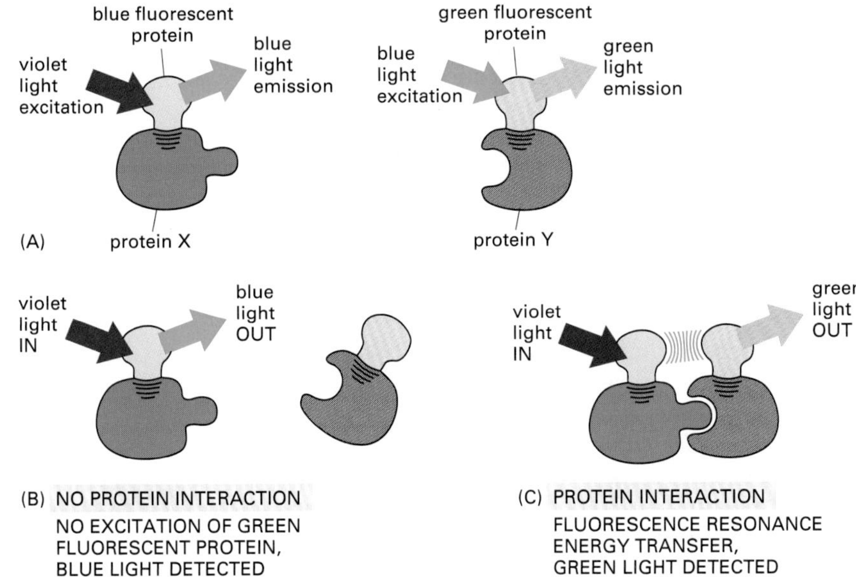

Figure 8–49 Fluorescence resonance energy transfer (FRET). To determine whether (and when) two proteins interact inside the cell, the proteins are first produced as fusion proteins attached to different variants of GFP. (A) In this example, protein X is coupled to a blue fluorescent protein, which is excited by violet light (370–440 nm) and emits blue light (440–480 nm); protein Y is coupled to a green fluorescent protein, which is excited by blue light and emits green light (510 nm). (B) If protein X and Y do not interact, illuminating the sample with violet light yields fluorescence from the blue fluorescent protein only. (C) When protein X and protein Y interact, FRET can now occur. Illuminating the sample with violet light excites the blue fluorescent protein, whose emission in turn excites the green fluorescent protein, resulting in an emission of green light. The fluorochromes must be quite close together—within about 1–10 nm of one another—for FRET to occur. Because not every molecule of protein X and protein Y is bound at all times, some blue light may still be detected. But as the two proteins begin to interact, emission from the donor GFP falls as the emission from the acceptor GFP rises.

The GFP fusion protein strategy has become a standard way to determine the distribution and dynamics of any protein of interest in living cells. We discuss its use further in Chapter 9.

GFP, and its derivatives of different color, can also be used to monitor protein–protein interactions. In this application, two proteins of interest are each labeled with a different fluorochrome, such that the emission spectrum of one fluorochrome overlaps the absorption spectrum of the second fluorochrome. If the two proteins—and their attached fluorochromes—come very close to each other (within about 1–10 nm), the energy of the absorbed light will be transferred from one fluorochrome to the other. The energy transfer, called **fluorescence resonance energy transfer (FRET)**, is determined by illuminating the first fluorochrome and measuring emission from the second (Figure 8–49). By using two different spectral variants of GFP as the fluorochromes in such studies, one can monitor the interaction of any two protein molecules inside a living cell.

Affinity Chromatography and Immunoprecipitation Allow Identification of Associated Proteins

Because most proteins in the cell function as part of a complex with other proteins, an important way to begin to characterize their biological roles is to identify their binding partners. If an uncharacterized protein binds to a protein whose role in the cell is understood, its function is likely to be related. For example, if a protein is found to be part of the proteasome complex, it is likely to be involved somehow in degrading damaged or misfolded proteins.

Protein affinity chromatography is one method that can be used to isolate and identify proteins that interact physically. To capture interacting proteins, a target protein is attached to polymer beads that are packed into a column. Cellular proteins are washed through the column and those proteins that interact with the target adhere to the affinity matrix (see Figure 8–11C). These proteins can then be eluted and their identity determined by mass spectrometry or another suitable method.

Perhaps the simplest method for identifying proteins that bind to one another tightly is **co-immunoprecipitation**. In this case, an antibody is used to recognize a specific target protein; affinity reagents that bind to the antibody and are coupled to a solid matrix are then used to drag the complex out of solution to the bottom of a test tube. If this protein is associated tightly enough with another protein when it is captured by the antibody, the partner precipitates as well. This method is useful for identifying proteins that are part of a complex inside cells, including those that interact only transiently—for example when cells are stimulated by signal molecules (discussed in Chapter 15).

Co-immunoprecipitation techniques require having a highly specific antibody against a known cellular protein target, which is not always available. One way to overcome this requirement is to use recombinant DNA techniques to add an epitope tag (see Figure 8–48) or to fuse the target protein to a well-characterized marker protein, such as the small enzyme glutathione S-transferase (GST). Commercially available antibodies directed against the epitope tag or the marker protein can then be used to precipitate the whole fusion protein, including any cellular proteins associated with the protein of interest. If the protein is fused to GST, antibodies may not be needed at all: the hybrid and its binding partners can be readily selected on beads coated with glutathione (Figure 8–50).

In addition to capturing protein complexes on columns or in test tubes, researchers are also developing high-density protein arrays for investigating protein function and protein interactions. These arrays, which contain thousands of different proteins or antibodies spotted onto glass slides or immobilized in tiny wells, allow one to examine the biochemical activities and binding profiles of a large number of proteins at once. To examine protein interactions with such an array, one incubates a labeled protein with each of the target proteins immobilized on the slide and then determines to which of the many proteins the labeled molecule binds.

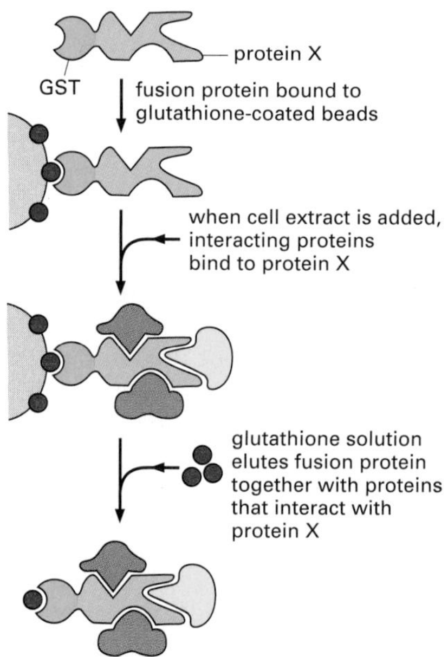
recombinant DNA techniques are used to make fusion between protein X and glutathione S-transferase (GST)
protein X
GST
fusion protein bound to glutathione-coated beads
when cell extract is added, interacting proteins bind to protein X
glutathione solution elutes fusion protein together with proteins that interact with protein X

Figure 8–50 Purification of protein complexes using a GST-tagged fusion protein. GST fusion proteins, generated by standard recombinant DNA techniques, can be captured on an affinity column containing beads coated with glutathione. To look for proteins that bind to protein X, cell extracts can be passed through this column. The hybrid protein and its binding partners can then be eluted with glutathione. The identities of these interacting proteins can be determined by mass spectrometry (see Figure 8–20). In an alternative approach, a cell extract can be made from a cell producing the GST fusion protein and passed directly through the glutathione affinity column. The GST fusion protein, along with proteins that have associated with it in the cell, are thereby retained. Affinity columns can also be made to contain antibodies against GST or another convenient small protein or epitope tag (see Figure 8–48).

Protein–Protein Interactions Can Be Identified by Use of the Two-Hybrid System

Methods such as co-immunoprecipitation and affinity chromatography allow the physical isolation of interacting proteins. A successful isolation yields a protein whose identity must then be ascertained by mass spectrometry, and whose gene must be retrieved and cloned before further studies characterizing its activity—or the nature of the protein–protein interaction—can be performed.

Other techniques allow the simultaneous isolation of interacting proteins along with the genes that encode them. The first method we discuss, called the **two-hybrid system**, uses a reporter gene to detect the physical interaction of a pair of proteins inside a yeast cell nucleus. This system has been designed so that when a target protein binds to another protein in the cell, their interaction brings together two halves of a transcriptional activator, which is then able to switch on the expression of the reporter gene.

The technique takes advantage of the modular nature of gene activator proteins (see Figure 7–42). These proteins both bind to DNA and activate transcription—activities that are often performed by two separate protein domains. Using recombinant DNA techniques, the DNA sequence that codes for a target protein is fused with DNA that encodes the DNA-binding domain of a gene activator protein. When this construct is introduced into yeast, the cells produce the target protein attached to this DNA-binding domain (Figure 8–51). This protein binds to the regulatory region of a reporter gene, where it serves as "bait" to fish for proteins that interact with the target protein inside a yeast cell. To prepare a set of potential binding partners, DNA encoding the activation domain of a gene activator protein is ligated to a large mixture of DNA fragments from a cDNA library. Members of this collection of genes—the "prey"—are introduced individually into yeast cells containing the bait. If the yeast cell has received a DNA clone that expresses a prey partner for the bait protein, the two halves of a transcriptional activator are united, switching on the reporter gene (see Figure 8–51). Cells that express this reporter are selected and grown, and the gene (or gene fragment) encoding the prey protein is retrieved and identified through nucleotide sequencing.

Although it sounds complex, the two-hybrid system is relatively simple to use in the laboratory. Although the protein–protein interactions occur in the yeast cell nucleus, proteins from every part of the cell and from any organism can be studied in this way. Of the thousands of protein–protein interactions that have been catalogued in yeast, half have been discovered with such two-hybrid screens.

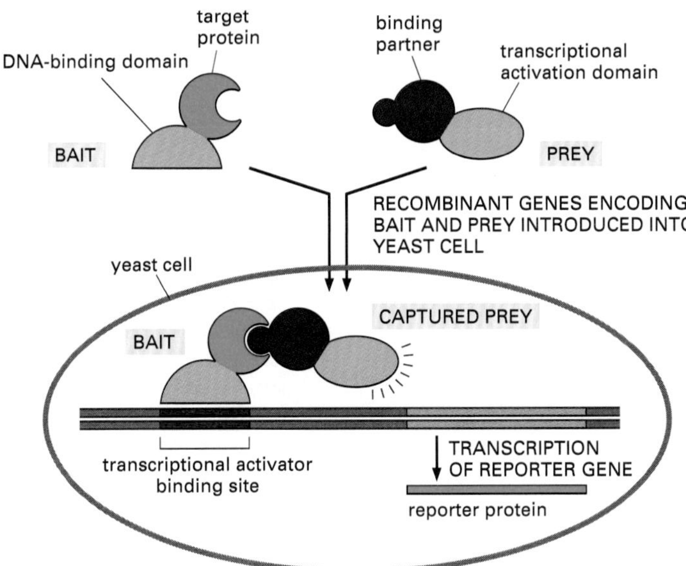

Figure 8–51 The yeast two-hybrid system for detecting protein–protein interactions. The target protein is fused to a DNA-binding domain that localizes it to the regulatory region of a reporter gene as "bait." When this target protein binds to another specially designed protein in the cell nucleus ("prey"), their interaction brings together two halves of a transcriptional activator, which then switches on the expression of the reporter gene. The reporter gene is often one that will permit growth on a selective medium. Bait and prey fusion proteins are generated by standard recombinant DNA techniques. In most cases, a single bait protein is used to fish for interacting partners among a large collection of prey proteins produced by ligating DNA encoding the activation domain of a transcriptional activator to a large mixture of DNA fragments from a cDNA library.

The two-hybrid system can be scaled up to map the interactions that occur among all of the proteins produced by an organism. In this case, a set of bait fusions is produced for every cellular protein, and each of these constructs is introduced into a separate yeast cell. These cells are then mated to yeast containing the prey library. Those rare cells that are positive for a protein–protein interaction are then characterized. In this way a protein linkage map has been generated for most of the 6,000 proteins in yeast (see Figure 3–78), and similar projects are underway to catalog the protein interactions in *C. elegans* and *Drosophila.*

A related technique, called a *reverse two-hybrid system,* can be used to identify mutations—or chemical compounds—that are able to disrupt specific protein–protein interactions. In this case the reporter gene can be replaced by a gene that kills cells in which the bait and prey proteins interact. Only those cells in which the proteins no longer bind—because an engineered mutation or a test compound prevents them from doing so—can survive. Like knocking out a gene (which we discuss shortly), eliminating a particular molecular interaction can reveal something about the role of the participating proteins in the cell. In addition, compounds that selectively interrupt protein interactions can be medically useful: a drug that prevents a virus from binding to its receptor protein on human cells could help people to avoid infections, for example.

Phage Display Methods Also Detect Protein Interactions

Another powerful method for detecting protein–protein interactions involves introducing genes into a virus that infects the *E. coli* bacterium (a bacteriophage, or "phage"). In this case the DNA encoding the protein of interest (or a smaller peptide fragment of this protein) is fused with a gene encoding one of the proteins that forms the viral coat. When this virus infects *E. coli,* it replicates, producing phage particles that display the hybrid protein on the outside of their coats (Figure 8–52A). This bacteriophage can then be used to fish for binding partners in a large pool of potential target proteins.

However, the most powerful use of this **phage display** method allows one to screen large collections of proteins or peptides for binding to selected targets. This approach requires first generating a library of fusion proteins, much like the prey library in the two-hybrid system. This collection of phage is then screened for binding to a purified protein of interest. For example, the phage library can be passed through an affinity column containing an immobilized target protein. Viruses that display a protein or peptide that binds tightly to the target are captured on the column and can be eluted with excess target protein. Those phage containing a DNA fragment that encodes an interacting protein or peptide are

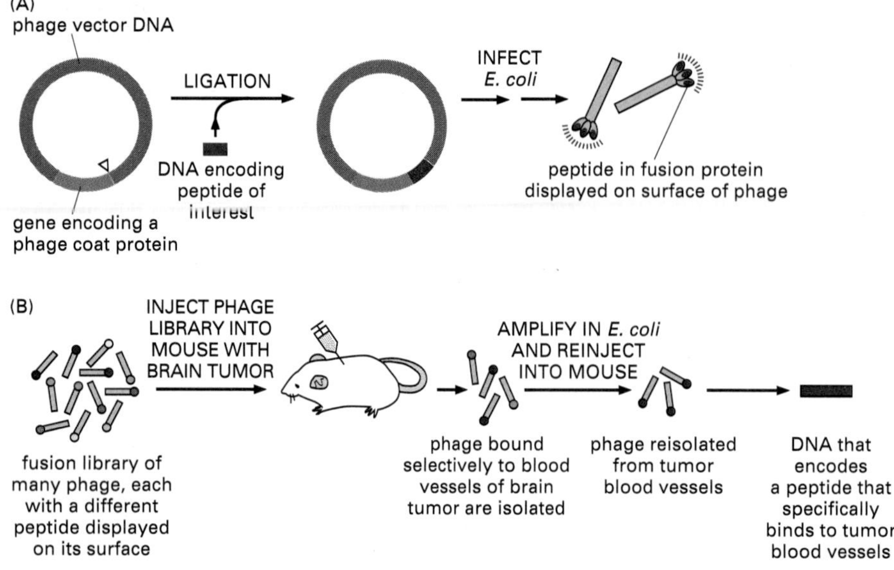

Figure 8–52 The phage display method for investigating protein interactions. (A) Preparation of the bacteriophage. DNA encoding the desired peptide is ligated into the phage vector, fused with the gene encoding the viral protein coat. The engineered phage are then introduced into *E. coli*, which produce phage displaying a hybrid coat protein that contains the peptide. (B) Phage libraries containing billions of different peptides can also be generated. In this example, the library is injected into a mouse with a brain tumor and phage that bind selectively to the blood vessels that supply the tumor are isolated and amplified. A peptide that binds specifically to tumor blood vessels can then be isolated from the purified phage. Such a peptide could be used to target drugs or toxins to the tumor.

collected and allowed to replicate in *E. coli*. The DNA from each phage can then be recovered and its nucleotide sequence determined to identify the protein or peptide partner that bound to the target protein. A similar technique has been used to isolate peptides that bind specifically to the inside of the blood vessels associated with human tumors. These peptides are presently being tested as agents for delivering therapeutic anti-cancer compounds directly to such tumors (Figure 8–52B).

Phage display has also been used to generate monoclonal antibodies that recognize a specific target molecule or cell. In this case, a library of phage expressing the appropriate parts of antibody molecules is screened for those phage that bind to a target antigen.

Protein Interactions Can Be Monitored in Real Time Using Surface Plasmon Resonance

Once two proteins—or a protein and a small molecule—are known to associate, it becomes important to characterize their interaction in more detail. Proteins can bind to one another permanently, or engage in transient encounters in which proteins remain associated only temporarily. These dynamic interactions are often regulated through reversible modifications (such as phosphorylation), through ligand binding, or through the presence or absence of other proteins that compete for the same binding site.

To begin to understand these intricacies, one must determine how tightly two proteins associate, how slowly or rapidly molecular complexes assemble and break down over time, and how outside influences can affect these parameters. As we have seen in this chapter, there are many different techniques available to study protein–protein interactions, each with its individual advantages and disadvantages. One particularly useful method for monitoring the dynamics of protein association is called **surface plasmon resonance (SPR)**. The SPR method has been used to characterize a wide variety of molecular interactions, including antibody–antigen binding, ligand–receptor coupling, and the binding of proteins to DNA, carbohydrates, small molecules, and other proteins.

SPR detects binding interactions by monitoring the reflection of a beam of light off the interface between an aqueous solution of potential binding molecules and a biosensor surface carrying immobilized bait protein. The bait protein is attached to a very thin layer of metal that coats one side of a glass prism (Figure 8–53). A light beam is passed through the prism; at a certain angle, called the resonance angle, some of the energy from the light interacts with the cloud of electrons in the metal film, generating a plasmon—an oscillation of the electrons at right angles to the plane of the film, bouncing up and down between its upper and lower surfaces like a weight on a spring. The plasmon, in turn, generates an electrical field that extends a short distance—about the wavelength of the light—above and below the metal surface. Any change in the composition of the environment within the range of the electrical field causes a measurable change in the resonance angle.

To measure binding, a solution containing proteins (or other molecules) that might interact with the immobilized bait protein is allowed to flow past the biosensor surface. When proteins bind to the bait, the composition of the molecular complexes on the metal surface change, causing a change in the resonance angle (see Figure 8–53). The changes in the resonance angle are monitored in real time and reflect the kinetics of the association—or dissociation—of molecules with the bait protein. The association rate (k_{on}) is measured as the molecules interact, and the dissociation rate (k_{off}) is determined as buffer washes the bound molecules from the sensor surface. A binding constant (K) is calculated by dividing k_{off} by k_{on}. In addition to determining the kinetics, SPR can be used to determine the number of molecules that are bound in each complex: the magnitude of the SPR signal change is proportional to the mass of the immobilized complex.

The SPR method is particularly useful because it requires only small amounts of proteins, the proteins do not have to be labeled in any way, and protein–protein interactions can be monitored in real time.

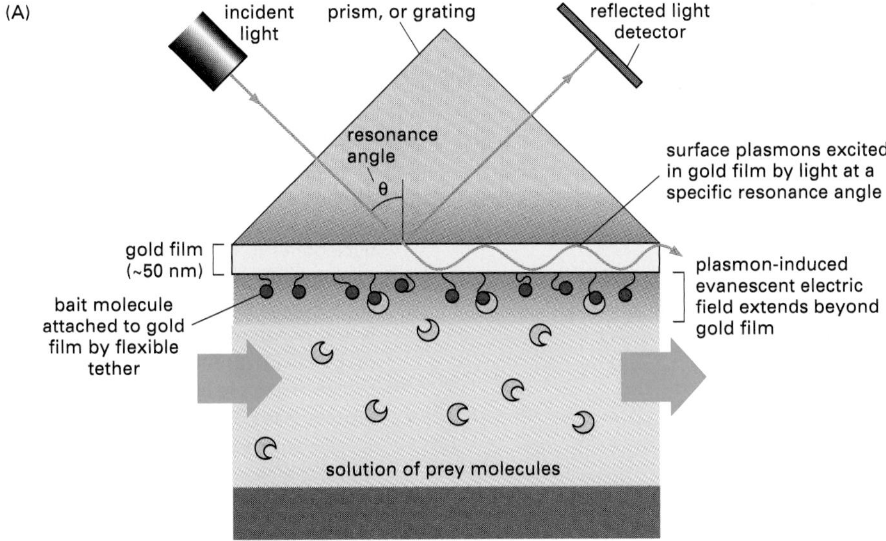

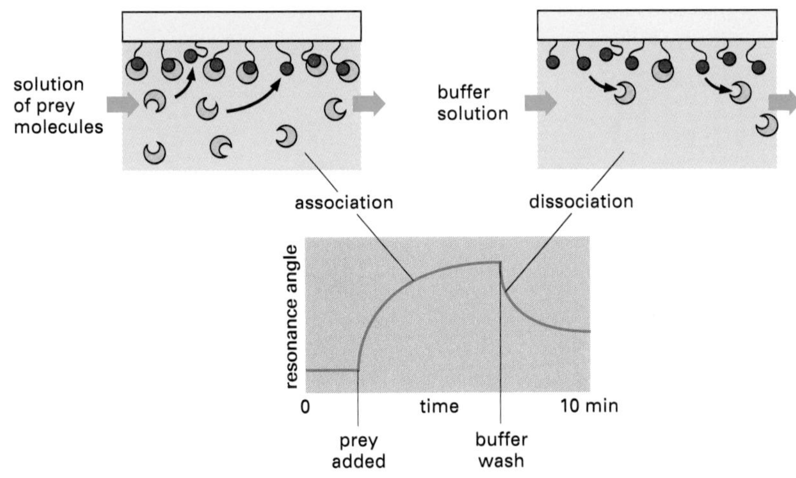

Figure 8–53 Surface plasmon resonance. (A) SPR can detect binding interactions by monitoring the reflection of a beam of light off the interface between an aqueous solution of potential binding molecules *(green)* and a biosensor surface coated with an immobilized bait protein *(red)*. (B) A solution of prey proteins is allowed to flow past the immobilized bait protein. Binding of prey molecules to bait proteins produces a measurable change in the resonance angle. These changes, monitored in real time, reflect the association and dissociation of the molecular complexes.

DNA Footprinting Reveals the Sites Where Proteins Bind on a DNA Molecule

So far we have concentrated on examining protein–protein interactions. But some proteins act by binding to DNA. Most of these proteins have a central role in determining which genes are active in a particular cell by binding to regulatory DNA sequences, which are usually located outside the coding regions of a gene.

In analyzing how such a protein functions, it is important to identify the specific nucleotide sequences to which it binds. A method used for this purpose is called **DNA footprinting**. First, a pure DNA fragment that is labeled at one end with ^{32}P is isolated (see Figure 8–24B); this molecule is then cleaved with a nuclease or a chemical that makes random single-stranded cuts in the DNA. After the DNA molecule is denatured to separate its two strands, the resultant fragments from the labeled strand are separated on a gel and detected by autoradiography. The pattern of bands from DNA cut in the presence of a DNA-binding protein is compared with that from DNA cut in its absence. When the protein is present, it covers the nucleotides at its binding site and protects their phosphodiester bonds from cleavage. As a result, the labeled fragments that terminate in the binding site are missing, leaving a gap in the gel pattern called a "footprint" (Figure 8–54). Similar methods can be used to determine the binding sites of proteins on RNA.

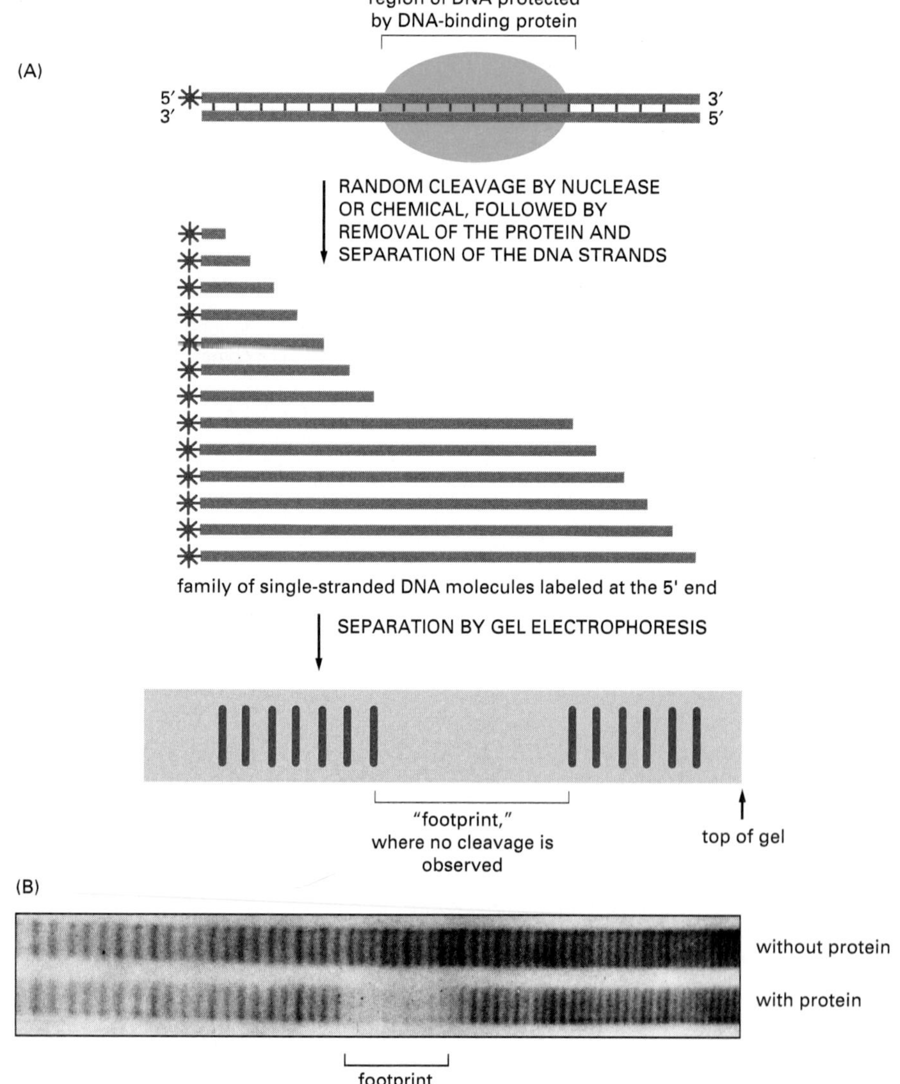

Figure 8–54 The DNA footprinting technique. (A) This technique requires a DNA molecule that has been labeled at one end (see Figure 8–24B). The protein shown binds tightly to a specific DNA sequence that is seven nucleotides long, thereby protecting these seven nucleotides from the cleaving agent. If the same reaction were performed without the DNA-binding protein, a complete ladder of bands would be seen on the gel (not shown). (B) An actual footprint used to determine the binding site for a human protein that stimulates the transcription of specific eucaryotic genes. These results locate the binding site about 60 nucleotides upstream from the start site for RNA synthesis. The cleaving agent was a small, iron-containing organic molecule that normally cuts at every phosphodiester bond with nearly equal frequency. (B, courtesy of Michele Sawadogo and Robert Roeder.)

Summary

Many powerful techniques are used to study the structure and function of a protein. To determine the three-dimensional structure of a protein at atomic resolution, large proteins have to be crystallized and studied by x-ray diffraction. The structure of small proteins in solution can be determined by nuclear magnetic resonance analysis. Because proteins with similar structures often have similar functions, the biochemical activity of a protein can sometimes be predicted by searching for known proteins that are similar in their amino acid sequences.

Further clues to the function of a protein can be derived from examining its subcellular distribution. Fusion of the protein with a molecular tag, such as the green fluorescent protein (GFP), allows one to track its movement inside the cell. Proteins that enter the nucleus and bind to DNA can be further characterized by footprint analysis, a technique used to determine which regulatory sequences the protein binds to as it controls gene transcription.

All proteins function by binding to other proteins or molecules, and many methods exist for studying protein–protein interactions and identifying potential protein partners. Either protein affinity chromatography or co-immunoprecipitation by antibodies directed against a target protein will allow physical isolation of interacting proteins. Other techniques, such as the two-hybrid system or phage display, permit the simultaneous isolation of interacting proteins and the genes that encode them. The identity of the proteins recovered from any of these approaches is then ascertained by determining the sequence of the protein or its corresponding gene.

STUDYING GENE EXPRESSION AND FUNCTION

Ultimately, one wishes to determine how genes—and the proteins they encode—function in the intact organism. Although it may sound counterintuitive, one of the most direct ways to find out what a gene does is to see what happens to the organism when that gene is missing. Studying mutant organisms that have acquired changes or deletions in their nucleotide sequences is a time-honored practice in biology. Because mutations can interrupt cellular processes, mutants often hold the key to understanding gene function. In the classical approach to the important field of **genetics**, one begins by isolating mutants that have an interesting or unusual appearance: fruit flies with white eyes or curly wings, for example. Working backward from the **phenotype**—the appearance or behavior of the individual—one then determines the organism's **genotype**, the form of the gene responsible for that characteristic (Panel 8–1).

Today, with numerous genome projects adding tens of thousands of nucleotide sequences to the public databases each day, the exploration of gene function often begins with a DNA sequence. Here the challenge is to translate sequence into function. One approach, discussed earlier in the chapter, is to search databases for well-characterized proteins that have similar amino acid sequences to the protein encoded by a new gene, and from there employ some of the methods described in the previous section to explore the gene's function further. But to tackle directly the problem of how a gene functions in a cell or organism, the most effective approach involves studying mutants that either lack the gene or express an altered version of it. Determining which cellular processes have been disrupted or compromised in such mutants will then frequently provide a window to a gene's biological role.

In this section, we describe several different approaches to determining a gene's function, whether one starts from a DNA sequence or from an organism with an interesting phenotype. We begin with the classical genetic approach to studying genes and gene function. These studies start with a *genetic screen* for isolating mutants of interest, and then proceed toward identification of the gene or genes responsible for the observed phenotype. We then review the collection of techniques that fall under the umbrella of *reverse genetics*, in which one begins with a gene or gene sequence and attempts to determine its function. This approach often involves some intelligent guesswork—searching for homologous sequences and determining when and where a gene is expressed—as well as generating mutant organisms and characterizing their phenotype.

GENES AND PHENOTYPES

Gene: a functional unit of inheritance, usually corresponding to the segment of DNA coding for a single protein.

Genome: an organism's set of genes.

locus: the site of the gene in the genome

alleles: alternative forms of a gene

Wild-type: the normal, naturally occurring type

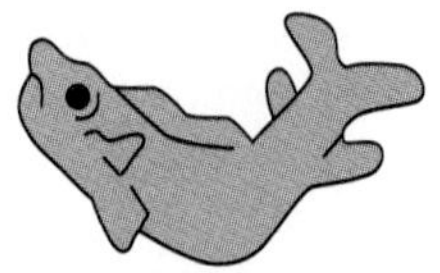

Mutant: differing from the wild-type because of a genetic change (a mutation)

GENOTYPE: the specific set of alleles forming the genome of an individual

homozygous A/A | heterozygous a/A | homozygous a/a

PHENOTYPE: the visible character of the individual

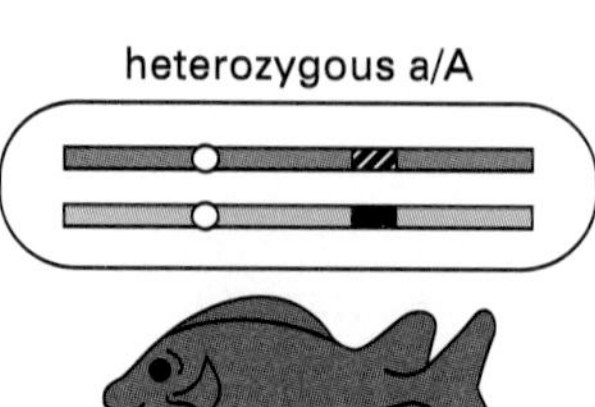

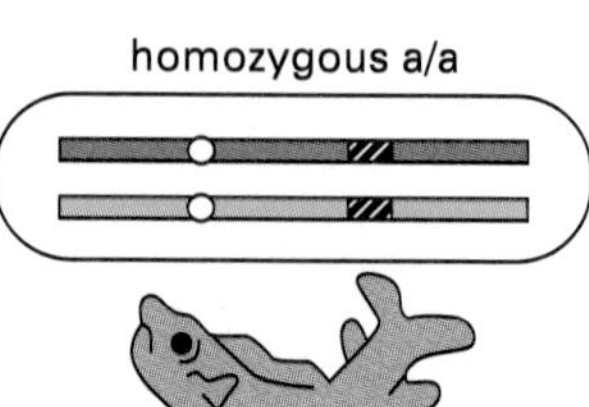

allele A is dominant (relative to a); allele a is recessive (relative to A)

In the example above, the phenotype of the heterozygote is the same as that of one of the homozygotes; in cases where it is different from both, the two alleles are said to be co-dominant.

CHROMOSOMES

a chromosome at the beginning of the cell cycle, in G_1 phase; the single long bar represents one long double helix of DNA

centromere

short "p" arm

long "q" arm

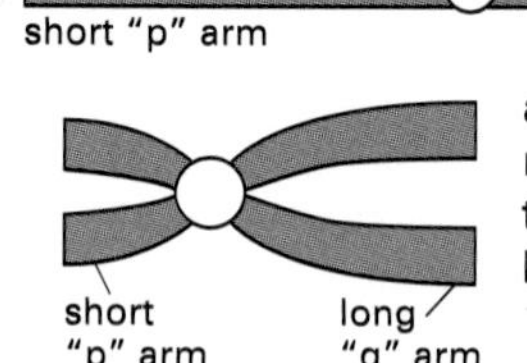

a chromosome at the end of the cell cycle, in metaphase; it is duplicated and condensed, consisting of two identical sister chromatids (each containing one DNA double helix) joined at the centromere.

pair of autosomes

maternal 1

paternal 1

maternal 3

paternal 3

paternal 2

maternal 2

X

Y

sex chromosomes

A normal diploid chromosome set, as seen in a metaphase spread, prepared by bursting open a cell at metaphase and staining the scattered chromosomes. In the example shown schematically here, there are three pairs of autosomes (chromosomes inherited symmetrically from both parents, regardless of sex) and two sex chromosomes—an X from the mother and a Y from the father. The numbers and types of sex chromosomes and their role in sex determination are variable from one class of organisms to another, as is the number of pairs of autosomes.

THE HAPLOID–DIPLOID CYCLE OF SEXUAL REPRODUCTION

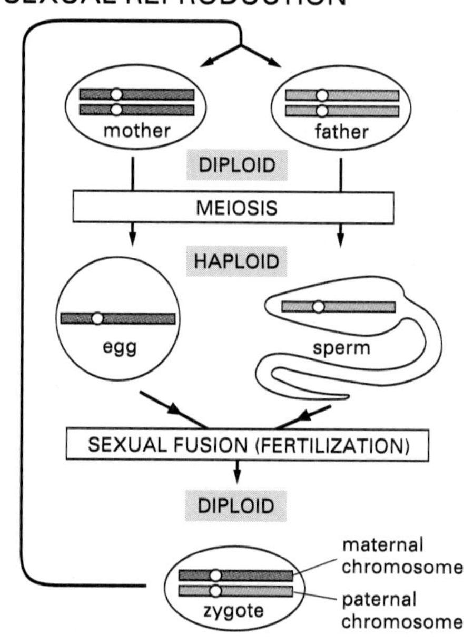

For simplicity, the cycle is shown for only one chromosome/chromosome pair.

MEIOSIS AND GENETIC RECOMBINATION

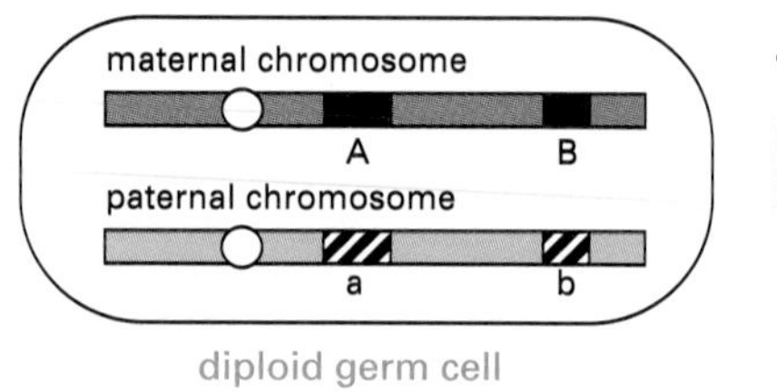

diploid germ cell

genotype $\frac{AB}{ab}$

MEIOSIS AND RECOMBINATION

genotype Ab

A b

site of crossing-over

genotype aB

a B

haploid gametes (eggs or sperm)

The greater the distance between two loci on a single chromosome, the greater is the chance that they will be separated by crossing-over occurring at a site between them. If two genes are thus reassorted in x% of gametes, they are said to be separated on a chromosome by a genetic map distance of x map units (or x centimorgans).

TYPES OF MUTATIONS

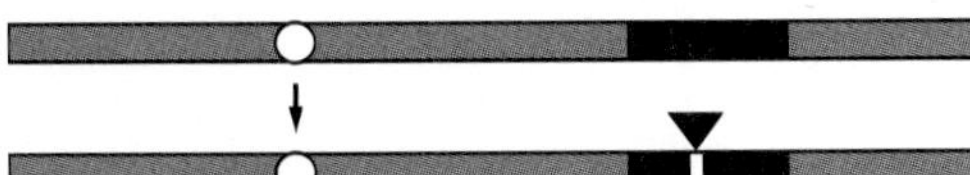

POINT MUTATION: maps to a single site in the genome, corresponding to a single nucleotide pair or a very small part of a single gene

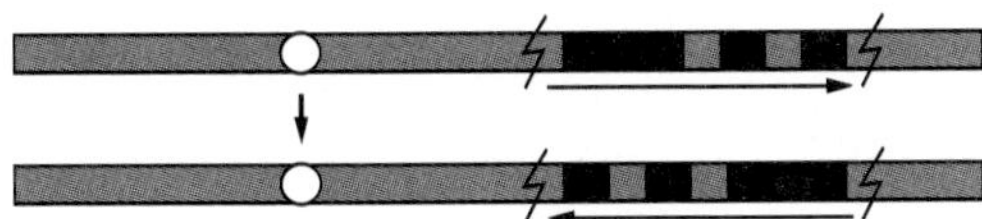

INVERSION: inverts a segment of a chromosome

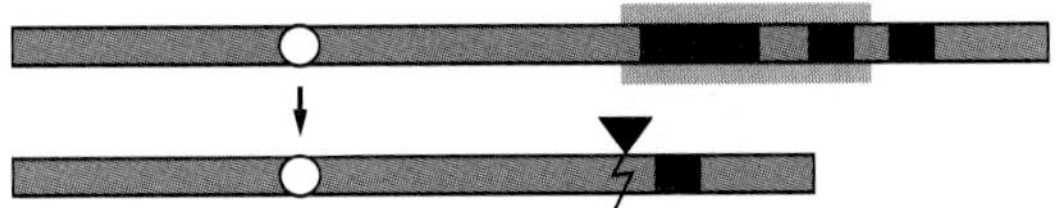

DELETION: deletes a segment of a chromosome

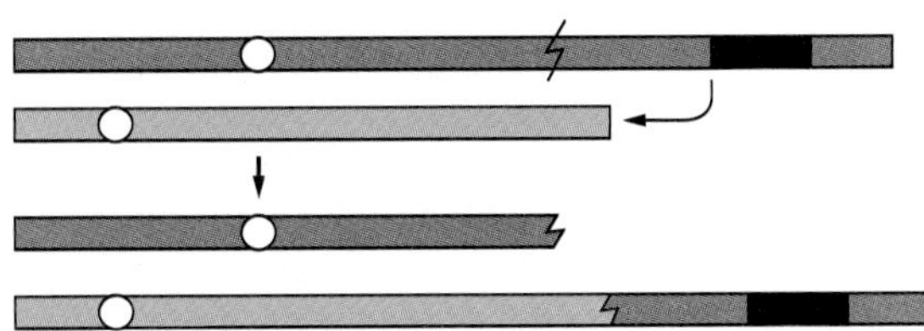

TRANSLOCATION: breaks off a segment from one chromosome and attaches it to another

lethal mutation: causes the developing organism to die prematurely.

conditional mutation: produces its phenotypic effect only under certain conditions, called the *restrictive* conditions. Under other conditions—the *permissive* conditions—the effect is not seen. For a *temperature-sensitive* mutation, the restrictive condition typically is high temperature, while the permissive condition is low temperature.

loss-of-function mutation: either reduces or abolishes the activity of the gene. These are the commonest class of mutations. Loss-of-function mutations are usually *recessive*—the organism can usually function normally as long as it retains at least one normal copy of the affected gene.

null mutation: a loss-of-function mutation that completely abolishes the activity of the gene.

gain-of-function mutation: increases the activity of the gene or makes it active in inappropriate circumstances; these mutations are usually *dominant.*

dominant negative mutation: dominant-acting mutation that blocks gene activity, causing a loss-of-function phenotype even in the presence of a normal copy of the gene. This phenomenon occurs when the mutant gene product interferes with the function of the normal gene product.

suppressor mutation: suppresses the phenotypic effect of another mutation, so that the double mutant seems normal. An *intragenic* suppressor mutation lies within the gene affected by the first mutation; an *extragenic* suppressor mutation lies in a second gene—often one whose product interacts directly with the product of the first.

TWO GENES OR ONE?

Given two mutations that produce the same phenotype, how can we tell whether they are mutations in the same gene? If the mutations are recessive (as they most often are), the answer can be found by a complementation test.

In the simplest type of complementation test, an individual who is homozygous for one mutation is mated with an individual who is homozygous for the other. The phenotype of the offspring gives the answer to the question.

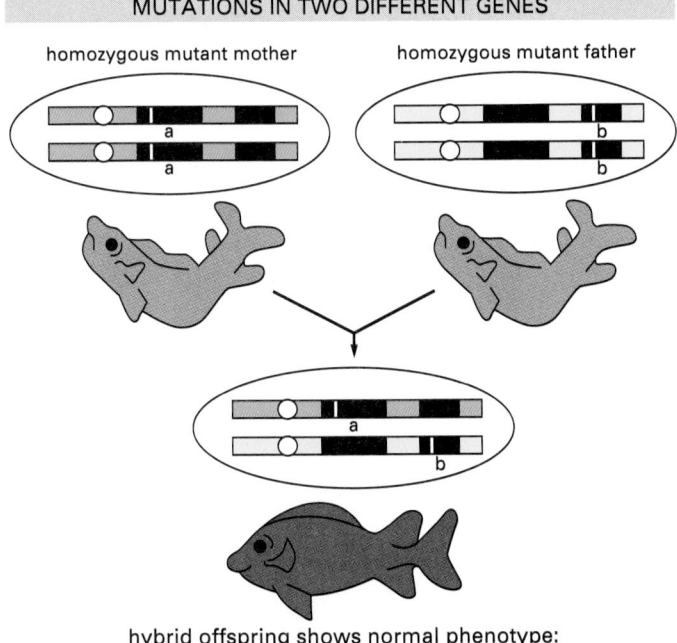

COMPLEMENTATION: MUTATIONS IN TWO DIFFERENT GENES

homozygous mutant mother

homozygous mutant father

a

a

b

b

a

b

hybrid offspring shows normal phenotype: one normal copy of each gene is present

NONCOMPLEMENTATION: TWO INDEPENDENT MUTATIONS IN THE SAME GENE

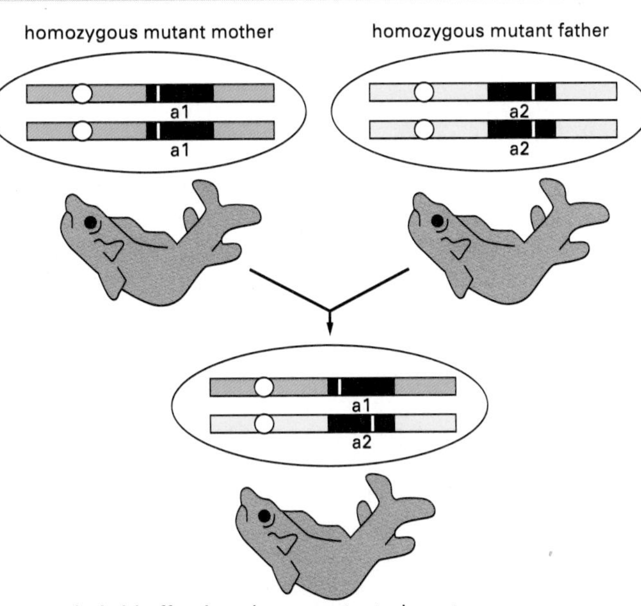

hybrid offspring shows mutant phenotype: no normal copies of the mutated gene are present

The Classical Approach Begins with Random Mutagenesis

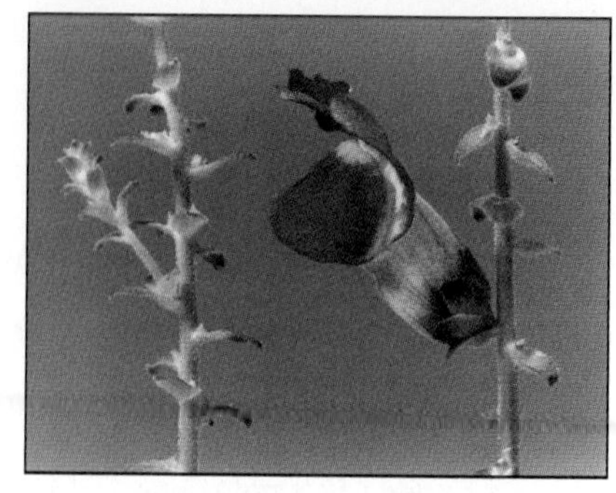

Figure 8–55 Insertional mutant of the snapdragon, *Antirrhinum*. A mutation in a single gene coding for a regulatory protein causes leafy shoots to develop in place of flowers. The mutation allows cells to adopt a character that would be appropriate to a different part of the normal plant. The mutant plant is on the left, the normal plant on the right. (Courtesy of Enrico Coen and Rosemary Carpenter.)

Before the advent of gene cloning technology, most genes were identified by the processes disrupted when the gene was mutated. This classical genetic approach—identifying the genes responsible for mutant phenotypes—is most easily performed in organisms that reproduce rapidly and are amenable to genetic manipulation, such as bacteria, yeasts, nematode worms, and fruit flies. Although spontaneous mutants can sometimes be found by examining extremely large populations—thousands or tens of thousands of individual organisms—the process of isolating mutants can be made much more efficient by generating mutations with agents that damage DNA. By treating organisms with mutagens, very large numbers of mutants can be created quickly and then screened for a particular defect of interest, as we will see shortly.

An alternative approach to chemical or radiation mutagenesis is called *insertional mutagenesis*. This method relies on the fact that exogenous DNA inserted randomly into the genome can produce mutations if the inserted fragment interrupts a gene or its regulatory sequences. The inserted DNA, whose sequence is known, then serves as a molecular tag that aids in the subsequent identification and cloning of the disrupted gene (Figure 8–55). In *Drosophila*, the use of the transposable P element to inactivate genes has revolutionized the study of gene function in the fruit fly. Transposable elements (see Table 5–3, p. 287) have also been used to generate mutants in bacteria, yeast, and in the flowering plant *Arabidopsis*. Retroviruses, which copy themselves into the host genome (see Figure 5–73), have been used to disrupt genes in zebrafish and in mice.

Such studies are well suited for dissecting biological processes in worms and flies, but how can we study gene function in humans? Unlike the organisms we have been discussing, humans do not reproduce rapidly, and they are not intentionally treated with mutagens. Moreover, any human with a serious defect in an essential process, such as DNA replication, would die long before birth.

There are two answers to the question of how we study human genes. First, because genes and gene functions have been so highly conserved throughout evolution, the study of less complex model organisms reveals critical information about similar genes and processes in humans. The corresponding human genes can then be studied further in cultured human cells. Second, many mutations that are not lethal—tissue-specific defects in lysosomes or in cell-surface receptors, for example—have arisen spontaneously in the human population. Analyses of the phenotypes of the affected individuals, together with studies of their cultured cells, have provided many unique insights into important human cell functions. Although such mutations are rare, they are very efficiently discovered because of a unique human property: the mutant individuals call attention to themselves by seeking special medical care.

Genetic Screens Identify Mutants Deficient in Cellular Processes

Once a collection of mutants in a model organism such as yeast or flies has been produced, one generally must examine thousands of individuals to find the altered phenotype of interest. Such a search is called a **genetic screen**. Because obtaining a mutation in a gene of interest depends on the likelihood that the gene will be inactivated or otherwise mutated during random mutagenesis, the larger the genome, the less likely it is that any particular gene will be mutated. Therefore, the more complex the organism, the more mutants must be examined to avoid missing genes. The phenotype being screened for can be simple or complex. Simple phenotypes are easiest to detect: a metabolic deficiency, for example, in which an organism is no longer able to grow in the absence of a particular amino acid or nutrient.

Phenotypes that are more complex, for example mutations that cause defects in learning or memory, may require more elaborate screens (Figure 8–56). But even genetic screens that are used to dissect complex physiological systems should be as simple as possible in design, and, if possible, should permit

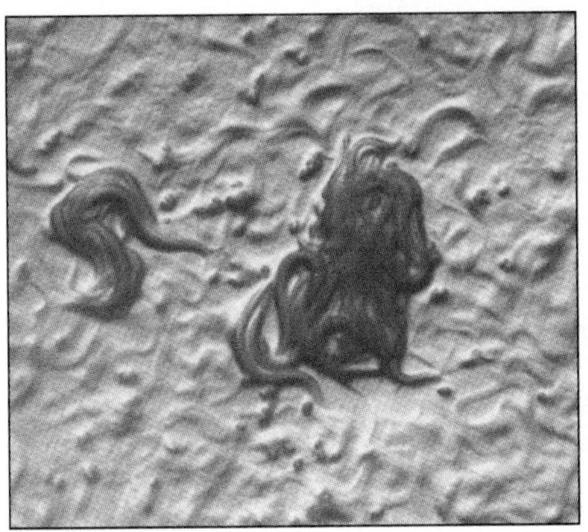

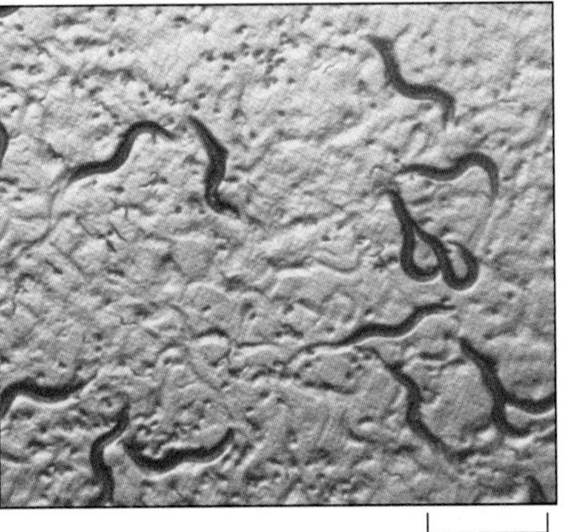
1 mm

Figure 8–56 Screens can detect mutations that affect an animal's behavior. (A) Wild-type *C. elegans* engage in social feeding. The worms swim around until they encounter their neighbors and commence feeding. (B) Mutant animals feed by themselves. (Courtesy of Cornelia Bargmann, *Cell* 94:cover, 1998. © Elsevier.)

the examination of large numbers of mutants simultaneously. As an example, one particularly elegant screen was designed to search for genes involved in visual processing in the zebrafish. The basis of this screen, which monitors the fishes' response to motion, is a change in behavior. Wild-type fish tend to swim in the direction of a perceived motion, while mutants with defects in their visual systems swim in random directions—a behavior that is easily detected. One mutant discovered in this screen is called *lakritz*, which is missing 80% of the retinal ganglion cells that help to relay visual signals from the eye to the brain. As the cellular organization of the zebrafish retina mirrors that of all vertebrates, the study of such mutants should also provide insights into visual processing in humans.

Because defects in genes that are required for fundamental cell processes—RNA synthesis and processing or cell cycle control, for example—are usually lethal, the functions of these genes are often studied in temperature-sensitive mutants. In these mutants the protein product of the mutant gene functions normally at a medium temperature, but can be inactivated by a small increase or decrease in temperature. Thus the abnormality can be switched on and off experimentally simply by changing the temperature. A cell containing a temperature-sensitive mutation in a gene essential for survival at a non-permissive temperature can nevertheless grow at the normal or permissive temperature (Figure 8–57). The temperature-sensitive gene in such a mutant usually contains a point mutation that causes a subtle change in its protein product.

Many temperature-sensitive mutants were isolated in the genes that encode the bacterial proteins required for DNA replication by screening populations of mutagen-treated bacteria for cells that stop making DNA when they are warmed from 30°C to 42°C. These mutants were later used to identify and characterize the corresponding DNA replication proteins (discussed in Chapter 5). Temperature-sensitive mutants also led to the identification of many proteins involved in regulating the cell cycle and in moving proteins through the secretory pathway in yeast (see Panel 13–1). Related screening approaches have demonstrated the function of enzymes involved in the principal metabolic pathways of bacteria and yeast (discussed in Chapter 2), as well as discovering many of the gene products

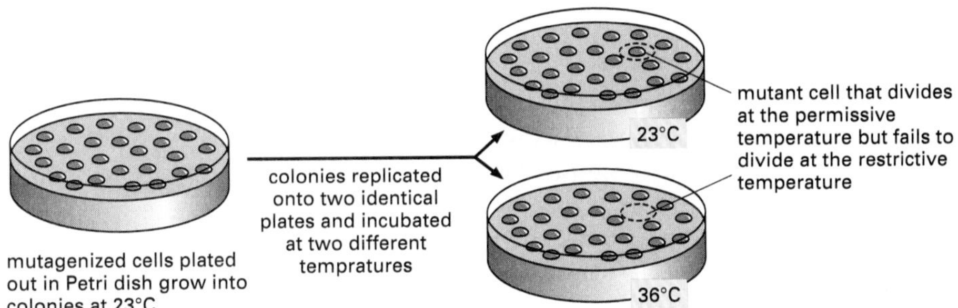

mutagenized cells plated out in Petri dish grow into colonies at 23°C
colonies replicated onto two identical plates and incubated at two different tempratures
23°C
36°C
mutant cell that divides at the permissive temperature but fails to divide at the restrictive temperature

Figure 8–57 Screening for temperature-sensitive bacterial or yeast mutants. Mutagenized cells are plated out at the permissive temperature. The resulting colonies are transferred to two identical Petri dishes by replica plating; one of these plates is incubated at the permissive temperature, the other at the non-permissive temperature. Cells containing a temperature-sensitive mutation in a gene essential for proliferation can divide at the normal, permissive temperature but fail to divide at the elevated, non-permissive temperature.

responsible for the orderly development of the *Drosophila* embryo (discussed in Chapter 21).

A Complementation Test Reveals Whether Two Mutations Are in the Same or in Different Genes

A large-scale genetic screen can turn up many different mutants that show the same phenotype. These defects might lie in different genes that function in the same process, or they might represent different mutations in the same gene. How can we tell, then, whether two mutations that produce the same phenotype occur in the same gene or in different genes? If the mutations are recessive—if, for example, they represent a loss of function of a particular gene—a complementation test can be used to ascertain whether the mutations fall in the same or in different genes. In the simplest type of complementation test, an individual that is homozygous for one mutation—that is, it possesses two identical alleles of the mutant gene in question—is mated with an individual that is homozygous for the other mutation. If the two mutations are in the same gene, the offspring show the mutant phenotype, because they still will have no normal copies of the gene in question (see Panel 8–1, pp. 526–527). If, in contrast, the mutations fall in different genes, the resulting offspring show a normal phenotype. They retain one normal copy (and one mutant copy) of each gene. The mutations thereby complement one another and restore a normal phenotype. Complementation testing of mutants identified during genetic screens has revealed, for example, that 5 genes are required for yeast to digest the sugar galactose; that 20 genes are needed for *E. coli* to build a functional flagellum; that 48 genes are involved in assembling bacteriophage T4 viral particles; and that hundreds of genes are involved in the development of an adult nematode worm from a fertilized egg.

Once a set of genes involved in a particular biological process has been identified, the next step is to determine in which order the genes function. Determining when a gene acts can facilitate the reconstruction of entire genetic or biochemical pathways, and such studies have been central to our understanding of metabolism, signal transduction, and many other developmental and physiological processes. In essence, untangling the order in which genes function requires careful characterization of the phenotype caused by mutations in each different gene. Imagine, for example, that mutations in a handful of genes all cause an arrest in cell division during early embryo development. Close examination of each mutant may reveal that some act extremely early, preventing the fertilized egg from dividing into two cells. Other mutations may allow early cell divisions but prevent the embryo from reaching the blastula stage.

To test predictions made about the order in which genes function, organisms can be made that are mutant in two different genes. If these mutations affect two different steps in the same process, such *double mutants* should have a phenotype identical to that of the mutation that acts earliest in the pathway. As an example, the pathway of protein secretion in yeast has been deciphered in this manner. Different mutations in this pathway cause proteins to accumulate aberrantly in the endoplasmic reticulum (ER) or in the Golgi apparatus. When a cell is engineered to harbor both a mutation that blocks protein processing in the ER *and* a mutation that blocks processing in the Golgi compartment, proteins accumulate in the ER. This indicates that proteins must pass through the ER before being sent to the Golgi before secretion (Figure 8–58).

Genes Can Be Located by Linkage Analysis

With mutants in hand, the next step is to identify the gene or genes that seem to be responsible for the altered phenotype. If insertional mutagenesis was used for the original mutagenesis, locating the disrupted gene is fairly simple. DNA fragments containing the insertion (a transposon or a retrovirus, for example) are collected and amplified, and the nucleotide sequence of the flanking DNA is determined. This sequence is then used to search a DNA database to identify the gene that was interrupted by insertion of the transposable element.

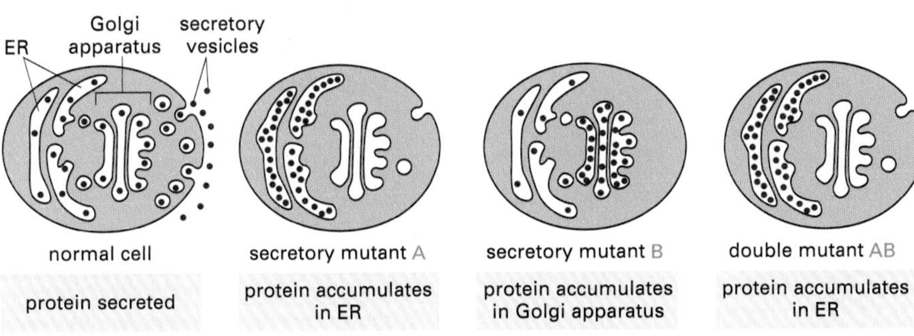

Figure 8–58 Using genetics to determine the order of function of genes. In normal cells, proteins are loaded into vesicles, which fuse with the plasma membrane and secrete their contents into the extracellular medium. In secretory mutant A, proteins accumulate in the ER. In a different secretory mutant B, proteins accumulate in the Golgi. In the double mutant AB, proteins accumulate in the ER; this indicates that the gene defective in mutant A acts before the gene defective in mutant B in the secretory pathway.

If a DNA-damaging chemical was used to generate the mutants, identifying the inactivated gene is often more laborious and can be accomplished by several different approaches. In one, the first step is to determine where on the genome the gene is located. To map a newly discovered gene, its rough chromosomal location is first determined by assessing how far the gene lies from other known genes in the genome. Estimating the distance between genetic loci is usually done by linkage analysis, a technique that relies on the fact that genes that lie near one another on a chromosome tend to be inherited together. The closer the genes are, the greater the likelihood they will be passed to offspring as a pair. Even closely linked genes, however, can be separated by recombination during meiosis. The larger the distance between two genetic loci, the greater the chance that they will be separated by a crossover (see Panel 8–1, pp. 526–527). By calculating the recombination frequency between two genes, the approximate distance between them can be determined.

Because genes are not always located close enough to one another to allow a precise pinpointing of their position, linkage analyses often rely on physical markers along the genome for estimating the location of an unknown gene. These markers are generally nucleotide fragments, with a known sequence and genome location, that can exist in at least two allelic forms. Single-nucleotide polymorphisms (SNPs), for example, are short sequences that differ by one or more nucleotides among individuals in a population. SNPs can be detected by hybridization techniques. Many such physical markers, distributed all along the length of chromosomes, have been collected for a variety of organisms, including more than 10^6 for humans. If the distribution of these markers is sufficiently dense, one can, through a linkage analysis that tests for the tight coinheritance of one or more SNPs with the mutant phenotype, narrow the potential location of a gene to a chromosomal region that may contain only a few gene sequences. These are then considered candidate genes, and their structure and function can be tested directly to determine which gene is responsible for the original mutant phenotype.

Linkage analysis can be used in the same way to identify the genes responsible for heritable human disorders. Such studies require that DNA samples be collected from a large number of families affected by the disease. These samples are examined for the presence of physical markers such as SNPs that seem to be closely linked to the disease gene—these sequences would always be inherited by individuals who have the disease, and not by their unaffected relatives. The disease gene is then located as described above (Figure 8–59). The genes for cystic fibrosis and Huntington's disease, for example, were discovered in this manner.

Searching for Homology Can Help Predict a Gene's Function

Once a gene has been identified, its function can often be predicted by identifying homologous genes whose functions are already known. As we discussed earlier, databases containing nucleotide sequences from a variety of organisms—including the complete genome sequences of many dozens of microbes, *C. elegans*, *A. thaliana*, *D. melanogaster*, and human—can be searched for sequences that are similar to those of the uncharacterized target gene.

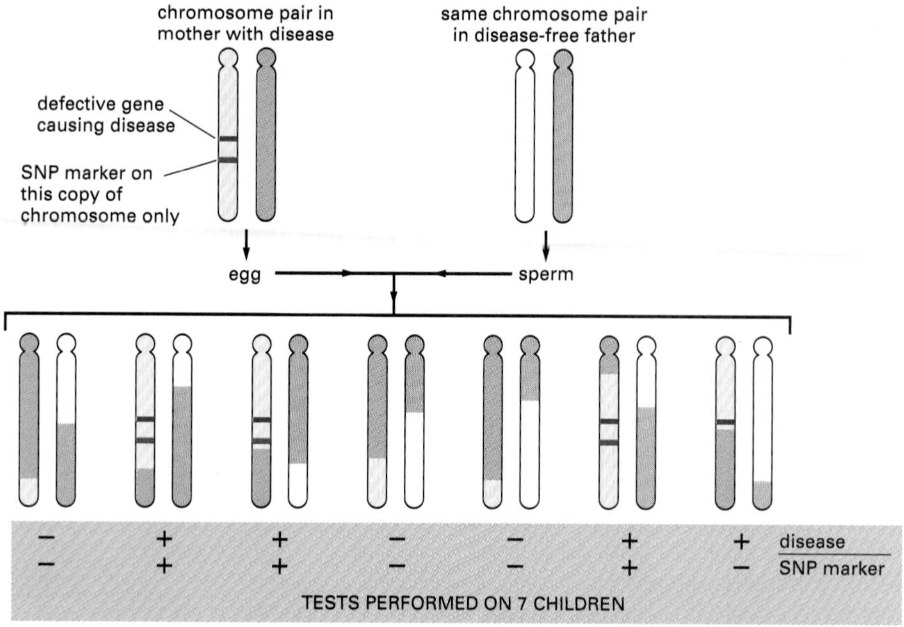

CONCLUSION: gene causing disease is coinherited with SNP marker from diseased mother in 75% of the diseased progeny. If this same correlation is observed in other families that have been examined, the gene causing disease is mapped to this chromosome close to the SNP. Note that a SNP that is either far away from the gene on the same chromosome, or located on a different chromosome than the gene of interest, will be coinherited only 50% of the time.

Figure 8–59 Genetic linkage analysis using physical markers on the DNA to find a human gene. In this example, one studies the coinheritance of a specific human phenotype (here a genetic disease) with a SNP marker. If individuals who inherit the disease nearly always inherit a particular SNP marker, then the gene causing the disease and the SNP are likely to be close together on the chromosome, as shown here. To prove that an observed linkage is statistically significant, hundreds of individuals may need to be examined. Note that the linkage will not be absolute unless the SNP marker is located in the gene itself. Thus, occasionally the SNP will be separated from the disease gene by meiotic crossing-over during the formation of the egg or sperm: this has happened in the case of the chromosome pair on the far right. When working with a sequenced genome, this procedure would be repeated with SNPs located on either side of the initial SNP, until a 100% coinheritance is found.

When analyzing a newly sequenced genome, such a search serves as a first-pass attempt to assign functions to as many genes as possible, a process called *annotation*. Further genetic and biochemical studies are then performed to confirm whether the gene encodes a product with the predicted function, as we discuss shortly. Homology analysis does not always reveal information about function: in the case of the yeast genome, 30% of the previously uncharacterized genes could be assigned a putative function by homology analysis; 10% had homologues whose function was also unknown; and another 30% had no homologues in any existing databases. (The remaining 30% of the genes had been identified before sequencing the yeast genome.)

In some cases, a homology search turns up a gene in organism A which produces a protein that, in a different organism, is fused to a second protein that is produced by an independent gene in organism A. In yeast, for example, two separate genes encode two proteins that are involved in the synthesis of tryptophan; in *E. coli*, however, these two genes are fused into one (Figure 8–60). Knowledge that these two proteins in yeast correspond to two domains in a single bacterial protein means that they are likely to be functionally associated, and probably work together in a protein complex. More generally, this approach is used to establish functional links between genes that, for most organisms, are widely separated in the genome.

Reporter Genes Reveal When and Where a Gene Is Expressed

Clues to gene function can often be obtained by examining when and where a gene is expressed in the cell or in the whole organism. Determining the pattern and timing of gene expression can be accomplished by replacing the coding

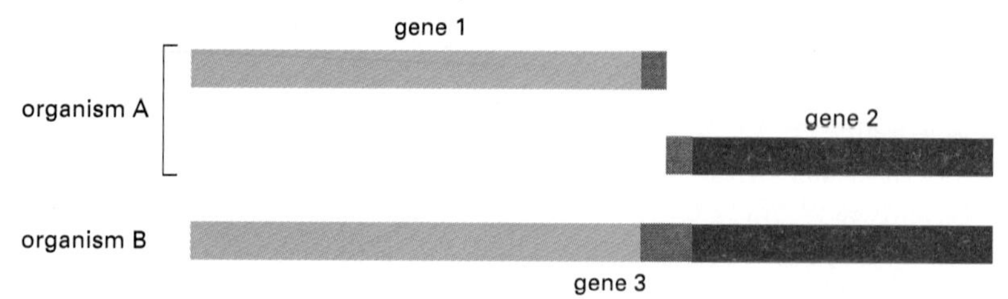

Figure 8–60 Domain fusions reveal relationships between functionally linked genes. In this example, the functional interaction of genes 1 and 2 in organism A is inferred by the fusion of homologous domains into a single gene (gene 3) in organism B.

portion of the gene under study with a reporter gene. In most cases, the expression of the reporter gene is then monitored by tracking the fluorescence or enzymatic activity of its protein product (pp. 518–519).

As discussed in detail in Chapter 7, gene expression is controlled by regulatory DNA sequences, located upstream or downstream of the coding region, which are not generally transcribed. These regulatory sequences, which control which cells will express a gene and under what conditions, can also be made to drive the expression of a reporter gene. One simply replaces the target gene's coding sequence with that of the reporter gene, and introduces these recombinant DNA molecules into cells. The level, timing, and cell specificity of reporter protein production reflect the action of the regulatory sequences that belong to the original gene (Figure 8–61).

Several other techniques, discussed previously, can also be used to determine the expression pattern of a gene. Hybridization techniques such as Northern analysis (see Figure 8–27) and *in situ* hybridization for RNA detection (see Figure 8–29) can reveal when genes are transcribed and in which tissue, and how much mRNA they produce.

Microarrays Monitor the Expression of Thousands of Genes at Once

So far we have discussed techniques that can be used to monitor the expression of only a single gene at a time. Many of these methods are fairly labor-intensive: generating reporter gene constructs or GFP fusions requires manipulating DNA and transfecting cells with the resulting recombinant molecules. Even Northern analyses are limited in scope by the number of samples that can be run on an agarose gel. Developed in the 1990s, **DNA microarrays** have revolutionized the way in which gene expression is now analyzed by allowing the RNA products of thousands of genes to be monitored at once. By examining the expression of so many genes simultaneously, we can now begin to identify and study the gene expression patterns that underlie cellular physiology: we can see which genes are switched on (or off) as cells grow, divide, or respond to hormones or to toxins.

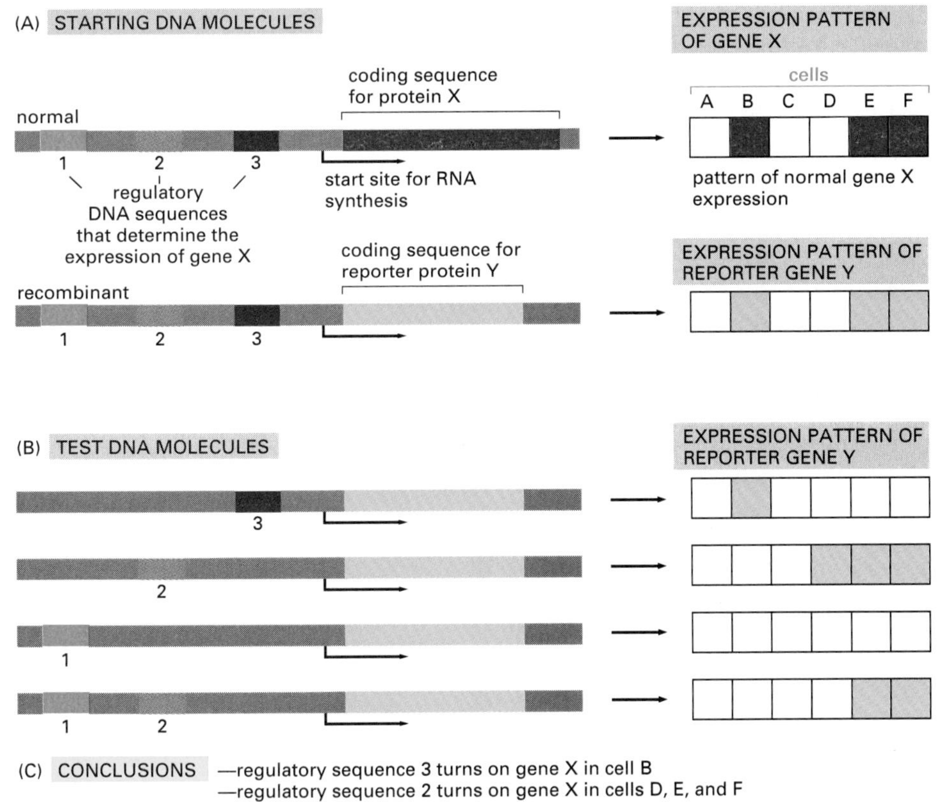

Figure 8–61 Using a reporter protein to determine the pattern of a gene's expression. (A) In this example the coding sequence for protein X is replaced by the coding sequence for protein Y. (B) Various fragments of DNA containing candidate regulatory sequences are added in combinations. The recombinant DNA molecules are then tested for expression after their transfection into a variety of different types of mammalian cells, and the results are summarized in (C). For experiments in eucaryotic cells, two commonly used reporter proteins are the enzymes β-galactosidase (*β-gal*) and green fluorescent protein or GFP (see Figure 9–44). Because these are bacterial enzymes, their presence can be monitored by simple and sensitive assays of enzyme activity, without any interference from host cell enzymes. Figure 7–39 shows an example in which the β-gal receptor gene is used to monitor the activity of the eve gene regulatory sequence in a *Drosophila* embryo.

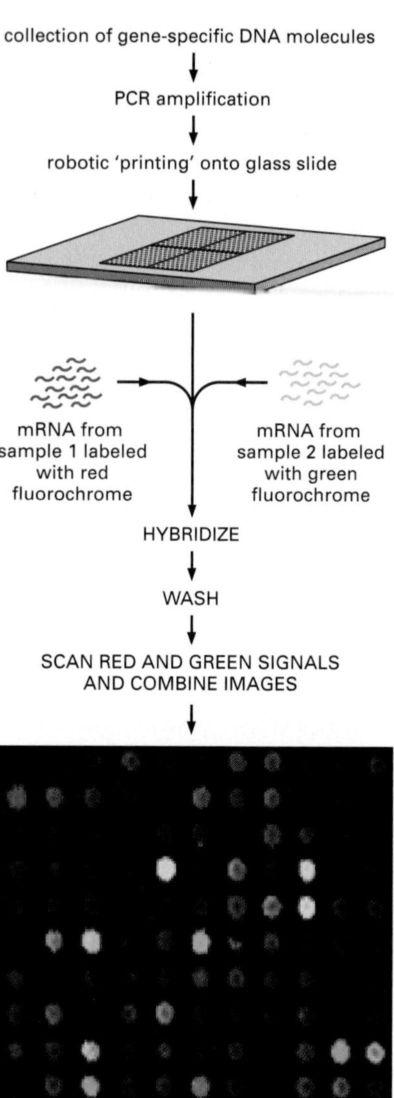

Figure 8–62 Using DNA microarrays to monitor the expression of thousands of genes simultaneously. To prepare the microarray, DNA fragments—each corresponding to a gene—are spotted onto a slide by a robot. Prepared arrays are also available commercially. In this example, mRNA is collected from two different cell samples for a direct comparison of their relative levels of gene expression. These samples are converted to cDNA and labeled, one with a red fluorochrome, the other with a green fluorochrome. The labeled samples are mixed and then allowed to hybridize to the microarray. After incubation, the array is washed and the fluorescence scanned. In the portion of a microarray shown, which represents the expression of 110 yeast genes, *red* spots indicate that the gene in sample 1 is expressed at a higher level than the corresponding gene in sample 2; *green* spots indicate that expression of the gene is higher in sample 2 than in sample 1. *Yellow* spots reveal genes that are expressed at equal levels in both cell samples. Dark spots indicate little or no expression in either sample of the gene whose fragment is located at that position in the array. For details see Figure 1–45. (Microarray courtesy of J.L. DeRisi et al., *Science* 278:680–686, 1997. © AAAS.)

DNA microarrays are little more than glass microscope slides studded with a large number of DNA fragments, each containing a nucleotide sequence that serves as a probe for a specific gene. The most dense arrays may contain tens of thousands of these fragments in an area smaller than a postage stamp, allowing thousands of hybridization reactions to be performed in parallel (Figure 8–62). Some microarrays are generated from large DNA fragments that have been generated by PCR and then spotted onto the slides by a robot. Others contain short oligonucleotides that are synthesized on the surface of the glass wafer with techniques similar to those that are used to etch circuits onto computer chips. In either case, the exact sequence—and position—of every probe on the chip is known. Thus any nucleotide fragment that hybridizes to a probe on the array can be identified as the product of a specific gene simply by detecting the position to which it is bound.

To use a DNA microarray to monitor gene expression, mRNA from the cells being studied is first extracted and converted to cDNA (see Figure 8–34). The cDNA is then labeled with a fluorescent probe. The microarray is incubated with this labeled cDNA sample and hybridization is allowed to occur (see Figure 8–62). The array is then washed to remove cDNA that is not tightly bound, and the positions in the microarray to which labeled DNA fragments have bound are identified by an automated scanning-laser microscope. The array positions are then matched to the particular gene whose sample of DNA was spotted in this location.

Typically the fluorescent DNA from the experimental samples (labeled, for example, with a red fluorescent dye) are mixed with a reference sample of cDNA fragments labeled with a differently colored fluorescent dye (green, for example). Thus, if the amount of RNA expressed from a particular gene in the cells of interest is increased relative to that of the reference sample, the resulting spot is red. Conversely, if the gene's expression is decreased relative to the reference sample, the spot is green. Using such an internal reference, gene expression profiles can be tabulated with great precision.

So far, DNA microarrays have been used to examine everything from the change in gene expression that make strawberries ripen to the gene expression "signatures" of different types of human cancer cells (see Figure 7–3). Arrays that contain probes representing all 6000 yeast genes have been used to monitor the changes that occur in gene expression as yeast shift from fermenting glucose to growing on ethanol; as they respond to a sudden shift to heat or cold; and as they proceed through different stages of the cell cycle. The first study showed that, as yeast use up the last glucose in their medium, their gene expression pattern changes markedly: nearly 900 genes are more actively transcribed, while another 1200 decrease in activity. About half of these genes have no known function, although this study suggests that they are somehow involved in the metabolic reprogramming that occurs when yeast cells shift from fermentation to respiration.

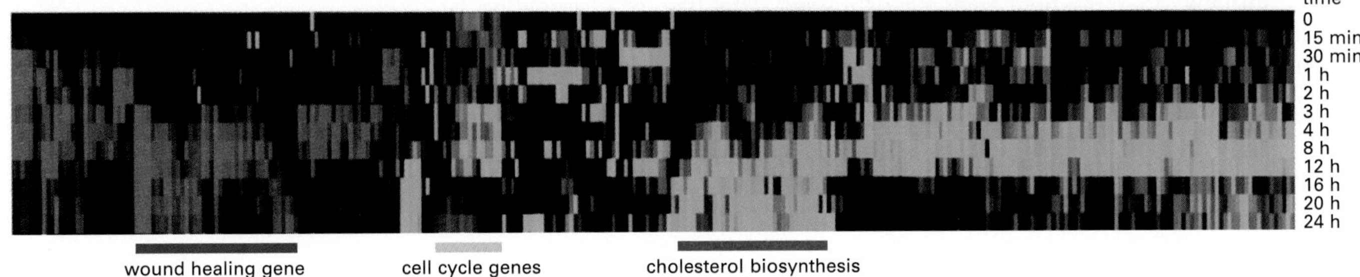

Figure 8–63 Using cluster analysis to identify sets of genes that are coordinately regulated. Genes that belong to the same cluster may be involved in common cellular pathways or processes. To perform a cluster analysis, microarray data are obtained from cell samples exposed to a variety of different conditions, and genes that show coordinate changes in their expression pattern are grouped together. In this experiment, human fibroblasts were deprived of serum for 48 hours; serum was then added back to the cultures at time 0 and the cells were harvested for microarray analysis at different time points. Of the 8600 genes analyzed on the DNA microarray, just over 300 showed threefold or greater variation in their expression patterns in response to serum reintroduction. Here, *red* indicates an increase in expression; *green* is a decrease in expression. On the basis of the results of many microarray experiments, the 8600 genes have been grouped in clusters based on similar patterns of expression. The results of this analysis show that genes involved in wound healing are turned on in response to serum, while genes involved in regulating cell cycle progression and cholesterol biosynthesis are shut down. (From M.B. Eisen et al., *Proc. Natl. Acad. Sci. USA* 95:14863–14868, 1998. © National Academy of Sciences.)

Comprehensive studies of gene expression also provide an additional layer of information that is useful for predicting gene function. Earlier we discussed how identifying a protein's interaction partners can yield clues about that protein's function. A similar principle holds true for genes: information about a gene's function can be deduced by identifying genes that share its expression pattern. Using a technique called *cluster analysis*, one can identify sets of genes that are coordinately regulated. Genes that are turned on or turned off together under a variety of different circumstances may work in concert in the cell: they may encode proteins that are part of the same multiprotein machine, or proteins that are involved in a complex coordinated activity, such as DNA replication or RNA splicing. Characterizing an unknown gene's function by grouping it with known genes that share its transcriptional behavior is sometimes called "guilt by association." Cluster analyses have been used to analyze the gene expression profiles that underlie many interesting biological processes, including wound healing in humans (Figure 8–63).

Targeted Mutations Can Reveal Gene Function

Although in rapidly reproducing organisms it is often not difficult to obtain mutants that are deficient in a particular process, such as DNA replication or eye development, it can take a long time to trace the defect to a particular altered protein. Recently, recombinant DNA technology and the explosion in genome sequencing have made possible a different type of genetic approach. Instead of beginning with a randomly generated mutant and using it to identify a gene and its protein, one can start with a particular gene and proceed to make mutations in it, creating mutant cells or organisms so as to analyze the gene's function. Because the new approach reverses the traditional direction of genetic discovery—proceeding from genes and proteins to mutants, rather than vice versa—it is commonly referred to as **reverse genetics**.

Reverse genetics begins with a cloned gene, a protein with interesting properties that has been isolated from a cell, or simply a genome sequence. If the starting point is a protein, the gene encoding it is first identified and, if necessary, its nucleotide sequence is determined. The gene sequence can then be altered *in vitro* to create a mutant version. This engineered mutant gene, together with an appropriate regulatory region, is transferred into a cell. Inside the cell, it can integrate into a chromosome, becoming a permanent part of the cell's genome. All of the descendants of the modified cell will now contain the mutant gene.

If the original cell used for the gene transfer is a fertilized egg, whole multicellular organisms can be obtained that contain the mutant gene, provided that the mutation does not cause lethality. In some of these animals, the altered gene will be incorporated into the germ cells—a germline mutation—allowing the mutant gene to be passed on to their progeny.

Genetic transformations of this kind are now routinely performed with organisms as complex as fruit flies and mammals. Technically, even humans could now be transformed in this way, although such procedures are not undertaken, even for therapeutic purposes, for fear of the unpredictable aberrations that might occur in such individuals.

Earlier in this chapter we discussed other approaches to discover a gene's function, including searching for homologous genes in other organisms and determining when and where a gene is expressed. This type of information is especially useful in suggesting what sort of phenotypes to look for in the mutant organisms. A gene that is expressed only in adult liver, for example, may have a role in degrading toxins, but is not likely to affect the development of the eye. All of these approaches can be used either to study single genes or to attempt a large-scale analysis of the function of every gene in an organism—a burgeoning field known as *functional genomics.*

Cells and Animals Containing Mutated Genes Can Be Made to Order

We have seen that searching for homologous genes and analyzing gene expression patterns can provide clues about gene function, but they do not reveal what exactly a gene does inside a cell. Genetics provides a powerful solution to this problem, because mutants that lack a particular gene may quickly reveal the function of the protein that it encodes. Genetic engineering techniques allow one to specifically produce such gene knockouts, as we will see. However, one can also generate mutants that express a gene at abnormally high levels (overexpression), in the wrong tissue or at the wrong time (misexpression), or in a slightly altered form that exerts a dominant phenotype. To facilitate such studies of gene function, the coding sequence of a gene and its regulatory regions can be engineered to change the functional properties of the protein product, the amount of protein made, or the particular cell type in which the protein is produced.

Altered genes are introduced into cells in a variety of ways, some of which are described in detail in Chapter 9. DNA can be microinjected into mammalian cells with a glass micropipette or introduced by a virus that has been engineered to carry foreign genes. In plant cells, genes are frequently introduced by a technique called particle bombardment: DNA samples are painted onto tiny gold beads and then literally shot through the cell wall with a specially modified gun. Electroporation is the method of choice for introducing DNA into bacteria and some other cells. In this technique, a brief electric shock renders the cell membrane temporarily permeable, allowing foreign DNA to enter the cytoplasm.

We will now examine how the study of such mutant cells and organisms allows the dissection of biological pathways.

The Normal Gene in a Cell Can Be Directly Replaced by an Engineered Mutant Gene in Bacteria and Some Lower Eucaryotes

Unlike higher eucaryotes (which are multicellular and diploid), bacteria, yeasts, and the cellular slime mold *Dictyostelium* generally exist as haploid single cells. In these organisms an artificially introduced DNA molecule carrying a mutant gene can, with a relatively high frequency, replace the single copy of the normal gene by homologous recombination (see p. 276), so that it is easy to produce cells in which the mutant gene has replaced the normal gene (Figure 8–64A). In

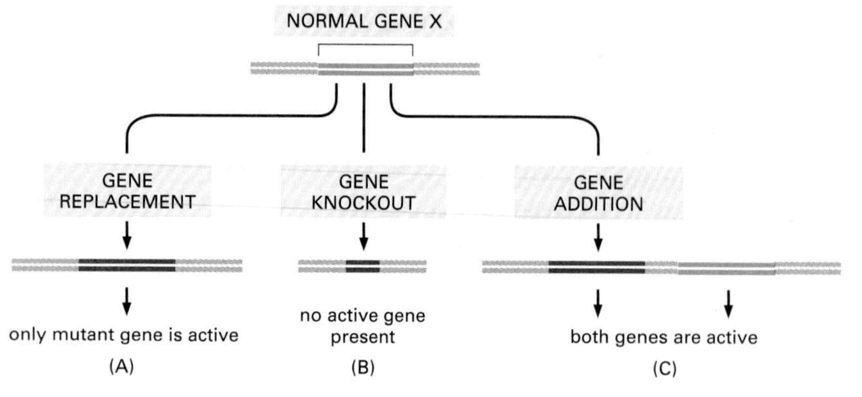

Figure 8–64 Gene replacement, gene knockout, and gene addition. A normal gene can be altered in several ways in a genetically engineered organism. (A) The normal gene *(green)* can be completely replaced by a mutant copy of the gene *(red)*, a process called gene replacement. This provides information on the activity of the mutant gene without interference from the normal gene, and thus the effects of small and subtle mutations can be determined. (B) The normal gene can be inactivated completely, for example, by making a large deletion in it; the gene is said to have suffered a knockout. (C) A mutant gene can simply be added to the genome. In some organisms this is the easiest type of genetic engineering to perform. This approach can provide useful information when the introduced mutant gene overrides the function of the normal gene.

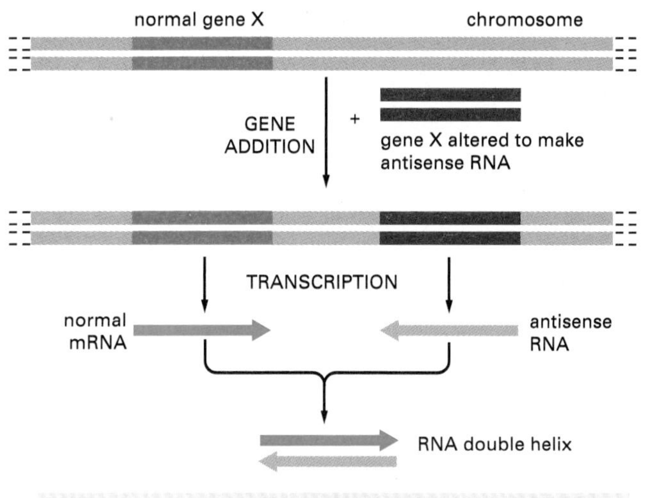

Figure 8–65 The antisense RNA strategy for generating dominant negative mutations. Mutant genes that have been engineered to produce antisense RNA, which is complementary in sequence to the RNA made by the normal gene X, can cause double-stranded RNA to form inside cells. If a large excess of the antisense RNA is produced, it can hybridize with—and thereby inactivate—most of the normal RNA produced by gene X. Although in the future it may become possible to inactivate any gene in this way, at present the technique seems to work for some genes but not others.

this way cells can be made to order that produce an altered form of any specific protein or RNA molecule instead of the normal form of the molecule. If the mutant gene is completely inactive and the gene product normally performs an essential function, the cell dies; but in this case a less severely mutated version of the gene can be used to replace the normal gene, so that the mutant cell survives but is abnormal in the process for which the gene is required. Often the mutant of choice is one that produces a temperature-sensitive gene product, which functions normally at one temperature but is inactivated when cells are shifted to a higher or lower temperature.

The ability to perform direct gene replacements in lower eucaryotes, combined with the power of standard genetic analyses in these haploid organisms, explains in large part why studies in these types of cells have been so important for working out the details of those processes that are shared by all eucaryotes. As we shall see, gene replacements are possible, but more difficult to perform in higher eucaryotes, for reasons that are not entirely understood.

Engineered Genes Can Be Used to Create Specific Dominant Negative Mutations in Diploid Organisms

Higher eucaryotes, such as mammals, fruit flies, or worms, are diploid and therefore have two copies of each chromosome. Moreover, transfection with an altered gene generally leads to gene addition rather than gene replacement: the altered gene inserts at a random location in the genome, so that the cell (or the organism) ends up with the mutated gene in addition to its normal gene copies.

Because gene addition is much more easily accomplished than gene replacement in higher eucaryotic cells, it is useful to create specific dominant negative mutations in which a mutant gene eliminates the activity of its normal counterparts in the cell. One ingenious approach exploits the specificity of hybridization reactions between two complementary nucleic acid chains. Normally, only one of the two DNA strands in a given portion of double helix is transcribed into RNA, and it is always the same strand for a given gene (see Figure 6–14). If a cloned gene is engineered so that the opposite DNA strand is transcribed instead, it will produce antisense RNA molecules that have a sequence complementary to the normal RNA transcripts. Such *antisense RNA*, when synthesized in large enough amounts, can often hybridize with the "sense" RNA made by the normal genes and thereby inhibit the synthesis of the corresponding protein (Figure 8–65). A related method involves synthesizing short antisense nucleic acid molecules chemically or enzymatically and then injecting (or otherwise delivering) them into cells, again blocking (although only temporarily) production of the corresponding protein. To avoid degradation of the

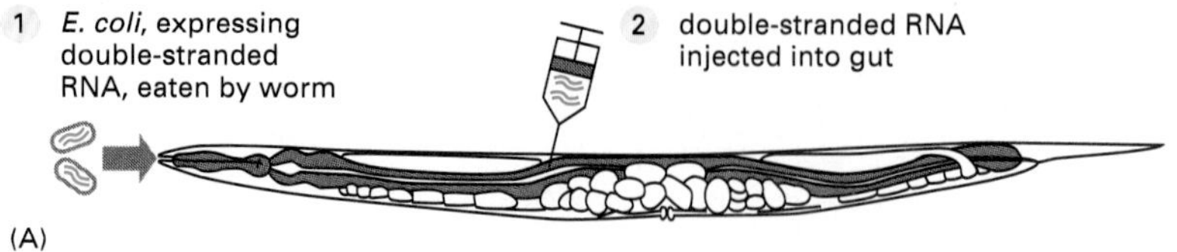

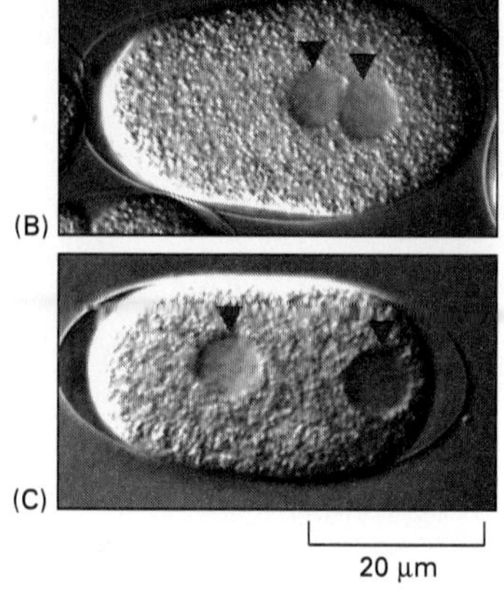

Figure 8–66 Dominant negative mutations created by RNA interference. (A) Double-stranded RNA (dsRNA) can be introduced into *C. elegans* (1) by feeding the worms with *E. coli* expressing the dsRNA or (2) by injecting dsRNA directly into the gut. (B) Wild-type worm embryo. (C) Worm embryo in which a gene involved in cell division has been inactivated by RNAi. The embryo shows abnormal migration of the two unfused nuclei of the egg and sperm. (B, C, from P. Gönczy et al., *Nature* 408:331–336, 2000. © Macmillan Magazines Ltd.)

injected nucleic acid, a stable synthetic RNA analog, called morpholino-RNA, is often used instead of ordinary RNA.

As investigators continued to explore the antisense RNA strategy, they made an interesting discovery. An antisense RNA strand can block gene expression, but a preparation of double-stranded RNA (dsRNA), containing both the sense and antisense strands of a target gene, inhibit the activity of target genes even more effectively (see Figure 7–107). This phenomenon, dubbed *RNA interference (RNAi),* has now been exploited for examining gene function in several organisms.

The RNAi technique has been widely used to study gene function in the nematode *C. elegans*. When working with worms, introducing the dsRNA is quite simple: RNA can be injected directly into the intestine of the animal, or the worm can be fed with *E. coli* expressing the target gene dsRNA (Figure 8–66A). The RNA is distributed throughout the body of the worm and is found to inhibit expression of the target gene in different tissue types. Further, as explained in Figure 7–107, the interference is frequently inherited by the progeny of the injected animal. Because the entire genome of *C. elegans* has been sequenced, RNAi is being used to help in assigning functions to the entire complement of worm genes. In one study, researchers were able to inhibit 96% of the approximately 2300 predicted genes on *C. elegans* chromosome III. In this way, they identified 133 genes involved in cell division in *C. elegans* embryos (Figure 8–66C). Of these, only 11 had been previously ascribed a function by direct experimentation.

For unknown reasons, RNA interference does not efficiently inactivate all genes. And interference can sometimes suppress the activity of a target gene in one tissue and not another. An alternative way to produce a dominant negative mutation takes advantage of the fact that most proteins function as part of a larger protein complex. Such complexes can often be inactivated by the inclusion of just one nonfunctional component. Therefore, by designing a gene that produces large quantities of a mutant protein that is inactive but still able to assemble into the complex, it is often possible to produce a cell in which all the complexes are inactivated despite the presence of the normal protein (Figure 8–67).

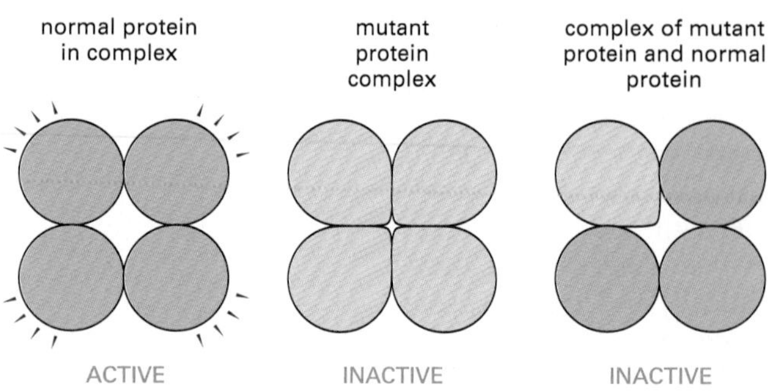

Figure 8–67 A dominant negative effect of a protein. Here a gene is engineered to produce a mutant protein that prevents the normal copies of the same protein from performing their function. In this simple example, the normal protein must form a multisubunit complex to be active, and the mutant protein blocks function by forming a mixed complex that is inactive. In this way a single copy of a mutant gene located anywhere in the genome can inactivate the normal products produced by other gene copies.

If a protein is required for the survival of the cell (or the organism), a dominant negative mutant dies, making it impossible to test the function of the protein. To avoid this problem, one can couple the mutant gene to control sequences that have been engineered to produce the gene product only on command—for example, in response to an increase in temperature or to the presence of a specific signaling molecule. Cells or organisms containing such a dominant mutant gene under the control of an *inducible promoter* can be deprived of a specific protein at a particular time, and the effect can then be followed. Inducible promoters also allow genes to be switched on or off in specific tissues, allowing one to examine the effect of the mutant gene in selected parts of the organism. In the future, techniques for producing dominant negative mutations to inactivate specific genes are likely to be widely used to determine the functions of proteins in higher organisms.

Gain-of-Function Mutations Provide Clues to the Role Genes Play in a Cell or Organism

In the same way that cells can be engineered to express a dominant negative version of a protein, resulting in a loss-of-function phenotype, they can also be engineered to display a novel phenotype through a *gain-of-function* mutation. Such mutations may confer a novel activity on a particular protein, or they may cause a protein with normal activity to be expressed at an inappropriate time or in the wrong tissue in an animal. Regardless of the mechanism, gain-of-function mutations can produce a new phenotype in a cell, tissue, or organism.

Often, gain-of-function mutants are generated by expressing a gene at a much higher level than normal in cells. Such overexpression can be achieved by coupling a gene to a powerful promoter sequence and placing it on a multicopy plasmid—or integrating it in multiple copies in the genome. In either case, the gene is present in many copies and each copy directs the transcription of unusually large numbers of mRNA molecules. Although the effect that such overexpression has on the phenotype of an organism must be interpreted with caution, this approach has provided invaluable insights into the activity of many genes. In an alternate type of gain-of-function mutation, the mutant protein is made in normal amounts, but is much more active than its normal counterpart. Such proteins are frequently found in tumors, and they have been exploited to study signal transduction pathways in cells (discussed in Chapter 15).

Genes can also be expressed at the wrong time or in the wrong place in an organism—often with striking results (Figure 8–68). Such misexpression is most often accomplished by re-engineering the genes themselves, thereby supplying them with the regulatory sequences needed to alter their expression.

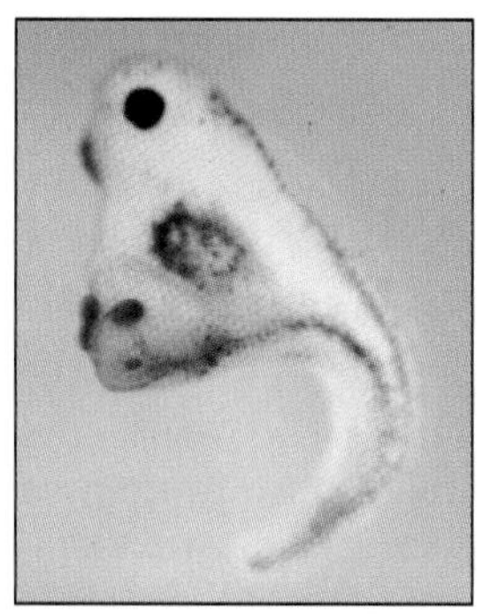

Figure 8–68 Ectopic misexpression of Wnt, a signaling protein that affects development of the body axis in the early *Xenopus* embryo. In this experiment, mRNA coding for Wnt was injected into the ventral vegetal blastomere, inducing a second body axis (discussed in Chapter 21). (From S. Sokol et al., *Cell* 67:741–752, 1992. © Elsevier.)

Genes Can Be Redesigned to Produce Proteins of Any Desired Sequence

In studying the action of a gene and the protein it encodes, one does not always wish to make drastic changes—flooding cells with huge quantities of hyperactive protein or eliminating a gene product entirely. It is sometimes useful to make slight changes in a protein's structure so that one can begin to dissect which portions of a protein are important for its function. The activity of an enzyme, for example, can be studied by changing a single amino acid in its active site. Special techniques are required to alter genes, and their protein products, in such subtle ways. The first step is often the chemical synthesis of a short DNA molecule containing the desired altered portion of the gene's nucleotide sequence. This synthetic DNA oligonucleotide is hybridized with single-stranded plasmid DNA that contains the DNA sequence to be altered, using conditions that allow imperfectly matched DNA strands to pair (Figure 8–69). The synthetic oligonucleotide will now serve as a primer for DNA synthesis by DNA polymerase, thereby generating a DNA double helix that incorporates the altered sequence into one of its two strands. After transfection, plasmids that carry the fully modified gene sequence are obtained. The appropriate DNA is

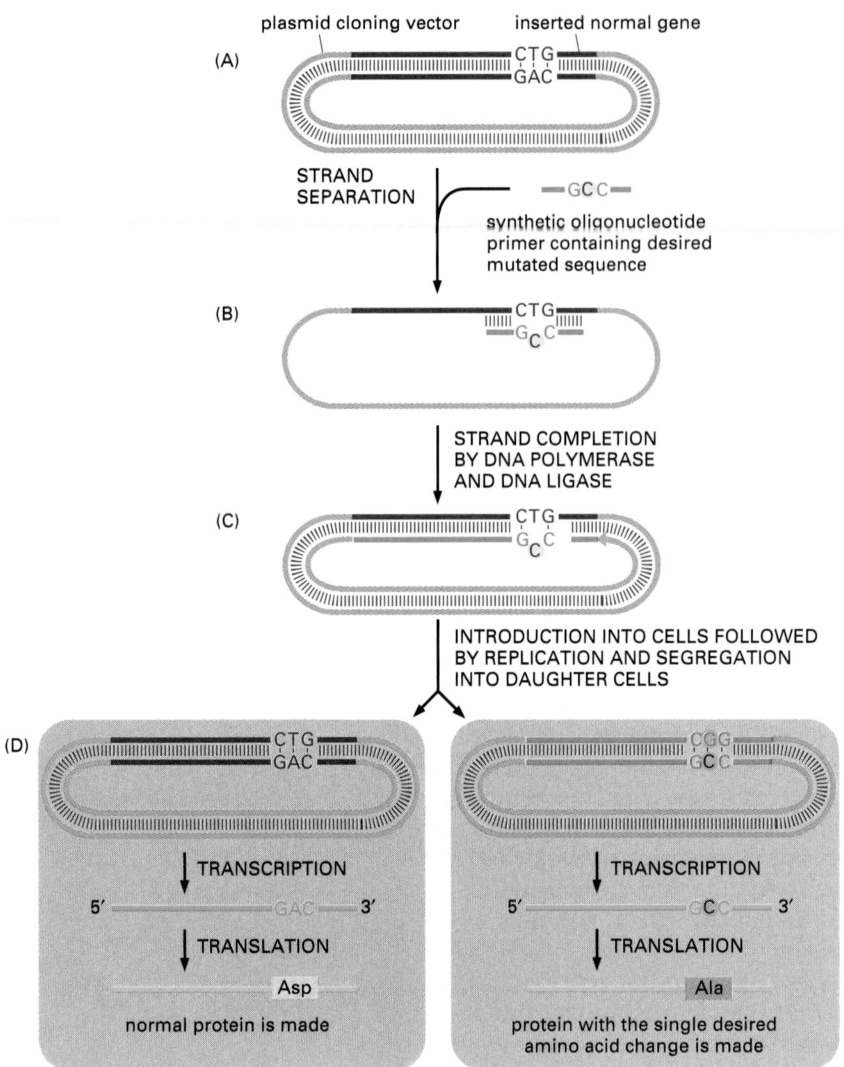

Figure 8–69 The use of a synthetic oligonucleotide to modify the protein-coding region of a gene by site-directed mutagenesis. (A) A recombinant plasmid containing a gene insert is separated into its two DNA strands. A synthetic oligonucleotide primer corresponding to part of the gene sequence but containing a single altered nucleotide at a predetermined point is added to the single-stranded DNA under conditions that permit less than perfect DNA hybridization. (B) The primer hybridizes to the DNA, forming a single mismatched nucleotide pair. (C) The recombinant plasmid is made double-stranded by *in vitro* DNA synthesis starting from the primer and sealed by DNA ligase. (D) The double-stranded DNA is introduced into a cell, where it is replicated. Replication using one strand of the template produces a normal DNA molecule, but replication using the other (the strand that contains the primer) produces a DNA molecule carrying the desired mutation. Only half of the progeny cells will end up with a plasmid that contains the desired mutant gene. However, a progeny cell that contains the mutated gene can be identified, separated from other cells, and cultured to produce a pure population of cells, all of which carry the mutated gene. Only one of the many changes that can be engineered in this way is shown here. With an oligonucleotide of the appropriate sequence, more than one amino acid substitution can be made at a time, or one or more amino acids can be inserted or deleted. Although not shown in this figure, it is also possible to create a site-directed mutation by using the appropriate oligonucleotides and PCR (instead of plasmid replication) to amplify the mutated gene.

then inserted into an expression vector so that the redesigned protein can be produced in the appropriate type of cells for detailed studies of its function. By changing selected amino acids in a protein in this way—a technique called **site-directed mutagenesis**—one can determine exactly which parts of the polypeptide chain are important for such processes as protein folding, interactions with other proteins, and enzymatic catalysis.

Engineered Genes Can Be Easily Inserted into the Germ Line of Many Animals

When engineering an organism that is to express an altered gene, ideally one would like to be able to replace the normal gene with the altered one so that the function of the mutant protein can be analyzed in the absence of the normal protein. As discussed above, this can be readily accomplished in some haploid, single-celled organisms. We shall see in the following section that much more complicated procedures have been developed that allow gene replacements of this type in mice. Foreign DNA can, however, be rather easily integrated into random positions of many animal genomes. In mammals, for example, linear DNA fragments introduced into cells are rapidly ligated end-to-end by intracellular enzymes to form long tandem arrays, which usually become integrated into a chromosome at an apparently random site. Fertilized mammalian eggs behave like other mammalian cells in this respect. A mouse egg injected with 200 copies of a linear DNA molecule often develops into a mouse containing, in many of its cells, a tandem array of copies of the injected gene integrated at a single random site in one of its chromosomes. If the modified chromosome is present in the germ line cells (eggs or sperm), the mouse will pass these foreign genes on to its progeny.

Animals that have been permanently reengineered by either gene insertion, gene deletion, or gene replacement are called **transgenic organisms**, and any foreign or modified genes that are added are called *transgenes*. When the normal gene remains present, only dominant effects of the alteration will show up in phenotypic analyses. Nevertheless, transgenic animals with inserted genes have provided important insights into how mammalian genes are regulated and how certain altered genes (called oncogenes) cause cancer.

It is also possible to produce transgenic fruit flies, in which single copies of a gene are inserted at random into the *Drosophila* genome. In this case the DNA fragment is first inserted between the two terminal sequences of a *Drosophila* transposon called the P element. The terminal sequences enable the P element to integrate into *Drosophila* chromosomes when the P element transposase enzyme is also present (see p. 288). To make transgenic fruit flies, therefore, the appropriately modified DNA fragment is injected into a very young fruit fly embryo along with a separate plasmid containing the gene encoding the transposase. When this is done, the injected gene often enters the germ line in a single copy as the result of a transposition event.

Gene Targeting Makes It Possible to Produce Transgenic Mice That Are Missing Specific Genes

If a DNA molecule carrying a mutated mouse gene is transferred into a mouse cell, it usually inserts into the chromosomes at random, but about once in a thousand times, it replaces one of the two copies of the normal gene by homologous recombination. By exploiting these rare "gene targeting" events, any specific gene can be altered or inactivated in a mouse cell by a direct gene replacement. In the special case in which the gene of interest is inactivated, the resulting animal is called a "knockout" mouse.

The technique works as follows: in the first step, a DNA fragment containing a desired mutant gene (or a DNA fragment designed to interrupt a target gene) is inserted into a vector and then introduced into a special line of embryo-derived mouse stem cells, called **embryonic stem cells** or ES cells, that grow in cell culture and are capable of producing cells of many different tissue types. After a period of cell proliferation, the rare colonies of cells in which a homologous recombination event is likely to have caused a gene replacement to occur are isolated. The correct colonies among these are identified by PCR or by Southern blotting: they contain recombinant DNA sequences in which the inserted fragment has replaced all or part of one copy of the normal gene. In the second step, individual cells from the identified colony are taken up into a fine micropipette and injected into an early mouse embryo. The transfected embryo-derived stem cells collaborate with the cells of the host embryo to produce a normal-looking mouse; large parts of this chimeric animal, including—in favorable cases—cells of the germ line, often derive from the artificially altered stem cells (Figure 8–70).

The mice with the transgene in their germ line are bred to produce both a male and a female animal, each heterozygous for the gene replacement (that is, they have one normal and one mutant copy of the gene). When these two mice are in turn mated, one-fourth of their progeny will be homozygous for the altered gene. Studies of these homozygotes allow the function of the altered gene—or the effects of eliminating a gene activity—to be examined in the absence of the corresponding normal gene.

The ability to prepare transgenic mice lacking a known normal gene has been a major advance, and the technique is now being used to dissect the functions of a large number of mammalian genes (Figure 8–71). Related techniques can be used to produce conditional mutants, in which a selected gene becomes disrupted in a specific tissue at a certain time in development. The strategy takes advantage of a site-specific recombination system to excise—and thus disable—

the target gene in a particular place or at a particular time. The most common of these recombination systems called **Cre/lox**, is widely used to engineer gene replacements in mice and in plants (see Figure 5–82). In this case the target gene in ES cells is replaced by a fully functional version of the gene that is flanked by a pair of the short DNA sequences, called lox sites, that are recognized by the Cre recombinase protein. The transgenic mice that result are phenotypically normal. They are then mated with transgenic mice that express the Cre recombinase gene under the control of an inducible promoter. In the specific cells or tissues in which Cre is switched on, it catalyzes recombination between the lox sequences—excising a target gene and eliminating its activity. Similar recombination systems are used to generate conditional mutants in *Drosophila* (see Figure 21–48).

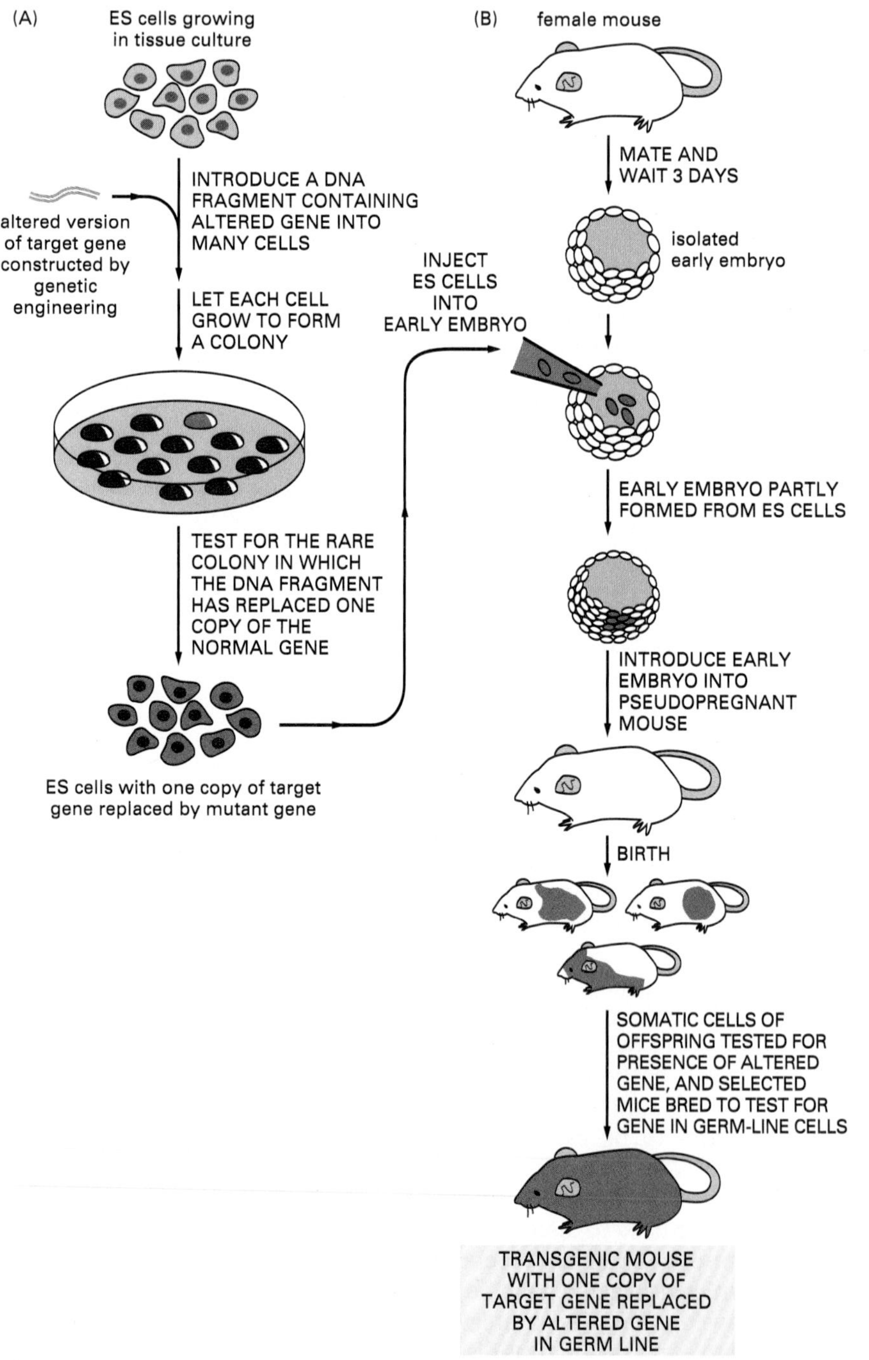

Figure 8–70 Summary of the procedures used for making gene replacements in mice. In the first step (A), an altered version of the gene is introduced into cultured ES (embryonic stem) cells. Only a few rare ES cells will have their corresponding normal genes replaced by the altered gene through a homologous recombination event. Although the procedure is often laborious, these rare cells can be identified and cultured to produce many descendants, each of which carries an altered gene in place of one of its two normal corresponding genes. In the next step of the procedure (B), these altered ES cells are injected into a very early mouse embryo; the cells are incorporated into the growing embryo, and a mouse produced by such an embryo will contain some somatic cells (indicated by *orange*) that carry the altered gene. Some of these mice will also contain germ-line cells that contain the altered gene. When bred with a normal mouse, some of the progeny of these mice will contain the altered gene in all of their cells. If two such mice are in turn bred (not shown), some of the progeny will contain two altered genes (one on each chromosome) in all of their cells.

If the original gene alteration completely inactivates the function of the gene, these mice are known as knockout mice. When such mice are missing genes that function during development, they often die with specific defects long before they reach adulthood. These defects are carefully analyzed to help decipher the normal function of the missing gene.

Transgenic Plants Are Important for Both Cell Biology and Agriculture

When a plant is damaged, it can often repair itself by a process in which mature differentiated cells "dedifferentiate," proliferate, and then redifferentiate into other cell types. In some circumstances the dedifferentiated cells can even form an apical meristem, which can then give rise to an entire new plant, including gametes. This remarkable plasticity of plant cells can be exploited to generate transgenic plants from cells growing in culture.

When a piece of plant tissue is cultured in a sterile medium containing nutrients and appropriate growth regulators, many of the cells are stimulated to proliferate indefinitely in a disorganized manner, producing a mass of relatively undifferentiated cells called a callus. If the nutrients and growth regulators are carefully manipulated, one can induce the formation of a shoot and then root apical meristems within the callus, and, in many species, a whole new plant can be regenerated.

Callus cultures can also be mechanically dissociated into single cells, which will grow and divide as a suspension culture. In several plants—including tobacco, petunia, carrot, potato, and *Arabidopsis*—a single cell from such a suspension culture can be grown into a small clump (a clone) from which a whole plant can be regenerated. Such a cell, which has the ability to give rise to all parts of the organism, is considered **totipotent**. Just as mutant mice can be derived by genetic manipulation of embryonic stem cells in culture, so transgenic plants can be created from single totipotent plant cells transfected with DNA in culture (Figure 8–72).

The ability to produce transgenic plants has greatly accelerated progress in many areas of plant cell biology. It has had an important role, for example, in isolating receptors for growth regulators and in analyzing the mechanisms of morphogenesis and of gene expression in plants. It has also opened up many new possibilities in agriculture that could benefit both the farmer and the consumer. It has made it possible, for example, to modify the lipid, starch, and protein storage reserved in seeds, to impart pest and virus resistance to plants, and to create modified plants that tolerate extreme habitats such as salt marshes or water-stressed soil.

Many of the major advances in understanding animal development have come from studies on the fruit fly *Drosophila* and the nematode worm *Caenorhabditis elegans*, which are amenable to extensive genetic analysis as well as to experimental manipulation. Progress in plant developmental biology has, in the past, been relatively slow by comparison. Many of the plants that have proved most amenable to genetic analysis—such as maize and tomato—have long life cycles and very large genomes, making both classical and molecular genetic analysis time-consuming. Increasing attention is consequently being paid to a fast-growing small weed, the common wall cress *(Arabidopsis thaliana)*, which has several major advantages as a "model plant" (see Figures 1–46 and 21–107). The relatively small *Arabidopsis* genome was the first plant genome to be completely sequenced.

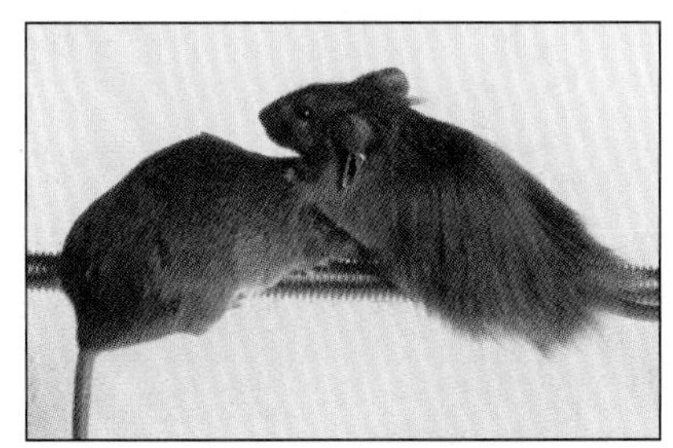

Figure 8–71 Mouse with an engineered defect in fibroblast growth factor 5 (FGF5). FGF5 is a negative regulator of hair formation. In a mouse lacking FGF5 *(right)*, the hair is long compared with its heterozygous littermate *(left)*. Transgenic mice with phenotypes that mimic aspects of a variety of human disorders, including Alzheimer's disease, atherosclerosis, diabetes, cystic fibrosis, and some type of cancers, have been generated. Their study may lead to the development of more effective treatments. (Courtesy of Gail Martin, from J.M. Hebert et al., *Cell* 78:1017–1025, 1994. © Elsevier.)

Large Collections of Tagged Knockouts Provide a Tool for Examining the Function of Every Gene in an Organism

Extensive collaborative efforts are underway to generate comprehensive libraries of mutations in several model organisms, including *S. cerevisiae, C. elegans, Drosophila, Arabidopsis*, and the mouse. The ultimate aim in each case is to produce a collection of mutant strains in which every gene in the organism has either been systematically deleted, or altered such that it can be conditionally disrupted. Collections of this type will provide an invaluable tool for investigating gene function on a genomic scale. In some cases, each of the individual mutants within the collection will sport a distinct molecular tag—a unique DNA sequence designed to make identification of the altered gene rapid and routine.

In *S. cerevisiae*, the task of generating a set of 6000 mutants, each missing

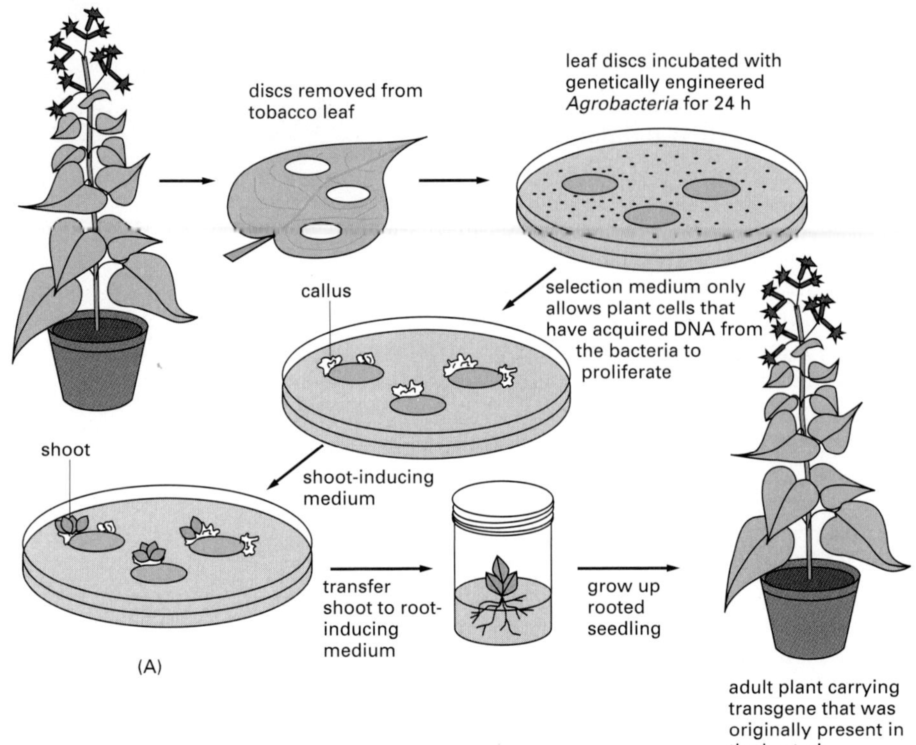
discs removed from
tobacco leaf
leaf discs incubated with
genetically engineered
Agrobacteria for 24 h
callus
selection medium only
allows plant cells that
have acquired DNA from
the bacteria to
proliferate
shoot
shoot-inducing
medium
transfer
shoot to root-
inducing
medium
grow up
rooted
seedling
(A)
adult plant carrying
transgene that was
originally present in
the bacteria

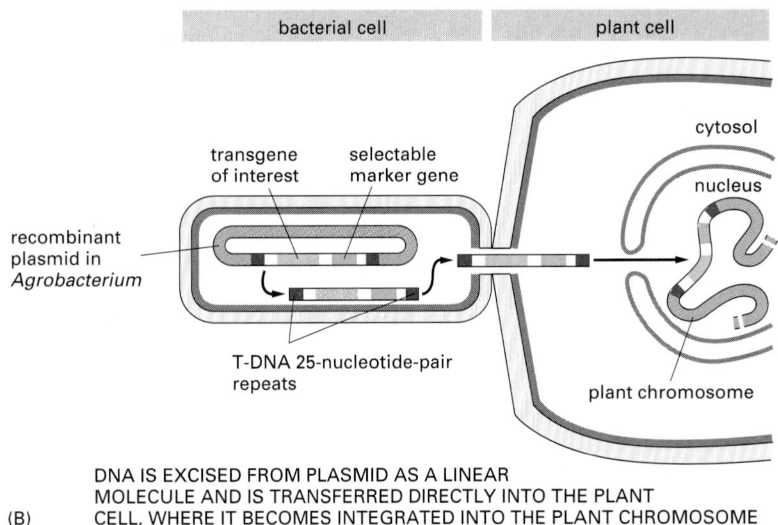
bacterial cell
plant cell
cytosol
transgene
of interest
selectable
marker gene
nucleus
recombinant
plasmid in
Agrobacterium
T-DNA 25-nucleotide-pair
repeats
plant chromosome
DNA IS EXCISED FROM PLASMID AS A LINEAR
MOLECULE AND IS TRANSFERRED DIRECTLY INTO THE PLANT
CELL, WHERE IT BECOMES INTEGRATED INTO THE PLANT CHROMOSOME
(B)

Figure 8–72 A procedure used to make a transgenic plant. (A) Outline of the process. A disc is cut out of a leaf and incubated in culture with *Agrobacteria* that carry a recombinant plasmid with both a selectable marker and a desired transgene. The wounded cells at the edge of the disc release substances that attract the *Agrobacteria* and cause them to inject DNA into these cells. Only those plant cells that take up the appropriate DNA and express the selectable marker gene survive to proliferate and form a callus. The manipulation of growth factors supplied to the callus induces it to form shoots that subsequently root and grow into adult plants carrying the transgene. (B) The preparation of the recombinant plasmid and its transfer to plant cells. An *Agrobacterium* plasmid that normally carries the T-DNA sequence is modified by substituting a selectable marker (such as the kanamycin-resistance gene) and a desired transgene between the 25-nucleotide-pair T-DNA repeats. When the *Agrobacterium* recognizes a plant cell, it efficiently passes a DNA strand that carries these sequences into the plant cell, using the special machinery that normally transfers the plasmid's T-DNA sequence.

only one gene, is made simpler by yeast's propensity for homologous recombination. For each gene, a "deletion cassette" is prepared. The cassette consists of a special DNA molecule that contains 50 nucleotides identical in sequence to each end of the targeted gene, surrounding a selectable marker. In addition, a special "barcode" sequence tag is embedded in this DNA molecule to facilitate the later rapid identification of each resulting mutant strain (Figure 8–73). A large mixture of such gene knockout mutants can then be grown under various selective test conditions—such as nutritional deprivation, temperature shift, or the presence of various drugs—and the cells that survive can be rapidly identified by their unique sequence tags. By assessing how well each mutant in the mixture fares, one can begin to assess which genes are essential, useful, or irrelevant for growth under various conditions.

The challenge in deriving information from the study of such yeast mutants lies in deducing a gene's activity or biological role based on a mutant phenotype.

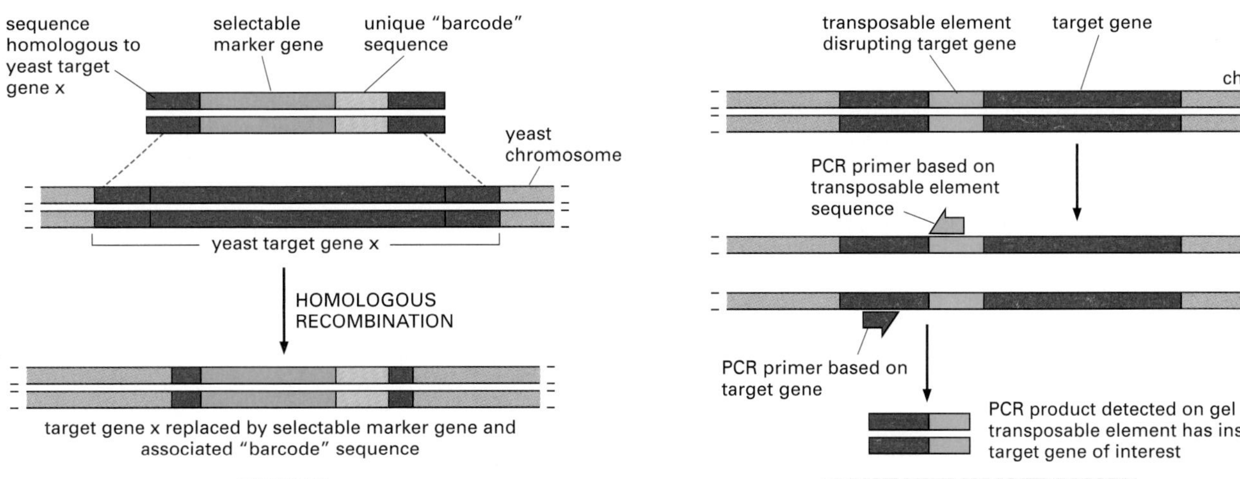

Figure 8–73 Making collections of mutant organisms. (A) A deletion cassette for use in yeast contains sequences homologous to each end of a target gene x *(red)*, a selectable marker *(blue)*, and a unique "barcode" sequence, approximately 20 nucleotide pairs in length *(green)*. This DNA is introduced into yeast, where it readily replaces the target gene by homologous recombination. (B) A similar approach can be taken to prepare tagged knockout mutants in *Arabidopsis* and *Drosophila*. In this case, mutants are generated by the accidental insertion of a transposable element into a target gene. The total DNA from the resulting organism can be collected and quickly screened for disruption of a gene of interest by using PCR primers that bind to the transposable element and to the target gene. A PCR product is detected on the gel only if the transposable element has inserted into the target gene.

Some defects—an inability to live without histidine, for example—point directly to the function of the wild-type gene. Other connections may not be so obvious. What might a sudden sensitivity to cold indicate about the role that a particular gene plays in the yeast cell? Such problems are even greater in organisms that are more complex than yeast. The loss of function of a single gene in the mouse, for example, can affect many different tissue types at different stages of development—whereas the loss of other genes is found to have no obvious effect. Adequately characterizing mutant phenotypes in mice often requires a thorough examination, along with extensive knowledge of mouse anatomy, histology, pathology, physiology, and complex behavior.

The insights generated by examination of mutant libraries, however, will be great. For example, studies of an extensive collection of mutants in *Mycoplasma genitalium*—the organism with the smallest known genome—have identified the minimum complement of genes essential for cellular life. Analysis of the mutant pool suggests that 265–350 of the 480 protein-coding genes in *M. genitalium* are required for growth under laboratory conditions. Approximately 100 of these essential genes are of unknown function, which suggests that a surprising number of the basic molecular mechanisms that underlie cellular life have yet to be discovered.

Summary

Genetics and genetic engineering provide powerful tools for the study of gene function in both cells and organisms. In the classical genetic approach, random mutagenesis is coupled with screening to identify mutants that are deficient in a particular biological process. These mutants are then used to locate and study the genes responsible for that process.

Gene function can also be ascertained by reverse genetic techniques. DNA engineering methods can be used to mutate any gene and to re-insert it into a cell's chromosomes so that it becomes a permanent part of the genome. If the cell used for this gene transfer is a fertilized egg (for an animal) or a totipotent plant cell in culture, transgenic organisms can be produced that express the mutant gene and pass it on to their progeny. Especially important for cell biology is the ability to alter cells and organisms in highly specific ways—allowing one to discern the effect on the cell or the organism of a designed change in a single protein or RNA molecule.

Many of these methods are being expanded to investigate gene function on a genome-wide scale. Technologies such as DNA microarrays can be used to monitor the expression of thousands of genes simultaneously, providing detailed, comprehensive snapshots of the dynamic patterns of gene expression that underlie complex cellular processes. And the generation of mutant libraries in which every gene in an organism has been systematically deleted or disrupted will provide an invaluable tool for exploring the role of each gene in the elaborate molecular collaboration that gives rise to life.

References

General

Ausubel FM, Brent R, Kingston RE et al. (eds) (1999) Short Protocols in Molecular Biology, 4th edn. New York: Wiley.

Brown TA (1999) Genomes. New York: Wiley-Liss.

Spector DL, Goldman RD, Leinwand LA (eds) (1998) Cells: A Laboratory Manual. Cold Spring Harbor, NY: Cold Spring Harbor Laboratory Press.

Watson JD, Gilman M, Witkowski J & Zoller M (1992) Recombinant DNA, 2nd edn. New York: WH Freeman.

Isolating Cells and Growing Them in Culture

Cohen S, Chang A, Boyer H & Helling R (1973) Construction of biologically functional bacterial plasmids *in vitro*. *Proc. Natl. Acad. Sci. USA* 70, 3240–3244.

Emmert-Buck, MR, Bonner RF, Smith PD et al. (1996) Laser capture microdissection. *Science* 274, 998–1001.

Freshney RI (2000) Culture of Animal Cells: A Manual of Basic Technique, 4th edn. New York: Wiley.

Ham RG (1965) Clonal growth of mammalian cells in a chemically defined, synthetic medium. *Proc. Natl. Acad. Sci. USA* 53, 288–293.

Harlow E & Lane D (1999) Using Antibodies: A Laboratory Manual. Cold Spring Harbor, NY: Cold Spring Harbor Laboratory Press.

Herzenberg LA, Sweet RG & Herzenberg LA (1976) Fluorescence-activated cell sorting. *Sci. Am.* 234(3), 108–116.

Jackson D, Symons R & Berg P (1972) Biochemical method for inserting new genetic information into DNA of simian virus 40: circular SV40 DNA molecules containing lambda phage genes and the galactose operon of *Escherichia coli*. *Proc. Natl. Acad. Sci. USA* 69, 2904–2909.

Levi-Montalcini R (1987) The nerve growth factor thirty-five years later. *Science* 237, 1154–1162.

Milstein C (1980) Monoclonal antibodies. *Sci. Am.* 243(4), 66–74.

Fractionation of Cells

de Duve C & Beaufay H (1981) A short history of tissue fractionation. *J. Cell Biol.* 91, 293s–299s.

Laemmli UK (1970) Cleavage of structural proteins during the assembly of the head of bacteriophage T4. *Nature* 227, 680–685.

Nirenberg MW & Matthaei JH (1961) The dependence of cell-free protein synthesis in *E. coli* on naturally occurring or synthetic polyribonucleotides. *Proc. Natl. Acad. Sci. USA* 47, 1588–1602.

O'Farrell PH (1975) High-resolution two-dimensional electrophoresis of proteins. *J. Biol. Chem.* 250, 4007–4021.

Palade G (1975) Intracellular aspects of the process of protein synthesis. *Science* 189, 347–358.

Pandey A & Mann M (2000) Proteomics to study genes and genomes. *Nature* 405, 837–846.

Scopes RK & Cantor CR (1993) Protein Purification: Principles and Practice, 3rd edn. New York: Springer-Verlag.

Yates JR (1998) Mass spectrometry and the age of the proteome. *J. Mass. Spectrom.* 33, 1–19.

Isolating, Cloning, and Sequencing DNA

Adams MD, Celniker SE, Holt RA et al. (2000) The genome sequence of *Drosophila melanogaster*. *Science* 287, 2185–2195.

Alwine JC, Kemp DJ & Stark GR (1977) Method for detection of specific RNAs in agarose gels by transfer to diabenzyloxymethyl-paper and hybridization with DNA probes. *Proc. Natl. Acad. Sci. USA* 74, 5350–5354.

Blattner FR, Plunkett G, Bloch CA et al. (1997) The complete genome sequence of *Escherichia coli* K-12. *Science* 277, 1453–1474.

Hunkapiller T, Kaiser RJ, Koop BK & Hood L Large-scale and automated DNA sequence determination. *Science* 254, 59–67.

International Human Genome Sequencing Consortium (2000) Initial sequencing and analysis of the human genome. *Nature* 409, 860–921.

Maniatis T et al. (1978) The isolation of structural genes from libraries of eucaryotic DNA. *Cell* 15, 687–701.

Saiki RK, Gelfand DH, Stoffel S et al. (1988) Primer-directed enzymatic amplification of DNA with a thermostable DNA polymerase. *Science* 239, 487–491.

Sambrook J, Russell D (2001) Molecular Cloning: A Laboratory Manual, 3rd edn. Cold Spring Harbor, NY: Cold Spring Harbor Laboratory Press.

Sanger F, Nicklen S & Coulson AR (1977) DNA sequencing with chain-terminating inhibitors. *Proc. Natl. Acad. Sci. USA* 74, 5463–5467.

Southern EM (1975) Detection of specific sequences among DNA fragments separated by gel electrophoresis. *J. Mol. Biol.* 98, 503–517.

The *Arabidopsis* Genome Initiative (2000) Analysis of the genome sequence of the flowering plant *Arabidopsis thaliana*. *Nature* 408, 796–815.

The *C. elegans* Sequencing Consortium (1998) Genome sequence of the nematode *C. elegans*: a platform for investigating biology. *Science* 282, 2012–2018.

Venter JC, Adams MA, Myers EW et al. (2000) The sequence of the human genome. *Science* 291, 1304–1351.

Analyzing Protein Structure and Function

Bamdad C (1997) Surface plasmon resonance for measurements of biological interest, in Current Protocols in Molecular Biology (FM Ausubel, R Brent, RE Kingston et al. eds), pp 20.4.1–20.4.12. New York: Wiley.

Branden C & Tooze J (1999) Introduction to Protein Structure, 2nd edn. New York: Garland Publishing.

Fields S & Song O (1989) A novel genetic system to detect protein–protein interactions. *Nature* 340, 245–246.

Kendrew JC (1961) The three-dimensional structure of a protein molecule. *Sci. Am.* 205(6), 96–111.

MacBeath G & Schreiber SL (2000) Printing proteins as microarrays for high-throughput function determination. *Science* 289, 1760–1763.

Miyawaki A, Tsien RY (2000) Monitoring protein conformations and interactions by fluorescence resonance energy transfer between mutants of green fluorescent protein. *Methods Enzymol.* 327, 472–500.

Sali A & Kuriyan J (1999) Challenges at the frontiers of structural biology. *Trends Genet.* 15, M20–M24.

Wuthrich K (1989) Protein structure determination in solution by nuclear magnetic resonance spectroscopy. *Science* 243, 45–50.

Studying Gene Expression and Function

Botstein D, White RL, Skolnick M & Davis RW (1980) Construction of a genetic linkage map in man using restriction fragment length polymorphisms. *Am. J. Hum. Genet.* 32, 314–331.

Brent R (2000) Genomic Biology. *Cell* 100, 169–182.

Capecchi MR (2001) Generating Mice with Targeted Mutations. *Nat. Med.* 7, 1086-1090.

Coelho PS, Kumar A & Snyder M (2000) Genome-wide mutant collections: toolboxes for functional genomics. *Curr. Opin. Microbiol.* 3, 309–315.

DeRisi JL, Iyer VR & Brown PO (1997) Exploring the metabolic and genetic control of gene expression on a genomic scale. *Science* 278, 680–686.

Eisenberg D, Marcotte EM, Xenarios I & Yeates TO (2000) Protein function in the post-genomic era. *Nature* 415, 823–826.

Enright AJ, Illiopoulos I, Kyrpides NC & Ouzounis CA (1999) Protein interaction maps for complete genomes based on gene fusion events. *Nature* 402, 86–90.

Hartwell LH, Hood L, Goldberg ML et al. (2000) Genetics: From Genes to Genomes. Boston: McGraw-Hill.

Lockhart DJ & Winzeler EA (2000) Genomics, gene expression and DNA arrays. *Nature* 405, 827–836, 2000.

Nusslein-Volhard C & Weischaus E (1980) Mutations affecting segment number and polarity in *Drosophila*. *Nature* 287, 795–801.

Palmiter RD & Brinster RL (1986) Germ line transformation of mice. *Annu. Rev. Genet.* 20, 465–499.

Rubin GM & Spradling AC (1982) Genetic transformation of *Drosophila* with transposable element vectors. *Science* 218, 348–353.

Tabara H, Grishok A & Mello CC (1998) RNAi in *C. elegans*: soaking in the genome sequence. *Science* 282, 430–431.

Weigel D & Glazebrook J (2001) *Arabidopsis:* A Laboratory Manual. Cold Spring Harbor, NY: Cold Spring Harbor Laboratory Press.

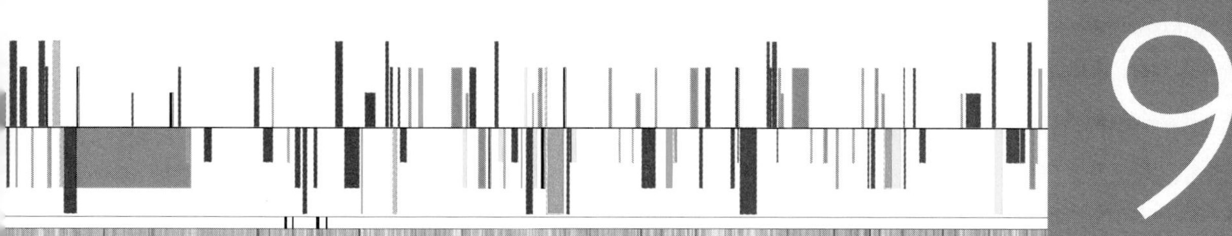

9

VISUALIZING CELLS

Cells are small and complex. It is hard to see their structure, hard to discover their molecular composition, and harder still to find out how their various components function. What we can learn about cells depends on the tools at our disposal, and major advances in cell biology have frequently sprung from the introduction of new techniques. To understand contemporary cell biology, therefore, it is necessary to know something of its methods.

In this chapter, we briefly review some of the principal methods in microscopy used to study cells. Understanding the structural organization of cells is an essential prerequisite for understanding how cells function. Optical microscopy will be our starting point because cell biology began with the light microscope, and it is still an essential tool. In recent years optical microscopy has become ever more important, largely owing to the development of methods for the specific labeling and imaging of individual cellular constituents and the reconstruction of their three-dimensional architecture. An important advantage of optical microscopy is that light is relatively nondestructive. By tagging specific cell components with fluorescent markers, such as green fluorescent protein (GFP), we can thus watch their movements and interactions in living cells and organisms.

Light microscopy is limited in the fineness of detail that it can reveal. Microscopes using other types of radiation—in particular, electron microscopes—can resolve much smaller structures than is possible with visible light. This comes at a cost: specimen preparation for electron microscopy is much more complex and it is harder to be sure that what we see in the image corresponds precisely to the actual structure being examined. It is now possible, however, to preserve structures faithfully for electron microscopy by very rapid freezing. Computerized image analysis can be used to reconstruct three-dimensional objects from multiple tilted views. Together these approaches are extending the resolution and scope of microscopy to the point where we can begin to image the structures of individual macromolecules.

Although methods are of basic importance, it is what we discover with them that makes them interesting. This chapter, is therefore, meant to be used for reference and to be read in conjunction with the later chapters of the book rather than as an introduction to them.

LOOKING AT THE STRUCTURE OF CELLS IN THE MICROSCOPE

A typical animal cell is 10–20 μm in diameter, which is about one-fifth the size of the smallest particle visible to the naked eye. It was not until good light microscopes became available in the early part of the nineteenth century that all plant and animal tissues were discovered to be aggregates of individual cells. This discovery, proposed as the **cell doctrine** by Schleiden and Schwann in 1838, marks the formal birth of cell biology.

Animal cells are not only tiny, they are also colorless and translucent. Consequently, the discovery of their main internal features depended on the development, in the latter part of the nineteenth century, of a variety of stains that provided sufficient contrast to make those features visible. Similarly, the introduction of the far more powerful electron microscope in the early 1940s required the development of new techniques for preserving and staining cells before the full complexities of their internal fine structure could begin to emerge. To this day, microscopy depends as much on techniques for preparing the specimen as on the performance of the microscope itself. In the discussions that follow, we therefore consider both instruments and specimen preparation, beginning with the light microscope.

Figure 9–1 shows a series of images illustrating an imaginary progression from a thumb to a cluster of atoms. Each successive image represents a tenfold increase in magnification. The naked eye could see features in the first two panels, the resolution of the light microscope would extend to about the fourth panel, and the electron microscope to about the seventh panel. Some of the landmarks in the development of light microscopy are outlined in Table 9–1. Figure 9–2 shows the sizes of various cellular and subcellular structures and the ranges of size that different types of microscopes can visualize.

The Light Microscope Can Resolve Details 0.2 μm Apart

In general, a given type of radiation cannot be used to probe structural details much smaller than its own wavelength. This is a fundamental limitation of all microscopes. The ultimate limit to the resolution of a light microscope is therefore set by the wavelength of visible light, which ranges from about 0.4 μm (for violet) to 0.7 μm (for deep red). In practical terms, bacteria and mitochondria, which are about 500 nm (0.5 μm) wide, are generally the smallest objects whose shape can be clearly discerned in the **light microscope**; details smaller than this are obscured by effects resulting from the wave nature of light. To understand why this occurs, we must follow what happens to a beam of light waves as it passes through the lenses of a microscope (Figure 9–3).

Because of its wave nature, light does not follow exactly the idealized straight ray paths predicted by geometrical optics. Instead, light waves travel through an optical system by a variety of slightly different routes, so that they interfere with one another and cause *optical diffraction* effects. If two trains of waves reaching the same point by different paths are precisely *in phase*, with crest matching crest and trough matching trough, they will reinforce each other so as to increase brightness. In contrast, if the trains of waves are *out of phase*, they will interfere with each other in such a way as to cancel each other partly or entirely (Figure 9–4). The interaction of light with an object changes the phase relationships of the light waves in a way that produces complex interference effects. At high magnification, for example, the shadow of a straight edge that is evenly illuminated with light of uniform wavelength appears as a set of parallel lines, whereas that of a circular spot appears as a set of concentric rings (Figure

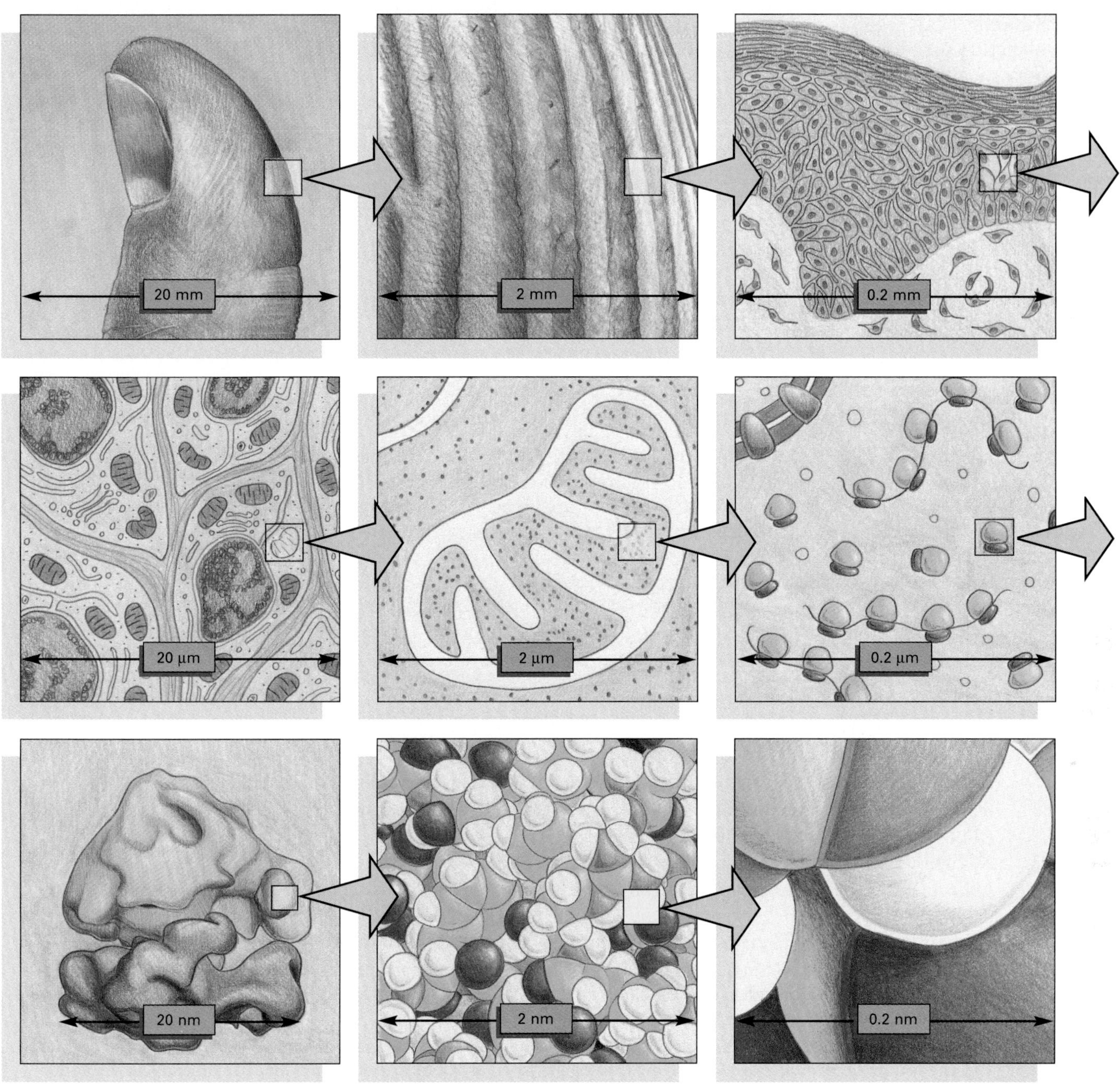

Figure 9–1 A sense of scale between living cells and atoms. Each diagram shows an image magnified by a factor of ten in an imaginary progression from a thumb, through skin cells, to a ribosome, to a cluster of atoms forming part of one of the many protein molecules in our body. Details of molecular structure, as shown in the last two panels, are beyond the power of the electron microscope.

9–5). For the same reason, a single point seen through a microscope appears as a blurred disc, and two point objects close together give overlapping images and may merge into one. No amount of refinement of the lenses can overcome this limitation imposed by the wavelike nature of light.

The limiting separation at which two objects can still be seen as distinct—the so-called **limit of resolution**—depends on both the wavelength of the light and the *numerical aperture* of the lens system used. This latter quantity is a measure of the width of the entry pupil of the microscope, scaled according to its distance from the object; the wider the microscope opens its eye, so to speak, the more sharply it can see (Figure 9–6). Under the best conditions, with violet light (wavelength = 0.4 μm) and a numerical aperture of 1.4, a limit of resolution of just under 0.2 μm can theoretically be obtained in the light microscope. This resolution was achieved by microscope makers at the end of the nineteenth century and is only rarely matched in contemporary, factory-produced microscopes. Although it is possible to *enlarge* an image as much as one wants—for example, by projecting it onto a screen—it is never possible to resolve two objects in the light microscope that are separated by less than about 0.2 μm; they will appear as a single object.

We see next how interference and diffraction can be exploited to study unstained cells in the living state. Later we discuss how permanent preparations of cells are made for viewing in the light microscope and how chemical stains are used to enhance the visibility of the cell structures in such preparations.

Living Cells Are Seen Clearly in a Phase-Contrast or a Differential-Interference-Contrast Microscope

The possibility that some components of the cell may be lost or distorted during specimen preparation has always challenged microscopists. The only certain way to avoid the problem is to examine cells while they are alive, without fixing or freezing. For this purpose, light microscopes with special optical systems are especially useful.

When light passes through a living cell, the phase of the light wave is changed according to the cell's refractive index: light passing through a relatively thick or

TABLE 9–1 Some Important Discoveries in the History of Light Microscopy

1611	**Kepler** suggests a way of making a compound microscope.
1655	**Hooke** uses a compound microscope to describe small pores in sections of cork he calls "cells".
1674	**Leeuwenhoek** reports his discovery of protozoa. He sees bacteria for the first time nine years later.
1833	**Brown** publishes his microscopic observations of orchids, clearly describing the cell nucleus.
1838	**Schleiden** and **Schwann** propose the cell theory, stating that the nucleated cell is the unit of structure and function in plants and animals.
1857	**Kolliker** describes mitrochondria in muscle cells.
1876	**Abbé** analyzes the effects of diffraction on image formation in the microscope and shows how to optimize microscope design.
1879	**Flemming** describes with great clarity chromosome behavior during mitosis in animals.
1881	**Retzius** describes many animal tissues with a detail that has not been surpassed by any other light microscopist. During the next two decades, he, Cajal, and other histologists develop staining methods and lay the foundations of microscopic anatomy.
1882	**Koch** uses aniline dyes to stain microorganisms and identifies the bacteria that cause tuberculosis and cholera. In the following two decades, other bacteriologists, such as **Klebs** and **Pasteur**, identify the causative agents of many other diseases by examining stained preparations under the microscope.
1886	**Zeiss** makes a series of lenses, to the design of **Abbé**, that enable microscopists to resolve structures at the theoretical limits of visible light.
1898	**Golgi** first sees and describes the Golgi apparatus by staining cells with silver nitrate.
1924	**Lacassagne** and collaborators develop the first autoradiographic method to localize radiographic polonium in biological specimens.
1930	**Lebedeff** designs and builds the first inference microscope. In 1932, **Zernicke** invents the phase-contrast microscope. These two developments allow unstained living cells to be seen in detail for the first time.
1941	**Coons** uses antibiotics coupled to fluorescent dyes to detect cellular antigens.
1952	**Nomarski** devises and patents the system of differential interference contrast for the light microscope that still bears his name.
1968	**Petran** and collaborators make the first confocal microscope.
1981	**Allen** and **Inoué** perfect video-enhanced light microscopy.
1984	**Agard and Sedat** use computer deconvolution to reconstruct *Drosophilia* polytene nuclei.
1988	Commercial confocal microscopes come into widespread use.
1994	**Chalfie** and collaborators introduce green fluorescent protein (GFP) as a marker in microscopy.

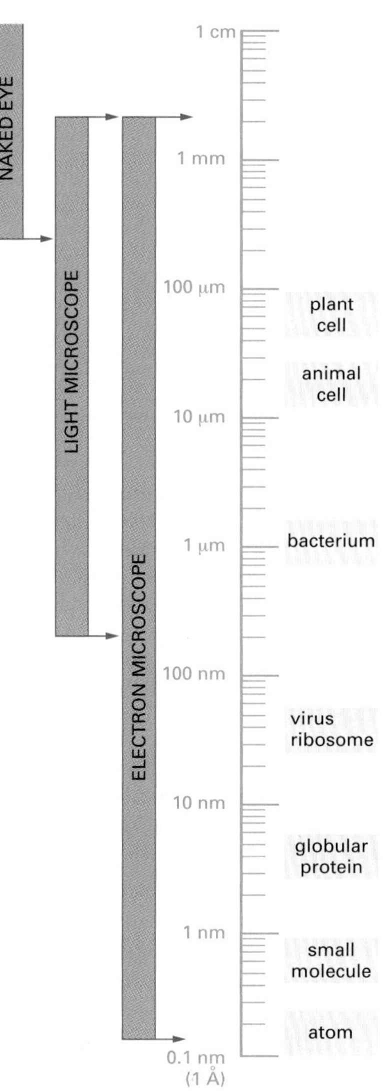

Figure 9–2 Resolving power. Sizes of cells and their components are drawn on a logarithmic scale, indicating the range of objects that can be readily resolved by the naked eye and in the light and electron microscopes. The following units of length are commonly employed in microscopy:

μm (micrometer) = 10^{-6} m
nm (nanometer) = 10^{-9} m
Å (Ångström unit) = 10^{-10} m

dense part of the cell, such as the nucleus, is retarded; its phase, consequently, is shifted relative to light that has passed through an adjacent thinner region of the cytoplasm. The **phase-contrast microscope** and, in a more complex way, the **differential-interference-contrast microscope**, exploit the interference effects produced when these two sets of waves recombine, thereby creating an image of the cell's structure (Figure 9–7). Both types of light microscopy are widely used to visualize living cells.

A simpler way to see some of the features of a living cell is to observe the light that is scattered by its various components. In the **dark-field microscope**, the illuminating rays of light are directed from the side so that only scattered light enters the microscope lenses. Consequently, the cell appears as a bright object against a dark background. With a normal **bright-field microscope**, the image is obtained by the simple transmission of light through a cell in culture. Images of the same cell obtained by four kinds of light microscopy are shown in Figure 9–8.

Phase-contrast, differential-interference-contrast, and dark-field microscopy make it possible to watch the movements involved in such processes as mitosis and cell migration. Since many cellular motions are too slow to be seen in real time, it is often helpful to take time-lapse motion pictures or video recordings. Here, successive frames separated by a short time delay are recorded, so that when the resulting picture series or videotape is played at normal speed, events appear greatly speeded up.

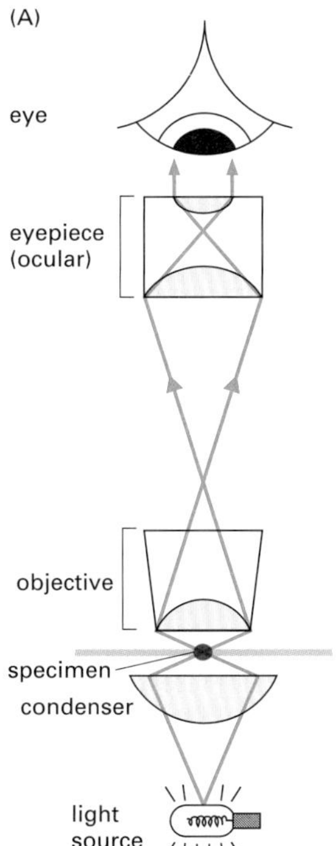

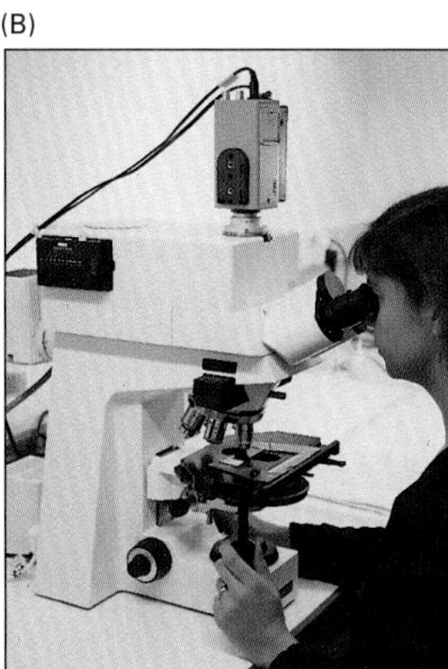

Figure 9–3 A light microscope. (A) Diagram showing the light path in a compound microscope. Light is focused on the specimen by lenses in the condensor. A combination of objective lenses and eyepiece lenses are arranged to focus an image of the illuminated specimen in the eye. (B) A modern research light microscope.

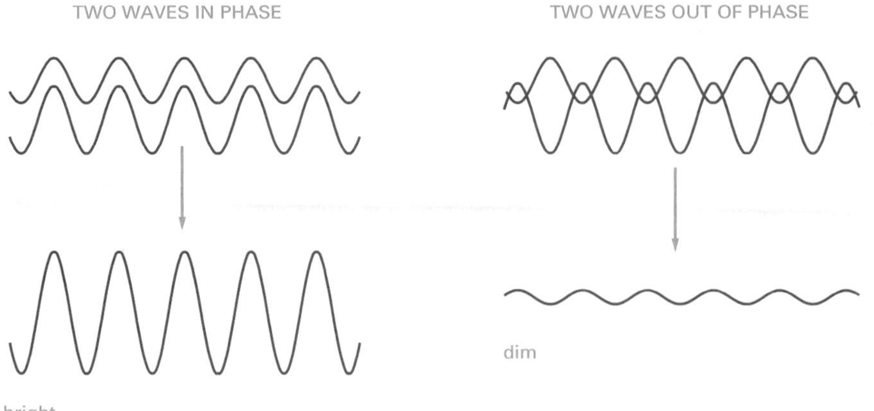
TWO WAVES IN PHASE
TWO WAVES OUT OF PHASE
bright
dim

Figure 9–4 Interference between light waves. When two light waves combine in phase, the amplitude of the resultant wave is larger and the brightness is increased. Two light waves that are out of phase cancel each other partly and produce a wave whose amplitude, and therefore brightness, is decreased.

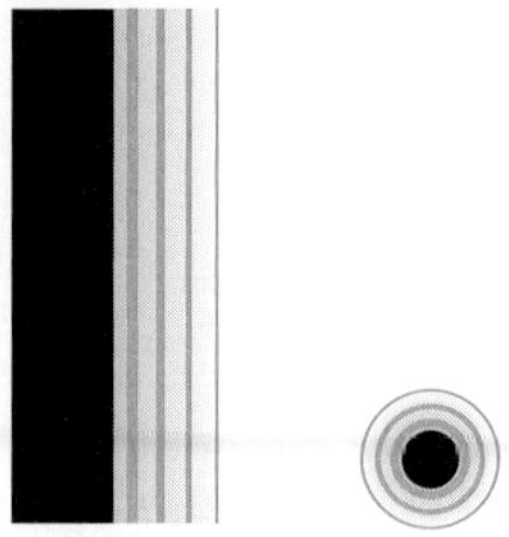

Figure 9–5 Edge effects. The interference effects observed at high magnification when light passes the edges of a solid object placed between the light source and the observer are shown here.

Images Can Be Enhanced and Analyzed by Electronic Techniques

In recent years electronic imaging systems and the associated technology of **image processing** have had a major impact on light microscopy. They have enabled certain practical limitations of microscopes (due to imperfections in the optical system) to be largely overcome. They have also circumvented two fundamental limitations of the human eye: the eye cannot see well in extremely dim light, and it cannot perceive small differences in light intensity against a bright background. The first limitation can be overcome by attaching highly sensitive video cameras (the kind used in night surveillance) to a microscope. It is then possible to observe cells for long periods at very low light levels, thereby avoiding the damaging effects of prolonged bright light (and heat). Such low-light cameras are especially important for viewing fluorescent molecules in living cells, as explained below.

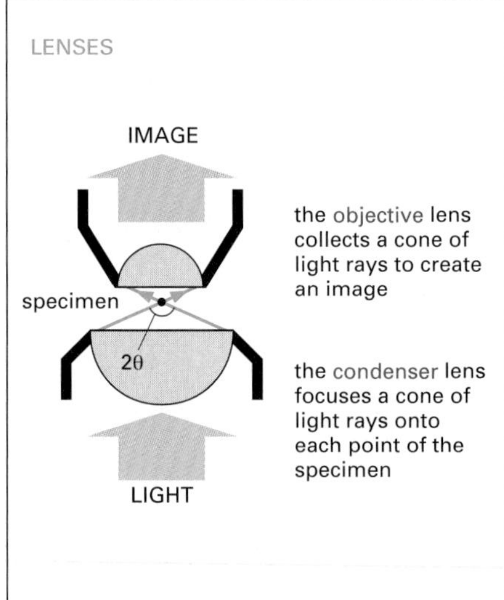
LENSES
IMAGE
the objective lens collects a cone of light rays to create an image
specimen
2θ
the condenser lens focuses a cone of light rays onto each point of the specimen
LIGHT

RESOLUTION: the resolving power of the microscope depends on the width of the cone of illumination and therefore on both the condenser and the objective lens. It is calculated using the formula

$$\text{resolution} = \frac{0.61\,\lambda}{n \sin \theta}$$

where:

θ = half the angular width of the cone of rays collected by the objective lens from a typical point in the specimen (since the maximum width is 180°, sin θ has a maximum value of 1)

n = the refractive index of the medium (usually air or oil) separating the specimen from the objective and condenser lenses

λ = the wavelength of light used (for white light a figure of 0.53 μm is commonly assumed)

NUMERICAL APERTURE: $n \sin \theta$ in the equation above is called the numerical aperture of the lens (NA) and is a function of its light-collecting ability. For dry lenses this cannot be more than 1, but for oil-immersion lenses it can be as high as 1.4. The higher the numerical aperture, the greater the resolution and the brighter the image (brightness is important in fluorescence microscopy). However, this advantage is obtained at the expense of very short working distances and a very small depth of field.

Figure 9–6 Numerical aperture. The path of light rays passing through a transparent specimen in a microscope illustrate the concept of numerical aperture and its relation to the limit of resolution.

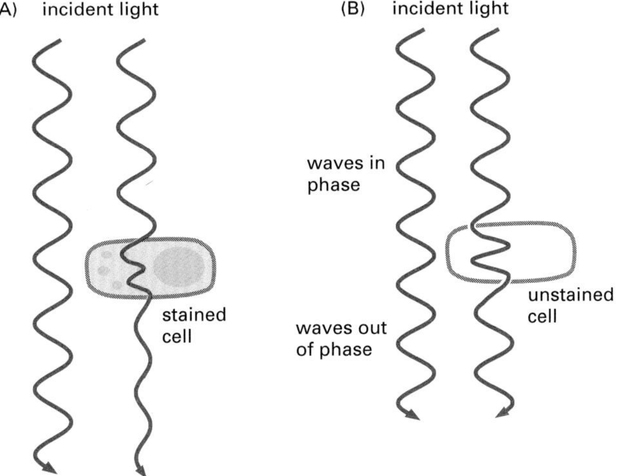

Figure 9–7 Two ways to obtain contrast in light microscopy. (A) The stained portions of the cell reduce the amplitude of light waves of particular wavelengths passing through them. A colored image of the cell is thereby obtained that is visible in the ordinary way. (B) Light passing through the unstained, living cell undergoes very little change in amplitude, and the structural details cannot be seen even if the image is highly magnified. The phase of the light, however, is altered by its passage through the cell, and small phase differences can be made visible by exploiting interference effects using a phase-contrast or a differential-interference-contrast microscope.

Because images produced by video cameras are in electronic form, they can be readily digitized, fed to a computer, and processed in various ways to extract latent information. Such image processing makes it possible to compensate for various optical faults in microscopes to attain the theoretical limit of resolution. Moreover, by electronic image processing, contrast can be greatly enhanced so that the eye's limitations in detecting small differences in light intensity are overcome. Although this processing also enhances the effects of random background irregularities in the optical system, such defects can be removed by electronically subtracting an image of a blank area of the field. Small transparent objects that were previously impossible to distinguish from the background then become visible.

The high contrast attainable by computer-assisted differential-interference-contrast microscopy makes it possible to see even very small objects such as single microtubules (Figure 9–9), which have a diameter of 0.025 µm, less than one-tenth the wavelength of light. Individual microtubules can also be seen in a fluorescence microscope if they are fluorescently labeled (see Figure 9–15). In both cases, however, the unavoidable diffraction effects badly blur the image so that the microtubules appear at least 0.2 µm wide, making it impossible to distinguish a single microtubule from a bundle of several microtubules.

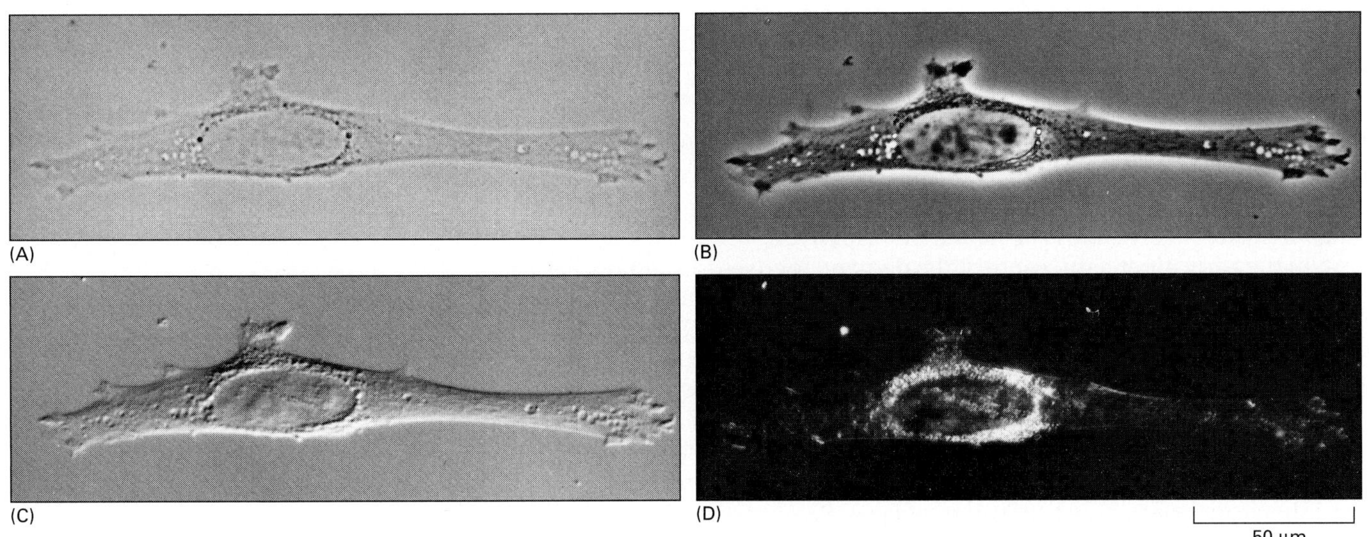

Figure 9–8 Four types of light microscopy. Four images are shown of the same fibroblast cell in culture. All four types of images can be obtained with most modern microscopes by interchanging optical components. (A) Bright-field microscopy. (B) Phase-contrast microscopy. (C) Nomarski differential-interference-contrast microscopy. (D) Dark-field microscopy.

Figure 9–9 Image processing. (A) Unstained microtubules are shown here in an unprocessed digital image, captured using differential-interference-contrast microscopy. (B) The image has now been processed, first by digitally subtracting the unevenly illuminated background, and second by digitally enhancing the contrast. The result of this image processing is a far more interpretable picture. Note that the microtubules are dynamic and some have changed length or position between the before-and-after images. (Courtesy of Viki Allan.)

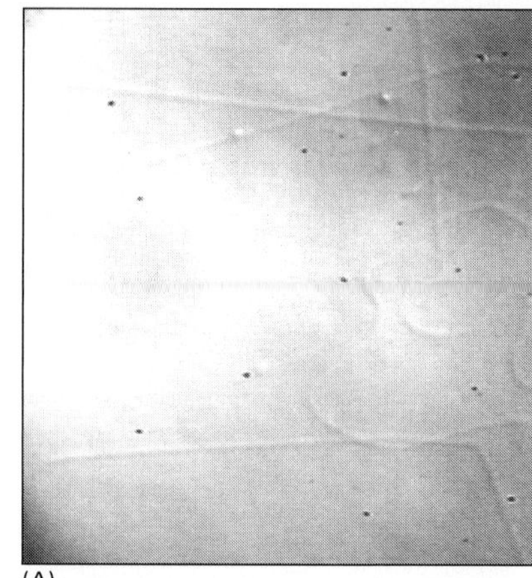

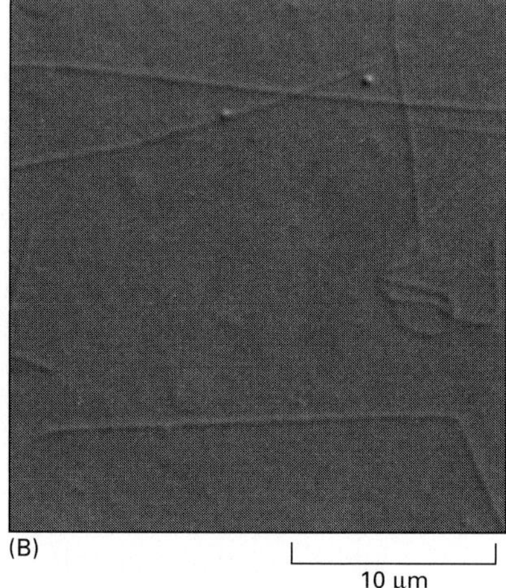

Tissues Are Usually Fixed and Sectioned for Microscopy

To make a permanent preparation that can be stained and viewed at leisure in the microscope, one first must treat cells with a fixative so as to immobilize, kill, and preserve them. In chemical terms, **fixation** makes cells permeable to staining reagents and cross-links their macromolecules so that they are stabilized and locked in position. Some of the earliest fixation procedures involved immersion in acids or in organic solvents, such as alcohol. Current procedures usually include treatment with reactive aldehydes, particularly formaldehyde and glutaraldehyde, which form covalent bonds with the free amino groups of proteins and thereby cross-link adjacent protein molecules.

Most tissue samples are too thick for their individual cells to be examined directly at high resolution. After fixation, therefore, the tissues are usually cut into very thin slices, or **sections**, with a *microtome,* a machine with a sharp blade that operates like a meat slicer (Figure 9–10). The sections (typically 1–10 μm thick) are then laid flat on the surface of a glass microscope slide.

Because tissues are generally soft and fragile, even after fixation, they need to be embedded in a supporting medium before sectioning. The usual embedding media are waxes or resins. In liquid form these media both permeate and surround the fixed tissue; they can then be hardened (by cooling or by polymerization) to form a solid block, which is readily sectioned by the microtome.

There is a serious danger that any treatment used for fixation and embedding may alter the structure of the cell or its constituent molecules in undesirable ways. Rapid freezing provides an alternative method of preparation that to some extent avoids this problem by eliminating the need for fixation and embedding. The frozen tissue can be cut directly with a special microtome that is maintained in a cold chamber. Although *frozen sections* produced in this way avoid some artifacts, they suffer from others: the native structures of individual molecules such as proteins are well preserved, but the fine structure of the cell is often disrupted by ice crystals.

Once sections have been cut, by whatever method, the next step is usually to stain them.

Different Components of the Cell Can Be Selectively Stained

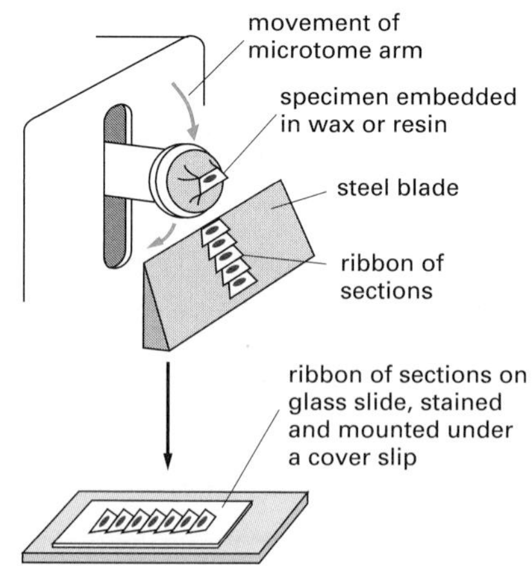

Figure 9–10 Making tissue sections. This illustration shows how an embedded tissue is sectioned with a microtome in preparation for examination in the light microscope.

There is little in the contents of most cells (which are 70% water by weight) to impede the passage of light rays. Thus, most cells in their natural state, even if fixed and sectioned, are almost invisible in an ordinary light microscope. One way to make them visible is to stain them with dyes.

In the early nineteenth century, the demand for dyes to stain textiles led to a fertile period for organic chemistry. Some of the dyes were found to stain biological tissues and, unexpectedly, often showed a preference for particular parts of the cell—the nucleus or mitochondria, for example—making these internal structures clearly visible. Today a rich variety of organic dyes is available, with such colorful names as *Malachite green, Sudan black,* and *Coomassie blue,* each of which has some specific affinity for particular subcellular components. The dye *hematoxylin,* for example, has an affinity for negatively charged molecules and therefore reveals the distribution of DNA, RNA, and acidic proteins in a cell (Figure 9–11). The chemical basis for the specificity of many dyes, however, is not known.

The relative lack of specificity of these dyes at the molecular level has stimulated the design of more rational and selective staining procedures and, in particular, of methods that reveal specific proteins or other macromolecules in cells. It is a problem, however, to achieve adequate sensitivity for this purpose. Since relatively few copies of most macromolecules are present in any given cell, one or two molecules of stain bound to each macromolecule are often invisible. One way to solve this problem is to increase the number of stain molecules associated with a single macromolecule. Thus, some enzymes can be located in cells through their catalytic activity: when supplied with appropriate substrate molecules, each enzyme molecule generates many molecules of a localized, visible reaction product. An alternative and much more generally applicable approach to the problem of sensitivity depends on using dyes that are fluorescent, as we explain next.

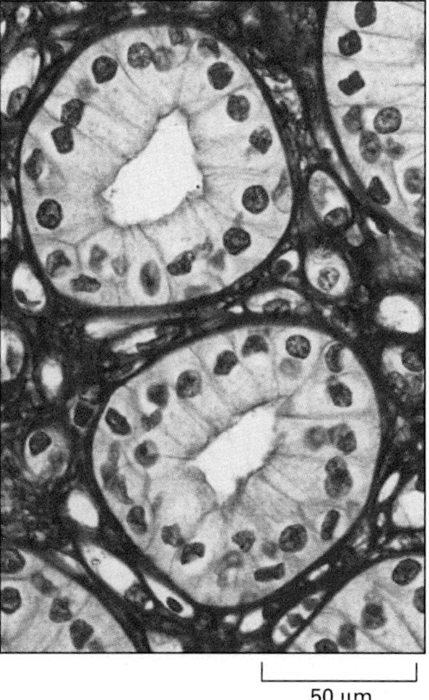
50 μm

Figure 9–11 A stained tissue section. This section of cells in the urine-collecting ducts of the kidney was stained with a combination of dyes, hematoxylin and eosin, commonly used in histology. Each duct is made of closely packed cells (with nuclei stained *red*) that form a ring. The ring is surrounded by extracellular matrix, stained *purple*. (From P.R. Wheater et al., Functional Histology, 2nd edn. London: Churchill Livingstone, 1987.)

Specific Molecules Can Be Located in Cells by Fluorescence Microscopy

Fluorescent molecules absorb light at one wavelength and emit it at another, longer wavelength. If such a compound is illuminated at its absorbing wavelength and then viewed through a filter that allows only light of the emitted wavelength to pass, it is seen to glow against a dark background. Because the background is dark, even a minute amount of the glowing fluorescent dye can be detected. The same number of molecules of an ordinary stain viewed conventionally would be practically invisible because they would give only the faintest tinge of color to the light transmitted through this stained part of the specimen.

The fluorescent dyes used for staining cells are detected by a **fluorescence microscope**. This microscope is similar to an ordinary light microscope except that the illuminating light, from a very powerful source, is passed through two sets of filters—one to filter the light before it reaches the specimen and one to filter the light obtained from the specimen. The first filter is selected so that it passes only the wavelengths that excite the particular fluorescent dye, while the second filter blocks out this light and passes only those wavelengths emitted when the dye fluoresces (Figure 9–12).

Fluorescence microscopy is most often used to detect specific proteins or other molecules in cells and tissues. A very powerful and widely used technique is to couple fluorescent dyes to antibody molecules, which then serve as highly specific and versatile staining reagents that bind selectively to the particular macromolecules they recognize in cells or in the extracellular matrix. Two fluorescent dyes that have been commonly used for this purpose are *fluorescein,* which emits an intense green fluorescence when excited with blue light, and *rhodamine,* which emits a deep red fluorescence when excited with green–yellow

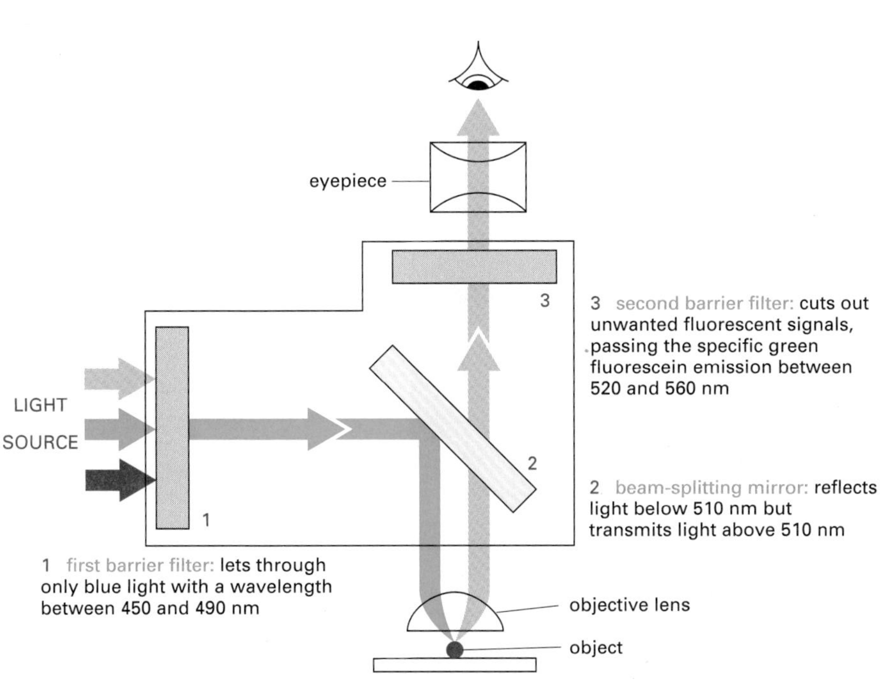
eyepiece
3
3 second barrier filter: cuts out unwanted fluorescent signals, passing the specific green fluorescein emission between 520 and 560 nm
LIGHT
SOURCE
2
2 beam-splitting mirror: reflects light below 510 nm but transmits light above 510 nm
1
1 first barrier filter: lets through only blue light with a wavelength between 450 and 490 nm
objective lens
object

Figure 9–12 The optical system of a fluorescence microscope. A filter set consists of two barrier filters (1 and 3) and a dichroic (beam-splitting) mirror (2). In this example, the filter set for detection of the fluorescent molecule fluorescein is shown. High-numerical-aperture objective lenses are especially important in this type of microscopy because, for a given magnification, the brightness of the fluorescent image is proportional to the fourth power of the numerical aperture (see also Figure 9–6).

light (Figure 9–13). By coupling one antibody to fluorescein and another to rhodamine, the distributions of different molecules can be compared in the same cell; the two molecules are visualized separately in the microscope by switching back and forth between two sets of filters, each specific for one dye. As shown in Figure 9–14, three fluorescent dyes can be used in the same way to distinguish between three types of molecules in the same cell. Many newer fluorescent dyes, such as Cy3, Cy5, and the Alexa dyes, have been specifically developed for fluorescence microscopy (see Figure 9–13).

Important methods, discussed later in the chapter, enable fluorescence microscopy to be used to monitor changes in the concentration and location of specific molecules inside *living* cells (see p. 574).

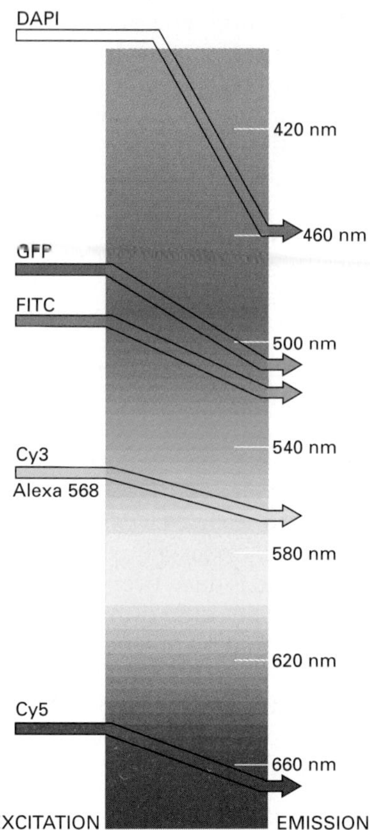

Figure 9–13 Fluorescent dyes. The maximum excitation and emission wavelengths of several commonly used fluorescent dyes are shown in relation to the corresponding colours of the spectrum. The photon emitted by a dye molecule is necessarily of lower energy (longer wavelength) than the photon absorbed and this accounts for the difference between the excitation and emission peaks. GFP is a green fluorescent protein, not a dye, and is discussed in detail later in the chapter. DAPI is widely used as a general fluorescent DNA dye, that absorbs UV light and fluoresces bright blue. The other dyes are all commonly used to fluorescently label antibodies and other proteins.

Antibodies Can Be Used to Detect Specific Molecules

Antibodies are proteins produced by the vertebrate immune system as a defense against infection (discussed in Chapter 24). They are unique among proteins because they are made in billions of different forms, each with a different binding site that recognizes a specific target molecule (or *antigen*). The precise antigen specificity of antibodies makes them powerful tools for the cell biologist. When labeled with fluorescent dyes, they are invaluable for locating specific molecules in cells by fluorescence microscopy (Figure 9–15); labeled with electron-dense particles such as colloidal gold spheres, they are used for similar purposes in the electron microscope (discussed below).

The sensitivity of antibodies as probes for detecting and assaying specific molecules in cells and tissues is frequently enhanced by chemical methods that amplify the signal. For example, although a marker molecule such as a fluorescent dye can be linked directly to an antibody used for specific recognition—the *primary antibody*—a stronger signal is achieved by using an unlabeled primary antibody and then detecting it with a group of labeled *secondary antibodies* that bind to it (Figure 9–16).

The most sensitive amplification methods use an enzyme as a marker molecule attached to the secondary antibody. The enzyme alkaline phosphatase, for example, in the presence of appropriate chemicals, produces inorganic phosphate and leads to the local formation of a colored precipitate. This reveals the location of the secondary antibody that is coupled to the enzyme and hence the location of the antibody–antigen complex to which the secondary

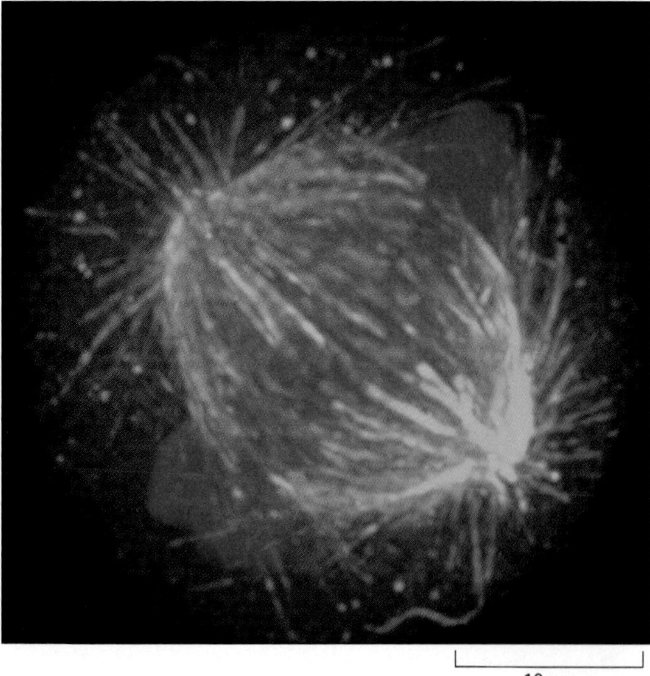

Figure 9–14 Multiple-fluorescent-probe microscopy. In this composite micrograph of a cell in mitosis, three different fluorescent probes have been used to stain three different cellular components. The spindle microtubules are revealed with a *green* fluorescent antibody, centromeres with a *red* fluorescent antibody and the DNA of the condensed chromosomes with the *blue* fluorescent dye DAPI. (Courtesy of Kevin F. Sullivan.)

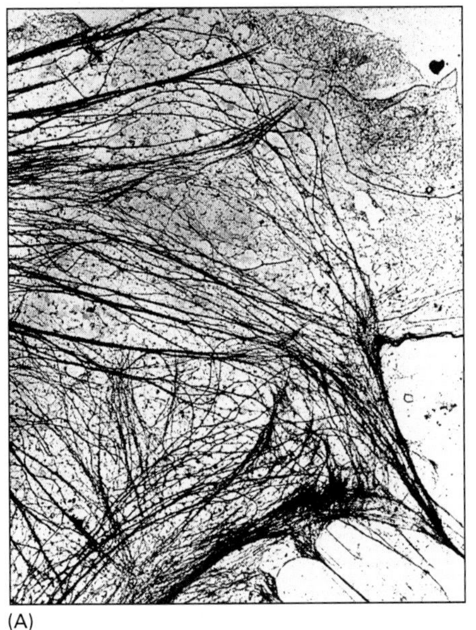

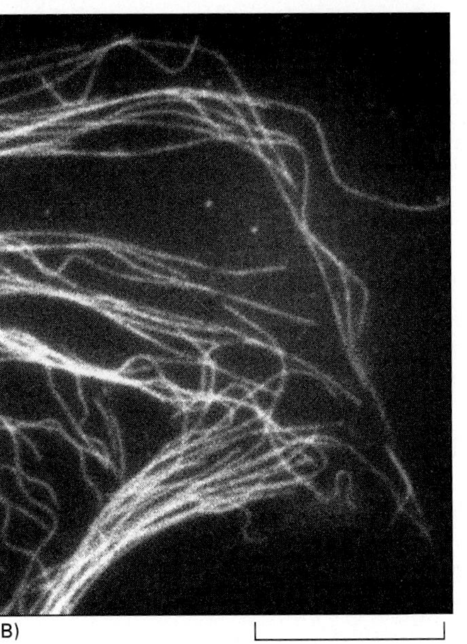

(A) (B)

Figure 9–15 Immunofluorescence. (A) A transmission electron micrograph of the periphery of a cultured epithelial cell showing the distribution of microtubules and other filaments. (B) The same area stained with fluorescent antibodies against tubulin, the protein subunit of microtubules, using the technique of indirect immunocytochemistry (see Figure 9–16). *Arrows* indicate individual microtubules that are readily recognizable in both images. Note that, because of diffraction effects, the microtubules in the light microscope appear 0.2 μm wide rather than their true width of 0.025 μm. (From M. Osborn, R. Webster, and K. Weber, *J. Cell Biol.* 77:R27–R34, 1978. © The Rockefeller University Press.)

antibody is bound. Since each enzyme molecule acts catalytically to generate many thousands of molecules of product, even tiny amounts of antigen can be detected. An enzyme-linked immunosorbent assay (ELISA) based on this principle is frequently used in medicine as a sensitive test—for pregnancy or for various types of infections, for example. Although the enzyme amplification makes enzyme-linked methods very sensitive, diffusion of the colored precipitate away from the enzyme means that the spatial resolution of this method for microscopy may be limited, and fluorescent labels are usually used for the most precise optical localization.

Antibodies are made most simply by injecting a sample of the antigen several times into an animal such as a rabbit or a goat and then collecting the antibody-rich serum. This *antiserum* contains a heterogeneous mixture of antibodies, each produced by a different antibody-secreting cell (a B lymphocyte). The different antibodies recognize various parts of the antigen molecule, as well as impurities in the antigen preparation. The specificity of an antiserum for a particular antigen can sometimes be sharpened by removing the unwanted antibody molecules that bind to other molecules; an antiserum produced against protein X, for example, can be passed through an affinity column of antigens Y and Z to remove any contaminating anti-Y and anti-Z antibodies. Even so, the heterogeneity of such antisera sometimes limits their usefulness. This problem is largely overcome by the use of monoclonal antibodies (see Figure 8–6). However, monoclonal antibodies can also have problems. Since they are single antibody protein species, they show almost perfect specificity for a single site or epitope on the antigen, but the accessibility of the epitope, and thus the usefulness of the antibody, may depend on the specimen preparation. For example, some monoclonal antibodies will react only with unfixed antigens, others only after the use of particular fixatives, and still others only with proteins denatured on SDS polyacrylamide gels, and not with the proteins in their native conformation.

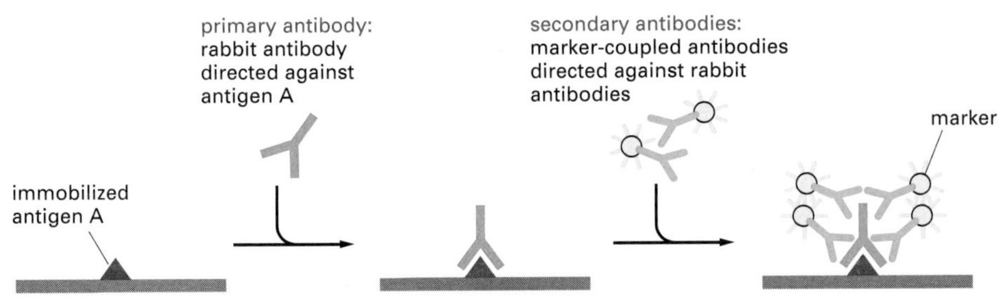

Figure 9–16 Indirect immunocytochemistry. This detection method is very sensitive because the primary antibody is itself recognized by many molecules of the secondary antibody. The secondary antibody is covalently coupled to a marker molecule that makes it readily detectable. Commonly used marker molecules include fluorescent dyes (for fluorescence microscopy), the enzyme horseradish peroxidase (for either conventional light microscopy or electron microscopy), colloidal gold spheres (for electron microscopy), and the enzymes alkaline phosphatase or peroxidase (for biochemical detection).

Imaging of Complex Three-dimensional Objects Is Possible with the Optical Microscope

For ordinary light microscopy, as we have seen, a tissue has to be sliced into thin sections to be examined; the thinner the section, the crisper the image. In the process of sectioning, information about the third dimension is lost. How, then, can one get a picture of the three-dimensional architecture of a cell or tissue, and how can one view the microscopic structure of a specimen that, for one reason or another, cannot first be sliced into sections? Although an optical microscope is focused on a particular focal plane within complex three-dimensional specimens, all the other parts of the specimen above and below the plane of focus are also illuminated, and the light originating from these regions contributes to the image as "out-of-focus" blur. This can make it very hard to interpret the image in detail, and can lead to fine image structure being obscured by the out-of-focus light.

Two approaches have been developed to solve this problem: one is computational, the other is optical. These three-dimensional microscopic imaging methods make it possible to focus on a chosen plane in a thick specimen while rejecting the light that comes from out-of-focus regions above and below that plane. Thus one sees a crisp, thin *optical section*. From a series of such optical sections taken at different depths and stored in a computer, it is easy to reconstruct a three-dimensional image. The methods do for the microscopist what the CT scanner does (by different means) for the radiologist investigating a human body: both machines give detailed sectional views of the interior of an intact structure.

The computational approach is often called *image deconvolution*. To understand how it works, remember how the wave nature of light means that the microscope lens system gives a small blurred disc as the image of a point light source, with increased blurring if the point source lies above or below the focal plane. This blurred image of a point source is called the *point spread function*. An image of a complex object can then be thought of as being built up by replacing each point of the specimen by a corresponding blurred disc, resulting in an image that is blurred overall. For deconvolution, we first obtain a series of (blurred) images, focusing the microscope in turn on a series of focal planes—in effect, a blurred three-dimensional image. The stack of images is then processed by computer to remove as much of the blur as possible. Essentially the computer program uses the microscope's point spread function to determine what the effect of the blurring would have been on the image, and then applies an equivalent "deblurring" (deconvolution), turning the blurred three-dimensional image into a series of clean optical sections. The computation required is quite complex, and used to be a serious limitation. However, with faster and cheaper computers, the image deconvolution method is gaining in power and popularity. An example is shown in Figure 9–17.

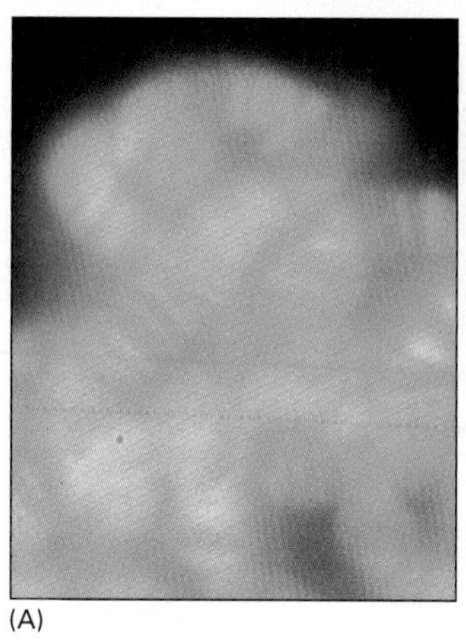

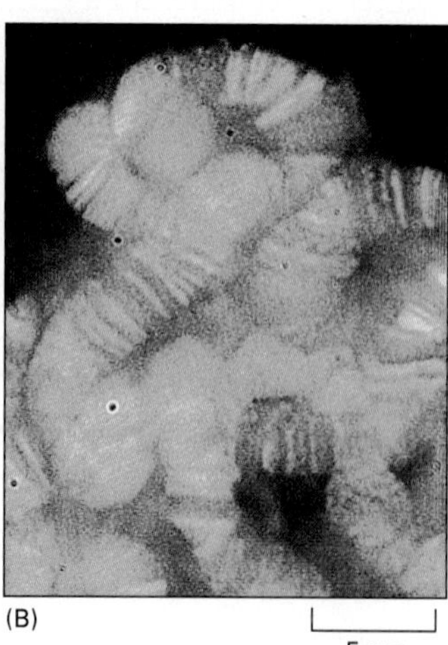

Figure 9–17 Image deconvolution. (A) A light micrograph of the large polytene chromosomes from *Drosophila*, stained with a fluorescent DNA-binding dye. (B) The same field of view after image deconvolution clearly reveals the banding pattern on the chromosomes. Each band is about 0.25 μm thick, approaching the resolution limit of the light microscope. (Courtesy of the John Sedat Laboratory.)

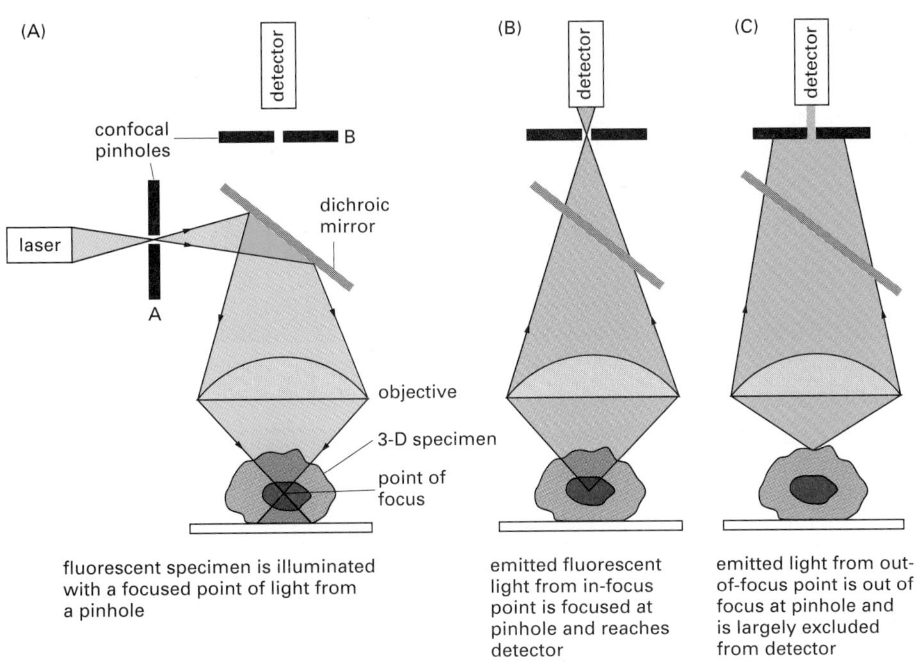
(A)
detector
confocal pinholes
B
laser
dichroic mirror
A
objective
3-D specimen
point of focus
fluorescent specimen is illuminated with a focused point of light from a pinhole
(B)
detector
emitted fluorescent light from in-focus point is focused at pinhole and reaches detector
(C)
detector
emitted light from out-of-focus point is out of focus at pinhole and is largely excluded from detector

Figure 9–18 The confocal fluorescence microscope. This simplified diagram shows that the basic arrangement of optical components is similar to that of the standard fluorescence microscope shown in Figure 9–12, except that a laser is used to illuminate a small pinhole whose image is focused at a single point in the specimen (A). Emitted fluorescence from this focal point in the specimen is focused at a second (confocal) pinhole (B). Emitted light from elsewhere in the specimen is not focused here and therefore does not contribute to the final image (C). By scanning the beam of light across the specimen, a very sharp two-dimensional image of the exact plane of focus is built up that is not significantly degraded by light from other regions of the specimen.

The Confocal Microscope Produces Optical Sections by Excluding Out-of-Focus Light

The confocal microscope achieves a result similar to that of deconvolution, but does so by manipulation of the light before it is measured; thus it is an analog technique rather than a digital one. The optical details of the **confocal microscope** are complex, but the basic idea is simple, as illustrated in Figure 9–18.

The microscope is generally used with fluorescence optics (see Figure 9–12), but instead of illuminating the whole specimen at once, in the usual way, the optical system at any instant focuses a spot of light onto a single point at a specific depth in the specimen. A very bright source of pinpoint illumination is required; this is usually supplied by a laser whose light has been passed through a pinhole. The fluorescence emitted from the illuminated material is collected and brought to an image at a suitable light detector. A pinhole aperture is placed in front of the detector, at a position that is *confocal* with the illuminating pinhole—that is, precisely where the rays emitted from the illuminated point in the specimen come to a focus. Thus, the light from this point in the specimen converges on this aperture and enters the detector.

By contrast, the light from regions out of the plane of focus of the spotlight is also out of focus at the pinhole aperture and is therefore largely excluded from the detector (Figure 9–19). To build up a two-dimensional image, data from each point in the plane of focus are collected sequentially by scanning across the field

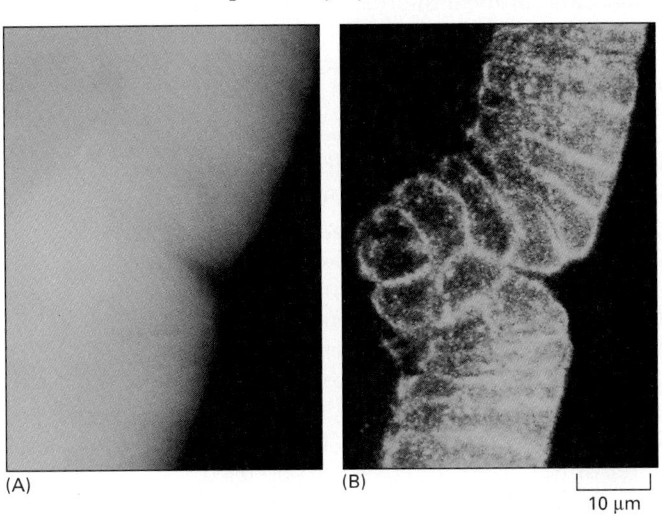
(A)
(B)
10 μm

Figure 9–19 Conventional and confocal fluorescence microscopy compared. These two micrographs are of the same intact gastrula-stage *Drosophila* embryo that has been stained with a fluorescent probe for actin filaments. (A) The conventional, unprocessed image is blurred by the presence of fluorescent structures above and below the plane of focus. (B) In the confocal image, this out-of-focus information is removed, resulting in a crisp optical section of the cells in the embryo. (Courtesy of Richard Warn and Peter Shaw.)

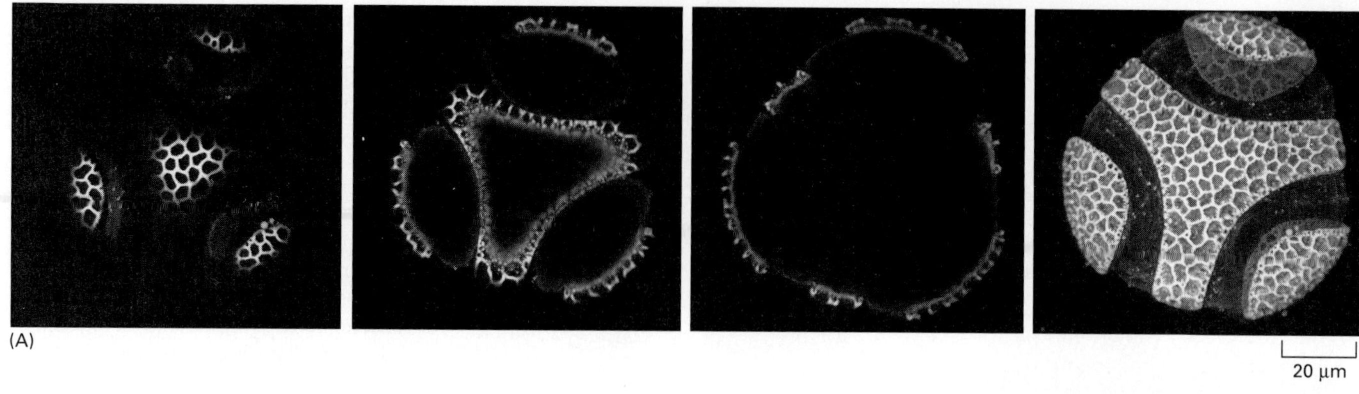

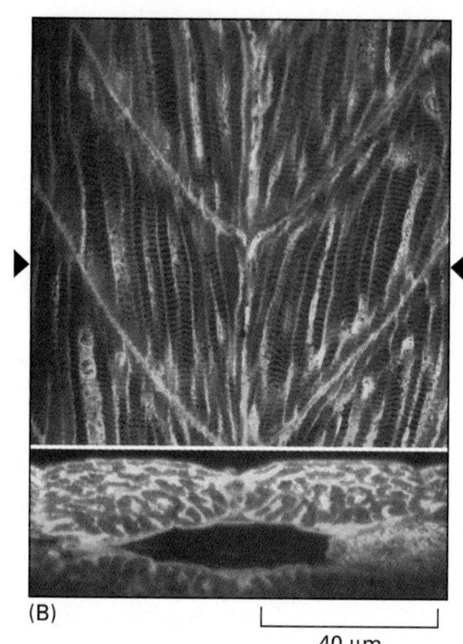

Figure 9–20 Three-dimensional reconstruction from confocal microscope images. (A) Pollen grains, in this case from a passion flower, have a complex sculptured cell wall that contains fluorescent compounds. Images obtained at different depths through the grain, using a confocal microscope, can be recombined to give a three-dimensional view of the whole grain, shown on the right. Three selected individual optical sections from the full set of 30, each of which shows little contribution from its neighbors, are shown on the left. (B) The tail region of this zebrafish embryo has been stained with two fluorescent dyes and imaged in a confocal microscope. The image below is a transverse optical section of the tail, digitally constructed by scanning the laser in a single line (indicated by arrowheads) while progressively varying the focus level. The dark cavity in the centre is the developing notochord. (A, courtesy of Brad Amos; B, courtesy of S. Reichelt and W.B. Amos.)

in a raster pattern (as on a television screen) and are displayed on a video screen. Although not shown in Figure 9–18, the scanning is usually done by deflecting the beam with an oscillating mirror placed between the dichroic mirror and the objective lens in such a way that the illuminating spotlight and the confocal pinhole at the detector remain strictly in register.

The confocal microscope has been used to resolve the structure of numerous complex three-dimensional objects (Figure 9–20), including the networks of cytoskeletal fibers in the cytoplasm and the arrangements of chromosomes and genes in the nucleus.

The relative merits of deconvolution methods and confocal microscopy for three-dimensional optical microscopy are still the subject of debate. Confocal microscopes are generally easier to use than deconvolution systems and the final optical sections can be seen quickly. On the other hand, modern, cooled CCD (charge-coupled device) cameras used for deconvolution systems are extremely efficient at collecting small amounts of light, and they can be used to make detailed three-dimensional images from specimens that are too weakly stained or too easily damaged by bright light for confocal microscopy.

The Electron Microscope Resolves the Fine Structure of the Cell

The relationship between the limit of resolution and the wavelength of the illuminating radiation (see Figure 9–6) holds true for any form of radiation, whether it is a beam of light or a beam of electrons. With electrons, however, the limit of resolution can be made very small. The wavelength of an electron decreases as its velocity increases. In an **electron microscope** with an accelerating voltage of 100,000 V, the wavelength of an electron is 0.004 nm. In theory the resolution of such a microscope should be about 0.002 nm, which is 10,000 times that of the light microscope. Because the aberrations of an electron lens are considerably harder to correct than those of a glass lens, however, the practical resolving power of most modern electron microscopes is, at best, 0.1 nm (1 Å) (Figure 9–21). This is because only the very center of the electron lenses can be used, and the effective numerical aperture is tiny. Furthermore, problems of specimen

Figure 9–21 The limit of resolution of the electron microscope. This transmission electron micrograph of a thin layer of gold shows the individual files of atoms in the crystal as bright spots. The distance between adjacent files of gold atoms is about 0.2 nm (2 Å). (Courtesy of Graham Hills.)

preparation, contrast, and radiation damage have generally limited the normal effective resolution for biological objects to 2 nm (20 Å). This is nonetheless about 100 times better than the resolution of the light microscope. Moreover, in recent years, the performance of electron microscopes has been improved by the development of electron illumination sources called field emission guns. These very bright and coherent sources can substantially improve the resolution achieved. The major landmarks in the development of electron microscopy are listed in Table 9–2.

In overall design the transmission electron microscope (TEM) is similar to a light microscope, although it is much larger and upside down (Figure 9–22). The

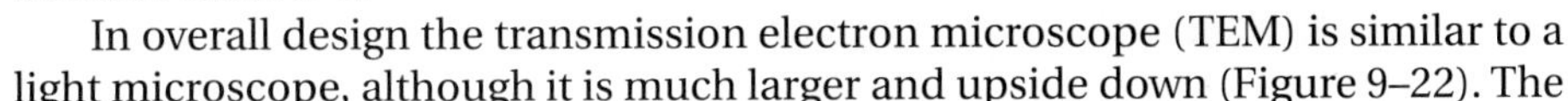

TABLE 9–2 Major Events in the Development of the Electron Microscope and Its Application to Cell Biology

1897	**Thomson** announces the existence of negatively charged particles, later termed electrons.
1924	**de Broglie** proposes that a moving electron has wavelike properties.
1926	**Busch** proves that it is possible to focus a beam of electrons with a cylindrical magnetic lens, laying the foundation of electron optics.
1931	**Ruska** and colleagues build the first transmission electron microscope.
1935	**Knoll** demonstrates the feasibility of the scanning electron microscope; three years later, a prototype instrument is built by **Von Ardenne**.
1939	**Siemens** produces the first commercial transmission electron microscope.
1944	**Williams** and **Wyckoff** introduce the metal shadowing technique.
1945	**Porter**, **Claude**, and **Fullam** use the electron microscope to examine cells in tissue culture, fixing and staining them with OsO_4.
1948	**Pease** and **Baker** reliably prepare thin sections (0.1–0.2 μm thick) of biological material.
1952	**Palade**, **Porter**, and **Sjöstrand** develop methods of fixation and thin sectioning that enable many intracellular structures to be seen for the first time. In one of the first applications of these techniques, **Huxley** shows that skeletal muscle contains overlapping arrays or protein filaments, supporting the sliding-filament hypothesis of muscle contraction.
1953	**Porter and Blum** develop the first widely accepted ultramicrotome, incorporating many features previously introduced by **Claude** and **Sjöstrand**.
1956	**Glauert** and colleagues show that the epoxy resin Araldite is a highly effective embedding agent for electron microscopy. **Luft** introduces another embedding resin, Epon, five years later.
1957	**Robertson** describes the trilaminar structure of the cell membrane, seen for the first time in the electron microscope.
1957	Freeze-fracture techniques, initially developed by **Steere**, are perfected by **Moor and Mühlethaler**. Later (1966), **Branton** demonstrates that freeze-fracture allows the interior of the membrane to be visualized.
1959	**Singer** uses antibodies coupled to ferritin to detect cellular molecules in the electron microscope.
1959	**Brenner and Horne** develop the negative staining technique, invented four years previously by **Hall**, into a generally useful technique for visualizing viruses, bacteria, and protein filaments.
1963	**Sabatini, Bensch, and Barrnett** introduce glutaraldeyhde (usually followed by OsO_4) as a fixative for electron microscopy.
1965	Cambridge Instruments produces the first commercial scanning electron microscope.
1968	**de Rosier and Klug** describe techniques for the reconstruction of three-dimensional structures from electron micrographs.
1975	**Henderson and Unwin** determine the first structure of a membrane protein by computer-based reconstruction from electron micrographs of unstained samples.
1979	**Heuser**, **Reese**, and colleagues develop a high-resolution, deep-etching technique using very rapidly frozen specimens.
1980s	**Dubochet** and colleagues introduce rapid freezing in thin films of vitreous ice for high-resolution electron microscopy.
1997–	**Crowther**, **Fuller**, **Frank**, and colleagues use single-particle reconstruction to determine structures of viruses and the ribosome at high resolution (8–10 Å).

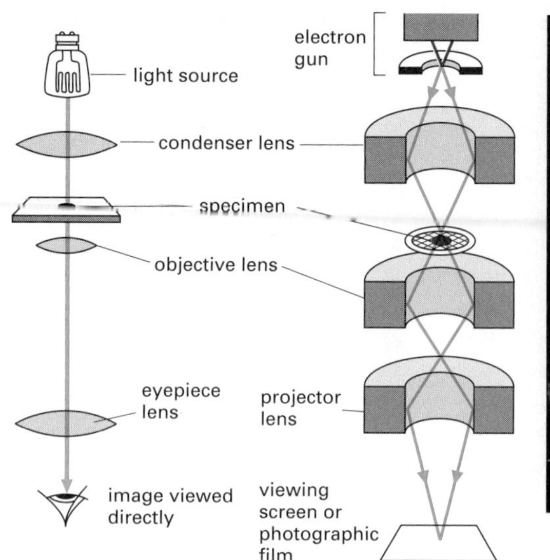

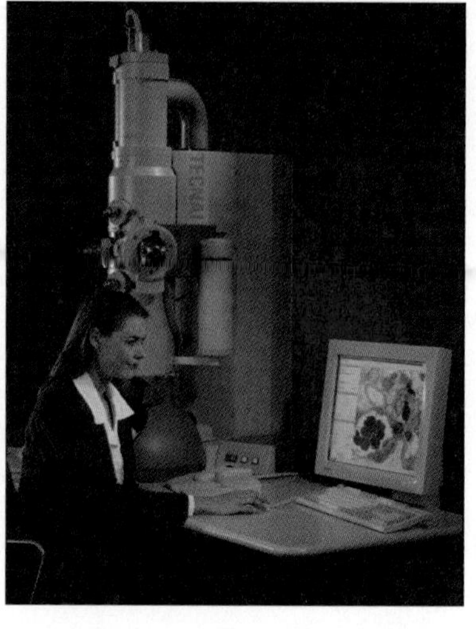

Figure 9–22 The principal features of a light microscope and a transmission electron microscope. These drawings emphasize the similarities of overall design. Whereas the lenses in the light microscope are made of glass, those in the electron microscope are magnetic coils. The electron microscope requires that the specimen be placed in a vacuum. The inset shows a transmission electron microscope in use. (Photograph courtesy of FEI Company Ltd.)

source of illumination is a filament or cathode that emits electrons at the top of a cylindrical column about 2 m high. Since electrons are scattered by collisions with air molecules, air must first be pumped out of the column to create a vacuum. The electrons are then accelerated from the filament by a nearby anode and allowed to pass through a tiny hole to form an electron beam that travels down the column. Magnetic coils placed at intervals along the column focus the electron beam, just as glass lenses focus the light in a light microscope. The specimen is put into the vacuum, through an airlock, into the path of the electron beam. As in light microscopy, the specimen is usually stained—in this case, with *electron-dense* material, as we see in the next section. Some of the electrons passing through the specimen are scattered by structures stained with the electron-dense material; the remainder are focused to form an image, in a manner analogous to the way an image is formed in a light microscope—either on a photographic plate or on a phosphorescent screen. Because the scattered electrons are lost from the beam, the dense regions of the specimen show up in the image as areas of reduced electron flux, which look dark.

Biological Specimens Require Special Preparation for the Electron Microscope

In the early days of its application to biological materials, the electron microscope revealed many previously unimagined structures in cells. But before these discoveries could be made, electron microscopists had to develop new procedures for embedding, cutting, and staining tissues.

Since the specimen is exposed to a very high vacuum in the electron microscope, there is no possibility of viewing it in the living, wet state. Tissues are usually preserved by fixation—first with *glutaraldehyde,* which covalently cross-links protein molecules to their neighbors, and then with *osmium tetroxide,* which binds to and stabilizes lipid bilayers as well as proteins (Figure 9–23). Because electrons have very limited penetrating power, the fixed tissues normally have to be cut into extremely thin sections (50–100 nm thick, about 1/200 the thickness of a single cell) before they are viewed. This is achieved by dehydrating the specimen and permeating it with a monomeric resin that polymerizes to form a solid block of plastic; the block is then cut with a fine glass or diamond knife on a special microtome. These *thin sections,* free of water and other volatile solvents, are placed on a small circular metal grid for viewing in the microscope (Figure 9–24).

The steps required to prepare biological material for viewing in the electron microscope have challenged electron microscopists from the beginning. How can we be sure that the image of the fixed, dehydrated, resin-embedded specimen finally seen bears any relation to the delicate aqueous biological system that was originally present in the living cell? The best current approaches to this

H–C(=O)–CH_2–CH_2–CH_2–C(=O)–H

OsO_4

glutaraldehyde osmium tetroxide

Figure 9–23 Two common chemical fixatives used for electron microscopy. The two reactive aldehyde groups of glutaraldehyde enable it to cross-link various types of molecules, forming covalent bonds between them. Osmium tetroxide is reduced by many organic compounds with which it forms cross-linked complexes. It is especially useful for fixing cell membranes, since it reacts with the C=C double bonds present in many fatty acids.

problem depend on rapid freezing. If an aqueous system is cooled fast enough to a low enough temperature, the water and other components in it do not have time to rearrange themselves or crystallize into ice. Instead, the water is supercooled into a rigid but noncrystalline state—a "glass"—called vitreous ice. This state can be achieved by slamming the specimen onto a polished copper block cooled by liquid helium, by plunging it into or spraying it with a jet of a coolant such as liquid propane, or by cooling it at high pressure.

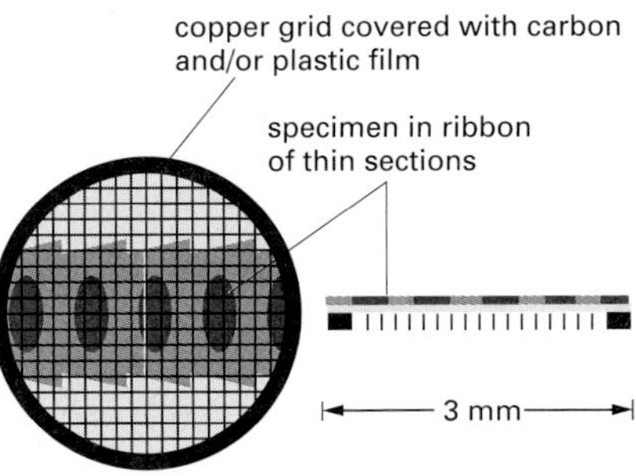

Figure 9–24 The copper grid that supports the thin sections of a specimen in a TEM.

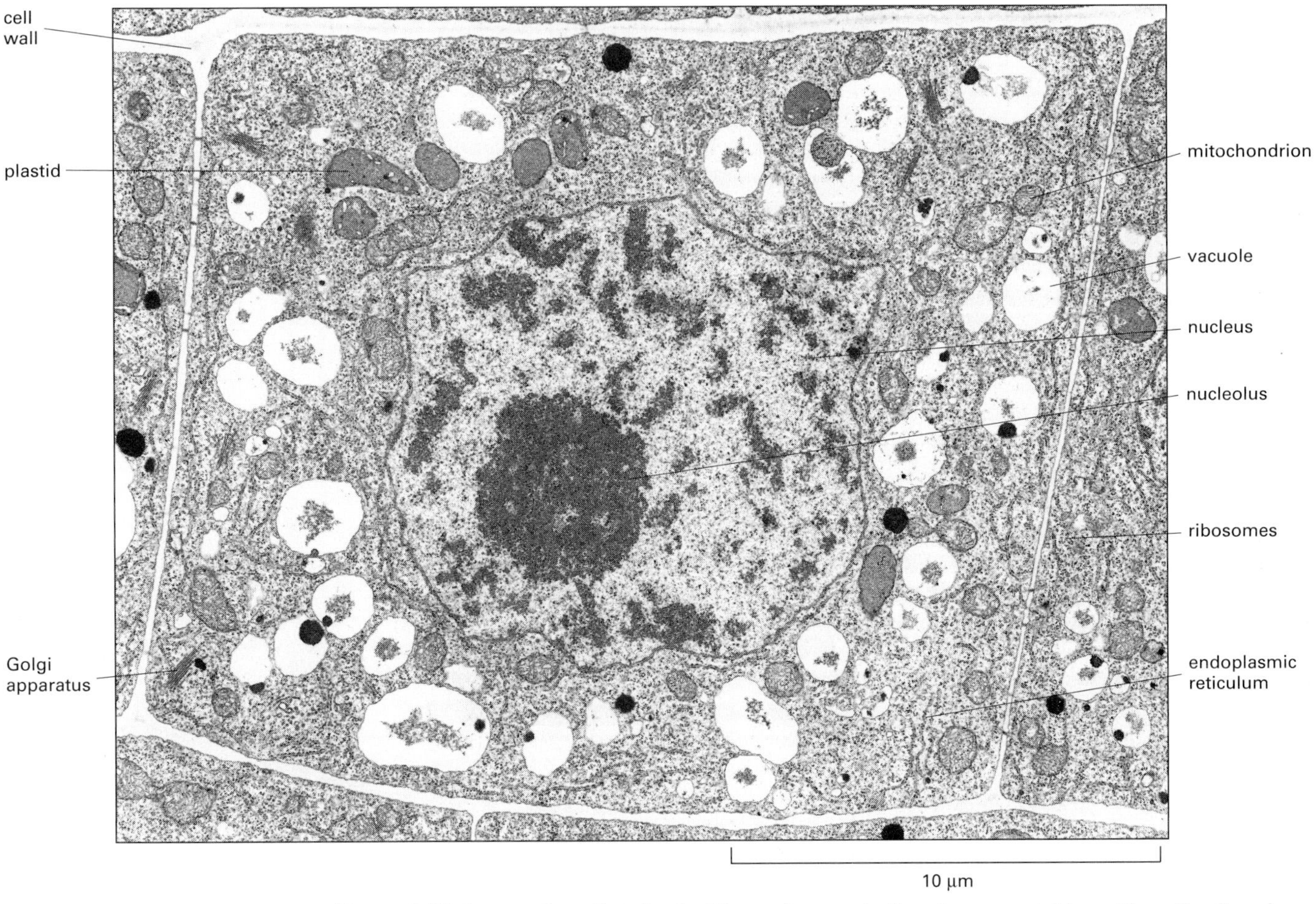

Some frozen specimens can be examined directly in the electron microscope using a special, cooled specimen holder. In other cases the frozen block can be fractured to reveal interior surfaces, or the surrounding ice can be sublimed away to expose external surfaces. However, we often want to examine thin sections, and to have them stained to yield adequate contrast in the electron microscope image (discussed further below). A compromise is therefore to rapid-freeze the tissue, then replace the water, maintained in the vitreous (glassy) state, by organic solvents, and finally embed the tissue in plastic resin, cut sections, and stain. Although technically still difficult, this approach stabilizes and preserves the tissue in a condition very close to its original living state.

Contrast in the electron microscope depends on the atomic number of the atoms in the specimen: the higher the atomic number, the more electrons are scattered and the greater the contrast. Biological tissues are composed of atoms of very low atomic number (mainly carbon, oxygen, nitrogen, and hydrogen). To make them visible, they are usually impregnated (before or after sectioning) with the salts of heavy metals such as uranium and lead. Different cellular constituents are revealed with various degrees of contrast according to their degree of impregnation, or "staining," with these salts. Lipids, for example, tend to stain darkly after osmium fixation, revealing the location of cell membranes (Figure 9–25).

cell wall
plastid
Golgi apparatus
mitochondrion
vacuole
nucleus
nucleolus
ribosomes
endoplasmic reticulum
10 μm

Figure 9–25 A root-tip cell stained with osmium and other heavy metal ions. The cell wall, nucleus, vacuoles, mitochondria, endoplasmic reticulum, Golgi apparatus, and ribosomes are easily visible in this transmission electron micrograph. (Courtesy of Brian Gunning.)

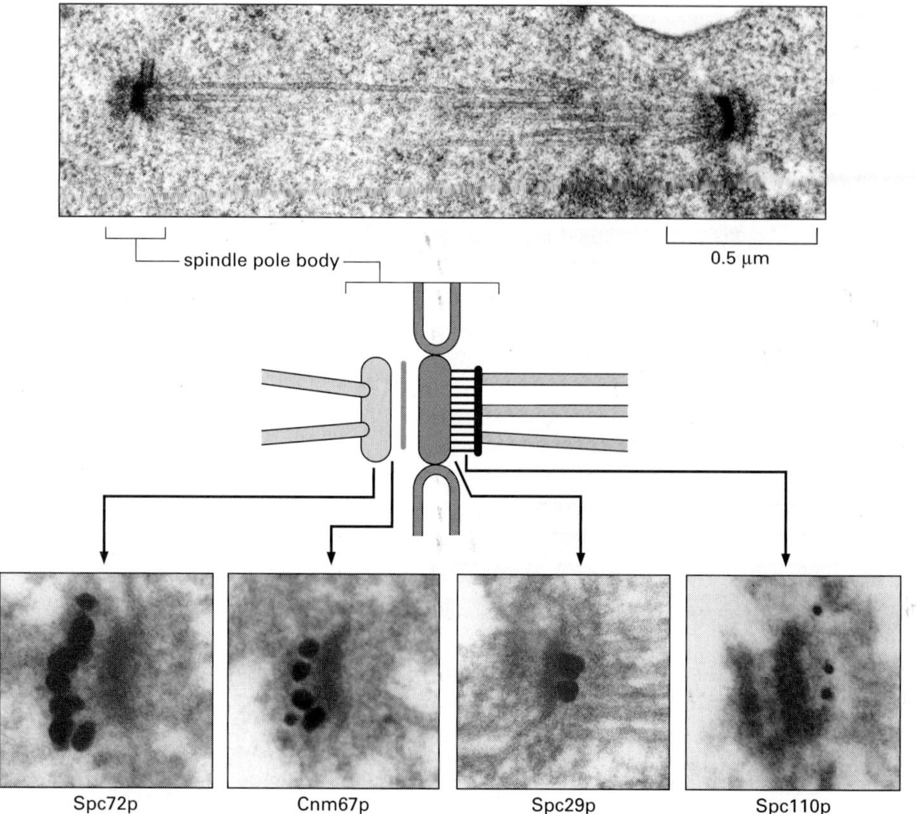

Figure 9–26 Localizing proteins in the electron microscope. Immunogold electron microscopy is used here to localize four different protein components to particular locations within the spindle pole body of yeast. At the top is a thin section of a yeast mitotic spindle showing the spindle microtubules that cross the nucleus, and connect at each end to spindle pole bodies embedded in the nuclear envelope. A diagram of the components of a single spindle pole body is shown below. Antibodies against four different proteins of the spindle pole body are used, together with colloidal gold particles (black dots), to reveal where within the complex structure each protein is located. (Courtesy of John Kilmartin.)

Specific Macromolecules Can Be Localized by Immunogold Electron Microscopy

We have seen how antibodies can be used in conjunction with fluorescence microscopy to localize specific macromolecules. An analogous method—**immunogold electron microscopy**—can be used in the electron microscope. The usual procedure is to incubate a thin section with a specific primary antibody, and then with a secondary antibody to which a colloidal gold particle has been attached. The gold particle is electron-dense and can be seen as a black dot in the electron microscope (Figure 9–26).

Thin sections often fail to convey the three-dimensional arrangement of cellular components in the TEM and can be very misleading: a linear structure such as a microtubule may appear in section as a pointlike object, for example, and a section through protruding parts of a single irregularly shaped solid body may give the appearance of two or more separate objects. The third dimension can be reconstructed from serial sections (Figure 9–27), but this is still a lengthy and tedious process.

Even thin sections, however, have a significant depth compared to the resolution of the electron microscope, so they can also be misleading in an opposite way. The optical design of the electron microscope—the very small aperture used—produces a large depth of field, so the image seen corresponds to a superimposition (a projection) of the structures at different depths. A further complication for immunogold labeling is that the antibodies and colloidal gold particles do not penetrate into the resin used for embedding; therefore, they only detect antigens right at the surface of the section. This means that first, the sensitivity of detection is low, since antigen molecules present in the deeper parts of

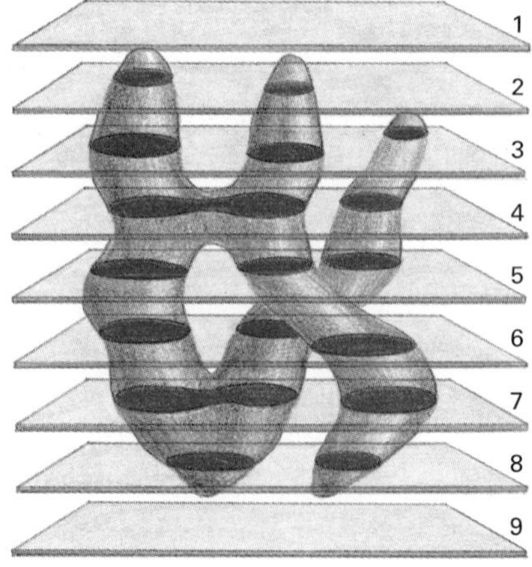

Figure 9–27 A three-dimensional reconstruction from serial sections. Single thin sections sometimes give misleading impressions. In this example, most sections through a cell containing a branched mitochondrion seem to contain two or three separate mitochondria. Sections 4 and 7, moreover, might be interpreted as showing a mitochondrion in the process of dividing. The true three-dimensional shape, however, can be reconstructed from serial sections.

the section are not detected, and second, one may get a false impression of which structures contain the antigen and which do not. A solution to this problem is to perform the labeling before embedding the specimen in plastic, when the cells and tissues are still fully accessible to labeling reagents. Extremely small gold particles, about 1 nm in diameter, work best for this procedure. Such small gold particles are usually not directly visible in the final sections, so additional silver or gold is nucleated around the 1 nm gold particles in a chemical process very much like photographic development.

Images of Surfaces Can Be Obtained by Scanning Electron Microscopy

A **scanning electron microscope (SEM)** directly produces an image of the three-dimensional structure of the surface of a specimen. The SEM is usually a smaller, simpler, and cheaper device than a transmission electron microscope. Whereas the TEM uses the electrons that have passed through the specimen to form an image, the SEM uses electrons that are scattered or emitted from the specimen's surface. The specimen to be examined is fixed, dried, and coated with a thin layer of heavy metal. Alternatively, it can be rapidly frozen, and then transferred to a cooled specimen stage for direct examination in the microscope. Often an entire plant or small animal can be put into the microscope with very little preparation (Figure 9–28). The specimen, prepared in any of these ways, is then scanned with a very narrow beam of electrons. The quantity of electrons scattered or emitted as this primary beam bombards each successive point of the metallic surface is measured and used to control the intensity of a second beam, which moves in synchrony with the primary beam and forms an image on a television screen. In this way, a highly enlarged image of the surface as a whole is built up (Figure 9–29).

The SEM technique provides great depth of field; moreover, since the amount of electron scattering depends on the angle of the surface relative to the beam, the image has highlights and shadows that give it a three-dimensional appearance (Figures 9–28 and 9–30). Only surface features can be examined, however, and in most forms of SEM, the resolution attainable is not very high (about 10 nm, with an effective magnification of up to 20,000 times). As a result, the technique is usually used to study whole cells and tissues rather than subcellular organelles. Very high-resolution SEMs have, however, been recently

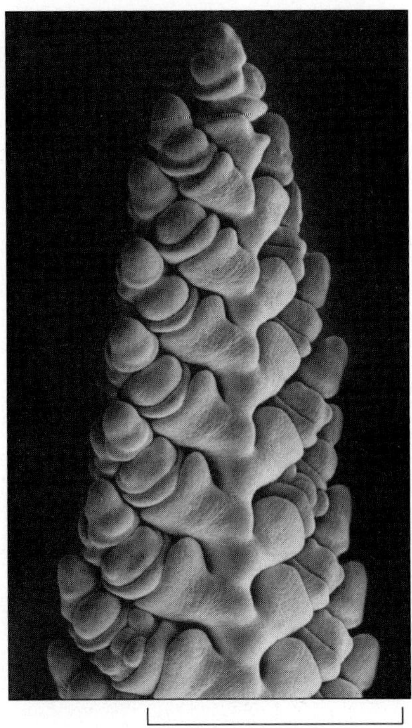

Figure 9–28 A developing wheat flower, or spike. This delicate flower spike was rapidly frozen, coated with a thin metal film, and examined in the frozen state in a SEM. This micrograph, which is at a low magnification, demonstrates the large depth of focus of the SEM. (Courtesy of Kim Findlay.)

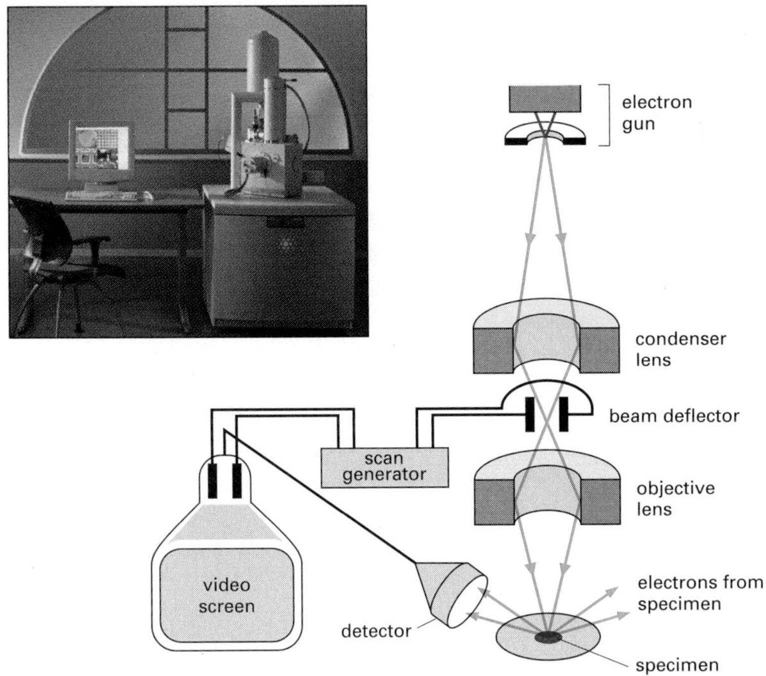

Figure 9–29 The scanning electron microscope. In a SEM, the specimen is scanned by a beam of electrons brought to a focus on the specimen by the electromagnetic coils that act as lenses. The quantity of electrons scattered or emitted as the beam bombards each successive point on the surface of the specimen is measured by the detector and is used to control the intensity of successive points in an image built up on a video screen. The SEM creates striking images of three-dimensional objects with great depth of focus and a resolution between 3 nm and 20 nm depending on the instrument. (Photograph courtesy of FEI Company Ltd.)

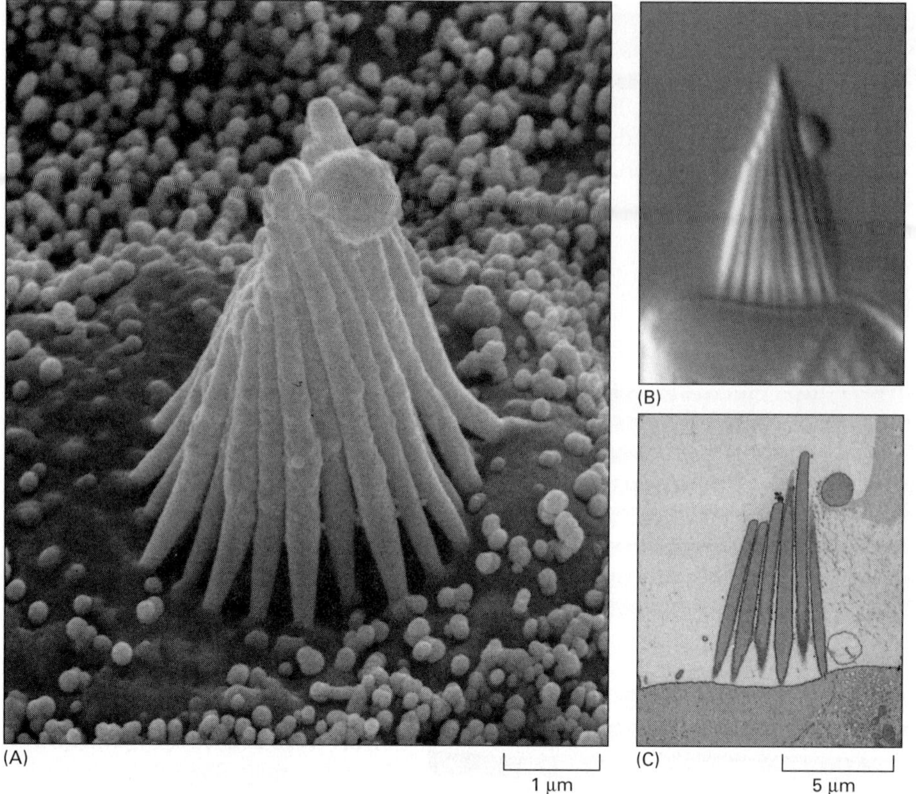

Figure 9–30 Scanning electron microscopy. (A) A scanning electron micrograph of the stereocilia projecting from a hair cell in the inner ear of a bullfrog. For comparison, the same structure is shown by (B) differential-interference-contrast light microscopy and (C) thin-section transmission electron microscopy. (Courtesy of Richard Jacobs and James Hudspeth.)

developed with a bright coherent-field emission gun as the electron source. This type of SEM can produce images that rival TEM images in resolution (Figure 9–31).

Metal Shadowing Allows Surface Features to Be Examined at High Resolution by Transmission Electron Microscopy

The TEM can also be used to study the surface of a specimen—and generally at a higher resolution than in the SEM—in such a way that individual macromolecules can be seen. As in scanning electron microscopy, a thin film of a heavy metal such as platinum is evaporated onto the dried specimen. The metal is sprayed from an oblique angle so as to deposit a coating that is thicker in some places than others—a process known as *shadowing* because a shadow effect is created that gives the image a three-dimensional appearance.

Some specimens coated in this way are thin enough or small enough for the electron beam to penetrate them directly. This is the case for individual molecules, viruses, and cell walls—all of which can be dried down, before shadowing, onto a flat supporting film made of a material that is relatively transparent to electrons, such as carbon or plastic. For thicker specimens, the organic material of the cell must be dissolved away after shadowing so that only the thin metal *replica* of the surface of the specimen is left. The replica is reinforced with a film of carbon so it can be placed on a grid and examined in the transmission electron microscope in the ordinary way (Figure 9–32).

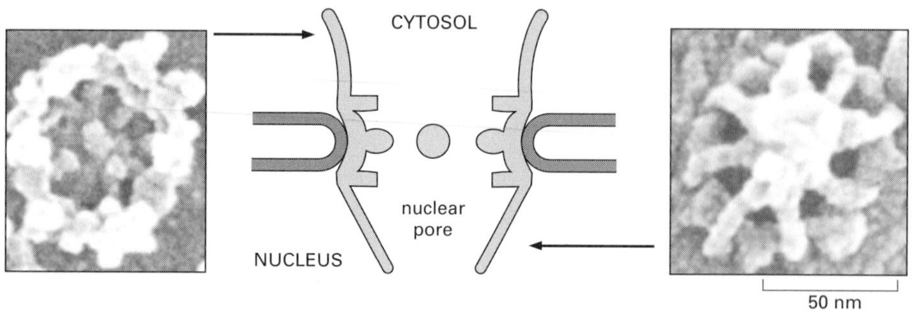

Figure 9–31 The nuclear pore. Rapidly frozen nuclear envelopes were imaged in a high-resolution SEM, equipped with a field emission gun as the source of electrons. These views of each side of a nuclear pore represent the limit of resolution of the SEM, and should be compared with Figure 12–10. (Courtesy of Martin Goldberg and Terry Allen.)

Freeze-Fracture and Freeze-Etch Electron Microscopy Provide Views of Surfaces Inside the Cell

Freeze-fracture electron microscopy provides a way of visualizing the interior of cell membranes. Cells are frozen (as described above) and then the frozen block is cracked with a knife blade. The fracture plane often passes through the hydrophobic middle of lipid bilayers, thereby exposing the interior of cell membranes. The resulting fracture faces are shadowed with platinum, the organic material is dissolved away, and the replicas are floated off and viewed in the electron microscope (see Figure 9–32). Such replicas are studded with small bumps, called *intramembrane particles,* which represent large transmembrane proteins. The technique provides a convenient and dramatic way to visualize the distribution of such proteins in the plane of a membrane (Figure 9–33).

Another related replica method is **freeze-etch electron microscopy**, which can be used to examine either the exterior or interior of cells. In this technique, the frozen block is cracked with a knife blade as described above. But now the ice level is lowered around the cells (and to a lesser extent within the cells) by the sublimation of ice in a vacuum as the temperature is raised—a process called *freeze-drying.* The parts of the cell exposed by this *etching* process are then shadowed as before to make a platinum replica. This technique exposes structures in the interior of the cell and can reveal their three-dimensional organization with exceptional clarity (Figure 9–34).

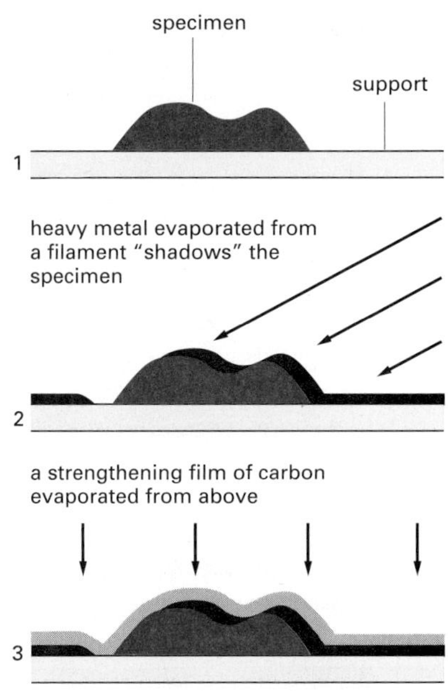

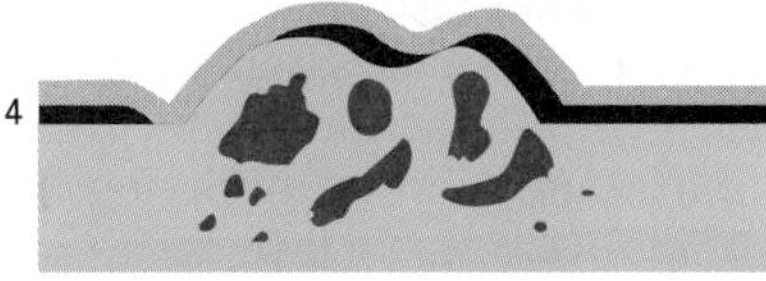

Figure 9–32 The preparation of a metal-shadowed replica of the surface of a specimen. Note that the thickness of the metal reflects the surface contours of the original specimen.

Negative Staining and Cryoelectron Microscopy Allow Macromolecules to Be Viewed at High Resolution

Although isolated macromolecules, such as DNA or large proteins, can be visualized readily in the electron microscope if they are shadowed with a heavy metal to provide contrast, finer detail can be seen by using **negative staining**. In this technique, the molecules, supported on a thin film of carbon, are washed with a concentrated solution of a heavy-metal salt such as uranyl acetate. After the sample has dried, a very thin film of metal salt covers the carbon film everywhere except where it has been excluded by the presence of an adsorbed macromolecule. Because the macromolecule allows electrons to pass much more readily than does the surrounding heavy-metal stain, a reversed or negative image of the molecule is created. Negative staining is especially useful for viewing large macromolecular aggregates such as viruses or ribosomes, and for seeing the subunit structure of protein filaments (Figure 9–35).

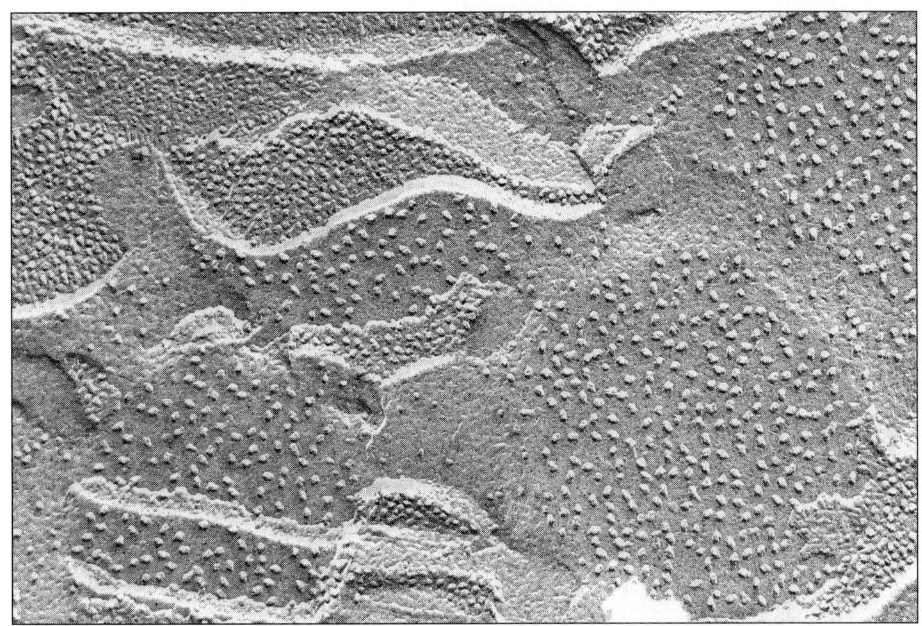

Figure 9–33 The thylakoid membranes from the chloroplast of a plant cell. In this freeze-fracture electron micrograph, the thylakoid membranes, which perform photosynthesis, are stacked up in multiple layers (see Figure 14–34). The plane of the fracture has moved from layer to layer, passing through the middle of each lipid bilayer and exposing transmembrane proteins that have sufficient bulk in the interior of the bilayer to cast a shadow and show up as intramembrane particles in this platinum replica. The largest particles seen in the membrane are the complete photosystem II—a complex of multiple proteins. (Courtesy of L.A. Staehelin.)

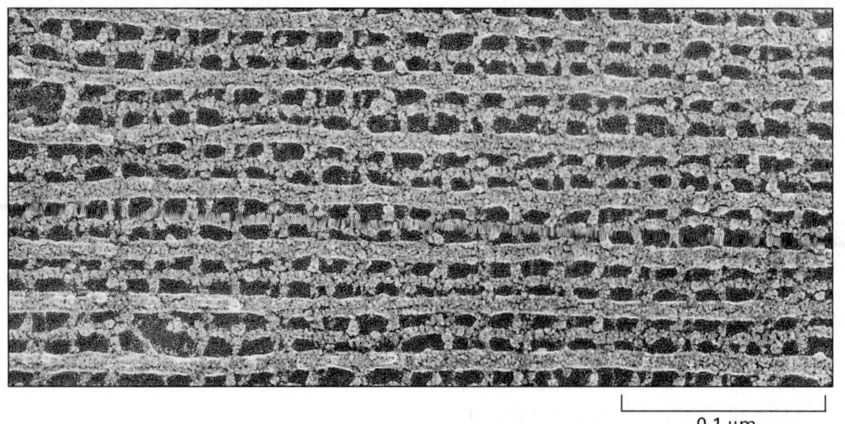

Figure 9–34 A regular array of protein filaments in an insect muscle. To obtain this image, the muscle cells were rapidly frozen to liquid helium temperature, fractured through the cytoplasm, and subjected to deep etching. A metal-shadowed replica was then prepared and examined at high magnification. (Courtesy of Roger Cooke and John Heuser.)

Shadowing and negative staining can provide high-contrast surface views of small macromolecular assemblies, but both techniques are limited in resolution by the size of the smallest metal particles in the shadow or stain used. Recent methods provide an alternative that has allowed even the interior features of three-dimensional structures such as viruses to be visualized directly at high resolution. In this technique, called **cryoelectron microscopy**, rapid freezing to form vitreous ice is again the key. A very thin (about 100 nm) film of an aqueous suspension of virus or purified macromolecular complex is prepared on a microscope grid. The specimen is then rapidly frozen by plunging it into a coolant. A special sample holder is used to keep this hydrated specimen at –160°C in the vacuum of the microscope, where it can be viewed directly without fixation, staining, or drying. Unlike negative staining, in which what is seen is the envelope of stain exclusion around the particle, hydrated cryoelectron microscopy produces an image from the macromolecular structure itself. However, to extract the maximum amount of structural information, special image-processing techniques must be used, as we describe next.

Multiple Images Can Be Combined to Increase Resolution

Any image, whether produced by an electron microscope or by an optical microscope, is made by particles—electrons or photons—striking a detector of some sort. But these particles are governed by quantum mechanics, so the numbers reaching the detector are predictable only in a statistical sense. In the limit of very large numbers of particles, the distribution at the detector is accurately determined by the imaged specimen. However, with smaller numbers of particles, this underlying structure in the image is obscured by the statistical fluctuations in the numbers of particles detected in each region. Random variability that confuses the underlying image of the specimen itself is referred to as noise.

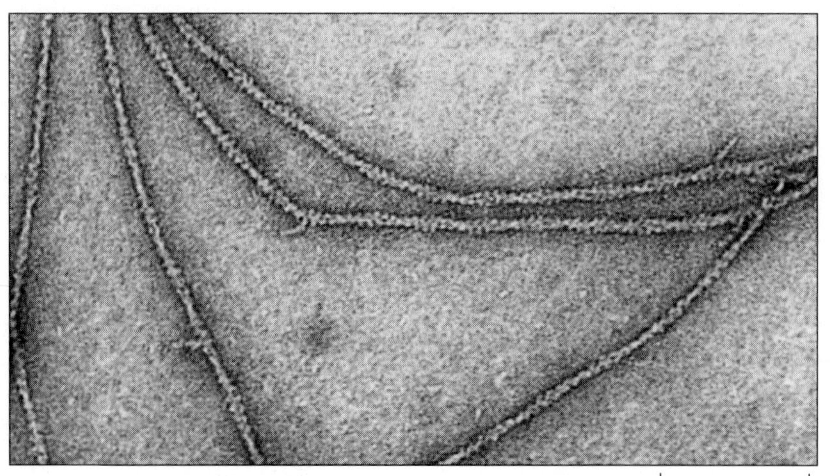

Figure 9–35 Negatively stained actin filaments. In this transmission electron micrograph, each filament is about 8 nm in diameter and is seen, on close inspection, to be composed of a helical chain of globular actin molecules. (Courtesy of Roger Craig.)

Noise is a particularly severe problem for electron microscopy of unstained macromolecules, but it is also important in light microscopy at low light levels. A protein molecule can tolerate a dose of only a few tens of electrons per square nanometer without damage, and this dose is orders of magnitude below what is needed to define an image at atomic resolution.

The solution is to obtain images of many identical molecules—perhaps tens of thousands of individual images—and combine them to produce an averaged image, revealing structural details that were hidden by the noise in the original images. Before the individual images can be combined, however, they must be aligned with each other. Sometimes it is possible to induce proteins and complexes to form crystalline arrays, in which each molecule is held in the same orientation in a regular lattice. In this case, the alignment problem is easily solved, and several protein structures have been determined at atomic resolution by this type of electron crystallography. In principle, however, crystalline arrays are not absolutely required. With the help of a computer, the images of randomly distributed molecules can be processed and combined to yield high-resolution reconstructions, as we now explain.

Views from Different Directions Can Be Combined to Give Three-dimensional Reconstructions

The detectors used to record images from electron microscopes produce two-dimensional pictures. Because of the large depth of field of the microscope, all the parts of the three-dimensional specimen are in focus, and the resulting image is a projection of the structure along the viewing direction. The lost information in the third dimension can be recovered if we have views of the same specimen from many different directions. The computational methods for this technique were worked out in the 1960s, and they are widely used in medical computed tomography (CT) scans. In a CT scan, the imaging equipment is moved around the patient to generate the different views. In **electron-microscope (EM) tomography**, the specimen holder is tilted in the microscope, which achieves the same result. In this way, one can arrive at a three-dimensional reconstruction, in a chosen standard orientation, by combining a set of views of many identical molecules in the microscope's field of view. Each view will be individually very noisy, but by combining them in three dimensions and taking an average, the noise can be largely eliminated, yielding a clear view of the molecular structure.

EM tomography is now widely applied for determining both molecular structures, using either crystalline or noncrystalline specimens, and larger objects such as thin sections of cells and organelles. It is a particularly successful technique for structures that have some intrinsic symmetry, such as helical or icosahedral viruses, because it makes the task of alignment easier and more accurate. Figure 9–36 shows the structure of an icosahedral virus that has been

Figure 9–36 EM tomography. Spherical protein shells of the hepatitis B virus are preserved in a thin film of ice (A) and imaged in the transmission electron microscope. Thousands of individual particles were combined by EM tomography to produce the three-dimensional map of the icosahedral particle shown in (B). The two views of a single protein dimer (C), that forms the spikes on the surface of the shell, show that the resolution of the reconstruction (7.4 Å) is sufficient to resolve the complete fold of the polypeptide chain. (A, courtesy of B. Böttcher, S.A. Wynne and R.A. Crowther; B and C, from B. Böttcher, S.A. Wynne and R.A. Crowther, *Nature* 386:88–91, 1997. © Macmillan Magazines Ltd.)

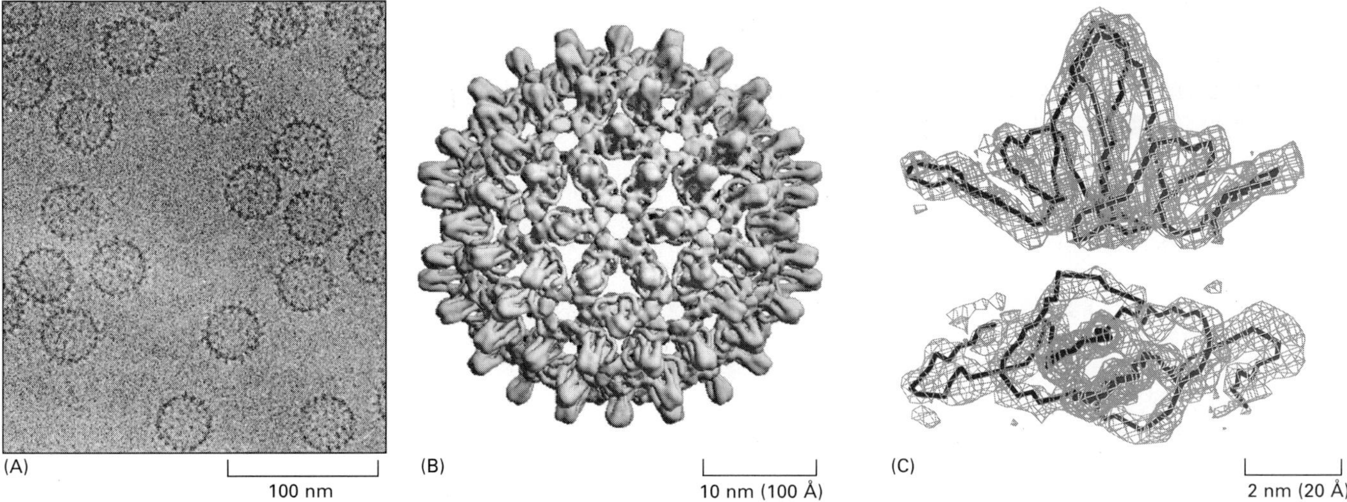

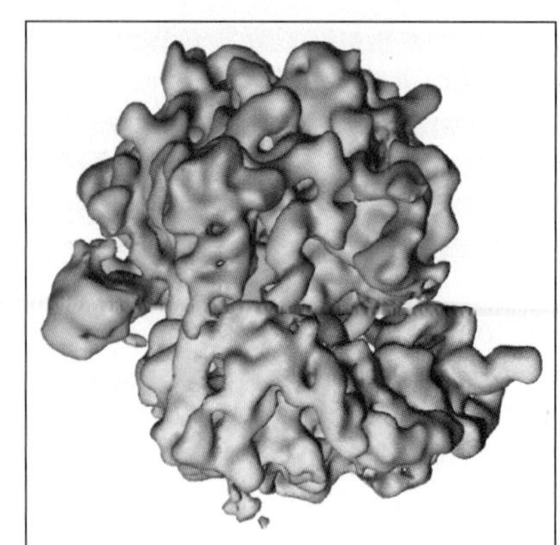

Figure 9–37 The three-dimensional structure of the 70S ribosome from *E. coli* determined by EM tomography. The small subunit is colored *yellow*, the large subunit *blue*. The overall resolution is 11.2 Å. (From I.S. Gabashvili et al., *Cell* 100: 537–549, 2000. © Elsevier.)

determined at high resolution by the combination of many particles and multiple views, and Figure 9–37 shows the structure of a ribosome determined in the same way.

With crystalline arrays, a resolution of 0.3 nm has been achieved by electron microscopy—enough to begin to see the internal atomic arrangements in a protein and to rival x-ray crystallography in resolution. With single-particle reconstruction, the limit at the moment is about 0.8 nm, enough to identify protein subunits and domains, and limited protein secondary structure. Although electron microscopy is unlikely to supersede x-ray crystallography (discussed in Chapter 8) as a method for macromolecular structure determination, it has some very clear advantages. First, it does not absolutely require crystalline specimens. Second, it can deal with extremely large complexes—structures that may be too large or too variable to crystallize satisfactorily. Electron microscopy provides a bridge between the scale of the single molecule and that of the whole cell.

Summary

Many light-microscope techniques are available for observing cells. Cells that have been fixed and stained can be studied in a conventional light microscope, while antibodies coupled to fluorescent dyes can be used to locate specific molecules in cells in a fluorescence microscope. Living cells can be seen with phase-contrast, differential-interference-contrast, dark-field, or bright-field microscopes. All forms of light microscopy are facilitated by electronic image-processing techniques, which enhance sensitivity and refine the image. Confocal microscopy and image deconvolution both provide thin optical sections and can be used to reconstruct three-dimensional images.

Determining the detailed structure of the membranes and organelles in cells requires the higher resolution attainable in a transmission electron microscope. Specific macromolecules can be localized with colloidal gold linked to antibodies. Three-dimensional views of the surfaces of cells and tissues are obtained by scanning electron microscopy. The shapes of isolated macromolecules that have been shadowed with a heavy metal or outlined by negative staining can also be readily determined by electron microscopy. Using computational methods, multiple images and views from different directions are combined to produce detailed reconstructions of macromolecules and molecular complexes through a technique known as electron-microscope tomography.

VISUALIZING MOLECULES IN LIVING CELLS

Even the most stable cellular structures must be assembled, disassembled, and reorganized during the cell's life cycle. Other structures, often enormous on the molecular scale, rapidly change, move, and reorganize themselves as the cell conducts its internal affairs and responds to its environment. Complex, highly organized pieces of molecular machinery move components around the cell, controlling traffic into and out of the nucleus, from one organelle to another, and into and out of the cell itself.

In this section we describe some of the methods that are used to study these dynamic processes in living cells. Most current methods use optical microscopy. All imaging requires the use of some form of radiation, and light is one of the least destructive types of radiation for living systems. Various techniques have been developed to make specific components of living cells visible in the microscope. Most of these methods are based on the use of fluorescent tags and indicators. The molecules that can be specifically imaged in this way range from small inorganic ions, such as Ca^{2+} or H^+, to large macromolecules, such as specific proteins, RNAs, or DNA sequences. Optical microscopy is not, however, the only possible approach to the problem, nor are microscopes the only equipment required.

Rapidly Changing Intracellular Ion Concentrations Can Be Measured with Light-emitting Indicators

One way to study the chemistry of a single living cell is to insert the tip of a fine, glass, ion-sensitive **microelectrode** directly into the cell interior through the plasma membrane. This technique is used to measure the intracellular concentrations of common inorganic ions, such as H^+, Na^+, K^+, Cl^- and Ca^{2+}. However, ion-sensitive microelectrodes reveal the ion concentration only at one point in a cell, and for an ion present at a very low concentration, such as Ca^{2+}, their responses are slow and somewhat erratic. Thus, these microelectrodes are not ideally suited to record the rapid and transient changes in the concentration of cytosolic Ca^{2+} that have an important role in allowing cells to respond to extracellular signals. Such changes can be analyzed with the use of **ion-sensitive indicators**, whose light emission reflects the local concentration of the ion. Some of these indicators are luminescent (emitting light spontaneously), while others are fluorescent (emitting light on exposure to light).

Aequorin is a luminescent protein isolated from a marine jellyfish; it emits light in the presence of Ca^{2+} and responds to changes in Ca^{2+} concentration in the range of 0.5–10 μM. If microinjected into an egg, for example, aequorin emits a flash of light in response to the sudden localized release of free Ca^{2+} into the cytoplasm that occurs when the egg is fertilized (Figure 9–38). Aequorin has also been expressed transgenically in plants and other organisms to provide a method of monitoring Ca^{2+} in all their cells without the need for microinjection, which can be a difficult procedure.

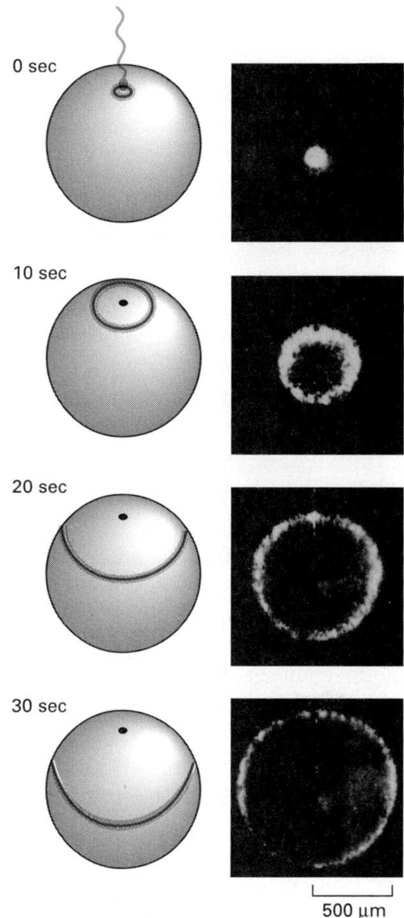

Figure 9–38 Aequorin, a luminescent protein. The luminescent protein aequorin emits light in the presence of free Ca^{2+}. Here, an egg of the medaka fish has been injected with aequorin, which has diffused throughout the cytosol, and the egg has then been fertilized with a sperm and examined with the help of an image intensifier. The four photographs were taken looking down on the site of sperm entry at intervals of 10 seconds and reveal a wave of release of free Ca^{2+} into the cytosol from internal stores just beneath the plasma membrane. This wave sweeps across the egg starting from the site of sperm entry, as indicated in the diagrams on the *left*. (Photographs reproduced from J.C. Gilkey, L.F. Jaffe, E.B. Ridgway, and G.T. Reynolds, *J. Cell Biol.* 76:448–466, 1978. © The Rockefeller University Press.)

Bioluminescent molecules like aequorin emit tiny amounts of light—at best, a few photons per indicator molecule—that are difficult to measure. Fluorescent indicators produce orders of magnitude more photons per molecule, they are therefore easier to measure and can give better spatial resolution. Fluorescent Ca^{2+} indicators have been synthesized that bind Ca^{2+} tightly and are excited or emit at slightly different wavelengths when they are free of Ca^{2+} than when they are in their Ca^{2+}-bound form. By measuring the ratio of fluorescence intensity at two excitation or emission wavelengths, the concentration ratio of the Ca^{2+}-bound indicator to the Ca^{2+}-free indicator can be determined, thereby providing an accurate measurement of the free Ca^{2+} concentration. Indicators of this type are widely used for second-by-second monitoring of changes in intracellular Ca^{2+} concentrations in the different parts of a cell viewed in a fluorescence microscope (Figure 9–39).

Similar fluorescent indicators are available for measuring other ions; some are used for measuring H^+, for example, and hence intracellular pH. Some of these indicators can enter cells by diffusion and thus need not be microinjected; this makes it possible to monitor large numbers of individual cells simultaneously in a fluorescence microscope. New types of indicators, used in conjunction with modern image-processing methods, are leading to similarly rapid and precise methods for analyzing changes in the concentrations of many types of small molecules in cells.

There Are Several Ways of Introducing Membrane-impermeant Molecules into Cells

It is often useful to be able to introduce membrane-impermeant molecules into a living cell, whether they are antibodies that recognize intracellular proteins, normal cell proteins tagged with a fluorescent label, or molecules that influence cell behavior. One approach is to microinject the molecules into the cell through a glass micropipette. An especially useful technique is called **fluorescent analog cytochemistry**, in which a purified protein is coupled to a fluorescent dye and

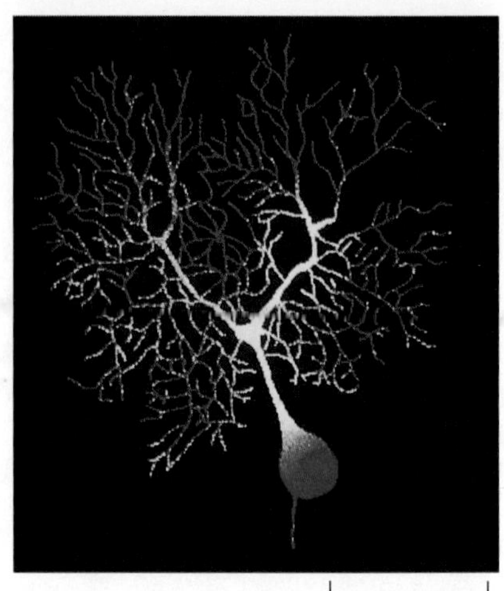

Figure 9–39 Visualizing intracellular Ca^{2+} concentrations by using a fluorescent indicator. The branching tree of dendrites of a Purkinje cell in the cerebellum receives more than 100,000 synapses from other neurons. The output from the cell is conveyed along the single axon seen leaving the cell body at the bottom of the picture. This image of the intracellular Ca^{2+} concentration in a single Purkinje cell (from the brain of a guinea pig) was taken with a low-light camera and the Ca^{2+}-sensitive fluorescent indictor fura-2. The concentration of free Ca^{2+} is represented by different colors, *red* being the highest and *blue* the lowest. The highest Ca^{2+} levels are present in the thousands of dendritic branches. (Courtesy of D.W. Tank, J.A. Connor, M. Sugimori, and R.R. Llinas.)

microinjected into a cell. In this way, the fate of the injected protein can be followed in a fluorescence microscope as the cell grows and divides. If tubulin (the subunit of microtubules) is labeled with a dye that fluoresces red, for example, microtubule dynamics can be followed second by second in a living cell (see Figures 18–21 and 16–12).

Antibodies can be microinjected into a cell to block the function of the molecule that the antibodies recognize. Anti-myosin-II antibodies injected into a fertilized sea urchin egg, for example, prevent the egg cell from dividing in two, even though nuclear division occurs normally. This observation demonstrates that this myosin has an essential role in the contractile process that divides the cytoplasm during cell division, but that it is not required for nuclear division.

Microinjection, although widely used, demands that each cell be injected individually; therefore, it is possible to study at most only a few hundred cells at a time. Other approaches allow large populations of cells to be permeabilized simultaneously. One can partly disrupt the structure of the cell plasma membrane, for example, to make it more permeable; this is usually accomplished by using a powerful electric shock or a chemical such as a low concentration of detergent. The electrical technique has the advantage of creating large pores in the plasma membrane without damaging intracellular membranes. The pores remain open for minutes or hours, depending on the cell type and the size of the electric shock, and allow even macromolecules to enter (and leave) the cytosol rapidly. This process of *electroporation* is valuable also in molecular genetics, as a way of introducing DNA molecules into cells. With a limited treatment, a large fraction of the cells repair their plasma membrane and survive.

A third method for introducing large molecules into cells is to cause membranous vesicles that contain these molecules to fuse with the cell's plasma membrane. To introduce new genes into the nucleus, gold particles coated with DNA can be shot into cells at high velocity. These methods are used widely in cell biology and are illustrated in Figure 9–40.

The Light-induced Activation of "Caged" Precursor Molecules Facilitates Studies of Intracellular Dynamics

The complexity and rapidity of many intracellular processes, such as the actions of signaling molecules or the movements of cytoskeletal proteins, make them difficult to study at a single-cell level. Ideally, one would like to be able to introduce any molecule of interest into a living cell at a precise time and location and follow its subsequent behavior, as well as the response of the cell. Microinjection is limited by the difficulty of controlling the place and time of delivery. A more powerful approach involves synthesizing an inactive form of the molecule of interest, introducing it into the cell and then activating it suddenly at a chosen site in the cell by focusing a spot of light on it. Inactive photosensitive precursors of this type, called **caged molecules**, have been made for a variety of small molecules, including Ca^{2+}, cyclic AMP, GTP, and inositol trisphosphate. The caged molecules can be introduced into living cells by any of the methods described in Figure 9–40 and then activated by a strong pulse of light from a laser (Figure 9–41). A microscope can be used to focus the light pulse on any tiny

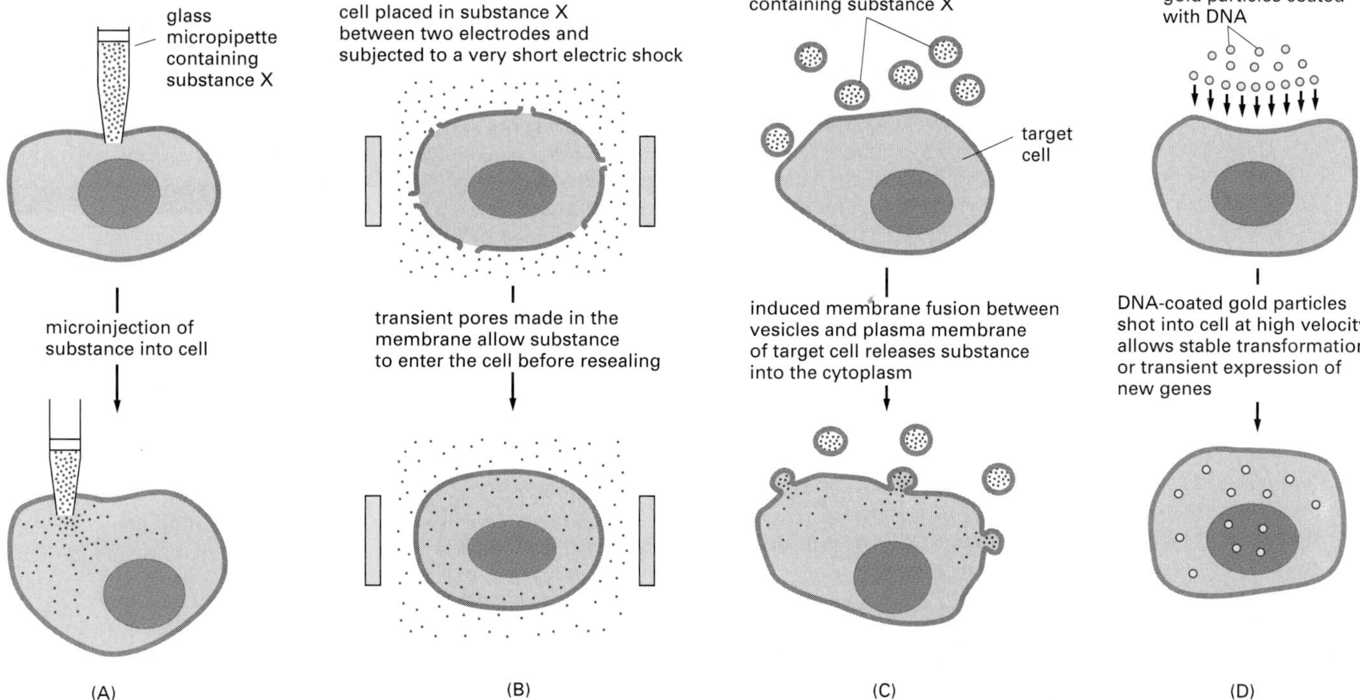

Figure 9–40 Methods of introducing a membrane-impermeant substance into a cell. (A) The substance is injected through a micropipette, either by applying pressure or, if the substance is electrically charged, by applying a voltage that drives the substance into the cell as an ionic current (a technique called *iontophoresis*). (B) The cell membrane is made transiently permeable to the substance by disrupting the membrane structure with a brief but intense electric shock (2000 V/cm for 200 μsec, for example). (C) Membrane-enclosed vesicles are loaded with the desired substance and then induced to fuse with the target cells. (D) Gold particles coated with DNA are used to introduce a novel gene into the nucleus.

region of the cell, so that the experimenter can control exactly where and when a molecule is delivered. In this way, for example, one can study the instantaneous effects of releasing an intracellular signaling molecule into the cytosol.

Fluorescent molecules are also valuable tools when caged. They are made by attaching a photoactivatable fluorescent dye to a purified protein. It is important that the modified protein remain biologically active: unlike labeling with radioisotopes (which changes only the number of neutrons in the nuclei of the labeled atoms), labeling with a caged fluorescent dye adds a large bulky group to the surface of a protein, which can easily change the protein's properties. A satisfactory labeling protocol is usually found by trial and error. Once a biologically active labeled protein has been produced, its behavior can be followed inside living cells. Tubulin labeled with caged fluorescein, for example, can be incorporated into microtubules of the mitotic spindle; when a small region of the spindle is illuminated with a laser, the labeled tubulin becomes fluorescent, so that its movement along the spindle microtubules can be readily followed (Figure 9–42). In principle, the same technique can be applied to any protein.

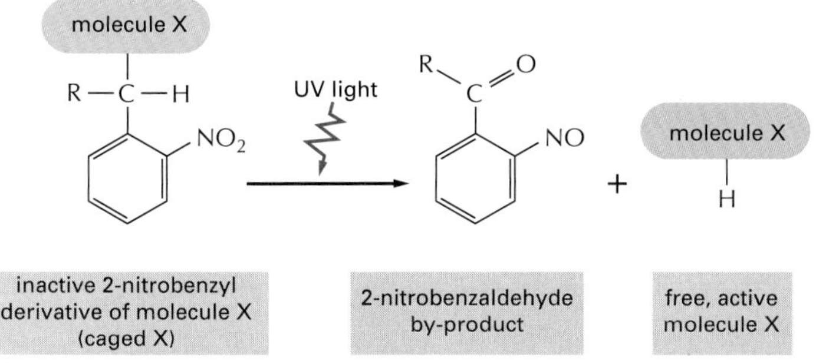

Figure 9–41 Caged molecules. This diagram shows how a light-sensitive caged derivative of a molecule (designated X) can be converted by a flash of UV light to its free, active form. Small molecules such as ATP can be caged in this way. Even ions like Ca^{2+} can be indirectly caged; in this case a Ca^{2+}-binding chelator is used, which is inactivated by photolysis, thus releasing its Ca^{2+}.

Figure 9–42 Determining microtubule flux in the mitotic spindle with caged fluorescein linked to tubulin. (A) A metaphase spindle formed *in vitro* from an extract of *Xenopus* eggs has incorporated three fluorescent markers: rhodamine-labeled tubulin (*red*) to mark all the microtubules, a *blue* DNA-binding dye that labels the chromosomes, and caged-fluorescein-labeled tubulin, which is also incorporated into all the microtubules but is invisible because it is nonfluorescent until activated by ultraviolet light. (B) A beam of UV light is used to uncage the caged-fluorescein-labeled tubulin locally, mainly just to the left side of the metaphase plate. Over the next few minutes (after 1.5 minutes in C, after 2.5 minutes in D), the uncaged fluorescein-tubulin signal is seen to move toward the left spindle pole, indicating that tubulin is continuously moving poleward even though the spindle (visualized by the *red* rhodamine-labeled tubulin fluorescence) remains largely unchanged. (From K.E. Sawin and T.J. Mitchison, *J. Cell Biol.* 112:941–954, 1991. © The Rockefeller University Press.)

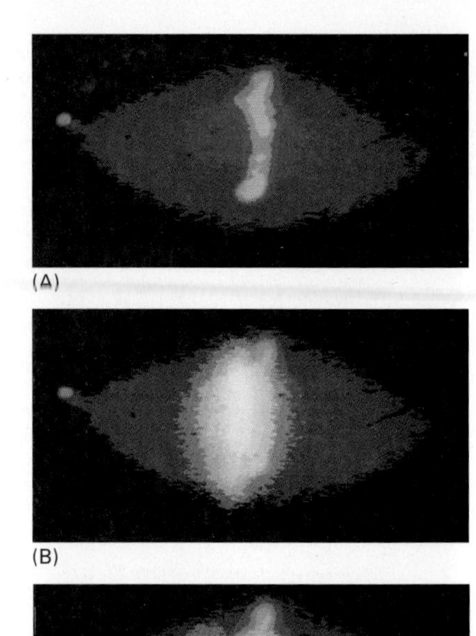

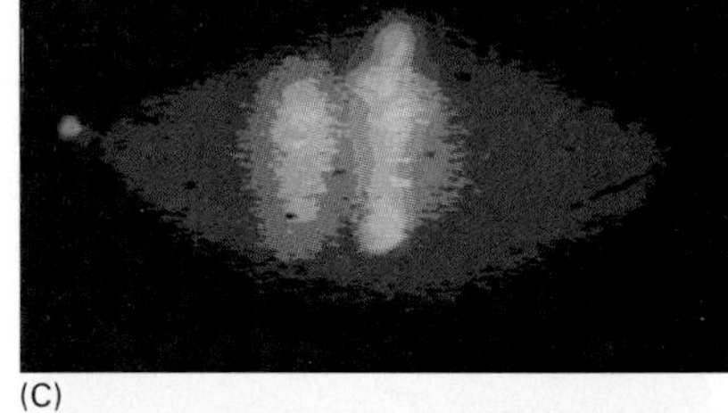

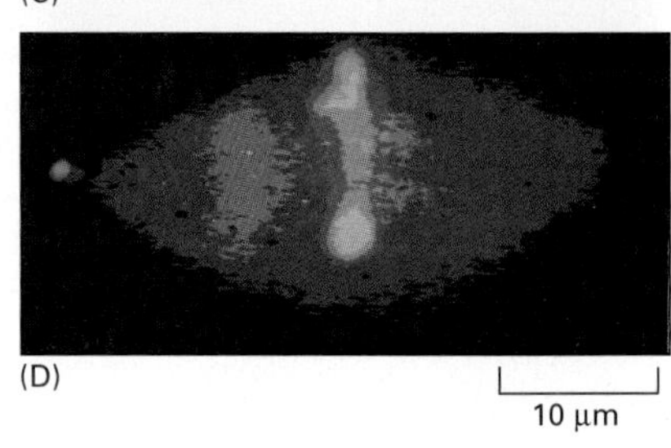

Green Fluorescent Protein Can Be Used to Tag Individual Proteins in Living Cells and Organisms

All of the fluorescent molecules discussed above have to be made outside the cell and then artificially introduced into it. A wealth of new opportunities has been opened up by the discovery of genes coding for protein molecules that are themselves inherently fluorescent. Genetic engineering then enables the creation of lines of cells that make their own visible tags and labels, without further interference. These cellular exhibitionists display their inner workings in glowing fluorescent color.

Foremost among the fluorescent proteins used for these purposes by cell biologists is the **green fluorescent protein (GFP)**, isolated (like aequorin) from the jellyfish *Aequoria victoria*. This protein is encoded in the normal way by a single gene that can be cloned and introduced into cells of other species. The freshly translated protein is not fluorescent, but within an hour or so (less for some alleles of the gene, more for others) it undergoes a self-catalyzed post-translational modification to generate an efficient and bright fluorescent center, shielded within the interior of a barrel-like protein (Figure 9–43). Extensive site-directed mutagenesis has been performed on the original gene sequence to obtain useful fluorescence in a wide range of organisms ranging from animals and plants to fungi and microbes. The fluorescence efficiency has also been improved, and variants have been generated with altered absorption and emission spectra in the blue–green–yellow range. Recently a family of related fluorescent proteins has been discovered in corals, thereby extending the range into the red region of the spectrum.

One of the simplest uses of GFP is as a reporter molecule to monitor gene expression. A transgenic organism can be made with the GFP-coding sequence placed under the transcriptional control of the promoter belonging to a gene of interest; this then gives a directly visible display of the gene's expression pattern in the living organism. In another application, a peptide location signal can be added to the GFP to direct it to a particular cellular compartment, such as the endoplasmic reticulum or a mitochondrion, lighting up these organelles so they can be observed in the living state.

The GFP DNA-coding sequence can also be inserted at the beginning or end of the gene for another protein, yielding a chimeric product consisting of that protein with a GFP domain attached. In many cases, this GFP-fusion protein behaves in the same way as the original protein, directly revealing its location and activities (Figure 9–44). It is often possible to prove that the GFP-fusion protein is functionally equivalent to the untagged protein, for example by using it to

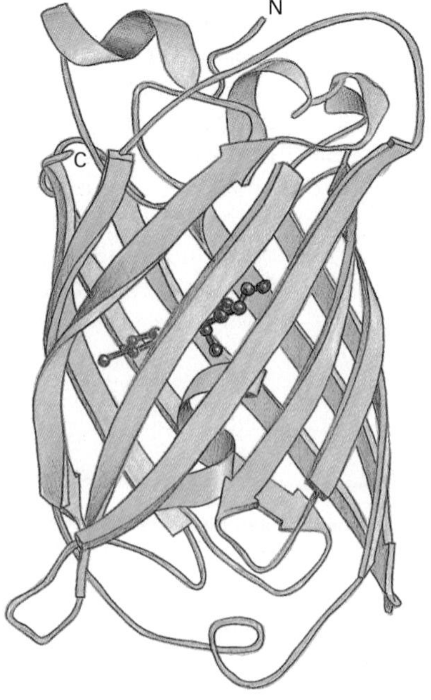

Figure 9–43 Green fluorescent protein (GFP). The structure of GFP, shown here schematically, highlights the eleven β strands that form the staves of a barrel. Buried within the barrel is the active chromaphore *(dark green)* that is formed post-translationally from the protruding side chains of three amino acid residues. (Adapted from M. Ormö et al., *Science* 273:1392–1395, 1995.)

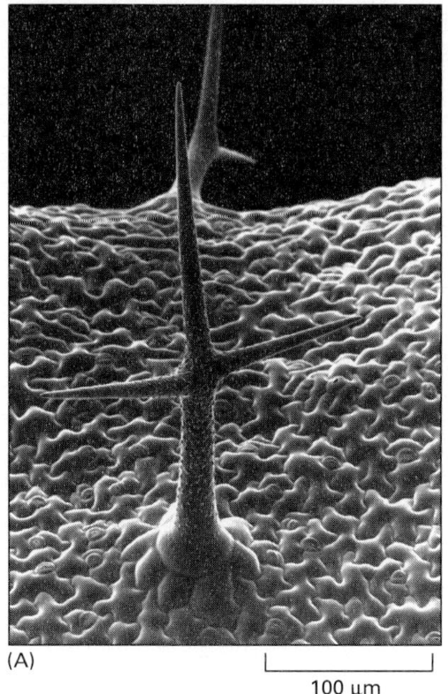

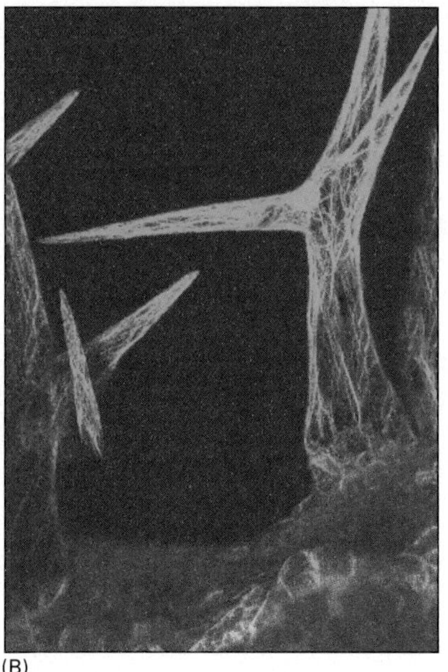

Figure 9–44 GFP tagging. (A) The upper surface of the leaves of *Arabidopsis* plants are covered with huge branched single-cell hairs that rise up from the surface of the epidermis. These hairs, or trichomes, can be imaged in the SEM. (B) If an *Arabidopsis* plant is transformed with a DNA sequence coding for talin (an actin-binding protein), fused to a DNA sequence coding for GFP, the fluorescent talin protein produced binds to actin filaments in all the living cells of the transgenic plant. Confocal microscopy can reveal the dynamics of the entire actin cytoskeleton of the trichome *(green)*. The red fluorescence arises from chlorophyl in cells within the leaf below the epidermis. (A, courtesy of Paul Linstead; B, courtesy of Jaideep Mathur.)

rescue a mutant lacking that protein. GFP tagging is the clearest and most unequivocal way of showing the distribution and dynamics of a protein in a living organism (Figure 9–45).

The uses to which GFP and its relatives can be put are multiplying rapidly. For example, DNA or RNA molecules can be marked by incorporating in them copies of an oligonucleotide sequence that is known to bind a specific protein, and expressing in the same cell a GFP-tagged version of that binding protein. Another use is in monitoring interactions between one protein and another by *fluorescence resonance energy transfer (FRET)* (see Figure 8–49). In this technique, the two molecules of interest are each labeled with a different fluorochrome, chosen so that the emission spectrum of one fluorochrome overlaps with the absorption spectrum of the other. If the two proteins bind so as to bring their fluorochromes into very close proximity (closer than about 2 nm), the energy of the absorbed light can be transferred directly from one fluorochrome to the other. Thus, when the complex is illuminated at the excitation wavelength of the first fluorochrome, light is emitted at the emission wavelength of the second. This method can be used with two different spectral variants of GFP as fluorochromes to monitor processes such as the interaction of signaling molecules with their receptors.

Light Can Be Used to Manipulate Microscopic Objects as Well as to Image Them

Photons carry a small amount of momentum. This means that an object that absorbs or deflects a beam of light experiences a small force. With ordinary light sources, this radiation pressure is too small to be significant. But it is important on a cosmic scale (helping prevent gravitational collapse inside stars), and, more modestly, in the cell biology lab, where an intense focused laser beam can exert large enough forces to push small objects around inside a cell. If the laser beam

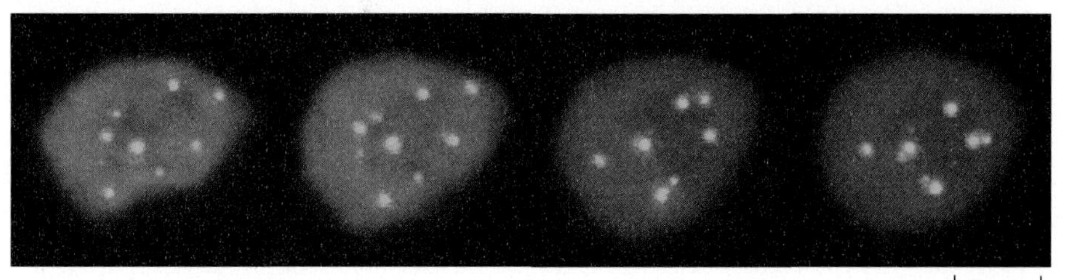

Figure 9–45 Dynamics of GFP tagging. This sequence of micrographs shows a set of three-dimensional images of a living nucleus taken over the course of an hour. Tobacco cells have been stably transformed with GFP fused to a spliceosomal protein that is concentrated in small nuclear bodies called Cajal Bodies (see Figure 6–48). The fluorescent Cajal bodies, easily visible in a living cell with confocal microscopy, are dynamic structures that move around within the nucleus. (Courtesy of Kurt Boudonck, Liam Dolan and Peter Shaw.)

is focused on an object having a higher refractive index than its surroundings, the beam is refracted, causing very large numbers of photons to change direction. The pattern of photon deflection holds the object at the focus of the beam; if it begins to drift away from this position, it is pushed back by radiation pressure acting more strongly on one side than the other. Thus, by steering a focused laser beam, usually an infrared laser, which is minimally absorbed by the cellular constituents, one can create "optical forceps" to move subcellular objects like organelles and chromosomes around. This method has been used to measure the force exerted by single actin–myosin molecules, by single microtubule motors, and by RNA polymerase.

Intense focused laser beams that are more strongly absorbed by biological material can also be used more straightforwardly as optical knives—to kill individual cells, to cut or burn holes in them, or to detach one intracellular component from another. In these ways, optical devices can provide a basic toolkit for cellular microsurgery.

Molecules Can Be Labeled with Radioisotopes

As we have seen, in cell biology it is often important to determine the quantities of specific molecules and to know where they are in the cell and how their level or location changes in response to extracellular signals. The molecules of interest range from small inorganic ions, such as Ca^{2+} or H^+, to large macromolecules, such as specific proteins, RNAs, or DNA sequences. We have so far described how sensitive fluorescence methods can be used for assaying these types of molecules, as well as for following the dynamic behavior of many of them in living cells. In ending this chapter, we describe how radioisotopes are used to trace the path of specific molecules through the cell.

Most naturally occurring elements are a mixture of slightly different *isotopes*. These differ from one another in the mass of their atomic nuclei, but because they have the same number of protons and electrons, they have the same chemical properties. In radioactive isotopes, or radioisotopes, the nucleus is unstable and undergoes random disintegration to produce a different atom. In the course of these disintegrations, either energetic subatomic particles, such as electrons, or radiations, such as gamma-rays, are given off. By using chemical synthesis to incorporate one or more radioactive atoms into a small molecule of interest, such as a sugar or an amino acid, the fate of that molecule can be traced during any biological reaction.

Although naturally occurring radioisotopes are rare (because of their instability), they can be produced in large amounts in nuclear reactors, where stable atoms are bombarded with high-energy particles. As a result, radioisotopes of many biologically important elements are readily available (Table 9–3). The radiation they emit is detected in various ways. Electrons (beta particles) can be detected in a *Geiger counter* by the ionization they produce in a gas, or they can be measured in a *scintillation counter* by the small flashes of light they induce in a scintillation fluid. These methods make it possible to measure accurately the quantity of a particular radioisotope present in a biological specimen. Using

TABLE 9–3 Some Radioisotopes in Common Use in Biological Research

ISOTOPE	HALF-LIFE
^{32}P	14 days
^{131}I	8.1 days
^{35}S	87 days
^{14}C	5570 years
^{45}Ca	164 days
^{3}H	12.3 years

The isotopes are arranged in decreasing order of the energy of the β radiation (electrons) they emit. ^{131}I also emits γ radiation. The half-life is the time required for 50% of the atoms of an isotope to disintegrate.

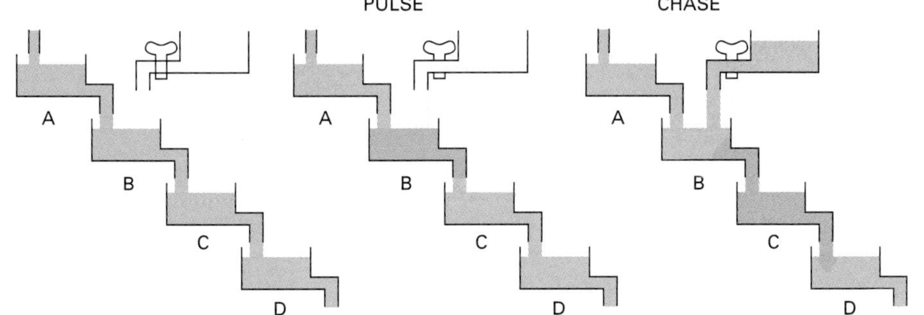

Figure 9–46 The logic of a typical pulse-chase experiment using radioisotopes. The chambers labeled A, B, C, and D represent either different compartments in the cell (detected by autoradiography or by cell-fractionation experiments) or different chemical compounds (detected by chromatography or other chemical methods). The results of a real pulse-chase experiment can be seen in Figure 9–47.

light or electron microscopy, it is also possible to determine the location of a radioisotope in a specimen by *autoradiography*, as we describe below. All of these methods of detection are extremely sensitive: in favorable circumstances, nearly every disintegration—and therefore every radioactive atom that decays—can be detected.

Radioisotopes Are Used to Trace Molecules in Cells and Organisms

One of the earliest uses of radioactivity in biology was to trace the chemical pathway of carbon during photosynthesis. Unicellular green algae were maintained in an atmosphere containing radioactively labeled CO_2 ($^{14}CO_2$), and at various times after they had been exposed to sunlight, their soluble contents were separated by paper chromatography. Small molecules containing ^{14}C atoms derived from CO_2 were detected by a sheet of photographic film placed over the dried paper chromatogram. In this way most of the principal components in the photosynthetic pathway from CO_2 to sugar were identified.

Radioactive molecules can be used to follow the course of almost any process in cells. In a typical experiment the cells are supplied with a precursor molecule in radioactive form. The radioactive molecules mix with the preexisting unlabeled ones; both are treated identically by the cell as they differ only in the weight of their atomic nuclei. Changes in the location or chemical form of the radioactive molecules can be followed as a function of time. The resolution of such experiments is often sharpened by using a **pulse-chase** labeling protocol, in which the radioactive material (the *pulse*) is added for only a very brief period and then washed away and replaced by nonradioactive molecules (the *chase*). Samples are taken at regular intervals, and the chemical form or location of the radioactivity is identified for each sample (Figure 9–46). Pulse-chase experiments, combined with autoradiography, have been important, for example, in elucidating the pathway taken by secreted proteins from the ER to the cell exterior (Figure 9–47).

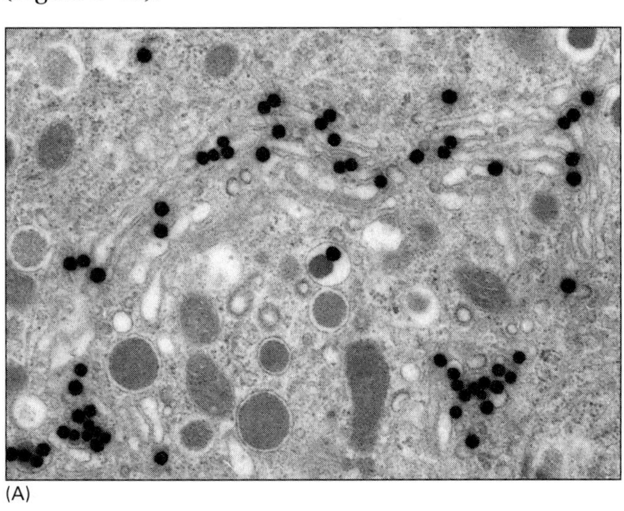

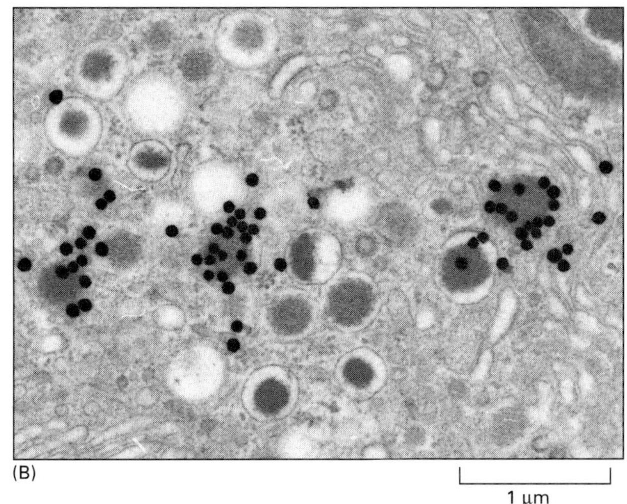

Figure 9–47 Electron-microscopic autoradiography. The results of a pulse-chase experiment in which pancreatic B cells were fed with ^3H-leucine for 5 minutes (the pulse) followed by excess unlabeled leucine (the chase). The amino acid is largely incorporated into insulin, which is destined for secretion. After a 10-minute chase the labeled protein has moved from the rough ER to the Golgi stacks (A), where its position is revealed by the black silver grains in the photographic emulsion. After a further 45-minute chase the labeled protein is found in electron-dense secretory granules (B). The small round silver grains seen here are produced by using a special photographic developer and should not be confused with the similar-looking black dots seen with immunogold labeling methods (e.g., Figure 9–26). Experiments similar to this were important in establishing the intracellular pathway taken by newly synthesized secretory proteins. (Courtesy of L. Orci, from *Diabetes* 31:538–565, 1982. © American Diabetes Association, Inc.)

Radioisotopic labeling is a uniquely valuable way of distinguishing between molecules that are chemically identical but have different histories—for example, those that differ in their time of synthesis. In this way, for example, it was shown that almost all of the molecules in a living cell are continually being degraded and replaced, even when the cell is not growing and is apparently in a steady state. This "turnover," which sometimes takes place very slowly, would be almost impossible to detect without radioisotopes.

Today, nearly all common small molecules are available in radioactive form from commercial sources, and virtually any biological molecule, no matter how complicated, can be radioactively labeled. Compounds can be made with radioactive atoms incorporated at particular positions in their structure, enabling the separate fates of different parts of the same molecule to be followed during biological reactions (Figure 9–48).

As mentioned previously, one of the important uses of radioactivity in cell biology is to localize a radioactive compound in sections of whole cells or tissues by **autoradiography**. In this procedure, living cells are briefly exposed to a pulse of a specific radioactive compound and then incubated for a variable period—to allow them time to incorporate the compound—before being fixed and processed for light or electron microscopy. Each preparation is then overlaid with a thin film of photographic emulsion and left in the dark for several days, during which the radioisotope decays. The emulsion is then developed, and the position of the radioactivity in each cell is indicated by the position of the developed silver grains (see Figure 5–33). If cells are exposed to ^{3}H-thymidine, a radioactive precursor of DNA, for example, it can be shown that DNA is made in the nucleus and remains there. By contrast, if cells are exposed to ^{3}H-uridine, a radioactive precursor of RNA, it is found that RNA is initially made in the nucleus and then moves rapidly into the cytoplasm. Radiolabeled molecules can also be

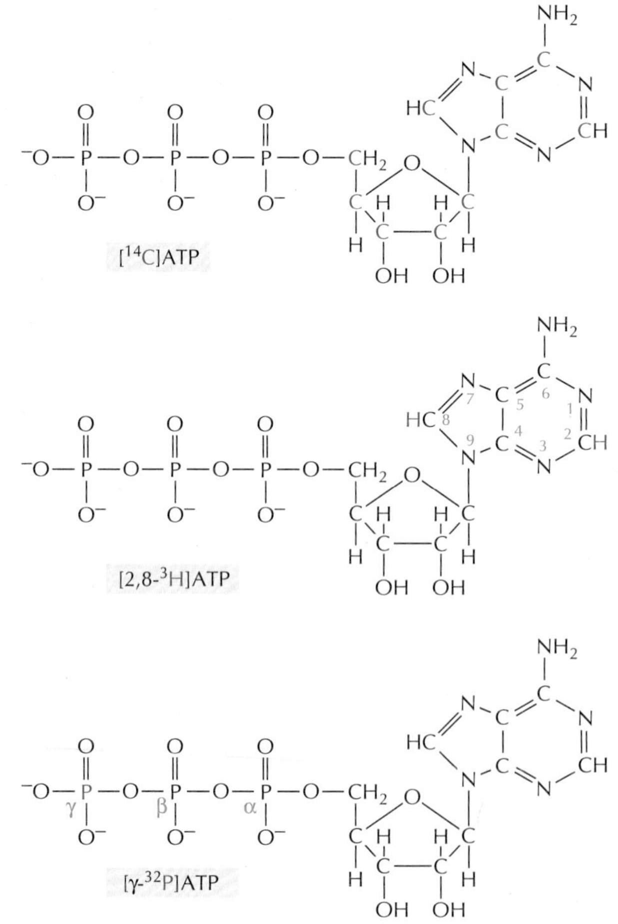

[^{14}C]ATP

[2,8-^{3}H]ATP

[γ-^{32}P]ATP

Figure 9–48 Radioisotopically labeled molecules. Three commercially available radioactive forms of ATP, with the radioactive atoms shown in *red*. The nomenclature used to identify the position and type of the radioactive atoms is also shown.

detected by autoradiography after they are separated from other molecules by gel electrophoresis: the positions of both proteins (see Figure 8–17) and nucleic acids (see Figure 8–23) are commonly detected on gels in this way.

Summary

Techniques are now available for detecting, measuring, and following almost any desired molecule in a living cell. For example, fluorescent indicator dyes can be introduced to measure the concentrations of specific ions in individual cells or in different parts of a cell. The dynamic behavior of many molecules can be followed in a living cell by constructing an inactive "caged" precursor, which can be introduced into a cell and then instantaneously activated in a selected region of the cell by a light-stimulated reaction. Green fluorescent protein (GFP) is an especially versatile probe that can be attached to other proteins by genetic manipulation. Virtually any protein of interest can be genetically engineered as a GFP-fusion protein, and then imaged in living cells by fluorescence microscopy. Radioactive isotopes of various elements can also be used to follow the fate of specific molecules both biochemically and microscopically.

References

General

Celis JE (ed) (1998) Cell Biology: A Laboratory Handbook, 2nd edn. San Diego: Academic Press.

Lacey AJ (ed) (1999) Light Microscopy in Biology: A Practical Approach, 2nd edn. Oxford: Oxford University Press.

Paddock SW (ed) (1999) Methods in Molecular Biology, vol 122: Confocal Microscopy Methods and Protocols. Totowa, NJ: Humana Press.

Watt IM (1997) The Principles and Practice of Electron Microscopy, 2nd edn, Cambridge: Cambridge University Press.

Looking at the Structure of Cells in the Microscope

Agard DA & Sedat JW (1983) Three-dimensional architecture of a polytene nucleus. *Nature* 302, 676–681.

Agard DA, Hiraoka Y, Shaw P & Sedat JW (1989) Fluorescence microscopy in three dimensions. In Methods in Cell Biology, vol 30: Fluorescence Microscopy of Living Cells in Culture, part B (DL Taylor, Y-L Wang eds). San Diego: Academic Press.

Allen RD (1985) New observations on cell architecture and dynamics by video-enhanced contrast optical microscopy. *Annu. Rev. Biophys. Chem.* 14, 265–290.

Allen TD & Goldberg MW (1993) High resolution SEM in cell biology. *Trends Cell Biol.* 3, 203–208.

Boon ME & Driver JS (1986) Routine Cytological Staining Methods. London: Macmillan.

Böttcher B, Wynne SA & Crowther RA (1997) Determination of the fold of the core protein of hepatits B virus by electron cryomicroscopy. *Nature* 386, 88–91.

Celis JE (ed) (1998) Cell Biology: A Laboratory Handbook, 2nd edn, vol 3, part 1—Histology and Histochemistry, pp 219–248. San Diego: Academic Press.

Celis JE (ed) (1998) Cell Biology: A Laboratory Handbook, 2nd edn, vol 2, part 6—Immunocytochemistry, pp 457–494. San Diego: Academic Press.

Celis JE (ed) (1998) Cell Biology: A Laboratory Handbook, 2nd edn, vol 3, part 9—Electron Microscopy, pp 249–362. San Diego: Academic Press.

Dubochet J, Adrian M, Chang J-J et al. (1988) Cryoelectron microscopy of vitrified specimens. *Q. Rev. Biophys.* 21, 129–228.

Everhart TE & Hayes TL (1972) The scanning electron microscope. *Sci. Am.* 226, 54–69.

Fowler WE & Aebi U (1983) Preparation of single molecules and supramolecular complexes for high resolution metal shadowing. *J. Ultrastruct. Res.* 83, 319–334.

Gabashvili IS, Agrawal RK, Spahn CMT et al. (2000) Solution structure of the *E. coli* 70S ribosome at 11.5 Å resolution. *Cell* 100, 537–549.

Griffiths G (1993) Fine Structure Immunocytochemistry. Heidelberg: Springer-Verlag.

Harlow E & Lane D (1988) Antibodies. A Laboratory Manual. Cold Spring Harbor: Cold Spring Harbor Laboratory.

Haugland RP (ed) (1996) Handbook of Fluorescent Probes and Research Chemicals, 8th edn. Eugene, OR: Molecular Probes, Inc. (Available online at http://www.probes.com/)

Hayat MA (1978) Introduction to Biological Scanning Electron Microscopy. Baltimore: University Park Press.

Hayat MA (1981) Fixation for Electron Microscopy. New York: Academic Press.

Hayat MA (1989) Cytochemical Methods. New York: Wiley-Liss.

Hayat MA (2000) Principles and Techniques of Electron Microscopy, 4th edn. Cambridge: Cambridge University Press.

Hecht E (2002) Optics, 4th edn. Reading, MA: Addison-Wesley.

Heitlinger E, Peter M, Haner M et al. (1991) Expression of chicken lamin B2 in *Escherichia coli*: characterization of its structure, assembly and molecular interactions. *J. Cell Biol.* 113, 485–495.

Heuser J (1981) Quick-freeze, deep-etch preparation of samples for 3-D electron microscopy. *Trends Biochem. Sci.* 6, 64–68.

Hoenger A & Aebi U (1996) Three-dimensional reconstructions from ice-embedded and negatively stained macromolecular assemblies: a critical comparison. *J. Struct. Biol.* 117, 99–116.

Inoué S & Spring HR (1997) Video Microscopy, 2nd edn. New York: Plenum Press.

Kuhlbrandt W, Wang DN & Fujiyoshi Y (1994) Atomic model of plant light-harvesting complex by electron crystallography. *Nature* 367, 614–621.

Mancini EJ, Clarke M, Gowen BE, Rutten T & Fuller SD (2000) Cryo-electron microscopy reveals the functional organization of an enveloped virus, Semliki Forest virus. *Mol. Cell* 5, 255–266.

Maunsbach AB (1998) Immunolabelling and staining of ultrathin sections in biological electron microscopy. In Cell Biology: A Laboratory Handbook, vol 3 (JE Celis ed), pp 268–276. San Diego: Academic Press.

Minsky M (1988) Memoir on inventing the confocal scanning microscope. *Scanning* 10, 128–138.

Pawley JB (ed) (1995) Handbook of Biological Confocal Microscopy, 2nd edn. New York: Plenum Press.

Pease DC & Porter KR (1981) Electron microscopy and ultramicrotomy. *J. Cell Biol.* 91, 287s–292s.

Petran M, Hadravsky M, Egger MD & Galambos R (1968) Tandem-scanning reflected light microscope. *J. Opt. Soc. Am.* 58, 661–664.

Ploem JS & Tanke HJ (1987) Introduction to Fluorescence Microscopy. Royal Microscopical Society Microscopy Handbook, No 10. Oxford: Oxford University Press.

Shotton DM (1998) Freeze fracture and freeze etching. In Cell Biology: A Laboratory Handbook, vol 3 (JE Celis ed), pp 310–322. San Diego: Academic Press.

Shotton DM (ed) (1990) Electronic Light Microscopy: Techniques in Modern Biomedical Microscopy. New York: Wiley-Liss.

Slayter EM & Slayter HS (1992) Light and Electron Microscopy. Cambridge: Cambridge University Press.

Taylor DL & Wang Y-L (eds) (1989) Methods in Cell Biology, vols 29 & 30: Fluorescence Microscopy of Living Cells in Culture, parts A & B. San Diego: Academic Press.

Unwin PNT & Henderson R (1975) Molecular structure determination by electron microscopy of unstained crystal specimens. *J. Mol. Biol.* 94, 425–440.

Weiss DG, Maile W, Wick RA & Steffen W (1999) Video microscopy. In Light Microscopy in Biology: A Practical Approach, 2nd edn (AJ Lacey ed). Oxford: Oxford University Press.

White JG, Amos WB & Fordham M (1987) An evaluation of confocal versus conventional imaging of biological structures by fluorescence light microscopy. *J. Cell Biol.* 105, 41–48.

Wigge PA, Jensen ON, Holmes S et al. (1998) Analysis of the Saccharomyces spindle pole by matrix-assisted laser desorption/ionization (MALDI) mass spectrometry. *J. Cell Biol.* 141, 967–977.

Wilson L & Matsudaira P (eds) (1993) Methods in Cell Biology, vol 37: Antibodies in Cell Biology. New York: Academic Press.

Zernike F (1955) How I discovered phase contrast. *Science* 121, 345–349.

Visualizing Molecules in Living Cells

Ammann D (1986) Ion-Selective Microelectrodes: Principles, Design and Application. Berlin: Springer-Verlag.

Celis JE (ed) (1998) Cell Biology: A Laboratory Handbook, 2nd edn, vol 4: part 13—Transfer of Macromolecules and Small Molecules, pp 3–183. San Diego: Academic Press. (This contains many descriptions of the variety of methods that are used in various different cell types for different purposes, including microinjection, electroporation, liposome delivery, biolistic bombardment and other methods.)

Celis JE (ed) (1998) Cell Biology: A Laboratory Handbook, 2nd edn, vol 3, part 10—Intracellular Measurements, pp 363–386. San Diego: Academic Press.

Chalfie M, Tu Y, Euskirchen G, Ward WW & Prasher DC (1994) Green fluorescent protein as a marker for gene expression. *Science* 263, 802–805.

Haseloff J, Dormand E-L & Brand A (1999) Live imaging with green fluorescent protein. In Methods in Molecular Biology, vol 122: Confocal Microscopy Methods and Protocols (SW Paddock ed). Totowa, NJ: Humana Press.

Mitchison TJ & Salmon ED (1992) Kinetochore fiber movement contributes to anaphase-A in newt lung cells. *J. Cell Biol.* 119, 569–582.

Neher E (1992) Ion channels for communication between and within cells. *Science* 256, 498–502.

Ormo M, Cubitt AB, Kallio et al. (1966) Crystal structure of the *Aequoria victoria* green fluorescent protein. *Science* 273, 1392–1395.

Parton RM & Read ND (1999) Calcium and pH imaging in living cells. In Light Microscopy in Biology. A Practical Approach, 2nd edn (Lacey AJ ed). Oxford: Oxford University Press.

Sakmann B (1992) Elementary steps in synaptic transmission revealed by currents through single ion channels. *Science* 256, 503–512.

Sawin KE, Theriot JA & Mitchison TJ (1999) Photoactivation of fluorescence as a probe for cytoskeletal dynamics in mitosis and cell motility. In Fluorescent and Luminescent Probes for Biological Activity, 2nd edn (Mason WT ed). San Diego: Academic Press.

Sheetz MP (ed) (1997) Methods in Cell Biology, vol 55: Laser Tweezers in Cell Biology. San Diego: Academic Press.

Theriot JA, Mitchison TJ, Tilney LG & Portnoy DA (1992) The rate of actin-based motility of intracellular *Listeria monocytogenes* equals the rate of actin polymerization. *Nature* 357, 257–260.

Tsien RY (1989) Fluorescent probes of cell signaling. *Ann. Rev. Neorosci.* 12, 227–253.

INTERNAL ORGANIZATION OF THE CELL

PART IV

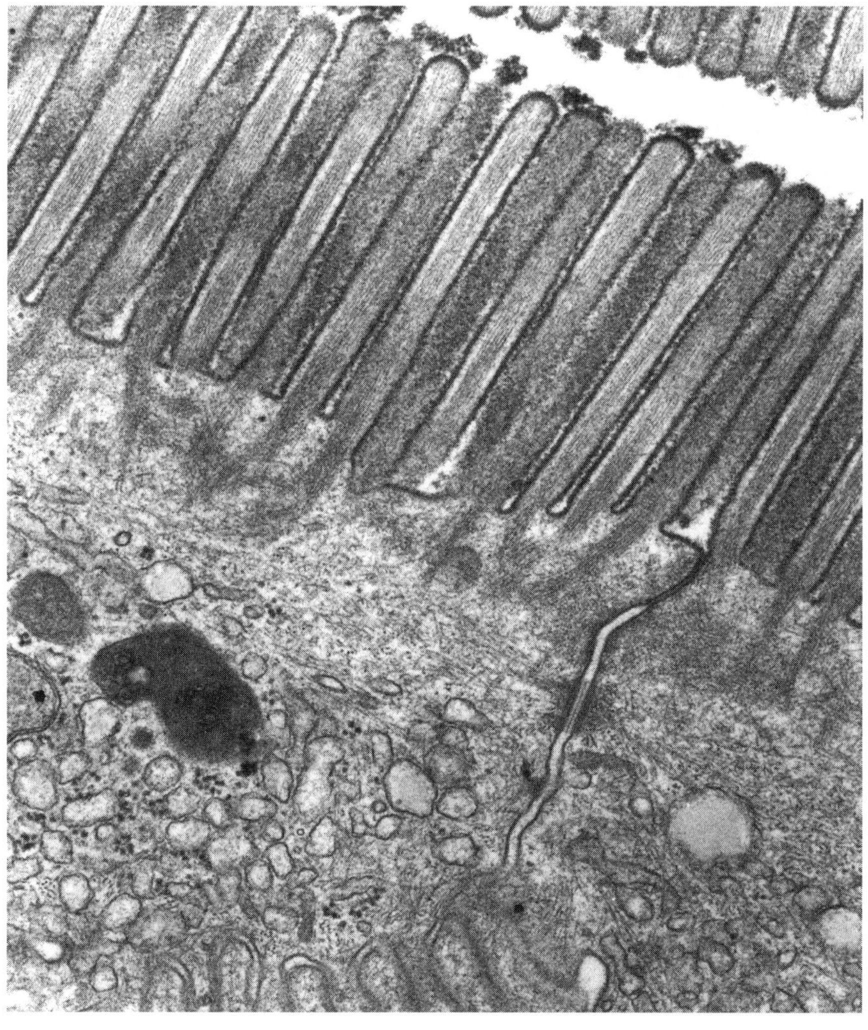

Cytoplasmic organization. This thin section through the microvilli found on the apical surface of intestinal epithelial cells illustrates the general ordering principles of eucaryotic cells. Cytoskeletal filaments (actin) provide a structural core for the microvilli, the cytosol is filled with membrane-enclosed vesicles and compartments, and adjacent cells are anchored to each other by junctions. (From P.T. Matsudaira and D.R. Burgess, *Cold Spring Harbor Symp. Quant. Biol.* 46:845–854, 1985.)

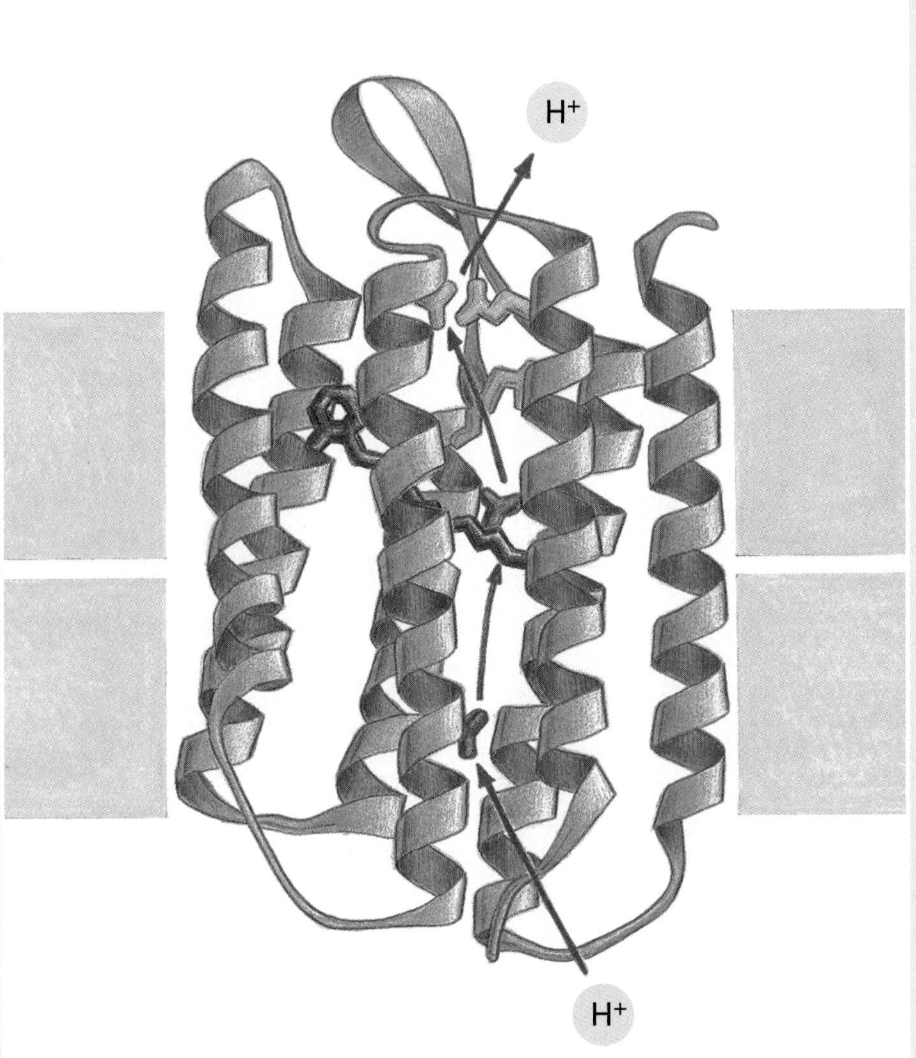

Membrane protein. Special proteins inserted in cellular membranes create pores that permit the passage of molecules across them. The bacterial protein shown here uses the energy from light (photons) to activate the pumping of protons across the plasma membrane. (Adapted from H. Luecke et al., *Science* 286:255–260, 1999.)

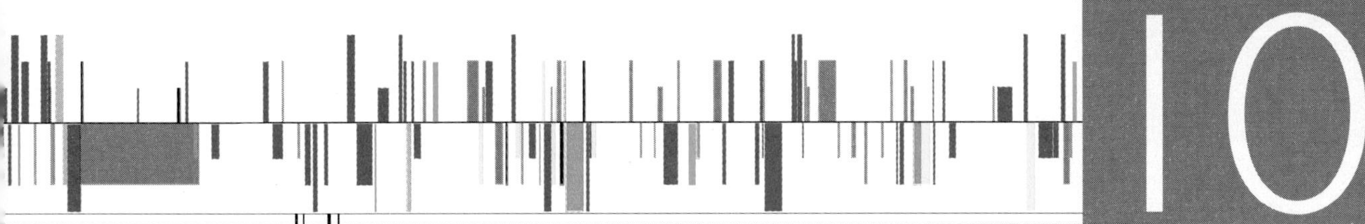

10

MEMBRANE STRUCTURE

Cell membranes are crucial to the life of the cell. The **plasma membrane** encloses the cell, defines its boundaries, and maintains the essential differences between the cytosol and the extracellular environment. Inside eucaryotic cells, the membranes of the endoplasmic reticulum, Golgi apparatus, mitochondria, and other membrane-enclosed organelles maintain the characteristic differences between the contents of each organelle and the cytosol. Ion gradients across membranes, established by the activities of specialized membrane proteins, can be used to synthesize ATP, to drive the transmembrane movement of selected solutes, or, in nerve and muscle cells, to produce and transmit electrical signals. In all cells, the plasma membrane also contains proteins that act as sensors of external signals, allowing the cell to change its behavior in response to environmental cues; these protein sensors, or *receptors,* transfer information—rather than ions or molecules—across the membrane.

Despite their differing functions, all biological membranes have a common general structure: each is a very thin film of lipid and protein molecules, held together mainly by noncovalent interactions. Cell membranes are dynamic, fluid structures, and most of their molecules are able to move about in the plane of the membrane. The lipid molecules are arranged as a continuous double layer about 5 nm thick (Figure 10–1). This *lipid bilayer* provides the basic fluid structure of the membrane and serves as a relatively impermeable barrier to the passage of most water-soluble molecules. Protein molecules that span the lipid bilayer mediate nearly all of the other functions of the membrane, transporting specific molecules across it, for example, or catalyzing membrane-associated

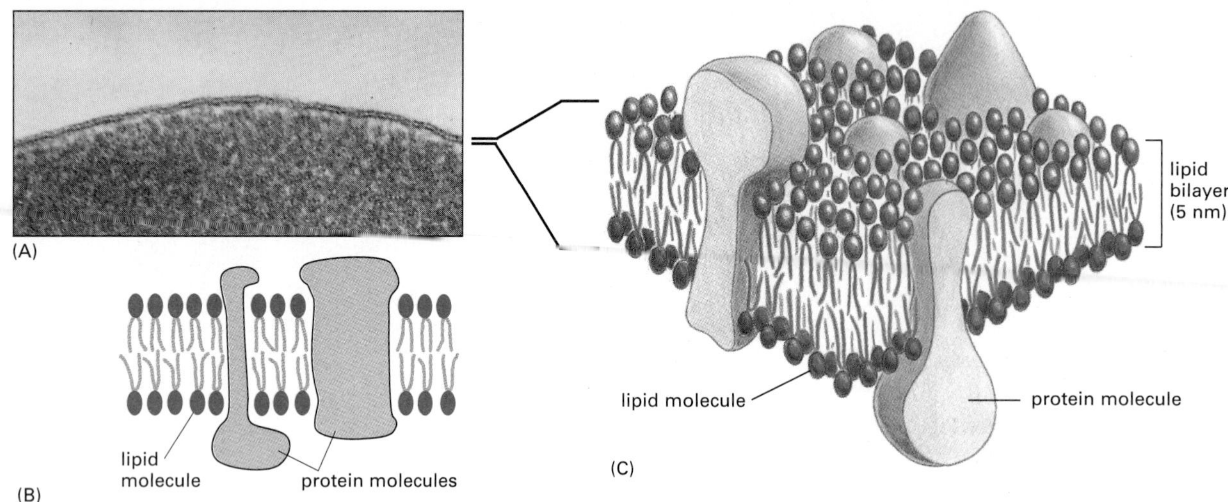

Figure 10–1 Three views of a cell membrane. (A) An electron micrograph of a plasma membrane (of a human red blood cell) seen in cross section. (B and C) These drawings show two-dimensional and three-dimensional views of a cell membrane. (A, courtesy of Daniel S. Friend.)

reactions, such as ATP synthesis. In the plasma membrane, some proteins serve as structural links that connect the cytoskeleton through the lipid bilayer to either the extracellular matrix or an adjacent cell, while others serve as receptors to detect and transduce chemical signals in the cell's environment. As would be expected, it takes many different membrane proteins to enable a cell to function and interact with its environment. In fact, it is estimated that about 30% of the proteins that are encoded in an animal cell's genome are membrane proteins.

In this chapter we consider the structure and organization of the two main constituents of biological membranes—the lipids and the membrane proteins. Although we focus mainly on the plasma membrane, most of the concepts discussed are applicable to the various internal membranes in cells as well. The functions of cell membranes are considered in later chapters. Their role in ATP synthesis, for example, is discussed in Chapter 14; their role in the transmembrane transport of small molecules, in Chapter 11; and their roles in cell signaling and cell adhesion in Chapters 15 and 19, respectively. In Chapters 12 and 13 we discuss the internal membranes of the cell and the protein traffic through and between them.

THE LIPID BILAYER

The **lipid bilayer** has been firmly established as the universal basis for cell-membrane structure. It is easily seen by electron microscopy, although specialized techniques, such as x-ray diffraction and freeze-fracture electron microscopy, are needed to reveal the details of its organization. The bilayer structure is attributable to the special properties of the lipid molecules, which cause them to assemble spontaneously into bilayers even under simple artificial conditions.

Membrane Lipids Are Amphipathic Molecules, Most of which Spontaneously Form Bilayers

Lipid—that is, fatty—molecules constitute about 50% of the mass of most animal cell membranes, nearly all of the remainder being protein. There are approximately 5×10^6 lipid molecules in a 1 μm × 1 μm area of lipid bilayer, or about 10^9 lipid molecules in the plasma membrane of a small animal cell. All of the lipid molecules in cell membranes are **amphipathic** (or amphiphilic)—that is, they have a **hydrophilic** ("water-loving") or *polar* end and a **hydrophobic** ("water-fearing") or *nonpolar* end.

The most abundant membrane lipids are the **phospholipids**. These have a polar head group and two hydrophobic *hydrocarbon tails*. The tails are usually fatty acids, and they can differ in length (they normally contain between 14 and 24 carbon atoms). One tail usually has one or more *cis*-double bonds (i.e., it is *unsaturated*), while the other tail does not (i.e., it is *saturated*). As shown in Figure 10–2, each double bond creates a small kink in the tail. Differences in the length and saturation of the fatty acid tails are important because they influence the ability of phospholipid molecules to pack against one another, thereby affecting the fluidity of the membrane (discussed below).

It is the shape and amphipathic nature of the lipid molecules that cause them to form bilayers spontaneously in aqueous environments. As discussed in Chapter 2, hydrophilic molecules dissolve readily in water because they contain charged groups or uncharged polar groups that can form either favorable electrostatic interactions or hydrogen bonds with water molecules. Hydrophobic molecules, by contrast, are insoluble in water because all, or almost all, of their atoms are uncharged and nonpolar and therefore cannot form energetically favorable interactions with water molecules. If dispersed in water, they force the adjacent water molecules to reorganize into icelike cages that surround the hydrophobic molecule (Figure 10–3). Because these cage structures are more ordered than the surrounding water, their formation increases the free energy. This free energy cost is minimized, however, if the hydrophobic molecules (or the hydrophobic portions of amphipathic molecules) cluster together so that the smallest number of water molecules is affected.

For the above reason, lipid molecules spontaneously aggregate to bury their hydrophobic tails in the interior and expose their hydrophilic heads to water. Depending on their shape, they can do this in either of two ways: they can form spherical *micelles*, with the tails inward, or they can form bimolecular sheets, or *bilayers*, with the hydrophobic tails sandwiched between the hydrophilic head groups (Figure 10–4).

Being cylindrical, phospholipid molecules spontaneously form bilayers in aqueous environments. In this energetically most-favorable arrangement, the hydrophilic heads face the water at each surface of the bilayer, and the hydrophobic tails are shielded from the water in the interior. The same forces

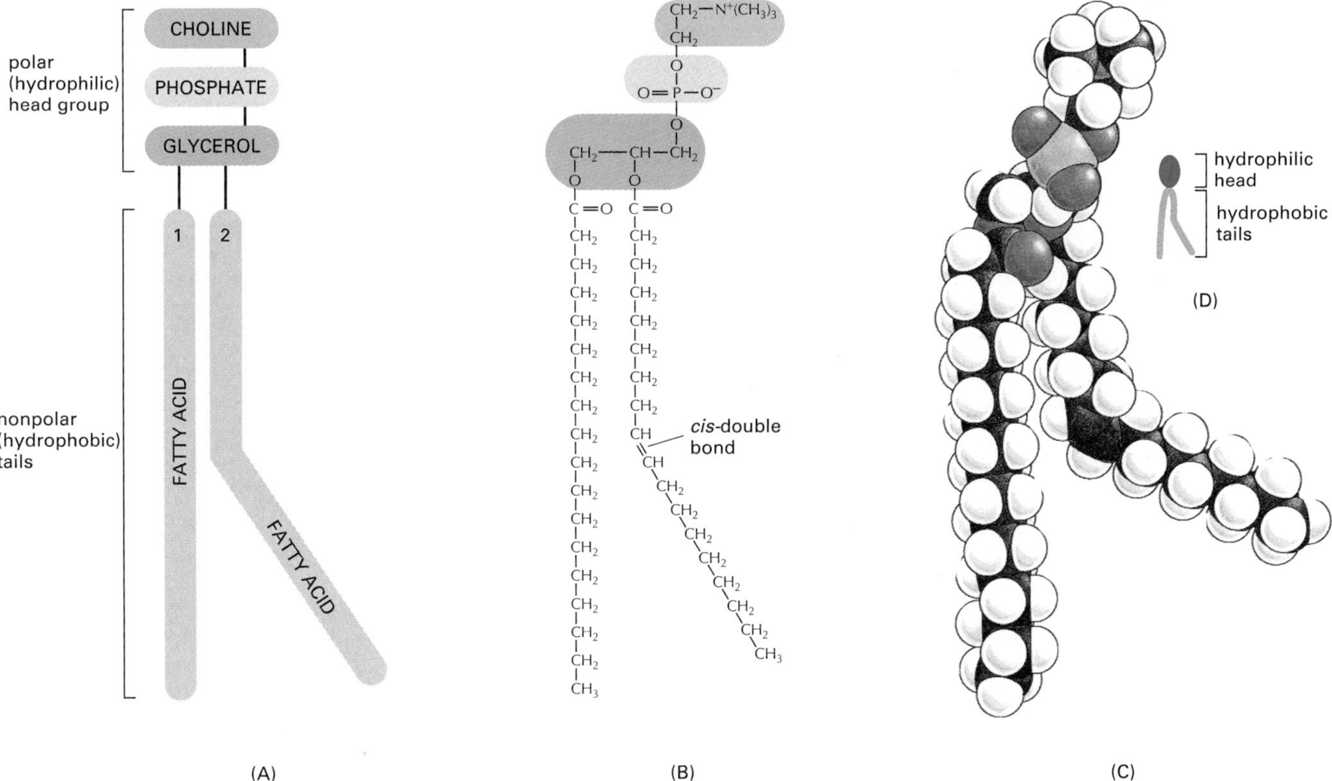

Figure 10–2 The parts of a phospholipid molecule. This example is phosphatidylcholine, represented (A) schematically, (B) by a formula, (C) as a space-filling model, and (D) as a symbol. The kink resulting from the *cis*-double bond is exaggerated for emphasis.

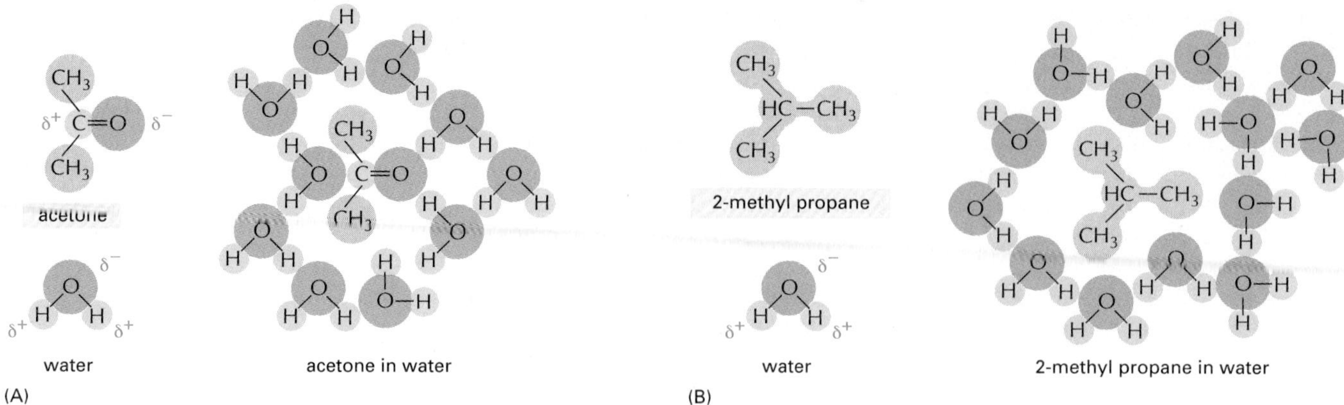

Figure 10–3 How hydrophilic and hydrophobic molecules interact differently with water. (A) Because acetone is polar, it can form favorable electrostatic interactions with water molecules, which are also polar. Thus, acetone readily dissolves in water. (B) By contrast, 2-methyl propane is entirely hydrophobic. It cannot form favorable interactions with water and it would force adjacent water molecules to reorganize into icelike cage structures, which increases the free energy. This compound therefore is virtually insoluble in water. The symbol δ^- indicates a partial negative charge, and δ^+ indicates a partial positive charge. Polar atoms are shown in color and nonpolar groups are shown in *gray.*

that drive phospholipids to form bilayers also provide a self-healing property. A small tear in the bilayer creates a free edge with water; because this is energetically unfavorable, the lipids spontaneously rearrange to eliminate the free edge. (In eucaryotic plasma membranes, larger tears are repaired by the fusion of intracellular vesicles.) The prohibition against free edges has a profound consequence: the only way for a bilayer to avoid having edges is by closing in on itself and forming a sealed compartment (Figure 10–5). This remarkable behavior, fundamental to the creation of a living cell, follows directly from the shape and amphipathic nature of the phospholipid molecule.

A lipid bilayer has other characteristics beside its self-sealing properties that make it an ideal structure for cell membranes. One of the most important of these is its fluidity, which is crucial to many membrane functions.

shape of lipid molecule

packing of lipid molecules

water

lipid micelle

lipid bilayer

(A)

(B)

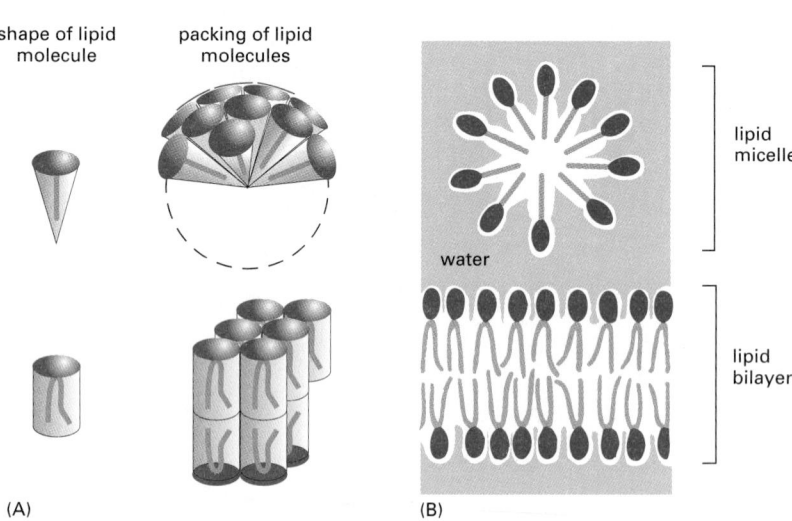

Figure 10–4 Packing arrangements of lipid molecules in an aqueous environment. (A) Wedge-shaped lipid molecules *(above)* form micelles, whereas cylinder-shaped phospholipid molecules *(below)* form bilayers. (B) A lipid micelle and a lipid bilayer seen in cross section. Lipid molecules spontaneously form one or other of these structures in water, depending on their shape.

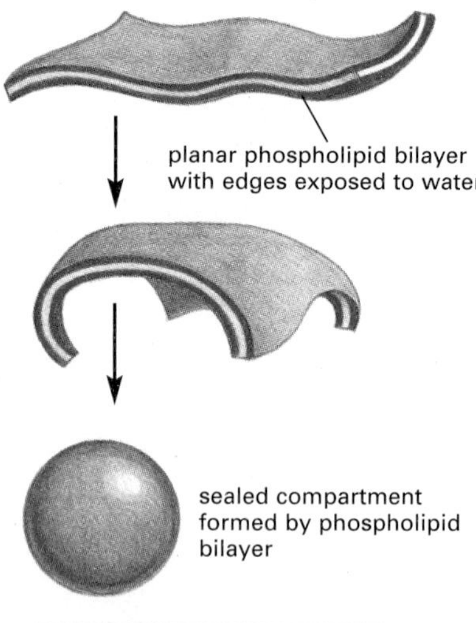

Figure 10–5 The spontaneous closure of a phospholipid bilayer to form a sealed compartment. The closed structure is stable because it avoids the exposure of the hydrophobic hydrocarbon tails to water, which would be energetically unfavorable.

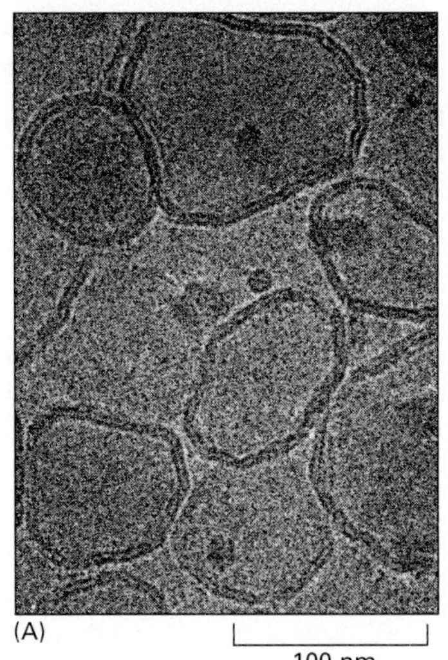

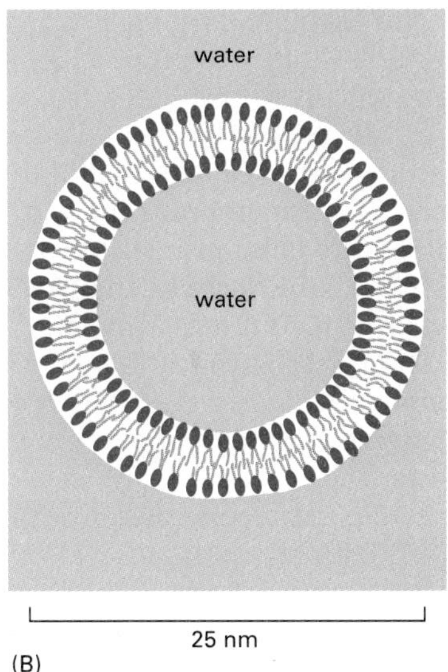

Figure 10–6 Liposomes. (A) An electron micrograph of unfixed, unstained phospholipid vesicles—liposomes—in water rapidly frozen to liquid nitrogen temperature. The bilayer structure of the liposomes is readily apparent. (B) A drawing of a small spherical liposome seen in cross section. Liposomes are commonly used as model membranes in experimental studies. (A, courtesy of Jean Lepault.)

The Lipid Bilayer Is a Two-dimensional Fluid

It was only around 1970 that researchers first recognized that individual lipid molecules are able to diffuse freely within lipid bilayers. The initial demonstration came from studies of synthetic lipid bilayers. Two types of preparations have been very useful in such studies: (1) bilayers made in the form of spherical vesicles, called **liposomes**, which can vary in size from about 25 nm to 1 μm in diameter depending on how they are produced (Figure 10–6); and (2) planar bilayers, called **black membranes**, formed across a hole in a partition between two aqueous compartments (Figure 10–7).

Various techniques have been used to measure the motion of individual lipid molecules and their different parts. One can construct a lipid molecule, for example, whose polar head group carries a "spin label," such as a nitroxyl group (>N–O); this contains an unpaired electron whose spin creates a paramagnetic signal that can be detected by electron spin resonance (ESR) spectroscopy. (The principles of this technique are similar to those of nuclear magnetic resonance, discussed in Chapter 8.) The motion and orientation of a spin-labeled lipid in a bilayer can be deduced from the ESR spectrum. Such studies show that phospholipid molecules in synthetic bilayers very rarely migrate from the monolayer (also called a *leaflet)* on one side to that on the other. This process, known as "flip-flop," occurs less than once a month for any individual molecule. In contrast, lipid molecules readily exchange places with their neighbors *within* a monolayer (~10^7 times per second). This gives rise to a rapid lateral diffusion, with a diffusion coefficient *(D)* of about 10^{-8} cm^2/sec, which means that an average lipid molecule diffuses the length of a large bacterial cell (~2 μm) in about 1 second. These studies have also shown that individual lipid molecules rotate very rapidly about their long axis and that their hydrocarbon chains are flexible (Figure 10–8).

Similar studies have been performed with labeled lipid molecules in isolated biological membranes and in living cells. The results are generally the same as for synthetic bilayers, and they demonstrate that the lipid component of a biological membrane is a two-dimensional liquid in which the constituent molecules are free to move laterally. As in synthetic bilayers, individual phospholipid molecules are normally confined to their own monolayer. This confinement creates a problem for their synthesis. Phospholipid molecules are made in only one monolayer of a membrane, mainly in the cytosolic monolayer of the endoplasmic reticulum (ER) membrane. If none of these newly made molecules could migrate reasonably promptly to the noncytosolic monolayer,

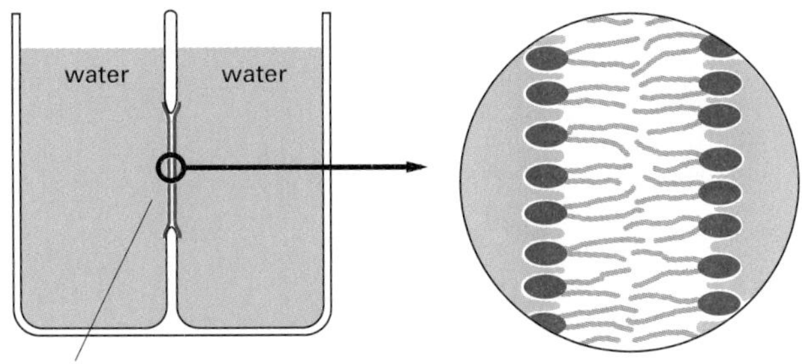

Figure 10–7 A cross-sectional view of a black membrane, a synthetic lipid bilayer. This planar bilayer appears black when it forms across a small hole in a partition separating two aqueous compartments. Black membranes are used to measure the permeability properties of synthetic membranes.

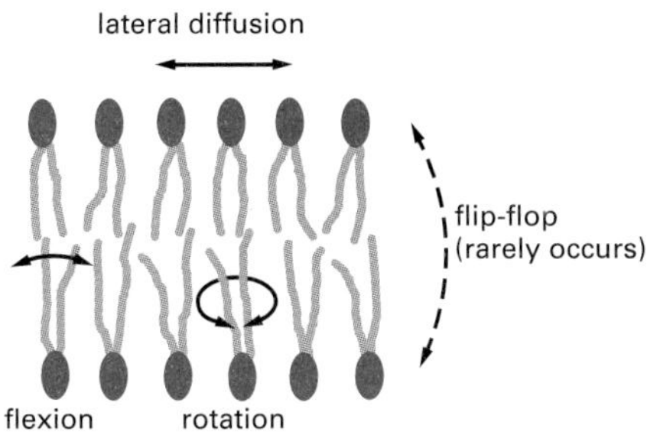

Figure 10–8 Phospholipid mobility. The types of movement possible for phospholipid molecules in a lipid bilayer.

new lipid bilayer could not be made. The problem is solved by a special class of membrane-bound enzymes called *phospholipid translocators,* which catalyze the rapid flip-flop of phospholipids from one monolayer to the other, as discussed in Chapter 12.

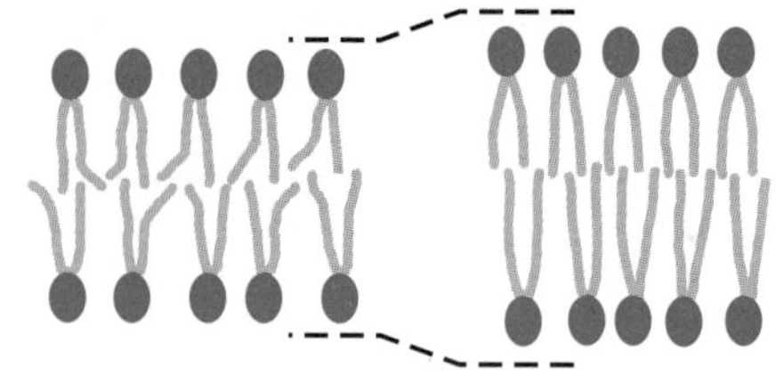

Figure 10–9 The influence of *cis*-double bonds in hydrocarbon chains. The double bonds make it more difficult to pack the chains together, thereby making the lipid bilayer more difficult to freeze. In addition, because the fatty acid chains of unsaturated lipids are more spread apart, lipid bilayers containing them are thinner than bilayers formed exclusively from saturated lipids.

The Fluidity of a Lipid Bilayer Depends on Its Composition

The fluidity of cell membranes has to be precisely regulated. Certain membrane transport processes and enzyme activities, for example, cease when the bilayer viscosity is experimentally increased beyond a threshold level.

The fluidity of a lipid bilayer depends on both its composition and its temperature, as is readily demonstrated in studies of synthetic bilayers. A synthetic bilayer made from a single type of phospholipid changes from a liquid state to a two-dimensional rigid crystalline (or gel) state at a characteristic freezing point. This change of state is called a *phase transition,* and the temperature at which it occurs is lower (that is, the membrane becomes more difficult to freeze) if the hydrocarbon chains are short or have double bonds. A shorter chain length reduces the tendency of the hydrocarbon tails to interact with one another, and *cis*-double bonds produce kinks in the hydrocarbon chains that make them more difficult to pack together, so that the membrane remains fluid at lower temperatures (Figure 10–9). Bacteria, yeasts, and other organisms whose temperature fluctuates with that of their environment adjust the fatty acid composition of their membrane lipids to maintain a relatively constant fluidity. As the temperature falls, for instance, fatty acids with more *cis*-double bonds are synthesized, so the decrease in bilayer fluidity that would otherwise result from the drop in temperature is avoided.

The lipid bilayer of many cell membranes is not composed exclusively of phospholipids, however; it often also contains *cholesterol* and *glycolipids.* Eucaryotic plasma membranes contain especially large amounts of **cholesterol** (Figure 10–10)—up to one molecule for every phospholipid molecule. The cholesterol molecules enhance the permeability-barrier properties of the lipid bilayer. They orient themselves in the bilayer with their hydroxyl groups close to the polar head groups of the phospholipid molecules. In this position, their rigid, platelike steroid rings interact with—and partly immobilize—those regions of the hydrocarbon chains closest to the polar head groups (Figure 10–11). By decreasing the mobility of the first few CH_2 groups of the hydrocarbon chains of the phospholipid molecules, cholesterol makes the lipid bilayer less deformable in this region and thereby decreases the permeability of the bilayer to small water-soluble molecules. Although cholesterol tends to make lipid bilayers less fluid, at the high concentrations found in most eucaryotic plasma membranes, it also prevents the hydrocarbon chains from coming together and crystallizing. In this way, it inhibits possible phase transitions.

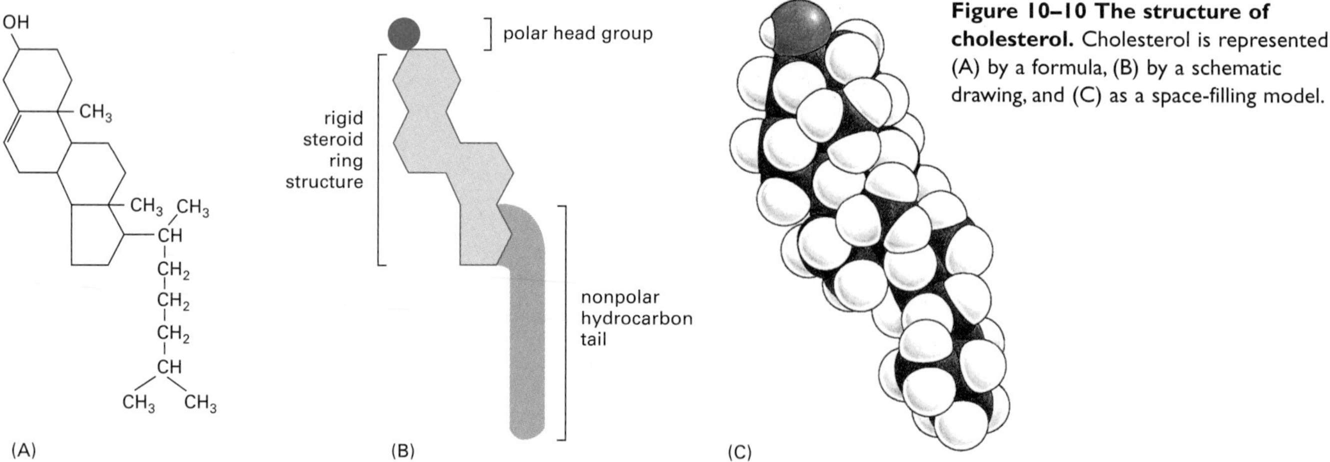

Figure 10–10 The structure of cholesterol. Cholesterol is represented (A) by a formula, (B) by a schematic drawing, and (C) as a space-filling model.

The lipid compositions of several biological membranes are compared in Table 10–1. Bacterial plasma membranes are often composed of one main type of phospholipid and contain no cholesterol; their mechanical stability is enhanced by an overlying cell wall (see Figure 11–17). The plasma membranes of most eucaryotic cells, by contrast, are more varied, not only in containing large amounts of cholesterol, but also in containing a mixture of different phospholipids.

Four major phospholipids predominate in the plasma membrane of many mammalian cells: *phosphatidylcholine, phosphatidylethanolamine, phosphatidylserine,* and *sphingomyelin.* The structures of these molecules are shown in Figure 10–12. Note that only phosphatidylserine carries a net negative charge, the importance of which we discuss later; the other three are electrically neutral at physiological pH, carrying one positive and one negative charge. Together these four phospholipids constitute more than half the mass of lipid in most membranes (see Table 10–1). Other phospholipids, such as the *inositol phospholipids,* are present in smaller quantities but are functionally very important. The inositol phospholipids, for example, have a crucial role in cell signaling, as discussed in Chapter 15.

One might wonder why eucaryotic membranes contain such a variety of phospholipids, with head groups that differ in size, shape, and charge. One can begin to understand why if one thinks of the membrane lipids as constituting a two-dimensional solvent for the proteins in the membrane, just as water constitutes a three-dimensional solvent for proteins in an aqueous solution. Some membrane proteins can function only in the presence of specific phospholipid head groups, just as many enzymes in aqueous solution require a particular ion for activity. Moreover, some cytosolic enzymes bind to specific lipid head groups exposed on the cytosolic face of a membrane and are thus recruited to and concentrated at specific membrane sites.

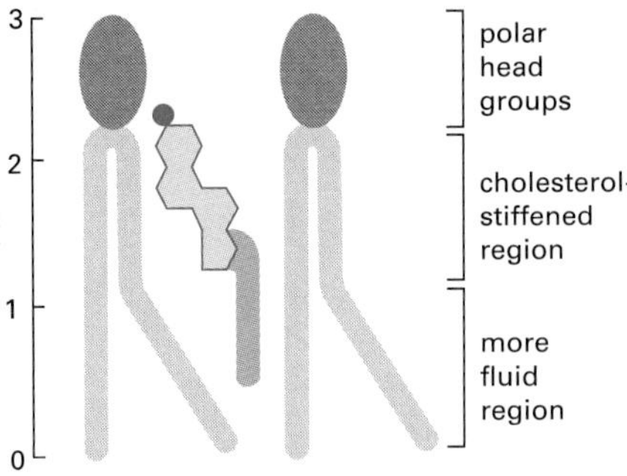

Figure 10–11 Cholesterol in a lipid bilayer. Schematic drawing of a cholesterol molecule interacting with two phospholipid molecules in one monolayer of a lipid bilayer.

The Plasma Membrane Contains Lipid Rafts That Are Enriched in Sphingolipids, Cholesterol, and Some Membrane Proteins

Most types of lipid molecules in cell membranes are randomly mixed in the lipid monolayer in which they reside. The van der Waals attractive forces between neighboring fatty acid tails are not selective enough to hold groups of molecules of this sort together. For some lipid molecules, however, such as the sphingolipids (discussed below), which tend to have long and saturated fatty hydrocarbon chains, the attractive forces can be just strong enough to hold the adjacent molecules together transiently in small microdomains. Such microdomains, or **lipid rafts**, can be thought of as transient phase separations in the fluid lipid bilayer where sphingolipids become concentrated.

The plasma membrane of animal cells is thought to contain many such tiny lipid rafts (~70 nm in diameter), which are rich in both sphingolipids and

TABLE 10–1 Approximate Lipid Compositions of Different Cell Membranes

	PERCENTAGE OF TOTAL LIPID BY WEIGHT					
LIPID	LIVER CELL PLASMA MEMBRANE	RED BLOOD CELL PLASMA MEMBRANE	MYELIN	MITOCHONDRION (INNER AND OUTER MEMBRANES)	ENDOPLASMIC RETICULUM	*E. COLI* BACTERIUM
Cholesterol	17	23	22	3	6	0
Phosphatidylethanolamine	7	18	15	25	17	70
Phosphatidylserine	4	7	9	2	5	trace
Phosphatidylcholine	24	17	10	39	40	0
Sphingomyelin	19	18	8	0	5	0
Glycolipids	7	3	28	trace	trace	0
Others	22	13	8	21	27	30

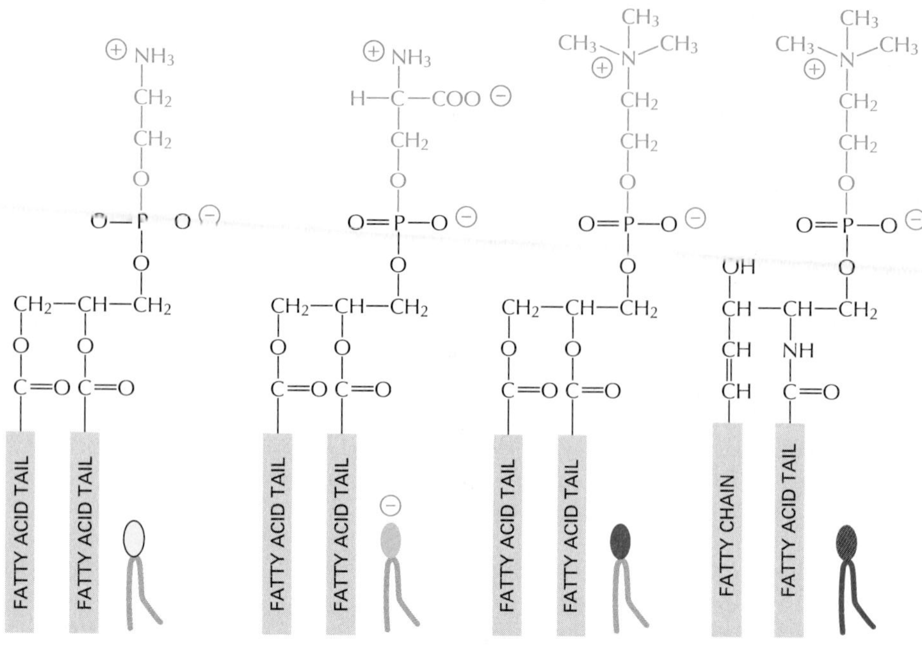

Figure 10–12 Four major phospholipids in mammalian plasma membranes. Note that different head groups are represented by different colors. All the lipid molecules shown are derived from glycerol except for sphingomyelin, which is derived from serine.

cholesterol. Because the hydrocarbon chains of the lipids concentrated there are longer and straighter than the fatty acid chains of most membrane lipids, the rafts are thicker than other parts of the bilayer (see Figure 10–9) and can better accommodate certain membrane proteins, which therefore tend to accumulate there (Figure 10–13). In this way, lipid rafts are thought to help organize such proteins—either concentrating them for transport in small vesicles or to enable the proteins to function together, as when they convert extracellular signals into intracellular ones (discussed in Chapter 15).

For the most part, lipid molecules in one monolayer of the bilayer move about independently of those in the other monolayer. In lipid rafts, however, the long hydrocarbon chains of the sphingolipids in one monolayer interact with those in the other monolayer. Thus, the two monolayers in a lipid raft communicate through their lipid tails.

The Asymmetry of the Lipid Bilayer Is Functionally Important

The lipid compositions of the two monolayers of the lipid bilayer in many membranes are strikingly different. In the human red blood cell membrane, for example, almost all of the lipid molecules that have choline—$(CH_3)_3N^+CH_2CH_2OH$— in their head group (phosphatidylcholine and sphingomyelin) are in the outer monolayer, whereas almost all of the phospholipid molecules that contain a terminal primary amino group (phosphatidylethanolamine and phosphatidylserine) are in the inner monolayer (Figure 10–14). Because the negatively charged phosphatidylserine is located in the inner monolayer, there is a significant difference in charge between the two halves of the bilayer. We discuss in Chapter 12 how lipid asymmetry is generated and maintained by membrane-bound phospholipid translocators.

Lipid asymmetry is functionally important. Many cytosolic proteins bind to specific lipid head groups found in the cytosolic monolayer of the lipid bilayer. The enzyme *protein kinase C (PKC)*, for example, is activated in response to various extracellular signals. It binds to the cytosolic face of the plasma membrane, where phosphatidylserine is concentrated, and requires this negatively charged phospholipid for its activity.

In other cases, the lipid head group must first be modified so that protein-binding sites are created at a particular time and place. *Phosphatidylinositol*, for instance, is a minor phospholipid that is concentrated in the cytosolic monolayer

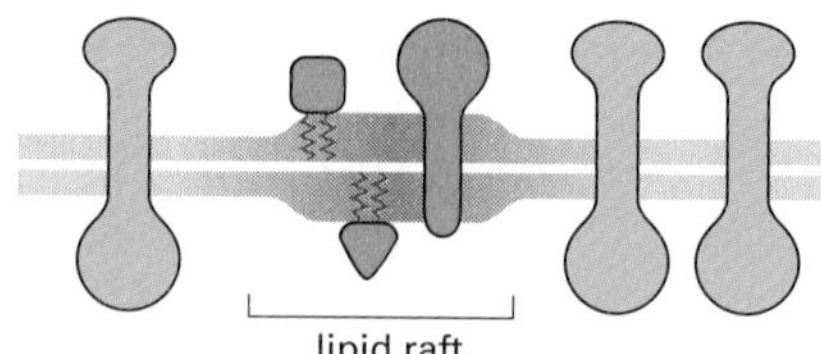

Figure 10–13 A lipid raft. Lipid rafts are small, specialized areas in membranes where some lipids (primarily sphingolipids and cholesterol) and proteins *(green)* are concentrated. Because the lipid bilayer is somewhat thicker in the rafts, certain membrane proteins accumulate. A more detailed view of a lipid raft is shown in Figure 13–63.

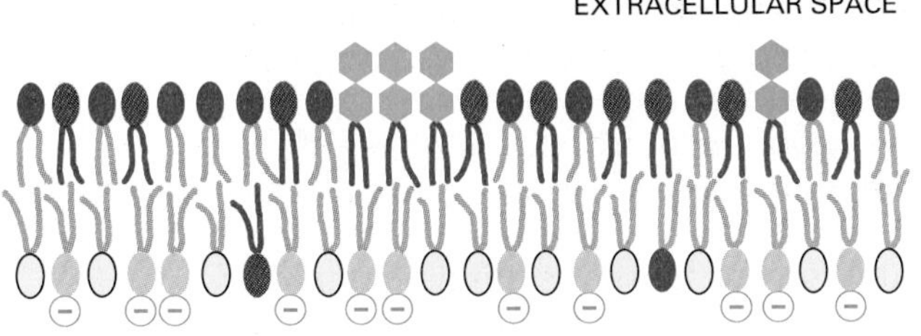

Figure 10–14 The asymmetrical distribution of phospholipids and glycolipids in the lipid bilayer of human red blood cells. The colors used for the phospholipid head groups are those introduced in Figure 10–12. In addition, glycolipids are drawn with hexagonal polar head groups *(blue)*. Cholesterol (not shown) is thought to be distributed about equally in both monolayers.

of cell membranes. A variety of lipid kinases can add phosphate groups at distinct positions in the inositol ring. The phosphorylated inositol phospholipids then act as binding sites that recruit specific proteins from the cytosol to the membrane. An important example of a lipid kinase is *phosphatidylinositol kinase (PI 3-kinase)*, which is activated in response to extracellular signals and helps to recruit specific intracellular signaling proteins to the cytosolic face of the plasma membrane (Figure 10–15A). Similar lipid kinases phosphorylate inositol phospholipids in intracellular membranes and thereby help to recruit proteins that guide membrane transport.

Phospholipids in the plasma membrane are used also in another way in the response to extracellular signals. The plasma membrane contains various *phospholipases* that are activated by extracellular signals to cleave specific phospholipid molecules, generating fragments of these molecules that act as short-lived intracellular mediators (Figure 10–15B). *Phospholipase C*, for example, cleaves an inositol phospholipid in the cytosolic monolayer of the plasma membrane to generate two fragments, one of which remains in the membrane and helps activate protein kinase C, while the other is released into the cytosol and stimulates the release of Ca^{2+} from the endoplasmic reticulum (see Figure 15–36).

Animals exploit the phospholipid asymmetry of their plasma membranes to distinguish between live and dead cells. When animal cells undergo programmed cell death, or apoptosis (discussed in Chapter 17), phosphatidylserine,

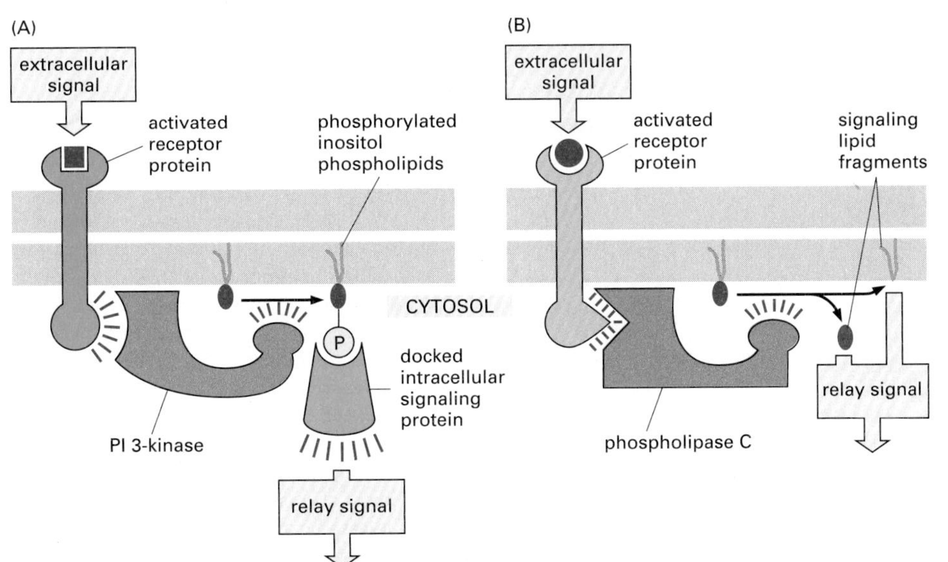

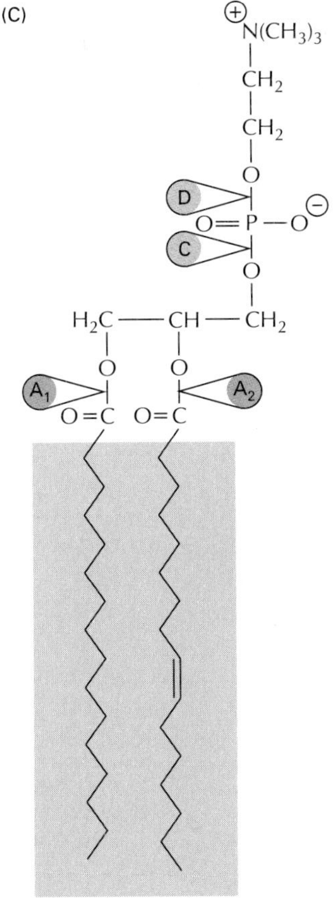

Figure 10–15 Some functions of membrane phospholipids in cell signaling. (A) Extracellular signals can activate PI 3-kinase, which phosphorylates inositol phospholipids in the plasma membrane. Various intracellular signaling molecules then bind to these phosphorylated lipids and are thus recruited to the membrane, where they can interact and help relay the signal into the cell. (B) Other extracellular signals activate phospholipases that cleave phospholipids. The lipid fragments then act as signaling molecules to relay the signal into the cell. (C) Illustration of the sites where different classes of phospholipases cleave phospholipids. As indicated, phospholipases A_1 and A_2 cleave ester bonds, whereas phospholipases C and D cleave at phosphoester bonds.

which is normally confined to the cytosolic monolayer of the plasma membrane lipid bilayer, rapidly translocates to the extracellular monolayer. The phosphatidylserine exposed on the cell surface serves as a signal to induce neighboring cells, such as macrophages, to phagocytose the dead cell and digest it. The translocation of the phosphatidylserine in apoptotic cells occurs by two mechanisms:

1. The phospholipid translocator that normally transports this lipid from the noncytosolic monolayer to the cytosolic monolayer is inactivated.
2. A "scramblase" that transfers phospholipids nonspecifically in both directions between the two monolayers is activated.

Glycolipids Are Found on the Surface of All Plasma Membranes

The lipid molecules with the most extreme asymmetry in their membrane distribution are the sugar-containing lipid molecules called **glycolipids**. These intriguing molecules are found exclusively in the noncytosolic monolayer of the lipid bilayer, where they are thought to partition preferentially into lipid rafts. The glycolipids tend to self-associate, partly through hydrogen bonds between their sugars and partly through van der Waals forces between their long and mainly saturated hydrocarbon chains. The asymmetric distribution of glycolipids in the bilayer results from the addition of sugar groups to the lipid molecules in the lumen of the Golgi apparatus, which is topologically equivalent to the exterior of the cell (discussed in Chapter 12). In the plasma membrane, the sugar groups are exposed at the cell surface (see Figure 10–14), where they have important roles in interactions of the cell with its surroundings.

Glycolipids probably occur in all animal cell plasma membranes, where they generally constitute about 5% of the lipid molecules in the outer monolayer. They are also found in some intracellular membranes. The most complex of the glycolipids, the **gangliosides**, contain oligosaccharides with one or more sialic acid residues, which give gangliosides a net negative charge (Figure 10–16). More than 40 different gangliosides have been identified. They are most abundant in the plasma membrane of nerve cells, where gangliosides constitute 5–10% of the total lipid mass; they are also found in much smaller quantities in other cell types.

Hints as to what the functions of glycolipids might be come from their localization. In the plasma membrane of epithelial cells, for example, glycolipids are

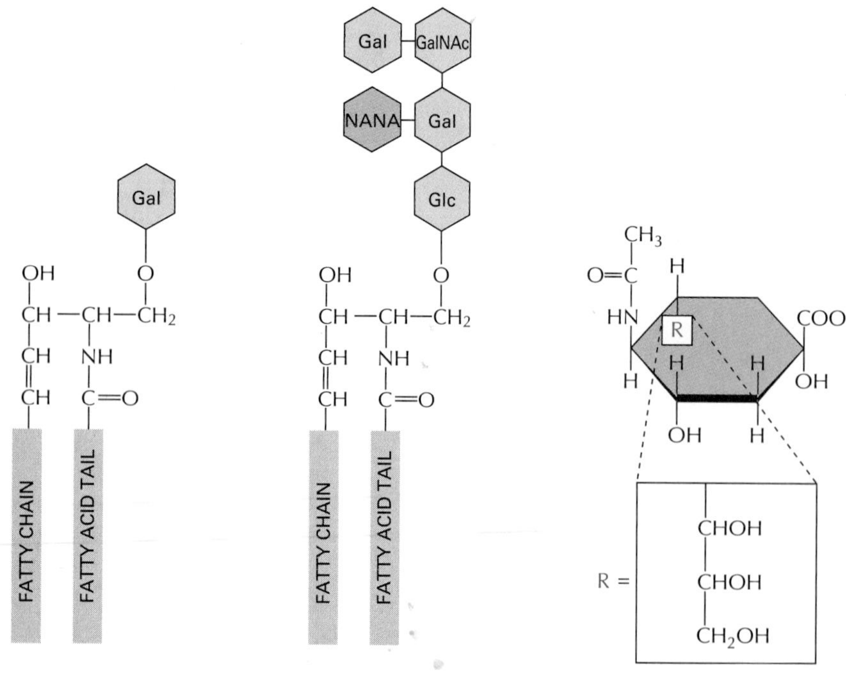

Figure 10–16 Glycolipid molecules. (A) Galactocerebroside is called a *neutral glycolipid* because the sugar that forms its head group is uncharged. (B) A ganglioside always contains one or more negatively charged sialic acid residues (also called *N*-acetylneuraminic acid, or NANA), whose structure is shown in (C). Whereas in bacteria and plants almost all glycolipids are derived from glycerol, as are most phospholipids, in animal cells they are almost always produced from serine, as is the case for the phospholipid sphingomyelin (see Figure 10–12). Gal = galactose; Glc = glucose, GalNAc = *N*-acetylgalactosamine; these three sugars are uncharged.

confined to the exposed apical surface, where they may help protect the membrane against the harsh conditions frequently found there (such as low pH and degradative enzymes). Charged glycolipids, such as gangliosides, may be important for their electrical effects: their presence alters the electrical field across the membrane and the concentrations of ions—especially Ca^{2+}—at the membrane surface. Glycolipids are also thought to function in cell-recognition processes, in which membrane-bound carbohydrate-binding proteins *(lectins)* bind to the sugar groups on both glycolipids and glycoproteins in the process of cell–cell adhesion (discussed in Chapter 19). Surprisingly, however, mutant mice that are deficient in all of their complex gangliosides show no obvious abnormalities, although the males cannot transport testosterone normally in the testes and are consequently sterile.

Whatever their normal function, some glycolipids provide entry points for certain bacterial toxins. The ganglioside G_{M1} (see Figure 10–16), for example, acts as a cell-surface receptor for the bacterial toxin that causes the debilitating diarrhea of cholera. Cholera toxin binds to and enters only those cells that have G_{M1} on their surface, including intestinal epithelial cells. Its entry into a cell leads to a prolonged increase in the concentration of intracellular cyclic AMP (discussed in Chapter 15), which in turn causes a large efflux of Na^+ and water into the intestine.

Summary

Biological membranes consist of a continuous double layer of lipid molecules in which membrane proteins are embedded. This lipid bilayer is fluid, with individual lipid molecules able to diffuse rapidly within their own monolayer. The membrane lipid molecules are amphipathic. The most numerous are the phospholipids. When placed in water they assemble spontaneously into bilayers, which form sealed compartments that reseal if torn.

There are three major classes of membrane lipid molecules—phospholipids, cholesterol, and glycolipids. The lipid compositions of the inner and outer monolayers are different, reflecting the different functions of the two faces of a cell membrane. Different mixtures of lipids are found in the membranes of cells of different types, as well as in the various membranes of a single eucaryotic cell. Some membrane-bound enzymes require specific lipid head groups in order to function. The head groups of some lipids form docking sites for specific cytosolic proteins. Some extracellular signals that act through membrane receptor proteins activate phospholipases that cleave selected phospholipid molecules in the plasma membrane, thereby generating fragments that act as intracellular signaling molecules.

MEMBRANE PROTEINS

Although the basic structure of biological membranes is provided by the lipid bilayer, membrane proteins perform most of the specific functions of membranes. It is the proteins, therefore, that give each type of membrane in the cell its characteristic functional properties. Accordingly, the amounts and types of proteins in a membrane are highly variable. In the myelin membrane, which serves mainly as electrical insulation for nerve cell axons, less than 25% of the membrane mass is protein. By contrast, in the membranes involved in ATP production (such as the internal membranes of mitochondria and chloroplasts), approximately 75% is protein. A typical plasma membrane is somewhere in between, with protein accounting for about 50% of its mass.

Because lipid molecules are small compared with protein molecules, there are always many more lipid molecules than protein molecules in membranes—about 50 lipid molecules for each protein molecule in a membrane that is 50% protein by mass. Like membrane lipids, membrane proteins often have oligosaccharide chains attached to them that face the cell exterior. Thus, the surface that the cell presents to the exterior is rich in carbohydrate, which forms a *cell coat*, as we discuss later.

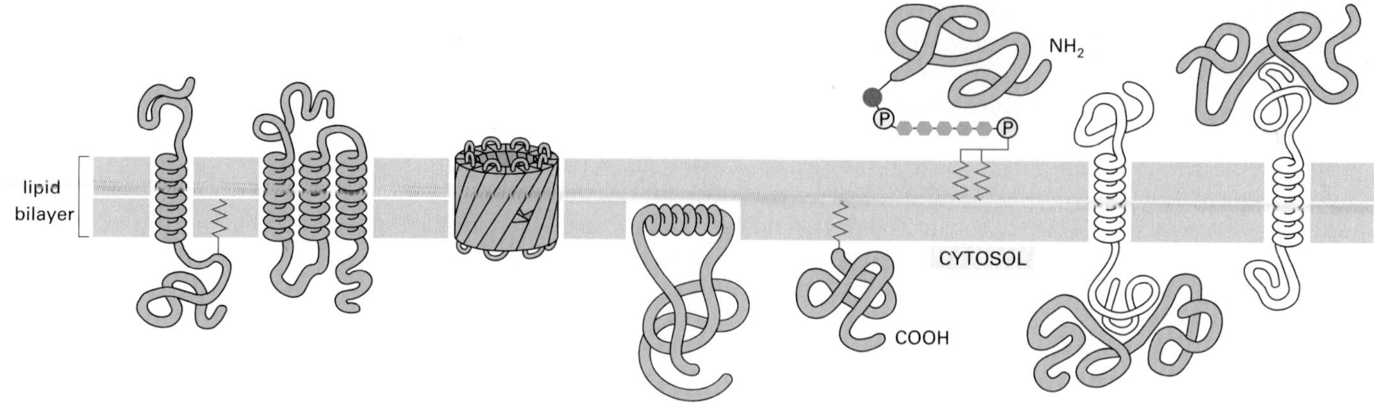

Figure 10–17 Various ways in which membrane proteins associate with the lipid bilayer. Most trans-membrane proteins are thought to extend across the bilayer as (1) a single α helix, (2) as multiple α helices, or (3) as a rolled-up β sheet (a β barrel). Some of these "single-pass" and "multipass" proteins have a covalently attached fatty acid chain inserted in the cytosolic lipid monolayer (1). Other membrane proteins are exposed at only one side of the membrane. (4) Some of these are anchored to the cytosolic surface by an amphipathic α helix that partitions into the cytosolic monolayer of the lipid bilayer through the hydrophobic face of the helix. (5) Others are attached to the bilayer solely by a covalently attached lipid chain—either a fatty acid chain or a prenyl group—in the cytosolic monolayer or, (6) via an oligosaccharide linker, to phosphatidylinositol in the noncytosolic monolayer. (7, 8) Finally, many proteins are attached to the membrane only by noncovalent interactions with other membrane proteins. The way in which the structure in (5) is formed is illustrated in Figure 10–18. The details of how membrane proteins become associated with the lipid bilayer are discussed in Chapter 12.

Membrane Proteins Can Be Associated with the Lipid Bilayer in Various Ways

Different membrane proteins are associated with the membranes in different ways, as illustrated in Figure 10–17. Many extend through the lipid bilayer, with part of their mass on either side (examples 1, 2, and 3 in Figure 10–17). Like their lipid neighbors, these **transmembrane proteins** are amphipathic, having regions that are hydrophobic and regions that are hydrophilic. Their hydrophobic regions pass through the membrane and interact with the hydrophobic tails of the lipid molecules in the interior of the bilayer, where they are sequestered away from water. Their hydrophilic regions are exposed to water on either side of the membrane. The hydrophobicity of some of these transmembrane proteins is increased by the covalent attachment of a fatty acid chain that inserts into the cytosolic monolayer of the lipid bilayer (example 1 in Figure 10–17).

Other membrane proteins are located entirely in the cytosol and are associated with the cytosolic monolayer of the lipid bilayer either by an amphipathic α helix exposed on the surface of the protein (example 4 in Figure 10–17) or by one or more covalently attached lipid chains, which can be fatty acid chains or *prenyl groups* (example 5 in Figure 10–17 and Figure 10–18). Yet other membrane proteins are entirely exposed at the external cell surface, being attached to the lipid bilayer only by a covalent linkage (via a specific oligosaccharide) to phosphatidylinositol in the outer lipid monolayer of the plasma membrane (example 6 in Figure 10–17).

The lipid-linked proteins in example 5 in Figure 10–17 are made as soluble proteins in the cytosol and are subsequently directed to the membrane by the covalent attachment of a lipid group (see Figure 10–18). The proteins in example 6, however, are made as single-pass transmembrane proteins in the ER. While still in the ER, the transmembrane segment of the protein is cleaved off and a **glycosylphosphatidylinositol (GPI) anchor** is added, leaving the protein bound to the noncytosolic surface of the membrane solely by this anchor (discussed in Chapter 12). Proteins bound to the plasma membrane by a GPI anchor can be readily distinguished by the use of an enzyme called phosphatidylinositol-specific phospholipase C. This enzyme cuts these proteins free from their anchors, thereby releasing them from the membrane.

Some membrane proteins do not extend into the hydrophobic interior of the lipid bilayer at all; they are instead bound to either face of the membrane by noncovalent interactions with other membrane proteins (examples 7 and 8 in Figure 10–17). Many of the proteins of this type can be released from the membrane by relatively gentle extraction procedures, such as exposure to solutions of very high or low ionic strength or of extreme pH, which interfere with protein–protein interactions but leave the lipid bilayer intact; these proteins are referred to as **peripheral membrane proteins**. Transmembrane proteins, many proteins held in the bilayer by lipid groups, and some proteins held on the membrane by unusually tight binding to other proteins cannot be released in these ways. These proteins are called **integral membrane proteins**.

How a membrane protein associates with the lipid bilayer reflects the function of the protein. Only transmembrane proteins can function on both sides of the bilayer or transport molecules across it. Cell-surface receptors are transmembrane proteins that bind signal molecules in the extracellular space and generate different intracellular signals on the opposite side of the plasma membrane. Proteins that function on only one side of the lipid bilayer, by contrast, are often associated exclusively with either the lipid monolayer or a protein domain on that side. Some of the proteins involved in intracellular signaling, for example, are bound to the cytosolic half of the plasma membrane by one or more covalently attached lipid groups.

In Most Transmembrane Proteins the Polypeptide Chain Crosses the Lipid Bilayer in an α-Helical Conformation

A transmembrane protein always has a unique orientation in the membrane. This reflects both the asymmetric manner in which it is synthesized and inserted into the lipid bilayer in the ER and the different functions of its cytosolic and noncytosolic domains. These domains are separated by the membrane-spanning segments of the polypeptide chain, which contact the hydrophobic environment of the lipid bilayer and are composed largely of amino acid residues with nonpolar side chains. Because the peptide bonds themselves are polar and because water is absent, all peptide bonds in the bilayer are driven to form

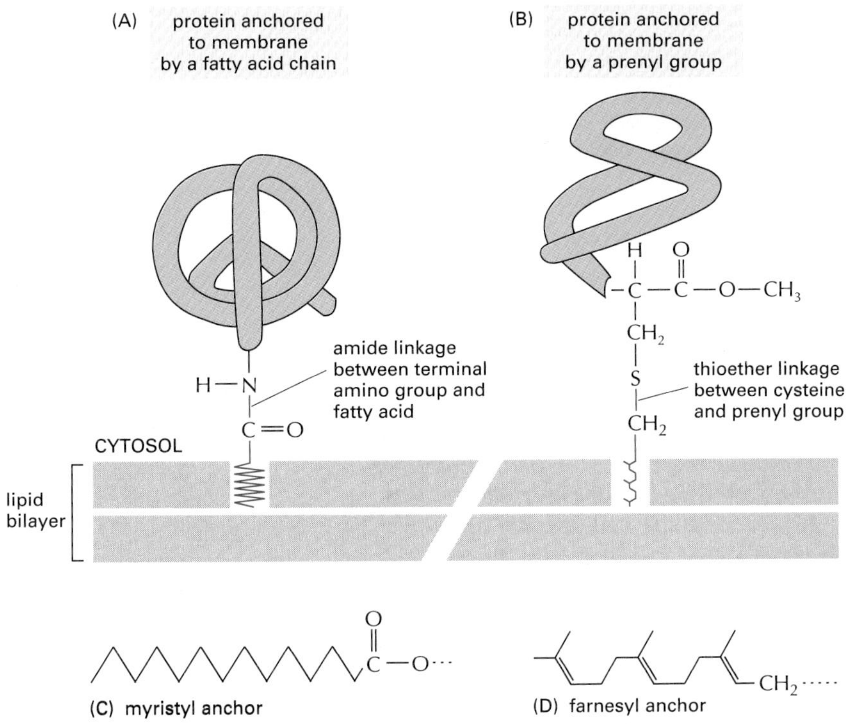

Figure 10–18 Membrane protein attachment by a fatty acid chain or a prenyl group. The covalent attachment of either type of lipid can help localize a water-soluble protein to a membrane after its synthesis in the cytosol. (A) A fatty acid chain (myristic acid) is attached via an amide linkage to an N-terminal glycine. (B) A prenyl group (either farnesyl or a longer geranylgeranyl group) is attached via a thioether linkage to a cysteine residue that is initially located four residues from the protein's C-terminus. After this prenylation, the terminal three amino acids are cleaved off, and the new C-terminus is methylated before insertion into the membrane. Palmitic acid, an 18 carbon saturated fatty acid, can also be attached to some proteins via thioester bonds formed with internal cysteine side chains. This modification is often reversible, allowing proteins to become recruited to membranes only when needed. The structures of two lipid anchors are shown below: (C) a myristyl anchor (a 14-carbon saturated fatty acid chain), and (D) a farnesyl anchor (a 15-carbon unsaturated hydrocarbon chain).

Figure 10–19 A segment of a transmembrane polypeptide chain crossing the lipid bilayer as an α helix. Only the α-carbon backbone of the polypeptide chain is shown, with the hydrophobic amino acids in *green* and *yellow*. The polypeptide segment shown is part of the bacterial photosynthetic reaction center illustrated in Figure 10–38, the structure of which was determined by x-ray diffraction. (Based on data from J. Deisenhofer et al., *Nature* 318:618–624, 1985, and H. Michel et al., *EMBO J.* 5:1149–1158, 1986.)

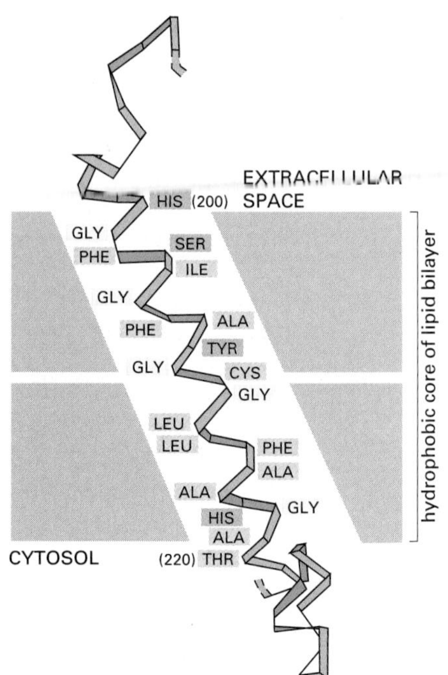

hydrogen bonds with one another. The hydrogen bonding between peptide bonds is maximized if the polypeptide chain forms a regular α helix as it crosses the bilayer, and this is how the great majority of the membrane-spanning segments of polypeptide chains are thought to traverse the bilayer (Figure 10–19).

In **single-pass transmembrane proteins**, the polypeptide crosses only once (see example 1 in Figure 10–17), whereas in **multipass transmembrane proteins**, the polypeptide chain crosses multiple times (see example 2 in Figure 10–17). An alternative way for the peptide bonds in the lipid bilayer to satisfy their hydrogen-bonding requirements is for multiple transmembrane strands of polypeptide chain to be arranged as a β sheet in the form of a closed barrel (a so-called *β barrel;* see example 3 in Figure 10–17). This form of multipass transmembrane structure is seen in *porin proteins,* which we discuss later. The strong drive to maximize hydrogen bonding in the absence of water also means that a polypeptide chain that enters the bilayer is likely to pass entirely through it before changing direction, since chain bending requires a loss of regular hydrogen-bonding interactions.

Because transmembrane proteins are notoriously difficult to crystallize, relatively few have been studied in their entirety by x-ray crystallography. The folded three-dimensional structures of almost all of the others are uncertain. DNA cloning and sequencing techniques, however, have revealed the amino acid sequences of large numbers of transmembrane proteins, and it is often possible to predict from an analysis of the protein's sequence which parts of the polypeptide chain extend across the lipid bilayer. Segments containing about 20–30 amino acids with a high degree of hydrophobicity are long enough to span a lipid bilayer as an α helix, and they can often be identified by means of a *hydropathy plot* (Figure 10–20). From such plots it is possible to predict what proportion of the proteins that an organism makes are transmembrane proteins. In budding yeast, for example, where the nucleotide sequence of the entire genome is known, about 20% of the proteins can be identified as transmembrane proteins, emphasizing the importance of membrane protein function. Hydropathy plots cannot identify the membrane-spanning segments of a β barrel, as 10 amino acids or fewer are sufficient to traverse a lipid bilayer as an extended β strand and only every other amino acid side chain is hydrophobic.

Some β Barrels Form Large Transmembrane Channels

Multipass transmembrane proteins that have their transmembrane segments arranged as a *β barrel* rather than as an α helix are comparatively rigid and tend to crystallize readily. Thus, the structures of a number of them have been determined by x-ray crystallography. The number of β strands varies widely, from as few as 8 strands to as many as 22 (Figure 10–21).

The β barrel proteins are abundant in the outer membrane of mitochondria, chloroplasts, and many bacteria. Some are pore-forming proteins, generating water-filled channels that allow selected hydrophilic solutes to cross the lipid bilayer of the bacterial outer membrane. The porins are well-studied examples (example 3 in Figure 10–21). The porin barrel is formed from a 16-stranded antiparallel β sheet, which is sufficiently large to roll up into a cylindrical structure. Polar side chains line the aqueous channel on the inside, while nonpolar side chains project from the outside of the barrel to interact with the hydrophobic core of the lipid bilayer. Loops of polypeptide chain often protrude into the lumen of the channel, narrowing it so that only certain solutes can pass. Some

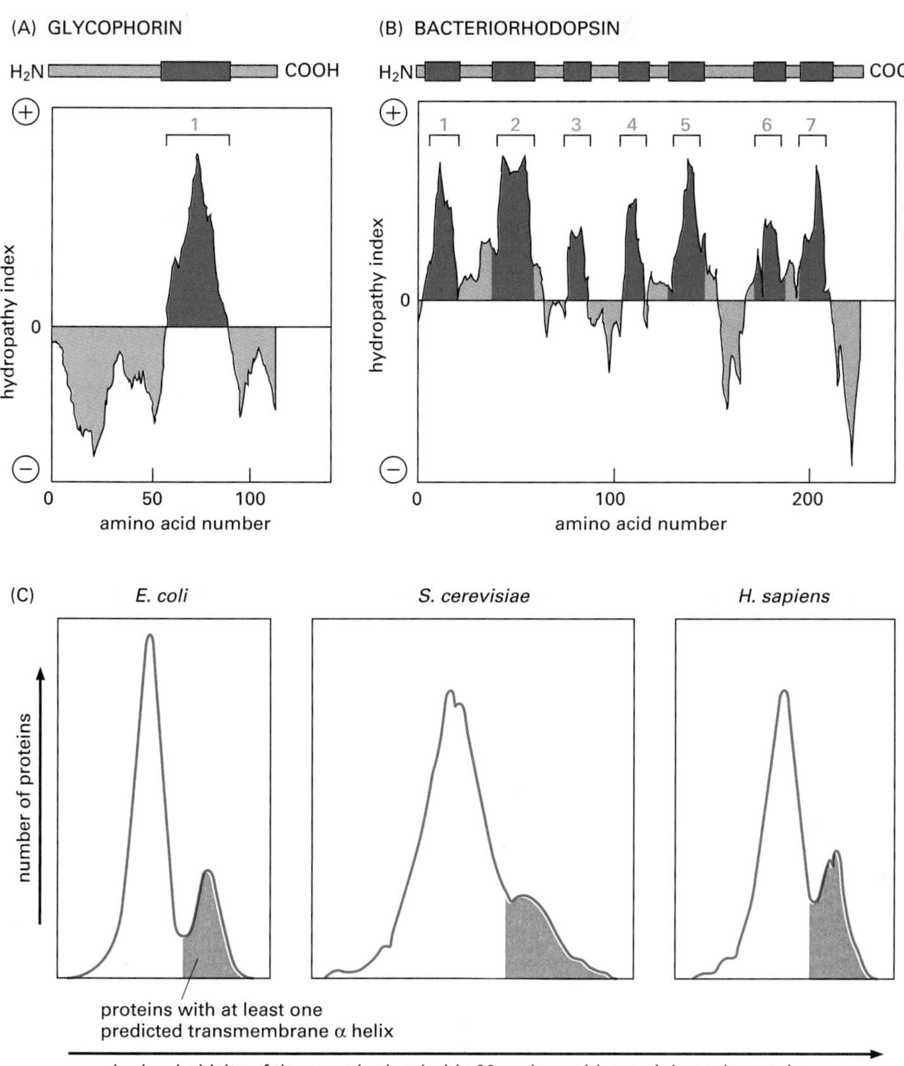

Figure 10–20 Using hydropathy plots to localize potential α-helical membrane-spanning segments in a polypeptide chain. The free energy needed to transfer successive segments of a polypeptide chain from a nonpolar solvent to water is calculated from the amino acid composition of each segment using data obtained with model compounds. This calculation is made for segments of a fixed size (usually around 10–20 amino acids), beginning with each successive amino acid in the chain. The "hydropathy index" of the segment is plotted on the y axis as a function of its location in the chain. A positive value indicates that free energy is required for transfer to water (i.e., the segment is hydrophobic), and the value assigned is an index of the amount of energy needed. Peaks in the hydropathy index appear at the positions of hydrophobic segments in the amino acid sequence. (A and B) Two examples of membrane proteins discussed later in this chapter are shown. Glycophorin (A) has a single membrane-spanning α helix and one corresponding peak in the hydropathy plot. Bacteriorhodopsin (B) has seven membrane-spanning α helices and seven corresponding peaks in the hydropathy plot. (C) The proportion of predicted membrane proteins in the genomes of *E. coli, S. cerevisiae,* and human. The area shaded in *green* indicates the fraction of proteins that contain at least one predicted transmembrane helix. The curves for *E. coli* and *S. cerevisiae* represent the whole genome; the curve for human proteins represents an incomplete set; in each case, the area under the curve is proportional to the number of genes analysed. (A, adapted from D. Eisenberg, *Annu. Rev. Biochem.* 53:595–624, 1984; C, adapted from D. Boyd et al., *Protein Sci.* 7:201–205, 1998.)

porins are therefore highly selective: *maltoporin,* for example, preferentially allows maltose and maltose oligomers to cross the outer membrane of *E. coli.*

The *FepA protein* is a more complex example of a transport protein of this kind (example 4 in Figure 10–21). It transports iron ions across the bacterial outer membrane. Its large barrel is constructed from 22 β strands, and a large globular domain completely fills the inside of the barrel. Iron ions bind to this domain, which is thought to undergo a large conformational change to transfer the iron across the membrane.

Not all β barrel proteins are transport proteins. Some form smaller barrels that are completely filled by amino acid side chains that project into the center of the barrel. These proteins function as receptors or enzymes (examples 1 and 2 in Figure 10–21); here the barrel is used primarily as a rigid anchor that holds the protein in the membrane and orients the cytosolic loop regions that form binding sites for specific intracellular molecules.

Although β barrels can serve many purposes, they are largely restricted to bacterial, mitochondrial, and chloroplast outer membranes. The vast majority of multipass transmembrane proteins in eucaryotic cells and in the bacterial plasma membrane are constructed from transmembrane α helices. The helices within these proteins can slide against each other, allowing the protein to undergo conformational changes that can be exploited to open and shut ion channels, transport solutes, or transduce extracellular signals into intracellular ones. In β barrel proteins, by contrast, each β strand is bound rigidly to its neighbors by hydrogen bonds, making conformational changes of the barrel itself unlikely.

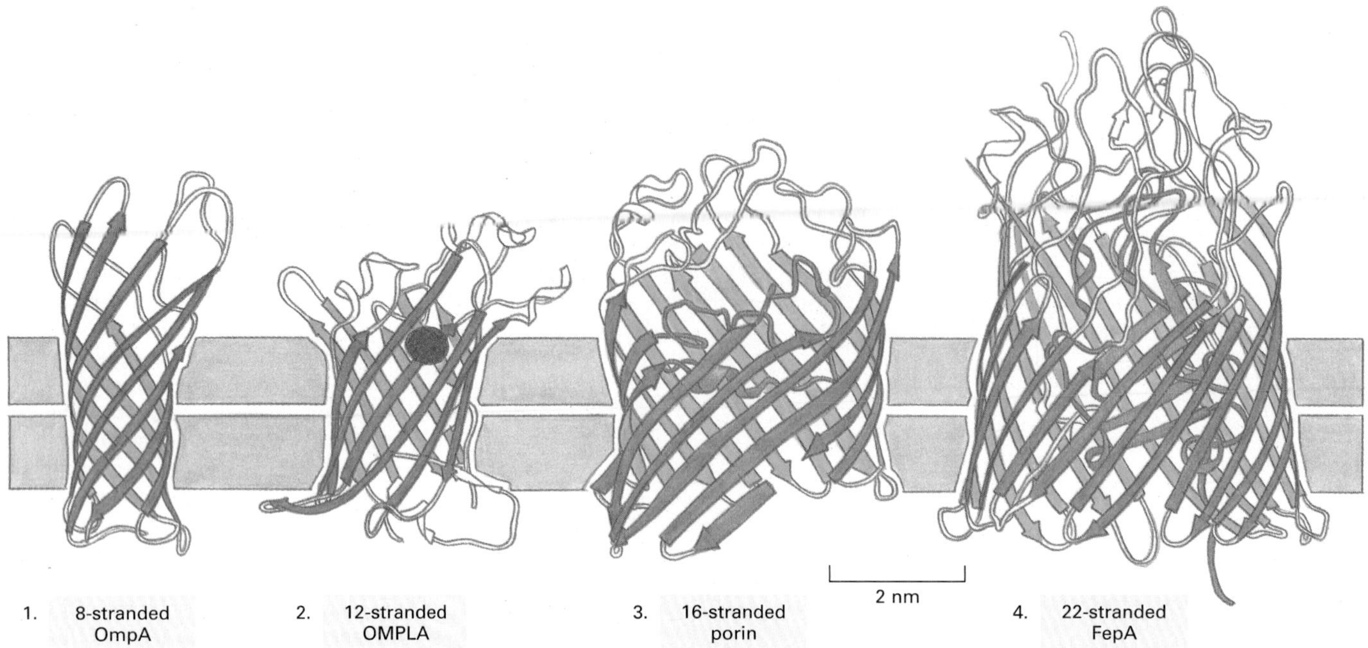

Figure 10–21 β barrels formed from different numbers of β strands. (1) The *E. coli* OmpA protein (8 β strands), which serves as a receptor for a bacterial virus. (2) The *E. coli* OMPLA protein (12 β strands), is a lipase that hydrolyses lipid molecules. The amino acids that catalyze the enzymatic reaction (shown in *red)* protrude from the outside surface of the barrel. (3) A porin from the bacterium *Rhodobacter capsulatus,* which forms water-filled pores across the outer membrane (16 β strands). The diameter of the channel is restricted by loops (shown in *blue)* that protrude into the channel. (4) The *E. coli* FepA protein (22 β strands), which transports iron ions. The inside of the barrel is completely filled by a globular protein domain (shown in *blue)* that contains an iron-binding site. This domain is thought to change its conformation to transport the bound iron, but the molecular details of the changes are not known.

Many Membrane Proteins Are Glycosylated

The great majority of transmembrane proteins in animal cells are glycosylated. As in glycolipids, the sugar residues are added in the lumen of the ER and Golgi apparatus (discussed in Chapters 12 and 13). For this reason, the oligosaccharide chains are always present on the noncytosolic side of the membrane. Another difference between proteins (or parts of proteins) on the two sides of the membrane results from the reducing environment of the cytosol. This environment decreases the likelihood that intrachain or interchain disulfide (S–S) bonds will form between cysteine residues on the cytosolic side of membranes. These intrachain and interchain bonds do form on the noncytosolic side, where they can have an important role in stabilizing either the folded structure of the polypeptide chain or its association with other polypeptide chains, respectively (Figure 10–22).

Membrane Proteins Can Be Solubilized and Purified in Detergents

In general, transmembrane proteins (and some other tightly bound membrane proteins) can be solubilized only by agents that disrupt hydrophobic associations and destroy the lipid bilayer. The most useful of these for the membrane biochemist are **detergents**, which are small amphipathic molecules that tend to

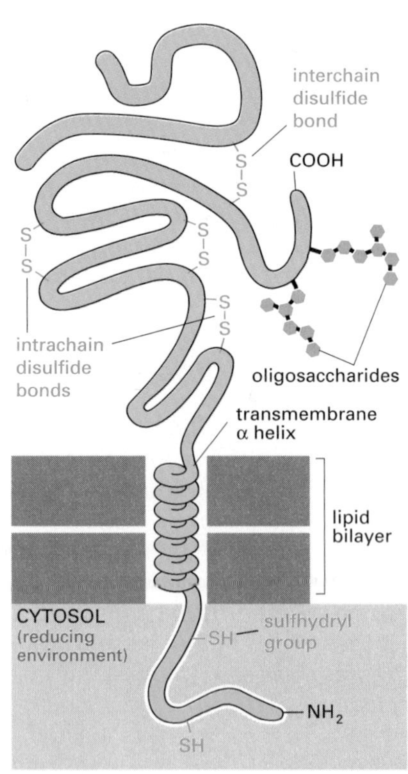

Figure 10–22 A single-pass transmembrane protein. Note that the polypeptide chain traverses the lipid bilayer as a right-handed α helix and that the oligosaccharide chains and disulfide bonds are all on the noncytosolic surface of the membrane. The sulfhydryl groups in the cytosolic domain of the protein do not normally form disulfide bonds because the reducing environment in the cytosol maintains these groups in their reduced (–SH) form.

form micelles in water (Figure 10–23). When mixed with membranes, the hydrophobic ends of detergents bind to the hydrophobic regions of the membrane proteins, thereby displacing the lipid molecules. Since the other end of the detergent molecule is polar, this binding tends to bring the membrane proteins into solution as detergent–protein complexes (although some lipid molecules may remain attached to the protein) (Figure 10–24). The polar (hydrophilic) ends of detergents can be either charged (ionic), as in *sodium dodecyl sulfate (SDS)*, or uncharged (nonionic), as in the *Triton* detergents. The structures of these two commonly used detergents are illustrated in Figure 10–25.

With strong ionic detergents, such as SDS, even the most hydrophobic membrane proteins can be solubilized. This allows them to be analyzed by *SDS polyacrylamide-gel electrophoresis* (discussed in Chapter 8), a procedure that has revolutionized the study of membrane proteins. Such strong detergents unfold (denature) proteins by binding to their internal "hydrophobic cores," thereby rendering the proteins inactive and unusable for functional studies. Nonetheless, proteins can be readily purified in their SDS-denatured form. In some cases the removal of the detergent allows the purified protein to renature, with recovery of functional activity.

Many hydrophobic membrane proteins can be solubilized and then purified in an active, if not entirely normal, form by the use of mild detergents, such as Triton X-100, which covers the membrane-spanning segments of the protein. In this way, functionally active membrane protein systems can be reconstituted from purified components, providing a powerful means of analyzing their activities (Figure 10–26).

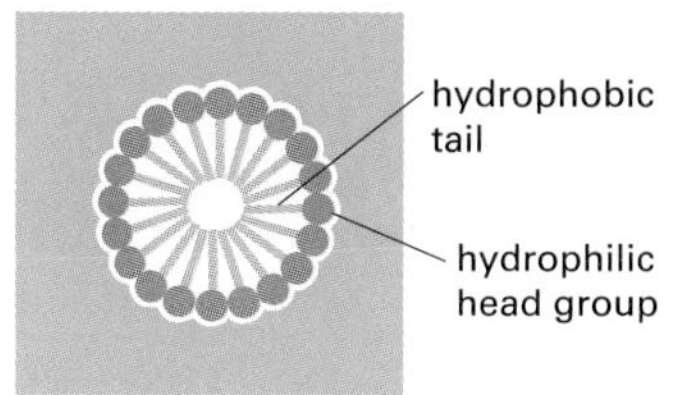

Figure 10–23 A detergent micelle in water, shown in cross section. Because they have both polar and nonpolar ends, detergent molecules are amphipathic. Because they are wedge-shaped, they form micelles rather than bilayers (see Figure 10–4).

The Cytosolic Side of Plasma Membrane Proteins Can Be Studied in Red Blood Cell Ghosts

More is known about the plasma membrane of the human red blood cell (Figure 10–27) than about any other eucaryotic membrane. There are several reasons for this. Red blood cells (also called erythrocytes) are available in large numbers (from blood banks, for example) relatively uncontaminated by other cell types.

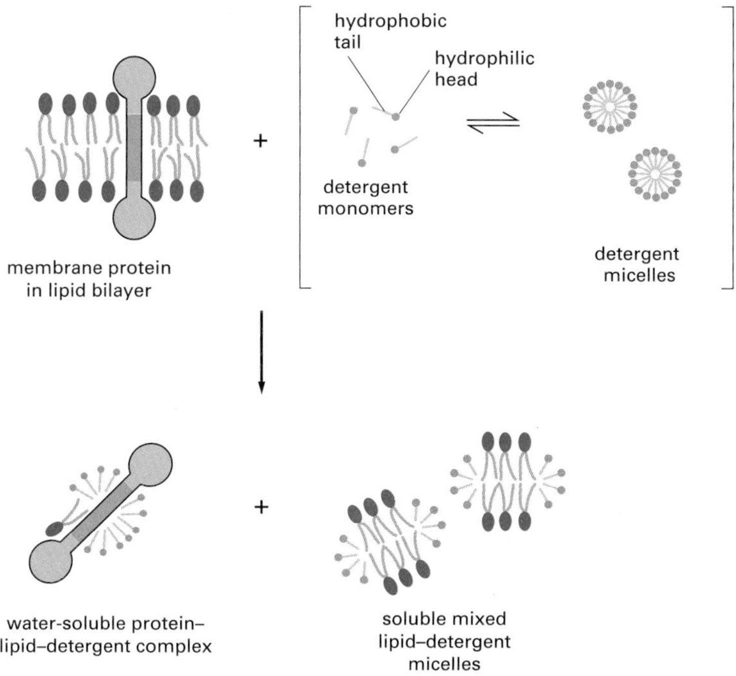

Figure 10–24 Solubilizing membrane proteins with a mild detergent. The detergent disrupts the lipid bilayer and brings the proteins into solution as protein–lipid–detergent complexes. The phospholipids in the membrane are also solubilized by the detergent.

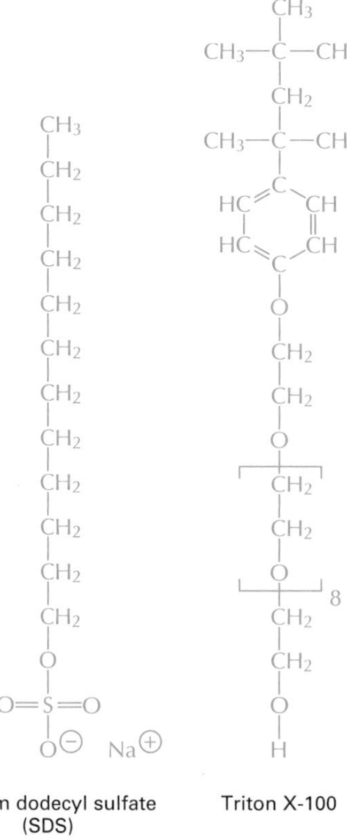

Figure 10–25 The structures of two commonly used detergents. Sodium dodecyl sulfate (SDS) is an anionic detergent, and Triton X-100 is a nonionic detergent. The hydrophobic portion of each detergent is shown in *green*, and the hydrophilic portion is shown in *blue*. The bracketed portion of Triton X-100 is repeated about eight times.

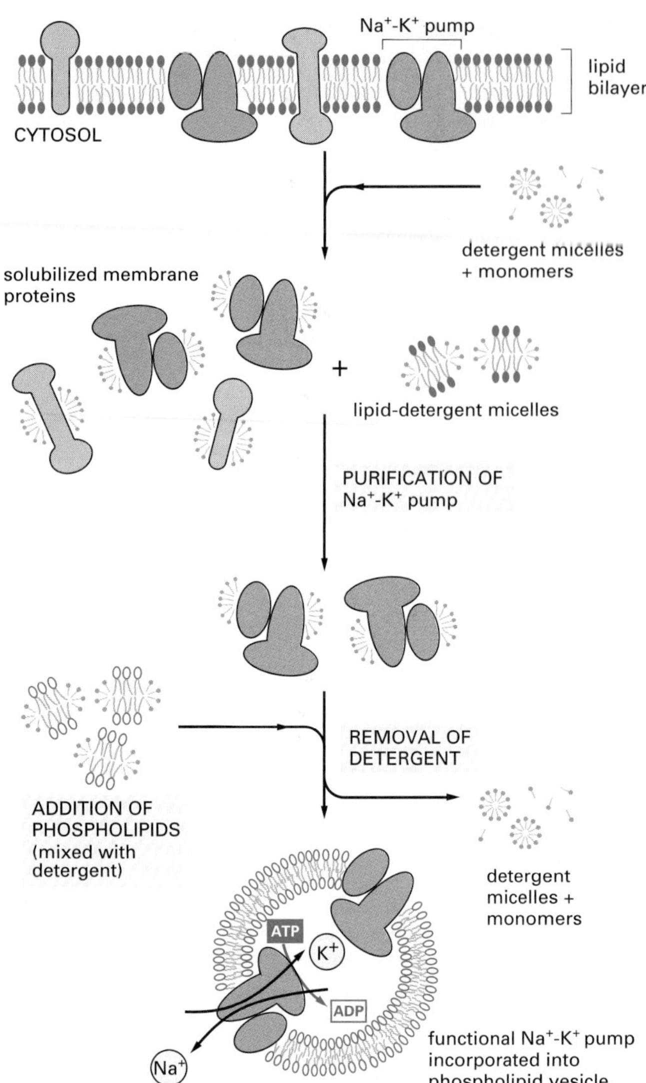

Figure 10–26 The use of mild detergents for solubilizing, purifying, and reconstituting functional membrane protein systems. In this example, functional Na^+-K^+ pump molecules are purified and incorporated into phospholipid vesicles. The Na^+-K^+ pump is an ion pump that is present in the plasma membrane of most animal cells; it uses the energy of ATP hydrolysis to pump Na^+ out of the cell and K^+ in, as discussed in Chapter 11.

Since they have no nucleus or internal organelles, the plasma membrane is their only membrane, and it can be isolated without contamination by internal membranes (thus avoiding a serious problem encountered in plasma membrane preparations from other eukaryotic cell types, in which the plasma membrane typically constitutes less than 5% of the cell's membrane).

It is easy to prepare empty red blood cell membranes, or "ghosts," by putting the cells in a medium with a lower salt concentration than the cell interior. Water then flows into the red cells, causing them to swell and burst (lyse) and release

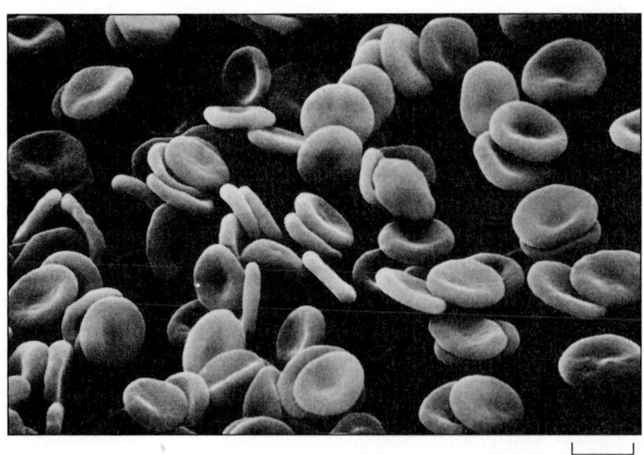

Figure 10–27 A scanning electron micrograph of human red blood cells. The cells have a biconcave shape and lack a nucleus and other organelles. (Courtesy of Bernadette Chailley.)

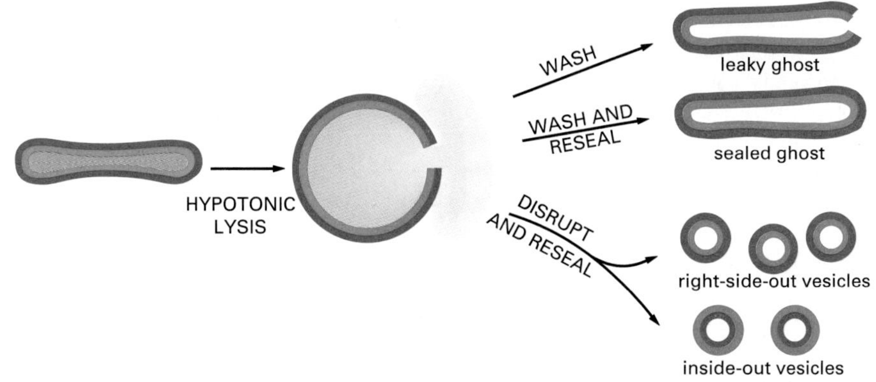

Figure 10–28 The preparation of sealed and unsealed red blood cell ghosts and of right-side-out and inside-out vesicles. As indicated, the red cells tend to rupture in only one place, giving rise to ghosts with a single hole in them. The smaller vesicles are produced by mechanically disrupting the ghosts; the orientation of the membrane in these vesicles can be either right-side-out or inside-out, depending on the ionic conditions used during the disruption and resealing procedures.

their hemoglobin and other soluble cytosolic proteins. Membrane ghosts can be studied while they are still leaky (in which case any reagent can interact with molecules on both faces of the membrane), or they can be allowed to reseal, so that water-soluble reagents cannot reach the internal face. Moreover, since sealed *inside-out* vesicles can also be prepared from red blood cell ghosts (Figure 10–28), the external side and internal (cytosolic) side of the membrane can be studied separately. The use of sealed and unsealed red cell ghosts led to the first demonstration that some membrane proteins extend across the lipid bilayer (discussed below) and that the lipid compositions of the two halves of the bilayer are different. Like most of the basic principles initially demonstrated in red blood cell membranes, these findings were later extended to the various membranes of nucleated cells and bacteria.

The "sidedness" of a membrane protein can be determined in several ways. One method is the use of a covalent labeling reagent (such as a radioactive or fluorescent marker) that is water-soluble and therefore cannot penetrate the lipid bilayer; such a marker attaches covalently only to the portion of the protein on the exposed side of the membrane. The membranes are then solubilized with detergent and the proteins are separated by SDS polyacrylamide-gel electrophoresis. The labeled proteins can be detected either by their radioactivity (by autoradiography of the gel) or by their fluorescence (by exposing the gel to ultraviolet light). By using such *vectorial labeling,* it is possible to determine how a particular protein, detected as a band on a gel, is oriented in the membrane: for example, if it is labeled from both the external side (when intact cells or sealed ghosts are labeled) and the internal (cytosolic) side (when sealed inside-out vesicles are labeled), then it must be a transmembrane protein. An alternative approach is to expose either the external or internal surface to proteolytic enzymes, which are membrane-impermeant: if a protein is partially digested from both surfaces, it must be a transmembrane protein. In addition, labeled antibodies that bind only to one part of a protein can be used to determine whether that part of a transmembrane protein is exposed on one side of the membrane or the other.

When the plasma membrane proteins of the human red blood cell are studied by SDS polyacrylamide-gel electrophoresis, approximately 15 major protein bands are detected, varying in molecular weight from 15,000 to 250,000 daltons. Three of these proteins—*spectrin, glycophorin,* and *band 3*—account for more than 60% (by weight) of the total membrane protein (Figure 10–29). Each of these proteins is arranged in the membrane in a different manner. We shall therefore use them as examples of three major ways in which proteins are associated with membranes, not only in red blood cells but in other cells as well.

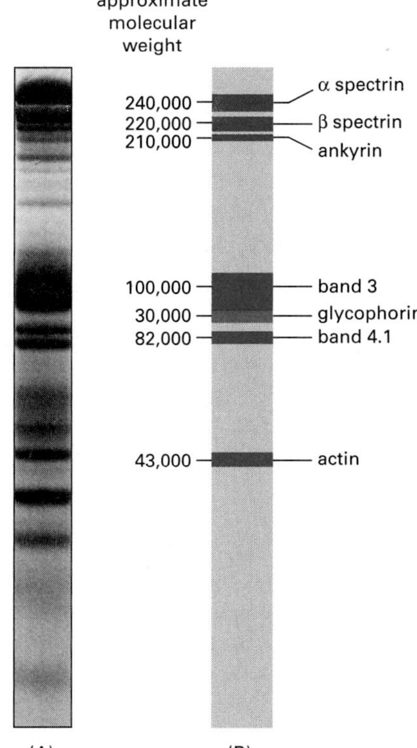

Figure 10–29 SDS polyacrylamide-gel electrophoresis pattern of the proteins in the human red blood cell membrane. (A) The gel is stained with Coomassie blue. (B) The positions of some of the major proteins in the gel are indicated in the drawing. Glycophorin is shown in *red* to distinguish it from band 3. Other bands in the gel are omitted from the drawing. The large amount of negatively charged carbohydrate in glycophorin molecules slows their migration so that they run almost as slowly as the much larger band 3 molecules. (A, courtesy of Ted Steck.)

Spectrin Is a Cytoskeletal Protein Noncovalently Associated with the Cytosolic Side of the Red Blood Cell Membrane

Most of the protein molecules associated with the human red blood cell membrane are peripheral membrane proteins bound to the cytosolic side of the lipid bilayer. The most abundant of these proteins is **spectrin**, a long, thin, flexible rod about 100 nm in length that constitutes around 25% of the membrane-associated protein mass (about 2.5×10^5 copies per cell). It is the principal component of the protein meshwork (the *cytoskeleton)* that underlies the red blood cell membrane, maintaining the structural integrity and biconcave shape of this membrane (see Figure 10–26). If the cytoskeleton is dissociated from red blood cell ghosts in low-ionic-strength solutions, the membrane fragments into small vesicles.

Spectrin is a heterodimer formed from two large, structurally similar subunits (Figure 10–30). The heterodimers self-associate head-to-head to form 200-nm-long tetramers. The tail ends of four or five tetramers are linked together by binding to short actin filaments and to other cytoskeletal proteins (including the *band 4.1 protein)* in a "junctional complex." The final result is a deformable, netlike meshwork that underlies the entire cytosolic surface of the membrane (Figure 10–31). It is this spectrin-based cytoskeleton that enables the red cell to withstand the stress on its membrane as it is forced through narrow capillaries. Mice and humans with genetic abnormalities in spectrin are anemic and have red cells that are spherical (instead of concave) and fragile; the severity of the anemia increases with the degree of the spectrin deficiency.

The protein mainly responsible for attaching the spectrin cytoskeleton to the red cell plasma membrane was identified by monitoring the binding of radiolabeled spectrin to red cell membranes from which spectrin and various other peripheral proteins had been removed. These experiments showed that the binding of spectrin depends on a large intracellular attachment protein called **ankyrin**, which attaches both to spectrin and to the cytosolic domain of the transmembrane protein *band 3* (see Figure 10–31). By connecting some of the band 3 molecules to spectrin, ankyrin links the spectrin network to the membrane; it also greatly reduces the rate of diffusion of the bound band 3 molecules in the lipid bilayer. The spectrin-based cytoskeleton is also attached to the membrane by a second mechanism, which depends on the band 4.1 protein mentioned above. This protein, which binds to spectrin and actin, also binds to the cytosolic domain of both band 3 and *glycophorin,* the other major transmembrane protein in red blood cells.

An analogous but much more elaborate and complicated cytoskeletal network exists beneath the plasma membrane of most other cells in our bodies. This network, which constitutes the cortical region (or *cortex)* of the cytosol, is rich in actin filaments, which are thought to be attached to the plasma membrane in numerous ways. Proteins that are structurally homologous to spectrin, ankyrin, and band 4.1 are present in the cortex of nucleated cells. We discuss the cortical cytoskeleton in nucleated cells and its interactions with the plasma membrane in Chapter 16.

Glycophorin Extends Through the Red Blood Cell Lipid Bilayer as a Single α Helix

Glycophorin is one of the two major proteins exposed on the outer surface of the human red blood cell and was the first membrane protein for which the complete amino acid sequence was determined. Like the model transmembrane protein shown in Figure 10–22, glycophorin is a small, single-pass transmembrane glycoprotein (131 amino acids) with most of its mass on the external surface of the membrane, where its hydrophilic N-terminal end is located. This part of the protein carries all of the carbohydrate (about 100 sugar residues in 16 separate oligosaccharide side chains), which accounts for 60% of the molecule's mass. In fact, the great majority of the total red blood cell surface carbohydrate (including more than 90% of the sialic acid and therefore most of the negative charge of the surface) is carried by glycophorin molecules. The hydrophilic C-terminal tail of glycophorin is exposed to the cytosol, while a hydrophobic α-helical segment 23 amino acids long spans the lipid bilayer (see Figure 10–20A).

Despite there being nearly a million glycophorin molecules per cell, their function remains unknown. Indeed, individuals whose red blood cells lack a major subset of these molecules seem to be perfectly healthy. Although glycophorin itself is found only in red blood cells, its structure is representative of

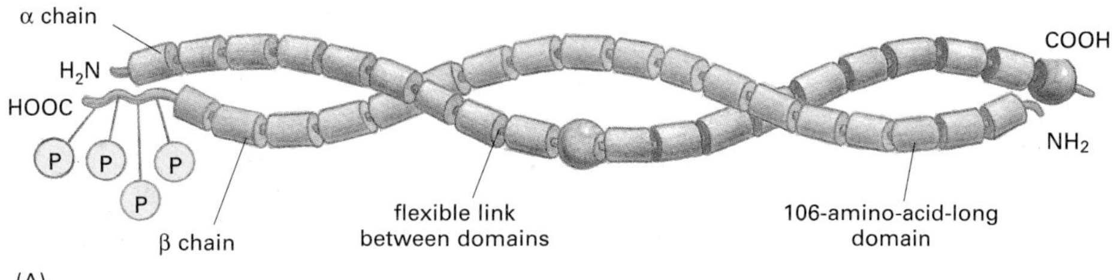

(B)

100 nm

Figure 10–30 Spectrin molecules from human red blood cells. The protein is shown (A) schematically and (B) in electron micrographs. Each spectrin heterodimer consists of two antiparallel, loosely intertwined, flexible polypeptide chains called α and β. These are attached noncovalently to each other at multiple points, including both ends. The phosphorylated "head" end, where two dimers associate to form a tetramer, is on the left. Both the α and the β chains are composed largely of repeating domains 106 amino acids long. In the micrographs, the spectrin molecules have been shadowed with platinum. (A, adapted from D.W. Speicher and V.T. Marchesi, *Nature* 311:177–180, 1984; B, courtesy of D.M. Shotton, with permission from D.M. Shotton, B.E. Burke, and D. Branton, *J. Mol. Biol.* 131:303–329, 1979. © Academic Press Inc. [London] Ltd.)

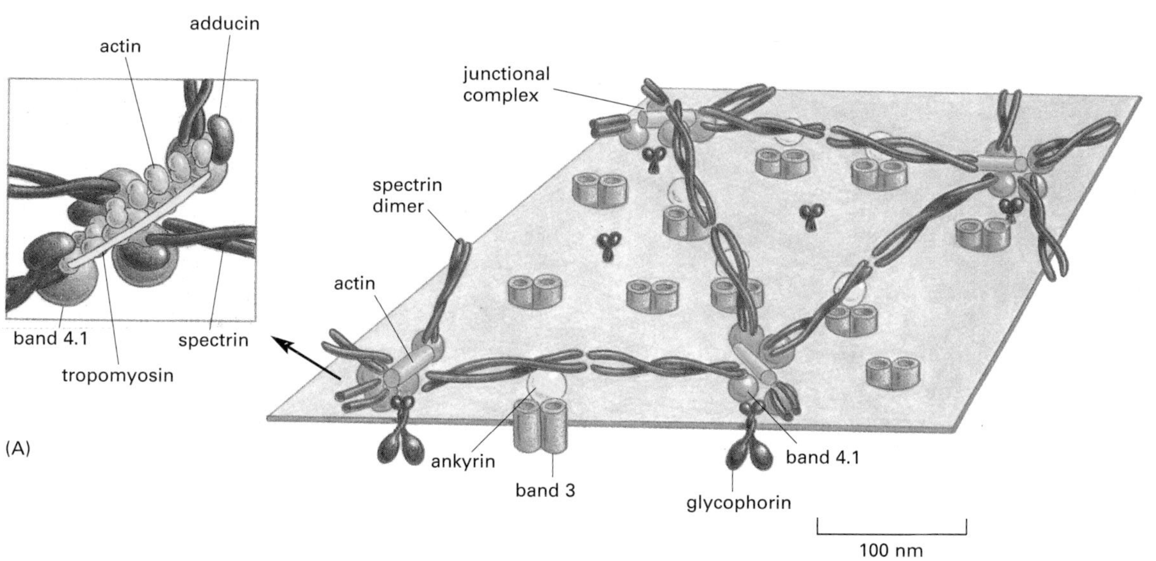

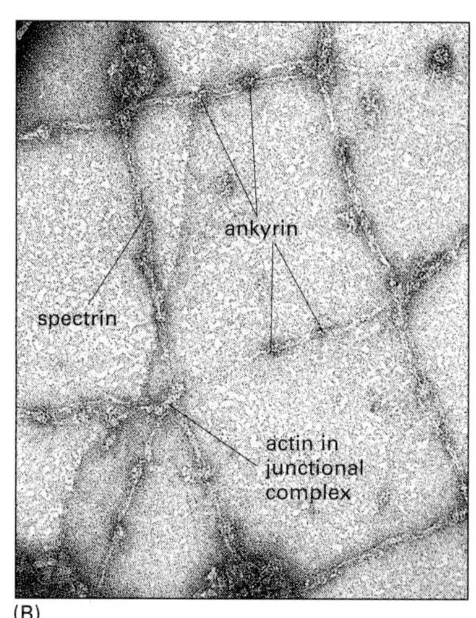

(B)

Figure 10–31 The spectrin-based cytoskeleton on the cytosolic side of the human red blood cell membrane. The structure is shown (A) schematically and (B) in an electron micrograph. The arrangement shown in the drawing has been deduced mainly from studies on the interactions of purified proteins *in vitro*. Spectrin dimers are linked together into a netlike meshwork by junctional complexes (enlarged in the box on the *left*) composed of short actin filaments (containing 13 actin monomers), band 4.1, adducin, and a tropomyosin molecule that probably determines the length of the actin filaments. The cytoskeleton is linked to the membrane by the indirect binding of spectrin tetramers to some band 3 proteins via ankyrin molecules, as well as by the binding of band 4.1 proteins to both band 3 and glycophorin (not shown). The electron micrograph shows the cytoskeleton on the cytosolic side of a red blood cell membrane after fixation and negative staining. The spectrin meshwork has been purposely stretched out to allow the details of its structure to be seen. In a normal cell, the meshwork shown would be much more crowded and occupy only about one-tenth of this area. (B, courtesy of T. Byers and D. Branton, *Proc. Natl. Acad. Sci. USA* 82:6153–6157, 1985. © National Academy of Sciences.)

a common class of membrane proteins that traverse the lipid bilayer as a single α helix. Many cell-surface receptors, for example, belong to this class.

Glycophorin normally exists as a homodimer, with its two identical chains linked primarily through noncovalent interactions between the transmembrane α helices. Thus, the transmembrane segment of a membrane protein is often more than just a hydrophobic anchor: the sequence of hydrophobic amino acids can contain information that mediates protein–protein interactions. Similarly, the individual transmembrane segments of a multipass membrane protein occupy defined positions in the folded protein structure that are determined by interactions between neighboring transmembrane α helices. Often, the cytosolic or noncytosolic loops of the polypeptide chain that link the transmembrane segments in multipass transmembrane proteins can be clipped with proteases and the resulting fragments stay together and function normally. In some cases, the separate pieces can be expressed in cells and they assemble properly to form a functional protein (Figure 10–32).

Band 3 of the Red Blood Cell Is a Multipass Membrane Protein That Catalyzes the Coupled Transport of Anions

Unlike glycophorin, the **band 3 protein** is known to have an important role in the function of red blood cells. It derives its name from its position relative to the other membrane proteins after electrophoresis in SDS polyacrylamide gels (see Figure 10–29). Like glycophorin, band 3 is a transmembrane protein, but it is a multipass membrane protein, traversing the membrane in a highly folded conformation. The polypeptide chain (about 930 amino acids long) is thought to extend across the bilayer 12 times.

The main function of red blood cells is to carry O_2 from the lungs to the tissues and to help in carrying CO_2 from the tissues to the lungs. The band 3 protein is crucial for the second of these functions. Because CO_2 is only sparingly soluble in water, it is carried in the blood plasma as bicarbonate (HCO_3^-), which is formed and broken down inside red blood cells by an enzyme that catalyzes the reaction $H_2O + CO_2 \rightleftharpoons HCO_3^- + H^+$. The band 3 protein acts as an *anion transporter*,

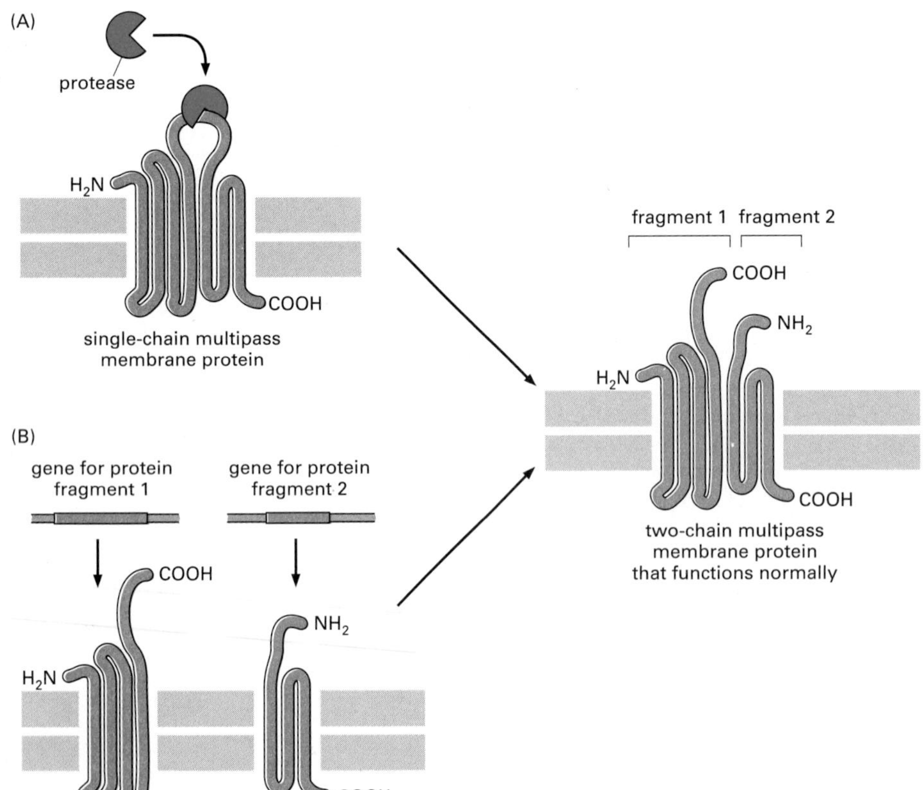

Figure 10–32 Converting a single-chain multipass protein into a two-chain multipass protein. (A) Proteolytic cleavage of one loop to create two fragments that stay together and function normally. (B) Expression of the same two fragments from separate genes gives rise to a similar protein that functions normally.

which allows HCO_3^- to cross the membrane in exchange for Cl^-. By making the red cell membrane permeable to HCO_3^-, this transporter increases the amount of CO_2 the blood can deliver to the lungs.

Band 3 proteins can be seen as distinct *intramembrane particles* by the technique of **freeze-fracture electron microscopy**. In this procedure, cells are frozen in liquid nitrogen, and the resulting block of ice is fractured with a knife. The fracture plane tends to pass through the hydrophobic middle of membrane lipid bilayers, separating them into their two monolayers (Figure 10–33). The exposed *fracture faces* are then shadowed with platinum, and the resulting platinum replica is examined with an electron microscope. When examined in this way, human red blood cell membranes appear to be studded with intramembrane particles that are relatively homogeneous in size (7.5 nm in diameter) and randomly distributed (Figure 10–34). The particles are thought to be principally band 3 molecules. When synthetic lipid bilayers are reconstituted with purified band 3 protein molecules, typical 7.5-nm intramembrane particles are observed when the bilayers are fractured. Figure 10–35 illustrates why band 3 molecules are seen in freeze-fracture electron microscopy of red blood cell membranes but glycophorin molecules probably are not.

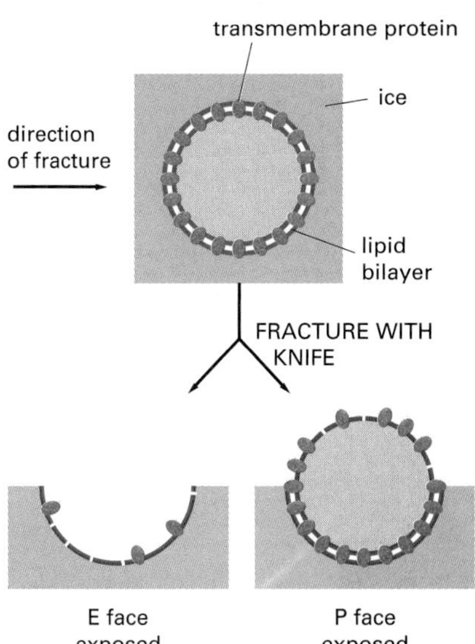

Figure 10–33 Freeze-fracture electron microscopy. This drawing shows how the technique provides images of both the hydrophobic interior of the cytosolic half of the bilayer (the P face) and the hydrophobic interior of the external half of the bilayer (the E face). After the fracturing process, the exposed fracture faces are shadowed with platinum and carbon, the organic material is digested away, and the resulting platinum replica is examined in an electron microscope (see also Figure 9–32).

To transfer small hydrophobic molecules across a membrane, a membrane transport protein must puncture the hydrophobic permeability barrier of the lipid bilayer and provide a path for the hydrophilic molecules to cross. As with the pore-forming β barrel proteins discussed earlier, the molecular architecture of multipass transmembrane proteins is ideally suited for this task. In many multipass transmembrane proteins, some of the transmembrane α helices contain both hydrophobic and hydrophilic amino acid side chains. The hydrophobic side chains lie on one side of the helix, exposed to the lipid of the membrane. The hydrophilic side chains are concentrated on the other side, where they form part of the lining of a hydrophilic pore created by packing several such amphipathic α helices side by side in a ring within the hydrophobic interior of the lipid bilayer.

In Chapter 11 we consider how multipass transmembrane proteins are thought to mediate the selective transport of small hydrophilic molecules across membranes. But a detailed understanding of how a membrane transport protein actually works requires precise information about its three-dimensional structure in the bilayer. A plasma membrane transport protein for which such detail is known is *bacteriorhodopsin*, a protein that serves as a light-activated proton (H^+) pump in the plasma membrane of certain archaea. The structure of bacteriorhodopsin is similar to that of many other membrane proteins, and it merits a brief digression here.

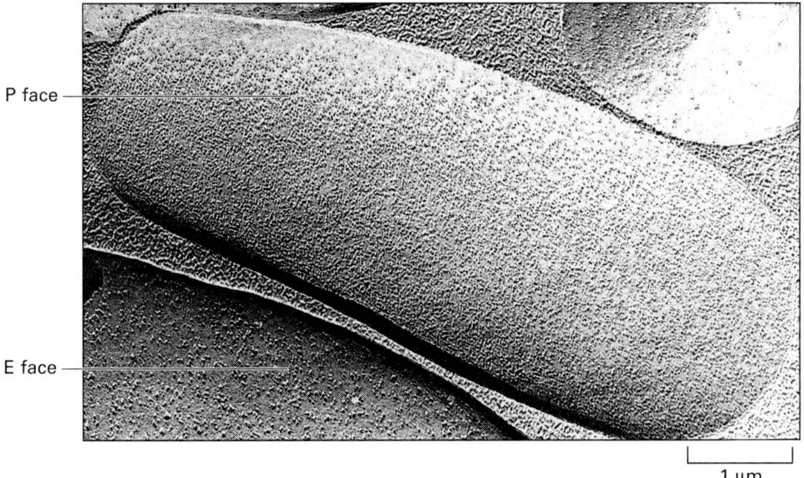

Figure 10–34 Freeze-fracture electron micrograph of human red blood cells. Note that the density of intramembrane particles on the cytosolic (P) face is higher than on the external (E) face. (Courtesy of L. Engstrom and D. Branton.)

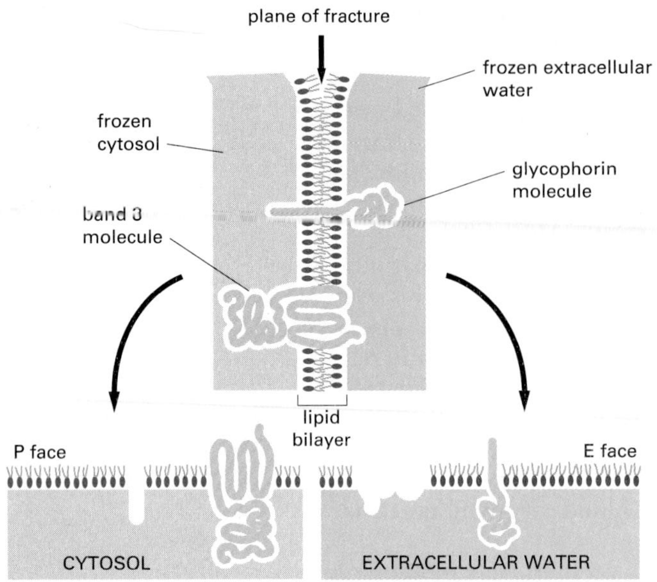

Figure 10–35 Probable fates of band 3 and glycophorin molecules in the human red blood cell membrane during freeze-fracture. When the lipid bilayer is split, either the inside or the outside half of each transmembrane protein is pulled out of the frozen monolayer with which it is associated; the protein tends to remain with the monolayer that contains the main bulk of the protein. For this reason, band 3 molecules usually remain with the inner (P) face; since they have sufficient mass above the fracture plane, they are readily seen as intramembrane particles. Glycophorin molecules usually remain with the outer (E) face, but it is thought that their cytosolic tails have insufficient mass to be seen.

Bacteriorhodopsin Is a Proton Pump That Traverses the Lipid Bilayer as Seven α Helices

The "purple membrane" of the archaean *Halobacterium salinarum* is a specialized patch in the plasma membrane that contains a single species of protein molecule, **bacteriorhodopsin** (Figure 10–36). Each bacteriorhodopsin molecule contains a single light-absorbing group, or chromophore (called *retinal),* which gives the protein its purple color. Retinal is vitamin A in its aldehyde form and is identical to the chromophore found in *rhodopsin* of the photoreceptor cells of the vertebrate eye (discussed in Chapter 15). Retinal is covalently linked to a lysine side chain of the bacteriorhodopsin protein. When activated by a single photon of light, the excited chromophore changes shape and causes a series of small conformational changes in the protein that results in the transfer of one H^+ from the inside to the outside of the cell (Figure 10–37). In bright light, each bacteriorhodopsin molecule can pump several hundred protons per second. The light-driven proton transfer establishes an H^+ gradient across the plasma membrane, which in turn drives the production of ATP by a second protein in the cell's plasma membrane, as well as other processes that use the energy stored in the H^+ gradient. Thus, bacteriorhodopsin is part of a solar energy transducer that provides energy to the bacterial cell.

To understand the function of a multipass transmembrane protein in molecular detail, it is necessary to locate each of its atoms precisely, which generally requires x-ray diffraction studies of well-ordered three-dimensional crystals of the protein. But because of their amphipathic nature, these proteins are soluble only in detergent solutions and are difficult to crystallize. The numerous bacteriorhodopsin molecules in the purple membrane, however, are arranged as a planar two-dimensional crystal. The regular packing has made it possible to determine the three-dimensional structure and orientation of bacteriorhodopsin in the membrane to high resolution (3 Å) by an alternative approach, which uses a combination of electron microscopy and electron diffraction analysis. This procedure, known as **electron crystallography**, is

Figure 10–36 Drawing of the archaean *Halobacterium salinarum* showing the patches of purple membrane that contain bacteriorhodopsin molecules. These archaeans, which live in saltwater pools where they are exposed to sunlight, have evolved a variety of light-activated proteins, including bacteriorhodopsin, which is a light-activated proton pump in the plasma membrane.

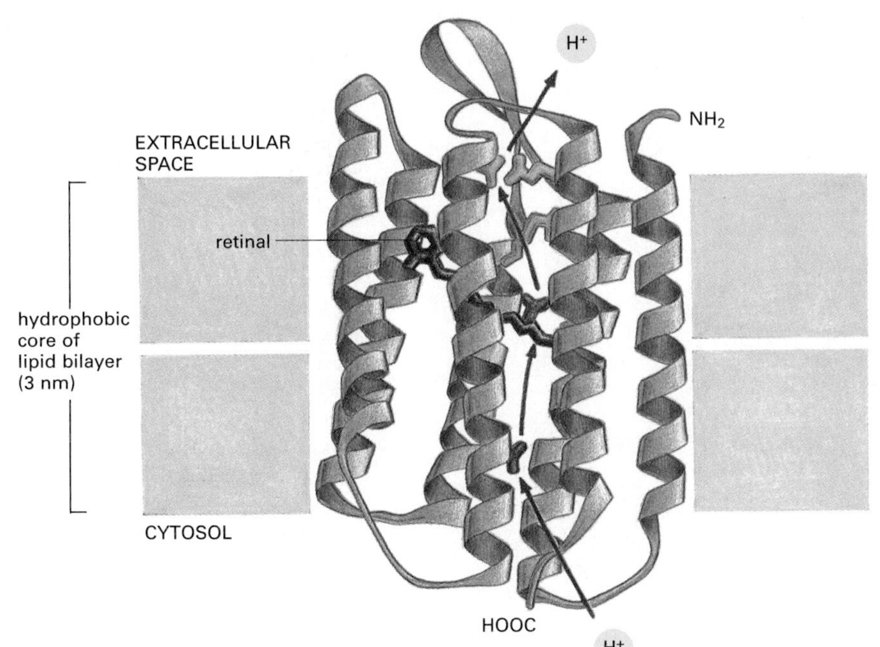

Figure 10–37 The three-dimensional structure of a bacteriorhodopsin molecule. The polypeptide chain crosses the lipid bilayer seven times as α helices. The location of the retinal chromophore *(purple)* and the probable pathway taken by protons during the light-activated pumping cycle are shown. The first and key step is the passing of a H^+ from the chromophore to the side chain of aspartic acid 85 *(red,* located next to the chromophore) that occurs upon absorption of a photon by the chromophore. Subsequently, other H^+ transfers—utilizing the hydrophilic amino acid side chains that line a path through the membrane—complete the pumping cycle and return the enzyme to its starting state. Color code: glutamic acid *(orange)*, aspartic acid *(red)*, arginine *(blue)*. (Adapted from H. Luecke et al., *Science* 286:255–260, 1999.)

analogous to the study of three-dimensional crystals of soluble proteins by x-ray diffraction analysis. The structure obtained by electron crystallography was later confirmed and extended to higher resolution by x-ray crystallography. These studies showed that each bacteriorhodopsin molecule is folded into seven closely packed α helices (each containing about 25 amino acids), which pass through the lipid bilayer at slightly different angles (see Figure 10–37). By freezing the protein crystals at very low temperatures, it has been possible to solve the structures of some of the intermediate conformations the protein goes through during its pumping cycle.

Bacteriorhodopsin is a member of a large superfamily of membrane proteins with similar structures but different functions. For example, rhodopsin in rod cells of the vertebrate retina and many cell-surface receptor proteins that bind extracellular signal molecules are also built from seven transmembrane α helices. These proteins function as signal transducers rather than as transporters: each responds to an extracellular signal by activating another protein inside the cell, which generates chemical signals in the cytosol, as we discuss in Chapter 15. Although the structures of bacteriorhodopsins and mammalian signaling receptors are strikingly similar, they show no sequence similarity and thus probably belong to two evolutionarily distant branches of an ancient protein family.

Membrane Proteins Often Function as Large Complexes

Some membrane proteins function as part of multicomponent complexes. A few of these have been studied by x-ray crystallography. One is a bacterial *photosynthetic reaction center,* which was the first transmembrane protein complex to be crystallized and analyzed by x-ray diffraction. The results of this analysis were of general importance to membrane biology because they showed for the first time how multiple polypeptides associate in a membrane to form a complex protein machine (Figure 10–38). In Chapter 14 we discuss how such photosynthetic complexes function to capture light energy and use it to pump H^+ across the membrane. In that chapter we also discuss the structure and function of membrane protein complexes that are even larger than the photosynthetic reaction center. Membrane proteins are often arranged in large complexes, not only for harvesting various forms of energy but also for transducing extracellular signals into intracellular ones (discussed in Chapter 15).

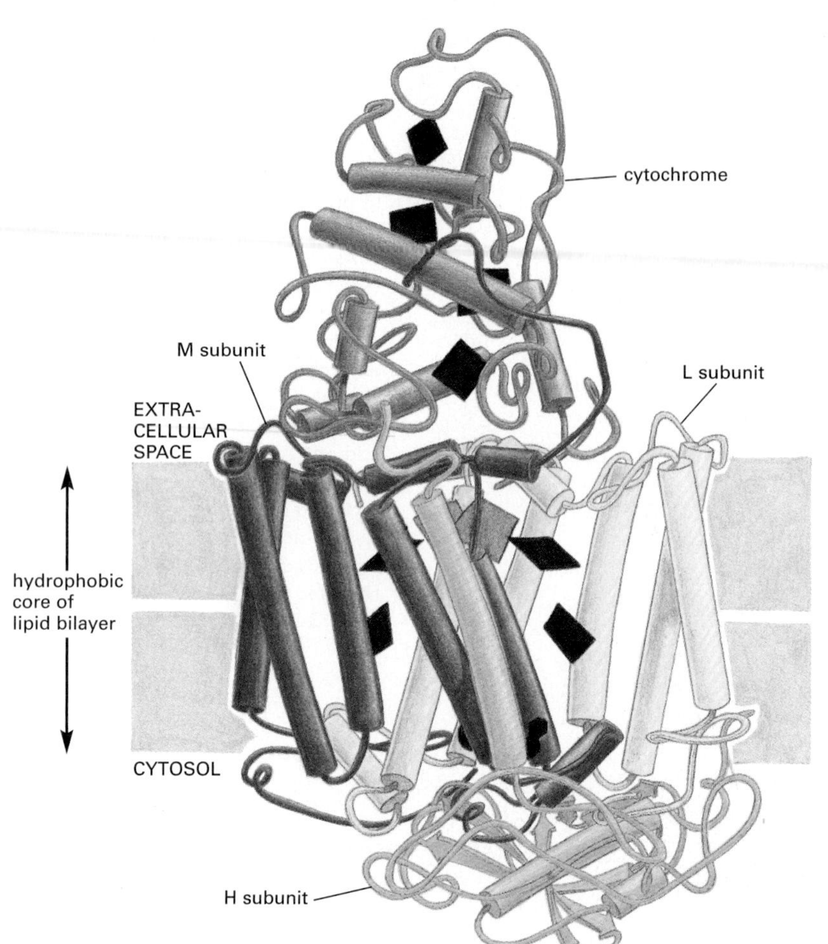

Figure 10–38 The three-dimensional structure of the photosynthetic reaction center of the bacterium *Rhodopseudomonas viridis*. The structure was determined by x-ray diffraction analysis of crystals of this transmembrane protein complex. The complex consists of four subunits L, M, H, and a cytochrome. The L and M subunits form the core of the reaction center, and each contains five α helices that span the lipid bilayer. The locations of the various electron carrier coenzymes are shown in *black*. Note that the coenzymes are arranged in the spaces between the helices. (Adapted from a drawing by J. Richardson based on data from J. Deisenhofer, O. Epp, K. Miki, R. Huber, and H. Michel, *Nature* 318:618–624, 1985.)

Many Membrane Proteins Diffuse in the Plane of the Membrane

Like membrane lipids, membrane proteins do not tumble *(flip-flop)* across the lipid bilayer, but they do rotate about an axis perpendicular to the plane of the bilayer *(rotational diffusion)*. In addition, many membrane proteins are able to move laterally within the membrane *(lateral diffusion)*. The first direct evidence that some plasma membrane proteins are mobile in the plane of the membrane was provided by an experiment in which mouse cells were artificially fused with human cells to produce hybrid cells *(heterocaryons)*. Two differently labeled antibodies were used to distinguish selected mouse and human plasma membrane proteins. Although at first the mouse and human proteins were confined to their own halves of the newly formed heterocaryon, the two sets of proteins diffused and mixed over the entire cell surface within half an hour or so (Figure 10–39).

The lateral diffusion rates of membrane proteins can be measured by using the technique of *fluorescence recovery after photobleaching (FRAP)*. The method usually involves marking the membrane protein of interest with a specific fluorescent group. This can be done either with a fluorescent ligand such as an antibody that binds to the protein or with recombinant DNA technology to express the protein fused to green fluorescent protein (GFP) (discussed in Chapter 9). The fluorescent group is then bleached in a small area by a laser beam, and the time taken for adjacent membrane proteins carrying unbleached ligand or GFP to diffuse into the bleached area is measured (Figure 10–40A). A complementary technique is *fluorescence loss in photobleaching (FLIP)*. Here, a laser beam continuously irradiates a small area to bleach all the fluorescent molecules that diffuse into it, thereby gradually depleting the surrounding membrane of fluorescently labeled molecules (Figure 10–40B). From such measurements one can calculate the diffusion coefficient for the particular cell-surface protein that was marked. The values of the diffusion coefficients for different membrane proteins in different cells are highly variable, because interactions with other proteins impede the diffusion of the proteins to varying degrees. Measurements of proteins that are minimally impeded in this way indicate that cell membranes have a viscosity comparable to that of olive oil.

Cells Can Confine Proteins and Lipids to Specific Domains Within a Membrane

The recognition that biological membranes are two-dimensional fluids was a major advance in understanding membrane structure and function. It has become clear, however, that the picture of a membrane as a lipid sea in which all proteins float freely is greatly oversimplified. Many cells have ways of confining membrane proteins to specific domains in a continuous lipid bilayer. In epithelial cells, such as those that line the gut or the tubules of the kidney, certain plasma membrane enzymes and transport proteins are confined to the apical surface of the cells, whereas others are confined to the basal and lateral surfaces (Figure 10–41). This asymmetric distribution of membrane proteins is often essential for the function of the epithelium, as we discuss in Chapter 19. The lipid compositions of these two membrane domains are also different, demonstrating that epithelial cells can prevent the diffusion of lipid as well as protein molecules between the domains. However, experiments with labeled lipids

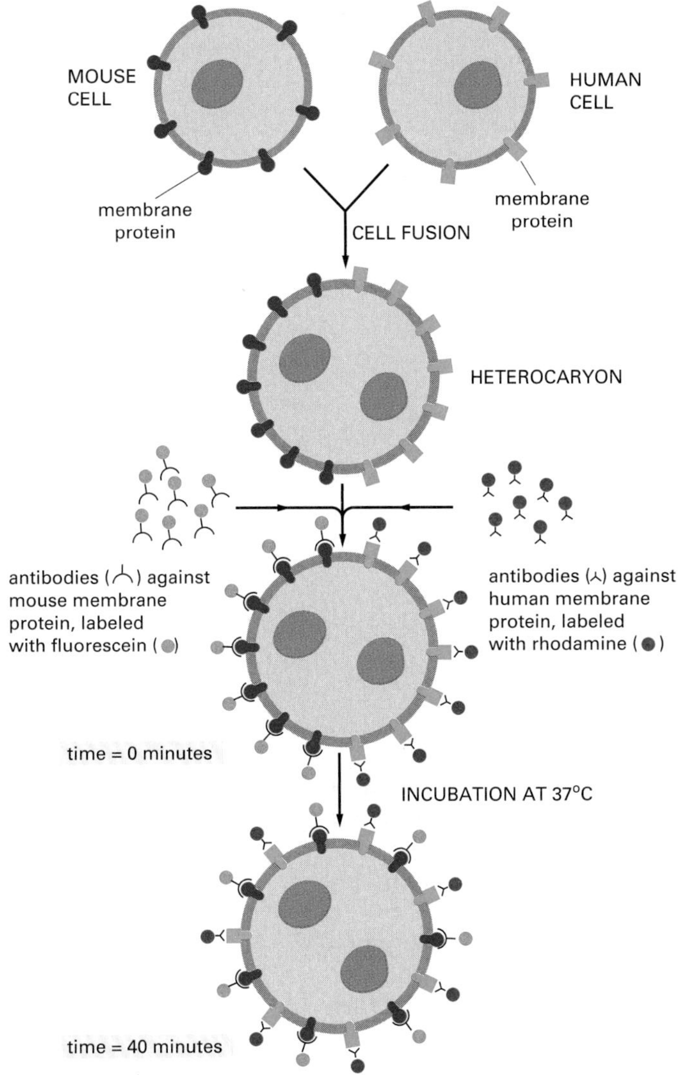

Figure 10–39 An experiment demonstrating the mixing of plasma membrane proteins on mouse–human hybrid cells. The mouse and human proteins are initially confined to their own halves of the newly formed heterocaryon plasma membrane, but they intermix with time. The two antibodies used to visualize the proteins can be distinguished in a fluorescence microscope because fluorescein is green whereas rhodamine is red. (Based on observations of L.D. Frye and M. Edidin, *J. Cell Sci.* 7:319–335, 1970. © The Company of Biologists.)

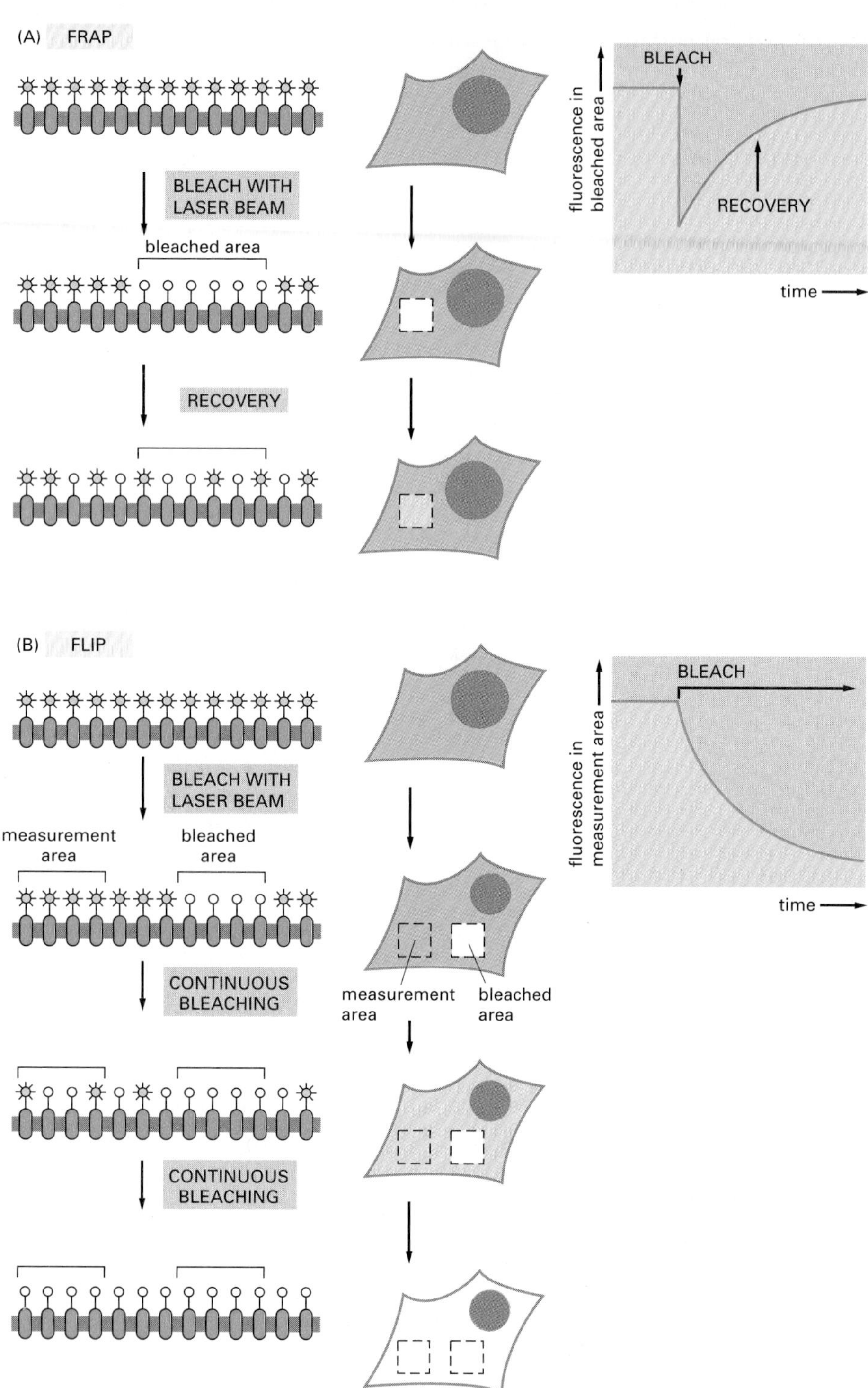

Figure 10–40 Measuring the rate of lateral diffusion of a membrane protein by photobleaching techniques. A specific protein of interest can be labeled with a fluorescent antibody (as shown here), or it can be expressed as a fusion protein with green fluorescent protein (GFP), which is intrinsically fluorescent. (A) In the FRAP technique, fluorescent molecules are bleached in a small area using a laser beam. The fluorescence intensity recovers as the bleached molecules diffuse away and unbleached molecules diffuse into the irradiated area (shown here in side and top views). The diffusion coefficient is calculated from a graph of the rate of recovery: the greater the diffusion coefficient of the membrane protein, the faster the recovery. (B) In the FLIP technique, an area in the membrane is irradiated continuously, and fluorescence is measured in a separate area. Fluorescence in the second area progressively decreases as fluorescent proteins diffuse out and bleached molecules diffuse in; eventually, all of the fluorescent protein molecules will be bleached, as long as they are mobile and not permanently anchored to the cytoskeleton or extracellular matrix.

suggest that only lipid molecules in the outer monolayer of the membrane are confined in this way. The separation of both protein and lipid molecules is thought to be maintained, at least in part, by the barriers set up by a specific type of intercellular junction (called a *tight junction,* discussed in Chapter 19). Clearly, the membrane proteins that form these intercellular junctions cannot be allowed to diffuse laterally in the interacting membranes.

A cell can also create membrane domains without using intercellular junctions. The mammalian spermatozoon, for instance, is a single cell that consists of several structurally and functionally distinct parts covered by a continuous plasma membrane. When a sperm cell is examined by immunofluorescence microscopy with a variety of antibodies, each of which reacts with a specific

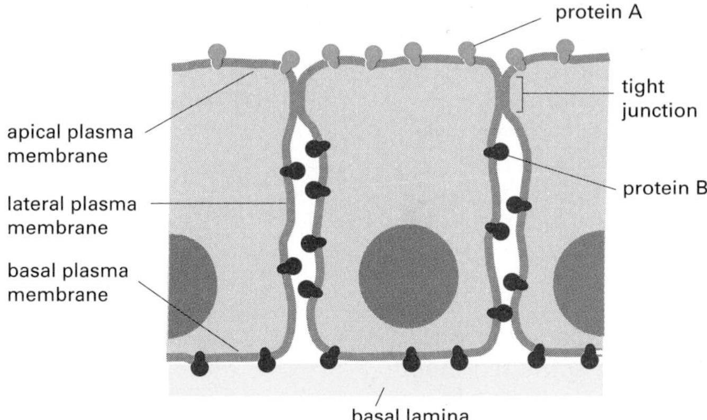

Figure 10–41 How a plasma membrane protein is restricted to a particular membrane domain. In this drawing of an epithelial cell, protein A (in the apical membrane) and protein B (in the basal and lateral membranes) can diffuse laterally in their own domains but are prevented from entering the other domain, at least partly by the specialized cell junction called a tight junction. Lipid molecules in the outer (noncytosolic) monolayer of the plasma membrane are likewise unable to diffuse between the two domains; lipids in the inner (cytosolic) monolayer, however, are able to do so (not shown).

cell-surface molecule, the plasma membrane is found to consist of at least three distinct domains (Figure 10–42). Some of the membrane molecules are able to diffuse freely within the confines of their own domain. The molecular nature of the "fence" that prevents the molecules from leaving their domain is not known. Many other cells have similar membrane fences that restrict membrane protein diffusion to certain membrane domains. The plasma membrane of nerve cells, for example, contains a domain enclosing the cell body and dendrites and another enclosing the axon. In this case, it is thought that a belt of actin filaments tightly associated with the plasma membrane at the cell-body–axon junction forms part of the barrier.

In all of these examples, the diffusion of protein and lipid molecules is confined to specialized domains within a continuous plasma membrane. Cells are known to have a variety of ways of immobilizing their membrane proteins. The bacteriorhodopsin molecules in the purple membrane of *Halobacterium* assemble into large two-dimensional crystals in which the individual protein molecules are relatively fixed in relationship to one another; large aggregates of

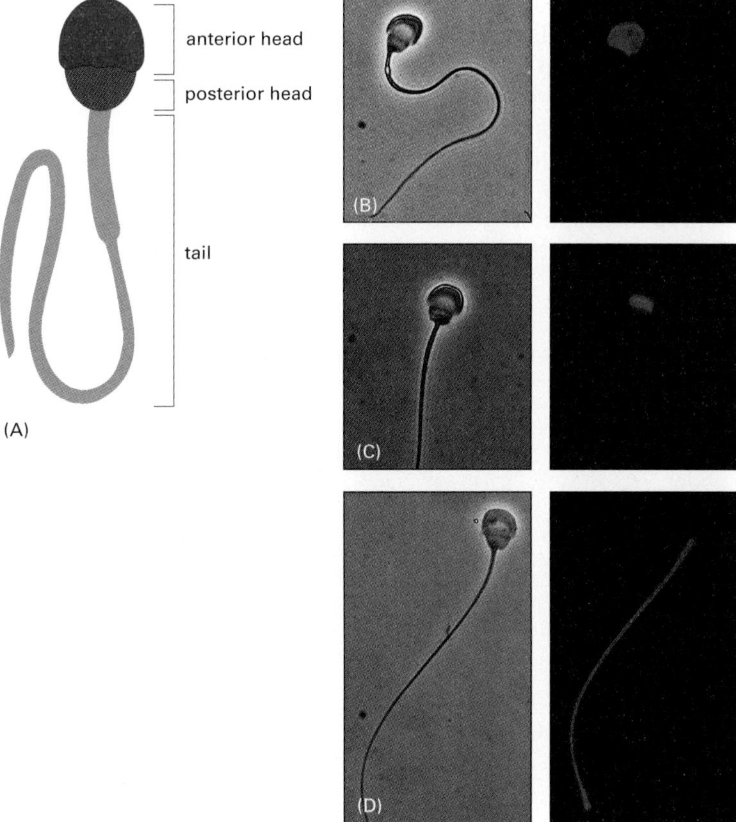

Figure 10–42 Three domains in the plasma membrane of a guinea pig sperm cell. (A) A basic drawing of a guinea pig sperm. In the three pairs of micrographs, phase-contrast micrographs are on the *left*, and the same cell is shown with cell-surface immunofluorescence staining on the *right*. Different monoclonal antibodies label selectively cell-surface molecules on (B) the anterior head, (C) the posterior head, and (D) the tail. (Micrographs courtesy of Selena Carroll and Diana Myles.)

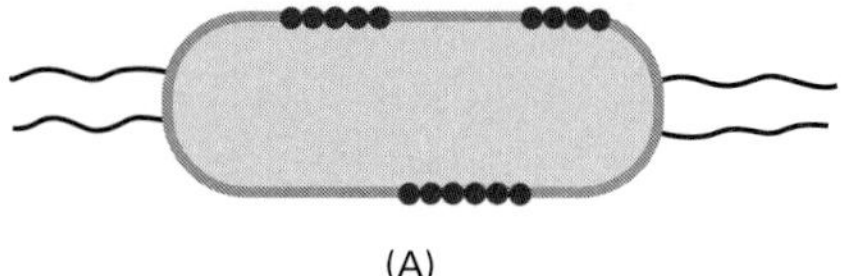

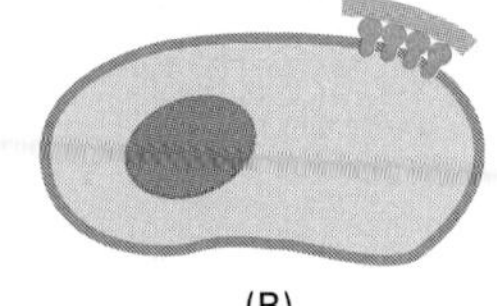

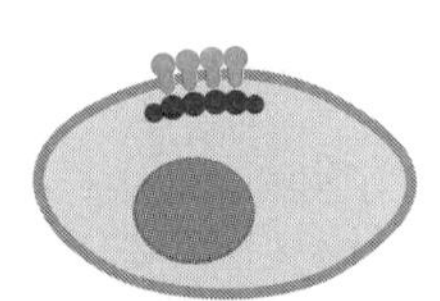

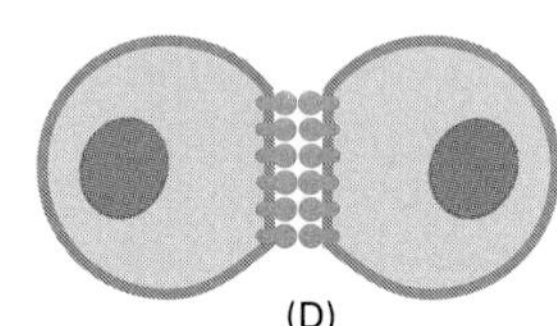

Figure 10–43 Four ways of restricting the lateral mobility of specific plasma membrane proteins. (A) The proteins can self-assemble into large aggregates (as seen for bacteriorhodopsin in the purple membrane of *Halobacterium*); they can be tethered by interactions with assemblies of macromolecules (B) outside or (C) inside the cell; or they can interact with proteins on the surface of another cell (D).

this kind diffuse very slowly. A more common way of restricting the lateral mobility of specific membrane proteins is to tether them to macromolecular assemblies either inside or outside the cell. We have seen that some red blood cell membrane proteins are anchored to the cytoskeleton inside. In other cell types, plasma membrane proteins can be anchored to the cytoskeleton, or to the extracellular matrix, or to both. The four known ways of immobilizing specific membrane proteins are summarized in Figure 10–43.

The Cell Surface Is Coated with Sugar Residues

Plasma membrane proteins, as a rule, do not protrude naked from the exterior of the cell. They are usually decorated by carbohydrates, which coat the surface of all eucaryotic cells. These carbohydrates occur as oligosaccharide chains covalently bound to membrane proteins (glycoproteins) and lipids (glycolipids). They also occur as the polysaccharide chains of integral membrane *proteoglycan* molecules. Proteoglycans, which consist of long polysaccharide chains linked covalently to a protein core, are found mainly outside the cell as part of the extracellular matrix (discussed in Chapter 19). But for some proteoglycans, the protein core either extends across the lipid bilayer or is attached to the bilayer by a glycosylphosphotidylinositol (GPI) anchor.

The term **cell coat** or **glycocalyx** is often used to describe the carbohydrate-rich zone on the cell surface. This zone can be visualized by a variety of stains, such as ruthenium red (Figure 10–44), as well as by its affinity for carbohydrate-binding proteins called **lectins**, which can be labeled with a fluorescent dye or some other visible marker. Although most of the carbohydrate is attached to intrinsic plasma membrane molecules, the glycocalyx also contains both glycoproteins and proteoglycans that have been secreted into the extracellular space and then adsorbed on the cell surface (Figure 10–45). Many of these adsorbed macromolecules are components of the extracellular matrix, so that where the plasma membrane ends and the extracellular matrix begins is largely a matter of semantics. One of the likely functions of the cell coat is to protect cells against mechanical and chemical damage and to keep foreign objects and other cells at a distance, preventing undesirable protein–protein interactions.

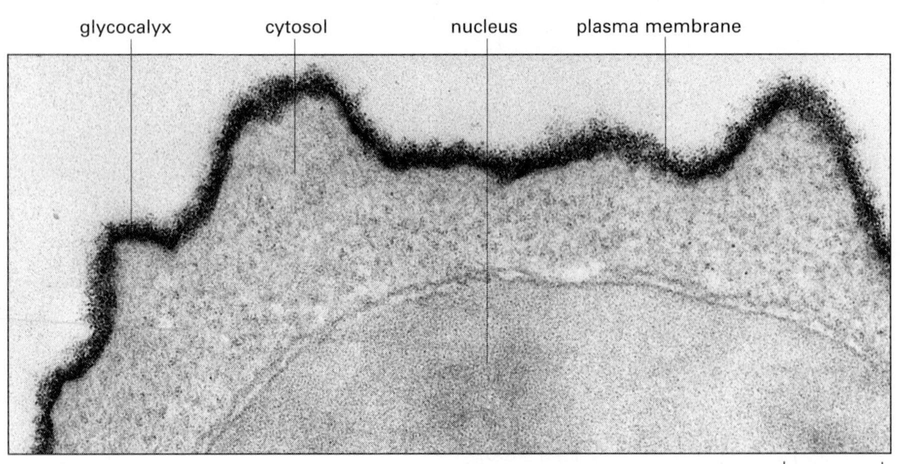

Figure 10–44 The cell coat, or glycocalyx. This electron micrograph of the surface of a lymphocyte stained with ruthenium red emphasizes the thick carbohydrate layer surrounding the cell. (Courtesy of Audrey M. Glauert and G.M.W. Cook.)

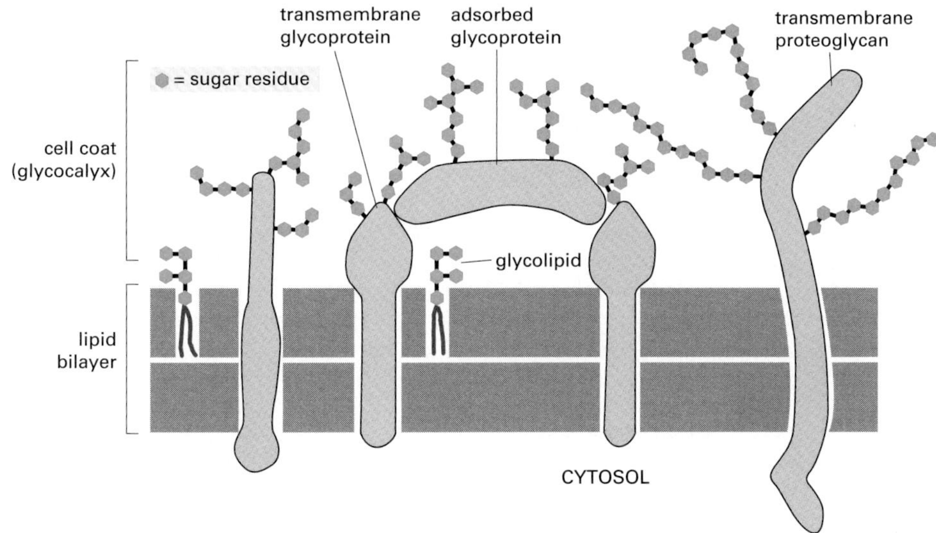

Figure 10–45 Simplified diagram of the cell coat (glycocalyx). The cell coat is made up of the oligosaccharide side chains of glycolipids and integral membrane glycoproteins and the polysaccharide chains on integral membrane proteoglycans. In addition, adsorbed glycoproteins and adsorbed proteoglycans (not shown) contribute to the glycocalyx in many cells. Note that all of the carbohydrate is on the noncytosolic surface of the membrane.

The oligosaccharide side chains of glycoproteins and glycolipids are enormously diverse in their arrangement of sugars. Although they usually contain fewer than 15 sugars, they are often branched, and the sugars can be bonded together by a variety of covalent linkages—unlike the amino acids in a polypeptide chain, which are all linked by identical peptide bonds. Even three sugars can be put together to form hundreds of different trisaccharides. In principle, both the diversity and the exposed position of the oligosaccharides on the cell surface make them especially well suited to a function in specific cell-recognition processes. For many years there was little evidence for this suspected function. More recently, however, plasma-membrane-bound lectins have been identified that recognize specific oligosaccharides on cell-surface glycolipids and glycoproteins. As we discuss in Chapter 19, these lectins are now known to mediate a variety of transient cell–cell adhesion processes, including those occurring in sperm–egg interactions, blood clotting, lymphocyte recirculation, and inflammatory responses.

Summary

Whereas the lipid bilayer determines the basic structure of biological membranes, proteins are responsible for most membrane functions, serving as specific receptors, enzymes, transport proteins, and so on. Many membrane proteins extend across the lipid bilayer. In some of these transmembrane proteins, the polypeptide chain crosses the bilayer as a single α helix (single-pass proteins). In others, including those responsible for the transmembrane transport of ions and other small water-soluble molecules, the polypeptide chain crosses the bilayer multiple times—either as a series of α helices or as a β sheet in the form of a closed barrel (multipass proteins). Other membrane-associated proteins do not span the bilayer but instead are attached to either side of the membrane. Many of these are bound by noncovalent interactions with transmembrane proteins, but others are bound via covalently attached lipid groups. Like the lipid molecules in the bilayer, many membrane proteins are able to diffuse rapidly in the plane of the membrane. However, cells have ways of immobilizing specific membrane proteins and of confining both membrane protein and lipid molecules to particular domains in a continuous lipid bilayer.

In the plasma membrane of all eucaryotic cells, most of the proteins exposed on the cell surface and some of the lipid molecules in the outer lipid monolayer have oligosaccharide chains covalently attached to them. Plasma membranes also contain integral proteoglycan molecules with surface-exposed polysaccharide chains. The resulting sugar coating is thought to protect the cell surface from mechanical and chemical damage. In addition, some of the oligosaccharide chains are recognized by cell-surface carbohydrate-binding proteins (lectins) that help mediate cell–cell adhesion events.

References

General

Bretscher MS (1985) The molecules of the cell membrane. *Sci. Am.* 253(4), 100–109.

Jacobson K, Sheets ED & Simson R (1995) Revisiting the fluid mosaic model of membranes. *Science* 268, 1441–1442.

Lipowsky R & Sackmann E (eds) (1995) The Structure and Dynamics of Membranes. Amsterdam: Elsevier.

Singer SJ & Nicolson GL (1972) The fluid mosaic model of the structure of cell membranes. *Science* 175, 720–731.

The Lipid Bilayer

Bevers EM, Comfurius P, Dekkers DW & Zwaal RF (1999) Lipid translocation across the plasma membrane of mammalian cells. *Biochim. Biophys. Acta* 1439, 317–330.

Chapman D & Benga G (1984) Biomembrane fluidity – studies of model and natural membranes. In Biological Membranes (Chapman D ed) Vol 5, pp 1–56. London: Academic Press.

Devaux PF (1993) Lipid transmembrane asymmetry and flip-flop in biological membranes and in lipid bilayers. *Curr. Opin. Struct. Biol.* 3, 489–494.

Dowhan W (1997) Molecular basis for membrane phospholipid diversity: why are there so many lipids? *Annu. Rev. Biochem.* 66, 199–232.

Hakomori S (1986) Glycosphingolipids. *Sci. Am.* 254(4), 44–53.

Harder T & Simons K (1997) Caveolae, DIGs, and the dynamics of sphingolipid-cholesterol microdomains. *Curr. Opin. Cell Biol.* 9, 534–542.

Ichikawa S & Hirabayashi Y (1998) Glucosylceramide synthase and glycosphingolipid synthesis. *Trends Cell Biol.* 8, 198–202.

Jain MK & White HBD (1977) Long-range order in biomembranes. *Adv. Lipid Res.* 15, 1–60.

Kornberg RD & McConnell HM (1971) Lateral diffusion of phospholipids in a vesicle membrane. *Proc. Natl. Acad. Sci. USA* 68, 2564–2568.

Menon AK (1995) Flippases. *Trends Cell Biol.* 5, 355–360.

Rothman J & Lenard J (1977) Membrane asymmetry. *Science* 195, 743–753.

Simons K & Ikonen E (1997) Functional rafts in cell membranes. *Nature* 387, 569–572.

Tanford C (1980) The Hydrophobic Effect: Formation of Micelles and Biological Membranes. New York: Wiley.

van Meer G (1989) Lipid traffic in animal cells. *Annu. Rev. Cell Biol.* 5, 247–275.

Membrane Proteins

Bennett V & Baines A (2001) Spectrin and ankyrin-based pathways: metazoan inventions for integrating cells into tissues. *Physiol. Rev.* 81, 1353–1392.

Branden C & Tooze J (1999) Introduction to Protein Structure, 2nd edn. New York: Garland.

Bretscher MS & Raff MC (1975) Mammalian plasma membranes. *Nature* 258, 43–49.

Buchanan SK (1999) β-Barrel proteins from bacterial outer membranes: structure, function and refolding. *Curr. Opin. Struct. Biol.* 9, 455–461.

Cross GA (1990) Glycolipid anchoring of plasma membrane proteins. *Annu. Rev. Cell Biol.* 6, 1–39.

Deisenhofer J & Michel H (1991) Structures of bacterial photosynthetic reaction centers. *Annu. Rev. Cell Biol.* 7, 1–23.

Drickamer K & Taylor ME (1993) Biology of animal lectins. *Annu. Rev. Cell Biol.* 9, 237–264.

Driscoll PC & Vuidepot AL (1999) Peripheral membrane proteins: FYVE sticky fingers. *Curr. Biol.* 9, R857–R860.

Edidin M (1992) Patches, posts and fences: proteins and plasma membrane domains. *Trends Cell Biol.* 2, 376–380.

Eisenberg D (1984) Three-dimensional structure of membrane and surface proteins. *Annu. Rev. Biochem.* 53, 595–623.

Frye LD & Edidin M (1970) The rapid intermixing of cell surface antigens after formation of mouse–human heterokaryons. *J. Cell Sci.* 7, 319–335.

Gahmberg CG & Tolvanen M (1996) Why mammalian cell surface proteins are glycoproteins. *Trends Biochem. Sci.* 21, 308–311.

Haupts U, Tittor J & Oesterhelt D (1999) Closing in on bacteriorhodopsin: progress in understanding the molecule. *Annu. Rev. Biophys. Biomol. Struct.* 28, 367–399.

Helenius A & Simons K (1975) Solubilization of membranes by detergents. *Biochim. Biophys. Acta* 415, 29–79.

Henderson R & Unwin PNT (1975) Three-dimensional model of purple membrane obtained by electron microscopy. *Nature* 257, 28–32.

Jacobson K & Vaz WLC (1992) Domains in biological membranes. *Comm. Mol. Cell. Biophys.* 8, 1–114.

Kyte J & Doolittle RF (1982) A simple method for displaying the hydropathic character of a protein. *J. Mol. Biol.* 157, 105–132.

Leevers SJ, Vanhaesebroeck B & Waterfield MD (1999) Signalling through phosphoinositide 3-kinases: the lipids take centre stage. *Curr. Opin. Cell Biol.* 11, 219–225.

Marchesi VT, Furthmayr H & Tomit M (1976) The red cell membrane. *Annu. Rev. Biochem.* 45, 667–698.

Pumplin DW & Bloch RJ (1993) The membrane skeleton. *Trends Cell Biol.* 3, 113–117.

Reithmeier RAF (1993) The erythrocyte anion transporter (band 3). *Curr. Opin. Cell Biol.* 3, 515–523.

Rodgers W & Glaser M (1993) Distributions of proteins and lipids in the erythrocyte membrane. *Biochemistry* 32, 12591–12598.

Sharon N & Lis H (1995) Lectins – proteins with a sweet tooth: functions in cell recognition. *Essays Biochem.* 30, 59–75.

Sheets ED, Simson R & Jacobson K (1995) New insights into membrane dynamics from the analysis of cell surface interactions by physical methods. *Curr. Opin. Cell Biol.* 7, 707–714.

Sheetz MP (1995) Cellular plasma membrane domains. *Mol. Membr. Biol.* 12, 89–91.

Silvius JR (1992) Solubilization and functional reconsititution of biomembrane components. *Annu. Rev. Biophys. Biomol. Struct.* 21, 323–348.

Steck TL (1974) The organization of proteins in the human red blood cell membrane. *J. Cell Biol.* 62, 1–19.

Stevens TJ & Arkin IT (1999) Are membrane proteins "inside-out" proteins? *Proteins: Struct. Funct. Genet.* 36, 135–143.

Subramaniam S (1999) The structure of bacteriorhodopsin: an emerging consensus. *Curr. Opin. Struct. Biol.* 9, 462–468.

Viel A & Branton D (1996) Spectrin: on the path from structure to function. *Curr. Opin. Cell Biol.* 8, 49–55.

Vince JW & Reithmeier RA (1996) Structure of the band 3 transmembrane domain. *Cell Mol. Biol. (Noisy-le-Grand)* 42, 1041–1051.

Wallin E & von Heijne G (1998) Genome-wide analysis of integral membrane proteins from eubacterial, archaean, and eukaryotic organisms. *Protein Sci.* 7, 1029–1038.

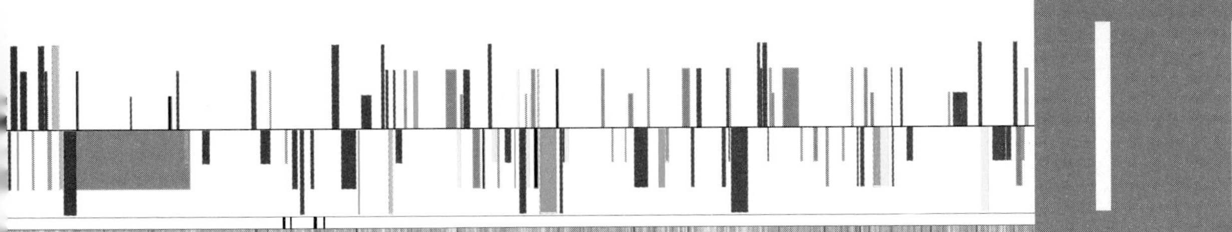

11

MEMBRANE TRANSPORT OF SMALL MOLECULES AND THE ELECTRICAL PROPERTIES OF MEMBRANES

Because of its hydrophobic interior, the lipid bilayer of cell membranes serves as a barrier to the passage of most polar molecules. This barrier function is crucially important because it allows the cell to maintain concentrations of solutes in its cytosol that are different from those in the extracellular fluid and in each of the intracellular membrane-enclosed compartments. To make use of this barrier, however, cells have had to evolve ways of transferring specific water-soluble molecules across their membranes in order to ingest essential nutrients, excrete metabolic waste products, and regulate intracellular ion concentrations. The transport of inorganic ions and small water-soluble organic molecules across the lipid bilayer is achieved by specialized transmembrane proteins, each of which is responsible for the transfer of a specific ion, molecule, or group of closely related ions or molecules. Cells can also transfer macromolecules and even large particles across their membranes, but the mechanisms involved in most of these cases are different from those used for transferring small molecules, and they are discussed in Chapters 12 and 13. The importance of membrane transport is indicated by the large number of genes in all organisms that code for transport proteins, which make up between 15 and 30% of the membrane proteins in all cells. Some specialized mammalian cells devote up to two-thirds of their total metabolic energy consumption to membrane transport processes.

We begin this chapter by considering some general principles of how small water-soluble molecules traverse cell membranes. We then consider, in turn, the two main classes of membrane proteins that mediate the transfer: *carrier proteins*, which have moving parts to shift specific molecules across the membrane, and *channel proteins*, which form a narrow hydrophilic pore, allowing the passive movement primarily of small inorganic ions. Carrier proteins can be coupled to a source of energy to catalyze active transport, and a combination of selective passive permeability and active transport creates large differences in the composition of the cytosol compared with either the extracellular fluid

TABLE 11-1 A Comparison of Ion Concentrations Inside and Outside a Typical Mammalian Cell

COMPONENT		INTRACELLULAR CONCENTRATION (mM)	EXTRACELLULAR CONCENTRATION (mM)
Cations			
	Na^+	5–15	145
	K^+	140	5
	Mg^{2+}	0.5	1–2
	Ca^{2+}	10^{-4}	1–2
	H^+	7×10^{-5} ($10^{-7.2}$ M or pH 7.2)	4×10^{-5} ($10^{-7.4}$ M or pH 7.4)
Anions*			
	Cl^-	5–15	110

*The cell must contain equal quantities of positive and negative charges (that is, be electrically neutral). Thus, in addition to Cl^-, the cell contains many other anions not listed in this table; in fact, most cellular constituents are negatively charged (HCO_3^-, PO_4^{3-}, proteins, nucleic acids, metabolites carrying phosphate and carboxyl groups, etc.). The concentrations of Ca^{2+} and Mg^{2+} given are for the free ions. There is a total of about 20 mM Mg^{2+} and 1–2 mM Ca^{2+} in cells, but this is mostly bound to proteins and other substances and, for Ca^{2+}, stored within various organelles.

(Table 11–1) or the fluid within membrane-enclosed organelles. By generating ionic concentration differences across the lipid bilayer, cell membranes can store potential energy in the form of electrochemical gradients, which are used to drive various transport processes, to convey electrical signals in electrically excitable cells, and (in mitochondria, chloroplasts, and bacteria) to make most of the cell's ATP. We focus our discussion mainly on transport across the plasma membrane, but similar mechanisms operate across the other membranes of the eucaryotic cell, as discussed in later chapters.

In the last part of the chapter, we concentrate mainly on the functions of ion channels in neurons (nerve cells). In these cells, channel proteins perform at their highest level of sophistication, enabling networks of neurons to carry out all the astonishing feats of which the human brain is capable.

PRINCIPLES OF MEMBRANE TRANSPORT

We begin this section by describing the permeability properties of protein-free, synthetic lipid bilayers. We then introduce some of the terms used to describe the various forms of membrane transport and some strategies for characterizing the proteins and processes involved.

Protein-free Lipid Bilayers Are Highly Impermeable to Ions

Given enough time, virtually any molecule will diffuse across a protein-free lipid bilayer down its concentration gradient. The rate at which it does so, however, varies enormously, depending partly on the size of the molecule, but mostly on its relative solubility in oil. In general, the smaller the molecule and the more soluble it is in oil (the more hydrophobic, or nonpolar, it is), the more rapidly it will diffuse across a lipid bilayer. Small nonpolar molecules, such as O_2 and CO_2, readily dissolve in lipid bilayers and therefore diffuse rapidly across them. Small uncharged polar molecules, such as water or urea, also diffuse across a bilayer, albeit much more slowly (Figure 11–1). By contrast, lipid bilayers are highly impermeable to charged molecules (ions), no matter how small: the charge and high degree of hydration of such molecules prevents them from entering the hydrocarbon phase of the bilayer. Thus, synthetic bilayers are 10^9 times more permeable to water than to even such small ions as Na^+ or K^+ (Figure 11–2).

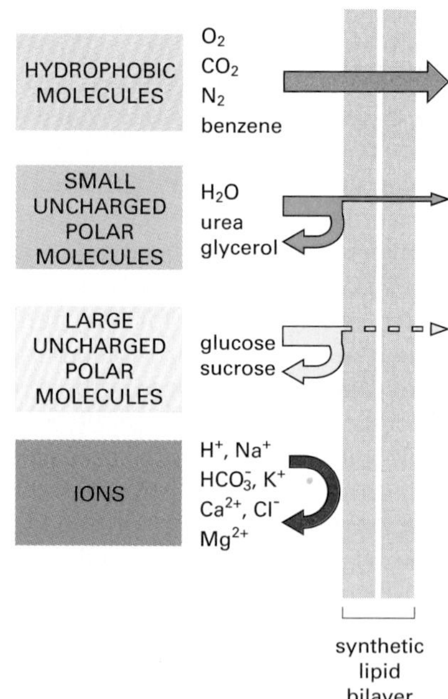

Figure 11–1 The relative permeability of a synthetic lipid bilayer to different classes of molecules. The smaller the molecule and, more importantly, the less strongly it associates with water, the more rapidly the molecule diffuses across the bilayer.

There Are Two Main Classes of Membrane Transport Proteins: Carriers and Channels

Like synthetic lipid bilayers, cell membranes allow water and nonpolar molecules to permeate by simple diffusion. Cell membranes, however, also have to allow the passage of various polar molecules, such as ions, sugars, amino acids, nucleotides, and many cell metabolites that cross synthetic lipid bilayers only very slowly. Special **membrane transport proteins** are responsible for transferring such solutes across cell membranes. These proteins occur in many forms and in all types of biological membranes. Each protein transports a particular class of molecule (such as ions, sugars, or amino acids) and often only certain molecular species of the class. The specificity of membrane transport proteins was first indicated in the mid-1950s by studies in which single gene mutations were found to abolish the ability of bacteria to transport specific sugars across their plasma membrane. Similar mutations have now been discovered in humans suffering from a variety of inherited diseases that affect the transport of a specific solute in the kidney, intestine, or many other cell types. Individuals with the inherited disease *cystinuria*, for example, are unable to transport certain amino acids (including cystine, the disulfide-linked dimer of cysteine) from either the urine or the intestine into the blood; the resulting accumulation of cystine in the urine leads to the formation of cystine stones in the kidneys.

All membrane transport proteins that have been studied in detail have been found to be multipass transmembrane proteins—that is, their polypeptide chains traverse the lipid bilayer multiple times. By forming a continuous protein pathway across the membrane, these proteins enable specific hydrophilic solutes to cross the membrane without coming into direct contact with the hydrophobic interior of the lipid bilayer.

Carrier proteins and channel proteins are the two major classes of membrane transport proteins. **Carrier proteins** (also called *carriers, permeases,* or *transporters)* bind the specific solute to be transported and undergo a series of conformational changes to transfer the bound solute across the membrane (Figure 11–3). **Channel proteins**, in contrast, interact with the solute to be transported much more weakly. They form aqueous pores that extend across the lipid bilayer; when these pores are open, they allow specific solutes (usually inorganic ions of appropriate size and charge) to pass through them and thereby cross the membrane (see Figure 11–3). Not surprisingly, transport through channel proteins occurs at a much faster rate than transport mediated by carrier proteins.

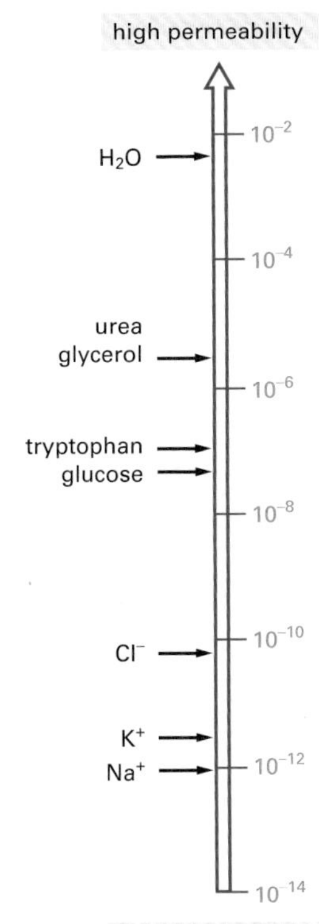

Figure 11–2 Permeability coefficients for the passage of various molecules through synthetic lipid bilayers. The rate of flow of a solute across the bilayer is directly proportional to the difference in its concentration on the two sides of the membrane. Multiplying this concentration difference (in mol/cm^3) by the permeability coefficient (in cm/sec) gives the flow of solute in moles per second per square centimeter of membrane. A concentration difference of tryptophan of 10^{-4} mol/cm^3 ($10^{-4}/10^{-3}$ L = 0.1 M), for example, would cause a flow of 10^{-4} mol/cm^3 $\times$ 10^{-7} cm/sec = 10^{-11} mol/sec through 1 cm^2 of membrane, or 6×10^4 molecules/sec through 1 μm^2 of membrane.

Active Transport Is Mediated by Carrier Proteins Coupled to an Energy Source

All channel proteins and many carrier proteins allow solutes to cross the membrane only passively ("downhill"), a process called **passive transport**, or **facilitated diffusion**. In the case of transport of a single uncharged molecule, it is simply the difference in its concentration on the two sides of the membrane—its *concentration gradient*—that drives passive transport and determines its direction (Figure 11–4A).

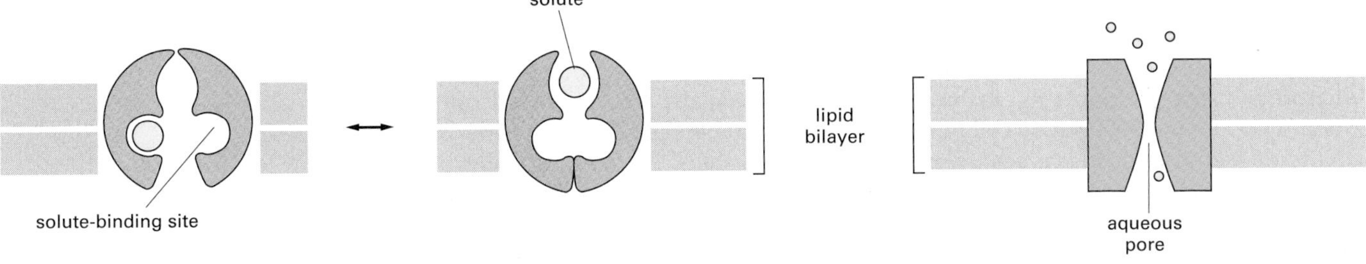

Figure 11–3 Carrier proteins and channel proteins. (A) A carrier protein alternates between two conformations, so that the solute-binding site is sequentially accessible on one side of the bilayer and then on the other. (B) In contrast, a channel protein forms a water-filled pore across the bilayer through which specific solutes can diffuse.

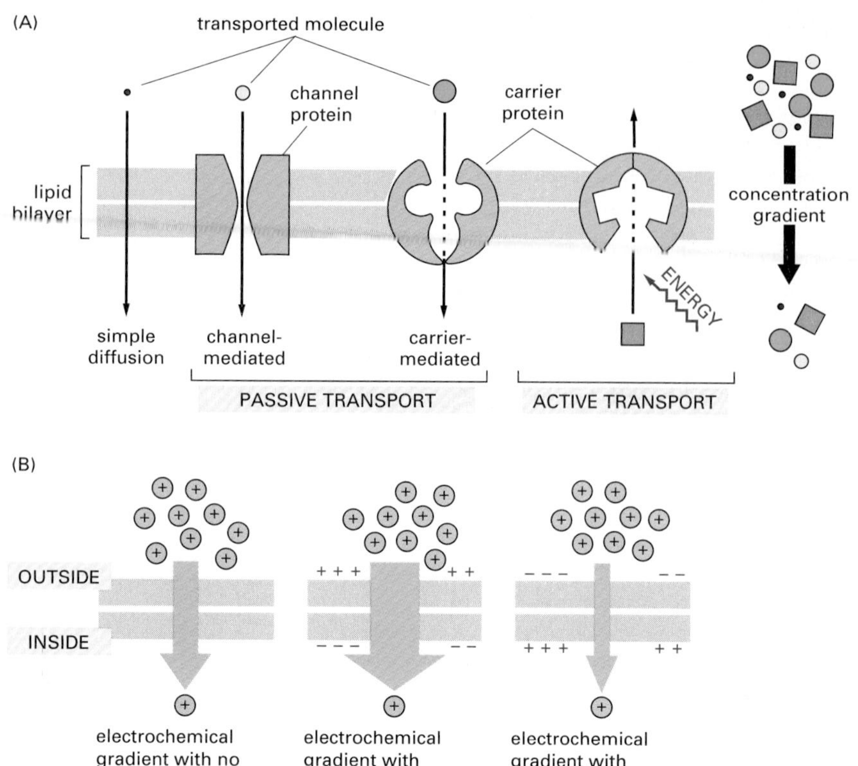

Figure 11–4 Passive and active transport compared. (A) Passive transport down an electrochemical gradient occurs spontaneously, either by simple diffusion through the lipid bilayer or by facilitated diffusion through channels and passive carriers. By contrast, active transport requires an input of metabolic energy and is always mediated by carriers that harvest metabolic energy to pump the solute against its electrochemical gradient. (B) An electrochemical gradient combines the membrane potential and the concentration gradient, which can work additively to increase the driving force on an ion across the membrane *(middle)* or can work against each other *(right).*

If the solute carries a net charge, however, both its concentration gradient and the electrical potential difference across the membrane, the *membrane potential,* influence its transport. The concentration gradient and the electrical gradient can be combined to calculate a net driving force, the **electrochemical gradient**, for each charged solute (Figure 11–4B). We discuss this in more detail in Chapter 14. In fact, almost all plasma membranes have an electrical potential difference (voltage gradient) across them, with the inside usually negative with respect to the outside. This potential difference favors the entry of positively charged ions into the cell but opposes the entry of negatively charged ions.

Cells also require transport proteins that will actively pump certain solutes across the membrane against their electrochemical gradient ("uphill"); this process, known as **active transport**, is mediated by carriers, which are also called *pumps.* In active transport, the pumping activity of the carrier protein is directional because it is tightly coupled to a source of metabolic energy, such as ATP hydrolysis or an ion gradient, as discussed later. Thus, transport by carriers can be either active or passive, whereas transport by channel proteins is always passive.

Ionophores Can Be Used as Tools to Increase the Permeability of Membranes to Specific Ions

Ionophores are small hydrophobic molecules that dissolve in lipid bilayers and increase their permeability to specific inorganic ions. Most are synthesized by microorganisms (presumably as biological weapons against competitors or prey). They are widely used by cell biologists as tools to increase the ion permeability of membranes in studies on synthetic bilayers, cells, or cell organelles. There are two classes of ionophores—**mobile ion carriers** and **channel formers** (Figure 11–5). Both types operate by shielding the charge of the transported ion so that it can penetrate the hydrophobic interior of the lipid bilayer. Since ionophores are not coupled to energy sources, they permit the net movement of ions only down their electrochemical gradients.

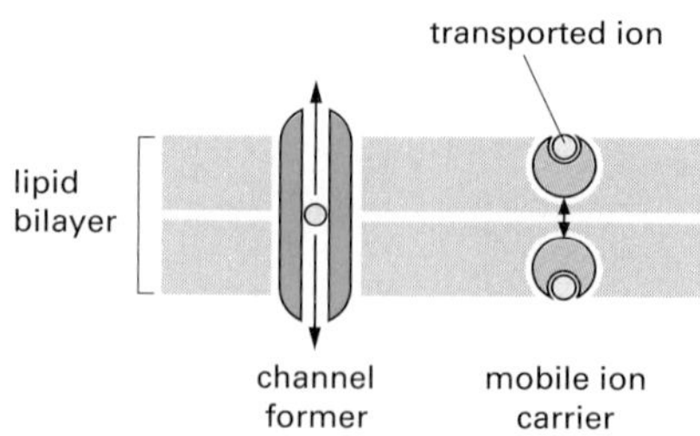

Figure 11–5 Ionophores: a channel-former and a mobile ion carrier. In both cases, net ion flow occurs only down an electrochemical gradient.

Valinomycin is an example of a mobile ion carrier. It is a ring-shaped polymer that transports K^+ down its electrochemical gradient by picking up K^+ on one side of the membrane, diffusing across the bilayer, and releasing K^+ on the other side. Similarly, *FCCP*, a mobile ion carrier that makes membranes selectively leaky to H^+, is often used to dissipate the H^+ electrochemical gradient across the mitochondrial inner membrane, thereby blocking mitochondrial ATP production. *A23187* is yet another example of a mobile ion carrier, only it transports divalent cations such as Ca^{2+} and Mg^{2+}. When cells are exposed to A23187, Ca^{2+} enters the cytosol from the extracellular fluid down a steep electrochemical gradient. Accordingly, this ionophore is widely used to increase the concentration of free Ca^{2+} in the cytosol, thereby mimicking certain cell-signaling mechanisms (discussed in Chapter 15).

Gramicidin A is an example of a channel-forming ionophore. It is a dimeric compound of two linear peptides (of 15 hydrophobic amino acids each), which wind around each other to form a double helix. Two gramicidin dimers are thought to come together end to end across the lipid bilayer to form what is probably the simplest of all transmembrane channels, which selectively allows monovalent cations to flow down their electrochemical gradients. Gramicidin is made by certain bacteria, perhaps to kill other microorganisms by collapsing the H^+, Na^+, and K^+ gradients that are essential for their survival, and it has been useful as an antibiotic.

Summary

Lipid bilayers are highly impermeable to most polar molecules. To transport small water-soluble molecules into or out of cells or intracellular membrane-enclosed compartments, cell membranes contain various membrane transport proteins, each of which is responsible for transferring a particular solute or class of solutes across the membrane. There are two classes of membrane transport proteins—carriers and channels. Both form continuous protein pathways across the lipid bilayer. Whereas transport by carriers can be either active or passive, solute flow through channel proteins is always passive. Ionophores, which are small hydrophobic molecules made by microorganisms, can be used as tools to increase the permeability of cell membranes to specific inorganic ions.

CARRIER PROTEINS AND ACTIVE MEMBRANE TRANSPORT

The process by which a carrier protein transfers a solute molecule across the lipid bilayer resembles an enzyme–substrate reaction, and in many ways carriers behave like enzymes. In contrast to ordinary enzyme–substrate reactions, however, the transported solute is not covalently modified by the carrier protein, but instead is delivered unchanged to the other side of the membrane.

Each type of carrier protein has one or more specific binding sites for its solute (substrate). It transfers the solute across the lipid bilayer by undergoing reversible conformational changes that alternately expose the solute-binding site first on one side of the membrane and then on the other. A schematic model of how such a carrier protein is thought to operate is shown in Figure 11–6. When

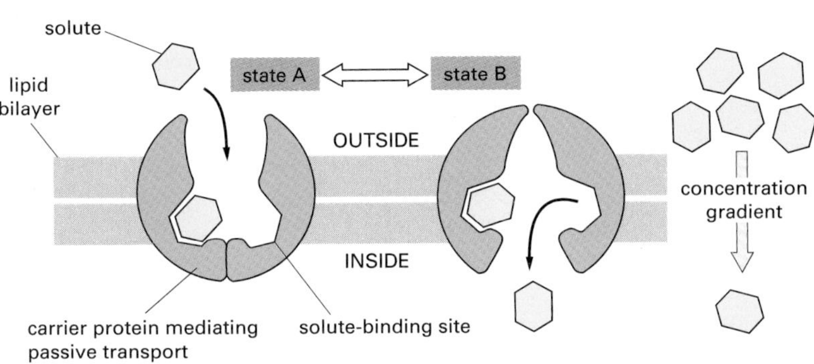

Figure 11–6 A model of how a conformational change in a carrier protein could mediate the passive transport of a solute. The carrier protein shown can exist in two conformational states: in state A, the binding sites for solute are exposed on the outside of the lipid bilayer; in state B, the same sites are exposed on the other side of the bilayer. The transition between the two states can occur randomly. It is completely reversible and does not depend on whether the solute binding site is occupied. Therefore, if the solute concentration is higher on the outside of the bilayer, more solute binds to the carrier protein in the A conformation than in the B conformation, and there is a net transport of solute down its concentration gradient (or, if the solute is an ion, down its electrochemical gradient).

the carrier is saturated (that is, when all solute-binding sites are occupied), the rate of transport is maximal. This rate, referred to as V_{max}, is characteristic of the specific carrier and reflects the rate with which the carrier can flip between its two conformational states. In addition, each transporter protein has a characteristic binding constant for its solute, K_m, equal to the concentration of solute when the transport rate is half its maximum value (Figure 11–7). As with enzymes, the binding of solute can be blocked specifically by either competitive inhibitors (which compete for the same binding site and may or may not be transported by the carrier) or noncompetitive inhibitors (which bind elsewhere and specifically alter the structure of the carrier).

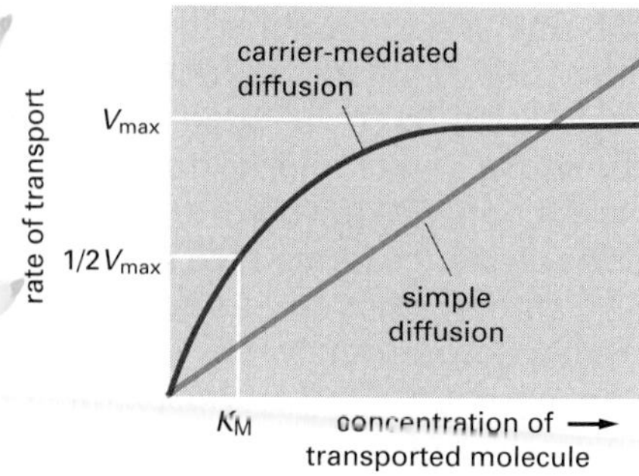

Figure 11–7 The kinetics of simple diffusion and carrier-mediated diffusion. Whereas the rate of the former is always proportional to the solute concentration, the rate of the latter reaches a maximum (V_{max}) when the carrier protein is saturated. The solute concentration when transport is at half its maximal value approximates the binding constant (K_m) of the carrier for the solute and is analogous to the K_m of an enzyme for its substrate. The graph applies to a carrier transporting a single solute; the kinetics of coupled transport of two or more solutes (see text) is more complex but shows basically similar phenomena.

As we discuss below, it requires only a relatively minor modification of the model shown in Figure 11–6 to link the carrier protein to a source of energy in order to pump a solute uphill against its electrochemical gradient. Cells carry out such active transport in three main ways (Figure 11–8):

1. *Coupled carriers* couple the uphill transport of one solute across the membrane to the downhill transport of another.
2. *ATP-driven pumps* couple uphill transport to the hydrolysis of ATP.
3. *Light-driven pumps,* which are found mainly in bacterial cells, couple uphill transport to an input of energy from light, as with bacteriorhodopsin (discussed in Chapter 10).

Amino acid sequence comparisons suggest that, in many cases, there are strong similarities in molecular design between carrier proteins that mediate active transport and those that mediate passive transport. Some bacterial carriers, for example, which use the energy stored in the H^+ gradient across the plasma membrane to drive the active uptake of various sugars, are structurally similar to the carriers that mediate passive glucose transport into most animal cells. This suggests an evolutionary relationship between various carrier proteins; and, given the importance of small metabolites and sugars as an energy source, it is not surprising that the superfamily of carriers is an ancient one.

We begin our discussion of active transport by considering carrier proteins that are driven by ion gradients. These proteins have a crucial role in the transport of small metabolites across membranes in all cells. We then discuss ATP-driven pumps, including the Na^+ pump that is found in the plasma membrane of almost all cells.

Active Transport Can Be Driven by Ion Gradients

Some carrier proteins simply transport a single solute from one side of the membrane to the other at a rate determined as above by V_{max} and K_m; they are called **uniporters**. Others, with more complex kinetics, function as *coupled carriers,* in which the transfer of one solute strictly depends on the transport of a second. Coupled transport involves either the simultaneous transfer of a second solute in the same direction, performed by **symporters**, or the transfer of a second solute in the opposite direction, performed by **antiporters** (Figure 11–9).

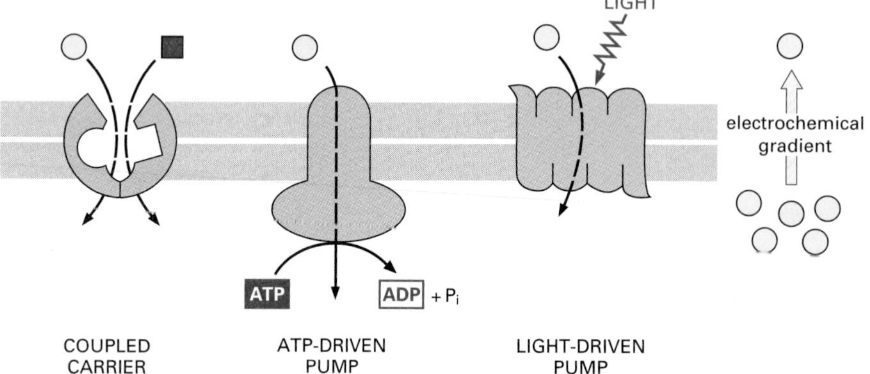

Figure 11–8 Three ways of driving active transport. The actively transported molecule is shown in *yellow,* and the energy source is shown in *red.*

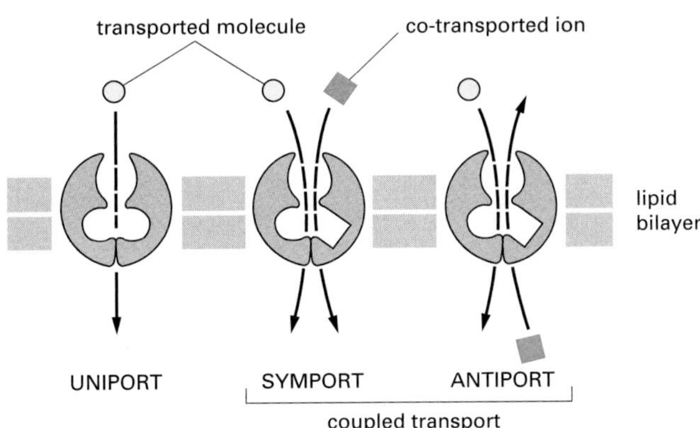

Figure 11–9 Three types of carrier-mediated transport. This schematic diagram shows carrier proteins functioning as uniporters, symporters, and antiporters.

The tight coupling between the transport of two solutes allows these carriers to harvest the energy stored in the electrochemical gradient of one solute, typically an ion, to transport the other. In this way, the free energy released during the movement of an inorganic ion down an electrochemical gradient is used as the driving force to pump other solutes uphill, against their electrochemical gradient. This principle can work in either direction; some coupled carriers function as symporters, others as antiporters. In the plasma membrane of animal cells, Na^+ is the usual co-transported ion whose electrochemical gradient provides a large driving force for the active transport of a second molecule. The Na^+ that enters the cell during transport is subsequently pumped out by an ATP-driven Na^+ pump in the plasma membrane (as we discuss later), which, by maintaining the Na^+ gradient, indirectly drives the transport. (For this reason ion-driven carriers are said to mediate *secondary active transport,* whereas ATP-driven carriers are said to mediate *primary active transport.)* Intestinal and kidney epithelial cells, for example, contain a variety of symport systems that are driven by the Na^+ gradient across the plasma membrane; each system is specific for importing a small group of related sugars or amino acids into the cell. In these systems, the solute and Na^+ bind to different sites on a carrier protein. Because the Na^+ tends to move into the cell down its electrochemical gradient, the sugar or amino acid is, in a sense, "dragged" into the cell with it. The greater the electrochemical gradient for Na^+, the greater the rate of solute entry; conversely, if the Na^+ concentration in the extracellular fluid is reduced, solute transport decreases (Figure 11–10).

Figure 11–10 One way in which a glucose carrier can be driven by a Na^+ gradient. As in the model shown in Figure 11–6, the carrier oscillates between two alternate states, A and B. In the A state, the protein is open to the extracellular space; in the B state, it is open to the cytosol. Binding of Na^+ and glucose is cooperative—that is, the binding of either ligand induces a conformational change that greatly increases the protein's affinity for the other ligand. Since the Na^+ concentration is much higher in the extracellular space than in the cytosol, glucose is more likely to bind to the carrier in the A state. Therefore, both Na^+ and glucose enter the cell (via an A $\rightarrow$ B transition) much more often than they leave it (via a B $\rightarrow$ A transition). The overall result is the net transport of both Na^+ and glucose into the cell. Note that because the binding is cooperative, if one of the two solutes is missing, the other fails to bind to the carrier. Thus, the carrier undergoes a conformational switch between the two states only if both solutes or neither are bound.

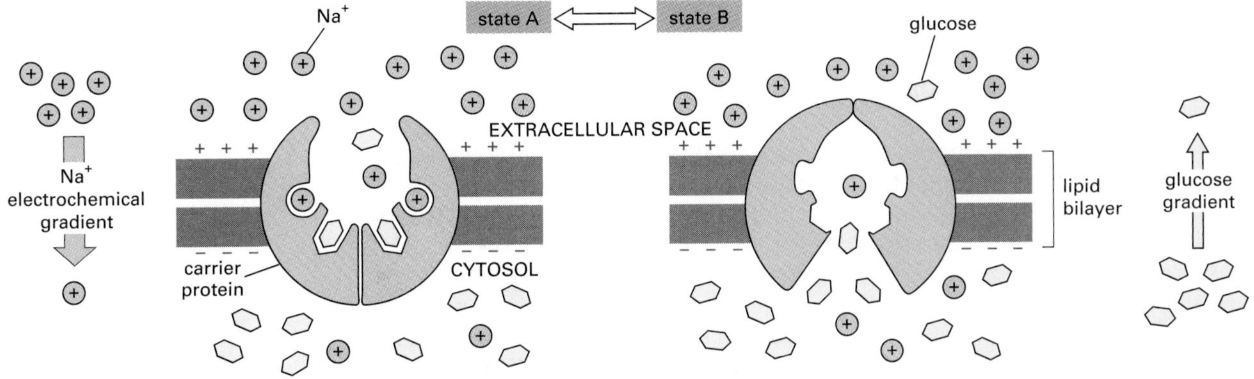

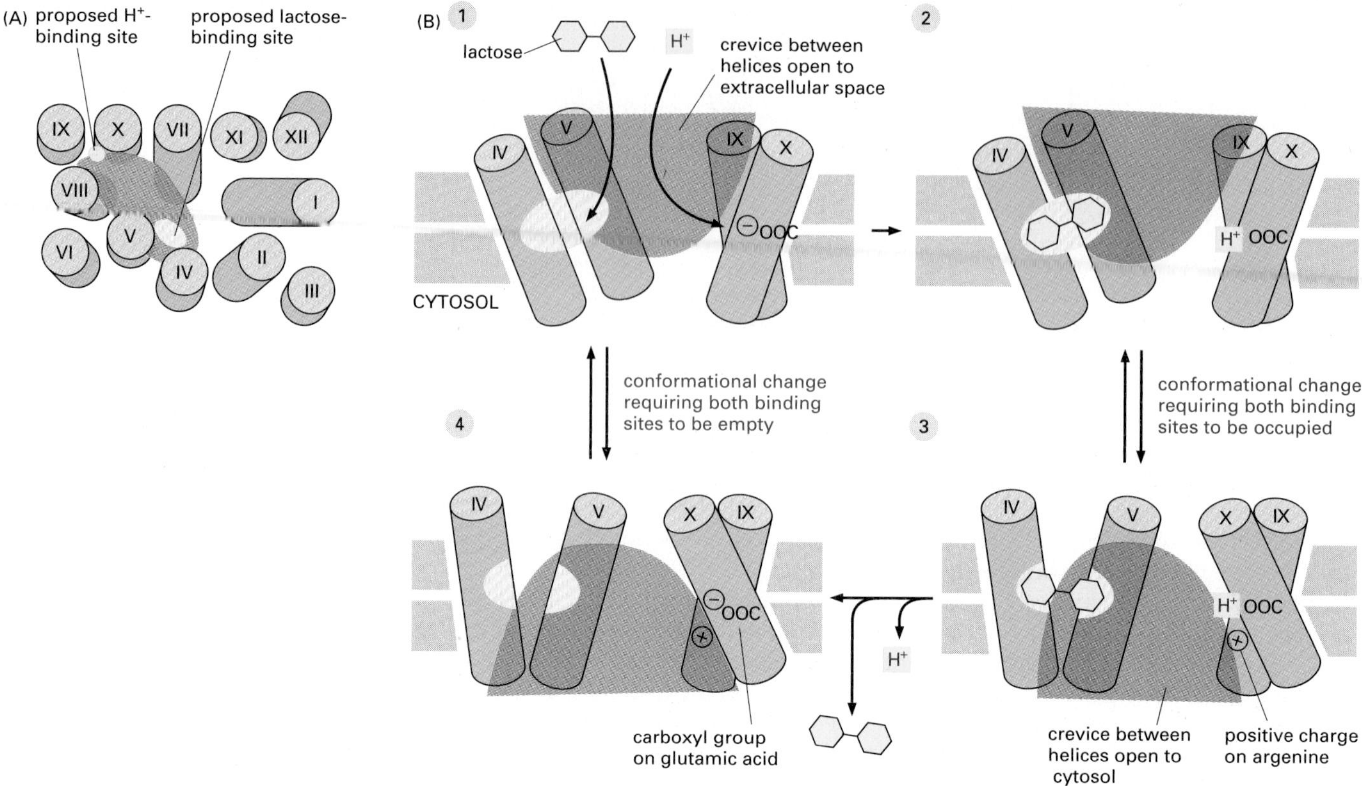

Figure 11–11 A model for the molecular mechanism of action of the bacterial lactose permease. (A) A view from the cytosol of the proposed arrangement of the 12 predicted transmembrane helices in the membrane. The loops that connect the helices on either side of the membrane are omitted for clarity. A glutamic acid on helix X binds H^+, and amino acids contributed by helices IV and V bind lactose. (B) During a transport cycle, the carrier flips between two conformational states: in one, the H^+- and lactose-binding sites are accessible to the extracellular space (1 and 2); in the other, they are exposed to the cytosol (3 and 4). Unloading of the solutes on the cytosolic face (3 → 4) is favored because the lactose-binding site is partly disrupted and a positive charge contributed by an arginine on helix IX displaces the H^+ from the glutamic acid on helix X. (A, adapted from H.R. Kaback and J. Wu, *Accts. Chem. Res.* 32:805–813, 1999.)

In bacteria and yeasts, as well as in many membrane-enclosed organelles of animal cells, most active transport systems driven by ion gradients depend on H^+ rather than Na^+ gradients, reflecting the predominance of H^+ pumps and the virtual absence of Na^+ pumps in these membranes. The active transport of many sugars and amino acids into bacterial cells, for example, is driven by the electrochemical H^+ gradient across the plasma membrane. One well-studied H^+-driven symport is **lactose permease**, which transports lactose across the plasma membrane of *E. coli*. Although the folded structure of the permease is unknown, biophysical studies and extensive analyses of mutant proteins have led to a detailed model of how the symport works. The permease consists of 12 loosely packed transmembrane α helices. During the transport cycle, some of the helices undergo sliding motions that cause them to tilt. These motions alternately open and close a crevice between the helices, exposing the binding sites for the solutes lactose and H^+, first on one side of the membrane and then on the other (Figure 11–11).

Na^+-driven Carrier Proteins in the Plasma Membrane Regulate Cytosolic pH

The structure and function of most macromolecules are greatly influenced by pH, and most proteins operate optimally at a particular pH. Lysosomal enzymes, for example, function best at the low pH (~5) found in lysosomes, whereas cytosolic enzymes function best at the close to neutral pH (~7.2) found in the cytosol. It is therefore crucial that cells be able to control the pH of their intracellular compartments.

Most cells have one or more types of Na^+-driven antiporters in their plasma membrane that help to maintain the cytosolic pH (pH_i), at about 7.2. These proteins use the energy stored in the Na^+ gradient to pump out excess H^+, which either leaks in or is produced in the cell by acid-forming reactions. Two mechanisms are used: either H^+ is directly transported out of the cell or HCO_3^- is brought into the cell to neutralize H^+ in the cytosol (according to the reaction $HCO_3^- + H^+ \rightarrow H_2O + CO_2$). One of the antiporters that uses the first mechanism is a *Na^+-H^+ exchanger*, which couples an influx of Na^+ to an efflux of H^+. Another, which uses a combination of the two mechanisms, is a *Na^+-driven Cl^-–HCO_3^- exchanger* that couples an influx of Na^+ and HCO_3^- to an efflux of Cl^- and H^+ (so that $NaHCO_3$ comes in and HCl goes out). The Na^+-driven Cl^-–HCO_3^- exchanger is twice as effective as the Na^+-H^+ exchanger, in the sense that it pumps out one H^+ and neutralizes another for each Na^+ that enters the cell. If HCO_3^- is available, as is usually the case, this antiporter is the most important carrier protein regulating pH_i. Both exchangers are regulated by pH_i and increase their activity as the pH in the cytosol falls.

An *Na^+-independent Cl^-–HCO_3^- exchanger* also has an important role in pH_i regulation. Like the Na^+-dependent transporters, the Cl^-–HCO_3^- exchanger is regulated by pH_i, but the movement of HCO_3^-, in this case, is normally out of the cell, down its electrochemical gradient. The rate of HCO_3^- efflux and Cl^- influx increases as pH_i rises, thereby decreasing pH_i whenever the cytosol becomes too alkaline. The Cl^-–HCO_3^- exchanger is similar to the band 3 protein in the membrane of red blood cells discussed in Chapter 10. In red blood cells, band 3 protein facilitates the quick discharge of CO_2 as the cells pass through capillaries in the lung.

ATP-driven H^+ pumps are also used to control the pH of many intracellular compartments. As discussed in Chapter 13, the low pH in lysosomes, as well as in endosomes and secretory vesicles, is maintained by such H^+ pumps, which use the energy of ATP hydrolysis to pump H^+ into these organelles from the cytosol.

An Asymmetric Distribution of Carrier Proteins in Epithelial Cells Underlies the Transcellular Transport of Solutes

In epithelial cells, such as those involved in absorbing nutrients from the gut, carrier proteins are distributed nonuniformly in the plasma membrane and thereby contribute to the **transcellular transport** of absorbed solutes. As shown in Figure 11–12, Na^+-linked symporters located in the apical (absorptive) domain of the plasma membrane actively transport nutrients into the cell, building up substantial concentration gradients for these solutes across the

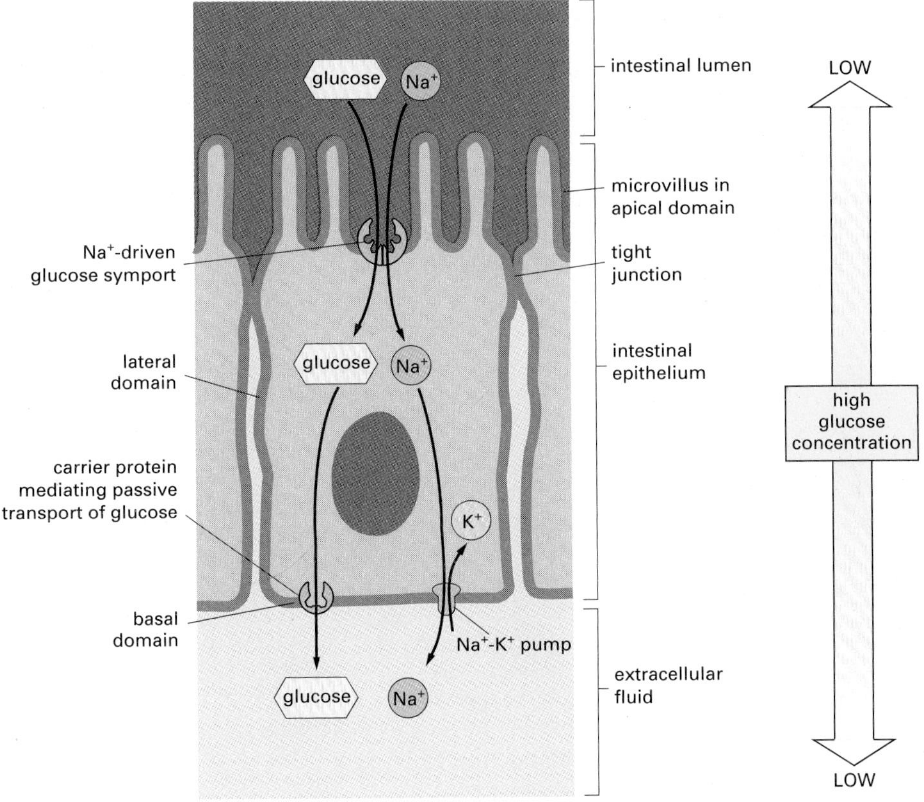

Figure 11–12 Transcellular transport. The transcellular transport of glucose across an intestinal epithelial cell depends on the nonuniform distribution of transport proteins in the cell's plasma membrane. The process shown here results in the transport of glucose from the intestinal lumen to the extracellular fluid (from where it passes into the blood). Glucose is pumped into the cell through the apical domain of the membrane by a Na^+-powered glucose symport. Glucose passes out of the cell (down its concentration gradient) by passive transport mediated by a different glucose carrier protein in the basal and lateral membrane domains. The Na^+ gradient driving the glucose symport is maintained by a Na^+ pump in the basal and lateral plasma membrane domains, which keeps the internal concentration of Na^+ low. Adjacent cells are connected by impermeable tight junctions, which have a dual function in the transport process illustrated: they prevent solutes from crossing the epithelium between cells, allowing a concentration gradient of glucose to be maintained across the cell sheet, and they also serve as diffusion barriers within the plasma membrane, which help confine the various carrier proteins to their respective membrane domains (see Figure 10–41).

plasma membrane. Na^+-independent transport proteins in the basal and lateral (basolateral) domain allow the nutrients to leave the cell passively down these concentration gradients.

In many of these epithelial cells, the plasma membrane area is greatly increased by the formation of thousands of microvilli, which extend as thin, fingerlike projections from the apical surface of each cell (see Figure 11–12). Such microvilli can increase the total absorptive area of a cell by as much as 25-fold, thereby enhancing its transport capabilities.

Although, as we have seen, ion gradients have a crucial role in driving many essential transport processes in cells, the ion pumps that use the energy of ATP hydrolysis are responsible mainly for establishing and maintaining these gradients, as we next discuss.

The Plasma Membrane Na^+-K^+ Pump Is an ATPase

The concentration of K^+ is typically 10 to 20 times higher inside cells than outside, whereas the reverse is true of Na^+ (see Table 11–1, p. 616). These concentration differences are maintained by a **Na^+-K^+ pump**, or **Na^+ pump**, found in the plasma membrane of virtually all animal cells. The pump operates as an antiporter, actively pumping Na^+ out of the cell against its steep electrochemical gradient and pumping K^+ in. Because the pump hydrolyzes ATP to pump Na^+ out and K^+ in, it is also known as a **Na^+-K^+ ATPase** (Figure 11–13).

We have seen that the Na^+ gradient produced by the pump drives the transport of most nutrients into animal cells and also has a crucial role in regulating cytosolic pH. As we discuss below, the pump also regulates cell volume through its osmotic effects; indeed, it keeps many animal cells from bursting. Almost one-third of the energy requirement of a typical animal cell is consumed in fueling this pump. In electrically active nerve cells, which, as we shall see, repeatedly gain small amounts of Na^+ and lose small amounts of K^+ during the propagation of nerve impulses, this number approaches two-thirds of the cell's energy requirement.

An essential characteristic of the Na^+-K^+ pump is that the transport cycle depends on autophosphorylation of the protein. The terminal phosphate group of ATP is transferred to an aspartic acid residue of the pump and is subsequently removed, as explained in Figure 11–14. Different states of the pump are thus distinguished by the presence or absence of the phosphate group. Ion pumps that phosphorylate themselves in this way are called *P-type transport ATPases*. They constitute a family of structurally and functionally related proteins, which includes a variety of Ca^{2+} pumps and H^+ pumps, as we discuss below.

Like any enzyme, the Na^+-K^+ pump can be driven in reverse, in this case to produce ATP. When the Na^+ and K^+ gradients are experimentally increased to such an extent that the energy stored in their electrochemical gradients is greater than the chemical energy of ATP hydrolysis, these ions move down their

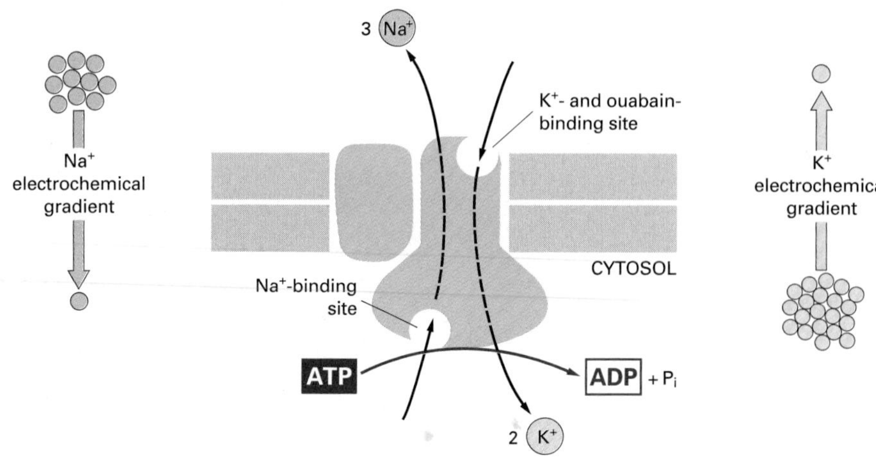

Figure 11–13 The Na^+-K^+ pump. This carrier protein actively pumps Na^+ out of and K^+ into a cell against their electrochemical gradients. For every molecule of ATP hydrolyzed inside the cell, three Na^+ are pumped out and two K^+ are pumped in. The specific inhibitor ouabain and K^+ compete for the same site on the extracellular side of the pump.

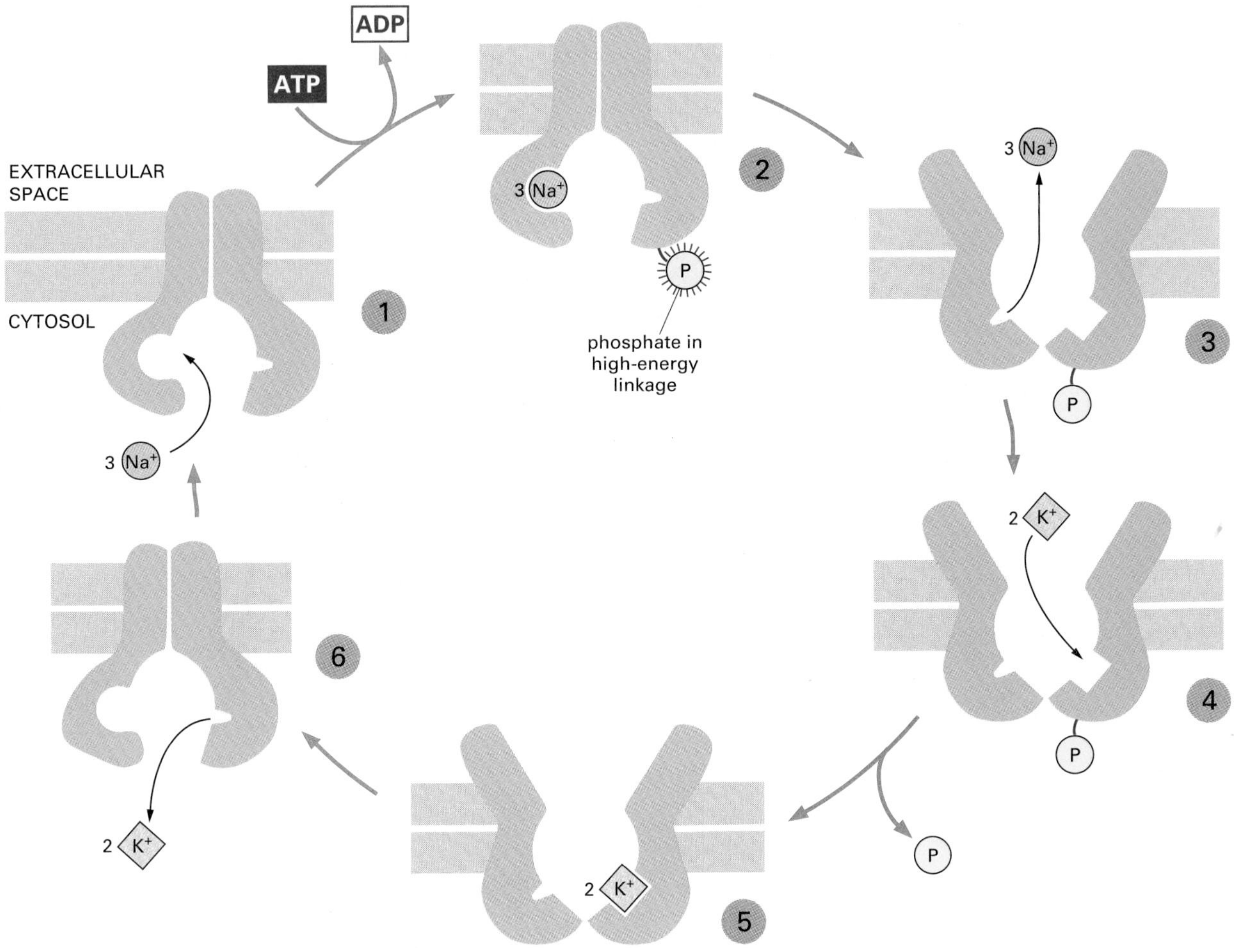

electrochemical gradients and ATP is synthesized from ADP and phosphate by the Na^+-K^+ pump. Thus, the phosphorylated form of the pump (step 2 in Figure 11–14) can relax by either donating its phosphate to ADP (step 2 to step 1) or changing its conformation (step 2 to step 3). Whether the overall change in free-energy is used to synthesize ATP or to pump Na^+ out of the cell depends on the relative concentrations of ATP, ADP, and phosphate, as well as on the electrochemical gradients for Na^+ and K^+.

Figure 11–14 A model of the pumping cycle of the Na^+-K^+ pump. (1) The binding of Na^+ and (2) the subsequent phosphorylation by ATP of the cytoplasmic face of the pump induce the protein to undergo a conformational change that (3) transfers the Na^+ across the membrane and releases it on the outside. (4) Then, the binding of K^+ on the extracellular surface and (5) the subsequent dephosphorylation return the protein to its original conformation, which (6) transfers the K^+ across the membrane and releases it into the cytosol. These changes in conformation are analogous to the A ↔ B transitions shown in Figure 11–6, except that here the Na^+-dependent phosphorylation and the K^+-dependent dephosphorylation of the protein cause the conformational transitions to occur in an orderly manner, enabling the protein to do useful work. Although for simplicity only one Na^+- and one K^+-binding site are shown, in the real pump there are thought to be three Na^+- and two K^+-binding sites. Moreover, although the pump is shown as alternating between two conformational states only, there is evidence that it goes through a more complex series of conformational changes during the pumping cycle.

Some Ca^{2+} and H^+ Pumps Are Also P-type Transport ATPases

In addition to the Na^+-K^+ pump, the P-type transport ATPase family includes *Ca^{2+} pumps* that remove Ca^{2+} from the cytosol after signaling events and the *H^+-K^+ pumps* that secrete acid from specialized epithelial cells in the lining of the stomach. The Ca^{2+} pumps are especially important. Eucaryotic cells maintain very low concentrations of free Ca^{2+} in their cytosol (~10^{-7} M) in the face of very much higher extracellular Ca^{2+} concentrations (~10^{-3} M). Even a small influx of Ca^{2+} significantly increases the concentration of free Ca^{2+} in the cytosol, and the flow of Ca^{2+} down its steep concentration gradient in response to extracellular signals is one means of transmitting these signals rapidly across the plasma membrane (discussed in Chapter 15). The maintenance of a steep Ca^{2+} gradient is therefore important to the cell. The Ca^{2+} gradient is maintained by Ca^{2+} transporters in the plasma membrane that actively move Ca^{2+} out of the cell. One of these is a P-type Ca^{2+} ATPase; the other is an antiporter (called a *Na^+-Ca^{2+} exchanger*) that is driven by the Na^+ electrochemical gradient (see Figure 15–38A).

The best-understood P-type transport ATPase is the **Ca^{2+} pump**, or **Ca^{2+} ATPase**, in the *sarcoplasmic reticulum* membrane of skeletal muscle cells. The sarcoplasmic reticulum is a specialized type of endoplasmic reticulum that

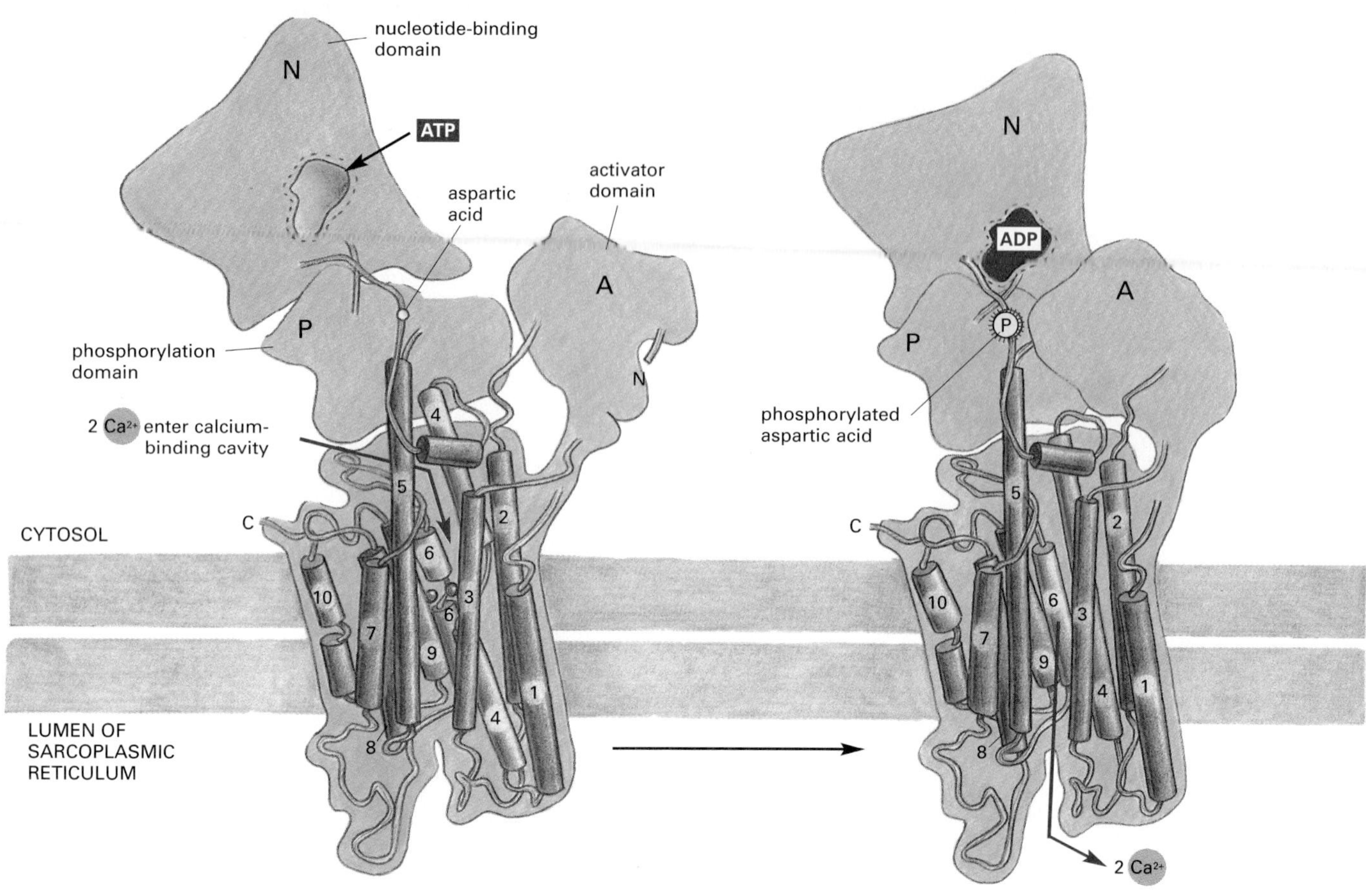

forms a network of tubular sacs in the muscle cell cytosol and serves as an intracellular store of Ca^{2+}. (When an action potential depolarizes the muscle cell membrane, Ca^{2+} is released from the sarcoplasmic reticulum through the Ca^{2+}-release channels into the cytosol, stimulating the muscle to contract, as discussed in Chapter 16.) The Ca^{2+} pump, which accounts for about 90% of the membrane protein of the organelle, is responsible for moving Ca^{2+} from the cytosol back into the sarcoplasmic reticulum. The endoplasmic reticulum of nonmuscle cells contains a similar Ca^{2+} pump, but in smaller quantities.

The three-dimensional structure of the sarcoplasmic reticulum Ca^{2+} pump has been determined at high resolution. This structure and the analysis of a related fungal H^+ pump have provided the first views of P-type transport ATPases, which are all thought to have similar structures. They contain 10 transmembrane α helices, three of which line a central channel that spans the lipid bilayer. In the unphosphorylated state of the Ca^{2+} pump, two of these helices are disrupted and form a cavity that is accessible from the cytosolic side of the membrane and binds two Ca^{2+} ions. The binding of ATP to a binding site on the same side of the membrane and the subsequent phosphorylation of an adjacent domain lead to a drastic rearrangement of the transmembrane helices. The rearrangement disrupts the Ca^{2+}-binding site and releases the Ca^{2+} ions on the other side of the membrane, into the lumen of the sarcoplasmic reticulum (Figure 11–15).

Figure 11–15 A model of how the sarcoplasmic reticulum Ca^{2+} pump moves Ca^{2+}. The structure of the unphosphorylated Ca^{2+}-bound state *(left)* is based on the X-ray crystallographic structure of the pump. The structure of the phosphorylated, Ca^{2+}-free state *(right)* is based on lower-resolution structures determined by electron microscopy, and so the arrangement of the transmembrane helices is speculative. Helices 4 and 6 are disrupted in the unphosphorylated state and form the Ca^{2+}-binding site on the cytosolic side of the membrane. ATP binding and hydrolysis cause drastic conformational changes, bringing the nucleotide-binding (N) and phosphorylation (P) domains into close proximity. This change is thought to cause a 90° rotation of the actuator domain (A), which leads to a rearrangement of the transmembrane helices. The rearrangement eliminates the breaks in helices 4 and 6 abolishing the Ca^{2+}-binding sites and releasing the Ca^{2+} ions on the other side of the membrane into the lumen of the sarcoplasmic reticulum. (Adapted from C. Toyoshima et al., *Nature* 405:647–655, 2000.)

The Na^+-K^+ Pump Is Required to Maintain Osmotic Balance and Stabilize Cell Volume

Since the Na^+-K^+ pump drives three positively charged ions out of the cell for every two it pumps in, it is *electrogenic*. It drives a net current across the membrane, tending to create an electrical potential, with the cell's inside negative relative to the outside. This electrogenic effect of the pump, however, seldom

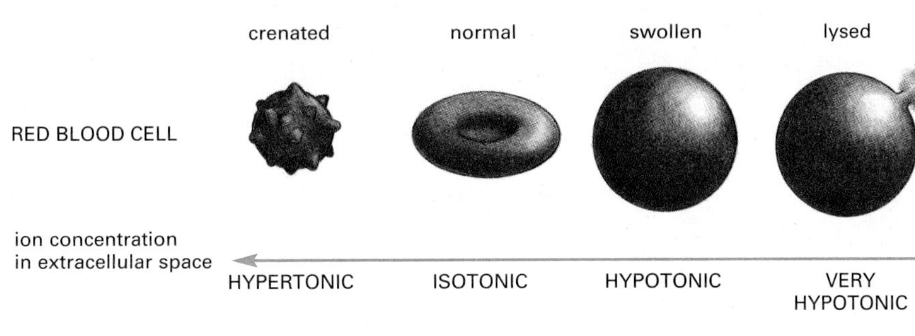

Figure 11–16 Response of a human red blood cell to changes in osmolarity of the extracellular fluid. The cell swells or shrinks as water moves into or out of the cell down its concentration gradient.

contributes more than 10% to the membrane potential. The remaining 90%, as we discuss later, depends on the Na^+-K^+ pump only indirectly.

On the other hand, the Na^+-K^+ pump does have a direct and crucial role in regulating cell volume. It controls the solute concentration inside the cell, thereby regulating the **osmolarity** (or *tonicity)* that can make a cell swell or shrink (Figure 11–16). Because the plasma membrane is weakly permeable to water, water moves slowly into or out of cells down its concentration gradient, a process called *osmosis.* If cells are placed in a *hypotonic solution* (that is, a solution having a low solute concentration and therefore a high water concentration), there is a net movement of water into the cells, causing them to swell and burst (lyse). Conversely, if cells are placed in a *hypertonic solution,* they shrink (see Figure 11–16). Many animal cells also contain specialized *water channels* in their plasma membrane to facilitate osmotic water flow called *aquaporins.*

The importance of the Na^+-K^+ pump in controlling cell volume is indicated by the observation that many animal cells swell, and often burst, if they are treated with *ouabain,* which inhibits the Na^+-K^+ pump. As explained in Panel 11–1, cells contain a high concentration of solutes, including numerous negatively charged organic molecules that are confined inside the cell (the so-called *fixed anions)* and their accompanying cations that are required for charge balance. This tends to create a large osmotic gradient that, unless balanced, would tend to "pull" water into the cell. For animal cells this effect is counteracted by an opposite osmotic gradient due to a high concentration of inorganic ions—chiefly Na^+ and Cl^-—in the extracellular fluid. The Na^+-K^+ pump maintains osmotic balance by pumping out the Na^+ that leaks in down its steep electrochemical gradient. The Cl^- is kept out by the membrane potential.

There are, of course, other ways for a cell to cope with its osmotic problems. Plant cells and many bacteria are prevented from bursting by the semirigid cell wall that surrounds their plasma membrane. In amoebae the excess water that flows in osmotically is collected in contractile vacuoles, which periodically discharge their contents to the exterior (see Panel 11–1). Bacteria have also evolved strategies that allow them to lose ions, and even macromolecules, quickly when subjected to an osmotic shock. But for most animal cells, the Na^+-K^+ pump is crucial.

Membrane-bound Enzymes That Synthesize ATP Are Transport ATPases Working in Reverse

The plasma membrane of bacteria, the inner membrane of mitochondria, and the thylakoid membrane of chloroplasts all contain transport ATPases. These, however, belong to the family of **V-type ATPases** and are structurally very different from P-type ATPases. They are turbine-like structures, constructed from multiple different protein subunits. We discuss them in more detail in Chapter 14.

Many V-type ATPases normally work in reverse: instead of ATP hydrolysis driving ion transport, H^+ gradients across these membranes drive the synthesis of ATP from ADP and phosphate. The H^+ gradients are generated during the electron-transport steps of oxidative phosphorylation (in aerobic bacteria and mitochondria) or photosynthesis (in chloroplasts), or by the light-activated H^+ pump (bacteriorhodopsin) in *Halobacterium.*

PANEL 11–1 Intracellular Water Balance: the Problem and Its Solution

SOURCES OF INTRACELLULAR OSMOLARITY

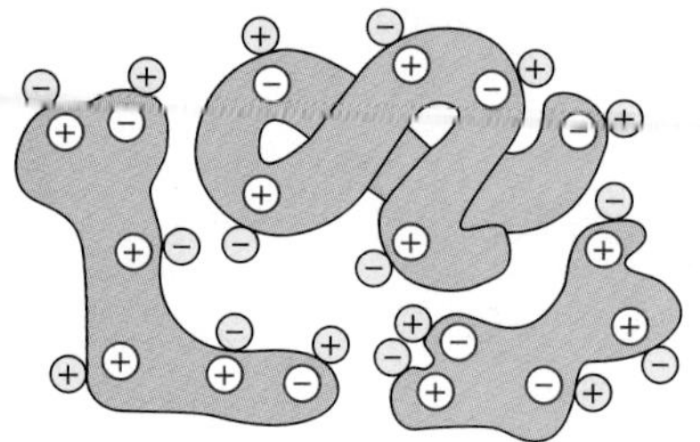

Macromolecules themselves contribute very little to the osmolarity of the cell interior since, despite their large size, each one counts only as a single molecule and there are relatively few of them compared to the number of small molecules in the cell. However, most biological macromolecules are highly charged, and they attract many inorganic ions of opposite charge. Because of their large numbers, these counterions make a major contribution to intracellular osmolarity.

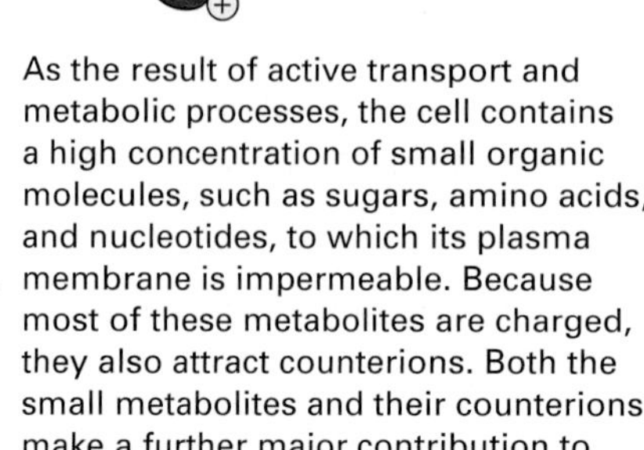

As the result of active transport and metabolic processes, the cell contains a high concentration of small organic molecules, such as sugars, amino acids, and nucleotides, to which its plasma membrane is impermeable. Because most of these metabolites are charged, they also attract counterions. Both the small metabolites and their counterions make a further major contribution to intracellular osmolarity.

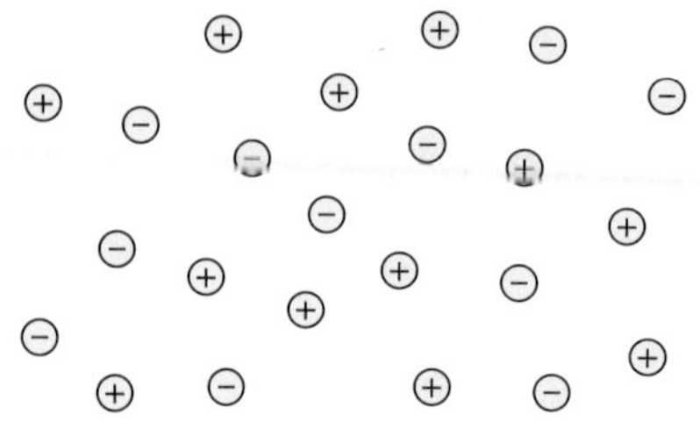

The osmolarity of the extracellular fluid is usually due mainly to small inorganic ions. These leak slowly across the plasma membrane into the cell. If they were not pumped out, and if there were no other molecules inside the cell that interacted with them so as to influence their distribution, they would eventually come to equilibrium with equal concentrations inside and outside the cell. However, the presence of charged macromolecules and metabolites in the cell that attract these ions gives rise to the Donnan effect: it causes the total concentration of inorganic ions (and therefore their contribution to the osmolarity) to be greater inside than outside the cell at equilibrium.

THE PROBLEM

Because of the above factors, a cell that does nothing to control its osmolarity will have a higher concentration of solutes inside than outside. As a result, water will be higher in concentration outside the cell than inside. This difference in water concentration across the plasma membrane will cause water to move continuously into the cell by osmosis, causing it to rupture.

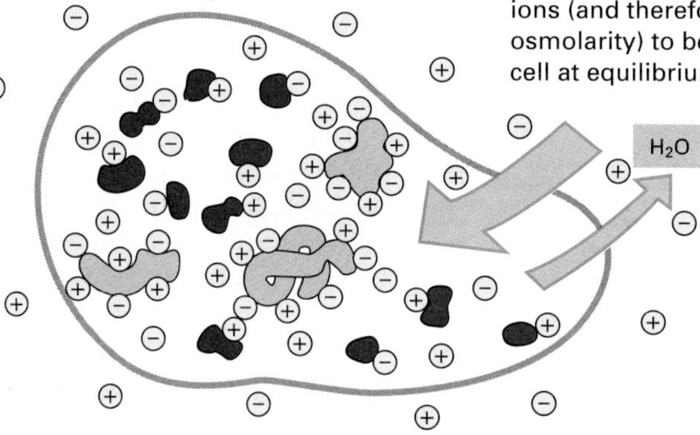

THE SOLUTION

Animal cells and bacteria control their intracellular osmolarity by actively pumping out inorganic ions, such as Na^+, so that their cytoplasm contains a lower total concentration of inorganic ions than the extracellular fluid, thereby compensating for their excess of organic solutes.

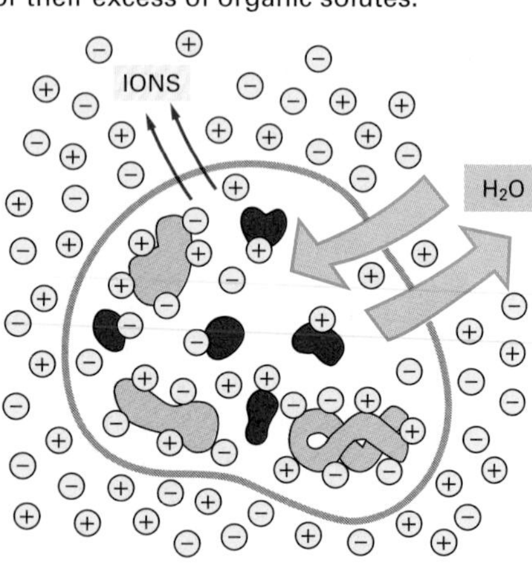

Plant cells are prevented from swelling by their rigid walls and so can tolerate an osmotic difference across their plasma membranes: an internal turgor pressure is built up, which at equilibrium forces out as much water as enters.

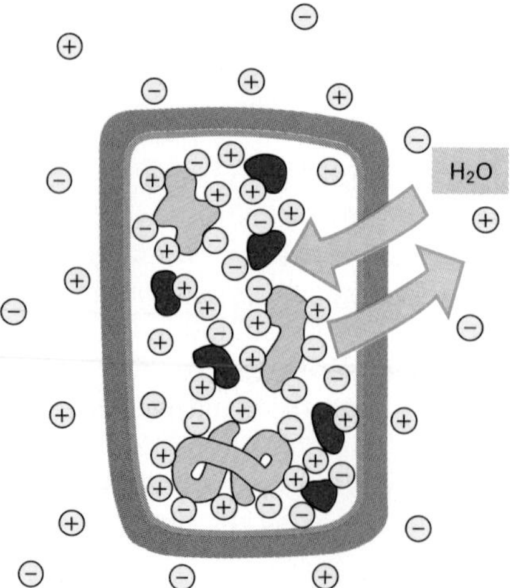

Many protozoa avoid becoming swollen with water, despite an osmotic difference across the plasma membrane, by periodically extruding water from special contractile vacuoles.

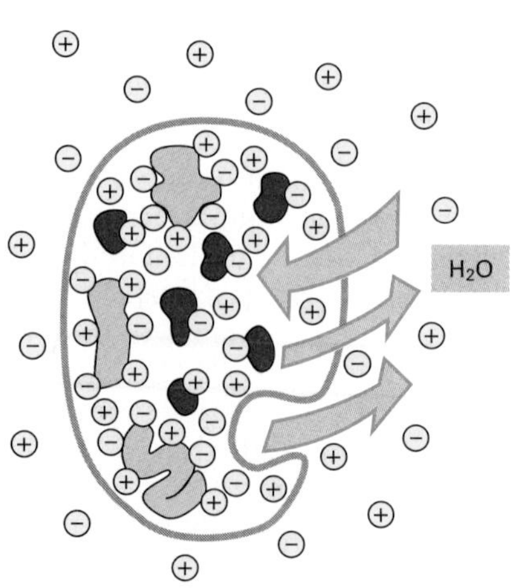

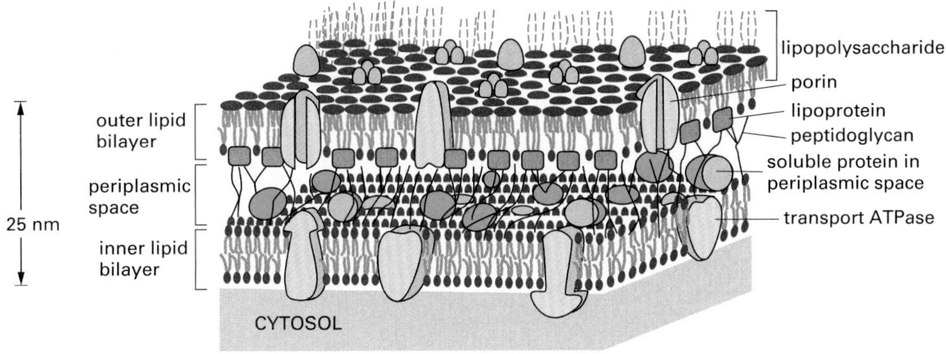

Figure 11–17 A small section of the double membrane of an *E. coli* bacterium. The inner membrane is the cell's plasma membrane. Between the inner and outer lipid bilayer membranes is a highly porous, rigid peptidoglycan, composed of protein and polysaccharide, that constitutes the bacterial cell wall. It is attached to lipoprotein molecules in the outer membrane and fills the *periplasmic space* (only a little of the peptidoglycan is shown). This space also contains a variety of soluble protein molecules. The dashed threads (shown in *green*) at the top represent the polysaccharide chains of the special lipopolysaccharide molecules that form the external monolayer of the outer membrane; for clarity, only a few of these chains are shown. Bacteria with double membranes are called *Gram-negative* because they do not retain the dark blue dye used in Gram staining. Bacteria with single membranes (but thicker cell walls), such as staphylococci and streptococci, retain the blue dye and therefore are called *Gram-positive;* their single membrane is analogous to the inner (plasma) membrane of Gram-negative bacteria.

Although many of these enzymes normally synthesize ATP and hence are also called *ATP synthases,* they can, like the P-type transport ATPases, work in either direction. Depending on the conditions, they can hydrolyze ATP and pump H^+ across the membrane, or they can synthesize ATP when H^+ flows through the enzyme in the reverse direction. Indeed, cells use V-type ATPases in both ways. Some organelles, such as lysosomes, synaptic vesicles, and plant vacuoles, contain V-type transport ATPases that pump H^+ and hence are responsible for acidification of the organelles.

ABC Transporters Constitute the Largest Family of Membrane Transport Proteins

The last type of carrier protein that we discuss is a family of transport ATPases that are of great clinical importance, even though their normal functions in eucaryotic cells are only just beginning to be discovered. The first of these proteins to be characterized was found in bacteria. We have already mentioned that the plasma membranes of all bacteria contain carrier proteins that use the H^+ gradient across the membrane to pump a variety of nutrients into the cell. Many also have transport ATPases that use the energy of ATP hydrolysis to import certain small molecules. In bacteria such as *E. coli,* which have double membranes (Figure 11–17), the transport ATPases are located in the inner membrane, and an auxiliary mechanism exists to capture the nutrients and deliver them to the transporters (Figure 11–18).

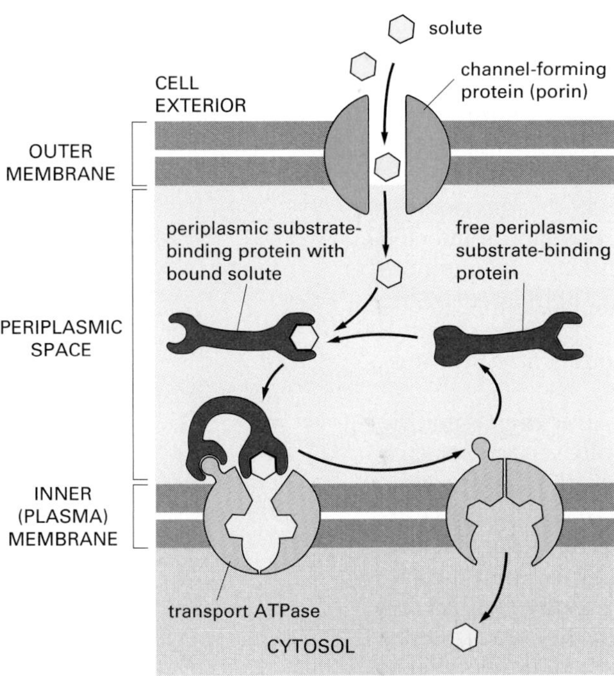

Figure 11–18 The auxiliary transport system associated with transport ATPases in bacteria with double membranes. The solute diffuses through channel-forming proteins (porins) in the outer membrane and binds to a *periplasmic substrate-binding protein.* As a result, the substrate-binding protein undergoes a conformational change that enables it to bind to an ABC transporter in the plasma membrane, which then picks up the solute and actively transfers it across the bilayer in a reaction driven by ATP hydrolysis. The peptidoglycan is omitted for simplicity; its porous structure allows the substrate-binding proteins and water-soluble solutes to move through it by simple diffusion.

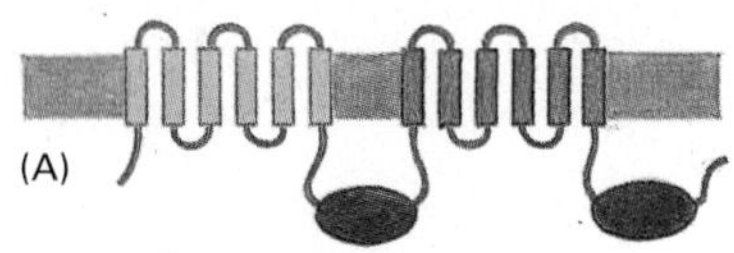

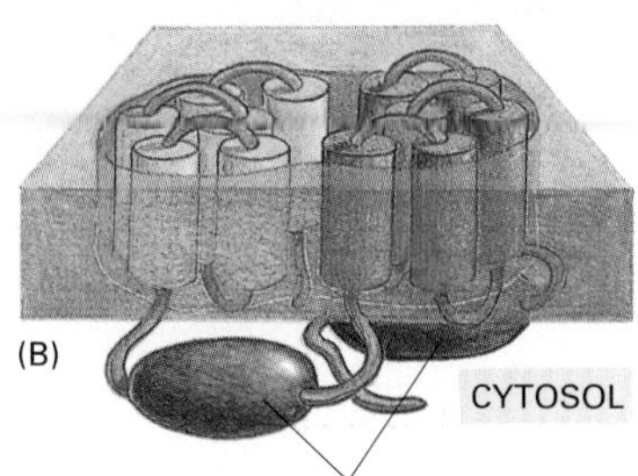

Figure 11–19 A typical ABC transporter. (A) A topology diagram. (B) A hypothetical arrangement of the polypeptide chain in the membrane. The transporter consists of four domains: two highly hydrophobic domains, each with six putative membrane-spanning segments that somehow form the translocation pathway, and two ATP-binding catalytic domains (or cassettes). In some cases the two halves of the transporter are formed by a single polypeptide (as shown), whereas in other cases they are formed by two or more separate polypeptides that assemble into a similar structure (see Figure 10–32).

The transport ATPases in the bacterial plasma membrane belong to the largest and most diverse family of transport proteins known. It is called the **ABC transporter superfamily** because each member contains two highly conserved ATP-binding cassettes (Figure 11–19). ATP binding leads to dimerization of the two ATP-binding domains, and ATP hydrolysis leads to their dissociation. These structural changes in the cytosolic domains are thought to be transmitted to the transmembrane segments, driving cycles of conformational changes that alternately expose substrate-binding sites to one or the other side of the membrane. In this way, ABC transporters use ATP binding and hydrolysis to transport molecules across the bilayer.

In *E. coli,* 78 genes (an amazing 5% of the bacterium's genes) encode ABC transporters, and animal cells contain many more. Although each is thought to be specific for a particular substrate or class of substrates, the variety of substrates transported by this superfamily is great and includes amino acids, sugars, inorganic ions, polysaccharides, peptides, and even proteins. Whereas bacterial ABC transporters are used for both import and export, those described in eucaryotes mostly seem specialized for export. ABC transporters also catalyze the flipping of lipids from one face of the lipid bilayer to the other, and thus have an important role in membrane biogenesis and maintenance, as discussed in Chapter 12. When the substrates are lipids or overall hydrophobic molecules, the binding sites for them must be exposed on the surface of the transporter that is in contact with the hydrophobic interior of the lipid bilayer.

Indeed, the first eucaryotic ABC transporters identified were discovered because of their ability to pump hydrophobic drugs out of the cytosol. One of these is the **multidrug resistance (MDR) protein**, whose overexpression in human cancer cells can make the cells simultaneously resistant to a variety of chemically unrelated cytotoxic drugs that are widely used in cancer chemotherapy. Treatment with any one of these drugs can result in the selection of cells that overexpress the MDR transport protein. The transporter pumps the drugs out of the cell, thereby reducing their toxicity and conferring resistance to a wide variety of therapeutic agents. Some studies indicate that up to 40% of human cancers develop multidrug resistance, making it a major hurdle in the battle against cancer.

A related and equally sinister phenomenon occurs in the protist *Plasmodium falciparum,* which causes malaria. More than 200 million people are infected with this parasite, which remains a major cause of human death, killing more than a million people every year. The control of malaria is hampered by the development of resistance to the antimalarial drug *chloroquine,* and resistant *P. falciparum* have been shown to have amplified a gene encoding an ABC transporter that pumps out the chloroquine.

In yeasts, an ABC transporter is responsible for exporting a mating pheromone (which is a peptide 12 amino acids long) across the yeast cell plasma membrane. In most vertebrate cells, an ABC transporter in the endoplasmic reticulum (ER) membrane actively transports a wide variety of peptides, produced by protein degradation, from the cytosol into the ER. This is the first step in a pathway of great importance in the surveillance of cells by the immune system (discussed in Chapter 24). The transported protein fragments, having entered the ER, are eventually carried to the cell surface, where they are displayed for scrutiny by cytotoxic T lymphocytes, which kill the cell if the fragments seem foreign (as they do if they derive from a virus or other microorganisms lurking in the cytosol).

Yet another member of the ABC family has been discovered through studies of the common genetic disease *cystic fibrosis*. This disease is caused by a mutation in a gene encoding an ABC transporter that functions as a regulator of a Cl^- channel in the plasma membrane of epithelial cells. One in 27 white persons carries a mutant gene encoding this protein; and, in 1 in 2500, both copies of the gene are mutant, causing the disease. It is still uncertain how the ABC transporter acts to regulate Cl^- conductance across the membrane.

Summary

Carrier proteins bind specific solutes and transfer them across the lipid bilayer by undergoing conformational changes that expose the solute-binding site sequentially on one side of the membrane and then on the other. Some carrier proteins simply transport a single solute "downhill," whereas others can act as pumps to transport a solute "uphill" against its electrochemical gradient, using energy provided by ATP hydrolysis, by a downhill flow of another solute (such as Na^+ or H^+), or by light to drive the requisite series of conformational changes in an orderly manner. Carrier proteins belong to a small number of families. Each family comprises proteins of similar amino acid sequence that are thought to have evolved from a common ancestral protein and to operate by a similar mechanism. The family of P-type transport ATPases, which includes the ubiquitous Na^+-K^+ pump, is an important example; each of these ATPases sequentially phosphorylates and dephosphorylates itself during the pumping cycle. The superfamily of ABC transporters is the largest family of membrane transport proteins and is especially important clinically. It includes proteins that are responsible for cystic fibrosis, as well as for drug resistance in cancer cells and in malaria-causing parasites.

ION CHANNELS AND THE ELECTRICAL PROPERTIES OF MEMBRANES

Unlike carrier proteins, channel proteins form hydrophilic pores across membranes. One class of channel proteins found in virtually all animals forms *gap junctions* between two adjacent cells; each plasma membrane contributes equally to the formation of the channel, which connects the cytoplasm of the two cells. These channels are discussed in Chapter 19 and will not be considered further here. Both gap junctions and *porins*, the channel-forming proteins of the outer membranes of bacteria, mitochondria, and chloroplasts (discussed in Chapter 10) have relatively large and permissive pores, which would be disastrous if they directly connected the inside of a cell to an extracellular space. Indeed, many bacterial toxins do exactly that to kill other cells (discussed in Chapter 25).

In contrast, most channel proteins in the plasma membrane of animal and plant cells that connect the cytosol to the cell exterior necessarily have narrow, highly selective pores that can open and close. Because these proteins are concerned specifically with inorganic ion transport, they are referred to as **ion channels**. For transport efficiency, channels have an advantage over carriers in that up to 100 million ions can pass through one open channel each second—a rate 10^5 times greater than the fastest rate of transport mediated by any known carrier protein. However, channels cannot be coupled to an energy source to perform active transport, so the transport that they mediate is always passive ("downhill"). Thus, the function of ion channels is to allow specific inorganic ions—primarily Na^+, K^+, Ca^{2+}, or Cl^-—to diffuse rapidly down their electrochemical gradients across the lipid bilayer. As we shall see, the ability to control ion fluxes through these channels is essential for many cell functions. Nerve cells (neurons), in particular, have made a specialty of using ion channels, and we shall consider how they use a diversity of such channels for receiving, conducting, and transmitting signals.

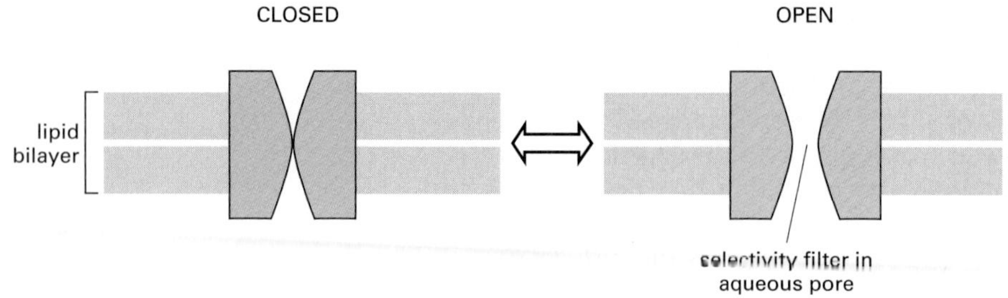

Figure 11–20 A typical ion channel, which fluctuates between closed and open conformations. The channel protein shown here in cross section forms a hydrophilic pore across the lipid bilayer only in the "open" conformational state. Polar groups are thought to line the wall of the pore, while hydrophobic amino acid side chains interact with the lipid bilayer (not shown). The pore narrows to atomic dimensions in one region (the selectivity filter), where the ion selectivity of the channel is largely determined.

Ion Channels Are Ion-Selective and Fluctuate Between Open and Closed States

Two important properties distinguish ion channels from simple aqueous pores. First, they show *ion selectivity,* permitting some inorganic ions to pass, but not others. This suggests that their pores must be narrow enough in places to force permeating ions into intimate contact with the walls of the channel so that only ions of appropriate size and charge can pass. The permeating ions have to shed most or all of their associated water molecules to pass, often in single file, through the narrowest part of the channel, which is called the *selectivity filter;* this limits their rate of passage. Thus, as ion concentrations are increased, the flux of ions through a channel increases proportionally but then levels off (saturates) at a maximum rate.

The second important distinction between ion channels and simple aqueous pores is that ion channels are not continuously open. Instead, they are *gated,* which allows them to open briefly and then close again (Figure 11–20). In most cases, the gate opens in response to a specific stimulus. The main types of stimuli that are known to cause ion channels to open are a change in the voltage across the membrane *(voltage-gated channels),* a mechanical stress *(mechanically gated channels),* or the binding of a ligand *(ligand-gated channels).* The ligand can be either an extracellular mediator—specifically, a neurotransmitter *(transmitter-gated channels)*—or an intracellular mediator, such as an ion *(ion-gated channels)* or a nucleotide *(nucleotide-gated channels)* (Figure 11–21). The activity of many ion channels is regulated, in addition, by protein phosphorylation and dephosphorylation; this type of channel regulation is discussed, together with nucleotide-gated ion channels, in Chapter 15. Moreover, with prolonged (chemical or electrical) stimulation, most channels go into a closed "desensitized" or "inactivated" state, in which they are refractory to further opening until the stimulus has been removed, as we discuss later.

More than 100 types of ion channels have been described thus far, and new ones are still being added to the list. They are responsible for the electrical

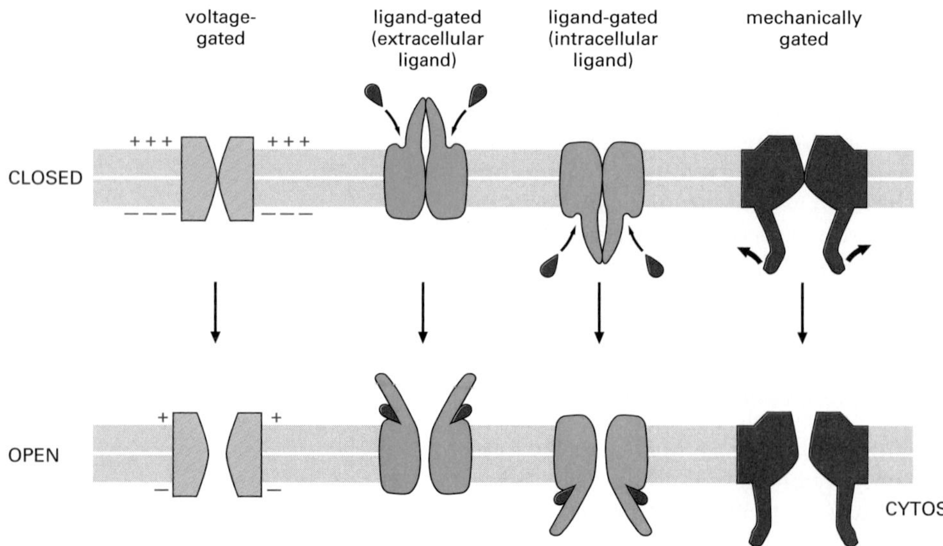

Figure 11–21 The gating of ion channels. This drawing shows different kinds of stimuli that open ion channels. Mechanically gated channels often have cytoplasmic extensions that link the channel to the cytoskeleton (not shown).

excitability of muscle cells, and they mediate most forms of electrical signaling in the nervous system. A single neuron might typically contain 10 kinds of ion channels or more, located in different domains of its plasma membrane. But ion channels are not restricted to electrically excitable cells. They are present in all animal cells and are found in plant cells and microorganisms: they propagate the leaf-closing response of the mimosa plant, for example, and allow the single-celled *Paramecium* to reverse direction after a collision.

Perhaps the most common ion channels are those that are permeable mainly to K^+. These channels are found in the plasma membrane of almost all animal cells. An important subset of K^+ channels are open even in an unstimulated or "resting" cell and are hence sometimes called **K^+ leak channels**. Although this term covers a variety of different K^+ channels, depending on the cell type, they serve a common purpose. By making the plasma membrane much more permeable to K^+ than to other ions, they have a crucial role in maintaining the membrane potential across all plasma membranes.

The Membrane Potential in Animal Cells Depends Mainly on K^+ Leak Channels and the K^+ Gradient Across the Plasma Membrane

A **membrane potential** arises when there is a difference in the electrical charge on the two sides of a membrane, due to a slight excess of positive ions over negative ones on one side and a slight deficit on the other. Such charge differences can result both from active electrogenic pumping (see p. 626) and from passive ion diffusion. As we discuss in Chapter 14, most of the membrane potential of the mitochondrion is generated by electrogenic H^+ pumps in the mitochondrial inner membrane. Electrogenic pumps also generate most of the electrical potential across the plasma membrane in plants and fungi. In typical animal cells, however, passive ion movements make the largest contribution to the electrical potential across the plasma membrane.

As explained earlier, the Na^+-K^+ pump helps maintain an osmotic balance across the animal cell membrane by keeping the intracellular concentration of Na^+ low. Because there is little Na^+ inside the cell, other cations have to be plentiful there to balance the charge carried by the cell's fixed anions—the negatively charged organic molecules that are confined inside the cell. This balancing role is performed largely by K^+, which is actively pumped into the cell by the Na^+-K^+ pump and can also move freely in or out through the *K^+ leak channels* in the plasma membrane. Because of the presence of these channels, K^+ comes almost to equilibrium, where an electrical force exerted by an excess of negative charges attracting K^+ into the cell balances the tendency of K^+ to leak out down its concentration gradient. The membrane potential is the manifestation of this electrical force, and its equilibrium value can be calculated from the steepness of the K^+ concentration gradient. The following argument may help to make this clear.

Suppose that initially there is no voltage gradient across the plasma membrane (the membrane potential is zero), but the concentration of K^+ is high inside the cell and low outside. K^+ will tend to leave the cell through the K^+ leak channels, driven by its concentration gradient. As K^+ moves out, it leaves behind an unbalanced negative charge, thereby creating an electrical field, or membrane potential, which will tend to oppose the further efflux of K^+. The net efflux of K^+ halts when the membrane potential reaches a value at which this electrical driving force on K^+ exactly balances the effect of its concentration gradient—that is, when the electrochemical gradient for K^+ is zero. Although Cl^- ions also equilibrate across the membrane, the membrane potential keeps most of these ions out of the cell because their charge is negative.

The equilibrium condition, in which there is no net flow of ions across the plasma membrane, defines the **resting membrane potential** for this idealized cell. A simple but very important formula, the **Nernst equation**, expresses the equilibrium condition quantitatively and, as explained in Panel 11–2, makes it possible to calculate the theoretical resting membrane potential if the ratio of internal and external ion concentrations is known. As the plasma membrane of

THE NERNST EQUATION AND ION FLOW

The flow of any ion through a membrane channel protein is driven by the electrochemical gradient for that ion. This gradient represents the combination of two influences: the voltage gradient and the concentration gradient of the ion across the membrane. When these two influences just balance each other the electrochemical gradient for the ion is zero and there is no *net* flow of the ion through the channel. The voltage gradient (membrane potential) at which this equilibrium is reached is called the equilibrium potential for the ion. It can be calculated from an equation that will be derived below, called the Nernst equation.

The Nernst equation is

$$V = \frac{RT}{zF} \ln \frac{C_o}{C_i}$$

where

V = the equilibrium potential in volts (internal potential minus external potential)

C_o and C_i = outside and inside concentrations of the ion, respectively

R = the gas constant (2 cal mol^{-1} K^{-1})

T = the absolute temperature (K)

F = Faraday's constant (2.3×10^4 cal V^{-1} mol^{-1})

z = the valence (charge) of the ion

ln = logarithm to the base e

The Nernst equation is derived as follows:

A molecule in solution (a solute) tends to move from a region of high concentration to a region of low concentration simply due to the random movement of molecules, which results in their equilibrium. Consequently, movement down a concentration gradient is accompanied by a favorable free-energy change ($\Delta G < 0$), whereas movement up a concentration gradient is accompanied by an unfavorable free-energy change ($\Delta G > 0$). (Free energy is introduced and discussed in Panel 14–1, p. 784.) The free-energy change per mole of solute moved across the plasma membrane (ΔG_{conc}) is equal to $-RT \ln C_o / C_i$. If the solute is an ion, moving it into a cell across a membrane whose inside is at a voltage V relative to the outside will cause an additional free-energy change (per mole of solute moved) of $\Delta G_{volt} = zFV$. At the point where the concentration and voltage gradients just balance, $\Delta G_{conc} + \Delta G_{volt} = 0$ and the ion distribution is at equlibrium across the membrane. Thus,

$$zFV - RT \ln \frac{C_o}{C_i} = 0$$

and, therefore,

$$V = \frac{RT}{zF} \ln \frac{C_o}{C_i} = 2.3 \frac{RT}{zF} \log_{10} \frac{C_o}{C_i}$$

For a univalent ion,

$$2.3 \frac{RT}{F} = 58 \text{ mV at } 20°\text{C} \quad \text{and} \quad 61.5 \text{ mV at } 37°\text{C}$$

Thus, for such an ion at 37°C, $V = +61.5$ mV for $C_o / C_i = 10$, whereas $V = 0$ for $C_o / C_i = 1$.

The K^+ equilibrium potential (V_K), for example, is $61.5 \log_{10}([K^+]_o / [K^+]_i)$ millivolts (–89 mV for a typical cell where $[K^+]_o = 5$ mM and $[K^+]_i = 140$ mM). At V_K, there is no net flow of K^+ across the membrane. Similarly, when the membrane potential has a value of $61.5 \log_{10}([Na^+]_o / [Na^+]_i)$, the Na^+ equilibrium potential (V_{Na}), there is no net flow of Na^+.

For any particular membrane potential, V_M, the net force tending to drive a particular type of ion out of the cell, is proportional to the difference between V_M and the equilibrium potential for the ion: hence, for K^+ it is $V_M - V_K$ and for Na^+ it is $V_M - V_{Na}$.

The number of ions that go to form the layer of charge adjacent to the membrane is minute compared with the total number inside the cell. For example, the movement of 6000 Na^+ ions across 1 μm^2 of membrane will carry sufficient charge to shift the membrane potential by about 100 mV. Because there are about 3×10^7 Na^+ ions in a typical cell (1 μm^3 of bulk cytoplasm), such a movement of charge will generally have a negligible effect on the ion concentration gradients across the membrane.

a real cell is not exclusively permeable to K^+ and Cl^-, however, the actual resting membrane potential is usually not exactly equal to that predicted by the Nernst equation for K^+ or Cl^-.

The Resting Potential Decays Only Slowly When the Na^+-K^+ Pump Is Stopped

The number of ions that must move across the plasma membrane to set up the membrane potential is minute. Thus, one can think of the membrane potential as arising from movements of charge that leave ion *concentrations* practically unaffected and result in only a very slight discrepancy in the number of positive and negative ions on the two sides of the membrane (Figure 11–22). Moreover, these movements of charge are generally rapid, taking only a few milliseconds or less.

It is illuminating to consider what happens to the membrane potential in a real cell if the Na^+-K^+ pump is suddenly inactivated. A slight drop in the membrane potential immediately occurs. This is because the pump is electrogenic and, when active, makes a small direct contribution to the membrane potential by pumping out three Na^+ for every two K^+ that it pumps in. However, switching off the pump does not abolish the major component of the resting potential, which is generated by the K^+ equilibrium mechanism outlined above. This component persists as long as the Na^+ concentration inside the cell stays low and the K^+ ion concentration high—typically for many minutes. But the plasma membrane is somewhat permeable to all small ions, including Na^+. Therefore, without the Na^+-K^+ pump, the ion gradients set up by pumping will eventually run down, and the membrane potential established by diffusion through the K^+ leak channels will fall as well. As Na^+ enters, the osmotic balance is upset, and water seeps into the cell (see Panel 11–1, pp. 628–629). If the cell does not burst, it eventually comes to a new resting state where Na^+, K^+, and Cl^- are all at equilibrium across the membrane. The membrane potential in this state is much less than it was in the normal cell with an active Na^+-K^+ pump.

The potential difference across the plasma membrane of an animal cell at rest varies between –20 mV and –200 mV, depending on the organism and cell type. Although the K^+ gradient always has a major influence on this potential, the gradients of other ions (and the disequilibrating effects of ion pumps) also have a significant effect: the more permeable the membrane for a given ion, the more strongly the membrane potential tends to be driven toward the equilibrium value for that ion. Consequently, changes in a membrane's permeability to ions can cause significant changes in the membrane potential. This is one of the key principles relating the electrical excitability of cells to the activities of ion channels.

To understand how ion channels select their ions and how they open and close, one needs to know their atomic structure. The first ion channel to be crystallized and studied by x-ray diffraction was a bacterial K^+ channel. The details of its structure revolutionized our understanding of ion channels.

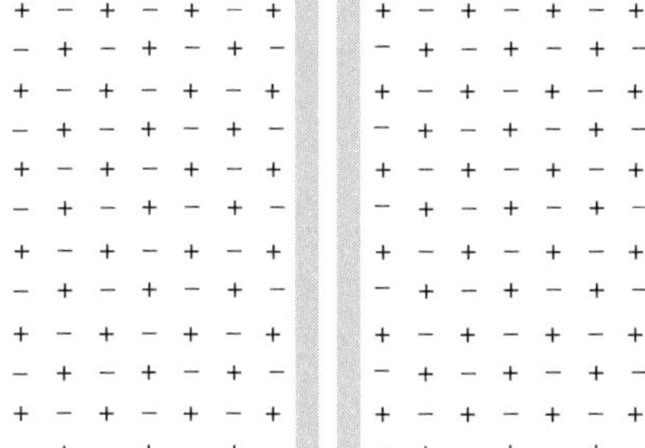
exact balance of charges on each side of the membrane; membrane potential = 0

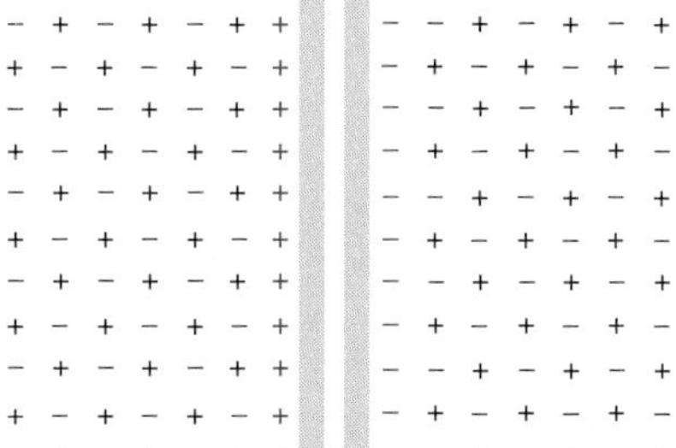
a few of the positive ions *(red)* cross the membrane from right to left, leaving their negative counterions *(red)* behind; this sets up a nonzero membrane potential

Figure 11–22 The ionic basis of a membrane potential. A small flow of ions carries sufficient charge to cause a large change in the membrane potential. The ions that give rise to the membrane potential lie in a thin (< 1 nm) surface layer close to the membrane, held there by their electrical attraction to their oppositely charged counterparts (counterions) on the other side of the membrane. For a typical cell, 1 microcoulomb of charge (6×10^{12} monovalent ions) per square centimeter of membrane, transferred from one side of the membrane to the other, changes the membrane potential by roughly 1 V. This means, for example, that in a spherical cell of diameter 10 μm, the number of K^+ ions that have to flow out to alter the membrane potential by 100 mV is only about 1/100,000 of the total number of K^+ ions in the cytosol.

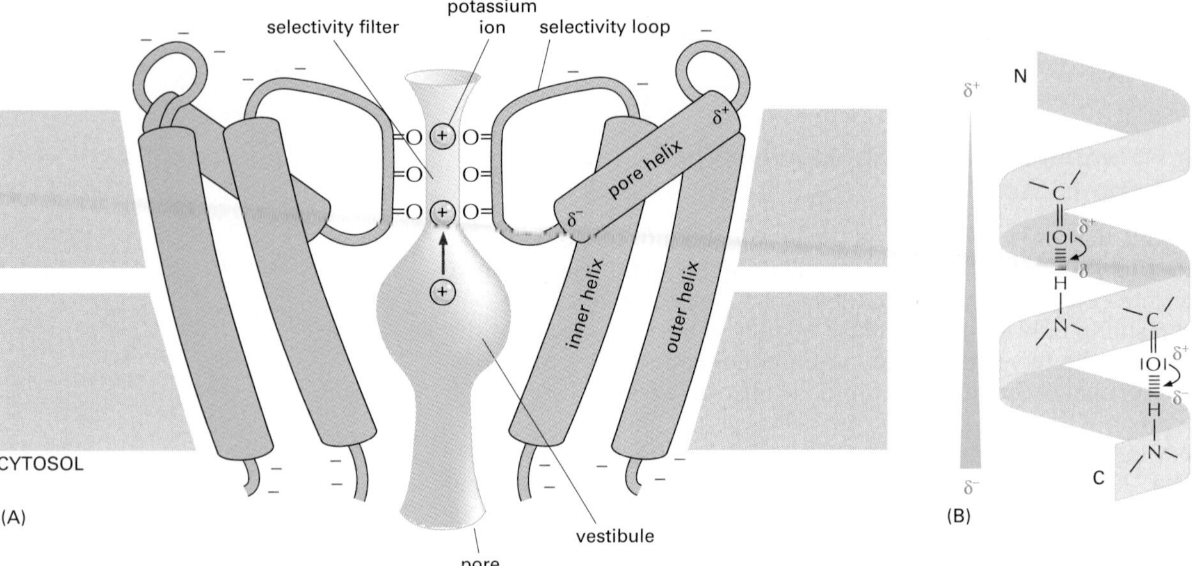

The Three-dimensional Structure of a Bacterial K^+ Channel Shows How an Ion Channel Can Work

The remarkable ability of ion channels to combine exquisite ion selectivity with a high conductance has long puzzled scientists. K^+ leak channels, for example, conduct K^+ 10,000-fold better than Na^+, yet the two ions are featureless spheres with similar diameters (0.133 nm and 0.095 nm, respectively). A single amino acid substitution in the pore of a K^+ channel can result in a loss of ion selectivity and cell death. The normal selectivity cannot be explained by pore size, because Na^+ is smaller than K^+. Moreover, the high conductance rate is incompatible with the channel's having selective, high-affinity K^+-binding sites, as the binding of K^+ ions to such sites would greatly slow their passage.

The puzzle was solved when the structure of a *bacterial K^+ channel* was determined by x-ray crystallography. The channel is made from four identical transmembrane subunits, which together form a central pore through the membrane (Figure 11–23). Negatively charged amino acids are concentrated at the cytosolic entrance to the pore and are thought to attract cations and repel anions, making the channel cation-selective. Each subunit contributes two transmembrane helices, which are tilted outward in the membrane and together form a cone, with its wide end facing the outside of the cell where K^+ ions exit the channel. The polypeptide chain that connects the two transmembrane helices forms a short α helix (the *pore helix)* and a crucial loop that protrudes into the wide section of the cone to form the **selectivity filter**. The selectivity loops from the four subunits form a short, rigid, narrow pore, which is lined by the carbonyl oxygen atoms of their polypeptide backbones. Because the selectivity loops of all known K^+ channels have similar amino acid sequences, it is likely that they form a closely similar structure. The crystal structure shows two K^+ ions in single file within the selectivity filter, separated by about 8 Å. Mutual repulsion between the two ions is thought to help move them through the pore into the extracellular fluid.

The structure of the selectivity filter explains the exquisite ion selectivity of the channel. For a K^+ ion to enter the filter, it must lose almost all of its bound water molecules and interact instead with the carbonyl oxygens lining the selectivity filter, which are rigidly spaced at the exact distance to accommodate a K^+ ion. A Na^+ ion, in contrast, cannot enter the filter because the carbonyl oxygens are too far away from the smaller Na^+ ion to compensate for the energy expense associated with the loss of water molecules required for entry (Figure 11–24).

Structural studies of the bacterial K^+ channel have indicated how these channels may open and close. The loops that form the selectivity filter are rigid

Figure 11–23 The structure of a bacterial K^+ channel. (A) Only two of the four identical subunits are shown. From the cytosolic side, the pore opens up into a vestibule in the middle of the membrane. The vestibule facilitates transport by allowing the K^+ ions to remain hydrated even though they are halfway across the membrane. The narrow selectivity filter links the vestibule to the outside of the cell. Carbonyl oxygens line the walls of the selectivity filter and form transient binding sites for dehydrated K^+ ions. The positions of the K^+ ions in the pore were determined by soaking crystals of the channel protein in a solution containing rubidium ions, which are more electron-dense but only slightly larger than K^+ ions; from the differences in the diffraction patterns with K^+ ions and with rubidium ions in the channel, the positions of the ions could be calculated. Two K^+ ions occupy sites in the selectivity filter, while a third K^+ ion is located in the center of the vestibule, where it is stabilized by electrical interactions with the more negatively charged ends of the pore helices. The ends of the four pore helices point precisely toward the center of the vestibule, thereby guiding K^+ ions into the selectivity filter. (B) Because of the polarity of the hydrogen bonds (in *red*) that link adjacent turns of an α helix, every α helix has an electric dipole along its axis, with a more negatively charged C-terminal end (δ^-) and a more positively charged N-terminal end (δ^+). (A, Adapted from D.A. Doyle et al., *Science* 280:69–77, 1998.)

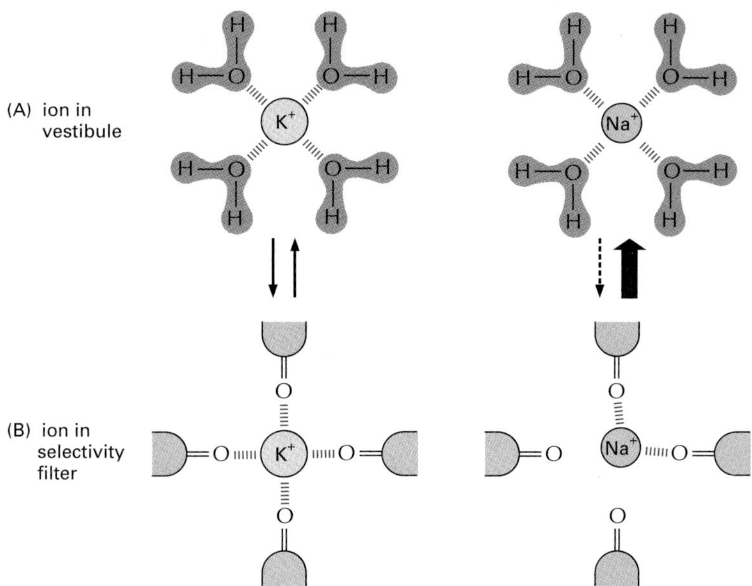

Figure 11–24 K^+ specificity of the selectivity filter in a K^+ channel. The drawing shows K^+ and Na^+ ions (A) in the vestibule and (B) in the selectivity filter of the pore, viewed in cross section. In the vestibule, the ions are hydrated. In the selectivity filter, the carbonyl oxygens are placed precisely to accommodate a dehydrated K^+ ion. The dehydration of the K^+ ion requires energy, which is precisely balanced by the energy regained by the interaction of the ion with the carbonyl oxygens that serve as surrogate water molecules. Because the Na^+ ion is too small to interact with the oxygens, it could enter the selectivity filter only at a great energetic expense. The filter therefore selects K^+ ions with high specificity. (Adapted from D.A. Doyle et al., *Science* 280:69–77, 1998.)

and do not change conformation when the channel opens or closes. In contrast, the inner and outer transmembrane helices that line the rest of the pore rearrange when the channel closes, causing the pore to constrict like a diaphragm at its cytosolic end (Figure 11–25). Although the pore does not close completely, the small opening that remains is lined by hydrophobic amino acid side chains, which block the entry of ions.

The cells that make most use of ion channels are neurons. Before discussing how they do so, we must digress to review briefly how a typical neuron is organized.

The Function of a Nerve Cell Depends on Its Elongated Structure

The fundamental task of a **neuron**, or **nerve cell**, is to receive, conduct, and transmit signals. To perform these functions, neurons are often extremely elongated. A single nerve cell in a human being, extending, for example, from the spinal cord to a muscle in the foot, may be as long as 1 meter. Every neuron consists of a cell body (containing the nucleus) with a number of thin processes radiating outward from it. Usually one long **axon** conducts signals away from the cell body toward distant targets, and several shorter branching **dendrites** extend from the cell body like antennae, providing an enlarged surface area to receive signals from the axons of other nerve cells (Figure 11–26). Signals are also received on the cell body itself. The typical axon divides at its far end into many branches, passing on its message to many target cells simultaneously. Likewise,

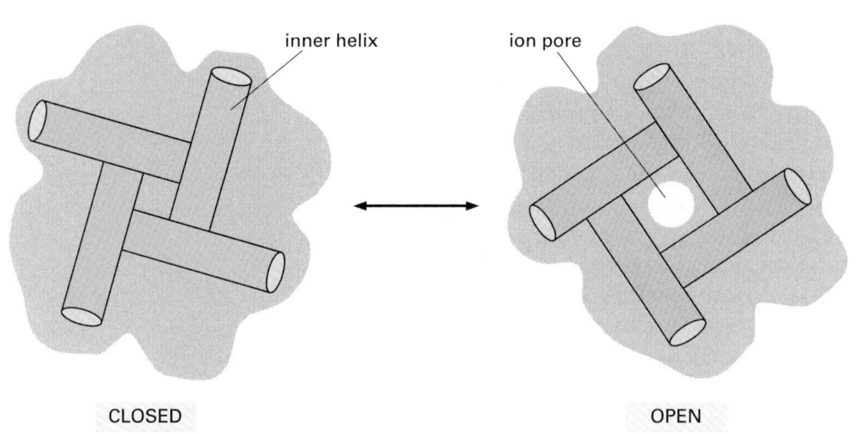

Figure 11–25 A model for the gating of a bacterial K^+ channel. The channel is viewed in cross section. To adopt the closed conformation, the four inner transmembrane helices that line the pore on the cytosolic side of the selectivity filter (see Figure 11–22) rearrange to close the cytosolic entrance to the channel. (Adapted from E. Perozo et al., *Science* 285:73–78, 1999.)

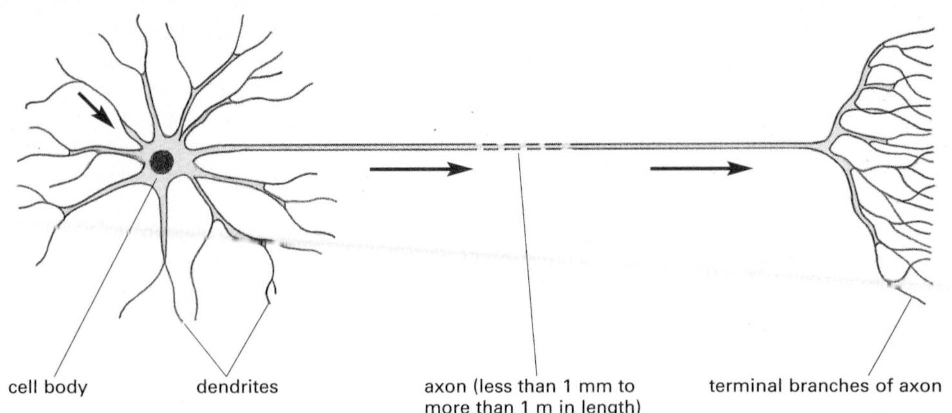

Figure 11–26 A typical vertebrate neuron. The arrows indicate the direction in which signals are conveyed. The single axon conducts signals away from the cell body, while the multiple dendrites receive signals from the axons of other neurons. The nerve terminals end on the dendrites or cell body of other neurons or on other cell types, such as muscle or gland cells.

the extent of branching of the dendrites can be very great—in some cases, sufficient to receive as many as 100,000 inputs on a single neuron.

Despite the varied significance of the signals carried by different classes of neurons, the form of the signal is always the same, consisting of changes in the electrical potential across the neuron's plasma membrane. Communication occurs because an electrical disturbance produced in one part of the cell spreads to other parts. Such a disturbance becomes weaker with increasing distance from its source, unless energy is expended to amplify it as it travels. Over short distances this attenuation is unimportant; in fact, many small neurons conduct their signals passively, without amplification. For long-distance communication, however, passive spread is inadequate. Thus, larger neurons employ an active signaling mechanism, which is one of their most striking features. An electrical stimulus that exceeds a certain threshold strength triggers an explosion of electrical activity that is propagated rapidly along the neuron's plasma membrane and is sustained by automatic amplification all along the way. This traveling wave of electrical excitation, known as an **action potential**, or *nerve impulse,* can carry a message without attenuation from one end of a neuron to the other at speeds as great as 100 meters per second or more. Action potentials are the direct consequence of the properties of voltage-gated cation channels, as we shall now see.

Voltage-gated Cation Channels Generate Action Potentials in Electrically Excitable Cells

The plasma membrane of all electrically excitable cells—not only neurons, but also muscle, endocrine, and egg cells—contains **voltage-gated cation channels**, which are responsible for generating the action potentials. An action potential is triggered by a *depolarization* of the plasma membrane—that is, by a shift in the membrane potential to a less negative value. (We shall see later how this can be caused by the action of a neurotransmitter.) In nerve and skeletal muscle cells, a stimulus that causes sufficient depolarization promptly causes **voltage-gated Na^+ channels** to open, allowing a small amount of Na^+ to enter the cell down its electrochemical gradient. The influx of positive charge depolarizes the membrane further, thereby opening more Na^+ channels, which admit more Na^+ ions, causing still further depolarization. This process continues in a self-amplifying fashion until, within a fraction of a millisecond, the electrical potential in the local region of membrane has shifted from its resting value of about –70 mV to almost as far as the Na^+ equilibrium potential of about +50 mV (see Panel 11–2, p. 634). At this point, when the net electrochemical driving force for the flow of Na^+ is almost zero, the cell would come to a new resting state, with all of its Na^+ channels permanently open, if the open conformation of the channel were stable. The cell is saved from such a permanent electrical spasm by two mechanisms that act in concert: inactivation of the Na^+ channels, and opening of voltage-gated K^+ channels.

The Na^+ channels have an automatic inactivating mechanism, which causes the channels to reclose rapidly even though the membrane is still depolarized. The Na^+ channels remain in this *inactivated* state, unable to reopen, until a few milliseconds after the membrane potential has returned to its initial negative value. The Na^+ channel can therefore exist in three distinct states—closed, open, and inactivated. How they contribute to the rise and fall of the action potential is shown in Figure 11–27.

The description just given of an action potential concerns only a small patch of plasma membrane. The self-amplifying depolarization of the patch, however, is sufficient to depolarize neighboring regions of membrane, which then go through the same cycle. In this way, the action potential spreads as a traveling wave from the initial site of depolarization to involve the entire plasma membrane, as shown in Figure 11–28.

Voltage-gated K^+ channels provide a second mechanism in most nerve cells to help bring the activated plasma membrane more rapidly back toward its original negative potential, ready to transmit a second impulse. These channels open, so that the transient influx of Na^+ is rapidly overwhelmed by an efflux of K^+, which quickly drives the membrane back toward the K^+ equilibrium potential, even before the inactivation of the Na^+ channels is complete. These K^+ channels respond to changes in membrane potential in much the same way as the Na^+ channels do, but with slightly slower kinetics; for this reason; they are sometimes called *delayed K^+ channels.*

Like the Na^+ channel, voltage-gated K^+ channels can inactivate. Studies of mutant voltage-gated K^+ channels show that the N-terminal 20 amino acids of the channel protein are required for rapid inactivation of the channel. If this region is altered, the kinetics of channel inactivation are changed, and if the region is entirely removed, inactivation is abolished. Amazingly, in the latter case, inactivation can be restored by exposing the cytoplasmic face of the plasma membrane to a small synthetic peptide corresponding to the missing amino terminus. These findings suggest that the amino terminus of each K^+ channel subunit acts like a tethered ball that occludes the cytoplasmic end of the pore soon after it opens, thereby inactivating the channel (Figure 11–29). A similar mechanism is thought to operate in the rapid inactivation of voltage-gated Na^+ channels (which we discuss later), although a different segment of the protein seems to be involved.

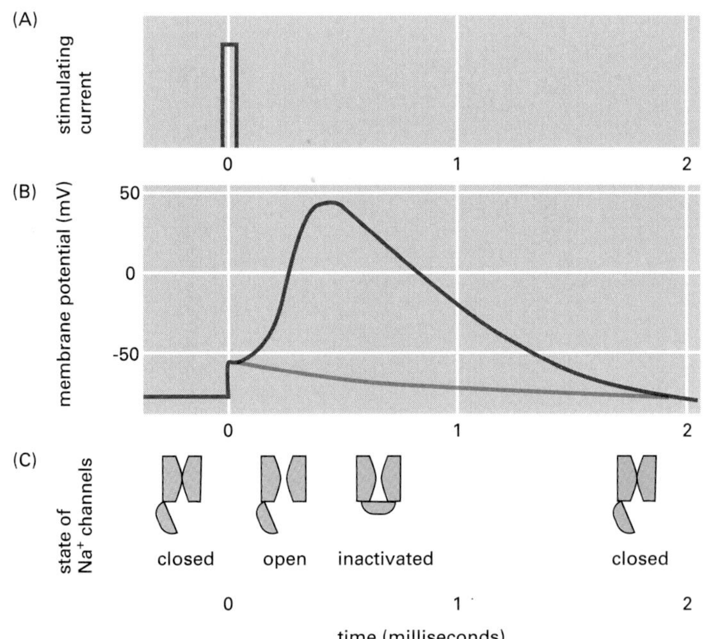

Figure 11–27 An action potential. (A) An action potential is triggered by a brief pulse of current, which (B) partially depolarizes the membrane, as shown in the plot of membrane potential versus time. The *green curve* shows how the membrane potential would have simply relaxed back to the resting value after the initial depolarizing stimulus if there had been no voltage-gated ion channels in the membrane; this relatively slow return of the membrane potential to its initial value of –70 mV in the absence of open Na^+ channels occurs because of the efflux of K^+ through K^+ channels, which open in response to membrane depolarization and drive the membrane back toward the K^+ equilibrium potential. The *red curve* shows the course of the action potential that is caused by the opening and subsequent inactivation of voltage-gated Na^+ channels, whose state is shown in (C). The membrane cannot fire a second action potential until the Na^+ channels have returned to the closed conformation; until then, the membrane is refractory to stimulation.

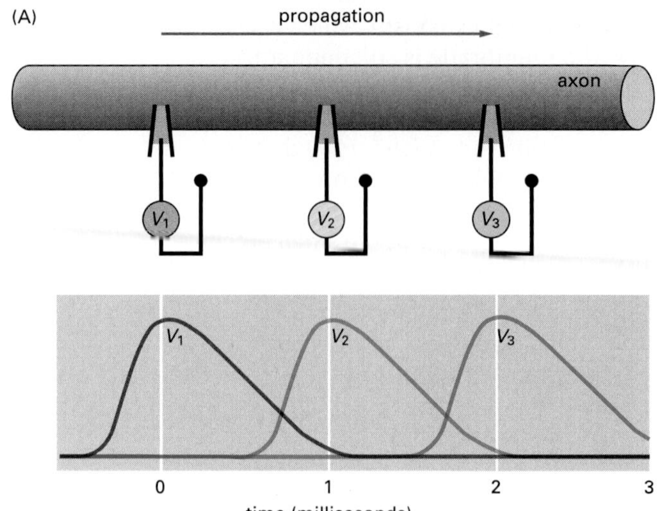

Figure 11–28 The propagation of an action potential along an axon. (A) The voltages that would be recorded from a set of intracellular electrodes placed at intervals along the axon. (B) The changes in the Na^+ channels and the current flows *(orange arrows)* that give rise to the traveling disturbance of the membrane potential. The region of the axon with a depolarized membrane is shaded in *blue*. Note that an action potential can only travel away from the site of depolarization, because Na^+-channel inactivation prevents the depolarization from spreading backward (see also Figure 11–30). On myelinated axons, clusters of Na^+ channels can be millimeters apart from each other.

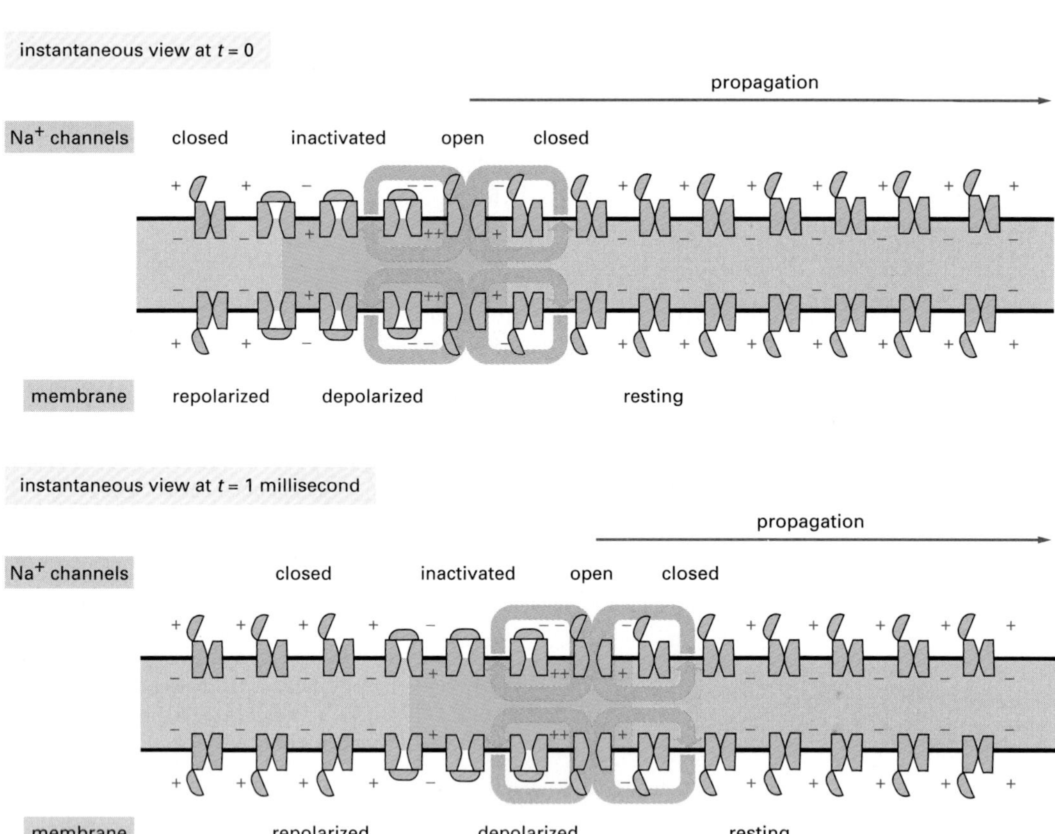

The electrochemical mechanism of the action potential was first established by a famous series of experiments carried out in the 1940s and 1950s. Because the techniques for studying electrical events in small cells had not yet been developed, the experiments exploited the giant neurons in the squid. Despite the many technical advances made since then, the logic of the original analysis continues to serve as a model for present-day work. Panel 11–3 outlines some of the key original experiments.

Myelination Increases the Speed and Efficiency of Action Potential Propagation in Nerve Cells

The axons of many vertebrate neurons are insulated by a **myelin sheath**, which greatly increases the rate at which an axon can conduct an action potential. The importance of myelination is dramatically demonstrated by the demyelinating disease *multiple sclerosis,* in which myelin sheaths in some regions of the central nervous system are destroyed; where this happens, the propagation of nerve impulses is greatly slowed, often with devastating neurological consequences.

Myelin is formed by specialized supporting cells called **glial cells**. **Schwann cells** myelinate axons in peripheral nerves and **oligodendrocytes** do so in the central nervous system. These glial cells wrap layer upon layer of their own plasma membrane in a tight spiral around the axon (Figure 11–30), thereby insulating the axonal membrane so that little current can leak across it. The myelin sheath is interrupted at regularly spaced *nodes of Ranvier,* where almost all the

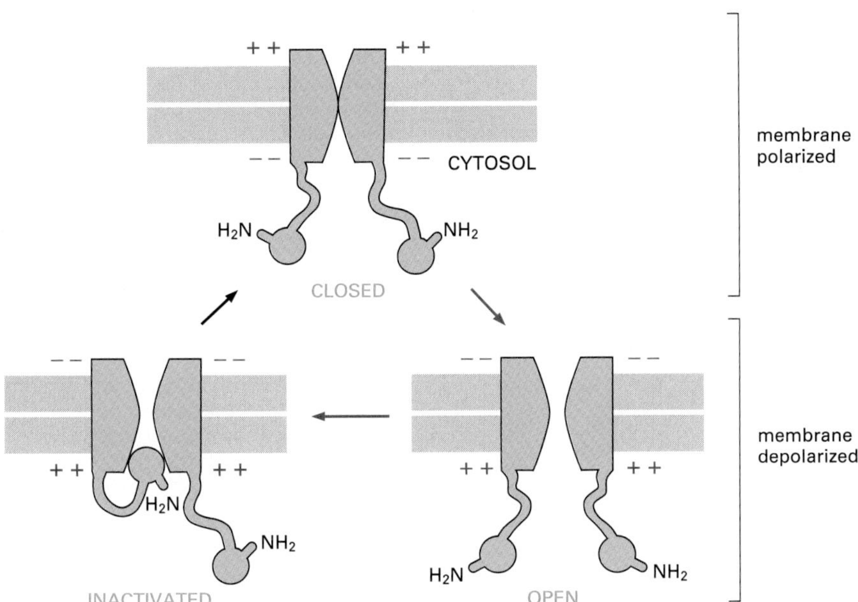

Figure 11–29 The "ball-and-chain" model of rapid inactivation for a voltage-gated K^+ channel. When the membrane potential is depolarized, the channel opens and begins to conduct ions. If the depolarization is maintained, the open channel adopts an inactive conformation, where the pore is occluded by the N-terminal 20 amino acid "ball," which is linked to the channel proper by a segment of unfolded polypeptide chain that serves as the "chain." For simplicity, only two balls are shown; in fact, there are four, one from each subunit. A similar mechanism, using a different segment of the polypeptide chain, is thought to operate in Na^+ channel inactivation. Internal forces stabilize each state against small disturbances, but a sufficiently violent collision with other molecules can cause the channel to flip from one of these states to another. The state of lowest energy depends on the membrane potential because the different conformations have different charge distributions. When the membrane is at rest (highly polarized), the closed conformation has the lowest free energy and is therefore most stable; when the membrane is depolarized, the energy of the *open* conformation is lower, so the channel has a high probability of opening. But the free energy of the *inactivated* conformation is lower still; therefore, after a randomly variable period spent in the open state, the channel becomes inactivated. Thus, the open conformation corresponds to a metastable state that can exist only transiently. The *red arrows* indicate the sequence that follows a sudden depolarization; the *black arrow* indicates the return to the original conformation as the lowest energy state after the membrane is repolarized.

1. Action potentials are recorded with an intracellular electrode

The squid giant axon is about 0.5–1 mm in diameter and several centimeters long. An electrode in the form of a glass capillary tube containing a conducting solution can be thrust down the axis of the axon so that its tip lies deep in the cytoplasm. With its help, one can measure the voltage difference between the inside and the outside of the axon—that is, the membrane potential—as an action potential sweeps past the electrode. The action potential is triggered by a brief electrical stimulus to one end of the axon. It does not matter which end, because the excitation can travel in either direction; and it does not matter how big the stimulus is, as long as it exceeds a certain threshold: the action potential is all or none.

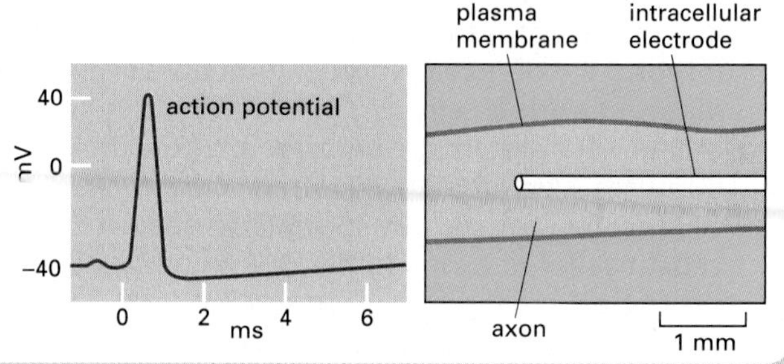

2. Action potentials depend only on the neuronal plasma membrane and on gradients of Na^+ and K^+ across it

The three most plentiful ions, both inside and outside the axon, are Na^+, K^+, and Cl^-. As in other cells, the Na^+-K^+ pump maintains a concentration gradient: the concentration of Na^+ is about 9 times lower inside the axon than outside, while the concentration of K^+ is about 20 times higher inside than outside. Which ions are important for the action potential?

The squid giant axon is so large and robust that it is possible to extrude the cytoplasm from it, like toothpaste from a tube, and then to perfuse it internally with pure artificial solutions of Na^+, K^+, and Cl^- or SO_4^{2-}. Remarkably, if (and only if) the concentrations of Na^+ and K^+ inside and outside approximate those found naturally, the axon will still propagate action potentials of the normal form. The important part of the cell for electrical signaling, therefore, must be the plasma membrane; the important ions are Na^+ and K^+; and a sufficient source of free energy to power the action potential must be provided by their concentration gradients across the membrane, because all other sources of metabolic energy have presumably been removed by the perfusion.

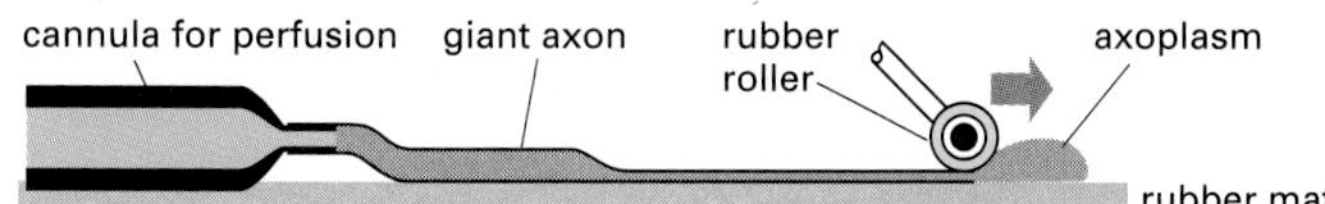

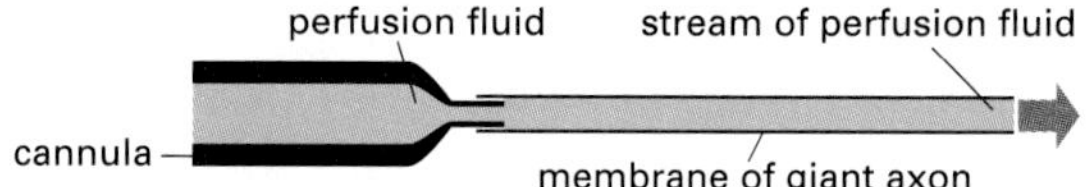

3. At rest, the membrane is chiefly permeable to K^+; during the action potential, it becomes transiently permeable to Na^+

At rest the membrane potential is close to the equilibrium potential for K^+. When the external concentration of K^+ is changed, the resting potential changes roughly in accordance with the Nernst equation for K^+ (see Panel 11–2). At rest, therefore, the membrane is chiefly permeable to K^+: K^+ leak channels provide the main ion pathway through the membrane.

If the external concentration of Na^+ is varied, there is no effect on the resting potential. However, the height of the peak of the action potential varies roughly in accordance with the Nernst equation for Na^+. During the action potential, therefore, the membrane appears to be chiefly permeable to Na^+: Na^+ channels have opened. In the aftermath of the action potential, the membrane potential reverts to a negative value that depends on the external concentration of K^+ and is even closer to the K^+ equilibrium potential than the resting potential is: the membrane has lost most of its permeability to Na^+ and has become even more permeable to K^+ than before—that is, Na^+ channels have closed, and additional K^+ channels have opened.

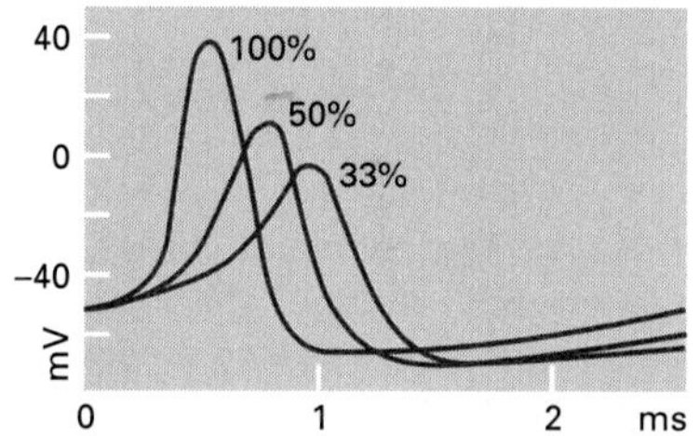

The form of the action potential when the external medium contains 100%, 50%, or 33% of the normal concentration of Na^+.

4. Voltage clamping reveals how the membrane potential controls opening and closing of ion channels

The membrane potential can be held constant ("voltage clamped") throughout the axon by passing a suitable current through a bare metal wire inserted along the axis of the axon while monitoring the membrane potential with another intracellular electrode. When the membrane is abruptly shifted from the resting potential and held in a depolarized state (A), Na^+ channels rapidly open until the Na^+ permeability of the membrane is much greater than the K^+ permeability; they then close again spontaneously, even though the membrane potential is clamped and unchanging. K^+ channels also open but with a delay, so that the K^+ permeability increases as the Na^+ permeability falls (B). If the experiment is now very promptly repeated, by returning the membrane briefly to the resting potential and then quickly depolarizing it again, the response is different: prolonged depolarization has caused the Na^+ channels to enter an inactivated state, so that the second depolarization fails to cause a rise and fall similar to the first. Recovery from this state requires a relatively long time—about 10 milliseconds—spent at the repolarized (resting) membrane potential.

In a normal unclamped axon, an inrush of Na^+ through the opened Na^+ channels produces the spike of the action potential; inactivation of Na^+ channels and opening of K^+ channels bring the membrane rapidly back down to the resting potential.

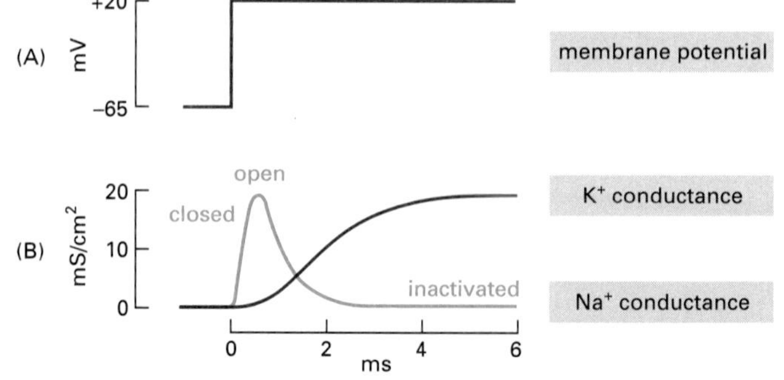

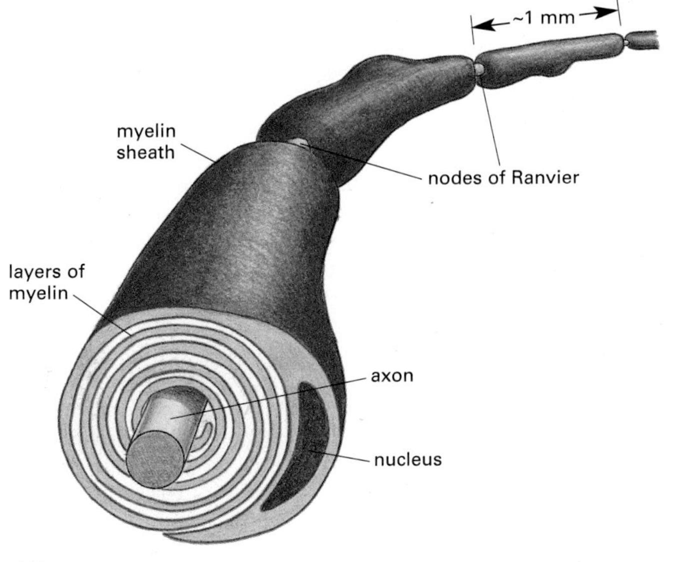

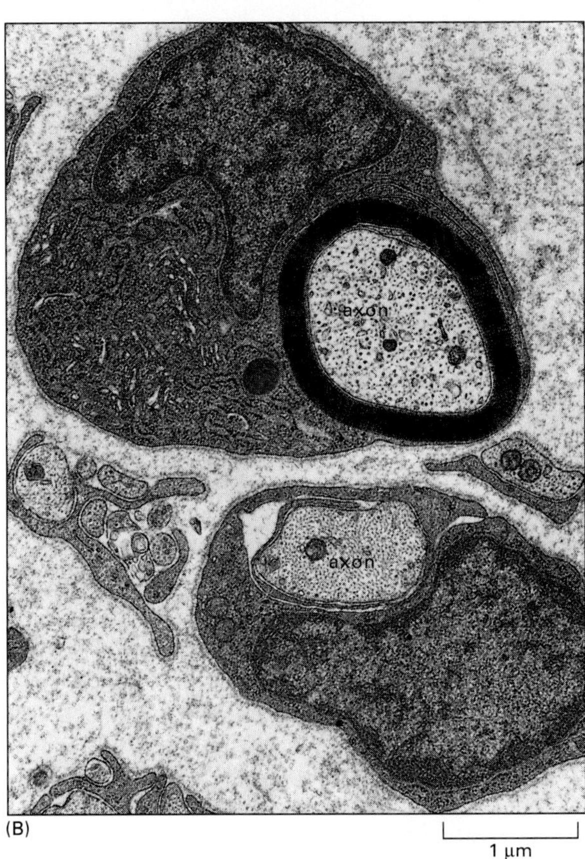

Figure 11–30 Myelination. (A) A myelinated axon from a peripheral nerve. Each Schwann cell wraps its plasma membrane concentrically around the axon to form a segment of myelin sheath about 1 mm long. For clarity, the layers of myelin in this drawing are not shown compacted together as tightly as they are in reality (see part B). (B) An electron micrograph of a section from a nerve in the leg of a young rat. Two Schwann cells can be seen: one near the bottom is just beginning to myelinate its axon; the one above it has formed an almost mature myelin sheath. (B, from Cedric S. Raine, in Myelin [P. Morell, ed.]. New York: Plenum, 1976.)

Na^+ channels in the axon are concentrated. Because the ensheathed portions of the axonal membrane have excellent cable properties (in other words, they behave electrically much like well-designed underwater telegraph cables), a depolarization of the membrane at one node almost immediately spreads passively to the next node. Thus, an action potential propagates along a myelinated axon by jumping from node to node, a process called *saltatory conduction.* This type of conduction has two main advantages: action potentials travel faster, and metabolic energy is conserved because the active excitation is confined to the small regions of axonal plasma membrane at nodes of Ranvier.

Patch-Clamp Recording Indicates That Individual Gated Channels Open in an All-or-Nothing Fashion

Neuron and skeletal muscle cell plasma membranes contain many thousands of voltage-gated Na^+ channels, and the current crossing the membrane is the sum of the currents flowing through all of these. This aggregate current can be recorded with an intracellular microelectrode, as shown in Figure 11–28. Remarkably, however, it is also possible to record current flowing through individual channels. This is achieved by means of **patch-clamp recording**, a method that has revolutionized the study of ion channels by allowing researchers to examine transport through a single molecule of channel protein in a small patch of membrane covering the mouth of a micropipette (Figure 11–31). With this simple but powerful technique, the detailed properties of ion channels can be studied in all sorts of cell types. This work has led to the discovery that even cells that are not electrically excitable usually have a variety of gated ion channels in their plasma membrane. Many of these cells, such as yeasts, are too small to be investigated by the traditional electrophysiologist's method of impalement with an intracellular microelectrode.

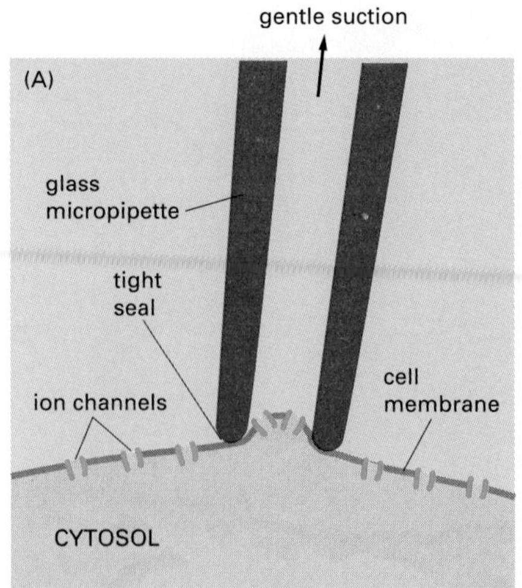

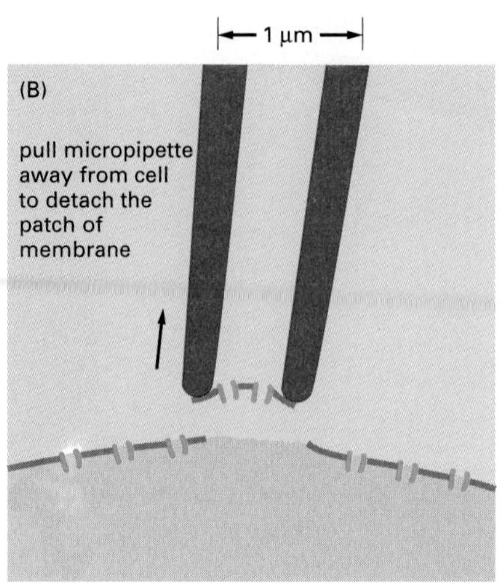

Figure 11–31 The technique of patch-clamp recording. Because of the extremely tight seal between the micropipette and the membrane, current can enter or leave the micropipette only by passing through the channels in the *patch* of membrane covering its tip. The term *clamp* is used because an electronic device is employed to maintain, or "clamp," the membrane potential at a set value while recording the ionic current through individual channels. Recordings of the current through these channels can be made with the patch still attached to the rest of the cell, as in (A), or detached, as in (B). The advantage of the detached patch is that it is easy to alter the composition of the solution on either side of the membrane to test the effect of various solutes on channel behavior. A detached patch can also be produced with the opposite orientation, so that the cytoplasmic surface of the membrane faces the inside of the pipette.

Patch-clamp recording indicates that individual voltage-gated Na^+ channels open in an all-or-nothing fashion. The times of a channel's opening and closing are random, but when open, the channel always has the same large conductance, allowing more than 1000 ions to pass per millisecond. Therefore, the aggregate current crossing the membrane of an entire cell does not indicate the *degree* to which a typical individual channel is open but rather the *total number* of channels in its membrane that are open at any one time (Figure 11–32).

The phenomenon of voltage gating can be understood in terms of simple physical principles. The interior of the resting neuron or muscle cell is at an electrical potential about 50–100 mV more negative than the external medium. Although this potential difference seems small, it exists across a plasma membrane only about 5 nm thick, so that the resulting voltage gradient is about 100,000 V/cm. Proteins in the membrane are thus subjected to a very large electrical field. These proteins, like all others, have a number of charged groups, as well as polarized bonds between their various atoms. The electrical field therefore exerts forces on the molecular structure. For many membrane proteins the effects of changes in the membrane electrical field are probably insignificant, but voltage-gated ion channels can adopt a number of alternative conformations whose stabilities depend on the strength of the field. Voltage-gated Na^+, K^+, and Ca^{2+} channels, for example, have characteristic positively charged amino acids in one of their transmembrane segments that respond to depolarization by moving outward, triggering conformational changes that open the channel. Each conformation can "flip" to another conformation if given a sufficient jolt by

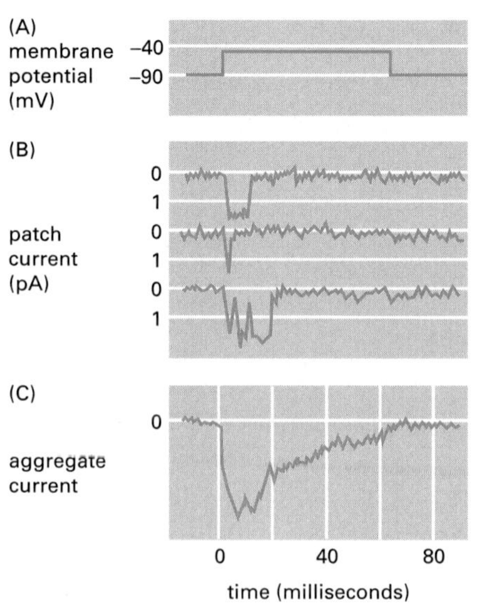

Figure 11–32 Patch-clamp measurements for a single voltage-gated Na^+ channel. A tiny patch of plasma membrane was detached from an embryonic rat muscle cell, as in Figure 11–31. (A) The membrane was depolarized by an abrupt shift of potential. (B) Three current records from three experiments performed on the same patch of membrane. Each major current step in (B) represents the opening and closing of a single channel. A comparison of the three records shows that, whereas the durations of channel opening and closing vary greatly, the rate at which current flows through an open channel is practically constant. The minor fluctuations in the current records arise largely from electrical noise in the recording apparatus. Current is measured in picoamperes (pA). (C) The sum of the currents measured in 144 repetitions of the same experiment. This aggregate current is equivalent to the usual Na^+ current that would be observed flowing through a relatively large region of membrane containing 144 channels. A comparison of (B) and (C) reveals that the time course of the aggregate current reflects the probability that any individual channel will be in the open state; this probability decreases with time as the channels in the depolarized membrane adopt their inactivated conformation. (Data from J. Patlak and R. Horn, *J. Gen. Physiol.* 79:333–351, 1982. © The Rockefeller University Press.)

the random thermal movements of the surroundings, and it is the relative stability of the closed, open, and inactivated conformations against flipping that is altered by changes in the membrane potential (see legend to Figure 11–29).

Voltage-gated Cation Channels Are Evolutionarily and Structurally Related

Na^+ channels are not the only kind of voltage-gated cation channel that can generate an action potential. The action potentials in some muscle, egg, and endocrine cells, for example, depend on *voltage-gated Ca^{2+} channels* rather than on Na^+ channels. Moreover, voltage-gated Na^+, K^+, or Ca^{2+} channels of unknown function are found in some cell types that are not normally electrically active.

There is a surprising amount of structural and functional diversity within each of these three classes, generated both by multiple genes and by the alternative splicing of RNA transcripts produced from the same gene. Nonetheless, the amino acid sequences of the known voltage-gated Na^+, K^+, and Ca^{2+} channels show striking similarities, suggesting that they all belong to a large superfamily of evolutionarily and structurally related proteins and share many of the design principles. Whereas the single-celled yeast *S. cerevisiae* contains a single gene that codes for a voltage-gated K^+ channel, the genome of the worm *C. elegans* contains 68 genes that encode different but related K^+ channels. This complexity indicates that even a simple nervous system made up of only 302 neurons uses a large number of different ion channels to compute its responses.

Humans who inherit mutant genes encoding ion channel proteins can suffer from a variety of nerve, muscle, brain, or heart diseases, depending on where the gene is expressed. Mutations in genes that encode voltage-gated Na^+ channels in skeletal muscle cells, for example, can cause *myotonia,* a condition in which muscle relaxation after voluntary contraction is greatly delayed, causing painful muscle spasms. In some cases this is because the abnormal channels fail to inactivate normally; as a result, Na^+ entry persists after an action potential finishes and repeatedly reinitiates membrane depolarization and muscle contraction. Similarly, mutations that affect Na^+ or K^+ channels in the brain can cause *epilepsy,* where excessive synchronized firing of large groups of nerve cells cause epileptic seizures (convulsions, or fits).

Transmitter-gated Ion Channels Convert Chemical Signals into Electrical Ones at Chemical Synapses

Neuronal signals are transmitted from cell to cell at specialized sites of contact known as **synapses**. The usual mechanism of transmission is indirect. The cells are electrically isolated from one another, the *presynaptic cell* being separated from the *postsynaptic cell* by a narrow *synaptic cleft.* A change of electrical potential in the presynaptic cell triggers it to release small signal molecules known as a **neurotransmitters**, which are stored in membrane-enclosed synaptic vesicles and released by exocytosis. The neurotransmitter diffuses rapidly across the synaptic cleft and provokes an electrical change in the postsynaptic cell by binding to *transmitter-gated ion channels* (Figure 11–33). After the neurotransmitter has been secreted, it is rapidly removed: it is either destroyed by specific enzymes in the synaptic cleft or taken up by the nerve terminal that released it or by surrounding glial cells. Reuptake is mediated by a variety of

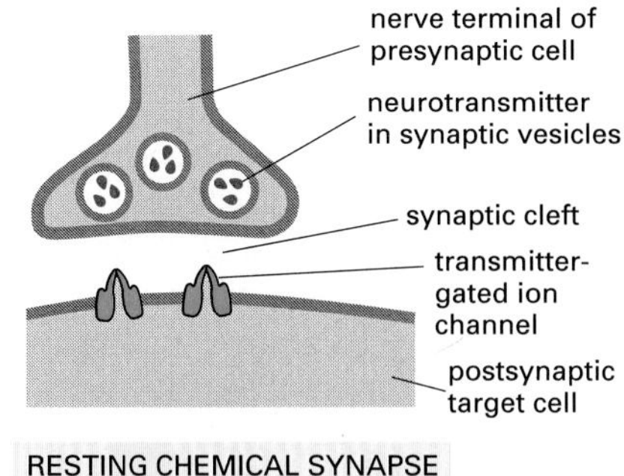

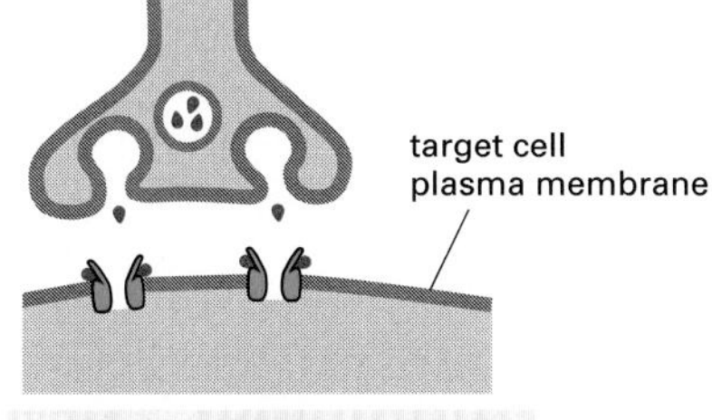

Figure 11–33 A chemical synapse. When an action potential reaches the nerve terminal in a presynaptic cell, it stimulates the terminal to release its neurotransmitter. The neurotransmitter molecules are contained in synaptic vesicles and are released to the cell exterior when the vesicles fuse with the plasma membrane of the nerve terminal. The released neurotransmitter binds to and opens the transmitter-gated ion channels concentrated in the plasma membrane of the postsynaptic target cell at the synapse. The resulting ion flows alter the membrane potential of the target cell, thereby transmitting a signal from the excited nerve.

Na^+-dependent neurotransmitter carrier proteins; in this way, neurotransmitters are recycled, allowing cells to keep up with high rates of release. Rapid removal ensures both spatial and temporal precision of signaling at a synapse. It decreases the chances that the neurotransmitter will influence neighboring cells, and it clears the synaptic cleft before the next pulse of neurotransmitter is released, so that the timing of repeated, rapid signaling events can be accurately communicated to the postsynaptic cell. As we shall see, signaling via such *chemical synapses* is far more versatile and adaptable than direct electrical coupling via gap junctions at *electrical synapses* (discussed in Chapter 19), which are also used by neurons but to a much smaller extent.

Transmitter-gated ion channels are specialized for rapidly converting extracellular chemical signals into electrical signals at chemical synapses. The channels are concentrated in the plasma membrane of the postsynaptic cell in the region of the synapse and open transiently in response to the binding of neurotransmitter molecules, thereby producing a brief permeability change in the membrane (see Figure 11–33). Unlike the voltage-gated channels responsible for action potentials, transmitter-gated channels are relatively insensitive to the membrane potential and therefore cannot by themselves produce a self-amplifying excitation. Instead, they produce local permeability changes, and hence changes of membrane potential, that are graded according to how much neurotransmitter is released at the synapse and how long it persists there. An action potential can be triggered from this site only if the local membrane potential increases enough to open a sufficient number of nearby voltage-gated cation channels that are present in the same target cell membrane.

Chemical Synapses Can Be Excitatory or Inhibitory

Transmitter-gated ion channels differ from one another in several important ways. First, as receptors, they have a highly selective binding site for the neurotransmitter that is released from the presynaptic nerve terminal. Second, as channels, they are selective as to the type of ions that they let pass across the plasma membrane; this determines the nature of the postsynaptic response. **Excitatory neurotransmitters** open cation channels, causing an influx of Na^+ that depolarizes the postsynaptic membrane toward the threshold potential for firing an action potential. **Inhibitory neurotransmitters**, by contrast, open either Cl^- channels or K^+ channels, and this suppresses firing by making it harder for excitatory influences to depolarize the postsynaptic membrane. Many transmitters can be either excitatory and inhibitory, depending on where they are released, what receptors they bind to, and the ionic conditions that they encounter. *Acetylcholine,* for example, can either excite or inhibit, depending on the type of acetylcholine receptors it binds to. Usually, however, acetylcholine, *glutamate,* and *serotonin* are used as excitatory transmitters, and *γ-aminobutyric acid (GABA)* and *glycine* are used as inhibitory transmitters. Glutamate, for instance, mediates most of the excitatory signaling in the vertebrate brain.

We have already discussed how the opening of cation channels depolarizes a membrane. The effect of opening Cl^- channels can be understood as follows. The concentration of Cl^- is much higher outside the cell than inside (see Table 11–1, p. 616), but its influx is opposed by the membrane potential. In fact, for many neurons, the equilibrium potential for Cl^- is close to the resting potential—or even more negative. For this reason, opening Cl^- channels tend to buffer the membrane potential; as the membrane starts to depolarize, more negatively charged Cl^- ions enter the cell and counteract the effect. Thus, the opening of Cl^- channels makes it more difficult to depolarize the membrane and hence to excite the cell. The opening of K^+ channels has a similar effect. The importance of inhibitory neurotransmitters is demonstrated by the effects of toxins that block their action: strychnine, for example, by binding to glycine receptors and blocking the action of glycine, causes muscle spasms, convulsions, and death.

However, not all chemical signaling in the nervous system operates through ligand-gated ion channels. Many of the signaling molecules that are secreted by nerve terminals, including a large variety of neuropeptides, bind to receptors that regulate ion channels only indirectly. These so-called *G-protein-linked receptors* and *enzyme-linked receptors* are discussed in detail in Chapter 15. Whereas signaling mediated by excitatory and inhibitory neurotransmitters binding to transmitter-gated ion channels is generally immediate, simple, and brief, signaling mediated by ligands binding to G-protein-linked receptors and enzyme-linked receptors tends to be far slower and more complex, and longer lasting in its consequences.

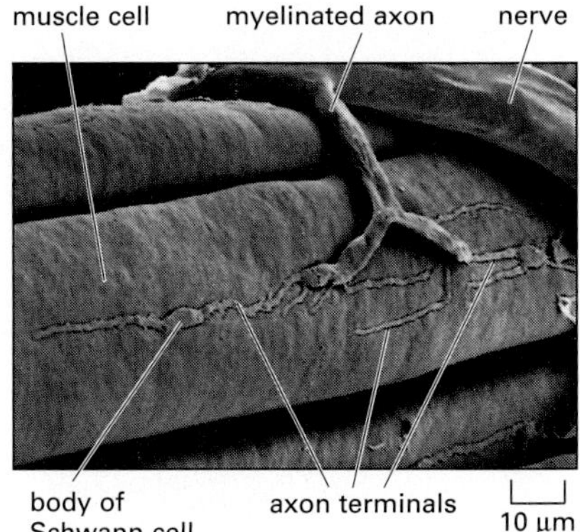

Figure 11–34 A low-magnification scanning electron micrograph of a neuromuscular junction in a frog. The termination of a single axon on a skeletal muscle cell is shown. (From J. Desaki and Y. Uehara, *J. Neurocytol.* 10:101–110, 1981. © Kluwer Academic Publishers.)

The Acetylcholine Receptors at the Neuromuscular Junction Are Transmitter-gated Cation Channels

The best-studied example of a transmitter-gated ion channel is the *acetylcholine receptor* of skeletal muscle cells. This channel is opened transiently by acetylcholine released from the nerve terminal at a **neuromuscular junction**—the specialized chemical synapse between a motor neuron and a skeletal muscle cell (Figure 11–34). This synapse has been intensively investigated because it is readily accessible to electrophysiological study, unlike most of the synapses in the central nervous system.

The **acetylcholine receptor** has a special place in the history of ion channels. It was the first ion channel to be purified, the first to have its complete amino acid sequence determined, the first to be functionally reconstituted in synthetic lipid bilayers, and the first for which the electrical signal of a single open channel was recorded. Its gene was also the first ion channel gene to be cloned and sequenced, and it is the only ligand-gated channel whose three-dimensional structure has been determined, albeit at moderate resolution. There were at least two reasons for the rapid progress in purifying and characterizing this receptor. First, an unusually rich source of the acetylcholine receptors exists in the electrical organs of electric fish and rays (these organs are modified muscles designed to deliver a large electrical shock to prey). Second, certain neurotoxins (such as *α-bungarotoxin)* in the venom of certain snakes bind with high affinity ($K_a = 10^9$ liters/mole) and specificity to the receptor and can therefore be used to purify it by affinity chromatography. Fluorescent or radiolabeled α-bungarotoxin can also be used to localize and count acetylcholine receptors. In this way, it has been shown that the receptors are densely packed in the muscle cell plasma membrane at a neuromuscular junction (about 20,000 such receptors per μm^2), with relatively few receptors elsewhere in the same membrane.

The acetylcholine receptor of skeletal muscle is composed of five transmembrane polypeptides, two of one kind and three others, encoded by four separate genes. The four genes are strikingly similar in sequence, implying that they evolved from a single ancestral gene. The two identical polypeptides in the pentamer each have binding sites for acetylcholine. When two acetylcholine molecules bind to the pentameric complex, they induce a conformational change that opens the channel. With ligand bound, the channel still flickers between open and closed states, but now has a 90% probability of being in the open state. This state continues until the concentration of acetylcholine is lowered sufficiently by hydrolysis by a specific enzyme (acetylcholinesterase) located in the neuromuscular junction. Once freed of its bound neurotransmitter, the acetylcholine receptor reverts to its initial resting state. If the presence of acetylcholine persists for a prolonged time as a result of excessive nerve stimulation, the channel inactivates (Figure 11–35).

The general shape of the acetylcholine receptor and the likely arrangement of its subunits have been determined by electron microscopy (Figure 11–36). The five subunits are arranged in a ring, forming a water-filled transmembrane channel that consists of a narrow pore through the lipid bilayer, which widens into vestibules at both ends. Clusters of negatively charged amino acids at either end of the pore help to exclude negative ions and encourage any positive ion of diameter less than 0.65 nm to pass through. The normal traffic consists chiefly of Na^+ and K^+, together with some Ca^{2+}. Thus, unlike voltage-gated cation channels, such as the K^+ channel discussed earlier, there is little selectivity among

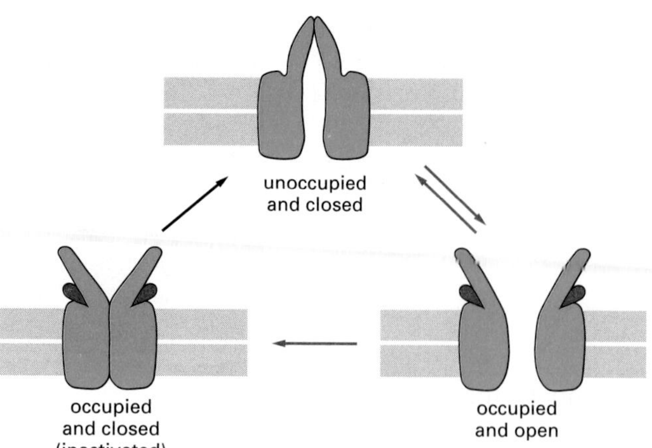

Figure 11–35 Three conformations of the acetylcholine receptor. The binding of two acetylcholine molecules opens this transmitter-gated ion channel. It then maintains a high probability of being open until the acetylcholine has been hydrolyzed. In the persistent presence of acetylcholine, however, the channel inactivates (desensitizes). Normally, the acetylcholine is rapidly hydrolyzed and the channel closes within about 1 millisecond, well before significant desensitization occurs. Desensitization would occur after about 20 milliseconds in the continued presence of acetylcholine.

cations, and the relative contributions of the different cations to the current through the channel depend chiefly on their concentrations and on the electrochemical driving forces. When the muscle cell membrane is at its resting potential, the net driving force for K^+ is near zero, since the voltage gradient nearly balances the K^+ concentration gradient across the membrane (see Panel 11–2, p. 634). For Na^+, in contrast, the voltage gradient and the concentration gradient both act in the same direction to drive the ion into the cell. (The same is true for Ca^{2+}, but the extracellular concentration of Ca^{2+} is so much lower than that of Na^+ that Ca^{2+} makes only a small contribution to the total inward current.) Therefore, the opening of the acetylcholine receptor channels leads to a large net influx of Na^+ (a peak rate of about 30,000 ions per channel each millisecond). This influx causes a membrane depolarization that signals the muscle to contract, as discussed below.

Transmitter-gated Ion Channels Are Major Targets for Psychoactive Drugs

The ion channels that open directly in response to the neurotransmitters acetylcholine, serotonin, GABA, and glycine contain subunits that are structurally similar, suggesting that they are evolutionarily related and probably form transmembrane pores in the same way, even though their neurotransmitter-binding specificities and ion selectivities are distinct. These channels seem to have a similar overall structure, in each case formed by homologous polypeptide subunits, which probably assemble as a pentamer resembling the acetylcholine receptor. Glutamate-gated ion channels are constructed from a distinct family of subunits and are thought to form tetramers, like the K^+ channels discussed earlier.

For each class of transmitter-gated ion channels, alternative forms of each type of subunit exist, either encoded by distinct genes or generated by alternative

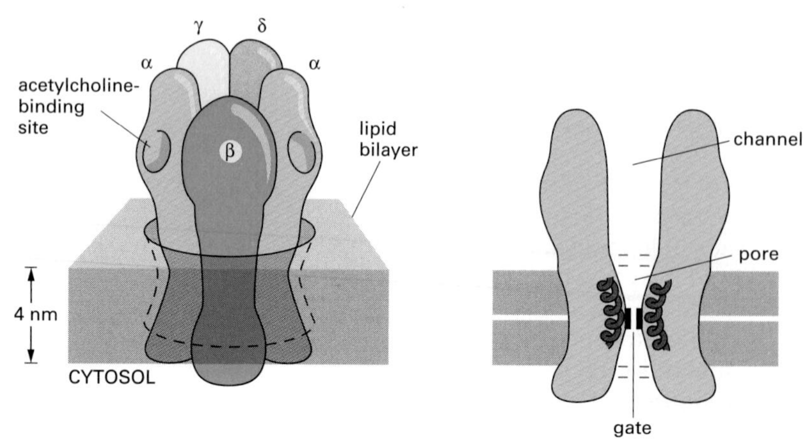

Figure 11–36 A model for the structure of the acetylcholine receptor. Five homologous subunits (α, α, β, γ, δ) combine to form a transmembrane aqueous pore. The pore is lined by a ring of five transmembrane α helices, one contributed by each subunit. In its closed conformation, the pore is thought to be occluded by the hydrophobic side chains of five leucines, one from each α helix, which form a gate near the middle of the lipid bilayer. The negatively charged side chains at either end of the pore ensure that only positively charged ions pass through the channel. Both of the α subunits contain an acetylcholine-binding site; when acetylcholine binds to both sites, the channel undergoes a conformational change that opens the gate, possibly by causing the leucines to move outward. (Adapted from N. Unwin, *Cell/Neuron* 72/10 [Suppl.]:31–41, 1993.)

RNA splicing of the same gene product. These combine in different variations to form an extremely diverse set of distinct channel subtypes, with different ligand affinities, different channel conductances, different rates of opening and closing, and different sensitivities to drugs and toxins. Vertebrate neurons, for example, have acetylcholine-gated ion channels that differ from those of muscle cells in that they are usually formed from two subunits of one type and three of another; but there are at least nine genes coding for different versions of the first type of subunit and at least three coding for different versions of the second, with further diversity due to alternative RNA splicing. Subsets of acetylcholine-sensitive neurons performing different functions in the brain are characterized by different combinations of these subunits. This, in principle and already to some extent in practice, makes it possible to design drugs targeted against narrowly defined groups of neurons or synapses, thereby influencing particular brain functions specifically.

Indeed, transmitter-gated ion channels have for a long time been important targets for drugs. A surgeon, for example, can make muscles relax for the duration of an operation by blocking the acetylcholine receptors on skeletal muscle cells with *curare*, a drug from a plant that was originally used by South American Indians to poison arrows. Most drugs used in the treatment of insomnia, anxiety, depression, and schizophrenia exert their effects at chemical synapses, and many of these act by binding to transmitter-gated channels. Both barbiturates and tranquilizers, such as Valium and Librium, for example, bind to GABA receptors, potentiating the inhibitory action of GABA by allowing lower concentrations of this neurotransmitter to open Cl^- channels. The new molecular biology of ion channels, by revealing both their diversity and the details of their structure, holds out the hope of designing a new generation of psychoactive drugs that will act still more selectively to alleviate the miseries of mental illness.

In addition to ion channels, many other components of the synaptic signaling machinery are potential targets for psychoactive drugs. As discussed earlier, many neurotransmitters are cleared from the synaptic cleft after release by Na^+-driven carriers. The inhibition of such a carrier prolongs the effect of the transmitter and thereby strengthens synaptic transmission. Many antidepressant drugs, including Prozac, for example, act by inhibiting the uptake of serotonin; others inhibit the uptake of both serotonin and norepinephrine.

Ion channels are the basic molecular components from which neuronal devices for signaling and computation are built. To provide a glimpse of how sophisticated the functions of these devices can be, we consider several examples that demonstrate how groups of ion channels work together in synaptic communication between electrically excitable cells.

Neuromuscular Transmission Involves the Sequential Activation of Five Different Sets of Ion Channels

The importance of ion channels to electrically excitable cells can be illustrated by following the process whereby a nerve impulse stimulates a muscle cell to contract. This apparently simple response requires the sequential activation of at least five different sets of ion channels, all within a few milliseconds (Figure 11–37).

1. The process is initiated when the nerve impulse reaches the nerve terminal and depolarizes the plasma membrane of the terminal. The depolarization transiently opens voltage-gated Ca^{2+} channels in this membrane. As the Ca^{2+} concentration outside cells is more than 1000 times greater than the free Ca^{2+} concentration inside, Ca^{2+} flows into the nerve terminal. The increase in Ca^{2+} concentration in the cytosol of the nerve terminal triggers the localized release of acetylcholine into the synaptic cleft.
2. The released acetylcholine binds to acetylcholine receptors in the muscle cell plasma membrane, transiently opening the cation channels associated with them. The resulting influx of Na^+ causes a localized membrane depolarization.
3. The local depolarization of the muscle cell plasma membrane opens voltage-gated Na^+ channels in this membrane, allowing more Na^+ to enter,

which further depolarizes the membrane. This, in turn, opens neighboring voltage-gated Na^+ channels and results in a self-propagating depolarization (an action potential) that spreads to involve the entire plasma membrane (see Figure 11–28).

4. The generalized depolarization of the muscle cell plasma membrane activates voltage-gated Ca^{2+} channels in specialized regions (the transverse [T] tubules—discussed in Chapter 16) of this membrane.
5. This, in turn, causes *Ca^{2+} release channels* in an adjacent region of the sarcoplasmic reticulum membrane to open transiently and release the Ca^{2+} stored in the sarcoplasmic reticulum into the cytosol. It is the sudden increase in the cytosolic Ca^{2+} concentration that causes the myofibrils in the muscle cell to contract. It is not certain how the activation of the voltage-gated Ca^{2+} channels in the T-tubule membrane leads to the opening of the Ca^{2+} release channels in the sarcoplasmic reticulum membrane. The two membranes are closely apposed, however, with the two types of channels joined together in a specialized structure (see Figure 16–74). It is possible, therefore, that a voltage-induced change in the conformation of the plasma membrane Ca^{2+} channel directly opens the Ca^{2+} release channel in the sarcoplasmic reticulum through a mechanical coupling.

Whereas the activation of muscle contraction by a motor neuron is complex, an even more sophisticated interplay of ion channels is required for a neuron to integrate a large number of input signals at synapses and compute an appropriate output, as we now discuss.

Single Neurons Are Complex Computation Devices

In the central nervous system, a single neuron can receive inputs from thousands of other neurons, and can in turn synapse on many thousands of other cells. Several thousand nerve terminals, for example, make synapses on an average motor neuron in the spinal cord; its cell body and dendrites are almost completely covered with them (Figure 11–38). Some of these synapses transmit signals from the brain or spinal cord; others bring sensory information from muscles or from the skin. The motor neuron must combine the information received from all these sources and react either by firing action potentials along its axon or by remaining quiet.

Of the many synapses on a neuron, some tend to excite it, others to inhibit it. Neurotransmitter released at an excitatory synapse causes a small depolarization in the postsynaptic membrane called an *excitatory postsynaptic potential (excitatory PSP),* while neurotransmitter released at an inhibitory synapse generally causes a small hyperpolarization called an *inhibitory PSP.* The membrane of the dendrites and cell body of most neurons contains a relatively low density

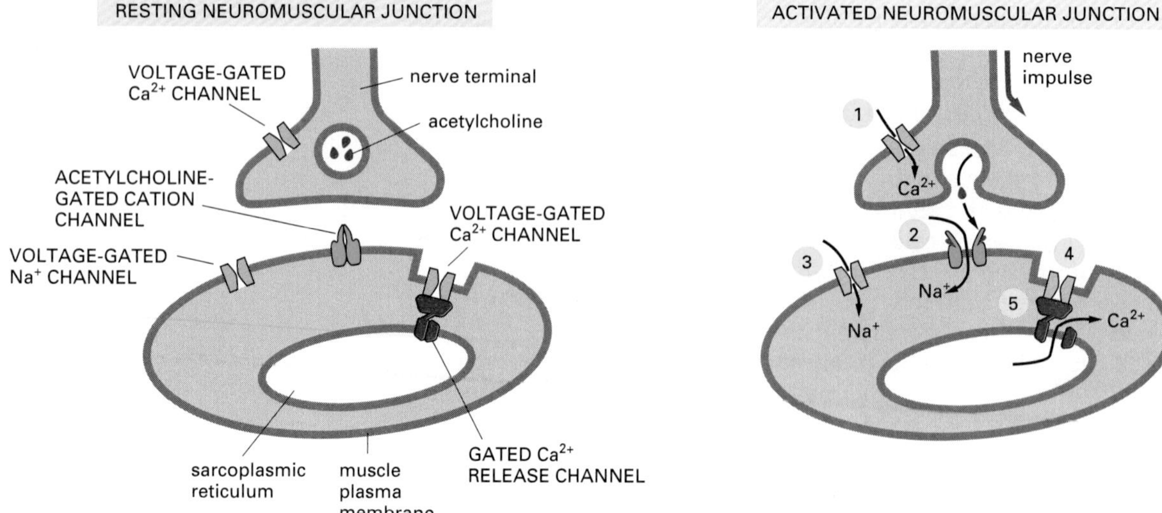

Figure 11–37 The system of ion channels at a neuromuscular junction. These gated ion channels are essential for the stimulation of muscle contraction by a nerve impulse. The various channels are numbered in the sequence in which they are activated, as described in the text.

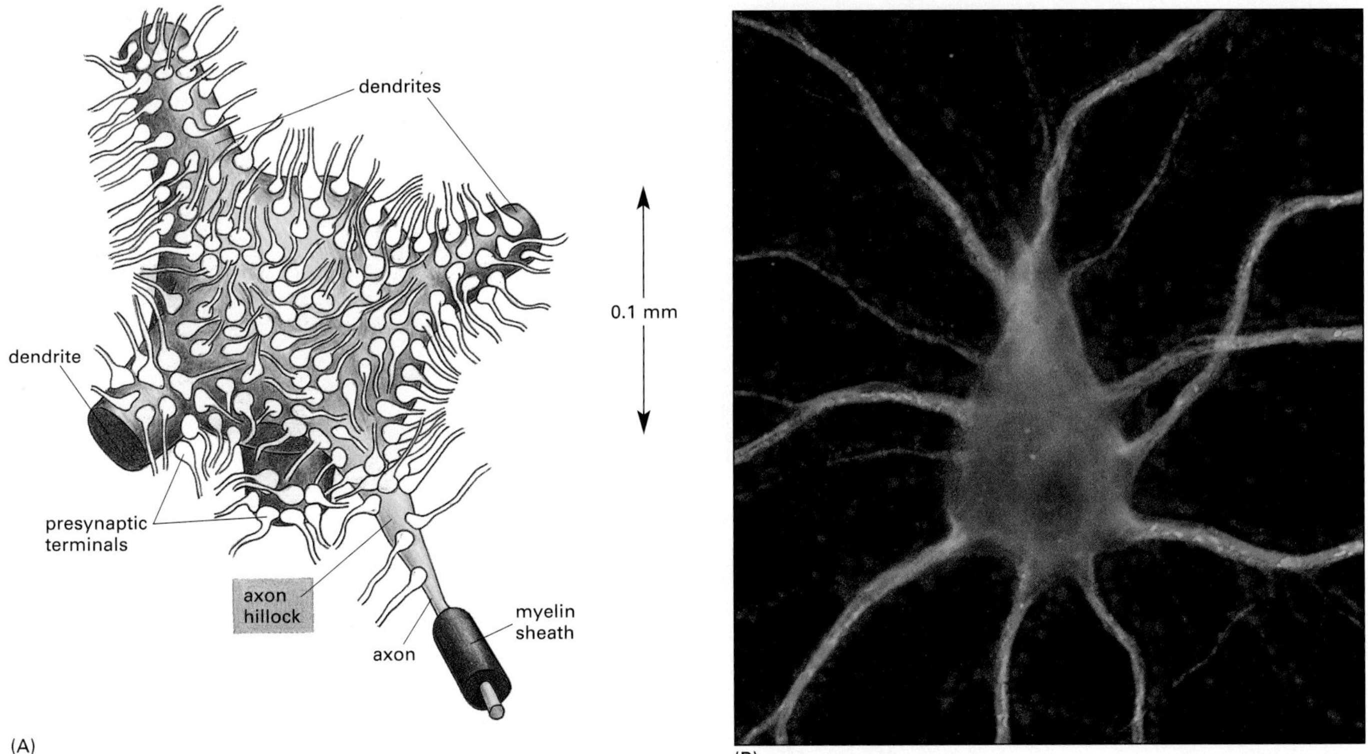

Figure 11–38 A motor neuron cell body in the spinal cord. (A) Many thousands of nerve terminals synapse on the cell body and dendrites. These deliver signals from other parts of the organism to control the firing of action potentials along the single axon of this large cell. (B) Micrograph showing a nerve cell body and its dendrites stained with a fluorescent antibody that recognizes a cytoskeletal protein *(green)*. Thousands of axon terminals *(red)* from other nerve cells (not visible) make synapses on the cell body and dendrites; they are stained with a fluorescent antibody that recognizes a protein in synaptic vesicles. (B, courtesy of Olaf Mundigl and Pietro de Camilli).

of voltage-gated Na^+ channels, and an individual excitatory PSP is generally too small to trigger an action potential. Instead, each incoming signal is reflected in a local PSP of graded magnitude, which decreases with distance from the site of the synapse. If signals arrive simultaneously at several synapses in the same region of the dendritic tree, the total PSP in that neighborhood will be roughly the sum of the individual PSPs, with inhibitory PSPs making a negative contribution to the total. The PSPs from each neighborhood spread passively and converge on the cell body. Because the cell body is small compared with the dendritic tree, its membrane potential is roughly uniform and is a composite of the effects of all the signals impinging on the cell, weighted according to the distances of the synapses from the cell body. The *combined PSP* of the cell body thus represents a **spatial summation** of all the stimuli being received. If excitatory inputs predominate, the combined PSP is a depolarization; if inhibitory inputs predominate, it is usually a hyperpolarization.

Whereas spatial summation combines the effects of signals received at different sites on the membrane, **temporal summation** combines the effects of signals received at different times. If an action potential arrives at a synapse and triggers neurotransmitter release before a previous PSP at the synapse has decayed completely, the second PSP adds to the remaining tail of the first. If many action potentials arrive in quick succession, each PSP adds to the tail of the preceding PSP, building up to a large sustained average PSP whose magnitude reflects the rate of firing of the presynaptic neuron (Figure 11–39). This is the essence of temporal summation: it translates the *frequency* of incoming signals into the *magnitude* of a net PSP.

Temporal and spatial summation together provide the means by which the rates of firing of many presynaptic neurons jointly control the membrane potential in the body of a single postsynaptic cell. The final step in the neuronal computation made by the postsynaptic cell is the generation of an output, usually in the form of action potentials, to relay a signal to other cells. The output signal reflects the magnitude of the PSP in the cell body. While the PSP is a continuously graded variable, however, action potentials are always all-or-nothing and uniform in size. The only variable in signaling by action potentials is the time interval between one action potential and the next. For long-distance transmission, the magnitude of the PSP is therefore translated, or *encoded,* into the

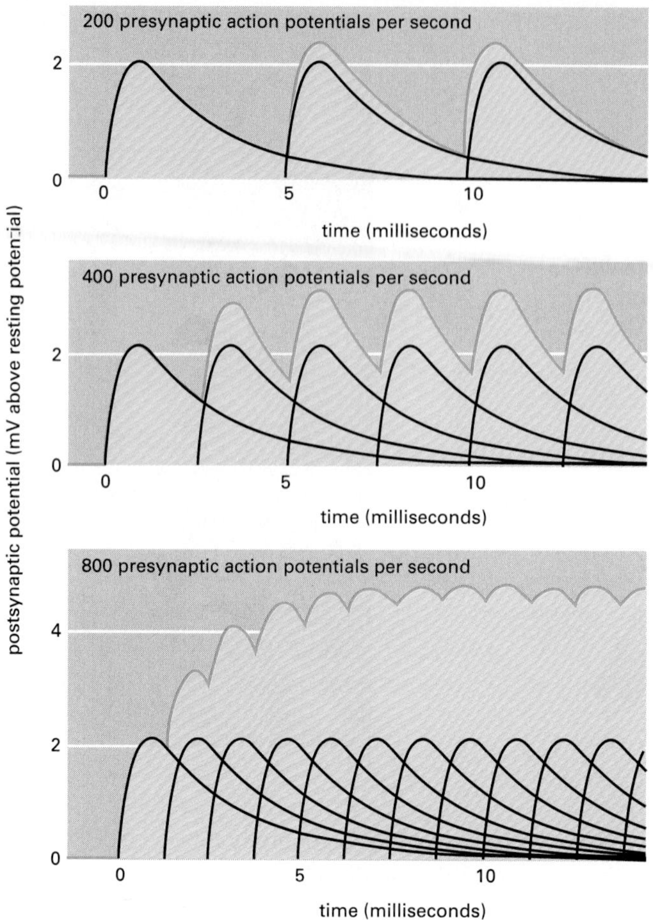

Figure 11–39 The principle of temporal summation. Each presynaptic action potential arriving at a synapse produces a small postsynaptic potential, or PSP *(black lines)*. When successive action potentials arrive at the same synapse, each PSP produced adds to the tail of the preceding one to produce a larger combined PSP *(green lines)*. The greater the frequency of incoming action potentials, the greater the size of the combined PSP.

frequency of firing of action potentials (Figure 11–40). This encoding is achieved by a special set of gated ion channels that are present at high density at the base of the axon, adjacent to the cell body, in a region known as the axon hillock (see Figure 11–38).

Neuronal Computation Requires a Combination of at Least Three Kinds of K^+ Channels

We have seen that the intensity of stimulation received by a neuron is determined for long-distance transmission by the frequency of action potentials that the neuron fires: the stronger the stimulation, the higher the frequency of action potentials. Action potentials are initiated at the axon hillock, a unique region of each neuron where voltage-gated Na^+ channels are plentiful. But to perform its special function of encoding, the membrane of the **axon hillock** also contains at least four other classes of ion channels—three selective for K^+ and one selective for Ca^{2+}. The three varieties of K^+ channels have different properties; we shall refer to them as the *delayed*, the *early*, and the *Ca^{2+}-activated K^+ channels.*

To understand the need for multiple types of channels, consider first what would happen if the only voltage-gated ion channels present in the nerve cell were the Na^+ channels. Below a certain threshold level of synaptic stimulation, the depolarization of the axon hillock membrane would be insufficient to trigger an action potential. With gradually increasing stimulation, the threshold would be crossed, the Na^+ channels would open, and an action potential would fire. The action potential would be terminated in the usual way by inactivation of the Na^+ channels. Before another action potential could fire, these channels would have to recover from their inactivation. But that would require a return of the membrane voltage to a very negative value, which would not occur as long as the strong depolarizing stimulus (from PSPs) was maintained. An additional channel type is needed, therefore, to repolarize the membrane after each action potential to prepare the cell to fire again.

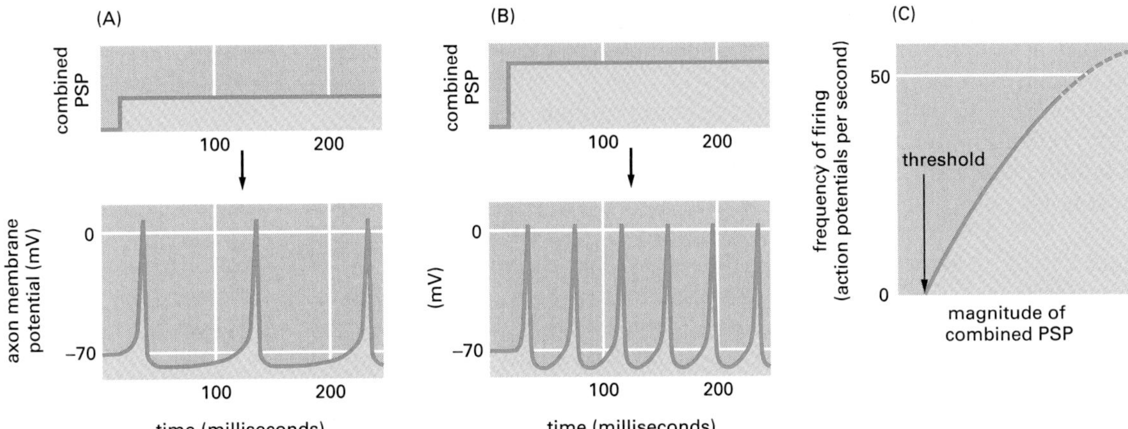

Figure 11–40 The encoding of the combined PSP in the form of the frequency of firing of action potentials by an axon. A comparison of (A) and (B) shows how the firing frequency of an axon increases with an increase in the combined PSP, while (C) summarizes the general relationship.

This task is performed by the **delayed K^+ channels** discussed previously in relation to the propagation of the action potential (see p. 639). They are voltage-gated, but because of their slower kinetics they open only during the falling phase of the action potential, when the Na^+ channels are inactive. Their opening permits an efflux of K^+ that drives the membrane back toward the K^+ equilibrium potential, which is so negative that the Na^+ channels rapidly recover from their inactivated state. Repolarization of the membrane also causes the delayed K^+ channels to close. The axon hillock is now reset so that the depolarizing stimulus from synaptic inputs can fire another action potential. In this way, sustained stimulation of the dendrites and cell body leads to repetitive firing of the axon.

Repetitive firing in itself, however, is not enough. The frequency of the firing has to reflect the intensity of the stimulation, and a simple system of Na^+ channels and delayed K^+ channels is inadequate for this purpose. Below a certain threshold level of steady stimulation, the cell will not fire at all; above that threshold level, it will abruptly begin to fire at a relatively rapid rate. The **early K^+ channels** solve the problem. These, too, are voltage-gated and open when the membrane is depolarized, but their specific voltage sensitivity and kinetics of inactivation are such that they act to reduce the rate of firing at levels of stimulation that are only just above the threshold required for firing. Thus, they remove the discontinuity in the relationship between the firing rate and the intensity of stimulation. The result is a firing rate that is proportional to the strength of the depolarizing stimulus over a very broad range (see Figure 11–40).

The process of encoding is usually further modulated by the two other types of ion channels in the axon hillock that were mentioned at the outset, voltage-gated Ca^{2+} channels and Ca^{2+}-activated K^+ channels. They act together to decrease the response of the cell to an unchanging, prolonged stimulation—a process called **adaptation**. These Ca^{2+} channels are similar to the Ca^{2+} channels that mediate the release of neurotransmitter from presynaptic axon terminals; they open when an action potential fires, transiently allowing Ca^{2+} into the axon hillock.

The **Ca^{2+}-activated K^+ channel** is both structurally and functionally different from any of the channel types described earlier. It opens in response to a raised concentration of Ca^{2+} at the cytoplasmic face of the nerve cell membrane. Suppose that a strong depolarizing stimulus is applied for a long time, triggering a long train of action potentials. Each action potential permits a brief influx of Ca^{2+} through the voltage-gated Ca^{2+} channels, so that the intracellular Ca^{2+} concentration gradually builds up to a level high enough to open the Ca^{2+}-activated K^+ channels. Because the resulting increased permeability of the membrane to K^+ makes the membrane harder to depolarize, it increases the delay between one action potential and the next. In this way, a neuron that is stimulated continuously for a prolonged period becomes gradually less responsive to the constant stimulus.

Such adaptation, which can also occur by other mechanisms, allows a neuron—indeed, the nervous system generally—to react sensitively to *change*, even against a high background level of steady stimulation. It is one of the strategies that help us, for example, to feel a light touch on the shoulder and yet ignore the constant pressure of our clothing. We discuss adaptation as a general feature in cell signaling processes in more detail in Chapter 15.

Other neurons do different computations, reacting to their synaptic inputs in myriad ways, reflecting the different assortments of members of the various ion channel families that reside in their membranes. There are, for example, at least five known types of voltage-gated Ca^{2+} channels in the vertebrate nervous system and at least four known types of voltage-gated K^+ channels. The multiplicity of genes evidently allows for a host of different types of neurons, whose electrical behavior is specifically tuned to the particular tasks that they must perform.

One of the crucial properties of the nervous system is its ability to learn and remember, which seems to depend largely on long-term changes in specific synapses. We end this chapter by considering a remarkable type of ion channel that is thought to have a special role in some forms of learning and memory. It is located at many synapses in the central nervous system, where it is gated by both voltage and the excitatory neurotransmitter glutamate. It is also the site of action of the psychoactive drug phencyclidine, or angel dust.

Long-term Potentiation (LTP) in the Mammalian Hippocampus Depends on Ca^{2+} Entry Through NMDA-Receptor Channels

Practically all animals can learn, but mammals seem to learn exceptionally well (or so we like to think). In a mammal's brain, the region called the *hippocampus* has a special role in learning. When it is destroyed on both sides of the brain, the ability to form new memories is largely lost, although previous long-established memories remain. Correspondingly, some synapses in the hippocampus show marked functional alterations with repeated use: whereas occasional single action potentials in the presynaptic cells leave no lasting trace, a short burst of repetitive firing causes **long-term potentiation (LTP)**, such that subsequent single action potentials in the presynaptic cells evoke a greatly enhanced response in the postsynaptic cells. The effect lasts hours, days, or weeks, according to the number and intensity of the bursts of repetitive firing. Only the synapses that were activated exhibit LTP; synapses that have remained quiet on the same postsynaptic cell are not affected. However, while the cell is receiving a burst of repetitive stimulation via one set of synapses, if a single action potential is delivered at *another* synapse on its surface, that latter synapse also will undergo LTP, even though a single action potential delivered there at another time would leave no such lasting trace.

The underlying rule in such synapses seems to be that *LTP occurs on any occasion when a presynaptic cell fires (once or more) at a time when the postsynaptic membrane is strongly depolarized* (either through recent repetitive firing of the same presynaptic cell or by other means). This rule reflects the behavior of a particular class of ion channels in the postsynaptic membrane. Glutamate is the main excitatory neurotransmitter in the mammalian central nervous system, and glutamate-gated ion channels are the most common of all transmitter-gated channels in the brain. In the hippocampus, as elsewhere, most of the depolarizing current responsible for excitatory PSPs is carried by glutamate-gated ion channels that operate in the standard way. But the current has, in addition, a second and more intriguing component, which is mediated by a separate subclass of glutamate-gated ion channels known as **NMDA receptors**, so named because they are selectively activated by the artificial glutamate analog N-methyl-D-aspartate. The NMDA-receptor channels are doubly gated, opening only when two conditions are satisfied simultaneously: glutamate must be bound to the receptor, and the membrane must be strongly depolarized. The second condition is required for releasing the Mg^{2+} that normally blocks the resting channel. This means that NMDA receptors are normally only activated when conventional glutamate-gated ion channels are activated as well and depolarize the membrane. The NMDA receptors are critical for LTP. When they are selectively blocked with a specific inhibitor, LTP does not occur, even though

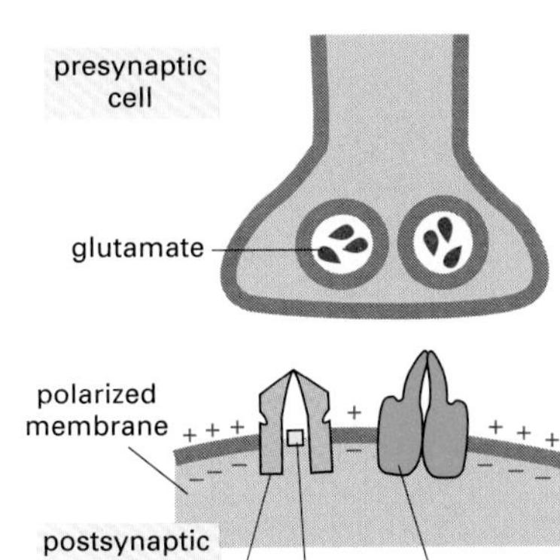
presynaptic cell
glutamate
polarized membrane
postsynaptic cell
Mg^{2+}
NMDA receptor
non-NMDA glutamate receptor

glutamate released by activated presynaptic nerve terminal opens non-NMDA glutamate receptor channels, allowing Na^+ influx that depolarizes the postsynaptic membrane

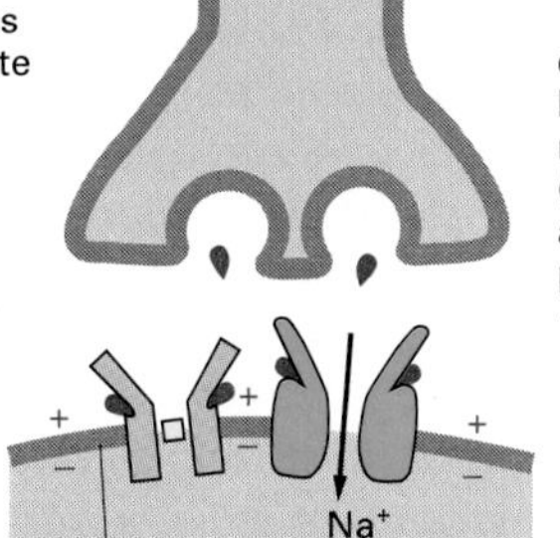
Na^+
depolarized membrane

depolarization removes Mg^{2+} block from NMDA-receptor channel, which (with glutamate bound) allows Ca^{2+} to enter the postsynaptic cell

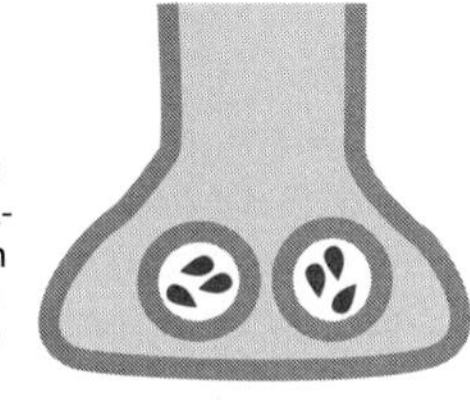

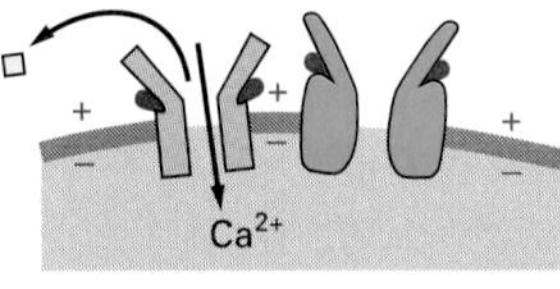
Ca^{2+}
increased Ca^{2+} in the cytosol induces postsynaptic cell to insert new non-NMDA glutamate receptors in the plasma membrane, increasing the cell's sensitivity to glutamate

Figure 11–41 The signaling events in long-term potentiation. Although not shown, evidence suggests that changes can also occur in the presynaptic nerve terminals in LTP, which may be stimulated by retrograde signals from the postsynaptic cell.

ordinary synaptic transmission continues. An animal treated with this inhibitor exhibits specific deficits in its learning abilities but behaves almost normally otherwise.

How do NMDA receptors mediate such a remarkable effect? The answer is that these channels, when open, are highly permeable to Ca^{2+}, which acts as an intracellular mediator in the postsynaptic cell, triggering a cascade of changes that are responsible for LTP. Thus, LTP is prevented when Ca^{2+} levels are held artificially low in the postsynaptic cell by injecting the Ca^{2+} chelator EGTA into it and can be induced by artificially raising intracellular Ca^{2+} levels. Among the long-term changes that increase the sensitivity of the postsynaptic cell to glutamate is the insertion of new conventional glutamate receptors into the plasma membrane (Figure 11–41). Evidence also indicates that changes can occur in the presynaptic cell as well, so that it releases more glutamate than normal when it is activated subsequently.

There is evidence that NMDA receptors have an important role in learning and related phenomena in other parts of the brain, as well as in the hippocampus. In Chapter 21 we see, moreover, that NMDA receptors have a crucial role in adjusting the anatomical pattern of synaptic connections in the light of experience during the development of the nervous system.

Thus, neurotransmitters released at synapses, besides relaying transient electrical signals, can also alter concentrations of intracellular mediators that bring about lasting changes in the efficacy of synaptic transmission. However, it is still uncertain how these changes endure for weeks, months, or a lifetime in the face of the normal turnover of cell constituents.

Some of the ion channel families that we have discussed are summarized in Table 11–2.

TABLE 11–2 Some Ion Channel Families

FAMILY*	REPRESENTATIVE SUBFAMILIES	
Voltage-gated cation channels	voltage-gated Na^+ channels	
	voltage-gated K^+ channels (including delayed and early)	
	voltage-gated Ca^{2+} channels	
Transmitter-gated ion channels	acetylcholine-gated cation channels	excitatory
	glutamate-gated Ca^{2+} channels	excitatory
	serotonin-gated cation channels	excitatory
	GABA-gated Cl^- channels	inhibitory
	glycine-gated Cl^- channels	inhibitory

*The members of a family are similar in amino acid sequence and are therefore thought to have derived from a common ancestor; within subfamilies, the resemblances are usually even closer.

Summary

Ion channels form aqueous pores across the lipid bilayer and allow inorganic ions of appropriate size and charge to cross the membrane down their electrochemical gradients at rates about 1000 times greater than those achieved by any known carrier. The channels are "gated" and usually open transiently in response to a specific perturbation in the membrane, such as a change in membrane potential (voltage-gated channels) or the binding of a neurotransmitter (transmitter-gated channels).

K^+-selective leak channels have an important role in determining the resting membrane potential across the plasma membrane in most animal cells. Voltage-gated cation channels are responsible for the generation of self-amplifying action potentials in electrically excitable cells, such as neurons and skeletal muscle cells. Transmitter-gated ion channels convert chemical signals to electrical signals at chemical synapses. Excitatory neurotransmitters, such as acetylcholine and glutamate, open transmitter-gated cation channels and thereby depolarize the postsynaptic membrane toward the threshold level for firing an action potential. Inhibitory neurotransmitters, such as GABA and glycine, open transmitter-gated Cl^- or K^+ channels and thereby suppress firing by keeping the postsynaptic membrane polarized. A subclass of glutamate-gated ion channels, called NMDA-receptor channels, are highly permeable to Ca^{2+}, which can trigger the long-term changes in synapses that are thought to be involved in some forms of learning and memory.

Ion channels work together in complex ways to control the behavior of electrically excitable cells. A typical neuron, for example, receives thousands of excitatory and inhibitory inputs, which combine by spatial and temporal summation to produce a postsynaptic potential (PSP) in the cell body. The magnitude of the PSP is translated into the rate of firing of action potentials by a mixture of cation channels in the membrane of the axon hillock.

References

General

Martonosi AN (ed) (1985) The Enzymes of Biological Membranes, vol 3: Membrane Transport, 2nd edn. New York: Plenum Press.

Stein WD (1990) Channels, Carriers and Pumps: An Introduction to Membrane Transport. San Deigo: Academic Press.

Principles of Membrane Transport

Higgins CF (1999) Membrane permeability transporters and channels: from disease to structure and back. *Curr. Opin. Cell Biol.* 11, 495.

Sansom MS (1998) Models and simulations of ion channels and related membrane proteins. *Curr. Opin. Struct. Biol.* 8, 237–244.

Tanford C (1983) Mechanism of free energy coupling in active transport. *Annu. Rev. Biochem.* 52, 379–409.

Carrier Proteins and Active Membrane Transport

Almers W & Stirling C (1984) Distribution of transport proteins over animal cells. *J. Membr. Biol.* 77, 169–186.

Ames GFL (1986) Bacterial periplasmic transport systems: structure, mechanism and evolution. *Annu. Rev. Biochem.* 55, 397–425.

Auer M, Scarborough GA & Kühlbrandt W (1998) Three-dimensional map of the plasma membrane H^+-ATPase in the open conformation. *Nature* 392, 840–843.

Boyer PD (1997) The ATP synthase—a splendid molecular machine. *Annu. Rev. Biochem.* 66, 717–749.

Carafoli E & Brini M (2000) Calcium pumps: structural basis for and mechanism of calcium transmembrane transport. *Curr. Opin. Chem. Biol.* 4, 152–161.

Dean M & Allikmets R (1995) Evolution of ATP-binding cassette transporter genes. *Curr. Opin. Genet. Dev.* 5, 779–785.

Graf J & Haussinger D (1996) Ion transport in hepatocytes: mechanisms and correlations to cell volume, hormone actions and metabolism. *J. Hepatol.* 24(Suppl 1), 53–77.

Henderson PJ (1993) The 12-transmembrane helix transporters. *Curr. Opin. Cell Biol.* 5, 708–721.

Higgins CF (1992) ABC transporters: from microorganisms to man. *Annu. Rev. Cell Biol.* 8, 67–113.

Junge W, Lill H & Engelbrecht S (1997) ATP synthase: an electrochemical transducer with rotatory mechanics. *Trends Biochem. Sci.* 22, 420–423.

Kaback HR, Voss J & Wu J (1997) Helix packing in polytopic membrane proteins: the lactose permease of *Escherichia coli*. *Curr. Opin. Struct. Biol.* 7, 537–542.

Kühlbrandt W, Auer M & Scarborough GA (1998) Structure of the P-type ATPases. *Curr. Opin. Struct. Biol.* 8, 510–516.

Linton KJ & Higgins CF (1998) The *Escherichia coli* ATP-binding cassette (ABC) proteins. *Mol. Microbiol.* 28, 5–13.

Lodish HF (1988) Anion-exchange and glucose transport proteins: structure, function and distribution. *Harvey Lect.* 82, 19–46.

Romero MF & Boron WF (1999) Electrogenic Na^+/HCO_3^- cotransporters: cloning and physiology. *Annu. Rev. Physiol.* 61, 699–723.

Saier MH, Jr (2000) Families of transmembrane sugar transport proteins. *Mol. Microbiol.* 35, 699–710.

Scarborough GA (1999) Structure and function of the P-type ATPases. *Curr. Opin. Cell Biol.* 11, 517–522.

Weber J & Senior AE (2000) ATP synthase: what we know about ATP hydrolysis and what we do not know about ATP synthesis. *Biochim. Biophys. Acta* 1458, 300–309.

Ion Channels and the Electrical Properties of Membranes

Armstrong C (1998) The vision of the pore. *Science* 280, 56–57.

Betz H (1990) Ligand-gated ion channels in the brain: the amino acid receptor family. *Neuron* 5, 383–392.

Biggin PC, Roosild T & Choe S (2000) Potassium channel structure: domain by domain. *Curr. Opin. Struct. Biol.* 10, 456–461.

Choe S, Kreusch A & Pfaffinger PJ (1999) Towards the three-dimensional structure of voltage-gated potassium channels. *Trends Biochem. Sci.* 24, 345–349.

Doyle DA, Cabral JM, Pfuetzner RA et al. (1998) The structure of the potassium channel: molecular basis of K^+ conduction and selectivity. *Science* 280, 69–77.

Franks NP & Lieb WR (1994) Molecular and cellular mechanisms of general anaesthesia. *Nature* 367, 607–614.

Hall ZW (1992) An Introduction to Molecular Neurobiology, pp 33–178. Sunderland, MA: Sinauer.

Hille B (1992) Ionic Channels of Excitable Membranes, 2nd edn. Sunderland, MA: Sinauer.

Hodgkin AL & Huxley AF (1952) A quantitative description of membrane current and its application to conduction and exictation in the nerve. *J. Physiol.* 117, 500–544.

Hodgkin AL & Huxley AF (1952) Currents carried by sodium and potassium ions through the membrane of the giant axon of Loligo. *J. Physiol. (Lond.)* 117, 449–472.

Jessell TM & Kandel ER (1993) Synaptic transmission: a bidirectional and self-modifiable form of cell-cell communication. *Cell* 72(Suppl), 1–30.

Kandel ER, Schwartz JH & Jessell TM (2000) Principles of Neural Science, 4th edn. New York: McGraw-Hill.

Karlin A (1991) Exploration of the nicotinic acetylcholine receptor. *Harvey Lect.* 85, 71–107.

Katz B (1966) Nerve, Muscle and Synapse. New York: McGraw-Hill.

Kim JH & Huganir RL (1999) Organization and regulation of proteins at synapses. *Curr. Opin. Cell Biol.* 11, 248–254.

MacKinnon R, Cohen SL, Kuo A et al. (1998) Structural conservation in prokaryotic and eukaryotic potassium channels. *Science* 280, 106–109.

Malenka RC & Nicoll RA (1999) Long-term potentiation—a decade of progress? *Science* 285, 1870–1874.

Neher E & Sakmann B (1992) The patch clamp technique. *Sci. Am.* 266(3), 28–35.

Nicholls JG, Fuchs PA, Martin AR & Wallace BG (2000) From Neuron to Brain, 4th edn. Sunderland, MA: Sinauer.

Numa S (1989) A molecular view of neurotransmitter receptors and ionic channels. *Harvey Lect.* 83, 121–165.

Perozo E, Cortes DM & Cuello LG (1999) Structural rearrangements underlying K^+-channel activation gating. *Science* 285, 73–78.

Roux B & MacKinnon R (1999) The cavity and pore helices in the KcsA K^+ channel: electrostatic stabilization of monovalent cations. *Science* 285, 100–102.

Sargent PB (1993) The diversity of neronal nicotinic acetylcholine receptors. *Annu. Rev. Neurosci.* 16, 403–443.

Sather WA, Yang J & Tsien RW (1994) Structural basis of ion channel permeation and selectivity. *Curr. Opin. Neurobiol.* 4, 313–323.

Seeberg PH (1993) The molecular biology of mammalian glutamate receptor channels. *Trends Neurosci.* 16, 359–365.

Snyder SH (1996) Drugs and the Brain. New York: WH Freeman/Scientific American Books.

Stevens CF & Sullivan J (1998) Synaptic plasticity. *Curr. Biol.* 8, R151–R153.

Tsien RW (1988) Multiple types of neuronal calcium channels and their selective modulation. *Trends Neurosci.* 11, 431–438.

Unwin N (1998) The nicotinic acetylcholine receptor of the Torpedo electric ray. *J. Struct. Biol.* 121, 181–190.

Utkin Yu N, Tsetlin VI & Hucho F (2000) Structural organization of nicotinic acetylcholine receptors. *Membr. Cell Biol.* 13, 143–164.

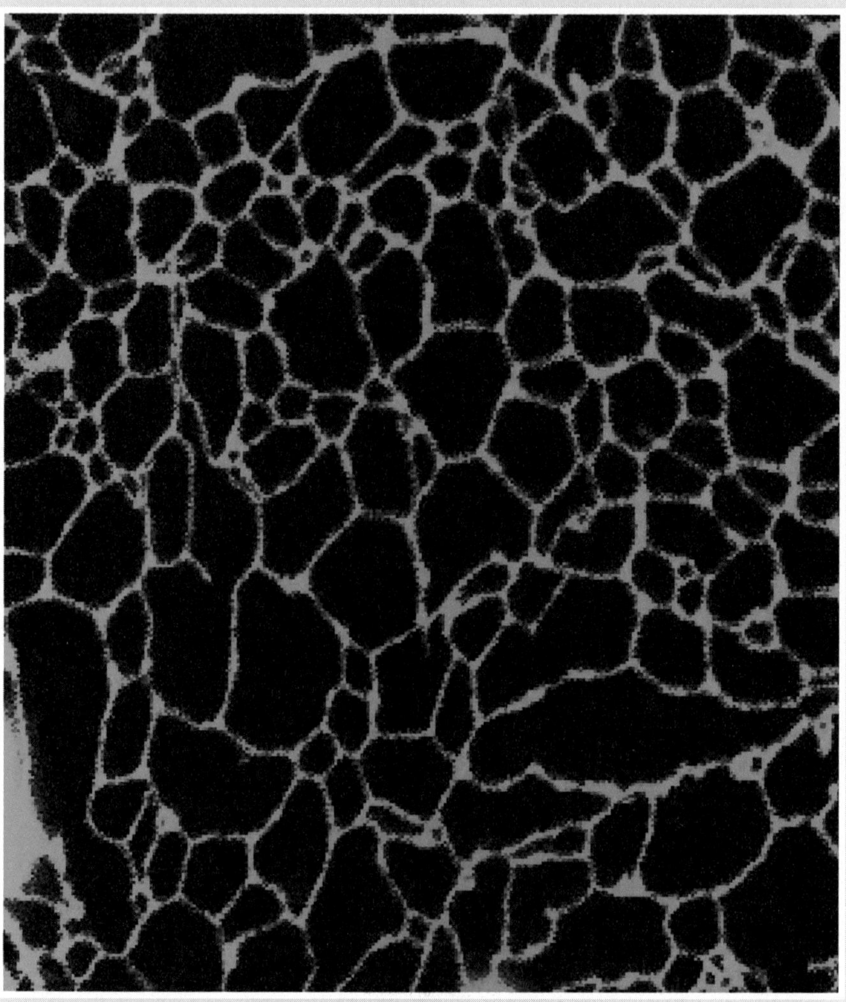

The endoplasmic reticulum (ER). In this tobacco cell, an ER resident protein, which has been coupled to a green fluorescent protein tag, reveals the complex cortical network of ER sheets and tubules within the cell. (Courtesy of Petra Boevink and Chris Hawes.)

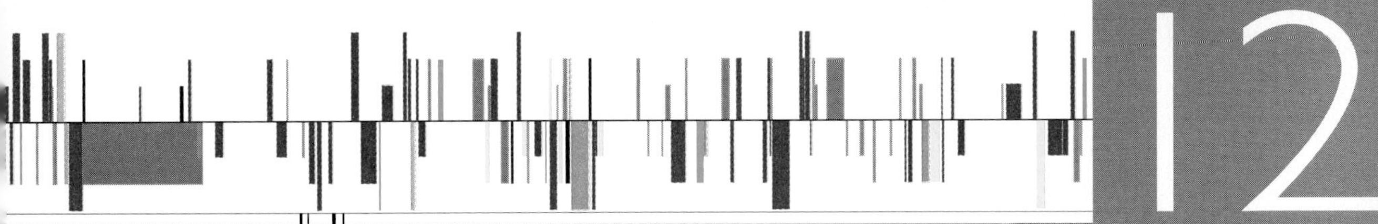

12

INTRACELLULAR COMPARTMENTS AND PROTEIN SORTING

Unlike a bacterium, which generally consists of a single intracellular compartment surrounded by a plasma membrane, a eucaryotic cell is elaborately subdivided into functionally distinct, membrane-enclosed compartments. Each compartment, or **organelle**, contains its own characteristic set of enzymes and other specialized molecules, and complex distribution systems transport specific products from one compartment to another. To understand the eucaryotic cell, it is essential to know what occurs in each of these compartments, how molecules move between them, and how the compartments themselves are created and maintained.

Proteins confer upon each compartment its characteristic structural and functional properties. They catalyze the reactions that occur in each organelle and selectively transport small molecules into and out of its interior, or **lumen**. Proteins also serve as organelle-specific surface markers that direct new deliveries of proteins and lipids to the appropriate organelle.

An animal cell contains about 10 billion (10^{10}) protein molecules of perhaps 10,000–20,000 kinds, and the synthesis of almost all of them begins in the cytosol. Each newly synthesized protein is then delivered specifically to the cell compartment that requires it. The intracellular transport of proteins is the central theme of both this chapter and the next. By tracing the protein traffic from one compartment to another, one can begin to make sense of the otherwise bewildering maze of intracellular membranes.

THE COMPARTMENTALIZATION OF CELLS

In this introductory section we present a brief overview of the compartments of the cell and the relationships between them. In doing so, we organize the organelles conceptually into a small number of discrete families and discuss how proteins are directed to specific organelles and how they cross organelle membranes.

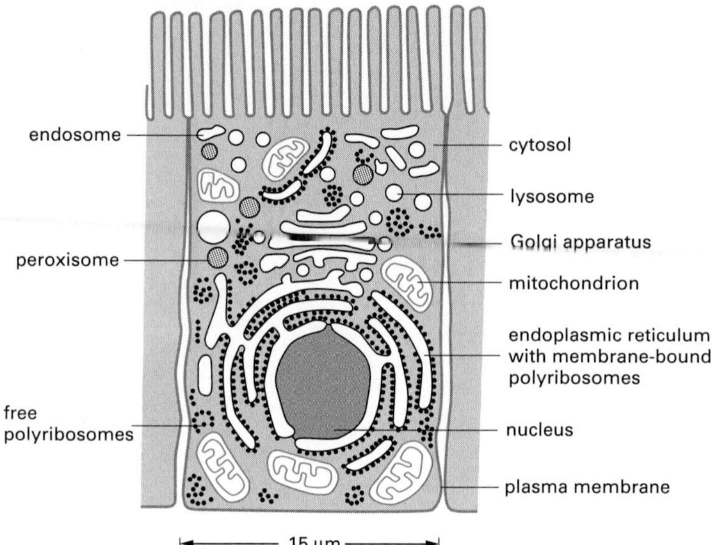

Figure 12–1 The major intracellular compartments of an animal cell. The cytosol *(gray)*, endoplasmic reticulum, Golgi apparatus, nucleus, mitochondrion, endosome, lysosome, and peroxisome are distinct compartments isolated from the rest of the cell by at least one selectively permeable membrane.

All Eucaryotic Cells Have the Same Basic Set of Membrane-enclosed Organelles

Many vital biochemical processes take place in or on membrane surfaces. Lipid metabolism, for example, is catalyzed mostly by membrane-bound enzymes, and oxidative phosphorylation and photosynthesis both require a membrane to couple the transport of H^+ to the synthesis of ATP. Intracellular membrane systems, however, do more for the cell than just provide increased membrane area: they create enclosed compartments that are separate from the cytosol, thus providing the cell with functionally specialized aqueous spaces. Because the lipid bilayer of organelle membranes is impermeable to most hydrophilic molecules, the membrane of each organelle must contain membrane transport proteins that are responsible for the import and export of specific metabolites. Each organelle membrane must also have a mechanism for importing, and incorporating into the organelle, the specific proteins that make the organelle unique.

The major intracellular compartments common to eucaryotic cells are illustrated in Figure 12–1. The *nucleus* contains the main genome and is the principal site of DNA and RNA synthesis. The surrounding **cytoplasm** consists of the cytosol and the cytoplasmic organelles suspended in it. The **cytosol**, constituting a little more than half the total volume of the cell, is the site of protein synthesis and degradation. It also performs most of the cell's intermediary metabolism—that is, the many reactions by which some small molecules are degraded and others are synthesized to provide the building blocks for macromolecules (discussed in Chapter 2).

About half the total area of membrane in a eucaryotic cell encloses the labyrinthine spaces of the *endoplasmic reticulum (ER)*. The ER has many ribosomes bound to its cytosolic surface; these are engaged in the synthesis of both soluble and integral membrane proteins, most of which are destined either for secretion to the cell exterior or for other organelles. We shall see that whereas proteins are translocated into other organelles only after their synthesis is complete, they are translocated into the ER as they are synthesized. This explains why the ER membrane is unique in having ribosomes tethered to it. The ER also produces most of the lipid for the rest of the cell and functions as a store for Ca^{2+} ions. The ER sends many of its proteins and lipids to the Golgi apparatus. The *Golgi apparatus* consists of organized stacks of disclike compartments called *Golgi cisternae;* it receives lipids and proteins from the ER and dispatches them to a variety of destinations, usually covalently modifying them *en route.*

Mitochondria and (in plants) *chloroplasts* generate most of the ATP used by cells to drive reactions that require an input of free energy; chloroplasts are a specialized version of *plastids,* which can also have other functions in plant cells, such as the storage of food or pigment molecules. *Lysosomes* contain digestive enzymes that degrade defunct intracellular organelles, as well as

TABLE 12–1 Relative Volumes Occupied by the Major Intracellular Compartments in a Liver Cell (Hepatocyte)

INTRACELLULAR COMPARTMENT	PERCENTAGE OF TOTAL CELL VOLUME
Cytosol	54
Mitochondria	22
Rough ER cisternae	9
Smooth ER cisternae plus Golgi cisternae	6
Nucleus	6
Peroxisomes	1
Lysosomes	1
Endosomes	1

macromolecules and particles taken in from outside the cell by endocytosis. On their way to lysosomes, endocytosed material must first pass through a series of organelles called *endosomes. Peroxisomes* are small vesicular compartments that contain enzymes utilized in a variety of oxidative reactions.

In general, each membrane-enclosed organelle performs the same set of basic functions in all cell types. But to serve the specialized functions of cells, these organelles will vary in abundance and can have additional properties that differ from cell type to cell type.

On average, the membrane-enclosed compartments together occupy nearly half the volume of a cell (Table 12–1), and a large amount of intracellular membrane is required to make them all. In liver and pancreatic cells, for example, the endoplasmic reticulum has a total membrane surface area that is, respectively, 25 times and 12 times that of the plasma membrane (Table 12–2). In terms of its area and mass, the plasma membrane is only a minor membrane in most eucaryotic cells (Figure 12–2).

Membrane-enclosed organelles often have characteristic positions in the cytosol. In most cells, for example, the Golgi apparatus is located close to the nucleus, whereas the network of ER tubules extends from the nucleus throughout the entire cytosol. These characteristic distributions depend on interactions

TABLE 12–2 Relative Amounts of Membrane Types in Two Kinds of Eucaryotic Cells

MEMBRANE TYPE	PERCENTAGE OF TOTAL CELL MEMBRANE	
	LIVER HEPATOCYTE*	PANCREATIC EXOCRINE CELL*
Plasma membrane	2	5
Rough ER membrane	35	60
Smooth ER membrane	16	<1
Golgi apparatus membrane	7	10
Mitochondria		
Outer membrane	7	4
Inner membrane	32	17
Nucleus		
Inner membrane	0.2	0.7
Secretory vesicle membrane	not determined	3
Lysosome membrane	0.4	not determined
Peroxisome membrane	0.4	not determined
Endosome membrane	0.4	not determined

*These two cells are of very different sizes: the average hepatocyte has a volume of about 5000 μm^3 compared with 1000 μm^3 for the pancreatic exocrine cell. Total cell membrane areas are estimated at about 110,000 μm^2 and 13,000 μm^2, respectively.

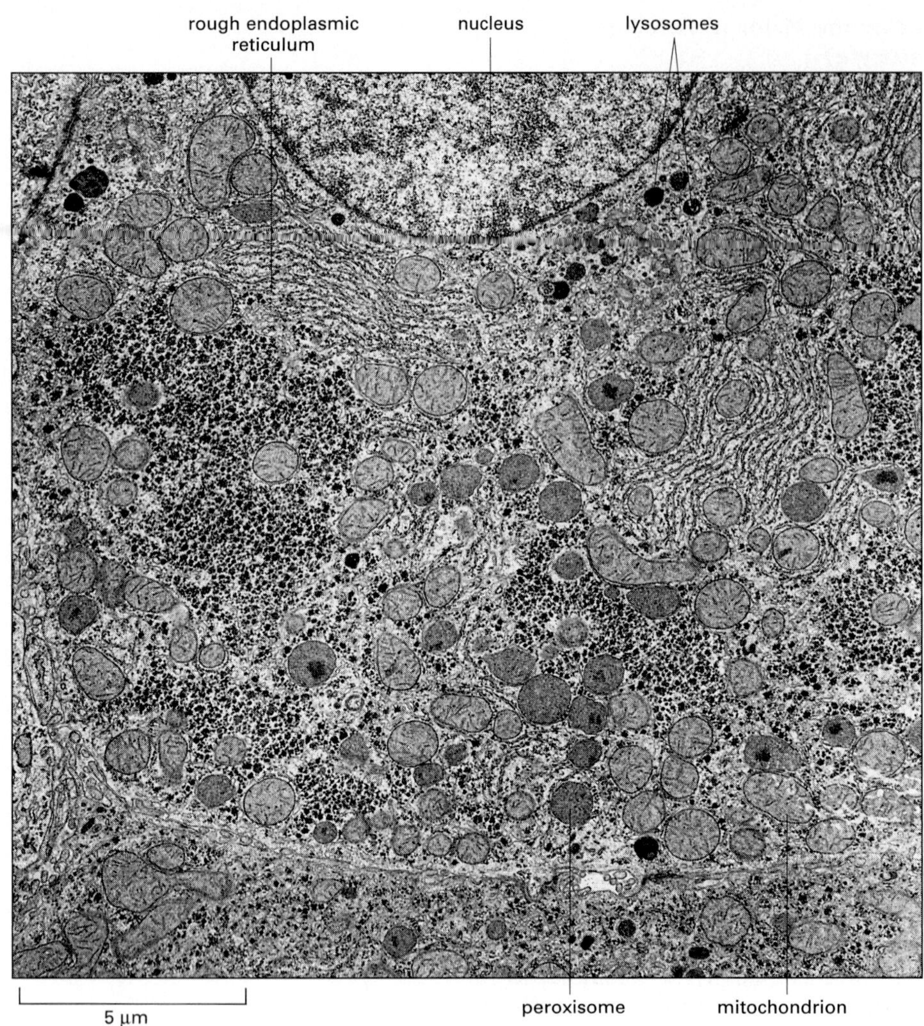

Figure 12–2 An electron micrograph of part of a liver cell seen in cross section. Examples of most of the major intracellular compartments are indicated. (Courtesy of Daniel S. Friend.)

of the organelles with the cytoskeleton. The localization of both the ER and the Golgi apparatus, for instance, depends on an intact microtubule array; if the microtubules are experimentally depolymerized with a drug, the Golgi apparatus fragments and disperses throughout the cell, and the ER network collapses toward the cell center (discussed in Chapter 16).

The Topological Relationships of Membrane-enclosed Organelles Can Be Interpreted in Terms of Their Evolutionary Origins

To understand the relationships between the compartments of the cell, it is helpful to consider how they might have evolved. The precursors of the first eucaryotic cells are thought to have been simple organisms that resembled bacteria, which generally have a plasma membrane but no internal membranes. The plasma membrane in such cells therefore provides all membrane-dependent functions, including the pumping of ions, ATP synthesis, protein secretion, and lipid synthesis. Typical present-day eucaryotic cells are 10–30 times larger in linear dimension and 1000–10,000 times greater in volume than a typical bacterium such as *E. coli*. The profusion of internal membranes can be seen in part as an adaptation to this increase in size: the eucaryotic cell has a much smaller ratio of surface area to volume, and its area of plasma membrane is presumably too small to sustain the many vital functions for which membranes are required. The extensive internal membrane systems of a eucaryotic cell alleviate this imbalance.

The evolution of internal membranes evidently accompanied the specialization of membrane function. Consider, for example, the generation of *thylakoid vesicles* in chloroplasts. These vesicles form during the development of chloroplasts from *proplastids* in the green leaves of plants. Proplastids are small precursor organelles that are present in all immature plant cells. They are surrounded by a double membrane and develop according to the needs of the differentiated cells: they develop into chloroplasts in leaf cells, for example, and into organelles that store starch, fat, or pigments in other cell types (Figure 12–3A). In the process of differentiating into chloroplasts, specialized membrane patches form and pinch off from the inner membrane of the proplastid. The vesicles that pinch off form a new specialized compartment, the *thylakoid,* that harbors all of the chloroplast's photosynthetic machinery (Figure 12–3B).

Other compartments in eucaryotic cells may have originated in a conceptually similar way (Figure 12–4A). Pinching off of specialized intracellular membrane structures from the plasma membrane, for example, would create organelles with an interior that is topologically equivalent to the exterior of the cell. We shall see that this topological relationship holds for all of the organelles involved in the secretory and endocytic pathways, including the ER, Golgi apparatus, endosomes, and lysosomes. We can therefore think of all of these organelles as members of the same family. As we discuss in detail in the next chapter, their interiors communicate extensively with one another and with the outside of the cell via *transport vesicles* that bud off from one organelle and fuse with another (Figure 12–5).

As described in Chapter 14, mitochondria and plastids differ from the other membrane-enclosed organelles in containing their own genomes. The nature of these genomes, and the close resemblance of the proteins in these organelles to those in some present-day bacteria, strongly suggest that mitochondria and plastids evolved from bacteria that were engulfed by other cells with which they initially lived in symbiosis (discussed in Chapters 1 and 14). According to the hypothetical scheme shown in Figure 12–4B, the inner membrane of mitochondria and plastids corresponds to the original plasma membrane of the bacterium, while the lumen of these organelles evolved from the bacterial cytosol. As might be expected from such an endocytic origin, these two organelles are surrounded by a double membrane, and they remain isolated from the extensive vesicular traffic that connects the interiors of most of the other membrane-enclosed organelles to each other and to the outside of the cell.

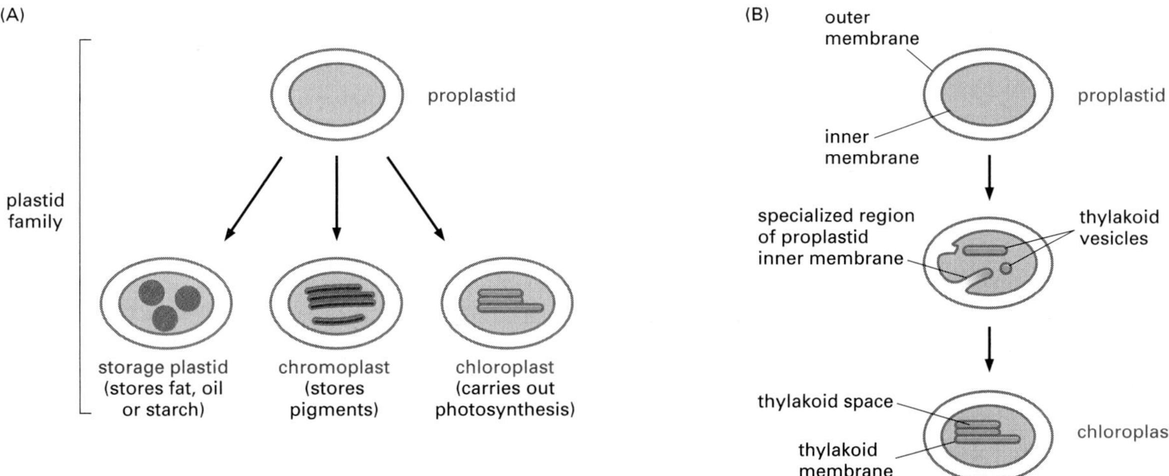

Figure 12–3 Development of plastids. (A) Proplastids are inherited with the cytoplasm of plant egg cells. As immature plant cells differentiate, the proplastids develop according to the needs of the specialized cell: they can become chloroplasts (in green leaf cells), storage plastids that accumulate starch (e.g., in potato tubers) or oil and lipid droplets (e.g., in fatty seeds), or chromoplasts that harbor pigments (e.g., in flower petals). (B) Development of the thylakoid. As chloroplasts develop, invaginated patches of specialized membrane from the proplastid inner membrane pinch off to form thylakoid vesicles, which then develop into the mature thylakoid. The thylakoid membrane forms a separate compartment, the thylakoid space, which is structurally and functionally distinct from the rest of the chloroplast. Thylakoids can grow and divide autonomously as chloroplasts proliferate.

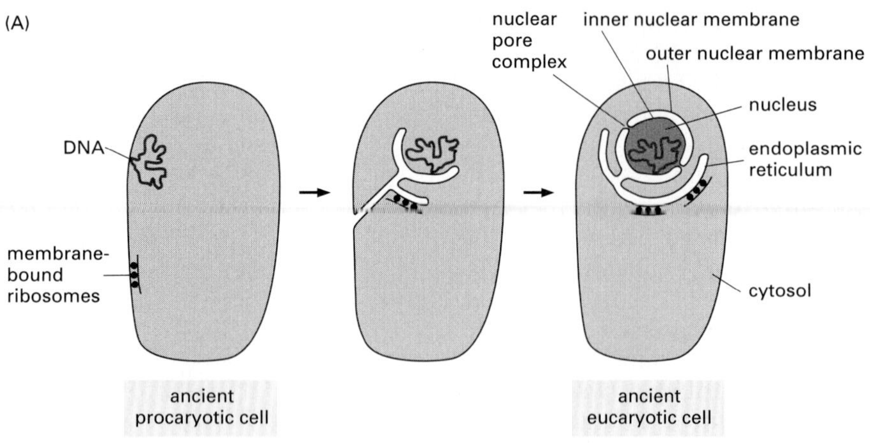

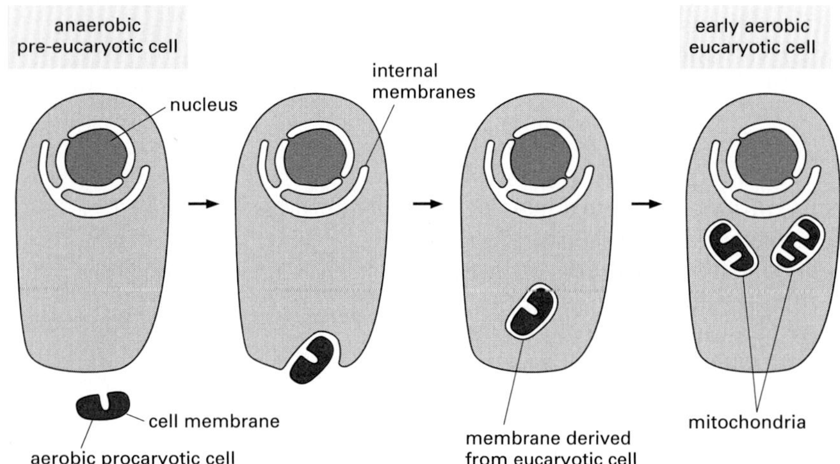

Figure 12–4 Hypothetical schemes for the evolutionary origins of some membrane-enclosed organelles. The origins of mitochondria, chloroplasts, ER, and the cell nucleus can explain the topological relationships of these intracellular compartments in eucaryotic cells.

(A) A possible pathway for the evolution of the cell nucleus and the ER. In some bacteria the single DNA molecule is attached to an invagination of the plasma membrane. Such an invagination in a very ancient procaryotic cell could have rearranged to form an envelope around the DNA, while still allowing the DNA access to the cell cytosol (as is required for DNA to direct protein synthesis). This envelope is presumed to have eventually pinched off completely from the plasma membrane, producing a nuclear compartment surrounded by a double membrane.

As illustrated, the nuclear envelope is penetrated by communicating channels called nuclear pore complexes. Because it is surrounded by two membranes that are in continuity where they are penetrated by these pores, the nuclear compartment is topologically equivalent to the cytosol; in fact, during mitosis the nuclear contents mix with the cytosol. The lumen of the ER is continuous with the space between the inner and outer nuclear membranes and topologically equivalent to the extracellular space.

(B) Mitochondria (and plastids) are thought to have originated when a bacterium was engulfed by a larger pre-eucaryotic cell. They retain their autonomy. This may explain why the lumens of these organelles remain isolated from the membrane traffic that interconnects the lumens of many other intracellular compartments.

The evolutionary scheme described above groups the intracellular compartments in eucaryotic cells into four distinct families: (1) the nucleus and the cytosol, which communicate through *nuclear pore complexes* and are thus topologically continuous (although functionally distinct); (2) all organelles that function in the secretory and endocytic pathways—including the ER, Golgi apparatus, endosomes, lysosomes, the numerous classes of transport intermediates such as transport vesicles, and possibly peroxisomes; (3) the mitochondria; and (4) the plastids (in plants only).

Proteins Can Move Between Compartments in Different Ways

All proteins begin being synthesized on ribosomes in the cytosol, except for the few that are synthesized on the ribosomes of mitochondria and plastids. Their subsequent fate depends on their amino acid sequence, which can contain **sorting signals** that direct their delivery to locations outside the cytosol. Most proteins

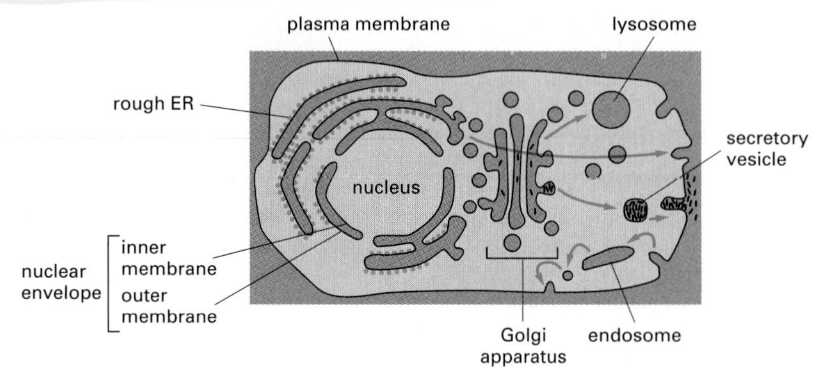

Figure 12–5 Topological relationships between compartments of the secretory and endocytic pathways in a eucaryotic cell. Topologically equivalent spaces are shown in *red*. In principle, cycles of membrane budding and fusion permit the lumen of any of these organelles to communicate with any other and with the cell exterior by means of transport vesicles. *Blue arrows* indicate the extensive network of outbound and inbound traffic routes, which we discuss in Chapter 13. Some organelles, most notably mitochondria and (in plant cells) plastids do not take part in this communication and are isolated from the traffic between organelles shown here.

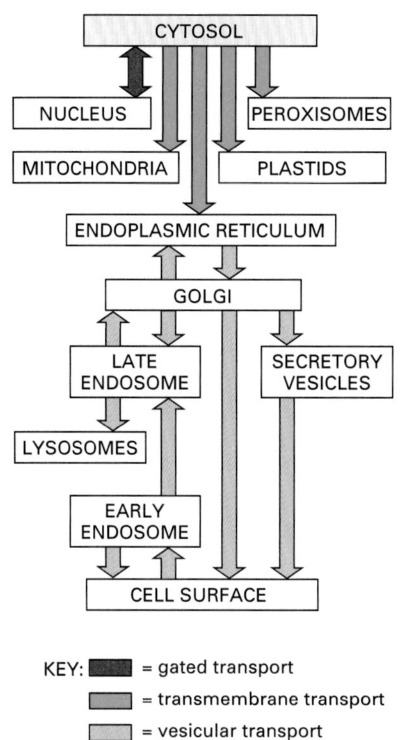

Figure 12–6 A simplified "roadmap" of protein traffic. Proteins can move from one compartment to another by gated transport *(red)*, transmembrane transport *(blue)*, or vesicular transport *(green)*. The signals that direct a given protein's movement through the system, and thereby determine its eventual location in the cell, are contained in each protein's amino acid sequence. The journey begins with the synthesis of a protein on a ribosome in the cytosol and terminates when the final destination is reached. At each intermediate station *(boxes)*, a decision is made as to whether the protein is to be retained in that compartment or transported further. In principle, a signal could be required for either retention in or exit from a compartment. We shall use this figure repeatedly as a guide throughout this chapter and the next, highlighting in color the particular pathway being discussed.

do not have a sorting signal and consequently remain in the cytosol as permanent residents. Many others, however, have specific sorting signals that direct their transport from the cytosol into the nucleus, the ER, mitochondria, plastids, or peroxisomes; sorting signals can also direct the transport of proteins from the ER to other destinations in the cell.

To understand the general principles by which sorting signals operate, it is important to distinguish three fundamentally different ways by which proteins move from one compartment to another. These three mechanisms are described below, and their sites of action in the cell are outlined in Figure 12–6. The first two mechanisms are detailed in this chapter, while the third *(green arrows* in Figure 12–6) is the subject of Chapter 13.

1. In **gated transport**, the protein traffic between the cytosol and nucleus occurs between topologically equivalent spaces, which are in continuity through the nuclear pore complexes. The nuclear pore complexes function as selective gates that actively transport specific macromolecules and macromolecular assemblies, although they also allow free diffusion of smaller molecules.
2. In **transmembrane transport**, membrane-bound *protein translocators* directly transport specific proteins across a membrane from the cytosol into a space that is topologically distinct. The transported protein molecule usually must unfold to snake through the translocator. The initial transport of selected proteins from the cytosol into the ER lumen or from the cytosol into mitochondria, for example, occurs in this way.
3. In **vesicular transport**, membrane-enclosed transport intermediates—which may be small, spherical transport vesicles or larger, irregularly shaped organelle fragments—ferry proteins from one compartment to another. The transport vesicles and fragments become loaded with a cargo of molecules derived from the lumen of one compartment as they pinch off from its membrane; they discharge their cargo into a second compartment by fusing with that compartment (Figure 12–7). The transfer of soluble proteins from the ER to the Golgi apparatus, for example, occurs in this way. Because the transported proteins do not cross a membrane, vesicular transport can move proteins only between compartments that are topologically equivalent (see Figure 12–5). We discuss vesicular transport in detail in Chapter 13.

Each of the three modes of protein transfer is usually guided by sorting signals in the transported protein that are recognized by complementary receptor proteins. If a large protein is to be imported into the nucleus, for example, it must possess a sorting signal that is recognized by receptor proteins that

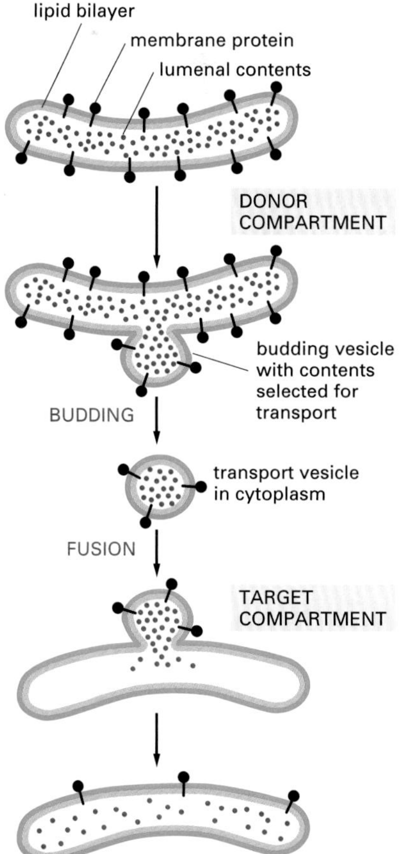

Figure 12–7 Vesicle budding and fusion during vesicular transport. Transport vesicles bud from one compartment (donor) and fuse with another (target) compartment. In the process, soluble components *(red dots)* are transferred from lumen to lumen. Note that membrane is also transferred, and that the original orientation of both proteins and lipids in the donor-compartment membrane is preserved in the target-compartment membrane. Thus, membrane proteins retain their asymmetric orientation, with the same domains always facing the cytosol.

guide it through the nuclear pore complex. If a protein is to be transferred directly across a membrane, it must possess a sorting signal that is recognized by the translocator in the membrane to be crossed. Likewise, if a protein is to be loaded into a certain type of vesicle or retained in certain organelles, its sorting signal must be recognized by a complementary receptor in the appropriate membrane.

Signal Sequences and Signal Patches Direct Proteins to the Correct Cellular Address

There are at least two types of sorting signals in proteins. One type resides in a continuous stretch of amino acid sequence, typically 15–60 residues long. Some of these **signal sequences** are removed from the finished protein by specialized **signal peptidases** once the sorting process has been completed. The other type consists of a specific three-dimensional arrangement of atoms on the protein's surface that forms when the protein folds up. The amino acid residues that comprise this **signal patch** can be distant from one another in the linear amino acid sequence, and they generally persist in the finished protein (Figure 12–8). Signal sequences are used to direct proteins from the cytosol into the ER, mitochondria, chloroplasts, and peroxisomes, and they are also used to transport proteins from the nucleus to the cytosol and from the Golgi apparatus to the ER. The sorting signals that direct proteins into the nucleus from the cytosol can be either short signal sequences or longer sequences that are likely to fold into signal patches. Signal patches also direct newly synthesized degradative enzymes into lysosomes.

Each signal sequence specifies a particular destination in the cell. Proteins destined for initial transfer to the ER usually have a signal sequence at their N terminus, which characteristically includes a sequence composed of about 5–10 hydrophobic amino acids. Many of these proteins will in turn pass from the ER to the Golgi apparatus, but those with a specific sequence of four amino acids at their C terminus are recognized as ER residents and are returned to the ER. Proteins destined for mitochondria have signal sequences of yet another type, in which positively charged amino acids alternate with hydrophobic ones. Finally, many proteins destined for peroxisomes have a signal peptide of three characteristic amino acids at their C terminus.

Some specific signal sequences are presented in Table 12–3. The importance of each of these signal sequences for protein targeting has been shown by experiments in which the peptide is transferred from one protein to another by genetic engineering techniques. Placing the N-terminal ER signal sequence at the beginning of a cytosolic protein, for example, redirects the protein to the ER. Signal sequences are therefore both necessary and sufficient for protein targeting. Even though their amino acid sequences can vary greatly, the signal sequences of all proteins having the same destination are functionally interchangeable, and physical properties, such as hydrophobicity, often seem to be more important in the signal-recognition process than the exact amino acid sequence.

Signal patches are far more difficult to analyze than signal sequences, so less is known about their structure. Because they often result from a complex

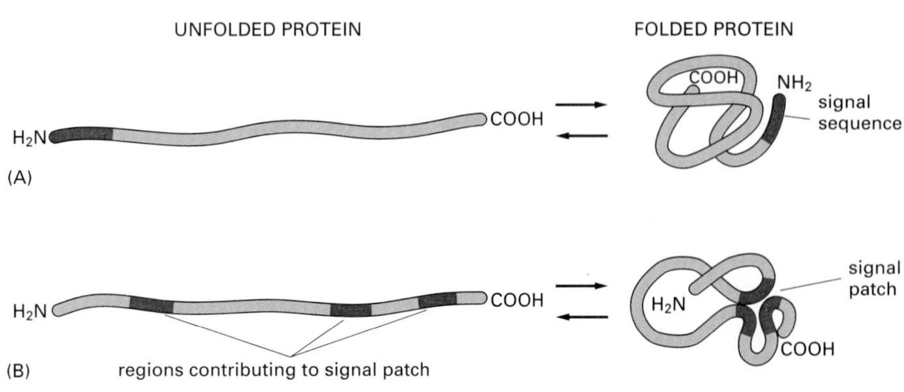

Figure 12–8 Two ways in which a sorting signal can be built into a protein. (A) The signal resides in a single discrete stretch of amino acid sequence, called a signal sequence, that is exposed in the folded protein. Signal sequences often occur at the end of the polypeptide chain (as shown), but they can also be located internally. (B) A signal patch can be formed by the juxtaposition of amino acids from regions that are physically separated before the protein folds (as shown). Alternatively, separate patches on the surface of the folded protein that are spaced a fixed distance apart can form the signal.

TABLE 12–3 Some Typical Signal Sequences

FUNCTION OF SIGNAL SEQUENCE	EXAMPLE OF SIGNAL SEQUENCE
Import into nucleus	-Pro-Pro-Lys-Lys-Lys-Arg-Lys-Val-
Export from nucleus	-Leu-Ala-Leu-Lys-Leu-Ala-Gly-Leu-Asp-Ile-
Import into mitochondria	$^{+}H_3N$-Met-Leu-Ser-Leu-Arg-Gln-Ser-Ile-Arg-Phe-Phe-Lys-Pro-Ala-Thr-Arg-Thr-Leu-Cys-Ser-Ser-Arg-Tyr-Leu-Leu-
Import into plastid	$^{+}H_3N$-Met-Val-Ala-Met-Ala-Met-Ala-Ser-Leu-Gln-Ser-Ser-Met-Ser-Ser-Leu-Ser-Leu-Ser-Ser-Asn-Ser-Phe-Leu-Gly-Gln-Pro-Leu-Ser-Pro-Ile-Thr-Leu-Ser-Pro-Phe-Leu-Gln-Gly-
Import into peroxisomes	-Ser-Lys-Leu-COO^{-}
Import into ER	$^{+}H_3N$-Met-Met-Ser-Phe-Val-Ser-Leu-Leu-Leu-Val-Gly-Ile-Leu-Phe-Trp-Ala-Thr-Glu-Ala-Glu-Gln-Leu-Thr-Lys-Cys-Glu-Val-Phe-Gln-
Return to ER	-Lys-Asp-Glu-Leu-COO^{-}

Some characteristic features of the different classes of signal sequences are highlighted in color. Where they are known to be important for the function of the signal sequence, positively charged amino acids are shown in *red* and negatively charged amino acids are shown in *green*. Similarly, important hydrophobic amino acids are shown in *yellow* and hydroxylated amino acids are shown in *blue*. $^{+}H_3N$ indicates the N-terminus of a protein; COO^{-} indicates the C-terminus.

three-dimensional protein-folding pattern, they cannot be easily transferred experimentally from one protein to another.

Both types of sorting signals are recognized by complementary *sorting receptors* that guide proteins to their appropriate destination, where the receptors unload their cargo. The receptors function catalytically: after completing one round of targeting, they return to their point of origin to be reused. Most sorting receptors recognize classes of proteins rather than just an individual protein species. They therefore can be viewed as public transportation systems dedicated to delivering groups of components to their correct location in the cell.

The main ways of studying how proteins are directed from the cytosol to a specific compartment and how they are translocated across membranes are illustrated in Panel 12–1.

Most Membrane-enclosed Organelles Cannot Be Constructed From Scratch: They Require Information in the Organelle Itself

When a cell reproduces by division, it has to duplicate its membrane-enclosed organelles. In general, cells do this by enlarging the existing organelles by incorporating new molecules into them; the enlarged organelles then divide and are distributed to the two daughter cells. Thus, each daughter cell inherits from its mother a complete set of specialized cell membranes. This inheritance is essential because a cell could not make such membranes from scratch. If the ER were completely removed from a cell, for example, how could the cell reconstruct it? As we shall discuss later, the membrane proteins that define the ER and perform many of its functions are themselves products of the ER. A new ER could not be made without an existing ER or, at the very least, a membrane that specifically contains the protein translocators required to import selected proteins into the ER from the cytosol (including the ER-specific translocators themselves). The same is true for mitochondria, plastids, and peroxisomes (see Figure 12–6).

Thus, it seems that the information required to construct a membrane-enclosed organelle does not reside exclusively in the DNA that specifies the organelle's proteins. *Epigenetic* information in the form of at least one distinct protein that preexists in the organelle membrane is also required, and this information is passed from parent cell to progeny cell in the form of the organelle itself. Presumably, such information is essential for the propagation of the cell's compartmental organization, just as the information in DNA is essential for the propagation of the cell's nucleotide and amino acid sequences.

As we discuss in more detail in Chapter 13, however, the ER sheds a constant stream of membrane vesicles that incorporate only specific proteins and therefore

A TRANSFECTION APPROACH FOR DEFINING SIGNAL SEQUENCES

One way to show that a signal sequence is required and sufficient to target a protein to a specific intracellular compartment is to create a fusion protein in which the signal sequence is attached by genetic engineering techniques to a protein that is normally resident in the cytosol. After the cDNA encoding this protein is transfected into cells, the location of the fusion protein is determined by immunostaining or by cell fractionation.

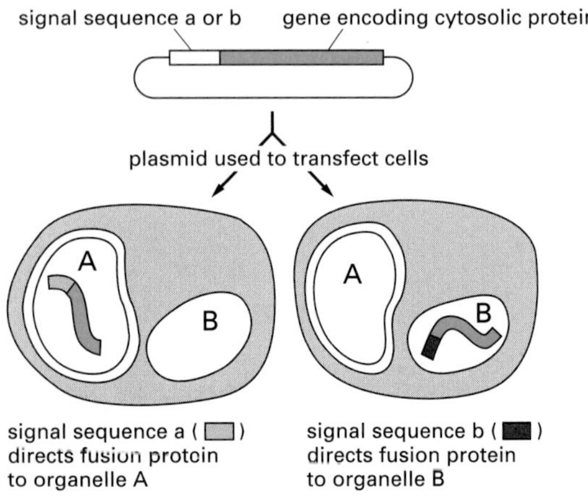

signal sequence a ( ) directs fusion protein to organelle A

signal sequence b ( ) directs fusion protein to organelle B

By altering the signal sequence using site-directed mutagenesis, one can determine which structural features are important for its function.

A BIOCHEMICAL APPROACH FOR STUDYING THE MECHANISM OF PROTEIN TRANSLOCATION

In this approach a labeled protein containing a specific signal sequence is transported into isolated organelles *in vitro*. The labeled protein is usually produced by cell-free translation of a purified mRNA encoding the protein; radioactive amino acids are used to label the newly synthesized protein so that it can be distinguished from the many other proteins that are present in the *in vitro* translation system.

Three methods are commonly used to test if the labeled protein has been translocated into the organelle:

1. The labeled protein co-fractionates with the organelle during centrifugation.

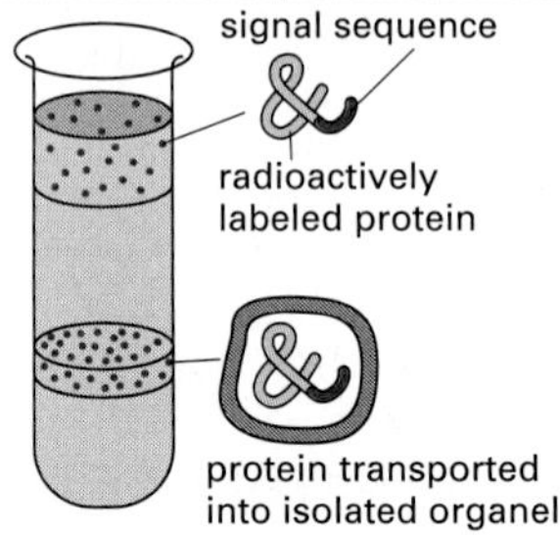

2. The signal sequence is removed by a specific protease that is present inside the organelle.

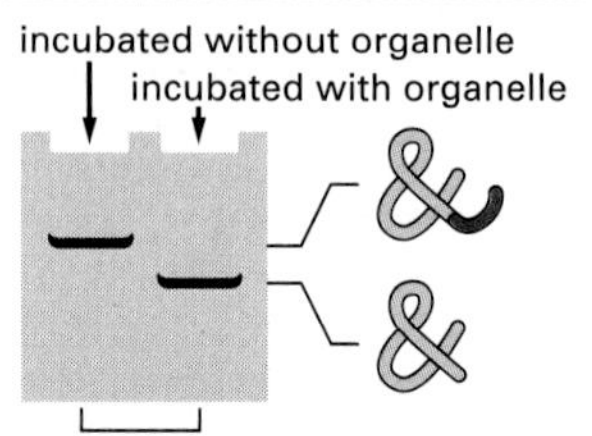

3. The protein is protected from digestion when proteases are added to the incubation medium but is susceptible if a detergent is first added to disrupt the organelle membrane.

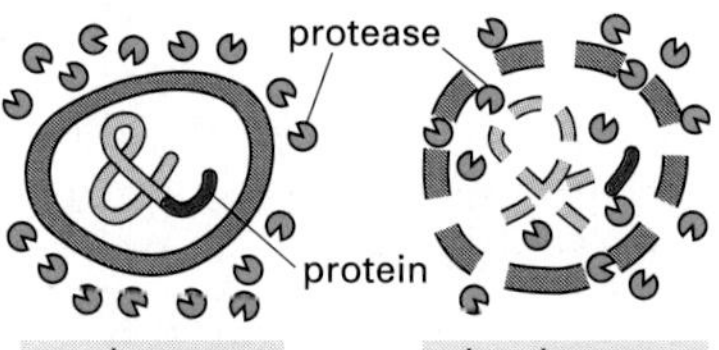

By exploiting such *in vitro* assays, one can determine what components (proteins, ATP, GTP, etc.) are required for the translocation process.

GENETIC APPROACHES FOR STUDYING THE MECHANISM OF PROTEIN TRANSLOCATION

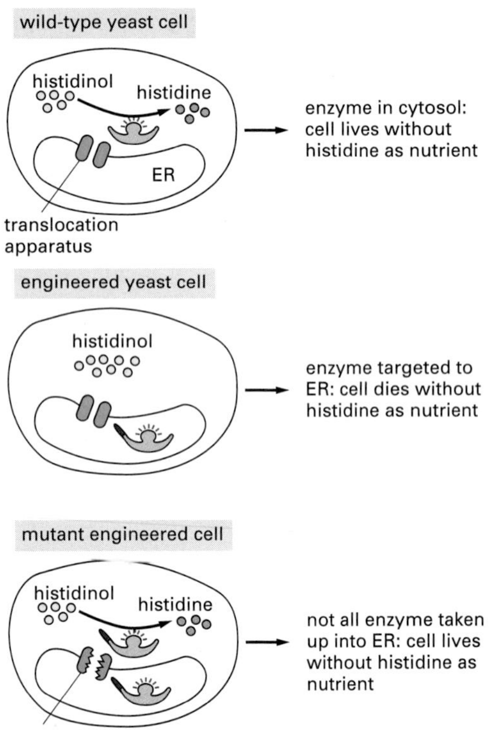

Yeast cells with mutations in genes that encode components of the translocation machinery have been useful for studying protein translocation. Because mutant cells that cannot translocate proteins across their membranes will die, the trick is to design a strategy that allows weak mutations that cause only a partial defect in protein translocation to be isolated.

One way uses genetic engineering to design special yeast cells. The enzyme histidinol dehydrogenase, for example, normally resides in the cytosol, where it is required to produce the essential amino acid histidine from its precursor histidinol. A yeast strain is constructed in which the histidinol dehydrogenase gene is replaced by a re-engineered gene encoding a fusion protein with an added signal sequence that misdirects the enzyme into the endoplasmic reticulum (ER). When such cells are grown without histidine, they die because all of the histidinol dehydrogenase is sequestered in the ER, where it is of no use. Cells with a mutation that partially inactivates the mechanism for translocating proteins from the cytosol to the ER, however, will survive because enough of the dehydrogenase will be retained in the cytosol to produce histidine. Often one obtains a cell in which the mutant protein still functions partially at normal temperature but is completely inactive at higher temperature. A cell carrying such a temperature-sensitive mutation dies at higher temperature, whether or not histidine is present, as it cannot transport any protein into the ER. This allows the normal gene that was disabled by the mutation to be identified by transfecting the mutant cells with a yeast plasmid vector into which random yeast genomic DNA fragments have been cloned: the specific DNA fragment that rescues the mutant cells when they are grown at high temperature should encode the wild-type version of the mutant gene.

have a different composition from the ER itself. Similarly, the plasma membrane constantly produces specialized endocytic vesicles. Thus, some membrane-enclosed compartments can form from other organelles and do not have to be inherited at cell division.

Summary

Eucaryotic cells contain intracellular membranes that enclose nearly half the cell's total volume in separate intracellular compartments called organelles. The main types of membrane-enclosed organelles present in all eucaryotic cells are the endoplasmic reticulum, Golgi apparatus, nucleus, mitochondria, lysosomes, endosomes, and peroxisomes; plant cells also contain plastids, such as chloroplasts. Each organelle contains a distinct set of proteins that mediate its unique functions.

Each newly synthesized organelle protein must find its way from a ribosome in the cytosol, where it is made, to the organelle where it functions. It does so by following a specific pathway, guided by signals in its amino acid sequence that function as signal sequences or signal patches. Signal sequences and patches are recognized by complementary sorting receptors that deliver the protein to the appropriate target organelle. Proteins that function in the cytosol do not contain sorting signals and therefore remain there after they are synthesized.

During cell division, organelles such as the ER and mitochondria are distributed intact to each daughter cell. These organelles contain information that is required for their construction so that they cannot be made from scratch.

THE TRANSPORT OF MOLECULES BETWEEN THE NUCLEUS AND THE CYTOSOL

The **nuclear envelope** encloses the DNA and defines the *nuclear compartment*. This envelope consists of two concentric membranes that are penetrated by nuclear pore complexes (Figure 12–9). Although the inner and outer nuclear membranes are continuous, they maintain distinct protein compositions. The **inner nuclear membrane** contains specific proteins that act as binding sites for chromatin and for the protein meshwork of the *nuclear lamina* that provides structural support for this membrane. The inner membrane is surrounded by the **outer nuclear membrane**, which is continuous with the membrane of the ER. Like the membrane of the ER that will be described later in this chapter, the outer nuclear membrane is studded with ribosomes engaged in protein synthesis. The proteins made on these ribosomes are transported into the space between the inner and outer nuclear membranes (the *perinuclear space)*, which is continuous with the ER lumen (see Figure 12–9).

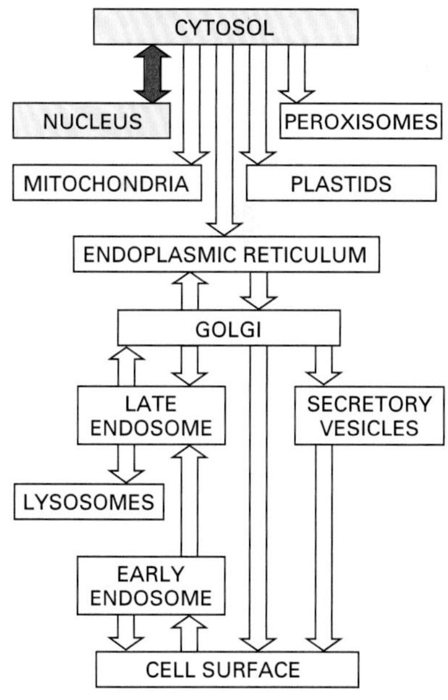

Bidirectional traffic occurs continuously between the cytosol and the nucleus. The many proteins that function in the nucleus—including histones, DNA and RNA polymerases, gene regulatory proteins, and RNA-processing proteins—are selectively imported into the nuclear compartment from the cytosol, where they are made. At the same time, tRNAs and mRNAs are synthesized in the nuclear compartment and then exported to the cytosol. Like the import process, the export process is selective; mRNAs, for example, are exported only after they have been properly modified by RNA-processing reactions in the nucleus. In some cases the transport process is complex: ribosomal proteins, for instance, are made in the cytosol, imported into the nucleus—where they assemble with newly made ribosomal RNA into particles—and are then exported again to the cytosol as part of a ribosomal subunit. Each of these steps requires selective transport across the nuclear envelope.

Nuclear Pore Complexes Perforate the Nuclear Envelope

The nuclear envelope of all eucaryotes is perforated by large, elaborate structures known as **nuclear pore complexes.** In animal cells, each complex has an estimated molecular mass of about 125 million and is thought to be composed

Figure 12–9 The nuclear envelope. The double-membrane envelope is penetrated by nuclear pore complexes and is continuous with the endoplasmic reticulum. The ribosomes that are normally bound to the cytosolic surface of the ER membrane and outer nuclear membrane are not shown. The nuclear lamina is a fibrous meshwork underlying the inner membrane.

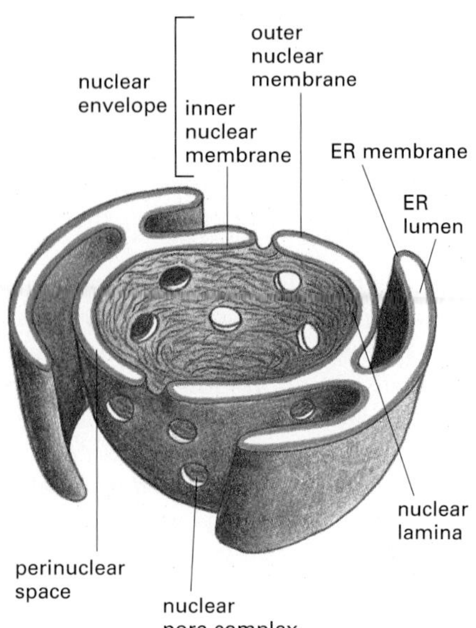

of more than 50 different proteins, called **nucleoporins**, that are arranged with a striking octagonal symmetry (Figure 12–10).

In general, the more active the nucleus is in transcription, the greater the number of pore complexes its envelope contains. The nuclear envelope of a typical mammalian cell contains 3000–4000 pore complexes. If the cell is synthesizing DNA, it needs to import about 10^6 histone molecules from the cytosol every 3 minutes to package the newly made DNA into chromatin, which means that, on average, each pore complex needs to transport about 100 histone molecules per minute. If the cell is growing rapidly, each complex also needs to transport about 6 newly assembled large and small ribosomal subunits per minute from

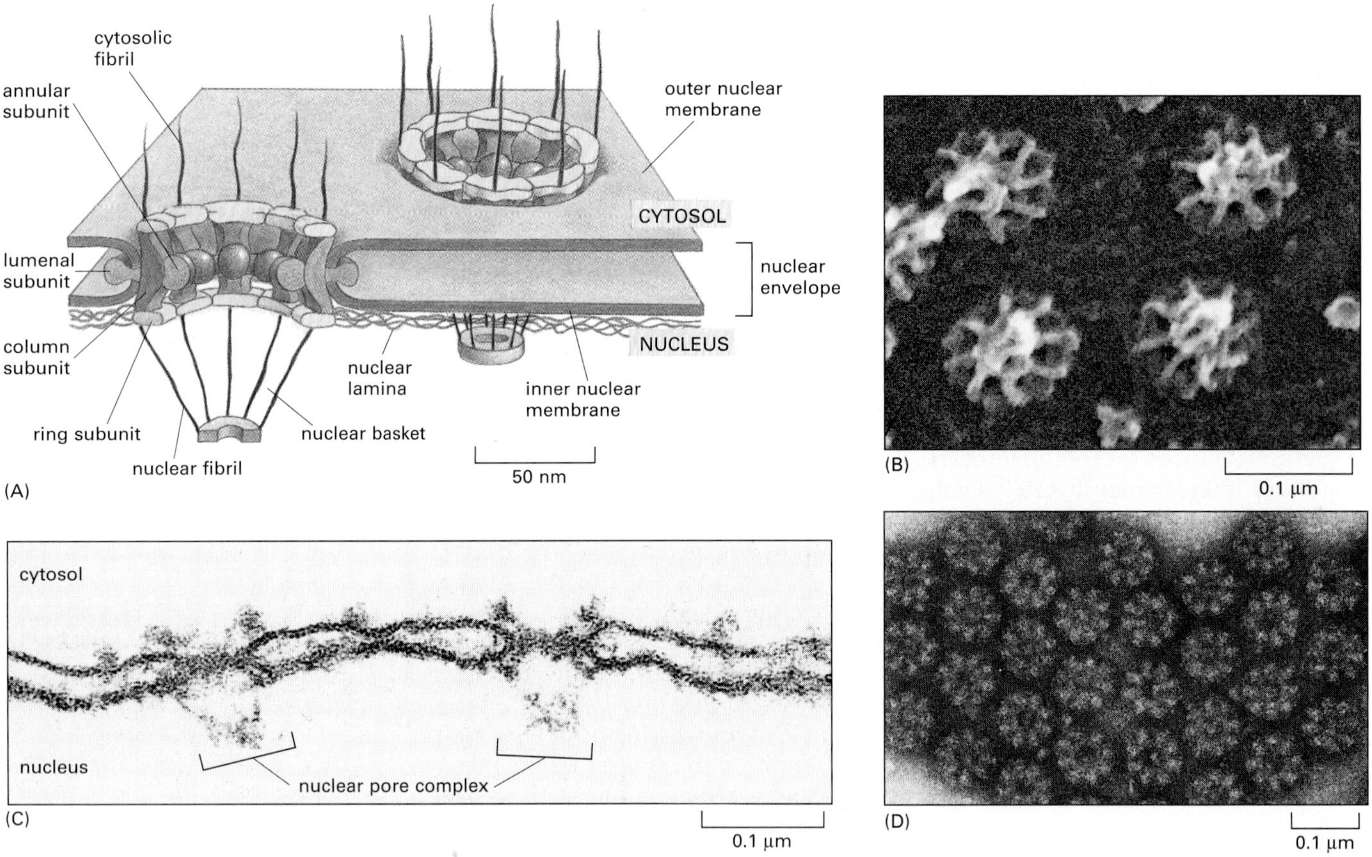

Figure 12–10 The arrangement of nuclear pore complexes in the nuclear envelope. (A) A small region of the nuclear envelope. In cross section, a nuclear pore complex seems to have four structural building blocks: column subunits, which form the bulk of the pore wall; annular subunits, which extend "spokes" (not shown) toward the center of the pore; lumenal subunits, which contain transmembrane proteins that anchor the complex to the nuclear membrane; and ring subunits, which form the cytosolic and nuclear faces of the complex. In addition, fibrils protrude from both the cytosolic and the nuclear sides of the complex. On the nuclear side, the fibrils converge to form basketlike structures. Localization studies using immunoelectron microscopy techniques showed that the proteins that make up the core of the nuclear pore complex are symmetrically distributed across the nuclear envelope so that the nuclear and cytosolic sides look identical. This is in contrast to proteins that make up the fibrils, which are different on each side of the cytosolic or the nuclear side. (B) A scanning electron micrograph of the nuclear side of the nuclear envelope of an oocyte. (C) The continuity of the inner and outer nuclear membrane at the pore is apparent in this thin section electron micrograph, showing a side view of two nuclear pore complexes *(brackets)*. (D) This electron micrograph shows face-on views of negatively stained nuclear pore complexes from which the membrane has been removed by detergent extraction. (B, from M.W. Goldberg and T.D. Allen, *J. Cell Biol.* 119:1429–1440, 1992. © The Rockefeller University Press; C, courtesy of Werner Franke and Ulrich Scheer; D, courtesy of Ron Milligan.)

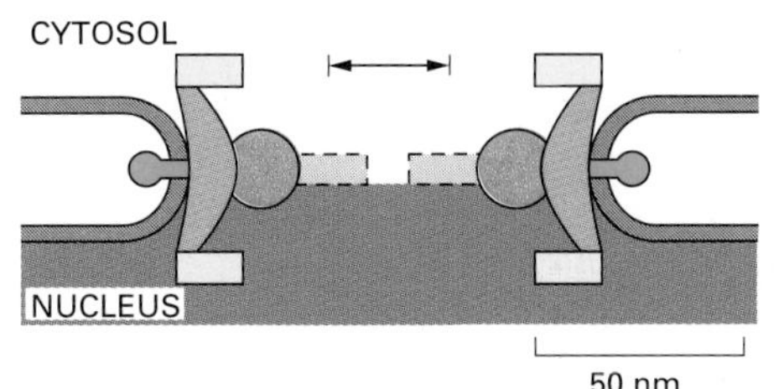

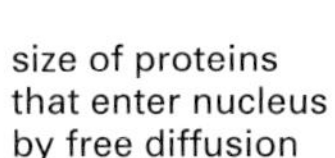

Figure 12–11 Possible paths for free diffusion through the nuclear pore complex. This drawing shows a hypothetical diaphragm *(gray)* inserted into the pore to restrict the size of the open channel to 9 nm, the pore size estimated from diffusion measurements. Nine nanometers is a much smaller diameter than that of the central opening apparent on the images of the nuclear pore complex derived from electron micrographs. It is also smaller than the opening estimated during active transport, when the pore dilates to allow the transport of particles of up to 26 nm in diameter *(arrow)*. Thus, it is likely that some pore components are lost during the preparation of specimens for electron microscopy, and that these normally restrict free diffusion through the central opening. Such components may form a diaphragm (or plug) that opens and closes to allow the passage of large objects during active transport, which depends on sorting signals (discussed below). Although plugs can be seen in some preparations, it is not clear whether they are components of the pore complex or material that is being transported through it. Three-dimensional computer reconstructions suggest that the channels permitting free diffusion might be located near the rim of the pore complex, between the column subunits, rather than at its center (see Figure 12–10A); this would mean that passive diffusion and active transport take place through different parts of the complex.

the nucleus, where they are produced, to the cytosol, where they are used. And that is only a very small part of the total traffic that passes through the pore complexes.

Each pore complex contains one or more open aqueous channels through which small water-soluble molecules can passively diffuse. The effective size of these channels has been determined by injecting labeled water-soluble molecules of different sizes into the cytosol and then measuring their rate of diffusion into the nucleus. Small molecules (5000 daltons or less) diffuse in so fast that the nuclear envelope can be considered to be freely permeable to them. A protein of 17,000 daltons takes 2 minutes to equilibrate between the cytosol and the nucleus, whereas proteins larger than 60,000 daltons are hardly able to enter the nucleus at all. A quantitative analysis of such data suggests that the nuclear pore complex contains a pathway for free diffusion equivalent to a water-filled cylindrical channel about 9 nm in diameter and 15 nm long; such a channel would occupy only a small fraction of the total volume of the pore complex (Figure 12–11).

Because many cell proteins are too large to pass by diffusion through the nuclear pore complexes, the nuclear envelope enables the nuclear compartment and the cytosol to maintain different complements of proteins. Mature cytosolic ribosomes, for example, are about 30 nm in diameter and thus cannot diffuse through the 9 nm channels; their exclusion from the nucleus ensures that protein synthesis is confined to the cytosol. But how does the nucleus export newly made ribosomal subunits or import large molecules, such as DNA and RNA polymerases, which have subunit molecular weights of 100,000–200,000 daltons? As we discuss next, these and many other protein and RNA molecules bind to specific receptor proteins that ferry them actively through nuclear pore complexes.

Nuclear Localization Signals Direct Nuclear Proteins to the Nucleus

When proteins are experimentally extracted from the nucleus and reintroduced into the cytosol (e.g., through experimentally induced perforations in the plasma membrane), even the very large ones reaccumulate efficiently in the nucleus. The selectivity of this nuclear import process resides in **nuclear localization signals (NLSs)**, which are present only in nuclear proteins. The signals have been precisely defined in numerous nuclear proteins by using recombinant DNA technology (Figure 12–12). As mentioned earlier, they can be either signal sequences or signal patches. In many nuclear proteins they consist of one or two short

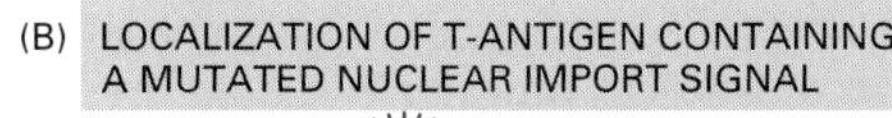

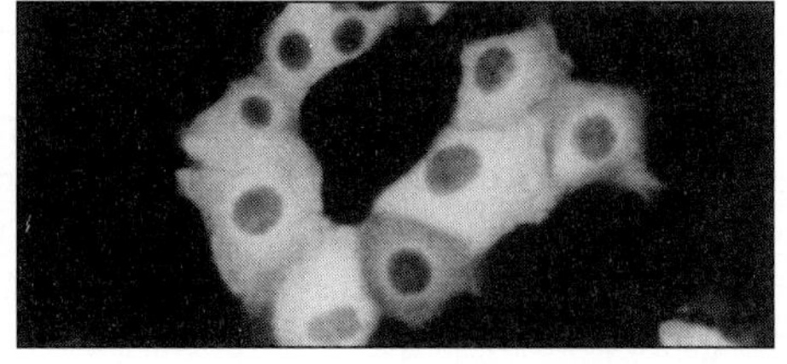

Figure 12–12 The function of a nuclear localization signal. Immunofluorescence micrographs showing the cellular location of SV40 virus T-antigen containing or lacking a short peptide that serves as a nuclear localization signal. (A) The normal T-antigen protein contains the lysine-rich sequence indicated and is imported to its site of action in the nucleus, as indicated by immunofluorescence staining with antibody against the T-antigen. (B) T-antigen with an altered nuclear localization signal (a threonine replacing a lysine) remains in the cytosol. (From D. Kalderon, B. Roberts, W. Richardson, and A. Smith, *Cell* 39:499–509, 1984. © Elsevier.)

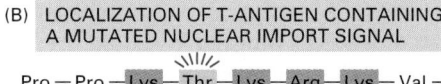

sequences that are rich in the positively charged amino acids lysine and arginine (see Table 12–3, p. 667), the precise sequence varying for different nuclear proteins. Other nuclear proteins contain different signals, some of which are not yet characterized.

The signals characterized this far can be located almost anywhere in the amino acid sequence and are thought to form loops or patches on the protein surface. Many function even when linked as short peptides to lysine side chains on the surface of a cytosolic protein, suggesting that the precise location of the signal within the amino acid sequence of a nuclear protein is not important.

The transport of nuclear proteins through nuclear pore complexes can be directly visualized by coating gold particles with a nuclear localization signal, injecting the particles into the cytosol, and then following their fate by electron microscopy (Figure 12–13). Studies with various sizes of gold beads indicate that the opening can dilate up to about 26 nm in diameter during the transport process. A structure in the center of the nuclear pore complex seems to function like a close-fitting diaphragm that opens just the right amount to let transport substrates pass (see Figure 12–11). The molecular basis of the gating mechanism remains a mystery.

The mechanism of macromolecular transport across nuclear pore complexes is fundamentally different from the transport mechanisms involved in protein transfer across the membranes of other organelles, because it occurs through a large aqueous pore rather than through a protein transporter spanning one or more lipid bilayers. For this reason, nuclear proteins can be transported through a pore complex while they are in a fully folded conformation. Likewise, a newly formed ribosomal subunit is transported out of the nucleus as an assembled particle. By contrast, proteins have to be extensively unfolded during their transport into most other organelles, as we discuss later. In the electron microscope, however, very large particles traversing the pore seem to become constricted as they squeeze through the nuclear pore complex, indicating that at least some of them must undergo restructuring during transport. This has been most extensively studied for the export of some very large mRNAs, as discussed in Chapter 6 (see Figure 6–39).

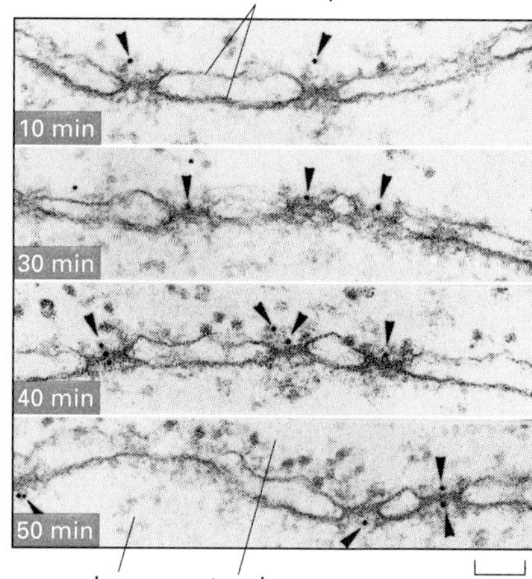

Figure 12–13 Visualizing active import through nuclear pores. This series of electron micrographs shows colloidal gold spheres *(arrowheads)* coated with peptides containing nuclear localization signals entering the nucleus by means of nuclear pore complexes. Gold particles were injected into living cells, which then were fixed and prepared for electron microscopy at various times after injection. At early time points (10 min), gold particles are seen in proximity to the cytosolic fibrils of the nuclear pore complexes. They then migrate to the center of the nuclear pore complexes, where they are first seen exclusively on the cytosolic face (30 and 40 min) and then appear on the nuclear face (50 min). These gold particles are much larger in diameter than the diffusion channel in the pore complex, which implies that the pores have been induced to widen to permit their passage. (From N. Panté and U. Aebi, *Science* 273:1729–1732, 1996. © AAAS.)

Nuclear Import Receptors Bind Nuclear Localization Signals and Nucleoporins

To initiate nuclear import, most nuclear localization signals must be recognized by **nuclear import receptors**, which are encoded by a family of related genes. Each family member encodes a receptor protein that is specialized for the transport of a group of nuclear proteins sharing structurally similar nuclear localization signals (Figure 12–14A).

The import receptors are soluble cytosolic proteins that bind both to the nuclear localization signal on the protein to be transported and to nucleoporins, some of which form the tentaclelike fibrils that extend into the cytosol from the rim of the nuclear pore complexes. The fibrils and many other nucleoporins contain a large number of short amino-acid repeats that contain phenylalanine and glycine and are therefore called *FG-repeats* (named after the one-letter code for amino acids, discussed in Chapter 5). FG-repeats serve as binding sites for the import receptors. They are thought to line the path through the nuclear pore complexes taken by the import receptors and their bound cargo proteins. These protein complexes move along the path by repeatedly binding, dissociating, and then re-binding to adjacent repeat sequences. Once in the nucleus, the import receptors dissociate from their cargo and are returned to the cytosol.

Nuclear import receptors do not always bind to nuclear proteins directly. Additional adaptor proteins are sometimes used that bridge between the import receptors and the nuclear localization signals on the proteins to be transported. Surprisingly, the adaptor proteins are structurally related to nuclear import receptors, suggesting a common evolutionary origin. The combined use of import receptors and adaptors allows a cell to recognize the broad repertoire of nuclear localization signals that are displayed on nuclear proteins.

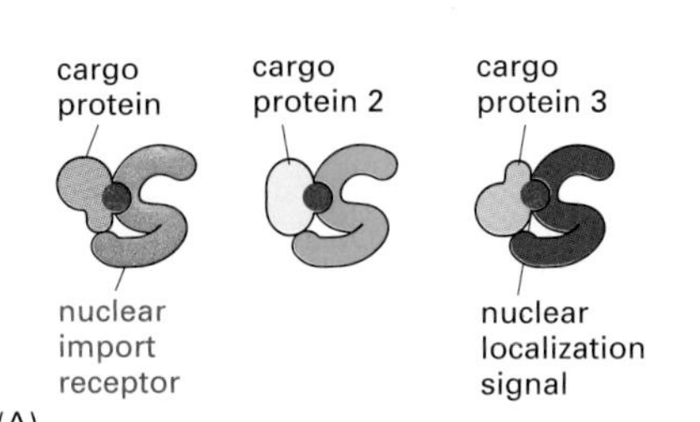

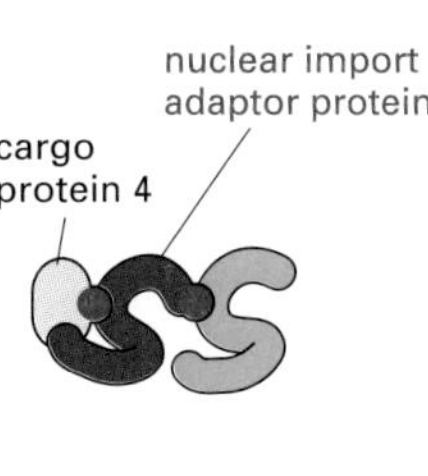

Figure 12–14 Nuclear import receptors. (A) Many nuclear import receptors bind both to nucleoporins and to a nuclear localization signal on the cargo proteins they transport. Cargo proteins 1, 2, and 3 in this example contain different nuclear localization signals, which causes each to bind to a different nuclear import receptor. (B) Cargo protein 4 shown here requires an adaptor protein to bind to its nuclear import receptor. The adaptors are structurally related to nuclear import receptors and recognize nuclear localization signals on cargo proteins. They also contain a nuclear localization signal that binds them to an import receptor.

Nuclear Export Works Like Nuclear Import, But in Reverse

The nuclear export of large molecules, such as new ribosomal subunits and RNA molecules, also occurs through nuclear pore complexes and depends on a selective transport system. The transport system relies on **nuclear export signals** on the macromolecules to be exported, as well as on complementary **nuclear export receptors**. These receptors bind both the export signal and nucleoporins to guide their cargo through the pore complex to the cytosol.

Nuclear export receptors are structurally related to nuclear import receptors, and they are encoded by the same gene family of **nuclear transport receptors**, or *karyopherins*. In yeast, there are 14 genes encoding members of this family; in animal cells the number is significantly larger. From their amino acid sequence alone, it is often not possible to distinguish whether a particular family member works as a nuclear import or nuclear export receptor. It comes as no surprise, therefore, that the import and export transport systems work in similar ways but in opposite directions: the import receptors bind their cargo molecules in the cytosol, release them in the nucleus, and are then exported to the cytosol for reuse, while the export receptors function in reverse.

If gold spheres similar to those used in the experiments shown in Figure 12–13 are coated with small RNA molecules (tRNA or ribosomal 5S RNA) and injected into the nucleus of a cultured cell, they are rapidly transported through the nuclear pore complexes into the cytosol. Using two sizes of gold particles, one coated with RNA and injected into the nucleus and the other coated with nuclear localization signals and injected into the cytosol, it can be shown that a single pore complex conducts traffic in both directions. How a pore complex coordinates the bidirectional flow of macromolecules to avoid congestion and head-on collisions is not known.

The Ran GTPase Drives Directional Transport Through Nuclear Pore Complexes

The import of nuclear proteins through the pore complex concentrates specific proteins in the nucleus, thereby increasing order in the cell, which must consume energy (discussed in Chapter 2). The energy is thought to be provided by the hydrolysis of GTP by the monomeric GTPase **Ran**. Ran is found in both the cytosol and the nucleus, and it is required for both the nuclear import and export systems.

Like other GTPases, Ran is a molecular switch that can exist in two conformational states, depending on whether GDP or GTP is bound (discussed in Chapter 3). Conversion between the two states is triggered by two Ran-specific regulatory proteins: a cytosolic *GTPase-activating protein (GAP)* that triggers GTP hydrolysis and thus converts Ran-GTP to Ran-GDP, and a nuclear *guanine exchange factor (GEF)* that promotes the exchange of GDP for GTP and thus converts Ran-GDP to Ran-GTP. Because *Ran-GAP* is located in the cytosol and *Ran-GEF* is located in the nucleus, the cytosol primarily contains Ran-GDP, and the nucleus primarily contains Ran-GTP (Figure 12–15).

This gradient of the two conformational forms of Ran drives nuclear transport in the appropriate direction (Figure 12–16). Docking of nuclear import receptors to FG-repeats on the cytosolic side of the nuclear pore complex, for

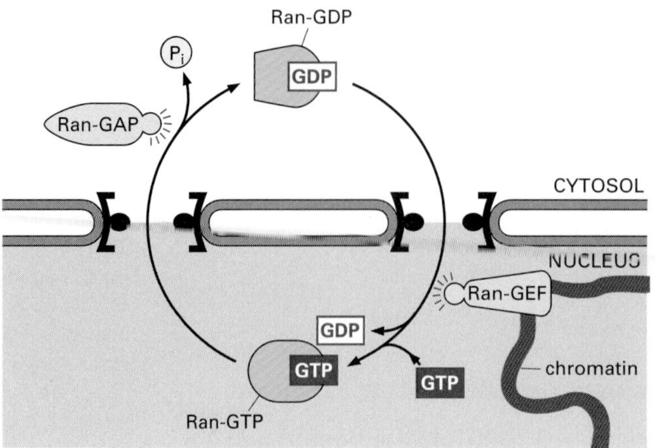

Figure 12–15 The compartmentalization of Ran-GDP and Ran-GTP. Localization of Ran-GDP to the cytosol and Ran-GTP to the nucleus results from the localization of two Ran regulatory proteins: Ran GTPase-activating protein (Ran-GAP) is located in the cytosol and Ran guanine nucleotide exchange factor (Ran-GEF) is bound to chromatin and is hence exclusively found in the nucleus. Another protein, called Ran Binding Protein (omitted here for clarity), collaborates with Ran-GAP in activating GTP hydrolysis.

example, occurs only when these receptors are loaded with an appropriate cargo. The import receptors with their bound cargo then move along tracks lined by FG-repeat sequences until they reach the nuclear side of the pore complex, where Ran-GTP binding causes the import receptors to release their cargo (Figure 12–17). By favoring cargo-dependent loading of import receptors onto the FG-repeat track in the cytosol and Ran-GTP-dependent cargo release in the nucleus, the nuclear localization of Ran-GTP imposes directionality.

Having discharged its cargo in the nucleus, the empty import receptor with Ran-GTP bound is transported back through the pore complex to the cytosol. There, two cytosolic proteins, *Ran Binding Protein* and Ran-GAP collaborate to convert Ran-GTP to Ran-GDP. The Ran Binding Protein first displaces Ran-GTP from the import receptor, which allows Ran-GAP to trigger Ran to hydrolyze its bound GTP. The Ran-GDP then dissociates from the Ran Binding Protein and is reimported into the nucleus, thereby completing the cycle.

Nuclear export occurs by a similar mechanism, except that Ran-GTP in the nucleus promotes cargo binding to the export receptor and the binding of the loaded receptor to the nuclear side of the pore complex. Once in the cytosol, Ran encounters Ran-GAP and Ran Binding Protein and hydrolyses its bound GTP. The export receptor then releases both its cargo and Ran-GDP in the cytosol and dissociates from the pore complex, and free export receptors are returned to the nucleus to complete the cycle (see Figure 12–16).

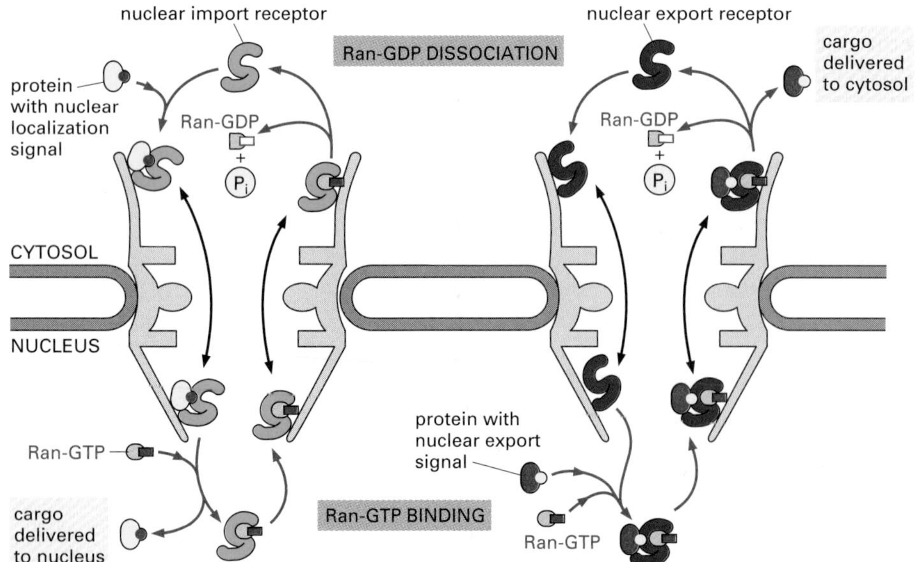

Figure 12–16 A model for how GTP hydrolysis by Ran provides directionality for nuclear transport. Movement through the pore complex of loaded nuclear transport receptors may occur by guided diffusion along the FG-repeats displayed by nucleoporins. The differential localization of Ran-GTP in the nucleus and Ran-GDP in the cytosol provides directionality *(red arrows)* to both nuclear import *(left)* and nuclear export *(right)*. The hydrolysis of GTP to produce Ran-GDP is mediated by Ran-GAP and Ran Binding Protein on the cytosolic side of the nuclear pore complex.

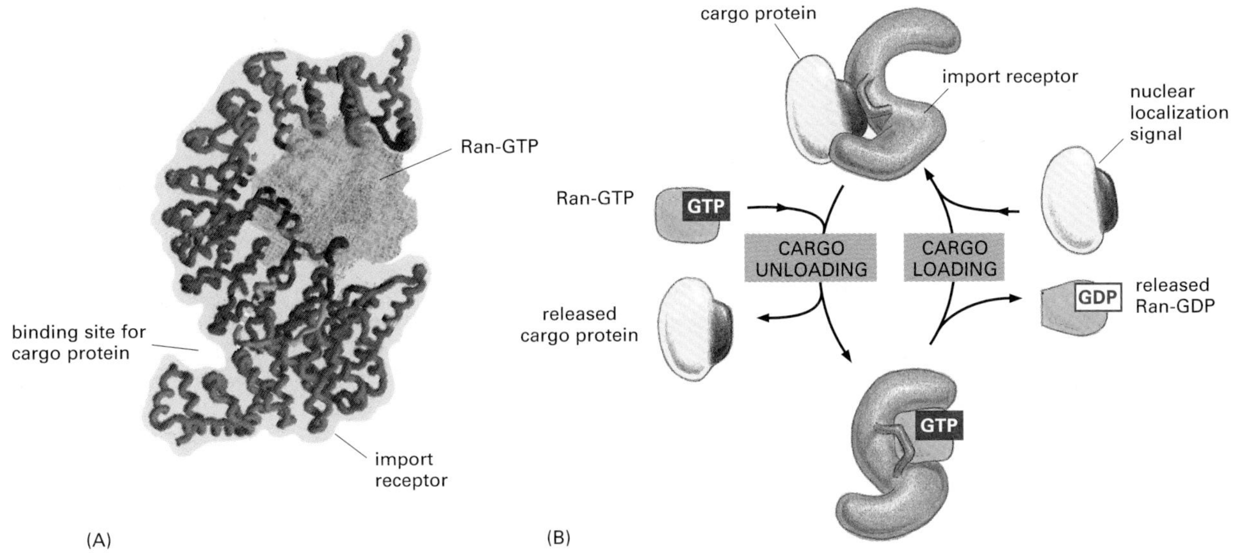

Figure 12–17 A model for how the binding of Ran-GTP might cause nuclear import receptors to release their cargo. (A) Nuclear transport receptors are composed of repeated α-helical motifs that stack into either large arches or snail-shaped coils, depending on the particular receptor or adaptor. Cargo proteins and Ran-GTP bind to different regions at the inside faces of the arches. In a co-crystal of a nuclear import receptor bound to Ran-GTP, a conserved loop *(red)* of the receptor becomes covered by bound Ran-GTP, which, in the Ran-free state of the receptor, is thought to be important for signal sequence binding. (B) The cycle of loading in the cytosol and unloading in the nucleus of a nuclear import receptor. (A, adapted from Y. M. Chook and G. Blobel, *Nature* 399:230–237, 1999.)

Transport Between the Nucleus and Cytosol Can Be Regulated by Controlling Access to the Transport Machinery

Some proteins, such as those that bind newly made mRNAs in the nucleus, contain both nuclear localization and nuclear export signals. These proteins continually shuttle between the nucleus and the cytosol. The steady-state localization of such *shuttling proteins* is determined by the relative rates of their import and export. If the rate of import exceeds the rate of export, a protein will be located primarily in the nucleus. Conversely, if the rate of export exceeds the rate of import, a protein will be located primarily in the cytosol. Thus, changing the rate of import, export, or both, can change the location of a protein.

Some shuttling proteins move continuously in and out of the nucleus. In other cases, however, the transport is stringently controlled. As discussed in Chapter 7, the activity of some gene regulatory proteins is controlled by keeping them out of the nuclear compartment until they are needed there (Figure 12–18). In many cases, this control depends on the regulation of nuclear localization and export signals; these can be turned on or off, often by phosphorylation of adjacent amino acids (Figure 12–19).

Other gene regulatory proteins are bound to inhibitory cytosolic proteins that either anchor them in the cytosol (through interactions with the cytoskeleton or with specific organelles), or mask their nuclear localization signals so that they are unable to interact with nuclear import receptors. When the cell receives an appropriate stimulus, the gene regulatory protein is released from its cytosolic anchor or mask and is transported into the nucleus. One important example is the latent gene regulatory protein that controls the expression of proteins involved in cholesterol metabolism. The protein is made and stored in an inactive form as a transmembrane protein in the ER. When deprived of cholesterol, the cell activates specific proteases that cleave the protein, releasing its cytosolic domain. This domain is then imported into the nucleus, where it activates the transcription of genes required for cholesterol import and synthesis.

Cells control the export of RNA from the nucleus in a similar way. Messenger RNAs become bound to proteins that are loaded onto the RNA as transcription and splicing proceed. These proteins contain nuclear export signals that are

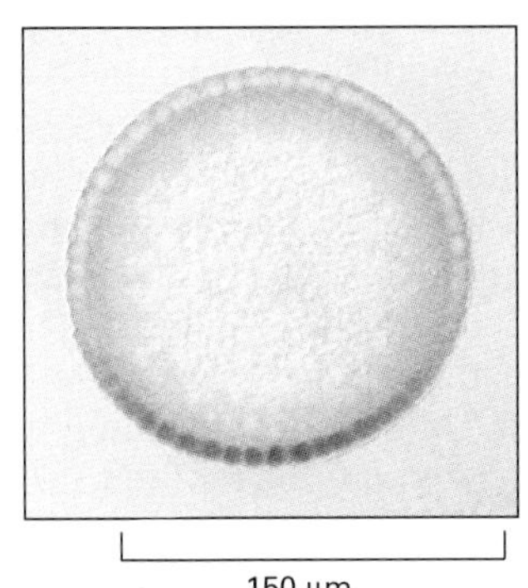

Figure 12–18 The control of fly embryo development by nuclear transport. The gene regulatory protein *dorsal* is expressed uniformly throughout this early *Drosophila* embryo, which is shown in cross section. It is active only in cells at the ventral side *(bottom)* of the embryo, where it is found in nuclei. The dorsal protein is visualized by staining with an enzyme-coupled antibody that yields a colored product. (Courtesy of Siegfried Roth.)

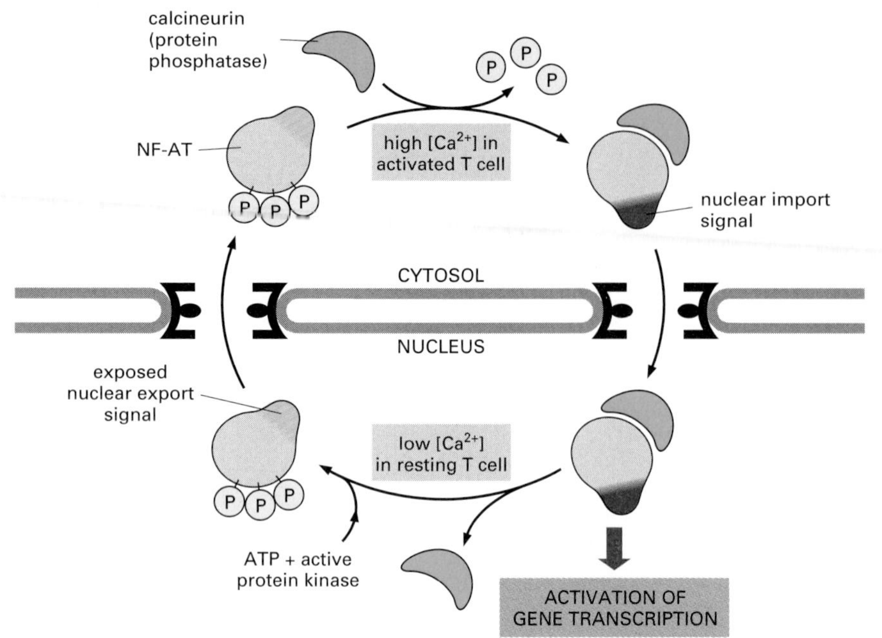

Figure 12–19 The control of nuclear import during T-cell activation. The nuclear factor of activated T cells (NF-AT) is a gene regulatory protein that, in the resting T cell, is found in the cytosol in a phosphorylated state. When T cells are activated, the intracellular Ca^{2+} concentration increases. In high Ca^{2+}, the protein phosphatase, calcineurin, binds to NF-AT. Binding of calcineurin dephosphorylates NF-AT, exposing one or more nuclear import signals, and it may also block a nuclear export signal. The complex of NF-AT bound to calcineurin is then imported into the nucleus, where NF-AT activates the transcription of numerous cytokine and cell-surface protein genes that are required for a proper immune response.

During the shut-off of the response, decreased Ca^{2+} levels lead to the release of calcineurin. Rephosphorylation of NF-AT inactivates the nuclear import signal, and it re-exposes the nuclear export signal of NF-AT causing NF-AT to relocate to the cytosol. Some of the most potent immunosuppressive drugs, such as cyclosporin A and FK506, inhibit the ability of calcineurin to dephosphorylate NF-AT; these drugs thereby block the nuclear accumulation of NF-AT.

recognized by export receptors that guide the RNA out of the nucleus through nuclear pore complexes. Upon entry into the cytosol, the proteins coating the RNA are stripped off and rapidly returned to the nucleus. Other RNAs, such as snRNAs and tRNAs, are exported by different sets of nuclear export receptors.

Incompletely processed pre-mRNAs are actively retained in the nucleus, anchored to the nuclear transcription and splicing machinery, which releases an RNA molecule only after its processing is completed. Genetic studies in yeast show that a mutant pre-mRNA that cannot properly engage with the splicing machinery is improperly exported as an unspliced molecule.

The Nuclear Envelope Is Disassembled During Mitosis

The **nuclear lamina** is a meshwork of interconnected protein subunits called **nuclear lamins**. The lamins are a special class of intermediate filament proteins (discussed in Chapter 16) that polymerize into a two-dimensional lattice (Figure 12–20). The nuclear lamina gives shape and stability to the nuclear envelope, to which it is anchored by attachment to both the nuclear pore complexes and integral membrane proteins of the inner nuclear membrane. The lamina also interacts directly with chromatin, which itself interacts with the integral membrane proteins of the inner nuclear membrane. Together with the lamina, these membrane proteins provide structural links between the DNA and the nuclear envelope.

When a nucleus disassembles during mitosis, the nuclear lamina depolymerizes. The disassembly is at least partly a consequence of direct phosphorylation of the nuclear lamins by the cyclin-dependent kinase activated at the onset of mitosis (discussed in Chapter 17). At the same time, proteins of the inner nuclear membrane are phosphorylated, and the nuclear pore complexes disassemble and disperse in the cytosol. Nuclear envelope membrane proteins—no longer tethered to the pore complexes, lamina, or chromatin—diffuse throughout the ER membrane. Together, these events break down the barriers that normally separate the nucleus and cytosol, and these nuclear proteins that are not bound to membranes or chromosomes intermix completely with the cytosol of the dividing cell (Figure 12–21).

Later in mitosis (in late anaphase), the nuclear envelope reassembles on the surface of the chromosomes, as inner nuclear membrane proteins and dephosphorylated lamins rebind to chromatin. ER membranes wrap around groups of chromosomes and continue fusing until a sealed nuclear envelope is reformed. During this process, the nuclear pore complexes also reassemble and start

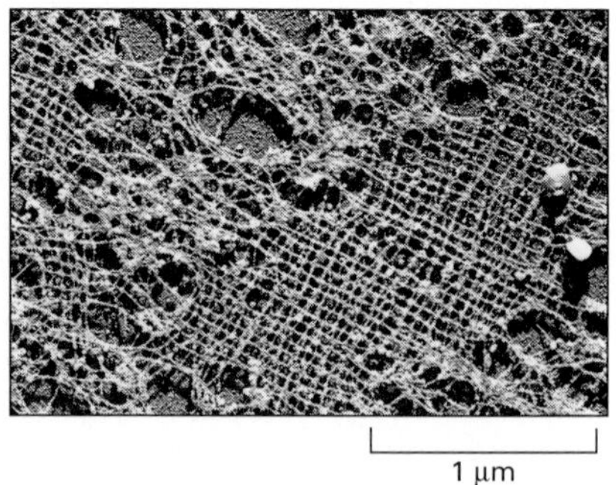

Figure 12–20 The nuclear lamina. An electron micrograph of a portion of the nuclear lamina in a *Xenopus* oocyte prepared by freeze-drying and metal shadowing. The lamina is formed by a regular lattice of specialized intermediate filaments. (Courtesy of Ueli Aebi.)

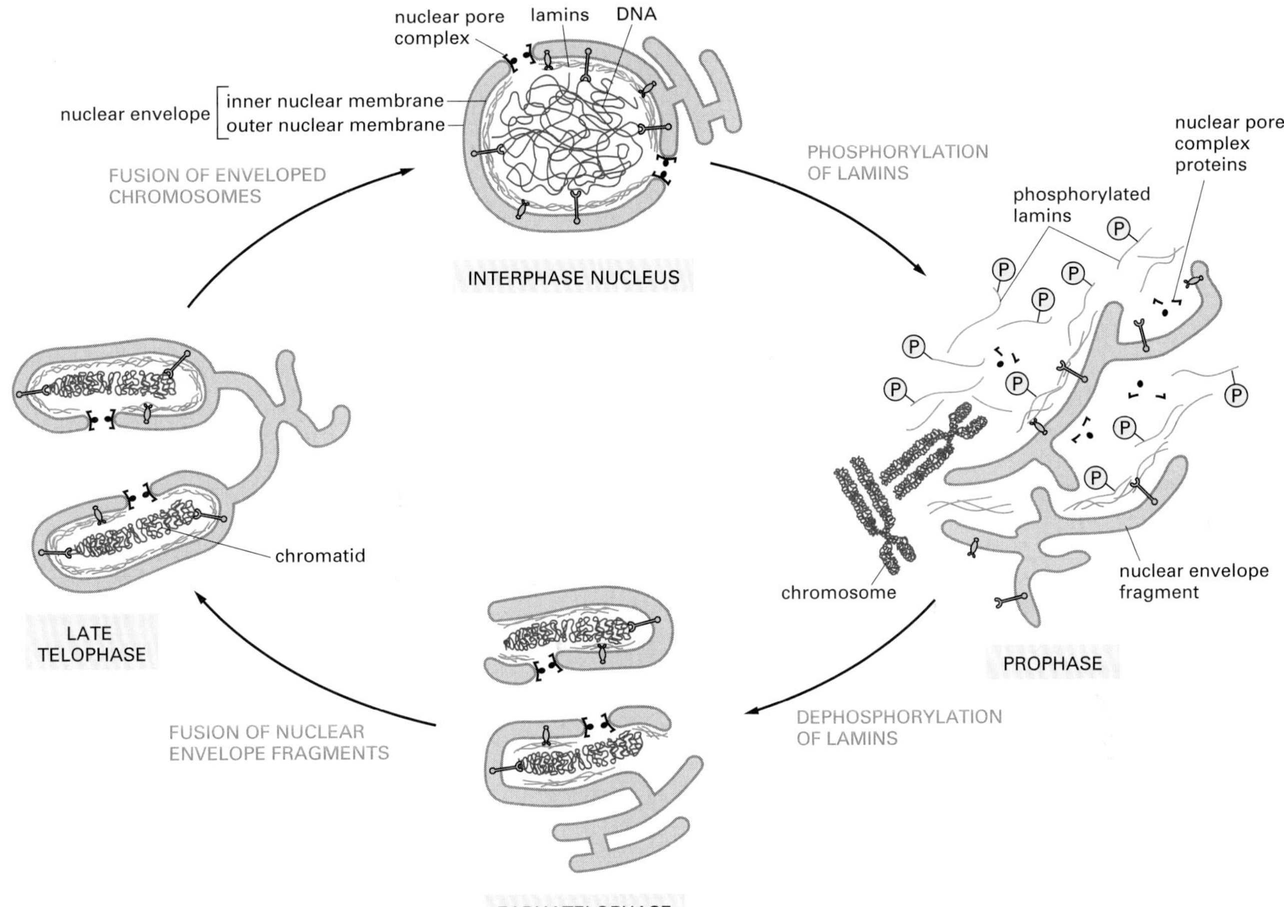

Figure 12–21 The breakdown and re-formation of the nuclear envelope during mitosis. The phosphorylation of the lamins is thought to trigger the disassembly of the nuclear lamina, which in turn causes the nuclear envelope to break up. Dephosphorylation of the lamins is thought to help to reverse the process.

actively reimporting proteins that contain nuclear localization signals. Because the nuclear envelope is initially closely applied to the surface of the chromosomes, the newly formed nucleus excludes all proteins except those initially bound to the mitotic chromosomes and those that are selectively imported through nuclear pore complexes. In this way, all other large proteins are kept out of the newly assembled nucleus.

Nuclear localization signals are not cleaved off after transport into the nucleus. This is presumably because nuclear proteins need to be imported repeatedly, once after every cell division. In contrast, once a protein molecule has been imported into any of the other membrane-enclosed organelles, it is passed on from generation to generation within that compartment and need never be translocated again; the signal sequence on these molecules is often removed after protein translocation.

Summary

The nuclear envelope consists of an inner and an outer nuclear membrane. The outer membrane is continuous with the ER membrane, and the space between it and the inner membrane is continuous with the ER lumen. RNA molecules, which are made in the nucleus, and ribosomal subunits, which are assembled there, are exported to the cytosol, while all the proteins that function in the nucleus are synthesized in the cytosol and are then imported. The extensive traffic of materials between the nucleus and cytosol occurs through nuclear pore complexes, which provide a direct passageway across the nuclear envelope.

Proteins containing nuclear localization signals are actively transported inward through the nuclear pore complexes, while RNA molecules and newly made ribosomal subunits contain nuclear export signals that direct their active transport outward through the pore complexes. Some proteins, including nuclear import and export receptors, continually shuttle between the cytosol and nucleus. The GTPase

Ran, provides directionality for nuclear transport. The transport of nuclear proteins and RNA molecules through the pore complexes can be regulated by denying these molecules access to the transport machinery. Because nuclear localization signals are not removed, nuclear proteins can be imported repeatedly, as is required each time that the nucleus reassembles after mitosis.

THE TRANSPORT OF PROTEINS INTO MITOCHONDRIA AND CHLOROPLASTS

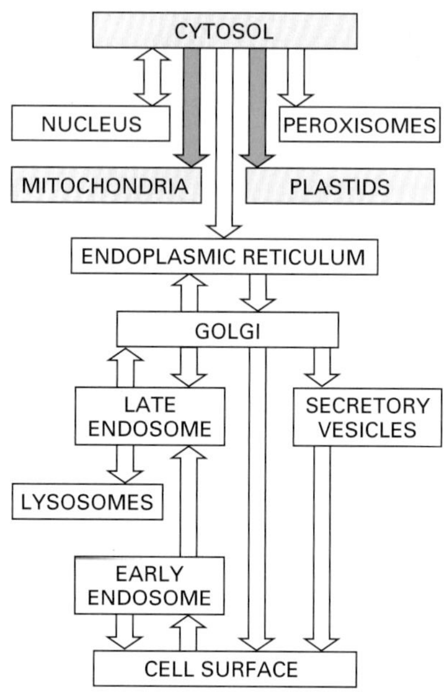

As discussed in Chapter 14, mitochondria and chloroplasts are double-membrane-enclosed organelles. They specialize in the synthesis of ATP, using energy derived from electron transport and oxidative phosphorylation in mitochondria and from photosynthesis in chloroplasts. Although both organelles contain their own DNA, ribosomes, and other components required for protein synthesis, most of their proteins are encoded in the cell nucleus and imported from the cytosol. Moreover, each imported protein must reach the particular organelle subcompartment in which it functions.

There are two subcompartments in mitochondria: the internal **matrix space** and the **intermembrane space**. These compartments are formed by the two concentric mitochondrial membranes: the **inner membrane**, which forms extensive invaginations, the *cristae*, and encloses the matrix space, and the **outer membrane**, which is in contact with the cytosol (Figure 12–22A). Chloroplasts have the same two subcompartments plus an additional subcompartment, the *thylakoid space,* which is surrounded by the *thylakoid membrane* (Figure 12–22B). Each of the subcompartments in mitochondria and chloroplasts contains a distinct set of proteins.

New mitochondria and chloroplasts are produced by the growth of preexisting organelles followed by fission (discussed in Chapter 14). Their growth depends mainly on the import of proteins from the cytosol. This requires that proteins be translocated across a number of membranes in succession and end up in the appropriate place. How this occurs is the subject of this section.

Translocation into the Mitochondrial Matrix Depends on a Signal Sequence and Protein Translocators

Proteins imported into the matrix of **mitochondria** are usually taken up from the cytosol within seconds or minutes of their release from ribosomes. Thus, in contrast to the protein translocation into the ER described later, mitochondrial proteins are first fully synthesized as precursor proteins in the cytosol and then translocated into mitochondria by a *posttranslational* mechanism. Most of the **mitochondrial precursor proteins** have a signal sequence at their N terminus that is rapidly removed after import by a protease (the *signal peptidase)* in the mitochondrial matrix. The signal sequences are both necessary and sufficient

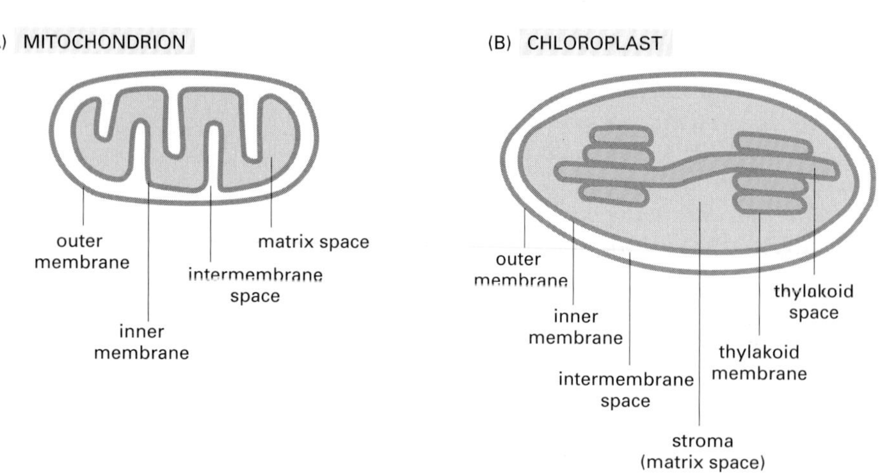

Figure 12–22 The subcompartments of mitochondria and chloroplasts. In contrast to the cristae of mitochondria (A), the thylakoids of chloroplasts (B) are not connected to the inner membrane and therefore form a compartment with a separate internal space (see Figure 12–3).

Figure 12–23 A signal sequence for mitochondrial protein import. Cytochrome oxidase is a large multiprotein complex located in the inner mitochondrial membrane, where it functions as the terminal enzyme in the electron-transport chain (discussed in Chapter 14). (A) The first 18 amino acids of the precursor to subunit IV of this enzyme serve as a signal sequence for import of the subunit into the mitochondrion. (B) When the signal sequence is folded as an α helix, the positively charged residues *(red)* are seen to be clustered on one face of the helix, while the nonpolar residues *(yellow)* are clustered primarily on the opposite face. Mitochondrial matrix-targeting sequences always have the potential to form such an amphipathic α helix, which is recognized by specific receptor proteins on the mitochondrial surface.

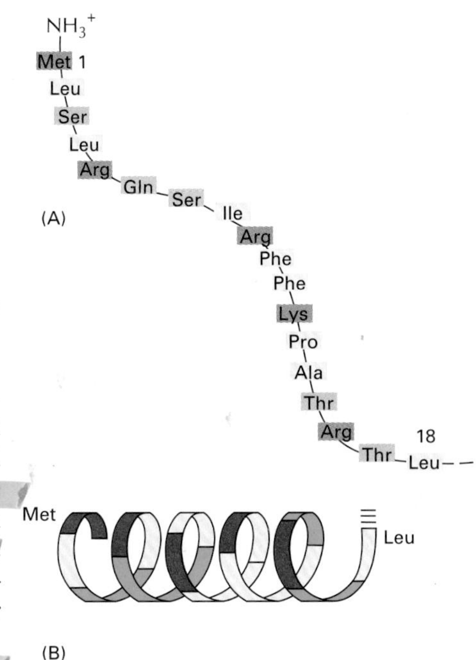

for import of the proteins that contain them: through the use of genetic engineering techniques, these signals can be linked to any cytosolic protein to direct the protein into the mitochondrial matrix. Sequence comparisons and physical studies of different matrix signal sequences suggest that their common feature is the propensity to fold into an amphipathic α helix, in which positively charged residues are clustered on one side of the helix, while uncharged hydrophobic residues are clustered on the opposite side (Figure 12–23). This configuration—rather than a precise amino acid sequence—is recognized by specific receptor proteins that initiate protein translocation.

Protein translocation across mitochondrial membranes is mediated by multisubunit protein complexes that function as **protein translocators:** the **TOM complex** functions across the outer membrane, and two **TIM complexes**, the TIM23 and TIM22 complexes, function across the inner membrane (Figure 12–24). TOM and TIM stand for translocase of the outer and inner mitochondrial membranes, respectively. These complexes contain some components that act as receptors for mitochondrial precursor proteins and other components that form the translocation channel. The TOM complex is required for the import of all nucleus-encoded mitochondrial proteins. It initially transports their signal sequences into the intermembrane space and helps to insert transmembrane proteins into the outer membrane. The TIM23 complex then transports some of these proteins into the matrix space, while helping to insert transmembrane proteins into the inner membrane. The TIM22 complex mediates the insertion of a subclass of inner membrane proteins, including the carrier protein that transports ADP, ATP, and phosphate. A third protein translocator in the inner mitochondrial membrane, the **OXA complex**, mediates the insertion of inner membrane proteins that are synthesized within the mitochondria. It also helps to insert some proteins that are initially transported into the matrix by the TOM and TIM complexes.

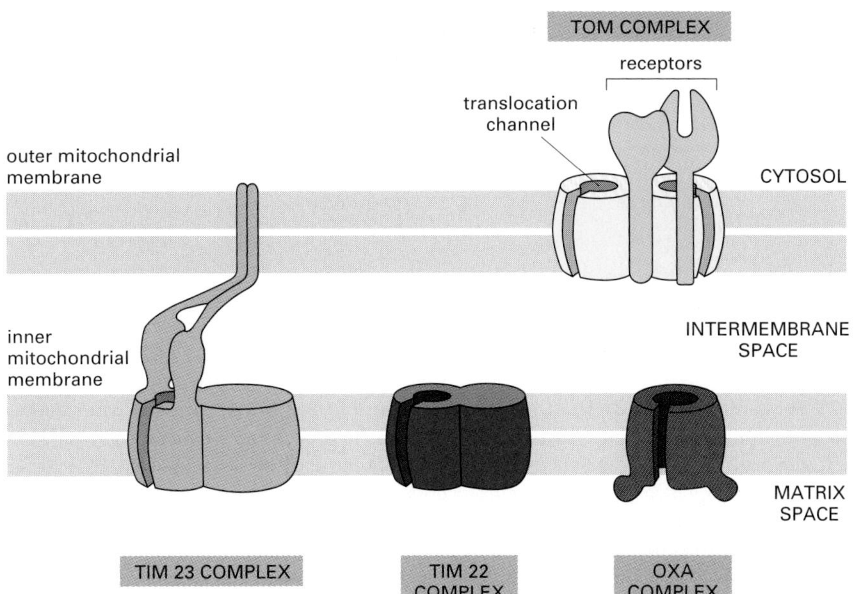

Figure 12–24 Three protein translocators in the mitochondrial membranes. The TOM and TIM complexes and the OXA complex are multimeric membrane protein assemblies that catalyze protein transport across mitochondrial membranes. The protein components of the TIM22 and TIM23 complexes that line the import channel are structurally related, suggesting a common evolutionary origin of both TIM complexes. As indicated, one of the core components of the TIM23 complex contains a hydrophobic α-helical extension that is inserted into the outer mitochondrial membrane; the complex is therefore unusual in that it simultaneously spans two membranes.

Mitochondrial Precursor Proteins Are Imported as Unfolded Polypeptide Chains

Almost everything we know about the molecular mechanism of protein import into mitochondria has been learned from analyses of cell-free, reconstituted transport systems. Mitochondria are first purified by differential centrifugation of homogenized cells and are then incubated with radiolabeled mitochondrial precursor proteins, which are generally taken up rapidly and efficiently. By changing the incubation conditions, it is possible to establish the biochemical requirements for the transport.

Mitochondrial precursor proteins do not fold into their native structures after they are synthesized; instead, they remain unfolded through interactions with other proteins in the cytosol. Some of these interacting proteins are general *chaperone proteins* belonging to the *hsp70 family* (discussed in Chapter 6), whereas others are dedicated to mitochondrial precursor proteins and bind directly to their signal sequences. All these interacting proteins help to prevent the precursor proteins from aggregating or folding up spontaneously before they engage with the TOM complex in the outer mitochondrial membrane. As a first step in the import process, the mitochondrial precursor proteins bind to import receptor proteins of the TOM complex, which recognize the mitochondrial signal sequences. The interacting proteins are then stripped off, and the unfolded polypeptide chain is fed—signal sequence first—into the translocation channel.

Mitochondrial Precursor Proteins Are Imported into the Matrix at Contact Sites That Join the Inner and Outer Membranes

In principle, a protein could reach the mitochondrial matrix by crossing the two membranes one at a time, or it could pass through both at once. To distinguish between these possibilities, a cell-free mitochondrial import system was cooled to a low temperature, arresting the proteins at an intermediate step in the translocation process. The proteins that accumulated at this step already had their N-terminal signal sequence removed by the matrix signal peptidase, indicating that their N terminus must be in the matrix space. Yet, the bulk of the protein could still be attacked from outside the mitochondria by externally added proteolytic enzymes (Figure 12–25). This result demonstrates that the precursor proteins can pass through both mitochondrial membranes at once to enter the matrix (Figure 12–26). It is thought that the TOM complex first transports the

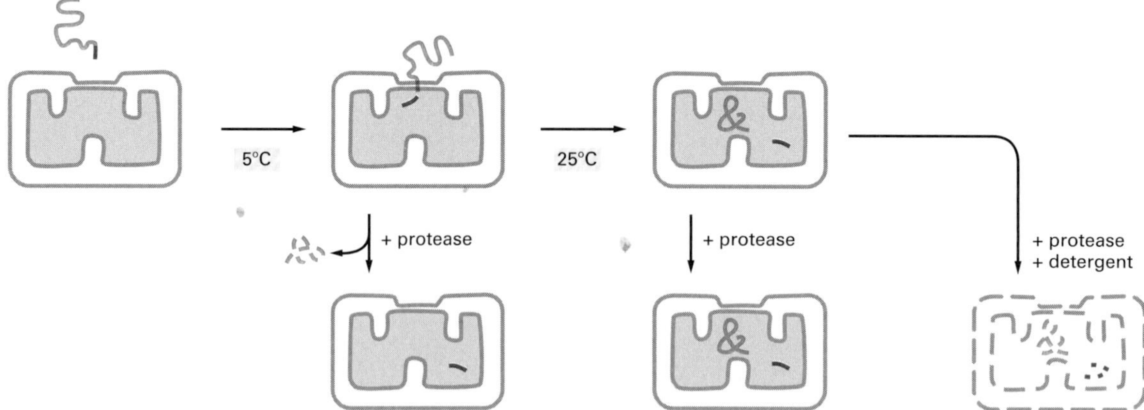

Figure 12–25 Proteins transiently spanning the inner and outer mitochondrial membranes during their translocation into the matrix. When isolated mitochondria are incubated with a precursor protein at 5°C, the precursor is only partly translocated. The N-terminal signal sequence *(red)* is cleaved off in the matrix; but most of the polypeptide chain remains outside the mitochondrion, where it is accessible to proteolytic enzymes. Upon warming to 25°C, the translocation is completed. Once inside the mitochondrion, the polypeptide chain is protected from externally added proteolytic enzymes. As a control, when detergents are added to disrupt the mitochondrial membranes, the imported proteins can be readily digested by the same protease treatment.

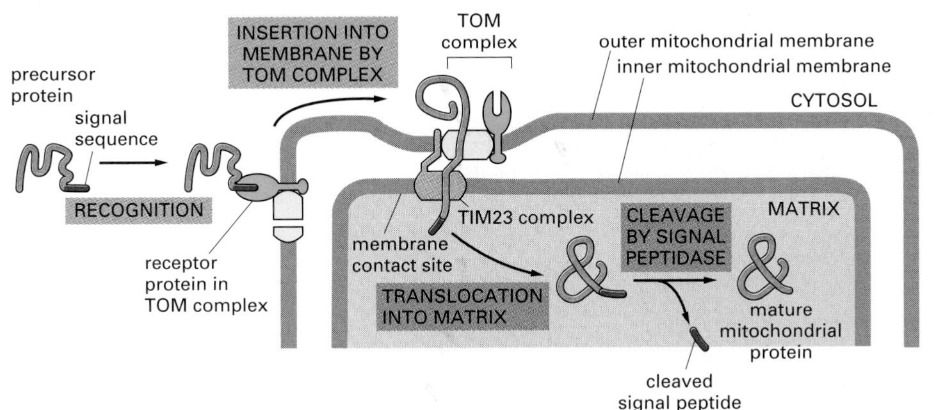

Figure 12–26 Protein import by mitochondria. The N-terminal signal sequence of the precursor protein is recognized by receptors of the TOM complex. The protein is thought to be translocated across both mitochondrial membranes at or near special contact sites. The signal sequence is cleaved off by a signal peptidase in the matrix to form the mature protein. The free signal sequence is then rapidly degraded (not shown).

mitochondrial targeting signal across the outer membrane. Once it reaches in the intermembrane space, the targeting signal binds to a TIM complex, opening the channel in the complex through which the polypeptide chain either enters the matrix or inserts into the inner membrane. Electron microscopists have noted numerous *contact sites* at which the inner and outer mitochondrial membranes are closely apposed, and it seems likely that translocation occurs at or near these sites.

Although the functions of the TOM and TIM complexes are usually coupled to allow protein transport across both membranes at the same time, both protein types of translocator can work independently. The TOM complex in isolated outer membranes, for example, can translocate the signal sequence of precursor proteins across the membrane. Similarly, mitochondria with experimentally disrupted outer membranes, and therefore with the TIM23 complex exposed on their surface, efficiently import precursor proteins into the matrix space. Despite the independent functional roles of the TOM and TIM translocators, the two mitochondrial membranes at contact sites may be permanently held together by the TIM23 complex, which spans both membranes (see Figure 12–24).

ATP Hydrolysis and a H^+ Gradient are Used to Drive Protein Import into Mitochondria

Directional transport requires energy. In most biological systems, energy is supplied by ATP hydrolysis. Mitochondrial protein import is fueled by ATP hydrolysis at two discrete sites, one outside the mitochondria and one in the matrix (Figure 12–27). In addition, another energy source is required: an electrochemical H^+ gradient across the inner mitochondrial membrane.

The first requirement for energy occurs at the initial stage of the translocation process, when the unfolded precursor protein, associated with chaperone proteins, interacts with the mitochondrial import receptors. As discussed in Chapter 6, the release of newly synthesized polypeptides from the hsp70 family of chaperone proteins requires ATP hydrolysis. Experimentally, the requirement

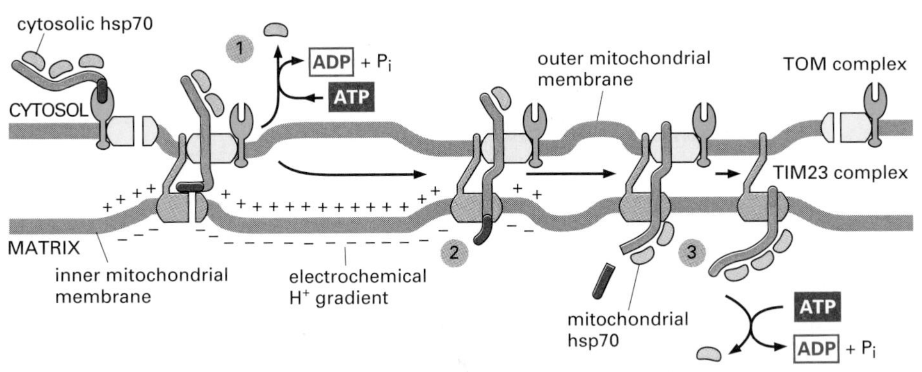

Figure 12–27 The role of energy in protein import into the mitochondrial matrix. (1) Bound *cytosolic hsp70* is released from the protein in a step that depends on ATP hydrolysis. After initial insertion of the signal sequence and of adjacent portions of the polypeptide chain into the TOM complex, the signal sequence interacts with a TIM complex. (2) The signal sequence is then translocated into the matrix in a process that requires an electrochemical H^+ gradient across the inner membrane, positioning the unfolded polypeptide chain so that it transiently spans both membranes. (3) *Mitochondrial hsp70* binds to regions of the polypeptide chain as they become exposed in the matrix, thereby "pulling" the protein into the matrix. ATP hydrolysis then removes the mitochondrial hsp70, allowing the imported protein to fold.

for hsp70 and ATP in the cytosol can be bypassed if the precursor protein is artificially unfolded prior to adding it to purified mitochondria.

Once the signal sequence has passed through the TOM complex and has become bound to either TIM complex, further translocation through the TIM requires an electrochemical H^+ gradient across the inner membrane. The electrochemical gradient is maintained by the pumping of H^+ from the matrix to the intermembrane space, driven by electron transport processes in the inner membrane. By contrast, the outer mitochondrial membrane, like that of Gram-negative bacteria (see Figure 11–17), contains a pore-forming protein called porin and is thus freely permeable to inorganic ions and metabolites (but not to most proteins), so that ion gradients cannot be maintained across it. The energy in the electrochemical H^+ gradient across the inner membrane is not only used to help drive most of the cell's ATP synthesis; it is also used to drive the translocation of the targeting signals through the TIM complexes. The precise mechanism by which this occurs is not known, but it is possible that the electrical components of the gradient (the membrane potential, see Figure 14–13) helps to drive the positively charged signal sequence into the matrix by electrophoresis.

Hsp70 chaperone proteins in the matrix space also have a role in the translocation process, and they are the third point in the import process at which ATP is consumed, as we discuss next.

Repeated Cycles of ATP Hydrolysis by Mitochondrial Hsp70 Complete the Import Process

We know that **mitochondrial hsp70** is crucial to the import process, because mitochondria containing mutant forms of the protein fail to import precursor proteins. Like its cytosolic cousin, mitochondrial hsp70 has a high affinity for unfolded polypeptide chains and it binds tightly to an imported protein as soon as it emerges from the translocator in the matrix. The hsp70 then releases the protein in an ATP-dependent step. This energy-driven cycle of binding and subsequent release is thought to provide the final driving force needed to complete protein import after a protein has initially inserted into the TIM23 complex (see Figure 12–27).

Two models have been proposed to explain how ATP hydrolysis by mitochondrial hsp70 drives protein import. In both models, hsp70 proteins are closely associated with the TIM23 complex, which deposits them onto the translocating polypeptide chain as it emerges into the matrix. In the *thermal ratchet model* (Figure 12–28A), the emerging chain slides back and forth in the TIM23 translocation channel by thermal motion. Each time a sufficiently long portion of the chain is exposed in the matrix, an hsp70 molecule binds to it, preventing further backsliding and thereby making the movement directional. Thus, a hand-over-hand binding of multiple hsp70 proteins translocates the polypeptide chain into the matrix. In the *cross-bridge ratchet model* (Figure 12–28B), the hsp70 proteins that bind to the emerging polypeptide chain undergo a conformational change, driven by ATP hydrolysis, that actively pulls a segment of the polypeptide chain into the matrix. A new hsp70 molecule can then bind to the segment just pulled in and repeat the cycle. In both models, therefore, hsp70 functions as a ratchet that prevents backsliding of the emerging polypeptide chain.

After the initial interaction with mitochondrial hsp70, many imported proteins are passed on to another chaperone protein, *mitochondrial hsp60*. As discussed in Chapter 6, hsp60 provides a chamber for the unfolded polypeptide chain that facilitates its folding by binding and releasing it through cycles of ATP hydrolysis.

Protein Transport into the Inner Mitochondrial Membrane and the Intermembrane Space Requires Two Signal Sequences

Proteins that are integrated into the inner mitochondrial membrane or that operate in the intermembrane space are initially transported from the cytosol by

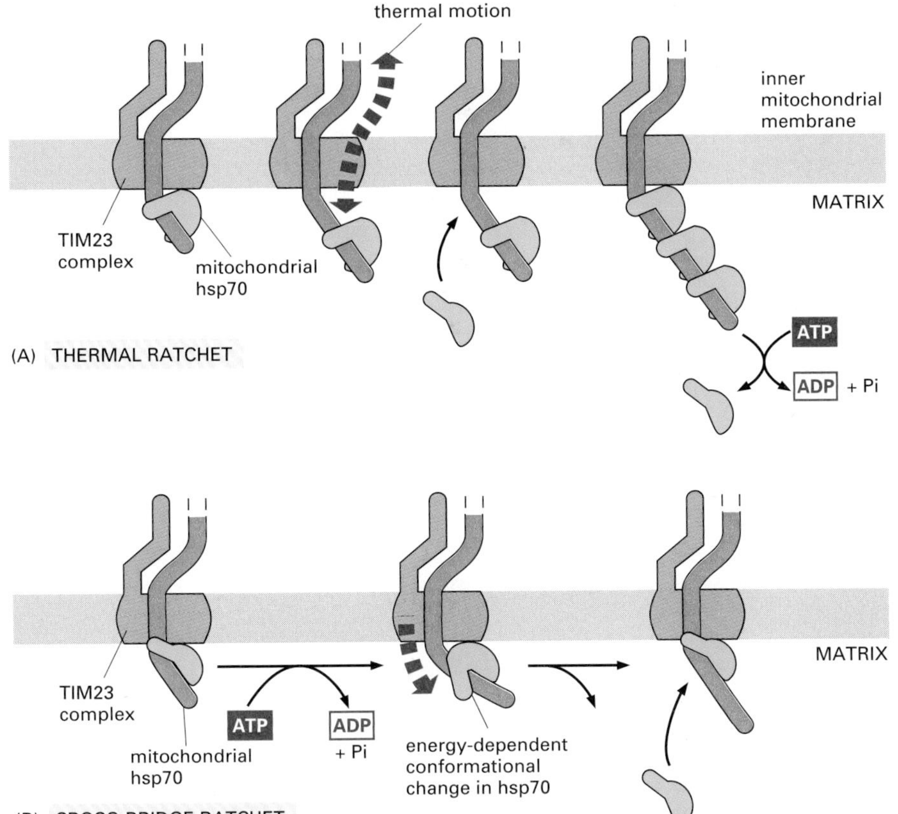

Figure 12–28 Two plausible models of how mitochondrial hsp70 could drive protein import. (A) In the thermal ratchet model, the translocating polypeptide chain slides back and forth, driven by thermal motion, and it is successively trapped in the matrix by hsp70 binding. (B) In the cross-bridge ratchet model, a conformational change in hsp70 actively pulls the chain into the matrix. In both models, hsp70 binds to the TIM23 complex, which loads the hsp70 onto the translocating polypeptide chain as it emerges from the complex into the matrix.

the same mechanism that transports proteins into the matrix. In some cases they are first transferred into the matrix (see Figure 12–26). A hydrophobic amino acid sequence, however, is strategically placed after the N-terminal signal sequence that guides import into the matrix. Once the N-terminal signal sequence has been removed by the matrix signal peptidase, the hydrophobic sequence functions as a new N-terminal signal sequence to translocate the protein from the matrix into or across the inner membrane, using the OXA complex as the translocator (see Figure 12–24). The OXA complex is also used to insert proteins encoded in the mitochondrion into the inner membrane (Figure 12–29A). Closely related translocators are found in the plasma membranes of bacteria and in the thylakoids of chloroplasts, where they are thought to help to insert membrane proteins by a similar mechanism.

An alternative route to the inner membrane avoids excursion into the matrix altogether (Figure 12–29B). In this case, the TIM23 translocator in the inner membrane binds to the hydrophobic sequence that follows the N-terminal signal sequence and initiates import, causing it to act as a *stop-transfer sequence* that prevents further translocation across the inner membrane. After the N-terminal signal sequence has been cleaved off, the remainder of the protein is pulled through the TOM complex in the outer membrane into the intermembrane space. Different proteins use one or the other of these two pathways to the inner membrane or intermembrane space.

Proteins destined for the intermembrane space are first inserted via their hydrophobic signal sequence into the inner membrane, and then cleaved by a signal peptidase in the intermembrane space to release the mature polypeptide chain as a soluble protein (Figure 12–29C). Many of these proteins attach as peripheral membrane proteins to the outer surface of the inner membrane, where they form subunits of protein complexes that also contain transmembrane proteins.

Mitochondria are the principal site of ATP synthesis in the cell, but they also contain many metabolic enzymes, such as those of the citric acid cycle. Thus, in addition to proteins, mitochondria must also transport small metabolites across their membranes. While the outer membrane contains porins that make the

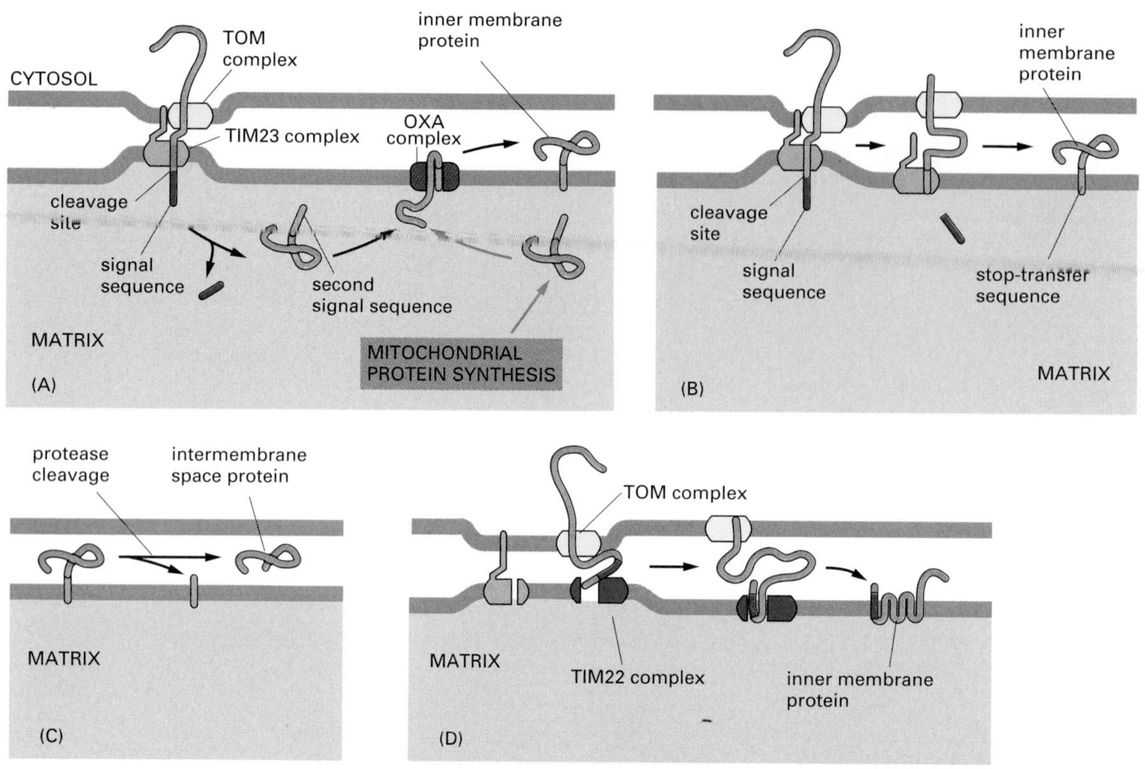

Figure 12–29 Protein import from the cytosol into the inner mitochondrial membrane or intermembrane space. (A) A pathway that requires two signal sequences and two translocation events is thought to be used to move some proteins from the cytosol to the inner membrane. The precursor protein is first imported into the matrix space (see Figure 12–26). Cleavage of the signal sequence *(red)* used for the initial translocation, however, unmasks an adjacent hydrophobic signal sequence *(orange)* at the new N terminus. This signal then directs the protein into the inner membrane, presumably by the same OXA-dependent pathway that is used to insert proteins encoded by the mitochondrial genome. (B) In some cases, the hydrophobic sequence that follows the matrix-targeting signal binds to the TIM23 translocator in the inner membrane and stops translocation. The remainder of the protein is then pulled into the intermembrane space through the TOM translocator in the outer membrane, and the hydrophobic sequence is released into the inner membrane. (C) Some soluble proteins of the intermembrane space may also use the pathways shown in (A) and (B) before they are released into the intermembrane space by a second signal peptidase, which has its active site in the intermembrane space and removes the hydrophobic signal sequence. (D) The import pathway used to insert metabolite carrier proteins into the inner mitochondrial membrane utilizes the TIM22 complex, which is specialized for the translocation of multipass membrane proteins.

membrane freely permeable to small molecules, the inner membrane does not. Instead, the transport of a vast number of small molecules across the inner membrane is mediated by a family of metabolite-specific carrier proteins. In yeast cells, these proteins comprise a family of 35 different proteins, of which the most abundant are those that transport ADP and ATP, or phosphate. These carrier proteins in the inner membrane are multipass transmembrane proteins, which do not have cleavable signal sequences at their N termini but, instead, contain internal signal sequences. These proteins cross the TOM complex in the outer membrane and are inserted into the inner membrane by the TIM22 complex (Figure 12–29D). Their integration into the inner membrane requires the electrochemical H^+ gradient, but not mitochondrial hsp70 or ATP. The energetically favorable partitioning of the hydrophobic transmembrane regions into the inner membrane is likely to help drive integration.

Two Signal Sequences Are Required to Direct Proteins to the Thylakoid Membrane in Chloroplasts

Protein transport into **chloroplasts** resembles transport into mitochondria in many respects. Both processes occur posttranslationally, use separate translocation complexes in each membrane, occur at contact sites, require energy, and use amphipathic N-terminal signal sequences that are removed after use. With the exception of some of the chaperone molecules, however, the protein components that form the translocation complexes are different. Moreover, whereas mitochondria harness the electrochemical H^+ gradient across their inner membrane to drive transport, chloroplasts, which have an electrochemical H^+ gradient across their thylakoid membrane but not their inner membrane, use the hydrolysis of GTP and ATP to power import across their double membrane. The functional similarities may thus result from convergent evolution, reflecting the common requirements for translocation across a double-membrane system.

Although the signal sequences for import into chloroplasts superficially resemble those for import into mitochondria, mitochondria and chloroplasts are both present in the same plant cells, so proteins must choose appropriately between them. In plants, for example, a bacterial enzyme can be directed specifically to mitochondria if it is experimentally joined to an N-terminal signal sequence of a mitochondrial protein; the same enzyme joined to an N-terminal signal sequence of a chloroplast protein ends up in chloroplasts. The different signal sequences can therefore be distinguished by the import receptors on each organelle.

Chloroplasts have an extra membrane-enclosed compartment, the **thylakoid**. Many chloroplast proteins, including the protein subunits of the photosynthetic system and of the ATP synthase (discussed in Chapter 14), are embedded in the thylakoid membrane. Like the precursors of some mitochondrial proteins, these proteins are transported from the cytosol to their final destination in two steps. First, they pass across the double membrane at contact sites into the matrix space of the chloroplast, called the **stroma**, and then they are translocated into the thylakoid membrane (or across this membrane into the thylakoid space) (Figure 12–30A). The precursors of these proteins have a hydrophobic thylakoid

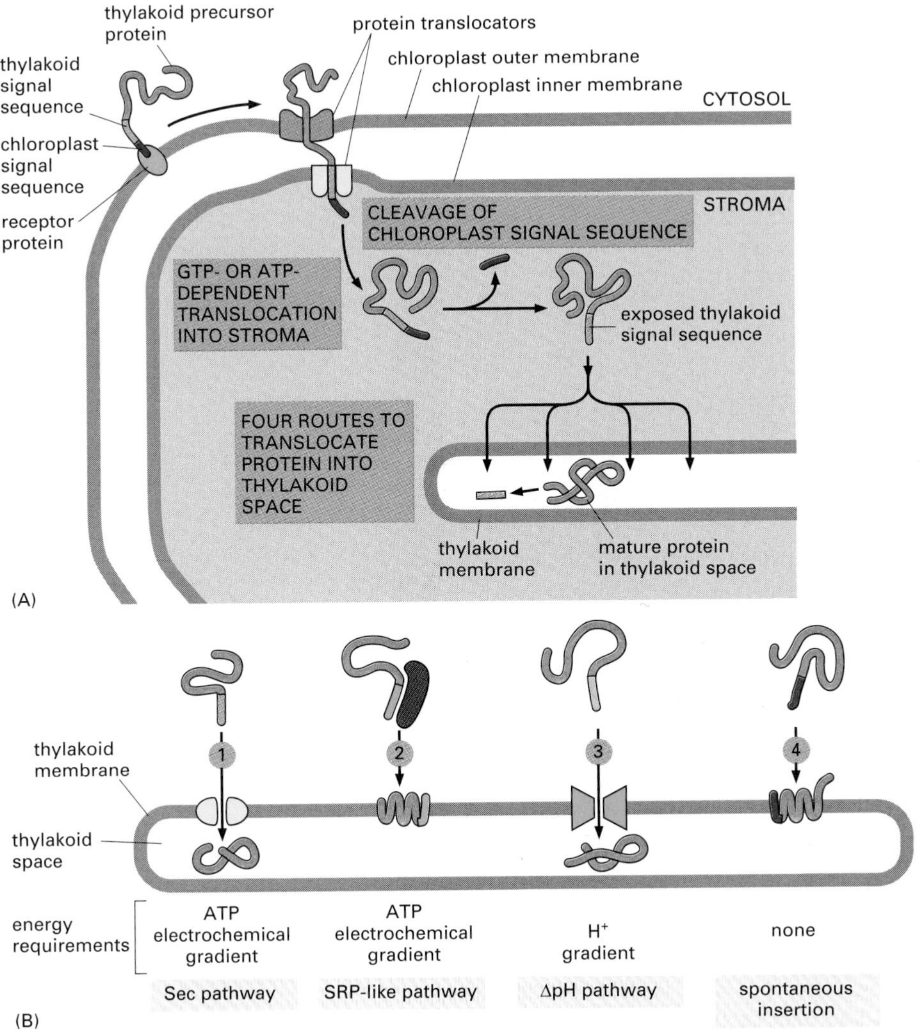

Figure 12–30 Translocation of a precursor protein into the thylakoid space of chloroplasts. (A) The precursor protein contains an N-terminal chloroplast signal sequence *(red)*, followed immediately by a thylakoid signal sequence *(orange)*. The chloroplast signal sequence initiates translocation into the stroma through a membrane contact site by a mechanism similar to that used for translocation into the mitochondrial matrix. The signal sequence is then cleaved off, unmasking the thylakoid signal sequence, which initiates translocation across the thylakoid membrane.
(B) Translocation into the thylakoid space or thylakoid membrane can occur by any one of at least four routes: (1) a Sec pathway, so called because it uses components that are homologs of Sec proteins, which mediate protein translocation across the bacterial plasma membrane (discussed later), (2) an SRP-like pathway, so called because it uses a chloroplast homolog of the signal recognition particle, or SRP (discussed later), (3) a ΔpH pathway, so called because it is driven by the H^+ gradient across the thylakoid membrane, and (4) a spontaneous insertion pathway that seems to require no protein translocator for membrane integration.

signal sequence following the N-terminal chloroplast signal sequence. After the N-terminal signal sequence has been used to import the protein into the stroma, it is removed by a stromal signal peptidase (analogous to the matrix signal peptidase in mitochondria). This cleavage unmasks the thylakoid signal sequence, which then initiates transport across the thylakoid membrane. There are at least four routes for proteins to cross or become integrated into the thylakoid membrane, distinguished by their need for different stromal chaperones and energy sources (Figure 12–30B).

Summary

Although mitochondria and chloroplasts have their own genetic systems, they produce only a small proportion of their own proteins. Instead, the two organelles import most of their proteins from the cytosol, using similar mechanisms. In both cases, proteins are imported in an unfolded state. Proteins are translocated into the mitochondrial matrix space by passing through the TOM and TIM complexes at sites of adhesion between the outer and inner membranes known as contact sites. Translocation into mitochondria is driven by both ATP hydrolysis and an electrochemical H^+ gradient across the inner membrane, whereas translocation into chloroplasts is driven solely by the hydrolysis of GTP and ATP.

Chaperone proteins of the cytosolic hsp70 family maintain the precursor proteins in an unfolded, translocation-competent state. A second set of hsp70 proteins in the matrix or stroma bind to the incoming polypeptide chain to pull it into the organelle. Only proteins that contain a specific signal sequence are translocated into mitochondria or chloroplasts. The signal sequence is usually located at the N terminus and is cleaved off after import. Some imported proteins also contain an internal signal sequence that guides their further transport. Transport across or into the inner membrane can occur as a second step if a hydrophobic signal sequence is unmasked when the first signal sequence is removed. In chloroplasts, import from the stroma into the thylakoid likewise requires a second signal sequence and can occur by one of several routes.

PEROXISOMES

Peroxisomes differ from mitochondria and chloroplasts in many ways. Most notably, they are surrounded by only a single membrane, and they do not contain DNA or ribosomes. Like mitochondria and chloroplasts, however, peroxisomes are thought to acquire their proteins by selective import from the cytosol. But because they have no genome, *all* of their proteins must be imported. Peroxisomes thus resemble the ER in being a self-replicating, membrane-enclosed organelle that exists without a genome of its own.

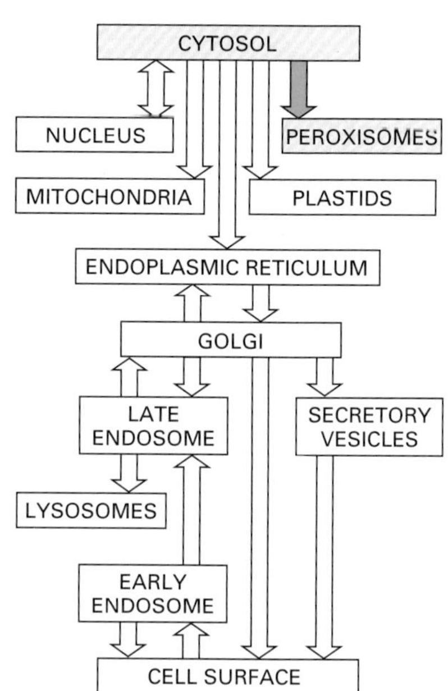

Because we do not discuss peroxisomes elsewhere, we shall digress to consider some of the functions of this diverse family of organelles, before discussing their biosynthesis. Peroxisomes are found in all eucaryotic cells. They contain oxidative enzymes, such as *catalase* and *urate oxidase*, at such high concentrations that in some cells the peroxisomes stand out in electron micrographs because of the presence of a crystalloid core (Figure 12–31).

Like mitochondria, peroxisomes are major sites of oxygen utilization. One hypothesis is that peroxisomes are a vestige of an ancient organelle that performed all the oxygen metabolism in the primitive ancestors of eucaryotic cells. When the oxygen produced by photosynthetic bacteria first began to accumulate in the atmosphere, it would have been highly toxic to most cells. Peroxisomes might have served to lower the intracellular concentration of oxygen, while also exploiting its chemical reactivity to perform useful oxidative reactions. According to this view, the later development of mitochondria rendered peroxisomes largely obsolete because many of the same reactions—which had formerly been carried out in peroxisomes without producing energy—were now coupled to ATP formation by means of oxidative phosphorylation. The oxidative reactions performed by peroxisomes in present-day cells would therefore be those that have important functions not taken over by mitochondria.

Peroxisomes Use Molecular Oxygen and Hydrogen Peroxide to Perform Oxidative Reactions

Peroxisomes are so named because they usually contain one or more enzymes that use molecular oxygen to remove hydrogen atoms from specific organic substrates (designated here as R) in an oxidative reaction that produces *hydrogen peroxide* (H_2O_2):

$$RH_2 + O_2 \rightarrow R + H_2O_2$$

Catalase utilizes the H_2O_2 generated by other enzymes in the organelle to oxidize a variety of other substrates—including phenols, formic acid, formaldehyde, and alcohol—by the "peroxidative" reaction: $H_2O_2 + R'H_2 \rightarrow R' + 2H_2O$. This type of oxidative reaction is particularly important in liver and kidney cells, where the peroxisomes detoxify various toxic molecules that enter the bloodstream. About 25% of the ethanol we drink is oxidized to acetaldehyde in this way. In addition, when excess H_2O_2 accumulates in the cell, catalase converts it to H_2O through the reaction:

$$2H_2O_2 \rightarrow 2H_2O + O_2$$

A major function of the oxidative reactions performed in peroxisomes is the breakdown of fatty acid molecules. In a process called *β oxidation*, the alkyl chains of fatty acids are shortened sequentially by blocks of two carbon atoms at a time, thereby converting the fatty acids to acetyl CoA. The acetyl CoA is then exported from the peroxisomes to the cytosol for reuse in biosynthetic reactions. In mammalian cells, β oxidation occurs in both mitochondria and peroxisomes; in yeast and plant cells, however, this essential reaction occurs exclusively in peroxisomes.

An essential biosynthetic function of animal peroxisomes is to catalyze the first reactions in the formation of *plasmalogens*, which are the most abundant class of phospholipids in myelin (Figure 12–32). Deficiency of plasmalogens causes profound abnormalities in the myelination of nerve cells, which is one reason why many peroxisomal disorders lead to neurological disease.

Peroxisomes are unusually diverse organelles, and even in the various cell types of a single organism they may contain different sets of enzymes. They can also adapt remarkably to changing conditions. Yeast cells grown on sugar, for example, have small peroxisomes. But when some yeasts are grown on methanol, they develop large peroxisomes that oxidize methanol; and when grown on fatty acids, they develop large peroxisomes that break down fatty acids to acetyl CoA by β oxidation.

Peroxisomes are also important in plants. Two different types have been studied extensively. One type is present in leaves, where it catalyzes the oxidation of a side product of the crucial reaction that fixes CO_2 in carbohydrate (Figure 12–33A). As discussed in Chapter 14, this process is called *photorespiration* because it uses up O_2 and liberates CO_2. The other type of peroxisome is present in germinating seeds, where it has an essential role in converting the fatty acids stored in seed lipids into the sugars needed for the growth of the young plant. Because this conversion of fats to sugars is accomplished by a series of reactions known as the *glyoxylate cycle*, these peroxisomes are also called *glyoxysomes* (Figure 12–33B). In the glyoxylate cycle, two molecules of acetyl CoA produced by fatty acid breakdown in the peroxisome are used to make succinic acid, which then leaves the peroxisome and is converted into glucose. The glyoxylate cycle does not occur in animal cells, and animals are therefore unable to convert the fatty acids in fats into carbohydrates.

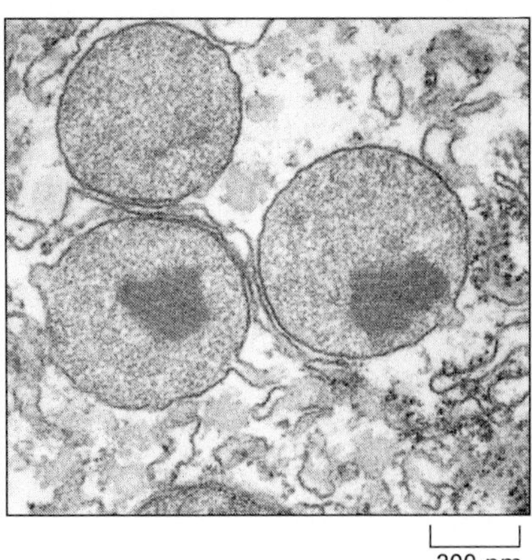

Figure 12–31 An electron micrograph of three peroxisomes in a rat liver cell. The paracrystalline electron-dense inclusions are composed of the enzyme urate oxidase. (Courtesy of Daniel S. Friend.)

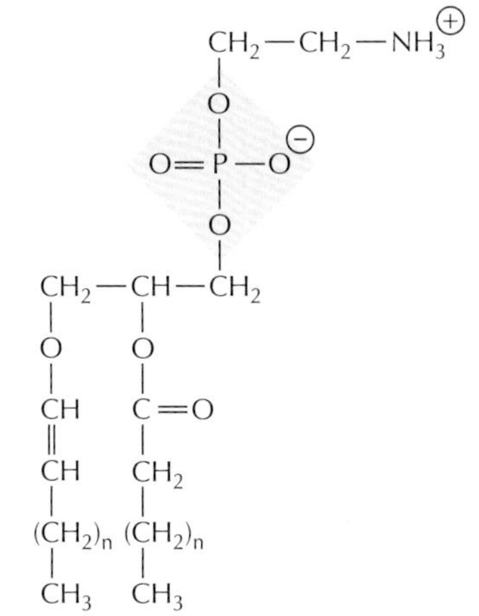

$CH_2—CH_2—NH_3^{\oplus}$
O
$O=P—O^{\ominus}$
O
$CH_2—CH—CH_2$
O O
CH C=O
CH CH_2
$(CH_2)_n$ $(CH_2)_n$
CH_3 CH_3

Figure 12–32 The structure of a plasmalogen. Plasmalogens are very abundant in the myelin sheaths that insulate the axons of nerve cells. They make up some 80–90% of the myelin membrane phospholipids. In addition to an ethanolamine head group and a long-chain fatty acid attached to the same glycerol phosphate backbone used for phospholipids, plasmalogens contain an unusual fatty alcohol that is attached through an ether linkage *(bottom left)*.

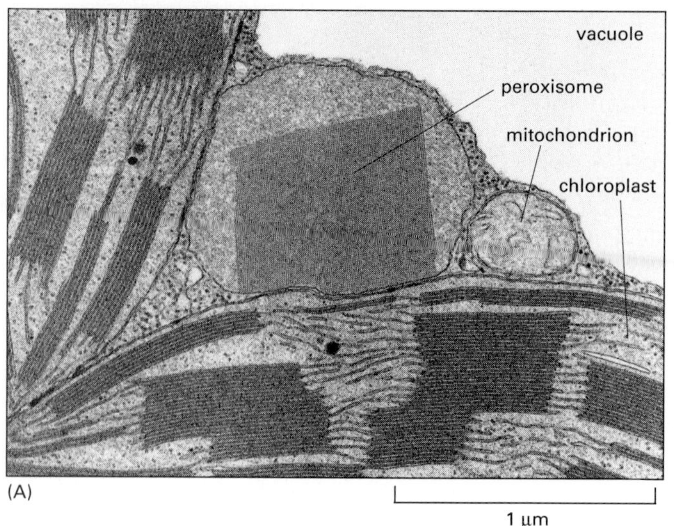

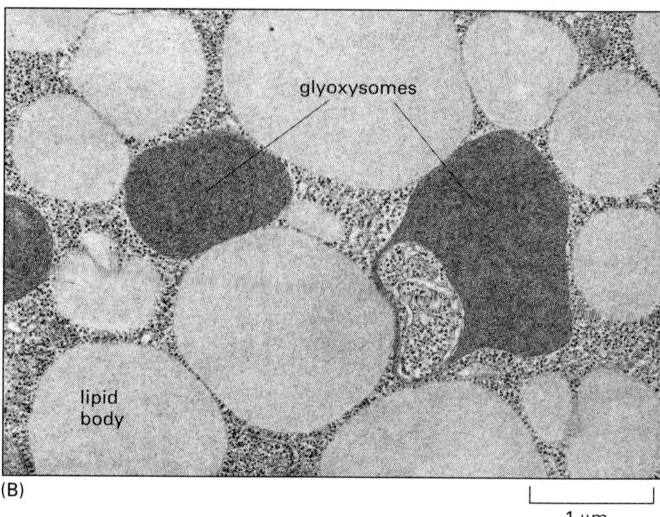

Figure 12–33 Electron micrographs of two types of peroxisomes found in plant cells. (A) A peroxisome with a paracrystalline core in a tobacco leaf mesophyll cell. Its close association with chloroplasts is thought to facilitate the exchange of materials between these organelles during photorespiration. (B) Peroxisomes in a fat-storing cotyledon cell of a tomato seed 4 days after germination. Here, the peroxisomes (glyoxysomes) are associated with the lipid bodies where fat is stored, reflecting their central role in fat mobilization and gluconeogenesis during seed germination. (A, from S.E. Frederick and E.H. Newcomb, *J. Cell Biol.* 43:343–353, 1969. © The Rockefeller Press; B, from W.P. Wergin, P.J. Gruber, and E.H. Newcomb, *J. Ultrastruct. Res.* 30:533–557, 1970. © Academic Press.)

A Short Signal Sequence Directs the Import of Proteins into Peroxisomes

A specific sequence of three amino acids located at the C terminus of many peroxisomal proteins functions as an import signal (see Table 12–3). Other peroxisomal proteins contain a signal sequence near the N terminus. If either of these sequences is experimentally attached to a cytosolic protein, the protein is imported into peroxisomes. The import process is still poorly understood, although it is known to involve soluble receptor proteins in the cytosol that recognize the targeting signals, as well as docking proteins on the cytosolic surface of the peroxisome. At least 23 distinct proteins, called **peroxins**, participate as components in the process, which is driven by ATP hydrolysis. Oligomeric proteins do not have to unfold to be imported into peroxisomes, indicating that the mechanism is distinct from that used by mitochondria and chloroplasts and at least one soluble import receptor, the peroxin Pex5, accompanies its cargo all the way into peroxisomes and, after cargo release, cycles back out into the cytosol. These aspects of peroxisomal protein import resemble protein tranport into the nucleus.

The importance of this import process and of peroxisomes is demonstrated by the inherited human disease *Zellweger syndrome,* in which a defect in importing proteins into peroxisomes leads to a severe peroxisomal deficiency. These individuals, whose cells contain "empty" peroxisomes, have severe abnormalities in their brain, liver, and kidneys, and they die soon after birth. One form of this disease has been shown to be due to a mutation in the gene encoding a peroxisomal integral membrane protein, the peroxin Pex2, involved in protein import. A milder inherited peroxisomal disease is caused by a defective receptor for the N-terminal import signal.

Most peroxisomal membrane proteins are made in the cytosol and then insert into the membrane of preexisting peroxisomes. Thus, new peroxisomes are thought to arise from preexisting ones, by organelle growth and fission—as mentioned earlier for mitochondria and plastids, and as we describe below for the ER (Figure 12–34).

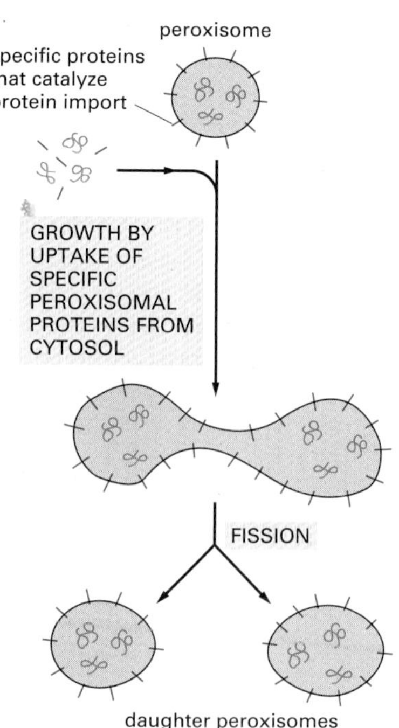

Figure 12–34 A model for how new peroxisomes are produced. The peroxisome membrane contains import receptor proteins. Peroxisomal proteins, including new copies of the import receptor, are synthesized by cytosolic ribosomes and then imported into the organelle. Presumably, the lipids required to make new peroxisome membrane are also imported. (We discuss later how lipids made in the ER can be transported through the cytosol to other organelles.) In this model, peroxisomes are thought to form only from preexisting peroxisomes by a process of growth and fission.

Summary

Peroxisomes are specialized for carrying out oxidative reactions using molecular oxygen. They generate hydrogen peroxide, which they use for oxidative purposes—destroying the excess by means of the catalase they contain. Peroxisomes also have an important role in the synthesis of specialized phospholipids required for nerve cell myelination. Like mitochondria and plastids, peroxisomes are thought to be self-replicating organelles. Because they contain no DNA or ribosomes, however, they have to import their proteins from the cytosol. A specific sequence of three amino acids near the C terminus of many of these proteins functions as a peroxisomal import signal. The mechanism of protein import is distinct from that of mitochondria and chloroplasts, and oligomeric proteins can be transported into peroxisomes without unfolding.

THE ENDOPLASMIC RETICULUM

All eucaryotic cells have an **endoplasmic reticulum (ER)**. Its membrane typically constitutes more than half of the total membrane of an average animal cell (see Table 12–2). The ER is organized into a netlike labyrinth of branching tubules and flattened sacs extending throughout the cytosol (Figure 12–35). The tubules and sacs are all thought to interconnect, so that the ER membrane forms a continuous sheet enclosing a single internal space. This highly convoluted space is called the **ER lumen** or the *ER cisternal space*, and it often occupies more than 10% of the total cell volume (see Table 12–1). The ER membrane separates the ER lumen from the cytosol, and it mediates the selective transfer of molecules between these two compartments.

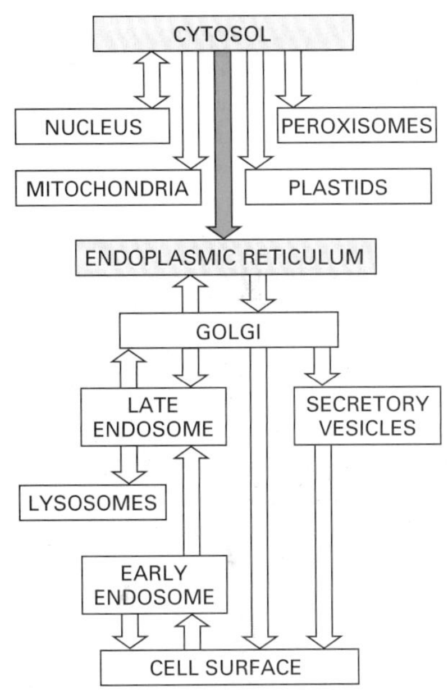

The ER has a central role in lipid and protein biosynthesis. Its membrane is the site of production of all the transmembrane proteins and lipids for most of the cell's organelles, including the ER itself, the Golgi apparatus, lysosomes, endosomes, secretory vesicles, and the plasma membrane. The ER membrane makes a major contribution to mitochondrial and peroxisomal membranes by producing most of their lipids. In addition, almost all of the proteins that will be secreted to the cell exterior—plus those destined for the lumen of the ER, Golgi apparatus, or lysosomes—are initially delivered to the ER lumen.

Membrane-bound Ribosomes Define the Rough ER

The ER captures selected proteins from the cytosol as they are being synthesized. These proteins are of two types: *transmembrane proteins*, which are only partly translocated across the ER membrane and become embedded in it, and *water-soluble proteins*, which are fully translocated across the ER membrane and are released into the ER lumen. Some of the transmembrane proteins function in the ER, but many are destined to reside in the plasma membrane or the membrane of another organelle. The water-soluble proteins are destined either

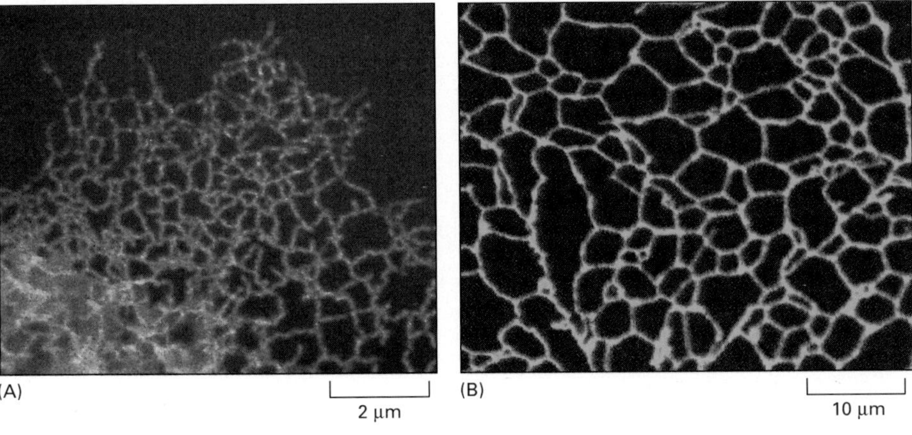

Figure 12–35 Fluorescent micrographs of the endoplasmic reticulum. (A) Part of the ER network in a cultured mammalian cell, stained with an antibody that binds to a protein retained in the ER. The ER extends as a network throughout the entire cytosol, so that all regions of the cytosol are close to some portion of the ER membrane. (B) Part of an ER network in a living plant cell that was genetically engineered to express a fluorescent protein in the ER. (A, courtesy of Hugh Pelham; B, courtesy of Petra Boevink and Chris Hawes.)

for the lumen of an organelle or for secretion. All of these proteins, regardless of their subsequent fate, are directed to the ER membrane by the same kind of signal sequence and are translocated across it by similar mechanisms.

In mammalian cells, the import of proteins into the ER begins before the polypeptide chain is completely synthesized—that is, import is a **co-translational** process. This distinguishes the process from the import of proteins into mitochondria, chloroplasts, nuclei, and peroxisomes, which are **posttranslational** processes. Since one end of the protein is usually translocated into the ER as the rest of the polypeptide chain is being made, the protein is never released into the cytosol and therefore is never in danger of folding up before reaching the translocator in the ER membrane. Thus, in contrast to the posttranslational import of proteins into mitochondria and chloroplasts, chaperone proteins are not required to keep the protein unfolded. The ribosome that is synthesizing the protein is directly attached to the ER membrane. These membrane-bound ribosomes coat the surface of the ER, creating regions termed **rough endoplasmic reticulum**, or **rough ER** (Figure 12–36A).

There are therefore two spatially separate populations of ribosomes in the cytosol. **Membrane-bound ribosomes**, attached to the cytosolic side of the ER membrane, are engaged in the synthesis of proteins that are being concurrently translocated into the ER. **Free ribosomes**, unattached to any membrane, synthesize all other proteins encoded by the nuclear genome. Membrane-bound and free ribosomes are structurally and functionally identical. They differ only in the proteins they are making at any given time. When a ribosome happens to be making a protein with an ER signal sequence, the signal directs the ribosome to the ER membrane.

Since many ribosomes can bind to a single mRNA molecule, a **polyribosome** is usually formed, which becomes attached to the ER membrane, directed there by the signal sequences on multiple growing polypeptide chains (Figure 12–36B). The individual ribosomes associated with such an mRNA molecule can return to the cytosol when they finish translation near the 3′ end of the mRNA molecule. The mRNA itself, however, remains attached to the ER membrane by a changing population of ribosomes, each transiently held at the membrane by the translocator. In contrast, if an mRNA molecule encodes a protein that lacks an ER signal sequence, the polyribosome that forms remains free in the cytosol, and its protein product is discharged there. Therefore, only those mRNA molecules that encode proteins with an ER signal sequence bind to rough ER membranes; those mRNA molecules that encode all other proteins remain free in the cytosol. Individual ribosomal subunits are thought to move randomly between these two segregated populations of mRNA molecules (Figure 12–37).

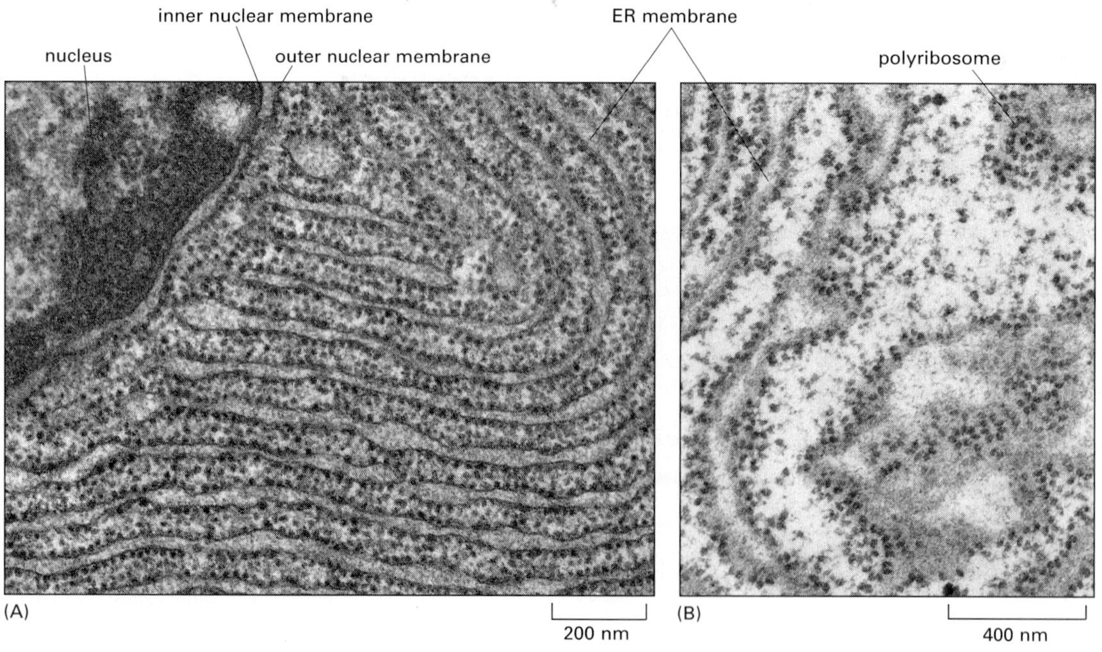

Figure 12–36 The rough ER. (A) An electron micrograph of the rough ER in a pancreatic exocrine cell that makes and secretes large amounts of digestive enzymes every day. The cytosol is filled with closely packed sheets of ER membrane studded with ribosomes. At the top left is a portion of the nucleus and its nuclear envelope; note that the outer nuclear membrane, which is continuous with the ER, is also studded with ribosomes. (B) A thin section electron micrograph of polyribosomes attached to the ER membrane. The plane of section in some places cuts through the ER roughly parallel to the membrane, giving a face-on view of the rosettelike pattern of the polyribosomes. (A, courtesy of Lelio Orci; B, courtesy of George Palade.)

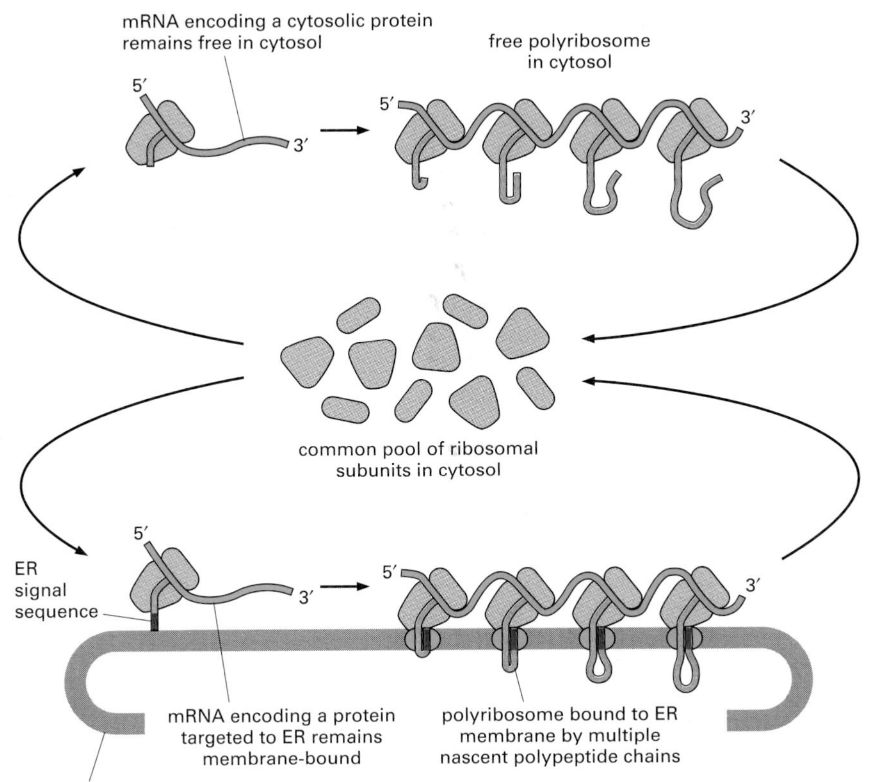

Figure 12–37 Free and membrane-bound ribosomes. A common pool of ribosomes is used to synthesize the proteins that stay in the cytosol and those that are transported into the ER. The ER signal sequence on a newly formed polypeptide chain directs the engaged ribosome to the ER membrane. The mRNA molecule remains permanently bound to the ER as part of a polyribosome, while the ribosomes that move along it are recycled; at the end of each round of protein synthesis, the ribosomal subunits are released and rejoin the common pool in the cytosol.

Smooth ER Is Abundant in Some Specialized Cells

Regions of ER that lack bound ribosomes are called **smooth endoplasmic reticulum**, or **smooth ER**. In the great majority of cells, such regions are scanty and are often partly smooth and partly rough. They are sometimes called *transitional ER* because they contain *ER exit sites* from which transport vesicles carrying newly synthesized proteins and lipids bud off for transport to the Golgi apparatus. In certain specialized cells, however, the smooth ER is abundant and has additional functions. In particular, it is usually prominent in cells that specialize in lipid metabolism. Cells that synthesize steroid hormones from cholesterol, for example, have an expanded smooth ER compartment to accommodate the enzymes needed to make cholesterol and to modify it to form the hormones (Figure 12–38A).

The main cell type in the liver, the *hepatocyte*, is another cell with an abundant smooth ER. It is the principal site of production of lipoprotein particles, which carry lipids via the bloodstream to other parts of the body. The enzymes that synthesize the lipid components of lipoproteins are located in the membrane of the smooth ER, which also contains enzymes that catalyze a series of reactions to detoxify both lipid-soluble drugs and various harmful compounds produced by metabolism. The most extensively studied of these *detoxification*

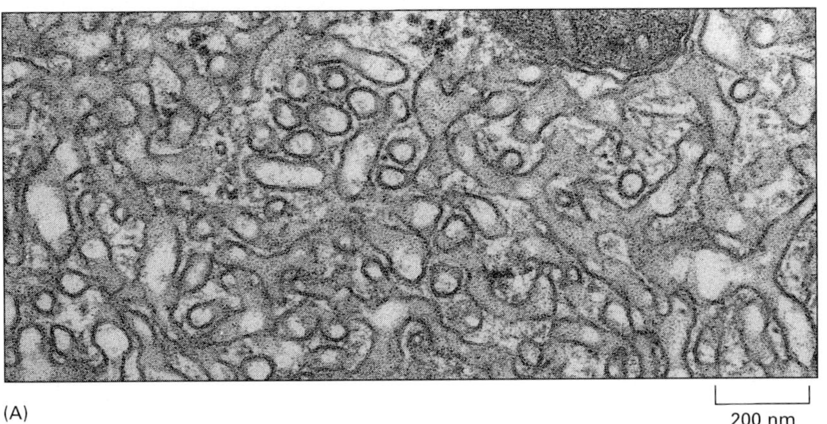

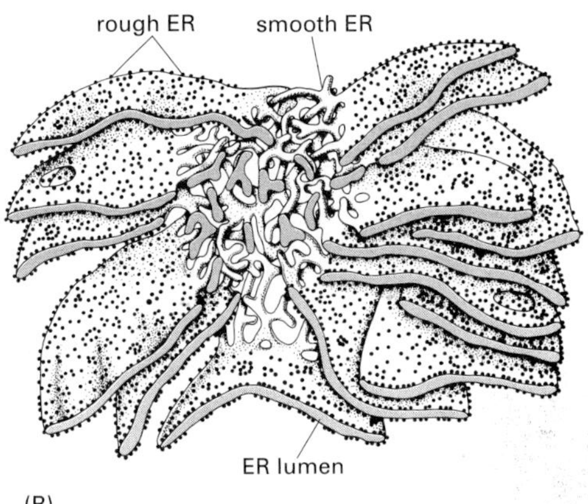

Figure 12–38 The smooth ER. (A) Abundant smooth ER in a steroid-hormone-secreting cell. This electron micrograph is of a testosterone-secreting Leydig cell in the human testis. (B) A three-dimensional reconstruction of a region of smooth ER and rough ER in a liver cell. The rough ER forms oriented stacks of flattened cisternae, each having a lumenal space 20–30 nm wide. The smooth ER membrane is connected to these cisternae and forms a fine network of tubules 30–60 nm in diameter. (A, courtesy of Daniel S. Friend; B, after R.V. Krstić, Ultrastructure of the Mammalian Cell. New York: Springer-Verlag, 1979.)

reactions are carried out by the *cytochrome P450* family of enzymes, which catalyze a series of reactions in which water-insoluble drugs or metabolites that would otherwise accumulate to toxic levels in cell membranes are rendered sufficiently water-soluble to leave the cell and be excreted in the urine. Because the rough ER alone cannot house enough of these and other necessary enzymes, a major portion of the membrane in a hepatocyte normally consists of smooth ER (Figure 12–38B; see Table 12–2).

When large quantities of certain compounds, such as the drug phenobarbital, enter the circulation, detoxification enzymes are synthesized in the liver in unusually large amounts, and the smooth ER doubles in surface area within a few days. Once the drug has disappeared, the excess smooth ER membrane is specifically and rapidly removed by a lysosome-dependent process called *autophagocytosis* (discussed in Chapter 13). It is not known how these dramatic changes are regulated.

Another function of the ER in most eucaryotic cells is to sequester Ca^{2+} from the cytosol. The release of Ca^{2+} into the cytosol from the ER, and its subsequent reuptake, is involved in many rapid responses to extracellular signals, as discussed in Chapter 15. The storage of Ca^{2+} in the ER lumen is facilitated by the high concentrations of Ca^{2+}-binding proteins there. In some cell types, and perhaps in most, specific regions of the ER are specialized for Ca^{2+} storage. Muscle cells, for example, have an abundant specialized smooth ER, called the *sarcoplasmic reticulum*, which sequesters Ca^{2+} from the cytosol by means of a Ca^{2+}-ATPase that pumps in Ca^{2+} into its lumen. The release and reuptake of Ca^{2+} by the sarcoplasmic reticulum trigger the contraction and relaxation, respectively, of the myofibrils during each round of muscle contraction (discussed in Chapter 16).

We now return to the two major roles of the ER: the synthesis and modification of proteins and the synthesis of lipids.

Rough and Smooth Regions of ER Can Be Separated by Centrifugation

To study the functions and biochemistry of the ER, it is necessary to isolate the ER membrane. This may seem like a hopeless task because the ER is intricately interleaved with other components of the cytosol. Fortunately, when tissues or cells are disrupted by homogenization, the ER breaks into fragments and reseals into many small (~100–200 nm in diameter) closed vesicles called **microsomes**, which are relatively easy to purify. Microsomes derived from rough ER are studded with ribosomes and are called *rough microsomes*. The ribosomes are always found on the outside surface, so the interior of the microsome is biochemically equivalent to the lumenal space of the ER (Figure 12–39). Because they can be readily purified in functional form, rough microsomes are especially useful for

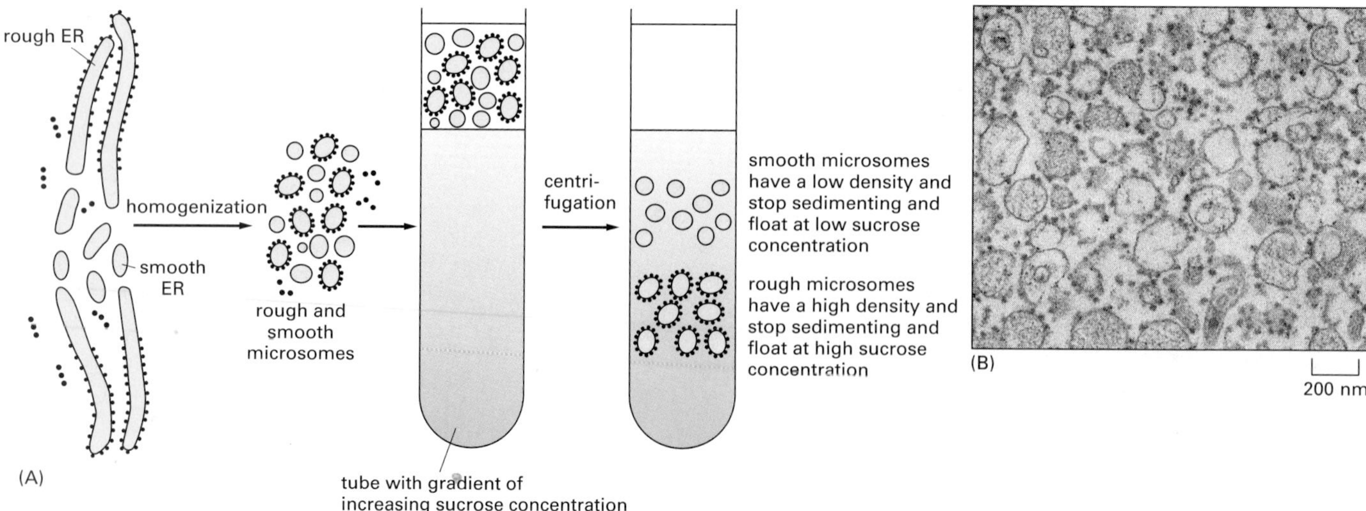

Figure 12–39 The isolation of purified rough and smooth microsomes from the ER. (A) When sedimented to equilibrium through a gradient of sucrose, the two types of microsomes separate from each other on the basis of their different densities. (B) A thin section electron micrograph of the purified rough ER fraction shows an abundance of ribosome-studded vesicles. (B, courtesy of George Palade.)

studying the many processes performed by the rough ER. To the biochemist they represent small authentic versions of the rough ER, still capable of protein synthesis, protein glycosylation, Ca^{2+} uptake, and lipid synthesis.

Many vesicles of a size similar to that of rough microsomes, but lacking attached ribosomes, are also found in these homogenates. Such *smooth microsomes* are derived in part from smooth portions of the ER and in part from vesiculated fragments of the plasma membrane, Golgi apparatus, endosomes, and mitochondria (the ratio depending on the tissue). Thus, whereas rough microsomes are derived from rough portions of ER, the origins of smooth microsomes cannot be as easily assigned. The microsomes of the liver are an exception. Because of the unusually large quantities of smooth ER in hepatocytes, most of the smooth microsomes in liver homogenates are derived from smooth ER.

The ribosomes attached to rough microsomes make them more dense than smooth microsomes (Figure 12–39B). As a result, the rough and smooth microsomes can be separated from each other by equilibrium centrifugation (see Figure 12–39A). When the separated rough and smooth microsomes of liver are compared with regard to such properties as enzyme activity or polypeptide composition, they are very similar, although not identical: apparently most of the components of the ER membrane can diffuse freely between the rough and smooth regions, as would be expected for a continuous, fluid membrane. The rough microsomes, however, contain more than 20 proteins that are not present in smooth microsomes, showing that some separation mechanism must exist for a subset of ER membrane proteins. Some of the proteins in this subset help to bind ribosomes to the rough ER, while others presumably produce the flattened shape of this part of the ER (see Figure 12–38B). It is not clear whether these membrane proteins are confined to the rough ER by forming large two-dimensional assemblies in the lipid bilayer, or whether they are instead held in place by interactions with a network of structural proteins on one or the other face of the rough ER membrane.

Signal Sequences Were First Discovered in Proteins Imported into the Rough ER

Signal sequences (and the signal sequence strategy of protein sorting) were first discovered in the early 1970s in secreted proteins that are translocated across the ER membrane as a first step toward their eventual discharge from the cell. In the key experiment, the mRNA encoding a secreted protein was translated by ribosomes *in vitro*. When microsomes were omitted from this cell-free system, the protein synthesized was slightly larger than the normal secreted protein, the extra length being the N-terminal leader peptide. In the presence of microsomes derived from the rough ER, however, a protein of the correct size was produced. These results were explained by the *signal hypothesis*, which postulated that the leader serves as an **ER signal sequence** that directs the secreted protein to the ER membrane and is then cleaved off by a *signal peptidase* in the ER membrane before the polypeptide chain has been completed (Figure 12–40).

According to the signal hypothesis, the secreted protein should be extruded into the lumen of the microsome during its synthesis *in vitro*. This can be demonstrated by treatment with a protease: a newly synthesized protein made in the absence of microsomes is degraded when the protease is added to the medium, whereas the same protein made in the presence of microsomes remains intact because it is protected by the microsomal membrane. When proteins without ER signal sequences are similarly synthesized *in vitro*, they are not imported into microsomes and are therefore degraded by protease treatment.

The signal hypothesis has been thoroughly tested by genetic and biochemical experiments and is found to apply to both plant and animal cells, as well as to protein translocation across the bacterial plasma membrane and, as we have seen, the membranes of mitochondria, chloroplasts, and peroxisomes. N-terminal ER signal sequences guide not only soluble secreted proteins, but also the precursors of all other proteins made by ribosomes bound to the rough ER membrane, including membrane proteins. The signaling function of these

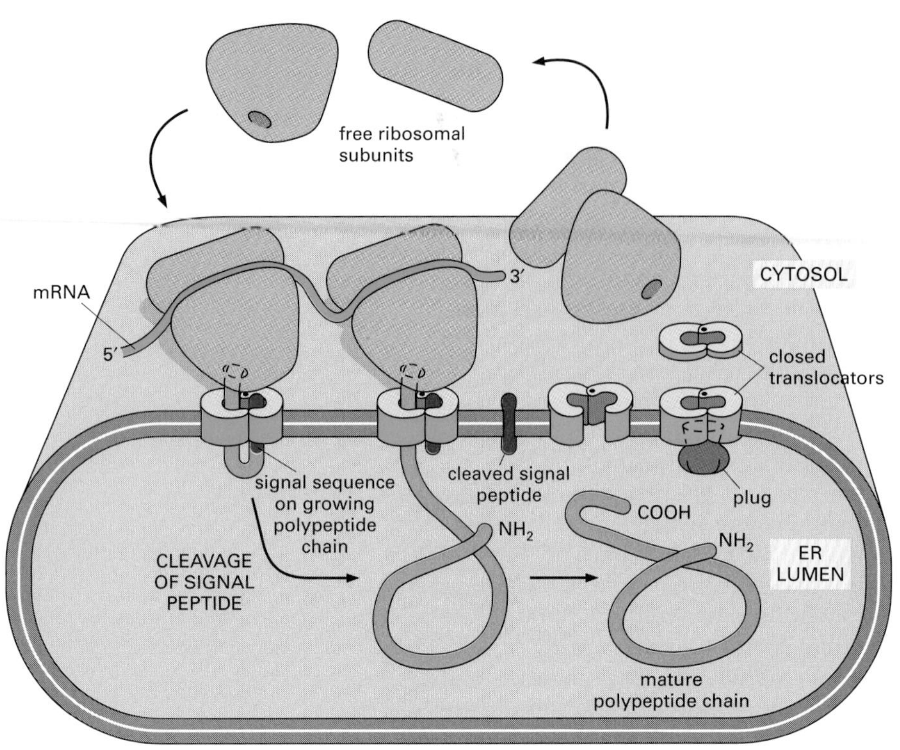

Figure 12–40 The signal hypothesis. A simplified view of protein translocation across the ER membrane, as originally proposed. When the ER signal sequence emerges from the ribosome, it directs the ribosome to a translocator on the ER membrane that forms a pore in the membrane through which the polypeptide is translocated. The signal sequence is clipped off during translation by a signal peptidase, and the mature protein is released into the lumen of the ER immediately after being synthesized. We now know that the hypothesis is correct in outline but that additional components besides those shown in this figure are required.

peptides has been demonstrated directly by using recombinant DNA techniques to attach ER signal sequences to proteins that do not normally have them; the resulting fusion proteins are directed to the ER.

Cell-free systems in which proteins are imported into microsomes have provided powerful assay procedures for identifying, purifying, and studying the various components of the molecular machinery responsible for the ER import process.

A Signal-Recognition Particle (SRP) Directs ER Signal Sequences to a Specific Receptor in the Rough ER Membrane

The ER signal sequence is guided to the ER membrane by at least two components: a **signal-recognition particle (SRP)**, which cycles between the ER membrane and the cytosol and binds to the signal sequence, and an **SRP receptor** in the ER membrane. The SRP is a complex particle consisting of six different polypeptide chains bound to a single small RNA molecule (Figure 12–41A). Homologs of the SRP and its receptor are found in all organisms that have been studied, indicating that this protein-targeting mechanism arose early in evolution and has been conserved.

ER signal sequences vary greatly in amino acid sequence, but each has eight or more nonpolar amino acids at its center (see Table 12–3, p. 667). How can the SRP bind specifically to so many different sequences? The answer has come

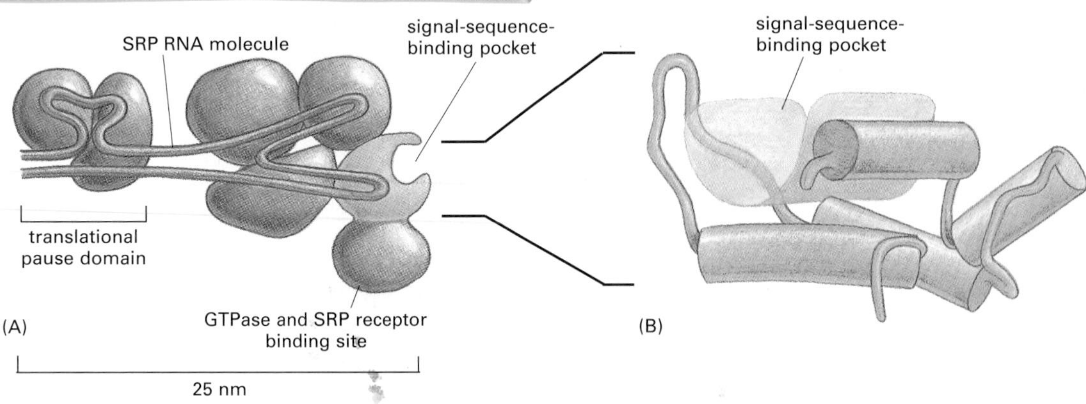

Figure 12–41 The signal-recognition particle (SRP). (A) A mammalian SRP is an elongated complex containing six protein subunits and one RNA molecule (SRP RNA). One end of the SRP binds to an ER signal sequence on a growing polypeptide chain, while the other end binds to the ribosome itself and pauses translation. The RNA in the particle may mediate an interaction with ribosomal RNA. (B) The crystal structure of the signal-sequence-binding domain of a bacterial SRP subunit. The domain contains a large, exposed binding pocket that is lined by hydrophobic amino acids, a large number of which are methionines. The outline of the pocket is shaded in *gray* to emphasize its location. The flexible side chains of methionine are ideal for building adaptable hydrophobic binding sites for other proteins. Calmodulin, for example (discussed in Chapter 15), binds to many different target proteins, and, like SRP, contains patches of methionines to clamp down on differently shaped targets (see Figure 15–40). (A, adapted from V. Siegel and P. Walter, *Nature* 320:82–84, 1986; B, adapted from Keenan et al., *Cell* 94:181–191, 1998.)

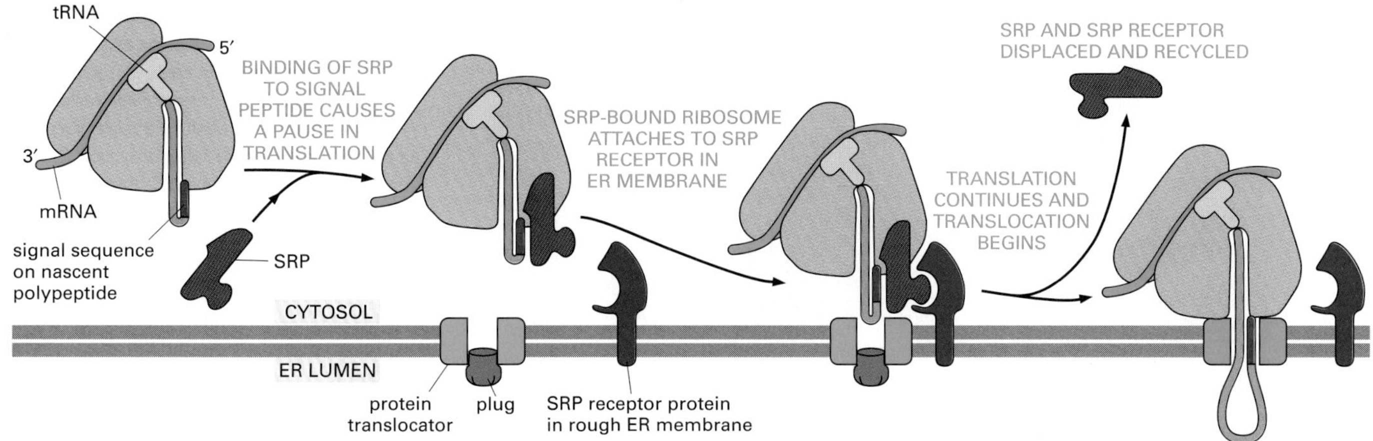

Figure 12–42 How ER signal sequences and SRP direct ribosomes to the ER membrane. The SRP and its receptor are thought to act in concert. The SRP binds to both the exposed ER signal sequence and the ribosome, thereby inducing a pause in translation. The SRP receptor in the ER membrane, which is composed of two different polypeptide chains, binds the SRP–ribosome complex and directs it to the translocator. In a poorly understood reaction, the SRP and SRP receptor are then released, leaving the ribosome bound to the translocator in the ER membrane. The translocator then inserts the polypeptide chain into the membrane and transfers it across the lipid bilayer. Because one of the SRP proteins and both chains of the SRP receptor contain GTP-binding domains, it is thought that conformational changes that occur during cycles of GTP binding and hydrolysis (discussed in Chapter 15) ensure that SRP release occurs only after the ribosome has become properly engaged with the translocator in the ER membrane. The translocator is closed (indicated schematically by the ER-lumenal plug) until the ribosome has bound, so that the permeability barrier of the ER membrane is maintained at all times.

from the crystal structure of the SRP protein, which shows that the signal-sequence-binding site is a large hydrophobic pocket lined by methionines (Figure 12–41B). Because methionines have an unbranched, flexible side chains, the pocket is sufficiently plastic to accommodate hydrophobic signal sequences of different sequences and shapes.

The SRP binds to the ER signal sequence as soon as the peptide has emerged from the ribosome. This causes a pause in protein synthesis, the pause presumably gives the ribosome enough time to bind to the ER membrane before the synthesis of the polypeptide chain is completed, thereby ensuring that the protein is not released into the cytosol. This safety device may be especially important for secreted and lysosomal hydrolases that could wreak havoc in the cytosol; however, cells that secrete large amounts of hydrolases take the added precaution of having high concentrations of hydrolase inhibitors in their cytosol.

Once formed, the SRP–ribosome complex binds to the SRP receptor, which is an integral membrane protein exposed only on the cytosolic surface of the rough ER membrane. This interaction brings the SRP–ribosome complex to a protein translocator. The SRP and SRP receptor are then released, and the growing polypeptide chain is transferred across the membrane (Figure 12–42).

The Polypeptide Chain Passes Through an Aqueous Pore in the Translocator

It has long been debated whether polypeptide chains are transferred across the ER membrane in direct contact with the lipid bilayer or through a pore in a protein translocator. The debate ended with the purification of the protein translocator, which was shown to form a water-filled pore in the membrane through which the polypeptide chain traverses the membrane. The translocator, called the **Sec61 complex**, consists of three or four protein complexes, each composed of three transmembrane proteins, that assemble into a donutlike structure.

When a ribosome binds, the central hole in the translocator lines up with a tunnel in the large ribosomal subunit through which the growing polypeptide chain exits from the ribosome (Figure 12–43). The bound ribosome forms a tight

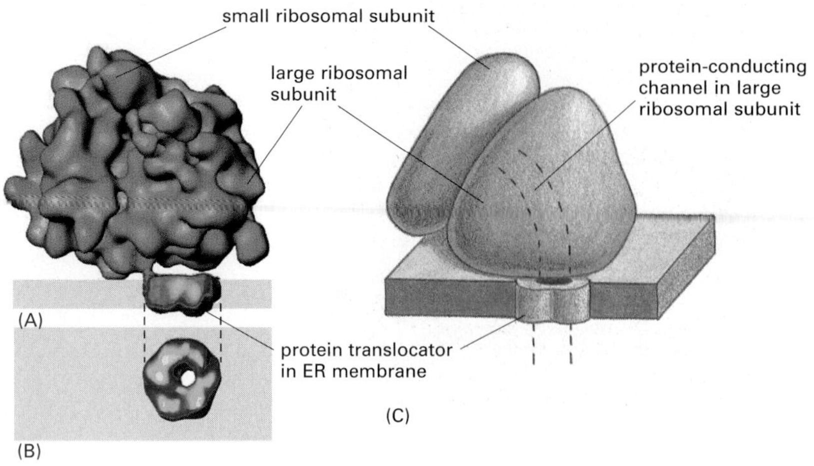

Figure 12–43 A ribosome bound to the Sec61 protein translocator. (A) A reconstruction of the complex from electron microscopic images viewed from the side. (B) A view of the translocator seen from the top (looking down on the membrane). (C) A schematic drawing of a membrane-bound ribosome attached to the translocator. The central pore in the translocator lines up with the tunnel in the large ribosomal subunit, through which the growing polypeptide chain exits from the ribosome (see Figure 6–68C). (A and B, from R. Beckmann et al., *Science* 278:2123–2126, 1997. © AAAS.)

seal with the translocator, such that the space inside the ribosome is continuous with the lumen of the ER and no molecules can escape from the ER (Figure 12–44). The pore in the translocator cannot be open permanently, however; if it were, Ca^{2+} would leak out of the ER when the ribosome detaches. It is thought that a lumenal ER protein serves as a plug or that the translocator itself can rearrange to close the pore when no ribosome is bound. Thus, the pore is a dynamic structure that opens only transiently when a ribosome with a growing polypeptide chain attaches to the ER membrane.

The signal sequence in the growing polypeptide chain is thought to trigger the opening of the pore: after the signal sequence is released from the SRP and the growing chain has reached a sufficient length, the signal sequence binds to a specific site inside the pore itself, thereby opening the pore. An ER signal sequence is therefore recognized twice: first, by an SRP in the cytosol, and then by a binding site in the ER protein translocator. This may help to ensure that only appropriate proteins enter the lumen of the ER.

Translocation Across the ER Membrane Does Not Always Require Ongoing Polypeptide Chain Elongation

As we have seen, translocation of proteins into mitochondria, chloroplasts, and peroxisomes occurs posttranslationally, after the protein has been made and released into the cytosol, whereas translocation across the ER membrane usually occurs during translation (co-translationally). This explains why ribosomes are bound to the ER but usually not to other organelles.

Some proteins, however, are imported into the ER after their synthesis has been completed, demonstrating that translocation does not always require ongoing translation. Posttranslational protein translocation is especially common across the ER membrane in yeast cells and across the bacterial plasma

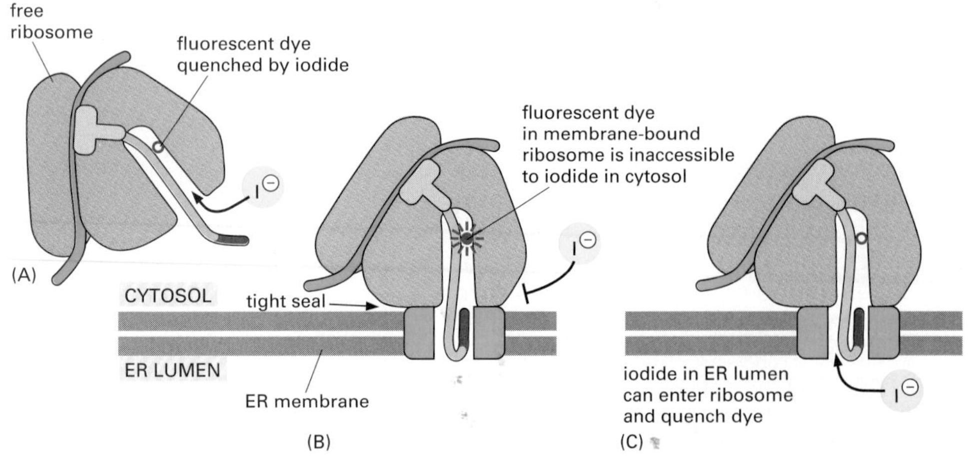

Figure 12–44 Evidence for a continuous aqueous pore joining the ER lumen and the interior of the ribosome. In this experiment, a fluorescent dye is attached to a portion of the growing polypeptide chain that is still contained within the ribosome. (A) In free ribosomes, the dye is accessible to iodide ions in solution in the cytosol. These ions quench the fluorescence when they come in contact with the dye. (B) In contrast, when a ribosome is membrane-bound, a tight seal is formed between the ribosome and the ER membrane that prevents access of the above iodide ions to the dye. (C) When iodide ions are added to the ER lumen, they can diffuse through the translocator all the way into the ribosome tunnel to quench the dye inside the membrane-bound ribosome.

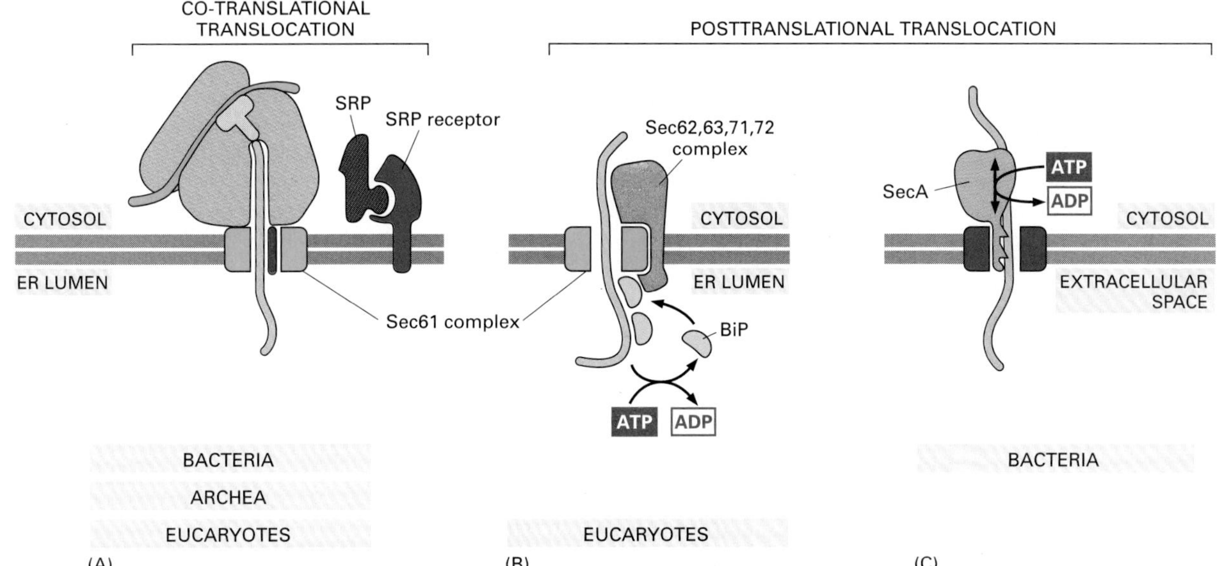

Figure 12–45 Three ways in which protein translocation can be driven through structurally similar translocators. (A) Co-translational translocation. The ribosome is brought to the membrane by the SRP and SRP receptor and forms a tight seal with the Sec61 protein translocator. The growing polypeptide chain is threaded across the membrane as it is made. No additional energy is needed, as the only path available to the growing chain is to cross the membrane. (B) Posttranslational translocation in eucaryotic cells. An additional complex composed of the Sec62, Sec63, Sec71, and Sec72 proteins is attached to the Sec61 translocator and deposits BiP molecules onto the translocating chain as it emerges into the ER lumen. ATP-driven cycles of BiP binding and release pull the protein into the lumen, a mechanism that closely resembles the thermal ratchet model for mitochondrial import in Figure 12–28. (C) Posttranslational translocation in bacteria. The completed polypeptide chain is fed from the cytosolic side into a translocator in the plasma membrane by the SecA ATPase. ATP-hydrolysis-driven conformational changes drive a pistonlike motion in SecA, each cycle pushing about 20 amino acids of the protein chain through the pore of the translocator. The Sec pathway used for protein translocation across the thylakoid membrane in chloroplasts uses a similar mechanism (see Figure 12–30B).

Whereas the Sec61 translocator, SRP, and SRP receptor are found in all organisms, SecA is found exclusively in bacteria, and the Sec62, Sec63, Sec71, and Sec72 proteins are found exclusively in eucaryotic cells. (Adapted from P. Walter and A.E. Johnson, *Annu. Rev. Cell Biol.* 10:87–119, 1994.)

membrane (which is thought to be evolutionarily related to the ER; see Figure 12–4). To function in posttranslational translocation, the translocator needs accessory proteins that feed the polypeptide chain into the pore and drive translocation (Figure 12–45). In bacteria, a translocation motor protein, the *SecA ATPase,* attaches to the cytosolic side of the translocator, where it undergoes cyclic conformational changes driven by ATP hydrolysis. Each time an ATP is hydrolyzed, a portion of the SecA protein inserts into the pore of the translocator, pushing a short segment of the passenger protein with it. As a result of this ratchet mechanism, the SecA protein pushes the polypeptide chain of the transported protein across the membrane.

Eucaryotic cells use a different set of accessory proteins that associate with the Sec61 complex. These proteins span the ER membrane and use a small domain on the lumenal side of the ER membrane to deposit an hsp70-like chaperone protein (called *BiP,* for *b*inding *p*rotein) onto the polypeptide chain as it emerges from the pore into the ER lumen. Unidirectional translocation is driven by cycles of BiP binding and release, as described earlier for the mitochondrial hsp70 proteins that pull proteins across mitochondrial membranes.

Proteins that are transported into the ER by a posttranslational mechanism are first released into the cytosol, where they are prevented from folding up by binding to chaperone proteins, as discussed earlier for proteins destined for mitochondria and chloroplasts. In all of these cases where translocation occurs without a ribosome sealing the pore, it remains a mystery how the polypeptide chain can slide through the pore in the translocator without allowing ions and other molecules to pass through.

The ER Signal Sequnce Is Removed from Most Soluble Proteins After Translocation

We have seen that in chloroplasts and mitochondria, the signal sequence is cleaved from precursor proteins once it has crossed the membrane. Similarly, N-terminal ER signal sequences are removed by a signal peptidase on the lumenal side of the ER membrane. The signal sequence by itself, however, is not sufficient for signal cleavage by the peptidase; this requires an adjacent cleavage site that is specifically recognized by the peptidase. We shall see below that ER signal sequences that occur within the polypeptide chain—rather than at the N-terminus—do not have these recognition sites and are never cleaved; instead, they can serve to retain transmembrane proteins in the lipid bilayer after the translocation process has been completed.

The N-terminal ER signal sequence of a soluble protein has two signaling functions. It directs the protein to the ER membrane, and it serves as a **start-transfer signal** (or start-transfer peptide) that opens the pore. Even after it is

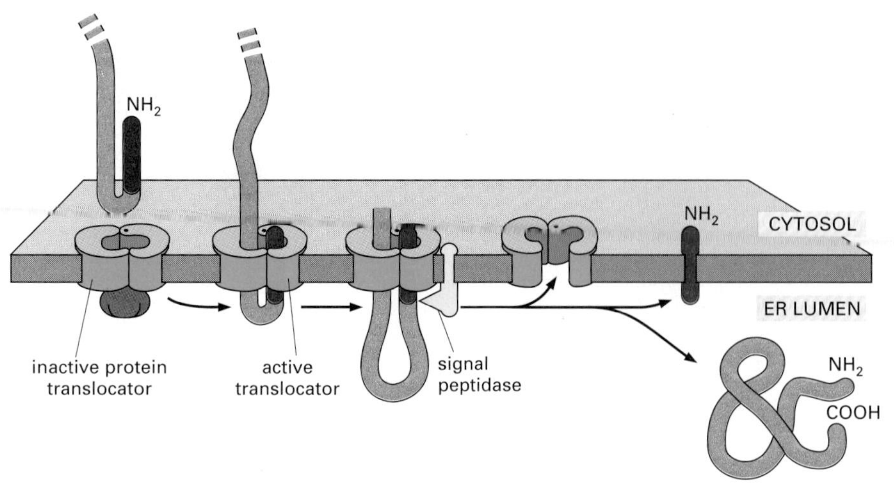

Figure 12–46 A model for how a soluble protein is translocated across the ER membrane. On binding an ER signal sequence (which acts as a start-transfer signal), the translocator opens its pore, allowing the transfer of the polypeptide chain across the lipid bilayer as a loop. After the protein has been completely translocated, the pore closes, but the translocator now opens laterally within the lipid bilayer, allowing the hydrophobic signal sequence to diffuse into the bilayer, where it is rapidly degraded. (In this figure and the three figures that follow, the ribosomes have been omitted for clarity.)

cleaved off by signal peptidase, the signal sequence is thought to remain bound to the translocator while the rest of the protein is threaded continuously through the membrane as a large loop. Once the C-terminus of the protein has passed through the membrane, the translocated protein is released into the ER lumen (Figure 12–46). The signal sequence is released from the pore and rapidly degraded to amino acids by other proteases in the ER.

While bound in the translocation pore, signal sequences are in contact not only with the Sec61 complex, which forms the walls of the pore, but also with the hydrophobic lipid core of the membrane. This was shown in chemical cross-linking experiments in which signal sequences and the hydrocarbon chains of lipids could be covalently linked together. To release the signal sequence into the membrane, the translocator has to open laterally. The translocator is therefore gated in two directions: it can open to form a pore across the membrane to let the hydrophilic portions of proteins cross the lipid bilayer, and it can open laterally within the membrane to let hydrophobic portions of proteins partition into the bilayer. This lateral gating mechanism is crucial for the insertion of transmembrane proteins into the lipid bilayer, as we discuss next.

In Single-Pass Transmembrane Proteins, a Single Internal ER Signal Sequence Remains in the Lipid Bilayer as a Membrane-spanning α Helix

The translocation process for proteins destined to remain in the membrane is more complex than it is for soluble proteins, as some parts of the polypeptide chain are translocated across the lipid bilayer whereas others are not. Nevertheless, all modes of insertion of membrane proteins can be considered as variants of the sequence of events just described for transferring a soluble protein into the lumen of the ER. We begin by describing the three ways in which **single-pass transmembrane proteins** (see Figure 10–17) become inserted into the ER.

In the simplest case, an N-terminal signal sequence initiates translocation, just as for a soluble protein, but an additional hydrophobic segment in the polypeptide chain stops the transfer process before the entire polypeptide chain is translocated. This **stop-transfer signal** anchors the protein in the membrane after the ER signal sequence (the start-transfer signal) has been released from the translocator and has been cleaved off (Figure 12–47). The stop-transfer sequence is transferred into the bilayer by the lateral gating mechanism, and it remains there as a single α-helical membrane-spanning segment, with the N-terminus of the protein on the lumenal side of the membrane and the C-terminus on the cytosolic side.

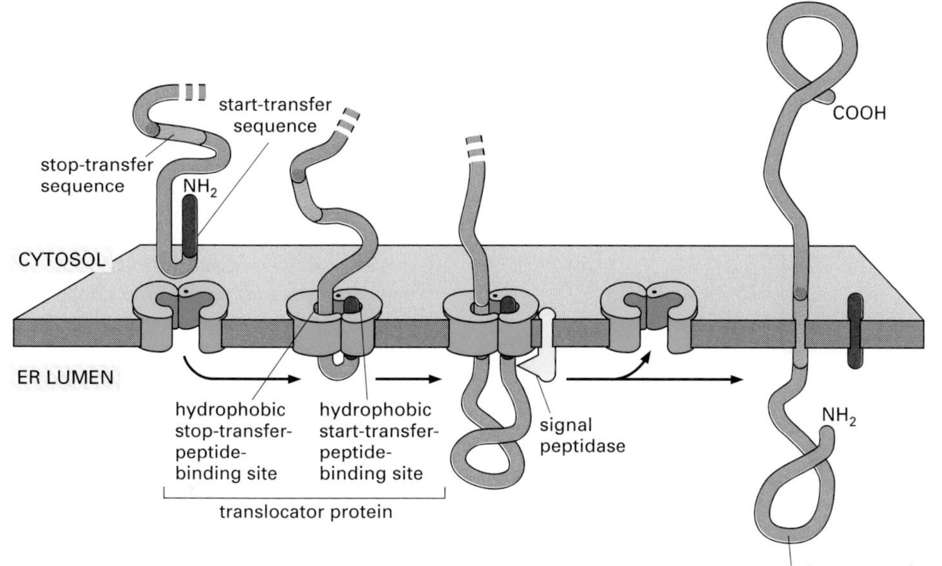

Figure 12–47 How a single-pass transmembrane protein with a cleaved ER signal sequence is integrated into the ER membrane. In this hypothetical protein the co-translational translocation process is initiated by an N-terminal ER signal sequence *(red)* that functions as a start-transfer signal, as in Figure 12–46. In addition to this start-transfer sequence, however, the protein also contains a stop-transfer sequence *(orange)*. When the stop-transfer sequence enters the translocator and interacts with a binding site, the translocator changes its conformation and discharges the protein laterally into the lipid bilayer.

In the other two cases, the signal sequence is internal, rather than at the N-terminal end of the protein. Like the N-terminal ER signal sequences, the internal signal sequence is recognized by an SRP, which brings the ribosome making the protein to the ER membrane and serves as a start-transfer signal that initiates the translocation of the protein. After release from the translocator, the internal start-transfer sequence remains in the lipid bilayer as a single membrane-spanning α helix.

Internal start-transfer sequences, can bind to the translocation apparatus in either of two orientations, and the orientation of the inserted start-transfer sequence, in turn, determines which protein segment (the one preceding or the one following the start-transfer sequence) is moved across the membrane into the ER lumen. In one case, the resulting membrane protein has its C-terminus on the lumenal side (Figure 12–48A), while in the other, it has its N-terminus on the lumenal side (Figure 12–48B). The orientation of the start-transfer sequence depends on the distribution of nearby charged amino acids, as described in the figure legend.

Combinations of Start-Transfer and Stop-Transfer Signals Determine the Topology of Multipass Transmembrane Proteins

In **multipass transmembrane proteins**, the polypeptide chain passes back and forth repeatedly across the lipid bilayer (see Figure 10–17). It is thought that an internal signal sequence serves as a start-transfer signal in these proteins to initiate translocation, which continues until a stop-transfer sequence is reached. In double-pass transmembrane proteins, for example, the polypeptide can then be released into the bilayer (Figure 12–49). In more complex multipass proteins, in which many hydrophobic α helices span the bilayer, a second start-transfer sequence reinitiates translocation further down the polypeptide chain until the next stop-transfer sequence causes polypeptide release, and so on for subsequent start-transfer and stop-transfer sequences (Figure 12–50).

Whether a given hydrophobic signal sequence functions as a start-transfer or stop-transfer sequence must depend on its location in a polypeptide chain, since its function can be switched by changing its location in the protein using recombinant DNA techniques. Thus, the distinction between start-transfer and stop-transfer sequences results mostly from their relative order in the growing polypeptide chain. It seems that the SRP begins scanning an unfolded polypeptide chain for hydrophobic segments at its N-terminus and proceeds toward the C-terminus, in the direction that the protein is synthesized. By recognizing the first appropriate hydrophobic segment to emerge from the ribosome, the SRP

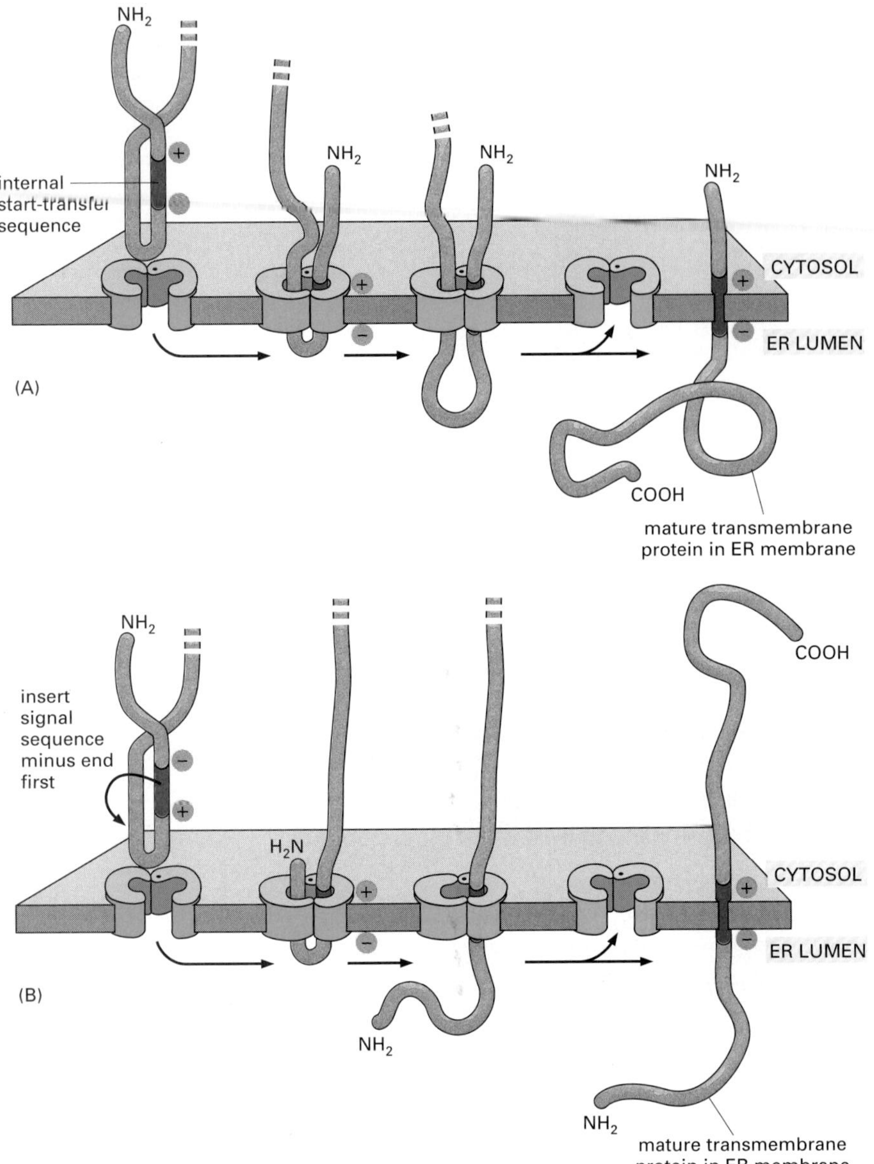

Figure 12–48 Integration of a single-pass membrane protein with an internal signal sequence into the ER membrane. In these hypothetical proteins, an internal ER signal sequence that functions as a start-transfer signal binds to the translocator in such a way that its more positively charged end remains in the cytosol. (A) If there are more positively charged amino acids immediately preceding the hydrophobic core of the start-transfer sequence than there are following it, the start-transfer sequence is inserted into the translocator in the orientation shown here. The part of the protein C-terminal to the start-transfer sequence will therefore be passed across the membrane. (B) If there are more positively charged amino acids immediately following the hydrophobic core of the start-transfer sequence than there are preceding it, the start-transfer sequence is inserted into the translocator in the orientation shown here. The part of the protein N-terminal to the start-transfer sequence will therefore be passed across the membrane. Because translocation cannot start before a start-transfer sequence appears outside the ribosome, translocation of the N-terminal portion of the protein shown in (B) can occur only after this portion has been fully synthesized.

Note that there are two ways to insert a single-pass membrane-spanning protein whose N-terminus is located in the ER lumen: that shown in Figure 12–47 and that shown in (B) here.

sets the "reading frame": if translocation is initiated, the next appropriate hydrophobic segment is recognized as a stop-transfer sequence, causing the region of the polypeptide chain in between to be threaded across the membrane. A similar scanning process continues until all of the hydrophobic regions in the protein have been inserted into the membrane.

Because membrane proteins are always inserted from the cytosolic side of the ER in this programmed manner, all copies of the same polypeptide chain will have the same orientation in the lipid bilayer. This generates an asymmetrical ER membrane in which the protein domains exposed on one side are different from those domains exposed on the other. This asymmetry is maintained during the many membrane budding and fusion events that transport the proteins made in the ER to other cell membranes (discussed in Chapter 13). Thus, the way in which a newly synthesized protein is inserted into the ER membrane determines the orientation of the protein in all of the other membranes as well.

When proteins are dissociated from a membrane and are then reconstituted into artificial lipid vesicles, a random mixture of right-side-out and inside-out protein orientations usually results. Thus, the protein asymmetry observed in cell membranes seems not to be an inherent property of the protein, but instead results solely from the process by which proteins are inserted into the ER membrane from the cytosol.

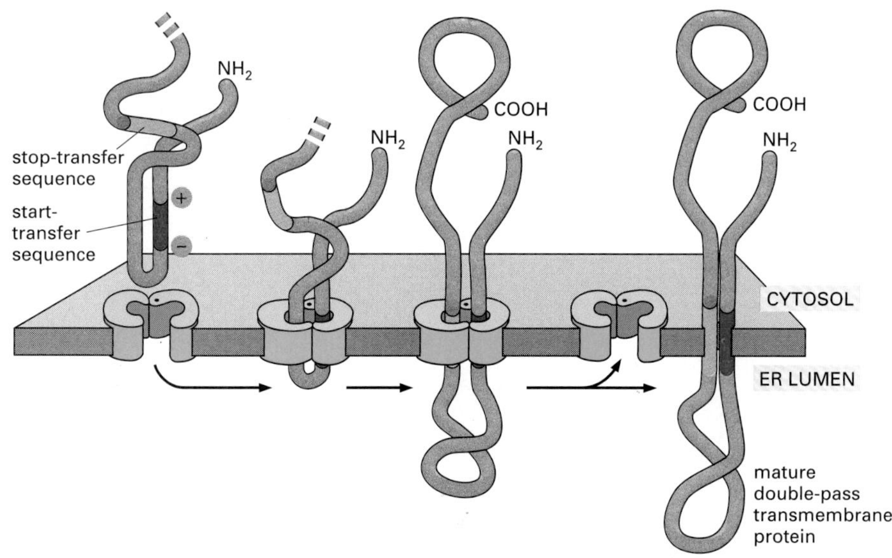

Figure 12–49 Integration of a double-pass membrane protein with an internal signal sequence into the ER membrane. In this hypothetical protein, an internal ER signal sequence acts as a start-transfer signal (as in Figure 12–48) and initiates the transfer of the C-terminal part of the protein. At some point after a stop-transfer sequence has entered the translocator, the translocator discharges the sequence laterally into the membrane.

Translocated Polypeptide Chains Fold and Assemble in the Lumen of the Rough ER

Many of the proteins in the lumen of the ER are in transit, *en route* to other destinations; others, however, are normally resident there and are present at high concentrations. These **ER resident proteins** contain an **ER retention signal** of four amino acids at their C terminus that is responsible for retaining the protein in the ER (see Table 12–3; discussed in Chapter 13). Some of these proteins function as catalysts that help the many proteins that are translocated into the ER to fold and assemble correctly.

One important ER resident protein is *protein disulfide isomerase (PDI)*, which catalyzes the oxidation of free sulfhydryl (SH) groups on cysteines to form disulfide (S–S) bonds. Almost all cysteines in protein domains exposed to either the extracellular space or the lumen of organelles in the secretory and endocytic pathways are disulfide-bonded; disulfide bonds do not form, however, in domains exposed to the cytosol because of the reducing environment there.

Another ER resident protein is the chaperone protein **BiP**. We have already discussed how BiP works to pull proteins posttranslationally into the ER through the ER translocator. Like other chaperones, BiP recognizes incorrectly folded proteins, as well as protein subunits that have not yet assembled into their final oligomeric complexes. To do so, it binds to exposed amino acid sequences that would normally be buried in the interior of correctly folded or assembled polypeptide chains. An example of a BiP-binding site is a stretch of alternating

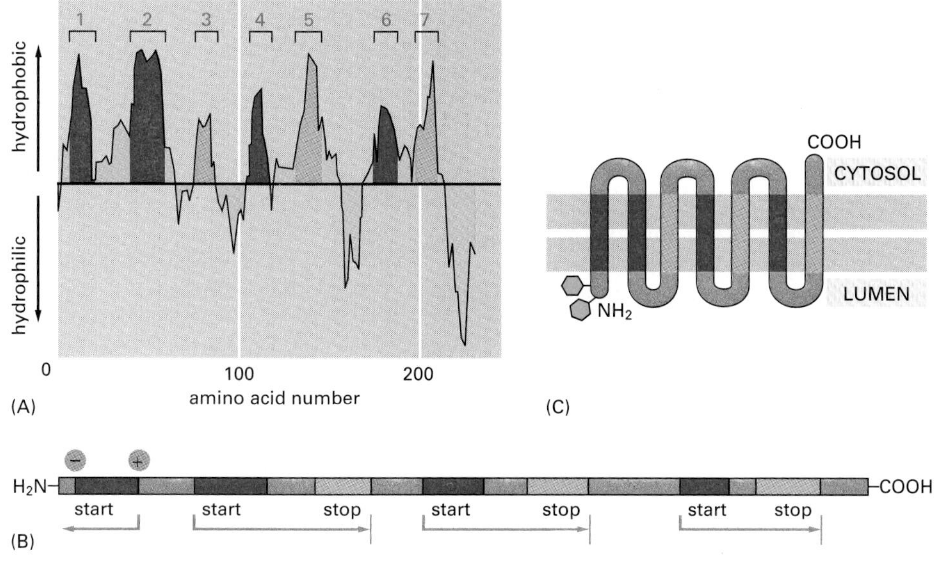

Figure 12–50 The insertion of the multipass membrane protein rhodopsin into the ER membrane. Rhodopsin is the light-sensitive protein in rod photoreceptor cells in the mammalian retina (discussed in Chapter 15). (A) A hydrophobicity plot identifies seven short hydrophobic regions in rhodopsin. (B) The most N-terminal region serves as a start-transfer sequence that causes the preceding N-terminal portion of the protein to be passed across the ER membrane. Subsequent hydrophobic sequences function in alternation as start-transfer and stop-transfer sequences. (C) The final integrated rhodopsin has its N-terminus located in the ER lumen and its C-terminus located in the cytosol. The *blue hexagons* represent covalently attached oligosaccharides. Arrows indicate the parts of the protein that are inserted into the translocator.

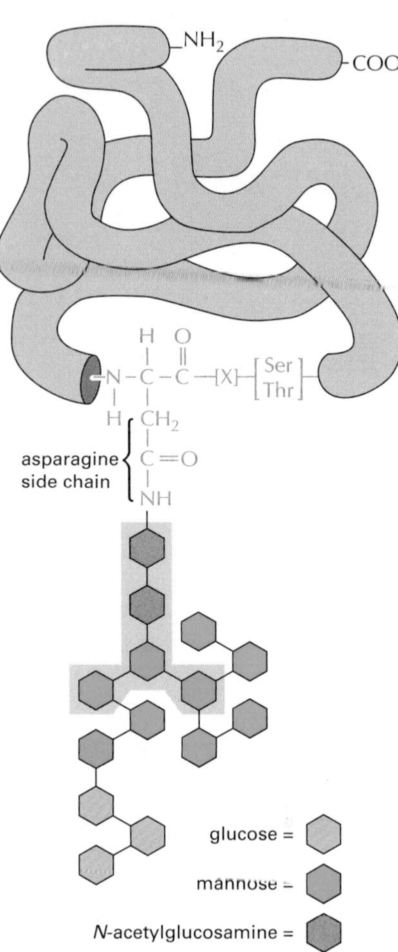

Figure 12–51 The asparagine-linked (*N*-linked) precursor oligosaccharide that is added to most proteins in the rough ER membrane. The five sugars in the *gray box* form the "core region" of this oligosaccharide. For many glycoproteins, only the core sugars survive the extensive oligosaccharide trimming process that takes place in the Golgi apparatus. Only asparagines in the sequences Asn-X-Ser and Asn-X-Thr (where X is any amino acid except proline) become glycosylated. These two sequences occur much less frequently in glycoproteins than in nonglycosylated cytosolic proteins; evidently there has been selective pressure against these sequences during protein evolution, presumably because glycosylation at too many sites would interfere with protein folding.

hydrophobic and hydrophilic amino acids that would normally be buried in a β sheet. The bound BiP both prevents the protein from aggregating and helps to keep it in the ER (and thus out of the Golgi apparatus and later parts of the secretory pathway). Like the hsp70 family of proteins, which bind unfolded proteins in the cytosol and facilitate their import into mitochondria and chloroplasts, BiP hydrolyzes ATP to provide the energy for its roles in protein folding and post-translational import into the ER.

Most Proteins Synthesized in the Rough ER Are Glycosylated by the Addition of a Common *N*-linked Oligosaccharide

The covalent addition of sugars to proteins is one of the major biosynthetic functions of the ER. Most of the soluble and membrane-bound proteins that are made in the ER—including those destined for transport to the Golgi apparatus, lysosomes, plasma membrane, or extracellular space—are **glycoproteins**. In contrast, very few proteins in the cytosol are glycosylated, and those that are carry a much simpler sugar modification, in which a single *N*-acetylglucosamine group is added to a serine or threonine residue of the protein.

An important advance in understanding the process of **protein glycosylation** was the discovery that a preformed *precursor oligosaccharide* (composed of *N*-acetylglucosamine, mannose, and glucose and containing a total of 14 sugars) is transferred *en bloc* to proteins in the ER. Because this oligosaccharide is transferred to the side-chain NH_2 group of an asparagine amino acid in the protein, it is said to be *N-linked* or *asparagine-linked* (Figure 12–51). The transfer is catalyzed by a membrane-bound enzyme, an *oligosaccharyl transferase*, which has its active site exposed on the lumenal side of the ER membrane; this explains why cytosolic proteins are not glycosylated in this way. The precursor oligosaccharide is held in the ER membrane by a special lipid molecule called **dolichol**, and it is transferred to the target asparagine in a single enzymatic step immediately after that amino acid has emerged into the ER lumen during protein translocation (Figure 12–52). Since most proteins are co-translationally imported into the ER, *N*-linked oligosaccharides are almost always added during protein synthesis.

The precursor oligosaccharide is linked to the dolichol lipid by a high-energy pyrophosphate bond, which provides the activation energy that drives the glycosylation reaction illustrated in Figure 12–52. The entire precursor oligosaccharide is built up sugar by sugar on this membrane-bound lipid molecule before its transfer to a protein. The sugars are first activated in the cytosol by the formation of *nucleotide-sugar intermediates,* which then donate their sugar (directly or indirectly) to the lipid in an orderly sequence. Partway through this process, the lipid-linked oligosaccharide is flipped from the cytosolic to the lumenal side of the ER membrane (Figure 12–53).

All of the diversity of the *N*-linked oligosaccharide structures on mature glycoproteins results from the later modification of the original precursor oligosaccharide. While still in the ER, three glucoses (see Figure 12–51) and one mannose are quickly removed from the oligosaccharides of most glycoproteins. We shall return to the importance of glucose trimming shortly. This oligosaccharide "trimming" or "processing" continues in the Golgi apparatus and is discussed in Chapter 13.

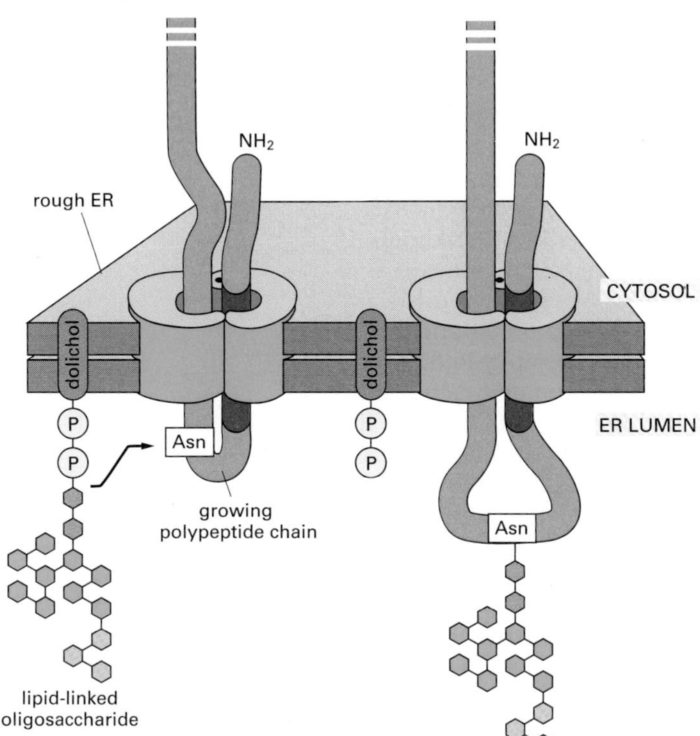

Figure 12–52 Protein glycosylation in the rough ER. Almost as soon as a polypeptide chain enters the ER lumen, it is glycosylated on target asparagine amino acids. The precursor oligosaccharide shown in Figure 12–51 is transferred to the asparagine as an intact unit in a reaction catalyzed by a membrane-bound *oligosaccharyl transferase* enzyme. As with signal peptidase, one copy of this enzyme is associated with each protein translocator in the ER membrane. (The ribosome is not shown for clarity.)

The *N*-linked oligosaccharides are by far the most common oligosaccharides found on glycoproteins. Less frequently, oligosaccharides are linked to the hydroxyl group on the side chain of a serine, threonine, or hydroxylysine amino acid. These *O-linked oligosaccharides* are formed in the Golgi apparatus by pathways that are not yet fully understood.

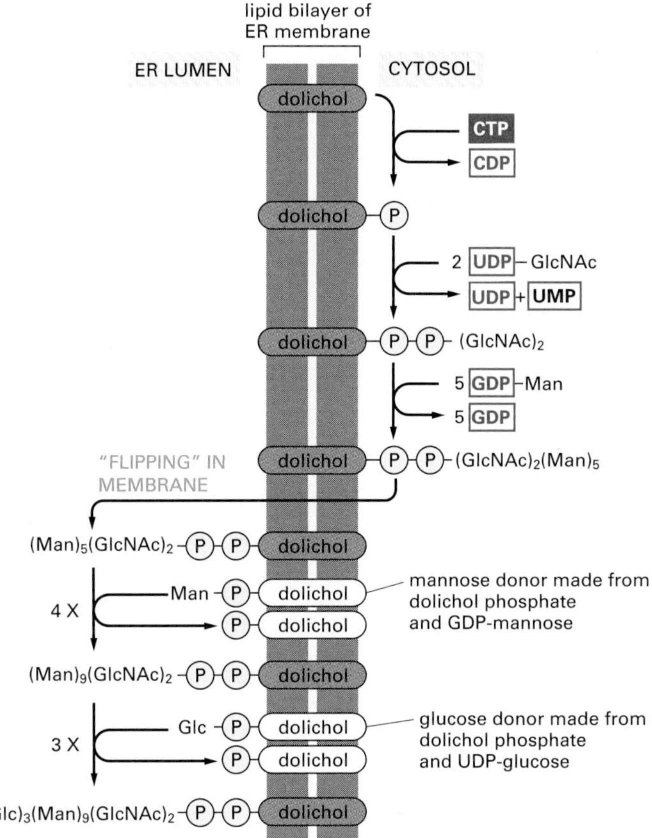

Figure 12–53 Synthesis of the lipid-linked precursor oligosaccharide in the rough ER membrane. The oligosaccharide is assembled sugar by sugar onto the carrier lipid dolichol (a polyisoprenoid; see Panel 2–5, pp. 118–119). Dolichol is long and very hydrophobic: its 22 five-carbon units can span the thickness of a lipid bilayer more than three times, so that the attached oligosaccharide is firmly anchored in the membrane. The first sugar is linked to dolichol by a pyrophosphate bridge. This high-energy bond activates the oligosaccharide for its eventual transfer from the lipid to an asparagine side chain of a nascent polypeptide on the lumenal side of the rough ER. As indicated, the synthesis of the oligosaccharide starts on the cytosolic side of the ER membrane and continues on the lumenal face after the $(Man)_5(GlcNAc)_2$ lipid intermediate is flipped across the bilayer by a transporter protein. All the subsequent glycosyl transfer reactions on the lumenal side of the ER involve transfers from dolichol-P-glucose and dolichol-P-mannose; these activated, lipid-linked monosaccharides are synthesized from dolichol phosphate and UDP-glucose or GDP-mannose (as appropriate) on the cytosolic side of the ER and are then thought to be flipped across the ER membrane. GlcNAc = *N*-acetylglucosamine; Man = mannose; Glc = glucose.

Oligosaccharides Are Used as Tags to Mark the State of Protein Folding

It has long been debated why glycosylation is such a common modification of proteins that enter the ER. One particularly puzzling observation has been that some proteins require *N*-linked glycosylation for proper folding in the ER, yet the precise location of the oligosaccharides attached to the protein's surface does not seem to matter. A clue to the role of glycosylation in protein folding came from studies of two ER chaperone proteins that are called **calnexin** and **calreticulin** because they require Ca^{2+} for their activities. These chaperones are lectins that bind to oligosaccharides on incompletely folded proteins and retain them in the ER. Like other chaperones, they prevent incompletely folded proteins from undergoing irreversible aggregation. Both calnexin and calreticulin also promote the association of incompletely folded protein with another ER chaperone, which binds to cysteines that have not yet formed disulfide bonds.

Calnexin and calreticulin recognize *N*-linked oligosaccharides that contain a single terminal glucose, and therefore bind proteins only after two of the three glucoses that are initially attached have been removed by ER glucosidases. When the third glucose is removed, the protein dissociates from its chaperone and can leave the ER.

How, then, do calnexin and calreticulin distinguish folded from incompletely folded proteins? The answer lies in yet another ER enzyme, a glucosyl transferase that keeps adding a glucose to those oligosaccharides that have lost their last glucose. It adds the glucose, however, only to oligosaccharides that are attached to unfolded proteins. Thus, an unfolded protein undergoes continuous cycles of glucose trimming (by glucosidase) and addition (by glycosyl transferase), and maintains an affinity for calnexin and calreticulin until it has achieved its fully folded state (Figure 12–54).

Improperly Folded Proteins Are Exported from the ER and Degraded in the Cytosol

Despite all the help from chaperones, many protein molecules (more than 80% for some proteins) translocated into the ER fail to achieve their properly folded or oligomeric state. Such proteins are exported from the ER back into the cytosol, where they are degraded. The retrotranslocation, also called *dislocation*, occurs via the same translocator (the Sec61 complex) through which the proteins entered the ER in the first place, although additional proteins help the

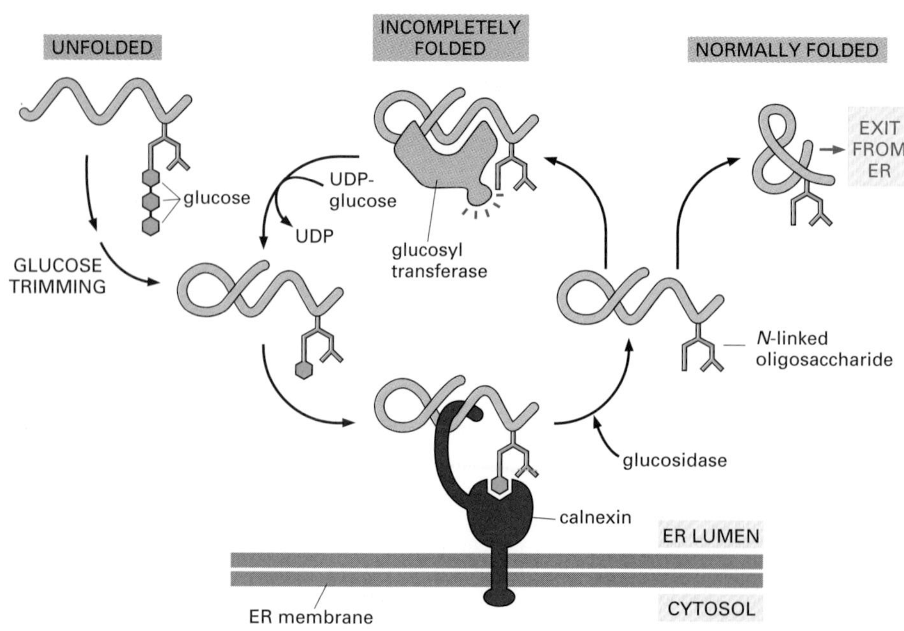

Figure 12–54 The role of *N*-linked glycosylation in ER protein folding. The ER-membrane-bound chaperone protein calnexin binds to incompletely folded proteins containing one terminal glucose on *N*-linked oligosaccharides, trapping the protein in the ER. Removal of the terminal glucose by a glucosidase releases the protein from calnexin. A glucosyl transferase is the crucial enzyme that determines whether the protein is folded properly or not: if the protein is still incompletely folded, the enzyme transfers a new glucose from UDP-glucose to the *N*-linked oligosaccharide, renewing the protein's affinity for calnexin and retaining it in the ER. The cycle repeats until the protein has folded completely. Calreticulin functions similarly, except that it is a soluble ER resident protein. Another ER chaperone, ERp57 (not shown), collaborates with calnexin and calreticulin in retaining an incompletely folded protein in the ER.

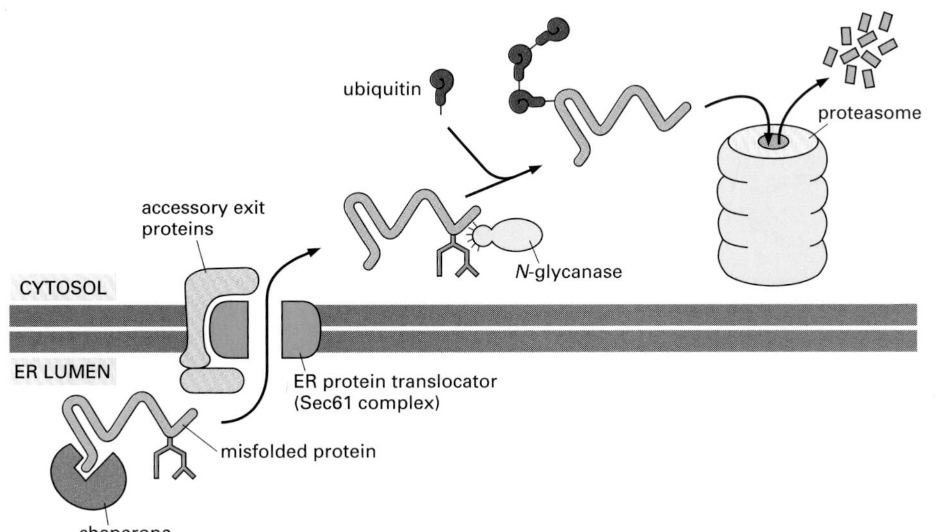

Figure 12–55 The export and degradation of misfolded ER proteins. Misfolded soluble proteins in the ER lumen are translocated back into the cytosol, where they are deglycosylated, ubiquitylated, and degraded in proteasomes. Misfolded membrane proteins follow a similar pathway. Misfolded proteins are exported through the same type of translocator that mediated their import; accessory proteins that are associated with the translocator allow it to operate in the export direction.

translocator to function in reverse. It is not known how such misfolded proteins, which no longer have their ER signal sequences, are recognized or transferred.

Once the misfolded protein has reached the cytosol, its oligosaccharides are removed. Deglycosylation is catalyzed by an *N*-glycanase, which removes the oligosaccharide chains by cleaving the amide bond between the carbonyl group and the amino group of the original asparagine to which the oligosaccharide was attached. The deglycosylated polypeptide is rapidly ubiquitylated by ER-bound ubiquitin-conjugating enzymes and is then fed into proteasomes (discussed in Chapter 6), where it is degraded (Figure 12–55).

Misfolded Proteins in the ER Activate an Unfolded Protein Response

Cells carefully monitor the amount of misfolded proteins they contain in various compartments. An accumulation of misfolded proteins in the cytosol, for example, triggers a *heat-shock response* (discussed in Chapter 6), which stimulates the transcription of genes encoding cytosolic chaperones that help to refold the proteins. Similarly, an accumulation of misfolded proteins in the ER triggers an **unfolded protein response**, which includes an increased transcription of genes encoding ER chaperones and enzymes involved in ER protein degradation.

How do misfolded proteins in the cytosol or ER signal to the nucleus? The pathway from the ER to the nucleus is especially well understood in yeast cells, and it is remarkable. A transmembrane protein kinase in the ER is activated by misfolded proteins, which cause its oligomerization and autophosphorylation. (Extracellular growth factors activate their receptors in the plasma membrane in a similar way, as discussed in Chapter 15). Oligomerization of the ER kinase leads to the activation of an endoribonuclease domain contained on the same molecule. This nuclease cleaves a specific, cytosolic RNA molecule at two positions, excising an intron. The separated exons are then joined by an RNA ligase, generating a spliced mRNA, which is translated on ribosomes to produce a gene regulatory protein. The protein migrates to the nucleus and activates the transcription of the genes encoding the proteins that mediate the unfolded protein response (Figure 12–56).

Some Membrane Proteins Acquire a Covalently Attached Glycosylphosphatidylinositol (GPI) Anchor

As discussed in Chapter 10, several cytosolic enzymes catalyze the covalent addition of a single fatty acid chain or prenyl group to selected proteins. The attached lipids help to direct these proteins to cell membranes. A related process

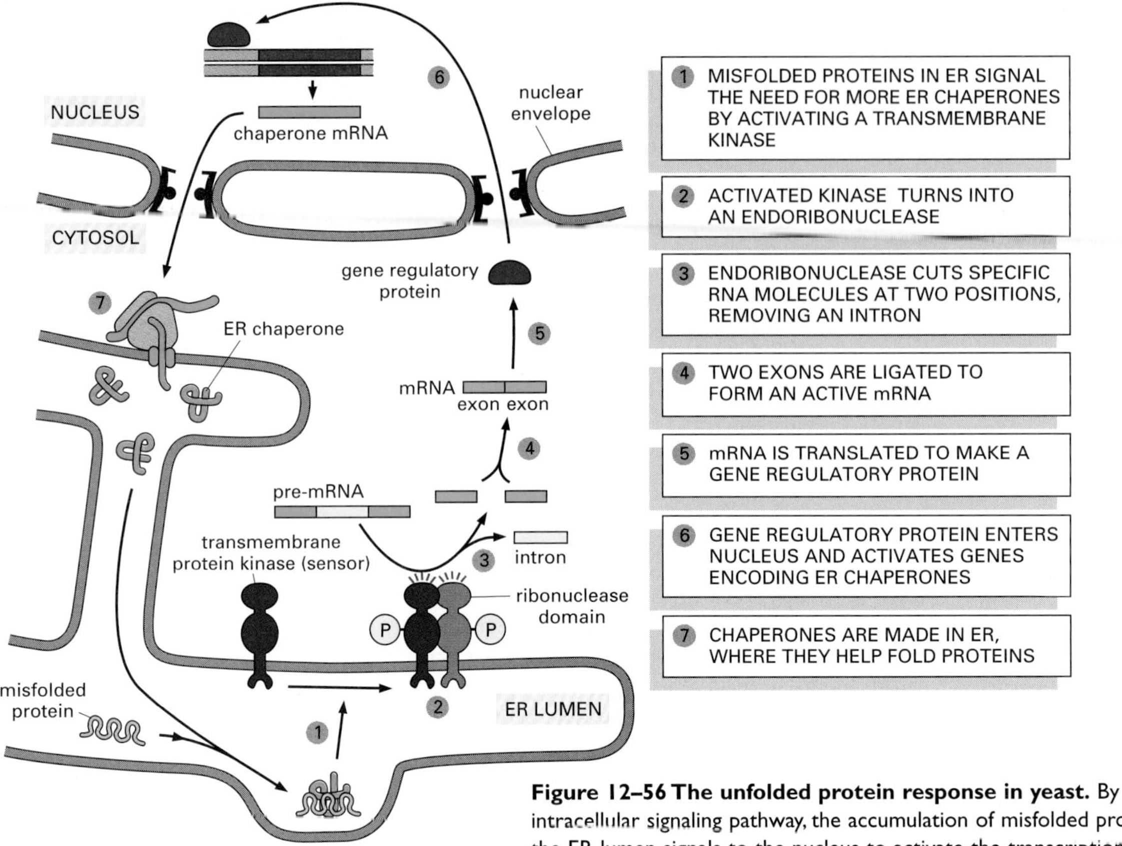
NUCLEUS
chaperone mRNA
6
nuclear envelope
CYTOSOL
gene regulatory protein
7
ER chaperone
5
mRNA
exon exon
4
pre-mRNA
transmembrane protein kinase (sensor)
3
intron
ribonuclease domain
P
P
misfolded protein
1
2
ER LUMEN
1 MISFOLDED PROTEINS IN ER SIGNAL THE NEED FOR MORE ER CHAPERONES BY ACTIVATING A TRANSMEMBRANE KINASE
2 ACTIVATED KINASE TURNS INTO AN ENDORIBONUCLEASE
3 ENDORIBONUCLEASE CUTS SPECIFIC RNA MOLECULES AT TWO POSITIONS, REMOVING AN INTRON
4 TWO EXONS ARE LIGATED TO FORM AN ACTIVE mRNA
5 mRNA IS TRANSLATED TO MAKE A GENE REGULATORY PROTEIN
6 GENE REGULATORY PROTEIN ENTERS NUCLEUS AND ACTIVATES GENES ENCODING ER CHAPERONES
7 CHAPERONES ARE MADE IN ER, WHERE THEY HELP FOLD PROTEINS

Figure 12–56 The unfolded protein response in yeast. By this novel intracellular signaling pathway, the accumulation of misfolded proteins in the ER lumen signals to the nucleus to activate the transcription of genes that encode proteins that help the cell to cope with the abundance of misfolded proteins in the ER.

is catalyzed by ER enzymes, which covalently attach a **glycosylphosphatidylinositol (GPI) anchor** to the C terminus of some membrane proteins destined for the plasma membrane. This linkage forms in the lumen of the ER, where, at the same time, the transmembrane segment of the protein is cleaved off (Figure 12–57). A large number of plasma membrane proteins are modified in this way. Since they are attached to the exterior of the plasma membrane only by their GPI anchors, they can in principle be released from cells in soluble form in response to signals that activate a specific phospholipase in the plasma membrane. Trypanosome parasites, for example, use this mechanism to shed their coat of GPI-anchored surface proteins if attacked by the immune system. GPI anchors are also used to direct plasma membrane proteins into *lipid rafts* and thus segregate the proteins from other membrane proteins, as we discuss in Chapter 13.

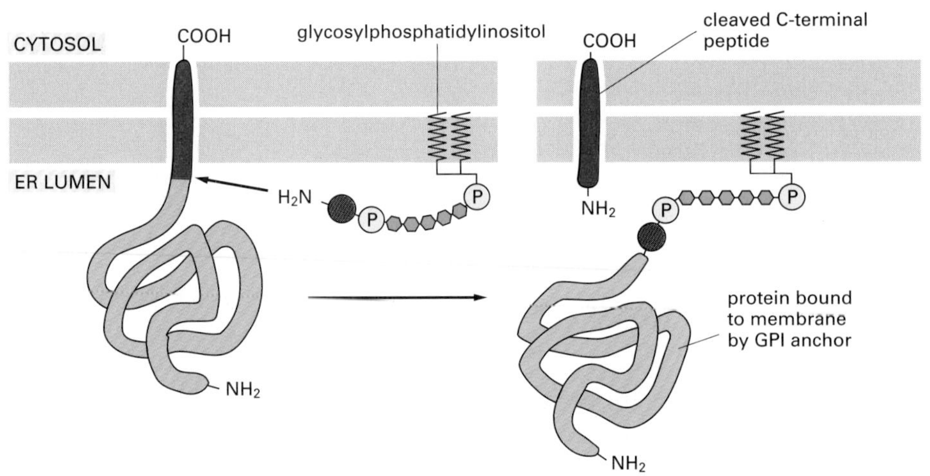
CYTOSOL
COOH
glycosylphosphatidylinositol
COOH
cleaved C-terminal peptide
ER LUMEN
H_2N
P
P
NH_2
P
P
NH_2
protein bound to membrane by GPI anchor
NH_2

Figure 12–57 The attachment of a GPI anchor to a protein in the ER. Immediately after the completion of protein synthesis, the precursor protein remains anchored in the ER membrane by a hydrophobic C-terminal sequence of 15–20 amino acids; the rest of the protein is in the ER lumen. Within less than a minute, an enzyme in the ER cuts the protein free from its membrane-bound C terminus and simultaneously attaches the new C terminus to an amino group on a preassembled GPI intermediate. The signal that specifies this modification is contained within the hydrophobic C-terminal sequence and a few amino acids adjacent to it on the lumenal side of the ER membrane; if this signal is added to other proteins, they too become modified in this way. Because of the covalently linked lipid anchor, the protein remains membrane-bound, with all of its amino acids exposed initially on the lumenal side of the ER and eventually on the cell exterior.

Most Membrane Lipid Bilayers Are Assembled in the ER

The ER membrane synthesizes nearly all of the major classes of lipids, including both phospholipids and cholesterol, required for the production of new cell membranes. The major phospholipid made is *phosphatidylcholine* (also called *lecithin*), which can be formed in three steps from choline, two fatty acids, and glycerol phosphate (Figure 12–58). Each step is catalyzed by enzymes in the ER membrane that have their active sites facing the cytosol, where all of the required metabolites are found. Thus, phospholipid synthesis occurs exclusively in the cytosolic leaflet of the ER membrane. In the first step, acyl transferases successively add two fatty acids to glycerol phosphate to produce phosphatidic acid, a compound sufficiently water-insoluble to remain in the lipid bilayer after it has been synthesized. It is this step that enlarges the lipid bilayer. The later steps determine the head group of a newly formed lipid molecule, and therefore the chemical nature of the bilayer, but they do not result in net membrane growth. The two other major membrane phospholipids—phosphatidylethanolamine and phosphatidylserine—as well as the minor phospholipid phosphatidylinositol (PI), are all synthesized in this way.

Because phospholipid synthesis takes place in the cytosolic half of the ER bilayer, there needs to be a mechanism that transfers some of the newly formed phospholipid molecules to the lumenal leaflet of the bilayer. In synthetic lipid bilayers, lipids do not "flip-flop" in this way. In the ER, however, phospholipids equilibrate across the membrane within minutes, which is almost 100,000 times faster than can be accounted for by spontaneous "flip-flop." This rapid transbilayer movement is thought to be mediated by a phospholipid translocator called a *scramblase* that equilibrates phospholipids between the two leaflets of

Figure 12–58 The synthesis of phosphatidylcholine. This phospholipid is synthesized from fatty acyl-coenzyme A (fatty acyl CoA), glycerol 3-phosphate, and cytidine-bisphosphocholine (CDP-choline).

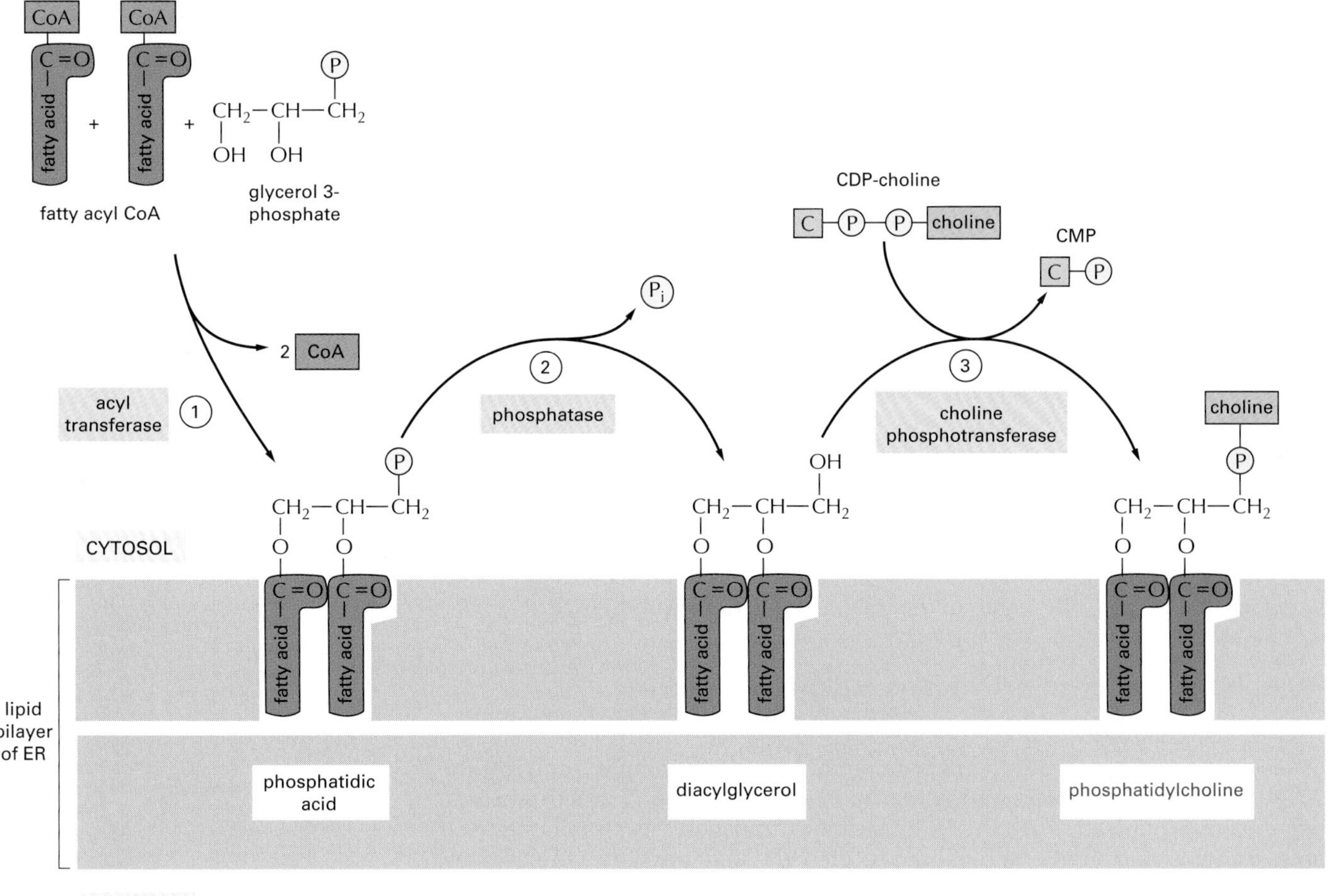

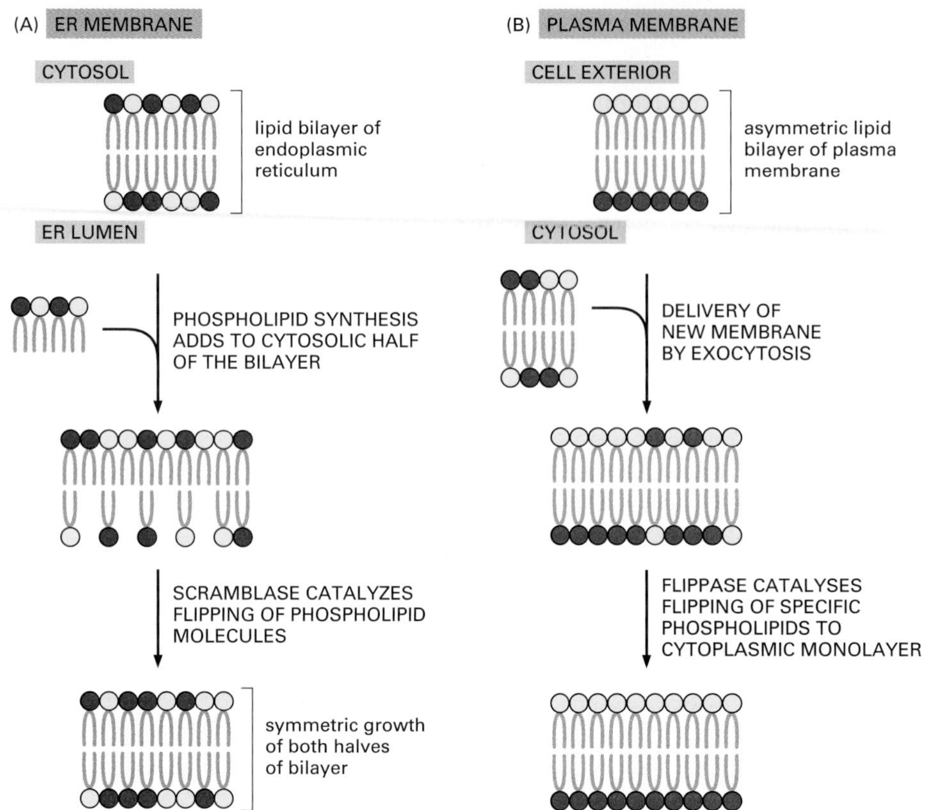

Figure 12–59 The role of phospholipid translocators in lipid bilayer synthesis. (A) Because new lipid molecules are added only to the cytosolic half of the bilayer and lipid molecules do not flip spontaneously from one monolayer to the other, a membrane-bound phospholipid translocator (called a scramblase) is required to transfer lipid molecules from the cytosolic half to the lumenal half so that the membrane grows as a bilayer. The scramblase is not specific for particular phospholipid head groups and therefore equilibrates the different phospholipids between the two monolayers. (B) Fueled by ATP hydrolysis, a head-group-specific flippase in the plasma membrane actively flips phosphatidylserine and phosphatidylethanolamine directionally from the extracellular to the cytosolic leaflet, creating the characteristically asymmetric lipid bilayer of the plasma membrane of animal cells (see Figure 10–14). A scramblase is also present in plasma membranes to ensure that both monolayers remain equally populated with lipids; the continuous action of the flippase is therefore necessary to maintain the phospholipid asymmetry.

the lipid bilayer (Figure 12–59). Thus, the different types of phospholipids are thought to be equally distributed between the two leaflets of the ER membrane. The plasma membrane contains, in addition to the scramblase, a different type of phospholipid translocator that belongs to the family of ABC transporters (discussed in Chapter 11). These *flippases* specifically remove phospholipids containing free amino groups (phosphatidylserine and phosphatidylethanolamine) from the extracellular leaflet and use the energy of ATP hydrolysis to flip them directionally into the leaflet facing the cytosol. The plasma membrane therefore has a highly asymmetric phospholipid composition, which is actively maintained by the flippases (see Figure 10–14).

The ER also produces cholesterol and ceramide. *Ceramide* is made by condensing the amino acid serine with a fatty acid to form the amino alcohol *sphingosine;* a second fatty acid is then added to form ceramide. The ceramide is exported to the Golgi apparatus, where it serves as a precursor for the synthesis of two types of lipids: oligosaccharide chains are added to form *glycosphingolipids* (glycolipids), and phosphocholine head groups are transferred from phosphatidylcholine to other ceramide molecules to form *sphingomyelin*. Thus, both glycolipids and sphingomyelin are produced relatively late in the process of membrane synthesis. Because they are produced by enzymes exposed to the Golgi lumen and are not substrates for lipid translocators, they are found exclusively in the noncytosolic leaflet of the lipid bilayers that contain them.

Phospholipid Exchange Proteins Help to Transport Phospholipids from the ER to Mitochondria and Peroxisomes

As discussed in Chapter 13, the plasma membrane and the membranes of the Golgi apparatus, lysosomes, and endosomes all form part of a membrane system that communicates with the ER by means of transport vesicles that transfer both proteins and lipids. Mitochondria, plastids, and possibly peroxisomes, however, do not belong to this system, and they therefore require different mechanisms for the import of proteins and lipids for growth. We have already seen that most of the proteins in these organelles are imported from the cytosol. Although mitochondria modify some of the lipids they import, they do not synthesize lipids from scratch; instead, their lipids have to be imported from the ER, either directly, or indirectly by way of other cell membranes. In either case, special mechanisms are required for the transfer.

Water-soluble carrier proteins called **phospholipid exchange proteins** (or *phospholipid transfer proteins)* transfer individual phospholipid molecules between membranes. Each exchange protein recognizes only specific types of phospholipids. It functions by "extracting" a molecule of the appropriate phospholipid from a membrane and diffusing away with the lipid buried within its lipid-binding site. When it encounters another membrane, the exchange protein tends to discharge the bound phospholipid molecule into the new lipid bilayer (Figure 12–60). It has been proposed that phosphatidylserine is imported into mitochondria in this way, where it is then decarboxylated to yield phosphatidylethanolamine. Phosphatidylcholine, by contrast, is imported intact.

Exchange proteins act to distribute phospholipids at random between all membranes present. In principle, such a random exchange process can result in a net transport of lipids from a lipid-rich to a lipid-poor membrane, allowing phosphatidylcholine and phosphatidylserine molecules, for example, to be transferred from the ER, where they are synthesized, to a mitochondrial or peroxisomal membrane. It might be that mitochondria and peroxisomes are the only "lipid-poor" organelles in the cytosol and that such an exchange process is sufficient. In electron micrographs, mitochondria are often seen in close juxtaposition to ER membranes, and there may be specific mechanisms of lipid transfer that operate at such regions of proximity.

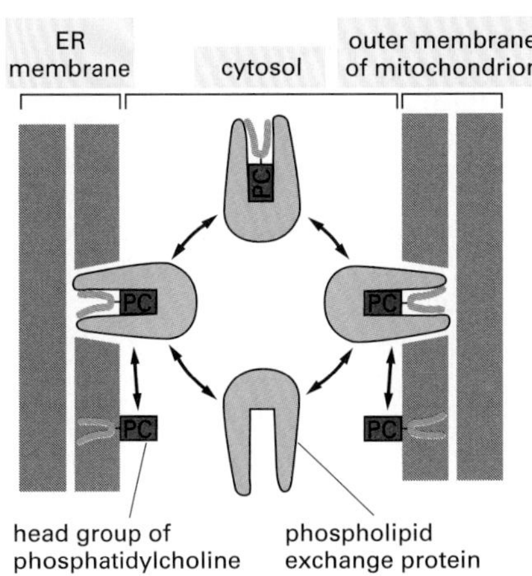

Figure 12–60 Phospholipid exchange proteins. Because phospholipids are insoluble in water, their passage between membranes requires carrier proteins. Phospholipid exchange proteins are water-soluble proteins that carry a single molecule of phospholipid at a time; they can pick up a lipid molecule from one membrane and release it at another, thereby redistributing phospholipids between membrane-enclosed compartments. The net transfer of phosphatidylcholine (PC) from the ER to mitochondria can occur without the input of additional energy, because the concentration of PC is high in the ER membrane (where it is made) and low in the outer mitochondrial membrane. One predicts a lipid translocator in the outer mitochondrial membrane to equilibrate the lipids between the two leaflets of its bilayer, and there must also be a mechanism to transfer lipids between the outer and inner mitochondrial membranes. These postulated pathways, however, remain to be discovered.

Summary

The extensive ER network serves as a factory for the production of almost all of the cell's lipids. In addition, a major portion of the cell's protein synthesis occurs on the cytosolic surface of the ER: all proteins destined for secretion and all proteins destined for the ER itself, the Golgi apparatus, the lysosomes, the endosomes, and the plasma membrane are first imported into the ER from the cytosol. In the ER lumen, the proteins fold and oligomerize, disulfide bonds are formed, and N-linked oligosaccharides are added. N-linked glycosylation is used to indicate the extent of protein folding, so that proteins leave the ER only when they are properly folded. Proteins that do not fold or oligomerize correctly are translocated back into the cytosol, where they are deglycosylated, ubiquitylated, and degraded in proteasomes. If misfolded proteins accumulate excessively in the ER, they trigger an unfolded protein response, which activates appropriate genes in the nucleus to help the ER to cope.

Only proteins that carry a special ER signal sequence are imported into the ER. The signal sequence is recognized by a signal recognition particle (SRP), which binds both the growing polypeptide chain and a ribosome and directs them to a receptor protein on the cytosolic surface of the rough ER membrane. This binding to the ER membrane initiates the translocation process by threading a loop of polypeptide chain across the ER membrane through the hydrophilic pore in a transmembrane protein translocator.

Soluble proteins—destined for the ER lumen, for secretion, or for transfer to the lumen of other organelles—pass completely into the ER lumen. Transmembrane proteins destined for the ER or for other cell membranes are translocated partway across the ER membrane and remain anchored there by one or more membrane-spanning α-helical regions in their polypeptide chains. These hydrophobic portions of the protein can act either as start-transfer or stop-transfer signals during the translocation process. When a polypeptide contains multiple, alternating start-transfer and stop-transfer signals, it will pass back and forth across the bilayer multiple times as a multipass transmembrane protein.

The asymmetry of protein insertion and glycosylation in the ER establishes the sidedness of the membranes of all of the other organelles that the ER supplies with membrane proteins.

References

General

Palade G (1975) Intracellular aspects of the process of protein synthesis. *Science* 189, 347–358.

The Compartmentalization of Cells

Blobel G (1980) Intracellular protein topogenesis. *Proc. Natl. Acad. Sci. USA* 77, 1496–1500.

Martoglio B & Dobberstein B (1998) Signal sequences: more than just greasy peptides. *Trends Cell Biol.* 8, 410–415.

von Heijne G (1990) Protein targeting signals. *Curr. Opin. Cell Biol.* 2, 604–608.

Warren G & Wickner W (1996) Organelle inheritance. *Cell* 84, 395–400.

The Transport of Molecules Between the Nucleus and the Cytosol

Adam SA (1999) Transport pathways of macromolecules between the nucleus and the cytoplasm. *Curr. Opin. Cell Biol.* 11, 402–406.

Arts G-J, Fornerod M & Mattaj IW (1998) Identification of a nuclear export receptor for tRNA. *Curr. Biol.* 8, 305–314.

Blobel G & Wozniak RW (2000) Proteomics for the pore. *Nature* 403, 835–836.

Gant TM & Wilson KL (1997) Nuclear assembly. *Annu. Rev. Cell Dev. Biol.* 13, 669–695.

Görlich D & Mattaj IW (1996) Nucleocytoplasmic transport. *Science* 271, 1513–1518.

Kaffman A & O'Shea EK (1999) Regulation of nuclear localization: a key to a door. *Annu. Rev. Cell Dev. Biol.* 15, 291–339.

Kalderon D, Roberts BL, Richardson WD & Smith AE (1984) A short amino acid sequence able to specify nuclear location. *Cell* 39, 499–509.

Lyman SK & Gerace L (2001) Nuclear pore complexes: dynamics in unexpected places. *J. Cell Biol.* 154, 17–20.

Mattaj IW & Conti E (1999) Snail mail to the nucleus. *Nature* 399, 208–210.

Politz JC & Pederson T (2000) Review: movement of mRNA from transcription site to nuclear pores. *J. Struct. Biol.* 129, 252–257.

Stoffler D, Fahrenkrog B & Aebi U (1999) The nuclear pore complex: from molecular architecture to functional dynamics. *Curr. Opin. Cell Biol.* 11, 391–401.

Weis K (1998) Importins and exportins: how to get in and out of the nucleus. *Trends Biochem. Sci.* 23, 185–189.

Wente SR (2000) Gatekeepers of the nucleus. *Science* 288, 1374–1377.

The Transport of Proteins Into Mitochondria and Chloroplasts

Chen X & Schnell DJ (1999) Protein import into chloroplasts. *Trends Cell Biol.* 9, 222–227.

Cline K & Henry R (1996) Import and routing of nucleus-encoded chloroplast proteins. *Annu. Rev. Cell Dev. Biol.* 12, 1–26.

Dalbey RE & Robinson C (1999) Protein translocation into and across the bacterial plasma membrane and the plant thylakoid membrane. *Trends Biochem. Sci.* 24, 17–22.

Haucke V & Schatz G (1997) Import of proteins into mitochondria and chloroplasts. *Trends Cell Biol.* 7, 103–106.

Jensen RE & Johnson AE (1999) Protein translocation: is Hsp70 pulling my chain? *Curr. Biol.* 9, R779–R782.

Neupert W (1997) Protein import into mitochondria. *Annu. Rev. Biochem.* 66, 863–917.

Rassow J & Pfanner N (2000) The protein import machinery of the mitochondrial membranes. *Traffic* 1, 457–464.

Schwartz MP & Matouschek A (1999) The dimensions of the protein import channels in the outer and inner mitochondrial membranes. *Proc. Natl. Acad. Sci. USA* 96, 13086–13090.

Peroxisomes

Purdue PE & Lazarow PB (2001) Peroxisome biogenesis. *Annu. Rev. Cell Dev. Biol.* 17, 701–752.

Tabak HF, Braakman I & Distel B (1999) Peroxisomes: simple in function but complex in maintenance. *Trends Cell Biol.* 9, 447–453.

The Endoplasmic Reticulum

Adelman MR, Sabatini DD & Blobel G (1973) Ribosome-membrane interaction: nondestructive dissembly of rat liver rough microsomes into ribosomal and membranous components. *J. Cell Biol.* 56, 206–229.

Bergeron JJ, Brenner MB, Thomas DY & Williams DB (1994) Calnexin: a membrane-bound chaperone of the endoplasmic reticulum. *Trends Biochem. Sci.* 19, 124–128.

Bishop WR & Bell RM (1988) Assembly of phospholipids into cellular membranes: biosynthesis, transmembrane movement and intracellular translocation. *Annu. Rev. Cell Biol.* 4, 579–610.

Blobel G & Dobberstein B (1975) Transfer of proteins across membranes. *J. Cell Biol.* 67, 835–851.

Borgese N, Mok W, Kreibich G & Sabatini DD (1974) Ribosomal-membrane interaction: in vitro binding of ribosomes to microsomal membranes. *J. Mol. Biol.* 88, 559–580.

Daleke DL & Lyles JV (2000) Identification and purification of aminophospholipid flippases. *Biochim. Biophys. Acta* 1486, 108–127.

deDuve C (1971) Tissue fractionation—past and present. *J. Cell Biol.* 50, 20d–55d.

Deshaies RJ, Sanders SL, Feldhiem DA & Schekman R (1991) Assembly of yeast Sec proteins involved in translocation into the endoplasmic reticulum into a membrane-bound multisubunit complex. *Nature* 349, 806–808.

Ellgaard L & Helenius A (2001) ER quality control: towards an understanding at the molecular level. *Curr. Opin. Cell Biol.* 13, 431–437.

Ferguson MA (1992) Colworth Medal Lecture. Glycosylphosphatidylinositol membrane anchors: the tale of a tail. *Biochem. Soc. Trans.* 20, 243–256.

Gething MJ (1999) Role and regulation of the ER chaperone BiP. *Semin. Cell Dev. Biol.* 10, 465–472.

Görlich D, Prehn S, Hartmann E et al. (1992) A mammalian homolog of SEC61p and SECYp is associated with ribosomes and nascent polypeptides during translocation. *Cell* 71, 489–503.

Johnson AE & van Waes MA (1999) The translocon: a dynamic gateway at the ER membrane. *Annu. Rev. Cell Dev. Biol.* 15, 799–842.

Keenan RJ, Freymann DM, Stroud RM & Walter P (2001) The signal recognition particle. *Annu. Rev. Biochem.* 10, 755–775.

Milstein C, Brownlee G, Harrision T & Mathews MB (1972) A possible precursor of immunoglobulin light chains. *Nature New Biol.* 239, 117–120.

Muniz M & Riezman H (2000) Intracellular transport of GPI-anchored proteins. *EMBO J.* 19, 10–15.

Parodi AJ (2000) Role of N-oligosaccharide endoplasmic reticulum processing reactions in glycoprotein folding and degradation. *Biochem. J.* 348, 1–13.

Plemper RK & Wolf DH (1999) Retrograde protein translocation: ERADication of secretory proteins in health and disease. *Trends Biochem. Sci.* 24, 266–270.

Rogers DP & Bankaitis VA (2000) Phospholipid transfer proteins and physiological functions. *Int. Rev. Cytol.* 197, 35–81.

Römisch K (1999) Surfing the Sec61 channel: bidirectional protein translocation across the ER membrane. *J. Cell Sci.* 112, 4185–4191.

Sidrauski C, Chapman R & Walter P (1998) The unfolded protein response: an intracellular signalling pathway with many surprising features. *Trends Cell Biol.* 8, 245–249.

Simon SM & Blobel G (1991) A protein-conducting channel in the endoplasmic reticulum. *Cell* 65, 371–380.

Staehelin LA (1997) The plant ER: a dynamic organelle composed of a large number of discrete functional domains. *The Plant Journal* 11, 1151–1165.

von Heijne G (1995) Membrane protein assembly: rules of the game. *Bioessays* 17, 25–30.

Yan Q & Lennarz WJ (1999) Oligosaccharyltransferase: a complex multisubunit enzyme of the endoplasmic reticulum. *Biochem. Biophys. Res. Commun.* 266, 684–689.

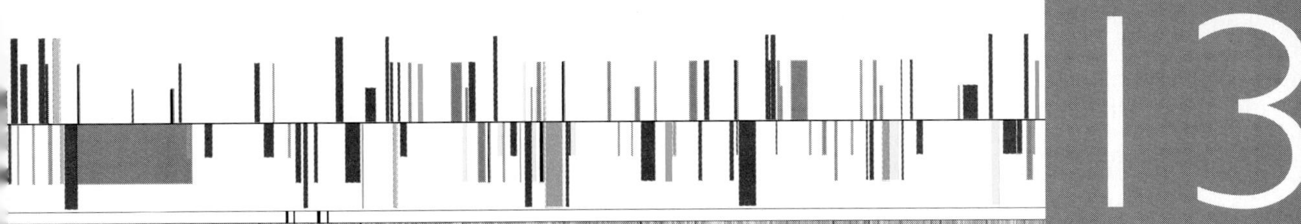

13

INTRACELLULAR VESICULAR TRAFFIC

Every cell must eat, and it must communicate with the world around it. In a procaryotic cell, all the eating and communicating takes place across the plasma membrane. The cell secretes digestive enzymes, for example, across the plasma membrane to the cell exterior. It then transports the small metabolites generated by digestion in the extracellular space across the same membrane into the cytosol. Eucaryotic cells, by contrast, have evolved an elaborate internal membrane system that allows them to take up macromolecules by the process of *endocytosis* and deliver them to digestive enzymes stored in lysosomes inside the cell. As a consequence, metabolites generated by the digestion of macromolecules are delivered directly from the lysosomes to the cytosol as they are produced. In addition to allowing the ingestion of macromolecules by the *endocytic pathway*, the internal membrane system allows eucaryotic cells to regulate the delivery of newly synthesized proteins, carbohydrates, and lipids to the cell exterior. The *biosynthetic–secretory pathway* allows the cell to modify the molecules it produces in a series of steps, store them until needed, and then deliver them to the exterior through a specific cell-surface domain by a process called *exocytosis*. An outline of the endocytic and biosynthetic–secretory pathways, which ultimately connect the plasma membrane to the endoplasmic reticulum (ER) deep within the cell, is shown in Figure 13–1.

The interior space, or *lumen*, of each membrane-enclosed compartment along the biosynthetic–secretory and endocytic pathways is topologically equivalent to the lumen of every other compartment. Moreover, these compartments are in constant communication, with molecules being passed from a donor compartment to a target compartment by means of numerous membrane-enclosed *transport packages*. Some of these packages are small spherical vesicles, while others are larger irregular vesicles or fragments of the donor compartment. We shall use the term *transport vesicle* to apply to all forms of these packages.

Vesicles continually bud off from one membrane and fuse with another, carrying membrane components and soluble molecules referred to as *cargo* (Figure

13–2). This membrane traffic flows along highly organized, directional routes, which allows the cell to secrete and eat. The biosynthetic–secretory pathway leads outward from the ER toward the Golgi apparatus and cell surface, with a side route leading to lysosomes, while the endocytic pathway leads inward from the plasma membrane (Figure 13–3). In each case, the flow of membrane between compartments is balanced, with retrieval pathways balancing the flow in the opposite direction, bringing membrane and selected proteins back to the compartment of origin.

To perform its function, each transport vesicle that buds from a compartment must be selective. It must take up only the appropriate proteins and must fuse only with the appropriate target membrane. A vesicle carrying cargo from the Golgi apparatus to the plasma membrane, for example, must exclude proteins that are to stay in the Golgi apparatus, and it must fuse only with the plasma membrane and not with any other organelle.

We begin this chapter by considering the molecular mechanisms of budding and fusion that underlie all transport. We then discuss the fundamental problem of how, in the face of this transport, the differences between the compartments are maintained. Finally, we consider the function of the Golgi apparatus, lysosomes, secretory vesicles, and endosomes, as we trace the pathways that connect these organelles.

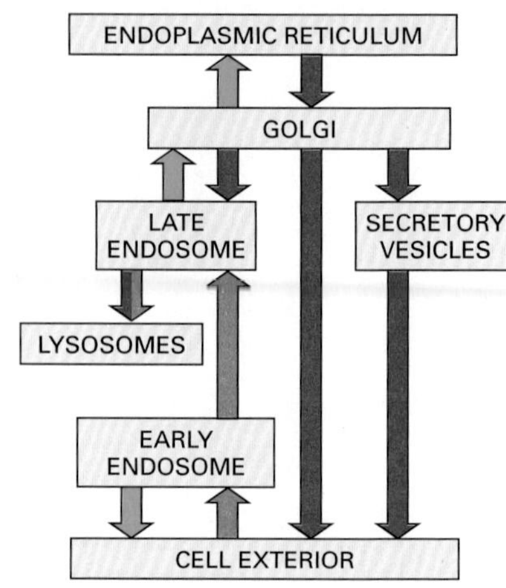

Figure 13–1 The endocytic and biosynthetic–secretory pathways. In this "road map" of biosynthetic protein traffic, which was introduced in Chapter 12, the endocytic and biosynthetic–secretory pathways are illustrated with *green* and *red* arrows, respectively. In addition, *blue* arrows are used to denote retrieval pathways by which the backflow of selected components is maintained (see also Figure 13–3).

THE MOLECULAR MECHANISMS OF MEMBRANE TRANSPORT AND THE MAINTENANCE OF COMPARTMENTAL DIVERSITY

Transport processes mediate a continual exchange of components between the ten or more chemically distinct, membrane-enclosed compartments that collectively comprise the biosynthetic–secretory and endocytic pathways. In the presence of this massive exchange, how can each compartment maintain its specialized character? To answer this question, we must first consider what defines the character of a compartment. Above all, it is the composition of the enclosing membrane: molecular markers displayed on the cytosolic surface of this membrane serve as guidance cues for incoming traffic and ensure that transport vesicles fuse only with the correct compartment, thereby dictating the pattern of traffic between one compartment and another. Many membrane markers, however, are found on more than one organelle, and thus it is the specific combination of marker molecules that gives each organelle its unique molecular address.

How are these membrane markers kept at high concentration on one compartment and at low concentration on another? To answer this question, we

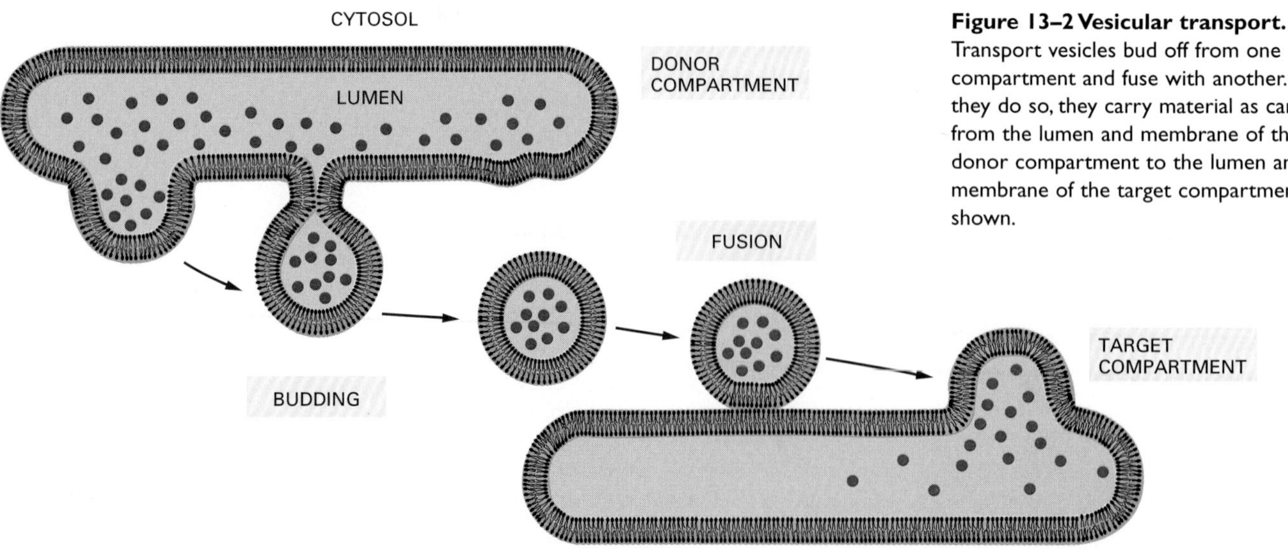

Figure 13–2 Vesicular transport. Transport vesicles bud off from one compartment and fuse with another. As they do so, they carry material as cargo from the lumen and membrane of the donor compartment to the lumen and membrane of the target compartment, as shown.

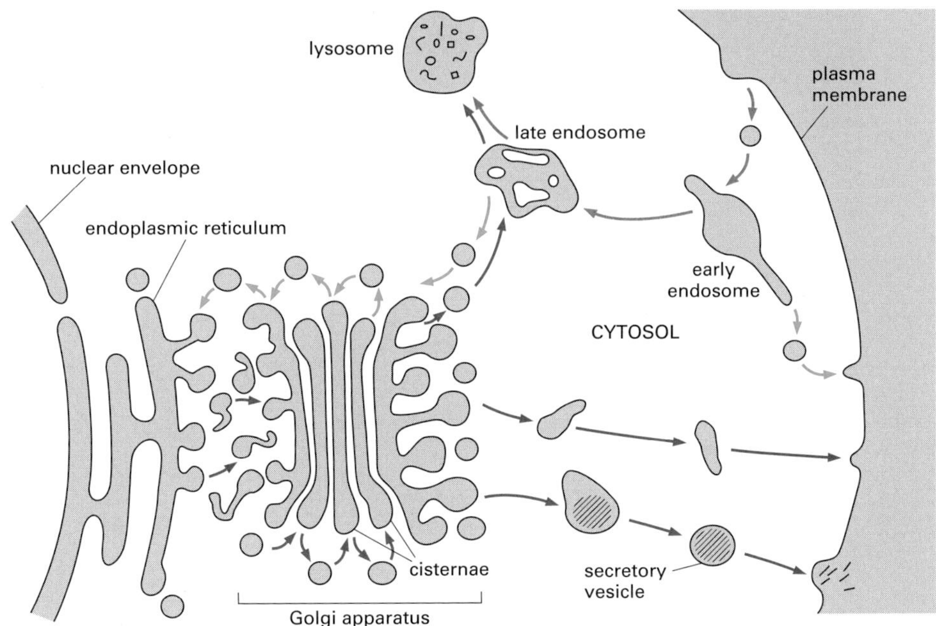

Figure 13–3 The intracellular compartments of the eucaryotic cell involved in the biosynthetic–secretory and endocytic pathways. Each compartment encloses a space, called a *lumen,* that is topologically equivalent to the outside of the cell, and all compartments shown communicate with one another and the outside of the cell by means of transport vesicles. In the biosynthetic–secretory pathway *(red arrows)* protein molecules are transported from the ER to the plasma membrane or (via late endosomes) to lysosomes. In the endocytic pathway *(green arrows)* molecules are ingested in vesicles derived from the plasma membrane and delivered to early endosomes and then (via late endosomes) to lysosomes. Many endocytosed molecules are retrieved from early endosomes and returned to the cell surface for reuse; similarly, some molecules are retrieved from the late endosome and returned to the Golgi apparatus, and some are retrieved from the Golgi apparatus and returned to the ER. All of these retrieval pathways are shown with *blue arrows* (see also Figure 13–1).

need to consider how patches of membrane, enriched or depleted in specific components, bud off from one compartment and transfer to another. In this section we describe how this is achieved. Some of the basic genetic and biochemical strategies that have been used to study the molecular machinery involved in vesicular transport are outlined in Panel 13–1.

We begin by discussing the sorting events that underlie the segregation of proteins into separate membrane domains. This sorting process depends on the assembly of a special protein coat on the cytosolic face of the donor membrane. We shall therefore consider how coats form, what they are made of, and how they enable specific components of a membrane to be extracted and delivered to another membrane. Finally, we discuss how transport vesicles dock at the appropriate target membrane and fuse with it to deliver the contents to their target organelle.

There Are Various Types of Coated Vesicles

Most transport vesicles form from specialized, coated regions of membranes. They bud off as **coated vesicles** that have a distinctive cage of proteins covering their cytosolic surface. Before the vesicle fuses with a target membrane, the coat is discarded, as is required to allow the two cytosolic membrane surfaces to interact directly and fuse.

The coat is thought to perform two principal functions. First, it concentrates specific membrane proteins in a specialized membrane patch that then gives rise to the vesicle membrane. It thus helps select the appropriate molecules for transport. Second, the assembly of the coat proteins into curved, basketlike lattices deforms the membrane patch and thereby molds the forming vesicles, which explains why vesicles with the same type of coat have a relatively uniform size.

There are three well-characterized types of coated vesicles, which differ in their coat proteins: *clathrin-coated, COPI-coated,* and *COPII-coated* vesicles (Figure 13–4). Each type is used for different transport steps in the cell. Clathrin-coated vesicles, for example, mediate transport from the Golgi apparatus and from the plasma membrane, whereas COPI- and COPII-coated vesicles most commonly mediate transport from the ER and the Golgi cisternae (Figure 13–5). There is, however, much more variety than this short list suggests. As we discuss below, there are at least three types of clathrin-coated vesicles, each specialized for a different transport step, and the COPI-coated vesicles may be similarly diverse. Moreover, still other coats have been seen in the electron microscope, whose molecular compositions and functions are not yet known.

CELL-FREE SYSTEMS FOR STUDYING THE COMPONENTS AND MECHANISM OF VESICULAR TRANSPORT

Vesicular transport can be reconstituted in cell-free systems. This was first achieved for the Golgi stack. When Golgi stacks are isolated from cells and incubated with cytosol and with ATP as a source of energy, transport vesicles bud from their rims and appear to transport proteins between cisternae. By following the progressive processing of the oligosaccharides on a glycoprotein as it moves from one Golgi compartment to the next, it is possible to follow the process of vesicular transport.

To follow the transport, two distinct populations of Golgi stacks are incubated together. The "donor" population is isolated from mutant cells that lack the enzyme *N*-acetylglucosamine (GlcNAc) transferase I and that have been infected with a virus; because of the mutation, the major viral glycoprotein fails to be modified with GlcNAc in the Golgi apparatus of the mutant cells. The "acceptor" Golgi stacks are isolated from uninfected wild-type cells and thus contain a good copy of GlcNAc transferase I, but lack the viral glycoprotein. In the mixture of Golgi stacks the viral glycoprotein acquires GlcNAc, indicating that it must have been transported between the Golgi stacks—presumably by vesicles that bud from the *cis* compartment of the donor Golgi and fuse with the *medial* compartment of the acceptor Golgi. This transport-dependent glycosylation is monitored by measuring the transfer of ^{3}H-GlcNAc from UDP-^{3}H-GlcNAc to the viral glycoprotein. Transport occurs only when ATP and cytosol are added. By fractionating the cytosol, a number of specific cytosolic proteins have been identified that are required for the budding and fusion of transport vesicles.

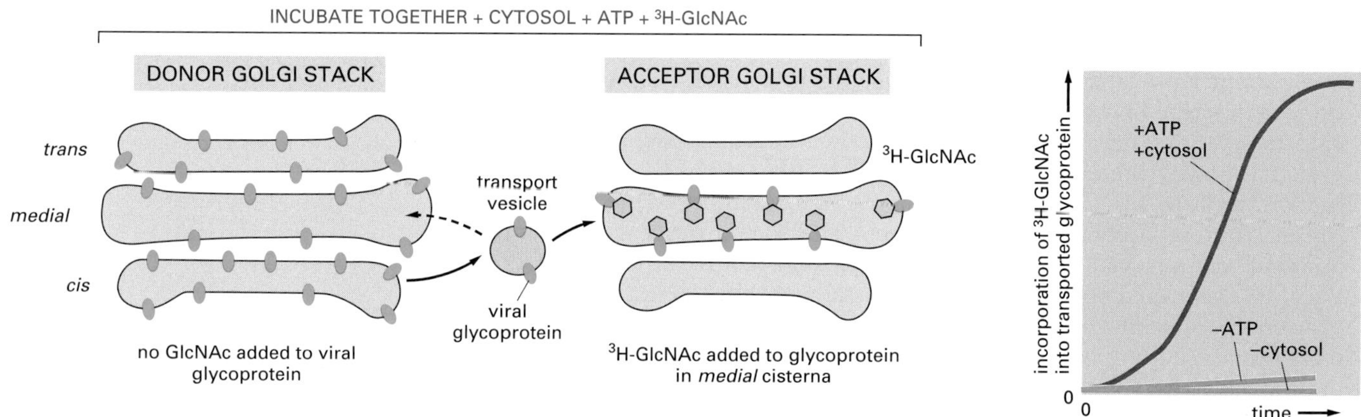

Similar cell-free systems have been used to study transport from the *medial* to the *trans* Golgi network, from the *trans* Golgi network to the plasma membrane, from endosomes to lysosomes, and from the *trans* Golgi network to late endosomes.

GENETIC APPROACHES FOR STUDYING VESICULAR TRANSPORT

Genetic studies of mutant yeast cells defective for secretion at high temperature have identified more than 25 genes that are involved in the secretory pathway. Many of the mutant genes encode *temperature-sensitive* proteins. These function normally at 25°C, but when the mutant cells are shifted to 35°C, some of them fail to transport proteins from the ER to the Golgi apparatus, others from one Golgi cisterna to another, and still others from the Golgi apparatus to the vacuole (the yeast lysosome) or to the plasma membrane.

Once a protein required for secretion has been identified in this way, one can identify genes that encode proteins that interact with it by making use of a phenomenon called *multicopy suppression.* A temperature-sensitive mutant protein at high temperature often has too low an affinity for the proteins it normally interacts with to bind to them. If the interacting proteins are produced at much higher concentration than normal, however, sufficient binding occurs to cure the defect. For this reason yeast cells with a temperature-sensitive mutation in a gene involved in vesicular transport are often transfected with a yeast plasmid vector into which random yeast genomic DNA fragments have been cloned. Because this plasmid is maintained in cells at high copy number, those that carry intact genes will overproduce the normal gene product, allowing rare cells to survive at the high temperature. The relevant DNA fragments, which presumably encode proteins that interact with the original mutant protein, can then be isolated from the surviving cell clones.

The genetic and biochemical approaches complement each other, and many of the proteins involved in vesicular transport have been identified independently by biochemical studies of mammalian cell-free systems and by genetic studies in yeast.

GFP-FUSION PROTEINS HAVE REVOLUTIONIZED THE STUDY OF INTRACELLULAR TRANSPORT

One way to follow the whereabouts of a protein in living cells is to construct fusion proteins, in which green fluorescent protein (GFP) is attached by genetic engineering techniques to the protein of interest. When a cDNA encoding such a fusion protein is expressed in a cell, the protein is readily visible in a fluorescent microscope, so that it can be followed in living cells in real time. Fortunately, for most proteins studied the addition of GFP to a protein does not perturb the protein's function.

GFP fusion proteins are widely used to study the location and movement of proteins in cells. GFP fused to proteins that shuttle in and out of the nucleus, for example, is used to study nuclear transport events and their regulation. GFP fused to mitochondrial or Golgi proteins is used to study the behavior of these organelles. GFP fused to plasma membrane proteins is used to measure the kinetics of their movement from the ER through the secretory pathway. Dramatic examples of such experiments can be seen as movies on the CD that accompanies this book.

The study of GFP fusion proteins is often combined with FRAP and FLIP techniques (discussed in Chapter 10), in which the GFP in selected regions of the cell is bleached by strong laser light. The rate of diffusion of unbleached GFP fusion proteins into that area can then be determined to provide measurement of the protein's diffusion or transport in the cell. In this way, for example, it was determined that many Golgi enzymes recycle between Golgi apparatus and the ER.

(A–D *right*, courtesy of Jennifer Lippincott-Schwartz Lab.)

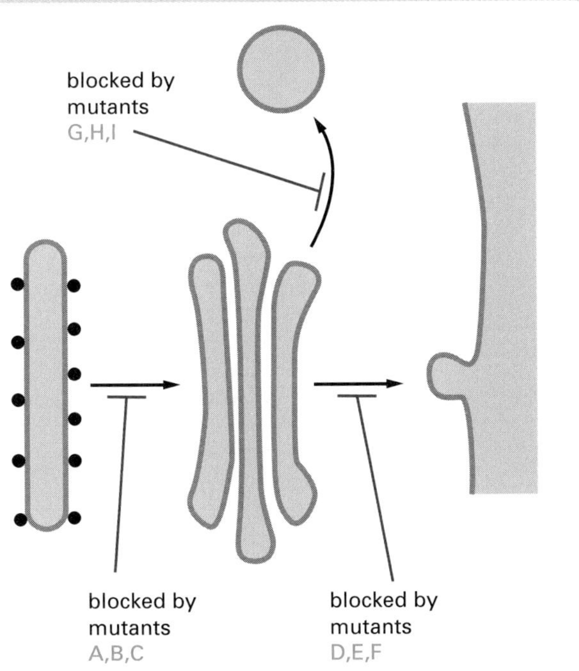

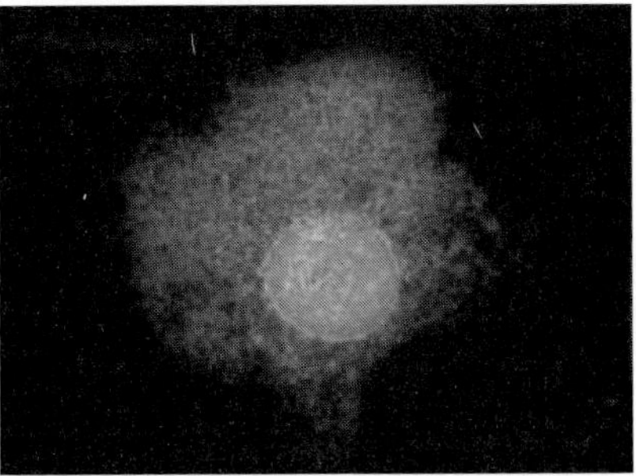

(A) In this experiment, GFP fused to the vesicular stomatitis virus coat protein was expressed in cultured cells. The viral protein is an integral membrane protein that normally moves through the secretory pathway from the ER to the cell surface, where the virus would be assembled if cells also expressed the other viral components. The viral protein contains a mutation that allows export from the ER only at a low temperature. Thus, at the high temperature shown, the fusion protein labels the ER.

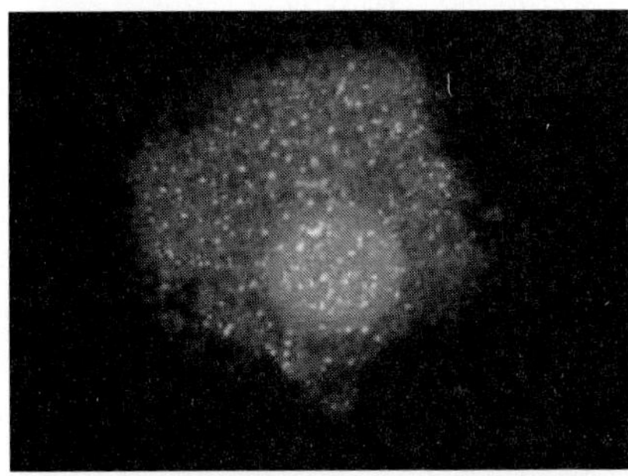

(B) As the temperature is lowered, the GFP fusion protein rapidly accumulates at ER exit sites.

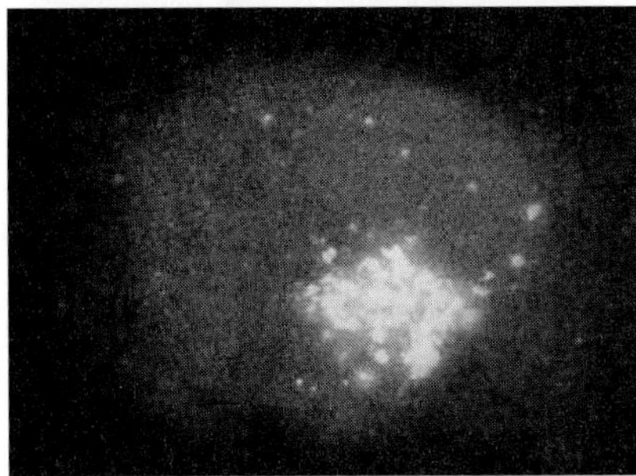

(C) The fusion protein then moves to the Golgi apparatus.

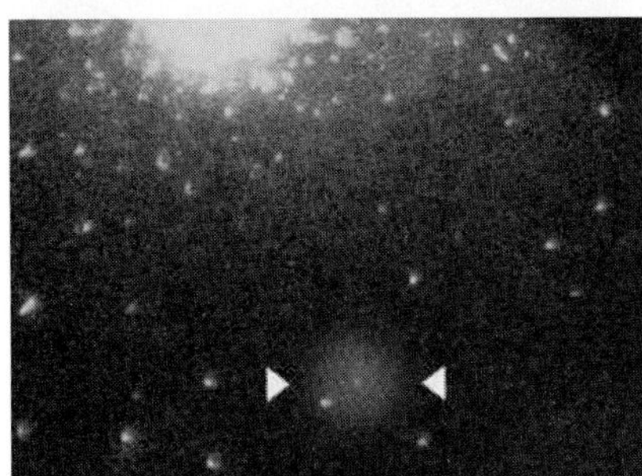

(D) Finally, the fusion protein is delivered to the plasma membrane. From such studies the kinetics of each step in the pathway can be determined.

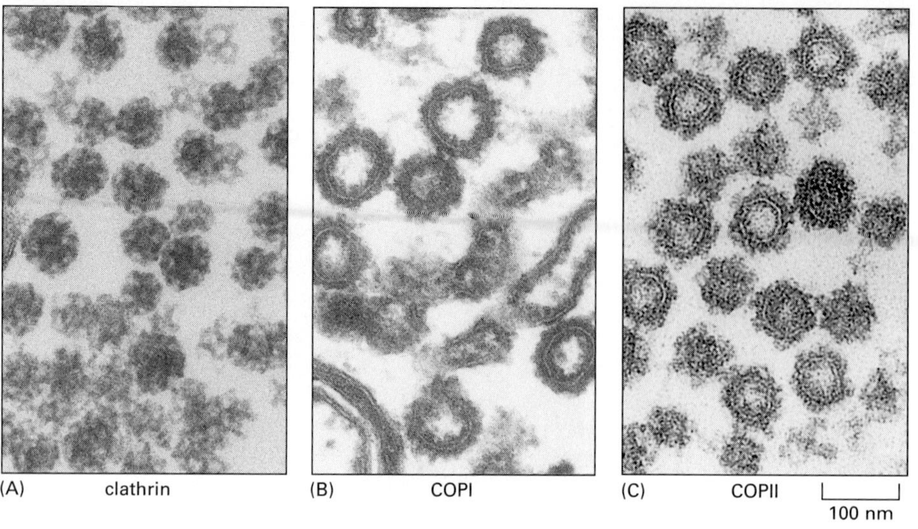

Figure 13–4 Electron micrograph of clathrin-coated, COPI-coated, and COPII-coated vesicles. All are shown in electron micrographs at the same scale. (A) Clathrin-coated vesicles. (B) Golgi cisternae from a cell-free system in which COPI-coated vesicles bud in the test tube. (C) COPII-coated vesicles. Note that the vesicles with clathrin coats have a more regular structure. (A and B, courtesy of Lelio Orci, from L. Orci, B. Glick, and J. Rothman, *Cell* 46:171–184, 1986. © Elsevier; C, courtesy of Charles Barlowe and Lelio Orci.)

The Assembly of a Clathrin Coat Drives Vesicle Formation

Clathrin-coated vesicles were the first coated vesicles discovered and have been the most thoroughly studied. They provide a good example of how vesicles form.

The major protein component of clathrin-coated vesicles is **clathrin** itself. Each clathrin subunit consists of three large and three small polypeptide chains that together form a three-legged structure called a *triskelion*. Clathrin triskelions assemble into a basketlike convex framework of hexagons and pentagons to form coated pits on the cytosolic surface of membranes (Figure 13–6). Under appropriate conditions, isolated triskelions spontaneously self-assemble into typical polyhedral cages in a test tube, even in the absence of the membrane vesicles that these baskets normally enclose (Figure 13–7). Thus, the geometry of the clathrin cage is determined by the clathrin triskelion alone.

A second major coat protein in clathrin-coated vesicles is a multisubunit complex called **adaptin**. It is required both to bind the clathrin coat to the membrane and to trap various transmembrane proteins, including transmembrane receptors that capture soluble cargo molecules inside the vesicle—so-called *cargo receptors*. In this way, a selected set of membrane proteins and the soluble

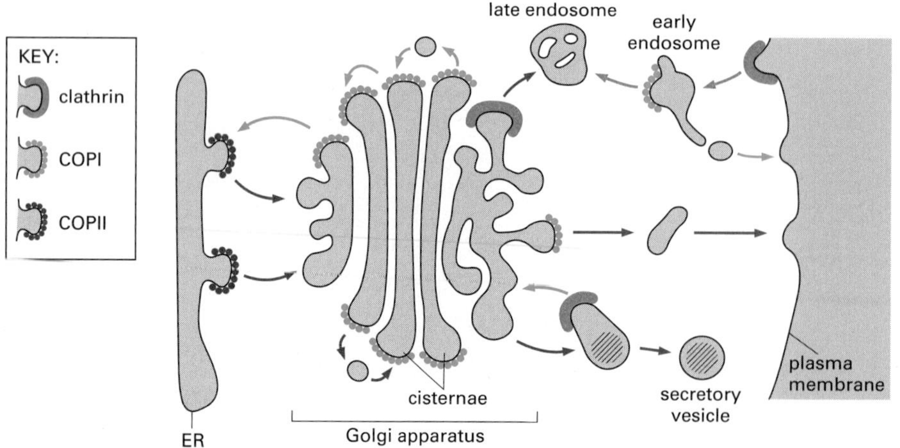

Figure 13–5 Utilization of different coats in vesicular traffic. Different coat proteins select different cargo and shape the transport vesicles that mediate the various steps in the biosynthetic–secretory and endocytic pathways. When the same coats function in different places in the cell, they can incorporate different coat protein subunits that modify their properties (not shown). Many differentiated cells have additional pathways beside those shown in this figure, including a sorting pathway from the *trans* Golgi network to the apical surface in polarized cells and a specialized recycling pathway for proteins of synaptic vesicles in the synapses of neurons.

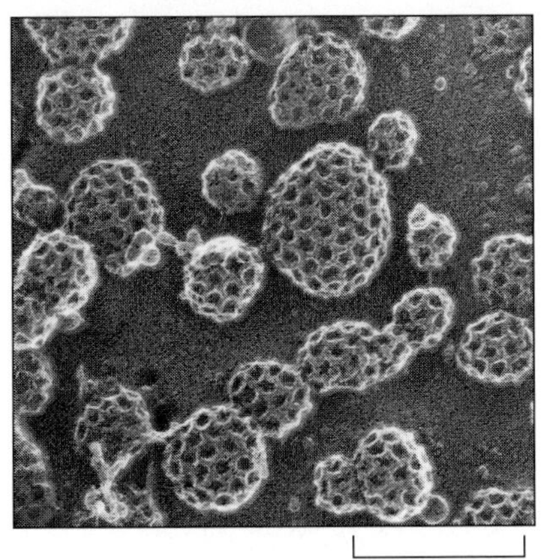

Figure 13–6 Clathrin-coated pits and vesicles. This rapid-freeze, deep-etch electron micrograph shows numerous clathrin-coated pits and vesicles on the inner surface of the plasma membrane of cultured fibroblasts. The cells were rapidly frozen in liquid helium, fractured, and deep-etched to expose the cytoplasmic surface of the plasma membrane. (From J. Heuser, *J. Cell Biol.* 84:560–583, 1980. © The Rockefeller University Press.)

proteins that interact with them are packaged into each newly formed clathrin-coated transport vesicle (Figure 13–8).

There are at least four types of adaptins, each specific for a different set of cargo receptors. Clathrin-coated vesicles budding from different membranes use different adaptins and thus package different receptors and cargo molecules. The formation of a clathrin-coated pit is driven by forces generated by the successive assembly of adaptins and the clathrin coat on the cytosolic surface of the membrane. The lateral interactions between adaptins and between clathrin molecules then aid in bud formation.

Both The Pinching-off and Uncoating of Coated Vesicles Are Regulated Processes

As a clathrin-coated bud grows, soluble cytoplasmic proteins, including **dynamin**, assemble as a ring around the neck of each bud. Dynamin is a GTPase,

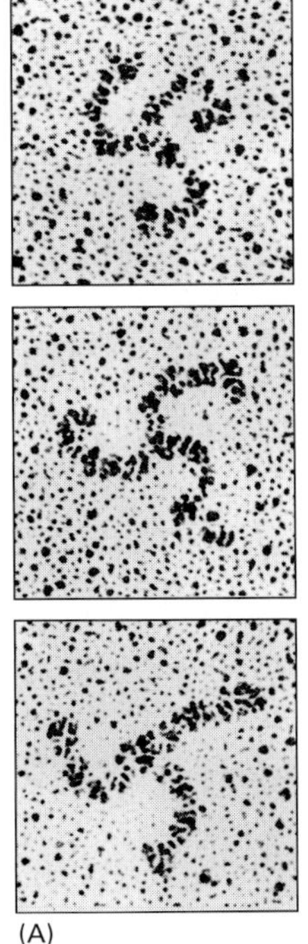

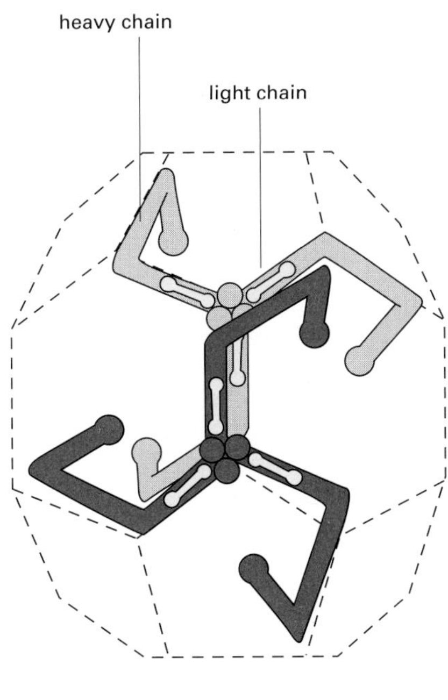

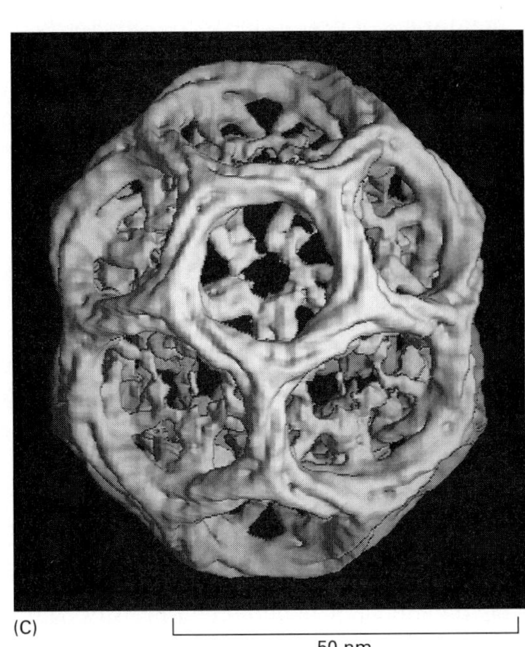

Figure 13–7 The structure of a clathrin coat. (A) Electron micrographs of clathrin triskelions shadowed with platinum. Although this feature cannot be seen in these micrographs, each triskelion is composed of 3 clathrin heavy chains and 3 clathrin light chains. (B) A schematic drawing of the probable arrangement of triskelions on the cytosolic surface of a clathrin-coated vesicle. Two triskelions are shown, with the heavy chains of one in *red* and those of the other in *gray;* the light chains are shown in *yellow.* The overlapping arrangement of the flexible triskelion arms provides both mechanical strength and flexibility. Note that the end of each leg of the triskelion turns inward, so that its N-terminal domain forms an intermediate shell. (C) A cryo electron micrograph taken of a clathrin coat composed of 36 triskelions organized in a network of 12 pentagons and 6 hexagons. The interwoven legs of the clathrin triskelions form an outer shell into which the N-terminal domains of the triskelions protrude to form an inner layer visible through the openings. It is this inner layer that contacts the adaptor proteins (adaptins) shown in the next figure. Although the coat shown is too small to enclose a membrane vesicle, the clathrin coats on vesicles are constructed in a similar way from 12 pentagons plus a larger number of hexagons, resembling the architecture of a soccer ball. (A, from E. Ungewickell and D. Branton, *Nature* 289:420–422, 1981. © Macmillan Magazines Ltd.; B, from I.S. Nathke et al., *Cell* 68:899–910, 1992. © Elsevier; C, courtesy of B.M.F. Pearse, from C.J. Smith et al., *EMBO J.* 17:4943–4953, 1998.)

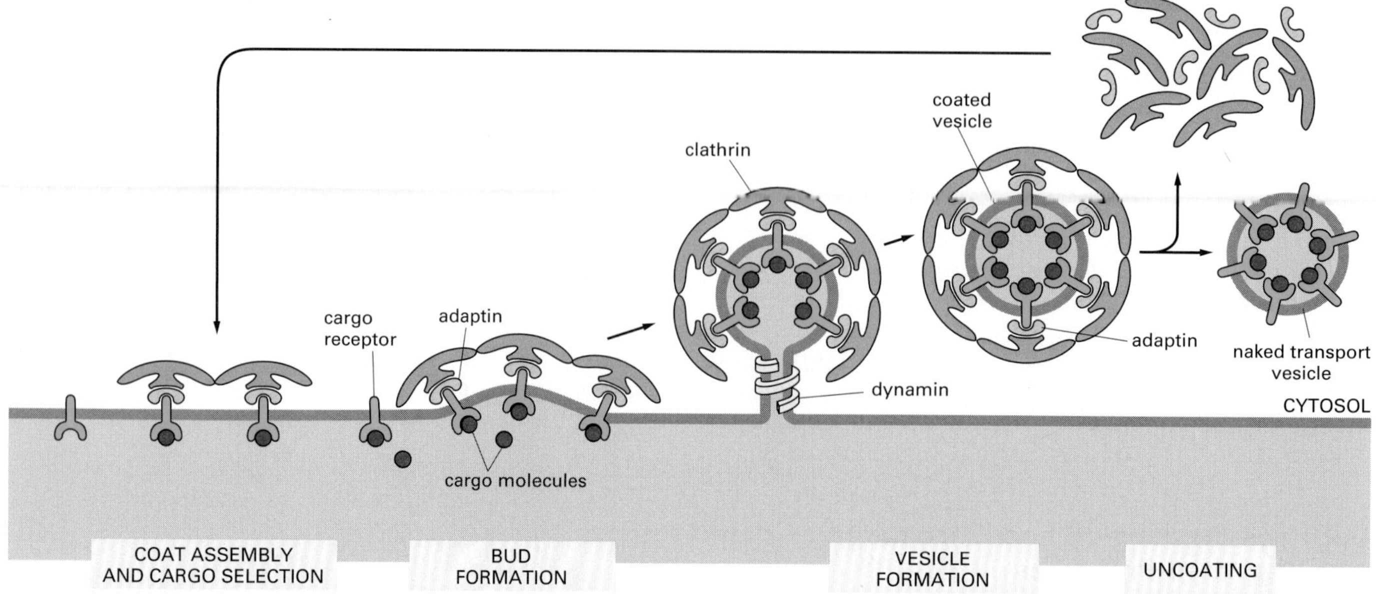

Figure 13–8 The assembly and disassembly of a clathrin coat. The assembly of the coat is thought to introduce curvature into the membrane, which leads in turn to the formation of uniformly sized coated buds. The adaptins bind both clathrin triskelions and membrane-bound cargo receptors, thereby mediating the selective recruitment of both membrane and cargo molecules into the vesicle. The pinching-off of the bud to form a vesicle involves membrane fusion; this is helped by the GTP-binding protein dynamin, which assembles around the neck of the bud. The coat of clathrin-coated vesicles is rapidly removed shortly after the vesicle forms.

which regulates the rate with which vesicles pinch off from the membrane. In the pinching-off process, the two noncytosolic leaflets of the membrane are brought into close proximity and fuse, sealing off the forming vesicle (Figure 13–9). To perform this task, dynamin recruits other proteins to the neck of the budding vesicle, which together with dynamin help to bend the membrane, either by directly distorting the bilayer structure locally or by changing the lipid composition, or both. A local change in lipid composition may result from the action of lipid-modifying enzymes that are recruited into the dynamin complex.

Once the vesicle is released from the membrane, the clathrin coat is rapidly lost. A chaperone protein of the hsp70 family functions as an uncoating ATPase, using the energy of ATP hydrolysis to peel off the coat. Another protein called *auxillin,* which is attached to the vesicle, is believed to activate the ATPase. Because the coated bud persists much longer than the coat on the vesicle, additional control mechanisms must somehow prevent the coat from being removed before it has formed a vesicle (discussed below).

Although there are many similarities in vesicle budding at various locations in the cell, each cell membrane poses its own special challenges. The plasma membrane, for example, is comparatively flat and stiff, owing to its cholesterol-rich lipid composition and underlying cortical cytoskeleton. Thus, clathrin coats have to produce considerable force to introduce curvature, especially at the neck of the bud where dynamin and its associated proteins facilitate the sharp bends required for the pinching-off of the vesicle. In contrast, vesicle budding from many intracellular membranes occurs preferentially at regions where the membranes are already curved, such as the rims of Golgi cisternae or membrane tubules.

COPI-coated vesicles and **COPII-coated vesicles** transport material early in the secretory pathway: COPII-coated packages bud from the ER, and COPI-coated packages bud from pre-Golgi compartments and Golgi cisternae (see Figure 13–5). The coats of COPI and COPII vesicles consist, in part, of large protein complexes that are composed of seven individual coat-protein subunits for COPI and four for COPII coats. Some COPI coat-protein subunits show sequence similarity to adaptins, suggesting a common evolutionary origin.

Not All Transport Vesicles are Spherical

Transport vesicles occur in various sizes and shapes. When living cells that have been genetically engineered to express fluorescent membrane components are observed under the microscope, endosomes and the *trans* Golgi network are

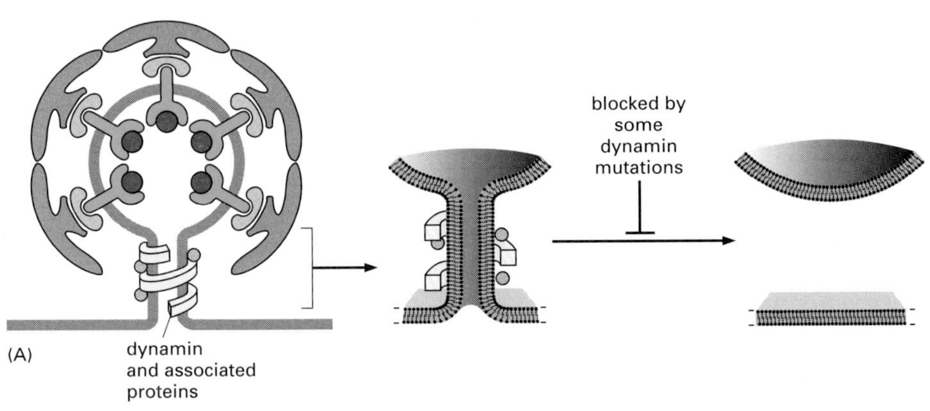

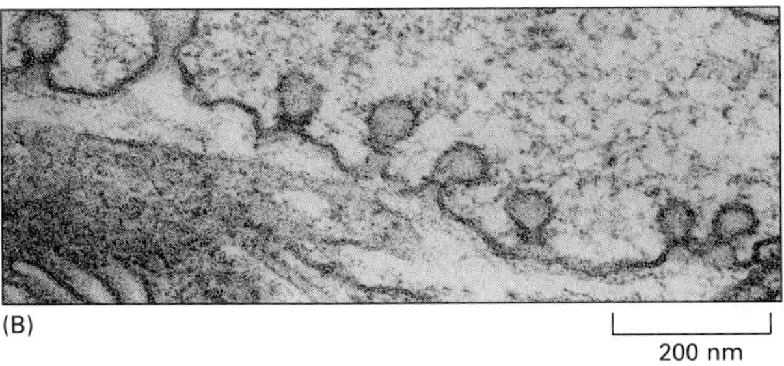

Figure 13–9 The role of dynamin in pinching off clathrin-coated vesicles from the membrane. (A) The dynamin binds to a forming bud on the membrane and assembles into a ring around the neck of the bud. The dynamin ring is thought to be a template that recruits other proteins to the vesicle neck, which together with dynamin destabilize the membrane so that the noncytoplasmic leaflets of the lipid bilayers flow together. The newly formed vesicle then pinches off the membrane. Specific mutations in dynamin can either enhance or block the pinching-off process. (B) Dynamin was discovered as the protein defective in the *shibire* mutant of *Drosophila*. These mutant flies become paralyzed because clathrin-mediated endocytosis stops, and the synaptic vesicle membrane fails to recycle, blocking synaptic signaling. Deeply invaginated clathrin-coated pits form in the fly's nerve cells, with a ring assembled around the neck, as shown in this thin-section electron micrograph, that is assumed to be mutant dynamin. The process then stops because membrane fusion does not take place. (B, from J.H. Koenig and K. Ikeda, *J. Neurosci.* 9:3844–3860, 1989. © Society of Neuroscience.)

seen to continually send out long tubules. Coat proteins assemble onto the tubules and help recruit specific cargo. The tubules then either withdraw, or they pinch off with the help of dynamin-like proteins and thus can serve as transport vesicles. Depending on the relative efficiencies of membrane tubulation and severing, differently sized portions of a donor organelle can pinch off.

Tubules have a much higher surface-to-volume ratio than the organelles from which they form. They are therefore relatively enriched in membrane proteins compared with soluble cargo proteins. As we discuss later, this property of tubules is used for sorting proteins in endosomes. Thus, vesicular transport does not necessarily occur only through uniformly sized spherical vesicles, but can involve larger portions of a donor organelle.

Monomeric GTPases Control Coat Assembly

The vesicular transport performed by both clathrin-coated and COP-coated vesicles depends on a variety of GTP-binding proteins that control both the spatial and the temporal aspects of membrane exchange. As discussed in Chapter 3, large families of GTP-binding proteins regulate diverse processes within cells. These proteins act as molecular switches that flip between an active state with GTP bound and an inactive state with GDP bound. Two classes of proteins regulate the flipping: *guanine-nucleotide-exchange factors (GEFs)* activate the proteins by catalyzing the exchange of GDP for GTP, and *GTPase-activating proteins (GAPs)* inactivate the proteins by triggering the hydrolysis of the bound GTP to GDP (see Figure 3–72). Although both monomeric GTP-binding proteins (monomeric GTPases) and trimeric GTP-binding proteins (G proteins) have essential roles in vesicular transport, the roles of the monomeric GTPases are better understood, and we focus our discussion on them.

To ensure that membrane traffic to and from an organelle is balanced, coat proteins must assemble only when and where they are needed. *Coat-recruitment GTPases,* which are members of a family of monomeric GTPases, usually serve this function. They include the **ARF proteins**, which are responsible for both COPI coat assembly and clathrin coat assembly at Golgi membranes, and the **Sar1 protein**, which is responsible for COPII coat assembly at the ER membrane. Clathrin coat assembly at the plasma membrane is also thought to involve a GTPase, but its identity is unknown.

Coat-recruitment GTPases are usually found in high concentration in the cytosol in an inactive, GDP-bound state. When a COPII-coated vesicle is to bud from the ER membrane, a specific GEF embedded in the ER membrane binds to cytosolic Sar1, causing the Sar1 to release its GDP and bind GTP in its place (recall that GTP is present in much higher concentration in the cytosol than GDP and therefore will spontaneously bind after GDP is released). In its GTP-bound state, Sar1 exposes a fatty acid tail, which inserts into the lipid bilayer of the ER membrane. The tightly bound Sar1 now recruits coat protein subunits to the ER membrane to initiate budding (Figure 13–10). Other GEFs and coat-recruitment GTPases operate in a similar way on other membranes.

Some coat protein subunits also interact, albeit more weakly, with the head groups of certain lipid molecules, in particular phosphatidic acid and phosphoinositides, as well as with the cytoplasmic tails of some of the membrane proteins they recruit into the bud. Activated coat-recruitment GTPases at sites of bud formation can locally activate phospholipase D, which converts some phospholipids to phosphatidic acid, thereby enhancing the binding of coat proteins. Together, these protein–protein and protein–lipid interactions tightly bind the coat to the membrane, causing the membrane to deform into a bud, which then pinches off as a coated vesicle.

The coat-recruitment GTPases also have a role in coat disassembly. The hydrolysis of bound GTP to GDP causes the GTPase to change its conformation so that its fatty acid tail pops out of the membrane, causing the vesicle's coat to disassemble. Although it is not known what triggers the ATP hydrolysis process, it has been proposed that the GTPases work like timers, which hydrolyze GTP at a slow but predictable rate. COPII coats, for example, accelerate GTP hydrolysis by Sar1, thereby triggering coat disassembly at a certain time after coat assembly has begun. Thus, a fully formed vesicle will be produced only when bud formation occurs faster than the timed disassembly process; otherwise, disassembly will be triggered before a vesicle pinches off, and the process will have to start again at a more appropriate time and place. Completion of coating or contact with the target membrane may also trigger coat disassembly.

SNARE Proteins and Targeting GTPases Guide Membrane Transport

To ensure that membrane traffic proceeds in an orderly way, transport vesicles must be highly selective in recognizing the correct target membrane with which to fuse. Because of the diversity of membrane systems, a vesicle is likely to encounter many potential target membranes before it finds the correct one. Specificity in targeting is ensured because all transport vesicles display surface markers that identify them according to their origin and type of cargo, while

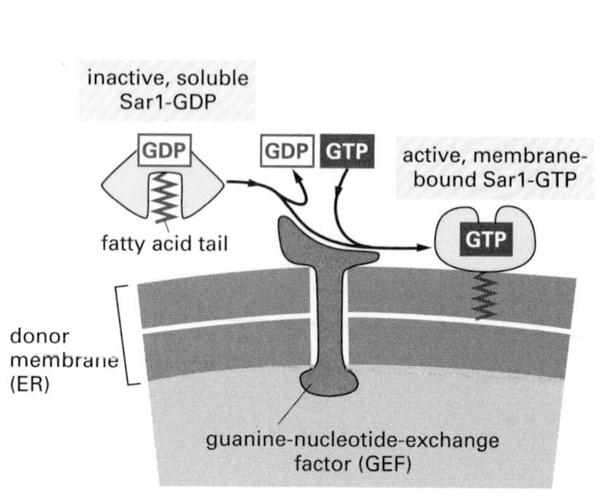

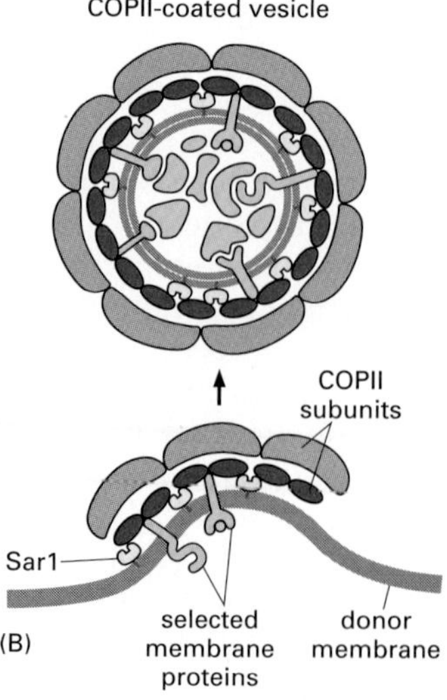

Figure 13–10 A current model of COPII-coated vesicle formation. (A) The Sar1 protein is a coat recruitment GTPase. Inactive, soluble Sar1-GDP binds to a GEF (called Sec12) in the ER membrane, causing the Sar1 to release its GDP and bind GTP. A GTP-triggered conformational change in Sar1 exposes its fatty acid chain, which inserts into the ER membrane. (B) Membrane-bound, active Sar1-GTP recruits COPII subunits to the membrane. This causes the membrane to form a bud, which includes selected membrane proteins. A subsequent membrane-fusion event pinches off and releases the coated vesicle. Other coated vesicles are thought to form in a similar way.

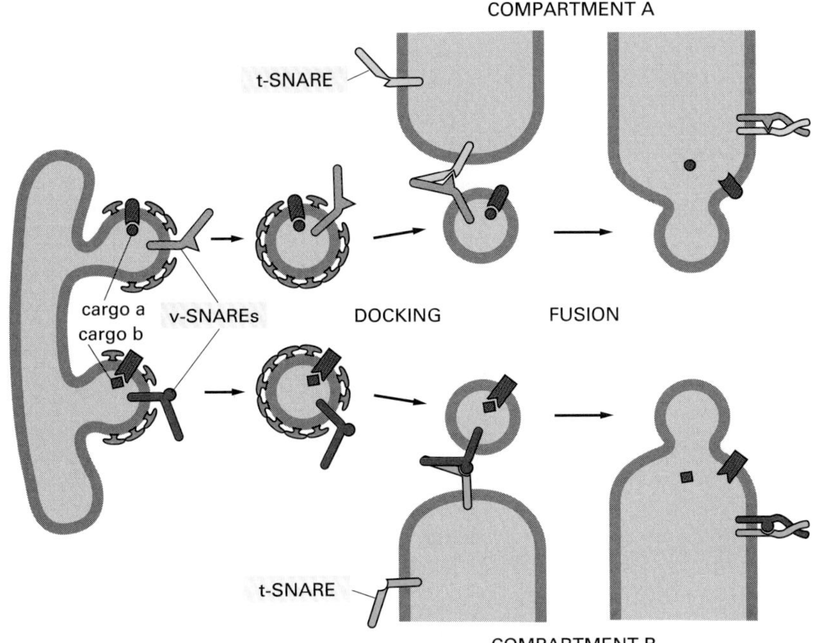

Figure 13–11 The postulated role of SNAREs in guiding vesicular transport. Complementary sets of vesicle SNAREs (v-SNAREs) and target membrane SNAREs (t-SNAREs) contribute to the selectivity of transport-vesicle docking and fusion. The v-SNAREs are packaged together with the coat proteins during the budding of transport vesicles from the donor membrane and bind to complementary t-SNAREs in the target membrane. After fusion, the v- and t-SNAREs remain associated in a tight complex. The complexes have to be dissociated before the t-SNAREs can accept a new vesicle or the v-SNAREs can be recycled to the donor compartment for participation in a new round of vesicular transport. As shown here, different v-SNAREs can be packaged with different cargo molecules (through association with other proteins; not shown) when leaving the donor compartment. In this case, the two sets of cargo will be delivered to different t-SNAREs and therefore to different target membranes.

target membranes display complementary receptors that recognize the appropriate markers. This crucial recognition step is thought to be controlled mainly by two classes of proteins: *SNAREs* and targeting GTPases called *Rabs*. SNARE proteins seem to have a central role both in providing specificity and in catalyzing the fusion of vesicles with the target membrane. Rabs seem to work together with other proteins to regulate the initial docking and tethering of the vesicle to the target membrane.

There are at least 20 different **SNAREs** in an animal cell, each associated with a particular membrane-enclosed organelle involved in the biosynthetic–secretory or endocytic pathway. These transmembrane proteins exist as complementary sets—vesicle membrane SNAREs, called **v-SNAREs**, and target membrane SNAREs, called **t-SNAREs**, (Figures 13–11 and 13–12). v-SNARESs and t-SNAREs have characteristic helical domains. When a v-SNARE interacts with a t-SNARE, the helical domains of one wrap around the helical domains of the other to form stable *trans-SNARE complexes*, which lock the two membranes together. We discuss later how the trans-SNARE complex is thought to contribute to membrane fusion. The specificity with which SNAREs interact determines the specificity of vesicle docking and fusion. In this way SNAREs specify compartment identity and govern the orderly transfer of material during vesicular transport.

SNAREs have been best characterized in nerve cells, where they mediate the docking and fusion of synaptic vesicles at the nerve terminal plasma membrane (see Figure 13–12). The SNARE complexes at neuron terminals are the targets of powerful neurotoxins that are secreted by the bacteria that cause tetanus and

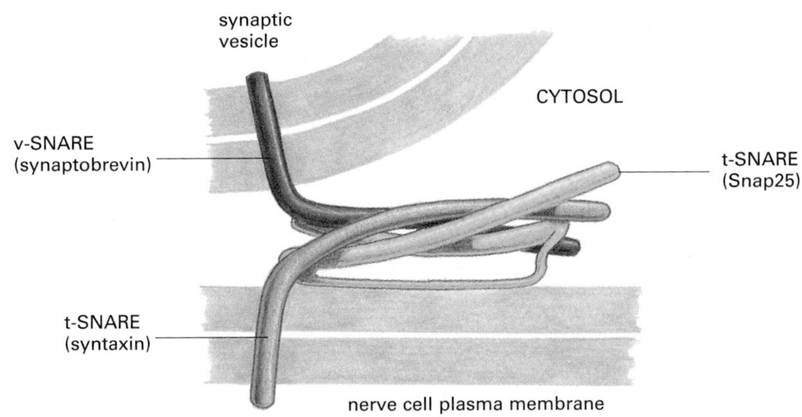

Figure 13–12 The structure of paired SNAREs. The SNAREs responsible for docking synaptic vesicles at the plasma membrane of nerve terminals consist of three proteins. The v-SNARE *synaptobrevin*, and the t-SNARE *syntaxin* are both transmembrane proteins and each contributes one α-helix to the complex. The t-SNARE *Snap25* is a peripheral membrane protein that contributes two α-helices to the four-helix bundle. Trans-SNARE complexes always consists of four tightly intertwined α-helices, three contributed by a t-SNARE and one by a v-SNARE.

The t-SNAREs are composed of multiple chains, one of which is always a transmembrane protein and contributes one helix, and one or two additional light chains that may or may not be transmembrane proteins and that contribute the remaining two helices to the four-helix bundle of the trans-SNARE complex. The crystal structure of a stable complex of the four intertwining α helices contributed by these proteins is modeled here in the context of the whole proteins. The α helices are shown as rods for simplicity. (Adapted from R.B. Sutton et al., *Nature* 395:347–353, 1998.)

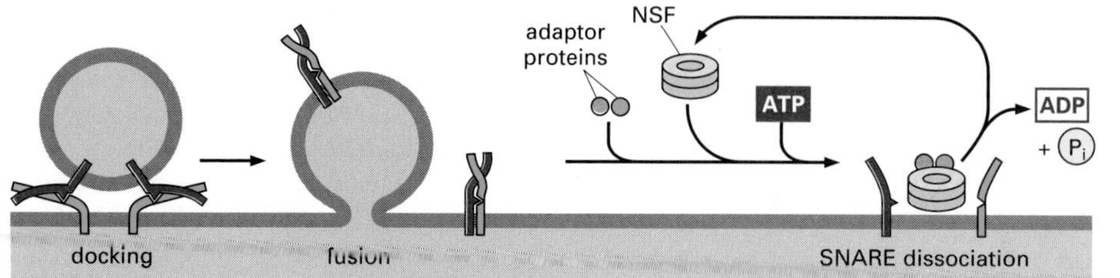

Figure 13–13 Dissociation of SNARE pairs by NSF after a membrane fusion cycle is completed. After the v-SNAREs and t-SNAREs have mediated the fusion of a vesicle on a target membrane, the NSF binds to the SNARE complex via adaptor proteins and hydrolyzes ATP to pry the SNAREs apart.

botulism. These toxins are highly specific proteases that enter specific neurons, cleave SNARE proteins in the nerve terminals and thereby block synaptic transmissions, often fatally.

Interacting SNAREs Need To Be Pried Apart Before They Can Function Again

Most SNARE proteins in cells have already participated in multiple rounds of membrane targeting and are sometimes present in a membrane as stable complexes with one or two partner SNAREs (see Figure 13–11). The complexes have to be disassembled before the SNAREs can mediate new rounds of transport. A crucial protein called **NSF** cycles between membranes and the cytosol and catalyzes the disassembly process. It is an ATPase that structurally resembles a minor class of cytosolic chaperone proteins that use the energy of ATP hydrolysis to solubilize and help refold denatured proteins. Similarly, NSF uses ATP to unravel the coiled-coil interaction between the helical domains of SNARE proteins, using several adaptor proteins to bind to the SNAREs (Figure 13–13).

The requirement for SNARE complex disassembly may help explain why membranes do not fuse indiscriminately in cells. If the t-SNAREs in a target membrane were always active, then any membrane containing an appropriate v-SNARE would fuse whenever the two membranes made contact. The requirement for NSF-mediated reactivation of SNAREs allows the cell to control when and where membranes fuse. In addition, t-SNAREs in target membranes are often associated with inhibitory proteins that must be released before the t-SNARE can function. This release step may be controlled by the targeting GTPases, as we discuss next.

Rab Proteins Help Ensure the Specificity of Vesicle Docking

Rab proteins make an important contribution to the specificity of vesicular transport. They are monomeric GTPases, and with over 30 known members, they are the largest subfamily of these GTPases. Like the SNAREs, each Rab protein has a characteristic distribution on cell membranes and every organelle has at least one Rab protein on its cytosolic surface (Table 13–1). Rab proteins are

TABLE 13–1 Subcellular Locations of Some Rab Proteins

PROTEIN	ORGANELLE
Rab1	ER and Golgi complex
Rab2	*cis* Golgi network
Rab3A	synaptic vesicles, secretory granules
Rab4	early endosomes
Rab5A	plasma membrane, clathrin-coated vesicles
Rab5C	early endosomes
Rab6	*medial* and *trans* Golgi cisternae
Rab7	late endosomes
Rab8	secretory vesicles (basolateral)
Rab9	late endosomes, *trans* Golgi network

thought to facilitate and regulate the rate of vesicle docking and the matching of v-SNAREs and t-SNAREs, as required for membrane fusion.

Like the coat-recruitment GTPases discussed earlier (see Figure 13–10), Rab proteins cycle between a membrane and the cytosol. In their GDP-bound state they are inactive and in the cytosol, and in their GTP-bound state they are active and associated with the membrane of an organelle or transport vesicle (Figure 13–14). Many transport vesicles only form if a proper complement of SNARE and Rab proteins are included in the membrane, so as to allow the vesicle to dock and fuse appropriately.

The amino acid sequences of Rab proteins are most dissimilar near their C-terminal tails. Tail-swapping experiments indicate that the tail determines the intracellular location of each family member, presumably by enabling the protein to bind to complementary proteins, including GEFs, on the surface of the appropriate organelle. Once in its GTP-bound state and membrane-bound through its lipid anchor, a Rab protein is thought to bind to other proteins (called Rab effectors) that facilitate the docking process.

In contrast to the highly conserved structure of Rab proteins, the structures of **Rab effectors** vary greatly from one Rab protein to the next. One Rab effector, for example, is a large protein complex that serves to direct vesicles to specific sites on the plasma membrane for exocytosis. Vesicle fusion is limited to the region where this complex resides, even though the required t-SNAREs are uniformly distributed in the membrane. Some Rab effectors are long, filamentous, tethering proteins, which may restrict the movement of vesicles between adjacent Golgi cisternae. Others bind to Rab proteins in their active GTP-bound state and prevent premature GTP hydrolysis. Yet others are motor proteins that propel vesicles along actin filaments or microtubules to their proper target.

Although the Rab proteins and their effectors use widely different molecular mechanisms to influence vesicular transport, they have a common function. They help concentrate and tether vesicles near their target site and trigger the

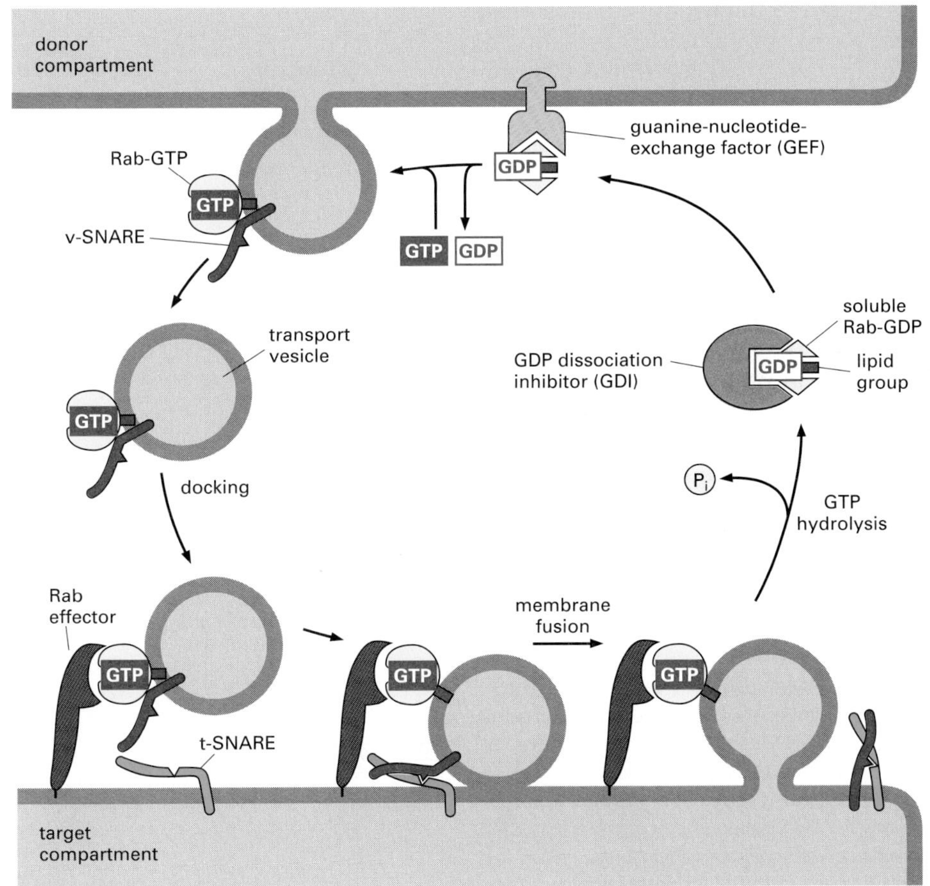

Figure 13–14 A postulated role of Rab proteins in facilitating the docking of transport vesicles. A GEF in the donor membrane recognizes a specific Rab protein and induces it to exchange GDP for GTP. GTP binding alters the conformation of the Rab protein, exposing its covalently attached lipid group, which helps anchor the protein in the membrane. Recall that an analogous mechanism helps to bind the coat-recruitment GTPases to these membranes, although a different GEF is involved (see Figure 13–10). The Rab-GTP remains bound to the surface of the transport vesicle after it pinches off from the donor membrane, and it then binds to varying Rab effector proteins on the target membrane. The Rab protein and its effectors help the vesicle dock and thereby facilitate the pairing of the appropriate v-SNAREs and t-SNAREs. After the vesicle has fused with the target membrane, the Rab protein hydrolyzes its bound GTP, releasing Rab-GDP into the cytosol, from where it can be reused in a new round of transport. As shown, Rab-GDP in the cytosol is bound to a GDP dissociation inhibitor (GDI), which prevents the Rab from releasing its bound GDP until it has interacted with appropriate proteins in the donor membrane. For clarity, we have omitted all of the proteins in the vesicle coats from this figure (see Figure 13–10).

release of SNARE control proteins. In this way Rab proteins speed up the process by which appropriate SNARE proteins in two membranes find each other. Some Rab proteins function on the vesicle, whereas others function on the target membrane. The pairing of v-SNAREs and t-SNAREs then locks the docked vesicle onto the target membrane, readying it for fusion, which we discuss next. After fusion, the Rab protein hydrolyzes its bound GTP and the inactive GDP-bound protein returns to the cytosol to participate in another cycle of transport.

SNAREs May Mediate Membrane Fusion

Once a transport vesicle has recognized its target membrane and docked there, it unloads its cargo by membrane fusion. Fusion does not always follow immediately, however. As we discuss later, in the process of regulated exocytosis, fusion is delayed until it is triggered by a specific extracellular signal.

Thus, docking and fusion are two distinct and separable processes. Docking requires only that the two membranes come close enough for proteins protruding from the lipid bilayers to interact and adhere. Fusion requires a much closer approach, bringing the lipid bilayers to within 1.5 nm of each other so that they can join. When the membranes are in such close apposition, lipids can flow from one bilayer to the other. For this close approach, water must be displaced from the hydrophilic surface of the membrane—a process that is energetically highly unfavorable. It seems likely that all membrane fusions in cells are catalyzed by specialized fusion proteins that provide a way to overcome this energy barrier. We have already discussed the role of dynamin in a related task during clathrin-coated vesicle budding (see Figure 13–9).

SNAREs are thought to have a central role in membrane fusion. The formation of the SNARE complex may work like a winch, using the energy that is freed when the interacting helices wrap around each other to pull the membrane faces together, while simultaneously squeezing out water molecules from the interface (Figure 13–15). When liposomes containing purified v-SNAREs are mixed with liposomes containing matching t-SNAREs, their membranes fuse, albeit slowly. In the cell, other proteins recruited to the fusion site presumably cooperate with SNAREs to initiate fusion. Moreover, inhibitory proteins may have to be released to allow the complete zipping-up of SNARE pairs. In some cases, such as in regulated exocytosis (discussed later), a localized influx of Ca^{2+} triggers the fusion process.

Viral Fusion Proteins and SNAREs May Use Similar Strategies

Membrane fusion is important in other processes beside vesicular transport. Examples are the fusion of the plasma membranes of sperm and egg that occurs at fertilization (discussed in Chapter 20) and the fusion of myoblasts during muscle cell development (discussed in Chapter 21). All cell membrane fusions require special proteins and are subject to tight controls, which ensure that only appropriate membranes fuse. The controls are crucial for maintaining both the identity of cells and the individuality of each type of intracellular compartment.

The membrane fusions catalyzed by viral fusion proteins are the best understood. These proteins have a crucial role in permitting the entry of enveloped viruses (which have a lipid-bilayer-based membrane coat) into the cells that they infect (discussed in Chapters 5 and 25, CD). For example, viruses—such as human immunodeficiency virus (HIV), which causes AIDS—bind to cell-surface

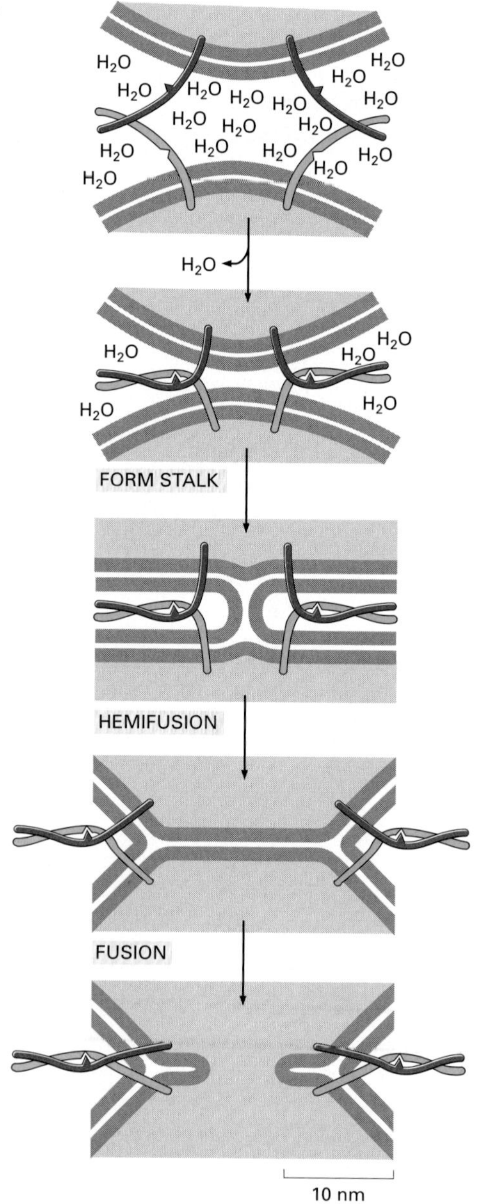

Figure 13–15 A model for how SNARE proteins may concentrate in membrane fusion. Bilayer fusion is proposed to occur in multiple steps. A tight SNARE pairing forces lipid bilayers into close apposition so that water molecules are expelled from the interface. Lipids of the two interacting leaflets of the bilayers then flow between the membranes to form a connecting stalk. Lipids of the other two leaflets then contact each other, forming a new bilayer, which widens the fusion zone (*hemifusion*, or half-fusion). Rupture of the new bilayer completes the fusion reaction.

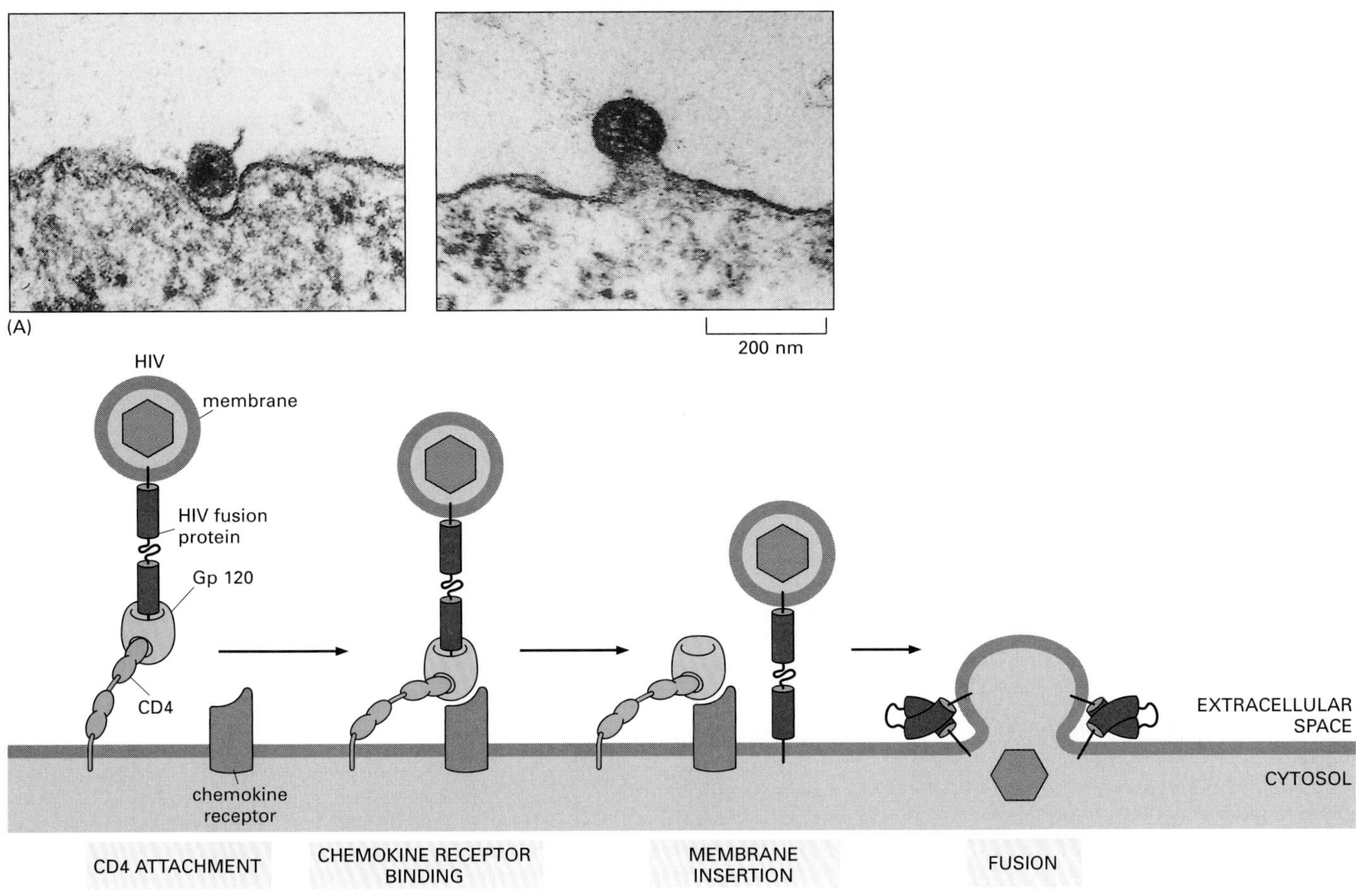

Figure 13–16 The entry of enveloped viruses into cells. (A) Electron micrographs showing how HIV enters a cell by fusing its membrane with the plasma membrane of the cell. (B) A model for the HIV membrane-fusion process. HIV binds first to the CD4 protein on the surface of the lymphocytes. This interaction is mediated by the viral gp120 protein bound to the HIV fusion protein. A second cell-surface protein on the host cell, which normally serves as a receptor for chemokines (discussed in Chapter 24), now interacts with gp120. This interaction releases the HIV fusion protein from gp120 allowing the previously buried hydrophobic fusion peptide, to insert into the plasma membrane. The fusion protein, which is a trimer (not shown), thus becomes transiently anchored as an integral membrane protein in two opposing membranes. The fusion protein then spontaneously rearranges, collapsing into a tightly packed six-helix bundle. The energy released by this rearrangement in multiple copies of the fusion protein is used to pull the two membranes together, overcoming the high activation energy barrier that normally prevents membrane fusion. Thus, like a mouse trap, the HIV fusion protein contains a reservoir of potential energy, which is released and harnessed to do mechanical work. (A, from B.S. Stein et al., *Cell* 49:659–668, 1987. © Elsevier; B, adapted from a drawing by Wayne Hendrickson.)

receptors, and then the viral and plasma membranes fuse (Figure 13–16). This fusion event allows the viral nucleic acid to enter the cytosol, where it replicates. Other viruses, such as influenza virus, first enter the cell by receptor-mediated endocytosis (discussed later) and are delivered to endosomes. In this case, the low pH in endosomes activates a fusion protein in the viral envelope that catalyzes the fusion of the viral and endosomal membranes. This likewise releases the viral nucleic acid into the cytosol.

The three-dimensional structures of the fusion proteins of influenza virus and HIV provide valuable insights into the molecular mechanism of the membrane fusion catalyzed by these proteins. An exposure of the influenza fusion protein to low pH, or an exposure of HIV fusion protein to receptors on the target cell membrane, uncovers previously buried hydrophobic regions. These regions, called fusion peptides, are observed to then insert directly into the hydrophobic core of lipid bilayer of the target membrane. Thus, the viral fusion proteins are, for a moment, integral membrane proteins in two separate lipid bilayers. Structural rearrangements in the fusion proteins then bring the two lipid bilayers into very close apposition and destabilize them so that the bilayers fuse (see Figure 13–16). For viral fusion, the fusion proteins are the only components required, supporting the possibility that SNAREs are also the central players in the process of bilayer fusion in cells.

Summary

The differences between the many different membrane-enclosed compartments in a eucaryotic cell are maintained by directed, selective transport of particular membrane components from one compartment to another. Transport vesicles, which can be spherical or tubular, bud from specialized coated regions of the donor membrane. The assembly of the coat helps to collect specific membrane and soluble cargo molecules for transport and to drive the formation of the vesicle.

Of the various types of coated vesicles, the best characterized are clathrin-coated vesicles, which mediate transport from the plasma membrane and the trans Golgi network, and COPI- and COPII-coated vesicles, which mediate transport between the ER and the Golgi apparatus and between Golgi cisternae. In clathrin-coated vesicles, adaptins link the clathrin to the vesicle membrane and also trap specific cargo molecules for packaging into the vesicle. The coat is shed rapidly after budding, which is necessary for a vesicle to fuse with its appropriate target membrane.

Monomeric GTPases help regulate various steps in vesicular transport, including both vesicle budding and docking. The coat-recruitment GTPases, including Sar1 and the ARF proteins, regulate coat assembly and disassembly. A family of Rab proteins functions as vesicle targeting GTPases. Being incorporated with v-SNAREs into budding transport vesicles, the Rab proteins help ensure that the vesicles deliver their contents only to the appropriate membrane-enclosed compartment: the one that displays complementary t-SNARE proteins. Complementary v-SNARE and t-SNARE proteins form stable trans-SNARE complexes, thereby bringing their membrane bilayers into close apposition for fusion.

TRANSPORT FROM THE ER THROUGH THE GOLGI APPARATUS

As discussed in Chapter 12, newly synthesized proteins enter the biosynthetic–secretory pathway in the ER by crossing the ER membrane from the cytosol. During their subsequent transport, from the ER to the Golgi apparatus and from the Golgi apparatus to the cell surface and elsewhere, these proteins pass through a series of compartments, where they are successively modified. Transfer from one compartment to the next involves a delicate balance between forward and backward (retrieval) transport pathways. Some transport vesicles select cargo molecules and move them to the next compartment in the pathway, while others retrieve escaped proteins and return them to a previous compartment where they normally function. Thus, the pathway from the ER to the cell surface involves many sorting steps, which continually select membrane and soluble lumenal proteins for packaging and transport—in vesicles or organelle fragments that bud from the ER and Golgi apparatus.

In this section we focus mainly on the **Golgi apparatus** (also called the **Golgi complex**). It is a major site of carbohydrate synthesis, as well as a sorting and dispatching station for the products of the ER. Many of the cell's polysaccharides are made in the Golgi apparatus, including the pectin and hemicellulose of the cell wall in plants and most of the glycosaminoglycans of the extracellular matrix in animals (discussed in Chapter 19). But the Golgi apparatus also lies on the exit route from the ER, and a large proportion of the carbohydrates that it makes are attached as oligosaccharide side chains to the many proteins and lipids that the ER sends to it. A subset of these oligosaccharide groups serve as tags to direct specific proteins into vesicles that then transport them to lysosomes. But most proteins and lipids, once they have acquired their appropriate oligosaccharides in the Golgi apparatus, are recognized in other ways for targeting into the transport vesicles going to other destinations.

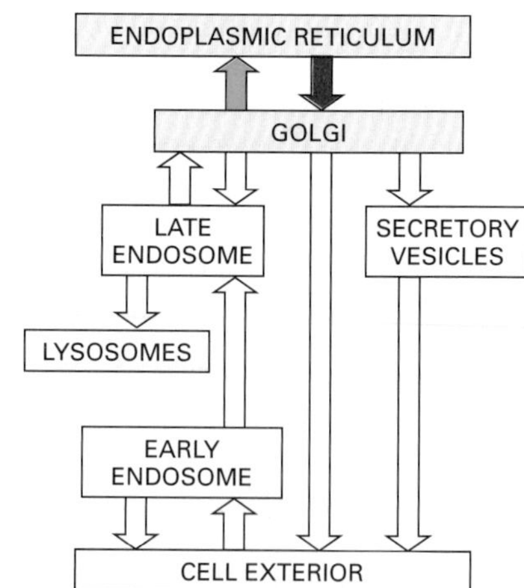

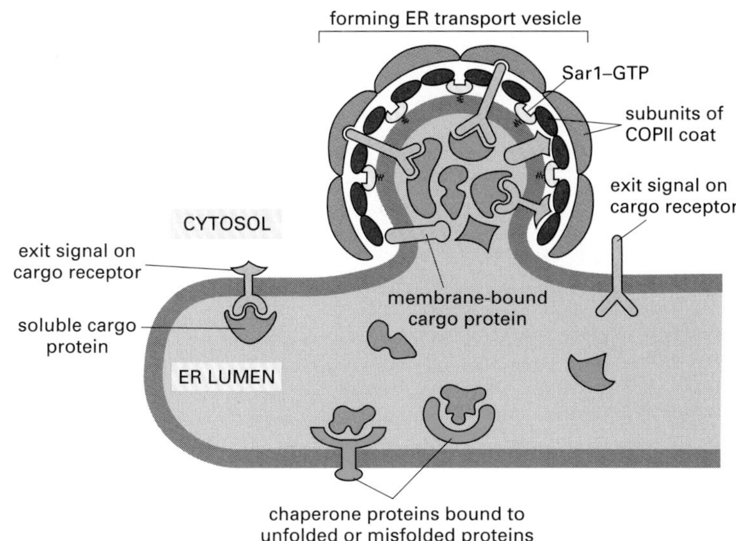

Figure 13–17 The recruitment of cargo molecules into ER transport vesicles. By binding to the COPII coat, membrane and cargo proteins become concentrated in the transport vesicles as they leave the ER. Membrane proteins are packaged into budding transport vesicles through the interactions of exit signals on their cytosolic tails with the COPII coat. Some of the membrane proteins trapped by the coat in turn function as cargo receptors, binding soluble proteins in the lumen and helping to package them into vesicles. A typical 50-nm transport vesicle contains about 200 membrane proteins, which can be of many different types. As indicated, unfolded or incompletely assembled proteins are bound to chaperones and are thereby retained in the ER compartment.

Proteins Leave the ER in COPII-coated Transport Vesicles

To initiate their journey along the biosynthetic–secretory pathway, proteins that have entered the ER and are destined for the Golgi apparatus or beyond are first packaged into small COPII-coated transport vesicles. These transport vesicles bud from specialized regions of the ER called *ER exit sites,* whose membrane lacks bound ribosomes. In most animal cells, ER exit sites seem to be randomly dispersed throughout the ER network.

Originally it was thought that all proteins that are not tethered in the ER enter transport vesicles by default. However, it is now clear that packaging into vesicles that leave the ER can also be a selective process. Some cargo proteins are actively recruited into such vesicles, where they become concentrated. It is thought that these cargo proteins display exit (transport) signals on their surface that are recognized by complementary receptor proteins that become trapped in the budding vesicle by interacting with components of the COPII coat (Figure 13–17). At a much lower rate, proteins without such exit signals can also get packaged in vesicles, so that even proteins that normally function in the ER (so-called *ER resident proteins)* slowly leak out of the ER. Similarly, secretory proteins that are made in high concentrations may leave the ER without the help of sorting receptors.

The exit signals that direct proteins out of the ER for transport to the Golgi and beyond are mostly not understood. There is one exception, however. The ERGIC53 protein seems to serve as a receptor for packaging some secretory proteins into COPII-coated vesicles. Its role in protein transport was identified because humans who lack it owing to an inherited mutation have lowered serum levels of two secreted blood-clotting factors (Factor V and Factor VIII) and therefore bleed excessively. The ERGIC53 protein is a lectin that binds mannose and is thought to recognize this sugar on Factor V and Factor VIII proteins, thereby packaging the proteins into transport vesicles in the ER.

Only Proteins That Are Properly Folded and Assembled Can Leave the ER

To exit from the ER, proteins must be properly folded and, if they are subunits of multimeric protein complexes, they may need to be completely assembled. Those that are misfolded or incompletely assembled are retained in the ER, where they are bound to chaperone proteins (see Chapter 6), such as *BiP* or *calnexin.* The chaperones may cover up the exit signals or somehow anchor the proteins in the ER (Figure 13–18). Such failed proteins are eventually transported back into the cytosol where they are degraded by proteasomes (discussed in Chapter 12). This quality-control step is important, as misfolded or misassembled proteins

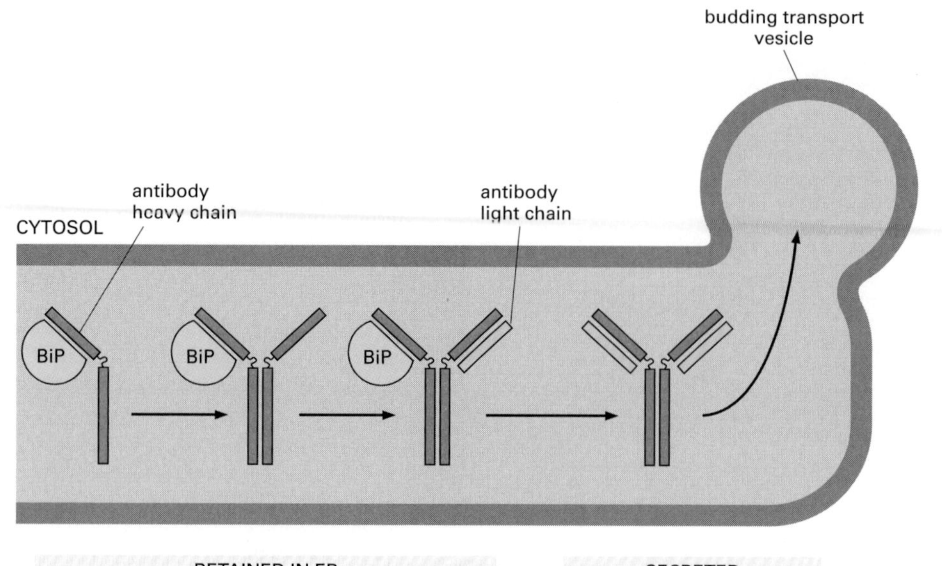

Figure 13–18 Retention of incompletely assembled antibody molecules in the ER. Antibodies are made up of two heavy and two light chains (discussed in Chapter 24), which assemble in the ER. The chaperone BiP is thought to bind to all incompletely assembled antibody molecules and to cover up an exit signal. Thus, only completely assembled antibodies leave the ER and are secreted.

could potentially interfere with the functions of normal proteins if they were transported onward. The amount of corrective action is surprisingly large. More than 90% of the newly synthesized subunits of the T cell receptor (discussed in Chapter 24) and of the acetylcholine receptor (discussed in Chapter 11), for example, are normally degraded in the cell without ever reaching the cell surface, where they function. Thus, cells must make a large excess of many protein molecules from which to select the few that fold and assemble properly.

Sometimes, however, this quality-control mechanism is detrimental. The predominant mutations that cause cystic fibrosis, a common inherited disease, produce a plasma membrane protein important for Cl^- transport that is only slightly misfolded. Although the mutant protein would function perfectly normally if it reached the plasma membrane, it is retained in the ER. The devastating disease thus results not because the mutation inactivates the protein, but because the active protein is discarded before it reaches the plasma membrane.

Transport from the ER to the Golgi Apparatus Is Mediated by Vesicular Tubular Clusters

After transport vesicles have budded from an ER exit site and have shed their coat, they begin to fuse with one another. This fusion of membranes from the same compartment is called *homotypic fusion,* to distinguish it from *heterotypic fusion,* in which a membrane from one compartment fuses with the membrane of a different compartment. As with heterotypic fusion, homotypic fusion requires a set of matching SNAREs. In this case, however, the interaction is symmetrical, with v-SNAREs and t-SNAREs contributed by both membranes (Figure 13–19).

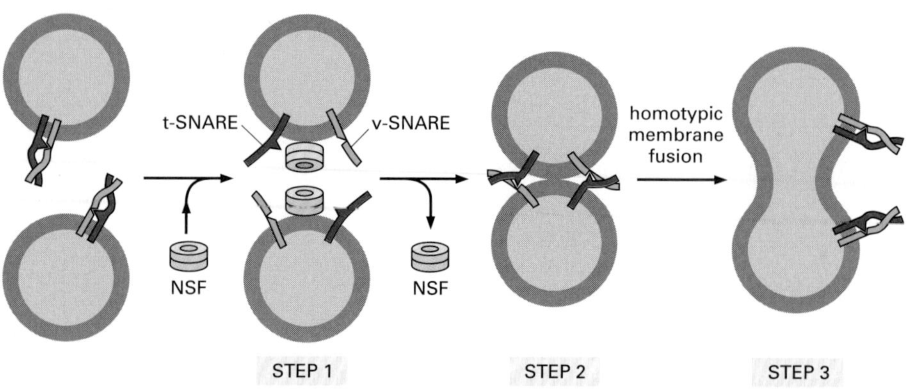

Figure 13–19 Homotypic membrane fusion. In step 1, identical pairs of v-SNAREs and t-SNAREs in both membranes are pried apart by NSF (see Figure 13–13). In steps 2 and 3, the separated matching SNAREs on adjacent identical membranes interact, which leads to membrane fusion and the formation of one continuous compartment. Subsequently, the compartment grows further by homotypic fusion with vesicles from the same kind of membrane, displaying matching SNAREs. Homotypic fusion is not restricted to the formation of vesicular tubular clusters; in a similar process, endosomes fuse to generate larger endosomes. Rab proteins help regulate the extent of homotypic fusion and hence the size of the compartments in a cell (not shown).

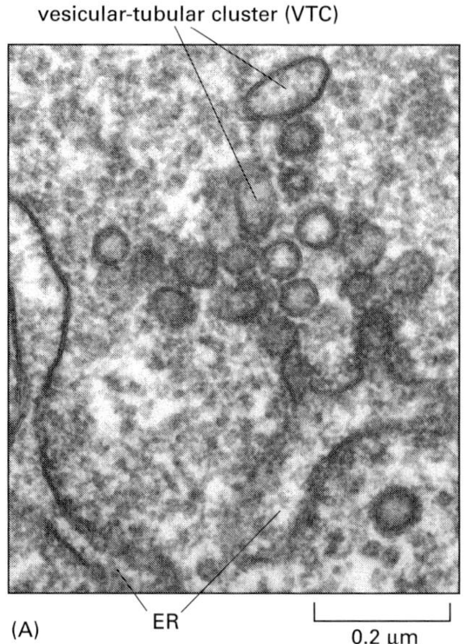

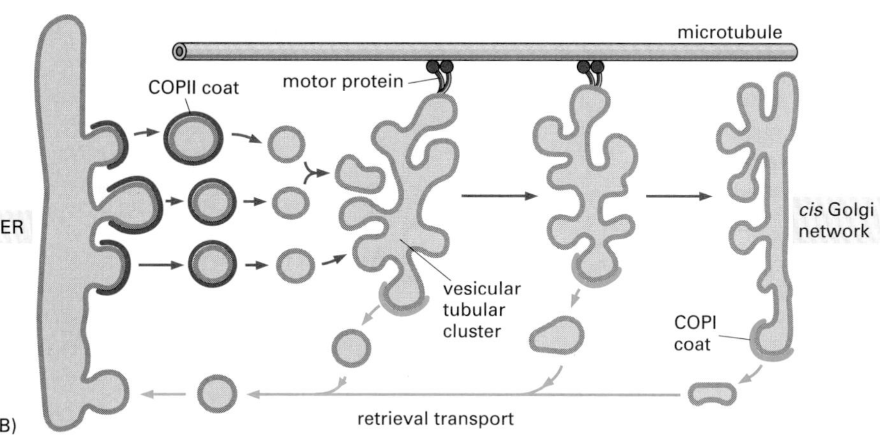

Figure 13–20 Vesicular tubular clusters. (A) An electron micrograph section of vesicular tubular clusters forming from the ER membrane. Many of the vesicle-like structures seen in the micrograph are cross sections of tubules that extend above and below the plane of this thin section and are interconnected. (B) Vesicular tubular clusters move along microtubules to carry proteins from the ER to the Golgi apparatus. COPI coats mediate the budding of vesicles that return to the ER from these clusters. As indicated, the coats quickly disassemble after the vesicles have formed. (A, courtesy of William Balch.)

The structures formed when ER-derived vesicles fuse with one another are called *vesicular tubular clusters,* on the basis of their convoluted appearance in the electron microscope (Figure 13–20A). These clusters constitute a new compartment that is separate from the ER and lacks many of the proteins that function in the ER. They are generated continually and function as transport packages that bring material from the ER to the Golgi apparatus. The clusters are relatively short-lived because they quickly move along microtubules to the Golgi apparatus, where they fuse and deliver their contents (Figure 13–20B).

As soon as vesicular tubular clusters form, they begin budding off vesicles of their own. Unlike the COPII-coated vesicles that bud from the ER, these vesicles are COPI-coated. They carry back to the ER resident proteins that have escaped, as well as proteins that participated in the ER budding reaction and are being returned. This retrieval process demonstrates the exquisite control mechanisms that regulate coat assembly reactions. The COPI coat assembly begins only seconds after the COPII coats have been shed. It remains a mystery how this switchover in coat assembly is controlled.

The *retrieval* (or *retrograde*) *transport* continues as the vesicular tubular clusters move to the Golgi apparatus. Thus, the clusters continuously mature, gradually changing their composition as selected proteins are returned to the ER. A similar retrieval process continues from the Golgi apparatus, after the vesicular tubular clusters have delivered their cargo.

The Retrieval Pathway to the ER Uses Sorting Signals

The retrieval pathway for returning escaped proteins back to the ER depends on *ER retrieval signals.* Resident ER membrane proteins, for example, contain signals that bind directly to COPI coats and are thus packaged into COPI-coated transport vesicles for retrograde delivery to the ER. The best-characterized signal of this type consists of two lysines, followed by any two other amino acids, at the extreme C-terminal end of the ER membrane protein. It is called a *KKXX sequence,* based on the single-letter amino acid code.

Soluble ER resident proteins, such as BiP, also contain a short retrieval signal at their C-terminal end, but it is different: it consists of a Lys-Asp-Glu-Leu or similar sequence. If this signal (called the *KDEL sequence*) is removed from BiP by genetic engineering, the protein is slowly secreted from the cell. If the signal is transferred to a protein that is normally secreted, the protein is now efficiently returned to the ER, where it accumulates.

Unlike the retrieval signals on ER membrane proteins that can interact directly with the COPI coat, soluble ER resident proteins must bind to specialized receptor proteins such as the *KDEL receptor*—a multipass transmembrane protein that binds to the KDEL sequence and packages any protein displaying it into COPI-coated retrograde transport vesicles. To accomplish this task, the KDEL receptor itself must cycle between the ER and the Golgi apparatus, and its affinity for the KDEL sequence must be different in these two compartments. The receptor must have a high affinity for the KDEL sequence in vesicular tubular clusters and the Golgi apparatus, so as to capture escaped ER resident proteins that are present there at low concentration. It must have a low affinity for the KDEL sequence in the ER, however, to unload its cargo in spite of the very high concentration of KDEL-containing resident proteins in the ER.

How can the affinity of the KDEL receptor change depending on the compartment in which it resides? The answer may be related to the different pH values established in the different compartments, regulated by H^+ pumps in their membrane. The KDEL receptor could bind the KDEL sequence under the slightly acidic conditions in vesicular tubular clusters and the Golgi compartment but release it at the neutral pH in the ER. As we discuss later, such pH-sensitive protein–protein interactions form the basis for many of the sorting steps in the cell (Figure 13–21).

Most membrane proteins that function at the interface between the ER and Golgi apparatus, including v- and t-SNARES and some cargo receptors, enter the retrieval pathway to the ER. Whereas the recycling of some of these proteins is signal-mediated as just described, for others no specific signal seems to be required. Thus, while retrieval signals increase the efficiency of the retrieval process, some proteins—including some Golgi enzymes—randomly enter budding vesicles destined for the ER and are returned to the ER at a slower rate. Such Golgi enzymes cycle constantly between the ER and the Golgi, but their rate of return to the ER is slow enough for most of the protein to be found in the Golgi apparatus.

Many Proteins are Selectively Retained in the Compartments in which they Function

The KDEL retrieval pathway only partly explains how ER resident proteins are maintained in the ER. As expected, cells that express genetically modified ER resident proteins, from which the KDEL sequence has been experimentally removed, secrete these proteins. But secretion occurs at a much slower rate than for a normal secretory protein. It seems that ER resident proteins are anchored

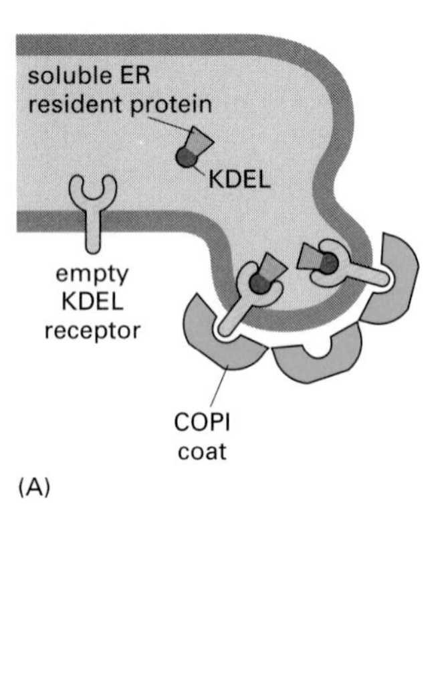

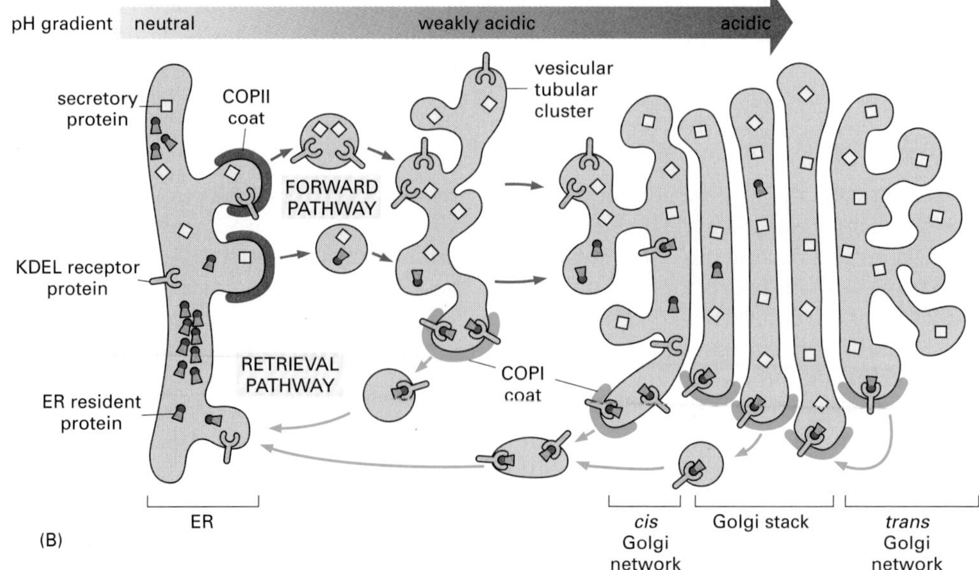

Figure 13–21 A model for the retrieval of ER resident proteins. Those ER resident proteins that escape from the ER are returned to the ER by vesicular transport. (A) The KDEL receptor present in vesicular tubular clusters and the Golgi apparatus, captures the soluble ER resident proteins and carries them in COPI-coated transport vesicles back to the ER. Upon binding its ligands in this low-pH environment, the KDEL receptor may change conformation, so as to facilitate its recruitment into budding COPI-coated vesicles. (B) The retrieval of ER proteins begins in vesicular tubular clusters and continues from all parts of the Golgi apparatus. In the neutral-pH environment of the ER, the ER proteins dissociate from the KDEL receptor, which is then returned to the Golgi for reuse.

in the ER by a mechanism that is independent of their KDEL signal and that only those proteins that escape retention are captured and returned via the KDEL receptor. A suggested mechanism of retention is that ER resident proteins bind to one another, thus forming complexes that are too big to enter transport vesicles. Because ER resident proteins are present in the ER at very high concentrations (estimated to be millimolar), relatively low-affinity interactions would suffice to have most of the proteins tied up in such complexes.

Aggregation of proteins that function in the same compartment—called *kin recognition*—is a general mechanism that compartments use to organize and retain their resident proteins. Golgi enzymes that function together, for example, also bind to each other and are thereby restrained from entering transport vesicles.

The Length of the Transmembrane Region of Golgi Enzymes Determines their Location in The Cell

Vesicles that leave the Golgi apparatus of animal cells destined for the plasma membrane are rich in cholesterol. The cholesterol fills the space between the kinked hydrocarbon chains of the lipids in the bilayer, forcing them into tighter alignment and increasing the separation between the lipid head groups of the two leaflets of the bilayer (see Figure 10–11). Thus the lipid bilayer of the cholesterol-derived vesicles is thicker than that of the Golgi membrane itself. Transmembrane proteins must have sufficiently long transmembrane segments to span this thickness if they are to enter the cholesterol-rich transport vesicle budding from the Golgi apparatus destined for the plasma membrane. Proteins with shorter transmembrane segments are excluded.

This exclusion is thought to explain why membrane proteins that normally reside in the Golgi and the ER have shorter transmembrane segments (around 15 amino acids) than do plasma membrane proteins (around 20–25 amino acids). When the transmembrane segments of Golgi proteins are extended by recombinant DNA techniques, the proteins are no longer efficiently retained in the Golgi apparatus and are transported to the plasma membrane instead. Thus, at least some Golgi proteins seem to be retained in the Golgi apparatus mainly because they cannot enter transport vesicles heading for the plasma membrane.

The Golgi Apparatus Consists of an Ordered Series of Compartments

Because of its large and regular structure, the Golgi apparatus was one of the first organelles described by early light microscopists. It consists of a collection of flattened, membrane-enclosed *cisternae,* somewhat resembling a stack of pancakes. Each of these Golgi stacks usually consists of four to six cisternae (Figure 13–22), although some unicellular flagellates can have up to 60. In animal cells, many stacks are linked by tubular connections between corresponding cisternae, thus forming a single complex, which is usually located near the cell nucleus and close to the centrosome (Figure 13–23A). This localization depends on microtubules. If microtubules are experimentally depolymerized, the Golgi apparatus reorganizes into individual stacks that are found throughout the cytoplasm, adjacent to ER exit sites. In some cells, including most plant cells, hundreds of individual Golgi stacks are normally dispersed throughout the cytoplasm (Figure 13–23B).

During their passage through the Golgi apparatus, transported molecules undergo an ordered series of covalent modifications. Each Golgi stack has two distinct faces: a ***cis* face** (or entry face) and a ***trans* face** (or exit face). Both *cis* and *trans* faces are closely associated with special compartments, each composed of a network of interconnected tubular and cisternal structures: the ***cis* Golgi network (CGN)** (also called the *intermediate compartment)* and the ***trans* Golgi network (TGN)**, respectively. Proteins and lipids enter the *cis* Golgi network in vesicular tubular clusters arriving from the ER and exit from the *trans* Golgi

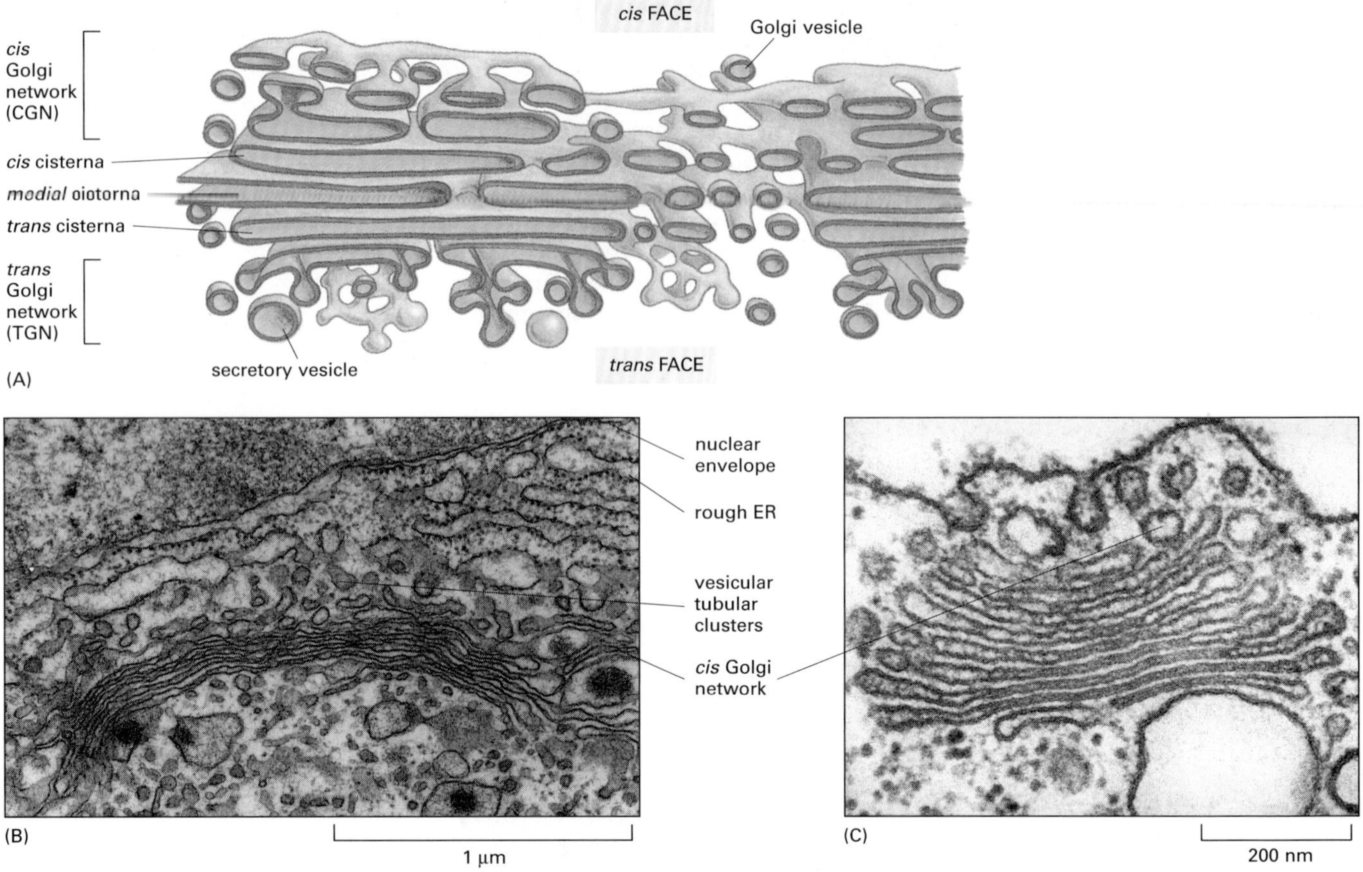

cis FACE
Golgi vesicle
cis Golgi network (CGN)
cis cisterna
medial cisterna
trans cisterna
trans Golgi network (TGN)
secretory vesicle
trans FACE
(A)

nuclear envelope
rough ER
vesicular tubular clusters
cis Golgi network
(B)
1 μm
(C)
200 nm

Figure 13–22 The Golgi apparatus. (A) Three-dimensional reconstruction from electron micrographs of the Golgi apparatus in a secretory animal cell. The *cis*-face of the Golgi stack is that closest to the ER. (B) A thin-section electron micrograph emphasizing the transitional zone between the ER and the Golgi apparatus in an animal cell. (C) An electron micrograph of a Golgi apparatus in a plant cell (the green alga *Chlamydomonas*) seen in cross section. In plant cells, the Golgi apparatus is generally more distinct and more clearly separated from other intracellular membranes than in animal cells. (A, redrawn from A. Rambourg and Y. Clermont, *Eur. J. Cell Biol.* 51:189–200, 1990; B, courtesy of Brij J. Gupta; C, courtesy of George Palade.)

network bound for the cell surface or another compartment. Both networks are thought to be important for protein sorting. As we have seen, proteins entering the CGN can either move onward in the Golgi apparatus or be returned to the ER. Similarly, proteins exiting from the TGN can either move onward and be sorted according to whether they are destined for lysosomes, secretory vesicles, or the cell surface, or be returned to an earlier compartment.

The Golgi apparatus is especially prominent in cells that are specialized for secretion, such as the goblet cells of the intestinal epithelium, which secrete large amounts of polysaccharide-rich mucus into the gut (Figure 13–24). In such cells, unusually large vesicles are found on the *trans* side of the Golgi apparatus, which faces the plasma membrane domain where secretion occurs.

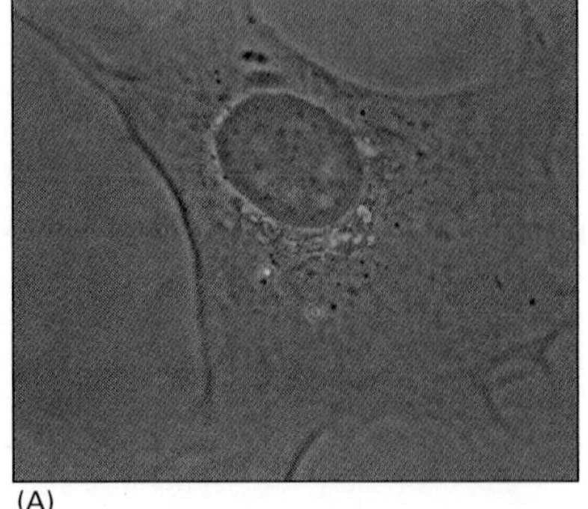
(A)

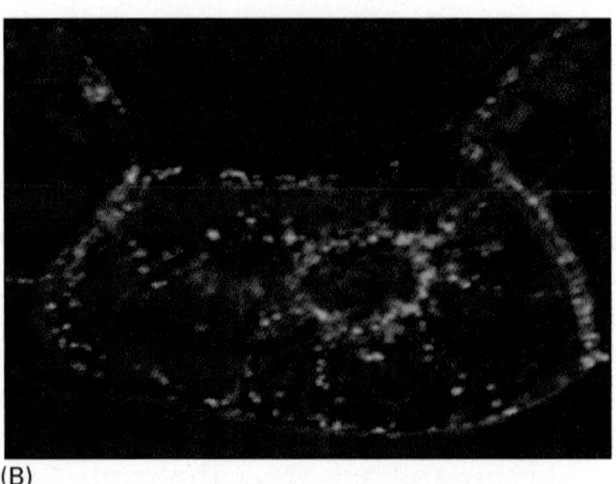
(B)

Figure 13–23 Light micrographs of the Golgi apparatus. (A) The Golgi apparatus in a cultured fibroblast stained with a fluorescent antibody that recognizes a Golgi resident protein. The Golgi apparatus is polarized, facing the direction in which the cell was crawling before fixation. (B) The Golgi apparatus in a plant cell that is expressing a fusion protein consisting of a resident Golgi enzyme fused to green fluorescent protein. The bright green spots are Golgi stacks. The faint green network is the ER, which contains some Golgi enzymes that are constantly returned to the ER from the Golgi via the retrieval pathway. (A, courtesy of John Henley and Mark McNiven; B, courtesy of Chris Hawes.)

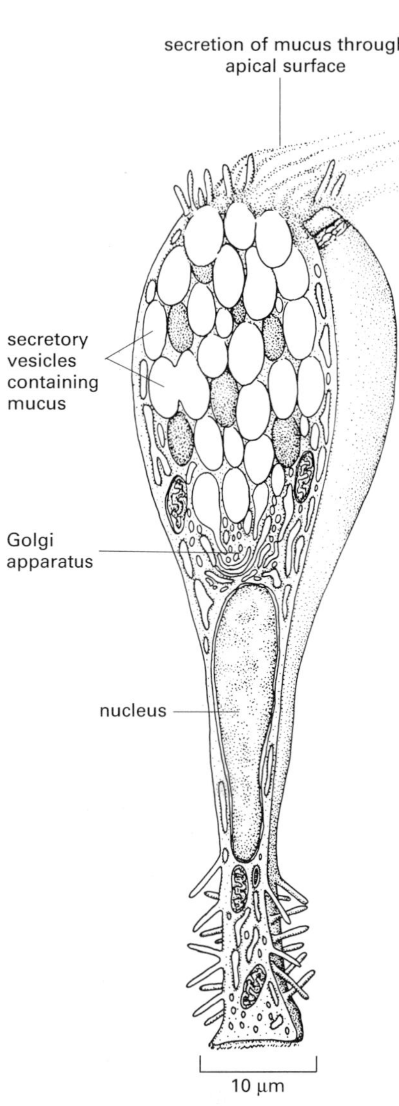

Figure 13–24 A goblet cell of the small intestine. This cell is specialized for secreting mucus, a mixture of glycoproteins and proteoglycans synthesized in the ER and Golgi apparatus. Like all epithelial cells, goblet cells are highly polarized, with the apical domain of their plasma membrane facing the lumen of the gut and the basolateral domain facing the basal lamina. The Golgi apparatus is also highly polarized, which facilitates the discharge of mucus by exocytosis at the apical domain of the plasma membrane. (After R.V. Krstić, Illustrated Encyclopedia of Human Histology. New York: Springer-Verlag, 1984.)

Oligosaccharide Chains Are Processed in the Golgi Apparatus

As described in Chapter 12, a single species of ***N*-linked oligosaccharide** is attached *en bloc* to many proteins in the ER and then trimmed while the protein is still in the ER. Further modifications and additions occur in the Golgi apparatus, depending on the protein. The outcome is that two broad classes of *N*-linked oligosaccharides, the *complex oligosaccharides* and the *high-mannose oligosaccharides,* are found attached to mammalian glycoproteins (Figure 13–25). Sometimes both types are attached (in different places) to the same polypeptide chain.

Complex oligosaccharides are generated by a combination of trimming the original *N*-linked oligosaccharide added in the ER and the addition of further sugars. By contrast, **high-mannose oligosaccharides** have no new sugars added

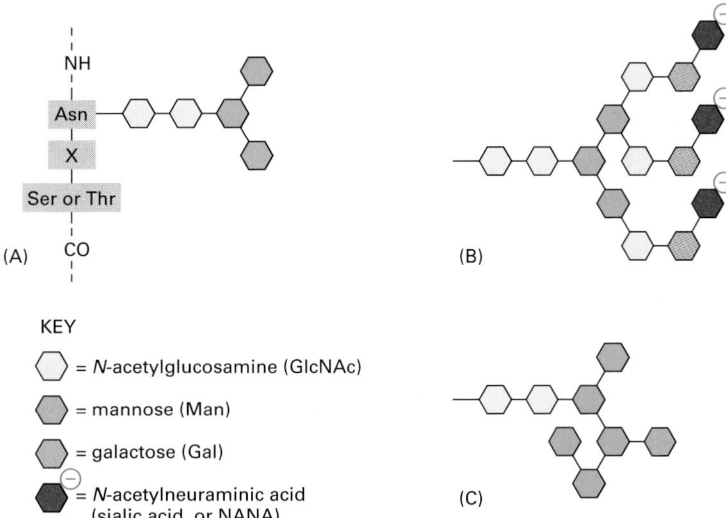

Figure 13–25 The two main classes of asparagine-linked (*N*-linked) oligosaccharides found in mature glycoproteins. (A) Both complex oligosaccharides and high-mannose oligosaccharides share a common *core region* derived from the original *N*-linked oligosaccharide added in the ER and typically containing two *N*-acetylglucosamines (GlcNAc) and three mannoses (Man). (B) Each complex oligosaccharide consists of a *core region,* together with a *terminal region* that contains a variable number of copies of a special trisaccharide unit *(N-acetylglucosamine–galactose–sialic acid)* linked to the core mannoses. Frequently, the terminal region is truncated and contains only GlcNAc and galactose (Gal) or just GlcNAc. In addition, a fucose residue may be added, usually to the core GlcNAc attached to the asparagine (Asn). Thus, although the steps of processing and subsequent sugar addition are rigidly ordered, complex oligosaccharides can be heterogeneous. Moreover, although the complex oligosaccharide shown has three terminal branches, two and four branches are also common, depending on the glycoprotein and the cell in which it is made. (C) High-mannose oligosaccharides are not trimmed back all the way to the core region and contain additional mannose residues. Hybrid oligosaccharides (not shown) with one Man branch and one GlcNAc and Gal branch are also found.

The three amino acids indicated in (A) constitute the sequence recognized by the oligosaccharyl transferase enzyme that adds the initial oligosaccharide to the protein. Ser = serine; Thr = threonine; X = any amino acid.

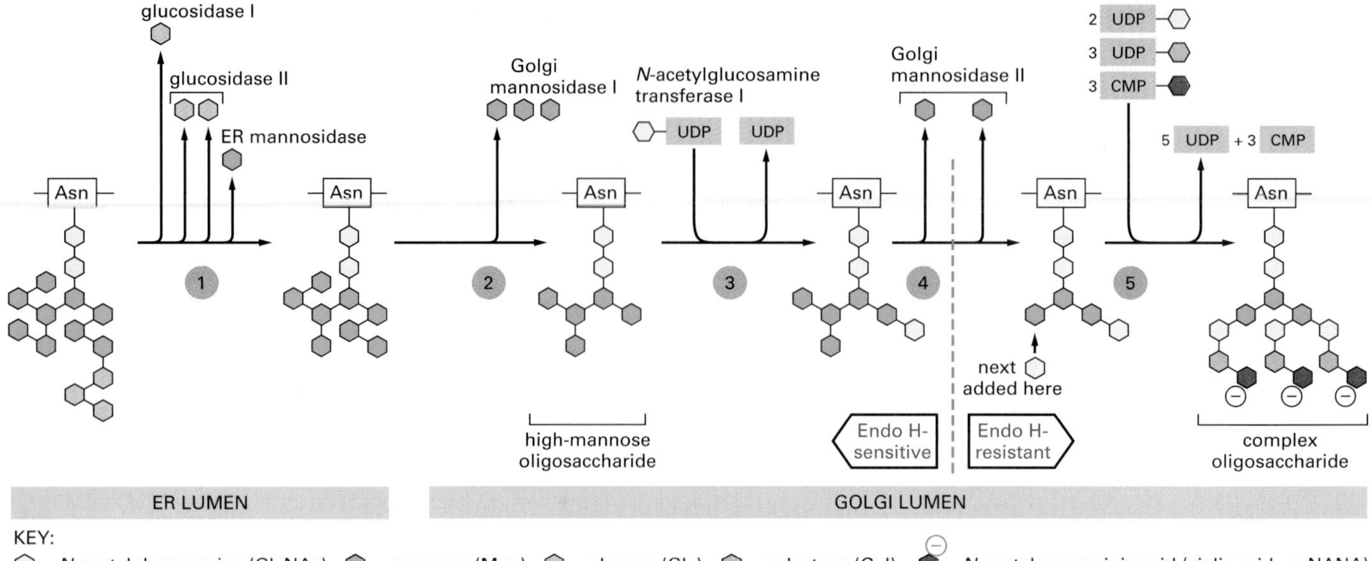

Figure 13–26 Oligosaccharide processing in the ER and the Golgi apparatus. The processing pathway is highly ordered, so that each step shown is dependent on the previous one. Processing begins in the ER with the removal of the glucoses from the oligosaccharide initially transferred to the protein. Then a mannosidase in the ER membrane removes a specific mannose. The remaining steps occur in the Golgi stack, where Golgi mannosidase I first removes three more mannoses and *N*-acetylglucosamine transferase I then adds an *N*-acetylglucosamine, which enables mannosidase II to remove two additional mannoses. This yields the final core of three mannoses that is present in a complex oligosaccharide. At this stage, the bond between the two *N*-acetylglucosamines in the core becomes resistant to attack by a highly specific endoglycosidase *(Endo H)*. Since all later structures in the pathway are also Endo H-resistant, treatment with this enzyme is widely used to distinguish complex from high-mannose oligosaccharides. Finally, as shown in Figure 13–24, additional *N*-acetylglucosamines, galactoses, and sialic acids are added. These final steps in the synthesis of a complex oligosaccharide occur in the cisternal compartments of the Golgi apparatus. Three types of glycosyl transferase enzymes act sequentially, using sugar substrates that have been activated by linkage to the indicated nucleotide. The membranes of the Golgi cisternae contain specific carrier proteins that allow each sugar nucleotide to enter in exchange for the nucleoside phosphates that are released after the sugar is attached to the protein on the lumenal face.

to them in the Golgi apparatus. They contain just two *N*-acetylglucosamines and many mannose residues, often approaching the number originally present in the lipid-linked oligosaccharide precursor added in the ER. Complex oligosaccharides can contain more than the original two *N*-acetylglucosamines as well as a variable number of galactose and sialic acid residues and, in some cases, fucose. Sialic acid is of special importance because it is the only sugar in glycoproteins that bears a negative charge. Whether a given oligosaccharide remains high-mannose or is processed is determined largely by its position on the protein. If the oligosaccharide is accessible to the processing enzymes in the Golgi apparatus, it is likely to be converted to a complex form; if it is inaccessible because its sugars are tightly held to the protein's surface, it is likely to remain in a high-mannose form. The processing that generates complex oligosaccharide chains follows the highly ordered pathway shown in Figure 13–26.

Proteoglycans Are Assembled in the Golgi Apparatus

It is not only the *N*-linked oligosaccharide chains on proteins that are altered as the proteins pass through the Golgi cisternae *en route* from the ER to their final destinations; many proteins are also modified in other ways. Some proteins have sugars added to the OH groups of selected serine or threonine side chains. This ***O*-linked glycosylation**, like the extension of *N*-linked oligosaccharide chains, is catalyzed by a series of glycosyl transferase enzymes that use the sugar nucleotides in the lumen of the Golgi apparatus to add sugar residues to a protein one at a time. Usually, *N*-acetylgalactosamine is added first, followed by a variable number of additional sugar residues, ranging from just a few to 10 or more.

The Golgi apparatus confers the heaviest glycosylation of all on *proteoglycan core proteins,* which it modifies to produce **proteoglycans.** As discussed in Chapter 19, this process involves the polymerization of one or more *glycosaminoglycan chains* (long unbranched polymers composed of repeating disaccharide units) via a xylose link onto serines on the core protein. Many proteoglycans are secreted and become components of the extracellular matrix, while others remain anchored to the plasma membrane. Still others form a major component of slimy materials, such as the mucus that is secreted to form a protective coating over many epithelia.

The sugars incorporated into glycosaminoglycans are heavily sulfated in the Golgi apparatus immediately after these polymers are made, thus adding a significant portion of their characteristically large negative charge. Some tyrosine residues in proteins also become sulfated shortly before they exit from the Golgi apparatus. In both cases, the sulfation depends on the sulfate donor 3′-phosphoadenosine-5′-phosphosulfate, or PAPS, that is transported from the cytosol into the lumen of the *trans* Golgi network.

What Is the Purpose of Glycosylation?

There is an important difference between the construction of an oligosaccharide and the synthesis of other macromolecules such as DNA, RNA, and protein. Whereas nucleic acids and proteins are copied from a template in a repeated series of identical steps using the same enzyme or set of enzymes, complex carbohydrates require a different enzyme at each step, each product being recognized as the exclusive substrate for the next enzyme in the series. Given the complicated pathways that have evolved to synthesize them, it seems likely that the oligosaccharides on glycoproteins and glycosphingolipids have important functions, but for the most part these functions are not known.

N-linked glycosylation, for example, is prevalent in all eucaryotes, including yeasts, but is absent from procaryotes. Because one or more *N*-linked oligosaccharides are present on most proteins transported through the ER and Golgi apparatus—a pathway that is unique to eucaryotic cells—one might suspect that they function to aid folding and the transport process. We have already discussed a number of instances for which this is so—the use of a carbohydrate as a marker during protein folding in the ER (see Chapter 12), for example, and the use of carbohydrate-binding lectins in guiding ER-to-Golgi transport. As we discuss later, lectins also participate in protein sorting in the *trans* Golgi network.

Because chains of sugars have limited flexibility, even a small *N*-linked oligosaccharide protrudes from the surface of a glycoprotein (Figure 13–27) and can thus limit the approach of other macromolecules to the protein surface. In this way, for example, the presence of oligosaccharides tends to make a glycoprotein more resistant to digestion by proteases. It may be that the

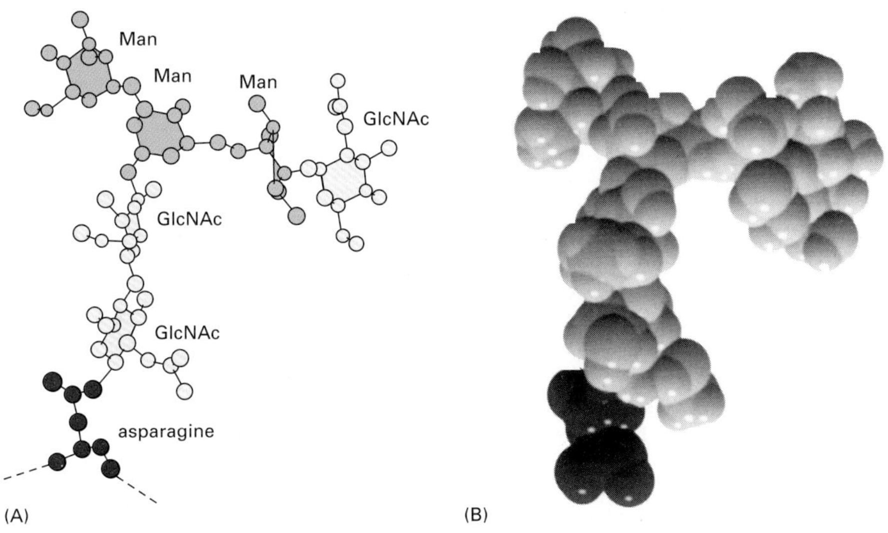

Figure 13–27 The three-dimensional structure of a small *N*-linked oligosaccharide. The structure was determined by x-ray crystallographic analysis of a glycoprotein. This oligosaccharide contains only 6 sugar residues, whereas there are 14 sugar residues in the *N*-linked oligosaccharide that is initially transferred to proteins in the ER (see Figure 12–51). (A) A backbone model showing all atoms except hydrogens. (B) A space-filling model, with the asparagine indicated by dark atoms. (B, courtesy of Richard Feldmann.)

oligosaccharides on cell-surface proteins originally provided an ancestral eucaryotic cell with a protective coat that, unlike the rigid bacterial cell wall, left the cell with the freedom to change shape and move. But if so, these sugar chains have since become modified to serve other purposes as well. The oligosaccharides attached to some cell-surface proteins, for example, are recognized by transmembrane lectins called *selectins*, which function in cell–cell adhesion processes, as discussed in Chapter 19.

Glycosylation can also have important regulatory roles. Signaling through the cell-surface signaling receptor Notch, for example, is important for proper cell fate determination in development. Notch is a transmembrane protein that is *O*-glycosylated by addition of a single fucose to some serines, threonines, and hydroxylysines. Some cell types express an additional glycosyltransferase that adds an *N*-acetylglucosamine to each of these fucoses in the Golgi apparatus. This addition sensitizes the Notch receptor, and thus allows these cells to respond selectively to activating stimuli. In this way, glycosylation has become important to the establishment of spatial boundaries in developing tissues.

The Golgi Cisternae Are Organized as a Series of Processing Compartments

Proteins exported from the ER enter the first of the Golgi processing compartments (the *cis* Golgi compartment), after having passed through the *cis* Golgi network. They then move to the next compartment (the *medial* compartment, consisting of the central cisternae of the stack) and finally to the *trans* compartment, where glycosylation is completed. The lumen of the *trans* compartment is thought to be continuous with the *trans* Golgi network, where proteins are segregated into different transport packages and dispatched to their final destinations—the plasma membrane, lysosomes, or secretory vesicles.

The oligosaccharide processing steps occur in a correspondingly organized sequence in the Golgi stack, with each cisterna containing a characteristic abundance of processing enzymes. Proteins are modified in successive stages as they move from cisterna to cisterna across the stack, so that the stack forms a multistage processing unit. This compartmentalization might seem unnecessary, since each oligosaccharide processing enzyme can accept a glycoprotein as a substrate only after it has been properly processed by the preceding enzyme. Nonetheless, it is clear that processing occurs in a spatial as well as a biochemical sequence: enzymes catalyzing early processing steps are concentrated in the cisternae toward the *cis* face of the Golgi stack, whereas enzymes catalyzing later processing steps are concentrated in the cisternae toward the *trans* face.

The functional differences between the *cis*, *medial*, and *trans* subdivisions of the Golgi apparatus were discovered by localizing the enzymes involved in processing *N*-linked oligosaccharides in distinct regions of the organelle, both by physical fractionation of the organelle and by labeling the enzymes in electron microscope sections with antibodies. The removal of mannose residues and the addition of *N*-acetylglucosamine, for example, were shown to occur in the *medial* compartment, while the addition of galactose and sialic acid was found to occur in the *trans* compartment and the *trans* Golgi network (Figure 13–28). The functional compartmentalization of the Golgi apparatus is summarized in diagrammatic form in Figure 13–29.

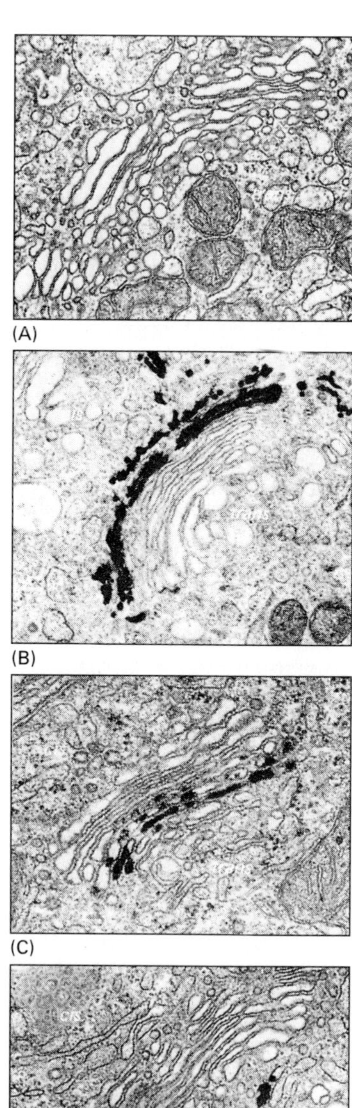

Figure 13–28 Histochemical stains demonstrating the biochemical compartmentalization of the Golgi apparatus. A series of electron micrographs shows the Golgi apparatus (A) unstained, (B) stained with osmium, which is preferentially reduced by the cisternae of the *cis* compartment, and (C and D) stained to reveal the location of a specific enzyme. The enzyme, nucleoside diphosphatase, is found in the *trans* Golgi cisternae (C), while acid phosphatase is found in the *trans* Golgi network (D). Note that usually more than one cisterna is stained. The enzymes are therefore thought to be highly enriched rather than precisely localized to a specific cisterna. (Courtesy of Daniel S. Friend.)

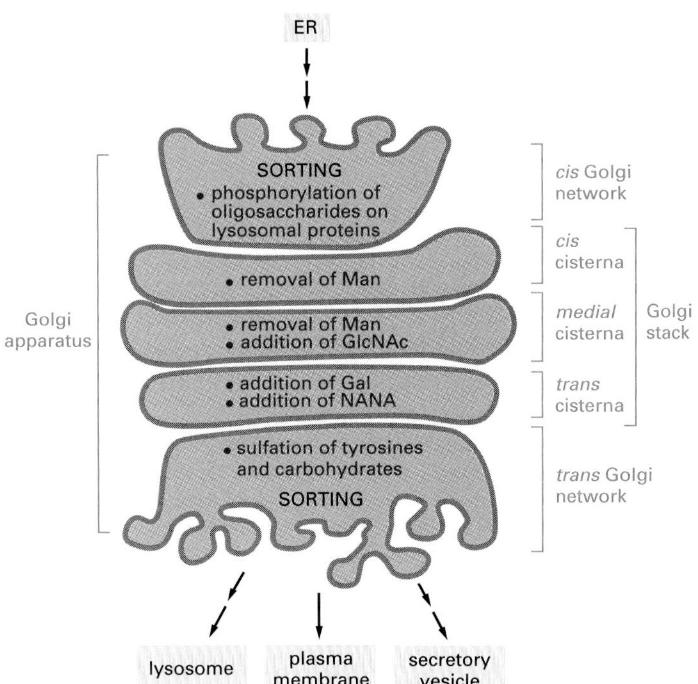

Figure 13–29 The functional compartmentalization of the Golgi apparatus. The localization of each processing step shown was determined by a combination of techniques, including biochemical subfractionation of the Golgi apparatus membranes and electron microscopy after staining with antibodies specific for some of the processing enzymes. The locations of many other processing reactions have not been determined. Although only three distinguishable cisternal compartments have so far been demonstrated, each of these sometimes consists of a group of two or more cisternae in sequence. It is likely that each processing enzyme is not completely restricted to a particular cisterna but that its distribution is graded across the stack—such that early acting enzymes are present mostly in the *cis* Golgi cisternae and later acting enzymes are mostly in the *trans* Golgi cisternae.

The functional and structural divisions of the Golgi stack pose two important questions. How are molecules transported from one Golgi cisterna to the next, and how are Golgi resident proteins retained in their appropriate places?

Transport Through the Golgi Apparatus May Occur by Vesicular Transport or Cisternal Maturation

It is still uncertain how the Golgi apparatus achieves and maintains its polarized structure and how molecules move from one cisterna to another. Functional evidence from *in vitro* transport assays and the finding of abundant transport vesicles in the vicinity of Golgi cisternae initially led to the view that these vesicles transport proteins between the cisternae, budding from one cisterna and fusing with the next. According to this **vesicular transport model**, the Golgi apparatus is a relatively static structure, with its enzymes held in place, while the molecules in transit are moved through the cisternae in sequence, carried by transport vesicles (Figure 13–30A). Retrograde flow retrieves escaped ER and Golgi proteins and returns them to preceding compartments. Directional flow is achieved as forward-moving cargo molecules are selectively packaged into forward-moving vesicles, whereas proteins to be retrieved are selectively packaged into retrograde vesicles. Although both types of vesicles are likely to be COPI-coated, the coats may contain different adaptor proteins to confer selectivity on the packaging of cargo molecules. Alternatively, transport vesicles that shuttle between Golgi cisternae may not be directional at all, transporting cargo material randomly back and forth; directional flow would then occur because of the continual input at the *cis* cisterna and output at the *trans* cisterna. In either case, the movement of vesicles from each cisterna to an adjacent one is helped by a neat trick: the budding vesicles remain tethered by filamentous proteins that restrict their movement, so that their fusion with the correct target membrane is facilitated.

According to an alternative hypothesis, called the **cisternal maturation model**, the Golgi is viewed as a dynamic structure in which the cisternae themselves move through the Golgi stack. The vesicular tubular clusters that arrive from the ER fuse with one another to become a *cis* Golgi network, and this network then progressively matures to become a *cis* cisterna, then a *medial* cisterna, and so on. Thus, at the *cis* face of a Golgi stack, new *cis* cisternae would continually form and then migrate through the stack as they mature (Figure 13–30B). This model is supported by microscopic observations demonstrating that large

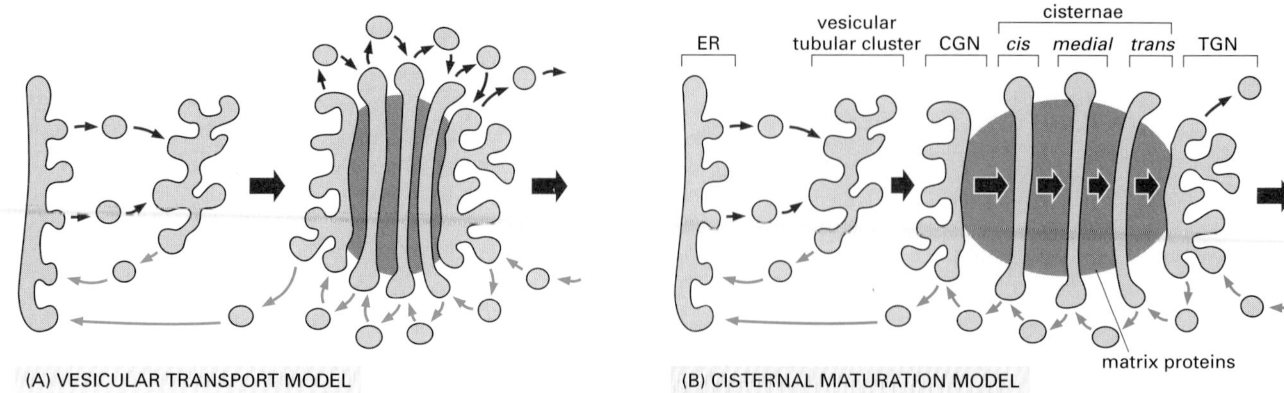

Figure 13–30 Two possible models explaining the organization of the Golgi apparatus and the transport of proteins from one cisterna to the next. It is likely that the transport through the Golgi apparatus in the forward direction *(red arrows)* involves elements of both of the views represented here. (A) In the vesicular transport model, Golgi cisternae are static organelles, which contain a characteristic complement of resident enzymes. The passing of molecules through the Golgi is accomplished by forward-moving transport vesicles, which bud from one cisterna and fuse with the next in a *cis*-to-*trans* direction. (B) According to the alternative cisternal maturation model, each Golgi cisterna matures as it migrates outwards through a stack. At each stage, the Golgi resident proteins that are carried forward in a cisterna are moved backward to an earlier compartment in COPI-coated vesicles. When a newly formed cisterna moves around to a *medial* position, for example, "left-over" *cis* Golgi enzymes would be extracted and transported backward to a new cis cisterna behind. Likewise, the *medial* enzymes would be received by retrograde transport from the cisternae just ahead. In this way, a *cis* cisterna would mature to a *medial* cisterna as it moves.

structures such as collagen rods in fibroblasts and scales in certain algae—which are much too large to fit into classical transport vesicles—move progressively through the Golgi stack.

In the maturation model, the characteristic distribution of Golgi enzymes is explained by retrograde flow. Everything moves continuously forward with the maturing cisterna, including the processing enzymes that belong in the early Golgi apparatus. But budding COPI-coated vesicles continually collect the appropriate enzymes, almost all of which are membrane proteins, and carry them back to the earlier cisterna where they function. A newly formed *cis* cisterna would therefore receive its normal complement of resident enzymes primarily from the cisterna just ahead of it and would later pass them back to the next *cis* cisterna that forms.

As we discuss later, when a cisterna finally moves up to become part of the *trans* Golgi network, various types of coated vesicles bud off of it until this network disappears, to be replaced by a maturing cisterna just behind. At the same time, other transport vesicles are continually retrieving membrane from post-Golgi compartments and returning this membrane to the *trans* Golgi network.

The vesicular transport and the cisternal maturation model are not mutually exclusive. Indeed, evidence suggests that transport may occur by a combination of the two mechanisms, in which some cargo is moved forward rapidly in transport vesicles, whereas other cargo is moved forward more slowly as the Golgi apparatus constantly renews itself through cisternal maturation.

Matrix Proteins Form a Dynamic Scaffold That Helps Organize the Apparatus

The unique architecture of the Golgi apparatus depends on both the microtubule cytoskeleton, as already discussed, and cytoplasmic Golgi matrix proteins, which form a scaffold between adjacent cisternae and give the Golgi stack its structural integrity. Some of the matrix proteins form long, filamentous tethers that are thought to help retain Golgi transport vesicles close to the organelle. When the cell prepares to divide, mitotic protein kinases phosphorylate the Golgi matrix proteins, causing the Golgi apparatus to fragment and disperse throughout the cytosol. During disassembly, Golgi enzymes are returned in vesicles to the ER, while other Golgi fragments are distributed to the two daughter cells. There, the matrix proteins are dephosphorylated, leading to the reassembly of the Golgi apparatus.

Remarkably, the Golgi matrix proteins can assemble into appropriately localized stacks near the centrosome even when Golgi membrane proteins are experimentally prevented from leaving the ER. This observation suggests that the matrix proteins are largely responsible for both the structure and location of the Golgi apparatus.

Summary

Correctly folded and assembled proteins in the ER are packaged into COPII-coated transport vesicles that pinch off from the ER membrane. Shortly thereafter the coat is shed and the vesicles fuse with one another to form vesicular tubular clusters, which move on microtubule tracks to the Golgi apparatus. Many resident ER proteins slowly escape, but they are returned to the ER from the vesicular tubular clusters and the Golgi apparatus by retrograde transport in COPI-coated vesicles.

The Golgi apparatus, unlike the ER, contains many sugar nucleotides, which are used by a variety of glycosyl transferase enzymes to perform glycosylation reactions on lipid and protein molecules as they pass through the Golgi apparatus. The N-*linked oligosaccharides that are added to proteins in the ER are often initially trimmed by the removal of mannoses, and further sugars are added. Moreover, the Golgi is the site where* O-*linked glycosylation occurs and where glycosaminoglycan chains are added to core proteins to form proteoglycans. Sulfation of the sugars in proteoglycans and of selected tyrosines on proteins also occurs in a late Golgi compartment.*

The Golgi apparatus distributes the many proteins and lipids that it receives from the ER and then modifies the plasma membrane, lysosomes, and secretory vesicles. It is a polarized structure consisting of one or more stacks of disc-shaped cisternae, each stack organized as a series of at least three functionally distinct compartments, termed cis, medial, and trans cisternae. The cis and trans cisternae are both connected to special sorting stations, called the cis Golgi network and the trans Golgi network, respectively. Proteins and lipids move through the Golgi stack in the cis-to-trans direction. This movement may occur by vesicular transport, by progressive maturation of the cis cisternae that migrate continuously through the stack, or by a combination of these two mechanisms. The enzymes that function in each particular region of the stack are thought to be kept there by continual retrograde vesicular transport from more distal cisternae. The finished new proteins end up in the trans Golgi network, which packages them in transport vesicles and dispatches them to their specific destinations in the cell.

TRANSPORT FROM THE *TRANS* GOLGI NETWORK TO LYSOSOMES

All of the proteins that pass through the Golgi apparatus, except those that are retained there as permanent residents, are sorted in the *trans* Golgi network according to their final destination. The mechanism of sorting is especially well understood for those proteins destined for the lumen of lysosomes, and in this section we consider this selective transport process. We begin with a brief account of lysosome structure and function.

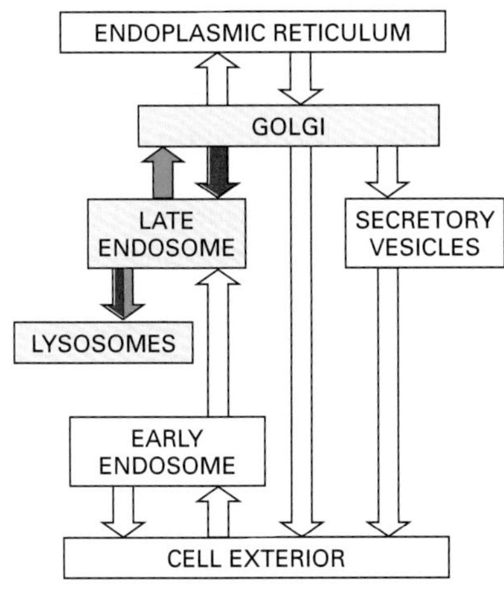

Lysosomes Are the Principal Sites of Intracellular Digestion

Lysosomes are membrane-enclosed compartments filled with hydrolytic enzymes that are used for the controlled intracellular digestion of macromolecules. They contain about 40 types of hydrolytic enzymes, including proteases, nucleases, glycosidases, lipases, phospholipases, phosphatases, and sulfatases. All are **acid hydrolases**. For optimal activity they require an acid environment, and the lysosome provides this by maintaining a pH of about 5.0

in its interior. In this way, the contents of the cytosol are doubly protected against attack by the cell's own digestive system. The membrane of the lysosome normally keeps the digestive enzymes out of the cytosol, but even if they should leak out, they can do little damage at the cytosolic pH of about 7.2.

Like all other intracellular organelles, the lysosome not only contains a unique collection of enzymes, but also has a unique surrounding membrane. Transport proteins in this membrane allow the final products of the digestion of macromolecules—such as amino acids, sugars, and nucleotides—to be transported to the cytosol, from where they can be either excreted or reutilized by the cell. An H^+ pump in the lysosomal membrane uses the energy of ATP hydrolysis to pump H^+ into the lysosome, thereby maintaining the lumen at its acidic pH (Figure 13–31). A similar or identical *vacuolar H^+ATPase* is thought to acidify all endocytic and exocytic organelles, including lysosomes, endosomes, selected compartments of the Golgi apparatus, and many transport and secretory vesicles. Most of the lysosomal membrane proteins are unusually highly glycosylated, which helps to protect them from the lysosomal proteases in the lumen.

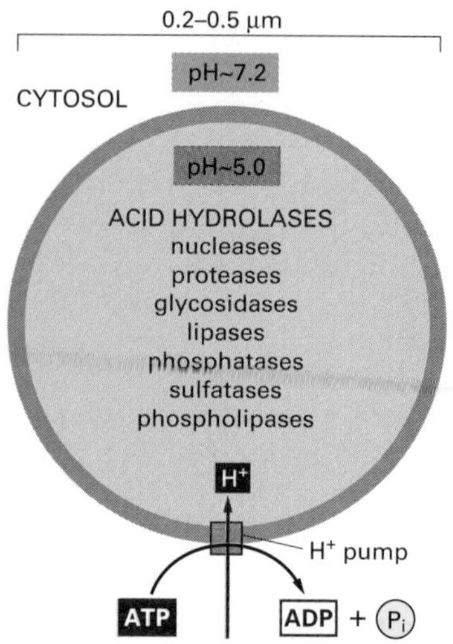

Figure 13–31 Lysosomes. The acid hydrolases are hydrolytic enzymes that are active under acidic conditions. The lumen is maintained at an acidic pH by an H^+ ATPase in the membrane that pumps H^+ into the lysosome.

Lysosomes Are Heterogeneous

Lysosomes were initially discovered by the biochemical fractionation of cell extracts; only later were they seen clearly in the electron microscope. Although extraordinarily diverse in shape and size, they can be identified as members of a single family of organelles by staining them with specific antibodies. They can also be identified by histochemistry, using the precipitate produced by the action of an acid hydrolase on its substrate to indicate which organelles contain the hydrolase (Figure 13–32). By this criterion, lysosomes are found in all eucaryotic cells.

The heterogeneity of lysosomal morphology contrasts with the relatively uniform structures of most other cell organelles. The diversity reflects the wide variety of digestive functions mediated by acid hydrolases, including the breakdown of intra- and extracellular debris, the destruction of phagocytosed microorganisms, and the production of nutrients for the cell. For this reason, lysosomes are sometimes viewed as a heterogeneous collection of distinct organelles whose common feature is a high content of hydrolytic enzymes. It is especially hard to apply a narrower definition than this in plant cells, as we see next.

Plant and Fungal Vacuoles Are Remarkably Versatile Lysosomes

Most plant and fungal cells (including yeasts) contain one or several very large, fluid-filled vesicles called **vacuoles**. They typically occupy more than 30% of the cell volume, and as much as 90% in some cell types (Figure 13–33). Vacuoles are related to the lysosomes of animal cells, containing a variety of hydrolytic enzymes, but their functions are remarkably diverse. The plant vacuole can act as a storage organelle for both nutrients and waste products, as a degradative compartment, as an economical way of increasing cell size (Figure 13–34), and as a controller of *turgor pressure* (the osmotic pressure that pushes outward on the cell wall and keeps the plant from wilting). Different vacuoles with distinct functions (e.g., digestion and storage) are often present in the same cell.

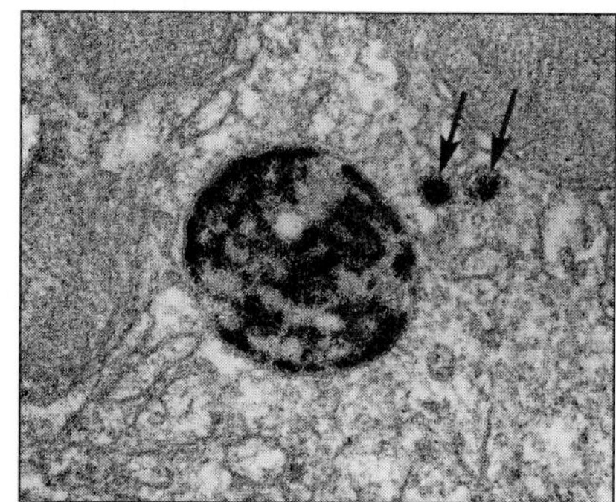

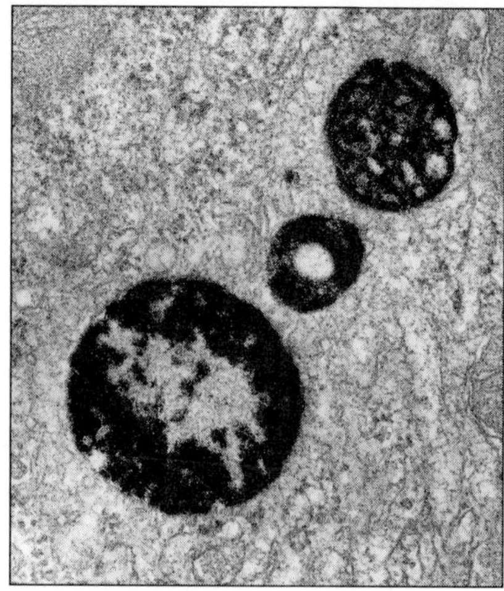

Figure 13–32 Histochemical visualization of lysosomes. These electron micrographs show two sections of a cell stained to reveal the location of acid phosphatase, a marker enzyme for lysosomes. The larger membrane-enclosed organelles, containing dense precipitates of lead phosphate, are lysosomes. Their diverse morphology reflects variations in the amount and nature of the material they are digesting. The precipitates are produced when tissue fixed with glutaraldehyde (to fix the enzyme in place) is incubated with a phosphatase substrate in the presence of lead ions. Two small vesicles thought to be carrying acid hydrolases from the Golgi apparatus are indicated by *red arrows* in the top panel. (Courtesy of Daniel S. Friend.)

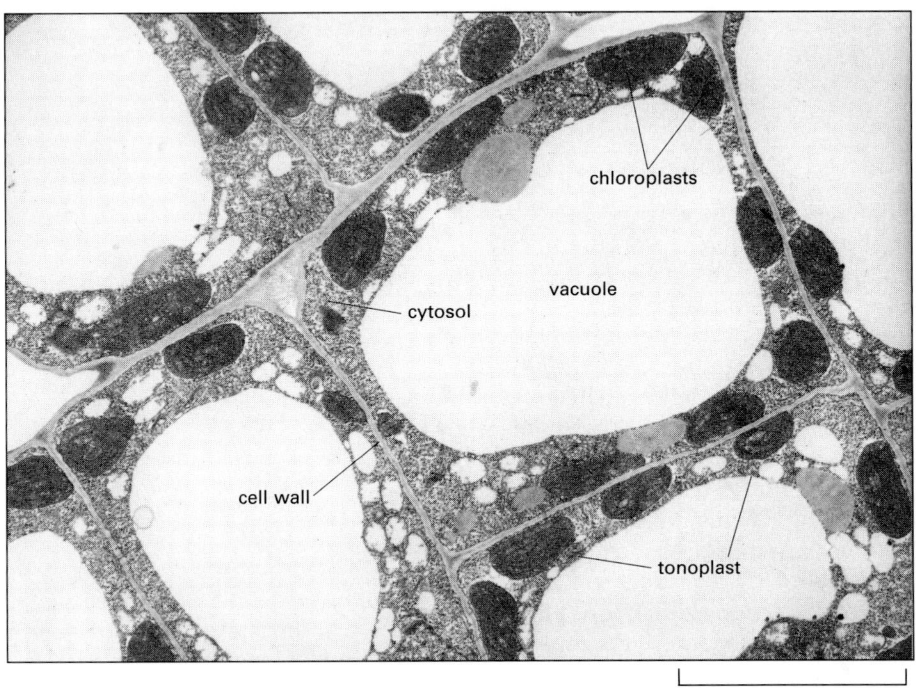

Figure 13–33 The plant cell vacuole. This electron micrograph of cells in a young tobacco leaf shows the cytosol as a thin layer, containing chloroplasts, pressed against the cell wall by the enormous vacuole. The membrane of the vacuole is called the tonoplast. (Courtesy of J. Burgess.)

The vacuole is important as a homeostatic device, enabling plant cells to withstand wide variations in their environment. When the pH in the environment drops, for example, the flux of H^+ into the cytosol is balanced, at least in part, by an increased transport of H^+ into the vacuole to keep the pH in the cytosol constant. Similarly, many plant cells maintain an almost constant turgor pressure in the face of large changes in the tonicity of the fluid in their immediate environment. They do so by changing the osmotic pressure of the cytosol and vacuole—in part by the controlled breakdown and resynthesis of polymers, such as polyphosphate, in the vacuole, and in part by altering the transport rates of sugars, amino acids, and other metabolites across the plasma membrane and the vacuolar membrane. The turgor pressure controls these fluxes by regulating the activities of the distinct sets of transporters in each membrane.

Substances stored in plant vacuoles are often harvested for human use: in different species, these range from rubber to opium to the flavoring of garlic. Many stored products have a metabolic function. Proteins, for example, can be preserved for years in the vacuoles of the storage cells of many seeds, such as those of peas and beans. When the seeds germinate, these proteins are hydrolyzed and the resulting amino acids provide a food supply for the developing embryo.

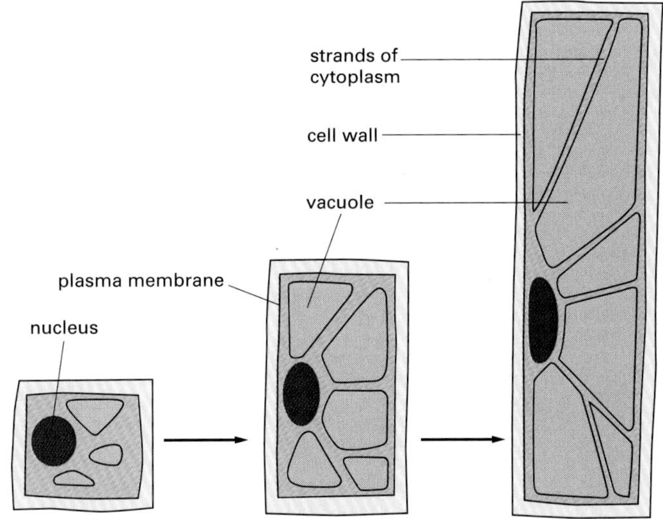

Figure 13–34 The role of the vacuole in controlling the size of plant cells. A large increase in cell volume can be achieved without increasing the volume of the cytosol. Localized weakening of the cell wall orients a turgor-driven cell enlargement that accompanies the uptake of water into an expanding vacuole. The cytosol is eventually confined to a thin peripheral layer, which is connected to the nuclear region by strands of cytosol, which are stabilized by bundles of actin filaments (not shown).

Anthocyanin pigments stored in vacuoles color the petals of many flowers so as to attract pollinating insects, while noxious molecules released from vacuoles when a plant is eaten or damaged provide a defense against predators.

Multiple Pathways Deliver Materials to Lysosomes

Lysosomes are usually meeting-places where several streams of intracellular traffic converge. Digestive enzymes are delivered to them by a route that leads outward from the ER via the Golgi apparatus, while substances to be digested are fed in by at least three paths, depending on their source.

The best studied of these paths to degradation in lysosomes is that followed by the macromolecules taken up from extracellular fluid by *endocytosis*. As discussed in detail later, endocytosed molecules are initially delivered in vesicles to small, irregularly shaped intracellular organelles called *early endosomes*. Some of these ingested molecules are selectively retrieved and recycled to the plasma membrane, while others pass on into **late endosomes**. It is here that endocytosed materials first meet the lysosomal hydrolases, which are delivered to the endosome from the Golgi apparatus. The interior of the late endosomes is mildly acidic (pH ~6), and it is the site where the hydrolytic digestion of the endocytosed molecules begins. Mature lysosomes form from the late endosomes, accompanied by a further decrease in internal pH. Lysosomes are thought to be produced by a gradual maturation process, during which endosomal membrane proteins are selectively retrieved from the developing lysosome by transport vesicles that deliver these proteins back to endosomes or the *trans* Golgi network.

A second pathway to degradation in lysosomes is used in all cell types for the disposal of obsolete parts of the cell itself—a process called **autophagy**. In a liver cell, for example, an average mitochondrion has a lifetime of about 10 days, and electron microscopic images of normal cells reveal lysosomes containing (and presumably digesting) mitochondria, as well as other organelles. The process seems to begin with the enclosure of an organelle by membranes of unknown origin, creating an *autophagosome,* which then fuses with a lysosome (or a late endosome). The process is highly regulated, and selected cell components can somehow be marked for lysosomal destruction during cell remodeling. The smooth ER that proliferates in a liver cell in response to the drug phenobarbital (discussed in Chapter 12), for example, is selectively removed by autophagy after the drug is withdrawn.

As we discuss later, the third pathway that brings materials to lysosomes for degradation is found mainly in cells specialized for the *phagocytosis* of large particles and microorganisms. Such professional phagocytes (macrophages and neutrophils in vertebrates) engulf objects to form a *phagosome,* which is then converted to a lysosome in the manner described for the autophagosome. The three pathways are summarized in Figure 13–35.

A Mannose 6-Phosphate Receptor Recognizes Lysosomal Proteins in the *Trans* Golgi Network

We now consider the pathway that delivers lysosomal hydrolases and membrane proteins to lysosomes. Both classes of proteins are synthesized in the rough ER and transported through the Golgi apparatus to the *trans* Golgi network. The transport vesicles that deliver these proteins to late endosomes (which later form lysosomes) bud from the *trans* Golgi network. The vesicles incorporate the lysosomal proteins and exclude the many other proteins being packaged into different transport vesicles for delivery elsewhere.

How are lysosomal proteins recognized and selected in the *trans* Golgi network with the required accuracy? For the lysosomal hydrolases the answer is known. They carry a unique marker in the form of *mannose 6-phosphate (M6P)* groups, which are added exclusively to the *N*-linked oligosaccharides of these soluble lysosomal enzymes as they pass through the lumen of the *cis* Golgi network (Figure 13–36). The M6P groups are recognized by transmembrane

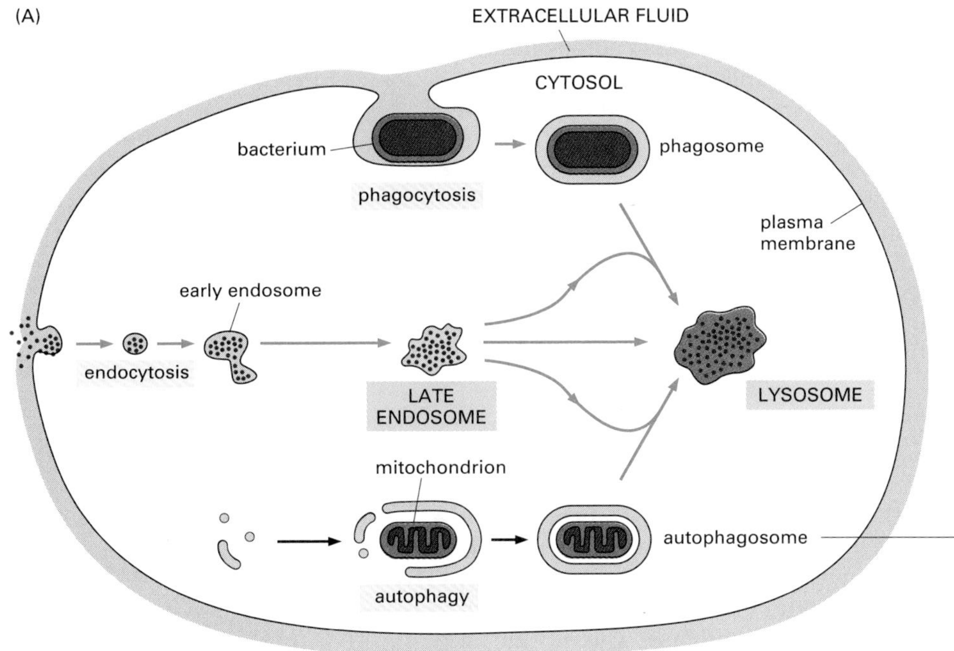

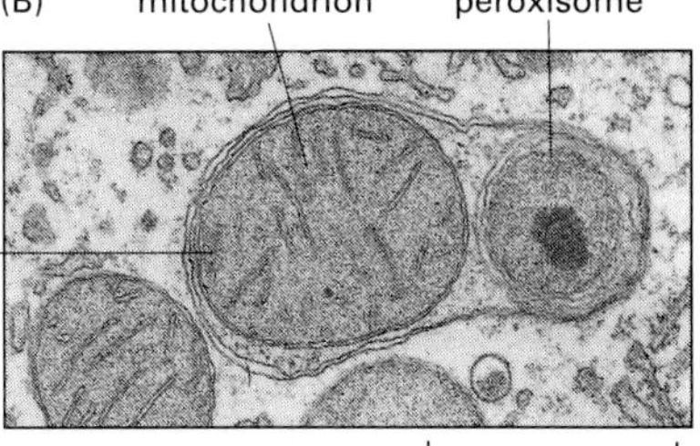

Figure 13–35 Three pathways to degradation in lysosomes. (A) Each pathway leads to the intracellular digestion of materials derived from a different source. (B) An electron micrograph of an autophagosome containing a mitochondrion and a peroxisome. (B, courtesy of Daniel S. Friend, from D.W. Fawcett, A Textbook of Histology, 12th edn. New York: Chapman and Hall, 1994.)

M6P receptor proteins, which are present in the *trans* Golgi network. The receptor proteins bind to lysosomal hydrolases on the lumenal side of the membrane and to adaptins in assembling clathrin coats on the cytosolic side. In this way, they help package the hydrolases into clathrin-coated vesicles that bud from the *trans* Golgi network. The vesicles subsequently deliver their contents to a late endosome.

The M6P Receptor Shuttles Between Specific Membranes

The M6P receptor protein binds its specific oligosaccharide at pH 6.5–6.7 in the *trans* Golgi network and releases it at pH 6, which is the pH in the interior of late endosomes. Thus, in the late endosomes, the lysosomal hydrolases dissociate from the M6P receptor. As the pH drops further during endosomal maturation, the hydrolases begin to digest the endocytosed material delivered from early endosomes. Having released their bound enzymes, the M6P receptors are retrieved into transport vesicles that bud from late endosomes; the receptors are then returned to the membrane of the *trans* Golgi network for reuse (Figure 13–37). Transport in either direction requires signal peptides in the cytoplasmic tail of the M6P receptor that specify transport of this protein to the late endosome or back to the Golgi apparatus. Thus, the recycling of the M6P receptor closely resembles the recycling of the KDEL receptor, discussed earlier.

Not all of the hydrolase molecules that are tagged with M6P for delivery to lysosomes get to their proper destination. Some escape the normal packaging process in the *trans* Golgi network and are transported "by default" to the cell surface, where they are secreted into the extracellular fluid. Some M6P receptors, however, also take a detour to the plasma membrane, where they recapture the escaped lysosomal hydrolases and return them by *receptor-mediated endocytosis* to lysosomes via early and late endosomes. As lysosomal hydrolases require an acidic milieu to work, they can do little harm in the extracellular fluids, which usually has a neutral pH.

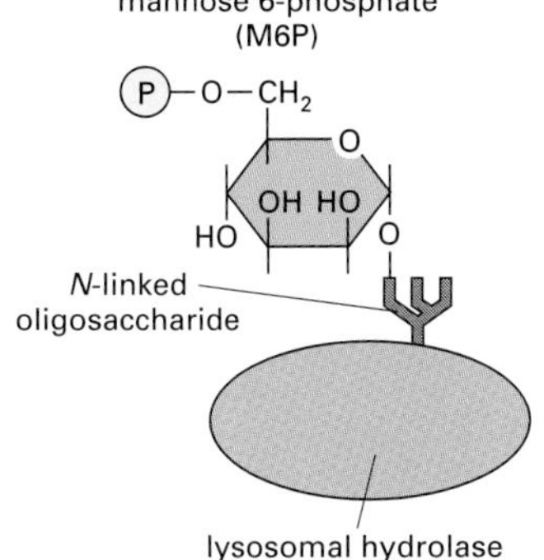

Figure 13–36 The structure of mannose 6-phosphate on a lysosomal enzyme.

A Signal Patch in the Hydrolyase Polypeptide Chain Provides the Cue for M6P Addition

The sorting system that segregates lysosomal hydrolases and dispatches them to late endosomes works because M6P groups are added only to the appropriate glycoproteins in the Golgi apparatus. This requires specific recognition of the hydrolases by the Golgi enzymes responsible for adding M6P. Since all glycoproteins leave the ER with identical *N*-linked oligosaccharide chains, the signal for adding the M6P units to oligosaccharides must reside somewhere in the polypeptide chain of each hydrolase. Genetic engineering experiments have revealed that the recognition signal is a cluster of neighboring amino acids on each protein's surface, known as a *signal patch.*

Two enzymes act sequentially to catalyze the addition of M6P groups to lysosomal hydrolases. The first is a GlcNAc phosphotransferase that specifically binds the hydrolase and adds GlcNAc-phosphate to one or two of the mannose residues on each oligosaccharide chain (Figure 13–38). A second enzyme then cleaves off the GlcNAc residue, leaving behind a newly created M6P marker. Since most lysosomal hydrolases contain multiple oligosaccharides, they acquire many M6P residues, providing a high affinity signal for the M6P receptor.

Defects in the GlcNAc Phosphotransferase Cause a Lysosomal Storage Disease in Humans

Lysosomal storage diseases are caused by genetic defects that affect one or more of the lysosomal hydrolases. The defect results in the accumulation of the undigested substrates in lysosomes, with severe pathological consequences, most often in the nervous system. In most cases there is a mutation in a structural gene that codes for an individual lysosomal hydrolase. This occurs in *Hurler's disease,* for example, in which the enzyme required for the breakdown of glycosaminoglycans is defective or missing. The most severe form of lysosomal storage disease, however, is a very rare disorder called *inclusion-cell disease* (*I-cell disease*). In this disease almost all of the hydrolytic enzymes are missing from the lysosomes of fibroblasts, and their undigested substrates accumulate in lysosomes, which consequently form large "inclusions" in the patients' cells.

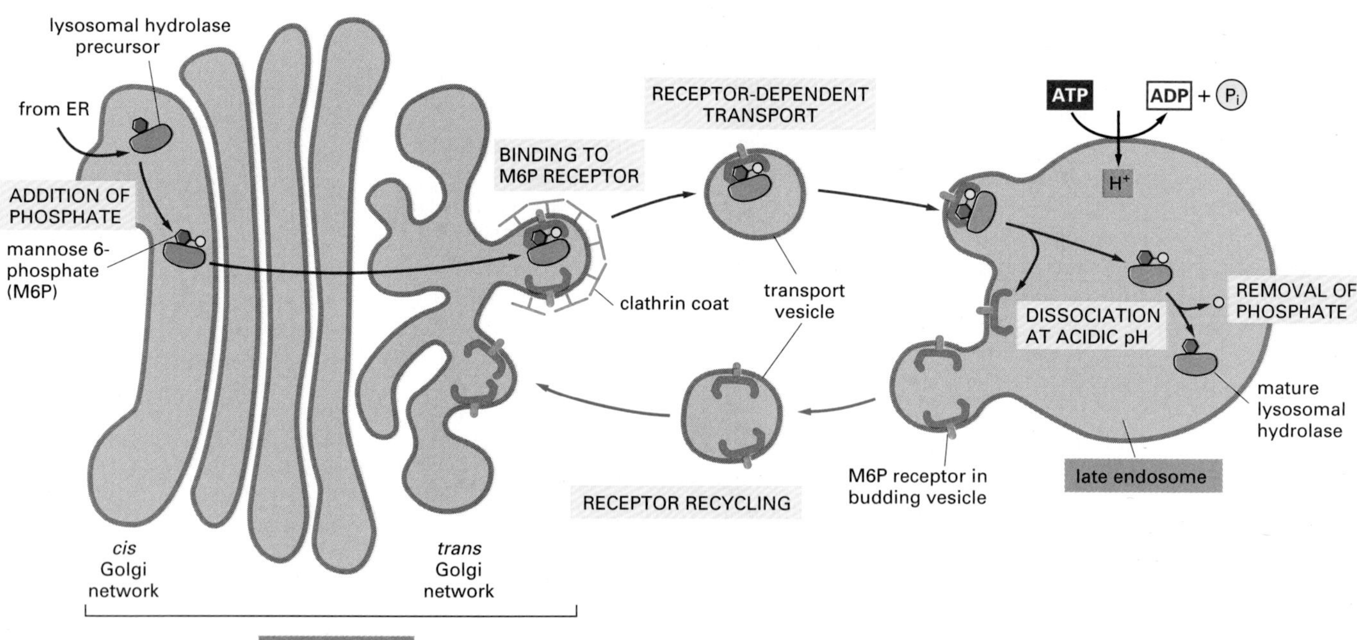

Figure 13–37 The transport of newly synthesized lysosomal hydrolases to lysosomes. The precursors of lysosomal hydrolases are covalently modified by the addition of mannose 6-phosphate (M6P) groups in the *cis* Golgi network. They then become segregated from all other types of proteins in the *trans* Golgi network because adaptins in the clathrin coat bind the M6P receptors, which, in turn, bind the modified lysosomal hydrolases. The clathrin-coated vesicles produced bud off from the *trans* Golgi network and fuse with late endosomes. At the low pH of the late endosome, the hydrolases dissociate from the M6P receptors, and the empty receptors are recycled to the Golgi apparatus for further rounds of transport. It is not known which type of coat mediates vesicle budding in the M6P receptor recycling pathway. In the late endosomes, the phosphate is removed from the mannose sugars attached to the hydrolases, further ensuring that the hydrolases do not return to the Golgi apparatus with the receptor.

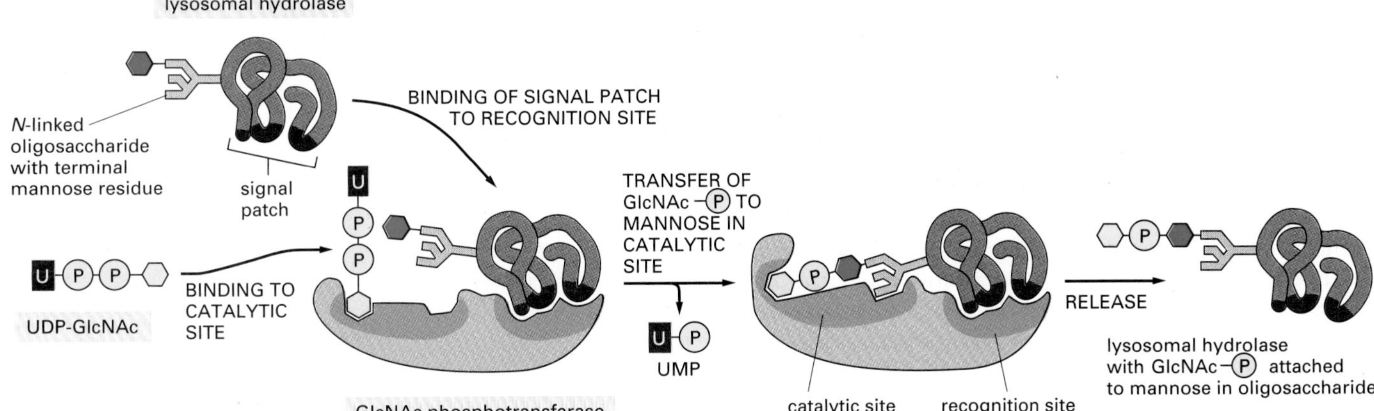

Figure 13–38 The recognition of a lysosomal hydrolase. The GlcNAc phosphotransferase enzyme that recognizes lysosomal hydrolases in the Golgi apparatus has separate catalytic and recognition sites. The catalytic site binds both high-mannose *N*-linked oligosaccharides and UDP-GlcNAc. The recognition site binds to a signal patch that is present only on the surface of lysosomal hydrolases. The GlcNAc is cleaved off by a second enzyme, leaving the M6P exposed (not shown).

I-cell disease is due to a single gene defect and, like most genetic enzyme deficiencies, it is recessive—that is, it is seen only in individuals in whom both copies of the gene are defective. In I-cell disease patients, all the hydrolases missing from lysosomes are found in the blood. Because they fail to be sorted properly in the Golgi apparatus, the hydrolases are secreted rather than transported to lysosomes. The missorting has been traced to a defective or missing GlcNAc-phosphotransferase. Because lysosomal enzymes are not phosphorylated in the *cis* Golgi network, they are not segregated by M6P receptors into the appropriate transport vesicles in the *trans* Golgi network. Instead, they are carried to the cell surface and secreted by a default pathway.

In I-cell disease the lysosomes in some cell types, such as hepatocytes, contain a normal complement of lysosomal enzymes, implying that there is another pathway for directing hydrolases to lysosomes that is used by some cell types but not others. The nature of this M6P-independent pathway is unknown. Similarly, the membrane proteins of lysosomes are sorted from the *trans* Golgi network to late endosomes by an M6P-independent pathway in all cells, and they are therefore normal in I-cell disease. These membrane proteins exit from the *trans* Golgi network in clathrin-coated vesicles distinct from those that transport the M6P-tagged hydrolases.

It is unclear why cells need more than one sorting pathway to construct a lysosome, although it is perhaps not surprising that different mechanisms operate for soluble and membrane-bound lysosomal proteins, especially since—unlike the M6P receptor—those membrane proteins are lysosomal residents and hence need not be returned to the *trans* Golgi network.

Some Lysosomes May Undergo Exocytosis

Targeting of material to lysosomes is not necessarily the end of the pathway. *Lysosomal secretion* (also called defecation) of their undigested content enables all cells to eliminate indigestible debris. For most cells, this seems to be a minor pathway, used only when cells are stressed. Some cell types, however, contain specialized lysosomes that have acquired the necessary machinery for fusion with the plasma membrane. *Melanocytes* in the skin, for example, produce and store pigments in their lysosomes. These pigment-containing *melanosomes* release their pigment into the extracellular space by exocytosis. The pigment is then taken up by keratinocytes, leading to normal skin pigmentation. In some genetic disorders, this transfer process is blocked owing to defects in melanosome exocytosis, leading to forms of hypopigmentation (albinism).

Summary

Lysosomes are specialized for the intracellular digestion of macromolecules. They contain unique membrane proteins and a wide variety of hydrolytic enzymes that operate best at pH 5, which is the internal pH of lysosomes. This low pH is maintained by an ATP-driven H^+ pump in the lysosomal membrane. Newly synthesized *lysosomal proteins are transferred into the lumen of the ER, transported through the Golgi apparatus, and then carried from the trans Golgi network to late endosomes by means of clathrin-coated transport vesicles.*

The lysosomal hydrolases contain N-linked oligosaccharides that are covalently modified in a unique way in the cis Golgi network so that their mannose residues are phosphorylated. These mannose 6-phosphate (M6P) groups are recognized by an M6P receptor protein in the trans Golgi network that segregates the hydrolases and helps package them into budding transport vesicles that deliver their contents to late endosomes (the organelle that matures into lysosomes). The M6P receptors shuttle back and forth between the trans Golgi network and these endosomes. The low pH in the late endosome dissociates the lysosomal hydrolases from these receptors, making the transport of the hydrolases unidirectional. A separate transport system uses clathrin-coated vesicles to deliver resident lysosomal membrane proteins from the trans Golgi network.

TRANSPORT INTO THE CELL FROM THE PLASMA MEMBRANE: ENDOCYTOSIS

The routes that lead inward from the cell surface to lysosomes start with the process of **endocytosis**, by which cells take up macromolecules, particulate substances, and, in specialized cases, even other cells. In this process, the material to be ingested is progressively enclosed by a small portion of the plasma membrane, which first invaginates and then pinches off to form an *endocytic vesicle* containing the ingested substance or particle. Two main types of endocytosis are distinguished on the basis of the size of the endocytic vesicles formed. One type is called *phagocytosis* ("cellular eating"), which involves the ingestion of large particles, such as microorganisms or dead cells via large vesicles called *phagosomes* (generally >250 nm in diameter). The other type is *pinocytosis* ("cellular drinking"), which involves the ingestion of fluid and solutes via small pinocytic vesicles (about 100 nm in diameter). Most eucaryotic cells are continually ingesting fluid and solutes by pinocytosis; large particles are most efficiently ingested by specialized phagocytic cells.

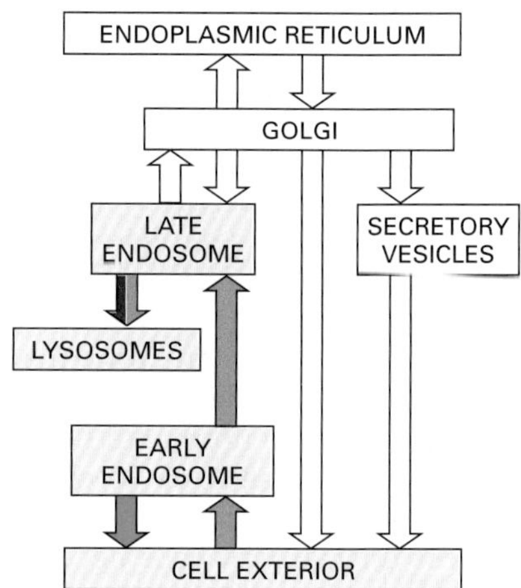

Specialized Phagocytic Cells Can Ingest Large Particles

Phagocytosis is a special form of endocytosis in which large particles such as microorganisms and dead cells are ingested via large endocytic vesicles called **phagosomes**. In protozoa, phagocytosis is a form of feeding: large particles taken up into phagosomes end up in lysosomes, and the products of the subsequent digestive processes pass into the cytosol to be utilized as food. However, few cells in multicellular organisms are able to ingest such large particles efficiently. In the gut of animals, for example, the particles of food are broken down extracellularly and their hydrolysis products are imported into cells.

Phagocytosis is important in most animals for purposes other than nutrition, and it is mainly carried out by specialized cells—so-called *professional phagocytes*. In mammals, three classes of white blood cells act as professional phagocytes—**macrophages**, **neutrophils**, and **dendritic cells**. These cells all develop from hemopoietic stem cells (discussed in Chapter 22), and they defend us against infection by ingesting invading microorganisms. Macrophages also have an important role in scavenging senescent cells and cells that have died by apoptosis (discussed in Chapter 17). In quantitative terms, the latter function is by far the most important: our macrophages phagocytose more than 10^{11} senescent red blood cells in each of us every day, for example.

Whereas the endocytic vesicles involved in pinocytosis are small and uniform, **phagosomes** have diameters that are determined by the size of the ingested particle, and they can be almost as large as the phagocytic cell itself (Figure 13–39). The phagosomes fuse with lysosomes inside the cell, and the ingested material is then degraded. Any indigestible substances will remain in lysosomes, forming *residual bodies*. Some of the internalized plasma membrane

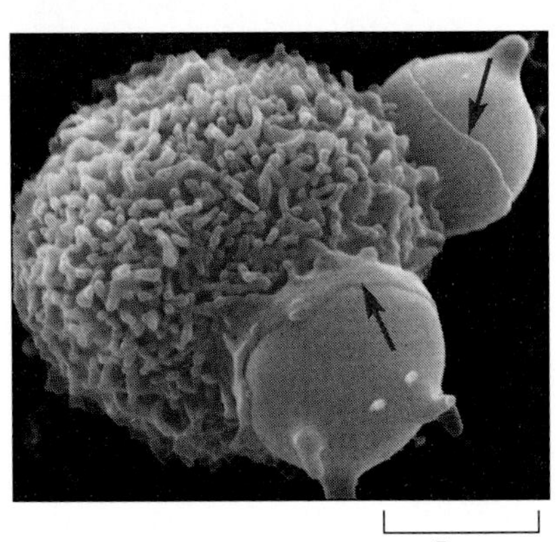

Figure 13–39 Phagocytosis by a macrophage. A scanning electron micrograph of a mouse macrophage phagocytosing two chemically altered red blood cells. The *red arrows* point to edges of thin processes (pseudopods) of the macrophage that are extending as collars to engulf the red cells. (Courtesy of Jean Paul Revel.)

components never reach the lysosome, because they are retrieved from the phagosome in transport vesicles and returned to the plasma membrane.

To be phagocytosed, particles must first bind to the surface of the phagocyte. However, not all particles that bind are ingested. Phagocytes have a variety of specialized surface receptors that are functionally linked to the phagocytic machinery of the cell. Unlike pinocytosis, which is a constitutive process that occurs continuously, phagocytosis is a triggered process, requiring that receptors be activated that transmit signals to the cell interior and initiate the response. The best-characterized triggers are antibodies, which protect us by binding to the surface of infectious microorganisms to form a coat in which the tail region of each antibody molecule, called the Fc region, is exposed on the exterior (discussed in Chapter 24). This antibody coat is recognized by specific *Fc receptors* on the surface of macrophages and neutrophils, whose binding induces the phagocytic cell to extend pseudopods that engulf the particle and fuse at their tips to form a phagosome (Figure 13–40).

Several other classes of receptors that promote phagocytosis have been characterized. Some recognize *complement* components, which collaborate with antibodies in targeting microbes for destruction (discussed in Chapter 25). Others directly recognize oligosaccharides on the surface of certain microorganisms. Still others recognize cells that have died by apoptosis. Apoptotic cells lose the asymmetric distribution of phospholipids in their plasma membrane. As a consequence, negatively charged phosphatidylserine, which is normally confined to the cytosolic leaflet of the lipid bilayer, is now exposed on the outside of the cell, where it triggers the phagocytosis of the dead cell.

Remarkably, macrophages will also phagocytose a variety of inanimate particles—such as glass, latex beads, or asbestos fibers—yet they do not phagocytose live animal cells. It seems that living animal cells display "don't-eat-me" signals in the form of cell-surface proteins that bind to inhibiting receptors on the surface of macrophages. The inhibitory receptors recruit tyrosine phosphatases that antagonize the intracellular signaling events required to initiate phagocytosis, thereby locally inhibiting the phagocytic process. Thus phagocytosis, like

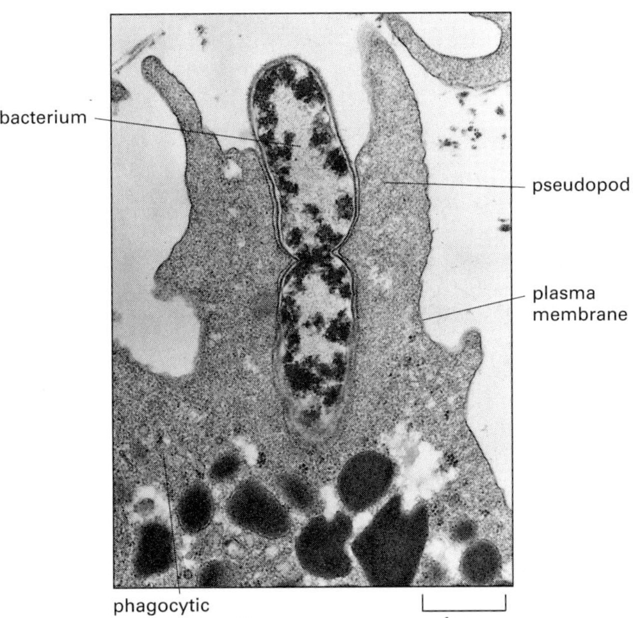

Figure 13–40 Phagocytosis by a neutrophil. An electron micrograph of a neutrophil phagocytosing a bacterium, which is in the process of dividing. (Courtesy of Dorothy F. Bainton, Phagocytic Mechanisms in Health and Disease. New York: Intercontinental Book Corporation, 1971.)

many other cell processes, depends on a balance between positive signals that activate the process and negative signals that inhibit it.

Pinocytic Vesicles Form from Coated Pits in the Plasma Membrane

Virtually all eucaryotic cells continually ingest bits of their plasma membrane in the form of small pinocytic (endocytic) vesicles, which are later returned to the cell surface. The rate at which plasma membrane is internalized in this process of **pinocytosis** varies between cell types, but it is usually surprisingly large. A macrophage, for example, ingests 25% of its own volume of fluid each hour. This means that it must ingest 3% of its plasma membrane each minute, or 100% in about half an hour. Fibroblasts endocytose at a somewhat lower rate (1% per minute), whereas some amoebae ingest their plasma membrane even more rapidly. Since a cell's surface area and volume remain unchanged during this process, it is clear that the same amount of membrane that is being removed by endocytosis is being added to the cell surface by *exocytosis*, the converse process, as we discuss later. In this sense, endocytosis and exocytosis are linked processes that can be considered to constitute an *endocytic–exocytic cycle.*

The endocytic part of the cycle often begins at **clathrin-coated pits**. These specialized regions typically occupy about 2% of the total plasma membrane area. The lifetime of a clathrin-coated pit is short: within a minute or so of being formed, it invaginates into the cell and pinches off to form a clathrin-coated vesicle (Figure 13–41). It has been estimated that about 2500 clathrin-coated vesicles leave the plasma membrane of a cultured fibroblast every minute. The coated vesicles are even more transient than the coated pits: within seconds of being formed, they shed their coat and are able to fuse with early endosomes. Since extracellular fluid is trapped in clathrin-coated pits as they invaginate to form coated vesicles, any substance dissolved in the extracellular fluid is internalized—a process called *fluid-phase endocytosis.*

Not All Pinocytic Vesicles Are Clathrin-coated

In addition to clathrin-coated pits and vesicles, there are other, less well-understood mechanisms by which cells can form pinocytic vesicles. One of these pathways initiates at **caveolae** (from the Latin for "little cavities"), originally recognized by their ability to transport molecules across endothelial cells, which form the inner lining of blood vessels. Caveolae are present in the plasma membrane of most cell types, and in some of these they are seen as deeply invaginated flasks in the electron microscope (Figure 13–42). They are thought to form from *lipid rafts,* which are patches of the plasma membrane that are especially rich in

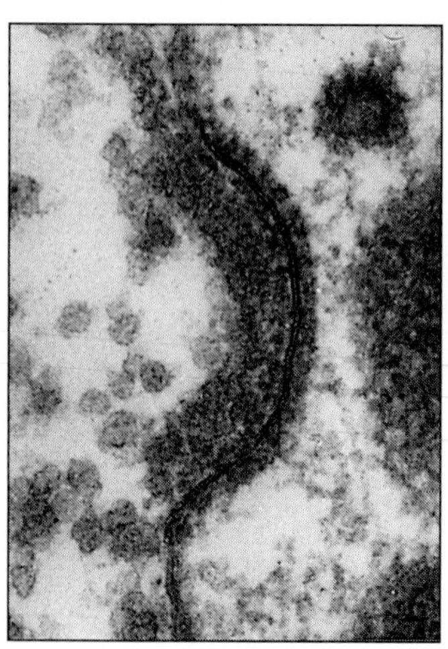

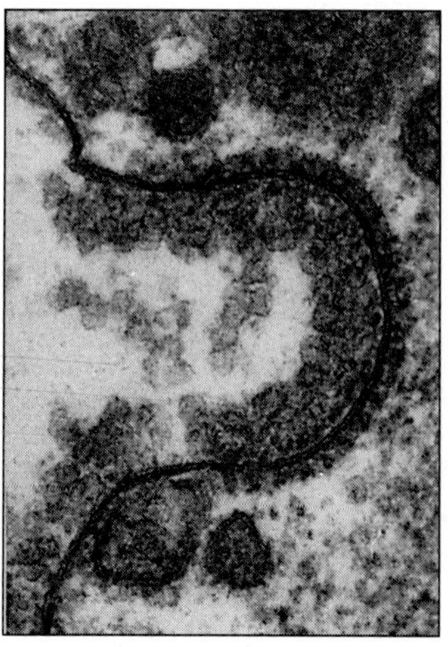

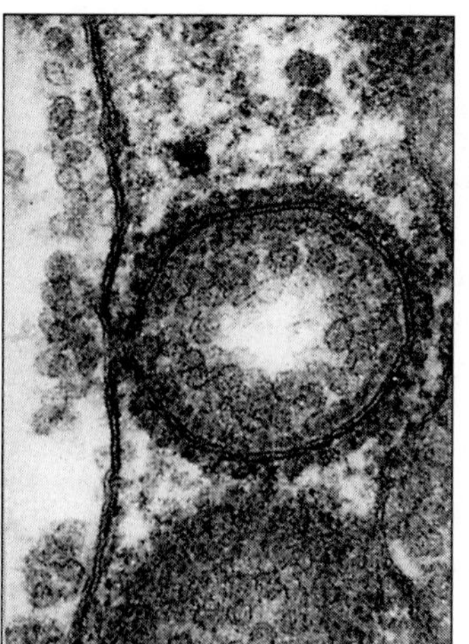

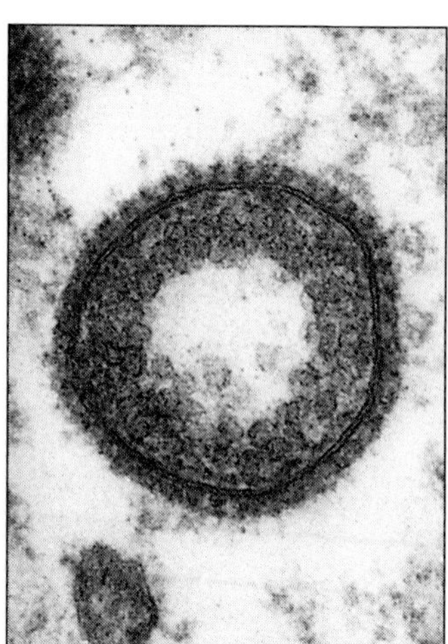

Figure 13–41 The formation of clathrin-coated vesicles from the plasma membrane. These electron micrographs illustrate the probable sequence of events in the formation of a clathrin-coated vesicle from a clathrin-coated pit. The clathrin-coated pits and vesicles shown are larger than those seen in normal-sized cells. They are involved in taking up lipoprotein particles into a very large hen oocyte to form yolk. The lipoprotein particles bound to their membrane-bound receptors can be seen as a dense, fuzzy layer on the extracellular surface of the plasma membrane—which is the inside surface of the vesicle. (Courtesy of M.M. Perry and A.B. Gilbert, *J. Cell Sci.* 39:257–272, 1979. © The Company of Biologists.)

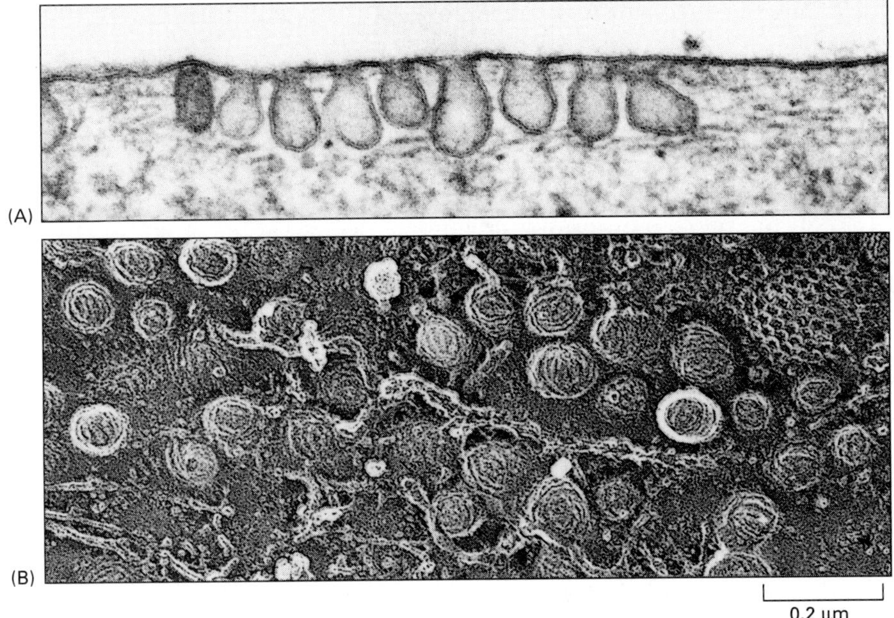

Figure 13–42 Caveolae in the plasma membrane of a fibroblast. (A) This electron micrograph shows a plasma membrane with a very high density of caveolae. Note that no cytosolic coat is visible. (B) This rapid-freeze deep-etch image demonstrates the characteristic "cauliflower" texture of the cytosolic face of the caveolae membrane. The regular texture is thought to result from aggregates of caveolin in the membrane. A clathrin-coated pit is also seen at the upper right. (Courtesy of R.G.W. Anderson, from K.G. Rothberg et al., *Cell* 68:673–682, 1992. © Elsevier)

cholesterol, glycosphingolipids, and GPI-anchored membrane proteins (see Figure 12–57). The major structural protein in caveolae is **caveolin**, a multipass integral membrane protein that is a member of a heterogeneous protein family.

In contrast to clathrin-coated and COPI- or COPII-coated vesicles, caveolae are thought to invaginate and collect cargo proteins by virtue of the lipid composition of the calveolar membrane, rather than by the assembly of a cytosolic protein coat. Caveolae pinch off from the plasma membrane and can deliver their contents either to endosome-like compartments or (in a process called transcytosis, which is discussed later) to the plasma membrane on the opposite side of a polarized cell. Some animal viruses also enter cells in vesicles derived from caveolae. The viruses are first delivered to an endosome-like compartment, from where they are moved to the ER. In the ER, they extrude their genome into the cytosol to start their infectious cycle. It remains a mystery how material endocytosed in caveolae-derived vesicles can end up in so many different locations in the cell.

Cells Import Selected Extracellular Macromolecules by Receptor-mediated Endocytosis

In most animal cells, clathrin-coated pits and vesicles provide an efficient pathway for taking up specific macromolecules from the extracellular fluid. In this process, called **receptor-mediated endocytosis**, the macromolecules bind to complementary transmembrane receptor proteins, accumulate in coated pits, and then enter the cell as receptor-macromolecule complexes in clathrin-coated vesicles (see Figure 13–41). Receptor-mediated endocytosis provides a selective concentrating mechanism that increases the efficiency of internalization of particular ligands more than a hundredfold, so that even minor components of the extracellular fluid can be specifically taken up in large amounts without taking in a correspondingly large volume of extracellular fluid. A particularly well-understood and physiologically important example is the process whereby mammalian cells take up cholesterol.

Many animals cells take up cholesterol through receptor-mediated endocytosis and, in this way, acquire most of the cholesterol they require to make new membrane. If the uptake is blocked, cholesterol accumulates in the blood and can contribute to the formation in blood vessel walls of *atherosclerotic plaques*, deposits of lipid and fibrous tissue that can cause strokes and heart attacks by blocking blood flow. In fact, it was through a study of humans with a strong genetic predisposition for *atherosclerosis* that the mechanism of receptor-mediated endocytosis was first clearly revealed.

Most cholesterol is transported in the blood as cholesteryl esters in the form of lipid–protein particles known as **low-density lipoproteins (LDL)** (Figure 13–43). When a cell needs cholesterol for membrane synthesis, it makes transmembrane receptor proteins for LDL and inserts them into its plasma membrane. Once in the plasma membrane, the LDL receptors diffuse until they associate with clathrin-coated pits that are in the process of forming (Figure 13–44A). Since coated pits constantly pinch off to form coated vesicles, any LDL particles bound to LDL receptors in the coated pits are rapidly internalized in coated vesicles. After shedding their clathrin coats, the vesicles deliver their contents to early endosomes, which are located near the cell periphery. Once the LDL and LDL receptors encounter the low pH in the endosomes, LDL is released from its receptor and is delivered via late endosomes to lysosomes. There the cholesteryl esters in the LDL particles are hydrolyzed to free cholesterol, which is now available to the cell for new membrane synthesis. If too much free cholesterol accumulates in a cell, the cell shuts off both its own cholesterol synthesis and the synthesis of LDL receptor proteins, so that it ceases either to make or to take up cholesterol.

This regulated pathway for the uptake of cholesterol is disrupted in individuals who inherit defective genes encoding LDL receptor proteins. The resulting high levels of blood cholesterol predispose these individuals to develop atherosclerosis prematurely, and many die at an early age of heart attacks resulting from coronary artery disease. In some cases, the receptor is lacking altogether. In others, the receptors are defective—in either the extracellular binding site for LDL or the intracellular binding site that attaches the receptor to the coat of a clathrin-coated pit (see Figure 13–44B). In the latter case, normal numbers of LDL-binding receptor proteins are present, but they fail to become localized in the clathrin-coated regions of the plasma membrane. Although LDL binds to the surface of these mutant cells, it is not internalized, directly demonstrating the importance of clathrin-coated pits in the receptor-mediated endocytosis of cholesterol.

More than 25 different receptors are known to participate in receptor-mediated endocytosis of different types of molecules, and they all apparently use the same clathrin-coated-pit pathway. Many of these receptors, like the LDL receptor, enter coated pits irrespective of whether they have bound their specific ligands. Others enter preferentially when bound to a specific ligand, suggesting that a ligand-induced conformational change is required for them to activate the signal sequence that guides them into the pits. Since most plasma membrane proteins fail to become concentrated in clathrin-coated pits, the pits must function as molecular filters, preferentially collecting certain plasma membrane proteins (receptors) over others.

Signal peptides guide transmembrane proteins into clathrin-coated pits by binding to the adaptins. Despite a common function, their amino acid sequences

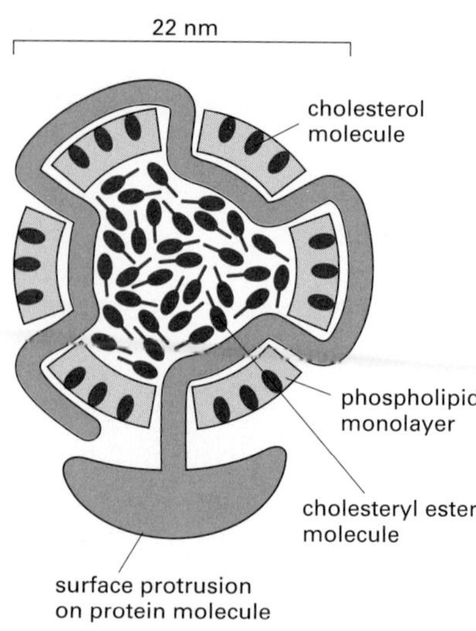

Figure 13–43 A low-density lipoprotein (LDL) particle. Each spherical particle has a mass of 3×10^6 daltons. It contains a core of about 1500 cholesterol molecules esterified to long-chain fatty acids that is surrounded by a lipid monolayer composed of about 800 phospholipid and 500 unesterified cholesterol molecules. A single molecule of a 500,000-dalton protein organizes the particle and mediates the specific binding of LDL to cell-surface receptor proteins.

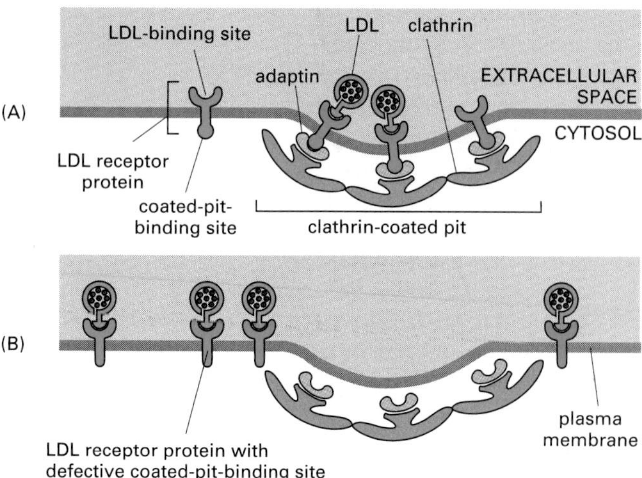

Figure 13–44 Normal and mutant LDL receptors. (A) LDL receptor proteins binding to a coated pit in the plasma membrane of a normal cell. The human LDL receptor is a single-pass transmembrane glycoprotein composed of about 840 amino acids, only 50 of which are on the cytoplasmic side of the membrane. (B) A mutant cell in which the LDL receptor proteins are abnormal and lack the site in the cytoplasmic domain that enables them to bind to adaptins in the clathrin-coated pits. Such cells bind LDL but cannot ingest it. In most human populations, 1 in 500 individuals inherits one defective LDL receptor gene and, as a result, has an increased risk of a heart attack caused by atherosclerosis.

vary. A common endocytosis signal consists of only four amino acids Y-X-X-Ψ, where Y is tyrosine, X any polar amino acid, and Ψ a hydrophobic amino acid. This short peptide, which is shared by many receptors, binds directly to one of the adaptins in clathrin-coated pits. By contrast, the cytosolic tail of the LDL receptor contains a unique signal (Asn-Pro-Val-Tyr) that apparently binds to the same adaptin protein.

Electron-microscope studies of cultured cells exposed simultaneously to different labeled ligands demonstrate that many kinds of receptors can cluster in the same coated pit. The plasma membrane of one clathrin-coated pit can probably accommodate up to 1000 receptors of assorted varieties. Although all of the receptor–ligand complexes that use this endocytic pathway are apparently delivered to the same endosomal compartment, the subsequent fates of the endocytosed molecules vary, as we discuss next.

Endocytosed Materials That Are Not Retrieved From Endosomes End Up in Lysosomes

The endosomal compartments of a cell can be complex. They can be made visible in the electron microscope by adding a readily detectable tracer molecule, such as the enzyme peroxidase, to the extracellular medium and leaving the cells for various lengths of time to take it up by endocytosis. The distribution of the molecule after its uptake reveals the endosomal compartments as a set of heterogeneous, membrane-enclosed tubes extending from the periphery of the cell to the perinuclear region, where it is often close to the Golgi apparatus. Two sequential sets of endosomes can be readily distinguished in such labeling experiments. The tracer molecule appears within a minute or so in early endosomes, just beneath the plasma membrane. After 5–15 minutes, it moves to late endosomes, close to the Golgi apparatus and near the nucleus. Early and late endosomes differ in their protein compositions; they are associated with different Rab proteins, for example.

As mentioned earlier, the interior of the endosomal compartment is kept acidic (pH ~6) by a vacuolar H^+ ATPase in the endosomal membrane that pumps H^+ into the lumen from the cytosol. In general, later endosomes are more acidic than early endosomes. This acidic environment has a crucial role in the function of these organelles.

We have already seen how endocytosed materials that reach the late endosomes become mixed with newly synthesized acid hydrolases and end up being degraded in lysosomes. Many molecules, however, are specifically diverted from this journey to destruction. They are recycled instead from the early endosomes back to the plasma membrane via transport vesicles. Only molecules that are not retrieved from endosomes in this way are delivered to lysosomes for degradation.

Specific Proteins Are Removed from Early Endosomes and Returned to the Plasma Membrane

The **early endosomes** form a compartment that acts as the main sorting station in the endocytic pathway, just as the *cis* and *trans* Golgi networks serve this function in the biosynthetic–secretory pathway. In the acidic environment of the early endosome, many internalized receptor proteins change their conformation and release their ligand, just as the M6P receptors unload their cargo of acid hydrolases in the even more acidic late endosomes. Those endocytosed ligands that dissociate from their receptors in the early endosome are usually doomed to destruction in lysosomes, along with the other soluble contents of the endosome. Some other endocytosed ligands, however, remain bound to their receptors, and thereby share the fate of the receptors.

The fates of the receptor proteins—and of any ligands remaining bound to them—vary according to the specific type of receptor. (1) Most receptors are recycled and return to the same plasma membrane domain from which they came; (2) some proceed to a different domain of the plasma membrane, thereby

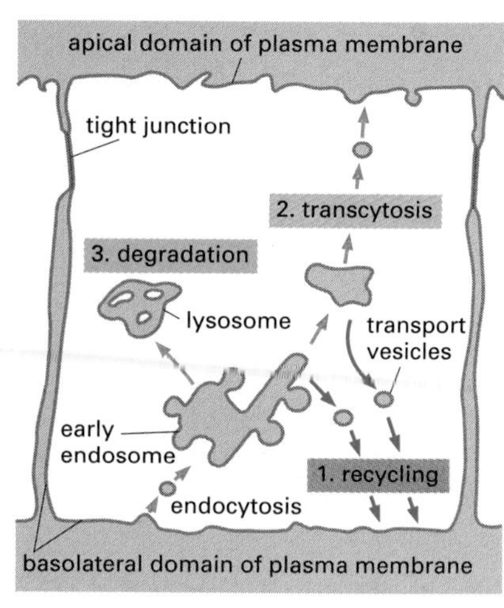

Figure 13–45 Possible fates for transmembrane receptor proteins that have been endocytosed. Three pathways from the endosomal compartment in an epithelial cell are shown. Retrieved receptors are returned (1) to the same plasma membrane domain from which they came *(recycling)* or (2) to a different domain of the plasma membrane *(transcytosis)*. (3) Receptors that are not specifically retrieved from endosomes follow the pathway from the endosomal compartment to lysosomes, where they are degraded *(degradation)*. The formation of oligomeric aggregates in the endosomal membrane may be one of the signals that guides receptors into the degradative pathway. If the ligand that is endocytosed with its receptor stays bound to the receptor in the acidic environment of the endosome, it follows the same pathway as the receptor; otherwise it is delivered to lysosomes.

mediating a process called *transcytosis*; and (3) some progress to lysosomes, where they are degraded (Figure 13–45).

The LDL receptor follows the first pathway. It dissociates from its ligand LDL in the early endosome and is recycled to the plasma membrane for reuse, leaving the discharged LDL to be carried to lysosomes (Figure 13–46). The recycling vesicles bud from long, narrow tubules that extend from the early endosomes. It is likely that the geometry of these tubules helps the sorting process. Because tubules have a large membrane area enclosing a small volume, membrane proteins tend to accumulate there. Transport vesicles that return material to the plasma membrane begin budding from the tubules, but tubular portions of the early endosome also pinch off and fuse with one another to form *recycling endosomes,* a way-station for the traffic between the early endosomes and the plasma membrane. During this process, the tubules and then the recycling endosome continuously shed vesicles that return to the plasma membrane.

The **transferrin receptor** follows a similar recycling pathway, but it also recycles its ligand. Transferrin is a soluble protein that carries iron in the blood. Cell-surface transferrin receptors deliver transferrin with its bound iron to early endosomes by receptor-mediated endocytosis. The low pH in the endosome induces transferrin to release its bound iron, but the iron-free transferrin itself (called apotransferrin) remains bound to its receptor. The receptor–apotransferrin complex enters the tubular extensions of the early endosome and from there is recycled back to the plasma membrane (Figure 13–47). When the apotransferrin returns to the neutral pH of the extracellular fluid, it dissociates from the

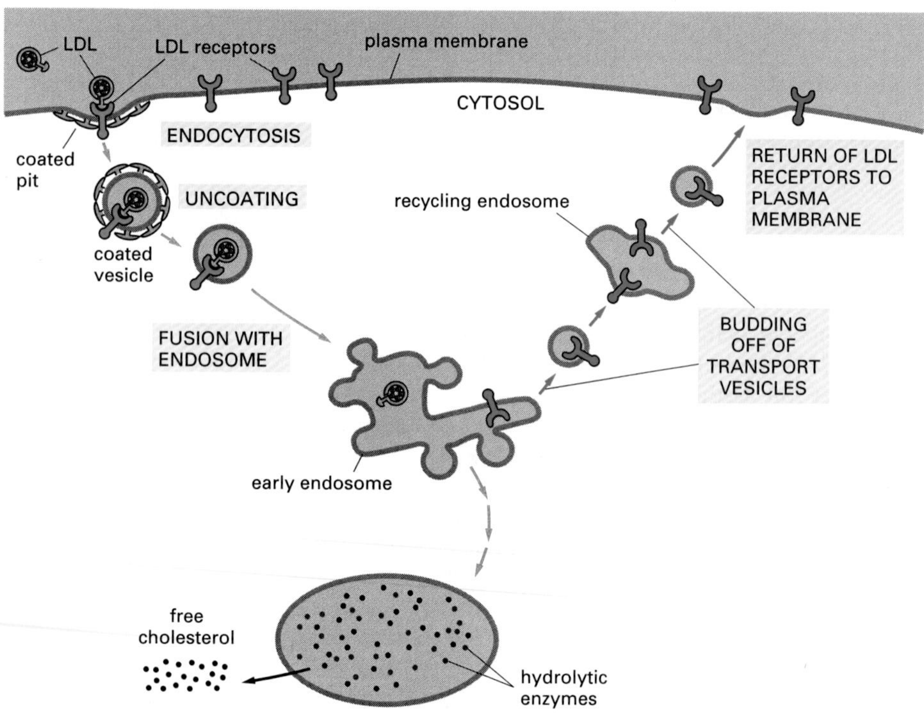

Figure 13–46 The receptor-mediated endocytosis of LDL. Note that the LDL dissociates from its receptors in the acidic environment of the endosome. After a number of steps (see Figure 13–48), the LDL ends up in lysosomes, where it is degraded to release free cholesterol. In contrast, the LDL receptor proteins are returned to the plasma membrane via clathrin-coated transport vesicles that bud off from the tubular region of the early endosome, as shown. For simplicity, only one LDL receptor is shown entering the cell and returning to the plasma membrane. Whether it is occupied or not, an LDL receptor typically makes one round trip into the cell and back to the plasma membrane every 10 minutes, making a total of several hundred trips in its 20-hour life-span.

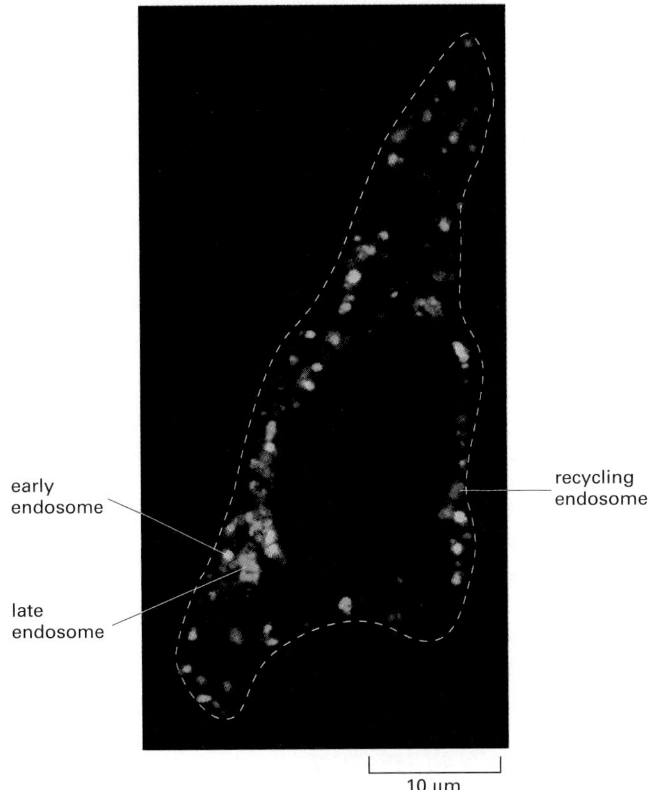

Figure 13–47 Sorting of membrane proteins in the endocytic pathway. Transferrin receptors mediate nutrient uptake and constitutively cycle between endosomes and the plasma membrane. By contrast, opioid receptors are signaling receptors that—after ligand binding—are down-regulated by endocytosis followed by degradation in lysosomes. Endocytosis of both receptors initiates in the same clathrin-coated pits. They are then delivered to the same early endosomes where their journeys part: transferrin receptors are sorted to the recycling endosomes, whereas opioid receptors are sorted to late endosomes. The micrograph shows both receptors—labeled with different fluorescent dyes—30 min after endocytosis (transferrin receptors labeled in *red* and opioid receptors labeled in *green*). At this time, some early endosomes still contain both receptors and are seen as *yellow* structures (yellow resulting from an overlap of red and green light emitted from the fluorescent dyes). By contrast, recycling and late endosomes are selectively enriched in either transferrin or opioid receptors, respectively—thus appearing as distinct red and green structures. (Courtesy of Mark von Zastrow.)

receptor and is thereby freed to pick up more iron and begin the cycle again. Thus, transferrin shuttles back and forth between the extracellular fluid and the endosomal compartment, avoiding lysosomes and delivering the iron that cells need to grow to the cell interior.

The second pathway that endocytosed receptors can follow from endosomes is taken both by opioid receptors (see Figure 13–47) and by the receptor that binds *epidermal growth factor (EGF)*. EGF is a small, extracellular signal protein that stimulates epidermal and various other cells to divide. Unlike LDL receptors, EGF receptors accumulate in clathrin-coated pits only after binding EGF, and most of them do not recycle but are degraded in lysosomes, along with the ingested EGF. EGF binding therefore first activates intracellular signaling pathways and then leads to a decrease in the concentration of EGF receptors on the cell surface, a process called *receptor down-regulation* that reduces the cell's subsequent sensitivity to EGF (discussed in Chapter 15).

Multivesicular Bodies Form on the Pathway to Late Endosomes

It is still uncertain how endocytosed molecules move from the early to the late endosomal compartment so as to end up in lysosomes. A current view is that portions of the early endosomes migrate slowly along microtubules toward the cell interior, shedding tubules of material to be recycled to the plasma membrane. The migrating endosomes enclose large amounts of invaginated membrane and internally pinched-off vesicles and are therefore called **multivesicular bodies** (Figure 13–48). It is unknown whether multivesicular bodies eventually fuse with a late endosomal compartment or if they fuse instead with each other to become late endosomes. At the end of this pathway, the late endosomes are converted to lysosomes as a result of their fusion with hydrolase-bearing transport vesicles from the *trans* Golgi network and their increased acidification (Figure 13–49).

The multivesicular bodies carry specific endocytosed membrane proteins that are to be degraded but exclude others that are to be recycled. As part of the protein-sorting process, specific proteins—for example, the occupied EGF

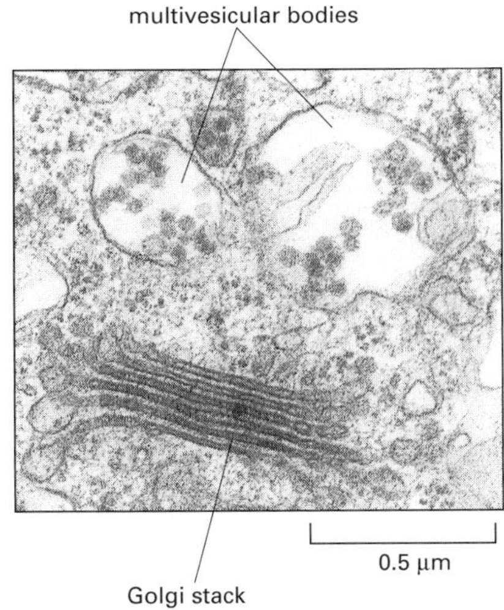

Figure 13–48 Electron micrograph of a multivesicular body in a plant cell. The large amount of internal membrane will be delivered to the vacuole, the plant equivalent of the lysosome, for digestion.

Figure 13–49 The endocytic pathway from the plasma membrane to lysosomes. Maturation from early to late endosomes occurs through the formation of multivesicular bodies, which contain large amounts of invaginated membrane and internal vesicles (hence their name). These bodies move inward along microtubules, and recycling of components to the plasma membrane continues as the bodies move. The multivesicular bodies gradually turn into late endosomes, either by fusing with each other or by fusing with preexisting late endosomes. The late endosomes no longer send vesicles to the plasma membrane but communicate with the *trans* Golgi network via transport vesicles, which deliver the proteins that will convert the late endosome into a lysosome.

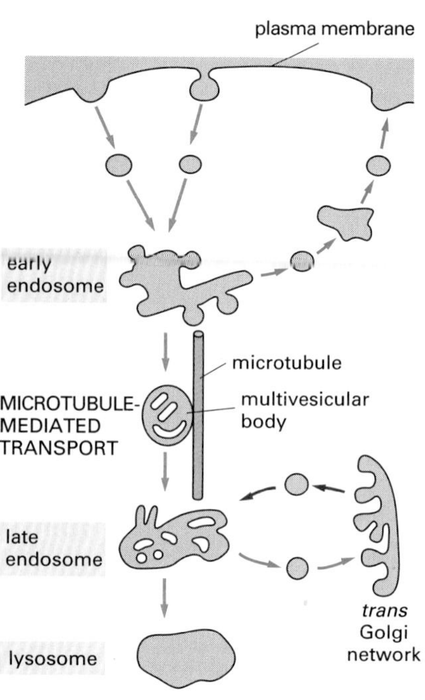

receptor described previously—selectively partition to the invaginating membrane of the multivesicular bodies (Figure 13–50). In this way, the receptors, as well as any signaling proteins strongly bound to them, are rendered fully accessible to the digestive enzymes that will degrade them (see Figure 13–50).

Membrane proteins that are sorted into the internal membrane vesicles of a multivesicular body are first covalently modified with the small protein ubiquitin. Unlike multi-ubiquitylation which typically targets substrate proteins for degradation in proteasomes (discussed in Chapter 6), ubiquitin tagging for sorting into the internal membrane vesicles of a multivesicular body requires the addition of only a single ubiquitin molecule that is added to activated receptors while still at the plasma membrane. The ubiquitin tag facilitates the uptake of the receptors into endocytic vesicles and is then recognized again by proteins that mediate the sorting process into the internal membrane vesicles of multivesicular bodies. In addition, membrane invagination in multivesicular bodies is regulated by a lipid kinase that phosphorylates phosphatidylinositol. The phosphorylated head groups of these lipids are thought to serve as docking sites for the proteins that mediate the invagination process. Local modification of lipid molecules is thus another way in which specific membrane patches can be induced to change shape and destiny.

In addition to endocytosed membrane proteins, multivesicular bodies also contain most of the soluble content of early endosomes destined for digestion in lysosomes.

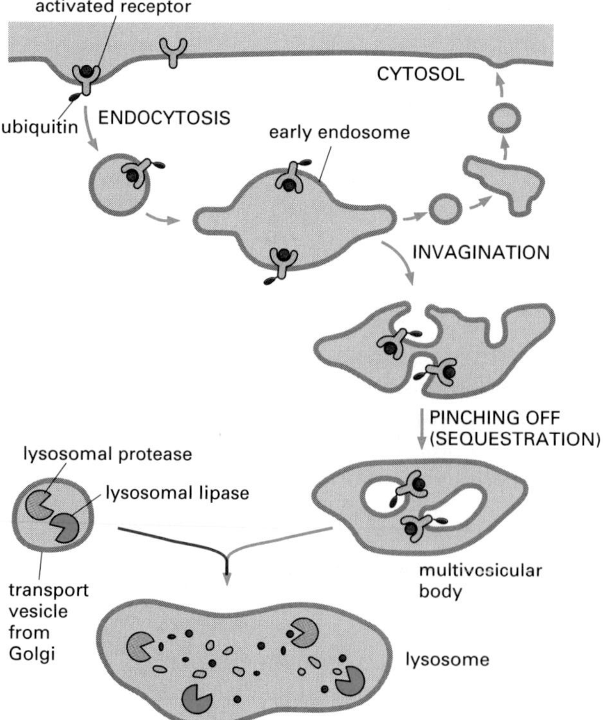

Figure 13–50 The sequestration of endocytosed proteins into internal membranes of multivesicular bodies. Eventually, all of the internal membranes produced by the invaginations shown are digested by proteases and lipases in lysosomes. The invagination is essential to achieve complete digestion of endocytosed membrane proteins. Because the outer membrane of the multivesicular body becomes continuous with the lysosomal membrane, lysosomal hydrolases could not digest the cytosolic domains of transmembrane proteins such as the EGF receptor shown here, if it were not for the invagination.

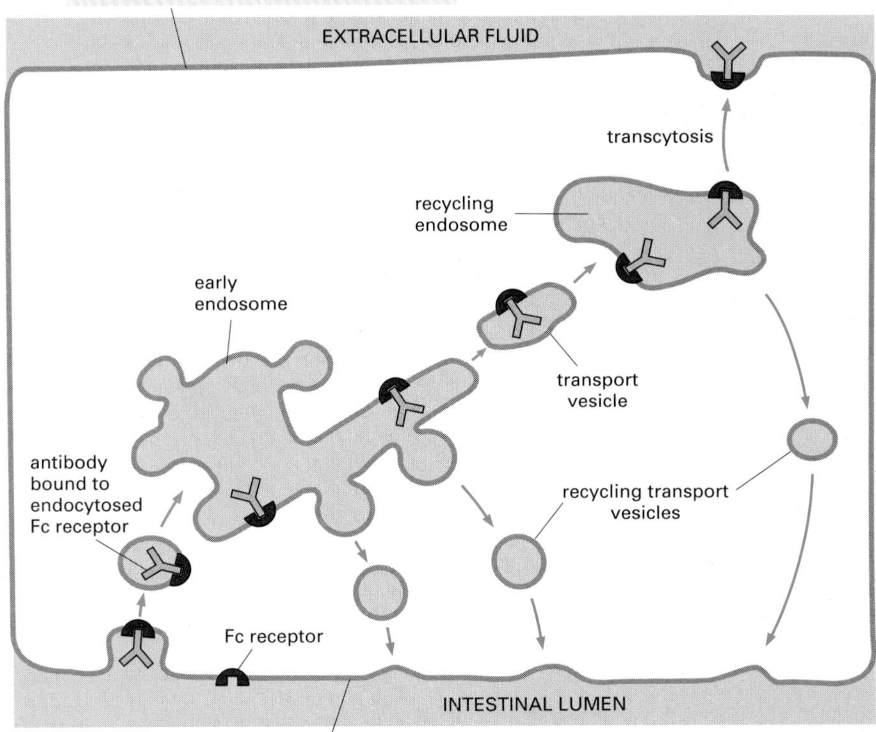

Figure 13–51 Transcytosis. Recycling endosomes form a way-station on the transcytotic pathway. In the example shown here, an antibody receptor on a gut epithelial cell binds antibody and is endocytosed, eventually carrying the antibody to the basolateral plasma membrane in contact with extracellular matrix that is permeated by blood vessels. The receptor is called an Fc receptor because it binds the Fc part of the antibody (discussed in Chapter 24).

Macromolecules Can Be Transferred Across Epithelial Cell Sheets by Transcytosis

Some receptors on the surface of polarized epithelial cells transfer specific macromolecules from one extracellular space to another by **transcytosis** (Figure 13–51). These receptors are endocytosed and then follow a pathway from endosomes to a different plasma membrane domain (see Figure 13–46). A newborn rat, for example, obtains antibodies from its mother's milk (which help protect it against infection) by transporting them across the epithelium of its gut. The lumen of the gut is acidic, and, at this low pH, the antibodies in the milk bind to specific receptors on the apical (absorptive) surface of the gut epithelial cells. The receptor–antibody complexes are internalized via clathrin-coated pits and vesicles and are delivered to early endosomes. The complexes remain intact and are retrieved in transport vesicles that bud from the early endosome and subsequently fuse with the basolateral domain of the plasma membrane. On exposure to the neutral pH of the extracellular fluid that bathes the basolateral surface of the cells, the antibodies dissociate from their receptors and eventually enter the newborn's bloodstream.

The transcytotic pathway from the early endosome to the plasma membrane is not direct. The receptors first move from the early endosome to an intermediate endosomal compartment, the **recycling endosome** described previously (see Figure 13–51). The variety of pathways that different receptors follow from endosomes implies that, in addition to binding sites for their ligands and binding sites for coated pits, many receptors also possess sorting signals that guide them into the appropriate type of transport vesicle leaving the endosome and thereby to the appropriate target membrane in the cell.

A unique property of a recycling endosomes is that the exit of membrane proteins from the compartment can be regulated. Thus, cells can adjust the flux of proteins through the transcytotic pathway according to need. Although the mechanism of regulation is uncertain, it allows recycling endosomes an important role in adjusting the concentration of specific plasma membrane proteins. Fat cells and muscle cells, for example, contain large intracellular pools of the glucose transporters that are responsible for the uptake of glucose across the

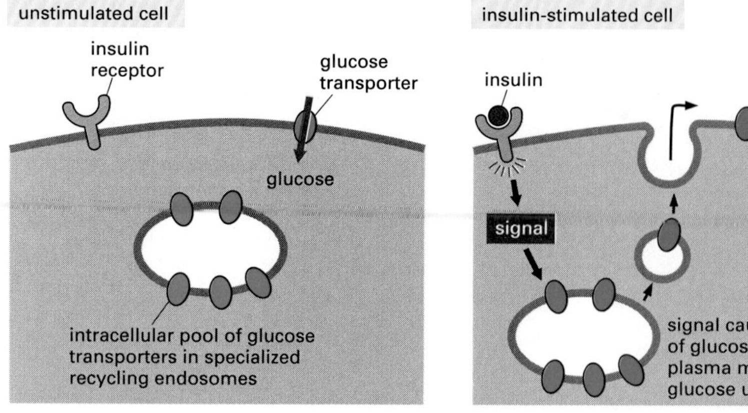

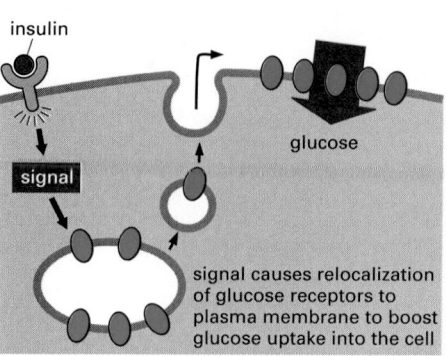

Figure 13–52 Storage of plasma membrane proteins in recycling endosomes. Recycling endosomes can serve as an intracellular pool for specialized plasma membrane proteins, enabling them to be mobilized when needed. In the example shown here, insulin binding to the insulin receptor triggers a signaling pathway that causes the rapid insertion of glucose transporters into the plasma membrane of a fat or muscle cell, greatly increasing glucose intake.

plasma membrane. These proteins are stored in specialized recycling endosomes until the cell is stimulated by the hormone insulin to increase its rate of glucose uptake. Then transport vesicles bud from the recycling endosome and deliver large numbers of glucose transporters to the plasma membrane, thereby greatly increasing the rate of glucose uptake into the cell (Figure 13–52).

Epithelial Cells Have Two Distinct Early Endosomal Compartments But a Common Late Endosomal Compartment

In polarized epithelial cells, endocytosis occurs from both the *basolateral domain* and the *apical domain* of the plasma membrane. Material endocytosed from either domain first enters an early endosomal compartment that is unique to that domain. This arrangement allows endocytosed receptors to be recycled back to their original membrane domain, unless they contain signals that mark them for transcytosis to the other domain. Molecules endocytosed from either plasma membrane domain that are not retrieved from the early endosomes end up in a common late endosomal compartment near the cell center and are eventually degraded in lysosomes (Figure 13–53).

Whether cells contain a few connected or many unconnected endosomal compartments seems to depend on the cell type and the physiological state of the cell. Like many other membrane-enclosed organelles, endosomes of the same type can readily fuse with one another (an example of homotypic fusion, discussed earlier) to create large continuous endosomes.

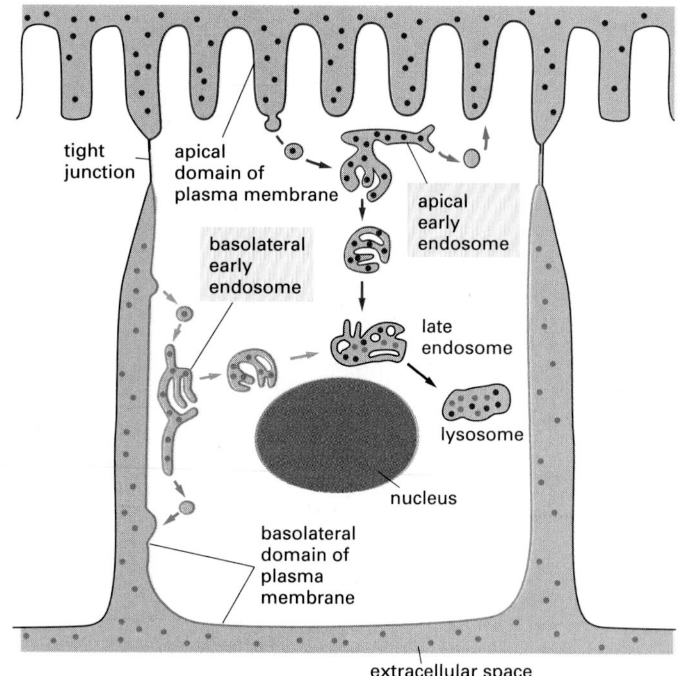

Figure 13–53 The two distinct early endosomal compartments in an epithelial cell. The basolateral and the apical domains of the plasma membrane communicate with separate early endosomal compartments. But endocytosed molecules from both domains that do not contain signals for recycling or transcytosis meet in a common late endosomal compartment before being digested in lysosomes.

Summary

Cells ingest fluid, molecules, and particles by endocytosis, in which localized regions of the plasma membrane invaginate and pinch off to form endocytic vesicles. Many of the endocytosed molecules and particles end up in lysosomes, where they are degraded. Endocytosis occurs both constitutively and as a triggered response to extracellular signals. Endocytosis is so extensive in many cells that a large fraction of the plasma membrane is internalized every hour. To make this possible, most of the plasma membrane components (proteins and lipid) that are endocytosed are continually returned to the cell surface by exocytosis. This large-scale endocytic–exocytic cycle is mediated largely by clathrin-coated pits and vesicles.

Many cell-surface receptors that bind specific extracellular macromolecules become localized in clathrin-coated pits. As a result, they and their ligands are efficiently internalized in clathrin-coated vesicles, a process called receptor-mediated endocytosis. The coated endocytic vesicles rapidly shed their clathrin coats and fuse with early endosomes.

Most of the ligands dissociate from their receptors in the acidic environment of the endosome and eventually end up in lysosomes, while most of the receptors are recycled via transport vesicles back to the cell surface for reuse. But receptor–ligand complexes can follow other pathways from the endosomal compartment. In some cases, both the receptor and the ligand end up being degraded in lysosomes, resulting in receptor down-regulation. In other cases, both are transferred to a different plasma membrane domain, and the ligand is thereby released by exocytosis at a surface of the cell different from that where it originated, a process called transcytosis. The transcytosis pathway includes recycling endosomes, where endocytosed plasma membrane proteins can be stored until they are needed.

TRANSPORT FROM THE *TRANS* GOLGI NETWORK TO THE CELL EXTERIOR: EXOCYTOSIS

Having considered the cell's internal digestive system and the various types of incoming membrane traffic that converge on lysosomes, we now return to the Golgi apparatus and examine the secretory pathways that lead out to the cell exterior. Transport vesicles destined for the plasma membrane normally leave the *trans* Golgi network in a steady stream. The membrane proteins and the lipids in these vesicles provide new components for the cell's plasma membrane, while the soluble proteins inside the vesicles are secreted to the extracellular space. The fusion of the vesicles with the plasma membrane is called **exocytosis**. In this way, for example, cells produce and secrete most of the proteoglycans and glycoproteins of the *extracellular matrix,* which is discussed in Chapter 19.

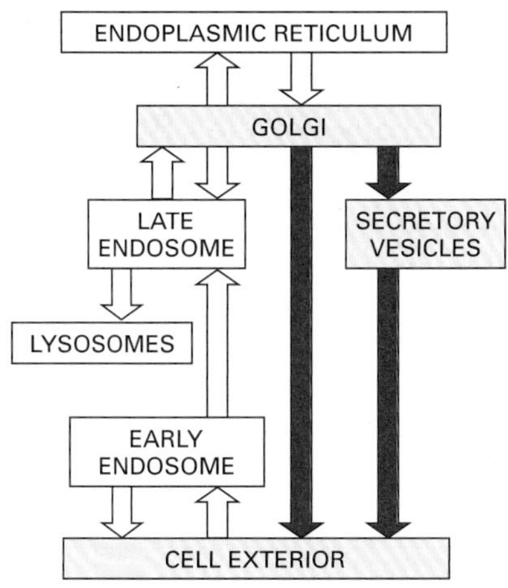

All cells require this **constitutive secretory pathway**. Specialized secretory cells, however, have a second secretory pathway in which soluble proteins and other substances are initially stored in *secretory vesicles* for later release. This is the **regulated secretory pathway**, found mainly in cells specialized for secreting products rapidly on demand—such as hormones, neurotransmitters, or digestive enzymes (Figure 13–54). In this section we consider the role of the Golgi apparatus in both of these secretory pathways and compare the two mechanisms of secretion.

Many Proteins and Lipids Seem to Be Carried Automatically from the Golgi Apparatus to the Cell Surface

In a cell capable of regulated secretion, at least three classes of proteins must be separated before they leave the *trans* Golgi network—those destined for lysosomes (via late endosomes), those destined for secretory vesicles, and those destined for immediate delivery to the cell surface. We have already noted that proteins destined for lysosomes are tagged for packaging into specific departing vesicles (by mannose-6-phosphate for lysosomal hydrolases), and analogous signals are thought to direct *secretory proteins* into secretory vesicles. Most other

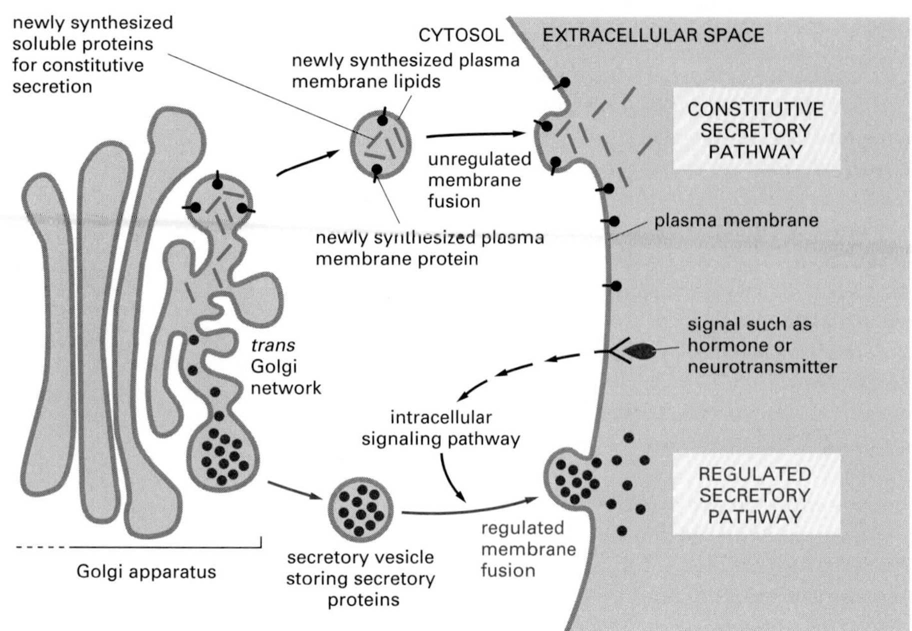

Figure 13–54 The constitutive and regulated secretory pathways. The two pathways diverge in the *trans* Golgi network. The constitutive secretory pathway operates in all cells. Many soluble proteins are continually secreted from the cell by this pathway, which also supplies the plasma membrane with newly synthesized lipids and proteins. Specialized secretory cells also have a regulated secretory pathway, by which selected proteins in the *trans* Golgi network are diverted into secretory vesicles, where the proteins are concentrated and stored until an extracellular signal stimulates their secretion. The regulated secretion of small molecules, such as histamine, occurs by a similar pathway; these molecules are actively transported from the cytosol into preformed secretory vesicles. There they are often complexed to specific macromolecules (proteoglycans, for histamine), so that they can be stored at high concentration without generating an excessively high osmotic pressure.

proteins are transported directly to the cell surface by the nonselective constitutive secretory pathway. Because entry into this pathway does not require a particular signal, it is also called the **default pathway** (Figure 13–55). Thus, in an unpolarized cell such as a white blood cell or a fibroblast, it seems that any protein in the lumen of the Golgi apparatus is automatically carried by the constitutive pathway to the cell surface unless it is either specifically returned to the ER, retained as a resident protein in the Golgi apparatus itself, or selected for the pathways that lead to regulated secretion or to lysosomes. In polarized cells, where different products have to be delivered to different domains of the cell surface, we shall see that the options are more complex.

Secretory Vesicles Bud from the *Trans* Golgi Network

Cells that are specialized for secreting some of their products rapidly on demand concentrate and store these products in **secretory vesicles** (often called *secretory granules* or *dense-core vesicles* because they have dense cores when viewed in the electron microscope). Secretory vesicles form from the *trans* Golgi network, and they release their contents to the cell exterior by exocytosis in response to extracellular signals. The secreted product can be either a small molecule (such as histamine) or a protein (such as a hormone or digestive enzyme).

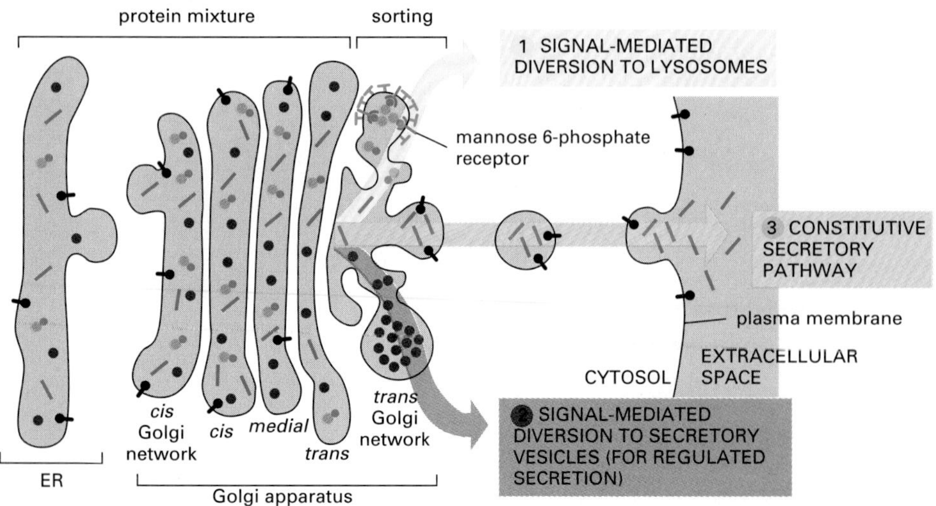

Figure 13–55 The three best-understood pathways of protein sorting in the *trans* Golgi network. (1) Proteins with the mannose 6-phosphate (M6P) marker are diverted to lysosomes (via late endosomes) in clathrin-coated transport vesicles (see Figure 13–37). (2) Proteins with signals directing them to secretory vesicles are concentrated in such vesicles as part of a regulated secretory pathway that is present only in specialized secretory cells. (3) In unpolarized cells, proteins with no special features are delivered to the cell surface by a constitutive secretory pathway. In polarized cells, however, secreted and plasma membrane proteins are selectively directed to either the apical or the basolateral plasma membrane domain, so that at least one of these two pathways must be mediated by a specific signal, as we discuss later.

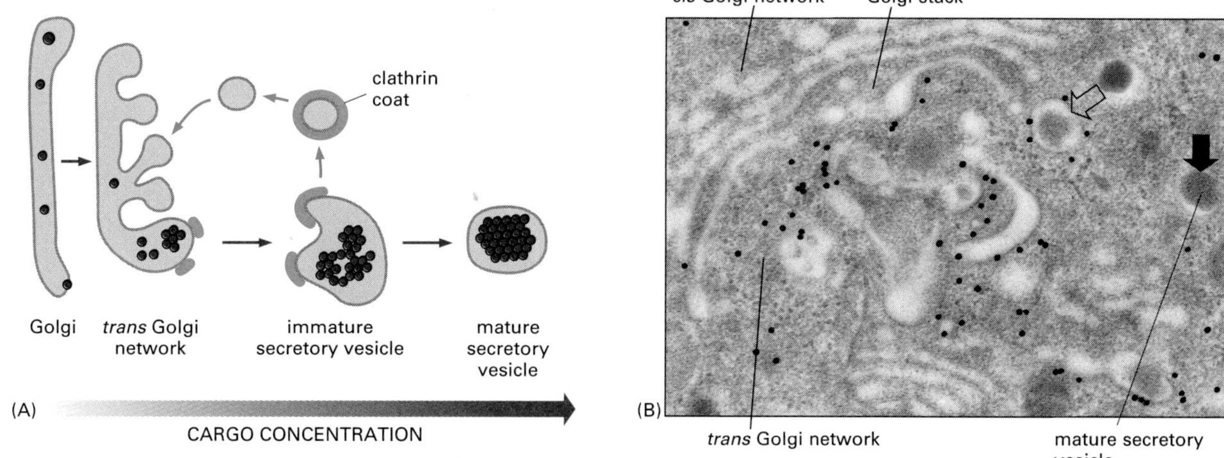

Figure 13–56 The formation of secretory vesicles. (A) Secretory proteins become segregated and highly concentrated in secretory vesicles by two mechanisms. First, they aggregate in the ionic environment of the *trans* Golgi network; often the aggregates become more condensed as secretory vesicles mature and their lumen becomes more acidic. Second, excess membrane and lumenal content present in immature secretory vesicles are retrieved in clathrin-coated vesicles as the secretory vesicles mature. (B) This electron micrograph shows secretory vesicles forming from the *trans* Golgi network in an insulin-secreting β-cell of the pancreas. An antibody conjugated to gold spheres *(black dots)* has been used to locate clathrin molecules. The immature secretory vesicles *(open arrow)*, which contain insulin precursor protein (proinsulin), contain clathrin patches. Clathrin coats are no longer seen on the mature secretory vesicle, which has a highly condensed core *(solid arrow)*. (Courtesy of Lelio Orci.)

Proteins destined for secretory vesicles (called *secretory proteins)* are packaged into appropriate vesicles in the *trans* Golgi network by a mechanism that is believed to involve the selective aggregation of the secretory proteins. Clumps of aggregated, electron-dense material can be detected by electron microscopy in the lumen of the *trans* Golgi network. The signal that directs secretory proteins into such aggregates is not known, but it is thought to be composed of signal patches that are common to proteins of this class. When a gene encoding a secretory protein is transferred to a secretory cell that normally does not make the protein, the foreign protein is appropriately packaged into secretory vesicles. This observation shows that although the proteins that an individual cell expresses and packages in secretory vesicles differ, they all contain common sorting signals, which function properly even when the proteins are expressed in cells that do not normally make them.

It is unclear how the aggregates of secretory proteins are segregated into secretory vesicles. Secretory vesicles have unique proteins in their membrane, some of which might serve as receptors for aggregated protein in the *trans* Golgi network. The aggregates are much too big, however, for each molecule of the secreted protein to be bound by its own cargo receptor, as proposed for transport of the lysosomal enzymes. The uptake of the aggregates into secretory vesicles may therefore more closely resemble the uptake of particles by phagocytosis at the cell surface, where the plasma membrane zippers up around large structures.

Initially, most of the membrane of the secretory vesicles that leave the *trans* Golgi network is only loosely wrapped around the clusters of aggregated secretory proteins. Morphologically, these **immature secretory vesicles** resemble dilated *trans* Golgi cisternae that have pinched off from the Golgi stack. As the vesicles mature, their contents become concentrated (Figure 13–56), probably as the result of both the continuous retrieval of membrane that is recycled back to late endosomes and the progressive acidification of the vesicle lumen that results from the progressive concentration of ATP-driven H^+ pumps in the vesicle membrane. The degree of concentration of proteins during the formation and maturation of secretory vesicles is small, however, compared with the total 200–400-fold concentration that occurs after they leave the ER. Secretory and membrane proteins become concentrated as they move from the ER through

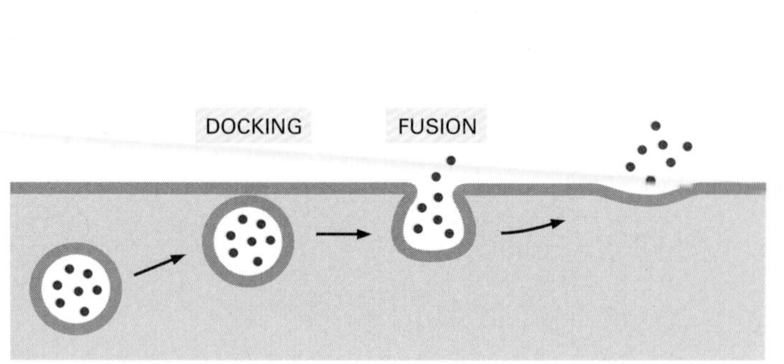

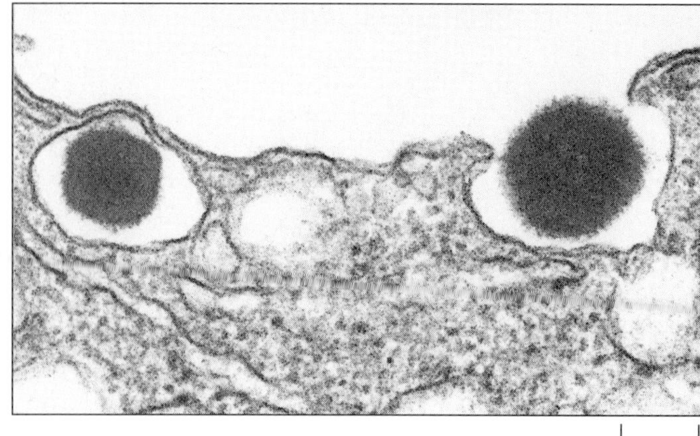

Figure 13–57 Exocytosis of secretory vesicles. The electron micrograph shows the release of insulin from a secretory vesicle of a pancreatic β-cell. (Courtesy of Lelio Orci, from L. Orci, J.-D. Vassali, and A. Perrelet, *Sci. Am.* 256:85–94, 1988.)

the Golgi apparatus because of an extensive retrograde retrieval process mediated by COPI-coated transport vesicles that exclude them (see Figure 13–21).

Membrane recycling is important for returning Golgi components to the Golgi apparatus, as well as for concentrating the contents of secretory vesicles. The vesicles that mediate this retrieval originate as clathrin-coated buds on the surface of immature secretory vesicles, often being seen even on budding secretory vesicles that have not yet severed from the Golgi stack (see Figure 13–56B).

Because the final mature secretory vesicles are so densely filled with contents, the secretory cell can disgorge large amounts of material promptly by exocytosis when triggered to do so (Figure 13–57).

Proteins Are Often Proteolytically Processed During the Formation of Secretory Vesicles

Condensation is not the only process to which secretory proteins are subject as the secretory vesicles mature. Many polypeptide hormones and neuropeptides, as well as many secreted hydrolytic enzymes, are synthesized as inactive protein precursors from which the active molecules have to be liberated by proteolysis. These cleavages begin in the *trans* Golgi network, and they continue in the secretory vesicles and sometimes in the extracellular fluid after secretion has occurred. Many secreted polypeptides have, for example, an N-terminal *propeptide* that is cleaved off to yield the mature protein. These proteins are thus synthesized as *pre-pro-proteins,* the *pre-peptide* consisting of the ER signal peptide that is cleaved off earlier in the rough ER (see Figure 12–40). In other cases, peptide-signaling molecules are made as *polyproteins* that contain multiple copies of the same amino acid sequence. In still more complex cases, a variety of peptide-signaling molecules are synthesized as parts of a single polyprotein that acts as a precursor for multiple end-products, which are individually cleaved from the initial polypeptide chain. The same polyprotein may be processed in various ways to produce different peptides in different cell types (Figure 13–58).

Why is proteolytic processing so common in the secretory pathway? Some of the peptides produced in this way, such as the *enkephalins* (five-amino-acid neuropeptides with morphine-like activity), are undoubtedly too short in their mature forms to be co-translationally transported into the ER lumen or to

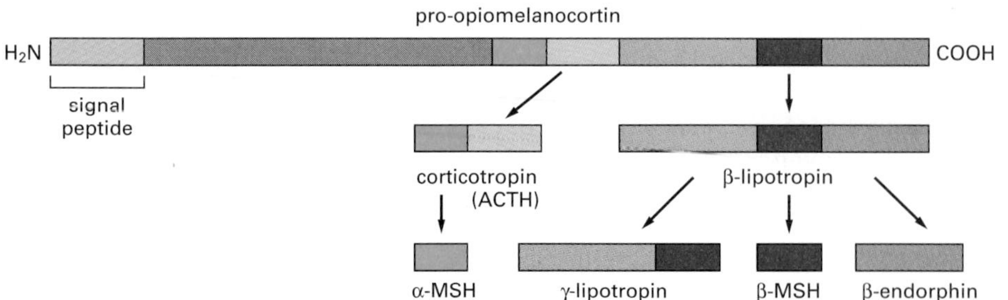

Figure 13–58 Alternative processing pathways for the prohormone pro-opiomelanocortin. The initial cleavages are made by proteases that cut next to pairs of positively charged amino acids (Lys-Arg, Lys-Lys, Arg-Lys, or Arg-Arg pairs). Trimming reactions then produce the final secreted products. Different cell types produce different concentrations of individual processing enzymes, so that the same prohormone precursor is cleaved to produce different peptide hormones. In the anterior lobe of the pituitary gland, for example, only corticotropin (ACTH) and β-lipotropin are produced from proopiomelanocortin, whereas in the intermediate lobe of the pituitary mainly α-melanocyte stimulating hormone (α-MSH), γ-lipotropin, β-MSH, and β-endorphin are produced.

include the necessary signal for packaging into secretory vesicles. In addition, for secreted hydrolytic enzymes—or any other protein whose activity could be harmful inside the cell that makes it—delaying activation of the protein until it reaches a secretory vesicle or until after it has been secreted has a clear advantage: it prevents it from acting prematurely inside the cell in which it is synthesized.

Secretory Vesicles Wait Near the Plasma Membrane Until Signaled to Release Their Contents

Once loaded, a secretory vesicle has to get to the site of secretion, which in some cells is far away from the Golgi apparatus. Nerve cells are the most extreme example. Secretory proteins, such as peptide neurotransmitters (neuropeptides) that are to be released from nerve terminals at the end of the axon, are made and packaged into vesicles in the cell body, where the ribosomes, ER, and Golgi apparatus are located. They must then travel along the axon to the nerve terminals, which can be a meter or more away. As discussed in Chapter 16, motor proteins propel the vesicles along axonal microtubules, whose uniform orientation guides the vesicles in the proper direction. Microtubules also guide vesicles to the cell surface for constitutive exocytosis.

Whereas vesicles containing materials for constitutive release fuse with the plasma membrane once they arrive there, secretory vesicles in the regulated pathway wait at the membrane until the cell receives a signal to secrete and then fuse. The signal is often a chemical messenger, such as a hormone, that binds to receptors on the cell surface. The resulting activation of the receptors generates intracellular signals, often including a transient increase in the concentration of free Ca^{2+} in the cytosol. In nerve terminals, the initial signal for exocytosis is usually an electrical excitation (an action potential) triggered by a chemical transmitter binding to receptors elsewhere on the same cell surface. When the action potential reaches the nerve terminals, it causes an influx of Ca^{2+} through voltage-gated Ca^{2+} channels. The binding of Ca^{2+} ions to specific sensors then triggers the secretory vesicles (called synaptic vesicles) to fuse with the plasma membrane and release their contents to the extracellular space.

The speed of transmitter release indicates that the proteins mediating the fusion reaction do not undergo complex, multistep rearrangements. After vesicles have been docked to the presynaptic plasma membrane, they undergo a priming step, which prepares them for rapid fusion. The SNAREs may be partly paired, but their helices are not fully wound into the final four-helix bundle required for fusion (see Figure 13–12). Other proteins are thought to keep the SNAREs from completing the fusion reaction until the Ca^{2+} influx releases this brake. At a typical synapse, only few of the docked vesicles seem to be primed and ready for exocytosis. The use of only a few vesicles at a time allows each synapse to fire over and over again in quick succession. With each firing, new synaptic vesicles become primed to replace those that have fused and released their contents.

Regulated Exocytosis Can Be a Localized Response of the Plasma Membrane and Its Underlying Cytoplasm

Histamine is a small molecule secreted by **mast cells**. It is released by the regulated pathway in response to specific ligands that bind to receptors on the mast cell surface. Histamine is responsible for many of the unpleasant symptoms that accompany allergic reactions, such as itching and sneezing. When mast cells are incubated in fluid containing a soluble stimulant, massive exocytosis occurs all over the cell surface (Figure 13–59). But if the stimulating ligand is artificially attached to a solid bead so that it can interact only with a localized region of the mast cell surface, exocytosis is now restricted to the region where the cell contacts the bead (Figure 13–60).

This experiment shows that individual segments of the plasma membrane can function independently in regulated exocytosis. As a result, the mast cell,

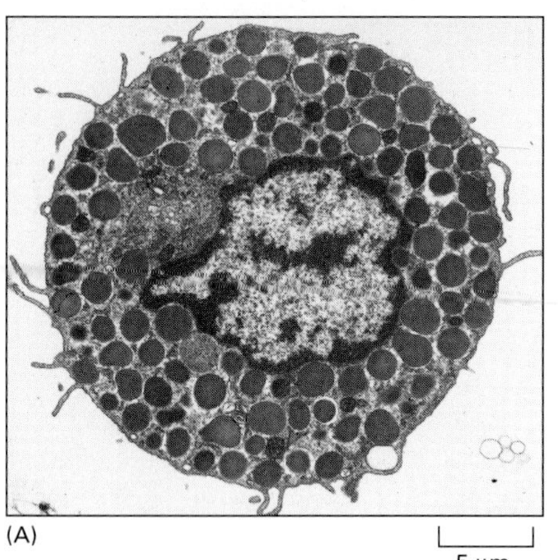

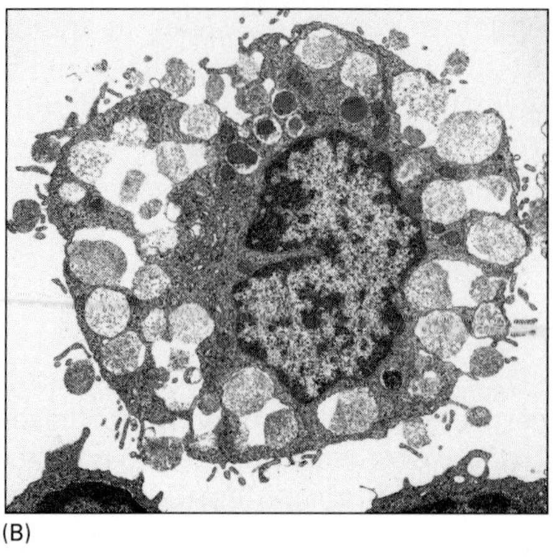

Figure 13–59 Electron micrographs of exocytosis in rat mast cells. (A) An unstimulated mast cell. (B) This cell has been activated to secrete its stored histamine by a soluble extracellular stimulant. Histamine-containing secretory vesicles are dark, while those that have released their histamine are light. The material remaining in the spent vesicles consists of a network of proteoglycans to which the stored histamine was bound. Once a secretory vesicle has fused with the plasma membrane, the secretory vesicle membrane often serves as a target to which other secretory vesicles fuse. Thus, the cell in (B) contains several large cavities lined by the fused membranes of many spent secretory vesicles, which are now in continuity with the plasma membrane. This continuity is not always apparent in one plane of section through the cell. (From D. Lawson, C. Fewtrell, B. Gomperts, and M. Raff, *J. Exp. Med.* 142:391–402, 1975. © The Rockefeller University Press.)

unlike a nerve cell, does not respond as a whole when it is triggered; the activation of receptors, the resulting intracellular signals, and the subsequent exocytosis are all localized in the particular region of the cell that has been excited. Such localized exocytosis enables a killer lymphocyte, for example, to deliver the proteins that induce the death of a single infected target cell precisely without endangering normal neighboring cells (see Figure 16–97).

Secretory Vesicle Membrane Components Are Quickly Removed from the Plasma Membrane

When a secretory vesicle fuses with the plasma membrane, its contents are discharged from the cell by exocytosis, and its membrane becomes part of the plasma membrane. Although this should greatly increase the surface area of the plasma membrane, it does so only transiently, because membrane components are removed from the surface by endocytosis almost as fast as they are added by exocytosis, reminiscent of the exocytosis–endocytosis cycle discussed earlier. After their removal from the plasma membrane, the proteins of the secretory vesicle membrane are thought to be shuttled to lysosomes for degradation. The amount of secretory vesicle membrane that is temporarily added to the plasma membrane can be enormous: in a pancreatic acinar cell discharging digestive enzymes for delivery to the gut lumen, about 900 μm^2 of vesicle membrane is inserted into the apical plasma membrane (whose area is only 30 μm^2) when the cell is stimulated to secrete.

Control of membrane traffic thus has a major role in maintaining the composition of the various membranes of the cell. To maintain each membrane-enclosed compartment in the secretory and endocytotic pathways at a constant size, the balance between the forward and retrograde flows of membrane needs to be precisely regulated. For cells to grow, the forward flow needs to be greater than the retrograde flow, so that the membrane can increase in area. For cells to maintain a constant size, the forward and retrograde flows must be equal. We still know very little about the mechanisms that coordinate these flows.

Polarized Cells Direct Proteins from the *Trans* Golgi Network to the Appropriate Domain of the Plasma Membrane

Most cells in tissues are *polarized* and have two (and sometimes more) distinct plasma membrane domains to which different types of vesicles must be directed. This raises the general problem of how the delivery of membrane from the Golgi apparatus is organized so as to maintain the differences between one cell-surface domain and another. A typical epithelial cell has an *apical domain,* which faces the lumen and often has specialized features such as cilia or a brush border of microvilli; it also has a *basolateral domain,* which covers the rest of the cell. The two domains are separated by a ring of *tight junctions* (see Figure 19–3), which prevent proteins and lipids (in the outer leaflet of the lipid bilayer) from

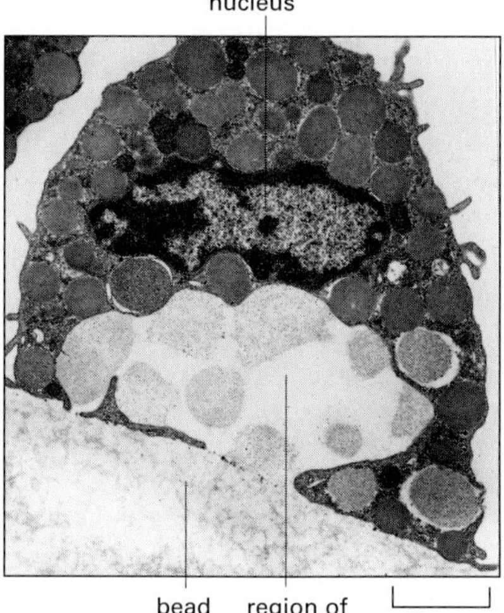

Figure 13–60 Exocytosis as a localized response. This electron micrograph shows a mast cell that has been activated to secrete histamine by a stimulant coupled to a large solid bead. Exocytosis has occurred only in the region of the cell that is in contact with the bead. (From D. Lawson, C. Fewtrell, and M. Raff, *J. Cell Biol.* 79:394–400, 1978. © The Rockefeller University Press.)

diffusing between the two domains, so that the compositions of the two domains are different.

A nerve cell is another example of a polarized cell. The plasma membrane of its axon and nerve terminals is specialized for signaling to other cells, whereas the plasma membrane of its cell body and dendrites is specialized to receive signals from other nerve cells. The two domains have distinct protein compositions. Studies of protein traffic in nerve cells in culture suggest that, with regard to vesicular transport from the *trans* Golgi network to the cell surface, the plasma membrane of the nerve cell body and dendrites resembles the basolateral membrane of a polarized epithelial cell, while the plasma membrane of the axon and its nerve terminals resembles the apical membrane of such a cell (Figure 13–61). Thus, some proteins that are targeted to a specific domain in the epithelial cell are also found to be targeted to the corresponding domain in the nerve cell.

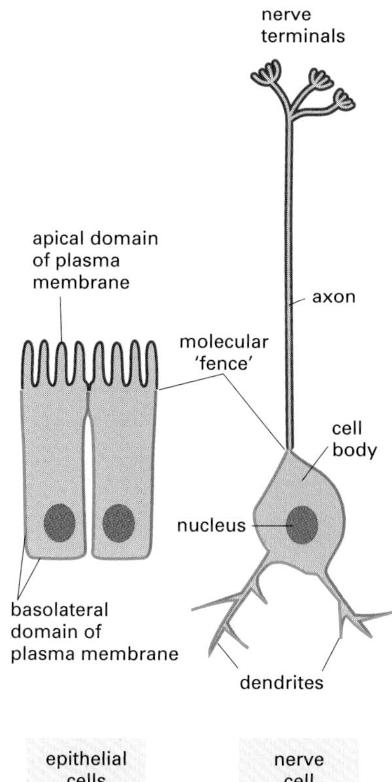

Figure 13–61 A comparison of two types of polarized cells. In terms of the mechanisms used to direct proteins to them, the plasma membrane of the nerve cell body and dendrites resembles the basolateral plasma membrane domain of a polarized epithelial cell, whereas the plasma membrane of the axon and its nerve terminals resembles the apical domain of an epithelial cell. The different membrane domains of both the epithelial cell and the nerve cell are separated by a molecular fence, consisting of a meshwork of membrane proteins tightly associated with the underlying actin cytoskeleton; this barrier—called a tight junction in the epithelial cell and an axonal hillock in neurons—keeps membrane proteins from diffusing between the two distinct domains.

Cytoplasmic Sorting Signals Guide Membrane Proteins Selectively to the Basolateral Plasma Membrane

In principle, differences between plasma membrane domains need not depend on the targeted delivery of the appropriate membrane components. Instead, membrane components could be delivered to all regions of the cell surface indiscriminately, but then be selectively stabilized in some locations and selectively eliminated in others. Although this strategy of random delivery followed by selective retention or removal seems to be used in certain cases, deliveries are often specifically directed to the appropriate membrane domain. Epithelial cells, for example, frequently secrete one set of products—such as digestive enzymes or mucus in cells lining the gut—at their apical surface, and another set of products—such as components of the basal lamina—at their basolateral surface. Thus, cells must have ways of directing vesicles carrying different cargoes to different plasma membrane domains.

By examining polarized epithelial cells in culture, it has been found that proteins from the ER destined for different domains travel together until they reach the *trans* Golgi network. Here they are separated and dispatched in secretory or transport vesicles to the appropriate plasma membrane domain (Figure 13–62).

Membrane proteins destined for delivery to the basolateral membrane contain sorting signals in their cytoplasmic tail. Two such signals are known, one containing a characteristic conserved tyrosine and the other two adjacent leucines. When present in an appropriate structural context, these amino acids are recognized by coat proteins that package them into appropriate transport vesicles in the *trans* Golgi network. The same basolateral signals that are recognized in the *trans* Golgi network also function in endosomes to redirect the proteins back to the basolateral plasma membrane after they have been endocytosed.

Lipid Rafts May Mediate Sorting of Glycosphingolipids and GPI-anchored Proteins to the Apical Plasma Membrane

The apical plasma membrane of most cells is greatly enriched in glycosphingolipids, which help protect this exposed surface from damage—by the digestive enzymes and low pH in such sites as the stomach or the lumen of the gut, for example. Plasma membrane proteins that are linked to the lipid bilayer by a glycosylphosphatidylinositol (GPI) anchor are also found exclusively in the apical plasma membrane. If recombinant DNA techniques are used to attach a GPI anchor to a protein that would normally be delivered to the basolateral surface, the protein is now delivered to the apical surface instead.

GPI-anchored proteins are thought to be directed to the apical membrane because they associate with the glycosphingolipids in **lipid rafts** that form in the membrane of the *trans* Golgi network. As discussed in Chapter 10, lipid rafts form in the *trans* Golgi network and plasma membrane when glycosphingolipids and cholesterol self-associate into microaggregates (see Figure 10–13). Membrane proteins with unusually long transmembrane domains

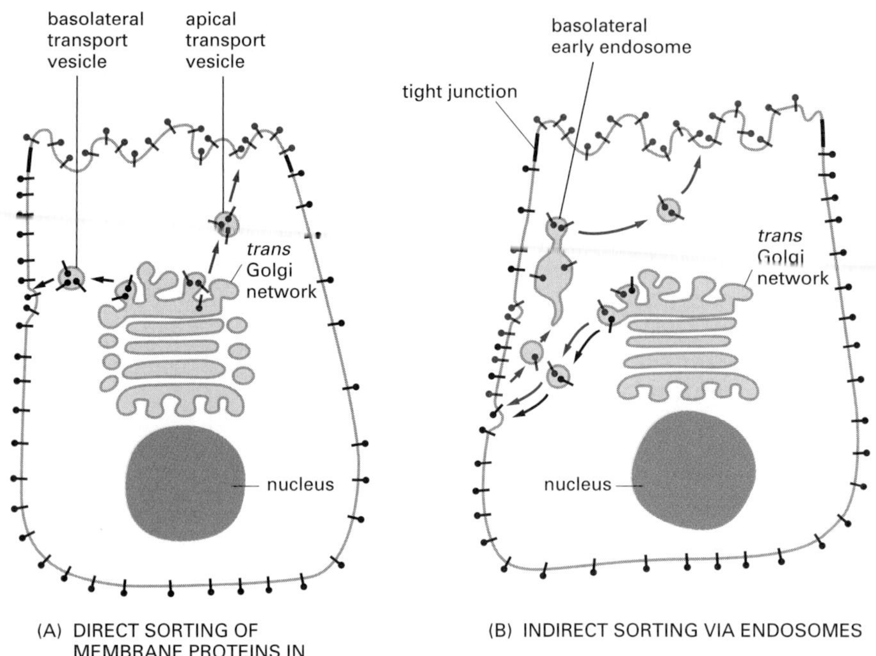

Figure 13–62 Two ways of sorting plasma membrane proteins in a polarized epithelial cell. Newly synthesized proteins can reach their proper plasma membrane domain by either (A) a direct pathway or (B) an indirect pathway. In the indirect pathway, a protein is retrieved from the inappropriate plasma membrane domain by endocytosis and then transported to the correct domain via early endosomes—that is, by transcytosis. The indirect pathway is known to be used in liver hepatocytes to deliver proteins to the apical domain that lines bile ducts. However, in other cases, the direct pathway is used, as described in the text for epithelial cells in the gut.

also accumulate in the rafts. In addition, the rafts preferentially contain GPI-anchored proteins and some carbohydrate-binding proteins (lectins) that may help stabilize the assemblies (Figure 13–63).

Having selected a unique set of cargo molecules, the rafts then bud from the *trans* Golgi network into transport vesicles destined for the apical plasma membrane.

Synaptic Vesicles Can Form Directly from Endocytic Vesicles

Nerve cells (and some endocrine cells) contain two types of secretory vesicles. As for all secretory cells, these cells package proteins and peptides in dense-cored secretory vesicles in the standard way for release by the regulated secretory pathway. In addition, however, they make use of another specialized class of tiny (~50-nm diameter) secretory vesicles, which are called **synaptic vesicles** and are generated in a different way. In nerve cells, these vesicles store small neurotransmitter molecules, such as acetylcholine, glutamate, glycine, and γ-aminobutyric acid (GABA), that mediate rapid signaling from cell to cell at chemical synapses. As discussed earlier, the vesicles are triggered to release their contents within a fraction of a millisecond when an action potential arrives at a nerve terminal. Some neurons fire more than 1000 times per second, releasing

Figure 13–63 Model of lipid rafts in the *trans* Golgi network. Glycosphingolipids and cholesterol are thought to form rafts in the lipid bilayer. Membrane proteins with long enough membrane-spanning segments preferentially partition into the lipid rafts and thus become sorted into transport vesicles. These rafts are subsequently packaged into transport vesicles that carry them to the apical domain of the plasma membrane. Carbohydrate-binding proteins (lectins) in the lumen of the *trans* Golgi network may help stabilize the rafts, as shown.

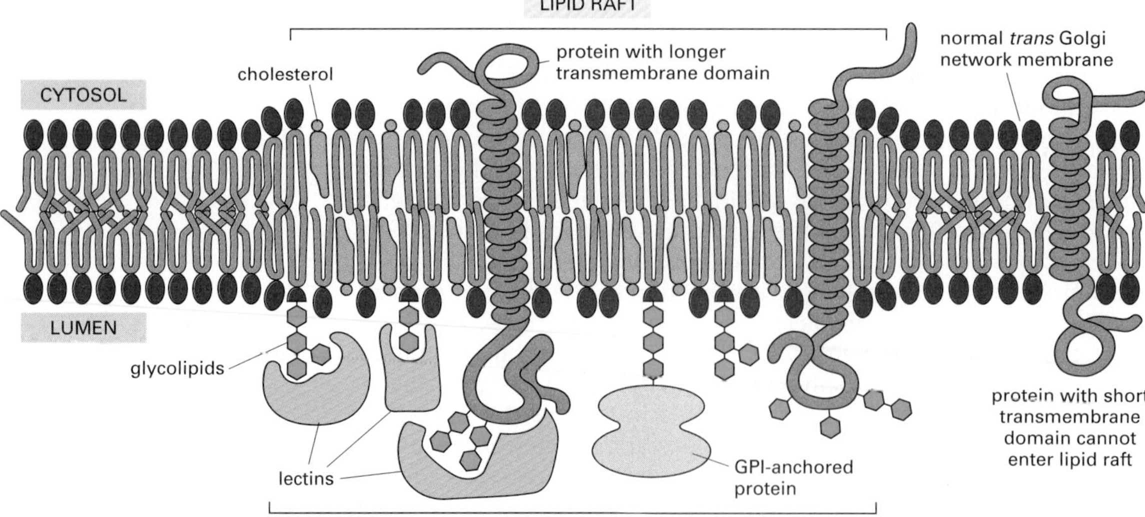

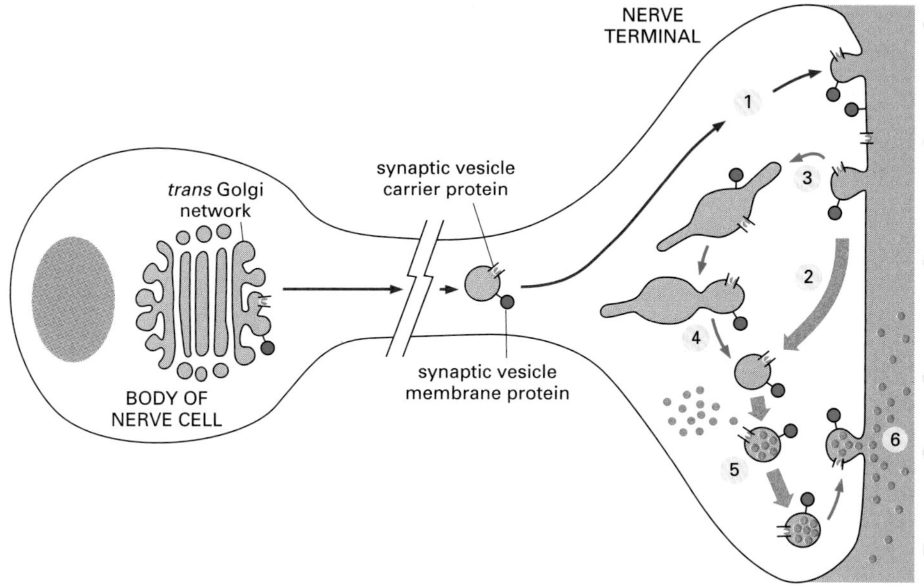

1 DELIVERY OF SYNAPTIC VESICLE COMPONENTS TO PLASMA MEMBRANE

2 ENDOCYTOSIS OF SYNAPTIC VESICLE COMPONENTS TO FORM NEW SYNAPTIC VESICLES DIRECTLY

3 ENDOCYTOSIS OF SYNAPTIC VESICLE COMPONENTS AND DELIVERY TO ENDOSOME

4 BUDDING OF SYNAPTIC VESICLE FROM ENDOSOME

5 LOADING OF NEUROTRANSMITTER INTO SYNAPTIC VESICLE

6 SECRETION OF NEUROTRANSMITTER BY EXOCYTOSIS IN RESPONSE TO AN ACTION POTENTIAL

Figure 13–64 The formation of synaptic vesicles. These tiny uniform vesicles are found only in nerve cells and in some endocrine cells, where they store and secrete small-molecule neurotransmitters. The import of neurotransmitter directly into the small endocytic vesicles that form from the plasma membrane is mediated by membrane carrier proteins that function as antiports, being driven by a H^+ gradient maintained by proton pumps in the vesicle membrane.

neurotransmitters each time. This rapid release is possible because some of the vesicles are docked and primed for fusion, which will occur only when an action potential causes an influx of Ca^{2+} into the terminal.

Only a small proportion of the synaptic vesicles in the nerve terminal fuse with the plasma membrane in response to each action potential. But for the nerve terminal to respond rapidly and repeatedly, the vesicles need to be replenished very quickly after they discharge. Thus, most synaptic vesicles are generated not from the Golgi membrane in the nerve cell body but by local recycling from the plasma membrane in the nerve terminals. It is thought that the membrane components of the synaptic vesicles are initially delivered to the plasma membrane by the constitutive secretory pathway and then retrieved by endocytosis. But instead of fusing with endosomes, most of the endocytic vesicles immediately fill with transmitter to become synaptic vesicles.

The membrane components of a synaptic vesicle include carrier proteins specialized for the uptake of neurotransmitter from the cytosol, where the small-molecule neurotransmitters that mediate fast synaptic signaling are synthesized. Once filled with neurotransmitter, the vesicles return to the plasma membrane, where they wait until the cell is stimulated. After they have released their contents, their membrane components are retrieved in the same way and used again (Figure 13–64).

Summary

Proteins can be secreted from cells by exocytosis in either a constitutive or a regulated fashion. In the regulated pathways, molecules are stored either in secretory vesicles or synaptic vesicles, which do not fuse with the plasma membrane to release their contents until an appropriate signal is received. Secretory vesicles bud from the trans Golgi network. The secretory proteins they contain condense during the formation and maturation of secretory vesicles. Synaptic vesicles, which are confined to nerve cells and some endocrine cells, form from endocytic vesicles and from endosomes, and they are responsible for the regulated secretion of small-molecule neurotransmitters. Whereas the regulated pathways operate only in specialized secretory cells, a constitutive secretory pathway operates in all eucaryotic cells, mediated by continual vesicular transport from the trans Golgi network to the plasma membrane.

Proteins are delivered from the trans Golgi network to the plasma membrane by the constitutive pathway unless they are diverted into other pathways or retained in the Golgi apparatus. In polarized cells, the transport pathways from the trans Golgi network to the plasma membrane operate selectively to ensure that different sets of membrane proteins, secreted proteins, and lipids are delivered to the different domains of the plasma membrane.

References

General

Mellman I & Warren G (2000) The road taken: past and future foundations of membrane traffic. *Cell* 100, 99–112.

Pelham HR (1999) The Croonian Lecture 1999. Intracellular membrane traffic: getting proteins sorted. *Philos. Trans. R. Soc. Lond. B. Biol. Sci.* 354(1388), 1471–1478.

Rothman JE & Wieland FT (1996) Protein sorting by transport vesicles. *Science* 272, 227–234.

Schekman RW (1996) Regulation of membrane traffic in the secretory pathway. In The Harvey Lectures Series 90 1994–95 (RM Evans, C Guthrie, LH Hartwell et al. eds), pp 41–57. New York: John Wiley.

The Molecular Mechanisms of Membrane Transport and the Maintenance of Compartmental Diversity

Chavrier P & Goud B (1999) The role of ARF and Rab GTPases in membrane transport. *Curr. Opin. Cell Biol.* 11, 466–475.

Kirchhausen T (1999) Adaptors for clathrin-mediated traffic. *Annu. Rev. Cell Dev. Biol.* 15, 705–732.

Kreis TE, Lowe M & Pepperkok R (1995) COPs regulating membrane traffic. *Annu. Rev. Cell Dev. Biol.* 11, 677–706.

Novick P & Zerial M (1997) The diversity of Rab proteins in vesicle transport. *Curr. Opin. Cell Biol.* 9, 496–504.

Nuoffer C & Balch WE (1994) GTPases: multifunctional molecular switches regulating vesicular traffic. *Annu. Rev. Biochem.* 63, 949–990.

Pelham HR (1999) SNAREs and the secretory pathway—lessons from yeast. *Exp. Cell Res.* 247, 1–8.

Pelham HR (2001) SNAREs and the specificity of membrane fusion. *Trends Cell Biol.* 11, 99–101.

Schekman R & Orci L (1996) Coat proteins and vesicle budding. *Science* 271, 1526–1533.

Schmid SL (1997) Clathrin-coated vesicle formation and protein sorting: an integrated process. *Annu. Rev. Biochem.* 66, 511–548.

Weber T, Zemelman BV, McNew JA et al. (1998) SNAREpins: minimal machinery for membrane fusion. *Cell* 92, 759–772.

Transport from the ER Through the Golgi Apparatus

Bannykh SI, Rowe T & Balch WE (1996) The organization of endoplasmic reticulum export complexes. *J. Cell Biol.* 135, 19–35.

Comer FI & Hart GW (1999) O-GlcNAc and the control of gene expression. *Biochim. Biophys. Acta* 1473, 161–171.

Ellgaard L, Molinari M & Helenius A (1999) Setting the standards: quality control in the secretory pathway. *Science* 286, 1882–1888.

Farquhar MG & Palade GE (1998) The Golgi apparatus: 100 years of progress and controversy. *Trends Cell Biol.* 8, 2–10. (Part of a special issue on the Golgi)

Fortoni ME (2000) Fringe benefits to carbohydrates. *Nature* 406, 357–358.

Glick BS (2000) Organization of the Golgi apparatus. *Curr. Opin. Cell Biol.* 12, 450–456.

Klumperman J (2000) Transport between ER and Golgi. *Curr. Opin. Cell Biol.* 12, 445–449.

Ladinsky MS, Mastronarde DN, McIntosh JR et al. (1999) Golgi structure in three dimensions: functional insights from the normal rat kidney cell. *J Cell Biol.* 144, 1135–1149.

Orci L, Stamnes M, Ravazzola M et al. (1997) Bidirectional transport by distinct populations of COPI-coated vesicles. *Cell* 90, 335–349.

Ruoslahti E (1988) Structure and biology of proteoglycans. *Annu. Rev. Cell Biol.* 4, 229–255.

Warren G & Malhotra V (1998) The organisation of the Golgi apparatus. *Curr. Opin. Cell Biol.* 10, 493–498.

Waters MG & Pfeffer SR (1999) Membrane tethering in intracellular transport. *Curr. Opin. Cell Biol.* 11, 453–459.

Transport from the *Trans* Golgi Network to Lysosomes

Amara J, Cheng SH & Smith A (1992) Intracellular protein trafficking defects in human disease. *Trends Cell Biol.* 2, 145–149.

Andrews NW (2000) Regulated secretion of conventional lysosomes. *Trends Cell Biol.* 10, 316–321.

Bainton D (1981) The discovery of lysosomes. *J. Cell Biol.* 91, 66s–76s.

Kornfeld S & Mellman I (1989) The biogenesis of lysosomes. *Annu. Rev. Cell Biol.* 5, 483–525.

Mizushima N, Yamamoto A, Hatano M et al. (2001) Dissection of autophagosome formation using Apg5-deficient mouse embryonic stem cells. *J. Cell Biol.* 152, 657–668.

Munier-Lehmann H, Mauxion F & Hoflack B (1996) Function of the two mannose 6-phosphate receptors in lysosomal enzyme transport. *Biochem. Soc. Trans.* 24, 133–136.

Storrie B & Desjardins M (1996) The biogenesis of lysosomes: is it a kiss and run, continuous fusion and fission process? *Bioessays* 18, 895–903.

Tjelle TE, Lovdal T & Berg T (2000) Phagosome dynamics and function. *Bioessays* 22, 255–263.

von Figura K (1991) Molecular recognition and targeting of lysosomal proteins. *Curr. Opin. Cell Biol.* 3, 642–646.

Transport into the Cell from the Plasma Membrane: Endocytosis

Anderson RG (1998) The caveolae membrane system. *Annu. Rev. Biochem.* 67, 199–225.

Béron W, Alvarez-Dominguez C, Mayorga L & Stahl PD (1995) Membrane trafficking along the phagocytic pathway. *Trends Cell Biol.* 5, 100–104. (Part of a special issue on phagocytosis)

Brown MS & Goldstein JL (1986) A receptor-mediated pathway for cholesterol homeostasis. *Science* 232, 34–47.

Fujita M, Reinhart F & Neutra M (1990) Convergence of apical and basolateral endocytic pathways at apical late endosomes in absorptive cells of suckling rat ileum *in vivo*. *J. Cell Sci.* 97, 385–394.

Hicke L (2001) A new ticket for entry into budding vesicles—ubiquitin. *Cell* 106, 527–530.

Marks MS, Ohno H, Kirchhausen T & Bonifacino JS (1997) Protein sorting by tyrosine-based signals: adapting to the Ys and wherefores. *Trends Cell Biol.* 7, 124–128

Mellman I (1996) Endocytosis and molecular sorting. *Annu. Rev. Cell Dev. Biol.* 12, 575–625.

Mostov KE, Verges M & Altschuler Y (2000) Membrane traffic in polarized epithelial cells. *Curr. Opin. Cell Biol.* 12, 483–490.

Odorizzi G, Babst M & Emr SD (2000) Phosphoinositide signaling and the regulation of membrane trafficking in yeast. *Trends Biochem. Sci.* 25, 229–235.

Transport from the *Trans* Golgi Network to the Cell Exterior: Exocytosis

Burgess TL & Kelly RB (1987) Constitutive and regulated secretion of proteins. *Annu. Rev. Cell Biol.* 3, 243–294.

Jacobson K & Dietrich C (1999) Looking at lipid rafts? *Trends Cell Biol.* 9, 87–91.

Jarousse N & Kelly RB (2001) Endocytotic mechanisms in synapses. *Curr. Opin. Cell Biol.* 13, 461–469.

Keller P & Simons K (1997) Post-Golgi biosynthetic trafficking. *J. Cell Sci.* 110, 3001–3009.

Klenchin VA & Martin TFJ (2000) Priming in exocytosis: attaining fusion-competence after vesicle docking. *Biochimie* 82, 399–407.

Lawson D, Fewtrell C & Raff M (1978) Localized mast cell degranulation induced by concanavalin A-sepharose beads, implications for the Ca^{2+} hypothesis of stimulus-secretion coupling. *J. Cell Biol.* 79, 394–400.

Martin TFJ (1997) Stages of regulated exocytosis. *Trends Cell Biol.* 7, 271–276.

Simons K & Ikonen E (1997) Functional rafts in cell membranes. *Nature* 387, 569–572.

Südhof TC (1995) The synaptic vesicle cycle: a cascade of protein–protein interactions. *Nature* 375, 645–653.

Traub LM & Kornfeld S (1997) The trans-Golgi network: a late secretory sorting station. *Curr. Opin. Cell Biol.* 9, 527–533.

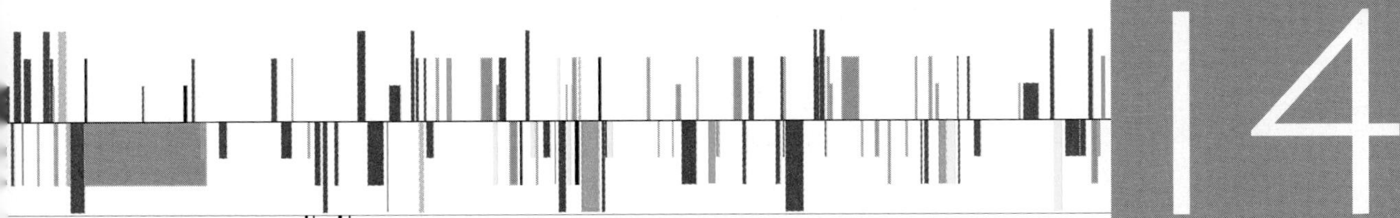

14

ENERGY CONVERSION: MITOCHONDRIA AND CHLOROPLASTS

Through a set of reactions that occur in the cytosol, energy derived from the partial oxidation of energy-rich carbohydrate molecules is used to form ATP, the chemical energy currency of cells (discussed in Chapter 2). But a much more efficient method of energy generation appeared very early in the history of life. This process is based on membranes, and it enables cells to acquire energy from a wide variety of sources. For example, it is central to the conversion of light energy into chemical bond energy in photosynthesis, as well as to the aerobic respiration that enables us to use oxygen to produce large amounts of ATP from food molecules.

The membrane that is used to produce ATP in procaryotes is the plasma membrane. But in eucaryotic cells, the plasma membrane is reserved for the transport processes described in Chapter 11. Instead, the specialized membranes inside *energy-converting organelles* are employed for the production of ATP. The membrane-enclosed organelles are **mitochondria**, which are present in the cells of virtually all eucaryotic organisms (including fungi, animals, and plants), and **plastids**—most notably **chloroplasts**—which occur only in plants. In electron micrographs the most striking morphological feature of mitochondria and chloroplasts is the large amount of internal membrane they contain. This internal membrane provides the framework for an elaborate set of electron-transport processes that produce most of the cell's ATP.

The common pathway used by mitochondria, chloroplasts, and procaryotes to harness energy for biological purposes operates by a process known as **chemiosmotic coupling**—reflecting a link between the chemical bond-forming reactions that generate ATP ("chemi") and membrane-transport processes ("osmotic"). The coupling process occurs in two linked stages, both of which are performed by protein complexes embedded in a membrane:

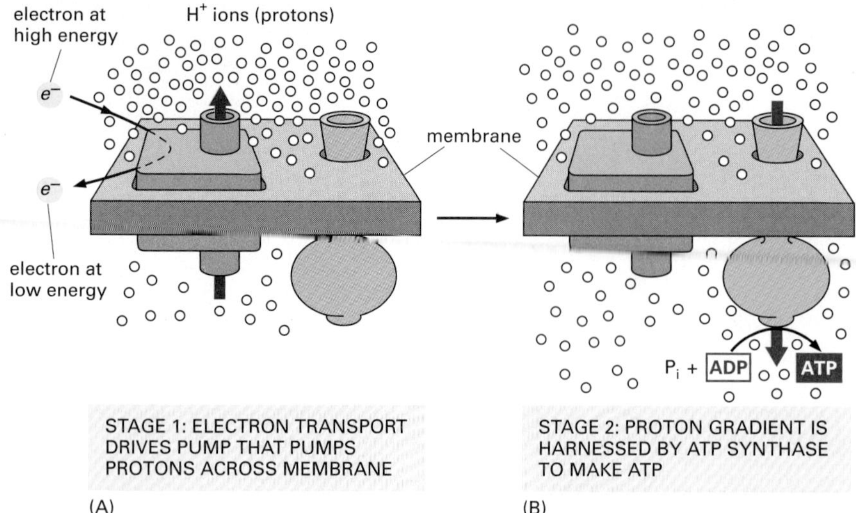

Figure 14–1 Harnessing energy for life. (A) The essential requirements for chemiosmosis are a membrane—in which are embedded a pump protein and an ATP synthase, plus a source of high-energy electrons (e^-). The protons (H^+) shown are freely available from water molecules. The pump harnesses the energy of electron transfer (details not shown here) to pump protons, creating a proton gradient across the membrane. (B) This proton gradient serves as an energy store that can be used to drive ATP synthesis by the ATP synthase enzyme. The *red arrow* shows the direction of proton movement at each stage.

Stage 1. High-energy electrons (derived from the oxidation of food molecules, from the action of sunlight, or from other sources discussed later) are transferred along a series of electron carriers embedded in the membrane. These electron transfers release energy that is used to pump protons (H^+, derived from the water that is ubiquitous in cells) across the membrane and thus generate an *electrochemical proton gradient.* As discussed in Chapter 11, an ion gradient across a membrane is a form of stored energy, which can be harnessed to do useful work when the ions are allowed to flow back across the membrane down their electrochemical gradient.

Stage 2. H^+ flows back down its electrochemical gradient through a protein machine called *ATP synthase,* which catalyzes the energy-requiring synthesis of ATP from ADP and inorganic phosphate (P_i). This ubiquitous enzyme plays the role of a turbine, permitting the proton gradient to drive the production of ATP (Figure 14–1).

The electrochemical proton gradient is also used to drive other membrane-embedded protein machines (Figure 14–2). In eucaryotes, special proteins couple the "downhill" H^+ flow to the transport of specific metabolites into and out of the organelles. In bacteria, the electrochemical proton gradient drives more than ATP synthesis and transport processes; as a store of directly usable energy, it also drives the rapid rotation of the bacterial flagellum, which enables the bacterium to swim.

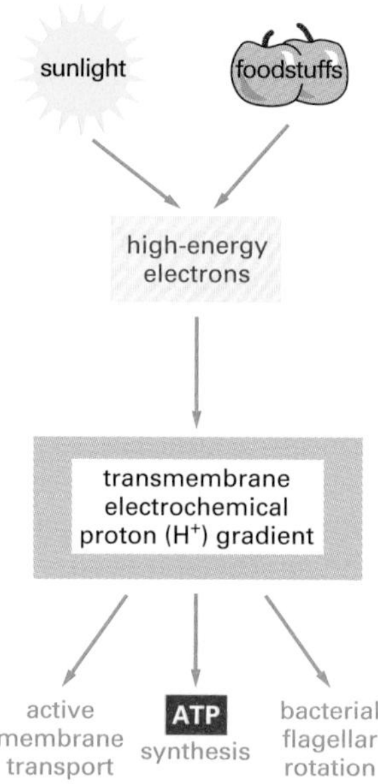

Figure 14–2 Chemiosmotic coupling. Energy from sunlight or the oxidation of foodstuffs is first used to create an electrochemical proton gradient across a membrane. This gradient serves as a versatile energy store and is used to drive a variety of energy-requiring reactions in mitochondria, chloroplasts, and bacteria.

It is useful to compare the electron-transport processes in mitochondria, which convert energy from chemical fuels, with those in chloroplasts, which convert energy from sunlight (Figure 14–3). In the mitochondrion, electrons—which have been released from a carbohydrate food molecule in the course of its degradation to CO_2—are transferred through the membrane by a chain of electron carriers, finally reducing oxygen gas (O_2) to form water. The free energy released as the electrons flow down this path from a high-energy state to a low-energy state is used to drive a series of three H^+ pumps in the inner mitochondrial membrane, and it is the third H^+ pump in the series that catalyzes the transfer of the electrons to O_2 (see Figure 14–3A).

The mechanism of electron transport can be compared to an electric cell driving a current through a set of electric motors. However, in biological systems, electrons are carried between one site and another not by conducting wires, but by diffusible molecules that can pick up electrons at one location and deliver them to another. For mitochondria, the first of these electron carriers is NAD^+, which takes up two electrons (plus an H^+) to become NADH, a water-soluble small molecule that ferries electrons from the sites where food molecules are degraded to the inner mitochondrial membrane. The entire set of proteins in the membrane, together with the small molecules involved in the orderly sequence of electron transfers, is called an **electron-transport chain**.

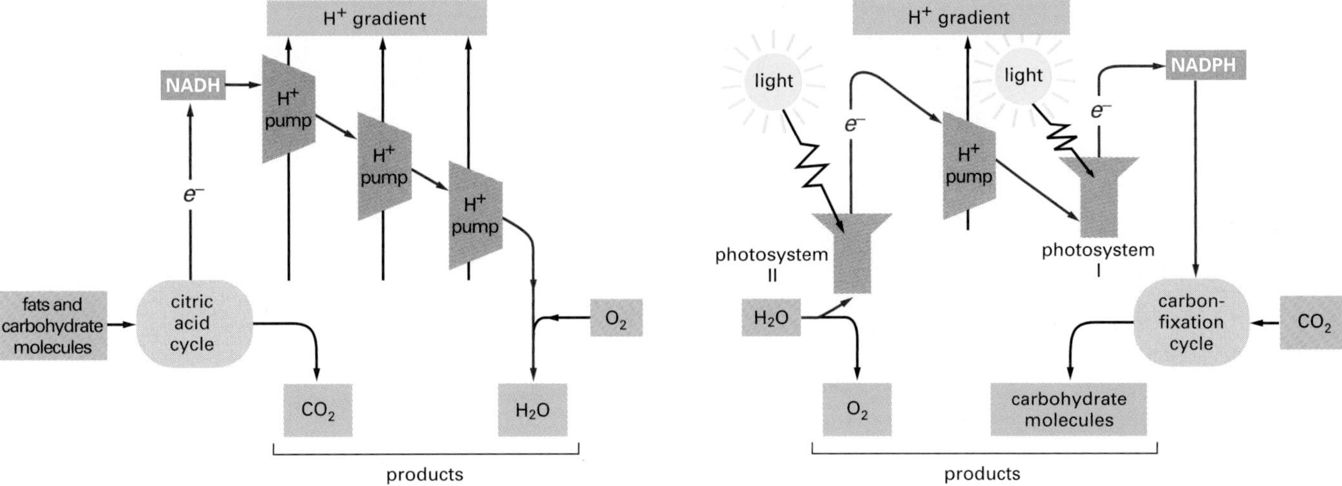

Although the chloroplast can be described in similar terms, and several of its main components are similar to those of the mitochondrion, the chloroplast membrane contains some crucial components not found in the mitochondrial membrane. Foremost among these are the *photosystems*, where light energy is captured by the green pigment chlorophyll and harnessed to drive the transfer of electrons, much as man-made photocells in solar panels absorb light energy and use it to drive an electric current. The electron-motive force generated by the chloroplast photosystems drives electron transfer in the direction opposite to that in mitochondria: electrons are taken *from* water to produce O_2, and they are donated (via NADPH, a compound closely related to NADH) *to* CO_2 to synthesize carbohydrate. Thus, the chloroplast generates O_2 and carbohydrate, whereas the mitochondrion consumes them (see Figure 14–3B).

It is generally believed that the energy-converting organelles of eucaryotes evolved from procaryotes that were engulfed by primitive eucaryotic cells and developed a symbiotic relationship with them. This would explain why mitochondria and chloroplasts contain their own DNA, which codes for some of their proteins. Since their initial uptake by a host cell, these organelles have lost much of their own genomes and have become heavily dependent on proteins that are encoded by genes in the nucleus, synthesized in the cytosol, and then imported into the organelle. Conversely, the host cells have become dependent on these organelles for much of the ATP they need for biosyntheses, ion pumping, and movement; they have also become dependent on selected biosynthetic reactions that occur inside these organelles.

Figure 14–3 Electron transport processes. (A) The mitochondrion converts energy from chemical fuels. (B) The chloroplast converts energy from sunlight. Inputs are *light green,* products are *blue,* and the path of electron flow is indicated by *red arrows.* Each of the protein complexes *(orange)* is embedded in a membrane. Note that the electron-motive force generated by the two chloroplast photosystems enables the chloroplast to drive electron transfer from H_2O to carbohydrate, and that this is *opposite* to the energetically favorable direction of electron transfer in a mitochondrion. Thus, whereas carbohydrate molecules and O_2 are inputs for the mitochondrion, they are products of the chloroplast.

THE MITOCHONDRION

Mitochondria occupy a substantial portion of the cytoplasmic volume of eucaryotic cells, and they have been essential for the evolution of complex animals. Without mitochondria, present-day animal cells would be dependent on anaerobic glycolysis for all of their ATP. When glucose is converted to pyruvate by glycolysis, only a very small fraction of the total free energy potentially available from the glucose is released. In mitochondria, the metabolism of sugars is completed: the pyruvate is imported into the mitochondrion and oxidized by O_2 to CO_2 and H_2O. This allows 15 times more ATP to be made than that produced by glycolysis alone.

Mitochondria are usually depicted as stiff, elongated cylinders with a diameter of 0.5–1 μm, resembling bacteria. Time-lapse microcinematography of living cells, however, shows that mitochondria are remarkably mobile and plastic organelles, constantly changing their shape (Figure 14–4) and even fusing with one another and then separating again. As they move about in the cytoplasm, they often seem to be associated with microtubules (Figure 14–5), which can

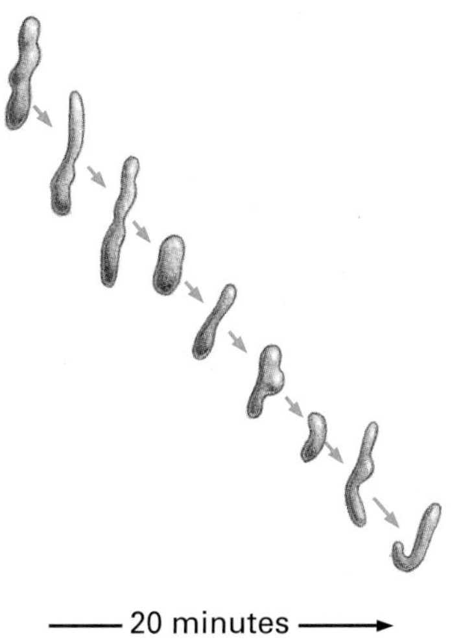

Figure 14–4 Mitochondrial plasticity. Rapid changes of shape are observed when an individual mitochondrion is followed in a living cell.

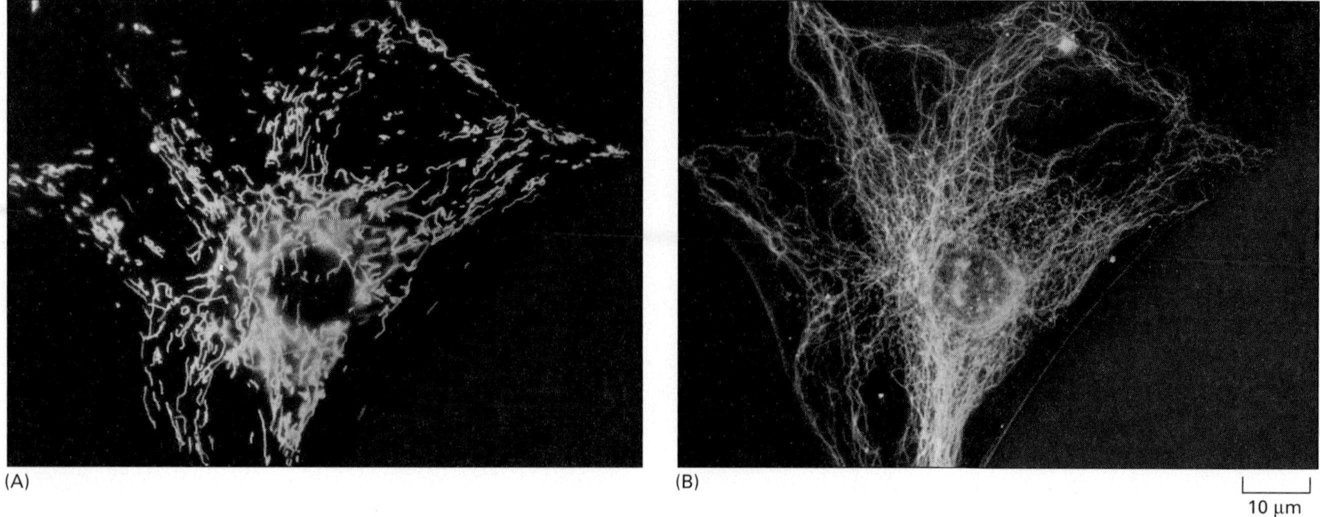

Figure 14–5 The relationship between mitochondria and microtubules. (A) A light micrograph of chains of elongated mitochondria in a living mammalian cell in culture. The cell was stained with a fluorescent dye (rhodamine 123) that specifically labels mitochondria in living cells. (B) An immunofluorescence micrograph of the same cell stained (after fixation) with fluorescent antibodies that bind to microtubules. Note that the mitochondria tend to be aligned along microtubules. (Courtesy of Lan Bo Chen.)

determine the unique orientation and distribution of mitochondria in different types of cells. Thus, the mitochondria in some cells form long moving filaments or chains. In others they remain fixed in one position where they provide ATP directly to a site of unusually high ATP consumption—packed between adjacent myofibrils in a cardiac muscle cell, for example, or wrapped tightly around the flagellum in a sperm (Figure 14–6).

Mitochondria are large enough to be seen in the light microscope, and they were first identified during the nineteenth century. Real progress in understanding their function, however, depended on procedures developed in 1948 for isolating intact mitochondria. For technical reasons, many of these biochemical studies have been performed with mitochondria purified from liver; each liver cell contains 1000–2000 mitochondria, which in total occupy about one-fifth of the cell volume.

The Mitochondrion Contains an Outer Membrane, an Inner Membrane, and Two Internal Compartments

Each mitochondrion is bounded by two highly specialized membranes, which have very different functions. Together they create two separate mitochondrial compartments: the internal **matrix** and a much narrower **intermembrane space**. If purified mitochondria are gently disrupted and then fractionated into separate components (Figure 14–7), the biochemical composition of each of the

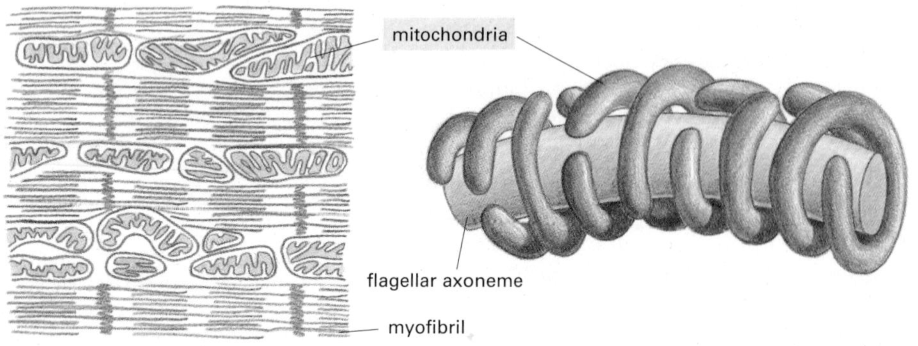

Figure 14–6 Localization of mitochondria near sites of high ATP utilization in cardiac muscle and a sperm tail. During the development of the flagellum of the sperm tail, microtubules wind helically around the axoneme, where they are thought to help localize the mitochondria in the tail; these microtubules then disappear, and the mitochondria fuse with one another to create the structure shown.

Figure 14–7 Biochemical fractionation of purified mitochondria into separate components. These techniques have made it possible to study the different proteins in each mitochondrial compartment. The method shown allows the processing of large numbers of mitochondria at the same time. It takes advantage of the fact that, in a solution of low osmotic strength, water flows into mitochondria and greatly expands the matrix space *(yellow)*. While the cristae of the inner membrane unfold to accommodate the expansion, the outer membrane—which has no folds—breaks, releasing a structure composed of only the inner membrane and the matrix.

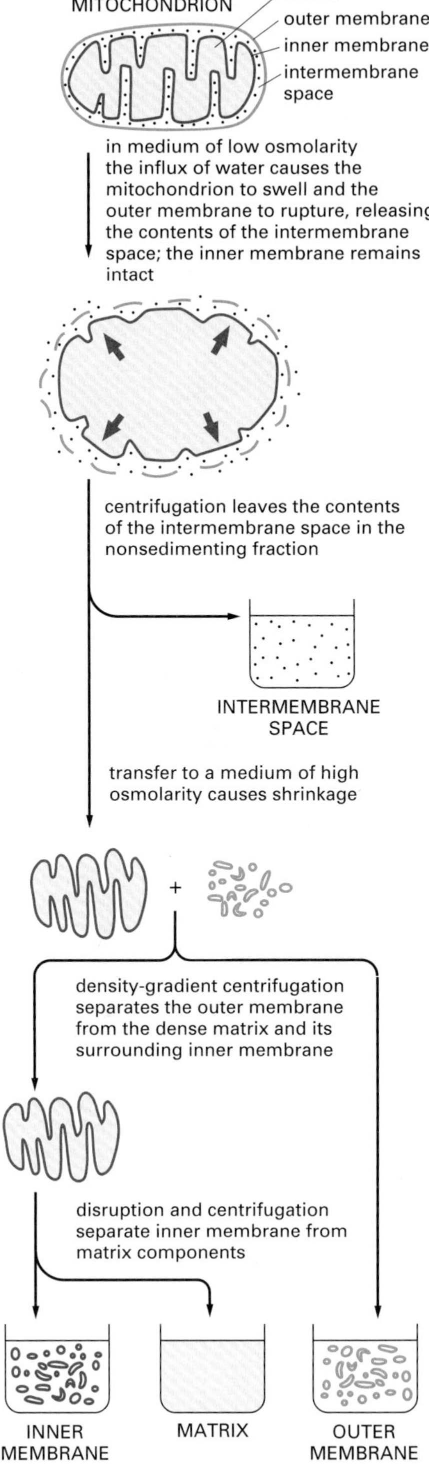

two membranes and of the spaces enclosed by them can be determined. As described in Figure 14–8, each contains a unique collection of proteins.

The **outer membrane** contains many copies of a transport protein called *porin* (discussed in Chapter 11), which forms large aqueous channels through the lipid bilayer. This membrane thus resembles a sieve that is permeable to all molecules of 5000 daltons or less, including small proteins. Such molecules can enter the intermembrane space, but most of them cannot pass the impermeable inner membrane. Thus, whereas the intermembrane space is chemically equivalent to the cytosol with respect to the small molecules it contains, the matrix contains a highly selected set of these molecules.

As we explain in detail later, the major working part of the mitochondrion is the matrix and the **inner membrane** that surrounds it. The inner membrane is highly specialized. Its lipid bilayer contains a high proportion of the "double" phospholipid *cardiolipin,* which has four fatty acids rather than two and may help to make the membrane especially impermeable to ions. This membrane also contains a variety of transport proteins that make it selectively permeable to those small molecules that are metabolized or required by the many mitochondrial enzymes concentrated in the matrix. The matrix enzymes include those that metabolize pyruvate and fatty acids to produce acetyl CoA and those that oxidize acetyl CoA in the *citric acid cycle.* The principal end-products of this oxidation are CO_2, which is released from the cell as waste, and NADH, which is the main source of electrons for transport along the **respiratory chain**—the name given to the electron-transport chain in mitochondria. The enzymes of the respiratory chain are embedded in the inner mitochondrial membrane, and they are essential to the process of *oxidative phosphorylation,* which generates most of the animal cell's ATP.

The inner membrane is usually highly convoluted, forming a series of infoldings, known as **cristae**, that project into the matrix. These convolutions greatly increase the area of the inner membrane, so that in a liver cell, for example, it constitutes about one-third of the total cell membrane. The number of cristae is three times greater in the mitochondrion of a cardiac muscle cell than in the mitochondrion of a liver cell, presumably because of the greater demand for ATP in heart cells. There are also substantial differences in the mitochondrial enzymes of different cell types. In this chapter, we largely ignore these differences and focus instead on the enzymes and properties that are common to all mitochondria.

High-Energy Electrons Are Generated via the Citric Acid Cycle

As previously mentioned, without mitochondria present-day eucaryotes would be dependent on the relatively inefficient process of glycolysis (described in Chapter 2) for all of their ATP production, and it seems unlikely that complex multicellular organisms could have been supported in this way. When glucose is converted to pyruvate by glycolysis, less than 10% of the total free energy potentially available from the glucose is released. In the mitochondria, the metabolism of sugars is completed, and the energy released is harnessed so efficiently that about 30 molecules of ATP are produced for each molecule of glucose oxidized. By contrast, only 2 molecules of ATP are produced per glucose molecule by glycolysis alone.

Mitochondria can use both pyruvate and fatty acids as fuel. Pyruvate comes from glucose and other sugars, whereas fatty acids come from fats. Both of these

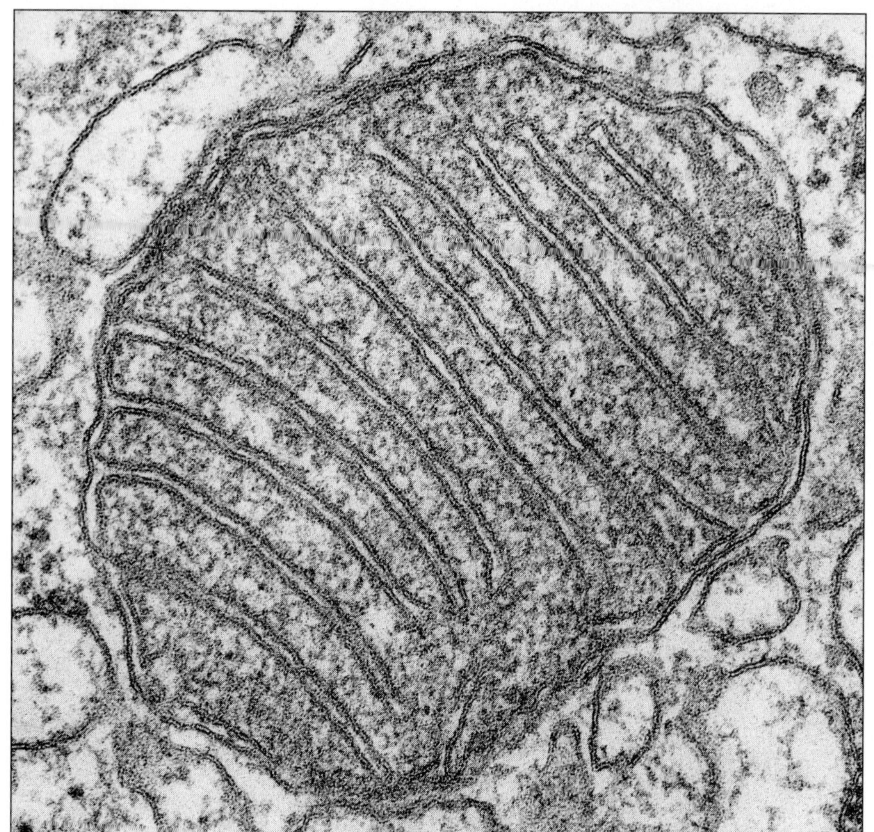

Figure 14–8 The general organization of a mitochondrion. In the liver, an estimated 67% of the total mitochondrial protein is located in the matrix, 21% is located in the inner membrane, 6% in the outer membrane, and 6% in the intermembrane space. As indicated below, each of these four regions contains a special set of proteins that mediate distinct functions. (Courtesy of Daniel S. Friend.)

Matrix. This large internal space contains a highly concentrated mixture of hundreds of enzymes, including those required for the oxidation of pyruvate and fatty acids and for the citric acid cycle. The matrix also contains several identical copies of the mitochondrial DNA genome, special mitochondrial ribosomes, tRNAs, and various enzymes required for expression of the mitochondrial genes.

Inner membrane. The inner membrane *(red)* is folded into numerous cristae, greatly increasing its total surface area. It contains proteins with three types of functions: (1) those that carry out the oxidation reactions of the electron-transport chain, (2) the ATP synthase that makes ATP in the matrix, and (3) transport proteins that allow the passage of metabolites into and out of the matrix. An electrochemical gradient of H^+, which drives the ATP synthase, is established across this membrane, so the membrane must be impermeable to ions and most small charged molecules.

Outer membrane. Because it contains a large channel-forming protein (called porin), the outer membrane is permeable to all molecules of 5000 daltons or less. Other proteins in this membrane include enzymes involved in mitochondrial lipid synthesis and enzymes that convert lipid substrates into forms that are subsequently metabolized in the matrix.

Intermembrane space. This space *(white)* contains several enzymes that use the ATP passing out of the matrix to phosphorylate other nucleotides.

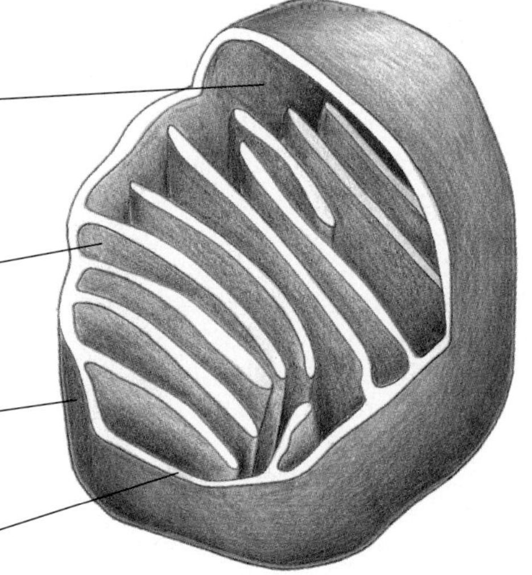

fuel molecules are transported across the inner mitochondrial membrane and then converted to the crucial metabolic intermediate *acetyl CoA* by enzymes located in the mitochondrial matrix. The acetyl groups in acetyl CoA are then oxidized in the matrix via the **citric acid cycle**, described in Chapter 2. The cycle converts the carbon atoms in acetyl CoA to CO_2, which is released from the cell as a waste product. Most importantly, the cycle generates high-energy electrons, carried by the activated carrier molecules NADH and $FADH_2$ (Figure 14–9). These high-energy electrons are then transferred to the inner mitochondrial membrane, where they enter the electron-transport chain; the loss of electrons from NADH and $FADH_2$ also regenerates the NAD^+ and FAD that is needed for continued oxidative metabolism. The entire sequence of reactions is outlined in Figure 14–10.

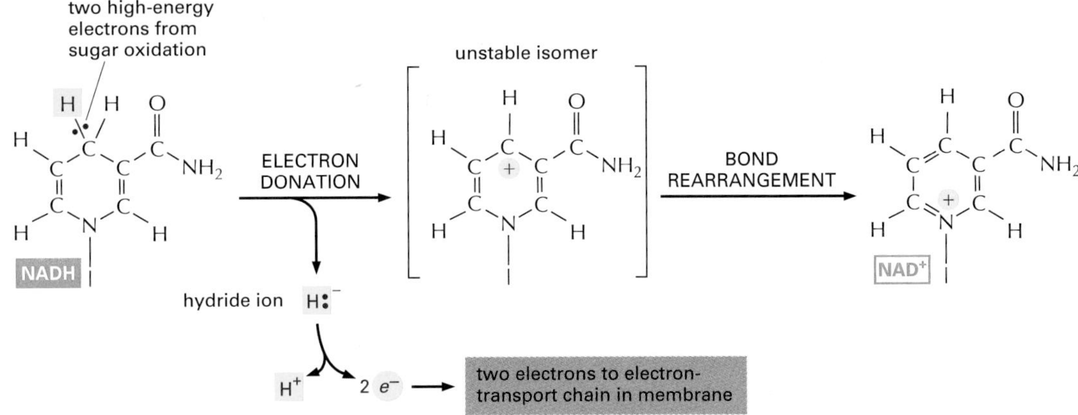

Figure 14–9 How electrons are donated by NADH. In this diagram, the high-energy electrons are shown as two *red dots* on a *yellow* hydrogen atom. A hydride ion (H^- a hydrogen atom and an extra electron) is removed from NADH and is converted into a proton and two high-energy electrons: $H^- \rightarrow H^+ + 2e^-$. Only the ring that carries the electrons in a high-energy linkage is shown; for the complete structure and the conversion of NAD^+ back to NADH, see the structure of the closely related NADPH in Figure 2–60. Electrons are also carried in a similar way by $FADH_2$, whose structure is shown in Figure 2–80.

A Chemiosmotic Process Converts Oxidation Energy into ATP

Although the citric acid cycle is considered to be part of aerobic metabolism, it does not itself use the oxygen. Only in the final catabolic reactions that take place on the inner mitochondrial membrane is molecular oxygen (O_2) directly consumed. Nearly all the energy available from burning carbohydrates, fats, and other foodstuffs in the earlier stages of their oxidation is initially saved in the form of high-energy electrons removed from substrates by NAD^+ and FAD. These electrons, carried by NADH and $FADH_2$, are then combined with O_2 by means of the respiratory chain embedded in the inner mitochondrial membrane. The large amount of energy released is harnessed by the inner membrane to drive

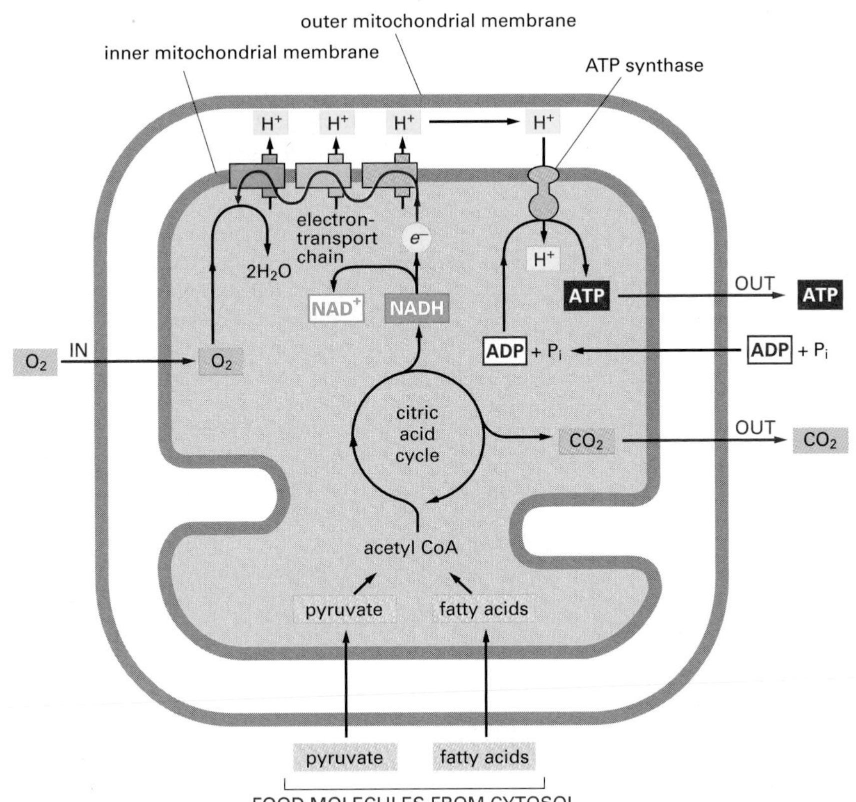

Figure 14–10 A summary of energy-generating metabolism in mitochondria. Pyruvate and fatty acids enter the mitochondrion *(bottom)* and are broken down to acetyl CoA. The acetyl CoA is then metabolized by the citric acid cycle, which reduces NAD^+ to NADH (and FAD to $FADH_2$, not shown). In the process of oxidative phosphorylation, high-energy electrons from NADH (and $FADH_2$) are then passed along the electron-transport chain in the inner membrane to oxygen (O_2). This electron transport generates a proton gradient across the inner membrane, which is used to drive the production of ATP by ATP synthase (see Figure 14–1).

The NADH generated by glycolysis in the cytosol also passes electrons to the respiratory chain (not shown). Since NADH cannot pass across the inner mitochondrial membrane, the electron transfer from cytosolic NADH must be accomplished indirectly by means of one of several "shuttle" systems that transport another reduced compound into the mitochondrion; after being oxidized, this compound is returned to the cytosol, where it is reduced by NADH again.

the conversion of ADP + P_i to ATP. For this reason, the term **oxidative phosphorylation** is used to describe this last series of reactions (Figure 14–11).

As previously mentioned, the generation of ATP by oxidative phosphorylation via the respiratory chain depends on a chemiosmotic process. When it was first proposed in 1961, this mechanism explained a long-standing puzzle in cell biology. Nonetheless, the idea was so novel that it was some years before enough supporting evidence accumulated to make it generally accepted.

In the remainder of this section we shall briefly outline the type of reactions that make oxidative phosphorylation possible, saving the details of the respiratory chain for later.

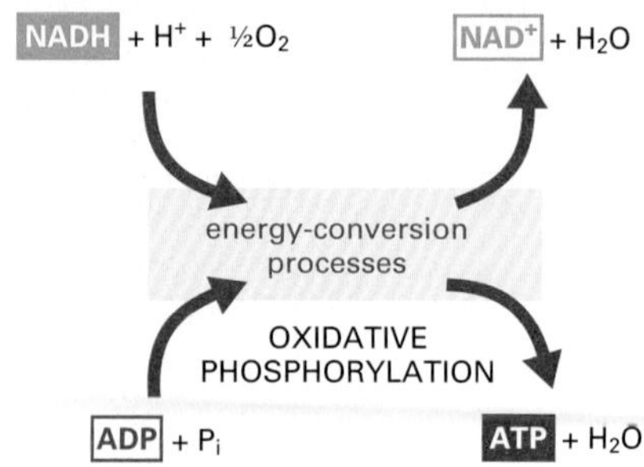

Figure 14–11 The major net energy conversion catalyzed by the mitochondrion. In this process of oxidative phosphorylation, the inner mitochondrial membrane serves as a device that changes one form of chemical bond energy to another, converting a major part of the energy of NADH (and $FADH_2$) oxidation into phosphate-bond energy in ATP.

Electrons Are Transferred from NADH to Oxygen Through Three Large Respiratory Enzyme Complexes

Although the mechanism by which energy is harvested by the respiratory chain differs from that in other catabolic reactions, the principle is the same. The energetically favorable reaction $H_2 + \frac{1}{2}O_2 \rightarrow H_2O$ is made to occur in many small steps, so that most of the energy released can be stored instead of being lost to the environment as heat. The hydrogen atoms are first separated into protons and electrons. The electrons pass through a series of electron carriers in the inner mitochondrial membrane. At several steps along the way, protons and electrons are transiently recombined. But only when the electrons reach the end of the electron-transport chain are the protons returned permanently, when they are used to neutralize the negative charges created by the final addition of the electrons to the oxygen molecule (Figure 14–12).

The process of electron transport begins when the hydride ion is removed from NADH (to regenerate NAD^+) and is converted into a proton and two electrons ($H^- \rightarrow H^+ + 2e^-$). The two electrons are passed to the first of the more than 15 different electron carriers in the respiratory chain. The electrons start with very high energy and gradually lose it as they pass along the chain. For the most part, the electrons pass from one metal ion to another, each of these ions being tightly bound to a protein molecule that alters the electron affinity of the metal ion (discussed in detail later). Most of the proteins involved are grouped into three large *respiratory enzyme complexes,* each containing transmembrane

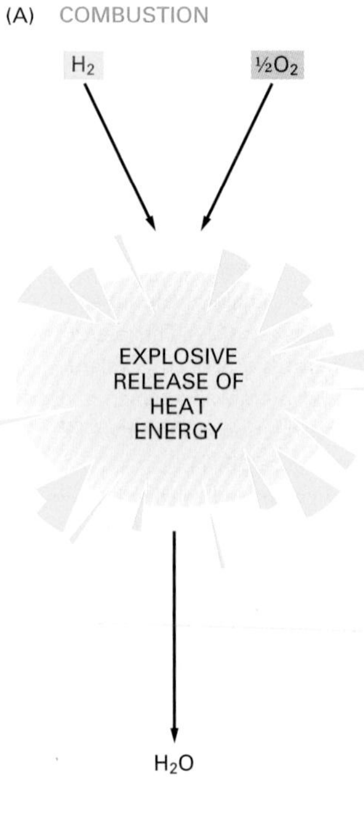

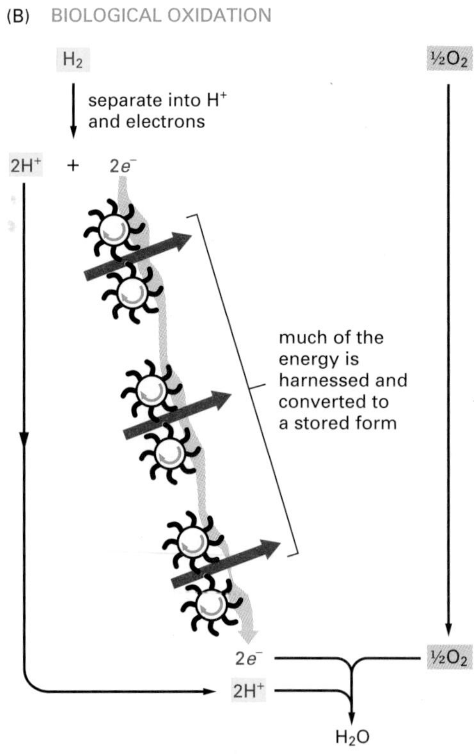

Figure 14–12 A comparison of biological oxidations with combustion. (A) Most of the energy would be released as heat if hydrogen were simply burned. (B) In biological oxidation by contrast, most of the released energy is stored in a form useful to the cell by means of the electron-transport chain in the inner mitochondrial membrane (the respiratory chain). The rest of the oxidation energy is released as heat by the mitochondrion. In reality, the protons and electrons shown are removed from hydrogen atoms that are covalently linked to NADH or $FADH_2$ molecules.

proteins that hold the complex firmly in the inner mitochondrial membrane. Each complex in the chain has a greater affinity for electrons than its predecessor, and electrons pass sequentially from one complex to another until they are finally transferred to oxygen, which has the greatest affinity of all for electrons.

As Electrons Move Along the Respiratory Chain, Energy Is Stored as an Electrochemical Proton Gradient Across the Inner Membrane

Oxidative phosphorylation is made possible by the close association of the electron carriers with protein molecules. The proteins guide the electrons along the respiratory chain so that the electrons move sequentially from one enzyme complex to another—with no short circuits. Most importantly, the transfer of electrons is coupled to oriented H^+ uptake and release, as well as to allosteric changes in energy-converting protein pumps. The net result is the pumping of H^+ across the inner membrane—from the matrix to the intermembrane space, driven by the energetically favorable flow of electrons. This movement of H^+ has two major consequences:

1. It generates a pH gradient across the inner mitochondrial membrane, with the pH higher in the matrix than in the cytosol, where the pH is generally close to 7. (Since small molecules equilibrate freely across the outer membrane of the mitochondrion, the pH in the intermembrane space is the same as in the cytosol.)
2. It generates a voltage gradient *(membrane potential)* across the inner mitochondrial membrane, with the inside negative and the outside positive (as a result of the net outflow of positive ions).

The pH gradient (ΔpH) drives H^+ back into the matrix and OH^- out of the matrix, thereby reinforcing the effect of the membrane potential (ΔV), which acts to attract any positive ion into the matrix and to push any negative ion out. Together, the ΔpH and the ΔV are said to constitute an **electrochemical proton gradient** (Figure 14–13).

The electrochemical proton gradient exerts a **proton-motive force**, which can be measured in units of millivolts (mV). Since each ΔpH of 1 pH unit has an effect equivalent to a membrane potential of about 60 mV, the total proton-motive force equals $\Delta V - 60(\Delta\text{pH})$. In a typical cell, the proton-motive force across the inner membrane of a respiring mitochondrion is about 200 mV and is made up of a membrane potential of about 140 mV and a pH gradient of about –1 pH unit.

How the Proton Gradient Drives ATP Synthesis

The electrochemical proton gradient across the inner mitochondrial membrane is used to drive ATP synthesis in the critical process of oxidative phosphorylation (Figure 14–14). This is made possible by the membrane-bound enzyme

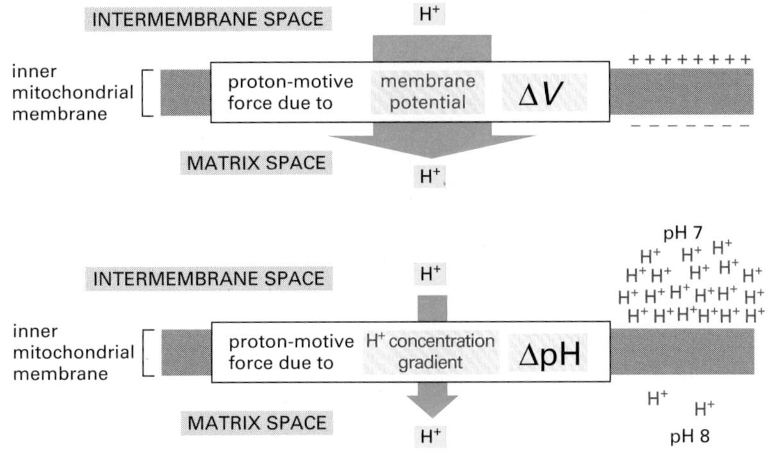

Figure 14–13 The two components of the electrochemical proton gradient. The total proton-motive force across the inner mitochondrial membrane consists of a large force due to the membrane potential (traditionally designated $\Delta\psi$ by experts, but designated ΔV in this text) and a smaller force due to the H^+ concentration gradient (ΔpH). Both forces act to drive H^+ into the matrix.

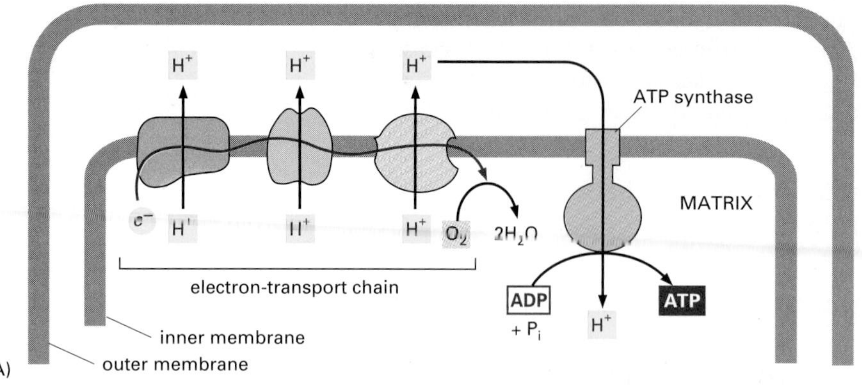

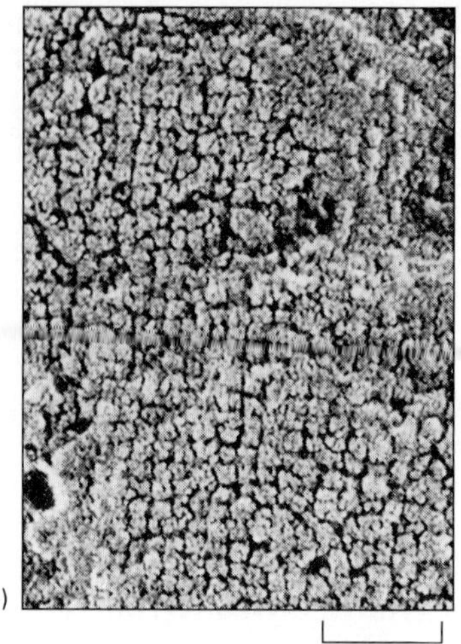

Figure 14–14 The general mechanism of oxidative phosphorylation. (A) As a high-energy electron is passed along the electron-transport chain, some of the energy released is used to drive the three respiratory enzyme complexes that pump H^+ out of the matrix. The resulting electrochemical proton gradient across the inner membrane drives H^+ back through the ATP synthase, a transmembrane protein complex that uses the energy of the H^+ flow to synthesize ATP from ADP and P_i in the matrix. (B) An electron micrograph of the inside surface of the inner mitochondrial membrane in a plant cell. Densely packed particles are visible, due to protruding portions of the ATP synthases and the respiratory enzyme complexes. (Micrograph courtesy of Brian Wells.)

ATP synthase, mentioned previously. This enzyme creates a hydrophilic pathway across the inner mitochondrial membrane that allows protons to flow down their electrochemical gradient. As these ions thread their way through the ATP synthase, they are used to drive the energetically unfavorable reaction between ADP and P_i that makes ATP (see Figure 2–27). The ATP synthase is of ancient origin; the same enzyme occurs in the mitochondria of animal cells, the chloroplasts of plants and algae, and in the plasma membrane of bacteria and archea.

The structure of ATP synthase is shown in Figure 14–15. Also called the F_0F_1 ATPase, it is a multisubunit protein with a mass of more than 500,000 daltons. A large enzymatic portion, shaped like a lollipop head and composed of a ring of 6 subunits, projects on the matrix side of the inner mitochondrial membrane. This head is held in place by an elongated arm that binds to the head, tying it to a group of transmembrane proteins that produce a "stator" in the membrane. This stator is in contact with a "rotor" that is formed by a ring of 10 to 14 identical transmembrane protein subunits. As protons pass through a narrow channel formed at the stator–rotor contact, their movement causes the rotor ring to spin. This spinning also turns a stalk attached to the rotor *(blue* in Figure 14–15B),

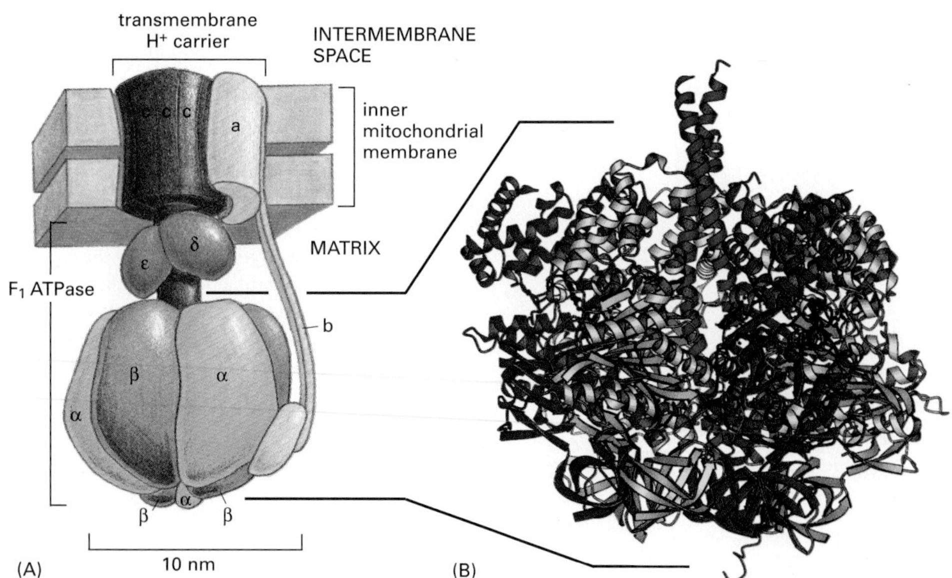

Figure 14–15 ATP synthase. (A) The enzyme is composed of a head portion, called the F_1 ATPase, and a transmembrane H^+ carrier, called F_0. Both F_1 and F_0 are formed from multiple subunits, as indicated. A rotating stalk turns with a rotor formed by a ring of 10 to 14 c subunits in the membrane *(red)*. The stator *(green)* is formed from transmembrane a subunits, tied to other subunits that create an elongated arm. This arm fixes the stator to a ring of 3α and 3β subunits that forms the head. (B) The three-dimensional structure of the F_1 ATPase, determined by x-ray crystallography. This part of the ATP synthase derives its name from its ability to carry out the reverse of the ATP synthesis reaction—namely, the hydrolysis of ATP to ADP and P_i, when detached from the transmembrane portion. (B, courtesy of John Walker, from J.P. Abrahams et al., *Nature* 370:621–628, 1994. © Macmillan Magazines Ltd.)

which is thereby made to turn rapidly inside the lollipop head. As a result, the energy of proton flow down a gradient has been converted into the mechanical energy of two sets of proteins rubbing against each other: rotating stalk proteins pushing against a stationary ring of head proteins.

Three of the six subunits in the head contain binding sites for ADP and inorganic phosphate. These are driven to form ATP as mechanical energy is converted into chemical bond energy through the repeated changes in protein conformation that the rotating stalk creates. In this way, the ATP synthase is able to produce more than 100 molecules of ATP per second. Three or four protons need to pass through this marvelous device to make each molecule of ATP.

How the Proton Gradient Drives Coupled Transport Across the Inner Membrane

The synthesis of ATP is not the only process driven by the electrochemical proton gradient. In mitochondria, many charged small molecules, such as pyruvate, ADP, and P_i, are pumped into the matrix from the cytosol, while others, such as ATP, must be moved in the opposite direction. Carrier proteins that bind these molecules can couple their transport to the energetically favorable flow of H^+ into the mitochondrial matrix. Thus, for example, pyruvate and inorganic phosphate (P_i) are co-transported inward with H^+ as the H^+ moves into the matrix.

In contrast, ADP is co-transported with ATP in opposite directions by a single carrier protein. Since an ATP molecule has one more negative charge than ADP, each nucleotide exchange results in a total of one negative charge being moved out of the mitochondrion. This ADP–ATP co-transport is thereby driven by the voltage difference across the membrane (Figure 14–16).

In eucaryotic cells, the proton gradient is thus used to drive both the formation of ATP and the transport of certain metabolites across the inner mitochondrial membrane. In bacteria, the proton gradient across the bacterial plasma membrane is harnessed for both types of functions. And in the plasma membrane of motile bacteria, the gradient also drives the rapid rotation of the bacterial flagellum, which propels the bacterium along (Figure 14–17).

Proton Gradients Produce Most of the Cell's ATP

As stated previously, glycolysis alone produces a net yield of 2 molecules of ATP for every molecule of glucose that is metabolized, and this is the total energy yield for the fermentation processes that occur in the absence of O_2 (discussed

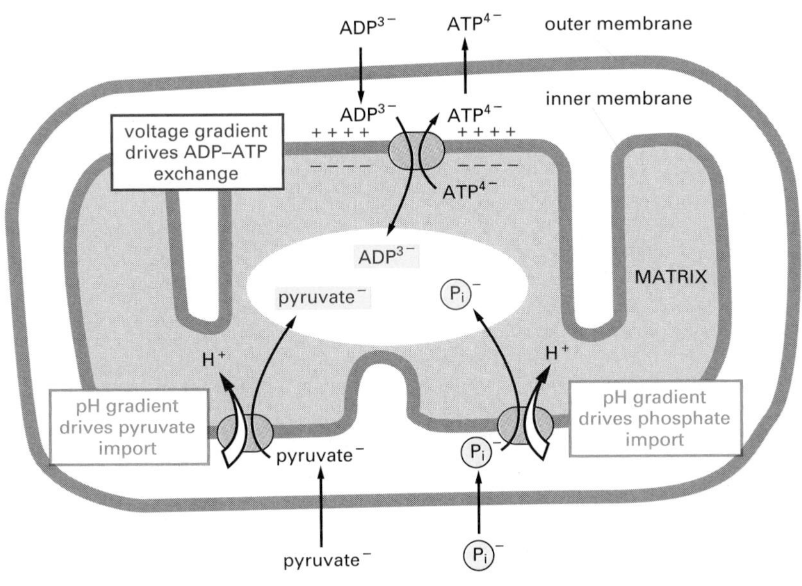

Figure 14–16 Some of the active transport processes driven by the electrochemical proton gradient across the inner mitochondrial membrane. Pyruvate, inorganic phosphate (P_i), and ADP are moved into the matrix, while ATP is pumped out. The charge on each of the transported molecules is indicated for comparison with the membrane potential, which is negative inside, as shown. The outer membrane is freely permeable to all of these compounds. The active transport of molecules across membranes by carrier proteins is discussed in Chapter 11.

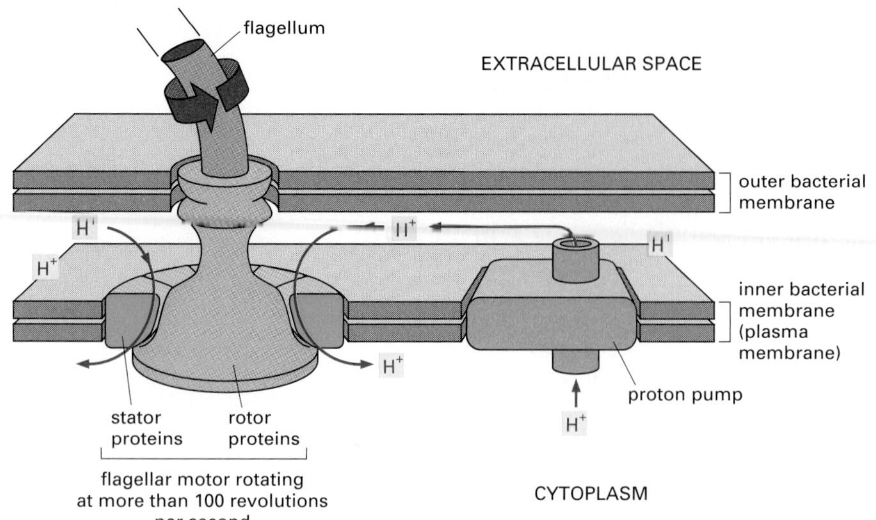

Figure 14–17 The rotation of the bacterial flagellum driven by H^+ flow. The flagellum is attached to a series of protein rings *(orange)*, which are embedded in the outer and inner membranes and rotate with the flagellum. The rotation is driven by a flow of protons through an outer ring of proteins (the stator) by mechanisms that may resemble those used by the ATP synthase, although they are not yet understood.

in Chapter 2). During oxidative phosphorylation, each pair of electrons donated by the NADH produced in mitochondria is thought to provide energy for the formation of about 2.5 molecules of ATP, after subtracting the energy needed for transporting this ATP to the cytosol. Oxidative phosphorylation also produces 1.5 ATP molecules per electron pair from $FADH_2$, or from the NADH molecules produced by glycolysis in the cytosol. From the product yields of glycolysis and the citric acid cycle summarized in Table 14–1A, one can calculate that the complete oxidation of one molecule of glucose—starting with glycolysis and ending with oxidative phosphorylation—gives a net yield of about 30 ATPs.

In conclusion, the vast majority of the ATP produced from the oxidation of glucose in an animal cell is produced by chemiosmotic mechanisms in the mitochondrial membrane. Oxidative phosphorylation in the mitochondrion also produces a large amount of ATP from the NADH and the $FADH_2$ that is derived from the oxidation of fats (Table 14–1B; see also Figure 2–78).

Mitochondria Maintain a High ATP:ADP Ratio in Cells

Because of the carrier protein in the inner mitochondrial membrane that exchanges ATP for ADP, the ADP molecules produced by ATP hydrolysis in the cytosol rapidly enter mitochondria for recharging, while the ATP molecules formed in the mitochondrial matrix by oxidative phosphorylation are rapidly

TABLE 14–1 Product Yields from the Oxidation of Sugars and Fats

A. NET PRODUCTS FROM OXIDATION OF ONE MOLECULE OF GLUCOSE

In cytosol (glycolysis)

1 glucose → 2 pyruvate + 2 NADH + 2 ATP

In mitochondrion (pyruvate dehydrogenase and citric acid cycle)

2 pyruvate → 2 acetyl CoA + 2 NADH

2 acetyl CoA → 6 NADH + 2 $FADH_2$ + 2 GTP

Net result in mitochondrion

2 pyruvate → 8 NADH + 2 $FADH_2$ + 2 GTP

B. NET PRODUCTS FROM OXIDATION OF ONE MOLECULE OF PALMITOYL COA (ACTIVATED FORM OF PALMITATE, A FATTY ACID)

In mitochondrion (fatty acid oxidation and citric acid cycle)

1 palmitoyl CoA → 8 acetyl CoA + 7 NADH + 7 $FADH_2$

8 acetyl CoA → 24 NADH + 8 $FADH_2$

Net result in mitochondrion

1 palmitoyl CoA → 31 NADH + 15 $FADH_2$

pumped into the cytosol, where they are needed. A typical ATP molecule in the human body shuttles out of a mitochondrion and back into it (as ADP) for recharging more than once per minute, keeping the concentration of ATP in the cell about 10 times higher than that of ADP.

As discussed in Chapter 2, biosynthetic enzymes often drive energetically unfavorable reactions by coupling them to the energetically favorable hydrolysis of ATP (see Figure 2–56). The ATP pool is therefore used to drive cellular processes in much the same way that a battery can be used to drive electric engines. If the activity of the mitochondria is blocked, ATP levels fall and the cell's battery runs down; eventually, energetically unfavorable reactions are no longer driven, and the cell dies. The poison cyanide, which blocks electron transport in the inner mitochondrial membrane, causes death in exactly this way.

It might seem that cellular processes would stop only when the concentration of ATP reaches zero; but, in fact, life is more demanding: it depends on cells maintaining a concentration of ATP that is high compared to the concentrations of ADP and P_i. To explain why, we must consider some elementary principles of thermodynamics.

A Large Negative Value of ΔG for ATP Hydrolysis Makes ATP Useful to the Cell

In Chapter 2, the concept of free energy (G) was discussed. The free energy change for a reaction, ΔG, determines whether this reaction will occur in a cell. We showed on p. 80 that the ΔG for a given reaction can be written as the sum of two parts: the first, called the *standard free-energy change*, $\Delta G°$, depends on the intrinsic characters of the reacting molecules; the second depends on their concentrations. For the simple reaction A → B,

$$\Delta G = \Delta G° + \mathrm{RT}\ln\frac{[\mathrm{B}]}{[\mathrm{A}]}$$

where [A] and [B] denote the concentrations of A and B, and ln is the natural logarithm. $\Delta G°$ is therefore only a reference value that equals the value of ΔG when the molar concentrations of A and B are equal (ln 1 = 0).

In Chapter 2, ATP was described as the major "activated carrier molecule" in cells. The large, favorable free-energy change (large negative ΔG) for its hydrolysis is used, through *coupled reactions*, to drive other chemical reactions that would otherwise not occur (see pp. 82–85). The ATP hydrolysis reaction produces two products, ADP and inorganic phosphate (P_i); it is therefore of the type A → B + C, where, as described in Figure 14–18,

$$\Delta G = \Delta G° + \mathrm{RT}\ln\frac{[\mathrm{B}][\mathrm{C}]}{[\mathrm{A}]}$$

When ATP is hydrolyzed to ADP and P_i under the conditions that normally exist in a cell, the free-energy change is roughly –11 to –13 kcal/mole. This extremely favorable ΔG depends on having a high concentration of ATP in the cell compared with the concentration of ADP and P_i. When ATP, ADP, and P_i are all present at the same concentration of 1 mole/liter (so-called standard conditions), the ΔG for ATP hydrolysis is the standard free-energy change ($\Delta G°$), which is only –7.3 kcal/mole. At much lower concentrations of ATP relative to ADP and P_i, ΔG becomes zero. At this point, the rate at which ADP and P_i will join to form ATP will be equal to the rate at which ATP hydrolyzes to form ADP and P_i. In other words, when $\Delta G = 0$, the reaction is at *equilibrium* (see Figure 14–18).

It is ΔG, not $\Delta G°$, that indicates how far a reaction is from equilibrium and determines whether it can be used to drive other reactions. Because the efficient conversion of ADP to ATP in mitochondria maintains such a high concentration of ATP relative to ADP and P_i, the ATP-hydrolysis reaction in cells is kept very far from equilibrium and ΔG is correspondingly very negative. Without this large disequilibrium, ATP hydrolysis could not be used to direct the reactions of the cell; for example, many biosynthetic reactions would run backward rather than forward at low ATP concentrations.

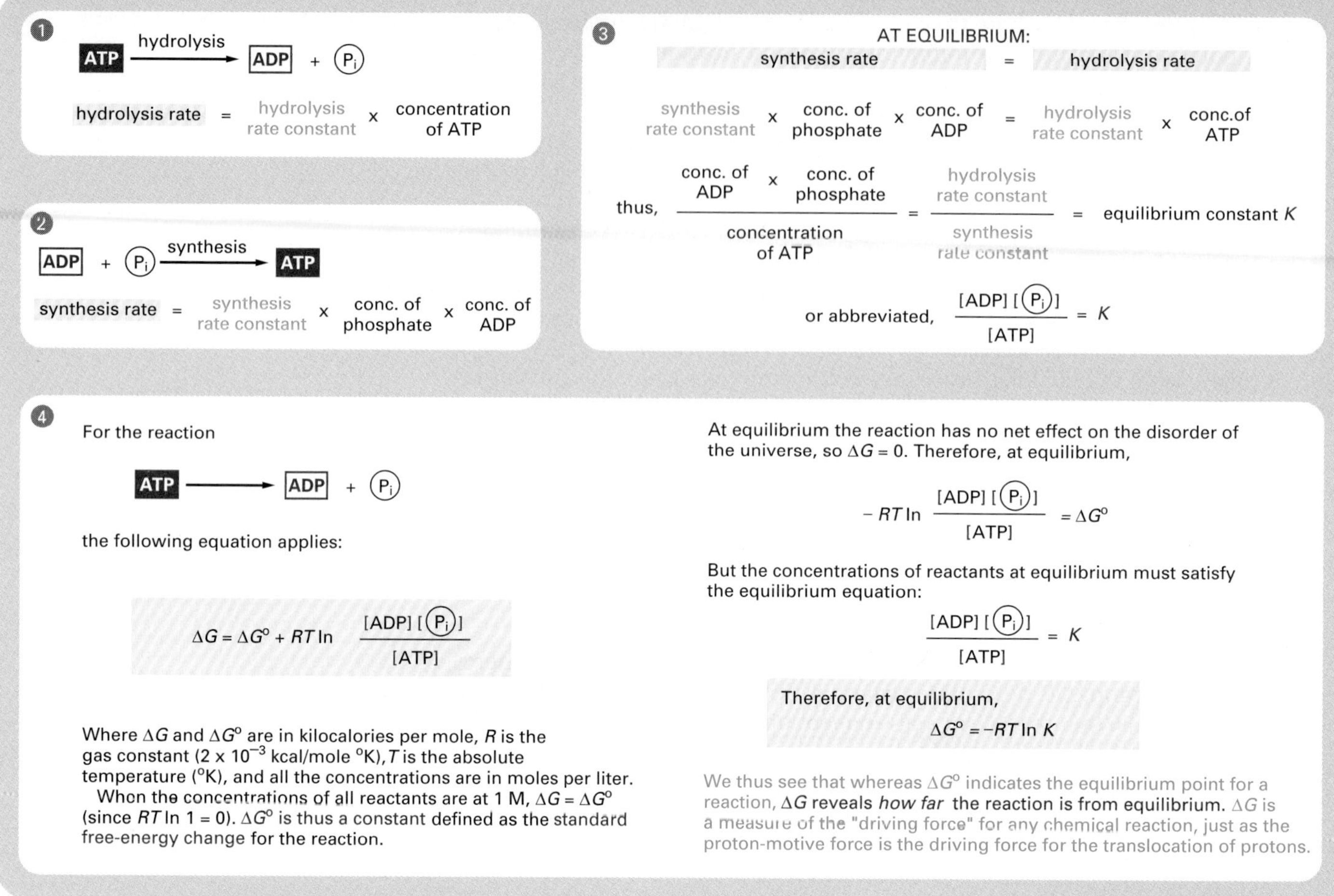

1
ATP hydrolysis ADP + Pi
hydrolysis rate = hydrolysis rate constant x concentration of ATP
2
ADP + Pi synthesis ATP
synthesis rate = synthesis rate constant x conc. of phosphate x conc. of ADP
3
AT EQUILIBRIUM:
synthesis rate = hydrolysis rate
synthesis rate constant x conc. of phosphate x conc. of ADP = hydrolysis rate constant x conc.of ATP
thus, (conc. of ADP x conc. of phosphate) / (concentration of ATP) = (hydrolysis rate constant) / (synthesis rate constant) = equilibrium constant K
or abbreviated, [ADP] [Pi] / [ATP] = K
4
For the reaction
ATP → ADP + Pi
the following equation applies:
ΔG = ΔG° + RT ln [ADP] [Pi] / [ATP]
Where ΔG and ΔG° are in kilocalories per mole, R is the gas constant (2 x 10⁻³ kcal/mole °K), T is the absolute temperature (°K), and all the concentrations are in moles per liter.
When the concentrations of all reactants are at 1 M, ΔG = ΔG° (since RT ln 1 = 0). ΔG° is thus a constant defined as the standard free-energy change for the reaction.
At equilibrium the reaction has no net effect on the disorder of the universe, so ΔG = 0. Therefore, at equilibrium,
– RT ln [ADP] [Pi] / [ATP] = ΔG°
But the concentrations of reactants at equilibrium must satisfy the equilibrium equation:
[ADP] [Pi] / [ATP] = K
Therefore, at equilibrium,
ΔG° = –RT ln K
We thus see that whereas ΔG° indicates the equilibrium point for a reaction, ΔG reveals how far the reaction is from equilibrium. ΔG is a measure of the "driving force" for any chemical reaction, just as the proton-motive force is the driving force for the translocation of protons.

Figure 14–18 The basic relationship between free-energy changes and equilibrium in the ATP hydrolysis reaction. The rate constants in boxes 1 and 2 are determined from experiments in which product accumulation is measured as a function of time. The equilibrium constant shown here, K, is in units of moles per liter. (See Panel 2–7, pp. 122–123, for a discussion of free energy and p. 160 for a discussion of the equilibrium constant.)

ATP Synthase Can Also Function in Reverse to Hydrolyze ATP and Pump H^+

In addition to harnessing the flow of H^+ down an electrochemical proton gradient to make ATP, the ATP synthase can work in reverse: it can use the energy of ATP hydrolysis to pump H^+ across the inner mitochondrial membrane (Figure 14–19). It thus acts as a *reversible coupling device,* interconverting electrochemical proton gradient and chemical bond energies. The direction of action at any instant depends on the balance between the steepness of the electrochemical proton gradient and the local ΔG for ATP hydrolysis, as we now explain.

Although the exact number of protons needed to make each ATP molecule is not known with certainty, we shall assume that one molecule of ATP is made by the ATP synthase for every 3 protons driven through it. Whether the ATP synthase works in its ATP-synthesizing or its ATP-hydrolyzing direction at any instant depends, in this case, on the exact balance between the favorable free-energy change for moving the three protons across the membrane into the matrix ΔG_{3H+} (which is less than zero) and the unfavorable free-energy change for ATP *synthesis* in the matrix $\Delta G_{\text{ATP synthesis}}$ (which is greater than zero). As just discussed, the value of $\Delta G_{\text{ATP synthesis}}$ depends on the exact concentrations of the three reactants ATP, ADP, and P_i in the mitochondrial matrix (see Figure 14–18). The value of ΔG_{3H+} in contrast, is directly proportional to the value of the proton-motive force across the inner mitochondrial membrane. The following example will help explain how the balance between these two free-energy changes affects the ATP synthase.

As explained in the legend to Figure 14–19, a single H^+ moving into the matrix down an electrochemical gradient of 200 mV liberates 4.6 kcal/mole of free energy, while the movement of three protons liberates three times this much free energy (ΔG_{3H+} = –13.8 kcal/mole). Thus, if the proton-motive force remains constant at 200 mV, the ATP synthase synthesizes ATP until a ratio of ATP to ADP

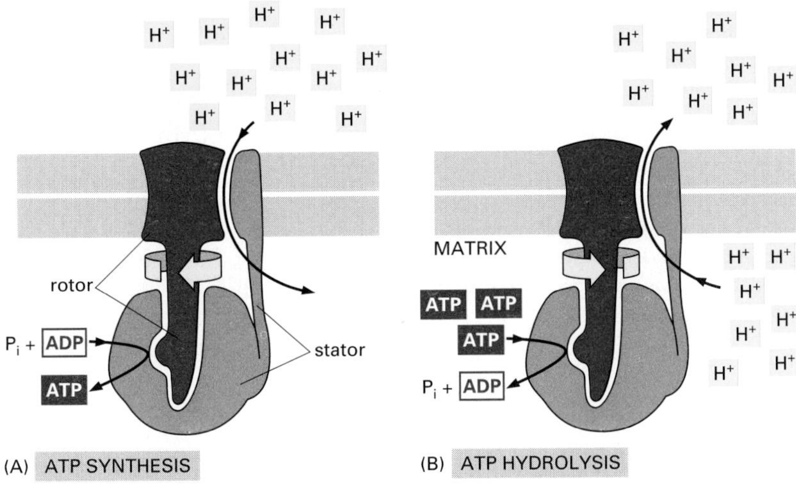

Figure 14–19 The ATP synthase is a reversible coupling device that can convert the energy of the electrochemical proton gradient into chemical-bond energy, or vice versa. The ATP synthase can either (A) synthesize ATP by harnessing the proton-motive force or (B) pump protons against their electrochemical gradient by hydrolyzing ATP. As explained in the text, the direction of operation at any given instant depends on the net free-energy change (ΔG) for the coupled processes of H^+ translocation across the membrane and the synthesis of ATP from ADP and P_i. Measurement of the torque that the ATP synthase can produce when hydrolyzing ATP reveals that the synthase can pump 60 times more strongly than a diesel-engine of equal weight.

We have previously shown how the free-energy change (ΔG) for ATP hydrolysis depends on the concentrations of the three reactants ATP, ADP, and P_i (see Figure 14–18); the ΔG for ATP synthesis is the negative of this value. The ΔG for proton translocation across the membrane is proportional to the proton-motive force. The conversion factor between them is the faraday. Thus, $\Delta G_{H^+} = -0.023$ (proton-motive force), where ΔG_{H^+} is in kcal/mole and the proton-motive force is in mV. For an electrochemical proton gradient (proton-motive force) of 200 mV, $\Delta G_{H^+} = -4.6$ kcal/mole.

and P_i is reached where $\Delta G_{\text{ATP synthesis}}$ is just equal to +13.8 kcal/mole (here $\Delta G_{\text{ATP synthesis}} + \Delta G_{3H^+} = 0$). At this point there is no further net ATP synthesis or hydrolysis by the ATP synthase.

Suppose a large amount of ATP is suddenly hydrolyzed by energy-requiring reactions in the cytosol, causing the ATP:ADP ratio in the matrix to fall. Now the value of $\Delta G_{\text{ATP synthesis}}$ will decrease (see Figure 14–18), and ATP synthase will begin to synthesize ATP again to restore the original ATP:ADP ratio. Alternatively, if the proton-motive force drops suddenly and is then maintained at a constant 160 mV, ΔG_{3H^+} will change to –11.0 kcal/mole. As a result, ATP synthase will start hydrolyzing some of the ATP in the matrix until a new balance of ATP to ADP and P_i is reached (where $\Delta G_{\text{ATP synthesis}} = +11.0$ kcal/mole), and so on.

In many bacteria, ATP synthase is routinely reversed in a transition between aerobic and anaerobic metabolism, as we shall see later. The same type of reversibility is shared by other membrane transport proteins that couple the transmembrane movement of an ion to ATP synthesis or hydrolysis. Both the Na^+-K^+ pump and the Ca^{2+} pump described in Chapter 11, for example, normally hydrolyze ATP and use the energy released to move their specific ions across a membrane. If either of these pumps is exposed to an abnormally steep gradient of the ions it transports, however, it will act in reverse—synthesizing ATP from ADP and P_i instead of hydrolyzing it. Thus, the ATP synthase is by no means unique in its ability to convert the electrochemical energy stored in a transmembrane ion gradient directly into phosphate-bond energy in ATP.

Summary

The mitochondrion performs most cellular oxidations and produces the bulk of the animal cell's ATP. The mitochondrial matrix contains a large variety of enzymes, including those that convert pyruvate and fatty acids to acetyl CoA and those that oxidize this acetyl CoA to CO_2 through the citric acid cycle. Large amounts of NADH (and $FADH_2$) are produced by these oxidation reactions.

The energy available from combining molecular oxygen with the reactive electrons carried by NADH and $FADH_2$ is harnessed by an electron-transport chain in the inner mitochondrial membrane called the respiratory chain. The respiratory chain pumps H^+ out of the matrix to create a transmembrane electrochemical proton (H^+) gradient, which includes contributions from both a membrane potential and a pH difference. The large amount of free energy released when H^+ flows back into the matrix (across the inner membrane) provides the basis for ATP production in the matrix by a remarkable protein machine—the ATP synthase. The transmembrane electrochemical gradient is also used to drive the active transport of selected metabolites across the mitochondrial inner membrane, including an efficient ATP–ADP exchange between the mitochondrion and the cytosol that keeps the cell's ATP pool highly charged. The resulting high ratio of ATP to its hydrolysis products makes the free-energy change for ATP hydrolysis extremely favorable, allowing this hydrolysis reaction to drive a large number of the cell's energy-requiring processes.

Figure 14–20 How protons behave in water. (A) Protons move very rapidly along a chain of hydrogen-bonded water molecules. In this diagram, proton jumps are indicated by *blue arrows*, and hydronium ions are indicated by *green shading*. As discussed in Chapter 2, naked protons rarely exist as such; they are instead associated with a water molecule in the form of a hydronium ion, H_3O^+. At a neutral pH (pH 7.0), the hydronium ions are present at a concentration of 10^{-7} M. However, for simplicity, one usually refers to this as an H^+ concentration of 10^{-7} M (see Panel 2–2, pp. 112–113). (B) Electron transfer can result in the transfer of entire hydrogen atoms, because protons are readily accepted from or donated to water inside cells. In this example, A picks up an electron plus a proton when it is reduced, and B loses an electron plus a proton when it is oxidized.

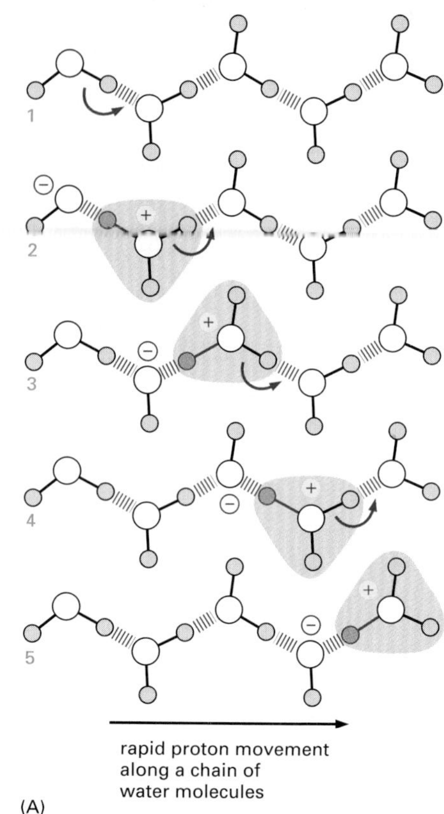

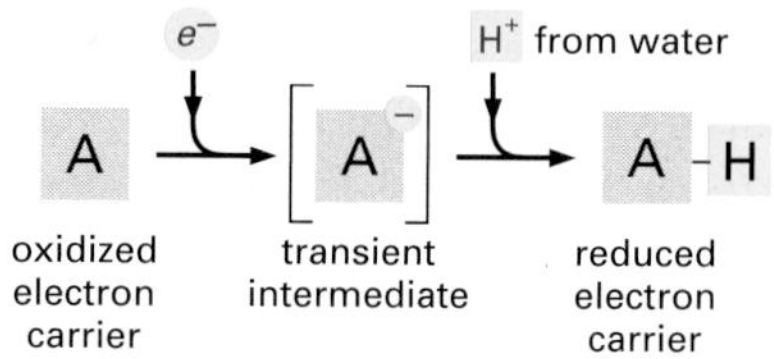

ELECTRON-TRANSPORT CHAINS AND THEIR PROTON PUMPS

Having considered in general terms how a mitochondrion uses electron transport to create an electrochemical proton gradient, we need to examine the mechanisms that underlie this membrane-based energy-conversion process. In doing so, we also accomplish a larger purpose. As emphasized at the beginning of this chapter, very similar chemiosmotic mechanisms are used by mitochondria, chloroplasts, archea, and bacteria. In fact, these mechanisms underlie the function of nearly all living organisms—including anaerobes that derive all their energy from electron transfers between two inorganic molecules. It is therefore rather humbling for scientists to remind themselves that the existence of chemiosmosis has been recognized for only about 40 years.

We begin with a look at some of the principles that underlie the electron-transport process, with the aim of explaining how it can pump protons across a membrane.

Protons Are Unusually Easy to Move

Although protons resemble other positive ions such as Na^+ and K^+ in their movement across membranes, in some respects they are unique. Hydrogen atoms are by far the most abundant type of atom in living organisms; they are plentiful not only in all carbon-containing biological molecules, but also in the water molecules that surround them. The protons in water are highly mobile, flickering through the hydrogen-bonded network of water molecules by rapidly dissociating from one water molecule to associate with its neighbor, as illustrated in Figure 14–20A. Protons are thought to move across a protein pump embedded in a lipid bilayer in a similar way: they transfer from one amino acid side chain to another, following a special channel through the protein.

Protons are also special with respect to electron transport. Whenever a molecule is reduced by acquiring an electron, the electron (e^-) brings with it a negative charge. In many cases, this charge is rapidly neutralized by the addition of a proton (H^+) from water, so that the net effect of the reduction is to transfer an entire hydrogen atom, $H^+ + e^-$ (Figure 14–20B). Similarly, when a molecule is oxidized, a hydrogen atom removed from it can be readily dissociated into its constituent electron and proton—allowing the electron to be transferred separately to a molecule that accepts electrons, while the proton is passed to the water. Therefore, in a membrane in which electrons are being passed along an electron-transport chain, pumping protons from one side of the membrane to another can be relatively simple. The electron carrier merely needs to be arranged in the membrane in a way that causes it to pick up a proton from one side of the membrane when it accepts an electron, and to release the proton on the other side of the membrane as the electron is passed to the next carrier molecule in the chain (Figure 14–21).

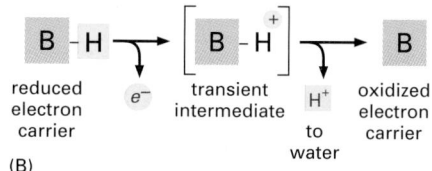

The Redox Potential Is a Measure of Electron Affinities

In biochemical reactions, any electrons removed from one molecule are always passed to another, so that whenever one molecule is oxidized, another is reduced. Like any other chemical reaction, the tendency of such oxidation–reduction reactions, or **redox reactions**, to proceed spontaneously depends on the free-energy change (ΔG) for the electron transfer, which in turn depends on the relative affinities of the two molecules for electrons.

Because electron transfers provide most of the energy for living things, it is worth spending the time to understand them. Many readers are already familiar with acids and bases, which donate and accept protons (see Panel 2–2, pp. 112–113). Acids and bases exist in conjugate acid–base pairs, in which the acid is readily converted into the base by the loss of a proton. For example, acetic acid (CH_3COOH) is converted into its conjugate base (CH_3COO^-) in the reaction:

$$CH_3COOH \rightleftharpoons CH_3COO^- + H^+$$

In exactly the same way, pairs of compounds such as NADH and NAD^+ are called **redox pairs**, since NADH is converted to NAD^+ by the loss of electrons in the reaction:

$$NADH \rightleftharpoons NAD^+ + H^+ + 2e^-$$

NADH is a strong electron donor: because its electrons are held in a high-energy linkage, the free-energy change for passing its electrons to many other molecules is favorable (see Figure 14–9). It is difficult to form a high-energy linkage. Therefore its redox partner, NAD^+, is of necessity a weak electron acceptor.

The tendency to transfer electrons from any redox pair can be measured experimentally. All that is required is the formation of an electrical circuit linking a 1:1 (equimolar) mixture of the redox pair to a second redox pair that has been arbitrarily selected as a reference standard, so the voltage difference can be measured between them (Panel 14–1, p. 784). This voltage difference is defined as the **redox potential**; as defined, electrons move spontaneously from a redox pair like $NADH/NAD^+$ with a low redox potential (a low affinity for electrons) to a redox pair like O_2/H_2O with a high redox potential (a high affinity for electrons). Thus, NADH is a good molecule for donating electrons to the respiratory chain, while O_2 is well suited to act as the "sink" for electrons at the end of the pathway. As explained in Panel 14–1, the difference in redox potential, $\Delta E_0'$, is a direct measure of the standard free-energy change ($\Delta G°$) for the transfer of an electron from one molecule to another.

Electron Transfers Release Large Amounts of Energy

As just discussed, those pairs of compounds that have the most negative redox potentials have the weakest affinity for electrons and therefore contain carriers with the strongest tendency to donate electrons. Conversely, those pairs that have the most positive redox potentials have the strongest affinity for electrons and therefore contain carriers with the strongest tendency to accept electrons. A 1:1 mixture of NADH and NAD^+ has a redox potential of –320 mV, indicating that NADH has a strong tendency to donate electrons; a 1:1 mixture of H_2O and $\frac{1}{2}O_2$ has a redox potential of +820 mV, indicating that O_2 has a strong tendency to accept electrons. The difference in redox potential is 1.14 volts (1140 mV), which means that the transfer of each electron from NADH to O_2 under these standard conditions is enormously favorable, where $\Delta G° = -26.2$ kcal/mole (–52.4 kcal/mole for the two electrons transferred per NADH molecule; see Panel 14–1). If we compare this free-energy change with that for the formation of the phosphoanhydride bonds in ATP ($\Delta G° = -7.3$ kcal/mole, see Figure 2–75), we see that more than enough energy is released by the oxidization of one NADH molecule to synthesize several molecules of ATP from ADP and P_i.

Living systems could certainly have evolved enzymes that would allow NADH to donate electrons directly to O_2 to make water in the reaction:

$$2H^+ + 2e^- + \tfrac{1}{2}O_2 \rightarrow H_2O$$

But because of the huge free-energy drop, this reaction would proceed with almost explosive force and nearly all of the energy would be released as heat.

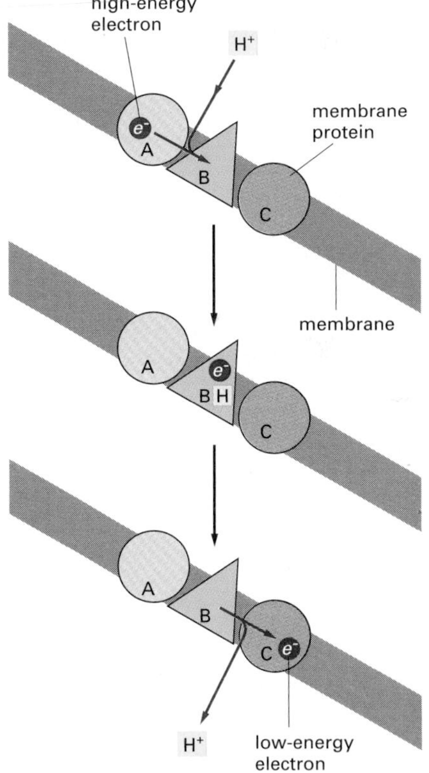

Figure 14–21 How protons can be pumped across membranes. As an electron passes along an electron-transport chain embedded in a lipid-bilayer membrane, it can bind and release a proton at each step. In this diagram, electron carrier B picks up a proton (H^+) from one side of the membrane when it accepts an electron (e^-) from carrier A; it releases the proton to the other side of the membrane when it donates its electron to carrier C.

HOW REDOX POTENTIALS ARE MEASURED

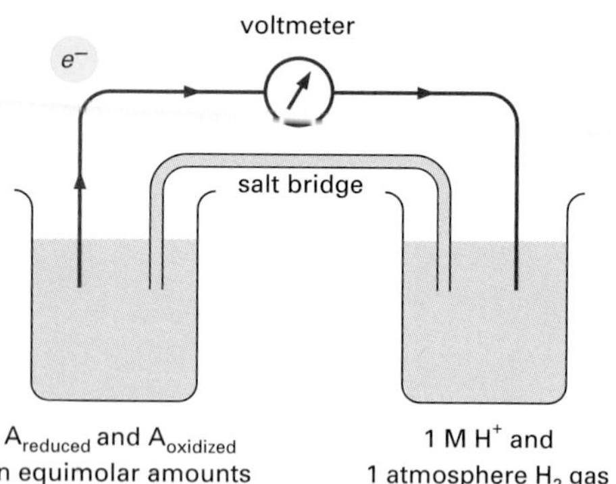

One beaker (*left*) contains substance A, with an equimolar mixture of the reduced ($A_{reduced}$) and oxidized ($A_{oxidized}$) members of its redox pair. The other beaker contains the hydrogen reference standard ($2H^+ + 2e^- \rightleftharpoons H_2$), whose redox potential is arbitrarily assigned as zero by international agreement. (A salt bridge formed from a concentrated KCl solution allows the ions K^+ and Cl^- to move between the two beakers, as required to neutralize the charges in each beaker when electrons flow between them.) The metal wire (*red*) provides a resistance-free path for electrons, and a voltmeter then measures the redox potential of substance A. If electrons flow from $A_{reduced}$ to H^+, as indicated here, the redox pair formed by substance A is said to have a negative redox potential. If they instead flow from H_2 to $A_{oxidized}$, the redox pair is said to have a positive redox potential.

SOME STANDARD REDOX POTENTIALS AT pH7

By convention, the redox potential for a redox pair is designated *E*. For the standard state, with all reactants at a concentration of 1 M, including H^+, one can determine a standard redox potential, designated E_0. Since biological reactions occur at pH 7, biologists use a different standard state in which $A_{reduced} = A_{oxidized}$ and $H^+ = 10^{-7}$ M. This standard redox potential is designated E'_0. A few examples of special relevance to oxidative phosphorylation are given here.

redox reactions	redox potential E'_0
$NADH \rightleftharpoons NAD^+ + H^+ + 2e^-$	–320 mV
reduced ubiquinone $\rightleftharpoons$ oxidized ubiquinone $+ 2H^+ + 2e^-$	+30 mV
reduced cytochrome c $\rightleftharpoons$ oxidized cytochrome c $+ e^-$	+230 mV
$H_2O \rightleftharpoons \frac{1}{2}O_2 + 2H^+ + 2e^-$	+820 mV

CALCULATION OF $\Delta G°$ FROM REDOX POTENTIALS

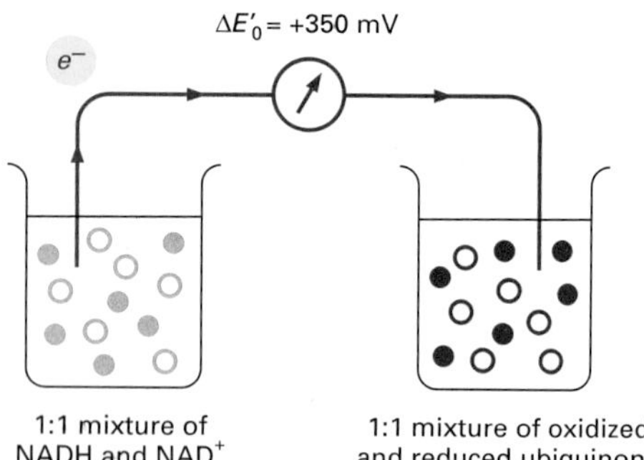

$\Delta G° = -8$ kcal/mole

$\Delta G° = -n(0.023)\ \Delta E'_0$, where *n* is the number of electrons transferred across a redox potential change of $\Delta E'_0$ millivolts (mV)

Example: The transfer of one electron from NADH to ubiquinone has a favorable $\Delta G°$ of –8.0 kcal/mole, whereas the transfer of one electron from ubiquinone to oxygen has an even more favorable $\Delta G°$ of –18.2 kcal/mole. The $\Delta G°$ value for the transfer of one electron from NADH to oxygen is the sum of these two values, –26.2 kcal/mole.

THE EFFECT OF CONCENTRATION CHANGES

The actual free-energy change for a reaction, ΔG, depends on the concentration of the reactants and generally is different from the standard free-energy change, $\Delta G°$. The standard redox potentials are for a 1:1 mixture of the redox pair. For example, the standard redox potential of –320 mV is for a 1:1 mixture of NADH and NAD^+. But when there is an excess of NADH over NAD^+, electron transfer from NADH to an electron acceptor becomes more favorable. This is reflected by a more negative redox potential and a more negative ΔG for electron transfer.

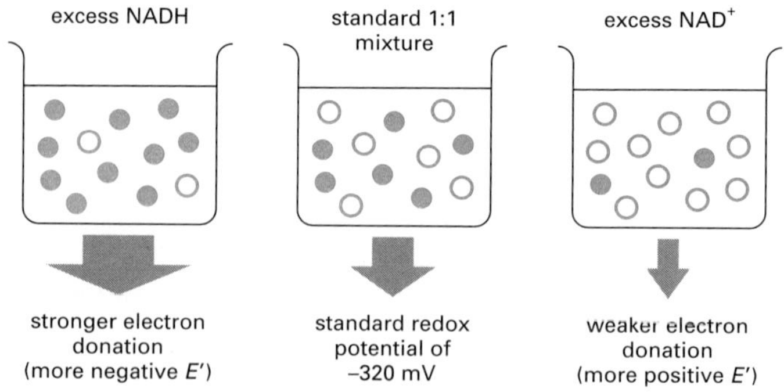

Cells do perform this reaction, but they make it proceed much more gradually by passing the high-energy electrons from NADH to O_2 via the many electron carriers in the electron-transport chain. Since each successive carrier in the chain holds its electrons more tightly, the highly energetically favorable reaction $2H^+ + 2e^- + \frac{1}{2}O_2 \rightarrow H_2O$ is made to occur in many small steps. This enables nearly half of the released energy to be stored, instead of being lost to the environment as heat.

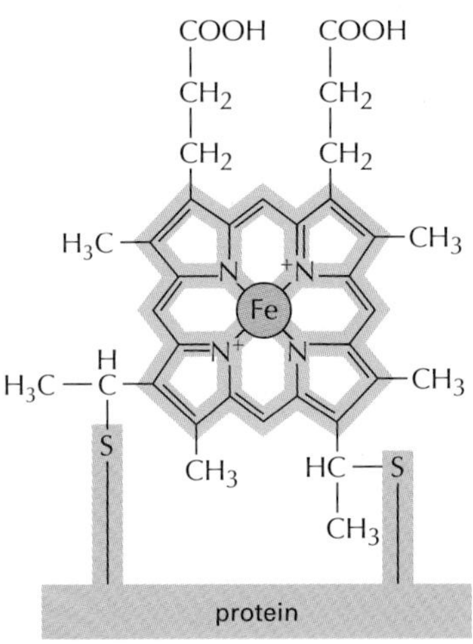

Figure 14–22 The structure of the heme group attached covalently to cytochrome c. The porphyrin ring is shown in *blue*. There are five different cytochromes in the respiratory chain. Because the hemes in different cytochromes have slightly different structures and are held by their respective proteins in different ways, each of the cytochromes has a different affinity for an electron.

Spectroscopic Methods Have Been Used to Identify Many Electron Carriers in the Respiratory Chain

Many of the electron carriers in the respiratory chain absorb visible light and change color when they are oxidized or reduced. In general, each has an absorption spectrum and reactivity that are distinct enough to allow its behavior to be traced spectroscopically, even in crude mixtures. It was therefore possible to purify these components long before their exact functions were known. Thus, the **cytochromes** were discovered in 1925 as compounds that undergo rapid oxidation and reduction in living organisms as disparate as bacteria, yeasts, and insects. By observing cells and tissues with a spectroscope, three types of cytochromes were identified by their distinctive absorption spectra and designated cytochromes *a, b,* and *c.* This nomenclature has survived, even though cells are now known to contain several cytochromes of each type and the classification into types is not functionally important.

The cytochromes constitute a family of colored proteins that are related by the presence of a bound *heme group,* whose iron atom changes from the ferric oxidation state (Fe^{3+}) to the ferrous oxidation state (Fe^{2+}) whenever it accepts an electron. The heme group consists of a *porphyrin ring* with a tightly bound iron atom held by four nitrogen atoms at the corners of a square (Figure 14–22). A similar porphyrin ring is responsible for the red color of blood and for the green color of leaves, being bound to iron in hemoglobin and to magnesium in chlorophyll, respectively.

Iron–sulfur proteins are a second major family of electron carriers. In these proteins, either two or four iron atoms are bound to an equal number of sulfur atoms and to cysteine side chains, forming an **iron–sulfur center** on the protein (Figure 14–23). There are more iron–sulfur centers than cytochromes in the respiratory chain. But their spectroscopic detection requires electron spin resonance (ESR) spectroscopy, and they are less completely characterized. Like the cytochromes, these centers carry one electron at a time.

The simplest of the electron carriers in the respiratory chain—and the only one that is not part of a protein—is a small hydrophobic molecule that is freely mobile in the lipid bilayer known as *ubiquinone,* or *coenzyme Q.* A **quinone (Q)** can pick up or donate either one or two electrons; upon reduction, it picks up a proton from the medium along with each electron it carries (Figure 14–24).

In addition to six different hemes linked to cytochromes, more than seven iron–sulfur centers, and ubiquinone, there are also two copper atoms and a flavin serving as electron carriers tightly bound to respiratory-chain proteins in the pathway from NADH to oxygen. This pathway involves more than 60 different proteins in all.

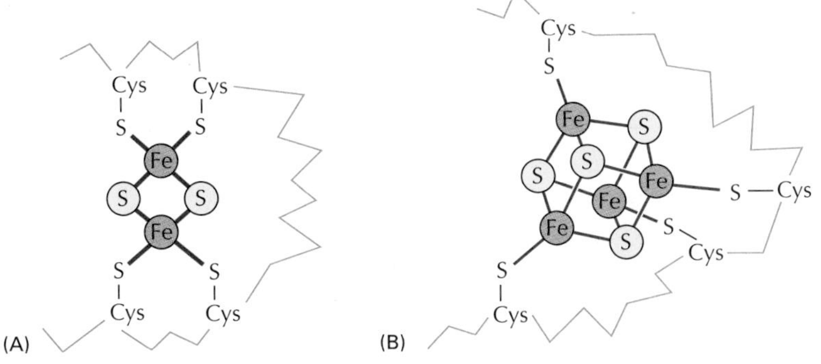

Figure 14–23 The structures of two types of iron–sulfur centers. (A) A center of the 2Fe2S type. (B) A center of the 4Fe4S type. Although they contain multiple iron atoms, each iron–sulfur center can carry only one electron at a time. There are more than seven different iron–sulfur centers in the respiratory chain.

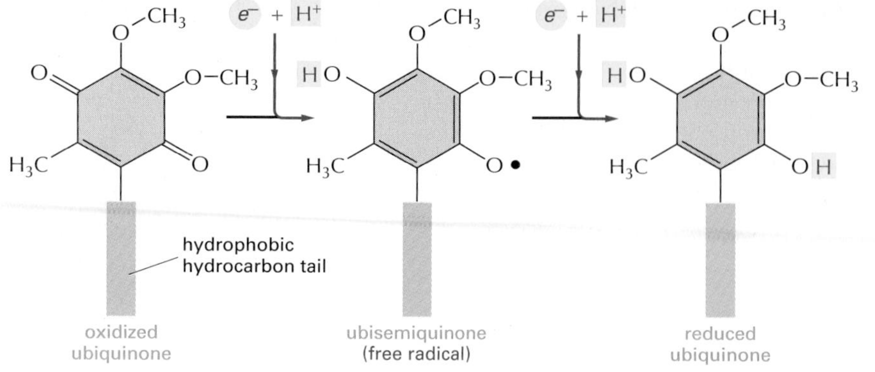

Figure 14–24 Quinone electron carriers. Ubiquinone in the respiratory chain picks up one H^+ from the aqueous environment for every electron it accepts, and it can carry either one or two electrons as part of a hydrogen atom *(yellow)*. When reduced ubiquinone donates its electrons to the next carrier in the chain, these protons are released. A long hydrophobic tail confines ubiquinone to the membrane and consists of 6–10 five-carbon isoprene units, the number depending on the organism. The corresponding electron carrier in the photosynthetic membranes of chloroplasts is plastoquinone, which is almost identical in structure. For simplicity, both ubiquinone and plastoquinone are referred to in this chapter as quinone (abbreviated as Q).

As one would expect, the electron carriers have higher and higher affinities for electrons (greater redox potentials) as one moves along the respiratory chain. The redox potentials have been fine-tuned during evolution by the binding of each electron carrier in a particular protein context, which can alter its normal affinity for electrons. However, because iron–sulfur centers have a relatively low affinity for electrons, they predominate in the early part of the respiratory chain; in contrast, the cytochromes predominate further down the chain, where a higher affinity for electrons is required.

The order of the individual electron carriers in the chain was determined by sophisticated spectroscopic measurements (Figure 14–25), and many of the proteins were initially isolated and characterized as individual polypeptides. A major advance in understanding the respiratory chain, however, was the later realization that most of the proteins are organized into three large enzyme complexes.

The Respiratory Chain Includes Three Large Enzyme Complexes Embedded in the Inner Membrane

Membrane proteins are difficult to purify as intact complexes because they are insoluble in aqueous solutions, and some of the detergents required to solubilize them can destroy normal protein–protein interactions. In the early 1960s, however, it was found that relatively mild ionic detergents, such as deoxycholate, can solubilize selected components of the inner mitochondrial membrane in their native form. This permitted the identification and purification of the three major membrane-bound **respiratory enzyme complexes** in the pathway from NADH to oxygen (Figure 14–26). As we shall see in this section, each of these complexes acts as an electron-transport-driven H^+ pump; however, they were initially characterized in terms of the electron carriers that they interact with and contain:

1. The **NADH dehydrogenase complex** (generally known as complex I) is the largest of the respiratory enzyme complexes, containing more than 40 polypeptide chains. It accepts electrons from NADH and passes them through a flavin and at least seven iron–sulfur centers to ubiquinone.

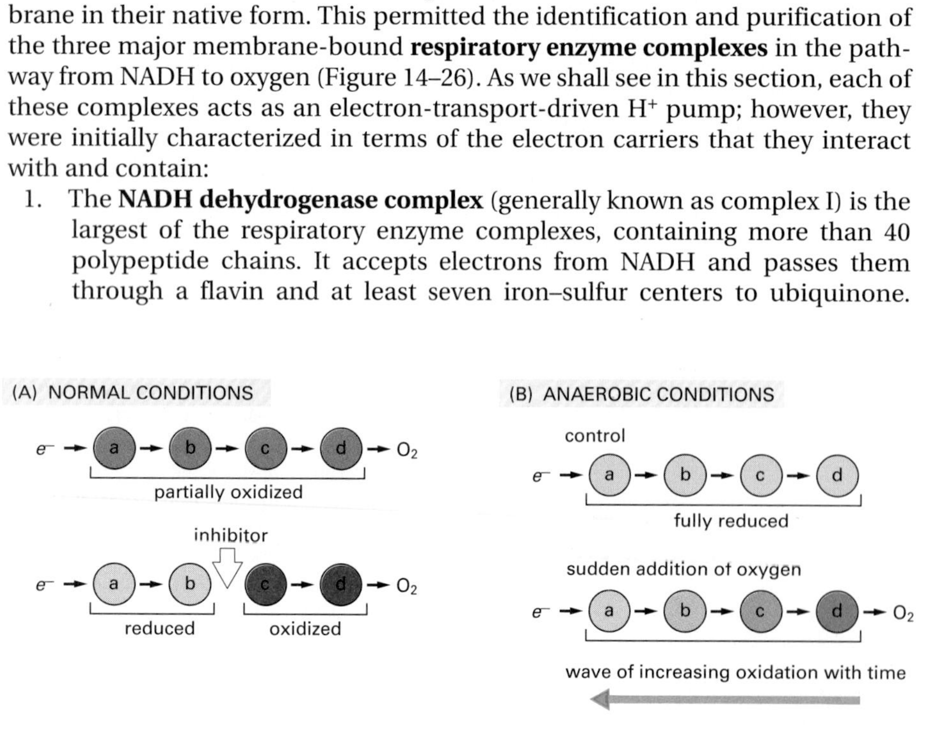

Figure 14–25 The general methods used to determine the path of electrons along an electron-transport chain. The extent of oxidation of electron carriers a, b, c, and d is continuously monitored by following their distinct spectra, which differ in their oxidized and reduced states. In this diagram an increased degree of oxidation is indicated by a *darker red*. (A) Under normal conditions, where oxygen is abundant, all carriers are in a partly oxidized state. The addition of a specific inhibitor causes the downstream carriers to become more oxidized *(red)* and the upstream carriers to become more reduced. (B) In the absence of oxygen, all carriers are in their fully reduced state *(gray)*. The sudden addition of oxygen converts each carrier to its partly oxidized form with a delay that is greatest for the most upstream carriers.

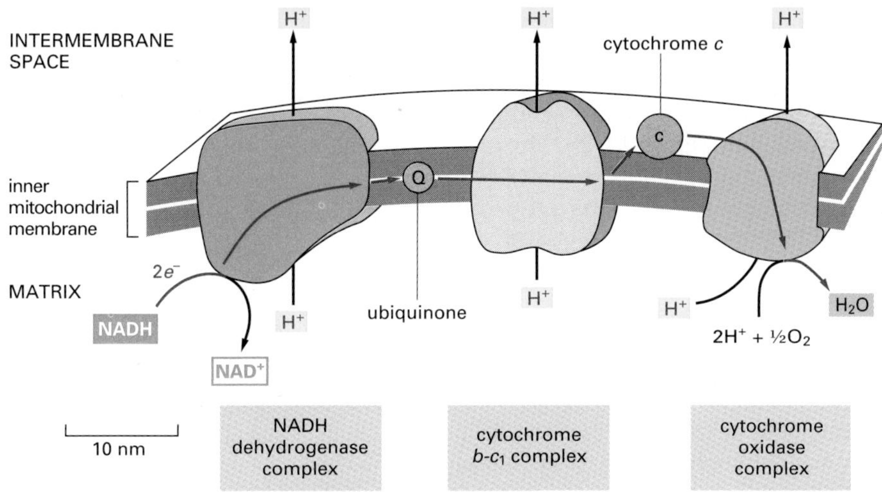

Figure 14–26 The path of electrons through the three respiratory enzyme complexes. The relative size and shape of each complex are shown. During the transfer of electrons from NADH to oxygen *(red lines)*, ubiquinone and cytochrome *c* serve as mobile carriers that ferry electrons from one complex to the next. As indicated, protons are pumped across the membrane by each of the respiratory enzyme complexes.

Ubiquinone then transfers its electrons to a second respiratory enzyme complex, the cytochrome b-c_1 complex.

2. The **cytochrome b-c_1 complex** contains at least 11 different polypeptide chains and functions as a dimer. Each monomer contains three hemes bound to cytochromes and an iron–sulfur protein. The complex accepts electrons from ubiquinone and passes them on to cytochrome *c*, which carries its electron to the cytochrome oxidase complex.
3. The **cytochrome oxidase complex** also functions as a dimer; each monomer contains 13 different polypeptide chains, including two cytochromes and two copper atoms. The complex accepts one electron at a time from cytochrome *c* and passes them four at a time to oxygen.

The cytochromes, iron–sulfur centers, and copper atoms can carry only one electron at a time. Yet each NADH donates two electrons, and each O_2 molecule must receive four electrons to produce water. There are several electron-collecting and electron-dispersing points along the electron-transport chain where these changes in electron number are accommodated. The most obvious of these is cytochrome oxidase.

An Iron–Copper Center in Cytochrome Oxidase Catalyzes Efficient O_2 Reduction

Because oxygen has a high affinity for electrons, it releases a large amount of free energy when it is reduced to form water. Thus, the evolution of cellular respiration, in which O_2 is converted to water, enabled organisms to harness much more energy than can be derived from anaerobic metabolism. This is presumably why all higher organisms respire. The ability of biological systems to use O_2 in this way, however, requires a very sophisticated chemistry. We can tolerate O_2 in the air we breathe because it has trouble picking up its first electron; this fact allows its initial reaction in cells to be controlled closely by enzymatic catalysis. But once a molecule of O_2 has picked up one electron to form a superoxide radical (O_2^-), it becomes dangerously reactive and rapidly takes up an additional three electrons wherever it can find them. The cell can use O_2 for respiration only because cytochrome oxidase holds onto oxygen at a special bimetallic center, where it remains clamped between a heme-linked iron atom and a copper atom until it has picked up a total of four electrons. Only then can the two oxygen atoms of the oxygen molecule be safely released as two molecules of water (Figure 14–27).

The cytochrome oxidase reaction is estimated to account for 90% of the total oxygen uptake in most cells. This protein complex is therefore crucial for all aerobic life. Cyanide and azide are extremely toxic because they bind tightly to the cell's cytochrome oxidase complexes to stop electron transport, thereby greatly reducing ATP production.

Although the cytochrome oxidase in mammals contains 13 different protein subunits, most of these seem to have a subsidiary role, helping to regulate either the activity or the assembly of the three subunits that form the core of the enzyme. The complete structure of this large enzyme complex has recently been determined by x-ray crystallography, as illustrated in Figure 14–28. The atomic resolution structures, combined with mechanistic studies of the effect of precisely tailored mutations introduced into the enzyme by genetic engineering of the yeast and bacterial proteins, are revealing the detailed mechanisms of this finely tuned protein machine.

Electron Transfers Are Mediated by Random Collisions in the Inner Mitochondrial Membrane

The two components that carry electrons between the three major enzyme complexes of the respiratory chain—ubiquinone and cytochrome *c*—diffuse rapidly in the plane of the inner mitochondrial membrane. The expected rate of random collisions between these mobile carriers and the more slowly diffusing enzyme complexes can account for the observed rates of electron transfer (each complex donates and receives an electron about once every 5–20 milliseconds). Thus, there is no need to postulate a structurally ordered chain of electron-transfer proteins in the lipid bilayer; indeed, the three enzyme complexes seem to exist as independent entities in the plane of the inner membrane, being present in different ratios in different mitochondria.

The ordered transfer of electrons along the respiratory chain is due entirely to the specificity of the functional interactions between the components of the chain: each electron carrier is able to interact only with the carrier adjacent to it in the sequence shown in Figure 14–26, with no short circuits.

Electrons move between the molecules that carry them in biological systems not only by moving along covalent bonds within a molecule, but also by jumping across a gap as large as 2 nm. The jumps occur by electron "tunneling," a

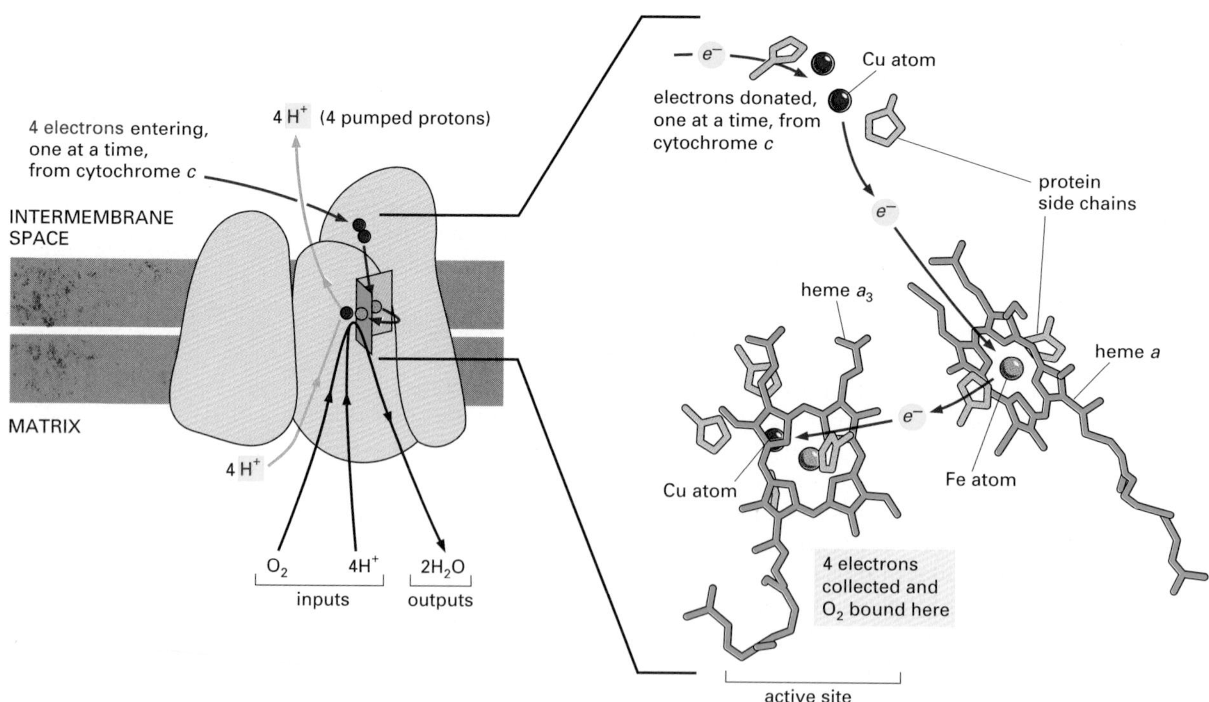

Figure 14–27 The reaction of O_2 with electrons in cytochrome oxidase. As indicated, the iron atom in heme *a* serves as an electron queuing point; this heme feeds four electrons into an O_2 molecule held at the bimetallic center active site, which is formed by the other heme-linked iron and a closely opposed copper atom. Note that four protons are pumped out of the matrix for each O_2 molecule that undergoes the reaction $4e^- + 4H^+ + O_2 \rightarrow 2H_2O$.

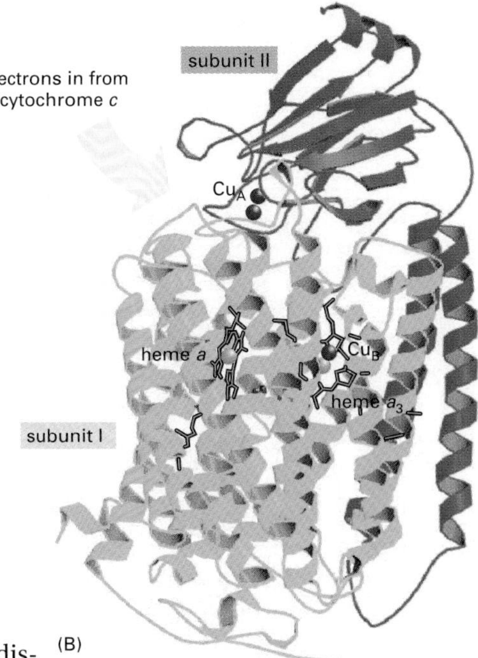

Figure 14–28 The molecular structure of cytochrome oxidase. This protein is a dimer formed from a monomer with 13 different protein subunits (monomer mass of 204,000 daltons). The three colored subunits are encoded by the mitochondrial genome, and they form the functional core of the enzyme. As electrons pass through this protein on the way to its bound O_2 molecule, they cause the protein to pump protons across the membrane (see Figure 14–27). (A) The entire protein is shown, positioned in the inner mitochondrial membrane. (B) The electron carriers are located in subunits I and II, as indicated.

quantum-mechanical property that is critical for the processes we are discussing. Insulation is needed to prevent short circuits that would otherwise occur when an electron carrier with a low redox potential collides with a carrier with a high redox potential. This insulation seems to be provided by carrying an electron deep enough inside a protein to prevent its tunneling interactions with an inappropriate partner.

How the changes in redox potential from one electron carrier to the next are harnessed to pump protons out of the mitochondrial matrix is the topic we discuss next.

A Large Drop in Redox Potential Across Each of the Three Respiratory Enzyme Complexes Provides the Energy for H^+ Pumping

We have previously discussed how the redox potential reflects electron affinities (see p. 783). An outline of the redox potentials measured along the respiratory chain is shown in Figure 14–29. These potentials drop in three large steps, one across each major respiratory complex. The change in redox potential between any two electron carriers is directly proportional to the free energy released when an electron transfers between them. Each enzyme complex acts as an energy-conversion device by harnessing some of this free-energy change to pump H^+ across the inner membrane, thereby creating an electrochemical proton gradient as electrons pass through that complex. This conversion can be demonstrated by purifying each respiratory enzyme complex and incorporating it separately into liposomes: when an appropriate electron donor and acceptor are added so that electrons can pass through the complex, H^+ is translocated across the liposome membrane.

The Mechanism of H^+ Pumping Will Soon Be Understood in Atomic Detail

Some respiratory enzyme complexes pump one H^+ per electron across the inner mitochondrial membrane, whereas others pump two. The detailed mechanism by which electron transport is coupled to H^+ pumping is different for the three different enzyme complexes. In the cytochrome b-c_1 complex, the quinones clearly have a role. As mentioned previously, a quinone picks up a H^+ from the aqueous medium along with each electron it carries and liberates it when it releases the electron (see Figure 14–24). Since ubiquinone is freely mobile in the lipid bilayer, it could accept electrons near the inside surface of the membrane

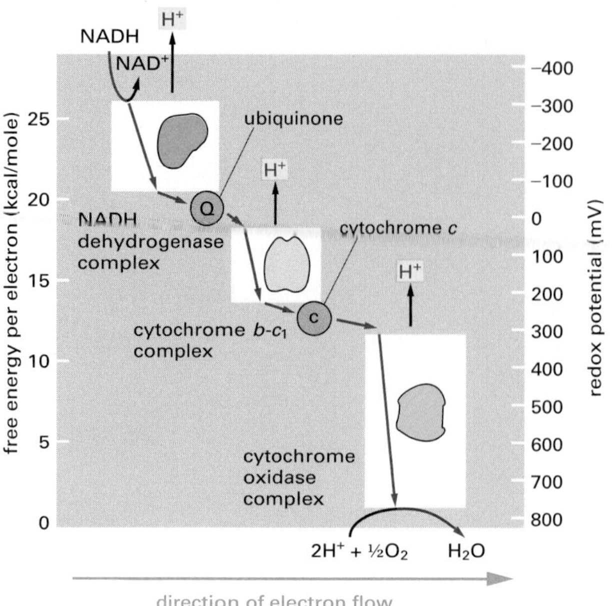

Figure 14–29 Redox potential changes along the mitochondrial electron-transport chain. The redox potential (designated E'_0) increases as electrons flow down the respiratory chain to oxygen. The standard free-energy change, $\Delta G°$, for the transfer of each of the two electrons donated by an NADH molecule can be obtained from the left-hand ordinate ($\Delta G = -n(0.023)\ \Delta E'_0$, where n is the number of electrons transferred across a redox potential change of $\Delta E'_0$ mV). Electrons flow through a respiratory enzyme complex by passing in sequence through the multiple electron carriers in each complex. As indicated, part of the favorable free-energy change is harnessed by each enzyme complex to pump H^+ across the inner mitochondrial membrane. It is thought that the NADH dehydrogenase and cytochrome b-c_1 complexes each pump two H^+ per electron, whereas the cytochrome oxidase complex pumps one.

It should be noted that NADH is not the only source of electrons for the respiratory chain. The flavin $FADH_2$ is also generated by fatty acid oxidation (see Figure 2–77) and by the citric acid cycle (see Figure 2–79). Its two electrons are passed directly to ubiquinone, bypassing NADH dehydrogenase; they therefore cause less H^+ pumping than the two electrons transported from NADH.

and donate them to the cytochrome b-c_1 complex near the outside surface, thereby transferring one H^+ across the bilayer for every electron transported. Two protons are pumped per electron in the cytochrome b-c_1 complex, however, and there is good evidence for a so-called *Q-cycle,* in which ubiquinone is recycled through the complex in an ordered way that makes this two-for-one transfer possible. Exactly how this occurs can now be worked out at the atomic level, because the complete structure of the cytochrome b-c_1 complex has been determined by x-ray crystallography (Figure 14–30).

Allosteric changes in protein conformations driven by electron transport can also pump H^+, just as H^+ is pumped when ATP is hydrolyzed by the ATP

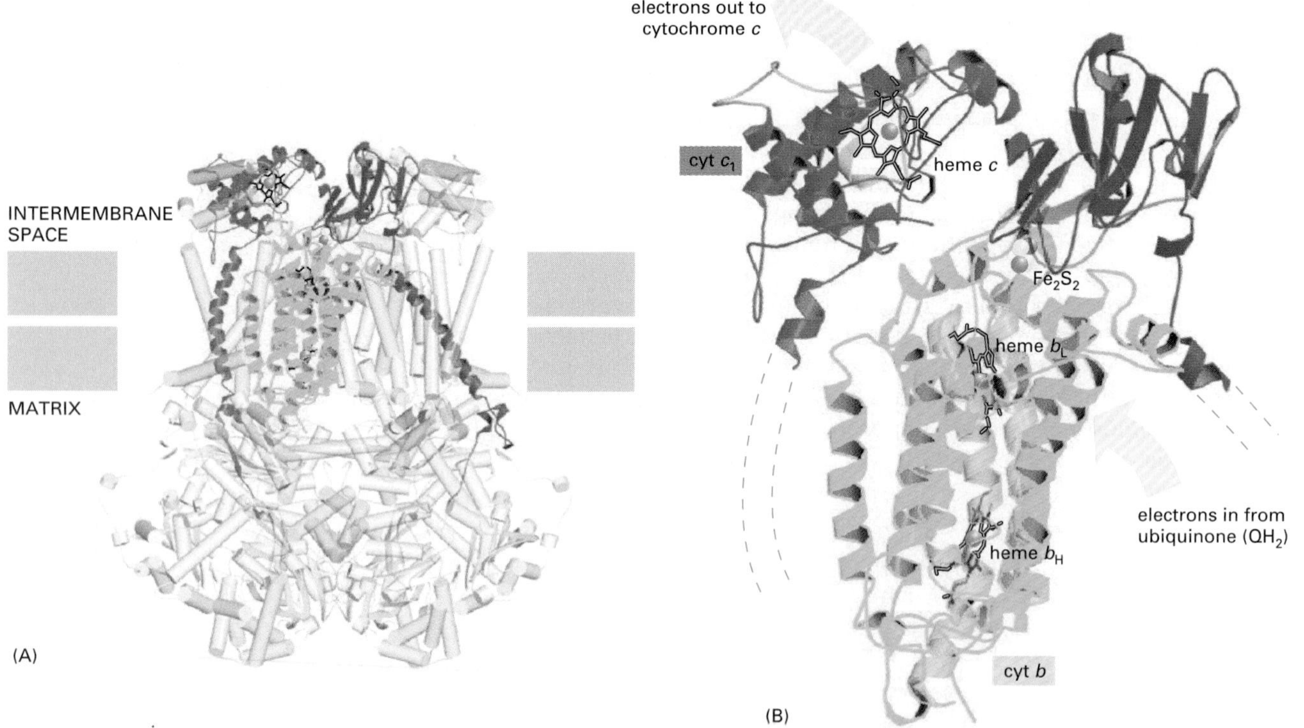

Figure 14–30 The atomic structure of cytochrome b-c_1. This protein is a dimer. The 240,000-dalton monomer is composed of 11 different protein molecules in mammals. The three colored proteins form the functional core of the enzyme: cytochrome b *(green),* cytochrome c_1 *(blue),* and the Rieske protein containing an iron–sulfur center *(purple).* (A) The interaction of these three proteins across the two monomers. (B) Their electron carriers, along with the entrance and exit sites for electrons.

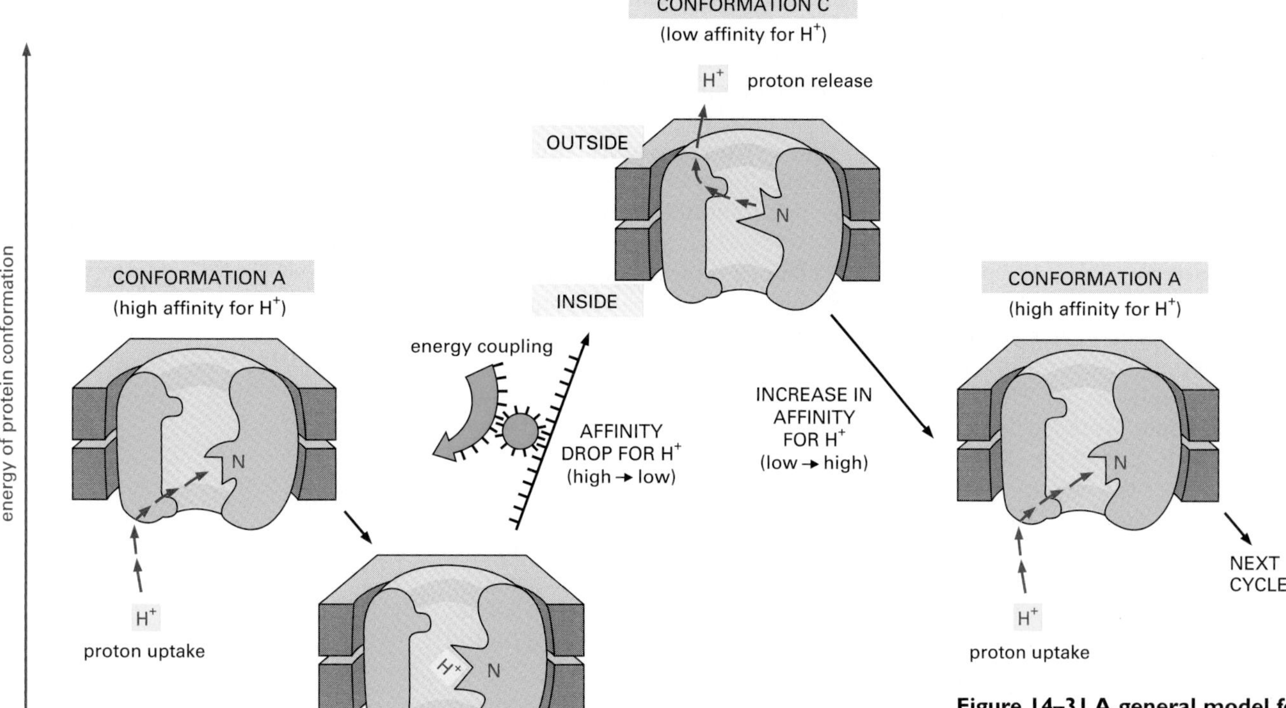

Figure 14–31 A general model for H^+ pumping. This model for H^+ pumping by a transmembrane protein is based on mechanisms that are thought to be used by both cytochrome oxidase and the light-driven procaryotic proton pump, bacteriorhodopsin. The protein is driven through a cycle of three conformations: A, B, and C. As indicated by their vertical spacing, these protein conformations have different energies. In conformation A, the protein has a high affinity for H^+, causing it to pick up a H^+ on the inside of the membrane. In conformation C, the protein has a low affinity for H^+, causing it to release a H^+ on the outside of the membrane. The transition from conformation B to conformation C that releases the H^+ is energetically unfavorable, and it occurs only because it is driven by being allosterically coupled to an energetically favorable reaction occurring elsewhere on the protein *(blue arrow)*. The other two conformational changes, A → B and C → A, lead to states of lower energy, and they proceed spontaneously.

Because the overall cycle A → B → C → A → B → C releases free energy, H^+ is pumped from the inside (the matrix in mitochondria) to the outside (the intermembrane space in mitochondria). For cytochrome oxidase, the energy required for the transition B → C is provided by electron transport, whereas for bacteriorhodopsin, this energy is provided by light (see Figure 10–37). For yet other proton pumps, the energy is derived from ATP hydrolysis.

synthase running in reverse. For both the NADH dehydrogenase complex and the cytochrome oxidase complex, it seems likely that electron transport drives sequential allosteric changes in protein conformation that cause a portion of the protein to pump H^+ across the mitochondrial inner membrane. A general mechanism for this type of H^+ pumping is presented in Figure 14–31.

H^+ Ionophores Uncouple Electron Transport from ATP Synthesis

Since the 1940s, several substances—such as 2,4-dinitrophenol—have been known to act as *uncoupling agents,* uncoupling electron transport from ATP synthesis. The addition of these low-molecular-weight organic compounds to cells stops ATP synthesis by mitochondria without blocking their uptake of oxygen. In the presence of an uncoupling agent, electron transport and H^+ pumping continue at a rapid rate, but no H^+ gradient is generated. The explanation for this effect is both simple and elegant: uncoupling agents are lipid-soluble weak acids that act as H^+ carriers (H^+ ionophores), and they provide a pathway for the flow of H^+ across the inner mitochondrial membrane that bypasses the ATP synthase. As a result of this short-circuiting, the proton-motive force is dissipated completely, and ATP can no longer be made.

Respiratory Control Normally Restrains Electron Flow Through the Chain

When an uncoupler such as dinitrophenol is added to cells, mitochondria increase their oxygen uptake substantially because of an increased rate of electron transport. This increase reflects the existence of **respiratory control**. The control is thought to act via a direct inhibitory influence of the electrochemical proton gradient on the rate of electron transport. When the gradient is collapsed by an uncoupler, electron transport is free to run unchecked at the maximal rate. As the gradient increases, electron transport becomes more difficult, and the process slows. Moreover, if an artificially large electrochemical proton gradient is experimentally created across the inner membrane, normal electron transport stops completely, and a *reverse electron flow* can be detected in some sections of the respiratory chain. This observation suggests that respiratory control reflects a simple balance between the free-energy change for electron-transport-linked proton pumping and the free-energy change for electron transport—that is, the magnitude of the electrochemical proton gradient affects both the rate and the direction of electron transport, just as it affects the directionality of the ATP synthase (see Figure 14–19).

Respiratory control is just one part of an elaborate interlocking system of feedback controls that coordinate the rates of glycolysis, fatty acid breakdown, the citric acid cycle, and electron transport. The rates of all of these processes are adjusted to the ATP:ADP ratio, increasing whenever an increased utilization of ATP causes the ratio to fall. The ATP synthase in the inner mitochondrial membrane, for example, works faster as the concentrations of its substrates ADP and P_i increase. As it speeds up, the enzyme lets more H^+ flow into the matrix and thereby dissipates the electrochemical proton gradient more rapidly. The falling gradient, in turn, enhances the rate of electron transport.

Similar controls, including feedback inhibition of several key enzymes by ATP, act to adjust the rates of NADH production to the rate of NADH utilization by the respiratory chain, and so on. As a result of these many control mechanisms, the body oxidizes fats and sugars 5–10 times more rapidly during a period of strenuous exercise than during a period of rest.

Natural Uncouplers Convert the Mitochondria in Brown Fat into Heat-generating Machines

In some specialized fat cells, mitochondrial respiration is normally uncoupled from ATP synthesis. In these cells, known as brown fat cells, most of the energy of oxidation is dissipated as heat rather than being converted into ATP. The inner membranes of the large mitochondria in these cells contain a special transport protein that allows protons to move down their electrochemical gradient, by-passing ATP synthase. As a result, the cells oxidize their fat stores at a rapid rate and produce more heat than ATP. Tissues containing brown fat serve as "heating pads," helping to revive hibernating animals and to protect sensitive areas of newborn human babies from the cold.

Bacteria Also Exploit Chemiosmotic Mechanisms to Harness Energy

Bacteria use enormously diverse energy sources. Some, like animal cells, are aerobic; they synthesize ATP from sugars they oxidize to CO_2 and H_2O by glycolysis, the citric acid cycle, and a respiratory chain in their plasma membrane that is similar to the one in the inner mitochondrial membrane. Others are strict anaerobes, deriving their energy either from glycolysis alone (by fermentation) or from an electron-transport chain that employs a molecule other than oxygen as the final electron acceptor. The alternative electron acceptor can be a nitrogen compound (nitrate or nitrite), a sulfur compound (sulfate or sulfite), or a carbon compound (fumarate or carbonate), for example. The electrons are transferred to these acceptors by a series of electron carriers in the plasma membrane that are comparable to those in mitochondrial respiratory chains.

Despite this diversity, the plasma membrane of the vast majority of bacteria contains an ATP synthase that is very similar to the one in mitochondria. In bacteria that use an electron-transport chain to harvest energy, the electron-transport pumps H^+ out of the cell and thereby establishes a proton-motive force across the plasma membrane that drives the ATP synthase to make ATP. In other bacteria, the ATP synthase works in reverse, using the ATP produced by glycolysis to pump H^+ and establish a proton gradient across the plasma membrane. The ATP used for this process is generated by fermentation processes (discussed in Chapter 2).

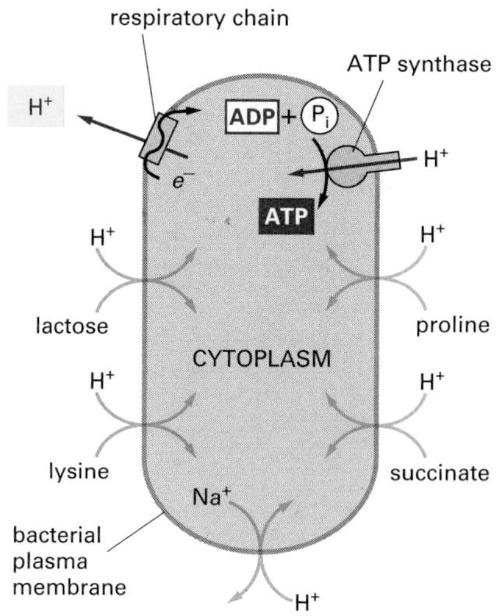

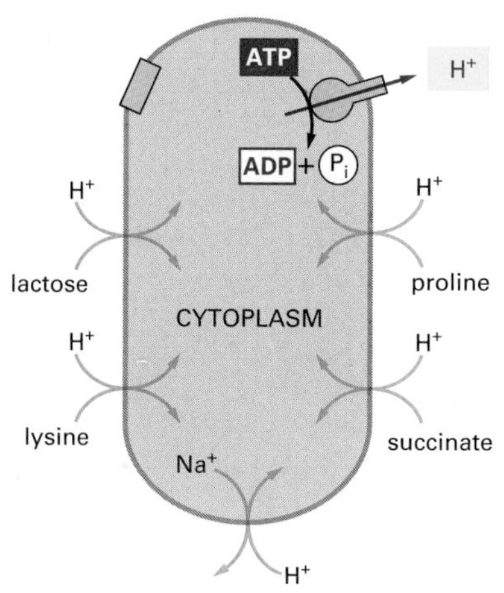

Figure 14–32 The importance of H^+-driven transport in bacteria. A proton-motive force generated across the plasma membrane pumps nutrients into the cell and expels Na^+. (A) In an aerobic bacterium, an electrochemical proton gradient across the plasma membrane is produced by a respiratory chain and is then used both to transport some nutrients into the cell and to make ATP. (B) The same bacterium growing under anaerobic conditions can derive its ATP from glycolysis. Part of this ATP is hydrolyzed by the ATP synthase to establish an electrochemical proton gradient that drives the same transport processes that depend on the respiratory chain in (A).

Thus, most bacteria, including the strict anaerobes, maintain a proton gradient across their plasma membrane. It can be harnessed to drive a flagellar motor, and it is used to pump Na^+ out of the bacterium via a Na^+-H^+ antiporter that takes the place of the Na^+-K^+ pump of eucaryotic cells. This gradient is also used for the active inward transport of nutrients, such as most amino acids and many sugars: each nutrient is dragged into the cell along with one or more H^+ through a specific symporter (Figure 14–32). In animal cells, by contrast, most inward transport across the plasma membrane is driven by the Na^+ gradient that is established by the Na^+-K^+ pump.

Some unusual bacteria have adapted to live in a very alkaline environment and yet must maintain their cytoplasm at a physiological pH. For these cells, any attempt to generate an electrochemical H^+ gradient would be opposed by a large H^+ concentration gradient in the wrong direction (H^+ higher inside than outside). Presumably for this reason, some of these bacteria substitute Na^+ for H^+ in all of their chemiosmotic mechanisms. The respiratory chain pumps Na^+ out of the cell, the transport systems and flagellar motor are driven by an inward flux of Na^+, and a Na^+-driven ATP synthase synthesizes ATP. The existence of such bacteria demonstrates that the principle of chemiosmosis is more fundamental than the proton-motive force on which it is normally based.

Summary

The respiratory chain in the inner mitochondrial membrane contains three respiratory enzyme complexes through which electrons pass on their way from NADH to O_2.

Each of these can be purified, inserted into synthetic lipid vesicles, and then shown to pump H^+ when electrons are transported through it. In the intact membrane, the mobile electron carriers ubiquinone and cytochrome c *complete the electron-transport chain by shuttling between the enzyme complexes. The path of electron flow is NADH → NADH dehydrogenase complex → ubiquinone → cytochrome* b-c$_1$ *complex → cytochrome* c *→ cytochrome oxidase complex → molecular oxygen (O_2).*

The respiratory enzyme complexes couple the energetically favorable transport of electrons to the pumping of H^+ out of the matrix. The resulting electrochemical proton gradient is harnessed to make ATP by another transmembrane protein complex, ATP synthase, through which H^+ flows back into the matrix. The ATP synthase is a reversible coupling device that normally converts a backflow of H^+ into ATP phosphate bond energy by catalyzing the reaction ADP + P_i → ATP, but it can also work in the opposite direction and hydrolyze ATP to pump H^+ if the electrochemical proton gradient is sufficiently reduced. Its universal presence in mitochondria, chloroplasts, and procaryotes testifies to the central importance of chemiosmotic mechanisms in cells.

CHLOROPLASTS AND PHOTOSYNTHESIS

All animals and most microorganisms rely on the continual uptake of large amounts of organic compounds from their environment. These compounds are used to provide both the carbon skeletons for biosynthesis and the metabolic energy that drives cellular processes. It is believed that the first organisms on the primitive Earth had access to an abundance of the organic compounds produced by geochemical processes, but that most of these original compounds were used up billions of years ago. Since that time, the vast majority of the organic materials required by living cells have been produced by *photosynthetic organisms,* including many types of photosynthetic bacteria.

The most advanced photosynthetic bacteria are the cyanobacteria, which have minimal nutrient requirements. They use electrons from water and the energy of sunlight when they convert atmospheric CO_2 into organic compounds—a process called *carbon fixation.* In the course of splitting water [in the overall reaction $n\text{H}_2\text{O} + n\text{CO}_2 \xrightarrow{\text{light}} (\text{CH}_2\text{O})_n + n\text{O}_2$], they also liberate into the atmosphere the oxygen required for oxidative phosphorylation. As we see in this section, it is thought that the evolution of cyanobacteria from more primitive photosynthetic bacteria eventually made possible the development of abundant aerobic life forms.

In plants and algae, which developed much later, photosynthesis occurs in a specialized intracellular organelle—the **chloroplast**. Chloroplasts perform photosynthesis during the daylight hours. The immediate products of photosynthesis, NADPH and ATP, are used by the photosynthetic cells to produce many organic molecules. In plants, the products include a low-molecular-weight sugar (usually sucrose) that is exported to meet the metabolic needs of the many nonphotosynthetic cells of the organism.

Biochemical and genetic evidence strongly suggest that chloroplasts are descendants of oxygen-producing photosynthetic bacteria that were endocytosed and lived in symbiosis with primitive eucaryotic cells. Mitochondria are also generally believed to be descended from an endocytosed bacterium. The many differences between chloroplasts and mitochondria are thought to reflect their different bacterial ancestors, as well as their subsequent evolutionary divergence. Nevertheless, the fundamental mechanisms involved in light-driven ATP synthesis in chloroplasts are very similar to those that we have already discussed for respiration-driven ATP synthesis in mitochondria.

The Chloroplast Is One Member of the Plastid Family of Organelles

Chloroplasts are the most prominent members of the **plastid** family of organelles. Plastids are present in all living plant cells, each cell type having its own characteristic complement. All plastids share certain features. Most notably, all plastids in a particular plant species contain multiple copies of the same relatively small genome. In addition, each is enclosed by an envelope composed of two concentric membranes.

As discussed in Chapter 12 (see Figure 12–3), all plastids develop from *proplastids,* small organelles in the immature cells of plant meristems (Figure 14–33A). Proplastids develop according to the requirements of each differentiated cell, and the type that is present is determined in large part by the nuclear genome. If a leaf is grown in darkness, its proplastids enlarge and develop into *etioplasts,* which have a semicrystalline array of internal membranes containing

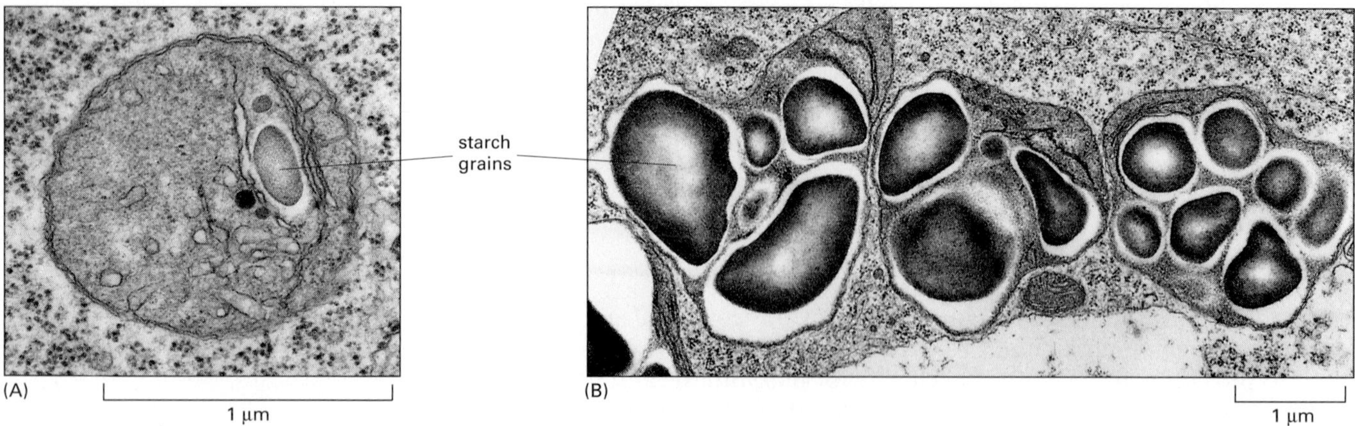

Figure 14–33 Plastid diversity. (A) A proplastid from a root tip cell of a bean plant. Note the double membrane; the inner membrane has also generated the relatively sparse internal membranes present. (B) Three amyloplasts (a form of leucoplast), or starch-storing plastids, in a root tip cell of soybean. (From B. Gunning and M. Steer, Plant Cell Biology: Structure and Function. Sudbury, MA: Jones & Bartlett, 1996. © Jones & Bartlett Publishers.)

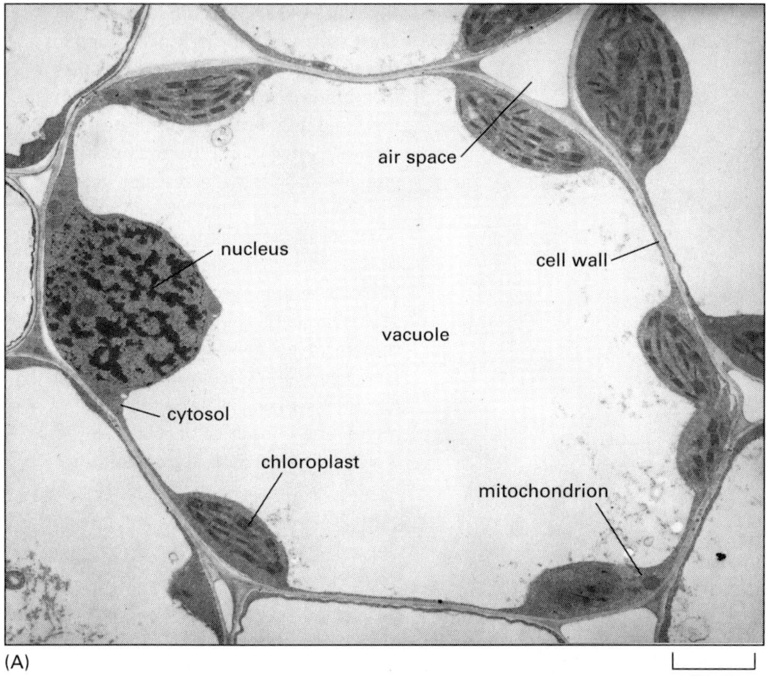

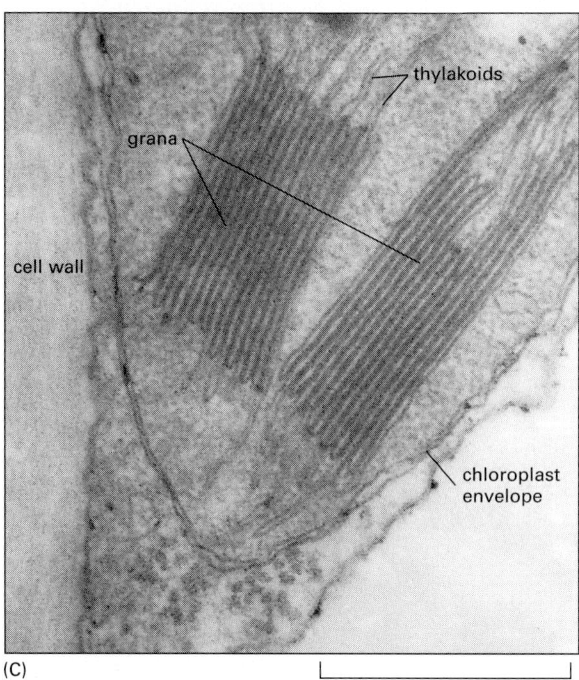

Figure 14–34 Electron micrographs of chloroplasts. (A) In a wheat leaf cell, a thin rim of cytoplasm—containing chloroplasts, the nucleus, and mitochondria—surrounds a large vacuole. (B) A thin section of a single chloroplast, showing the chloroplast envelope, starch granules, and lipid (fat) droplets that have accumulated in the stroma as a result of the biosyntheses occurring there. (C) A high-magnification view of two grana. A granum is a stack of thylakoids. (Courtesy of K. Plaskitt.)

a yellow chlorophyll precursor instead of chlorophyll. When exposed to light, the etioplasts rapidly develop into chloroplasts by converting this precursor to chlorophyll and by synthesizing new membrane pigments, photosynthetic enzymes, and components of the electron-transport chain.

Leucoplasts are plastids present in many epidermal and internal tissues that do not become green and photosynthetic. They are little more than enlarged proplastids. A common form of leucoplast is the *amyloplast* (Figure 14–33B), which accumulates the polysaccharide starch in storage tissues—a source of sugar for future use. In some plants, such as potatoes, the amyloplasts can grow to be as large as an average animal cell.

It is important to realize that plastids are not just sites for photosynthesis and the deposition of storage materials. Plants have also used their plastids to compartmentalize their intermediary metabolism. Purine and pyrimidine synthesis, most amino acid synthesis, and all of the fatty acid synthesis of plants takes place in the plastids, whereas in animal cells these compounds are produced in the cytosol.

Chloroplasts Resemble Mitochondria But Have an Extra Compartment

Chloroplasts carry out their energy interconversions by chemiosmotic mechanisms in much the same way that mitochondria do. Although much larger (Figure 14–34A), they are organized on the same principles. They have a highly

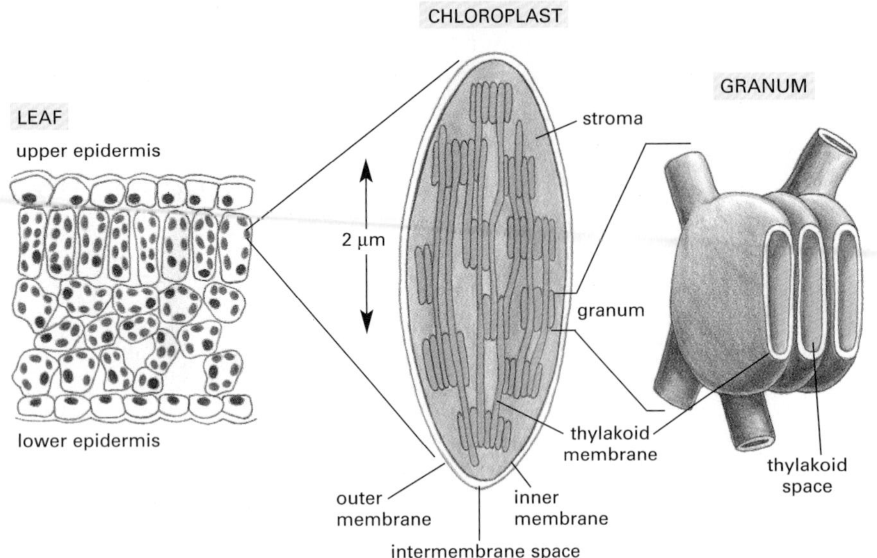

Figure 14–35 The chloroplast. This photosynthetic organelle contains three distinct membranes (the outer membrane, the inner membrane, and the thylakoid membrane) that define three separate internal compartments (the intermembrane space, the stroma, and the thylakoid space). The thylakoid membrane contains all the energy-generating systems of the chloroplast, including its chlorophyll. In electron micrographs, this membrane seems to be broken up into separate units that enclose individual flattened vesicles (see Figure 14–34), but these are probably joined into a single, highly folded membrane in each chloroplast. As indicated, the individual thylakoids are interconnected, and they tend to stack to form grana.

permeable outer membrane; a much less permeable inner membrane, in which membrane transport proteins are embedded; and a narrow intermembrane space in between. Together, these membranes form the chloroplast envelope (Figure 14–34B,C). The inner membrane surrounds a large space called the **stroma**, which is analogous to the mitochondrial matrix and contains many metabolic enzymes. Like the mitochondrion, the chloroplast has its own genome and genetic system. The stroma therefore also contains a special set of ribosomes, RNAs, and the chloroplast DNA.

There is, however, an important difference between the organization of mitochondria and that of chloroplasts. The inner membrane of the chloroplast is not folded into cristae and does not contain electron-transport chains. Instead, the electron-transport chains, photosynthetic light-capturing systems, and ATP synthase are all contained in the *thylakoid membrane,* a third distinct membrane that forms a set of flattened disclike sacs, the *thylakoids* (Figure 14–35). The lumen of each thylakoid is thought to be connected with the lumen of other thylakoids, thereby defining a third internal compartment called the *thylakoid space,* which is separated by the thylakoid membrane from the stroma that surrounds it.

The structural similarities and differences between mitochondria and chloroplasts are illustrated in Figure 14–36. The head of the chloroplast ATP synthase, where ATP is made, protrudes from the thylakoid membrane into the stroma, whereas it protrudes into the matrix from the inner mitochondrial membrane.

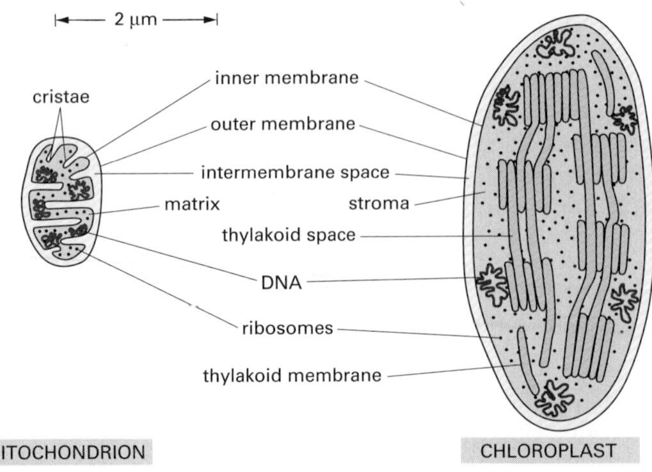

Figure 14–36 A mitochondrion and chloroplast compared. A chloroplast is generally much larger than a mitochondrion and contains, in addition to an outer and inner membrane, a thylakoid membrane enclosing a thylakoid space. Unlike the chloroplast inner membrane, the inner mitochondrial membrane is folded into cristae to increase its surface area.

Chloroplasts Capture Energy from Sunlight and Use It to Fix Carbon

The many reactions that occur during photosynthesis in plants can be grouped into two broad categories:

1. In the **photosynthetic electron-transfer** reactions (also called the "light reactions"), energy derived from sunlight energizes an electron in the green organic pigment *chlorophyll,* enabling the electron to move along an electron-transport chain in the thylakoid membrane in much the same way that an electron moves along the respiratory chain in mitochondria. The chlorophyll obtains its electrons from water (H_2O), producing O_2 as a by-product. During the electron-transport process, H^+ is pumped across the thylakoid membrane, and the resulting electrochemical proton gradient drives the synthesis of ATP in the stroma. As the final step in this series of reactions, high-energy electrons are loaded (together with H^+) onto $NADP^+$, converting it to NADPH. All of these reactions are confined to the chloroplast.
2. In the **carbon-fixation reactions** (also called the "dark reactions"), the ATP and the NADPH produced by the photosynthetic electron-transfer reactions serve as the source of energy and reducing power, respectively, to drive the conversion of CO_2 to carbohydrate. The carbon-fixation reactions, which begin in the chloroplast stroma and continue in the cytosol, produce sucrose and many other organic molecules in the leaves of the plant. The sucrose is exported to other tissues as a source of both organic molecules and energy for growth.

Thus, the formation of ATP, NADPH, and O_2 (which requires light energy directly) and the conversion of CO_2 to carbohydrate (which requires light energy only indirectly) are separate processes (Figure 14–37), although elaborate feedback mechanisms interconnect the two. Several of the chloroplast enzymes required for carbon fixation, for example, are inactivated in the dark and reactivated by light-stimulated electron-transport processes.

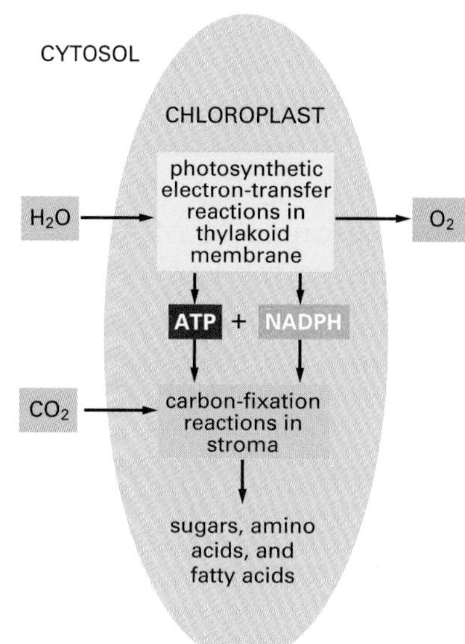

Figure 14–37 The reactions of photosynthesis in a chloroplast. Water is oxidized and oxygen is released in the photosynthetic electron-transfer reactions, while carbon dioxide is assimilated (fixed) to produce sugars and a variety of other organic molecules in the carbon-fixation reactions.

Carbon Fixation Is Catalyzed by Ribulose Bisphosphate Carboxylase

We have seen earlier in this chapter how cells produce ATP by using the large amount of free energy released when carbohydrates are oxidized to CO_2 and H_2O. Clearly, therefore, the reverse reaction, in which CO_2 and H_2O combine to make carbohydrate, must be a very unfavorable one that can only occur if it is coupled to other, very favorable reactions that drive it.

The central reaction of **carbon fixation**, in which an atom of inorganic carbon is converted to organic carbon, is illustrated in Figure 14–38: CO_2 from the atmosphere combines with the five-carbon compound ribulose 1,5-bisphosphate plus water to yield two molecules of the three-carbon compound 3-phosphoglycerate. This "carbon-fixing" reaction, which was discovered in 1948, is catalyzed in the chloroplast stroma by a large enzyme called *ribulose bisphosphate carboxylase*. Since each molecule of the complex works sluggishly (processing

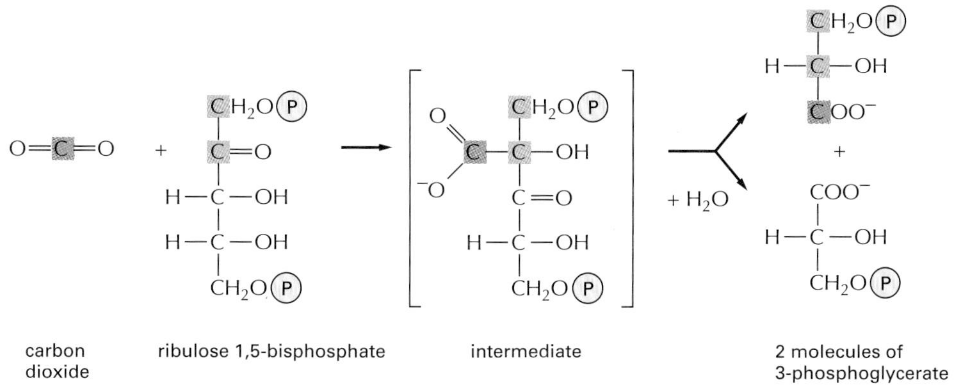

carbon dioxide + ribulose 1,5-bisphosphate → [intermediate] + H_2O → 2 molecules of 3-phosphoglycerate

Figure 14–38 The initial reaction in carbon fixation. This reaction, in which carbon dioxide is converted into organic carbon, is catalyzed in the chloroplast stroma by the abundant enzyme ribulose bisphosphate carboxylase. The product is 3-phosphoglycerate, which is also an intermediate in glycolysis. The two carbon atoms shaded in *blue* are used to produce phosphoglycolate when the same enzyme adds oxygen instead of CO_2 (see text).

only about 3 molecules of substrate per second compared to 1000 molecules per second for a typical enzyme), many enzyme molecules are needed. Ribulose bisphosphate carboxylase often constitutes more than 50% of the total chloroplast protein, and it is thought to be the most abundant protein on Earth.

Three Molecules of ATP and Two Molecules of NADPH Are Consumed for Each CO_2 Molecule That Is Fixed

The actual reaction in which CO_2 is fixed is energetically favorable because of the reactivity of the energy-rich compound *ribulose 1,5-bisphosphate,* to which each molecule of CO_2 is added (see Figure 14–38). The elaborate metabolic pathway that produces ribulose 1,5-bisphosphate requires both NADPH and ATP; it was worked out in one of the first successful applications of radioisotopes as tracers in biochemistry. This **carbon-fixation cycle** (also called the **Calvin cycle**) is outlined in Figure 14–39. It starts when 3 molecules of CO_2 are fixed by ribulose bisphosphate carboxylase to produce 6 molecules of 3-phosphoglycerate (containing $6 \times 3 = 18$ carbon atoms in all: 3 from the CO_2 and 15 from ribulose 1,5-bisphosphate). The 18 carbon atoms then undergo a cycle of reactions that regenerates the 3 molecules of ribulose 1,5-bisphosphate used in the initial carbon-fixation step (containing $3 \times 5 = 15$ carbon atoms). This leaves 1 molecule of *glyceraldehyde 3-phosphate* (3 carbon atoms) as the net gain.

A total of 3 molecules of ATP and 2 molecules of NADPH are consumed for each CO_2 molecule converted into carbohydrate. The net equation is:

$$3CO_2 + 9ATP + 6NADPH + \text{water} \rightarrow$$
$$\text{glyceraldehyde 3-phosphate} + 8P_i + 9ADP + 6NADP^+$$

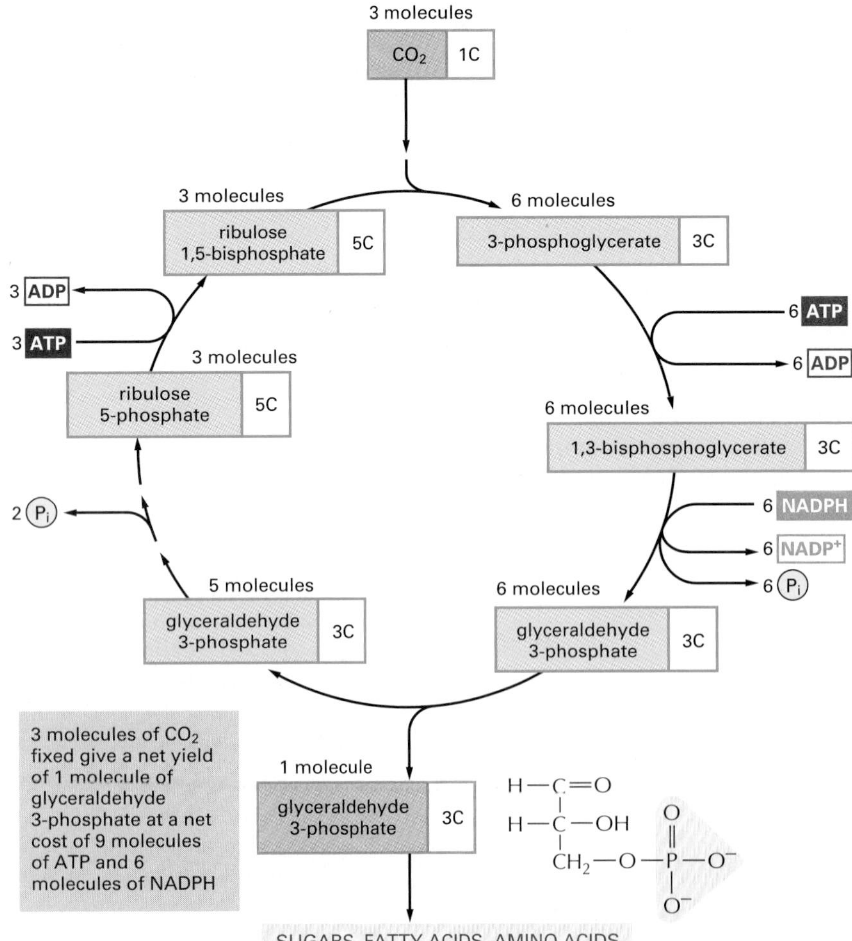

Figure 14–39 The carbon-fixation cycle, which forms organic molecules from CO_2 and H_2O. The number of carbon atoms in each type of molecule is indicated in the *white box*. There are many intermediates between glyceraldehyde 3-phosphate and ribulose 5-phosphate, but they have been omitted here for clarity. The entry of water into the cycle is also not shown.

Thus, both *phosphate-bond energy* (as ATP) and *reducing power* (as NADPH) are required for the formation of organic molecules from CO_2 and H_2O. We return to this important point later.

The glyceraldehyde 3-phosphate produced in chloroplasts by the carbon-fixation cycle is a three-carbon sugar that also serves as a central intermediate in glycolysis. Much of it is exported to the cytosol, where it can be converted into fructose 6-phosphate and glucose 1-phosphate by the reversal of several reactions in glycolysis (see Panel 2–8, pp. 124–125). The glucose 1-phosphate is then converted to the sugar nucleotide UDP-glucose, and this combines with the fructose 6-phosphate to form sucrose phosphate, the immediate precursor of the disaccharide sucrose. **Sucrose** is the major form in which sugar is transported between plant cells: just as glucose is transported in the blood of animals, sucrose is exported from the leaves via vascular bundles, providing the carbohydrate required by the rest of the plant.

Most of the glyceraldehyde 3-phosphate that remains in the chloroplast is converted to *starch* in the stroma. Like glycogen in animal cells, **starch** is a large polymer of glucose that serves as a carbohydrate reserve (see Figure 14–33B). The production of starch is regulated so that it is produced and stored as large grains in the chloroplast stroma during periods of excess photosynthetic capacity. This occurs through reactions in the stroma that are the reverse of those in glycolysis: they convert glyceraldehyde 3-phosphate to glucose 1-phosphate, which is then used to produce the sugar nucleotide ADP-glucose, the immediate precursor of starch. At night the starch is broken down to help support the metabolic needs of the plant. Starch provides an important part of the diet of all animals that eat plants.

Carbon Fixation in Some Plants Is Compartmentalized to Facilitate Growth at Low CO_2 Concentrations

Although ribulose bisphosphate carboxylase preferentially adds CO_2 to ribulose 1,5-bisphosphate, it can use O_2 as a substrate in place of CO_2, and if the concentration of CO_2 is low, it will add O_2 to ribulose 1,5-bisphosphate instead (see Figure 14–38). This is the first step in a pathway called **photorespiration**, whose ultimate effect is to use up O_2 and liberate CO_2 without the production of useful energy stores. In many plants, about one-third of the CO_2 fixed is lost again as CO_2 because of photorespiration.

Photorespiration can be a serious liability for plants in hot, dry conditions, which cause them to close their stomata (the gas exchange pores in their leaves) to avoid excessive water loss. This in turn causes the CO_2 levels in the leaf to fall precipitously, thereby favoring photorespiration. A special adaptation, however, occurs in the leaves of many plants, such as corn and sugar cane that live in hot, dry environments. In these plants, the carbon-fixation cycle occurs only in the chloroplasts of specialized *bundle-sheath cells,* which contain all of the plant's ribulose bisphosphate carboxylase. These cells are protected from the air and are surrounded by a specialized layer of *mesophyll cells* that use the energy harvested by their chloroplasts to "pump" CO_2 into the bundle-sheath cells. This supplies the ribulose bisphosphate carboxylase with a high concentration of CO_2, thereby greatly reducing photorespiration.

The CO_2 pump is produced by a reaction cycle that begins in the cytosol of the mesophyll cells. A CO_2-fixation step is catalyzed by an enzyme that binds carbon dioxide (as bicarbonate) and combines it with an activated three-carbon molecule to produce a four-carbon molecule. The four-carbon molecule diffuses into the bundle-sheath cells, where it is broken down to release the CO_2 and generate a molecule with three carbons. The pumping cycle is completed when this three-carbon molecule is returned to the mesophyll cells and converted back to its original activated form. Because the CO_2 is initially captured by converting it into a compound containing four carbons, the CO_2-pumping plants are called *C_4 plants.* All other plants are called *C_3 plants* because they capture CO_2 into the three-carbon compound 3-phosphoglycerate (Figure 14–40).

As for any vectorial transport process, pumping CO_2 into the bundle-sheath

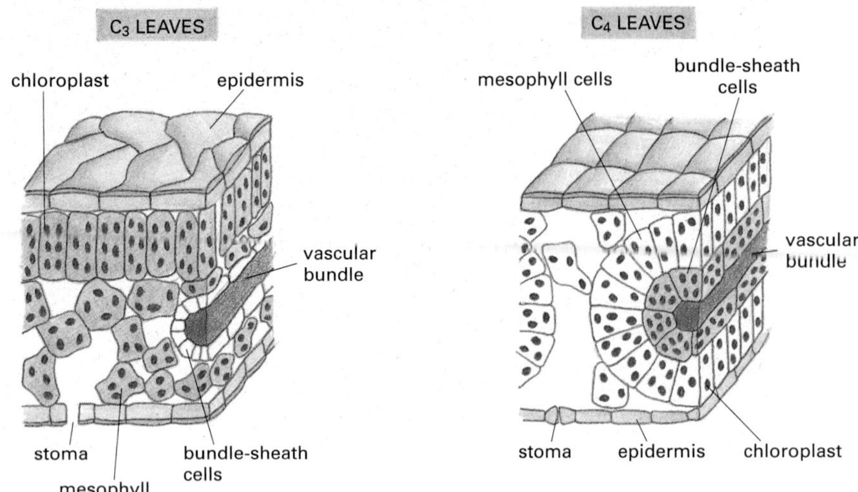

Figure 14–40 Comparative leaf anatomy in a C_3 plant and a C_4 plant. The cells with *green* cytosol in the leaf interior contain chloroplasts that perform the normal carbon-fixation cycle. In C_4 plants, the mesophyll cells are specialized for CO_2 pumping rather than for carbon fixation, and they thereby create a high ratio of CO_2 to O_2 in the bundle-sheath cells, which are the only cells in these plants where the carbon-fixation cycle occurs. The vascular bundles carry the sucrose made in the leaf to other tissues.

cells in C_4 plants costs energy. In hot, dry environments, however, this cost can be much less than the energy lost by photorespiration in C_3 plants, so C_4 plants have a potential advantage. Moreover, because C_4 plants can perform photosynthesis at a lower concentration of CO_2 inside the leaf, they need to open their stomata less often and therefore can fix about twice as much net carbon as C_3 plants per unit of water lost. Although the vast majority of plant species are C_3 plants, C_4 plants such as corn and sugar cane are much more effective at converting sunlight energy into biomass than C_3 plants such as cereal grains. They are therefore of special importance in world agriculture.

Photosynthesis Depends on the Photochemistry of Chlorophyll Molecules

Having discussed the carbon-fixation reactions, we now return to the question of how the photosynthetic electron-transfer reactions in the chloroplast generate the ATP and the NADPH needed to drive the production of carbohydrates from CO_2 and H_2O. The required energy is derived from sunlight absorbed by **chlorophyll** molecules (Figure 14–41). The process of energy conversion begins when a chlorophyll molecule is excited by a quantum of light (a photon) and an electron is moved from one molecular orbital to another of higher energy. As illustrated in Figure 14–42, such an excited molecule is unstable and tends to return to its original, unexcited state in one of three ways:

1. By converting the extra energy into heat (molecular motions) or to some combination of heat and light of a longer wavelength (fluorescence), which is what happens when light energy is absorbed by an isolated chlorophyll molecule in solution.
2. By transferring the energy—but not the electron—directly to a neighboring chlorophyll molecule by a process called *resonance energy transfer*.
3. By transferring the high-energy electron to another nearby molecule, an *electron acceptor*, and then returning to its original state by taking up a low-energy electron from some other molecule, an *electron donor*.

The last two mechanisms are exploited in the process of photosynthesis.

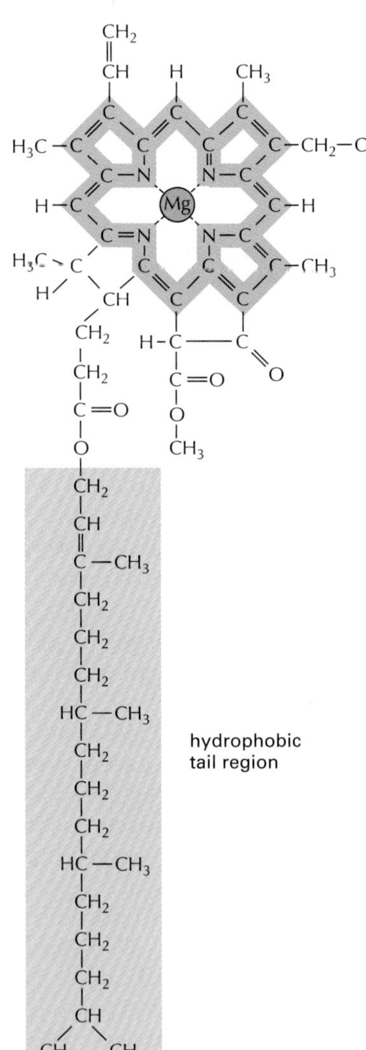

Figure 14–41 The structure of chlorophyll. A magnesium atom is held in a porphyrin ring, which is related to the porphyrin ring that binds iron in heme (see Figure 14–22). Electrons are delocalized over the bonds shown in *blue*.

A Photosystem Consists of a Reaction Center Plus an Antenna Complex

Multiprotein complexes called **photosystems** catalyze the conversion of the light energy captured in excited chlorophyll molecules to useful forms. A photosystem consists of two closely linked components: an antenna complex, consisting of a large set of pigment molecules that capture light energy and feed it to the reaction center; and a photochemical reaction center, consisting of a

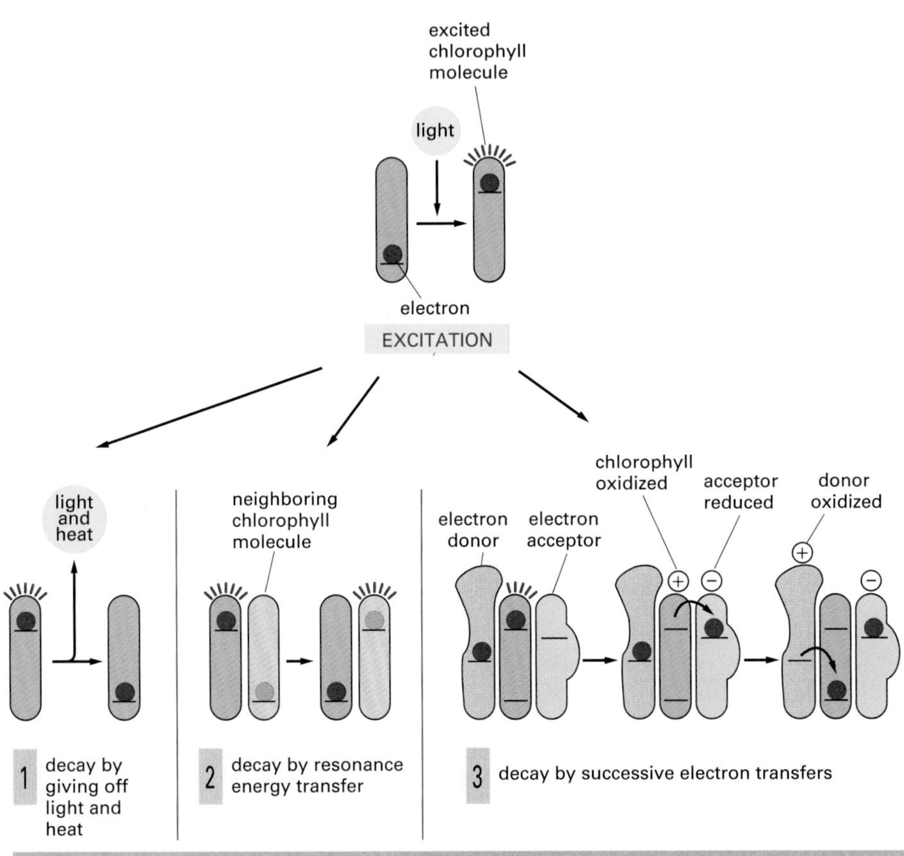

Figure 14–42 Three ways for an excited chlorophyll molecule to return to its original, unexcited state. The light energy absorbed by an isolated chlorophyll molecule is completely released as light and heat by process 1. In photosynthesis, by contrast, chlorophylls undergo process 2 in the antenna complex and process 3 in the reaction center, as described in the text.

complex of proteins and chlorophyll molecules that enable light energy to be converted into chemical energy (Figure 14–43).

The **antenna complex** is important for capturing light. In chloroplasts it consists of a number of distinct membrane protein complexes (known as light-harvesting complexes); together, these proteins bind several hundred chlorophyll molecules per reaction center, orienting them precisely in the thylakoid membrane. Depending on the plant, different amounts of accessory pigments called *carotenoids,* which protect the chlorophylls from oxidation and can help collect light of other wavelengths, are also located in each complex. When a chlorophyll molecule in the antenna complex is excited, the energy is rapidly transferred from one molecule to another by resonance energy transfer until it reaches a special pair of chlorophyll molecules in the photochemical reaction

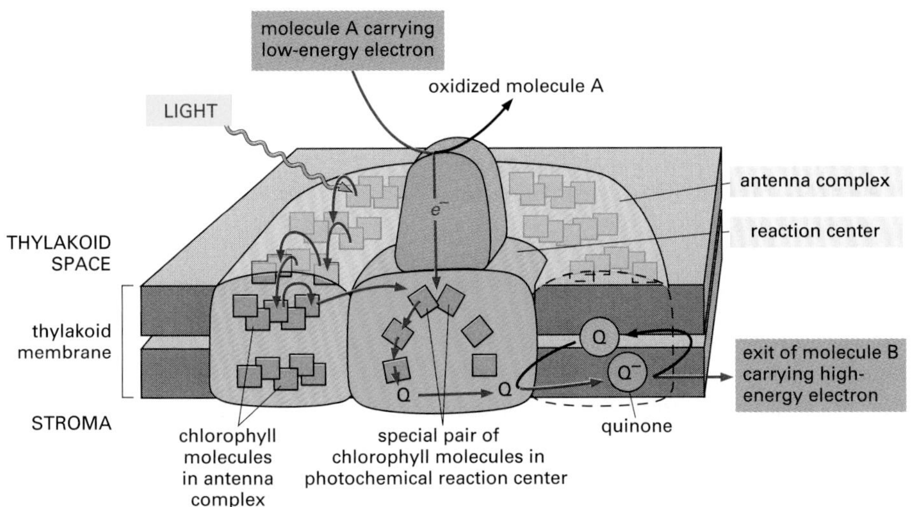

Figure 14–43 The antenna complex and photochemical reaction center in a photosystem. The antenna complex is a collector of light energy in the form of excited electrons. The energy of the excited electrons is funneled, through a series of resonance energy transfers, to a special pair of chlorophyll molecules in the photochemical reaction center. The reaction center then produces a high-energy electron that can be passed rapidly to the electron-transport chain in the thylakoid membrane, via a quinone.

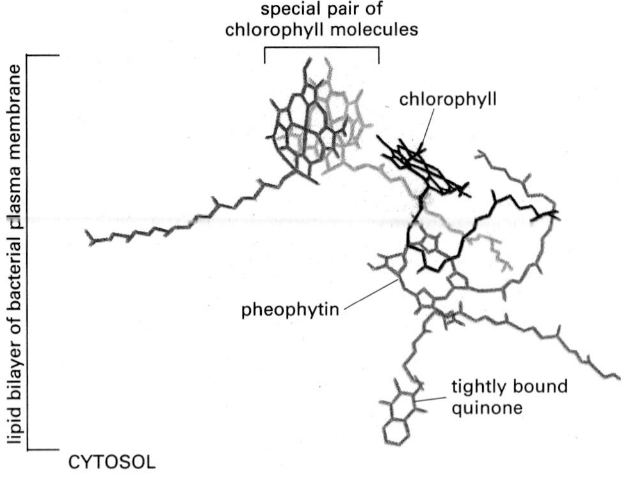

Figure 14–44 The arrangement of the electron carriers in a bacterial photochemical reaction center, as determined by x-ray crystallography. The pigment molecules shown are held in the interior of a transmembrane protein and are surrounded by the lipid bilayer of the bacterial plasma membrane. An electron in the special pair of chlorophyll molecules is excited by resonance from an antenna complex chlorophyll, and the excited electron is then transferred stepwise from the special pair to the quinone (see also Figure 14–45). A similar arrangement of electron carriers is present in the reaction centers of plants.

center. Each antenna complex thereby acts as a funnel, collecting light energy and directing it to a specific site where it can be used effectively (see Figure 14–43).

The **photochemical reaction center** is a transmembrane protein–pigment complex that lies at the heart of photosynthesis. It is thought to have evolved more than 3 billion years ago in primitive photosynthetic bacteria. The special pair of chlorophyll molecules in the reaction center acts as an irreversible trap for excitation quanta because its excited electron is immediately passed to a chain of electron acceptors that are precisely positioned as neighbors in the same protein complex (Figure 14–44). By moving the high-energy electron rapidly away from the chlorophylls, the photochemical reaction center transfers it to an environment where it is much more stable. The electron is thereby suitably positioned for subsequent reactions, which require more time to complete.

In a Reaction Center, Light Energy Captured by Chlorophyll Creates a Strong Electron Donor from a Weak One

The electron transfers involved in the photochemical reactions just outlined have been analyzed extensively by rapid spectroscopic methods. An enormous amount of detailed information is available for the photosystem of purple bacteria, which is somewhat simpler than the evolutionarily related photosystems in chloroplasts. The reaction center in this photosystem is a large protein–pigment complex that can be solubilized with detergent and purified in active form. In 1985, its complete three-dimensional structure was determined by x-ray crystallography (see Figure 10–38). This structure, combined with kinetic data, provides the best picture we have of the initial electron-transfer reactions that underlie photosynthesis.

The sequence of electron transfers that take place in the reaction center of purple bacteria is shown in Figure 14–45. As outlined previously for the general case (see Figure 14–43), light causes a net electron transfer from a weak electron donor (a molecule with a strong affinity for electrons) to a molecule that is a strong electron donor in its reduced form. The excitation energy in chlorophyll that would normally be released as fluorescence or heat is thereby used instead to create a strong electron donor (a molecule carrying a high-energy electron) where none had been before. In the purple bacterium, the weak electron donor used to fill the electron-deficient hole created by a light-induced charge separation is a cytochrome (see *orange box* in Figure 14–45); the strong electron donor produced is a quinone. In the chloroplasts of higher plants, a quinone is similarly produced. However, as we discuss next, water serves as the initial weak electron donor, which is why oxygen gas is released by photosynthesis in plants.

Noncyclic Photophosphorylation Produces Both NADPH and ATP

Photosynthesis in plants and cyanobacteria produces both ATP and NADPH directly by a two-step process called **noncyclic photophosphorylation**. Because two photosystems—called photosystems I and II—are used in series to energize an electron, the electron can be transferred all the way from water to NADPH. As the high-energy electrons pass through the coupled photosystems to generate NADPH, some of their energy is siphoned off for ATP synthesis.

The first of the two photosystems—paradoxically called *photosystem II* for historical reasons—has the unique ability to withdraw electrons from water. The oxygens of two water molecules bind to a cluster of manganese atoms in a poorly understood water-splitting enzyme. This enzyme enables electrons to be removed one at a time from the water, as required to fill the electron-deficient holes created by light in chlorophyll molecules in the reaction center. As soon as four electrons have been removed from the two water molecules (requiring four quanta of light), O_2 is released. Photosystem II thus catalyzes the reaction $2H_2O + 4\text{ photons} \rightarrow 4H^+ + 4e^- + O_2$. As we discussed for the electron-transport chain in mitochondria, which uses O_2 and produces water, the mechanism ensures that no partly oxidized water molecules are released as dangerous, highly reactive oxygen radicals. Essentially all the oxygen in the Earth's atmosphere has been produced in this way.

The core of the reaction center in photosystem II is homologous to the bacterial reaction center just described, and it likewise produces strong electron donors in the form of reduced quinone molecules dissolved in the lipid bilayer of the membrane. The quinones pass their electrons to a H^+ pump called the *cytochrome b_6-f complex,* which resembles the cytochrome b-c_1 complex in the respiratory chain of mitochondria. The cytochrome b_6-f complex pumps H^+ into the thylakoid space across the thylakoid membrane (or out of the cytosol across the plasma membrane in cyanobacteria), and the resulting electrochemical gradient drives the synthesis of ATP by an ATP synthase (Figure 14–46).

The final electron acceptor in this electron-transport chain is the second photosystem, *photosystem I,* which accepts an electron into the electron-deficient hole created by light in the chlorophyll molecule in its reaction center. Each electron that enters photosystem I is finally boosted to a very high-energy

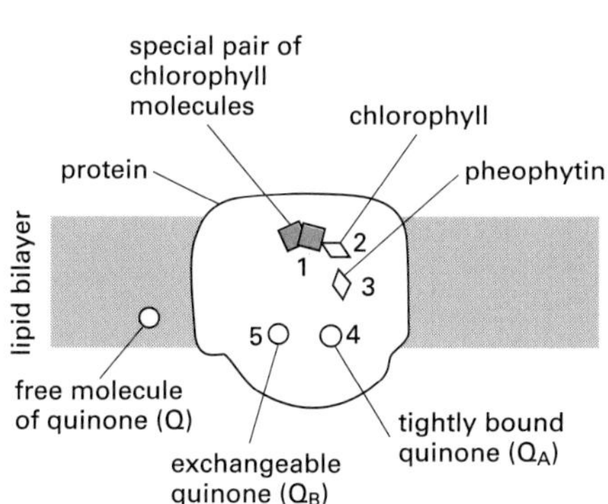

Figure 14–45 The electron transfers that occur in the photochemical reaction center of a purple bacterium. A similar set of reactions occurs in the evolutionarily related photosystem II in plants. At the top left is an orientating diagram showing the molecules that carry electrons, which are those in Figure 14–44, plus an exchangeable quinone (Q_B) and a freely mobile quinone (Q) dissolved in the lipid bilayer. Electron carriers 1–5 are each bound in a specific position on a 596-amino-acid transmembrane protein formed from two separate subunits (see Figure 10–38). After excitation by a photon of light, a high-energy electron passes from pigment molecule to pigment molecule, very rapidly creating a *charge separation*, as shown below in the sequence of steps A–C, in which the pigment molecule carrying a high-energy electron is colored *red*. Steps D and E then occur progressively. After a second photon has repeated this sequence with a second electron, the exchangeable quinone is released into the bilayer. This quinone quickly loses its charge by picking up two protons (see Figure 14–24).

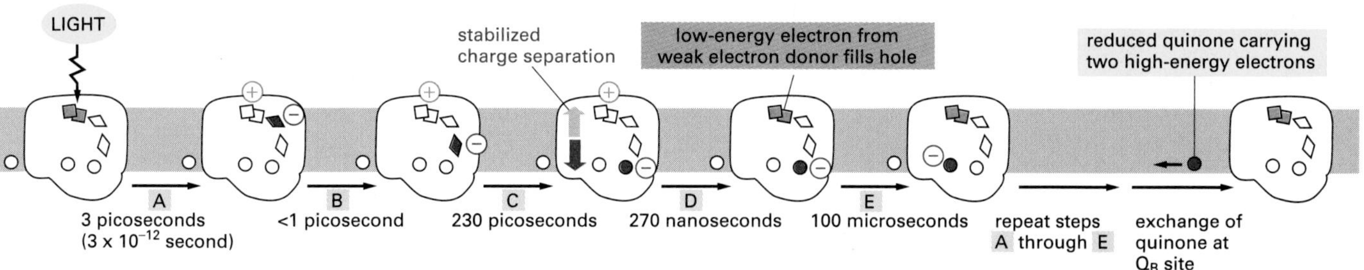

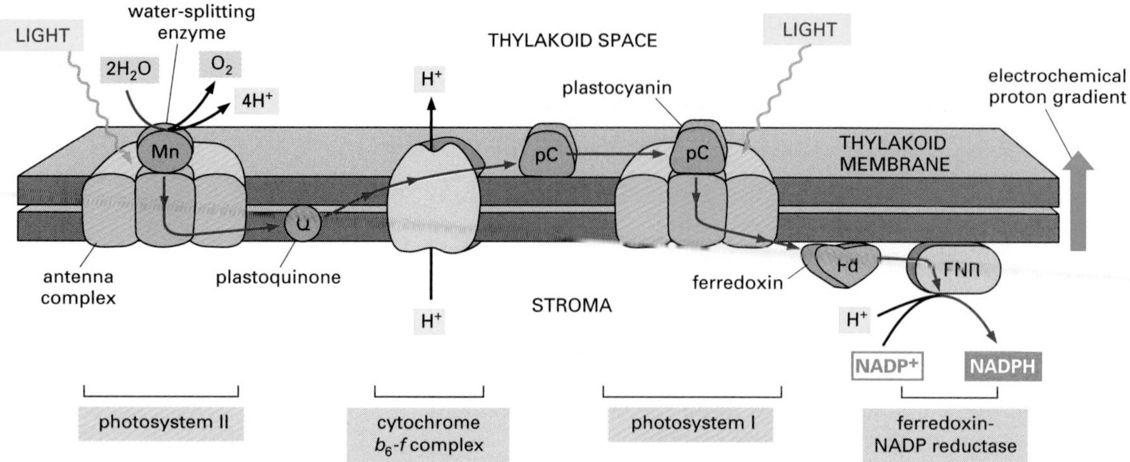

Figure 14–46 Electron flow during photosynthesis in the thylakoid membrane. The mobile electron carriers in the chain are plastoquinone (which closely resembles the ubiquinone of mitochondria), plastocyanin (a small copper-containing protein), and ferredoxin (a small protein containing an iron–sulfur center). The cytochrome b_6-f complex resembles the b-c_1 complex of mitochondria and the b-c complex of bacteria (see Figure 14–71): all three complexes accept electrons from quinones and pump H^+ across the membrane. The H^+ released by water oxidation into the thylakoid space, and the H^+ consumed during NADPH formation in the stroma, also contribute to the generation of the electrochemical H^+ gradient. This gradient drives ATP synthesis by an ATP synthase present in this same membrane (not shown here).

level that allows it to be passed to the iron–sulfur center in ferredoxin and then to $NADP^+$ to generate NADPH (Figure 14–47).

The scheme for photosynthesis just discussed is known as the *Z scheme*. By means of its two electron-energizing steps, one catalyzed by each photosystem, an electron is passed from water, which normally holds on to its electrons very tightly (redox potential = +820 mV), to NADPH, which normally holds on to its electrons loosely (redox potential = –320 mV). There is not enough energy in a single quantum of visible light to energize an electron all the way from the bottom

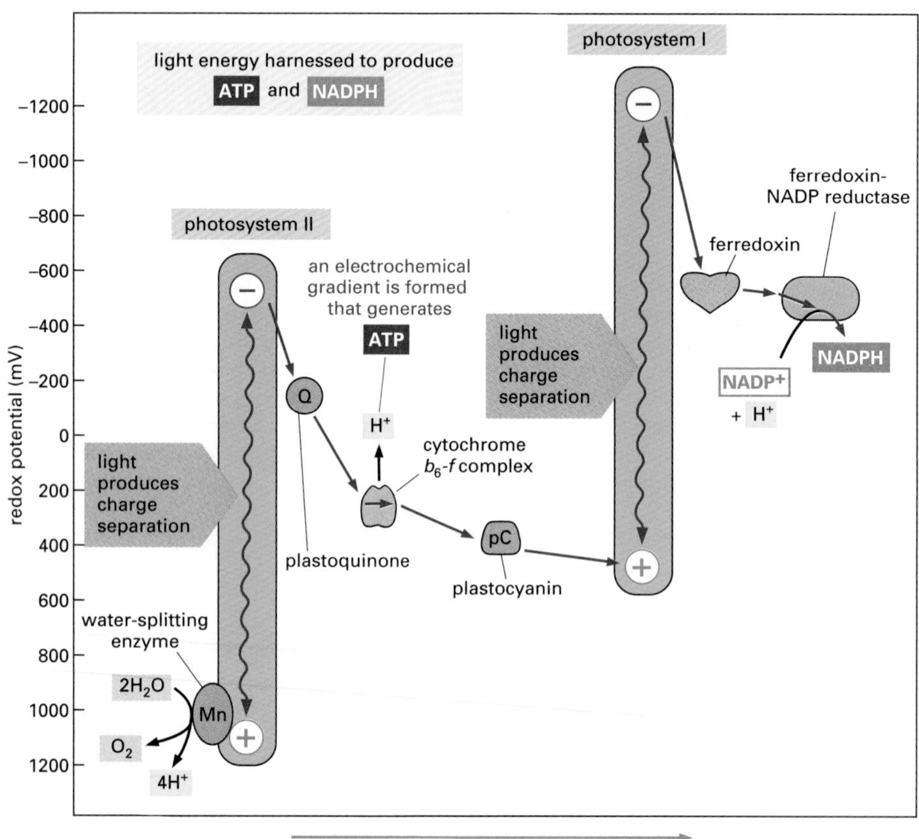

Figure 14–47 Changes in redox potential during photosynthesis. The redox potential for each molecule is indicated by its position along the vertical axis. In photosystem II, the excited reaction center chlorophyll has a redox potential high enough to withdraw electrons from water, by means of a specially organized cluster of four manganese atoms. Photosystem II closely resembles the reaction center in purple bacteria, and it passes electrons from its excited chlorophyll to an electron-transport chain that leads to photosystem I. Photosystem I then passes electrons from its excited chlorophyll through a series of tightly bound iron–sulfur centers. The net electron flow through the two photosystems in series is from water to $NADP^+$, and it produces NADPH as well as ATP.

The ATP is synthesized by an ATP synthase that harnesses the electrochemical proton gradient produced by the three sites of H^+ activity that are highlighted in Figure 14–46. This Z scheme for ATP production is called noncyclic photophosphorylation, to distinguish it from a cyclic scheme that utilizes only photosystem I (see the text).

of photosystem II to the top of photosystem I, which is presumably the energy change required to pass an electron efficiently from water to $NADP^+$. The use of two separate photosystems in series means that the energy from two quanta of light is available for this purpose. In addition, there is enough energy left over to enable the electron-transport chain that links the two photosystems to pump H^+ across the thylakoid membrane (or the plasma membrane of cyanobacteria), so that the ATP synthase can harness some of the light-derived energy for ATP production.

Chloroplasts Can Make ATP by Cyclic Photophosphorylation Without Making NADPH

In the noncyclic photophosphorylation scheme just discussed, high-energy electrons leaving photosystem II are harnessed to generate ATP and are passed on to photosystem I to drive the production of NADPH. This produces slightly more than 1 molecule of ATP for every pair of electrons that passes from H_2O to $NADP^+$ to generate a molecule of NADPH. But 1.5 molecules of ATP per NADPH are needed for carbon fixation (see Figure 14–39). To produce extra ATP, the chloroplasts in some species of plants can switch photosystem I into a cyclic mode so that it produces ATP instead of NADPH. In this process, called *cyclic photophosphorylation,* the high-energy electrons from photosystem I are transferred to the cytochrome b_6-f complex rather than being passed on to $NADP^+$. From the b_6-f complex, the electrons are passed back to photosystem I at a low energy. The only net result, besides the conversion of some light energy to heat, is that H^+ is pumped across the thylakoid membrane by the b_6-f complex as electrons pass through it, thereby increasing the electrochemical proton gradient that drives the ATP synthase. (This is analogous to the right side of the diagram for purple nonsulfur bacteria in Figure 14–71, below.)

To summarize, cyclic photophosphorylation involves only photosystem I, and it produces ATP without the formation of either NADPH or O_2. The relative activities of cyclic and noncyclic electron flows can be regulated by the cell to determine how much light energy is converted into reducing power (NADPH) and how much into high-energy phosphate bonds (ATP).

Photosystems I and II Have Related Structures, and Also Resemble Bacterial Photosystems

The mechanisms of fundamental cell processes such as DNA replication or respiration generally turn out to be the same in eucaryotic cells and in bacteria, even though the number of protein components involved is considerably greater in eucaryotes. Eucaryotes evolved from procaryotes, and the additional proteins presumably were selected for during evolution because they provided an extra degree of efficiency and/or regulation that was useful to the cell.

Photosystems provide a clear example of this type of evolution. Photosystem II, for example, is formed from more than 25 different protein subunits, creating a large assembly in the thylakoid membrane with a mass of about 1 million daltons. The atomic structures of the eucaryotic photosystems are being revealed by a combination of electron and x-ray crystallography. The task is difficult because the complexes are large and embedded in the lipid bilayer. Nevertheless, as illustrated in Figure 14–48, the close relationship of photosystem I, photosystem II, and the photochemical reaction center of purple bacteria has been clearly demonstrated from these atomic-level analyses.

The Proton-Motive Force Is the Same in Mitochondria and Chloroplasts

The presence of the thylakoid space separates a chloroplast into three rather than the two internal compartments of a mitochondrion. The net effect of H^+ translocation in the two organelles is, however, similar. As illustrated in Figure

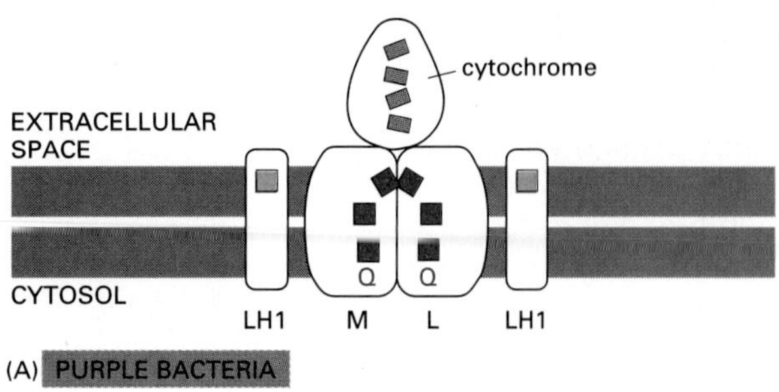

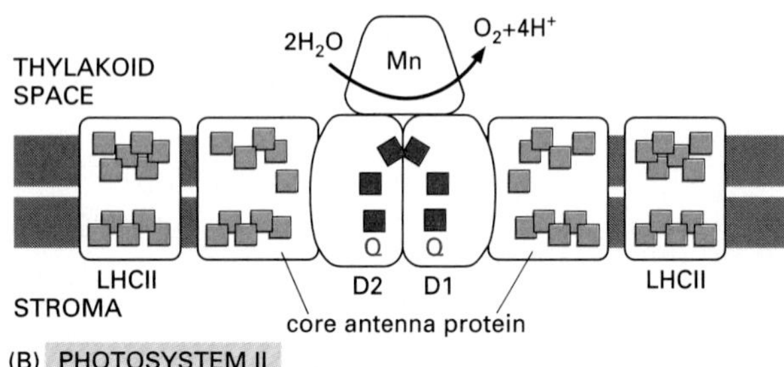

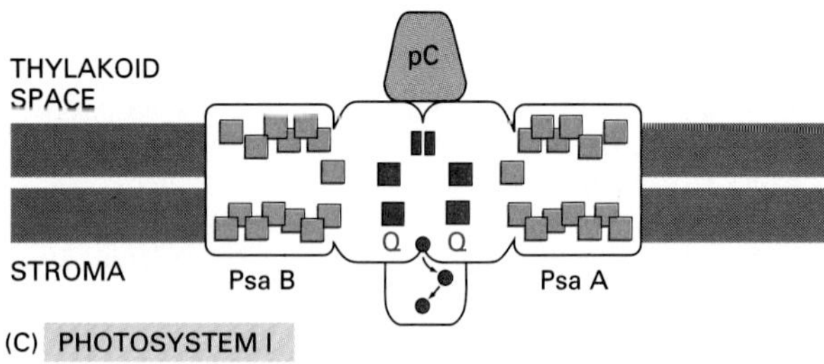

Figure 14–48 Three types of photosynthetic reaction centers compared. Pigments involved in light harvesting are colored *green;* those involved in the central photochemical events are colored *red.* (A) The photochemical reaction center of purple bacteria, whose detailed structure is illustrated in Figure 10–38, contains two related protein subunits, L and M, that bind the pigments involved in the central process illustrated in Figure 14–45. Low-energy electrons are fed into the excited chlorophylls by a cytochrome. LH1 is a protein–pigment complex involved in light harvesting. (B) Photosystem II contains the D_1 and D_2 proteins, which are homologous to the L and M subunits in (A). Low-energy electrons from water are fed into the excited chlorophylls by a manganese cluster. LHCII is a light-harvesting complex that feeds energy into the core antenna proteins. (C) Photosystem I contains the Psa A and Psa B proteins, each of which is equivalent to a fusion of the D_1 or D_2 protein to a core antenna protein of photosystem II. Low-energy electrons are fed into the excited chlorophylls by loosely bound plastocyanin (pC). As indicated, in photosystem I, high-energy electrons are passed from a nonmobile quinone (Q) through a series of three iron-sulfur centers *(red circles).* (Modified from K. Rhee, E. Morris, J. Barber, and W. Kühlbrandt, *Nature* 396:283–286, 1998; W. Kühlbrandt, *Nature* 411:896–899, 2001.)

14–49, in chloroplasts, H^+ is pumped out of the stroma (pH 8) into the thylakoid space (pH ~5), creating a gradient of 3–3.5 pH units. This represents a proton-motive force of about 200 mV across the thylakoid membrane, and it drives ATP synthesis by the ATP synthase embedded in this membrane. The force is the same as that across the inner mitochondrial membrane, but nearly all of it is contributed by the pH gradient rather than by a membrane potential, unlike the case in mitochondria.

Like the stroma, the mitochondrial matrix has a pH of about 8. This is created by pumping H^+ out of the mitochondrion into the cytosol (pH ~7) rather than into an interior space in the organelle. Thus, the pH gradient is relatively small, and most of the proton-motive force across the inner mitochondrial membrane is instead caused by the resulting membrane potential (see Figure 14–13).

For both mitochondria and chloroplasts, the catalytic site of the ATP synthase is at a pH of about 8 and is located in a large organelle compartment (matrix or stroma) that is packed full of soluble enzymes. Consequently, it is here that all of the organelle's ATP is made (see Figure 14–49).

Carrier Proteins in the Chloroplast Inner Membrane Control Metabolite Exchange with the Cytosol

If chloroplasts are isolated in a way that leaves their inner membrane intact, this membrane can be shown to have a selective permeability, reflecting the presence of specific carrier proteins. Most notably, much of the glyceraldehyde 3-phosphate produced by CO_2 fixation in the chloroplast stroma is transported out of the chloroplast by an efficient antiport system that exchanges three-carbon sugar phosphates for an inward flux of inorganic phosphate.

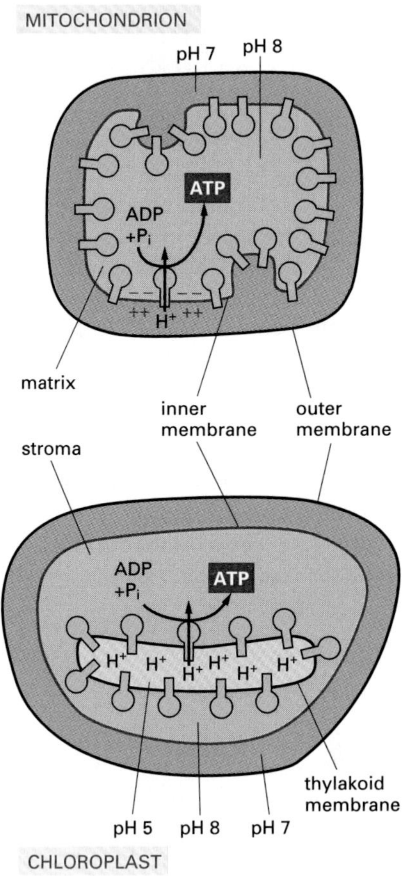

Figure 14–49 A comparison of the flow of H^+ and the orientation of the ATP synthase in mitochondria and chloroplasts. Those compartments with similar pH values have been colored the same. The proton-motive force across the thylakoid membrane consists almost entirely of the pH gradient; a high permeability of this membrane to Mg^{2+} and Cl^- ions allows the flow of these ions to dissipate most of the membrane potential. Mitochondria presumably need a large membrane potential because they could not tolerate having their matrix at pH 10, as would be required to generate their proton-motive force without one.

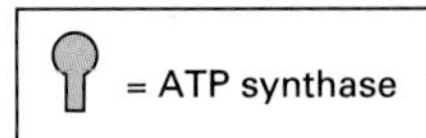

Glyceraldehyde 3-phosphate normally provides the cytosol with an abundant source of carbohydrate, which is used by the cell as the starting point for many other biosyntheses—including the production of sucrose for export. But this is not all that this molecule provides. Once the glyceraldehyde 3-phosphate reaches the cytosol, it is readily converted (by part of the glycolytic pathway) to 1,3-phosphoglycerate and then 3-phosphoglycerate (see p. 97), generating one molecule of ATP and one of NADH. (A similar two-step reaction, but working in reverse, forms glyceraldehyde 3-phosphate in the carbon-fixation cycle; see Figure 14–39.) As a result, the export of glyceraldehyde 3-phosphate from the chloroplast provides not only the main source of fixed carbon to the rest of the cell, but also the reducing power and ATP needed for metabolism outside the chloroplast.

Chloroplasts Also Perform Other Crucial Biosyntheses

The chloroplast performs many biosyntheses in addition to photosynthesis. All of the cell's fatty acids and a number of amino acids, for example, are made by enzymes in the chloroplast stroma. Similarly, the reducing power of light-activated electrons drives the reduction of nitrite (NO_2^-) to ammonia (NH_3) in the chloroplast; this ammonia provides the plant with nitrogen for the synthesis of amino acids and nucleotides. The metabolic importance of the chloroplast for plants and algae therefore extends far beyond its role in photosynthesis.

Summary

Chloroplasts and photosynthetic bacteria obtain high-energy electrons by means of photosystems that capture the electrons that are excited when sunlight is absorbed by chlorophyll molecules. Photosystems are composed of an antenna complex that funnels energy to a photochemical reaction center, where a precisely ordered complex of proteins and pigments allows the energy of an excited electron in chlorophyll to be captured by electron carriers. The best-understood photochemical reaction center is that of purple photosynthetic bacteria, which contain only a single photosystem. In contrast, there are two distinct photosystems in chloroplasts and cyanobacteria. The two photosystems are normally linked in series, and they transfer electrons from water to NADP$^+$ to form NADPH, with the concomitant production of a transmembrane electrochemical proton gradient. In these linked photosystems, molecular oxygen (O_2) is generated as a by-product of removing four low-energy electrons from two specifically positioned water molecules.

Compared with mitochondria, chloroplasts have an additional internal membrane (the thylakoid membrane) and a third internal space (the thylakoid space). All electron-transport processes occur in the thylakoid membrane: to make ATP, H$^+$ is pumped into the thylakoid space, and a backflow of H$^+$ through an ATP synthase then produces the ATP in the chloroplast stroma. This ATP is used in conjunction with the NADPH made by photosynthesis to drive a large number of biosynthetic reactions in the chloroplast stroma, including the all-important carbon-fixation cycle, which creates carbohydrate from CO_2. Along with some other important chloroplast products, this carbohydrate is exported to the cell cytosol, where—as glyceraldehyde 3-phosphate—it provides organic carbon, ATP, and reducing power to the rest of the cell.

THE GENETIC SYSTEMS OF MITOCHONDRIA AND PLASTIDS

It is widely accepted that mitochondria and plastids evolved from bacteria that were engulfed by nucleated ancestral cells. As a relic of this evolutionary past, both types of organelles contain their own genomes, as well as their own biosynthetic machinery for making RNA and organelle proteins. Mitochondria and plastids are never made from scratch, but instead arise by the growth and division of an existing mitochondrion or plastid. On average, each organelle must double in mass in each cell generation and then be distributed into each daughter cell. Even nondividing cells must replenish organelles that are degraded as part of the continual process of organelle turnover, or produce additional organelles as the need arises. The process of organelle growth and proliferation is complicated because mitochondrial and plastid proteins are encoded in two places: the nuclear genome and the separate genomes harbored in the organelles themselves (Figure 14–50). In Chapter 12, we discuss how selected proteins and lipids are imported into mitochondria and chloroplasts from the cytosol. Here we describe how the organelle genomes are maintained and the contributions they make to organelle biogenesis.

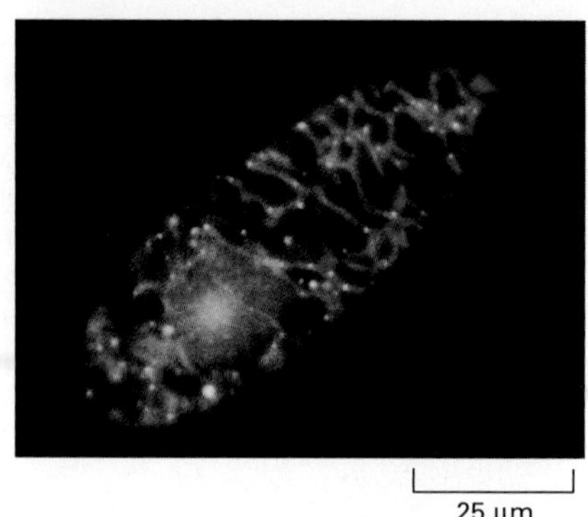

Figure 14–50 Mitochondrial and nuclear DNA stained with a fluorescent dye. This micrograph shows the distribution of the nuclear genome *(red)* and the multiple small mitochondrial genomes (bright *yellow* spots) in a *Euglena gracilis* cell. The DNA is stained with ethidium bromide, a fluorescent dye that emits red light. In addition, the mitochondrial matrix space is stained with a green fluorescent dye that reveals the mitochondria as a branched network extending throughout the cytosol. The superposition of the *green* matrix and the *red* DNA gives the mitochondrial genomes their *yellow* color. (Courtesy of Y. Hayashi and K. Ueda, *J. Cell Sci.* 93:565–570, 1989. © The Company of Biologists.)

Mitochondria and Chloroplasts Contain Complete Genetic Systems

The biosynthesis of mitochondria and plastids requires contributions from two separate genetic systems. Most of the proteins in mitochondria and chloroplasts are encoded by special genes devoted to this purpose in nuclear DNA. These proteins are imported into the organelle from the cytosol after they have been synthesized on cytosolic ribosomes. Other organelle proteins are encoded by organelle DNA and synthesized on ribosomes within the organelle, using organelle-produced mRNA to specify their amino acid sequence (Figure 14–51). The protein traffic between the cytosol and these organelles seems to be unidirectional, as no known proteins are exported from mitochondria or chloroplasts to the cytosol. An exception occurs under special conditions when a cell is about to undergo apoptosis. The release of intermembrane space proteins (including cytochrome *c*) from mitochondria through the outer mitochondrial membrane is part of a signaling pathway that is triggered in cells undergoing programmed cell death (discussed in Chapter 17).

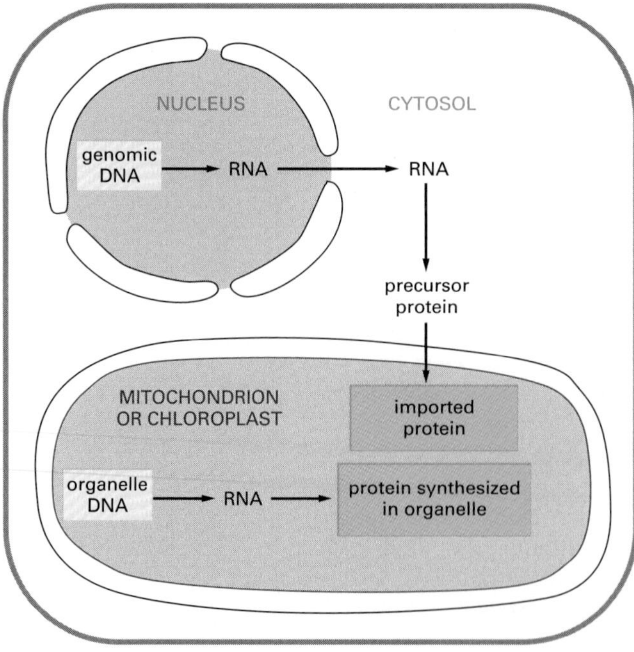

Figure 14–51 The production of mitochondrial and chloroplast proteins by two separate genetic systems. Most of the proteins in these organelles are encoded by the nucleus and must be imported from the cytosol.

The processes of organelle DNA transcription, protein synthesis, and DNA replication (Figure 14–52) take place where the genome is located: in the matrix of mitochondria and the stroma of chloroplasts. Although the proteins that mediate these genetic processes are unique to the organelle, most of them are encoded in the nuclear genome. This is all the more surprising because the protein-synthesis machinery of the organelles resembles that of bacteria rather than that of eucaryotes. The resemblance is particularly close in chloroplasts. For example, chloroplast ribosomes are very similar to *E. coli* ribosomes, both in their structure and in their sensitivity to various antibiotics (such as chloramphenicol, streptomycin, erythromycin, and tetracycline). In addition, protein synthesis in chloroplasts starts with *N*-formyl methionine, as in bacteria, and not with the methionine used for this purpose in the cytosol of eucaryotic cells. Although mitochondrial genetic systems are much less similar to those of present-day bacteria than are the genetic systems of chloroplasts, their ribosomes are also sensitive to antibacterial antibiotics, and protein synthesis in mitochondria also starts with *N*-formyl methionine.

Figure 14–52 An electron micrograph of an animal mitochondrial DNA molecule caught during the process of DNA replication. The circular DNA genome has replicated only between the two points marked by *red* arrows. The newly synthesized DNA is colored *yellow*. (Courtesy of David A. Clayton.)

Organelle Growth and Division Determine the Number of Mitochondria and Plastids in a Cell

Mitochondria and plastids are large enough to be observed by light microscopy in living cells. For example, mitochondria can be visualized by expressing a genetically engineered fusion of a mitochondrial protein linked to the green fluorescent protein (GFP) in cells, or cells can be incubated with a fluorescent dye that is specifically taken up by mitochondria because of the electrochemical gradient across their membranes. From such images, the mitochondria in living cells are seen to be very dynamic—frequently dividing, fusing, and changing shape (Figure 14–53), as mentioned previously. Division (fission) and fusion of these organelles are topologically complex processes, because the organelles are enclosed by a double membrane and the integrity of the separate mitochondrial compartments must be maintained (Figure 14–54).

The copy number and shape of mitochondria vary dramatically in different cell types and can change in the same cell type under different physiological conditions, ranging from multiple spherical organelles to a single organelle with a branched structure (or reticulum). The arrangement is controlled by the relative rates of mitochondrial division and fusion, which are regulated by dedicated GTPases that reside on mitochondrial membranes. The regulation of mitochondrial morphology and distribution is important for cell differentiation

Figure 14–53 Dynamic mitochondrial reticulum. (A) In yeast cells, mitochondria form a continuous reticulum underlying the plasma membrane. (B) A balance between fission and fusion determines the arrangement of the mitochondria in different cells. (C) Time-lapse fluorescent microscopy shows the dynamic behavior of the mitochondrial network in a yeast cell. In addition to shape changes, the network is constantly remodeled by fission and fusion *(red arrows)*. The pictures were taken at 3-minute intervals. (A and C, from J. Nunnari et al., *Mol. Biol. Cell* 8:1233–1242, 1997, with permission by the American Society for Cell Biology.)

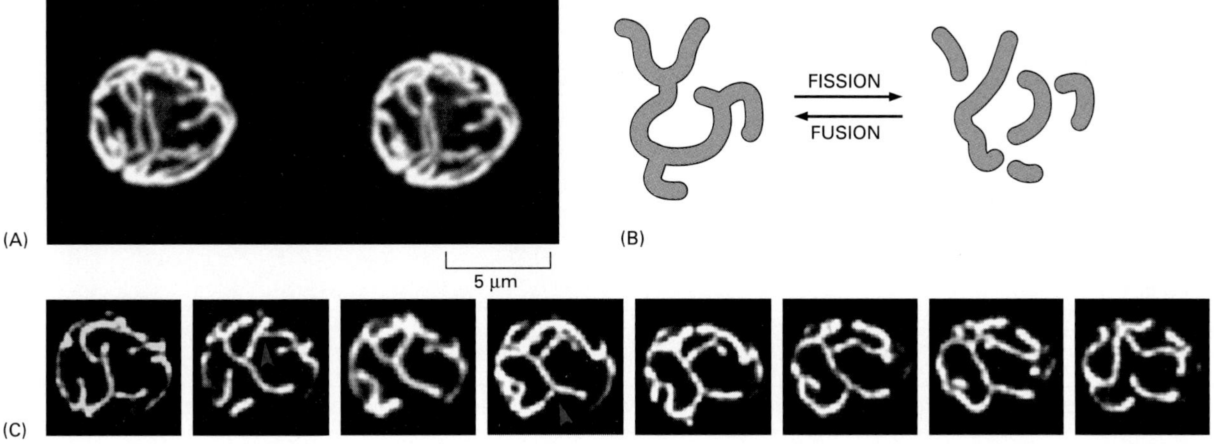

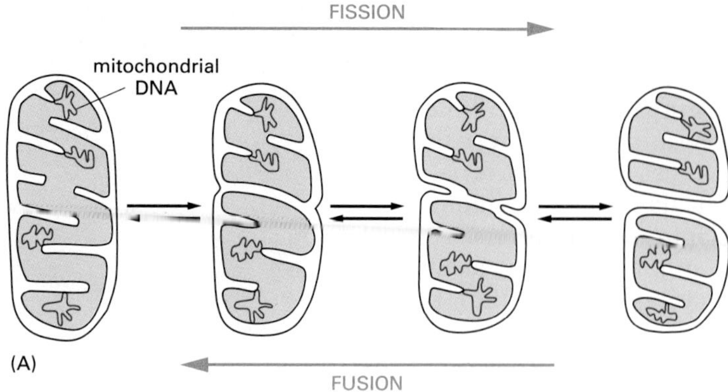

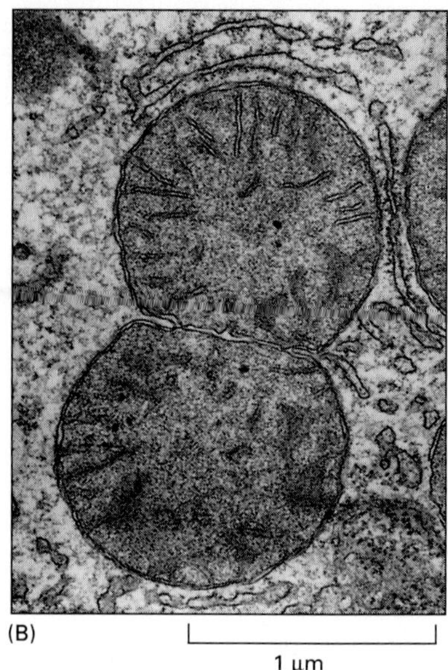

Figure 14–54 Mitochondrial fission and fusion. These processes involve both outer and inner mitochondrial membranes. (A) During fusion and fission, both matrix and intermembrane space compartments are maintained. Different membrane fusion machines are thought to operate at the outer and inner membranes. Conceptually, the fission process resembles that of bacterial cell division (discussed in Chapter 18). The pathway shown has been postulated from static views such as that shown in (B). (B) An electron micrograph of a dividing mitochondrion in a liver cell. (B, courtesy of Daniel S. Friend.)

and function. As an example, mutations in *Drosophila* that impair mitochondrial fusion, and hence cause extensive mitochondrial fragmentation, block sperm development and produce infertility.

There can be many copies of the mitochondrial and plastid genomes in the space enclosed by each organelle's inner membrane. How many of these genomes are present in a single organelle depends on the degree of organelle fragmentation; frequently, many genomes are housed in the same compartment (Table 14–2). In most cells, the replication of the organelle DNA is not limited to the S phase of the cell cycle, when the nuclear DNA replicates, but occurs throughout the cell cycle—out of phase with cell division. Individual organelle DNA molecules seem to be selected at random for replication, so that in a given cell cycle, some may replicate more than once and others not at all. Nonetheless, under constant conditions, the process is regulated to ensure that the total number of organelle DNA molecules doubles in every cell cycle, as required if each cell type is to maintain a constant amount of organelle DNA. When conditions change, the total organelle mass per cell can be regulated according to need. A large increase in mitochondria (as much as five- to tenfold), for example, is observed if a resting skeletal muscle is repeatedly stimulated to contract for a prolonged period.

In special circumstances, organelle division can be precisely controlled by the cell. In some algae that contain only one or a few chloroplasts, the organelle divides just before the cell does, in a plane that is identical to the future plane of cell division.

The Genomes of Mitochondria and Chloroplasts Are Diverse

The multiple copies of mitochondrial and chloroplast DNA contained within the matrix or stroma of these organelles are usually distributed in several clusters,

TABLE 14–2 Relative Amounts of Organelle DNA in Some Cells and Tissues

ORGANISM	TISSUE OR CELL TYPE	DNA MOLECULES PER ORGANELLE	ORGANELLES PER CELL	ORGANELLE DNA AS PERCENTAGE OF TOTAL CELLULAR DNA
MITOCHONDRIAL DNA				
Rat	liver	5–10	1000	1
Yeast*	vegetative	2–50	1–50	15
Frog	egg	5–10	10^7	99
CHLOROPLAST DNA				
Chlamydomonas	vegetative	80	1	7
Maize	leaves	20–40	20–40	15

*The large variation in the number and size of mitochondria per cell in yeasts is due to mitochondrial fusion and fission.

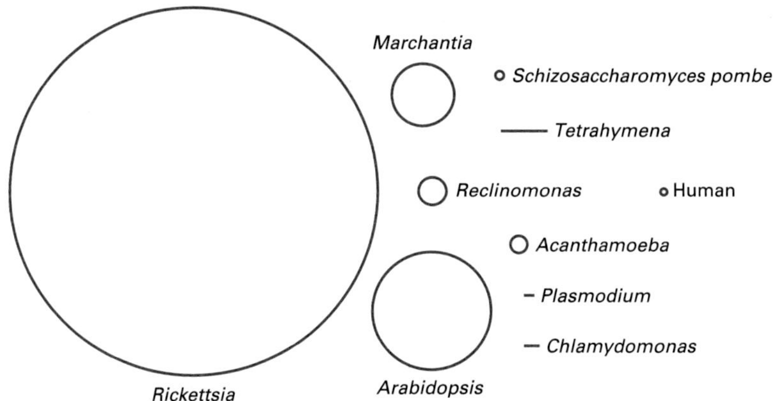

Figure 14–55 Various sizes of mitochondrial genomes. The complete DNA sequences for more than 200 mitochondrial genomes have been determined. The lengths of a few of these mitochondrial DNAs are shown to scale as *circles* for circular genomes and *lines* for linear genomes. The largest circle represents the genome of *Rickettsia prowazekii*, a small pathogenic bacterium whose genome most closely resembles that of mitochondria. The size of mitochondrial genomes does not correlate well with the number of proteins encoded in them: while human mitochondrial DNA encodes 13 proteins, the 22-fold larger mitochondrial DNA of *Arabidopsis* encodes only 32 proteins—that is, about 2.5-fold as many as human mitochondrial DNA. The extra DNA that is found in *Arabidopsis*, *Marchantia*, and other plant mitochondria may be "junk DNA". The mitochondrial DNA of the protozoan *Reclinomonas americana* has 97 genes, more than the mitochondrion of any other organism analyzed so far. (Adapted from M.W. Gray et al., *Science* 283:1476–1481, 1999.)

called *nucleoids*. Nucleoids are thought to be attached to the inner mitochondrial membrane. Although it is not known how the DNA is packaged, the DNA structure in nucleoids is likely to resemble that in bacteria rather than that in eucaryotic chromatin. As in bacteria, for example, there are no histones.

The size range of organelle DNAs is similar to that of viral DNAs. Mitochondrial DNA molecules range in size from less than 6000 nucleotide pairs in *Plasmodium falciparum* (the human malaria parasite) to more than 300,000 nucleotide pairs in some land plants (Figure 14–55). Like a typical bacterial genome, most mitochondrial DNAs are circular molecules, although linear mitochondrial DNA exists as well. In mammals, the mitochondrial genome is a DNA circle of about 16,500 base pairs (less than 0.001% of the size of the nuclear genome). It is nearly the same size in animals as diverse as *Drosophila* and sea urchins. The chloroplast genome of land plants ranges in size from 70,000 to 200,000 nucleotide pairs, and it is circular in all organisms examined thus far.

In mammalian cells, mitochondrial DNA makes up less than 1% of the total cellular DNA. In other cells, however, such as the leaves of higher plants or the very large egg cells of amphibians, a much larger fraction of the cellular DNA may be present in mitochondria or chloroplasts (see Table 14–2), and a large fraction of RNA and protein synthesis takes place there.

Mitochondria and Chloroplasts Probably Both Evolved from Endosymbiotic Bacteria

The procaryotic character of the organelle genetic systems, especially striking in chloroplasts, suggests that mitochondria and chloroplasts evolved from bacteria that were endocytosed more than 1 billion years ago. According to this *endosymbiont hypothesis*, eucaryotic cells started out as anaerobic organisms without mitochondria or chloroplasts and then established a stable endosymbiotic relation with a bacterium, whose oxidative phosphorylation system they subverted for their own use (Figure 14–56). The endocytic event that led to the development of mitochondria is presumed to have occurred when oxygen entered the atmosphere in substantial amounts, about 1.5×10^9 years ago, before animals and plants separated (see Figure 14–69).

Plant and algal chloroplasts seem to have been derived later from an endocytic event involving an oxygen-producing photosynthetic bacterium. To explain the different pigments and properties of the chloroplasts found in present-day higher plants and algae, it is usually assumed that at least three independent endosymbiotic events occurred.

Most of the genes encoding present-day mitochondrial and chloroplast proteins are in the cell nucleus. Thus, an extensive transfer of genes from organelle to nuclear DNA must have occurred during eucaryote evolution. In contrast, present organelle genomes are stable, indicating that a successful transfer is a rare evolutionary process. This is expected, because a gene moved from organelle DNA needs to change to become a functional nuclear gene: it must adapt to the nuclear and cytoplasmic transcription and translation requirements, and also acquire a signal sequence so that the encoded protein can be delivered to the organelle after its synthesis in the cytosol.

The gene transfer hypothesis explains why many of the nuclear genes encoding mitochondrial and chloroplast proteins resemble bacterial genes. The amino acid sequence of the chicken mitochondrial enzyme *superoxide dismutase,* for example, resembles the corresponding bacterial enzyme much more than it resembles the superoxide dismutase found in the cytosol of the same eucaryotic cell. Further evidence that such DNA transfers have occurred during evolution comes from the discovery of some noncoding DNA sequences in nuclear DNA that seem to be of recent mitochondrial origin; they have apparently integrated into the nuclear genome as "junk DNA."

Gene transfer seems to have been a gradual process. When mitochondrial genomes encoding different numbers of proteins are compared, a pattern of sequential reduction of encoded mitochondrial functions emerges (Figure 14–57). The smallest and presumably most highly evolved mitochondrial genomes, for example, encode only a few inner-membrane proteins involved in electron-transport reactions, plus ribosomal RNAs and tRNAs. Mitochondrial genomes that have remained more complex contain this same subset of genes, plus others. The most complex genomes are characterized by the presence of many extra genes compared with animal and yeast mitochondrial genomes. Many of these genes encode components of the mitochondrial genetic system, such as RNA polymerase subunits and ribosomal proteins; these genes are instead found in the cell nucleus in organisms that have reduced their mitochondrial DNA content.

What type of bacterium gave rise to the mitochondrion? From sequence comparisons, it seems that mitochondria are descendants of a particular type of purple photosynthetic bacterium that had previously lost its ability to perform photosynthesis and was left with only a respiratory chain. It is not certain that all mitochondria have originated from the same endosymbiotic event, however.

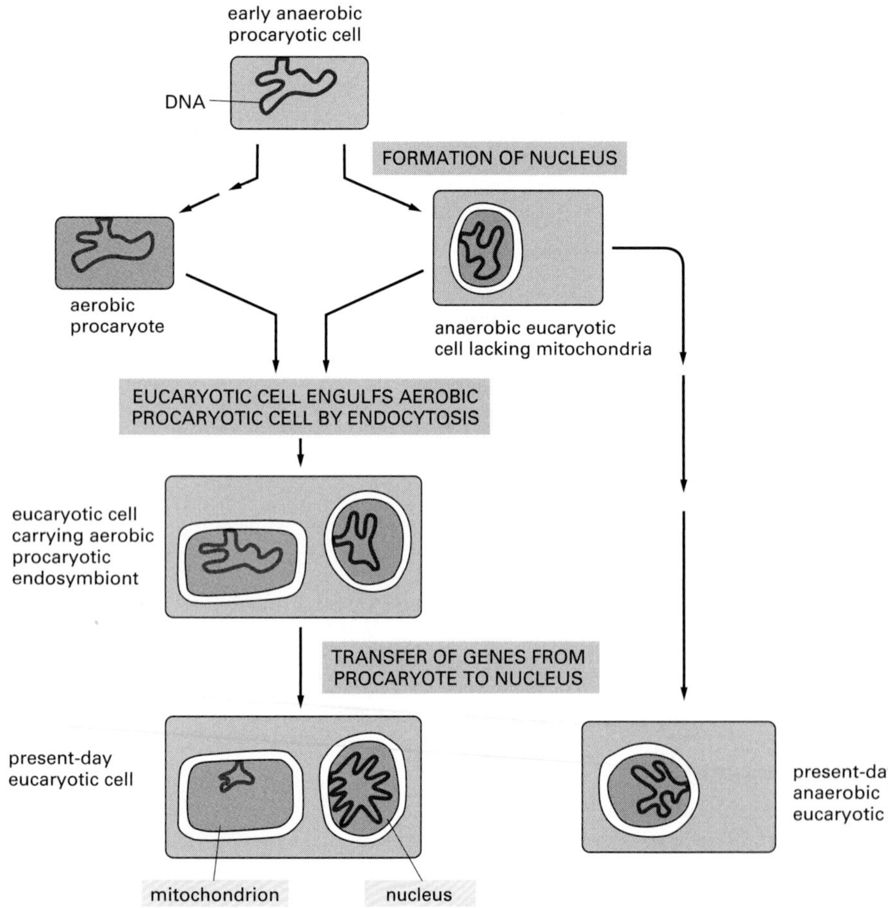

Figure 14–56 A suggested evolutionary pathway for the origin of mitochondria. *Microsporidia* and *Giardia* are two present-day anaerobic single-celled eucaryotes (protozoans) without mitochondria. Because they have an rRNA sequence that suggests a great deal of evolutionary distance from all other known eucaryotes, it has been postulated that their ancestors were also anaerobic and resembled the eucaryote that first engulfed the precursors of mitochondria.

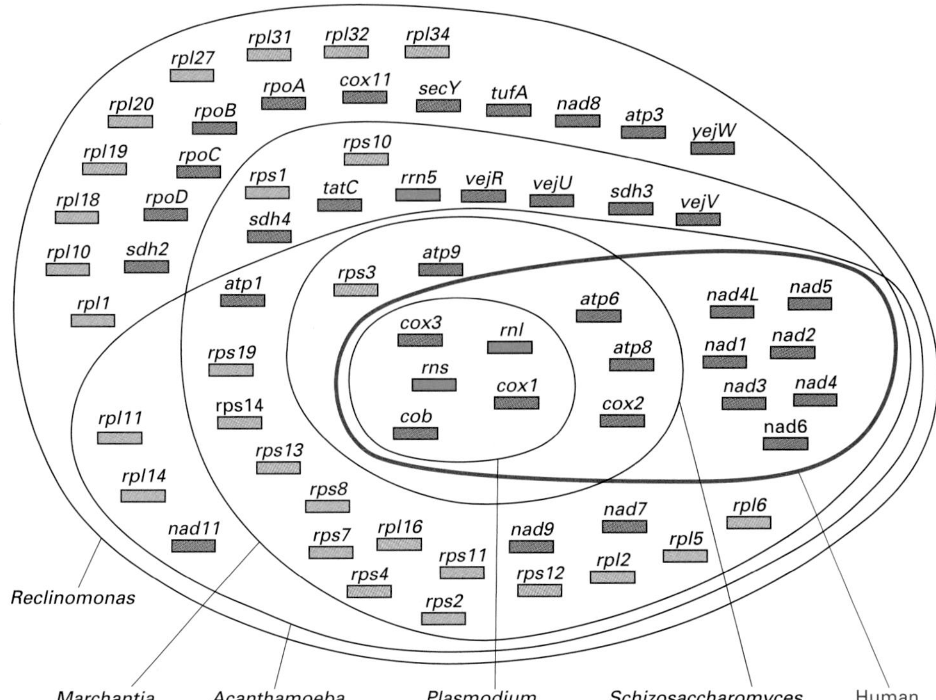

Figure 14–57 Comparison of mitochondrial genomes. Less complex mitochondrial genomes encode subsets of the proteins and ribosomal RNAs that are encoded by larger mitochondrial genomes. The five genes present in all known mitochondrial genomes encode ribosomal RNAs *(rns* and *rnl)*, cytochrome *b (cob)*, and two cytochrome oxidase subunits *(cox1* and *cox3)*. (Adapted from M.W. Gray et al., *Science* 283:1476–1481, 1999.)

Mitochondrial Genomes Have Several Surprising Features

The relatively small size of the human mitochondrial genome made it a particularly attractive target for early DNA-sequencing projects, and in 1981, the complete sequence of its 16,569 nucleotides was published. By comparing this sequence with known mitochondrial tRNA sequences and with the partial amino acid sequences available for proteins encoded by the mitochondrial DNA, all of the human mitochondrial genes were mapped on the circular DNA molecule (Figure 14–58).

Compared with nuclear, chloroplast, and bacterial genomes, the human mitochondrial genome has several surprising features:

1. *Dense gene packing.* Unlike other genomes, nearly every nucleotide seems to be part of a coding sequence, either for a protein or for one of the rRNAs or tRNAs. Since these coding sequences run directly into each other, there is very little room left for regulatory DNA sequences.
2. *Relaxed codon usage.* Whereas 30 or more tRNAs specify amino acids in the cytosol and in chloroplasts, only 22 tRNAs are required for mitochondrial protein synthesis. The normal codon-anticodon pairing rules are relaxed in mitochondria, so that many tRNA molecules recognize any one of the four nucleotides in the third (wobble) position. Such "2 out of 3" pairing allows one tRNA to pair with any one of four codons and permits protein synthesis with fewer tRNA molecules.
3. *Variant genetic code.* Perhaps most surprising, comparisons of mitochondrial gene sequences and the amino acid sequences of the corresponding proteins indicate that the genetic code is different: 4 of the 64 codons have different "meanings" from those of the same codons in other genomes (Table 14–3).

The observation that the genetic code is nearly the same in all organisms provides strong evidence that all cells have evolved from a common ancestor. How, then, does one explain the few differences in the genetic code in many mitochondria? A hint comes from the finding that the mitochondrial genetic code is different in different organisms. In the mitochondrion with the largest number of genes in Figure 14–57, that of the protozoan *Reclinomonas*, the genetic code is unchanged from the standard genetic code of the cell nucleus. Yet UGA, which is a stop codon elsewhere, is read as tryptophan in mitochondria of mammals, fungi, and invertebrates. Similarly, the codon AGG normally codes for arginine, but it codes for *stop* in the mitochondria of mammals and codes for

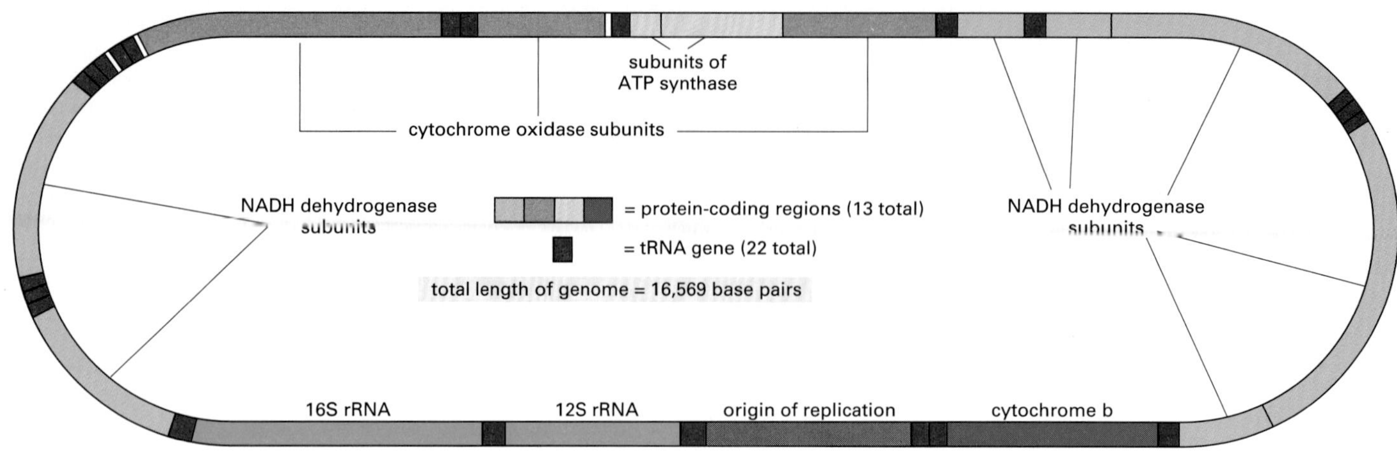

Figure 14–58 The organization of the human mitochondrial genome. The genome contains 2 rRNA genes, 22 tRNA genes, and 13 protein-coding sequences. The DNAs of many other animal mitochondrial genomes have also been completely sequenced. Most of these animal mitochondrial DNAs encode precisely the same genes as humans, with the gene order being identical for animals that range from mammals to fish.

serine in the mitochondria of *Drosophila* (see Table 14–3). Such variation suggests that a random drift can occur in the genetic code in mitochondria. Presumably, the unusually small number of proteins encoded by the mitochondrial genome makes an occasional change in the meaning of a rare codon tolerable, whereas such a change in a large genome would alter the function of many proteins and thereby destroy the cell.

Animal Mitochondria Contain the Simplest Genetic Systems Known

Comparisons of DNA sequences in different organisms reveal that the rate of nucleotide substitution during evolution has been 10 times greater in mitochondrial genomes than in nuclear genomes, which presumably is due to a reduced fidelity of mitochondrial DNA replication, inefficient DNA repair, or both. Because only about 16,500 DNA nucleotides need to be replicated and expressed as RNAs and proteins in animal cell mitochondria, the error rate per nucleotide copied by DNA replication, maintained by DNA repair, transcribed by RNA polymerase, or translated into protein by mitochondrial ribosomes can be relatively high without damaging one of the relatively few gene products. This could explain why the mechanisms that perform these processes are relatively simple compared with those used for the same purpose elsewhere in cells. The presence of only 22 tRNAs and the unusually small size of the rRNAs (less than two-thirds the size of the *E. coli* rRNAs), for example, would be expected to reduce the fidelity of protein synthesis in mitochondria, although this has not yet been tested adequately.

The relatively high rate of evolution of mitochondrial genes makes a comparison of mitochondrial DNA sequences especially useful for estimating the dates of relatively recent evolutionary events, such as the steps in primate evolution.

TABLE 14–3 Some Differences Between the "Universal" Code and Mitochondrial Genetic Codes*

		MITOCHONDRIAL CODES			
CODON	"UNIVERSAL" CODE	MAMMALS	INVERTEBRATES	YEASTS	PLANTS
UGA	STOP	*Trp*	*Trp*	*Trp*	STOP
AUA	Ile	*Met*	*Met*	*Met*	Ile
CUA	Leu	Leu	Leu	*Thr*	Leu
AGA, AGG	Arg	*STOP*	*Ser*	Arg	Arg

*Italics and color shading indicate that the code differs from the "Universal" code.

Some Organelle Genes Contain Introns

The processing of precursor RNAs has an important role in the two mitochondrial systems studied in most detail—human and yeast. In human cells, both strands of the mitochondrial DNA are transcribed at the same rate from a single promoter region on each strand, producing two different giant RNA molecules, each containing a full-length copy of one DNA strand. Transcription is therefore completely symmetric. The transcripts made on one strand are extensively processed by nuclease cleavage to yield the two rRNAs, most of the tRNAs, and about 10 poly-A-containing RNAs. In contrast, the transcript of the other strand is processed to produce only 8 tRNAs and 1 small poly-A-containing RNA; the remaining 90% of this transcript apparently contains no useful information (being complementary to coding sequences synthesized on the other strand) and is degraded. The poly-A-containing RNAs are the mitochondrial mRNAs: although they lack a cap structure at their 5′ end, they carry a poly-A tail at their 3′ end that is added posttranscriptionally by a mitochondrial poly-A polymerase.

Unlike human mitochondrial genes, some plant and fungal (including yeast) mitochondrial genes contain *introns,* which must be removed by RNA splicing. Introns have also been found in plant chloroplast genes. Many of the introns in organelle genes consist of a family of related nucleotide sequences that are capable of splicing themselves out of the RNA transcripts by RNA-mediated catalysis (discussed in Chapter 6), although these self-splicing reactions are generally aided by proteins. The presence of introns in organelle genes is surprising, as introns are not common in the genes of the bacteria whose ancestors are thought to have given rise to mitochondria and plant chloroplasts.

In yeasts, the same mitochondrial gene may have an intron in one strain but not in another. Such "optional introns" seem to be able to move in and out of genomes like transposable elements. In contrast, introns in other yeast mitochondrial genes have also been found in a corresponding position in the mitochondria of *Aspergillus* and *Neurospora,* implying that they were inherited from a common ancestor of these three fungi. It is possible that these intron sequences are of ancient origin—tracing back to a bacterial ancestor—and that, although they have been lost from many bacteria, they have been preferentially retained in some organelle genomes where RNA splicing is regulated to help control gene expression.

The Chloroplast Genome of Higher Plants Contains About 120 Genes

More than 20 chloroplast genomes have now been sequenced. The genomes of even distantly related plants (such as tobacco and liverwort) are nearly identical, and even those of green algae are closely related (Figure 14–59). Chloroplast genes are involved in four main types of processes: transcription, translation, photosynthesis, and the biosynthesis of small molecules such as amino acids, fatty acids, and pigments. Plant chloroplast genes also encode at least 40 proteins whose functions are as yet unknown; in addition, about twice that many genes of unknown function are present in the chloroplasts of some algae. Paradoxically, all of the known proteins encoded in the chloroplast are part of larger protein complexes that also contain one or more subunits encoded in the nucleus. We discuss possible reasons for this paradox later.

The similarities between the genomes of chloroplasts and bacteria are striking. The basic regulatory sequences, such as transcription promoters and terminators, are virtually identical in the two cases. The amino acid sequences of the proteins encoded in chloroplasts are clearly recognizable as bacterial, and several clusters of genes with related functions (such as those encoding ribosomal proteins) are organized in the same way in the genomes of chloroplasts, *E. coli,* and cyanobacteria.

Further comparisons of large numbers of homologous nucleotide sequences should help clarify the exact evolutionary pathway from bacteria to chloroplasts, but several conclusions can already be drawn:

1. Chloroplasts in higher plants arose from photosynthetic bacteria.

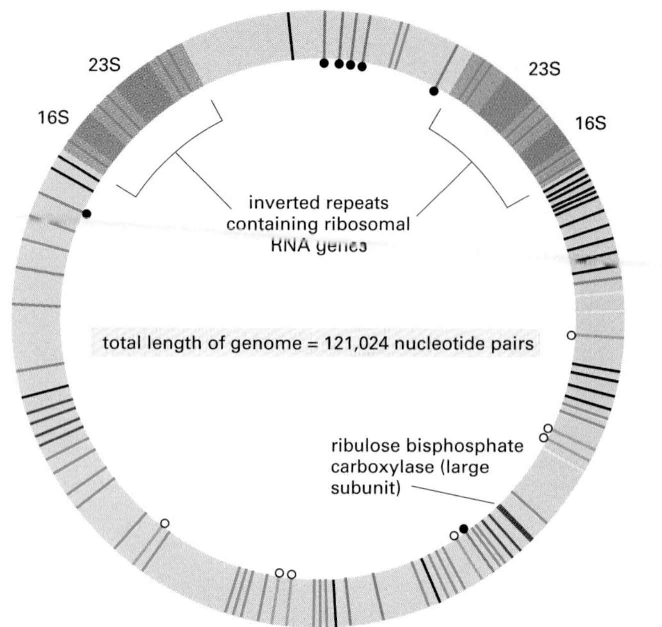

Figure 14–59 The organization of the liverwort chloroplast genome. The chloroplast genome organization is very similar in all higher plants, although the size varies from species to species—depending on how much of the DNA surrounding the genes encoding the chloroplast's 16S and 23S ribosomal RNAs is present in two copies.

2. The chloroplast genome has been stably maintained for at least several hundred million years, the estimated time of divergence of liverwort and tobacco.
3. Many of the genes of the original bacterium are now present in the nuclear genome, where they have been integrated and are stably maintained. In higher plants, for example, two-thirds of the 60 or so chloroplast ribosomal proteins are encoded in the cell nucleus; these genes have a clear bacterial ancestry, and the chloroplast ribosomes retain their original bacterial properties.

Mitochondrial Genes Are Inherited by a Non-Mendelian Mechanism

Many experiments on the mechanisms of mitochondrial biogenesis have been performed with *Saccharomyces cerevisiae* (baker's yeast). There are several reasons for this preference. First, when grown on glucose, this yeast has an ability to live by glycolysis alone and can therefore survive with defective mitochondria that cannot perform oxidative phosphorylation. This makes it possible to grow cells with mutations in mitochondrial or nuclear DNA that interfere with mitochondrial function; such mutations are lethal in many other eucaryotes. Second, yeasts are simple unicellular eucaryotes that are easy to grow and characterize biochemically. Finally, these yeast cells normally reproduce asexually by budding, but they can also reproduce sexually. During sexual reproduction two haploid cells mate and fuse to form a diploid zygote, which can either grow mitotically or divide by meiosis to produce new haploid cells.

The ability to control the alternation between asexual and sexual reproduction in the laboratory greatly facilitates genetic analyses. Mutations in mitochondrial genes are not inherited in accordance with the Mendelian rules that govern the inheritance of nuclear genes. Therefore, long before the mitochondrial genome could be sequenced, genetic studies revealed which of the genes involved in yeast mitochondrial function are located in the nucleus and which in the mitochondria. An example of non-Mendelian (cytoplasmic) inheritance of mitochondrial genes in a haploid yeast cell is shown in Figure 14–60. In this example, we follow the inheritance of a mutant gene that makes mitochondrial protein synthesis resistant to chloramphenicol.

When a chloramphenicol-resistant haploid cell mates with a chloramphenicol-sensitive wild-type haploid cell, the resulting diploid zygote contains a mixture of mutant and wild-type genomes. The two mitochondrial networks fuse in the zygote, creating one continuous reticulum that contains genomes of both parental cells. When the zygote undergoes mitosis, copies of both mutant and wild-type mitochondrial DNA are segregated to the diploid daughter cell. In the case of nuclear DNA, each daughter cell receives exactly two copies of each chromosome, one from each parent. By contrast, in the case of mitochondrial DNA, the daughter cell may inherit either more copies of the mutant DNA or more copies of the wild-type DNA. Successive mitotic divisions can further enrich for either DNA, so that subsequently many cells will arise that contain mitochondrial DNA of only one genotype. This stochastic process is called *mitotic segregation*.

When diploid cells that have segregated their mitochondrial genomes in this way undergo meiosis to form four haploid daughter cells, each of the four daughters receives the same mitochondrial genes. This type of inheritance is called non-Mendelian, or *cytoplasmic inheritance,* to contrast it with the Mendelian inheritance of nuclear genes (see Figure 14–60). When non-Mendelian inheritance occurs, it demonstrates that the gene in question is located outside the nuclear chromosomes.

Although clusters of mitochondrial DNA molecules (nucleoids) are relatively immobile in the mitochondrial reticulum because of their anchorage to the inner membrane, individual nucleoids occasionally come together. This occurs frequently, for example, at sites where the two parental mitochondrial networks fuse during zygote formation. When different DNAs are present in the same nucleoid, genetic recombination can occur. This recombination can result in mitochondrial genomes that contain DNA from both parent cells, which are stably inherited after their mitotic segregation.

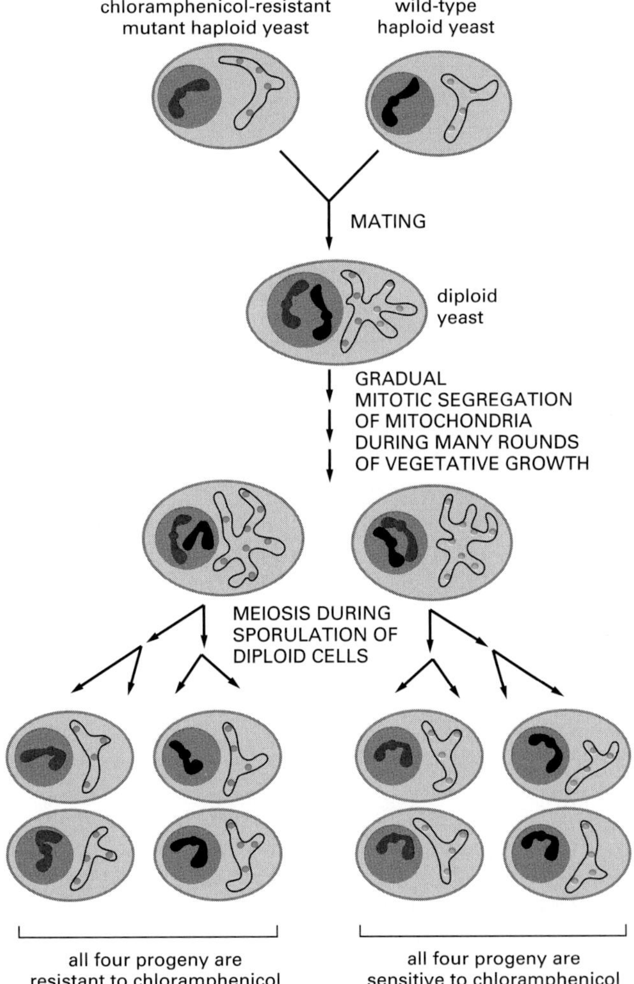

Figure 14–60 The difference in the patterns of inheritance between mitochondrial and nuclear genes of yeast cells. For nuclear genes (Mendelian inheritance), two of the four cells that result from meiosis inherit the gene from one of the original haploid parent cells, and the remaining two cells inherit the gene from the other. By contrast, for mitochondrial genes (non-Mendelian inheritance), it is possible for all four of the cells that result from meiosis to inherit their mitochondrial genes from only one of the two original haploid cells.

In this example, the mitochondrial gene is one that, in its mutant form (mitochondrial DNA denoted by *blue dots),* makes protein synthesis in the mitochondrion resistant to chloramphenicol—a protein synthesis inhibitor that acts specifically on the procaryotic-like ribosomes in mitochondria and chloroplasts. Yeast cells that contain the mutant gene can be detected by their ability to grow in the presence of chloramphenicol on a substrate, such as glycerol, that cannot be used for glycolysis. With glycolysis blocked, ATP must be provided by functional mitochondria, and therefore the cells that carry the normal (wild-type) mitochondrial DNA *(green dots)* cannot grow.

Organelle Genes Are Maternally Inherited in Many Organisms

The consequences of cytoplasmic inheritance are more profound for some organisms, including ourselves, than they are for yeasts. In yeasts, when two haploid cells mate, they are equal in size and contribute equal amounts of mitochondrial DNA to the zygote (see Figure 14–60). Mitochondrial inheritance in yeasts is therefore *biparental:* both parents contribute equally to the mitochondrial gene pool of the progeny (although, as we have just seen, after several generations of vegetative growth, the *individual* progeny often contain mitochondria from only one parent). In higher animals, by contrast, the egg cell always contributes much more cytoplasm to the zygote than does the sperm. One would therefore expect mitochondrial inheritance in higher animals to be nearly *uniparental*—or, more precisely, maternal. Such *maternal inheritance* has been demonstrated in laboratory animals. When animals carrying type A mitochondrial DNA are crossed with animals carrying type B, the progeny contain only the maternal type of mitochondrial DNA. Similarly, by following the distribution of variant mitochondrial DNA sequences in large families, it has been shown that human mitochondrial DNA is maternally inherited.

Figure 14–61 A variegated leaf. In the white patches, the plant cells have inherited a defective chloroplast. (Courtesy of John Innes Foundation.)

In about two-thirds of higher plants, the chloroplasts from the male parent (contained in pollen grains) do not enter the zygote, so that chloroplast as well as mitochondrial DNA is maternally inherited. In other plants, the pollen chloroplasts enter the zygote, making chloroplast inheritance biparental. In such plants, defective chloroplasts are a cause of *variegation:* a mixture of normal and defective chloroplasts in a zygote may sort out by mitotic segregation during plant growth and development, thereby producing alternating green and white patches in leaves. The green patches contain normal chloroplasts, while the white patches contain defective chloroplasts (Figure 14–61).

A fertilized human egg carries perhaps 2000 copies of the human mitochondrial genome, all but one or two inherited from the mother. A human in which all of these genomes carried a deleterious mutation would generally not survive. But some mothers carry a mixed population of both mutant and normal mitochondrial genomes. Their daughters and sons inherit this mixture of normal and mutant mitochondrial DNAs and are healthy unless the process of mitotic segregation by chance results in a majority of defective mitochondria in a particular tissue. Muscle and nervous tissues are most at risk, because of their need for particularly large amounts of ATP.

An inherited disease in humans caused by a mutation in mitochondrial DNA can be recognized by its passage from affected mothers to both their daughters and their sons, with the daughters but not the sons producing grandchildren with the disease. As expected from the random nature of mitotic segregation, the symptoms of these diseases vary greatly between different family members—including not only the severity and age of onset, but also which tissue is affected.

Consider, for example, the inherited disease *myoclonic epilepsy and ragged red fiber disease (MERRF)*, which can be caused by a mutation in one of the mitochondrial transfer RNA genes. This disease appears when, by chance, a particular tissue inherits a threshold amount of defective mitochondrial DNA genomes. Above this threshold, the pool of defective tRNA causes a decrease in the synthesis of the mitochondrial proteins required for electron transport and production of ATP. The result may be muscle weakness or heart problems (from effects on heart muscle), forms of epilepsy or dementia (from effects on nerve cells), or other symptoms. Not surprisingly, a similar variability in phenotypes is found for many other mitochondrial diseases.

Because of the unusually high rate of mutation observed in mitochondria, it has also been suggested that mutations that accumulate in mitochondrial DNAs may contribute to many of the medical problems of old age.

Petite Mutants in Yeasts Demonstrate the Overwhelming Importance of the Cell Nucleus for Mitochondrial Biogenesis

Genetic studies of yeasts have had a crucial role in the analysis of mitochondrial biogenesis. A striking example is provided by studies of yeast mutants that

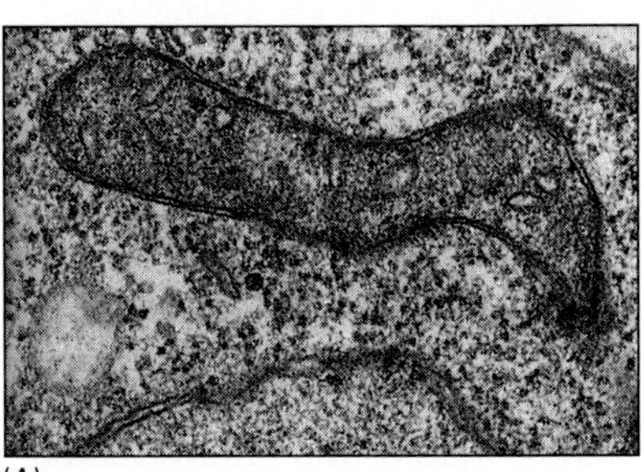
(A)

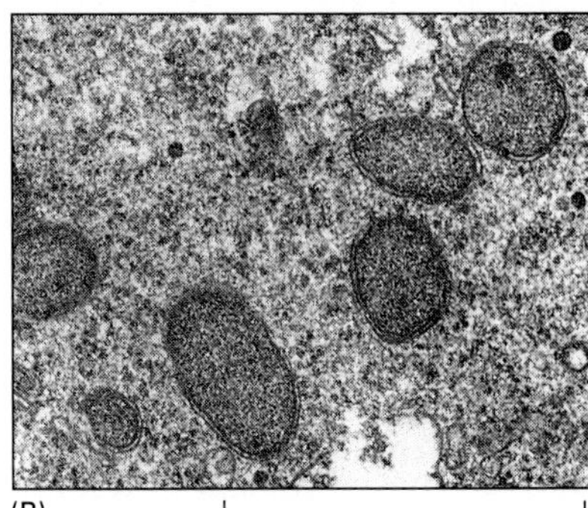
(B)

Figure 14–62 Electron micrographs of yeast cells. (A) The structure of normal mitochondria. (B) Mitochondria in a petite mutant. In petite mutants, all the mitochondrion-encoded gene products are missing, and the organelle is constructed entirely from nucleus-encoded proteins. (Courtesy of Barbara Stevens.)

contain large deletions in their mitochondrial DNA, so that all mitochondrial protein synthesis is abolished. Not surprisingly, these mutants cannot make respiring mitochondria. Some of these mutants lack mitochondrial DNA altogether. Because they form unusually small colonies when grown in media with low glucose, all mutants with such defective mitochondria are called *cytoplasmic petite mutants.*

Although petite mutants cannot synthesize proteins in their mitochondria and therefore cannot make mitochondria that produce ATP, they nevertheless contain mitochondria. These mitochondria have a normal outer membrane and an inner membrane with poorly developed cristae (Figure 14–62). They contain virtually all the mitochondrial proteins that are specified by nuclear genes and imported from the cytosol—including DNA and RNA polymerases, all of the citric acid cycle enzymes, and most inner membrane proteins—demonstrating the overwhelming importance of the nucleus in mitochondrial biogenesis. Petite mutants also show that an organelle that divides by fission can replicate indefinitely in the cytoplasm of proliferating eucaryotic cells, even in the complete absence of its own genome. It is possible that peroxisomes normally replicate in this way (see Figure 12–34).

For chloroplasts, the nearest equivalent to yeast mitochondrial petite mutants are mutants of unicellular algae such as *Euglena.* Mutant algae in which no chloroplast protein synthesis occurs still contain chloroplasts and are perfectly viable if oxidizable substrates are provided. If the development of mature chloroplasts is blocked in higher plants, however, either by raising the plants in the dark or because chloroplast DNA is defective or absent, the plants die.

Mitochondria and Plastids Contain Tissue-specific Proteins that Are Encoded in the Cell Nucleus

Mitochondria can have specialized functions in particular types of cells. The *urea cycle,* for example, is the central metabolic pathway in mammals for disposing of cellular breakdown products that contain nitrogen. These products are excreted in the urine as urea. Nucleus-encoded enzymes in the mitochondrial matrix perform several steps in the cycle. Urea synthesis occurs in only a few tissues, such as the liver, and the required enzymes are synthesized and imported into mitochondria only in these tissues.

The respiratory enzyme complexes in the mitochondrial inner membrane of mammals contain several tissue-specific, nucleus-encoded subunits that are thought to act as regulators of electron transport. Thus, some humans with a genetic muscle disease have a defective subunit of cytochrome oxidase; since the subunit is specific to skeletal muscle cells, their other cells, including their heart muscle cells, function normally, allowing the individuals to survive. As would be expected, tissue-specific differences are also found among the nucleus-encoded proteins in chloroplasts.

Mitochondria Import Most of Their Lipids; Chloroplasts Make Most of Theirs

The biosynthesis of new mitochondria and chloroplasts requires lipids in addition to nucleic acids and proteins. Chloroplasts tend to make the lipids they require. In spinach leaves, for example, all cellular fatty acid synthesis takes place in the chloroplast, although desaturation of the fatty acids occurs elsewhere. The major glycolipids of the chloroplast are also synthesized locally.

Mitochondria, in contrast, import most of their lipids. In animal cells, the phospholipids phosphatidylcholine and phosphatidylserine are synthesized in the endoplasmic reticulum and then transferred to the outer membrane of mitochondria. In addition to decarboxylating imported phosphatidylserine to phosphatidylethanolamine, the main reaction of lipid biosynthesis catalyzed by the mitochondria themselves is the conversion of imported lipids to cardiolipin (bisphosphatidylglycerol). Cardiolipin is a "double" phospholipid that contains four fatty acid tails (Figure 14–63). It is found mainly in the mitochondrial inner membrane, where it constitutes about 20% of the total lipid.

We discuss the important question of how specific cytosolic proteins are imported into mitochondria and chloroplasts in Chapter 12.

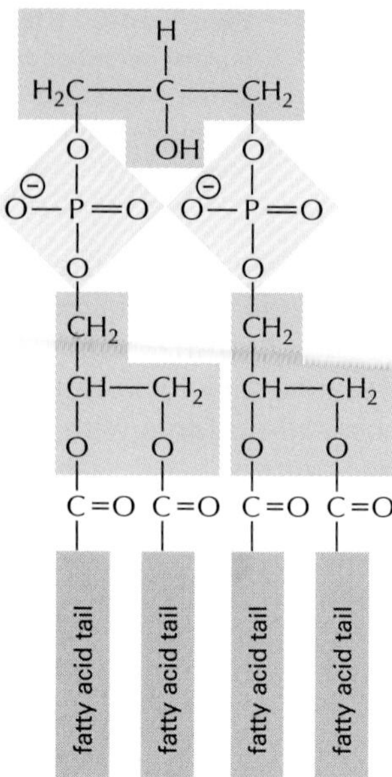

Figure 14–63 The structure of cardiolipin. Cardiolipin is an unusual lipid in the inner mitochondrial membrane.

Why Do Mitochondria and Chloroplasts Have Their Own Genetic Systems?

Why do mitochondria and chloroplasts require their own separate genetic systems, when other organelles that share the same cytoplasm, such as peroxisomes and lysosomes, do not? The question is not trivial, because maintaining a separate genetic system is costly: more than 90 proteins—including many ribosomal proteins, aminoacyl-tRNA synthases, DNA and RNA polymerases, and RNA-processing and RNA-modifying enzymes—must be encoded by nuclear genes specifically for this purpose (Figure 14–64). The amino acid sequences of most of these proteins in mitochondria and chloroplasts differ from those of their counterparts in the nucleus and cytosol, and it appears that these organelles have relatively few proteins in common with the rest of the cell. This means that the nucleus must provide at least 90 genes just to maintain each organelle's genetic system. The reason for such a costly arrangement is not clear, and the hope that the nucleotide sequences of mitochondrial and chloroplast

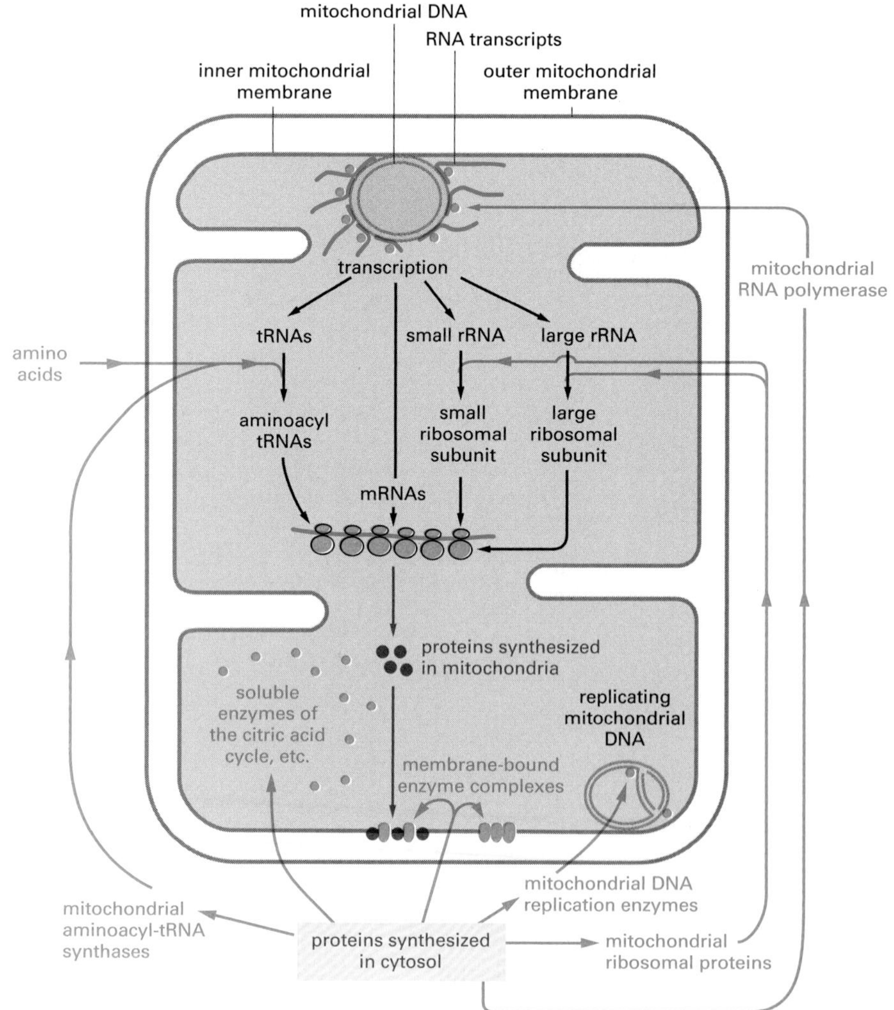

Figure 14–64 The origins of mitochondrial RNAs and proteins. The proteins encoded in the nucleus and imported from the cytosol have a major role in creating the genetic system of the mitochondrion, in addition to contributing most of the organelle's other proteins. Not indicated in this diagram are the additional nucleus-encoded proteins that regulate the expression of individual mitochondrial genes at posttranscriptional levels. The mitochondrion itself contributes only mRNAs, rRNAs, and tRNAs to its genetic system.

genomes would provide the answer has proved to be unfounded. We cannot think of compelling reasons why the proteins made in mitochondria and chloroplasts should be made there rather than in the cytosol.

At one time, it was suggested that some proteins have to be made in the organelle because they are too hydrophobic to get to their site in the membrane from the cytosol. More recent studies, however, make this explanation implausible. In many cases, even highly hydrophobic subunits are synthesized in the cytosol. Moreover, although the individual protein subunits in the various mitochondrial enzyme complexes are highly conserved in evolution, their site of synthesis is not (see Figure 14–57). The diversity in the location of the genes coding for the subunits of functionally equivalent proteins in different organisms is difficult to explain by any hypothesis that postulates a specific evolutionary advantage of present-day mitochondrial or chloroplast genetic systems.

Perhaps the organelle genetic systems are an evolutionary dead-end. In terms of the endosymbiont hypothesis, this would mean that the process whereby the endosymbionts transferred most of their genes to the nucleus stopped before it was complete. Further transfers may have been ruled out, for mitochondria, by recent alterations in the mitochondrial genetic code that made the remaining mitochondrial genes nonfunctional if they were transferred to the nucleus.

Summary

Mitochondria and chloroplasts grow in a coordinated process that requires the contribution of two separate genetic systems—one in the organelle and one in the cell nucleus. Most of the proteins in these organelles are encoded by nuclear DNA, synthesized in the cytosol, and then imported individually into the organelle. Some organelle proteins and RNAs are encoded by the organelle DNA and are synthesized in the organelle itself. The human mitochondrial genome contains about 16,500 nucleotides and encodes 2 ribosomal RNAs, 22 transfer RNAs, and 13 different polypeptide chains. Chloroplast genomes are about 10 times larger and contain about 120 genes. But partly functional organelles form in normal numbers even in mutants that lack a functional organelle genome, demonstrating the overwhelming importance of the nucleus for the biogenesis of both organelles.

The ribosomes of chloroplasts closely resemble bacterial ribosomes, while mitochondrial ribosomes show both similarities and differences that make their origin more difficult to trace. Protein similarities, however, suggest that both organelles originated when a primitive eucaryotic cell entered into a stable endosymbiotic relationship with a bacterium. A purple bacterium is thought to have given rise to the mitochondrion, and (later) a relative of a cyanobacterium is thought to have given rise to the plant chloroplast. Although many of the genes of these ancient bacteria still function to make organelle proteins, most of them have become integrated into the nuclear genome, where they encode bacterialike enzymes that are synthesized on cytosolic ribosomes and then imported into the organelle.

THE EVOLUTION OF ELECTRON-TRANSPORT CHAINS

Much of the structure, function, and evolution of cells and organisms can be related to their need for energy. We have seen that the fundamental mechanisms for harnessing energy from such disparate sources as light and the oxidation of glucose are the same. Apparently, an effective method for synthesizing ATP arose early in evolution and has since been conserved with only small variations. How did the crucial individual components—ATP synthase, redox-driven H^+ pumps, and photosystems—first arise? Hypotheses about events occurring on an evolutionary time scale are difficult to test. But clues abound, both in the many different primitive electron-transport chains that survive in some present-day bacteria, and in geological evidence about the environment of the Earth billions of years ago.

The Earliest Cells Probably Produced ATP by Fermentation

As explained in Chapter 1, the first living cells on Earth are thought to have arisen more than 3.5×10^9 years ago, when the Earth was not more than about 10^9 years old. The environment lacked oxygen but was presumably rich in geochemically produced organic molecules, and some of the earliest metabolic pathways for producing ATP may have resembled present-day forms of fermentation.

In the process of fermentation, ATP is made by a phosphorylation event that harnesses the energy released when a hydrogen-rich organic molecule, such as glucose, is partly oxidized (see Figure 2–72). The electrons lost from the oxidized organic molecules are transferred (via NADH or NADPH) to a different organic molecule (or to a different part of the same molecule), which thereby becomes more reduced. At the end of the fermentation process, one or more of the organic molecules produced are excreted into the medium as metabolic waste products; others, such as pyruvate, are retained by the cell for biosynthesis.

The excreted end-products are different in different organisms, but they tend to be organic acids (carbon compounds that carry a COOH group). Among the most important of such products in bacterial cells are lactic acid (which also accumulates in anaerobic mammalian glycolysis) and formic, acetic, propionic, butyric, and succinic acids.

Electron-transport Chains Enabled Anaerobic Bacteria to Use Nonfermentable Molecules as Their Major Source of Energy

The early fermentation processes would have provided not only the ATP but also the reducing power (as NADH or NADPH) required for essential biosyntheses. Thus, many of the major metabolic pathways could have evolved while fermentation was the only mode of energy production. With time, however, the metabolic activities of these procaryotic organisms must have changed the local environment, forcing organisms to evolve new biochemical pathways. The accumulation of waste products of fermentation, for example, might have resulted in the following series of changes:

Stage 1. The continuous excretion of organic acids lowered the pH of the environment, favoring the evolution of proteins that function as transmembrane H^+ pumps that can pump H^+ out of the cell to protect it from the dangerous effects of intracellular acidification. One of these pumps may have used the energy available from ATP hydrolysis and could have been the ancestor of the present-day ATP synthase.

Stage 2. At the same time as nonfermentable organic acids were accumulating in the environment and favoring the evolution of an ATP-consuming H^+ pump, the supply of geochemically generated fermentable nutrients, which provided the energy for the pumps and for all other cellular processes, was dwindling. This favored bacteria that could excrete H^+ without hydrolyzing ATP, allowing the ATP to be conserved for other cellular activities. Selective pressures of this kind might have led to the first membrane-bound proteins that could use electron transport between molecules of different redox potentials as the energy source for transporting H^+ across the plasma membrane. Some of these proteins would have found their electron donors and electron acceptors among the nonfermentable organic acids that had accumulated. Many such electron-transport proteins can be found in present-day bacteria; some bacteria that grow on formic acid, for example, pump H^+ by using the small amount of redox energy derived from the transfer of electrons from formic acid to fumarate (Figure 14–65). Others have similar electron-transport components devoted solely to the oxidation and reduction of inorganic substrates (see Figure 14–67, for example).

Stage 3. Eventually some bacteria developed H^+-pumping electron-transport systems that were efficient enough to harness more redox energy than they needed just to maintain their internal pH. Now, bacteria

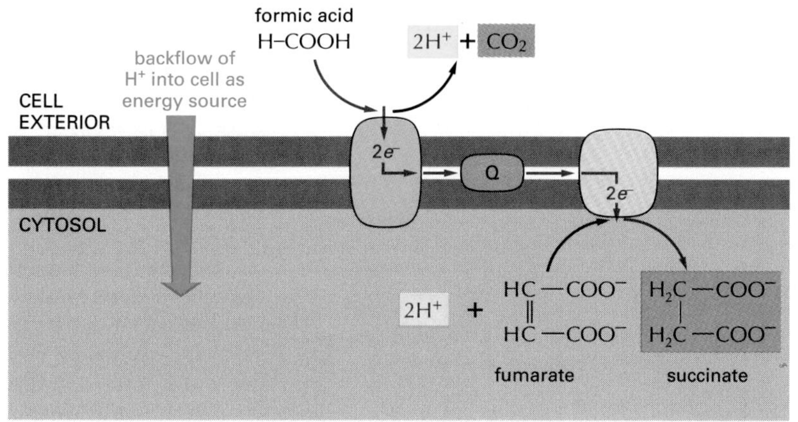

Figure 14–65 The oxidation of formic acid in some present-day bacteria. In such anaerobic bacteria, including *E. coli*, the oxidation is mediated by an energy-conserving electron-transport chain in the plasma membrane. As indicated, the starting materials are formic acid and fumarate, and the products are succinate and CO_2. Note that H^+ is consumed inside the cell and generated outside the cell, which is equivalent to pumping H^+ to the cell exterior. Thus, this membrane-bound electron-transport system can generate an electrochemical proton gradient across the plasma membrane. The redox potential of the formic acid–CO_2 pair is –420 mV, while that of the fumarate–succinate pair is +30 mV.

that carried both types of H^+ pumps were at an advantage. In these cells, a large electrochemical proton gradient generated by excessive H^+ pumping allowed protons to leak back into the cell through the ATP-driven H^+ pumps, thereby running them in reverse, so that they functioned as ATP synthases to make ATP. Because such bacteria required much less of the increasingly scarce supply of fermentable nutrients, they proliferated at the expense of their neighbors.

These three hypothetical stages in the evolution of oxidative phosphorylation mechanisms are summarized in Figure 14–66.

By Providing an Inexhaustible Source of Reducing Power, Photosynthetic Bacteria Overcame a Major Evolutionary Obstacle

The evolutionary steps just outlined would have solved the problem of maintaining both a neutral intracellular pH and an abundant store of energy, but these steps would not have solved another problem that was equally serious. The depletion of organic nutrients from the environment meant that organisms had to find some alternative source of carbon to make the sugars that served as the precursors for so many other cellular molecules. Although the CO_2 in the atmosphere provided an abundant potential carbon source, to convert it into an organic molecule such as a carbohydrate requires that the fixed CO_2 be reduced by a strong electron donor, such as NADH or NADPH, which can provide the high-energy electrons needed to generate each (CH_2O) unit from CO_2 (see Figure 14–39). Early in cellular evolution, strong reducing agents (electron donors) would have been plentiful as products of fermentation. But as the supply of fermentable nutrients dwindled and a membrane-bound ATP synthase began to produce most of the ATP, the plentiful supply of NADH and other reducing agents would have disappeared. It thus became imperative for cells to evolve a new way of generating strong reducing agents.

Presumably, the main reducing agents still available were the organic acids produced by the anaerobic metabolism of carbohydrates, inorganic molecules such as hydrogen sulfide (H_2S) generated geochemically, and water. But the reducing power of these molecules is far too weak to be useful for CO_2 fixation. An early supply of strong electron donors could have been generated by using the electrochemical proton gradient across the plasma membrane to drive a *reverse electron flow.* This would have required the evolution of membrane-bound enzyme complexes resembling an NADH dehydrogenase, and mechanisms of this kind survive in the anaerobic metabolism of some present-day bacteria (Figure 14–67).

The major evolutionary breakthrough in energy metabolism, however, was almost certainly the development of photochemical reaction centers that could use the energy of sunlight to produce molecules such as NADH. It is thought that this occurred early in the process of cellular evolution—more than 3×10^9 years

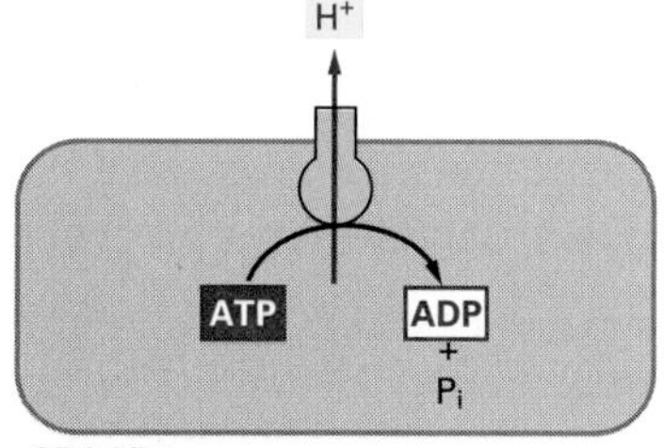

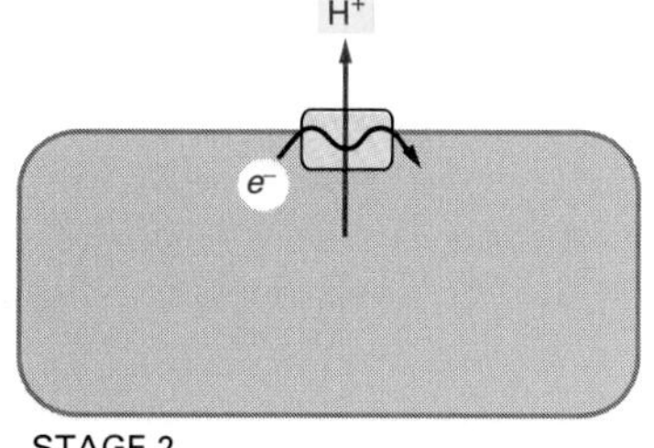

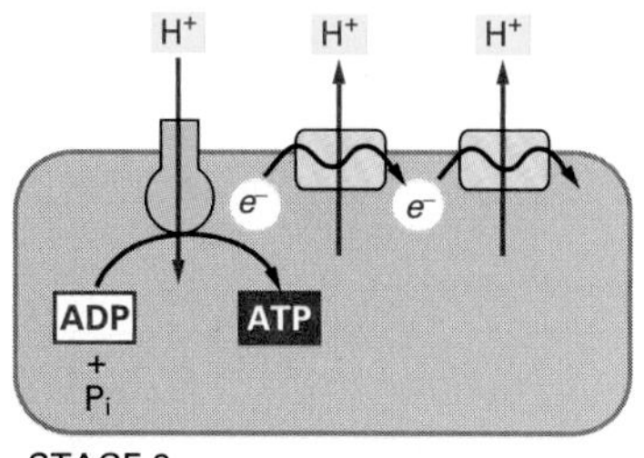

Figure 14–66 The evolution of oxidative phosphorylation mechanisms. One possible sequence is shown; the stages are described in the text.

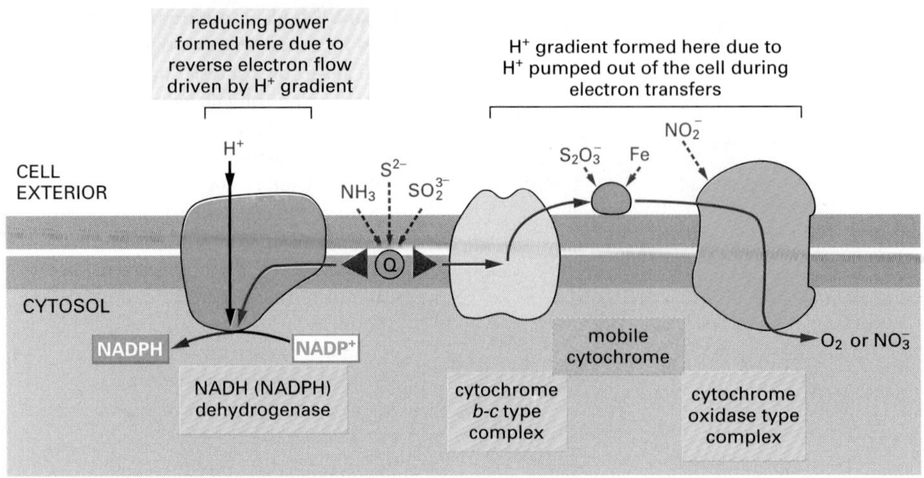

Figure 14–67 Some of the electron-transport pathways in present-day bacteria. These pathways generate all the cell's ATP and reducing power from the oxidation of inorganic molecules, such as iron, ammonia, nitrite, and sulfur compounds. As indicated, some species can grow anaerobically by substituting nitrate for oxygen as the terminal electron acceptor. Most use the carbon-fixation cycle and synthesize their organic molecules entirely from carbon dioxide.

Both forward and reverse electron flows occur from the quinone (Q). As in the respiratory chain, the forward electron flows cause H^+ to be pumped out of the cell, and the resulting H^+ gradient drives the production of ATP by an ATP synthase (not shown). The NADPH required for carbon fixation is produced by an energy-requiring reverse electron flow; this flow is also driven by the H^+ gradient, as indicated.

ago, in the ancestors of the green sulfur bacteria. Present-day green sulfur bacteria use light energy to transfer hydrogen atoms (as an electron plus a proton) from H_2S to NADPH, thereby creating the strong reducing power required for carbon fixation (Figure 14–68). Because the electrons removed from H_2S are at a much more negative redox potential than those of H_2O (–230 mV for H_2S compared +820 mV for H_2O), one quantum of light absorbed by the single photosystem in these bacteria is sufficient to achieve a high enough redox potential to generate NADPH via a relatively simple photosynthetic electron-transport chain.

The Photosynthetic Electron-transport Chains of Cyanobacteria Produced Atmospheric Oxygen and Permitted New Life-Forms

The next step, which is thought to have occurred with the development of the cyanobacteria at least 3×10^9 years ago, was the evolution of organisms capable of using water as the electron source for CO_2 reduction. This entailed the evolution of a water-splitting enzyme and also required the addition of a second photosystem, acting in series with the first, to bridge the enormous gap in redox potential between H_2O and NADPH. Present-day structural homologies between photosystems suggest that this change involved the cooperation of a photosystem derived from green bacteria (photosystem I) with a photosystem derived from purple bacteria (photosystem II). The biological consequences of this evolutionary step were far-reaching. For the first time, there were organisms that made only very minimal chemical demands on their environment. These

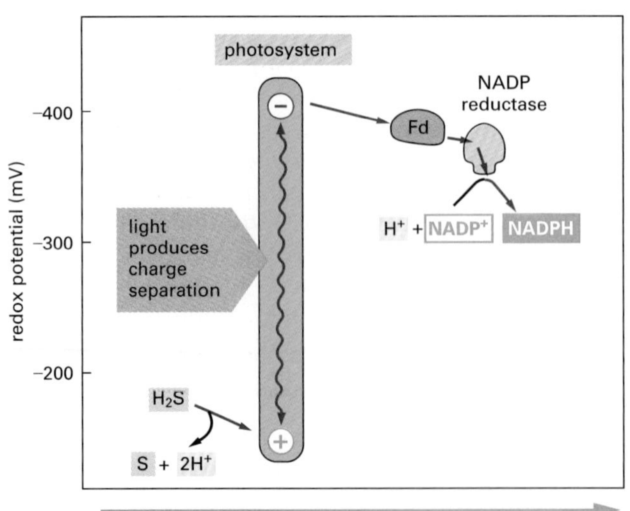

Figure 14–68 The general flow of electrons in a relatively primitive form of photosynthesis observed in present-day green sulfur bacteria. The photosystem in green sulfur bacteria resembles photosystem I in plants and cyanobacteria. Both photosystems use a series of iron–sulfur centers as the electron acceptors that eventually donate their high-energy electrons to ferredoxin (Fd). An example of a bacterium of this type is *Chlorobium tepidum*, which can thrive at high temperatures and low light intensities in hot springs.

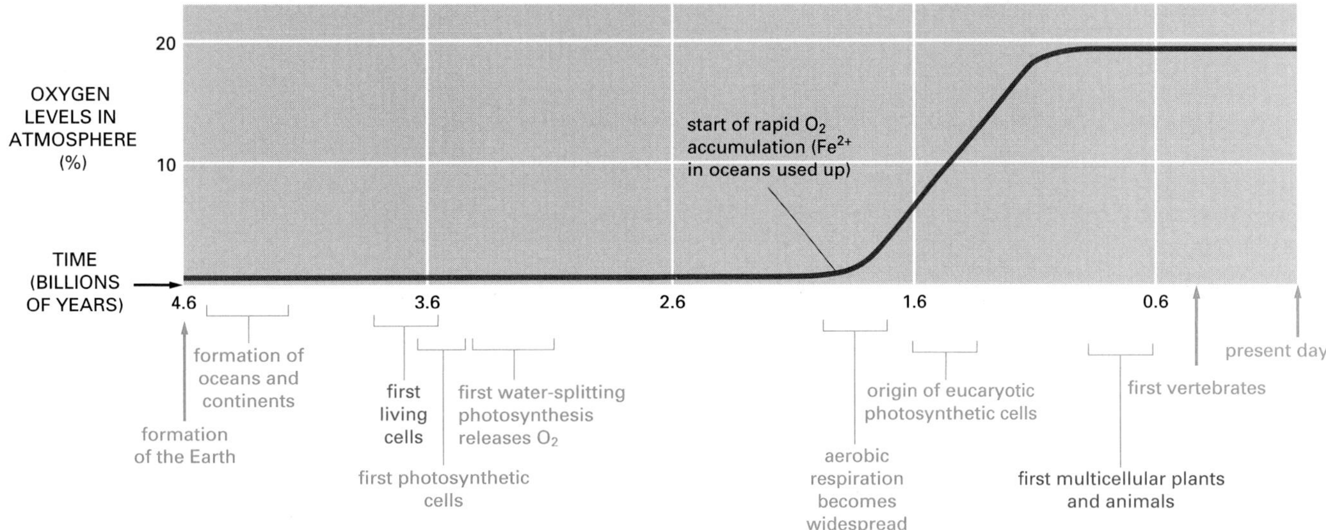

Figure 14–69 Some major events that are believed to have occurred during the evolution of living organisms on Earth. With the evolution of the membrane-based process of photosynthesis, organisms could make their own organic molecules from CO_2 gas. As explained in the text, the delay of more than 10^9 years between the appearance of bacteria that split water and released O_2 during photosynthesis and the accumulation of high levels of O_2 in the atmosphere is thought to be due to the initial reaction of the oxygen with abundant ferrous iron (Fe^{2+}) dissolved in the early oceans. Only when the iron was used up would oxygen have started to accumulate in the atmosphere. In response to the rising levels of oxygen in the atmosphere, nonphotosynthetic oxygen-using organisms appeared, and the concentration of oxygen in the atmosphere leveled out.

cells could spread and evolve in ways denied to the earlier photosynthetic bacteria, which needed H_2S or organic acids as a source of electrons. Consequently, large amounts of biologically synthesized, reduced organic materials accumulated. Moreover, oxygen entered the atmosphere for the first time.

Oxygen is highly toxic because the oxidation reactions it brings about can randomly alter biological molecules. Many present-day anaerobic bacteria, for example, are rapidly killed when exposed to air. Thus, organisms on the primitive Earth would have had to evolve protective mechanisms against the rising O_2 levels in the environment. Late evolutionary arrivals, such as ourselves, have numerous detoxifying mechanisms that protect our cells from the ill effects of oxygen. Even so, an accumulation of oxidative damage to our macromolecules is postulated to be a major cause of human aging.

The increase in atmospheric O_2 was very slow at first and would have allowed a gradual evolution of protective devices. The early seas contained large amounts of iron in its ferrous oxidation state (Fe^{2+}), and nearly all the O_2 produced by early photosynthetic bacteria was utilized in converting Fe^{2+} to Fe^{3+}. This conversion caused the precipitation of huge amounts of ferric oxides, and the extensive banded iron formations in sedimentary rocks beginning about 2.7×10^9 years ago help to date the spread of the cyanobacteria. By about 2×10^9 years ago, the supply of ferrous iron was exhausted, and the deposition of further iron precipitates ceased. Geological evidence suggests that O_2 levels in the atmosphere then began to rise, reaching current levels between 0.5 and 1.5×10^9 years ago (Figure 14–69).

The availability of O_2 made possible the development of bacteria that relied on aerobic metabolism to make their ATP. As explained previously, these organisms could harness the large amount of energy released by breaking down carbohydrates and other reduced organic molecules all the way to CO_2 and H_2O. Components of preexisting electron-transport complexes were modified to produce a cytochrome oxidase, so that the electrons obtained from organic or inorganic substrates could be transported to O_2 as the terminal electron acceptor. Depending on the availability of light and O_2, many present-day purple photosynthetic bacteria can switch between photosynthesis and respiration, requiring only relatively minor reorganizations of their electron-transport chains.

As organic materials accumulated on Earth as a result of photosynthesis, some photosynthetic bacteria (including the precursors of *E. coli*) lost their ability to survive on light energy alone and came to rely entirely on respiration. As described previously (see Figure 14–56), it has been suggested that mitochondria first arose some 1.5×10^9 years ago, when a primitive eucaryotic cell endocytosed such a respiration-dependent bacterium. Plants are believed to have evolved somewhat later, when a descendant of this early aerobic eucaryotic cell endocytosed a photosynthetic bacterium that became the precursor of chloroplasts. Present-day chloroplasts are so different in some types of algae that chloroplasts probably evolved separately in different algal lineages. Figure 14–70 relates these postulated pathways to the various types of bacteria discussed in this chapter.

Evolution is always conservative, taking parts of the old and building on them to create something new. Thus, parts of the electron-transport chains that were derived to service anaerobic bacteria $3–4 \times 10^9$ years ago probably survive, in altered form, in the mitochondria and chloroplasts of today's higher eucaryotes. Consider, for example, the striking homology in structure and function between the enzyme complex that pumps H^+ in the central segment of the mitochondrial respiratory chain (the cytochrome b-c_1 complex) and its analogs in the electron-transport chains of both bacteria and chloroplasts (Figure 14–71).

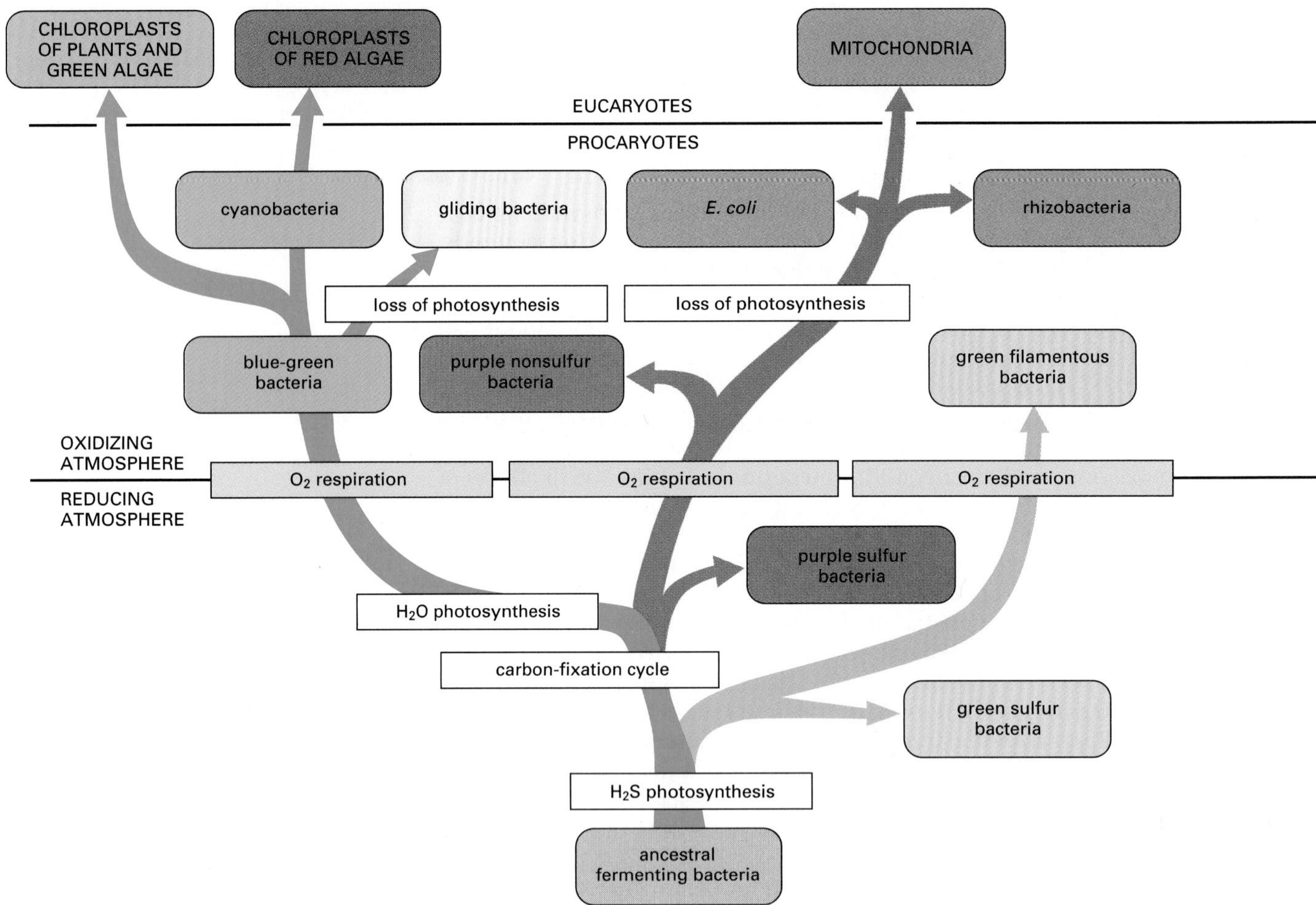

Figure 14–70 A phylogenetic tree of the proposed evolution of mitochondria and chloroplasts and their bacterial ancestors. Oxygen respiration is thought to have begun developing about 2×10^9 years ago. As indicated, it seems to have evolved independently in the green, purple, and blue-green (cyanobacterial) lines of photosynthetic bacteria. It is thought that an aerobic purple bacterium that had lost its ability to photosynthesize gave rise to the mitochondrion, while several different blue-green bacteria gave rise to chloroplasts. Nucleotide sequence analyses suggest that mitochondria arose from purple bacteria that resembled the rhizobacteria, agrobacteria, and rickettsias—three closely related species known to form intimate associations with present-day eucaryotic cells. Archea are not known to contain the type of photosystems described in this chapter, and they are not included here.

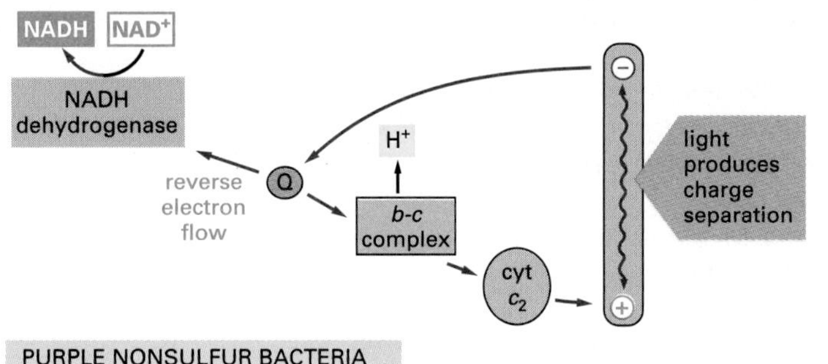

NADP+
NADPH
H+
Q
b-f complex
pC
light produces charge separation
light produces charge separation
H_2O
PLANT CHLOROPLASTS AND CYANOBACTERIA

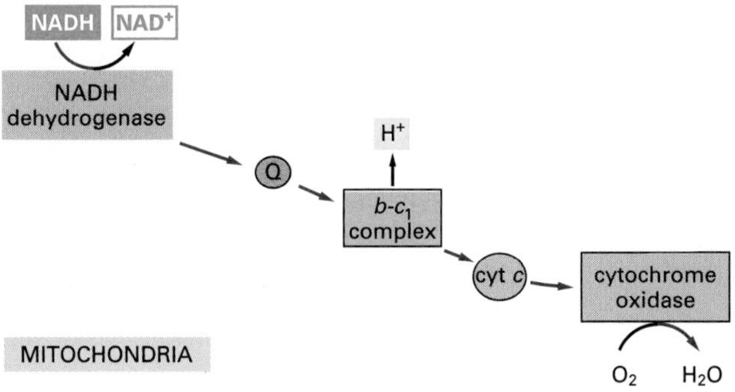

Figure 14–71 A comparison of three electron-transport chains discussed in this chapter. Bacteria, chloroplasts, and mitochondria all contain a membrane-bound enzyme complex that resembles the cytochrome *b-c*$_1$ complex of mitochondria. These complexes all accept electrons from a quinone carrier (Q) and pump H^+ across their respective membranes. Moreover, in reconstituted *in vitro* systems, the different complexes can substitute for one another, and the amino acid sequences of their protein components reveal that they are evolutionarily related.

Summary

Early cells are believed to have been bacteriumlike organisms living in an environment rich in highly reduced organic molecules that had been formed by geochemical processes over the course of hundreds of millions of years. They may have derived most of their ATP by converting these reduced organic molecules to a variety of organic acids, which were then released as waste products. By acidifying the environment, these fermentations may have led to the evolution of the first membrane-bound H^+ pumps, which could maintain a neutral pH in the cell interior by pumping out H^+. The properties of present-day bacteria suggest that an electron-transport-driven H^+ pump and an ATP-driven H^+ pump first arose in this anaerobic environment. Reversal of the ATP-driven pump would have allowed it to function as an ATP synthase. As more effective electron-transport chains developed, the energy released by redox reactions between inorganic molecules and/or accumulated non-fermentable compounds produced a large electrochemical proton gradient, which could be harnessed by the ATP-driven pump for ATP production.

Because preformed organic molecules were replenished only very slowly by geochemical processes, the proliferation of bacteria that used them as the source of both carbon and reducing power could not go on forever. The depletion of fermentable organic nutrients presumably led to the evolution of bacteria that could use CO_2 to make carbohydrates. By combining parts of the electron-transport chains that had developed earlier, light energy was harvested by a single photosystem in

photosynthetic bacteria to generate the NADPH required for carbon fixation. The subsequent appearance of the more complex photosynthetic electron-transport chains of the cyanobacteria allowed H_2O to be used as the electron donor for NADPH formation, rather than the much less abundant electron donors required by other photosynthetic bacteria. Life could then proliferate over large areas of the Earth, so that reduced organic molecules accumulated again.

About 2×10^9 years ago, the O_2 released by photosynthesis in cyanobacteria began to accumulate in the atmosphere. Once both organic molecules and O_2 had become abundant, electron-transport chains became adapted for the transport of electrons from NADH to O_2, and efficient aerobic metabolism developed in many bacteria. Exactly the same aerobic mechanisms operate today in the mitochondria of eucaryotes, and there is increasing evidence that both mitochondria and chloroplasts evolved from aerobic bacteria that were endocytosed by primitive eucaryotic cells.

References

General

Cramer WA & Knaff DB (1990) Energy Transduction in Biological Membranes: A Textbook of Bioenergetics. New York: Springer-Verlag.

Harold FM (1986) The Vital Force: A Study of Bioenergetics. New York: WH Freeman.

Mathews CK, van Holde KE & Ahern K-G (2000) Biochemistry, 3rd edn. San Francisco: Benjamin Cummings.

Nicholls DG & Ferguson SJ (1992) Bioenergetics 2. London: Academic Press.

Zubay GI (1998) Biochemistry, 4th edn. Dubuque, IO: William C Brown.

The Mitochondrion

Abrahams JP, Leslie AG, Lutter R & Walker JE (1994) Structure at 2.8Å resolution of F1-ATPase from bovine heart mitochondria. *Nature* 370, 621–628.

Baldwin JE & Krebs H (1981) The evolution of metabolic cycles. *Nature* 291, 381–382.

Bereiter-Hahn J (1990) Behavior of mitochondria in the living cell. *Int. Rev. Cytol.* 122, 1–63.

Berg HC (2000) Constraints on models for the flagellar rotary motor. *Philos. Trans. R. Soc. Lond. B. Biol. Sci.* 355, 491–501.

Boyer PD (1997) The ATP synthase—a splendid molecular machine. *Annu. Rev. Biochem.* 66, 717–749.

Ernster L & Schatz G (1981) Mitochondria: a historical review. *J. Cell Biol.* 91, 227s–255s.

Frey TG & Mannella CA (2000) The internal structure of mitochondria. *Trends Biochem. Sci.* 25, 319–324.

Harold FM (2001) Gleanings of a chemiosmotic eye. *BioEssays* 23, 848–855.

Hatefi Y (1985) The mitochondrial electron transport and oxidative phosphorylation system. *Annu. Rev. Biochem.* 54, 1015–1070.

Klingenberg M & Nelson DR (1994) Structure–function relationships of the ADP/ATP carrier. *Biochim. Biophys. Acta* 1187, 241–244.

Krstic RV (1979) Ultrastructure of the Mammalian Cell, pp 28–57. New York/Berlin: Springer-Verlag.

Kuan J & Saier MH, Jr (1993) The mitochondrial carrier family of transport proteins: structural, functional, and evolutionary relationships. *Crit. Rev. Biochem. Mol. Biol.* 28, 209–233.

Mitchell P (1961) Coupling of phosphorylation to electron and hydrogen transfer by a chemi-osmotic type of mechanism. *Nature* 191, 144–148.

Racker E & Stoeckenius W (1974) Reconstitution of purple membrane vesicles catalyzing light-driven proton uptake and adenosine triphosphate formation. *J. Biol. Chem.* 249, 662–663.

Racker E (1980) From Pasteur to Mitchell: a hundred years of bioenergetics. *Fed. Proc.* 39, 210–215.

Rastogi VK & Girvin ME (1999) Structural changes linked to proton translocation by subunit *c* of the ATP synthase. *Nature* 402, 263–268.

Sambongi Y, Iko Y, Tanabe M et al. (1999) Mechanical rotation of the *c* subunit oligomer in ATP synthase (F_oF_1): direct observation. *Science* 286, 1722–1724.

Saraste M (1999) Oxidative phosphorylation at the fin de siecle. *Science* 283, 1488–1493.

Scheffler IE (1999) Mitochondria. New York/Chichester: Wiley-Liss.

Stock D, Gibbons C, Arechaga I et al. (2000) The rotary mechanism of ATP synthase. *Curr. Opin. Struct. Biol.* 10, 672–679.

Electron-Transport Chains and Their Proton Pumps

Beinert H, Holm RH & Munck E (1997) Iron-sulfur clusters: nature's modular, multipurpose structures. *Science* 277, 653–659.

Berry EA, Guergova-Kuras M, Huang LS & Crofts AR (2000) Structure and function of cytochrome *bc* complexes. *Annu. Rev. Biochem.* 69, 1005–1075.

Brand MD & Murphy MP (1987) Control of electron flux through the respiratory chain in mitochondria and cells. *Biol. Rev. Camb. Philos. Soc.* 62, 141–193.

Brandt U (1997) Proton-translocation by membrane-bound NADH: ubiquinone-oxidoreductase (complex I) through redox-gated ligand conduction. *Biochim. Biophys. Acta* 1318, 79–91.

Chance B & Williams GR (1955) A method for the localization of sites for oxidative phosphorylation. *Nature* 176, 250–254.

Edman K, Nollert P, Royant A et al. (1999) High-resolution X-ray structure of an early intermediate in the bacteriorhodopsin photocycle. *Nature* 401, 822–826.

Gottschalk G (1986) Bacterial Metabolism, 2nd edn. New York: Springer.

Gray HB & Winkler JR (1996) Electron transfer in proteins. *Annu. Rev. Biochem.* 65, 537–561.

Grigorieff N (1999) Structure of the respiratory NADH:ubiquinone oxidoreductase (complex I) from *Escherichia coli*. *Curr. Opin. Struct. Biol.* 9, 476–483.

Keilin D (1966) The History of Cell Respiration and Cytochromes. Cambridge: Cambridge University Press.

Lowell BB & Spiegelman BM (2000) Towards a molecular understanding of adaptive thermogenesis. *Nature* 404, 652–660.

Michel H, Behr J, Harrenga A & Kannt A (1998) Cytochrome *c* oxidase: structure and spectroscopy. *Annu. Rev. Biophys. Biomol. Struct.* 27, 329–356.

Schultz BE & Chan SI (2001) Structures and proton-pumping strategies of mitochondrial respiratory enzymes. *Annu. Rev. Biophys. Biomol. Struct.* 30, 23–65.

Thauer R, Jungermann K & Decker K (1977) Energy conservation in chemotrophic anaerobic bacteria. *Bacteriol. Rev.* 41, 100–180.

Tsukihara T, Aoyama H, Yamashita E et al. (1996) The whole structure of the 13-subunit oxidized cytochrome *c* oxidase at 2.8Å. *Science* 272, 1136–1144.

Wikstrom M (1998) Proton translocation by bacteriorhodopsin and heme-copper oxidases. *Curr. Opin. Struct. Biol.* 8, 480–488.

Chloroplasts and Photosynthesis

Allen JF & Pfannschmidt T (2000) Balancing the two photosystems: photosynthetic electron transfer governs transcription of reaction centre genes in chloroplasts. *Philos. Trans. R. Soc. Lond. B. Biol. Sci.* 355, 1351–1359.

Allen JP & Williams JC (1998) Photosynthetic reaction centers. *FEBS Lett.* 438, 5–9.

Barber J & Kuhlbrandt W (1999) Photosystem II. *Curr. Opin. Struct. Biol.* 9, 469–475.

Bassham JA (1962) The path of carbon in photosynthesis. *Sci. Am.* 206(6), 88–100.

Bogorad L & Vasil IK (1991) The Molecular Biology of Plastids. San Diego: Academic Press.

Buchanan BB, Gruissem W & Jones RL (2000) Biochemistry and Molecular Biology of Plants. Rockville, MD: American Society of Plant Physiologists.

Deisenhofer J & Michel H (1989) Nobel lecture. The photosynthetic reaction centre from the purple bacterium *Rhodopseudomonas viridis. EMBO J.* 8, 2149–2170.

Edwards GE, Furbank RT, Hatch MD & Osmond CB (2001) What does it take to be c(4)? lessons from the evolution of c(4) photosynthesis. *Plant Physiol.* 125, 46–49.

Flugge UI (1998) Metabolite transporters in plastids. *Curr. Opin. Plant Biol.* 1, 201–206.

Golbeck JH (1993) Shared thematic elements in photochemical reaction centers. *Proc. Natl. Acad. Sci. USA* 90, 1642–1646.

Grossman AR, Bhaya D, Apt KE & Kehoe DM (1995) Light-harvesting complexes in oxygenic photosynthesis: diversity, control, and evolution. *Annu. Rev. Genet.* 29, 231–288.

Hartman FC & Harpel MR (1994) Structure, function, regulation, and assembly of D-ribulose-1,5-bisphosphate carboxylase/oxygenase. *Annu. Rev. Biochem.* 63, 197–234.

Hatch MD (1987) C4 photosynthesis: a unique blend of modifed biochemistry, anatomy and ultrastructure. *Biochim. Biophys. Acta* 895, 81–106.

Jordan P, Fromme P, Witt HT et al. (2001) Three-dimensional structure of cyanobacterial photosystem I at 2.5Å resolution. *Nature* 411, 909–917.

Langdale JA & Nelson T (1991) Spatial regulation of photosynthetic development in C4 plants. *Trends Genet.* 7, 191–196.

Nugent JH (1996) Oxygenic photosynthesis. Electron transfer in photosystem I and photosystem II. *Eur. J. Biochem.* 237, 519–531.

Rao K & Hall DO (1999) Photosynthesis. Cambridge: Cambridge University Press.

Rhee KH, Morris EP, Barber J & Kuhlbrandt W (1998) Three-dimensional structure of the plant photosystem II reaction centre at 8Å resolution. *Nature* 396, 283–286.

Rogner M, Boekema EJ & Barber J (1996) How does photosystem 2 split water? The structural basis of efficient energy conversion. *Trends Biochem. Sci.* 21, 44–49.

Zouni A, Witt HT, Kern J et al. (2001) Crystal structure of photosystem II from Synechococcus elongatus at 3.8Å resolution. *Nature* 409, 739–743.

The Genetic Systems of Mitochondria and Plastids

Anderson S, Bankier AT, Barrell BG et al. (1981) Sequence and organization of the human mitochondrial genome. *Nature* 290, 457–465.

Birky CW, Jr (1995) Uniparental inheritance of mitochondrial and chloroplast genes: mechanisms and evolution. *Proc. Natl. Acad. Sci. USA* 92, 11331–11338.

Clayton DA (2000) Vertebrate mitochondrial DNA—a circle of surprises. *Exp. Cell Res.* 255, 4–9.

Contamine V & Picard M (2000) Maintenance and integrity of the mitochondrial genome: a plethora of nuclear genes in the budding yeast. *Microbiol. Mol. Biol. Rev.* 64, 281–315.

de Duve C (1996) The birth of complex cells. *Sci. Am.* 274(4), 50–57.

Douglas SE (1998) Plastid evolution: origins, diversity, trends. *Curr. Opin. Genet. Dev.* 8, 655–661.

Gray MW, Burger G & Lang BF (1999) Mitochondrial evolution. *Science* 283, 1476–1481.

Hermann GJ & Shaw JM (1998) Mitochondrial dynamics in yeast. *Annu. Rev. Cell Dev. Biol.* 14, 265–303.

Kurland CG (1992) Evolution of mitochondrial genomes and the genetic code. *BioEssays* 14, 709–714.

Kuroiwa T, Kuroiwa H, Sakai A et al. (1998) The division apparatus of plastids and mitochondria. *Int. Rev. Cytol.* 181, 1–41.

Martin W & Schnarrenberger C (1997) The evolution of the Calvin cycle from prokaryotic to eukaryotic chromosomes: a case study of functional redundancy in ancient pathways through endosymbiosis. *Curr. Genet.* 32, 1–18.

Ohlrogge J & Jaworski J (1997) Regulation of plant fatty acid biosynthesis. *Annu. Rev. Plant Physiol. Plant Mol. Biol.* 48, 109–136.

Saccone C, Gissi C, Lanave C et al. (2000) Evolution of the mitochondrial genetic system: an overview. *Gene* 261, 153–159.

Sugita M & Sugiura M (1996) Regulation of gene expression in chloroplasts of higher plants. *Plant Mol. Biol.* 32, 315–326.

Wallace DC (1999) Mitochondrial diseases in man and mouse. *Science* 283, 1482–1488.

Yaffe MP (1999) The machinery of mitochondrial inheritance and behavior. *Science* 283, 1493–1497.

The Evolution of Electron-Transport Chains

Blankenship RE & Bauer CE (eds) (1995) Anoxygenic Photosynthetic Bacteria. Dordrecht: Kluwer.

Blankenship RE & Hartman H (1998) The origin and evolution of oxygenic photosynthesis. *Trends Biochem. Sci.* 23, 94–97.

Blankenship RE (1994) Protein structure, electron transfer and evolution of prokaryotic photosynthetic reaction centers. *Antonie Van Leeuwenhoek* 65, 311–329.

Nisbet EG & Sleep NH (2001) The habitat and nature of early life. *Nature* 409, 1083–1091.

Orgel LE (1998) The origin of life—a review of facts and speculations. *Trends Biochem. Sci.* 23, 491–495.

Schafer G, Purschke W & Schmidt CL (1996) On the origin of respiration: electron transport proteins from archaea to man. *FEMS Microbiol. Rev.* 18, 173–188.

Schopf JW (1993) Microfossils of the Early Archean Apex chert: new evidence of the antiquity of life. *Science* 260, 640–646.

Skulachev VP (1994) Bioenergetics: the evolution of molecular mechanisms and the development of bioenergetic concepts. *Antonie Van Leeuwenhoek* 65, 271–284.

Xiong J, Fischer WM, Inoue K et al. (2000) Molecular evidence for the early evolution of photosynthesis. *Science* 289, 1724–1730.

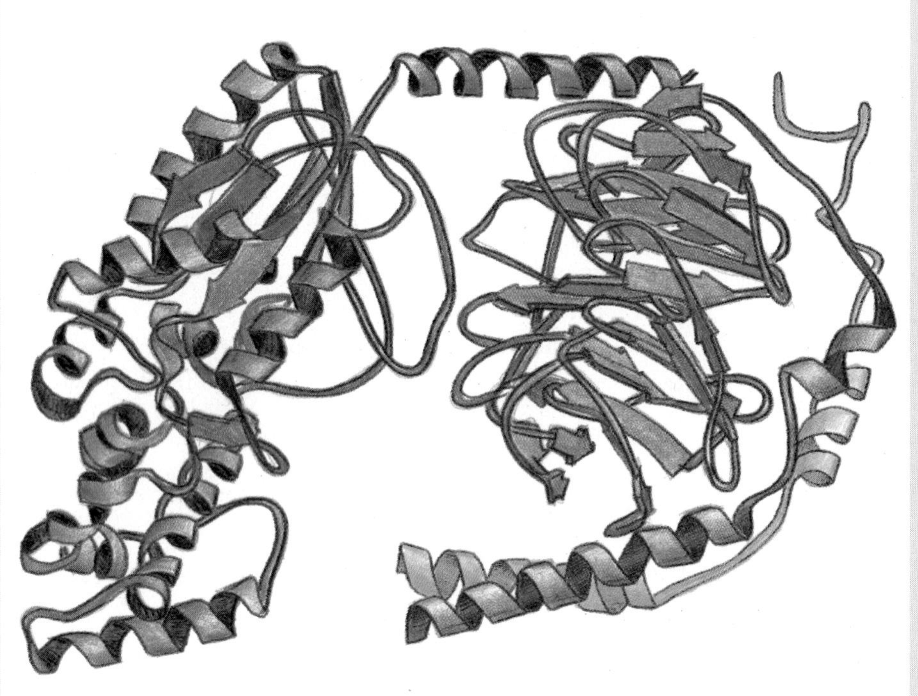

A trimeric GTP-binding protein, or G protein. This type of G protein functionally couples transmembrane receptors to either enzymes or ion channels in the plasma membrane. (Based on D.G. Lombright et al., *Nature* 379:311–319, 1996.)

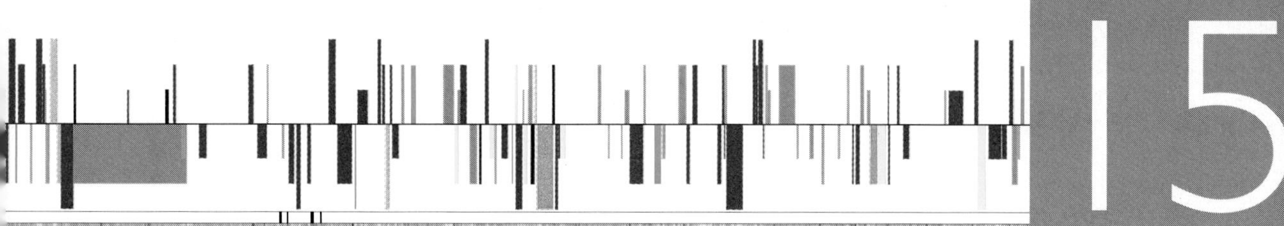

15

CELL COMMUNICATION

According to the fossil record, sophisticated unicellular organisms resembling present-day bacteria were present on Earth for about 2.5 billion years before the first multicellular organisms appeared. One reason why multicellularity was so slow to evolve may have been related to the difficulty of developing the elaborate cell communication mechanisms that a multicellular organism needs. Its cells have to be able to communicate with one another in complex ways if they are to be able to govern their own behavior for the benefit of the organism as a whole.

These communication mechanisms depend heavily on extracellular **signal molecules**, which are produced by cells to signal to their neighbors or to cells further away. They also depend on elaborate systems of proteins that each cell contains to enable it to respond to a particular subset of these signals in a cell-specific way. These proteins include cell-surface *receptor proteins*, which bind the signal molecule, plus a variety of *intracellular signaling proteins* that distribute the signal to appropriate parts of the cell. Among the intracellular signaling proteins are kinases, phosphatases, GTP-binding proteins, and many other proteins with which they interact. At the end of each intracellular signaling pathway are *target proteins*, which are altered when the pathway is active and change the behavior of the cell. Depending on the signal's effect, these target proteins can be gene regulatory proteins, ion channels, components of a metabolic pathway, parts of the cytoskeleton, and so on (Figure 15–1).

We begin this chapter by discussing the general principles of cell communication. We then consider, in turn, each of the main families of cell-surface receptor proteins and the intracellular signaling pathways they activate. The main focus of the chapter is on animal cells, but we end by considering the special features of cell communication in plants.

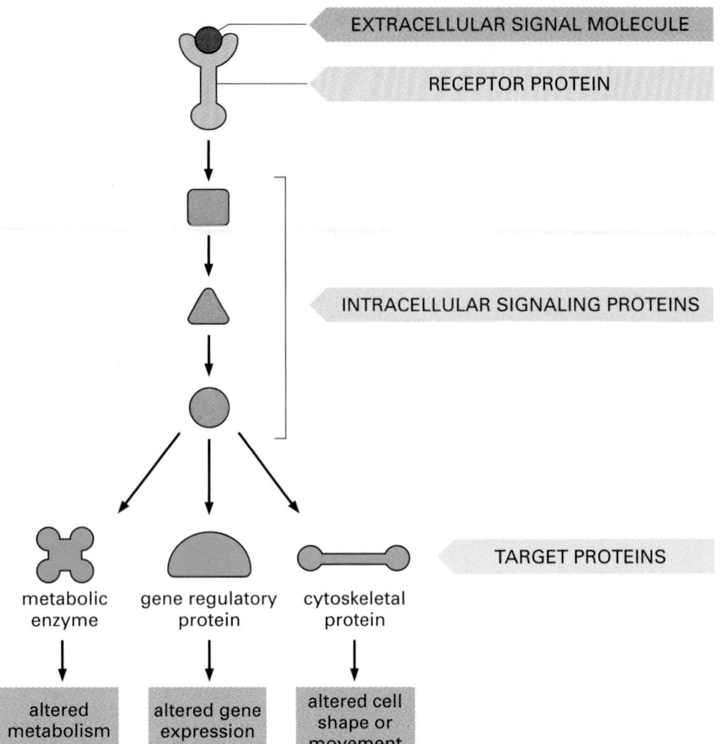

Figure 15–1 A simple intracellular signaling pathway activated by an extracellular signal molecule. The signal molecule binds to a receptor protein (which is usually embedded in the plasma membrane), thereby activating an intracellular signaling pathway that is mediated by a series of signaling proteins. Finally, one or more of these intracellular signaling proteins interacts with a target protein, altering the target protein so that it helps to change the behavior of the cell.

GENERAL PRINCIPLES OF CELL COMMUNICATION

Mechanisms enabling one cell to influence the behavior of another almost certainly existed in the world of unicellular organisms long before multicellular organisms appeared on Earth. Evidence comes from studies of present-day unicellular eucaryotes such as yeasts. Although these cells normally lead independent lives, they can communicate and influence one another's behavior in preparation for sexual mating. In the budding yeast *Saccharomyces cerevisiae,* for example, when a haploid individual is ready to mate, it secretes a peptide *mating factor* that signals cells of the opposite mating type to stop proliferating and prepare to mate (Figure 15–2). The subsequent fusion of two haploid cells of opposite mating types produces a diploid cell, which can then undergo meiosis and sporulate, generating haploid cells with new assortments of genes.

Studies of yeast mutants that are unable to mate have identified many proteins that are required in the signaling process. These proteins form a signaling network that includes cell-surface receptor proteins, GTP-binding proteins, and protein kinases, each of which has close relatives among the proteins that carry out signaling in animal cells. Through gene duplication and divergence, however, the signaling systems in animals have become much more elaborate than those in yeasts.

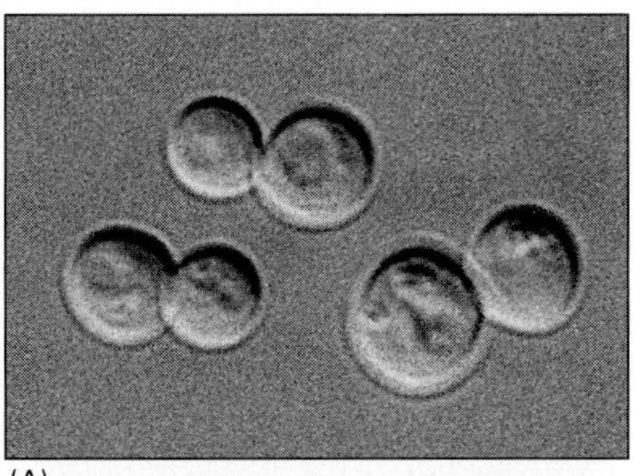

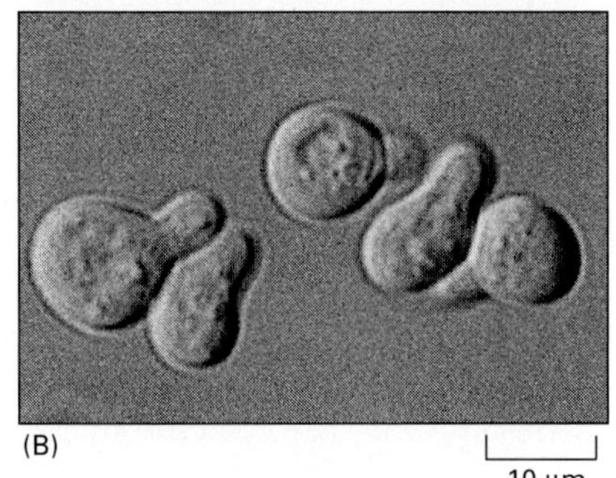

Figure 15–2 Budding yeast cells responding to mating factor. (A) The cells are normally spherical. (B) In response to mating factor secreted by neighboring yeast cells, they put out a protrusion toward the source of the factor in preparation for mating. (Courtesy of Michael Snyder.)

Extracellular Signal Molecules Bind to Specific Receptors

Yeast cells communicate with one another for mating by secreting a few kinds of small peptides. In contrast, cells in higher animals communicate by means of hundreds of kinds of signal molecules. These include proteins, small peptides, amino acids, nucleotides, steroids, retinoids, fatty acid derivatives, and even dissolved gases such as nitric oxide and carbon monoxide. Most of these signal molecules are secreted from the *signaling cell* into the extracellular space by exocytosis (discussed in Chapter 13). Others are released by diffusion through the plasma membrane, and some are exposed to the extracellular space while remaining tightly bound to the signaling cell's surface.

Figure 15–3 The binding of extracellular signal molecules to either cell-surface receptors or intracellular receptors. Most signal molecules are hydrophilic and are therefore unable to cross the plasma membrane directly; instead, they bind to cell-surface receptors, which in turn generate one or more signals inside the target cell. Some small signal molecules, by contrast, diffuse across the plasma membrane and bind to receptors inside the target cell—either in the cytosol or in the nucleus (as shown here). Many of these small signal molecules are hydrophobic and nearly insoluble in aqueous solutions; they are therefore transported in the bloodstream and other extracellular fluids after binding to carrier proteins, from which they dissociate before entering the target cell.

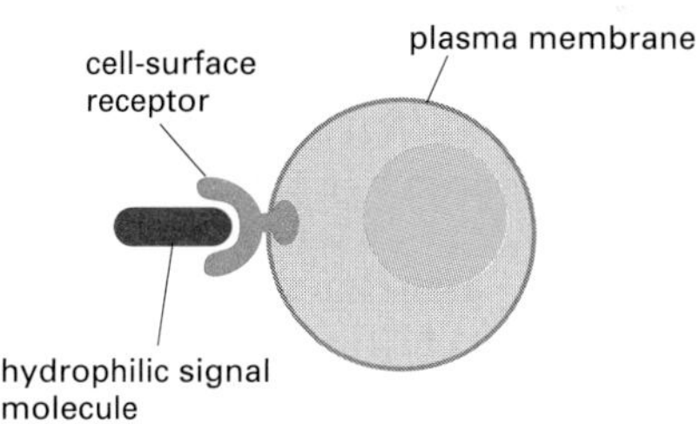

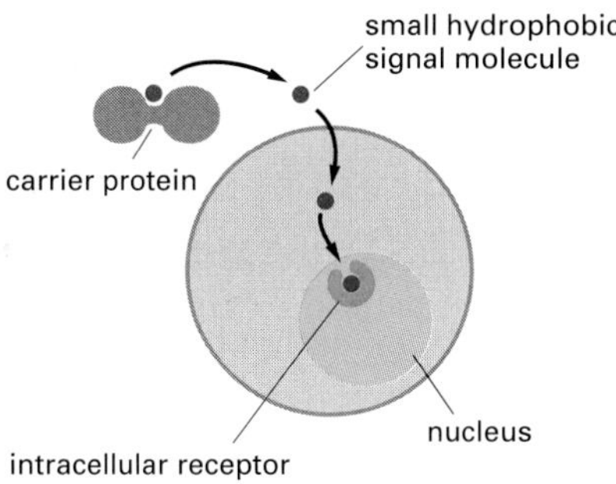

Regardless of the nature of the signal, the *target cell* responds by means of a specific protein called a **receptor**, which specifically binds the signal molecule and then initiates a response in the target cell. The extracellular signal molecules often act at very low concentrations (typically $\leq 10^{-8}$ M), and the receptors that recognize them usually bind them with high affinity (affinity constant $K_\alpha \geq 10^8$ liters/mole; see Figure 3–44). In most cases, these receptors are transmembrane proteins on the target cell surface. When they bind an extracellular signal molecule *(a ligand),* they become activated and generate a cascade of intracellular signals that alter the behavior of the cell. In other cases, the receptors are inside the target cell, and the signal molecule has to enter the cell to activate them: this requires that the signal molecules be sufficiently small and hydrophobic to diffuse across the plasma membrane (Figure 15–3).

Extracellular Signal Molecules Can Act Over Either Short or Long Distances

Many signal molecules remain bound to the surface of the signaling cell and influence only cells that contact it (Figure 15–4A). Such **contact-dependent signaling** is especially important during development and in immune responses. In most cases, however, signal molecules are secreted. The secreted molecules may be carried far afield to act on distant targets, or they may act as **local**

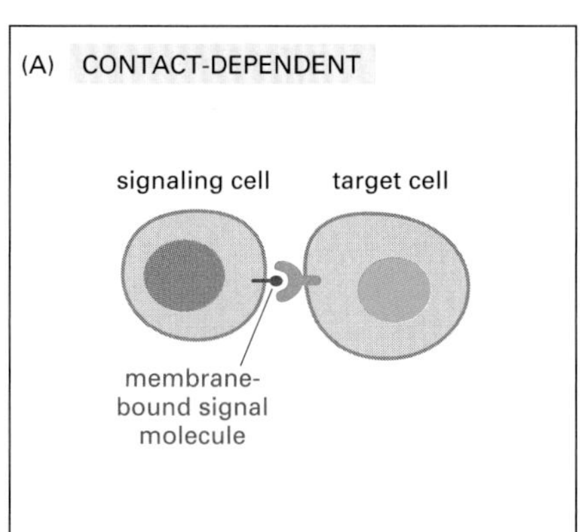

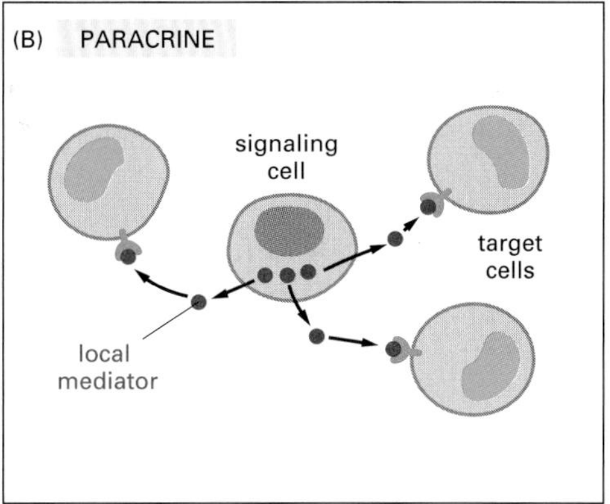

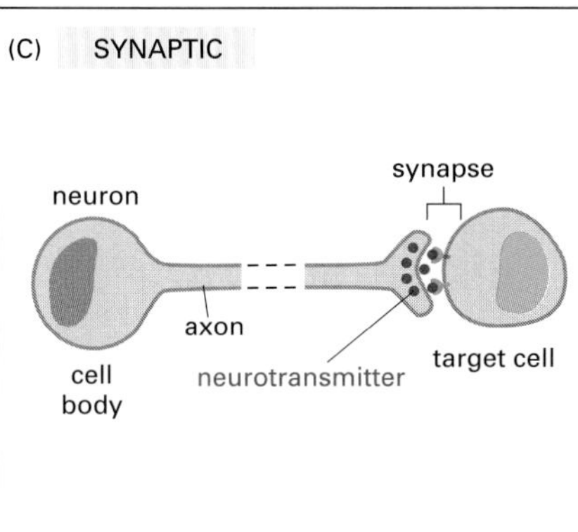

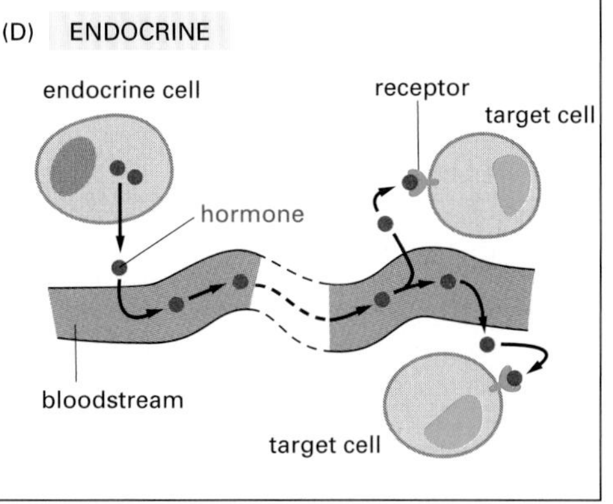

Figure 15–4 Forms of intracellular signaling. (A) Contact-dependent signaling requires cells to be in direct membrane–membrane contact. (B) Paracrine signaling depends on signals that are released into the extracellular space and act locally on neighboring cells. (C) Synaptic signaling is performed by neurons that transmit signals electrically along their axons and release neurotransmitters at synapses, which are often located far away from the cell body. (D) Endocrine signaling depends on endocrine cells, which secrete hormones into the bloodstream that are then distributed widely throughout the body. Many of the same types of signaling molecules are used in paracrine, synaptic, and endocrine signaling; the crucial differences lie in the speed and selectivity with which the signals are delivered to their targets.

mediators, affecting only cells in the immediate environment of the signaling cell. This latter process is called **paracrine signaling** (Figure 15–4B). For paracrine signals to be delivered only to their proper target cells, the secreted molecules must not be allowed to diffuse too far; for this reason they are often rapidly taken up by neighboring target cells, destroyed by extracellular enzymes, or immobilized by the extracellular matrix.

For a large, complex multicellular organism, short-range signaling is not sufficient on its own to coordinate the behavior of its cells. In these organisms, sets of specialized cells have evolved with a specific role in communication between widely separate parts of the body. The most sophisticated of these are nerve cells, or neurons, which typically extend long processes (axons) that enable them to contact target cells far away. When activated by signals from the environment or from other nerve cells, a neuron sends electrical impulses (action potentials) rapidly along its axon; when such an impulse reaches the end of the axon, it causes the nerve terminals located there to secrete a chemical signal called a **neurotransmitter**. These signals are secreted at specialized cell junctions called *chemical synapses*, which are designed to ensure that the neurotransmitter is delivered specifically to the postsynaptic target cell (Figure 15–4C). The details of this **synaptic signaling** process are discussed in Chapter 11.

A second type of specialized signaling cell that controls the behavior of the organism as a whole is an **endocrine cell**. These cells secrete their signal molecules, called **hormones**, into the bloodstream, which carries the signal to target cells distributed widely throughout the body (Figure 15–4D).

The mechanisms that allow endocrine cells and nerve cells to coordinate cell behavior in animals are compared in Figure 15–5. Because endocrine signaling relies on diffusion and blood flow, it is relatively slow. Synaptic signaling, by contrast, can be much faster, as well as more precise. Nerve cells can transmit information over long distances by electrical impulses that travel at rates of up to 100 meters per second; once released from a nerve terminal, a neurotransmitter has to diffuse less than 100 nm to the target cell, a process that takes less than a millisecond. Another difference between endocrine and synaptic signaling is that, whereas hormones are greatly diluted in the bloodstream and interstitial fluid and therefore must be able to act at very low concentrations (typically $< 10^{-8}$ M), neurotransmitters are diluted much less and can achieve high local concentrations. The concentration of *acetylcholine* in the synaptic cleft of an active neuromuscular junction, for example, is about 5×10^{-4} M. Correspondingly, neurotransmitter receptors have a relatively low affinity for their ligand, which means that the neurotransmitter can dissociate rapidly from the receptor to terminate a response. Moreover, after its release from a nerve terminal, a neurotransmitter is quickly removed from the synaptic cleft, either by specific hydrolytic enzymes that destroy it or by specific membrane transport proteins that pump it back into either the nerve terminal or neighboring glial cells. Thus, synaptic signaling is much more precise than endocrine signaling, both in time and in space.

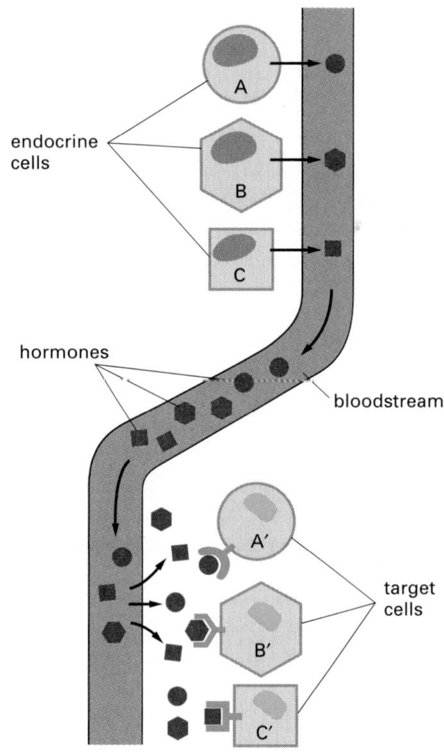

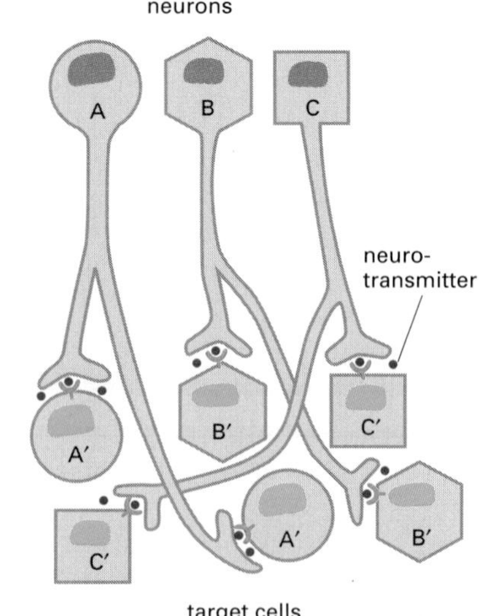

Figure 15–5 The contrast between endocrine and synaptic signaling. In complex animals, endocrine cells and nerve cells work together to coordinate the diverse activities of the billions of cells. Whereas different endocrine cells must use different hormones to communicate specifically with their target cells, different nerve cells can use the same neurotransmitter and still communicate in a highly specific manner. (A) Endocrine cells secrete hormones into the blood, which signal only the specific target cells that recognize them. These target cells have receptors for binding a specific hormone, which the cells "pull" from the extracellular fluid. (B) In synaptic signaling, by contrast, specificity arises from the synaptic contacts between a nerve cell and the specific target cells it signals. Usually, only a target cell that is in synaptic communication with a nerve cell is exposed to the neurotransmitter released from the nerve terminal (although some neurotransmitters act in a paracrine mode, serving as local mediators that influence multiple target cells in the area).

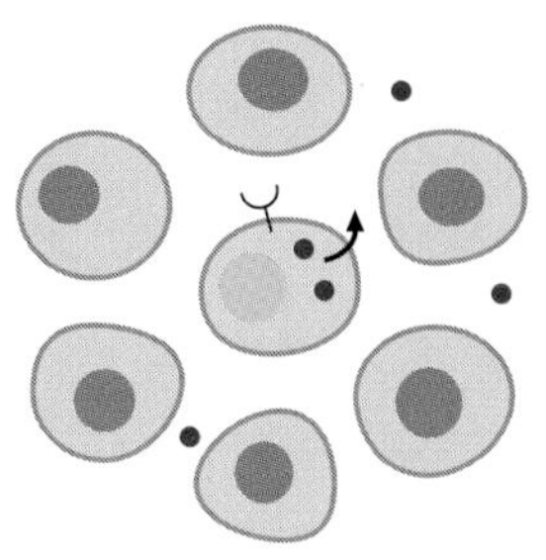
A SINGLE SIGNALING CELL RECEIVES A WEAK AUTOCRINE SIGNAL

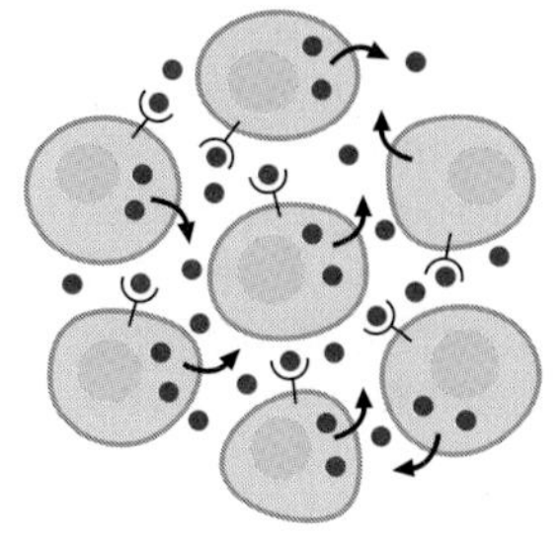
IN A GROUP OF IDENTICAL SIGNALING CELLS, EACH CELL RECEIVES A STRONG AUTOCRINE SIGNAL

Figure 15–6 Autocrine signaling. A group of identical cells produces a higher concentration of a secreted signal than does a single cell. When this signal binds back to a receptor on the same cell type, it encourages the cells to respond coordinately as a group.

The speed of a response to an extracellular signal depends not only on the mechanism of signal delivery, but also on the nature of the response in the target cell. Where the response requires only changes in proteins already present in the cell, it can occur in seconds or even milliseconds. When the response involves changes in gene expression and the synthesis of new proteins, however, it usually requires hours, irrespective of the mode of signal delivery.

Autocrine Signaling Can Coordinate Decisions by Groups of Identical Cells

All of the forms of signaling discussed so far allow one cell to influence another. Often, the signaling cell and target are different cell types. Cells, however, can also send signals to other cells of the same type, as well as to themselves. In such **autocrine signaling**, a cell secretes signal molecules that can bind back to its own receptors. During development, for example, once a cell has been directed along a particular pathway of differentiation, it may begin to secrete autocrine signals to itself that reinforce this developmental decision.

Autocrine signaling is most effective when performed simultaneously by neighboring cells of the same type, and it is likely to be used to encourage groups of identical cells to make the same developmental decisions. Thus, autocrine signaling is thought to be one possible mechanism underlying the "community effect" that is observed in early development, during which a group of identical cells can respond to a differentiation-inducing signal but a single isolated cell of the same type cannot (Figure 15–6).

Unfortunately, cancer cells often use autocrine signaling to overcome the normal controls on cell proliferation and survival that we discuss later. By secreting signals that act back on the cell's own receptors, cancer cells can stimulate their own survival and proliferation and thereby survive and proliferate in places where normal cells of the same type could not. How this dangerous perturbation of normal cell behavior comes about is discussed in Chapter 23.

Gap Junctions Allow Signaling Information to Be Shared by Neighboring Cells

Another way to coordinate the activities of neighboring cells is through **gap junctions**. These are specialized cell–cell junctions that can form between closely apposed plasma membranes and directly connect the cytoplasms of the joined cells via narrow water-filled channels (see Figure 19–15). The channels allow the exchange of small intracellular signaling molecules *(intracellular mediators)*, such as Ca^{2+} and cyclic AMP (discussed later), but not of macromolecules, such as proteins or nucleic acids. Thus, cells connected by gap junctions can communicate with each other directly, without having to surmount the barrier presented by the intervening plasma membranes (Figure 15–7).

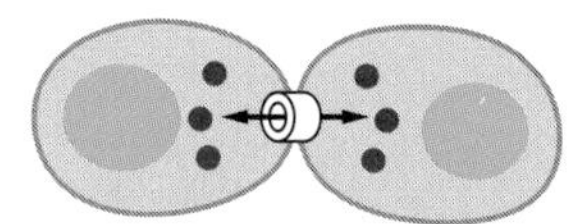

Figure 15–7 Signaling via gap junctions. Cells connected by gap junctions share small molecules, including small intracellular signaling molecules, and can therefore respond to extracellular signals in a coordinated way.

As discussed in Chapter 19, the pattern of gap-junction connections in a tissue can be revealed either electrically, with intracellular electrodes, or visually, after the microinjection of small water-soluble dyes. Studies of this kind indicate that the cells in a developing embryo make and break gap-junction connections in specific and interesting patterns, strongly suggesting that these junctions have an important role in the signaling processes that occur between these cells. Mice and humans that are deficient in one particular gap-junction protein (connexin 43), for example, have severe defects in heart development. Like the autocrine signaling described above, gap-junction communication helps adjacent cells of a similar type to coordinate their behavior. It is still not known, however, which particular small molecules are important as carriers of signals through gap junctions, and the specific functions of gap-junction communication in animal development remain uncertain.

Each Cell Is Programmed to Respond to Specific Combinations of Extracellular Signal Molecules

A typical cell in a multicellular organism is exposed to hundreds of different signals in its environment. These signals can be soluble, bound to the extracellular matrix, or bound to the surface of a neighboring cell, and they can act in many millions of combinations. The cell must respond to this babel of signals selectively, according to its own specific character, which it has acquired through progressive cell specialization in the course of development. A cell may be programmed to respond to one combination of signals by differentiating, to another combination by multiplying, and to yet another by performing some specialized function such as contraction or secretion.

Most of the cells in a complex animal are also programmed to depend on a specific combination of signals simply to survive. When deprived of these signals (in a culture dish, for example), a cell activates a suicide program and kills itself—a process called *programmed cell death*, or *apoptosis* (Figure 15–8). Because different types of cells require different combinations of survival signals, each cell type is restricted to different environments in the body. The ability to undergo apoptosis is a fundamental property of animal cells, and it is discussed in Chapter 17.

In principle, the hundreds of signal molecules that animals make can be used to create an almost unlimited number of signaling combinations. The use of these combinations to control cell behavior enables an animal to control its cells in highly specific ways by using a limited diversity of signal molecules.

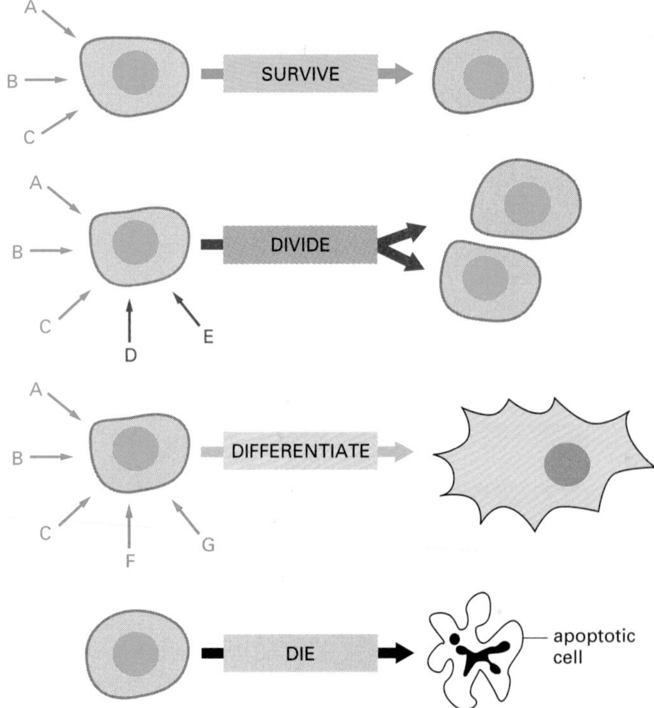

Figure 15–8 An animal cell's dependence on multiple extracellular signals. Each cell type displays a set of receptors that enables it to respond to a corresponding set of signal molecules produced by other cells. These signal molecules work in combinations to regulate the behavior of the cell. As shown here, an individual cell requires multiple signals to survive *(blue arrows)* and additional signals to divide *(red arrow)* or differentiate *(green arrows)*. If deprived of appropriate survival signals, a cell will undergo a form of cell suicide known as programmed cell death, or apoptosis.

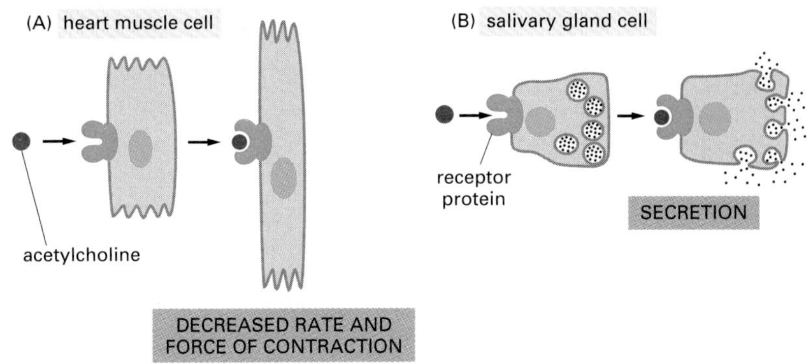

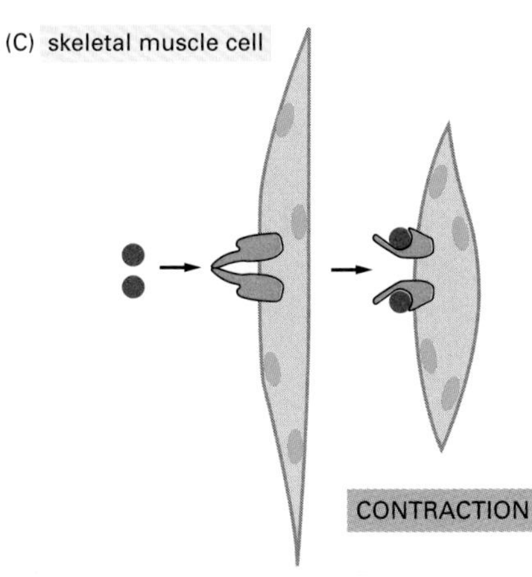

(D) acetylcholine

$$H_3C-\overset{\displaystyle O}{\overset{\|}{C}}-O-CH_2-CH_2-\underset{\displaystyle CH_3}{\overset{\displaystyle CH_3}{N^+}}-CH_3$$

Figure 15–9 Various responses induced by the neurotransmitter acetylcholine. Different cell types are specialized to respond to acetylcholine in different ways. (A and B) For these two cell types, acetylcholine binds to similar receptor proteins, but the intracellular signals produced are interpreted differently in cells specialized for different functions. (C) This muscle cell produces a distinct type of receptor protein for acetylcholine, which generates different intracellular signals from the receptor shown in (A) and (B), and results in a different effect. (D) The chemical structure of acetylcholine.

Different Cells Can Respond Differently to the Same Extracellular Signal Molecule

The specific way in which a cell reacts to its environment varies. It varies according to the set of receptor proteins the cell possesses, which determines the particular subset of signals it can respond to, and it varies according to the intracellular machinery by which the cell integrates and interprets the signals it receives (see Figure 15–1). Thus, a single signal molecule often has different effects on different target cells. The neurotransmitter acetylcholine, for example, stimulates the contraction of skeletal muscle cells, but it decreases the rate and force of contraction in heart muscle cells. This is because the acetylcholine receptor proteins on skeletal muscle cells are different from those on heart muscle cells. But receptor differences are not always the explanation for the different effects. In many cases, the same signal molecule binds to identical receptor proteins yet produces very different responses in different types of target cells, reflecting differences in the internal machinery to which the receptors are coupled (Figure 15–9).

The Concentration of a Molecule Can Be Adjusted Quickly Only If the Lifetime of the Molecule Is Short

It is natural to think of signaling systems in terms of the changes produced when a signal is delivered. But it is just as important to consider what happens when a signal is withdrawn. During development, transient signals often produce lasting effects: they can trigger a change in the cell's development that persists indefinitely, through cell memory mechanisms such as those discussed in Chapters 7 and 21. In most cases in adult tissues, however, the response fades when a signal ceases. The effect is transitory because the signal exerts its effects by altering a set of molecules that are unstable, undergoing continual turnover. Thus, once the signal is shut off, the replacement of the old molecules by new ones wipes out all traces of its action. It follows that the speed with which a cell responds to signal removal depends on the rate of destruction, or turnover, of the molecules the signal affects.

It is also true, although much less obvious, that this turnover rate also determines the promptness of the response when a signal is turned on. Consider, for example, two intracellular signaling molecules X and Y, both of which are normally maintained at a concentration of 1000 molecules per cell. Molecule Y is synthesized and degraded at a rate of 100 molecules per second, with each molecule having an average lifetime of 10 seconds. Molecule X has a turnover rate that is 10 times slower than that of Y: it is both synthesized and degraded at a rate of 10 molecules per second, so that each molecule has an average lifetime in the cell of 100 seconds. If a signal acting on the cell boosts the rates of synthesis of both X and Y tenfold without any change in the molecular lifetimes, at the end of 1 second the concentration of Y will have increased by nearly 900 molecules per cell ($10 \times 100 - 100$), while the concentration of X will have increased by only 90 molecules per cell. In fact, after a molecule's synthesis rate has been either increased or decreased abruptly, the time required for the

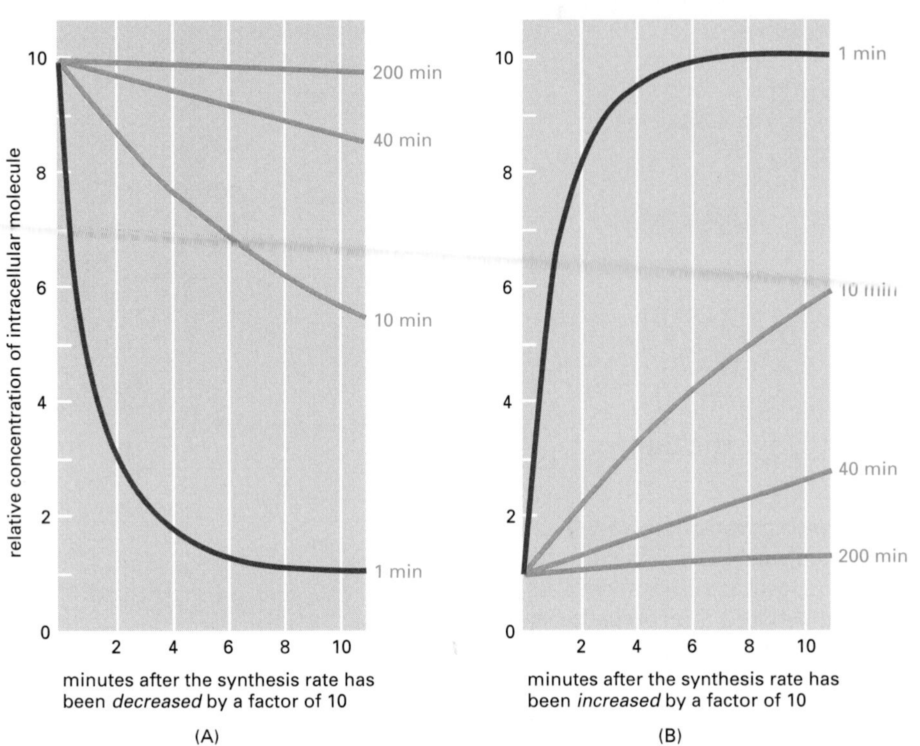

Figure 15–10 The importance of rapid turnover. The graphs show the predicted relative rates of change in the intracellular concentrations of molecules with differing turnover times when their synthesis rates are either (A) decreased or (B) increased suddenly by a factor of 10. In both cases, the concentrations of those molecules that are normally being rapidly degraded in the cell *(red lines)* change quickly, whereas the concentrations of those that are normally being slowly degraded *(green lines)* change proportionally more slowly. The numbers (in *blue*) on the right are the half-lives assumed for each of the different molecules.

molecule to shift halfway from its old to its new equilibrium concentration is equal to its normal half-life—that is, equal to the time that would be required for its concentration to fall by half if all synthesis were stopped (Figure 15–10).

The same principles apply to proteins and small molecules, and to molecules in the extracellular space and inside cells. Many intracellular proteins have short half-lives, some surviving for less than 10 minutes. In most cases, these are proteins with key regulatory roles, whose concentrations are rapidly regulated in the cell by changes in their rates of synthesis. Likewise, any covalent modifications of proteins that occur as part of a rapid signaling process—most commonly, the addition of a phosphate group to an amino acid side chain—must be continuously removed at a rapid rate to make rapid signaling possible.

We shall discuss some of these molecular events in detail later for signaling pathways that operate via cell-surface receptors. But the principles apply quite generally, as the next example illustrates.

Nitric Oxide Gas Signals by Binding Directly to an Enzyme Inside the Target Cell

Although most extracellular signals are hydrophilic molecules that bind to receptors on the surface of the target cell, some signal molecules are hydrophobic enough and/or small enough to pass readily across the target-cell plasma membrane. Once inside, they directly regulate the activity of a specific intracellular protein. An important and remarkable example is the gas **nitric oxide (NO)**, which acts as a signal molecule in both animals and plants. In mammals, one of its functions is to regulate smooth muscle contraction. Acetylcholine, for example, is released by autonomic nerves in the walls of a blood vessel, and it causes smooth muscle cells in the vessel wall to relax. The acetylcholine acts indirectly by inducing the nearby endothelial cells to make and release NO, which then signals the underlying smooth muscle cells to relax. This effect of NO on blood vessels provides an explanation for the mechanism of action of nitroglycerine, which has been used for about 100 years to treat patients with angina (pain resulting from inadequate blood flow to the heart muscle). The nitroglycerine is converted to NO, which relaxes blood vessels. This reduces the workload on the heart and, as a consequence, it reduces the oxygen requirement of the heart muscle.

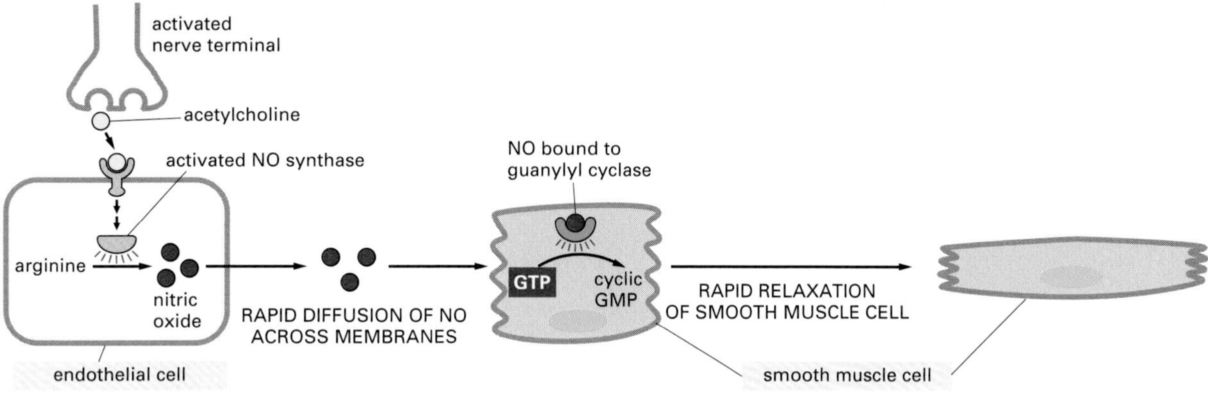

Figure 15–11 The role of nitric oxide (NO) in smooth muscle relaxation in a blood vessel wall. Acetylcholine released by nerve terminals in the blood vessel wall activates NO synthase in endothelial cells lining the blood vessel, causing the endothelial cells to produce NO. The NO diffuses out of the endothelial cells and into the underlying smooth muscle cells, where it binds to and activates guanylyl cyclase to produce cyclic GMP. The cyclic GMP triggers a response that causes the smooth muscle cells to relax, enhancing blood flow through the blood vessel.

Many types of nerve cells use NO gas to signal to their neighbors. The NO released by autonomic nerves in the penis, for example, causes the local blood vessel dilation that is responsible for penile erection. NO is also produced as a local mediator by activated macrophages and neutrophils to help them to kill invading microorganisms. In plants, NO is involved in the defensive responses to injury or infection.

NO gas is made by the deamination of the amino acid arginine, catalyzed by the enzyme *NO synthase*. Because it passes readily across membranes, dissolved NO rapidly diffuses out of the cell where it is produced and into neighboring cells. It acts only locally because it has a short half-life—about 5–10 seconds—in the extracellular space before it is converted to nitrates and nitrites by oxygen and water. In many target cells, including endothelial cells, NO binds to iron in the active site of the enzyme *guanylyl cyclase*, stimulating this enzyme to produce the small intracellular mediator *cyclic GMP*, which we discuss later (Figure 15–11). The effects of NO can occur within seconds, because the normal rate of turnover of cyclic GMP is high: a rapid degradation to GMP by a phosphodiesterase constantly balances the production of cyclic GMP from GTP by guanylyl cyclase. The drug Viagra inhibits this cyclic GMP phosphodiesterase in the penis, thereby increasing the amount of time that cyclic GMP levels remain elevated after NO production is induced by local nerve terminals. The cyclic GMP, in turn, keeps blood vessels relaxed and the penis erect.

Carbon monoxide (CO) is another gas that is used as an intercellular signal. It can act in the same way as NO, by stimulating guanylyl cyclase. These gases are not the only signal molecules that can pass directly across the target-cell plasma membrane. A group of small, hydrophobic, nongaseous hormones and local mediators also enter target cells in this way. But instead of binding to enzymes, they bind to intracellular receptor proteins that directly regulate gene transcription, as we discuss next.

Nuclear Receptors Are Ligand-activated Gene Regulatory Proteins

A number of small hydrophobic signal molecules diffuse directly across the plasma membrane of target cells and bind to intracellular receptor proteins. These signal molecules include *steroid hormones, thyroid hormones, retinoids,* and *vitamin D*. Although they differ greatly from one another in both chemical structure (Figure 15–12) and function, they all act by a similar mechanism. When these signal molecules bind to their receptor proteins, they activate the receptors, which bind to DNA to regulate the transcription of specific genes. The receptors are all structurally related, being part of the **nuclear receptor superfamily**. This very large superfamily also includes some receptor proteins that are activated by intracellular metabolites rather than by secreted signal molecules. Many family members have been identified by DNA sequencing only, and their ligand is not yet known; these proteins are therefore referred to as *orphan*

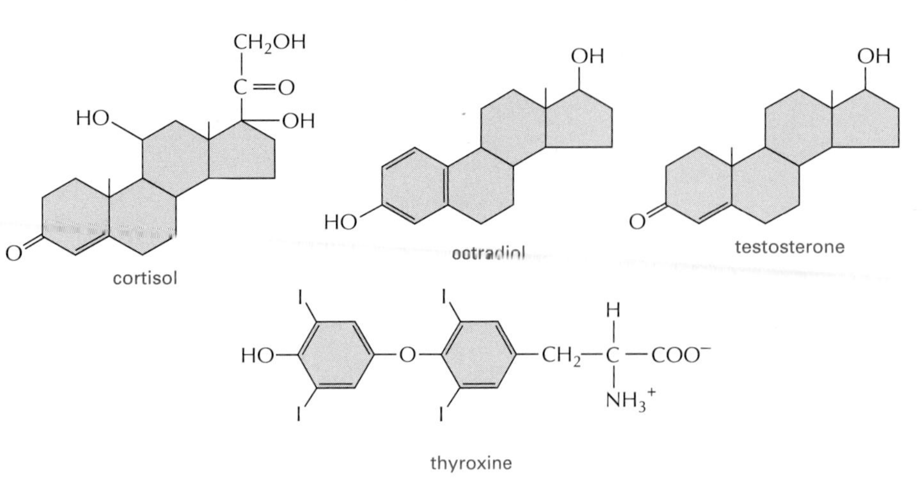

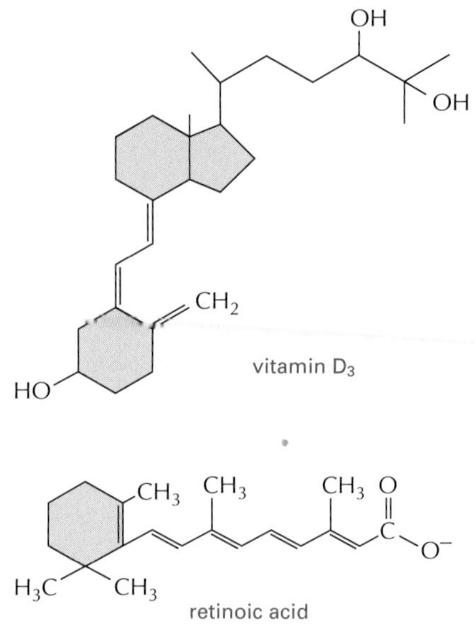

Figure 15–12 Some signaling molecules that bind to nuclear receptors. Note that all of them are small and hydrophobic. The active, hydroxylated form of vitamin D3 is shown. Estradiol and testosterone are steroid sex hormones.

nuclear receptors. The importance of such nuclear receptors in some animals is indicated by the fact that 1–2% of the genes in the nematode *C. elegans* code for them, although there are fewer than 50 in humans (see Figure 7–114).

Steroid hormones—which include cortisol, the steroid sex hormones, vitamin D (in vertebrates), and the moulting hormone ecdysone (in insects)—are all made from cholesterol. *Cortisol* is produced in the cortex of the adrenal glands and influences the metabolism of many types of cells. The steroid sex hormones are made in the testes and ovaries, and they are responsible for the secondary sex characteristics that distinguish males from females. Vitamin D is synthesized in the skin in response to sunlight; after it has been converted to its active form in the liver or kidneys, it regulates Ca^{2+} metabolism, promoting Ca^{2+} uptake in the gut and reducing its excretion in the kidneys. The thyroid hormones, which are made from the amino acid tyrosine, act to increase the metabolic rate in a wide variety of cell types, while the retinoids, such as retinoic acid, are made from vitamin A and have important roles as local mediators in vertebrate development. Although all of these signal molecules are relatively insoluble in water, they are made soluble for transport in the bloodstream and other extracellular fluids by binding to specific carrier proteins, from which they dissociate before entering a target cell (see Figure 15–3).

Beside a fundamental difference in the way they signal their target cells, most water-insoluble signal molecules differ from water-soluble ones in the length of time they persist in the bloodstream or tissue fluids. Most water-soluble hormones are removed and/or broken down within minutes of entering the blood, and local mediators and neurotransmitters are removed from the extracellular space even faster—within seconds or milliseconds. Steroid hormones, by contrast, persist in the blood for hours and thyroid hormones for days. Consequently, water-soluble signal molecules usually mediate responses of short duration, whereas water-insoluble ones tend to mediate responses that are longer lasting.

The intracellular receptors for the steroid and thyroid hormones, retinoids, and vitamin D all bind to specific DNA sequences adjacent to the genes the ligand regulates. Some receptors, such as those for cortisol, are located primarily in the cytosol and enter the nucleus after ligand binding; others, such as the thyroid and retinoid receptors, are bound to DNA in the nucleus even in the absence of ligand. In either case, the inactive receptors are bound to inhibitory protein complexes, and ligand binding alters the conformation of the receptor protein, causing the inhibitory complex to dissociate. The ligand binding also causes the receptor to bind to coactivator proteins that induce gene transcription

(Figure 15–13). The transcriptional response usually takes place in successive steps: the direct activation of a small number of specific genes occurs within about 30 minutes and constitutes the *primary response;* the protein products of these genes in turn activate other genes to produce a delayed, *secondary response;* and so on. In this way, a simple hormonal trigger can cause a very complex change in the pattern of gene expression (Figure 15–14).

The responses to steroid and thyroid hormones, vitamin D, and retinoids, like responses to extracellular signals in general, are determined as much by the nature of the target cell as by the nature of the signal molecule. Many types of cells have the identical intracellular receptor, but the set of genes that the receptor regulates is different in each cell type. This is because more than one type of gene regulatory protein generally must bind to a eucaryotic gene to activate its transcription. An intracellular receptor can therefore activate a gene only if there is the right combination of other gene regulatory proteins, and many of these are cell-type specific. Thus, each of these hormones induces a characteristic set of responses in an animal for two reasons. First, only certain types of cells have receptors for it. Second, each of these cell types contains a different combination of other cell-type-specific gene regulatory proteins that collaborate with the activated receptor to influence the transcription of specific sets of genes.

The molecular details of how nuclear receptors and other gene regulatory proteins control specific gene transcription are discussed in Chapter 7.

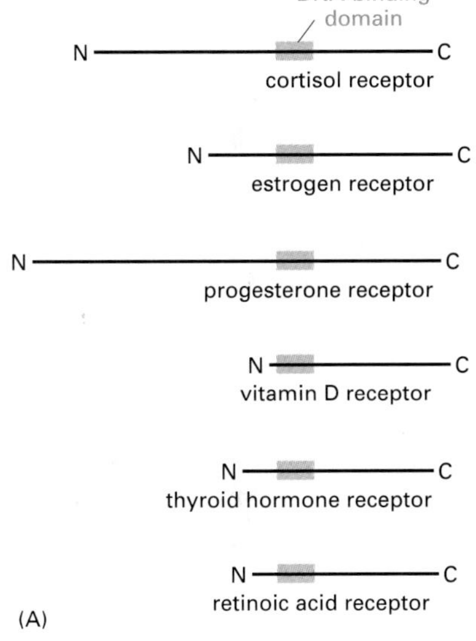

Figure 15–13 The nuclear receptor superfamily. All nuclear hormone receptors bind to DNA as either homodimers or heterodimers, but for simplicity we show them as monomers here. (A) The receptors all have a related structure. The short DNA-binding domain in each receptor is shown in *green.* (B) A receptor protein in its inactive state is bound to inhibitory proteins. Domain-swap experiments suggest that many of the ligand-binding, transcription-activating, and DNA-binding domains in these receptors can function as interchangeable modules. (C) The binding of ligand to the receptor causes the ligand-binding domain of the receptor to clamp shut around the ligand, the inhibitory proteins to dissociate, and coactivator proteins to bind to the receptor's transcription-activating domain, thereby increasing gene transcription. (D) The three-dimensional structure of a ligand-binding domain with *(right)* and without *(left)* ligand bound. Note that the *blue* α helix acts as a lid that snaps shut when the ligand (shown in *red*) binds, trapping the ligand in place.

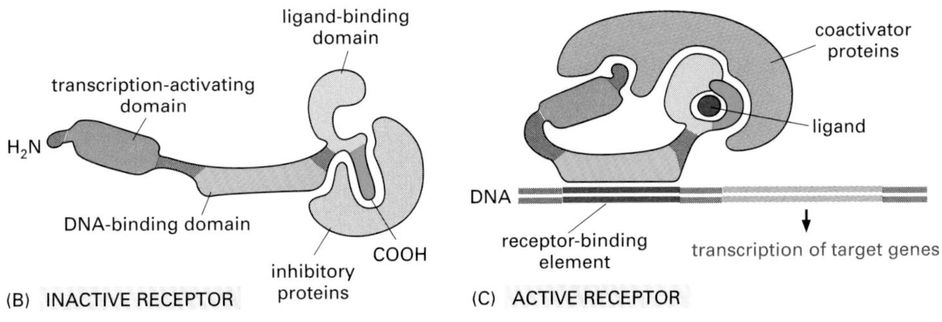

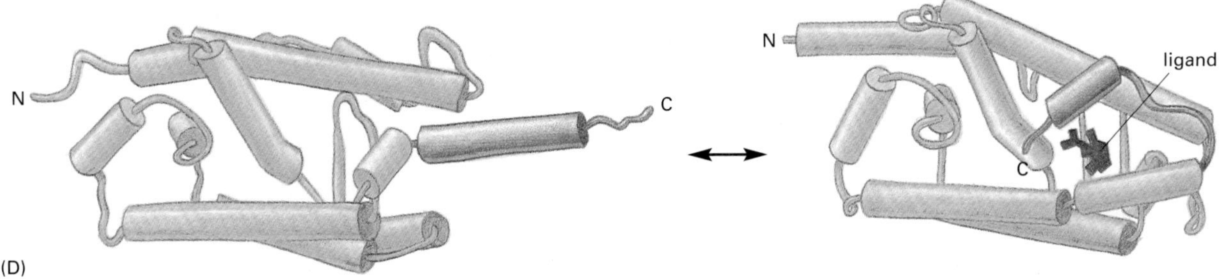

The Three Largest Classes of Cell-Surface Receptor Proteins Are Ion-Channel-linked, G-Protein-linked, and Enzyme-linked Receptors

As mentioned previously, all water-soluble signal molecules (including neurotransmitters and all signal proteins) bind to specific receptor proteins on the surface of the target cells that they influence. These cell-surface receptor proteins act as *signal transducers*. They convert an extracellular ligand-binding event into intracellular signals that alter the behavior of the target cell.

Most cell-surface receptor proteins belong to one of three classes, defined by the transduction mechanism they use. **Ion-channel-linked receptors**, also known as *transmitter-gated ion channels* or *ionotropic receptors*, are involved in rapid synaptic signaling between electrically excitable cells (Figure 15–15A). This type of signaling is mediated by a small number of neurotransmitters that transiently open or close an ion channel formed by the protein to which they bind, briefly changing the ion permeability of the plasma membrane and thereby the excitability of the postsynaptic cell. The ion-channel-linked receptors belong to a large family of homologous, multipass transmembrane proteins. Because they are discussed in detail in Chapter 11, we shall not consider them further here.

G-protein-linked receptors act indirectly to regulate the activity of a separate plasma-membrane-bound target protein, which can be either an enzyme or an ion channel. The interaction between the receptor and this target protein is mediated by a third protein, called a *trimeric GTP-binding protein (G protein)* (Figure 15–15B). The activation of the target protein can change the concentration of one or more intracellular mediators (if the target protein is an enzyme), or it can change the ion permeability of the plasma membrane (if the target protein is an ion channel). The intracellular mediators affected act in turn to alter the behavior of yet other signaling proteins in the cell. All of the G-protein-linked receptors belong to a large family of homologous, seven-pass transmembrane proteins.

Enzyme-linked receptors, when activated, either function directly as enzymes or are directly associated with enzymes that they activate (Figure 15–15C). They are formed by single-pass transmembrane proteins that have their ligand-binding site outside the cell and their catalytic or enzyme-binding site inside. Enzyme-linked receptors are heterogeneous in structure compared with the other two classes. The great majority, however, are protein kinases, or are associated with protein kinases, and ligand binding to them causes the phosphorylation of specific sets of proteins in the target cell.

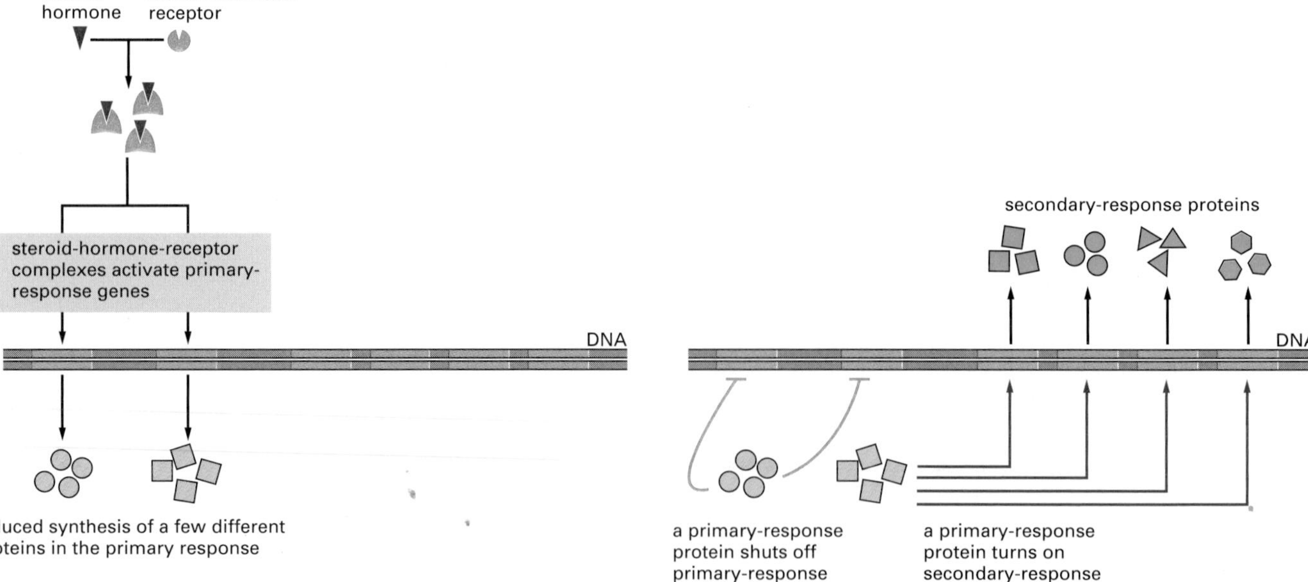

Figure 15–14 Responses induced by the activation of a nuclear hormone receptor. (A) Early primary response and (B) delayed secondary response. The figure shows the responses to a steroid hormone, but the same principles apply for all ligands that activate this family of receptor proteins. Some of the primary-response proteins turn on secondary-response genes, whereas others turn off the primary-response genes. The actual number of primary- and secondary-response genes is greater than shown. As expected, drugs that inhibit protein synthesis suppress the transcription of secondary-response genes but not primary-response genes, allowing these two classes of gene transcription responses to be readily distinguished.

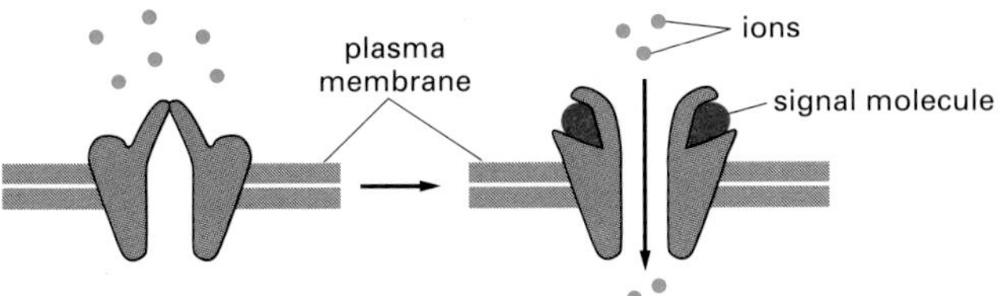

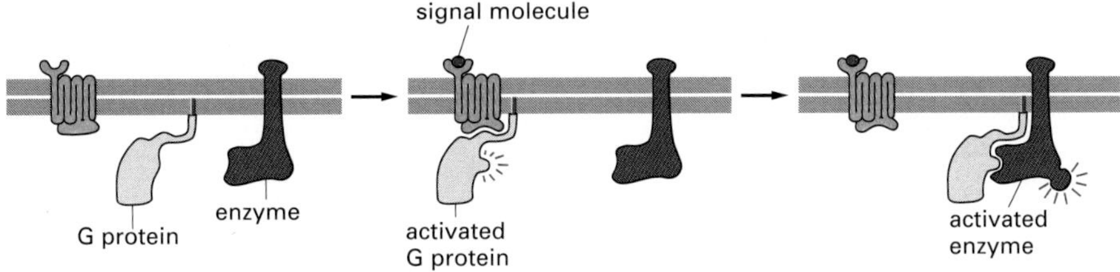

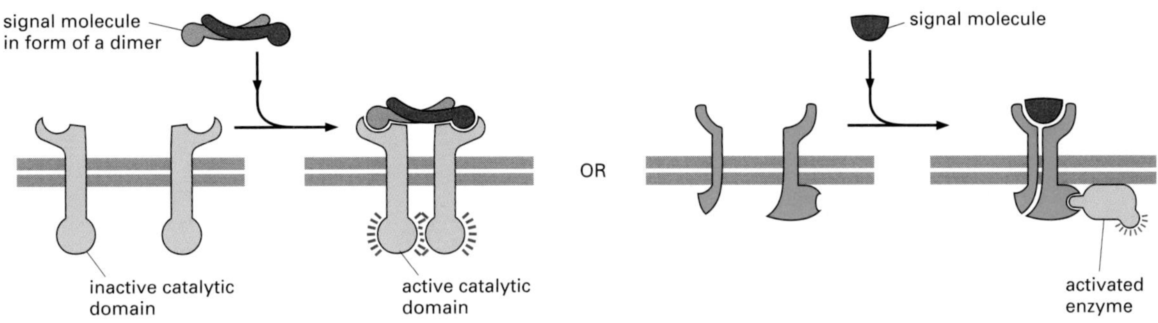

Figure 15–15 Three classes of cell-surface receptors. (A) Ion-channel-linked receptors, (B) G-protein-linked receptors, and (C) enzyme-linked receptors. Although many enzyme-linked receptors have intrinsic enzyme activity, as shown on the left, many others rely on associated enzymes, as shown on the right.

There are some cell-surface receptors that do not fit into any of the above classes. Some of these depend on intracellular proteolytic events to signal the cell, and we discuss them only after we explain in detail how G-protein-linked receptors and enzyme-linked receptors operate. We start with some general principles of signaling via cell-surface receptors.

Most Activated Cell-Surface Receptors Relay Signals Via Small Molecules and a Network of Intracellular Signaling Proteins

Signals received at the surface of a cell by either G-protein-linked or enzyme-linked receptors are relayed into the cell interior by a combination of small and large *intracellular signaling molecules*. The resulting chain of intracellular signaling events ultimately alters *target proteins*, and these altered target proteins are responsible for modifying the behavior of the cell (see Figure 15–1).

The small intracellular signaling molecules are called **small intracellular mediators**, or **second messengers** (the "first messengers" being the extracellular signals). They are generated in large numbers in response to receptor activation and rapidly diffuse away from their source, broadcasting the signal to other parts of the cell. Some, such as *cyclic AMP* and *Ca^{2+}*, are water-soluble and diffuse in the cytosol, while others, such as *diacylglycerol,* are lipid-soluble and diffuse in the plane of the plasma membrane. In either case, they pass the signal on by binding to and altering the behavior of selected signaling proteins or target proteins.

The large intracellular signaling molecules are **intracellular signaling proteins.** Many of these relay the signal into the cell by either activating the next signaling protein in the chain or generating small intracellular mediators. These proteins can be classified according to their particular function, although many fall into more than one category (Figure 15–16):

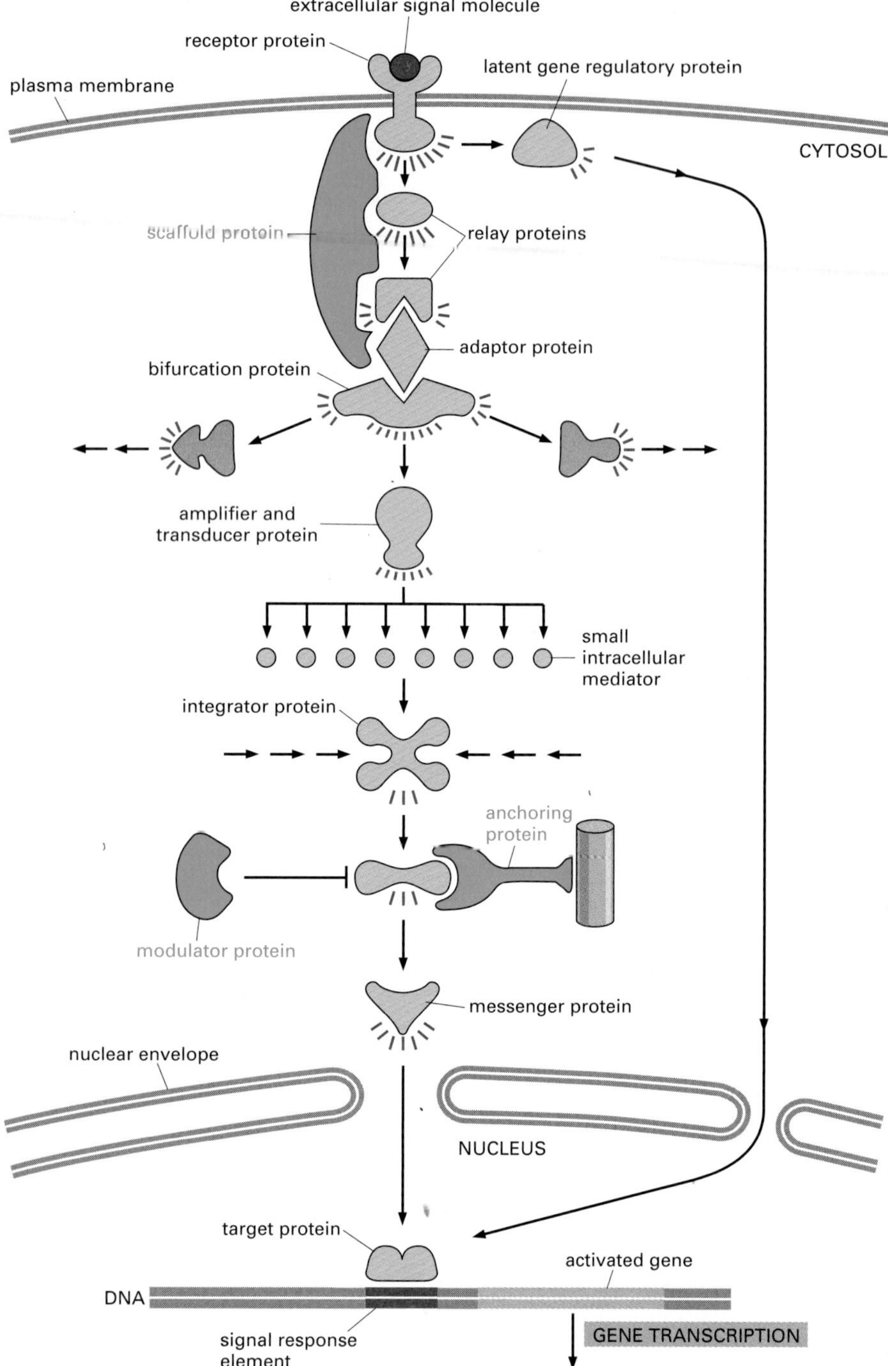

Figure 15–16 Different kinds of intracellular signaling proteins along a signaling pathway from a cell-surface receptor to the nucleus. In this example, a series of signaling proteins and small intracellular mediators relay the extracellular signal into the cell, causing a change in gene expression. The signal is amplified, altered (transduced), and distributed *en route*. Many of the steps can be modulated by other extracellular and intracellular signals, so that the final result of one signal depends on other factors affecting the cell (see Figure 15–8). Ultimately, the signaling pathway activates (or inactivates) target proteins that alter cell behavior. In this example, the target is a gene regulatory protein.

1. *Relay proteins* simply pass the message to the next signaling component in the chain.
2. *Messenger proteins* carry the signal from one part of the cell to another, such as from the cytosol to the nucleus.
3. *Adaptor proteins* link one signaling protein to another, without themselves conveying a signal.
4. *Amplifier proteins,* which are usually either enzymes or ion channels, greatly increase the signal they receive, either by producing large amounts of small intracellular mediators or by activating large numbers of downstream intracellular signaling proteins. When there are multiple amplification steps in a relay chain, the chain is often referred to as a **signaling cascade**.
5. *Transducer proteins* convert the signal into a different form. The enzyme that makes cyclic AMP is an example: it both converts the signal and amplifies it, thus acting as both a transducer and an amplifier.

6. *Bifurcation proteins* spread the signal from one signaling pathway to another.
7. *Integrator proteins* receive signals from two or more signaling pathways and integrate them before relaying a signal onward.
8. *Latent gene regulatory proteins* are activated at the cell surface by activated receptors and then migrate to the nucleus to stimulate gene transcription.

As shown in *blue* in Figure 15–16, other types of intracellular proteins also have important roles in intracellular signaling. *Modulator proteins* modify the activity of intracellular signaling proteins and thereby regulate the strength of signaling along the pathway. *Anchoring proteins* maintain specific signaling proteins at a precise location in the cell by tethering them to a membrane or the cytoskeleton. *Scaffold proteins* are adaptor and/or anchoring proteins that bind multiple signaling proteins together in a functional complex and often hold them at a specific location.

Some Intracellular Signaling Proteins Act as Molecular Switches

Many intracellular signaling proteins behave like **molecular switches:** on receipt of a signal they switch from an inactive to an active state, until another process switches them off. As we discussed earlier, the switching off is just as important as the switching on. If a signaling pathway is to recover after transmitting a signal so that it can be ready to transmit another, every activated molecule in the pathway must be returned to its original inactivated state.

The molecular switches fall into two main classes that operate in different ways, although in both cases it is the gain or loss of phosphate groups that determines whether the protein is active or inactive. The largest class consists of proteins that are activated or inactivated by phosphorylation (discussed in Chapter 3). For these proteins, the switch is thrown in one direction by a protein kinase, which adds one or more phosphate groups to the signaling protein, and in the other direction by a protein phosphatase, which removes the phosphate groups from the protein (Figure 15–17A). It is estimated that one-third of the proteins in a eucaryotic cell are phosphorylated at any given time.

Many of the signaling proteins controlled by phosphorylation are themselves protein kinases, and these are often organized into **phosphorylation cascades.** One protein kinase, activated by phosphorylation, phosphorylates the next protein kinase in the sequence, and so on, relaying the signal onward and, in the process, amplifying it and sometimes spreading it to other signaling pathways. Two main types of protein kinases operate as intracellular signaling proteins. The great majority are *serine/threonine kinases*, which phosphorylate proteins on serines and (less often) threonines. Others are *tyrosine kinases,* which phosphorylate proteins on tyrosines. An occasional kinase can do both. Genome sequencing reveals that about 2% of our genes encode protein kinases,

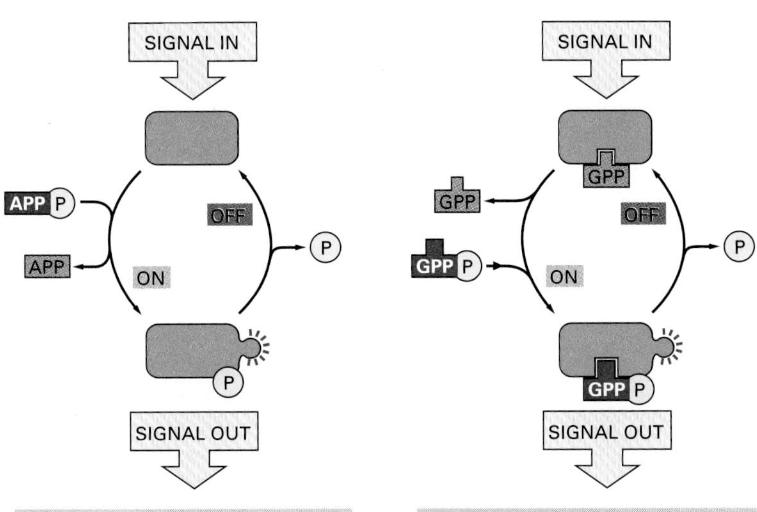

Figure 15–17 Two types of intracellular signaling proteins that act as molecular switches. In both cases, a signaling protein is activated by the addition of a phosphate group and inactivated by the removal of the phosphate. (A) The phosphate is added covalently to the signaling protein by a protein kinase. (B) A signaling protein is induced to exchange its bound GDP for GTP. To emphasize the similarity in the two mechanisms, ATP is shown as APPP, ADP as APP, GTP as GPPP, and GDP as GPP.

and it is thought that hundreds of distinct types of protein kinases are present in a typical mammalian cell.

The other main class of molecular switches involved in signaling are **GTP-binding proteins** (discussed in Chapter 3). These switch between an active state when GTP is bound and an inactive state when GDP is bound. Once activated, they have intrinsic GTPase activity and shut themselves off by hydrolyzing their bound GTP to GDP (Figure 15–17B). There are two major types of GTP-binding proteins—large *trimeric GTP-binding proteins* (also called *G proteins*), which relay the signals from G-protein-linked receptors (see Figure 15–15B), and small *monomeric GTPases* (also called *monomeric GTP-binding proteins*). The latter also help to relay intracellular signals, but in addition they are involved in regulating vesicular traffic and many other processes in eucaryotic cells.

As discussed earlier, complex cell behaviors, such as cell survival and cell proliferation, are generally stimulated by specific combinations of extracellular signals rather than by a single signal acting alone (see Figure 15–8). The cell therefore has to integrate the information coming from separate signals so as to make an appropriate response—to live or die, to divide or not, and so on. This integration usually depends on integrator proteins (see Figure 15–16), which are equivalent to the microprocessors in a computer: they require multiple signal inputs to produce an output that causes the desired biological effect. Two examples that show how such integrator proteins can operate are illustrated in Figure 15–18.

Intracellular Signaling Complexes Enhance the Speed, Efficiency, and Specificity of the Response

Even a single type of extracellular signal acting through a single type of G-protein-linked or enzyme-linked receptor usually activates multiple parallel signaling pathways and can thereby influence multiple aspects of cell behavior—such as shape, movement, metabolism, and gene expression. Indeed, these two main classes of cell-surface receptors often activate some of the same signaling pathways, and there is usually no obvious reason why a particular extracellular signal utilizes one class of receptors rather than the other.

The complexity of these signal-response systems, with multiple interacting relay chains of signaling proteins, is daunting. It is not clear how an individual cell manages to display specific responses to so many different extracellular signals, many of which bind to the same class of receptor and activate many of the same signaling pathways. One strategy that the cell uses to achieve specificity involves **scaffold proteins** (see Figure 15–16), which organize groups of interacting signaling proteins into *signaling complexes* (Figure 15–19A). Because the scaffold guides the interactions between the successive components in such a

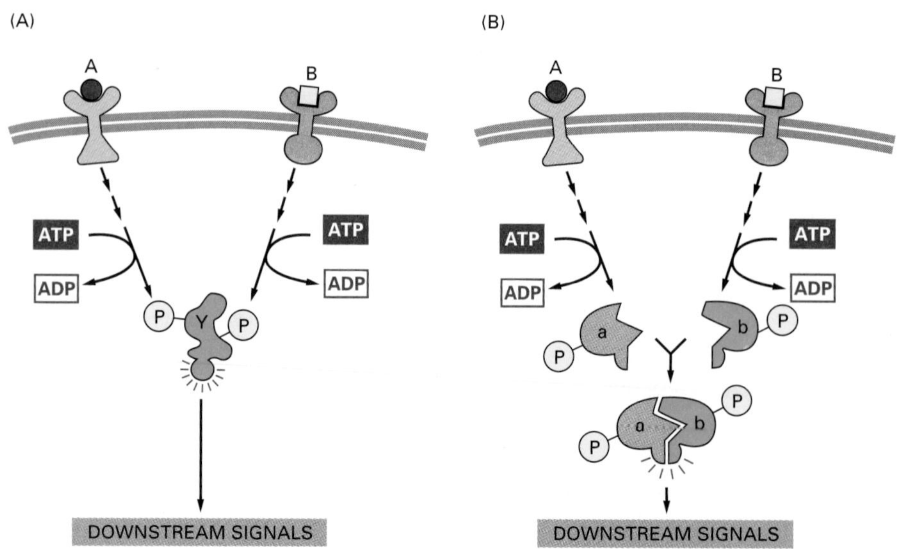

Figure 15–18 Signal integration. (A) Extracellular signals A and B both activate a different series of protein phosphorylations, each of which leads to the phosphorylation of protein Y but at different sites on the protein. Protein Y is activated only when both of these sites are phosphorylated, and therefore it becomes active only when signals A and B are simultaneously present. For this reason, integrator proteins are sometimes called coincidence detectors. (B) Extracellular signals A and B lead to the phosphorylation of two proteins, a and b, which then bind to each other to create the active protein. In both of the examples illustrated, the proteins themselves are phosphorylated. An equivalent form of control can also occur, however, by the exchange of GTP for GDP on GTP-binding proteins (see Figure 15–17).

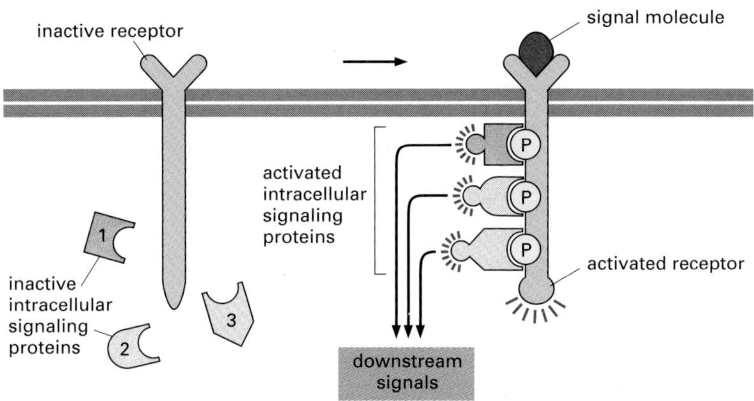

Figure 15–19 Two types of intracellular signaling complexes. (A) A receptor and some of the intracellular signaling proteins it activates in sequence are preassembled into a signaling complex by a large scaffold protein. (B) A large signaling complex is assembled after a receptor has been activated by the binding of an extracellular signal molecule; here the activated receptor phosphorylates itself at multiple sites, which then act as docking sites for intracellular signaling proteins.

complex, the signal can be relayed with precision, speed, and efficiency; moreover, unwanted cross-talk between signaling pathways is avoided. In order to amplify a signal, however, and spread it to other parts of the cell, at least some of the components in most signaling pathways are likely to be freely diffusible.

In other cases, signaling complexes form only transiently, as when signaling proteins assemble around a receptor after an extracellular signal molecule has activated it. In some of these cases, the cytoplasmic tail of the activated receptor is phosphorylated during the activation process, and the phosphorylated amino acids then serve as docking sites for the assembly of other signaling proteins (Figure 15–19B). In yet other cases, receptor activation leads to the production of modified phospholipid molecules in the adjacent plasma membrane, and these lipids then recruit specific intracellular signaling proteins to this region of membrane. All such signaling complexes form only transiently and rapidly disassemble after the extracellular ligand dissociates from the receptor.

Interactions Between Intracellular Signaling Proteins Are Mediated by Modular Binding Domains

The assembly of both stable and transient signaling complexes depends on a variety of highly conserved, small **binding domains** that are found in many intracellular signaling proteins. Each of these compact protein modules binds to a particular structural motif in the protein (or lipid) with which the signaling protein interacts. Because of these modular domains, signaling proteins bind to one another in multiple combinations, like Lego bricks, with the proteins often forming a three-dimensional network of interactions that determines the route followed by the signaling pathway. By joining existing domains together in novel

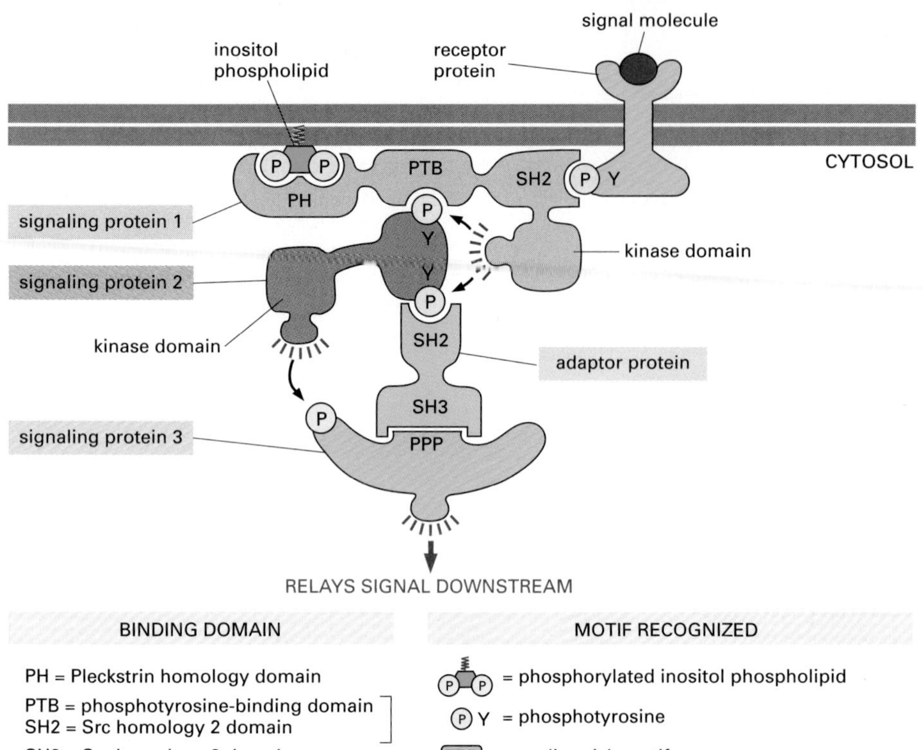

Figure 15–20 A hypothetical signaling pathway using modular binding domains. Signaling protein 1 contains three different binding domains, plus a catalytic protein kinase domain. It moves to the plasma membrane when extracellular signals lead to the creation of various phosphorylated docking sites on the cytosolic face of the membrane. Its SH2 domain binds to phosphorylated tyrosines on the receptor protein, and its PH domain binds to phosphorylated inositol phospholipids in the inner leaflet of the lipid bilayer. Protein 1 then phosphorylates signaling protein 2 on tyrosines, which allows protein 2 to bind to the PTB domain on protein 1 and to the SH2 domain on an adaptor protein. The adaptor protein then links protein 2 to protein 3, causing the phosphorylation of protein 3 by protein 2. The adaptor protein shown consists of two binding domains—an SH2 domain, which binds to a phosphotyrosine on protein 2, and an SH3 domain, which binds to a proline-rich motif on protein 3.

combinations, the use of such modular binding domains has presumably facilitated the rapid evolution of new signaling pathways.

Src homology 2 (SH2) domains and *phosphotyrosine-binding (PTB) domains*, for example, bind to phosphorylated tyrosines in a particular peptide sequence on activated receptors or intracellular signaling proteins. *Src homology 3 (SH3)* domains bind to a short proline-rich amino acid sequence. *Pleckstrin homology (PH)* domains (first described in the Pleckstrin protein in blood platelets) bind to the charged headgroups of specific phosphorylated inositol phospholipids that are produced in the plasma membrane in response to an extracellular signal; they thereby enable the protein they are part of to dock on the membrane and interact with other recruited signaling proteins. Some signaling proteins function only as adaptors to link two other proteins together in a signaling pathway, and they consist solely of two or more binding domains (Figure 15–20).

Scaffold proteins often contain multiple *PDZ domains* (originally found in a region of a synapse called the postsynaptic density), each of which binds to a specific motif on a receptor or signaling protein. The *InaD* scaffold protein in *Drosophila* photoreceptor cells is a striking example. It contains five PDZ domains, one of which binds a light-activated ion channel, while the others each bind to a different signaling protein involved in the response of the cell to light. If any of these PDZ domains are missing, the corresponding signaling protein fails to assemble in the complex, and the fly's vision is defective.

Some cell-surface receptors and intracellular signaling proteins are thought to cluster together transiently in specific microdomains in the lipid bilayer of the plasma membrane that are enriched in cholesterol and glycolipids. Some of the proteins are directed to these **lipid rafts** by covalently attached lipid molecules. Like scaffold proteins, these lipid scaffolds may promote speed and efficiency in the signaling process by serving as sites where signaling molecules can assemble and interact (see Figure 10–13).

Cells Can Respond Abruptly to a Gradually Increasing Concentration of an Extracellular Signal

Some cellular responses to extracellular signal molecules are smoothly graded in simple proportion to the concentration of the molecule. The primary responses to steroid hormones (see Figure 15–14) often follow this pattern, presumably because the nuclear hormone receptor protein binds a single molecule of hormone and each specific DNA recognition sequence in a steroid-hormone-responsive gene acts independently. As the concentration of hormone increases, the concentration of activated receptor–hormone complexes increases proportionally, as does the number of complexes bound to specific recognition sequences in the responsive genes; the response of the cell is therefore a gradual and linear one.

Many responses to extracellular signal molecules, however, begin more abruptly as the concentration of the molecule increases. Some may even occur in a nearly all-or-none manner, being undetectable below a threshold concentration of the molecule and then reaching a maximum as soon as this concentration is exceeded. What might be the molecular basis for such steep or even switchlike responses to graded signals?

One mechanism for sharpening the response is to require that more than one intracellular effector molecule or complex bind to some target macromolecule to induce a response. In some steroid-hormone-induced responses, for example, it seems that more than one activated receptor–hormone complex must bind simultaneously to specific regulatory sequences in the DNA to activate a particular gene. As a result, as the hormone concentration rises, gene activation begins more abruptly than it would if only one bound complex were sufficient for activation (Figure 15–21). A similar cooperative mechanism often operates in the signaling cascades activated by cell-surface receptors. As we discuss later, four molecules of the small intracellular mediator cyclic AMP, for example, must bind to each molecule of cyclic-AMP-dependent protein kinase to activate the kinase. Such responses become sharper as the number of cooperating molecules increases, and if the number is large enough, responses approaching the all-or-none type can be achieved (Figures 15–22 and 15–23).

Responses are also sharpened when an intracellular signaling molecule activates one enzyme and, at the same time, inhibits another enzyme that catalyzes the opposite reaction. A well-studied example of this common type of regulation is the stimulation of glycogen breakdown in skeletal muscle cells induced by the hormone *adrenaline* (epinephrine). Adrenaline's binding to a G-protein-linked cell-surface receptor leads to an increase in intracellular cyclic AMP concentration, which both activates an enzyme that promotes glycogen breakdown and inhibits an enzyme that promotes glycogen synthesis.

All of these mechanisms can produce responses that are very steep but, nevertheless, always smoothly graded according to the concentration of the

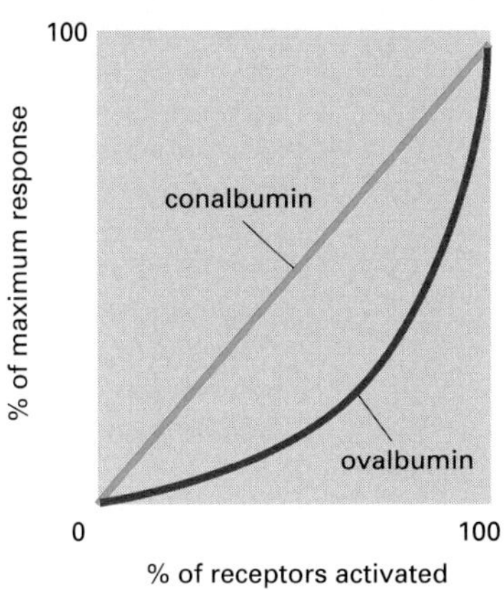

Figure 15–21 The primary response of chick oviduct cells to the steroid sex hormone estradiol. When activated, estradiol receptors turn on the transcription of several genes. Dose–response curves for two of these genes are shown, one coding for the egg protein conalbumin and the other coding for the egg protein ovalbumin. The linear response curve for conalbumin indicates that each activated receptor molecule that binds to the conalbumin gene increases the activity of the gene by the same amount. In contrast, the lag followed by the steep increase in the response curve for ovalbumin suggests that more than one activated receptor (in this case, two receptors) must bind simultaneously to the ovalbumin gene to initiate its transcription. (Adapted from E.R. Mulvihill and R.D. Palmiter, *J. Biol. Chem.* 252:2060–2068, 1977.)

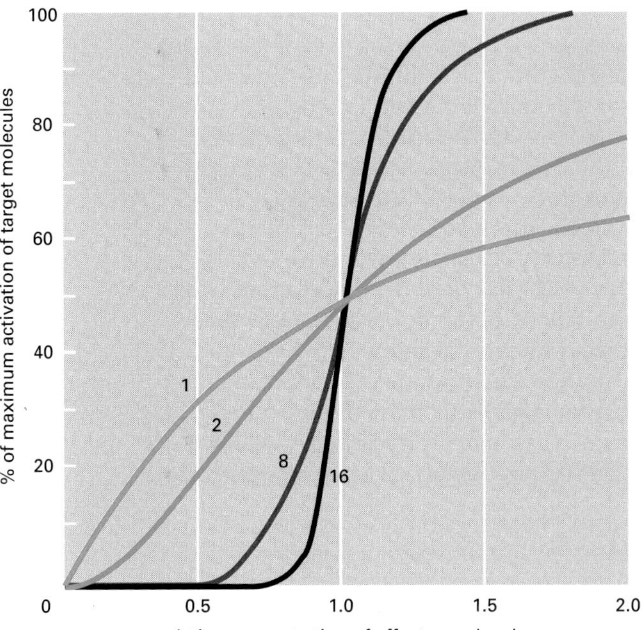

Figure 15–22 Activation curves as a function of signal-molecule concentration. The curves show how the sharpness of the response increases with an increase in the number of effector molecules that must bind simultaneously to activate a target macromolecule. The curves shown are those expected if the activation requires the simultaneous binding of 1, 2, 8, or 16 effector molecules.

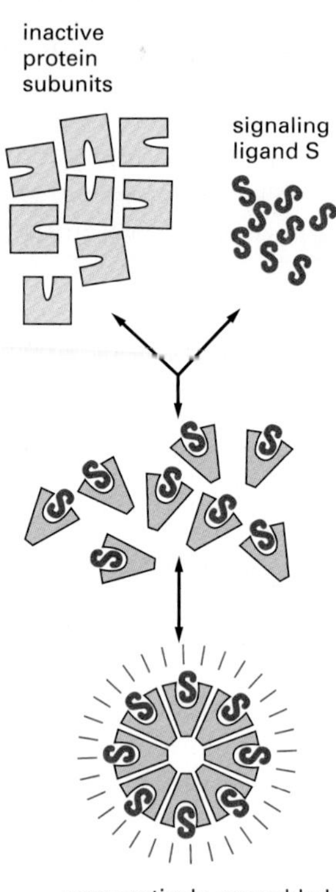

Figure 15–23 One type of signaling mechanism expected to show a steep thresholdlike response. Here, the simultaneous binding of eight molecules of a signaling ligand to a set of eight protein subunits is required to form an active protein complex. The ability of the subunits to assemble into the active complex depends on an allosteric conformational change that the subunits undergo when they bind their ligand. The binding of the ligand in the formation of such a complex is generally a cooperative process, causing a steep response as the ligand concentration is changed, as explained in Chapter 3. At low ligand concentrations, the number of active complexes increases roughly in proportion to the eighth power of the ligand concentration.

extracellular signal molecule. Another mechanism, however, can produce true all-or-none responses, such that raising the signal above a critical threshold level trips a sudden switch in the responding cell. All-or-none threshold responses of this type generally depend on *positive feedback;* by this mechanism, nerve and muscle cells generate all-or-none *action potentials* in response to neurotransmitters (discussed in Chapter 11). The activation of ion-channel-linked acetylcholine receptors at a neuromuscular junction, for example, results in a net influx of Na^+ that locally depolarizes the muscle plasma membrane. This causes voltage-gated Na^+ channels to open in the same membrane region, producing a further influx of Na^+, which further depolarizes the membrane and thereby opens more Na^+ channels. If the initial depolarization exceeds a certain threshold value, this positive feedback has an explosive "runaway" effect, producing an action potential that propagates to involve the entire muscle membrane.

An accelerating positive feedback mechanism can also operate through signaling proteins that are enzymes rather than ion channels. Suppose, for example, that a particular intracellular signaling ligand activates an enzyme located downstream in a signaling pathway and that two or more molecules of the product of the enzymatic reaction bind back to the same enzyme to activate it further (Figure 15–24). The consequence is a very low rate of synthesis of the product in the absence of the ligand. The rate increases slowly with the concentration of ligand until, at some threshold level of ligand, enough of the product has been synthesized to activate the enzyme in a self-accelerating, runaway fashion. The concentration of the product then suddenly increases to a much higher level. Through these and a number of other mechanisms not discussed here, the cell will often translate a gradual change in the concentration of a signaling ligand into a switchlike change, creating an all-or-none response by the cell.

A Cell Can Remember The Effect of Some Signals

The effect of an extracellular signal on a target cell can, in some cases, persist well after the signal has disappeared. The enzymatic accelerating positive feedback system just described represents one type of mechanism that displays this kind of persistence. If such a system has been switched on by raising the concentration of intracellular activating ligand above threshold, it will generally remain switched on even when the extracellular signal disappears; instead of faithfully reflecting the current level of signal, the response system displays a memory. We shall encounter a specific example of this later, when we discuss a protein kinase that is activated by Ca^{2+} to phosphorylate itself and other proteins; the autophosphorylation keeps the kinase active long after Ca^{2+} levels return to normal, providing a memory trace of the initial signal.

Transient extracellular signals often induce much longer-term changes in cells during the development of a multicellular organism. Some of these changes can persist for the lifetime of the organism. They usually depend on self-activating memory mechanisms that operate further downstream in a signaling

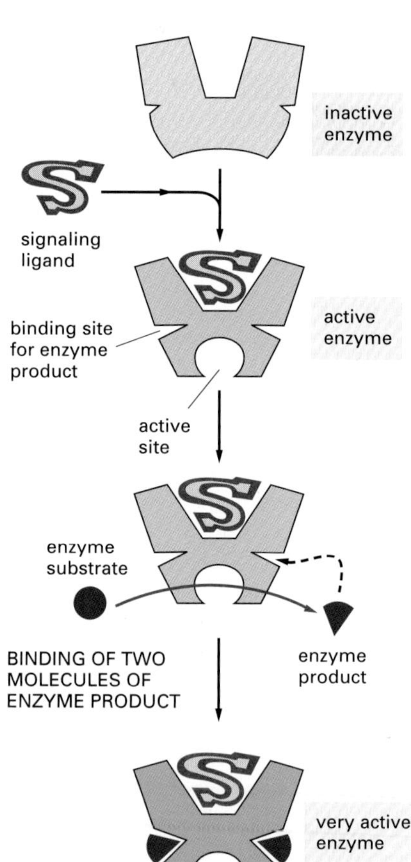

Figure 15–24 An accelerating positive feedback mechanism. In this example, the initial binding of the signaling ligand activates the enzyme to generate a product that binds back to the enzyme, further increasing the enzyme's activity.

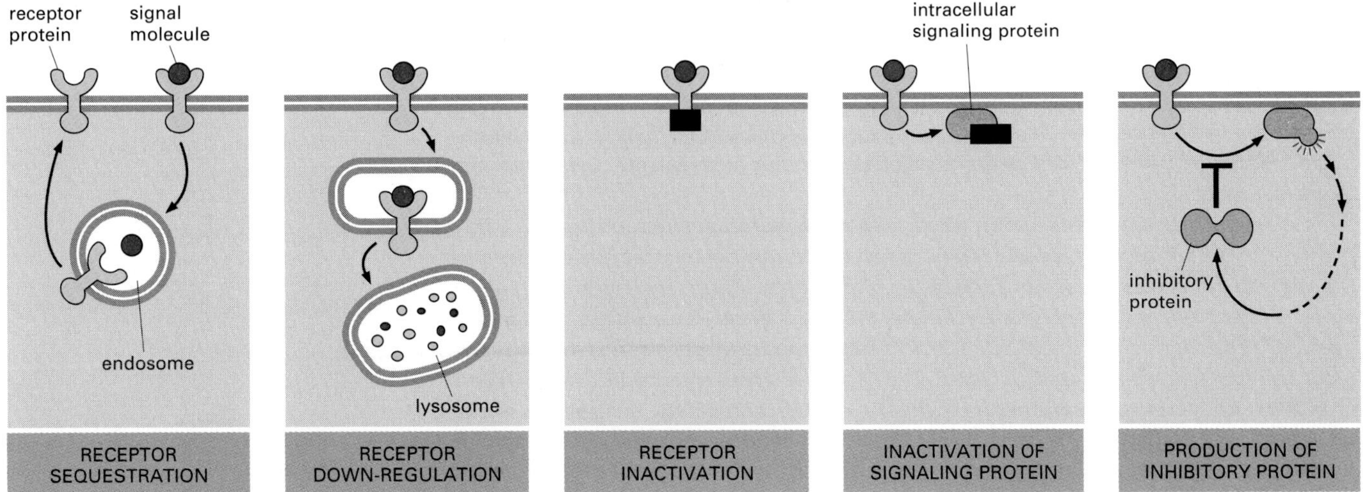

Figure 15–25 Five ways in which target cells can become desensitized to a signal molecule. The inactivation mechanisms shown here for both the receptor and the intracellular signaling protein often involve phosphorylation of the protein that is inactivated, although other types of modification are also known to occur. In bacterial chemotaxis, which we discuss later, desensitization depends on methylation of the receptor protein.

pathway, at the level of gene transcription. The signals that trigger muscle cell determination, for example, turn on a series of muscle-specific gene regulatory proteins that stimulate the transcription of their own genes, as well as genes producing many other muscle cell proteins. In this way, the decision to become a muscle cell is made permanent (see Figure 7–72B).

Cells Can Adjust Their Sensitivity to a Signal

In responding to many types of stimuli, cells and organisms are able to detect the same percentage of change in a signal over a very wide range of stimulus intensities. This requires that the target cells undergo a reversible process of **adaptation**, or **desensitization**, whereby a prolonged exposure to a stimulus decreases the cells' response to that level of exposure. In chemical signaling, adaptation enables cells to respond to *changes* in the concentration of a signaling ligand (rather than to the absolute concentration of the ligand) over a very wide range of ligand concentrations. The general principle is one of a negative feedback that operates with a delay. A strong response modifies the machinery for making that response, such that the machinery resets itself to an off position. Owing to the delay, however, a sudden change in the stimulus is able to make itself felt strongly for a short period before the negative feedback has time to kick in.

Desensitization to a signal molecule can occur in various ways. Ligand binding to cell-surface receptors, for example, may induce their endocytosis and temporary sequestration in endosomes. Such ligand-induced receptor endocytosis can lead to the destruction of the receptors in lysosomes, a process referred to as *receptor down-regulation.* In other cases, desensitization results from a rapid inactivation of the receptors—for example, as a result of a receptor phosphorylation that follows its activation, with a delay. Desensitization can also be caused by a change in a protein involved in transducing the signal or by the production of an inhibitor that blocks the transduction process (Figure 15–25).

Having discussed some of the general principles of cell signaling, we now turn to the G-protein-linked receptors. These are by far the largest class of cell-surface receptors, and they mediate the responses to the great majority of extracellular signals. This superfamily of receptor proteins not only mediates intercellular communication; it is also central to vision, smell, and taste perception.

Summary

Each cell in a multicellular animal has been programmed during development to respond to a specific set of extracellular signals produced by other cells. These signals act in various combinations to regulate the behavior of the cell. Most of the signals mediate a form of signaling in which local mediators are secreted, but then are

rapidly taken up, destroyed, or immobilized, so that they act only on neighboring cells. Other signals remain bound to the outer surface of the signaling cell and mediate contact-dependent signaling. Centralized control is exerted both by endocrine signaling, in which hormones secreted by endocrine cells are carried in the blood to target cells throughout the body, and by synaptic signaling, in which neurotransmitters secreted by nerve cell axons act locally on the postsynaptic cells that the axons contact.

Cell signaling requires not only extracellular signal molecules, but also a complementary set of receptor proteins in each cell that enable it to bind and respond to the signal molecules in a characteristic way. Some small hydrophobic signal molecules, including steroid and thyroid hormones, diffuse across the plasma membrane of the target cell and activate intracellular receptor proteins that directly regulate the transcription of specific genes. The dissolved gases nitric oxide and carbon monoxide act as local mediators by diffusing across the plasma membrane of the target cell and activating an intracellular enzyme—usually guanylyl cyclase, which produces cyclic GMP in the target cell. But most extracellular signal molecules are hydrophilic and can activate receptor proteins only on the surface of the target cell; these receptors act as signal transducers, converting the extracellular binding event into intracellular signals that alter the behavior of the target cell.

There are three main families of cell-surface receptors, each of which transduces extracellular signals in a different way. Ion-channel-linked receptors are transmitter-gated ion channels that open or close briefly in response to the binding of a neurotransmitter. G-protein-linked receptors indirectly activate or inactivate plasma-membrane-bound enzymes or ion channels via trimeric GTP-binding proteins (G proteins). Enzyme-linked receptors either act directly as enzymes or are associated with enzymes; these enzymes are usually protein kinases that phosphorylate specific proteins in the target cell.

Once activated, enzyme- and G-protein-linked receptors relay a signal into the cell interior by activating chains of intracellular signaling proteins; some transduce, amplify, or spread the signal as they relay it, while others integrate signals from different signaling pathways. Many of these signaling proteins function as switches that are transiently activated by phosphorylation or GTP binding. Functional signaling complexes are often formed by means of modular binding domains in the signaling proteins; these domains allow complicated protein assemblies to function in signaling networks.

Target cells can use a variety of intracellular mechanisms to respond abruptly to a gradually increasing concentration of an extracellular signal or to convert a short-lasting signal into a long-lasting response. In addition, through adaptation, they can often reversibly adjust their sensitivity to a signal to allow the cells to respond to changes in the concentration of a particular signal molecule over a large range of concentrations.

SIGNALING THROUGH G-PROTEIN-LINKED CELL-SURFACE RECEPTORS

G-protein-linked receptors form the largest family of cell-surface receptors and are found in all eucaryotes. About 5% of the genes in the nematode *C. elegans*, for example, encode such receptors, and thousands have already been defined in mammals; in mice, there are about 1000 concerned with the sense of smell alone. G-protein-linked receptors mediate the responses to an enormous diversity of signal molecules, including hormones, neurotransmitters, and local mediators. These signal molecules that activate them are as varied in structure as they are in function: the list includes proteins and small peptides, as well as derivatives of amino acids and fatty acids. The same ligand can activate many different receptor family members; at least 9 distinct G-protein-linked receptors are activated by adrenaline, for example, another 5 or more by acetylcholine, and at least 15 by the neurotransmitter serotonin.

Despite the chemical and functional diversity of the signal molecules that bind to them, all G-protein-linked receptors have a similar structure. They consist of a single polypeptide chain that threads back and forth across the lipid bilayer seven times and are therefore sometimes called *serpentine receptors* (Figure 15–26). In addition to their characteristic orientation in the plasma membrane, they have the same functional relationship to the G proteins they use to signal the cell interior that an extracellular ligand is present.

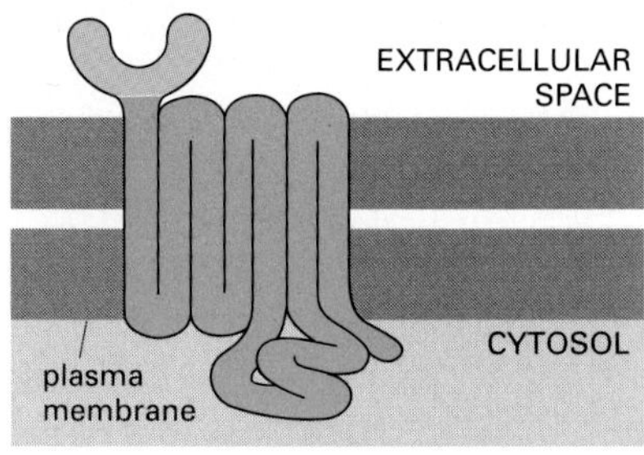

Figure 15–26 A G-protein-linked receptor. Receptors that bind protein ligands have a large extracellular domain formed by the part of the polypeptide chain shown in *light green.* This domain, together with some of the transmembrane segments, binds the protein ligand. Receptors for small ligands such as adrenaline have small extracellular domains, and the ligand usually binds deep within the plane of the membrane to a site that is formed by amino acids from several transmembrane segments.

As we discuss later, this superfamily of seven-pass transmembrane proteins includes *rhodopsin,* the light-activated protein in the vertebrate eye, as well as the large number of olfactory receptors in the vertebrate nose. Other family members are found in unicellular organisms: the receptors in yeasts that recognize secreted mating factors are an example. In fact, it is thought that the G-protein-linked receptors that mediate cell–cell signaling in multicellular organisms evolved from sensory receptors that were possessed by their unicellular eucaryotic ancestors.

It is remarkable that about half of all known drugs work through G-protein-linked receptors. Genome sequencing projects are revealing vast numbers of new family members, many of which are likely targets for new drugs that remain to be discovered.

Trimeric G Proteins Disassemble to Relay Signals from G-Protein-linked Receptors

When extracellular signaling molecules bind to serpentine receptors, the receptors undergo a conformational change that enables them to activate **trimeric GTP-binding proteins (G proteins)**. These G proteins are attached to the cytoplasmic face of the plasma membrane, where they serve as relay molecules, functionally coupling the receptors to enzymes or ion channels in this membrane. There are various types of G proteins, each specific for a particular set of serpentine receptors and for a particular set of downstream target proteins in the plasma membrane. All have a similar structure, however, and they operate in a similar way.

G proteins are composed of three protein subunits—α, β, and γ. In the unstimulated state, the α subunit has GDP bound and the G protein is inactive (Figure 15–27). When stimulated by an activated receptor, the α subunit releases its bound GDP, allowing GTP to bind in its place. This exchange causes the trimer to dissociate into two activated components—an *α subunit* and a *βγ complex* (Figure 15–28).

The dissociation of the trimeric G protein activates its two components in different ways. GTP binding causes a conformational change that affects the surface of the α subunit that associates with the βγ complex in the trimer. This change causes the release of the βγ complex, but it also causes and the α subunit to adopt a new shape that allows it to interact with its target proteins. The βγ complex does not change its conformation, but the surface previously masked by the α subunit is now available to interact with a second set of target proteins.

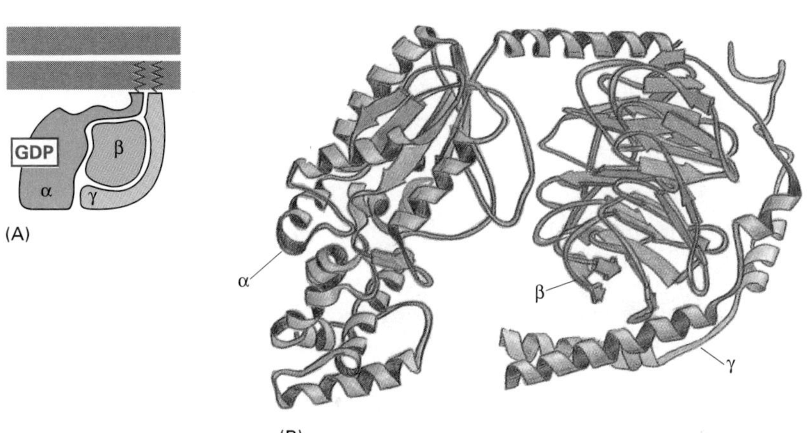

Figure 15–27 The structure of an inactive G protein. (A) Note that both the α and the γ subunits have covalently attached lipid molecules *(red)* that help to bind them to the plasma membrane, and the α subunit has GDP bound. (B) The three-dimensional structure of an inactive G protein, based on transducin, the G protein in visual transduction (discussed later). The α subunit contains the GTPase domain and binds to one side of the β subunit, which locks the GTPase domain in an inactive conformation that binds GDP. The γ subunit binds to the opposite side of the β subunit. (B, based on D.G. Lombright et al., *Nature* 379:311–319, 1996.)

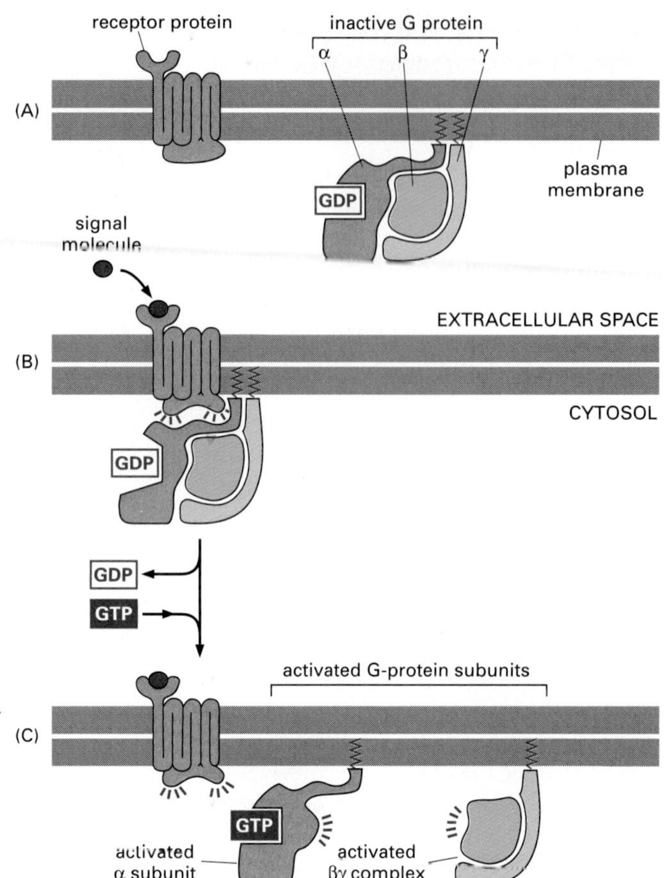

Figure 15–28 The disassembly of an activated G-protein into two signaling components. (A) In the unstimulated state, the receptor and the G protein are both inactive. Although they are shown here as separate entities in the plasma membrane, in some cases, at least, they are associated in a preformed complex. (B) Binding of an extracellular signal to the receptor changes the conformation of the receptor, which in turn alters the conformation of the G protein that is bound to the receptor. (C) The alteration of the α subunit of the G protein allows it to exchange its GDP for GTP. This causes the G protein to break up into two active components—an α subunit and a βγ complex, both of which can regulate the activity of target proteins in the plasma membrane. The receptor stays active while the external signal molecule is bound to it, and it can therefore catalyze the activation of many molecules of G protein.

The targets of the dissociated components of the G protein are either enzymes or ion channels in the plasma membrane, and they relay the signal onward.

The α subunit is a GTPase, and once it hydrolyzes its bound GTP to GDP, it reassociates with a βγ complex to re-form an inactive G protein, reversing the activation process (Figure 15–29). The time during which the α subunit and βγ complex remain apart and active is usually short, and it depends on how quickly the α subunit hydrolyzes its bound GTP. An isolated α subunit is an inefficient GTPase, and, left to its own devices, the subunit would inactivate only after several minutes. Its activation is usually reversed much faster than this, however, because the GTPase activity of the α subunit is greatly enhanced by the binding of a second protein, which can be either its target protein or a specific modulator known as a *regulator of G protein signaling (RGS)*. **RGS proteins** act as α-subunit-specific *GTPase activating proteins (GAPs),* and they are thought to have a crucial role in shutting off G-protein-mediated responses in all eucaryotes. There are about 25 RGS proteins encoded in the human genome, each of which is thought to interact with a particular set of G proteins.

The importance of the GTPase activity in shutting off the response can be easily demonstrated in a test tube. If cells are broken open and exposed to an analogue of GTP (GTPγS) in which the terminal phosphate cannot be hydrolyzed, the activated α subunits remain active for a very long time.

Some G Proteins Signal By Regulating the Production of Cyclic AMP

Cyclic AMP (cAMP) was first identified as a small intracellular mediator in the 1950s. It has since been found to act in this role in all procaryotic and animal cells that have been studied. The normal concentration of cyclic AMP inside the cell is about 10^{-7} M, but an extracellular signal can cause cyclic AMP levels to change by more than twentyfold in seconds (Figure 15–30). As explained earlier (see Figure 15–10), such a rapid response requires that a rapid synthesis of the molecule be balanced by its rapid breakdown or removal. In fact, cyclic AMP is synthesized from ATP by a plasma-membrane-bound enzyme **adenylyl cyclase**, and it is rapidly and continuously destroyed by one or more **cyclic AMP phosphodiesterases** that hydrolyze cyclic AMP to adenosine 5′-monophosphate (5′-AMP) (Figure 15–31).

Many extracellular signal molecules work by increasing cyclic AMP content, and they do so by increasing the activity of adenylyl cyclase rather than decreasing the activity of phosphodiesterase. Adenylyl cyclase is a large multipass transmembrane protein with its catalytic domain on the cytosolic side of the plasma membrane. There are at least eight isoforms in mammals, most of which are regulated by both G proteins and Ca^{2+}. All receptors that act via cyclic AMP are coupled to a **stimulatory G protein (G_s)**, which activates adenylyl cyclase and thereby increases cyclic AMP concentration. Another G protein, called **inhibitory**

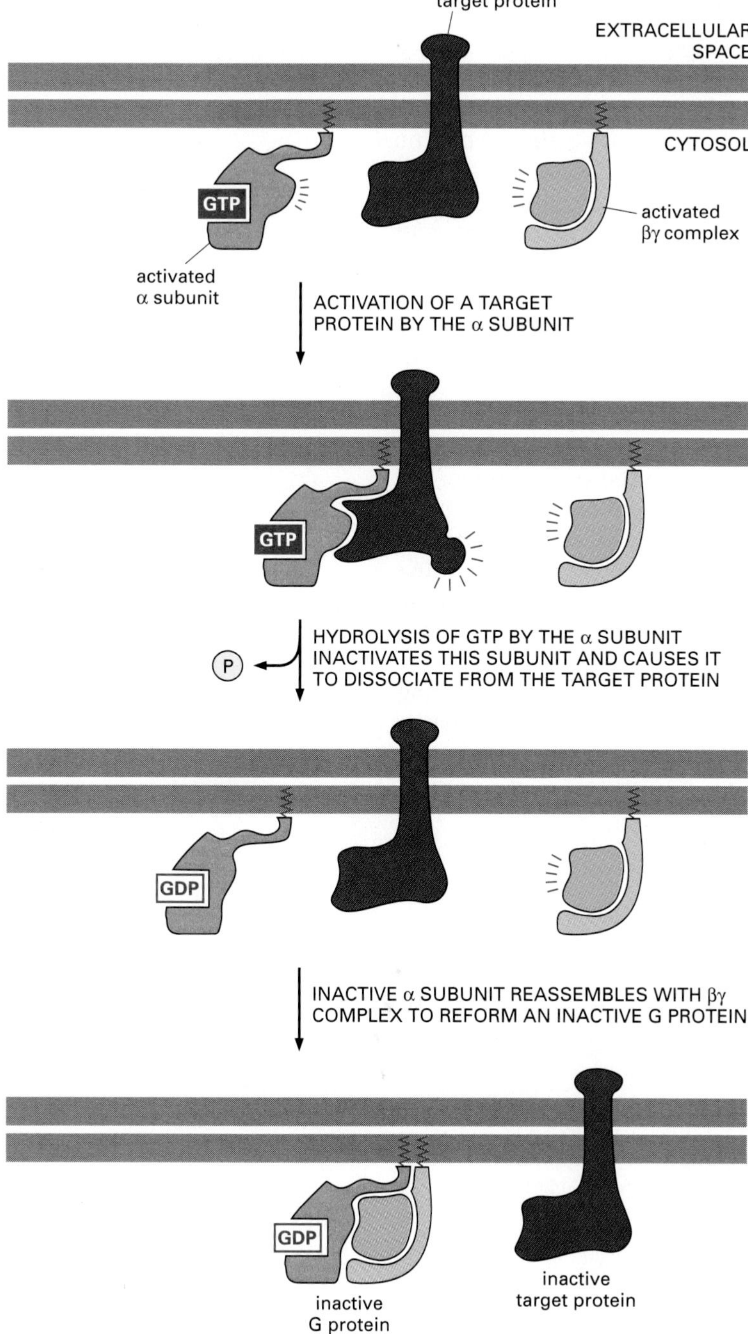

target protein
EXTRACELLULAR SPACE
CYTOSOL
GTP
activated α subunit
activated βγ complex
ACTIVATION OF A TARGET PROTEIN BY THE α SUBUNIT
GTP
HYDROLYSIS OF GTP BY THE α SUBUNIT INACTIVATES THIS SUBUNIT AND CAUSES IT TO DISSOCIATE FROM THE TARGET PROTEIN
P
GDP
INACTIVE α SUBUNIT REASSEMBLES WITH βγ COMPLEX TO REFORM AN INACTIVE G PROTEIN
GDP
inactive G protein
inactive target protein

Figure 15–29 The switching off of the G-protein α subunit by the hydrolysis of its bound GTP. After a G-protein α subunit activates its target protein, it shuts itself off by hydrolyzing its bound GTP to GDP. This inactivates the α subunit, which dissociates from the target protein and reassociates with a βγ complex to re-form an inactive G protein. Binding to the target protein or to a membrane-bound RGS protein (not shown) usually stimulates the GTPase activity of the α subunit; this stimulation greatly speeds up the inactivation process shown here.

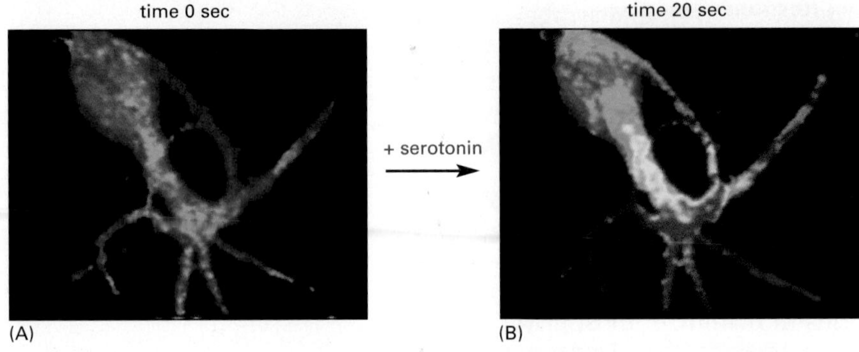

Figure 15–30 An increase in cyclic AMP in response to an extracellular signal. This nerve cell in culture is responding to the neurotransmitter serotonin, which acts through a G-protein-linked receptor to cause a rapid rise in the intracellular concentration of cyclic AMP. To monitor the cyclic AMP level, the cell has been loaded with a fluorescent protein that changes its fluorescence when it binds cyclic AMP. *Blue* indicates a low level of cyclic AMP, *yellow* an intermediate level, and *red* a high level. (A) In the resting cell, the cyclic AMP level is about 5×10^{-8} M. (B) Twenty seconds after the addition of serotonin to the culture medium, the intracellular level of cyclic AMP has increased to more than 10^{-6} M, an increase of more than twentyfold. (From Brian J. Bacskai et al., *Science* 260:222–226, 1993. © AAAS.)

G protein (G_i), inhibits adenylyl cyclase, but it mainly acts by directly regulating ion channels (as we discuss later) rather than by decreasing cyclic AMP content. Although it is usually the α subunit that regulates the cyclase, the βγ complex sometimes does so as well, either increasing or decreasing the enzyme's activity, depending on the particular βγ complex and the isoform of the cyclase.

Both G_s and G_i are targets for some medically important bacterial toxins. *Cholera toxin,* which is produced by the bacterium that causes cholera, is an enzyme that catalyzes the transfer of ADP ribose from intracellular NAD^+ to the α subunit of G_s. This ADP ribosylation alters the α subunit so that it can no longer hydrolyze its bound GTP, causing it to remain in an active state that stimulates adenylyl cyclase indefinitely. The resulting prolonged elevation in cyclic AMP levels within intestinal epithelial cells causes a large efflux of Cl^- and water into the gut, thereby causing the severe diarrhea that characterizes cholera. *Pertussis toxin,* which is made by the bacterium that causes pertussis (whooping cough), catalyzes the ADP ribosylation of the α subunit of G_i, preventing the subunit from interacting with receptors; as a result, this α subunit retains its bound GDP and is unable to regulate its target proteins. These two toxins are widely used as tools to determine whether a cell's response to a signal is mediated by G_s or by G_i.

Some of the responses mediated by a G_s-stimulated increase in cyclic AMP concentration are listed in Table 15–1. It is clear that different cell types respond differently to an increase in cyclic AMP concentration, and that any one cell type usually responds in the same way, even if different extracellular signals induce the increase. At least four hormones activate adenylyl cyclase in fat cells, for example, and all of them stimulate the breakdown of triglyceride (the storage form of fat) to fatty acids.

Individuals who are genetically deficient in a particular G_s α subunit show decreased responses to certain hormones. As a consequence, they display metabolic abnormalities, have abnormal bone development, and are mentally retarded.

Cyclic-AMP-dependent Protein Kinase (PKA) Mediates Most of the Effects of Cyclic AMP

Although cyclic AMP can directly activate certain types of ion channels in the plasma membrane of some highly specialized cells, in most animal cells it exerts its effects mainly by activating **cyclic-AMP-dependent protein kinase (PKA)**. This enzyme catalyzes the transfer of the terminal phosphate group from ATP to specific serines or threonines of selected target proteins, thereby regulating their activity.

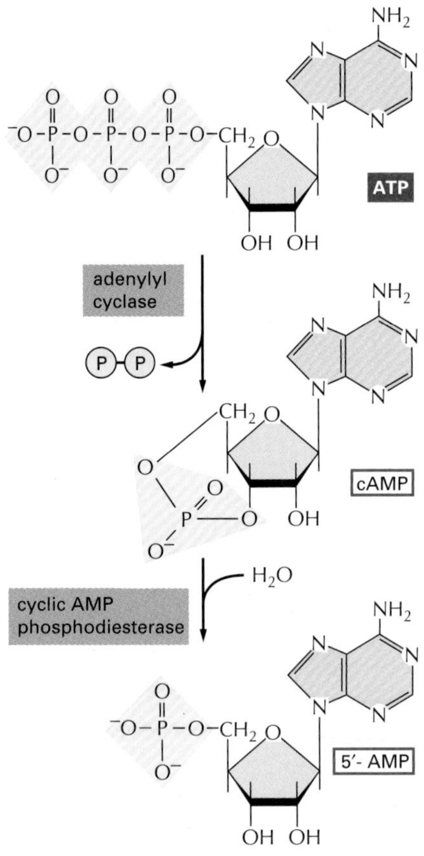

Figure 15–31 The synthesis and degradation of cyclic AMP. In a reaction catalyzed by the enzyme adenylyl cyclase, cyclic AMP (cAMP) is synthesized from ATP through a cyclization reaction that removes two phosphate groups as pyrophosphate (Ⓟ – Ⓟ); a pyrophosphatase drives this synthesis by hydrolyzing the released pyrophosphate to phosphate (not shown). Cyclic AMP is unstable in the cell, because it is itself hydrolyzed by a specific phosphodiesterase to form 5′-AMP, as indicated.

TABLE 15–1 Some Hormone-induced Cell Responses Mediated by Cyclic AMP

TARGET TISSUE	HORMONE	MAJOR RESPONSE
Thyroid gland	thyroid-stimulating hormone (TSH)	thyroid hormone synthesis and secretion
Adrenal cortex	adrenocorticotrophic hormone (ACTH)	cortisol secretion
Ovary	luteinizing hormone (LH)	progesterone secretion
Muscle	adrenaline	glycogen breakdown
Bone	parathormone	bone resorption
Heart	adrenaline	increase in heart rate and force of contraction
Liver	glucagon	glycogen breakdown
Kidney	vasopressin	water resorption
Fat	adrenaline, ACTH, glucagon, TSH	triglyceride breakdown

PKA is found in all animal cells and is thought to account for the effects of cyclic AMP in most of these cells. The substrates for PKA differ in different cell types, which explains why the effects of cyclic AMP vary so markedly depending on the cell type.

In the inactive state, PKA consists of a complex of two catalytic subunits and two regulatory subunits. The binding of cyclic AMP to the regulatory subunits alters their conformation, causing them to dissociate from the complex. The released catalytic subunits are thereby activated to phosphorylate specific substrate protein molecules (Figure 15–32). The regulatory subunits of PKA also are important for localizing the kinase inside the cell: special *PKA anchoring proteins* bind both to the regulatory subunits and to a membrane or a component of the cytoskeleton, thereby tethering the enzyme complex to a particular subcellular compartment. Some of these anchoring proteins also bind other kinases and some phosphatases, creating a signaling complex.

Some responses mediated by cyclic AMP are rapid while others are slow. In skeletal muscle cells, for example, activated PKA phosphorylates enzymes involved in glycogen metabolism, which simultaneously triggers the breakdown of glycogen to glucose and inhibits glycogen synthesis, thereby increasing the amount of glucose available to the muscle cell within seconds (see also Figure 15–30). At the other extreme are responses that take hours to develop fully and involve changes in the transcription of specific genes. In cells that secrete the peptide hormone *somatostatin,* for example, cyclic AMP activates the gene that encodes this hormone. The regulatory region of the somatostatin gene contains a short DNA sequence, called the *cyclic AMP response element (CRE),* that is also found in the regulatory region of many other genes activated by cyclic AMP. A

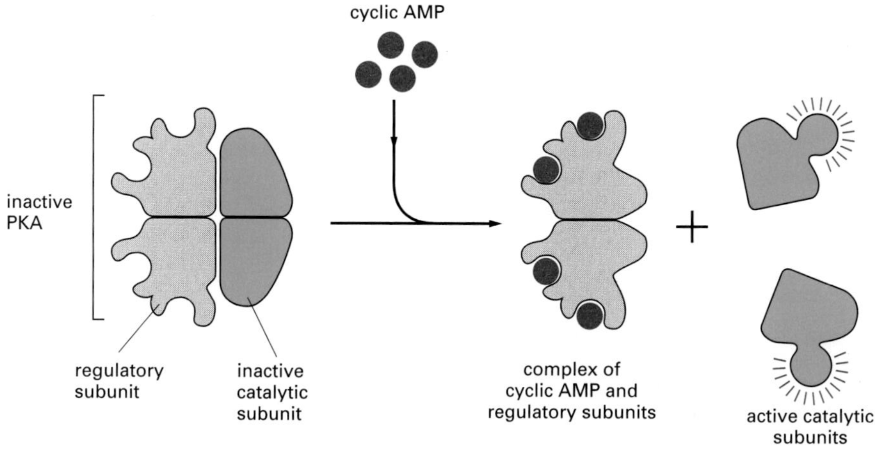

Figure 15–32 The activation of cyclic-AMP-dependent protein kinase (PKA). The binding of cyclic AMP to the regulatory subunits induces a conformational change, causing these subunits to dissociate from the catalytic subunits, thereby activating the kinase activity of the catalytic subunits. The release of the catalytic subunits requires the binding of more than two cyclic AMP molecules to the regulatory subunits in the tetramer. This requirement greatly sharpens the response of the kinase to changes in cyclic AMP concentration, as discussed earlier. Mammalian cells have at least two types of PKAs: type I is mainly in the cytosol, whereas type II is bound via its regulatory subunit and special anchoring proteins to the plasma membrane, nuclear membrane, mitochondrial outer membrane, and microtubules. In all cases, however, once the catalytic subunits are freed and active, they can migrate into the nucleus (where they can phosphorylate gene regulatory proteins), while the regulatory subunits remain in the cytoplasm. The three-dimensional structure of the protein kinase domain of the PKA catalytic subunit is shown in Figure 3–64.

specific gene regulatory protein called **CRE-binding (CREB) protein** recognizes this sequence. When CREB is phosphorylated by PKA on a single serine, it recruits a transcriptional coactivator called *CREB-binding protein (CBP),* which stimulates the transcription of these genes (Figure 15–33). If this serine is mutated, CREB cannot recruit CBP, and it no longer stimulates gene transcription in response to a rise in cyclic AMP levels.

Protein Phosphatases Make the Effects of PKA and Other Protein Kinases Transitory

Since the effects of cyclic AMP are usually transient, cells must be able to dephosphorylate the proteins that have been phosphorylated by PKA. Indeed, the activity of any protein regulated by phosphorylation depends on the balance at any instant between the activities of the kinases that phosphorylate it and the phosphatases that are constantly dephosphorylating it. In general, the dephosphorylation of phosphorylated serines and threonines is catalyzed by four types of **serine/threonine phosphoprotein phosphatases**—protein phosphatases I, IIA, IIB, and IIC. Except for protein phosphatase-IIC (which is a minor phosphatase, unrelated to the others), all of these phosphatases are composed of a homologous catalytic subunit complexed with one or more of a large set of regulatory subunits; the regulatory subunits help to control the phosphatase activity and enable the enzyme to select specific targets. *Protein phosphatase I* is responsible for dephosphorylating many of the proteins phosphorylated by PKA. It inactivates CREB, for example, by removing its activating phosphate, thereby turning off the transcriptional response caused by a rise in cyclic AMP

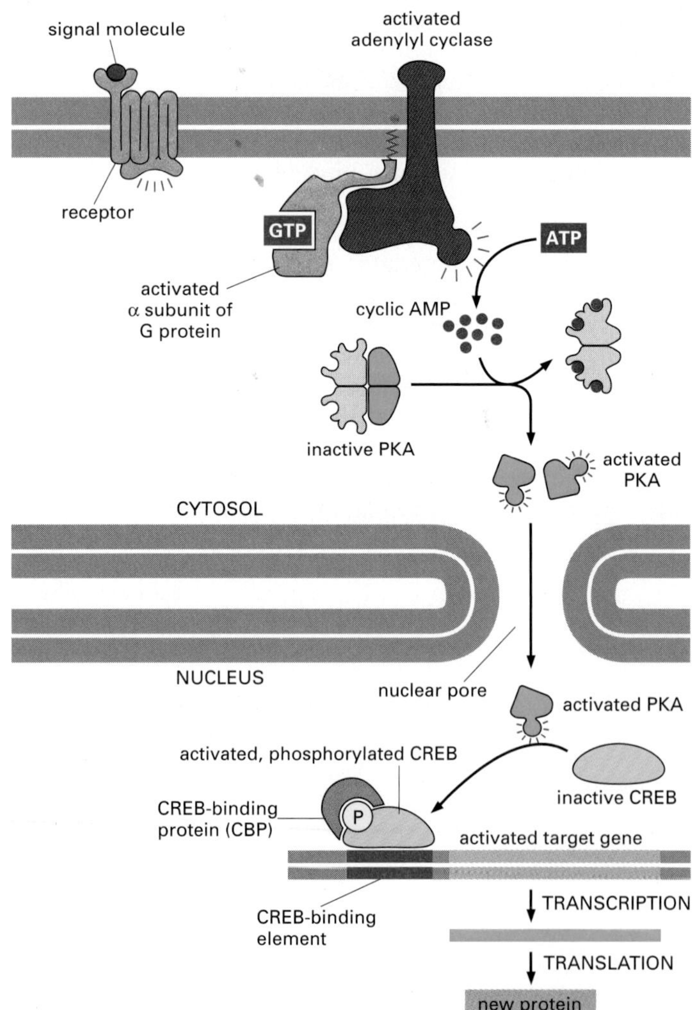

Figure 15–33 How gene transcription is activated by a rise in cyclic AMP concentration. The binding of an extracellular signal molecule to its G-protein-linked receptor leads to the activation of adenylyl cyclase and a rise in cyclic AMP concentration. The increase in cyclic AMP concentration activates PKA in the cytosol, and the released catalytic subunits then move into the nucleus, where they phosphorylate the CREB gene regulatory protein. Once phosphorylated, CREB recruits the coactivator CBP, which stimulates gene transcription. This signaling pathway controls many processes in cells, ranging from hormone synthesis in endocrine cells to the production of proteins required for long-term memory in the brain. We shall see later that some kinases that are activated by a rise in intracellular Ca^{2+} can also phosphorylate and thereby activate CREB.

TABLE 15–2 Some Cell Responses in Which G-Protein-linked Receptors Activate the Inositol-Phospholipid Signaling Pathway

TARGET TISSUE	SIGNALING MOLECULE	MAJOR RESPONSE
Liver	vasopressin	glycogen breakdown
Pancreas	acetylcholine	amylase secretion
Smooth muscle	acetylcholine	contraction
Blood platelets	thrombin	aggregation

concentration. *Protein phosphatase IIA* has a broad specificity and seems to be the main phosphatase responsible for reversing many of the phosphorylations catalyzed by serine/threonine kinases. *Protein phosphatase IIB,* also called *calcineurin,* is activated by Ca^{2+} and is especially abundant in the brain.

Having discussed how trimeric G proteins link activated receptors to adenylyl cyclase, we now consider how they couple activated receptors to another crucial enzyme, *phospholipase C*. The activation of this enzyme leads to an increase in the concentration of Ca^{2+} in the cytosol, which helps to relay the signal onward. Ca^{2+} is even more widely used as an intracellular mediator than is cyclic AMP.

Some G Proteins Activate the Inositol Phospholipid Signaling Pathway by Activating Phospholipase C-β

Many G-protein-linked receptors exert their effects mainly via G proteins that activate the plasma-membrane-bound enzyme **phospholipase C-β**. Several examples of responses activated in this way are listed in Table 15–2. The phospholipase acts on an inositol phospholipid (a *phosphoinositide*) called **phosphatidylinositol 4,5-bisphosphate [PI(4,5)P_2]**, which is present in small amounts in the inner half of the plasma membrane lipid bilayer (Figure 15–34). Receptors that operate through this **inositol phospholipid signaling pathway** mainly activate a G protein called $\mathbf{G_q}$, which in turn activates phospholipase C-β,

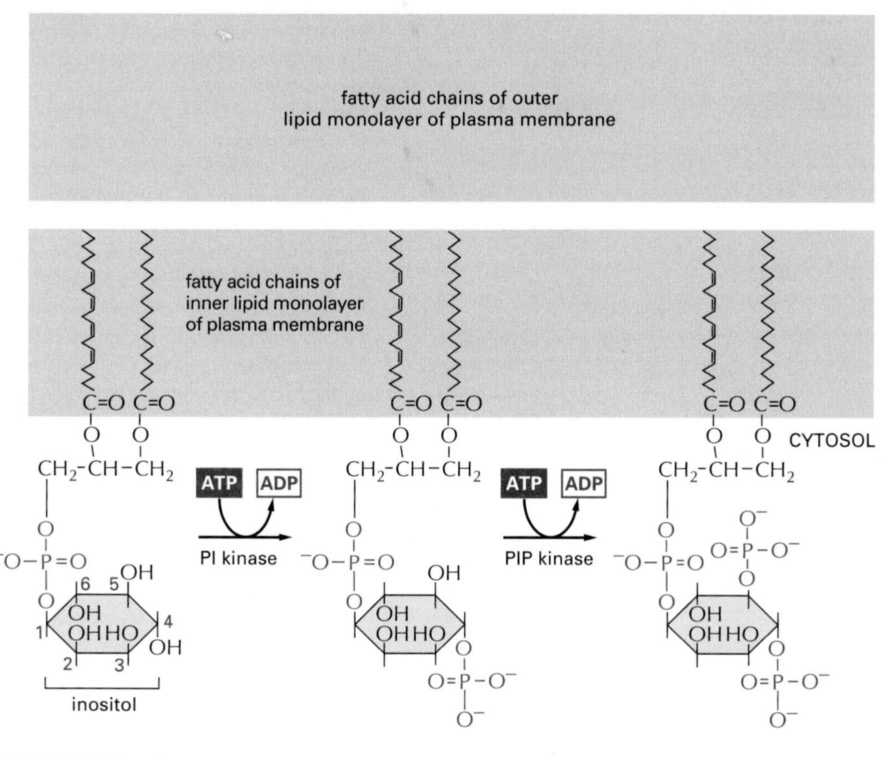

Figure 15–34 Three types of inositol phospholipids (phosphoinositides). The polyphosphoinositides—PI(4)P and PI(4,5)P_2—are produced by the phosphorylation of phosphatidylinositol (PI) and PI(4)P, respectively. Although all three inositol phospholipids may be broken down in a signaling response, it is the breakdown of PI(4,5)P_2 that is most critical because it generates two intracellular mediators, as shown in the next two figures. Nevertheless, PI(4,5)P_2 is the least abundant, constituting less than 10% of the total inositol lipids and less than 1% of the total phospholipids in a cell. The conventional numbering of the carbon atoms in the inositol ring is shown in *red numbers* on the PI molecule.

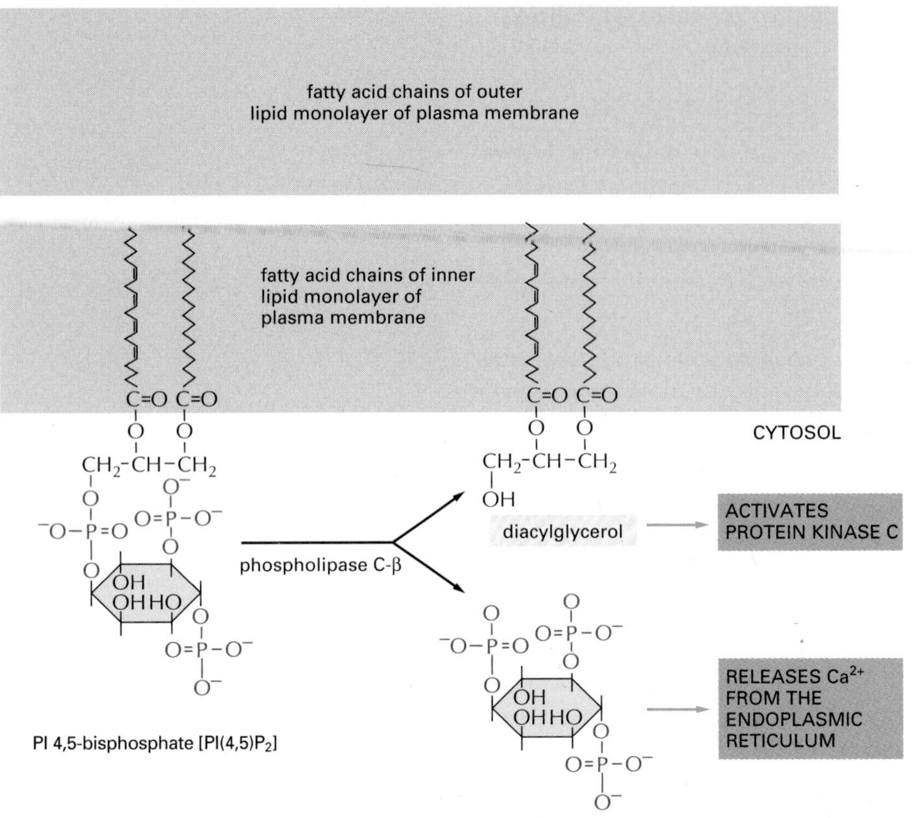

Figure 15–35 The hydrolysis of $PI(4,5)P_2$ by phospholipase C-β. Two intracellular mediators are produced when $PI(4,5)P_2$ is hydrolyzed: inositol 1,4,5-trisphosphate (IP_3), which diffuses through the cytosol and releases Ca^{2+} from the ER, and diacylglycerol, which remains in the membrane and helps to activate the enzyme protein kinase C (see Figure 15–36). There are at least three classes of phospholipase C—β, γ, and σ—and it is the β class that is activated by G-protein-linked receptors. We shall see later that the γ class is activated by a second class of receptors, called receptor tyrosine kinases, that activate the inositol phospholipid signaling pathway without an intermediary G protein.

in much the same way that G_s activates adenylyl cyclase. The activated phospholipase cleaves $PI(4,5)P_2$ to generate two products: *inositol 1,4,5-trisphosphate and diacylglycerol* (Figure 15–35). At this step, the signaling pathway splits into two branches.

Inositol 1,4,5-trisphosphate (IP_3) is a small, water-soluble molecule that leaves the plasma membrane and diffuses rapidly through the cytosol. When it reaches the endoplasmic reticulum (ER), it binds to and opens *IP_3-gated Ca^{2+}-release channels* in the ER membrane. Ca^{2+} stored in the ER is released through the open channels, quickly raising the concentration of Ca^{2+} in the cytosol (Figure 15–36). We discuss later how Ca^{2+} acts to propagate the signal. Several mechanisms operate to terminate the initial Ca^{2+} response: (1) IP_3 is rapidly dephosphorylated by specific phosphatases to form IP_2; (2) IP_3 is phosphorylated

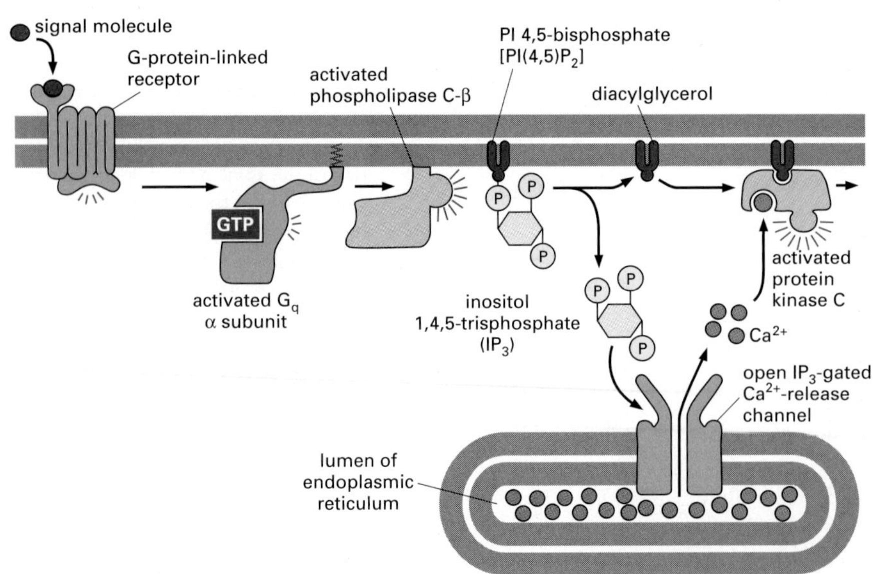

Figure 15–36 The two branches of the inositol phospholipid pathway. The activated receptor stimulates the plasma-membrane-bound enzyme phospholipase C-β via a G protein. Depending on the isoform of the enzyme, it may be activated by the α subunit of G_q as shown, by the βγ complex of another G protein, or by both. Two intracellular messenger molecules are produced when $PI(4,5)P_2$ is hydrolyzed by the activated phospholipase C-β. Inositol 1,4,5-trisphosphate (IP_3) diffuses through the cytosol and releases Ca^{2+} from the endoplasmic reticulum by binding to and opening IP_3-gated Ca^{2+}-release channels in the endoplasmic reticulum membrane. The large electrochemical gradient for Ca^{2+} across this membrane causes Ca^{2+} to escape into the cytosol. Diacylglycerol remains in the plasma membrane and, together with phosphatidylserine (not shown) and Ca^{2+}, helps to activate the enzyme protein kinase C, which is recruited from the cytosol to the cytosolic face of the plasma membrane. Of the 11 or more distinct isoforms of PKC in mammals, at least four are activated by diacylglycerol.

to IP_4 (which may function as another intracellular mediator); and (3) Ca^{2+} that enters the cytosol is rapidly pumped out, mainly to the exterior of the cell.

At the same time that the IP_3 produced by the hydrolysis of PI(4,5)P_2 is increasing the concentration of Ca^{2+} in the cytosol, the other cleavage product of PI(4,5)P_2—**diacylglycerol**—is exerting different effects. Diacylglycerol remains embedded in the membrane, where it has two potential signaling roles. First, it can be further cleaved to release arachidonic acid, which can either act as a messenger in its own right or be used in the synthesis of other small lipid messengers called *eicosanoids.* Eicosanoids, such as the *prostaglandins,* are made by most vertebrate cell types and have a wide variety of biological activities. They participate in pain and inflammatory responses, for example, and most anti-inflammatory drugs (such as aspirin, ibuprofen, and cortisone) act—in part, at least—by inhibiting their synthesis.

The second, and more important, function of diacylglycerol is to activate a crucial serine/threonine protein kinase called **protein kinase C (PKC),** so named because it is Ca^{2+}-dependent. The initial rise in cytosolic Ca^{2+} induced by IP_3 alters the PKC so that it translocates from the cytosol to the cytoplasmic face of the plasma membrane. There it is activated by the combination of Ca^{2+}, diacylglycerol, and the negatively charged membrane phospholipid phosphatidylserine (see Figure 15–36). Once activated, PKC phosphorylates target proteins that vary depending on the cell type. The principles are the same as discussed earlier for PKA, although most of the target proteins are different.

Each of the two branches of the inositol phospholipid signaling pathway can be mimicked by the addition of specific pharmacological agents to intact cells. The effects of IP_3 can be mimicked by using a *Ca^{2+} ionophore,* such as A23187 or ionomycin, which allows Ca^{2+} to move into the cytosol from the extracellular fluid (discussed in Chapter 11). The effects of diacylglycerol can be mimicked by *phorbol esters,* plant products that bind to PKC and activate it directly. Using these reagents, it has been shown that the two branches of the pathway often collaborate in producing a full cellular response. Some cell types, such as lymphocytes, for example, can be stimulated to proliferate in culture when treated with both a Ca^{2+} ionophore and a PKC activator, but not when they are treated with either reagent alone.

Ca^{2+} Functions as a Ubiquitous Intracellular Messenger

Many extracellular signals induce an increase in cytosolic Ca^{2+} level, not just those that work via G proteins. In egg cells, for example, a sudden rise in cytosolic Ca^{2+} concentration upon fertilization by a sperm triggers a Ca^{2+} wave that is responsible for the onset of embryonic development (Figure 15–37). In muscle cells, Ca^{2+} triggers contraction, and in many secretory cells, including nerve cells, it triggers secretion. Ca^{2+} can be used as a signal in this way because its concentration in the cytosol is normally kept very low (~10^{-7} M), whereas its concentration in the extracellular fluid (~10^{-3} M) and in the ER lumen is high. Thus, there is a large gradient tending to drive Ca^{2+} into the cytosol across both the plasma membrane and the ER membrane. When a signal transiently opens Ca^{2+} channels in either of these membranes, Ca^{2+} rushes into the cytosol, increasing the local Ca^{2+} concentration by 10–20-fold and triggering Ca^{2+}-responsive proteins in the cell.

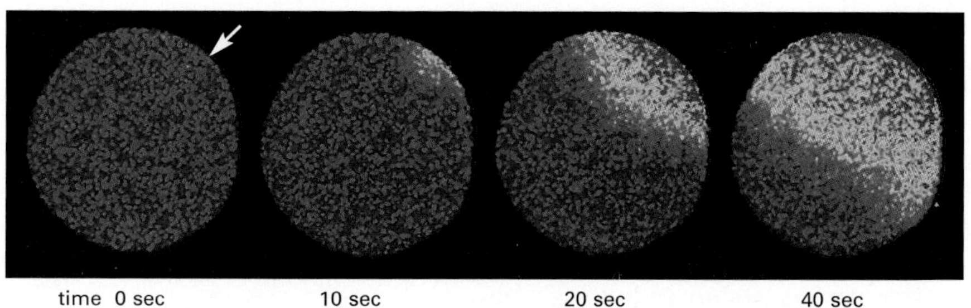

Figure 15–37 Fertilization of an egg by a sperm triggering an increase in cytosolic Ca^{2+}. This starfish egg was injected with a Ca^{2+}-sensitive fluorescent dye before it was fertilized. A wave of cytosolic Ca^{2+} *(red),* released from the endoplasmic reticulum, is seen to sweep across the egg from the site of sperm entry *(arrow).* This Ca^{2+} wave provokes a change in the egg cell membrane, preventing the entry of other sperm, and it also initiates embryonic development (discussed in Chapter 20). (Courtesy of Stephen A. Stricker.)

Figure 15–38 The main ways eucaryotic cells maintain a very low concentration of free Ca^{2+} in their cytosol. (A) Ca^{2+} is actively pumped out of the cytosol to the cell exterior. (B) Ca^{2+} is pumped into the ER and mitochondria, and various molecules in the cell bind free Ca^{2+} tightly.

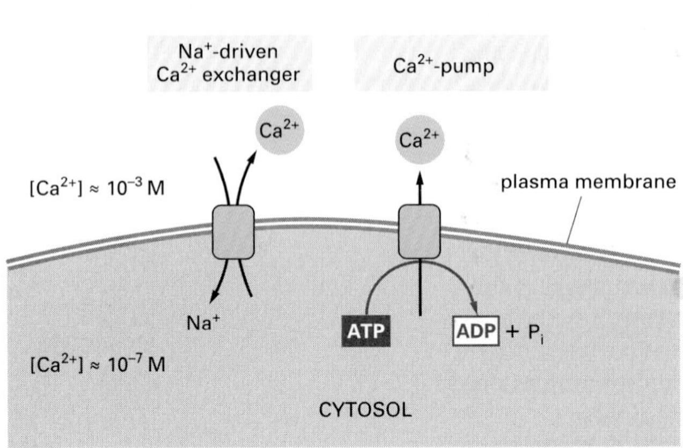

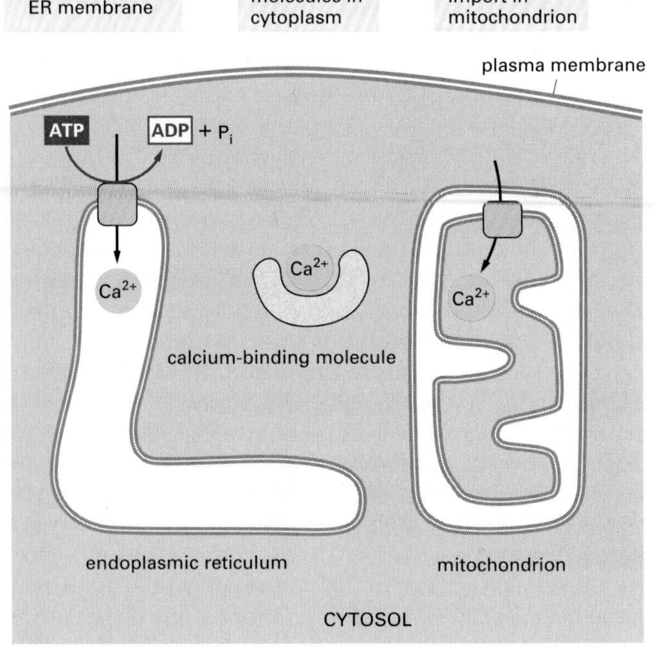

Three main types of Ca^{2+} channels can mediate this Ca^{2+} signaling:

1. *Voltage-dependent Ca^{2+} channels* in the plasma membrane open in response to membrane depolarization and allow, for example, Ca^{2+} to enter activated nerve terminals and trigger neurotransmitter secretion.
2. *IP_3-gated Ca^{2+}-release channels* allow Ca^{2+} to escape from the ER when the inositol phospholipid signaling pathway is activated, as just discussed (see Figure 15–36).
3. *Ryanodine receptors* (so called because they are sensitive to the plant alkaloid ryanodine) react to a change in plasma membrane potential to release Ca^{2+} from the sarcoplasmic reticulum and thereby stimulate the contraction of muscle cells; they are also present in the ER of many nonmuscle cells, including neurons, where they can contribute to Ca^{2+} signaling.

The concentration of Ca^{2+} in the cytosol is kept low in resting cells by several mechanisms (Figure 15–38). Most notably, all eucaryotic cells have a Ca^{2+}-pump in their plasma membrane that uses the energy of ATP hydrolysis to pump Ca^{2+} out of the cytosol. Cells such as muscle and nerve cells, which make extensive use of Ca^{2+} signaling, have an additional Ca^{2+} transport protein (exchanger) in their plasma membrane that couples the efflux of Ca^{2+} to the influx of Na^+. A Ca^{2+} pump in the ER membrane also has an important role in keeping the cytosolic Ca^{2+} concentration low: this Ca^{2+}-pump enables the ER to take up large amounts of Ca^{2+} from the cytosol against a steep concentration gradient, even when Ca^{2+} levels in the cytosol are low. In addition, a low-affinity, high-capacity Ca^{2+} pump in the inner mitochondrial membrane has an important role in returning the Ca^{2+} concentration to normal after a Ca^{2+} signal; it uses the electrochemical gradient generated across this membrane during the electron-transfer steps of oxidative phosphorylation to take up Ca^{2+} from the cytosol.

The Frequency of Ca^{2+} Oscillations Influences a Cell's Response

Ca^{2+}-sensitive fluorescent indicators, such as aequorin or fura-2 (discussed in Chapter 9), are often used to monitor cytosolic Ca^{2+} in individual cells after the inositol phospholipid signaling pathway has been activated. When viewed in this way, the initial Ca^{2+} signal is often seen to be small and localized to one or more discrete regions of the cell. These signals have been called Ca^{2+} blips, quarks, puffs, or sparks, and they are thought to reflect the local opening of individual (or small groups of) Ca^{2+}-release channels in the ER and to represent elementary Ca^{2+} signaling units. If the extracellular signal is sufficiently strong and persistent, this localized signal can propagate as a regenerative Ca^{2+} wave through the cytosol, much like an action potential in an axon (see Figure 15–37). Such a Ca^{2+} "spike" is often followed by a series of further spikes, each usually lasting seconds (Figure 15–39). These Ca^{2+} oscillations can persist for as long as receptors are activated at the cell surface. Both the waves and the oscillations are thought to depend, in part at least, on a combination of positive and negative feedback by Ca^{2+} on both the IP_3-gated Ca^{2+}-release channels and the ryanodine receptors: the released Ca^{2+} initially stimulates more Ca^{2+} release, a process known as Ca^{2+}-induced Ca^{2+} release. But then, as its concentration gets high enough, the Ca^{2+} inhibits further release.

The frequency of the Ca^{2+} oscillations reflects the strength of the extracellular stimulus (see Figure 15–39), and it can be translated into a frequency-dependent cell response. In some cases, the frequency-dependent response itself is also oscillatory. In hormone-secreting pituitary cells, for example, stimulation by an extracellular signal induces repeated Ca^{2+} spikes, each of which is associated with a burst of hormone secretion. The frequency-dependent response can also be nonoscillatory. In some types of cells, for instance, one frequency of Ca^{2+} spikes activates the transcription of one set of genes, while a higher frequency activates the transcription of a different set. How do cells sense the frequency of Ca^{2+} spikes and change their response accordingly? The mechanism presumably depends on Ca^{2+}-sensitive proteins that change their activity as a function of Ca^{2+} spike frequency. A protein kinase that acts as a molecular memory device seems to have this remarkable property, as we discuss next.

Ca^{2+}/Calmodulin-dependent Protein Kinases (CaM-Kinases) Mediate Many of the Actions of Ca^{2+} in Animal Cells

Ca^{2+}-binding proteins serve as transducers of the cytosolic Ca^{2+} signal. The first such protein to be discovered was *troponin C* in skeletal muscle cells; its role in muscle contraction is discussed in Chapter 16. A closely related Ca^{2+}-binding protein, known as **calmodulin**, is found in all eucaryotic cells, where it can constitute as much as 1% of the total protein mass. Calmodulin functions as a multipurpose intracellular Ca^{2+} receptor, mediating many Ca^{2+}-regulated processes. It consists of a highly conserved, single polypeptide chain with four high-affinity

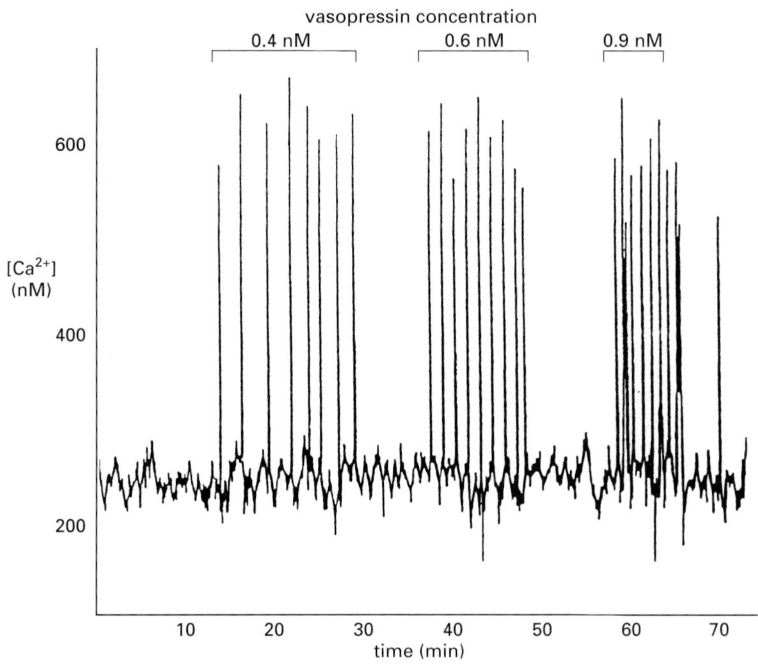

Figure 15–39 Vasopressin-induced Ca^{2+} oscillations in a liver cell. The cell was loaded with the Ca^{2+}-sensitive protein aequorin and then exposed to increasing concentrations of vasopressin. Note that the frequency of the Ca^{2+} spikes increases with an increasing concentration of vasopressin, but that the amplitude of the spikes is not affected. (Adapted from N.M. Woods, K.S.R. Cuthbertson, and P.H. Cobbold, *Nature* 319:600–602, 1986.)

Ca^{2+}-binding sites (Figure 15–40A). When activated by binding Ca^{2+}, it undergoes a conformational change. Because two or more Ca^{2+} ions must bind before calmodulin adopts its active conformation, the protein responds in a switchlike manner to increasing concentrations of Ca^{2+} (see Figure 15–22): a tenfold increase in Ca^{2+} concentration, for example, typically causes a fiftyfold increase in calmodulin activation.

The allosteric activation of calmodulin by Ca^{2+} is analogous to the allosteric activation of PKA by cyclic AMP, except that Ca^{2+}/calmodulin has no enzymic activity itself but instead acts by binding to other proteins. In some cases, calmodulin serves as a permanent regulatory subunit of an enzyme complex, but mostly the binding of Ca^{2+} enables calmodulin to bind to various target proteins in the cell to alter their activity.

When an activated molecule of Ca^{2+}/calmodulin binds to its target protein, it undergoes a marked change in conformation (Figure 15–40B). Among the targets regulated by calmodulin binding are many enzymes and membrane transport proteins. As one example, Ca^{2+}/calmodulin binds to and activates the plasma membrane Ca^{2+}-pump that pumps Ca^{2+} out of cells. Thus, whenever the concentration of Ca^{2+} in the cytosol rises, the pump is activated, which helps to return the cytosolic Ca^{2+} level to normal.

Many effects of Ca^{2+}, however, are more indirect and are mediated by protein phosphorylations catalyzed by a family of **Ca^{2+}/calmodulin-dependent protein kinases (CaM-kinases)**. These kinases, just like PKA and PKC, phosphorylate serines or threonines in proteins, and, as with PKA and PKC, the response of a target cell depends on which CaM-kinase-regulated target proteins are present in the cell. The first CaM-kinases to be discovered—*myosin light-chain kinase,* which activates smooth muscle contraction, and *phosphorylase kinase,* which activates glycogen breakdown—have narrow substrate specificities. A number of CaM-kinases, however, have much broader specificities, and these seem to be responsible for mediating many of the actions of Ca^{2+} in animal cells. Some phosphorylate gene regulatory proteins, such as the CREB protein discussed earlier, and in this way activate or inhibit the transcription of specific genes.

The best-studied example of such a *multifunctional CaM-kinase* is **CaM-kinase II**, which is found in all animal cells but is especially enriched in the nervous system. It constitutes up to 2% of the total protein mass in some regions of the brain, and it is highly concentrated in synapses. CaM-kinase II has at least two remarkable properties that are related. First, it can function as a molecular memory device, switching to an active state when exposed to Ca^{2+}/calmodulin and then remaining active even after the Ca^{2+} signal has decayed. This is because the kinase phosphorylates itself (a process called *autophosphorylation*) as well as other cell proteins when it is activated by Ca^{2+}/calmodulin. In its autophosphorylated state, the enzyme remains active even in the absence of

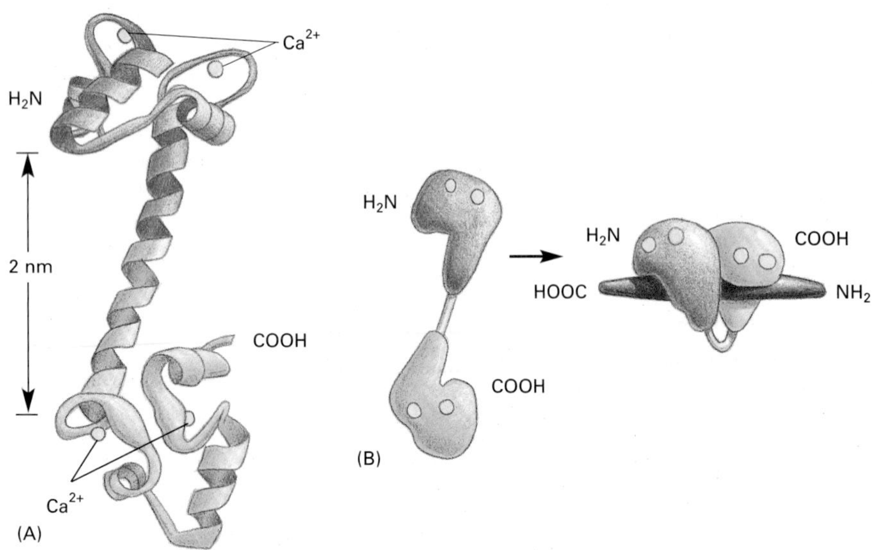

Figure 15–40 The structure of Ca^{2+}/calmodulin based on x-ray diffraction and NMR studies. (A) The molecule has a "dumbbell" shape, with two globular ends connected by a long, exposed α helix. Each end has two Ca^{2+}-binding domains, each with a loop of 12 amino acids, in which aspartic acid and glutamic acid side chains form ionic bonds with Ca^{2+}. The two Ca^{2+}-binding sites in the carboxyl-terminal part of the molecule have a tenfold higher affinity for Ca^{2+} than the two in the amino-terminal part. In solution, the molecule is flexible, displaying a range of forms, from extended (as shown) to more compact. (B) The major structural change in Ca^{2+}/calmodulin that occurs when it binds to a target protein (in this example, a peptide that consists of the Ca^{2+}/calmodulin-binding domain of a Ca^{2+}/calmodulin-dependent protein kinase). Note that the Ca^{2+}/calmodulin has "jack-knifed" to surround the peptide. (A, based on x-ray crystallographic data from Y.S. Babu et al., *Nature* 315:37–40, 1985; B, based on x-ray crystallographic data from W.E. Meador, A.R. Means, and F.A. Quiocho, *Science* 257:1251–1255, 1992, and on NMR data from M. Ikura et al., *Science* 256:632–638, 1992. © AAAS.)

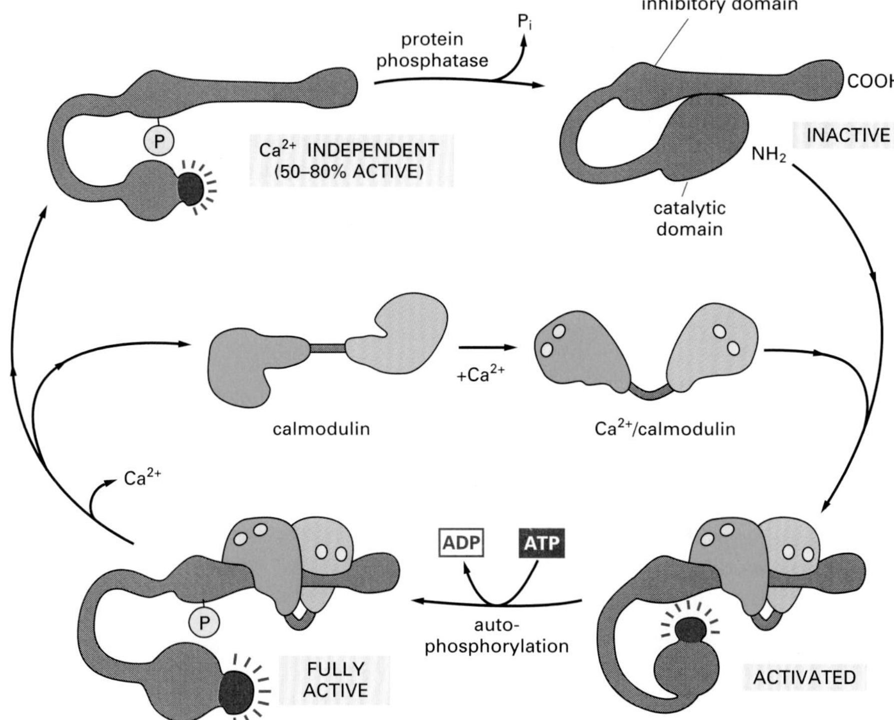

Figure 15–41 The activation of CaM-kinase II. The enzyme is a large protein complex of about 12 subunits, although, for simplicity, only one subunit is shown. The subunits are of four homologous kinds (α, β, γ, and σ), which are expressed in different proportions in different cell types. In the absence of Ca^{2+}/calmodulin, the enzyme is inactive as the result of an interaction between the inhibitory domain and the catalytic domain. The binding of Ca^{2+}/calmodulin alters the conformation of the protein, allowing the catalytic domain to phosphorylate the inhibitory domain of neighboring subunits in the complex, as well as other proteins in the cell (not shown). The autophosphorylation of the enzyme complex (by mutual phosphorylation of its subunits) prolongs the activity of the enzyme in two ways. First, it traps the bound Ca^{2+}/calmodulin so that it does not dissociate from the enzyme complex until cytosolic Ca^{2+} levels return to basal values for at least 10 seconds (not shown). Second, it converts the enzyme to a Ca^{2+}-independent form so that the kinase remains active even after the Ca^{2+}/calmodulin dissociates from it. This activity continues until the autophosphorylation process is overridden by a protein phosphatase.

Ca^{2+}, thereby prolonging the duration of the kinase activity beyond that of the initial activating Ca^{2+} signal. This activity is maintained until phosphatases overwhelm the autophosphorylating activity of the enzyme and shut it off (Figure 15–41). CaM-kinase II activation can thereby serve as a memory trace of a prior Ca^{2+} pulse, and it seems to have an important role in some types of memory and learning in the vertebrate nervous system. Mutant mice that lack the brain-specific subunit illustrated in Figure 15–41 have specific defects in their ability to remember where things are in space. A point mutation in CaM-kinase II that removes its autophosphorylation site, but otherwise leaves the kinase activity intact, produces the same learning defect, revealing that the autophosphorylation is critical in these animals.

The second remarkable property of CaM-kinase II is that it can use its memory mechanism to act as a frequency decoder of Ca^{2+} oscillations. This property is thought to be especially important at a nerve cell synapse, where changes in intracellular Ca^{2+} levels in an activated postsynaptic cell can lead to long-term changes in the subsequent effectiveness of that synapse (discussed in Chapter 11). When CaM-kinase II is immobilized on a solid surface and exposed to both a protein phosphatase and repetitive pulses of Ca^{2+}/calmodulin at different frequencies that mimic those observed in stimulated cells, the enzyme's activity increases steeply as a function of pulse frequency (Figure 15–42). Moreover, the frequency response of this multisubunit enzyme depends on its exact subunit composition, so that a cell can tailor its response to Ca^{2+} oscillations to particular needs by adjusting the composition of the CaM-kinase II enzyme that it makes.

Some G Proteins Directly Regulate Ion Channels

G proteins do not act exclusively by regulating the activity of membrane-bound enzymes that alter the concentration of cyclic AMP or Ca^{2+} in the cytosol. The α subunit of one type of G protein (called G_{12}), for example, activates a protein that converts a monomeric GTPase of the Rho family (discussed in Chapter 16) into its active form, which then alters the actin cytoskeleton. In some other cases, G proteins directly activate or inactivate ion channels in the plasma membrane of the target cell, thereby altering the ion permeability—and hence the excitability of the membrane. Acetylcholine released by the vagus nerve, for

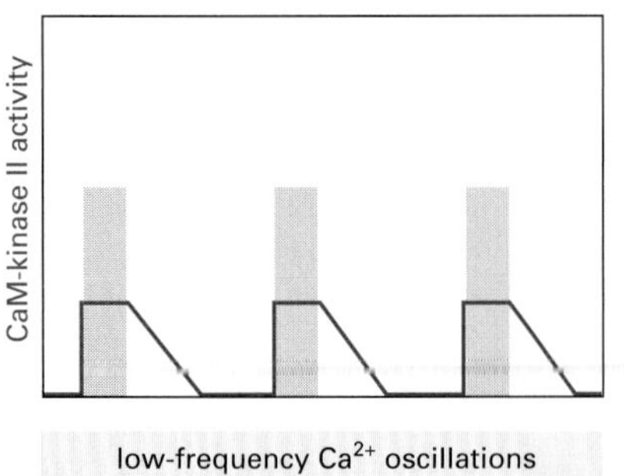

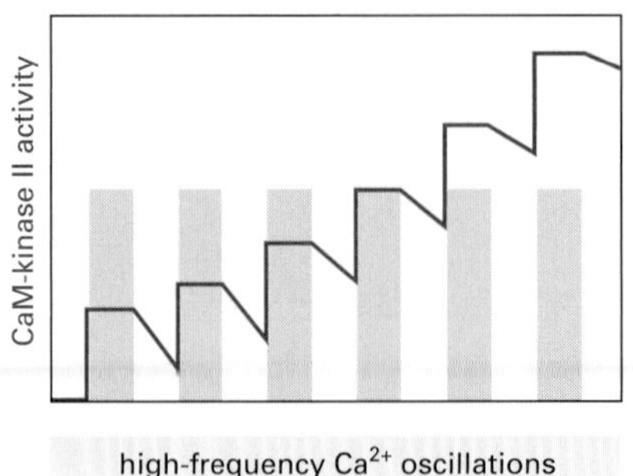

Figure 15–42 CaM-kinase II as a frequency decoder of Ca^{2+} oscillations. (A) At low frequencies of Ca^{2+} spikes *(gray bars)*, the enzyme becomes inactive after each spike, as the autophosphorylation induced by Ca^{2+}/calmodulin binding does not maintain the enzyme's activity long enough for the enzyme to remain active until the next Ca^{2+} spike arrives. (B) At higher spike frequencies, however, the enzyme fails to inactivate completely between Ca^{2+} spikes, so its activity ratchets up with each spike. If the spike frequency is high enough, this progressive increase in enzyme activity would continue until the enzyme is autophosphorylated on all subunits and is therefore maximally activated. Once enough of its subunits are autophosphorylated, the enzyme can be maintained in a highly active state even with a relatively low frequency of Ca^{2+} spikes (a form of cell memory). The binding of Ca^{2+}/calmodulin to the enzyme is enhanced by the CaM-kinase II autophosphorylation (a form of positive feedback), causing the response of the enzyme to repeated Ca^{2+} spikes to exhibit a steep threshold in its frequency response, as discussed earlier.

example, reduces both the rate and strength of heart muscle cell contraction (see Figure 15–9A). A special class of acetylcholine receptors that activate the G_i protein discussed earlier mediates this effect. Once activated, the α subunit of G_i inhibits adenylyl cyclase (as described previously), while the $\beta\gamma$ complex binds to K^+ channels in the heart muscle cell plasma membrane to open them. The opening of these K^+ channels makes it harder to depolarize the cell, which contributes to the inhibitory effect of acetylcholine on the heart. (These acetylcholine receptors, which can be activated by the fungal alkaloid muscarine, are called *muscarinic acetylcholine receptors* to distinguish them from the very different *nicotinic acetylcholine receptors,* which are ion-channel-linked receptors on skeletal muscle and nerve cells that can be activated by the binding of nicotine, as well as by acetylcholine.)

Other trimeric G proteins regulate the activity of ion channels less directly, either by stimulating channel phosphorylation (by PKA, PKC, or CaM-kinase, for example) or by causing the production or destruction of cyclic nucleotides that directly activate or inactivate ion channels. The *cyclic-nucleotide-gated ion channels* have a crucial role in both smell (olfaction) and vision, as we now discuss.

Smell and Vision Depend on G-Protein-linked Receptors That Regulate Cyclic-Nucleotide-gated Ion Channels

Humans can distinguish more than 10,000 distinct smells, which are detected by specialized olfactory receptor neurons in the lining of the nose. These cells recognize odors by means of specific G-protein-linked **olfactory receptors**, which are displayed on the surface of the modified cilia that extend from each cell (Figure 15–43). The receptors act through cyclic AMP. When stimulated by odorant binding, they activate an olfactory-specific G protein (known as G_{olf}), which in turn activates adenylyl cyclase. The resulting increase in cyclic AMP opens *cyclic-AMP-gated cation channels,* thereby allowing an influx of Na^+, which depolarizes the olfactory receptor neuron and initiates a nerve impulse that travels along its axon to the brain.

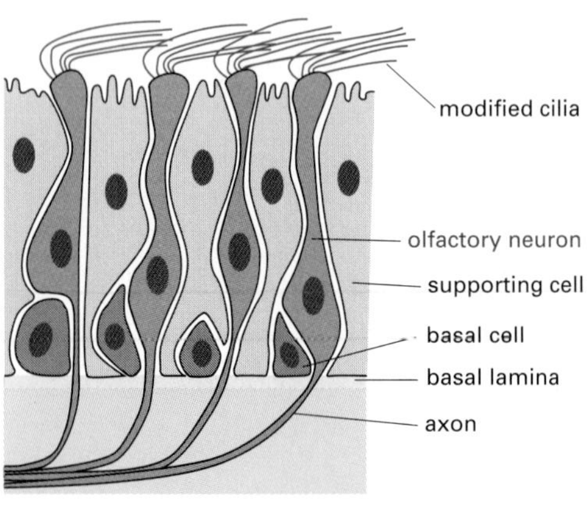

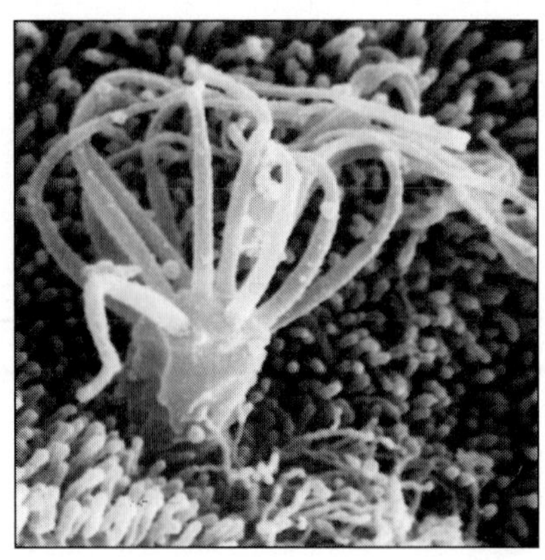

Figure 15–43 Olfactory receptor neurons. (A) This drawing shows a section of olfactory epithelium in the nose. Olfactory receptor neurons possess modified cilia, which project from the surface of the epithelium and contain the olfactory receptors, as well as the signal transduction machinery. The axon, which extends from the opposite end of the receptor neuron, conveys electrical signals to the brain when the cell is activated by an odorant to produce an action potential. The basal cells act as stem cells, producing new receptor neurons throughout life, to replace the neurons that die. (B) A scanning electron micrograph of the cilia on the surface of an olfactory neuron. (B, from E.E. Morrison and R.M. Costanzo, *J. Comp. Neurol.* 297:1–13, 1990. © Wiley-Liss, Inc.)

There are about 1000 different olfactory receptors in a mouse, each encoded by a different gene and each recognizing a different set of odorants. All of these receptors belong to the G-protein-linked receptor superfamily. Each olfactory receptor neuron produces only one of these 1000 receptors, and the neuron responds to a specific set of odorants by means of the specific receptor it displays. The same receptor also has a crucial role in directing the elongating axon of each developing olfactory neuron to the specific target neurons that it will connect to in the brain. A different set of more than 100 G-protein-linked receptors acts in a similar way to mediate a mouse's responses to *pheromones,* chemical signals detected in a different part of the nose that are used in communication between members of the same species.

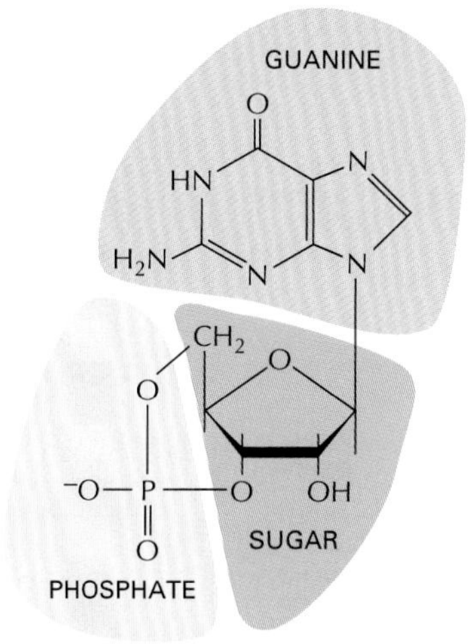

Figure 15–44 Cyclic GMP.

Vertebrate vision involves a similarly elaborate, highly sensitive, signal-detection process. Cyclic-nucleotide-gated ion channels are also involved, but the crucial cyclic nucleotide is **cyclic GMP** (Figure 15–44) rather than cyclic AMP. As with cyclic AMP, a continuous rapid synthesis (by *guanylyl cyclase*) and rapid degradation (by *cyclic GMP phosphodiesterase*) controls the concentration of cyclic GMP in cells.

In visual transduction responses, which are the fastest G-protein-mediated responses known in vertebrates, the receptor activation caused by light leads to a fall rather than a rise in the level of the cyclic nucleotide. The pathway has been especially well studied in **rod photoreceptors (rods)** in the vertebrate retina. Rods are responsible for noncolor vision in dim light, whereas *cone photoreceptors (cones)* are responsible for color vision in bright light. A rod photoreceptor is a highly specialized cell with outer and inner segments, a cell body, and a synaptic region where the rod passes a chemical signal to a retinal nerve cell; this nerve cell in turn relays the signal along the visual pathway (Figure 15–45). The phototransduction apparatus is in the outer segment, which contains a stack of *discs,* each formed by a closed sac of membrane in which many photosensitive **rhodopsin** molecules are embedded. The plasma membrane surrounding the outer segment contains *cyclic-GMP-gated* Na^+ *channels.* These channels are kept open in the dark by cyclic GMP that has bound to them. Paradoxically, light causes a hyperpolarization (which inhibits synaptic signaling) rather than a depolarization of the plasma membrane (which could stimulate synaptic signaling). Hyperpolarization (an increase in the membrane potential—discussed in Chapter 11) results because the activation by light of rhodopsin molecules in the disc membrane leads to a fall in cyclic GMP concentration and the *closure* of the special Na^+ channels in the surrounding plasma membrane (Figure 15–46).

Rhodopsin is a seven-pass transmembrane molecule homologous to other members of the G-protein-linked receptor family, and, like its cousins, it acts through a trimeric G protein. The activating extracellular signal, however, is not a molecule but a photon of light. Each rhodopsin molecule contains a covalently attached chromophore, 11-*cis* retinal, which isomerizes almost instantaneously to all-*trans* retinal when it absorbs a single photon. The isomerization alters the shape of the retinal, forcing a conformational change in the protein (opsin). The activated rhodopsin molecule then alters the G-protein *transducin* (G_t), causing its α subunit to dissociate and activate **cyclic GMP phosphodiesterase**. The phosphodiesterase then hydrolyzes cyclic GMP, so that cyclic GMP levels in the cytosol fall. This drop in cyclic GMP concentration leads to a decrease in the amount of cyclic GMP bound to the plasma membrane Na^+ channels, allowing more of these highly cyclic-GMP-sensitive channels to close. In this way, the signal quickly passes from the disc membrane to the plasma membrane, and a light signal is converted into an electrical one.

A number of mechanisms operate in rods to allow the cells to revert quickly to a resting, dark state in the aftermath of a flash of light—a requirement for perceiving the shortness of the flash. A *rhodopsin-specific kinase (RK)* phosphorylates the cytosolic tail of activated rhodopsin on multiple serines, partially

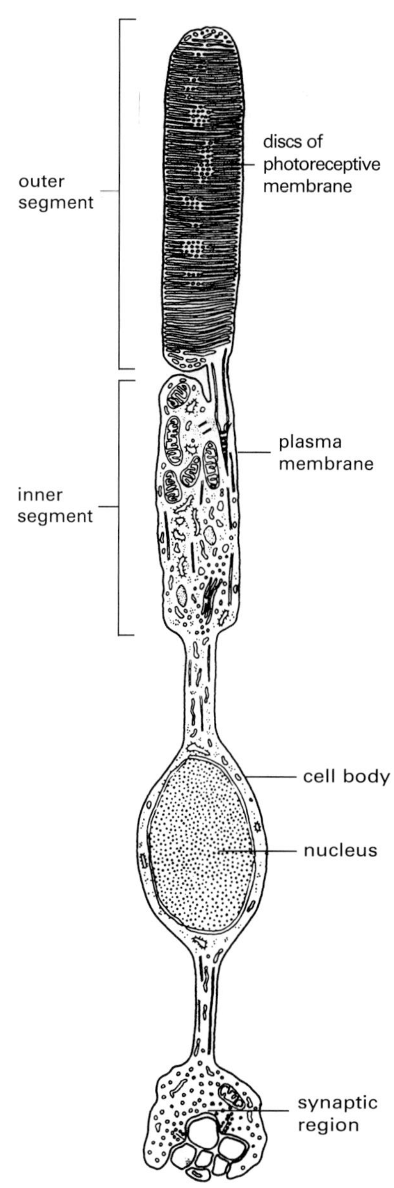

Figure 15–45 A rod photoreceptor cell. There are about 1000 discs in the outer segment. The disc membranes are not connected to the plasma membrane.

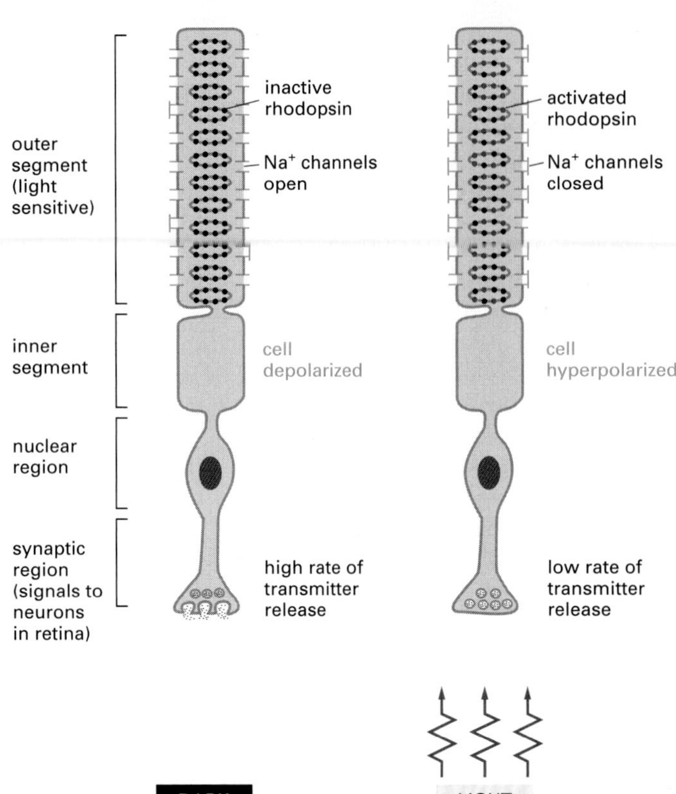

Figure 15–46 The response of a rod photoreceptor cell to light. Rhodopsin molecules in the outer-segment discs absorb photons. Photon absorption leads to the closure of Na^+ channels in the plasma membrane, which hyperpolarizes the membrane and reduces the rate of neurotransmitter release from the synaptic region. Because the neurotransmitter acts to inhibit many of the postsynaptic retinal neurons, illumination serves to free the neurons from inhibition and thus, in effect, excites them.

inhibiting the ability of the rhodopsin to activate transducin. An inhibitory protein called *arrestin* then binds to the phosphorylated rhodopsin, further inhibiting rhodopsin's activity. If the gene encoding RK is inactivated by mutation in mice or humans, the light response of rods is greatly prolonged, and the rods eventually die.

At the same time as rhodopsin is being shut off, an RGS protein (see p. 854) binds to activated transducin, stimulating the transducin to hydrolyze its bound GTP to GDP, which returns transducin to its inactive state. In addition, the Na^+ channels that close in response to light are also permeable to Ca^{2+}, so that when they close, the normal influx of Ca^{2+} is inhibited, causing the Ca^{2+} concentration in the cytosol to fall. The decrease in Ca^{2+} concentration stimulates guanylyl cyclase to replenish the cyclic GMP, rapidly returning its level to where it was before the light was switched on. A specific Ca^{2+}-sensitive protein mediates the activation of guanylyl cyclase in response to a fall in Ca^{2+} levels. In contrast to calmodulin, this protein is inactive when Ca^{2+} is bound to it and active when it is Ca^{2+}-free. It therefore stimulates the cyclase when Ca^{2+} levels fall following a light response.

These shut-off mechanisms do more than just return the rod to its resting state after a light flash; they also help to enable the photoreceptor to *adapt*, stepping down the response when it is exposed to light continuously. Adaptation, as we discussed earlier, allows the receptor cell to function as a sensitive detector of *changes* in stimulus intensity over an enormously wide range of baseline levels of stimulation.

The various trimeric G proteins we have discussed in this chapter are summarized in Table 15–3.

Extracellular Signals Are Greatly Amplified by the Use of Small Intracellular Mediators and Enzymatic Cascades

Despite the differences in molecular details, the signaling systems that are triggered by G-protein-linked receptors share certain features and are governed by similar general principles. They depend on relay chains of intracellular signaling

TABLE 15–3 Three Major Families of Trimeric G Proteins*

FAMILY	SOME FAMILY MEMBERS	ACTION MEDIATED BY	FUNCTIONS
I	G_s	α	activates adenylyl cyclase; activates Ca^{2+} channels
	G_{olf}	α	activates adenylyl cyclase in olfactory sensory neurons
II	G_i	α	inhibits adenylyl cyclase
		$\beta\gamma$	activates K^+ channels
	G_o	$\beta\gamma$	activates K^+ channels; inactivates Ca^{2+} channels
		α and $\beta\gamma$	activates phospholipase C-β
	G_t (transducin)	α	activates cyclic GMP phosphodiesterase in vertebrate rod photoreceptors
III	G_q	α	activates phospholipase C-β

*Families are determined by amino acid sequence relatedness of the α subunits. Only selected examples are shown. About 20 α subunits and at least 4 β subunits and 7 γ subunits have been described in mammals.

proteins and small intracellular mediators. In contrast to the more direct signaling pathways used by nuclear receptors discussed earlier, and by ion-channel-linked receptors discussed in Chapter 11, these relay chains provide numerous opportunities for amplifying the responses to extracellular signals. In the visual transduction cascade just described, for example, a single activated rhodopsin molecule catalyzes the activation of hundreds of molecules of transducin at a rate of about 1000 transducin molecules per second. Each activated transducin molecule activates a molecule of cyclic GMP phosphodiesterase, each of which hydrolyzes about 4000 molecules of cyclic GMP per second. This catalytic cascade lasts for about 1 second and results in the hydrolysis of more than 10^5 cyclic GMP molecules for a single quantum of light absorbed, and the resulting drop in the concentration of cyclic GMP in turn transiently closes hundreds of Na^{2+} channels in the plasma membrane (Figure 15–47). As a result, a rod cell can respond to a single photon of light, in a way that is highly reproducible in its timing and magnitude.

Likewise, when an extracellular signal molecule binds to a receptor that indirectly activates adenylyl cyclase via G_s, each receptor protein may activate many molecules of G_s protein, each of which can activate a cyclase molecule. Each cyclase molecule, in turn, can catalyze the conversion of a large number of ATP molecules to cyclic AMP molecules. A similar amplification operates in the inositol-phospholipid pathway. A nanomolar (10^{-9} M) change in the concentration of an extracellular signal can thereby induce micromolar (10^{-6} M) changes in the concentration of a small intracellular mediator such as cyclic AMP or Ca^{2+}. Because these mediators function as allosteric effectors to activate specific enzymes or ion channels, a single extracellular signal molecule can cause many thousands of protein molecules to be altered within the target cell.

Any such amplifying cascade of stimulatory signals requires that there be counterbalancing mechanisms at every step of the cascade to restore the system to its resting state when stimulation ceases. Cells therefore have efficient mechanisms for rapidly degrading (and resynthesizing) cyclic nucleotides and for buffering and removing cytosolic Ca^{2+}, as well as for inactivating the responding enzymes and ion channels once they have been activated. This is not only essential for turning a response off, it is also important for defining the resting state from which a response begins. As we saw earlier, in general, the response to stimulation can be rapid only if the inactivating mechanisms are also rapid. Each protein in the relay chain of signals can be a separate target for regulation, including the receptor, as we discuss next.

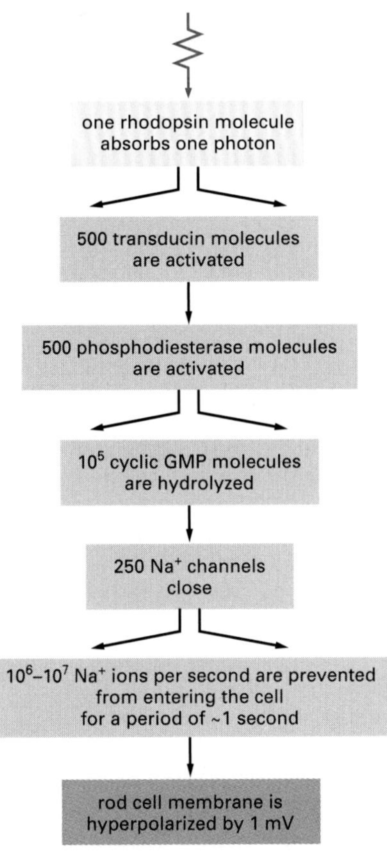
one rhodopsin molecule absorbs one photon
500 transducin molecules are activated
500 phosphodiesterase molecules are activated
10^5 cyclic GMP molecules are hydrolyzed
250 Na^+ channels close
10^6–10^7 Na^+ ions per second are prevented from entering the cell for a period of ~1 second
rod cell membrane is hyperpolarized by 1 mV

Figure 15–47 Amplification in the light-induced catalytic cascade in vertebrate rods. The divergent arrows indicate the steps where amplification occurs.

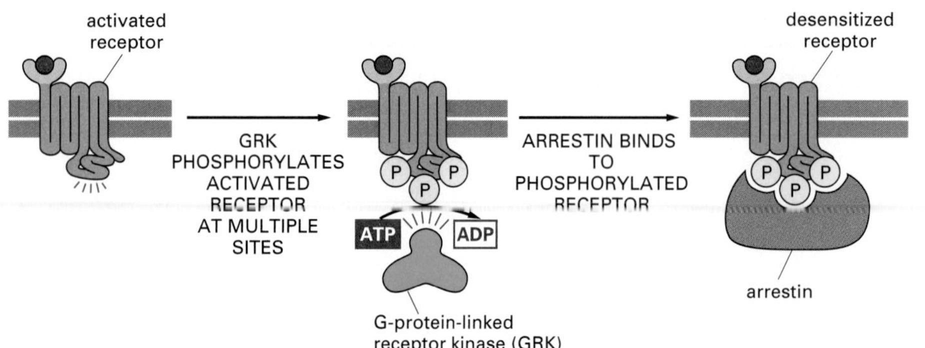

Figure 15–48 The roles of G-protein-linked receptor kinases (GRKs) and arrestins in receptor desensitization. The binding of an arrestin to the phosphorylated receptor prevents the receptor from binding to its G protein and can direct its endocytosis. Mice that are deficient in one form of arrestin fail to desensitize in response to morphine, for example, attesting to the importance of arrestins for desensitization.

G-Protein-linked Receptor Desensitization Depends on Receptor Phosphorylation

As discussed earlier, target cells use a variety of mechanisms to *desensitize*, or *adapt*, when they are exposed to a high concentration of stimulating ligand for a prolonged period (see Figure 15–25). We discuss here only those mechanisms that involve an alteration in G-protein-linked receptors themselves.

These receptors can desensitize in three general ways:

1. They can become altered so that they can no longer interact with G proteins *(receptor inactivation).*
2. They can be temporarily moved to the interior of the cell (internalized) so that they no longer have access to their ligand *(receptor sequestration).*
3. They can be destroyed in lysosomes after internalization *(receptor down-regulation).*

In each case, the desensitization process depends on phosphorylation of the receptor, by PKA, PKC, or a member of the family of **G-protein-linked receptor kinases (GRKs)**. (The GRKs include the rhodopsin-specific kinase involved in rod photoreceptor desensitization discussed earlier.) The GRKs phosphorylate multiple serine and threonines on a receptor, but they do so only after the receptor has been activated by ligand binding. As with rhodopsin, once a receptor has been phosphorylated in this way, it binds with high affinity to a member of the **arrestin** family of proteins (Figure 15–48).

The bound arrestin can contribute to the desensitization process in at least two ways. First, it inactivates the receptor by preventing it from interacting with G proteins, an example of receptor uncoupling. Second, it can serve as an adaptor protein to couple the receptor to clathrin-coated pits (discussed in Chapter 13), inducing receptor-mediated endocytosis. Endocytosis results in either the sequestration or degradation (down-regulation) of the receptor, depending on the specific receptor and cell type, the concentration of the stimulating ligand, and the duration of the ligand's presence.

Summary

G-protein-linked receptors can indirectly activate or inactivate either plasma-membrane-bound enzymes or ion channels via G proteins. When stimulated by an activated receptor, a G protein disassembles into an α subunit and a $\beta\gamma$ complex, both of which can directly regulate the activity of target proteins in the plasma membrane. Some G-protein-linked receptors either activate or inactivate adenylyl cyclase, thereby altering the intracellular concentration of the intracellular mediator cyclic AMP. Others activate a phosphoinositide-specific phospholipase C (phospholipase C-β), which hydrolyzes phosphatidylinositol 4,5-bisphosphate [PI(4,5)P_2] to generate two small intracellular mediators. One is inositol 1,4,5-trisphosphate (IP_3), which releases Ca^{2+} from the ER and thereby increases the concentration of Ca^{2+} in the cytosol. The other is diacylglycerol, which remains in the plasma membrane and activates protein kinase C (PKC). A rise in cyclic AMP or Ca^{2+} levels affects cells mainly by stimulating protein kinase A (PKA) and Ca^{2+}/calmodulin-dependent protein kinases (CaM-kinases), respectively.

PKC, PKA, and CaM-kinases phosphorylate specific target proteins on serines or threonines and thereby alter the activity of the proteins. Each type of cell has characteristic sets of target proteins that are regulated in these ways, enabling the cell to make its own distinctive response to the small intracellular mediators. The intracellular signaling cascades activated by G-protein-linked receptors allow the responses to be greatly amplified, so that many target proteins are changed for each molecule of extracellular signaling ligand bound to its receptor.

The responses mediated by G-protein-linked receptors are rapidly turned off when the extracellular signaling ligand is removed. Thus, the G-protein α subunit is induced to inactivate itself by hydrolyzing its bound GTP to GDP, IP_3 is rapidly dephosphorylated by a phosphatase (or phosphorylated by a kinase), cyclic nucleotides are hydrolyzed by phosphodiesterases, Ca^{2+} is rapidly pumped out of the cytosol, and phosphorylated proteins are dephosphorylated by protein phosphatases. Activated G-protein-linked receptors themselves are phosphorylated by GRKs, thereby trigging arrestin binding, which uncouples the receptors from G proteins and promotes receptor endocytosis.

SIGNALING THROUGH ENZYME-LINKED CELL-SURFACE RECEPTORS

Enzyme-linked receptors are a second major type of cell-surface receptor. They were recognized initially through their role in responses to extracellular signal proteins that promote the growth, proliferation, differentiation, or survival of cells in animal tissues. These signal proteins are often collectively called *growth factors*, and they usually act as local mediators at very low concentrations (about 10^{-9}–10^{-11} M). The responses to them are typically slow (on the order of hours) and usually require many intracellular signaling steps that eventually lead to changes in gene expression. Enzyme-linked receptors have since been found also to mediate direct, rapid effects on the cytoskeleton, controlling the way a cell moves and changes its shape. The extracellular signals that induce these rapid responses are often not diffusible but are instead attached to surfaces over which the cell is crawling. Disorders of cell proliferation, differentiation, survival, and migration are fundamental events that can give rise to cancer, and abnormalities of signaling through enzyme-linked receptors have major roles in this class of disease.

Like G-protein-linked receptors, enzyme-linked receptors are transmembrane proteins with their ligand-binding domain on the outer surface of the plasma membrane. Instead of having a cytosolic domain that associates with a trimeric G protein, however, their cytosolic domain either has an intrinsic enzyme activity or associates directly with an enzyme. Whereas a G-protein-linked receptor has seven transmembrane segments, each subunit of an enzyme-linked receptor usually has only one.

Six classes of enzyme-linked receptors have thus far been identified:

1. *Receptor tyrosine kinases* phosphorylate specific tyrosines on a small set of intracellular signaling proteins.
2. *Tyrosine-kinase-associated receptors* associate with intracellular proteins that have tyrosine kinase activity.
3. *Receptorlike tyrosine phosphatases* remove phosphate groups from tyrosines of specific intracellular signaling proteins. (They are called "receptorlike" because the presumptive ligands have not yet been identified, and so their receptor function has not been directly demonstrated.)
4. *Receptor serine/threonine kinases* phosphorylate specific serines or threonines on associated latent gene regulatory proteins.
5. *Receptor guanylyl cyclases* directly catalyze the production of cyclic GMP in the cytosol.
6. *Histidine-kinase-associated receptors* activate a "two-component" signaling pathway in which the kinase phosphorylates itself on histidine and then immediately transfers the phosphate to a second intracellular signaling protein.

We begin our discussion with the receptor tyrosine kinases, the most numerous of the enzyme-linked receptors. We then consider the other classes in turn.

Activated Receptor Tyrosine Kinases Phosphorylate Themselves

The extracellular signal proteins that act through **receptor tyrosine kinases** consist of a large variety of secreted growth factors and hormones. Notable examples discussed elsewhere in this book include *epidermal growth factor (EGF), platelet-derived growth factor (PDGF), fibroblast growth factors (FGFs), hepatocyte growth factor (HGF), insulin, insulinlike growth factor-1 (IGF-1), vascular endothelial growth factor (VEGF), macrophage-colony-stimulating factor (M-CSF),* and all the *neurotrophins,* including *nerve growth factor (NGF).*

Many cell-surface-bound signal proteins also act through these receptors. The largest class of these membrane-bound ligands is the **ephrins**, which regulate the cell adhesion and repulsion responses that guide the migration of cells and axons along specific pathways during animal development (discussed in Chapter 21). The receptors for ephrins, called **Eph receptors**, are also the most numerous receptor tyrosine kinases. The ephrins and Eph receptors are unusual in that they can simultaneously act as both ligand and receptor: on binding to an Eph receptor, some ephrins not only activate the Eph receptor but also become activated themselves to transmit signals into the interior of the ephrin-expressing cell. In this way, an interaction between an ephrin protein on one cell and an Eph protein on another cell can lead to bidirectional reciprocal signaling that changes the behavior of both cells. Such *bidirectional signaling* between ephrins and Eph receptors is required, for example, to keep cells in particular parts of the developing brain from mixing with cells in neighboring parts.

Receptor tyrosine kinases can be classified into more than 16 structural subfamilies, each dedicated to its complementary family of protein ligands. Several of these families that operate in mammals are shown in Figure 15–49, and some of their ligands and functions are given in Table 15–4. In all cases, the binding of a signal protein to the ligand-binding domain on the outside of the cell activates the intracellular tyrosine kinase domain. Once activated, the kinase domain transfers a phosphate group from ATP to selected tyrosine side chains, both on the receptor proteins themselves and on intracellular signaling proteins that subsequently bind to the phosphorylated receptors.

How does the binding of an extracellular ligand activate the kinase domain on the other side of the plasma membrane? For a G-protein-linked receptor, ligand binding is thought to change the relative orientation of several of the transmembrane α helices, thereby shifting the position of the cytoplasmic loops relative to each other. It is difficult to imagine, however, how a conformational

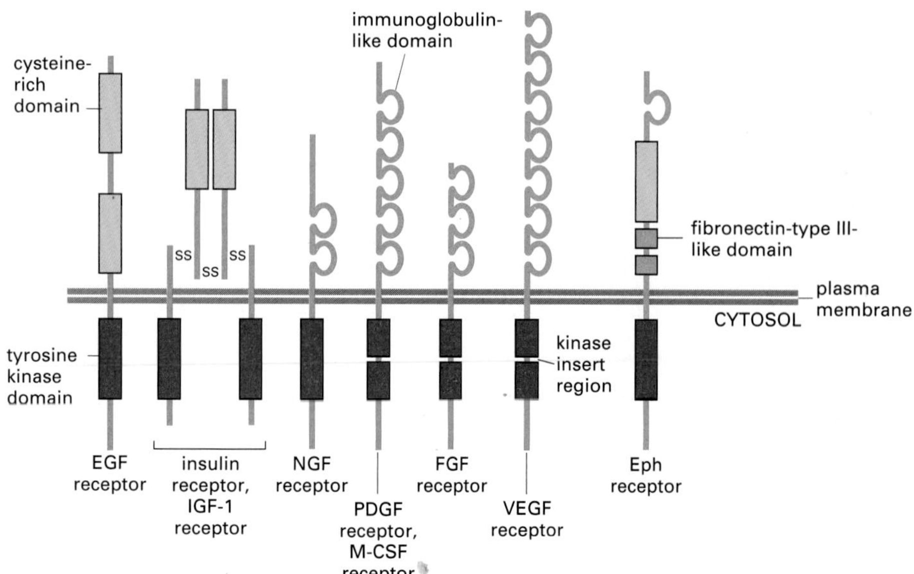

Figure 15–49 Seven subfamilies of receptor tyrosine kinases. Only one or two members of each subfamily are indicated. Note that the tyrosine kinase domain is interrupted by a "kinase insert region" in some of the subfamilies. The functional roles of most of the cysteine-rich, immunoglobulin-like, and fibronectin-type III-like domains are not known. Some of the ligands and responses for the receptors shown are listed in Table 15–4.

TABLE 15–4 Some Signaling Proteins That Act Via Receptor Tyrosine Kinases

SIGNALING LIGAND	RECEPTORS	SOME RESPONSES
Epidermal growth factor (EGF)	EGF receptor	stimulates proliferation of various cell types
Insulin	insulin receptor	stimulates carbohydrate utilization and protein synthesis
Insulin-like growth factors (IGF-1 and IGF-2)	IGF receptor-1	stimulate cell growth and survival
Nerve growth factor (NGF)	Trk A	stimulates survival and growth of some neurons
Platelet-derived growth factors (PDGF AA, BB, AB)	PDGF receptors (α and β)	stimulate survival, growth, and proliferation of various cell types
Macrophage-colony-stimulating factor (M-CSF)	M-CSF receptor	stimulates monocyte/macrophage proliferation and differentiation
Fibroblast growth factors (FGF-1 and FGF-24)	FGR receptors (FGF-R1–FGF-R4, plus multiple isoforms of each)	stimulate proliferation of various cell types; inhibit differentiation of some precursor cells; inductive signals in development
Vascular endothelial factor (VEGF)	VEGF receptor	stimulates angiogenesis
Ephrins (A and B types)	Eph receptors (A and B types)	stimulate angiogenesis; guide cell and axon migration

change could propagate across the lipid bilayer through a single transmembrane α helix. Instead, for the enzyme-linked receptors, two or more receptor chains come together in the membrane, forming a dimer or higher oligomer. In some cases, ligand binding induces the oligomerization. In other cases, the oligomerization occurs before ligand binding, and the ligand causes a reorientation of the receptor chains in the membrane. In either case, the rearrangement induced in cytosolic tails of the receptors initiates the intracellular signaling process. For receptor tyrosine kinases, the rearrangement enables the neighboring kinase domains of the receptor chains to cross-phosphorylate each other on multiple tyrosines, a process referred to as *autophosphorylation*.

To activate a receptor tyrosine kinase the ligand usually has to bind simultaneously to two adjacent receptor chains. PDGF, for example, is a dimer, which cross-links two receptors together (Figure 15–50A). Even some monomeric ligands, such as EGF, bind to two receptors simultaneously and cross-link them directly. By contrast, FGFs, which are also monomers, first form multimers by binding to heparan sulfate proteoglycans, either on the target cell surface or in the extracellular matrix. In this way, they are able to cross-link adjacent receptors (Figure 15–50B). In contact-dependent signaling, the ligands form clusters in the plasma membrane of the signaling cell and can thereby cross-link the receptors on the target cell (Figure15–50C); thus, whereas membrane-bound ephrins activate Eph receptors, soluble ephrins will do so only if they are aggregated.

Because of the requirement for receptor oligomerization, it is relatively easy to inactivate a specific receptor tyrosine kinase to determine its importance for a cell response. For this purpose, cells are transfected with DNA encoding a mutant form of the receptor that oligomerizes normally but has an inactive kinase domain. When coexpressed at a high level with normal receptors, the mutant receptor acts in a *dominant-negative* way, disabling the normal receptors by forming inactive dimers with them (Figure 15–51).

Autophosphorylation of the cytosolic tail of receptor tyrosine kinases contributes to the activation process in two ways. First, phosphorylation of tyrosines within the kinase domain increases the kinase activity of the enzyme. Second, phosphorylation of tyrosines outside the kinase domain creates high-affinity docking sites for the binding of a number of intracellular signaling proteins in the target cell. Each type of signaling protein binds to a different phosphorylated site on the activated receptor because it contains a specific phosphotyrosine-binding domain that recognizes surrounding features of the polypeptide chain in addition to the phosphotyrosine. Once bound to the activated kinase, the

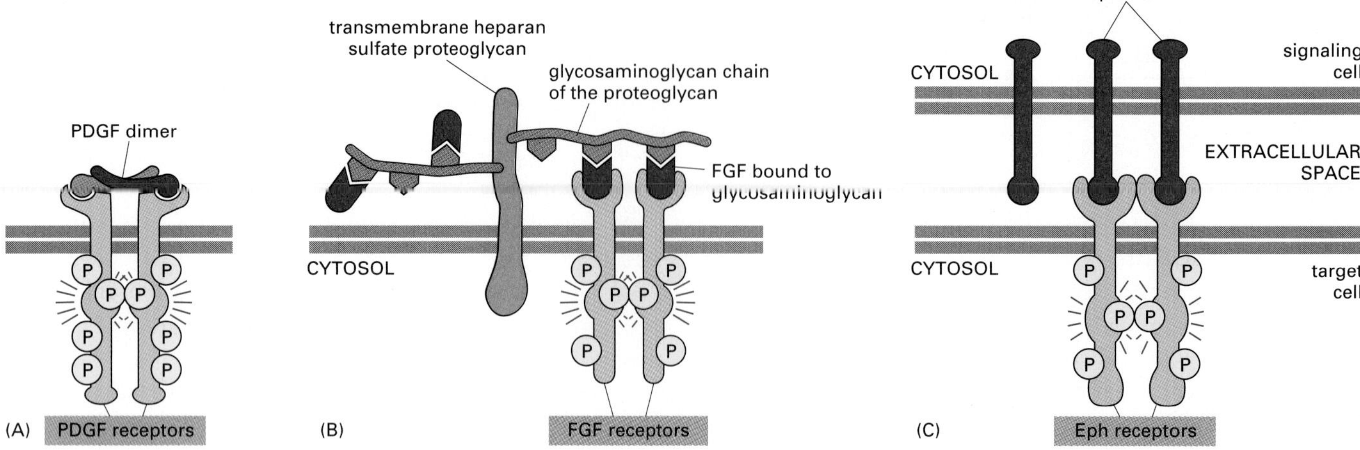

Figure 15–50 Three ways in which signaling proteins can cross-link receptor chains. When the receptor chains are cross-linked, the kinase domains of adjacent receptors cross-phosphorylate each other, stimulating the kinase activity of the receptor and creating docking sites for intracellular proteins. (A) Platelet-derived growth factor (PDGF) is a covalently linked dimer with two receptor-binding sites, so it can directly cross-link adjacent receptors to initiate the intracellular signaling process. The PDGF dimers are composed of A and B chains in different combinations, and they can activate different combinations of two types of PDGF receptor chains (α and β), which have somewhat different signaling properties. (B) Some monomeric ligands, such as fibroblast growth factors (FGFs), bind in clusters to proteoglycans, enabling the ligands to cross-link their receptors. The proteoglycans can be in the extracellular matrix or, as shown here, on the cell surface. There are more than 20 types of FGFs and more than 4 types of FGF receptors. (C) Membrane-bound signaling proteins, such as ephrins, can cross-link their receptors even though they are monomeric, because they cluster in the plasma membrane of the signaling cell. Some ephrins (B-type) are transmembrane proteins (as shown), whereas others (A-type) are linked to the membrane by a glycosylphosphatidylinositol (GPI) anchor (see Figure 12–56). For the most part, A-type ephrins activate A-type Eph receptors, and B-type ephrins activate B-type Eph receptors.

signaling protein may itself become phosphorylated on tyrosines and thereby activated; alternatively, the binding alone may be sufficient to activate the docked signaling protein. In summary, autophosphorylation serves as a switch to trigger the transient assembly of a large intracellular signaling complex, which then broadcasts signals along multiple routes to many destinations in the cell (Figure 15–52). Because different receptor tyrosine kinases bind different combinations of these signaling proteins, they activate different responses.

The receptors for insulin and IGF-1 act in a slightly different way. They are tetramers to start with (see Figure 15–49), and ligand binding is thought to induce a rearrangement of the transmembrane receptor chains, so that the two kinase domains come close together. Most of the phosphotyrosine docking sites generated by ligand binding are not on the receptor itself, but on a specialized docking protein called *insulin receptor substrate-1 (IRS-1)*. The activated receptor first autophosphorylates its kinase domains, which then phosphorylate IRS-1 on multiple tyrosines, thereby creating many more docking sites than could be accommodated on the receptor alone. Other docking proteins are used in a similar way by some other receptor tyrosine kinases to enlarge the size of the signaling complex.

Phosphorylated Tyrosines Serve as Docking Sites For Proteins With SH2 Domains

A whole menagerie of intracellular signaling proteins can bind to the phosphotyrosines on activated receptor tyrosine kinases (or on special docking proteins such as IRS-1) to help to relay the signal onward. Some docked proteins are enzymes, such as **phospholipase C-γ (PLC-γ)**, which functions in the same way as phospholipase C-β—activating the inositol phospholipid signaling pathway discussed earlier in connection with G-protein-linked receptors. Through this pathway, receptor tyrosine kinases can increase cytosolic Ca^{2+} levels. Much more often, these receptors depend more on relay chains of protein–protein

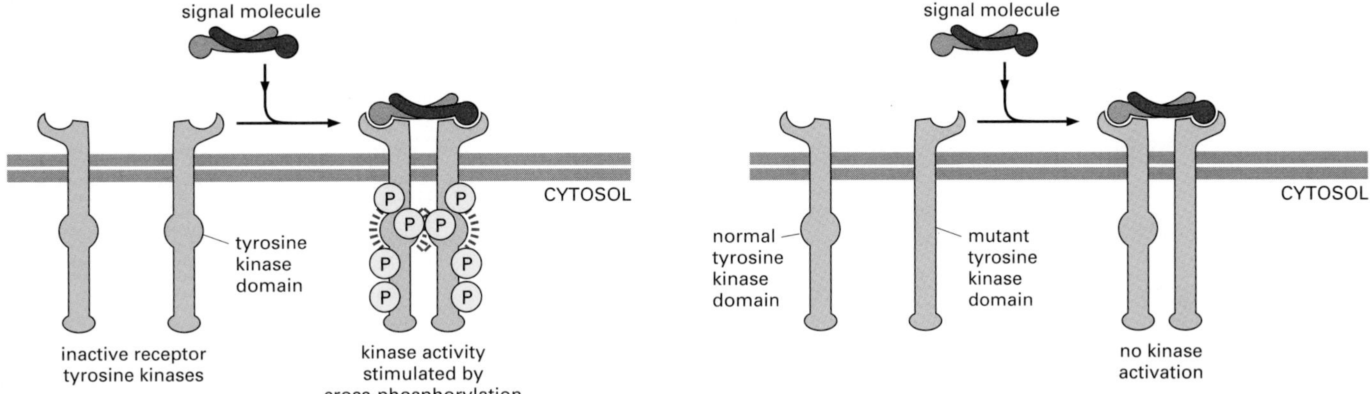

Figure 15–51 Inhibition of signaling through normal receptor tyrosine kinases by an excess of mutant receptors. (A) In this example, the normal receptors dimerize in response to ligand binding. The two kinase domains cross-phosphorylate each other, increasing the activity of the kinase domains, which can now further activate the receptor dimer by phosphorylating other sites on the receptors. (B) The mutant receptor with an inactivated kinase domain can dimerize normally, but it cannot cross-phosphorylate a normal receptor in a dimer. For this reason, the mutant receptors, if present in excess, will block signaling by the normal receptors—a process called *dominant-negative inhibition.* Cell biologists frequently use this strategy to inhibit a specific type of receptor tyrosine kinase in a cell to determine its normal function.

interactions. For example, another enzyme that docks on these receptors is the cytoplasmic tyrosine kinase Src, which phosphorylates other signaling proteins on tyrosines. Yet another is *phosphatidylinositol 3′-kinase (PI 3-kinase),* which, as we discuss later, generates specific lipid molecules in the plasma membrane to attract other signaling proteins there.

Although the intracellular signaling proteins that bind to phosphotyrosines on activated receptor tyrosine kinases and docking proteins have varied structures and functions, they usually share highly conserved phosphotyrosine-binding domains. These can be either **SH2 domains** (for *Src homology region,* because it was first found in the Src protein) or, less commonly, *PTB domains* (for *phosphotyrosine-binding*). By recognizing specific phosphorylated tyrosines, these small domains serve as modules that enable the proteins that contain them to bind to activated receptor tyrosine kinases, as well as to many other intracellular signaling proteins that have been transiently phosphorylated on tyrosines (Figure 15–53). Many signaling proteins also contain other protein modules that allow them to interact specifically with other proteins as part of the signaling process. These include the *SH3 domain* (again, so named because it was first discovered in Src), which binds to proline-rich motifs in intracellular proteins (see Figure 15–20).

Not all proteins that bind to activated receptor tyrosine kinases via SH2 domains help to relay the signal onward. Some act to decrease the signaling process, providing negative feedback. One example is the *c-Cbl* protein, which can dock on some activated receptors and catalyze their conjugation with ubiquitin. This ubiquitylation promotes the internalization and degradation of the receptors—a process called receptor down-regulation (see Figure 15–25).

Some signaling proteins are composed almost entirely of SH2 and SH3 domains and function as adaptors to couple tyrosine-phosphorylated proteins to other proteins that do not have their own SH2 domains (see Figure 15–20). Such adaptor proteins help to couple activated receptors to the important downstream signaling protein *Ras.* As we discuss next, Ras acts as a transducer and bifurcation signaling protein, changing the nature of the signal and broad-

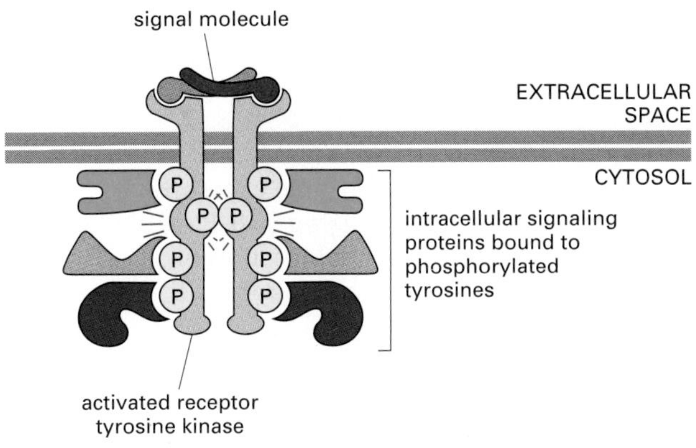

Figure 15–52 The docking of intracellular signaling proteins on an activated receptor tyrosine kinase. The activated receptor and its bound signaling proteins form a signaling complex that can then broadcast signals along multiple signaling pathways.

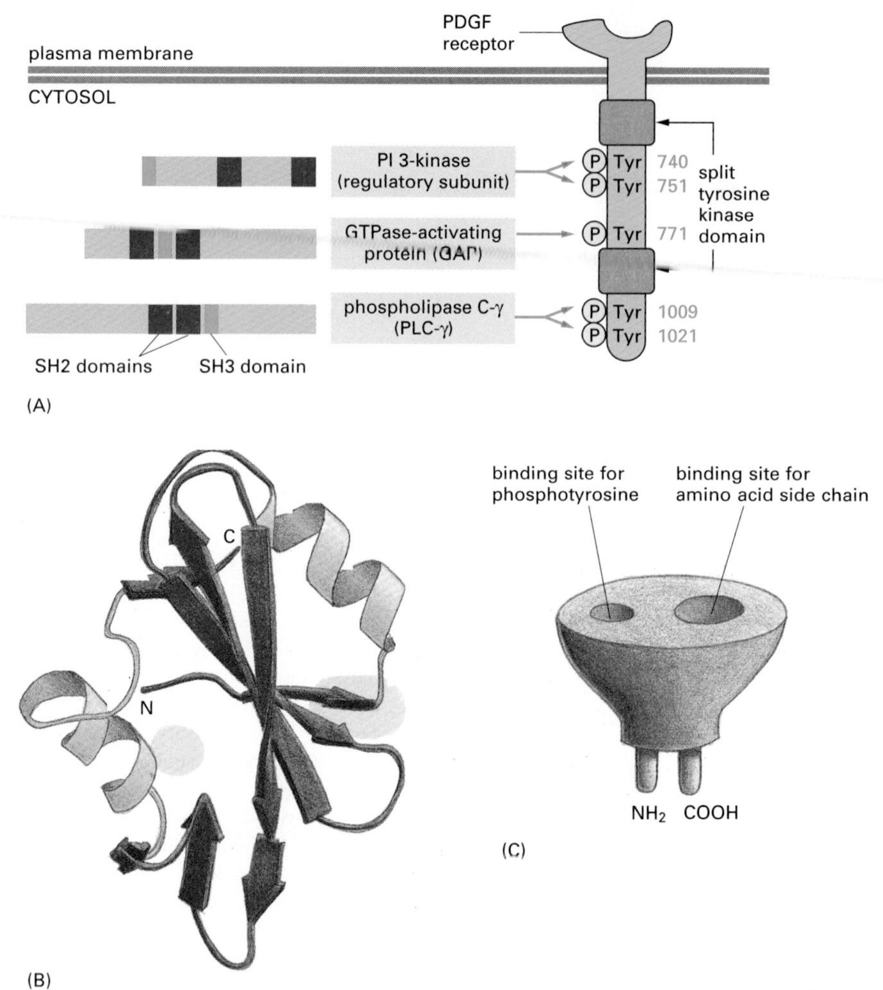

Figure 15–53 The binding of SH2-containing intracellular signaling proteins to an activated PDGF receptor. (A) This drawing of a PDGF receptor shows five of the tyrosine autophosphorylation sites, three in the kinase insert region and two on the C-terminal tail, to which the three signaling proteins shown bind as indicated. The numbers on the right indicate the positions of the tyrosines in the polypeptide chain. These binding sites have been identified by using recombinant DNA technology to mutate specific tyrosines in the receptor. Mutation of tyrosines 1009 and 1021, for example, prevents the binding and activation of PLC-γ, so that receptor activation no longer stimulates the inositol phospholipid signaling pathway. The locations of the SH2 *(red)* and SH3 *(blue)* domains in the three signaling proteins are indicated. (Additional autophosphorylation sites on this receptor are not shown, including those that serve as binding sites for the cytoplasmic tyrosine kinase Src and the adaptor proteins Grb2 and Shc, discussed later.) (B) The three-dimensional structure of an SH2 domain, as determined by x-ray crystallography. The binding pocket for phosphotyrosine is shown in *yellow* on the right, and a pocket for binding a specific amino acid side chain (isoleucine, in this case) is shown in *yellow* on the left (see also Figure 3–40). (C) The SH2 domain is a compact, "plug-in" module, which can be inserted almost anywhere in a protein without disturbing the protein's folding or function (see Figure 3–19). Because each domain has distinct sites for recognizing phosphotyrosine and for recognizing a particular amino acid side chain, different SH2 domains recognize phosphotyrosine in the context of different flanking amino acid sequences. (B, based on data from G. Waksman et al., *Cell* 72:1–20, 1993. © Elsevier.)

casting it along multiple downstream pathways, including a major signaling pathway that can help stimulate cells to proliferate or differentiate. Mutations that activate this pathway, and thereby stimulate cell division inappropriately, are a causative factor in many types of cancer.

Ras Is Activated by a Guanine Nucleotide Exchange Factor

The **Ras proteins** belong to the large *Ras superfamily of monomeric GTPases*. The family also contains two other subfamilies: the *Rho family,* involved in relaying signals from cell-surface receptors to the actin cytoskeleton and elsewhere (discussed in Chapter 16), and the *Rab family,* involved in regulating the traffic of intracellular transport vesicles (discussed in Chapter 13). Like almost all of these monomeric GTPases, the Ras proteins contain a covalently attached lipid group that helps to anchor the protein to a membrane—in this case, to the cytoplasmic face of the plasma membrane where the protein functions. There are multiple Ras proteins, and different ones act in different cell types. Because they all seem to work in much the same way, we shall refer to them simply as Ras.

Ras helps to broadcast signals from the cell surface to other parts of the cell. It is often required, for example, when receptor tyrosine kinases signal to the nucleus to stimulate cell proliferation or differentiation by altering gene expression. If Ras function is inhibited by the microinjection of neutralizing anti-Ras antibodies or a dominant-negative mutant form of Ras, the cell proliferation or differentiation responses normally induced by the activated receptor tyrosine kinases do not occur. Conversely, if a hyperactive mutant Ras protein is introduced into some cell lines, the effect on cell proliferation or differentiation is sometimes the same as that induced by the binding of ligands to cell-surface receptors. In fact, Ras was first discovered as the hyperactive product of a mutant *ras* gene that promoted the development of cancer; we now know that about 30% of human tumors have a hyperactive *ras* mutation.

Like other GTP-binding proteins, Ras functions as a switch, cycling between two distinct conformational states—active when GTP is bound and inactive when GDP is bound (see Figure 15–17). Two classes of signaling proteins regulate Ras activity by influencing its transition between active and inactive states. **Guanine nucleotide exchange factors (GEFs)** promote the exchange of bound nucleotide by stimulating the dissociation of GDP and the subsequent uptake of GTP from the cytosol, thereby activating Ras. **GTPase-activating proteins (GAPs)** increase the rate of hydrolysis of bound GTP by Ras, thereby inactivating Ras (Figure 15–54). Hyperactive mutant forms of Ras are resistant to GAP-mediated GTPase stimulation and are locked permanently in the GTP-bound active state, which is why they promote the development of cancer.

In principle, receptor tyrosine kinases could activate Ras either by activating a GEF or by inhibiting a GAP. Even though some GAPs bind directly (via their SH2 domains) to activated receptor tyrosine kinases (see Figure 15–53), whereas GEFs bind only indirectly, it is the indirect coupling of the receptor to a GEF that is responsible for driving Ras into its active state. In fact, the loss of function of a Ras-specific GEF has a similar effect to the loss of function of that Ras. The activation of the other Ras-like proteins, including those of the Rho family, is also thought to occur through the activation of GEFs.

Genetic studies in flies and worms, and biochemical studies in mammalian cells, indicate that adaptor proteins link receptor tyrosine kinases to Ras. The **Grb-2 protein** in mammalian cells, for example, binds through its SH2 domain to specific phosphotyrosines on activated receptor tyrosine kinases and through its SH3 domains to proline-rich motifs on a GEF called **Sos**. Some activated receptor tyrosine kinases, however, do not display the specific phosphotyrosines required for Grb-2 docking; these receptors recruit another adaptor protein called *Shc*, which binds both to the activated receptor and to Grb-2, thereby coupling the receptor to Sos by a more indirect route. The assembly of the complex of receptor–Grb-2–Sos (or receptor–Shc–Grb-2–Sos) brings Sos into position to activate neighboring Ras molecules by stimulating it to exchange its bound GDP for GTP (Figure 15–55). The importance of Grb-2 is indicated by the finding that Grb-2-deficient mice die early in embryogenesis. Very similar sets of proteins are thought to operate in all animals to activate Ras.

This pathway from receptor tyrosine kinases is not the only means of activating Ras. Other Ras GEFs are activated independently of Sos. One that is found mainly in the brain, for example, is activated by Ca^{2+} and diacylglycerol and can couple G-protein-linked receptors to Ras activation.

Once activated, Ras in turn activates various other signaling proteins to relay the signal downstream along several pathways. One of the signaling pathways Ras activates is a serine/threonine phosphorylation cascade that is highly conserved in eucaryotic cells from yeasts to humans. As we discuss next, a crucial component in this cascade is a novel type of protein kinase called *MAP-kinase*.

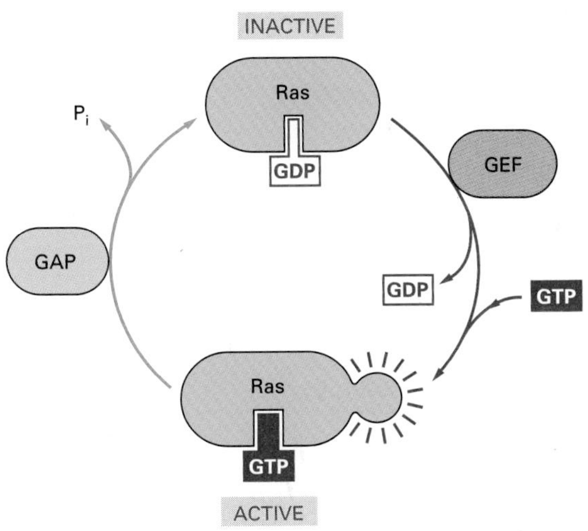

Figure 15–54 The regulation of Ras activity. GTPase-activating proteins (GAPs) inactivate Ras by stimulating it to hydrolyze its bound GTP; the inactivated Ras remains tightly bound to GDP. Guanine nucleotide exchange factors (GEFs) activate Ras by stimulating it to give up its GDP; the concentration of GTP in the cytosol is 10 times greater than the concentration of GDP, and Ras rapidly binds GTP once GDP has been ejected. Several Ras-regulating GAPs (Ras GAPs) have been characterized in mammalian cells, including *p120*^{GAP} and *neurofibromin* (so named because it is encoded by the gene that is mutated in the common human genetic disease neurofibromatosis, which is associated with tumors of nerves). The Ras GAPs maintain most of the Ras protein (~95%) in unstimulated cells in an inactive GDP-bound state.

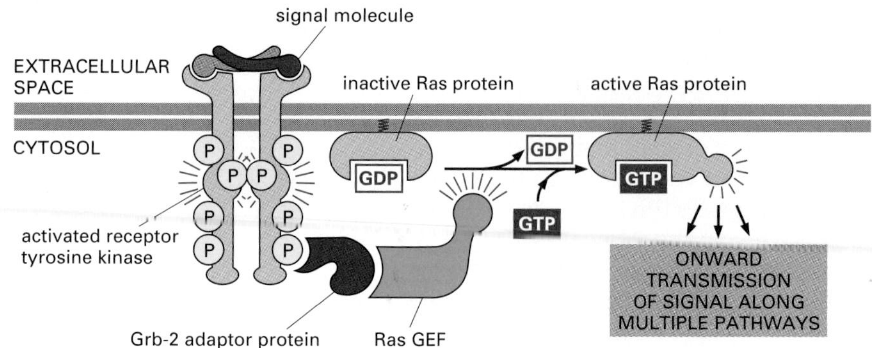

Figure 15–55 The activation of Ras by an activated receptor tyrosine kinase. Most of the signaling proteins bound to the activated receptor are omitted for simplicity. The Grb-2 adaptor protein binds to a specific phosphotyrosine on the receptor and to the Ras guanine nucleotide exchange factor (GEF), which stimulates Ras to exchange its bound GDP for GTP. The activated Ras then activates several downstream signaling pathways, one of which is shown in Figure 15–56.

Ras Activates a Downstream Serine/Threonine Phosphorylation Cascade That Includes a MAP-Kinase

Both the tyrosine phosphorylations and the activation of Ras triggered by activated receptor tyrosine kinases are short-lived. Tyrosine-specific protein phosphatases (discussed later) quickly reverse the phosphorylations, and GAPs induce activated Ras to inactivate itself by hydrolyzing its bound GTP to GDP. To stimulate cells to proliferate or differentiate, these short-lived signaling events must be converted into longer-lasting ones that can sustain the signal and relay it downstream to the nucleus to alter the pattern of gene expression. Activated Ras triggers this conversion by initiating a series of downstream serine/threonine phosphorylations, which are much longer-lived than tyrosine phosphorylations. Many serine/threonine kinases participate in this phosphorylation cascade, but three of them constitute the core module of the cascade. The last of the three is called a **mitogen-activated protein kinase (MAP-kinase)**.

An unusual feature of a MAP-kinase is that its full activation requires the phosphorylation of both a threonine and a tyrosine, which are separated in the protein by a single amino acid. The protein kinase that catalyzes both of these phosphorylations is called a *MAP-kinase-kinase,* which in the mammalian Ras signaling pathway is called MEK. The requirement for both a tyrosine and a threonine phosphorylation ensures that the MAP-kinase is kept inactive unless specifically activated by a MAP-kinase-kinase, whose only known substrate is a MAP-kinase. MAP-kinase-kinase is itself activated by phosphorylation catalyzed by the first kinase in the three-component module, *MAP-kinase-kinase-kinase,* which in the mammalian Ras signaling pathway is called **Raf**. The Raf kinase is activated by activated Ras.

Once activated, the MAP-kinase relays the signal downstream by phosphorylating various proteins in the cell, including gene regulatory proteins and other protein kinases (Figure 15–56). It enters the nucleus, for example, and phosphorylates one or more components of a gene regulatory complex. This activates the transcription of a set of *immediate early genes,* so named because they turn on within minutes of the time that cells are stimulated by an extracellular signal, even if protein synthesis is experimentally blocked with drugs. Some of these genes encode other gene regulatory proteins that turn on other genes, a process that requires both protein synthesis and more time. In this way the Ras–MAP-kinase signaling pathway conveys signals from the cell surface to the nucleus and alters the pattern of gene expression in significant ways. Among the genes activated by this pathway are those required for cell proliferation, such as the genes encoding *G_1 cyclins* (discussed in Chapter 17).

MAP-kinases are usually activated only transiently in response to extracellular signals, and the period of time they remain active can profoundly influence the nature of the response. When EGF activates its receptors on a neural precursor cell line, for example, MAP-kinase activity peaks at 5 minutes and rapidly declines, and the cells later go on to divide. By contrast, when NGF activates its receptors on the same cells, MAP-kinase activity remains high for many hours, and the cells stop proliferating and differentiate into neurons.

MAP-kinases are inactivated by dephosphorylation, and the specific removal of phosphate from either the tyrosine or the threonine is enough to inactivate the enzyme. In some cases, stimulation by an extracellular signal

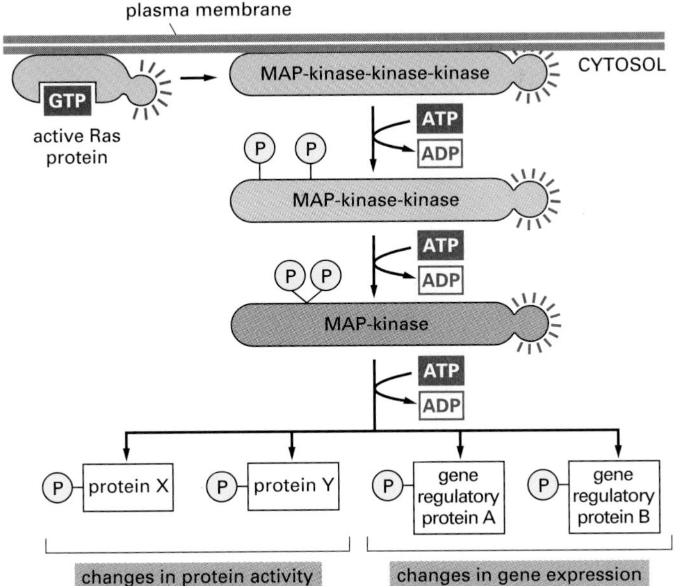

Figure 15–56 The MAP-kinase serine/threonine phosphorylation pathway activated by Ras. Multiple such pathways involving structurally and functionally related proteins operate in all eucaryotes, each coupling an extracellular stimulus to a variety of cell outputs. The pathway activated by Ras begins with a MAP-kinase-kinase-kinase called *Raf,* which activates the MAP-kinase-kinase *Mek,* which then activates the MAP-kinase called *Erk.* Erk in turn phosphorylates a variety of downstream proteins, including other kinases, as well as gene regulatory proteins in the nucleus. The resulting changes in gene expression and protein activity cause complex changes in cell behavior.

induces the expression of a dual-specificity phosphatase that removes both phosphates and inactivates the kinase, providing a form of negative feedback. In other cases, stimulation causes the kinase to be switched off more rapidly by phosphatases that are already present.

Three-component MAP-kinase signaling modules operate in all animal cells, as well as in yeasts, with different ones mediating different responses in the same cell. In budding yeast, for example, one such module mediates the mating pheromone response via the $\beta\gamma$ complex of a G protein, another the response to starvation, and yet another the response to osmotic shock. Some of these three-component MAP-kinase modules use one or more of the same kinases and yet manage to activate different effector proteins and hence different responses. How do cells avoid cross talk between the different parallel signaling pathways to ensure that each response is specific? One way is to use scaffold proteins that bind all or some of the kinases in a specific module to form a complex, as illustrated in Figure 15–57 and discussed earlier (see Figure 15–19A).

Mammalian cells also use this strategy to prevent cross talk between MAP-kinase signaling pathways. At least 5 parallel MAP-kinase modules can operate in a mammalian cell. These modules are composed of at least 12 MAP-kinases, 7 MAP-kinase-kinases, and 7 MAP-kinase-kinase-kinases. Several of these modules are activated by different kinds of cell stresses, such as UV irradiation, heat shock, osmotic stress, and stimulation by inflammatory cytokines. The three kinases in at least some of these stress-activated modules are held together by binding to a common scaffold protein, just as in yeast. The scaffold strategy

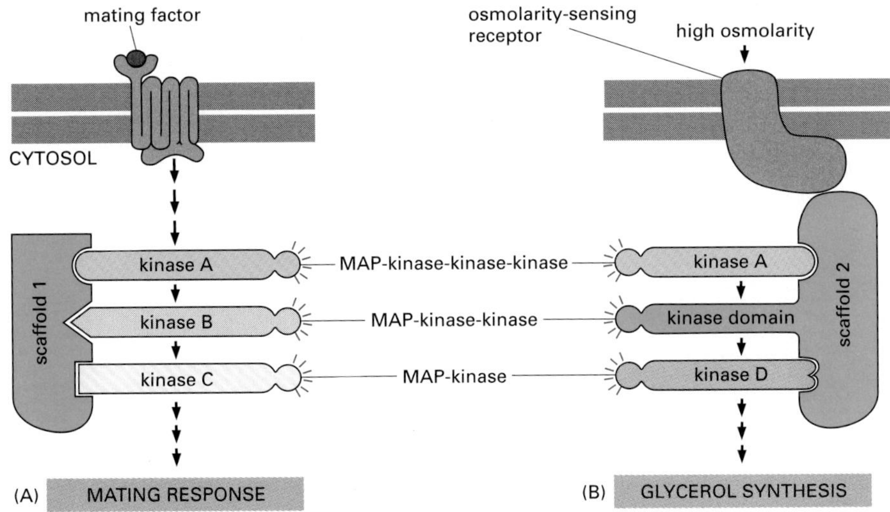

Figure 15–57 The organization of MAP-kinase pathways by scaffold proteins in budding yeast. Budding yeast have at least six three-component MAP-kinase modules involved in a variety of biological processes, including the two responses illustrated here—a mating response and the response to high osmolarity. (A) The mating response is triggered when a mating factor secreted by a yeast of opposite mating type binds to a G-protein-linked receptor. This activates a G protein, and the $\beta\gamma$ complex of the G protein indirectly activates the MAP-kinase-kinase-kinase (kinase A), which then relays the response onward. Once activated, the MAP-kinase (kinase C) phosphorylates and thereby activates several proteins that mediate the mating response, in which the yeast cell stops dividing and prepares for fusion. The three kinases in this module are bound to scaffold protein 1. (B) In a second response, a yeast cell exposed to a high-osmolarity environment is induced to synthesize glycerol to increase its internal osmolarity. This response is mediated by a transmembrane, osmolarity-sensing, receptor protein and a different MAP-kinase module bound to a second scaffold protein. (Note that the kinase domain of scaffold 2 provides the MAP-kinase-kinase activity of this module.) Although both pathways use the same MAP-kinase-kinase-kinase (kinase A, *green*), there is no cross talk between them, because the kinases in each module are tightly bound to different scaffold proteins, and the osmosensor is bound to the same scaffold protein as the particular kinase it activates.

provides precision, helps to create a large change in MAP-kinase activity in response to small changes in signal molecule concentration, and avoids cross-talk. However, it reduces the opportunities for amplification and spreading of the signal to different parts of the cell, which require at least some of the components to be diffusible (see Figure 15–16).

When Ras is activated by receptor tyrosine kinases, it usually activates more than just the MAP-kinase signaling pathway. It also usually helps activate *PI3-kinase,* which can signal cells to survive and grow.

PI 3-Kinase Produces Inositol Phospholipid Docking Sites in the Plasma Membrane

Extracellular signal proteins stimulate cells to divide, in part by activating the Ras–MAP-kinase pathway just discussed. If cells continually divided without growing, however, they would get progressively smaller and would eventually disappear. Thus, to proliferate, most cells need to be stimulated to enlarge (grow), as well as to divide. In some cases, one signal protein does both; in others one signal protein (a *mitogen)* mainly stimulates cell division, while another (a *growth factor)* mainly stimulates cell growth. One of the major intracellular signaling pathways leading to cell growth involves **phosphatidylinositol 3-kinase (PI 3-kinase)**. This kinase principally phosphorylates inositol phospholipids rather than proteins; it can be activated by receptor tyrosine kinases, as well as by many other types of cell-surface receptors, including some that are G-protein-linked.

Phosphatidylinositol (PI) is unique among membrane lipids because it can undergo reversible phosphorylation at multiple sites to generate a variety of distinct inositol phospholipids. When activated, PI 3-kinase catalyzes the phosphorylation of inositol phospholipids at the 3 position of the inositol ring to generate lipids called *PI(3,4)P$_2$* or *PI(3,4,5)P$_3$* (Figure 15–58). The PI(3,4)P$_2$ and PI(3,4,5)P$_3$ then serve as docking sites for intracellular signaling proteins, bringing these proteins together into signaling complexes, which relay the signal into the cell from the cytosolic face of the plasma membrane.

It is important to distinguish this use of inositol phospholipids from their use we discussed earlier. We considered earlier how PI(4,5)P$_2$ is cleaved by PLC-β (in the case of G-protein-linked receptors) or PLC-γ (in the case of receptor tyrosine kinases) to generate soluble IP$_3$ and membrane-bound diacylglycerol. The IP$_3$ releases Ca^{2+} from the ER, while the diacylglycerol activates PKC (see Figures 15–58 and 15–35). By contrast, PI(3,4)P$_2$ and PI(3,4,5)P$_3$ are not cleaved

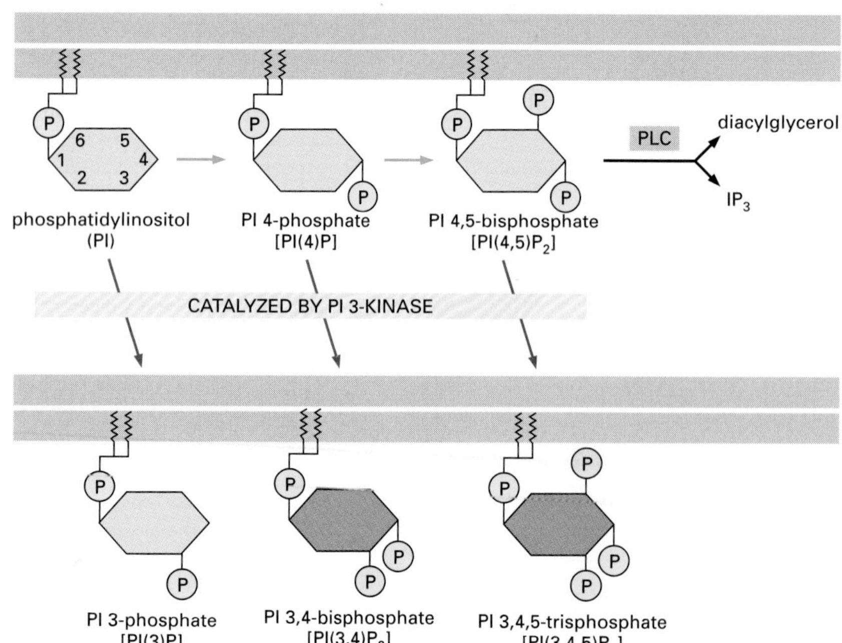

Figure 15–58 The generation of inositol phospholipid docking sites by PI 3-kinase. PI 3-kinase phosphorylates the inositol ring on carbon atom 3 to generate the inositol phospholipids shown at the bottom of the figure; the two lipids shown in *red* can serve as docking sites for signaling proteins with PH domains. The phosphorylations indicated by the *green* arrows are catalyzed by other inositol phospholipid kinases. As discussed earlier, phospholipase C (PLC-β or PLC-γ) can cleave PI(4,5)P$_2$ to produce the two small signaling molecules diacylglycerol and inositol 1,4,5-trisphosphate (IP$_3$).

by PLC. They remain in the plasma membrane until they are dephosphorylated by specific *inositol phospholipid phosphatases* that remove phosphate from the 3 position of the inositol ring. Mutations that inactivate one such phosphatase (called PTEN), and thereby prolong signaling by PI 3-kinase, promote the development of cancer, and they are found in many human cancers. The mutations result in prolonged cell survival, indicating that signaling through PI 3-kinase normally promotes cell survival, as well as cell growth.

There are various types of PI 3-kinases. The one that is activated by receptor tyrosine kinases consists of a catalytic and regulatory subunit. The regulatory subunit is an adaptor protein that binds to phosphotyrosines on activated receptor tyrosine kinases through its SH2 domains (see Figure 15–53). Another PI 3-kinase has a different regulatory subunit and is activated by the $\beta\gamma$ complex of a trimeric G protein when G-protein-linked receptors are activated by their extracellular ligand. The catalytic subunit, which is similar in both cases, also has a binding site for activated Ras, which allows Ras to directly stimulate PI 3-kinases.

Intracellular signaling proteins bind to the PI(3,4)P_2 and PI(3,4,5)P_3 that are produced by activated PI 3-kinase mainly through their **Pleckstrin homology (PH) domain**, first identified in the platelet protein Pleckstrin. PH domains are found in about 200 human proteins, including Sos (the GEF discussed earlier that activates Ras), and some atypical PKCs that do not depend on Ca^{2+} for their activation. The importance of these domains is illustrated dramatically by certain genetic immunodeficiency diseases in both humans and mice, where the PH domain in a cytoplasmic tyrosine kinase called **BTK** is inactivated by mutation. Normally, when antigen receptors on B lymphocytes (B cells) activate PI 3-kinase, the resulting inositol lipid docking sites recruit both BTK and PLC-γ to the cytoplasmic face of the plasma membrane. There, the two proteins interact: BTK phosphorylates and activates PLC-γ, which then cleaves PI(4,5)P_2 to generate IP_3 and diacylglycerol to relay the signal onward (Figure 15–59). Because the mutant BTK cannot bind to the lipid docking sites produced after receptor activation, the receptors cannot signal the B cells to proliferate or survive, resulting in a severe deficiency in antibody production.

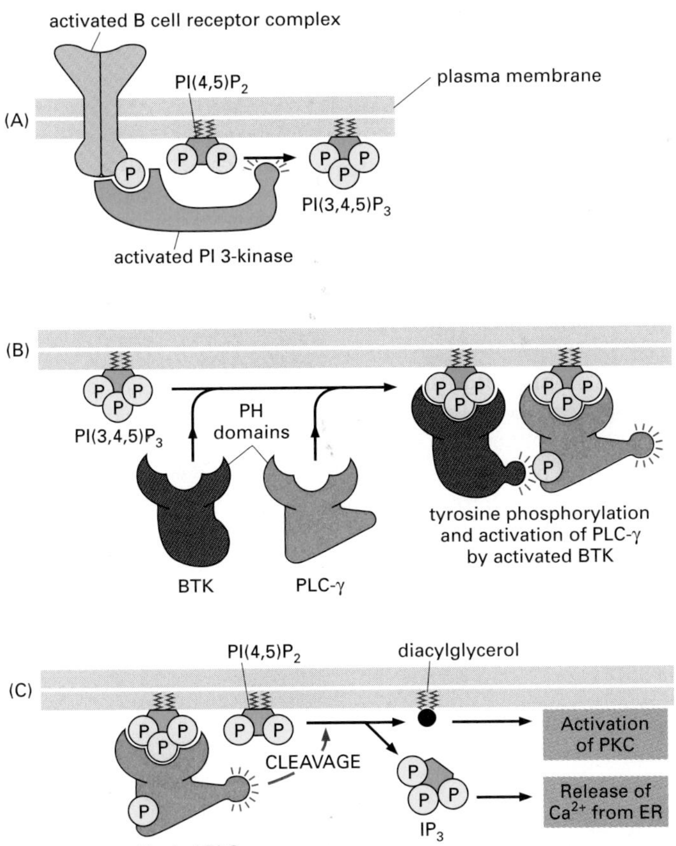

Figure 15–59 The recruitment of signaling proteins with PH domains to the plasma membrane during B cell activation. (A) PI 3-kinase binds to a phosphotyrosine on the activated B cell receptor complex and is thereby activated to phosphorylate the inositol phospholipid PI 4,5-bisphosphate [PI(4,5)P_2], generating PI 3,4,5-trisphosphate [PI(3,4,5)P_3]. (B) The PI 3,4,5-trisphosphate serves as a docking site for two signaling proteins with PH domains that then interact with each other. This causes the cytoplasmic tyrosine kinase BTK to phosphorylate and thereby activate PLC-γ at the membrane. (C) The activated PLC-γ then cleaves PI(4,5)P_2 to generate diacylglycerol and IP_3, which relay the signal onward.

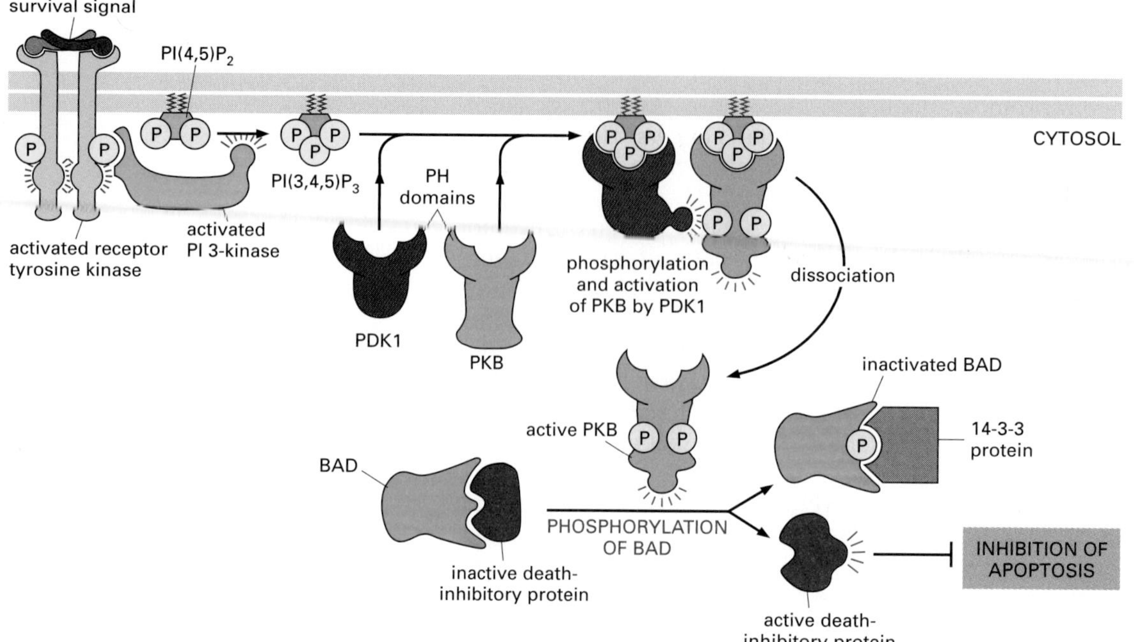

Figure 15–60 One way in which signaling through PI 3-kinase promotes cell survival. An extracellular survival signal activates a receptor tyrosine kinase, which recruits and activates PI 3-kinase. The PI 3-kinase produces PI(3,4,5)P$_3$ and PI(3,4)P$_2$ (not shown), both of which serve as docking sites for two serine/threonine kinases with PH domains—protein kinase B (PKB) and the phosphoinositol-dependent kinase PDK1. The binding of PKB to the inositol lipids alters its conformation so that the protein can be phosphorylated and activated by PDK1. The activated PKB now dissociates from the plasma membrane and phosphorylates the BAD protein, which, when unphosphorylated, holds one or more death-inhibitory proteins in an inactive state. Once phosphorylated, BAD releases the inhibitory proteins, which now can block programmed cell death (apoptosis) and thereby promote cell survival.

As shown, once phosphorylated, BAD binds to a ubiquitous cytosolic protein called *14-3-3*, which keeps BAD out of action. There are about 20 14-3-3 proteins in human cells, all of which bind to specific phosphoserine-containing motifs in proteins. The activation of other signaling pathways can also lead to BAD phosphorylation and the promotion of cell survival (not shown).

The PI 3-Kinase/Protein Kinase B Signaling Pathway Can Stimulate Cells to Survive and Grow

One way in which PI 3-kinase signals cells to survive is by indirectly activating **protein kinase B (PKB)** (also called **Akt**). This kinase contains a PH domain, which directs it to the plasma membrane when PI 3-kinase is activated there by an extracellular survival signal. After binding to PI(3,4,5)P$_3$ on the cytosolic face of the membrane, the PKB alters its conformation so that it can now be activated in a process that requires phosphorylation by a phosphatidylinositol-dependent protein kinase called **PDK1**, which is recruited to the membrane in the same way. Once activated, the PKB returns to the cytoplasm and phosphorylates a variety of target proteins. One of these, called *BAD,* is a protein that normally encourages cells to undergo programmed cell death, or apoptosis (mentioned earlier and discussed in detail in Chapter 17). By phosphorylating BAD, PKB inactivates it, thereby promoting cell survival (Figure 15–60). PKB also promotes cell survival by inhibiting other cell death activators, in some cases by inhibiting the transcription of the genes that encode them.

The pathways by which PI 3-kinase signals cells to grow (and increase their metabolism generally) are complex and still poorly understood. One way in which growth factors stimulate cell growth is by increasing the rate of protein synthesis through enhancing the efficiency with which ribosomes translate certain mRNAs into protein. A protein kinase called **S6 kinase** is part of one of the signaling pathways from PI 3-kinase to the ribosome. It phosphorylates and thereby activates the S6 subunit of ribosomes, which helps to increase the translation of a subset of mRNAs that encode ribosomal proteins and other components of the translational apparatus. The activation of S6 kinase is itself a complex process that depends on PDK1 and the phosphorylation of many sites on the protein. PDK1 may phosphorylate one of these sites in response to PI 3-kinase activation.

Figure 15–61 summarizes the five parallel intracellular signaling pathways we have discussed so far—one triggered by G-protein-linked receptors, two triggered by receptor tyrosine kinases, and two triggered by both kinds of receptors.

Tyrosine-Kinase-associated Receptors Depend on Cytoplasmic Tyrosine Kinases for Their Activity

Many cell-surface receptors depend on tyrosine phosphorylation for their activity and yet lack an obvious tyrosine kinase domain. These receptors act through

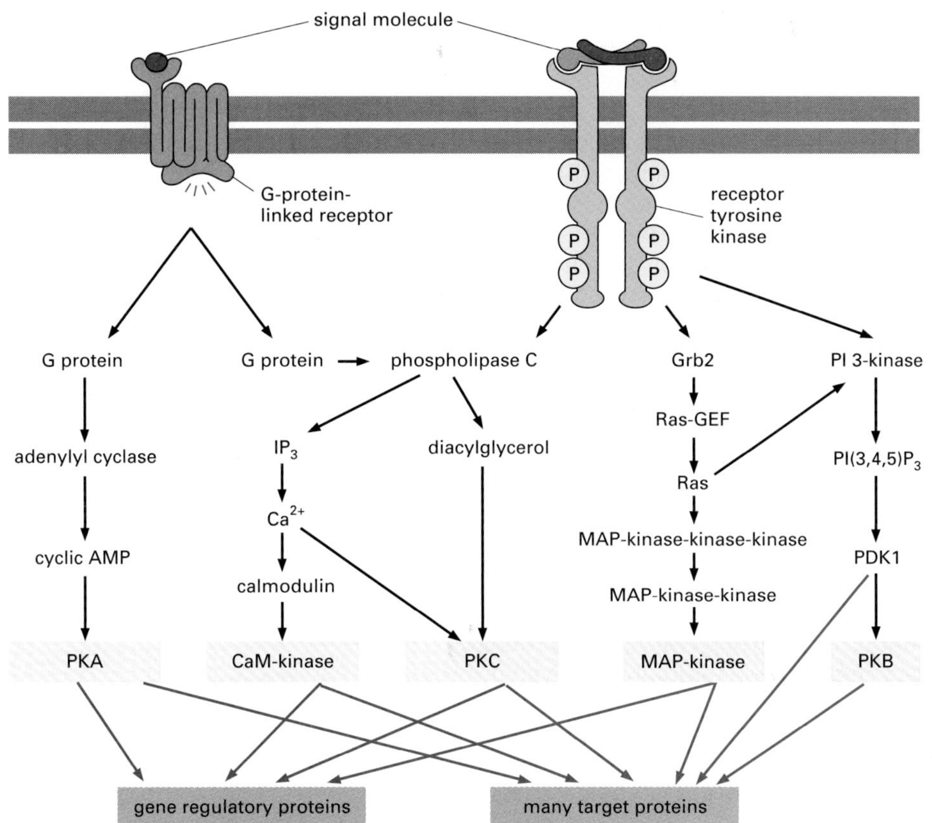

Figure 15–61 Five parallel intracellular signaling pathways activated by G-protein-linked receptors, receptor tyrosine kinases, or both. In this schematic example, the five kinases (shaded *yellow*) at the end of each pathway phosphorylate target proteins (shaded *red*), some of which are phosphorylated by more than one of the kinases. The specific phospholipase C activated by the two types of receptors is different: G-protein-linked receptors activate PLC-β, whereas receptor tyrosine kinases activate PLC-γ (not shown).

cytoplasmic tyrosine kinases, which are associated with the receptors and phosphorylate various target proteins, often including the receptors themselves, when the receptors bind their ligand. The receptors thus function in much the same way as receptor tyrosine kinases, except that their kinase domain is encoded by a separate gene and is noncovalently associated with the receptor polypeptide chain. As with receptor tyrosine kinases, these receptors must oligomerize to function (Figure 15–62).

Many of these receptors depend on members of the largest family of mammalian cytoplasmic tyrosine kinases, the **Src family** of protein kinases (see Figure 3–68). This family includes the following members: *Src, Yes, Fgr, Fyn, Lck, Lyn, Hck,* and *Blk*. These protein kinases all contain SH2 and SH3 domains and are located on the cytoplasmic side of the plasma membrane, held there partly by their interaction with transmembrane receptor proteins and partly by covalently attached lipid chains. Different family members are associated with different receptors and phosphorylate overlapping but distinct sets of target proteins. Lyn, Fyn, and Lck, for example, are each associated with different sets of receptors in lymphocytes. In each case the kinase is activated when an extracellular ligand binds to the appropriate receptor protein. Src itself, as well as several other family members, can also bind to activated receptor tyrosine kinases; in these cases, the receptor and cytoplasmic kinases mutually stimulate each other's catalytic activity, thereby strengthening and prolonging the signal.

Another type of cytoplasmic tyrosine kinase associates with *integrins,* the main family of receptors that cells use to bind to the extracellular matrix (discussed in Chapter 19). The binding of matrix components to integrins can activate intracellular signaling pathways that influence the behavior of the cell. When integrins cluster at sites of matrix contact, they help trigger the assembly of cell–matrix junctions called *focal adhesions.* Among the many proteins recruited into these junctions is the cytoplasmic tyrosine kinase called **focal adhesion kinase (FAK)**, which binds to the cytosolic tail of one of the integrin subunits with the assistance of other cytoskeletal protein. The clustered FAK molecules cross-phosphorylate each other, creating phosphotyrosine docking sites where the Src kinase can bind. Src and FAK now phosphorylate each other

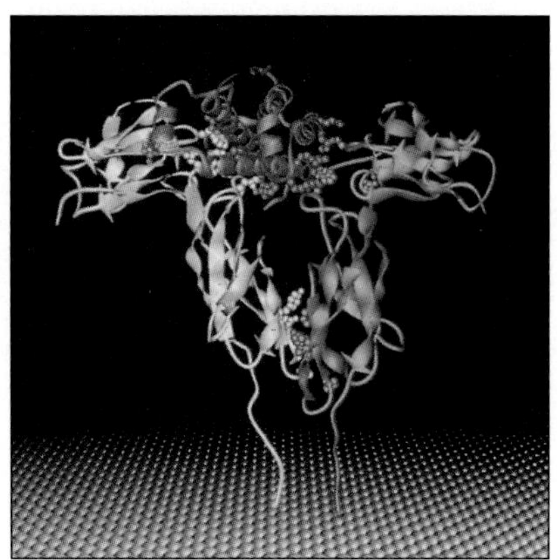

Figure 15–62 The three-dimensional structure of human growth hormone bound to its receptor. The hormone (*red*) has cross-linked two identical receptors (one shown in *green* and the other in *blue*). Hormone binding activates cytoplasmic tyrosine kinases that are tightly bound to the cytosolic tails of the receptors (not shown). The structures shown were determined by x-ray crystallographic studies of complexes formed between the hormone and extracellular receptor domains produced by recombinant DNA technology. It was entirely unexpected that a monomeric ligand such as growth hormone would cross-link its receptors, as it requires that the two identical receptors recognize different parts of the hormone. As mentioned earlier, EGF does the same thing. (From A.M. deVos, M. Ultsch, and A.A. Kossiakoff, *Science* 255:306–312, 1992. © AAAS.)

and other proteins that assemble in the junction, including many of the signaling proteins used by receptor tyrosine kinases. In this way, the two kinases signal to the cell that it has adhered to a suitable substratum, where the cell can now survive, grow, divide, migrate, and so on. Mice deficient in FAK die early in development, and their cells do not migrate normally in a culture dish.

Cytokine receptors are the subfamily of enzyme-linked receptors that we discuss next. They constitute the largest and most diverse class of receptors that rely on cytoplasmic kinases to relay signals into the cell. They include receptors for many kinds of local mediators (collectively called *cytokines*), as well as receptors for some hormones, such as growth hormone (see Figure 15–62) and prolactin. As we discuss next, these receptors are stably associated with a class of cytoplasmic tyrosine kinases called *Jaks,* which activate latent gene regulatory proteins called *STATs*. The STAT proteins are normally inactive, being located at the cell surface; cytokine or hormone binding causes them to migrate to the nucleus and activate gene transcription.

Cytokine Receptors Activate the Jak–STAT Signaling Pathway, Providing a Fast Track to the Nucleus

Many intracellular signaling pathways lead from cell-surface receptors to the nucleus, where they alter gene transcription. The **Jak-STAT signaling pathway**, however, provides one of the most direct routes. It was initially discovered in studies on the effects of *interferons,* which are cytokines secreted by cells (especially white blood cells) in response to viral infection. Interferons bind to receptors on noninfected neighboring cells and induce the cells to produce proteins that increase their resistance to viral infection. When activated, interferon receptors activate a novel class of cytoplasmic tyrosine kinases called **Janus kinases (Jaks)** (after the two-faced Roman god). The Jaks then phosphorylate and activate a set of latent gene regulatory proteins called **STATs** (**s**ignal **t**ransducers and **a**ctivators of **t**ranscription), which move into the nucleus and stimulate the transcription of specific genes. More than 30 cytokines and hormones activate the Jak–STAT pathway by binding to cytokine receptors, some of which are listed in Table 15–5.

All STATs also have an SH2 domain that enables them to dock onto specific phosphotyrosines on some activated receptor tyrosine kinase receptors. These receptors can directly activate the bound STAT, independently of Jaks. In fact, the nematode *C. elegans* uses STATs for signaling but does not make any Jaks or cytokine receptors, suggesting that STATs evolved before Jaks and cytokine receptors.

Cytokine receptors are composed of two or more polypeptide chains. Some cytokine receptor chains are specific to a particular cytokine receptor, while others are shared among several such receptors. All cytokine receptors, however, are associated with one or more Jaks. There are four known Jaks—Jak1, Jak2, Jak3,

TABLE 15–5 Some Signaling Proteins That Act Through Cytokine Receptors and the Jak–STAT Signaling Pathway

SIGNALING LIGAND	RECEPTOR-ASSOCIATED JAKS	STATS ACTIVATED	SOME RESPONSES
γ-interferon	Jak1 and Jak2	STAT1	activates macrophages; increases MHC protein expression
α-interferon	Tyk2 and Jak2	STAT1 and STAT2	increases cell resistance to viral infection
Erythropoietin	Jak2	STAT5	stimulates production of erythrocytes
Prolactin	Jak1 and Jak2	STAT5	stimulates milk production
Growth hormone	Jak2	STAT1 and STAT5	stimulates growth by inducing IGF-1 production
GM-CSF	Jak2	STAT5	stimulates production of granulocytes and macrophages
IL-3	Jak2	STAT5	stimulates early blood cell production

and Tyk2—and each is associated with particular cytokine receptors. The receptors for α-interferon, for example, are associated with Jak1 and Tyk2, whereas the receptors for γ-interferon are associated with Jak1 and Jak2 (see Table 15–5). As expected, mice that lack Jak1 do not respond to either of these interferons. The receptor for the hormone *erythropoietin*, which stimulates erythrocyte precursor cells to survive, proliferate, and differentiate, is associated with only Jak2. In Jak2-deficient mice, erythrocyte development fails, and the mice die early in development.

Cytokine binding either induces the receptor chains to oligomerize or reorients the chains in a preformed oligomer. In either case, the binding brings the associated Jaks close enough together for them to cross-phosphorylate each other, thereby increasing the activity of their tyrosine kinase domains. The Jaks then phosphorylate tyrosines on the cytokine receptors, creating phosphotyrosine docking sites for STATs and other signaling proteins.

There are seven known STATs, each with an SH2 domain that performs two functions. First, it mediates the binding of the STAT protein to a phosphotyrosine docking site on an activated cytokine receptor (or receptor tyrosine kinase); once bound, the Jaks phosphorylate the STAT on tyrosines, causing it to dissociate from the receptor. Second, the SH2 domain on the released STAT now mediates its binding to a phosphotyrosine on another STAT molecule, forming either a STAT homodimer or heterodimer. The STAT dimer then moves into the nucleus, where, in combination with other gene regulatory proteins, it binds to a specific DNA response element in various genes and stimulates their transcription (Figure 15–63). In response to the hormone prolactin, for example, which stimulates breast cells to produce milk, activated STAT5 stimulates the transcription of genes that encode milk proteins.

Cytokine receptors activate the appropriate STAT proteins because the SH2 domain of these STATs recognizes only the specific phosphotyrosine docking sites on these receptors. Activated receptors for α-interferon, for example, recruit both STAT1 and STAT2, whereas activated receptors for γ-interferon recruit only STAT1. If the SH2 domain of the α-interferon receptor is replaced with the SH2 domain of the γ-interferon receptor, the activated hybrid receptor recruits both STAT1 and STAT2, just like the α-interferon receptor itself.

Figure 15–63 The Jak–STAT signaling pathway activated by α-interferon. The binding of interferon either causes two separate receptor polypeptide chains to dimerize (as shown) or reorients the receptor chains in a preformed dimer. In either case, the associated Jaks are brought together so that they can cross-phosphorylate each other on tyrosines, starting the signaling process. The two different receptor chains are associated with different Jaks (Tyk2 and Jak1), and they recruit different STATs (STAT1 and STAT2). The STATs dissociate from the receptors and form heterodimers when activated by phosphorylation, and they bind to specific DNA sequences in the cell nucleus, where, together with other gene regulatory proteins, they induce the transcription of adjacent genes.

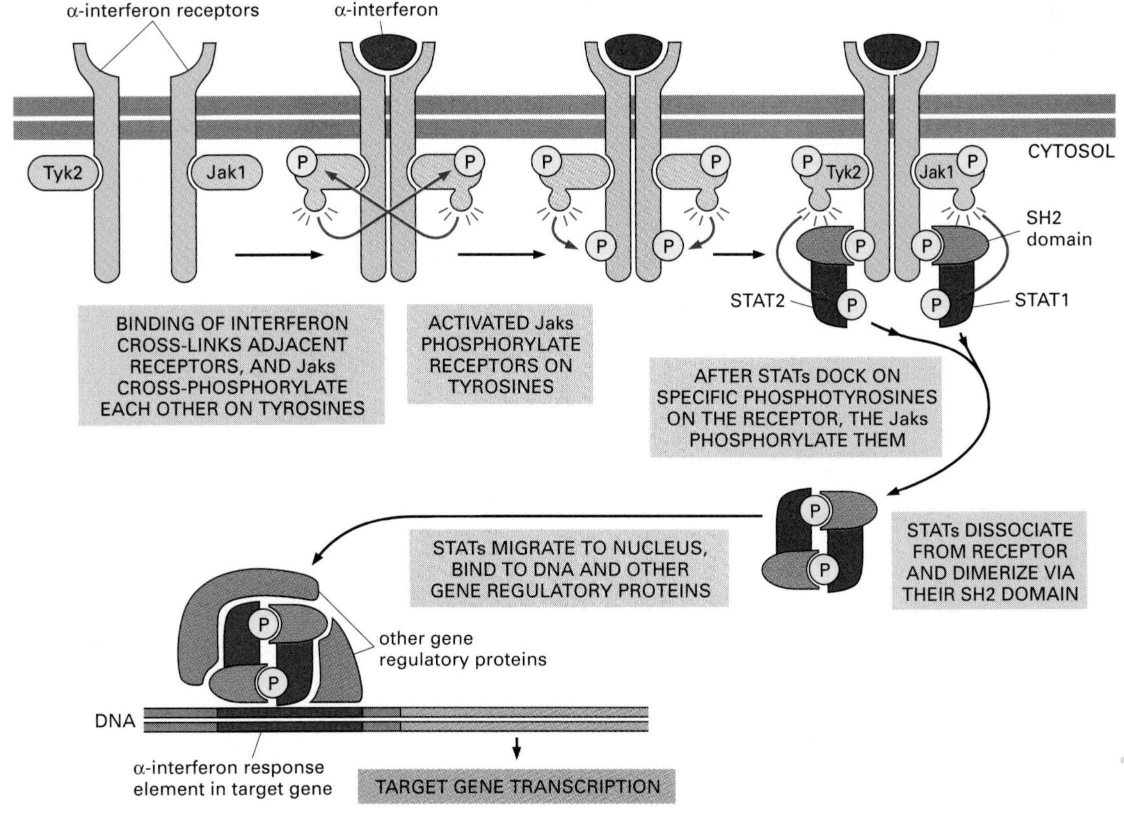

α-interferon receptors
α-interferon
Tyk2
Jak1
P
CYTOSOL
SH2 domain
STAT2
STAT1
BINDING OF INTERFERON CROSS-LINKS ADJACENT RECEPTORS, AND Jaks CROSS-PHOSPHORYLATE EACH OTHER ON TYROSINES
ACTIVATED Jaks PHOSPHORYLATE RECEPTORS ON TYROSINES
AFTER STATs DOCK ON SPECIFIC PHOSPHOTYROSINES ON THE RECEPTOR, THE Jaks PHOSPHORYLATE THEM
STATs DISSOCIATE FROM RECEPTOR AND DIMERIZE VIA THEIR SH2 DOMAIN
STATs MIGRATE TO NUCLEUS, BIND TO DNA AND OTHER GENE REGULATORY PROTEINS
other gene regulatory proteins
DNA
α-interferon response element in target gene
TARGET GENE TRANSCRIPTION

The responses mediated by STATs are often regulated by negative feedback. In addition to activating genes that encode proteins mediating the cytokine-induced response, the STAT dimers may also activate genes that encode inhibitory proteins. In some cases, the inhibitor binds to both the activated cytokine receptors and STAT proteins, which blocks further STAT activation and helps to shut off the response; in other cases, the inhibitor achieves the same result by blocking Jak function.

Such negative feedback mechanisms, however, are not enough on their own to turn off the response. The activated Jaks and STATs also have to be inactivated by dephosphorylation of their phosphotyrosines. As in all signaling pathways that use tyrosine phosphorylation, the dephosphorylation is performed by *protein tyrosine phosphatases,* which are as important in the signaling process as the protein tyrosine kinases that add the phosphates.

Some Protein Tyrosine Phosphatases May Act as Cell-Surface Receptors

As discussed earlier, only a small number of serine/threonine phosphatase catalytic subunits are responsible for removing phosphate groups from phosphorylated serines and threonines on proteins. By contrast, there are about 30 **protein tyrosine phosphatases (PTPs)** encoded in the human genome. Like tyrosine kinases, they occur in both cytoplasmic and transmembrane forms, none of which are structurally related to serine/threonine protein phosphatases. Individual protein tyrosine phosphatases display exquisite specificity for their substrates, removing phosphate groups from only selected phosphotyrosines on a subset of tyrosine-phosphorylated proteins. Together, these phosphatases ensure that tyrosine phosphorylations are short-lived and that the level of tyrosine phosphorylation in resting cells is very low. They do not, however, simply continuously reverse the effects of protein tyrosine kinases; they are regulated to act only at the appropriate time in a signaling response or in the cell-division cycle (discussed in Chapter 17).

Two cytoplasmic tyrosine phosphatases in vertebrates have SH2 domains and are therefore called **SHP-1** and **SHP-2** (Figure 15–64). SHP-1 helps to terminate some cytokine responses in blood cells by dephosphorylating activated Jaks: mutant erythropoietin receptors that cannot recruit SHP-1, for example, activate Jak2 for much longer than normal. Moreover, SHP-1-deficient mice have abnormalities in almost all blood cell lineages, emphasizing the importance of SHP-1 in blood cell development. Both SHP-1 and SHP-2 also help terminate responses mediated by some receptor tyrosine kinases.

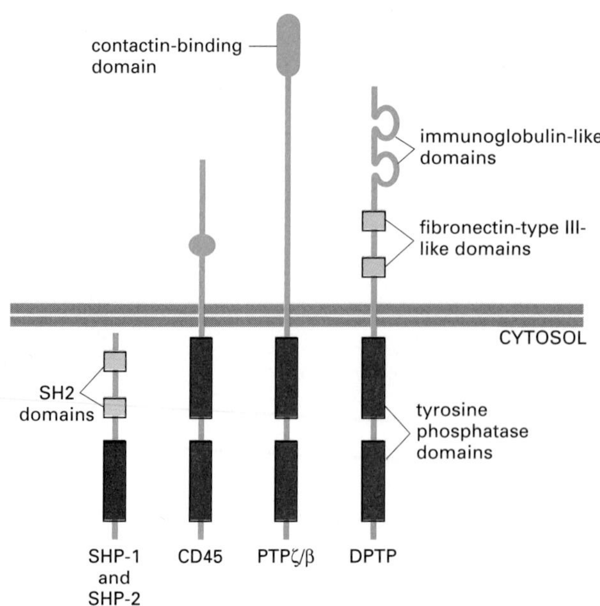

Figure 15–64 Some protein tyrosine phosphatases. The cytoplasmic tyrosine phosphatases SHP-1 and SHP-2 have similar structures, with two SH2 domains. The three transmembrane receptorlike tyrosine phosphatases have two tandemly arranged intracellular phosphatase domains, with the one closest to the membrane providing most or all of the catalytic activity. DPTP is a *Drosophila* protein; the others in the figure are mammalian proteins.

There are a large number of transmembrane protein tyrosine phosphatases, but the functions of most of them are unknown. At least some are thought to function as receptors; as this has not been directly demonstrated, however, they are referred to as **receptorlike tyrosine phosphatases**. They all have a single transmembrane segment and usually possess two tyrosine phosphatase domains on the cytosolic side of the plasma membrane. An important example is the **CD45 protein** (see Figure 15–64), which is found on the surface of all white blood cells and has an essential role in the activation of both T and B lymphocytes by foreign antigens. The ligand that is presumed to bind to the extracellular domain of the CD45 protein has not been identified. However, the role of CD45 in signal transduction has been studied by using recombinant DNA techniques to construct a hybrid protein with an extracellular EGF-binding domain and intracellular CD45 tyrosine phosphatase domains. The surprising result is that EGF binding seems to inactivate the phosphatase activity of the hybrid protein rather than activating it.

This finding raises the possibility that some receptor tyrosine kinases and receptor tyrosine phosphatases may collaborate when they bind their respective cell-surface-bound ligands—with the kinases adding more phosphates and the phosphatase removing fewer—to maximally stimulate the tyrosine phosphorylation of selected intracellular signaling proteins. The significance of ligand-induced inhibition of CD45 phosphatase is still uncertain, however, and it seems unlikely to be the whole story; CD45 requires its phosphatase activity to function in lymphocyte activation.

Some receptorlike tyrosine phosphatases display features of cell-adhesion proteins and can even mediate homophilic cell–cell binding in cell adhesion assays (see Figure 19–26). In the developing nervous system, for example, they may have an important role in guiding the growing tips of developing nerve cell axons to their targets. In *Drosophila,* the genes encoding several receptorlike tyrosine phosphatases are expressed exclusively in the nervous system, and when some of them are inactivated by mutation, the axons of certain developing neurons fail to find their way to their normal targets. In some cases at least, the phosphatase activity of the protein is required to counteract the action of a cytoplasmic tyrosine kinase for normal axon guidance.

Transmembrane tyrosine phosphatases can also serve as signaling ligands that activate receptors on a neighboring cell. An example is the *protein tyrosine phosphatase ζ/β* (see Figure 15–64), which is expressed on the surface of certain glial cells in the mammalian brain. It binds to a receptor protein (called *contactin*) on developing nerve cells, stimulating the cells to extend long processes. It is possible that the phosphatase also conveys a signal to the glial cell in this interaction, but such bidirectional signaling has not been directly demonstrated for transmembrane tyrosine phosphatases.

Having discussed the crucial role of tyrosine phosphorylation and dephosphorylation in the intracellular signaling pathways activated by many enzyme-linked receptors, we now turn to a class of enzyme-linked receptors that rely entirely on serine/threonine phosphorylation. These transmembrane serine/threonine kinases activate an even more direct signaling pathway to the nucleus than does the Jak–STAT pathway discussed earlier. They directly phosphorylate latent gene regulatory proteins called *Smads,* which then migrate into the nucleus to activate gene transcription.

Signal Proteins of the TGF-β Superfamily Act Through Receptor Serine/Threonine Kinases and Smads

The **transforming growth factor-β (TGF-β) superfamily** consists of a large number of structurally related, secreted, dimeric proteins. They act either as hormones or, more commonly, as local mediators to regulate a wide range of biological functions in all animals. During development, they regulate pattern formation and influence various cell behaviors, including proliferation, differentiation, extracellular matrix production, and cell death. In adults, they are involved in tissue repair and in immune regulation, as well as in many other processes. The superfamily includes the *TGF-βs* themselves, the *activins*, and the *bone morphogenetic proteins (BMPs)*. The BMPs constitute the largest family.

All of these proteins act through enzyme-linked receptors that are single-pass transmembrane proteins with a serine/threonine kinase domain on the cytosolic side of the plasma membrane. There are two classes of these **receptor serine/threonine kinases**—*type I* and *type II*—which are structurally similar. Each member of the TGF-β superfamily binds to a characteristic combination of type-I and type-II receptors, both of which are required for signaling. Typically, the ligand first binds to and activates a type-II receptor homodimer, which recruits, phosphorylates, and activates a type-I receptor homodimer, forming an active tetrameric receptor complex.

Once activated, the receptor complex uses a strategy for rapidly relaying the signal to the nucleus that is very similar to the Jak–STAT strategy used by cytokine receptors. The route to the nucleus, however, is even more direct. The type-I receptor directly binds and phosphorylates a latent gene regulatory protein of the **Smad family** (named after the first two identified, Sma in *C. elegans* and Mad in *Drosophila*). Activated TGF-β receptors and activin receptors phosphorylate Smad2 or Smad3, while activated BMP receptors phosphorylate Smad1, Smad5, or Smad8. Once one of these Smads has been phosphorylated, it dissociates from the receptor and binds to Smad4, which can form a complex with any of the above five *receptor-activated Smads*. The Smad complex then moves into the nucleus, where it associates with other gene regulatory proteins, binds to specific sites in DNA, and activates a particular set of target genes (Figure 15–65).

Some TGF-β family members serve as graded morphogens during development, inducing different responses in a developing cell depending on their concentration (discussed in Chapter 21). The different responses can be reproduced by experimentally altering the amount of active Smad complexes in the nucleus, suggesting that the level of these complexes may provide a direct readout of the level of receptor activation. If the DNA-binding sites in different target genes have different affinities for the complexes, then the particular genes activated would reflect the cell's position in the concentration gradient of the morphogen.

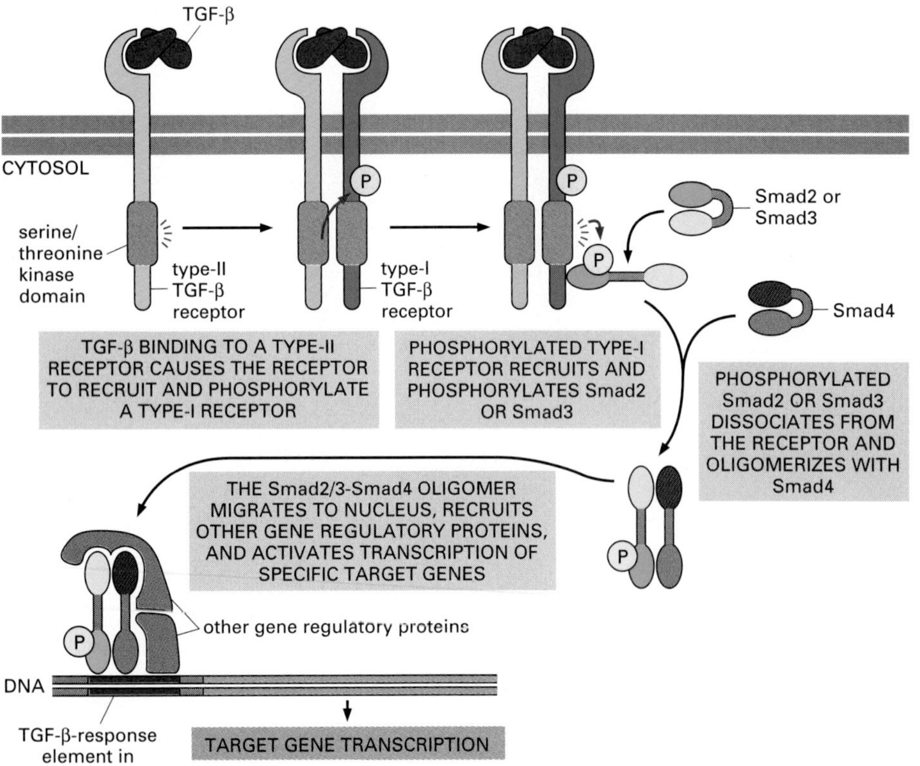

Figure 15–65 A model for the Smad-dependent signaling pathway activated by TGF-β. Note that TGF-β is a dimer and that Smads open up to expose a dimerization surface when they are phosphorylated. Several features of the pathway have been omitted for simplicity, including the following: (1) The type-I and type-II receptor proteins are both thought to be dimers. (2) The type-I receptors are normally associated with an inhibitory protein, which dissociates when the type-I receptor is phosphorylated by a type-II receptor. (3) The individual Smads are thought to be trimers. (4) An anchoring protein (called SARA, for *Smad anchor for receptor activation*) helps to recruit Smad2 or Smad3 to the activated type I receptor by binding to the receptor, to the Smad, and to inositol phospholipid molecules in the plasma membrane. (5) The function of certain Smads is regulated by enzymes that enhance their ubiquitylation and thereby their degradation.

As with the Jak–STAT pathway, the Smad pathway is also often regulated by feedback inhibition. Among the target genes activated by Smad complexes are those that encode *inhibitory Smads,* including Smad6 and Smad7. These Smads act as decoys. They bind to activated type-I receptors and prevent other Smads from binding there. This blocks the formation of active Smad complexes and shuts off the response to the TGF-β family ligand. Other types of extracellular ligands can also stimulate the production of inhibitory Smads to antagonize signaling by a TGF-β ligand; γ-interferon, for example, activates the Jak–STAT pathway, and the resulting activated STAT dimers induce the production of Smad7, which inhibits signaling by TGF-β.

In addition to these intracellular inhibitors, a number of secreted extracellular inhibitory proteins can also neutralize signaling mediated by TGF-β family members. They directly bind to the signal molecules and prevent them from activating their receptors on target cells. *Noggin* and *chordin,* for example, inhibit BMPs, and *follistatin* inhibits activins. Noggin and chordin help to induce the development of the vertebrate nervous system by preventing BMPs from inhibiting this development (discussed in Chapter 21). The TGF-β family members, as well as some of their inhibitors, are usually secreted as inactive precursors that are subsequently activated by proteolytic cleavage.

We turn now to enzyme-linked receptors that are neither kinases nor associated with kinases. We saw earlier that nitric oxide is widely used as a signaling molecule, diffusing through the plasma membrane of a target cell and stimulating a cytoplasmic guanylyl cyclase to produce the intracellular mediator cyclic GMP. The receptors we now consider are transmembrane proteins with guanylyl cyclase activity.

Receptor Guanylyl Cyclases Generate Cyclic GMP Directly

Receptor guanylyl cyclases are single-pass transmembrane proteins with an extracellular binding site for a signal molecule and an intracellular guanylyl cyclase catalytic domain. The binding of the signal molecule activates the cyclase domain to produce cyclic GMP, which in turn binds to and activates a *cyclic GMP-dependent protein kinase (PKG),* which phosphorylates specific proteins on serine or threonine. Thus, receptor guanylyl cyclases use cyclic GMP as an intracellular mediator in the same way that some G-protein-linked receptors use cyclic AMP, except that the linkage between ligand binding and cyclase activity is a direct one.

Among the signal molecules that use receptor guanylyl cyclase receptors are the *natriuretic peptides (NPs),* a family of structurally related secreted signal peptides that regulate salt and water balance and dilate blood vessels. There are several types of NPs, including *atrial natriuretic peptide (ANP)* and *brain natriuretic peptide (BNP).* Muscle cells in the atrium of the heart secrete ANP when blood pressure rises. The ANP stimulates the kidneys to secrete Na^+ and water and induces the smooth muscle cells in blood vessels walls to relax. Both of these effects tend to lower the blood pressure. When gene targeting is used to inactivate the ANP receptor guanylyl cyclase in mice, the mice have chronically elevated blood pressure, resulting in progressive heart enlargement.

An increasing number of receptor guanylyl cyclases are being discovered, but in most cases they are orphan receptors, where the ligand that normally activates them is unknown. The genome of the nematode *C. elegans,* for example, encodes 26 of these receptors. Most of those that have been studied are expressed in specific subsets of sensory neurons, suggesting that they may be involved in detecting particular molecules in the worm's environment. Some of the orphan receptors in mammals are found in sensory neurons in the part of the nose involved in detecting pheromones.

All the signaling pathways activated by G-protein-linked and enzyme-linked receptors we have discussed so far depend on serine/threonine-specific protein kinases, tyrosine-specific protein kinases, or both. These kinases are all structurally related, as reviewed in Figure 15–66. Some enzyme-linked receptors, however, depend on an entirely unrelated type of protein kinase, as we now discuss.

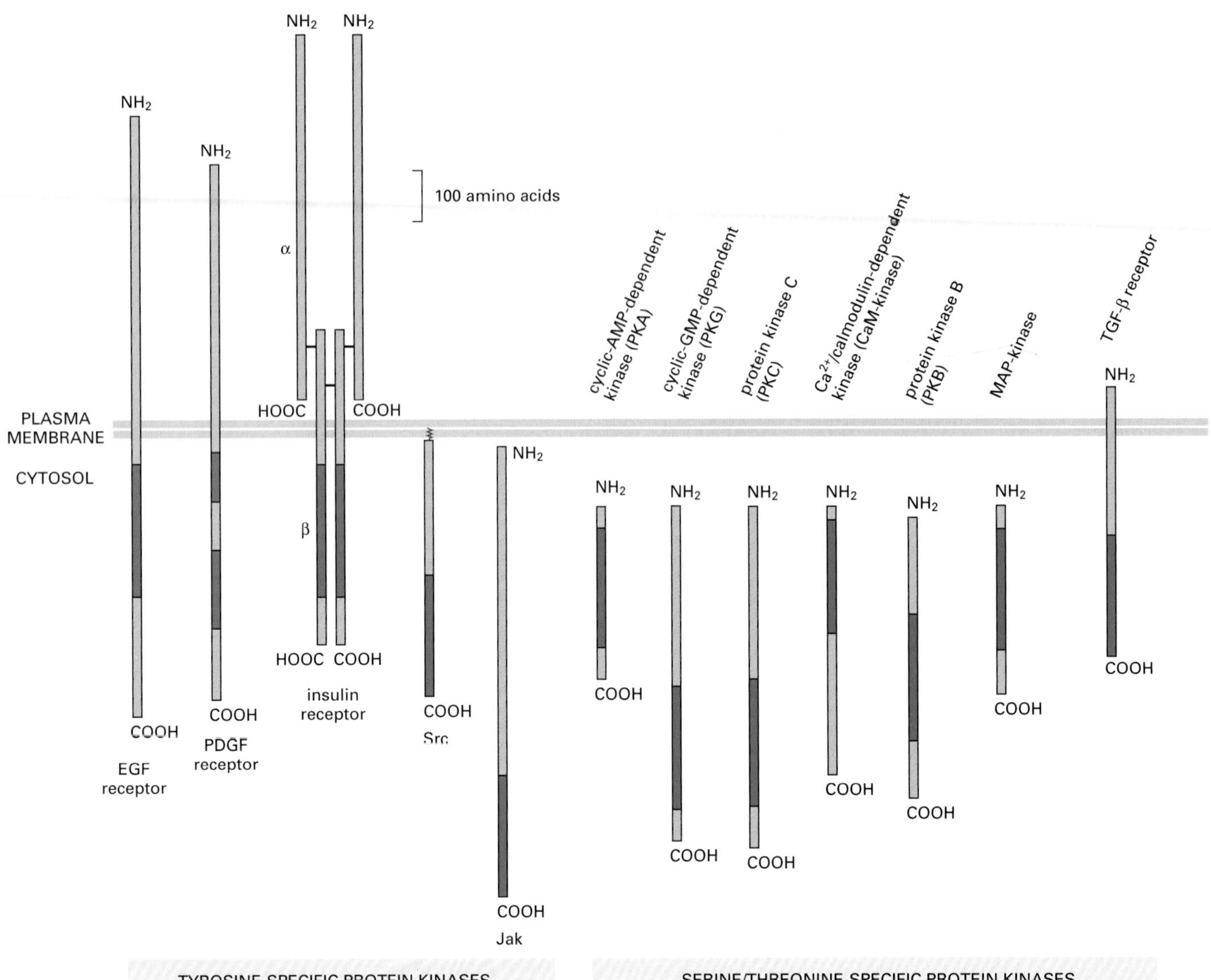

Figure 15–66 Some of the protein kinases discussed in this chapter. The size and location of their catalytic domains *(dark green)* are shown. In each case the catalytic domain is about 250 amino acids long. These domains are all similar in amino acid sequence, suggesting that they have all evolved from a common primordial kinase (see also Figure 3–65). Note that all of the tyrosine kinases shown are bound to the plasma membrane (Jaks are bound by their association with cytokine receptors), whereas most of the serine/threonine kinases are in the cytosol.

Bacterial Chemotaxis Depends on a Two-Component Signaling Pathway Activated by Histidine–Kinase-associated Receptors

As pointed out earlier, many of the mechanisms involved in chemical signaling between cells in multicellular animals are thought to have evolved from mechanisms used by unicellular organisms to respond to chemical changes in their environment. In fact, some of the same intracellular mediators, such as cyclic nucleotides and Ca^{2+}, are used by both types of organisms. Among the best-studied reactions of unicellular organisms to extracellular signals are their chemotactic responses, in which cell movement is oriented toward or away from a source of some chemical in the environment. We conclude this section on enzyme-linked receptors with a brief account of bacterial chemotaxis, which depends on a **two-component signaling pathway**, involving **histidine-kinase-associated receptors**. The same type of signaling pathway is used by yeasts and plants, although apparently not by animals.

Motile bacteria will swim toward higher concentrations of nutrients *(attractants)*, such as sugars, amino acids, and small peptides, and away from higher concentrations of various noxious chemicals *(repellents)*. They swim by means of flagella, each of which is attached by a short, flexible hook at its base to a small protein disc embedded in the bacterial membrane. This disc is part of a tiny motor that uses the energy stored in the transmembrane H^+ gradient to rotate rapidly and turn the helical flagellum (Figure 15–67). Because the flagella on the bacterial surface have an intrinsic "handedness," different directions of rotation have different effects on movement. Counterclockwise rotation allows all the

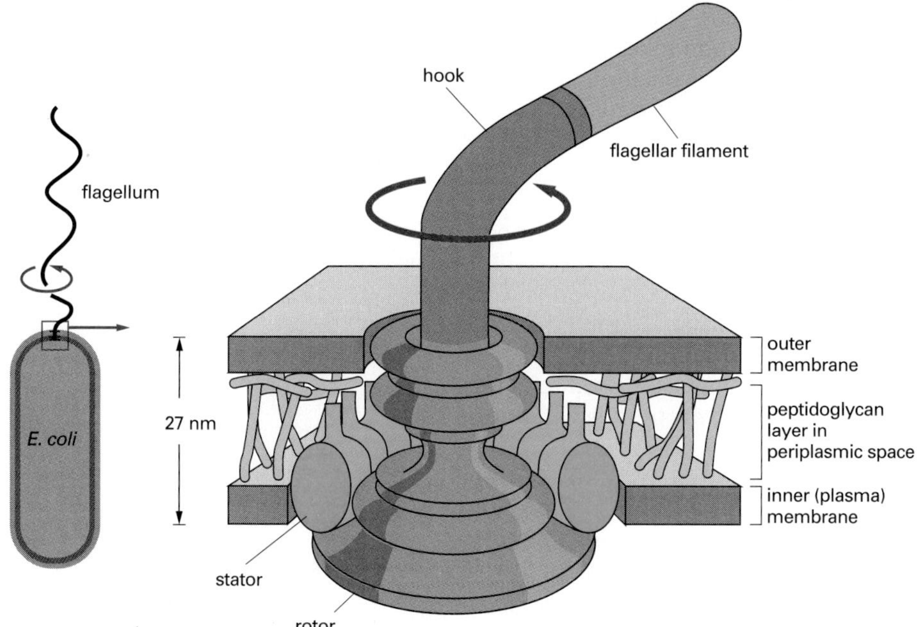

Figure 15–67 The bacterial flagellar motor. The flagellum is linked to a flexible hook. The hook is attached to a series of protein rings (shown in *red*), which are embedded in the outer and inner (plasma) membranes. The rings form a rotor, which rotates with the flagellum at more than 100 revolutions per second. The rotation is driven by a flow of protons through an outer ring of proteins (see Figure 14–17), the stator, which also contains the proteins responsible for switching the direction of rotation. (Based on data from T. Kubori et al., *J. Mol. Biol.* 226:433–446, 1992, and N.R. Francis et al., *Proc. Natl. Acad. Sci. USA* 89:6304–6308, 1992.)

flagella to draw together into a coherent bundle, so that the bacterium swims uniformly in one direction. Clockwise rotation causes them to fly apart, so that the bacterium tumbles chaotically without moving forward (Figure 15–68). In the absence of any environmental stimulus, the direction of rotation of the disc reverses every few seconds, producing a characteristic pattern of movement in which smooth swimming in a straight line is interrupted by abrupt, random changes in direction caused by tumbling.

The normal swimming behavior of bacteria is modified by chemotactic attractants or repellents, which bind to specific receptor proteins and affect the frequency of tumbling by increasing or decreasing the time that elapses between successive changes in direction of flagellar rotation. When bacteria are swimming in a favorable direction (toward a higher concentration of an attractant or away from a higher concentration of a repellent), they tumble less frequently than when they are swimming in an unfavorable direction (or when no gradient is present). Since the periods of smooth swimming are longer when a bacterium is traveling in a favorable direction, it will gradually progress in that direction—toward an attractant or away from a repellent.

These responses are mediated by histidine-kinase-associated **chemotaxis receptors**, which typically are dimeric transmembrane proteins that bind specific attractants and repellents on the outside of the plasma membrane. The cytoplasmic tails of the receptors are stably associated with an adaptor protein *CheW* and a histidine kinase *CheA*, which help to couple the receptors to the flagellar motor. Repellent binding activates the receptors, whereas attractant binding inactivates them; a single receptor can bind either type of molecule, with opposite consequences. The binding of a repellent to the receptor activates *CheA*, which phosphorylates itself on a histidine and almost immediately transfers the phosphate to an aspartic acid on a messenger protein *CheY*. The phosphorylated *CheY* dissociates from the receptor, diffuses through the cytosol, binds to the flagellar motor, and causes the motor to rotate clockwise, so that the bacterium tumbles. CheY has intrinsic phosphatase activity and dephosphorylates itself in a process that is greatly accelerated by the *CheZ* protein (Figure 15–69).

The response to an increase in the concentration of an attractant or repellent is only transient, even if the higher level of ligand is maintained, as the

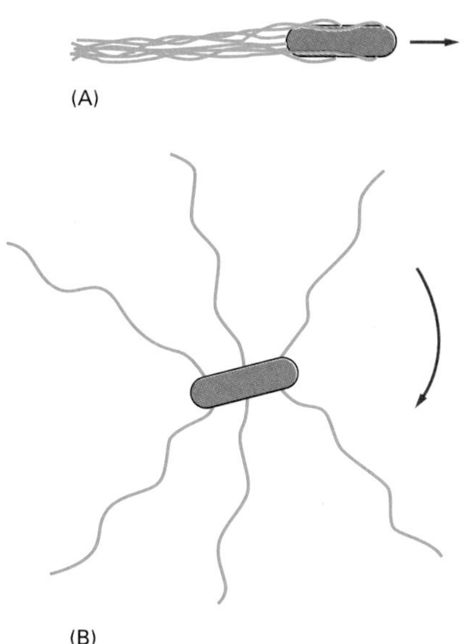

Figure 15–68 Positions of the flagella on *E. coli* during swimming. (A) When the flagella rotate counterclockwise, they are drawn together into a single bundle, which acts as a propeller to produce smooth swimming. (B) When the flagella rotate clockwise, they fly apart and produce tumbling.

bacteria *desensitize,* or *adapt,* to the increased stimulus. Whereas the initial effect on tumbling occurs in less than a second, adaptation takes minutes. The adaptation is a crucial part of the response, as it enables the bacteria to respond to *changes* in concentration of ligand rather than to steady-state levels. It is mediated by the covalent methylation (catalyzed by a methyl transferase) and demethylation (catalyzed by a methylase) of the chemotaxis receptors, which change their responsiveness to ligand binding when methylated.

All of the genes and proteins involved in this highly adaptive behavior have now been identified. It therefore seems likely that bacterial chemotaxis will be the first signaling system to be completely understood in molecular terms. Even in this relatively simple signaling network, computer-based simulations are required to comprehend how the system works as an integrated network. Cell signaling pathways will provide an especially rich area of investigation for a new generation of computational biologists, as their network properties will not be understandable without powerful computational tools.

There are some cell-surface receptor proteins that do not fit into the three major classes we have discussed thus far—ion-channel-linked, G-protein-linked, and enzyme-linked. In the next section, we consider cell-surface receptors that activate signaling pathways that depend on proteolysis. These pathways have especially important roles in animal development.

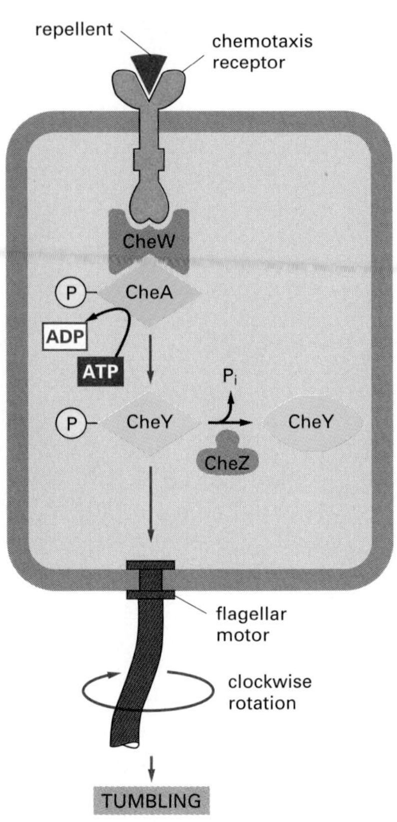

Figure 15–69 The two-component signaling pathway that enables chemotaxis receptors to control the flagellar motor during bacterial chemotaxis. The histidine kinase CheA is stably bound to the receptor via the adaptor protein CheW. The binding of a repellent increases the activity of the receptor, which stimulates CheA to phosphorylate itself on histidine. CheA quickly transfers its covalently bound, high-energy phosphate directly to CheY to generate CheY-phosphate, which then diffuses away, binds to the flagellar motor, and causes the motor to rotate clockwise, which results in tumbling. The binding of an attractant has the opposite effect: it decreases the activity of the receptor and therefore decreases the phosphorylation of CheA and CheY, which results in counterclockwise flagellar rotation and smooth swimming. CheZ accelerates the autodephosphorylation of CheY-phosphate, thereby inactivating it. Each of the phosphorylated intermediates decays in about 10 seconds, enabling the bacterium to respond very quickly to changes in its environment (see Figure 15–10).

Summary

There are five known classes of enzyme-linked receptors: (1) receptor tyrosine kinases, (2) tyrosine-kinase-associated receptors, (3) receptor serine/threonine kinases, (4) transmembrane guanylyl cyclases, and (5) histidine-kinase-associated receptors. In addition, some transmembrane tyrosine phosphatases, which remove phosphate from phosphotyrosine side chains of specific proteins, are thought to function as receptors, although for the most part their ligands are unknown. The first two classes of receptors are by far the most numerous.

Ligand binding to receptor tyrosine kinases induces the receptors to cross-phosphorylate their cytoplasmic domains on multiple tyrosines. The autophosphorylation activates the kinases, as well as producing a set of phosphotyrosines that then serve as docking sites for a set of intracellular signaling proteins, which bind via their SH2 (or PTB) domains. Some of the docked proteins serve as adaptors to couple the receptors to the small GTPase Ras, which, in turn, activates a cascade of serine/threonine phosphorylations that converge on a MAP-kinase, which relays the signal to the nucleus by phosphorylating gene regulatory proteins there. Ras can also activate another protein that docks on activated receptor tyrosine kinases—PI 3-kinase—which generates specific inositol phospholipids that serve as docking sites in the plasma membrane for signaling proteins with PH domains, including protein kinase B (PKB).

Tyrosine-kinase-associated receptors depend on various cytoplasmic tyrosine kinases for their action. These kinases include members of the Src family, which associate with many kinds of receptors, and the focal adhesion kinase (FAK), which associates with integrins at focal adhesions. The cytoplasmic tyrosine kinases then phosphorylate a variety of signaling proteins to relay the signal onward. The largest family of receptors in this class is the cytokine receptors family. When stimulated by ligand binding, these receptors activate Jak cytoplasmic tyrosine kinases, which phosphorylate STATs. The STATs then dimerize, migrate to the nucleus, and activate the transcription of specific genes. Receptor serine/threonine kinases, which are activated by signaling proteins of the TGF-β superfamily, act similarly: they directly phosphorylate and activate Smads, which then oligomerize with another Smad, migrate to the nucleus, and activate gene transcription.

Bacterial chemotaxis is mediated by histidine-kinase-associated chemotaxis receptors. When activated by the binding of a repellent, the receptors stimulate their associated protein kinase to phosphorylate itself on histidine and then transfer that phosphate to a messenger protein, which relays the signal to the flagellar motor to alter the bacterium's swimming behavior. Attractants have the opposite effect on this kinase and therefore on swimming.

SIGNALING PATHWAYS THAT DEPEND ON REGULATED PROTEOLYSIS

The need for intercellular signaling is never greater than during animal development. Each cell in the embryo has to be guided along one developmental pathway or another according to its history, its position, and the character of its neighbors. At each step in the pathway, it must exchange signals with its neighbors to coordinate its behavior with theirs. Most of the signaling pathways already discussed are widely used for these purposes. But there are also others that relay signals in other ways from cell-surface receptors to the interior of the cell. These additional signaling pathways all depend, in part at least, on regulated proteolysis. Although most of them first came to light through genetic studies in *Drosophila,* they have been highly conserved in evolution and are used over and over again during animal development. As we discuss in Chapter 22, they also have a crucial role in the many developmental processes that continue in adult tissues.

We discuss four of these signaling pathways in this section: the pathway mediated by the receptor protein *Notch,* the pathway activated by secreted *Wnt* proteins, the pathway activated by secreted *Hedgehog* proteins, and the pathway that depends on activation of the latent gene regulatory protein *NF-κB.* All of these pathways have crucial roles in animal development. If any one of them is inactivated in a mouse, for example, development is seriously disturbed, and the mouse dies as an embryo or at birth. (We discuss the roles of Notch, Wnt, and Hedgehog signaling in embryonic development in Chapter 21.)

The Receptor Protein Notch Is Activated by Cleavage

Signaling through the **Notch** receptor protein may be the most widely used signaling pathway in animal development. As discussed in Chapter 21, it has a general role in controlling cell fate choices during development, mainly by amplifying and consolidating molecular differences between adjacent cells. Although Notch signaling is involved in the development of most tissues, it is best known for its role in nerve cell production in *Drosophila.* The nerve cells usually arise as isolated single cells within an epithelial sheet of precursor cells. During the process, each future nerve cell or committed nerve-cell precursor signals to its immediate neighbors not to develop in the same way at the same time, a process known as *lateral inhibition.* In a fly embryo, for example, the inhibited cells around the future nerve-cell precursors develop into epidermal cells. Lateral inhibition depends on a contact-dependent signaling mechanism that is mediated by a signal protein called **Delta**, displayed on the surface of the future neural cell. By binding to Notch on a neighboring cell, Delta signals to the neighbor not to become neural (Figure 15–70). When this signaling process is defective in flies, the neighbors of neural cells also develop as neural cells, producing a huge excess of neurons at the expense of epidermal cells, which is lethal. Signaling between adjacent cells via Notch and Delta (or the Deltalike

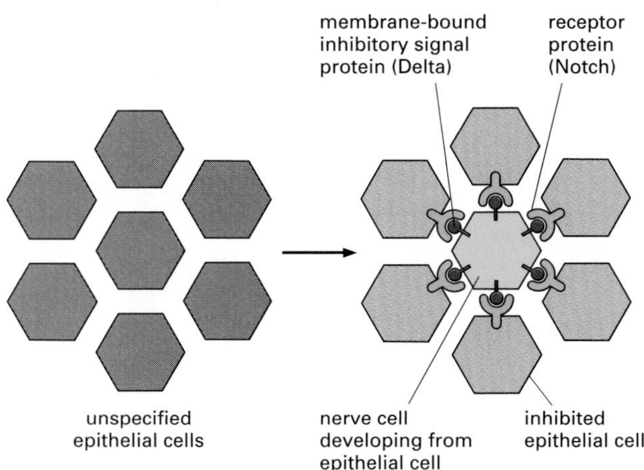

Figure 15–70 Lateral inhibition mediated by Notch and Delta during nerve cell development in *Drosophila*. When individual cells in the epithelium begin to develop as neural cells, they signal to their neighbors not to do the same. This inhibitory, contact-dependent signaling is mediated by the ligand Delta that appears on the surface of the future nerve cell and binds to Notch proteins on the neighboring cells. In many tissues, all the cells in a cluster initially express Delta and Notch, and a competition occurs, with one cell emerging as winner, expressing Delta strongly and inhibiting its neighbors from doing likewise. In other cases, additional factors interact with Delta or Notch to make some cells susceptible to the lateral inhibition signal and others deaf to it.

ligand *Serrate*) regulates cell fate choices in a wide variety of tissues and animals, helping to create fine-grained patterns of distinct cell types. The Notch-mediated signal can have other effects beside lateral inhibition; in some tissues, for example, it works in the opposite way, causing neighboring cells to behave similarly.

Both Notch and Delta are single-pass transmembrane proteins, and both require proteolytic processing to function. Although it is still unclear why Delta has to be cleaved, the cleavage of Notch is central to how Notch activation alters gene expression in the nucleus. When activated by the binding of Delta on another cell, an intracellular protease cleaves off the cytoplasmic tail of Notch, and the released tail moves into the nucleus to activate the transcription of a set of Notch-response genes. The Notch tail acts by binding to a gene regulatory protein called *CSL* (so named because it is called CBF1 in mammals, Suppressor of Hairless in flies, and Lag-1 in worms); this converts CSL from a transcriptional repressor into a transcriptional activator. The products of the main genes directly activated by Notch signaling are themselves gene regulatory proteins, but with an inhibitory action: they block the expression of genes required for neural differentiation (in the nervous system), and of various other genes in other tissues.

The Notch receptor undergoes three proteolytic cleavages, but only the last two depend on Delta. As part of its normal biosynthesis, a protease called *furin* acts in the Golgi apparatus to cleave the newly synthesized Notch protein in its future extracellular domain. This cleavage converts Notch into a heterodimer, which is then transported to the cell surface as the mature receptor. The binding of Delta to Notch induces a second cleavage in the extracellular domain, mediated by a different protease. A final cleavage quickly follows, cutting free the cytoplasmic tail of the activated receptor (Figure 15–71).

Figure 15–71 The processing and activation of Notch by proteolytic cleavage. The *numbered red arrowheads* indicate the sites of proteolytic cleavage. The first proteolytic processing step occurs within the *trans* Golgi network to generate the mature heterodimeric Notch receptor that is then displayed on the cell surface. The binding of Delta, which is displayed on a neighboring cell, triggers the next two proteolytic steps. Note that Notch and Delta interact through their repeated EGF-like domains. Some evidence suggests that the tension exerted on Notch by the endocytic machinery of the interacting cells triggers the cleavage at site 2.

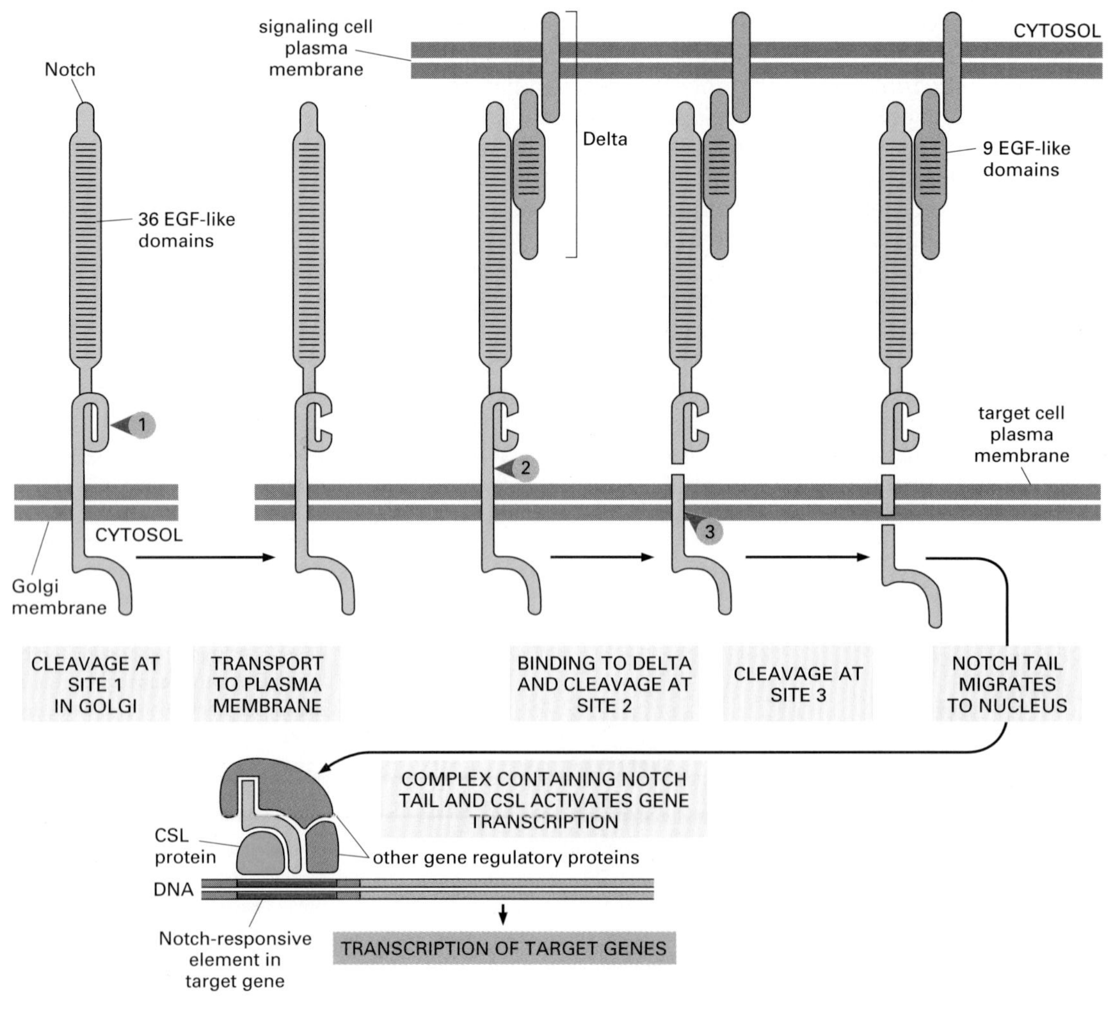

The cleavage of the Notch tail occurs very close to the plasma membrane, just within the transmembrane segment. In this respect it resembles the cleavage of another, more sinister transmembrane protein—the *β-amyloid precursor protein (APP)*, which is expressed in neurons and is implicated in Alzheimer's disease. APP is cleaved within its transmembrane segment, releasing one peptide fragment into the extracellular space of the brain and another into the cytosol of the neuron. In Alzheimer's disease, the extracellular fragments accumulate in excessive amounts and aggregate into filaments that form amyloid plaques, which are believed to injure nerve cells and contribute to their loss. The most frequent genetic cause of early-onset Alzheimer's disease is a mutation in the *presenilin-1 (PS-1)* gene, which encodes an 8-pass transmembrane protein that participates in the cleavage of APP. The mutations in PS-1 cause cleavage of APP into amyloid-plaque-forming fragments at an increased rate. Genetic evidence in *C. elegans, Drosophila,* and mice indicates that the PS-1 protein is a required component of the Notch signaling pathway, helping to perform the final cleavage that activates Notch. Indeed, Notch signaling and cleavage are greatly impaired in PS-1-deficient cells.

Remarkably, Notch signaling is regulated by glycosylation. The *Fringe family* of glycosyltransferases adds extra sugars to the O-linked oligosaccharide (discussed in Chapter 13) on Notch, which alters the specificity of Notch for its ligands. This has provided the first example of the modulation of ligand-receptor signaling by differential receptor glycosylation.

Wnt Proteins Bind to Frizzled Receptors and Inhibit the Degradation of β-Catenin

Wnt proteins are secreted signal molecules that act as local mediators to control many aspects of development in all animals that have been studied. They were discovered independently in flies and in mice: in *Drosophila,* the *wingless (wg)* gene originally came to light because of its role in wing development, while in mice, the *Int-1* gene was found because it promoted the development of breast tumors when activated by the integration of a virus next to it. The cell-surface receptors for the Wnts belong to the **Frizzled** family of seven-pass transmembrane proteins. They resemble G-protein-linked receptors in structure, and some of them can signal through G proteins and the inositol phospholipid pathway discussed earlier. They mainly signal, however, through G-protein-independent pathways, which require a cytoplasmic signaling protein called *Dishevelled.*

The best characterized of the Dishevelled-dependent pathways acts by regulating the proteolysis of a multifunctional protein called **β-catenin** (or Armadillo in flies), which functions both in cell–cell adhesion and as a latent gene regulatory protein. Wnts activate this pathway by binding to both a Frizzled protein and a co-receptor protein. The co-receptor protein is related to the low density lipoprotein (LDL) receptor protein (discussed in Chapter 13) and is therefore called *LDL-receptor-related protein (LRP).* It is uncertain how Frizzled and LRP activate Dishevelled, which relays the signal onward.

In the absence of Wnt signaling, most of a cell's β-catenin is located at cell–cell adherens junctions, where it is associated with *cadherins,* which are transmembrane adhesion proteins. As discussed in Chapter 19, the β-catenin in these junctions helps link the cadherins to the actin cytoskeleton. Any β-catenin not associated with cadherins is rapidly degraded in the cytoplasm. This degradation depends on a large degradation complex, which recruits β-catenin and contains at least three other proteins (Figure 15–72A):

1. A serine/threonine kinase called **glycogen synthase kinase-3β (GSK-3β)** phosphorylates β-catenin, thereby marking the protein for ubiquitylation and rapid degradation in proteasomes.
2. The tumor-suppressor protein **adenomatous polyposis coli (APC)** is so named because the gene encoding it is often mutated in a type of benign tumor (adenoma) of the colon. The tumor projects into the lumen as a polyp, which can eventually become malignant. APC helps promote the

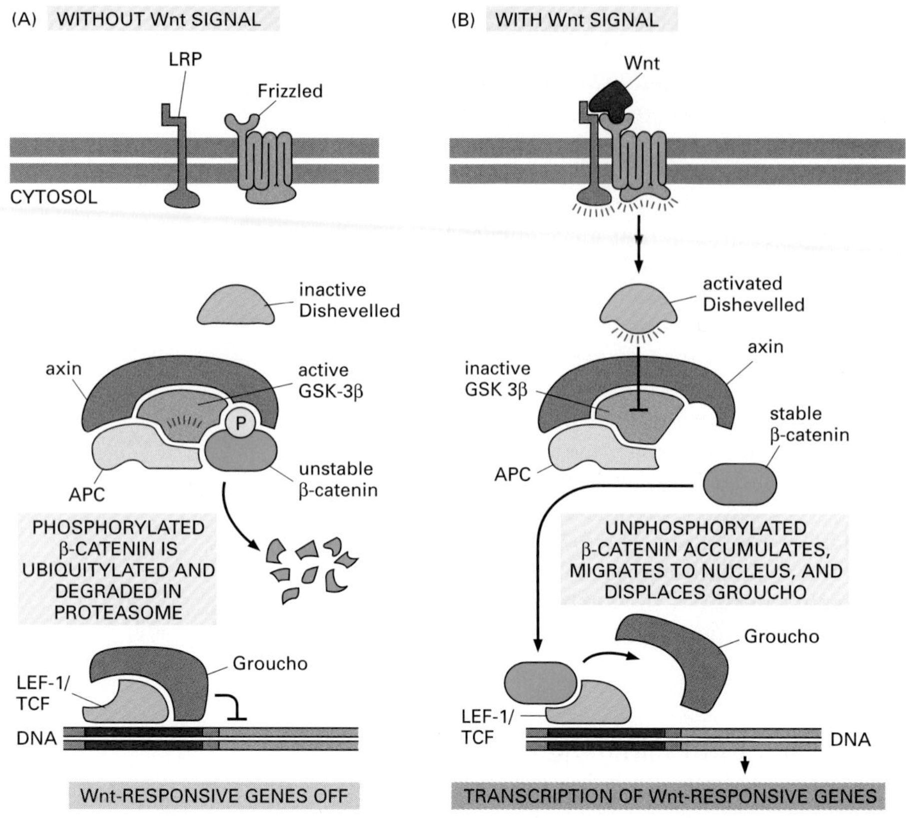

Figure 15–72 A model for the Wnt activation of the β-catenin signaling pathway. (A) In the absence of a Wnt signal, some β-catenin is bound to the cytosolic tail of cadherin proteins (not shown) and any cytosolic β-catenin becomes bound by the APC–axin–GSK-3β degradation complex. In this complex, β-catenin is phosphorylated by GSK-3β, triggering its ubiquitylation and degradation in proteasomes. Wnt-responsive genes are kept inactive by the Groucho corepressor protein bound to the gene regulatory protein LEF-1/TCF. (B) Wnt binding to Frizzled and LRP activates Dishevelled by an unknown mechanism. By an equally mysterious mechanism, which requires casein kinase I (not shown), this leads to the inactivation of GSK-β3 in the degradation complex. As a result, the phosphorylation and degradation of β-catenin is inhibited, and β-catenin accumulates in the cytoplasm and nucleus. In the nucleus, β-catenin binds to LEF-1/TCF, displaces Groucho, and acts as a coactivator to stimulate the transcription of Wnt target genes.

degradation of β-catenin by increasing the affinity of the degradation complex for β-catenin, as required for effective phosphorylation of β-catenin by GSK-3β.

3. A scaffold protein called *axin* holds the protein complex together.

The binding of a Wnt protein to Frizzled and LRP leads to the inhibition of β-catenin phosphorylation and degradation. The mechanism is not understood in detail, but it requires Dishevelled and several other signaling proteins that bind to Dishevelled, including the serine/threonine kinase called *casein kinase 1*. As a result, unphosphorylated β-catenin accumulates in the cytoplasm and nucleus (Figure 15–72B).

In the nucleus, the target genes for Wnt signaling are normally kept silent by an inhibitory complex of gene regulatory proteins, which includes proteins of the *LEF-1/TCF* family bound to the corepressor protein *Groucho* (see Figure 15–72A). The increase in undegraded β-catenin caused by Wnt signaling allows β-catenin to enter the nucleus and bind to LEF-1/TCF, displacing Groucho. The β-catenin now functions as a coactivator, inducing the transcription of the Wnt target genes (see Figure 15–72B).

Among the genes activated by β-catenin is *c-myc,* which encodes a protein (c-Myc) that is a powerful stimulator of cell growth and proliferation (discussed in Chapter 17). Mutations of the *APC* gene occur in 80% of human colon cancers. These mutations inhibit the protein's ability to bind β-catenin, so that β-catenin accumulates in the nucleus and stimulates the transcription of *c-myc* and other Wnt target genes, even in the absence of Wnt signaling. The resulting uncontrolled cell proliferation promotes the development of cancer.

Hedgehog Proteins Act Through a Receptor Complex of Patched and Smoothened, Which Oppose Each Other

Like Wnt proteins, the **Hedgehog** proteins are a family of secreted signal molecules that act as local mediators in many developmental processes in both invertebrates and vertebrates. Abnormalities in the Hedgehog pathway during development can be lethal and in adult cells can also lead to cancer. The Hedgehog proteins were discovered in *Drosophila,* where a mutation in the only gene encoding such a protein produces a larva with spiky processes (denticles) resembling a hedgehog. At least three genes encode Hedgehog proteins in vertebrates—*sonic, desert,* and *indian hedgehog*. The active form of all Hedgehog proteins is unusual in that it is covalently coupled to cholesterol, which helps to restrict its diffusion following secretion. The cholesterol is added during a remarkable processing step, in which the protein cleaves itself. The proteins are also modified by the addition of a fatty acid chain, which, for unknown reasons, can be required for their signaling activity.

Two transmembrane proteins, Patched and Smoothened, mediate the responses to all Hedgehog proteins. **Patched** is predicted to cross the plasma membrane 12 times, and it is the receptor that binds the Hedgehog protein. In the absence of a Hedgehog signal, Patched inhibits the activity of **Smoothened**, which is a 7-pass transmembrane protein with a structure similar to a Frizzled protein. This inhibition is relieved when a Hedgehog protein binds to Patched, allowing Smoothened to relay the signal into the cell. Most of what we know about the downstream signaling pathway activated by Smoothened comes from genetic studies in flies, and it is the fly pathway that we summarize here.

In some respects the Hedgehog signaling pathway in *Drosophila* operates similarly to the Wnt pathway. In the absence of a Hedgehog signal, a gene regulatory protein called **Cubitus interruptus (Ci)** is proteolytically cleaved in proteasomes. Instead of being completely degraded, however, it is processed to form a smaller protein that accumulates in the nucleus, where it acts as a transcriptional repressor, helping to keep some Hedgehog-responsive genes silent. The proteolytic processing of the Ci protein depends on a large multiprotein complex. The complex contains a serine/threonine kinase (called Fused) of unknown function, an anchoring protein (called Costal) that binds the complex to microtubules (keeping Ci out of the nucleus), and an adaptor protein (called Suppressor of Fused) (Figure 15–73A). When Hedgehog binds to Patched to

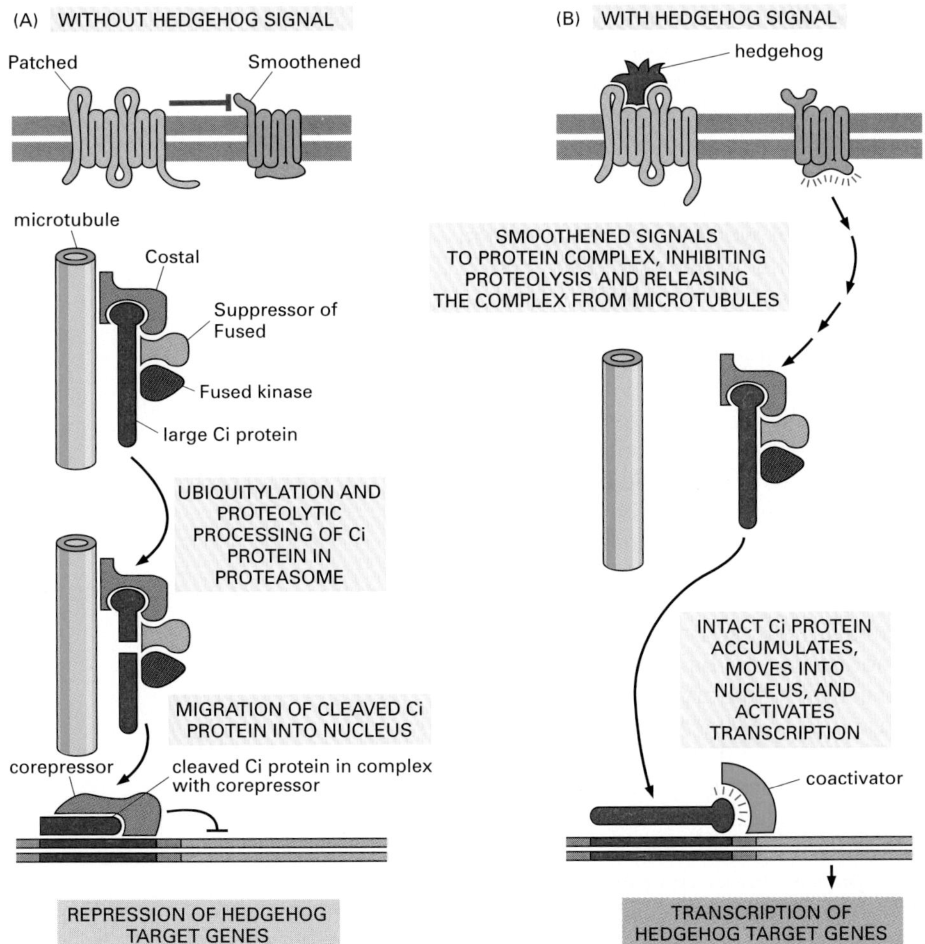

Figure 15–73 A model for Hedgehog signaling in *Drosophila*. (A) In the absence of Hedgehog, the Patched receptor inhibits Smoothened probably by promoting the degradation or intracellular sequestration of Smoothened. The Ci protein is located in a protein complex and is cleaved to form a transcriptional repressor, which accumulates in the nucleus to help keep Hedgehog target genes inactive. The protein complex includes the serine/threonine kinase Fused, the anchoring protein Costal (which binds the complex to microtubules), and the adaptor protein Suppressor of Fused. (B) Hedgehog binding to Patched relieves the inhibition of Smoothened, which now signals to the protein complex to stop processing Ci, to dissociate from microtubules, and to release the unprocessed Ci so it can accumulate in the nucleus and activate the transcription of Hedgehog-responsive genes. Most of the molecular events in the pathway are unknown.

activate the signaling pathway, Ci processing is suppressed, and the unprocessed Ci protein is released from its complex and enters the nucleus, where it activates the transcription of Hedgehog target genes (Figure 15–73B).

Among the genes activated by Ci is the gene that encodes the Wnt protein Wingless, which helps pattern tissues in the fly embryo (discussed in Chapter 21). Another target gene is *patched* itself; the resulting increase in Patched protein on the cell surface inhibits further Hedgehog signaling—a form of negative feedback.

Many gaps in the Hedgehog signaling pathway still remain to be filled in. It is not known, for example, how Patched inhibits Smoothened, how Smoothened activates the pathway, how the proteolysis of Ci is regulated (although it is known that Ci phosphorylation by PKA is required for the processing), or how the release of the complex from microtubules and unprocessed Ci from the complex is controlled.

Even less is known about the Hedgehog pathway in vertebrate cells. In addition to there being at least three types of vertebrate Hedgehog proteins, there are two forms of Patched and three Ci-like proteins (*Gli1, Gli2,* and *Gli3*). Unlike in flies, Hedgehog signaling stimulates the transcription of the *Gli* genes, and it is unclear whether all of the Gli proteins undergo proteolytic processing, although there is evidence that Gli3 does. Inactivating mutations in one of the human *patched* genes, which leads to excessive Hedgehog signaling, occur frequently in the most common form of skin cancer *(basal cell carcinoma),* suggesting that Patched normally helps to keep skin cell proliferation in check.

Multiple Stressful and Proinflammatory Stimuli Act Through an NF-κB-Dependent Signaling Pathway

The **NF-κB proteins** are latent gene regulatory proteins that lie at the heart of most inflammatory responses. These responses occur as a reaction to infection or injury and help protect the animal and its cells from these stresses. When excessive or inappropriate, however, inflammatory responses can also damage tissue and cause severe pain, as happens in joints in rheumatoid arthritis, for example. NF-κB proteins also have an important role in intercellular signaling during normal vertebrate development, although the extracellular signals that activate NF-κB in these circumstances are unknown. In *Drosophila,* however, genetic studies have identified both the extracellular and the intracellular proteins that activate the NF-κB family member *Dorsal,* which has a crucial role in specifying the dorsal–ventral axis of the developing fly embryo (discussed in Chapter 21). The same intracellular signaling pathway is also involved in defending the fly from infection, just as in vertebrates.

Two vertebrate cytokines are especially important in inducing inflammatory responses—*tumor necrosis factor α (TNF-α)* and *interleukin-1 (IL-1).* Both are made by cells of the innate immune system, such as macrophages, in response to infection or tissue injury. These proinflammatory cytokines bind to cell-surface receptors and activate NF-κB, which is normally sequestered in an inactive form in the cytoplasm of almost all of our cells. Once activated, NF-κB turns on the transcription of more than 60 known genes that participate in inflammatory responses. Although TNF-α receptors and IL-1 receptors are structurally unrelated, they operate in much the same way.

There are five NF-κB proteins in mammals (RelA, RelB, c-Rel, NF-κB1, and NF-κB2), and they form a variety of homodimers and heterodimers, each of which activates its own characteristic set of genes. Inhibitory proteins called **IκB** bind tightly to the dimers and hold them in an inactive state within large protein complexes in the cytoplasm. Signals such as TNF-α or IL-1 activate the dimers by triggering a signaling pathway that leads to the phosphorylation, ubiquitylation, and consequent degradation of IκB. The degradation of IκB exposes a nuclear localization signal on the NF-κB proteins, which now move into the nucleus and stimulate the transcription of specific genes. The phosphorylation of IκB is performed by a specific serine/threonine kinase called **IκB kinase (IKK)**.

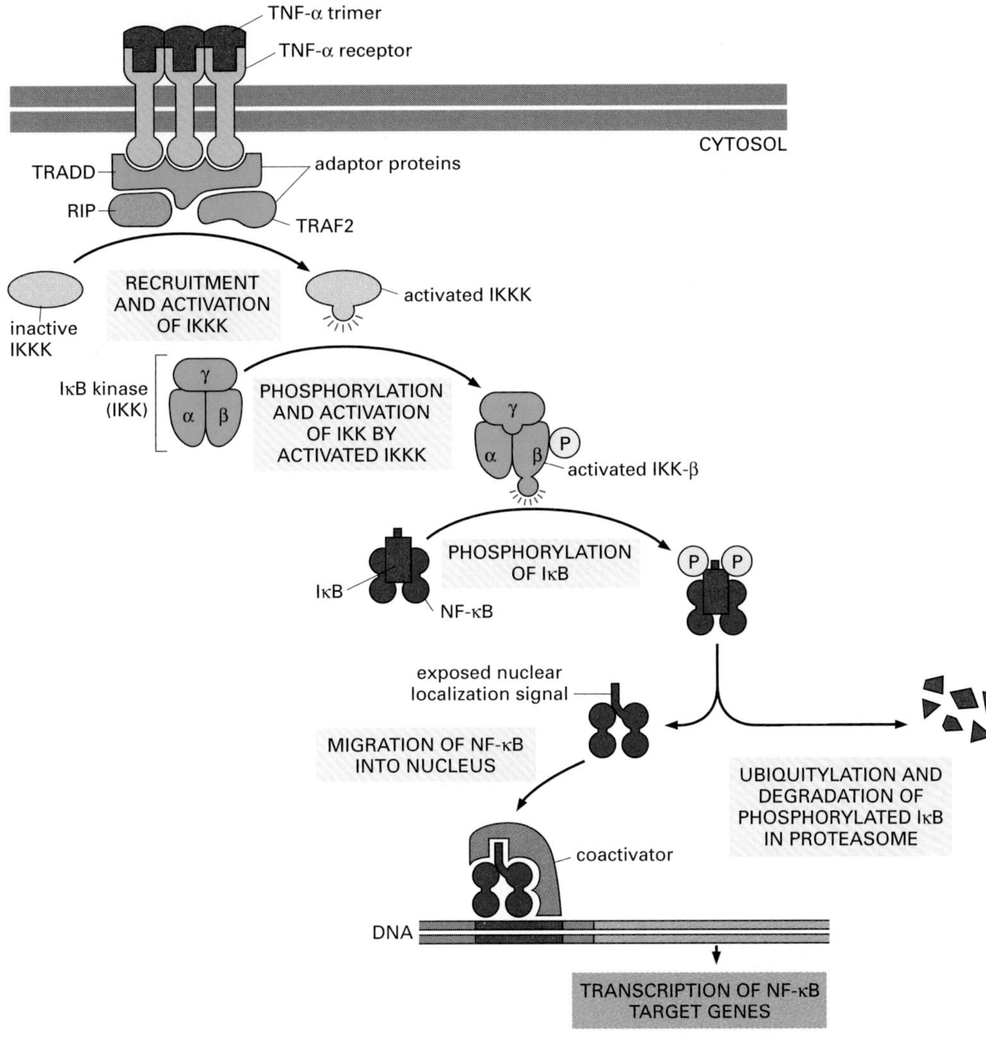

Figure 15–74 The activation of NF-κB by TNF-α. Both TNF-α and its receptors are trimers. The binding of TNF-α causes a rearrangement of the clustered cytosolic tails of the receptors, which now recruit a number of intracellular signaling proteins, including the *receptor-interacting protein kinase (RIP)* and two adaptor proteins, *TNF-associated death-domain protein (TRADD)* and *TNF-receptor-associated factor 2 (TRAF2)*. These then recruit and activate an unidentified kinase, IκB kinase kinase (IKKK), which phosphorylates and activates IκB kinase (IKK). IKK is a heterotrimer composed of two kinase subunits (IKK-α and IKK-β) and a regulatory adaptor subunit called IKK-γ. The IKK-β then phosphorylates IκB on two serines, which marks the protein for ubiquitylation and degradation in proteasomes. The nuclear localization signal on the free NF-κB now directs the transport of this protein into the nucleus where, in collaboration with coactivator proteins, it stimulates the transcription of its target genes. In addition to target genes involved in the inflammatory response, NF-κB also activates the *IκB* gene, providing negative feedback (not shown). The role of RIP is unclear: although it is required for TNF signaling, its kinase domain is not.

The mechanism by which the binding of a proinflammatory cytokine to its cell-surface receptors activates IκB kinase is illustrated for the TNF-α receptor in Figure 15–74. Ligand binding causes the cytosolic tails of the clustered receptors to recruit various adaptor proteins and cytoplasmic serine/threonine kinases. One of the recruited kinases is thought to be an *IκB kinase kinase (IKKK)* that directly phosphorylates and activates the IκB kinase (IKK).

Not all of the signaling proteins recruited to the cytosolic tail of the TNF-α receptor contribute to NF-κB activation, however. Some can trigger a MAP-kinase cascade, while others can activate a proteolytic cascade that leads to apoptosis (discussed in Chapter 17).

Thus far, we have discussed cell signaling mainly in animals, with a few diversions into yeasts and bacteria. But intercellular signaling is just as important for plants as it is for animals, although the mechanisms and molecules used are mainly different, as we discuss next.

Summary

Some signaling pathways that are especially important in animal development depend on proteolysis for at least part of their action. Notch receptors are activated by cleavage when Delta (or a related ligand) on another cell binds to them; the cleaved cytosolic tail of Notch migrates into the nucleus, where it stimulates gene transcription. In the Wnt signaling pathway, by contrast, the proteolysis of the latent gene regulatory protein β-catenin is inhibited when secreted Wnt proteins bind to their receptors; as a result, β-catenin accumulates in the nucleus and activates the transcription of Wnt target genes.

Hedgehog signaling in flies works much like Wnt signaling: in the absence of a signal, a bifunctional, cytoplasmic gene regulatory protein Ci is proteolytically cleaved to form a transcriptional repressor that keeps Hedgehog target genes silenced. The binding of Hedgehog to its receptor inhibits the proteolytic processing of Ci; as a result, the larger form of Ci accumulates in the nucleus and activates the transcription of Hedgehog-responsive genes. Signaling through the latent gene regulatory protein NF-κB also depends on proteolysis. NF-κB is normally held in an inactive state by the inhibitory protein IκB within a multiprotein complex in the cytoplasm. A variety of extracellular stimuli, including proinflammatory cytokines, trigger a phosphorylation cascade that ultimately phosphorylates IκB, marking it for degradation; this enables the freed NF-κB to enter the nucleus and activate the transcription of its target genes.

SIGNALING IN PLANTS

In plants, as in animals, cells are in constant communication with one another. Plant cells communicate to coordinate their activities in response to the changing conditions of light, dark, and temperature that guide the plant's cycle of growth, flowering, and fruiting. Plant cells also communicate to coordinate what goes on in their roots, stems, and leaves. In this final section, we consider how plant cells signal to one another and how they respond to light. Much less is known about the receptors and intracellular signaling mechanisms involved in cell communication in plants, and we shall concentrate mainly on how these differ from those used by animals. We discuss some of the details of plant development in Chapter 21.

Multicellularity and Cell Communication Evolved Independently in Plants and Animals

Although plants and animals are both eucaryotes, they have had separate evolutionary histories for more than a billion years. Their last common ancestor was a unicellular eucaryote that had mitochondria but no chloroplasts. The plant lineage acquired chloroplasts after plants and animals diverged. The earliest fossils of multicellular animals and plants date from almost 600 million years ago. Thus, it seems that plants and animals evolved multicellularity independently, each starting from a different unicellular eucaryote, sometime between 1.6 and 0.6 billion years ago (Figure 15–75).

If multicellularity evolved independently in plants and animals, the molecules and mechanisms used for cell communication will have evolved separately and would be expected to be different. Some degree of resemblance is expected, however, as both plant and animal genes diverged from the set of

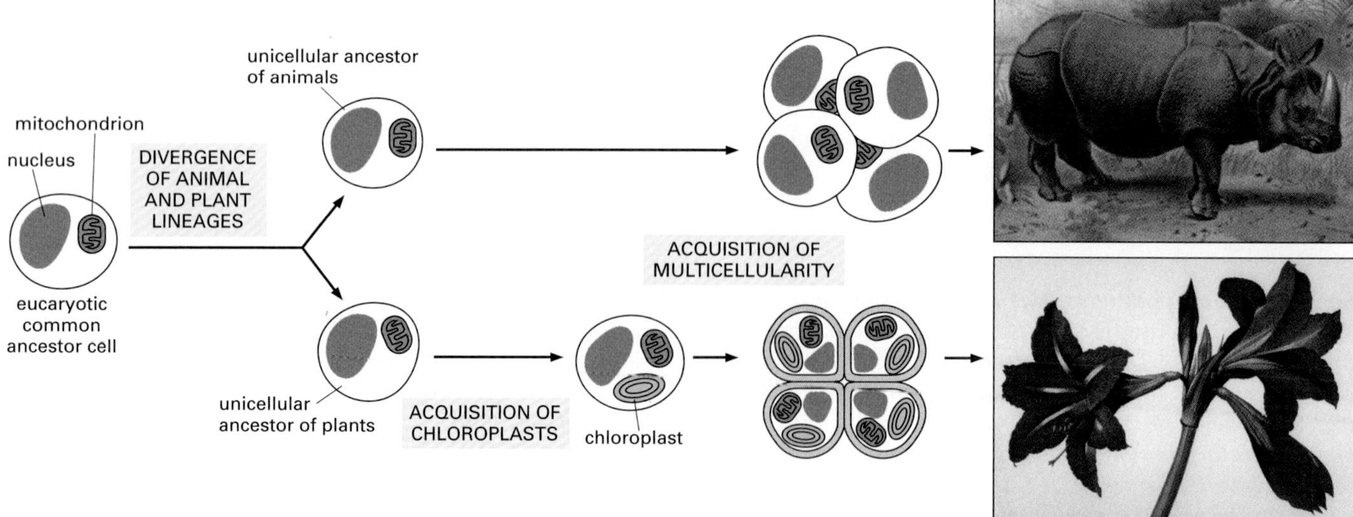

Figure 15–75 The proposed divergence of plant and animal lineages from a common unicellular eucaryotic ancestor. The plant lineage acquired chloroplasts after the two lineages diverged. Both lineages independently gave rise to multicellular organisms—plants and animals. (Paintings courtesy of John Innes Foundation.)

genes contained by the unicellular eucaryote that was the last common ancestor of plants and animals. Nitric oxide and Ca^{2+} are widely used for signaling in both plants and animals. However, because the genome of *Arabidopsis thaliana*, a widely studied small flowering plant, has been completely sequenced, we know that there are no homologs of Wnt, Hedgehog, Notch, Jak/STAT, TGF-β, Ras, or the nuclear receptor family in this organism. Similarly, cyclic AMP has not been definitively implicated in intracellular signaling in plants, although cyclic GMP has.

Much of what is known about the molecular mechanisms involved in signaling in plants has come from genetic studies on *Arabidopsis*. Although the specific molecules used in cell communication in plants often differ from those used in animals, the general strategies are frequently very similar. Enzyme-linked cell-surface receptors, for example, are used in both lineages, as we now discuss.

Receptor Serine/Threonine Kinases Function as Cell-Surface Receptors in Plants

Like animals, plants make extensive use of cell-surface receptors. Whereas most cell-surface receptors in animals are G-protein-linked, most found so far in plants are enzyme-linked. Moreover, whereas the largest class of enzyme-linked receptors in animals is receptor tyrosine kinases, this type of receptor is extremely rare in plants, even though they contain many cytoplasmic tyrosine kinases, and tyrosine phosphorylation and dephosphorylation have important roles in plant cell signaling. Instead, plants seem to rely on a great diversity of transmembrane *receptor serine/threonine kinases*, which are distinct from this type of receptor used by animal cells. Like the animal receptors, however, they have a typical serine/threonine kinase cytoplasmic domain and an extracellular ligand-binding domain. The most abundant types identified so far have a tandem array of extracellular leucine-rich repeats (Figure 15–76) and are therefore called **leucine-rich repeat (LRR) proteins**.

There are about 80 LRR receptor kinases encoded by the *Arabidopsis* genome. One of the best-studied examples is **CLAVATA 1 (CLV1)**, which was originally identified in genetic studies. Mutations that inactivate the protein cause the production of flowers with extra floral organs and a progressive enlargement of both the shoot and floral *meristems*, which are groups of self-renewing stem cells producing the cells that give rise to stems, leaves, and flowers (discussed in Chapter 21). The extracellular signal molecule for the receptor is thought to be a small protein called CLV3, which is secreted by neighboring cells. The binding of CLV3 to its receptor, CLV1, suppresses meristem growth, either by inhibiting cell division there or, more probably, by stimulating cell differentiation (Figure 15–77A). The intracellular signaling pathway from CLV1 to the cell response is largely unknown, but it includes a serine/threonine protein phosphatase that inhibits CLV1 signaling; also involved is a small GTP-binding protein of the Rho class and a nuclear gene regulatory protein that is distantly related to homeodomain proteins. Mutations that inactivate this gene regulatory

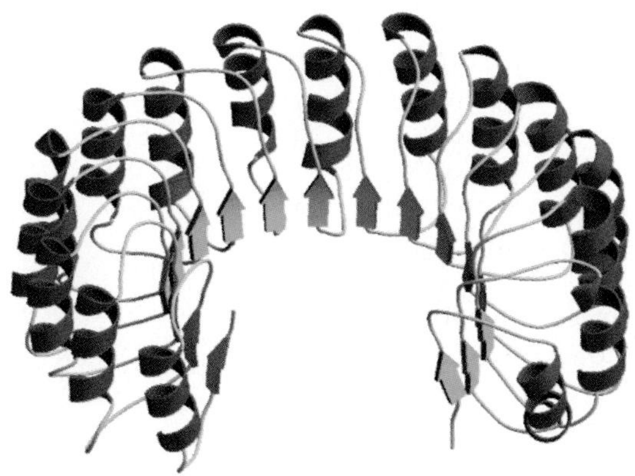

Figure 15–76 The three-dimensional structure of leucine-rich repeats, similar to those found in the LRR serine/threonine receptor kinases. (Courtesy of David Lawson.)

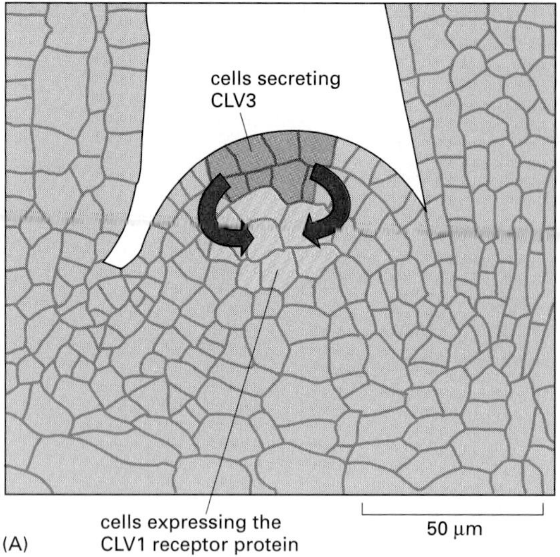

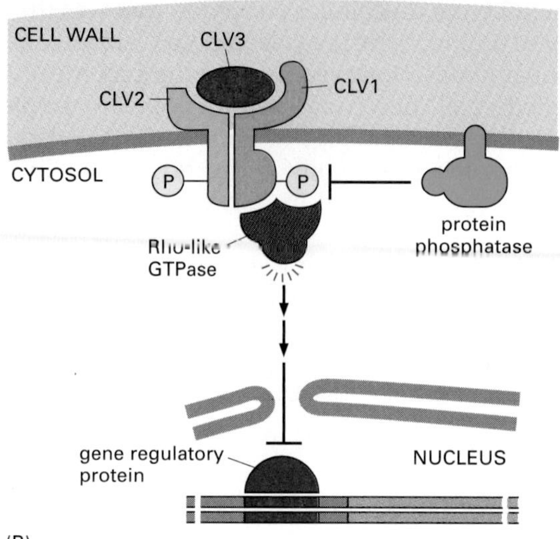

Figure 15–77 A hypothetical model for CLV1 receptor serine/threonine kinase regulation of cell proliferation and/or differentiation in the shoot meristem. (A) Cells in the outer layer of the meristem secrete CLV3 protein, which binds to CLV1 receptor proteins on target cells in an adjacent, more central region of the meristem, presumably stimulating the differentiation of the target cells. (B) Some parts of the intracellular signaling pathway activated by CLV3 binding. The CLV receptor protein is thought to be a homodimer or heterodimer, which phosphorylates itself on serines and threonines, thereby activating the receptor and leading to the activation of a Rho-like GTPase. The signaling pathway after this point is unclear, but it leads to the inhibition of a gene regulatory protein in the nucleus, thereby blocking the transcription of genes that otherwise might inhibit differentiation. The phosphatase dephosphorylates the receptor and thereby negatively regulates the signaling pathway.

protein have the opposite effect of mutations that inactivate CLV1: cell division is greatly decreased in the shoot meristem, and the plant produces flowers with too few organs. Thus, the intracellular signaling pathway activated by CLV1 is thought to normally stimulate cell differentiation by inhibiting the gene regulatory protein that normally inhibits cell differentiation (Figure 15–77B).

A different LRR receptor kinase called *BRI1* acts as a cell-surface steroid hormone receptor in *Arabidopsis*. Plants synthesize a class of steroids called *brassinosteroids,* because they were originally identified in the mustard family Brassicaceae, which includes *Arabidopsis*. During development, these plant growth regulators stimulate cell expansion and help mediate responses to darkness. Mutant plants that are deficient in the BRI1 receptor kinase are insensitive to brassinosteroids. Normally, *Arabidopsis* plants grown in darkness are white and gangly as a result of brassinosteroid signaling; in the absence of brassinosteroid signaling, they become green, as though they were growing in light, and the mature plant is severely dwarfed. As for the other known LRR receptor kinases in plants, the nature of the signal transduction pathway that leads from the receptor to the response remains a mystery.

The LRR receptor kinases are only one of many classes of transmembrane receptor serine/threonine kinases in plants. There are at least six additional families, each with its own characteristic set of extracellular domains. The *lectin receptor kinases,* for example, have extracellular domains that bind carbohydrate signal molecules. The *Arabidopsis* genome encodes over 300 receptor serine/threonine kinases, which makes this family of receptors the largest one known in plants. Many of these are involved in defense responses against pathogens.

Ethylene Activates a Two-Component Signaling Pathway

Various **growth regulators** (also called *plant hormones*) help to coordinate plant development. They include *ethylene, auxin, cytokinins, gibberellins,* and *abscisic acid*. Growth regulators are all small molecules made by most plant cells. They diffuse readily through cell walls and they can act locally or be transported to influence cells further away. Each growth regulator can have multiple effects. The specific effect depends on which other growth regulators are acting, on environmental conditions, on the nutritional state of the plant, and on the responsiveness of the target cell.

Ethylene is an important example. This small gas molecule can influence plant development in various ways, including the promotion of fruit ripening, leaf abscission, and plant senescence. It also functions as a stress signal in response to wounding, infection, flooding, and so on. When the shoot of a germinating seedling, for example, encounters an obstacle, such as a piece of

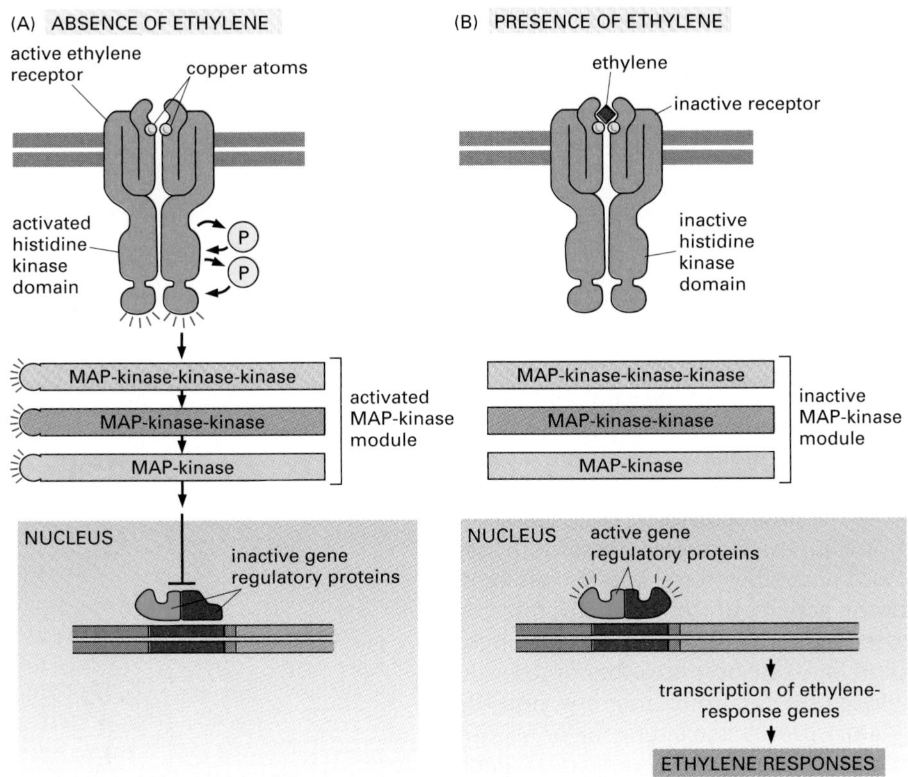

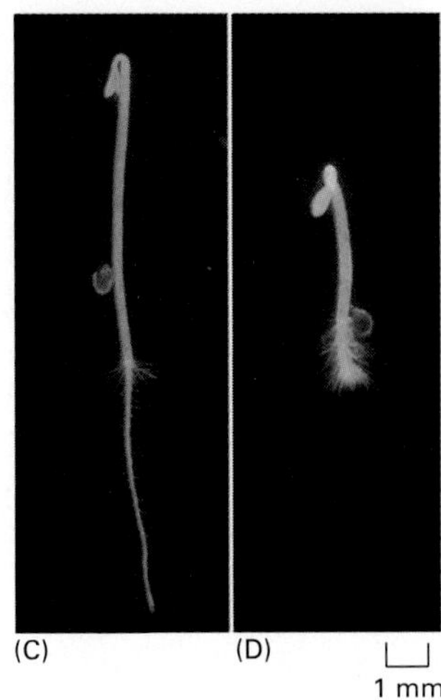

Figure 15–78 A current view of the ethylene two-component signaling pathway. (A) In the absence of ethylene, the receptors and the MAP-kinase module are active, leading to inhibition of the gene regulatory proteins in the nucleus that are responsible for the transcription of ethylene-responsive genes. (B) In the presence of ethylene, the receptors and the MAP-kinase module are inactive, so the ethylene-responsive genes are transcribed. (C and D) The ethylene-mediated "triple response" that occurs when the growing shoot of a germinating seedling encounters an obstacle underground. After such an encounter (D), the shoot thickens, and the protective hook (at *top*) increases its curvature to protect the tip of the shoot. (C and D, courtesy of Melanie Webb.)

gravel underground in the soil, the seedling responds to the encounter in three ways. First, it thickens its stem, which can then exert more force on the obstacle. Second, it shields the tip of the shoot by increasing the curvature of a specialized hook structure. Third, it reduces the shoot's tendency to grow away from the direction of gravity to avoid the obstacle. This "triple response" is controlled by ethylene (Figure 15–78C, D).

Plants have a number of ethylene receptors, which are all structurally related. They are dimeric transmembrane proteins that are thought to function as *histidine kinases*. Ethylene receptors have an extracellular domain, which contains a copper atom that binds ethylene, and an intracellular histidine-kinase-like domain. In a manner similar to the two-component signaling pathway involved in bacterial chemotaxis discussed earlier, the kinase domain, when active, phosphorylates itself on histidine and then is believed to transfer the phosphate to an aspartic acid in another domain of the receptor. In bacterial chemotaxis, the binding of an attractant inactivates the receptor. Similarly, the binding of ethylene inactivates ethylene receptors, inhibiting the kinase domain and the downstream signaling pathway emanating from it. In its unbound, active state, the receptor activates the first component of a MAP-kinase signaling module (see Figure 15–56). The activation of this MAP-kinase cascade leads to the *inactivation* of gene regulatory proteins in the nucleus that are responsible for stimulating the transcription of ethylene-responsive genes. The binding of ethylene to the receptors inactivates this signaling pathway, thereby turning these genes on (Figure 15–78A, B).

Two-component signaling systems operate in bacteria and fungi, as well as in plants, but apparently not in animals. Why animals should have given up this way of signaling remains a mystery.

Plant development is greatly influenced by environmental conditions. Unlike animals, plants cannot move on when conditions become unfavorable; they have to adapt, or they die. The most important environmental influence is light, which is their energy source and has a major role throughout their entire life cycle—from germination, through seedling development, to flowering and senescence. Plants have evolved a large set of light-sensitive proteins to monitor the quantity, quality, direction, and duration of light, as we now discuss.

Phytochromes Detect Red Light, and Cryptochromes Detect Blue Light

Both plants and animals use a variety of light-responsive proteins to sense light of different wavelengths. In plants, these are usually referred to as *photoreceptors*. However, because the term photoreceptor is also used for light-sensitive cells in the animal retina (see p. 867), we shall use the term *photoprotein* instead. All photoproteins sense light by means of a covalently attached light-absorbing chromophore, which changes its shape in response to light and then induces a change in the protein's conformation. Animals use some of the same photoprotein families used by plants. The most extensively studied animal photoproteins are the rhodopsins, which are membrane-bound, G-protein-linked proteins that regulate ion channels in the light-sensitive cells of the retina, as discussed earlier.

The best-known plant photoproteins are the **phytochromes**, which are present in all plants and in some algae. These are dimeric, cytoplasmic serine/threonine kinases that respond differentially and reversibly to red and far-red light: whereas red light usually activates the kinase activity of the phytochrome, far-red light inactivates it. When activated by red light, the phytochrome is thought to phosphorylate itself and then to phosphorylate one or more other proteins in the cell. In some light responses, the activated phytochrome migrates into the nucleus, where it interacts with gene regulatory proteins to alter gene transcription (Figure 15–79). In other cases, the activated phytochrome activates a gene regulatory protein in the cytoplasm, which then migrates into the nucleus to regulate gene transcription. In still other cases, the photoprotein triggers signaling pathways in the cytosol that alter the cell's behavior without involving the nucleus.

Although the phytochromes possess serine/threonine kinase activity, parts of their structure resemble the histidine kinases involved in bacterial chemotaxis. This finding suggests that the plant phytochromes originally descended from bacterial histidine kinases and only later in evolution altered their substrate specificity from histidine to serine and threonine.

Plants sense blue light using two types of photoproteins, phototropin and cryptochromes. **Phototropin** is associated with the plasma membrane and is

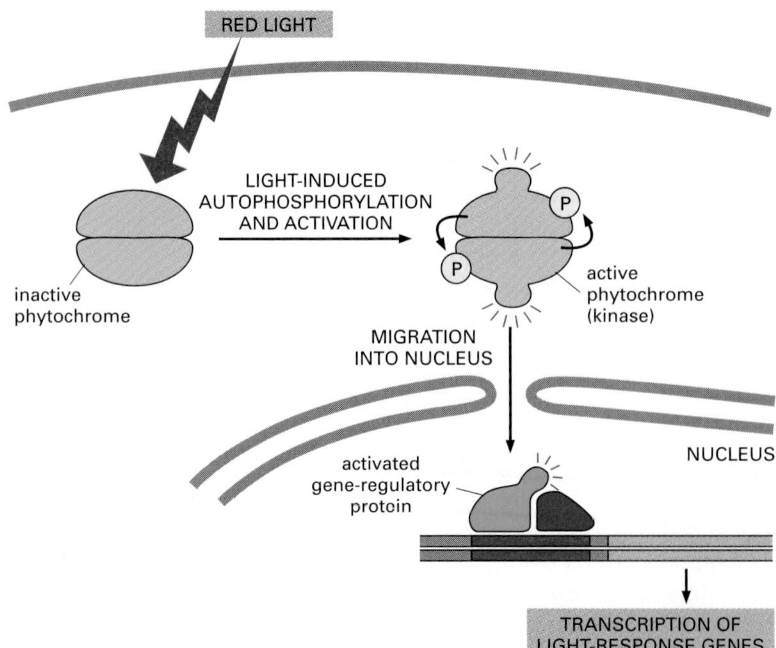

Figure 15–79 A current view of one way in which phytochromes mediate a light response in plant cells. When activated by light, the phytochrome, which is a dimer, phosphorylates itself and then moves into the nucleus, where it activates gene regulatory proteins to stimulate the transcription of specific genes.

partly responsible for *phototropism,* the tendency of plants to grow toward light. Phototropism occurs by directional cell elongation, which is stimulated by the growth regulator auxin, but the links between phototropin and auxin are unknown.

Cryptochromes are flavoproteins that are sensitive to blue light. They are structurally related to blue-light-sensitive enzymes called *photolyases,* which are involved in the repair of ultraviolet-induced DNA damage in all organisms, except most mammals. Unlike phytochromes, cryptochromes are also found in animals, where they have an important role in circadian clocks that operate in most cells and cycle with a 24-hour rhythm (discussed in Chapter 7). The cryptochromes do not have a DNA repair activity, but they are thought to have evolved from the photolyases.

In this chapter, we have discussed how extracellular signals can influence cell behavior. One crucial intracellular target of these signals is the cytoskeleton, which determines cell shape and is responsible for cell movements, as we discuss in the next chapter.

Summary

Plants and animals are thought to have evolved multicellularity and cell communication mechanisms independently, each starting from a different unicellular eucaryote, which in turn evolved from a common unicellular eucaryotic ancestor. Not surprisingly, therefore, the mechanisms of signaling between cells in animals and plants have both similarities and differences. Whereas animals rely mainly on G-protein-linked surface receptors, for example, plants rely mainly on enzyme-linked receptors of the receptor serine/threonine type, especially ones with extracellular leucine-rich repeats. A number of growth regulators, including ethylene, help coordinate plant development. Ethylene acts through receptor histidine kinases in a two-component signaling pathway that resembles the pathway used in bacterial chemotaxis. Light has an important role in regulating plant development. These light responses are mediated by a variety of light-sensitive photoproteins, including phytochromes, which are responsive to red light, and cryptochromes and phototropin, which are sensitive to blue light.

References

General

Ari S (ed) (1999) Introduction to Cellular Signal Transduction. New York: Birkhauser.

Heldin C-H & Purton M (eds) (1996) Signal Transduction. Cheltenham, UK: Nelson Thornes.

Krauss G (1999) Biochemistry of Signal Transduction and Regulation. New York: John Wiley & Sons.

General Principles of Cell Communication

Cohen P (1992) Signal integration at the level of protein kinases, protein phosphatases and their substrates. *Trends Biochem. Sci.* 17, 408–413.

Forman BM & Evans RM (1995) Nuclear hormone receptors activate direct, inverted, and everted repeats. *Ann. N.Y. Acad. Sci.* 761, 29–37.

Gurdon JB, Lemaire P & Kato K (1993) Community effects and related phenomena in development. *Cell* 75, 831–834.

Murad F (1998) Nitric oxide signaling: would you believe that a simple free radical could be a second messenger, autacoid, paracrine substance, neurotransmitter, and hormone? *Recent Prog. Horm. Res.* 53, 43–59; discussion 59–60.

Pawson T & Nash P (2000) Protein–protein interactions define specificity in signal transduction. *Genes Dev.* 14, 1027–1047.

Pawson T & Scott JD (1997) Signalling through scaffold, anchoring, and adaptor proteins. *Science* 278, 2075–2080.

Signaling Through G-protein-linked Cell-surface Receptors

Baylor D (1996) How photons start vision. *Proc. Natl. Acad. Sci. USA* 93, 560–565.

Berridge MJ (1997) Elementary and global aspects of calcium signalling. *J. Physiol.* (Lond). 499, 291–306.

Bourne HR (1997) How receptors talk to trimeric G proteins. *Curr. Opin. Cell Biol.* 9, 134–142.

Buck LB (2000) The molecular architecture of odor and pheromone sensing in mammals. *Cell* 100, 611–618.

Chin D & Means AR (2000) Calmodulin: a prototypical calcium sensor. *Trends Cell Biol.* 10, 322–328.

Cohen PT (1997) Novel protein serine/threonine phosphatases: variety is the spice of life. *Trends Biochem. Sci.* 22, 245–251.

Collins S, Caron MG & Lefkowitz RJ (1992) From ligand binding to gene expression: new insights into the regulation of G-protein-coupled receptors. *Trends Biochem. Sci.* 17, 37–39.

Gilman AG (1995) Nobel Lecture. G proteins and regulation of adenylyl cyclase. *Biosci. Rep.* 15, 65–97.

Irvine RF (1992) Inositol lipids in cell signalling. *Curr. Opin. Cell Biol.* 4, 212–219.

Koninck PD & Schulman H (1998) Sensitivity of CaM kinase II to the frequency of Ca2+ oscillations. *Science* 279, 227–229.

Montminy M (1997) Transcriptional regulation by cyclic AMP. *Annu. Rev. Biochem.* 66, 807–822.

Nishizuka Y (1992) Intracellular signaling by hydrolysis of phospholipids and activation of protein kinase C. *Science* 258, 607–614.

Pitcher JA, Freedman NJ & Lefkowitz RJ (1998) G protein-coupled receptor kinases. *Annu. Rev. Biochem.* 67, 653–692.

Signaling Through Enzyme-linked Cell-surface Receptors

Darnell JE, Jr, Kerr IM & Stark GR (1994) Jak-STAT pathways and transcriptional activation in response to IFNs and other extracellular signaling proteins. *Science* 264, 1415–1421.

Downward J (1996) Control of ras activation. *Cancer Surv.* 27, 87–100.

Downward J (1998) Mechanisms and consequences of activation of protein kinase B/Akt. *Curr. Opin. Cell Biol.* 10, 262–267.

Falke JJ, Bass RB, Butler SL et al. (1997) The two-component signaling pathway of bacterial chemotaxis: a molecular view of signal transduction by receptors, kinases, and adaptation enzymes. *Annu. Rev. Cell Dev. Biol.* 13, 457–512.

Gale NW & Yancopoulos GD (1997) Ephrins and their receptors: a repulsive topic? *Cell Tissue Res.* 290, 227–241.

Leevers SJ, Vanhaesebroeck B & Waterfield MD (1999) Signalling through phosphoinositide 3-kinases: the lipids take centre stage. *Curr. Opin. Cell Biol.* 11, 219–225.

Massagué J (1998) TGF-β signal transduction. *Annu. Rev. Biochem.* 67, 753–791.

Neel BG & Tonks NK (1997) Protein tyrosine phosphatases in signal transduction. *Curr. Opin. Cell Biol.* 9, 192–204.

Pawson T (1995) Protein modules and signalling networks. *Nature* 373, 573–580.

Potter LR & Hunter T (2001) Guanylyl cyclase-linked natriuretic peptide receptors: structure and regulation. *J. Biol. Chem.* 276, 6057–6060.

Schlessinger J (2000) Cell signaling by receptor tyrosine kinases. *Cell* 103, 211–225.

Thomas SM & Brugge JS (1997) Cellular functions regulated by Src family kinases. *Annu. Rev. Cell Dev. Biol.* 13, 513–609.

Widmann C, Gibson S, Jarpe B & Johnson GI (1999) Mitogen-activation protein kinase: conservation of a three-kinase module from yeast to human. *Physiol. Rev.* 79, 143–180.

Signaling Pathways That Depend on Regulated Proteolysis

Chan YM & Jan YN (1998) Roles for proteolysis and trafficking in notch maturation and signal transduction. *Cell* 94, 423–426.

Ghosh S, May M & Kopp E (1998) NF-κB and Rel proteins: evolutionarily conserved mediators of immune responses. *Annu. Rev. Immunol.* 16, 225–260.

Hlsken J & Behrens J (2000) The Wnt signalling pathway. *J. Cell Sci.* 113, 3545–3546.

Ingham PW (1998) Transducing Hedgehog: the story so far. *EMBO J.* 17, 3505–3511.

Signaling in Plants

Ahmad M (1999) Seeing the world in red and blue: insight into plant vision and photoreceptors. *Curr. Opin. Plant Biol.* 2, 230–235.

Buchanan SG & Gay NJ (1996) Structural and functional diversity in the leucine-rich repeat family of proteins. *Prog. Biophys. Mol. Biol.* 65, 1–44.

Chang C & Shockey JA (1999) The ethylene-response pathway: signal perception to gene regulation. *Curr. Opin. Plant Biol.* 2, 352–358.

Iten M, Hoffmann T & Grill E (1999) Receptors and signalling components of plant hormones. *J. Recept. Signal Transduct. Res.* 19, 41–58.

Nagy F & Schäfer E (2000) Nuclear and cytosolic events of light-induced, phytochrome-regulated signaling in higher plants. *EMBO J.* 19, 157–163.

Schumacher K & Chory J (2000) Brassinosteroid signal transduction: still casting the actors. *Curr. Opin. Plant Biol.* 3, 79–84.

Shiu SH & Bleecker AB (2001) Receptor-like kinases from *Arabidopsis* form a monophyletic gene family related to animal receptor kinases. *Proc. Natl. Acad. Sci. USA* 98, 10763–10768.

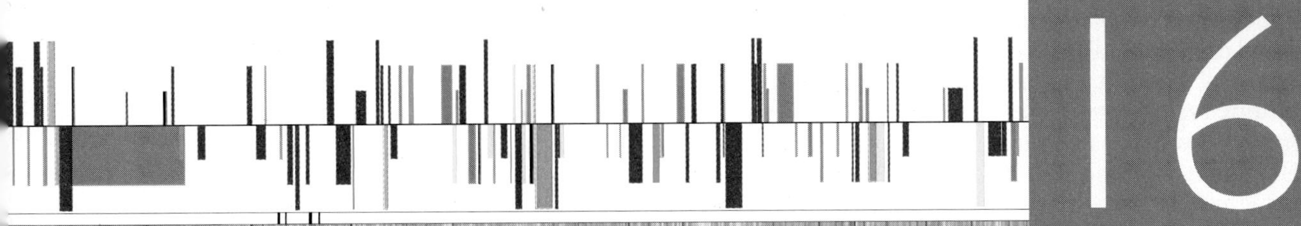

16

THE CYTOSKELETON

Cells have to organize themselves in space and interact mechanically with their environment. They have to be correctly shaped, physically robust, and properly structured internally. Many of them also have to be able to change their shape and move from place to place. All of them have to be able to rearrange their internal components as they grow, divide, and adapt to changing circumstances. All these spatial and mechanical functions are developed to a very high degree in eucaryotic cells, where they depend on a remarkable system of filaments called the **cytoskeleton** (Figure 16–1).

The cytoskeleton pulls the chromosomes apart at mitosis and then splits the dividing cell into two. It drives and guides the intracellular traffic of organelles, ferrying materials from one part of the cell to another. It supports the fragile plasma membrane and provides the mechanical linkages that let the cell bear stresses and strains without being ripped apart as the environment shifts and changes. It enables some cells, such as sperm, to swim, and others, such as fibroblasts and white blood cells, to crawl across surfaces. It provides the machinery in the muscle cell for contraction and in the neuron to extend an axon and dendrites. It guides the growth of the plant cell wall and controls the amazing diversity of eucaryotic cell shapes.

The varied functions of the cytoskeleton center on the behavior of three families of protein molecules, which assemble to form three main types of filaments. Each type of filament has distinct mechanical properties and dynamics, but certain fundamental principles are common to them all. These principles provide the basis for a general understanding of how the cytoskeleton works.

In this chapter, we begin by describing the three main types of filaments, the basic principles underlying their assembly and disassembly, and their individual peculiarities. We then describe how other proteins interact with the three main filament systems, enabling the cell to establish and maintain internal order, to shape and remodel its surface, to move organelles in a directed manner from one place to another, and—when appropriate—to move itself to new locations. The wonderfully coordinated actions of the cytoskeleton in cell division are discussed separately, in Chapter 18.

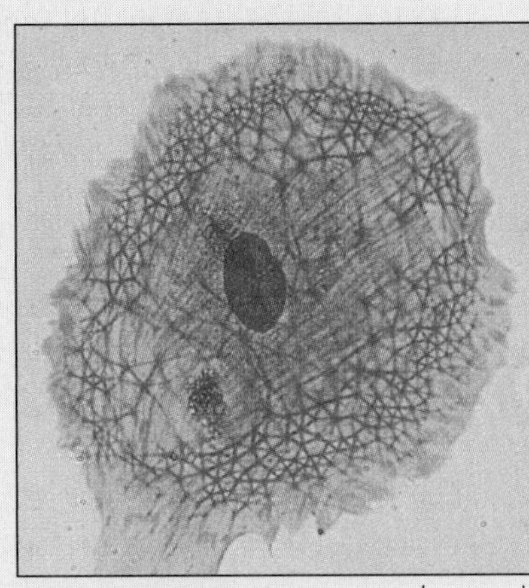

Figure 16–1 The cytoskeleton. A cell in culture has been fixed and stained with Coomassie blue, a general stain for proteins. Note the variety of filamentous structures that extend throughout the cell. (Courtesy of Colin Smith.)

THE SELF-ASSEMBLY AND DYNAMIC STRUCTURE OF CYTOSKELETAL FILAMENTS

Three types of cytoskeletal filaments are common to many eucaryotic cells and are fundamental to the spatial organization of these cells. *Intermediate filaments* provide mechanical strength and resistance to shear stress. *Microtubules* determine the positions of membrane-enclosed organelles and direct intracellular transport. *Actin filaments* determine the shape of the cell's surface and are necessary for whole-cell locomotion (Panel 16–1). But these cytoskeletal filaments would be ineffective on their own. Their usefulness to the cell depends on a large number of accessory proteins that link the filaments to other cell components, as well as to each other. This set of *accessory proteins* is essential for the controlled assembly of the cytoskeletal filaments in particular locations, and it includes the *motor proteins* that either move organelles along the filaments or move the filaments themselves.

Cytoskeletal systems are dynamic and adaptable, organized more like ant trails than interstate highways. A single trail of ants may persist for many hours, extending from the ant nest to a delectable picnic site, but the individual ants within the trail are anything but static. If the ant scouts find a new and better source of food, or if the picnickers clean up and leave, the dynamic structure rearranges itself with astonishing rapidity, to deal with the new situation. In a similar way, large-scale cytoskeletal structures can change or persist, according to need, lasting for lengths of time ranging from less than a minute up to the lifetime of the cell. The individual macromolecular components that make up these structures are in a constant state of flux. Thus, like the alteration of an ant trail, a structural rearrangement in a cell requires little extra energy when conditions change. In this section, we discuss the remarkable mechanisms that cause cytoskeletal filaments to be dynamic, and thereby able to respond rapidly to any eventuality.

Regulation of the dynamic behavior and assembly of the cytoskeletal filaments allows eucaryotic cells to build an enormous range of structures from the three basic filament systems. Micrographs that reveal some of these structures are shown in Panel 16–1. Microtubules, which are frequently found in a star-like cytoplasmic array emanating from the center of an interphase cell, can quickly rearrange themselves to form a bipolar *mitotic spindle* during cell division. They can also form motile whips called *cilia* and *flagella* on the surface of the cell, or tightly aligned bundles that serve as tracks for the transport of materials down long neuronal axons. Actin filaments form many types of cell-surface projections. Some of these are dynamic structures, such as the *lamellipodia* and *filopodia* that cells use to explore territory and pull themselves around. Others are stable structures such as the regular bundles of *stereocilia* on the surface of hair cells in the inner ear, which tilt as rigid rods in response to sound. Inside cells, actin filaments can also form either transient or stable structures: the *contractile ring*, for example, assembles transiently to divide cells in two during cytokinesis; more stable arrays allow cells to brace themselves against an underlying substratum and enable muscle to contract. Intermediate filaments line the inner face of the nuclear envelope, forming a protective cage for the cell's DNA; in the cytosol, they are twisted into strong cables that can hold epithelial cell sheets together, help neuronal cells to extend long and robust axons, or allow us to form tough appendages such as hair and fingernails.

Each Type of Cytoskeletal Filament Is Constructed from Smaller Protein Subunits

Cytoskeletal structures frequently reach all the way from one end of the cell to the other, spanning tens or even hundreds of micrometers. Yet the individual protein molecules of the cytoskeleton are generally only a few nanometers in size. The cell is able to build the large structures by the repetitive assembly of large numbers of the small subunits, like building a skyscraper out of bricks. Because these subunits are small, they can diffuse rapidly within cytoplasm, whereas the assembled filaments cannot. In this way, cells can undergo rapid

PANEL 16–1 The Three Major Types of Protein Filaments that Form the Cytoskeleton

ACTIN FILAMENTS

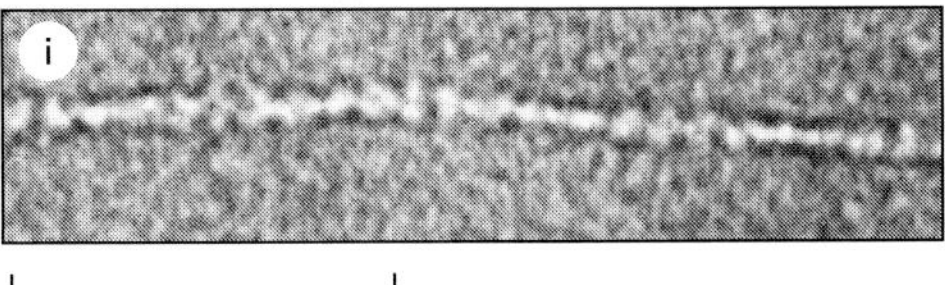

100 nm

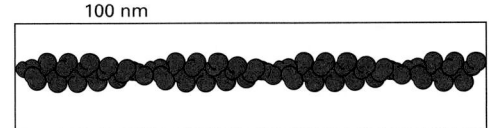

25 nm

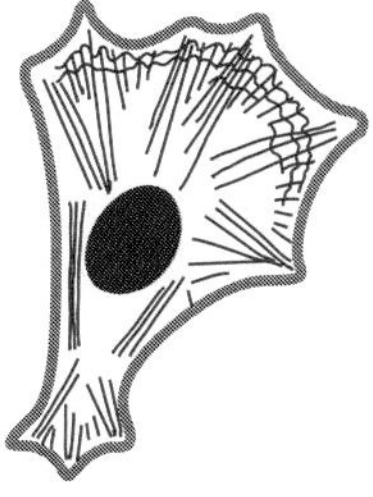

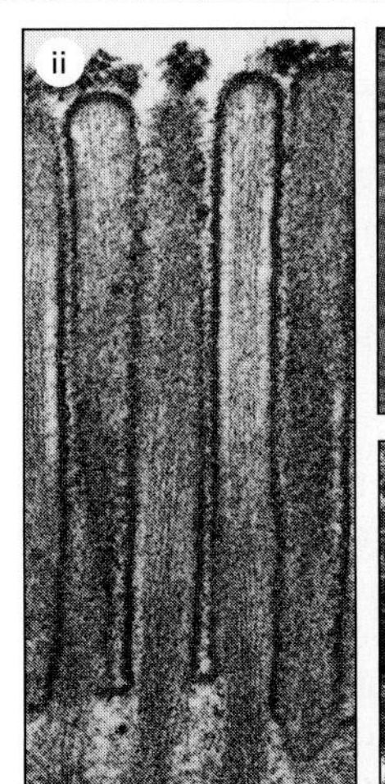

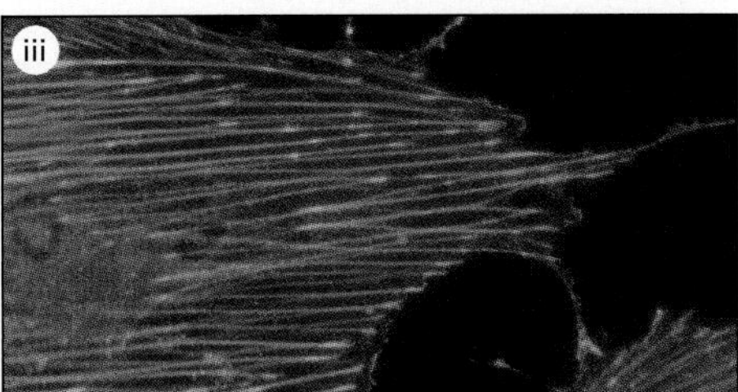

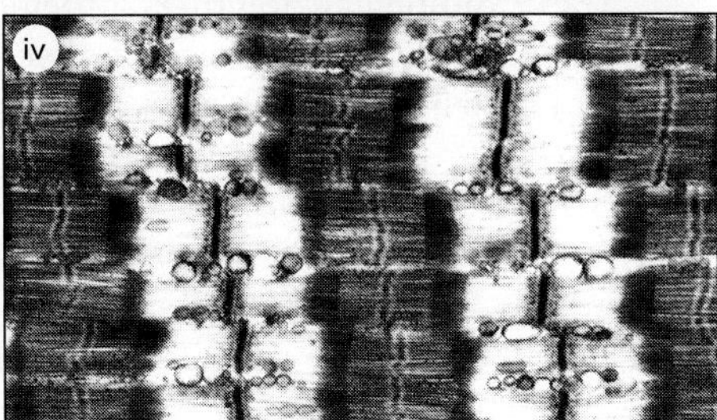

Actin filaments (also known as *microfilaments*) are two-stranded helical polymers of the protein actin. They appear as flexible structures, with a diameter of 5–9 nm, and they are organized into a variety of linear bundles, two-dimensional networks, and three-dimensional gels. Although actin filaments are dispersed throughout the cell, they are most highly concentrated in the *cortex*, just beneath the plasma membrane.

Micrographs courtesy of Roger Craig (i and iv); P.T. Matsudaira and D.R. Burgess (ii); Keith Burridge (iii).

MICROTUBULES

100 nm

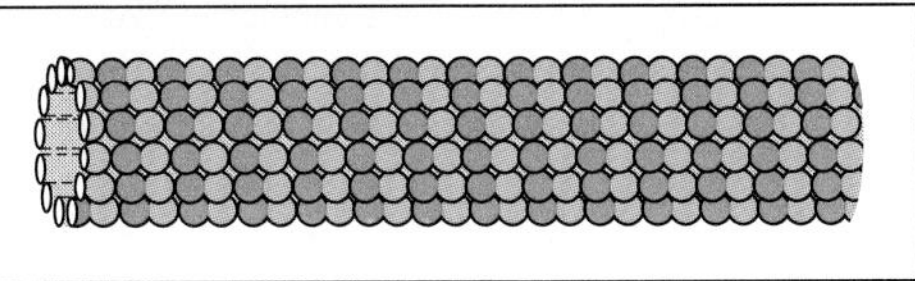

25 nm

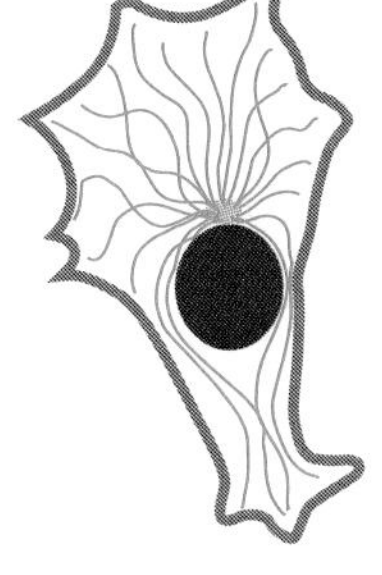

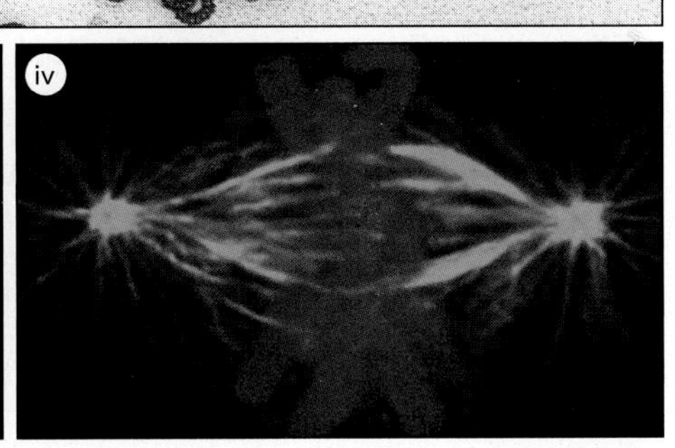

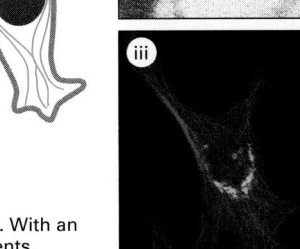

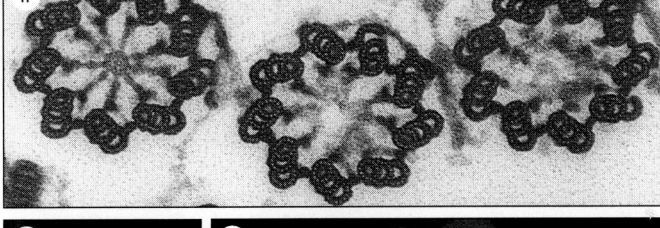

Microtubules are long, hollow cylinders made of the protein tubulin. With an outer diameter of 25 nm, they are much more rigid than actin filaments. Microtubules are long and straight and typically have one end attached to a single microtubule-organizing center (MTOC) called a *centrosome*, as shown here.

Micrographs courtesy of Richard Wade (i); D.T. Woodrow and R.W. Linck (ii); David Shima (iii); A. Desai (iv).

INTERMEDIATE FILAMENTS

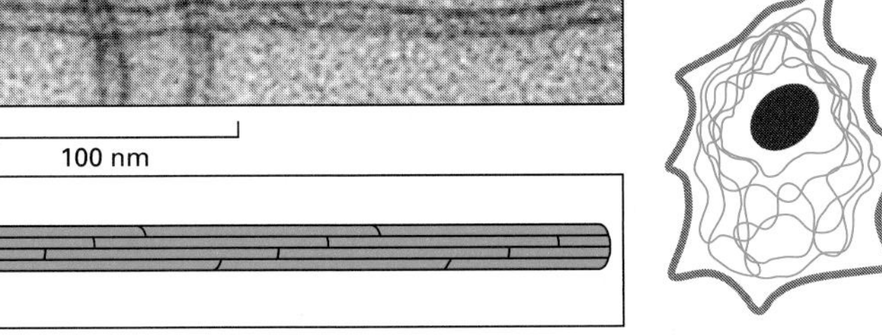

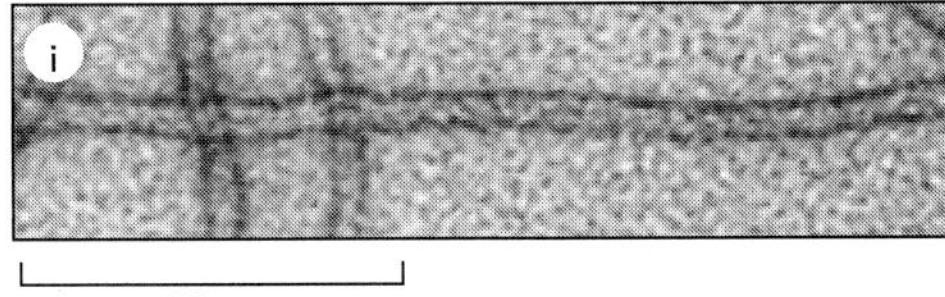

100 nm

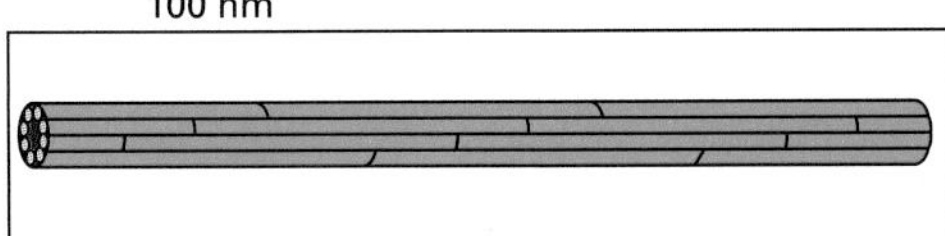

25 nm

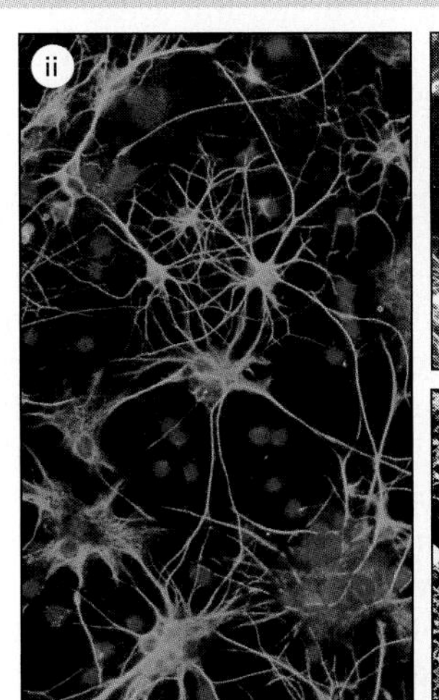

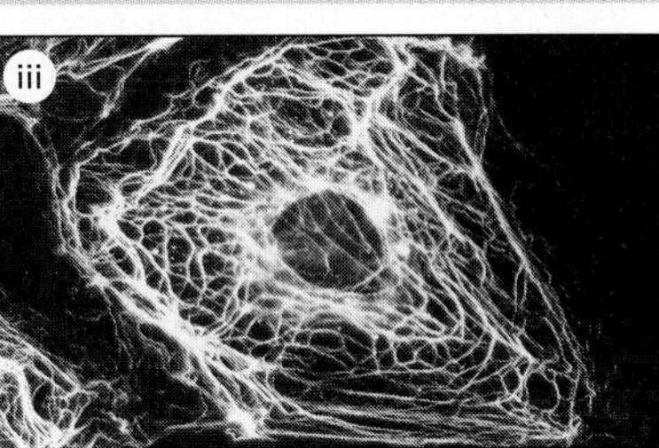

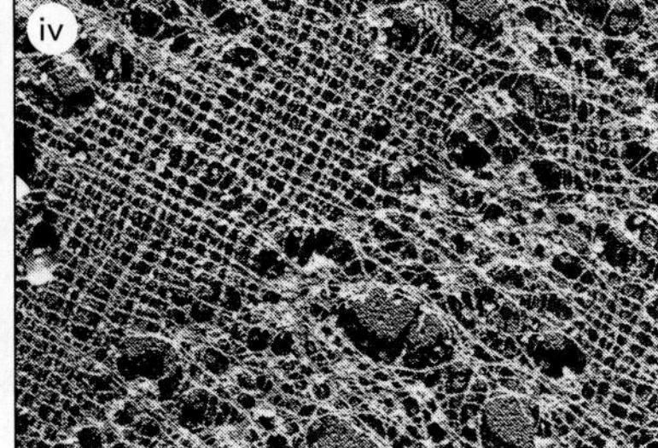

Intermediate filaments are ropelike fibers with a diameter of around 10 nm; they are made of intermediate filament proteins, which constitute a large and heterogeneous family. One type of intermediate filament forms a meshwork called the nuclear lamina just beneath the inner nuclear membrane. Other types extend across the cytoplasm, giving cells mechanical strength. In an epithelial tissue, they span the cytoplasm from one cell-cell junction to another, thereby strengthening the entire epithelium.

Micrographs courtesy of Roy Quinlan (i); Nancy L. Kedersha (ii); Mary Osborn (iii); Ueli Aebi (iv).

structural reorganizations, disassembling filaments at one site and reassembling them at another site far away (Figure 16–2).

Intermediate filaments are made up of smaller subunits that are themselves elongated and fibrous, whereas actin filaments and microtubules are made of subunits that are compact and globular—*actin subunits* for actin filaments, *tubulin subunits* for microtubules. All three types of cytoskeletal filaments form as helical assemblies of subunits (see Figure 3–27), which self-associate, using a combination of end-to-end and side-to-side protein contacts. Differences in the structures of the subunits and the strengths of the attractive forces between them produce critical differences in the stability and mechanical properties of each type of filament.

Many biological polymers—including DNA, RNA, and proteins—are held together by covalent linkages between their subunits. In contrast, the three types of cytoskeletal "polymers" are held together by weak noncovalent interactions, which means that their assembly and disassembly can occur rapidly, without covalent bonds being formed or broken.

Within the cell, hundreds of different cytoskeleton-associated accessory proteins regulate the spatial distribution and the dynamic behavior of the filaments. These accessory proteins bind to the filaments or their subunits to determine the sites of assembly of new filaments, to regulate the partitioning of polymer proteins between filament and subunit forms, to change the kinetics of filament assembly and disassembly, to harness energy to generate force, and to link filaments to one another or to other cellular structures such as organelles and the plasma membrane. In these processes, the accessory proteins bring cytoskeletal structure under the control of extracellular and intracellular signals, including those that trigger the dramatic transformations of the cytoskeleton that occur during the cell cycle. Acting together, the accessory proteins enable a eucaryotic cell to maintain a highly organized but flexible internal structure and, in many cases, to move.

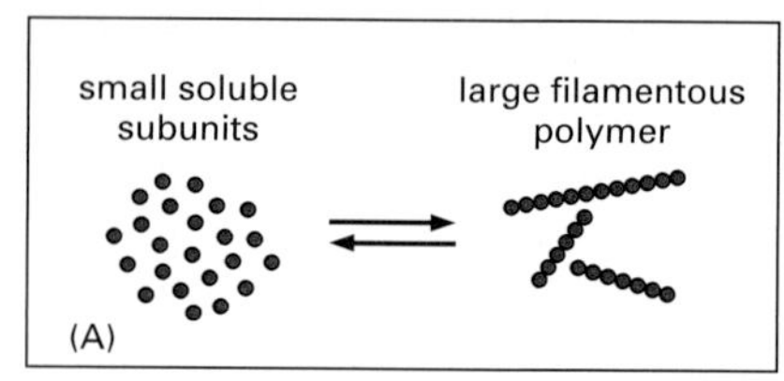

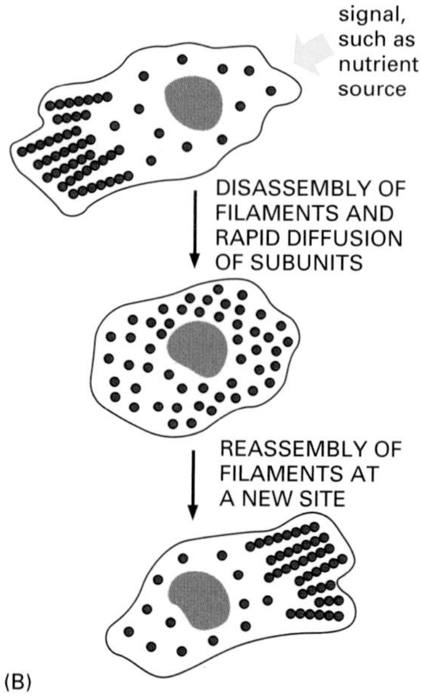

Figure 16–2 The cytoskeleton and changes in cell shape. The formation of protein filaments from much smaller protein subunits allows regulated filament assembly and disassembly to reshape the cytoskeleton. (A) Filament formation from a small protein. (B) Rapid reorganization of the cytoskeleton in a cell in response to an external signal.

Filaments Formed from Multiple Protofilaments Have Advantageous Properties

In general, the linking of protein subunits together to form a filament can be thought of as a simple association reaction. A free subunit binds to the end of a filament that contains n subunits to generate a filament of length $n + 1$. The addition of each subunit to the end of the polymer creates a new end to which yet another subunit can bind. However, the robust cytoskeletal filaments in living cells are not built by simply stringing subunits together in this way in a single straight file. A thousand tubulin monomers, for example, lined up end to end, would be enough to span the diameter of a small eucaryotic cell, but a filament formed in this way would not have enough strength to avoid breakage by ambient thermal energy, unless each subunit were bound very tightly to its neighbor. Such tight binding would limit the rate at which the filaments could disassemble, making the cytoskeleton a static and less useful structure.

Cytoskeletal polymers combine strength with adaptability because they are built out of multiple **protofilaments**—long linear strings of subunits joined end to end—that associate with one another laterally. Typically, the protofilaments twist around one another in a helical lattice. The addition or loss of a subunit at the end of one protofilament makes or breaks one set of longitudinal bonds and either one or two sets of lateral bonds. In contrast, breakage of the composite filament in the middle requires breaking sets of longitudinal bonds in several protofilaments all at the same time (Figure 16–3). The large energy difference between these two processes allows most cytoskeletal filaments to resist thermal breakage, while leaving the filament ends as dynamic structures in which addition and loss of subunits occur rapidly.

As with other specific protein–protein interactions, the subunits in a cytoskeletal filament are held together by a large number of hydrophobic interactions and weak noncovalent bonds (see Figure 3–5). The locations and types of subunit–subunit contacts are different for the different cytoskeletal filaments.

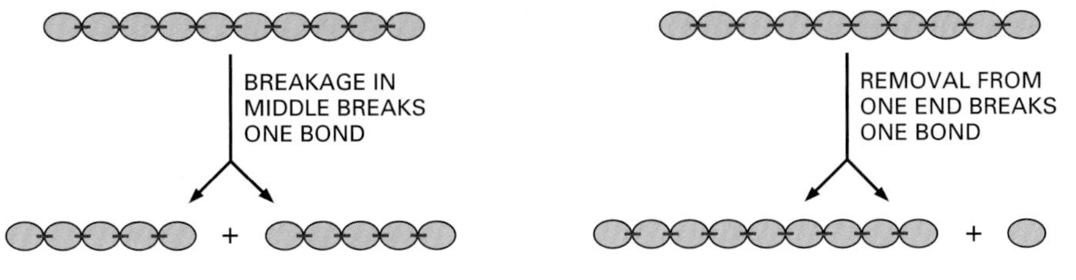

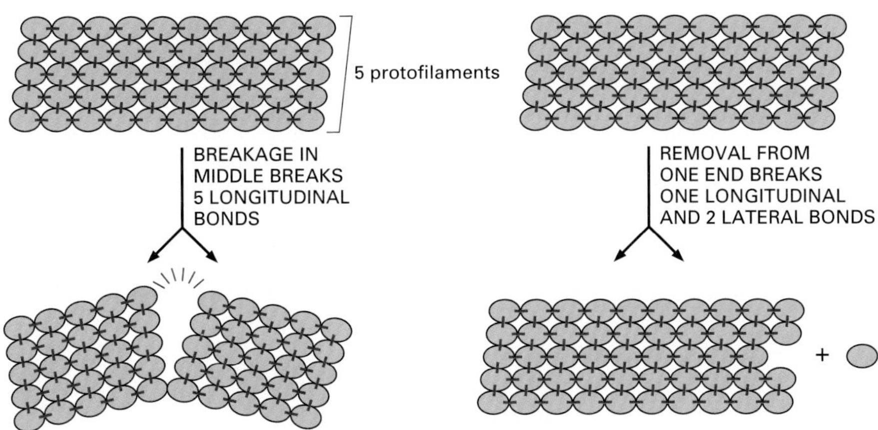

Figure 16–3 The thermal stability of cytoskeletal filaments with dynamic ends. Formation of a cytoskeletal filament from more than one protofilament allows the ends to be dynamic, while the filaments themselves are resistant to thermal breakage. In this hypothetical example, the stable filament is formed from five protofilaments. The bonds holding the subunits together in the filaments are shown in *red*.

Intermediate filaments, for example, assemble by forming strong lateral contacts between α-helical coiled coils, which extend over most of the length of each elongated fibrous subunit. Because the individual subunits are staggered in the filament, intermediate filaments tolerate stretching and bending, forming strong rope-like structures (Figure 16–4). Microtubules, by contrast, are built from globular subunits held together primarily by longitudinal bonds, and the lateral bonds holding the 13 protofilaments together are comparatively weak. For this reason, microtubules break much more easily than do intermediate filaments.

Nucleation Is the Rate-limiting Step in the Formation of a Cytoskeletal Polymer

There is an important additional consequence of the multiple-protofilament organization of cytoskeletal polymers. Short oligomers composed of a few subunits can assemble spontaneously, but they are unstable and disassemble readily because each monomer is bonded only to a few other monomers. For a new large filament to form, subunits must assemble into an initial aggregate, or

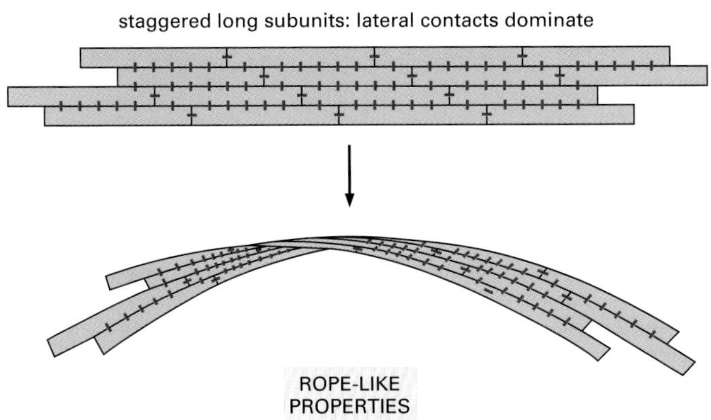

Figure 16–4 A strong filament formed from elongated fibrous subunits with strong lateral contacts. Intermediate filaments are formed in this way and are consequently especially resistant to bending or stretching forces.

ON RATES AND OFF RATES

A linear polymer of protein molecules, such as an actin filament or a microtubule, assembles (polymerizes) and disassembles (depolymerizes) by the addition and removal of subunits at the ends of the polymer. The rate of addition of these subunits (called monomers) is given by the rate constant k_{on}, which has units of M^{-1} sec^{-1}. The rate of loss is given by k_{off} (units of sec^{-1}).

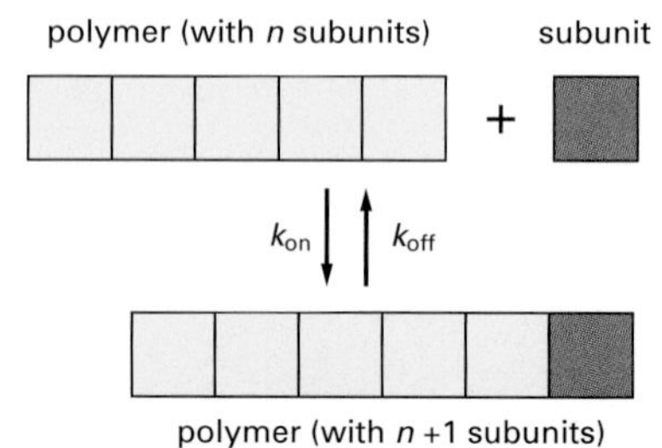

NUCLEATION

A helical polymer is stabilized by multiple contacts between adjacent subunits. In the case of actin, two actin molecules bind relatively weakly to each other, but addition of a third actin monomer to form a trimer makes the entire group more stable.

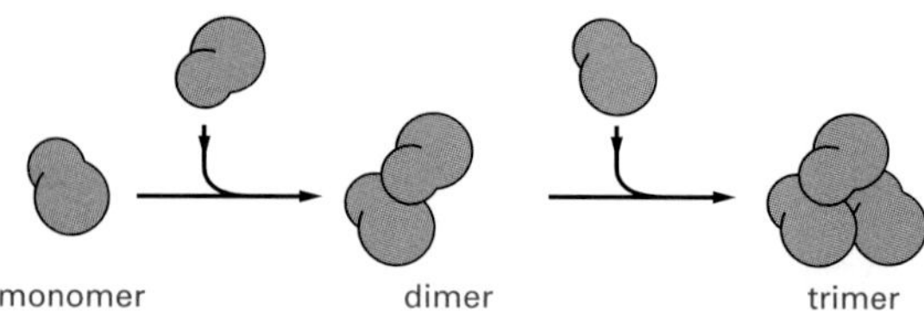

Further monomer addition can take place onto this trimer, which therefore acts as a nucleus for polymerization. For tubulin, the nucleus is larger and has a more complicated structure (possibly a ring of 13 or more tubulin molecules)—but the principle is the same.

The assembly of a nucleus is relatively slow, which explains the lag phase seen during polymerization. The lag phase can be reduced or abolished entirely if premade nuclei, such as fragments of already polymerized microtubules or actin filaments, are added.

THE CRITICAL CONCENTRATION

The number of monomers that add to the polymer (actin filament or microtubule) per second will be proportional to the concentration of the free subunit ($k_{on}C$), but the subunits will leave the polymer end at a constant rate (k_{off}) that does not depend on C. As the polymer grows, subunits are used up, and C is observed to drop until it reaches a constant value, called the critical concentration (C_c). At this concentration the rate of subunit addition equals the rate of subunit loss.

At this equilibrium,

$$k_{on}\, C = k_{off}$$

so that

$$C_c = \frac{k_{off}}{k_{on}} = \frac{1}{K}$$

(where K is the equilibrium constant for subunit addition; see Figure 3–44).

TIME COURSE OF POLYMERIZATION

The assembly of a protein into a long helical polymer such as a cytoskeletal filament or a bacterial flagellum typically shows the following time course:

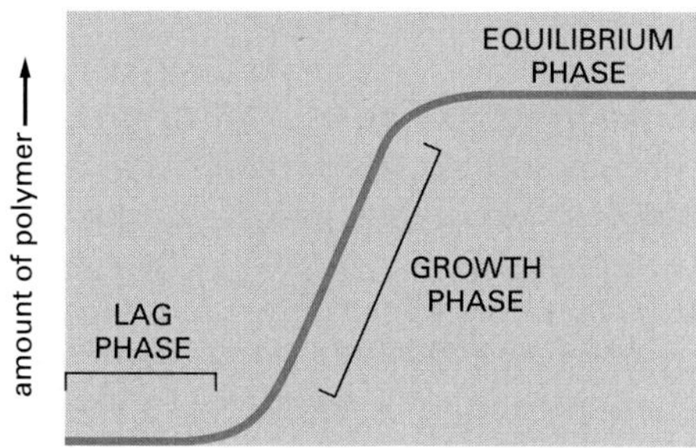

The lag phase corresponds to time taken for nucleation.

The growth phase occurs as monomers add to the exposed ends of the growing filament, causing filament elongation.

The equilibrium phase, or steady state, is reached when the growth of the polymer due to monomer addition is precisely balanced by the shrinkage of the polymer due to disassembly back to monomers.

PLUS AND MINUS ENDS

The two ends of an actin filament or microtubule polymerize at different rates. The fast-growing end is called the plus end, whereas the slow-growing end is called the minus end. The difference in the rates of growth at the two ends is made possible by changes in the conformation of each subunit as it enters the polymer.

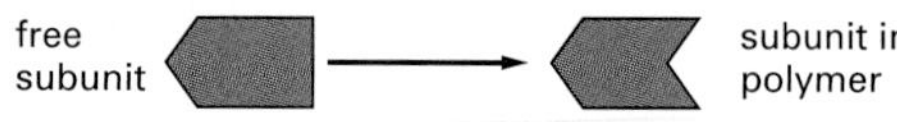

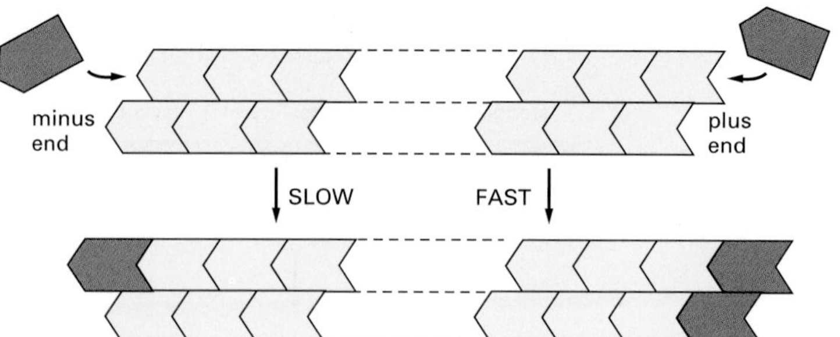

This conformational change affects the rates at which subunits add to the two ends.

Even though k_{on} and k_{off} will have different values for the plus and minus ends of the polymer, their ratio k_{off}/k_{on}—and hence C_c—must be the same at both ends for a simple polymerization reaction (no ATP or GTP hydrolysis). This is because exactly the same subunit interactions are broken when a subunit is lost at either end, and the final state of the subunit after dissociation is identical. Therefore, the ΔG for subunit loss, which determines the equilibrium constant for its association with the end, is identical at both ends: if the plus end grows four times faster than the minus end, it must also shrink four times faster. Thus, for $C > C_c$, both ends grow; for $C < C_c$, both ends shrink.

The nucleoside triphosphate hydrolysis that accompanies actin and tubulin polymerization removes this constraint.

NUCLEOTIDE HYDROLYSIS

Each actin molecule carries a tightly bound ATP molecule that is hydrolyzed to a tightly bound ADP molecule soon after its assembly into polymer. Similarly, each tubulin molecule carries a tightly bound GTP that is converted to a tightly bound GDP molecule soon after the molecule assembles into the polymer.

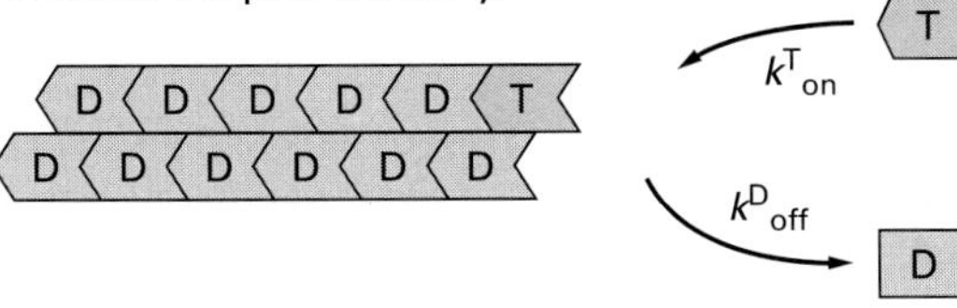

(T = monomer carrying ATP or GTP)
(D = monomer carrying ADP or GDP)

Hydrolysis of the bound nucleotide reduces the binding affinity of the subunit for neighboring subunits and makes it more likely to dissociate from each end of the filament (see Figure 16–11 for a possible mechanism). It is usually the T form that adds to the filament and the D form that leaves.

Considering events at the plus end only:

D D D D D T
D D D D D D
T
k^{T}_{on}
k^{D}_{off}
D

As before, the polymer will grow until $C = C_c$. For illustrative purposes, we can ignore k^{D}_{on} and k^{T}_{off} since they are usually very small, so that polymer growth ceases when

$$k^{T}_{on}\, C = k^{D}_{off} \qquad \text{or} \qquad C_c = \frac{k^{D}_{off}}{k^{T}_{on}}$$

This is a steady state and not a true equilibrium, because the ATP or GTP that is hydrolyzed must be replenished by a nucleotide exchange reaction of the free subunit (D → T).

ATP CAPS AND GTP CAPS

The rate of addition of subunits to a growing actin filament or microtubule can be faster than the rate at which their bound nucleotide is hydrolyzed. Under such conditions, the end has a "cap" of subunits containing the nucleoside triphosphate—an ATP cap on an actin filament or a GTP cap on a microtubule.

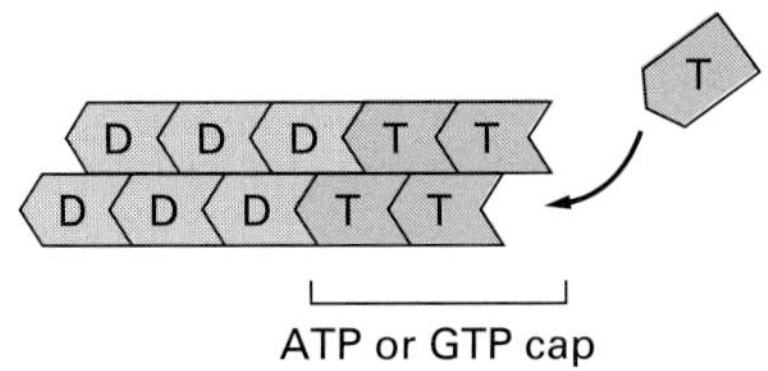

ATP or GTP cap

DYNAMIC INSTABILITY and TREADMILLING are two behaviors observed in cytoskeletal polymers. Both are associated with nucleoside triphosphate hydrolysis. Dynamic instability is believed to predominate in microtubules, whereas treadmilling may predominate in actin filaments.

TREADMILLING

One consequence of the nucleotide hydrolysis that accompanies polymer formation is to change the critical concentration at the two ends of the polymer. Since k^{D}_{off} and k^{T}_{on} refer to different reactions, their ratio k^{D}_{off}/k^{T}_{on} need not be the same at both ends of the polymer, so that:

$$C_c \text{ (minus end)} > C_c \text{ (plus end)}$$

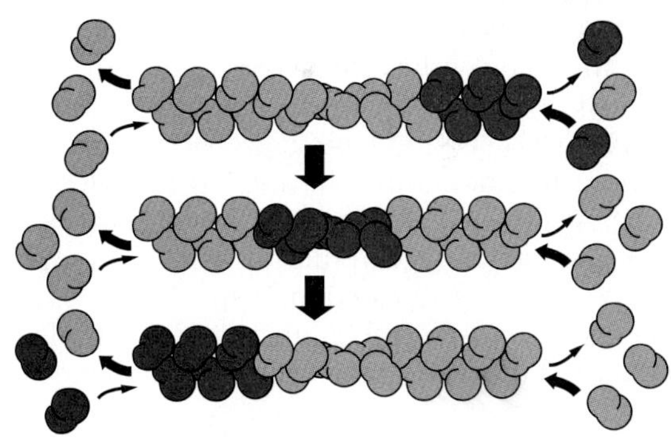

Thus, if both ends of a polymer are exposed, polymerization proceeds until the concentration of free monomer reaches a value that is above C_c for the plus end but below C_c for the minus end. At this steady state, subunits undergo a net assembly at the plus end and a net disassembly at the minus end at an identical rate. The polymer maintains a constant length, even though there is a net flux of subunits through the polymer, known as treadmilling.

DYNAMIC INSTABILITY

Microtubules depolymerize about 100 times faster from an end containing GDP tubulin than from one containing GTP tubulin. A GTP cap favors growth, but if it is lost, then depolymerization ensues.

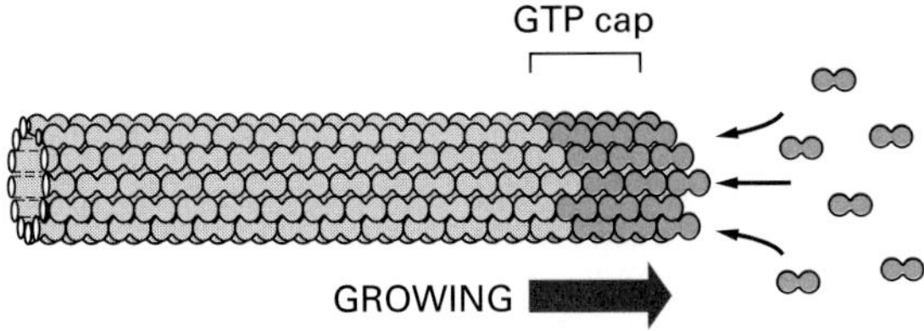

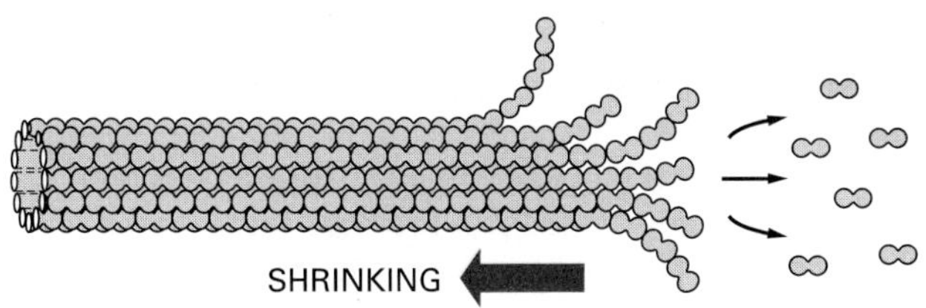

Individual microtubules can therefore alternate between a period of slow growth and a period of rapid disassembly, a phenomenon called dynamic instability.

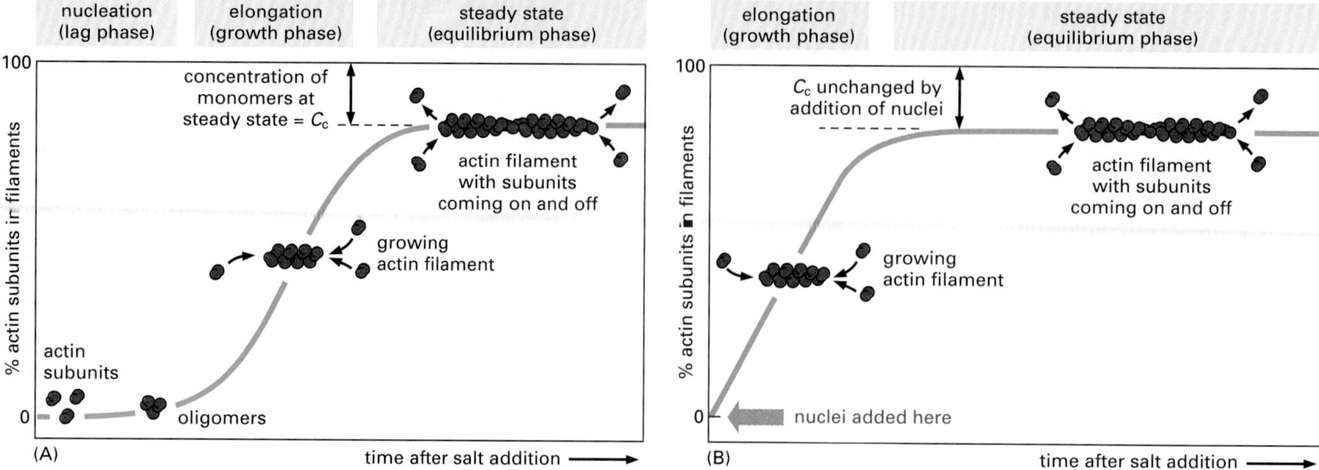

Figure 16–5 The time course of actin polymerization in a test tube. (A) Polymerization is begun by raising the salt concentration in a solution of pure actin subunits. (B) Polymerization is begun in the same way, but with preformed fragments of actin filaments present to act as nuclei for filament growth.

nucleus, that is stabilized by many subunit–subunit contacts and can then elongate rapidly by addition of more subunits. The initial process of nucleus assembly is called filament *nucleation*, and it can take quite a long time, depending on how many subunits must come together to form the nucleus. The instability of smaller aggregates creates a kinetic barrier to nucleation, which is easily observed in a solution of pure actin or tubulin—the subunits of actin filaments and microtubules, respectively. When polymerization is initiated in a test tube containing a solution of pure individual subunits (by raising the temperature or raising the salt concentration), there is an initial lag phase, during which no filaments are observed. During this lag phase, however, nuclei are assembling slowly, so that the lag phase is followed by a phase of rapid filament elongation, during which subunits add quickly onto the ends of the nucleated filaments (Figure 16–5A). Finally, the system approaches a steady state at which the rate of addition of new subunits to the filament ends is exactly balanced by the rate of subunit dissociation from the ends. The concentration of free subunits left in solution at this point is called the *critical concentration*, C_c. As explained in Panel 16–2, the value of the critical concentration is equal to the rate constant for subunit loss divided by the rate constant for subunit addition—that is, $C_c = k_{off} / k_{on}$.

The lag phase in filament growth is eliminated if preexisting nuclei (such as filament fragments that have been chemically cross-linked) are added to the solution at the beginning of the polymerization reaction (Figure 16–5B). The cell takes great advantage of this nucleation requirement: it uses special proteins to catalyze filament nucleation at specific sites, thereby determining the location where new cytoskeletal filaments are assembled. Indeed, this type of regulation of filament nucleation is a primary way that cells control their shape and movement.

The Tubulin and Actin Subunits Assemble Head-to-Tail, Creating Filaments that Are Polar

Microtubules are formed from protein subunits of **tubulin**. The tubulin subunit is itself a heterodimer formed from two closely related globular proteins called *α-tubulin* and *β-tubulin,* tightly bound together by noncovalent bonds (Figure 16–6). These two tubulin proteins are found only in this complex. Each α or β monomer has a binding site for one molecule of GTP. The GTP that is bound to the α-tubulin monomer is physically trapped at the dimer interface and is never hydrolyzed or exchanged; it can therefore be considered to be an integral part of the tubulin heterodimer structure. The nucleotide on the β-tubulin, in contrast, may be in either the GTP or the GDP form, and it is exchangeable. As we see, the hydrolysis of GTP at this site to produce GDP has an important effect on microtubule dynamics.

A microtubule is a stiff, hollow cylindrical structure built from 13 parallel protofilaments, each composed of alternating α-tubulin and β-tubulin

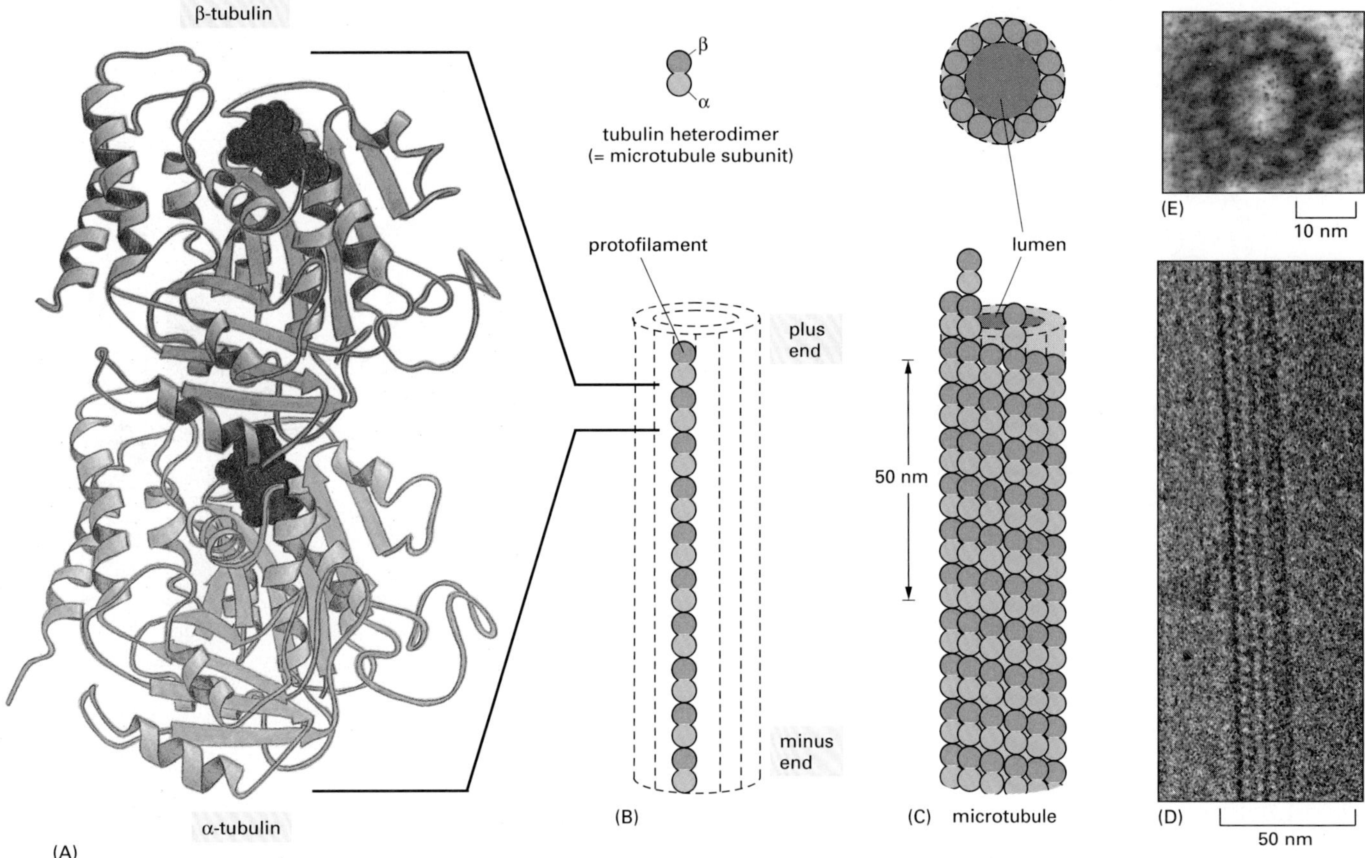

molecules. When the tubulin heterodimers assemble to form the hollow cylindrical microtubule, they generate two new types of protein–protein contacts. Along the longitudinal axis of the microtubule, the "top" of one β-tubulin molecule forms an interface with the "bottom" of the α-tubulin molecule in the adjacent dimer subunit. This interface is very similar to the interface holding the α and β monomers together in the dimer subunit, and the binding energy is strong. Perpendicular to these interactions, lateral contacts are formed between neighboring protofilaments. In this dimension, the main lateral contacts are between monomers of the same type (α–α and β–β). Together, the longitudinal and lateral contacts are repeated in the regular helical lattice of the microtubule. Because most of the subunits in a microtubule are held in place by multiple contacts within the lattice, the addition and loss of subunits occurs almost exclusively at the microtubule ends (see Figure 16–3).

Each protofilament in a microtubule is assembled from subunits that all point in the same direction, and the protofilaments themselves are aligned in parallel (in Figure 16–6, for example, the α-tubulin is down and the β-tubulin up in each heterodimer). Therefore, the microtubule itself has a distinct structural polarity, with α-tubulins exposed at one end and β-tubulins exposed at the other end.

The actin subunit is a single globular polypeptide chain and is thus a monomer rather than a dimer. Like tubulin, each actin subunit has a binding site for a nucleotide, but for actin the nucleotide is ATP (or ADP) rather than GTP (or GDP) (Figure 16–7). As for tubulin, the actin subunits assemble head-to-tail to generate filaments with a distinct structural polarity. The actin filament can be considered to consist of two parallel protofilaments that twist around each other in a right-handed helix. Actin filaments are relatively flexible compared with the hollow cylindrical microtubules. But in a living cell, actin filaments are cross-linked and bundled together by a variety of accessory proteins (see below), making these large-scale actin structures much stronger than an individual actin filament.

Figure 16–6 The structure of a microtubule and its subunit. (A) The subunit of each protofilament is a tubulin heterodimer, formed from a very tightly linked pair of α- and β-tubulin monomers. The GTP molecule in the α-tubulin monomer is so tightly bound that it can be considered an integral part of the protein. The GTP molecule in the β-tubulin monomer, however, is less tightly bound and has an important role in filament dynamics. Both nucleotides are shown in *red*. (B) One tubulin subunit (α–β heterodimer) and one protofilament are shown schematically. Each protofilament consists of many adjacent subunits with the same orientation. (C) The microtubule is a stiff hollow tube formed from 13 protofilaments aligned in parallel. (D) A short segment of a microtubule viewed in an electron microscope. (E) Electron micrograph of a cross section of a microtubule showing a ring of 13 distinct protofilaments. (D, courtesy of Richard Wade; E, courtesy of Richard Linck.)

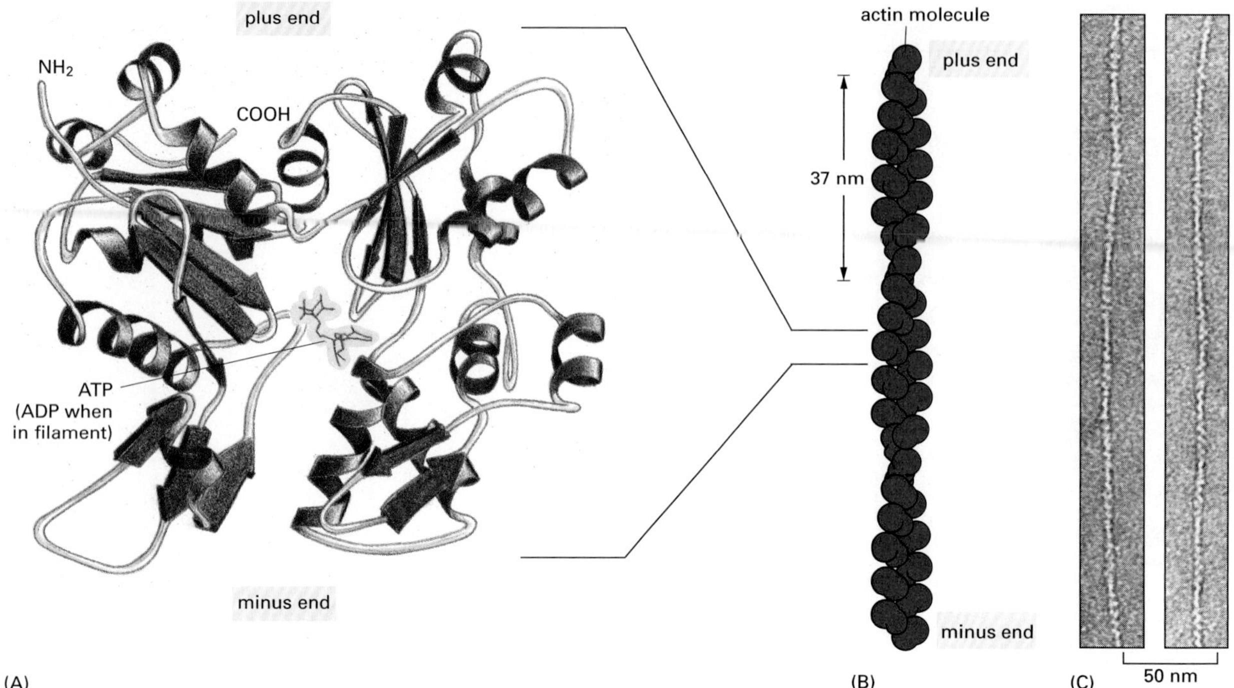

Figure 16–7 The structures of an actin monomer and actin filament. (A) The actin monomer has a nucleotide (either ATP or ADP) bound in a deep cleft in the center of the molecule. (B) Arrangement of monomers in a filament. Although the filament is often described as a single helix of monomers, it can also be thought of as consisting of two protofilaments, held together by lateral contacts, which wind around each other as two parallel strands of a helix, with a twist repeating every 37 nm. All the subunits within the filament have the same orientation. (C) Electron micrographs of negatively stained actin filaments. (C, courtesy of Roger Craig.)

The Two Ends of a Microtubule and of an Actin Filament Are Distinct and Grow at Different Rates

The structural polarity of actin filaments and microtubules is created by the regular, parallel orientation of all of their subunits. This orientation makes the two ends of each polymer different in ways that have a profound effect on filament growth rates. Addition of a subunit to either end of a filament of n subunits results in a filament of $n + 1$ subunits. In the absence of ATP or GTP hydrolysis, the free energy difference, and therefore the equilibrium constant (and the critical concentration), must be the same for addition of subunits at either end of the polymer. In this case, the ratio of the forward and backward rate constants, k_{on}/k_{off}, must be identical at the two ends, even though the absolute values of these rate constants may be very different at each end.

In a structurally polar filament, the kinetic rate constants for association and dissociation—k_{on} and k_{off}, respectively—are often much greater at one end than at the other. Thus, if an excess of purified subunits is allowed to assemble onto marked fragments of preformed filaments, one end of each fragment elongates much faster than the other (Figure 16–8). If filaments are rapidly diluted so that the free subunit concentration drops below the critical concentration, the fast-growing end also depolymerizes fastest. The more dynamic of the two ends of a filament, where both growth and shrinkage are fast, is called the **plus end**, and the other end is called the **minus end**.

On microtubules, α subunits are exposed at the minus end, and β subunits are exposed at the plus end. On actin filaments, the ATP-binding cleft on the monomer points toward the minus end. (For historical reasons, the plus ends of actin filaments are usually referred to as "barbed" ends, and minus ends as "pointed" ends, because of the arrowhead appearance of myosin heads when bound along the filament.)

Filament elongation proceeds spontaneously when the free energy change (ΔG) for addition of the soluble subunit is less than zero. This is the case when the concentration of subunits in solution exceeds the critical concentration. Likewise, filament depolymerization proceeds spontaneously when this free energy change is greater than zero. A cell can couple an energetically unfavorable process to these spontaneous processes; thus, the free energy released during spontaneous filament polymerization or depolymerization can be used to do mechanical work—in particular, to push or pull an attached load. For example, elongating microtubules can help push out membranes, and shrinking microtubules can help pull mitotic chromosomes away from their sisters during anaphase. Similarly, elongating actin filaments help protrude the leading edge of motile cells, as we discuss later.

Filament Treadmilling and Dynamic Instability Are Consequences of Nucleotide Hydrolysis by Tubulin and Actin

Thus far, our discussion of filament dynamics has ignored a critical fact that applies to both actin filaments and microtubules. In addition to their ability to form noncovalent polymers, the actin and tubulin subunits are both enzymes that can catalyze the hydrolysis of a nucleoside triphosphate, ATP or GTP, respectively. For the free subunits, this hydrolysis proceeds very slowly; however, it is accelerated when the subunits are incorporated into filaments. Shortly after incorporation of an actin or tubulin subunit into a filament, nucleotide hydrolysis occurs; the free phosphate group is released from each subunit, but the nucleoside diphosphate remains trapped in the filament structure. (On tubulin, the nucleotide-binding site lies at the interface between two neighboring subunits—see Figure 16–6, whereas in actin, the nucleotide is deep in a cleft near the center of the subunit—see Figure 16–7.) Thus, two different types of filament structures can exist, one with the "T form" of the nucleotide bound (ATP for actin, GTP for tubulin), and one with the "D form" bound (ADP for actin, GDP for tubulin).

When the nucleotide is hydrolyzed, much of the free energy released by cleavage of the high-energy phosphate–phosphate bond is stored in the polymer lattice, making the free energy change upon dissociation of the D-form polymer higher than the free energy change upon dissociation of the T-form polymer. Consequently, the equilibrium constant for dissociation $K_D = k_{off} / k_{on}$ for the D-form polymer, which is numerically equal to its critical concentration [C_c(D)], is larger than the corresponding equilibrium constant for the T-form polymer. Thus, C_c(D) is greater than C_c(T). For certain concentrations of free subunits, D-form polymers will therefore shrink while T-form polymers grow.

In living cells, most of the free subunits are in the T form, as the free concentration of both ATP and GTP is much higher than that of ADP and GDP. The longer the time that subunits have been in the polymer lattice, the more likely they are to have hydrolyzed their bound nucleotide. Whether the subunit at the very end of a filament is in the T or the D form will depend on the relative rates of hydrolysis and subunit addition. If the rate of subunit addition is high, that is if the filament is growing rapidly, then it is likely that a new subunit will add on to the polymer before the nucleotide in the previously added subunit has been hydrolyzed, so that the tip of the polymer remains in the T form, forming an *ATP cap* or *GTP cap*. However, if the rate of subunit addition is low, hydrolysis may occur before the next subunit is added, and the tip of the filament will then be in the D form.

The rate of subunit addition at the end of a filament is the product of the free subunit concentration and the rate constant k_{on}. The k_{on} is much faster for the plus end of a filament than for the minus end because of a structural difference between the two ends (see Panel 16–2, pp. 912–913). At an intermediate concentration of free subunits, it is therefore possible for the rate of subunit addition to be faster than nucleotide hydrolysis at the plus end, but slower than nucleotide hydrolysis at the minus end. In this case, the plus end of the filament remains in the T conformation, while the minus end adopts the D conformation. As just explained, the D form has a higher critical concentration than the T form; in other words, the D form leans more readily toward disassembly, while the T form leans more readily toward assembly. If the concentration of free subunits in solution is in an intermediate range—higher than the critical concentration of the T form (that is, the plus end), but lower than the critical concentration of the D form (that is, the minus end)—the filament adds subunits at the plus end, and simultaneously loses subunits from the minus end. This leads to the remarkable property of filament **treadmilling** (Figure 16–9 and Panel 16–2).

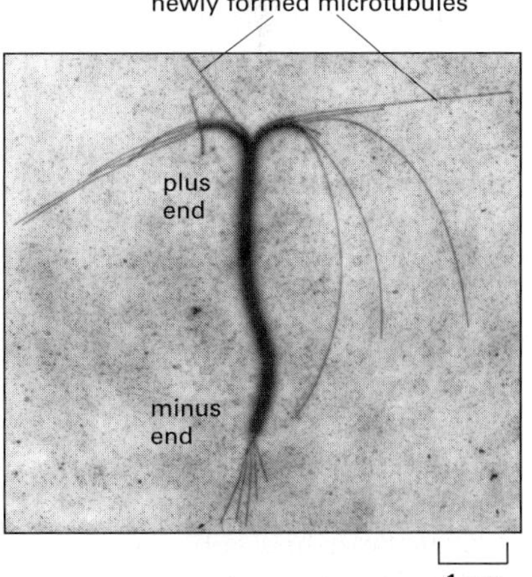

Figure 16–8 The preferential growth of microtubules at the plus end. Both microtubules and actin filaments grow faster at one end than at the other. In this case, a stable bundle of microtubules obtained from the core of a cilium (discussed later) was incubated with tubulin subunits under polymerizing conditions. Microtubules grow fastest from the plus end of the microtubule bundle, the end at the *top* in this micrograph. (Courtesy of Gary Borisy.)

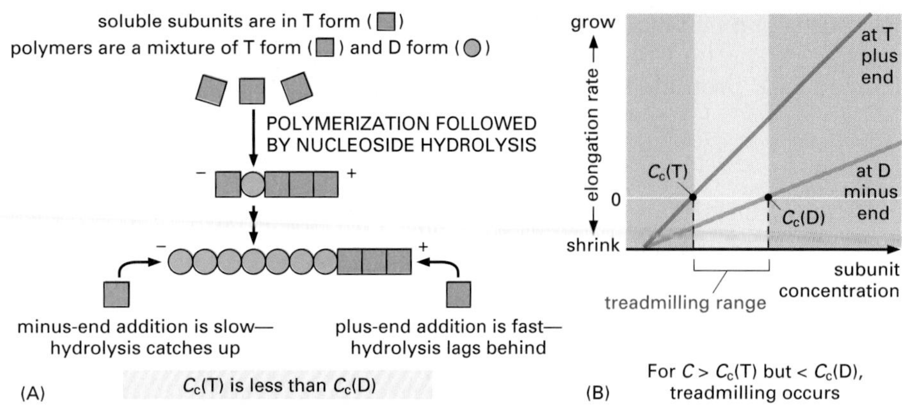
soluble subunits are in T form
polymers are a mixture of T form and D form
POLYMERIZATION FOLLOWED BY NUCLEOSIDE HYDROLYSIS
minus-end addition is slow—hydrolysis catches up
plus-end addition is fast—hydrolysis lags behind
(A)
C_c(T) is less than C_c(D)
grow
elongation rate
shrink
at T plus end
at D minus end
C_c(T)
C_c(D)
subunit concentration
treadmilling range
For $C > C_c$(T) but $< C_c$(D), treadmilling occurs
(B)

Figure 16–9 The treadmilling of an actin filament or microtubule, made possible by the nucleoside triphosphate hydrolysis that follows subunit addition. (A) Explanation for the different critical concentrations (C_c) at the plus and minus ends. Subunits with bound nucleoside triphosphate (T-form subunits) polymerize at both ends of a growing filament, and then undergo nucleotide hydrolysis in the filament lattice. As the filament grows, elongation is faster than hydrolysis at the plus end in this example, and the terminal subunits at this end are therefore always in the T form. However, hydrolysis is faster than elongation at the minus end, and so terminal subunits at this end are in the D form. (B) Treadmilling occurs at intermediate concentrations of free subunits. The critical concentration for polymerization on a filament end in the T form is lower than for a filament end in the D form. If the actual subunit concentration is somewhere between these two values, the plus end grows while the minus end shrinks, resulting in treadmilling.

During treadmilling, subunits are recruited at the plus end of the polymer in the T form and shed from the minus end in the D form. The ATP or GTP hydrolysis that occurs along the way gives rise to the difference in the free energy of the association/dissociation reactions at the plus and minus ends of the actin filament or microtubule and thereby makes treadmilling possible. At a particular intermediate subunit concentration, the filament growth at the plus end is exactly balanced by the filament shrinkage at the minus end. Now, the subunits cycle rapidly between the free and filamentous states, while the total length of the filament remains unchanged. This "steady-state treadmilling" requires a constant consumption of energy in the form of nucleoside triphosphate hydrolysis. While the extent of treadmilling inside the cell is uncertain, the treadmilling of single filaments has been observed *in vitro* for actin, and a phenomenon that looks like treadmilling can be observed in live cells for individual microtubules (Figure 16–10).

The kinetic differences between the behavior of the T form and the D form have another important consequence for the behaviors of filaments. If the rate of subunit addition at one end is similar in magnitude to the rate of hydrolysis, there is a finite probability that this end will start out in a T form, but that hydrolysis will eventually "catch up" with the addition and transform the end to a D form. This transformation is sudden and random, with a certain probability per unit time. Thus, in a population of microtubules, at any instant some of the ends are in the T form, and some are in the D form, with the ratio depending on the hydrolysis rate and the free subunit concentration.

Suppose that the concentration of free subunits is intermediate between the critical concentration for a T-form end and the critical concentration for a D-form end (that is, in the same range of concentrations where treadmilling is observed). Now, any end that happens to be in the T form will grow, whereas any end that happens to be in the D form will shrink. On a single filament, an end might grow for a certain length of time in a T form, but then suddenly change to the D form and begin to shrink rapidly, even while the free subunit concentration is held constant. At some later time, it might then regain a T-form and begin to grow again. This rapid interconversion between a growing and shrinking state, at a uniform free subunit concentration, is called **dynamic instability** (Figure 16–11). The change to rapid shrinkage is called a *catastrophe,* while the change to

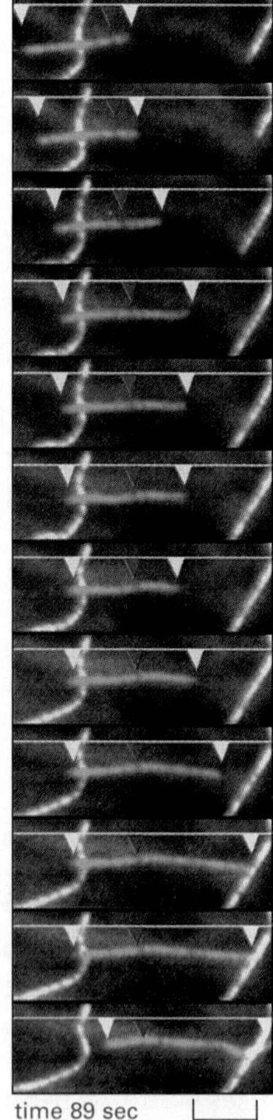
time 0 sec
time 89 sec
2 μm

Figure 16–10 Treadmilling behavior of a microtubule, as observed in a living cell. A cell was injected with tubulin that had been covalently linked to the fluorescent dye rhodamine, so that approximately 1 tubulin subunit in 20 was fluorescent. The fluorescence of individual microtubules was then observed with a sensitive electronic camera. The microtubule shown appears to be sliding from *left* to *right*, but, in fact, the microtubule lattice remains stationary (as shown by the dark mark indicated by the *red* arrowhead), while the plus end (on the *right*) grows and the minus end (on the *left*) shrinks. At the same time, the plus end is displaying dynamic instability. (From C.M. Waterman-Storer and E.D. Salmon, *J. Cell Biol.* 139:417–434, 1997. © The Rockefeller University Press.)

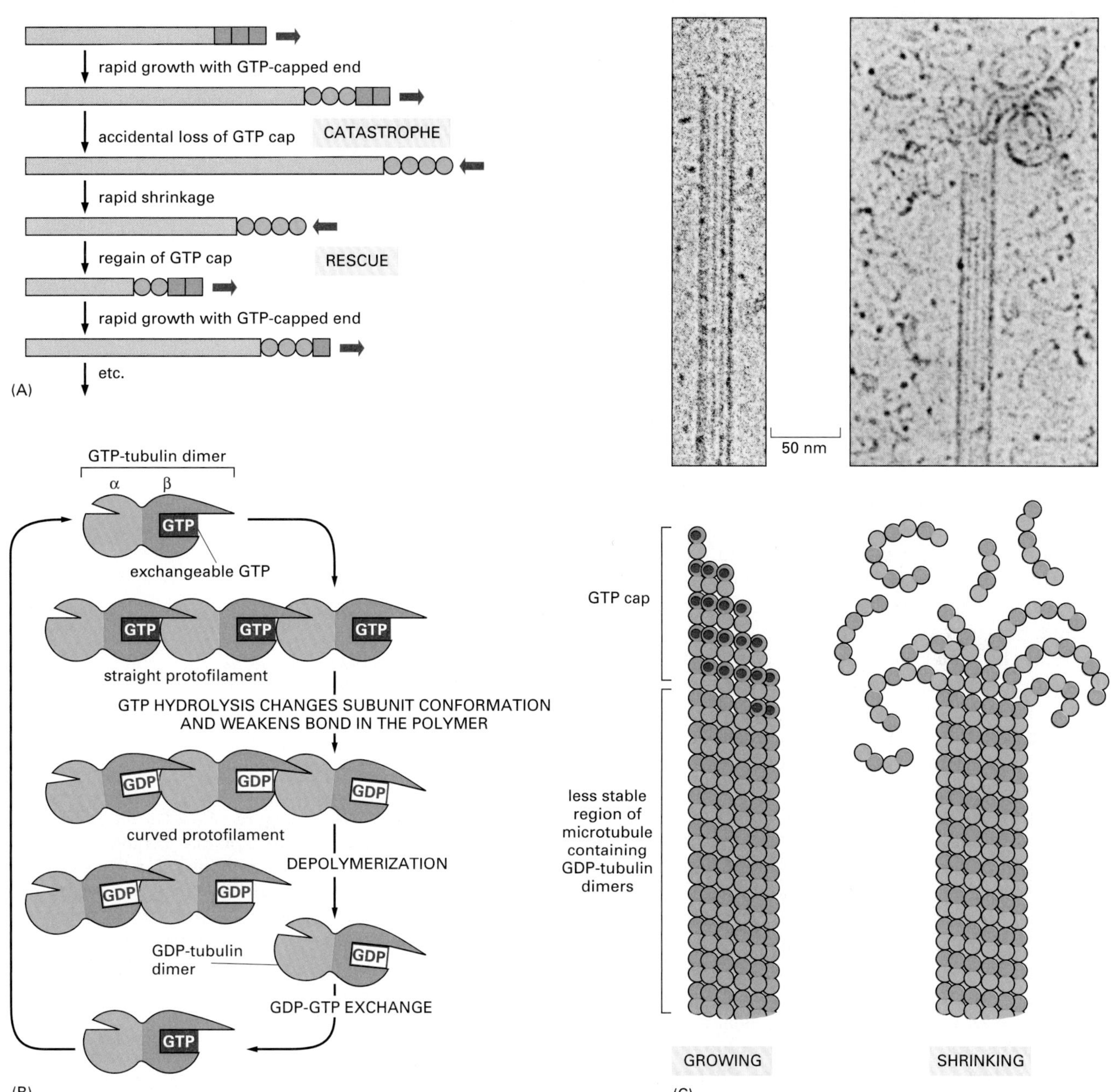

Figure 16–11 Dynamic instability due to the structural differences between a growing and a shrinking microtubule end. (A) If the free tubulin concentration in solution is between the critical values indicated in Figure 16–9, a single microtubule end may undergo transitions between a growing state and a shrinking state. A growing microtubule has GTP-containing subunits at its end, forming a GTP cap. If nucleotide hydrolysis proceeds more rapidly than subunit addition, the cap is lost and the microtubule begins to shrink, an event called a "catastrophe." But GTP-containing subunits may still add to the shrinking end, and if enough add to form a new cap, then microtubule growth resumes, an event called "rescue." (B) Model for the structural consequences of GTP hydrolysis in the microtubule lattice. The addition of GTP-containing tubulin subunits to the end of a protofilament causes the end to grow in a linear conformation that can readily pack into the cylindrical wall of the microtubule. Hydrolysis of GTP after assembly changes the conformation of the subunits and tends to force the protofilament into a curved shape that is less able to pack into the microtubule wall. (C) In an intact microtubule, protofilaments made from GDP-containing subunits are forced into a linear conformation by the many lateral bonds within the microtubule wall, given a stable cap of GTP-containing subunits. Loss of the GTP cap, however, allows the GDP-containing protofilaments to relax into their more curved conformation. This leads to a progressive disruption of the microtubule. Above the drawings of a growing and a shrinking microtubule, electron micrographs show actual microtubules in each of these two states, as observed in preparations in vitreous ice. Note particularly the curling, disintegrating GDP-containing protofilaments at the end of the shrinking microtubule. (C, courtesy of E.M. Mandelkow, E. Mandelkow and R.A. Milligan, *J. Cell Biol.* 114:977–991, 1991. © The Rockefeller University Press.)

growth is called a *rescue.* In cells, dynamic instability is thought to predominate in microtubules, whereas treadmilling may predominate in actin filaments.

For microtubules, the structural difference between a T-form end and a D-form end is dramatic. Tubulin subunits with GTP bound to the β-monomer produce straight protofilaments that make strong and regular lateral contacts with one another. But the hydrolysis of GTP to GDP is associated with a subtle conformational change in the protein, which makes the protofilaments curved. On a rapidly growing microtubule, the GTP cap constrains the curvature of the protofilaments, and the ends appear straight. But when the terminal subunits have hydrolyzed their nucleotides, this constraint is released, and the curved protofilaments spring apart. This cooperative release of the energy of hydrolysis stored in the microtubule lattice results in the curled protofilaments peeling off rapidly, and rings and curved oligomers of GDP-containing tubulin are seen near the ends of depolymerizing microtubules (Figure 16–11C).

Treadmilling and Dynamic Instability Require Energy but Are Useful

Both dynamic instability and treadmilling allow a cell to maintain the same overall filament content, while individual subunits constantly recycle between the filaments and the cytosol. How dynamic are the microtubules and actin filaments inside a living cell? Typically, a microtubule, with major structural differences between its growing and shrinking ends, switches between growth and shrinkage a few times per minute. The ends of individual microtubules can therefore be seen in real time to exhibit such dynamic instability (Figure 16–12). For actin filaments, by contrast, the difference between the two types of ends is not so extreme, and the dynamic instability of an individual filament cannot be observed directly with the light microscope. With appropriate techniques based on fluorescence microscopy, however, one can show that actin filament turnover is typically rapid, with individual filaments persisting for only a few minutes.

At first glance, this dynamic behavior of filaments seems like a waste of energy. To maintain a constant concentration of actin filaments and microtubules, most of which are undergoing a process of either treadmilling or dynamic instability, the cell must hydrolyze large amounts of nucleoside triphosphate. As we explained with our ant-trail analogy at the beginning of the chapter, the advantage to the cell seems to lie in the spatial and temporal flexibility that is inherent in a structural system with constant turnover. Individual subunits are small and can diffuse very rapidly; an actin or tubulin subunit can diffuse across the diameter of a typical eucaryotic cell in a second or two. As noted above, the rate-limiting step in the formation of a new filament is nucleation, so these rapidly diffusing subunits tend to assemble either on the ends of preexisting filaments or at special sites where the nucleation step is catalyzed by special proteins. The new filaments in either case are highly dynamic, and

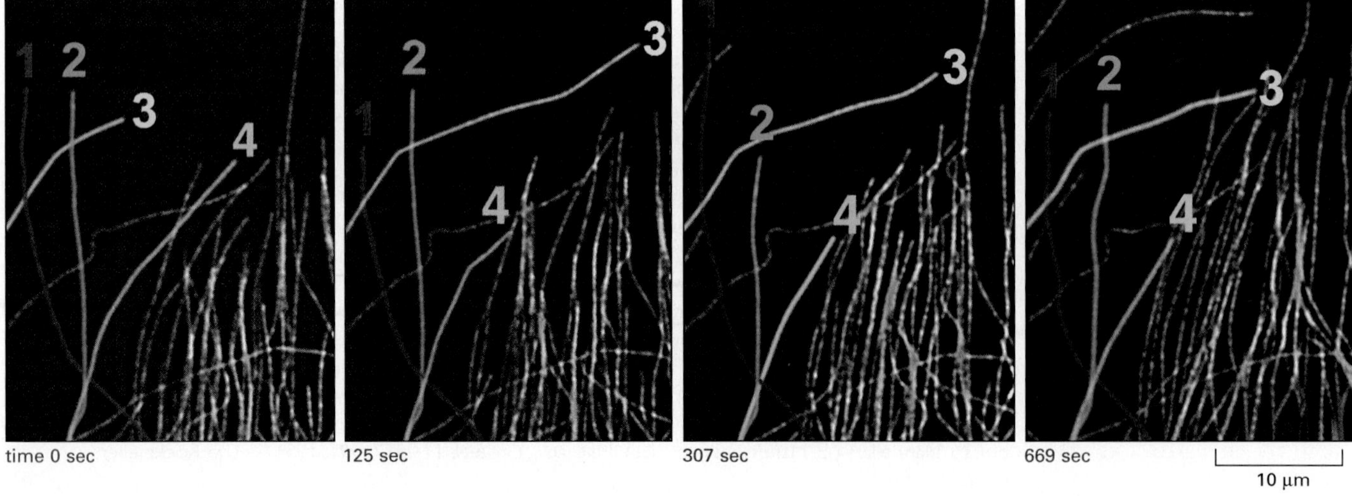

Figure 16–12 Direct observation of the dynamic instability of microtubules in a living cell. Microtubules in a newt lung epithelial cell were observed after the cell was injected with a small amount of rhodamine labeled tubulin, as in Figure 16–10. The dynamic instability of microtubules at the edge of the cell can be readily observed. Four individual microtubules are highlighted for clarity; each of these shows alternating shrinkage and growth. (Courtesy of Wendy C. Salmon and Clare Waterman-Storer.)

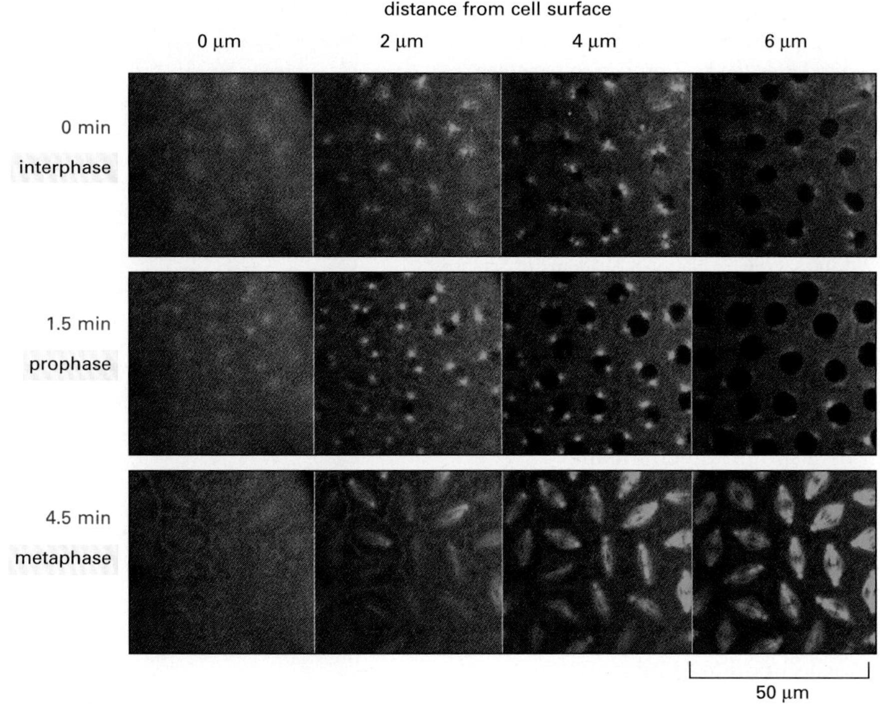

Figure 16–13 The rapid changes in cytoskeletal organization observed during the development of a *Drosophila* early embryo. In this giant multinuclear cell, the early nuclear divisions occur every 10 minutes or so in a common cytoplasm. The rapid rearrangements of the actin filaments *(red)* and microtubules *(green)* seen here in a living embryo, are required to separate the chromosomes at mitosis, while keeping each nucleus from colliding with its neighbors. (Courtesy of William Sullivan.)

unless specifically stabilized, they have only a fleeting existence. By controlling where filaments are nucleated and selectively stabilized, a cell can control the location of its filament systems, and hence its structure. It seems that the cell is continually testing a wide variety of internal structures and only preserving those that are useful. When external conditions change, or when internal signals arise (as during the transitions in the cell cycle), the cell is poised to change its structure rapidly (Figure 16–13).

In certain specialized structures, particularly in various specialized cells in a multicellular organism, parts of the cytoskeleton become less dynamic. In a terminally differentiated cell such as a neuron, for example, it is desirable to maintain a consistent structure over time, and many of the actin filaments and microtubules are stabilized by association with other proteins. However, when new connections are made in the brain, as when the information you are reading here is transferred into long-term memory, even a cell as stable as a neuron can grow new processes to make new synapses. To do this, a neuron requires the inherently dynamic and exploratory activities of its cytoskeletal filaments.

Other Polymeric Proteins Also Use Nucleotide Hydrolysis to Couple a Conformational Change to Cell Movements

It is remarkable that actin and tubulin have both evolved nucleoside triphosphate hydrolysis for the same basic reason—to enable them to depolymerize readily after they have polymerized. Actin and tubulin are completely unrelated in amino acid sequence: actin is distantly related in structure to the glycolytic enzyme hexokinase, whereas tubulin is distantly related to a large family of GTPases that includes the heterotrimeric G proteins and monomeric GTPases such as Ras, structures that are discussed in Chapter 3.

Several other types of protein polymers bind and hydrolyze nucleoside triphosphates. For example, *dynamin* assembles in helical coils around the base of membrane invaginations, helping to pinch off the invaginations to form vesicles (see Figure 13–9). Hydrolysis of GTP by dynamin generates a cooperative conformational change in the dynamin coil and provides the energy for the constriction. Bacteria and archaea have a tubulin homolog called *FtsZ* that is essential for cell division. A band of FtsZ protein forms at the site of septation, where the new cell wall is to form. Constriction and disassembly of the FtsZ band then help to pinch the two daughter cells apart. *In vitro*, FtsZ forms protofilaments

Figure 16–14 The bacterial FtsZ protein, a tubulin homolog in procaryotes. (A) A band of FtsZ protein forms a ring in a dividing bacterial cell. This ring has been labeled by fusing the FtsZ protein to the green fluorescent protein (GFP) and observing the living *E. coli* cells with a fluorescence microscope. *Top*, side view shows the ring as a bar in the middle of the dividing cell. *Bottom*, rotated view showing the ring structure. (B) FtsZ filaments and rings, formed *in vitro*, as visualized using electron microscopy. Compare this image with that of the microtubule shown on the *right* in Figure 16–11. (A, from X. Ma, D.W. Ehrhardt and W. Margolin, *Proc. Natl. Acad. Sci. USA* 93:12999–13003, 1996. © National Academy of Sciences; B, from H.A. Erickson et al., *Proc. Natl. Acad. Sci. USA* 93:519–523, 1996. © National Academy of Sciences.)

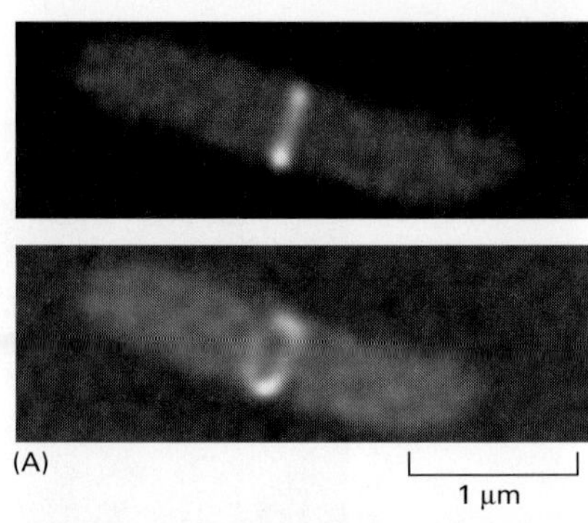

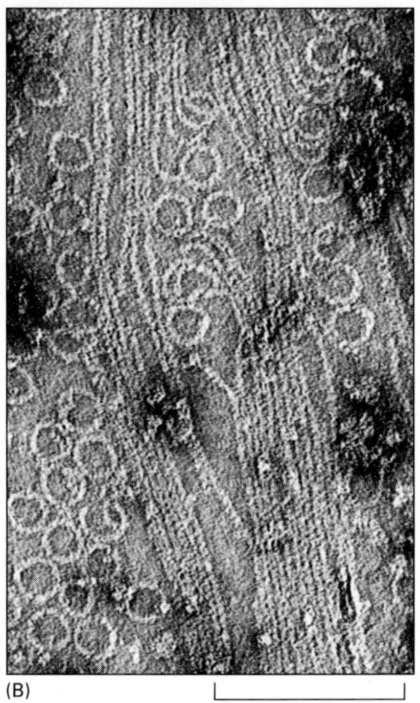

and rings reminiscent of tubulin, and it also binds and hydrolyzes GTP (Figure 16–14). The increase in ring curvature and disassembly of the FtsZ filaments is similar to the disassembly of curved GDP-tubulin protofilaments described earlier (see Figure 16–11), and it may occur by a similar mechanism.

These examples point to a second general way in which protein polymers can use the energy of nucleotide hydrolysis to affect cell structure and dynamics. In a linear protein polymer, a modest hydrolysis-linked conformational change by a few tenths of a nanometer in each subunit can be amplified by the thousands of subunits acting in parallel, so as to drive movements over tens or hundreds of nanometers. For dynamin and FtsZ, apparently, the effect of the GTP-dependent conformational changes is to bend lipid bilayers.

Tubulin and Actin Have Been Highly Conserved During Eucaryotic Evolution

Tubulin is found in all eucaryotic cells, and FtsZ has been found in all bacteria and archaea examined. Rod-shaped and spiral-shaped bacteria also contain an actin homolog, *MreB,* that forms filaments and regulates bacterial length. Thus, the principle of organizing cell structure by a self-association of nucleotide-binding proteins seems to be used in all cells.

Tubulin molecules themselves come in multiple isoforms. In mammals, there are at least six forms of α-tubulin and a similar number of forms of β-tubulin, each encoded by a different gene. The different forms of tubulin are very similar, and they generally copolymerize into mixed microtubules in the test tube, although they can have distinct locations in the cell and perform subtly different functions. As a particularly striking example, the microtubules in six specialized touch-sensitive neurons in the nematode *Caenorhabditis elegans* contain a specific form of β-tubulin, and mutations in the gene for this protein result in the loss of touch-sensitivity with no apparent defect in other cell functions.

Yeast and human tubulins are 75% identical in amino acid sequence. Most of the variation among different isoforms of tubulin is found in the amino acids near the C-terminus, which form a ridge on the surface of the microtubule. Thus, variations among isoforms are expected to affect primarily the association of accessory proteins with the surface of the microtubule, rather than microtubule polymerization *per se.*

Like tubulin, actin is found in all eucaryotic cells. Most organisms have multiple genes encoding actin; humans have six. Actin is extraordinarily well conserved among eucaryotes. The amino acid sequences of actins from different species are usually about 90% identical. But, again like tubulin, small variations in actin amino acid sequence can create significant functional differences. In vertebrates, there are three subtly different isoforms of actin, termed α, β, and γ, that differ slightly in their amino acid sequences. The α-actin is expressed only in muscle cells, while β and γ are found together in almost all nonmuscle cells. Yeast actin and *Drosophila* muscle actin are 89% identical, yet the expression of yeast actin in *Drosophila* results in a fly that looks normal but is unable to fly.

What is the explanation for the unusually strict conservation of actin and tubulin in eucaryotic evolution? Most other cytoskeletal proteins, including intermediate filament proteins and the large families of accessory proteins that bind to actin or tubulin, are not particularly well conserved at the level of amino acid sequence. The likely explanation is that the structure of the entire surface of an actin filament or microtubule is constrained because so many other proteins must be able to interact with these two ubiquitous and abundant cell components. A mutation in actin that could result in a desirable change in its interaction with one other protein might cause undesirable changes in its interactions with a number of other proteins that bind at or near the same site. Genetic and biochemical studies in the yeast *Saccharomyces cerevisiae* have demonstrated that actin interacts directly with dozens of other proteins, and indirectly with even more (Figure 16–15). Over time, evolving organisms have found it more profitable to leave actin and tubulin alone, and alter their binding partners instead.

Intermediate Filament Structure Depends on The Lateral Bundling and Twisting of Coiled Coils

All eucaryotic cells contain actin and tubulin. But the third major type of cytoskeletal filament, the *intermediate filament,* is found in only some metazoans, including vertebrates, nematodes, and molluscs. Even in these organisms, intermediate filaments are not required in every cell type. The specialized glial cells (called oligodendrocytes) that make myelin in the vertebrate central nervous system, for example, do not contain intermediate filaments. Cytoplasmic intermediate filaments are closely related to their ancestors, the much more widely used *nuclear lamins.* The nuclear lamins are filamentous proteins that form a meshwork lining the inner membrane of the eucaryotic nuclear envelope, where they provide anchorage sites for chromosomes and nuclear pores (their dynamic behavior during cell division is discussed in Chapter 12). Several times during metazoan evolution, lamin genes have apparently duplicated, and the duplicates have evolved to produce rope-like, cytoplasmic intermediate filaments. These filaments have mechanical properties that are especially useful to soft-bodied animals such as nematodes and vertebrates that do not have an exoskeleton. Intermediate filaments are particularly prominent in the cytoplasm of cells that are subject to mechanical stress, and their major function seems to be to impart physical strength to cells and tissues.

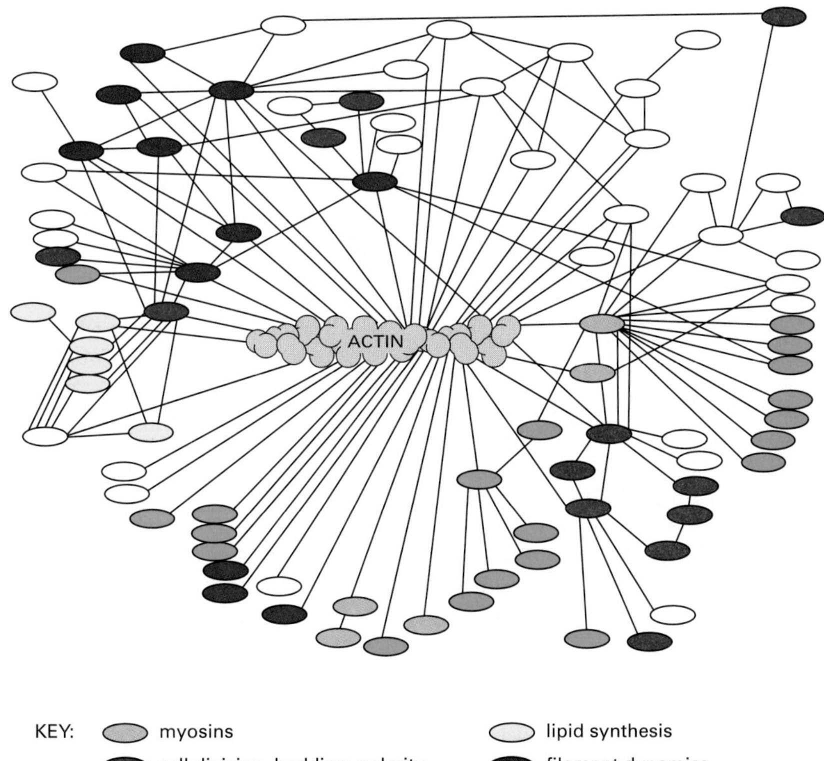

Figure 16–15 Actin at the crossroads. Actin binds to a very large variety of accessory proteins in all eucaryotic cells. This diagram shows most of the interactions that have been demonstrated, using either genetic or biochemical techniques, in the yeast *Saccharomyces cerevisiae.* Accessory proteins that operate in the same intracellular process are shown in the same color, as indicated in the key. (Adapted from D. Botstein et al., in The Molecular and Cellular Biology of the Yeast *Saccharomyces* [J.R. Broach, J.R. Pringle, E.W. Jones, eds.], Cold Spring Harbor, NY: Cold Spring Harbor Laboratory Press, 1991.)

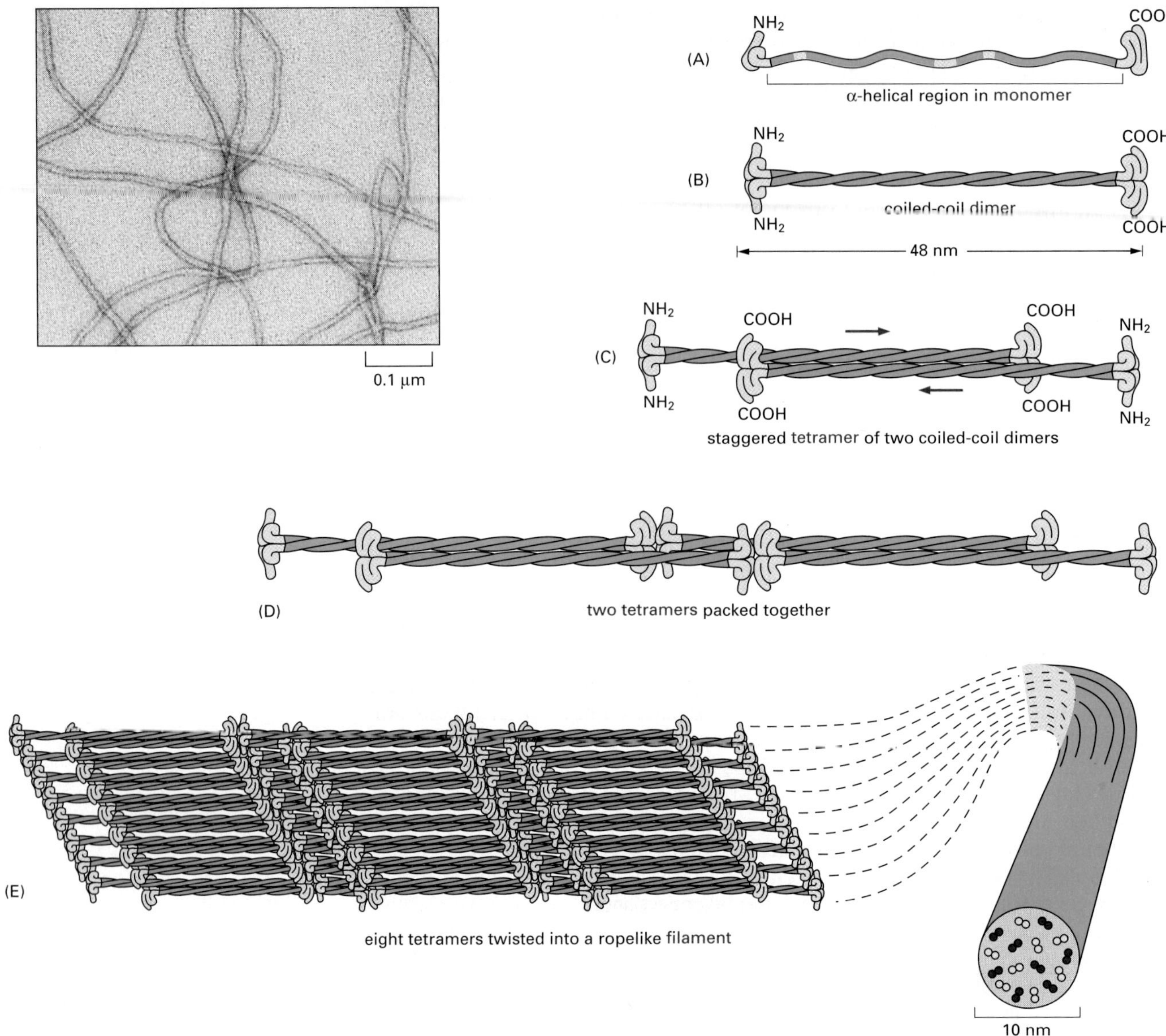

Figure 16–16 A model of intermediate filament construction. The monomer shown in (A) pairs with an identical monomer to form a dimer (B), in which the conserved central rod domains are aligned in parallel and wound together into a coiled coil. (C) Two dimers then line up side by side to form an antiparallel tetramer of four polypeptide chains. The tetramer is the soluble subunit of intermediate filaments. (D) Within each tetramer, the two dimers are offset with respect to one another, thereby allowing it to associate with another tetramer. (E) In the final 10-nm rope-like filament, tetramers are packed together in a helical array, which has 16 dimers in cross-section. Half of these dimers are pointing in each direction. An electron micrograph of intermediate filaments are shown on the upper left. (Electron micrograph courtesy of Roy Quinlan.)

The individual polypeptides of **intermediate filaments** are elongated molecules with an extended central α-helical domain that forms a parallel coiled coil with another monomer. A pair of parallel dimers then associates in an antiparallel fashion to form a staggered tetramer. This tetramer represents the soluble subunit that is analogous to the αβ-tubulin dimer, or the actin monomer (Figure 16–16).

Since the tetrameric subunit is made up of two dimers pointing in opposite directions, its two ends are the same. The assembled intermediate filament therefore lacks the overall structural polarity that is critical for actin filaments and microtubules. The tetramers pack together laterally to form the filament, which includes eight parallel protofilaments made up of tetramers. Each individual intermediate filament therefore has a cross section of 32 individual α-helical coils. This large number of polypeptides all lined up together, with the strong lateral hydrophobic interactions typical of coiled-coil proteins, gives intermediate filaments a rope-like character. They can be easily bent but are extremely difficult to break (Figure 16–17).

Less is understood about the mechanism of assembly and disassembly of intermediate filaments than of actin filaments and microtubules, but they are clearly highly dynamic structures in most cell types. Under normal conditions, protein phosphorylation probably regulates their disassembly, in much the same way that phosphorylation regulates the disassembly of nuclear lamins in

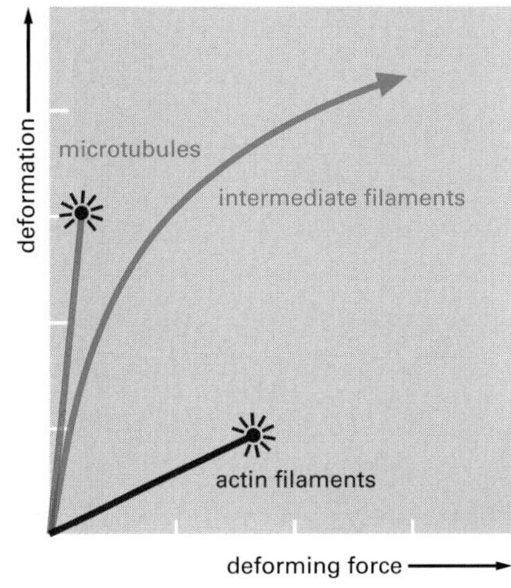

Figure 16–17 Mechanical properties of actin, tubulin, and intermediate filament polymers. Networks composed of microtubules, actin filaments, or a type of intermediate filament called vimentin, all at equal concentration, were exposed to a shear force in a viscometer, and the resulting degree of stretch was measured. The results show that microtubule networks are easily deformed but that they rupture (indicated by *red starburst)* and begin to flow without limit when stretched beyond 150% of their original length. Actin filament networks are much more rigid, but they also rupture easily. Intermediate filament networks, by contrast, are not only easily deformed, but they withstand large stresses and strains without rupture; they are thereby well suited to maintain cell integrity. (Adapted from P. Janmey et al., *J. Cell Biol.* 113:155–160, 1991.)

mitosis (see Figure 12–20). As evidence for a rapid turnover, labeled subunits microinjected into tissue culture cells rapidly add themselves onto the existing intermediate filaments within a few minutes, while an injection of peptides derived from a conserved helical region of the subunit induces the rapid disassembly of the intermediate filament network. Interestingly, the latter injection can induce the disassembly of the microtubule and actin filament networks in some cells that contain all three networks revealing that there is a fundamental mechanical integration among the different cytoskeletal systems in these cells.

Intermediate Filaments Impart Mechanical Stability to Animal Cells

Intermediate filaments come in a wide variety of types, with substantially more sequence variation in the subunit isoforms than occurs in the isoforms of actin or tubulin. A central α-helical domain has 40 or so heptad repeat motifs that form an extended coiled-coil (see Figure 3–11). This domain is similar in the different isoforms, but the N- and C-terminal globular domains can vary a great deal.

Different families of intermediate filaments are expressed in different cell types (Table 16–1). The most diverse intermediate filament family is that of the **keratins**: there are about 20 found in different types of human epithelial cells, and about 10 more that are specific to hair and nails. A single epithelial cell may produce multiple types of keratins, and these copolymerize into a single network (Figure 16–18). Every keratin filament is made up of an equal mixture of type I (acidic) and type II (neutral/basic) keratin chains; these form heterodimers, two of which then join to form the fundamental tetrameric subunit (see Figure 16–16). Cross-linked keratin networks held together by disulfide bonds may survive even the death of their cells, forming tough coverings for animals, as in the outer layer of skin and in hair, nails, claws, and scales. The diversity in

TABLE 16–1 Major Types of Intermediate Filament Proteins in Vertebrate Cells

TYPES OF IF	COMPONENT POLYPEPTIDES	CELLULAR LOCATION
Nuclear	lamins A, B, and C	nuclear lamina (inner lining of nuclear envelope)
Vimentin-like	vimentin	many cells of mesenchymal origin
	desmin	muscle
	glial fibrillary acidic protein	glial cells (astrocytes and some Schwann cells)
	peripherin	some neurons
Epithelial	type I keratins (acidic) type II keratins (basic)	epithelial cells and their derivatives (e.g., hair and nails)
Axonal	neurofilament proteins (NF-L, NF-M, and NF-H)	neurons

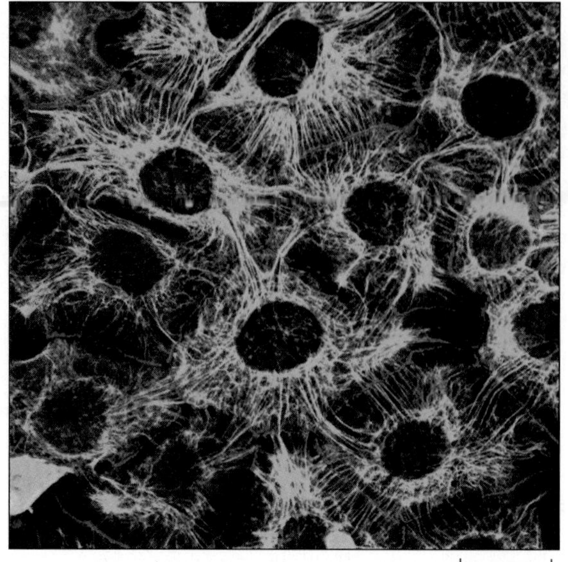

Figure 16–18 Keratin filaments in epithelial cells. Immunofluorescence micrograph of the network of keratin filaments *(green)* in a sheet of epithelial cells in culture. The filaments in each cell are indirectly connected to those of its neighbors by desmosomes (discussed in Chapter 19). A second protein *(blue)* has been stained to reveal the location of the cell boundaries. (Courtesy of Kathleen Green and Evangeline Amargo.)

keratins is clinically useful in the diagnosis of epithelial cancers (carcinomas), as the particular set of keratins expressed gives an indication of the epithelial tissue in which the cancer originated and thus can help to guide the choice of treatment.

Mutations in keratin genes cause several human genetic diseases. For example, when defective keratins are expressed in the basal cell layer of the epidermis, they produce a disorder called *epidermolysis bullosa simplex,* in which the skin blisters in response to even very slight mechanical stress, which ruptures the basal cells (Figure 16–19). Other types of blistering diseases, including disorders of the mouth, esophageal lining, and the cornea of the eye, are caused by mutations in the different keratins whose expression is specific to those tissues. All of these maladies are typified by cell rupture as a consequence of mechanical trauma and a disorganization or clumping of the keratin filament cytoskeleton. Many of the specific mutations that cause these diseases alter the ends of the central rod domain, underlining the importance of this particular part of the protein for correct filament assembly.

A second family of intermediate filaments, called **neurofilaments**, is found in high concentrations along the axons of vertebrate neurons (Figure 16–20). Three types of neurofilament proteins (NF-L, NF-M, NF-H) coassemble *in vivo,* forming heteropolymers that contain NF-L plus one of the others. The NF-H and NF-M proteins have lengthy C-terminal tail domains that bind to neighboring filaments, generating aligned arrays with a uniform interfilament spacing. During axonal growth, new neurofilament subunits are incorporated all along the

Figure 16–19 Blistering of the skin caused by a mutant keratin gene. A mutant gene encoding a truncated keratin protein (lacking both the N- and C-terminal domains) was expressed in a transgenic mouse. The defective protein assembles with the normal keratins and thereby disrupts the keratin filament network in the basal cells of the skin. Light micrographs of cross sections of normal (A) and mutant (B) skin show that the blistering results from the rupturing of cells in the basal layer of the mutant epidermis (small *red* arrows). (C) A sketch of three cells in the basal layer of the mutant epidermis, as observed by electron microscopy. As indicated by the *red* arrow, the cells rupture between the nucleus and the hemidesmosomes (discussed in Chapter 19), which connect the keratin filaments to the underlying basal lamina. (From P.A. Coulombe et al., *J. Cell Biol.* 115:1661–1674, 1991. © The Rockefeller University Press.)

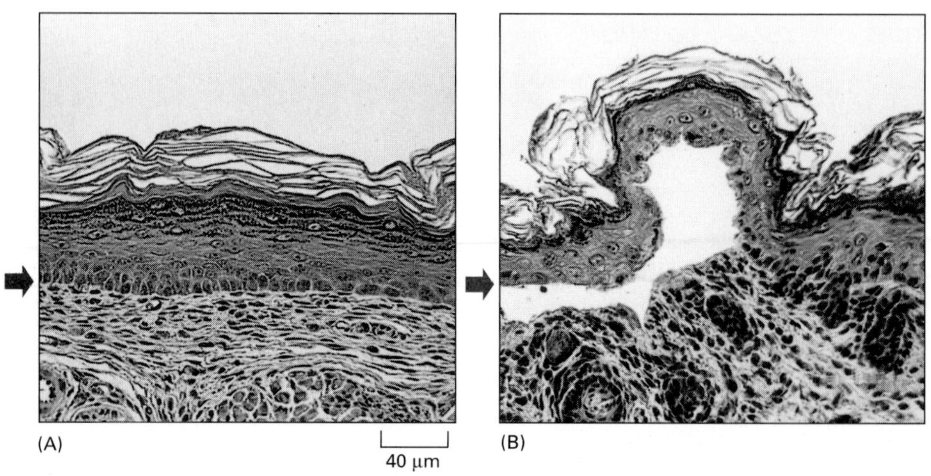

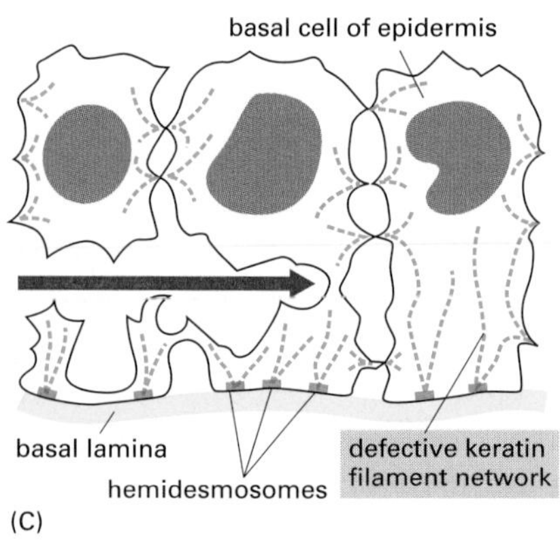

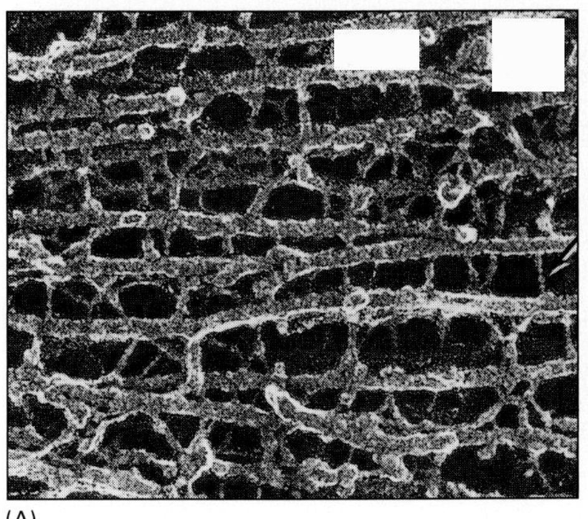

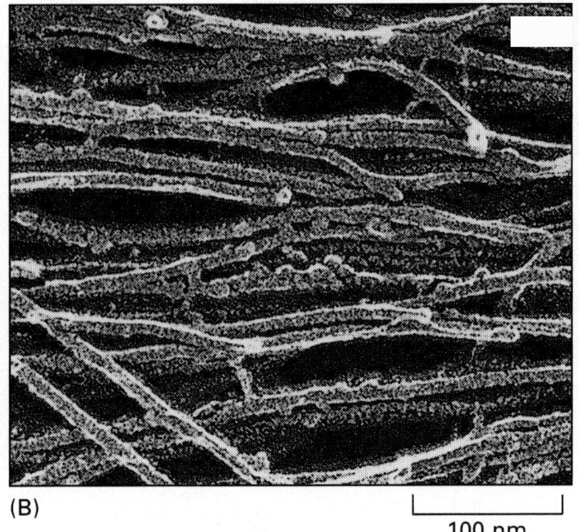

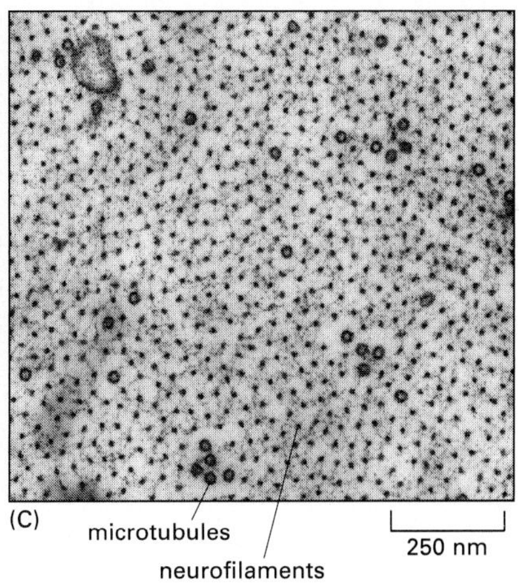

Figure 16–20 Two types of intermediate filaments in cells of the nervous system. (A) Freeze-etch electron microscopic image of neurofilaments in a nerve cell axon, showing the extensive cross-linking through protein cross-bridges—an arrangement believed to give this long cell process great tensile strength. The cross-bridges are formed by the long, nonhelical extensions at the C-terminus of the largest neurofilament protein (NF-H). (B) Freeze-etch image of glial filaments in glial cells, showing that these intermediate filaments are smooth and have few cross-bridges. (C) Conventional electron micrograph of a cross section of an axon showing the regular side-to-side spacing of the neurofilaments, which greatly outnumber the microtubules. (A and B, courtesy of Nobutaka Hirokawa; C, courtesy of John Hopkins.)

axon in a dynamic process that involves the addition of subunits along the filament length, as well as the addition of subunits at the filament ends. After an axon has grown and connected with its target cell, the diameter of the axon may increase as much as fivefold. The level of neurofilament gene expression seems to directly control axonal diameter, which in turn controls how fast electrical signals travel down the axon.

The neurodegenerative disease amyotrophic lateral sclerosis (ALS, or Lou Gehrig's Disease) is associated with an accumulation and abnormal assembly of neurofilaments in motor neuron cell bodies and in the axon, which may interfere with normal axonal transport. The degeneration of the axons leads to muscle weakness and atrophy, which is usually fatal. The overexpression of human NF-L or NF-H in mice results in mice that have an ALS-like disease.

The vimentin-like filaments are a third family of intermediate filaments. Desmin, a member of this family, is expressed in skeletal, cardiac, and smooth muscle. Mice lacking desmin show normal initial muscle development, but adults have a variety of muscle cell abnormalities, including misaligned muscle fibers.

Thus, intermediate filaments in general seem to serve as the ligaments of the cell, whereas microtubules and actin filaments seem to serve, respectively, as the bones and muscles of the cell. Just as we require our ligaments, bones, and muscles to work together, so all three cytoskeletal filament systems must normally function collectively to give a cell its strength, its shape, and its ability to move.

Filament Polymerization Can Be Altered by Drugs

Because the survival of eucaryotic cells depends on a balanced assembly and disassembly of the highly conserved cytoskeletal filaments formed from actin and tubulin, the two types of filaments are frequent targets for natural toxins. These toxins are produced in self-defense by plants, fungi, or sponges that do not wish to be eaten but cannot run away from predators, and they generally perturb the filament polymerization reaction. The toxin binds tightly to either the filament form or the free subunit form of a polymer, driving the assembly reaction in the direction that favors the form to which the toxin binds. For example, the drug *latrunculin*, extracted from the sea sponge *Latrunculia magnifica*, binds to and stabilizes actin monomers; it thereby causes a net depolymerization of actin filaments. In contrast, *phalloidin*, from the fungus *Amanita phalloides* (death cap), binds to and stabilizes actin filaments, causing a net increase in actin polymerization. (One remedy for *Amanita* mushroom poisoning is to eat a large quantity of raw meat: the high concentration of actin filaments in the muscle tissue binds the phalloidin and thereby reduces phalloidin's toxicity.) Either change in actin filaments is very toxic for cells. Similarly, *colchicine*, from the meadow saffron (or autumn crocus), binds to and stabilizes free tubulin, causing microtubule depolymerization. In contrast, *taxol*, extracted from the bark of a rare species of yew tree, binds to and stabilizes microtubules, causing

TABLE 16–2 Drugs That Affect Actin Filaments and Microtubules

ACTIN-SPECIFIC DRUGS	
Phalloidin	binds and stabilizes filaments
Cytochalasin	caps filament plus ends
Swinholide	severs filaments
Latrunculin	binds subunits and prevents their polymerization
MICROTUBULE-SPECIFIC DRUGS	
Taxol	binds and stabilizes microtubules
Colchine, colcemid	binds subunits and prevents their polymerization
Vinblastine, vincristine	binds subunits and prevents their polymerization
Nocodazole	binds subunits and prevents their polymerization

a net increase in tubulin polymerization. These and some other natural products that are commonly used by cell biologists to manipulate the cytoskeleton are listed in Table 16–2.

Drugs like these have a rapid and profound effect on the organization of the cytoskeleton in living cells (Figure 16–21). They provided early evidence that the cytoskeleton is a dynamic structure, maintained by a rapid and continual exchange of subunits between the soluble and filamentous forms and that this subunit flux is necessary for normal cytoskeletal function.

The drugs listed in Table 16–2 have been useful to cell biologists trying to probe the roles of actin and microtubules in various cell processes. Some of them are also used in the treatment of cancer. Both microtubule-depolymerizing drugs such as vinblastine and microtubule-polymerizing drugs such as taxol

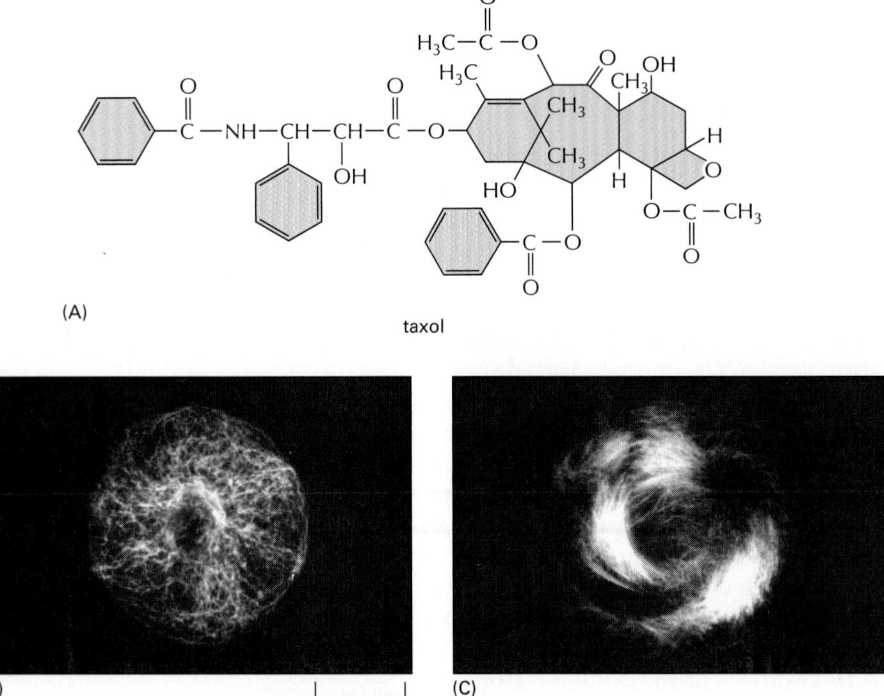

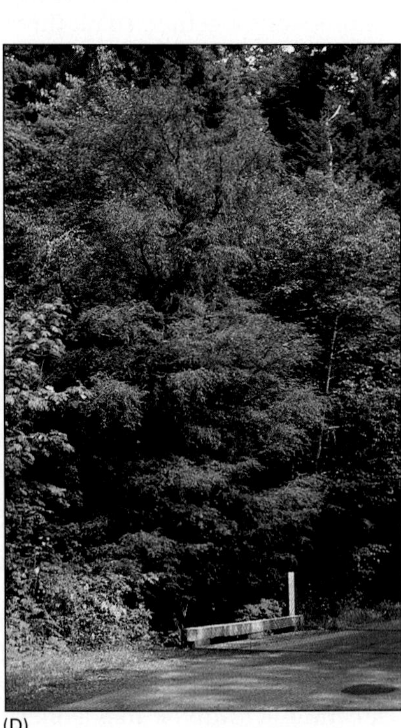

Figure 16–21 Effect of the drug taxol on microtubule organization. (A) Molecular structure of taxol. Recently, organic chemists have succeeded in synthesizing this complex molecule, which is widely used for cancer treatment. (B) Immunofluorescence micrograph showing the microtubule organization in a liver epithelial cell before the addition of taxol. (C) Microtubule organization in the same type of cell after taxol treatment. Note the thick circumferential bundles of microtubules around the periphery of the cell. (D) A Pacific yew tree, the natural source of taxol. (B, C from N.A. Gloushankova et al., *Proc. Natl. Acad. Sci. USA* 91:8597–8601, 1994. © National Academy of Sciences; D, courtesy of A.K. Mitchell 2001. © Her Majesty the Queen in Right of Canada, Canadian Forest Service.)

preferentially kill dividing cells, since both microtubule assembly and disassembly are crucial for correct function of the mitotic spindle (discussed in Chapter 18). These drugs efficiently kill certain types of tumor cells in a human patient, although not without toxicity to rapidly dividing normal cells, including those in the bone marrow, intestine, and hair follicles. Taxol in particular has been widely used to treat some specific cancers that are resistant to other chemotherapeutic agents.

The common laboratory reagent acrylamide, used as the precursor in making polyacrylamide gels for size separation of proteins and nucleic acids (see Figure 8–14), is also a cytoskeletal toxin. By an unknown mechanism, it causes the disassembly or rearrangement of intermediate filament networks. Acrylamide can be absorbed through the skin and acts as a potent neurotoxin by dismantling the neurofilament bundles in peripheral nerve axons. This is why it causes an unpleasant tingling sensation when spilled on bare skin and why it should always be handled cautiously.

Summary

The cytoplasm of eucaryotic cells is spatially organized by a network of protein filaments known as the cytoskeleton. This network contains three principal types of filaments: microtubules, actin filaments, and intermediate filaments. All three types of filaments form as helical assemblies of subunits that self-associate using a combination of end-to-end and side-to-side protein contacts. Differences in the structure of the subunits and the manner of their self-assembly give the filaments different mechanical properties. Intermediate filaments are rope-like and easy to bend but hard to break. Microtubules are strong, rigid hollow tubes. Actin filaments are the thinnest of the three and are hard to stretch but easy to break.

In living cells, all three types of cytoskeletal filaments undergo constant remodeling through the assembly and disassembly of their subunits. Microtubules and actin filaments add and lose subunits only at their ends, with one end (the plus end) growing faster than the other. Tubulin and actin (the subunits of microtubules and actin filaments, respectively) bind and hydrolyze nucleoside triphosphates (tubulin binds GTP and actin binds ATP). Nucleotide hydrolysis underlies the characteristic dynamic behavior of these two filaments. Actin filaments in cells seem to predominantly undergo treadmilling, where a filament assembles at one end while simultaneously disassembling at the other end. Microtubules in cells predominantly display dynamic instability, where a microtubule end undergoes alternating bouts of growth and shrinkage.

Whereas tubulin and actin have been strongly conserved in eucaryotic evolution, the family of intermediate filaments is very diverse. There are a variety of tissue-specific forms, including keratin filaments in epithelial cells, neurofilaments in nerve cells, and desmin filaments in muscle cells. In all these cells, the primary job of intermediate filaments is to provide mechanical strength.

HOW CELLS REGULATE THEIR CYTOSKELETAL FILAMENTS

Microtubules, actin filaments, and intermediate filaments are much more dynamic in cells than they are in the test tube. The cell regulates the length and stability of its cytoskeletal filaments, as well as their number and the geometry. It does so largely by regulating their attachments to one another and to other components of the cell, so that the filaments can form a wide variety of higher-order structures. Some filament properties are regulated by direct covalent modification of the filament subunits, but most of the regulation is performed by accessory proteins that bind to either the filaments or their free subunits.

This section focuses on how these accessory proteins modify the dynamics and structure of cytoskeletal filaments. We begin with a discussion of how microtubules and actin filaments are nucleated in cells, as this plays a major part in determining the overall organization of the cell's interior.

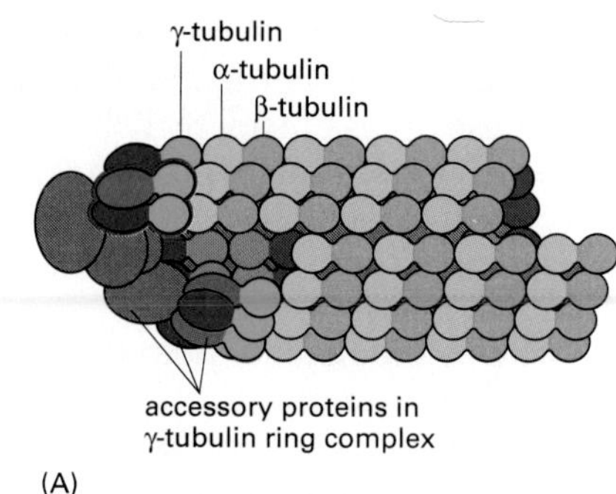

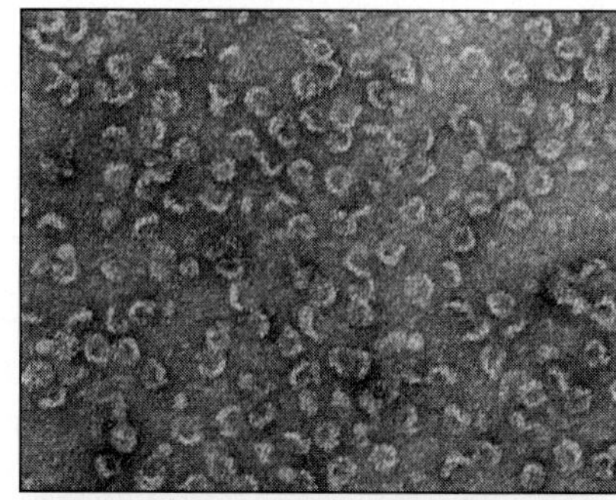

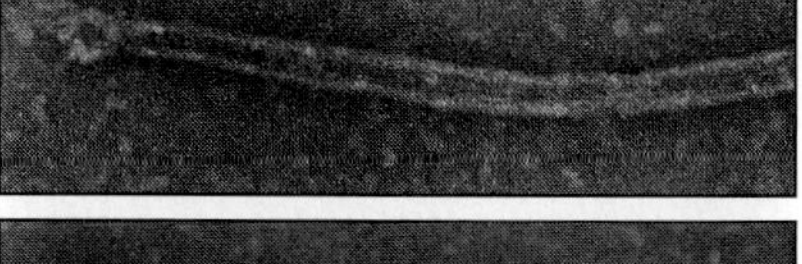

Figure 16–22 Polymerization of tubulin nucleated by γ-tubulin ring complexes. (A) Model for the nucleation of microtubule growth by the γ-TuRC. The red outline indicates a pair of proteins bound to two molecules of γ-tubulin; this group can be isolated as a separate subcomplex of the larger ring. (B) Electron micrographs of purified γ-tubulin ring complexes *(top)* and single microtubules nucleated from the purified γ-tubulin ring complexes *(middle* and *bottom)*. (A, modified from M. Moritz et al., *Nature Cell Biol.* 2:365–370, 2000; B, courtesy of Y. Zheng et al., *Nature* 378:578–583, 1994. © Macmillan Magazines Ltd.)

Microtubules Are Nucleated by a Protein Complex Containing γ-tubulin

While α- and β-tubulins are the regular building blocks of microtubules, another type of tubulin, called *γ-tubulin*, has a more specialized role. Present in much smaller amounts than α- and β-tubulin, this protein is involved in the nucleation of microtubule growth in organisms ranging from yeasts to humans. Microtubules are generally nucleated from a specific intracellular location known as a **microtubule-organizing center (MTOC)**. Antibodies against γ-tubulin stain the MTOC in virtually all species and cell types thus far examined.

Microtubules are nucleated at their minus end, with the plus end growing outward from each MTOC to create various types of microtubule arrays. A **γ-tubulin ring complex (γ-TuRC)** has been isolated from both insect and vertebrate cells and is an impressively efficient nucleator of microtubule growth in a test tube. Two proteins, conserved from yeasts to humans, bind directly to the γ-tubulin, along with several other proteins that help create the ring that can be seen at the minus ends of the microtubules nucleated by γ-TuRC. This ring of γ-tubulin molecules is therefore thought to serve as a template that nucleates a microtubule with 13 protofilaments (Figure 16–22).

Microtubules Emanate from the Centrosome in Animal Cells

In most animal cells, there is a single, well-defined MTOC called the **centrosome**, located near the nucleus. From this focal point, the cytoplasmic microtubules emanate in a star-like, "astral" conformation. Microtubules are nucleated at the centrosome at their minus ends, so the plus ends point outward and grow toward the cell periphery. A centrosome is composed of a fibrous *centrosome matrix* that contains more than fifty copies of γ-TuRC. Most of the proteins that form this matrix, remain to be discovered, and it is not yet known how they recruit and activate the γ-TuRC.

Embedded in the centrosome is a pair of cylindrical structures arranged at right angles to each other in an L-shaped configuration (Figure 16–23). These are the **centrioles**, which become the basal bodies of cilia and flagella in motile cells

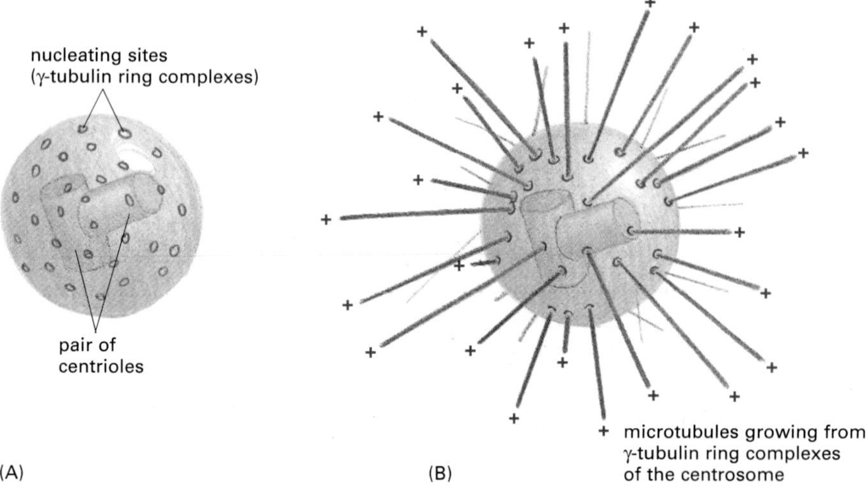

Figure 16–23 The centrosome. (A) The centrosome is the major MTOC of animal cells. Located in the cytoplasm next to the nucleus, it consists of an amorphous matrix of protein containing the γ-tubulin ring complexes that nucleate microtubule growth. This matrix is organized by a pair of centrioles, as described in the text. (B) A centrosome with attached microtubules. The minus end of each microtubule is embedded in the centrosome, having grown from a γ-tubulin ring complex, whereas the plus end of each microtubule is free in the cytoplasm.

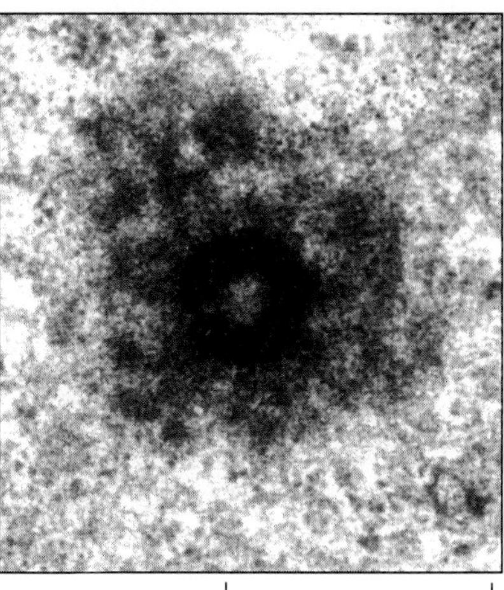

Figure 16–24 A centriole in the centrosome. An electron micrograph of a thick section of a centrosome showing an end-on view of a centriole. The ring of modified microtubules of the centriole is visible, surrounded by the fibrous centrosome matrix. (Courtesy of P. Witt and G.G. Borisy.)

(described later). The centrioles organize the centrosome matrix (also called the pericentriolar material), ensuring its duplication during each cell cycle as the centrioles themselves duplicate. As described in Chapter 18, the centrosome duplicates and splits into two equal parts during interphase, each half containing a duplicated centriole pair (see Figure 18–6). These two daughter centrosomes move to opposite sides of the nucleus when mitosis begins, and they form the two poles of the mitotic spindle (see Figure 18–7). A centriole consists of a short cylinder of modified microtubules, plus a large number of accessory proteins (Figure 16–24). The molecular basis for its duplication is not known.

In fungi and diatoms, microtubules are nucleated at an MTOC that is embedded in the nuclear envelope as a small plaque called the *spindle pole body.* Higher-plant cells seem to nucleate microtubules at sites distributed all around the nuclear envelope. Neither fungi nor most plant cells contain centrioles. Despite these differences, all these cells contain γ-tubulin and seem to use it to nucleate their microtubules.

In animal cells, the astral configuration of microtubules is very robust, with dynamic plus ends pointing outward toward the cell periphery and stable minus ends collected near the nucleus. The system of microtubules radiating from the centrosome acts as a device to survey the outlying regions of the cell and position the centrosome at its center, and it does this even in artificial enclosures (Figure 16–25). Even in an isolated cell fragment lacking the centrosome, dynamic microtubules interacting with membraneous organelles and motor proteins arrange themselves into a star-shaped array with the microtubule minus ends clustered at the center (Figure 16–26). This ability of the microtubule cytoskeleton to find the center of the cell establishes a general coordinate system, which is then used to position many organelles within the cell.

Actin Filaments Are Often Nucleated at the Plasma Membrane

In contrast to microtubule nucleation, which occurs primarily deep within the cytoplasm near the nucleus, actin filament nucleation most frequently occurs at the plasma membrane (Figure 16–27). Consequently, the highest density of actin filaments in most cells is at the cell periphery. These actin filaments in the layer underlying the plasma membrane, called the **cell cortex**, determine the shape and movement of the cell surface. For example, depending on their attachments to one another and to the plasma membrane, actin structures can form many strikingly different types of cell surface projections. These include spiky bundles such as *microvilli* or *filopodia*, flat protrusive veils called *lamellipodia* that help move cells over solid substrates, and the phagocytic cups in macrophages.

The nucleation of actin filaments at the plasma membrane is frequently regulated by external signals, allowing the cell to change its shape and stiffness rapidly in response to changes in its external environment. This nucleation is catalyzed by a complex of proteins that includes two *actin-related proteins*, or *ARPs*, each of which is about 45% identical to actin. Analogous to the function of

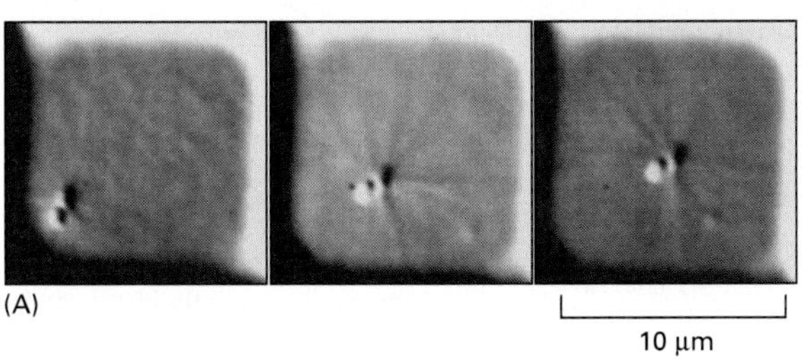

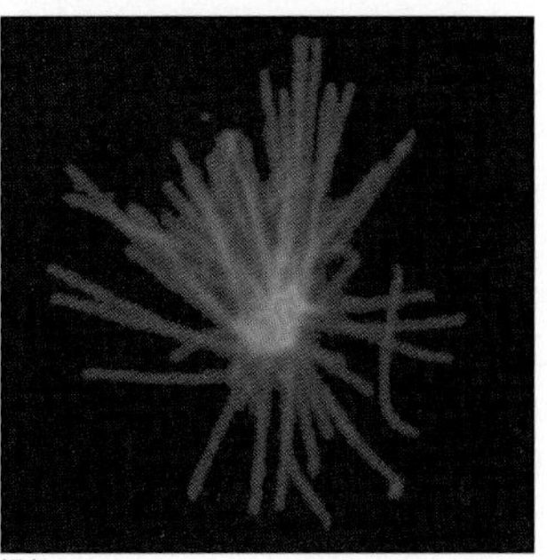

Figure 16–25 The center-seeking behavior of a centrosome. (A) Small square wells were micromachined into a plastic substrate. A single centrosome was placed into one of these wells, along with tubulin subunits in solution. As the microtubules polymerize, nucleated by the centrosome, they push against the walls of the well. The requirement for equal pushing in all directions if the position is to be stabilized forces the centrosome to the center of the well. The three pictures were taken at three-minute intervals. (B) A similar self-centered centrosome, fixed and stained to show the distribution of the microtubules pushing on all four walls of the enclosure. (From T.E. Holy et al., *Proc. Natl. Acad. Sci. USA* 94:6228–6231, 1997. © National Academy of Sciences.)

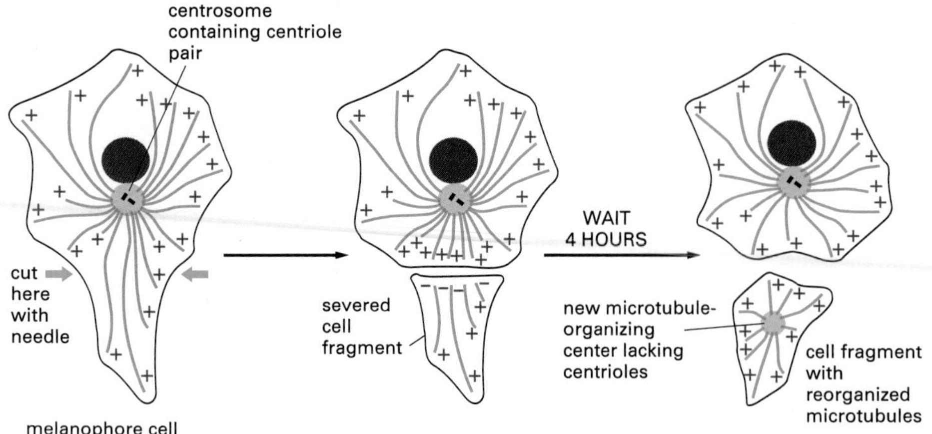

Figure 16–26 A microtubule array can find the center of a cell. After the arm of a fish pigment cell is cut off with a needle, the microtubules in the detached cell fragment reorganize so that their minus ends end up near the center of the fragment, buried in a new microtubule-organizing center.

the γ-TuRC, the **ARP complex** (also known as the *Arp 2/3 complex*) nucleates actin filament growth from the minus end, allowing rapid elongation at the plus end. However, the complex can also attach to the side of another actin filament while remaining bound to the minus end of the filament that it has nucleated, thereby building individual filaments into a treelike web (Figure 16–28). The ARP complex is localized in regions of rapid actin filament growth such as lamellipodia, and its nucleating activity is regulated by intracellular signaling molecules and components at the cytosolic face of the plasma membrane.

Both γ-tubulin and ARPs are evolutionarily ancient, and they are conserved among a wide variety of eucaryotic species. Their genes seem to have arisen by early duplication of the gene for the microtubule or actin filament subunit, respectively, followed by divergence and specialization of the gene copies so that they encode proteins with a special nucleating function. That a similar strategy has evolved for two separate cytoskeletal systems underlines the central importance of regulated filament nucleation as a general organizing principle in cells.

Figure 16–27 The tip of the leading edge of a cell nucleates actin filaments. Fibroblasts in culture were gently permeabilized using a nonionic detergent and were then incubated with rhodamine-labeled actin molecules to visualize newly formed actin filaments *(red)*. After 5 minutes, the cells were fixed and stained with fluorescein-labeled phalloidin to stain all actin filaments *(green)*. (A) All of the actin filaments, most of which were formed before the cells were permeabilized, are shown in *green*. (B) The location of the newly formed actin filaments *(red)* show that the leading edge is the predominant site of actin filament assembly in the cell. Further studies show that this is because new actin filaments are constantly being nucleated here. (From M.H. Symons and T.J. Mitchison, *J. Cell Biol.* 114:503–513, 1991. © The Rockefeller University Press.)

Filament Elongation Is Modified by Proteins That Bind to the Free Subunits

Once cytoskeletal filaments have been nucleated, they generally elongate by the addition of soluble subunits. In most nonmuscle vertebrate cells, approximately 50% of the actin is in filaments and 50% is soluble. The soluble monomer concentration is typically 50–200 μM (2–8 mg/ml), which is surprisingly high, given the low critical concentration observed for pure actin in a test tube (less than 1 μM). Why does the soluble actin not polymerize into filaments? The reason is that the abundant subunit pool contains special proteins that bind to the actin monomers, thereby making polymerization much less favorable (the action is similar to that of latrunculin). A small protein called *thymosin* is the most abundant of these proteins. Actin monomers bound to thymosin are in a locked state, where they cannot associate with either the plus or minus ends of actin filament and cannot hydrolyze or exchange their bound nucleotide.

How do cells recruit actin monomers from this sequestered pool and use them for polymerization? One might imagine that the thymosin itself could be regulated by signal transduction pathways, but this has not been found to be the case. Instead, recruitment depends on another monomer-binding protein, *profilin*. Profilin binds to the face of the actin monomer opposite the ATP-binding cleft, blocking the side of the monomer that would normally associate with the filament minus end (Figure 16–29). However, the profilin–actin complex can readily add onto a free plus end. As soon as this addition occurs, a conformational change is induced in the actin that reduces its affinity for profilin, so the profilin falls off, leaving the actin filament one subunit longer. Profilin competes with thymosin in binding to individual actin monomers, and a local activation of profilin molecules moves actin subunits from the sequestered thymosin-bound pool onto filament plus ends (Figure 16–30).

Several types of intracellular mechanisms regulate the activity of profilin, including profilin phosphorylation and profilin binding to inositol phospholipids. These mechanisms can define the sites where profilin acts. Profilin's ability to move sequestered actin subunits onto the growing ends of filaments is critical for filament assembly at the plasma membrane, for example. Profilin is localized at the cytosolic face of the plasma membrane because it binds to acidic membrane phospholipids there. At this location, extracellular signals can produce explosive local actin polymerization and the extension of actin-rich motile structures such as filopodia and lamellipodia (see below). Besides binding to

Figure 16–28 Nucleation and actin web formation by the ARP complex. (A) The structures of Arp2 and Arp3, compared to the structure of actin. Although the face of the molecule equivalent to the plus end *(top)* in both Arp2 and Arp3 is very similar to the plus end of actin itself, differences on the sides and minus end *(bottom)* prevent these actin-related proteins from forming filaments on their own or coassembling into filaments with actin. (B) A model for actin filament nucleation by the ARP complex. Arp2 and Arp3 may be held by their accessory proteins in an orientation that resembles the plus end of an actin filament. Actin subunits can then assemble onto this structure, bypassing the rate-limiting step of filament nucleation (see Figure 16–5). (C) The ARP complex nucleates filaments more efficiently when it is bound to the side of a preexisting actin filament. The result is a filament branch that grows at a 70° angle relative to the original filament. Repeated rounds of branching nucleation result in a treelike web of actin filaments. (D) Electron micrographs of branched actin filaments formed by mixing purified actin subunits with purified ARP complexes. (D, courtesy of R.D. Mullins et al., *Proc. Natl. Acad. Sci. USA* 95:6181–6186, 1998. © National Academy of Sciences.)

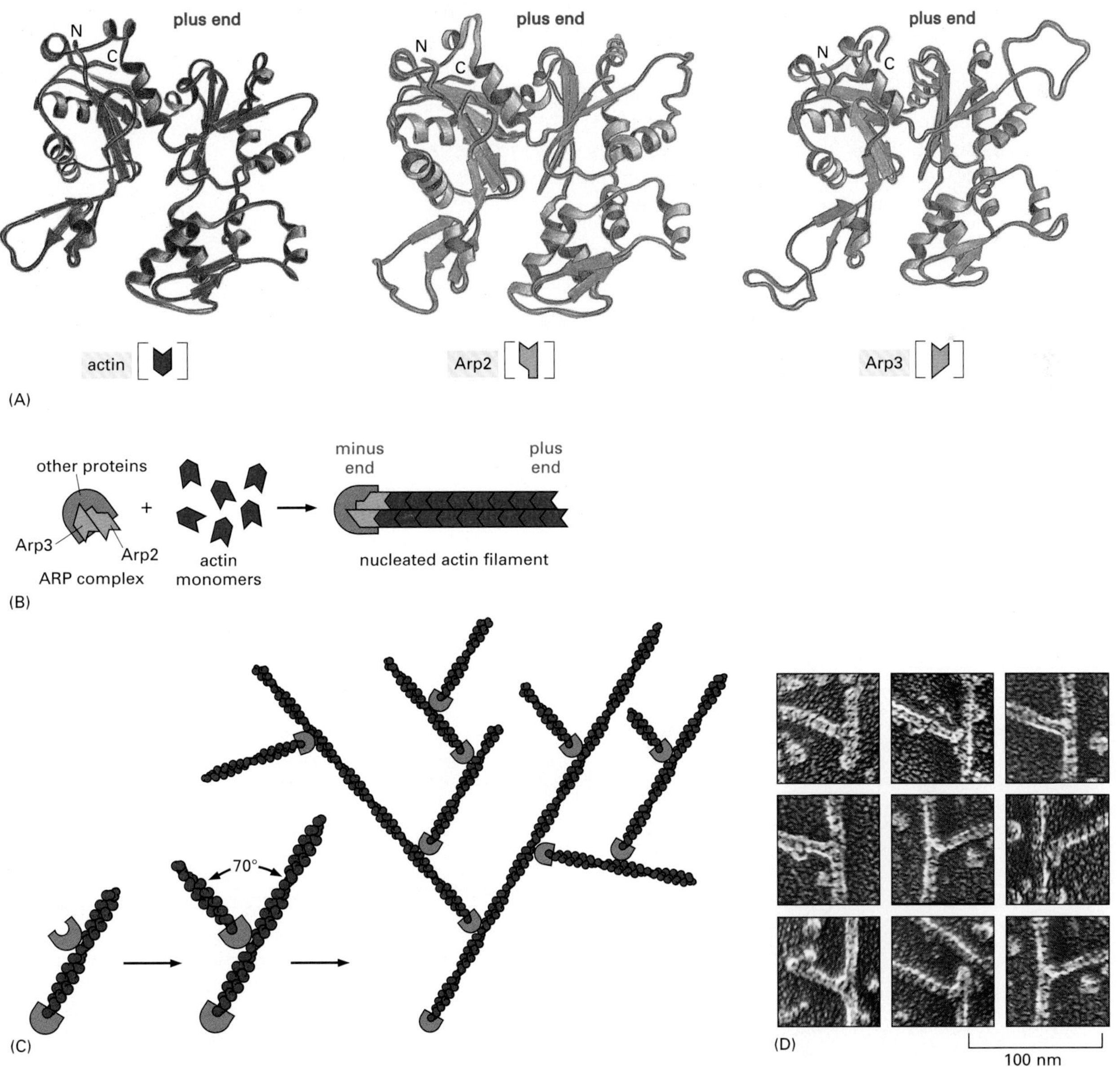

Figure 16–29 Profilin bound to an actin monomer. The profilin protein molecule is shown in *blue*, and the actin in *red*. ATP is shown in *green*. Profilin binds to the face of actin opposite the ATP-binding cleft. This profilin–actin heterodimer can therefore bind to and elongate the plus end of an actin filament, but it is sterically prevented from binding to the minus end. (Courtesy of Michael Rozycki and Clarence E. Schutt.)

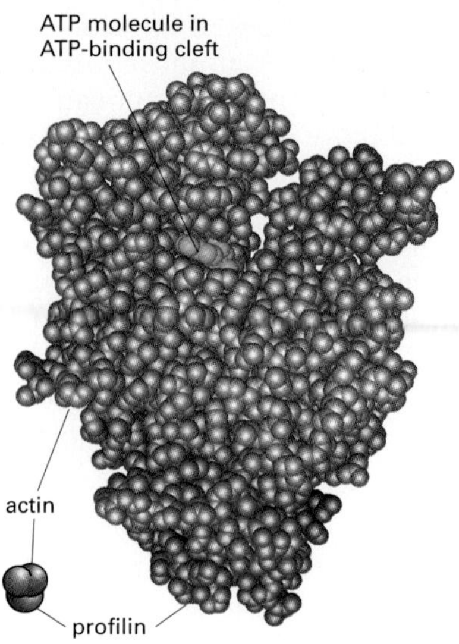

actin and phospholipids, profilin also binds to various other intracellular proteins that have domains rich in proline; these proteins may also help to localize profilin to sites where rapid actin assembly may be required.

Like actin monomers, unpolymerized tubulin subunits are sequestered in the cell to maintain the subunit pool at a level substantially higher than the critical concentration. One molecule of the small protein *stathmin* binds to two tubulin heterodimers and prevents their addition onto the ends of microtubules. Stathmin thus decreases the effective concentration of tubulin subunits that are available for polymerization (the action is analogous to that of colchicine). High levels of active stathmin in a cell decrease the elongation rate of microtubules, since the elongation rate is just the product of the concentration of available tubulin subunits and the rate constant k_{on}. The slower elongation rate also has a second remarkable effect. Since the transition from the growing state to the shrinking state for a microtubule undergoing dynamic instability depends on the race between GTP hydrolysis and filament elongation, slowing the elongation rate by sequestering tubulin subunits can increase the frequency of microtubule shrinkage. Thus, a protein that inhibits tubulin polymerization can have the secondary effect of dramatically increasing the dynamic turnover of microtubules in living cells (Figure 16–31).

Stathmin's binding to tubulin is inhibited by the phosphorylation of stathmin, and signals that result in stathmin phosphorylation can increase the rate of microtubule elongation and suppress dynamic instability.

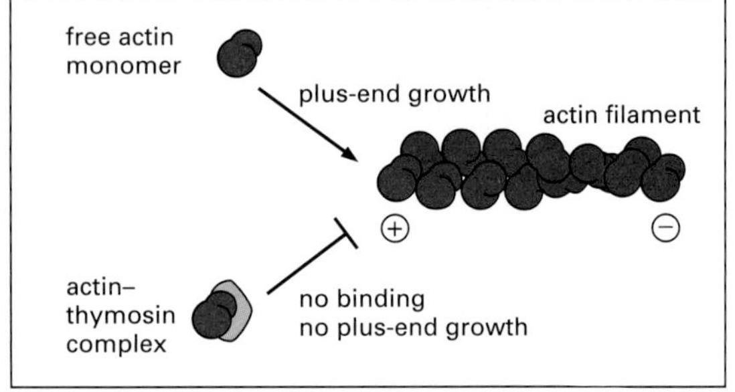

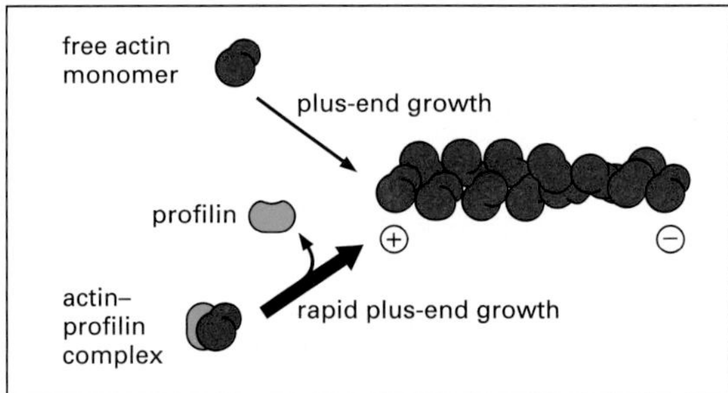

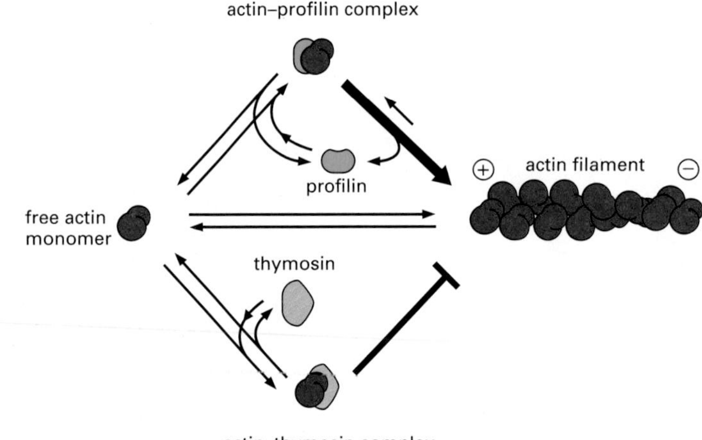

Figure 16–30 Effects of thymosin and profilin on actin polymerization. An actin monomer bound to thymosin is sterically prevented from binding to and elongating the plus end of an actin filament. An actin monomer bound to profilin, on the other hand, is capable of elongating a filament. Thymosin and profilin cannot both bind to a single actin monomer at the same time. In a cell in which most of the actin monomer is bound to thymosin, the activation of a small amount of profilin can produce rapid filament assembly. As indicated, profilin binds to actin monomers that are transiently released from the thymosin-bound monomer pool, shuttles them onto the plus ends of actin filaments, and is then released and recycled for further rounds of filament elongation.

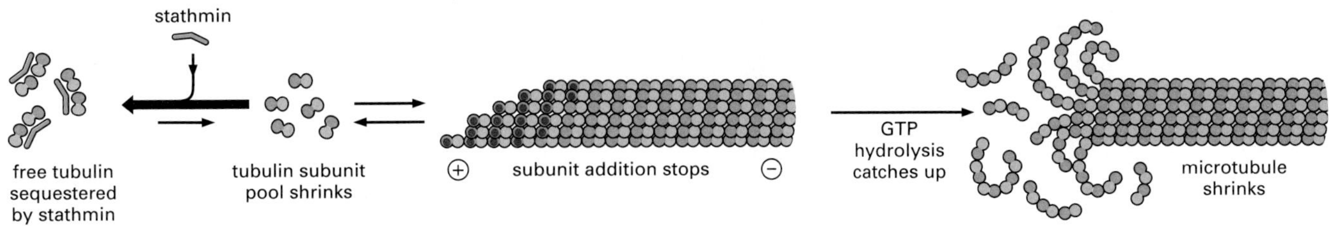
stathmin
free tubulin sequestered by stathmin
tubulin subunit pool shrinks
subunit addition stops
GTP hydrolysis catches up
microtubule shrinks

Proteins That Bind Along the Sides of Filaments Can Either Stabilize or Destabilize Them

Once a cytoskeletal filament is formed by nucleation and elongated from the subunit pool, its stability and mechanical properties are often altered by a set of proteins that bind along the sides of the polymer. Different filament-associated proteins use their binding energy to either lower or raise the free energy of the polymer state, and they thereby either stabilize or destabilize the polymer, respectively.

Proteins that bind along the sides of microtubules are collectively called **microtubule-associated proteins**, or **MAPs**. Like the drug taxol, MAPs can stabilize microtubules against disassembly. A subset of MAPs can also mediate the interaction of microtubules with other cellular components. This subset is prominent in neurons, where stabilized microtubule bundles form the core of the axons and dendrites that extend from the cell body (Figure 16–32). These MAPs have at least one domain that binds to the microtubule surface and another that projects outward. The length of the projecting domain can determine how closely MAP-coated microtubules pack together, as demonstrated in cells engineered to overproduce different MAPs. Cells overexpressing *MAP2*, which has a long projecting domain, form bundles of stable microtubules that are kept widely spaced, while cells overexpressing *tau*, a MAP with a much shorter projecting domain, form bundles of more closely packed microtubules (Figure 16–33).

The microtubule-binding domain of several MAPs, including tau and MAP2, includes multiple copies of a tubulin-binding motif. When such MAPs are added to a solution of pure unpolymerized tubulin, they greatly accelerate nucleation, presumably because they stabilize the small tubulin oligomers that form early in polymerization. MAPs are the targets of several protein kinases, and the resulting phosphorylation of a MAP can have a primary role in controlling both its activity and localization inside cells.

Whereas MAP2 and tau are confined to selected cell types in vertebrates, there are other MAPs that seem to have a central role in microtubule dynamics in nearly all eucaryotic cells. In particular, a ubiquitous protein called *XMAP215* has close homologs in organisms that range from yeast to humans (XMAP stands for *Xenopus* microtubule-associated protein, and the number refers to its molecular weight). This protein binds along the sides of microtubules, but, as discussed later, it also has a special ability to stabilize free microtubule ends and inhibit their switch from a growing to a shrinking state. The phosphorylation of XMAP215 during mitosis inhibits this activity, making a substantial contribution to the tenfold increase in the dynamic instability of microtubules observed during mitosis (see Figure 18–12).

Figure 16–31 Effects of stathmin on microtubule polymerization. Polymerization of free tubulin subunits to form microtubules is an energetically favorable reaction. As long as abundant free tubulin subunits are available, microtubule elongation continues. Stathmin sequesters free tubulin subunits, binding two tubulin αβ-heterodimers per stathmin molecule. As the pool of free tubulin subunits shrinks, microtubule elongation slows. This makes it more likely that the rate of GTP hydrolysis will catch up with the rate of subunit addition, and the GTP cap will be lost, resulting in a transition of the microtubule from the growing state to the shrinking state. Stathmin has a second activity that promotes this transition, which involves a direct interaction with a microtubule plus end (not shown). Stathmin is also called "oncoprotein 18," since its expression level is frequently increased in tumor cells, resulting in increased turnover of microtubules.

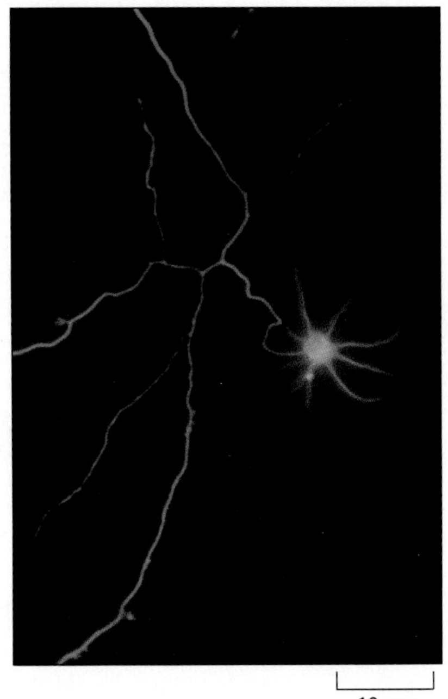
10 μm

Figure 16–32 Localization of MAPs in axons and dendrites of a neuron. This immunofluorescence micrograph shows the distribution of tau staining *(green)* and MAP2 staining *(orange)* in a hippocampal neuron in culture. Whereas tau staining is confined to the axon (long and branched in this neuron), MAP2 staining is confined to the cell body and its dendrites. The antibody used here to detect tau binds only to unphosphorylated tau; phosphorylated tau is also present in dendrites. (Courtesy of James W. Mandell and Gary A. Banker.)

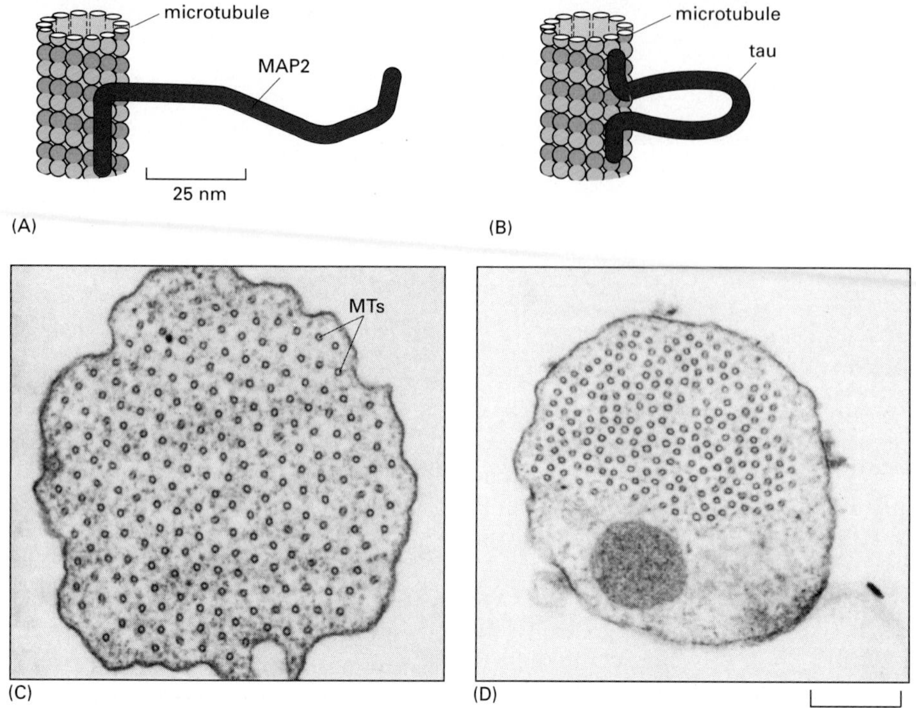

Figure 16–33 Organization of microtubule bundles by MAPs. (A) MAP2 binds along the microtubule lattice at one end and extends a long projecting arm with a second microtubule-binding domain at the other end. (B) Tau binds to the microtubule lattice at both its N- and C-termini, with a short projecting loop. (C) Electron micrograph showing a cross section through a microtubule bundle in a cell overexpressing MAP2. The regular spacing of the microtubules (MTs) in this bundle result from the constant length of the projecting arms of the MAP2. (D) Similar cross section through a microtubule bundle in a cell overexpressing tau. Here the microtubules are spaced more closely together than they are in (C) because of tau's relatively short projecting arm. (C and D, courtesy of V. Chen et al., *Nature* 360:674–647, 1992. © Macmillan Magazines Ltd.)

Actin filaments are likewise strongly affected by the binding of accessory proteins along their sides. Selected actin filaments in most cells are stabilized by the binding of *tropomyosin*, an elongated protein that binds simultaneously to seven adjacent actin subunits in one protofilament. The binding of tropomyosin along an actin filament can prevent the filament from interacting with other proteins; for this reason, the regulation of tropomyosin binding is an important step in muscle contraction, as we discuss later (see Figure 16–74).

Another important actin-filament binding protein present in all eucaryotic cells is *cofilin*, which destabilizes actin filaments. Also called *actin depolymerizing factor*, cofilin is unusual in that it binds to actin in both the filament and free subunit forms. Cofilin binds along the length of the actin filament, forcing the filament to twist a little more tightly (Figure 16–34). This mechanical stress weakens the contacts between actin subunits in the filament, making the filament brittle and more easily severed. In addition, it makes it easier for an ADP-actin subunit to dissociate from the minus end of the filament. Because the rate of actin filament treadmilling is normally limited by the slow dissociation rate at the minus end, cofilin binding causes a large increase in the rate of actin filament treadmilling. As a result, most of the actin filaments inside cells are much shorter-lived than are filaments formed from pure actin in a test tube.

Cofilin binds preferentially to ADP-containing actin filaments rather than to ATP-containing filaments. Since ATP hydrolysis is usually slower than filament assembly, the newest actin filaments in the cell still contain mostly ATP and are resistant to depolymerization by cofilin. Cofilin therefore efficiently dismantles the older filaments in the cell, ensuring that all actin filaments turn over rapidly.

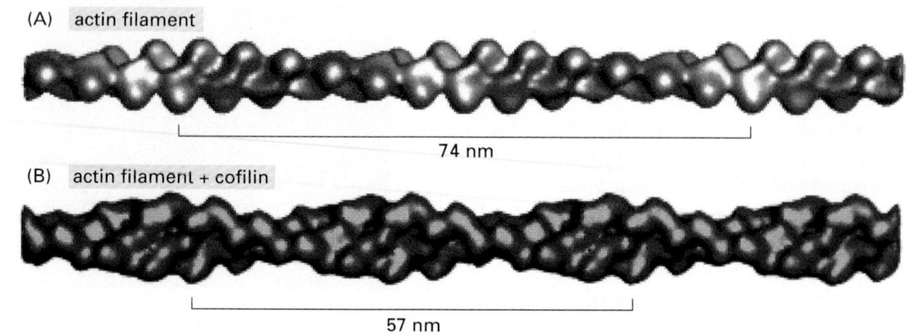

Figure 16–34 Twisting of an actin filament induced by cofilin. (A) Three-dimensional reconstruction from cryo-electron micrographs of filaments made of pure actin. The bracket shows the span of two turns of the actin helix. (B) Reconstruction of an actin filament coated with cofilin, which binds in a 1:1 stoichiometry to actin subunits all along the filament. Cofilin is a small protein (14 kilodaltons) compared to actin (43 kilodaltons), and so the filament appears only slightly thicker. The energy of cofilin binding serves to deform the actin filament lattice, twisting it more tightly so that the distance spanned by two turns of the helix is reduced. (From A. McGough et al., *J. Cell Biol.* 138:771–781, 1997. © The Rockefeller University Press.)

Proteins That Interact with Filament Ends Can Dramatically Change Filament Dynamics

As we have just seen, proteins that bind along the side of a filament can change the filament's dynamic behavior. For maximum effect, however, these proteins often need to coat the filament completely, and this means they have to be present at fairly high stoichiometries (for example, about one tropomyosin for every seven actin subunits, one tau for every four tubulin subunits, or one cofilin for every actin subunit). In contrast, proteins that bind preferentially to the ends of filaments can have dramatic effects on filament dynamics even when they are present at very low levels. Since subunit addition and loss occur primarily at filament ends, one molecule of such a protein per actin filament (typically one per about 200–500 actin monomers) can be enough to transform the architecture of an actin filament network.

As previously discussed, an actin filament that ceases elongation and is not specifically stabilized by the cell can depolymerize rapidly: it can lose subunits from either its plus or its minus end, once the actin molecules at that end have hydrolyzed their ATP to convert to the D form. The most rapid changes, however, occur at the plus end. An actin filament can be stabilized at its plus end by the binding of a plus end *capping protein*, which greatly slows the rates of both filament growth and filament depolymerization by making the plus end inactive (Figure 16–35). Indeed, most of the actin filaments in a cell are capped at their plus end by proteins such as *CapZ* (named for its location in the muscle Z band, see below; it is also called Capping Protein). At the minus end, an actin filament may be capped by remaining bound to the ARP complex that was responsible for its nucleation, although it is possible that many of the actin filament minus ends in typical cells are released from the ARP complex and are uncapped.

A local regulation of the level of capping helps cells to assemble and dismantle specific parts of their actin cytoskeleton. The association of capping proteins with actin filament ends is regulated by various localized intracellular signals. The regulation of plus end capping proteins by the inositol phospholipid PIP_2 is especially important: an increase in PIP_2 in the cytosolic leaflet of the plasma membrane caused by activation of an appropriate cell surface receptor (discussed in Chapter 15) uncaps plus ends. The uncapping makes the plus ends available for elongation, thereby promoting actin filament polymerization at the cell surface.

In muscle cells, where actin filaments are exceptionally long-lived, the filaments are known to be specially capped at both ends—by CapZ at the plus end and by *tropomodulin* at the minus end. Tropomodulin binds only to the minus end of actin filaments that have been coated by tropomyosin and have thereby already been somewhat stabilized.

The end of a microtubule, with thirteen protofilaments in a hollow ring (see Figure 16–6), is a much larger and more complex structure than the end of an actin filament, with many more possibilities for accessory protein action. We have already discussed an important microtubule capper: the γ-tubulin ring complex (γ-TuRC), which both nucleates the growth of microtubules at an organizing

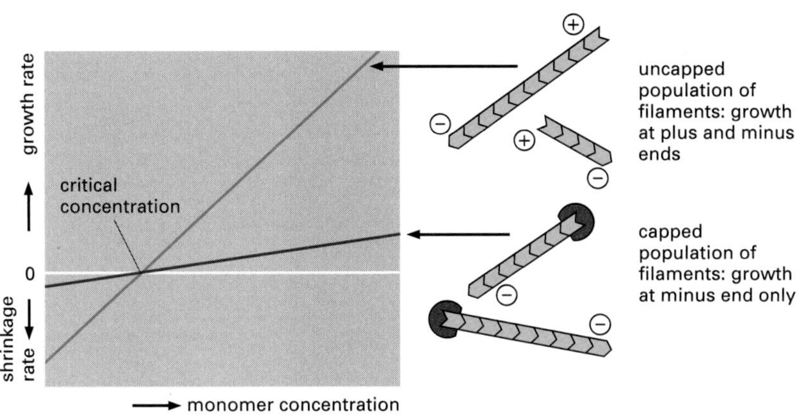

Figure 16–35 Filament capping and its effects on filament dynamics. A population of uncapped filaments adds and loses subunits at both the plus and minus ends, resulting in rapid growth or shrinkage, depending on the concentration of available free monomers (*green* line). In the presence of a protein that caps the plus end (*red* line), only the minus end is able to add or lose subunits; consequently, filament growth will be slower at all monomer concentrations above the critical concentration, and filament shrinkage will be slower at all monomer concentrations below the critical concentration. Plus-end capping of this type is widely used for actin filaments but not for microtubules.

center and caps their minus ends. Another true capping protein for microtubules is the special protein complex found at the ends of the microtubules in cilia (discussed later), where microtubules are both stable and uniform in length.

Some proteins that act at the ends of microtubules have crucial roles beyond those expected for a simple capping protein. In particular, they can have dramatic effects on the dynamic instability of microtubules (see Figure 16–11). They can influence the rate at which a microtubule switches from a growing to a shrinking state (the frequency of catastrophes) or from a shrinking to a growing state (the frequency of rescues). For example, a family of kinesin-related proteins known as *catastrophins* induce catastrophes. They bind specifically to ends and seem to pry protofilaments apart, lowering the normal activation energy barrier that prevents a microtubule from springing apart into the curved protofilament characteristic of the shrinking state (see Figure 16–11C). Opposing their actions are MAPs that bind preferentially to ends to favor continued microtubule growth (Figure 16–36A). Both groups of proteins are regulated, and a greatly altered balance between their activities causes the major increase in microtubule turnover rates observed during mitosis (see Figure 18–12).

Proteins that bind to the ends of microtubules also serve to control microtubule positioning. In addition to γ-TuRC at the minus ends, there are several groups of plus-end-binding proteins that help localize the growing microtubule end to specific target proteins in the cell cortex. An end-binding protein present in both yeasts and humans, for example, is essential for yeast mitotic spindle positioning, directing the growing plus ends of yeast spindle microtubules to a specific docking region in the yeast bud and then helping to anchor them there (Figure 16–36B).

Filaments Are Organized into Higher-order Structures in Cells

So far, we have described how cells use accessory proteins to regulate the location and dynamic behavior of cytoskeletal filaments. These proteins can nucleate

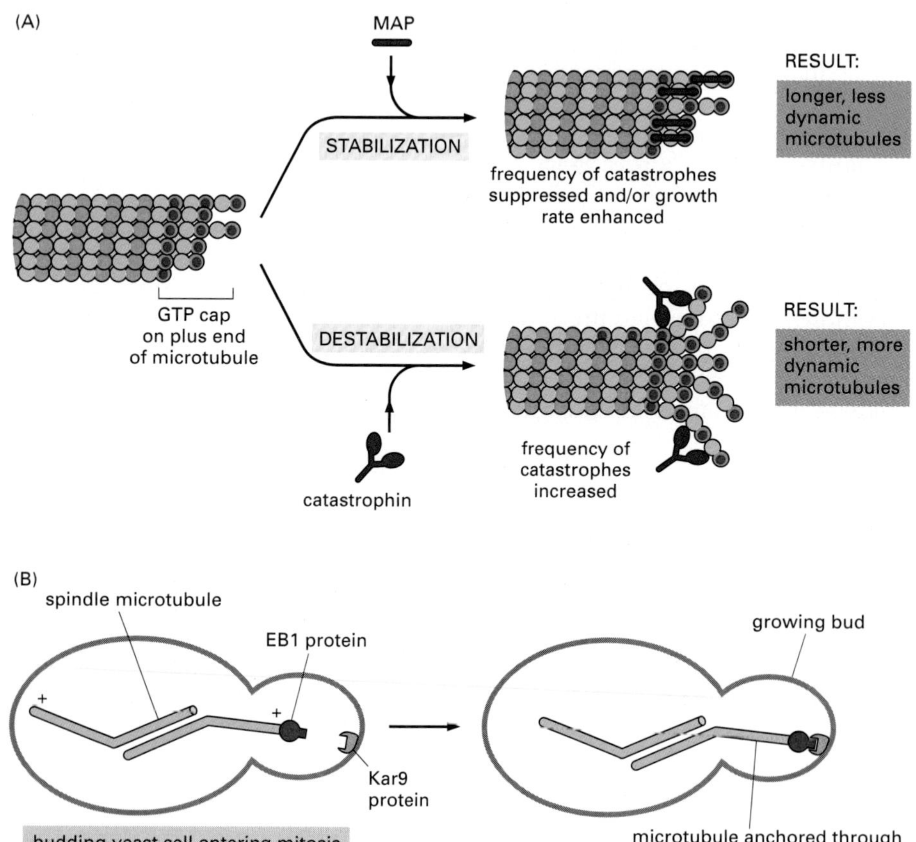

Figure 16–36 The effects of proteins that bind to microtubule ends. (A) The transition between microtubule growth and microtubule shrinking is controlled in cells by special proteins. A MAP such as XMAP215 stabilizes the end of a growing microtubule by its preferential binding there. Opposing its action are members of the kinesin superfamily (discussed later) designated here as catastrophins, but called by a variety of other names (such as, the Kin I family that includes KIF2 in Figure 16–55). (B) Capping proteins help to localize microtubules. Here, the EB1 microtubule capping protein interacts with localized Kar9 proteins in the budding yeast *Saccharomyces cerevisiae*. This interaction directs spindle microtubules into the growing bud during mitosis. A mutation in either protein causes the indicated spindle microtubules to lose their way. The EB1 protein is also found on the plus ends of microtubules in human cells.

filament assembly, bind to the ends or sides of the filaments, or bind to the free subunits of filaments. But in order for the cytoskeletal filaments to form a useful intracellular scaffold that gives the cell mechanical integrity and determines its shape, the individual filaments must be organized and tied together into larger-scale structures. The centrosome is one example of such a cytoskeletal organizer; in addition to nucleating the growth of microtubules, it holds them together in a defined geometry, with all of the minus ends buried in the centrosome and the plus ends pointing outward. In this way, the centrosome creates the astral array of microtubules that is able to find the center of each cell (see Figure 16–25).

Another mechanism that is used to organize filaments into large structures is filament cross-linking. Any tendency of filaments to stick to one another causes them to align in parallel arrays so as to maximize interfilament contacts. As described earlier, some MAPs can bundle microtubules together: they have two domains—one that binds along the microtubule side (and thereby stabilizes the filament) and another that projects outward to contact other MAP-coated microtubules. In the actin cytoskeleton, the stabilizing and cross-linking functions are separated. Tropomyosin binds along the sides of actin filaments, but it does not have an outward projecting domain. Filament cross-linking is instead mediated by a second group of actin-binding proteins that have only this function. Intermediate filaments are different yet again; they are organized both by a lateral self-association of the filaments themselves and by the cross-linking activity of accessory proteins.

Intermediate Filaments Are Cross-linked and Bundled Into Strong Arrays

Each individual intermediate filament forms as a long bundle of tetrameric subunits (see Figure 16–16). Many intermediate filaments further bundle themselves by self-association; for example, the neurofilament proteins NF-M and NF-H contain a C-terminal domain that extends outward from the surface of the assembled intermediate filament and binds to a neighboring filament. Thus groups of neurofilaments form robust parallel arrays that are held together by multiple lateral contacts, giving strength and stability to the long cell processes of neurons (see Figure 16–20).

Other types of intermediate filament bundles are held together by accessory proteins, such as *filaggrin,* which bundles keratin filaments in differentiating cells of the epidermis to give the outermost layers of the skin their special toughness. *Plectin,* which makes bundles of vimentin, is a particularly interesting cross-linking protein. Besides bundling intermediate filaments, it also links the intermediate filaments to microtubules, actin filament bundles, and filaments of the motor protein myosin II (discussed below), as well as helping to attach intermediate filament bundles to adhesive structures at the plasma membrane (Figure 16–37). Mutations in the gene for plectin cause a devastating human disease that combines epidermolysis bullosa (caused by disruption of skin keratin filaments), muscular dystrophy (caused by disruption of desmin filaments), and neurodegeneration (caused by disruption of neurofilaments). Mice lacking a functional plectin gene die within a few days of birth, with blistered skin and abnormal skeletal and heart muscles. Thus, although plectin may

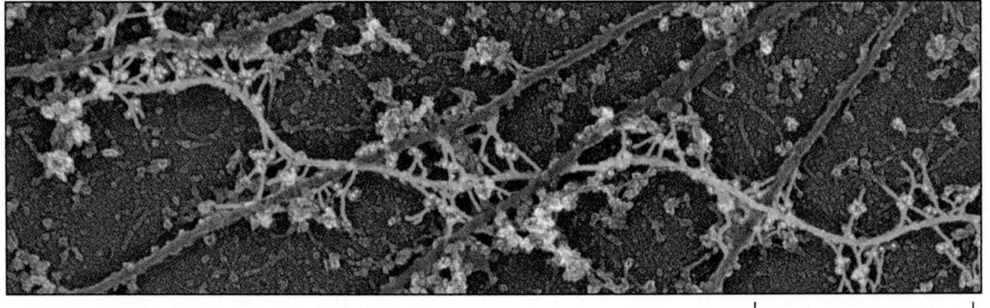

Figure 16–37 Plectin cross-linking of diverse cytoskeletal elements. Plectin *(green)* makes cross-links from intermediate filaments *(blue)* to other intermediate filaments, to microtubules *(red)*, and to myosin thick filaments. In this electron micrograph, the dots *(yellow)* are gold particles linked to anti-plectin antibodies. The entire actin filament network was removed to reveal these proteins. (From T.M. Svitkina and G.G. Borisy, *J. Cell Biol.* 135:991–1007, 1996. © The Rockefeller University Press.)

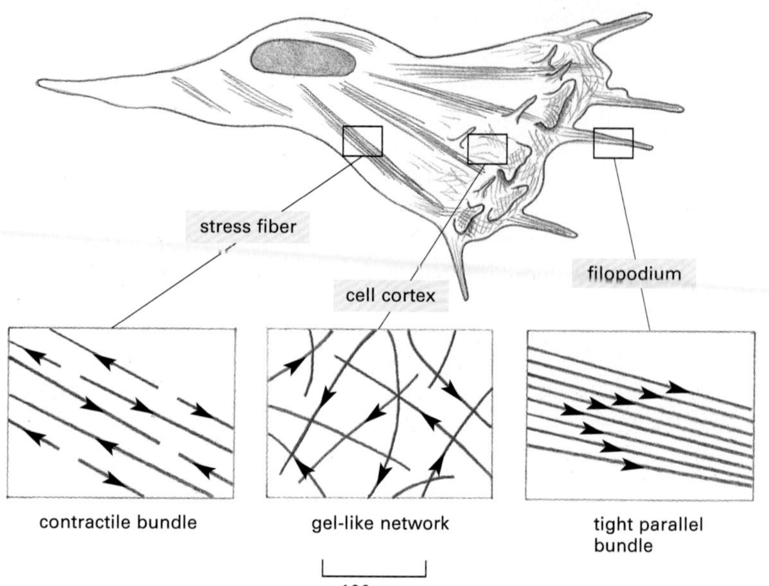

Figure 16–38 Actin arrays in a cell. A crawling cell, drawn to scale, is shown with three areas enlarged to show the arrangement of actin filaments. The actin filaments are shown in *red*, with arrowheads pointing toward the plus end. Stress fibers are contractile and exert tension. Filopodia are spike-like projections of the plasma membrane that allow a cell to explore its environments. The cortex underlies the plasma membrane.

not be necessary for the initial formation and assembly of intermediate filaments, its cross-linking action is required to provide cells with the strength they need to withstand the mechanical stresses inherent to vertebrate life.

Cross-linking Proteins with Distinct Properties Organize Different Assemblies of Actin Filaments

Actin filaments in animal cells are organized into two types of arrays: bundles and weblike (gel-like) networks (Figure 16–38). Actin filament cross-linking proteins are accordingly divided into two classes, *bundling proteins* and *gel-forming proteins*. Bundling proteins cross-link actin filaments into a parallel array, while gel-forming proteins hold two actin filaments together at a large angle to each other, thereby creating a looser meshwork. Both types of cross-linking protein generally have two similar actin-filament-binding sites, which can either be part of a single polypeptide chain or contributed by each of two polypeptide chains held together in a dimer. The spacing and arrangement of these two filament-binding domains determines the type of actin structure that a given cross-linking protein forms (Figure 16–39).

Fimbrin and *α-actinin* are widely distributed actin-bundling proteins. Fimbrin is a small crosslinker, with two actin-binding domains close together in a single polypeptide chain. It is enriched in the parallel actin filament bundles in filopodia at the leading edge of cells, and it is presumably responsible for the tight association of these actin filaments (Figure 16–40). α-actinin contains two actin-binding domains that are further apart (see Figure 16–40A and B); it is concentrated in stress fibers, where it is responsible for the relatively loose cross-linking of actin filaments in these contractile bundles. It also helps form the structure that holds stress fiber ends in *focal contacts* at the plasma membrane (see below).

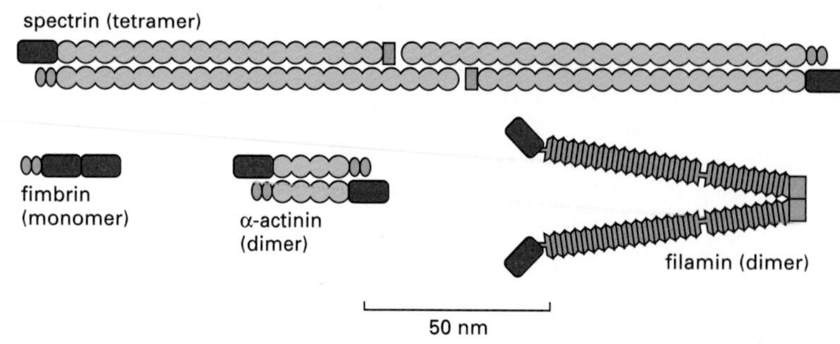

Figure 16–39 The modular structures of four actin-cross-linking proteins. Each of the proteins shown has two actin-binding sites *(red)* that are related in sequence. Fimbrin has two directly adjacent actin-binding sites, so that it holds its two actin filaments very close together (14 nm apart), aligned with the same polarity (see Figure 16–40A). The two actin-binding sites in α-actinin are separated by a spacer around 30 nm long, so that it forms more loosely packed actin bundles (see Figure 16–40). Filamin has two actin-binding sites with a V-shaped linkage between them, so that it cross-links actin filaments into a network with the filaments oriented almost at right angles to one another (see Figure 16–42). Spectrin is a tetramer of two α and two β subunits, and the tetramer has two actin-binding sites spaced about 200 nm apart (see Figure 10–31).

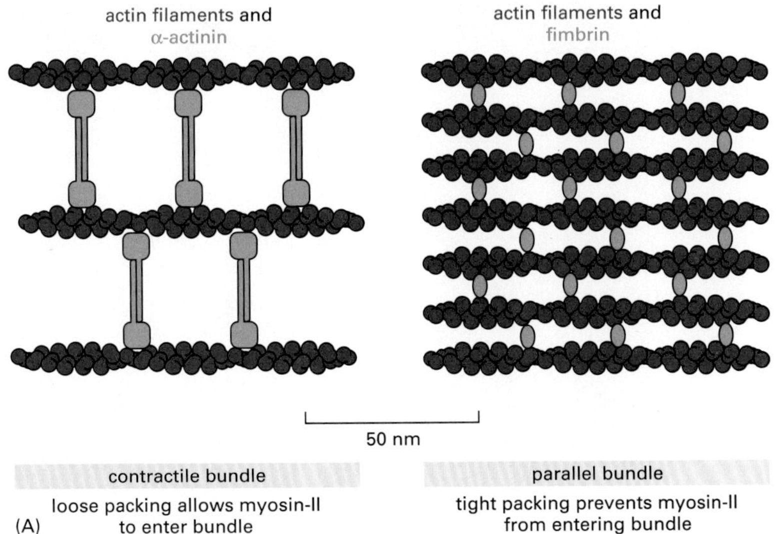

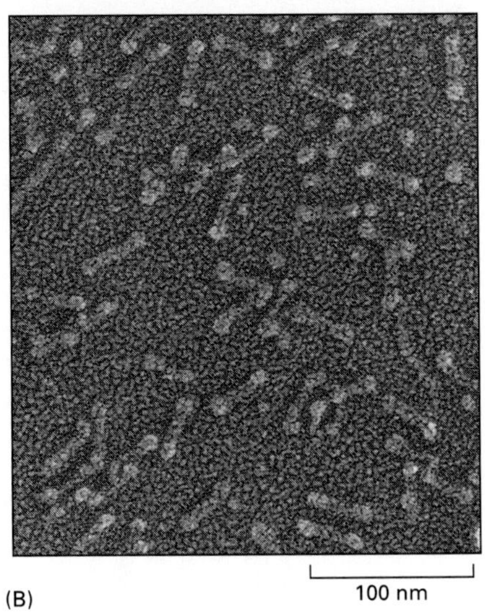

Figure 16–40 The formation of two types of actin filament bundles. (A) α-actinin, which is a homodimer, cross-links actin filaments into loose bundles, which allow the motor protein myosin II (not shown) to participate in the assembly. Fimbrin cross-links actin filaments into tight bundles, which exclude myosin. Fimbrin and α-actinin tend to exclude one another because of the very different spacing of the actin filament bundles that they form. (B) Electron micrograph of purified α-actinin molecules. (B, courtesy of John Heuser.)

Each type of bundling protein determines which other molecules can interact with an actin filament. Myosin II (discussed later) is the protein in stress fibers and other contractile arrays that is responsible for their ability to contract. The very close packing of actin filaments caused by fimbrin apparently excludes myosin, and thus filopodia are not contractile; on the other hand, the looser packing caused by α-actinin allows myosin molecules to enter, making stress fibers contractile (see Figure 16–40A). Because of the very different spacing between the actin filaments, bundling by fimbrin automatically discourages bundling by α-actinin, and vice-versa, so that the two types of bundling protein are themselves mutually exclusive.

Villin is another bundling protein that, like fimbrin, has two actin-filament-binding sites very close together in a single polypeptide chain. Villin (together with fimbrin) helps cross-link the 20 to 30 tightly bundled actin filaments found in microvilli, the finger-like extensions of the plasma membrane on the surface of many epithelial cells (Figure 16–41). A single absorptive epithelial cell in the human small intestine, for example, has several thousand microvilli on its apical surface. Each is about 0.08 μm wide and 1 μm long, making the cell's absorptive surface area about 20 times greater than it would be without microvilli. When villin is introduced into cultured fibroblasts, which do not normally contain villin and have only a few small microvilli, the existing microvilli become greatly elongated and stabilized, and new ones are induced. The actin filament core of the microvillus is attached to the plasma membrane along its sides by lateral sidearms made of *myosin I* (discussed later), which has a binding site for filamentous actin on one end and a domain that binds lipids on the other end. These two types of cross-linkers, one binding actin filaments to each other and the other binding these filaments to the membrane, seem to be sufficient to form microvilli on cells.

The various bundling proteins that we have discussed so far have straight, stiff connections between their two actin-filament-binding domains, and they tend to align filaments in parallel bundles. In contrast, those actin cross-linking proteins that have either a flexible or a stiff, bent connection between their two binding domains form actin filament webs or gels, rather than actin bundles.

A well-studied web-forming protein is *spectrin*, which was first identified in red blood cells. Spectrin is a long, flexible protein made out of four elongated polypeptide chains (two α subunits and two β subunits), arranged so that the two actin-filament-binding sites are about 200 nm apart (compared with 14 nm for fimbrin and about 30 nm for α-actinin, see Figure 16–39). In the red blood cell, spectrin is concentrated just beneath the plasma membrane, where it forms a two-dimensional web held together by short actin filaments; spectrin links this web to the plasma membrane because it has separate binding sites for peripheral membrane proteins, which are themselves positioned near the lipid bilayer by integral membrane proteins (see Figure 10–31). The resulting network creates

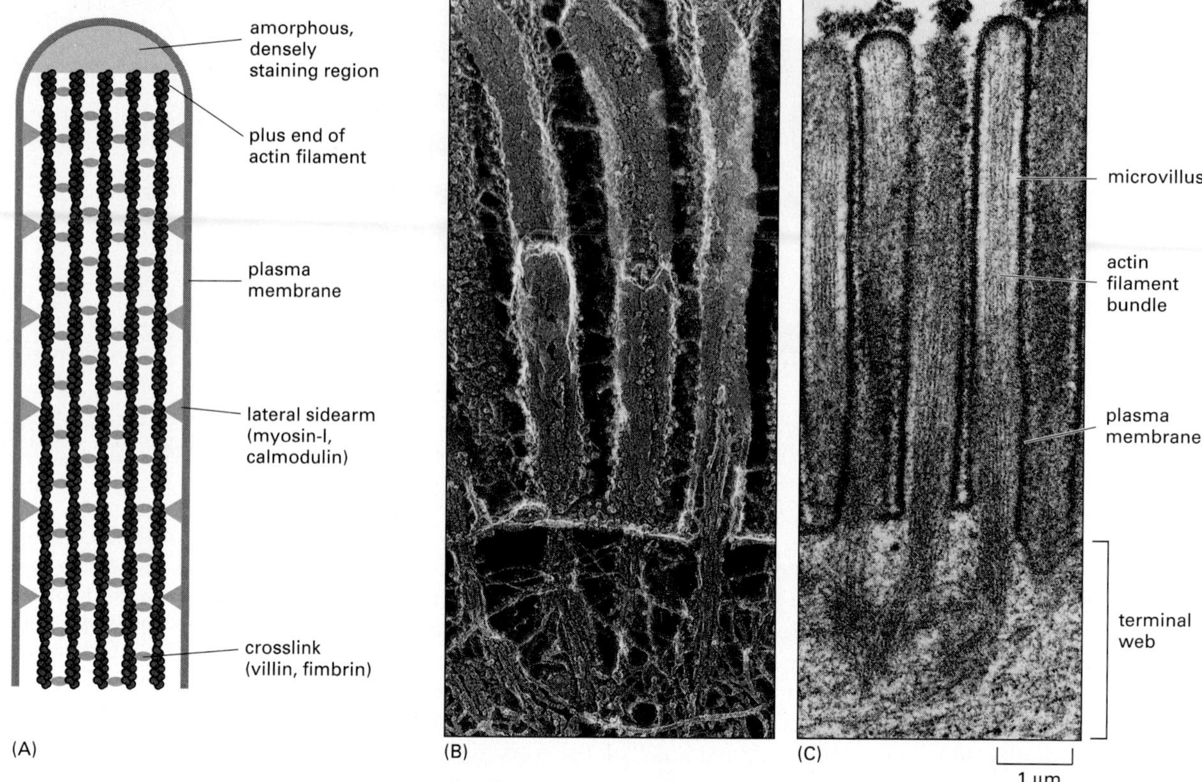

Figure 16–41 A microvillus. (A) A bundle of parallel actin filaments cross-linked by the actin-bundling proteins villin and fimbrin forms the core of a microvillus. Lateral sidearms (composed of myosin I and the Ca^{2+}-binding protein calmodulin) connect the sides of the actin filament bundle to the overlying plasma membrane. All the plus ends of the actin filaments are at the tip of the microvillus, where they are embedded in an amorphous, densely staining substance of unknown composition. (B) Freeze-fracture electron micrograph of the apical surface of an intestinal epithelial cell, showing microvilli. Actin bundles from the microvilli extend down into the cell and are rooted in the terminal web, where they are linked together by a complex set of proteins that includes spectrin and myosin II. Below the terminal web is a layer of intermediate filaments. (C) Thin section electron micrograph of microvilli. (B, courtesy of John Heuser; C, from P.T. Matsudaira and D.R. Burgess, *Cold Spring Harbor Symp. Quant. Biol.* 46:845–854, 1985.)

a stiff cell cortex that provides mechanical support for the overlying plasma membrane, allowing the red blood cell to spring back to its original shape after squeezing through a capillary. Close relatives of spectrin are found in the cortex of most other vertebrate cell types, where they also help to shape and stiffen the surface membrane.

Any cross-linking protein that has its two actin-binding domains joined by a long bent linkage can form three-dimensional actin gels. *Filamin* (see Figure 16–39) promotes the formation of a loose and highly viscous gel by clamping together two actin filaments roughly at right angles (Figure 16–42). The actin gels formed by filamin are required for cells to extend the thin sheet-like membrane

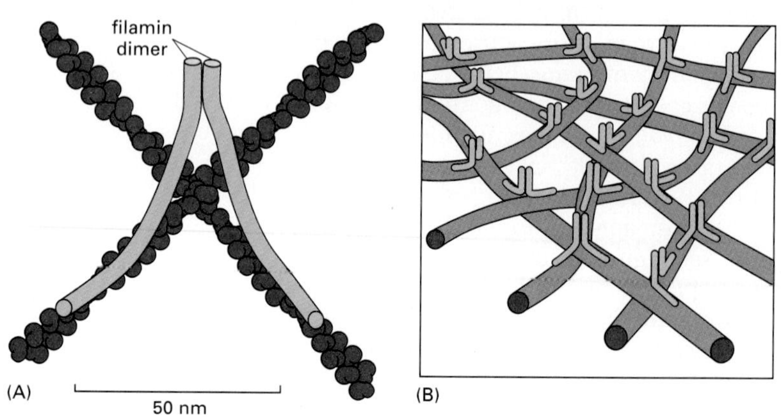

Figure 16–42 Filamin cross-links actin filaments into a three-dimensional network with the physical properties of a gel. (A) Each filamin homodimer is about 160 nm long when fully extended and forms a flexible, high-angle link between two adjacent actin filaments. (B) A set of actin filaments cross-linked by filamin forms a mechanically strong web or gel.

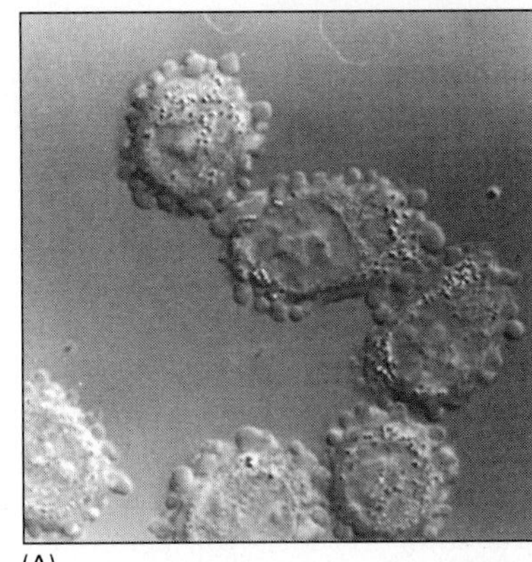

(A)

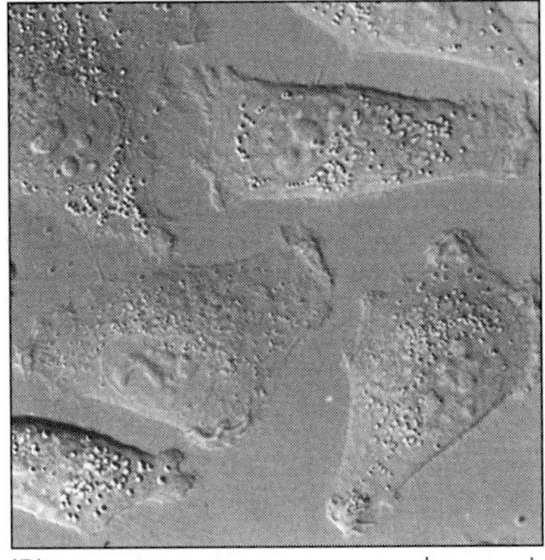

(B)

10 μm

Figure 16–43 Loss of filamin causes abnormal cell motility. (A) A group of melanoma cells that have an abnormally low level of filamin. These cells are not able to make normal lamellipodia and instead are covered with membrane "blebs." As a result, they crawl poorly and tend not to metastasize. (B) The same melanoma cells in which filamin expression has been artificially restored. The cells now make normal lamellipodia and are highly metastatic. This example is one of many demonstrating the profound effect that the presence or absence of a single structural protein can have on cell morphology and motility. (From C. Cunningham et al., *J. Cell Biol.* 136:845–857, 1997. © The Rockefeller University Press.)

projections called *lamellipodia* that help them to crawl across solid surfaces. Filamin is lacking in some types of cancer cells, especially some malignant melanomas (pigment-cell cancers). These cells cannot crawl properly, and instead they protrude disorganized membrane blebs (Figure 16–43). Losing filamin is bad news for the melanoma cells but good news for the melanoma patient; because of the cells' inability to crawl, melanoma cells that have lost filamin expression are less invasive than similar melanoma cells that still express filamin, and, as a result, the cancer is much less likely to metastasize.

Severing Proteins Regulate the Length and Kinetic Behavior of Actin Filaments and Microtubules

In some situations, a cell may break an existing long filament into many smaller filaments. This generates a large number of new filament ends: one long filament with just one plus end and one minus end might be broken into dozens of short filaments, each with its own minus end and plus end. Under some intracellular conditions, these newly formed ends nucleate filament elongation, and in this case severing accelerates the assembly of new filament structures. Under other conditions, severing promotes the depolymerization of old filaments, speeding up the depolymerization rate by tenfold or more (Figure 16–44). In addition, severing filaments changes the physical and mechanical properties of the cytoplasm: stiff, large bundles and gels become more fluid when the filaments are severed.

To sever a microtubule, thirteen longitudinal bonds must be broken, one for each protofilament. The protein *katanin,* named after the Japanese word for "sword," accomplishes this demanding task (Figure 16–45). Katanin is made up of two subunits, a smaller subunit that hydrolyzes ATP and performs the actual severing, and a larger one that directs katanin to the centrosome. Katanin releases microtubules from their attachment to a microtubule organizing center, and it is thought to have a particularly crucial role in the rapid microtubule depolymerization observed at the poles of a mitotic spindle during mitosis. It is also found in proliferating cells in interphase and in postmitotic cells such as neurons, where it may also be involved in microtubule release and depolymerization.

In contrast to microtubule severing by katanin, which requires ATP, the severing of actin filaments does not require an extra energy input. Most actin-severing proteins are members of the *gelsolin superfamily,* whose severing activity is activated by high levels of cytosolic Ca^{2+}. Gelsolin has subdomains that

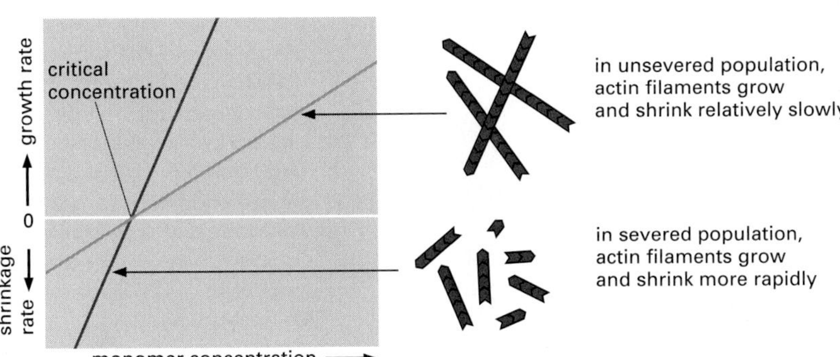

Figure 16–44 Filament severing and its effects on filament dynamics. Normally, filaments are able to gain and lose subunits only at their ends, limiting the maximum rates of growth and shrinkage of the total amount of actin in filamentous form *(green* line). If the filaments are severed into smaller pieces, the number of ends increases, so the increase in filament mass in the population can be faster at all monomer concentrations above the critical concentration, and the decrease in filament mass in the population can also be faster at all monomer concentrations below the critical concentration *(red* line). Thus, severing enhances the rate of filament dynamics.

bind to two different sites on the actin subunit, one exposed on the surface of the filament and one that is normally hidden in the longitudinal bond to the next subunit in the protofilament. According to one model for gelsolin severing, gelsolin binds on the side of an actin filament and waits until a thermal fluctuation happens to create a small gap between neighboring subunits in the protofilament; gelsolin then insinuates its subdomain into the gap, breaking the filament (Figure 16–46). Once gelsolin has severed an actin filament, it remains bound to the plus end and acts as an effective capping protein. However, like several other actin filament capping proteins, it can be removed from the filament end by a local rise in PIP_2 concentration.

The process of platelet activation shows how a cell can regulate its actin accessory proteins that mediate severing, capping, and cross-linking to generate rapid and dramatic morphological changes. Platelets are tiny cells without a nucleus that circulate in the blood and help to form clots at sites of injury. The resting platelet is discoid in shape, and it contains short actin filaments capped by CapZ, surrounded by a large pool of actin monomer bound to profilin. When the platelet is activated by physical contact with the edge of a damaged blood vessel or by a chemical clotting signal such as thrombin, a rapid, intracellular, signal transduction cascade results in a massive influx of Ca^{2+} into the platelet cytosol. The Ca^{2+} activates gelsolin, which cleaves the capped filaments into tiny fragments, each now capped by gelsolin. With slower kinetics, the same signaling pathway causes a rise in PIP_2 levels, which inactivates both gelsolin and CapZ, removing them from the filament plus ends. The large numbers of free plus ends generated by severing and uncapping are then rapidly elongated by the monomeric actin pool, forming many long filaments. Some of these long actin filaments are cross-linked into a gel by filamin, while others are bundled by α-actinin and fimbrin. This causes the activated platelet to extend lamellipodia and filopodia and to spread itself across the clot (Figure 16–47), attaching to the clot by transmembrane adhesion proteins called integrins. Once the PIP_2 signal subsides, the CapZ returns to the ends of the filaments, rendering them stable against depolymerization and locking the platelet into its spread form. Finally, myosin II uses ATP hydrolysis to slide the long actin filaments relative to one another, causing a contraction of the platelet that pulls the edges of the wound together.

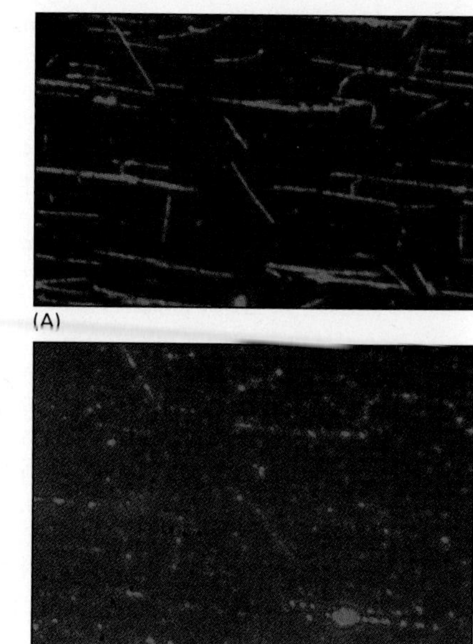

Figure 16–45 Microtubule severing by katanin. Taxol-stabilized, rhodamine-labeled microtubules were adsorbed on the surface of a glass slide, and purified katanin was added along with ATP. (A) There are a few breaks in the microtubules 30 seconds after the addition of katanin. (B) The same field 3 minutes after the addition of katanin. The filaments have been severed in many places, leaving a series of small fragments at the previous locations of the long microtubules. (From J.J. Hartman et al., *Cell* 93:277–287, 1998. © Elsevier.)

Cytoskeletal Elements Can Attach to the Plasma Membrane

A common function of actin cytoskeletal structures is to stiffen or to change the shape of the plasma membrane. We have already encountered at least two examples: the spectrin-actin web that underlies the red blood cell plasma membrane and the villin-actin bundles in microvilli that enlarge the absorptive surface area of epithelial cells. The effectiveness of these structures depends on both the bundling and cross-linking of actin filaments and the specific attachments between the actin filament structures and proteins or lipids of the plasma membrane. In many cases, the cytoskeletal assemblies have the additional function of helping to connect the internal structure of a cell to its surrounding

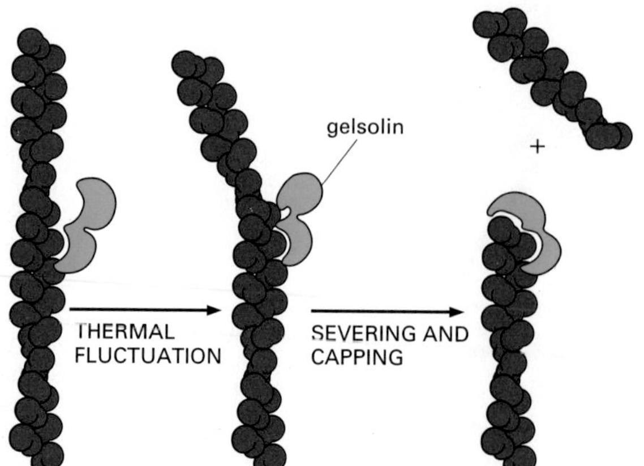

Figure 16–46 A model for actin filament severing by gelsolin. There are two actin-binding sites in the gelsolin protein. These sites allow gelsolin to bind both along the side and at the end of an actin filament. Filaments that have been severed by gelsolin remain capped at their plus ends. Villin, an actin-bundling protein (see Figure 16–41A), is a member of the same protein family as gelsolin, but it has gained another actin-binding site and lost its ability to sever filaments except at very high (nonphysiological) Ca^{2+} concentrations.

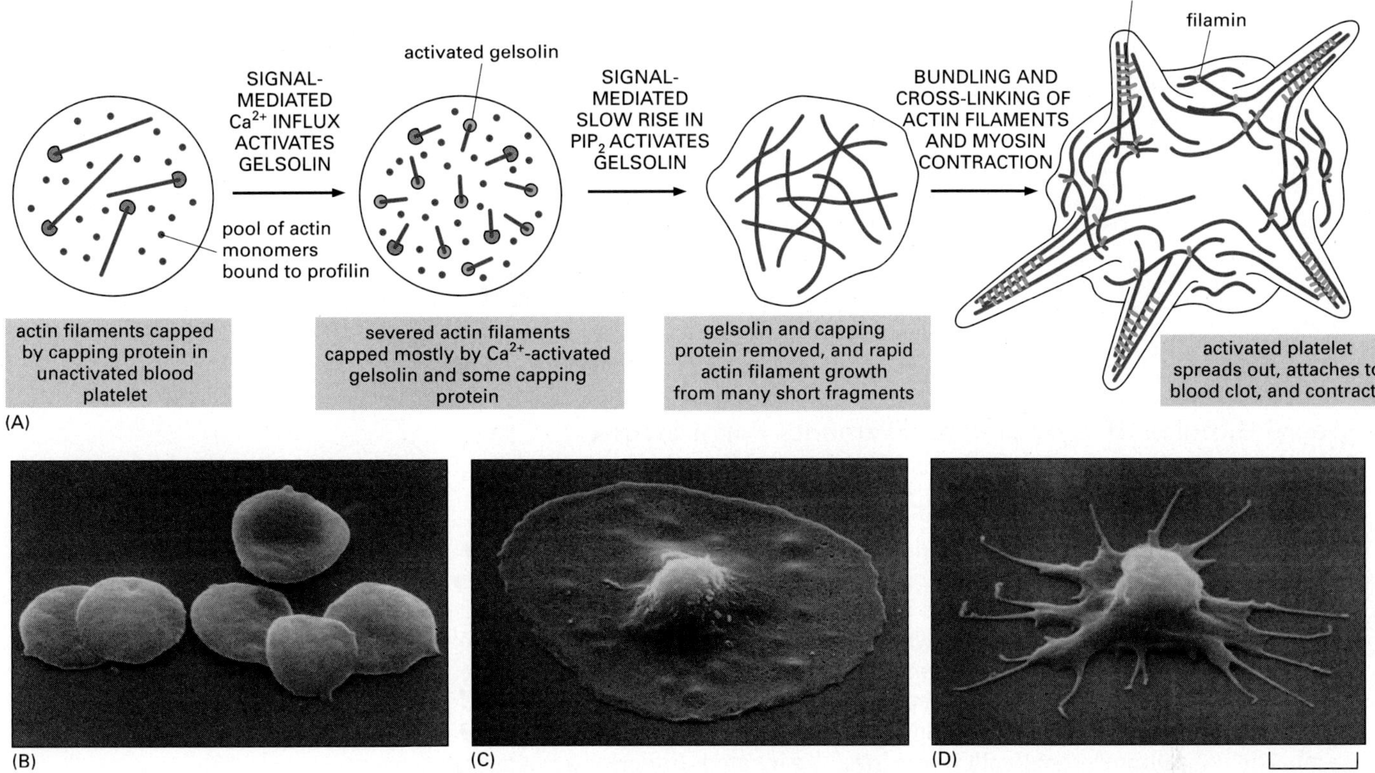

Figure 16–47 Platelet activation. (A) Platelet activation is a controlled sequence of actin filament severing, uncapping, elongation, recapping, and cross-linking that creates a dramatic shape change in the platelet. (B) Scanning electron micrograph of platelets prior to activation. (C) An activated platelet with its large spread lamellipodium. (D) An activated platelet at a later stage than the one shown in C, after myosin II-mediated contraction. (Courtesy of James G. White.)

environment, including other cells and the extracellular matrix. Both actin filaments and intermediate filaments are critical for these connections. Here, we describe several types of specific interactions between cytoskeletal filaments and the transmembrane proteins that mediate cell adhesion. We consider only the intracellular side of these interactions; the extracellular interactions are discussed in Chapter 19.

A widespread family of closely related intracellular proteins, the *ERM* family (named for its first three members, ezrin, radixin, and moesin), acts to attach actin filaments to the plasma membrane in many cell types. The C-terminal domain of ERM proteins binds directly to the sides of actin filaments. The N-terminal domain binds to the cytoplasmic face of several transmembrane glycoproteins, including CD44, the receptor for the extracellular matrix component hyaluronan. The functional importance of the ERM proteins is indicated by the consequences of mutations that lead to a loss of one of the members of the family, called merlin. This results in one form of the human genetic disease called *neurofibromatosis,* in which multiple benign tumors develop in the auditory nerves and certain other parts of the nervous system.

Unlike the attachments between actin and the plasma membrane mediated by spectrin or myosin I, those mediated by ERM proteins are regulated by both intracellular and extracellular signals. ERM proteins can exist in two conformations, an active extended conformation that oligomerizes and binds actin and CD44, and an inactive folded conformation, in which the N- and C-termini are held together by an intramolecular interaction. Switching between the two conformations can be triggered by phosphorylation or binding to PIP_2, either of which can occur, for example, in response to extracellular signals. Thus, the strength of ERM-mediated contacts between the actin cytoskeleton and the extracellular matrix is sensitive to a variety of signals received by the cell (Figure 16–48).

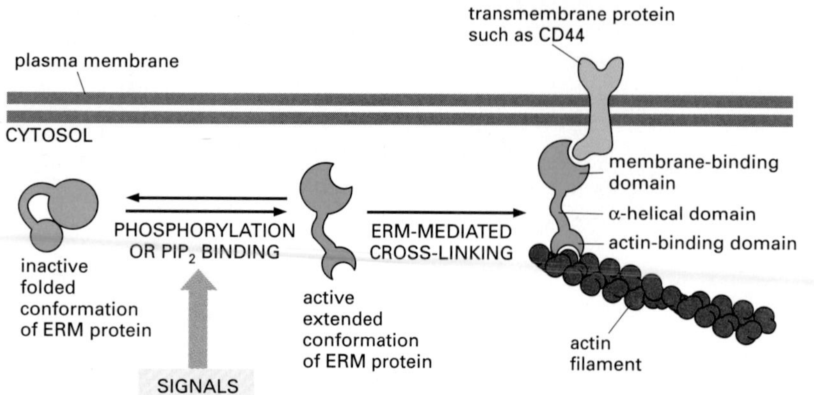

Figure 16–48 The role of ERM-family proteins in attaching actin filaments to the plasma membrane. Regulated unfolding of an ERM-family protein, caused by phosphorylation or by binding to PIP_2, exposes two binding sites, one for an actin filament and one for a transmembrane protein. Activation of ERM-family proteins can thereby generate and stabilize cell-surface protrusions that form in response to extracellular signals.

Special Bundles of Cytoskeletal Filaments Form Strong Attachments Across the Plasma Membrane: Focal Contacts, Adhesion Belts, and Desmosomes

Focal contacts are a highly specialized type of attachment between actin filaments and the extracellular matrix that allows cells to pull on the substratum to which they are bound. Focal contacts are particularly easy to see in cultured fibroblasts, because they create spots where the normal 50 nm gap between the bottom of the cell and the substratum is reduced to only 10–15 nm (Figure 16–49). At these sites, stress fibers, consisting of contractile bundles of actin and myosin II filaments, terminate at the plasma membrane, where clusters of transmembrane adhesion proteins, called *integrins,* are located. Integrins are a large family of heterodimeric proteins that bind to various components of the extracellular matrix (discussed in Chapter 19), and their linkage to the bundle of actin filaments is indirect, being mediated by an elaborate complex composed of multiple intracellular anchor proteins.

Besides serving as anchors for the cell, focal contacts can also relay signals from the extracellular matrix to the inside of the cell. Thus, the clustering of integrins at focal contacts activates a tyrosine kinase, *focal adhesion kinase (FAK)* (discussed in Chapter 19). Its activity is sensitive to the type of substratum on which the cell rests, and it is also regulated by the amount of tension at the cell attachment site. How this occurs is uncertain, but FAK is a target of the Src cytoplasmic tyrosine kinase (discussed in Chapter 5), which is also enriched at focal contacts. Once activated, FAK phosphorylates numerous targets, including many components of the focal contact complex itself, and it thereby helps to regulate the survival, growth, proliferation, morphology, movement, and differentiation of cells in response to the extracellular environment. Although we know few of the details, it is possible that the growth and development of a cell are influenced as much by signals received by means of its adhesion proteins as by the more extensively studied receptors for soluble, extracellular signal molecules described in Chapter 15.

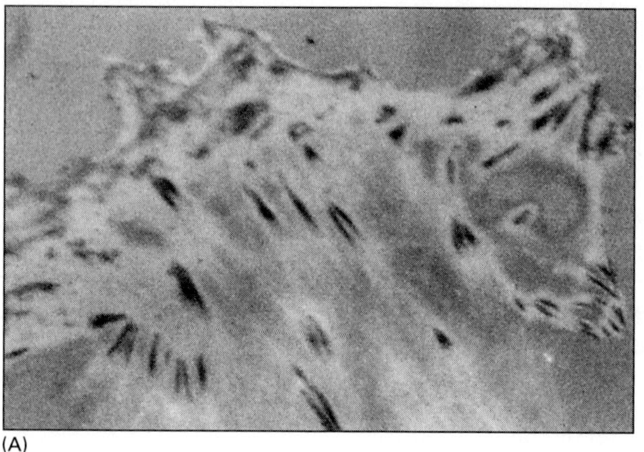
(A)

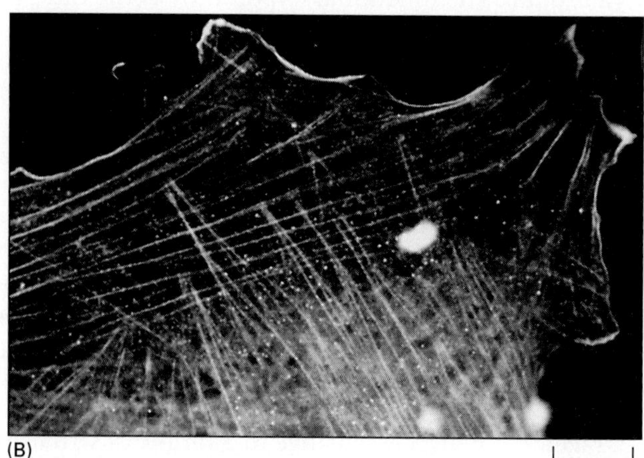
(B)
10 μm

Figure 16–49 Focal contacts and stress fibers in a cultured fibroblast. (A) Focal contacts are best seen in living cells by reflection-interference microscopy. In this technique, light is reflected from the lower surface of a cell attached to a glass slide, and the focal contacts appear as dark patches. (B) Immunofluorescence staining of the same cell (after fixation) with antibodies to actin shows that most of the cell's actin filament bundles (or stress fibers) terminate at or close to a focal contact. (Courtesy of Grenham Ireland.)

Besides forming mechanically strong attachments to the extracellular matrix, cytoskeletal elements anchored to the plasma membrane also provide attachments to other cells. For example, the cell–cell contacts formed in epithelia are mediated largely by the interactions of transmembrane *cadherin* proteins (see Figure 19–24). The cytoplasmic tail of a cadherin molecule binds a complex of proteins called catenins, which in turn bind to actin filaments (see Figure 19–29). Clusters of these actin-reinforced, cadherin-mediated cell–cell contacts form the *adherens junctions* described in Chapter 19.

The general principle that we see in focal contacts and adherens junctions, in which bundles of cytoskeletal filaments are indirectly attached through elaborate multiprotein complexes to transmembrane adhesion proteins, is recapitulated in the attachment of intermediate filaments to the plasma membrane. Intermediate filaments are anchored to the plasma membrane at structures called *desmosomes* (at cell–cell junctions) and *hemidesmosomes* (at cell–extracellular matrix junctions). These cell junctions are especially important in maintaining the strength of epithelial tissues (discussed in Chapter 19).

Extracellular Signals Can Induce Major Cytoskeletal Rearrangements

In the preceding sections, we have seen how accessory proteins associated with cytoskeletal filament systems can regulate filament length, location, organization, and dynamic behavior. Extracellular signals can alter the activity of these accessory proteins, thereby changing the cytoskeleton and cell behavior.

For the actin cytoskeleton, global structural rearrangements in response to external signals are triggered through diverse cell-surface receptors. But all of these signals seem to converge inside the cell on a group of closely related monomeric GTPases that are members of the **Rho protein family**—*Cdc42, Rac,* and *Rho.* Like other members of the Ras superfamily, these Rho proteins act as molecular switches to control cellular processes by cycling between an active, GTP-bound state and an inactive, GDP-bound state (see Figure 3–70). Activation of Cdc42 triggers actin polymerization and bundling to form either filopodia or shorter cell protrusions called microspikes. Activation of Rac promotes actin polymerization at the cell periphery leading to the formation of sheet-like lamellipodial extensions and membrane ruffles. Activation of Rho promotes both the bundling of actin filaments with myosin II filaments into stress fibers and the clustering of integrins and associated proteins to form focal contacts (Figure 16–50). These dramatic and complex structural changes occur because each of these three molecular switches has numerous downstream target proteins that affect actin organization and dynamics (at least eight have been found for each).Other target proteins affect gene transcription.

Some key targets of activated Cdc42 are members of the WASp protein family. **WASp proteins**, like ERM family proteins, can exist in an inactive folded conformation and an activated open conformation. Association with Cdc42-GTP stabilizes the open form of WASp, enabling it to bind to the ARP complex. This strongly enhances the actin-nucleating activity of the ARP complex (see Figure 16–28). Thus, activation of Cdc42 causes an increase in actin nucleation. Rac-GTP also activates WASp family members, but in addition it activates PI(4)P-5 kinase, which generates PI(4,5)P_2 (a form of PIP_2). As we have seen previously, PIP_2 can cause the uncapping of filaments whose plus end is bound by gelsolin or CapZ, thereby providing even more sites for actin assembly near the plasma membrane, resulting in the formation of large lamellipodia and ruffles. Rho-GTP activates a protein kinase that inhibits a phosphatase acting on myosin light chains (see Figure 16–67). The consequent increase in the net amount of myosin light chain phosphorylation increases the amount of contractile myosin activity in the cell, enhancing the formation of tension-dependent structures such as stress fibers.

The mechanisms that turn on the three Rho protein family members are similarly complex. Their activation, through an exchange of GTP for GDP, is promoted by guanine nucleotide exchange factors (GEFs), of which more than 20

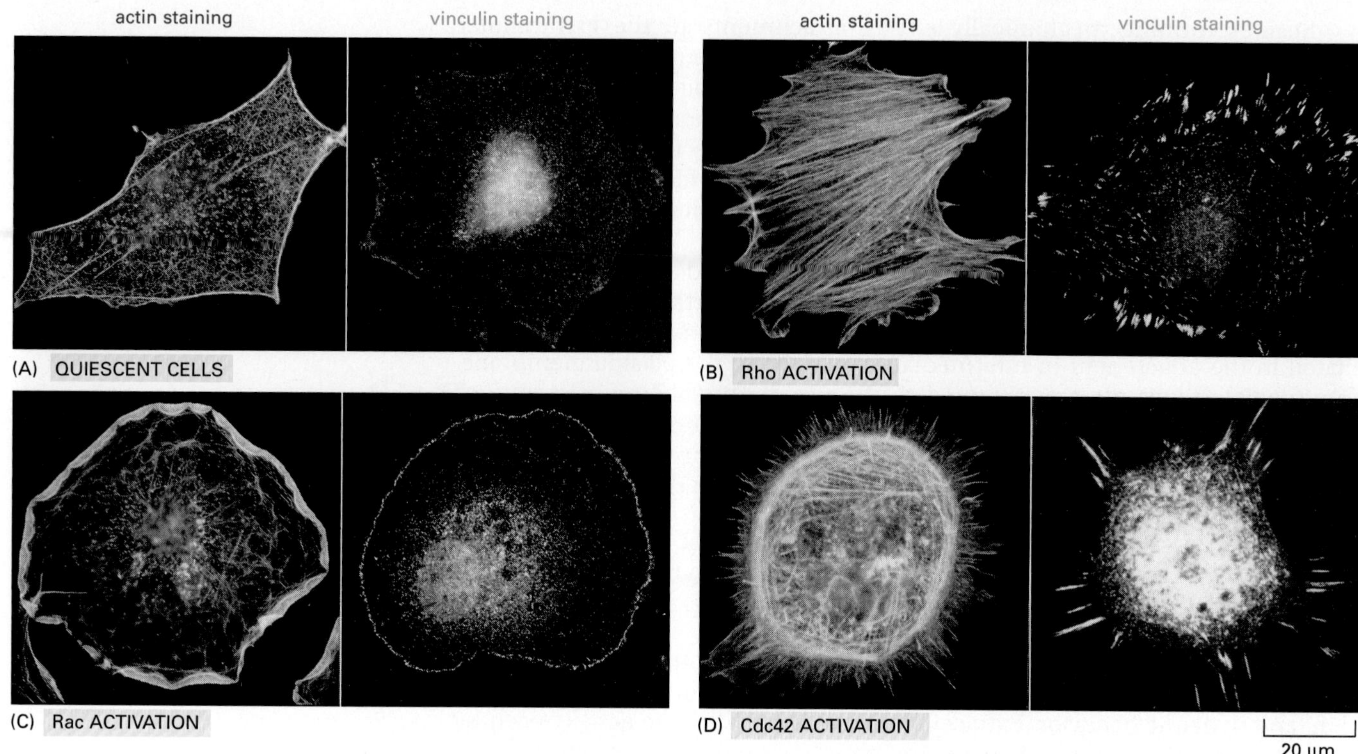

Figure 16–50 The dramatic effects of Rac, Rho, and Cdc42 on actin organization in fibroblasts. In each case, the actin filaments have been labeled with fluorescent phalloidin, and focal contacts have been located with an antibody against vinculin. (A) Serum-starved fibroblasts have actin filaments primarily in the cortex, and relatively few focal contacts. (B) Microinjection of a constitutively activated form of Rho causes the rapid assembly of many prominent stress fibers and focal contacts. (C) Microinjection of a constitutively activated form of Rac, a closely related monomeric GTPase, causes the formation of an enormous lamellipodium that extends from the entire circumference of the cell. (D) Microinjection of a constitutively activated form of Cdc42, another Rho family member, causes the protrusion of many long filopodia at the cell periphery that form adhesive contacts with the substratum. The distinct global effects of these three GTPases on the organization of the actin cytoskeleton are mediated by the actions of dozens of other protein molecules that are regulated by the GTPases. Many of these target proteins resemble the various actin-associated proteins that we have discussed in this chapter. (From A. Hall, *Science* 279:509–514, 1998. © AAAS.)

have been identified. Some are specific for an individual Rho family GTPase, whereas others seem to act on all three. Upstream from the GEFs are various cell-surface receptors. In most cases, the mechanisms that couple the receptors to GEF activation are unknown.

Summary

The varied forms and functions of cytoskeletal filament structures in eucaryotic cells depend on a versatile repertoire of accessory proteins. Each of the three major filament classes (microtubules, intermediate filaments, and actin filaments) has a large dedicated subset of such accessory proteins. A primary determinant of the sites of cytoskeletal structures is the regulation of the processes that initiate the nucleation of new filaments. In most animal cells, microtubules are nucleated at the centrosome, a complex assembly located near the center of the cell. In contrast, most actin filaments are nucleated near the plasma membrane.

The kinetics of filament assembly and disassembly can be either slowed or accelerated by accessory proteins that bind to either the free subunits or the filaments themselves. Some of these proteins alter filament dynamics by binding to the ends of filaments or by severing the filaments into smaller fragments. Another class of accessory proteins assembles the filaments into larger ordered structures by cross-linking them to one another in geometrically defined ways. Yet other accessory proteins determine the shape and adhesive properties of cells by attaching filaments to the plasma membrane. Cytoskeletal rearrangements are mediated by controlling the activities of various accessory proteins.

MOLECULAR MOTORS

Perhaps the most fascinating proteins that associate with the cytoskeleton are the molecular motors called **motor proteins**. These remarkable proteins bind to a polarized cytoskeletal filament and use the energy derived from repeated cycles of ATP hydrolysis to move steadily along it. Dozens of different motor proteins coexist in every eucaryotic cell. They differ in the type of filament they bind to (either actin or microtubules), the direction in which they move along the filament, and the "cargo" they carry. Many motor proteins carry membrane-enclosed organelles—such as mitochondria, Golgi stacks, or secretory vesicles—to their appropriate locations in the cell. Other motor proteins cause cytoskeletal filaments to slide against each other, generating the force that drives such phenomena as muscle contraction, ciliary beating, and cell division.

Cytoskeletal motor proteins that move unidirectionally along an oriented polymer track are reminiscent of some other proteins and protein complexes discussed elsewhere in this book, such as DNA and RNA polymerases, helicases, and ribosomes. All of these have the ability to use chemical energy to propel themselves along a linear track, with the direction of sliding dependent on the structural polarity of the track. All of them generate motion by coupling nucleoside triphosphate hydrolysis to a large-scale conformational change in a protein, as explained in Chapter 3.

The cytoskeletal motor proteins associate with their filament tracks through a "head" region, or *motor domain*, that binds and hydrolyzes ATP. Coordinated with their cycle of nucleotide hydrolysis and conformational change, the proteins cycle between states in which they are bound strongly to their filament tracks and states in which they are unbound. Through a mechanochemical cycle of filament binding, conformational change, filament release, conformational relaxation, and filament rebinding, the motor protein and its associated cargo move one step at a time along the filament (typically a distance of a few nanometers). The identity of the track and the direction of movement along it are determined by the motor domain (head), while the identity of the cargo (and therefore the biological function of the individual motor protein) is determined by the tail of the motor protein.

In this section, we begin by describing the three groups of cytoskeletal motor proteins. We then describe how they work to transport membrane-enclosed organelles or to change the shape of structures built from cytoskeletal filaments. We end by describing their action in muscle contraction and in powering the whiplike motion of structures formed from microtubules.

Actin-based Motor Proteins Are Members of the Myosin Superfamily

The first motor protein identified was skeletal muscle **myosin**, which is responsible for generating the force for muscle contraction. This myosin, called *myosin II* (see below) is an elongated protein that is formed from two heavy chains and two copies of each of two light chains. Each of the heavy chains has a globular head domain at its N-terminus that contains the force-generating machinery, followed by a very long amino acid sequence that forms an extended coiled-coil that mediates heavy chain dimerization (Figure 16–51). The two light chains bind close to the N-terminal head domain, while the long coiled-coil tail bundles itself with the tails of other myosin molecules. These tail–tail interactions result in the formation of large bipolar "thick filaments" that have several hundred myosin heads, oriented in opposite directions at the two ends of the thick filament (Figure 16–52).

Each myosin head binds and hydrolyses ATP, using the energy of ATP hydrolysis to walk toward the plus end of an actin filament. The opposing orientation of the heads in the thick filament makes the filament efficient at sliding pairs of oppositely oriented actin filaments past each other. In skeletal muscle, in which carefully arranged actin filaments are aligned in "thin filament" arrays surrounding the myosin thick filaments, the ATP-driven sliding of actin filaments

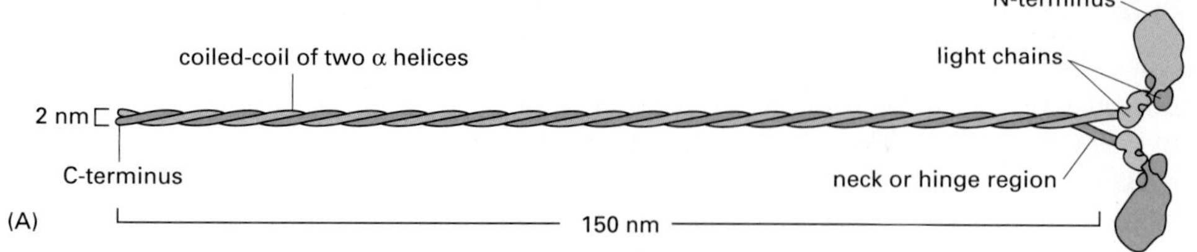

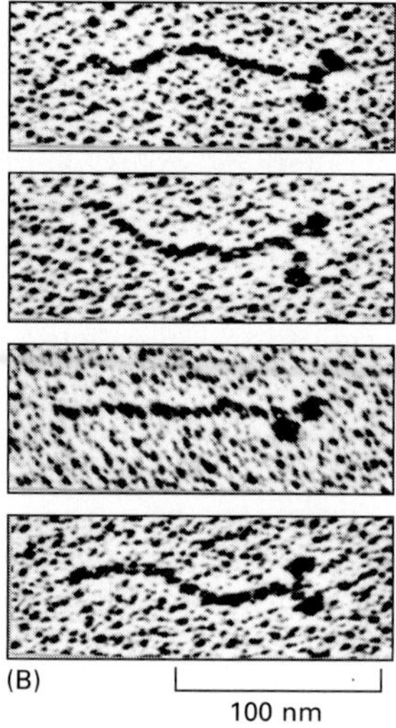

Figure 16–51 Myosin II. (A) A myosin II molecule is composed of two heavy chains (each about 2000 amino acids long *(green)* and four light chains *(blue)*. The light chains are of two distinct types, and one copy of each type is present on each myosin head. Dimerization occurs when the two α helices of the heavy chains wrap around each other to form a coiled-coil, driven by the association of regularly spaced hydrophobic amino acids (see Figure 3–11). The coiled-coil arrangement makes an extended rod in solution, and this part of the molecule is called the tail. (B) The two globular heads and the tail can be clearly seen in electron micrographs of myosin molecules shadowed with platinum. (B, courtesy of David Shotton.)

results in muscle contraction (discussed later). Cardiac and smooth muscle contain myosins that are similarly arranged, although they are encoded by different genes.

When a muscle myosin is digested by chymotrypsin and papain, the head domain is released as an intact fragment (called S1). The S1 fragment alone can generate filament sliding *in vitro,* proving that the motor activity is contained completely within the head (Figure 16–53).

It was initially thought that myosin was present only in muscle, but in the 1970's, researchers found that a similar two-headed myosin protein was also present in nonmuscle cells, including protozoan cells. At about the same time, other researchers found a myosin in the freshwater amoeba *Acanthamoeba castellanii* that was unconventional in having a motor domain similar to the head of muscle myosin but a completely different tail. This molecule seemed to function as a monomer and was named *myosin I* (for one-headed); the conventional myosin was renamed *myosin II* (for two-headed).

Subsequently, many other myosin types were discovered. The heavy chains generally start with a recognizable myosin motor domain at the N-terminus, and then diverge widely with a variety of C-terminal tail domains (Figure 16–54). The new types of myosins include a number of one-headed and two-headed varieties that are approximately equally related to myosin I and myosin II, and the nomenclature now reflects their approximate order of discovery (myosin III through at least myosin XVIII). The myosin tails (and the tails of motor proteins generally) have apparently diversified during evolution to permit the proteins to dimerize with other subunits and to interact with different cargoes.

Some myosins (such as VIII and XI) have been found only in plants, and some have been found only in vertebrates (IX). Most, however, are found in all eucaryotes, suggesting that myosins arose early in eucaryotic evolution. The

Figure 16–52 The myosin II bipolar thick filament. (A) Electron micrograph of a myosin II thick filament isolated from frog muscle. Note the central bare zone, which is free of head domains.
(B) Schematic diagram, not drawn to scale. The myosin II molecules aggregate by means of their tail regions, with their heads projecting to the outside of the filament. The bare zone in the center of the filament consists entirely of myosin II tails. (C) A small section of a myosin II filament as reconstructed from electron micrographs. An individual myosin molecule is highlighted in *green*.
(A, courtesy of Murray Stewart; C, based on R.A. Crowther, R. Padron and R. Craig, *J. Mol. Biol.* 184:429–439, 1985. © Academic Press.)

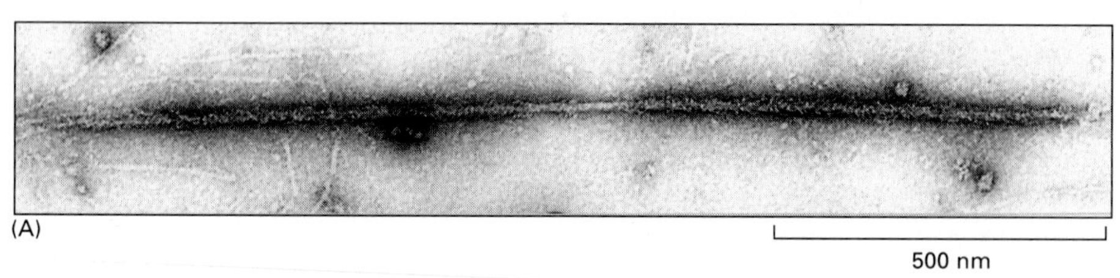

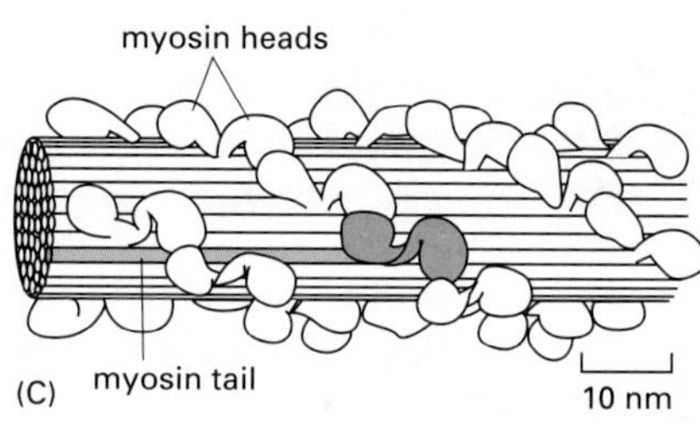

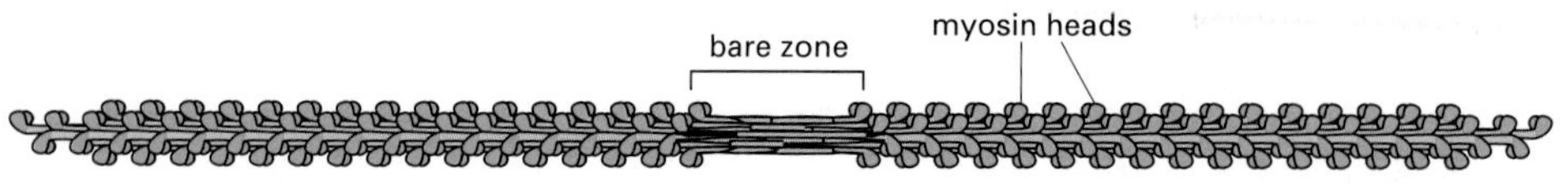

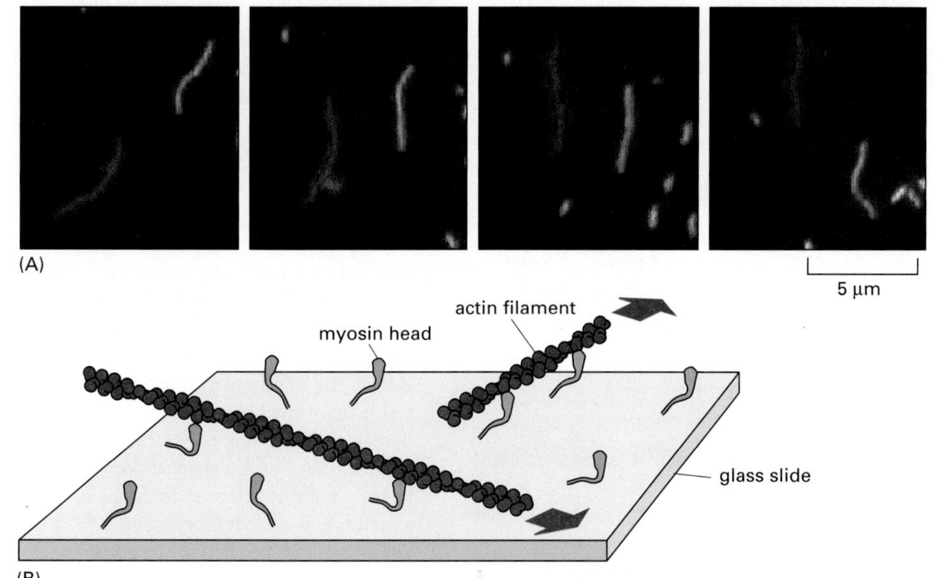

Figure 16–53 Direct evidence for the motor activity of the myosin head. In this experiment, purified S1 myosin heads were attached to a glass slide, and then actin filaments labeled with fluorescent phalloidin were added and allowed to bind to the myosin heads. (A) When ATP was added, the actin filaments began to glide along the surface, owing to the many individual steps taken by each of the dozens of myosin heads bound to each filament. The video frames shown in this sequence were recorded about 0.6 second apart; the two actin filaments shown (one *red* and one *green*) were moving in opposite directions at a rate of about 4 μm/sec. (B) Diagram of the experiment. The large *red* arrows indicate the direction of actin filament movement. (A, courtesy of James Spudich.)

yeast *Saccharomyces cerevisiae* contains five myosins: two myosin Is, one myosin II, and two myosin Vs. One can speculate that these three types of myosins are necessary for a eucaryotic cell to survive and that other myosins perform more specialized functions in multicellular organisms. The nematode *C. elegans*, for example, has at least 15 myosin genes, representing at least seven structural classes; the human genome includes about 40 myosin genes.

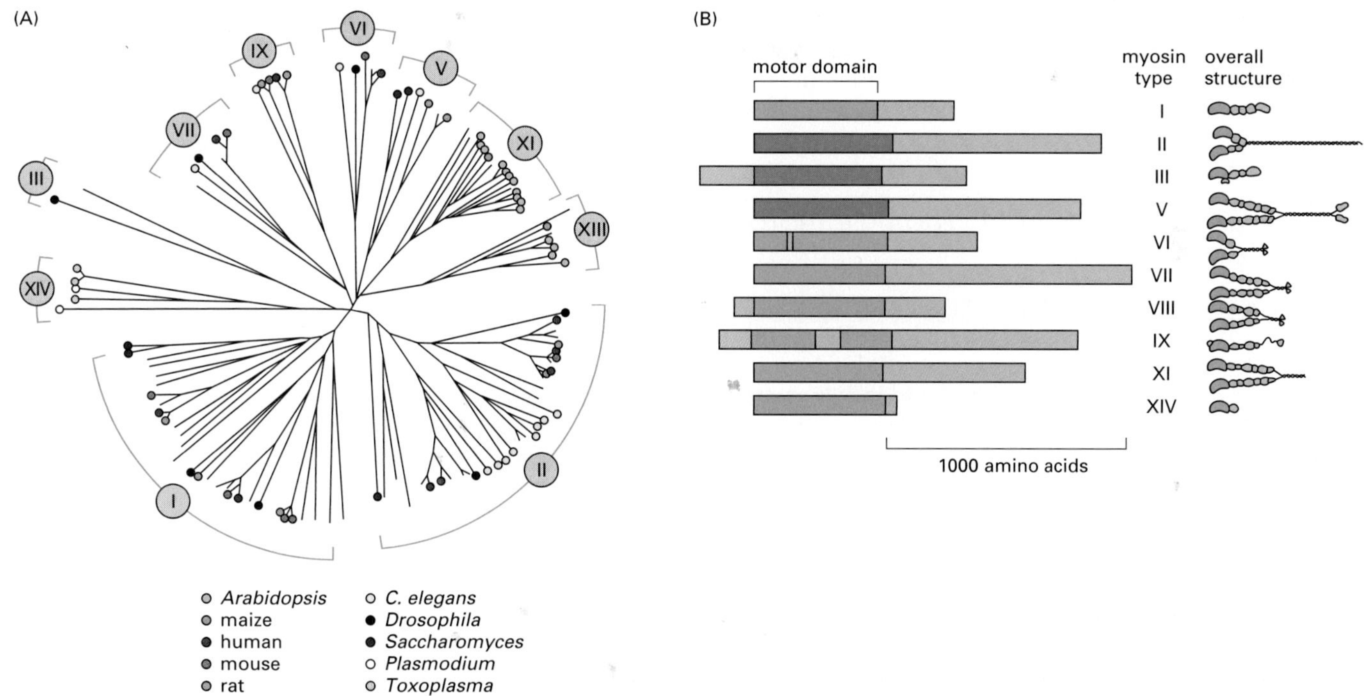

Figure 16–54 Myosin superfamily tree. (A) A family tree for a few of the many known members of the myosin superfamily. The length of the lines separating individual family members indicates the amount of difference in the amino acid sequence of the motor domain. Groups of related myosins that share a similar structure are assigned a Roman numeral. The species of origin of some of the myosins shown are indicated with colored dots. Some myosin types (such as I and II) have members in many eucaryotic species, ranging from animals to plants to parasitic protozoa; others are found only in particular groups of eucaryotes (XIV, for example, has been found so far only in parasitic protozoa such as *Toxoplasma* and *Plasmodium*). (B) Comparison of the domain structure of the heavy chains of some myosin types. All myosins share similar motor domains (shown in *dark green*), but their C-terminal tails *(light green)* and N-terminal extensions *(light blue)* are very diverse. On the right are depictions of the molecular structure for these family members. Many myosins form dimers, with two motor domains per molecule, but a few (such as I, IX, and XIV) seem to function as monomers, with just one motor domain. Myosin VI, despite its overall structural similarity to other family members, is unique in moving toward the minus end (instead of the plus end) of an actin filament. The small insertion within its motor head domain, not found in other myosins, is probably responsible for this change in direction.

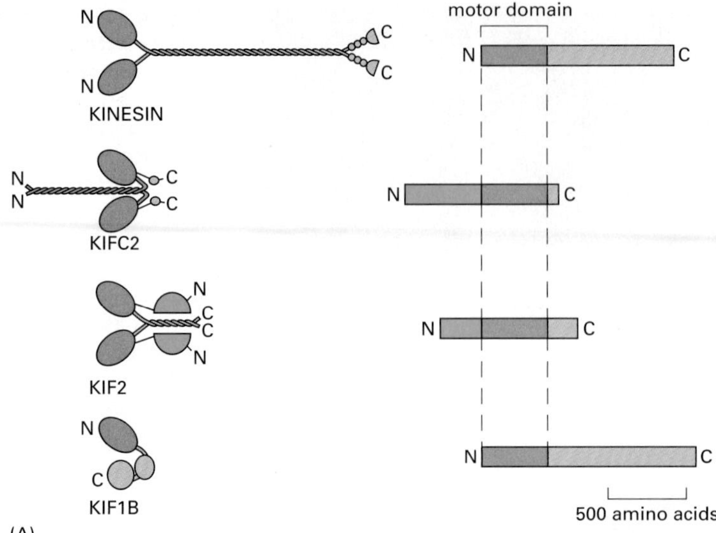

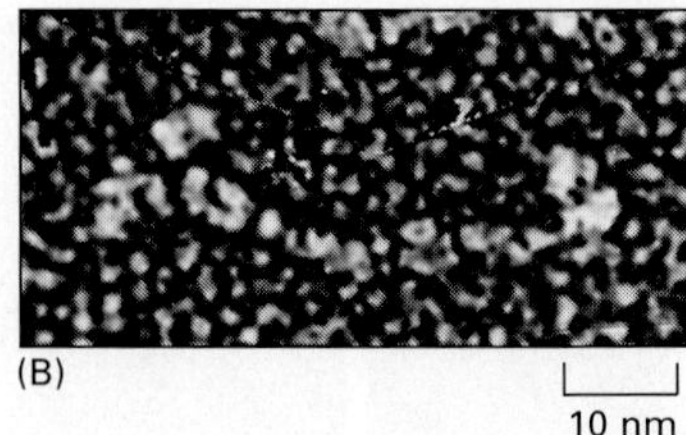

Figure 16–55 Kinesin and kinesin-related proteins. (A) Structures of four kinesin superfamily members. As in the myosin superfamily, only the motor domains are conserved. Conventional kinesin has the motor domain at the N-terminus of the heavy chain. The middle domain forms a long coiled-coil, mediating dimerization. The C-terminal domain forms a tail that attaches to cargo, such as a membrane-enclosed organelle. KIFC2 is a member of a family of C-terminal kinesins that includes the *Drosophila* protein Ncd and the yeast protein KAR3. These kinesins generally travel in the opposite direction from the majority of kinesins, toward the minus end instead of the plus end of a microtubule. KIF2 has its motor domain located in the middle of the heavy chain. It is a member of a family of kinesins that have lost typical motor activity and instead bind to microtubule ends to increase dynamic instability of microtubules; they are therefore called catastrophins. KIF1B is a member of the unusual class of kinesins that seem to function as monomers and move membrane-enclosed organelles along microtubules. Not shown is the BimC family of kinesins, which form tetramers. (B) Freeze-etch electron micrograph of a kinesin molecule with the head domains on the left. (B, courtesy of John Heuser.)

All of the myosins except one move toward the plus end of an actin filament, although they do so at different speeds. The exception is myosin VI, which moves toward the minus end.

The exact functions for most of the myosins remain to be determined. Myosin II is always associated with contractile activity in muscle and nonmuscle cells. It is also generally required for cytokinesis, the pinching apart of a dividing cell into two daughters (discussed in Chapter 18), as well as for the forward translocation of the body of a cell during cell migration. The myosin I proteins contain a second actin-binding site or a membrane-binding site in their tails, and they are generally involved in intracellular organization and the protrusion of actin-rich structures at the cell surface. Myosin V is involved in vesicle and organelle transport. Myosin VII is found in the inner ear in vertebrates, and certain mutations in the gene coding for myosin VII cause deafness in mice and humans.

There Are Two Types of Microtubule Motor Proteins: Kinesins and Dyneins

Kinesin is a motor protein that moves along microtubules. It was first identified in the giant axon of the squid, where it carries membrane-enclosed organelles away from the neuronal cell body toward the axon terminal by walking toward the plus end of microtubules. Kinesin is similar structurally to myosin II in having two heavy chains and two light chains per active motor, two globular head motor domains, and an elongated coiled-coil responsible for heavy chain dimerization. Like myosin, kinesin is a member of a large protein superfamily, for which the motor domain is the only common element (Figure 16–55). The yeast *Saccharomyces cerevisiae* has six distinct kinesins. The nematode *C. elegans* has 16 kinesins, and humans have about 40.

There are at least ten families of **kinesin-related proteins**, or **KRPs**, in the kinesin superfamily. Most of them have the motor domain at the N-terminus of the heavy chain and walk toward the plus end of the microtubule. A particularly interesting family has the motor domain at the C-terminus and walks in the opposite direction, toward the minus end of the microtubule. Some KRP heavy chains lack a coiled-coil sequence and seem to function as monomers, analogous to myosin I. Some others are homodimers, and yet others are heterodimers. At least one KRP (BimC) can self-associate through the tail domain, forming a bipolar motor that slides oppositely oriented microtubules past one another, much as a myosin II thick filament does for actin filaments. Most kinesins carry a binding site in the tail for either a membrane-enclosed organelle or another microtubule. Many of the kinesin superfamily members have specific roles in mitotic and meiotic spindle formation and chromosome separation during cell division.

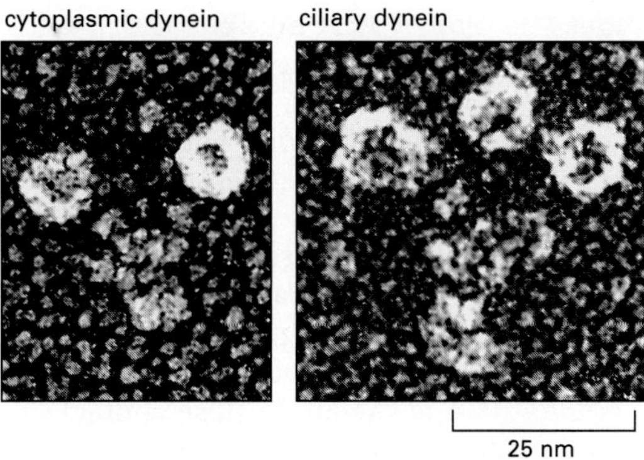

Figure 16–56 Dyneins. Freeze-etch electron micrographs of a molecule of cytoplasmic dynein and a molecule of ciliary (axonemal) dynein. Like myosin II and kinesin, cytoplasmic dynein is a two-headed molecule. The ciliary dynein shown has three heads. Note that the dynein head is very large compared with the head of either myosin or kinesin. (Courtesy of John Heuser.)

The **dyneins** are a family of minus-end-directed microtubule motors, but they are unrelated to the kinesin superfamily. They are composed of two or three heavy chains (that include the motor domain) and a large and variable number of associated light chains. The dynein family has two major branches (Figure 16–56). The most ancient branch contains the *cytoplasmic dyneins*, which are typically heavy-chain homodimers, with two large motor domains as heads. Cytoplasmic dyneins are probably found in all eucaryotic cells, and they are important for vesicle trafficking, as well as for localization of the Golgi apparatus near the center of the cell. *Axonemal dyneins*, the other large branch, include heterodimers and heterotrimers, with two or three motor-domain heads, respectively. They are highly specialized for the rapid and efficient sliding movements of microtubules that drive the beating of cilia and flagella (discussed later). A third, minor, branch shares greater sequence similarity with cytoplasmic than with axonemal dyneins but seems to be involved in the beating of cilia.

Dyneins are the largest of the known molecular motors, and they are also among the fastest: axonemal dyneins can move microtubules in a test tube at the remarkable rate of 14 μm/sec. In comparison, the fastest kinesins can move their microtubules at about 2–3 μm/sec.

The Structural Similarity of Myosin and Kinesin Indicates a Common Evolutionary Origin

The motor domain of myosins is substantially larger than that of kinesins, about 850 amino acids compared with about 350. The two classes of motor proteins track along different filaments and have different kinetic properties, and they have no identifiable amino acid sequence similarities. However, determination of the three-dimensional structure of the motor domains of myosin and kinesin has revealed that these two motor domains are built around nearly identical cores (Figure 16–57). The central force-generating element that the two types of

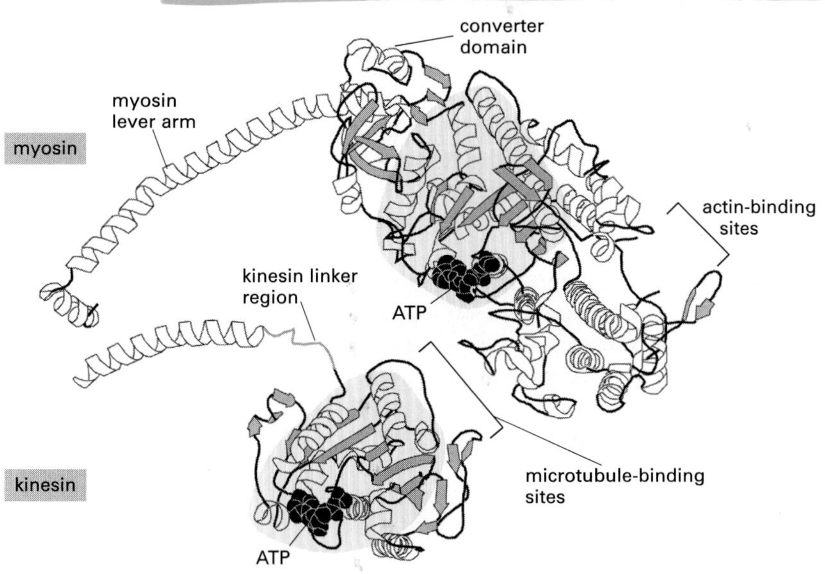

Figure 16–57 X-ray crystal structures of myosin and kinesin heads. The central nucleotide-binding domains of myosin and kinesin (shaded in *yellow*) are structurally very similar. The very different sizes and functions of the two motors are due to major differences in the polymer-binding and force-transduction portions of the motor domain. (Adapted from L.A. Amos and R.A. Cross, *Curr. Opin. Struct. Biol.* 7:239–246, 1997.)

motor proteins have in common includes the site of ATP binding and the machinery necessary to translate ATP hydrolysis into an allosteric conformational change. The differences in domain size and in the choice of track can be attributed to large loops extending outward from this central core. These loops include the actin-binding and microtubule-binding sites, respectively.

An important clue to how the central core is involved in force generation has come from the observation that the motor core also bears some structural resemblance to the nucleotide binding site of the small GTPases of the Ras superfamily. As discussed in Chapter 3 (see Figure 3–74), these proteins exhibit distinct conformations in their GTP-bound (active) and GDP-bound (inactive) forms: mobile "switch" loops in the nucleotide-binding site are in close contact with the γ-phosphate in the GTP-bound state, but these loops swing out when the hydrolyzed γ-phosphate is released. Although the details of the movement are different for the two motor proteins, and ATP rather than GTP is hydrolyzed, the relatively small structural change in the active site—the presence or absence of a terminal phosphate—is similarly amplified to cause a rotation of a different part of the protein. In kinesin and myosin, a switch loop interacts extensively with those regions of the protein involved in microtubule and actin binding, respectively, allowing the structural transitions caused by the ATP hydrolysis cycle to be relayed to the polymer-binding interface. The relay of structural changes between the polymer-binding site and the nucleotide hydrolysis site seems to work in both directions, since the ATPase activity of motor proteins is strongly activated by binding to their filament tracks.

Motor Proteins Generate Force by Coupling ATP Hydrolysis to Conformational Changes

Although the cytoskeletal motor proteins and GTP-binding proteins both use structural changes in their nucleoside-triphosphate-binding sites to produce cyclic interactions with a partner protein, the motor proteins have a further requirement: each cycle of binding and release must propel them forward in a single direction along a filament to a new binding site on the filament. For such unidirectional motion, a motor protein must use the energy derived from ATP binding and hydrolysis to force a large movement in part of the protein molecule. For myosin, each step of the movement along actin is generated by the swinging of an 8.5-nm-long α helix, or *lever arm* (see Figure 16–57), which is structurally stabilized by the binding of light chains. At the base of this lever arm next to the head, there is a piston-like helix that connects movements at the ATP-binding cleft in the head to small rotations of the so-called converter domain. A small change at this point can swing the helix like a long lever, causing the far end of the helix to move by about 5.0 nm. These changes in the conformation of the myosin are coupled to changes in its binding affinity for actin, allowing the myosin head to release its grip on the actin filament at one point and snatch hold of it again at another. The full mechanochemical cycle of nucleotide binding, nucleotide hydrolysis, and phosphate release (which causes the "power stroke") produces a single step of movement (Figure 16–58). In the myosin VI subfamily of myosins, which move backward (toward the minus end of the actin filament), the converter domain probably lies in a different orientation, so that the same piston-like movement of the small helix causes the lever arm to rotate in the opposite direction.

In kinesin, instead of the rocking of a lever arm, the small movements of switch loops at the nucleotide-binding site regulate the docking and undocking of the motor head domain to a long linker region that connects this motor head at one end to the coiled-coil dimerization domain at the other end. When the front (leading) kinesin head is bound to a microtubule before the power stroke, its linker region is relatively unstructured. On the binding of ATP to this bound head, its linker region docks along the side of the head, which throws the second head forward to a position where it will be able to bind a new attachment site on the protofilament, 8 nm closer to the microtubule plus end than the binding site for the first head. The nucleotide hydrolysis cycles in the two heads are closely coordinated, so that this cycle of linker docking and undocking can allow the two-headed motor to move in a hand-over-hand (or head-over-head) stepwise manner (Figure 16–59A).

The coiled-coil domain seems both to coordinate the mechanochemical cycles of the two heads (motor domains) of the kinesin dimer and to determine its directionality of movement. Recall that whereas most members of the kinesin superfamily, with their motor domains at the N-terminus, move toward the plus end of the microtubule, a few superfamily members have their motor domains at the C-terminus and move toward the minus end. Since the motor domains of these two types of kinesins are essentially identical, how can they move in opposite directions? The answer seems to lie in the way in which the heads are connected. In high-resolution images of forward-walking and backward-walking members of the kinesin superfamily bound to microtubules, the heads that are attached to the microtubule are essentially indistinguishable, but the second,

Figure 16–58 The cycle of structural changes used by myosin to walk along an actin filament. (Based on I. Rayment et al., *Science* 261:50–58, 1993. © AAAS.)

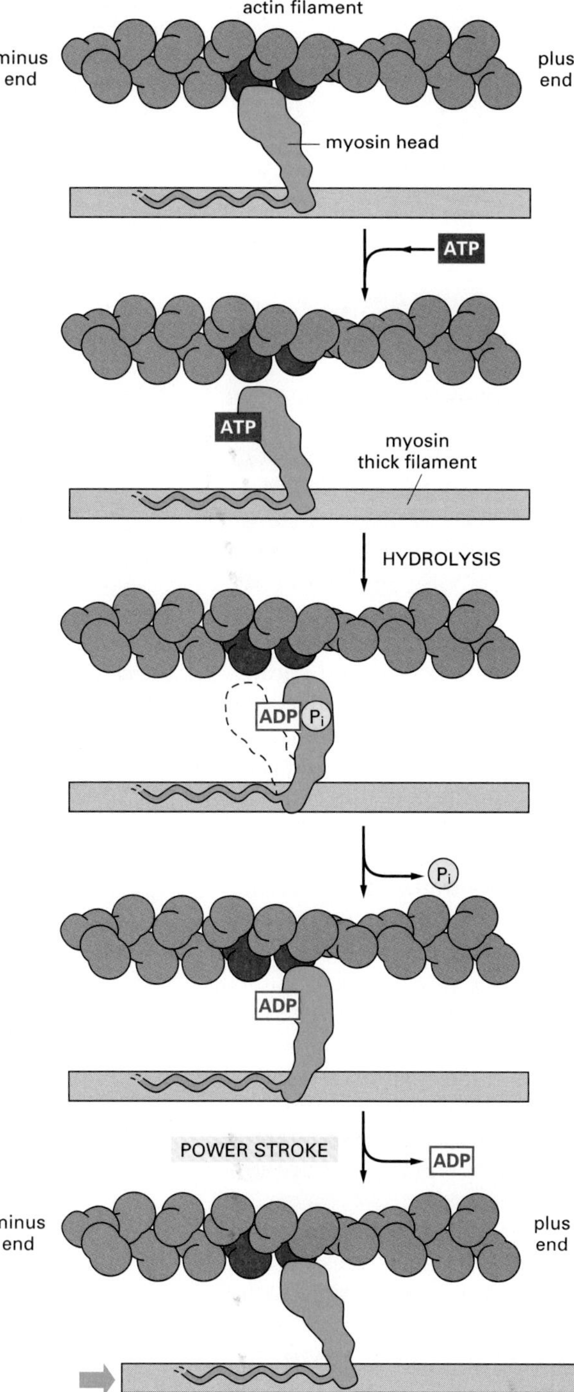

ATTACHED At the start of the cycle shown in this figure, a myosin head lacking a bound nucleotide is locked tightly onto an actin filament in a *rigor* configuration (so named because it is responsible for *rigor mortis*, the rigidity of death). In an actively contracting muscle, this state is very short-lived, being rapidly terminated by the binding of a molecule of ATP.

RELEASED A molecule of ATP binds to the large cleft on the "back" of the head (that is, on the side furthest from the actin filament) and immediately causes a slight change in the conformation of the domains that make up the actin-binding site. This reduces the affinity of the head for actin and allows it to move along the filament. (The space drawn here between the head and actin emphasizes this change, although in reality the head probably remains very close to the actin.)

COCKED The cleft closes like a clam shell around the ATP molecule, triggering a large shape change that causes the head to be displaced along the filament by a distance of about 5 nm. Hydrolysis of ATP occurs, but the ADP and inorganic phosphate (P_i) produced remain tightly bound to the protein.

FORCE-GENERATING A weak binding of the myosin head to a new site on the actin filament causes release of the inorganic phosphate produced by ATP hydrolysis, concomitantly with the tight binding of the head to actin. This release triggers the power stroke—the force-generating change in shape during which the head regains its original conformation. In the course of the power stroke, the head loses its bound ADP, thereby returning to the start of a new cycle.

ATTACHED At the end of the cycle, the myosin head is again locked tightly to the actin filament in a rigor configuration. Note that the head has moved to a new position on the actin filament.

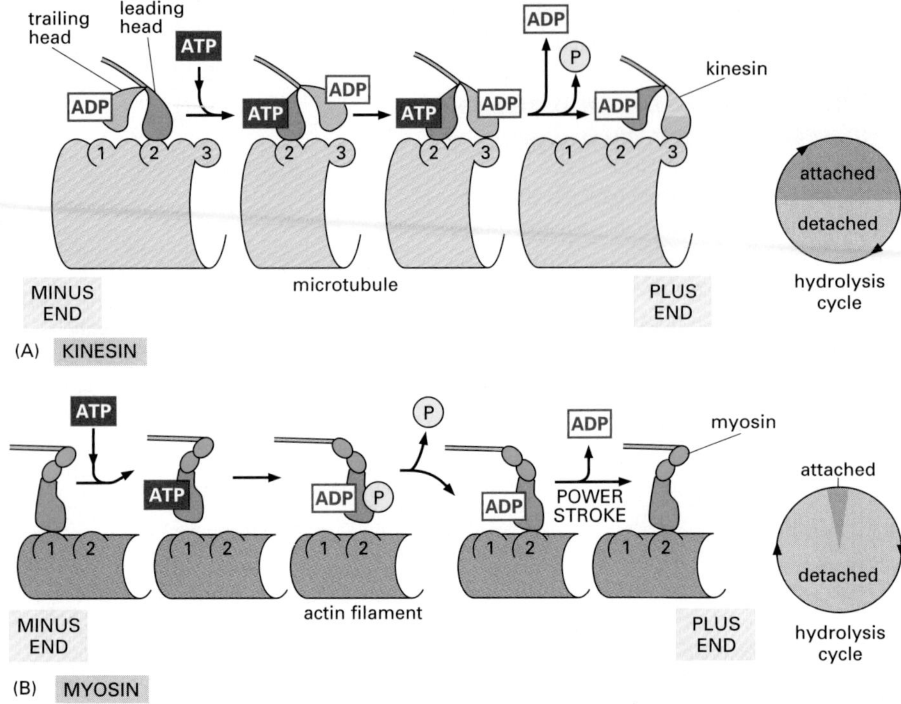

Figure 16–59 Comparison of the mechanochemical cycles of kinesin and myosin II. The shading in the two circles representing the hydrolysis cycle indicates the proportion of the cycle spent in attached and detached states for each motor protein. (A) Summary of the coupling between ATP hydrolysis and conformational changes for kinesin. At the start of the cycle, one of the two kinesin heads, the front or leading head *(dark green)* is bound to the microtubule, with the rear or trailing head *(light green)* detached. Binding of ATP to the front head causes the rear head to be thrown forward, past the binding site of the attached head, to another binding site further toward the plus end of the microtubule. Release of ADP from the second head (now in the front) and hydrolysis of ATP on the first head (now in the rear) brings the dimer back to the original state, but the two heads have switched their relative positions, and the motor protein has moved one step along the microtubule protofilament. In this cycle, each head spends about 50% of its time attached to the microtubule and 50% of its time detached. (B) Summary of the coupling between ATP hydrolysis and conformational changes for myosin II. Myosin begins its cycle tightly bound to the actin filament, with no associated nucleotide, the so-called "rigor" state. ATP binding releases the head from the filament. ATP hydrolysis occurs while the myosin head is detached from the filament, causing the head to assume a cocked conformation, although both ADP and inorganic phosphate remain tightly bound to the head. When the head rebinds to the filament, the release of phosphate, followed by the release of ADP, trigger the power stroke that moves the filament relative to the motor protein. ATP binding releases the head to allow the cycle to begin again. In the myosin cycle, the head remains bound to the actin filament for only about 5% of the entire cycle time, allowing many myosins to work together to move a single actin filament.

unattached heads are oriented very differently. This difference in tilt apparently biases the next binding site for the second head, and thereby determines the directionality of motor movement (Figure 16–60).

Although both myosin and kinesin undergo analogous mechanochemical cycles, the exact nature of the coupling between the mechanical and chemical cycles is different in the two cases (see Figure 16–60). For example, myosin without any nucleotide is tightly bound to its actin track, in a so-called "rigor" state, and it is released from this track by the association of ATP. In contrast, kinesin forms a rigor-like tight association with a microtubule when ATP is bound to the kinesin, and it is hydrolysis of ATP that promotes release of the motor from its track.

Thus, cytoskeletal motor proteins work in a manner highly analogous to GTP-binding proteins, except that in motor proteins the small protein conformational changes (a few tenths of a nanometer) associated with nucleotide hydrolysis are amplified by special protein domains—the lever arm in the case of myosin and the linker in the case of kinesin—to generate large-scale (several nanometers) conformational changes that move the motor proteins stepwise along their filament tracks. The analogy between the GTPases and the cytoskeletal motor proteins has recently been extended by the observation that one of the GTP-binding proteins—the bacterial elongation factor G—translates the chemical energy of GTP hydrolysis into directional movement of the mRNA molecule on the ribosome.

Motor Protein Kinetics Are Adapted to Cell Functions

The motor proteins in the myosin and kinesin superfamilies exhibit a remarkable diversity of motile properties, well beyond their choice of different polymer tracks. Most strikingly, a single dimer of conventional kinesin moves in a highly processive fashion, traveling for hundreds of ATPase cycles along a microtubule without dissociating. Skeletal muscle myosin II, in contrast, cannot move processively and makes just one or a few steps along an actin filament before letting go. These differences are critical for the motors' various biological roles. A small number of kinesin molecules must be able to transport a mitochondrion all the way down a nerve cell axon, and therefore require a high level of processivity. Skeletal muscle myosin, in contrast, never operates as a single molecule but rather as part of a huge array of myosin II molecules. Here processivity would actually inhibit biological function, since efficient muscle contraction requires that each myosin head perform its power stroke and then quickly get out of the way, to avoid interfering with the actions of the other heads attached to the same actin filament.

There are two reasons for the high degree of processivity of kinesin movement. The first is that the mechanochemical cycles of the two motor heads in a kinesin dimer are coordinated with each other, so that one kinesin head does not let go until the other is poised to bind. This coordination allows the motor protein to operate in a hand-over-hand fashion, never allowing the organelle cargo to diffuse away from the microtubule track. There is no apparent coordination between the myosin heads in a myosin II dimer. The second reason for the high processivity of kinesin movement is that kinesin spends a relatively large fraction of its ATPase cycle tightly bound to the microtubule. For both kinesin and myosin, the conformational change that produces the force-generating working stroke must occur while the motor protein is tightly bound to its polymer, and the recovery stroke in preparation for the next step must occur while the motor is unbound. But as we have seen in Figure 16–59, myosin spends only about 5% of its ATPase cycle in the tightly bound state and is unbound the rest of the time.

What myosin loses in processivity it gains in speed; in an array in which many motor heads are interacting with the same actin filament, a set of linked myosins can move their filament a total distance equivalent to 20 steps during a single cycle time, while kinesins can move only two. Thus, myosins can typically drive filament sliding much more rapidly than kinesins, even though they hydrolyze ATP at comparable rates and take molecular steps of comparable length.

Within each motor protein class, movement speeds vary widely, from about 0.2 to 60 μm/sec for myosins, and from about 0.02 to 2 μm/sec for kinesins. These differences arise from a fine-tuning of the mechanochemical cycle. The number of steps that an individual motor molecule can take in a given time, and thereby the velocity, can be increased by either increasing the motor protein's intrinsic ATPase rate or decreasing the proportion of cycle time spent bound to the filament track. Moreover, the size of each step can be changed by either

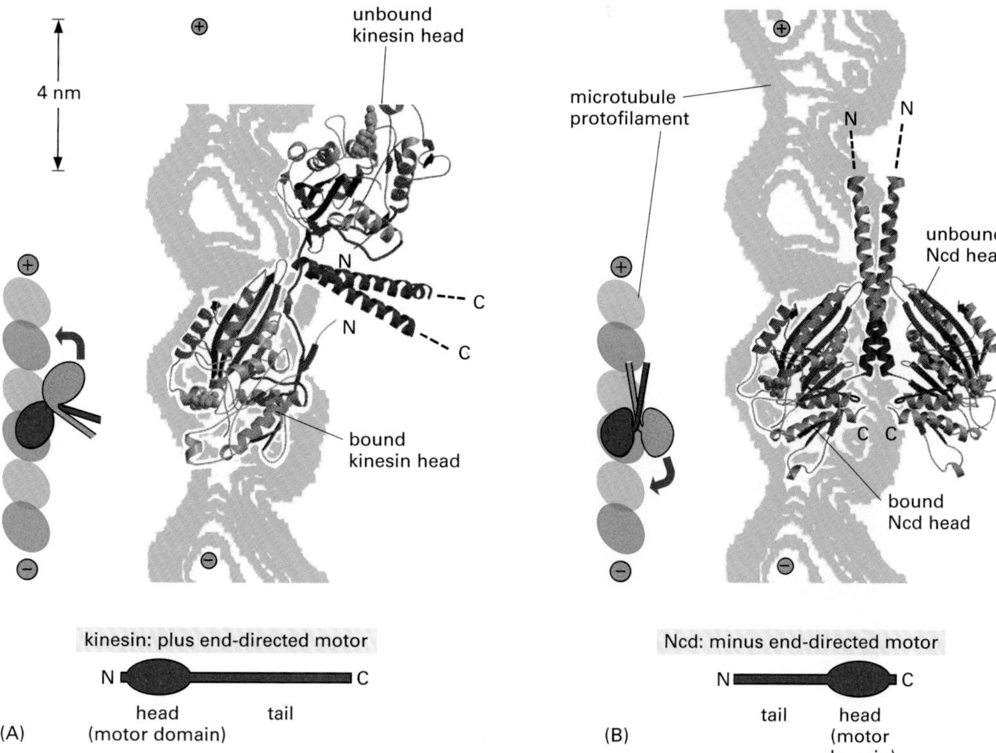

Figure 16–60 Orientation of forward- and backward-walking kinesin superfamily proteins bound to microtubules. These images were generated by fitting the structures of the free motor-protein dimers (determined by x-ray crystallography) onto a lower resolution image of the dimers attached to microtubules (determined by cryoelectron microscopy). (A) Conventional kinesin has its motor domain at the protein's N-terminus and moves toward the plus end of the microtubule. When one head of the dimer is bound to the microtubule in a post-stroke state (with ATP in the nucleotide binding site), the second, unbound head is pointing toward the microtubule plus end, poised to take the next step. (B) Ncd, a minus-end-directed motor with the motor domain at the C-terminus, forms dimers with the opposite orientation. (From E. Sablin et al., *Nature* 395:813–816, 1998. © Macmillan Magazines Ltd.)

Figure 16–61 The effect of lever arm length on the step size for a motor protein. The lever arm of myosin II is much shorter than the lever arm of myosin V. The power stroke in the head swings their lever arms through the same angle, so myosin V is able to take a bigger step than myosin II.

Myosin II
5 to 10 nm swing of lever arm
head
minus end
actin filament
plus end

Myosin V
30 to 40 nm swing of lever arm
head
minus end
plus end

changing the length of the lever arm (for example, the lever arm of myosin V is about three times longer than the lever arm of myosin II) or the angle through which the helix swings (Figure 16–61). Each of these parameters varies slightly among different members of the myosin and kinesin families, corresponding to slightly different protein sequences and structures. It is assumed that the behavior of each motor protein, whose function is determined by the identity of the cargo attached through its tail-domain, has been fine-tuned during evolution for speed and processivity according to the specific needs of the cell.

Motor Proteins Mediate the Intracellular Transport of Membrane-enclosed Organelles

A major function of cytoskeletal motors in interphase cells is the transport and positioning of membrane-enclosed organelles. Kinesin was originally identified as the protein responsible for fast axonal transport, the rapid movement of mitochondria, secretory vesicle precursors, and various synapse components down the microtubule highways of the axon to the distant nerve terminals. Although organelles in most cells need not cover such long distances, their polarized transport is equally necessary. A typical microtubule array in an interphase cell is oriented with the minus ends near the center of the cell at the centrosome, and the plus ends extending to the cell periphery. Thus, centripetal movements of organelles toward the cell center require the action of minus-end-directed motor proteins such as cytoplasmic dynein, whereas centrifugal movements toward the periphery require plus-end-directed motors such as kinesins.

The role of microtubules and microtubule motors in the behavior of intracellular membranes is best exemplified by the part they play in organizing the endoplasmic reticulum (ER) and the Golgi apparatus. The network of ER membrane tubules aligns with microtubules and extends almost to the edge of the cell, whereas the Golgi apparatus is located near the centrosome. When cells are treated with a drug that depolymerizes microtubules, such as colchicine or nocodazole, the ER collapses to the center of the cell, while the Golgi apparatus fragments and disperses throughout the cytoplasm (Figure 16–62). *In vitro*, kinesins can tether ER-derived membranes to preformed microtubule tracks, and walk toward the microtubule plus ends, dragging the ER membranes out into tubular protrusions and forming a membranous web very much like the ER in cells. Likewise, the outward movement of ER tubules toward the cell periphery is associated with microtubule growth in living cells. Conversely, dyneins are required for positioning the Golgi apparatus near the cell center, moving Golgi vesicles along microtubule tracks toward minus ends at the centrosome.

The different tails and their associated light chains on specific motor proteins allow the motors to attach to their appropriate organelle cargo. For example,

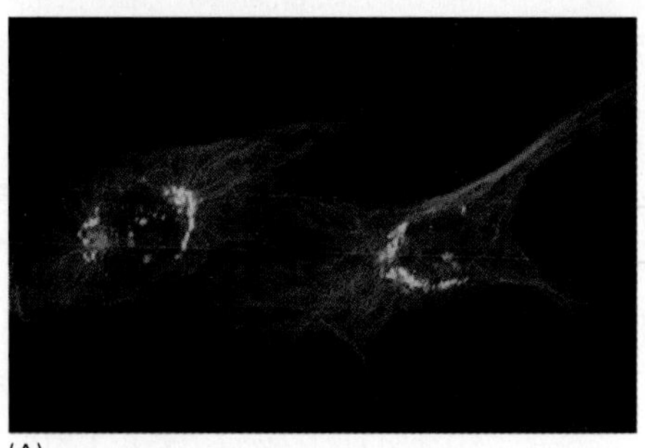
(A)

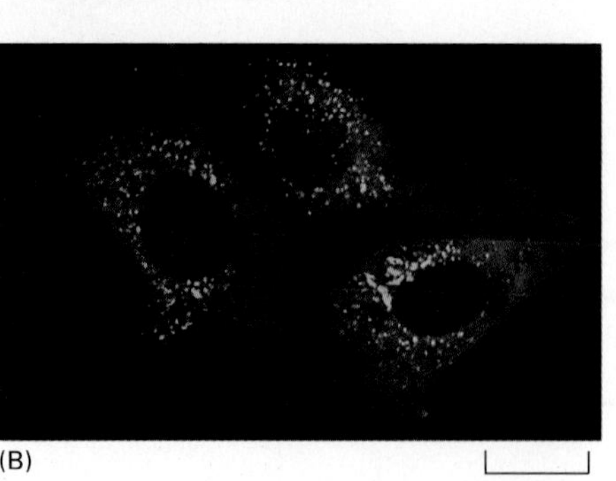
(B)

10 μm

Figure 16–62 Effect of depolymerizing microtubules on the Golgi apparatus. (A) In this endothelial cell, the microtubules are labeled in *red*, and the Golgi apparatus is labeled in *green* (using an antibody against a Golgi protein). As long as the system of microtubules remains intact, the Golgi is localized near the centrosome, close to the nucleus at the center of the cell. The cell on the *right* is in interphase, with a single centrosome. The cell on the *left* is in prophase, and the duplicated centrosomes have moved to opposite sides of the nucleus. (B) After exposure to nocodazole, which causes microtubules to depolymerize (see Table 16–2), the Golgi apparatus fragments and is dispersed throughout the cell cytoplasm. (Courtesy of David Shima.)

there is evidence for membrane-associated motor receptors, sorted to specific membrane-enclosed compartments, that interact directly or indirectly with the tails of the appropriate kinesin family members. One of these receptors seems to be the amyloid precursor protein, APP, which binds directly to a light chain on the tail of kinesin-I and is proposed to be a transmembrane motor protein receptor molecule in nerve-cell axons. It is the abnormal processing of this protein that gives rise to Alzheimer's disease, as discussed in Chapter 15.

For dynein, attachment to membranes is known to be mediated by a large macromolecular assembly. Cytoplasmic dynein is itself a huge protein complex, and it requires association with a second large protein complex called *dynactin* to translocate organelles effectively. The dynactin complex includes a short actinlike filament that is made of the actin-related protein Arp1 (distinct from Arp2 and Arp3, the components of the ARP complex involved in the nucleation of conventional actin filaments). Membranes of the Golgi apparatus are coated with the proteins ankyrin and spectrin, which have been proposed to associate with the Arp1 filament in the dynactin complex to form a planar cytoskeletal array reminiscent of the erythrocyte membrane cytoskeleton (see Figure 10–31). The spectrin array probably gives structural stability to the Golgi membrane, and—via the Arp1 filament—it may mediate the regulatable attachment of dynein to the organelle (Figure 16–63).

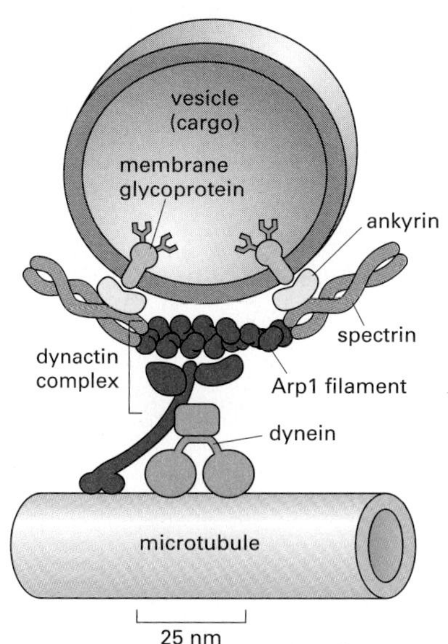

Figure 16–63 A model for the attachment of dynein to a membrane-enclosed organelle. Dynein requires the presence of a large number of accessory proteins to associate with membrane-enclosed organelles. Dynactin is a large complex *(red)* that includes components that bind weakly to microtubules, components that bind to dynein itself, and components that form a small actinlike filament made of the actin-related protein Arp1. It is thought that the Arp1 filament may mediate attachment of this large complex to membrane-enclosed organelles through a network of spectrin and ankyrin, similar to the membrane-associated cytoskeleton of the red blood cell (see Figure 10–31).

Motor proteins also have a significant role in organelle transport along actin filaments. The first myosin shown to mediate organelle motility was myosin V, a two-headed myosin with a large step size (see Figure 16–61). In mice, mutations in the myosin V gene result in a "dilute" phenotype, in which fur color looks faded. In mice (and humans), membrane-enclosed pigment granules, called *melanosomes,* are synthesized in cells called *melanocytes* beneath the skin surface. These melanosomes move out to the ends of dendritic processes in the melanocytes, from where they are delivered to the overlying keratinocytes that form the skin and fur. Myosin V is associated with the surface of melanosomes, and it is able to mediate their actin-based movement in a test tube (Figure 16–64). In dilute mutant mice, the melanosomes are not delivered to the keratinocytes efficiently, and pigmentation is defective. Other myosins, including myosin I, are associated with endosomes and a variety of other organelles.

Motor Protein Function Can Be Regulated

The cell can regulate the activity of motor proteins, allowing it to change either the positioning of its membrane-enclosed organelles or its whole-cell movements. One of the most dramatic examples is provided by fish melanocytes. These giant cells, which are responsible for rapid changes in skin coloration in several species of fish, contain large pigment granules that can alter their location in response to neuronal or hormonal stimulation (Figure 16–65). These pigment granules aggregate or disperse by moving along an extensive network of microtubules. The minus ends of these microtubules are nucleated by the centrosome and are located in the center of the cell, while the plus ends are distributed around the cell periphery. The tracking of individual pigment granules reveals that the inward movement is rapid and smooth, while the outward movement is jerky, with frequent backward steps (Figure 16–66). Both dynein and kinesin are associated with the pigment granules. The jerky outward movements apparently result from a tug-of-war between the two motor proteins, with the stronger kinesin winning out overall. When the kinesin light chains become phosphorylated after a hormonal stimulation that signals skin color change, kinesin is inactivated, leaving dynein free to drag the pigment granules rapidly

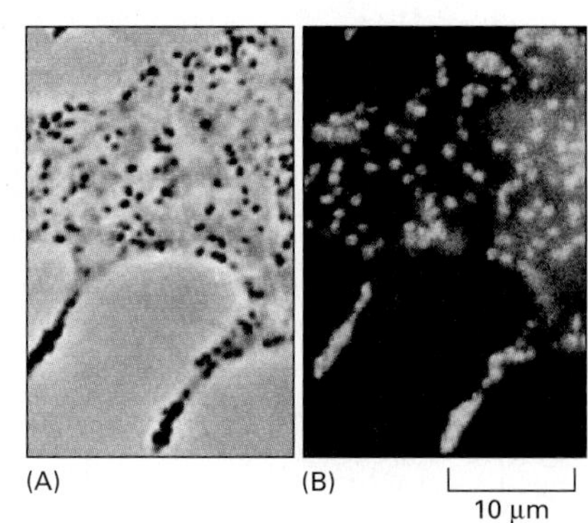

Figure 16–64 Myosin V on melanosomes. (A) Phase-contrast image of a portion of a melanocyte isolated from a mouse. The black spots are melanosomes, which are membrane-enclosed organelles filled with the skin pigment melanin. (B) The same cell labeled with a fluorescent antibody against myosin V. Every melanosome is associated with a large number of copies of this motor protein. (From X. Wu et al., *J. Cell Sci.* 110:847–859, 1997. © The Company of Biologists.)

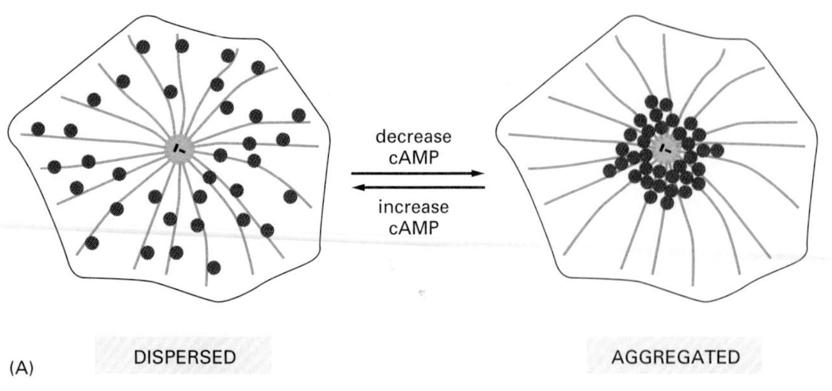

Figure 16–65 Regulated melanosome movements in fish pigment cells. These giant cells, which are responsible for changes in skin coloration in several species of fish, contain large pigment granules, or melanosomes *(brown)*. The melanosomes can change their location in the cell in response to a hormonal or neuronal stimulus. (A) Schematic view of a pigment cell, showing the dispersal and aggregation of melanosomes in response to an increase or decrease in intracellular cyclic AMP (cAMP), respectively. Both redistributions of melanosomes occur along microtubules. (B) Bright-field images of a single cell in a scale of an African cichlid fish, showing its melanosomes either dispersed throughout the cytoplasm *(left)* or aggregated in the center of the cell *(right)*. (B, courtesy of Leah Haimo.)

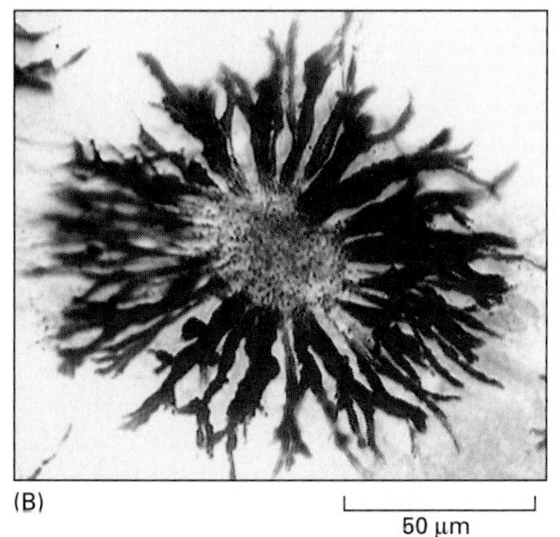

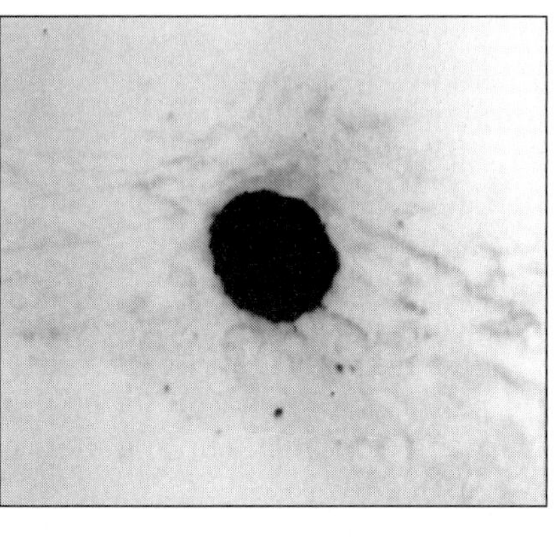

toward the cell center, changing the fish's color. In a similar way, the movement of other membrane organelles coated with particular motor proteins is controlled by a complex balance of competing signals that regulate both motor protein attachment and activity.

Myosin activity can also be regulated by phosphorylation. In nonmuscle cells, myosin II can be phosphorylated on a variety of sites on both heavy and light chains, affecting both motor activity and thick filament assembly. The myosin II can exist in two different conformational states in such cells, an extended state that is capable of forming bipolar filaments, and a bent state in which the tail domain apparently interacts with the motor head. Phosphorylation of the regulatory light chain by the calcium-dependent *myosin light-chain kinase (MLCK)* causes the myosin II to preferentially assume the extended state, which promotes its assembly into a bipolar filament and leads to cell contraction (Figure 16–67). Myosin light-chain phosphorylation is an indirect target of activated Rho, the small GTPase discussed previously whose activation causes a

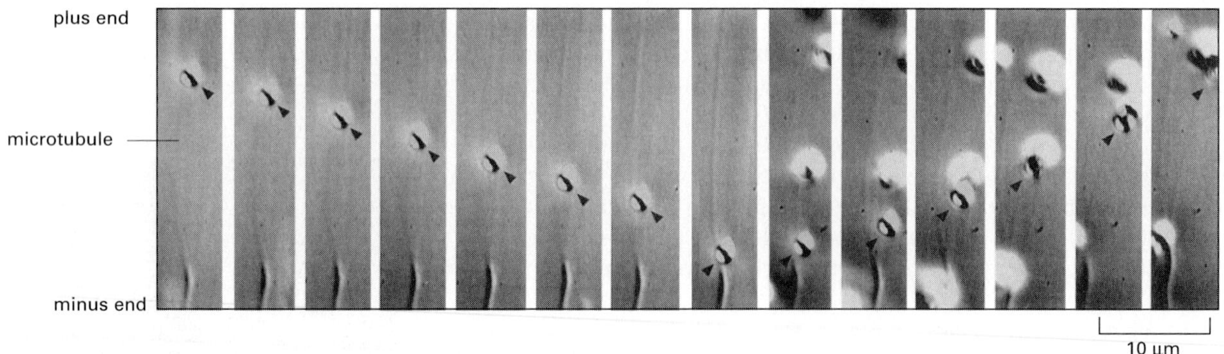

Figure 16–66 Bidirectional movement of a melanosome on a microtubule. An isolated melanosome *(yellow)* moves along a microtubule on a glass slide, from the plus end toward the minus end. Halfway through the video sequence, it abruptly switches direction and moves from the minus end toward the plus end. (From S.L. Rogers et al., *Proc. Natl. Acad. Sci. USA* 94:3720–3725, 1997. © National Academy of Sciences.)

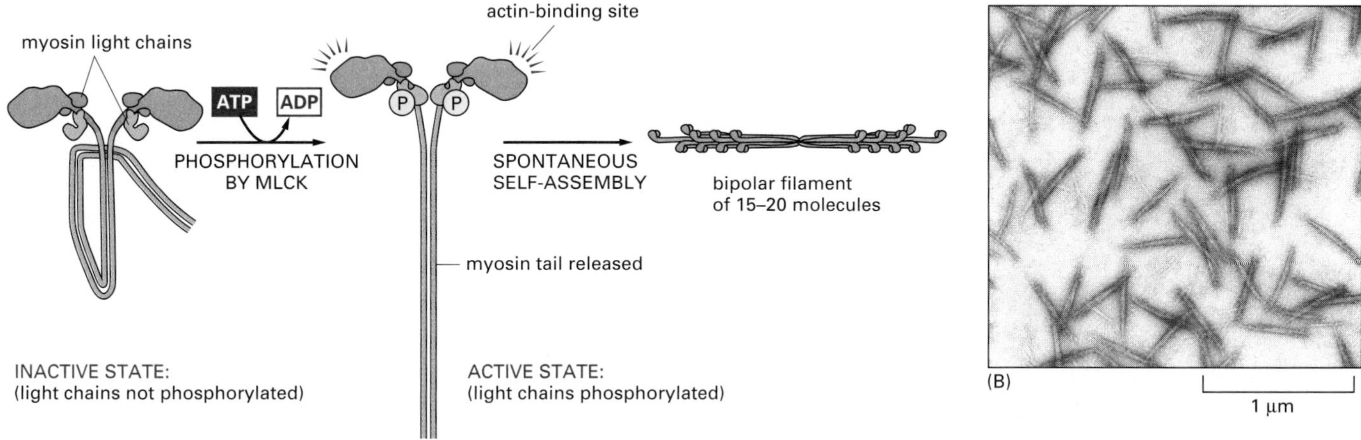

Figure 16–67 Light-chain phosphorylation and the regulation of the assembly of myosin II into thick filaments. (A) The controlled phosphorylation by the enzyme myosin light-chain kinase (MLCK) of one of the two light chains (the so-called regulatory light chain, shown in *light blue)* on nonmuscle myosin II in a test tube has at least two effects: it causes a change in the conformation of the myosin head, exposing its actin-binding site, and it releases the myosin tail from a "sticky patch" on the myosin head, thereby allowing the myosin molecules to assemble into short, bipolar, thick filaments. (B) Electron micrograph of negatively stained short filaments of myosin II that have been induced to assemble in a test tube by phosphorylation of their light chains. These myosin II filaments are much smaller than those found in skeletal muscle cells (see Figure 16–52). (B, courtesy of John Kendrick-Jones.)

reorganization of the actin cytoskeleton into contractile stress fibers. The MLCK is also activated during mitosis, causing myosin II to assemble into the contractile ring that is responsible for dividing the mitotic cell into two. Regulation of other members of the myosin superfamily is not as well understood, but control of these myosins is also likely to involve site-specific phosphorylations.

Muscle Contraction Depends on the Sliding of Myosin II and Actin Filaments

Muscle contraction is the most familiar and the best understood form of movement in animals. In vertebrates, running, walking, swimming, and flying all depend on the rapid contraction of skeletal muscle on its scaffolding of bone, while involuntary movements such as heart pumping and gut peristalsis depend on the contraction of cardiac muscle and smooth muscle, respectively. All these forms of muscle contraction depend on the ATP-driven sliding of highly organized arrays of actin filaments against arrays of myosin II filaments.

Muscle was a relatively late evolutionary development, and muscle cells are highly specialized for rapid and efficient contraction. The long thin muscle fibers of skeletal muscle are actually huge single cells that form during development by the fusion of many separate cells, as discussed in Chapter 22. The many nuclei of the contributing cells are retained in this large cell and lie just beneath the plasma membrane, but the bulk of the cytoplasm inside is made up of myofibrils, which is the name given to the basic contractile elements of the muscle cell (Figure 16–68). A **myofibril** is a cylindrical structure 1–2 μm in diameter that is often as long as the muscle cell itself. It consists of a long repeated chain of tiny contractile units—called *sarcomeres,* each about 2.2 μm long, which give the vertebrate myofibril its striated appearance (Figure 16–69).

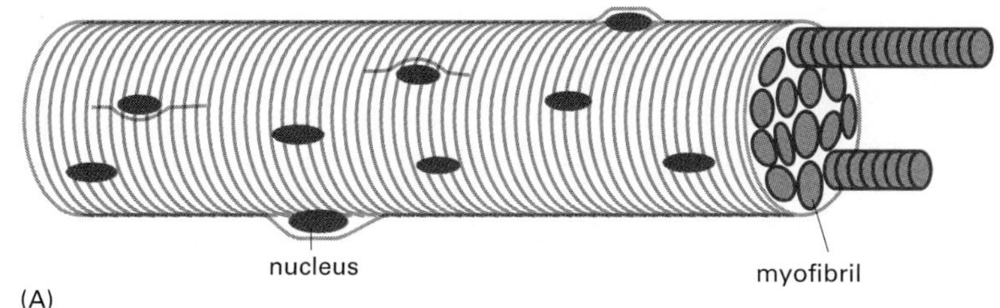

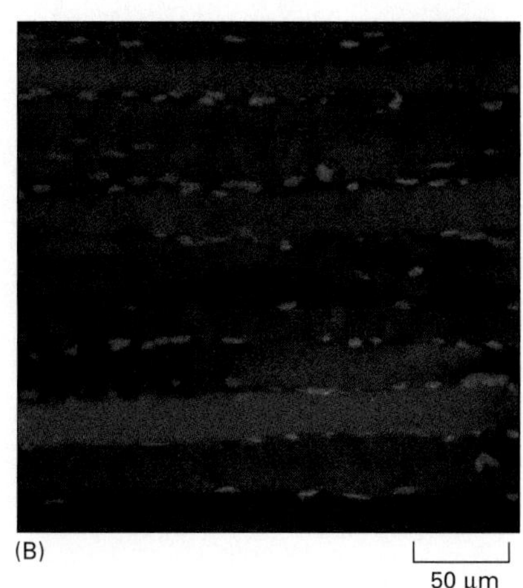

Figure 16–68 Skeletal muscle cells (also called muscle fibers). (A) These huge multinucleated cells form by the fusion of many muscle cell precursors, called myoblasts. In an adult human, a muscle cell is typically 50 μm in diameter and can be up to several centimeters long. (B) Fluorescence micrograph of rat muscle, showing the peripherally located nuclei *(blue)* in these giant cells. (B, courtesy of Nancy L. Kedersha.)

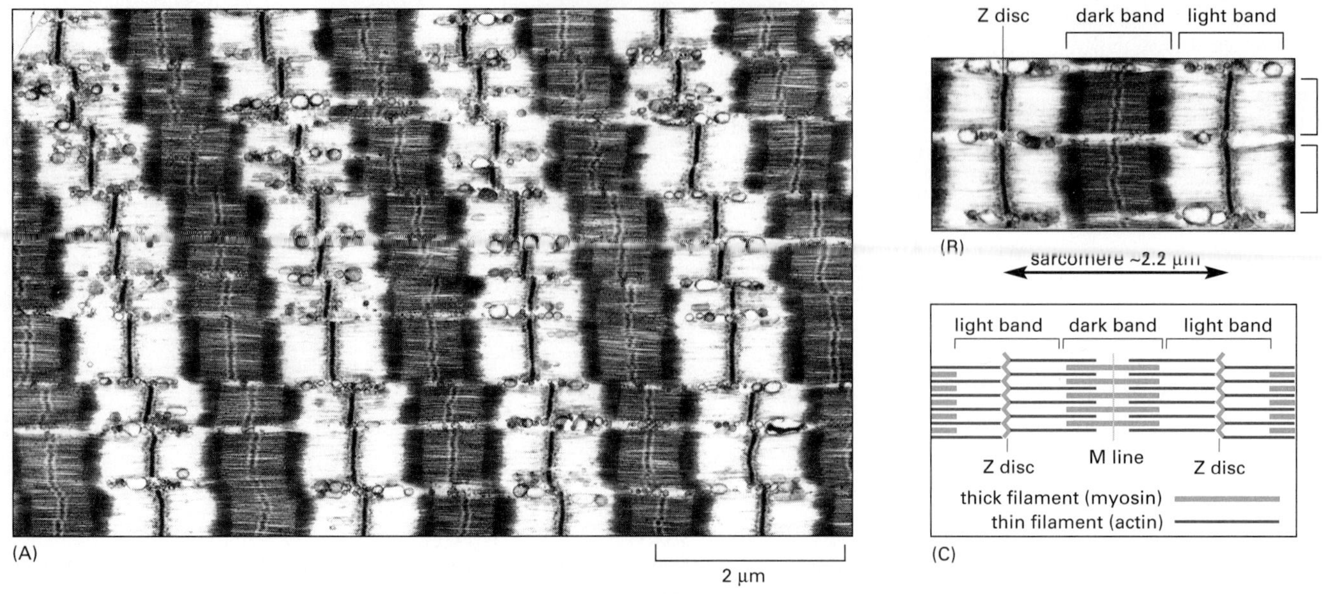

Figure 16–69 Skeletal muscle myofibrils. (A) Low-magnification electron micrograph of a longitudinal section through a skeletal muscle cell of a rabbit, showing the regular pattern of cross-striations. The cell contains many myofibrils aligned in parallel (see Figure 16–68). (B) Detail of the skeletal muscle shown in (A), showing portions of two adjacent myofibrils and the definition of a sarcomere. (C) Schematic diagram of a single sarcomere, showing the origin of the dark and light bands seen in the electron micrographs. The Z discs, at each end of the sarcomere, are attachment sites for the plus ends of actin filaments (thin filaments); the M line, or midline, is the location of proteins that link adjacent myosin II filaments (thick filaments) to one another. The dark bands, which mark the location of the thick filaments, are sometimes called A bands because they appear anisotropic in polarized light (that is, their refractive index changes with the plane of polarization). The light bands, which contain only thin filaments and therefore have a lower density of protein, are relatively isotropic in polarized light and are sometimes called I bands. (A and B, courtesy of Roger Craig.)

Each sarcomere is formed from a miniature, precisely ordered array of parallel and partly overlapping thin and thick filaments. The *thin filaments* are composed of actin and associated proteins, and they are attached at their plus ends to a *Z disc* at each end of the sarcomere. The capped minus ends of the actin filaments extend in toward the middle of the sarcomere, where they overlap with *thick filaments,* the bipolar assemblies formed from specific muscle isoforms of myosin II (see Figure 16–52). When this region of overlap is examined in cross section by electron microscopy, the myosin filaments are seen to be arranged in a regular hexagonal lattice, with the actin filaments evenly spaced between them (Figure 16–70). Cardiac muscle and smooth muscle also contain sarcomeres, although the organization is not as regular as that in skeletal muscle.

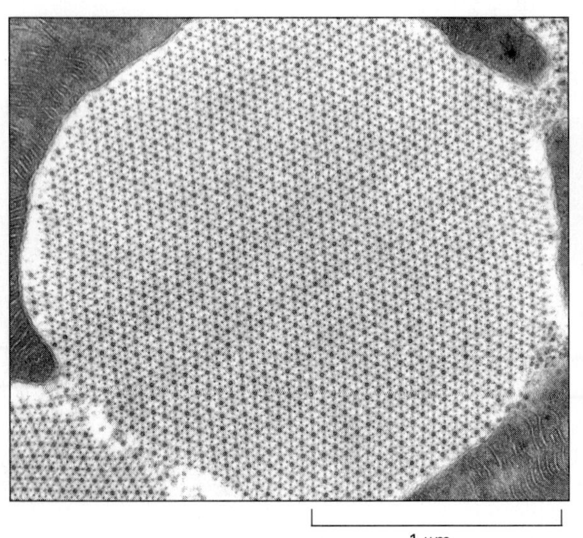

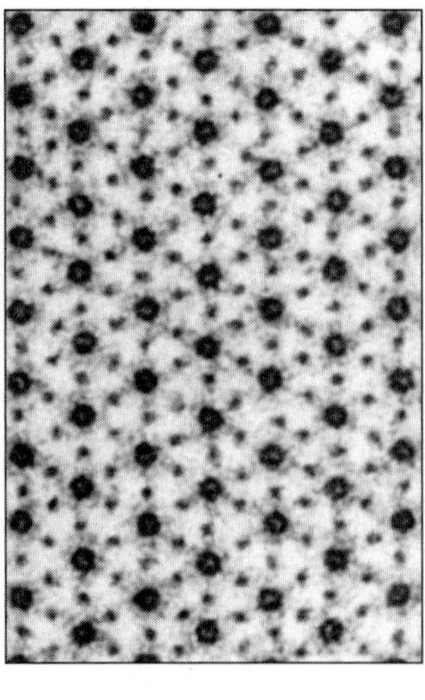

Figure 16–70 Electron micrographs of an insect flight muscle viewed in cross section. The myosin and actin filaments are packed together with almost crystalline regularity. Unlike their vertebrate counterparts, these myosin filaments have a hollow center, as seen in the enlargement on the right. The geometry of the hexagonal lattice is slightly different in vertebrate muscle. (From J. Auber, *J. de Microsc.* 8:197–232, 1969.)

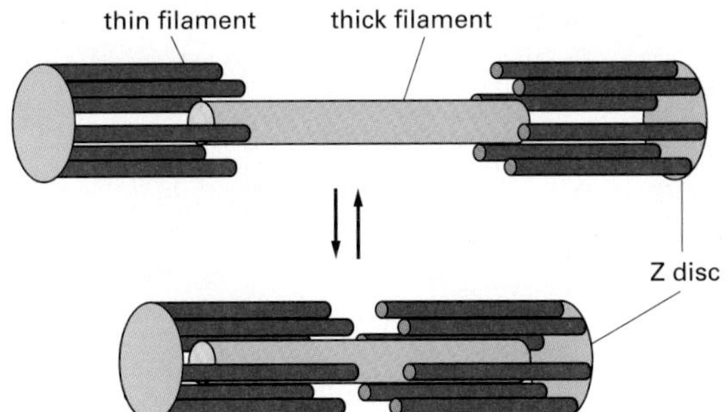

Figure 16–71 The sliding-filament model of muscle contraction. The actin *(red)* and myosin *(green)* filaments in a sarcomere slide past one another without shortening.

Sarcomere shortening is caused by the myosin filaments sliding past the actin thin filaments, with no change in the length of either type of filament (Figure 16–71). Bipolar thick filaments walk toward the plus ends of two sets of thin filaments of opposite orientations, driven by dozens of independent myosin heads that are positioned to interact with each thin filament. There is no coordination among the movements of the myosin heads, so it is critical that they operate with a low processivity, remaining tightly bound to the actin filament for only a small fraction of each ATPase cycle so that they do not hold one another back. Each myosin thick filament has about 300 heads (294 in frog muscle), and each head cycles about five times per second in the course of a rapid contraction—sliding the myosin and actin filaments past one another at rates of up to 15 μm/sec and enabling the sarcomere to shorten by 10% of its length in less than 1/50th of a second. The rapid synchronized shortening of the thousands of sarcomeres lying end-to-end in each myofibril gives skeletal muscle the ability to contract rapidly enough for running and flying, and even for playing the piano.

Accessory proteins govern the remarkable uniformity in filament organization, length, and spacing in the sarcomere (Figure 16–72). As mentioned previously, the actin filament plus ends are anchored in the Z disc, which is built from CapZ and α-actinin; the Z disc caps the filaments (preventing depolymerization), while holding them together in a regularly spaced bundle. The precise length of each filament is determined by a template protein of enormous size, called *nebulin*, which consists almost entirely of a repeating 35-amino-acid actin-binding motif. Nebulin stretches from the Z disc to the minus end of each thin filament and acts as a "molecular ruler" to dictate the length of the filament. The minus ends of the thin filaments are capped and stabilized by tropomodulin. Thus, the actin filaments in sarcomeres are remarkably stable, unlike the dynamic actin filaments characteristic of most other cell types.

The thick filaments are positioned midway between the Z discs by opposing pairs of an even longer template protein, called *titin*. Titin acts as a molecular

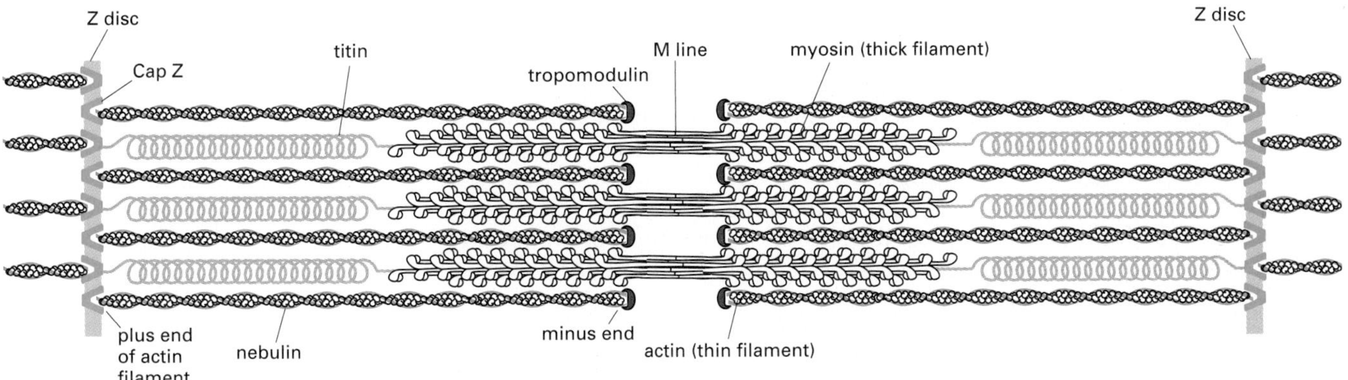

Figure 16–72 Organization of accessory proteins in a sarcomere. Each giant titin molecule extends from the Z disc to the M line—a distance of over 1 μm. Part of each titin molecule is closely associated with a myosin thick filament (which switches polarity at the M line); the rest of the titin molecule is elastic and changes length as the sarcomere contracts and relaxes. Each nebulin molecule is exactly the length of a thin filament. The actin filaments are also coated with tropomyosin and troponin (not shown; see Figure 16–74) and are capped at both ends. Tropomodulin caps the minus end of the actin filaments, and CapZ anchors the plus end at the Z disc, which also contains α-actinin.

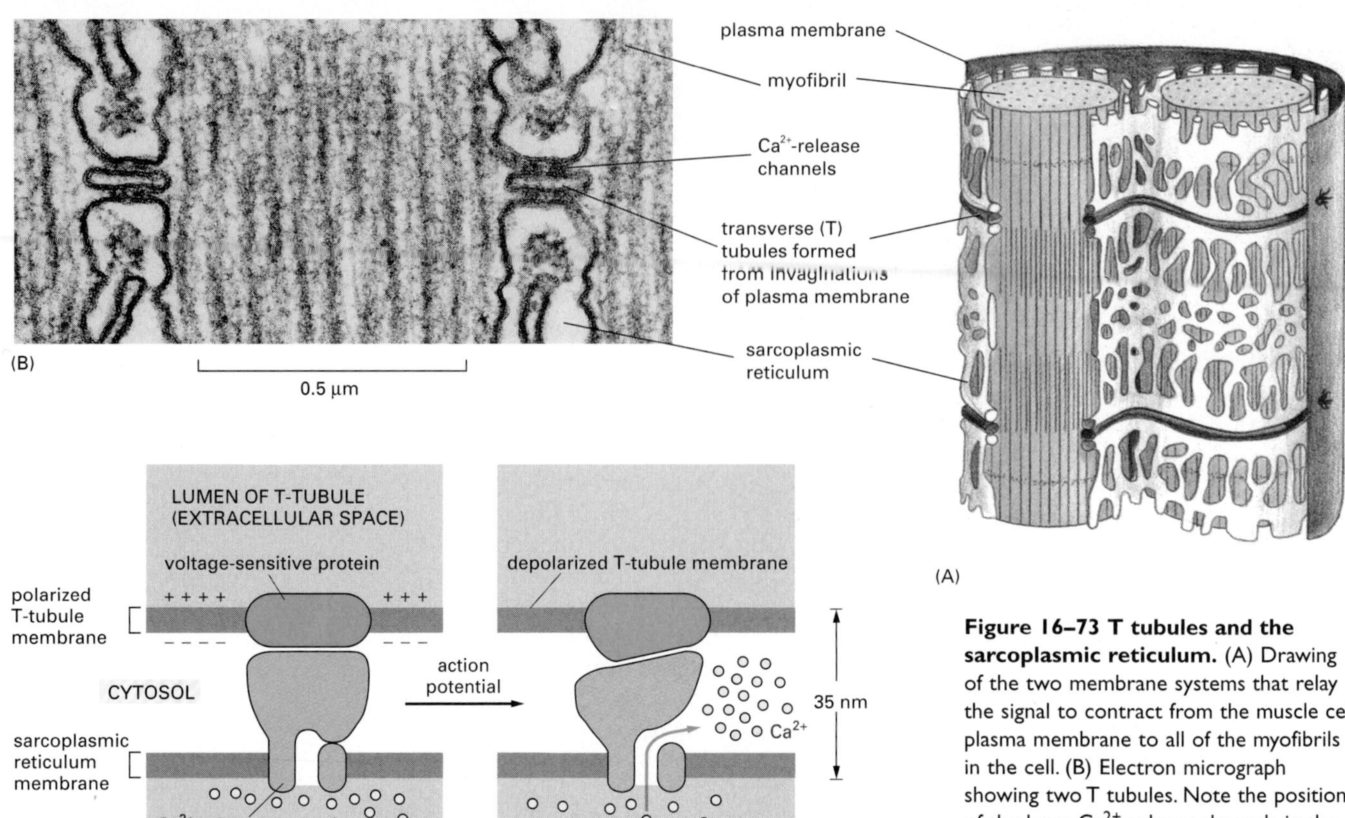

Figure 16–73 T tubules and the sarcoplasmic reticulum. (A) Drawing of the two membrane systems that relay the signal to contract from the muscle cell plasma membrane to all of the myofibrils in the cell. (B) Electron micrograph showing two T tubules. Note the position of the large Ca^{2+}-release channels in the sarcoplasmic reticulum membrane; they look like square-shaped "feet" that connect to the adjacent T-tubule membrane. (C) Schematic diagram showing how a Ca^{2+}-release channel in the sarcoplasmic reticulum membrane is thought to be opened by a voltage-sensitive transmembrane protein in the adjacent T-tubule membrane. (B, courtesy of Clara Franzini-Armstrong.)

spring, with a long series of immunoglobulin-like domains that can unfold one by one as stress is applied to the protein. A springlike unfolding and refolding of these domains keeps the thick filaments poised in the middle of the sarcomere and allows the muscle fiber to recover after being overstretched. In *C. elegans*, whose sarcomeres are longer than those in vertebrates, titin is also longer, suggesting that it too serves as a moleculer ruler, determining in this case the overall length of each sarcomere (see Figure 3–34).

Muscle Contraction Is Initiated by a Sudden Rise in Cytosolic Ca^{2+} Concentration

The force-generating molecular interaction between myosin thick filaments and actin thin filaments takes place only when a signal passes to the skeletal muscle from its motor nerve. The signal from the nerve triggers an action potential in the muscle cell plasma membrane (discussed in Chapter 11), and this electrical excitation spreads rapidly into a series of membraneous folds, the transverse tubules, or *T tubules*, that extend inward from the plasma membrane around each myofibril. The signal is then relayed across a small gap to the *sarcoplasmic reticulum*, an adjacent web-like sheath of modified endoplasmic reticulum that surrounds each myofibril like a net stocking (Figure 16–73A, B).

When voltage-sensitive proteins in the T-tubule membrane are activated by the incoming action potential, they trigger the opening of Ca^{2+}-release channels in the sarcoplasmic reticulum (Figure 16–73C). Ca^{2+} flooding into the cytosol then initiates the contraction of each myofibril. Because the signal from the muscle-cell plasma membrane is passed within milliseconds (via the T tubules and sarcoplasmic reticulum) to every sarcomere in the cell, all of the myofibrils in the cell contract at the same time. The increase in Ca^{2+} concentration is transient because the Ca^{2+} is rapidly pumped back into the sarcoplasmic reticulum

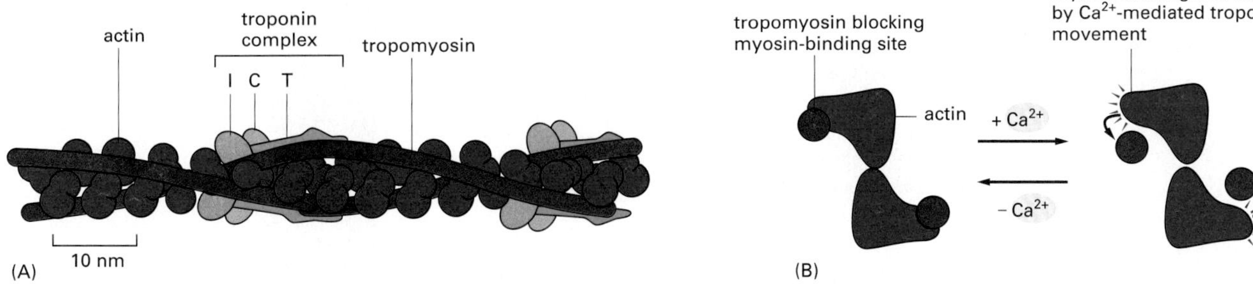

Figure 16–74 The control of skeletal muscle contraction by troponin. (A) A skeletal muscle cell thin filament, showing the positions of tropomyosin and troponin along the actin filament. Each tropomyosin molecule has seven evenly spaced regions with similar amino acid sequences, each of which is thought to bind to an actin subunit in the filament. (B) A thin filament shown end-on, illustrating how Ca^{2+} (binding to troponin) is thought to relieve the tropomyosin blockage of the interaction between actin and the myosin head. (A, adapted from G.N. Phillips, J.P. Fillers and C. Cohen, *J. Mol. Biol.* 192:111–131, 1986. © Academic Press.)

by an abundant, ATP-dependent Ca^{2+}-pump (also called a Ca^{2+}-ATPase), in its membrane (see Figure 3–77). Typically, the cytoplasmic Ca^{2+} concentration is restored to resting levels within 30 msec, allowing the myofibrils to relax. Thus, muscle contraction depends on two processes that consume enormous amounts of ATP: filament sliding, driven by the ATPase of the myosin motor domain, and Ca^{2+} pumping, driven by the Ca^{2+}-pump.

The Ca^{2+} dependence of vertebrate skeletal muscle contraction, and hence its dependence on motor commands transmitted via nerves, is due entirely to a set of specialized accessory proteins that are closely associated with the actin thin filaments. One of these accessory proteins is a muscle form of *tropomyosin*, an elongated molecule that binds along the groove of the actin helix. The other is *troponin*, a complex of three polypeptides, troponins T, I, and C (named for their tropomyosin-binding, inhibitory, and Ca^{2+}-binding activities, respectively). Troponin I binds to actin as well as to troponin T. In a resting muscle, the troponin I–T complex pulls the tropomyosin out of its normal binding groove into a position along the actin filament that interferes with the binding of myosin heads, thereby preventing any force-generating interaction. When the level of Ca^{2+} is raised, troponin C—which binds up to four molecules of Ca^{2+}—causes troponin I to release its hold on actin. This allows the tropomyosin molecules to slip back into their normal position so that the myosin heads can walk along the actin filaments (Figure 16–74). Troponin C is closely related to the ubiquitous Ca^{2+}-binding protein calmodulin (see Figure 15–40); it can be thought of as a specialized form of calmodulin that has acquired binding sites for troponin I and troponin T, thereby ensuring that the myofibril responds extremely rapidly to an increase in Ca^{2+} concentration.

Heart Muscle Is a Precisely Engineered Machine

The heart is the most heavily worked muscle in the body, contracting about 3 billion (3×10^9) times during the course of a human lifetime. This number is about the same as the average number of revolutions in the lifetime of an automobile's internal combustion engine. Several specific isoforms of cardiac muscle myosin and cardiac muscle actin are expressed in heart cells. Even subtle changes in contractile proteins expressed in the heart—changes that would not cause any noticeable consequences in other tissues—can cause serious heart disease (Figure 16–75).

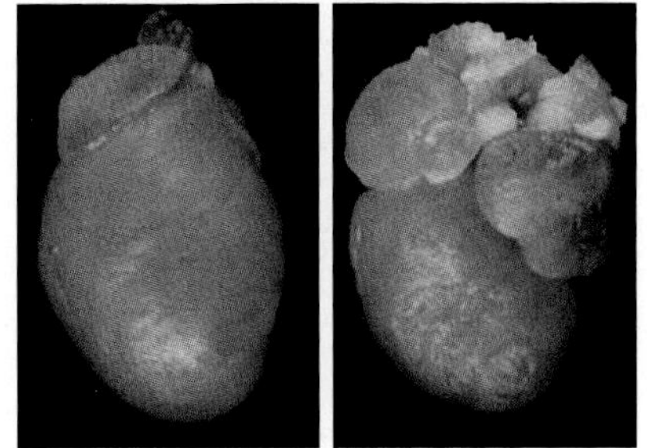

Figure 16–75 Effect on the heart of a subtle mutation in cardiac myosin. *Left*, normal heart from a 6-day old mouse pup. *Right*, heart from a pup with a point mutation in both copies of its cardiac myosin gene, changing Arg 403 to Gln. Both atria are greatly enlarged (hypertrophic), and the mice die within a few weeks of birth. (From D. Fatkin et al., *J. Clin. Invest.* 103:147, 1999.)

Familial hypertrophic cardiomyopathy is a frequent cause of sudden death in young athletes. It is an inherited condition that affects about two out of every thousand people, and it is associated with heart enlargement, abnormally small coronary vessels, and disturbances in heart rhythm (cardiac arrhythmias). Over 40 subtle point mutations in the genes encoding cardiac β myosin heavy chain (almost all causing changes in or near the motor domain), as well as about a dozen in other genes encoding contractile proteins, including myosin light chains, cardiac troponin, and tropomyosin, have been found that cause this condition. Minor missense mutations in the cardiac actin gene can cause another type of heart condition, called *dilated cardiomyopathy*, that also frequently results in early heart failure. The normal cardiac contractile apparatus seems to be such a highly tuned machine that a tiny abnormality anywhere in the works can be enough to gradually wear it down over years of repetitive motion.

Figure 16–76 The contrasting motions of flagella and cilia. (A) The wave-like motion of the flagellum of a sperm cell from a tunicate. The cell was photographed with stroboscopic illumination at 400 flashes per second. Note that waves of constant amplitude move continuously from the base to the tip of the flagellum. (B) The beat of a cilium, which resembles the breast stroke in swimming. A fast power stroke *(red* arrows), in which fluid is driven over the surface of the cell, is followed by a slow recovery stroke. Each cycle typically requires 0.1–0.2 sec and generates a force perpendicular to the axis of the axoneme. (A, courtesy of C.J. Brokaw.)

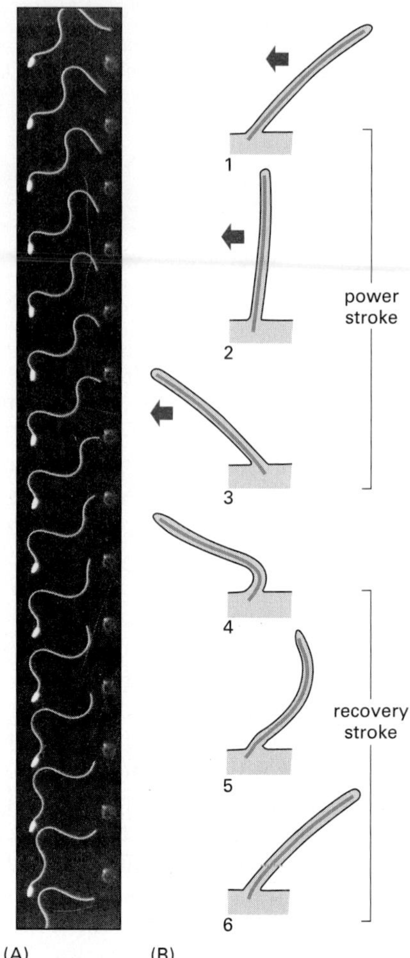

Cilia and Flagella Are Motile Structures Built from Microtubules and Dyneins

Just as myofibrils are highly specialized and efficient motility machines built from actin and myosin filaments, cilia and flagella are highly specialized and efficient motility structures built from microtubules and dynein. Both cilia and flagella are hair-like cellular appendages that have a bundle of microtubules at their core. **Flagella** are found on sperm and many protozoa. By their undulating motion, they enable the cells to which they are attached to swim through liquid media (Figure 16–76A). **Cilia** tend to be shorter than flagella and are organized in a similar fashion, but they beat with a whip-like motion that resembles the breast stroke in swimming (Figure 16–76B). The cycles of adjacent cilia are almost but not quite in synchrony, creating the wave-like patterns that can be seen in fields of beating cilia under the microscope. Ciliary beating can either propel single cells through a fluid (as in the swimming of the protozoan *Paramecium*) or can move fluid over the surface of a group of cells in a tissue. In the human body, huge numbers of cilia (10^9/cm^2 or more) line our respiratory tract, sweeping layers of mucus, trapped particles of dust, and bacteria up to the mouth where they are swallowed and ultimately eliminated. Likewise, cilia along the oviduct help to sweep eggs toward the uterus.

The movement of a cilium or a flagellum is produced by the bending of its core, which is called the **axoneme**. The axoneme is composed of microtubules and their associated proteins, arranged in a distinctive and regular pattern. Nine special doublet microtubules (comprising one complete and one partial microtubule fused together so that they share a common tubule wall) are arranged in a ring around a pair of single microtubules (Figure 16–77). This characteristic

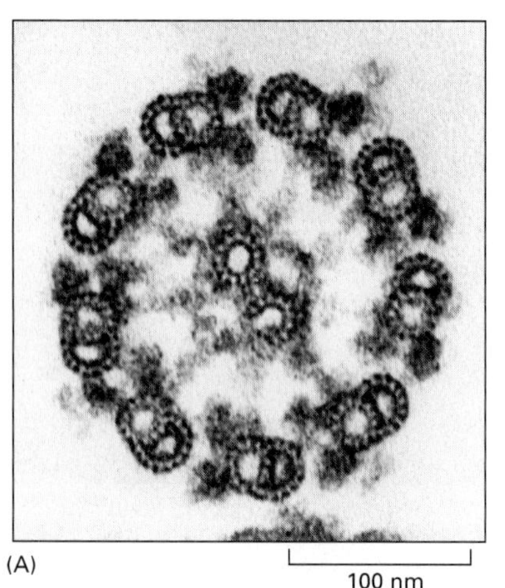

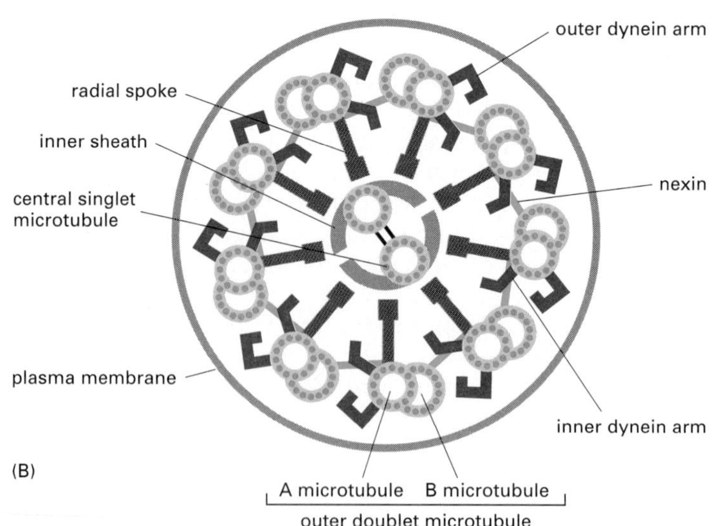

Figure 16–77 The arrangement of microtubules in a flagellum or cilium. (A) Electron micrograph of the flagellum of a green-alga cell *(Chlamydomonas)* shown in cross section, illustrating the distinctive "9 + 2" arrangement of microtubules. (B) Diagram of the parts of a flagellum or cilium. The various projections from the microtubules link the microtubules together and occur at regular intervals along the length of the axoneme. (A, courtesy of Lewis Tilney.)

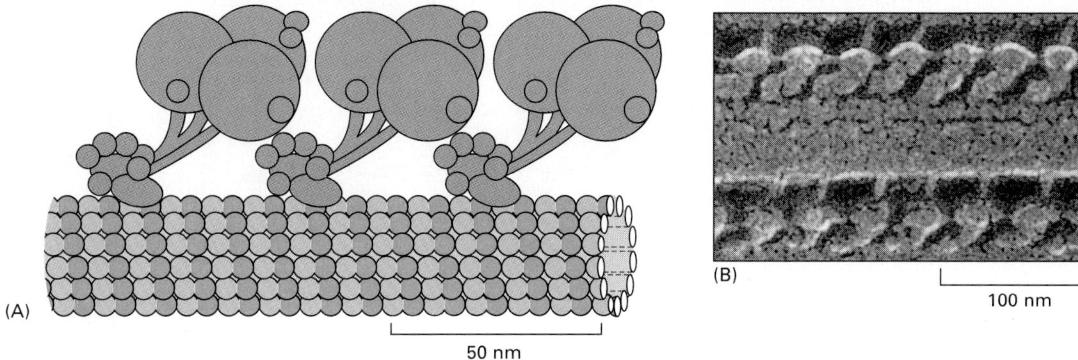

Figure 16–78 Ciliary dynein. Ciliary (axonemal) dynein is a large protein assembly (nearly 2 million daltons) composed of 9–12 polypeptide chains, the largest of which is the heavy chain of more than 500,000 daltons. (A) The heavy chains are believed to form the major portion of the globular head and stem domains, and many of the smaller chains are clustered around the base of the stem (see Figure 16–56 for a view of the isolated molecule). The base of the molecule binds tightly to an A microtubule in an ATP-independent manner, while the large globular heads have an ATP-dependent binding site for a B microtubule (see Figure 16–77). When the heads hydrolyze their bound ATP, they move toward the minus end of the B microtubule, thereby producing a sliding force between the adjacent microtubule doublets in a cilium or flagellum. The three-headed form of ciliary dynein, formed from three heavy chains, is illustrated here. (B) Freeze-etch electron micrograph of a cilium showing the dynein arms projecting at regular intervals from the doublet microtubules; only the A microtubule is shown. (B, courtesy of John Heuser.)

arrangement is found in almost all forms of eucaryotic flagella and cilia from protozoans to humans. The microtubules extend continuously for the length of the axoneme, which can be 10–200 μm. At regular positions along the length of the microtubules, accessory proteins cross-link the mircotubules together. Molecules of *ciliary dynein* form bridges between the neighboring doublet microtubules around the circumference of the axoneme (Figure 16–78). When the motor domain of this dynein is activated, the dynein molecules attached to one microtubule doublet attempt to walk along the adjacent microtubule doublet, tending to force the adjacent doublets to slide relative to one another, much as actin thin filaments slide during muscle contraction. However, the presence of other links between the microtubule doublets prevents this sliding, and the dynein force is instead converted into a bending motion (Figure 16–79).

Bacteria also swim using cell surface structures called flagella, but these do not contain microtubules or dynein and do not wave or beat. Instead, *bacterial flagella* are long, rigid helical filaments, made up of repeating subunits of the

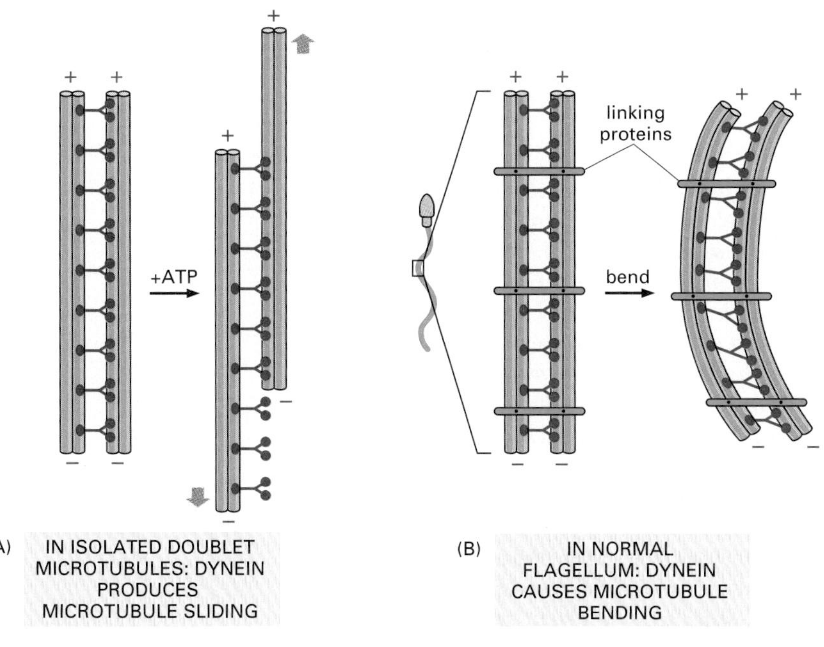

Figure 16–79 The bending of an axoneme. (A) When axonemes are exposed to the proteolytic enzyme trypsin, the linkages holding neighboring doublet microtubules together are broken. In this case, the addition of ATP allows the motor action of the dynein heads to slide one pair of doublet microtubules against the other pair. (B) In an intact axoneme (such as in a sperm), sliding of the doublet microtubules is prevented by flexible protein links. The motor action therefore causes a bending motion, creating waves or beating motions, as seen in Figure 16–76.

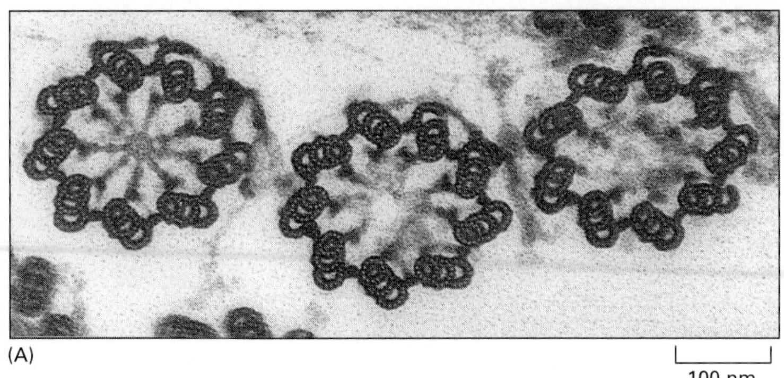

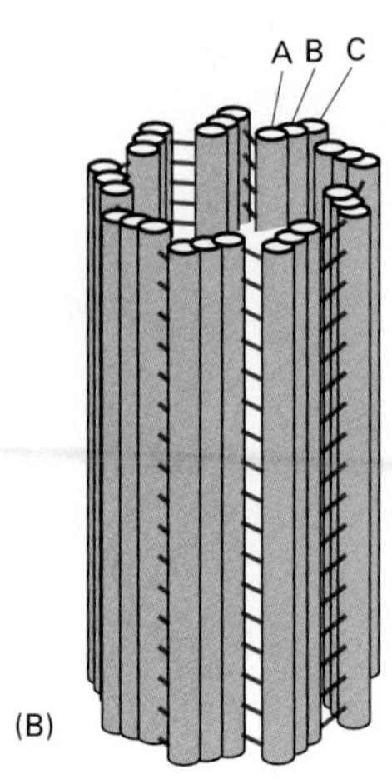

Figure 16–80 Basal bodies. (A) Electron micrograph of a cross section through three basal bodies in the cortex of a protozoan. (B) Diagram of a basal body viewed from the side. Each basal body forms the lower portion of a ciliary axoneme and is composed of nine sets of triplet microtubules, each triplet containing one complete microtubule (the A microtubule) fused to two incomplete microtubules (the B and C microtubules). Other proteins (shown in *red* in B) form links that hold the cylindrical array of microtubules together. The arrangement of microtubules in a centriole is essentially the same (see Figure 16–24). (A, courtesy of D.T. Woodrow and R.W. Linck.)

protein flagellin. The flagella rotate like propellers, driven by a special rotary motor embedded in the bacterial cell wall (see Figure 15–67). The use of the same name to denote these two very different types of swimming apparatus is an unfortunate historical accident.

Structures called *basal bodies* firmly root eucaryotic cilia and flagella at the cell surface. The basal bodies have the same form as the centrioles that are found embedded at the center of animal centrosomes, with nine groups of fused triplet microtubules arranged in a cartwheel (Figure 16–80). Indeed, in some organisms, basal bodies and centrioles are functionally interconvertible: during each mitosis in the unicellular alga *Chlamydomonas*, for example, the flagella are resorbed, and the basal bodies move into the cell interior and become part of the spindle poles. New centrioles and basal bodies arise by a curious replication process, in which a smaller daughter is formed perpendicular to the original structure by a still mysterious mechanism (see Figure 18–6).

In humans, hereditary defects in ciliary dynein cause Kartagener's syndrome. The syndrome is characterized by male sterility due to immotile sperm, a high susceptibility to lung infections owing to the paralyzed cilia in the respiratory tract that fail to clear debris and bacteria, and defects in determination of the left–right axis of the body during early embryonic development (discussed in Chapter 21).

Summary

Motor proteins use the energy of ATP hydrolysis to move along microtubules or actin filaments. They mediate the sliding of filaments relative to one another and the transport of membrane-enclosed organelles along filament tracks. All known motor proteins that move on actin filaments are members of the myosin superfamily. The motor proteins that move on microtubules are members of either the kinesin superfamily or the dynein family. The myosin and kinesin superfamilies are diverse, with about 40 genes encoding each type of protein in humans. The only structural element shared among all members of each superfamily is the motor "head" domain. These heads can be attached to a wide variety of "tails," which attach to different types of cargo and enable the various family members to perform different functions in the cell. Although myosin and kinesin walk along different tracks and use different mechanisms to produce force and movement by ATP hydrolysis, they share a common structural core, suggesting that they are derived from a common ancestor.

Two types of specialized motility structures in eucaryotic cells consist of highly ordered arrays of motor proteins that move on stabilized filament tracks. The myosin–actin system of the sarcomere powers the contraction of various types of muscle, including skeletal, smooth, and cardiac muscle. The dynein–microtubule system of the axoneme powers the beating of cilia and the undulations of flagella.

THE CYTOSKELETON AND CELL BEHAVIOR

A central challenge in all areas of cell biology is to understand how the functions of many individual molecular components combine to produce complex cell behaviors. Thus far in this chapter, we have seen how various basic cytoskeletal mechanisms, such as muscle contraction, can be understood in terms of the activities of just a few types of molecules. In this final section, we examine more complex cell behaviors and consider how they are built up from these basic cytoskeletal mechanisms.

The cell behaviors that we describe all rely on coordinated deployment of the same components and processes that we have explored in the first three sections of the chapter: the dynamic assembly and disassembly of cytoskeletal polymers, the regulation and modification of their structure by polymer-associated proteins, and the actions of motor proteins moving along the polymers. How are all these activities coordinated to define a cell's shape, to enable it to crawl, or to divide it neatly into two at mitosis? These problems of cytoskeletal coordination will challenge scientists for many years to come. We shall focus on the control of cell shape and locomotion here, leaving the role of the cytoskeleton in cell division for discussion in Chapter 18.

Mechanisms of Cell Polarization Can Be Readily Analyzed in Yeast Cells

Compared with most animal cells, the budding yeast *Saccharomyces cerevisiae* has a very simple structure. The most notable feature is the marked asymmetry of the yeast cell, evident in the asymmetrical way it divides by budding to create a small daughter cell and a large mother cell. This asymmetry derives from the polar orientation of its actin cytoskeleton. Because genetic analysis in yeast is relatively easy, it has been possible to use genetic screens to identify many of the molecules involved in generating and maintaining this cell polarity. As in other fields where budding yeast has proved its utility as a model organism for cell biologists, most of these molecules have turned out to have homologs with similar functions in more complex cells.

Budding yeast cells are unusual among eucaryotes in having very few cytoplasmic microtubules, so that most aspects of its polarity depend on actin. There are two types of actin filament assemblies in these cells: actin cables (long bundles of actin filaments) and actin patches (small assemblies of filaments associated with the cell cortex, probably marking sites of endocytosis and exocytosis). Proliferating budding yeast cells must be highly polarized to allow the mother cell to grow a bud from a single site on the cell surface. In this process, the actin patches are highly concentrated at the growing tip of the bud, with the actin cables aligned and pointing toward them. This actin organization presumably directs the secretion of new cell wall material to the site of budding (Figure 16–81).

On starvation, yeasts, like many other unicellular organisms, sporulate. But sporulation can occur only in diploid budding yeast cells, whereas budding yeasts mainly proliferate as haploid cells. A starving haploid individual must therefore locate a partner of the opposite mating type, woo it, and mate with it before sporulating. Yeast cells are unable to swim and, instead, reach their mates by polarized growth. The haploid form of budding yeast comes in two mating types, a and α, which secrete mating factors known as a-factor and α-factor, respectively. These secreted signal molecules act by binding to cell-surface receptors that belong to the G-protein-linked receptor superfamily discussed in Chapter 15. One consequence of the binding of α-factor to its receptor is to cause the recipient cell to become polarized, adopting a shape known as a

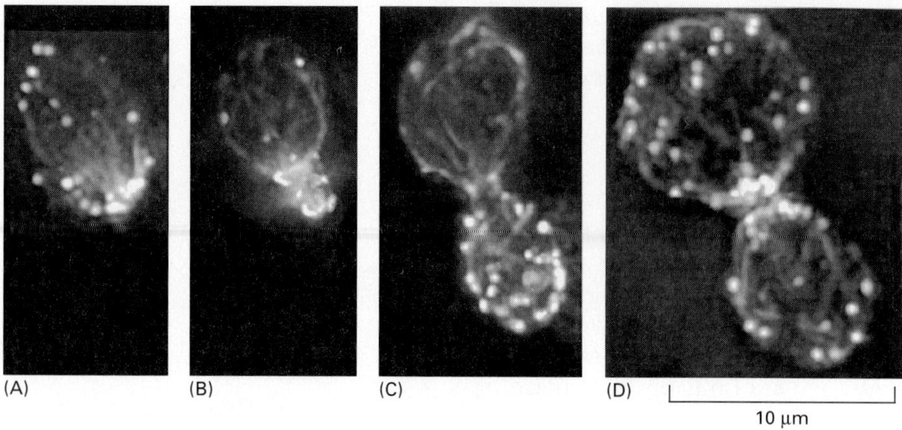

Figure 16–81 Polarity of actin patches and cables throughout the yeast cell cycle. Filamentous actin structures in the yeast cell, labeled here with fluorescent phalloidin, include actin patches (round bright spots) and actin cables (extended lines). (A) In a mother cell, before the formation of the bud, most of the patches are clustered at one end. The cables are lined up and point toward the cluster of patches, which is the site where the bud will emerge. (B) As the small bud grows, most patches remain within it. Cables in the mother cell continue to point toward this site of new cell wall growth. (C) Patches are almost uniformly distributed over the surface of a full-sized bud. Cables in the mother cell remain polarized. (D) Immediately after cell division, mother and daughter cells form new patches, which are concentrated near the division site, although both cells have randomly oriented cables. (From T.S. Karpova et al., *J. Cell Biol.* 142:1501–1517, 1998. © The Rockefeller University Press.)

"shmoo" (Figure 16–82). In the presence of an α-factor gradient, the a-cell shmoo tip is directed toward the highest concentration of the signal molecule, which under normal circumstances would direct it toward an amorous α cell located nearby.

This polarized cell growth requires alignment of the actin cytoskeleton in response to the mating factor signal. When the signal binds to its receptor, the receptor activates Cdc42, the Rho-family GTPase that is also responsible for some types of actin rearrangements in animal cells (see Figure 16–50D). The signals that the actin cytoskeleton receives from the outside world are then relayed to the microtubule cytoskeleton, causing polarized growth. In addition, the microtubule-organizing center (which in yeast is embedded in the nuclear membrane and is called the *spindle pole body)* rotates to the side of the nucleus that is closest to the shmoo tip, so that the nucleus becomes oriented along an axis that will guide it to meet and fuse with the nucleus of its mating partner. Extensive genetic studies have identified most of the genes involved in receiving and transducing this spatial signal. As might be expected, many of the proteins that they encode are involved in regulating actin filament and microtubule rearrangements (Figure 16–83).

Haploid budding yeast cells use this same polarization machinery during vegetative growth. To form the bud that will grow out to become a daughter cell, the yeast must direct new plasma membrane and cell wall material primarily to a single site. As with shmoo formation, this requires an initial cytoskeletal polarity, with most actin patches in the growing bud and actin cables oriented along the bud axis. In haploid cells, a new bud site is always constructed immediately adjacent to the previous bud site. In this case, the spatial cues that set up cytoskeletal polarity are intrinsic to the cell, left behind from previous rounds of cell division. Cdc42 is once again involved in transducing the signal from the destined bud site to the cytoskeleton, and most of the proteins involved in the upstream and downstream pathways have been identified through genetic experiments. Subsequent to their identification in yeast, many of these proteins have been found to have homologs in other organisms, where they are often likewise involved in the establishment of cell polarity.

Specific RNA Molecules Are Localized by the Cytoskeleton

The cytoskeleton is used to position RNA molecules as well as proteins in cells. The yeast mother and daughter cells retain distinct identities, as revealed by

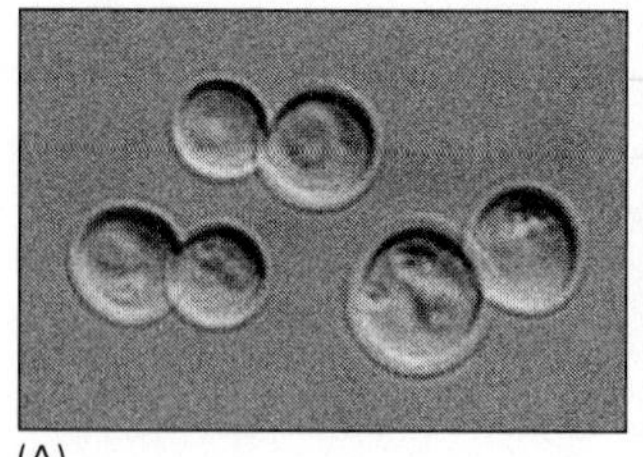

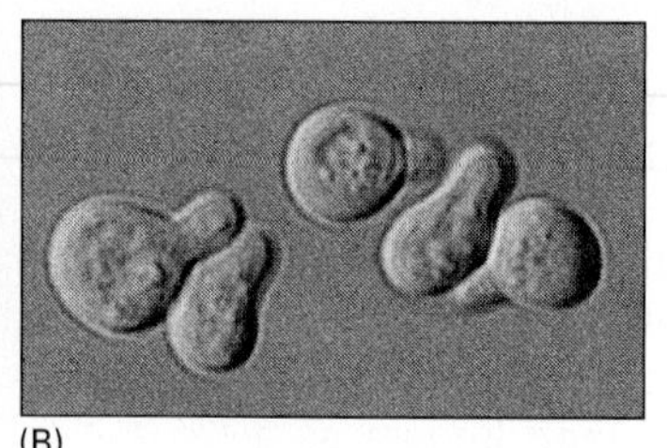

Figure 16–82 Morphological polarization of yeast cells in response to mating factor. (A) Cells of *Saccharomyces cerevisiae* are usually spherical. (B) They become polarized when treated with mating factor from cells of the opposite mating type. The polarized cells are called "shmoos." (C) Al Capp's famous cartoon character, the original Shmoo. (A and B, courtesy of Michael Snyder; C, © 1948 Capp Enterprises, Inc., all rights reserved.)

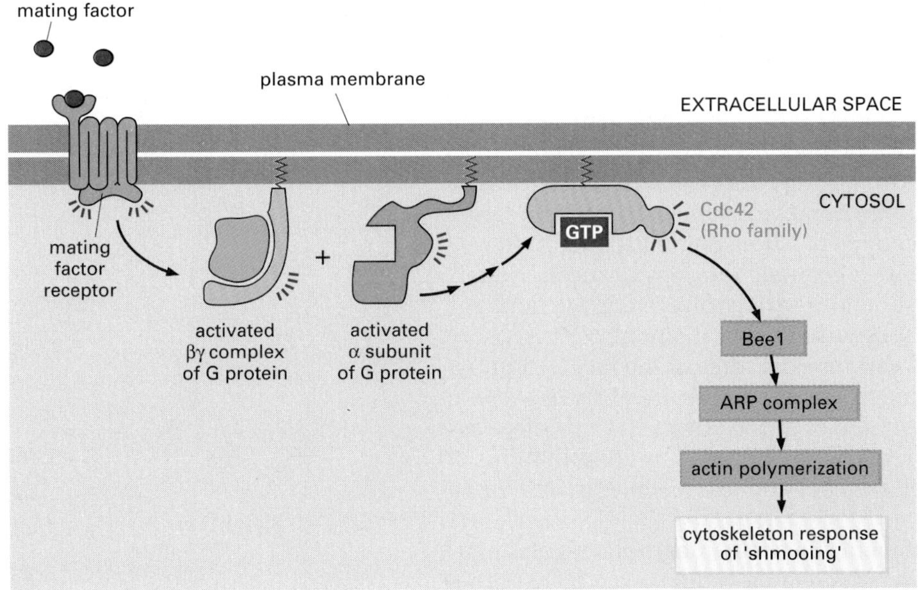

Figure 16–83 The signaling pathway in the yeast mating factor response. The extracellular mating factor binds to a G-protein-linked receptor in the plasma membrane. Activation of the receptor triggers dissociation of the GTP-bound Gα subunit from a heterotrimeric G-protein (discussed in Chapter 15). This in turn activates the Rho family GTP-binding protein, Cdc42, the yeast homolog of the mammalian small GTPase whose activation triggers the formation of filopodia (see Figure 16–50D). As in mammalian cells, Cdc42 activates a protein (Bee1, the homolog of WASp) that activates the ARP complex, leading to local actin nucleation at the site of mating factor binding. The local actin nucleation and filament growth lead to polarized growth and acquisition of a shmoo shape. In addition, receptor activation triggers other responses through a MAP kinase cascade (discussed in Chapter 15), preparing the haploid cell for mating (not shown).

major differences in their subsequent ability to undergo mating-type switching (discussed in Chapter 7) and in the choice of their next bud site. Many of these differences are caused by a gene regulatory protein called *Ash1*. Both *ash1* mRNA and protein are localized exclusively to the growing bud and therefore end up only in the daughter cell. One of the two type V myosins found in yeast, MyoV, is required for this asymmetric distribution of *ash1* mRNA. A genetic screen for other mutations that disrupt the mother/daughter difference has revealed that at least six other gene products that are associated with the cytoskeleton are required for normal polarity; these include tropomyosin, profilin, and actin itself, as well as a complex of two proteins that form a direct link between a specific sequence in the *ash1* mRNA and the myosin V protein (Figure 16–84).

This type of selective positioning of RNA represents yet another way in which the cytoskeleton is involved in cell polarity. For example, in the developing *Drosophila* embryo (discussed in Chapter 21), a group of mRNAs that encode genes necessary for proper development of the posterior region of the embryo are localized posteriorly by their association with actin filaments, but only after they are transported there by motor proteins moving along oriented microtubules. Identifying the molecules and organizational principles that determine cell polarity in these more complex systems is an active area of research.

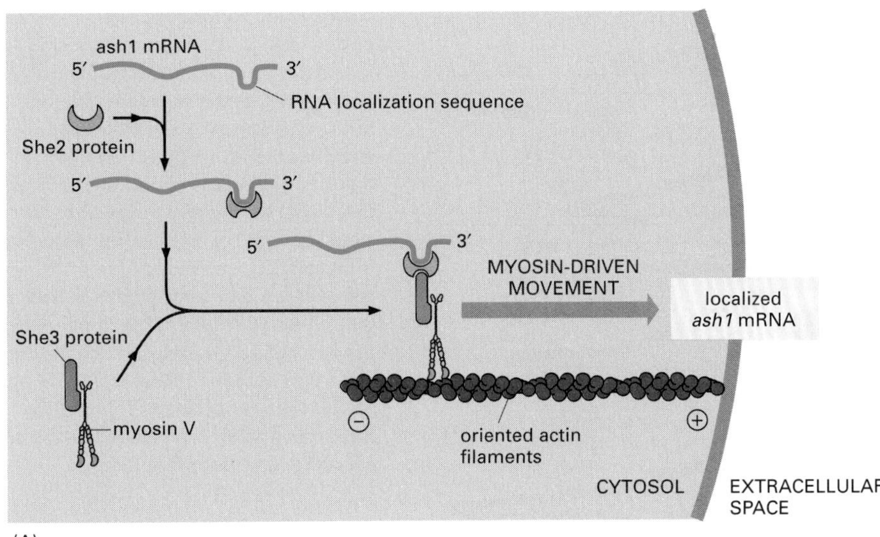

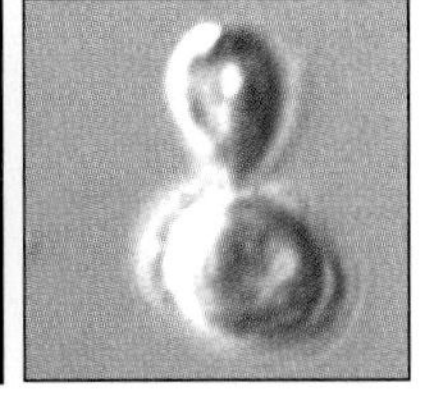
(B)

Figure 16–84 Polarized mRNA localization in the yeast bud tip. (A) The molecular mechanism of *ash1* mRNA localization, as determined by genetics and biochemistry. (B) Fluorescent *in situ* hybridization (FISH) was used to localize the *ash1* mRNA (*red*) in this dividing yeast cell. The mRNA is confined to the far tip of the daughter cell (here, still a large bud). Ash1 protein, transcribed from this localized mRNA, is also confined to the daughter cell. (B, courtesy of Peter Takizawa and Ron Vale.)

Many Cells Can Crawl Across A Solid Substratum

Many cells move by crawling over surfaces rather than by using cilia or flagella to swim. Predatory amoebae crawl continuously in search of food, and they can easily be observed to attack and devour smaller ciliates and flagellates in a drop of pond water. In animals, almost all cell locomotion occurs by crawling, with the notable exception of swimming sperm. During embryogenesis, the structure of an animal is created by the migrations of individual cells to specific target locations and by the coordinated movements of whole epithelial sheets. In vertebrates, *neural crest cells* are remarkable for their long-distance migrations from their site of origin in the neural tube to a variety of sites throughout the embryo. These cells have diverse fates, becoming skin pigment cells, sensory and sympathetic neurons and glia, and various structures of the face. Long-distance crawling is fundamental to the construction of the entire nervous system: it is in this way that the actin-rich growth cones at the advancing tips of developing axons travel to their eventual synaptic targets, guided by combinations of soluble signals and signals bound to cell surfaces and extracellular matrix along the way.

The adult animal is also seething with crawling cells. Macrophages and neutrophils crawl to sites of infection and engulf foreign invaders as a critical part of the innate immune response. Osteoclasts tunnel into bone, forming channels that are filled in by the osteoblasts that follow after them, in a continuous process of bone remodeling and renewal. Similarly, fibroblasts can migrate through connective tissues, remodeling them where necessary and helping to rebuild damaged structures at sites of injury. In an ordered procession, the cells in the epithelial lining of the intestine travel up the sides of the intestinal villi, replacing absorptive cells lost at the tip of the villus. Cell crawling also has a role in many cancers, when cells in a primary tumor invade neighboring tissues and crawl into blood vessels or lymph vessels and are thereby carried to other sites in the body to form metastases.

Cell crawling is a highly complex integrated process, dependent on the actin-rich cortex beneath the plasma membrane. Three distinct activities are involved: *protrusion,* in which actin-rich structures are pushed out at the front of the cell; *attachment,* in which the actin cytoskeleton connects across the plasma membrane to the substratum; and *traction,* in which the bulk of the trailing cytoplasm is drawn forward (Figure 16–85). In some crawling cells, such as keratocytes from the fish epidermis, these activities are closely coordinated, and the cells seem to glide forward smoothly without changing shape. In other cells, such as fibroblasts, these activities are more independent, and the locomotion is jerky and irregular.

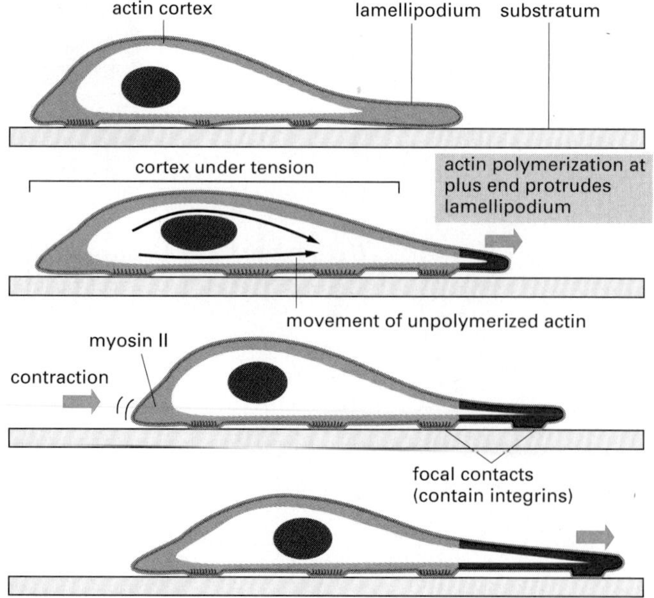

Figure 16–85 A model of how forces generated in the actin-rich cortex move a cell forward. The actin-polymerization-dependent protrusion and firm attachment of a lamellipodium at the leading edge of the cell moves the edge forward (*green* arrows at front) and stretches the actin cortex. Contraction at the rear of the cell propels the body of the cell forward (*green* arrow at back) to relax some of the tension (traction). New focal contacts are made at the front, and old ones are disassembled at the back as the cell crawls forward. The same cycle can be repeated, moving the cell forward in a stepwise fashion. Alternatively, all steps can be tightly coordinated, moving the cell forward smoothly. The newly polymerized cortical actin is shown in *red*.

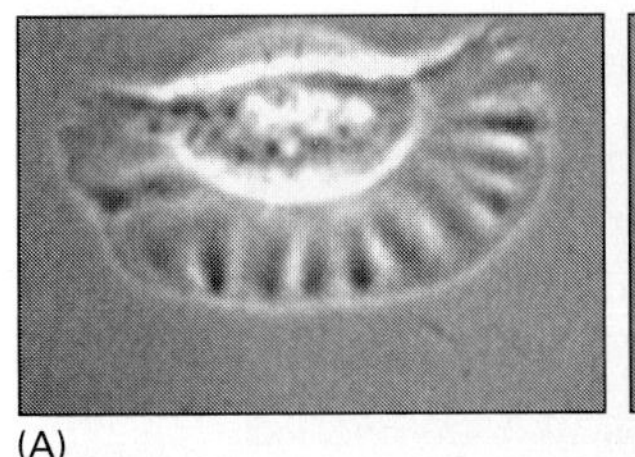
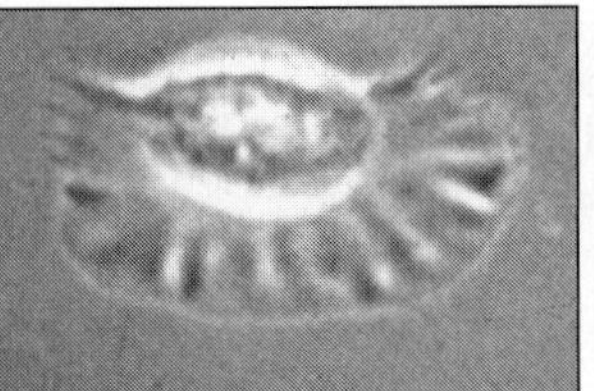
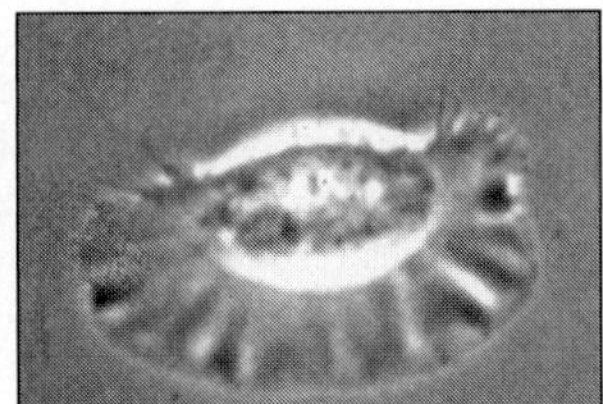
(A)

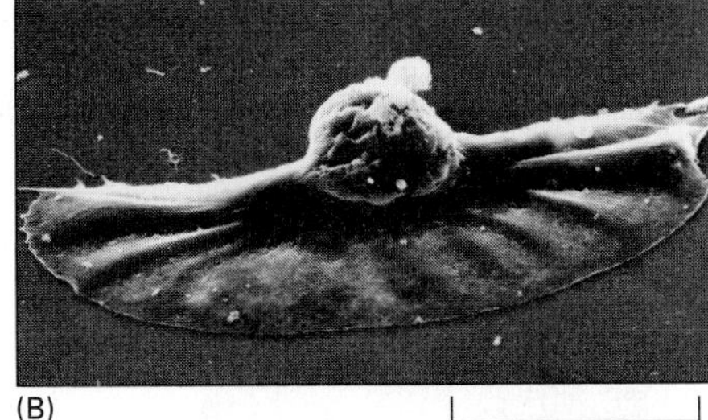
(B) 10 μm

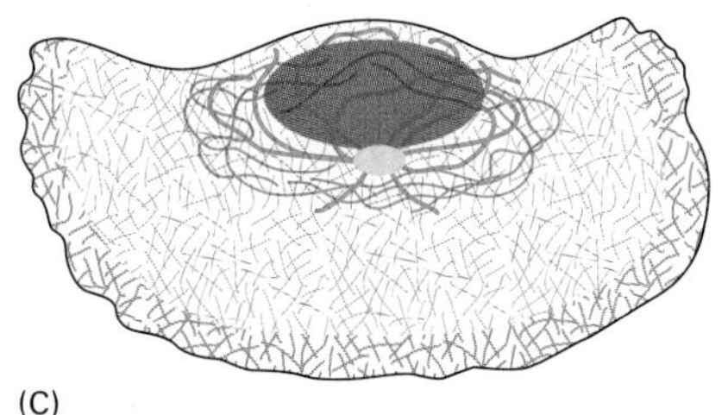
(C)

Figure 16–86 Migratory keratocytes from a fish epidermis. (A) Light micrographs of a keratocyte in culture, taken about 15 sec apart. This cell is moving at about 15 μm/sec. (B) Keratocyte seen by scanning electron microscopy, showing its broad, flat lamellipodium and small cell body, including the nucleus, carried up above the substratum at the rear. (C) Distribution of cytoskeletal filaments in this cell. Actin filaments *(red)* fill the large lamellipodium and are responsible for the cell's rapid movement. Microtubules *(green)* and intermediate filaments *(blue)* are restricted to the regions close to the nucleus. (A and B, courtesy of Juliet Lee.)

Plasma Membrane Protrusion Is Driven by Actin Polymerization

The first step in locomotion, protrusion of a leading edge, seems to rely primarily on forces generated by actin polymerization pushing the plasma membrane outward. Different cell types generate different types of protrusive structures, including filopodia (also known as microspikes), lamellipodia, and pseudopodia. All are filled with a dense core of filamentous actin, which excludes membrane-enclosed organelles. The three structures differ primarily in the way in which the actin is organized—in one, two, or three dimensions, respectively—and we have already discussed how this results from the presence of different actin-associated proteins.

Filopodia, formed by migrating growth cones and some types of fibroblasts, are essentially one-dimensional. They contain a core of long, bundled actin filaments, which are reminiscent of those in microvilli but longer and thinner, as well as more dynamic. **Lamellipodia**, formed by epithelial cells and fibroblasts, as well as by some neurons, are two-dimensional, sheet-like structures. They contain an orthogonally cross-linked mesh of actin filaments, most of which lie in a plane parallel to the solid substratum. **Pseudopodia**, formed by amoebae and neutrophils, are stubby three-dimensional projections filled with an actin-filament gel. Perhaps because their two-dimensional geometry is most convenient for examination with the light microscope, we know more about the dynamic organization and protrusion mechanism of lamellipodia than we do for either filopodia or pseudopodia.

Lamellipodia contain all of the machinery that is required for cell motility. They have been especially well studied in the epithelial cells of the epidermis of fish and frogs, which are known as "keratocytes" because of their abundant keratin filaments. These cells normally cover the animal by forming an epithelial sheet, and they are specialized to close wounds very rapidly, moving at rates up to 30 μm/min. When cultured as individual cells, keratocytes assume a distinctive shape with a very large lamellipodium and a small, trailing cell body that is not attached to the substratum (Figure 16–86). Fragments of this lamellipodium can be sliced off with a micropipette. Although the fragments generally lack microtubules and membrane-enclosed organelles, they continue to crawl normally, looking like tiny keratocytes (Figure 16–87).

The dynamic behavior of actin filaments can be studied in keratocyte lamellipodia by marking a small patch of actin and examining its fate. This reveals that, while the lamellipodia crawl forward, the actin filaments remain stationary with respect to the substrate. The actin filaments in the meshwork are mostly oriented with their plus ends facing forward. The minus ends are frequently attached to the sides of other actin filaments by ARP complexes (see Figure 16–28), helping to form the two-dimensional web (Figure 16–88). The web as a

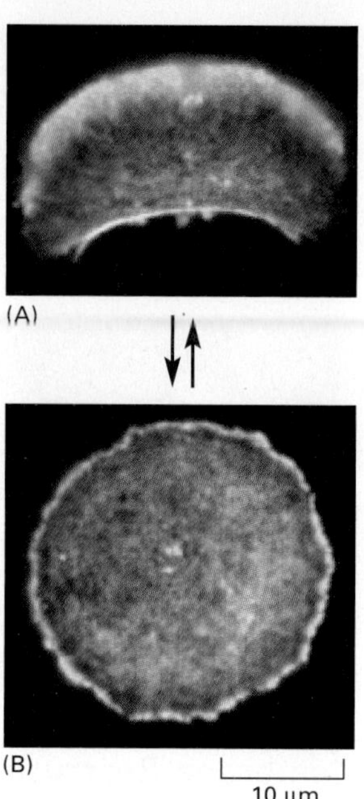

Figure 16–87 Behavior of lamellipodial fragments. Fragments of the large lamellipodium of a keratocyte can be separated from the main cell, either by surgery with a micropipette or by treating the cell with certain drugs. (A) Many of these fragments continue to move rapidly, with the same overall cytoskeletal organization as the intact keratocytes. Actin *(blue)* forms a protrusive meshwork at the front of the fragment. Myosin II *(pink)* is gathered into a band at the rear. This cell fragment was moving toward the top of the photo at the time that it was fixed and stained. (B) Some of the fragments stop moving and assume a symmetric, circular shape. Actin *(blue)* is distributed uniformly around the circumference, and myosin *(pink)* is found in spots throughout the cytoplasm. These circular fragments are trying to protrude uniformly in all directions and are therefore paralyzed. Only by coordinating protrusion at the front with contraction at the rear can steady movement be maintained. (From A. Verkovsky et al., *Curr. Biol.* 9:11–20, 1999. © Elsevier.)

whole seems to be undergoing treadmilling, assembling at the front and disassembling at the back, reminiscent of the treadmilling that occurs in individual actin filaments and microtubules discussed previously (see Figure 16–9).

Maintenance of unidirectional motion by lamellipodia is thought to require the cooperation and mechanical integration of several factors. Filament nucleation is localized at the leading edge, with new actin filament growth occurring primarily in that location to push the plasma membrane forward. Most filament depolymerization occurs at sites located well behind the leading edge. Because *cofilin* (see Figure 16–34) binds cooperatively and preferentially to actin filaments containing ADP-actin (the D form), the new T-form filaments generated at the leading edge should be resistant to depolymerization by cofilin (Figure 16–89). As the filaments age and ATP hydrolysis proceeds, cofilin can efficiently disassemble the older filaments. Thus, the delayed ATP hydrolysis by filamentous actin is thought to provide the basis for a mechanism that maintains an efficient, unidirectional treadmilling process in the lamellipodium (Figure 16–90). Finally, bipolar myosin II filaments seem to associate with the actin filaments in the web and pull them into a new orientation—from nearly perpendicular to the leading edge to an orientation almost parallel to the leading edge. This contraction prevents protrusion and it pinches in the sides of the locomoting lamellipodium, helping to gather in the sides of the cell as it moves forward.

Cell Adhesion and Traction Allow Cells to Pull Themselves Forward

Current evidence suggests that lamellipodia of all cells share this basic, simple type of dynamic organization, but the interactions between the cell and its physical environment usually make the situation considerably more complex than for fish keratocytes crawling on a culture dish. Particularly important in locomotion is the intimate crosstalk between the cytoskeleton and cell adhesion. Although some degree of adhesion to the substratum is necessary for any form of cell crawling, adhesion and locomotion rate seem generally to be inversely related, with highly adhesive cells moving more slowly than weakly adhesive

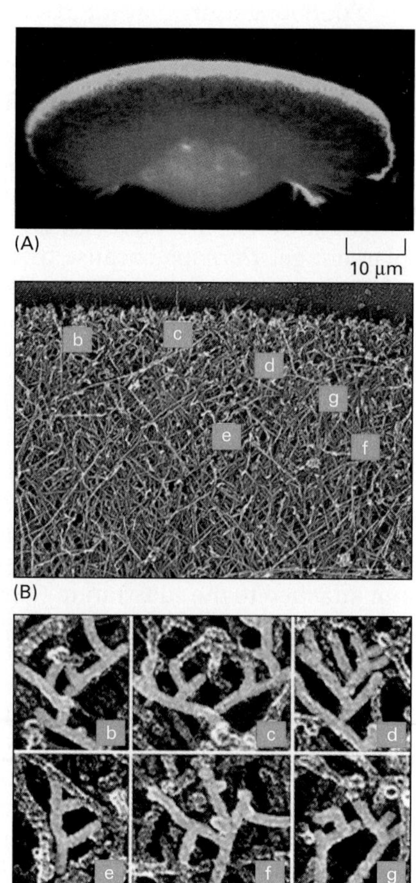

Figure 16–88 Actin filament nucleation and web formation by the ARP complex in lamellipodia. (A) A keratocyte with actin filaments labeled in *red* by fluorescent phalloidin, and the ARP complex labeled in *green* with an antibody raised against one of its component proteins. The regions where the two overlap appear *yellow*. The ARP complex is highly concentrated near the front of the lamellipodium, where actin nucleation is most active. (B) Electron micrograph of a platinum-shadowed replica of the leading edge of a keratocyte, showing the dense actin filament meshwork. The labels denote areas enlarged in C. (C) Closeup views of the marked regions of the actin web at the leading edge shown in B. Numerous branched filaments can be seen, with the characteristic 70° angle formed when the ARP complex nucleates a new actin filament off the side of a preexisting filament (see Figure 16–29). (From T. Svitkina and G. Borisy, *J. Cell Biol.* 145:1009–1026, 1999. © The Rockefeller University Press.)

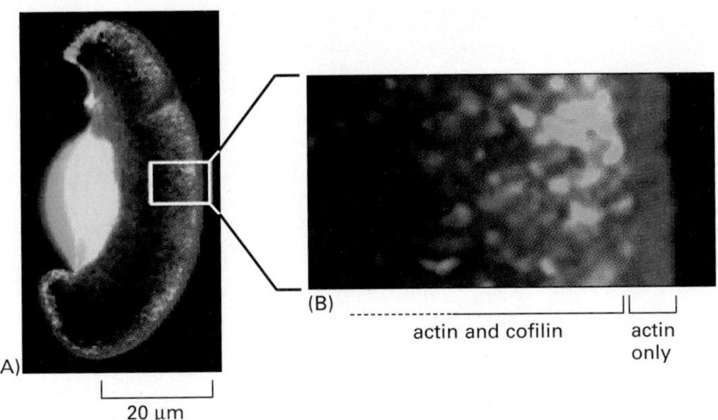

Figure 16–89 Cofilin in lamellipodia. (A) A keratocyte with actin filaments labeled in *red* by fluorescent phalloidin and cofilin labeled in *green* with a fluorescent antibody. The regions where the two overlap appear *yellow*. Although the dense actin meshwork reaches all the way through the lamellipodium, cofilin is not found at the very leading edge. (B) Closeup view of the region marked with the *white* rectangle in A. The actin filaments closest to the leading edge, which are also the ones that have formed most recently and that are most likely to contain ATP actin (rather than ADP actin) in the filament lattice are generally not associated with cofilin. (From T. Svitkina and G. Borisy, *J. Cell Biol.* 145:1009–1026, 1999. © The Rockefeller University Press.)

ones. Keratocytes are so weakly adhesive to the substratum that the force of actin polymerization can push the leading edge forward very rapidly. In contrast, when neurons from the sea slug *Aplysia* are cultured on a sticky substratum, they form large lamellipodia that become stuck too tightly to move forward. In these lamellipodia, the same cycle of localized nucleation of new actin filaments, depolymerization of old filaments, and myosin-dependent contraction continues to operate. But because the leading edge is prevented physically from moving forward, the entire actin mesh moves backward toward the cell body instead (Figure 16–91). The adhesion of most cells lies somewhere between these two extremes, and most lamellipodia exhibit some combination of forward actin filament protrusion (like keratocytes) and rearward actin flux (like the *Aplysia* neurons).

As a lamellipodium, filopodium, or pseudopodium extends forward over a substratum, it can form new attachment sites. These sites can be seen by *interference reflection microscopy,* as sites where the cell plasma membrane is extremely close to the substratum. In crawling cells, the individual attachment sites, which contain much of the same molecular machinery found in focal contacts (see Figure 16–49), generally form at the cell front and remain stationary as the cell moves forward over them, persisting until the rear of the cell catches up with them. When an individual lamellipodium fails to adhere to the substratum, it is usually lifted up onto the dorsal surface of the cell and rapidly carried backward as a "ruffle" (Figure 16–92).

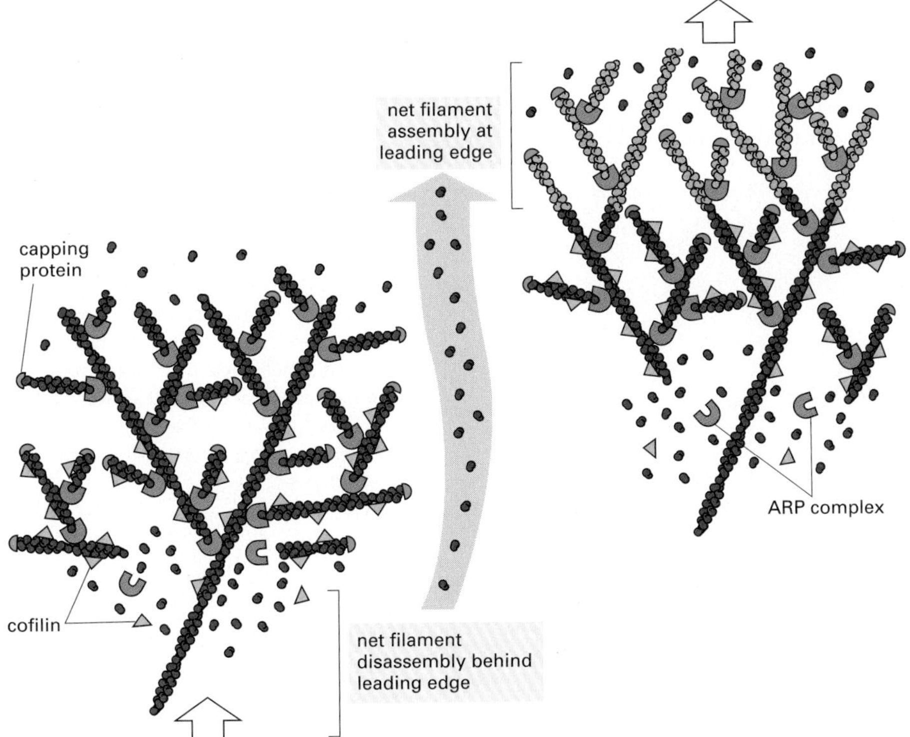

Figure 16–90 A model for protrusion of the actin meshwork at the leading edge. Two time points during advance of the lamellipodium are illustrated, with newly assembled structures at the later time point shown in a lighter color. Nucleation is mediated by the ARP complex at the front. Newly nucleated actin filaments are attached to the sides of preexisting filaments, primarily at a 70° angle. Filaments elongate, pushing the plasma membrane forward because of some sort of anchorage of the array behind. At a steady rate, actin filament plus ends become capped. After newly polymerized actin subunits hydrolyze their bound ATP in the filament lattice, the filaments become susceptible to depolymerization by cofilin. This cycle causes a spatial separation between net filament assembly at the front and net filament disassembly at the rear, so that the actin filament network as a whole can move forward, even though the individual filaments within it remain stationary with respect to the substratum.

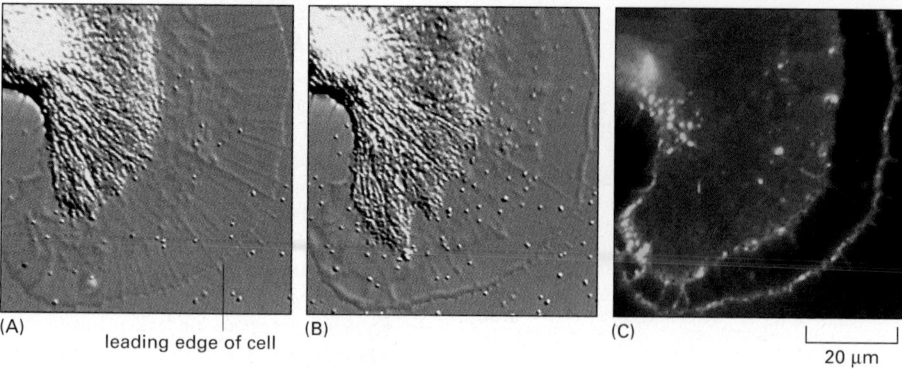

(D)

cytochalasin B

Figure 16–91 Rearward movement of the actin network in a growth-cone lamellipodium. (A) A growth cone from a neuron of the sea slug *Aplysia* is cultured on a highly adhesive substratum and viewed by differential-interference-contrast microscopy. Microtubules and membrane-enclosed organelles are confined to the bright, rear area of the growth cone (to the *left*), while a meshwork of actin filaments fills the lamellipodium (on the *right*). (B) After brief treatment with the drug cytochalasin, which caps the plus ends of actin filaments (see Table 16–2), the actin meshwork has detached from the front edge of the lamellipodium and has been pulled backward. (C) At the time point shown in B, the cell was fixed and labeled with fluorescent phalloidin to show the distribution of the actin filaments. Some actin filaments persist at the leading edge, but the region behind the leading edge is devoid of filaments. Note the sharp boundary of the rearward-moving actin meshwork. (D) The complex cyclic structure of cytochalasin B. (A, B, and C courtesy of Paul Forscher.)

The attachment sites established at the leading edge serve as anchorage points, which allow the cell to generate traction on the substratum and pull its body forward. Traction forces seem to be generated by myosin motor proteins, especially myosin II. In many locomoting cells, myosin II is highly concentrated at the posterior of the cell (Figure 16–93). *Dictyostelium* amoebae that are deficient in myosin II are able to protrude pseudopodia at normal speeds, but the translocation of their cell body is much slower than that of wild-type amoebae, indicating the importance of myosin II contraction in this part of the cell locomotion cycle. It is not clear how the activity of myosin II pulls the cell body forward. One might imagine that contractile bundles of actin and myosin linking the back of the cell to the front would be responsible; but the observed distribution of myosin II, in some cells at least, suggests a somewhat different strategy, which is depicted in Figure 16–85. Contraction of the actin-rich cortex at the rear of the cell may also selectively weaken the older adhesive interactions that tend

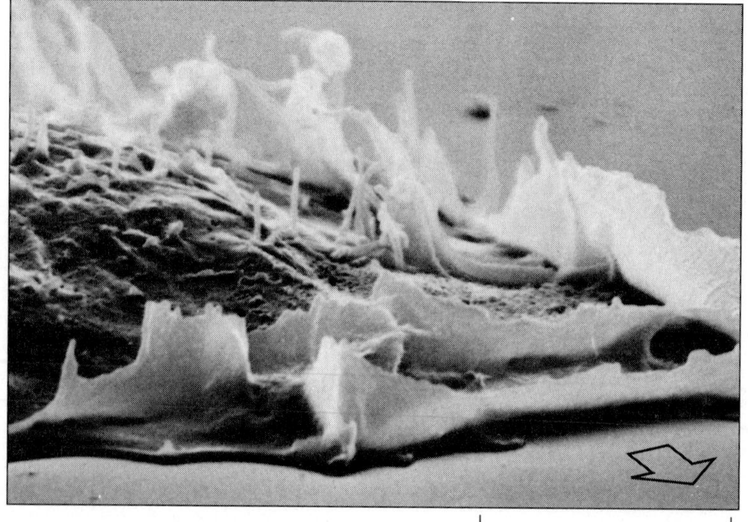

Figure 16–92 Lamellipodia and ruffles at the leading edge of a human fibroblast migrating in culture. The *arrow* in this scanning electron micrograph shows the direction of cell movement. As the cell moves forward, lamellipodia that fail to attach to the substratum are swept backward over the dorsal surface of the cell, a movement known as ruffling. (Courtesy of Julian Heath.)

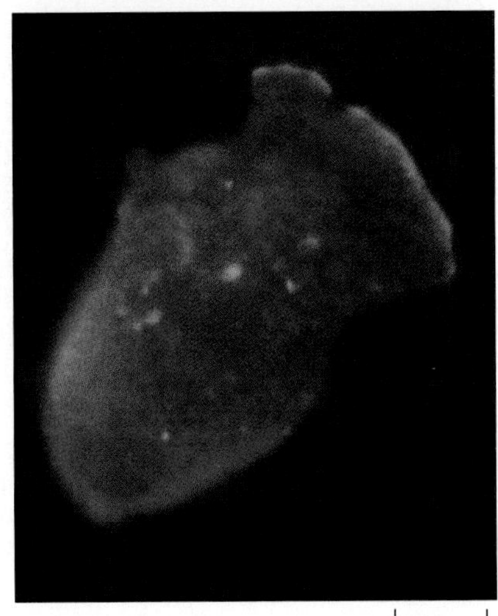

Figure 16–93 The localization of myosin I and myosin II in a normal crawling *Dictyostelium* amoeba. This cell was crawling toward the upper right at the time that it was fixed and labeled with antibodies specific for two myosin isoforms. Myosin I *(green)* is mainly restricted to the leading edge of pseudopodia at the front of the cell. Myosin II *(red)* is highest in the posterior, actin-rich cortex. Contraction of the cortex at the posterior of the cell by myosin II may help to push the cell body forward. (Courtesy of Yoshio Fukui.)

to hold the cell back. In addition, myosin II may transport cell body components forward over a polarized array of actin filaments.

The traction forces generated by locomoting cells exert a significant pull on the substratum (Figure 16–94). In a living animal, most crawling cells move across a semiflexible substratum made of extracellular matrix, which can be deformed and rearranged by these cell forces. In culture, movement of fibroblasts through a gel of collagen fibrils aligns the collagen, generating an organized extracellular matrix that in turn affects the shape and direction of locomotion of the fibroblasts within it (Figure 16–95). This two-way mechanical interaction between cells and their physical environment is thought to be a major way that vertebrate tissues organize themselves.

External Signals Can Dictate the Direction of Cell Migration

Cell locomotion requires an initial polarization of the cell to set it off in a particular direction. Carefully controlled cell polarization processes are also required for oriented cell divisions in tissues and for formation of a coherent, organized multicellular structure. The molecular mechanisms that generate cell polarity are only beginning to be understood for vertebrates. Most of what we know so far about the topic has been learned through genetic studies in yeasts, flies, and worms. In all known cases, however, the cytoskeleton has a central role, and many of the molecular components have been evolutionarily conserved.

Polarized cell movement is of little use unless it is responsive to cues in the environment, and these can be both chemical and physical. *Chemotaxis* is defined as cell movement in a direction controlled by a gradient of a diffusible chemical. One well-studied example is the chemotactic movement of a class of white blood cells, called *neutrophils,* toward a source of bacterial infection. Receptor proteins on the surface of neutrophils enable them to detect the very low concentrations of the *N*-formylated peptides that are derived from bacterial proteins (only procaryotes begin protein synthesis with *N*-formylmethionine). Using these receptors, neutrophils are guided to bacterial targets by their ability to detect a difference of only 1% in the concentration of these diffusible peptides

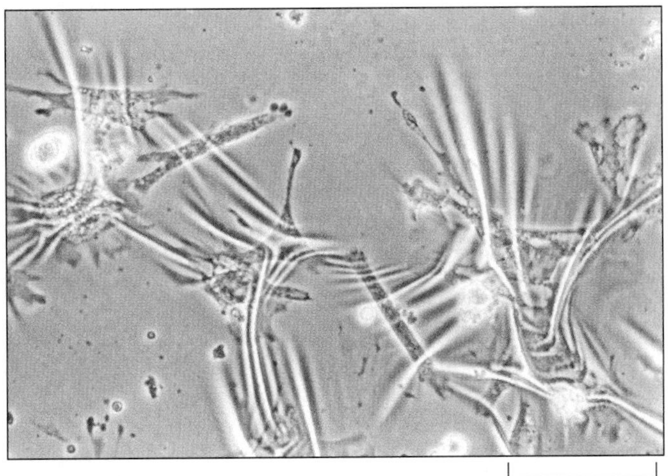

Figure 16–94 Adhesive cells exert traction forces on the substratum. These fibroblasts have been cultured on a very thin sheet of silicon rubber. Attachment of the cells, followed by contraction of their cytoskeleton, has caused the rubber substratum to wrinkle. (From A.K. Harris, P. Wild, and D. Stopak, *Science* 208:177–179, 1980. © AAAS.)

on one side of the cell versus the other (Figure 16–96). Both in this case and in the similar chemotaxis of *Dictyostelium* amoebae toward a source of cyclic AMP, a local polymerization of actin near the receptors is stimulated when the receptors bind their ligands. This actin polymerization response depends on the monomeric Rho-family GTPases discussed earlier. As in the shmooing yeast (see Figure 16–83), the responding cell extends a protrusion toward the signal. The preferential localization of protrusive activity toward one side of the cell indirectly causes reorientation of the traction-generating machinery, and the cell body then follows its "nose" and moves toward the attractive signal.

The direction of cell migration can also be influenced by nondiffusible chemical cues attached to the extracellular matrix or to the surface of cells. Receptor activation by these signals can cause increased cell adhesion, in addition to directed actin polymerization. Most long-distance cell migrations in animals, including neural-crest-cell migration and the travels of neuronal growth cones, depend on a combination of diffusible and non-diffusible signals to steer the locomoting cells or growth cones to their proper destinations (see Figure 21–98).

To help organize persistent movement in a particular direction cells use their microtubules. In many locomoting cells, the position of the centrosome is influenced by the location of protrusive actin polymerization, being found on the forward side of the nucleus. The mechanism of centrosome reorientation is not clear. It is thought that the activation of receptors on one edge of a cell might not only stimulate actin polymerization there (and therefore local protrusion) but also locally activate dynein-like motor proteins that move the centrosome by pulling on its microtubules. The centrosome nucleates a large number of dynamic microtubules, and its repositioning means that many of these have their plus ends extending from the centrosome into the protrusive region of the cell. The dynamic microtubule plus ends may indirectly modulate local adhesion and also activate the Rac GTPase to further increase actin polymerization in the protrusive region. The increased concentration of microtubules would thereby encourage further protrusion, creating a positive feedback loop that enables protrusive motility to persist in the same direction for a prolonged period. Regardless of the exact mechanism, the orientation of the centrosome seems to reinforce the polarity information that the actin cytoskeleton receives from the outside world, allowing a sensitive response to weak signals.

A similar cooperative feedback loop seems to operate in other instances of cell polarization. A particularly interesting example is the killing of specific target cells by cytotoxic T lymphocytes. These cells kill other cells that carry foreign antigens on their surface and are a critical component of the vertebrate's adaptive immune response to infection. When the receptors on the surface of the T cell recognize antigen on the surface of the target cell, Rho-family GTPases are activated and cause actin polymerization under the zone of contact between the two cells, creating a specialized region of the cortex. This specialized site causes the centrosome to reorient, moving with its microtubules to the zone of T-cell-target contact (Figure 16–97). The microtubules, in turn, position the Golgi apparatus right under the contact zone, focusing the killing machinery onto the target cell. The mechanism of killing is discussed in Chapter 24.

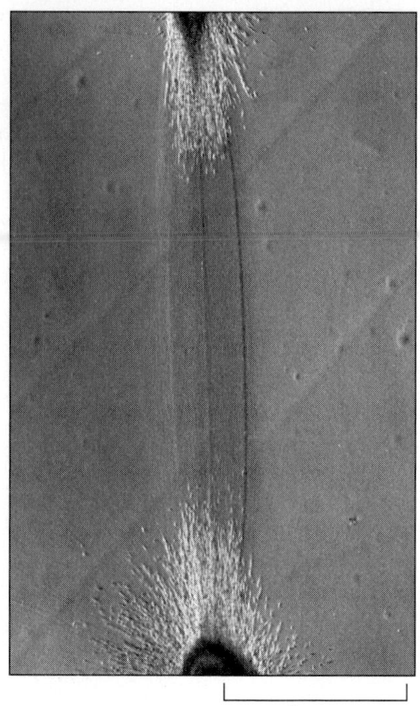

Figure 16–95 Shaping of the extracellular matrix by cell pulling. This micrograph shows a region between two pieces of embryonic chick heart (tissue explants rich in fibroblasts and heart muscle cells) that were grown in culture on a collagen gel for 4 days. A dense tract of aligned collagen fibers has formed between the two explants, apparently as a result of fibroblasts tugging on the collagen. (From D. Stopak and A.K. Harris, *Dev. Biol.* 90:383–398, 1982. © Academic Press.)

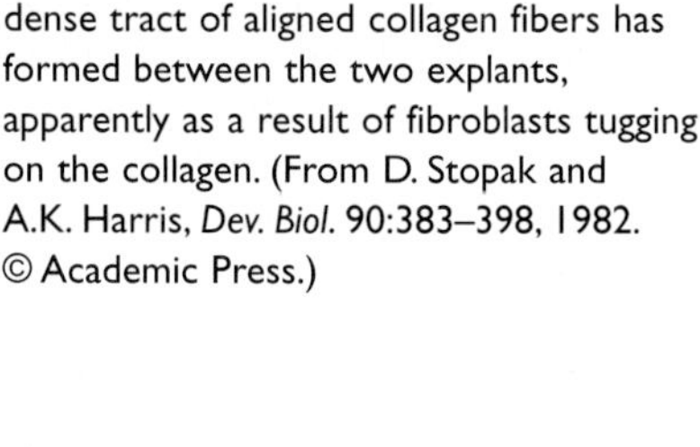

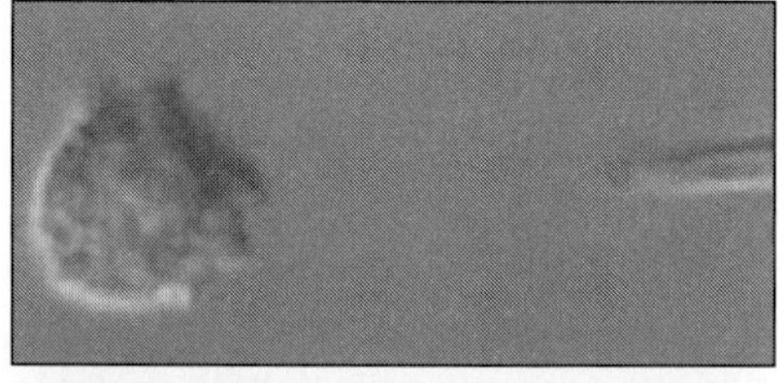

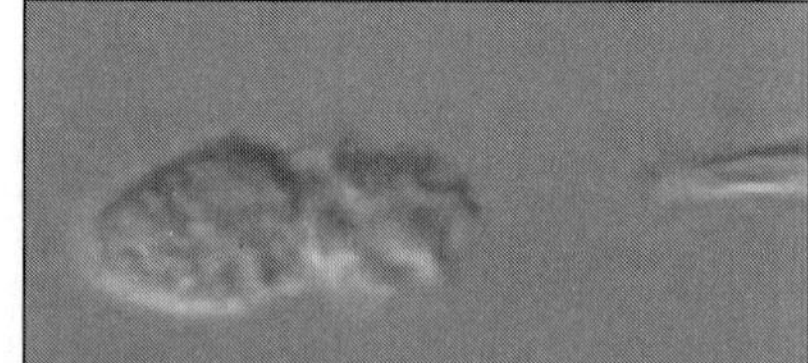

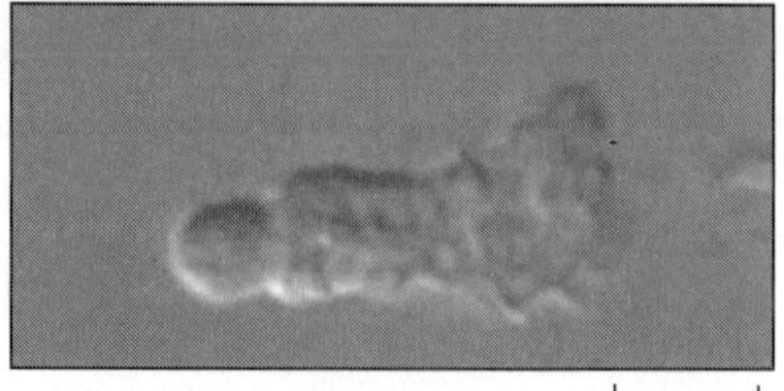

Figure 16–96 Neutrophil polarization and chemotaxis. The pipette tip at the right is leaking a small amount of the peptide formyl-Met-Leu-Phe. Only bacterial proteins have formylated methionine residues, so the human neutrophil recognizes this peptide as the product of a foreign invader (discussed in Chapter 25). The neutrophil quickly extends a new lamellipodium toward the source of the chemoattractant peptide *(top)*. It then extends this lamellipodium and polarizes its cytoskeleton so that contractile myosin II is located primarily at the rear, opposite the position of the lamellipodium *(middle)*. Finally, the cell crawls toward the source of this peptide *(bottom)*. If a real bacterium were the source of the peptide, rather than an investigator's pipette, the neutrophil would engulf the bacterium and destroy it. (From O.D. Weiner et al., *Nature Cell Biol.* 1:75–81, 1999. © Macmillan Magazines Ltd.)

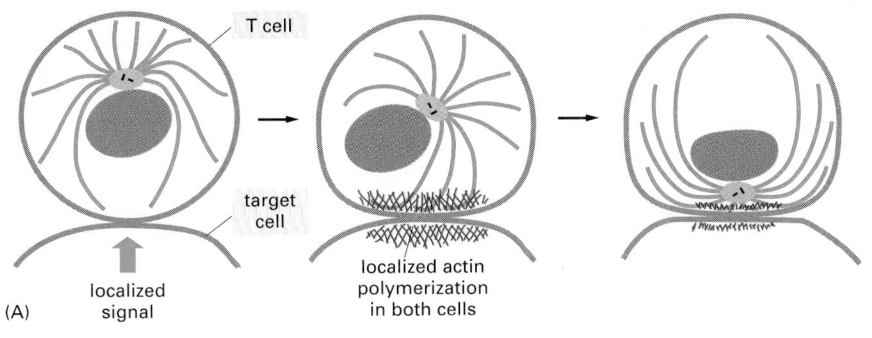

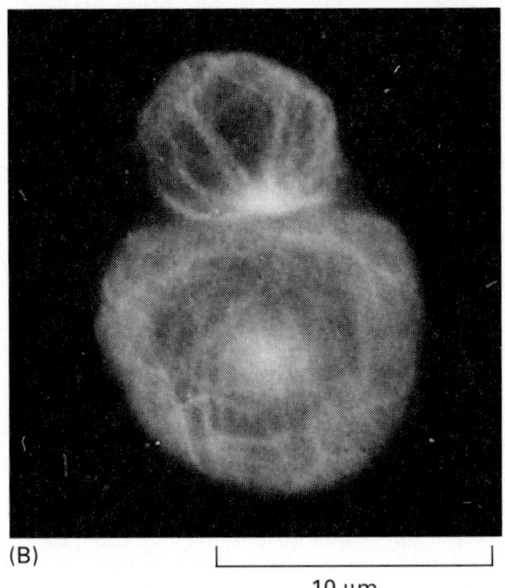

Figure 16–97 The polarization of a cytotoxic T cell after target-cell recognition. (A) Changes in the cytoskeleton of a cytotoxic T cell after it has made contact with a target cell. The initial recognition event results in signals that cause actin polymerization in both cells at the site of contact. In the T cell, interactions between the actin-rich contact zone and microtubules emanating from the centrosome result in reorientation of the centrosome, so that the associated Golgi apparatus is directly apposed to the target cell. (B) Immunofluorescence micrograph in which both the T cell *(top)* and its target cell *(bottom)* have been stained with an antibody against microtubules. The centrosome and the microtubules radiating from it in the T cell are oriented toward the point of cell–cell contact. In contrast, the microtubule array in the target cell is not polarized. (B, from B. Geiger, D. Rosen and G. Berke, *J. Cell Biol.* 95:137–143, 1982. © The Rockefeller University Press.)

The Complex Morphological Specialization of Neurons Depends on The Cytoskeleton

Many of the most striking differences between eucaryotic and prokaryotic cells are consequences of the eucaryotic cytoskeleton. The regulated, directed transport of intracellular components by motor proteins enables the eucaryotic cell to enlarge beyond the limits set by the speed of diffusion, which restrict procaryotic cell growth. At the same time, the cytoskeleton provides the framework for the elaborate architectures so characteristic of eucaryotic cells (Figure 1–27; Panel 16–1). No cell exemplifies these two features of size and architectural complexity better than a neuron.

Neurons begin life in the embryo as unremarkable cells, which use actin-based motility to migrate to specific locations. Once there, however, they send out a series of long specialized processes that will either receive electrical signals *(dendrites)* or transmit these electrical signals to their target cells *(axons).* Both types of processes (collectively called *neurites)* are filled with bundles of microtubules, which are critical to both their structure and their function.

In axons, all the microtubules are oriented in the same direction, with their minus end pointing back toward the cell body and their plus end pointing forward toward the axon terminals (Figure 16–98). The microtubules do not reach from the cell body all the way to the axon terminals; each is typically only a few micrometers in length, but large numbers are staggered in an overlapping array. This set of perfectly aligned microtubule tracks acts as a highway to transport many specific proteins and protein-containing vesicles. These are needed at the axon terminals, where synapses must be constructed and maintained, but they are made only in the cell body and dendrites where the cell keeps all its ribosomes. The longest axon in the human body reaches from the base of the spinal cord to the tip of the big toe, being up to a meter in length. Mitochondria, large numbers of specific proteins in transport vesicles, and synaptic vesicle precursors make the long journey in the forward (anterograde) direction. They are carried there by plus-end-directed kinesin-family motor proteins that can move them a meter in about a day, which is a great improvement over diffusion, which would take approximately 8 years to move a mitochondrion this distance. Many members of the kinesin superfamily contribute to this *anterograde axonal transport,* most carrying specific subsets of membrane-enclosed organelles along the microtubules. The great diversity of the kinesin family motor proteins used in axonal transport suggests that they are involved in specific targeting, as well as in movement. At the same time, old components from the axon terminals are carried back to the cell body for degradation and recycling; this *retrograde axonal transport* occurs along the same set of oriented microtubules, and it relies on cytoplasmic dynein, which is a minus-end-directed motor protein.

Axonal structure depends on these microtubules, as well as on the contributions of the other two major cytoskeletal systems—actin filaments and intermediate filaments. Actin filaments line the cortex of the axon, just beneath the plasma membrane, and actin-based motor proteins such as myosin V are also

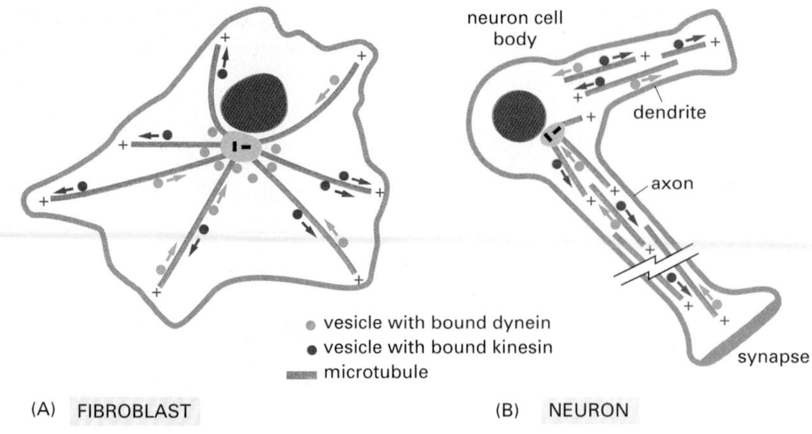

Figure 16–98 Microtubule organization in fibroblasts and neurons. (A) In a fibroblast, microtubules emanate outward from the centrosome in the middle of the cell. Vesicles with plus-end-directed kinesin attached move outward, and vesicles with minus-end-directed dynein attached move inward. (B) In a neuron, microtubule organization is more complex. In the axon, all microtubules share the same polarity, with the plus ends pointing outward toward the axon terminus. No one microtubule stretches the entire length of the axon; instead, short overlapping segments of parallel microtubules make the tracks for fast axonal transport. In dendrites, the microtubules are of mixed polarity, with some plus ends pointing outward and some pointing inward.

abundant in the axon, perhaps to help move materials, although their exact function is still unclear. Neurofilaments, the specialized intermediate filaments of nerve cells, provide the most important structural support in the axon. A disruption in neurofilament structure, or in the cross-linking proteins that attach the neurofilaments to the microtubules and actin filaments distributed along the axon, can result in axonal disorganization and eventually axonal degeneration.

The construction of the elaborate branching architecture of the neuron during development depends on actin-based motility. As mentioned earlier, the tips of growing axons and dendrites extend by means of a *growth cone,* a specialized motile structure rich in actin (Figure 16–99). Most neuronal growth cones produce filopodia, and some make lamellipodia as well. The protrusion and stabilization of growth-cone filopodia are exquisitely sensitive to environmental cues. Some cells secrete soluble proteins such as netrin to attract or repel growth cones. In addition, there are fixed guidance markers along the way, attached to the extracellular matrix or to the surfaces of cells. When a filopodium encounters such a "guidepost" in its exploration, it quickly forms adhesive contacts. It is thought that a myosin-dependent collapse of the actin meshwork in the unstabilized part of the growth cone then causes the developing axon to turn toward the guidepost. Other extracellular directional signals act as repellents and cause growth cone collapse at the site of signal action. A complex combination of positive and negative signals, both soluble and insoluble, accurately guide the growth cone to its final destination (discussed in Chapter 21).

Microtubules reinforce the directional decisions made by the actin-rich protrusive structures at the leading edge of the growth cone. Microtubules from the

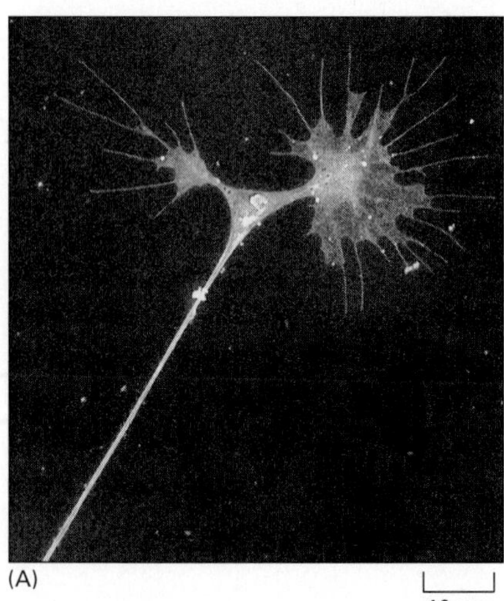

Figure 16–99 Neuronal growth cones. (A) Scanning electron micrograph of two growth cones at the end of a neurite, put out by a chick sympathetic neuron in culture. Here, a previously single growth cone has recently split into two. Note the many filopodia and the large lamellipodia. The taut appearance of the neurite is due to tension generated by the forward movement of the growth cones, which are often the only firm points of attachment of the axon to the substratum. (B) Scanning electron micrograph of the growth cone of a sensory neuron crawling over the inner surface of the epidermis of a *Xenopus* tadpole. (A, from D. Bray, in Cell Behaviour [R. Bellairs, A. Curtis, and G. Dunn, eds.]. Cambridge, UK: Cambridge University Press, 1982; B, from A. Roberts, *Brain Res.* 118:526–530, 1976. © Elsevier.)

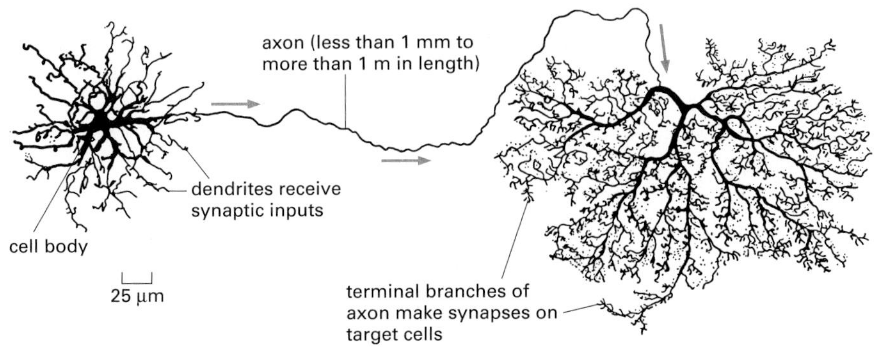

Figure 16–100 The complex architecture of a vertebrate neuron. The neuron shown is from the retina of a monkey. The arrows indicate the direction of travel of the electrical signal along the axon. The longest and largest neurons in the human body extend for a distance of about 1 m (1 million μm), from the base of the spinal cord to the tip of the big toe, and have an axon diameter of 15 μm. (Adapted from B.B. Boycott, in Essays on the Nervous System [R. Bellairs and E.G. Gray, eds.]. Oxford, UK: Clarendon Press, 1974.)

axonal parallel array just behind the growth cone are constantly growing into the growth cone and shrinking back by dynamic instability. Adhesive guidance signals are somehow relayed to the dynamic microtubule ends, so that microtubules growing in the correct direction are stabilized against disassembly. In this way, a microtubule-rich axon is left behind, marking the path that the growth cone has traveled.

Dendrites are generally much shorter projections than axons, and they function to receive signals rather than to send them. The microtubules in dendrites all lie parallel to one another but their polarities are mixed, with some pointing their plus ends toward the dendrite tip, while others point back toward the cell body. Nevertheless, dendrites also form as the result of growth-cone activity (see Figure 21–97). Travelling along their complex paths, the growth cones at the tips of both axons and dendrites create the intricate and highly individual morphology of each mature neuronal cell (Figure 16–100). In this way, the cytoskeleton provides the engine for construction of the entire nervous system, as well as the supporting structures that strengthen, stabilize, and maintain its parts.

Summary

Whole-cell movements and the large-scale shaping and structuring of cells require the coordinated activities of all three basic filament systems along with a large variety of cytoskeletal accessory proteins, including motor proteins. Cell crawling—a widespread behavior important in embryonic development and also in wound healing, tissue maintenance, and immune system function in the adult animal—is a prime example of such complex coordinated cytoskeletal action. For a cell to crawl, it must generate and maintain an overall structural polarity, which is influenced by external cues. Interactions between the microtubule and actin cytoskeletons reinforce this polarity. In addition, the cell must coordinate protrusion at the leading edge (by assembly of new actin filaments), adhesion of the newly protruded part of the cell to the substratum, traction via molecular motors to bring the cell body forward, and disassembly of old cell–substratum contacts.

Nowhere is the elaborate potential of eucaryotic cell morphology better illustrated than in the neuron. Neurons may be many millimeters long, with exceedingly complex dendrites and axons. The elaboration and maintenance of these beautiful and astonishingly complex cells requires the coordinated assembly of microtubules, neurofilaments, and actin filaments, as well as the actions of dozens of highly specialized molecular motors to transport subcellular components to their required destinations.

At the other end of the complexity spectrum, the budding yeast Saccharomyces cerevisiae *uses its cytoskeletal plasticity to change shape during vegetative growth and during mating. Through genetic analyses, many of the molecular components that are required for these processes were first identified in yeast. Quite surprisingly, it has turned out that most of the same proteins also mediate the polarity and movements of animal cells, including those of humans.*

References

General

Bray D (2001) Cell Movements: From Molecules to Motility, 2nd edn. New York: Garland Publishing.

Howard J (2001) Mechanics of Motor Proteins and the Cytoskeleton. Sunderland, MA: Sinauer.

The Self-Assembly and Dynamic Structure of Cytoskeletal Filaments

Cooper JA (1987) Effects of cytochalasin and phalloidin on actin. *J. Cell Biol.* 105, 1473–1478.

Desai A & Mitchison TJ (1997) Microtubule polymerization dynamics. *Annu. Rev. Cell Dev. Biol.* 13, 83–117.

Downing KH (2000) Structural basis for the interaction of tubulin with proteins and drugs that affect microtubule dynamics. *Annu. Rev. Cell Dev. Biol.* 16, 89–111.

Fuchs E (1996) The cytoskeleton and disease: genetic disorders of intermediate filaments. *Annu. Rev. Genet.* 30, 197–231.

Gilson PR & Beech PL (2001) Cell division protein FtsZ: running rings around bacteria, chloroplasts and mitochondria. *Res. Microbiol.* 152, 3–10.

Herrmann H & Aebi U (1998) Intermediate filament assembly: fibrillogenesis is driven by decisive dimer–dimer interactions. *Curr. Opin. Struct. Biol.* 8, 177–185.

Hill TL & Kirschner MW (1982) Subunit treadmilling of microtubules or actin in the presence of cellular barriers: possible conversion of chemical free energy into mechanical work. *Proc. Natl. Acad. Sci. USA* 79, 490–494.

Holmes KC, Popp D, Gebhard W & Kabsch W (1990) Atomic model of the actin filament. *Nature* 347, 44–49.

Lee MK & Cleveland DW (1996) Neuronal intermediate filaments. *Annu. Rev. Neurosci.* 19, 187–217.

Mitchison TJ (1995) Evolution of a dynamic cytoskeleton. *Philos. Trans. R. Soc. Lond. B. Biol. Sci.* 349, 299–304.

Nogales E, Wolf SG & Downing KH (1998) Structure of the αβ-tubulin dimer by electron crystallography. *Nature* 391, 199–203.

Oosawa F & Asakura S (1975) Thermodynamics of the Polymerization of Protein, pp 41–55, 90–108. New York: Academic Press.

van der Bliek AM (1999) Functional diversity in the dynamin family. *Trends Cell Biol.* 9, 96–102.

How Cells Regulate Their Cytoskeletal Filaments

Bretscher A, Chambers D, Nguyen R & Reczek D (2000) ERM-Merlin and EBP50 protein families in plasma membrane organization and function. *Annu. Rev. Cell Dev. Biol.* 16, 113–143.

Burridge K & Chrzanowska-Wodnicka M (1996) Focal adhesions, contractility, and signaling. *Annu. Rev. Cell Dev. Biol.* 12, 463–518.

Furukawa R & Fechheimer M (1997) The structure, function, and assembly of actin filament bundles. *Int. Rev. Cytol.* 175, 29–90.

Garcia ML & Cleveland DW (2001) Going new places using an old MAP: tau, microtubules and human neurodegenerative disease. *Curr. Opin. Cell Biol.* 13, 41–48.

Hall A (1998) Rho GTPases and the actin cytoskeleton. *Science* 279, 509–514.

Janmey PA (1998) The cytoskeleton and cell signaling: component localization and mechanical coupling. *Physiol. Rev.* 78, 763–781.

Machesky LM & Gould KL (1999) The Arp2/3 complex: a multifunctional actin organizer. *Curr. Opin. Cell Biol.* 11, 117–121.

Mullins RD (2000) How WASP-family proteins and the Arp2/3 complex convert intracellular signals into cytoskeletal structures. *Curr. Opin. Cell Biol.* 12, 91–96.

Quarmby L (2000) Cellular Samurai: katanin and the severing of microtubules. *J. Cell Sci.* 113, 2821–2827.

Sun HQ, Yamamoto M, Mejillano M & Yin HL (1999) Gelsolin, a multifunctional actin regulatory protein. *J. Biol. Chem.* 274, 33179–33182.

Troyanovsky SM (1999) Mechanism of cell–cell adhesion complex assembly. *Curr. Opin. Cell Biol.* 11, 561–566.

Weber A (1999) Actin binding proteins that change extent and rate of actin monomer-polymer distribution by different mechanisms. *Mol. Cell. Biochem.* 190, 67–74.

Zheng Y, Wong ML, Alberts B & Mitchison T (1995) Nucleation of microtubule assembly by a γ-tubulin-containing ring complex. *Nature* 378, 578–583.

Zimmerman W, Sparks CA & Doxsey SJ (1999) Amorphous no longer: the centrosome comes into focus. *Curr. Opin. Cell Biol.* 11, 122–128.

Molecular Motors

Amos LA & Cross RA (1997) Structure and dynamics of molecular motors. *Curr. Opin. Struct. Biol.* 7, 239–246.

Berg JS, Powell BC & Cheney RE (2001) A millennial myosin census. *Mol. Biol. Cell* 12, 780–794.

Cooke R (1986) The mechanism of muscle contraction. *CRC Crit. Rev. Biochem.* 21, 53–118.

Hackney DD (1996) The kinetic cycles of myosin, kinesin, and dynein. *Annu. Rev. Physiol.* 58, 731–750.

Hirokawa N, Noda Y & Okada Y (1998) Kinesin and dynein superfamily proteins in organelle transport and cell division. *Curr. Opin. Cell Biol.* 10, 60–73.

Holmes KC (1997) The swinging lever-arm hypothesis of muscle contraction. *Curr. Biol.* 7, R112–R118.

Howard J (1997) Molecular motors: structural adaptations to cellular functions. *Nature* 389, 561–567.

Kamal A & Goldstein LS (2000) Connecting vesicle transport to the cytoskeleton. *Curr. Opin. Cell Biol.* 12, 503–508.

Porter ME (1996) Axonemal dyneins: assembly, organization, and regulation. *Curr. Opin. Cell Biol.* 8, 10–17.

Reilein AR, Rogers SL, Tuma MC & Gelfand VI (2001) Regulation of molecular motor proteins. *Int. Rev. Cytol.* 204, 179–238.

Squire JM & Morris EP (1998) A new look at thin filament regulation in vertebrate skeletal muscle. *FASEB J.* 12, 761–771.

Vale RD & Milligan RA (2000) The way things move: looking under the hood of molecular motor proteins. *Science* 288, 88–95.

Vikstrom KL & Leinwand LA (1996) Contractile protein mutations and heart disease. *Curr. Opin. Cell Biol.* 8, 97–105.

The Cytoskeleton and Cell Behavior

Abercrombie M (1980) The crawling movement of metazoan cells. *Proc. Roy. Soc. B* 207, 129–147.

Baas PW (1998) The role of motor proteins in establishing the microtubule arrays of axons and dendrites. *J. Chem. Neuroanat.* 14, 175–180.

Condeelis J (1993) Life at the leading edge: the formation of cell protrusions. *Annu. Rev. Cell Biol.* 9, 411–444.

Devreotes PN & Zigmond SH (1988) Chemotaxis in eukaryotic cells: a focus on leukocytes and Dictyostelium. *Annu. Rev. Cell Biol.* 4, 649–686.

Drubin DG & Nelson WJ (1996) Origins of cell polarity. *Cell* 84, 335–344.

Harris AK, Wild P & Stopak D (1980) Silicone rubber substrata: a new wrinkle in the study of cell locomotion. *Science* 208, 177–179.

Heath JP & Holifield BF (1991) Cell locomotion: new research tests old ideas on membrane and cytoskeletal flow. *Cell Motil. Cytoskeleton* 18, 245–257.

Madden K & Snyder M (1998) Cell polarity and morphogenesis in budding yeast. *Annu. Rev. Microbiol.* 52, 687–744.

Nasmyth K & Jansen RP (1997) The cytoskeleton in mRNA localization and cell differentiation. *Curr. Opin. Cell Biol.* 9, 396–400.

Pantaloni D, Le Clainche C & Carlier MF (2001) Mechanism of actin-based motility. *Science* 292, 1502–1506.

Parent CA & Devreotes PN (1999) A cell's sense of direction. *Science* 284, 765–770.

Suter DM & Forscher P (2000) Substrate-cytoskeletal coupling as a mechanism for the regulation of growth cone motility and guidance. *J. Neurobiol.* 44, 97–113.

Svitkina TM & Borisy GG (1999) Arp2/3 complex and actin depolymerizing factor/cofilin in dendritic organization and treadmilling of actin filament array in lamellipodia. *J. Cell Biol.* 145, 1009–1026.

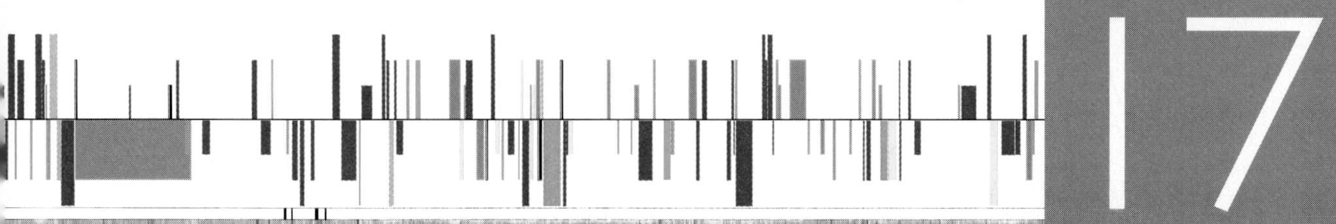

17

THE CELL CYCLE AND PROGRAMMED CELL DEATH

"Where a cell arises, there must be a previous cell, just as animals can only arise from animals and plants from plants." This *cell doctrine*, proposed by the German pathologist Rudolf Virchow in 1858, carried with it a profound message for the continuity of life. Cells are generated from cells, and the only way to make more cells is by division of those that already exist. All living organisms, from the unicellular bacterium to the multicellular mammal, are products of repeated rounds of cell growth and division extending back in time to the beginnings of life on Earth over three billion years ago.

A cell reproduces by performing an orderly sequence of events in which it duplicates its contents and then divides in two. This cycle of duplication and division, known as the **cell cycle**, is the essential mechanism by which all living things reproduce. In unicellular species, such as bacteria and yeasts, each cell division produces a complete new organism. In multicellular species, long and complex sequences of cell divisions are required to produce a functioning organism. Even in the adult body, cell division is usually needed to replace cells that die. In fact, each of us must manufacture many millions of cells every second simply to survive: if all cell division were stopped—by exposure to a very large dose of x-rays, for example—we would die within a few days.

The details of the cell cycle vary from organism to organism and at different times in an organism's life. Certain characteristics, however, are universal. The minimum set of processes that a cell has to perform are those that allow it to accomplish its most fundamental task: the passing on of its genetic information to the next generation of cells. To produce two genetically identical daughter cells, the DNA in each chromosome must first be faithfully replicated to produce two complete copies, and the replicated chromosomes must then be accurately distributed *(segregated)* to the two daughter cells, so that each receives a copy of the entire genome (Figure 17–1).

Eucaryotic cells have evolved a complex network of regulatory proteins, known as the **cell-cycle control system**, that governs progression through the cell cycle. The core of this system is an ordered series of biochemical switches that control the main events of the cycle, including DNA replication and the

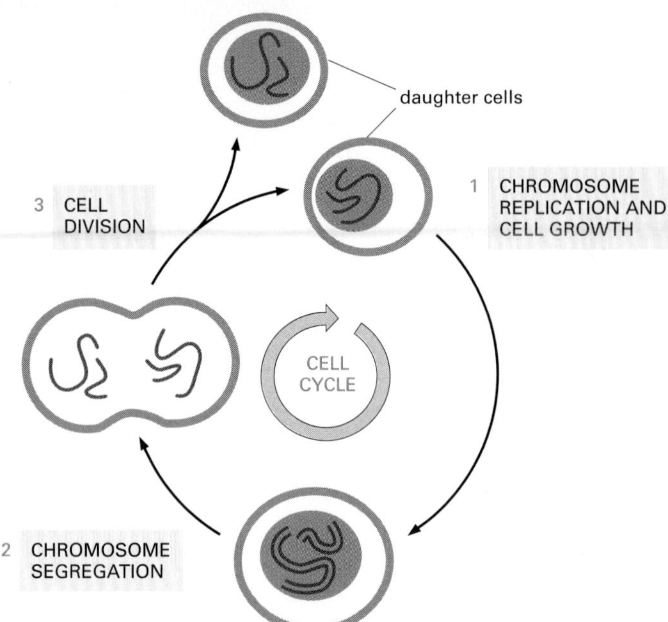

Figure 17–1 The cell cycle. The division of a hypothetical eucaryotic cell with two chromosomes is shown to illustrate how two genetically identical daughter cells are produced in each cycle. Each of the daughter cells will often divide again by going through additional cell cycles.

segregation of the replicated chromosomes. In most cells, additional layers of regulation enhance the fidelity of cell division and allow the control system to respond to various signals from both inside and outside the cell. Inside the cell, the control system monitors progression through the cell cycle and delays later events until earlier events have been completed. Preparations for the segregation of replicated chromosomes, for example, are not permitted until DNA replication is complete. The control system also monitors conditions outside the cell. In a multicellular animal, the system is highly responsive to signals from other cells, stimulating cell division when more cells are needed and blocking it when they are not. The cell-cycle control system therefore has a central role in regulating cell numbers in the tissues of the body. When the system malfunctions, excessive cell divisions can result in cancer.

In addition to duplicating their genome, most cells also duplicate their other organelles and macromolecules; otherwise, they would get smaller with each division. To maintain their size, dividing cells must coordinate their growth (i.e., their increase in cell mass) with their division; it is still not clear how this coordination is achieved.

This chapter is concerned primarily with how the various events of the cell cycle are controlled and coordinated. We begin with a brief overview of these events, the molecular details of which are discussed in other chapters (DNA replication in Chapter 5; chromosome segregation and cell division in Chapter 18). We then describe the cell-cycle control system, examining how it organizes the sequence of cell-cycle events and how it responds to intracellular signals to regulate cell division. We next discuss how multicellular organisms eliminate unwanted cells by the process of *programmed cell death,* or *apoptosis,* in which a cell commits suicide when the interests of the organism demand it. Finally, we consider how animals regulate cell numbers and cell size—using extracellular signals to control cell survival, cell growth, and cell division.

AN OVERVIEW OF THE CELL CYCLE

The most basic function of the cell cycle is to duplicate accurately the vast amount of DNA in the chromosomes and then segregate the copies precisely into two genetically identical daughter cells. These processes define the two major *phases* of the cell cycle. DNA duplication occurs during *S phase* (S for synthesis), which requires 10–12 hours and occupies about half of the cell-cycle time in a typical mammalian cell. After S phase, chromosome segregation and

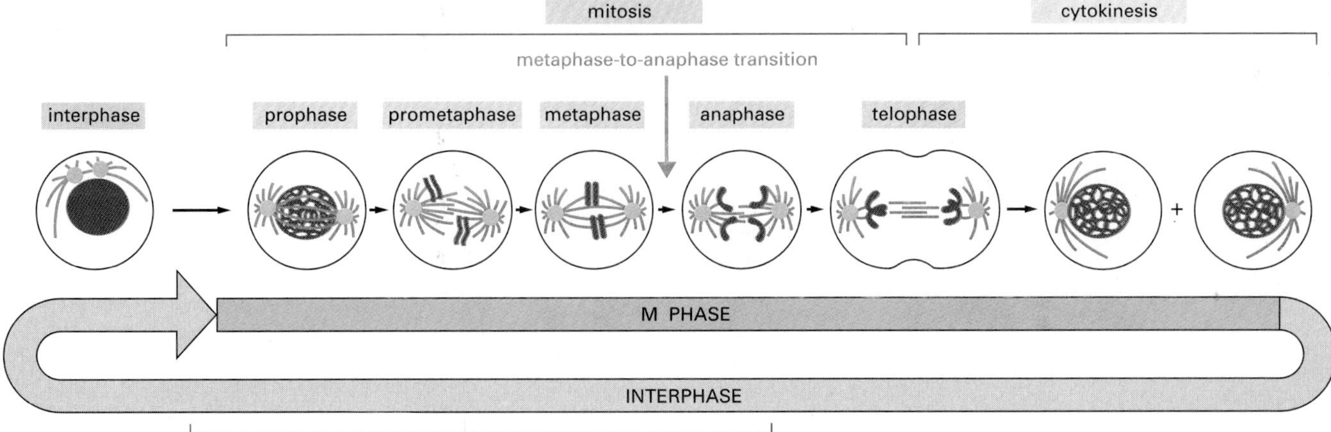

Figure 17–2 The events of eucaryotic cell division as seen under a microscope. The easily visible processes of nuclear division (mitosis) and cell division (cytokinesis), collectively called M phase, typically occupy only a small fraction of the cell cycle. The other, much longer, part of the cycle is known as interphase. The five stages of mitosis are shown: an abrupt change in the biochemical state of the cell occurs at the transition from metaphase to anaphase. A cell can pause in metaphase before this transition point, but once the point has been passed, the cell carries on to the end of mitosis and through cytokinesis into interphase. Note that DNA replication occurs in interphase. The part of interphase where DNA is replicated is called S phase (not shown).

cell division occur in *M phase* (M for *m*itosis), which requires much less time (less than an hour in a mammalian cell). M phase involves a series of dramatic events that begin with nuclear division, or mitosis. As discussed in detail in Chapter 18, mitosis begins with chromosome condensation: the duplicated DNA strands, packaged into elongated chromosomes, condense into the much more compact chromosomes required for their segregation. The nuclear envelope then breaks down, and the replicated chromosomes, each consisting of a pair of *sister chromatids,* become attached to the microtubules of the *mitotic spindle.* As mitosis proceeds, the cell pauses briefly in a state called *metaphase,* when the chromosomes are aligned at the equator of the mitotic spindle, poised for segregation. The sudden separation of sister chromatids marks the beginning of *anaphase,* during which the chromosomes move to opposite poles of the spindle, where they decondense and reform intact nuclei. The cell is then pinched in two by cytoplasmic division, or *cytokinesis,* and cell division is complete (Figure 17–2).

Most cells require much more time to grow and double their mass of proteins and organelles than they require to replicate their DNA and divide. Partly to allow more time for growth, extra *gap phases* are inserted in most cell cycles—a **G_1 phase** between M phase and S phase and a **G_2 phase** between S phase and mitosis. Thus, the eucaryotic cell cycle is traditionally divided into four sequential phases: G_1, S, G_2, and M (Figure 17–3). G_1, S, and G_2 together are called *interphase*. In a typical human cell proliferating in culture, interphase might occupy 23 hours of a 24 hour cycle, with 1 hour for M phase.

The two gap phases serve as more than simple time delays to allow cell growth. They also provide time for the cell to monitor the internal and external environment to ensure that conditions are suitable and preparations are complete before the cell commits itself to the major upheavals of S phase and mitosis. The G_1 phase is especially important in this respect. Its length can vary greatly depending on external conditions and extracellular signals from other

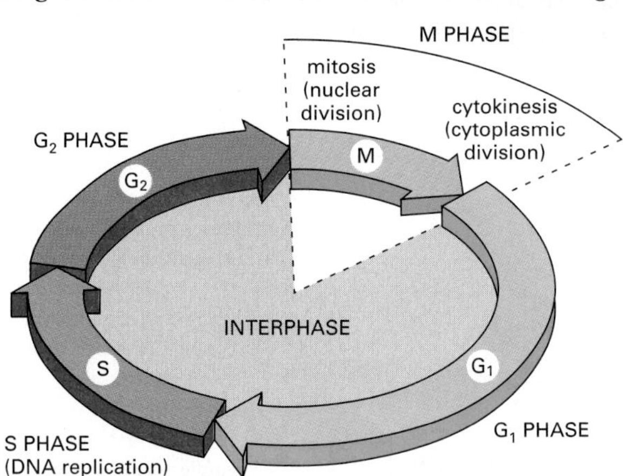

Figure 17–3 The phases of the cell cycle. The cell grows continuously in interphase, which consists of three phases: DNA replication is confined to S phase; G_1 is the gap between M phase and S phase, while G_2 is the gap between S phase and M phase. In M phase, the nucleus and then the cytoplasm divide.

cells. If extracellular conditions are unfavorable, for example, cells delay progress through G_1 and may even enter a specialized resting state known as *G_0* (G zero), in which they can remain for days, weeks, or even years before resuming proliferation. Indeed, many cells remain permanently in G_0 until they or the organism dies. If extracellular conditions are favorable and signals to grow and divide are present, cells in early G_1 or G_0 progress through a commitment point near the end of G_1 known as **Start** (in yeasts) or the **restriction point** (in mammalian cells). After passing this point, cells are committed to DNA replication, even if the extracellular signals that stimulate cell growth and division are removed.

The Cell-Cycle Control System Is Similar in All Eucaryotes

Some features of the cell cycle, including the time required to complete certain events, vary greatly from one cell type to another, even in the same organism. The basic organization of the cycle and its control system, however, are essentially the same in all eucaryotic cells. The proteins of the control system first appeared over a billion years ago. Remarkably, they have been so well conserved over the course of evolution that many of them function perfectly when transferred from a human cell to a yeast cell. We can therefore study the cell cycle and its regulation in a variety of organisms and use the findings from all of them to assemble a unified picture of how eucaryotic cells divide. In the following section, we briefly review the three eucaryotic systems in which cell-cycle control is commonly studied—yeasts, frog embryos, and cultured mammalian cells.

The Cell-Cycle Control System Can Be Dissected Genetically in Yeasts

Yeasts are tiny, single-celled fungi whose mechanisms of cell-cycle control are remarkably similar to our own. Two species are generally used in studies of the cell cycle. The **fission yeast** *Schizosaccharomyces pombe* is named after the African beer it is used to produce. It is a rod-shaped cell that grows by elongation at its ends. Division occurs by the formation of a septum, or cell plate, in the center of the rod (Figure 17–4A). The **budding yeast** *Saccharomyces cerevisiae* is used by brewers, as well as by bakers. It is an oval cell that divides by forming a bud, which first appears during G_1 and grows steadily until it separates from the mother cell after mitosis (Figure 17–4B).

Despite their outward differences, the two yeast species share a number of features that are extremely useful for genetic studies. They reproduce almost as rapidly as bacteria and have a genome size less than 1% that of a mammal. They are amenable to rapid molecular genetic manipulation, whereby genes can be

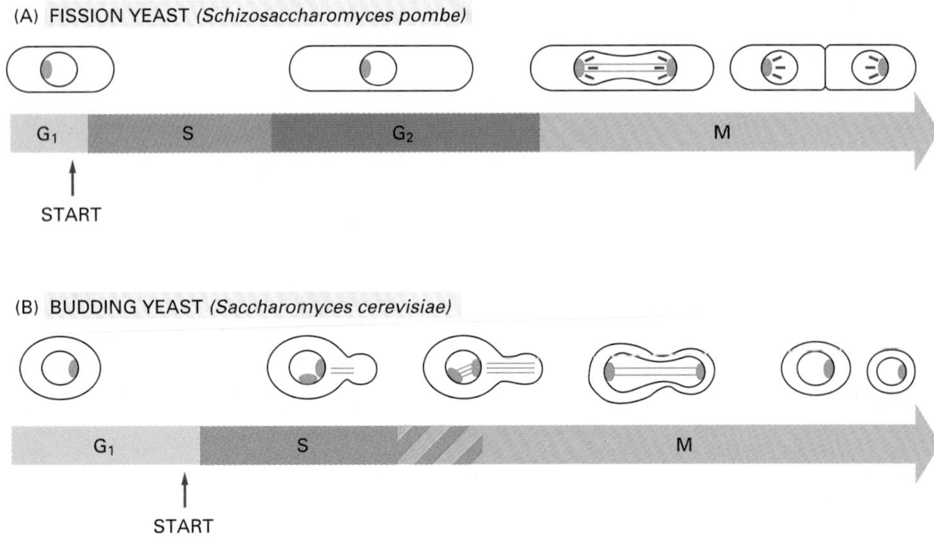

Figure 17–4 A comparison of the cell cycles of fission yeasts and budding yeasts. (A) The fission yeast has a typical eucaryotic cell cycle with G_1, S, G_2, and M phases. In contrast with what happens in higher eucaryotic cells, however, the nuclear envelope of the yeast cell does not break down during M phase. The microtubules of the mitotic spindle *(light green)* form inside the nucleus and are attached to spindle pole bodies *(dark green)* at its periphery. The cell divides by forming a partition (known as the cell plate) and splitting in two. The condensed mitotic chromosomes *(red)* are readily visible in fission yeast, but are less easily seen in budding yeasts. (B) The budding yeast has normal G_1 and S phases but does not have a normal G_2 phase. Instead, a microtubule-based spindle begins to form inside the nucleus early in the cycle, during S phase. In contrast with a fission yeast cell, the cell divides by budding. As in fission yeasts, but in contrast with higher eucaryotic cells, the nuclear envelope remains intact during mitosis, and the spindle forms within the nucleus.

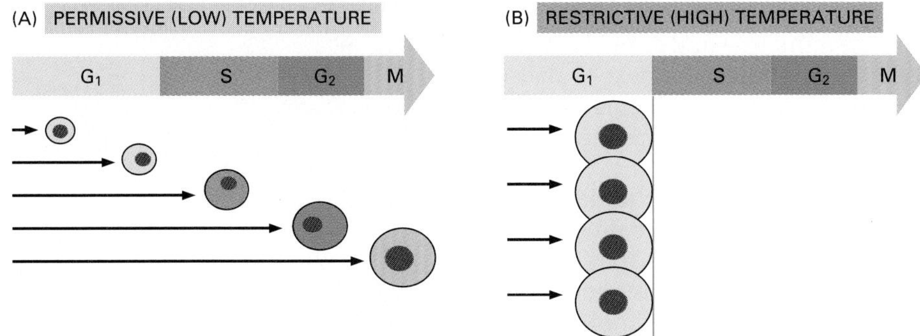

Figure 17–5 The behavior of a temperature-sensitive *cdc* mutant. (A) At the permissive (low) temperature, the cells divide normally and are found in all phases of the cycle (the phase of the cell is indicated by its color). (B) On warming to the restrictive (high) temperature, at which the mutant gene product functions abnormally, the mutant cells continue to progress through the cycle until they come to the specific step that they are unable to complete (initiation of S phase, in this example). Because the *cdc* mutants still continue to grow, they become abnormally large. By contrast, non-*cdc* mutants, if deficient in a process that is necessary throughout the cycle for biosynthesis and growth (such as ATP production), halt haphazardly at any stage of the cycle—depending on when their biochemical reserves run out (not shown).

deleted, replaced, or altered. Most importantly, they have the unusual ability to proliferate in a *haploid* state, in which only a single copy of each gene is present in the cell. When cells are haploid, it is easy to isolate and study mutations that inactivate a gene, as one avoids the complication of having a second copy of the gene in the cell.

Many important discoveries about cell-cycle control have come from systematic searches for mutations in yeasts that inactivate genes encoding essential components of the cell-cycle control system. The genes affected by these mutations are known as **cell-division-cycle genes**, or ***cdc* genes**. Many of these mutations cause cells to arrest at a specific point in the cell cycle, suggesting that the normal gene product is required to get the cell past this point.

A mutant that cannot complete the cell cycle cannot be propagated. Thus, *cdc* mutants can be selected and maintained only if their phenotype is *conditional*—that is, if the gene product fails to function only in certain specific conditions. Most conditional cell-cycle mutations are *temperature-sensitive mutations*, in which the mutant protein fails to function at high temperatures but functions well enough to allow cell division at low temperatures. A temperature-sensitive *cdc* mutant can be propagated at a low temperature (the *permissive condition)* and then raised to a higher temperature (the *restrictive condition)* to switch off the function of the mutant gene. At the higher temperature, the cells continue through the cell cycle until they reach the point where the function of the mutant gene is required for further progress, and at this point they halt (Figure 17–5). In budding yeasts, a uniform cell-cycle arrest of this type can be detected by just looking at the cells: the presence or absence of a bud, and bud size, indicate the point in the cycle at which the mutant is arrested (Figure 17–6).

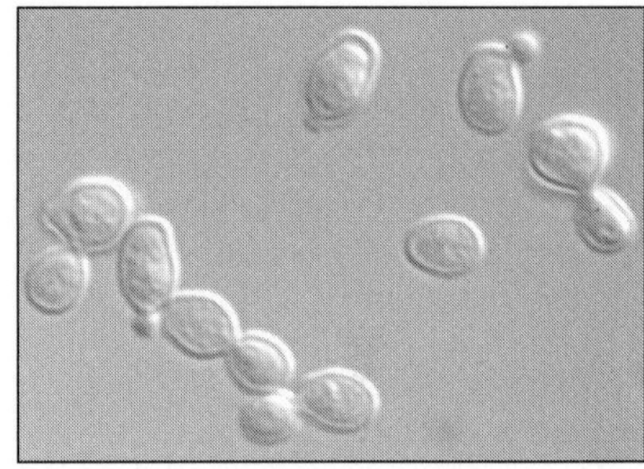
(A)

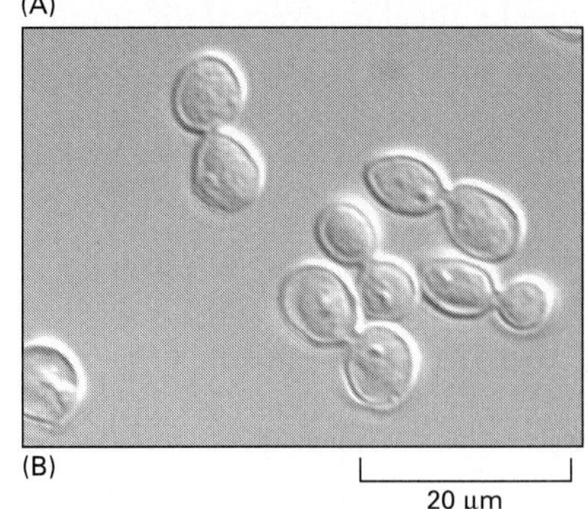
(B)

Figure 17–6 The morphology of budding yeast cells arrested by a *cdc* mutation. (A) In a normal population of proliferating yeast cells, buds vary in size according to the cell-cycle stage. (B) In a *cdc15* mutant grown at the restrictive temperature, cells complete anaphase but cannot complete the exit from mitosis and cytokinesis. As a result, they arrest uniformly with the large buds, which are characteristic of late M phase. (Courtesy of Jeff Ubersax.)

The Cell-Cycle Control System Can Be Analyzed Biochemically in Animal Embryos

While yeasts are ideal for studying the genetics of the cell cycle, the biochemistry of the cycle is most easily analyzed in the giant fertilized eggs of many animals, which carry large stockpiles of the proteins needed for cell division. The egg of the frog *Xenopus*, for example, is over 1 mm in diameter and carries 100,000 times more cytoplasm than an average cell in the human body (Figure 17–7). Fertilization of the *Xenopus* egg triggers an astonishingly rapid sequence of cell divisions, called *cleavage divisions*, in which the single giant cell divides, without growing, to generate an embryo containing thousands of smaller cells

(Figure 17–8). In this process, almost the only macromolecules synthesized are DNA—required to produce the thousands of new nuclei—and a small amount of protein. After a first division that takes about 90 minutes, the next 11 divisions occur, more or less synchronously, at 30-minute intervals, producing about 4096 (2^{12}) cells within 7 hours. Each cycle is divided into S and M phases of about 15 minutes each, without detectable G_1 or G_2 phases.

The cells in early embryos of *Xenopus,* as well as those of the clam *Spisula* and the fruit fly *Drosophila,* are thus capable of exceedingly rapid division in the absence of either growth or many of the control mechanisms that operate in more complex cell cycles. These *early embryonic cell cycles* therefore reveal the workings of the cell-cycle control system stripped down and simplified to the minimum needed to achieve the most fundamental requirements—the duplication of the genome and its segregation into two daughter cells. Another advantage of these early embryos for cell-cycle analysis is their large size. It is relatively easy to inject test substances into an egg to determine their effect on cell-cycle progression. It is also possible to prepare almost pure cytoplasm from *Xenopus* eggs and reconstitute many events of the cell cycle in a test tube (Figure 17–9). In such cell extracts, one can observe and manipulate cell-cycle events under highly simplified and controllable conditions.

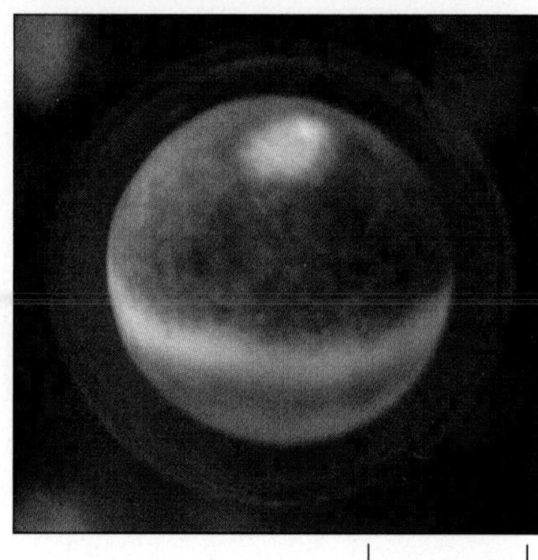

Figure 17–7 A mature *Xenopus* egg, ready for fertilization. The pale spot near the top shows the site of the nucleus, which has displaced the *brown* pigment in the surface layer of the egg cytoplasm. Although this cannot be seen in the picture, the nuclear envelope has broken down during the process of egg maturation. (Courtesy of Tony Mills.)

The Cell-Cycle Control System of Mammals Can Be Studied in Culture

It is not easy to observe individual cells in an intact mammal. Most studies on mammalian cell-cycle control therefore use cells that have been isolated from normal tissues or tumors and grown in plastic culture dishes in the presence of essential nutrients and other factors (Figure 17–10). There is a complication, however. When cells from normal mammalian tissues are cultured in standard conditions, they often stop dividing after a limited number of division cycles. Human fibroblasts, for example, permanently cease dividing after 25–40 divisions, a process called *replicative cell senescence,* which we discuss later.

Mammalian cells occasionally undergo mutations that allow them to proliferate readily and indefinitely in culture as "immortalized" *cell lines.* Although they are not normal, such cell lines are used widely for cell-cycle studies—and for cell biology generally—because they provide an unlimited source of genetically homogeneous cells. In addition, these cells are sufficiently large to allow detailed cytological observations of cell-cycle events, and they are amenable to biochemical analysis of the proteins involved in cell-cycle control.

Figure 17–8 Oocyte growth and egg cleavage in *Xenopus*. The oocyte grows without dividing for many months in the ovary of the mother frog and finally matures into an egg. Upon fertilization, the egg cleaves very rapidly—initially at a rate of one division cycle every 30 minutes—forming a multicellular tadpole within a day or two. The cells get progressively smaller with each division, and the embryo remains the same size. Growth starts only when the tadpole begins feeding. The drawings in the top row are all on the same scale (but the frog below is not).

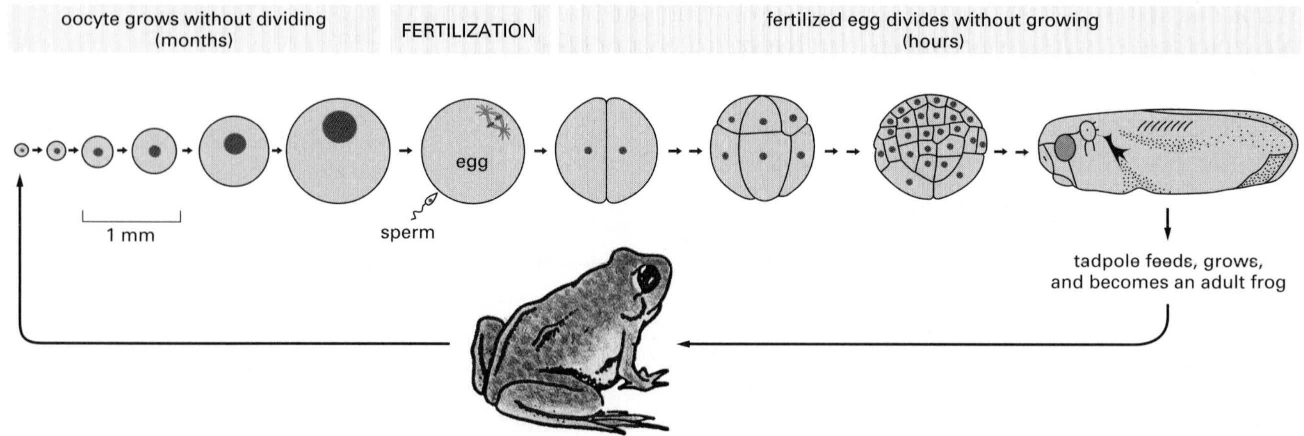

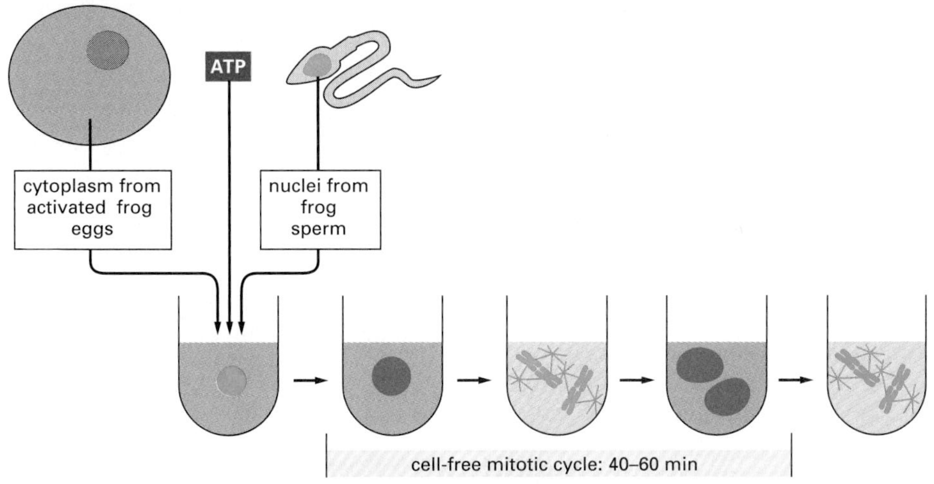

Figure 17–9 Studying the cell cycle in a cell-free system. A large batch of activated frog eggs is broken open by gentle centrifugation, which also separates the cytoplasm from other cell components. The undiluted cytoplasm is collected, and sperm nuclei are added to it, together with ATP. The sperm nuclei decondense and then go through repeated cycles of DNA replication and mitosis, indicating that the cell-cycle control system is operating in this cell-free cytoplasmic extract.

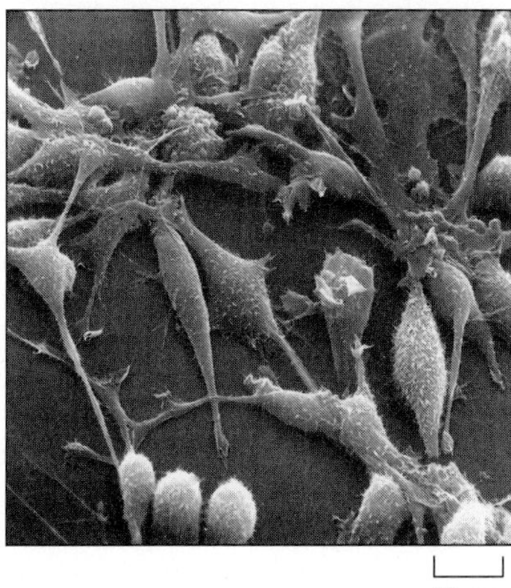

Figure 17–10 Mammalian cells proliferating in culture. The cells in this scanning electron micrograph are rat fibroblasts. (Courtesy of Guenter Albrecht-Buehler.)

Studies of cultured mammalian cells have been especially useful for examining the molecular mechanisms governing the control of cell proliferation in multicellular organisms. Such studies are important not only for understanding the normal controls of cell numbers in tissues but also for understanding the loss of these controls in cancer (discussed in Chapter 23).

Cell-Cycle Progression Can Be Studied in Various Ways

How can one tell at what stage an animal cell is in the cell cycle? One way is to simply look at living cells with a microscope. A glance at a population of mammalian cells proliferating in culture reveals that a fraction of the cells have rounded up and are in mitosis. Others can be observed in the process of cytokinesis. The S-phase cells, however, cannot be detected by simple observation. They can be recognized, however, by supplying them with visualizable molecules that are incorporated into newly synthesized DNA, such as ^{3}H-thymidine or the artificial thymidine analog bromo-deoxyuridine (BrdU). Cell nuclei that have incorporated ^{3}H-thymidine are visualized by autoradiography (Figure 17–11A), whereas those that have incorporated BrdU are visualized by staining with anti-BrdU antibodies (Figure 17–11B).

Typically, in a population of cells that are all proliferating rapidly but asynchronously, about 30–40% will be in S phase at any instant and become labeled by a brief pulse of ^{3}H-thymidine or BrdU. From the proportion of cells in such a population that are labeled (the *labeling index),* one can estimate the duration

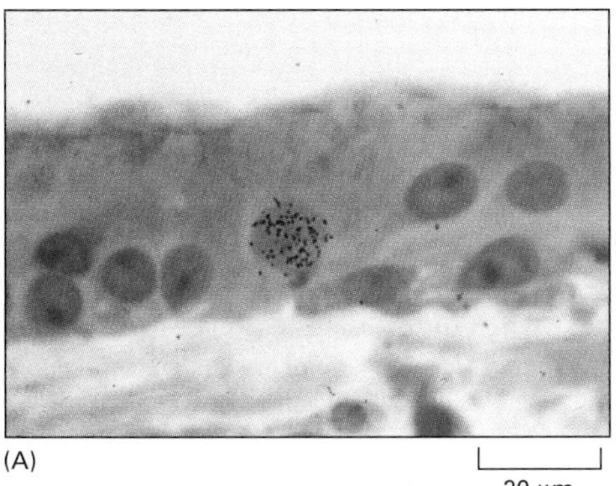

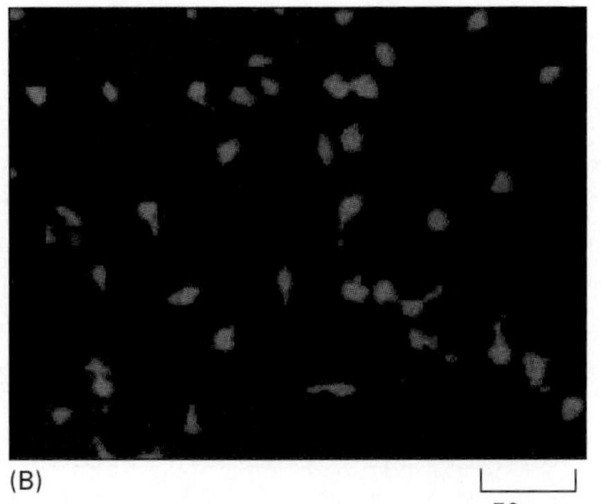

Figure 17–11 Labeling S-phase cells. (A) The tissue has been exposed for a short period to ^{3}H-thymidine and the labeled cells have been visualized by autoradiography. Silver grains *(black dots)* in the photographic emulsion over a nucleus indicate that the cell incorporated ^{3}H-thymidine into its DNA and thus was in S phase some time during the labeling period. In this specimen, showing the sensory epithelium from the inner ear of a chicken, the presence of an S-phase cell is evidence of cell proliferation occurring in response to damage. (B) An immunofluorescence micrograph of BrdU-labeled glial precursor cells in culture. The cells were exposed to BrdU for 4 h and were then fixed and labeled with fluorescent anti-BrdU antibodies *(red).* All the cells are stained with a *blue* fluorescent dye. (A, courtesy of Mark Warchol and Jeffrey Corwin; B, from D. Tang, Y. Tokumoto, and M. Raff, *J. Cell Biol.* 148:971–984, 2000. © The Rockefeller University Press.)

of S phase as a fraction of the whole cell cycle duration. Similarly, from the proportion of these cells in mitosis (the *mitotic index*), one can estimate the duration of M phase. In addition, by giving a pulse of ^{3}H-thymidine or BrdU and allowing the cells to continue around the cycle for measured lengths of time, one can determine how long it takes for an S-phase cell to progress through G_2 into M phase, through M phase into G_1, and finally through G_1 back into S phase.

Another way to assess the stage that a cell has reached in the cell cycle is by measuring its DNA content, which doubles during S phase. This approach is greatly facilitated by the use of DNA-binding fluorescent dyes and a *flow cytometer*, which allows large numbers of cells to be analyzed rapidly and automatically (Figure 17–12). One can also use flow cytometry to determine the lengths of G_1, S, and G_2 + M phases, by following over time a population of cells that have been preselected to be in one particular phase of the cell cycle: DNA content measurements on such a synchronized population of cells reveal how the cells progress through the cycle.

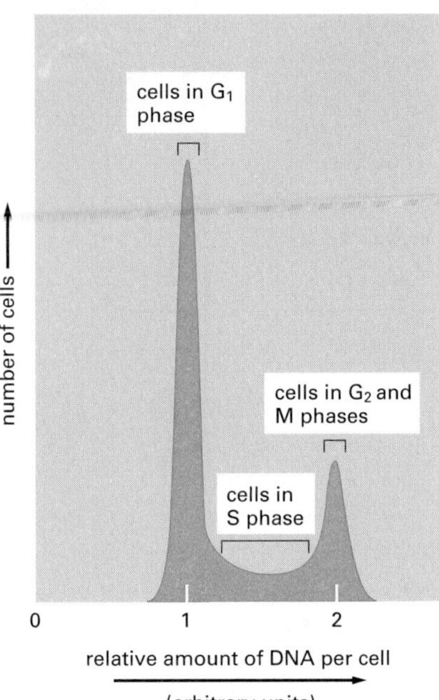

Figure 17–12 Analysis of DNA content with a flow cytometer. This graph shows typical results obtained for a proliferating cell population when the DNA content of its individual cells is determined in a flow cytometer. (A flow cytometer, also called a fluorescence-activated cell sorter, or FACS, can also be used to sort cells according to their fluorescence—see Figure 8–2). The cells analyzed here were stained with a dye that becomes fluorescent when it binds to DNA, so that the amount of fluorescence is directly proportional to the amount of DNA in each cell. The cells fall into three categories: those that have an unreplicated complement of DNA and are therefore in G_1 phase, those that have a fully replicated complement of DNA (twice the G_1 DNA content) and are in G_2 or M phase, and those that have an intermediate amount of DNA and are in S phase. The distribution of cells in the case illustrated indicates that there are greater numbers of cells in G_1 phase than in G_2 + M phase, showing that G_1 is longer than G_2 + M in this population.

Summary

Cell reproduction begins with duplication of the cell's contents, followed by distribution of those contents into two daughter cells. Chromosome duplication occurs during S phase of the cell cycle, whereas most other cell components are duplicated continuously throughout the cycle. During M phase, the replicated chromosomes are segregated into individual nuclei (mitosis), and the cell then splits in two (cytokinesis). S phase and M phase are usually separated by gap phases called G_1 and G_2, when cell-cycle progression can be regulated by various intracellular and extracellular signals. Cell-cycle organization and control have been highly conserved during evolution, and studies in a wide range of systems—including yeasts, frog embryos, and mammalian cells in culture—have led to a unified view of eucaryotic cell-cycle control.

COMPONENTS OF THE CELL-CYCLE CONTROL SYSTEM

For many years cell biologists watched the puppet show of DNA synthesis, mitosis, and cytokinesis but had no idea of what lay behind the curtain controlling these events. The cell-cycle control system was simply a black box inside the cell. It was not even clear whether there was a separate control system, or whether the processes of DNA synthesis, mitosis, and cytokinesis somehow controlled themselves. A major breakthrough came in the late 1980s with the identification of the key proteins of the control system, along with the realization that they are distinct from the proteins that perform the processes of DNA replication, chromosome segregation, and so on.

We first consider the basic principles upon which the cell-cycle control system operates. Then we discuss the protein components of the system and how they work together to activate the different phases of the cell cycle.

The Cell-Cycle Control System Triggers the Major Processes of the Cell Cycle

The cell-cycle control system operates much like the control system of an automatic clothes-washing machine. The washing machine functions in a series of stages: it takes in water, mixes it with detergent, washes the clothes, rinses them, and spins them dry. These essential processes of the wash cycle are analogous to the essential processes of the cell cycle—DNA replication, mitosis, and so on. In both cases, a central controller triggers each process in a set sequence (Figure 17–13).

How might one design a control system that safely guides the cell through the events of the cell cycle (or a wash cycle, for that matter)? In principle, one can imagine that the most basic control system should possess the following features:

- A clock, or timer, that turns on each event at a specific time, thus providing a fixed amount of time for the completion of each event.
- A mechanism for initiating events in the correct order; entry into mitosis, for example, must always come after DNA replication.
- A mechanism to ensure that each event is triggered only once per cycle.
- Binary (on/off) switches that trigger events in a complete, irreversible fashion. It would clearly be disastrous, for example, if events like chromosome condensation or nuclear envelope breakdown were initiated but not completed.
- Robustness: backup mechanisms to ensure that the cycle can work properly even when parts of the system malfunction.
- Adaptability, so that the system's behavior can be modified to suit specific cell types or environmental conditions.

We shall see in this chapter that the cell-cycle control system possesses all of these features, and that we are now beginning to understand the molecular mechanisms involved.

The Control System Can Arrest the Cell Cycle at Specific Checkpoints

We can illustrate the importance of an adjustable cell-cycle control system by extending our washing machine analogy. The control system of simple embryonic cell cycles, like the controller in a simple washing machine, is based on a clock. The clock is unaffected by the events it regulates and will progress through the whole sequence of events even if one of those events has not been successfully completed. In contrast, the control system of most cell cycles (and sophisticated washing machines) is responsive to information received back from the processes it is controlling. Sensors, for example, detect the completion of DNA synthesis (or the successful filling of the washtub), and, if some malfunction prevents the successful completion of this process, signals are sent to the control system to delay progression to the next phase. These delays provide time for the machinery to be repaired and also prevent the disaster that might result if the cycle progressed prematurely to the next stage.

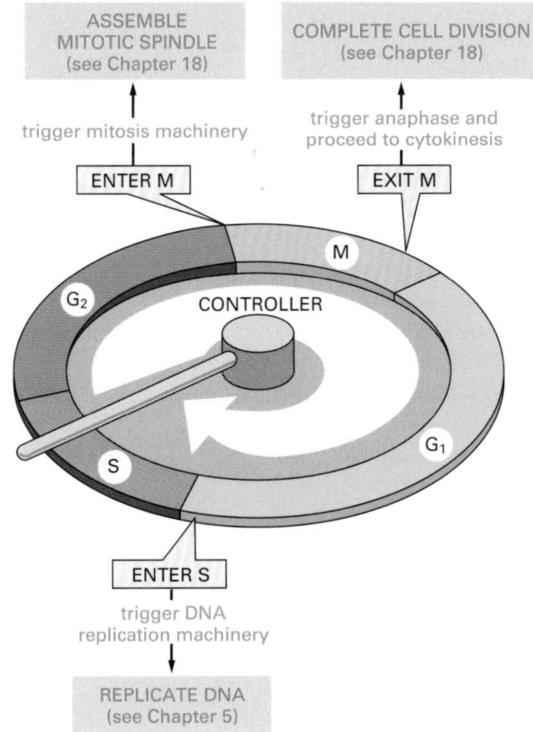

Figure 17–13 The control of the cell cycle. The essential processes of the cell cycle—such as DNA replication, mitosis, and cytokinesis—are triggered by a cell-cycle control system. By analogy with a washing machine, the cell-cycle control system is shown here as a central arm—the controller—that rotates clockwise, triggering essential processes when it reaches specific points on the outer dial.

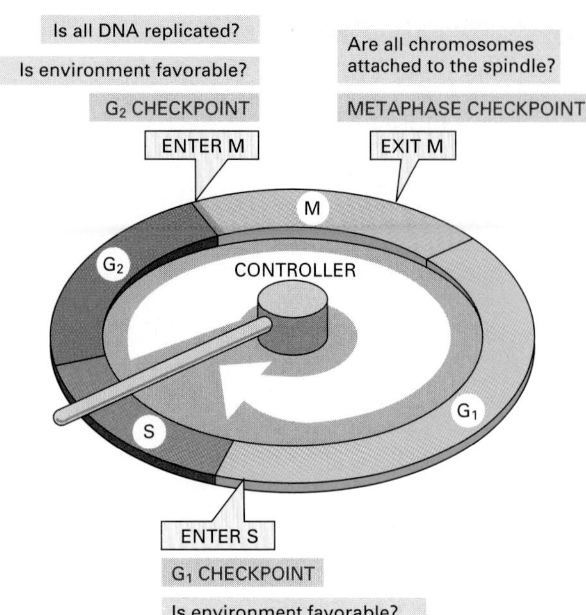

Figure 17–14 Checkpoints in the cell-cycle control system. Information about the completion of cell-cycle events, as well as signals from the environment, can cause the control system to arrest the cycle at specific checkpoints. The most prominent checkpoints occur at locations marked with *yellow boxes*.

In most cells there are several points in the cell cycle, called **checkpoints**, at which the cycle can be arrested if previous events have not been completed (Figure 17–14). Entry into mitosis is prevented, for example, when DNA replication is not complete, and chromosome separation in mitosis is delayed if some chromosomes are not properly attached to the mitotic spindle.

Progression through G_1 and G_2 is delayed by braking mechanisms if the DNA in the chromosomes is damaged by radiation or chemicals. Delays at these *DNA damage checkpoints* provide time for the damaged DNA to be repaired, after which the cell-cycle brakes are released and progress resumes.

Checkpoints are important in another way as well. They are points in the cell cycle at which the control system can be regulated by extracellular signals from other cells. These signals—which can either promote or inhibit cell proliferation—tend to act by regulating progression through a G_1 checkpoint, using mechanisms discussed later in the chapter.

Checkpoints Generally Operate Through Negative Intracellular Signals

Checkpoint mechanisms like those just described tend to act through negative intracellular signals that arrest the cell cycle, rather than through the removal of positive signals that normally stimulate cell-cycle progression. The following argument suggests why this is so.

Consider, for example, the checkpoint that monitors the attachment of chromosomes to the mitotic spindle. If a cell proceeds into anaphase and starts to segregate its chromosomes into separate daughter cells before all chromosomes are appropriately attached, one daughter receives an incomplete chromosome set, while the other daughter receives a surplus. The cell therefore needs to be able to detect the attachment of the last unattached chromosome to the microtubules of the spindle. In a cell with many chromosomes, if each chromosome sends a positive signal to the cell-cycle control system once it is attached, the attachment of the last chromosome will be hard to detect, as it will be signaled by only a small fractional change in the total intensity of the "go" signal. On the other hand, if each unattached chromosome sends a negative signal to inhibit progress through the cell cycle, the attachment of the last chromosome will be easily detected because it will cause a change from some "stop" signal to none. A similar argument would imply that unreplicated DNA inhibits the initiation of mitosis, creating a stop signal that persists until the completion of DNA replication.

The most convincing evidence that checkpoints operate through negative signals comes from studies of cells in which a checkpoint is inactivated by either mutation or chemical treatment. In these cells, the cell cycle continues to progress even if DNA replication or spindle assembly is incomplete, indicating that checkpoints are generally not essential for cell-cycle progression. Checkpoints are best viewed as accessory braking systems that have been added to the cell-cycle control system to provide a more sophisticated form of regulation.

Although most checkpoints are not essential for normal cell-cycle progression under ideal conditions, populations of cells with checkpoint defects often accumulate mutations due to occasional malfunctions in DNA replication, DNA repair, or spindle assembly. Some of these mutations can promote the development of cancer, as we discuss later and in Chapter 23.

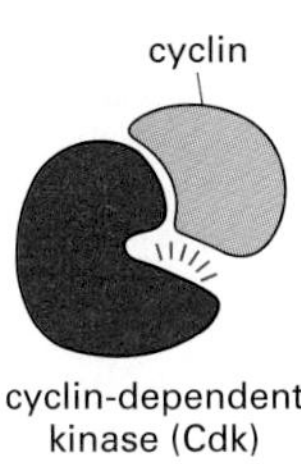

Figure 17–15 Two key components of the cell-cycle control system. A complex of cyclin with Cdk acts as a protein kinase to trigger specific cell-cycle events. Without cyclin, Cdk is inactive.

The Cell-Cycle Control System Is Based on Cyclically Activated Protein Kinases

At the heart of the cell-cycle control system is a family of protein kinases known as **cyclin-dependent kinases (Cdks)**. The activity of these kinases rises and falls as the cell progresses through the cycle. The oscillations lead directly to cyclical changes in the phosphorylation of intracellular proteins that initiate or regulate the major events of the cell cycle—DNA replication, mitosis, and cytokinesis. An increase in Cdk activity at the beginning of mitosis, for example, leads to increased phosphorylation of proteins that control chromosome condensation, nuclear envelope breakdown, and spindle assembly.

Cyclical changes in Cdk activity are controlled by a complex array of enzymes and other proteins. The most important of these Cdk regulators are proteins known as **cyclins**. Cdks, as their name implies, are dependent on cyclins for their activity: unless they are tightly bound to a cyclin, they have no protein kinase activity (Figure 17–15). Cyclins were originally named as such because they undergo a cycle of synthesis and degradation in each cell cycle. Cdk levels, by contrast, are constant, at least in the simplest cell cycles. Cyclical changes in cyclin levels result in the cyclic assembly and activation of the **cyclin–Cdk complexes**; this activation in turn triggers cell-cycle events (Figure 17–16).

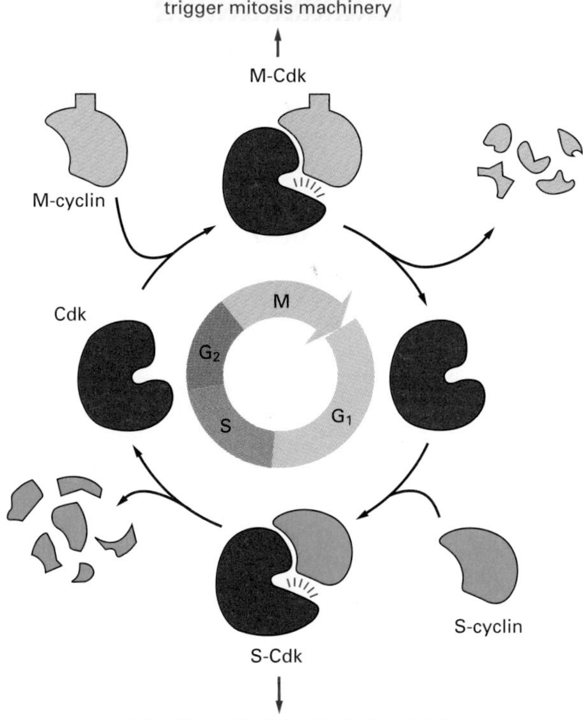

Figure 17–16 A simplified view of the core of the cell-cycle control system. Cdk associates successively with different cyclins to trigger the different events of the cycle. Cdk activity is usually terminated by cyclin degradation. For simplicity, only the cyclins that act in S phase (S-cyclin) and M phase (M-cyclin) are shown, and they interact with a single Cdk; as indicated, the resulting cyclin–Cdk complexes are referred to as S-Cdk and M-Cdk, respectively.

TABLE 17–1 The Major Cyclins and Cdks of Vertebrates and Budding Yeast

CYCLIN–CDK COMPLEX	VERTEBRATES CYCLIN	VERTEBRATES CDK PARTNER	BUDDING YEAST CYCLIN	BUDDING YEAST CDK PARTNER
G_1-Cdk	cyclin D*	Cdk4, Cdk6	Cln3	Cdk1**
G_1/S-Cdk	cyclin E	Cdk2	Cln1, 2	Cdk1
S-Cdk	cyclin A	Cdk2	Clb5, 6	Cdk1
M-Cdk	cyclin B	Cdk1**	Clb1, 2, 3, 4	Cdk1

* There are three D cyclins in mammals (cyclins D1, D2, and D3).
** The original name of Cdk1 was Cdc2 in both vertebrates and fission yeast, and Cdc28 in budding yeast.

There are four classes of cyclins, each defined by the stage of the cell cycle at which they bind Cdks and function. Three of these classes are required in all eucaryotic cells:

1. **G_1/S-cyclins** bind Cdks at the end of G_1 and commit the cell to DNA replication.
2. **S-cyclins** bind Cdks during S phase and are required for the initiation of DNA replication.
3. **M-cyclins** promote the events of mitosis.

In most cells, a fourth class of cyclins, the **G_1-cyclins**, helps promote passage through Start or the restriction point in late G_1.

In yeast cells, a single Cdk protein binds all classes of cyclins and drives all cell-cycle events by changing cyclin partners at different stages of the cycle. In vertebrate cells, by contrast, there are four Cdks. Two interact with G_1-cyclins, one with G_1/S- and S-cyclins, and one with M-cyclins. In this chapter, we simply refer to the different cyclin–Cdk complexes as **G_1-Cdk, G_1/S-Cdk, S-Cdk,** and **M-Cdk.** The names of the individual Cdks and cyclins are given in Table 17–1.

How do different cyclin–Cdk complexes drive different cell-cycle events? The answer, at least in part, seems to be that the cyclin protein does not simply activate its Cdk partner but also directs it to specific target proteins. As a result, each cyclin–Cdk complex phosphorylates a different set of substrate proteins. The same cyclin–Cdk complex can also induce different effects at different times in the cycle, probably because the accessibility of some Cdk substrates changes during the cell cycle. Certain proteins that function in mitosis, for example, may become available for phosphorylation only in G_2.

Studies of the three-dimensional structures of Cdk and cyclin proteins have revealed that, in the absence of cyclin, the active site in the Cdk protein is partly obscured by a slab of protein, like a stone blocking the entrance to a cave (Figure 17–17A). Cyclin binding causes the slab to move away from the active site, resulting in partial activation of the Cdk enzyme (Figure 17–17B). Full activation of the cyclin–Cdk complex then occurs when a separate kinase, the **Cdk-activating kinase (CAK)**, phosphorylates an amino acid near the entrance of the

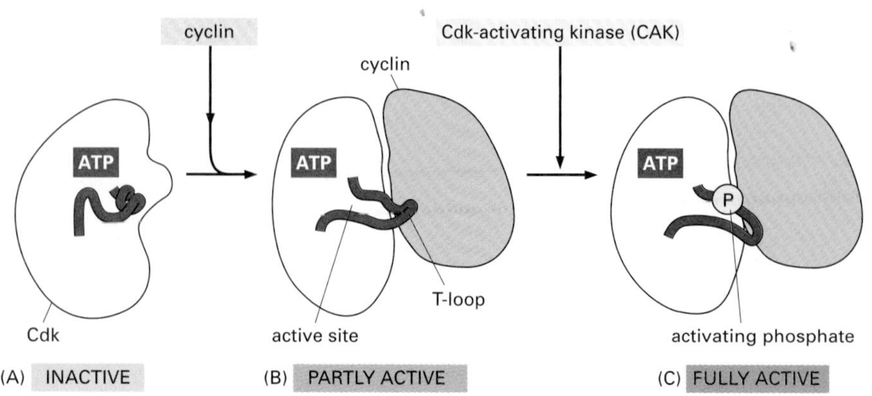

Figure 17–17 The structural basis of Cdk activation. These drawings are based on three-dimensional structures of human Cdk2, as determined by x-ray crystallography. The location of the bound ATP is indicated. The enzyme is shown in three states. (A) In the inactive state, without cyclin bound, the active site is blocked by a region of the protein called the T-loop (*red*). (B) The binding of cyclin causes the T-loop to move out of the active site, resulting in partial activation of the Cdk2. (C) Phosphorylation of Cdk2 (by CAK) at a threonine residue in the T-loop further activates the enzyme by changing the shape of the T-loop, improving the ability of the enzyme to bind its protein substrates.

Cdk active site. This causes a small conformational change that further increases the activity of the Cdk, allowing the kinase to phosphorylate its target proteins effectively and thereby induce specific cell-cycle events (Figure 17–17C).

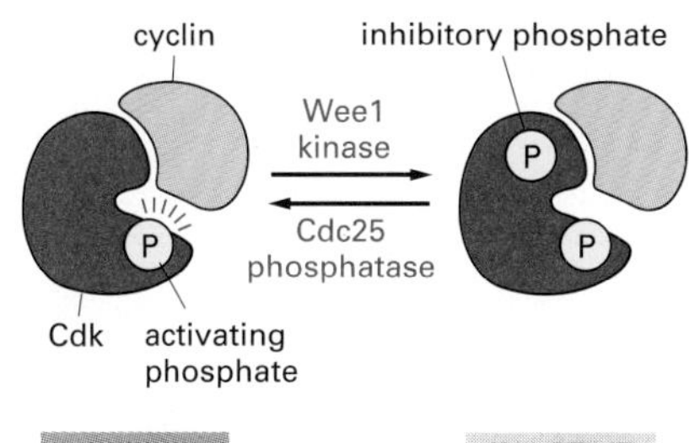

Figure 17–18 The regulation of Cdk activity by inhibitory phosphorylation. The active cyclin–Cdk complex is turned off when the kinase Wee1 phosphorylates two closely spaced sites above the active site. Removal of these phosphates by the phosphatase Cdc25 results in activation of the cyclin–Cdk complex. For simplicity, only one inhibitory phosphate is shown. The activating phosphate is added by CAK, as shown in Figure 17–17.

Cdk Activity Can Be Suppressed Both by Inhibitory Phosphorylation and by Inhibitory Proteins

The rise and fall of cyclin levels is the primary determinant of Cdk activity during the cell cycle. Several additional mechanisms, however, are important for fine-tuning Cdk activity at specific stages in the cell cycle.

The activity of a cyclin–Cdk complex can be inhibited by phosphorylation at a pair of amino acids in the roof of the active site. Phosphorylation of these sites by a protein kinase known as **Wee1** inhibits Cdk activity, while dephosphorylation of these sites by a phosphatase known as **Cdc25** increases Cdk activity (Figure 17–18). We see later that this regulatory mechanism is particularly important in the control of M-Cdk activity at the onset of mitosis.

Cyclin–Cdk complexes can also be regulated by the binding of **Cdk inhibitor proteins (CKIs)**. There are a variety of CKI proteins, and they are primarily employed in the control of G_1 and S phase. The three-dimensional structure of a cyclin–Cdk–CKI complex reveals that CKI binding dramatically rearranges the structure of the Cdk active site, rendering it inactive (Figure 17–19).

The Cell-Cycle Control System Depends on Cyclical Proteolysis

Cell-cycle control depends crucially on at least two distinct enzyme complexes that act at different times in the cycle to cause the proteolysis of key proteins of the cell-cycle control system thereby, inactivating them. Most notably, cyclin–Cdk complexes are inactivated by regulated proteolysis of cyclins at certain cell-cycle stages. This cyclin destruction occurs by a ubiquitin-dependent mechanism, like that involved in the proteolysis of many other intracellular proteins (discussed in Chapter 6). An activated enzyme complex recognizes specific amino-acid sequences on the cyclin and attaches multiple copies of ubiquitin to it, marking the protein for complete destruction in proteasomes.

The rate-limiting step in cyclin destruction is the final ubiquitin-transfer reaction catalyzed by enzymes known as **ubiquitin ligases** (see Figure 6–87B). Two ubiquitin ligases are important in the destruction of cyclins and other cell-cycle regulators. In G_1 and S phase, an enzyme complex called *SCF* (after its three main protein subunits) is responsible for the ubiquitylation and destruction of G_1/S-cyclins and certain CKI proteins that control S-phase initiation. In M phase, the *anaphase-promoting complex* (*APC*) is responsible for the ubiquitylation and proteolysis of M-cyclins and other regulators of mitosis.

These two large, multisubunit complexes contain some related components, but they are regulated in different ways. SCF activity is constant during the cell cycle. Ubiquitylation by SCF is controlled by changes in the phosphorylation state of its target proteins: only specifically phosphorylated proteins are recognized, ubiquitylated, and destroyed (Figure 17–20A). APC activity, by contrast, changes at different stages of the cell cycle. APC is turned on mainly by the addition of activating subunits to the complex (Figure 17–20B). We discuss the functions of SCF and APC in more detail later.

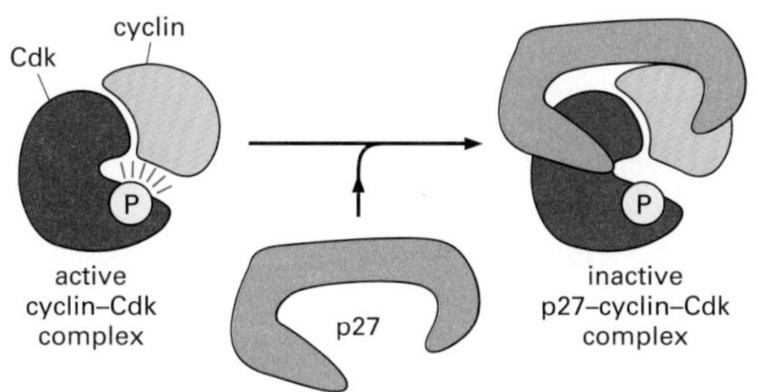

Figure 17–19 The inhibition of a cyclin–Cdk complex by a CKI. This drawing is based on the three-dimensional structure of the human cyclin A–Cdk2 complex bound to the CKI p27, as determined by x-ray crystallography. The p27 binds to both the cyclin and Cdk in the complex, distorting the active site of the Cdk. It also inserts into the ATP-binding site, further inhibiting the enzyme activity.

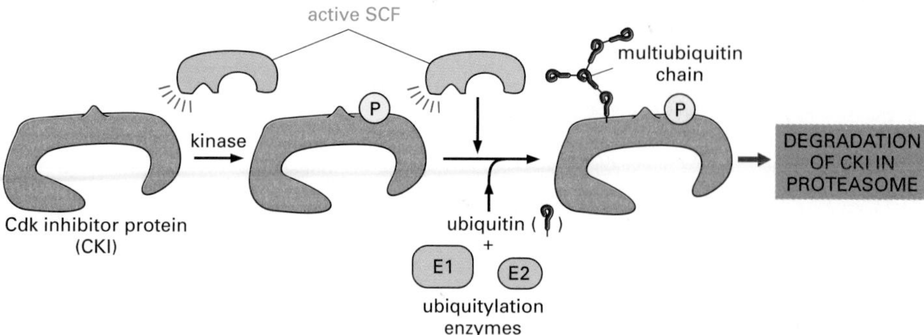

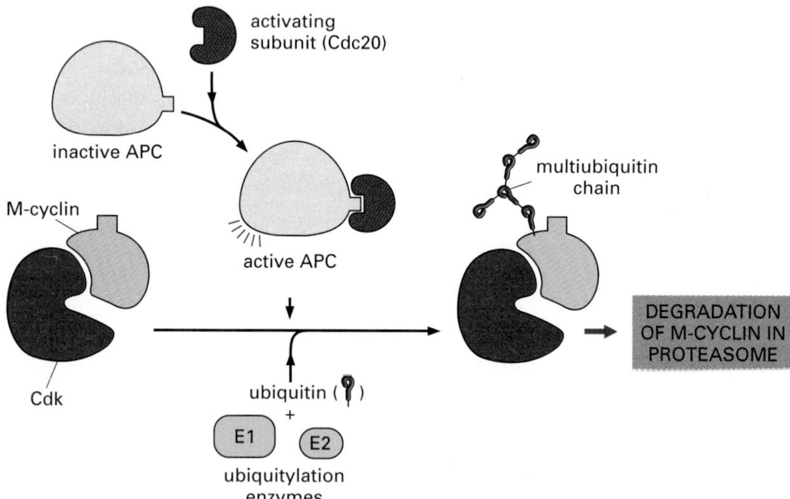

Figure 17–20 The control of proteolysis by SCF and APC during the cell cycle. (A) The phosphorylation of a target protein, such as the CKI shown, allows the protein to be recognized by SCF, which is constitutively active. With the help of two additional proteins called E1 and E2, SCF serves as a ubiquitin ligase that transfers multiple ubiquitin molecules onto the CKI protein. The ubiquitylated CKI protein is then immediately recognized and degraded in a proteasome. (B) M-cyclin ubiquitylation is performed by APC, which is activated in late mitosis by the addition of an activating subunit to the complex. Both SCF and APC contain binding sites that recognize specific amino acid sequences of the target protein.

Cell-Cycle Control Also Depends on Transcriptional Regulation

In the frog embryonic cell cycle discussed earlier, gene transcription does not occur. Cell-cycle control depends exclusively on post-transcriptional mechanisms that involve the regulation of Cdk activity by phosphorylation and the binding of regulatory proteins such as cyclins, which are themselves regulated by proteolysis. In the more complex cell cycles of most cell types, however, transcriptional control provides an added level of regulation. Cyclin levels in most cells, for example, are controlled not only by changes in cyclin degradation but also by changes in cyclin gene transcription and cyclin synthesis.

In certain organisms, such as budding yeasts, one can use DNA arrays (discussed in Chapter 8) to analyze changes in the expression of all of the genes in the genome as the cell progresses through the cell cycle. The results of these studies are surprising. About 10% of the yeast genes encode mRNAs whose levels oscillate during the cell cycle. Some of these genes encode proteins with known cell-cycle functions, but the functions of many others are unknown. It seems likely that these oscillations in gene expression are controlled by the cyclin–Cdk-dependent phosphorylation of gene regulatory proteins, but the details of this regulation remain unknown.

Summary

Events of the cell cycle are triggered by an independent cell-cycle control system, which ensures that the events are properly timed, occur in the correct order, and occur only once per cell cycle. The control system is responsive to various intracellular and extracellular signals, so that cell-cycle progression can be arrested when the cell either fails to complete an essential cell-cycle process or encounters unfavorable environmental conditions.

The central components of the cell-cycle control system are cyclin-dependent protein kinases (Cdks), whose activity depends on association with regulatory subunits called cyclins. Oscillations in the activities of various cyclin–Cdk complexes leads to the initiation of various cell-cycle events. Thus, activation of S-phase cyclin–Cdk complexes initiates S phase, while activation of M-phase cyclin–Cdk complexes triggers mitosis. The activities of cyclin–Cdk complexes are influenced by several mechanisms, including phosphorylation of the Cdk subunit, the binding of special inhibitory proteins (CKIs), proteolysis of cyclins, and changes in the transcription of genes encoding Cdk regulators. Two enzyme complexes, SCF and APC, are also crucial components of the cell-cycle control system; they induce the proteolysis of specific cell-cycle regulators by ubiquitylating them and thereby trigger several critical events in the cycle.

INTRACELLULAR CONTROL OF CELL-CYCLE EVENTS

Each of the different cyclin–Cdk complexes serves as a molecular switch that triggers a specific cell-cycle event. We now consider how these switches initiate such events and how the cell-cycle control system ensures that the switches fire in the correct order and only once per cell cycle. We begin with the two central events of the cell cycle: the replication of DNA during S phase and the chromosome segregation and cell division of M phase. We then discuss how crucial regulatory mechanisms in G_1 phase control whether or not a cell proliferates.

S-Phase Cyclin–Cdk Complexes (S-Cdks) Initiate DNA Replication Once Per Cycle

A cell must solve several problems in controlling the initiation and completion of DNA replication. Not only must replication occur with extreme accuracy to minimize the risk of mutations in the next cell generation, but every nucleotide in the genome must be copied once, and only once, to prevent the damaging effects of gene amplification. In Chapter 5, we discuss the sophisticated protein machinery that performs DNA replication with astonishing speed and accuracy. In this chapter, we consider the elegant mechanisms by which the cell-cycle control system initiates the replication process and, at the same time, prevents it from happening more than once per cycle.

Early clues about the regulation of S phase came from studies in which human cells at various cell-cycle stages were fused to form single cells with two nuclei. These experiments revealed that when a G_1 cell is fused with an S-phase cell, DNA replication occurs in the G_1 nucleus (presumably triggered by S-Cdk activity in the S-phase cell). Fusion of a G_2 cell with an S-phase cell, however, does not cause DNA synthesis in the G_2 nucleus (Figure 17–21). These studies provided a clear hint that only G_1 cells are competent to initiate DNA replication and that cells that have completed S phase (i.e. G_2 cells) are not able to rereplicate their

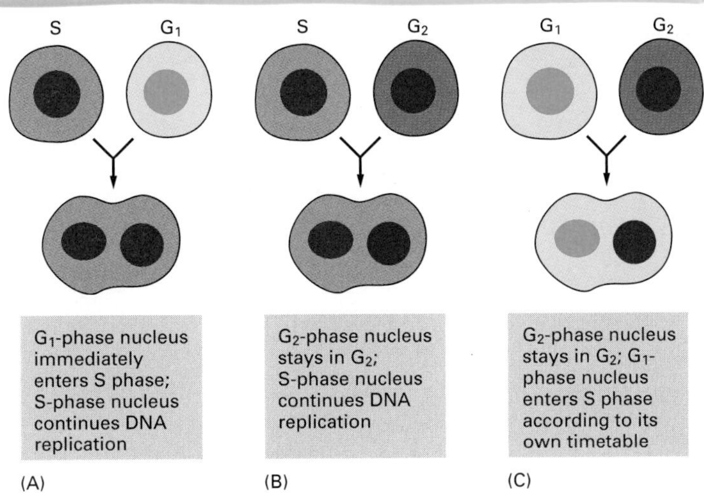

Figure 17–21 Evidence from cell-fusion experiments for a rereplication block. These experiments were carried out in 1970 in cultured mammalian cells. (A) The results show that S-phase cytoplasm contains factors that drive a G_1 nucleus directly into DNA synthesis. (B) A G_2 nucleus, having already replicated its DNA, is refractory to these factors. (C) Fusion of a G_2 cell with a G_1 cell does not drive the G_1 nucleus into DNA synthesis, indicating that the cytoplasmic factors for DNA replication that were present in the S-phase cell disappear when the cell moves from S phase into G_2. (Adapted from R.T. Johnson and P.N. Rao, *Nature* 226:717–722, 1970.)

DNA, even when provided with S-Cdk activity. Apparently, passage through mitosis is required for the cell to regain the ability to undergo S phase.

We have begun to decipher the molecular basis of these cell fusion experiments only recently. DNA replication begins at *origins of replication*, which are scattered at various locations in the chromosome. Replication origins are simple and well defined in the budding yeast *S. cerevisiae*, and most of our understanding of the initiation machinery comes from studies of this organism. Analyses of proteins that bind to the yeast replication origin have identified a large, multi-protein complex known as the **origin recognition complex (ORC)**. These complexes bind to replication origins throughout the cell cycle and serve as landing pads for several additional regulatory proteins.

One of these regulatory proteins is **Cdc6**. It is present at low levels during most of the cell cycle but increases transiently in early G_1. It binds to ORC at replication origins in early G_1, where it is required for the binding of a complex composed of a group of closely related proteins, the **Mcm proteins**. The resulting large protein complex formed at an origin is known as the **pre-replicative complex**, or **pre-RC** (Figure 17–22).

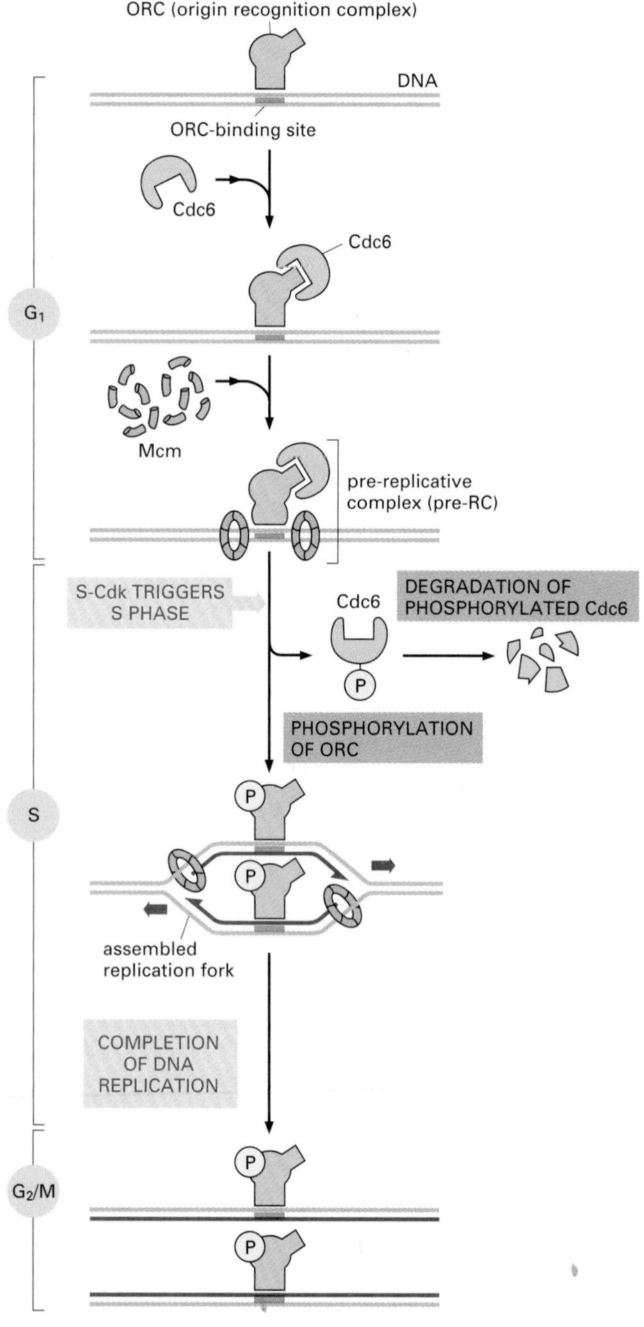

Figure 17–22 The initiation of DNA replication once per cell cycle. The ORC remains associated with a replication origin throughout the cell cycle. In early G_1, Cdc6 associates with ORC. Aided by Cdc6, Mcm ring complexes then assemble on the adjacent DNA, resulting in the formation of the pre-replicative complex. The S-Cdk (with assistance from another protein kinase, not shown) then triggers origin firing, assembling DNA polymerase and other replication proteins and activating the Mcm protein rings to migrate along DNA strands as DNA helicases. The S-Cdk also blocks rereplication by causing the dissociation of Cdc6 from origins, its degradation, and the export of all excess Mcm out of the nucleus. Cdc6 and Mcm cannot return to reset an ORC-containing origin for another round of DNA replication until M-Cdk has been inactivated at the end of mitosis (see text).

Once the pre-RC has been assembled in G_1, the replication origin is ready to fire. The activation of S-Cdk in late G_1 pulls the trigger and initiates DNA replication. The initiation of replication also requires the activity of a second protein kinase, which collaborates with S-Cdk to cause the phosphorylation of ORC.

The S-Cdk not only initiates origin firing, but also helps to prevent rereplication in several ways. First, it causes the Cdc6 protein to dissociate from ORC after an origin has fired. This results in the disassembly of the pre-RC, which prevents replication from occurring again at the same origin. Second, it prevents the Cdc6 and Mcm proteins from reassembling at any origin. By phosphorylating Cdc6, it triggers Cdc6 ubiquitylation by the SCF enzyme complex discussed earlier. As a result, any Cdc6 protein that is not bound to an origin is rapidly degraded in proteasomes. S-Cdk also phosphorylates certain Mcm proteins, which triggers their export from the nucleus, further ensuring that the Mcm protein complex cannot bind to a replication origin (see Figure 17–22).

S-Cdk activity remains high during G_2 and early mitosis, preventing rereplication from occurring after the completion of S phase. M-Cdk also helps ensure that rereplication does not occur during mitosis by phosphorylating the Cdc6 and Mcm proteins. The G_1/S-Cdks help as well, by inducing Mcm export from the nucleus, ensuring that excess Mcm proteins that have not bound to origins in late G_1 are taken out of action before replication begins.

Thus, several cyclin–Cdk complexes cooperate to restrain pre-RC assembly and prevent DNA rereplication after S phase. How, then, is the cell-cycle control system reset to allow replication to occur in the next cell cycle? The answer is simple. At the end of mitosis, all Cdk activity in the cell is reduced to zero. The resulting dephosphorylation of the Cdc6 and Mcm proteins allows pre-RC assembly to occur once again, readying the chromosomes for a new round of replication.

The Activation of M-Phase Cyclin–Cdk Complexes (M-Cdks) Triggers Entry into Mitosis

The completion of DNA replication leaves the G_2 cell with two accurate copies of the entire genome, with each replicated chromosome consisting of two identical *sister chromatids* glued together along their length. The cell then undergoes the dramatic upheaval of M phase, in which the duplicated chromosomes and other cell contents are distributed equally to the two daughter cells. The events of mitosis are triggered by M-Cdk, which is activated after S phase is complete.

The activation of M-Cdk begins with the accumulation of M-cyclin (cyclin B in vertebrate cells, see Table 17–1). In embryonic cell cycles, the synthesis of M-cyclin is constant throughout the cell cycle, and M-cyclin accumulation results from a decrease in its degradation. In most cell types, however, M-cyclin synthesis increases during G_2 and M, owing primarily to an increase in *M-cyclin* gene transcription. This increase in M-cyclin protein leads to a gradual accumulation of M-Cdk (the complex of Cdk1 and M-cyclin) as the cell approaches mitosis. Although the Cdk in these complexes is phosphorylated at an activating site by the enzyme CAK discussed earlier, it is held in an inactive state by inhibitory phosphorylation at two neighboring sites by the protein kinase Wee1 (see Figure 17–18). Thus, by the time the cell reaches the end of G_2, it contains an abundant stockpile of M-Cdk that is primed and ready to act, but the M-Cdk activity is repressed by the presence of two phosphate groups that block the active site of the kinase.

What, then, triggers the activation of the M-Cdk stockpile? The crucial event is the activation in late G_2 of the protein phosphatase Cdc25, which removes the inhibitory phosphates that restrain M-Cdk (Figure 17–23). At the same time, the activity of the inhibitory kinase Wee1 is also suppressed, further ensuring that M-Cdk activity increases abruptly. Two protein kinases activate Cdc25. One, known as **Polo kinase**, phosphorylates Cdc25 at one set of sites. The other activating kinase is M-Cdk itself, which phosphorylates a different set of sites on Cdc25. M-Cdk also phosphorylates and inhibits Wee1.

The ability of M-Cdk to activate its own activator (Cdc25) and inhibit its own

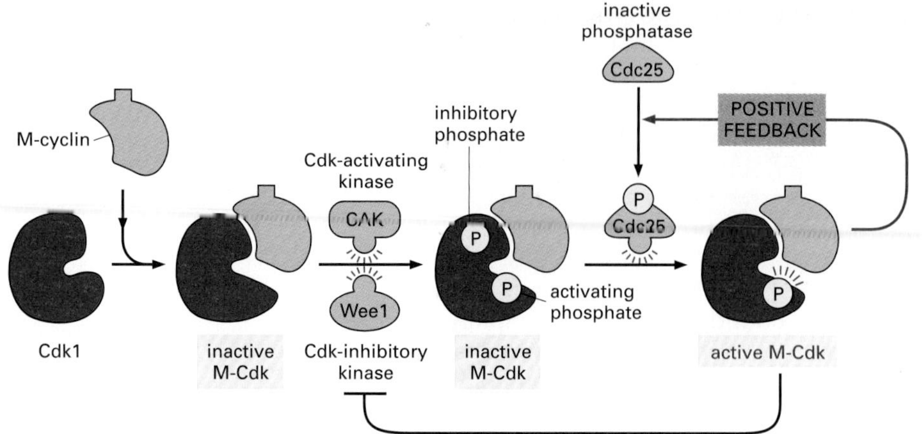

Figure 17–23 The activation of M-Cdk. Cdk1 associates with M-cyclin as the levels of M-cyclin gradually rise. The resulting M-Cdk complex is phosphorylated on an activating site by the Cdk-activating kinase (CAK) and on a pair of inhibitory sites by the Wee1 kinase. The resulting inactive M-Cdk complex is then activated at the end of G_2 by the phosphatase Cdc25. Cdc25 is stimulated in part by Polo kinase, which is not shown for simplicity. Cdc25 is further stimulated by active M-Cdk, resulting in positive feedback. This feedback is enhanced by the ability of M-Cdk to inhibit Wee1.

inhibitor (Wee1) suggests that M-Cdk activation in mitosis involves a positive feedback loop (see Figure 17–23). According to this attractive model, the partial activation of Cdc25, perhaps by Polo kinase, leads to the partial activation of a subpopulation of M-Cdk complexes, which then phosphorylate more Cdc25 and Wee1 molecules. This leads to more M-Cdk dephosphorylation and activation, and so on. Such a mechanism would quickly promote the complete activation of all the M-Cdk complexes in the cell, converting a gradual increase in M-cyclin levels into a switchlike, abrupt rise in M-Cdk activity. As mentioned earlier, similar molecular switches operate at various points in the cell cycle to ensure that events such as entry into mitosis occur in an all-or-none fashion.

Entry into Mitosis Is Blocked by Incomplete DNA Replication: The DNA Replication Checkpoint

If a cell is driven into mitosis before it has finished replicating its DNA, it will pass on broken or incomplete sets of chromosomes to its daughter cells. This disaster is avoided in most cells by a **DNA replication checkpoint** mechanism, which ensures that the initiation of mitosis cannot occur until the last nucleotide in the genome has been copied. Sensor mechanisms, of unknown molecular nature, detect either the unreplicated DNA or the corresponding unfinished replication forks and send a negative signal to the cell-cycle control system, blocking the activation of M-Cdk. Thus, normal cells treated with chemical inhibitors of DNA synthesis, such as hydroxyurea, do not progress into mitosis. If the checkpoint mechanism is defective, however, as in yeast cells with certain mutations or in mammalian cells treated with high doses of caffeine, the cells plunge into a suicidal mitosis despite the failure to complete DNA replication (Figure 17–24).

The final targets of the negative checkpoint signal are the enzymes that control M-Cdk activation. The negative signal activates a protein kinase that inhibits the Cdc25 protein phosphatase (see Figures 17–18 and 17–23). As a result, M-Cdk remains phosphorylated and inactive until DNA replication is complete.

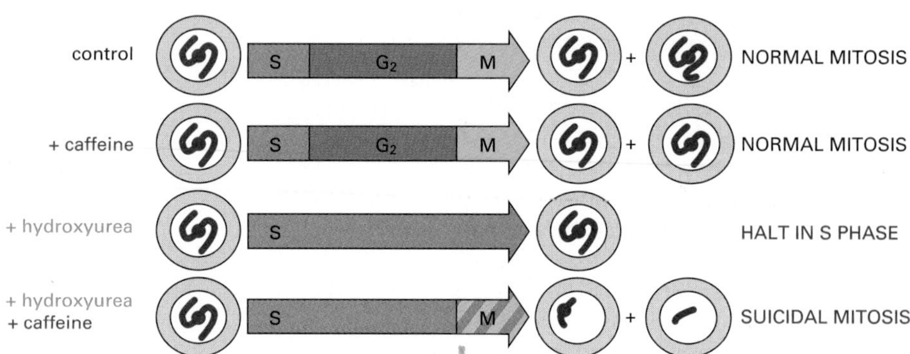

Figure 17–24 The DNA replication checkpoint. In the experiments diagrammed here, mammalian cells in culture were treated with caffeine and hydroxyurea, either alone or in combination. Hydroxyurea blocks DNA synthesis. This block activates a checkpoint mechanism that arrests the cells in S phase, delaying mitosis. But if caffeine is added as well as hydroxyurea, the checkpoint mechanism fails, and the cells proceed into mitosis according to their normal schedule, with incompletely replicated DNA. As a result, the cells die.

M-Cdk Prepares the Duplicated Chromosomes for Separation

One of the most remarkable features of cell-cycle control is that a single protein kinase, M-Cdk, is able to bring about all of the diverse and complex rearrangements that occur in the early stages of mitosis (discussed in Chapter 18). At a minimum, M-Cdk must induce the assembly of the mitotic spindle and ensure that replicated chromosomes attach to the spindle. In many organisms, M-Cdk also triggers chromosome condensation, nuclear envelope breakdown, actin cytoskeleton rearrangement, and the reorganization of the Golgi apparatus and endoplasmic reticulum. Each of these events is thought to be triggered when M-Cdk phosphorylates specific structural or regulatory proteins involved in the event, although most of these proteins have not yet been identified.

The breakdown of the nuclear envelope, for example, requires the disassembly of the *nuclear lamina*—the underlying shell of polymerized lamin filaments that gives the nuclear envelope its structural rigidity. Direct phosphorylation of lamin proteins by M-Cdk results in their depolymerization, which is an essential first step in the dismantling of the envelope (see Figure 12–21).

Chromosome condensation also seems to be a direct consequence of phosphorylation by M-Cdk. A complex of five proteins, known as the **condensin complex**, is required for chromosome condensation in *Xenopus* embryos. After M-Cdk has phosphorylated several subunits in the complex, two of the subunits are able to change the coiling of DNA molecules in a test tube. It is thought that this coiling activity is important for chromosome condensation during mitosis (see Figure 4–56).

Phosphorylation by M-Cdk also triggers the complex microtubule rearrangements and other events that lead to the assembly of the mitotic spindle. As discussed in Chapter 18, M-Cdk is known to phosphorylate a number of proteins that regulate microtubule behavior, causing the increase in microtubule instability that is required for spindle assembly.

Sister Chromatid Separation Is Triggered by Proteolysis

After M-Cdk has triggered the complex rearrangements that occur in early mitosis, the cell cycle reaches its culmination with the separation of the **sister chromatids** at the *metaphase-to-anaphase transition*. Although M-Cdk activity sets the stage for this event, an entirely different enzyme complex—the **anaphase-promoting complex (APC)** introduced earlier—throws the switch that initiates sister-chromatid separation. The APC is a highly regulated ubiquitin ligase that promotes the destruction of several mitotic regulatory proteins (see Figure 17–20B).

The attachment of the two sister chromatids to opposite poles of the mitotic spindle early in mitosis results in forces tending to pull the two chromatids apart. These pulling forces are initially resisted because the sister chromatids are bound tightly together, both at their centromeres and all along their arms. This sister-chromatid cohesion depends on a complex of proteins, the **cohesin complex**, that is deposited along the chromosomes as they are duplicated in S phase. The cohesin proteins *(cohesins)* are closely related to the proteins of the condensin complex involved in chromosome condensation, suggesting a common evolutionary origin for the two processes (see Figure 18–3).

Anaphase begins with a sudden disruption of the cohesion between sister chromatids, which allows them to separate and move to opposite poles of the spindle. This process is initiated by a remarkable cascade of signaling events. The sister-chromatid separation requires the activation of the APC enzyme complex, suggesting that proteolysis is central to the process (Figure 17–25). The relevant target of the APC is the protein **securin**. Before anaphase, securin binds to and inhibits the activity of a protease called **separase**. The destruction of securin at the end of metaphase releases separase, which is then free to cleave one of the subunits of the cohesin complex. In an instant, the cohesin complex falls away from the chromosomes, and the sister chromatids separate (Figure 17–26).

If the APC triggers anaphase, what triggers the APC? The answer is only partly known. APC activation requires the protein **Cdc20**, which binds to and

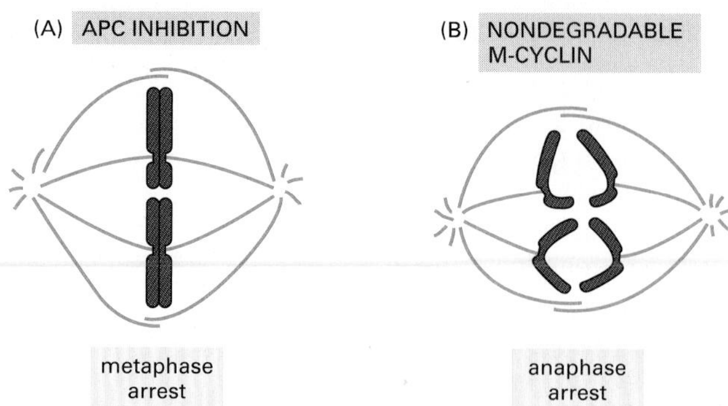

Figure 17–25 Two experiments that demonstrate the requirement for protein degradation to exit from mitosis. (A) An APC inhibitor was added to frog egg extracts undergoing mitosis *in vitro* (see Figure 17–9). The inhibitor arrested mitosis at metaphase, indicating that proteolysis is required for the separation of sister chromatids at the metaphase-to-anaphase transition. A similar arrest occurs in budding yeasts with mutations in components of the APC. (B) A nondegradable mutant form of M-cyclin was added to mitotic frog egg extracts. This addition arrested mitosis after sister-chromatid separation, indicating that destruction of M-cyclin is not required for sister-chromatid separation but is required for the subsequent exit from mitosis. (Based on S.L. Holloway et al., *Cell* 73:1393–1402, 1993.)

activates the APC at mitosis (see Figures 17–26 and 17–20B). At least two processes regulate Cdc20 and its association with the APC. First, Cdc20 synthesis increases as the cell approaches mitosis, owing to an increase in the transcription of its gene. Second, phosphorylation of the APC helps Cdc20 bind to the APC, thereby helping to create an active complex.

It is not clear what kinases phosphorylate and activate the Cdc20–APC complex. M-Cdk activity is required for the activity of these kinases, but there is a significant delay, or *lag phase*, between M-Cdk activation and the activation of the Cdc20–APC complex. The molecular basis of this delay is still mysterious, but it is likely to hold the key to how anaphase is initiated at the correct time in M phase.

Unattached Chromosomes Block Sister-Chromatid Separation: The Spindle-Attachment Checkpoint

The cell does not commit itself to the momentous events of anaphase before it is fully prepared. In most cell types, a **spindle-attachment checkpoint** mechanism operates to ensure that all chromosomes are properly attached to the spindle before sister-chromatid separation occurs. The checkpoint depends on a sensor mechanism that monitors the state of the *kinetochore*, the specialized region of the chromosome that attaches to microtubules of the spindle. Any

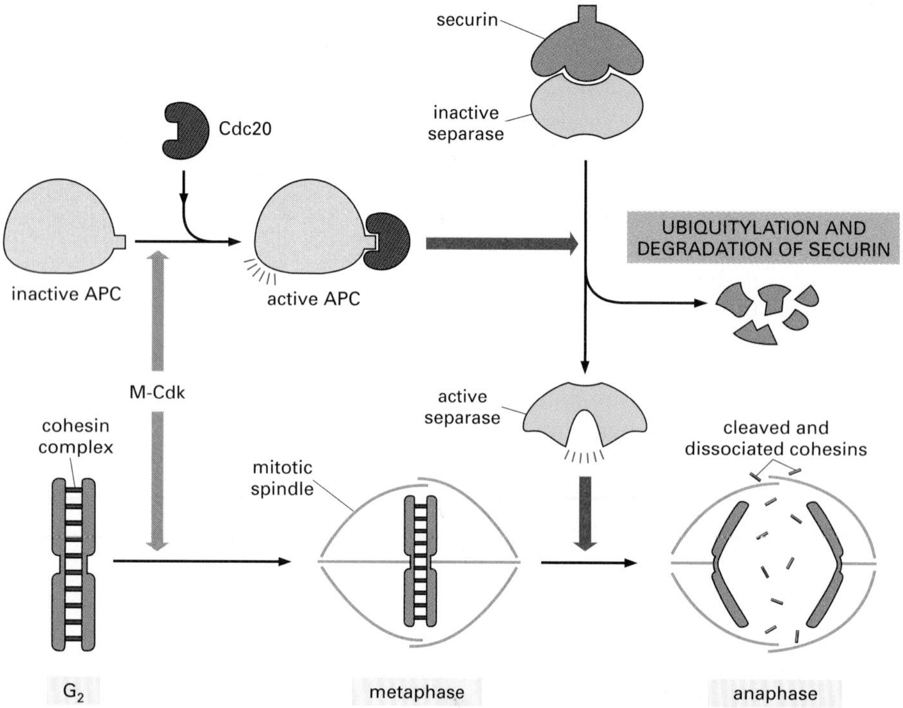

Figure 17–26 The triggering of sister-chromatid separation by the APC. The activation of APC by Cdc20 leads to the ubiquitylation and destruction of securin, which normally holds separase in an inactive state. The destruction of securin allows separase to cleave a subunit of the cohesin complex holding the sister chromatids together. The pulling forces of the mitotic spindle then pull the sister chromatids apart. In budding yeasts at least, cohesin cleavage by separase is facilitated by the phosphorylation of the cohesin complex adjacent to the cleavage site, just before anaphase begins. The phosphorylation is mediated by Polo kinase and provides an additional control on the timing of the metaphase-to-anaphase transition.

kinetochore that is not properly attached to the spindle sends out a negative signal to the cell-cycle control system, blocking Cdc20–APC activation and sister-chromatid separation.

The nature of the signal generated by an unattached kinetochore is not clear, although several proteins, including *Mad2,* are recruited to unattached kinetochores and are required for the spindle-attachment checkpoint to function. Even a single unattached kinetochore in the cell results in Mad2 binding and the inhibition of Cdc20–APC activity and Securin destruction (Figure 17–27). Thus, sister-chromatid separation cannot occur until the last kinetochore is attached.

Surprisingly, the normal timing of anaphase does not require a functional spindle-attachment checkpoint, at least in frog embryos and yeasts. Mutant yeast cells with a defective checkpoint undergo anaphase with normal timing, indicating that some other mechanism normally determines the timing of anaphase in these cells. In mammalian cells, however, a defect in the spindle-attachment checkpoint causes anaphase to occur slightly earlier than normal. This finding suggests that, in our cells, the checkpoint has evolved from a useful accessory to an essential component of the cell-cycle control system.

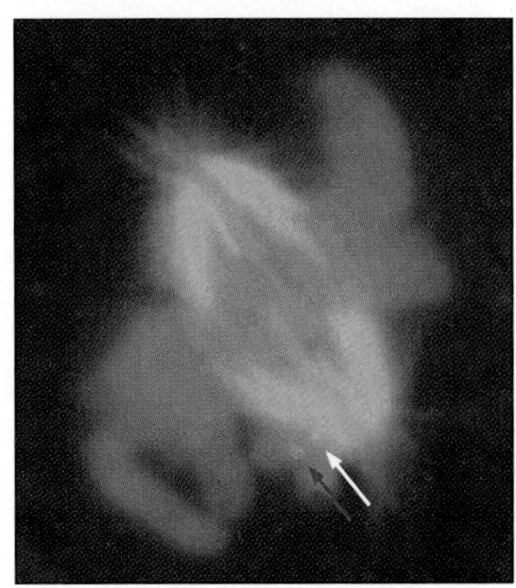

Figure 17–27 Mad2 protein on unattached kinetochores. This fluorescence micrograph shows a mammalian cell in prometaphase, with the mitotic spindle in *green* and the sister chromatids in *blue*. One sister chromatid pair is not yet attached to the spindle. The presence of Mad2 on the kinetochore of the unattached chromosome is revealed by the binding of anti-Mad2 antibodies *(red dot,* indicated by *red arrow).* Another chromosome has just attached to the spindle, and its kinetochore has a low level of Mad2 still associated with it *(pale dot,* indicated by *white arrow).* (From J.C. Waters et al., *J. Cell Biol.* 141:1181–1191, 1998. © The Rockefeller University Press.)

Exit from Mitosis Requires the Inactivation of M-Cdk

After the chromosomes have been segregated to the poles of the spindle, the cell must reverse the complex changes of early mitosis. The spindle must be disassembled, the chromosomes decondensed, and the nuclear envelope reformed. Because the phosphorylation of various proteins is responsible for getting cells into mitosis in the first place, it is not surprising that the dephosphorylation of these same proteins is required to get them out. In principle, these dephosphorylations and the exit from mitosis could be triggered by the inactivation of M-Cdk, the activation of phosphatases, or both. Evidence suggests that M-Cdk inactivation is primarily responsible.

M-Cdk inactivation occurs mainly by ubiquitin-dependent proteolysis of M-cyclins. Ubiquitylation of the cyclin is usually triggered by the same Cdc20–APC complex that promotes the destruction of Securin at the metaphase-to-anaphase transition (see Figure 17–20B). Thus, the activation of the Cdc20–APC complex leads not only to anaphase, but also to M-Cdk inactivation—which in turn leads to all of the other events that take the cell out of mitosis.

The G_1 Phase Is a State of Stable Cdk Inactivity

In early animal embryos, the inactivation of M-Cdk in late mitosis is due almost entirely to the action of Cdc20–APC. Recall, however, that M-Cdk stimulates Cdc20–APC activity (see Figure 17–26). Thus, the destruction of M-cyclin in late mitosis soon leads to the inactivation of all APC activity in an embryonic cell. This is a useful arrangement in rapid embryonic cell cycles, as APC inactivation immediately after mitosis allows the cell to quickly begin accumulating new M-cyclin for the next cycle (Figure 17–28A).

Rapid cyclin accumulation immediately after mitosis is not useful, however, in cell cycles containing a G_1 phase. In these cycles, progression into the next S phase is delayed in G_1 to allow for cell growth and for the cycle to be regulated by extracellular signals. Thus, most cells employ several mechanisms to ensure that Cdk reactivation is prevented after mitosis. One mechanism makes use of

(A) embryonic cells with no G_1 phase

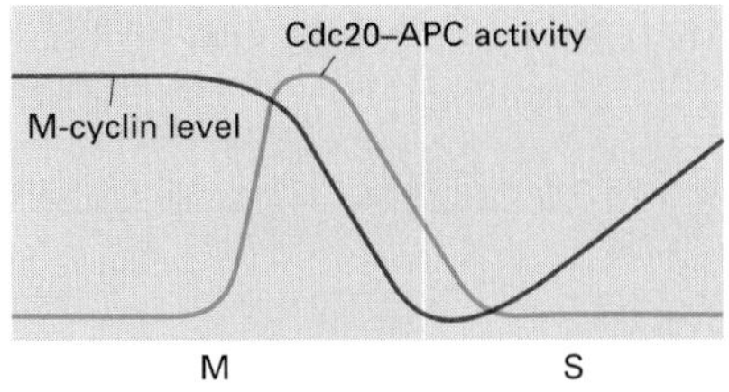

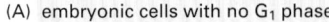

(B) cells with G_1 phase

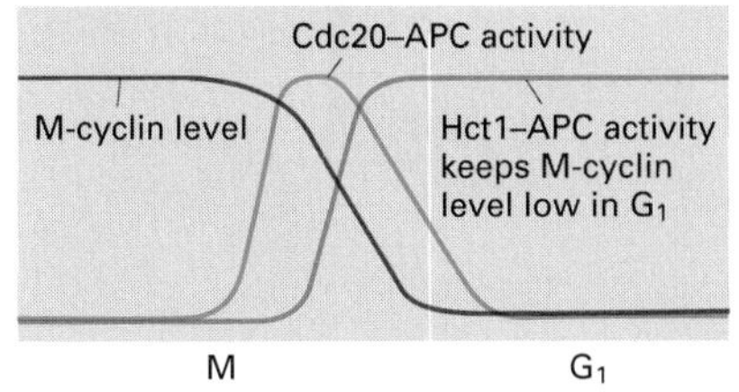

Figure 17–28 The creation of a G_1 phase by stable Cdk inhibition after mitosis. (A) In early embryonic cell cycles, Cdc20–APC activity rises at the end of metaphase, triggering M-cyclin destruction. Because M-Cdk activity stimulates Cdc20–APC activity, the loss of M-cyclin leads to APC inactivation after mitosis, which allows M-cyclins to begin accumulating again. (B) In cells containing a G_1 phase, the drop in M-Cdk activity in late mitosis leads to the activation of Hct1–APC (as well as to the accumulation of CKI proteins, not shown). This ensures a continued suppression of Cdk activity after mitosis, as required for a G_1 phase.

another APC-activating protein called *Hct1*, a close relative of Cdc20. Although both Hct1 and Cdc20 bind and activate the APC, they differ in one important respect. Whereas the Cdc20–APC complex is activated by M-Cdk, the Hct1–APC complex is inhibited by M-Cdk, which directly phosphorylates Hct1. As a result of this relationship, Hct1–APC activity increases in late mitosis after the Cdc20–APC complex has initiated the destruction of M-cyclin. M-cyclin destruction therefore continues after mitosis: although Cdc20–APC activity has declined, Hct1–APC activity is high (Figure 17–28B).

A second mechanism that suppresses Cdk activity in G_1 depends on the increased production of CKIs, the Cdk inhibitory proteins discussed earlier. Budding yeast cells, in which this mechanism is best understood, contain a CKI protein called *Sic1*, which binds to and inactivates M-Cdk in late mitosis and G_1. Like Hct1, Sic1 is inhibited by M-Cdk, which phosphorylates Sic1 during mitosis. M-Cdk also phosphorylates and inhibits a gene regulatory protein required for Sic1 synthesis, resulting in decreased Sic1 production. Thus, Sic1 and M-Cdk, like Hct1 and M-Cdk, mutually inhibit each other. As a result, the decline in M-Cdk activity that occurs in late mitosis triggers the rapid accumulation of Sic1 protein, and this CKI helps ensure that M-Cdk activity is stably inhibited after mitosis.

In most cells, M-Cdk inactivation in late mitosis also results from decreased transcription of M-cyclin genes. In budding yeast, for example, M-Cdk promotes the expression of these genes, resulting in a positive feedback loop. This loop is turned off as cells exit from mitosis: the inactivation of M-Cdk by Hct1 and Sic1 leads to decreased M-cyclin gene transcription and thus decreased M-cyclin synthesis.

In summary Hct1–APC activation, CKI accumulation, and decreased cyclin production act together to ensure that the early G_1 phase is a time when essentially all Cdk activity is suppressed. As in many other aspects of cell-cycle control, the use of multiple regulatory mechanisms makes the suppression system robust, so that it still operates with reasonable efficiency even if one mechanism fails.

How does the cell escape from this stable G_1 state to initiate S phase? As we describe later, escape usually occurs through the accumulation of G_1-cyclins. In budding yeast, for example, these cyclins are not targeted for destruction by Hct1–APC and are not inhibited by Sic1. As a result, the accumulation of G_1 cyclins leads to an unopposed increase in G_1-Cdk activity (Figure 17–29). In animal cells, the accumulation of G_1-cyclins is stimulated by the extracellular signals that promote cell proliferation, as we discuss later.

In budding yeast, G_1-Cdk activity triggers the transcription of G_1/S-cyclin genes, leading to increased synthesis of G_1/S-cyclins and the formation of G_1/S-Cdk complexes, which are also resistant to Hct1–APC and Sic1. The increased G_1/S-Cdk activity initiates the events that commit the cell to enter S phase. It stimulates the transcription of S-cyclin genes, leading to the synthesis of S-cyclins and the formation of S-Cdk complexes. These complexes are inhibited by Sic1, but G_1/S-Cdk phosphorylates and inactivates Sic1, thereby causing S-Cdk activation. G_1/S-Cdk and S-Cdk also phosphorylate and inactivate Hct1–APC.

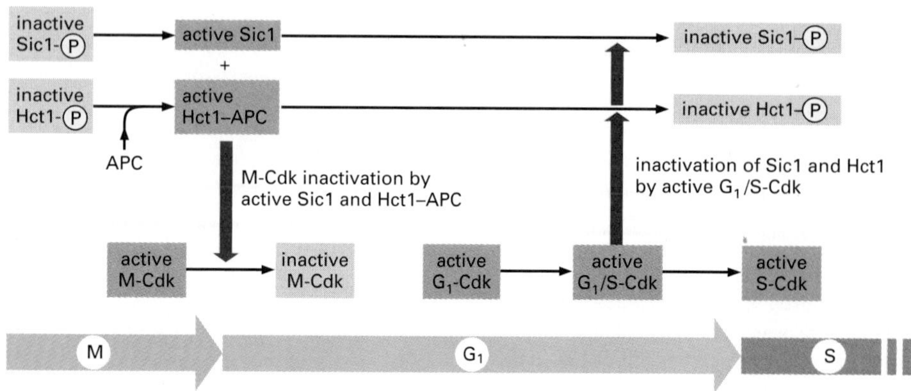

Figure 17–29 The control of G_1 progression by Cdk activity in budding yeast. As cells exit from mitosis and inactivate M-Cdk, the resulting increase in Hct1 and Sic1 activities results in stable Cdk inactivation during G_1. When conditions are right for entering a new cell cycle, the increase in G_1-Cdk and G_1/S-Cdk activities leads to the inhibition of Sic1 and Hct1 by phosphorylation, allowing S-Cdk activity to increase.

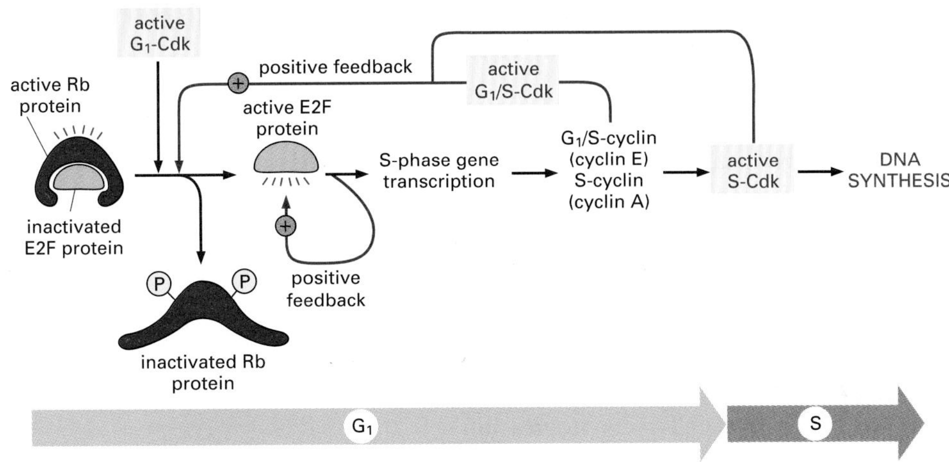

Figure 17–30 Mechanisms controlling S-phase initiation in animal cells. G_1-Cdk activity (cyclin D–Cdk4) initiates Rb phosphorylation. This inactivates Rb, freeing E2F to activate the transcription of S-phase genes, including the genes for a G_1/S-cyclin (cyclin E) and S-cyclin (cyclin A). The resulting appearance of G_1/S-Cdk and S-Cdk activities further enhances Rb phosphorylation, forming a positive feedback loop. E2F acts back to stimulate the transcription of its own gene, forming another positive feedback loop.

Thus, the same feedback loops that trigger rapid M-Cdk inactivation in late mitosis now work in reverse at the end of G_1 to ensure the rapid and complete activation of S-Cdk activity.

The Rb Protein Acts as a Brake in Mammalian G_1 Cells

The control of G_1 progression and S-phase initiation is often disrupted in cancer cells, leading to unrestrained cell-cycle entry and cell proliferation (discussed in Chapter 23). To develop improved methods for controlling cancer growth, we need a better understanding of the proteins that control G_1 progression in mammalian cells.

Animal cells suppress Cdk activity in G_1 by the same three mechanisms mentioned earlier for budding yeast: Hct1 activation, the accumulation of a CKI protein (p27 in mammalian cells), and the inhibition of cyclin gene transcription. As in yeasts, the activation of G_1-Cdk complexes reverses all three inhibitory mechanisms in late G_1.

The best understood effects of G_1-Cdk activity in animal cells are mediated by a gene regulatory protein called **E2F**. It binds to specific DNA sequences in the promoters of many genes that encode proteins required for S-phase entry, including G_1/S-cyclins and S-cyclins. E2F function is controlled primarily by an interaction with the **retinoblastoma protein (Rb)**, an inhibitor of cell-cycle progression. During G_1, Rb binds to E2F and blocks the transcription of S-phase genes. When cells are stimulated to divide by extracellular signals, active G_1-Cdk accumulates and phosphorylates Rb, reducing its affinity for E2F. The Rb then dissociates, allowing E2F to activate S-phase gene expression (Figure 17–30).

This transcriptional control system, like so many other control systems that regulate the cell cycle, includes feedback loops that sharpen the G_1/S transition (see Figure 17–30):

- The liberated E2F increases the transcription of its own gene.
- E2F-dependent transcription of G_1/S-cyclin and S-cyclin genes leads to increased G_1/S-Cdk and S-Cdk activities, which in turn increase Rb phosphorylation and promote further E2F release.
- The increase in G_1/S-Cdk and S-Cdk activities enhances the phosphorylation of Hct1 and p27, leading to their inactivation or destruction.

As in yeast cells, the result of all these interactions is the rapid and complete activation of the S-Cdk complexes required for S-phase initiation.

The Rb protein was identified originally through studies of an inherited form of eye cancer in children, known as *retinoblastoma* (discussed in Chapter 23). The loss of both copies of the Rb gene leads to excessive cell proliferation in the immature retina, suggesting that the Rb protein is particularly important for restraining the rate of cell division in the developing retina. The complete loss of Rb does not immediately cause increased proliferation of other cell types, in part because Hct1 and p27 provide assistance in G_1 control, and in part because other cell types contain Rb-related proteins that provide backup support in the absence of Rb. It is also likely that other proteins, unrelated to Rb, help to regulate the activity of E2F.

Cell-Cycle Progression Is Somehow Coordinated With Cell Growth

For proliferating cells to maintain a relatively constant size, the length of the cell cycle must match the time it takes the cell to double in size. If the cycle time is shorter than this, the cells will get smaller with each division; if it is longer, the cells will get bigger with each division. Because cell growth depends on nutrients and growth signals in the environment, the length of the cell cycle has to be able to adjust to varying environmental conditions (Figure 17–31). It is not clear how proliferating cells coordinate their growth with the rate of cell-cycle progression to maintain their size.

There is evidence that budding yeasts coordinate their growth and cell-cycle progression by monitoring the total amount of a G_1 cyclin called **Cln3** (see Table 17–1, p. 994). Because Cln3 is synthesized in parallel with cell growth, its concentration remains constant while its total amount increases as the cell grows. If the amount of Cln3 is artificially increased, the cells divide at a smaller size than normal, whereas if it is artificially decreased, the cells divide at a larger size than normal. These experiments are consistent with the idea that the cells commit themselves to division when the total amount of Cln3 reaches some threshold value. How, then, can the cell monitor the total amount of Cln3, rather than its concentration? One possibility is that cells inherit a fixed amount of an inhibitor that can bind to Cln3 and block its activity. When the amount of Cln3 exceeds the amount of this inhibitor, the extra Cln3 triggers G_1-Cdk activation and a new cell cycle. Since all cells receive a fixed and equal quantity of DNA, it has been speculated that the Cln3 inhibitor could be DNA itself, or some protein bound to DNA (Figure 17–32). Such a mechanism would also explain why cell size in all organisms is proportional to ploidy (the number of copies of the nuclear genome per cell).

Whereas yeast cells grow and proliferate constitutively if nutrients are plentiful, animal cells generally grow and proliferate only when they are stimulated to do so by signals from other cells. The size at which an animal cell divides depends, at least in part, on these extracellular signals, which can regulate cell growth and proliferation independently. Animal cells can also completely uncouple cell growth and division so as to grow without dividing or to divide without growing. The eggs of many animals, for example, grow to an extremely large size without dividing. After fertilization, this relationship is reversed, and many rounds of division occur without growth (see Figure 17–8). Thus, although

Figure 17–31 Cell size control through control of the cell cycle in yeasts. These graphs show the relationship between growth rate, cell size, and cell cycle time. (A) If cell division continued at an unchanged rate when cells were starved and stopped growing, the daughter cells produced at each division would become progressively smaller. (B) Yeast cells respond to some forms of nutritional deprivation by slowing the rate of progress through the cell cycle so that the cells have more time to grow. As a result, cell size remains unchanged or is reduced slightly. (A unit of time is the cycle time observed when nutrients are in excess.)

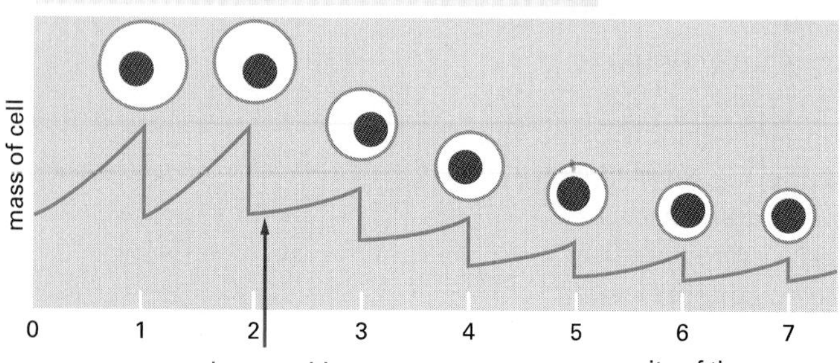

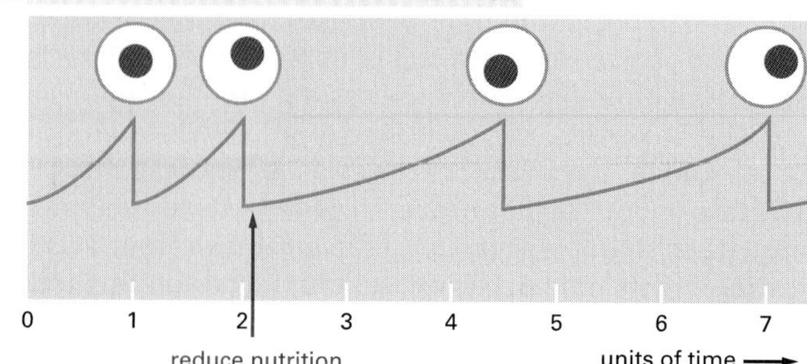

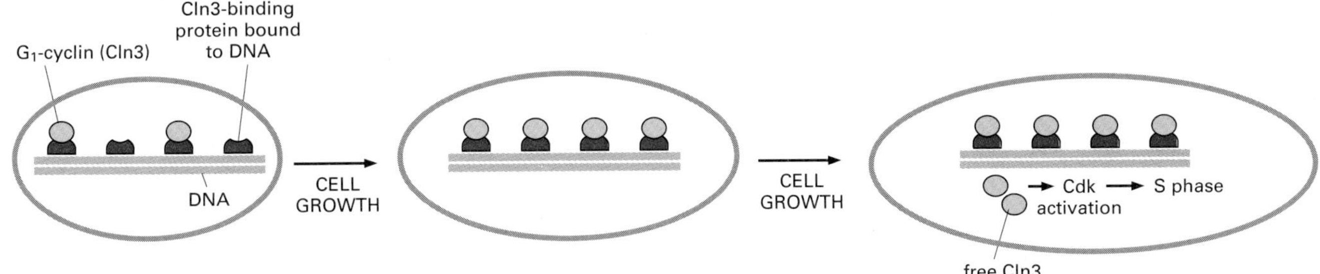

Figure 17–32 A hypothetical model of how budding yeast cells might coordinate cell growth and cell-cycle progression. The cell contains a fixed number of proteins *(red)* that are bound to DNA and bind and inhibit Cln3 molecules *(green)*. As the cell grows, the total number of Cln3 molecules increases in parallel with total cell protein. When the cell is small *(left)*, all of the Cln3 is inactivated by the excess of Cln-3-binding protein. As the cell grows, however, it reaches a threshold size at which the number of Cln3 molecules equals the number of Cln-3-binding proteins *(middle)*. When the cell exceeds this size, free Cln3 can bind to Cdk, which can now trigger the next cell cycle *(right)*.

cell growth and cell division are usually coordinated, they can be regulated independently. Cell growth does not depend on cell-cycle progression. Yeast cells continue to grow when cell-cycle progression is blocked by a mutation; and many animal cells, including neurons and muscle cells, grow large after they have withdrawn permanently from the cell cycle.

Cell-Cycle Progression is Blocked by DNA Damage and p53: DNA Damage Checkpoints

When chromosomes are damaged, as can occur after exposure to radiation or certain chemicals, it is essential that they be repaired before the cell attempts to duplicate or segregate them. The cell-cycle control system can readily detect DNA damage and arrest the cycle at **DNA damage checkpoints**. Most cells have at least two such checkpoints—one in late G_1, which prevents entry into S phase, and one in late G_2, which prevents entry into mitosis.

The G_2 checkpoint depends on a mechanism similar to the one discussed earlier that delays entry into mitosis in response to incomplete DNA replication. When cells in G_2 are exposed to damaging radiation, for example, the damaged DNA sends a signal to a series of protein kinases that phosphorylate and inactivate the phosphatase Cdc25. This blocks the dephosphorylation and activation of M-Cdk, thereby blocking entry into mitosis. When the DNA damage is repaired, the inhibitory signal is turned off, and cell-cycle progression resumes.

The G_1 checkpoint blocks progression into S phase by inhibiting the activation of G_1/S-Cdk and S-Cdk complexes. In mammalian cells, for example, DNA damage leads to the activation of the gene regulatory protein **p53**, which stimulates the transcription of several genes. One of these genes encodes a CKI protein called **p21**, which binds to G_1/S-Cdk and S-Cdk and inhibits their activities, thereby helping to block entry into S phase.

DNA damage activates p53 by an indirect mechanism. In undamaged cells, p53 is highly unstable and is present at very low concentrations. This is because it interacts with another protein, **Mdm2**, that acts as a ubiquitin ligase that targets p53 for destruction by proteasomes. DNA damage activates protein kinases that phosphorylate p53 and thereby reduce its binding to Mdm2. This decreases p53 degradation, which results in a marked increase in p53 concentration in the cell. In addition, the decreased binding to Mdm2 enhances the ability of p53 to stimulate gene transcription (Figure 17–33).

Like many other checkpoints, DNA damage checkpoints are not essential for normal cell division if environmental conditions are ideal. Conditions are rarely ideal, however: a low level of DNA damage occurs in the normal life of any cell, and this damage accumulates in the cell's progeny if the damage checkpoints are not functioning. Over the long term, the accumulation of genetic damage in cells lacking checkpoints leads to an increased frequency of cancer-promoting

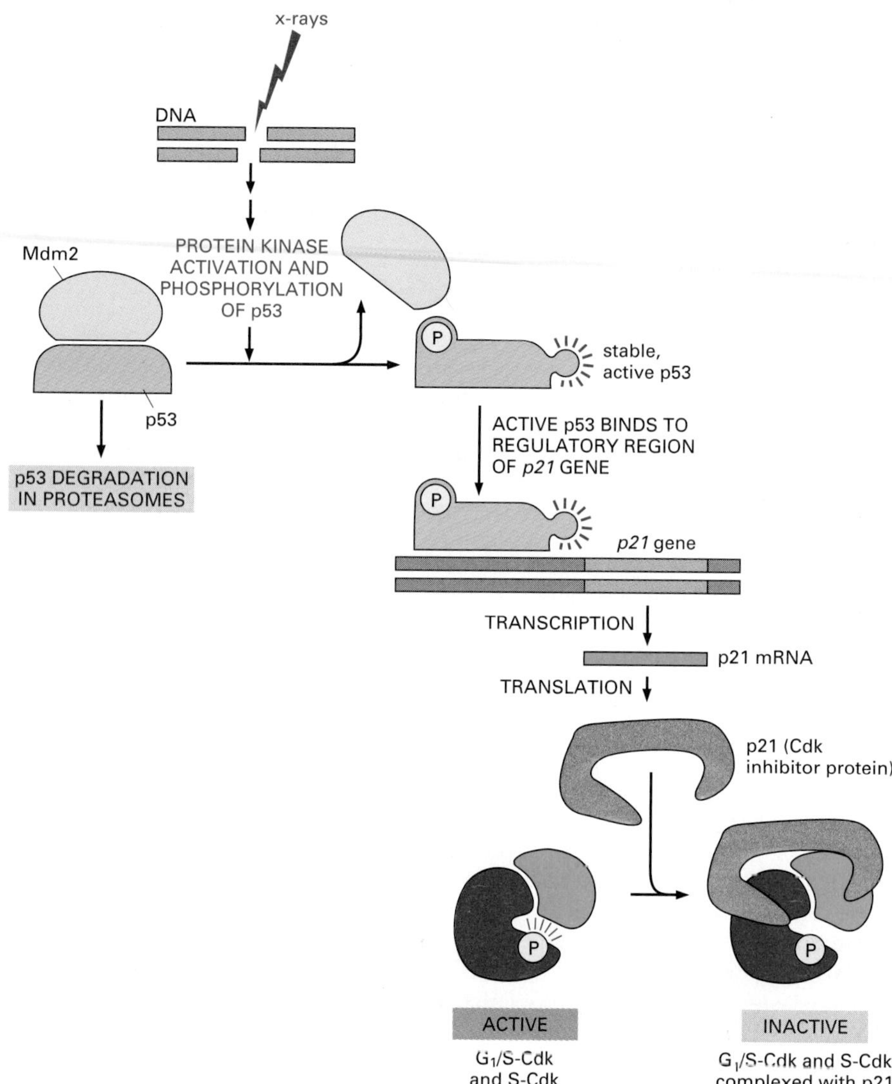

Figure 17–33 How DNA damage arrests the cell cycle in G_1. When DNA is damaged, protein kinases that phosphorylate p53 are activated. Mdm2 normally binds to p53 and promotes its ubiquitylation and destruction in proteasomes. Phosphorylation of p53 blocks its binding to Mdm2; as a result, p53 accumulates to high levels and stimulates transcription of the gene that encodes the CKI protein p21. The p21 binds and inactivates G_1/S-Cdk and S-Cdk complexes, arresting the cell in G_1. In some cases, DNA damage also induces either the phosphorylation of Mdm2 or a decrease in Mdm2 production, which causes an increase in p53 (not shown).

mutations. Indeed, mutations in the p53 gene occur in at least half of all human cancers (discussed in Chapter 23). This loss of p53 function allows the cancer cell to accumulate mutations more readily. Similarly, a rare genetic disease known as *ataxia telangiectasia* is caused by a defect in one of the protein kinases that phosphorylates and activates p53 in response to x-ray-induced DNA damage; patients with this disease are very sensitive to x-rays due to the loss of the DNA damage checkpoints, and they consequently suffer from increased rates of cancer.

What if DNA damage is so severe that repair is not possible? In this case, the response is different in different organisms. Unicellular organisms such as budding yeast transiently arrest their cell cycle to repair the damage. If repair cannot be completed, the cycle resumes despite any damage. For a single-celled organism, life with mutations is apparently better than no life at all. In multicellular organisms, however, the health of the organism takes precedence over the life of an individual cell. Cells that divide with severe DNA damage threaten the life of the organism, since genetic damage can often lead to cancer and other lethal defects. Thus, animal cells with severe DNA damage do not attempt to continue division, but instead commit suicide by undergoing programmed cell death, or apoptosis, as we discuss in the next section. The decision to die in this way also depends on the activation of p53, and it is this function of p53 that is apparently most important in protecting us aganist cancer.

As a review, the major cell-cycle regulatory proteins are summarized in Table 17–2, with the general structure of the cell-cycle control system shown in Figure 17–34.

TABLE 17–2 Summary of the Major Cell-cycle Regulatory Proteins

GENERAL NAME	FUNCTIONS AND COMMENTS
Protein kinases and protein phosphatases that modify Cdks	
Cdk-activating kinase (CAK)	phosphorylates an activating site in Cdks
Wee1 kinase	phosphorylates inhibitory sites in Cdks; primarily involved in controlling entry into mitosis
Cdc25 phosphatase	removes inhibitory phosphates from Cdks; three family members (Cdc25A, B, C) in mammals; Cdc25C is the activator of Cdk1 at the onset of mitosis
Cdk inhibitory proteins (CKIs)	
Sic1 (budding yeast)	suppresses Cdk activity in G_1; phosphorylation by Cdk1 triggers its destruction
p27 (mammals)	suppresses G_1/S-Cdk and S-Cdk activities in G_1; helps cells to withdraw from cell cycle when they terminally differentiate; phosphorylation by Cdk2 triggers its ubiquitylation by SCF
p21 (mammals)	suppresses G_1/S-Cdk and S-Cdk activities following DNA damage in G_1; transcriptionally activated by p53
p16 (mammals)	suppresses G_1-Cdk activity in G_1; frequently inactivated in cancer
Ubiquitin ligases and their activators	
SCF	catalyzes ubiquitylation of regulatory proteins involved in G_1 control, including CKIs (Sic1 in budding yeast, p27 in mammals); phosphorylation of target protein usually required for this activity
APC	catalyzes ubiquitylation of regulatory proteins involved primarily in exit from mitosis, including Securin and M-cyclins; regulated by association with activating subunits
Cdc20	APC-activating subunit in all cells; triggers initial activation of APC at metaphase-to-anaphase transition; stimulated by M-Cdk activity
Hct1	maintains APC activity after anaphase and throughout G_1; inhibited by Cdk activity
Gene regulatory proteins	
E2F	promotes transcription of genes required for G_1/S progression, including genes encoding G_1/S cyclins, S-cyclins, and proteins required for DNA synthesis; stimulated when G_1-Cdk phosphorylates Rb in response to extracellular mitogens
p53	promotes transcription of genes that induce cell cycle arrest (especially p21) or apoptosis in response to DNA damage or other cell stress; regulated by association with Mdm2, which promotes p53 degradation

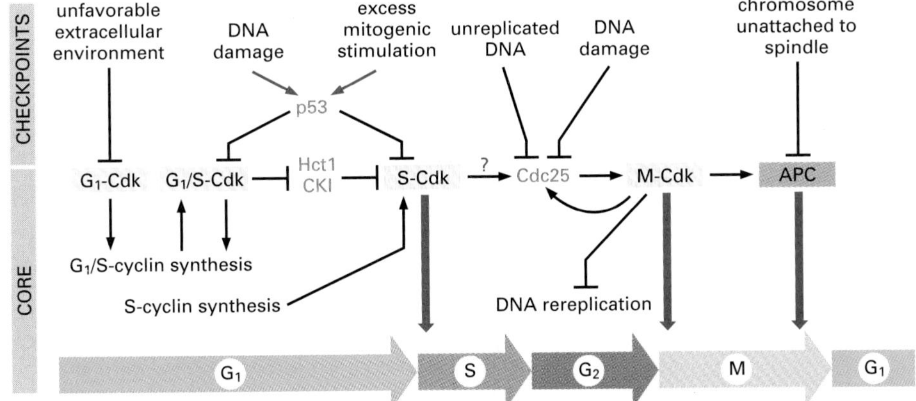

Figure 17–34 An overview of the cell-cycle control system. The core of the cell-cycle control system consists of a series of cyclin-Cdk complexes *(yellow)*. The activity of each complex is also influenced by various inhibitory checkpoint mechanisms, which provide information about the extracellular environment, cell damage, and incomplete cell-cycle events *(top)*. These mechanisms are not present in all cell types; many are missing in early embryonic cell cycles, for example.

Summary

An ordered sequence of cyclin–Cdk activities triggers most of the events of the cell cycle. During G_1 phase, Cdk activity is reduced to a minimum by Cdk inhibitors (CKIs), cyclin proteolysis, and decreased cyclin gene transcription. When environmental conditions are favorable, G_1- and G_1/S-Cdks increase in concentration, overcoming these inhibitory barriers in late G_1 and triggering the activation of S-Cdk. The S-Cdk phosphorylates proteins at DNA replication origins, initiating DNA synthesis through a mechanism that ensures that the DNA is duplicated only once per cell cycle.

Once S phase is completed, the activation of M-Cdk leads to the events of early mitosis, whereby the cell assembles a mitotic spindle and prepares for segregation of the duplicated chromosomes—which consist of sister chromatids glued together. Anaphase is triggered by the destruction of the proteins that hold the sisters together. The M-Cdk is then inactivated by cyclin proteolysis, which leads to cytokinesis and the end of M phase. Progression through the cell cycle is regulated precisely by various inhibitory mechanisms that arrest the cell cycle at specific checkpoints when events are not completed successfully, when DNA damage occurs, or when extracellular conditions are unfavorable.

PROGRAMMED CELL DEATH (APOPTOSIS)

The cells of a multicellular organism are members of a highly organized community. The number of cells in this community is tightly regulated—not simply by controlling the rate of cell division, but also by controlling the rate of cell death. If cells are no longer needed, they commit suicide by activating an intracellular death program. This process is therefore called **programmed cell death**, although it is more commonly called **apoptosis** (from a Greek word meaning "falling off," as leaves from a tree).

The amount of apoptosis that occurs in developing and adult animal tissues can be astonishing. In the developing vertebrate nervous system, for example, up to half or more of the nerve cells normally die soon after they are formed. In a healthy adult human, billions of cells die in the bone marrow and intestine every hour. It seems remarkably wasteful for so many cells to die, especially as the vast majority are perfectly healthy at the time they kill themselves. What purposes does this massive cell death serve?

In some cases, the answers are clear. Mouse paws, for example, are sculpted by cell death during embryonic development: they start out as spadelike structures, and the individual digits separate only as the cells between them die (Figure 17–35). In other cases, cells die when the structure they form is no longer needed. When a tadpole changes into a frog, the cells in the tail die, and the tail, which is not needed in the frog, disappears (Figure 17–36). In many other cases, cell death helps regulate cell numbers. In the developing nervous system, for example, cell death adjusts the number of nerve cells to match the number of target cells that require innervation. In all these cases, the cells die by apoptosis.

In adult tissues, cell death exactly balances cell division. If this were not so, the tissue would grow or shrink. If part of the liver is removed in an adult rat, for example, liver cell proliferation increases to make up the loss. Conversely, if a rat

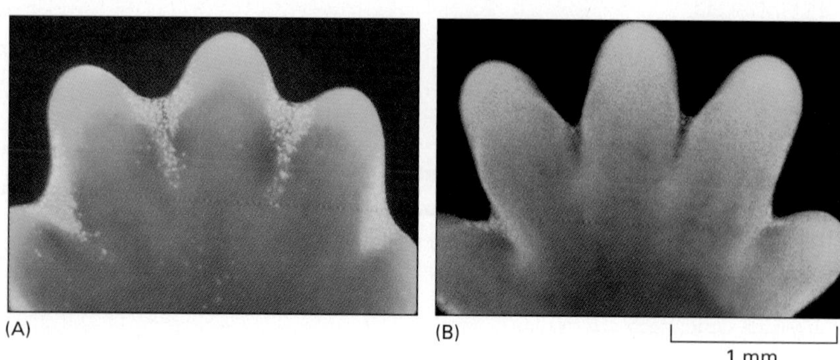

Figure 17–35 Sculpting the digits in the developing mouse paw by apoptosis. (A) The paw in this mouse embryo has been stained with a dye that specifically labels cells that have undergone apoptosis. The apoptotic cells appear as *bright green* dots between the developing digits. (B) This interdigital cell death eliminates the tissue between the developing digits, as seen one day later, when few, if any, apoptotic cells can be seen. (From W. Wood et al., *Development* 127:5245–5252, 2000. © The Company of Biologists.)

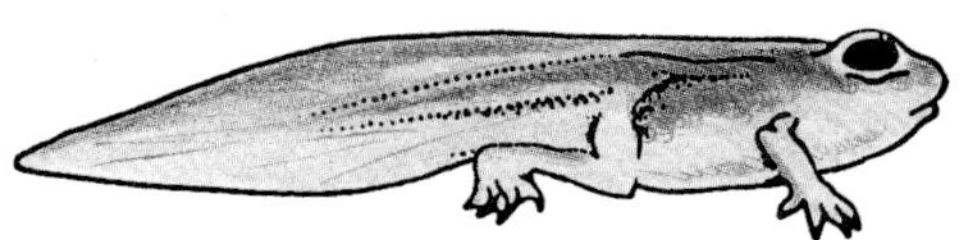

Figure 17–36 Apoptosis during the metamorphosis of a tadpole into a frog. As a tadpole changes into a frog, the cells in the tadpole tail are induced to undergo apoptosis; as a consequence, the tail is lost. All the changes that occur during metamorphosis, including the induction of apoptosis in the tail, are stimulated by an increase in thyroid hormone in the blood.

is treated with the drug phenobarbital—which stimulates liver cell division (and thereby liver enlargement)—and then the phenobarbital treatment is stopped, apoptosis in the liver greatly increases until the liver has returned to its original size, usually within a week or so. Thus, the liver is kept at a constant size through the regulation of both the cell death rate and the cell birth rate.

In this short section, we describe the molecular mechanisms of apoptosis and its control. In the final section, we consider how the extracellular control of cell proliferation and cell death contributes to the regulation of cell numbers in multicellular organisms.

Apoptosis Is Mediated by an Intracellular Proteolytic Cascade

Cells that die as a result of acute injury typically swell and burst. They spill their contents all over their neighbors—a process called *cell necrosis*—causing a potentially damaging inflammatory response. By contrast, a cell that undergoes apoptosis dies neatly, without damaging its neighbors. The cell shrinks and condenses. The cytoskeleton collapses, the nuclear envelope disassembles, and the nuclear DNA breaks up into fragments. Most importantly, the cell surface is altered, displaying properties that cause the dying cell to be rapidly phagocytosed, either by a neighboring cell or by a macrophage (a specialized phagocytic cell, discussed in Chapter 24), before any leakage of its contents occurs (Figure 17–37). This not only avoids the damaging consequences of cell necrosis but also allows the organic components of the dead cell to be recycled by the cell that ingests it.

The intracellular machinery responsible for apoptosis seems to be similar in all animal cells. This machinery depends on a family of proteases that have a cysteine at their active site and cleave their target proteins at specific aspartic acids. They are therefore called **caspases**. Caspases are synthesized in the cell as inactive precursors, or *procaspases,* which are usually activated by cleavage at aspartic acids by other caspases (Figure 17–38A). Once activated, caspases cleave, and thereby activate, other procaspases, resulting in an amplifying proteolytic

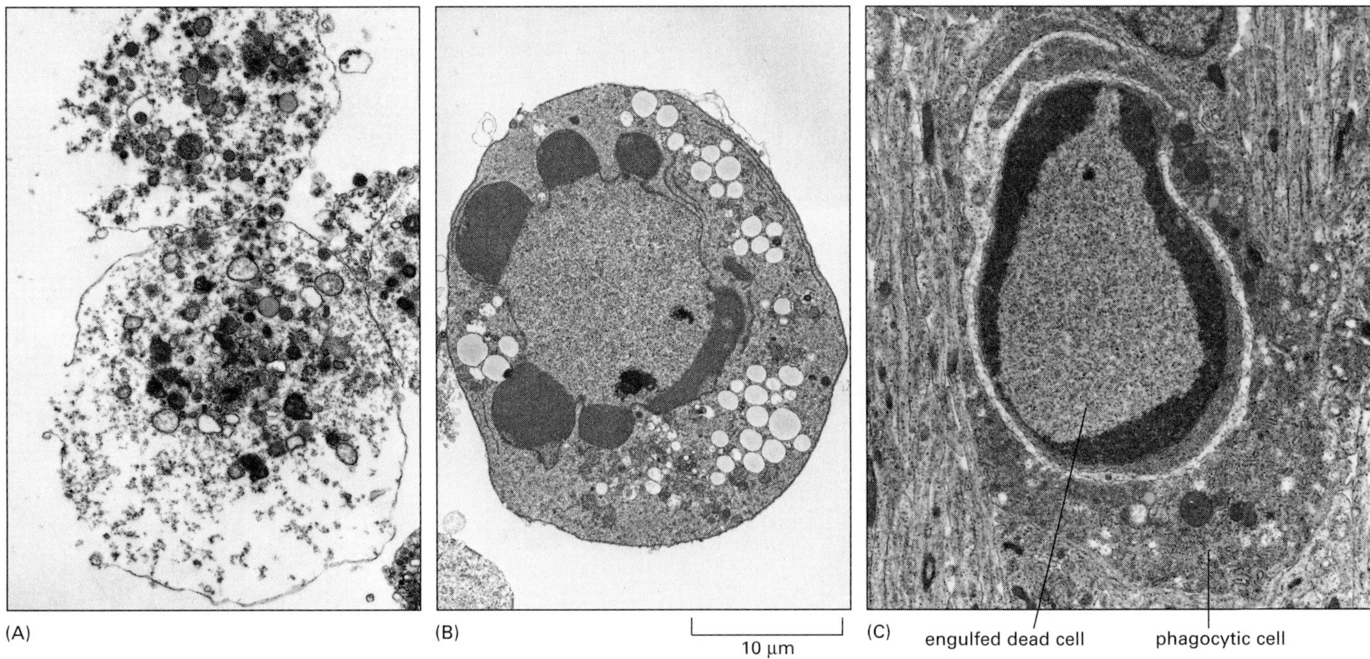

Figure 17–37 Cell death. These electron micrographs show cells that have died by (A) necrosis or (B and C) apoptosis. The cells in (A) and (B) died in a culture dish, whereas the cell in (C) died in a developing tissue and has been engulfed by a neighboring cell. Note that the cell in (A) seems to have exploded, whereas those in (B) and (C) have condensed but seem relatively intact. The large vacuoles visible in the cytoplasm of the cell in (B) are a variable feature of apoptosis. (Courtesy of Julia Burne.)

cascade (Figure 17–38B). Some of the activated caspases then cleave other key proteins in the cell. Some cleave the nuclear lamins, for example, causing the irreversible breakdown of the nuclear lamina; another cleaves a protein that normally holds a DNA-degrading enzyme (a DNAse) in an inactive form, freeing the DNAse to cut up the DNA in the cell nucleus. In this way, the cell dismantles itself quickly and neatly, and its corpse is rapidly taken up and digested by another cell.

Activation of the intracellular cell death pathway, like entry into a new stage of the cell cycle, is usually triggered in a complete, all-or-none fashion. The protease cascade is not only destructive and self-amplifying but also irreversible, so that once a cell reaches a critical point along the path to destruction, it cannot turn back.

Procaspases Are Activated by Binding to Adaptor Proteins

All nucleated animal cells contain the seeds of their own destruction, in the form of various inactive procaspases that lie waiting for a signal to destroy the cell. It is therefore not surprising that caspase activity is tightly regulated inside the cell to ensure that the death program is held in check until needed.

How are procaspases activated to initiate the caspase cascade? A general principle is that the activation is triggered by **adaptor proteins** that bring multiple copies of specific procaspases, known as *initiator procaspases,* close together in a complex or aggregate. In some cases, the initiator procaspases have a small amount of protease activity, and forcing them together into a complex causes them to cleave each other, triggering their mutual activation. In other cases, the aggregation is thought to cause a conformational change that activates the procaspase. Within moments, the activated caspase at the top of the cascade cleaves downstream procaspases to amplify the death signal and spread it throughout the cell (see Figure 17–38B).

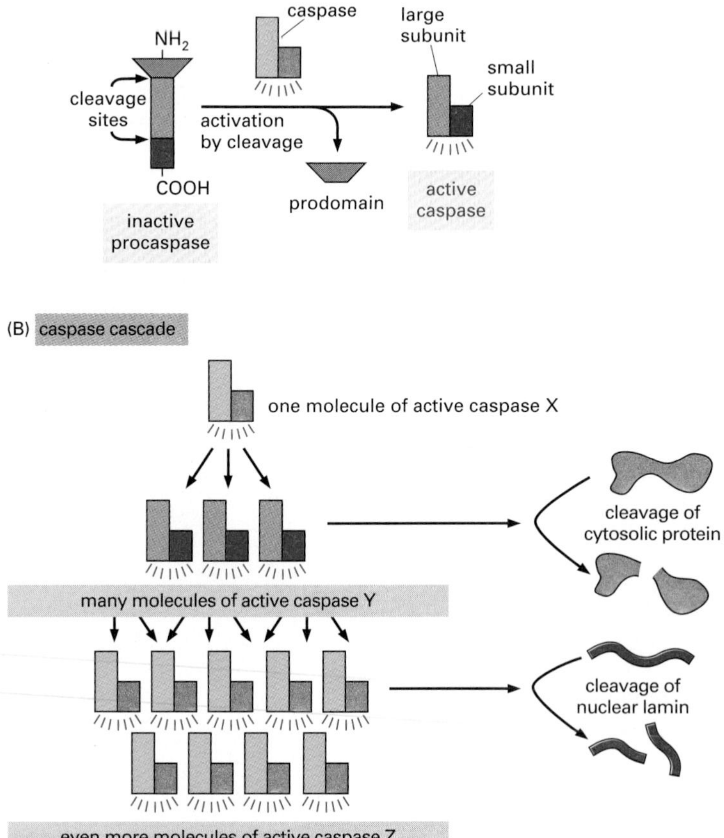

Figure 17–38 The caspase cascade involved in apoptosis. (A) Each suicide protease is made as an inactive proenzyme (procaspase), which is usually activated by proteolytic cleavage by another member of the caspase family. As indicated, two of the cleaved fragments associate to form the active site of the caspase. The active enzyme is thought to be a tetramer of two of these units (not shown). (B) Each activated caspase molecule can cleave many procaspase molecules, thereby activating them, and these can then activate even more procaspase molecules. In this way, an initial activation of a small number of procaspase molecules (called initiator caspases) can lead, via an amplifying chain reaction (a cascade), to the explosive activation of a large number of procaspase molecules. Some of the activated caspases (called effector caspases) then cleave a number of key proteins in the cell, including specific cytosolic proteins and nuclear lamins, leading to the controlled death of the cell.

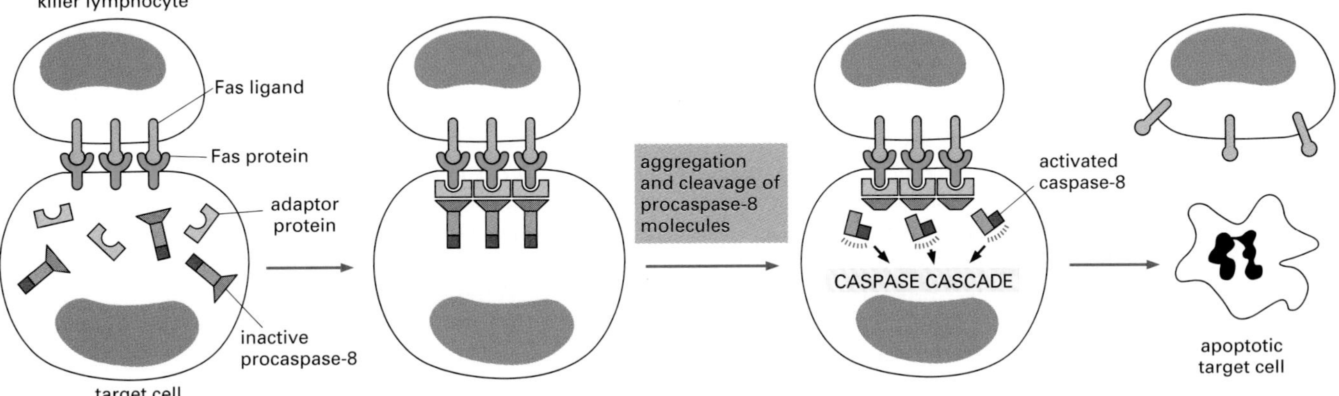

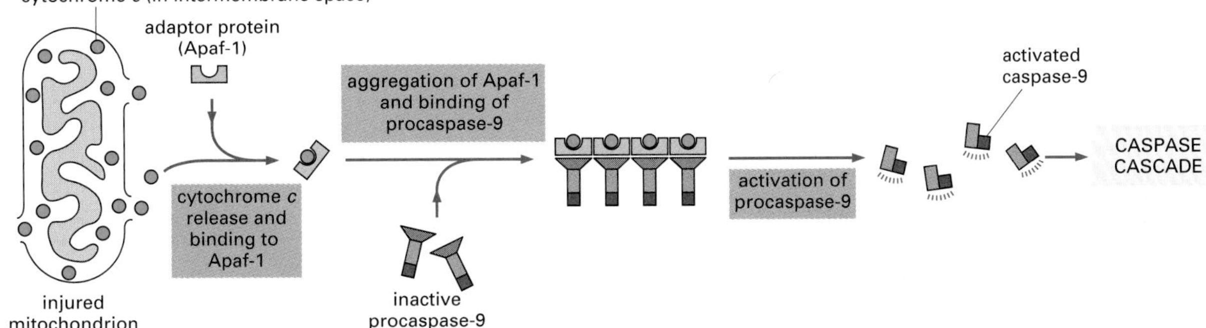

Figure 17–39 Induction of apoptosis by either extracellular or intracellular stimuli. (A) Extracellular activation. A killer lymphocyte carrying the Fas ligand binds and activates Fas proteins on the surface of the target cell. Adaptor proteins bind to the intracellular region of aggregated Fas proteins, causing the aggregation of procaspase-8 molecules. These then cleave one another to initiate the caspase cascade. (B) Intracellular activation. Mitochondria release cytochrome *c*, which binds to and causes the aggregation of the adaptor protein Apaf-1. Apaf-1 binds and aggregates procaspase-9 molecules, which leads to the cleavage of these molecules and the triggering of a caspase cascade. Other proteins that contribute to apoptosis are also released from the mitochondrial intermembrane space (not shown).

Procaspase activation can be triggered from outside the cell by the activation of *death receptors* on the cell surface. Killer lymphocytes (discussed in Chapter 24), for example, can induce apoptosis by producing a protein called **Fas ligand**, which binds to the death receptor protein **Fas** on the surface of the target cell. The clustered Fas proteins then recruit intracellular adaptor proteins that bind and aggregate procaspase-8 molecules, which cleave and activate one another. The activated caspase-8 molecules then activate downstream procaspases to induce apoptosis (Figure 17–39A). Some stressed or damaged cells kill themselves by producing both the Fas ligand and the Fas protein, thereby triggering an intracellular caspase cascade.

When cells are damaged or stressed, they can also kill themselves by triggering procaspase aggregation and activation from within the cell. In the best understood pathway, mitochondria are induced to release the electron carrier protein *cytochrome c* (see Figure 14–26) into the cytosol, where it binds and activates an adaptor protein called **Apaf-1** (Figure 17–39B). This mitochondrial pathway of procaspase activation is recruited in most forms of apoptosis to initiate or to accelerate and amplify the caspase cascade. DNA damage, for example, as discussed earlier, can trigger apoptosis. This response usually requires p53, which can activate the transcription of genes that encode proteins that promote the release of cytochrome *c* from mitochondria. These proteins belong to the Bcl-2 family.

Bcl-2 Family Proteins and IAP Proteins Are the Main Intracellular Regulators of the Cell Death Program

The **Bcl-2 family** of intracellular proteins helps regulate the activation of procaspases. Some members of this family, like *Bcl-2* itself or *Bcl-X*$_L$, inhibit apoptosis, at least partly by blocking the release of cytochrome *c* from mitochondria. Other members of the Bcl-2 family are not death inhibitors, but instead promote procaspase activation and cell death. Some of these apoptosis promoters, such as *Bad*, function by binding to and inactivating the death-inhibiting members of the family, whereas others, like *Bax* and *Bak*, stimulate the release of cytochrome *c* from mitochondria. If the genes encoding Bax and Bak are both inactivated, cells are remarkably resistant to most apoptosis-inducing stimuli, indicating the crucial importance of these proteins in apoptosis induction. Bax and Bak are themselves activated by other apoptosis-promoting members of the Bcl-2 family such as Bid.

Another important family of intracellular apoptosis regulators is the **IAP (inhibitor of apoptosis) family**. These proteins are thought to inhibit apoptosis in two ways: they bind to some procaspases to prevent their activation, and they bind to caspases to inhibit their activity. IAP proteins were originally discovered as proteins produced by certain insect viruses, which use them to prevent the infected cell from killing itself before the virus has had time to replicate. When mitochondria release cytochrome *c* to activate Apaf-1, they also release a protein that blocks IAPs, thereby greatly increasing the efficiency of the death activation process.

The intracellular cell death program is also regulated by extracellular signals, which can either activate apoptosis or inhibit it. These signal molecules mainly act by regulating the levels or activity of members of the Bcl-2 and IAP families. We see in the next section how these signal molecules help multicellular organisms regulate their cell numbers.

Summary

In multicellular organisms, cells that are no longer needed or are a threat to the organism are destroyed by a tightly regulated cell suicide process known as programmed cell death, or apoptosis. Apoptosis is mediated by proteolytic enzymes called caspases, which trigger cell death by cleaving specific proteins in the cytoplasm and nucleus. Caspases exist in all cells as inactive precursors, or procaspases, which are usually activated by cleavage by other caspases, producing a proteolytic caspase cascade. The activation process is initiated by either extracellular or intracellular death signals, which cause intracellular adaptor molecules to aggregate and activate procaspases. Caspase activation is regulated by members of the Bcl-2 and IAP protein families.

EXTRACELLULAR CONTROL OF CELL DIVISION, CELL GROWTH, AND APOPTOSIS

A fertilized mouse egg and a fertilized human egg are similar in size, yet they produce animals of very different sizes. What factors in the control of cell behavior in humans and mice are responsible for these size differences? The same fundamental question can be asked for each organ and tissue in an animal's body. What factors in the control of cell behavior explain the length of an elephant's trunk or the size of its brain or its liver? These questions are largely unanswered, at least in part because they have received relatively little attention compared with other questions in cell and developmental biology. It is nevertheless possible to say what the ingredients of an answer must be.

The size of an organ or organism depends mainly on its total cell mass, which depends on both the total number of cells and the size of the cells. Cell number, in turn, depends on the amounts of cell division and cell death. Organ and body size are therefore determined by three fundamental processes: cell growth, cell division, and cell death. Each is independently regulated—both by intracellular programs and by extracellular signal molecules that control these programs.

The extracellular signal molecules that regulate cell size and cell number are generally either soluble secreted proteins, proteins bound to the surface of cells, or components of the extracellular matrix. The factors that promote organ or organism growth can be operationally divided into three major classes:

1. *Mitogens*, which stimulate cell division, primarily by relieving intracellular negative controls that otherwise block progress through the cell cycle.

2. *Growth factors,* which stimulate cell growth (an increase in cell mass) by promoting the synthesis of proteins and other macromolecules and by inhibiting their degradation.
3. *Survival factors,* which promote cell survival by suppressing apoptosis.

Some extracellular signal molecules promote all of these processes, while others promote one or two of them. Indeed, the term *growth factor* is often used inappropriately to describe a factor that has any of these activities. Even worse, the term *cell growth* is often used to mean an increase in cell number, or *cell proliferation.*

In this section, we first discuss how these extracellular signals stimulate cell division, cell growth, and cell survival, thereby promoting the growth of an animal and its organs. We then consider how other extracellular signals can act in the opposite way, to inhibit cell growth or cell division or to stimulate apoptosis, thereby inhibiting organ growth.

Mitogens Stimulate Cell Division

Unicellular organisms tend to grow and divide as fast as they can, and their rate of proliferation depends largely on the availability of nutrients in the environment. The cells of a multicellular organism, however, divide only when more cells are needed by the organism. Thus, for an animal cell to proliferate, nutrients are not enough. It must also receive stimulatory extracellular signals, in the form of **mitogens**, from other cells, usually its neighbors. Mitogens act to overcome intracellular braking mechanisms that block progress through the cell cycle.

One of the first mitogens to be identified was **platelet-derived growth factor (PDGF)**, and it is typical of many others discovered since. The path to its isolation began with the observation that fibroblasts in a culture dish proliferate when provided with *serum* but not when provided with *plasma.* Plasma is prepared by removing the cells from blood without allowing clotting to occur; serum is prepared by allowing blood to clot and taking the cell-free liquid that remains. When blood clots, platelets incorporated in the clot are triggered to release the contents of their secretory vesicles (Figure 17–40). The superior ability of serum to support cell proliferation suggested that platelets contain one or more mitogens. This hypothesis was confirmed by showing that extracts of platelets could serve instead of serum to stimulate fibroblast proliferation. The crucial factor in the extracts was shown to be a protein, which was subsequently purified and named PDGF. In the body, PDGF liberated from blood clots probably has a major role in stimulating cell division during wound healing.

PDGF is only one of over 50 proteins that are known to act as mitogens. Most of these proteins are broad-specificity factors, like PDGF and *epidermal growth factor (EGF),* that can stimulate many types of cells to divide. Thus, PDGF acts on a range of cell types, including fibroblasts, smooth muscle cells, and neuroglial cells. Similarly, EGF acts not only on epidermal cells but also on many other cell types, including both epithelial and nonepithelial cells. At the opposite extreme lie narrow-specificity factors such as *erythropoietin,* which induces the proliferation of red blood cell precursors only.

In addition to mitogens that stimulate cell division, there are factors, such as some members of the *transforming growth factor-β (TGF-β)* family, that act on some cells to stimulate cell proliferation and others to inhibit it, or that stimulate at one concentration and inhibit at another. Indeed, like PDGF, many mitogens have other actions beside the stimulation of cell division: they can stimulate cell growth, survival, differentiation, or migration, depending on the circumstances and the cell type.

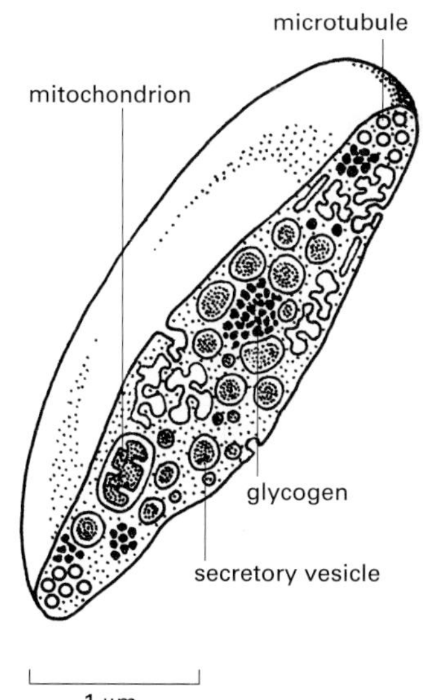

Figure 17–40 A platelet. Platelets are miniature cells without a nucleus. They circulate in the blood and help stimulate blood clotting at sites of tissue damage, thereby preventing excessive bleeding. They also release various factors that stimulate healing. The platelet shown here has been cut in half to show its secretory vesicles, some of which contain platelet-derived growth factor (PDGF). See also Figure 16–47B–D.

Cells Can Delay Division by Entering a Specialized Nondividing State

In the absence of a mitogenic signal to proliferate, Cdk inhibition in G_1 is maintained, and the cell cycle arrests. In some cases, cells partly disassemble their cell-cycle control system and exit from the cycle to a specialized, nondividing state called $\mathbf{G_0}$.

Most cells in our body are in G_0, but the molecular basis and reversibility of this state vary in different cell types. Neurons and skeletal muscle cells, for example, are in a *terminally differentiated* G_0 state, in which their cell-cycle control system is completely dismantled: the expression of the genes encoding various Cdks and cyclins are permanently turned off, and cell division never occurs. Other cell types withdraw from the cell cycle only transiently and retain the ability to reassemble the cell-cycle control system quickly and reenter the cycle. Most liver cells, for example, are in G_0, but they can be stimulated to divide if the liver is damaged. Still other types of cells, including some lymphocytes, withdraw from and re-enter the cell cycle repeatedly throughout their lifetime.

Almost all the variation in cell-cycle length in the adult body occurs during the time the cell spends in G_1 or G_0. By contrast, the time taken for a cell to progress from the beginning of S phase through mitosis is usually brief (typically 12–24 hours in mammals) and relatively constant, regardless of the interval from one division to the next.

Mitogens Stimulate G_1-Cdk and G_1/S-Cdk Activities

For the vast majority of animal cells, mitogens control the rate of cell division by acting in the G_1 phase of the cell cycle. As discussed earlier, multiple mechanisms act during G_1 to suppress Cdk activity and thereby hinder entry into S phase. Mitogens act to release the brakes on Cdk activity, thereby allowing S phase to begin. They do so by binding to cell-surface receptors to initiate a complex array of intracellular signals that penetrate deep into the cytoplasm and nucleus (discussed in Chapter 15). The ultimate result is the activation of G_1-Cdk and G_1/S-Cdk complexes, which overcome the inhibitory barriers that normally block progression into S phase.

As we discuss in Chapter 15, an early step in mitogen signaling is often the activation of the small GTPase **Ras**, which leads to the activation of a *MAP kinase cascade*. By uncertain mechanisms, this leads to increased levels of the gene regulatory protein **Myc**. Myc promotes cell-cycle entry by several overlapping mechanisms (Figure 17–41). It increases the transcription of genes that encode G_1 cyclins (D cyclins), thereby increasing G_1-Cdk (cyclin D–Cdk4) activity. In addition, Myc increases the transcription of a gene that encodes a component of the SCF ubiquitin ligase. This mechanism promotes the degradation of the CKI protein p27, leading to increased G_1/S-Cdk (cyclin E–Cdk2) activity. As discussed earlier, increased G_1-Cdk and G_1/S-Cdk activities stimulate phosphorylation of the inhibitory protein Rb, which then leads to activation of the gene regulatory protein E2F. Myc may also stimulate the transcription of the gene encoding E2F, further promoting E2F activity in the cell. The end result is the increased transcription of genes required for entry into S phase (see Figure 17–30). As we discuss later, Myc also has a major role in stimulating the transcription of genes that increase cell growth.

Abnormal Proliferation Signals Cause Cell-Cycle Arrest or Cell Death

As we discuss in Chapter 23, many of the components of intracellular signaling pathways are encoded by genes that were originally identified as cancer-promoting genes, or *oncogenes*, because mutations in them contribute to the development of cancer. The mutation of a single amino acid in Ras, for example, causes the protein to become permanently overactive, leading to constant stimulation of Ras-dependent signaling pathways, even in the absence of mitogenic stimulation. Similarly, mutations that cause an overexpression of Myc promote excessive cell growth and proliferation and thereby promote the development of cancer.

Surprisingly, however, when Ras or Myc is experimentally hyperactivated in most normal cells, the result is not excessive proliferation but the opposite: the activation of checkpoint mechanisms causes the cells to undergo either cell-cycle arrest or apoptosis. The normal cell seems able to detect abnormal mitogenic stimulation, and it responds by preventing further division. Such

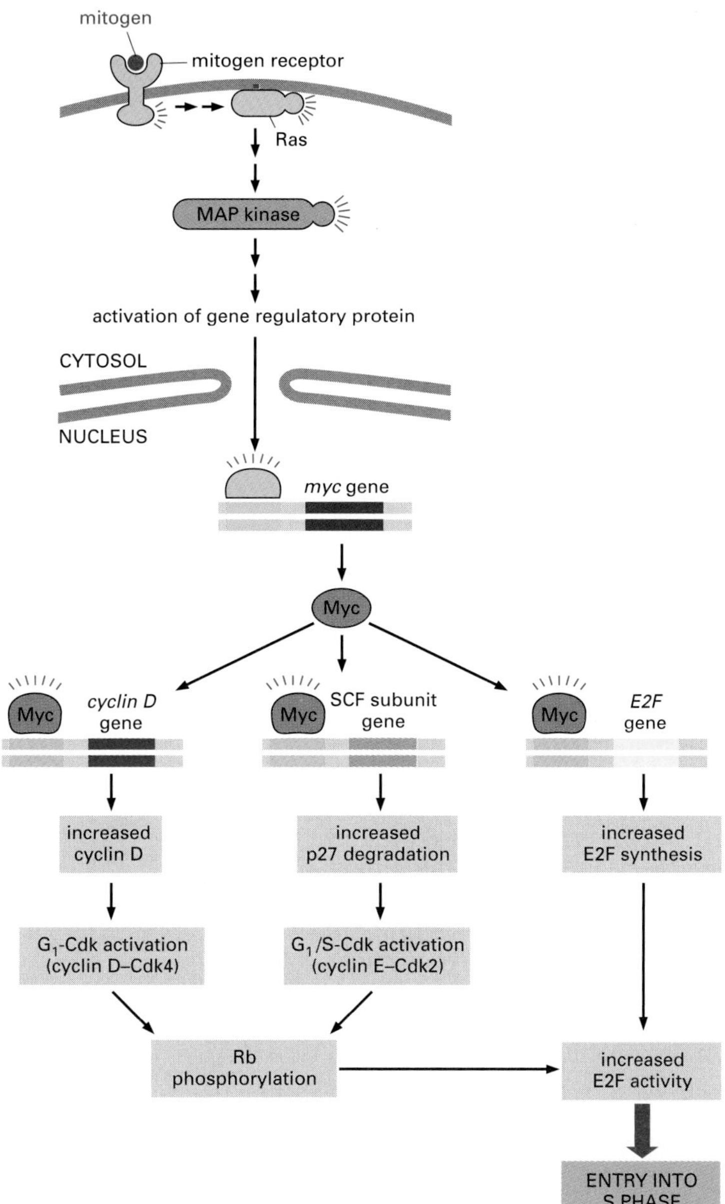

Figure 17–41 A simplified model of one way that mitogens stimulate cell division. The binding of mitogens to cell-surface receptors leads to the activation of Ras and a MAP kinase cascade. One effect of this pathway is the increased production of the gene regulatory protein Myc. Myc increases the transcription of several genes, including the gene encoding cyclin D and a gene encoding a subunit of the SCF ubiquitin ligase. The resulting increase in G_1-Cdk and G_1/S-Cdk activities promotes Rb phosphorylation and activation of the gene regulatory protein E2F, resulting in S-phase entry (see Figure 17–30). Myc may also promote E2F activity directly by stimulating the transcription of the *E2F* gene. Although, for simplicity, Myc is shown as a monomer, it functions as a heterodimer with another protein called Max.

checkpoint responses help prevent the survival and proliferation of cells with various cancer-promoting mutations.

Although it is not known how a cell detects excessive mitogenic stimulation, such stimulation often leads to the production of a cell-cycle inhibitor protein called **p19$^{\mathbf{ARF}}$**, which binds and inhibits Mdm2. As discussed earlier, Mdm2 normally promotes p53 degradation. Activation of p19^{ARF} therefore causes p53 levels to increase, thereby inducing either cell-cycle arrest or apoptosis (Figure 17–42).

How do cancer cells ever arise if these mechanisms block the division or survival of mutant cells with overactive proliferation signals? The answer is that the protective system is often inactivated in cancer cells by mutations in the genes that encode essential components of the checkpoint responses, such as p19^{ARF} or p53.

Human Cells Have a Built-in Limitation on the Number of Times They Can Divide

Cell division is controlled not only by extracellular mitogens but also by intracellular mechanisms that can limit cell proliferation. Many animal precursor cells, for example, divide a limited number of times before they stop and terminally

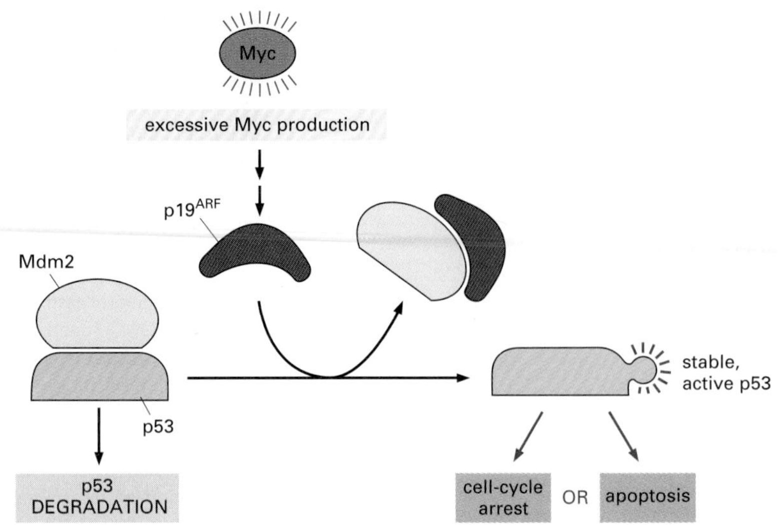

Figure 17–42 Cell-cycle arrest or apoptosis induced by excessive stimulation of mitogenic pathways. Abnormally high levels of Myc cause the activation of $p19^{ARF}$, which binds and inhibits Mdm2 and thereby causes increased p53 levels (see Figure 17–33). Depending on the cell type and extracellular conditions, p53 then causes either cell-cycle arrest or apoptosis.

differentiate into permanently arrested, specialized cells. Although the stopping mechanisms are poorly understood, a progressive increase in CKI proteins probably contributes in some cases. Mice that are deficient in the CKI p27, for example, have more cells than normal in all of their organs because the stopping mechanisms are apparently defective.

The best-understood intracellular mechanism that limits cell proliferation occurs in human fibroblasts. Fibroblasts taken from a normal human tissue go through only about 25–50 population doublings when cultured in a standard mitogenic medium. Toward the end of this time, proliferation slows down and finally halts, and the cells enter a nondividing state from which they never recover. This phenomenon is called **replicative cell senescence**, although it is unlikely to be responsible for the senescence (aging) of the organism. Organism senescence is thought to depend, in part at least, on progressive oxidative damage to macromolecules, in as much as strategies that reduce metabolism (such as reduced food intake), and thereby reduce the production of reactive oxygen species, can extend the lifespan of experimental animals.

Replicative cell senescence in human fibroblasts seems to be caused by changes in the structure of the **telomeres**, the repetitive DNA sequences and associated proteins at the ends of chromosomes. As discussed in Chapter 5, when a cell divides, telomeric DNA sequences are not replicated in the same manner as the rest of the genome but instead are synthesized by the enzyme **telomerase**. By mechanisms that remain unclear, telomerase also promotes the formation of protein cap structures that protect the chromosome ends. Because human fibroblasts, and many other human somatic cells, are deficient in telomerase, their telomeres become shorter with every cell division, and their protective protein caps progressively deteriorate. Eventually, DNA damage occurs at chromosome ends. The damage activates a p53-dependent cell-cycle arrest that resembles the arrest caused by other types of DNA damage (see Figure 17–33).

The lack of telomerase in most somatic cells has been proposed to help protect humans from the potentially damaging effects of runaway cell proliferation, as occurs in cancer. Unfortunately, most cancer cells have regained the ability to produce telomerase and therefore maintain telomere function as they proliferate; as a result, they do not undergo replicative cell senescence (discussed in Chapter 23). The forced expression of telomerase in normal human fibroblasts, using genetic engineering techniques, has the same effect (Figure 17–43).

Normal rodent cells, by contrast, usually maintain telomerase activity and telomere function as they proliferate and therefore do not undergo this type of replicative senescence. When overstimulated to proliferate in culture, however, they frequently activate the $p19^{ARF}$-dependent checkpoint mechanism described earlier and eventually stop dividing. Mutations that inactivate these checkpoints make it easier for rodent cells to proliferate indefinitely in culture. Such mutant cells are often described as "immortalized". If cultured in optimal

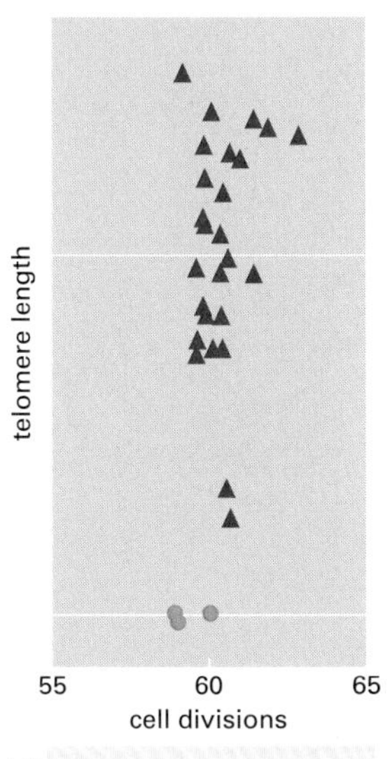

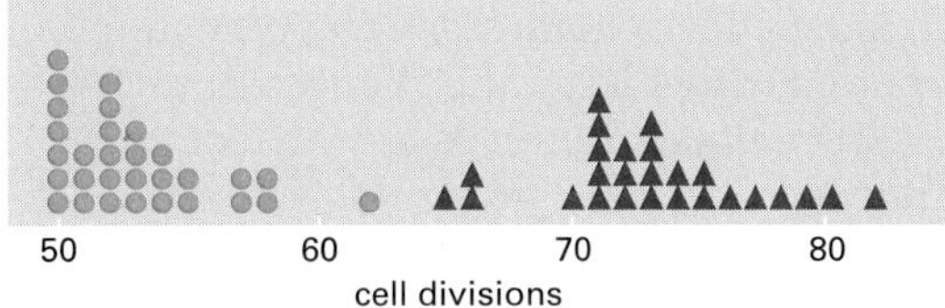

Figure 17–43 Overcoming replicative cell senescence by the forced expression of telomerase. (A) Normal human fibroblasts do not contain telomerase, and so their telomeres gradually shorten and lose their normal cap structure as the cells proliferate. Cells forced to express telomerase, however, maintain telomere length (and normal cap structures) after many divisions. (B) The normal human fibroblasts stopped dividing after about 50–60 divisions in these experiments, where as the cells expressing telomerase were still dividing at the end of the experiment. (Based on A. Bodnar et al., *Science* 279:349–352, 1998.)

conditions that avoid the activation of checkpoint responses, however, at least some normal rodent cells also seem able to proliferate indefinitely. Nevertheless, rodents age much more rapidly than humans.

Extracellular Growth Factors Stimulate Cell Growth

The growth of an organism or organ depends on cell growth: cell division alone cannot increase total cell mass without cell growth. In single-celled organisms such as yeasts, cell growth (like cell division) requires only nutrients. In animals, by contrast, cell growth and cell division both depend on signals from other cells.

The extracellular **growth factors** that stimulate cell growth bind to receptors on the cell surface and activate intracellular signaling pathways. These pathways stimulate the accumulation of proteins and other macromolecules, and they do so by both increasing their rate of synthesis and decreasing their rate of degradation.

One of the most important intracellular signaling pathways activated by growth factor receptors involves the enzyme **PI 3-kinase**, which adds a phosphate from ATP to the 3 position of inositol phospholipids in the plasma membrane. As discussed in Chapter 15, the activation of PI 3-kinase leads to the activation of several protein kinases, including **S6 kinase**. The S6 kinase phosphorylates ribosomal protein S6, increasing the ability of ribosomes to translate a subset of mRNAs, most of which encode ribosomal components. Protein synthesis therefore increases. When the gene encoding S6 kinase is inactivated in *Drosophila*, the mutant flies are small; whereas cell numbers are normal, cell size is abnormally small. Growth factors also activate a translation initiation factor called *eIF4E*, further increasing protein synthesis and cell growth (Figure 17–44).

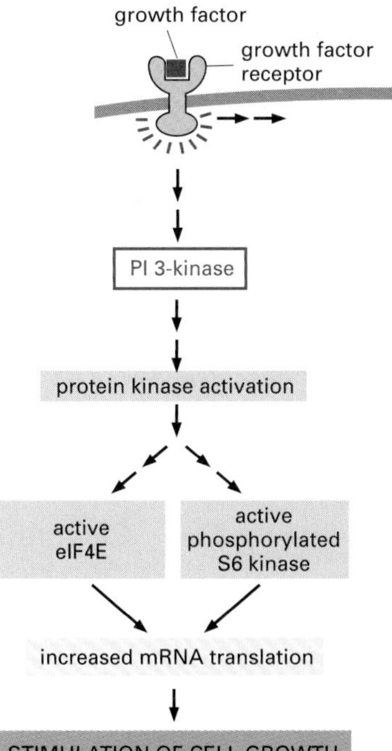

Figure 17–44 One way in which growth factors promote cell growth. In this simplified scheme, activation of cell-surface receptors leads to the activation of PI 3-kinase, which promotes protein synthesis, at least partly through the activation of eIF4E and S6 kinase. Growth factors also inhibit protein breakdown (not shown) by poorly understood pathways.

Growth factor stimulation also leads to increased production of the gene regulatory protein Myc, which also plays an important part in signaling by mitogens (see Figure 17–41). Myc increases the transcription of a number of genes that encode proteins involved in cell metabolism and macromolecular synthesis. In this way, it stimulates both cell metabolism and cell growth.

Some extracellular signal proteins, including PDGF, can act as both growth factors and mitogens, stimulating both cell growth and cell-cycle progression. This functional overlap is achieved in part by overlaps in the intracellular signaling pathways that control these two processes. The signaling protein Ras, for example, is activated by both growth factors and mitogens. It can stimulate the PI3-kinase pathway to promote cell growth and the MAP-kinase pathway to trigger cell-cycle progression. Similarly, as described above, Myc stimulates both cell growth and cell-cycle progression. Extracellular factors that act as both

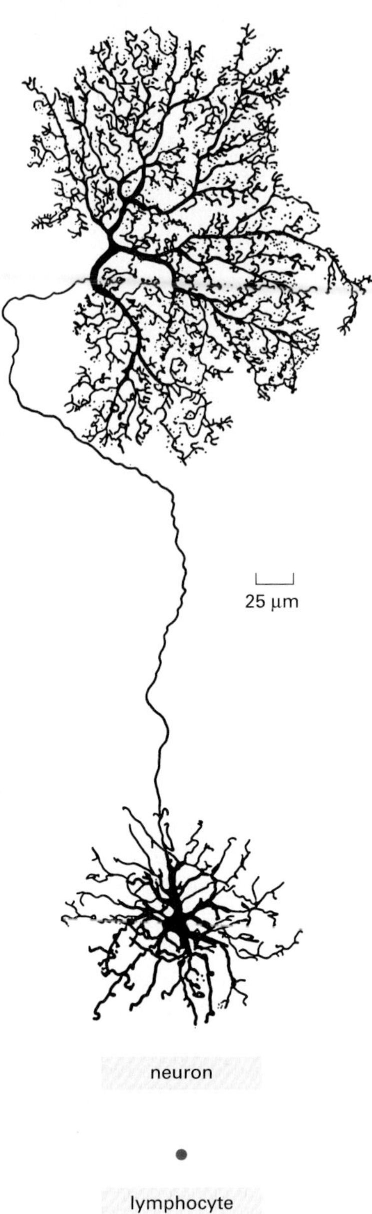

Figure 17–45 The size difference between a neuron (from the retina) and a lymphocyte in a mammal. Both cells contain the same amount of DNA. A neuron grows progressively larger after it has permanently withdrawn from the cell cycle. During this time, the ratio of cytoplasm to DNA increases enormously (by a factor of more than 10^5 for some neurons). (Neuron from B.B. Boycott, in Essays on the Nervous System [R. Bellairs and E.G. Gray, eds]. Oxford, UK: Clarendon Press, 1974.)

growth factors and mitogens help ensure that cells maintain their appropriate size as they proliferate.

Cell growth and division, however, can be controlled by separate extracellular signal proteins in some cell types. Such independent control may be particularly important during embryonic development, when dramatic changes in the size of certain cell types can occur. Even in adult animals, however, growth factors can stimulate cell growth without affecting cell division. The size of a sympathetic neuron, for example, which has permanently withdrawn from the cell cycle, depends on the amount of **nerve growth factor (NGF)** secreted by the target cells it innervates. The greater the amount of NGF the neuron has access to, the larger it becomes. It remains a mystery, however, how different cell types in the same animal grow to be so different in size (Figure 17–45).

Extracellular Survival Factors Suppress Apoptosis

Animal cells need signals from other cells—not only to grow and proliferate, but also to survive. If deprived of such **survival factors**, cells activate their intracellular death program and die by apoptosis. This arrangement ensures that cells survive only when and where they are needed. Nerve cells, for example, are produced in excess in the developing nervous system and then compete for limited amounts of survival factors that are secreted by the target cells they contact. Nerve cells that receive enough survival factor live, while the others die by apoptosis (Figure 17–46). A similar dependence on survival signals from neighboring cells is thought to control cell numbers in other tissues, both during development and in adulthood.

Survival factors, just like mitogens and growth factors, usually bind to cell-surface receptors. Binding activates signaling pathways that keep the death program suppressed, often by regulating members of the Bcl-2 family of proteins. Some factors, for example, stimulate the increased production of apoptosis-suppressing members of this family. Others act by inhibiting the function of apoptosis-promoting members of the family (Figure 17–47A). In *Drosophila,* and probably in vertebrates as well, some survival factors also act by stimulating the activity of IAPs, which suppress apoptosis (Figure 17–47B).

Neighboring Cells Compete for Extracellular Signal Proteins

When most types of mammalian cells are cultured in a dish in the presence of serum, they adhere to the bottom of the dish, spread out, and divide until a confluent monolayer is formed. Each cell is attached to the dish and contacts its neighbors on all sides. At this point, normal cells, unlike cancer cells, stop

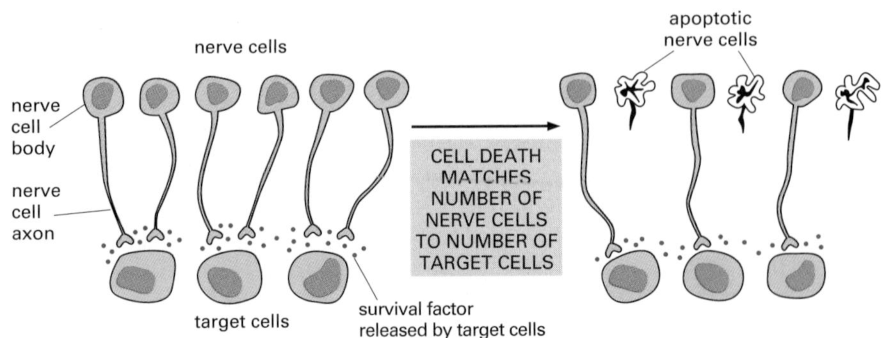

Figure 17–46 The function of cell death in matching the number of developing nerve cells to the number of target cells they contact. More nerve cells are produced than can be supported by the limited amount of survival factors released by the target cells. Therefore, some cells receive an insufficient amount of survival factors to keep their suicide program suppressed and, as a consequence, undergo apoptosis. This strategy of overproduction followed by culling ensures that all target cells are contacted by nerve cells and that the extra nerve cells are automatically eliminated.

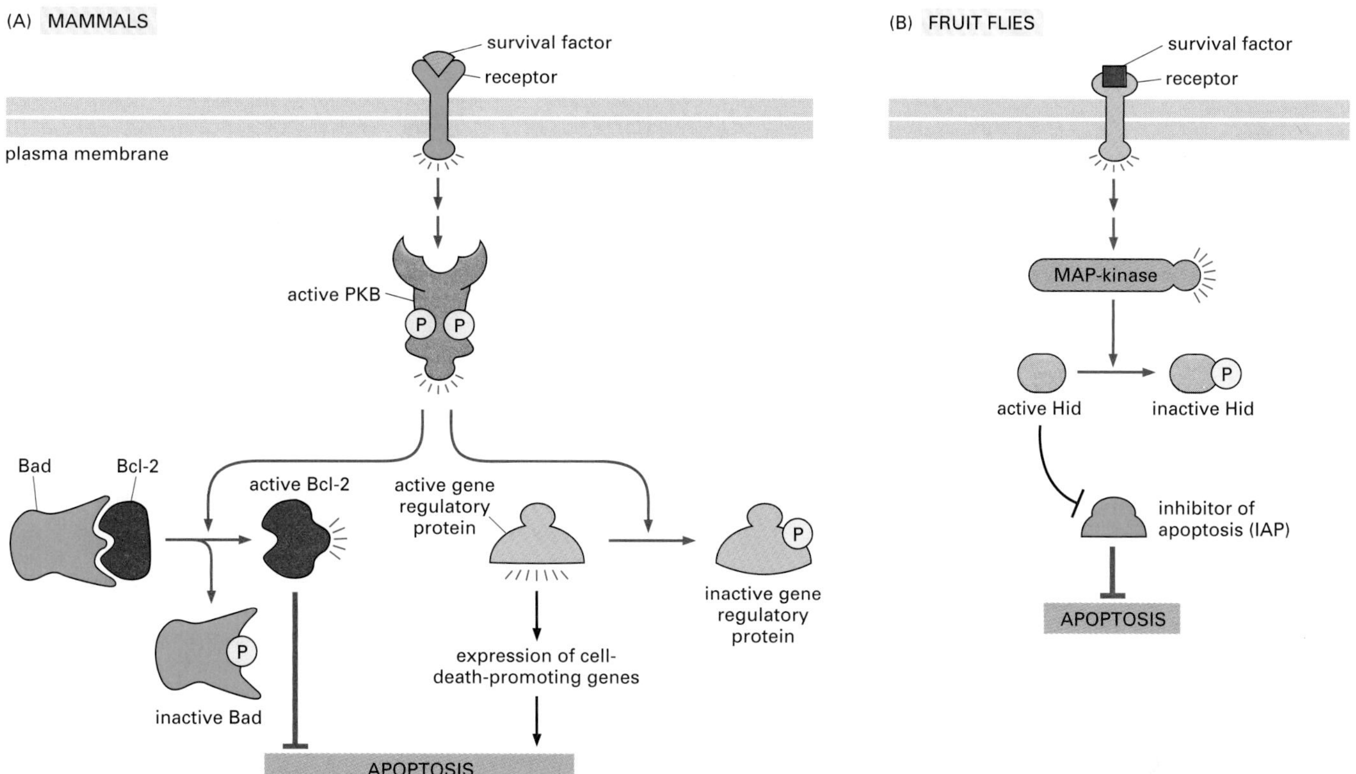

Figure 17–47 Two ways in which survival factors suppress apoptosis. (A) In mammalian cells, the binding of some survival factors to cell-surface receptors leads to the activation of various protein kinases, including protein kinase B (PKB), that phosphorylate and inactivate the Bcl-2 family member Bad. When not phosphorylated, Bad promotes apoptosis by binding and inhibiting Bcl-2. Once phosphorylated, Bad dissociates, freeing Bcl-2 to suppress apoptosis. As indicated, PKB also suppresses death by phosphorylating and thereby inhibiting gene regulatory proteins of the Forkhead family that stimulate the transcription of genes that encode proteins that promote apoptosis. (B) In *Drosophila*, some survival factors inhibit apoptosis by stimulating the phosphorylation of the Hid protein. When not phosphorylated, Hid promotes cell death by inhibiting IAPs. Once phosphorylated, Hid no longer inhibits IAPs, which become active and block cell death.

proliferating—a phenomenon known as *density-dependent inhibition of cell division*. This phenomenon was originally described in terms of "contact inhibition" of cell division, but it is unlikely that cell–cell contact interactions are solely responsible. The cell population density at which cell proliferation ceases in the confluent monolayer increases with increasing concentration of serum in the medium. Moreover, passing a stream of fresh culture medium over a confluent layer of fibroblasts reduces the diffusional limitation to the supply of mitogens, and it induces the cells under the stream of medium to divide at densities at which they would normally be inhibited from doing so (Figure 17–48). Thus, density-dependent inhibition of cell proliferation seems to reflect, in part at least, the ability of a cell to deplete the medium locally of extracellular mitogens, thereby depriving its neighbors.

This type of competition could be important for cells in tissues as well as in culture, because it prevents them from proliferating beyond a certain population density, determined by the available amounts of mitogens, growth factors, and survival factors. The amounts of these factors in tissues is usually limited, and increasing their amounts results in an increase in cell number, cell size, or both. Thus, the concentrations of these factors in tissues have important roles in determining cell size and number.

Many Types of Normal Animal Cells Need Anchorage to Grow and Proliferate

The shape of a cell changes as it spreads and crawls out over a substratum to occupy vacant space, and this can have a major impact on cell growth, cell division,

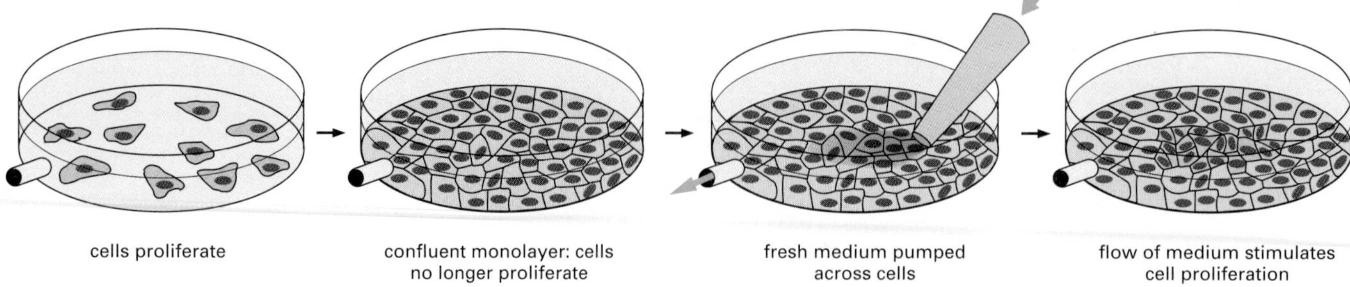

Figure 17–48 The effect of fresh medium on a confluent cell monolayer. Cells in a confluent monolayer do not divide *(gray)*. The cells resume dividing *(green)* when exposed directly to fresh culture medium. Apparently, in the undisturbed confluent monolayer, proliferation has halted because the medium close to the cells is depleted of mitogens, for which the cells compete.

and cell survival. When normal fibroblasts or epithelial cells, for example, are cultured in suspension, unattached to any solid surface and therefore rounded up, they almost never divide—a phenomenon known as *anchorage dependence* of cell division (Figure 17–49). But when these cells are allowed to settle and adhere to a sticky substrate, they rapidly form focal adhesions at sites of attachment, and then begin to grow and proliferate.

How are the growth and proliferation signals generated by cell attachments? *Focal adhesions* are places where extracellular matrix molecules, such as laminin or fibronectin, interact with cell-surface matrix receptors called *integrins*, which are linked to the actin cytoskeleton (discussed in Chapter 19). The binding of extracellular matrix molecules to integrins leads to the local activation of protein kinases, including *focal adhesion kinase (FAK)*, which in turn leads to the activation of intracellular signaling pathways that can promote the survival, growth, and division of cells (Figure 17–50).

Like other controls on cell division, anchorage control operates in G_1. Cells require anchorage to progress through G_1 into S phase, but anchorage is not required for completing the cycle. In fact, cells commonly loosen their attachments and round up as they pass through M phase. This cycle of attachment and detachment presumably allows cells in tissues to rearrange their contacts with other cells and with the extracellular matrix. In this way, tissues can accommodate the daughter cells produced by cell division and then bind them securely into the tissue before they are allowed to begin the next division cycle.

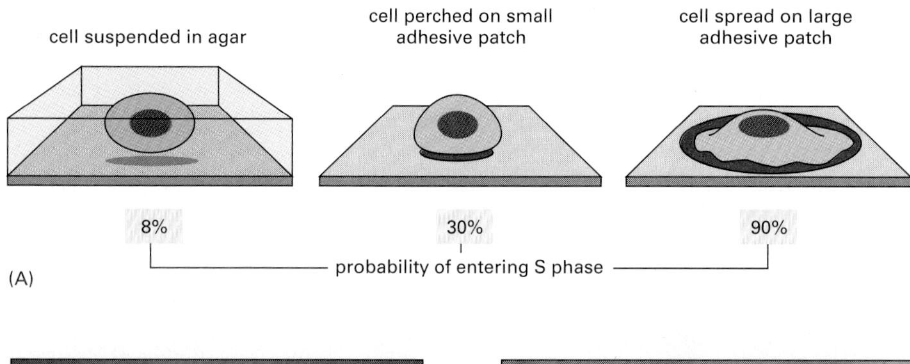

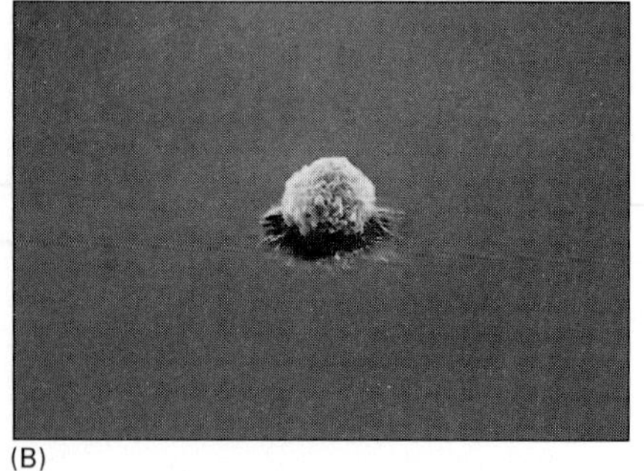

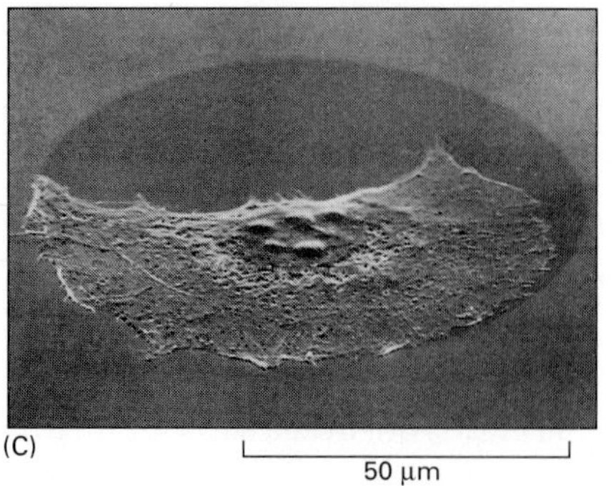

Figure 17–49 The dependence of cell division on cell shape and anchorage. In this experiment, cells are either held in suspension or allowed to settle on patches of an adhesive material (palladium) on a nonadhesive substratum. The patch diameter, which is variable, determines the extent to which an individual cell spreads and the probability that it will progress into S phase. ^{3}H-thymidine is added to the culture medium, and after 1 or 2 days, the culture is fixed and autoradiographed to determine the percentage of cells that have entered S phase (see Figure 17–11A). (A) Few cells of the 3T3 cell line enter S phase when held rounded up in suspension, but adherence even to a very tiny patch—one that is too small to allow spreading—enables many of them to enter S phase. (B and C) These scanning electron micrographs show a cell perched on a small patch compared with a cell spread on a large patch.

In contrast to fibroblasts and epithelial cells, some cell types in the body (including lymphocytes and blood cell precursors) can divide readily in suspension (see also Figure 19–62). (B and C, from C. O'Neill, P. Jordan, and G. Ireland, *Cell* 44:489–496, 1986. © Elsevier.)

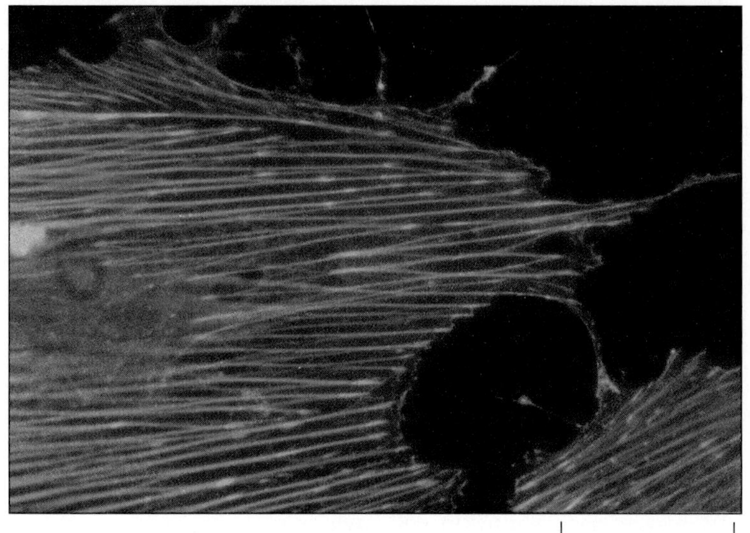

Figure 17–50 Focal adhesions as production sites of intracellular signals. This fluorescence micrograph shows a fibroblast cultured on a substratum coated with the extracellular matrix molecule fibronectin. Actin filaments have been labeled to fluoresce *green*, while activated proteins that contain phosphotyrosine have been labeled with an antibody that is tagged to fluoresce *red*. Where the two components overlap, the resulting color is *orange*. The actin filaments terminate at focal adhesions, where the cell attaches to the substratum. Proteins containing phosphotyrosine are also concentrated at these sites. This is thought to reflect the local activation of focal adhesion kinase (FAK) and other protein kinases stimulated by transmembrane integrin proteins that bind to fibronectin extracellularly and (indirectly) to actin filaments intracellularly. Signals generated at such adhesion sites help regulate cell division, growth, and survival, in both fibroblasts and epithelial cells. (Courtesy of Keith Burridge.)

Some Extracellular Signal Proteins Inhibit Cell Growth, Cell Division, and Survival

The extracellular signal proteins discussed in this chapter—mitogens, growth factors and survival factors—are positive regulators of cell-cycle progression, cell growth, and cell survival, respectively. They therefore tend to increase the size of organs and organisms. In some tissues, however, cell and tissue size also is influenced by inhibitory extracellular signal proteins that oppose the positive regulators and thereby inhibit organ growth.

The best-understood inhibitory signal proteins are TGF-β and its relatives. TGF-β inhibits the proliferation of several cell types, either by blocking cell-cycle progression in G_1 or by stimulating apoptosis. As discussed in Chapter 15, TGF-β binds to cell-surface receptors and initiates an intracellular signaling pathway that leads to changes in the activities of gene regulatory proteins called Smads. This results in complex and poorly understood changes in the transcription of genes encoding regulators of cell division and cell death.

One example of an apoptosis-inducing extracellular signal is *bone morphogenetic protein (BMP),* a TGF-β family member. BMP helps trigger the apoptosis that removes the tissue between the developing digits in the mouse paw (see Figure 17–35). Like TGF-β, BMP stimulates changes in the transcription of genes that regulate cell death, although the nature of these genes remains unclear.

The overall size of an organ may be limited in some cases by inhibitory signaling proteins. *Myostatin,* for example, is a TGF-β family member that normally inhibits the proliferation of myoblasts that fuse to form skeletal muscle cells. When the gene that encodes myostatin is deleted in mice, muscles grow to be several times larger than normal (see Figure 22–43). Both the number and the size of muscle cells increase. Remarkably, two breeds of cattle that were bred for large muscles have both turned out to have mutations in the gene encoding myostatin (Figure 17–51).

Intricately Regulated Patterns of Cell Division Generate and Maintain Body Form

The life of multicellular organisms begins with a series of division cycles that are controlled according to intricate rules. This is strikingly illustrated by the nematode *Caenorhabditis elegans.* The fertilized egg of *C. elegans* divides to produce an adult worm with precisely 959 somatic cell nuclei (in the male), each of which is generated by its own characteristic and absolutely predictable sequence of cell divisions. (The initial cell number is greater than this, but more than 100 cells die by apoptosis during development.) In general, the controls that generate

Figure 17–51 The effects of a myostatin mutation on muscle size. The mutation leads to a dramatic increase in the mass of muscle tissue, as illustrated in this Belgian Blue bull. The Belgium Blue was produced by cattle breeders and was only recently found to have a mutation in the *myostatin* gene. Mice purposely made deficient in the same gene also have remarkably big muscles (see Figure 22–43). (From A.C. McPherron and S.-J. Lee, *Proc. Natl. Acad. Sci. USA* 94:12457–12461, 1997. © National Academy of Sciences.)

such precise cell numbers do not operate by merely counting cell divisions according to a clocklike schedule. Instead, the organism seems mainly to control total cell mass, which depends not only on cell numbers but also on cell size. Salamanders of different ploidies, for example, are the same size but have different numbers of cells. Individual cells in a pentaploid salamander are about five times the volume of those in a haploid salamander, and in each organ the pentaploids have generated only one-fifth as many cells as their haploid cousins, so that the organs are about the same size in the two animals (Figures 17–52 and 17–53). Evidently, in this case (and in many others) the size of organs and organisms depends on mechanisms that can somehow measure total cell mass.

The development of limbs and organs of specific size and shape depends on complex positional controls, as well as on local concentrations of extracellular signal proteins that stimulate or inhibit cell growth, division, and survival. As we

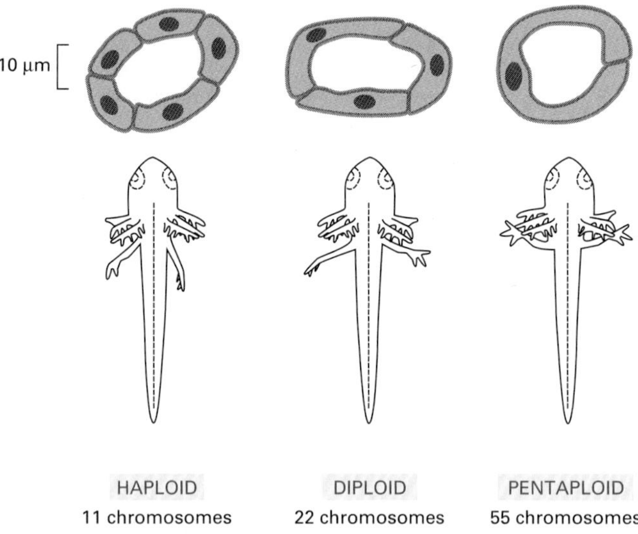

Figure 17–52 Sections of kidney tubules from salamander larvae of different ploidies. In all organisms, from bacteria to humans, cell size is proportional to ploidy. Pentaploid salamanders, for example, have cells that are much larger than those of haploid salamanders. The animals and their individual organs, however, are the same size because each tissue in the pentaploid animal contains fewer cells. This indicates that the size of an organism or organ is not controlled simply by counting cell divisions or cell numbers; total cell mass must somehow be regulated. (Adapted from G. Fankhauser, in Analysis of Development [B.H. Willier, P.A. Weiss, and V. Hamburger, eds.], pp. 126–150. Philadelphia: Saunders, 1955.)

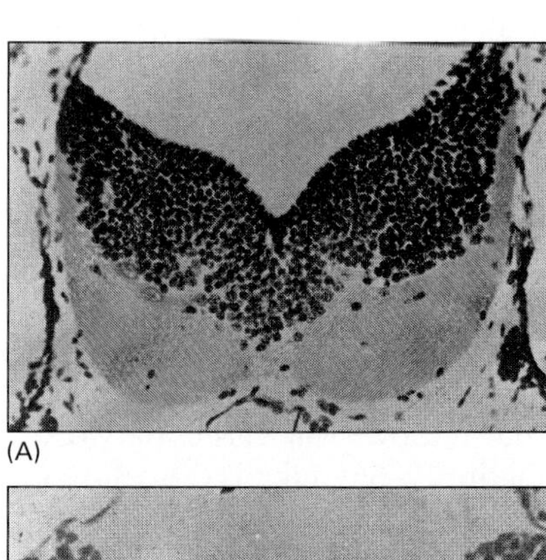

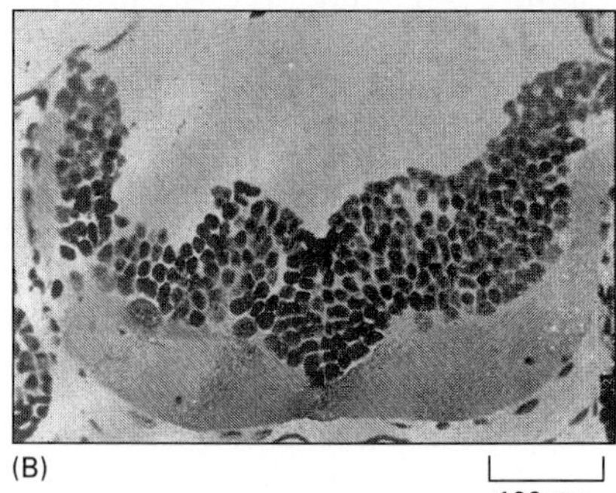

Figure 17–53 The hindbrain in a haploid and in a tetraploid salamander. (A) This light micrograph shows a cross section of the hindbrain of a haploid salamander. (B) A corresponding cross section of the hindbrain of a tetraploid salamander, revealing how reduced cell numbers compensate for increased cell size. (From G. Fankhauser, *Int. Rev. Cytol.* 1:165–193, 1952.)

discuss in Chapter 21, some of the genes that help pattern these processes in the embryo are now known. A great deal remains to be learned, however, about how these genes regulate cell growth, division, survival, and differentiation to generate a complex organism (discussed in Chapter 21).

The controls that govern these processes in an adult body are also poorly understood. When a skin wound heals in a vertebrate, for example, about a dozen cell types, ranging from fibroblasts to Schwann cells, must be regenerated in appropriate numbers and in appropriate positions to reconstruct the lost tissue. The mechanisms that control cell proliferation in tissues are likewise central to the understanding of cancer, a disease in which the controls go wrong, as discussed in Chapter 23.

Summary

In multicellular animals, cell size, cell division, and cell death are carefully controlled to ensure that the organism and its organs achieve and maintain an appropriate size. Three classes of extracellular signal proteins contribute to this control, although many of them affect two or more of these processes. Mitogens stimulate the rate of cell division by removing intracellular molecular brakes that restrain cell-cycle progression in G_1. Growth factors promote an increase in cell mass by stimulating the synthesis and inhibiting the degradation of macromolecules. Survival factors increase cell numbers by inhibiting apoptosis. Extracellular signals that inhibit cell division or cell growth, or induce cells to undergo apoptosis, also contribute to size control.

References

General

Baserga R (1985) The Biology of Cell Reproduction. Cambridge, MA: Harvard University Press.

Mitchison JM (1971) The Biology of the Cell Cycle. Cambridge: Cambridge University Press.

Murray AW & Hunt T (1993) The Cell Cycle: An Introduction. New York: WH Freeman.

An Overview of the Cell Cycle

Hartwell LH, Culotti J, Pringle JR & Reid BJ (1974) Genetic control of the cell division cycle in yeast. *Science* 183, 46–51.

Kirschner M, Newport J & Gerhart J (1985) The timing of early developmental events in Xenopus. *Trends Genet.* 1, 41–47.

Lohka MJ, Hayes MK & Maller JL (1988) Purification of maturation-promoting factor, an intracellular regulator of early mitotic events. *Proc. Natl. Acad. Sci. USA* 85, 3009–3013.

Masui Y & Markert CL (1971) Cytoplasmic control of nuclear behavior during meiotic maturation of frog oocytes. *J. Exp. Zool.* 177, 129–146.

Nurse P, Thuriaux P & Nasmyth K (1976) Genetic control of the cell division cycle in the fission yeast *Schizosaccharomyces pombe*. *Mol. Gen. Genet.* 146, 167–178.

Components of the Cell-Cycle Control System

Deshaies R (1999) SCF and cullin/Ring H2-based ubiquitin ligases. *Ann. Rev. Cell Dev. Biol.* 15, 435–467.

Hartwell LH & Weinert TA (1989) Checkpoints: controls that ensure the order of cell cycle events. *Science* 246, 629–634.

Koch C & Nasmyth K (1994) Cell cycle regulated transcription in yeast. *Curr. Opin. Cell Biol.* 6, 451–459.

Morgan DO (1997) Cyclin-dependent kinases: engines, clocks, and microprocessors. *Annu. Rev. Cell Dev. Biol.* 13, 261–291.

Pavletich NP (1999) Mechanisms of cyclin-dependent kinase regulation: structures of Cdks, their cyclin activators, and CIP and Ink4 inhibitors. *J. Mol. Biol.* 287, 821–828.

Spellman PT, Sherlock G, Zhang MQ et al. (1988) Comprehensive identification of cell cycle-regulated genes of the yeast *Saccharomyces cerevisiae* by microarray hybridization. *Mol. Biol. Cell* 9, 3273–3297.

Zachariae W & Nasmyth K (1999) Whose end is destruction: cell division and the anaphase-promoting complex. *Genes Dev.* 13, 2039–2058.

Intracellular Control of Cell-Cycle Events

Caspari T (2000) How to activate p53. *Curr. Biol.* 10, R315–R317.

Donaldson AD & Blow JJ (1999) The regulation of replication origin activation. *Curr. Opin. Genet. Dev.* 9, 62–68.

Dunphy WG (1994) The decision to enter mitosis. *Trends Cell Biol.* 4, 202–207.

Elledge SJ (1996) Cell cycle checkpoints: preventing an identity crisis. *Science* 274, 1664–1672.

Harbour JW & Dean DC (2000) The Rb/E2F pathway: expanding roles and emerging paradigms. *Genes Dev.* 14, 2393–2409.

Hirano T (2000) Chromosome cohesion, condensation, and separation. *Annu. Rev. Biochem.* 69, 115–144.

Johnston GC, Pringle JR & Hartwell LH (1977) Coordination of growth with cell division in the yeast *Saccharomyces cerevisiae*. *Exp. Cell Res.* 105, 79–98.

Kelly TJ & Brown GW (2000) Regulation of chromosome replication. *Annu. Rev. Biochem.* 69, 829–880.

Nasmyth K, Peters JM & Uhlmann F (2000) Splitting the chromosome: cutting the ties that bind sister chromatids. *Science* 288, 1379–1385.

Neufeld TP & Edgar BA (1998) Connections between growth and the cell cycle. *Curr. Opin. Cell Biol.* 10, 784–790.

Rao PN & Johnson RT (1970) Mammalian cell fusion: studies on the regulation of DNA synthesis and mitosis. *Nature* 225, 159–164.

Shah JV & Cleveland DW (2000) Waiting for anaphase: Mad2 and the spindle assembly checkpoint. *Cell* 103, 997–1000.

Stillman B (1996) Cell cycle control of DNA replication. *Science* 274, 1659–1664.

Zhou BB & Elledge SJ (2000) The DNA damage response: putting checkpoints in perspective. *Nature* 408, 433–439.

Programmed Cell Death (Apoptosis)

Adams JM & Cory S (1988) The Bcl-2 protein family: arbiters of cell survival. *Science* 281, 1322–1326.

Chao DT & Korsmeyer SJ (1998) BCL-2 family: regulators of cell death. *Annu. Rev. Immunol.* 16, 395–419.

Ekert PG, Silke J & Vaux DL (1999) Caspase inhibitors. *Cell Death Differ.* 6, 1081–1086.

Ellis RE, Yuan JY & Horvitz RA (1991) Mechanisms and functions of cell death *Annu. Rev. Cell Biol.* 7, 663–698.

Kerr JF, Wyllie AH & Currie AR (1972) Apoptosis: a basic biological phenomenon with wide-ranging implications in tissue kinetics. *Br. J. Cancer* 26, 239–257.

Li H & Yuan J (1999) Deciphering the pathways of life and death. *Curr. Opin. Cell Biol.* 11, 261–266.

Nicholson DW & Thornberry NA (1977) Caspases: killer proteases. *TIBS* 22, 299–306.

Extracellular Control of Cell Division, Cell Growth, and Apoptosis

Assoian RK (1997) Anchorage-dependent cell cycle progression. *J. Cell Biol.* 136, 1–4.

Blackburn EH (2000) Telomere states and cell fates. *Nature* 408, 53–56.

Conlon I & Raff M (1999) Size control in animal development. *Cell* 96, 235–244.

Datta SR, Brunet A & Greenberg ME (1999) Cellular survival: a play in three Akts. *Genes Dev.* 13, 2905–2927.

Raff MC (1992) Social controls on cell survival and cell death. *Nature* 356, 397–400.

Sherr CJ (1994) G1 phase progression: cycling on cue. *Cell* 79, 551–555.

Sherr CJ (1998) Tumor surveillance via the ARF-p53 pathway. *Genes Dev.* 12, 2984–2991.

Sherr CJ & DePinho RA (2000) Cellular senescence: mitotic clock or culture shock? *Cell* 102, 407–410.

Stocker H & Hafen E (2000) Genetic control of cell size. *Curr. Opin. Genet. Dev.* 10, 529–535.

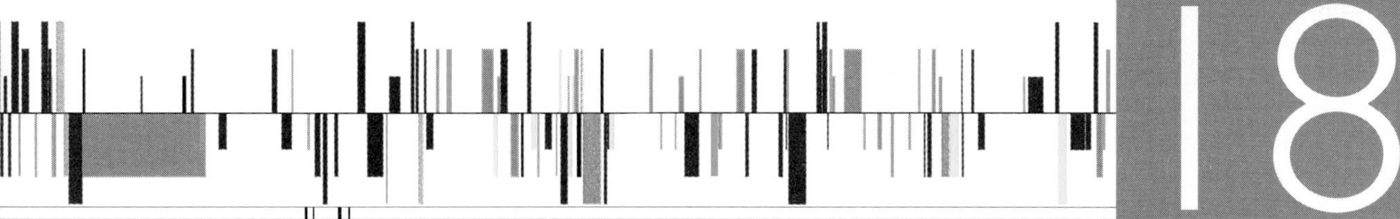

18

THE MECHANICS OF CELL DIVISION

Cells reproduce by duplicating their contents and dividing in two. This cycle of duplication and division, called the **cell cycle**, is discussed in Chapter 17. In this chapter, we consider the mechanical events of the **M phase** of the cycle, which is the culmination of the cycle and includes the various stages of nuclear division *(mitosis)* and cytoplasmic division *(cytokinesis)*. In a comparatively brief period, the contents of the parent cell, which were doubled during earlier phases of the cycle, are partitioned into two daughter cells. The period between one M phase and the next is called *interphase*, and in most rapidly proliferating cells, it is divided into three phases: *S phase*, in which DNA is replicated (discussed in Chapter 5), and two gap phases, G_1 and G_2, which provide additional time for the cell to grow (Figure 18–1).

As discussed in detail in Chapter 17, the events of the cell cycle are controlled by the *cell-cycle control system*. The core of the control system consists of various **cyclin-dependent kinases (Cdks)**, which are activated in sequence to trigger various steps of the cycle. The Cdks are activated by the binding of *cyclin* regulatory proteins, as well as by phosphorylation and dephosphorylation of the kinase. They are inactivated by various Cdk inhibitory proteins (CKIs) and by the degradation of the cyclin subunits at specific stages of the cycle.

The **M-phase Cdk (M-Cdk)** triggers a cascade of protein phosphorylation that initiates M phase. These phosphorylations are responsible for the many morphological changes that occur during mitosis in animal cells. The chromosomes condense, the nuclear envelope breaks down, the endoplasmic reticulum and Golgi apparatus reorganize, the cell loosens its adhesions both to other cells and to the extracellular matrix, and the cytoskeleton radically reorganizes to bring about the highly ordered movements that will segregate the replicated chromosomes and divide the cell in two.

Targetted protein degradation by the *anaphase-promoting complex (APC)* (discussed in Chapter 17) has an equally important regulatory role in mitosis. It initiates the separation and segregation of the replicated chromosomes, and it inactivates M-Cdk at the end of mitosis.

We begin this chapter with an overview of M phase. We then discuss mitosis and cytokinesis in turn, focusing mainly on animal cells. We end by considering how M phase may have evolved. We discuss the special features of meiotic cell division in Chapter 20, where we describe the development of germ cells.

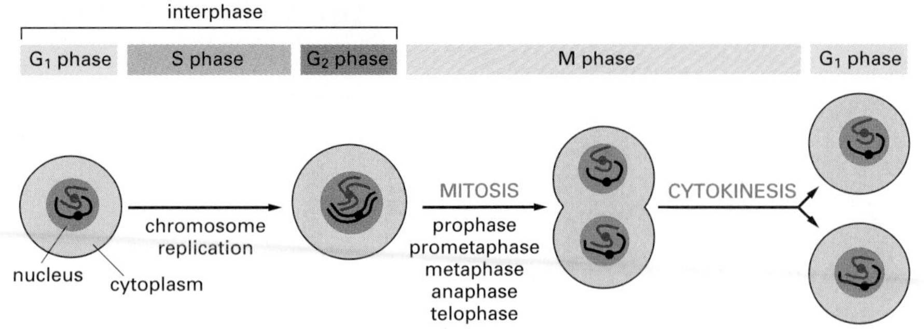

Figure 18–1 The M phase of the cell cycle. M phase starts at the end of G_2 and ends at the start of the next G_1 phase. It includes the five stages of nuclear division (mitosis), as well as cytoplasmic division (cytokinesis).

AN OVERVIEW OF M PHASE

The central problem for a mitotic cell in M phase is how to accurately separate and distribute *(segregate)* its chromosomes, which were replicated in the preceding S phase, so that each new daughter cell receives an identical copy of the genome (see Figure 18–1). With minor variations, all eucaryotes solve this problem in a similar way: they assemble specialized cytoskeletal machines—first to pull the duplicated chromosome sets apart and then to split the cytoplasm into two halves. Before the duplicated chromosomes can be separated and distributed equally to the two daughter cells during mitosis, however, they must be appropriately configured, and this process begins in S phase.

Cohesins and Condensins Help Configure Replicated Chromosomes for Segregation

When the chromosomes are duplicated in S phase, the two copies of each replicated chromosome remain tightly bound together as identical **sister chromatids**. The sister chromatids are glued together by multisubunit protein complexes called **cohesins**, which are deposited along the length of each sister chromatid as the DNA is replicated. This cohesion between sister chromatids is crucial to the chromosome segregation process and is broken only late in mitosis (at the start of *anaphase*) to allow the sisters to be pulled apart.

The first readily visible sign that a cell is about to enter M phase is the progressive compaction of the replicated chromosomes, which become visible as threadlike structures—a process called **chromosome condensation**. In humans, for example, each interphase chromosome becomes compacted after replication into a mitotic chromosome that is about 50 times shorter (Figure 18–2). As discussed in Chapter 4, proteins called **condensins** do the work of chromosome condensation. Activated M-Cdk phosphorylates some of the condensin subunits, triggering the assembly of condensin complexes on DNA and, thereby, the

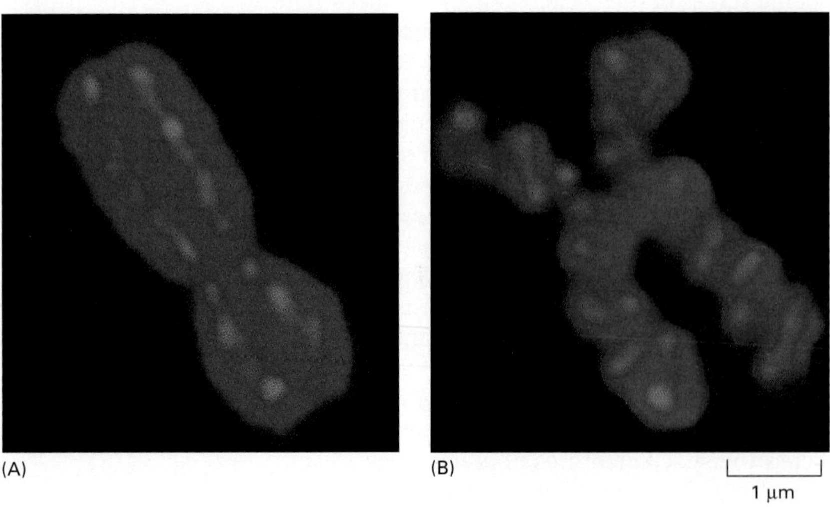

Figure 18–2 Human mitotic chromosomes stained to reveal a scaffold-like structure along the chromosome axis. In these confocal fluorescence micrographs, the DNA has been stained with a *blue* dye, and the axis has been stained *red* with a fluorescent antibody against a protein in the condensin complex. Only part of the scaffolding is visible in these optical sections. (A) A typical mitotic chromosome, which has a gently coiled scaffold along each of the two chromatids. (B) A metaphase chromosome from a cell artificially blocked in metaphase; in the chromosomes of these cells, the scaffold has condensed by further helical folding. (Courtesy of Ulrich Laemmli and Kazuhiro Maeshima).

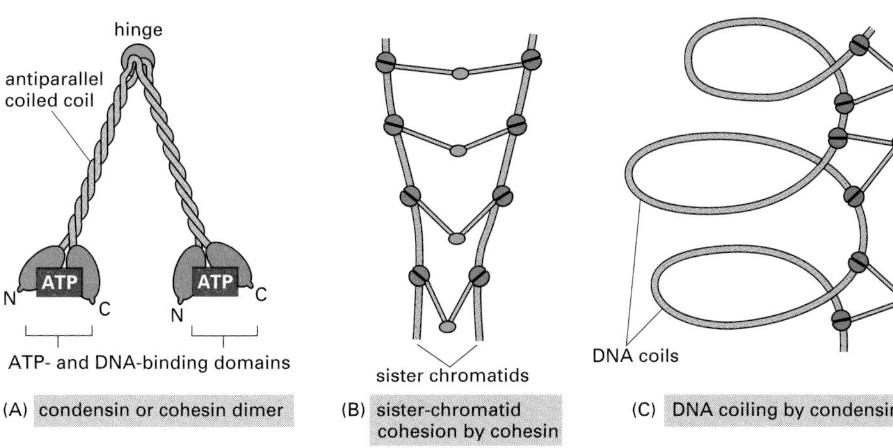

Figure 18–3 The related structure and function of cohesins and condensins. (A) Both proteins have two identical DNA- and ATP-binding domains at one end and a hinge region at the other, joined by two, long, coiled-coil regions. This flexible structure is well suited for their role as DNA cross-linkers. (B) Cohesins cross-link two adjacent sister chromatids, gluing them together. (C) Condensins mediate intramolecular cross-linking to coil DNA in the process of chromosome condensation. (Adapted from T. Hirano, *Genes and Dev.* 13:11–19, 1999.)

progressive condensation of the chromosomes. The condensins can use the energy of ATP hydrolysis to promote DNA coiling in a test tube, and they are thought to do the same in cells during chromosome condensation. By a mechanism that is still poorly understood, they eventually produce fully condensed mitotic chromosomes, with each of the two sister chromatids organized around a linear central axis, where the condensin complexes are concentrated (see Figure 18–2).

The cohesins and condensins are structurally related, and they work together to configure the replicated chromosomes in preparation for mitosis. Genetic studies show that, if chromatid cohesion is not established properly in S phase, full condensation cannot occur in M phase and chromosomes are abnormally segregated at anaphase. A model for how cohesins can glue two DNA molecules together, while the closely related condensins can bind to a single DNA molecule and induce its supercoiling and condensation, is illustrated in Figure 18–3.

In budding yeast, the sudden degradation and release of the cohesin complexes allows sister chromatids to separate at anaphase. In vertebrate cells, by contrast, most of the cohesin is released from the chromosomes at the start of mitosis, when condensins bind to drive condensation. The small amount of cohesin that remains, however, is enough to hold sister chromatids together until anaphase, when the residual cohesins are degraded, allowing the chromatids to separate, as we discuss later.

Cytoskeletal Machines Perform Both Mitosis and Cytokinesis

After the chromosomes have condensed, two distinct cytoskeletal machines are assembled in sequence to perform the mechanical processes of mitosis and cytokinesis. Both machines disassemble rapidly after they have completed their tasks.

To produce two genetically identical daughter cells, the cell has to separate its replicated chromosomes and allocate one copy to each daughter cell. In all eucaryotic cells, this task is performed during mitosis by a bipolar *mitotic spindle*, which is composed of microtubules and various proteins that interact with them, including *microtubule-dependent motor proteins* (discussed in Chapter 16).

Different cytoskeletal structures are responsible for cytokinesis. In animal cells and many unicellular eucaryotes, it is the *contractile ring*; in most plant cells, it is the *phragmoplast*. The contractile ring contains both actin and myosin filaments and forms around the equator of the cell, just under the plasma membrane; as the ring contracts, it pulls the membrane inward, thereby dividing the cell in two (Figure 18–4). Plant cells, which have a cell wall to contend with, divide their cytoplasm by a very different mechanism. As we discuss later, instead of using a contractile process that acts on the plasma membrane, the phragmoplast constructs a new cell wall from within the cell, between the two sets of replicated chromosomes.

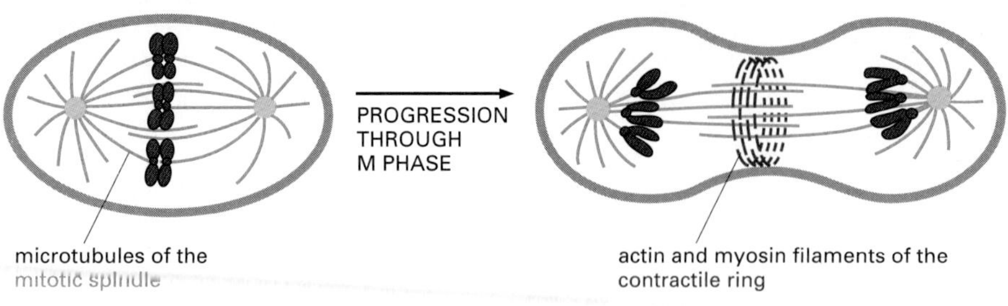

Figure 18–4 Two cytoskeletal machines that operate in M phase. The mitotic spindle assembles first and segregates the chromosomes. The contractile ring assembles later and divides the cell in two. Plant cells use a very different mechanism to divide the cytoplasm, as we discuss later.

Two Mechanisms Help Ensure That Mitosis Always Precedes Cytokinesis

In most animal cells, M phase takes only about an hour—a small fraction of the total cell-cycle time, which often lasts 12–24 hours. The rest of the cycle is occupied by **interphase**. Under the microscope, interphase appears as a deceptively uneventful interlude, in which the cell simply continues to grow in size. Other techniques, however, reveal that interphase is actually a busy time for a proliferating cell, during which elaborate preparations for cell division are occurring in a tightly ordered sequence. Two critical preparatory events that are completed during interphase are DNA replication and duplication of the *centrosome*.

As discussed in Chapter 17, cyclical oscillations in the activities of the Cdks and of proteolytic complexes drive the cell cycle forward. Cdks trigger various steps of the cycle either by directly phosphorylating structural or regulatory proteins or by activating other protein kinases to do so. The proteolytic complexes activate specific steps in the cycle by degrading key cell-cycle proteins such as cyclins and Cdk inhibitor proteins. Like throwing switches, the activation of Cdks and proteolytic complexes triggers cell-cycle transitions that are normally points of no return. Thus, a green light from M-Cdk to enter M phase results in chromosome condensation, nuclear envelope breakdown, and a dramatic change in microtubule dynamics, all triggered by the phosphorylation of regulatory proteins that control these processes.

It is crucial that the two major events of M phase—nuclear division (mitosis) and cytoplasmic division (cytokinesis)—occur in the correct sequence (see Figure 18–1). It would be catastrophic if cytokinesis occurred before all of the chromosomes had segregated during mitosis. At least two mechanisms seem to prevent this catastrophe. First, the cell-cycle control system that activates proteins required for mitosis is thought to inactivate some of the proteins required for cytokinesis; presumably for this reason, cytokinesis cannot occur until M-Cdk is inactivated at the end of mitosis. Second, after the mitotic spindle has segregated the two sets of chromosomes to opposite poles of the cell, the residual central region of the spindle is required to maintain a functional contractile ring (see Figure 18–4); thus, until the spindle has separated the chromosomes and formed a *central spindle,* the ring cannot divide the cytoplasm in two.

M Phase in Animal Cells Depends on Centrosome Duplication in the Preceding Interphase

Two critical events must be completed in interphase before M phase begins—replication of the DNA and, in animal cells, duplication of the centrosome. DNA is duplicated so that each new daughter cell inherits an identical copy of the genome, while the centrosome is duplicated to help initiate the formation of the two poles of the mitotic spindle and to supply each daughter cell with its own centrosome. As we discuss later, after the chromosomes have been segregated in late mitosis, the microtubules that emanate from the two centrosomes signal to the cell cortex to help establish the plane of cytoplasmic division. This ensures that the division occurs exactly midway between the two separated groups of chromosomes (see Figure 18–4).

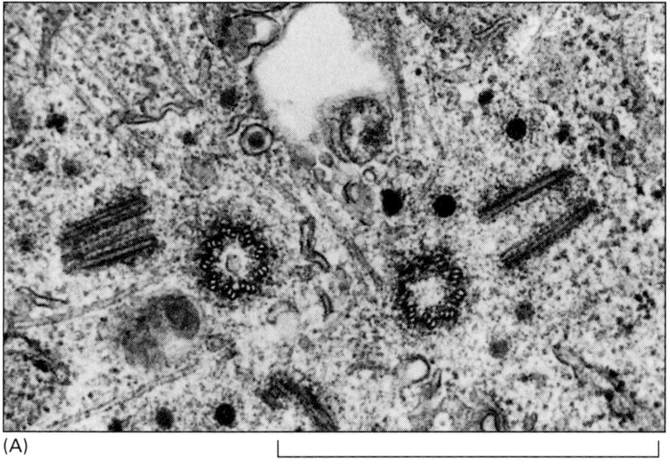

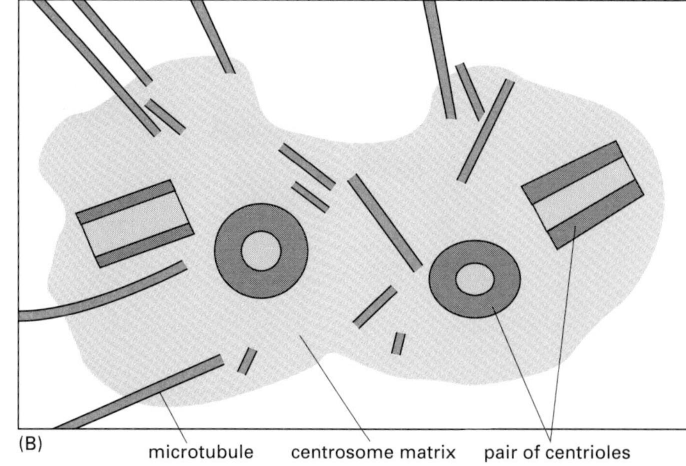

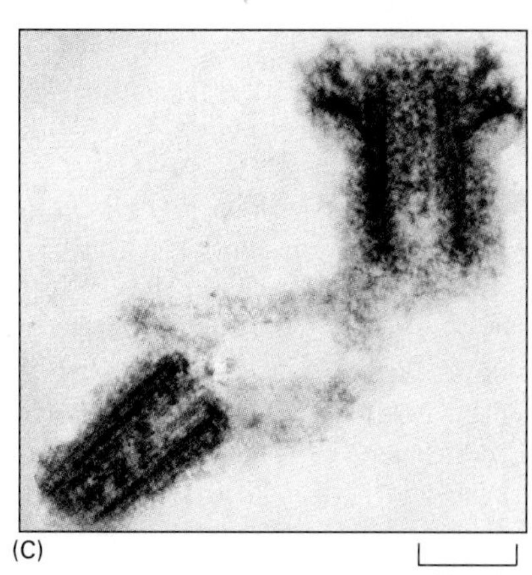

Figure 18–5 Centrioles. (A) Electron micrograph of an S-phase mammalian cell in culture, showing a duplicated centrosome. Each centrosome contains a pair of centrioles; although the centrioles have duplicated, they remain together in a single complex, as shown in the drawing of the micrograph in (B). One centriole of each centriole pair has been cut in cross-section, while the other is cut in longitudinal section, indicating that the two members of each pair are aligned at right angles to each other. The two halves of the replicated centrosome, each consisting of a centriole pair surrounded by matrix, will split and migrate apart to initiate the formation of the two poles of the mitotic spindle when the cell enters M phase (see Figure 18–7). (C) Electron micrograph of a centriole pair that has been isolated from a cell. The two centrioles have partly separated during the isolation procedure but remain tethered together by fine fibers, which keep the centriole pair together until it is time for them to separate (see Figure 18–6). Both centrioles are cut longitudinally, and it can now be seen that the two have different structures: the mother centriole is larger and more complex than the daughter centriole, and, as shown in Figure 18–6, only the mother centriole is associated with matrix that nucleates microtubules. Each daughter centriole will mature during the next cell cycle, when it will replicate to give rise to its own daughter centriole. (A, from M. McGill, D.P. Highfield, T.M. Monahan, and B.R. Brinkley, *J. Ultrastruct. Res.* 57:43–53, 1976. © Academic Press; C, from M. Paintrand et al., *J. Struct. Biol.* 108:107–128, 1992. © Academic Press.)

The **centrosome** is the principal *microtubule-organizing center* in animal cells (discussed in Chapter 16). It consists of a cloud of amorphous material (called the *centrosome matrix* or *pericentriolar material)* that surrounds a pair of *centrioles* (see Figure 16–24). During interphase, the centrosome matrix nucleates a *cytoplasmic array* of microtubules, with their fast-growing plus ends projecting outward toward the cell perimeter and their minus ends associated with the centrosome. The matrix contains a great variety of proteins, including microtubule-dependent motor proteins, coiled-coil proteins that are thought to link the motors to the centrosome, structural proteins, and components of the cell-cycle control system. Most important, it contains the **γ-tubulin ring complex**, which is the component mainly responsible for nucleating microtubules (see Figure 16–22).

The process of centrosome duplication and separation is known as the **centrosome cycle**. During interphase of each animal cell cycle, the centrioles and other components of the centrosome are duplicated (by an unknown mechanism) but remain together as a single complex on one side of the nucleus (Figure 18–5). As mitosis begins, this complex splits in two, and each centriole pair becomes part of a separate microtubule organizing center that nucleates a radial array of microtubules called an **aster** (Figure 18–6). The two asters move to opposite sides of the nucleus to initiate the formation of the two poles of the

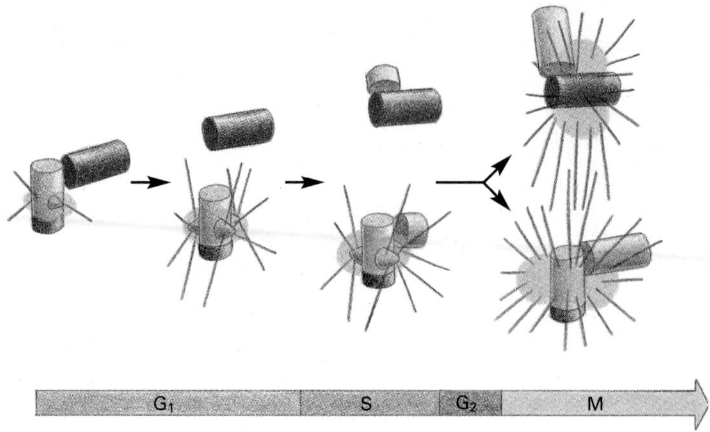

Figure 18–6 Centriole replication. The centrosome consists of a centriole pair and associated matrix *(green)*. At a certain point in G_1, the two centrioles of the pair separate by a few micrometers. During S phase, a daughter centriole begins to grow near the base of each mother centriole and at a right angle to it. The elongation of the daughter centriole is usually completed by G_2. The two centriole pairs remain close together in a single centrosomal complex until the beginning of M phase (see Figure 18–7), when the complex splits in two and the two halves begin to separate. Each centrosome now nucleates its own radial array of microtubules called an aster.

mitotic spindle. When the nuclear envelope breaks down (at *prometaphase),* the spindle captures the chromosomes; it will separate them toward the end of mitosis (Figure 18–7). As mitosis ends and the nuclear envelope re-forms around the separated chromosomes, each daughter cell receives a centrosome in association with its chromosomes.

In early embryonic cell cycles, the centrosome cycle can operate even if the nucleus is physically removed or nuclear DNA replication is blocked by a drug that inhibits DNA synthesis. Cycles of centrosome duplication and separation proceed almost normally, first yielding two centrosomes, then four, and then eight, and so on. Egg cell extracts from the frog *Xenopus* (see Figure 17–9) support multiple rounds of centrosome duplication in a test tube. This system has been used to test the individual protein components of the cell-cycle control system for their ability to stimulate centrosome duplication. Such experiments show that the G_1/S-Cdk (a complex of cyclin E and Cdk2) that initiates DNA replication in S phase (discussed in Chapter 17) also stimulates centrosome duplication, presumably explaining why centrosome duplication begins at the start of S phase.

M Phase Is Traditionally Divided into Six Stages

The first five stages of M phase constitute **mitosis**, which was originally defined as the period in which the chromosomes are visibly condensed. *Cytokinesis* occurs in the sixth stage, which overlaps with the end of mitosis. These six stages form a dynamic sequence, in which many independent cycles, involving the chromosomes, cytoskeleton, and centrosomes, have to be coordinated in order to produce two genetically identical daughter cells. The stages that occur during M phase are summarized in Panel 18–1. The complexity and beauty of M phase, however, are hard to appreciate from written descriptions or from a set of static pictures.

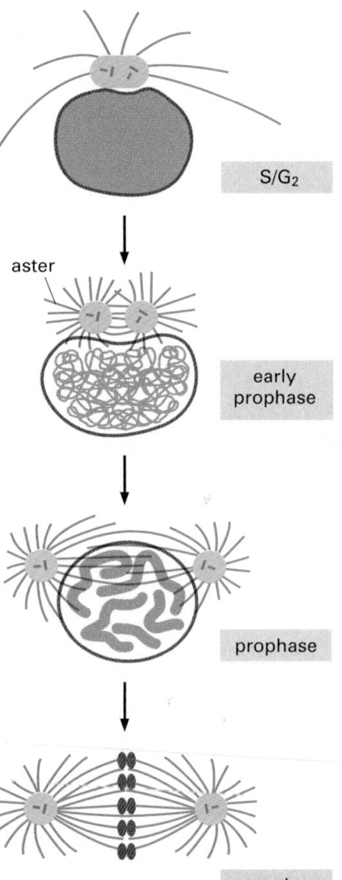

Figure 18–7 The centrosome cycle. The centrosome in a proliferating animal cell duplicates in interphase in preparation for mitosis. In most animal cells, a centriole pair (shown here as a pair of *dark green* bars) is associated with the centrosome matrix *(light green)* that nucleates microtubule outgrowth. (The volume of centrosome matrix is exaggerated in this diagram for clarity; Figure 18–6 gives a more accurate representation.) Centriole duplication begins in G_1 and is completed by G_2 (see Figure 18–6). Initially the two centriole pairs and associated centrosome matrix remain together as a single complex. In early M phase, this complex separates into two, each of which nucleates its own aster. The two asters, which initially lie side by side and close to the nuclear envelope, move apart. By late prophase, the microtubules that interact between the two asters preferentially elongate as the two asters move apart along the outside of the nucleus. In this way, a bipolar mitotic spindle is rapidly formed. At metaphase, the nuclear envelope breaks down, enabling the spindle microtubules to interact with the chromosomes.

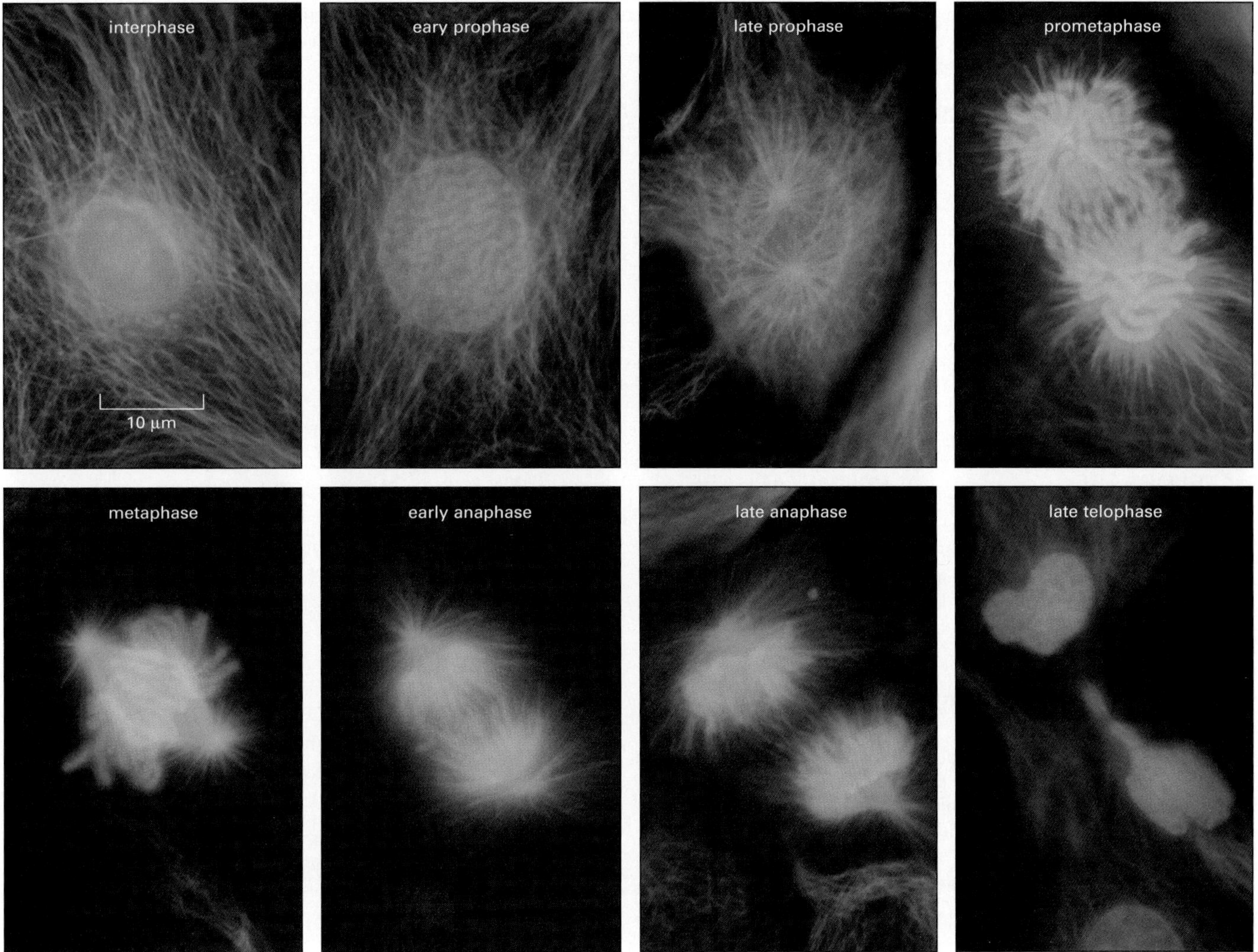

Figure 18–8 The course of mitosis in a typical animal cell. In these micrographs of cultured newt lung cells, the microtubules *(green)* have been visualized by immunofluorescence, while the chromatin is stained with a *blue* fluorescent dye. During *interphase,* the centrosome (not visible) forms the focus for the interphase microtubule array. By *early prophase,* the single centrosome contains two centriole pairs (not visible). At *late prophase,* the centrosome divides, and the resulting two asters can be seen to have moved apart. At *prometaphase,* the nuclear envelope breaks down, allowing the spindle microtubules to interact with the fully condensed chromosomes. At *metaphase,* the bipolar structure of the spindle is clear, and all the chromosomes are aligned at the equator of the spindle. At *early anaphase,* the sister chromatids all separate synchronously and, under the influence of the microtubules, the daughter chromosomes begin to move toward the poles. By *late anaphase,* the spindle poles have moved farther apart, increasing the separation of the two groups of chromosomes. At *telophase,* the daughter nuclei re-form, and by *late telophase,* cytokinesis is almost complete, with the midbody (discussed later) persisting between the daughter cells. (Photographs courtesy of C.L. Rieder, J.C. Waters, and R.W. Cole.)

The five stages of mitosis—*prophase, prometaphase, metaphase, anaphase, and telophase*—occur in strict sequential order, while cytokinesis begins in anaphase and continues through telophase. Light micrographs of cell division in a typical animal cell and a typical plant cell are shown in Figures 18–8 and 18–9, respectively. During prophase, the replicated chromosomes condense in step with the reorganization of the cytoskeleton. In metaphase, the chromosomes are aligned at the equator of the mitotic spindle, and in anaphase they are segregated to the two poles of the spindle. Cytoplasmic division is complete by the end of telophase, and the nucleus and cytoplasm of each of the daughter cells then return to interphase, signaling the end of M phase.

PANEL 18–1 The Principal Stages of M Phase (Mitosis and Cytokinesis) in an Animal Cell

CELL DIVISION AND THE CELL CYCLE

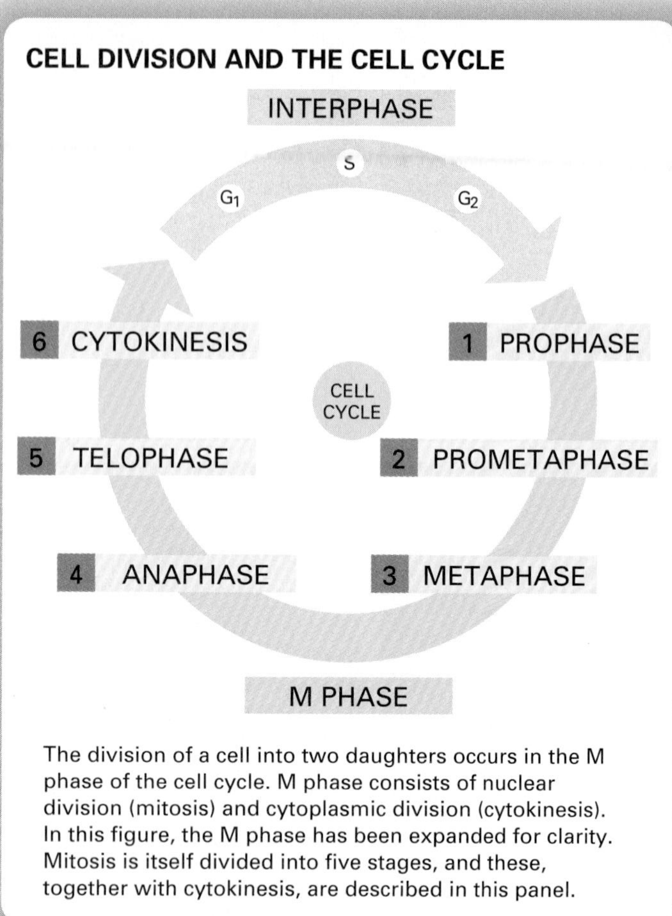

The division of a cell into two daughters occurs in the M phase of the cell cycle. M phase consists of nuclear division (mitosis) and cytoplasmic division (cytokinesis). In this figure, the M phase has been expanded for clarity. Mitosis is itself divided into five stages, and these, together with cytokinesis, are described in this panel.

INTERPHASE

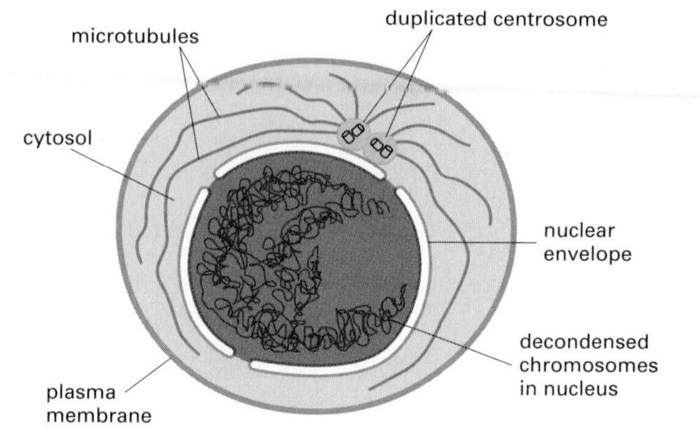

During interphase, the cell increases in size. The DNA of the chromosomes is replicated, and the centrosome is duplicated.

The light micrographs shown in this panel are of a living cell from the lung epithelium of a newt. The same cell has been photographed when viewed by differential-interference-contrast microscopy at different times during its division into two daughter cells. (Courtesy of Conly L. Rieder.)

1 PROPHASE

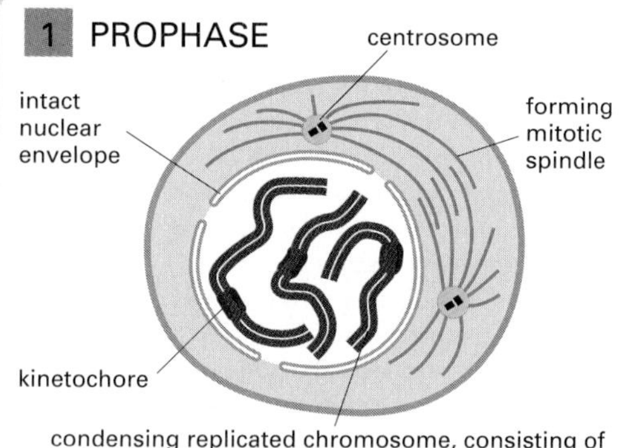

At prophase, the replicated chromosomes, each consisting of two closely associated sister chromatids, condense. Outside the nucleus, the mitotic spindle assembles between the two centrosomes, which have replicated and moved apart. For simplicity, only three chromosomes are shown. In diploid cells, there would be two copies of each chromosome present.

time = 0 min

2 PROMETAPHASE

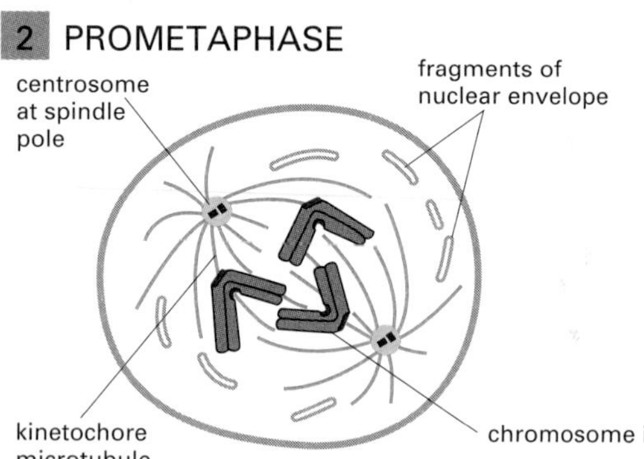

Prometaphase starts abruptly with the breakdown of the nuclear envelope. Chromosomes can now attach to spindle microtubules via their kinetochores and undergo active movement.

time = 79 min

3 METAPHASE

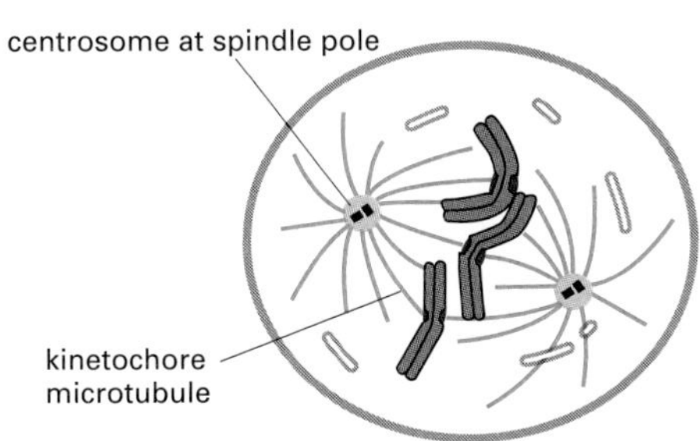

At metaphase, the chromosomes are aligned at the equator of the spindle, midway between the spindle poles. The kinetochore microtubules attach sister chromatids to opposite poles of the spindle.

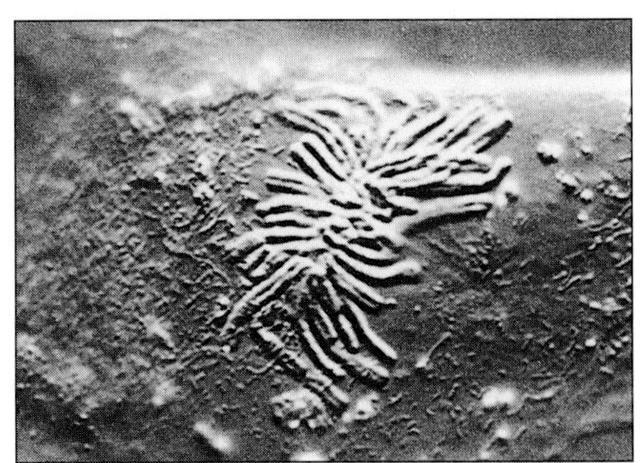

time = 250 min

4 ANAPHASE

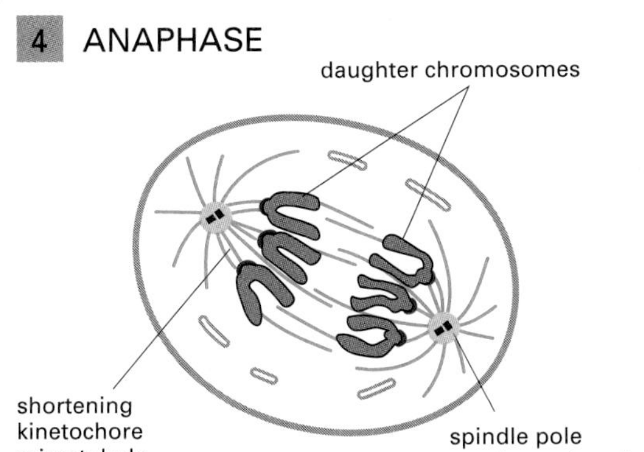

At anaphase, the sister chromatids synchronously separate to form two daughter chromosomes, and each is pulled slowly toward the spindle pole it faces. The kinetochore microtubules get shorter, and the spindle poles also move apart; both processes contribute to chromosome separation.

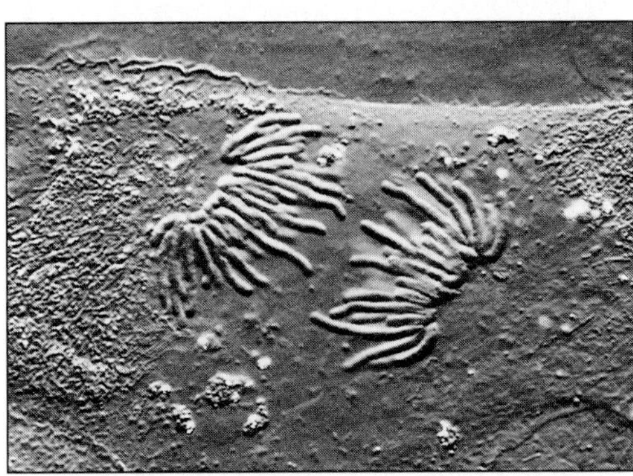

time = 279 min

5 TELOPHASE

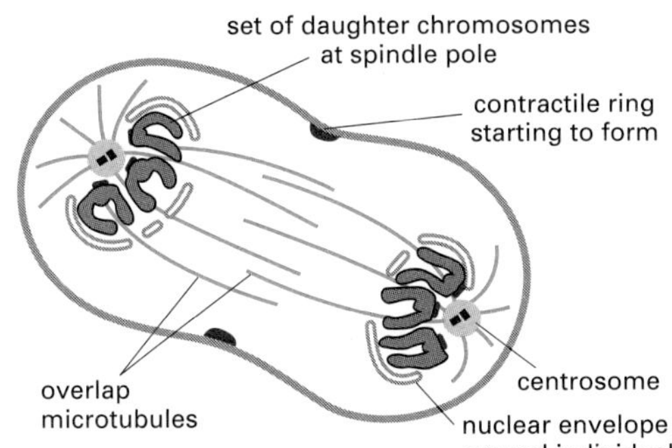

During telophase, the two sets of daughter chromosomes arrive at the poles of the spindle and decondense. A new nuclear envelope reassembles around each set, completing the formation of two nuclei and marking the end of mitosis. The division of the cytoplasm begins with the assembly of the contractile ring.

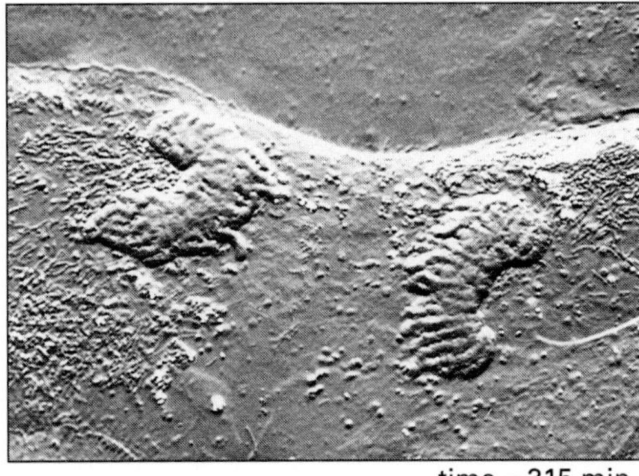

time = 315 min

6 CYTOKINESIS

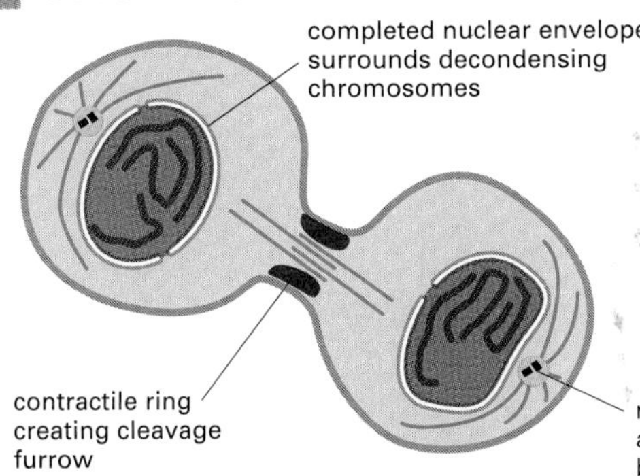

During cytokinesis, the cytoplasm is divided in two by a contractile ring of actin and myosin filaments, which pinches the cell in two to create two daughters, each with one nucleus.

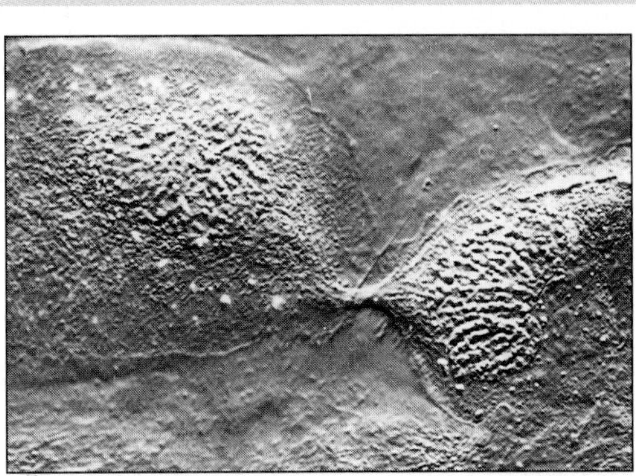

time = 362 min

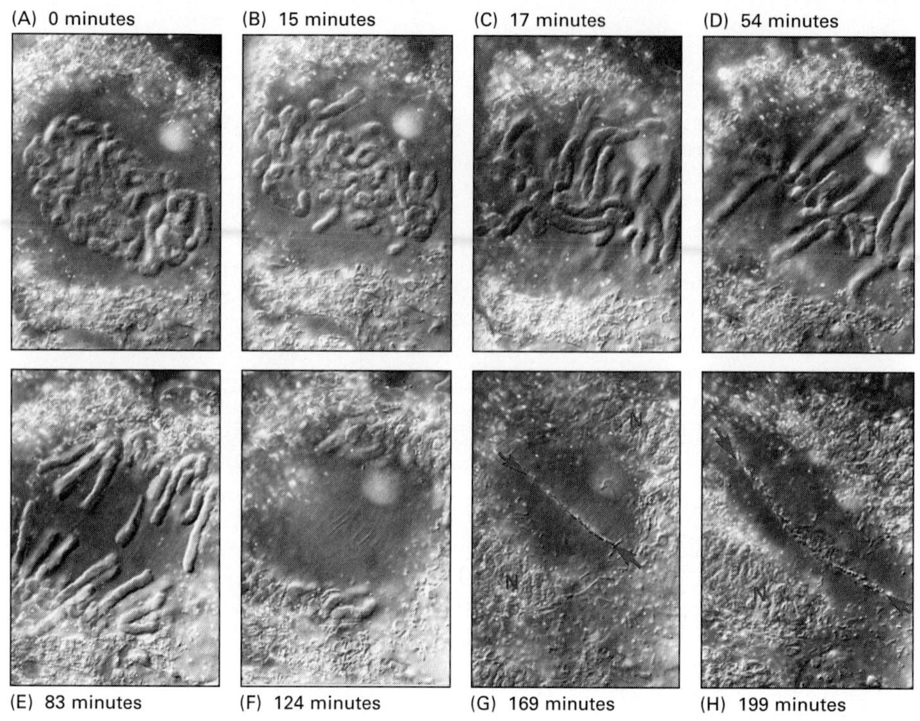

Figure 18–9 The course of mitosis in a plant cell. These light micrographs of a living *Haemanthus* (lily) cell were taken at the times indicated, using differential-interference-contrast microscopy. The cell has unusually large chromosomes that are easy to see. (A) At *prophase*, the chromosomes condense and are clearly visible in the cell nucleus. (B and C) At *prometaphase*, the nuclear envelope breaks down and the chromosomes interact with the microtubules that emanate from the two spindle poles. Plants do not have centrosomes, but their spindle poles contain proteins related to those found in the centrosomal matrix of animal cells. (D) At *metaphase*, the chromosomes line up at the equator of the spindle. (E) At *anaphase*, the daughter chromosomes separate and start moving to opposite poles. (F) At *telophase*, the chromosomes decondense and daughter nuclei re-form (not seen). (G and H) During *cytokinesis*, a new cell wall (the cell plate, *red arrows*) forms between the two nuclei (N). (Courtesy of Andrew Bajer.)

Summary

Cell division occurs during M phase, which consists of nuclear division (mitosis) followed by cytoplasmic division (cytokinesis). The DNA is replicated in the preceding S phase; the two copies of each replicated chromosome (called sister chromatids) remain glued together by cohesins. At the start of M phase, cohesin-related proteins called condensins bind to the replicated chromosomes and progressively condense them. A microtubule-based mitotic spindle is responsible for chromosome segregation in all eucaryotic cells. The mitotic spindle in animal cells develops from the microtubule asters that form around each of the two centrosomes produced when the centrosome duplicates, beginning in S phase; at the onset of M phase, the duplicated centrosomes separate and move to opposite sides of the nucleus to initiate the formation of the two poles of the spindle. An actin and myosin-based contractile ring is responsible for cytoplasmic division in animal cells and in many unicellular eucaryotes, but not in plant cells.

MITOSIS

The segregation of the replicated chromosomes is brought about by a complex cytoskeletal machine with many moving parts—the **mitotic spindle**. It is constructed from microtubules and their associated proteins, which both pull the daughter chromosomes toward the poles of the spindle and move the poles apart.

As we have seen, the spindle starts to form outside the nucleus while the chromosomes are condensing during prophase. When the nuclear envelope breaks down at prometaphase, the microtubules of the spindle are able to capture the chromosomes, which eventually become aligned at the spindle equator, forming the *metaphase plate* (see Panel 18–1). At anaphase, the sister chromatids abruptly separate and are drawn to opposite poles of the spindle; at about the same time, the spindle elongates, increasing the separation between the poles. The spindle continues to elongate during telophase, as the chromosomes arriving at the poles are released from the spindle microtubules and the nuclear envelope re-forms around them.

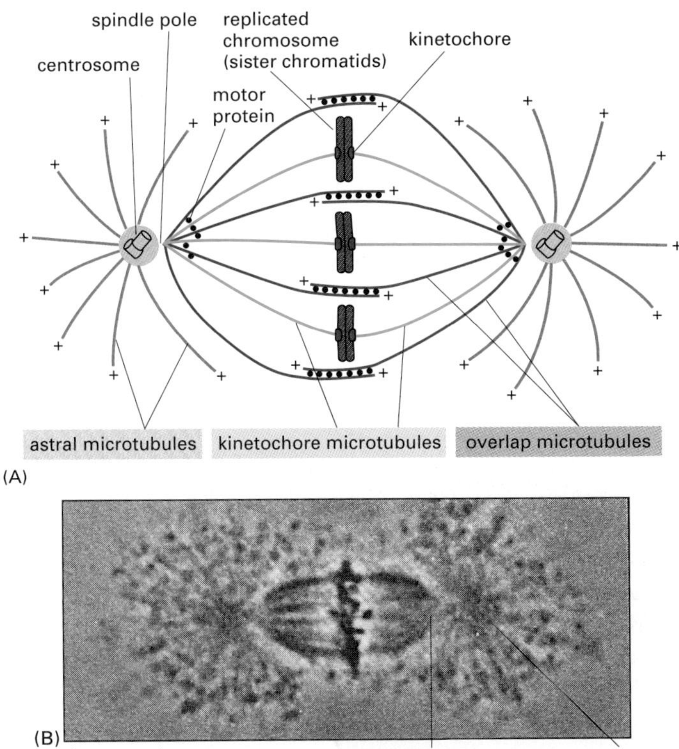

Figure 18–10 The three classes of microtubules of the fully formed mitotic spindle in an animal cell. (A) In reality, the chromosomes are proportionally much larger than shown in this drawing, and multiple microtubules are attached to each kinetochore. Note that the plus ends of the microtubules project away from the centrosomes, while the minus ends are anchored at the spindle poles. Although the centrosomes initiate the assembly of the spindle poles, most of the kinetochore and overlap microtubules that they nucleate are released from the centrosomes and are then held and organized at the poles by motor proteins. For simplicity, in the other figures in this chapter, we draw all the spindle microtubules at the poles emanating from the centrosomes. (B) A phase-contrast micrograph of an isolated mitotic spindle at metaphase, with the chromosomes aligned at the spindle equator. (B, from E.D. Salmon and R.R. Segall, *J. Cell Biol.* 86:355–365, 1980. © The Rockefeller University Press.)

Both the assembly and the function of the mitotic spindle depend on **microtubule-dependent motor proteins.** As discussed in Chapter 16, these proteins belong to two families—the **kinesin-related proteins**, which usually move toward the plus end of microtubules, and the **dyneins**, which move toward the minus end. In the mitotic spindle, the motor proteins operate at or near the ends of the microtubules. These ends are not only sites of microtubule assembly and disassembly; they are also sites of force production. The assembly and dynamics of the mitotic spindle rely on the shifting balance between opposing plus-end-directed and minus-end-directed motor proteins.

Three classes of spindle microtubules can be distinguished in mitotic animal cells (Figure 18–10). **Astral microtubules** radiate in all directions from the centrosomes and are thought to contribute to the forces that separate the poles. They also act as “handles” for orienting and positioning the spindle in the cell. **Kinetochore microtubules** attach end-on to the *kinetochore,* which forms at the *centromere* of each duplicated chromosome. They serve to attach the chromosomes to the spindle. **Overlap microtubules** interdigitate at the equator of the spindle and are responsible for the symmetrical, bipolar shape of the spindle. All three classes of microtubules have their plus ends projecting away from their centrosome. The behavior of each class is thought to be different because of the different protein complexes that are associated with their plus and minus ends.

Microtubule Instability Increases Greatly at M Phase

The mitotic spindle begins to self-assemble in the cytoplasm during **prophase**. In animal cells, each of the replicated centrosomes nucleates its own array of microtubules, and the two sets of microtubules interact to form the mitotic spindle. We see later that the self-assembly process depends on a balance between opposing forces that originate within the spindle itself and are generated by motor proteins associated with the spindle microtubules.

Many animal cells in interphase contain a cytoplasmic array of microtubules radiating out from the single centrosome. As discussed in Chapter 16, the microtubules of this *interphase array* are in a state of **dynamic instability**, in which individual microtubules are either growing or shrinking and stochastically

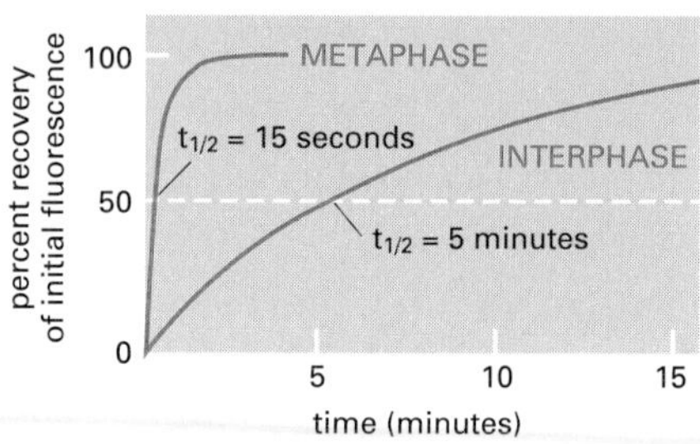

Figure 18–11 The half-life of microtubules in mitosis. Microtubules in an M-phase cell are much more dynamic, on average, than the microtubules at interphase. Mammalian cells in culture were injected with tubulin that had been covalently linked to a fluorescent dye. After the fluorescent tubulin had become incorporated into the cell's microtubules, an intense laser beam was used to bleach all the fluorescence in a small region. The recovery of fluorescence in the bleached region of microtubules, caused by their replacement by microtubules formed from unbleached fluorescent tubulin from the soluble pool, was then monitored as a function of time. The time for 50% recovery of fluorescence ($t_{1/2}$) is thought to be equal to the time required for half of the microtubules in the region to depolymerize and re-form. (Data from W.M. Saxton et al., *J. Cell Biol.* 99:2175–2187, 1984. © The Rockefeller University Press.)

switch between the two states. The switch from growth to shrinkage is called a *catastrophe*, and the switch from shrinkage to growth is called a *rescue* (see Figure 16–11). New microtubules are continually being created to balance the loss of those that disappear completely by depolymerization.

Prophase signals an abrupt change in the cell's microtubules. The relatively few, long microtubules of the interphase array rapidly convert to a larger number of shorter and more dynamic microtubules surrounding each centrosome, which will begin to form the mitotic spindle. During prophase, the half-life of microtubules decreases dramatically. This can be seen by labeling the microtubules in living cells with fluorescent tubulin subunits (Figure 18–11). As the instability of microtubules increases, the number of microtubules radiating from the centrosomes greatly increases as well, apparently because of an alteration in the centrosomes themselves that increases the rate at which they nucleate new microtubules. How does the cell-cycle control system trigger these dramatic changes in the cell's microtubules at the onset of mitosis?

M-Cdk initiates the changes by causing the phosphorylation of two classes of proteins that control microtubule dynamics (discussed in Chapter 16). These include microtubule motor proteins and **microtubule-associated proteins (MAPs)**. The roles of these regulators in controlling microtubule dynamics have been revealed by experiments using *Xenopus* egg extracts, which reproduce many of the changes that occur in intact cells during M phase. If centrosomes and fluorescent tubulin are mixed with extracts made from either M-phase or interphase cells, fluorescent microtubules nucleate from the centrosomes, permitting the behavior of individual microtubules to be analyzed by time-lapse fluorescence video microscopy. The microtubules in mitotic extracts differ from those in interphase extracts primarily by the increased rate of catastrophes, where they switch abruptly from slow growth to rapid shortening.

Proteins called **catastrophins** destabilize microtubule arrays by increasing the frequency of catastrophes (see Figure 16–36A). Among the catastrophins is a kinesin-related protein that does not function as a motor. In general, MAPs have the opposite effect of catastrophins, stabilizing microtubules in various ways: they can increase the frequency of rescues, in which microtubules switch from shrinkage to growth, or they can either increase the growth rate or decrease the shrinkage rate of microtubules. Thus, in principle, changes in catastrophins and MAPs can make microtubules much more dynamic in M phase by increasing total microtubule depolymerization rates, decreasing total microtubule polymerization rates, or both.

In *Xenopus* egg extracts, the balance between a single type of catastrophin and a single type of MAP can be shown to determine the catastrophe rate and the steady-state length of microtubules. This balance, in turn, governs the assembly of the mitotic spindle, as microtubules that are either too long or too short are incapable of assembling into a spindle (Figure 18–12). One way in which M-Cdk may control microtubule length is by phosphorylating this MAP and reducing its ability to stabilize microtubules. Even if the activity of the catastrophin remained constant throughout the cell cycle, the balance between the two opposing activities of the MAP and catastrophin would shift, increasing the dynamic instability of the microtubules.

The cell contains a variety of MAPs, catastrophins, and motor proteins, each with subtly different activities. It is the balance between the opposing activities of these proteins that is responsible for the dynamic behavior of the mitotic spindle. We see later how changes in this balance help the spindle to segregate the chromosomes at anaphase.

Interactions Between Opposing Motor Proteins and Microtubules of Opposite Polarity Drive Spindle Assembly

While a shift in the balance between MAPs and catastrophins early in M phase creates more dynamic microtubules, a different sort of balance, between minus-end-directed and plus-end-directed motor proteins, helps assemble the mitotic spindle. Because some of these motor proteins form oligomers that can cross-link adjacent microtubules, they can move one microtubule relative to the other, with the direction of movement dependent on the polarity of both the motor protein and the microtubules. In this way, these motor proteins can form foci by bringing together a group of microtubule ends (Figure 18–13A). Alternatively, such motor proteins can slide antiparallel microtubules past each other (Figure 18–13B). These two different motor protein functions play a crucial part in the

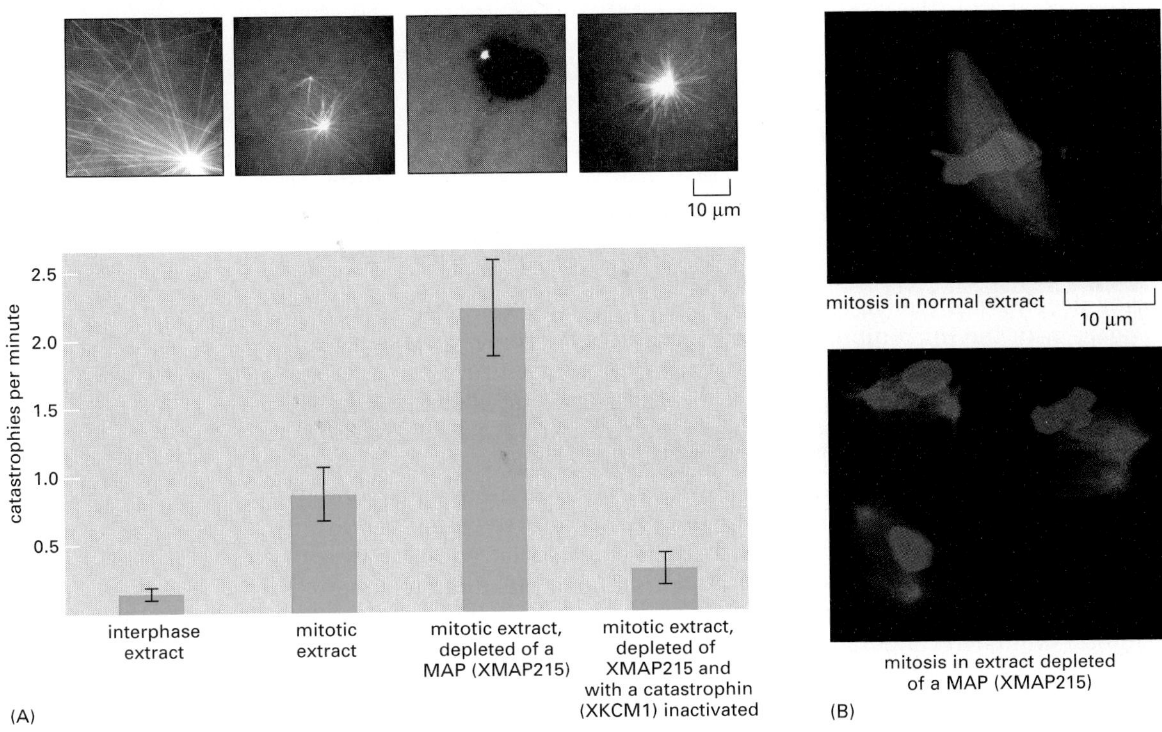

Figure 18–12 Experimental evidence that the balance between catastrophins and MAPs influences the frequency of microtubule catastrophes and microtubule length. (A) Interphase or mitotic *Xenopus* egg extracts were incubated with centrosomes, and the behavior of individual microtubules nucleated from the centrosomes was followed by fluorescence video microscopy. As expected, the catastrophe rate is higher in mitotic than in interphase extracts. The depletion of a specific MAP (XMAP215) from the mitotic extracts greatly increases the catastrophe rate, indicating that this MAP normally inhibits catastrophes in mitotic extracts. The addition of antibodies that block the function of a specific catastrophin (the kinesin-related protein XKCM1) greatly reduces the catastrophe rate in the XMAP215-depleted mitotic extracts, indicating that this catastrophin is normally responsible for stimulating catastrophes in mitotic extracts and that the catastrophe rate depends on the balance between the MAP and the catastrophin. Fluorescence micrographs of the asters formed in the different experimental conditions are shown in the top panels; note that the higher the catastrophe rates, the shorter the microtubules. (B) Mitotic spindle formation in mitotic extracts when both centrosomes and sperm nuclei are added. Microtubules are shown in *red* and chromosomes in *blue*. Whereas normal spindles form in normal mitotic extracts, very abnormal spindles form when XMAP215 is depleted from the extracts, presumably because the microtubules nucleated by the centrosomes are too short. Remarkably, microtubules formed in a test tube from a mixture of purified tubulin, XMAP215, and XKCM1 exhibit normal dynamic instability. (From R. Tournebize et al., *Nature Cell Biol.* 2:13–19, 2000. © Macmillan Magazines Ltd.)

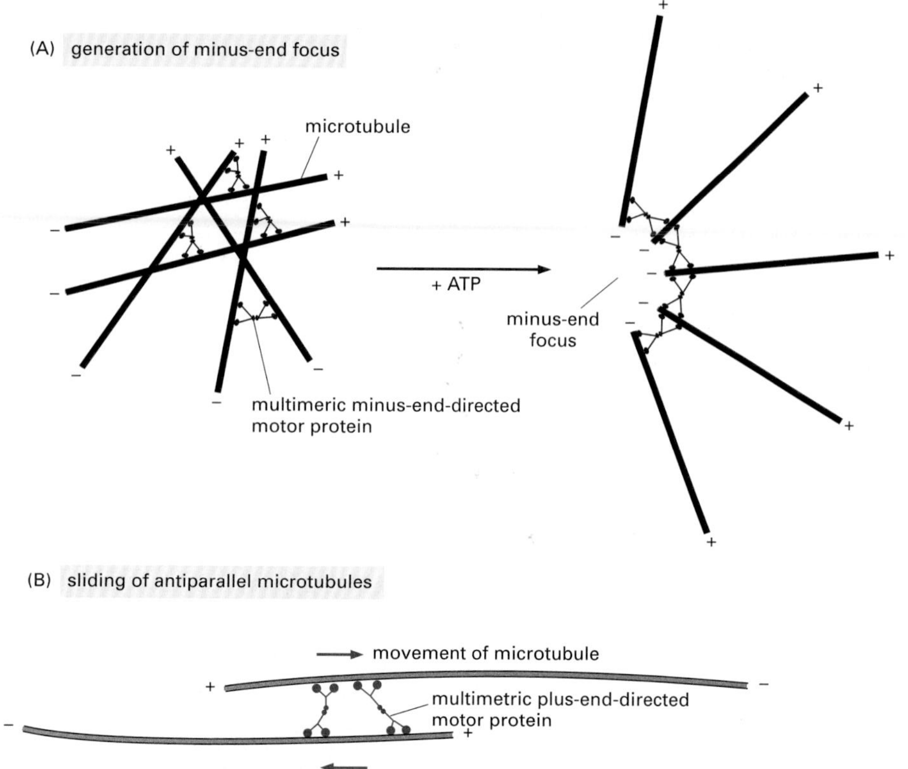

Figure 18–13 Two functions of multimeric motor proteins that are important for mitotic spindle assembly and function. Microtubule-dependent motor proteins hydrolyze ATP and move along a microtubule toward either its plus or its minus end. If the motor protein is multimeric, as in these examples, it can cross-link two adjacent microtubules and move them relative to one another. (A) Some minus-end-directed multimeric motor proteins rearrange microtubules to form a focus of minus ends, where the motor proteins accumulate. (B) If microtubules are aligned so that they are antiparallel (that is, their plus ends are facing in opposite directions), a cross-linking motor protein can slide the microtubules past each other, as shown here for a plus-end-directed motor protein that could elongate the spindle. (Adapted from A.A. Hyman and E. Karsenti, *Cell* 84:406–410, 1996.)

assembly and function of the spindle: they create the foci of microtubule minus ends that form the two spindle poles, and they slide antiparallel microtubules past each other in the overlap zone of the spindle (see Figure 18–10).

During prophase in animal cells, microtubules growing from one centrosome engage with the microtubules of the adjacent centrosome. Because the plus ends of the microtubules are oriented away from the centrosomes, these two sets of microtubules have opposite polarities. Plus-end-directed motor proteins cross-link the two sets of microtubules and help push the centrosomes apart to begin to form the two poles of the mitotic spindle (Figure 18–14). A balance between plus-end-directed motor proteins and minus-end-directed motor proteins is crucial for spindle assembly: whereas plus-end-directed motor proteins operating on overlap microtubules tend to push the two halves of the spindle apart, some minus-end-directed motor proteins tend to pull them together.

At least seven families of kinesin-related motor proteins have been localized to the mitotic spindle in vertebrate cells. In the budding yeast *S. cerevisiae*, five such motor proteins have been shown to work together in the spindle. Increasing the level of one of the plus-end-directed motor proteins produces abnormally long spindles, whereas increasing the level of one of the minus-end-directed motor proteins produces abnormally short spindles (Figure 18–15). Thus, the balance between plus-end-directed and minus-end-directed motor proteins seems to determine spindle length. A similar balance between motor proteins of opposite polarities occurs in human mitotic cells. At least, one of the motor proteins in human cells has to be phosphorylated by M-Cdk to bind to the spindle, suggesting one way in which M-Cdk might control the balance between opposing motor proteins.

Kinetochores Attach Chromosomes to the Mitotic Spindle

Prometaphase in animal cells begins abruptly with the breakdown of the nuclear envelope. The breakdown is triggered when M-Cdk directly phosphorylates the nuclear lamina that underlies the nuclear envelope (see Figure 12–20). The disassembly of the nuclear envelope allows the microtubules access to the condensed chromosomes for the first time. Now, the assembly of a mature mitotic spindle can begin (see Panel 18–1, pp. 1034–1035).

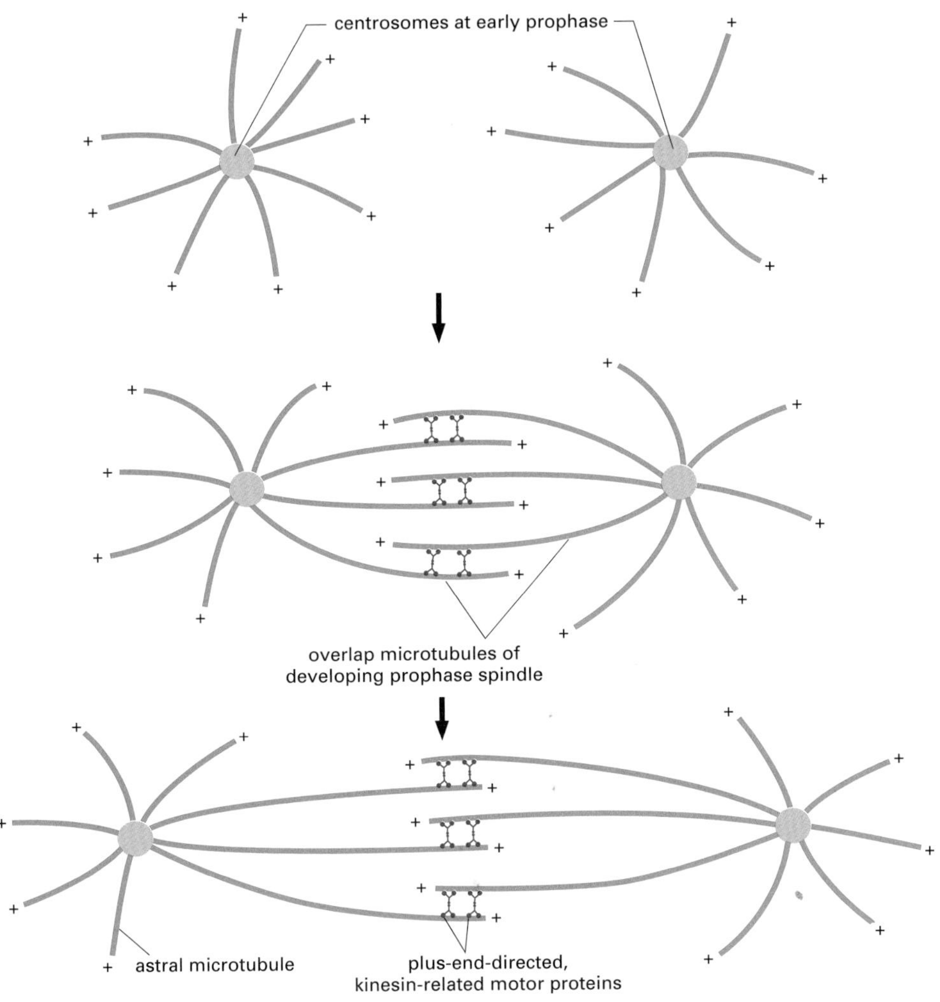

Figure 18–14 Separation of the two spindle poles in prophase in an animal cell. In this model, plus-end-directed motor proteins operating on interacting antiparallel microtubules help separate the two poles of a forming mitotic spindle. New microtubules grow out in random directions from two nearby centrosomes. The microtubules are anchored at the centrosome by their minus ends, while their plus ends extend outward. When two microtubules from opposite centrosomes interact in an overlap zone, plus-end-directed, kinesin-related motor proteins cross-link the microtubules together and tend to drive the microtubules in the direction that will push the centrosomes apart (see Figure 18–13B). Minus-end-directed dynein motors associated with the nuclear envelope are also thought to help separate the two centrosomes by pulling on the two sets of astral microtubules (not shown).

The attachment of the chromosomes to the spindle is a dynamic process. When viewed by video microscopy, it seems to involve a "search and capture" mechanism, in which microtubules nucleated from each of the rapidly separating centrosomes grow outward toward the chromosomes. Microtubules that attach to a chromosome become stabilized, so that they no longer undergo catastrophes. They eventually end up attached end-on at the **kinetochore**, a complex protein machine that assembles onto the highly condensed DNA at the centromere (discussed in Chapter 4) during late prophase. The end-on attachment to the kinetochore is through the plus end of the microtubule, which is now called a *kinetochore microtubule* (Figure 18–16).

In newt lung cells, where the initial capture event can be readily visualized, the kinetochore is seen first to bind to the side of the microtubule and then to slide rapidly along it toward one of the centrosomes. The lateral attachment to the chromosome is rapidly converted to an end-on attachment. At the same time, microtubules growing from the opposite spindle pole attach to the kinetochore

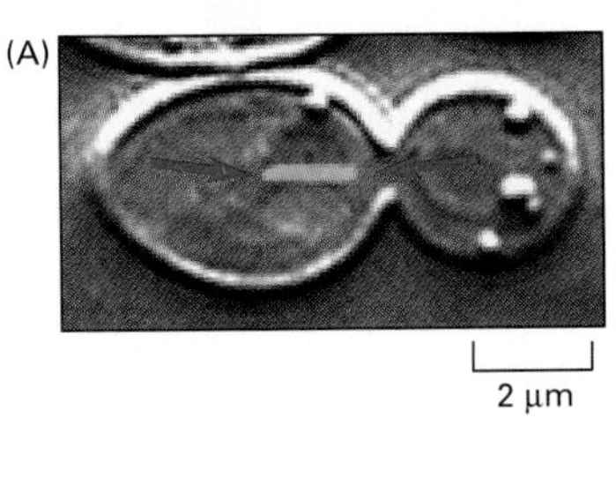

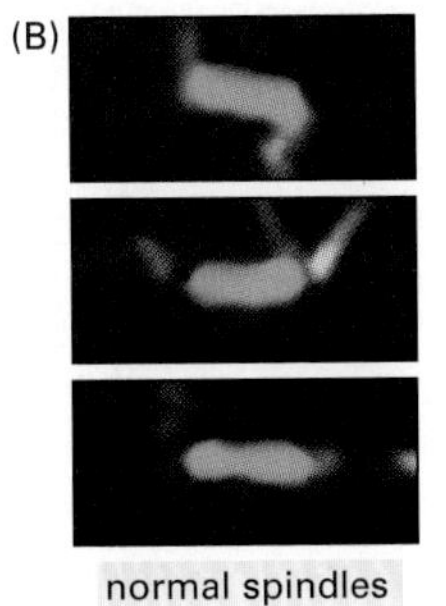

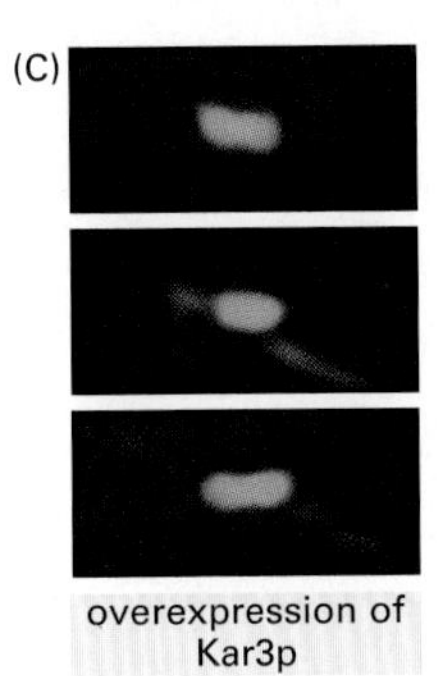

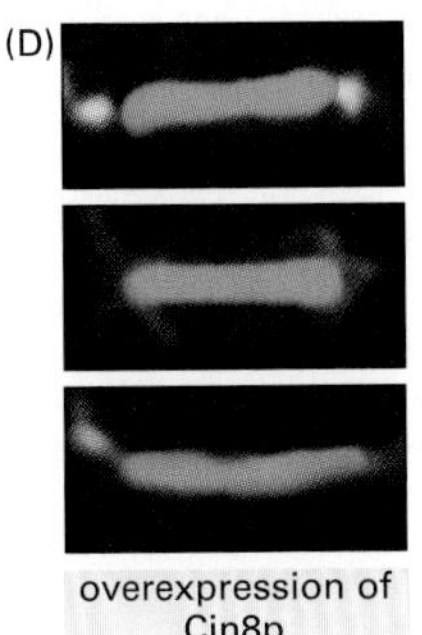

Figure 18–15 The influence of opposing motor proteins on spindle length in budding yeast. (A) A differential-interference-contrast micrograph of a mitotic yeast cell. The spindle is highlighted in *green*, and the position of the spindle poles is indicated by *red arrows*. The nuclear envelope does not break down during mitosis in yeasts, and the spindle forms inside the nucleus (see Figure 18–41). In (B–D), the mitotic spindles have been stained with fluorescent anti-tubulin antibodies. (B) Normal yeast cells. (C) Overexpression of the minus-end-directed motor protein Kar3p leads to abnormally short spindles. (D) Overexpression of the plus-end-directed motor protein Cin8p leads to abnormally long spindles. Thus, it seems that a balance between opposing motor proteins determines spindle length in these cells. (A, courtesy of Kerry Bloom; B–D, from W. Saunders, V. Lengyel, and M.A. Hoyt, *Mol. Biol. Cell* 8:1025–1033, 1997. © American Society for Cell Biology.)

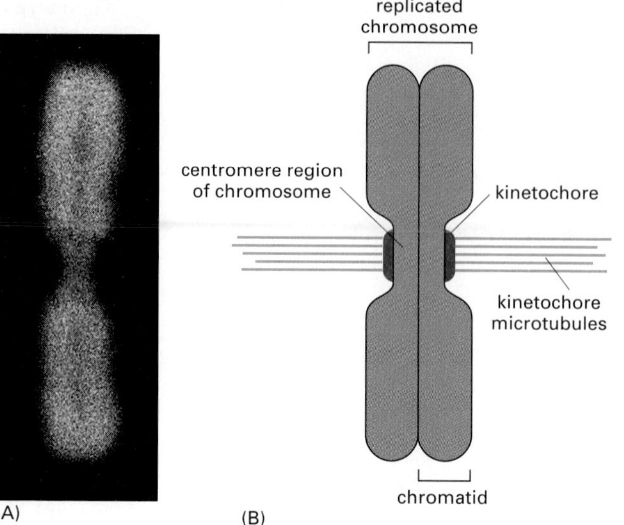
replicated
chromosome
centromere region
of chromosome
kinetochore
kinetochore
microtubules
chromatid
(A)
(B)

Figure 18–16 Kinetochore microtubules. (A) A fluorescence micrograph of a metaphase chromosome stained with a DNA-binding fluorescent dye and with human autoantibodies that react with specific kinetochore proteins. The two kinetochores, one associated with each chromatid, are stained *red.* (B) A drawing of a metaphase chromosome showing its two sister chromatids attached to kinetochore microtubules, which bind by their plus ends. Each kinetochore forms a plaque on the surface of the centromere. The number of microtubules bound to a metaphase kinetochore varies from 1 in budding yeast to over 40 in some mammalian cells. (A, courtesy of B.R. Brinkley.)

on the opposite side of the chromosome, forming a *bipolar attachment* (Figure 18–17). Then begins a truly mesmerizing stage of mitosis. First, the chromosomes are tugged back and forth, eventually assuming a position equidistant between the two spindle poles, a position called the **metaphase plate** (Figure 18–18). In vertebrate cells, the chromosomes then oscillate gently at the metaphase plate, awaiting the signal to separate. The signal is produced after a predictable lag time after the bipolar attachment of the last of the chromosomes (discussed in Chapter 17).

As we discuss later, kinetochores play a crucial part in moving chromosomes on the spindle. They have a platelike organization when viewed in the electron microscope (Figure 18–19), and they are associated with both plus-end-directed and minus-end-directed microtubule motor proteins. But it remains a mystery how the plus ends of microtubules are attached to the kinetochore, especially because these ends are continuously polymerizing or depolymerizing, depending on the stage of mitosis.

Microtubules Are Highly Dynamic in the Metaphase Spindle

The **metaphase** spindle is a complex and beautiful assembly, suspended in a state of dynamic equilibrium and tensed for action that will begin in anaphase. All of the spindle microtubules, except the kinetochore microtubules, are in a state of dynamic instability, with their free plus ends shifting stochastically

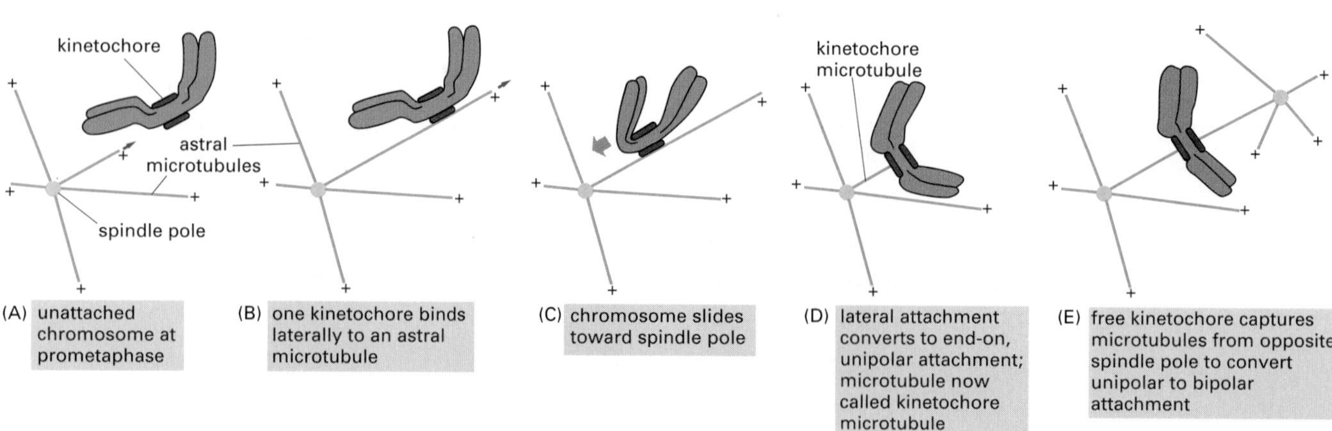
kinetochore
astral
microtubules
spindle pole
kinetochore
microtubule
(A) unattached chromosome at prometaphase
(B) one kinetochore binds laterally to an astral microtubule
(C) chromosome slides toward spindle pole
(D) lateral attachment converts to end-on, unipolar attachment; microtubule now called kinetochore microtubule
(E) free kinetochore captures microtubules from opposite spindle pole to convert unipolar to bipolar attachment

Figure 18–17 The capture of microtubules by kinetochores. The *red arrow* in (A) indicates the direction of microtubule growth, while the *gray arrow* in (C) indicates the direction of chromosome sliding.

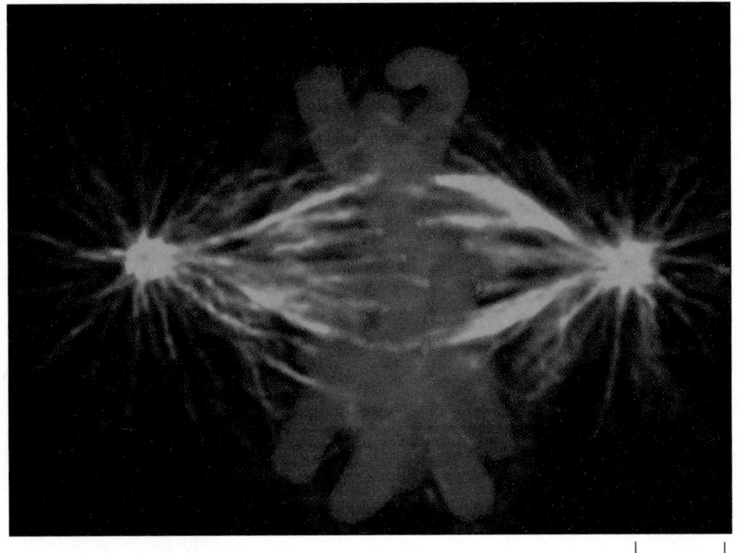

Figure 18–18 Chromosomes at the metaphase plate of a mitotic spindle. In this fluorescence micrograph, kinetochores are labeled in *red,* microtubules in *green,* and chromosomes in *blue.* (From A. Desai, *Curr. Biol.* 10:R508, 2000. © Elsevier.)

between slow growth and rapid shrinkage. In addition, the kinetochore and overlap microtubules exhibit a behavior called *poleward flux,* with a net addition of tubulin subunits at their plus end, balancing a net loss at their minus ends, near the spindle poles.

The poleward flux in kinetochore and overlap microtubules in metaphase spindles has been studied directly by allowing the microtubules to incorporate tubulin that has been covalently coupled to photoactivatable, "caged" fluorescein. When such spindles are marked with a beam of UV light from a laser, the marks move continuously toward the spindle poles (Figure 18–20). The fluorescent marks get dimmer with time, indicating that many of the overlap and kinetochore microtubules depolymerize completely and are replaced. The dynamics of individual spindle microtubules of all classes (astral, kinetochore, and overlap) can be studied by an ingenious method in which very low amounts of fluorescent tubulin are injected into living cells (Figure 18–21). In these studies, a poleward flux is seen in both kinetochore and overlap microtubules, but not in astral microtubules. The function of the poleward flux, which does not occur in simple spindles such as those in yeasts, is unknown, although it might aid chromosome movement in anaphase.

One of the most striking aspects of metaphase in vertebrate cells is the continuous oscillatory movement of the chromosomes at the metaphase plate. These movements have been studied by video microscopy in newt lung cells and are seen to switch between two states—a poleward (P) state, which is a minus-end-directed pulling movement, and an away-from-the-pole (AP) state, which is a plus-end-directed movement. Kinetochores are thought to pull the

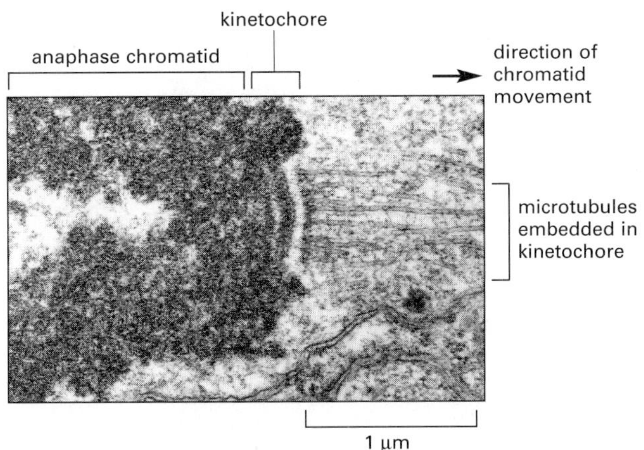

Figure 18–19 The kinetochore. Electron micrograph of an anaphase chromatid with microtubules attached to its kinetochore. While most kinetochores have a trilaminar structure, the one shown here (from a green alga) has an unusually complex structure with additional layers. (From J.D. Pickett-Heaps and L.C. Fowke, *Aust. J. Biol. Sci.* 23:71–92, 1970. Reproduced by permission of CSIRO.)

Figure 18–20 **The dynamic behavior of microtubules in the metaphase spindle studied by photoactivation of fluorescence.** A metaphase spindle formed *in vitro* by adding *Xenopus* sperm to an extract of *Xenopus* eggs (see Figure 17–9) has incorporated three fluorescent markers: rhodamine-labeled tubulin *(red)* to mark all of the microtubules, a *blue* DNA-binding dye that labels the chromosomes, and caged-fluorescein-labeled tubulin, which is also incorporated into all of the microtubules but is invisible because it is nonfluorescent until activated by ultraviolet light. (A) The distribution of the chromosomes and microtubules in the spindle. (B) A beam of ultraviolet light was used to uncage the caged-fluorescein-labeled tubulin locally, mainly just to the left side of the metaphase plate. Over the next few minutes (after 1.5 minutes in C, after 2.5 minutes in D), the uncaged fluorescein-tubulin signal is seen to move toward the left spindle pole, indicating that tubulin is continuously moving poleward, even though the spindle (visualized by the *red* rhodamine-tubulin fluorescence) remains largely unchanged. The caged fluorescein signal also diminishes in intensity, indicating that the individual microtubules are continually depolymerizing and being replaced. (From K.E. Sawin and T.J. Mitchison, *J. Cell Biol.* 112:941–954, 1991. © The Rockefeller University Press.)

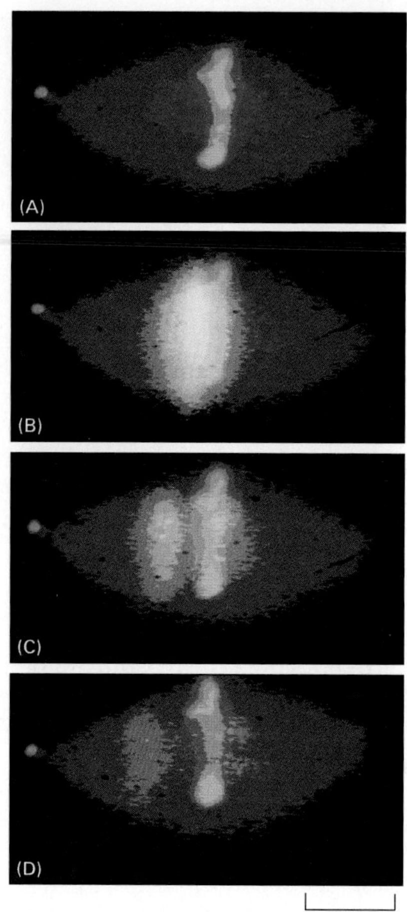

chromosomes toward the poles, while an *astral ejection force* is thought to push the chromosomes away from the poles, toward the spindle equator (Figure 18–22A). Plus-end-directed motor proteins located on the chromosome arms are believed to interact with the astral microtubules to produce the ejection force (Figure 18–22B). Interestingly, spindles without centrosomes, including those in higher plants and some meiotic spindles, do not display these oscillations, which might reflect the absence of astral microtubules and, consequently, the absence of the astral ejection force.

Functional Bipolar Spindles Can Assemble Around Chromosomes in Cells Without Centrosomes

Chromosomes are not just passive passengers in the process of spindle assembly. By creating a local environment that favors both microtubule nucleation and microtubule stabilization, they play an active part in spindle formation. The influence of the chromosomes can be demonstrated by using a fine glass needle to reposition them after the spindle has formed. For some cells in metaphase, if a single chromosome is tugged out of alignment, a mass of new spindle microtubules rapidly appears around the newly positioned chromosome, while the spindle microtubules at the chromosome's former position depolymerize. This property of the chromosomes seems to depend on a guanine-nucleotide exchange factor (GEF) that is bound to chromatin; it stimulates a small GTPase in the cytosol called *Ran*, inducing Ran to bind GTP in place of GDP. The activated Ran–GTP, which is also involved in nuclear transport (discussed in Chapter 12),

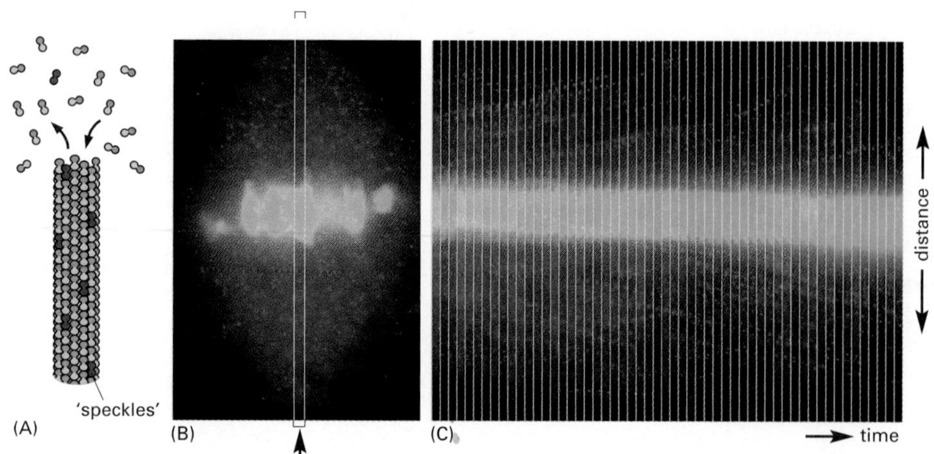

Figure 18–21 Visualizing the dynamics of individual microtubules by fluorescence speckle microscopy. (A) The principle of the method. A very low amount of fluorescent tubulin is injected into living cells so that individual microtubules form with a very small proportion of fluorescent tubulin. Such microtubules have a speckled appearance when viewed by fluorescence microscopy. (B) Fluorescence micrographs of a mitotic spindle in a living newt lung epithelial cell. The chromosomes are stained *green,* and the tubulin speckles are *red.* (C) The movement of individual speckles can be readily followed by time-lapse video microscopy. Images of the long, thin, rectangular, boxed region *(arrow)* in (B) were taken at sequential times and pasted side by side to make a montage of the region over time. Individual speckles can be seen to move toward the poles (representing poleward flux) at a rate of about 0.75 μm/min. (From T.J. Mitchison and E.D. Salmon, *Nature Cell Biol.* 3:E17–21, 2001.)

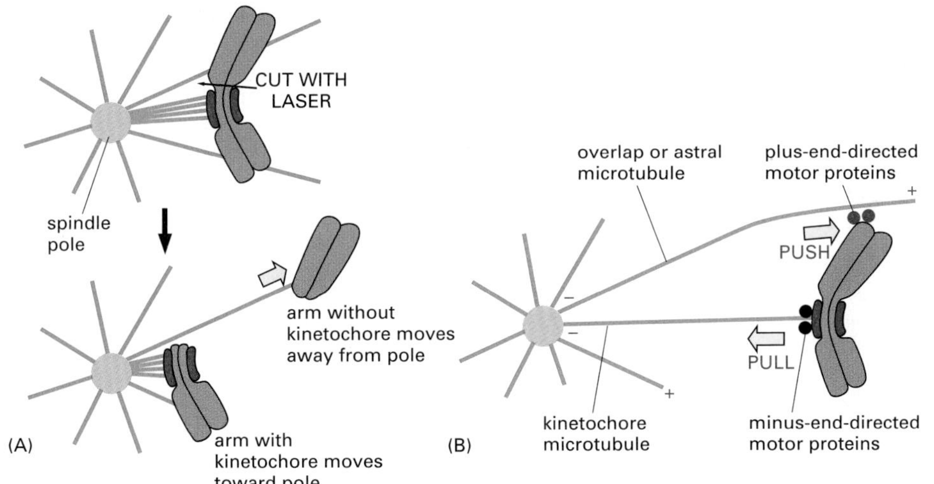

Figure 18–22 How opposing forces may drive chromosomes to the metaphase plate. (A) Evidence for an astral ejection force that pushes chromosomes away from the spindle poles toward the spindle equator. In this experiment, a prometaphase chromosome that is temporarily attached to a single pole by kinetochore microtubules is cut in half with a laser beam. The half that is freed from the kinetochore is pushed rapidly away from the pole, whereas the half that remains attached to the kinetochore moves toward the pole, reflecting a decreased repulsion.
(B) A model of how two opposing forces may cooperate to move chromosomes to the metaphase plate. Plus-end-directed motor proteins on the chromosome arms are thought to interact with astral microtubules to generate the astral ejection force, which pushes chromosomes toward the spindle equator. Minus-end-directed motor proteins at the kinetochore are thought to interact with kinetochore microtubules to pull chromosomes toward the pole.

releases microtubule-stabilizing proteins from protein complexes in the cytosol, thereby stimulating the local nucleation of microtubules around chromosomes.

In cells without centrosomes, the chromosomes direct the assembly of a functional bipolar spindle. This is how spindles form in cells of higher plants, as well as in many meiotic cells. It is also how they assemble in certain insect embryos that have been induced to develop from eggs without fertilization (i.e., *parthenogenetically*); as the sperm normally provides the centrosome when it fertilizes an egg (discussed in Chapter 20), the mitotic spindles in these parthenogenic embryos develop without centrosomes (Figure 18–23). Remarkably, in artificial systems, spindles can self-assemble without either centrosomes or centromeres. When beads coated with DNA that lack centromere sequences (and therefore lack kinetochore complexes) are added to *Xenopus* egg extracts in the absence of centrosomes, bipolar spindles assemble around the beads (Figure 18–24A).

The centrosome-independent spindle assembly process is different from the assembly that is directed by centrosomes. In the DNA-coated bead model, for example, the microtubules first nucleate near the surface of the DNA, and then microtubule motor proteins sort the microtubules into bundles of uniform polarity, push the minus ends of the microtubules apart, and focus them into spindle poles (Figure 18–24B).

Even vertebrate cells can use such a centrosome-independent pathway to construct a functional bipolar spindle if the centrosomes are destroyed with a laser beam. Although the resulting acentrosomal spindle can segregate chromosomes normally, it lacks astral microtubules, which are responsible for positioning the spindle in animal cells; as a result, the spindle is often mispositioned, resulting in abnormalities in cytokinesis. If present, however, centrosomes normally direct spindle assembly because they are more efficient at nucleating microtubule polymerization than are chromosomes.

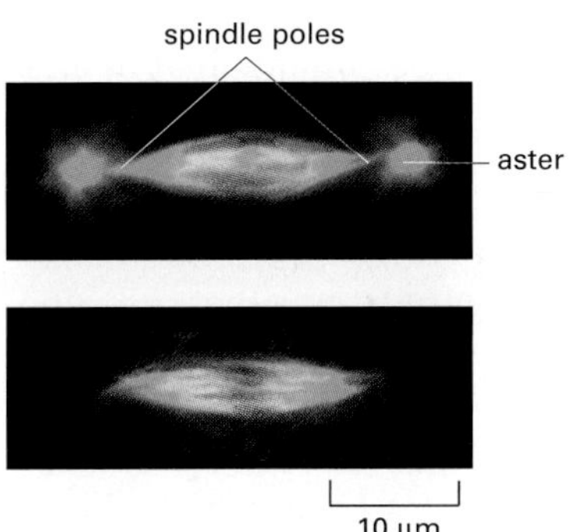

Figure 18–23 Bipolar spindle assembly without centrosomes in parthenogenetic embryos of the insect *Sciara*. The microtubules are stained *green*, the chromosomes *red*. The *top* fluorescence micrograph shows a normal spindle formed with centrosomes in a normally fertilized *Sciara* embryo. The *bottom* micrograph shows a spindle formed without centrosomes in an embryo that initiated development without fertilization. Note that the spindle with centrosomes has an aster at each pole of the spindle, whereas the spindle formed without centrosomes does not. Both types of spindles are able to segregate the replicated chromosomes. (From B. de Saint Phalle and W. Sullivan, *J. Cell Biol.* 141:1383–1391, 1998. © The Rockefeller University Press.)

Anaphase Is Delayed Until All Chromosomes Are Positioned at the Metaphase Plate

Mitotic cells usually spend about half of M phase in metaphase, with the chromosomes aligned on the metaphase plate, jostling about, awaiting the signal that induces sister chromatids to separate to begin anaphase. Treatment with drugs that destabilize microtubules, such as colchicine or vinblastine (discussed in Chapter 16), arrests mitosis for hours or even days. This observation led to the identification of a **spindle-attachment checkpoint**, which is activated by the drug treatment and arrests progress in mitosis. The checkpoint mechanism is used by the cell-cycle control system to ensure that cells do not enter anaphase until all chromosomes are attached to both poles of the spindle (discussed in

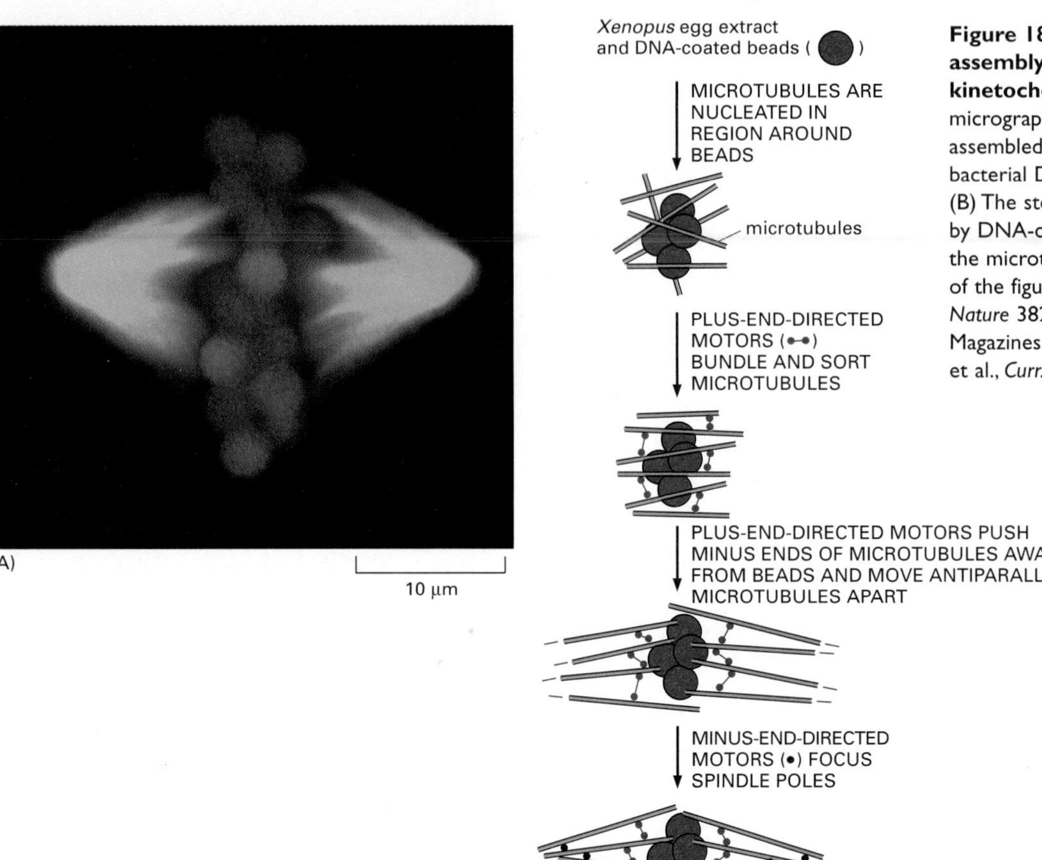
(A)
10 μm
Xenopus egg extract and DNA-coated beads ( )
MICROTUBULES ARE NUCLEATED IN REGION AROUND BEADS
microtubules
PLUS-END-DIRECTED MOTORS (•–•) BUNDLE AND SORT MICROTUBULES
PLUS-END-DIRECTED MOTORS PUSH MINUS ENDS OF MICROTUBULES AWAY FROM BEADS AND MOVE ANTIPARALLEL MICROTUBULES APART
MINUS-END-DIRECTED MOTORS (•) FOCUS SPINDLE POLES
(B)

Figure 18–24 Bipolar spindle assembly without centrosomes or kinetochores. (A) A fluorescence micrograph of a spindle *(green)* that self-assembled around beads *(red)* coated with bacterial DNA in *Xenopus* egg extracts. (B) The steps in spindle assembly directed by DNA-coated beads. The minus ends of the microtubules are indicated in one part of the figure. (A, from R. Heald et al., *Nature* 382:420–425, 1996. © Macmillan Magazines Ltd; B, based on C.E. Walczak et al., *Curr. Biol.* 8:903–913, 1998.)

Chapter 17). If one of the protein components of the checkpoint mechanism is inactivated by mutation or by an intracellular injection of antibodies against the component, the cells initiate anaphase prematurely.

The spindle-attachment checkpoint monitors the attachment of the chromosomes to the mitotic spindle. It is thought to detect either unattached kinetochores or kinetochores that are not under the tension that results from bipolar attachment. In either case, unattached kinetochores emit a signal that delays anaphase until they all are properly attached to the spindle (see Figure 17–27). Drugs that destabilize microtubules prevent such attachment and therefore maintain the signal and delay anaphase. The inhibitory signaling role of the kinetochore can be demonstrated in mammalian cells in culture, where a single unattached kinetochore can block anaphase; destruction of this kinetochore with a laser causes the cell to enter anaphase.

Sister Chromatids Separate Suddenly at Anaphase

Anaphase begins abruptly with the release of the cohesin linkage that holds the sister chromatids together at the metaphase plate. As discussed in Chapter 17, this *metaphase-to-anaphase transition* is triggered by the activation of the *anaphase promoting complex (APC)*. Once this proteolytic complex is activated, it has at least two crucial functions: (1) it cleaves and inactivates the M-phase cyclin (M-cyclin), thereby inactivating M-Cdk; and (2) it cleaves an inhibitory protein (securin), thereby activating a protease called *separase*. Separase then cleaves a subunit in the cohesin complex to unglue the sister chromatids (see Figure 17–26). The sisters immediately separate—and are now called daughter chromosomes—and move to opposite poles (Figure 18–25).

The chromosomes move by two independent and overlapping processes. The first, referred to as **anaphase A**, is the initial poleward movement of the chromosomes. It is accompanied by shortening of the kinetochore microtubules at their attachment to the chromosome and, to a lesser extent, by the

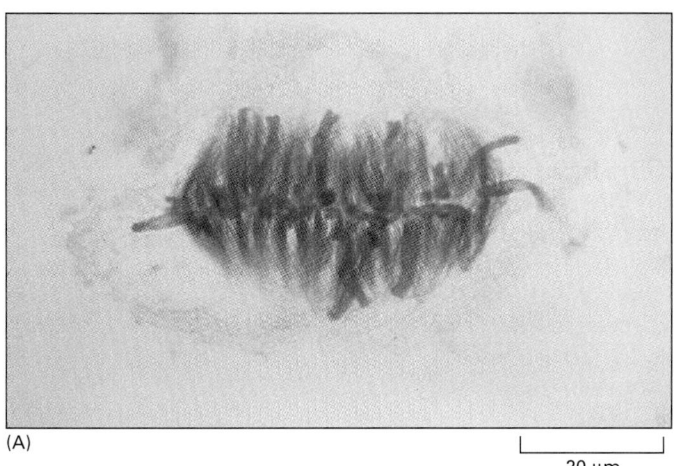
(A)
20 μm

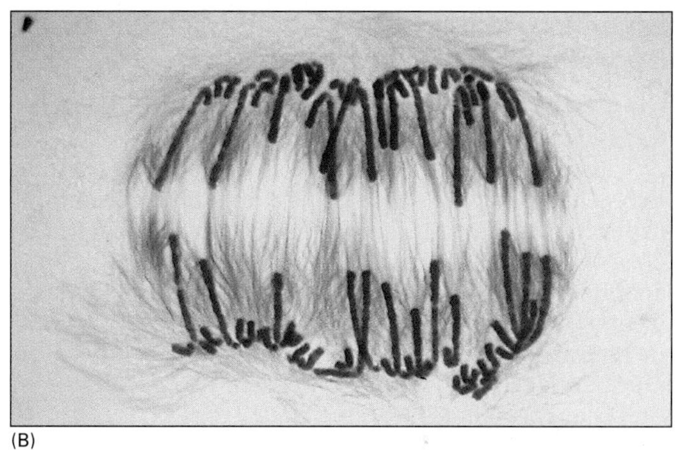
(B)

Figure 18–25 Chromatid separation at anaphase. In the transition from metaphase (A) to anaphase (B), sister chromatids suddenly separate and move toward opposite poles—as shown in these light micrographs of *Haemanthus* (lily) endosperm cells that were stained with gold-labeled antibodies against tubulin. (Courtesy of Andrew Bajer.)

depolymerization of spindle microtubules at the two spindle poles. The second process, referred to as **anaphase B**, is the separation of the poles themselves, which begins after the sister chromatids have separated and the daughter chromosomes have moved some distance apart. Anaphase A depends on motor proteins at the kinetochore. Anaphase B depends on motor proteins at the poles that pull the poles apart, as well as on motor proteins at the *central spindle* (the bundles of antiparallel overlap microtubules between the separating chromosomes) that push the poles apart (Figure 18–26). Originally, anaphase A and anaphase B were distinguished by their different sensitivities to drugs. These differences are now thought to reflect differences in the sensitivities of the microtubule motor proteins that mediate the two processes.

Figure 18–26 The major forces that separate daughter chromosomes at anaphase in mammalian cells. Anaphase A depends on motor proteins operating at the kinetochores that, together with the depolymerization of the kinetochore microtubules, pull the daughter chromosomes toward the nearest pole. In anaphase B, the two spindle poles move apart. Two separate forces are thought to be responsible for anaphase B. The elongation and sliding of the overlap microtubules past one another in the central spindle push the two poles apart, and outward forces exerted by the astral microtubules at each spindle pole act to pull the poles away from each other, toward the cell surface.

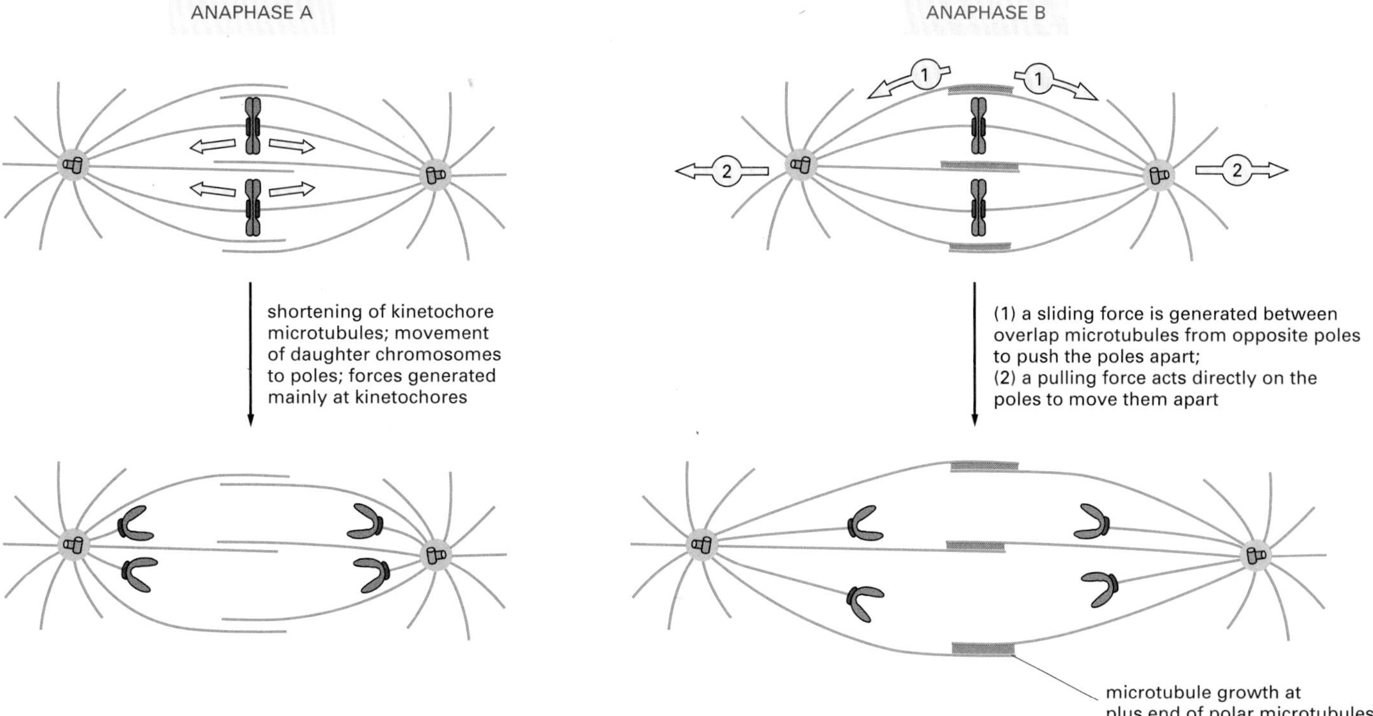
ANAPHASE A
ANAPHASE B
1
1
2
2
shortening of kinetochore microtubules; movement of daughter chromosomes to poles; forces generated mainly at kinetochores
(1) a sliding force is generated between overlap microtubules from opposite poles to push the poles apart;
(2) a pulling force acts directly on the poles to move them apart
microtubule growth at plus end of polar microtubules

Kinetochore Microtubules Disassemble at Both Ends During Anaphase A

As each daughter chromosome moves poleward, its kinetochore microtubules depolymerize, so that they have nearly disappeared at telophase. We can see this process by fluorescence video microscopy, in which labeled tubulin is injected into cells so that the sites of recent tubulin incorporation can be seen. In such experiments, the kinetochore ends of the kinetochore microtubules are observed to be the primary sites of tubulin addition during metaphase. In anaphase A, however, the kinetochore microtubules shorten mainly by the loss of tubulin from their kinetochore ends. It is not known how this switch from polymerization to depolymerization at kinetochores occurs at anaphase, but it may be triggered by the loss of tension that occurs when the cohesion between the sister chromatids is destroyed.

The poleward flux discussed earlier, with the continuous loss of tubulin subunits from both the overlap and kinetochore microtubules at the poles (see Figure 18–21), continues through anaphase. Thus, the kinetochore microtubules disassemble from both ends in anaphase.

Although it is clear that both microtubule motor proteins and microtubule depolymerization at the kinetochores contribute to chromosome movement during anaphase A, the exact molecular mechanism that drives the movement is still unknown. It is also unclear how kinetochores can remain attached to a microtubule that is losing tubulin subunits at its kinetochore (plus) end. There are two main ideas about how chromosomes move in anaphase A. One is that motor proteins at the kinetochores use the energy of ATP hydrolysis to pull the chromosomes along the kinetochore microtubules, which depolymerize as a consequence. Another is that the depolymerization itself drives the movement, without using ATP (Figure 18–27). The second possibility might seem implausible at first, but it has been shown that purified kinetochores in a test tube, with no ATP present, can remain attached to depolymerizing microtubules and thereby move. The energy that drives the movement is stored in the microtubule and is released when the microtubule depolymerizes; it ultimately comes from the hydrolysis of GTP that occurs after a tubulin subunit adds to the end of a microtubule (discussed in Chapter 16). How motor proteins and microtubule depolymerization at the kinetochore combine to drive chromosome movement remains one of the fundamental mysteries of mitosis.

Both Pushing and Pulling Forces Contribute to Anaphase B

In anaphase B, the spindle elongates, pulling the two sets of chromosomes farther apart. In contrast to anaphase A, where the depolymerization of kinetochore microtubules is coupled to chromosome movement toward the poles, in

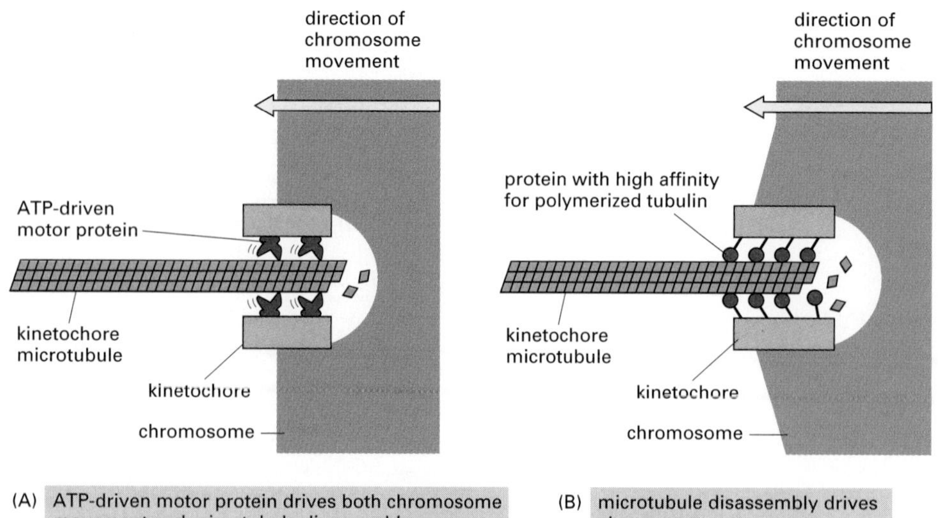

Figure 18–27 Two alternative models of how the kinetochore may generate a poleward force on its chromosome during anaphase A. (A) Microtubule motor proteins at the kinetochore use the energy of ATP hydrolysis to pull the chromosome along its bound microtubules. Depolymerization of the kinetochore microtubules follows as a consequence. (B) Chromosome movement is driven by kinetochore microtubule disassembly: as tubulin subunits dissociate, the kinetochore is obliged to slide poleward to maintain its binding to the walls of the microtubule. As motor proteins are required for anaphase A, in this model, they would be required mainly for the microtubules to remain attached to the kinetochore.

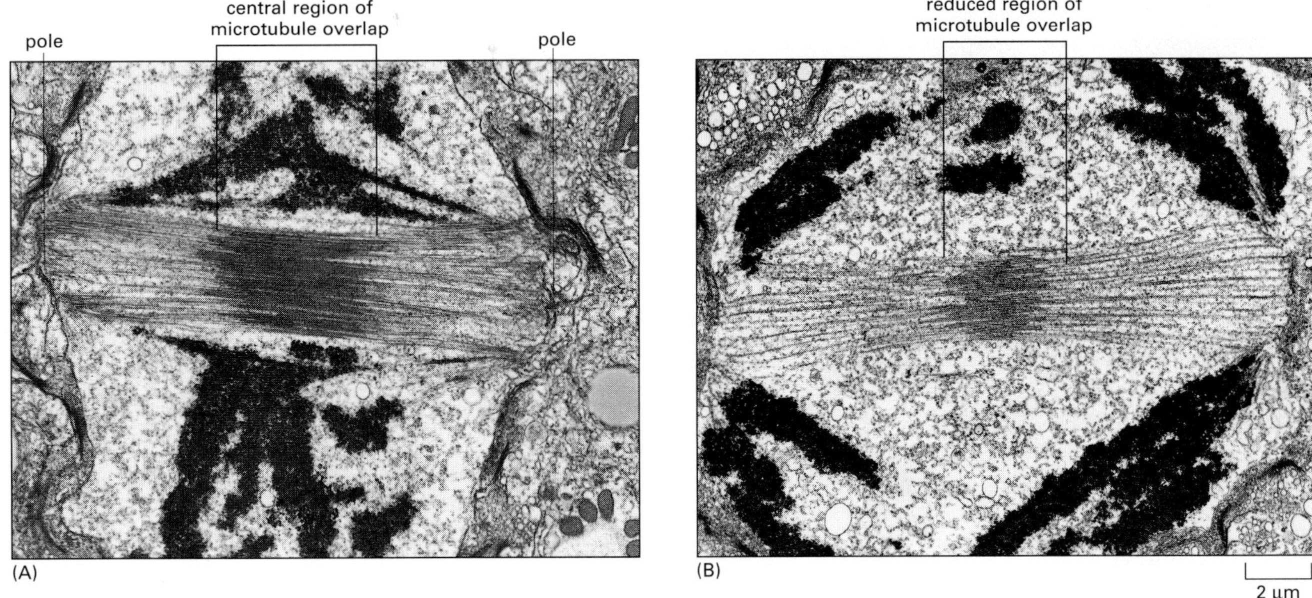

Figure 18–28 The sliding of overlap microtubules at anaphase. These electron micrographs show the reduction in the degree of microtubule overlap in the central spindle during mitosis in a diatom. (A) Metaphase. (B) Late anaphase. (Courtesy of Jeremy D. Pickett-Heaps.)

anaphase B, the overlap microtubules actually elongate, helping to push the spindle poles apart. Anaphase B is driven by two distinct forces (see Figure 18–26). The first depends on plus-end-directed microtubule motor proteins in the central spindle that form a bridge between the overlapping microtubules of opposite polarities; the translocation of these motors toward the microtubule plus ends slides the microtubules past one another, pushing the poles apart. As a result, the bundle of overlap microtubules in the central spindle thins out (Figure 18–28). This mechanism is similar to that described earlier, in which motor proteins push the poles apart during spindle assembly in prophase (see Figure 18–14). The second force contributing to anaphase B depends on minus-end-directed motor proteins that interact with both the astral microtubules and the cell cortex and pull the two poles of the spindle apart (Figure 18–29).

The relative contributions of anaphase A and anaphase B to chromosome segregation vary greatly, depending on the cell type. In mammalian cells, anaphase B begins shortly after anaphase A and stops when the spindle is about twice its metaphase length; in contrast, the spindles of yeasts and certain protozoa primarily use anaphase B to separate the chromosomes at anaphase, and their spindles elongate to up to 15 times the metaphase length in the process.

At Telophase, the Nuclear Envelope Re-forms Around Individual Chromosomes

By the end of anaphase, the daughter chromosomes have separated into two equal groups at opposite ends of the cell and have begun to decondense. In

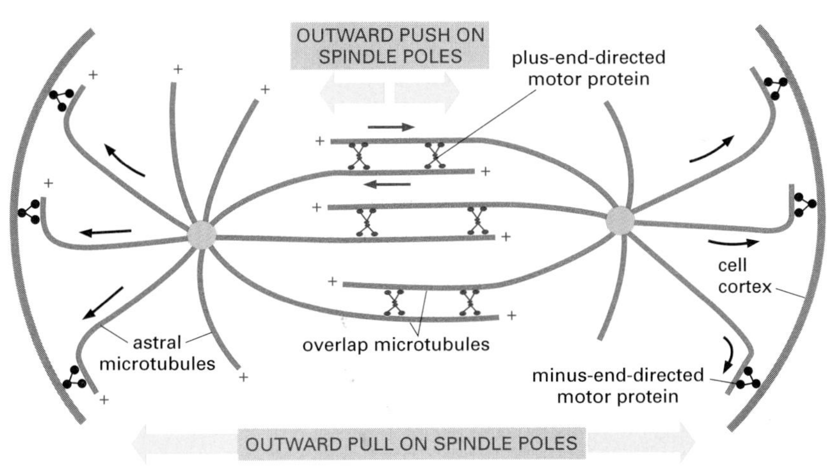

Figure 18–29 A model for how motor proteins may act in anaphase B. Plus-end-directed motor proteins cross-link the overlapping, antiparallel, overlap microtubules and slide the microtubules past each other, thereby pushing the spindle poles apart. The *red arrows* indicate the direction of microtubule sliding. Minus-end-directed motor proteins bind to the cell cortex and to those astral microtubules that point away from the spindle and pull the poles apart. These astral microtubules shorten as the spindle poles are pulled toward the cortex.

telophase, the final stage of mitosis, a nuclear envelope reassembles around each group of chromosomes to form the two daughter interphase nuclei.

The sudden transition from metaphase to anaphase initiates the dephosphorylation of the many proteins that were phosphorylated at prophase. Although the relevant phosphatases are active throughout mitosis, it is not until M-Cdk is switched off that the phosphatases can act unopposed. Shortly thereafter, at telophase, nuclear membrane fragments associate with the surface of individual chromosomes and fuse to re-form the nuclear envelope. Initially, the fused membrane fragments partly enclose clusters of chromosomes; the fragments then coalesce to re-form the complete nuclear envelope (see Figure 12–21). During this process, the nuclear pore complexes are incorporated into the envelope, and the dephosphorylated lamins reassociate to form the nuclear lamina. The nuclear envelope once again becomes continuous with the extensive membrane sheets of the endoplasmic reticulum. Once the nuclear envelope has re-formed, the pore complexes pump in nuclear proteins, the nucleus expands, and the condensed mitotic chromosomes decondense into their interphase state, thereby allowing gene transcription to resume. A new nucleus has been created, and mitosis is complete. All that remains is for the cell to complete its division into two.

Summary

Mitosis begins with prophase, which is marked by an increase in microtubule instability, triggered by M-Cdk. In animal cells, an unusually dynamic microtubule array (an aster) forms around each of the duplicated centrosomes, which separate to initiate the formation of the two spindle poles. Interactions between the asters and a balance between minus-end-directed and plus-end-directed microtubule-dependent motor proteins result in the self-assembly of the bipolar spindle. In higher-plant cells and other cells that lack centrosomes, a functional, bipolar spindle self-assembles instead around the replicated chromosomes. Prometaphase begins with the breakdown of the nuclear envelope, which allows the kinetochores on the condensed chromosomes to capture and stabilize microtubules from each spindle pole. The kinetochore microtubules from opposite spindle poles pull in opposite directions on each duplicated chromosome, creating, together with a polar ejection force, a tension that helps bring the chromosomes to the spindle equator to form the metaphase plate. The spindle microtubules at metaphase are highly dynamic and undergo a continuous poleward flux of tubulin subunits. Anaphase begins with the sudden proteolytic cleavage of the cohesin linkage holding sister chromatids together. The breakage of this linkage allows the chromosomes to be pulled to opposite poles (the anaphase A movement). At about the same time, the two spindle poles move apart (the anaphase B movement). In telophase, the nuclear envelope re-forms on the surface of each group of separated chromosomes as the proteins phosphorylated at the onset of M phase are dephosphorylated.

CYTOKINESIS

The cell cycle culminates in the division of the cytoplasm by **cytokinesis**. In a typical cell, cytokinesis accompanies every mitosis, although some cells, such as *Drosophila* embryos (discussed later) and vertebrate osteoclasts (discussed in Chapter 22), undergo mitosis without cytokinesis and become multinucleate. Cytokinesis begins in anaphase and ends in telophase, reaching completion as the next interphase begins.

The first visible change of cytokinesis in an animal cell is the sudden appearance of a pucker, or *cleavage furrow*, on the cell surface. The furrow rapidly deepens and spreads around the cell until it completely divides the cell in two. In animal cells and many unicellular eucaryotes, the structure that accomplishes cytokinesis is the *contractile ring*—a dynamic assembly composed of actin filaments, myosin II filaments, and many structural and regulatory proteins. The ring assembles just beneath the plasma membrane and contracts to constrict the cell into two (see Figure 18–4). At the same time, new membrane is inserted into the plasma membrane adjacent to the contractile ring by the fusion of intracellular vesicles. This addition of membrane is required to compensate for the increase in surface area that accompanies cytoplasmic division. Thus, cytokinesis can be considered to occur in four stages—initiation, contraction, membrane insertion, and completion.

The central problem for a cell undergoing cytokinesis is to ensure that it occurs at the right time and in the right place. Cytokinesis must not occur too early in M-phase, or it will disrupt the path of the separating chromosomes. It must also occur at the right place to separate the two segregating sets of chromosomes properly so that each daughter cell receives a complete set.

The Microtubules of the Mitotic Spindle Determine the Plane of Animal Cell Division

The mitotic spindle in animal cells not only separates the daughter chromosomes, it also specifies the location of the contractile ring, and thereby the plane of cell division. The contractile ring invariably forms in the plane of the metaphase plate, at right angles to the long axis of the mitotic spindle, thereby ensuring that division occurs between the two sets of separated chromosomes. The part of the spindle that specifies the division plane varies depending on the cell type: in some cells, it is the astral microtubules; in others, it is the overlapping antiparallel microtubules in the central spindle.

The relationship between the spindle microtubules and the placement of the contractile ring has been studied by manipulating fertilized eggs of marine invertebrates. After fertilization, these embryos undergo a series of rapid cleavage divisions, without intervening periods of growth. In this way, the original egg is progressively divided up into smaller and smaller cells. During cytokinesis, the cleavage furrow appears suddenly on the surface of the cell and deepens rapidly (Figure 18–30). Because the cytoplasm is clear, the spindle can be observed in real time through a microscope. If the spindle is tugged into a new position with a fine glass needle in early anaphase, the incipient cleavage furrow disappears, and a new one develops in accord with the new spindle site.

How does the mitotic spindle control the plane of division? Ingenious experiments in large embryonic cells demonstrate that a cleavage furrow forms midway between the asters originating from the two centrosomes, even when the two centrosomes are not connected to each other by a mitotic spindle (Figure 18–31). Thus, in these cells, the microtubule asters—not the chromosomes or other parts of the spindle—signal to the cell cortex to specify where the contractile ring should assemble. In other cells, the central spindle, rather than the astral microtubules, is apparently responsible for this specification. In either case, it has been speculated that the overlapping microtubules may provide tracks for motor proteins to deliver contractile ring regulators, and perhaps new

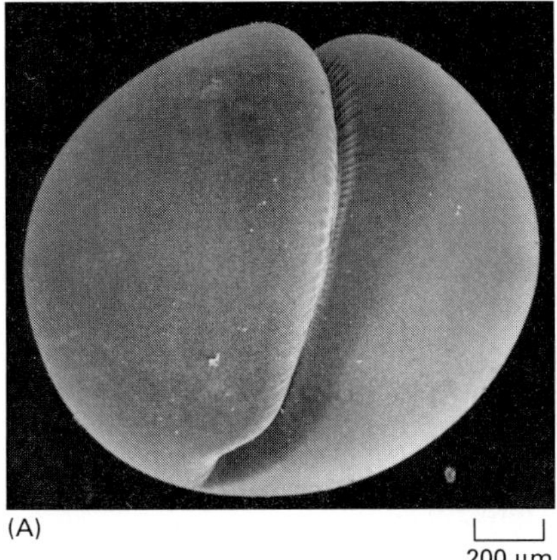

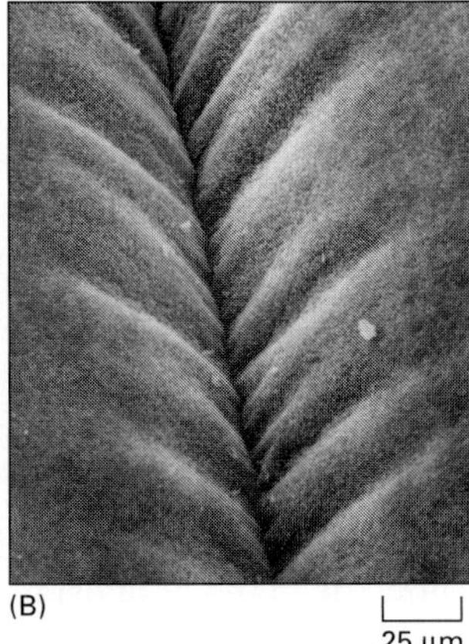

Figure 18–30 Cleavage in a fertilized frog egg. In these scanning electron micrographs, the cleavage furrow is especially obvious and well defined, as the cell is unusually large. The furrowing of the cell membrane is caused by the activity of the contractile ring underneath it. (A) A low-magnification view of the cleaving egg surface. (B) The surface of a furrow at higher magnification. (From H.W. Beams and R.G. Kessel, *Am. Sci.* 64:279–290, 1976.)

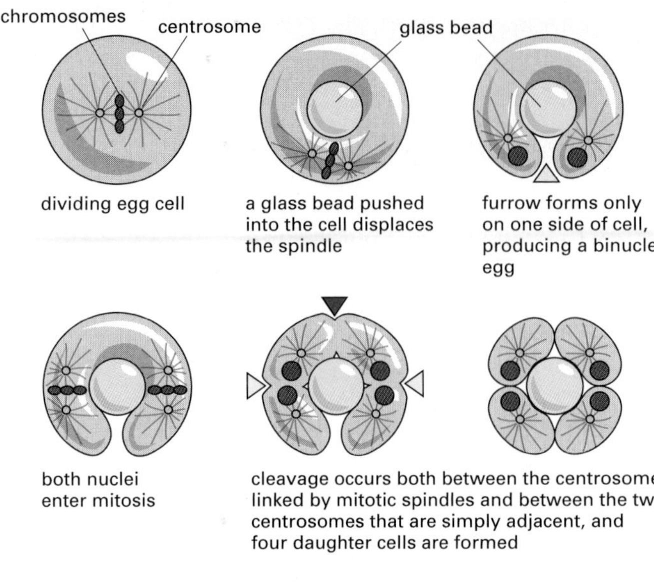
chromosomes
centrosome
glass bead
dividing egg cell
a glass bead pushed into the cell displaces the spindle
furrow forms only on one side of cell, producing a binucleate egg
both nuclei enter mitosis
cleavage occurs both between the centrosomes linked by mitotic spindles and between the two centrosomes that are simply adjacent, and four daughter cells are formed

Figure 18–31 An experiment demonstrating the influence of the position of microtubule asters on the subsequent plane of cleavage in a large egg cell. If the mitotic spindle is mechanically pushed to one side of the cell with a glass bead, the membrane furrowing is incomplete, failing to occur on the opposite side of the cell. Subsequent cleavages occur not only in the conventional relation to each of the two subsequent mitotic spindles *(yellow arrowheads)*, but also between the two adjacent asters that are not linked by a mitotic spindle—but in this abnormal cell share the same cytoplasm *(red arrowhead)*. Apparently, the contractile ring that produces the cleavage furrow in these cells always forms in the region midway between two asters, suggesting that the asters somehow alter the adjacent region of cell cortex to induce furrow formation between them.

membrane, to the appropriate region of the dividing cell. But, in fact, the molecular mechanism by which the spindle positions the cleavage furrow remains a mystery.

In some cells, the site of ring assembly is chosen before mitosis, according to a landmark placed in the cortex during a previous cell cycle. In budding yeasts, for example, a ring of proteins called *septins* assembles before mitosis, adjacent to a bud scar left on the cell surface as the mother and daughter cells separated in the previous division. The septins are thought to form a scaffold onto which other components of the contractile ring, including myosin II, assemble. As we discuss later, in plant cells, an organized band of microtubules and actin filaments assembles just before mitosis and marks the site where the cell wall will assemble and divide the cell in two.

Some Cells Reposition Their Spindle to Divide Asymmetrically

Most cells divide symmetrically. In most animal cells, for example, the contractile ring forms around the equator of the parent cell, so that the two daughter cells produced are of equal size and have similar properties. This symmetry results from the placement of the mitotic spindle, which in most cases tends to center itself in the cytoplasm. The centering process depends both on astral microtubules and on motor proteins that either push or pull on the astral microtubules to center the spindle.

There are many instances in development, however, when cells divide asymmetrically to produce two cells that differ in size, in the cytoplasmic contents they inherit, or in both. Usually, the two daughter cells are destined to develop along different pathways. To create daughter cells with different fates, the mother cell must first segregate some components (called *fate determinants)* to one side of the cell and then position the plane of division so that the appropriate daughter cell inherits these components (Figure 18–32). To position the plane of division asymmetrically, the spindle has to be moved in a controlled manner within the dividing cell. It seems likely that such spindle movements are directed by changes in local regions of the cell cortex and that motor proteins localiazed there pull one of the spindle poles, via its astral microtubules, to the appropriate region (Figure 18–33). Some of the proteins required for such asymmetrical divisions have been identified through genetic analyses in *C. elegans* and *Drosophila* (discussed in Chapter 21), and some of these seem to have a similar role in vertebrates.

Asymmetric division is particularly important in plant cells. As these cells cannot move after division, the selection of division planes is crucial for controlling tissue morphology. We discuss later how the plane of division is determined in these cells.

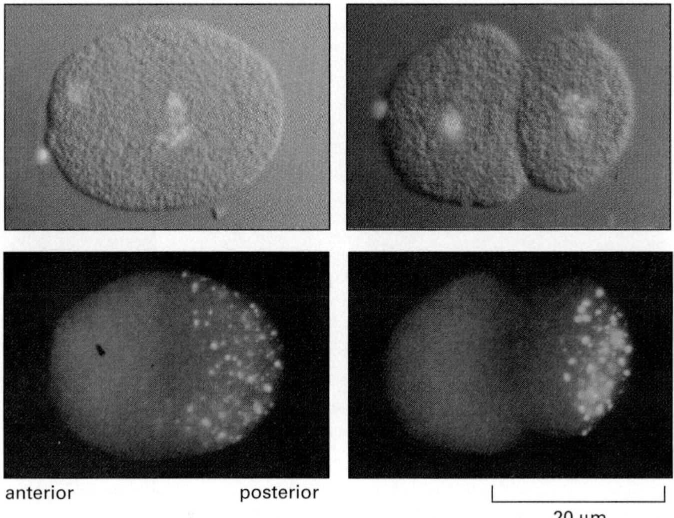

Figure 18–32 An asymmetric cell division segregating cytoplasmic components to only one daughter cell. These light micrographs illustrate the controlled asymmetric segregation of specific cytoplasmic components to one daughter cell during the first division of a fertilized egg of the nematode *C. elegans*. The cells *above* have been stained with a *blue*, DNA-binding, fluorescent dye to show the nucleus (and polar bodies—see Figure 20–22); they are viewed by both differential-interference-contrast and fluorescence microscopy. The cells *below* are the same cells stained with an antibody against P-granules and viewed by fluorescence microscopy. These small granules of unknown function are distributed randomly throughout the cytoplasm of the unfertilized egg (not shown) but become segregated to the posterior pole of the fertilized egg, as shown on the *left*. The cleavage plane is oriented to ensure that only the posterior daughter cell receives the P-granules when the egg divides, as shown on the *right*. The same segregation process is repeated in several subsequent cell divisions, so that the P-granules end up only in the cells that give rise to eggs and sperm. (Courtesy of Susan Strome.)

Actin and Myosin II in the Contractile Ring Generate the Force for Cytokinesis

As the astral microtubules in anaphase become longer lived and less dynamic in response to the loss of M-Cdk activity, the **contractile ring** begins to assemble beneath the plasma membrane. Much of the preparation for cytokinesis, however, happens earlier in mitosis, before the division of the cytoplasm actually begins. In interphase cells, actin and myosin filaments are assembled into a cortical network and, in some cells, also into large cytoplasmic bundles called *stress fibers* (discussed in Chapter 16). As cells enter mitosis, these arrays disassemble; much of the actin is reorganized, and myosin II filaments are released. As the

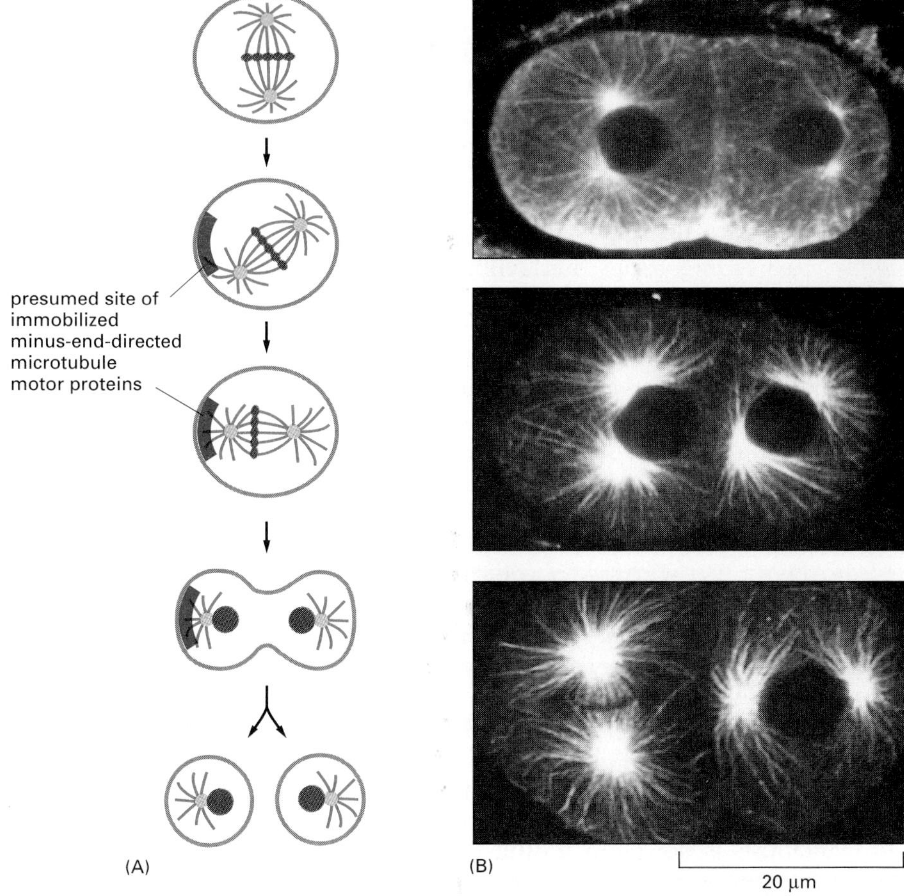

Figure 18–33 Spindle rotation. (A) A possible mechanism underlying the controlled rotation of a mitotic spindle. The *red bar* represents a specialized region of cell cortex toward which one spindle pole is pulled by its astral microtubules. (B) Fluorescence micrographs showing a precisely programmed rotation of a mitotic spindle in a *C. elegans* embryo at the two-cell stage in preparation for cleavage to form four cells in a specific pattern. The microtubules are stained with an antibody against tubulin. The spindle in the cell on the right rotates almost 90° clockwise, as diagrammed in (A). (B, courtesy of Tony Hyman and John White.)

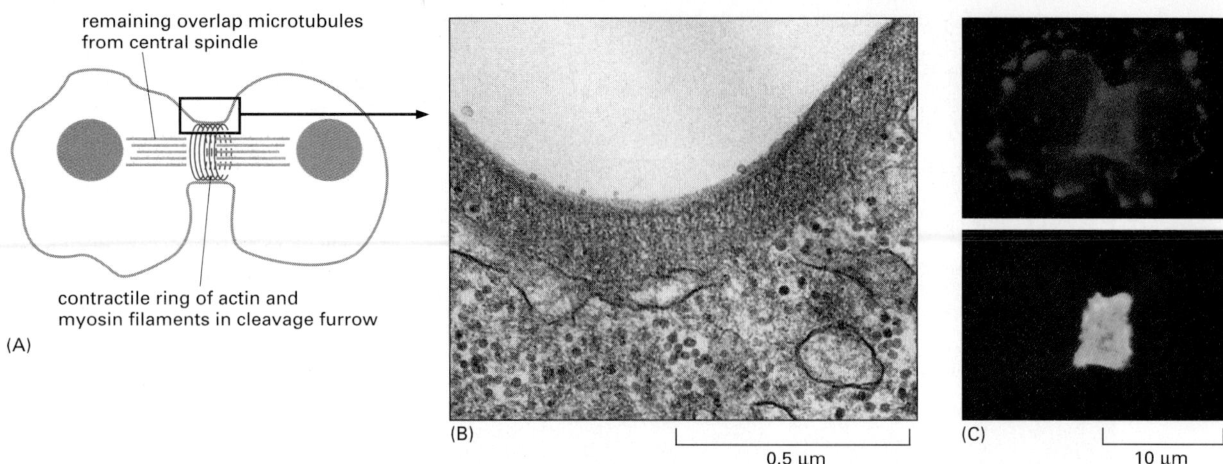

Figure 18–34 The contractile ring. (A) A drawing of the cleavage furrow in a dividing cell. (B) An electron micrograph of the ingrowing edge of a cleavage furrow of a dividing animal cell. (C) Fluorescence micrographs of a dividing slime mold amoeba stained for actin *(red)* and myosin II *(green)*. Whereas all of the visible myosin II has redistributed to the contractile ring, only some of the actin has done so; the rest remains in the cortex of the nascent daughter cells. (B, from H.W. Beams and R.G. Kessel, *Am. Sci.* 64:279–290, 1976; C, courtesy of Yoshio Fukui.)

chromatids separate in anaphase, myosin II begins to accumulate in the rapidly assembling contractile ring (Figure 18–34).

In many cells, cytokinesis requires the activation of one or more members of the **polo-like family** of protein kinases. These kinases regulate the assembly of both the mitotic spindle and the contractile ring and are therefore thought to help coordinate mitosis and cytokinesis, but it is uncertain how they do so. The fully assembled contractile ring contains many proteins in addition to actin and myosin II. The overlapping arrays of actin filaments and bipolar myosin II filaments, however, generate the force that divides the cytoplasm in two. They are thought to contract by a mechanism that is biochemically similar to that used by smooth muscle cells; in both cases, for example, the contraction begins when Ca^{2+}-calmodulin activates myosin light-chain kinase to phosphorylate myosin II. Once contraction has been stimulated, the ring develops a force large enough to bend a fine glass needle that is inserted in the path of the constricting ring.

How the contractile ring constricts is still a mystery. It seems not to operate by a simple "purse-string" mechanism, with actin and myosin II filaments sliding past each other as in skeletal muscle (see Figure 16–71). As the ring constricts, the ring maintains the same thickness in cross-section, suggesting that its total volume and the number of filaments it contains decrease steadily. Moreover, unlike in muscle, the actin filaments in the ring are highly dynamic, and their arrangement changes extensively during cytokinesis.

In addition to specifying the site of contractile ring assembly in early anaphase, in many cells, microtubules also work continuously during anaphase and telophase to stabilize the advancing cleavage furrow. Drugs that depolymerize microtubules, for example, cause the actin filaments in the contractile ring to become less organized. Moreover, if a needle is used to tear microtubules away from the cell cortex, the contractile ring disassembles and the cleavage furrow regresses. It is not known how the microtubules stabilize the ring, although it has been shown that growing microtubules can activate some members of the Rho family of small GTPases, which in turn stimulate actin polymerization (discussed in Chapter 16). One member of this family, Rho A, is required for cytokinesis.

The contractile ring is finally dispensed with altogether when cleavage ends, as the plasma membrane of the cleavage furrow narrows to form the **midbody**. The midbody persists as a tether between the two daughter cells and contains the remains of the central spindle, which now consists of the two sets of antiparallel overlap microtubules packed tightly together within a dense matrix material (Figure 18–35). Remarkably, in some cells, before cytokinesis has been completed, the mother centriole from one or both daughter cells separates from its daughter centriole (see Figure 18–5C) and migrates into the midbody, where it lingers for minutes, before returning to its daughter cell. Only then do the two daughter cells separate to complete cytokinesis. What the centriole might do in the midbody to trigger the final steps of cytokinesis is not known. After the daughter cells separate completely, some of the components of the residual midbody often remain on the inside of the plasma membrane of each cell, where they may serve as a mark on the cortex that helps to orient the spindle in the subsequent cell division.

Membrane-enclosed Organelles Must Be Distributed to Daughter Cells During Cytokinesis

The process of mitosis ensures that each daughter cell receives a full complement of chromosomes. But when a eucaryotic cell divides, each daughter cell must also inherit all of the other essential cell components, including the membrane-enclosed organelles. As discussed in Chapter 12, organelles like mitochondria and chloroplasts cannot assemble spontaneously from their individual components; they can arise only from the growth and division of the preexisting organelles. Similarly, cells cannot make a new endoplasmic reticulum (ER) unless some part of it is already present.

How, then, are the various membrane-enclosed organelles segregated when a cell divides? Organelles such as mitochondria and chloroplasts are usually present in large enough numbers to be safely inherited if, on average, their numbers roughly double once each cycle. The ER in interphase cells is continuous with the nuclear membrane and is organized by the microtubule cytoskeleton. Upon entry into M phase, the reorganization of the microtubules releases the ER, which fragments as the nuclear envelope breaks down. The Golgi apparatus probably fragments as well, although in some cells it seems to redistribute transiently into the ER, only to re-emerge at telophase. Some of the organelle fragments associate with the spindle microtubules via motor proteins, thereby hitching a ride into the daughter cells as the spindle elongates in anaphase.

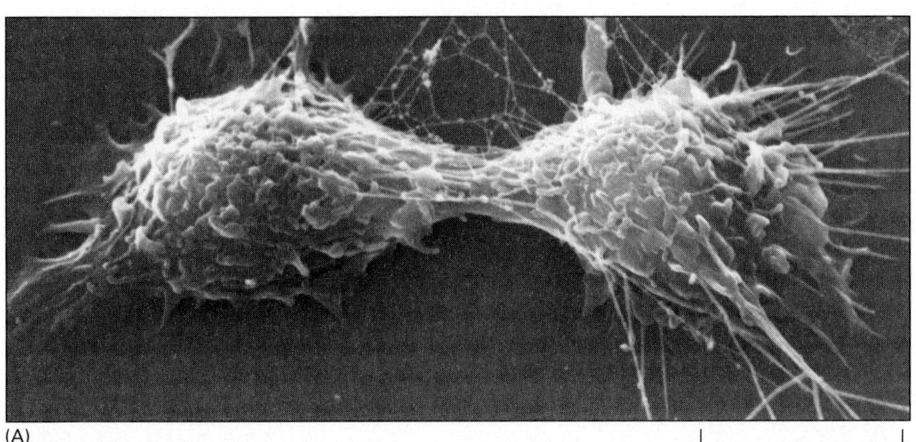

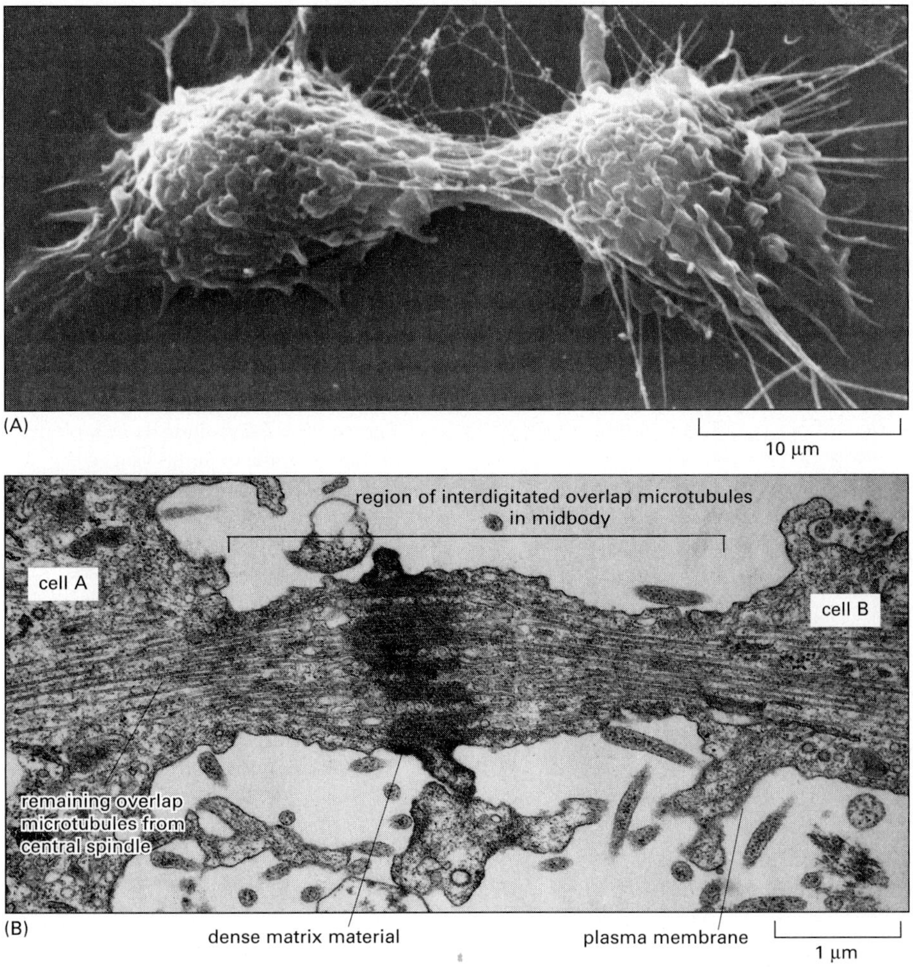

(B)

region of interdigitated overlap microtubules in midbody

cell A

cell B

remaining overlap microtubules from central spindle

dense matrix material

plasma membrane

1 μm

Figure 18–35 The midbody. (A) A scanning electron micrograph of an animal cell in culture in the process of dividing; the midbody still joins the two daughter cells. (B) A conventional electron micrograph of the midbody of a dividing animal cell. Cleavage is almost complete, but the daughter cells remain attached by this thin strand of cytoplasm containing the remains of the central spindle. See also the late telophase panel in Figure 18–8. (A, courtesy of Guenter Albrecht-Buehler; B, courtesy of J.M. Mullins.)

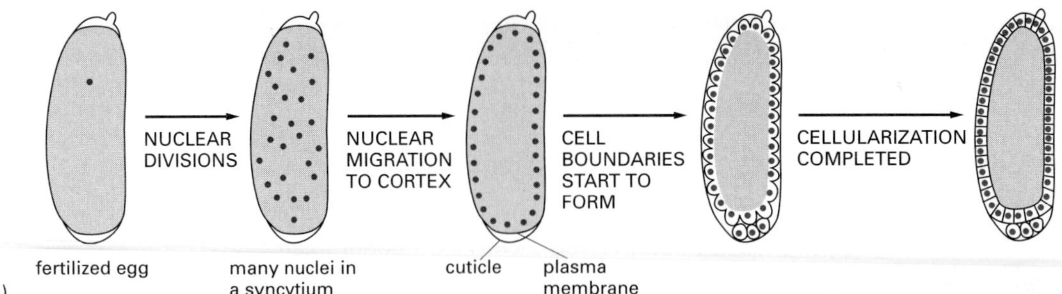

Figure 18–36 Mitosis without cytokinesis in the *Drosophila* embryo. (A) The first 13 nuclear divisions occur synchronously and without cytoplasmic division to create a large syncytium. Most of the nuclei then migrate to the cortex, and the plasma membrane extends inward and pinches off to surround each nucleus to form individual cells in a process called cellularization. (B) Fluorescence micrograph of multiple mitotic spindles at metaphase in a *Drosophila* embryo before cellularization. The microtubules are stained *green* and the centrosomes *red*. (B, courtesy of William Sullivan.)

(B) 10 μm

Mitosis Can Occur Without Cytokinesis

Although nuclear division is usually followed by cytoplasmic division, there are exceptions. Some cells undergo multiple rounds of nuclear division without intervening cytoplasmic division. In the early *Drosophila* embryo, for example, the first 13 rounds of nuclear division occur without cytoplasmic division, resulting in the formation of a single large cell containing 6000 nuclei, arranged in a monolayer near the surface (Figure 18–36). This arrangement greatly speeds up early development, as the cells do not have to take the time to go through all the steps of cytokinesis for each division. After these rapid nuclear divisions, cells are created around each nucleus in one round of coordinated cytokinesis called **cellularization**. Contractile rings form at the cell surface, and the plasma membrane extends inward and pinches off to enclose each nucleus.

Nuclear division without cytokinesis also occurs in some types of mammalian cells. Osteoclasts, trophoblasts, and some hepatocytes and heart muscle cells, for example, become multinucleated in this way.

The Phragmoplast Guides Cytokinesis in Higher Plants

Most higher-plant cells are enclosed by a semirigid *cell wall,* and their mechanism of cytokinesis is different from that just described for animal cells. Rather than a contractile ring dividing the cytoplasm from the outside in, the cytoplasm of the plant cell is partitioned from the inside out by the construction of a new cell wall, called the **cell plate**, between the two daughter nuclei (Figure 18–37).

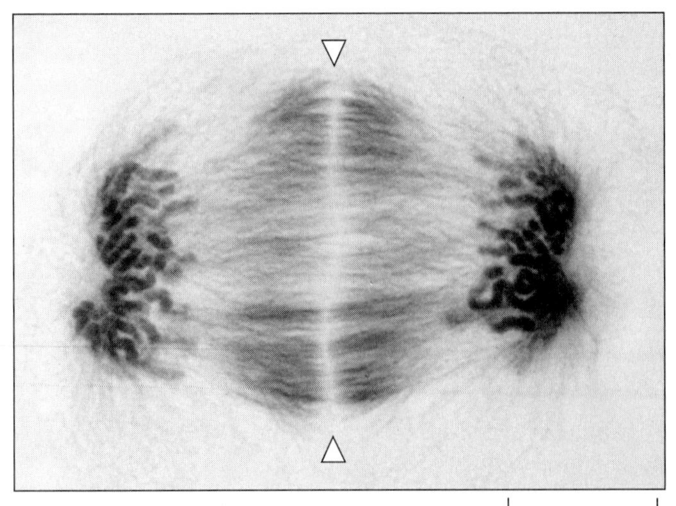

Figure 18–37 Cytokinesis in a plant cell in telophase. In this light micrograph, the early cell plate (between the two *arrowheads)* is forming in a plane perpendicular to the plane of the page. The microtubules of the spindle are stained with gold-labeled antibodies against tubulin, while the DNA in the two sets of daughter chromosomes is stained with a fluorescent dye. Note that there are no astral microtubules, because there are no centrosomes in higher-plant cells. (Courtesy of Andrew Bajer.)

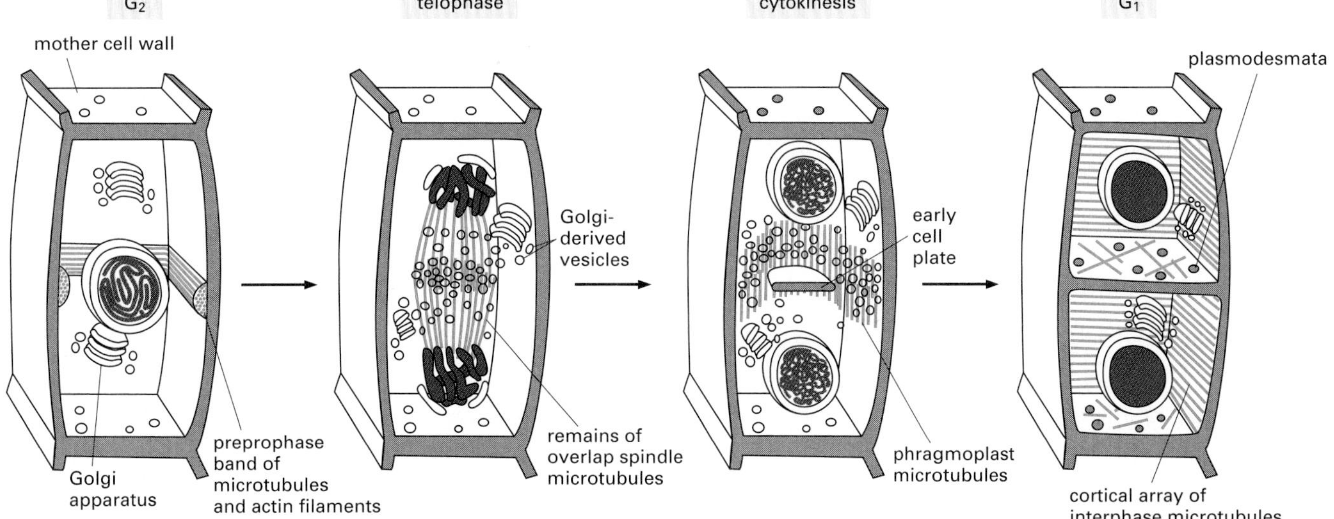

At some point in G_2, the cortical microtubules and actin filaments rearrange to form a band that encircles the cell, just below the plasma membrane. This preprophase band gradually narrows and exactly predicts where the new cell wall will join the mother cell wall when the cell divides.

The phragmoplast is formed by the overlap spindle microtubules at telophase. Golgi-derived vesicles carrying cell-wall precursors associate with microtubules, accumulate in the equatorial region, and fuse to form the early cell plate.

Phragmoplast microtubules re-form at the periphery of the early cell plate. New Golgi-derived vesicles are recruited into this region, fuse with edge of the cell plate, and extend it outward.

The membrane of the extending cell plate fuses with the plasma membrane of the mother cell, completing the new cell wall. The cortical array of interphase microtubules is reestablished in each of the two daughter cells.

Figure 18–38 The special features of cytokinesis in a higher plant cell. The division plane is established before M phase by a band of microtubules and actin filaments (the preprophase band) at the cell cortex. At the beginning of telophase, after the chromosomes have segregated, a new cell wall starts to assemble inside the cell at the equator of the old spindle. The overlap microtubules of the mitotic spindle remaining at telophase form the phragmoplast and guide vesicles derived from the Golgi apparatus toward the center of the spindle. The vesicles are filled with cell-wall material and fuse to form the growing new cell wall, which grows outward to reach the plasma membrane and original cell wall at the site determined earlier by the preprophase band. The plasma membrane and the membrane surrounding the new cell wall fuse, completely separating the two daughter cells.

The orientation of the cell plate determines the positions of the two daughter cells relative to neighboring cells. It follows that altering the planes of cell division, together with enlargement of the cells by expansion or growth, leads to different cell and tissue shapes that help determine the form of the plant.

The mitotic spindle by itself is not sufficient to determine the exact position and orientation of the cell plate. The first visible sign that a higher-plant cell has become committed to divide in a particular plane is seen in G_2, when the cortical array of microtubules disappears in preparation for mitosis. At this time, a circumferential band of microtubules and actin filaments forms a ring around the entire cell just beneath the plasma membrane. Because this cytoskeletal array appears before prophase begins, it is called the **preprophase band**. The band becomes thinner as the cell progresses to prophase, and it disappears completely before metaphase is reached. Yet, the division plane has somehow been established: when the new cell plate forms later during cytokinesis, it grows outward to fuse with the parental wall precisely at the zone formerly occupied by the preprophase band. Even if the cell contents are displaced by centrifugation after the preprophase band has disappeared, the growing cell plate tends to find its way back to the plane defined by the former preprophase band.

The assembly of the cell plate begins in late anaphase and is guided by a structure called the **phragmoplast**, which contains the remaining overlap microtubules of the mitotic spindle that interdigitate at their growing plus ends. This region of overlap is similar in structure to the central spindle in animal cells in late anaphase. Small vesicles, largely derived from the Golgi apparatus and filled with polysaccharide and glycoproteins required for the synthesis of the new cell-wall matrix, are transported along the microtubules to the equator of the phragmoplast, apparently by the action of microtubule-dependent motor proteins. Here, the vesicles fuse to form a disclike, membrane-enclosed structure called the *early cell plate* (see Figure 18–9G). The plate expands outward by further vesicle fusion until it reaches the plasma membrane and the original cell wall and divides the cell in two. Later, cellulose microfibrils are laid down within the matrix of the cell plate to complete the construction of the new cell wall (Figure 18–38).

Figure 18–39 Cell division in the bacterium *E. coli*. The single, circular chromosome contains an origin of replication called *oriC*. Before division, the chromosome is polarized, so that *oriC* is at one pole of the bacterium. As soon as the *oriC* sequence is copied, one of the copies is actively translocated to the other pole, before the rest of the chromosome is replicated. Cell growth occurs continuously, and when the cell reaches an appropriate size, the plasma membrane and cell wall grow inward to divide the cell in two, exactly between the two daughter chromosomes.

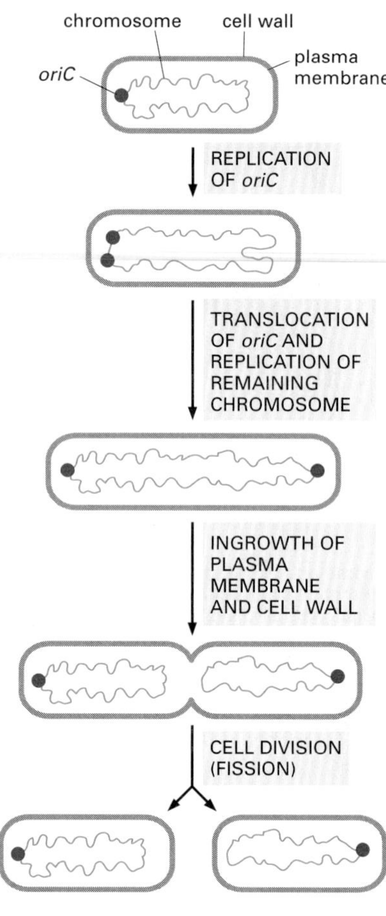

The Elaborate M Phase of Higher Organisms Evolved Gradually from Procaryotic Fission Mechanisms

Procaryotic cells divide by a process called **binary fission**. The single, circular DNA molecule replicates and division occurs by the invagination of the plasma membrane and the laying down of new cell wall between the two chromosomes to produce two separate daughter cells. In *E. coli*, before the chromosome replicates, the single origin of replication *(oriC)* is located at one pole of the rod-shaped bacterium. As soon as *oriC* is replicated, one copy of the sequence is immediately translocated to the opposite pole of the cell, after which the rest of the chromosome is replicated. Like the two spindle-pole asters in an animal cell, the bacterial daughter chromosomes at the cell poles somehow determine the location of the plane of cell division, ensuring that fission takes place at the cell equator, so that each daughter cell inherits one chromosome (Figure 18–39). Although a number of genes and proteins involved have been identified, the mechanisms responsible for the active translocation of *oriC* and the inhibition of fission everywhere but at the equator remain unknown.

Binary fission in procaryotes depends on filaments made of the **FtsZ protein**. FtsZ is a cytoskeletal GTPase that is structurally related to tubulin and assembles into a ring at the equator of the cell (Figure 18–40A, and see Figure 16–17). The FtsZ filaments are essential for the recruitment of all the other cell division proteins to the division site. Together, these proteins guide the inward growth of the cell wall and membrane, leading to the formation of a septum that divides the cell into two. Bacteria in which the *ftsZ* gene is inactivated by mutation cannot divide. A FtsZ-based mechanism is also used in the division of chloroplasts in plant cells (Figure 18–40B) and mitochondria in protists. In fungi and animal cells, another self-assembling GTPase called *dynamin* (discussed in Chapter 13) has apparently taken over the function of FtsZ in mitochondrial division.

With the evolution of the eucaryotes, the genome increased in complexity, and the chromosomes increased in both number and size. For these organisms, a more elaborate mechanism for dividing the chromosomes between daughter cells was apparently required. Clearly, the mitotic apparatus could not have evolved all at once. In many primitive eucaryotes, such as the dinoflagellate *Crypthecodinium cohnii*, mitosis depends on a membrane-attachment mechanism, in which the chromosomes have to bind to the inner nuclear membrane for segregation. The intermediate status of this large, single-celled alga is reflected in the composition of its chromosomes, which, like those of procaryotes, have

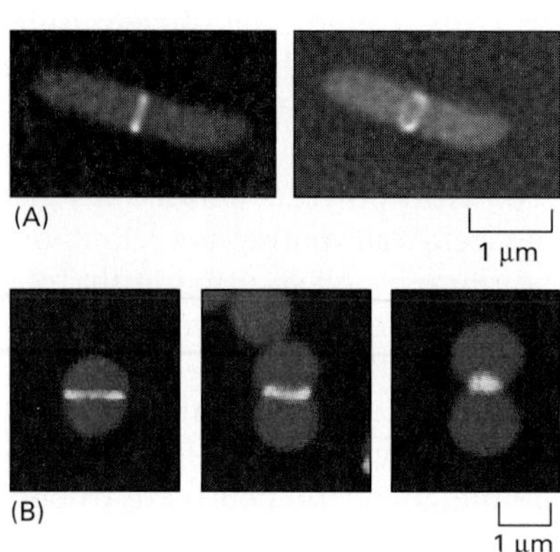

Figure 18–40 The FtsZ protein. (A) Fluorescence micrographs showing the location of the FtsZ protein during binary fission in *E. coli*. The protein assembles into a ring at the center of the cell, where it helps orchestrate cell division. The bacteria here have been genetically engineered to produce a fluorescent form of the protein (FtsZ fused to green fluorescent protein). (B) Dividing chloroplasts *(red)* from a red alga also make use of a FtsZ protein ring *(green)* for cleavage. (A, from X. Ma, D.W. Ehrhardt, and W. Margolin, *Proc. Natl. Acad. Sci. USA* 93:12999–13003, 1996. © National Academy of Sciences; B, from S. Miyagishima et al., *Plant Cell* 13:2257–2268, 2001. © American Society of Plant Biologists.)

relatively little associated protein. The nuclear membrane in *C. cohnii* remains intact throughout mitosis, and the spindle microtubules remain entirely outside the nucleus. Where these spindle microtubules press on the outside of the nuclear envelope, the envelope becomes indented in a series of parallel channels (Figure 18–41). The chromosomes become attached to the inner membrane

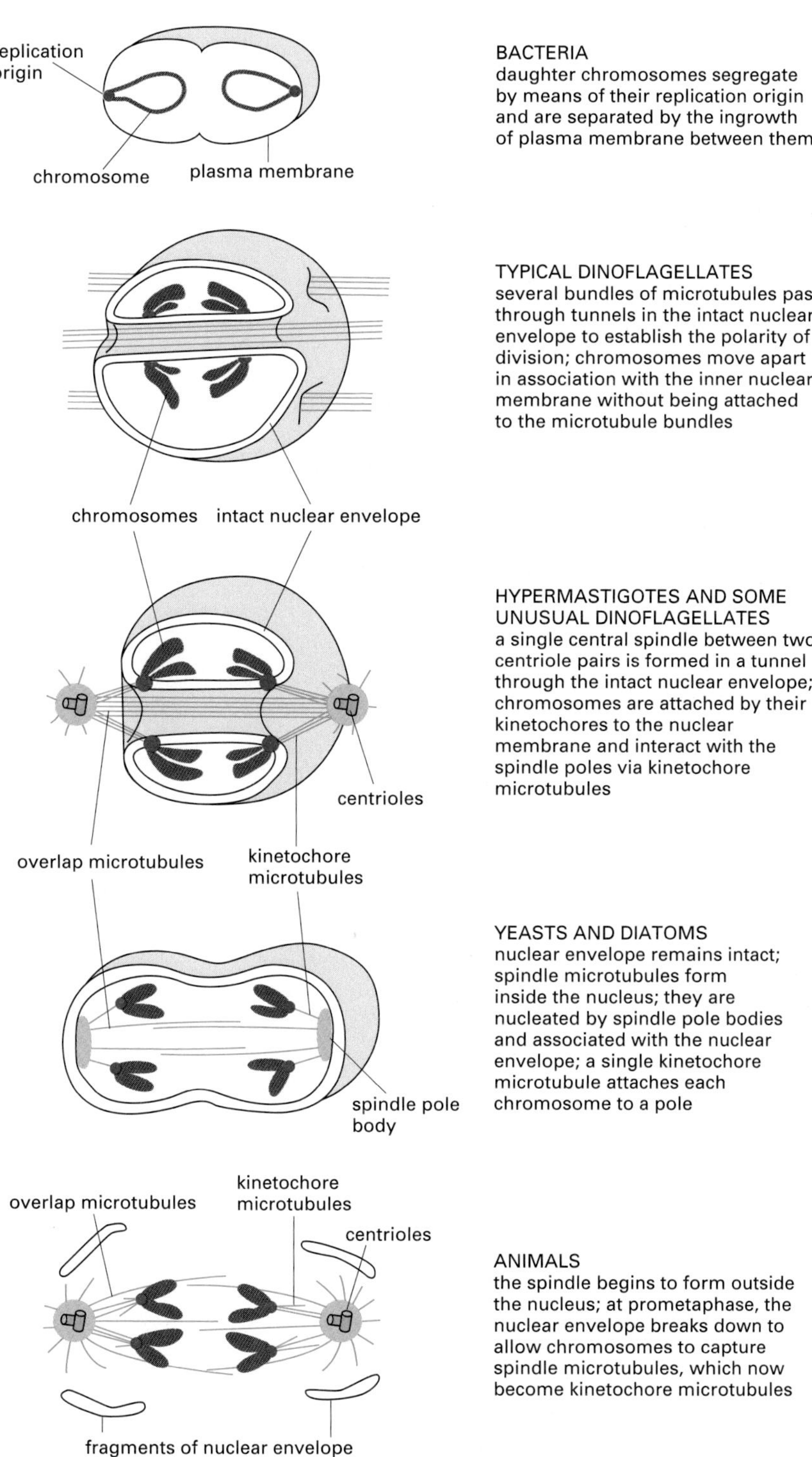

BACTERIA
daughter chromosomes segregate by means of their replication origin and are separated by the ingrowth of plasma membrane between them

TYPICAL DINOFLAGELLATES
several bundles of microtubules pass through tunnels in the intact nuclear envelope to establish the polarity of division; chromosomes move apart in association with the inner nuclear membrane without being attached to the microtubule bundles

HYPERMASTIGOTES AND SOME UNUSUAL DINOFLAGELLATES
a single central spindle between two centriole pairs is formed in a tunnel through the intact nuclear envelope; chromosomes are attached by their kinetochores to the nuclear membrane and interact with the spindle poles via kinetochore microtubules

YEASTS AND DIATOMS
nuclear envelope remains intact; spindle microtubules form inside the nucleus; they are nucleated by spindle pole bodies and associated with the nuclear envelope; a single kinetochore microtubule attaches each chromosome to a pole

ANIMALS
the spindle begins to form outside the nucleus; at prometaphase, the nuclear envelope breaks down to allow chromosomes to capture spindle microtubules, which now become kinetochore microtubules

Figure 18–41 The use of different chromosome separation mechanisms by different organisms. Some of these may have been intermediate stages in the evolution of the mitotic spindle of higher organisms. For all examples except bacteria, only the central nuclear region of the cell is shown.

of the nuclear envelope opposite these channels, and chromosome segregation occurs on the inside of this channeled nuclear membrane. Thus, the extranuclear "spindle" is used to order the nuclear membrane and thereby define the plane of division. Kinetochores in this species seem to be integrated into the nuclear membrane and may therefore have evolved from some membrane component.

Eucaryotic tubulin and procaryotic FtsZ clearly have a common evolutionary history. But, microtubules are important for chromosome segregation in even the most primitive eucaryotes, where they are also present in flagellar axonemes (discussed in Chapter 16). Whether the flagellum or the spindle evolved first is unclear.

A somewhat more advanced, although still extranuclear, spindle is seen in *hypermastigotes,* in which the nuclear envelope again remains intact throughout mitosis. These large protozoa from the guts of insects provide a particularly clear illustration of the independence of spindle elongation and the chromosome movements that separate the chromatids. The sister kinetochores become separated by the growth of the nuclear membrane (to which they are attached) before becoming attached to the spindle. Only when the kinetochores are near the poles of the spindle do they acquire the kinetochore microtubules needed to attach them to the spindle. Because the spindle microtubules remain separated from the chromosomes by the nuclear envelope, the kinetochore microtubules, which are formed outside the nucleus, must somehow attach to the chromosomes through the nuclear membranes. After this attachment has occurred, the kinetochores are drawn poleward in a conventional manner (see Figure 18–41).

Organisms that form spindles inside an intact nucleus may represent a further stage in the evolution of mitotic mechanisms. In both yeasts and diatoms, the spindle is attached to chromosomes by their kinetochores, and the chromosomes are segregated in a way loosely similar to that described for animal cells—except that the entire process generally occurs within the confines of the nuclear envelope (see Figure 18–41). It is thought that the "open" mitosis of higher organisms and the "closed" mitosis of yeasts and diatoms evolved separately from a common ancestor resembling the modern hypermastigote spindle. At present, there is no convincing explanation for why higher plants and animals have evolved a mitotic apparatus that requires the controlled and reversible dissolution of the nuclear envelope.

Summary

Cell division ends as the cytoplasm divides into two by the process of cytokinesis. Except for plants, cytokinesis in eucaryotic cells is mediated by a contractile ring, which is composed of actin and myosin filaments and a variety of other proteins. By an unknown mechanism, the mitotic spindle determines when and where the contractile ring assembles and, thereby, when and where the cell divides. Most cells divide symmetrically to produce two cells of the same content and size. Some cells, however, specifically position their spindle to divide asymmetrically, producing two daughter cells that differ in size, content, or both. Cytokinesis occurs by a special mechanism in higher-plant cells—in which the cytoplasm is partitioned by the construction of a new cell wall, the cell plate, inside the cell. The position of the cell plate is determined by the position of a preprophase band of microtubules and actin filaments. The organization of mitosis in fungi and some protozoa differs from that in animals and plants, suggesting how the complex process of eucaryotic cell division may have evolved.

References

General

Hyams JS & Brinkely BR (eds) (1989) Mitosis: Molecules and Mechanisms. San Diego: Academic Press.

Wilson EB (1987) The Cell in Development and Heredity, 3rd edn. New York: Garland.

An Overview of M Phase

Hinchcliffe EH & Sluder G (2001) "It takes two to tango": understanding how centrosome duplication is regulated throughout the cell cycle. *Genes Dev.* 15, 1167–1181.

Hirano T (2000) Chromosome cohesion, condensation, and separation. *Annu. Rev. Biochem.* 69, 115–144.

Hyman AA (2000) Centrosomes: Sic transit gloria centri. *Curr Biol.* 10, R276–R278.

Losada A & Hirano T (2001) Shaping the metaphase chromosome: coordination of cohesion and condensation. *Bioessays* 23, 924–935.

Mitchison TJ & Salmon ED (2001) Mitosis: a history of division. *Nature Cell Biol.* 3, E28–E34.

Morgan DO (1997) Cyclin-dependent kinases: engines, clocks, and microprocessors. *Annu. Rev. Cell Dev. Biol.* 13, 261–291.

Moritz M & Agard DA (2001) Gamma-tubulin complexes and microtubule nucleation. *Curr. Opin. Struct. Biol.* 11, 174–181.

Oakley BR (2000) Gamma-tubulin. *Curr. Top. Dev. Biol.* 49, 27–54.

Pines J & Rieder CL (2001) Re-staging mitosis: a contemporary view of mitotic progression. *Nat. Cell Biol.* 3, E3–E6.

Raff JW (2001) Centrosomes: Central no more? *Curr Biol.* 6, 11, R159–R161.

Rieder CL, Faruki S & Khodjakov A (2001) The centrosome in vertebrates: more than a microtubule-organizing center. *Trends Cell Biol.* 11, 413–419.

Sadler KC & do Carmo Avides M (2000) Orchestrating cell division. *Trends Cell Biol.* 10, 447–450.

Stearns T (2001) Centrosome duplication: a centriolar pas de deux. *Cell* 105, 417–420.

Mitosis

Amon A (1999) The spindle assembly checkpoint. *Curr. Opin. Genet. Dev.* 9, 69–75.

Belmont LD, Hyman AA, Sawin KE & Mitchison TJ (1990) Real-time visualization of cell cycle-dependent changes in microtubule dynamics in cytoplasmic extracts. *Cell* 62, 579–589.

Brunet S & Vernos I (2001) Chromosome motors on the move: From motion to spindle checkpoint activity. *EMBO Rep.* 2, 669–673.

Cande WZ & Hogan CJ (1989) The mechanism of anaphase spindle elongation. *Bioessays* 11, 5–9.

Compton DA (2000) Spindle assembly in animal cells. *Annu. Rev. Biochem.* 69, 95–114.

Desai A & Mitchison TJ (1997) Microtubule polymerization dynamics. *Annu. Rev. Cell Dev. Biol.* 13, 83–117.

Desai A, Maddox PS, Mitchison TJ & Salmon ED (1998) Anaphase A chromosome movement and poleward spindle microtubule flux occur at similar rates in Xenopus extract spindles. *J. Cell Biol.* 141, 703–713.

Desai A, Verma S, Mitchison TJ & Walczak CE (1999) Kinl kinesins are microtubule-destabilizing enzymes. *Cell* 96, 69–78.

Heald R, Tournebize R, Blank T et al. (1996) Self-organization of microtubules into bipolar spindles around artificial chromosomes in Xenopus egg extracts. *Nature* 382, 420–425.

Hoyt MA (2000) Exit from mitosis: spindle pole power. *Cell* 102, 267–270.

Hunter AW & Wordeman L (2000) How motor proteins influence microtubule polymerization dynamics. *J Cell Sci.* 24, 4379–4389.

Hyman AA & Karsenti E (1996) Morphogenetic properties of microtubules and mitotic spindle assembly. *Cell* 84, 401–410.

Mitchison TJ (1989) Polewards microtubule flux in the mitotic spindle: evidence from photoactivation of fluorescence. *J. Cell Biol.* 109, 637–652.

Mitchison TJ & Kirschner MW (1984) Dynamic instability of microtubule growth. *Nature* 312, 237–242.

Mitchison T, Evans L, Schulze E & Kirschner M (1986) Sites of microtubule assembly and disassembly in the mitotic spindle. *Cell* 45, 515–527.

Nasmyth K, Peters JM & Uhlmann F (2000) Splitting the chromosome: cutting the ties that bind sister chromatids. *Science* 288, 1379–1385.

Nislow C, Lombillo VA, Kuriyama R & McIntosh JR (1992) A plus-end-directed motor enzyme that moves antiparallel microtubules in vitro localizes to the interzone of mitotic spindles. *Nature* 359, 543–547.

Pines J & Rieder CL (2001) Re-staging mitosis: a contemporary view of mitotic progression. *Nat Cell Biol.* 3, E3–E6.

Rieder CL & Salmon ED (1994) Motile kinetochores and polar ejection forces dictate chromosome position on the vertebrate mitotic spindle. *J. Cell Biol.* 124, 223–233.

Rieder CL & Salmon ED (1998) The vertebrate cell kinetochore and its roles during mitosis. *Trends Cell Biol.* 8, 310–318.

Rudner AD & Murray AW (1996) The spindle assembly checkpoint. *Curr. Opin. Cell Biol.* 8, 773–780.

Saunders WS, Lengyel B & Hoyt MA (1997) Mitotic spindle function in *Saccharomyces cerevisiae* requires a balance between different types of kinesin-related motors. *Mol. Biol. Cell* 8, 1025–1033.

Sharp DJ, Rogers GC & Scholey JM (2000) Microtubule motors in mitosis. *Nature* 407, 41–47.

Tournebize R, Popov A, Kinoshita K, Ashford AJ, Fybina S, Pozniakovsky A, Mayer TU, Walczak CE, Karsenti E, Hyman AA (2000) Control of microtubule dynamics by the antagonistic activities of XMAP215 and XKCM1 in *Xenopus* egg extracts. *Nat. Cell Biol.* 2, 13–19.

Walczak CE, Mitchison TJ & Desai A (1996) XKCM1: A Xenopus kinesin-related protein that regulates microtubule dynamics during mitotic spindle assembly. *Cell* 84, 37–47.

Waterman-Storer CM, Desai A, Bulinski JG & Salmon ED (1998) Fluorescent speckle microscopy, a method to visualize the dynamics of protein assemblies in living cells. *Curr. Biol.* 8, 1227–1230.

Wittmann T, Hyman A & Desai A (2001) The mitotic spindle: a dynamic assembly of microtubules and motors. *Nature Cell Biol.* 3, E28–E34.

Zhang D & Nicklas RB (1995) The impact of chromosomes and centrosomes on spindle assembly as observed in living cells. *J. Cell Biol.* 129, 1287–1300.

Cytokinesis

Conrad GW & Schroeder TE (eds) (1990) Cytokinesis: Mechanisms of Furrow Formation During Cell Division. Ann. NY. Acad. Sci., vol. 582. New York: Academy of Sciences.

Field D, Li R & Oegema K (1999) Cytokinesis in eukaryotes: a mechanistic comparison. *Curr. Opin. Cell Biol.* 11, 68–80.

Glotzer M (1997) The mechanism and control of cytokinesis. *Curr. Opin. Cell Biol.* 9, 815–823.

Harry EJ (2001) Bacterial cell division: regulating Z-ring formation. *Mol. Microbiol.* 40, 795–803.

Heese M, Mayer U & Jurgens G (1998) Cytokinesis in flowering plants: cellular process and developmental integration. *Curr. Opin. Plant Biol.* 1, 486–491.

Horvitz HR & Herskowitz I (1992) Mechanisms of asymmetric cell division: two Bs or not two Bs, that is the question. *Cell* 68, 237–255.

Jacobs C & Shapiro L (1999) Bacterial cell division: a moveable feast. *Proc. Natl. Acad. Sci. USA* 96, 5891–5893.

Knoblich JA (2001) Asymmetric cell division during animal development. *Nat. Rev. Mol. Cell Biol.* 2, 11–20.

Lecuit T & Wieschaus E (2000) Polarized insertion of new membrane from a cytoplasmic reservoir during cleavage of the *Drosophila* embryo. *J. Cell Biol.* 150, 849–860.

Margolin W (2000) Themes and variations in prokaryotic cell division. *FEMS Microbiol. Rev.* 24, 531–548.

Nanninga N (2001) Cytokinesis in prokaryotes and eukaryotes: common principles and different solutions. *Microbiol. Mol. Biol. Rev.* 65, 319–333.

Rappaport R (1996) Cytokinesis in Animal Cells. Cambridge, New York: Cambridge University Press.

Robinson DN & Spudich JA (2000) Towards a molecular understanding of cytokinesis. *Trends Cell Biol.* 10, 228–237.

Segal M & Bloom K (2001) Control of spindle polarity and orientation in *Saccharomyces cerevisiae*. *Trends Cell Biol.* 11, 160–166.

Sharpe ME (1999) Upheaval in the bacterial nucleoid: an active chromosome segregation mechanism. *Trends Genet.* 15, 70–74.

Shima DT, Cabrera-Poch N, Pepperkok R & Warren G (1998) An ordered inheritance strategy for the Golgi apparatus: visualization of mitotic disassembly reveals a role for the mitotic spindle. *J. Cell Biol.* 141, 955–966.

Smith L (1999) Divide and conquer: cytokinesis in plant cells. *Curr. Opin. Plant Biol.* 2, 447–453.

Sylvester AW (2000) Division decisions and the spatial regulation of cytokinesis. *Curr. Opin. Plant Biol.* 3, 58–66.

Wolf WA, Chew TL & Chisholm RL (1999) Regulation of cytokinesis. *Cell Mol. Life Sci.* 55, 108–120.

CELLS IN THEIR SOCIAL CONTEXT

PART V

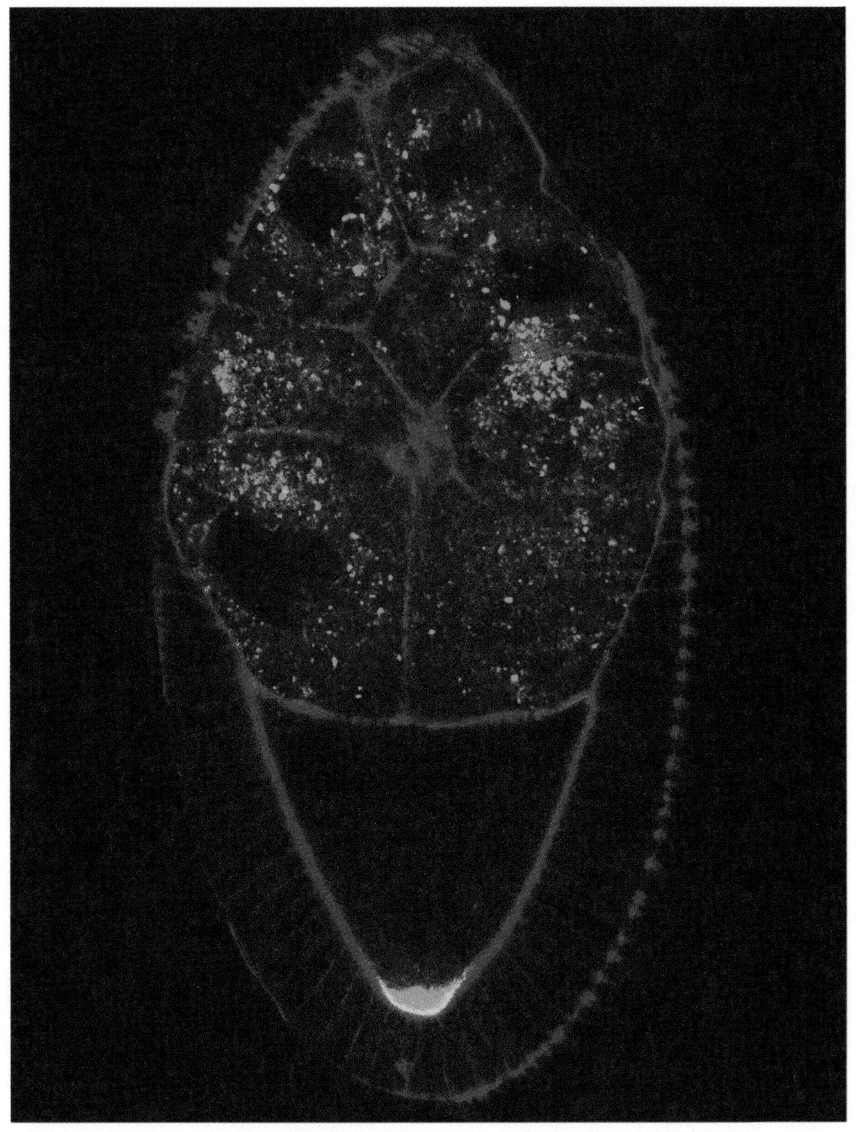

Multicellular organisms start life as a fertilized egg cell. In this egg chamber of the fly *Drosophila*, the outlines of the component cells are revealed by the actin (stained *red*) that underlies their plasma membranes. The large cell at the bottom is the egg cell, which contains an asymmetrically localized protein called Staufen *(green)*. Surrounding the egg cell is a row of follicle cells and at the top is a group of large nurse cells. (Courtesy of Daniel St Johnston.)

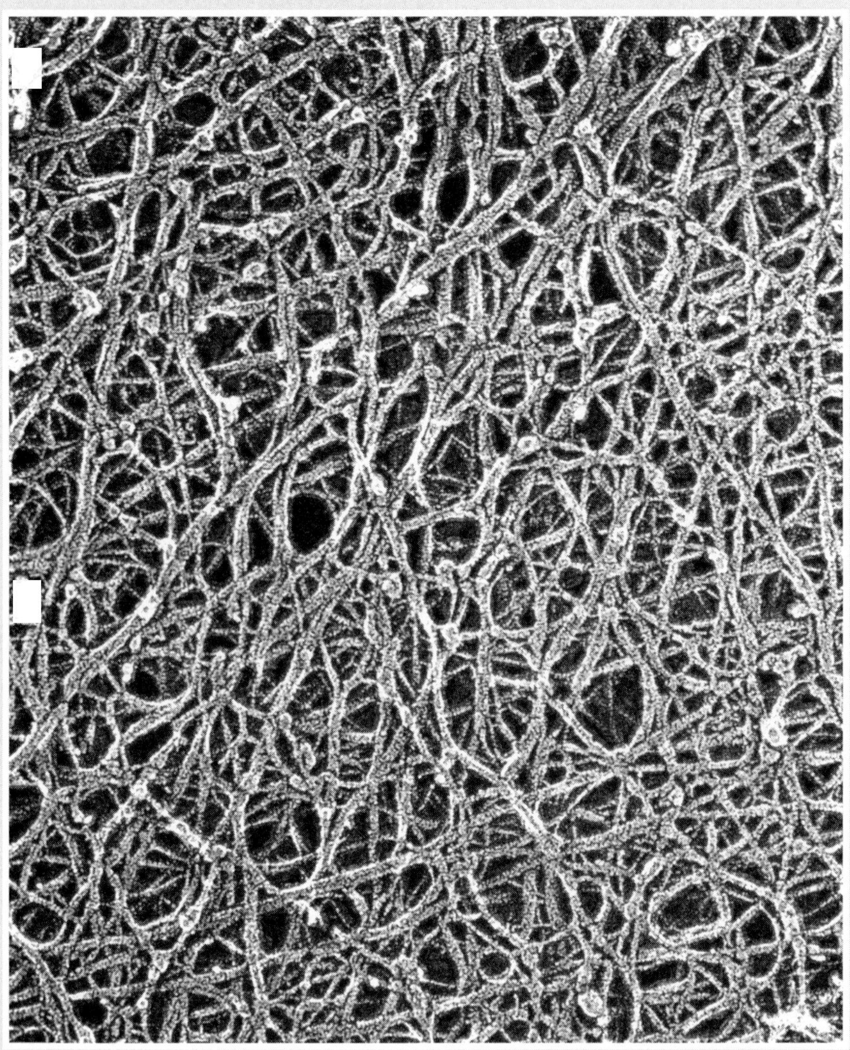

The plant cell wall. The extracellular matrix of both plants and animals consist of tough fibers embedded in a jelly-like matrix. An onion cell wall has been extracted to reveal the underlying meshwork of tough, load-bearing cellulose fibers. This structure is mimicked in paper manufacture. (Courtesy of Maureen McCann.)

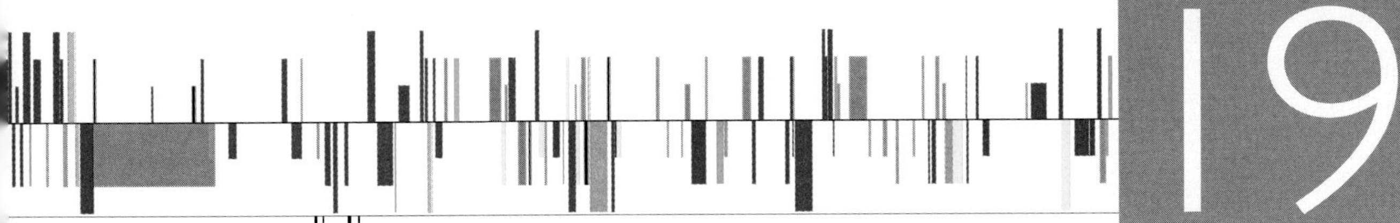

19

CELL JUNCTIONS, CELL ADHESION, AND THE EXTRACELLULAR MATRIX

In this chapter, we consider what holds the cells of a multicellular organism together. Cells are small, deformable, and often motile objects, filled with an aqueous medium and enclosed in a flimsy plasma membrane; yet they can combine in their millions to form a structure as massive, as strong, and as strictly ordered as a horse or a tree. We have to explain what gives such multicellular assemblies their strength and keeps the cells in their proper places.

The building technologies of animals and plants are different, and each type of organism is formed of many types of *tissues,* in which the cells are assembled and bound together in different ways. In both animals and plants, however, an essential part is played in most tissues by the *extracellular matrix.* This complex network of secreted extracellular macromolecules has many functions, but first and foremost it forms a supporting framework. It helps hold cells and tissues together, and, in animals, it provides an organized environment within which migratory cells can move and interact with one another in orderly ways. The extracellular matrix, however, is only half the story. In animals especially, the cells of most tissues are bound directly to one another by *cell–cell junctions.* These too are of many types, serving many purposes in addition to mechanical attachment; but without them, our bodies would disintegrate.

In vertebrates, the major tissue types are nerve, muscle, blood, lymphoid, epithelial, and connective tissues. Connective tissue and epithelial tissue represent two extremes of organization (Figure 19–1). In **connective tissue**, the extracellular matrix is plentiful, and cells are sparsely distributed within it. The matrix is rich in fibrous polymers, especially *collagen,* and it is the matrix—rather than the cells—that bears most of the mechanical stress to which the tissue is subjected. Direct attachments between one cell and another are relatively rare.

In **epithelial tissue**, by contrast, cells are tightly bound together into sheets called **epithelia**. The extracellular matrix is scanty, consisting mainly of a thin mat called the *basal lamina,* which underlies the epithelium. The cells are attached to each other by cell–cell adhesions, which bear most of the mechanical stresses. For this purpose, strong intracellular protein filaments (components of the cytoskeleton) cross the cytoplasm of each epithelial cell and attach to specialized junctions in the plasma membrane. The junctions, in turn, tie the surfaces of adjacent cells either to each other or to the underlying basal lamina.

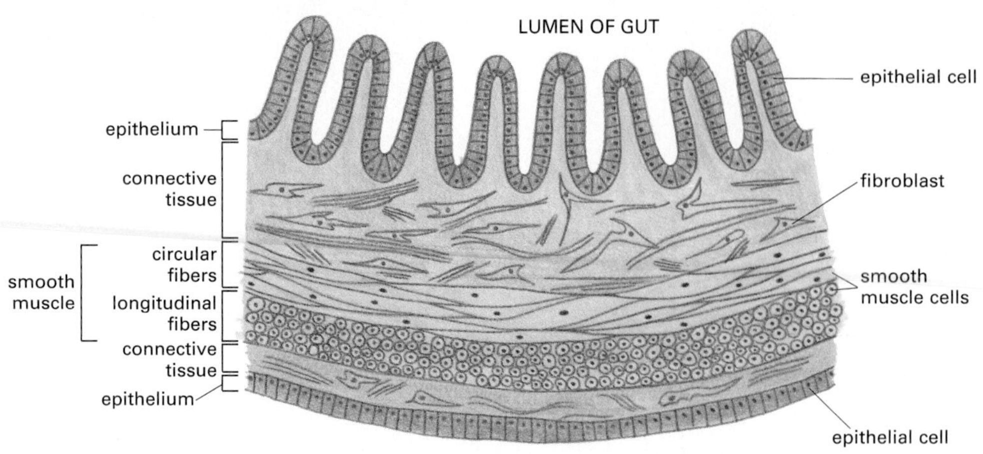

Figure 19–1 A cross-sectional view of part of the wall of the intestine. This long, tubelike organ is constructed from epithelial tissue *(red)*, connective tissue *(green)*, and muscle tissue *(yellow)*. Each tissue is an organized assembly of cells held together by cell–cell adhesions, extracellular matrix, or both.

Epithelial cell sheets line all the cavities and free surfaces of the body. The specialized junctions between the cells enable epithelia to form barriers that inhibit the movement of water, solutes, and cells from one body compartment to another. As illustrated in Figure 19–1, epithelia almost always rest on a supporting bed of connective tissue. This supporting bed may in turn attach them to other tissues, such as the muscle shown in the figure. In this way, tissues join together in various combinations to form larger functional units called *organs*.

In this chapter, we first discuss the structure and function of the specialized cell–cell and cell–matrix junctions, which are collectively called *cell junctions*. We next consider how animal cells recognize and bind to one another as they move and assemble into tissues and organs—a process called *cell adhesion*. We then discuss the structure and organization of the extracellular matrix in animals, as well as the main cell-surface receptors that animal cells use to bind to the matrix. Finally, we consider the special extracellular matrix that surrounds each cell in a plant—the plant cell wall.

CELL JUNCTIONS

Specialized **cell junctions** occur at points of cell–cell and cell–matrix contact in all tissues, and they are particularly plentiful in epithelia. Cell junctions are best visualized using either conventional or freeze-fracture electron microscopy (discussed in Chapter 9), which reveals that the interacting plasma membranes (and often the underlying cytoplasm and the intervening intercellular space as well) are highly specialized in these regions.

Cell junctions can be classified into three functional groups:

1. **Occluding junctions** seal cells together in an epithelium in a way that prevents even small molecules from leaking from one side of the sheet to the other.
2. **Anchoring junctions** mechanically attach cells (and their cytoskeletons) to their neighbors or to the extracellular matrix.
3. **Communicating junctions** mediate the passage of chemical or electrical signals from one interacting cell to its partner.

The major kinds of intercellular junctions within each group are listed in Table 19–1. We discuss each of them in turn, except for chemical synapses, which are formed exclusively by nerve cells and are considered in Chapters 11 and 15.

Occluding Junctions Form a Selective Permeability Barrier Across Epithelial Cell Sheets

All epithelia have at least one important function in common: they serve as selective permeability barriers, separating fluids on either side that have a different chemical composition. This function requires that the adjacent cells be sealed together by occluding junctions. **Tight junctions** have this barrier role in vertebrates, as we illustrate by considering the epithelium of the mammalian small intestine, or gut.

The epithelial cells lining the small intestine form a barrier that keeps the gut contents in the gut cavity, the *lumen*. At the same time, however, the cells must transport selected nutrients across the epithelium from the lumen into the extracellular fluid that permeates the connective tissue on the other side (see Figure 19–1). From there, these nutrients diffuse into small blood vessels to provide nourishment to the organism. This *transcellular transport* depends on two sets of membrane-bound membrane transport proteins. One set is confined to the *apical surface* of the epithelial cell (the surface facing the lumen) and actively transports selected molecules into the cell from the gut. The other set is confined to the *basolateral* (basal and lateral) *surfaces* of the cell, and it allows the same molecules to leave the cell by facilitated diffusion into the extracellular fluid on the other side of the epithelium. To maintain this directional transport, the apical set of transport proteins must not be allowed to migrate to the basolateral surface of the cell, and the basolateral set must not be allowed to migrate to the apical surface. Furthermore, the spaces between epithelial cells must be tightly sealed, so that the transported molecules cannot diffuse back into the gut lumen through these spaces (Figure 19–2).

The tight junctions between epithelial cells are thought to have both of these roles. First, they function as barriers to the diffusion of some membrane proteins (and lipids) between apical and basolateral domains of the plasma membrane (see Figure 19–2). Mixing of such proteins and lipids occurs if tight junctions are disrupted, for example, by removing the extracellular Ca^{2+} that is required for tight junction integrity. Second, tight junctions seal neighboring cells together so that, if a low-molecular-weight tracer is added to one side of an epithelium, it will generally not pass beyond the tight junction (Figure 19–3). This seal is not absolute, however. Although all tight junctions are impermeable to macromolecules, their permeability to small molecules varies greatly in different epithelia. Tight junctions in the epithelium lining the small intestine, for example, are 10,000 times more permeable to inorganic ions, such as Na^+, than the tight junctions in the epithelium lining the urinary bladder. These differences reflect differences in tight junction proteins that form the junctions.

Epithelial cells can transiently alter their tight junctions to permit an increased flow of solutes and water through breaches in the junctional barriers. Such *paracellular transport* is especially important in the absorption of amino acids and monosaccharides from the lumen of the intestine, where their concentration can increase enough after a meal to drive passive transport in the desired direction.

TABLE 19–1 A Functional Classification of Cell Junctions

OCCLUDING JUNCTIONS	
1.	tight junctions (vertebrates only)
2.	septate junctions (invertebrates mainly)
ANCHORING JUNCTIONS	
Actin filament attachment sites	
1.	cell–cell junctions (adherens junctions)
2.	cell–matrix junctions (focal adhesions)
Intermediate filament attachment sites	
1.	cell–cell junctions (desmosomes)
2.	cell–matrix junctions (hemidesmosomes)
COMMUNICATING JUNCTIONS	
1.	gap junctions
2.	chemical synapses
3.	plasmodesmata (plants only)

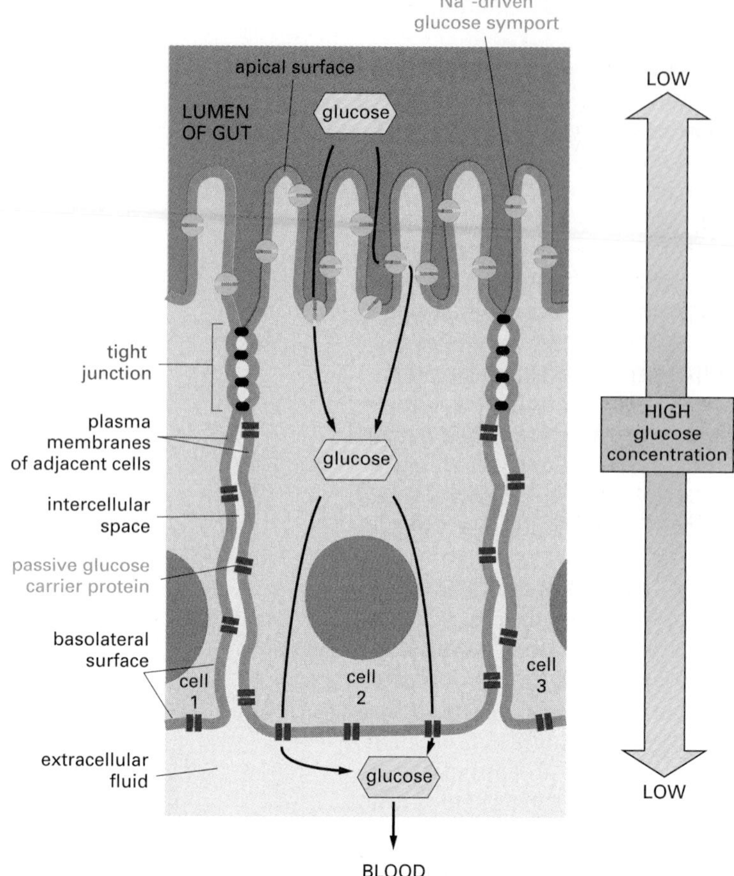

Figure 19–2 The role of tight junctions in transcellular transport. Transport proteins are confined to different regions of the plasma membrane in epithelial cells of the small intestine. This segregation permits a vectorial transfer of nutrients across the epithelium from the gut lumen to the blood. In the example shown, glucose is actively transported into the cell by Na^+-driven glucose symports at the apical surface, and it diffuses out of the cell by facilitated diffusion mediated by glucose carriers in the basolateral membrane. Tight junctions are thought to confine the transport proteins to their appropriate membrane domains by acting as diffusion barriers within the lipid bilayer of the plasma membrane; these junctions also block the backflow of glucose from the basal side of the epithelium into the gut lumen.

When tight junctions are visualized by freeze-fracture electron microscopy, they seem to be composed of a branching network of *sealing strands* that completely encircles the apical end of each cell in the epithelial sheet (Figure 19–4A and B). In conventional electron micrographs, the outer leaflets of the two interacting plasma membranes are seen to be tightly apposed where sealing strands are present (Figure 19–4C). The ability of tight junctions to restrict the passage of ions through the spaces between cells is found to increase logarithmically with increasing numbers of strands in the network, suggesting that each strand acts as an independent barrier to ion flow.

Each tight junction sealing strand is composed of a long row of transmembrane adhesion proteins embedded in each of the two interacting plasma membranes. The extracellular domains of these proteins join directly to one another to occlude the intercellular space (Figure 19–5). The major transmembrane proteins in a tight junction are the *claudins,* which are essential for tight junction formation and function and differ in different tight junctions. A specific claudin

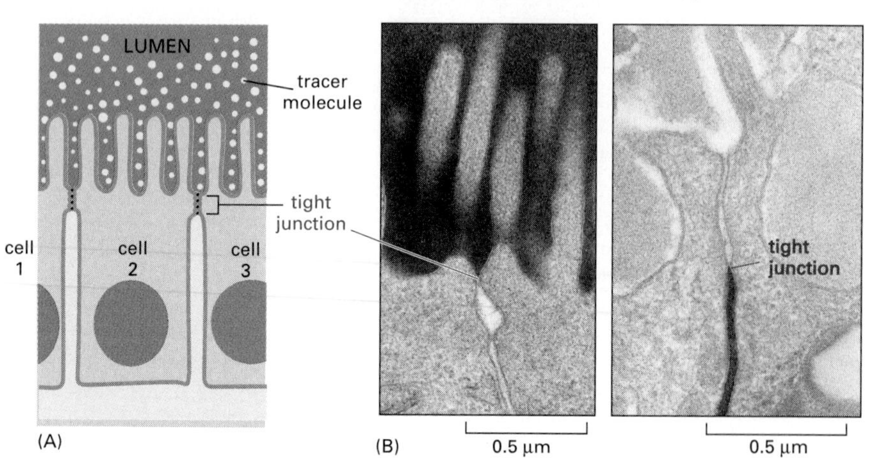

Figure 19–3 The role of tight junctions in allowing epithelia to serve as barriers to solute diffusion. (A) The drawing shows how a small extracellular tracer molecule added on one side of an epithelium cannot traverse the tight junctions that seal adjacent cells together. (B) Electron micrographs of cells in an epithelium in which a small, extracellular, electron-dense tracer molecule has been added to either the apical side (on the *left*) or the basolateral side (on the *right*). In both cases, the tracer is stopped by the tight junction. (B, courtesy of Daniel Friend.)

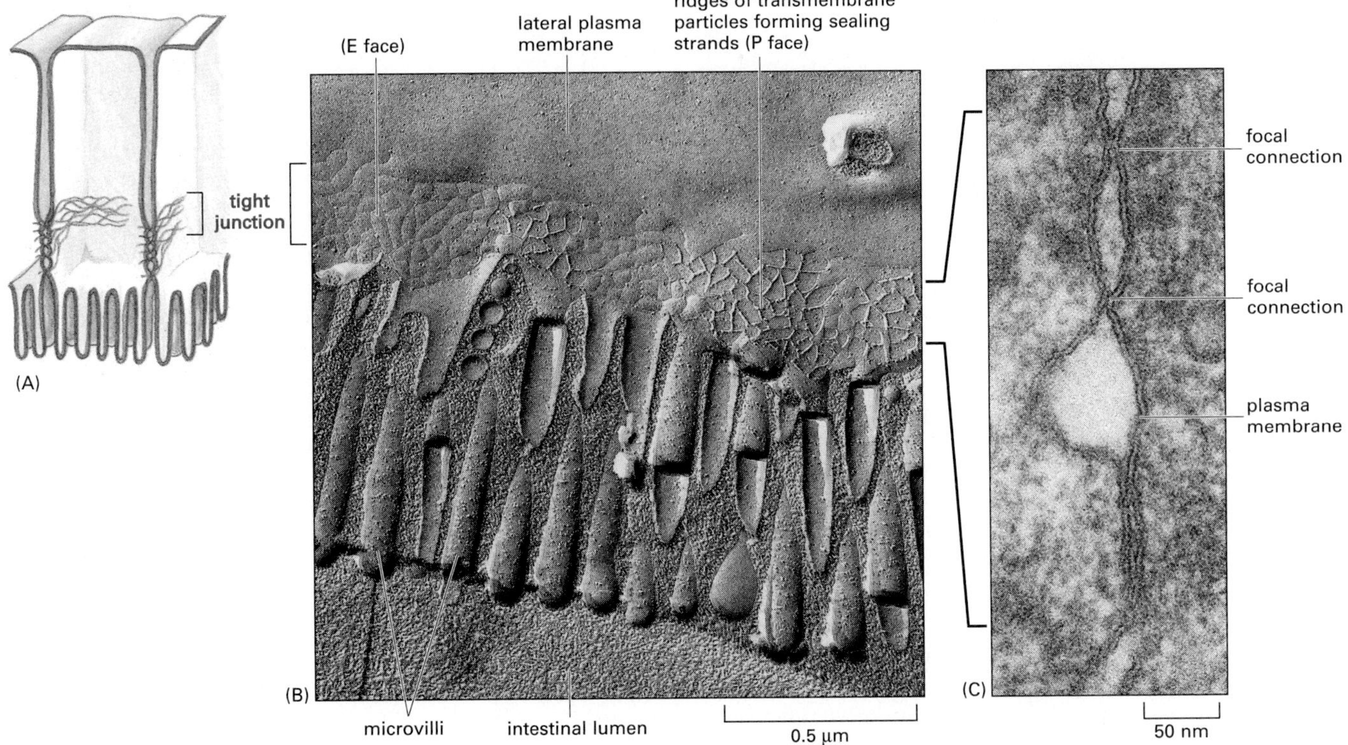

Figure 19–4 The structure of a tight junction between epithelial cells of the small intestine. The junctions are shown (A) schematically, (B) in a freeze-fracture electron micrograph, and (C) in a conventional electron micrograph. Note that the cells are oriented with their apical ends down. In (B), the plane of the micrograph is parallel to the plane of the membrane, and the tight junction appears as a beltlike band of branching sealing strands that encircle each cell in the epithelium. The sealing strands are seen as ridges of intramembrane particles on the cytoplasmic fracture face of the membrane (the P face) or as complementary grooves on the external face of the membrane (the E face) (see Figure 19–5A). In (C), the junction is seen in cross section as a series of focal connections between the outer leaflets of the two interacting plasma membranes, each connection corresponding to a sealing strand in cross section. (B and C, from N.B. Gilula, in Cell Communication [R.P. Cox, ed.], pp. 1–29. New York: Wiley, 1974. Reprinted by permission of John Wiley & Sons, Inc.)

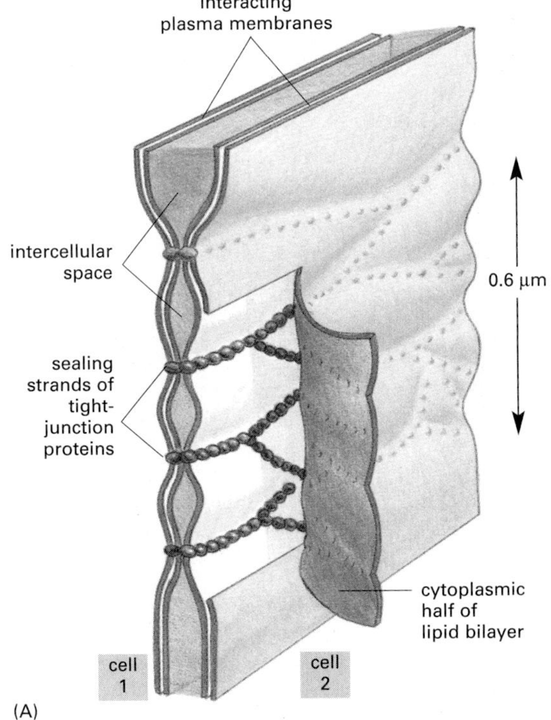

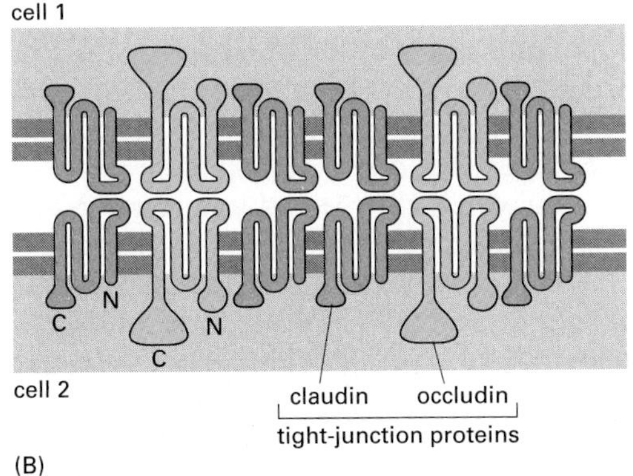

Figure 19–5 A current model of a tight junction. (A) This drawing shows how the sealing strands hold adjacent plasma membranes together. The strands are composed of transmembrane proteins that make contact across the intercellular space and create a seal. (B) This drawing shows the transmembrane claudin and occludin proteins in a tight junction. The claudins are the main components of the sealing strands; the function of the occludins is uncertain.

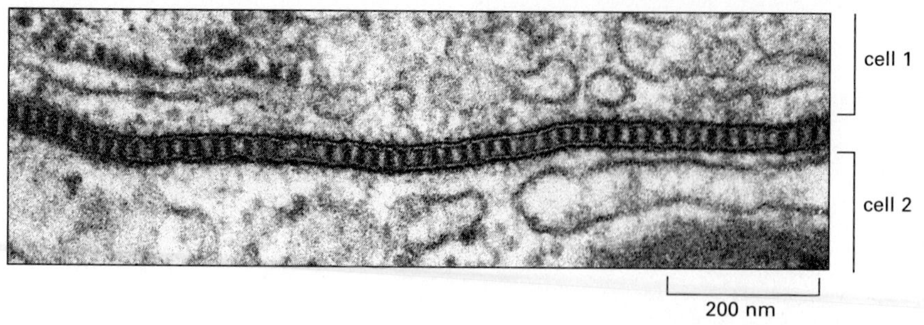

Figure 19–6 A septate junction. A conventional electron micrograph of a septate junction between two epithelial cells in a mollusk. The interacting plasma membranes, seen in cross section, are connected by parallel rows of junctional proteins. The rows, which have a regular periodicity, are seen as dense bars, or septa. (From N.B. Gilula, in Cell Communication [R.P. Cox, ed.], pp. 1–29. New York: Wiley, 1974. Reprinted by permission of John Wiley & Sons, Inc.)

found in kidney epithelial cells, for example, is required for Mg^{2+} to be resorbed from the urine into the blood. A mutation in the gene encoding this claudin results in excessive loss of Mg^{2+} in the urine. A second major transmembrane protein in tight junctions is *occludin,* the function of which is uncertain. Claudins and occludins associate with intracellular peripheral membrane proteins called *ZO proteins* (a tight junction is also known as a *zonula occludens*), which anchor the strands to the actin cytoskeleton.

In addition to claudins, occludins, and ZO proteins, several other proteins can be found associated with tight junctions. These include some that regulate epithelial cell polarity and others that help guide the delivery of components to the appropriate domain of the plasma membrane. Thus, the tight junction may serve as a regulatory center to help in coordinating multiple cell processes.

In invertebrates, **septate junctions** are the main occluding junction. More regular in structure than a tight junction, they likewise form a continuous band around each epithelial cell. But their morphology is distinct because the interacting plasma membranes are joined by proteins that are arranged in parallel rows with a regular periodicity (Figure 19–6). A protein called *Discs-large,* which is required for the formation of septate junctions in *Drosophila,* is structurally related to the ZO proteins found in vertebrate tight junctions. Mutant flies that are deficient in this protein not only lack septate junctions but also develop epithelial tumors. This observation suggests that the normal regulation of cell proliferation in epithelial tissues may depend, in part, on intracellular signals that emanate from occluding junctions.

Anchoring Junctions Connect the Cytoskeleton of a Cell Either to the Cytoskeleton of Its Neighbors or to the Extracellular Matrix

The lipid bilayer is flimsy and cannot by itself transmit large forces from cell to cell or from cell to extracellular matrix. Anchoring junctions solve the problem by forming a strong membrane-spanning structure that is tethered inside the cell to the tension-bearing filaments of the cytoskeleton (Figure 19–7).

Anchoring junctions are widely distributed in animal tissues and are most abundant in tissues that are subjected to severe mechanical stress, such as heart, muscle, and epidermis. They are composed of two main classes of proteins (Figure 19–8). *Intracellular anchor proteins* form a distinct plaque on the cytoplasmic face of the plasma membrane and connect the junctional complex to either actin filaments or intermediate filaments. *Transmembrane adhesion proteins* have a cytoplasmic tail that binds to one or more intracellular anchor proteins and an extracellular domain that interacts with either the extracellular matrix or the extracellular domains of specific transmembrane adhesion proteins on another cell. In addition to anchor proteins and adhesion proteins, many anchoring junctions contain intracellular signaling proteins that enable the junctions to signal to the cell interior.

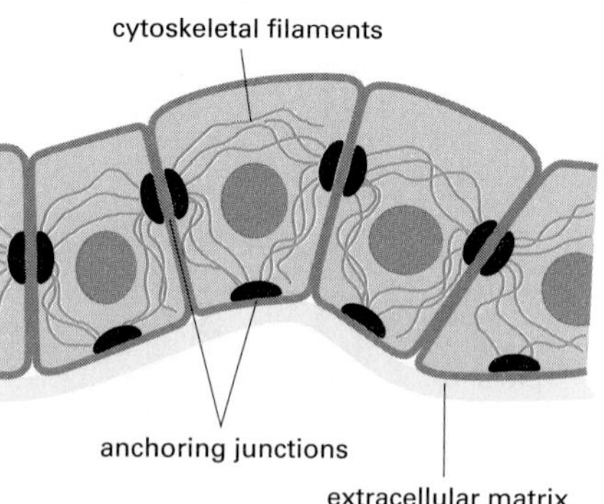

Figure 19–7 Anchoring junctions in an epithelium. This drawing illustrates, in a very general way, how anchoring junctions join cytoskeletal filaments from cell to cell and from cells to the extracellular matrix.

Anchoring junctions occur in two functionally different forms:

1. Adherens junctions and desmosomes hold cells together and are formed by transmembrane adhesion proteins that belong to the *cadherin family.*
2. Focal adhesions and hemidesmosomes bind cells to the extracellular matrix and are formed by transmembrane adhesion proteins of the *integrin family.*

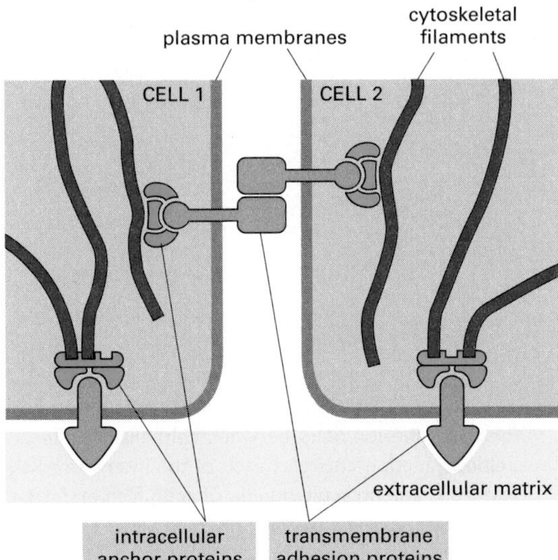

Figure 19–8 The construction of an anchoring junction from two classes of proteins. This drawing shows how intracellular anchor proteins and transmembrane adhesion proteins form anchoring junctions.

On the intracellular side of the membrane, adherens junctions and focal adhesions serve as connection sites for actin filaments, while desmosomes and hemidesmosomes serve as connection sites for intermediate filaments (see Table 19–1, p. 1067).

Adherens Junctions Connect Bundles of Actin Filaments from Cell to Cell

Adherens junctions occur in various forms. In many nonepithelial tissues, they take the form of small punctate or streaklike attachments that indirectly connect the cortical actin filaments beneath the plasma membranes of two interacting cells. But the prototypical examples of adherens junctions occur in epithelia, where they often form a continuous **adhesion belt** (or *zonula adherens)* just below the tight junctions, encircling each of the interacting cells in the sheet. The adhesion belts are directly apposed in adjacent epithelial cells, with the interacting plasma membranes held together by the cadherins that serve here as transmembrane adhesion proteins.

Within each cell, a contractile bundle of actin filaments lies adjacent to the adhesion belt, oriented parallel to the plasma membrane. The actin is attached to this membrane through a set of intracellular anchor proteins, including *catenins, vinculin,* and *α-actinin,* which we consider later. The actin bundles are thus linked, via the cadherins and anchor proteins, into an extensive transcellular network (Figure 19–9). This network can contract with the help of myosin motor proteins (discussed in Chapter 16), and it is thought to help in mediating a fundamental process in animal morphogenesis—the folding of epithelial cell sheets into tubes and other related structures (Figure 19–10).

The assembly of tight junctions between epithelial cells seems to require the prior formation of adherens junctions. Anti-cadherin antibodies that block the formation of adherens junctions, for example, also block the formation of tight junctions.

Desmosomes Connect Intermediate Filaments from Cell to Cell

Desmosomes are buttonlike points of intercellular contact that rivet cells together (Figure 19–11A). Inside the cell, they serve as anchoring sites for ropelike intermediate filaments, which form a structural framework of great tensile strength (Figure 19–11B). Through desmosomes, the intermediate filaments of adjacent cells are linked into a net that extends throughout the many cells of a tissue. The particular type of intermediate filaments attached to the desmosomes

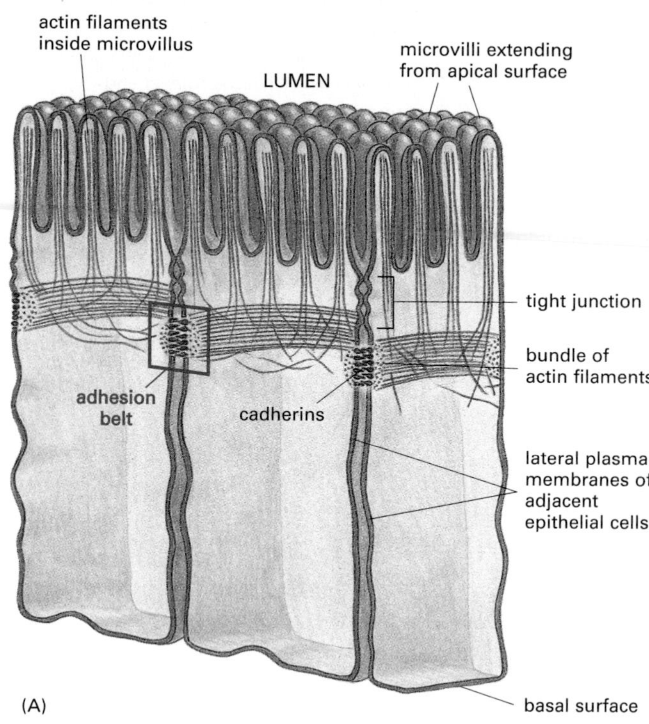

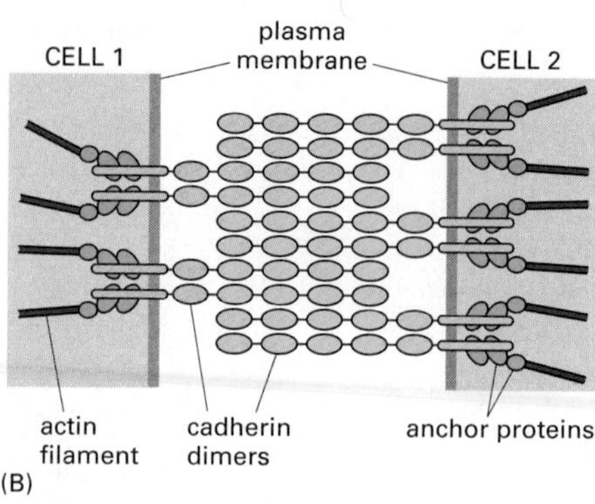

Figure 19–9 Adherens junctions. (A) Adherens junctions, in the form of adhesion belts, between epithelial cells in the small intestine. The beltlike junction encircles each of the interacting cells. Its most obvious feature is a contractile bundle of actin filaments running along the cytoplasmic surface of the junctional plasma membrane. (B) Some of the molecules that form an adherens junction. The actin filaments are joined from cell to cell by transmembrane adhesion proteins called cadherins. The cadherins form homodimers in the plasma membrane of each interacting cell. The extracellular domain of one cadherin dimer binds to the extracellular domain of an identical cadherin dimer on the adjacent cell. The intracellular tails of the cadherins bind to anchor proteins that tie them to actin filaments. These anchor proteins include α-catenin, β-catenin, γ-catenin (also called plakoglobin), α-actinin, and vinculin.

depends on the cell type: they are *keratin filaments* in most epithelial cells, for example, and *desmin filaments* in heart muscle cells.

The general structure of a desmosome is illustrated in Figure 19–11C, and some of the proteins that form it are shown in Figure 19–11D. The junction has a dense cytoplasmic plaque composed of a complex of intracellular anchor proteins (*plakoglobin* and *desmoplakin*) that are responsible for connecting the cytoskeleton to the transmembrane adhesion proteins. These adhesion proteins (*desmoglein* and *desmocollin*), like those at an adherens junction, belong to the cadherin family. They interact through their extracellular domains to hold the adjacent plasma membranes together.

The importance of desmosome junctions is demonstrated by some forms of the potentially fatal skin disease *pemphigus*. Affected individuals make antibodies against one of their own desmosomal cadherin proteins. These antibodies bind to and disrupt the desmosomes that hold their skin epithelial cells

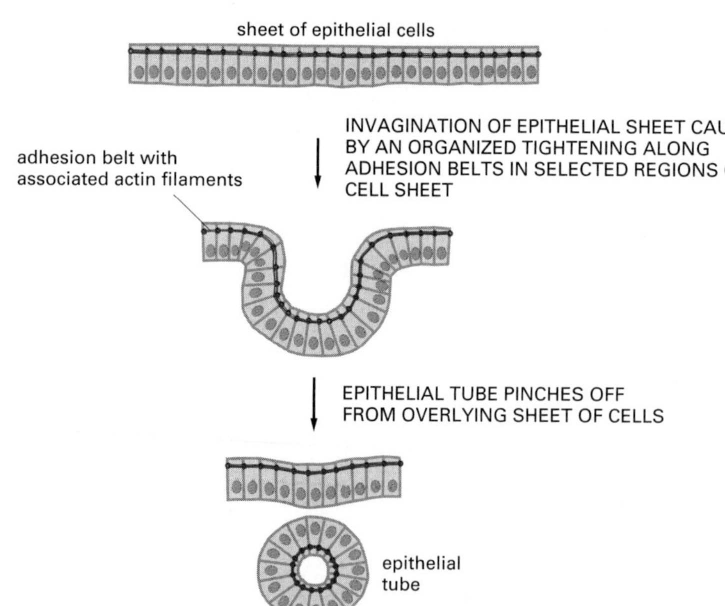

Figure 19–10 The folding of an epithelial sheet to form an epithelial tube. The oriented contraction of the bundles of actin filaments running along adhesion belts causes the epithelial cells to narrow at their apex and helps the epithelial sheet to roll up into a tube. An example is the formation of the neural tube in early vertebrate development (discussed in Chapter 21). Although not shown here, rearrangements of the cells within the epithelial sheet are also thought to have an important role in the process.

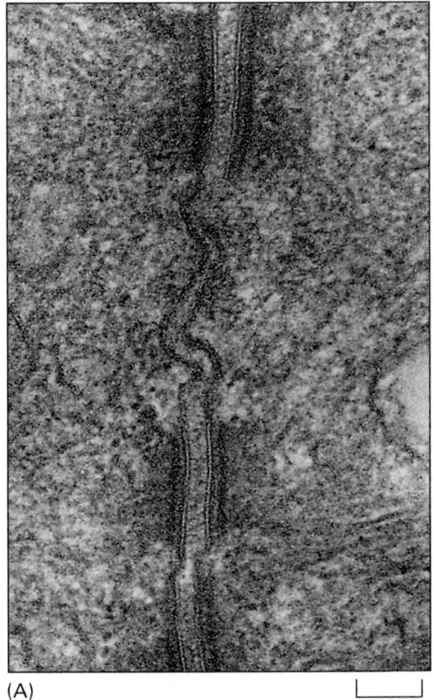
(A) 0.1 μm

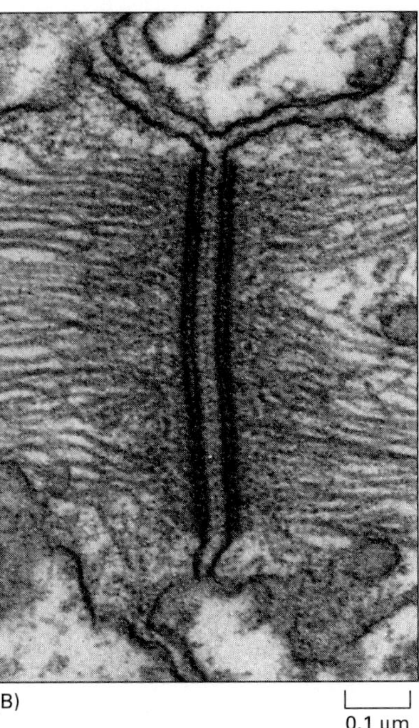
(B) 0.1 μm

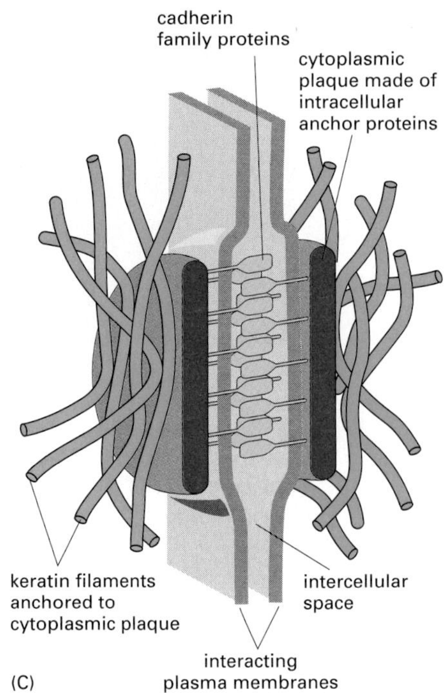

(C)

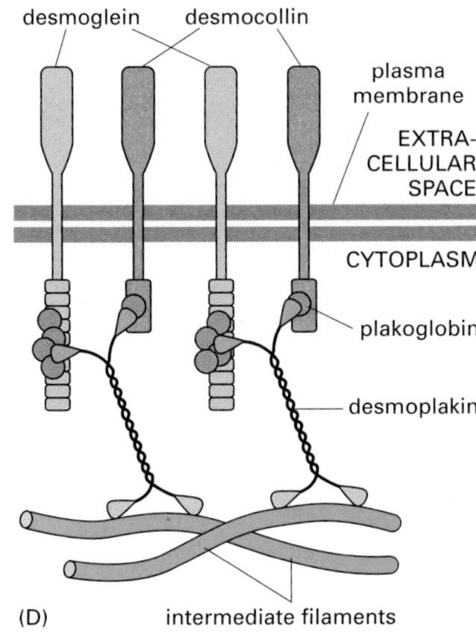

(D)

Figure 19–11 Desmosomes. (A) An electron micrograph of three desmosomes between two epithelial cells in the intestine of a rat. (B) An electron micrograph of a single desmosome between two epidermal cells in a developing newt, showing clearly the attachment of intermediate filaments. (C) The structural components of a desmosome. On the cytoplasmic surface of each interacting plasma membrane is a dense plaque composed of a mixture of intracellular anchor proteins. A bundle of keratin intermediate filaments is attached to the surface of each plaque. Transmembrane adhesion proteins of the cadherin family bind to the plaques and interact through their extracellular domains to hold the adjacent membranes together by a Ca^{2+}-dependent mechanism. (D) Some of the molecular components of a desmosome. Desmoglein and desmocollin are members of the cadherin family of adhesion proteins. Their cytoplasmic tails bind plakoglobin (γ-catenin), which in turn binds to desmoplakin. Desmoplakin also binds to the sides of intermediate filaments, thereby tying the desmosome to these filaments. (A, from N.B. Gilula, in Cell Communication [R.P. Cox, ed.], pp. 1–29. New York: Wiley, 1974. Reprinted by permission of John Wiley & Sons, Inc.; B, from D.E. Kelly, *J. Cell Biol.* 28:51–59, 1966. © The Rockefeller University Press.)

(keratinocytes) together. This results in a severe blistering of the skin, with leakage of body fluids into the loosened epithelium.

Anchoring Junctions Formed by Integrins Bind Cells to the Extracellular Matrix: Focal Adhesions and Hemidesmosomes

Some anchoring junctions bind cells to the extracellular matrix rather than to other cells. The transmembrane adhesion proteins in these cell–matrix junctions are *integrins*—a large family of proteins distinct from the cadherins. **Focal adhesions** enable cells to get a hold on the extracellular matrix through integrins that link intracellularly to actin filaments. In this way, muscle cells, for example, attach to their tendons at the *myotendinous junction*. Likewise, when cultured fibroblasts migrate on an artificial substratum coated with extracellular matrix molecules, they also grip the substratum at focal adhesions, where bundles of actin filaments terminate. At all such adhesions, the extracellular domains of transmembrane integrin proteins bind to a protein component of the extracellular matrix, while their intracellular domains bind indirectly to bundles of actin filaments via the intracellular anchor proteins talin, α-actinin, filamin, and vinculin (Figure 19–12B).

Hemidesmosomes, or half-desmosomes, resemble desmosomes morphologically and in connecting to intermediate filaments, and, like desmosomes, they act as rivets to distribute tensile or shearing forces through an epithelium.

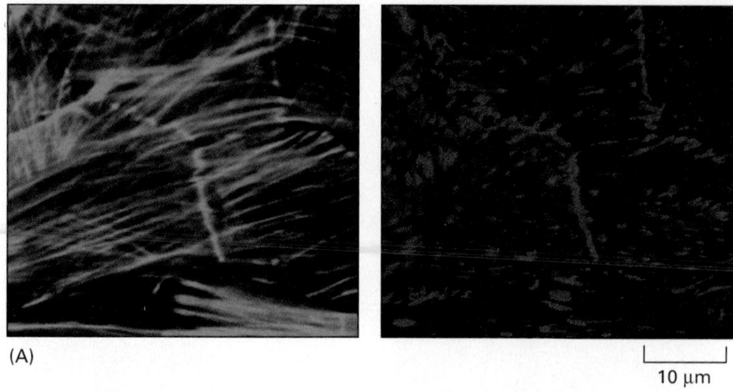

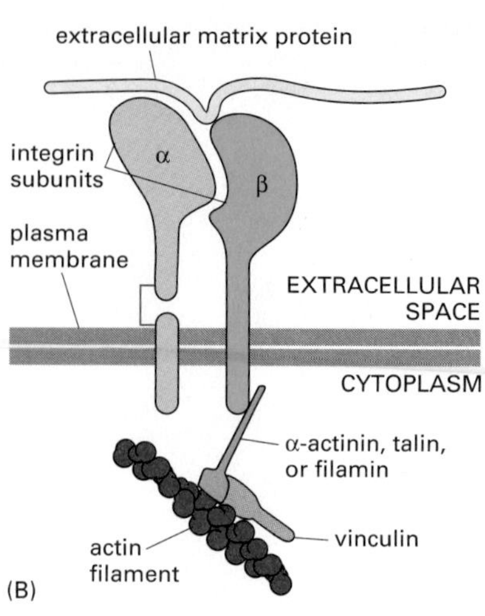

Figure 19–12 Focal adhesions. (A) In these immunofluorescence micrographs, cells in culture have been labeled with antibodies against both actin *(green)* and the intracellular anchor protein vinculin *(red)*. Note that vinculin is located at focal adhesions, which is also where bundles of actin filaments terminate at the plasma membrane. (B) Some of the proteins that form focal adhesions. The transmembrane adhesion protein is an integrin heterodimer, composed of an α and a β subunit. Its extracellular domains bind to components of the extracellular matrix, while the cytoplasmic tail of the β subunit binds indirectly to actin filaments via several intracellular anchor proteins. (A, from B. Geiger, E. Schmid, and W. Franke, *Differentiation* 23:189–205, 1983.)

Instead of joining adjacent epithelial cells, however, hemidesmosomes connect the basal surface of an epithelial cell to the underlying basal lamina (Figure 19–13). The extracellular domains of the integrins that mediate the adhesion bind to a *laminin* protein (discussed later) in the basal lamina, while an intracellular domain binds via an anchor protein *(plectin)* to keratin intermediate filaments. Whereas the keratin filaments associated with desmosomes make lateral attachments to the desmosomal plaques (see Figure 19–11C and D), many keratin filaments associated with hemidesmosomes have their ends buried in the plaque (see Figure 19–13).

Although the terminology for the various anchoring junctions can be confusing, the molecular principles (for vertebrates, at least) are relatively simple (Table 19–2). Integrins in the plasma membrane anchor a cell to extracellular matrix molecules; cadherin family members in the plasma membrane anchor it to the plasma membrane of an adjacent cell. In both cases, there is an intracellular coupling to cytoskeletal filaments, either actin filaments or intermediate filaments, depending on the types of intracellular anchor proteins involved.

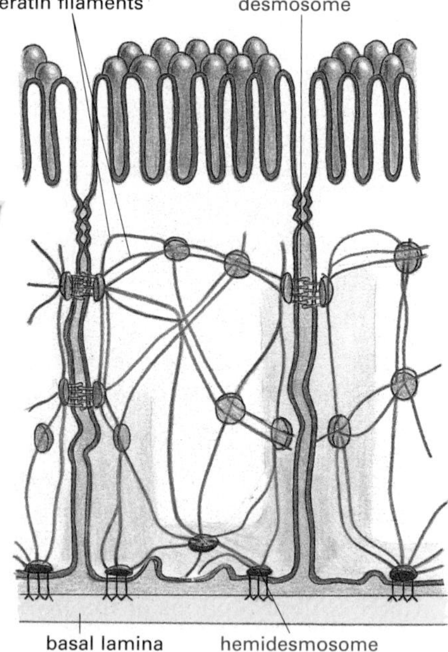

Figure 19–13 Desmosomes and hemidesmosomes. The distribution of desmosomes and hemidesmosomes in epithelial cells of the small intestine. The keratin intermediate filament networks of adjacent cells are indirectly connected to one another through desmosomes and to the basal lamina through hemidesmosomes.

Gap Junctions Allow Small Molecules to Pass Directly from Cell to Cell

With the exception of a few terminally differentiated cells such as skeletal muscle cells and blood cells, most cells in animal tissues are in communication with their neighbors via **gap junctions**. Each gap junction appears in conventional electron micrographs as a patch where the membranes of two adjacent cells are separated by a uniform narrow gap of about 2–4 nm. The gap is spanned by channel-forming proteins *(connexins)*. The channels they form *(connexons)* allow inorganic ions and other small water-soluble molecules to pass directly from the cytoplasm of one cell to the cytoplasm of the other, thereby coupling the cells both electrically and metabolically. Dye-injection experiments suggest a maximal functional pore size for the connecting channels of about 1.5 nm, implying that coupled cells share their small molecules (such as inorganic ions, sugars, amino acids, nucleotides, vitamins, and the intracellular mediators cyclic AMP and inositol trisphosphate) but not their macromolecules (proteins, nucleic acids, and polysaccharides) (Figure 19–14). This cell coupling has important functional implications, many of which are only beginning to be understood.

TABLE 19–2 Anchoring Junctions

JUNCTION	TRANSMEMBRANE ADHESION PROTEIN	EXTRACELLULAR LIGAND	INTRACELLULAR CYTOSKELETAL ATTACHMENT	INTRACELLULAR ANCHOR PROTEINS
Cell–Cell				
Adherens junction	cadherin (E-cadherin)	cadherin in neighboring cell	actin filaments	α- and β-catenins, vinculin, α-actinin, plakoglobin (γ-catenin)
Desmosome	cadherin (desmoglein, desmocollin)	desmogleins and desmocollins in neighboring cell	intermediate filaments	desmoplakins, plakoglobin (γ-catenin)
Cell–Matrix				
Focal adhesion	integrin	extracellular matrix proteins	actin filaments	talin, vinculin, α-actinin, filamin
Hemidesmosome	integrin $\alpha_6\beta_4$, BP180	extracellular matrix proteins	intermediate filaments	plectin, BP230

Evidence that gap junctions mediate electrical and chemical coupling has come from many experiments. When, for example, connexin mRNA is injected into either frog oocytes or gap-junction-deficient cultured cells, channels with the properties expected of gap-junction channels can be demonstrated electrophysiologically where pairs of injected cells make contact.

The mRNA injection approach has been useful for identifying new gap-junction proteins. Genetic studies in the fruit fly *Drosophila* identified the gene *shaking B*, which, when mutated, resulted in flies that failed to jump in response to a visual stimulus. Although these flies had defective gap junctions, the sequence of the Shaking B protein did not resemble a connexin, and the function of the protein was unclear. An injection of the *shaking B* mRNA into frog oocytes, however, led to the formation of functional gap-junction channels, just like those formed by connexins. Shaking B thus became the first member of a new family of invertebrate gap-junction proteins called *innexins*. There are more than 15 *innexin* genes in *Drosophila* and 25 in the nematode *C. elegans*.

A Gap-Junction Connexon Is Made Up of Six Transmembrane Connexin Subunits

Connexins are four-pass transmembrane proteins, six of which assemble to form a channel, a **connexon**. When the connexons in the plasma membranes of two cells in contact are aligned, they form a continuous aqueous channel that connects the two cell interiors (Figure 19–15A). The connexons hold the interacting plasma membranes at a fixed distance apart—hence the gap.

Gap junctions in different tissues can have different properties. The permeability of their individual channels can vary, reflecting differences in the connexins that form the junctions. In humans, for instance, there are 14 distinct connexins, each encoded by a separate gene and each having a distinctive, but sometimes overlapping, tissue distribution. Most cell types express more than one type of connexin, and two different connexin proteins can assemble into a heteromeric connexon, the properties of which differ from those of a homomeric connexon constructed from a single type of connexin. Moreover, adjacent cells expressing different connexins can form intercellular channels in which the two aligned half-channels are different (Figure 19–15B). Each gap junction can contain a cluster of a few to many thousands of connexons (Figure 19–16B).

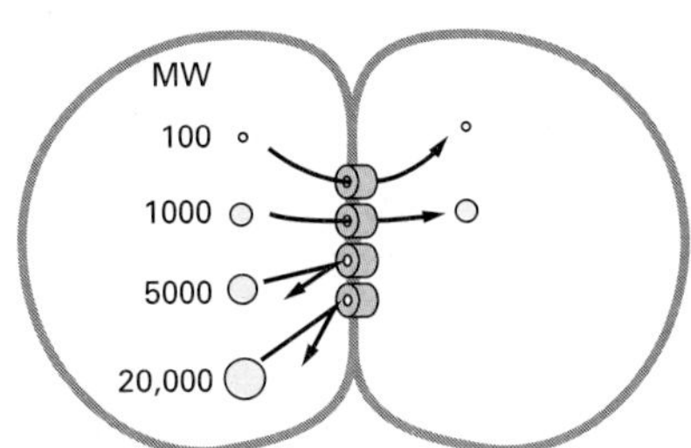

Figure 19–14 Determining the size of a gap-junction channel. When fluorescent molecules of various sizes are injected into one of two cells coupled by gap junctions, molecules with a mass of less than about 1000 daltons can pass into the other cell, but larger molecules cannot.

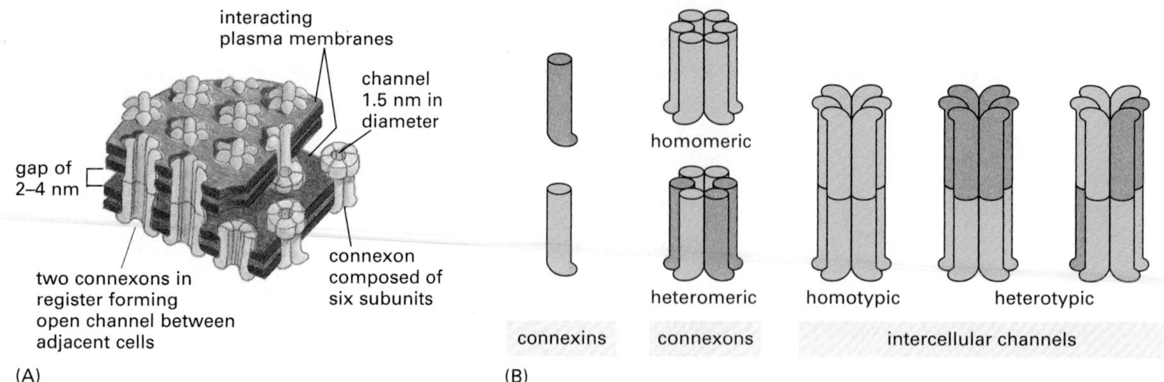

Figure 19–15 Gap junctions. (A) A three-dimensional drawing showing the interacting plasma membranes of two adjacent cells connected by gap junctions. The apposed lipid bilayers *(red)* are penetrated by protein assemblies called connexons *(green)*, each of which is formed by six connexin subunits. Two connexons join across the intercellular gap to form a continuous aqueous channel connecting the two cells. (B) The organization of connexins into connexons and connexons into intercellular channels. The connexons can be homomeric or heteromeric, and the intercellular channels can be homotypic or heterotypic.

Gap Junctions Have Diverse Functions

In tissues containing electrically excitable cells, coupling via gap junctions serves an obvious purpose. Some nerve cells, for example, are electrically coupled, allowing action potentials to spread rapidly from cell to cell, without the delay that occurs at chemical synapses. This is advantageous when speed and reliability are crucial, as in certain escape responses in fish and insects. Similarly, in vertebrates, electrical coupling through gap junctions synchronizes the contractions of both heart muscle cells and the smooth muscle cells responsible for the peristaltic movements of the intestine.

Gap junctions also occur in many tissues that do not contain electrically excitable cells. In principle, the sharing of small metabolites and ions provides a mechanism for coordinating the activities of individual cells in such tissues and for smoothing out random fluctuations in small molecule concentrations in different cells. In the liver, for example, the release of noradrenaline from sympathetic nerve endings in response to a fall in blood glucose levels stimulates hepatocytes to increase glycogen breakdown and release glucose into the blood. Not all the hepatocytes are innervated by sympathetic nerves, however. By means of the gap junctions that connect hepatocytes, the signal is transmitted from the innervated hepatocytes to the noninnervated ones. Thus, mice with a mutation in the major connexin gene expressed in the liver fail to mobilize glucose normally when blood glucose levels fall.

The normal development of ovarian follicles also depends on gap-junction-mediated communication—in this case, between the oocyte and the surrounding granulosa cells. A mutation in the gene that encodes the connexin that normally couples these two cell types causes infertility (Figure 19–17).

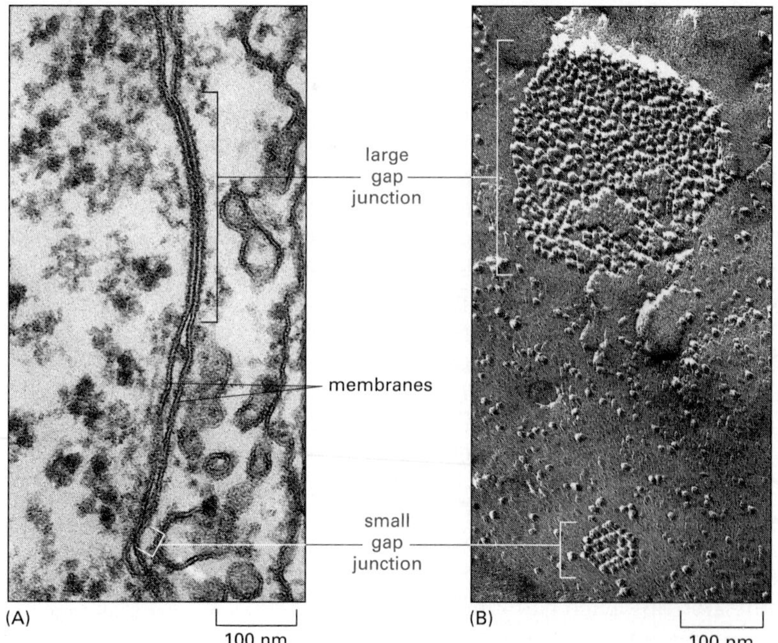

Figure 19–16 Gap junctions as seen in the electron microscope. (A) Thin-section and (B) freeze-fracture electron micrographs of a large and a small gap junction between fibroblasts in culture. In (B), each gap junction is seen as a cluster of homogeneous intramembrane particles associated exclusively with the cytoplasmic fracture face (P face) of the plasma membrane. Each intramembrane particle corresponds to a connexon. (From N.B. Gilula, in Cell Communication [R.P. Cox, ed.], pp. 1–29. New York: Wiley, 1974. Reprinted by permission of John Wiley & Sons, Inc.)

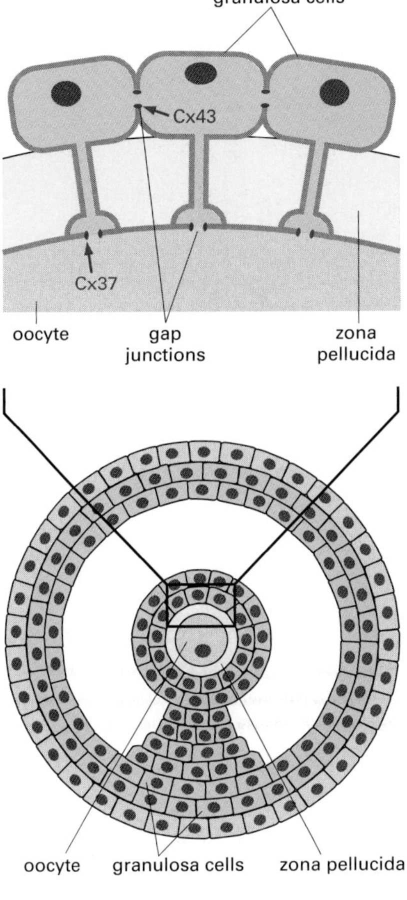

Figure 19–17 Gap junction coupling in the ovarian follicle. The oocyte is surrounded by a thick layer of extracellular matrix called the zona pellucida (discussed in Chapter 20). The surrounding granulosa cells are coupled to each other by gap junctions formed by connexin 43 (Cx43). In addition, the granulosa cells extend processes through the zona pellucida and make gap junctions with the oocyte. These gap junctions contain a different connexin (Cx37). Mutations in the gene encoding Cx37 cause infertility by disrupting the development of both the granulosa cells and the oocyte.

Cell coupling via gap junctions also seems to be important in embryogenesis. In early vertebrate embryos, beginning with the late eight-cell stage in mouse embryos, most cells are electrically coupled to one another. As specific groups of cells in the embryo develop their distinct identities and begin to differentiate, they commonly uncouple from surrounding tissue. As the neural plate folds up and pinches off to form the neural tube, for instance (see Figure 19–10), its cells uncouple from the overlying ectoderm. Meanwhile, the cells within each group remain coupled with one another and therefore tend to behave as a cooperative assembly, all following a similar developmental pathway in a coordinated fashion.

The Permeability of Gap Junctions Can Be Regulated

Like conventional ion channels (discussed in Chapter 11), individual gap-junction channels do not remain continuously open; instead, they flip between open and closed states. Moreover, the permeability of gap junctions is rapidly (within seconds) and reversibly reduced by experimental manipulations that decrease the cytosolic pH or increase the cytosolic concentration of free Ca^{2+} to very high levels. Thus, gap-junction channels are dynamic structures that can undergo a reversible conformational change that closes the channel in response to changes in the cell.

The purpose of the pH regulation of gap-junction permeability is unknown. In one case, however, the purpose of Ca^{2+} control seems clear. When a cell is damaged, its plasma membrane can become leaky. Ions present at high concentration in the extracellular fluid, such as Ca^{2+} and Na^+, then move into the cell, and valuable metabolites leak out. If the cell were to remain coupled to its healthy neighbors, these too would suffer a dangerous disturbance of their internal chemistry. But the large influx of Ca^{2+} into the damaged cell causes its gap-junction channels to close immediately, effectively isolating the cell and preventing the damage from spreading to other cells.

Gap-junction communication can also be regulated by extracellular signals. The neurotransmitter *dopamine*, for example, reduces gap-junction communication between a class of neurons in the retina in response to an increase in light intensity (Figure 19–18). This reduction in gap-junction permeability helps the retina switch from using rod photoreceptors, which are good detectors of low light, to cone photoreceptors, which detect color and fine detail in bright light.

Figure 19–19 summarizes the various types of junctions formed by vertebrate cells in an epithelium. In the most apical portion of the cell, the relative

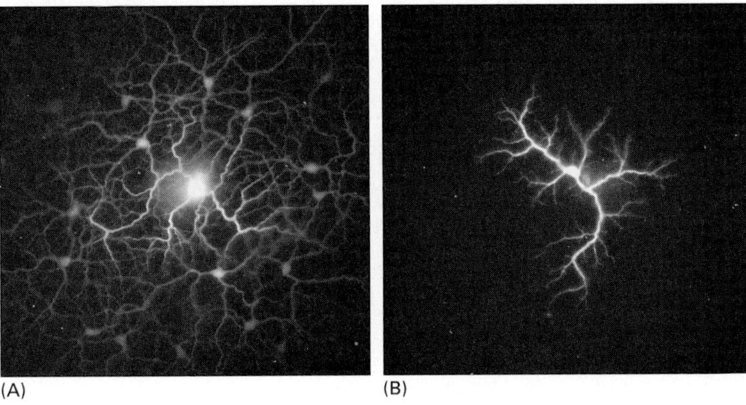

Figure 19–18 The regulation of gap-junction coupling by a neurotransmitter. (A) A neuron in a rabbit retina was injected with the dye Lucifer yellow, which passes readily through gap junctions and labels other neurons of the same type that are connected to the injected cell by gap junctions. (B) The retina was first treated with the neurotransmitter dopamine, before the neuron was injected with dye. As can be seen, the dopamine treatment greatly decreased the permeability of the gap junctions. Dopamine acts by increasing intracellular cyclic AMP levels. (Courtesy of David Vaney.)

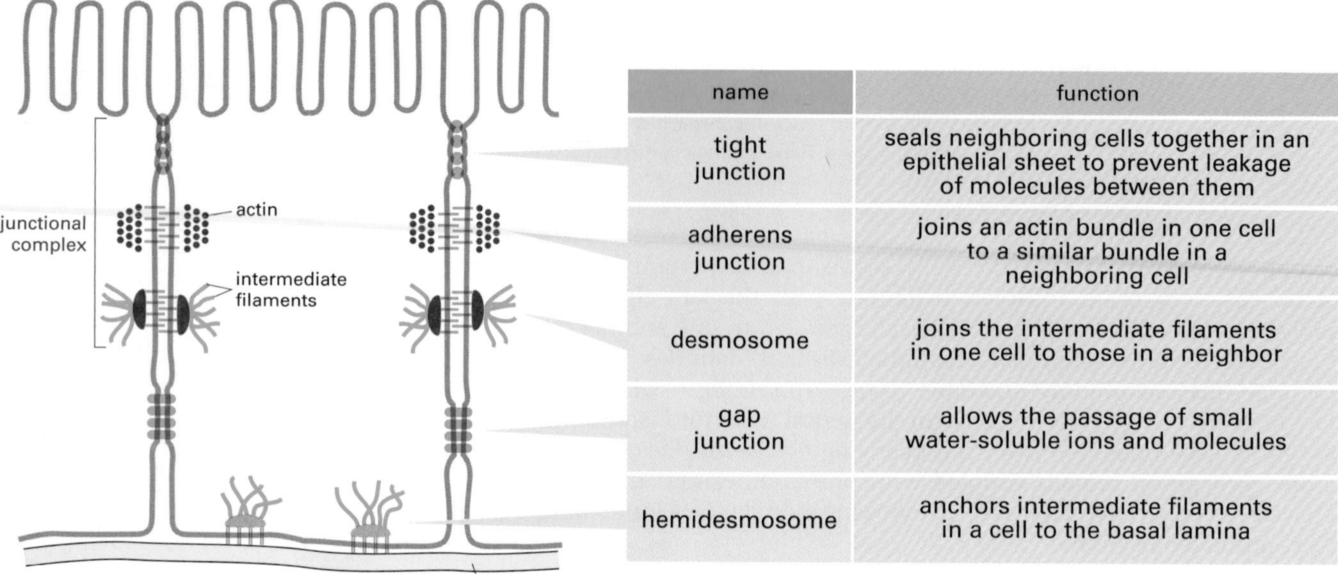

name	function
tight junction	seals neighboring cells together in an epithelial sheet to prevent leakage of molecules between them
adherens junction	joins an actin bundle in one cell to a similar bundle in a neighboring cell
desmosome	joins the intermediate filaments in one cell to those in a neighbor
gap junction	allows the passage of small water-soluble ions and molecules
hemidesmosome	anchors intermediate filaments in a cell to the basal lamina

Figure 19–19 A summary of the various cell junctions found in a vertebrate epithelial cell. The drawing is based on epithelial cells of the small intestine.

positions of the junctions are the same in nearly all vertebrate epithelia. The tight junction occupies the most apical position, followed by the adherens junction (adhesion belt) and then by a special parallel row of desmosomes; together these form a structure called a *junctional complex.* Gap junctions and additional desmosomes are less regularly organized.

In Plants, Plasmodesmata Perform Many of the Same Functions as Gap Junctions

The tissues of a plant are organized on different principles from those of an animal. This is because plant cells are imprisoned within rigid *cell walls* composed of an extracellular matrix rich in cellulose and other polysacharides, as we discuss later. The cell walls of adjacent cells are firmly cemented to those of their neighbors, which eliminates the need for anchoring junctions to hold the cells in place. But a need for direct cell–cell communication remains. Thus, plant cells have only one class of intercellular junctions, **plasmodesmata** (singular, plasmodesma). Like gap junctions, they directly connect the cytoplasms of adjacent cells.

In plants, however, the cell wall between a typical pair of adjacent cells is at least 0.1 μm thick, and so a structure very different from a gap junction is required to mediate communication across it. Plasmodesmata solve the problem. With a few specialized exceptions, every living cell in a higher plant is connected to its living neighbors by these structures, which form fine cytoplasmic channels through the intervening cell walls. As shown in Figure 19–20A, the

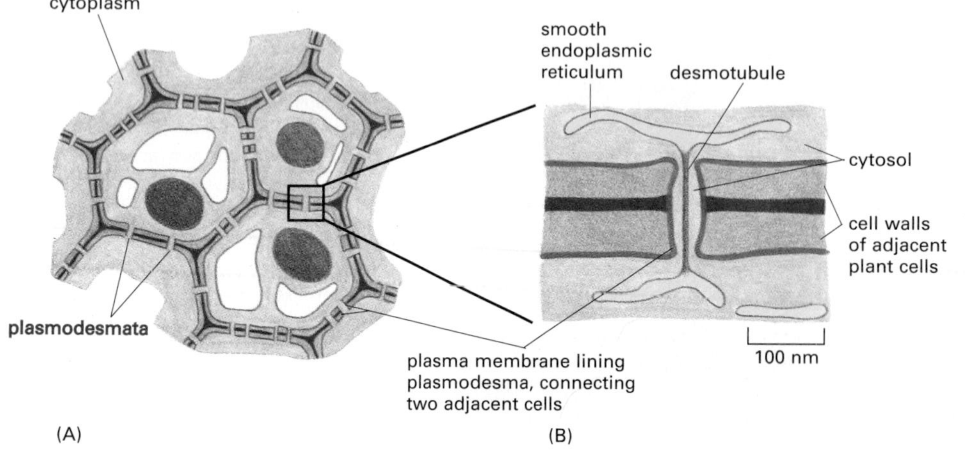

Figure 19–20 Plasmodesmata. (A) The cytoplasmic channels of plasmodesmata pierce the plant cell wall and connect all cells in a plant together. (B) Each plasmodesma is lined with plasma membrane that is common to two connected cells. It usually also contains a fine tubular structure, the desmotubule, derived from smooth endoplasmic reticulum.

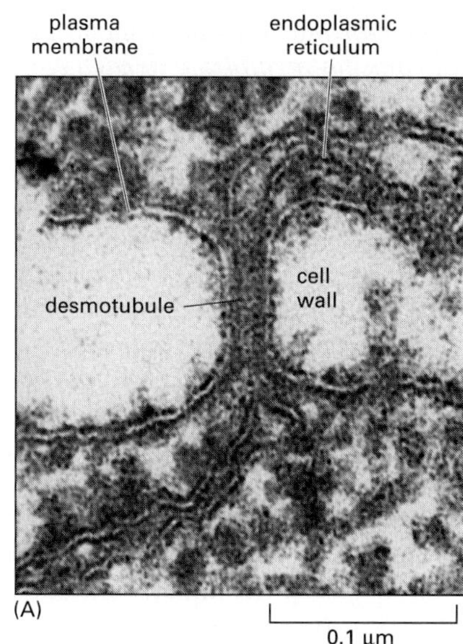

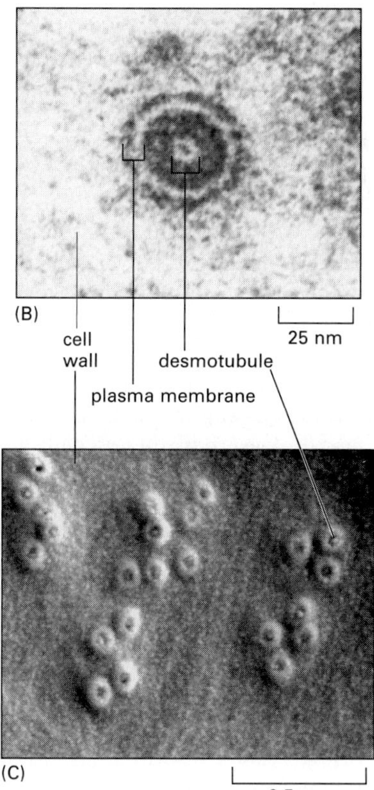

Figure 19–21 Various views of plasmodesmata. (A) Electron micrograph of a longitudinal section of a plasmodesma from a water fern. The plasma membrane lines the pore and is continuous from one cell to the next. Endoplasmic reticulum and its association with the central desmotubule can be seen. (B) A similar plasmodesma seen in cross section. (C) Small pit fields of plasmodesmata in a cell wall isolated from a tobacco leaf. In this SEM they are seen in face view and in many of them the central desmotubule can be seen. (A and B, from R. Overall, J. Wolfe and B.E.S. Gunning, in Protoplasm 9, pp. 137 and 140. Heidelberg: Springer-Verlag, 1982; C, courtesy of Kim Findlay.)

plasma membrane of one cell is continuous with that of its neighbor at each plasmodesma, and the cytoplasm of the two cells is connected by a roughly cylindrical channel with a diameter of 20–40 nm. Thus, the cells of a plant can be viewed as forming a syncytium, in which many cell nuclei share a common cytoplasm.

Running through the center of the channel in most plasmodesmata is a narrower cylindrical structure, the *desmotubule,* which is continuous with elements of the smooth endoplasmic reticulum in each of the connected cells (Figures 19–20B and 19–21A and B). Between the outside of the desmotubule and the inner face of the cylindrical channel formed by plasma membrane is an annulus of cytosol through which small molecules can pass from cell to cell. As each new cell wall is assembled during the cytokinesis phase of cell division, plasmadesmata are created within it. They form around elements of smooth ER that become trapped across the developing cell plate (discussed in Chapter 18). They can also be inserted *de novo* through pre-existing cell walls, where they are commonly found in dense clusters called *pit fields* (Figure 19–21C). When no longer required, plasmadesmata can be readily removed.

In spite of the radical difference in structure between plasmodesmata and gap junctions, they seem to function in remarkably similar ways. Evidence obtained by injecting tracer molecules of different sizes suggests that plasmodesmata allow the passage of molecules with a molecular weight of less than about 800, which is similar to the molecular-weight cutoff for gap junctions. As with gap junctions, transport through plasmodesmata is regulated. Dye-injection experiments, for example, show that there can be barriers to the movement of even low-molecular-weight molecules between certain cells, or groups of cells, that are connected by apparently normal plasmodesmata; the mechanisms that restrict communication in these cases are not understood.

During plant development, groups of cells within the shoot and root meristems signal to one another in the process of defining their future fates (discussed in Chapter 21). Some gene regulatory proteins involved in this process of cell fate determination pass from cell to cell through plasmodesmata. They bind to components of the plasmodesmata and override the size exclusion mechanism that would otherwise prevent their passage. In some cases, the mRNA that encodes the protein can also pass through. Some plant viruses also exploit this route: infectious viral RNA, or even intact virus particles, can pass from cell to cell in this way. These viruses produce proteins that bind to components of the plasmodesmata to increase dramatically the effective pore size of the channel. As the functional components of plasmodesmata are unknown, it is unclear how endogenous or viral macromolecules regulate the transport properties of the channel to pass through it.

Summary

Many cells in tissues are linked to one another and to the extracellular matrix at specialized contact sites called cell junctions. Cell junctions fall into three functional classes: occluding junctions, anchoring junctions, and communicating junctions. Tight junctions are occluding junctions that are crucial in maintaining the concentration differences of small hydrophilic molecules across epithelial cell sheets. They do so in two ways. First, they seal the plasma membranes of adjacent cells together to create a continuous impermeable, or semipermeable, barrier to diffusion across the cell sheet. Second, they act as barriers in the lipid bilayer to restrict the diffusion

of membrane transport proteins between the apical and the basolateral domains of the plasma membrane in each epithelial cell. Septate junctions serve as occluding junctions in invertebrate tissues.

The main types of anchoring junctions in vertebrate tissues are adherens junctions, desmosomes, focal adhesions, and hemidesmosomes. Adherens junctions and desmosomes connect cells together and are formed by cadherins, while focal adhesions and hemidesmosomes connect cells to the extracellular matrix and are formed by integrins. Adherens junctions and focal adhesions are connecting sites for bundles of actin filaments, whereas desmosomes and hemidesmosomes are connecting sites for intermediate filaments.

Gap junctions are communicating junctions composed of clusters of connexons that allow molecules smaller than about 1000 daltons to pass directly from the inside of one cell to the inside of the next. Cells connected by gap junctions share many of their inorganic ions and other small molecules and are therefore chemically and electrically coupled. Gap junctions are important in coordinating the activities of electrically active cells, and they have a coordinating role in other groups of cells as well. Plasmodesmata are the only intercellular junctions in plants. Although their structure is entirely different, and they can sometimes transport informational macromolecules, in general, they function like gap junctions.

CELL–CELL ADHESION

To form an anchoring junction, cells must first adhere. A bulky cytoskeletal apparatus must then be assembled around the molecules that directly mediate the adhesion. The result is a well-defined structure—a desmosome, a hemidesmosome, a focal adhesion, or an adherens junction—that is easily identified in the electron microscope. Indeed, electron microscopy provided the basis for the original classification of cell junctions. In the early stages of cell junction development, however, before the cytoskeletal apparatus has assembled, cells often adhere to one another without clearly displaying these characteristic structures; in the electron microscope, one may simply see two plasma membranes separated by a small gap of a definite width. Functional tests show, nevertheless, that the two cells are stuck to each other, and biochemical analysis can reveal the molecules responsible for the adhesion.

The study of cell–cell junctions and the study of cell–cell adhesion were once quite distinct endeavors, originating from two different experimental approaches—junctions through electron microscopic description, and adhesion through functional tests and biochemistry. Only in recent years have these two approaches begun to converge in a unified view of the molecular basis of cell junctions and cell adhesion. In the previous section, we concentrated on the structures of mature cell junctions. In this section, we turn to functional and biochemical studies of the cell–cell adhesion mechanisms that operate when cells migrate over other cells and when they assemble into tissues—mechanisms that precede the construction of mature cell–cell anchoring junctions. We begin with a critical question for embryonic development: what mechanisms ensure that a cell attaches to appropriate neighbors at the right time?

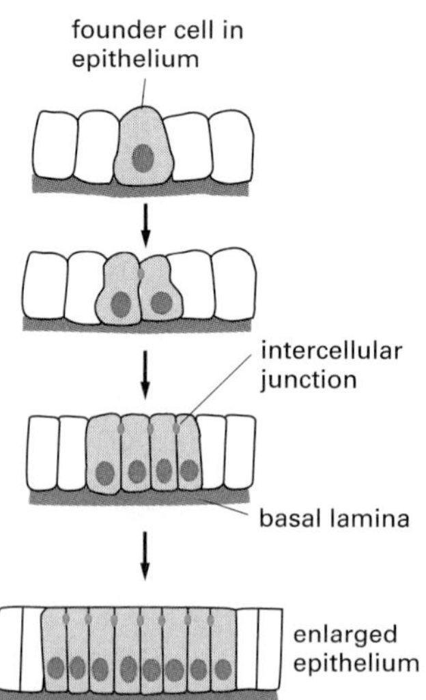

Figure 19–22 The simplest mechanism by which cells assemble to form a tissue. The progeny of the founder cell are retained in the epithelium by the basal lamina and by cell–cell adhesion mechanisms, including the formation of intercellular junctions.

Animal Cells Can Assemble into Tissues Either in Place or After They Migrate

Many simple tissues, including most epithelial tissues, derive from precursor cells whose progeny are prevented from wandering away by being attached to the extracellular matrix, to other cells, or to both (Figure 19–22). But the accumulating cells do not simply remain passively stuck together; instead, the tissue architecture is generated and actively maintained by selective adhesions that the cells make and progressively adjust.

Selective adhesion is even more essential for the development of tissues that have more complex origins involving cell migration. In these tissues, one population of cells invades another and assembles with it, and perhaps with other

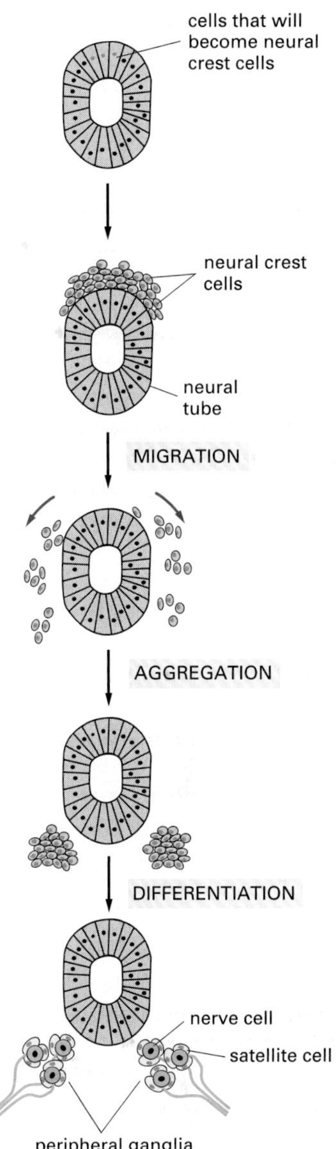

Figure 19–23 An example of a more complex mechanism by which cells assemble to form a tissue. Some cells that are initially part of the epithelial neural tube alter their adhesive properties and disengage from the epithelium to form the neural crest on the upper surface of the neural tube. The cells then migrate away and form a variety of cell types and tissues throughout the embryo. Here they are shown assembling and differentiating to form two collections of nerve cells, called ganglia, in the peripheral nervous system. Other neural crest cells differentiate in the ganglion to become supporting (satellite) cells surrounding the neurons. The crest cells tend to migrate in clusters, and they proliferate rapidly as they migrate.

migrant cells, to form an orderly structure. In vertebrate embryos, for example, cells from the *neural crest* break away from the epithelial neural tube, of which they are initially a part, and migrate along specific paths to many other regions (discussed in Chapter 21). There they assemble with other cells and with one another to differentiate into a variety of tissues, including those of the peripheral nervous system (Figure 19–23).

Cell motility and cell adhesion combine to bring about these kinds of morphogenetic events. The process requires some mechanism for directing the cells to their final destination. This may involve *chemotaxis* or *chemorepulsion,* the secretion of a soluble chemical that attracts or repels migrating cells, respectively, or *pathway guidance,* the laying down of adhesive or repellent molecules in the extracellular matrix or on cell surfaces to guide the migrating cells along the right paths. Then, once a migrating cell has reached its destination, it must recognize and join other cells of the appropriate type to assemble into a tissue. How this latter process occurs can be studied if cells of different embryonic tissues are artificially mingled, after which they often spontaneously sort out to restore a more normal arrangement, as we discuss next.

Dissociated Vertebrate Cells Can Reassemble into Organized Tissues Through Selective Cell–Cell Adhesion

Unlike adult vertebrate tissues, which are difficult to dissociate, embryonic vertebrate tissues are easily dissociated. This is usually done by treating the tissue with low concentrations of a proteolytic enzyme such as trypsin, sometimes combined with the removal of extracellular Ca^{2+} and Mg^{2+} with a divalent-cation chelator (such as EDTA). These reagents disrupt the protein–protein interactions (many of which are divalent-cation-dependent) that hold cells together. Remarkably, the dissociated cells often reassemble *in vitro* into structures that resemble the original tissue. Such findings reveal that tissue structure is not just a product of history; it is actively maintained and stabilized by the system of affinities that cells have for one another and for the extracellular matrix.

A striking example of this phenomenon is seen when dissociated cells from two embryonic vertebrate organs, such as the liver and the retina, are mixed together and artificially formed into a pellet: the mixed aggregates gradually sort out according to their organ of origin. More generally, disaggregated cells are found to adhere more readily to aggregates of their own organ than to aggregates of other organs. Evidently there are cell–cell recognition systems that make cells of the same differentiated tissue preferentially adhere to one another; these adhesive preferences are presumably important in stabilizing tissue architecture.

Cells adhere to each other and to the extracellular matrix through cell-surface proteins called **cell adhesion molecules (CAMs)**—a category that includes the transmembrane adhesion proteins we have already discussed. CAMs can be *cell–cell adhesion molecules* or *cell–matrix adhesion molecules.* Some CAMs are Ca^{2+}-dependent, whereas others are Ca^{2+}-independent. The Ca^{2+}-dependent CAMs seem to be primarily responsible for the tissue-specific cell–cell adhesion seen in early vertebrate embryos, explaining why these cells can be disaggregated with Ca^{2+}-chelating agents.

CAMs were initially identified by making antibodies against cell-surface molecules and then testing the antibodies for their ability to inhibit cell–cell adhesion in a test tube. Those rare antibodies that inhibit the adhesion were then used to characterize and isolate the adhesion molecule recognized by the antibodies.

Cadherins Mediate Ca^{2+}-dependent Cell–Cell Adhesion

The **cadherins** are the major CAMs responsible for Ca^{2+}-dependent cell–cell adhesion in vertebrate tissues. The first three cadherins that were discovered were named according to the main tissues in which they were found: *E-cadherin* is present on many types of epithelial cells; *N-cadherin* on nerve, muscle, and lens cells; and *P-cadherin* on cells in the placenta and epidermis. All are also found in various other tissues; N-cadherin, for example, is expressed in fibroblasts, and E-cadherin is expressed in parts of the brain. These and other **classical cadherins** are related in sequence throughout their extracellular and intracellular domains. There are also a large number of **nonclassical cadherins**, with more than 50 expressed in the brain alone. The nonclassical cadherins include proteins with known adhesive function, such as the desmosomal cadherins discussed earlier and the diverse *protocadherins* found in the brain. They also include proteins that appear to have nonadhesive functions, such as *T-cadherin,* which lacks a transmembrane domain and is attached to the plasma membrane of nerve and muscle cells by a glycosylphosphatidylinositol (GPI) anchor, and the *Fat protein,* which was first identified as the product of a tumor-suppressor gene in *Drosophila*. Together, the classical and nonclassical cadherin proteins constitute the **cadherin superfamily** (Table 19–3).

Cadherins are expressed in both invertebrates and vertebrates. Virtually all vertebrate cells seem to express one or more cadherins, according to the cell type. They are the main adhesion molecules holding cells together in early embryonic tissues. In culture, the removal of extracellular Ca^{2+} or treatment with anti-cadherin antibodies disrupts embryonic tissues, and, if cadherin-mediated adhesion is left intact, antibodies against other adhesion molecules have little effect. Mutations that inactivate the function of E-cadherin cause mouse embryos to fall apart and die early in development.

Most cadherins are single-pass transmembrane glycoproteins about 700–750 amino acids long. Structural studies suggest that they associate in the plasma membrane to form dimers or larger oligomers. The large extracellular part of the polypeptide chain is usually folded into five or six *cadherin repeats,*

TABLE 19–3 Some Members of the Cadherin Superfamily

NAME	MAIN LOCATION	JUNCTION ASSOCIATION	PHENOTYPE WHEN INACTIVATED IN MICE
Classical cadherins			
E-cadherin	epithelia	adherens junctions	die at blastocyst stage; embryos fail to undergo compaction
N-cadherin	neurons, heart, skeletal muscle, lens, and fibroblasts	adherens junctions and chemical synapses	embryos die from heart defects
P-cadherin	placenta, epidermis, breast epithelium	adherens junctions	abnormal mammary gland development
VE-cadherin	endothelial cells	adherens junctions	abnormal vascular development (apoptosis of endothelial cells)
Nonclassical cadherins			
Desmocollin	skin	desmosomes	unknown
Desmoglein	skin	desmosomes	blistering skin disease due to loss of keratinocyte cell–cell adhesion
T-cadherin	neurons, muscle	none	unknown
Fat (in *Drosophila*)	epithelia and CNS	none	enlarged imaginal discs and tumors
Protocadherins	neurons	chemical synapses	unknown

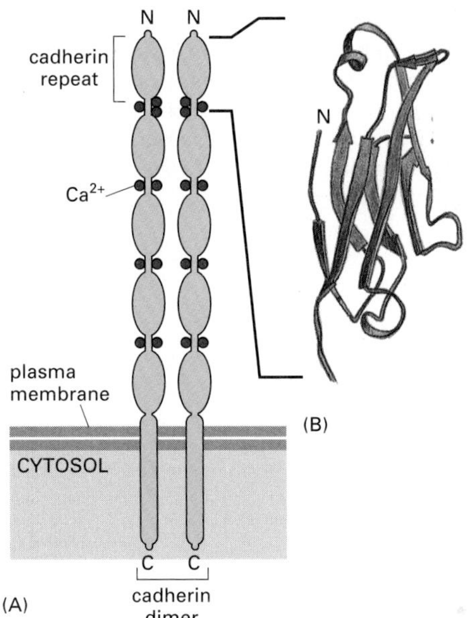

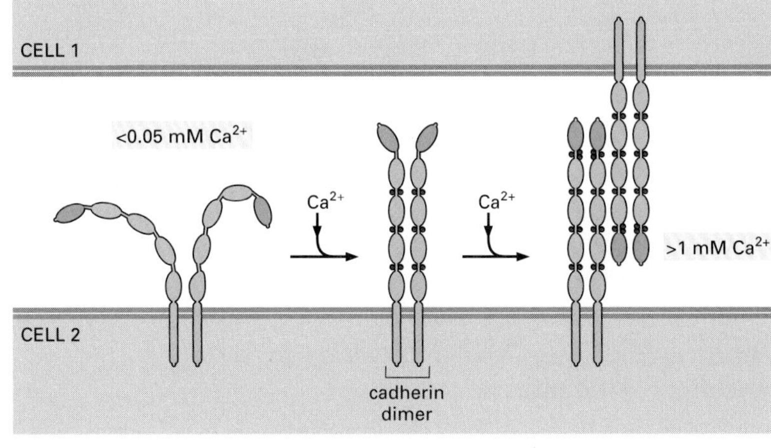

Figure 19–24 The structure and function of cadherins. (A) A classical cadherin molecule. The protein is a homodimer, with the extracellular part of each polypeptide folded into five cadherin repeats. There are Ca^{2+}-binding sites between each pair of repeats. (B) The crystal structure of a single cadherin repeat, which resembles an immunoglobulin (Ig) domain. (C) The influence of extracellular Ca^{2+}. As the amount of Ca^{2+} increases, the extracellular parts of the cadherin chains become more rigid. When enough Ca^{2+} is bound, the cadherin dimer extends from the surface, where it can bind to a cadherin dimer on a neighboring cell. If Ca^{2+} is removed, the extracellular part of the protein becomes floppy and is degraded by proteolytic enzymes.

which are structurally related to immunoglobulin (Ig) domains (Figure 19–24A and B). The crystal structures of E- and N-cadherin have helped to explain the importance of Ca^{2+} binding for cadherin function. The Ca^{2+} ions are positioned between each pair of cadherin repeats, locking the repeats together to form a stiff, rodlike structure: the more Ca^{2+} ions that are bound, the more rigid the structure is. If Ca^{2+} is removed, the extracellular part of the protein becomes floppy and is rapidly degraded by proteolytic enzymes (Figure 19–24C).

Cadherins Have Crucial Roles in Development

E-cadherin is the best-characterized cadherin. It is usually concentrated in adherens junctions in mature epithelial cells, where it helps connect the cortical actin cytoskeletons of the cells it holds together (see Figure 19–9B). E-cadherin is also the first cadherin expressed during mammalian development. It helps cause compaction, an important morphological change that occurs at the eight-cell stage of mouse embryo development. During compaction, the loosely attached cells, called *blastomeres,* become tightly packed together and joined by intercellular junctions. Antibodies against E-cadherin block blastomere compaction, whereas antibodies that react with various other cell-surface molecules on these cells do not.

It seems likely that cadherins are also crucial in later stages of vertebrate development, since their appearance and disappearance correlate with major morphogenetic events in which tissues segregate from one another. As the neural tube forms and pinches off from the overlying ectoderm, for example, neural tube cells lose E-cadherin and acquire other cadherins, including N-cadherin, while the cells in the overlying ectoderm continue to express E-cadherin (Figure 19–25). Then, when the neural crest cells migrate away from the neural tube, these cadherins become scarcely detectable, and another cadherin (cadherin-7) appears that helps hold the migrating cells together as loosely associated cell groups. Finally, when the cells aggregate to form a ganglion, they re-express N-cadherin (see Figure 19–23).

If N-cadherin is overexpressed in the emerging neural crest cells, the cells fail to escape from the neural tube. Thus, not only do cell groups that originate

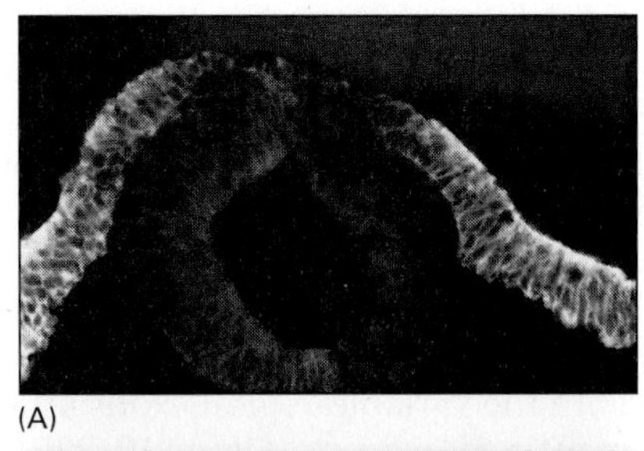
(A)

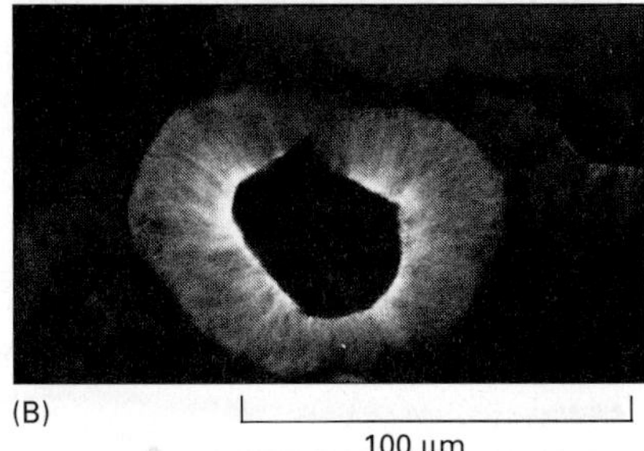

(B)

Figure 19–25 The distribution of E-cadherin and N-cadherin in the developing nervous system. Immunofluorescence micrographs of a cross section of a chick embryo showing the developing neural tube labeled with antibodies against (A) E-cadherin and (B) N-cadherin. Note that the overlying ectoderm cells express only E-cadherin, while the cells in the neural tube have lost E-cadherin and have acquired N-cadherin. See also Figure 19–10. (Courtesy of Kohei Hatta and Masatoshi Takeichi.)

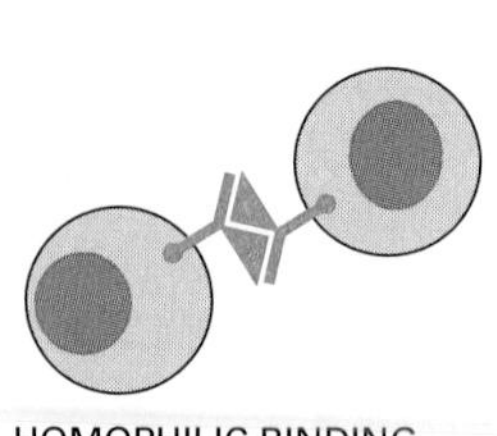

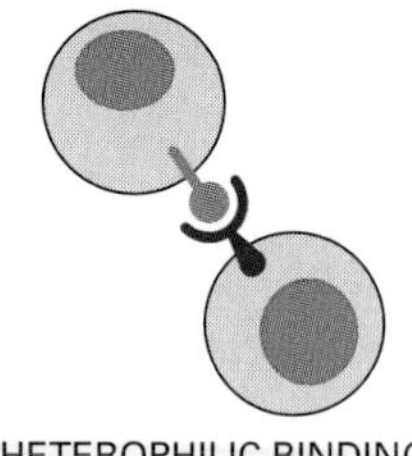

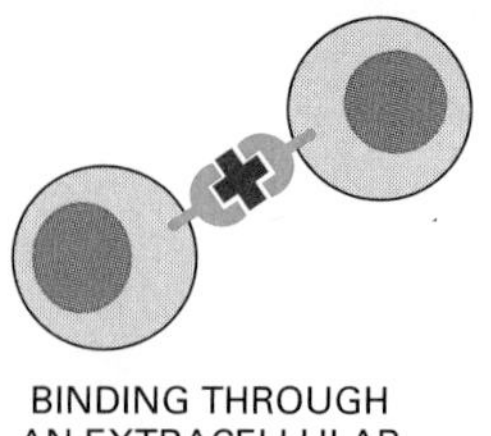

Figure 19–26 Three mechanisms by which cell-surface molecules can mediate cell–cell adhesion. Although all of these mechanisms can operate in animals, the one that depends on an extracellular linker molecule seems to be the least common.

from one cell layer exhibit distinct patterns of cadherin expression when separating from one another, but these switches in cadherin expression seem to be intimately involved in the separation process.

Cadherins Mediate Cell–Cell Adhesion by a Homophilic Mechanism

How do cell–cell adhesion molecules such as the cadherins bind cells together? Three possibilities are illustrated in Figure 19–26: (1) in *homophilic binding,* molecules on one cell bind to other molecules of the same kind on adjacent cells; (2) in *heterophilic binding,* the molecules on one cell bind to molecules of a different kind on adjacent cells; (3) in *linker-dependent binding,* cell-surface receptors on adjacent cells are linked to one another by secreted multivalent linker molecules. Although all three mechanisms have been found to operate in animals, cadherins usually link cells by the homophilic mechanism. In a line of cultured fibroblasts called *L cells,* for example, the cells neither express cadherins nor adhere to one another. When L cells are transfected with DNA encoding E-cadherin, the transfected cells become adherent to one another by a Ca^{2+}-dependent mechanism, and the adhesion is inhibited by anti-E-cadherin antibodies. Since cadherin proteins can bind directly to one another and the transfected cells do not bind to untransfected L cells, one can conclude that E-cadherin binds cells together through the interaction of two E-cadherin molecules on different cells.

If L cells expressing different cadherins are mixed together, they sort out and aggregate separately, indicating that different cadherins preferentially bind to their own type (Figure 19–27A), mimicking what happens when cells derived from tissues that express different cadherins are mixed together. A similar segregation of cells occurs if L cells expressing different amounts of the same cadherin are mixed together (Figure 19–27B). It therefore seems likely that both qualitative and quantitative differences in the expression of cadherins have a role in organizing tissues.

In the nervous system especially, there are many different cadherins, each with a distinct but overlapping pattern of expression (Figure 19–28A). As they are concentrated at synapses, they are thought to have a role in synapse formation and stabilization. Some of the nonclassical cadherins, such as the protocadherins, are strong candidates for helping to determine the specificity of synaptic connections. Like antibodies, they differ in their N-terminal (variable) regions but are identical in their C-terminal (constant) regions. The extracellular variable region and intracellular constant region are encoded by separate exons, with the variable-region exons arranged in tandem arrays upstream of the constant-region exons (Figure 19–28B). The diversity of protocadherins is generated by a combination of differential promoter usage and alternative RNA splicing, rather than by site-specific recombination as occurs in antibody diversification (discussed in Chapter 24).

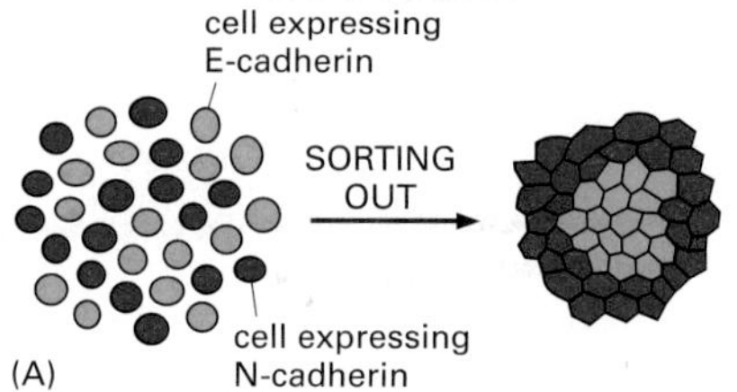

Figure 19–27 Cadherin-dependent cell sorting. Cells in culture can sort themselves out according to the type and level of cadherins they express. This can be visualized by labeling different populations of cells with dyes of different colors. (A) Cells expressing N-cadherin sort out from cells expressing E-cadherin. (B) Cells expressing high levels of E-cadherin sort out from cells expressing low levels of E-cadherin.

Cadherins Are Linked to the Actin Cytoskeleton by Catenins

Most cadherins, including all classical and some nonclassical ones, function as transmembrane adhesion proteins that indirectly link the actin cytoskeletons of the cells they join together. This arrangement occurs in adherens junctions (see

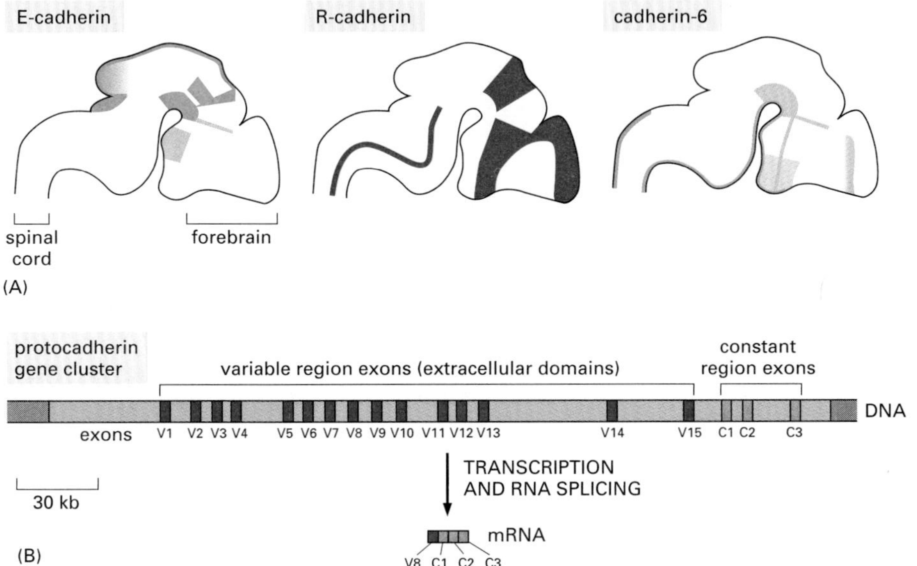

Figure 19–28 Cadherin diversity in the central nervous system. (A) Expression patterns for three classical cadherins in the embryonic mouse brain. (B) The arrangement of exons that encode the members of one of the three known protocadherin families of nonclassical cadherins in humans. Each of the variable-region exons encodes the extracellular region of one type of transmembrane protocadherin protein, but all of the protocadherins encoded by this cluster share the same cytoplasmic tail, which is encoded by the three constant-region exons C1, C2, and C3. A diversity of proteins is generated from the gene cluster by the use of different promoters and by alternative RNA splicing.

Figure 19–9B). The highly conserved cytoplasmic tail of these cadherins interacts indirectly with actin filaments by means of a group of intracellular anchor proteins called **catenins** (Figure 19–29). This interaction is essential for efficient cell–cell adhesion, as classical cadherins that lack their cytoplasmic domain cannot hold cells strongly together.

As discussed earlier, the nonclassical cadherins that form desmosomes interact with intermediate filaments, rather than with actin filaments. Their cytoplasmic domain binds to a different set of intracellular anchor proteins, which in turn bind to intermediate filaments (see Figure 19–11D).

Some cells can regulate the adhesive activity of their cadherins. This regulation may be important for the cellular rearrangements that occur within epithelia when these cell sheets change their shape and organization during animal development (see Figure 19–10). The molecular basis of this regulation is uncertain but may involve the phosphorylation of anchor proteins attached to the cytoplasmic tail of the cadherins.

Some cadherins can help transmit signals to the cell interior. Vascular endothelial cadherin (VE-cadherin), for example, not only mediates adhesion between endothelial cells but also is required for endothelial cell survival. Although endothelial cells that do not express VE-cadherin still adhere to one another via N-cadherin, they do not survive (see Table 19–3, p. 1082). Their survival depends on an extracellular signal protein called *vascular endothelial growth factor (VEGF),* which binds to a receptor tyrosine kinase (discussed in Chapter 15) that uses VE-cadherin as a co-receptor.

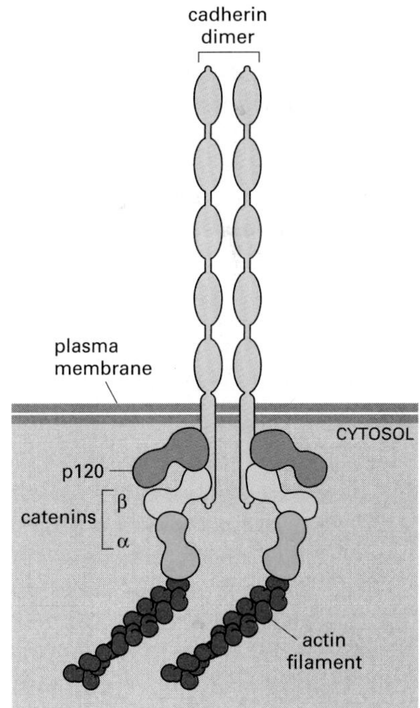

Figure 19–29 The linkage of classical cadherins to actin filaments. The cadherins are coupled indirectly to actin filaments by the anchor proteins α-catenin and β-catenin. A third intracellular protein, called p120, also binds to the cadherin cytoplasmic tail and regulates cadherin function. β-catenin has a second, and very important, function in intracellular signaling, as we discuss in Chapter 15 (see Figure 15–72).

Selectins Mediate Transient Cell–Cell Adhesions in the Bloodstream

White blood cells lead a nomadic life, moving to and fro between the bloodstream and the tissues, and this necessitates special adhesive properties. These properties depend on **selectins**. Selectins are cell-surface carbohydrate-binding proteins *(lectins)* that mediate a variety of transient, Ca^{2+}-dependent, cell–cell adhesion interactions in the bloodstream. There are at least three types: *L-selectin* on white blood cells, *P-selectin* on blood platelets and on endothelial cells that have been locally activated by an inflammatory response, and *E-selectin* on activated endothelial cells. Each selectin is a transmembrane protein with a highly conserved lectin domain that binds to a specific oligosaccharide on another cell (Figure 19–30A).

Selectins have an important role in binding white blood cells to endothelial cells lining blood vessels, thereby enabling the blood cells to migrate out of the

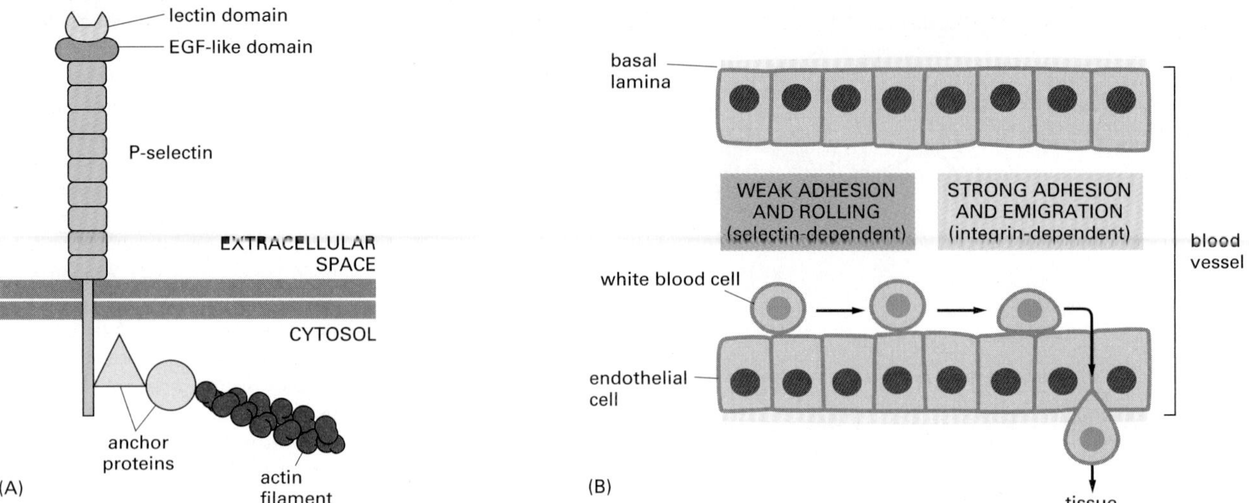

Figure 19–30 The structure and function of selectins. (A) The structure of P-selectin. The selectin attaches to the actin cytoskeleton through anchor proteins that are still poorly characterized. (B) How selectins and integrins mediate the cell–cell adhesions required for a white blood cell to migrate out of the bloodstream into a tissue.

bloodstream into a tissue. In a lymphoid organ, the endothelial cells express oligosaccharides that are recognized by L-selectin on lymphocytes, causing the lymphocytes to loiter and become trapped. Conversely, at sites of inflammation, the endothelial cells switch on expression of selectins, which recognize the oligosaccharides on white blood cells and platelets, flagging the cells down to help deal with the local emergency. Selectins do not act alone, however; they collaborate with integrins, which strengthen the binding of the blood cells to the endothelium. The cell–cell adhesions mediated by both selectins and integrins are heterophilic (see Figure 19–26): selectins bind to specific oligosaccharides on glycoproteins and glycolipids, while integrins bind to specific proteins.

Selectins and integrins act in sequence to let white blood cells leave the bloodstream and enter tissues. The selectins mediate a weak adhesion because the binding of the lectin domain of the selectin to its carbohydrate ligand is of low affinity. This allows the white blood cell to adhere weakly and reversibly to the endothelium, rolling along the surface of the blood vessel propelled by the flow of blood. The rolling continues until the blood cell activates its integrins (discussed later), now causing the cell to bind strongly to the endothelial cell surface and to crawl out of the blood vessel between adjacent endothelial cells (Figure 19–30B).

Members of the Immunoglobulin Superfamily of Proteins Mediate Ca^{2+}-independent Cell–Cell Adhesion

Cadherins, selectins, and integrins all depend on extracellular Ca^{2+} (or Mg^{2+} for some integrins) to function in cell adhesion. The molecules responsible for Ca^{2+}-independent cell–cell adhesion belong mainly to the large and ancient *immunoglobulin (Ig) superfamily* of proteins. These proteins contain one or more Ig-like domains that are characteristic of antibody molecules (discussed in Chapter 24). One of the best-studied examples is the **neural cell adhesion molecule (N-CAM)**, which is expressed by a variety of cell types, including most nerve cells. N-CAM is the most prevalent of the Ca^{2+}-independent cell–cell adhesion molecules in vertebrates, and, like cadherins, it is thought to bind cells together by a homophilic mechanism (between N-CAM molecules on adjacent cells). Some Ig-like cell–cell adhesion proteins, however, use a heterophilic mechanism. *Intercellular adhesion molecules (ICAMs)* on endothelial cells, for example, bind to integrins on blood cells when blood cells migrate out of the bloodstream, as just discussed.

There are at least 20 forms of N-CAM, all generated by alternative splicing of an RNA transcript produced from a single gene. In all forms, the large extracellular part of the polypeptide chain is folded into five Ig-like domains (Figure 19–31). Some forms of N-CAM carry an unusually large quantity of sialic acid (with chains containing hundreds of repeating sialic acid units). By virtue of their negative charge, these long polysialic acid chains hinder cell adhesion, and there is increasing evidence that N-CAM heavily loaded with sialic acid serves to prevent adhesion, rather than cause it.

Although cadherins and Ig family members are frequently expressed on the same cells, the adhesions mediated by cadherins are much stronger, and they are largely responsible for holding cells together, segregating cell collectives into discrete tissues, and maintaining tissue integrity. N-CAM and other members of the Ig family seem to contribute more to the fine-tuning of these adhesive interactions during development and regeneration. In the developing rodent pancreas, for example, the formation of the islets of Langerhans requires cell aggregation, followed by cell sorting. Whereas inhibition of cadherin function prevents cell aggregation and islet formation, loss of N-CAM only impairs the cell sorting process, so that disorganized islets form.

Similarly, whereas mutant mice that lack N-cadherin die early in development, mutant mice that lack N-CAM develop relatively normally, although they do have some defects in neural development. Mutations in other genes that encode Ig-like cell adhesion proteins, however, can cause more severe neural defects. *L1* gene mutations in humans, for example, cause mental retardation and other neurological defects resulting from abnormalities in the migration of nerve cells and their axons.

The importance of Ig-like cell adhesion proteins in connecting the neurons of the developing nervous system has been demonstrated dramatically in *Drosophila*. An N-CAM-like protein called *fasciclin III (FAS3)* is expressed transiently on some motor neurons, as well as on the muscle cells they normally innervate. If FAS3 is genetically removed from these neurons, they fail to recognize their muscle targets and do not make synapses with them. Conversely, if motor neurons that normally do not express FAS3 are made to express this protein, they now synapse with FAS3-expressing muscle cells to which they normally do not connect. It seems that FAS3 mediates these synaptic connections by a homophilic "matchmaking" mechanism.

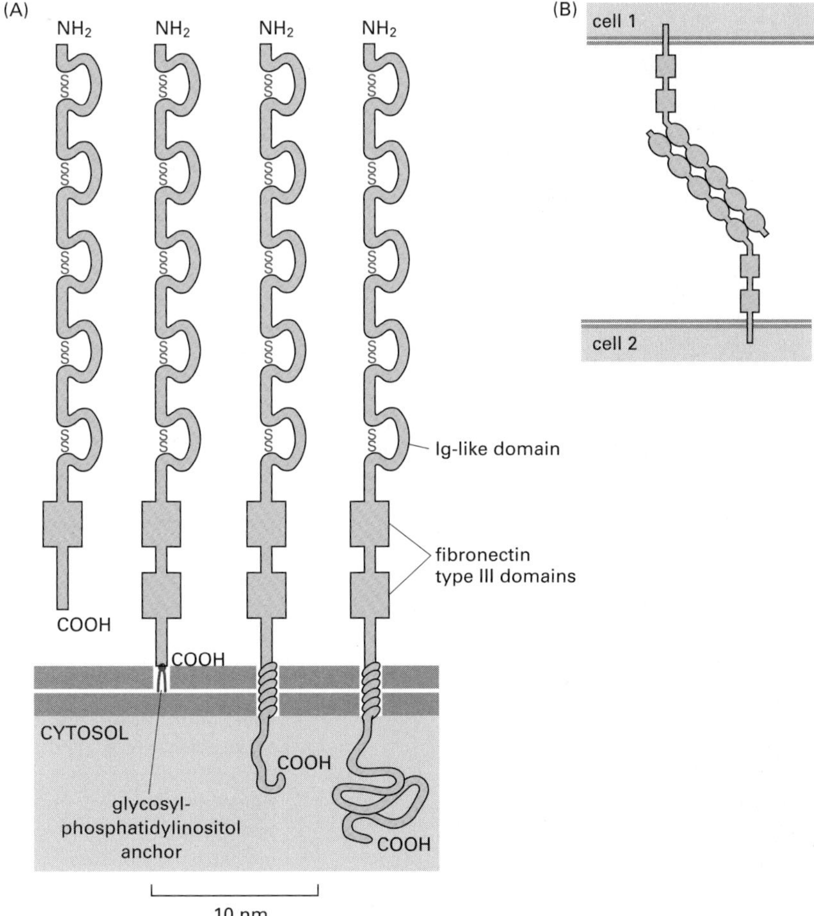

Figure 19–31 The cell adhesion protein N-CAM. (A) Four forms of N-CAM. The extracellular part of the polypeptide chain in each case is folded into five Ig-like domains (and one or two other domains called fibronectin type III repeats). Disulfide bonds (*red*) connect the ends of each loop that forms an Ig-like domain. (B) A model for the homophilic interaction that allows N-CAM to mediate cell–cell adhesion.

Like the cadherins, some Ig-like proteins do more than just bind cells together. They can also transmit signals to the cell interior. Some forms of N-CAM in nerve cells, for example, associate with *Src family* cytoplasmic tyrosine kinases (discussed in Chapter 15), which relay signals onward by phosphorylating intracellular proteins on tyrosines. Other Ig family members are transmembrane tyrosine phosphatases (discussed in Chapter 15) that help guide growing axons to their target cells, presumably by dephosphorylating specific intracellular proteins.

Multiple Types of Cell-Surface Molecules Act in Parallel to Mediate Selective Cell–Cell Adhesion

A single type of cell utilizes multiple molecular mechanisms in adhering to other cells. Some of these mechanisms involve organized cell junctions, while others do not (Figure 19–32). Each cell in a multicellular animal contains an assortment of cell-surface receptors that enables the cell to respond specifically to a complementary set of soluble extracellular signal molecules, such as hormones and growth factors. Likewise, each cell in a tissue has a particular combination (and concentration) of cell-surface adhesion molecules that enables it to bind in its own characteristic way to other cells and to the extracellular matrix. And just as receptors for soluble extracellular signal molecules generate intracellular signals that alter the cell's behavior, so too can cell adhesion molecules, although the signaling mechanisms they use are generally not as well understood.

Unlike receptors for soluble signal molecules, which bind their specific ligand with high affinity, the receptors that bind to molecules on cell surfaces or in the extracellular matrix usually do so with relatively low affinity. These low-affinity receptors rely on the enormous increase in binding strength gained through the simultaneous binding of multiple receptors to multiple ligands on an opposing cell or in the adjacent matrix. One could call this the "Velcro principle."

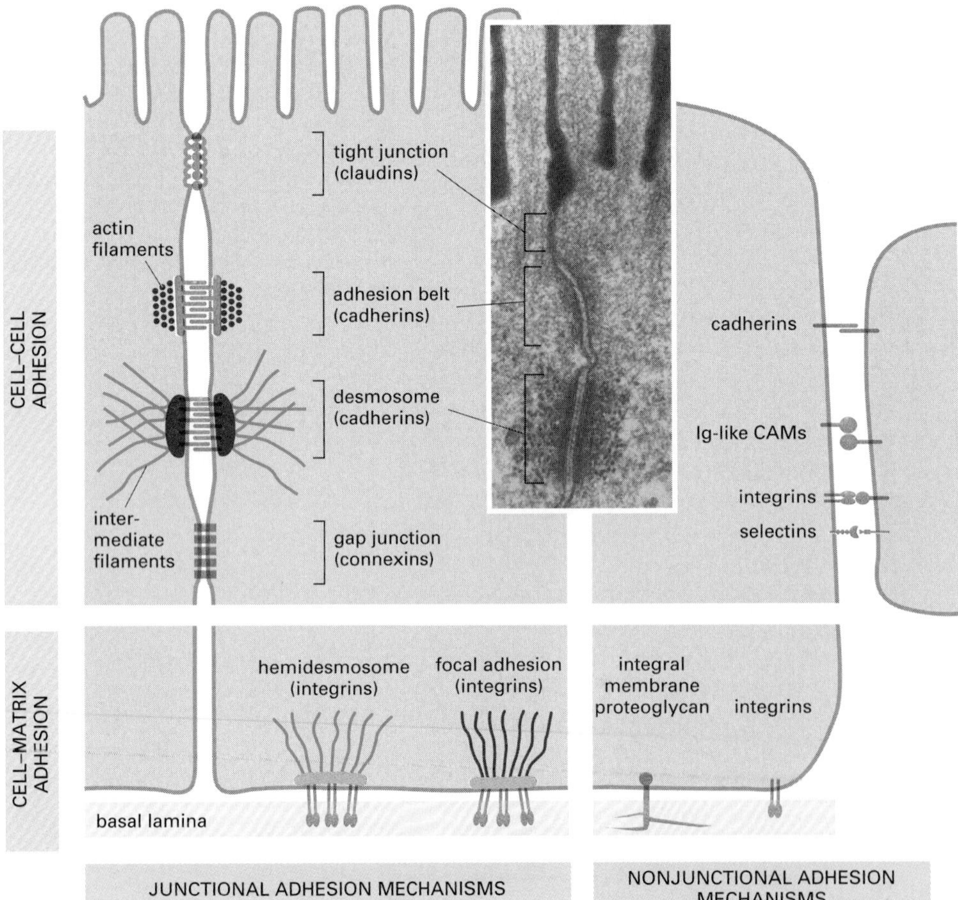

Figure 19–32 A summary of the junctional and nonjunctional adhesive mechanisms used by animal cells in binding to one another and to the extracellular matrix. The junctional mechanisms are shown in epithelial cells, while the nonjunctional mechanisms are shown in nonepithelial cells. A junctional adhesion is operationally defined as one that can be seen as a specialized region of contact by conventional or freeze-fracture electron microscopy. A nonjunctional adhesion shows no such obvious specialized structure. Note that the integrins and cadherins are involved in both nonjunctional and junctional cell–cell (cadherins) and cell–matrix (integrins) contacts. Cadherins generally mediate homophilic interactions, whereas integrins mediate heterophilic interactions (see Figure 19–26). The cadherins, integrins, and selectins act as transmembrane adhesion molecules and depend on extracellular divalent cations to function; for this reason, most cell–cell and cell–matrix contacts are divalent-cation-dependent. On blood cells, selectins and integrins can also act as heterophilic cell–cell adhesion molecules: the selectins bind to carbohydrate, while the cell-binding integrins bind to members of the Ig superfamily. The integrins and integral membrane proteoglycans that mediate nonjunctional adhesion to the extracellular matrix are discussed later. (Insert courtesy of Daniel S. Friend.)

We have seen, however, that the interaction of the extracellular binding domains of these cell-surface molecules is not enough to ensure cell adhesion. At least in the case of cadherins and, as we shall see, integrins, the adhesion molecules must also attach (via anchor proteins) to the cytoskeleton inside the cell. The cytoskeleton is thought to assist and stabilize the lateral clustering of the adhesion molecules to facilitate multipoint binding. The cytoskeleton is also required to enable the adhering cell to exert traction on the adjacent cell or matrix (and vice versa). Thus, the mixture of specific types of cell–cell adhesion molecules present on any two cells, as well as their concentration, cytoskeletal linkages, and distribution on the cell surface, determine the total affinity with which the two cells bind to each other.

Nonjunctional Contacts May Initiate Cell–Cell Adhesions That Junctional Contacts Then Orient and Stabilize

We have seen that adhesive contacts between cells play a crucial part in organizing the formation of tissues and organs in developing embryos or in adult tissues undergoing repair after injury. Most often, these contacts do not involve the formation of organized intercellular junctions that show up as specialized structures in the electron microscope. The interacting plasma membranes are simply seen to come close together and run parallel, separated by a space of 10–20 nm. This type of "nonjunctional" contact may be optimal for cell locomotion—close enough to provide traction and to allow transmembrane adhesion proteins to interact, but not so tight, or so solidly anchored to the cytoskeleton, as to immobilize the cell.

A reasonable hypothesis is that nonjunctional cell–cell adhesion proteins initiate cell–cell adhesions, which are then oriented and stabilized by the assembly of full-blown intercellular junctions. Many of the transmembrane proteins involved can diffuse in the plane of the plasma membrane and, in this or other ways, can be recruited to sites of cell–cell (and cell–matrix) contact, enabling nonjunctional adhesions to enlarge and mature into junctional adhesions. This has been demonstrated for some integrins and cadherins, which help initiate cell adhesion and then later become integral parts of cell junctions. The migrating tip of an axon, for example, has an even distribution of cadherins on its surface, which helps it adhere to other cells along the migration pathway. It also has an intracellular pool of cadherins in vesicles just under the plasma membrane. When the axon reaches its target cell, it is thought to release the intracellular cadherin molecules onto the cell surface, where they help form a stable contact, which matures into a chemical synapse.

As discussed earlier, antibodies against adherens junction proteins block the formation of tight junctions, as well as adherens junctions, suggesting that the assembly of one type of junction can be a prerequisite for the formation of another. An increasing number of monoclonal antibodies and peptide fragments have been produced that can block a single type of cell adhesion molecule. Moreover, an increasing number of genes encoding these cell-surface proteins have been identified, creating new opportunities for manipulating the adhesive machinery of cells in culture and in experimental animals. It is now possible, therefore, to inactivate the various cell–cell adhesion proteins in combinations—a requirement for deciphering the rules of cell–cell recognition and binding used to build complex tissues.

Summary

Cells dissociated from various tissues of vertebrate embryos preferentially reassociate with cells from the same tissue when they are mixed together. This tissue-specific recognition process in vertebrates is mediated mainly by a family of Ca^{2+}-dependent cell–cell adhesion proteins called cadherins, which hold cells together by a homophilic interaction between these transmembrane proteins on adjacent cells. For this interaction to be effective, the cytoplasmic part of the cadherins must be linked to the cytoskeleton by cytoplasmic anchor proteins called catenins.

Two other families of transmembrane adhesion proteins have major roles in cell–cell adhesion. Selectins function in transient Ca^{2+}-dependent cell–cell adhesions in the bloodstream by binding to specific oligosaccharides on the surface of another cell. Members of the immunoglobulin superfamily, including N-CAM, mediate Ca^{2+}-independent cell–cell adhesion processes that are especially important during neural development.

Even a single cell type uses multiple molecular mechanisms in adhering to other cells (and to the extracellular matrix). Thus, the specificity of cell–cell (and cell–matrix) adhesion seen in embryonic development must result from the integration of several different adhesion systems, of which some are associated with specialized cell junctions, while others are not.

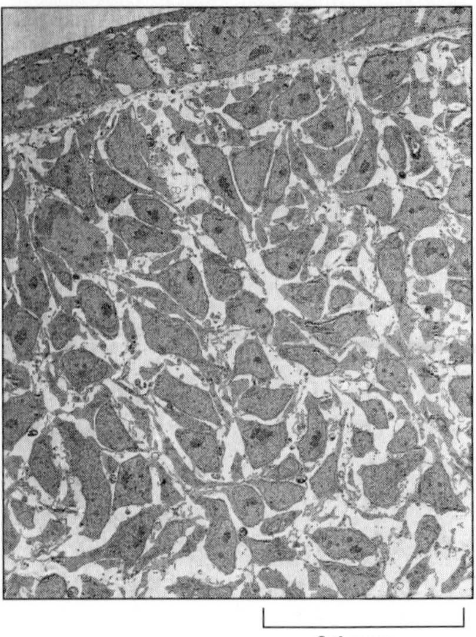

Figure 19–33 Cells surrounded by spaces filled with extracellular matrix. The particular cells shown in this low-power electron micrograph are those in an embryonic chick limb bud. The cells have not yet acquired their specialized characteristics. (Courtesy of Cheryll Tickle.)

THE EXTRACELLULAR MATRIX OF ANIMALS

Tissues are not made up solely of cells. A substantial part of their volume is *extracellular space,* which is largely filled by an intricate network of macromolecules constituting the **extracellular matrix** (Figure 19–33). This matrix is composed of a variety of proteins and polysaccharides that are secreted locally and assembled into an organized meshwork in close association with the surface of the cell that produced them.

Whereas we have discussed cell junctions chiefly in the context of epithelial tissues, our account of the extracellular matrix concentrates on connective tissues (Figure 19–34). The extracellular matrix in connective tissue is frequently more plentiful than the cells it surrounds, and it determines the tissue's physical properties. Connective tissues form the framework of the vertebrate body, but the amounts found in different organs vary greatly—from cartilage and bone, in which they are the major component, to brain and spinal cord, in which they are only minor constituents.

Variations in the relative amounts of the different types of matrix macromolecules and the way in which they are organized in the extracellular matrix give rise to an amazing diversity of forms, each adapted to the functional requirements of the particular tissue. The matrix can become calcified to form the rock-hard structures of bone or teeth, or it can form the transparent matrix of the cornea, or it can adopt the ropelike organization that gives tendons their enormous tensile strength. At the interface between an epithelium and connective tissue, the matrix forms a basal lamina (see Figure 19–34), which is important in controlling cell behavior.

The vertebrate extracellular matrix was once thought to serve mainly as a relatively inert scaffold to stabilize the physical structure of tissues. But now it is clear that the matrix has a far more active and complex role in regulating the behavior of the cells that contact it, influencing their survival, development, migration, proliferation, shape, and function. The extracellular matrix has a correspondingly complex molecular composition. Although our understanding of

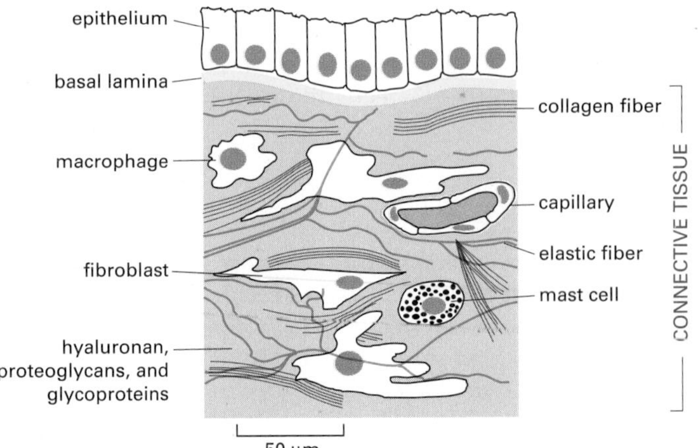

Figure 19–34 The connective tissue underlying an epithelium. This tissue contains a variety of cells and extracellular matrix components. The predominant cell type is the fibroblast, which secretes abundant extracellular matrix.

its organization is still incomplete, there has been rapid progress in characterizing many of its major components.

We focus on the extracellular matrix of vertebrates, but the origins of the extracellular matrix are very ancient and virtually all multicellular organisms, make it; examples include the cuticles of worms and insects, the shells of mollusks, and, as we discuss later, the cell walls of plants.

The Extracellular Matrix Is Made and Oriented by the Cells Within It

The macromolecules that constitute the extracellular matrix are mainly produced locally by cells in the matrix. As we discuss later, these cells also help to organize the matrix: the orientation of the cytoskeleton inside the cell can control the orientation of the matrix produced outside. In most connective tissues, the matrix macromolecules are secreted largely by cells called **fibroblasts** (Figure 19–35). In certain specialized types of connective tissues, such as cartilage and bone, however, they are secreted by cells of the fibroblast family that have more specific names: *chondroblasts,* for example, form cartilage, and *osteoblasts* form bone.

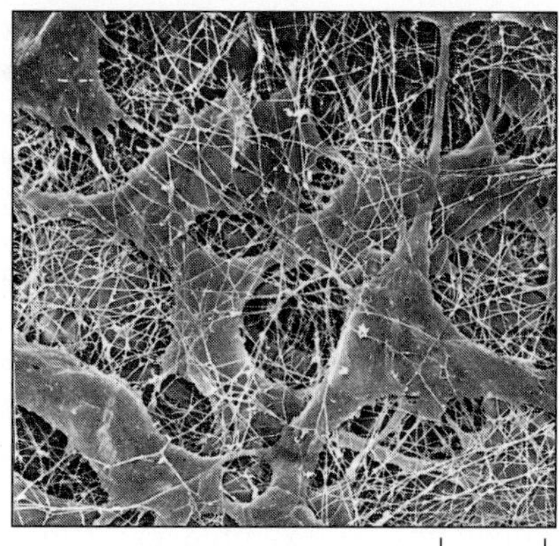

Figure 19–35 Fibroblasts in connective tissue. This scanning electron micrograph shows tissue from the cornea of a rat. The extracellular matrix surrounding the fibroblasts is composed largely of collagen fibrils (there are no elastic fibers in the cornea). The glycoproteins, hyaluronan, and proteoglycans, which normally form a hydrated gel filling the interstices of the fibrous network, have been removed by enzyme and acid treatment. (From T. Nishida et al., *Invest. Ophthalmol. Vis. Sci.* 29:1887–1890, 1988. © Association for Research in Vision and Opthalmology.)

Two main classes of extracellular macromolecules make up the matrix: (1) polysaccharide chains of the class called *glycosaminoglycans* (*GAGs*), which are usually found covalently linked to protein in the form of *proteoglycans,* and (2) fibrous proteins, including *collagen, elastin, fibronectin,* and *laminin,* which have both structural and adhesive functions. We shall see that the members of both classes come in a great variety of shapes and sizes.

The proteoglycan molecules in connective tissue form a highly hydrated, gel-like "ground substance" in which the fibrous proteins are embedded. The polysaccharide gel resists compressive forces on the matrix while permitting the rapid diffusion of nutrients, metabolites, and hormones between the blood and the tissue cells. The collagen fibers both strengthen and help organize the matrix, and rubberlike elastin fibers give it resilience. Finally, many matrix proteins help cells attach in the appropriate locations.

Glycosaminoglycan (GAG) Chains Occupy Large Amounts of Space and Form Hydrated Gels

Glycosaminoglycans (GAGs) are unbranched polysaccharide chains composed of repeating disaccharide units. They are called GAGs because one of the two sugars in the repeating disaccharide is always an amino sugar (*N*-acetylglucosamine or *N*-acetylgalactosamine), which in most cases is sulfated. The second sugar is usually a uronic acid (glucuronic or iduronic). Because there are sulfate or carboxyl groups on most of their sugars, GAGs are highly negatively charged (Figure 19–36). Indeed, they are the most anionic molecules produced by animal cells. Four main groups of GAGs are distinguished according to their

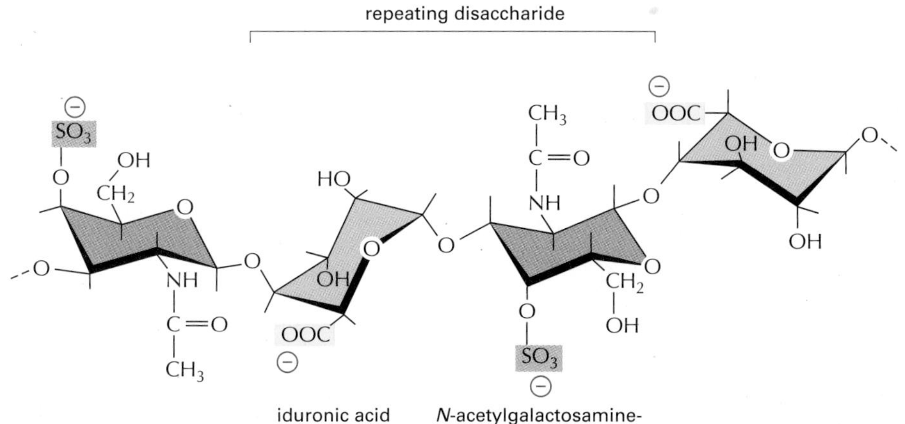

Figure 19–36 The repeating disaccharide sequence of a dermatan sulfate glycosaminoglycan (GAG) chain. These chains are typically 70–200 sugars long. There is a high density of negative charges along the chain resulting from the presence of both carboxyl and sulfate groups.

sugars, the type of linkage between the sugars, and the number and location of sulfate groups: (1) *hyaluronan,* (2) *chondroitin sulfate* and *dermatan sulfate,* (3) *heparan sulfate,* and (4) *keratan sulfate.*

Polysaccharide chains are too stiff to fold up into the compact globular structures that polypeptide chains typically form. Moreover, they are strongly hydrophilic. Thus, GAGs tend to adopt highly extended conformations that occupy a huge volume relative to their mass (Figure 19–37), and they form gels even at very low concentrations. Their high density of negative charges attracts a cloud of cations, most notably Na^+, that are osmotically active, causing large amounts of water to be sucked into the matrix. This creates a swelling pressure, or turgor, that enables the matrix to withstand compressive forces (in contrast to collagen fibrils, which resist stretching forces). The cartilage matrix that lines the knee joint, for example, can support pressures of hundreds of atmospheres in this way.

The GAGs in connective tissue usually constitute less than 10% of the weight of the fibrous proteins. But, because they form porous hydrated gels, the GAG chains fill most of the extracellular space, providing mechanical support to the tissue. In one rare human genetic disease, there is a severe deficiency in the synthesis of the dermatan sulfate disaccharide shown in Figure 19–36. The affected individuals have a short stature, prematurely aged appearance, and generalized defects in their skin, joints, muscles, and bones.

It should be emphasized, however, that, in invertebrates and plants, other types of polysaccharides often dominate the extracellular matrix. Thus, in higher plants, as we discuss later, cellulose (polyglucose) chains are packed tightly together in ribbonlike crystalline arrays to form the microfibrillar component of the cell wall. In insects, crustaceans, and other arthropods, chitin (poly-*N*-acetylglucosamine) similarly forms the main component of the exoskeleton. Together, cellulose and chitin are the most abundant biopolymers on Earth.

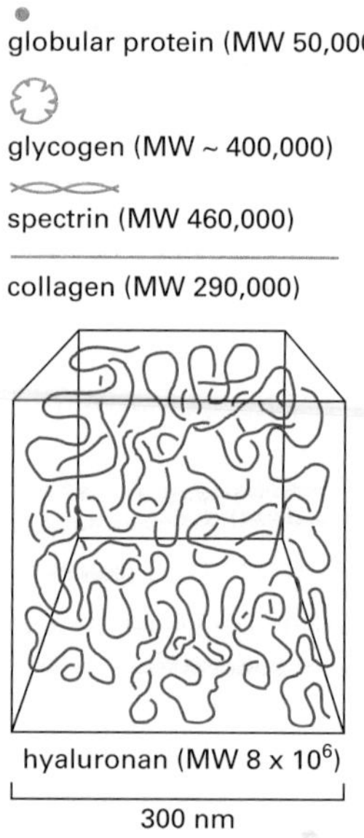

globular protein (MW 50,000)
glycogen (MW ~ 400,000)
spectrin (MW 460,000)
collagen (MW 290,000)
hyaluronan (MW 8 x 10^6)
300 nm

Figure 19–37 The relative dimensions and volumes occupied by various macromolecules. Several proteins, a glycogen granule, and a single hydrated molecule of hyaluronan are shown.

Hyaluronan Is Thought to Facilitate Cell Migration During Tissue Morphogenesis and Repair

Hyaluronan (also called *hyaluronic acid* or *hyaluronate*) is the simplest of the GAGs (Figure 19–38). It consists of a regular repeating sequence of up to 25,000 nonsulfated disaccharide units, is found in variable amounts in all tissues and fluids in adult animals, and is especially abundant in early embryos. Hyaluronan is not typical of the majority of GAGs. In contrast with all of the others, it contains no sulfated sugars, all its disaccharide units are identical, its chain length is enormous (thousands of sugar monomers), and it is not generally linked covalently to any core protein. Moreover, whereas other GAGs are synthesized inside the cell and released by exocytosis, hyaluronan is spun out directly from the cell surface by an enzyme complex embedded in the plasma membrane.

Hyaluronan is thought to have a role in resisting compressive forces in tissues and joints. It is also important as a space filler during embryonic development, where it can be used to force a change in the shape of a structure, as a small quantity expands with water to occupy a large volume (see Figure 19–37). Hyaluronan synthesized from the basal side of an epithelium, for example, often

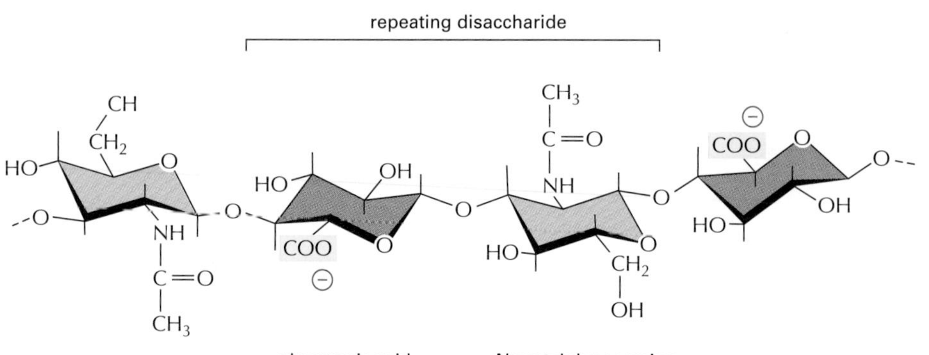

Figure 19–38 The repeating disaccharide sequence in hyaluronan, a relatively simple GAG. This ubiquitous molecule in vertebrates consists of a single long chain of up to 25,000 sugars. Note the absence of sulfate groups.

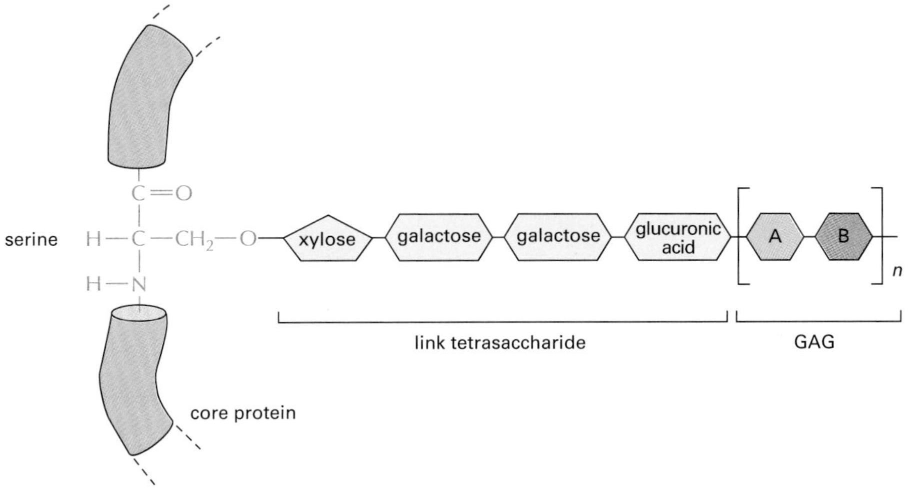

serine $H-C-CH_2-O-$ xylose — galactose — galactose — glucuronic acid — [A — B]$_n$

link tetrasaccharide GAG

core protein

Figure 19–39 The linkage between a GAG chain and its core protein in a proteoglycan molecule. A specific link tetrasaccharide is first assembled on a serine side chain. In most cases, it is unclear how the particular serine is selected, but it seems that a specific local conformation of the polypeptide chain, rather than a specific linear sequence of amino acids, is recognized. The rest of the GAG chain, consisting mainly of a repeating disaccharide unit, is then synthesized, with one sugar being added at a time. In chondroitin sulfate, the disaccharide is composed of D-glucuronic acid and *N*-acetyl-D-galactosamine; in heparan sulfate, it is D-glucosamine (or L-iduronic acid) and *N*-acetyl-D-glucosamine; in keratan sulfate, it is D-galactose and *N*-acetyl-D-glucosamine.

serves to create a cell-free space into which cells subsequently migrate; this occurs in the formation of the heart, the cornea, and several other organs. When cell migration ends, the excess hyaluronan is generally degraded by the enzyme *hyaluronidase*. Hyaluronan is also produced in large quantities during wound healing, and it is an important constituent of joint fluid, where it serves as a lubricant.

Many of the functions of hyaluronan depend on specific interactions with other molecules, including both proteins and proteoglycans—molecules consisting of GAG chains covalently linked to a protein. Some of these molecules that bind to hyaluronan are constituents of the extracellular matrix, while others are integral components of the surface of cells.

Proteoglycans Are Composed of GAG Chains Covalently Linked to a Core Protein

Except for hyaluronan, all GAGs are found covalently attached to protein in the form of **proteoglycans**, which are made by most animal cells. The polypeptide chain, or *core protein*, of a proteoglycan is made on membrane-bound ribosomes and threaded into the lumen of the endoplasmic reticulum. The polysaccharide chains are mainly assembled on this core protein in the Golgi apparatus. First, a special *link tetrasaccharide* is attached to a serine side chain on the core protein to serve as a primer for polysaccharide growth; then, one sugar at a time is added by specific glycosyl transferases (Figure 19–39). While still in the Golgi apparatus, many of the polymerized sugars are covalently modified by a sequential and coordinated series of reactions. Epimerizations alter the configuration of the substituents around individual carbon atoms in the sugar molecule; sulfations increase the negative charge.

Proteoglycans are usually easily distinguished from other glycoproteins by the nature, quantity, and arrangement of their sugar side chains. By definition, at least one of the sugar side chains of a proteoglycan must be a GAG. Whereas glycoproteins contain 1–60% carbohydrate by weight in the form of numerous relatively short, branched oligosaccharide chains, proteoglycans can contain as much as 95% carbohydrate by weight, mostly in the form of long, unbranched GAG chains, each typically about 80 sugars long. Proteoglycans can be huge. The proteoglycan *aggrecan*, for example, which is a major component of cartilage, has a mass of about 3×10^6 daltons with over 100 GAG chains. Other proteoglycans are much smaller and have only 1–10 GAG chains; an example is *decorin*, which is secreted by fibroblasts and has a single GAG chain (Figure 19–40).

In principle, proteoglycans have the potential for almost limitless heterogeneity. Even a single type of core protein can vary greatly in the number and types of attached GAG chains. Moreover, the underlying repeating pattern of disaccharides in each GAG can be modified by a complex pattern of sulfate groups. The heterogeneity of these GAGs makes it difficult to identify and

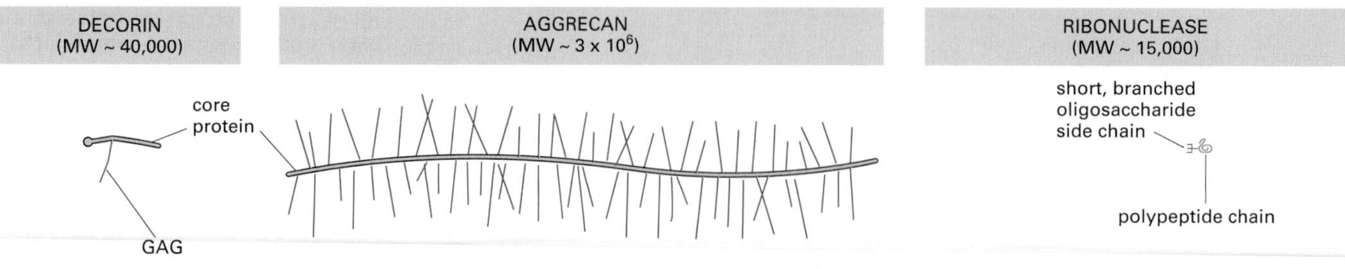

Figure 19–40 Examples of a small (decorin) and a large (aggrecan) proteoglycan found in the extracellular matrix. These two proteoglycans are compared with a typical secreted glycoprotein molecule, pancreatic ribonuclease B. All three are drawn to scale. The core proteins of both aggrecan and decorin contain oligosaccharide chains as well as the GAG chains, but these are not shown. Aggrecan typically consists of about 100 chondroitin sulfate chains and about 30 keratan sulfate chains linked to a serine-rich core protein of almost 3000 amino acids. Decorin "decorates" the surface of collagen fibrils, hence its name.

classify proteoglycans in terms of their sugars. The sequences of many core proteins have been determined with the aid of recombinant DNA techniques, and they, too, are extremely diverse. Although a few small families have been recognized, no common structural feature clearly distinguishes proteoglycan core proteins from other proteins, and many have one or more domains that are homologous to domains found in other proteins of the extracellular matrix or plasma membrane. Thus, it is probably best to regard proteoglycans as a diverse group of highly glycosylated glycoproteins whose functions are mediated by both their core proteins and their GAG chains.

Proteoglycans Can Regulate the Activities of Secreted Proteins

Given the great abundance and structural diversity of proteoglycan molecules, it would be surprising if their function were limited to providing hydrated space around and between cells. Their GAG chains, for example, can form gels of varying pore size and charge density; one possible function, therefore, is to serve as selective sieves to regulate the traffic of molecules and cells according to their size, charge, or both. Evidence suggests that a heparan sulfate proteoglycan called *perlecan* has this role in the basal lamina of the kidney glomerulus, which filters molecules passing into the urine from the bloodstream (discussed below).

Proteoglycans are thought to have a major role in chemical signaling between cells. They bind various secreted signal molecules, such as certain protein growth factors, and can enhance or inhibit their signaling activity. For example, the heparan sulfate chains of proteoglycans bind to *fibroblast growth factors* (*FGFs*), which stimulate a variety of cell types to proliferate; this interaction oligomerizes the growth factor molecules, enabling them to cross-link and activate their cell-surface receptors, which are transmembrane tyrosine kinases (see Figure 15–50B). Whereas in most cases the signal molecules bind to the GAG chains of the proteoglycan, this is not always so. Some members of the *transforming growth factor β (TGF-β)* family bind to the core proteins of several matrix proteoglycans, including decorin; binding to decorin inhibits the activity of the growth factors.

Proteoglycans also bind, and regulate the activities of, other types of secreted proteins, including proteolytic enzymes (proteases) and protease inhibitors. Binding to a proteoglycan could control the activity of a secreted protein in any of the following ways: (1) it could immobilize the protein close to the site where it is produced, thereby restricting its range of action; (2) it could sterically block the activity of the protein; (3) it could provide a reservoir of the protein for delayed release; (4) it could protect the protein from proteolytic degradation, thereby prolonging its action; (5) it could alter or concentrate the protein for more effective presentation to cell-surface receptors.

Proteoglycans are thought to act in all these ways to help regulate the activities of secreted proteins. An example of the last function occurs in inflammatory responses, in which heparan sulfate proteoglycans immobilize secreted chemotactic attractants called *chemokines* (discussed in Chapter 24) on the endothelial surface of a blood vessel at an inflammatory site. In this way, the chemokines remain there for a prolonged period, stimulating white blood cells to leave the bloodstream and migrate into the inflamed tissue.

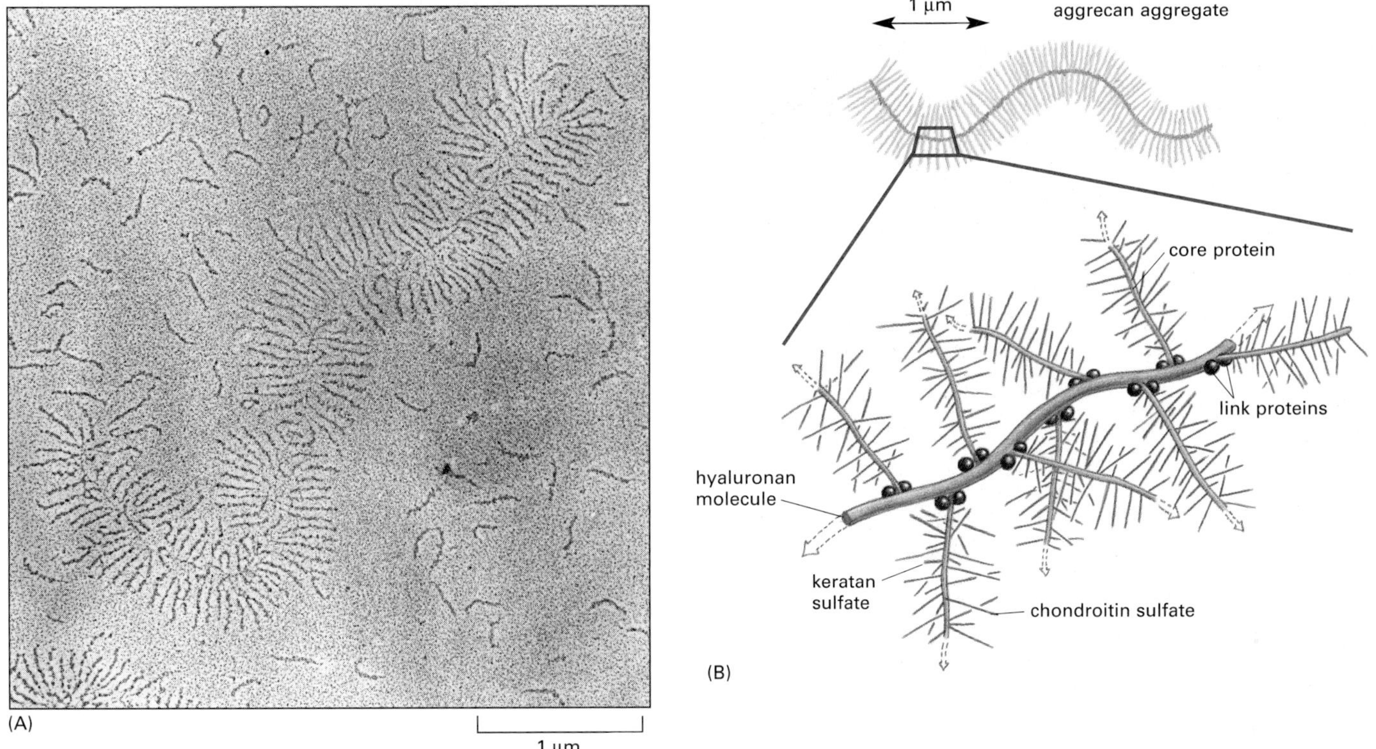

Figure 19–41 An aggrecan aggregate from fetal bovine cartilage. (A) An electron micrograph of an aggrecan aggregate shadowed with platinum. Many free aggrecan molecules are also visible. (B) A drawing of the giant aggrecan aggregate shown in (A). It consists of about 100 aggrecan monomers (each like the one shown in Figure 19–40) noncovalently bound to a single hyaluronan chain through two link proteins that bind both to the core protein of the proteoglycan and to the hyaluronan chain, thereby stabilizing the aggregate. The link proteins are members of a family of hyaluronan-binding proteins, some of which are cell-surface proteins. The molecular weight of such a complex can be 10^8 or more, and it occupies a volume equivalent to that of a bacterium, which is about 2×10^{-12} cm^3. (A, courtesy of Lawrence Rosenberg.)

GAG Chains May Be Highly Organized in the Extracellular Matrix

GAGs and proteoglycans can associate to form huge polymeric complexes in the extracellular matrix. Molecules of aggrecan, for example, the major proteoglycan in cartilage (see Figure 19–40), assemble with hyaluronan in the extracellular space to form aggregates that are as big as a bacterium (Figure 19–41).

Moreover, besides associating with one another, GAGs and proteoglycans associate with fibrous matrix proteins such as collagen and with protein meshworks such as the basal lamina, creating extremely complex structures. Decorin, which binds to collagen fibrils, is essential for collagen fiber formation; mice that cannot make decorin have fragile skin that has reduced tensile strength. The arrangement of proteoglycan molecules in living tissues is generally hard to determine. As the molecules are highly water-soluble, they may be washed out of the extracellular matrix when tissue sections are exposed to aqueous solutions during fixation. In addition, changes in pH, ionic, or osmotic conditions can drastically alter their conformation. Thus, specialized methods must be used to visualize them in tissues (Figure 19–42).

Cell-Surface Proteoglycans Act as Co-receptors

Not all proteoglycans are secreted components of the extracellular matrix. Some are integral components of plasma membranes and have their core protein either inserted across the lipid bilayer or attached to the lipid bilayer by a glycosylphosphatidylinositol (GPI) anchor. Some of these plasma membrane proteoglycans act as *co-receptors* that collaborate with conventional cell-surface receptor proteins, in both binding cells to the extracellular matrix and initiating

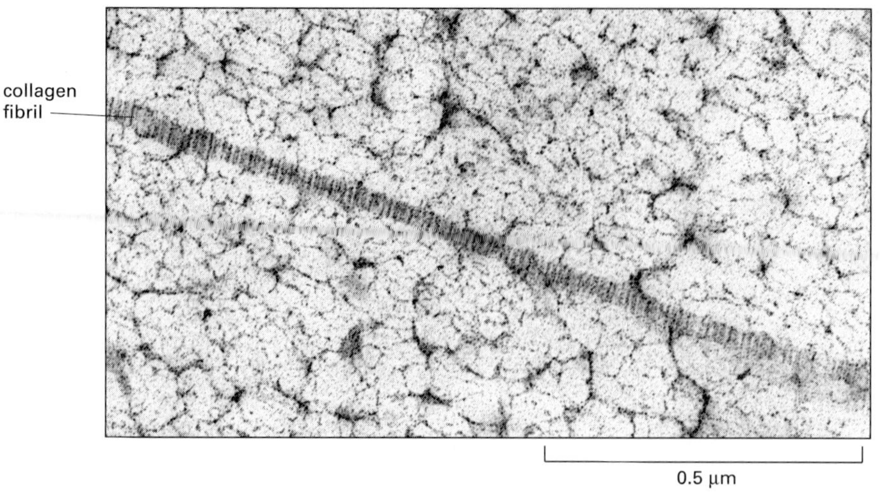

Figure 19–42 Proteoglycans in the extracellular matrix of rat cartilage. The tissue was rapidly frozen at –196°C, and fixed and stained while still frozen (a process called freeze substitution) to prevent the GAG chains from collapsing. In this electron micrograph, the proteoglycan molecules are seen to form a fine filamentous network in which a single striated collagen fibril is embedded. The more darkly stained parts of the proteoglycan molecules are the core proteins; the faintly stained threads are the GAG chains. (Reproduced from E.B. Hunziker and R.K. Schenk, *J. Cell Biol.* 98:277–282, 1985. © The Rockefeller University Press.)

the response of cells to some extracellular signal proteins. In addition, some conventional receptors have one or more GAG chains and are therefore proteoglycans themselves.

Among the best-characterized plasma membrane proteoglycans are the *syndecans,* which have a membrane-spanning core protein. The extracellular domains of these transmembrane proteoglycans carry up to three chondroitin sulfate and heparan sulfate GAG chains, while their intracellular domains are thought to interact with the actin cytoskeleton in the cell cortex.

Syndecans are located on the surface of many types of cells, including fibroblasts and epithelial cells, where they serve as receptors for matrix proteins. In fibroblasts, syndecans can be found in focal adhesions, where they modulate integrin function by interacting with fibronectin on the cell surface and with cytoskeletal and signaling proteins inside the cell. Syndecans also bind FGFs and present them to FGF receptor proteins on the same cell. Similarly, another plasma membrane proteoglycan, called *betaglycan,* binds TGF-β and may present it to TGF-β receptors.

The importance of proteoglycans as co-receptors is illustrated by the severe developmental defects that can occur when specific proteoglycans are inactivated by mutation. In *Drosophila,* for example, signaling by the secreted signal protein *Wingless* depends on the protein's binding to a specific heparan sulfate proteoglycan co-receptor called *Dally* on the target cell. In mutant flies deficient in Dally, Wingless signaling fails, and the severe developmental defects that result are similar to those that result from mutations in the *wingless* gene itself. In some tissues, inactivation of Dally also inhibits signaling by a secreted protein of the TGF-β family called *Decapentaplegic (DPP).*

Some of the proteoglycans discussed in this chapter are summarized in Table 19–4.

Collagens Are the Major Proteins of the Extracellular Matrix

The **collagens** are a family of fibrous proteins found in all multicellular animals. They are secreted by connective tissue cells, as well as by a variety of other cell types. As a major component of skin and bone, they are the most abundant proteins in mammals, constituting 25% of the total protein mass in these animals.

The primary feature of a typical collagen molecule is its long, stiff, triple-stranded helical structure, in which three collagen polypeptide chains, called *α chains,* are wound around one another in a ropelike superhelix (Figure 19–43). Collagens are extremely rich in proline and glycine, both of which are important in the formation of the triple-stranded helix. Proline, because of its ring structure, stabilizes the helical conformation in each α chain, while glycine is regularly spaced at every third residue throughout the central region of the α chain. Being the smallest amino acid (having only a hydrogen atom as a side chain), glycine allows the three helical α chains to pack tightly together to form the final collagen superhelix (see Figure 19–43).

TABLE 19–4 Some Common Proteoglycans

PROTEOGLYCAN	APPROXIMATE MOLECULAR WEIGHT OF CORE PROTEIN	TYPE OF GAG CHAINS	NUMBER OF GAG CHAINS	LOCATION	FUNCTIONS
Aggrecan	210,000	chondroitin sulfate + keratan sulfate	~130	cartilage	mechanical support; forms large aggregates with hyaluronan
Betaglycan	36,000	chondroitin sulfate/ dermatan sulfate	1	cell surface and matrix	binds TGF-β
Decorin	40,000	chondroitin sulfate/ dermatan sulfate	1	widespread in connective tissues	binds to type I collagen fibrils and TGF-β
Perlecan	600,000	heparan sulfate	2–15	basal laminae	structural and filtering function in basal lamina
Syndecan-1	32,000	chondroitin sulfate + heparan sulfate	1–3	epithelial cell surface	cell adhesion; binds FGF and other growth factors
Dally (in *Drosophila*)	60,000	heparan sulfate	1–3	cell surface	co-receptor for Wingless and Decapentaplegic signaling proteins

So far, about 25 distinct collagen α chains have been identified, each encoded by a separate gene. Different combinations of these genes are expressed in different tissues. Although in principle more than 10,000 types of triple-stranded collagen molecules could be assembled from various combinations of the 25 or so α chains, only about 20 types of collagen molecules have been found. The main types of collagen found in connective tissues are types I, II, III, V, and XI, type I being the principal collagen of skin and bone and by far the most common. These are the **fibrillar collagens**, or fibril-forming collagens, with the ropelike structure illustrated in Figure 19–43. After being secreted into the extracellular space, these collagen molecules assemble into higher-order polymers called *collagen fibrils,* which are thin structures (10–300 nm in diameter) many hundreds of micrometers long in mature tissues and clearly visible in electron micrographs (Figure 19–44; see also Figure 19–42). Collagen fibrils often aggregate into larger, cablelike bundles, several micrometers in diameter, which can be seen in the light microscope as *collagen fibers.*

Collagen types IX and XII are called *fibril-associated collagens* because they decorate the surface of collagen fibrils. They are thought to link these fibrils to one another and to other components in the extracellular matrix. Types IV and VII are *network-forming collagens.* Type IV molecules assemble into a feltlike sheet or meshwork that constitutes a major part of mature basal laminae, while type VII molecules form dimers that assemble into specialized structures called *anchoring fibrils.* Anchoring fibrils help attach the basal lamina of multilayered epithelia to the underlying connective tissue and therefore are especially abundant in the skin.

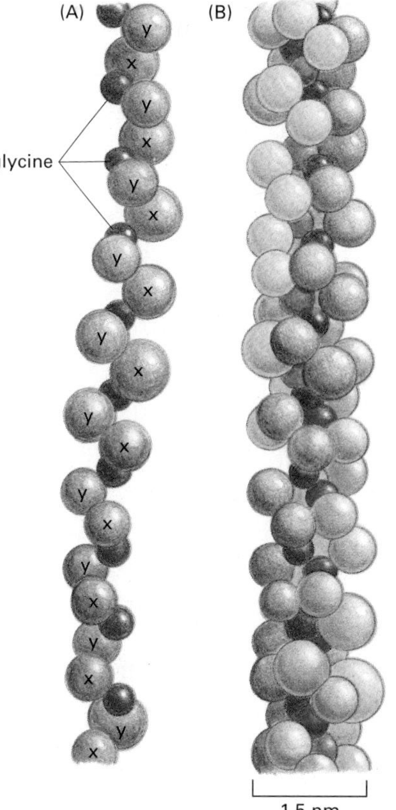

Figure 19–43 The structure of a typical collagen molecule. (A) A model of part of a single collagen α chain in which each amino acid is represented by a sphere. The chain is about 1000 amino acids long. It is arranged as a left-handed helix, with three amino acids per turn and with glycine as every third amino acid. Therefore, an α chain is composed of a series of triplet Gly-X-Y sequences, in which X and Y can be any amino acid (although X is commonly proline and Y is commonly hydroxyproline). (B) A model of part of a collagen molecule in which three α chains, each shown in a different color, are wrapped around one another to form a triple-stranded helical rod. Glycine is the only amino acid small enough to occupy the crowded interior of the triple helix. Only a short length of the molecule is shown; the entire molecule is 300 nm long. (From model by B.L. Trus.)

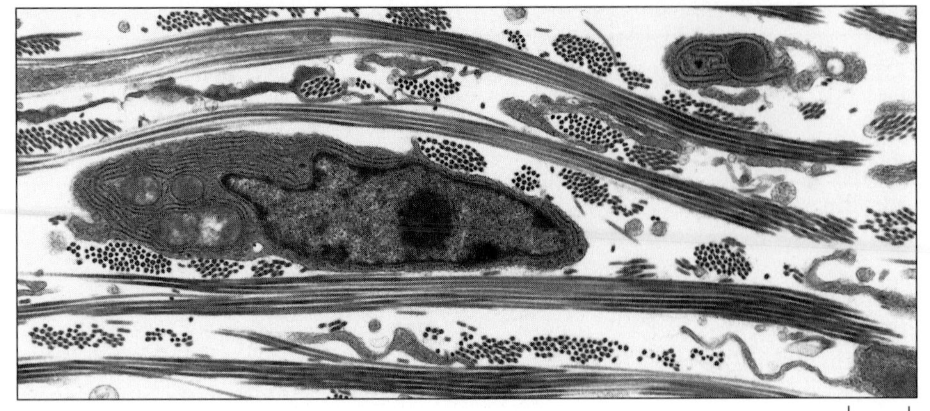

Figure 19–44 Fibroblast surrounded by collagen fibrils in the connective tissue of embryonic chick skin. In this electron micrograph, the fibrils are organized into bundles that run approximately at right angles to one another. Therefore, some bundles are oriented longitudinally, whereas others are seen in cross section. The collagen fibrils are produced by the fibroblasts, which contain abundant endoplasmic reticulum, where secreted proteins such as collagen are synthesized. (From C. Ploetz, E.I. Zycband, and D.E. Birk, *J. Struct. Biol.* 106:73–81, 1991. © Academic Press.)

There are also a number of "collagen-like" proteins, including type XVII, which has a transmembrane domain and is found in hemidesmosomes, and type XVIII, which is located in the basal laminae of blood vessels. Cleavage of the C-terminal domain of type XVIII collagen yields a peptide called *endostatin,* which inhibits new blood vessel formation and is therefore being investigated as an anticancer drug. Some of the collagen types discussed in this chapter are listed in Table 19–5.

Many proteins that contain a repeated pattern of amino acids have evolved by duplications of DNA sequences. The fibrillar collagens apparently arose in this way. Thus, the genes that encode the α chains of most of these collagens are very large (up to 44 kilobases in length) and contain about 50 exons. Most of the exons are 54, or multiples of 54, nucleotides long, suggesting that these collagens arose by multiple duplications of a primordial gene containing 54 nucleotides and encoding exactly 6 Gly-X-Y repeats (see Figure 19–43).

Collagens Are Secreted with a Nonhelical Extension at Each End

Individual collagen polypeptide chains are synthesized on membrane-bound ribosomes and injected into the lumen of the endoplasmic reticulum (ER) as

TABLE 19–5 Some Types of Collagen and Their Properties

	TYPE	MOLECULAR FORMULA	POLYMERIZED FORM	TISSUE DISTRIBUTION
Fibril-forming (fibrillar)	I	$[a1(I)]_2\alpha2$ (I)	fibril	bone, skin, tendons, ligaments, cornea, internal organs (accounts for 90% of body collagen)
	II	$[\alpha1(II)]_3$	fibril	cartilage, invertebral disc, notochord, vitreous humor of the eye
	III	$[\alpha1(III)]_3$	fibril	skin, blood vessels, internal organs
	V	$[\alpha1(V)]_2\alpha2(V)$ and $\alpha1(V)\alpha2(V)\alpha3(V)$	fibril (with type I)	as for type I
	XI	$\alpha1(XI)\alpha2(IX)\alpha3(XI)$	fibril (with type II)	as for type II
Fibril-associated	IX	$\alpha1(IX)\alpha2(IX)\alpha3(IX)$	lateral association with type II fibrils	cartilage
	XII	$[\alpha1(XII)]_3$	lateral association with some type I fibrils	tendons, ligaments, some other tissues
Network-forming	IV	$[\alpha1(IV)]_2\alpha2(IV)$	sheetlike network	basal lamina
	VII	$[\alpha1(VII)]_3$	anchoring fibrils	beneath stratified squamous epithelia
Transmembrane	XVII	$[\alpha1(XVII)]_3$	not known	hemidesmosomes
Others	XVIII	$[\alpha1(XVIII)]_3$	not known	basal lamina around blood vessels

Note that types I, IV, V, IX, and XI are each composed of two or three types of α chains, whereas types II, III, VII, XII, XVII, and XVIII are composed of only one type of α chain each. Only 11 types of collagen are shown, but about 20 types of collagen and about 25 types of α chains have been identified so far.

larger precursors, called *pro-α chains*. These precursors not only have the short amino-terminal signal peptide required to direct the nascent polypeptide to the ER, they also have additional amino acids, called *propeptides*, at both their N- and C-terminal ends. In the lumen of the ER, selected prolines and lysines are hydroxylated to form hydroxyproline and hydroxylysine, respectively, and some of the hydroxylysines are glycosylated. Each pro-α chain then combines with two others to form a hydrogen-bonded, triple-stranded, helical molecule known as *procollagen*.

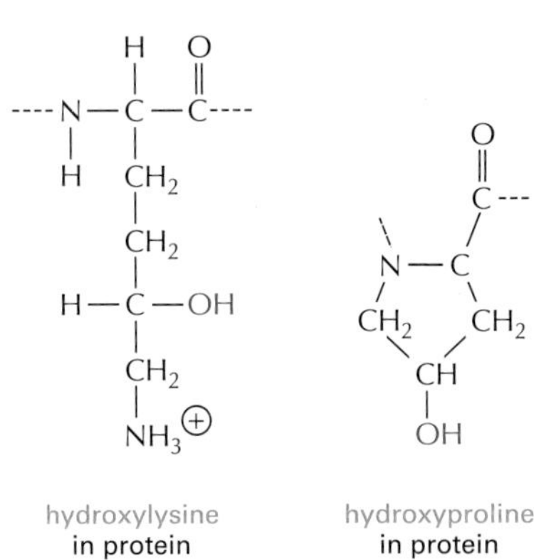

Figure 19–45 Hydroxylysine and hydroxyproline. These modified amino acids are common in collagen. They are formed by enzymes that act after the lysine and proline have been incorporated into procollagen molecules.

Hydroxylysines and *hydroxyprolines* (Figure 19–45) are infrequently found in other animal proteins, although hydroxyproline is abundant in some proteins in the plant cell wall. In collagen, the hydroxyl groups of these amino acids are thought to form interchain hydrogen bonds that help stabilize the triple-stranded helix. Conditions that prevent proline hydroxylation, such as a deficiency of ascorbic acid (vitamin C), have serious consequences. In *scurvy*, the disease caused by a dietary deficiency of vitamin C that was common in sailors until the nineteenth century, the defective pro-α chains that are synthesized fail to form a stable triple helix and are immediately degraded within the cell. Consequently, with the gradual loss of the preexisting normal collagen in the matrix, blood vessels become extremely fragile and teeth become loose in their sockets, implying that in these particular tissues the degradation and replacement of collagen occur relatively rapidly. In many other adult tissues, however, the turnover of collagen (and other extracellular matrix macromolecules) is thought to be very slow. In bone, to take an extreme example, collagen molecules persist for about 10 years before they are degraded and replaced. By contrast, most cell proteins have half-lives of hours or days.

After Secretion, Fibrillar Procollagen Molecules Are Cleaved to Collagen Molecules, Which Assemble into Fibrils

After secretion, the propeptides of the fibrillar procollagen molecules are removed by specific proteolytic enzymes outside the cell. This converts the procollagen molecules to collagen molecules, which assemble in the extracellular space to form much larger **collagen fibrils**. The propeptides have at least two functions. First, they guide the intracellular formation of the triple-stranded collagen molecules. Second, because they are removed only after secretion, they prevent the intracellular formation of large collagen fibrils, which could be catastrophic for the cell.

The process of fibril formation is driven, in part, by the tendency of the collagen molecules, which are more than a thousandfold less soluble than procollagen molecules, to self-assemble. The fibrils begin to form close to the cell surface, often in deep infoldings of the plasma membrane formed by the fusion of secretory vesicles with the cell surface. The underlying cortical cytoskeleton can therefore influence the sites, rates, and orientation of fibril assembly.

When viewed in an electron microscope, collagen fibrils have characteristic cross-striations every 67 nm, reflecting the regularly staggered packing of the individual collagen molecules in the fibril. After the fibrils have formed in the extracellular space, they are greatly strengthened by the formation of covalent cross-links between lysine residues of the constituent collagen molecules (Figure 19–46). The types of covalent bonds involved are found only in collagen and elastin. If cross-linking is inhibited, the tensile strength of the fibrils is drastically reduced; collagenous tissues become fragile, and structures such as skin, tendons, and blood vessels tend to tear. The extent and type of cross-linking vary

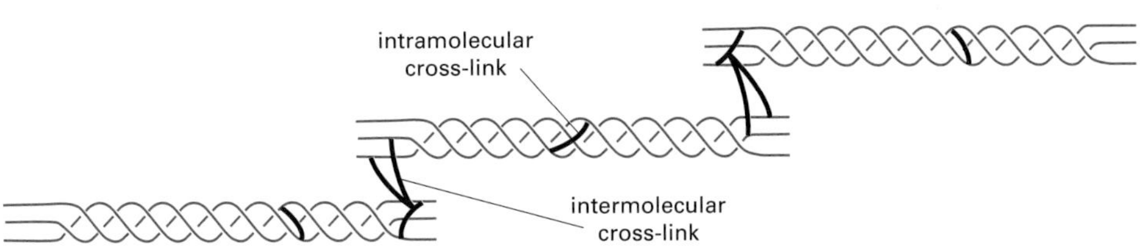

Figure 19–46 Cross-links formed between modified lysine side chains within a collagen fibril. Covalent intramolecular and intermolecular cross-links are formed in several steps. First, certain lysines and hydroxylysines are deaminated by the extracellular enzyme lysyl oxidase to yield highly reactive aldehyde groups. The aldehydes then react spontaneously to form covalent bonds with each other or with other lysines or hydroxylysines. Most of the cross-links form between the short nonhelical segments at each end of the collagen molecules.

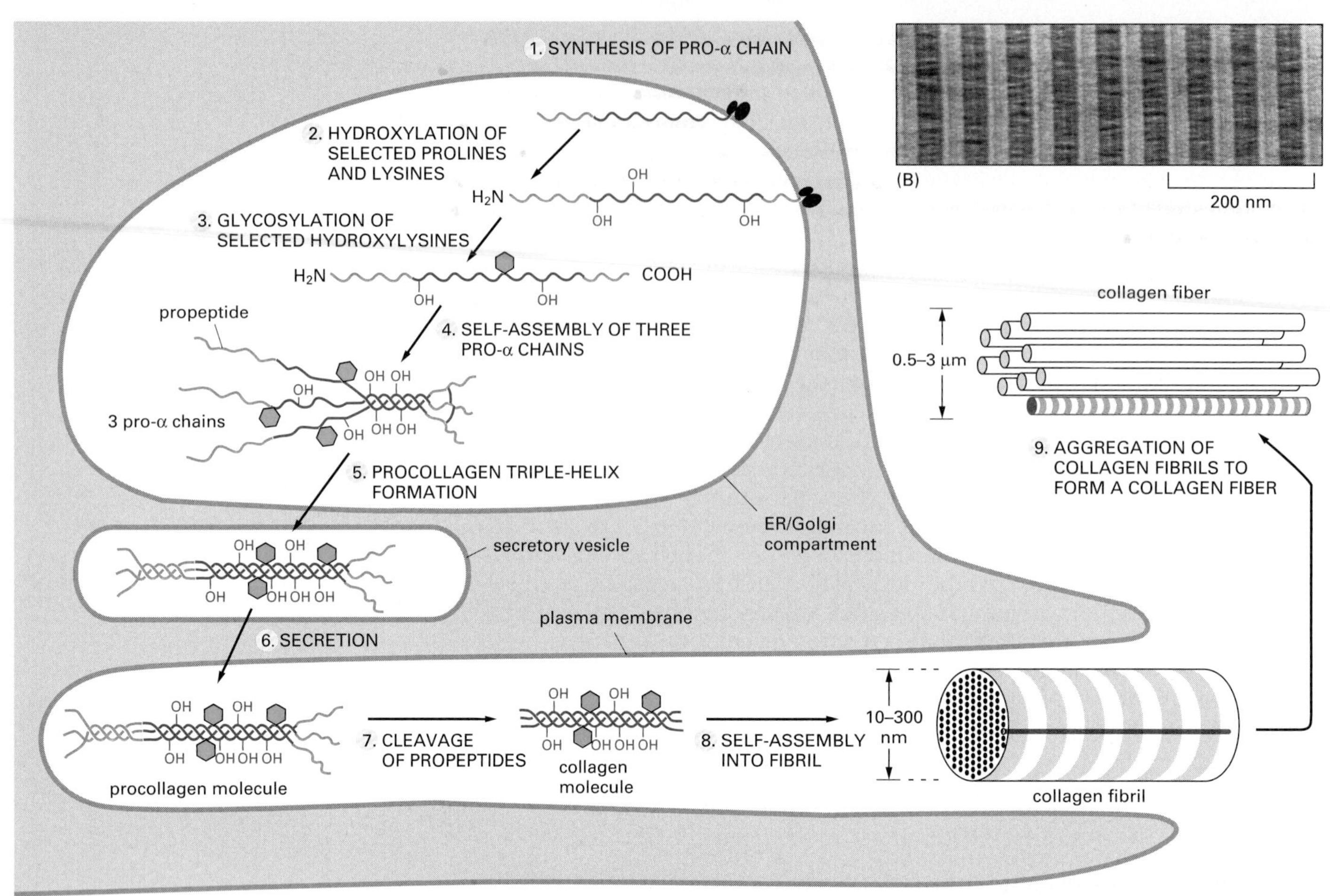

Figure 19–47 The intracellular and extracellular events in the formation of a collagen fibril. (A) Note that collagen fibrils are shown assembling in the extracellular space contained within a large infolding in the plasma membrane. As one example of how collagen fibrils can form ordered arrays in the extracellular space, they are shown further assembling into large collagen fibers, which are visible in the light microscope. The covalent cross-links that stabilize the extracellular assemblies are not shown. (B) Electron micrograph of a negatively stained collagen fibril reveals its typical striated appearance. (B, courtesy of Robert Horne.)

from tissue to tissue. Collagen is especially highly cross-linked in the Achilles tendon, for example, where tensile strength is crucial.

Figure 19–47 summarizes the various steps in the synthesis and assembly of collagen fibrils. Given the large number of enzymatic steps involved, it is not surprising that there are many human genetic diseases that affect fibril formation. Mutations affecting type I collagen cause *osteogenesis imperfecta,* characterized by weak bones that fracture easily. Mutations affecting type II collagen cause *chondrodysplasias*, characterized by abnormal cartilage, which leads to bone and joint deformities. Mutations affecting type III collagen cause *Ehlers–Danlos syndrome,* characterized by fragile skin and blood vessels and hypermobile joints.

Fibril-associated Collagens Help Organize the Fibrils

In contrast to GAGs, which resist compressive forces, collagen fibrils form structures that resist tensile forces. The fibrils have various diameters and are organized in different ways in different tissues. In mammalian skin, for example, they are woven in a wickerwork pattern so that they resist tensile stress in multiple directions. In tendons, they are organized in parallel bundles aligned along the major axis of tension. In mature bone and in the cornea, they are arranged in orderly plywoodlike layers, with the fibrils in each layer lying parallel to one another but nearly at right angles to the fibrils in the layers on either side. The same arrangement occurs in tadpole skin (Figure 19–48).

The connective tissue cells themselves must determine the size and arrangement of the collagen fibrils. The cells can express one or more genes for the different types of fibrillar procollagen molecules. But even fibrils composed of the same mixture of fibrillar collagen molecules have different arrangements in different tissues. How is this achieved? Part of the answer is that cells can regulate the disposition of the collagen molecules after secretion by guiding collagen fibril formation in close association with the plasma membrane (see Figure 19–46). In addition, as the spatial organization of collagen fibrils at least partly reflects their interactions with other molecules in the matrix, cells can influence this organization by secreting, along with their fibrillar collagens, different kinds and amounts of other matrix macromolecules.

Fibril-associated collagens, such as types IX and XII collagens, are thought to be especially important in this regard. They differ from fibrillar collagens in several ways.

1. Their triple-stranded helical structure is interrupted by one or two short nonhelical domains, which makes the molecules more flexible than fibrillar collagen molecules.
2. They are not cleaved after secretion and therefore retain their propeptides.
3. They do not aggregate with one another to form fibrils in the extracellular space. Instead, they bind in a periodic manner to the surface of fibrils formed by the fibrillar collagens. Type IX molecules bind to type-II-collagen-containing fibrils in cartilage, the cornea, and the vitreous of the eye (Figure 19–49), whereas type XII molecules bind to type-I-collagen-containing fibrils in tendons and various other tissues.

Fibril-associated collagens are thought to mediate the interactions of collagen fibrils with one another and with other matrix macromolecules. In this way, they have a role in determining the organization of the fibrils in the matrix.

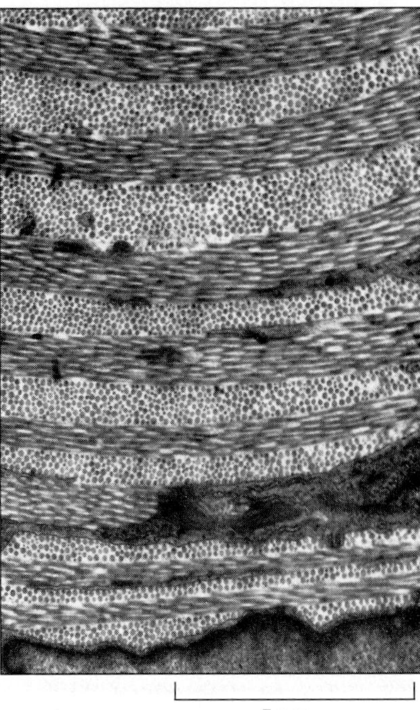

Figure 19–48 Collagen fibrils in the tadpole skin. This electron micrograph shows the plywoodlike arrangement of the fibrils. Successive layers of fibrils are laid down nearly at right angles to each other. This organization is also found in mature bone and in the cornea. (Courtesy of Jerome Gross.)

Cells Help Organize the Collagen Fibrils They Secrete by Exerting Tension on the Matrix

Cells interact with the extracellular matrix mechanically as well as chemically, with dramatic effects on the architecture of the tissue. Thus, for example, fibroblasts work on the collagen they have secreted, crawling over it and tugging on it—helping to compact it into sheets and draw it out into cables. When fibroblasts are mixed with a meshwork of randomly oriented collagen fibrils that form a gel in a culture dish, the fibroblasts tug on the meshwork, drawing in collagen from their surroundings and thereby causing the gel to contract to a small fraction of its initial volume. By similar activities, a cluster of fibroblasts surrounds itself with a capsule of densely packed and circumferentially oriented collagen fibers.

If two small pieces of embryonic tissue containing fibroblasts are placed far apart on a collagen gel, the intervening collagen becomes organized into a compact band of aligned fibers that connect the two explants (Figure 19–50). The

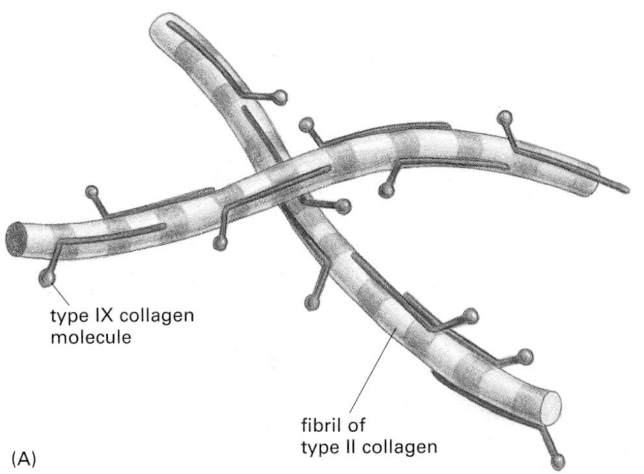

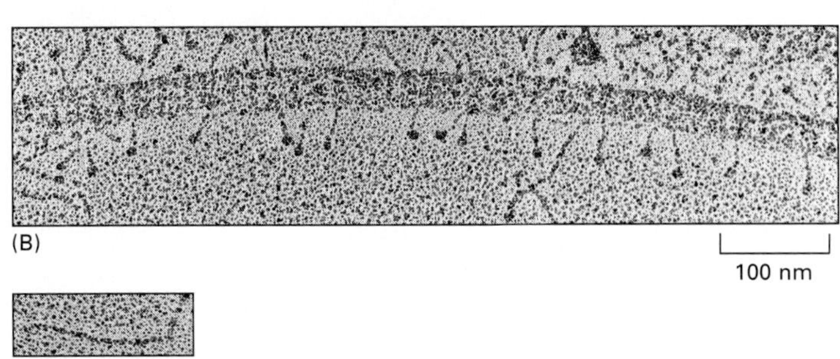

Figure 19–49 Type IX collagen. (A) Type IX collagen molecules binding in a periodic pattern to the surface of a fibril containing type II collagen. (B) Electron micrograph of a rotary-shadowed type-II-collagen-containing fibril in cartilage, sheathed in type IX collagen molecules. (C) An individual type IX collagen molecule. (B and C, from L. Vaughan et al., *J. Cell Biol.* 106:991–997, 1988. © The Rockefeller University Press.)

fibroblasts subsequently migrate out from the explants along the aligned collagen fibers. Thus, the fibroblasts influence the alignment of the collagen fibers, and the collagen fibers in turn affect the distribution of the fibroblasts. Fibroblasts presumably have a similar role in generating long-range order in the extracellular matrix inside the body—in helping to create tendons and ligaments, for example, and the tough, dense layers of connective tissue that ensheathe and bind together most organs.

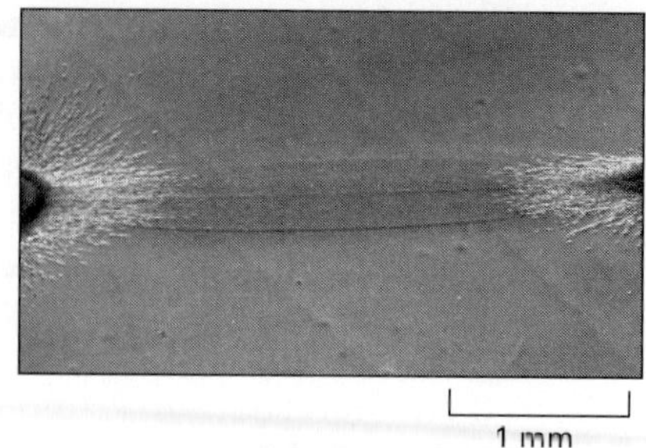

Figure 19–50 The shaping of the extracellular matrix by cells. This micrograph shows a region between two pieces of embryonic chick heart (rich in fibroblasts as well as heart muscle cells) that were cultured on a collagen gel for 4 days. A dense tract of aligned collagen fibers has formed between the explants, presumably as a result of the fibroblasts in the explants tugging on the collagen. (From D. Stopak and A.K. Harris, *Dev. Biol.* 90:383–398, 1982. © Academic Press.)

Elastin Gives Tissues Their Elasticity

Many vertebrate tissues, such as skin, blood vessels, and lungs, need to be both strong and elastic in order to function. A network of **elastic fibers** in the extracellular matrix of these tissues gives them the required resilience so that they can recoil after transient stretch (Figure 19–51). Elastic fibers are at least five times more extensible than a rubber band of the same cross-sectional area. Long, inelastic collagen fibrils are interwoven with the elastic fibers to limit the extent of stretching and prevent the tissue from tearing.

The main component of elastic fibers is **elastin**, a highly hydrophobic protein (about 750 amino acids long), which, like collagen, is unusually rich in proline and glycine but, unlike collagen, is not glycosylated and contains some hydroxyproline but no hydroxylysine. Soluble *tropoelastin* (the biosynthetic precursor of elastin) is secreted into the extracellular space and assembled into elastic fibers close to the plasma membrane, generally in cell-surface infoldings. After secretion, the tropoelastin molecules become highly cross-linked to one another, generating an extensive network of elastin fibers and sheets. The cross-links are formed between lysines by a mechanism similar to the one discussed earlier that operates in cross-linking collagen molecules.

The elastin protein is composed largely of two types of short segments that alternate along the polypeptide chain: hydrophobic segments, which are responsible for the elastic properties of the molecule; and alanine- and lysine-rich α-helical segments, which form cross-links between adjacent molecules. Each segment is encoded by a separate exon. There is still controversy, however, concerning the conformation of elastin molecules in elastic fibers and how the structure of these fibers accounts for their rubberlike properties. In one view, the elastin polypeptide chain, like the polymer chains in ordinary rubber, adopts a loose "random coil" conformation, and it is the random coil structure of the component molecules cross-linked into the elastic fiber network that allows the network to stretch and recoil like a rubber band (Figure 19–52).

Elastin is the dominant extracellular matrix protein in arteries, comprising 50% of the dry weight of the largest artery—the aorta. Mutations in the elastin gene causing a deficiency of the protein in mice or humans result in narrowing of the aorta or other arteries as a result of excessive proliferation of smooth

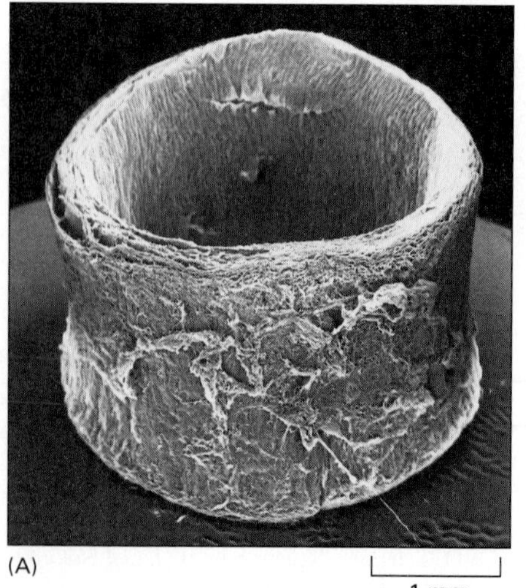

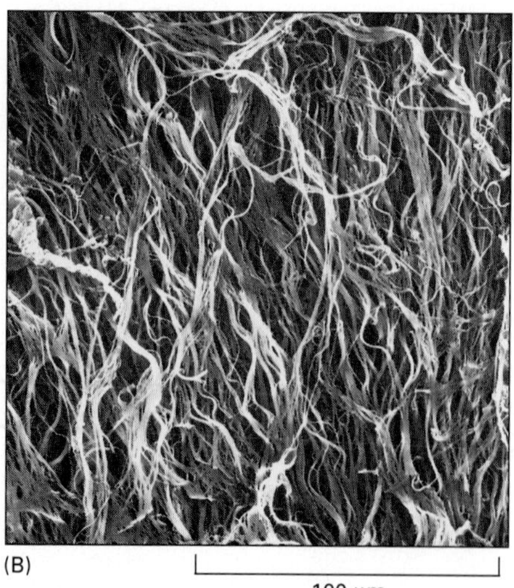

Figure 19–51 Elastic fibers. These scanning electron micrographs show (A) a low-power view of a segment of a dog's aorta and (B) a high-power view of the dense network of longitudinally oriented elastic fibers in the outer layer of the same blood vessel. All the other components have been digested away with enzymes and formic acid. (From K.S. Haas, S.J. Phillips, A.J. Comerota, and J.W. White, *Anat. Rec.* 230:86–96, 1991. © Wiley-Liss, Inc.)

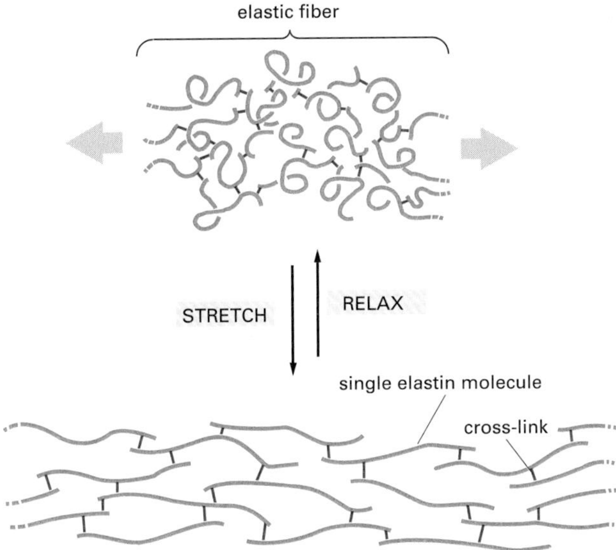

Figure 19–52 Stretching a network of elastin molecules. The molecules are joined together by covalent bonds *(red)* to generate a cross-linked network. In this model, each elastin molecule in the network can expand and contract as a random coil, so that the entire assembly can stretch and recoil like a rubber band.

muscle cells in the arterial wall. Apparently, the normal elasticity of an artery is required to restrain the proliferation of these cells.

Elastic fibers are not composed solely of elastin. The elastin core is covered with a sheath of *microfibrils,* each of which has a diameter of about 10 nm. Microfibrils are composed of a number of distinct glycoproteins, including the large glycoprotein *fibrillin,* which binds to elastin and is essential for the integrity of elastic fibers. Mutations in the fibrillin gene result in *Marfan's syndrome,* a relatively common human genetic disease affecting connective tissues that are rich in elastic fibers; in the most severely affected individuals, the aorta is prone to rupture. Microfibrils are thought to be important in the assembly of elastic fibers. They appear before elastin in developing tissues and seem to form a scaffold on which the secreted elastin molecules are deposited. As the elastin is deposited, the microfibrils become displaced to the periphery of the growing fiber.

Fibronectin Is an Extracellular Protein That Helps Cells Attach to the Matrix

The extracellular matrix contains a number of noncollagen proteins that typically have multiple domains, each with specific binding sites for other matrix macromolecules and for receptors on the surface of cells. These proteins therefore contribute to both organizing the matrix and helping cells attach to it. The first of them to be well characterized was **fibronectin**, a large glycoprotein found in all vertebrates. Fibronectin is a dimer composed of two very large subunits joined by disulfide bonds at one end. Each subunit is folded into a series of functionally distinct domains separated by regions of flexible polypeptide chain (Figure 19–53A and B). The domains in turn consist of smaller modules, each of which is serially repeated and usually encoded by a separate exon, suggesting that the fibronectin gene, like the collagen genes, evolved by multiple exon duplications. All forms of fibronectin are encoded by a single large gene that contains about 50 exons of similar size. Transcription produces a single large RNA molecule that can be alternatively spliced to produce the various isoforms of fibronectin. The main type of module, called the **type III fibronectin repeat**, binds to integrins. It is about 90 amino acids long and occurs at least 15 times in each subunit. The type III fibronectin repeat is among the most common of all protein domains in vertebrates.

One way to analyze a complex multifunctional protein molecule like fibronectin is to chop it into pieces and determine the function of its individual domains. When fibronectin is treated with a low concentration of a proteolytic enzyme, the polypeptide chain is cut in the connecting regions between the domains, leaving the domains themselves intact. One can then show that one of

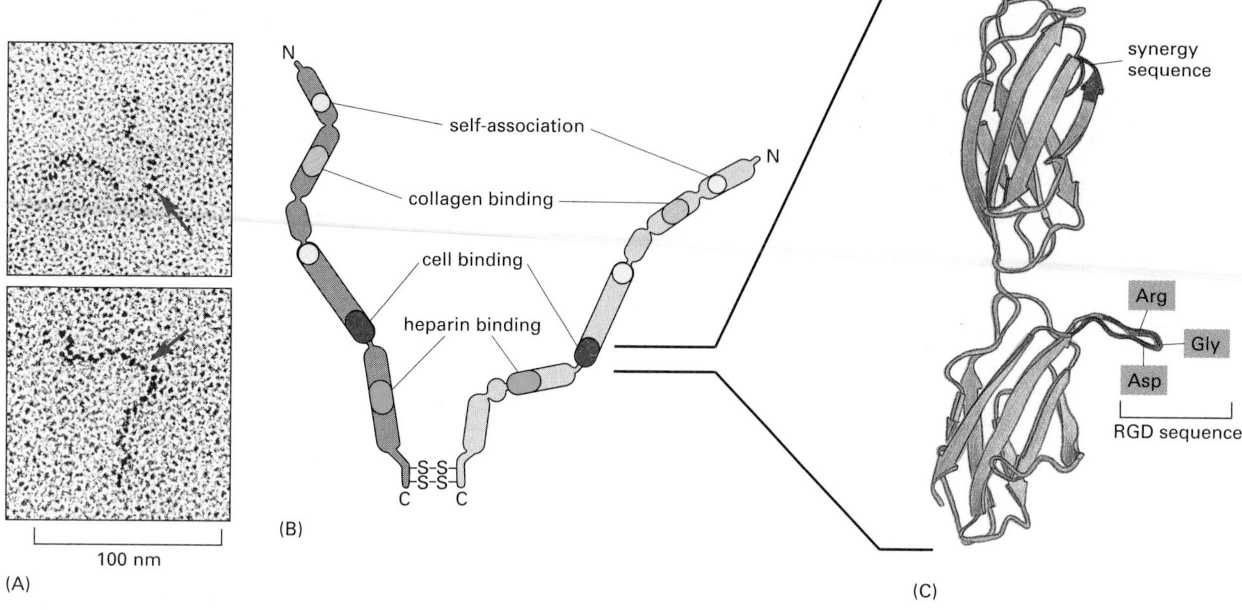

its domains binds to collagen, another to heparin, another to specific receptors on the surface of various types of cells, and so on (see Figure 19–53B). Synthetic peptides corresponding to different segments of the cell-binding domain have been used to identify a specific tripeptide sequence (*Arg-Gly-Asp,* or *RGD*), which is found in one of the type III repeats (see Figure 19–53C), as a central feature of the binding site. Even very short peptides containing this **RGD sequence** can compete with fibronectin for the binding site on cells, thereby inhibiting the attachment of the cells to a fibronectin matrix. If these peptides are coupled to a solid surface, they cause cells to adhere to it.

The RGD sequence is not confined to fibronectin. It is found in a number of extracellular proteins, including, for example, the blood-clotting factor *fibrinogen.* Fibrinogen peptides containing this RGD sequence have been useful in the development of anti-clotting drugs that mimic these peptides. Snakes use a similar strategy to cause their victims to bleed: they secrete RGD-containing anti-clotting proteins called *disintegrins* into their venom.

RGD sequences are recognized by several members of the integrin family of cell-surface matrix receptors. Each integrin, however, specifically recognizes its own small set of matrix molecules, indicating that tight binding requires more than just the RGD sequence.

Figure 19–53 The structure of a fibronectin dimer. (A) Electron micrographs of individual fibronectin dimer molecules shadowed with platinum; *red arrows* mark the C-termini. (B) The two polypeptide chains are similar but generally not identical (being made from the same gene but from differently spliced mRNAs). They are joined by two disulfide bonds near the C-termini. Each chain is almost 2500 amino acids long and is folded into five or six domains connected by flexible polypeptide segments. Individual domains are specialized for binding to a particular molecule or to a cell, as indicated for five of the domains. For simplicity, not all of the known binding sites are shown (there are other cell-binding sites, for example). (C) The three-dimensional structure of two type III fibronectin repeats as determined by x-ray crystallography. The type III repeat is the main repeating module in fibronectin. Both the Arg-Gly-Asp (RGD) and the "synergy" sequences shown in *red* form part of the major cell-binding site (shown *blue* in B). (A, from J. Engel et al., *J. Mol. Biol.* 150:97–120, 1981. © Academic Press; C, from Daniel J. Leahy, *Annu. Rev. Cell Dev. Biol.* 13:363–393, 1997. © Annual Reviews.)

Fibronectin Exists in Both Soluble and Fibrillar Forms

There are multiple isoforms of fibronectin. One, called *plasma fibronectin,* is soluble and circulates in the blood and other body fluids, where it is thought to enhance blood clotting, wound healing, and phagocytosis. All of the other forms assemble on the surface of cells and are deposited in the extracellular matrix as highly insoluble *fibronectin fibrils.* In these cell-surface and matrix forms, fibronectin dimers are cross-linked to one another by additional disulfide bonds.

Unlike fibrillar collagen molecules, which can be made to self-assemble into fibrils in a test tube, fibronectin molecules assemble into fibrils only on the surface of certain cells. This is because additional proteins are needed for fibril formation, especially fibronectin-binding integrins. In the case of fibroblasts, fibronectin fibrils are associated with integrins at sites called *fibrillar adhesions.* These are distinct from focal adhesions, in that they are more elongated and contain different intracellular anchor proteins. The fibronectin fibrils on the cell surface are highly stretched and under tension. The tension is exerted by the cell and is essential for fibril formation, as we discuss below. Some secreted proteins function to prevent fibronectin assembly in inappropriate places. *Uteroglobin,* for example, binds to fibronectin and prevents it from forming fibrils in the kidney. Mice that have a mutation in the uteroglobin gene accumulate insoluble fibronectin fibrils in their kidneys.

The importance of fibronectin in animal development is dramatically demonstrated by gene inactivation experiments. Mutant mice that are unable to make fibronectin die early in embryogenesis because their endothelial cells fail to form proper blood vessels. This defect is thought to result from abnormalities in the interactions of these cells with the surrounding extracellular matrix, which normally contains fibronectin.

Intracellular Actin Filaments Regulate the Assembly of Extracellular Fibronectin Fibrils

The fibronectin fibrils that form on or near the surface of fibroblasts are usually aligned with adjacent intracellular actin stress fibers (Figure 19–54). In fact, intracellular actin filaments promote the assembly of secreted fibronectin molecules into fibrils and influence fibril orientation. If cells are treated with the drug cytochalasin, which disrupts actin filaments, the fibronectin fibrils dissociate from the cell surface (just as they do during mitosis when a cell rounds up).

The interactions between extracellular fibronectin fibrils and intracellular actin filaments across the fibroblast plasma membrane are mediated mainly by integrin transmembrane adhesion proteins. The contractile actin and myosin cytoskeleton thereby pulls on the fibronectin matrix to generate tension. As a result, the fibronectin fibrils are stretched, exposing a cryptic (hidden) binding site in the fibronectin molecules that allows them to bind directly to one another. In addition, the stretching exposes more binding sites for integrins. In this way, the actin cytoskeleton promotes fibronectin polymerization and matrix assembly.

Extracellular signals can regulate the assembly process by altering the actin cytoskeleton and thereby the tension on the fibrils. Many other extracellular matrix proteins have multiple repeats similar to the type III fibronectin repeat, and it is possible that tension exerted on these proteins also uncovers cryptic binding sites and thereby influences their polymerization.

Glycoproteins in the Matrix Help Guide Cell Migration

Fibronectin is important not only for cell adhesion to the matrix but also for guiding cell migrations in vertebrate embryos. Large amounts of fibronectin, for example, are found along the pathway followed by migrating prospective mesodermal cells during amphibian gastrulation (discussed in Chapter 21). Although all cells of the early embryo can attach to fibronectin, only these migrating cells

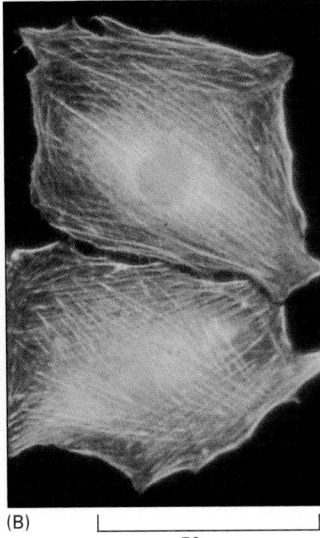

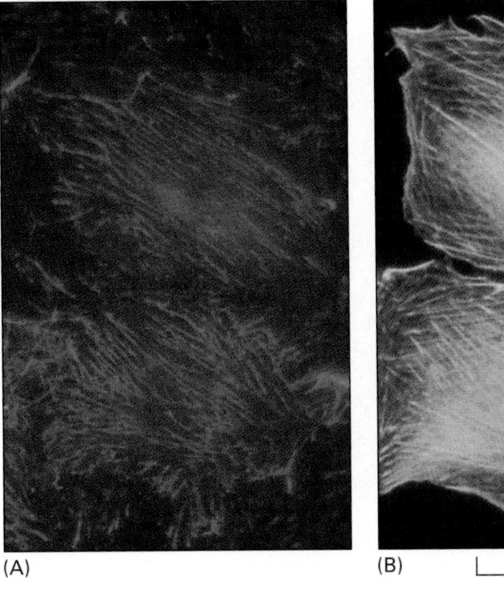

Figure 19–54 Coalignment of extracellular fibronectin fibrils and intracellular actin filament bundles. (A) The fibronectin is revealed in two rat fibroblasts in culture by the binding of rhodamine-coupled anti-fibronectin antibodies. (B) The actin is revealed by the binding of fluorescein-coupled anti-actin antibodies. (From R.O. Hynes and A.T. Destree, *Cell* 15:875–886, 1978. © Elsevier.)

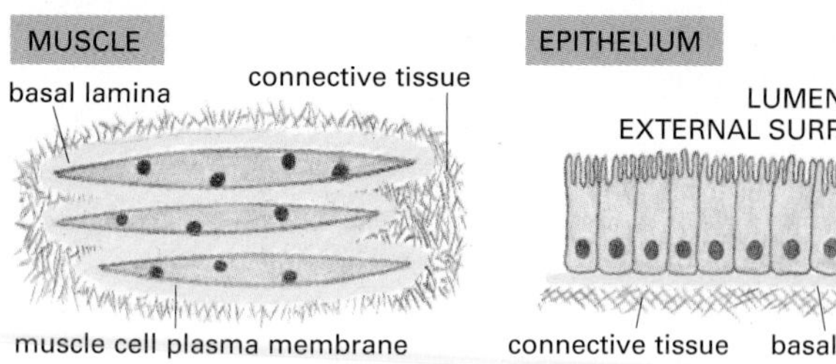

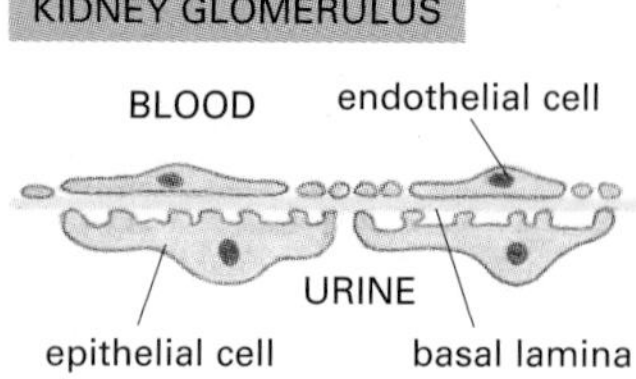

Figure 19–55 Three ways in which basal laminae are organized. Basal laminae *(yellow)* surround certain cells (such as skeletal muscle cells), underlie epithelia, and are interposed between two cell sheets (as in the kidney glomerulus). Note that, in the kidney glomerulus, both cell sheets have gaps in them, so that the basal lamina serves as the permeability barrier determining which molecules will pass into the urine from the blood.

can spread and migrate on fibronectin. The migration is inhibited by an injection into the developing amphibian embryo of various ligands that disrupt the ability of the cells to bind to fibronectin.

Many matrix proteins are believed to have a role in guiding cell movements during development. The *tenascins* and *thrombospondins,* for example, are composed of several types of short amino acid sequences that are repeated many times and form functionally distinct domains. They can either promote or inhibit cell adhesion, depending on the cell type. Indeed, anti-adhesive interactions are as important as adhesive ones in guiding cell migration, as we discuss in Chapter 21.

Basal Laminae Are Composed Mainly of Type IV Collagen, Laminin, Nidogen, and a Heparan Sulfate Proteoglycan

As mentioned earlier, **basal laminae** are flexible, thin (40–120 nm thick) mats of specialized extracellular matrix that underlie all epithelial cell sheets and tubes. They also surround individual muscle cells, fat cells, and Schwann cells (which wrap around peripheral nerve cell axons to form myelin). The basal lamina thus separates these cells and epithelia from the underlying or surrounding connective tissue. In other locations, such as the kidney glomerulus, a basal lamina lies between two cell sheets and functions as a highly selective filter (Figure 19–55). Basal laminae have more than simple structural and filtering roles, however. They are able to determine cell polarity, influence cell metabolism, organize the proteins in adjacent plasma membranes, promote cell survival, proliferation, or differentiation, and serve as specific highways for cell migration.

The basal lamina is synthesized largely by the cells that rest on it (Figure 19–56). In some multilayered epithelia, such as the stratified squamous epithelium

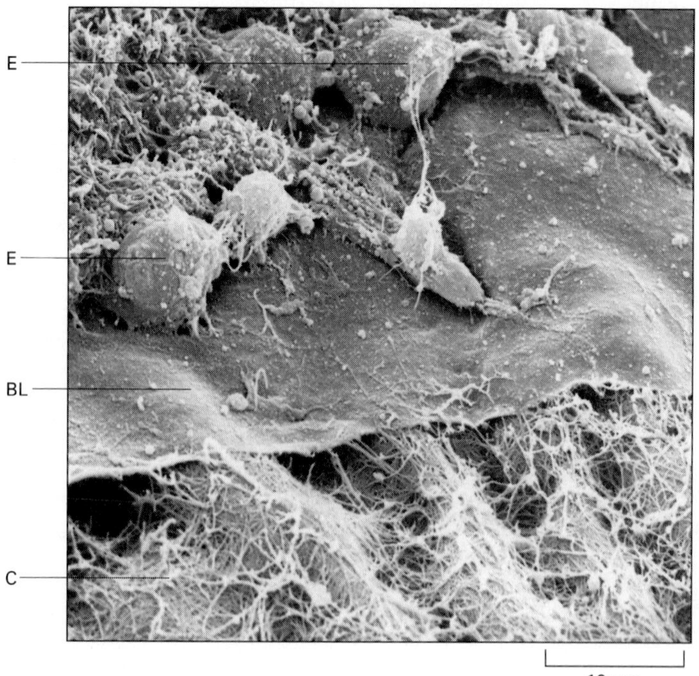

Figure 19–56 The basal lamina in the cornea of a chick embryo. In this scanning electron micrograph, some of the epithelial cells (E) have been removed to expose the upper surface of the matlike basal lamina (BL). A network of collagen fibrils (C) in the underlying connective tissue interacts with the lower face of the lamina. (Courtesy of Robert Trelstad.)

that forms the epidermis of the skin, the basal lamina is tethered to the underlying connective tissue by specialized anchoring fibrils made of type VII collagen molecules. The term *basement membrane* is often used to describe the composite of the basal lamina and this layer of collagen fibrils. In one type of skin disease, these connections are either absent or destroyed, and the epidermis and its basal lamina become detached from the underlying connective tissue, causing blistering.

Although its precise composition varies from tissue to tissue and even from region to region in the same lamina, most mature basal laminae contain *type IV collagen*, the large heparan sulfate proteoglycan perlecan, and the glycoproteins laminin and nidogen (also called entactin).

Type IV collagens exist in several isoforms. They all have a more flexible structure than the fibrillar collagens; their triple-stranded helix is interrupted in 26 regions, allowing multiple bends. They are not cleaved after secretion, but interact via their uncleaved terminal domains to assemble extracellularly into a flexible, sheetlike, multilayered network.

Early in development, basal laminae contain little or no type IV collagen and consist mainly of laminin molecules. **Laminin-1** (classical laminin) is a large, flexible protein composed of three very long polypeptide chains (α, β, and γ) arranged in the shape of an asymmetric cross and held together by disulfide bonds (Figure 19–57). Several isoforms of each type of chain can associate in different combinations to form a large family of laminins. The laminin γ-1 chain is a component of most laminin heterotrimers, and mice lacking it die during embryogenesis because they are unable to make a basal lamina. Like many other proteins in the extracellular matrix, the laminin in basement membranes consists of several functional domains: one binds to *perlecan*, one to *nidogen*, and two or more to laminin receptor proteins on the surface of cells.

Like type IV collagen, laminins can self-assemble *in vitro* into a feltlike sheet, largely through interactions between the ends of the laminin arms. As nidogen and perlecan can bind to both laminin and type IV collagen, it is thought that they connect the type IV collagen and laminin networks (Figure 19–58). In tissues, laminins and type IV collagen preferentially polymerize while bound to receptors on the surface of the cells producing the proteins. Many of the cell-surface receptors for type IV collagen and laminin are members of the integrin family. Another important type of laminin receptor is the

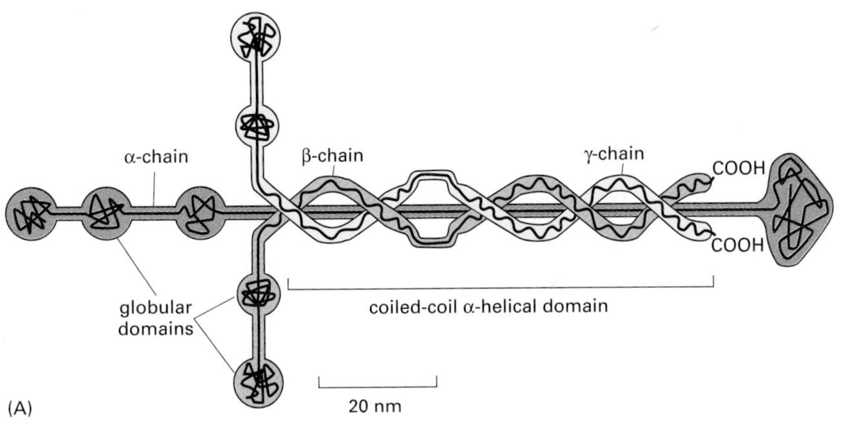

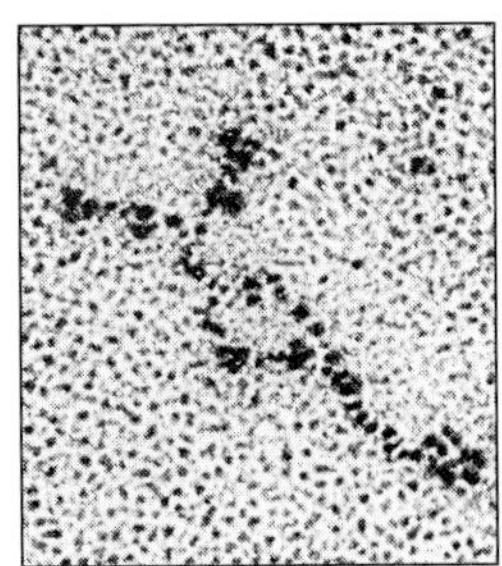

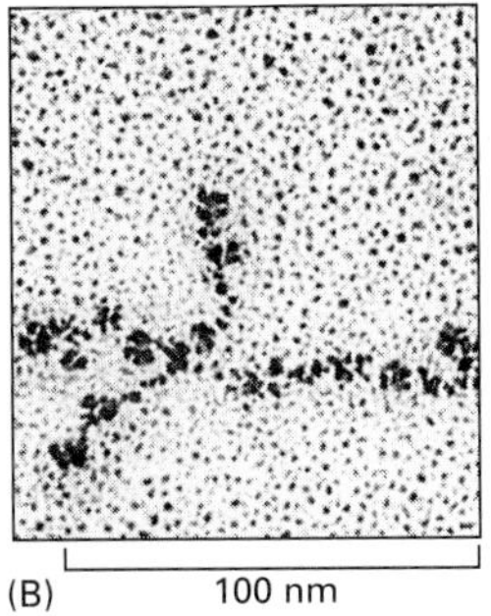

Figure 19–57 The structure of laminin. (A) The subunits of a laminin-1 molecule. This multidomain glycoprotein is composed of three polypeptides (α, β, and γ) that are disulfide-bonded into an asymmetric crosslike structure. Each of the polypeptide chains is more than 1500 amino acids long. Five types of α chains, three types of β chains, and three types of γ chains are known; in principle, they can assemble to form 45 ($5 \times 3 \times 3$) laminin isoforms. Several such isoforms have been found, each with a characteristic tissue distribution. (B) Electron micrographs of laminin molecules shadowed with platinum. (B, from J. Engel et al., *J. Mol. Biol.* 150:97–120, 1981. © Academic Press.)

transmembrane protein *dystroglycan,* which, together with integrins, may organize the assembly of the basal lamina.

The shapes and sizes of some of the extracellular matrix molecules discussed in this chapter are compared in Figure 19–59.

Basal Laminae Perform Diverse Functions

As we have mentioned, in the kidney glomerulus, an unusually thick basal lamina acts as a molecular filter, preventing the passage of macromolecules from the blood into the urine as urine is formed (see Figure 19–55). The heparan sulfate proteoglycan in the basal lamina seems to be important for this function: when its GAG chains are removed by specific enzymes, the filtering properties of the lamina are destroyed. Type IV collagen also has a role, as a human hereditary kidney disorder *(Alport syndrome)* results from mutations in type IV collagen α-chain genes.

The basal lamina can also act as a selective barrier to the movement of cells. The lamina beneath an epithelium, for example, usually prevents fibroblasts in the underlying connective tissue from making contact with the epithelial cells. It does not, however, stop macrophages, lymphocytes, or nerve processes from

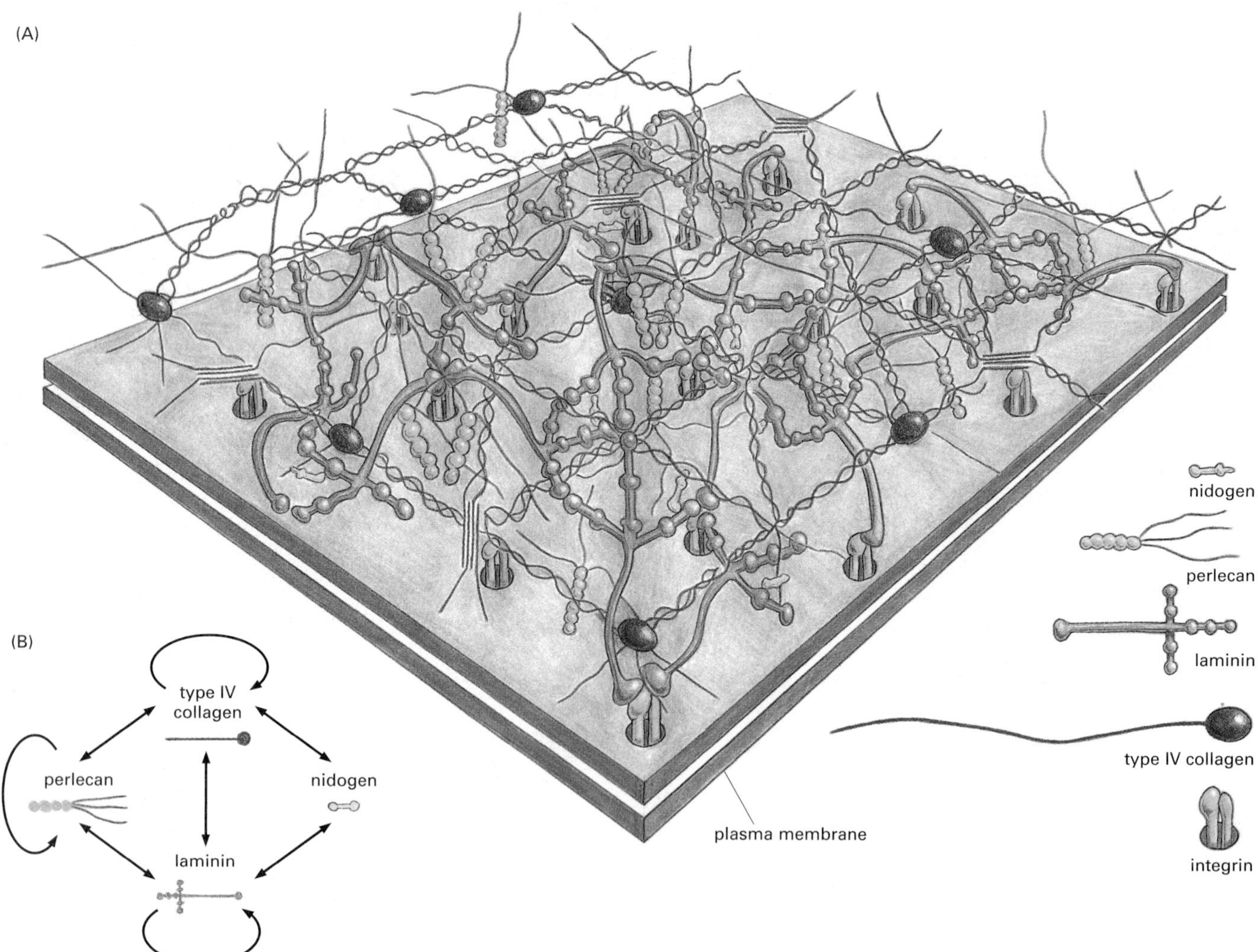

Figure 19–58 A model of the molecular structure of a basal lamina. (A) The basal lamina is formed by specific interactions (B) between the proteins type IV collagen, laminin, and nidogen, and the proteoglycan perlecan. Arrows in (B) connect molecules that can bind directly to each other. There are various isoforms of type IV collagen and laminin, each with a distinctive tissue distribution. Transmembrane laminin receptors (integrins and dystroglycan) in the plasma membrane are thought to organize the assembly of the basal lamina; only the integrins are shown. (Based on H. Colognato and P.D. Yurchenco, *Dev. Dynamics* 218:213–234, 2000.)

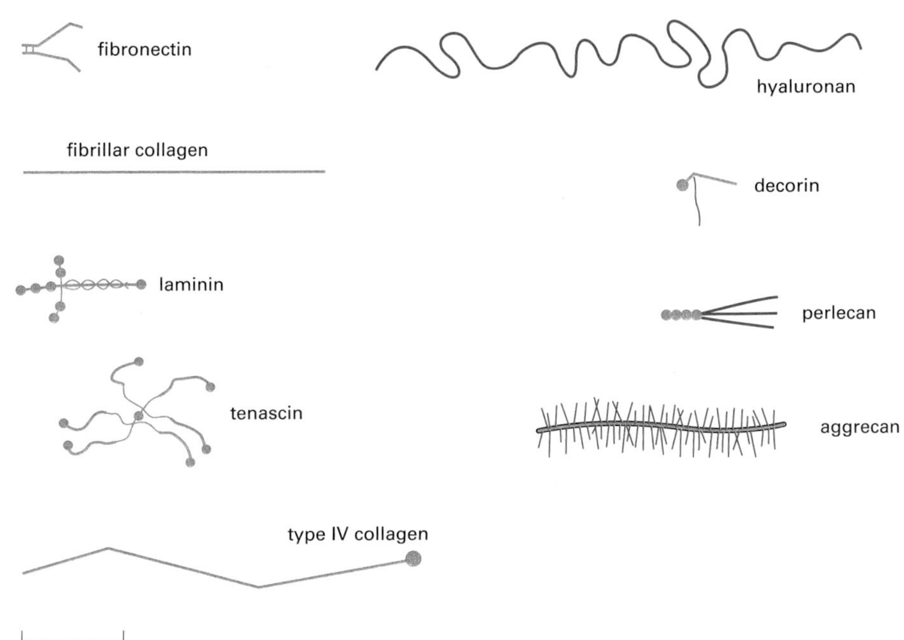

Figure 19–59 The comparative shapes and sizes of some of the major extracellular matrix macromolecules. Protein is shown in *green*, and glycosaminoglycan in *red*.

passing through it. The basal lamina is also important in tissue regeneration after injury. When tissues such as muscles, nerves, and epithelia are damaged, the basal lamina survives and provides a scaffold along which regenerating cells can migrate. In this way, the original tissue architecture is readily reconstructed. In some cases, as in the skin or cornea, the basal lamina becomes chemically altered after injury—for example, by the addition of fibronectin, which promotes the cell migration required for wound healing.

A particularly striking example of the instructive role of the basal lamina in regeneration comes from studies on the *neuromuscular junction*, the site where the nerve terminals of a motor neuron form a chemical synapse with a skeletal muscle cell (discussed in Chapter 11). The basal lamina that surrounds the muscle cell separates the nerve and muscle cell plasma membranes at the synapse, and the synaptic region of the lamina has a distinctive chemical character, with special isoforms of type IV collagen and laminin and a heparan sulfate proteoglycan called **agrin**.

This basal lamina at the synapse has a central role in reconstructing the synapse after nerve or muscle injury. If a frog muscle and its motor nerve are destroyed, the basal lamina around each muscle cell remains intact and the sites of the old neuromuscular junctions are still recognizable. If the motor nerve, but not the muscle, is allowed to regenerate, the nerve axons seek out the original synaptic sites on the empty basal lamina and differentiate there to form normal-looking nerve terminals. Thus, the junctional basal lamina by itself can guide the regeneration of motor nerve terminals.

Similar experiments show that the basal lamina also controls the localization of the acetylcholine receptors that cluster in the muscle cell plasma membrane at a neuromuscular junction. If the muscle and nerve are both destroyed, but now the muscle is allowed to regenerate while the nerve is prevented from doing so, the acetylcholine receptors synthesized by the regenerated muscle localize predominantly in the region of the old junctions, even though the nerve is absent (Figure 19–60). Thus, the junctional basal lamina apparently coordinates the local spatial organization of the components in each of the two cells that form a neuromuscular junction. Some of the matrix proteins have been identified. Motor neuron axons, for example, deposit agrin in the junctional basal lamina, where it triggers the assembly of acetylcholine receptors and other proteins in the junctional plasma membrane of the muscle cell. Conversely, muscle cells deposit a particular isoform of laminin in the junctional basal lamina. Both agrin and this isoform of laminin are essential for the formation of normal neuromuscular junctions.

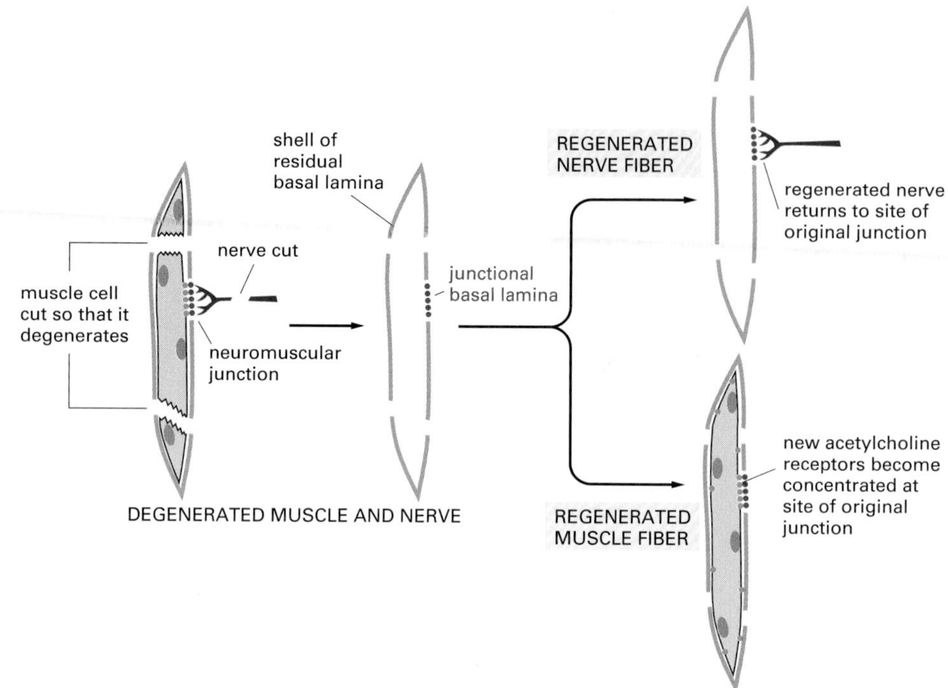

Figure 19–60 Regeneration experiments demonstrating the special character of the junctional basal lamina at a neuromuscular junction. When the nerve, but not the muscle, is allowed to regenerate after both the nerve and muscle have been damaged (*upper* part of figure), the junctional basal lamina directs the regenerating nerve to the original synaptic site. When the muscle, but not the nerve, is allowed to regenerate (*lower* part of figure), the junctional basal lamina causes newly made acetylcholine receptors *(blue)* to accumulate at the original synaptic site. The muscle regenerates from satellite cells (discussed in Chapter 22) located between the basal lamina and the original muscle cell (not shown). These experiments show that the junctional basal lamina controls the localization of synaptic components on both sides of the lamina.

The Extracellular Matrix Can Influence Cell Shape, Cell Survival, and Cell Proliferation

The extracellular matrix can influence the organization of a cell's cytoskeleton. This can be vividly demonstrated by using transformed (cancerlike) fibroblasts in culture (discussed in Chapter 23). Transformed cells often make less fibronectin than normal cultured cells and behave differently. They adhere poorly to the culture substratum, for example, and fail to flatten out or develop the organized intracellular bundles of actin filaments known as *stress fibers*. The decrease in fibronectin production and adhesion may contribute to the tendency of cancer cells to break away from the primary tumor and spread to other parts of the body.

In some cases, fibronectin deficiency seems also to be at least partly responsible for this abnormal morphology of cancer cells: if the cells are grown on a matrix of organized fibronectin fibrils, they flatten out and assemble intracellular stress fibers that are aligned with the extracellular fibronectin fibrils. This interaction between the extracellular matrix and the cytoskeleton is reciprocal in that intracellular actin filaments can promote the assembly and influence the orientation of fibronectin fibrils, as described earlier. Since the cytoskeleton can exert forces that orient the matrix macromolecules the cell secretes and the matrix macromolecules can in turn organize the cytoskeleton of the cells they contact, the extracellular matrix can in principle propagate order from cell to cell (Figure 19–61), creating large-scale oriented structures, as described earlier (see Figure 19–50). The integrins serve as the main adaptors in this ordering process, mediating the interactions between cells and the matrix around them.

Most cells need to attach to the extracellular matrix to grow and proliferate—and, in many cases, even to survive. This dependence of cell growth, proliferation, and survival on attachment to a substratum is known as **anchorage dependence**, and it is mediated mainly by integrins and the intracellular signals they generate. The physical spreading of a cell on the matrix also has a strong influence on intracellular events. Cells that are forced to spread over a large surface area survive better and proliferate faster than cells that are not so spread

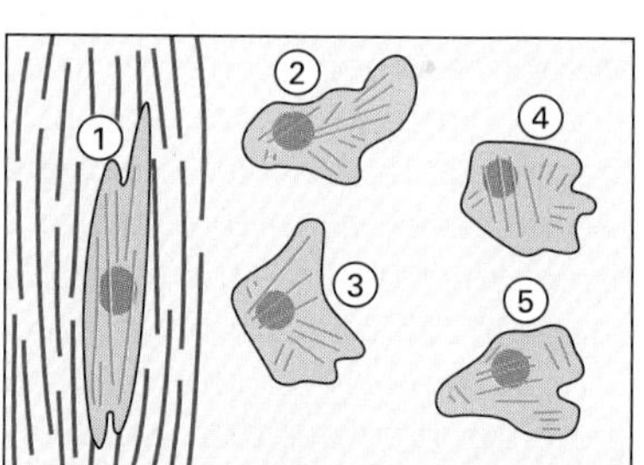

orientation of cytoskeleton in cell (1) orients the assembly of secreted extracellular matrix molecules in the vicinity

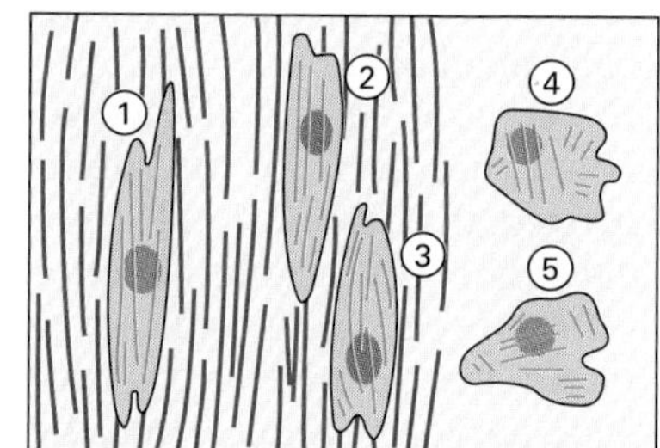

the oriented extracellular matrix reaches cells (2) and (3) and orients the cytoskeleton of those cells

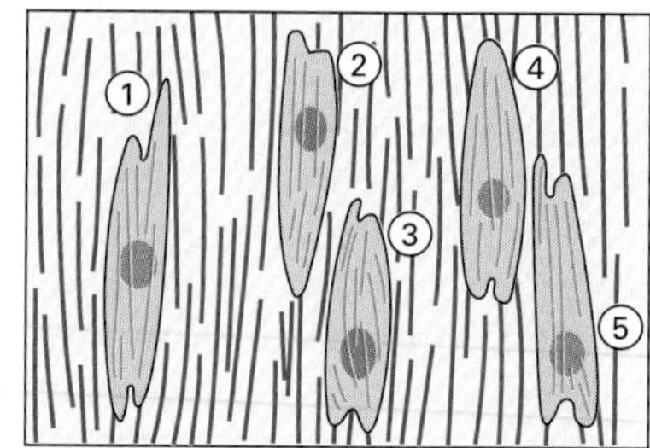

cells (2) and (3) now secrete an oriented matrix in their vicinity; in this way the ordering of cytoskeletons is propagated to cells (4) and (5)

Figure 19–61 How the extracellular matrix could, in principle, propagate order from cell to cell within a tissue. For simplicity, the figure represents a hypothetical scheme in which one cell influences the orientation of its neighboring cells. It is more likely, however, that the cells would mutually affect one another's orientation.

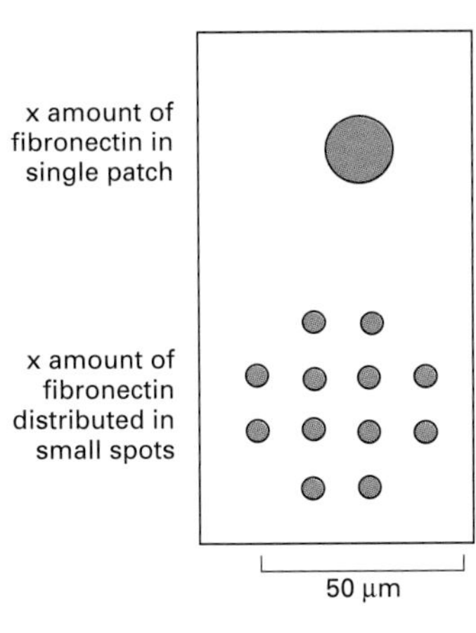

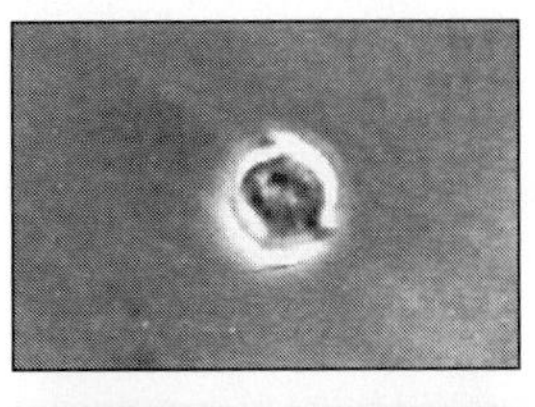

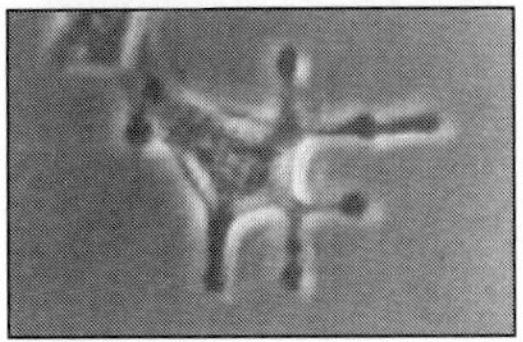

Figure 19–62 Anchorage dependence and the importance of cell spreading. For many cells, contact with the extracellular matrix is essential for survival, growth, and proliferation. In this experiment, the extent of cell spreading on a substratum, rather than the number of matrix molecules the cell contacts, influences cell survival (see also Figure 17–49). (Based on C.S. Chen et al., *Science* 276:1425–1428, 1997. © AAAS.)

out, even if in both cases the cells have the same area making contact with the matrix directly (Figure 19–62). This stimulatory effect of cell spreading presumably helps tissues to regenerate after injury. If cells are lost from an epithelium, for example, the spreading of the remaining cells into the vacated space will stimulate them to proliferate until they fill the gap. It is still uncertain, however, how a cell senses its extent of spreading so as to adjust its behavior accordingly.

The Controlled Degradation of Matrix Components Helps Cells Migrate

The regulated turnover of extracellular matrix macromolecules is crucial to a variety of important biological processes. Rapid degradation occurs, for example, when the uterus involutes after childbirth, or when the tadpole tail is resorbed during metamorphosis (see Figure 17–36). A more localized degradation of matrix components is required when cells migrate through a basal lamina. This occurs when white blood cells migrate across the basal lamina of a blood vessel into tissues in response to infection or injury, and when cancer cells migrate from their site of origin to distant organs via the bloodstream or lymphatic vessels—the process known as *metastasis*. Even in the seemingly static extracellular matrix of adult animals, there is a slow, continuous turnover, with matrix macromolecules being degraded and resynthesized.

In each of these cases, matrix components are degraded by extracellular proteolytic enzymes (proteases) that are secreted locally by cells. Thus, antibodies that recognize the products of proteolytic cleavage stain matrix only around cells. Many of these proteases belong to one of two general classes. Most are **matrix metalloproteases**, which depend on bound Ca^{2+} or Zn^{2+} for activity; the others are **serine proteases**, which have a highly reactive serine in their active site. Together, metalloproteases and serine proteases cooperate to degrade matrix proteins such as collagen, laminin, and fibronectin. Some metalloproteases, such as the *collagenases*, are highly specific, cleaving particular proteins at a small number of sites. In this way, the structural integrity of the matrix is largely retained, but cell migration can be greatly facilitated by the small amount of proteolysis. Other metalloproteases may be less specific, but, because they are anchored to the plasma membrane, they can act just where they are needed.

The importance of proteolysis in cell migration can be shown by using protease inhibitors, which often block migration. Moreover, cells that migrate readily on type I collagen in culture can no longer do so if the collagen is made resistant to proteolysis by mutating the collagenase-sensitive cleavage sites. The proteolysis of matrix proteins can contribute to cell migration in several ways: (1) it can simply clear a path through the matrix; (2) it can expose cryptic sites on the cleaved proteins that promote cell binding, cell migration, or both; (3) it can promote cell detachment so that a cell can move onward, or (4) it can release extracellular signal proteins that stimulate cell migration.

Three basic mechanisms operate to ensure that the proteases that degrade the matrix components are tightly controlled.

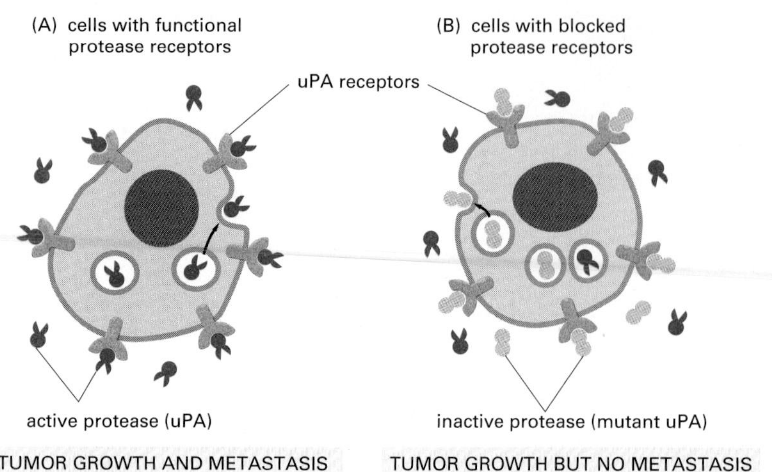

Figure 19–63 The importance of proteases bound to cell-surface receptors. (A) Human prostate cancer cells make and secrete the serine protease uPA, which binds to cell-surface uPA receptor proteins. (B) The same cells have been transfected with DNA that encodes an excess of an inactive form of uPA, which binds to the uPA receptors but has no protease activity. By occupying most of the uPA receptors, the inactive uPA prevents the active protease from binding to the cell surface. Both types of cells secrete active uPA, grow rapidly, and produce tumors when injected into experimental animals. But the cells in (A) metastasize widely, whereas the cells in (B) do not. To metastasize via the blood, tumor cells have to crawl through basal laminae and other extracellular matrices on the way into and out of the bloodstream. This experiment suggests that proteases must be cell-surface bound to mediate migration through the matrix.

Local activation: Many proteases are secreted as inactive precursors that can be activated locally when needed. An example is *plasminogen,* an inactive protease precursor that is abundant in the blood. It is cleaved locally by other proteases called *plasminogen activators* to yield the active serine protease *plasmin,* which helps break up blood clots. *Tissue-type plasminogen activator (tPA)* is often given to patients who have just had a heart attack or thrombotic stroke; it helps dissolve the arterial clot that caused the attack, thereby restoring bloodflow to the tissue.

Confinement by cell-surface receptors: Many cells have receptors on their surface that bind proteases, thereby confining the enzyme to the sites where it is needed. A second type of plasminogen activator called *urokinase-type plasminogen activator (uPA)* is an example. It is found bound to receptors on the growing tips of axons and at the leading edge of some migrating cells, where it may serve to clear a pathway for their migration. Receptor-bound uPA may also help some cancer cells metastasize (Figure 19–63).

Secretion of inhibitors: The action of proteases is confined to specific areas by various secreted protease inhibitors, including the *tissue inhibitors of metalloproteases (TIMPs)* and the serine protease inhibitors known as *serpins.* These inhibitors are protease-specific and bind tightly to the activated enzyme, blocking its activity. An attractive idea is that the inhibitors are secreted by cells at the margins of areas of active protein degradation in order to protect uninvolved matrix; they may also protect cell-surface proteins required for cell adhesion and migration. The overexpression of TIMPs inhibits the migration of some cell types, indicating the importance of metalloproteases for the migration.

Summary

Cells in connective tissues are embedded in an intricate extracellular matrix that not only binds the cells together but also influences their survival, development, shape, polarity, and behavior. The matrix contains various protein fibers interwoven in a hydrated gel composed of a network of glycosaminoglycan (GAG) chains.

GAGs are a heterogeneous group of negatively charged polysaccharide chains that (except for hyaluronan) are covalently linked to protein to form proteoglycan molecules. They occupy a large volume and form hydrated gels in the extracellular space. Proteoglycans are also found on the surface of cells, where they function as co-receptors to help cells respond to secreted signal proteins.

Fiber-forming proteins strengthen the matrix and give it form. They also provide surfaces for cells to adhere to. Elastin molecules form an extensive cross-linked network of fibers and sheets that can stretch and recoil, imparting elasticity to the matrix. The fibrillar collagens (types I, II, III, V, and XI) are ropelike, triple-stranded helical molecules that aggregate into long fibrils in the extracellular space. The fibrils in turn can assemble into a variety of highly ordered arrays. Fibril-associated collagen molecules, such as types IX and XII, decorate the surface of collagen fibrils and influence the interactions of the fibrils with one another and with other matrix components.

In contrast, type IV collagen molecules assemble into a sheetlike meshwork that is a crucial component of all mature basal laminae. All basal laminae are based on a mesh of laminin molecules. The collagen and laminin networks in mature basal laminae are bridged by the protein nidogen and the large heparan sulfate proteoglycan perlecan. Fibronectin and laminin are examples of large, multidomain matrix glycoproteins. By means of their multiple binding domains, such proteins help organize the matrix and help cells adhere to it.

Matrix proteins such as collagens, laminins, and fibronectin are assembled into fibrils or networks on the surface of the cells that produce them by a process that depends on the underlying actin cortex. The organization of the matrix can reciprocally influence the organization of the cell's cytoskeleton and can mechanically influence cell spreading. The matrix also influences cell behavior by binding to cell-surface receptors that activate intracellular signaling pathways.

Matrix components are degraded by extracellular proteolytic enzymes. Most of these are matrix metalloproteases, which depend on bound Ca^{2+} or Zn^{2+} for activity, while others are serine proteases, which have a reactive serine in their active site. Various mechanisms operate to ensure that the degradation of matrix components is tightly controlled. Cells can, for example, cause a localized degradation of matrix components to clear a path through the matrix.

INTEGRINS

The linkage of the extracellular matrix to the cell requires transmembrane cell adhesion proteins that act as *matrix receptors* and tie the matrix to the cell's cytoskeleton. Although we have seen that some transmembrane proteoglycans function as co-receptors for matrix components, the principal receptors on animal cells for binding most extracellular matrix proteins—including collagens, fibronectin, and laminins—are the **integrins**. These constitute a large family of homologous transmembrane, cell–matrix adhesion receptors. In blood cells, as we have seen, integrins also serve as cell–cell adhesion molecules, helping the cells bind to other cells, as well as to the extracellular matrix.

Integrins, like other cell adhesion molecules, differ from cell-surface receptors for hormones and for other extracellular soluble signal molecules in that they usually bind their ligand with lower affinity and are usually present at about tenfold to a hundredfold higher concentration on the cell surface. If the binding were too tight, cells would presumably become irreversibly glued to the matrix and would be unable to move—a problem that does not arise if attachment depends on large numbers of weak adhesions. This is an example of the "Velcro principle" mentioned earlier. Like other transmembrane cell adhesion proteins, however, integrins do more than just attach a cell to its surroundings. They also activate intracellular signaling pathways that communicate to the cell the character of the extracellular matrix that is bound.

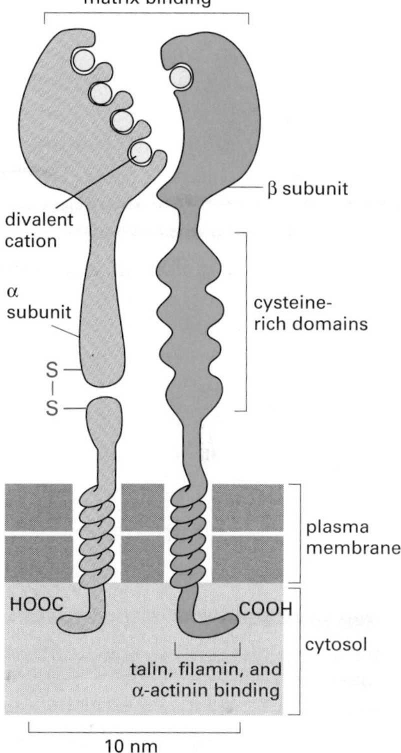

Figure 19–64 The subunit structure of an integrin cell-surface matrix receptor. Electron micrographs of isolated receptors suggest that the molecule has approximately the shape shown here, with the globular head projecting more than 20 nm from the lipid bilayer. By binding to a matrix protein outside the cell and to the actin cytoskeleton (via the anchor proteins indicated) inside the cell (see Figure 19–12B), the protein serves as a transmembrane linker. The α and β subunits are held together by noncovalent bonds. In the fibronectin receptor shown, the α subunit is made initially as a single 140,000-dalton polypeptide chain, which is then cleaved into one small transmembrane domain and one large extracellular domain that contains four divalent-cation-binding sites; the two domains remain held together by a disulfide bond. The extracellular part of the β subunit contains a single divalent-cation-binding site, as well as a repeating cysteine-rich region, where intrachain disulfide bonding occurs.

Integrins Are Transmembrane Heterodimers

Integrins are crucially important because they are the main receptor proteins that cells use to both bind to and respond to the extracellular matrix. An integrin molecule is composed of two noncovalently associated transmembrane glycoprotein subunits called α and β (Figure 19–64; see also Figure 19–12B). Because the same integrin molecule in different cell types can have different ligand-binding specificities, it seems that additional cell-type-specific factors can interact with integrins to modulate their binding activity.

The binding of integrins to their ligands depends on extracellular divalent cations (Ca^{2+} or Mg^{2+}, depending on the integrin), reflecting the presence of divalent-cation-binding domains in the extracellular part of the α and β subunits. The type of divalent cation can influence both the affinity and the specificity of the binding of an integrin to its ligands.

Many matrix proteins in vertebrates are recognized by multiple integrins. At least 8 integrins bind fibronectin, for example, and at least 5 bind laminin. A variety of human integrin heterodimers are formed from 9 types of β subunits and 24 types of α subunits. This diversity is further increased by alternative splicing of some integrin RNAs. Some of the best-studied integrins are listed in Table 19–6.

β_1 subunits form dimers with at least 12 distinct α subunits. They are found on almost all vertebrate cells: $\alpha_5\beta_1$, for example, is a fibronectin receptor and $\alpha_6\beta_1$ a laminin receptor on many types of cells. Mutant mice that cannot make any β_1 integrins die at implantation, whereas mice that are only unable to make the α_7 subunit (the partner for β_1 in muscle) survive but develop muscular dystrophy (as do mice that cannot make the laminin ligand for the $\alpha_7\beta_1$ integrin).

The β_2 subunits form dimers with at least four types of α subunit. They are expressed exclusively on the surface of white blood cells, where they have an essential role in enabling these cells to fight infection. The β_2 integrins mainly mediate cell–cell rather than cell–matrix interactions, binding to specific ligands on another cell, such as an endothelial cell. The ligands, sometimes referred to as *counterreceptors*, are members of the Ig superfamily of cell–cell adhesion molecules discussed earlier. The β_2 integrins enable white blood cells, for example, to attach firmly to endothelial cells at sites of infection and migrate out of the bloodstream into the infected site (see Figure 19–30B). Humans with the genetic disease called *leucocyte adhesion deficiency* are unable to synthesize β_2 subunits. As a consequence, their white blood cells lack the entire family of β_2 receptors, and they suffer repeated bacterial infections.

The β_3 integrins are found on a variety of cells, including blood platelets. They bind several matrix proteins, including fibrinogen. Platelets interact with fibrinogen during blood clotting, and humans with *Glanzmann's disease,* who are genetically deficient in β_3 integrins, bleed excessively.

Integrins Must Interact with the Cytoskeleton to Bind Cells to the Extracellular Matrix

Integrins function as transmembrane linkers (or "integrators"), mediating the interactions between the cytoskeleton and the extracellular matrix that are required for cells to grip the matrix. Most integrins are connected to bundles of actin filaments. The $\alpha_6\beta_4$ integrin found in hemidesmosomes is an exception: it is connected to intermediate filaments. After the binding of a typical integrin to its ligand in the matrix, the cytoplasmic tail of the β subunit binds to several intracellular anchor proteins, including talin, α-actinin, and filamin. These anchor proteins can bind directly to actin or to other anchor proteins such as vinculin, thereby linking the integrin to actin filaments in the cell cortex. Given the right conditions, this linkage leads to a clustering of the integrins and the formation of focal adhesions between the cell and the extracellular matrix, as discussed earlier.

If the cytoplasmic domain of the β subunit is deleted using recombinant DNA techniques, the shortened integrins still bind to their ligands, but they no longer mediate robust adhesion, and they fail to cluster at focal adhesions. It seems that integrins must interact with the cytoskeleton to bind cells strongly to

TABLE 19–6 Some Types of Integrins

INTEGRIN	LIGAND*	DISTRIBUTION
$\alpha_5\beta_1$	fibronectin	ubiquitous
$\alpha_6\beta_1$	laminin	ubiquitous
$\alpha_7\beta_1$	laminin	muscle
$\alpha_L\beta_2$ (LFA-1, see p. 1411)	Ig superfamily counterreceptors	white blood cells
$\alpha_2\beta_3$	fibrinogen	platelets
$\alpha_6\beta_4$	laminin	epithelial hemidesmosomes

*Not all ligands are listed.

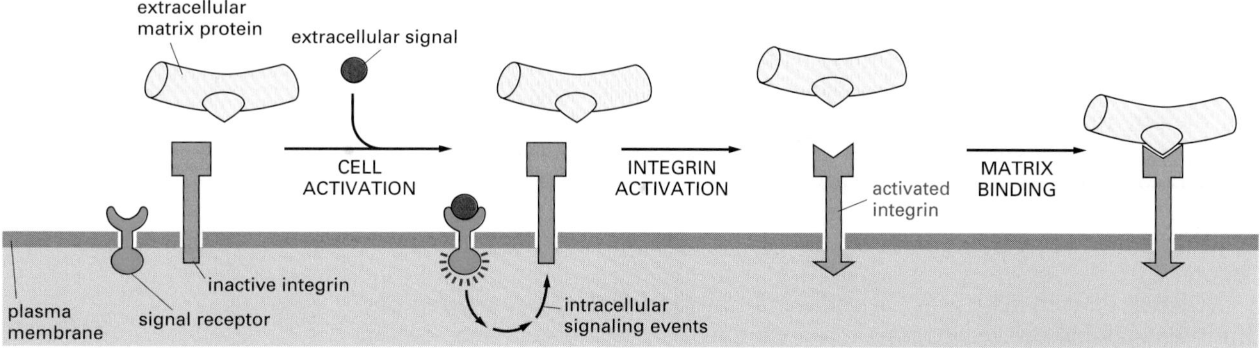

Figure 19–65 The regulation of the extracellular binding activity of a cell's integrins from within. In this example, an extracellular signal activates an intracellular signaling cascade that alters the integrin so that its extracellular binding site can now mediate cell adhesion. The molecular nature of the alteration is still poorly understood.

the matrix, just as cadherins must interact with the cytoskeleton to hold cells together efficiently. The cytoskeletal attachment may help cluster the integrins, providing a stronger aggregate bond; it may also lock the integrin in a conformation that allows the integrin to bind its ligand more tightly.

Just as cadherins can promote cell–cell adhesion without forming mature adherens junctions, integrins can mediate cell–matrix adhesion without forming mature focal adhesions. In both cases, however, the transmembrane adhesion proteins may still bind to the cytoskeleton. For integrins, this kind of adhesion occurs when cells are spreading or migrating, and it results in the formation of *focal complexes.* For such focal complexes to mature into the focal adhesions that are typical of many well-spread cells, the activation of the small GTPase Rho is required. The activation of Rho leads to the recruitment of more actin filaments and integrins to the contact site (discussed in Chapter 16).

Cells Can Regulate the Activity of Their Integrins

We discuss below how integrin clustering activates intracellular signaling pathways. But signaling also operates in the opposite direction: signals generated inside the cell can either enhance or inhibit the ability of integrins to bind to their ligand outside the cell (Figure 19–65). This regulation is poorly understood, but it may involve the phosphorylation of the cytoplasmic tails of the integrins, the association of the tails with activating cytoplasmic proteins, or both.

The ability of a cell to control integrin–ligand interactions from within is termed *inside-out signaling.* It is particularly important in platelets and white blood cells, where integrins usually have to be activated before they can mediate adhesion. In most other cells, integrins are usually maintained in an adhesion-competent state. Regulated adhesion allows white blood cells to circulate unimpeded until they are activated by an appropriate stimulus. Because the integrins do not need to be synthesized *de novo,* the signaled adhesion response can be rapid. Platelets, for example, are activated either by contact with a damaged blood vessel or by various soluble signal molecules. In either case, the stimulus triggers intracellular signaling pathways that rapidly activate a β_3 integrin in the platelet membrane. This induces a conformational change in the extracellular domain of the integrin that enables the protein to bind the blood-clotting protein fibrinogen with high affinity. The fibrinogen links platelets together to form a platelet plug, which helps stop bleeding.

Similarly, the weak binding of a T lymphocyte to its specific antigen on the surface of an antigen-presenting cell (discussed in Chapter 24) triggers intracellular signaling pathways in the T cell that activate its β_2 integrins. The activated integrins then enable the T cell to adhere strongly to the antigen-presenting cell so that it remains in contact long enough to become stimulated fully. The integrins may then return to an inactive state, allowing the T cell to disengage.

There are occasions, especially during development, when cells other than blood cells also regulate the activity of their integrins. If a constitutively active integrin (made by deleting the cytoplasmic tail of the α subunit) is expressed in a developing *Drosophila* embryo, for example, it disrupts normal muscle development. The muscle precursor cells expressing the activated integrin cannot disengage from the extracellular matrix and therefore cannot migrate normally.

Integrins Activate Intracellular Signaling Pathways

We have already discussed how integrins function as transmembrane linkers that connect extracellular matrix molecules to actin filaments in the cell cortex and thereby regulate the shape, orientation, and movement of cells. But the clustering of integrins at the sites of contact with the matrix (or with another cell) can also activate intracellular signaling pathways. Signaling is initiated by the assembly of signaling complexes at the cytoplasmic face of the plasma membrane, much as in signaling by conventional signaling receptors (discussed in Chapter 15).

Whereas activated integrins, like activated conventional signaling receptors, can induce global cell responses, often including changes in gene expression, activated integrins are especially adept at stimulating localized changes in the cytoplasm close to the cell–matrix contact. This may be a fundamental feature of signaling by transmembrane cell adhesion proteins in general. In the developing nervous system, for example, the growing tip of an axon is guided mainly by its responses to local adhesive (and repellent) cues in the environment that are recognized by transmembrane cell adhesion proteins. The primary effects of the adhesion proteins are thought to result from the activation of intracellular signaling pathways that act locally in the axon tip, rather than through cell–cell adhesion itself or signals conveyed to the cell body.

Many of the signaling functions of integrins depend on a cytoplasmic protein tyrosine kinase called **focal adhesion kinase (FAK)**. Focal adhesions are often the most prominent sites of tyrosine phosphorylation in cells in culture (see Figure 17–50), and FAK is one of the major tyrosine-phosphorylated proteins found in focal adhesions (although it can also associate with conventional signaling receptors). When integrins cluster at sites of cell–matrix contact, FAK is recruited to focal adhesions by intracellular anchor proteins such as *talin,* which binds to the integrin β subunit, or *paxillin,* which binds to one type of integrin α subunit. The clustered FAK molecules cross-phosphorylate each other on a specific tyrosine, creating a phosphotyrosine docking site for members of the Src family of cytoplasmic tyrosine kinases. These kinases then phosphorylate FAK on additional tyrosines, creating docking sites for a variety of intracellular signaling proteins; they also phosphorylate other proteins at focal adhesions. In this way, the signal is relayed into the cell (as discussed in Chapter 15).

One way to analyze the function of FAK is to examine focal adhesions in cells from mutant mice that lack the protein. FAK-deficient fibroblasts still adhere to fibronectin and form focal adhesions. Suprisingly, they form too many focal adhesions rather than too few; as a result, cell spreading and migration are slowed (Figure 19–66). This unexpected finding suggests that FAK normally helps disassemble focal adhesions and that this loss of adhesions is required for normal cell migration. By interacting with both conventional signaling receptors and focal adhesions, FAK can couple migratory signals to changes in cell adhesion. Many cancer cells have elevated levels of FAK, which may help explain why they are often more motile than their normal counterparts.

Integrins and conventional signaling receptors can work together in several ways. The signaling pathways activated by conventional signaling receptors can increase the expression of integrins or extracellular matrix molecules, while those activated by integrins can increase the expression of conventional signaling receptors or the ligands that bind to them. The intracellular signaling pathways

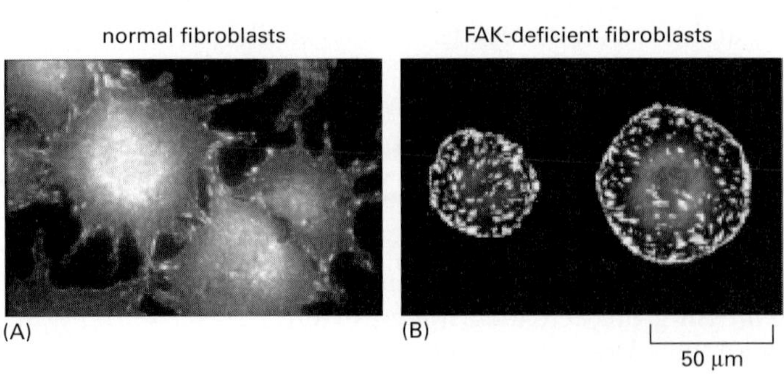

Figure 19–66 Excessive numbers of focal adhesions in FAK-deficient fibroblasts. Normal and FAK-deficient fibroblasts were stained with antibodies against vinculin to reveal the location of focal adhesions (see Figure 19–12). (A) The normal fibroblasts have fewer focal adhesions and have spread after 2 hours in culture. (B) At the same time point, the FAK-deficient fibroblasts have more focal adhesions and have not spread. (From D. Ilic et al., *Nature* 377:539–544, 1995. © Macmillan Magazines Ltd.)

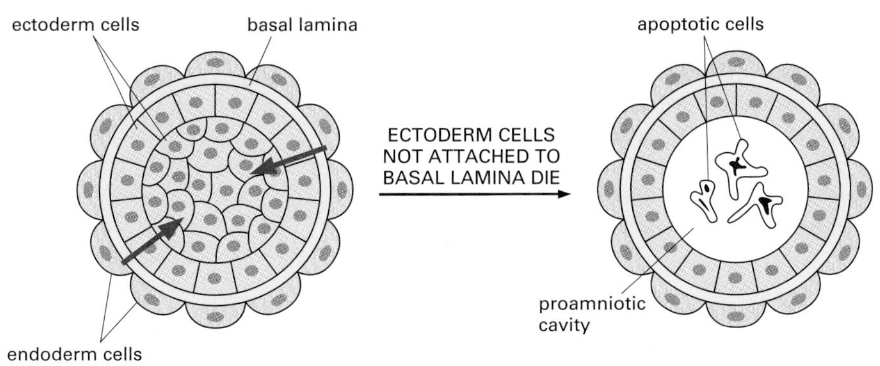

Figure 19–67 Matrix-dependent cell survival in the formation of the proamniotic cavity. Endoderm cells are thought to produce a signal (indicated by the *red arrows)* that causes ectoderm cells to die by apoptosis. The ectoderm cells in contact with the basal lamina, however, are saved by the survival-promoting action of the matrix molecules in the lamina, while the other ectoderm cells die by apoptosis, forming the proamniotic cavity. (Based on E. Coucouvanis and G.R. Martin, *Cell* 83:279–287, 1995.)

themselves can also interact and reinforce one another. While some conventional signaling receptors and integrins activate the Ras/MAP kinase pathway (see Figure 15–56) independently, for example, they often act together to sustain the activation of this pathway long enough to induce cell proliferation. Integrins and conventional signaling receptors cooperate to stimulate many types of cell response. Many cells in culture, for example, will not grow or proliferate in response to extracellular growth factors unless the cells are attached via integrins to extracellular matrix molecules. For some cell types, including epithelial, endothelial, and muscle cells, even cell survival depends on signaling through integrins. When these cells lose contact with the extracellular matrix, they undergo programmed cell death, or apoptosis. This dependence on attachment to the extracellular matrix for survival and proliferation may help ensure that the cells survive and proliferate only when they are in their appropriate location, which may protect animals against the spread of cancer cells. Attachment-dependent cell survival is exploited for special purposes in embryonic development, as shown in Figure 19–67. The signaling pathways that integrins activate to promote cell survival are similar to those activated by conventional signaling receptors, as discussed in Chapters 15 and 17.

The cell adhesion molecules discussed in this chapter are summarized in Table 19–7.

TABLE 19–7 Cell Adhesion Molecule Families

	SOME FAMILY MEMBERS	Ca^{2+} OR Mg^{2+} DEPENDENCE	HOMOPHILIC OR HETEROPHILIC	CYTOSKELETON ASSOCIATIONS	CELL JUNCTION ASSOCIATIONS
Cell–Cell Adhesion					
Classical cadherins	E, N, P, VE	yes	homophilic	actin filaments (via catenins)	adherens junctions
Desmosomal cadherins	desmoglein	yes	homophilic	intermediate filaments (via desmoplakin, plakoglobin, and other proteins)	desmosomes
Ig family members	N-CAM	no	both	unknown	no
Selectins (blood cells and endothelial cells only)	L-, E-, and P-selectins	yes	heterophilic	actin filaments	no
Integrins on blood cells	$\alpha_L\beta_2$ (LFA-1)	yes	heterophilic	actin filaments	no
Cell–Matrix Adhesion					
Integrins	many types	yes	heterophilic	actin filaments (via talin, filamin, α-actinin, and vinculin)	focal adhesions
	$\alpha_6\beta_4$	yes	heterophilic	intermediate filaments (via plectin)	hemidesmosomes
Transmembrane proteoglycans	syndecans	no	heterophilic	actin filaments	no

Summary

Integrins are the principal receptors used by animal cells to bind to the extracellular matrix. They are heterodimers and function as transmembrane linkers between the extracellular matrix and the actin cytoskeleton. A cell can regulate the adhesive activity of its integrins from within. Integrins also function as signal transducers, activating various intracellular signaling pathways when activated by matrix binding. Integrins and conventional signaling receptors often cooperate to promote cell growth, cell survival, and cell proliferation.

THE PLANT CELL WALL

The plant cell wall is an elaborate extracellular matrix that encloses each cell in a plant. It was the thick cell walls of cork, visible in a primitive microscope, that in 1663 enabled Robert Hooke to distinguish and name cells for the first time. The walls of neighboring plant cells, cemented together to form the intact plant (Figure 19–68), are generally thicker, stronger, and, most important of all, more rigid than the extracellular matrix produced by animal cells. In evolving relatively rigid walls, which can be up to many micrometers thick, early plant cells forfeited the ability to crawl about and adopted a sedentary life-style that has persisted in all present-day plants.

The Composition of the Cell Wall Depends on the Cell Type

All cell walls in plants have their origin in dividing cells, as the cell plate forms during cytokinesis to create a new partition wall between the daughter cells (discussed in Chapter 18). The new cells are usually produced in special regions called *meristems* (discussed in Chapter 21), and they are generally small in comparison with their final size. To accommodate subsequent cell growth, their

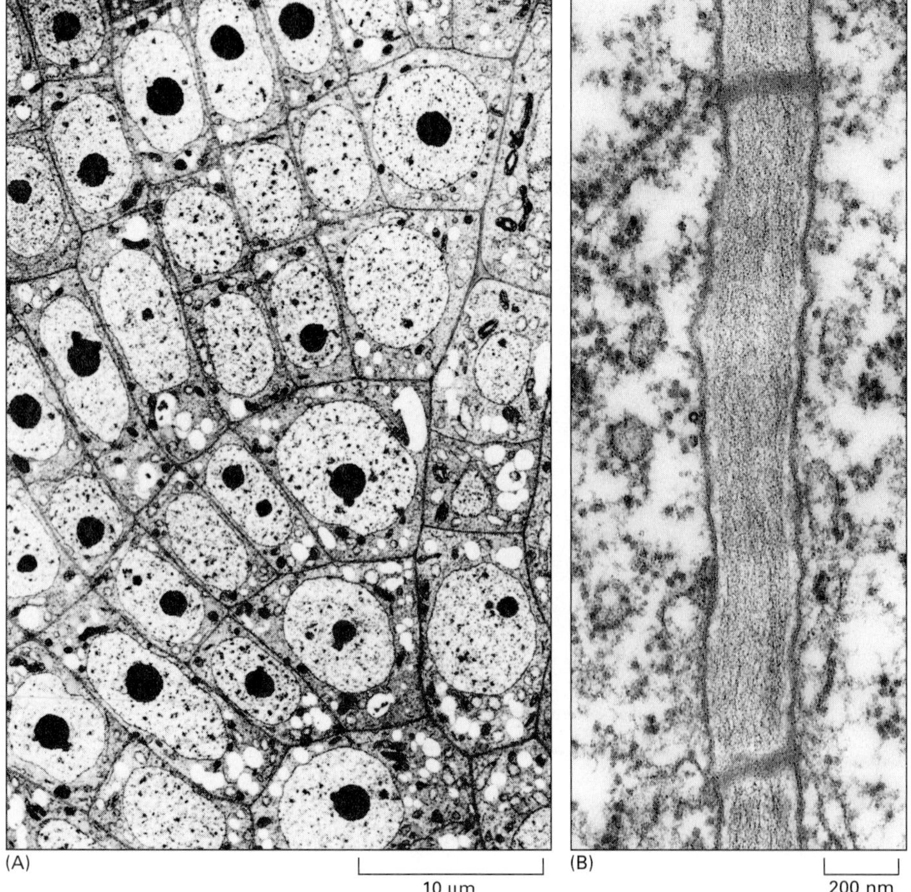

Figure 19–68 Plant cell walls. (A) Electron micrograph of the root tip of a rush, showing the organized pattern of cells that results from an ordered sequence of cell divisions in cells with relatively rigid cell walls. In this growing tissue, the cell walls are still relatively thin, appearing as fine black lines between the cells in the micrograph. (B) Section of a typical cell wall separating two adjacent plant cells. The two dark transverse bands correspond to plasmodesmata that span the wall (see Figure 19–20). (A, courtesy of C. Busby and B. Gunning, *Eur. J. Cell Biol.* 21:214–233, 1980; B, courtesy of Jeremy Burgess.)

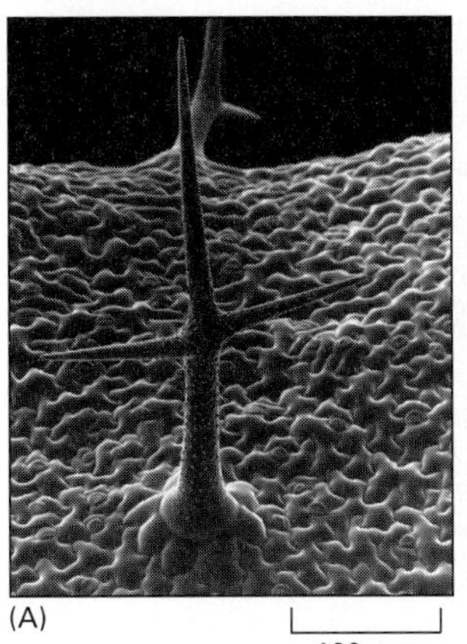

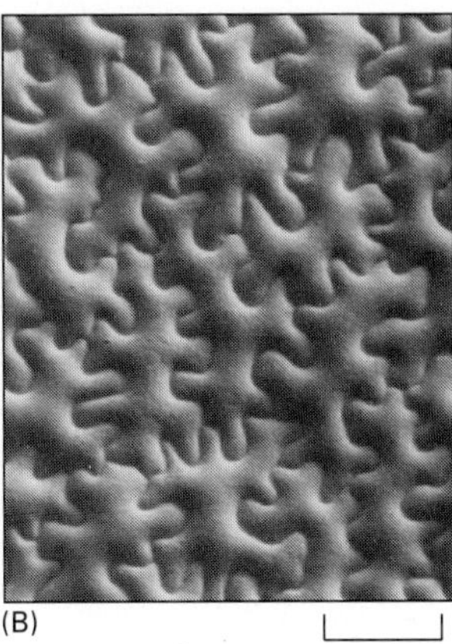

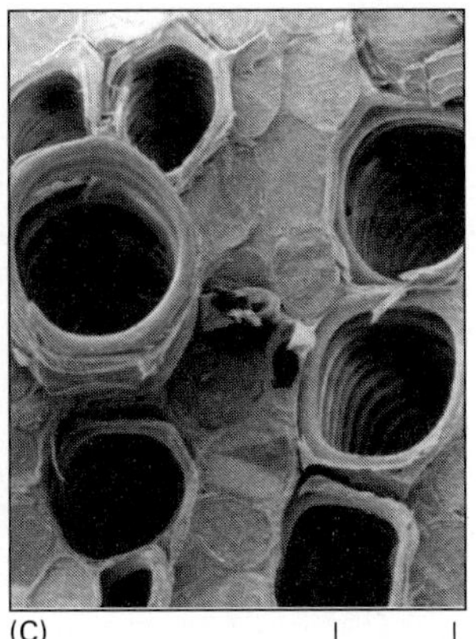

Figure 19–69 Specialized cell types with appropriately modified cell walls. (A) A trichome, or hair, on the upper surface of an *Arabidopsis* leaf. This spiky, protective single cell is shaped by the local deposition of a tough, cellulose-rich wall. (B) Surface view of tomato leaf epidermal cells. The cells fit together snugly like the pieces of a jigsaw puzzle, providing a strong outer covering for the leaf. The outer cell wall is reinforced with a cuticle and waxes that waterproof the leaf and help defend it against pathogens. (C) This view into young xylem elements shows the thick, lignified, hoop-reinforced secondary cell wall that creates robust tubes for the transport of water throughout the plant. (A, courtesy of Paul Linstead; B and C, courtesy of Kim Findlay.)

walls, called **primary cell walls**, are thin and extensible, although tough. Once growth stops, the wall no longer needs to be extensible: sometimes the primary wall is retained without major modification, but, more commonly, a rigid, **secondary cell wall** is produced by depositing new layers inside the old ones. These may either have a composition similar to that of the primary wall or be markedly different. The most common additional polymer in secondary walls is **lignin**, a complex network of phenolic compounds found in the walls of the xylem vessels and fiber cells of woody tissues.The plant cell wall thus has a "skeletal" role in supporting the structure of the plant as a whole, a protective role as an enclosure for each cell individually, and a transport role, helping to form channels for the movement of fluid in the plant. When plant cells become specialized, they generally adopt a specific shape and produce specially adapted types of walls, according to which the different types of cells in a plant can be recognized and classified (Figure 19–69; see also Panel 21–3).

Although the cell walls of higher plants vary in both composition and organization, they are all constructed, like animal extracellular matrices, using a structural principle common to all fiber-composites, including fibreglass and reinforced concrete. One component provides tensile strength, while another, in which the first is embedded, provides resistance to compression. While the principle is the same in plants and animals, the chemistry is different. Unlike the animal extracellular matrix, which is rich in protein and other nitrogen-containing polymers, the plant cell wall is made almost entirely of polymers that contain no nitrogen, including *cellulose* and lignin. Trees make a huge investment in the cellulose and lignin that comprise the bulk of their biomass. For a sedentary organism that depends on CO_2, H_2O and sunlight, these two abundant biopolymers represent "cheap," carbon-based, structural materials, helping to conserve the scarce fixed nitrogen available in the soil that generally limits plant growth.

In the cell walls of higher plants, the tensile fibers are made from the polysaccharide cellulose, the most abundant organic macromolecule on Earth, tightly linked into a network by *cross-linking glycans.* In primary cell walls, the matrix in which the cellulose network is embedded is composed of *pectin,* a highly hydrated network of polysaccharides rich in galacturonic acid. Secondary cell walls contain additional components, such as lignin, which is hard and occupies the interstices between the other components, making the walls rigid and permanent. All of these molecules are held together by a combination of covalent and noncovalent bonds to form a highly complex structure, whose composition, thickness and architecture depends on the cell type.

We focus here on the primary cell wall and the molecular architecture that underlies its remarkable combination of strength, resilience, and plasticity, as seen in the growing parts of a plant.

The Tensile Strength of the Cell Wall Allows Plant Cells to Develop Turgor Pressure

The aqueous extracellular environment of a plant cell consists of the fluid contained in the walls that surround the cell. Although the fluid in the plant cell wall contains more solutes than does the water in the plant's external milieu (for example, soil), it is still hypotonic in comparison with the cell interior. This osmotic imbalance causes the cell to develop a large internal hydrostatic pressure, or **turgor pressure**, that pushes outward on the cell wall, just as an inner tube pushes outward on a tire. The turgor pressure increases just to the point where the cell is in osmotic equilibrium, with no net influx of water despite the salt imbalance (see Panel 11–1, pp. 628–629). This pressure is vital to plants because it is the main driving force for cell expansion during growth, and it provides much of the mechanical rigidity of living plant tissues. Compare the wilted leaf of a dehydrated plant, for example, with the turgid leaf of a well-watered one. It is the mechanical strength of the cell wall that allows plant cells to sustain this internal pressure.

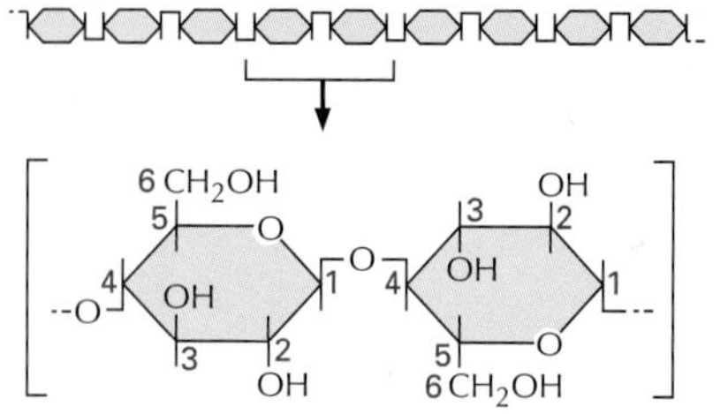

Figure 19–70 Cellulose. Cellulose molecules are long, unbranched chains of β1,4-linked glucose units. Each glucose is inverted with respect to its neighbors, and the resulting disacchride repeat occurs hundreds of times in a single cellulose molecule.

The Primary Cell Wall Is Built from Cellulose Microfibrils Interwoven with a Network of Pectic Polysaccharides

The cellulose molecules provide tensile strength to the primary cell wall. Each molecule consists of a linear chain of at least 500 glucose residues that are covalently linked to one another to form a ribbonlike structure, which is stabilized by hydrogen bonds within the chain (Figure 19–70). In addition, intermolecular hydrogen bonds between adjacent cellulose molecules cause them to adhere strongly to one another in overlapping parallel arrays, forming a bundle of about 40 cellulose chains, all of which have the same polarity. These highly ordered crystalline aggregates, many micrometers long, are called **cellulose microfibrils**, and they have a tensile strength comparable to steel. Sets of microfibrils are arranged in layers, or lamellae, with each microfibril about 20–40 nm from its neighbors and connected to them by long cross-linking glycan molecules that are bound by hydrogen bonds to the surface of the microfibrils. The primary cell wall consists of several such lamellae arranged in a plywoodlike network (Figure 19–71).

The **cross-linking glycans** are a heterogeneous group of branched polysaccharides that bind tightly to the surface of each cellulose microfibril and thereby help to cross-link microfibrils into a complex network. Their function is analogous to that of the fibril-associated collagens discussed earlier (see Figure 19–49). There are many classes of cross-linking glycans, but they all have a long linear backbone composed of one type of sugar (glucose, xylose, or mannose) from which short side chains of other sugars protrude. It is the backbone sugar molecules that form hydrogen bonds with the surface of cellulose microfibrils,

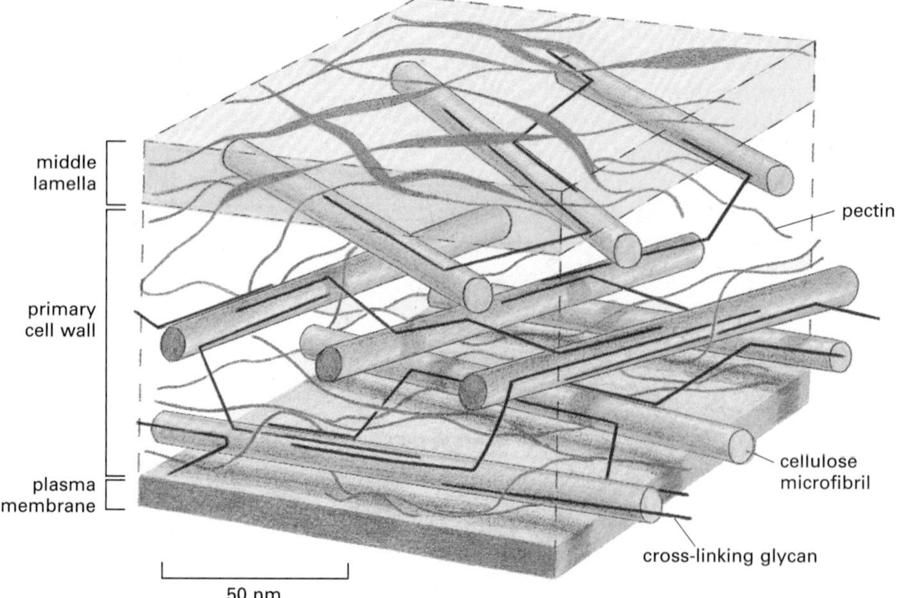

Figure 19–71 Scale model of a portion of a primary cell wall showing the two major polysaccharide networks. The orthogonally arranged layers of cellulose microfibrils *(green)* are tied into a network by cross-linking glycans *(red)* that form hydrogen bonds with the microfibrils. This network is coextensive with a network of pectin polysaccharides *(blue)*. The network of cellulose and cross-linking glycans provides tensile strength, while the pectin network resists compression. Cellulose, cross-linking glycans, and pectin are typically present in roughly equal amounts in a primary cell wall. The middle lamella is rich in pectin and cements adjacent cells together.

TABLE 19–8 The Polymers of the Plant Cell Wall

POLYMER	COMPOSITION	FUNCTIONS
Cellulose	linear polymer of glucose	fibrils confer tensile strength to all walls
Cross-linking glycans	xyloglucan, glucuronoarabinoxylan, and mannans	cross-link cellulose fibrils into robust network
Pectin	homogalacturonans and rhamnogalacturonans	forms negatively charged, hydrophilic network that gives compressive strength to primary walls; cell–cell adhesion
Lignin	cross-linked coumaryl, coniferyl, and sinapyl alcohols	forms strong waterproof polymer that reinforces secondary cell walls
Proteins and glycoproteins	enzymes, hydroxyproline-rich proteins	responsible for wall turnover and remodeling helps defend against pathogens

cross-linking them in the process. Both the backbone and the side-chain sugars vary according to the plant species and its stage of development.

Coextensive with this network of cellulose microfibrils and cross-linking glycans is another cross-linked polysaccharide network based on pectins (see Figure 19–71). **Pectins** are a heterogeneous group of branched polysaccharides that contain many negatively charged galacturonic acid units. Because of their negative charge, pectins are highly hydrated and associated with a cloud of cations, resembling the glycosaminoglycans of animal cells in the large amount of space they occupy (see Figure 19–37). When Ca^{2+} is added to a solution of pectin molecules, it cross-links them to produce a semirigid gel (it is pectin that is added to fruit juice to make jelly). Certain pectins are particularly abundant in the *middle lamella*, the specialized region that cements together the walls of adjacent cells (see Figure 19–71); here, Ca^{2+} cross-links are thought to help hold cell-wall components together. Although covalent bonds also play a part in linking the components together, very little is known about their nature. Regulated separation of cells at the middle lamella underlies such processes as the ripening of tomatoes and the abscission (detachment) of leaves in the fall.

In addition to the two polysaccharide-based networks that are present in all plant primary cell walls, proteins can contribute up to about 5% of the wall's dry mass. Many of these proteins are enzymes, responsible for wall turnover and remodelling, particularly during growth. Another class of wall proteins contains high levels of hydroxyproline, as in collagen. These proteins are thought to strengthen the wall, and they are produced in greatly increased amounts as a local response to attack by pathogens. From the genome sequence of *Arabidopsis*, it has been estimated that more than 700 genes are required to synthesize, assemble, and remodel the plant cell wall. Some of the main polymers found in the primary and secondary cell wall are listed in Table 19–8.

For a plant cell to grow or change its shape, the cell wall has to stretch or deform. Because of their crystalline structure, however, individual cellulose microfibrils are unable to stretch. Thus, stretching or deformation of the cell wall must involve either the sliding of microfibrils past one another, the separation of adjacent microfibrils, or both. As we discuss next, the direction in which the growing cell enlarges depends in part on the orientation of the cellulose microfibrils in the primary wall, which in turn depends on the orientation of microtubules in the underlying cell cortex at the time the wall was deposited.

Microtubules Orient Cell-Wall Deposition

The final shape of a growing plant cell, and hence the final form of the plant, is determined by controlled cell expansion. Expansion occurs in response to turgor pressure in a direction that depends in part on the arrangement of the cellulose microfibrils in the wall. Cells, therefore, anticipate their future morphology by controlling the orientation of microfibrils that they deposit in the wall. Unlike most other matrix macromolecules, which are made in the endoplasmic

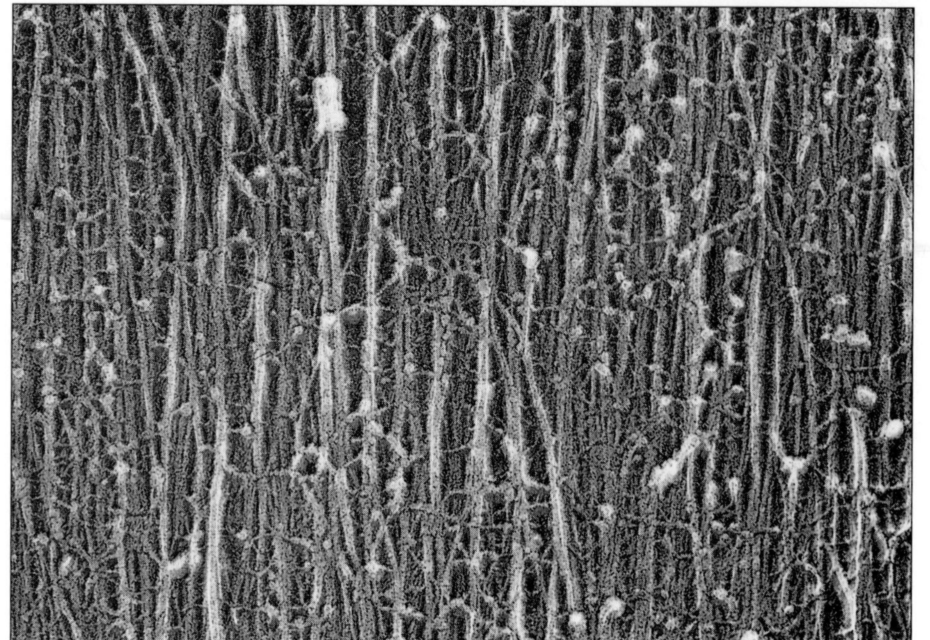

Figure 19–72 The orientation of cellulose microfibrils in the primary cell wall of an elongating carrot cell. This electron micrograph of a shadowed replica from a rapidly frozen and deep-etched cell wall shows the largely parallel arrangements of cellulose microfibrils, oriented perpendicular to the axis of cell elongation. The microfibrils are cross-linked by, and interwoven with, a complex web of matrix molecules (compare with Figure 19–71). (Courtesy of Brian Wells and Keith Roberts.)

reticulum and Golgi apparatus and are secreted, cellulose, like hyaluronan, is spun out from the surface of the cell by a plasma-membrane-bound enzyme complex (cellulose synthase), which uses as its substrate the sugar nucleotide UDP-glucose supplied from the cytosol. As they are being synthesized, the nascent cellulose chains assemble spontaneously into microfibrils that form on the extracellular surface of the plasma membrane—forming a layer, or lamella, in which all the microfibrils have more or less the same alignment (see Figure 19–71). Each new lamella forms internally to the previous one, so that the wall consists of concentrically arranged lamellae, with the oldest on the outside. The most recently deposited microfibrils in elongating cells commonly lie perpendicular to the axis of cell elongation (Figure 19–72). Although the orientation of the microfibrils in the outer lamellae that were laid down earlier may be different, it is the orientation of these inner lamellae that is thought to have a dominant influence on the direction of cell expansion (Figure 19–73).

An important clue to the mechanism that dictates this orientation came from observations of the microtubules in plant cells. These are arranged in the cortical cytoplasm with the same orientation as the cellulose microfibrils that are currently being deposited in the cell wall in that region. These cortical microtubules form a *cortical array* close to the cytosolic face of the plasma membrane, held there by poorly characterized proteins (Figure 19–74). The congruent orientation of the cortical array of microtubules (lying just inside the plasma membrane) and cellulose microfibrils (lying just outside) is seen in many types and shapes of plant cells and is present during both primary and secondary cell-wall deposition, suggesting a causal relationship.

If the entire system of cortical microtubules is disassembled by treating a plant tissue with a microtubule-depolymerizing drug, the consequences for sub-

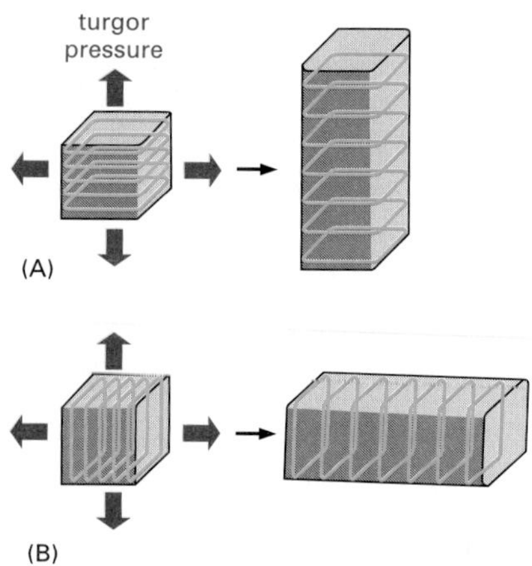

Figure 19–73 How the orientation of cellulose microfibrils within the cell wall influences the direction in which the cell elongates. The cells in (A) and (B) start off with identical shapes (shown here as cubes) but with different orientations of cellulose microfibrils in their walls. Although turgor pressure is uniform in all directions, cell-wall weakening causes each cell to elongate in a direction perpendicular to the orientation of the microfibrils, which have great tensile strength. The final shape of an organ, such as a shoot, is determined by the direction in which its cells expand.

sequent cellulose deposition are not as straightforward as might be expected. The drug treatment has no effect on the production of new cellulose microfibrils, and in some cases cells can continue to deposit new microfibrils in the preexisting orientation. Any developmental change in the microfibril pattern that would normally occur between successive lamellae, however, is invariably blocked. It seems that a preexisting orientation of microfibrils can be propagated even in the absence of microtubules, but any change in the deposition of cellulose microfibrils requires that intact microtubules be present to determine the new orientation.

These observations are consistent with the following model. The cellulose-synthesizing complexes embedded in the plasma membrane are thought to spin out long cellulose molecules. As the synthesis of cellulose molecules and their self-assembly into microfibrils proceeds, the distal end of each microfibril presumably forms indirect cross-links to the previous layer of wall material as it becomes integrated into the texture of the wall. At the growing, proximal end of each microfibril, the synthesizing complexes would therefore need to move through the membrane in the direction of synthesis. Since the growing cellulose microfibrils are stiff, each layer of microfibrils would tend to be spun out from the membrane in the same orientation as the previously laid down layer, with the cellulose synthase complex following along the preexisting tracks of oriented microfibrils outside the cell. Oriented microtubules inside the cell, however, can change this predetermined direction in which the synthase complexes move: they can create boundaries in the plasma membrane that act like the banks of a canal to constrain movement of the synthase complexes (Figure 19–75). In this view, cellulose synthesis can occur independently of microtubules but is constrained spatially when cortical microtubules are present to define membrane domains within which the enzyme complex can move.

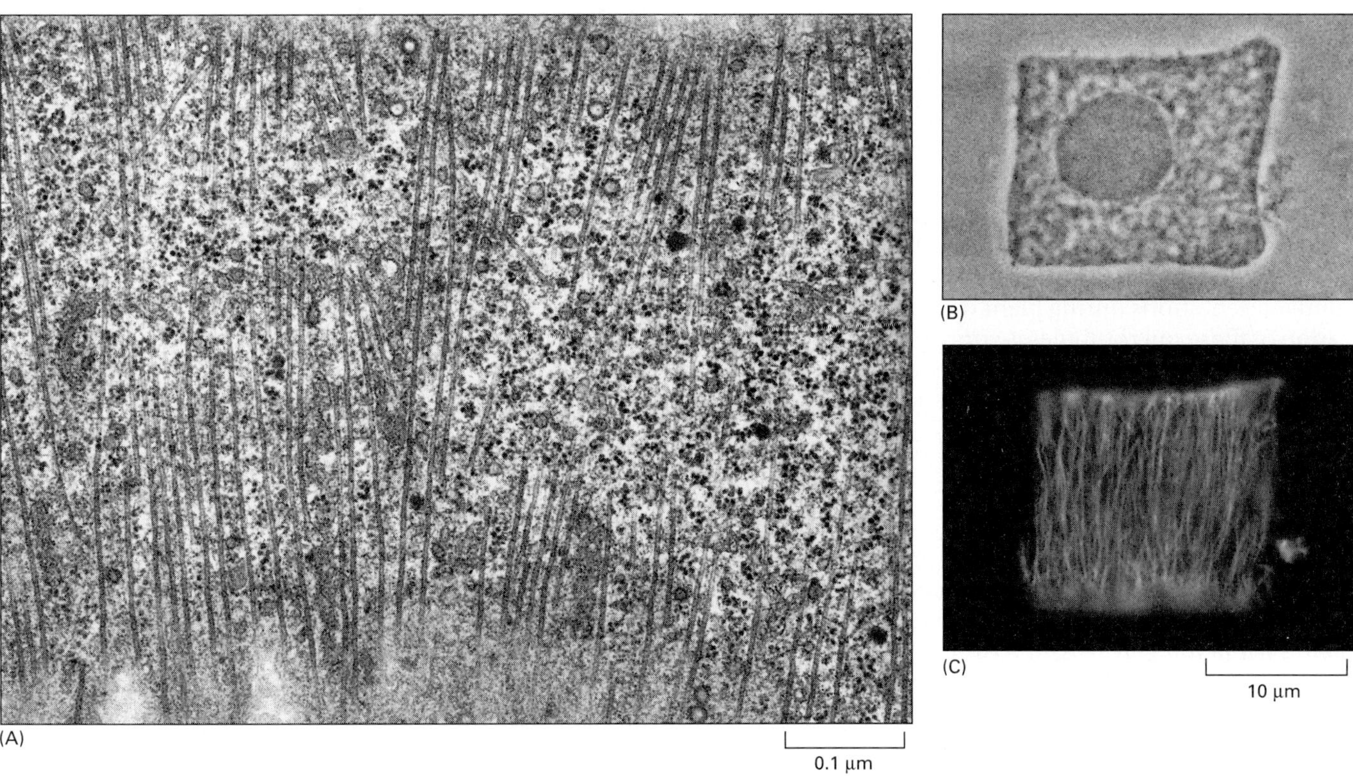

Figure 19–74 The cortical array of microtubules in a plant cell. (A) A grazing section of a root-tip cell from Timothy grass, showing a cortical array of microtubules lying just below the plasma membrane. These microtubules are oriented perpendicularly to the long axis of the cell. (B) An isolated onion root-tip cell. (C) The same cell shown in (B) stained by immunofluorescence to show the transverse cortical array of microtubules. (A, courtesy of Brian Gunning; B and C, courtesy of Kim Findlay.)

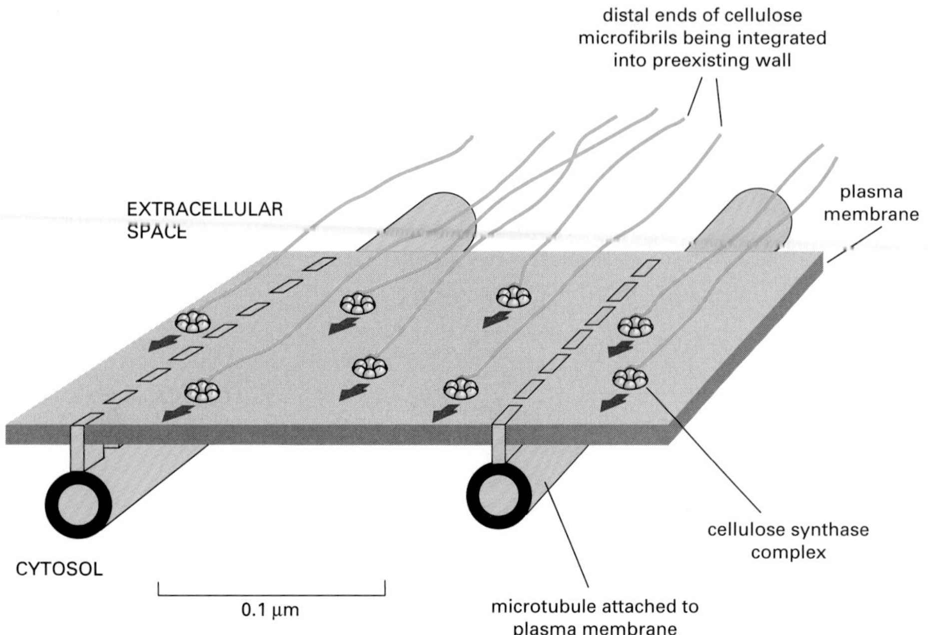

Figure 19–75 One model of how the orientation of newly deposited cellulose microfibrils might be determined by the orientation of cortical microtubules. The large cellulose synthase complexes are integral membrane proteins that continuously synthesize cellulose microfibrils on the outer face of the plasma membrane. The distal ends of the stiff microfibrils become integrated into the texture of the wall, and their elongation at the proximal end pushes the synthase complex along in the plane of the membrane. Because the cortical array of microtubules is attached to the plasma membrane in a way that confines this complex to defined membrane channels, the orientation of these microtubules—when they are present—determines the axis along which the new microfibrils are laid down.

Plant cells can change their direction of expansion by a sudden change in the orientation of their cortical array of microtubules. Because plant cells cannot move (being constrained by their walls), the entire morphology of a multicellular plant depends on the coordinated, highly patterned control of cortical microtubule orientations during plant development. It is not known how the organization of these microtubules is controlled, although it has been shown that they can reorient rapidly in response to extracellular stimuli, including low-molecular-weight plant growth regulators such as ethylene and gibberellic acid (see Figure 21–113).

Summary

Plant cells are surrounded by a tough extracellular matrix in the form of a cell wall, which is responsible for many of the unique features of a plant's life style. The cell wall is composed of a network of cellulose microfibrils and cross-linking glycans embedded in a highly cross-linked matrix of pectin polysaccharides. In secondary cell walls, lignin may be deposited. A cortical array of microtubules can determine the orientation of newly deposited cellulose microfibrils, which in turn determines directional cell expansion and therefore the final shape of the cell and, ultimately, of the plant as a whole.

References

General

Ayad S, Boot-Handford R, Humphries M et al. (1998) Extracellular Matrix Factsbook. London: Academic Press.

Beckerle M (ed) (2002) Cell Adhesion. Oxford: Oxford University Press.

Howlett A (ed) (1999) Integrins in Biological Processes. Totowa, NJ: Humana Press.

Kreis T & Vale R (ed) (1999) Guidebook to the Extracellular Matrix, Anchor, and Adhesion Proteins, 2nd edn. Oxford: Oxford University Press.

Cell Junctions

Green KJ & Gaudry CA (2000) Are desmosomes more than tethers for intermediate filaments? *Nature Rev. Mol. Cell Biol.* 1, 208–216.

Jockusch BM, Bubeck P, Giehl K et al. (1995) The molecular architecture of focal adhesions. *Annu. Rev. Cell Dev. Biol.* 11, 379–416.

Madara JL (1998) Regulation of the movement of solutes across tight junctions. *Annu. Rev. Physiol.* 60, 143–159.

Nievers MG, Schaapveld RQ & Sonnenberg A (1999) Biology and function of hemidesmosomes. *Matrix Biol.* 18, 5–17.

Saitou M, Furuse M, Sasaki H, Schulzke JD, Fromm M, Takano H, Noda T & Tsukita S (2000) Complex phenotype of mice lacking occludin, a component of tight junction strands. *Mol. Biol. Cell* 11, 4131–4142.

Sastry SK & Burridge K (2000) Focal adhesions: a nexus for intracellular signaling and cytoskeletal dynamics. *Exp. Cell Res.* 261, 25–36.

Simon AM & Goodenough DA (1988) Diverse functions of vertebrate gap junctions. *Trends Cell Biol.* 8, 477–483.

Tsukita S & Furuse M (1999) Occludin and claudins in tight-junction strands: leading or supporting players?, *Trends Cell Biol.* 9, 268–273.

Tsukita S & Furuse M (2000) Pores in the wall: claudins constitute tight junction strands containing aqueous pores. *J. Cell Biol.* 149, 13–16.

Yap AS, Brieher WM & Gumbiner BM (1977) Molecular and functional analysis of cadherin-based adherens junctions. *Annu. Rev. Cell Dev. Biol.* 13, 119–146.

Yeager M, Unger VM & Falk MM (1998) Synthesis, assembly and structure of gap junction intercellular channels. *Curr. Opin. Struct. Biol.* 8, 517–524.

Zambryski P & Crawford K (2000) Plasmodesmata: gatekeepers for cell-to-cell transport of developmental signals in plants. *Annu. Rev. Cell Dev. Biol.* 16, 393–421.

Cell–cell Adhesion

Crossin KL & Krushel LA (2000) Cellular signaling by neural cell adhesion molecules of the immunoglobulin superfamily. *Dev. Dyn.* 218, 260–279.

Gumbiner BM (1996) Cell adhesion: the molecular basis of tissue architecture and morphogenesis. *Cell* 84, 345–357.

Hynes RO & Zhao Q (2000) The evolution of cell adhesion. *J. Cell Biol.* 150, F89–F96.

Koch AW, Bozic D, Pertz O & Engel J (1999) Homophilic adhesion by cadherins. *Curr. Opin. Struct. Biol.* 9, 275–281.

Salzer JL & Colman DR (1989) Mechanisms of cell adhesion in the nervous system: role of the immunoglobulin gene superfamily. *Dev. Neurosci.* 11, 377–390.

Takeichi M, Nakagawa S, Aono S et al. (2000) Patterning of cell assemblies regulated by adhesion receptors of the cadherin superfamily. *Philos. Trans. R. Soc. Lond. B. Biol. Sci.* 355, 885–890.

Tepass U, Truong K, Godt D et al. (2000) Cadherins in embryonic and neuronal morphogenesis. *Nature Rev. Mol. Cell Biol.* 1, 91–100.

Vestweber D & Blanks JE (1999) Mechanisms that regulate the function of the selectins and their ligands. *Physiol. Rev.* 79, 181–213.

Vleminckx K & Kemler R (1999) Cadherins and tissue formation: integrating adhesion and signaling. *Bioessays* 21, 211–220.

Yagi T & Takeichi M (2000) Cadherin superfamily genes: functions, genomic organization, and neurologic diversity. *Genes Dev.* 14, 1169–1180.

The Extracellular Matrix of Animals

Bernfield M, Gotte M, Park PW et al. (1999) Functions of cell surface heparan sulfate proteoglycans. *Annu. Rev. Biochem.* 68, 729–777.

Chiquet-Ehrismann R (1995) Inhibition of cell adhesion by anti-adhesive molecules. *Curr. Opin. Cell Biol.* 7, 715–719.

Colognato H & Yurchenco PD (2000) Form and function: the laminin family of heterotrimers. *Dev. Dyn.* 18, 213–234.

Debelle L & Tamburro AM (1999) Elastin: molecular description and function. *Int. J. Biochem. Cell Biol.* 31, 261–72.

Geiger B, Bershadsky A, Pankov R & Yamada KM (2001) Transmembrane crosstalk between the extracellular matrix and the cytoskeleton. *Nat. Rev. Mol. Cell Biol.* 2, 793–805.

Henry MD & Campbell KP (1999) Dystroglycan inside and out. *Curr. Opin. Cell Biol.* 11, 602–607.

Huang S & Ingber DE (1999) The structural and mechanical complexity of cell-growth control. *Nature Cell Biol.* 1, E131–E138.

Hutter H, Vogel BE, Plenefisch JD et al. (2000) Conservation and novelty in the evolution of cell adhesion and extracellular matrix genes. *Science* 287, 989–894.

Iozzo RV (1998) Matrix proteoglycans: from molecular design to cellular function. *Annu. Rev. Biochem.* 67, 609–652.

Leahy DJ (1997) Implications of atomic-resolution structures for cell adhesion. *Annu. Rev. Cell Dev. Biol.* 13, 363–393.

Prockop DJ & Kivirikko KI (1995) Collagens: molecular biology, diseases, and potentials for therapy. *Annu. Rev. Biochem.* 64, 403–434.

Sanes JR & Lichtman JW (1999) Development of the vertebrate neuromuscular junction. *Annu. Rev. Neurosci.* 22, 389–442.

Schwarzbauer JE & Sechler JL (1999) Fibronectin fibrillogenesis: a paradigm for extracellular matrix assembly. *Curr. Opin. Cell Biol.* 11, 622–627.

Tessier-Lavigne M & Goodman CS (1996) The molecular biology of axon guidance. *Science* 274, 1123–1133.

Toole BP (2000) Hyaluronan is not just a goo! *J. Clin. Invest.* 106, 335–336.

Vu TH & Werb Z (2000) Matrix metalloproteinases: effectors of development and normal physiology. *Genes Dev.* 14. 2123–2133.

Integrins

Brown NH, Gregory SL & Martin-Bermudo MD (2000) Integrins as mediators of morphogenesis in Drosophila. *Dev. Biol.* 223, 1–16.

Calderwood DA, Shattil SJ & Ginsberg MH (2000) Integrins and actin filaments: reciprocal regulation of cell adhesion and signaling. *J. Biol.Chem.* 275, 22607–22610.

Giancotti FG & Ruoslahti E (1999) Integrin signaling. *Science* 285, 1028–1032.

Harris ES, McIntyre TM, Prescott SM & Zimmerman GA (2000) The leukocyte integrins. *J. Biol. Chem.* 275, 23409–23412.

Hynes RO (1992) Integrins: versatility, modulation, and signaling in cell adhesion. *Cell* 69, 11–25.

Hynes RO (1996) Targeted mutations in cell adhesion genes: what have we learned from them? *Dev. Biol.* 180, 402–412.

Parsons JT, Martin KH & Slack JK (2000) Focal adhesion kinase: a regulator of focal adhesion dynamics and cell movement. *Oncogene* 19, 5606–5613.

Plow EF, Haas TA, Zhang L et al. (2000) Ligand binding to integrins. *J. Biol. Chem.* 275, 21785–21788.

The Plant Cell Wall

Carpita NC & McCann M (2000) The Cell Wall. In Biochemistry and Molecular Biology of Plants (Buchanan BB, Gruissem W & Jones RL eds), pp 52–108. Rockville, MD: ASPB.

Carpita NC, Campbell M & Tierney M (eds) (2001) Plant cell walls. *Plant Mol. Biol.* 47, 1–340. (special issue)

Cosgrove DJ (2001) Wall structure and wall loosening. A look backwards and forwards. *Plant Physiol.* 125, 131–134.

Dhugga KS (2001) Building the wall: genes and enzyme complexes for polysaccharide synthases. *Curr. Opin. Plant Biol.* 4, 488–493.

McCann MC & Roberts K (1991) Architecture of the primary cell wall. In The Cytoskeletal Basis of Plant Growth and Form (Lloyd CW ed), pp 109–129. London: Academic Press.

Reiter W-D (1998) *Arabidopsis thaliana* as a model system to study synthesis, structure and function of the plant cell wall. *Plant Physiol. Biochem.* 36, 167–176.

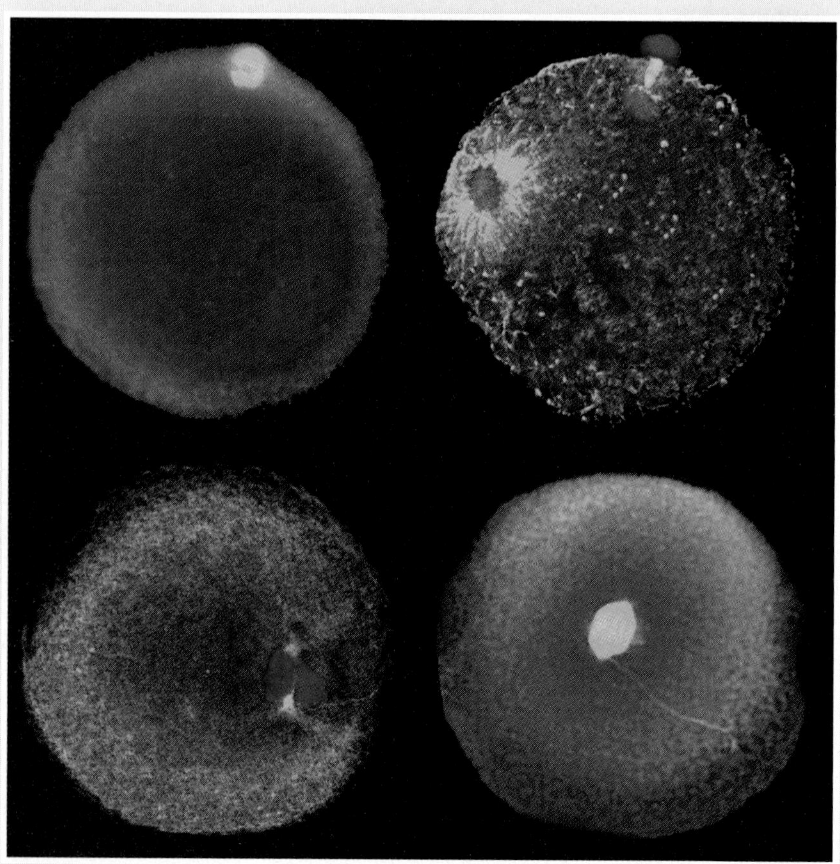

Human fertilization *in vitro*. Four sequential stages of fertilization are shown, beginning with a mature, but unfertilized human egg *(top left)*. Next comes the post-fertilization incorporation of a sperm pronucleus, which in this example is moving inward from the left *(top right)*. This stage is followed by the fusion of the egg and sperm pronuclei *(bottom left)*, which in turn leads to the alignment of the mixed chromosomes from both parents on the first mitotic spindle *(bottom right)*. (From C. Simerly et al., *Nat. Med.* 1:47–53. © Macmillan Magazines Ltd.)

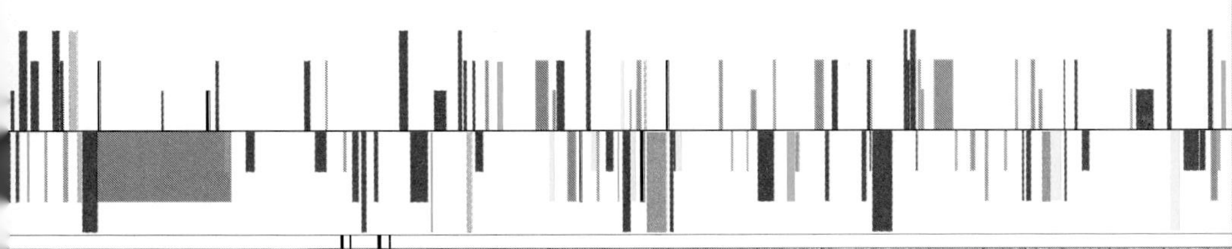

20

GERM CELLS AND FERTILIZATION

Sex is not absolutely necessary. Single-celled organisms can reproduce by simple mitotic division, and many plants propagate vegetatively, by forming multicellular offshoots that later detach from the parent. Likewise, in the animal kingdom, a solitary multicellular *Hydra* can produce offspring by budding (Figure 20–1). Sea anemones and marine worms can split into two half-organisms, each of which then regenerates its missing half. There are even species of lizards that consist only of females and reproduce without mating. Although such **asexual reproduction** is simple and direct, it gives rise to offspring that are genetically identical to the parent organism. In **sexual reproduction**, on the other hand, the genomes from two individuals are mixed to produce offspring that differ genetically from one another and from both their parents. This mode of reproduction apparently has great advantages, as the vast majority of plants and animals have adopted it. Even many procaryotes and other organisms that normally reproduce asexually engage in occasional bouts of sexual reproduction, thereby creating offspring with new combinations of genes. This chapter describes the cellular machinery of sexual reproduction. Before discussing in detail how the machinery works, however, we shall pause briefly to consider why it exists and what benefits it brings.

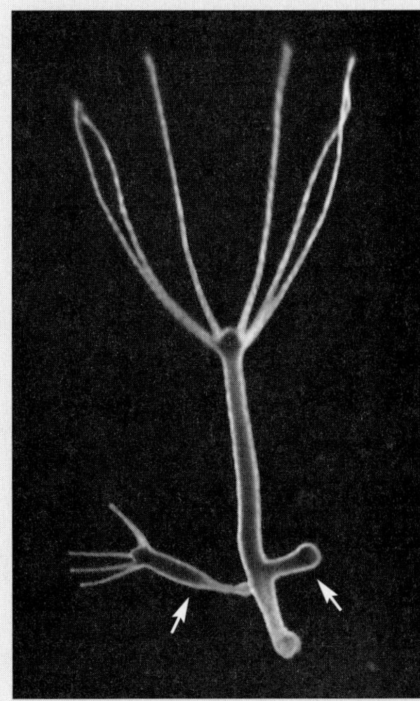

Figure 20–1 Photograph of a *Hydra* from which two new organisms are budding *(arrows)*. The offspring, which are genetically identical to their parent, will eventually detach and live independently. (Courtesy of Amata Hornbruch.)

THE BENEFITS OF SEX

In the sexual reproductive cycle, **haploid** generations of cells, each carrying a single set of chromosomes, alternate with **diploid** generations of cells, each carrying a double set of chromosomes (Figure 20–2). Genomes mix when two haploid cells fuse to form a diploid cell. Later, new haploid cells are generated when a descendant of this diploid cell divides by the process of *meiosis*. During meiosis, the chromosomes of the double chromosome set exchange DNA by *genetic recombination* before being shared out, in new combinations, into single chromosome sets. Because each single chromosome set will contain genes originating from one ancestral cell of the previous haploid generation mixed with genes from the other ancestral cell, each cell of the new haploid generation will receive a novel assortment of genes. Thus, through cycles of haploidy, cell fusion, diploidy, and meiosis, old combinations of genes are broken up and new combinations are created.

In Multicellular Animals and Most Plants, the Diploid Phase Is Complex and Long, the Haploid Simple and Fleeting

Cells proliferate by mitotic division. In most organisms that reproduce sexually, this proliferation occurs during the diploid phase. Some primitive organisms, such as fission yeasts, are exceptional in that the haploid cells proliferate mitotically and the diploid cells, once formed, proceed directly to meiosis. A less extreme exception occurs in plants, where mitotic cell divisions occur in both the haploid and the diploid phases. In all but the most primitive plants, such as mosses and ferns, however, the haploid phase is very brief and simple, while the diploid phase is extended into a long period of development and proliferation. For almost all multicellular animals, including vertebrates, practically the whole of the life cycle is spent in the diploid state: the haploid cells exist only briefly, do not divide at all, and are highly specialized for sexual fusion (Figure 20–3).

Haploid cells that are specialized for sexual fusion are called **gametes**. Typically, two types of gametes are formed: one is large and nonmotile and is referred to as the *egg* (or *ovum);* the other is small and motile and is referred to as the *sperm* (or *spermatozoon)* (Figure 20–4). During the diploid phase that follows the fusion of gametes, the cells proliferate and diversify to form a complex multicellular organism. In most animals, a useful distinction can be drawn between the cells of the **germ line**, from which the next generation of gametes will be derived, and the **somatic cells**, which form the rest of the body and ultimately leave no progeny. In a sense, the somatic cells exist only to help the cells of the germ line (the **germ cells**) survive and propagate.

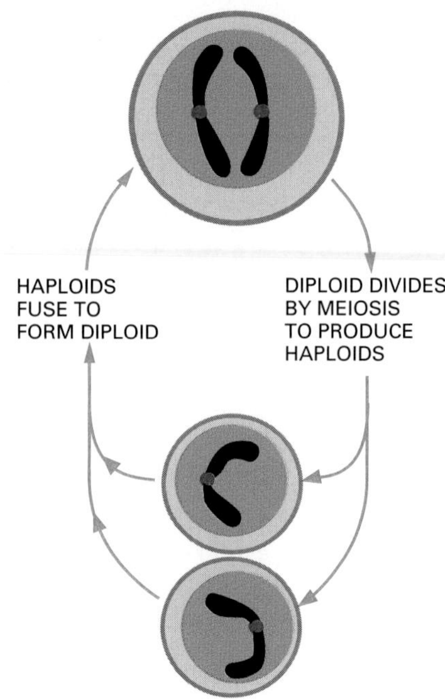

Figure 20–2 The sexual reproductive cycle. It involves an alternation of haploid and diploid generations of cells.

Sexual Reproduction Gives a Competitive Advantage to Organisms in an Unpredictably Variable Environment

The machinery of sexual reproduction is elaborate, and the resources spent on it are large (Figure 20–5). What benefits does it bring, and why did it evolve? Through genetic recombination, sexual individuals produce unpredictably dissimilar offspring, whose haphazard genotypes are at least as likely to represent a

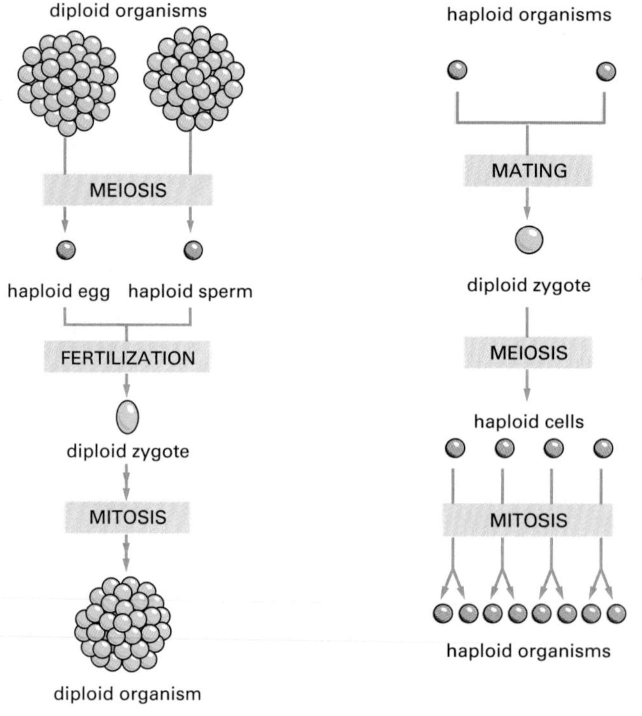

Figure 20–3 Haploid and diploid cells in the life cycle of higher and some lower eucaryotes. The haploid cells are shown in *red* and the diploid cells in *blue*. Cells in higher eucaryotic organisms usually proliferate in the diploid phase to form a multicellular organism; only the gametes are haploid, and they fuse at fertilization to form a diploid zygote, which develops into a new individual. In some lower eucaryotes, by contrast, the haploid cells proliferate, and the only diploid cell is the zygote, which exists transiently after mating.

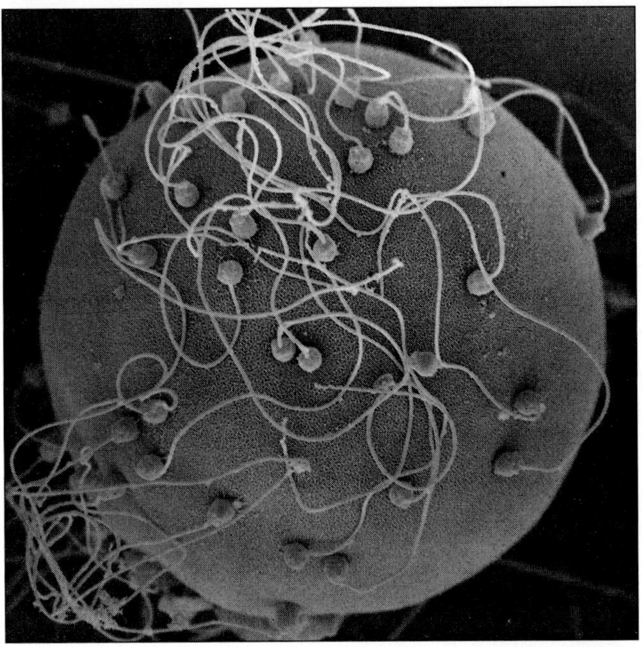

Figure 20–4 Scanning electron micrograph of a clam egg with sperm bound to its surface. Although many sperm are bound to the egg, only one will fertilize it, as we discuss later. (Courtesy of David Epel.)

change for the worse as a change for the better. Why, then, should sexual individuals have a competitive advantage over individuals that breed true, by an asexual process? This problem continues to perplex population geneticists, but the general conclusion seems to be that the reshuffling of genes in sexual reproduction helps a species to survive in an unpredictably variable environment. If a parent produces many offspring with a wide variety of gene combinations, there is a better chance that at least one of the offspring will have the assortment of features necessary for survival. Sexual reproduction also allows the many deleterious mutations that accumulate randomly to be eliminated, while permitting the rare advantageous mutations that arise in separate individuals to be combined in a single individual.

Whatever the benefits of sex may be, it is striking that practically all complex present-day organisms have evolved largely through generations of sexual, rather than asexual, reproduction. Asexual organisms, although plentiful, seem mostly to have remained simple and primitive.

We now turn to the cellular mechanisms of sex, beginning with the events of meiosis, which segregates the chromosomes into new sets, as the diploid cells in the germ line divide to produce haploid gametes. We then focus our discussion on mammals. We consider the diploid cells of the germ line that give rise to the gametes and how the sex of a mammal is determined. Finally, we discuss the nature of the gametes themselves, as well as the process of *fertilization,* in which two gametes fuse to form a *zygote,* which develops into a new diploid organism.

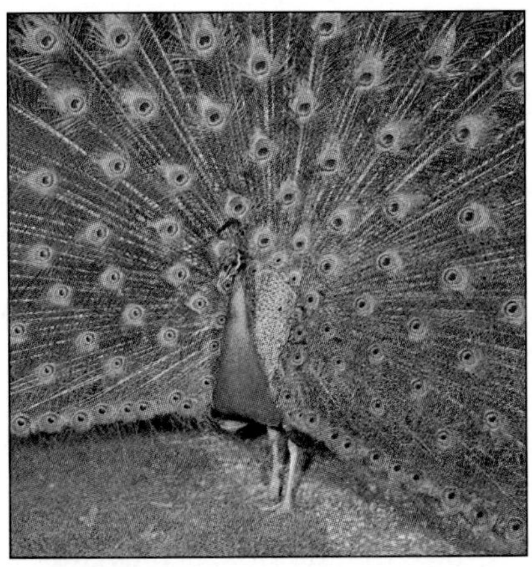

Figure 20–5 A peacock displaying his elaborate tail. This extravagant plumage serves solely to attract females for the purpose of sexual reproduction. (© Cyril Laubscher.)

Summary

Sexual reproduction has been favored by evolution probably because the random recombination of genetic information improves the chances of producing at least some offspring that will survive in an unpredictably variable environment. The sexual reproductive cycle involves an alternation of diploid and haploid states: diploid cells divide by meiosis to form haploid cells, and the haploid cells from two individuals fuse in pairs at fertilization to form new diploid cells. In the process, genomes are mixed and recombined to produce individuals that inherit novel assortments of genes. Most of the life cycle of higher plants and animals is spent in the diploid phase; only a small proportion of the diploid cells (those in the germ line) undergo meiosis to produce haploid cells (the gametes), and the haploid phase is very brief.

MEIOSIS

The realization that gametes are haploid, and must therefore be produced by a special type of cell division, came from an observation that was also among the first to suggest that chromosomes carry genetic information. In 1883, it was discovered that, whereas the fertilized egg of a roundworm contains four chromosomes, the nucleus of the egg and that of the sperm each contain only two chromosomes. The chromosome theory of heredity therefore explained the long-standing paradox that the maternal and paternal contributions to the character of the progeny seem to be equal, despite the enormous difference in size between the egg and sperm (see Figure 20–4).

The finding also implied that germ cells must be formed by a special kind of nuclear division in which the chromosome complement is precisely halved. This type of division is called **meiosis**, from the Greek, meaning diminution. *(Mitosis,* which refers to the nuclear division that occurs during an ordinary mitotic cell division (discussed in Chapter 18), is from the Greek word *mitos,* meaning "a thread." The term refers to the threadlike appearance of the chromosomes as they condense during nuclear division—a process that occurs in both meiotic and mitotic divisions.) The behavior of the chromosomes during meiosis turned out to be considerably more complex than expected. Consequently, it was not until the early 1930s, as a result of painstaking cytological and genetic studies, that the essential events of meiosis were finally established. More recent genetic and molecular studies have begun to identify the meiosis-specific proteins that cause the chromosomes to behave in a special way and mediate the genetic recombination events that occur in meiosis.

Duplicated Homologous Chromosomes Pair During Meiosis

The set of chromosomes of a typical sexually-reproducing organism consists of *autosomes,* which are common to all members of the species, and *sex chromosomes,* which are differently allocated according to the sex of the individual. A diploid nucleus contains two closely similar versions of each chromosome. For each of the autosomal chromosome pairs, one member was initially inherited from the male parent (a paternal chromosome) and the other was initially inherited from the female parent (a maternal chromosome). The two versions, which are very similar but not identical in DNA sequence, are called **homologs**, and in most cells they maintain a completely separate existence as independent chromosomes.

After a chromosome is duplicated by DNA replication, the twin copies of the fully replicated chromosome at first remain tightly linked along their length and are called **sister chromatids**. In a mitotic cell division, the sister chromatids line up at the equator of the spindle with their *kinetochores* (protein complexes associated with the centromeres, discussed in Chapter 18) and attached microtubules pointing toward opposite poles. The sister chromatids then separate completely from each other at anaphase to become individual chromosomes. In this manner each daughter cell formed by a mitotic cell division inherits one copy of each paternal chromosome and one copy of each maternal chromosome and is therefore unchanged in its genetic composition from the parent cell.

In contrast, a haploid gamete produced from a diploid cell through meiosis must contain half the original number of chromosomes. It must contain only one chromosome in place of each homologous pair of chromosomes, so it is endowed with either the maternal or the paternal copy of each gene but not both. This requirement makes an extra demand on the machinery for cell division. The mechanism that has evolved to accomplish the additional sorting requires that homologs recognize each other and become physically connected side-by-side along their entire length before they line up on the spindle. How the maternal and the paternal copy of each chromosome recognize each other is still uncertain. In many organisms, the initial association (a process called **pairing**) seems to be mediated by complementary DNA base-pair interactions at numerous and widely dispersed sites along the chromosomes.

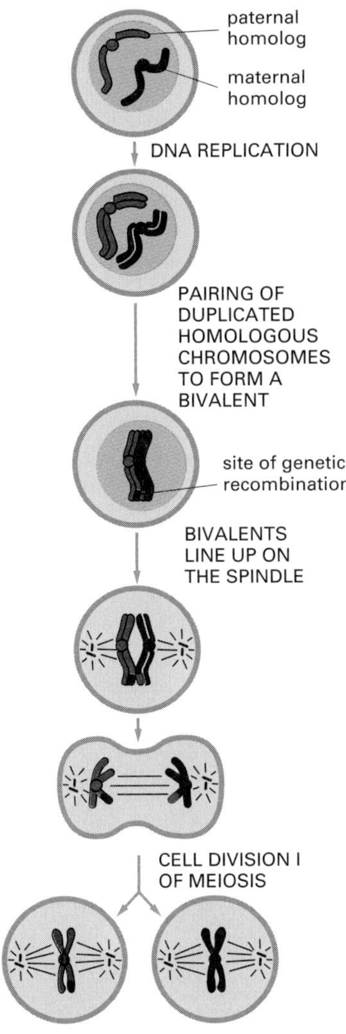

Figure 20–6 Events through the first cell division of meiosis. For clarity, only one pair of homologous chromosomes is shown. Each chromosome has been duplicated and exists as attached sister chromatids before pairing with its homologous chromosome (homolog), thereby forming a structure containing four chromatids known as a bivalent. As shown by the formation of chromosomes that are part *red* and part *black*, chromosome pairing in meiosis leads to genetic recombination between homologous chromosomes, as explained later.

Before the homologs pair, each chromosome in the diploid cell replicates to produce two sister chromatids, just as in a mitotic cell division. It is only after DNA replication has been completed that the special features of meiosis become evident. Each duplicated chromosome pairs with its duplicated homolog, forming a structure called a **bivalent**, which contains four chromatids. The pairing occurs during a long meiotic prophase, which often lasts for days and can last for years. As we shall see, pairing allows genetic recombination to occur, whereby a fragment of a maternal chromatid may be exchanged for a corresponding fragment of a homologous paternal chromatid. At the subsequent metaphase all of the bivalents line up on the spindle, and at anaphase the two duplicated homologs (each consisting of two sister chromatids) separate from each other and move to opposite poles of the spindle, and the cell divides (Figure 20–6). To produce haploid gametes, however, another cell division is required.

Gametes Are Produced by Two Meiotic Cell Divisions

The meiotic cell division just described—referred to as **division I of meiosis**—does not produce cells with a haploid amount of DNA. Because the sister chromatids behave as a unit, each daughter cell of this division inherits two copies of one of the two homologs. The two copies are identical except where genetic recombination has occurred. The two daughter cells therefore contain a haploid number of chromosomes but a diploid amount of DNA. They differ from normal diploid cells in two ways. First, the two DNA copies of each chromosome derive from only one of the two homologous chromosomes in the original cell (except for the bits exchanged by genetic recombination). Second, these two DNA copies are inherited as joined sister chromatids (see Figure 20–6).

Formation of the actual gamete nuclei can now proceed simply through a second cell division, **division II of meiosis**, without further DNA replication. The duplicated chromosomes align on a second spindle, and the sister chromatids separate to produce cells with a haploid DNA content. Meiosis thus consists of a single phase of DNA replication followed by two cell divisions. Four haploid cells are therefore produced from each cell that enters meiosis. Meiosis and mitosis are compared in Figure 20–7.

Occasionally during meiosis, chromosomes fail to separate normally into the four haploid cells, a phenomenon known as *nondisjunction*. In such abnormal meiotic divisions some of the haploid cells that are produced lack a chromosome, while others have more than one copy. The resulting gametes form abnormal embryos, most of which die. Some survive, however: *Down syndrome* in humans, for example, is caused by an extra copy of chromosome 21, resulting from nondisjunction during meiotic division I or II. The vast majority of such segregation errors occur during meiosis in females, and the error rate increases with advancing maternal age. The frequency of missegregation in human oocytes is remarkably high (about 10% of meioses), and this is thought to be one reason for the high rate of miscarriages (spontaneous abortions) in early pregnancy.

Genetic Reassortment Is Enhanced by Crossing-over Between Homologous Nonsister Chromatids

Unless they are identical twins, which develop from a single zygote, no two offspring of the same parents are genetically the same. This is because, long before

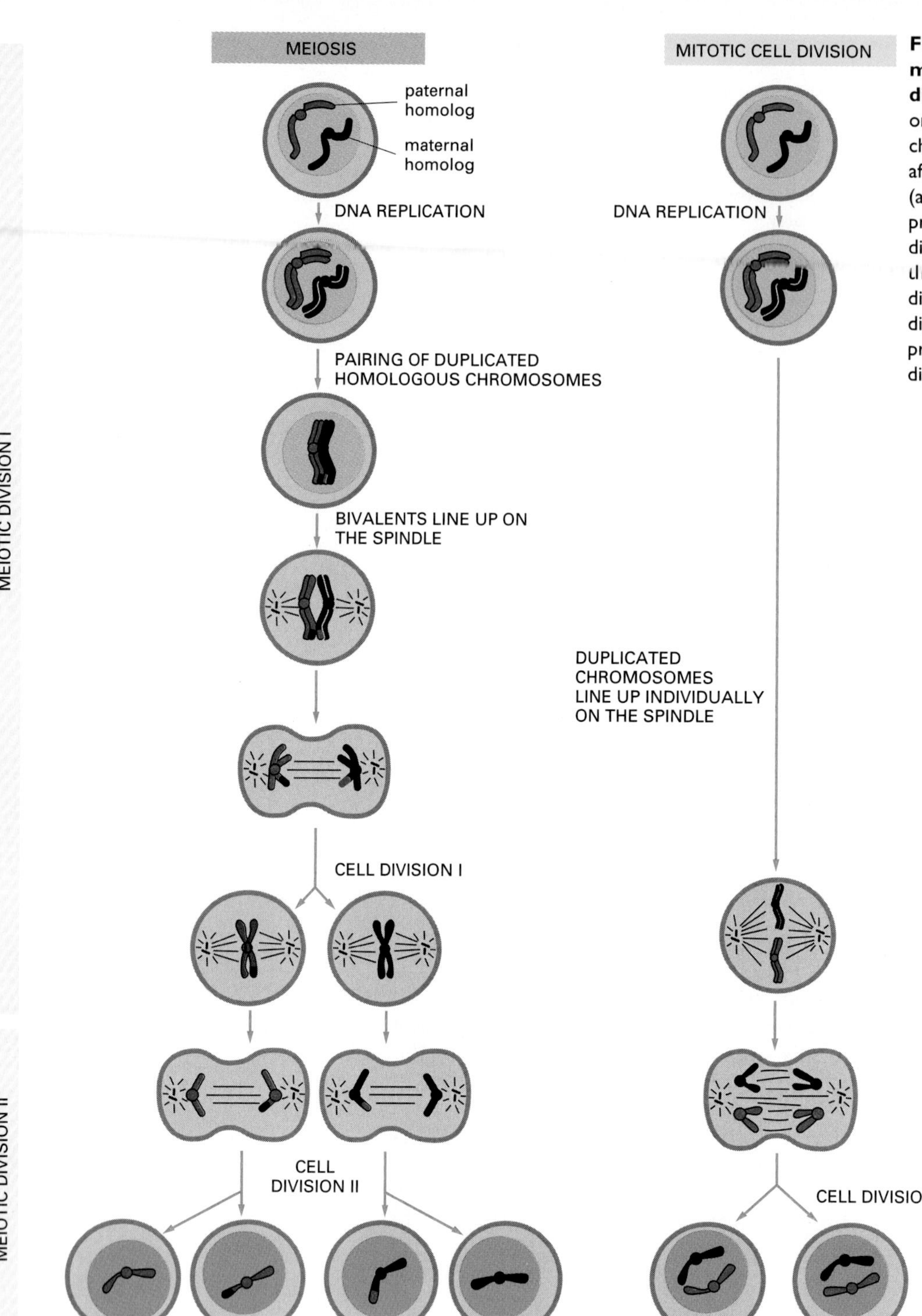

Figure 20–7 Comparison of meiosis and mitotic cell division. As in the previous figure, only one pair of homologous chromosomes is shown. In meiosis, after DNA replication, two nuclear (and cell) divisions are required to produce the haploid gametes. Each diploid cell that enters meiosis therefore produces four genetically different haploid cells, whereas each diploid cell that divides by mitosis produces two genetically identical diploid cells.

the two gametes fuse at fertilization, two kinds of randomizing genetic reassortment have occurred during meiosis.

One kind of reassortment is a consequence of the random distribution of the maternal and paternal homologs between the daughter cells at meiotic division I, as a result of which each gamete acquires a different mixture of maternal and paternal chromosomes. From this process alone, one individual could, in principle, produce 2^n genetically different gametes, where n is the haploid number of chromosomes (Figure 20–8A). In humans, for example, each individual can produce at least $2^{23} = 8.4 \times 10^6$ genetically different gametes. But the actual number of variants is very much greater than this because a second type of

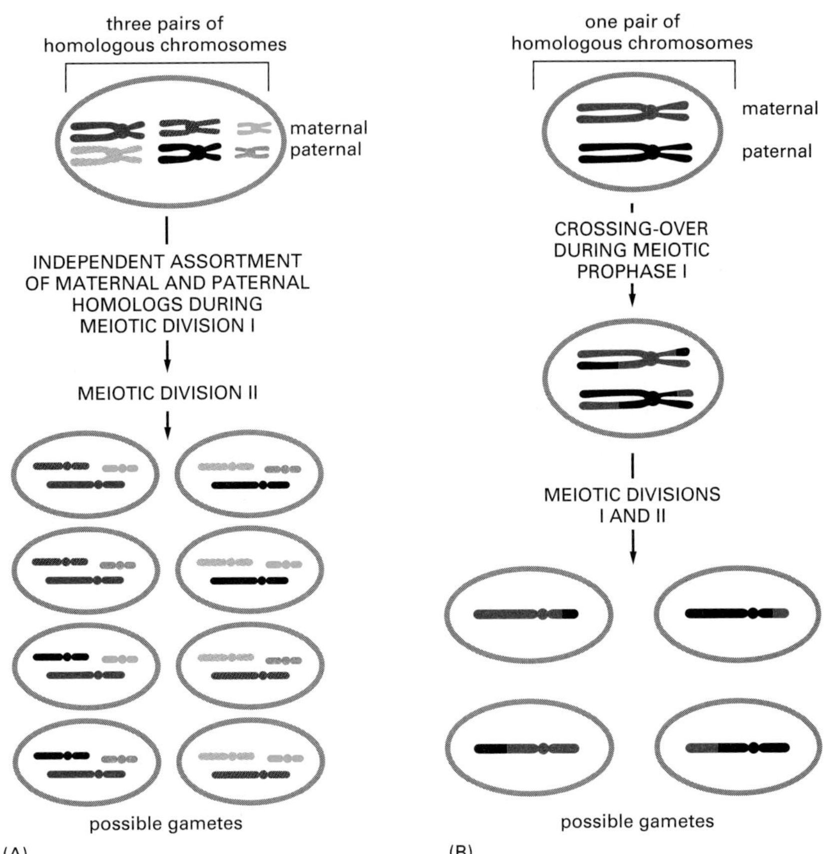

Figure 20–8 Two major contributions to the reassortment of genetic material that occurs in the production of gametes during meiosis. (A) The independent assortment of the maternal and paternal homologs during the first meiotic division produces 2^n different haploid gametes for an organism with *n* chromosomes. Here *n* = 3, and there are eight different possible gametes. (B) Crossing-over during meiotic prophase I exchanges segments of homologous chromosomes and thereby reassorts genes on individual chromosomes. Because of the many small differences in DNA sequence that always exist between any two homologs, both mechanisms increase the genetic variability of organisms that reproduce sexually.

reassortment, called **chromosomal crossing-over**, occurs during meiosis. It takes place during the long prophase of meiotic division I (prophase I), in which parts of homologous chromosomes are exchanged. On average, between two and three crossover events occur on each pair of human chromosomes during meiotic division I. This process scrambles the genetic constitution of each of the chromosomes in gametes, as illustrated in Figure 20–8B.

During chromosomal crossing-over, the DNA double helix is broken in both a maternal chromatid and a homologous paternal chromatid, so as to exchange fragments between the two nonsister chromatids in a reciprocal fashion by a process known as **genetic recombination**. The molecular details of this process are discussed in Chapter 5. The consequences of each crossover event can be observed in the microscope at the latest stages of prophase I, when the chromosomes in the bivalents are highly condensed. At this stage, the sister chromatids are tightly apposed along their entire length, and the two duplicated homologs (maternal and paternal) that form each bivalent are seen to be physically connected at specific points. Each connection, called a **chiasma** (plural **chiasmata**), corresponds to a crossover between two nonsister chromatids (Figure 20–9). Each of the two chromatids of a duplicated chromosome can cross over with either of the two chromatids of the other chromosome in the bivalent, as illustrated in Figure 20–10.

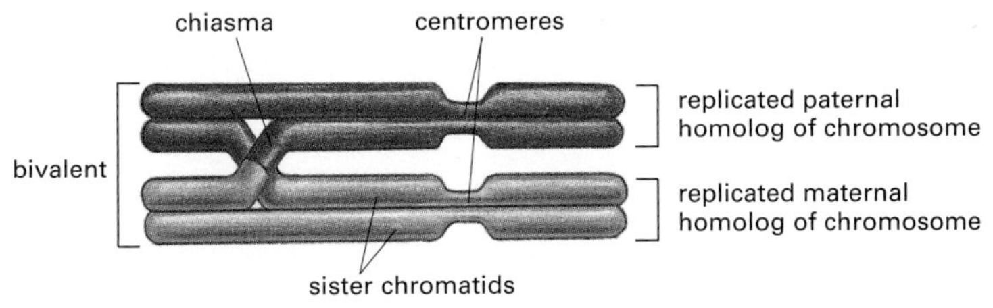

Figure 20–9 Paired homologous chromosomes during the transition to metaphase of meiotic division I. A single crossover event has occurred earlier in prophase to create one chiasma. Note that the four chromatids are arranged as two distinct pairs of sister chromatids. As in mitosis, the sister chromatids in each pair are tightly connected along their entire lengths, as well as at their centromeres, by proteins called cohesins. The entire unit of four chromatids is referred to as a bivalent. The combination of the chiasma and the tight attachment of the sister chromatids holds the two duplicated homologs together.

At this stage of meiosis, each pair of duplicated homologs is held together by at least one chiasma. Many bivalents contain more than one chiasma, indicating that multiple crossovers can occur between homologs.

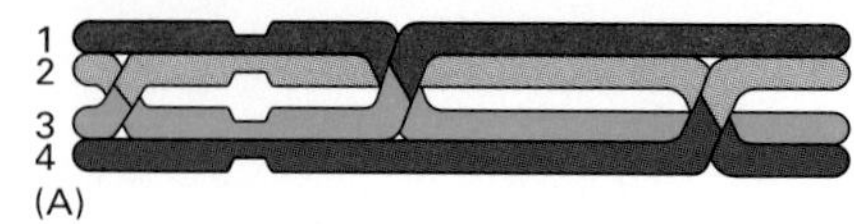

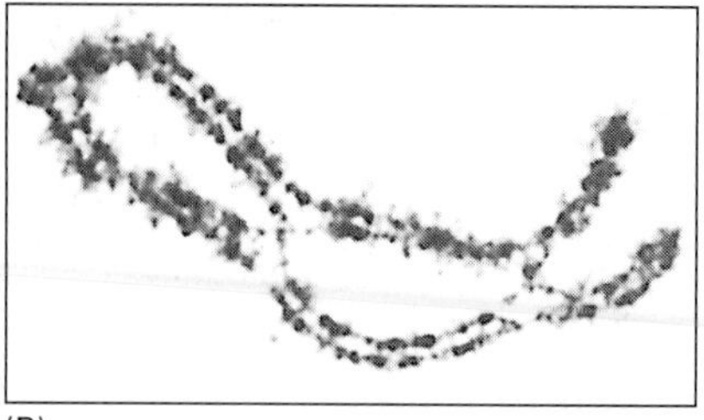

Figure 20–10 Bivalents with three chiasmata resulting from separate crossover events. (A) In this drawing, chromatid 1 has undergone an exchange with chromatid 3, and chromatid 2 has undergone exchanges with chromatid 3 and 4. Note that the sister chromatids of the same chromosome do not exchange with each other. (B) Light micrograph of a grasshopper bivalent with three chiasmata. (B, courtesy of Bernard John.)

Chiasmata Have an Important Role in Chromosome Segregation in Meiosis

In addition to reassorting genes, chromosomal crossing-over is crucial in most organisms for the correct segregation of the two duplicated homologs to separate daughter nuclei. This is because the chiasmata created by crossover events have a crucial role in holding the maternal and paternal homologs together until the spindle separates them at anaphase I (see Figure 20–9). Before anaphase I, the two poles of the spindle pull on the duplicated homologs in opposite directions, and the chiasmata resist this pulling. In mutant organisms that have a reduced frequency of meiotic chromosome crossing-over, some of the chromosome pairs lack chiasmata. These pairs fail to segregate normally, and many of the resulting gametes contain too many or too few chromosomes.

The duplicated homologs are held together at chiasmata only because the arms of sister chromatids are glued together along their length by proteins called *cohesins* (discussed in Chapter 18; see Figure 20–9). In *Drosophila,* for example, if a meiosis-specific cohesin is defective, sister chromatids separate prior to metaphase I and, as a consequence, the homologs segregate abnormally.

As illustrated in Figure 20–11, the arms of sister chromatids suddenly become unglued at the start of anaphase I, when the cohesins holding the arms together are degraded, allowing the duplicated homologs to separate and be

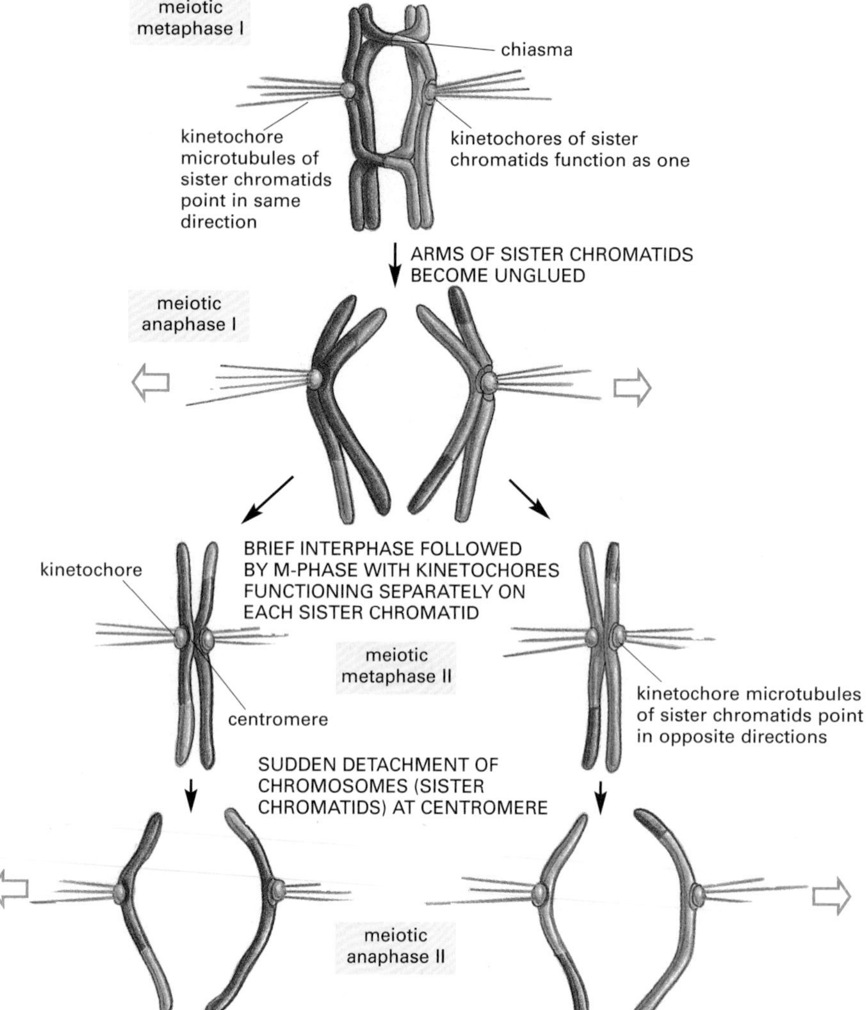

Figure 20–11 Comparison of the mechanisms of chromosome alignment (at metaphase) and separation (at anaphase) in meiotic division I and meiotic division II. The ungluing of the sister chromatid arms allows the duplicated homologs to separate at anaphase I, while an ungluing of the chromosomes at their centromeres allows the sister chromatids to separate at anaphase II. In contrast, at anaphase in mitosis, both the arms and the centromeres come apart at the same time (discussed in Chapter 18).

pulled to opposite poles of the spindle. The sister chromatids of each duplicated homolog remain attached at the centromere by meiosis-specific cohesins, which are degraded at anaphase of meiotic division II (anaphase II); only then can the sister chromatids separate.

In meiotic division II, as in a mitotic division, the kinetochores on each sister chromatid have attached kinetochore microtubules pointing in opposite directions, so that the chromatids are drawn into different daughter cells at anaphase. In meiotic division I, by contrast, the kinetochores on both sister chromatids behave as a single functional unit, as their attached kinetochore microtubules all point in the same direction so that the sister chromatids stay together when the duplicated homologs separate (see Figure 20–11). In budding yeasts, a meiosis-specific protein located at the kinetochores of meiosis I chromosomes has been shown to be required for this special behavior.

Pairing of the Sex Chromosomes Ensures That They Also Segregate

We have seen that duplicated homologous chromosomes must pair and form at least one chiasma during the first meiotic division if they are to segregate accurately between the daughter cells. But what happens to the **sex chromosomes**? Female mammals have two X chromosomes, which can pair and segregate like other homologs. But males have one X and one Y chromosome, and these chromosomes are not homologous. Yet, they must pair and then cross over during the first metaphase of meiosis if the sperm are to contain either one Y or one X chromosome and not both or neither. The crossovers are possible because of a small region of homology between the X and the Y at one end of these chromosomes. The two chromosomes pair and cross over in this region during prophase I. The chiasmata resulting from this genetic recombination keep the X and Y chromosomes connected on the spindle so that only two types of sperm are normally produced: sperm containing one Y chromosome, which will give rise to male embryos, and sperm containing one X chromosome, which will give rise to female embryos.

Having considered the general way in which chromosomes behave and segregate during meiosis, we now return to the process of genetic recombination that occurs during the long prophase of meiotic division I and has such an important role in reassorting genes during gamete formation.

Meiotic Chromosome Pairing Culminates in the Formation of the Synaptonemal Complex

A series of complex events occurs during the long prophase of meiotic division I: duplicated homologous chromosomes pair, genetic recombination is initiated between nonsister chromatids, and each pair of duplicated homologs assembles into an elaborate structure called the **synaptonemal complex**. In some organisms, genetic recombination begins before the synaptonemal complex assembles and is required for the complex to form; in others, the complex can form in the absence of recombination. In all organisms, however, the recombination process is completed while the DNA is held in the synaptonemal complex, which serves to space out the crossover events along each chromosome.

The prophase of meiotic division I is traditionally divided into five sequential stages—*leptotene, zygotene, pachytene, diplotene,* and *diakinesis*—defined by the morphological changes associated with the assembly *(synapsis)* and disassembly *(desynapsis)* of the synaptonemal complex. Prophase begins with **leptotene**, when the duplicated paired homologs condense. At **zygotene**, the synaptonemal complex begins to develop between the two sets of sister chromatids in each bivalent. **Pachytene** begins when synapsis is complete, and it generally persists for days, until desynapsis begins the **diplotene** stage, in which the chiasmata are first seen (Figure 20–12).

The synaptonemal complex consists of a long, ladderlike protein core, on opposite sides of which the two duplicated homologs are aligned to form a long

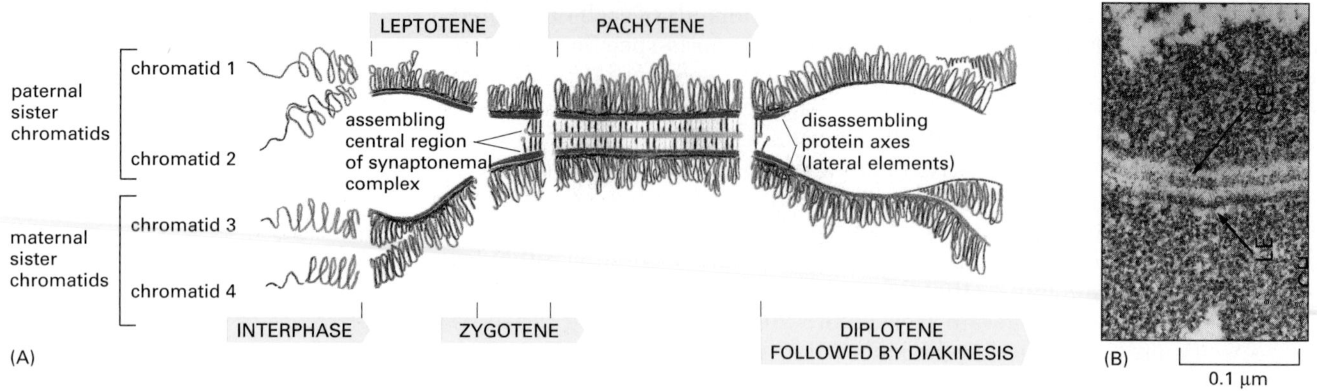

Figure 20–12 Chromosome synapsis and desynapsis during the different stages of meiotic prophase I. (A) A single bivalent is shown. The pachytene stage is defined as the period during which a fully formed synaptonemal complex exists. At leptotene, the two sister chromatids condense, and their chromatin loops each extend from a common protein axis *(red)*. As meiosis progresses, the two homologs become tightly connected by proteins that form the central region of the synaptonemal complex, composed of a central element *(blue)*, transverse filaments *(thin black lines)*, and the lateral elements *(red)* that anchor the chromatin loops. In the gametes of many female animals, but not those of mammals, the subsequent diplotene stage is an enormously prolonged period of cell growth, during which the chromosomes are decondensed and very active in transcription. Diplotene ends with diakinesis—the stage of transition to metaphase—in which the chromosomes recondense and transcription halts. In male gametes, diplotene and diakinesis are briefer and less distinct. (B) An electron micrograph of a synaptonemal complex from a meiotic cell at pachytene in a lily flower. (B, courtesy of Brian Wells.)

linear chromosome pair (Figure 20–13). The sister chromatids in each homolog are kept tightly packed together, with their DNA extending from their own side of the protein ladder in a series of loops. In the central region, a central element is connected by transverse filaments to lateral elements that run along each pair of sister chromatids, forming the sides of the ladder.

Several protein components of the synaptonemal complex have been identified and localized to specific structures of the complex. Yeast mutants that lack specific components have provided insights into the functions of the complex and some of its proteins. One yeast protein, for example, seems to nucleate the assembly of the lateral elements: if this protein is defective, these elements fail to form. Another yeast protein helps to form the transverse filaments: if this protein is absent, homolog pairing occurs without intimate synapsis, while an abnormally long mutant form of the protein creates a larger than normal separation between the two lateral elements of the synaptonemal complex.

Recombination Nodules Mark the Sites of Genetic Recombination

The crossover events that take place during the prophase of meiotic division I can occur nearly anywhere along a chromosome. They are not distributed uniformly, however: there are recombination "hot spots," where double-stranded DNA breaks seem to be preferentially induced by the meiotic endonuclease called *Spo*11. Moreover, both genetic and cytological experiments indicate that the occurrence of one crossover event decreases the probability of a second occurring at a nearby chromosomal site. This "interference" seems to ensure that the limited number of crossovers are spread out so that even small chromosomes get at least one, as required for the homologs to segregate normally. Although the molecular basis of the interference is unknown, the synaptonemal complex is thought to mediate the process.

There is strong indirect evidence that the general genetic recombination events in meiosis are catalyzed by **recombination nodules**. These are very large protein complexes that sit at intervals on the synaptonemal complex, placed like basketballs on a ladder between the two homologous chromosomes (see Figure 20–13). These nodules contain Rad51, which is the eucaryotic version of the

RecA protein, which mediates general recombination in *E. coli* (discussed in Chapter 5). They seem to mark the site of a multienzyme "recombination machine" that interacts with local regions of DNA on the maternal and paternal chromatids across the 100-nm-wide synaptonemal complex.

There are two main types of recombination nodule. *Early nodules* are present before pachytene and are thought to mark the sites of the initial DNA-strand-exchange events of the recombination process. *Late nodules* are less numerous, are present during pachytene, and are thought to mark the sites where the initial strand-exchange events are being resolved as stable crossovers. Proteins known to be involved in general recombination have been identified in recombination nodules, and there is a strong correspondence between the number and distribution of late nodules and the number and distribution of crossovers. Moreover, meiosis-specific versions of proteins involved in mismatch DNA repair (discussed in Chapter 5) are also located in late nodules, where they help to resolve recombination intermediates as stable crossovers.

The occurrence of crossovers has enabled geneticists to map the relative positions of genes on chromosomes, as we now explain. Such maps have been crucial in the cloning of human disease genes.

Genetic Maps Reveal Favored Sites for Crossovers

On average, a human chromosome participates in two or three crossover events during meiosis, and every chromosome participates in at least one. Thus, whereas two genes very close to each other on a chromosome almost always end up together in the same gamete after meiosis, two genes located at the opposite ends of a chromosome are no more likely to end up together than are genes located on different chromosomes. One can therefore determine whether two genes—a gene with a mutant form causing congenital deafness, for example, and a second gene with a mutant form causing muscular dystrophy—are located close together on the same chromosome. This is done by measuring the frequency with which a child inherits the mutant forms of both genes from a parent that carries one mutant and one nonmutant version of each of them. If the two mutant genes are on different chromosomes, one will be inherited without the other 50% of the time, as chromosomes are independently segregated at meiosis. The same result is expected, however, if the two mutant genes are far apart on the same chromosome, as one or more crossover events will separate them at meiosis. To determine whether genes are on the same chromosome and, if so, how close they are to one another, human geneticists measure the frequency of coinheritance of many genes in large numbers of families. In this way, they can discover not only the neighbors of a particular gene but also the neighbors of the neighbors and thereby work their way down an entire chromosome. By this means, they have defined 24 *linkage groups,* one corresponding to each human chromosome (22 autosome pairs plus 2 sex chromosomes).

Using such measurements, geneticists have constructed detailed **genetic maps** of the entire human genome, in which the distance between each pair of neighboring genes is displayed as the percentage recombination between them. The standard unit of genetic distance is the *centimorgan (cM),* which corresponds to a 1% probability that two genes will be separated by a crossover event during meiosis. A typical human chromosome is more than 100 centimorgans long, indicating that more than one crossover is likely to occur on a typical human chromosome.

Another way to construct a genetic map is to measure the coinheritance of short DNA sequences (called *DNA markers)* that differ between individuals in the population—that is, that are *polymorphic* (see p. 464). Genetic maps constructed in this way have two advantages over genetic maps constructed by tracing the phenotypes of individuals that inherit mutant genes. First, they can be more detailed, as there are large numbers of DNA markers that can be measured. Second, they can reveal the real distance in nucleotide pairs between the markers, so that genetic distances in centimorgans can be compared directly with true physical distances along a chromosome.

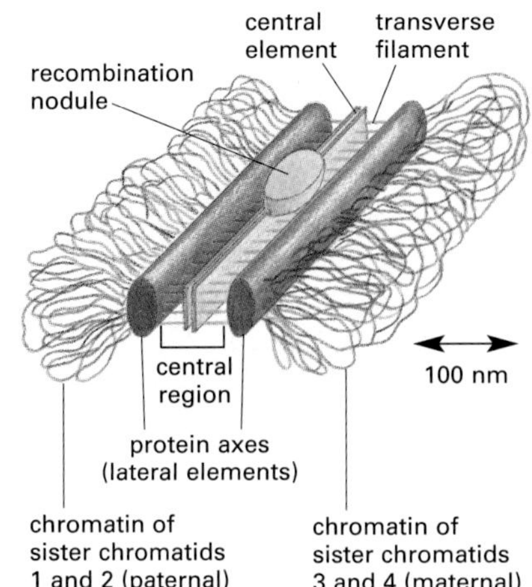

Figure 20–13 A mature synaptonemal complex. Only a short section of the long ladderlike complex is shown. A similar synaptonemal complex is present in organisms as diverse as yeasts and humans.

A direct comparison of genetic and physical distances on part of a budding yeast chromosome is shown in Figure 20–14. As the entire DNA sequence of this organism's genome is known, the physical map indicates the true distances between the DNA markers. The regions of the genetic map that are expanded in comparison with the physical map indicate recombination "hotspots," where crossovers during meiosis occur with an unusually high frequency. Regions that are contracted indicate recombination "coldspots," where crossovers occur with unusually low frequency. Human genetic maps show similar expansions and contractions. A likely explanation for the hotspots is that they contain an abundance of sites where the DNA helix is cut by the meiotic endonuclease (spo11) that creates the double-strand DNA breaks that begin the recombination process (see Figure 5–56).

Meiosis Ends with Two Successive Cell Divisions Without DNA Replication

Prophase I can occupy 90% or more of the time taken by meiosis. Although it is traditionally called prophase, it actually resembles the G_2 phase of a mitotic cell division. The nuclear envelope remains intact and disappears only when the meiotic spindle begins to form, as prophase I gives way to metaphase I. After prophase I is completed, two successive cell divisions follow without an intervening period of DNA synthesis. These divisions produce four cells from one and bring meiosis to an end (see Figure 20–7).

Meiotic division I is far more complex and requires much more time than either mitosis or meiotic division II. Even the preparatory DNA replication during meiotic division I tends to take much longer than an ordinary S phase, and cells can then spend days, months, or even years in prophase I, depending on the species and on the gamete being formed (Figure 20–15).

When meiotic division I ends, nuclear membranes re-form around the two daughter nuclei, and the brief interphase of division II begins. During this period, the chromosomes may decondense somewhat, but usually they soon recondense and prophase II begins. (Because there is no DNA synthesis during this interval, in some organisms the chromosomes seem to pass almost directly from one division phase into the next.) Prophase II is brief: the nuclear envelope breaks down as the new spindle forms, after which metaphase II, anaphase II, and telophase II usually follow in quick succession. After nuclear envelopes have formed around the four haploid nuclei produced at telophase II, cytokinesis occurs, and meiosis is complete (see Figure 20–7).

As in mitosis, a separate set of kinetochore microtubules is present on each sister chromatid at metaphase II, and these two sets of microtubules extend in opposite directions (see Figure 20–11). In mitosis, however, the sister chromatids are glued together along their length, as well as at the centromere, and both types of contact are released at the start of anaphase. In meiosis, by contrast, the sister chromatids come apart in two steps—their arms have separated at anaphase I, while their centromeres remain attached, separating only at anaphase II (see Figures 20–7 and 20–11).

The principles of meiosis are the same in plants and animals and in males and females. But the production of gametes involves more than just meiosis, and the other processes required vary widely among organisms and are very different for eggs and sperm. We shall focus our discussion of gametogenesis mainly on mammals. As we shall see, by the end of meiosis a mammalian egg is fully mature, whereas a sperm that has completed meiosis has only just begun its differentiation. Before discussing these gametes, however, we consider how certain

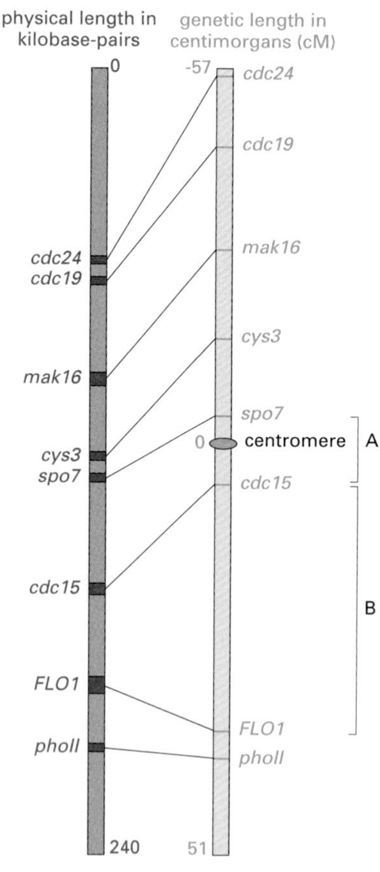

Figure 20–14 Comparison of the physical and genetic maps of part of chromosome I in budding yeast. The DNA markers shown are various genes. A indicates a region where the genetic map is contracted owing to decreased frequency of crossing-over. B indicates a region where the genetic map is expanded owing to increased frequency of crossing-over.

cells in the mammalian embryo become committed to developing into germ cells and how these cells then become committed to developing into either sperm or eggs, depending on the sex of the individual.

Summary

The formation of both eggs and sperm begins in a similar way, with meiosis. In this process two successive cell divisions following one round of DNA replication give rise to four haploid cells from a single diploid cell. Meiosis is dominated by prophase of meiotic division I, which can occupy 90% or more of the total meiotic period. As it enters this prophase, each chromosome consists of two tightly joined sister chromatids. The two replicated homologs present in each diploid nucleus then pair to form a bivalent, consisting of four chromatids. Chromosomal crossover events occur during this time. Each results in the formation of a chiasma, which helps hold each pair of homologs together during metaphase I. Crossing-over has an important role in reassorting genes during gamete formation, and it allows geneticists to map the relative positions of genes on chromosomes. The pairing of homologs culminates in the formation of a synaptonemal complex, which somehow serves to spread out the crossover events along the chromosomes. At anaphase of the first meiotic cell division, the arms of the sister chromatids suddenly become unglued, causing one member of each chromosome pair, still composed of a pair of sister chromatids linked at their centromeres, to be distributed to each daughter nucleus. A second cell division cycle, without DNA replication, then rapidly ensues; in anaphase II, each sister chromatid separates from its sister and is segregated into a separate haploid nucleus.

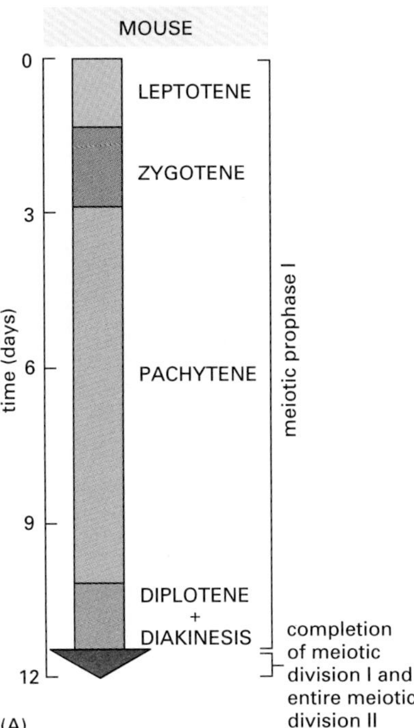

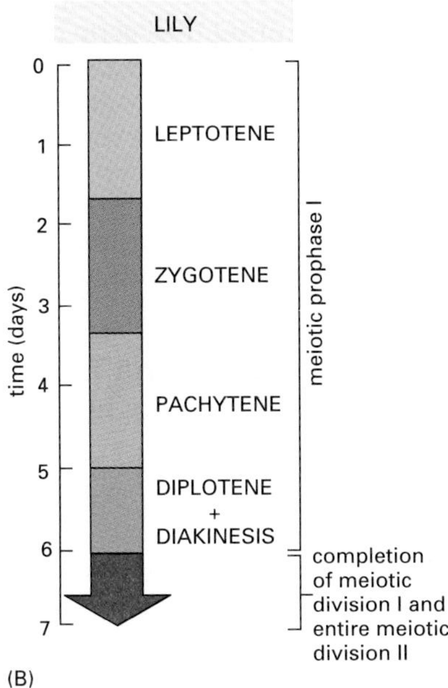

Figure 20–15 Comparison of times required for each of the stages of meiosis. (A) Approximate times for a male mammal (mouse). (B) Approximate times for the male tissue of a plant (lily). Times differ for male and female gametes (sperm and eggs, respectively) of the same species, as well as for the same gametes of different species. Meiosis in a human male, for example, lasts for 24 days, compared with 12 days in the mouse. In all species, however, meiotic prophase I is always much longer than all the other meiotic stages combined.

PRIMORDIAL GERM CELLS AND SEX DETERMINATION IN MAMMALS

Sexual reproductive strategies can vary enormously between different organisms. In the rest of this chapter, we focus mainly on the strategies used by mammals.

In all vertebrate embryos, certain cells are singled out early in development as progenitors of the gametes. These **primordial germ cells** migrate to the developing gonads, which will form the *ovaries* in females and the *testes* in males. After a period of mitotic proliferation, the primordial germ cells undergo meiosis and differentiate into mature gametes—either eggs or sperm. Later, the fusion of egg and sperm after mating initiates embryogenesis. The subsequent production in this embryo of new primordial germ cells begins the cycle again.

In this section, we consider how mammalian primordial germ cells arise, how the sex of a mammal is determined, and how sex determination dictates whether the primordial germ cells develop into sperm or eggs.

Primordial Germ Cells Migrate into the Developing Gonad

In most animals, including many vertebrates, the unfertilized egg is asymmetrical, with different regions of cytoplasm containing different sets of mRNA and protein molecules (discussed in Chapter 21). When the egg is fertilized and divides repeatedly to produce the cells of the early embryo, the cells that inherit specific molecules localized in a particular region of the egg cytoplasm become primordial germ cells. In mammals, by contrast, the egg is more symmetrical, and the cells produced by the first few divisions of the fertilized egg are all *totipotent*—that is, they can give rise to any of the cell types in the body, including germ cells. A small group of cells in the early mammalian embryo is induced to become primordial germ cells by signals produced by neighboring cells. In mice, for example, 1 week after fertilization, about 50 cells in tissue lying outside the embryo proper are induced by their neighbors to become primordial germ cells. In the next few days, these cells proliferate and are swept back into the embryo proper along with the invaginating hindgut. They then actively migrate through the gut to their final destination in the developing gonads (Figure 20–16). As the

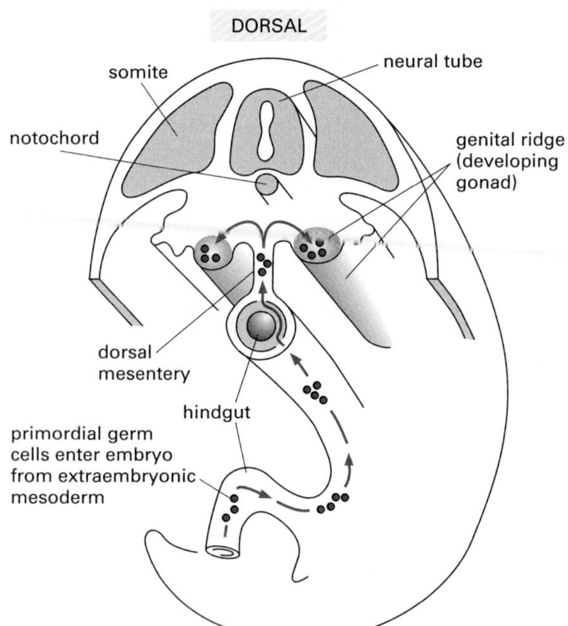

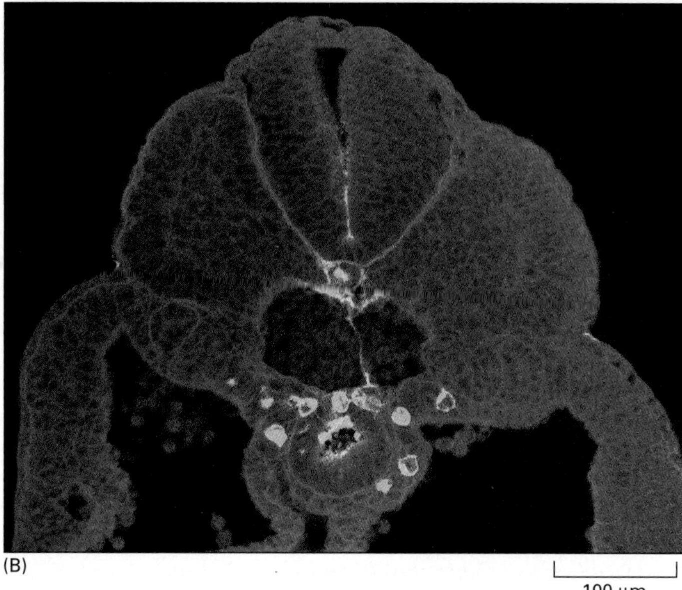

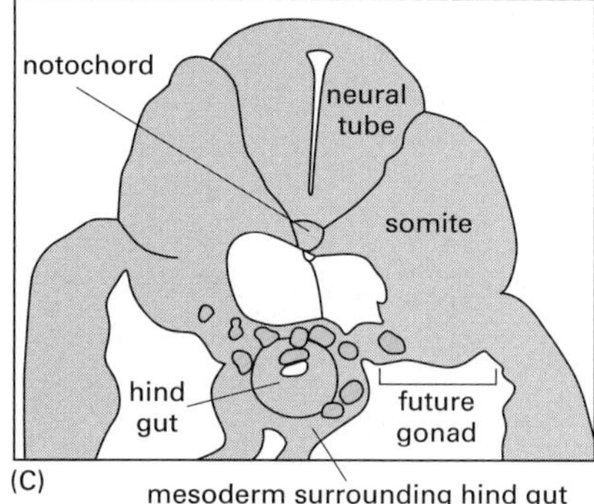

Figure 20–16 Migration of mammalian primordial germ cells. (A) Drawing showing the final stages of migration through the hindgut into the two genital ridges, each of which will develop into a gonad—either an ovary or a testis. (B) Micrograph showing migrating primordial germ cells in an early mouse embryo. The primordial germ cells are stained with a monoclonal antibody (in *green)* that specifically labels these cells at this stage of embryogenesis. The remaining cells in the embryo are stained with a lectin that binds to sialic acid, which is found on the surface of all cells. (C) Drawing corresponding to the micrograph shown in (B). (B, courtesy of Robert Anderson and Chris Wylie.)

primordial germ cells migrate through the embryo, they are signaled to survive, proliferate, and migrate by various extracellular proteins produced by adjacent somatic cells.

After the primordial germ cells enter the developing mouse gonad, which at this stage is called the *genital ridge,* they continue to proliferate for 2 or 3 more days. At this point, they commit to a developmental pathway that will lead them to become either eggs or sperm, depending not on their own sex chromosome constitution but on whether the genital ridge has begun to develop into an ovary or a testis, respectively. The sex chromosomes in the somatic cells of the genital ridge determine which type of gonad the ridge becomes. A single gene on the Y chromosome has an especially important role in this decision.

The *Sry* Gene on the Y Chromosome Can Redirect a Female Embryo to Become a Male

Aristotle believed that the temperature of the male during sexual intercourse determined the sex of offspring: the higher the temperature, the greater the chance of producing a male. We now know that the sex of a mammal is determined by its sex chromosomes, rather than by the environment (although for some animals, such as crocodiles and many fish, the opposite is true). Female mammals have two X chromosomes in all of their somatic cells, whereas males have one X and one Y. The Y chromosome is the determining factor. Individuals with a Y chromosome develop as males no matter how many X chromosomes they have, whereas individuals without a Y chromosome develop as females, even if they have only one X chromosome. The sperm that fertilizes the egg determines the sex of the resulting zygote: eggs have a single X chromosome, whereas the sperm can have either an X or a Y.

The Y chromosome influences the sex of the individual by inducing the somatic cells of the genital ridge to develop into a testis instead of an ovary. The crucial gene on the Y chromosome that has this *testis determining* function is called ***Sry***, for "sex-determining region of Y." Remarkably, when this gene is introduced into the genome of an XX mouse zygote, the transgenic embryo produced develops as a male, even though it lacks all of the other genes on the Y chromosome (Figure 20–17). Such mice, however, cannot produce sperm, in part, at least, because the presence of two X chromosomes suppresses sperm development.

Sry is expressed only in a subset of the somatic cells of the developing gonad, and it causes these cells to differentiate into **Sertoli cells**, which are the main type of supporting cells found in the testis. The Sertoli cells direct sexual development along a male pathway by affecting other cells in the genital ridge in at least four ways:

1. They stimulate the newly arriving primordial germ cells to develop along a pathway that produces sperm.
2. They secrete *anti-Müllerian hormone,* which suppresses the development of the female reproductive tract by causing the Müllerian duct to regress (this duct otherwise gives rise to the oviduct, uterus, and upper part of the vagina).
3. They stimulate particular somatic cells that lie adjacent to the developing gonad to migrate into the gonad and form critical connective tissue structures that are required for normal sperm production.
4. They help to induce other somatic cells in the developing gonad to become *Leydig cells,* which secrete the male sex hormone *testosterone;* this hormone is responsible for inducing all male secondary sexual characteristics. These include the structures of the male reproductive tract, such as the prostate and seminal vesicles, which develop from another duct, called the Wolffian duct system. This duct system degenerates in the developing female because it requires testosterone to survive and develop. The testosterone also masculinizes the early developing brain and thereby plays a major part in determining male sexual identity and orientation, and thereby behavior: female rats that are treated with testosterone around birth, for example, later display malelike sexual behavior.

Sry encodes a gene-regulatory protein (Sry) that activates the transcription of other gene-regulatory proteins required for Sertoli cell development, including the Sry-related protein *Sox9.* In the absence of either Sry or Sox9, the genital ridge develops into an ovary. The supporting cells become *follicle cells* instead of Sertoli cells. Other somatic cells become *theca cells* instead of Leydig cells and, beginning at puberty secrete the female sex hormone *estrogen* instead of testosterone. The primordial germ cells develop into eggs instead of sperm (Figure 20–18, and see Figure 20–28), and the animal develops as a female.

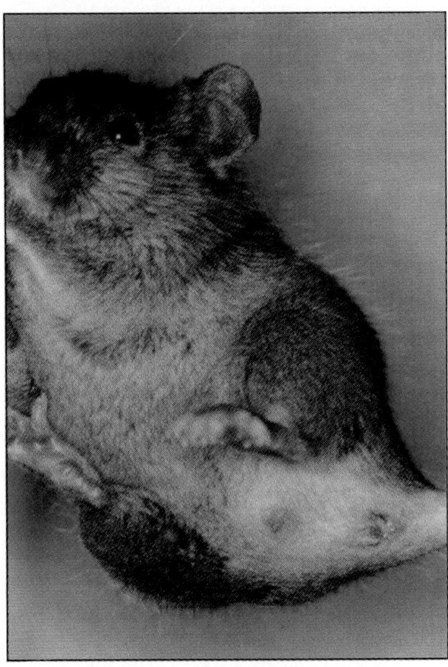

Figure 20–17 *Sry*-induced reprogramming of a female mouse embryo to develop into a male. The *Sry* gene, injected into the nucleus of an XX female zygote, caused the transgenic embryo produced to develop into a male. The external genitalia of the transgenic mouse are indistinguishable from those of a normal XY male mouse. (From P. Koopman et al., *Nature* 351:117–121, 1991. © Macmillan Magazines Ltd.)

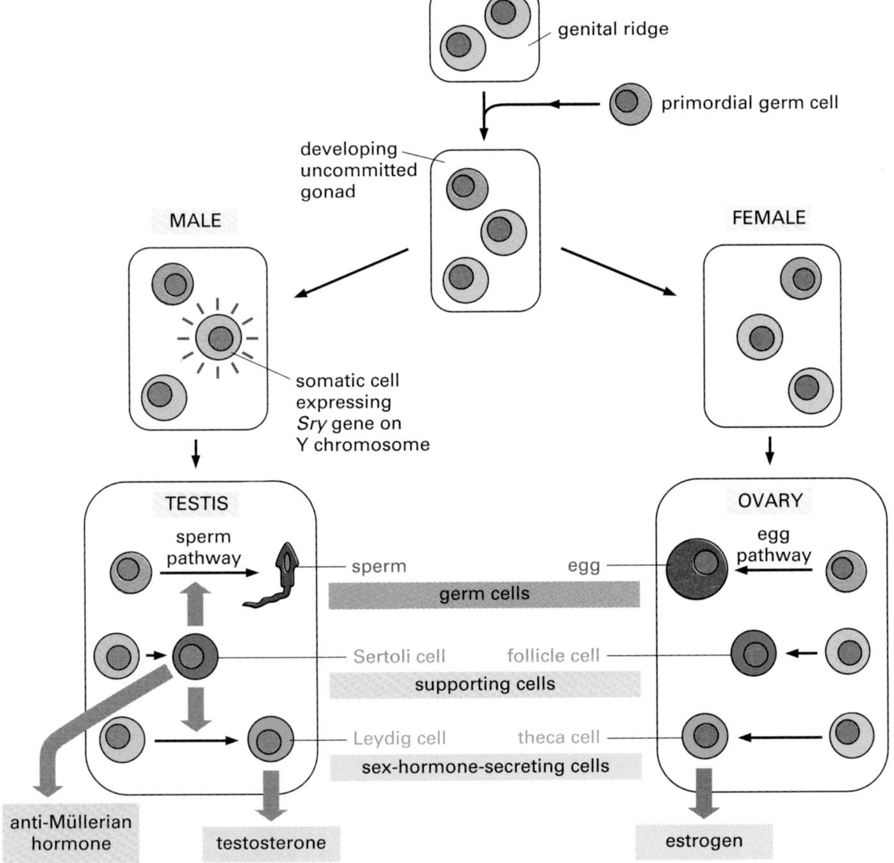

Figure 20–18 Influence of *Sry* on gonad development. The germ line cells are shaded in *red,* and the somatic cells are shaded in *green* and *blue.* The change from light to darker color indicates that the cell has matured or differentiated. The *Sry* gene acts in a subpopulation of somatic cells in the developing gonad to direct them to differentiate into Sertoli cells instead of into follicle cells. The Sertoli cells then induce primordial germ cells to commit to sperm development. They also secrete anti-Müllerian hormone, which causes the Müllerian duct to regress, and they help to induce other somatic cells to differentiate into Leydig cells, which secrete testosterone (see Figure 20–28). In the absence of *Sry,* the primordial germ cells commit to egg development, and the somatic cells develop into either follicle cells, which support egg development, or theca cells, which secrete estrogen. Whereas Leydig cells begin secreting testosterone in the fetus, theca cells do not begin secreting estrogen until puberty.

If the genital ridges are removed before they have started to develop into testes or ovaries, a mammal develops into a female, regardless of the sex chromosomes it carries. It seems that female development is the "default" pathway of sexual development in mammals.

Summary

A small number of cells in the gastrulating mammalian embryo are signaled by their neighbors to become primordial germ cells. These cells migrate into the genital ridges, which develop into the gonads. Here, the primordial germ cells start to develop into either eggs, if the gonad is becoming an ovary, or sperm, if the gonad is becoming a testis. A developing gonad will develop into an ovary unless its somatic cells contain a Y chromosome, in which case it develops into a testis. The Sry gene on the Y chromosome is responsible for this testis-determining function: it is expressed in a subset of somatic cells in the developing gonad, and it induces these cells to differentiate into Sertoli cells. The Sertoli cells in turn produce the signal molecules that promote the development of male characteristics, suppress the development of female characteristics, and induce the primordial germ cells to commit to sperm development.

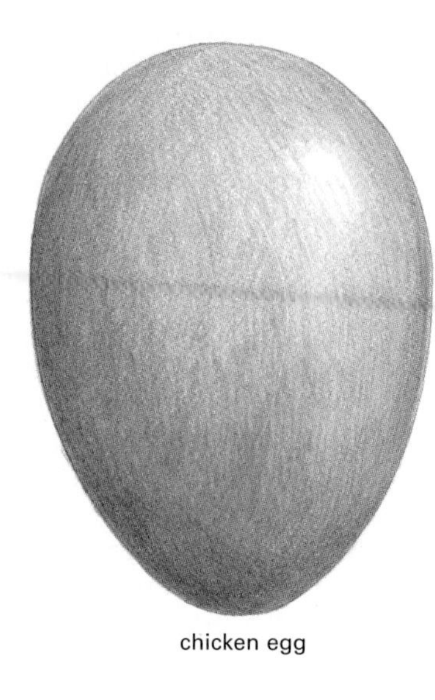

Figure 20–19 The actual sizes of three eggs. The human egg is 0.1 mm in diameter.

EGGS

In one respect at least, eggs are the most remarkable of animal cells: once activated, they can give rise to a complete new individual within a matter of days or weeks. No other cell in a higher animal has this capacity. Activation is usually the consequence of *fertilization*—fusion of a sperm with the egg. In some organisms, however, the sperm itself is not strictly required, and an egg can be activated artificially by a variety of nonspecific chemical or physical treatments. Indeed, some organisms, including a few vertebrates such as some lizards, normally reproduce from eggs that become activated in the absence of sperm—that is, **parthenogenetically**.

Although an egg can give rise to every cell type in the adult organism, it is itself a highly specialized cell, uniquely equipped for the single function of generating a new individual. The cytoplasm of an egg can even reprogram a somatic cell nucleus so that the nucleus can direct the development of a new individual. That is how the famous sheep Dolly was produced. The nucleus of an unfertilized sheep egg was destroyed and replaced with the nucleus of an adult somatic cell. An electric shock was used to activate the egg, and the resulting embryo was implanted in the uterus of a surrogate mother. The resulting normal adult sheep had the genome of the donor somatic cell and was therefore a clone of the donor sheep.

In this section, we briefly consider some of the specialized features of an egg before discussing how it develops to the point of being ready for fertilization.

An Egg Is Highly Specialized for Independent Development, with Large Nutrient Reserves and an Elaborate Coat

The eggs of most animals are giant single cells, containing stockpiles of all the materials needed for initial development of the embryo through to the stage at which the new individual can begin feeding. Before the feeding stage, the giant cell cleaves into many smaller cells, but no net growth occurs. The mammalian embryo is an exception. It can start to grow early by taking up nutrients from the mother via the placenta. Thus, a mammalian egg, although still a large cell, does not have to be as large as a frog or bird egg, for example. In general, eggs are typically spherical or ovoid, with a diameter of about 0.1 mm in humans and sea urchins (whose feeding larvae are tiny), 1 mm to 2 mm in frogs and fishes, and many centimeters in birds and reptiles (Figure 20–19). A typical somatic cell, by contrast, has a diameter of only about 10 or 20 μm (Figure 20–20).

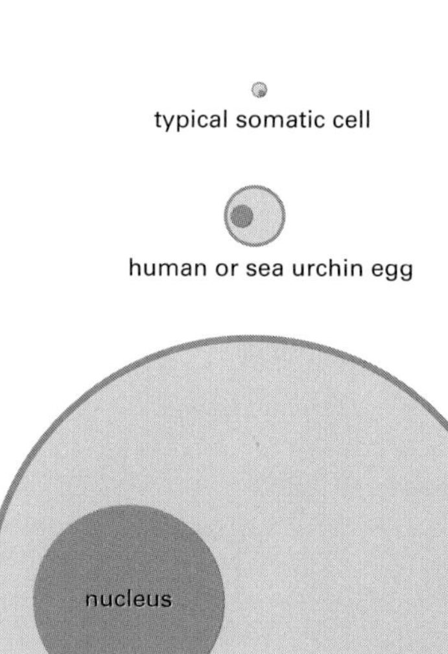

Figure 20–20 The relative sizes of various eggs. Sizes are compared with that of a typical somatic cell.

The egg cytoplasm contains nutritional reserves in the form of **yolk**, which is rich in lipids, proteins, and polysaccharides and is usually contained within

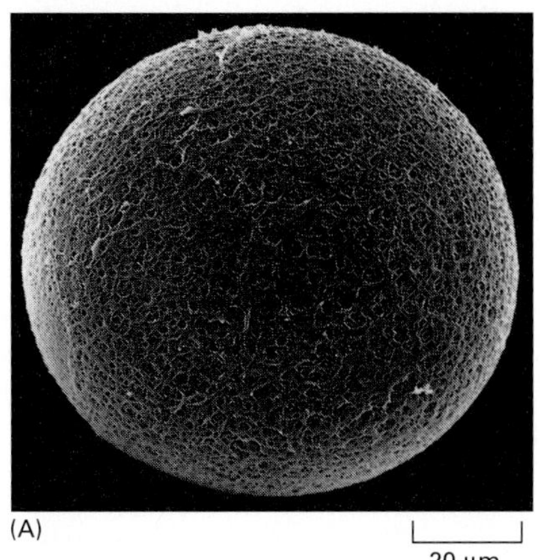

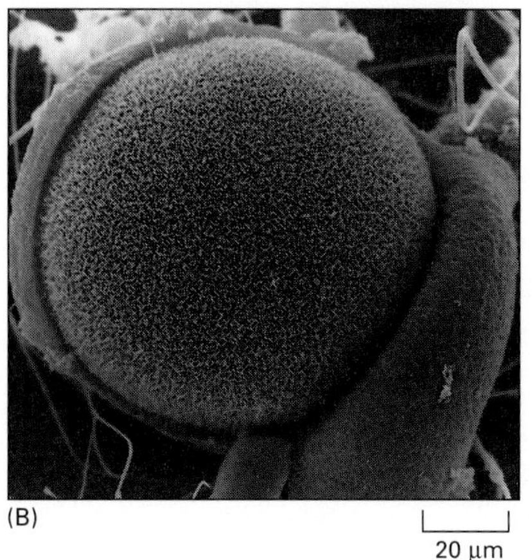

Figure 20–21 The zona pellucida. (A) Scanning electron micrograph of a hamster egg, showing the zona pellucida. (B) A scanning electron micrograph of a similar egg in which the zona (to which many sperm are attached) has been peeled back to reveal the underlying plasma membrane, which contains numerous microvilli. The zona is made entirely by the developing oocyte. (From D.M. Phillips, *J. Ultrastruct. Res.* 72:1–12, 1980.)

discrete structures called *yolk granules.* In some species, each yolk granule is membrane-enclosed, whereas in others it is not. In eggs that develop into large animals outside the mother's body, yolk can account for more than 95% of the volume of the cell. In mammals, whose embryos are largely nourished by their mothers, there is little, if any, yolk.

The **egg coat** is another peculiarity of eggs. It is a specialized form of extracellular matrix consisting largely of glycoprotein molecules, some secreted by the egg and others deposited on it by surrounding cells. In many species, the major coat is a layer immediately surrounding the egg plasma membrane; in nonmammalian eggs, such as those of sea urchins or chickens, it is called the *vitelline layer,* whereas in mammalian eggs it is called the *zona pellucida* (Figure 20–21). This layer protects the egg from mechanical damage, and in many eggs it also acts as a species-specific barrier to sperm, admitting only those of the same or closely related species.

Many eggs (including those of mammals) contain specialized secretory vesicles just under the plasma membrane in the outer region, or *cortex,* of the egg cytoplasm. When the egg is activated by a sperm, these **cortical granules** release their contents by exocytosis; the contents of the granules act to alter the egg coat so as to prevent more than one sperm from fusing with the egg (discussed below).

Cortical granules are usually distributed evenly throughout the egg cortex, but in some organisms other cytoplasmic components have a strikingly asymmetrical distribution. Some of these localized components later serve to help establish the polarity of the embryo, as discussed in Chapter 21.

Eggs Develop in Stages

A developing egg is called an **oocyte**. Its differentiation into a mature egg (or *ovum)* involves a series of changes whose timing is geared to the steps of meiosis in which the germ cells go through their two final, highly specialized divisions. Oocytes have evolved special mechanisms for arresting progress through meiosis: they remain suspended in prophase I for a prolonged period while the oocyte grows in size, and in many cases they later arrest in metaphase II while awaiting fertilization (although they can arrest at various other points, depending on the species).

While the details of oocyte development (**oogenesis**) vary from species to species, the general stages are similar, as outlined in Figure 20–22. Primordial germ cells migrate to the forming gonad to become *oogonia,* which proliferate by mitosis for a period before differentiating into *primary oocytes.* At this stage (usually before birth in mammals), the first meiotic division begins: the DNA replicates so that each chromosome consists of two sister chromatids, the duplicated homologous chromosomes pair along their long axes, and crossing-over occurs between nonsister chromatids of these paired chromosomes. After these events, the cell remains arrested in prophase of division I of meiosis (in a state equivalent, as we previously pointed out, to a G_2 phase of a mitotic division

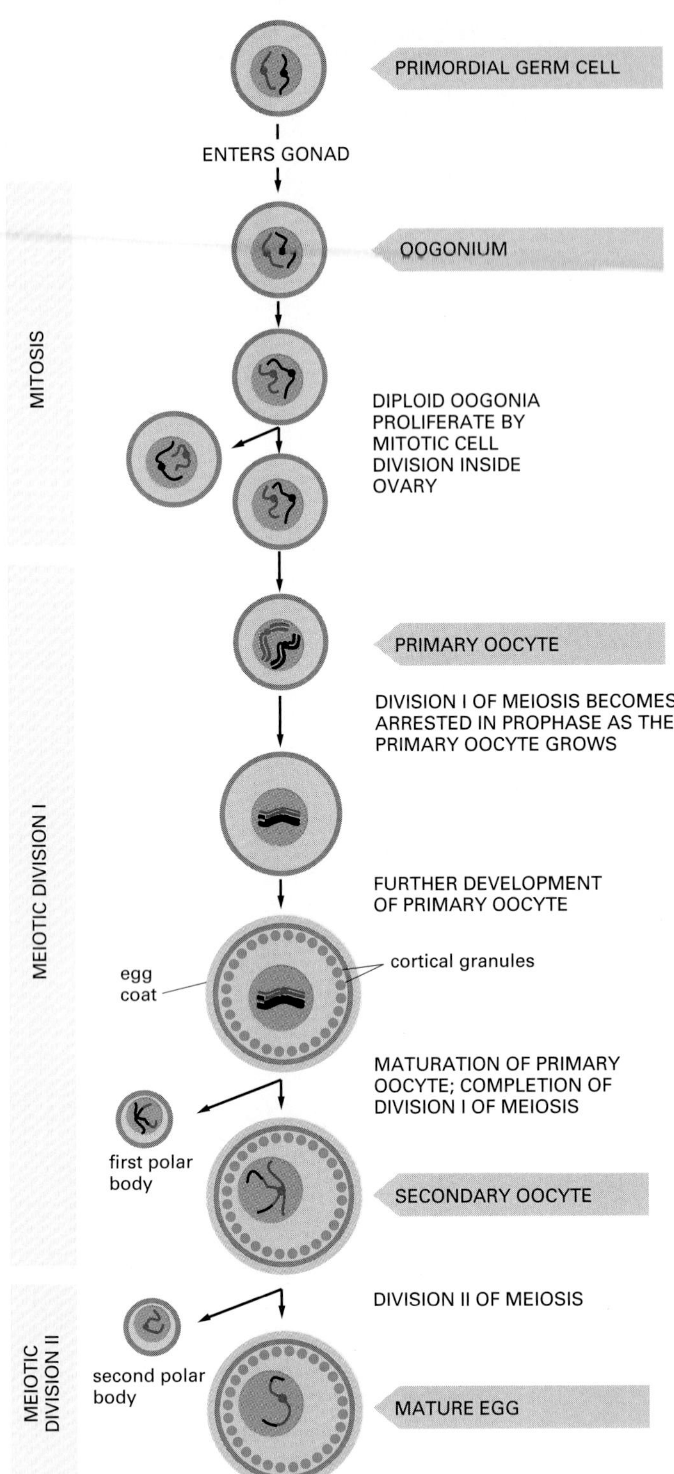

Figure 20–22 The stages of oogenesis. Oogonia develop from primordial germ cells that migrate into the developing gonad early in embryogenesis. After a number of mitotic divisions, oogonia begin meiotic division I, after which they are called primary oocytes. In mammals, primary oocytes are formed very early (between 3 and 8 months of gestation in the human embryo) and remain arrested in prophase of meiotic division I until the female becomes sexually mature. At this point, a small number periodically mature under the influence of hormones, completing meiotic division I to become secondary oocytes, which eventually undergo meiotic division II to become mature eggs (ova). The stage at which the egg or oocyte is released from the ovary and is fertilized varies from species to species. In most vertebrates, oocyte maturation is arrested at metaphase of meiosis II and the secondary oocyte completes meiosis II only after fertilization. All of the polar bodies eventually degenerate. In most animals, the developing oocyte is surrounded by specialized accessory cells that help to isolate and nourish it (not shown).

cycle) for a period lasting from a few days to many years, depending on the species. During this long period (or, in some cases, at the onset of sexual maturity), the primary oocytes synthesize a coat and cortical granules. In the case of large nonmammalian oocytes, they also accumulate ribosomes, yolk, glycogen, lipid, and the mRNA that will later direct the synthesis of proteins required for early embryonic growth and the unfolding of the developmental program. In many oocytes, the intensive biosynthetic activities are reflected in the structure of the chromosomes, which decondense and form lateral loops, taking on a characteristic "lampbrush" appearance, signifying that they are very busily engaged in RNA synthesis (see Figures 4–36 and 4–37).

The next phase of oocyte development is called *oocyte maturation.* It usually does not occur until sexual maturity, when the oocyte is stimulated by hormones.

Under these hormonal influences, the cell resumes its progress through division I of meiosis. The chromosomes recondense, the nuclear envelope breaks down (this is generally taken to mark the beginning of maturation), and the replicated homologous chromosomes segregate at anaphase I into two daughter nuclei, each containing half the original number of chromosomes. To end division I, the cytoplasm divides asymmetrically to produce two cells that differ greatly in size: one is a small *polar body,* and the other is a large **secondary oocyte**, the precursor of the egg. At this stage, each of the chromosomes is still composed of two sister chromatids. These chromatids do not separate until division II of meiosis, when they are partitioned into separate cells, as previously described. After this final chromosome separation at anaphase II, the cytoplasm of the large secondary oocyte again divides asymmetrically to produce the mature **egg** (or **ovum**) and a second small polar body, each with a haploid set of single chromosomes (see Figure 20–22). Because of these two asymmetrical divisions of their cytoplasm, oocytes maintain their large size despite undergoing the two meiotic divisions. Both of the polar bodies are small, and they eventually degenerate.

In most vertebrates, oocyte maturation proceeds to metaphase of meiosis II and then arrests until fertilization. At **ovulation**, the arrested secondary oocyte is released from the ovary and undergoes a rapid maturation step that transforms it into an egg that is prepared for fertilization. If fertilization occurs, the egg is stimulated to complete meiosis.

Oocytes Use Special Mechanisms to Grow to Their Large Size

A somatic cell with a diameter of 10–20 μm typically takes about 24 hours to double its mass in preparation for cell division. At this rate of biosynthesis, such a cell would take a very long time to reach the thousand-fold greater mass of a mammalian egg with a diameter of 100 μm. It would take even longer to reach the million-fold greater mass of an insect egg with a diameter of 1000 μm. Yet some insects live only a few days and manage to produce eggs with diameters even greater than 1000 μm. It is clear that eggs must have special mechanisms for achieving their large size.

One simple strategy for rapid growth is to have extra gene copies in the cell. Thus, the oocyte delays completion of the first meiotic division so as to grow while it contains the diploid chromosome set in duplicate. In this way, it has twice as much DNA available for RNA synthesis as does an average somatic cell in the G_1 phase of the cell cycle. The oocytes of some species go to even greater lengths to accumulate extra DNA: they produce many extra copies of certain genes. We discuss in Chapter 6 how the somatic cells of most organisms require 100 to 500 copies of the ribosomal RNA genes in order to produce enough ribosomes for protein synthesis. Eggs require even greater numbers of ribosomes to support protein synthesis during early embryogenesis, and in the oocytes of many animals the ribosomal RNA genes are specifically amplified; some amphibian eggs, for example, contain 1 or 2 million copies of these genes.

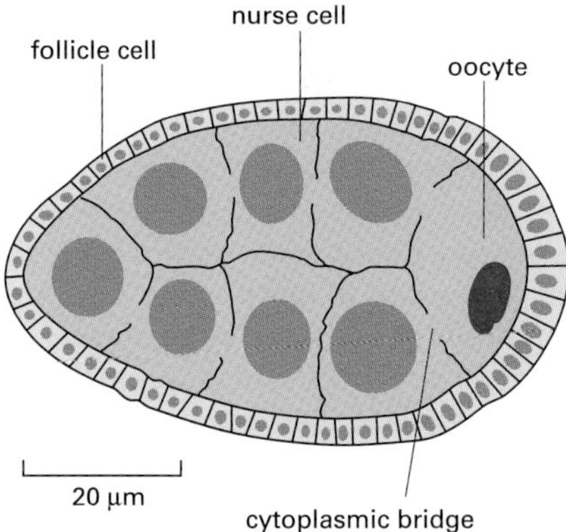

Figure 20–23 Nurse cells and follicle cells associated with a *Drosophila* oocyte. The nurse cells and the oocyte arise from a common oogonium, which gives rise to one oocyte and 15 nurse cells (only 7 of which are seen in this plane of section). These cells remain joined by cytoplasmic bridges, which result from incomplete cell division. Eventually the nurse cells dump their cytoplasmic contents into the developing oocyte and then kill themselves. The follicle cells develop independently (from mesodermal cells).

Oocytes may also depend partly on the synthetic activities of other cells for their growth. Yolk, for example, is usually synthesized outside the ovary and imported into the oocyte. In birds, amphibians, and insects, yolk proteins are made by liver cells (or their equivalents), which secrete these proteins into the blood. Within the ovaries, oocytes take up the yolk proteins from the extracellular fluid by receptor-mediated endocytosis (see Figure 13–41). Nutritive help can also come from neighboring accessory cells in the ovary. These can be of two types. In some invertebrates, some of the progeny of the oogonia become **nurse cells** instead of becoming oocytes. These cells usually are connected to the oocyte by cytoplasmic bridges through which macromolecules can pass directly into the oocyte cytoplasm (Figure 20–23). For the insect oocyte, the nurse cells manufacture many of the products—ribosomes, mRNA, protein, and so on—that vertebrate oocytes have to manufacture for themselves.

The other accessory cells in the ovary that help to nourish developing oocytes are ordinary somatic cells called **follicle cells**, which are found in both invertebrates and vertebrates. They are arranged as an epithelial layer around

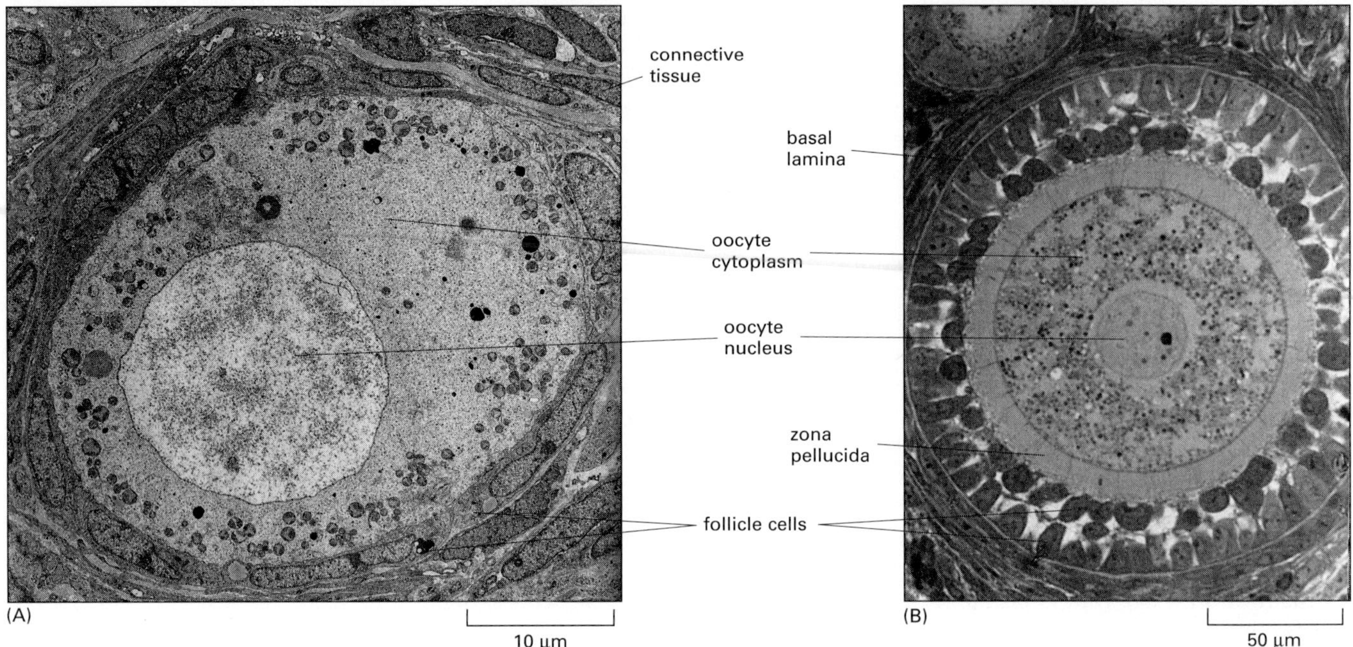

Figure 20–24 Electron micrographs of developing primary oocytes in the rabbit ovary. (A) An early stage of primary oocyte development. Neither a zona pellucida nor cortical granules have developed, and the oocyte is surrounded by a single layer of flattened follicle cells. (B) A more mature primary oocyte, which is shown at a sixfold lower magnification because it is much larger than the oocyte in (A). This oocyte has acquired a thick zona pellucida and is surrounded by several layers of follicle cells and a basal lamina, which isolate the oocyte from the other cells in the ovary. The primary oocyte together with its surrounding follicle cells is called a primary follicle. The follicle cells are connected to one another and to the oocyte by gap junctions. (From The Cellular Basis of Mammalian Reproduction [J. Van Blerkom and P. Motta eds.]. Baltimore–Munich: Urban & Schwarzenberg, 1979.)

the oocyte (Figure 20–24, and see Figure 20–23), to which they are connected only by gap junctions, which permit the exchange of small molecules but not macromolecules. While these cells are unable to provide the oocyte with preformed macromolecules through these communicating junctions, they may help to supply the smaller precursor molecules from which macromolecules are made. In addition, follicle cells frequently secrete macromolecules that contribute to the egg coat, or are taken up by receptor-mediated endocytosis into the growing oocyte, or act on egg cell-surface receptors to control the spatial patterning and axial asymmetries of the egg (discussed in Chapter 21).

Summary

Eggs develop in stages from primordial germ cells that migrate into the developing gonad early in development to become oogonia. After mitotic proliferation, oogonia become primary oocytes, which begin meiotic division I and then arrest at prophase I for days to years, depending on the species. During this prophase-I arrest period, primary oocytes grow, synthesize a coat, and accumulate ribosomes, mRNAs, and proteins, often enlisting the help of other cells, including surrounding accessory cells. In the process of maturation, primary oocytes complete meiotic division I to form a small polar body and a large secondary oocyte, which proceeds into metaphase of meiotic division II. There, in many species, the oocyte is arrested until stimulated by fertilization to complete meiosis and begin embryonic development.

SPERM

In most species, there are just two types of gamete, and they are radically different. The egg is among the largest cells in an organism, while the **sperm** (**spermatozoon**, plural **spermatozoa**) is often the smallest. The egg and the sperm are optimized in opposite ways for the propagation of the genes they carry. The egg is nonmotile and aids the survival of the maternal genes by providing large stocks of raw materials for growth and development, together with an effective protective wrapping. The sperm, by contrast, is optimized to propagate the paternal genes by exploiting this maternal investment: it is usually highly motile and streamlined for speed and efficiency in the task of fertilization. Competition between sperm is fierce, and the vast majority fail in their mission: of the billions of sperm released during the reproductive life of a human male, only a few ever manage to fertilize an egg.

Sperm Are Highly Adapted for Delivering Their DNA to an Egg

Typical sperm are "stripped-down" cells, equipped with a strong flagellum to propel them through an aqueous medium but unencumbered by cytoplasmic organelles such as ribosomes, endoplasmic reticulum, or Golgi apparatus, which are unnecessary for the task of delivering the DNA to the egg. Sperm, however, contain many mitochondria strategically placed where they can most efficiently power the flagellum. Sperm usually consist of two morphologically and functionally distinct regions enclosed by a single plasma membrane: the *tail,* which propels the sperm to the egg and helps it to burrow through the egg coat, and the *head,* which contains a condensed haploid nucleus (Figure 20–25). The DNA in the nucleus is extremely tightly packed, so that its volume is minimized for transport, and transcription is shut down. The chromosomes of many sperm have dispensed with the histones of somatic cells and are packed instead with simple, highly positively charged proteins called *protamines.*

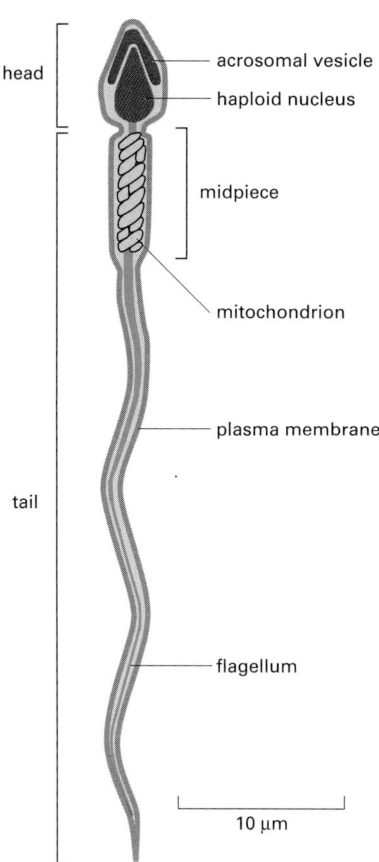

Figure 20–25 A human sperm. It is shown in longitudinal section.

In the head of most animal sperm, closely apposed to the anterior end of the nuclear envelope, is a specialized secretory vesicle called the **acrosomal vesicle** (see Figure 20–25). This vesicle contains hydrolytic enzymes that may help the sperm to penetrate the egg's outer coat. When a sperm contacts an egg, the contents of the vesicle are released by exocytosis in the so-called *acrosome reaction;* in some sperm, this reaction also exposes or releases specific proteins that help bind the sperm tightly to the egg coat.

The motile tail of a sperm is a long flagellum, whose central axoneme emanates from a basal body situated just posterior to the nucleus. As described in Chapter 16, the axoneme consists of two central singlet microtubules surrounded by nine evenly spaced microtubule doublets. The flagellum of some sperm (including those of mammals) differs from other flagella in that the usual 9 + 2 pattern of the axoneme is further surrounded by nine outer dense fibers (Figure 20–26). These dense fibers are stiff and noncontractile, and it is not known what role they have in the active bending of the flagellum, which is caused by the sliding of adjacent microtubule doublets past one another. Flagellar movement is driven by dynein motor proteins, which use the energy of ATP hydrolysis to slide the microtubules, as discussed in Chapter 16. The ATP is generated by highly specialized mitochondria in the anterior part of the sperm tail (called the *midpiece),* where the ATP is needed (see Figures 20–25 and 20–26).

Sperm Are Produced Continuously in Most Mammals

In mammals, there are major differences in the way in which eggs are produced (oogenesis) and the way in which sperm are produced (**spermatogenesis**). In human females, for example, oogonia proliferate only in the fetus, enter meiosis before birth, and become arrested as oocytes in the first meiotic prophase, in which state they may remain for up to 50 years. Individual oocytes mature from this strictly limited stock and are ovulated at intervals, generally one at a time, beginning at puberty. In human males, by contrast, meiosis and spermatogenesis do not begin in the testes until puberty and then go on continuously in the epithelial lining of very long, tightly coiled tubes, called *seminiferous tubules.* Immature germ cells, called *spermatogonia* (singular, *spermatogonium),* are located around the outer edge of these tubes next to the basal lamina, where they proliferate continuously by mitosis. Some of the daughter cells stop proliferating and differentiate into *primary spermatocytes.* These cells enter the first

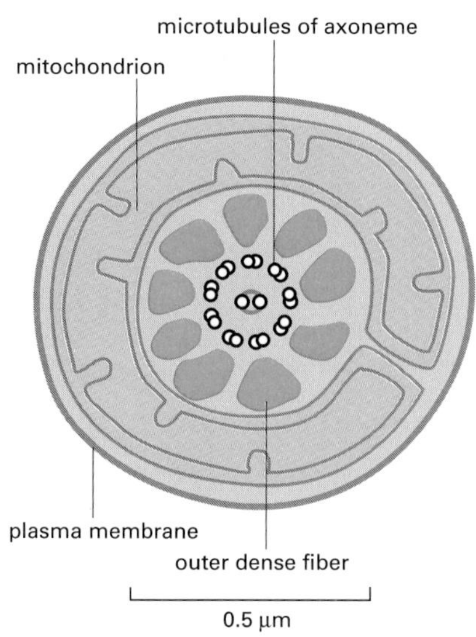

Figure 20–26 Drawing of the midpiece of a mammalian sperm as seen in cross section in an electron microscope. The core of the flagellum is composed of an axoneme surrounded by nine dense fibers. The axoneme consists of two singlet microtubules surrounded by nine microtubule doublets. The mitochondrion (shown in *green)* is well placed for providing the ATP required for flagellar movement; its unusual spiral structure (see Figure 20–25) results from the fusion of individual mitochondria during spermatid differentiation.

meiotic prophase, in which their paired homologous chromosomes participate in crossing-over, and then proceed with division I of meiosis to produce two *secondary spermatocytes,* each containing 22 duplicated autosomal chromosomes and either a duplicated X or a duplicated Y chromosome. The two secondary spermatocytes derived from each primary spermatocyte proceed through meiotic division II to produce four *spermatids,* each with a haploid number of single chromosomes. These haploid spermatids then undergo morphological differentiation into sperm (Figure 20–27), which escape into the lumen of the

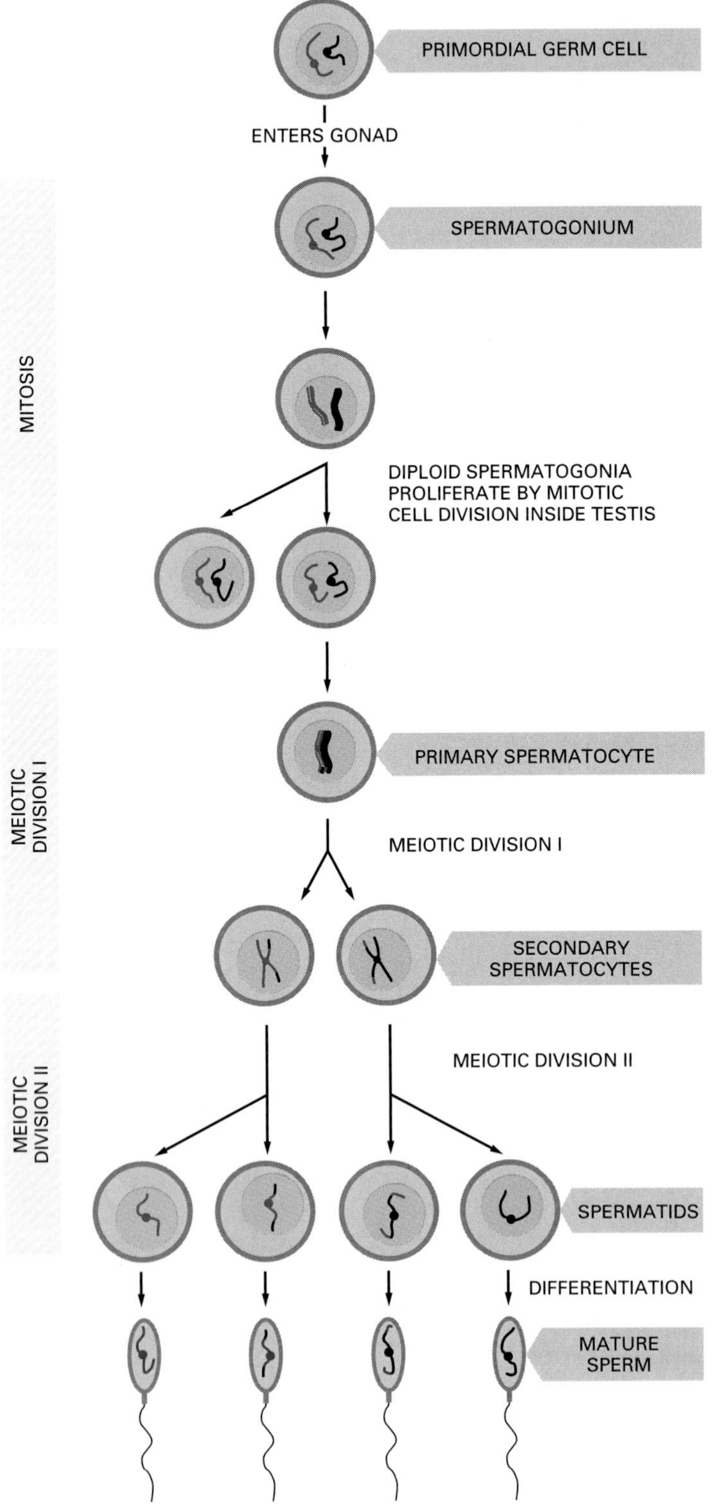

Figure 20–27 The stages of spermatogenesis. Spermatogonia develop from primordial germ cells that migrate into the testis early in embryogenesis. When the animal becomes sexually mature, the spermatogonia begin to proliferate rapidly, generating some progeny that retain the capacity to continue dividing indefinitely (as stem-cell spermatogonia) and other progeny (maturing spermatogonia) that will, after a limited number of further mitotic division cycles, embark on meiosis to become primary spermatocytes. The primary spermatocytes continue through meiotic division I to become secondary spermatocytes. After they complete meiotic division II, the secondary spermatocytes produce haploid spermatids, which differentiate into mature sperm (spermatozoa). Spermatogenesis differs from oogenesis (see Figure 20–22) in several ways. (1) New cells enter meiosis continually from the time of puberty. (2) Each cell that begins meiosis gives rise to four mature gametes rather than one. (3) Mature sperm form by an elaborate process of cell differentiation that begins after meiosis is complete. (4) About twice as many cell divisions occur in the production of a sperm as in the production of an egg; in a mouse, for example, it is estimated that on average about 56 divisions occur from zygote to mature sperm, and about 27 divisions occur from zygote to mature egg.

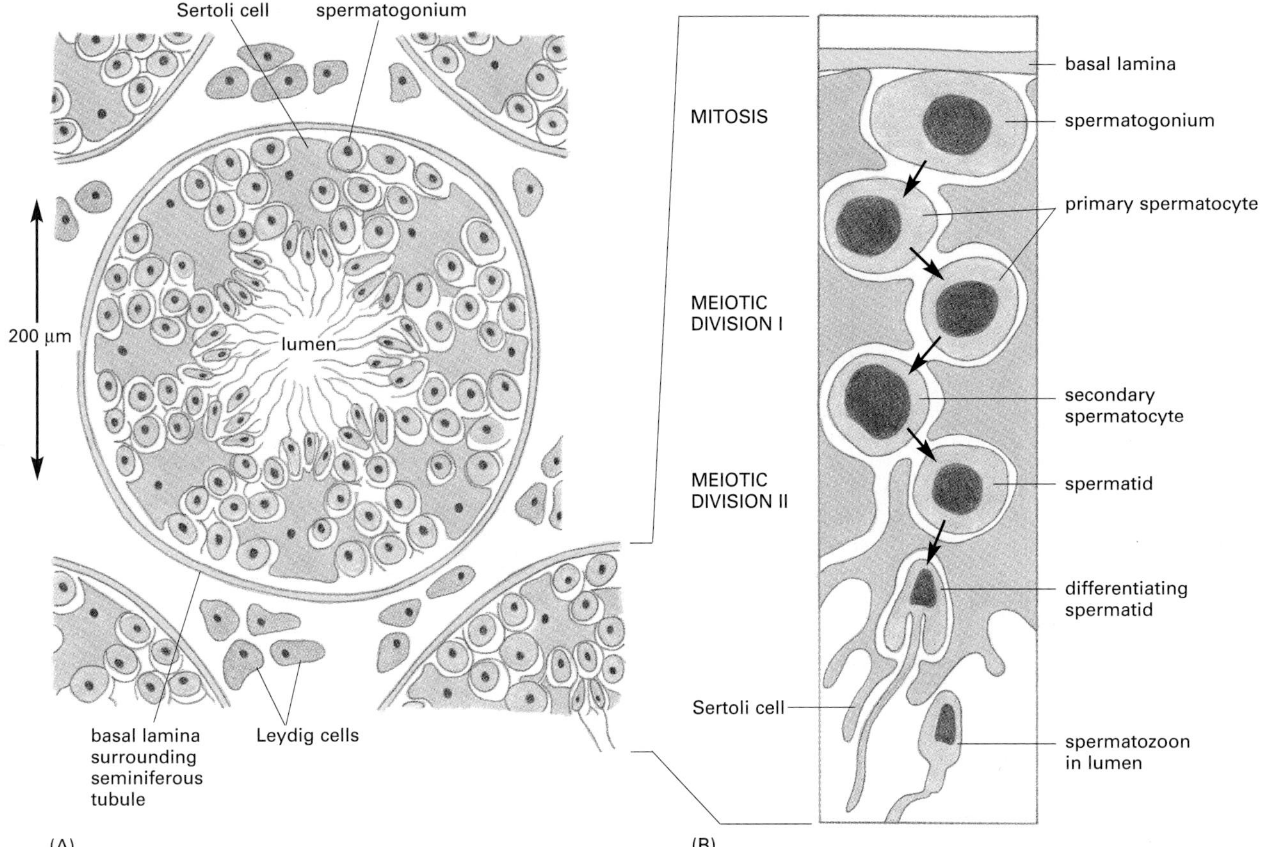

Figure 20–28 Highly simplified drawing of a cross section of a seminiferous tubule in a mammalian testis. (A) All of the stages of spermatogenesis shown take place while the developing gametes are in intimate association with Sertoli cells. These large cells extend from the basal lamina to the lumen of the seminiferous tubule; they are required for the survival of the germ cells and are analogous to follicle cells in the ovary (see Figure 20–18). Spermatogenesis also depends on testosterone secreted by Leydig cells, located between the seminiferous tubules. (B) Some of these cells are self-renewing stem-cell spermatogonia, whereas others are maturing spermatogonia; after a number of mitotic divisions, the maturing spermatogonia stop dividing by mitosis and enter meiosis to become primary spermatocytes. Eventually, sperm are released into the lumen. In man, it takes about 24 days for a spermatocyte to complete meiosis to become a spermatid and another 5 weeks for a spermatid to develop into a sperm. Sperm undergo further maturation and become motile in the epididymis; only then are they fully mature sperm.

seminiferous tubule (Figure 20–28). The sperm subsequently pass into the *epididymis,* a coiled tube overlying the testis, where they undergo further maturation and are stored.

An intriguing feature of spermatogenesis is that the developing male germ cells fail to complete cytoplasmic division (cytokinesis) during mitosis and meiosis. Consequently, large clones of differentiating daughter cells that have descended from one maturing spermatogonium remain connected by cytoplasmic bridges, forming a syncytium (Figure 20–29). The cytoplasmic bridges persist until the very end of sperm differentiation, when individual sperm are released into the tubule lumen. This accounts for the observation that mature sperm arise synchronously in any given area of a seminiferous tubule. But what is the function of the syncytial arrangement?

Unlike oocytes, sperm undergo most of their differentiation after their nuclei have completed meiosis to become haploid. The presence of cytoplasmic bridges between them, however, means that each developing haploid sperm shares a common cytoplasm with its neighbors. In this way, it can be supplied with all the products of a complete diploid genome. Developing sperm that carry a Y chromosome, for example, can be supplied with essential proteins encoded by genes on the X chromosome. Thus, the diploid genome directs sperm differentiation just as it directs egg differentiation.

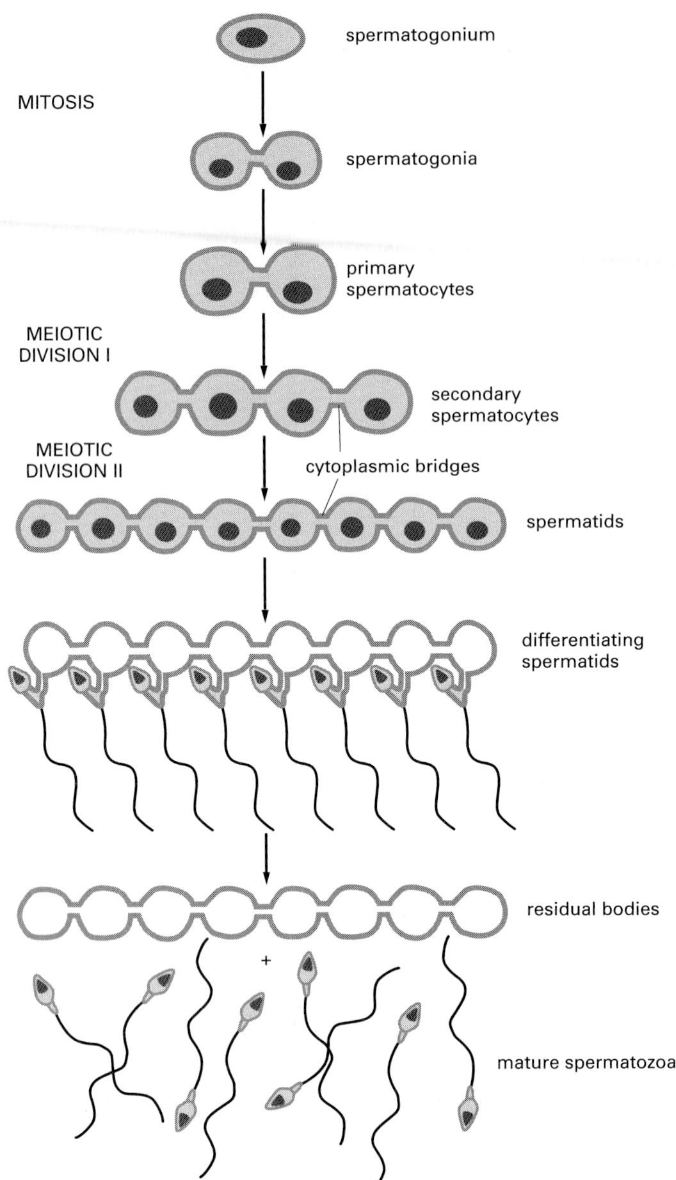

Figure 20–29 Cytoplasmic bridges in developing sperm cells and their precursors. The progeny of a single maturing spermatogonium remain connected to one another by cytoplasmic bridges throughout their differentiation into mature sperm. For the sake of simplicity, only two connected maturing spermatogonia are shown entering meiosis, eventually to form eight connected haploid spermatids. In fact, the number of connected cells that go through two meiotic divisions and differentiate synchronously is very much larger than shown here. Note that in the process of differentiating, most of the spermatid cytoplasm is discarded as residual bodies.

Some of the genes that regulate spermatogenesis have been conserved in evolution from flies to humans. The *DAZ* gene, for example, which encodes an RNA-binding protein and is located on the Y chromosome, is deleted in many infertile men, many of whom cannot make sperm. Two *Drosophila* genes that are homologous to *DAZ* are essential for spermatogenesis in the fly. RNA-binding proteins are especially important in spermatogenesis, because many of the genes expressed in the sperm lineage are regulated at the level of RNA translation.

Summary

A sperm is usually a small, compact cell, highly specialized for the task of fertilizing an egg. Whereas in human females the total pool of oocytes is produced before birth, in males new germ cells enter meiosis continually from the time of sexual maturation, with each diploid primary spermatocyte giving rise to four haploid mature sperm. The process of sperm differentiation occurs after meiosis is complete, requiring five weeks in humans. Because the maturing spermatogonia and spermatocytes fail to complete cytokinesis, however, the progeny of a single spermatogonium develop as a large syncytium. Sperm differentiation is therefore directed by the products from both parental chromosomes, even though each nucleus is haploid.

FERTILIZATION

Once released, egg and sperm alike are destined to die within minutes or hours unless they find each other and fuse in the process of **fertilization**. Through fertilization, the egg and sperm are saved: the egg is activated to begin its developmental program, and the haploid nuclei of the two gametes come together to form the genome of a new diploid organism. The mechanism of fertilization has been most intensively studied in marine invertebrates, especially sea urchins. In these organisms fertilization occurs in sea water, into which huge numbers of both sperm and eggs are released. Such *external fertilization* has been more accessible to study than the internal fertilization of mammals, which normally occurs in the female reproductive tract after mating.

In the late 1950s, however, it became possible to fertilize mammalian eggs *in vitro,* opening the way to an analysis of the cellular and molecular events in mammalian fertilization. Progress in understanding mammalian fertilization has brought substantial medical benefit: mammalian eggs that have been fertilized *in vitro* can develop into normal individuals when transplanted into the uterus; in this way many previously infertile women have been able to produce normal children. As mentioned earlier, it is possible to use *in vitro* fertilization to produce a clone of a sheep, a pig, or a mouse by transferring the nucleus of one of its somatic cells into an unfertilized egg that has had its own nucleus removed or destroyed. There is no reason to doubt that a human could be cloned in the same way, although there are serious ethical arguments about whether this should ever be done, especially as the liklihood of producing an abnormal child is very high. In this section, we focus our discussion on the fertilization of mammalian eggs.

Species-Specific Binding to the Zona Pellucida Induces the Sperm to Undergo an Acrosome Reaction

Of the 300,000,000 human sperm ejaculated during coitus, only about 200 reach the site of fertilization in the oviduct. There is evidence that chemical signals released by the follicle cells that surround the ovulated egg attract the sperm to the egg, but the nature of the chemoattractant molecules is unknown. Once it finds an egg, the sperm must first migrate through the layer of follicle cells and then bind to and cross the egg coat—the *zona pellucida.* Finally, the sperm must bind to and fuse with the egg plasma membrane. To become competent to accomplish these tasks, ejaculated mammalian sperm must normally be modified by conditions in the female reproductive tract, a process called **capacitation**, which requires about 5–6 hours in humans. Capacitation is triggered by bicarbonate ions (HCO_3^-) in the vagina, which enter the sperm and directly activate a soluble adenylyl cyclase enzyme in the cytosol. The cyclase produces cyclic AMP (discussed in Chapter 15), which helps to initiate the changes associated with capacitation. Capacitation alters the lipid and glycoprotein composition of the sperm plasma membrane, increases sperm metabolism and motility, and markedly decreases the membrane potential (that is, the membrane potential moves to a more negative value so that the membrane becomes hyperpolarized).

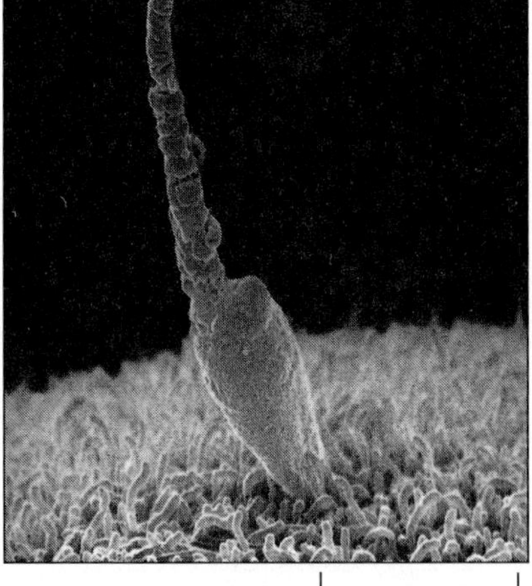

Figure 20–30 Scanning electron micrograph of a human sperm contacting a hamster egg. The zona pellucida of the egg has been removed, exposing the plasma membrane, which contains numerous microvilli. The ability of an individual's sperm to penetrate hamster eggs is used as an assay of male fertility; penetration of more than 10–25% of the eggs is considered to be normal. (Courtesy of David M. Phillips.)

Once a capacitated sperm has penetrated the layer of follicle cells, it binds to the **zona pellucida** (see Figure 20–21). The zona usually acts as a barrier to fertilization across species, and removing it often eliminates this barrier. Human sperm, for example, will fertilize hamster eggs from which the zona has been removed with specific enzymes; not surprisingly, such hybrid zygotes fail to develop. Zona-free hamster eggs, however, are sometimes used in infertility clinics to assess the fertilizing capacity of human sperm *in vitro* (Figure 20–30).

The zona pellucida of mammalian eggs is composed mainly of three glycoproteins, all of which are produced exclusively by the growing oocyte. Two of them, ZP2 and ZP3, assemble into long filaments, while the other, ZP1, cross-links the filaments into a three-dimensional network. The protein ZP3 is crucial: female mice with an inactivated *ZP3* gene produce eggs lacking a zona and are infertile. ZP3 is responsible for the species-specific binding of sperm to the zona,

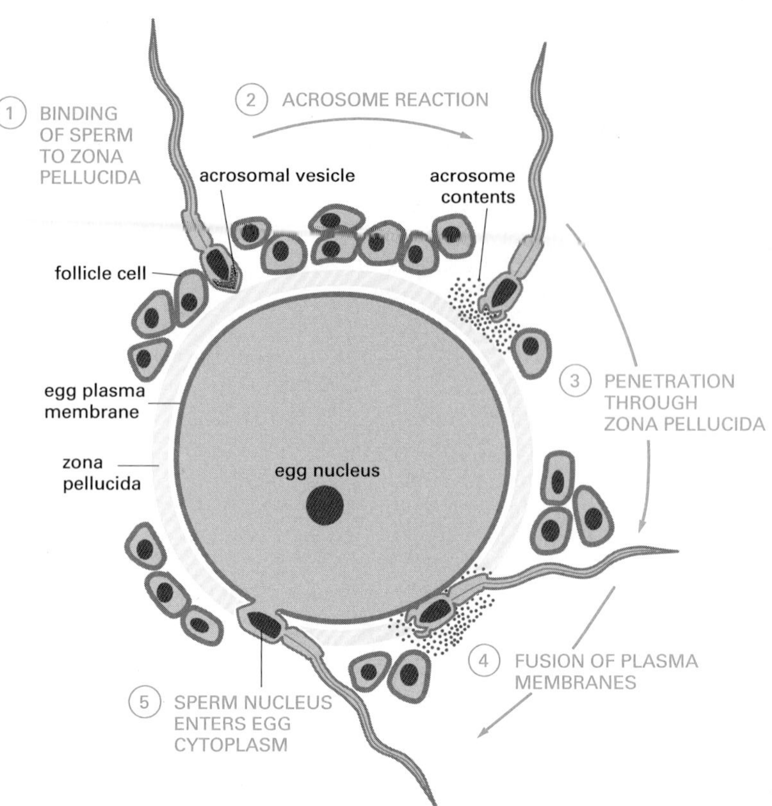

Figure 20–31 The acrosome reaction that occurs when a mammalian sperm fertilizes an egg. In mice, a single glycoprotein in the zona pellucida, ZP3, is thought to be responsible for both binding the sperm and inducing the acrosome reaction. Note that a mammalian sperm interacts tangentially with the egg plasma membrane so that fusion occurs at the equator, rather than at the tip, of the sperm head. In mice, the zona pellucida is about 6 μm thick, and sperm cross it at a rate of about 1 μm/min.

at least in mice. Several proteins on the sperm surface that bind to specific O-linked oligosaccharides on ZP3 have been implicated as ZP3 receptors, but the contribution of each is uncertain. On binding to the zona, the sperm is induced to undergo the **acrosome reaction**, in which the contents of the acrosome are released by exocytosis (Figure 20–31). In the mouse, at least, the trigger for the acrosome reaction is ZP3 in the zona, which induces an influx of Ca^{2+} into the sperm cytosol; this in turn initiates exocytosis. An increase in cytosolic Ca^{2+} seems to be necessary and sufficient to trigger the acrosome reaction in all animals.

The acrosome reaction is required for fertilization. It exposes various hydrolytic enzymes that help the sperm tunnel through the zona pellucida, and it exposes other proteins on the sperm surface that bind to the ZP2 protein and thereby help the sperm maintain its tight binding to the zona while burrowing through it. In addition, the acrosome reaction exposes proteins in the sperm plasma membrane that mediate the binding and fusion of this membrane with that of the egg, as we discuss below. Although fertilization normally occurs by sperm–egg fusion, it can also be achieved artificially, by injecting the sperm into the egg cytoplasm; this is sometimes done in infertility clinics when there is a problem with sperm–egg fusion.

The Egg Cortical Reaction Helps to Ensure That Only One Sperm Fertilizes the Egg

Although many sperm can bind to an egg, normally only one fuses with the egg plasma membrane and injects its nucleus and other organelles into the egg cytoplasm. If more than one sperm fuses—a condition called *polyspermy*—multipolar or extra mitotic spindles are formed, resulting in faulty segregation of chromosomes during cell division; nondiploid cells are produced, and development usually stops. Two mechanisms can operate to ensure that only one sperm fertilizes the egg. In many cases, a rapid depolarization of the egg plasma membrane, which is caused by the fusion of the first sperm, prevents further sperm from fusing and thereby acts as a fast *primary block to polyspermy.* But the membrane potential returns to normal soon after fertilization, so that a second mechanism is required to ensure a longer-term, *secondary block to polyspermy.* This is provided by the egg **cortical reaction**.

When the sperm fuses with the egg plasma membrane, it causes a local increase in cytosolic Ca^{2+}, which spreads through the cell in a wave. In some mammalian eggs, the initial increase in Ca^{2+} is followed by prolonged Ca^{2+} oscillations. There is evidence that the Ca^{2+} wave or oscillations are induced by a protein that is introduced into the egg by the sperm, but the nature of the protein is unknown.

The Ca^{2+} wave or oscillations activate the egg to begin development, and they initiate the cortical reaction, in which the cortical granules release their contents by exocytosis. If the cytosolic concentration of Ca^{2+} is increased artificially—either directly by an injection of Ca^{2+} or indirectly by the use of Ca^{2+}-carrying ionophores (discussed in Chapter 11)—the eggs of all animals so far tested, including mammals, are activated. Conversely, preventing the increase in Ca^{2+} by injecting the Ca^{2+} chelator EGTA inhibits activation of the egg in response to fertilization. The contents of the cortical granules include various enzymes that are released by the cortical reaction and change the structure of the zona pellucida. The altered zona becomes "hardened," so that sperm no longer bind to it, and it therefore provides a block to polyspermy. Among the changes that occur in the zona is the proteolytic cleavage of ZP2 and the hydrolysis of sugar groups on ZP3 (Figure 20–32).

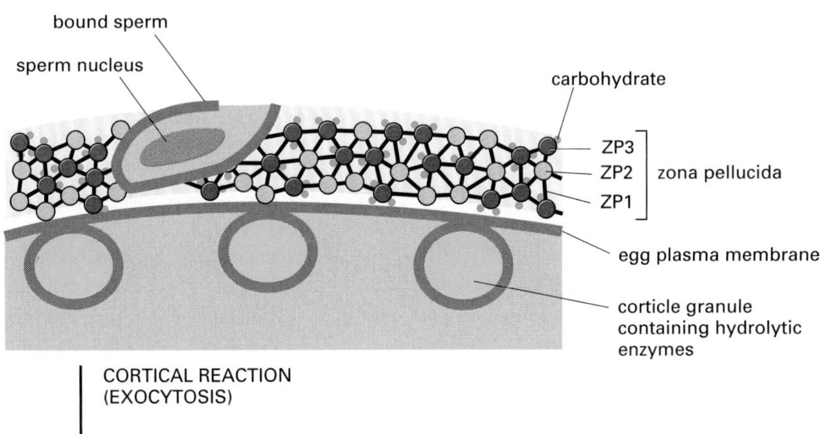

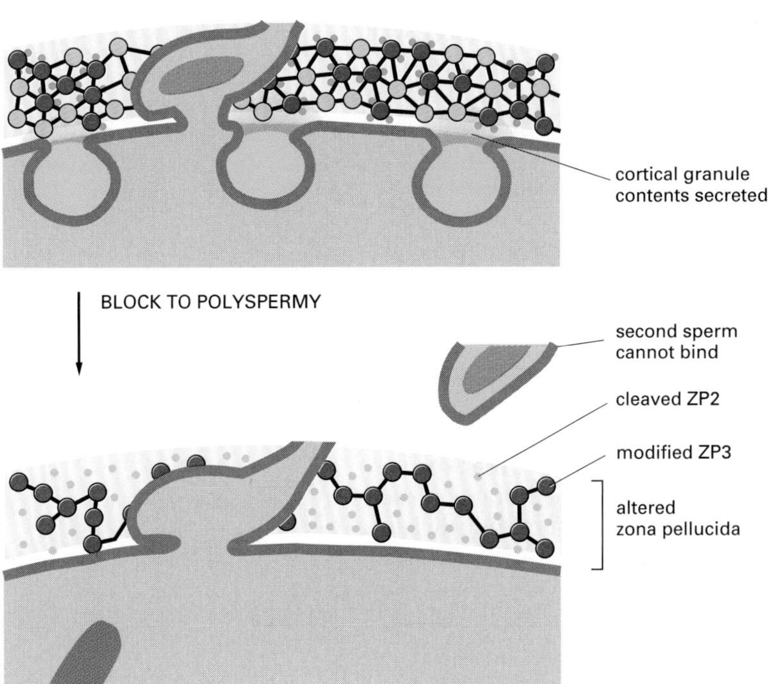

Figure 20–32 How the cortical reaction in a mouse egg is thought to prevent additional sperm from entering the egg. The released contents of the cortical granules both remove carbohydrate from ZP3 so it no longer can bind to the sperm plasma membrane and partly cleave ZP2, hardening the zona pellucida. Together these changes provide a block to polyspermy.

The Mechanism of Sperm–Egg Fusion Is Still Unknown

After a sperm has penetrated the extracellular coat of the egg, it interacts with the egg plasma membrane overlying the tips of microvilli on the egg surface (see Figure 20–30). Neighboring microvilli then rapidly elongate and cluster around the sperm to ensure that it is held firmly so that it can fuse with the egg. After fusion, the entire sperm is drawn head-first into the egg as the microvilli are resorbed. In mouse sperm, a transmembrane protein called *fertilin,* which becomes exposed on the sperm surface during the acrosome reaction, helps the sperm bind to the egg plasma membrane and may also have a role in the fusion of the two plasma membranes.

Fertilin is composed of two glycosylated transmembrane subunits called α and β, which are held together by noncovalent bonds (Figure 20–33). The extracellular N-terminal domain of the fertilin subunits is thought to bind to integrins in the egg plasma membrane and thereby help the sperm adhere to the egg membrane in preparation for fusion. The integrin in the egg plasma membrane is associated with a member of the *tetraspan* family of membrane proteins—so-called because they have four membrane-spanning segments. Female mice that are deficient in this protein are infertile, as their eggs cannot fuse with sperm. The extracellular domain of the α subunit of fertilin contains a hydrophobic region that resembles the fusogenic region of viral fusion proteins, which mediates the fusion of enveloped viruses with the cells that they infect (discussed in Chapter 13). Synthetic peptides corresponding to this region of the fertilin α chain can induce membrane fusion in a test-tube, consistent with the possibility that fertilin helps to mediate sperm–egg fusion.

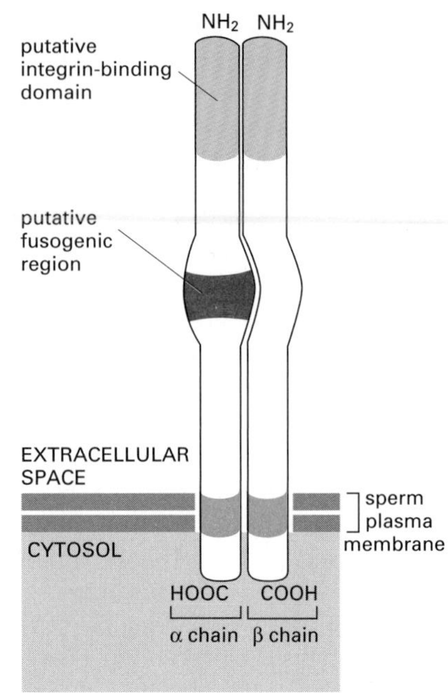

Figure 20–33 The fertilin protein in the sperm plasma membrane. The α and β subunits, which are both glycosylated (not shown), are noncovalently associated. Both subunits belong to the ADAM family of proteins, which includes proteins thought to function in either cell adhesion or the proteolytic processing of other transmembrane proteins (such as Notch, which is discussed in Chapter 15). The proteolytic domain that is normally present at the amino terminus of these proteins is removed from the fertilin protein during sperm maturation.

Male mice that are fertilin-deficient are infertile, and their sperm are eight-fold less efficient than normal sperm in binding to the egg plasma membrane but only 50% less efficient in fusing with it. Surprisingly, these defects do not seem to be the main cause of the infertility. The fertilin-deficient sperm are even more impaired in their ability to bind to the zona pellucida and to migrate out of the uterus into the oviduct, where the egg is normally fertilized. Clearly, fertilin's roles in fertilization are more complex than originally suspected and are still not completely understood. The finding that fertilin-deficient sperm can still fertilize eggs in a test tube, albeit inefficiently, suggests that other sperm proteins normally help to mediate sperm binding and fusion to the egg plasma membrane.

As the cell biology of mammalian fertilization becomes better understood and the molecules that mediate the various steps in the process are defined, new strategies for contraception become possible. One approach currently being investigated, for example, is to immunize males or females with molecules that are required for reproduction in the hope that the antibodies produced will inhibit the activities of these molecules. In addition to the various hormones and hormone receptors involved in reproduction, ZP3 and fertilin might be appropriate target molecules. An alternative approach would be to administer oligosaccharides or peptides corresponding to ligands that operate in fertilization, such as the postulated integrin-binding domain of fertilin. Small molecules of this type block fertilization in a test-tube by competing with the normal ligand for its receptor.

The Sperm Provides a Centriole for the Zygote

Once fertilized, the egg is called a **zygote**. Fertilization is not complete, however, until the two haploid nuclei (called *pronuclei)* have come together and combined their chromosomes into a single diploid nucleus. In fertilized mammalian eggs, the two pronuclei do not fuse directly as they do in many other species. They approach each other but remain distinct until after the membrane of each pronucleus has broken down in preparation for the zygote's first mitotic division (Figure 20–34).

In most animals, including humans, the sperm contributes more than DNA to the zygote. It also donates a centriole—an organelle that is lacking in unfertilized human eggs. The sperm centriole enters the egg along with the sperm nucleus and tail and a centrosome forms around it. In humans, it replicates and

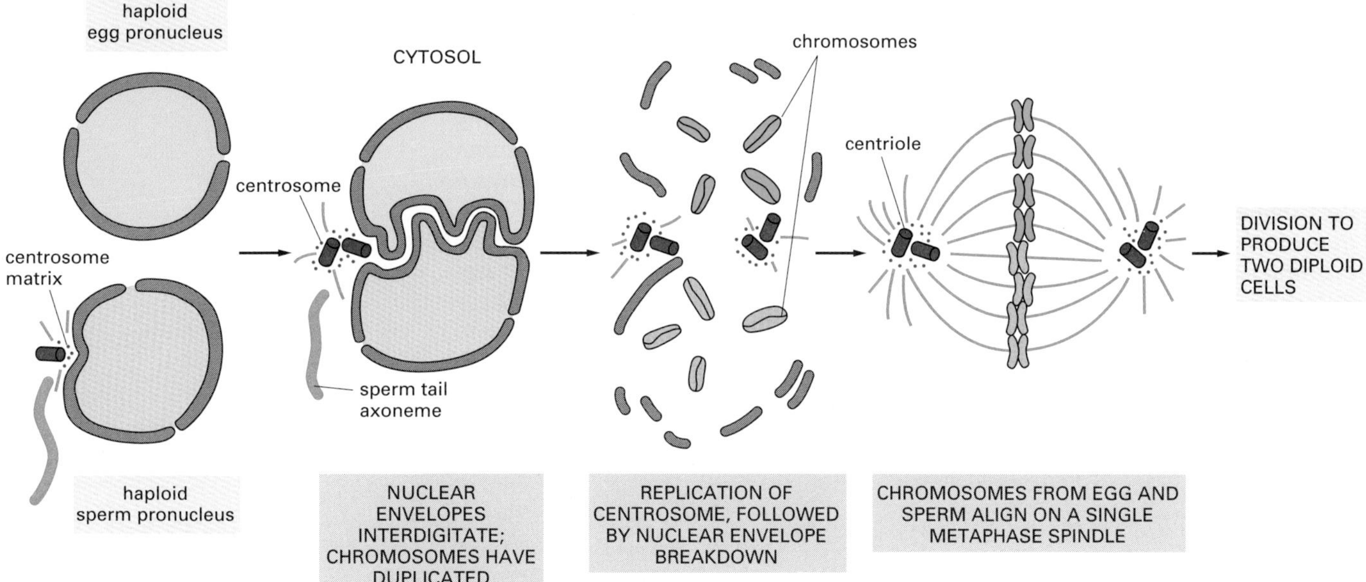

Figure 20–34 The coming together of the sperm and egg pronuclei after mammalian fertilization. The pronuclei migrate toward the center of the egg. When they come together, their nuclear envelopes interdigitate. The centrosome replicates, the nuclear envelopes break down, and the chromosomes of both gametes are eventually integrated into a single mitotic spindle, which mediates the first cleavage division of the zygote. (Adapted from drawings and electron micrographs provided by Daniel Szöllösi.)

helps organize the assembly of the first mitotic spindle in the zygote (Figure 20–35). This explains why multipolar or extra mitotic spindles form in cases of polyspermy, where several sperm contribute their centrioles to the egg.

Fertilization marks the beginning of one of the most remarkable phenomena in all of biology—the process of embryogenesis, in which the zygote develops into a new individual. This is the subject of the next chapter.

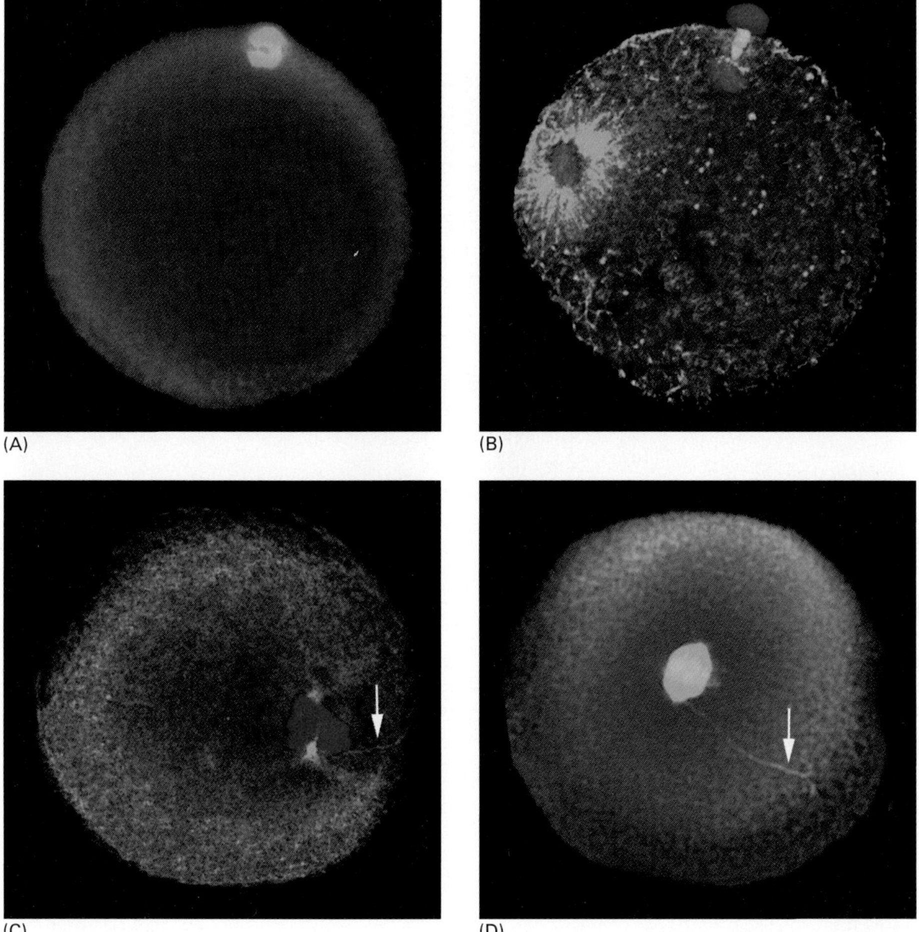

Figure 20–35 Immunofluorescence micrographs of human sperm and egg pronuclei coming together after *in vitro* fertilization. Spindle microtubules are stained in *green* with anti-tubulin antibodies, and DNA is labeled in *blue* with a DNA stain. (A) A meiotic spindle in a mature, unfertilized oocyte. (B) This fertilized egg is extruding its second polar body and is shown about 5 hours after fusion with a sperm. The sperm head *(left)* has nucleated an array of microtubules. (C) The two pronuclei have come together. (D) By 16 hours after fusion with a sperm, the centrosome that entered the egg with the sperm has duplicated, and the daughter centrosomes have organized a bipolar mitotic spindle. The chromosomes of both pronuclei are aligned at the metaphase plate of the spindle. As indicated by the arrows in (C) and (D), the sperm tail is associated with one of the centrosomes. (From C. Simerly et al., *Nat. Med.* 1:47–53, 1995. © Macmillan Magazines Ltd.)

Summary

Mammalian fertilization begins when the head of a sperm binds in a species-specific manner to the zona pellucida surrounding the egg. This induces the acrosome reaction in the sperm, which releases the contents of its acrosomal vesicle, exposing enzymes that help the sperm to digest its way through the zona to the egg plasma membrane in order to fuse with it. The fusion of the sperm with the egg induces a Ca^{2+} signal in the egg. The Ca^{2+} signal activates the egg to undergo the cortical reaction, in which cortical granules release their contents, including enzymes that alter the zona pellucida and thereby prevent the fusion of additional sperm. The Ca^{2+} signal also triggers the development of the zygote, which begins after sperm and egg haploid pronuclei have come together, and their chromosomes have aligned on a single mitotic spindle, which mediates the first division of the zygote.

References

General

Austin CR & Short RV (eds) (1982) Reproduction in Mammals Vol I, Germ Cells and Fertilization, 2nd edn. Cambridge, UK: Cambridge University Press.

Gilbert SF (2000) Developmental Biology, 6th edn, pp 185–221. Sunderland, MA: Sinauer.

Knobil E, Neill JD, Greenwald GS et al. (eds) (1994) Vol 1 and 2, The Physiology of Reproduction, 2nd edn. New York: Raven Press.

The Benefits of Sex

Barton NH & Charlesworth B (1998) Why sex and recombination? *Science* 281, 1986–1990.

Maynard Smith J (1978) Evolution of Sex. Cambridge, UK: Cambridge University Press.

Meiosis

Carpenter AT (1994) Chiasma function. *Cell* 77, 957–962.

Cherry JM, Ball C, Weng S et al. (1997) Genetic and physical maps of *Saccharomyces cerevisiae*. *Nature* 387(Suppl), 67–73.

Dej KJ & Orr-Weaver TL (2000) Separation anxiety at the centromere. *Trends Cell Biol.* 10, 392–399.

Eichenlaub-Ritter U (1994) Mechanisms of nondisjunction in mammalian meiosis. *Curr. Top. Dev. Biol.* 29, 281–324.

Kleckner N (1996) Meiosis: how could it work? *Proc. Natl Acad. Sci. USA* 93, 8167–8174.

Roeder GS (1997) Meiotic chromosomes: it takes two to tango. *Genes Dev.* 11, 2600–2621.

Smith KN & Nicolas A (1998) Recombination at work for meiosis. *Curr. Opin. Genet. Dev.* 8, 200–211.

Zickler D & Kleckner N (1999) Meiotic chromosomes: integrating structure and function. *Annu. Rev. Genet.* 33, 603–754.

Primordial Germ Cells and Sex Determination in Mammals

Goodfellow PN & Lovell-Badge R (1993) SRY and sex determination in mammals. *Annu. Rev. Genet.* 27, 71–92.

McLaren A (1999) Signaling for germ cells. *Genes Dev.* 13, 373–376.

Swain A & Lovell-Badge R (1999) Mammalian sex determination: a molecular drama. *Genes Dev.* 13, 755–767.

Williamson A & Lehmann R (1996) Germ cell development in Drosophila. *Annu. Rev. Cell Dev. Biol.* 12, 365–391.

Wylie C (1999) Germ cells. *Cell* 96, 165–174.

Eggs

Baker TG (1982) Oogenesis and ovulation. In Reproduction in Mammals: Vol I, Germ Cells and Fertilization (Austin CR & Short RV eds), 2nd edn, pp 17–45. Cambridge, UK: Cambridge University Press.

Campbell KH, McWhir J, Ritchie WA & Wilmut I (1996) Sheep cloned by nuclear transfer from a cultured cell line. *Nature* 380, 64–66.

Dietl J (ed) (1989) The Mammalian Egg Coat: Structure and Function. Berlin: Springer-Verlag.

Goodenough DA, Simon AM & Paul DL (1999) Gap junctional intercellular communication in the mouse ovarian follicle. *Novartis Found. Symp.* 219, 226–235; discussion 235–240.

Gosden R, Krapez J & Briggs D (1997) Growth and development of the mammalian oocyte. *Bioessays* 19, 875–882.

Sathananthan AH (1997) Ultrastructure of the human egg. *Hum. Cell* 10, 21–38.

Solter D (1996) Lambing by nuclear transfer. *Nature* 380, 24–25.

Spradling AC, de Cuevas M, Drummond-Barbosa D et al. (1997) The *Drosophila* germarium: stem cells, germ line cysts, and oocytes. *Cold Spring Harb. Symp. Quant. Biol.* 62, 25–34.

Sperm

Clermont Y (1972) Kinetics of spermatogenesis in mammals: seminiferous epithelium cycle and spermatogonial renewal. *Physiol. Rev.* 52, 198–236.

Fawcett DW & Bedford JM (eds) (1979) The Spermatozoon. Baltimore: Urban & Schwarzenberg.

Metz CB & Monroy A (eds) (1985) Biology of Fertilization Vol 2, Biology of the Sperm. Orlando, FL: Academic Press.

Ward WS & Coffey DS (1991) DNA packaging and organization in mammalian spermatozoa: comparison with somatic cells. *Biol. Reprod.* 44, 569–574.

Fertilization

Chen MS, Tung KS, Coonrod SA et al. (1999) Role of the integrin-associated protein CD9 in binding between sperm ADAM 2 and the egg integrin $\alpha6\beta1$: implications for murine fertilization. *Proc. Natl Acad. Sci. USA* 96, 11830–11835.

Cho C, O'Dell Bunch D, Faure J-E et al. (1998) Fertilization defects in sperm from mice lacking fertilin β. *Science* 281, 1857–1859.

Longo FJ (1997) Fertilization, 2nd edn. Cheltenham, UK: Nelson Thornes.

McLeskey SB, Dowds C, Carballada R et al. (1998) Molecules involved in mammalian sperm–egg interaction. *Int. Rev. Cytol.* 177, 57–113.

Myles DG & Primakoff P (1997) Why did the sperm cross the cumulus? To get to the oocyte. Functions of the sperm surface proteins PH-20 and fertilin in arriving at, and fusing with, the egg. *Biol. Reprod.* 56, 320–327.

Snell WJ & White JM (1996) The molecules of mammalian fertilization. *Cell* 85, 629–637.

Vacquier VD (1998) Evolution of gamete recognition proteins. *Science* 281, 1995–1998.

Wassarman PM (1999) Mammalian fertilization: molecular aspects of gamete adhesion, exocytosis, and fusion. *Cell* 96, 175–183.

Wassarman PM, Jovine L & Litscher ES (2001) A profile of fertilization in mammals. *Nat. Cell Biol.* 3, E59–E64.

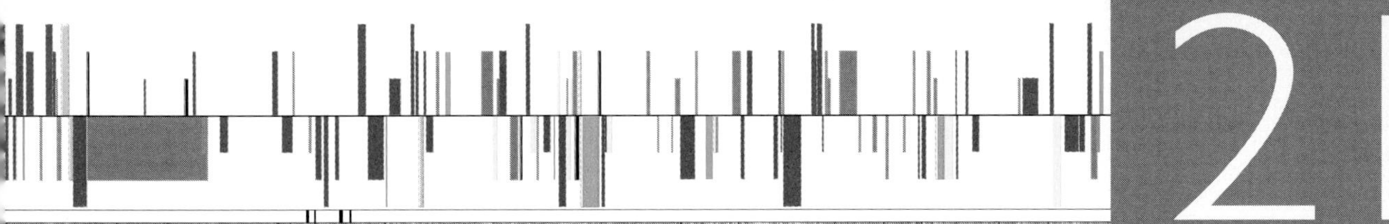

21

DEVELOPMENT OF MULTICELLULAR ORGANISMS

An animal or plant starts its life as a single cell—a fertilized egg. During development, this cell divides repeatedly to produce many different cells in a final pattern of spectacular complexity and precision. Ultimately, the genome determines the pattern, and the puzzle of developmental biology is to understand how it does so.

The genome is normally identical in every cell; the cells differ not because they contain different genetic information, but because they express different sets of genes. This selective gene expression controls the four essential processes by which the embryo is constructed: (1) *cell proliferation,* producing many cells from one, (2) *cell specialization,* creating cells with different characteristics at different positions, (3) *cell interactions,* coordinating the behavior of one cell with that of its neighbors, and (4) *cell movement,* rearranging the cells to form structured tissues and organs (Figure 21–1).

In a developing embryo, all these processes are happening at once, in a kaleidoscopic variety of different ways in different parts of the organism. To understand the basic strategies of development, we have to narrow our focus. In particular, we must understand the course of events from the standpoint of the individual cell and the way the genome acts within it. There is no commanding officer standing above the fray to direct the troops; each of the millions of cells in the embryo has to make its own decisions, according to its own copy of the genetic instructions and its own particular circumstances.

The complexity of animals and plants depends on a remarkable feature of the genetic control system. Cells have a memory: the genes a cell expresses and the way it behaves depend on the cell's past as well as its present environment. The cells of your body—the muscle cells, the neurons, the skin cells, the gut cells, and so on—maintain their specialized characters not because they continually receive the same instructions from their surroundings, but because they retain a record of signals their ancestors received in early embryonic development. The molecular mechanisms of cell memory have been introduced in Chapter 7. In this chapter we shall encounter its consequences.

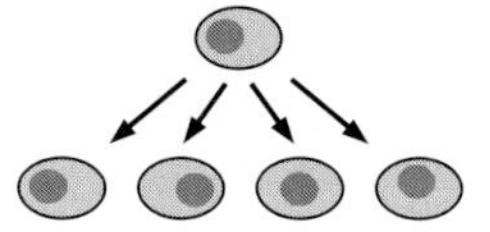

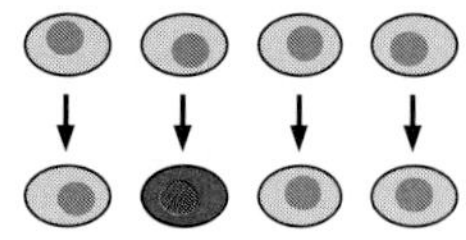

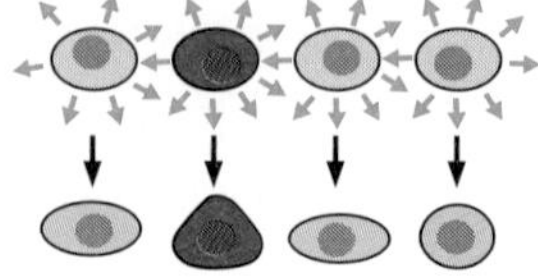

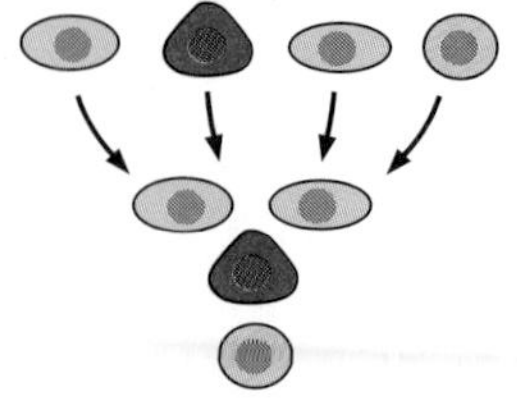

Figure 21–1 The four essential processes by which a multicellular organism is made: cell proliferation, cell specialization, cell interaction, and cell movement.

UNIVERSAL MECHANISMS OF ANIMAL DEVELOPMENT

There are about ten million species of animals, and they are fantastically varied. One would no more expect the worm, the flea, the eagle and the giant squid all to be generated by the same developmental mechanisms, than one would suppose that the same methods were used to make a shoe and an airplane. Some similar abstract principles might be involved, perhaps, but surely not the same specific molecules?

One of the most astonishing revelations of the past ten or twenty years has been that our initial suspicions are wrong. In fact, much of the basic machinery of development is essentially the same, not just in all vertebrates but in all the major phyla of invertebrates too. Recognizably similar, evolutionarily related molecules define our specialized cell types, mark the differences between body regions, and help create the body's pattern. Homologous proteins are often functionally interchangeable between very different species. A mouse protein produced artificially in a fly can often perform the same function as the fly's own version of that protein, and vice versa, successfully controlling the development of an eye, for example, or the architecture of the brain (Figure 21–2). Thanks to this underlying unity of mechanism, as we shall see, developmental biologists are now well on their way toward a coherent understanding of animal development.

Plants are a separate kingdom: they have evolved their multicellular organization independently of animals. For their development too, a unified account can be given, but it is different from that for animals. Animals will be our main concern in this chapter, but we shall return to plants briefly at the end.

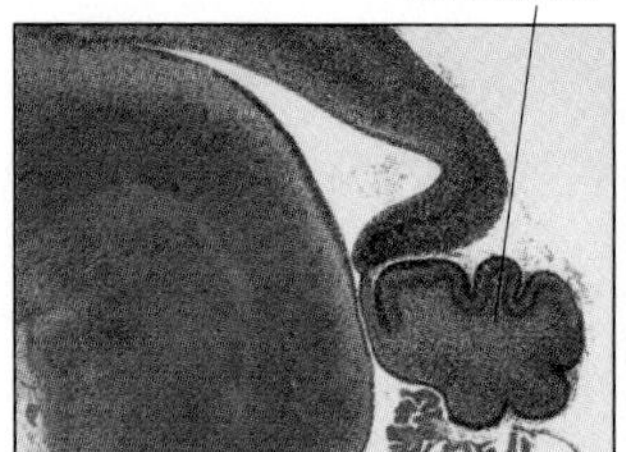

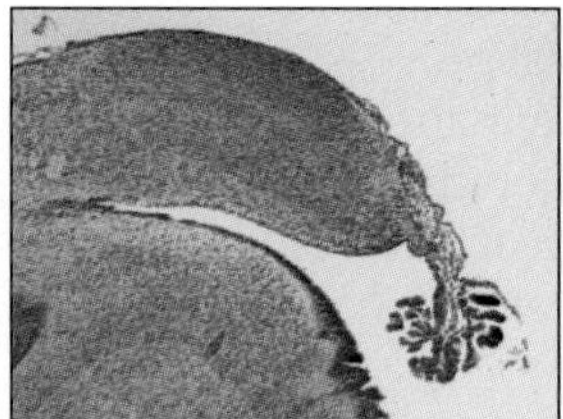

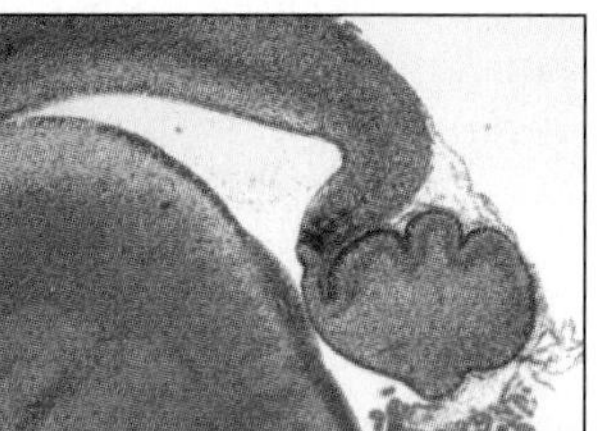

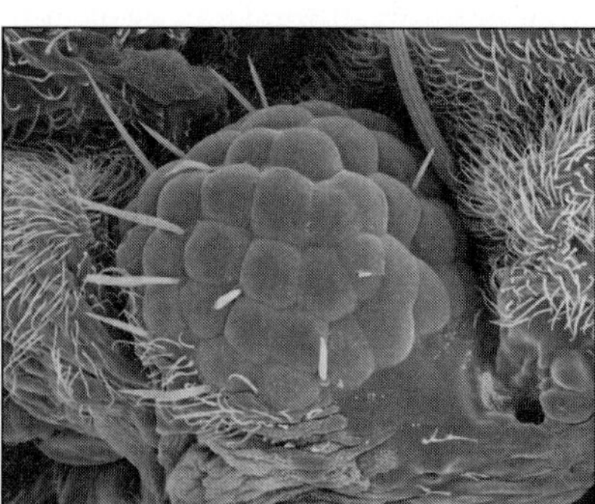

Figure 21–2 Homologous proteins functioning interchangeably in the development of mice and flies. (A) A fly protein used in a mouse. The DNA sequence from *Drosophila* coding for the Engrailed protein (a gene regulatory protein) can be substituted for the corresponding sequence coding for the Engrailed-1 protein of the mouse. Loss of Engrailed-1 in the mouse causes a defect in its brain (the cerebellum fails to develop); the *Drosophila* protein acts as an efficient substitute, rescuing the transgenic mouse from this deformity. (B) A mollusc protein used in a fly. The Eyeless protein controls eye development in *Drosophila*, and when misexpressed can cause an eye to develop in an abnormal site, such as a leg. The homologous protein, Pax-6, from a mouse, a squid, or practically any animal possessing eyes, when similarly misexpressed in a transgenic fly, has the same effect. The scanning electron micrographs show a patch of eye tissue on the leg of a fly resulting from misexpression of *Drosophila* Eyeless *(above)* and of squid Pax-6 *(below)*. (A, from M.C. Hanks et al., *Development* 125:4521–30, 1998. © The Company of Biologists; B, from S. I. Tomarev et al., *Proc. Natl. Acad. Sci. USA* 94:2421–2426, 1997. © National Academy of Sciences.)

We begin by reviewing some of the basic general principles of animal development and by introducing the seven animal species that developmental biologists have adopted as their chief model organisms.

Animals Share Some Basic Anatomical Features

The similarities between animal species in the genes that control development reflect the evolution of animals from a common ancestor in which these genes were already present. Although we do not know what it looked like, the common ancestor of worms, molluscs, insects, vertebrates, and other complex animals must have had many differentiated cell types that would be recognizable to us: epidermal cells, for example, forming a protective outer layer; gut cells to absorb nutrients from ingested food; muscle cells to move; neurons and sensory cells to control the movements. The body must have been organized with a sheet of skin covering the exterior, a mouth for feeding and a gut tube to contain and process the food—with muscles, nerves and other tissues arranged in the space between the external sheet of skin and the internal gut tube.

These features are common to almost all animals, and they correspond to a common basic anatomical scheme of development. The egg cell—a giant storehouse of materials—divides, or **cleaves**, to form many smaller cells. These cohere to create an epithelial sheet facing the external medium. Much of this sheet remains external, constituting the **ectoderm**—the precursor of the epidermis and of the nervous system. A part of the sheet becomes tucked into the interior to form **endoderm**—the precursor of the gut and its appendages, such as lung and liver. Another group of cells move into the space between ectoderm and endoderm, and form the **mesoderm**—the precursor of muscles, connective tissues, and various other components. This transformation of a simple ball or hollow sphere of cells into a structure with a gut is called **gastrulation** (from the Greek word for a belly), and in one form or another it is an almost universal feature of animal development. Figure 21–3 illustrates the process as it is seen in the sea urchin.

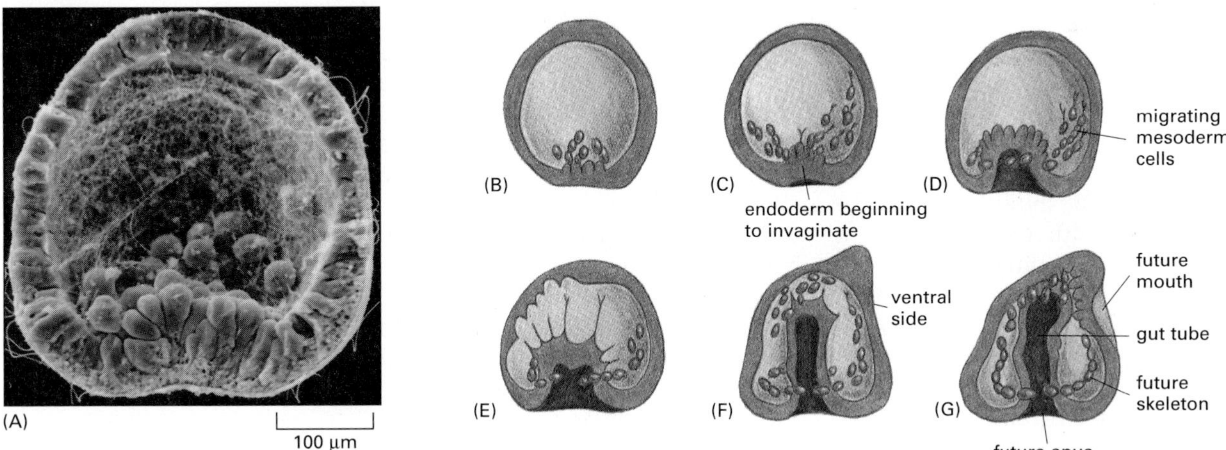

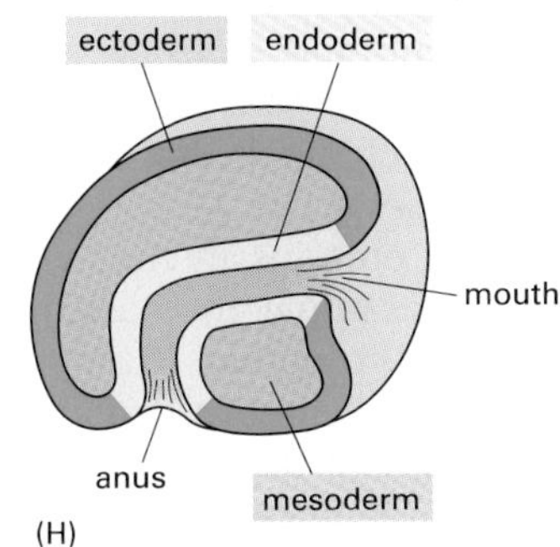

Figure 21–3 Sea urchin gastrulation. A fertilized egg divides to produce a *blastula*—a hollow sphere of epithelial cells surrounding a cavity. Then, in the process of gastrulation, some of the cells tuck into the interior to form the gut and other internal tissues. (A) Scanning electron micrograph showing the initial intucking of the epithelium. (B) Drawing showing how a group of cells break loose from the epithelium to become mesoderm. (C) These cells then crawl over the inner face of the wall of the blastula. (D) Meanwhile epithelium is continuing to tuck inward to become endoderm. (E and F) The invaginating endoderm extends into a long gut tube. (G) The end of the gut tube makes contact with the wall of the blastula at the site of the future mouth opening. Here the ectoderm and endoderm will fuse and a hole will form. (H) The basic animal body plan, with a sheet of ectoderm on the outside, a tube of endoderm on the inside, and mesoderm sandwiched between them. (A, from R.D. Burke et al., *Dev. Biol.* 146:542–557, 1991. © Academic Press; B–G, after L. Wolpert and T. Gustafson, *Endeavour* 26:85–90, 1967.)

Evolution has diversified upon the molecular and anatomical fundamentals that we describe in this chapter to produce the wonderful variety of present-day species. But the underlying conservation of genes and mechanisms means that studying the development of one animal very often leads to general insights into the development of many other types of animals. As a result, developmental biologists today, like cell biologists, have the luxury of addressing fundamental questions in whatever species offers the easiest path to an answer.

Multicellular Animals Are Enriched in Proteins Mediating Cell Interactions and Gene Regulation

Genome sequencing reveals the extent of molecular similarities between species. The nematode worm *Caenorhabditis elegans*, the fly *Drosophila melanogaster*, and the vertebrate *Homo sapiens* are the first three animals for which a complete genome sequence was obtained. In the family tree of animal evolution, they are very distant from one another: the lineage leading to the vertebrates is thought to have diverged from that leading to the nematodes, insects and molluscs more than 600 million years ago. Nevertheless, when the 19,000 genes of *C. elegans*, the 14,000 genes of *Drosophila*, and the 30,000 genes of the human are systematically compared with one another, it is found that about 50% of the genes in each of these species have clearly recognizable homologs in one or both of the other two species. In other words, recognizable versions of at least 50% of all human genes were already present in the common ancestor of worms, flies, and humans.

Of course, not everything is conserved: there are some genes with key roles in vertebrate development that have no homologs in the genome of *C. elegans* or *Drosophila*, and vice versa. However, a large proportion of the 50% of genes that lack identifiable homologs in other phyla may do so simply because their functions are of minor importance. Although these nonconserved genes are transcribed and well represented in cDNA libraries, studies of DNA and amino acid sequence variability in and between natural populations indicate that these genes are unusually free to mutate without seriously harming fitness. Because they are free to evolve so rapidly, a few tens of millions of years may be enough to obliterate any family resemblance or to permit loss from the genome.

The genomes of different classes of animals differ also because, as discussed in Chapter 1, there are substantial variations in the extent of gene duplication: the amount of gene duplication in the evolution of the vertebrates has been particularly large, with the result that a mammal or a fish often has several homologs corresponding to a single gene in a worm or a fly.

Despite such differences, to a first approximation we can say that all these animals have a similar set of proteins at their disposal for their key functions. In other words, they construct their bodies using roughly the same molecular kit of parts.

What genes, then, are needed to produce a multicellular animal, beyond those necessary for a solitary cell? Comparison of animal genomes with that of budding yeast—a unicellular eucaryote—suggests that two classes of proteins are especially important for multicellular organization. The first class is that of the transmembrane molecules used for cell adhesion and cell signaling. As many as 2000 *C. elegans* genes encode cell surface receptors, cell adhesion proteins, and ion channels that are either not present in yeast or present in much smaller numbers. The second class is that of gene regulatory proteins: these DNA-binding proteins are much more numerous in the *C. elegans* genome than in yeast. For example, the basic helix–loop–helix family has 41 members in *C. elegans*, 84 in *Drosophila*, 131 in humans and only 7 in yeast, and other families of regulators of gene expression are also dramatically overrepresented in animals as compared to yeast. Not surprisingly, these two classes of proteins are central to developmental biology: as we shall see, the development of multicellular animals is dominated by cell–cell interactions and by differential gene expression.

Regulatory DNA Defines the Program of Development

The fundamental similarity in the gene sets of different animals amazed developmental biologists when it was first discovered. A worm, a fly, a mollusc and a mammal do indeed share many of the same essential cell types, and they do all have a mouth, a gut, a nervous system and a skin; but beyond a few such basic features they seem radically different in their body structure. If the genome determines the structure of the body and these animals all have such a similar collection of genes, how can they be so different?

The proteins encoded in the genome can be viewed as the components of a construction kit. Many things can be built with this kit, just as a child's construction kit can be used to make trucks, houses, bridges, cranes, and so on by assembling the components in different combinations. Some components necessarily go together—nuts with bolts, wheels with tires and axles—but the large-scale organization of the final object is not defined by these substructures. Rather, it is defined by the instructions that accompany the components and prescribe how they are to be assembled.

To a large extent, the instructions needed to produce a multicellular animal are contained in the non-coding, regulatory DNA that is associated with each gene. As discussed in Chapter 4, each gene in a multicellular organism is associated with thousands or tens of thousands of nucleotides of noncoding DNA. This DNA may contain, scattered within it, dozens of separate regulatory elements or *enhancers*—short DNA segments that serve as binding sites for specific complexes of gene regulatory proteins. Roughly speaking, as explained in Chapter 7, the presence of a given regulatory module of this sort leads to expression of the gene whenever the complex of proteins recognizing that segment of DNA is appropriately assembled in the cell (in some cases, an inhibition or a more complicated effect on gene expression is produced instead). If we could decipher the full set of regulatory modules associated with a gene, we would understand all the different molecular conditions under which the product of that gene is to be made. This regulatory DNA can therefore be said to define the sequential program of development: the rules for stepping from one state to the next, as the cells proliferate and read their positions in the embryo by reference to their surroundings, switching on new sets of genes according to the activities of the proteins that they currently contain (Figure 21–4).

When we compare animal species with similar body plans—different vertebrates such as a fish, a bird and a mammal, for example—we find that corresponding genes usually have similar sets of regulatory modules: the DNA sequences of many of the individual modules have been well conserved and are recognizably homologous in the different animals. The same is true if we compare different species of nematode worm, or different species of insect. But when we compare vertebrate regulatory regions with those of worm or fly, it is hard to see any such resemblance. The protein-coding sequences are unmistakably similar, but the corresponding regulatory DNA sequences appear very different. This is the expected result if different body plans are produced mainly by changing the program embodied in the regulatory DNA, while retaining most of the same kit of proteins.

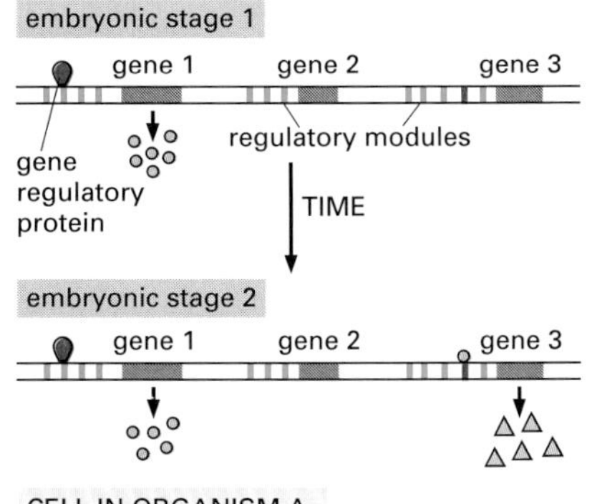

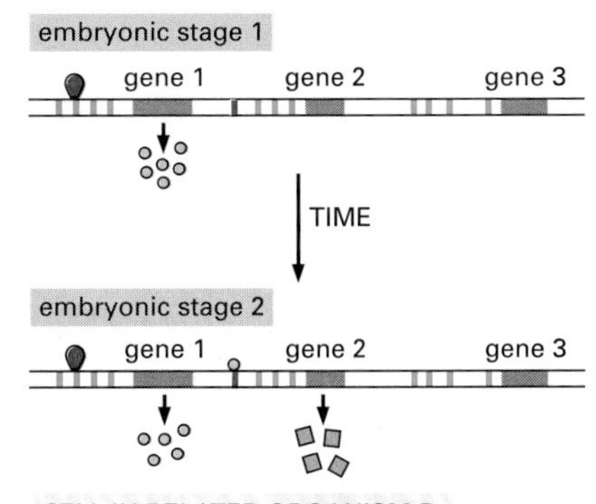

Figure 21–4 How regulatory DNA defines the succession of gene expression patterns in development. The genomes of organisms A and B code for the same set of proteins but have different regulatory DNA. The two cells in the cartoon start in the same state, expressing the same proteins at stage 1, but step to quite different states at stage 2 because of their different arrangements of regulatory modules.

Manipulation of the Embryo Reveals the Interactions Between its Cells

Confronted with an adult animal, in all its complexity, how does one begin to analyze the process that brought it into being? The first essential step is to describe the anatomical changes—the patterns of cell division, growth, and movement—that convert the egg into the mature organism. This is the job of *descriptive embryology*, and it is harder than one might think. To explain development in terms of cell behavior, we need to be able to track the individual cells through all their divisions, transformations, and migrations in the embryo. The foundations of descriptive embryology were laid in the 19th century, but the fine-grained task of *cell lineage tracing* continues to tax the ingenuity of developmental biologists (Figure 21–5)

Given a description, how can one go on to discover the causal mechanisms? Traditionally, *experimental embryologists* have tried to understand development in terms of the ways in which cells and tissues interact to generate the multicellular structure. *Developmental geneticists*, meanwhile, have tried to analyze development in terms of the actions of genes. These two approaches are complementary, and they have converged to produce our present understanding.

In experimental embryology, cells and tissues from developing animals are removed, rearranged, transplanted, or grown in isolation, in order to discover how they influence one another. The results are often startling: an early embryo cut in half, for example, may yield two complete and perfectly formed animals or a small piece of tissue transplanted to a new site may reorganize the whole structure of the developing body (Figure 21–6). Observations of this type can be extended and refined to decipher the underlying cell–cell interactions and rules of cell behavior. The experiments are easiest to perform in large embryos that are readily accessible for microsurgery. Thus, the most widely used species have been birds—especially the chick—and amphibians—particularly the African frog *Xenopus laevis*.

Studies of Mutant Animals Identify the Genes That Control Developmental Processes

Developmental genetics begins with the isolation of mutant animals whose development is abnormal. This typically involves a *genetic screen*, as described in Chapter 8. Parent animals are treated with a chemical mutagen or ionizing radiation to induce mutations in their germ cells, and large numbers of their progeny are examined. Those rare mutant individuals that show some interesting developmental abnormality—altered development of the eye, for example—are picked out for further study. In this way, it is possible to discover genes that are required specifically for the normal development of any chosen feature. By cloning and sequencing a gene found in this way, it is possible to identify its

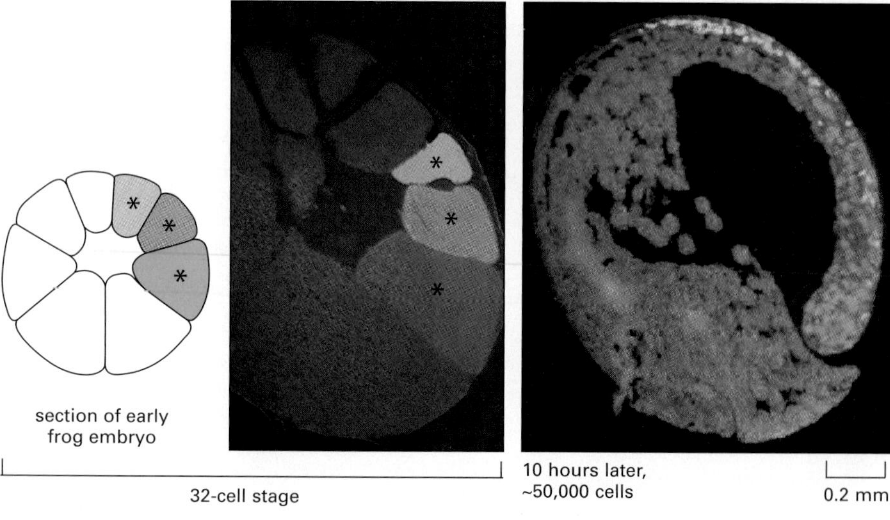

Figure 21–5 A cell lineage tracing experiment in the *Xenopus* embryo. Different fluorescent dyes are injected into three cells at an early stage (cells with asterisks at the 32-cell stage), and the embryo is then left to develop for 10 hours before being fixed and sectioned. The pattern of fluorescent labeling reveals how the descendants of the three cells have moved. (From M.A. Vodicka and J.C. Gerhart, *Development* 121:3505–3518, 1995. © The Company of Biologists.)

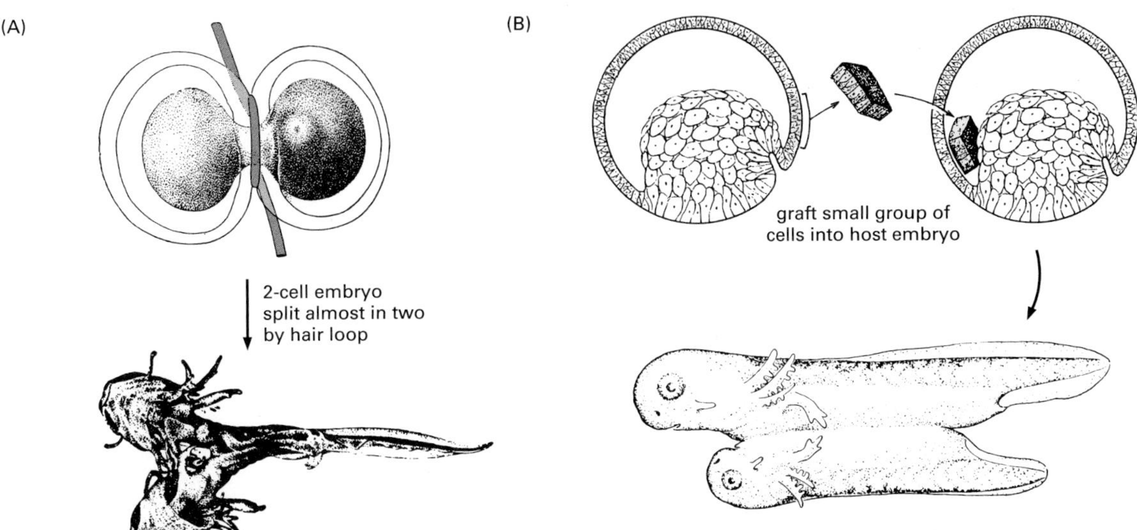

Figure 21–6 Some striking results obtained by experimental embryology. In (A), an early amphibian embryo is split almost into two parts with a hair loop. In (B), an amphibian embryo at a somewhat later stage receives a graft of a small cluster of cells from another embryo at that stage. The two quite different operations both cause a single embryo to develop into a pair of conjoined (Siamese) twins. It is also possible in experiment (A) to split the early embryo into two completely separate halves; two entire separate well-formed tadpoles are then produced. (A, after H. Spemann, Embryonic Development and Induction. New Haven: Yale University Press, 1938; B, after J. Holtfreter and V. Hamburger, in Analysis of Development [B.H. Willier, P.A. Weiss, and V. Hamburger, eds.], pp. 230–296. Philadelphia: Saunders, 1955.)

protein product, to investigate how it works, and to begin an analysis of the regulatory DNA that controls its expression.

The genetic approach is easiest in small animals with short generation times that can be grown in the laboratory. The first animal to be studied in this way was the fruit fly *Drosophila melanogaster,* which will be discussed at length below. But the same approach has been successful in the nematode worm, *Caenorhabditis elegans,* the zebrafish, *Danio rerio,* and the mouse, *Mus musculus.* Although humans are not intentionally mutagenized, they get screened for abnormalities in enormous numbers through the medical care system. Many mutations have arisen in humans that cause abnormalities compatible with life, and analyses of the affected individuals and of their cells have provided important insights into developmental processes.

A Cell Makes Developmental Decisions Long Before It Shows a Visible Change

By simply watching closely, or with the help of tracer dyes and other cell-marking techniques, one can discover what the fate of a given cell in an embryo will be if that embryo is left to develop normally. The cell may be fated to die, for example, or to become a neuron, to form part of an organ such as the foot, or to give progeny cells scattered all over the body. To know the **cell fate**, in this sense, however, is to know next to nothing about the cell's intrinsic character. At one extreme, the cell that is fated to become, say, a neuron may be already specialized in a way that guarantees that it will become a neuron no matter how its surroundings are disturbed; such a cell is said to be **determined** for its fate. At the opposite extreme, the cell may be biochemically identical to other cells destined for other fates, the only difference between them being the accident of position, which exposes the cells to different future influences.

A cell's state of determination can be tested by transplanting it to altered environments (Figure 21–7). One of the key conclusions of experimental embryology has been that, thanks to cell memory, a cell can become determined long before it shows any obvious outward sign of differentiation.

Between the extremes of the fully determined and the completely undetermined cell, there is a whole spectrum of possibilities. A cell may, for example, be already somewhat specialized for its normal fate, with a strong tendency to develop in that direction, but still able to change and undergo a different fate if it is put in a sufficiently coercive environment. (Some developmental biologists would describe such a cell as *specified* or *committed,* but not yet determined.) Or the cell may be determined, say, as a brain cell, but not yet determined as to whether it is to be a neuronal or a glial component of the brain. And often, it seems, adjacent cells of the same type interact and depend on mutual support to maintain their specialized character, so that they will behave as determined if

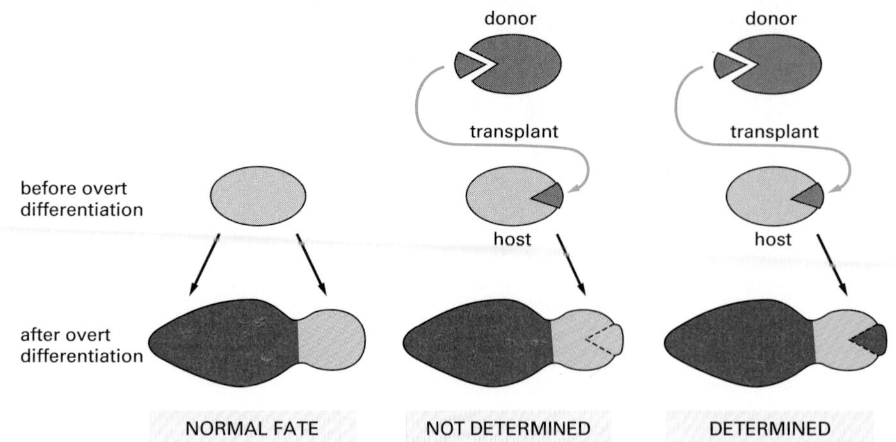

Figure 21–7 The standard test for cell determination.

kept together in a cluster, but not if taken singly and isolated from their usual companions.

Cells Have Remembered Positional Values That Reflect Their Location in the Body

In many systems, long before cells become committed to differentiating as a specific cell type, they become *regionally determined*: that is, they switch on and maintain expression of genes that can best be regarded as markers of position or region in the body. This position-specific character of a cell is called its **positional value**, and it shows its effects in the way the cell behaves in subsequent steps of pattern formation.

The development of the chick leg and wing provides a striking example. The leg and the wing of the adult both consist of muscle, bone, skin, and so on—almost exactly the same range of differentiated tissues. The difference between the two limbs lies not in the types of tissues, but in the way in which those tissues are arranged in space. So how does the difference come about?

In the chick embryo the leg and the wing originate at about the same time in the form of small tongue-shaped buds projecting from the flank. The cells in the two pairs of limb buds appear similar and uniformly undifferentiated at first. But a simple experiment shows that this appearance of similarity is deceptive. A small block of undifferentiated tissue at the base of the leg bud, from the region that would normally give rise to part of the thigh, can be cut out and grafted into the tip of the wing bud. Remarkably, the graft forms not the appropriate part of the wing tip, nor a misplaced piece of thigh tissue, but a toe (Figure 21–8). This experiment shows that the early leg-bud cells are already determined as leg but are not yet irrevocably committed to form a particular part of the leg: they can still respond to cues in the wing bud so that they form structures appropriate to the tip of the limb rather than the base. The signaling system that controls the differences between the parts of the limb is apparently the same for leg and wing. The difference between the two limbs results from a difference in the internal states of their cells at the outset of limb development.

The difference of positional value between vertebrate forelimb cells and hindlimb cells appears to be a reflection of differential expression of a class of gene regulatory proteins called T-box (Tbx) proteins. The cells of the hindlimb bud express the *Tbx4* gene while those of the forelimb bud express *Tbx5*, and this is thought to control their subsequent behavior (Figure 21–9). Later in this chapter we shall explain how the next, more detailed level of patterning is set up inside an individual limb bud.

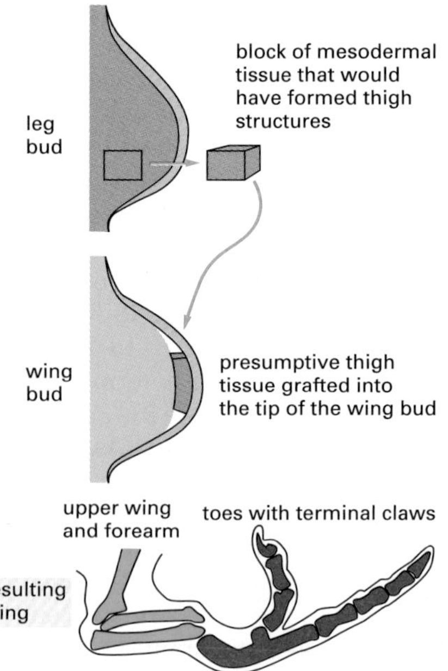

Figure 21–8 Prospective thigh tissue grafted into the tip of a chick wing bud forms toes. (After J.W. Saunders et al., *Dev. Biol.* 1:281–301, 1959.)

Sister Cells Can Be Born Different by an Asymmetric Cell Division

At each stage in its development, a cell in an embryo is presented with a limited set of options according to the state it has attained: the cell travels along a

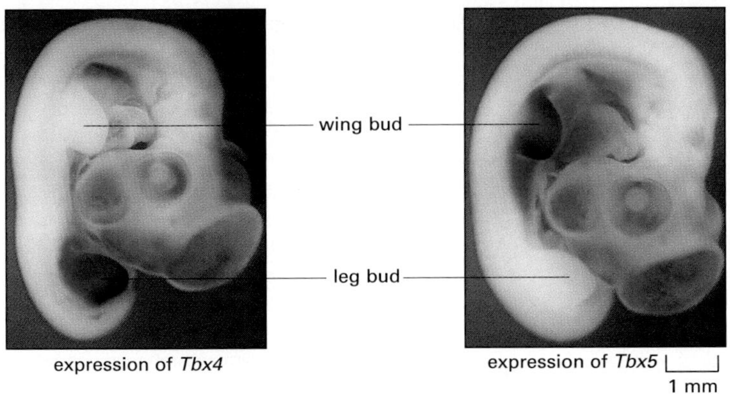

Figure 21–9 Chick embryos at 4 days of incubation, showing the limb buds stained by *in situ* hybridization with probes to detect expression of the *Tbx4* and *Tbx5* genes. The cells expressing *Tbx5* will form a wing; those expressing *Tbx4* will form a leg. *Tbx4* and *Tbx5* code for related gene regulatory proteins, which are thought to dictate which type of limb develops. (From H. Ohuchi et al., *Development* 125:51–60, 1998. © The Company of Biologists.)

developmental pathway that branches repeatedly. At each branch in the pathway it has to make a choice, and its sequence of choices determines its final destiny. In this way, a complicated array of different cell types is produced.

To understand development, we need to know how each choice between options is controlled, and how those options depend on the choices made previously. To reduce the question to its simplest form: how do two cells with the same genome come to be different?

When a cell undergoes mitosis, both of the resulting daughter cells receive a precise copy of the mother cell's genome. Yet those daughters will often have different specialized fates, and, at some point, they or their progeny must acquire different characters.

In some cases, the two sister cells are born different as a result of an **asymmetric cell division**, in which some significant set of molecules is divided unequally between the two daughter cells at the time of division. This asymmetrically segregated molecule (or set of molecules) then acts as a *determinant* for one of the cell fates by directly or indirectly altering the pattern of gene expression within the daughter cell that receives it (Figure 21–10).

Asymmetric divisions are particularly common at the beginning of development, when the fertilized egg divides to give daughter cells with different fates, but they also occur at later stages—in the genesis of nerve cells, for example.

Inductive Interactions Can Create Orderly Differences Between Initially Identical Cells

An alternative and by far the commonest way to make cells different is by exposing them to different environments, and the most important environmental

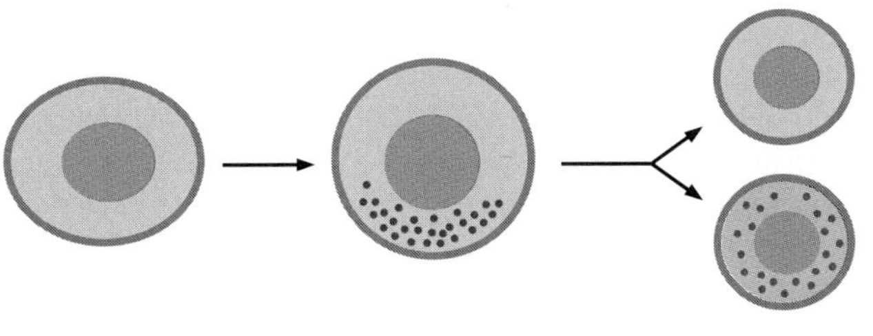

1. asymmetric division : sister cells born different

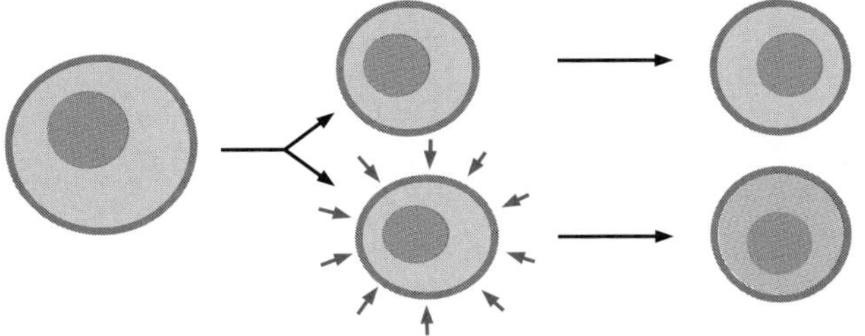

2. symmetric division : sister cells become different as result of influences acting on them after their birth

Figure 21–10 Two ways of making sister cells different.

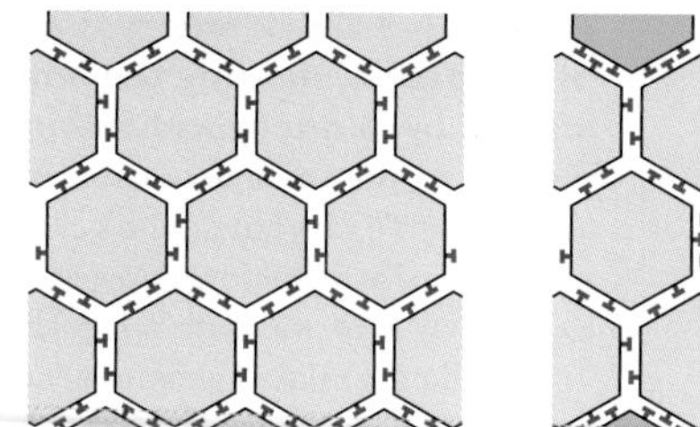
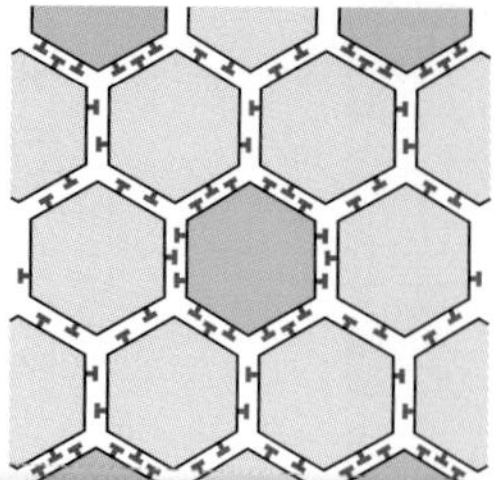
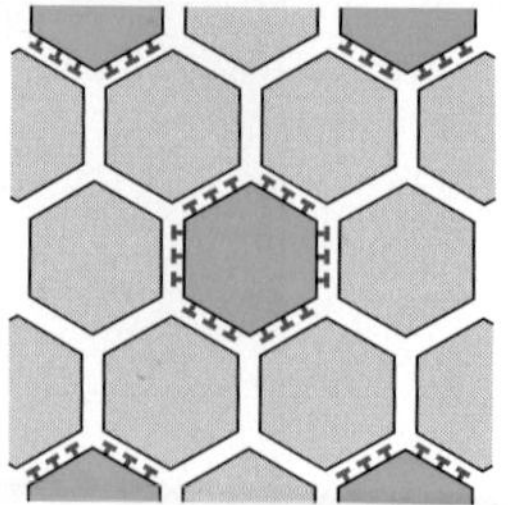

Figure 21–11 Lateral inhibition and cell diversification. Adjacent cells compete to adopt the primary character *(blue)*, by delivering inhibitory signals to one another. At first, all cells in the patch are similar. Any cell that gains an advantage in the competition *(darker blue)* delivers a stronger inhibitory signal (more *red* inhibition signals) to its neighbors, inhibiting them from delivering inhibitory signals themselves in return. This effect is self-reinforcing, and it leads to creation of a fine-grained mixture in which the cells finally adopting the primary character *(deep blue)* are surrounded by inhibited cells that adopt a different character *(gray)*.

cues acting on cells in an embryo are signals from neighboring cells.

In some cases, adjacent initially similar cells exchange signals that drive them to become different from one another, as in a competition between identical twins. A sort of shouting match occurs, from which one cell or group of cells emerges as winner—not only specializing in a particular way but also delivering a signal to neighboring cells that inhibits them from doing likewise—a phenomenon called *lateral inhibition* (Figure 21–11). Very often, this process is based on an exchange of signals at cell–cell contacts via the Notch pathway (discussed in Chapter 15).

In another strategy, perhaps the widely used of all, a group of cells start out all having the same developmental potential, and a signal from cells outside the group then drives one or more of the members of the group into a different developmental pathway, leading to a changed character. This process is called an **inductive interaction**. Generally, the signal is limited in time and space so that only a subset of the competent cells—those closest to the source of the signal—take on the induced character (Figure 21–12).

Some inductive signals are short-range—notably those transmitted via cell–cell contacts; others are long-range, mediated by molecules that can diffuse through the extracellular medium. The group of initially similar cells competent to respond to the signal is sometimes called an *equivalence group* or a *morphogenetic field.* It can consist of as few as two cells or as many as thousands, and any number of the total can be induced depending on the amount and distribution of the signal.

In principle, any kind of signal molecule could serve as an inducer. In practice, most of the known inductive events in animal development are governed by just a handful of highly conserved families of signal proteins, which are used over and over again in different contexts. The discovery of this limited vocabulary that cells use for developmental communications has emerged over the last ten or twenty years as one of the great simplifying discoveries of developmental biology. In Table 21–1, we briefly review five major families of signal proteins that serve repeatedly as inducers in animal development. Details of the intracellular mechanisms through which these molecules act are given in Chapter 15.

The ultimate result of most inductive events is a change in DNA transcription in the responding cell: some genes are turned on and others are turned off. Different signaling molecules activate different kinds of gene regulatory proteins. Moreover, the effect of activating a given gene regulatory protein will depend on which other gene regulatory proteins are also present in the cell, since these generally function in combinations. As a result, different types of cells will generally respond differently to the same signal. The response will depend both on the other gene regulatory proteins that are present before the signal arrives—reflecting the cell's memory of signals received previously—and on the other signals that the cell is receiving at the same time.

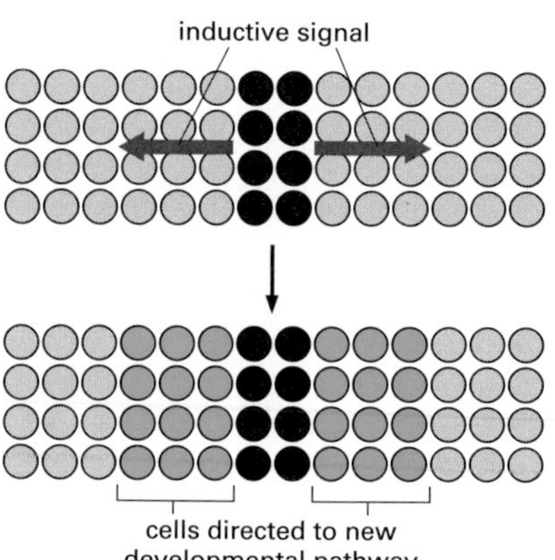

Figure 21–12 Inductive signaling.

Morphogens Are Long-Range Inducers That Exert Graded Effects

So far, we have spoken of signal molecules as though they governed a simple yes–no choice: one effect in their presence, another in their absence. In many cases, however, responses are more finely graded: a high concentration may, for example, direct target cells into one developmental pathway, an intermediate concentration into another, and a low concentration into yet another. An

TABLE 21–1 Some Signal Proteins That Are Used Over and Over Again as Inducers in Animal Development

SIGNALING PATHWAY	LIGAND FAMIILY	RECEPTOR FAMILY	EXTRACELLULAR INHIBITORS/MODULATORS
Receptor tyrosine kinase (RTK)	EGF	EGF receptors	Argos
	FGF (Branchless)	FGF receptors (Breathless)	
	ephrins	Eph receptors	
TGFβ superfamily	TGFβ	TGFβ receptors	chordin (Sog), noggin
	BMP (Dpp)	BMP receptors	
	Nodal		
Wnt	Wnt (Wingless)	Frizzled	Dickkopf, Cerberus
Hedgehog	Hedgehog	Patched, Smoothened	
Notch	Delta	Notch	Fringe

Only a few representatives of each class of proteins are listed—mainly those mentioned in this chapter. Names peculiar to *Drosophila* are shown in parentheses. Many of the listed components have several homologs distinguished by numbers (FGF1, FGF2, etc.) or by forenames (Sonic hedgehog, Lunatic fringe). For further details, see Chapter 15.

important case is that in which the signaling molecule diffuses out from a localized source, creating a signal concentration gradient. Cells at different distances from the source are driven to behave in a variety of different ways, according to the signal concentration that they experience.

A signaling molecule that imposes a pattern on a whole field of cells in this way is called a **morphogen**. Vertebrate limbs provide a striking example: a specialized group of cells at one side of the embryonic limb bud secrete Sonic hedgehog protein—a member of the Hedgehog family of signal molecules—and this protein spreads out from its source, forming a *morphogen gradient* that controls the characters of the cells along the thumb-to-little-finger axis of the limb bud. If an additional group of signaling cells is grafted into the opposite side of the bud, a mirror duplication of the pattern of digits is produced (Figure 21–13).

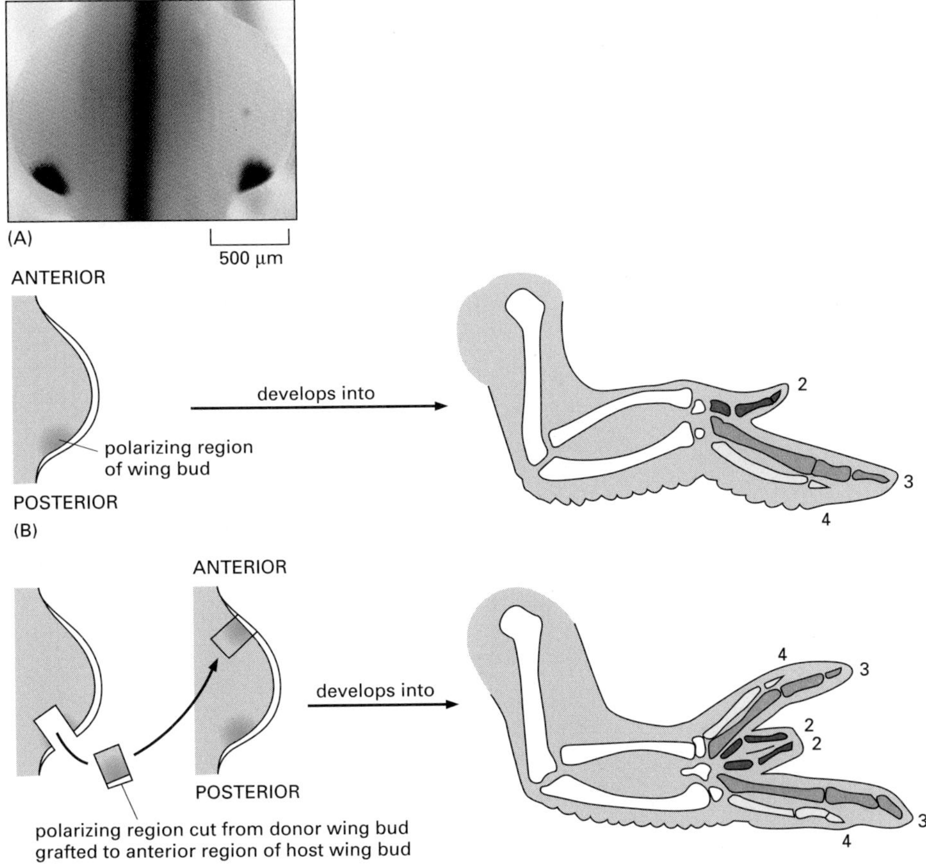

Figure 21–13 Sonic hedgehog as a morphogen in chick limb development. (A) Expression of the *Sonic hedgehog* gene in a 4-day chick embryo, shown by *in situ* hybridization (dorsal view of the trunk at the level of the wing buds). The gene is expressed in the midline of the body and at the posterior border (the polarizing region) of each of the two wing buds. Sonic hedgehog protein spreads out from these sources. (B) Normal wing development. (C) A graft of tissue from the polarizing region causes a mirror-image duplication of the pattern of the host wing. The type of digit that develops is thought to be dictated by the local concentration of Sonic hedgehog protein; different types of digit (labeled 2, 3, and 4) therefore form according to their distance from a source of Sonic hedgehog. (A, courtesy of Randall S. Johnson and Robert D. Riddle.)

Extracellular Inhibitors of Signal Molecules Shape the Response to the Inducer

Especially for molecules that can act at a distance, it is important to limit the action of the signal, as well as to produce it. Most developmental signal proteins have extracellular antagonists that can inhibit their function. These antagonists are generally proteins that bind to the signal or its receptor, preventing a productive interaction from taking place.

A surprisingly large number of developmental decisions are actually regulated by inhibitors rather than by the primary signal molecule. The nervous system in a frog embryo arises from a field of cells that is competent to form either neural or epidermal tissue. An inducing tissue releases the protein chordin, which favors the formation of neural tissue. Chordin does not have its own receptor. Instead it is an inhibitor of signal proteins of the BMP/TGFβ family, which induce epidermal development and are present throughout the neuroepithelial region where neurons and epidermis form. The induction of neural tissue is thus due to an inhibitory gradient of an antagonistic signal (Figure 21–14).

Programs That Are Intrinsic to a Cell Often Define the Time-Course of its Development

Signals such as those we have just discussed play a large part in controlling the timing of events in development, but it would be wrong to imagine that every developmental change needs an inductive signal to trigger it. Many of the mechanisms that alter cell character are intrinsic to the cell and require no cue from the cell's surroundings: the cell will step through its developmental program even when kept in a constant environment. There are numerous cases where one might suspect that something of this sort is occurring to control the duration of a developmental process. For example, in a mouse, the neural progenitor cells in the spinal cord carry on dividing and generating neurons for just 11 cell cycles, and those in the cerebral cortex of the brain for just 23 cycles, after which they stop. Moreover, different kinds of neurons are generated at different stages in this program, suggesting that as the progenitor cell ages, it changes the specifications that it supplies to the differentiating progeny cells.

It is difficult to prove in the context of the intact embryo that such a course of events is strictly the result of a cell-autonomous timekeeping process, since the cell environment is changing. Experiments on cells in culture, however, give clear-cut evidence. For example, glial progenitor cells isolated from the optic nerve of a 7-day postnatal rat and cultured under constant conditions in an appropriate medium will carry on proliferating for a strictly limited time (corresponding to a maximum of about eight cell division cycles) and then differentiate into oligodendrocytes (the glial cells that form myelin sheaths around axons

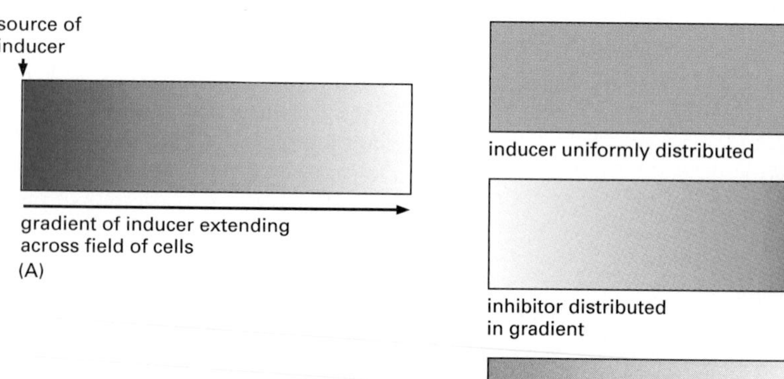

Figure 21–14 Two ways to create a morphogen gradient. (A) By localized production of an inducer—a morphogen—that diffuses away from its source, (B) By localized production of an inhibitor that diffuses away from its source and blocks the action of a uniformly distributed inducer.

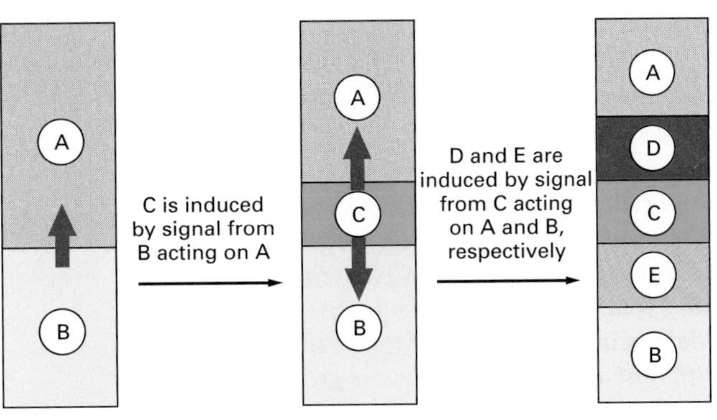

Figure 21–15 Patterning by sequential induction. A series of inductive interactions can generate many types of cells, starting from only a few.

in the brain), obeying a timetable similar to the one that they would have followed if they had been left in place in the embryo.

Initial Patterns Are Established in Small Fields of Cells and Refined by Sequential Induction as the Embryo Grows

The signals that organize the spatial pattern of an embryo generally act over short distances and govern relatively simple choices. A morphogen, for example, typically acts over a distance of less than 1 mm—an effective range for diffusion—and directs choices between no more than a handful of developmental options for the cells on which it acts. But the organs that eventually develop are much larger and more complex than this.

The cell proliferation that follows the initial specification accounts for the size increase, while the refinement of the initial pattern is explained by a series of local inductions that embroider successive levels of detail on an initially simple sketch. As soon as two sorts of cells are present, one of them can produce a factor that induces a subset of the neighboring cells to specialize in a third way. The third cell type can in turn signal back to the other two cell types nearby, generating a fourth and a fifth cell type, and so on (Figure 21–15).

This strategy for generating a progressively more complicated pattern is called **sequential induction**. It is chiefly through sequential inductions that the body plan of a developing animal, after being first roughed out in miniature, becomes elaborated with finer and finer details as development proceeds.

In the sections that follow, we focus on a small selection of model organisms to see how the principles that we have outlined in this first section operate in practice. We begin with the nematode worm, *Caenorhabditis elegans*.

Summary

The obvious changes of cell behavior that we see as a multicellular organism develops are the outward signs of a complex molecular computation, dependent on cell memory, that is taking place inside the cells as they receive and process signals from their neighbours and emit signals in return. The final pattern of differentiated cell types is thus the outcome of a more hidden program of cell specialization—a program played out in the changing patterns of expression of gene regulatory proteins, giving one cell different potentialities from another long before terminal differentiation begins. Developmental biologists seek to decipher the hidden program and to relate it, through genetic and microsurgical experiments, to the signals the cells exchange as they proliferate, interact, and move.

Animals as different as worms, flies, and humans use remarkably similar sets of proteins to control their development, so that what we discover in one organism very often gives insight into the others. A handful of evolutionarily conserved cell–cell signaling pathways are used repeatedly, in different organisms and at different times, to regulate the creation of an organized multicellular pattern. Differences of body plan seem to arise to a large extent from differences in the regulatory DNA associated with each gene. This DNA has a central role in defining the sequential program of development, calling genes into action at specific times and places according to the pattern of gene expression that was present in each cell at the previous developmental stage.

Differences between cells in an embryo arise in various ways. Sister cells can be born different, as a result of an asymmetric cell division. Alternatively, cells that are born similar can become different through a competitive interaction with one another, as in lateral inhibition. Or a group of initially similar cells may receive different exposures to inductive signals from cells outside the group; long-range inducers with graded effects, called morphogens, can organize a complex pattern. Through cell memory, such transient signals can have a lasting effect on the internal state of a cell, causing it, for example, to become determined for specific fate. In these ways, sequences of simple signals acting at different times and places in growing cell arrays give rise to the intricate and varied multicellular organisms that fill the world around us.

CAENORHABDITIS ELEGANS: DEVELOPMENT FROM THE PERSPECTIVE OF THE INDIVIDUAL CELL

The nematode worm *Caenorhabditis elegans* is a small, relatively simple, and precisely structured organism. The anatomy of its development has been described in extraordinary detail, and one can map out the exact lineage of every cell in the body. Its complete genome sequence is also known, and large numbers of mutant phenotypes have been analyzed to determine gene functions. If there is any multicellular animal whose development we should be able to understand in terms of genetic control, this is it.

There are, it is true, some factors that make such an understanding difficult. Anatomically simple though it is, it has more genes (about 18,000) and more cells (about 1000) than one can easily keep track of without a computer. It is not the most directly instructive model organism if one wants to understand how animals such as ourselves form organs such as eyes or limbs, which the worm lacks, and it is hard to do transplantation experiments. Moreover, DNA sequence comparisons indicate that, while the lineages leading to nematodes, insects, and vertebrates diverged from one another at about the same time, the rate of evolutionary change in the nematode lineage has been substantially greater: both its genes and its body structure are more divergent from our own than are those of *Drosophila*.

In spite of all this, the worm presents us with an excellent introductory example: it poses the basic general questions of animal development in a relatively simple form, and it lets us answer them in terms of the behavior of individual, identified cells.

Caenorhabditis elegans Is Anatomically Simple

As an adult, *C. elegans* consists of only about 1000 somatic cells and 1000–2000 germ cells (exactly 959 somatic cell nuclei plus about 2000 germ cells are counted in one sex; exactly 1031 somatic cell nuclei plus about 1000 germ cells in the other) (Figure 21–16). The anatomy has been reconstructed, cell by cell, by

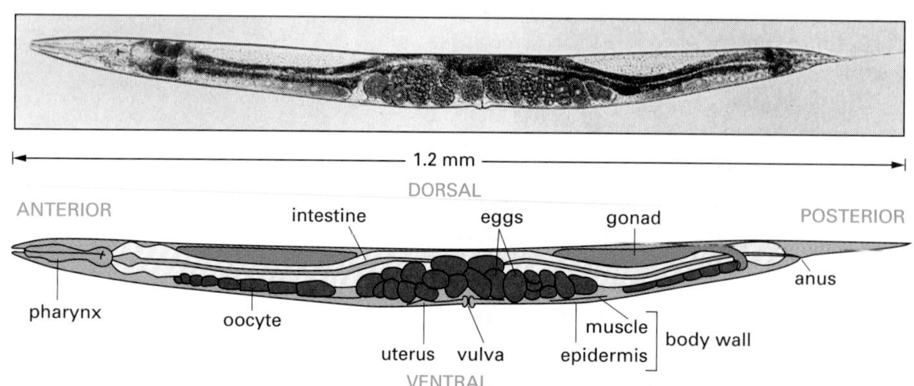

Figure 21–16 *Caenorhabditis elegans*. A side view of an adult hermaphrodite is shown. (From J.E. Sulston and H.R. Horvitz, *Dev. Biol.* 56:110–156, 1977. © Academic Press.)

electron microscopy of serial sections. The body plan of the worm is simple: it has a roughly bilaterally symmetrical, elongate body composed of the same basic tissues as in other animals (nerve, muscle, gut, skin), organized with mouth and brain at the anterior end and anus at the posterior. The outer body wall is composed of two layers: the protective epidermis, or "skin," and the underlying muscular layer. A tube of endodermal cells forms the intestine. A second tube, located between the intestine and the body wall, constitutes the gonad; its wall is composed of somatic cells, with the germ cells inside it.

C. elegans has two sexes—a hermaphrodite and a male. The hermaphrodite can be viewed most simply as a female that produces a limited number of sperm: she can reproduce either by self-fertilization, using her own sperm, or by cross-fertilization after transfer of male sperm by mating. Self-fertilization allows a single heterozygous worm to produce homozygous progeny. This is an important feature that helps to make *C. elegans* an exceptionally convenient organism for genetic studies.

Cell Fates in the Developing Nematode Are Almost Perfectly Predictable

C. elegans begins life as a single cell, the fertilized egg, which gives rise, through repeated cell divisions, to 558 cells that form a small worm inside the egg shell. After hatching, further divisions result in the growth and sexual maturation of the worm as it passes through four successive larval stages separated by molts. After the final molt to the adult stage, the hermaphrodite worm begins to produce its own eggs. The entire developmental sequence, from egg to egg, takes only about three days.

The lineage of all of the cells from the single-cell egg to the multicellular adult was mapped out by direct observation of the developing animal. In the nematode, a given precursor cell follows the same pattern of cell divisions in every individual, and with very few exceptions the fate of each descendant cell can be predicted from its position in the lineage tree (Figure 21–17).

This degree of stereotyped precision is not seen in the development of larger animals. At first sight, it might seem to suggest that each cell lineage in the nematode embryo is rigidly and independently programmed to follow a set pattern of cell division and cell specialization, making the worm a woefully unrepresentative model organism for development. We shall see that this is far from true: as in other animals, development depends on cell–cell interactions as well as on processes internal to the individual cells. The outcome in the nematode is almost perfectly predictable simply because the pattern of cell–cell interactions is highly reproducible and is accurately correlated with the sequence of cell divisions.

Products of Maternal-Effect Genes Organize the Asymmetric Division of the Egg

In the developing worm, as in other animals, most cells do not become restricted to a single pathway of differentiation until quite late in development, and cells of a particular type, such as muscle, usually derive from several spatially dispersed precursors that also give rise to other types of cells. The exceptions, in the worm, are the gut and the gonad, each of which forms from a single dedicated *founder cell*, born at the 8-cell stage of development for the gut-cell lineage and at the 16-cell stage for the germ-cell lineage, or *germ line*. Other cells in the embryo, meanwhile, are also becoming different from one another, even though they are not yet committed with regard to their terminal mode of differentiation. How do these early differences between cells arise?

The worm is typical of most animals in the early specification of the cells that will eventually give rise to the germ cells (eggs or sperm). The worm's germ line is produced by a strict series of asymmetric cell divisions of the fertilized egg. The asymmetry originates with a cue from the egg's environment: the sperm entry point defines the future posterior pole of the elongated egg. The proteins

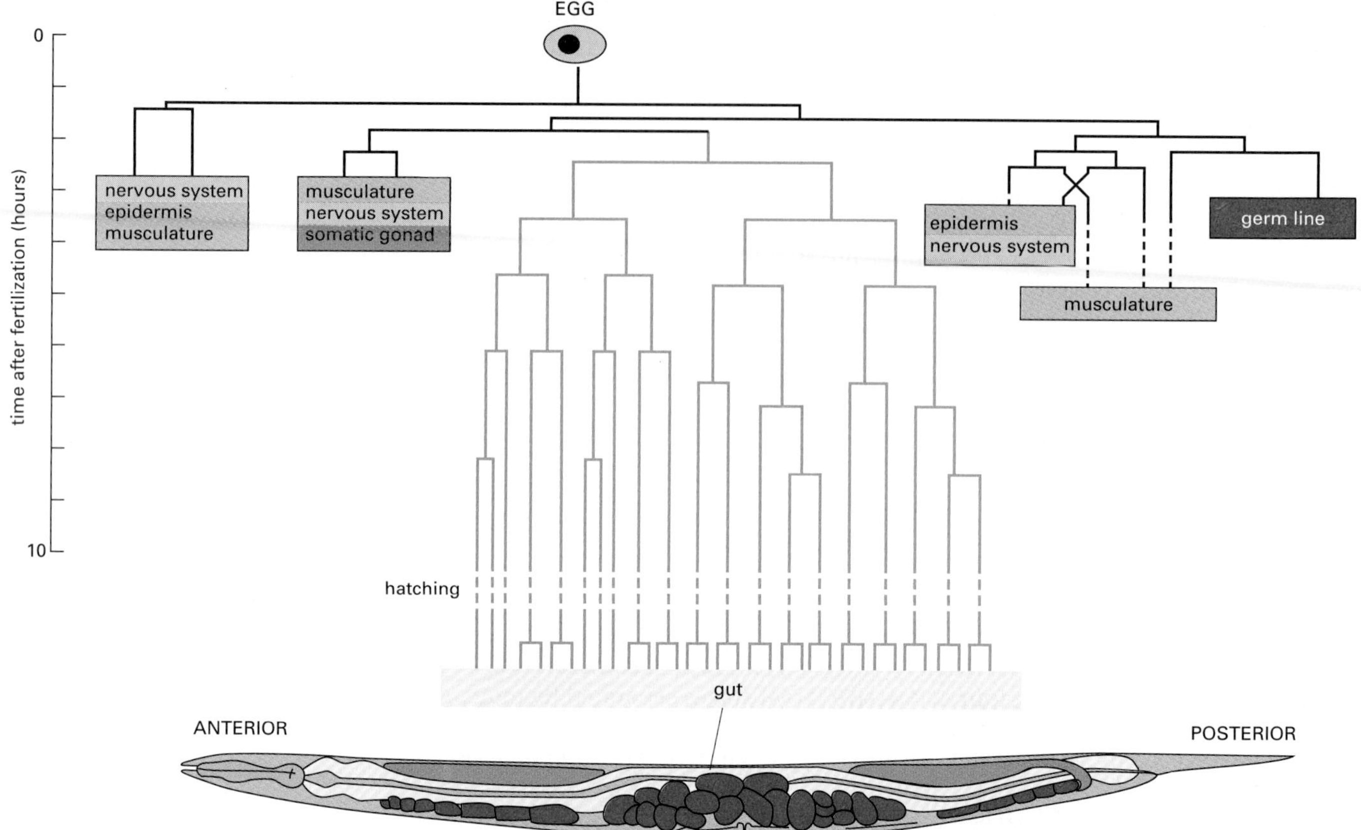

Figure 21–17 The lineage tree for the cells that form the gut (the intestine) of *C. elegans*. Note that although the intestinal cells form a single clone (as do the germ-line cells), the cells of most other tissues do not. Nerve cells (not shown in the drawing of the adult at the bottom) are mainly clustered in ganglia near the anterior and posterior ends of the animal and in a ventral nerve cord that runs the length of the body.

in the egg then interact with one another and organize themselves in relation to this point so as to create a more elaborate and extreme asymmetry in the interior of the cell. The proteins involved are mainly translated from the accumulated mRNA products of the genes of the mother. Because this RNA is made before the egg is laid, it is only the mother's genotype that dictates what happens in the first steps of development. Genes acting in this way are called **maternal-effect genes.**

A subset of maternal-effect genes are specifically required to organize the asymmetric pattern of the nematode egg. These are called *par* (*partitioning-defective*) genes, and at least six have been identified, through genetic screens for mutants where this pattern is disrupted. The *par* genes have homologs in insects and vertebrates, where they are also involved in organizing cell asymmetry, coordinating the polarization of the cytoskeleton with the distribution of other cell components

In the nematode egg, the Par proteins (the products of the *par* genes) serve to bring a set of ribonucleoprotein particles called *P granules* to the posterior pole, so that the posterior daughter cell inherits P granules and the anterior daughter cell does not. Throughout the next few cell divisions, the Par proteins operate in a similar way, orienting the mitotic spindle and segregating the P granules to one daughter cell at each mitosis, until, at the 16-cell stage, there is just one cell that contains the P granules (Figure 21–18). This one cell gives rise to the germ line.

The specification of the germ-cell precursors as distinct from somatic-cell precursors is a key event in the development of practically every type of animal, and the process has common features even in phyla with very different body plans. Thus in *Drosophila*, particles similar to P granules are also segregated into one end of the egg, and become incorporated into the germ-line precursor cells to determine their fate. Similar phenomena occur in fish and frogs. In all these species, one can recognise at least some of the same proteins in the germ-cell-determining material, including homologs of an RNA-binding protein called Vasa. How Vasa and its associated proteins and RNA molecules act to define the germ line is still unknown.

Progressively More Complex Patterns Are Created by Cell–Cell Interactions

The egg, in *C. elegans* as in other animals, is an unusually big cell, with room for complex internal patterning. In addition to the P granules, other factors become distributed in an orderly way along its anteroposterior axis under the control of the Par proteins, and thus are allocated to different cells as the egg goes through its first few cell-division cycles. These divisions occur without growth (since feeding cannot begin until a mouth and a gut have formed) and therefore subdivide the egg into progressively smaller cells. Several of the localized factors are gene regulatory proteins, which act directly in the cell that inherits them to either drive or block the expression of specific genes, adding to the differences between that cell and its neighbors and committing it to a specialized fate.

While the first few differences between cells along the anteroposterior axis of *C. elegans* result from asymmetric divisions, further patterning, including the pattern of cell types along the other axes, depends on interactions between one cell and another. The cell lineages in the embryo are so reproducible that individual cells can be assigned names and identified in every animal (Figure 21–19); the cells at the four-cell stage, for example, are called ABa and ABp (the two anterior sister cells), and EMS and P_2 (the two posterior sister cells). As a result of the asymmetric divisions we have just described, the P_2 cell expresses a signal protein on its surface—a nematode homolog of the Notch ligand Delta—while the ABa and ABp cells express the corresponding transmembrane receptor—a homolog of Notch. The elongated shape of the eggshell forces these cells into an arrangement such that the most anterior cell, ABa, and the most posterior cell, P_2, are no longer in contact with one another. Thus only the ABp cell and the EMS cell are exposed to the signal from P_2. This signal acts on ABp, making it different from ABa and defining the future dorsal–ventral axis of the worm (Figure 21–20).

At the same time, P_2 also expresses another signaling molecule, a Wnt protein, which acts on a Wnt receptor (a Frizzled protein) in the membrane of the EMS cell. This signal polarizes the EMS cell in relation to its site of contact with P_2, controlling the orientation of the mitotic spindle. The EMS cell then divides to give two daughters that become committed to different fates as a result of the Wnt signal from P_2. One daughter, the MS cell, will give rise to muscles and various other body parts; the other daughter, the E cell, is the founder cell for the

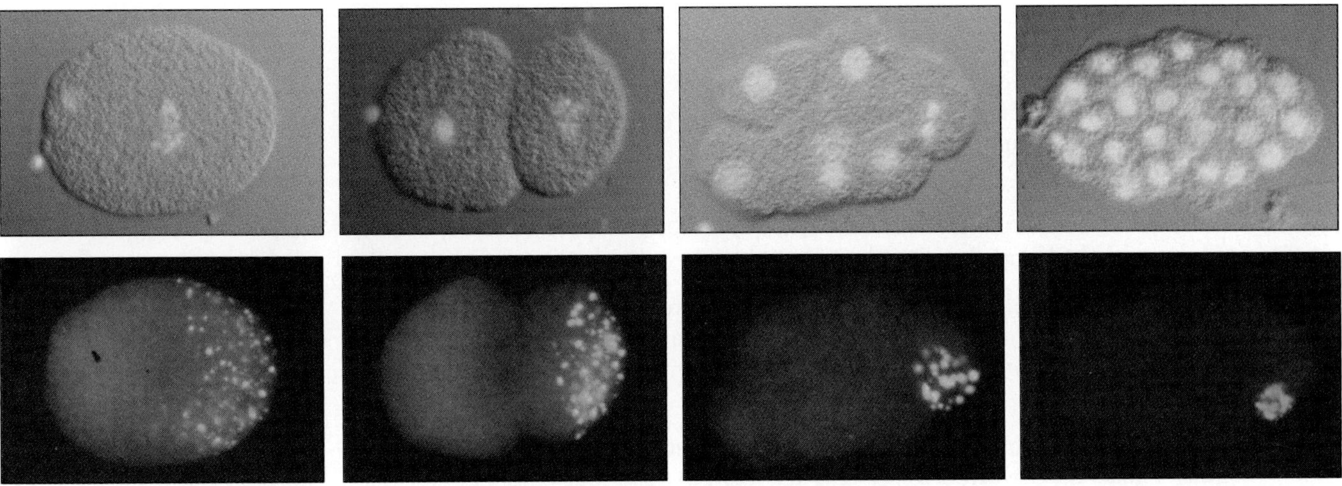

Figure 21–18 Asymmetric divisions segregating P granules into the founder cell of the *C. elegans* germ line. The micrographs in the upper row show the pattern of cell divisions, with cell nuclei stained blue with a DNA-specific fluorescent dye; below are the same cells stained with an antibody against P granules. These small granules (0.5–1 μm in diameter) are distributed randomly throughout the cytoplasm in the unfertilized egg (not shown). After fertilization, at each cell division up to the 16-cell stage, both they and the intracellular machinery that localizes them asymmetrically are segregated into a single daughter cell. (Courtesy of Susan Strome.)

gut, committed to give rise to all the cells of the gut and to no other tissues (see Figure 21–20).

Having described the chain of cause and effect in early nematode development, we now examine some of the methods that have been used to decipher it.

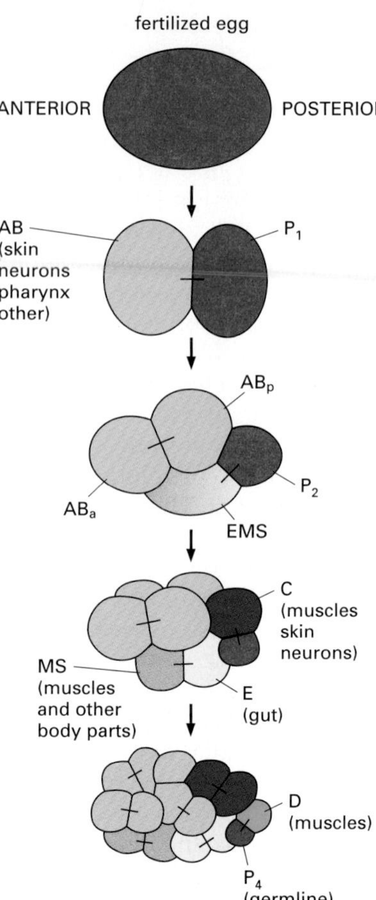

Figure 21–19 The pattern of cell divisions in the early nematode embryo, indicating the names and fates of the individual cells. Cells that are sisters are shown linked by a short *black* line. (After K. Kemphues, *Cell* 101:345–348, 2000.)

Microsurgery and Genetics Reveal the Logic of Developmental Control; Gene Cloning and Sequencing Reveal Its Molecular Mechanisms

To discover the causal mechanisms, we need to know the developmental potential of the individual cells in the embryo. At what points in their lives do they undergo decisive internal changes that determine them for a particular fate, and at what points do they depend on signals from other cells? In the nematode, laser microbeam microsurgery can be conveniently used to kill one or more of a cell's neighbors and then to observe directly how the cell behaves in the altered circumstances. Alternatively, cells of the early embryo can be pushed around and rearranged inside the eggshell using a fine needle. For example, the relative positions of ABa and ABp can be flipped at the four-cell stage of development. The ABa cell then undergoes what would normally be the fate of the ABp cell, and vice versa, showing that the two cells initially have the same developmental potential and depend on signals from their neighbors to make them different. A third tactic is to remove the eggshell of an early *C. elegans* embryo by digesting it with enzymes, and then to manipulate the cells in culture. The existence of a polarizing signal from P_2 to EMS was demonstrated in this way.

Genetic screens were used to identify the genes involved in the P_2–EMS cell interaction. A search was made for mutant strains of worms in which no gut cells were induced (called *mom* mutants, because they had *mo*re *m*esoderm—mesoderm being the fate of both of the EMS cell daughters when induction fails). Cloning and sequencing the *mom* genes revealed that one encodes a Wnt signal protein that is expressed in the P_2 cell, while another encodes a Frizzled protein (a Wnt receptor) that is expressed in the EMS cell. A second genetic screen was conducted for mutant strains of worms with the opposite phenotype, in which extra gut cells are induced (called *pop* mutants, because they have *p*lenty *o*f *p*harynx as a result of the extra gut). One of the *pop* genes (*pop-1*) turns out to encode a gene regulatory protein (a LEF-1/TCF homolog) whose activity is down-regulated by Wnt signaling in *C. elegans*. When Pop-1 activity is absent, both daughters of the EMS cell behave as though they have received the Wnt signal from P_2. Similar genetic methods were used to identify the genes whose products mediate the Notch-dependent signaling from P_2 to ABa.

Continuing in this way, it is possible to build up a detailed picture of the decisive events in nematode development, and of the genetically specified machinery that controls them.

Cells Change Over Time in Their Responsiveness to Developmental Signals

The complexity of the adult nematode body is achieved through repeated use of a handful of patterning mechanisms, including those we have just seen in action in the early embryo. For example, asymmetric cell divisions dependent on the

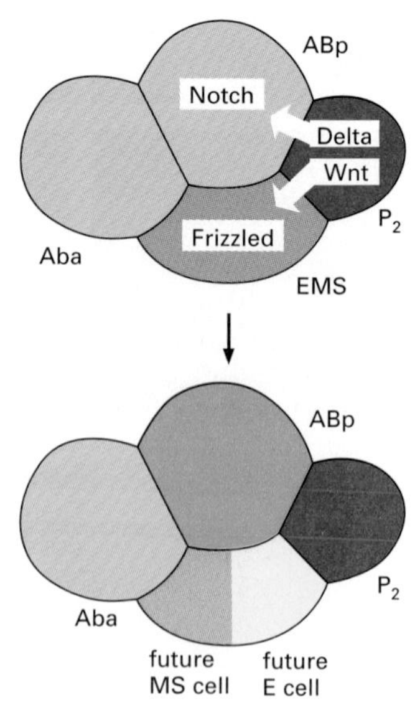

Figure 21–20 Cell-signaling pathways controlling assignment of different characters to the cells in a four-cell nematode embryo. The P_2 cell uses the Notch signaling pathway to send an inductive signal to the ABp cell, causing this to adopt a specialized character. The ABa cell has all the molecular apparatus to respond in the same way to the same signal, but it does not do so because it is out of contact with P_2. Meanwhile, a Wnt signal from the P_2 cell causes the EMS cell to orient its mitotic spindle and generate two daughters that become committed to different fates as a result of their different exposure to Wnt protein—the MS cell and the E cell (the founder cell of the gut).

Pop-1 gene regulatory proteins occur throughout *C. elegans* development, creating differences between anterior and posterior sister cells.

As emphasized earlier, while the same few types of signals act repeatedly at different times and places, the effects they have are different because the cells are programmed to respond differently according to their age and their past history. We have seen, for example, that at the four-cell stage of development, one cell, ABp, changes its developmental potential because of a signal received via the Notch pathway. At the 12-cell stage of development, the granddaughters of the ABp cell and the granddaughters of the ABa cell both encounter another Notch signal, this time from a daughter of the EMS cell. The ABa granddaughter changes its internal state in response to this signal and begins to form the pharynx. The ABp granddaughter does no such thing—the earlier exposure to a Notch signal has made it unresponsive. Thus, at different times in their history, both ABa lineage cells and ABp lineage cells respond to Notch, but the outcomes are different. Somehow a Notch signal at the 12-cell stage induces pharynx, but a Notch signal at the 4-cell stage has other effects—which include the prevention of pharynx induction by Notch at a later stage.

Heterochronic Genes Control the Timing of Development

A cell does not have to receive an external cue in order to change: one set of regulatory molecules inside the cell can provoke production of another, and the cell can thus step through a series of different states autonomously. These states differ not only in their responsiveness to external signals, but also in other aspects of their internal chemistry, including proteins that stop or start the cell-division cycle. In this way, the internal mechanisms of the cell, together with the past and present signals received, dictate both the sequence of biochemical changes in the cell and the timing of its cell divisions.

The specific molecular details of the mechanisms governing the temporal program of development are still mysterious. Remarkably little is known, even in the nematode embryo with its rigidly predictable pattern of cell divisions, about how the sequence of cell divisions is controlled. However, for the later stages, when the larva feeds and grows and moults to become an adult, it has been possible to identify some of the genes that control the timing of cellular events. Mutations in these genes cause *heterochronic* phenotypes: the cells in a larva of one stage behave as though they belonged to a larva of a different stage, or cells in the adult carry on dividing as though they belonged to a larva (Figure 21–21).

Through genetic analyses, one can determine that the products of the heterochronic genes act in series, forming regulatory cascades. Curiously, two genes

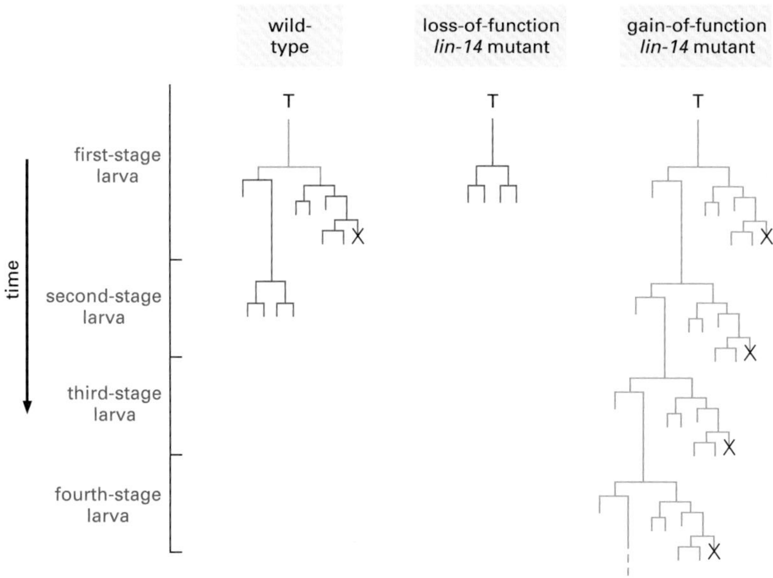

Figure 21–21 Heterochronic mutations in the *lin-14* gene of *C. elegans*. The effects on only one of the many affected lineages are shown. The loss-of-function (recessive) mutation in *lin-14* causes premature occurrence of the pattern of cell division and differentiation characteristic of a late larva, so that the animal reaches its final state prematurely and with an abnormally small number of cells. The gain-of-function (dominant) mutation has the opposite effect, causing cells to reiterate the patterns of cell divisions characteristic of the first larval stage, continuing through as many as five or six molt cycles and persisting in the manufacture of an immature type of cuticle. The cross denotes a programmed cell death. *Green* lines represent cells that contain Lin-14 protein (which binds to DNA), *red* lines those that do not. In normal development the disappearance of Lin-14 is triggered by the beginning of larval feeding. (After V. Ambros and H.R. Horvitz, *Science* 226:409–416, 1984; P. Arasu, B. Wightman, and G. Ruvkun, *Growth Dev. Aging* 5:1825–1833, 1991.)

at the top of their respective cascades, called *lin-4* and *let-7*, do not code for proteins but for short untranslated RNA molecules (21 or 22 nucleotides long). These act by binding to complementary sequences in the non-coding regions of mRNA molecules transcribed from other heterochronic genes, thereby controlling their rate of translation or possibly their degradation (perhaps by a mechanism similar to that of RNAi—see p. 451). Increasing levels of *lin-4* RNA govern the progression from larval stage-1 cell behavior to larval stage-3 cell behavior, increasing levels of *let-7* RNA govern the progression from late larva to adult.

RNA molecules that are identical or almost identical to the *let-7* RNA are found in many other species, including *Drosophila*, zebrafish, and human. Moreover, these RNAs appear to act in a similar way to regulate the level of their target mRNA molecules, and the targets themselves are homologous to the targets of *let-7* RNA in the nematode. The evidence therefore suggests that this system has a universal role in governing the timing of a switch from an early to a late style of cell behavior.

Cells Do Not Count Cell Divisions in Timing Their Internal Programs

Since the steps of cell specialization have to be coordinated with cell divisions, it is often suggested that the cell division cycle might serve as a clock to control the tempo of other events in development. In this view, changes of internal state would be locked to passage through each division cycle: the cell would click to the next state as it went through mitosis, so to speak. There is a substantial amount of evidence that this idea is wrong. Cells in developing embryos, whether they be worms, flies, or vertebrates, usually carry on with their standard timetable of determination and differentiation even when progress through the cell-division cycle is artificially blocked. Necessarily, there are some abnormalities, if only because a single undivided cell cannot differentiate in two ways at once. But in most cases that have been studied, it seems clear that the cell changes its state with time regardless of cell division, and that this changing state controls both the decision to divide and the decision as to when and how to specialize.

Selected Cells Die by Apoptosis as Part of the Program of Development

The control of cell numbers in development depends on cell death as well as cell division. A *C. elegans* hermaphrodite generates 1030 somatic cell nuclei in the course of its development, but 131 of the cells die. These programmed cell deaths occur in an absolutely predictable pattern. In *C. elegans*, they can be chronicled in detail, because one can trace the fate of each individual cell and see who dies, watching as each suicide victim undergoes apoptosis and is rapidly engulfed and digested by neighboring cells (Figure 21–22). In other organisms, where close observation is harder, such deaths easily go unnoticed; but cell death by apoptosis is probably the fate of a substantial fraction of the cells produced in most animals, playing an essential part in generating an individual with the right cell types in the right numbers and places, as discussed in Chapter 17.

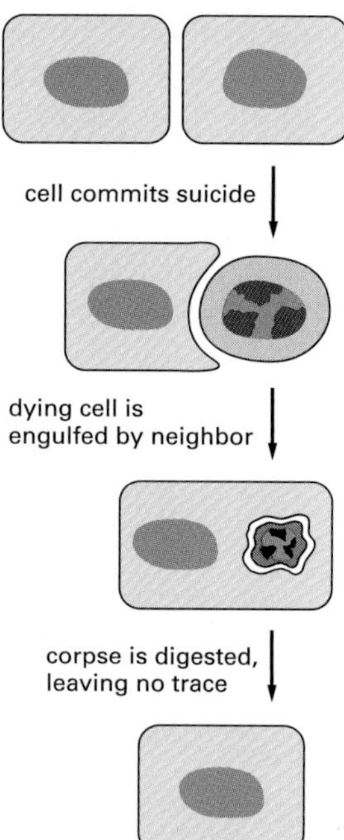
cell commits suicide
dying cell is
engulfed by neighbor
corpse is digested,
leaving no trace

Figure 21–22 Apoptotic cell death in *C. elegans*. Death depends on expression of the *ced-3* and *ced-4* genes in the absence of *ced-9* expression—all in the dying cell itself. The subsequent engulfment and disposal of the remains depend on expression of other genes in the neighboring cells.

Genetic screens in *C. elegans* have been crucial in identifying the genes that bring about apoptosis and in highlighting its importance in development. Three genes, called *ced-3*, *ced-4*, and *egl-1*, (*ced* stands for *cell death abnormal*), are found to be required for the 131 normal cell deaths to occur. If these genes are inactivated by mutation, cells that are normally fated to die survive instead, differentiating as recognizable cell types such as neurons. Conversely, over-expression or misplaced expression of the same genes causes many cells to die that would normally survive, and the same effect results from mutations that inactivate another gene, *ced-9*, which normally represses the death program.

All these genes code for conserved components of the cell-death machinery. As described in Chapter 17, *ced-3* codes for a caspase homolog, while *ced-4*, *ced-9* and *egl-1* are respectively homologs of *Apaf-1*, *Bcl-2*, and *Bad*. Without the insights that came from detailed analysis of the development of the transparent, genetically tractable nematode worm, it would have been very much harder to discover these genes and understand the cell-death process in vertebrates.

Summary

The development of the small, relatively simple, transparent nematode worm **Caenorhabditis elegans** *is extraordinarily reproducible and has been chronicled in detail, so that a cell at any given position in the body has the same lineage in every individual, and this lineage is fully known. Also, the genome has been completely sequenced. Thus powerful genetic and microsurgical approaches can be combined to decipher developmental mechanisms. As in other organisms, development depends on an interplay of cell–cell interactions and cell-autonomous processes. Development begins with an asymmetric division of the fertilized egg, dividing it into two smaller cells containing different cell-fate determinants. The daughters of these cells interact via the Notch and Wnt cell signaling pathways to create a more diverse array of cell states. Meanwhile, through further asymmetric divisions one cell inherits materials from the egg that determine it at an early stage as progenitor of the germ line.*

Genetic screens identify the sets of genes responsible for these and later steps in development, including, for example, cell-death genes that control the apoptosis of a specific subset of cells as part of the normal developmental program. Heterochronic genes that govern the timing of developmental events have also been found, although in general our understanding of temporal control of development is still very poor. There is good evidence, however, that the tempo of development is not set by the counting of cell divisions.

DROSOPHILA AND THE MOLECULAR GENETICS OF PATTERN FORMATION: GENESIS OF THE BODY PLAN

It is the fly *Drosophila melanogaster* (Figure 21–23), more than any other organism, that has transformed our understanding of how genes govern the patterning of the body. The anatomy of *Drosophila* is more complex than that of *C. elegans*, with more than 100 times as many cells, and it shows more obvious parallels with our own body structure. Surprisingly, the fly has fewer genes than the worm—about 14,000 as compared with 19,000. On the other hand, it

(A)

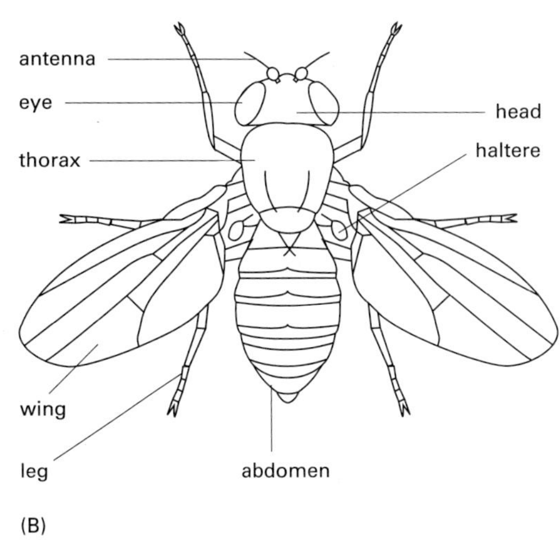

Figure 21–23 *Drosophila melanogaster.* Dorsal view of a normal adult fly. (A) Photograph. (B) Labeled drawing. (Photograph courtesy of E.B. Lewis.)

has almost twice as much noncoding DNA per gene (about 10,000 nucleotides on average, as compared with about 5000). The molecular construction kit has fewer types of parts, but the assembly instructions—as specified by the regulatory sequences in the non-coding DNA—seem to be more voluminous.

Decades of genetic study, culminating in massive systematic genetic screens, have yielded a catalogue of the developmental control genes that define the spatial pattern of cell types and body structures of the fly, and molecular biology has given us the tools to watch these genes in action. By *in situ* hybridization using DNA or RNA probes on whole embryos, or by staining with labeled antibodies to reveal the distribution of specific proteins, one can observe directly how the internal states of the cells are defined by the sets of regulatory genes that they express at different times of development. Moreover, by analyzing animals that are a patchwork of mutant and nonmutant cells, one can discover how each gene operates as part of a system to specify the organization of the body.

Most of the genes controlling the pattern of the body in *Drosophila* turn out to have close counterparts in higher animals, including ourselves. In fact, many of the basic devices for defining the body plan and patterning individual organs and tissues are astonishingly similar. Thus, quite surprisingly, the fly has provided the key to understanding the molecular genetics of our own development.

Flies, like nematode worms, are ideal for genetic studies: cheap to breed, easy to mutagenize, and rapid in their reproductive cycle. But there is a more fundamental reason why they have been so important for developmental geneticists. As emphasized earlier, as a result of gene duplications, vertebrate genomes often contain two or three homologous genes corresponding to a single gene in the fly. A mutation that disrupts one of these genes very often fails to reveal the gene's core function, because the other homologs share this function and remain active. In the fly, with its more economical gene set, this phenomenon of genetic redundancy is less prevalent. The phenotype of a single mutation in the fly therefore more often directly uncovers the function of the mutant gene.

The Insect Body Is Constructed as a Series of Segmental Units

The timetable of *Drosophila* development, from egg to adult, is summarized in Figure 21–24. The period of *embryonic development* begins at fertilization and takes about a day, at the end of which the embryo hatches out of the egg shell to become a *larva*. The larva then passes through three stages, or *instars*, separated by molts in which it sheds its old coat of cuticle and lays down a larger one. At the end of the third instar it pupates. Inside the *pupa*, a radical remodeling of the body takes place—a process called *metamorphosis*. Eventually, about nine days after fertilization, an adult fly, or *imago*, emerges.

The fly consists of a head, with mouth, eyes, and antennae, followed by three thoracic segments (numbered T1 to T3), and eight or nine abdominal segments (numbered A1 to A9). Each segment, although different from the others, is built according to a similar plan. Segment T1, for example, carries a pair of legs, T2 carries a pair of legs plus a pair of wings, and T3 carries a pair of legs plus a pair of halteres—small knob-shaped balancers important in flight, evolved from the second pair of wings that more primitive insects possess. The quasi-repetitive segmentation develops in the early embryo during the first few hours after fertilization (Figure 21–25), but it is more obvious in the larva (Figure 21–26), where the segments look more similar than in the adult. In the embryo it can be seen that the rudiments of the head, or at least the future adult mouth parts, are likewise segmental. At the two ends of the animal, however, there are highly specialized terminal structures that are not segmentally derived.

The boundaries between segments are traditionally defined by visible anatomical markers; but in discussing gene expression patterns it is often convenient to draw a different set of segmental boundaries, defining a series of segmental units called *parasegments*, half a segment out of register with the traditional divisions (see Figure 21–26).

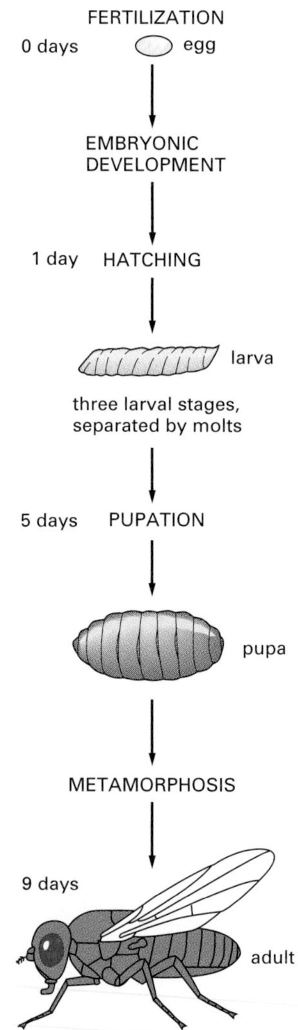

Figure 21–24 Synopsis of *Drosophila* development from egg to adult fly.

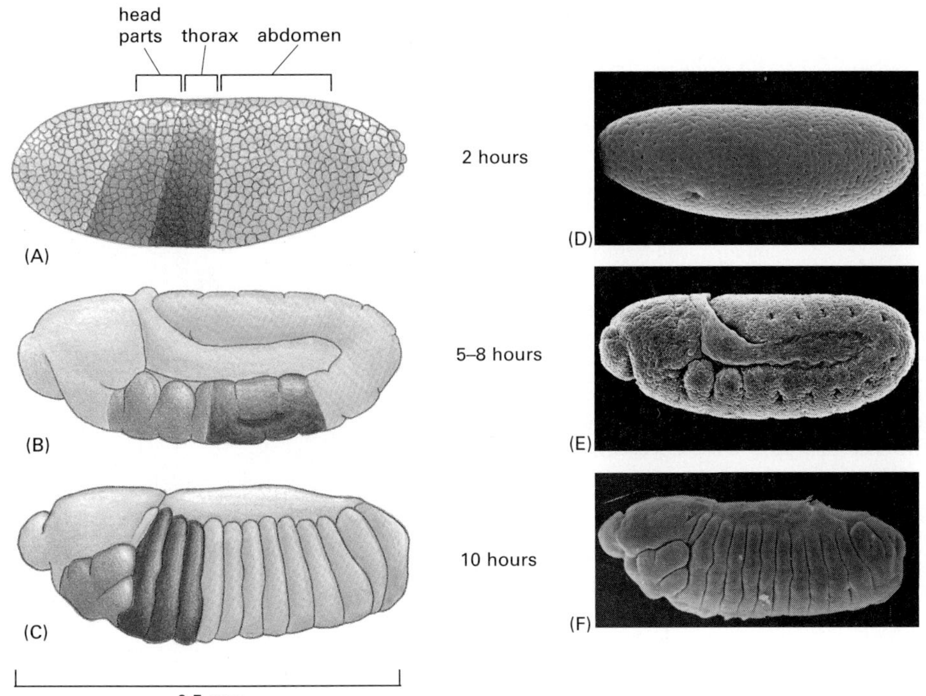

Figure 21–25 The origins of the *Drosophila* body segments during embryonic development. The embryos are seen in side view in drawings (A–C) and corresponding scanning electron micrographs (D–F). (A and D) At 2 hours the embryo is at the *syncytial blastoderm* stage (see Figure 21–51) and no segmentation is visible, although a fate map can be drawn showing the future segmented regions (*color* in A). (B and E) At 5–8 hours the embryo is at the *extended germ band* stage: gastrulation has occurred, segmentation has begun to be visible, and the segmented axis of the body has lengthened, curving back on itself at the tail end so as to fit into the egg shell. (C and F) At 10 hours the body axis has contracted and become straight again, and all the segments are clearly defined. The head structures, visible externally at this stage, will subsequently become tucked into the interior of the larva, to emerge again only when the larva goes through pupation to become an adult. (D and E, courtesy of F.R. Turner and A.P. Mahowald, *Dev. Biol.* 50:95–108, 1976. © Academic Press; F, from J.P. Petschek, N. Perrimon, and A.P. Mahowald, *Dev. Biol.* 119:175–189, 1987. © Academic Press.)

Drosophila Begins Its Development as a Syncytium

The egg of *Drosophila* is about 0.5 mm long and 0.15 mm in diameter, with a clearly defined polarity. Like the eggs of other insects, but unlike vertebrates, it begins its development in an unusual way: a series of nuclear divisions without cell division creates a syncytium. The early nuclear divisions are synchronous and extremely rapid, occurring about every 8 minutes. The first nine divisions generate a cloud of nuclei, most of which migrate from the middle of the egg toward the surface, where they form a monolayer called the *syncytial blastoderm*. After another four rounds of nuclear division, plasma membranes grow inward from the egg surface to enclose each nucleus, thereby converting the syncytial blastoderm into a *cellular blastoderm* consisting of about 6000 separate cells (Figure 21–27). About 15 of the nuclei populating the extreme posterior end of the egg are segregated into cells a few cycles earlier; these *pole cells* are the germ-line precursors (primordial germ cells) that will give rise to eggs or sperm.

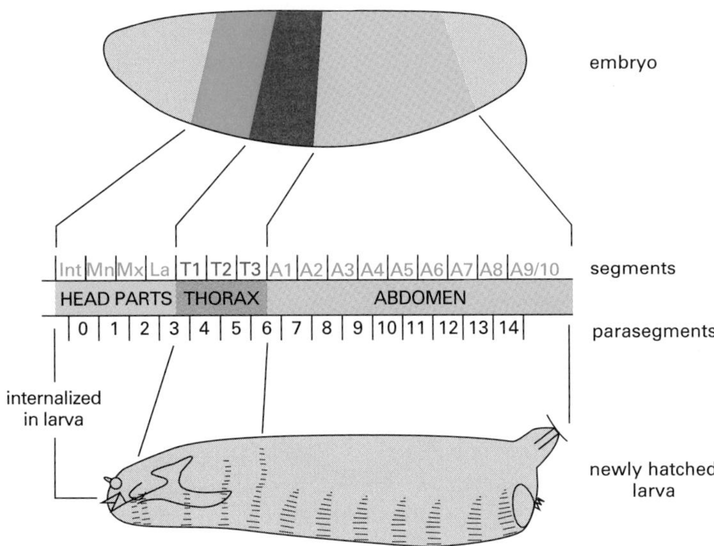

Figure 21–26 The segments of the *Drosophila* larva and their correspondence with regions of the blastoderm. The parts of the embryo that become organized into segments are shown in color. The two ends of the embryo, shaded *gray*, are not segmented and become tucked into the interior of the body to form the internal structures of the head and gut. (The future external, segmental structures of the adult head are also transiently tucked into the interior in the larva.) Segmentation in *Drosophila* can be described in terms of either segments or parasegments: the relationship is shown in the middle part of the figure. Parasegments often correspond more simply to patterns of gene expression. The exact number of abdominal segments is debatable: eight are clearly defined, and a ninth is present vestigially in the larva, but absent in the adult.

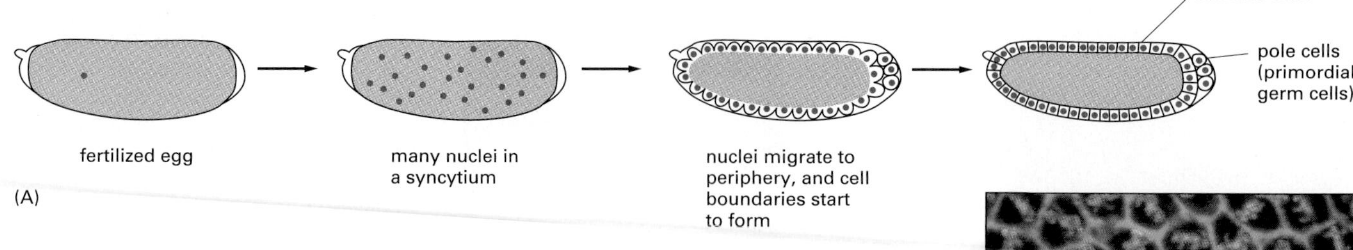

(B)

Figure 21–27 Development of the *Drosophila* egg from fertilization to the cellular blastoderm stage. (A) Schematic drawings. (B) Surface view—an optical-section photograph of blastoderm nuclei undergoing mitosis at the transition from the syncytial to the cellular blastoderm stage. Actin is stained *green,* chromosomes *orange.* (A, after H.A. Schneiderman, in Insect Development [P.A. Lawrence, ed.], pp. 3–34. Oxford, UK: Blackwell, 1976; B, courtesy of William Sullivan.)

Up to the cellular blastoderm stage, development depends largely—although not exclusively—on stocks of maternal mRNA and protein that accumulated in the egg before fertilization. The frantic rate of DNA replication and nuclear division evidently gives little opportunity for transcription. After cellularization, cell division continues in a more conventional way, asynchronously and at a slower rate, and the rate of transcription increases dramatically. Gastrulation begins a little while before cellularization is complete, when parts of the sheet of cells forming the exterior of the embryo start to tuck into the interior to form the gut, the musculature, and associated internal tissues. A little later and in another region of the embryo, a separate set of cells move from the surface epithelium into the interior to form the central nervous system. By marking and following the cells through these various movements, one can draw a fate map for the monolayer of cells on the surface of the blastoderm (Figure 21–28).

As gastrulation nears completion, a series of indentations and bulges appear in the surface of the embryo, marking the subdivision of the body into segments along its anteroposterior axis (see Figure 21–25). Soon a fully segmented larva emerges, ready to start eating and growing. Within the body of the larva, small groups of cells remain apparently undifferentiated, forming structures called *imaginal discs*. These will grow as the larva grows, and eventually they will give rise to most of the structures of the adult body, as we shall see later.

A head end and a tail end, a ventral (belly) side and a dorsal (back) side, a gut, a nervous system, a series of body segments—these are all features of the basic body plan that *Drosophila* shares with many other animals, including ourselves. We begin our account of the mechanisms of *Drosophila* development by considering how this body plan is set up.

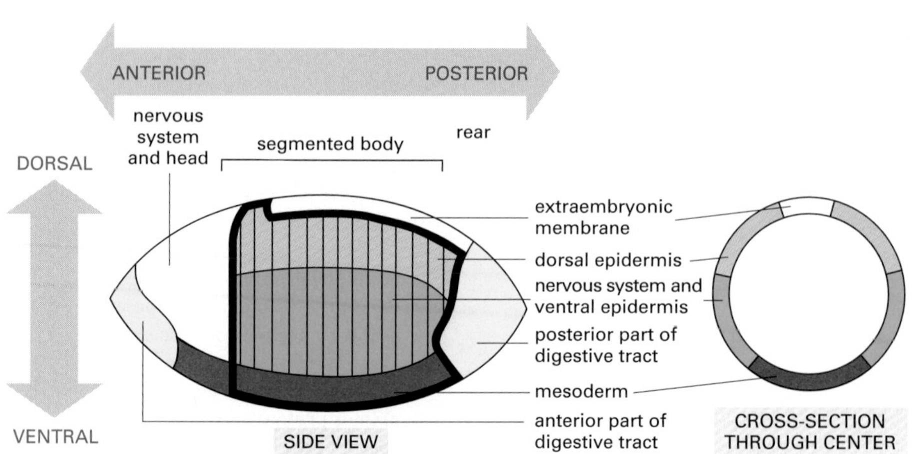

Figure 21–28 Fate map of a *Drosophila* embryo at the cellular blastoderm stage. The embryo is shown in side view and in cross section, displaying the relationship between the dorsoventral subdivision into future major tissue types and the anteroposterior pattern of future segments. A heavy line encloses the region that will form segmental structures. During gastrulation the cells along the ventral midline invaginate to form mesoderm, while the cells fated to form the gut invaginate near each end of the embryo. (After V. Hartenstein, G.M. Technau, and J.A. Campos-Ortega, *Wilhelm Roux'Arch. Dev. Biol.* 194:213–216, 1985.)

Genetic Screens Define Groups of Genes Required for Specific Aspects of Early Patterning

By carrying out a series of genetic screens based on saturation mutagenesis (see Chapter 8), it has been possible to amass a collection of *Drosophila* mutants that appears to include changes in a large proportion of the genes affecting development. Independent mutations in the same gene can be distinguished from mutations in separate genes by a complementation test (see Panel 8–1, p. 527), leading to a catalog of genes classified according to their mutant phenotypes. In such a catalog, a group of genes with very similar mutant phenotypes will often code for a set of proteins that work together to perform a particular function.

Sometimes the developmental functions revealed by mutant phenotypes are those that one would expect; sometimes they are a surprise. A large-scale genetic screen focusing on early *Drosophila* development revealed that the key genes fall into a relatively small set of functional classes defined by their mutant phenotypes. Some—*the egg-polarity genes* (Figure 21–29)—are required to define the anteroposterior and dorsoventral axes of the embryo and mark out its two ends for special fates, by mechanisms involving interactions between the oocyte and surrounding cells in the ovary. Others, the *gap genes*, are required in specific broad regions along the anteroposterior axis of the early embryo to allow their proper development. A third category, the *pair-rule genes*, are

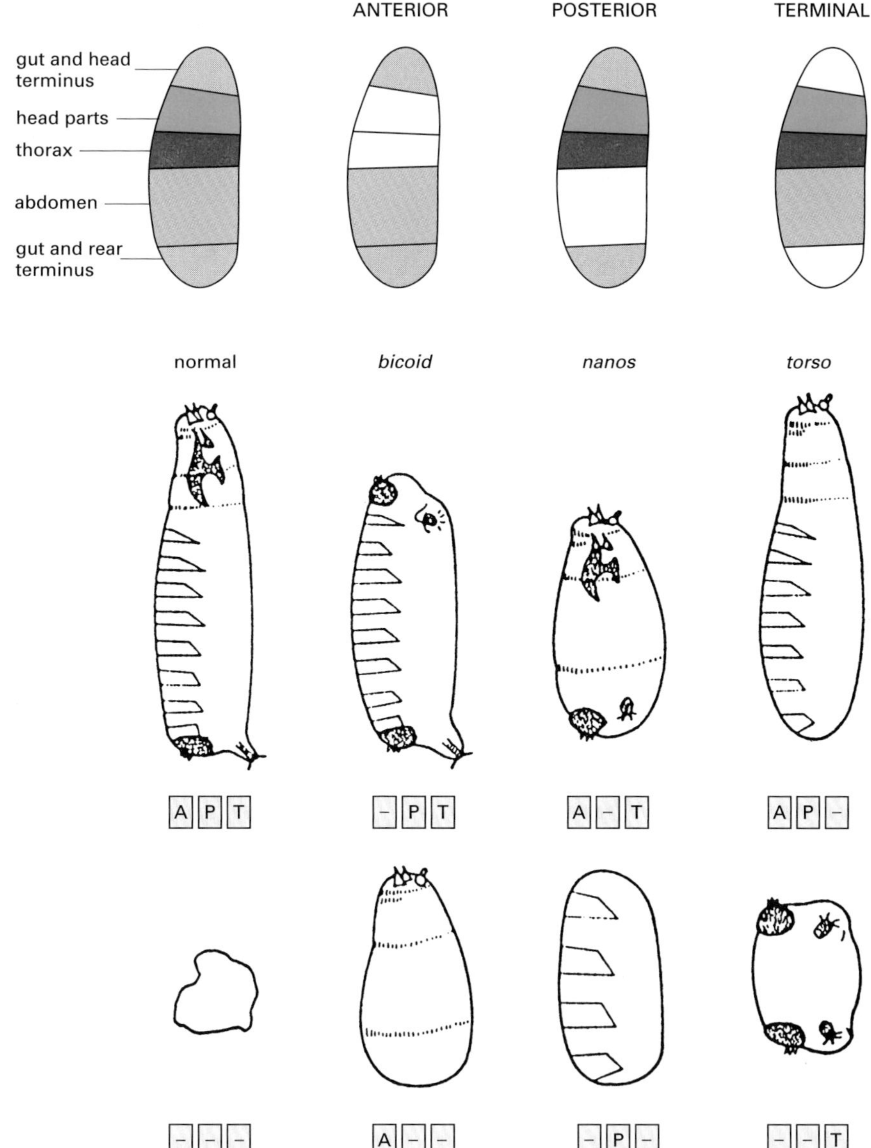

Figure 21–29 The domains of the anterior, posterior, and terminal systems of egg-polarity genes. The *upper* diagrams show the fates of the different regions of the egg/early embryo and indicate (in *white*) the parts that fail to develop if the anterior, posterior, or terminal system is defective. The *middle* row shows schematically the appearance of a normal larva and of mutant larvae that are defective in a gene of the anterior system (for example, *bicoid*), of the posterior system (for example, *nanos*), or of the terminal system (for example, *torso*). The *bottom* row of drawings shows the appearances of larvae in which none or only one of the three gene systems is functional. The lettering beneath each larva specifies which systems are intact (A P T for a normal larva, –P T for a larva where the anterior system is defective but the posterior and terminal systems are intact, and so on).

Inactivation of a particular gene system causes loss of the corresponding set of body structures; the body parts that form correspond to the gene systems that remain functional. Note that larvae with a defect in the anterior system can still form terminal structures at their anterior end, but these are of the type normally found at the rear end of the body rather than the front of the head. (Slightly modified from D. St. Johnston and C. Nüsslein-Volhard, *Cell* 68:201–219, 1992.)

required, more surprisingly, for development of alternate body segments. A fourth category, the *segment polarity genes,* are responsible for organizing the anteroposterior pattern of each individual segment.

The discovery of these four systems of genes and the subsequent analysis of their functions (an enterprise that still continues) was a famous tour-de-force of developmental genetics. It has had a revolutionary impact on all of developmental biology by showing the way toward a systematic, comprehensive account of the genetic control of embryonic development. In this section, we shall summarize only briefly the conclusions relating to the earliest phases of *Drosophila* development, because these are insect-specific; we dwell at greater length on the parts of the process that illustrate more general principles.

Interactions of the Oocyte With Its Surroundings Define the Axes of the Embryo: the Role of the Egg-Polarity Genes

Surprisingly, the earliest steps of animal development are among the most variable, even within a phylum. A frog, a chicken, and a mammal, for example, even though they develop in similar ways later, make eggs that differ radically in size and structure, and they begin their development with different sequences of cell divisions and cell specialization events.

The style of early development that we have described for *C. elegans* is typical of many classes of animals. In contrast, the early development of *Drosophila* represents a rather extreme variant. The main axes of the future insect body are defined before fertilization by a complex exchange of signals between the unfertilized egg, or oocyte, and the follicle cells that surround it in the ovary (Figure 21–30). Then, in the syncytial phase following fertilization, an exceptional amount of patterning occurs in the array of rapidly dividing nuclei, before the first partitioning of the egg into separate cells. Here, there is no need for the usual forms of cell–cell communication involving transmembrane signaling; neighboring regions of the early *Drosophila* embryo can communicate by means of gene regulatory proteins and mRNA molecules that diffuse or are actively transported through the cytoplasm of the giant multinuclear cell.

In the stages before fertilization, the anteroposterior axis of the future embryo becomes defined by three systems of molecules that create landmarks in the oocyte (Figure 21–31). Following fertilization, each landmark serves as a beacon, providing a signal, in the form of a morphogen gradient, that organizes the developmental process in its neighborhood. Two of these signals are generated from localized deposits of specific mRNA molecules. The future anterior end of the embryo contains a high concentration of mRNA for a gene regulatory protein called Bicoid; this mRNA is translated to produce Bicoid protein, which diffuses away from its source to form a concentration gradient with its maximum at the anterior end of the egg. The future posterior end of the embryo contains a high concentration of mRNA for a regulator of translation called Nanos, which sets up a posterior gradient in the same way. The third signal is generated symmetrically at both ends of the egg, by local activation of a transmembrane tyrosine kinase receptor called Torso. The activated receptor exerts its effects over a shorter range, marking the sites of specialized terminal structures that will

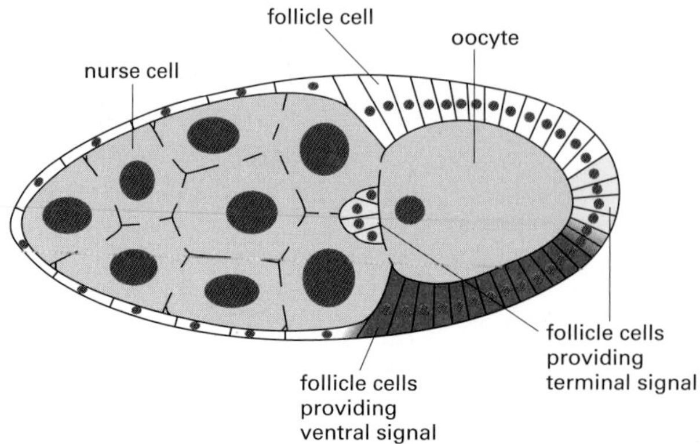

Figure 21–30 A *Drosophila* oocyte in its follicle. The oocyte is derived from a germ cell that divides four times to give a family of 16 cells that remain in communication with one another via cytoplasmic bridges *(gray)*. One member of the family group becomes the oocyte, while the others become nurse cells, which make many of the components required by the oocyte and pass them into it via the cytoplasmic bridges. The follicle cells that partially surround the oocyte have a separate ancestry. As indicated, they are the sources of terminal and ventral egg-polarizing signals.

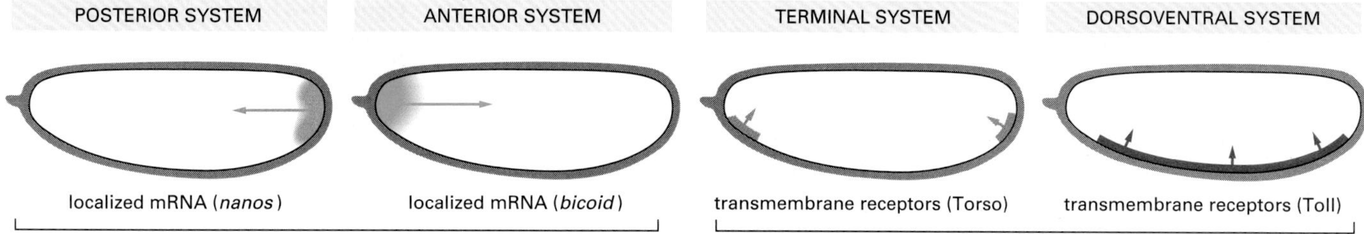

Figure 21–31 The organization of the four egg-polarity gradient systems. The receptors Toll and Torso are distributed all over the membrane; the coloring in the diagrams on the right indicates where they become activated by extracellular ligands.

form at the head and tail ends of the future larva and also defining the rudiments of the future gut. The three sets of genes responsible for these localized determinants are referred to as the **anterior**, **posterior**, and **terminal** sets of **egg-polarity** genes.

A fourth landmark defines the dorsoventral axis (see Figure 21–31): a protein that is produced by follicle cells underneath the future ventral region of the embryo leads to localized activation of another transmembrane receptor, called Toll, in the oocyte membrane. The genes required for this function are called **dorsoventral** egg-polarity genes.

All the egg-polarity genes in these four classes are maternal-effect genes: it is the mother's genome, not the zygotic genome, that is critical. Thus, a fly whose chromosomes are mutant in both copies of the *bicoid* gene but who is born from a mother carrying one normal copy of *bicoid* develops perfectly normally, without any defects in the head pattern. However, if that daughter fly is a female no functional *bicoid* mRNA can be deposited into the anterior part of her own eggs, and all of these will develop into headless embryos regardless of the father's genotype.

Each of the four egg-polarity signals—provided by Bicoid, Nanos, Torso, and Toll—exerts its effect by regulating (directly or indirectly) the expression of genes in the nuclei of the blastoderm. The use of these particular molecules to organize the egg is not a general feature of early animal development—indeed, only *Drosophila* and closely related insects possess a *bicoid* gene. And Toll has been coopted here for dorsoventral patterning; its more ancient and universal function is in the innate immune response.

Nevertheless, the egg-polarity system shows some highly conserved features. For example, the localization of *nanos* mRNA at one end of the egg is linked to, and dependent on, the localization of germ-cell determinants at that site, just as it is in *C. elegans*. Later in development, as the zygotic genome comes into play under the influence of the egg-polarity system, more similarities with other animal species become apparent. We shall use the dorsoventral system to illustrate this point.

The Dorsoventral Signaling Genes Create a Gradient of a Nuclear Gene Regulatory Protein

Localized activation of the Toll receptor on the ventral side of the egg controls the distribution of Dorsal, a gene regulatory protein inside the egg. The Dorsal protein belongs to the same family as the NF-κB gene regulatory protein of vertebrates (discussed in Chapter 15). Its Toll-regulated activity, like that of NF-κB, depends on its translocation from the cytoplasm, where it is held in an inactive form, to the nucleus, where it regulates gene expression. In the newly laid egg, both the *dorsal* mRNA (detected by *in situ* hybridization) and the protein it encodes (detected with antibodies) are distributed uniformly in the cytoplasm. After the nuclei have migrated to the surface of the embryo to form the blastoderm, however, a remarkable redistribution of the Dorsal protein occurs: dorsally the protein remains in the cytoplasm, but ventrally it is concentrated in the nuclei, with a smooth gradient of nuclear localization between these two extremes (Figure 21–32). The signal transmitted by the Toll protein controls this

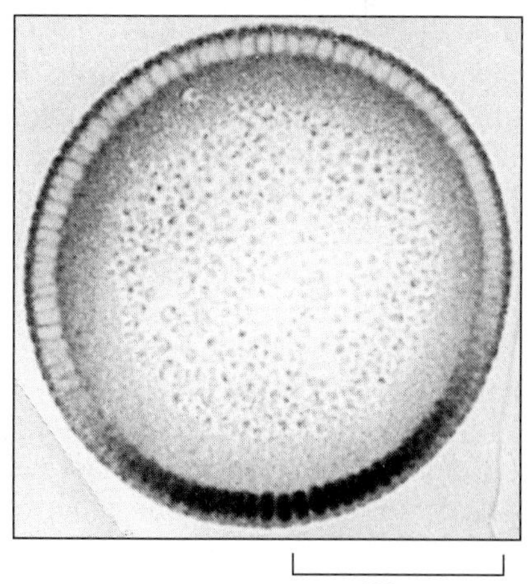

Figure 21–32 The concentration gradient of Dorsal protein in the nuclei of the blastoderm, as revealed by an antibody. Dorsally, the protein is present in the cytoplasm and absent from the nuclei; ventrally, it is depleted in the cytoplasm and concentrated in the nuclei. (From S. Roth, D. Stein, and C. Nüsslein-Volhard, *Cell* 59:1189–1202, 1989. © Elsevier.)

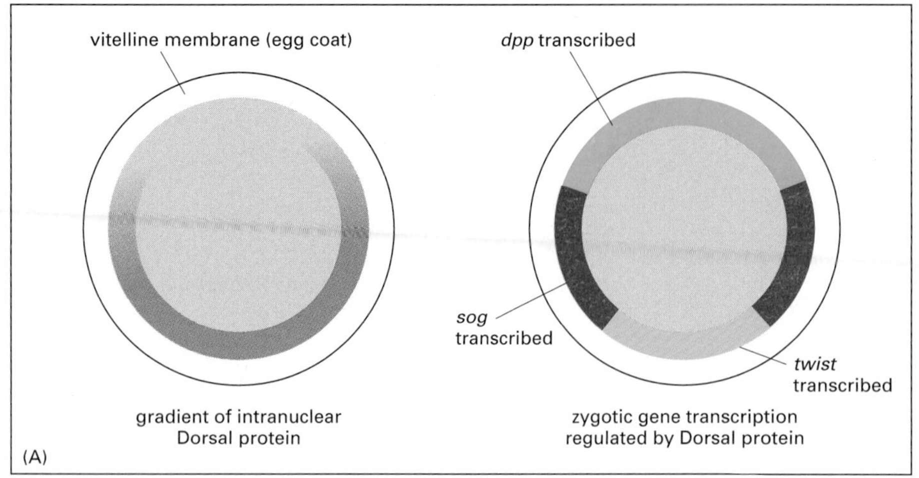

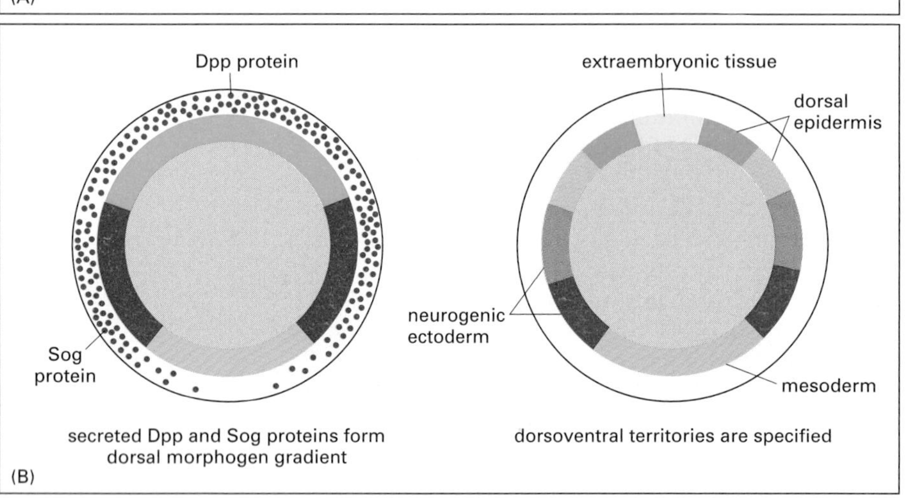

Figure 21–33 Morphogen gradients patterning the dorsoventral axis of the embryo. (A) The gradient of Dorsal protein defines three broad territories of gene expression, marked here by the expression of three representative genes—*dpp*, *sog*, and *twist*. (B) Slightly later, the cells expressing *dpp* and *sog* secrete, respectively, the signal proteins Dpp (a TGFβ family member) and Sog (an antagonist of Dpp). These two proteins diffuse and interact with one another (and with certain other factors) to set up a gradient of Dpp activity that guides a more detailed patterning process.

redistribution of Dorsal through a signaling pathway that is essentially the same as the Toll-dependent pathway involved in innate immunity.

Once inside the nucleus, the Dorsal protein turns on or off the expression of different sets of genes depending on its concentration. The expression of each responding gene depends on its regulatory DNA—specifically, on the number and affinity of the binding sites that this DNA contains for Dorsal and other regulatory proteins. In this way, the regulatory DNA can be said to *interpret* the positional signal provided by the Dorsal protein gradient, so as to define a dorsoventral series of territories—distinctive bands of cells that run the length of the embryo (Figure 21–33A). Most ventrally—where the concentration of Dorsal protein is highest—it switches on, for example, the expression of a gene called *twist* that is specific for mesoderm (Figure 21–34). Most dorsally, where the concentration of Dorsal protein is lowest, the cells switch on *decapentaplegic (dpp)*. And in an intermediate region, where the concentration of Dorsal protein is high enough to repress *dpp* but too low to activate *twist,* the cells switch on another set of genes, including one called *short gastrulation (sog).*

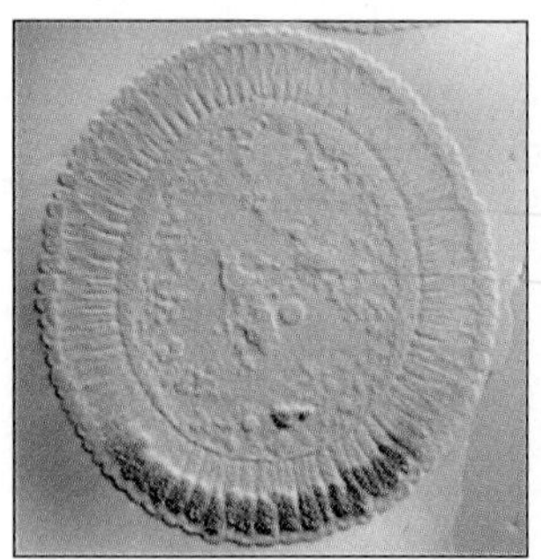
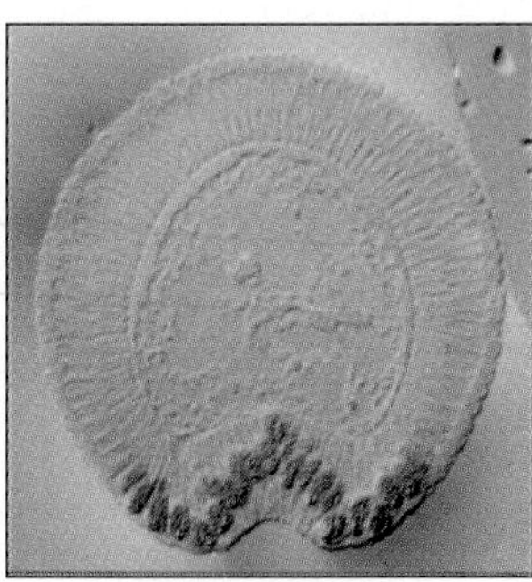
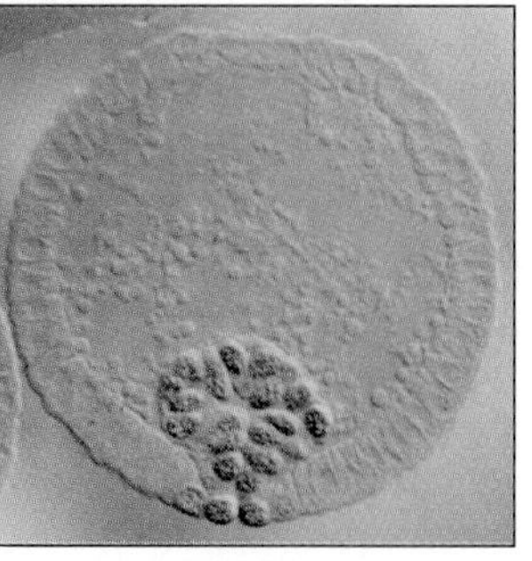

Figure 21–34 Origin of the mesoderm from cells expressing *twist*. Embryos were fixed at successive stages, cross-sectioned, and stained with an antibody against the Twist protein, a gene regulatory protein of the bHLH family. The cells that express Twist move into the interior of the embryo to form mesoderm. (From M. Leptin, J. Casal, B. Grunewald, and R. Reuter, *Development Suppl.* 23–31, 1992. © The Company of Biologists.)

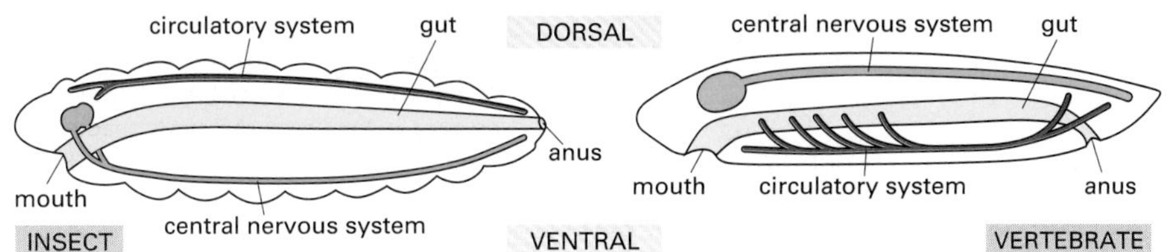

Figure 21–35 The vertebrate body plan as a dorsoventral inversion of the insect body plan. The mechanism of dorsoventral patterning in a vertebrate embryo is discussed in more detail later in this chapter. Note the correspondence with regard to the circulatory system as well as the gut and nervous system. In insects, the circulatory system is represented by a tubular heart and a main dorsal blood vessel, which pumps blood out into the tissue spaces through one set of apertures and receives blood back from the tissues through another set. In contrast with vertebrates, there is no system of capillary vessels to contain the blood as it percolates through the tissues. Nevertheless, heart development depends on homologous genes in vertebrates and insects, reinforcing the relationship between the two body plans. (After E.L. Ferguson, *Curr. Opin. Genet. Dev.* 6:424–431, 1996.)

Dpp and Sog Set Up a Secondary Morphogen Gradient to Refine the Pattern of the Dorsal Part of the Embryo

Products of the genes directly regulated by the Dorsal protein generate in turn more local signals that define finer subdivisions of the dorsoventral axis. These signals act after cellularization, and take the form of conventional extracellular signaling molecules. In particular, *dpp* codes for the secreted Dpp protein, which forms a local morphogen gradient in the dorsal part of the embryo. The gene *sog*, meanwhile, codes for another secreted protein that is produced in the neurogenic ectoderm and acts as an antagonist of Dpp. The opposing diffusion gradients of these two proteins create a steep gradient of Dpp activity. The highest Dpp activity levels, in combination with certain other factors, cause development of the most dorsal tissue of all—extraembryonic membrane; intermediate levels cause development of dorsal ectoderm; and very low levels allow development of neurogenic ectoderm (Figure 21–33B).

The Insect Dorsoventral Axis Corresponds to the Vertebrate Ventrodorsal Axis

Dpp is a member of the TGFβ superfamily of signaling molecules that is also important in vertebrates; Sog is a homolog of the vertebrate protein chordin. It is striking that a Dpp homolog, BMP4, and chordin work together in vertebrates in the same way as do Dpp and Sog in *Drosophila.* These two proteins control the dorsoventral pattern of the ectoderm, with high levels of chordin defining the region that is neurogenic and high levels of BMP4 activity defining the region that is not. This, combined with other molecular parallels, strongly suggests that this part of the body plan, has been conserved between insects and vertebrates. However, the axis is inverted, so that dorsal in the fly corresponds to ventral in the vertebrate (Figure 21–35). At some point in its evolutionary history, it seems, the ancestor of one of these classes of animals took to living life upside down.

Three Classes of Segmentation Genes Refine the Anterior–Posterior Maternal Pattern and Subdivide the Embryo

After the initial gradients of Bicoid and Nanos are created to define the anteroposterior axis, the **segmentation genes** refine the pattern. Mutations in any one of the segmentation genes alter the number of segments or their basic internal organization without affecting the global polarity of the embryo. Segmentation genes are expressed by subsets of cells in the embryo, so their products are the first components that the embryo's own genome, rather than the maternal genome, contributes to embryonic development. They are therefore called *zygotic-effect genes* to distinguish them from the earlier maternal-effect genes.

The segmentation genes fall into three groups according to their mutant phenotypes and the stages at which they act (Figure 21–36). First come a set of at least six **gap genes**, whose products mark out coarse subdivisions of the embryo. Mutations in a gap gene eliminate one or more groups of adjacent segments, and mutations in different gap genes cause different but partially overlapping defects. In the mutant *Krüppel,* for example, the larva lacks eight segments, from T1 to A5 inclusive.

The next segmentation genes to act are a set of eight **pair-rule genes**. Mutations in these cause a series of deletions affecting alternate segments, leaving the embryo with only half as many segments as usual. While all the pair-rule

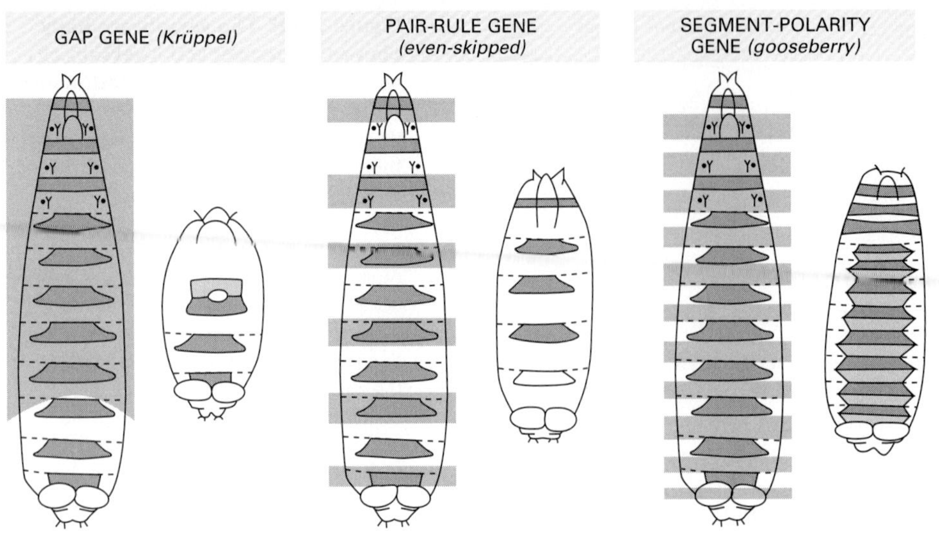

Figure 21–36 Examples of the phenotypes of mutations affecting the three types of segmentation genes. In each case the areas shaded in *green* on the normal larva *(left)* are deleted in the mutant or are replaced by mirror-image duplicates of the unaffected regions. By convention, dominant mutations are written with an initial capital letter and recessive mutations are written with a lower-case letter. Several of the patterning mutations of *Drosophila* are classed as dominant because they have a perceptible effect on the phenotype of the heterozygote, even though the characteristic major, lethal effects are recessive—that is, visible only in the homozygote. (Modified from C. Nüsslein-Volhard and E. Wieschaus, *Nature* 287:795–801, 1980.)

mutants display this two-segment periodicity, they differ in the precise positioning of the deletions relative to the segmental or parasegmental borders. The pair-rule mutant *even-skipped (eve)*, for example, which is discussed in Chapter 9, lacks the whole of each odd-numbered parasegment, while the pair-rule mutant *fushi tarazu (ftz)* lacks the whole of each even-numbered parasegment, and the pair-rule mutant *hairy* lacks a series of regions that are of similar width but out of register with the parasegmental units.

Finally, there are at least 10 **segment-polarity genes**. Mutations in these genes produce larvae with a normal number of segments but with a part of each segment deleted and replaced by a mirror-image duplicate of all or part of the rest of the segment. In *gooseberry* mutants, for example, the posterior half of each segment (that is, the anterior half of each parasegment) is replaced by an approximate mirror image of the adjacent anterior half-segment (see Figure 21–36).

We see later that, in parallel with the segmentation process, a further set of genes, the *homeotic selector genes*, serve to define and preserve the differences between one segment and the next.

The phenotypes of the various segmentation mutants suggest that the segmentation genes form a coordinated system that subdivides the embryo progressively into smaller and smaller domains along the anteroposterior axis, distinguished by different patterns of gene expression. Molecular genetics has helped to reveal how this system works.

The Localized Expression of Segmentation Genes Is Regulated by a Hierarchy of Positional Signals

About three-quarters of the segmentation genes, including all of the gap genes and pair-rule genes, code for gene regulatory proteins. Their actions on one another and on other genes can therefore be observed by comparing gene expression in normal and mutant embryos. By using appropriate probes to detect the gene transcripts or their protein products, one can, in effect, take snapshots as genes switch on and off in changing patterns. Repeating the process with mutants that lack a particular segmentation gene, one can begin to dissect the logic of the entire gene control system.

The products of the egg-polarity genes provide the global positional signals in the early embryo. These cause particular gap genes to be expressed in particular regions. The products of the gap genes then provide a second tier of positional signals that act more locally to regulate finer details of patterning through the expression of yet other genes, including the pair-rule genes (Figure 21–37). The pair-rule genes in turn collaborate with one another and with the gap genes to set up a regular periodic pattern of expression of segment-polarity genes, and the segment-polarity genes collaborate with one another to define the internal pattern of each individual segment. The strategy, therefore, is one of sequential induction (see Figure 21–15). By the end of the process, the global gradients produced by the egg-polarity genes have triggered the creation of a fine-grained pattern through a hierarchy of sequential, progressively more local, positional controls. Because the global positional signals that start the process do not have to directly specify fine details, the individual cell nuclei do not have to be governed with extreme precision by small differences in the concentration of these signals. Instead, at each step in the sequence, new signals come into play, providing substantial localized differences of concentration to define new details. Sequential induction is thus a robust strategy. It works reliably to produce fly embryos that all have the same pattern, despite the essential imprecision of biological control systems, and despite variations in conditions such as the temperature at which the fly develops.

The Modular Nature of Regulatory DNA Allows Genes to Have Multiple Independently Controlled Functions

The elaborate patterning process just described depends on the long stretches of noncoding DNA sequence that control the expression of each of the genes involved. These regulatory regions bind multiple copies of the gene regulatory proteins produced by the patterning genes expressed earlier. Like an input-output logic device, an individual gene is thus turned on and off according to the particular combination of proteins bound to its regulatory regions at each stage of development. In Chapter 7 we describe one particular segmentation gene—the pair-rule gene *even-skipped (eve)*—and discuss how the decision whether to transcribe the gene is made on the basis of all these inputs (see Figure 7–55). This example can be taken further to illustrate some important principles of developmental patterning.

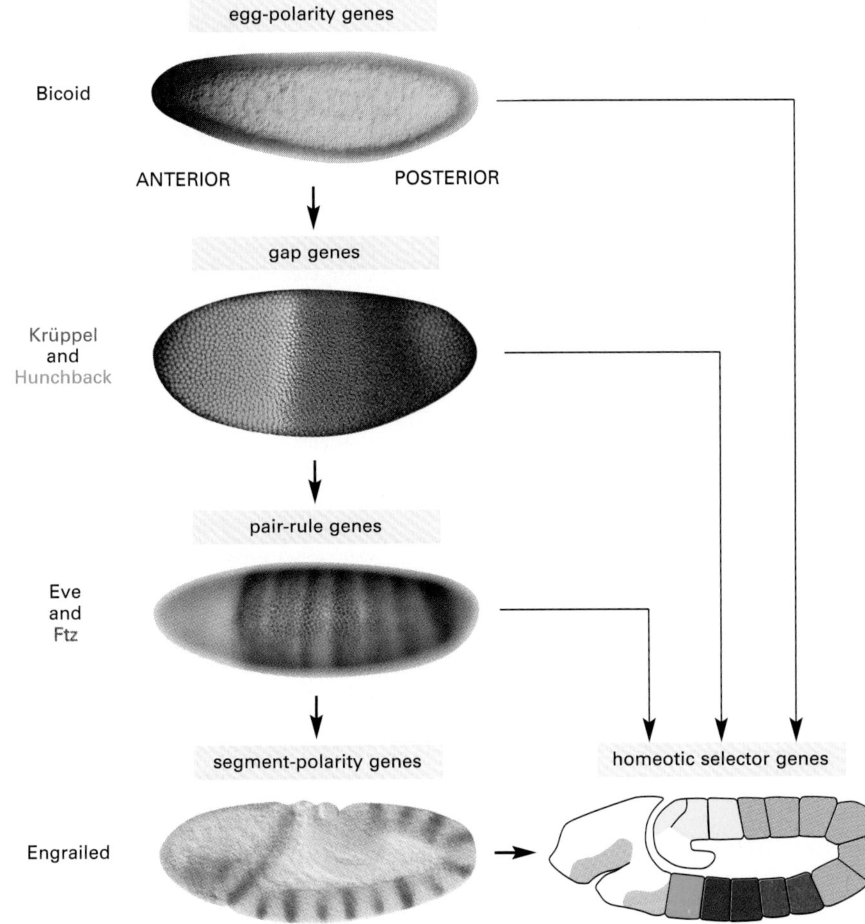

Figure 21–37 The regulatory hierarchy of egg-polarity, gap, segmentation, and homeotic selector genes. The photographs show expression patterns of representative examples of genes in each category, revealed by staining with antibodies against the protein products. The homeotic selector genes, discussed below, define the lasting differences between one segment and the next. (Photographs (i) from W. Driever and C. Nüsslein-Volhard, *Cell* 54:83–104, 1988. © Elsevier; (ii) courtesy of Jim Langeland, Steve Paddock, Sean Carroll, and the Howard Hughes Medical Institute; (iii) from P.A. Lawrence, The Making of a Fly. Oxford, UK: Blackwell, 1992; (iv) from C. Hama, Z. Ali, and T.B. Kornberg, *Genes Dev.* 4:1079–1093, 1990. © Cold Spring Harbor Press.)

Individual stripes of *eve* expression depend on separate regulatory modules in the *eve* regulatory DNA. Thus, one regulatory module is responsible for driving *eve* expression in stripes 1 + 5, another for stripe 2, another for stripes 3 + 7, and yet another for stripes 4 + 6 (Figure 21–38). Each regulatory module defines a different set of requirements for gene expression according to the concentrations of the products of the egg-polarity and gap genes. In this way, the *eve* regulatory DNA serves to translate the complex nonrepetitive pattern of egg-polarity and gap proteins into the periodic pattern of expression of a pair-rule gene.

The modular organization of the *eve* regulatory DNA just described is typical of gene regulation in multicellular animals and plants, and it has profound implications. By stringing together sequences of modules that respond to different combinations of regulatory proteins, it is possible to generate almost any pattern of gene expression on the basis of almost any other. Modularity, moreover, allows the regulatory DNA to define patterns of gene expression that are not merely complex, but whose parts are independently adjustable. A change in one of the regulatory modules can alter one part of the expression pattern without affecting the rest, and without requiring changes in regulatory proteins that would have repercussions for the expression of other genes in the genome. As described in Chapter 7, it is such regulatory DNA that contains the key to the complex organization of multicellular plants and animals, and its properties make possible the independent adaptability of each part of an organism's body structure in the course of evolution.

Most of the segmentation genes also have important functions at other times and places in the development of *Drosophila*. The *eve* gene, for example, is expressed in subsets of neurons, in muscle precursor cells, and in various other sites, under the control of additional enhancers (see Figure 21–38). By addition of new modules to its regulatory DNA, any gene can be coopted during evolution for new purposes at new sites in the body, without detriment to its other functions.

Egg-Polarity, Gap, and Pair-Rule Genes Create a Transient Pattern That Is Remembered by Other Genes

Within the first few hours after fertilization, the gap genes and the pair-rule genes are activated one after another. Their mRNA products appear first in patterns that only approximate the final picture; then, within a short time—through a series of interactive adjustments—the fuzzy initial distribution of gene products resolves itself into a regular, crisply defined system of stripes (Figure 21–39). But this system itself is unstable and transient. As the embryo proceeds through gastrulation and beyond, the regular segmental pattern of gap and pair-rule

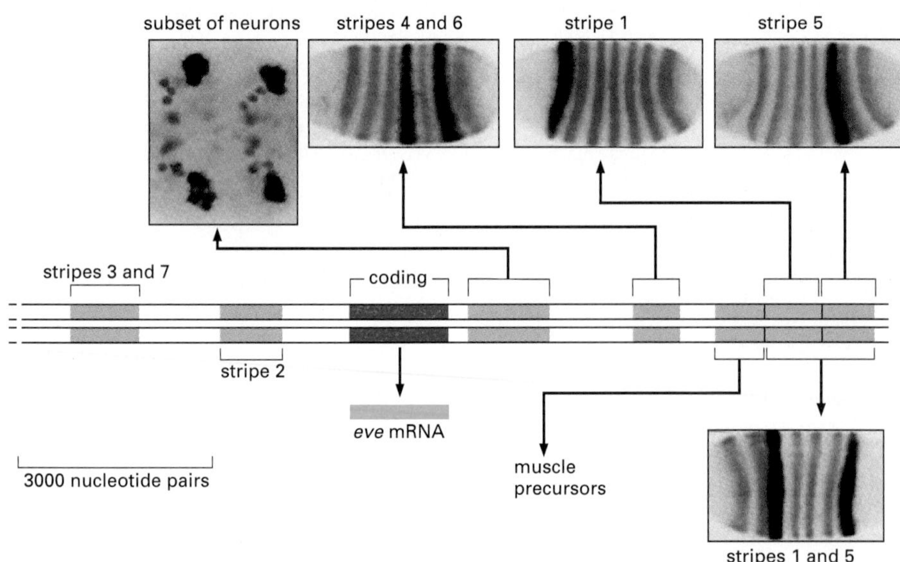

Figure 21–38 Modular organization of the regulatory DNA of the *eve* gene. In the experiment shown, cloned fragments of the regulatory DNA were linked to a *LacZ* reporter (a bacterial gene). Transgenic embryos containing these constructs were then stained by *In situ* hybridization to reveal the pattern of expression of *LacZ (blue/black)*, and counterstained with an anti-Eve antibody *(orange)* to show the positions of the normal *eve* expression stripes. Different segments of the *eve* regulatory DNA *(ochre)* are thus found to drive gene expression in regions corresponding to different parts of the normal *eve* expression pattern.

Two segments in tandem drive expression in a pattern that is the sum of the patterns generated by each of them individually. Separate regulatory modules are responsible for different times of gene expression, as well as different locations: the leftmost panel shows the action of a module that comes into play later than the others illustrated and drives expression in a subset of neurons. (From M. Fujioka et al., *Development* 126:2527–538, 1999. © The Company of Biologists.)

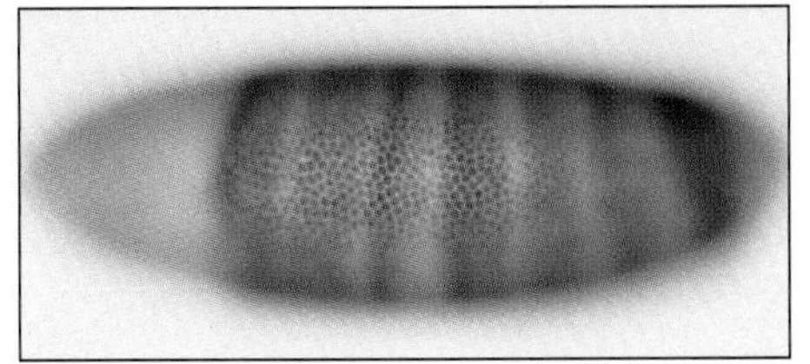

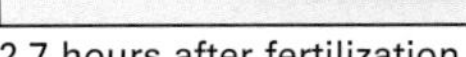

2.7 hours after fertilization

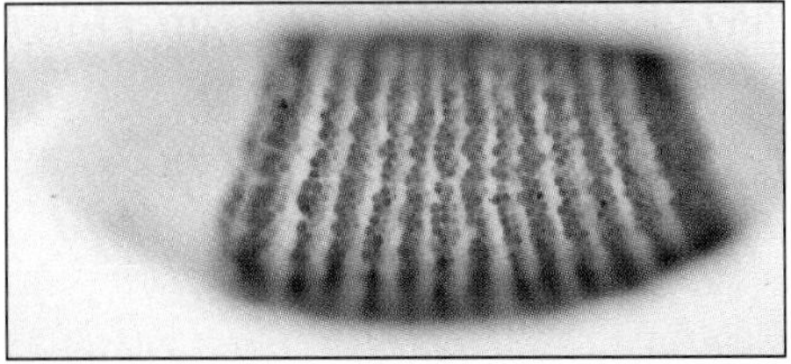

3.5 hours after fertilization

Figure 21–39 The formation of *ftz* and *eve* stripes in the *Drosophila* blastoderm. *ftz* and *eve* are both pair-rule genes. Their expression patterns (shown in *brown* for *ftz* and in *gray* for *eve*) are at first blurred but rapidly resolve into sharply defined stripes. (From P.A. Lawrence, The Making of a Fly. Oxford, UK: Blackwell, 1992.)

gene products disintegrates. Their actions, however, have stamped a permanent set of labels—positional values—on the cells of the blastoderm. These positional labels are recorded in the persistent activation of certain of the segment-polarity genes and of the homeotic selector genes, which serve to maintain the segmental organization of the larva and adult. The segment-polarity gene *engrailed* provides a good example. Its RNA transcripts are seen in the cellular blastoderm in a series of 14 bands, each approximately one cell wide, corresponding to the anteriormost portions of the future parasegments (Figure 21–40).

The segment-polarity genes are expressed in patterns that repeat from one parasegment to the next, and their bands of expression appear in a fixed relationship to the bands of expression of the pair-rule genes that help to induce them. However, the production of this pattern within each parasegment depends on interactions among the segment-polarity genes themselves. These interactions occur at stages when the blastoderm has already become fully partitioned into separate cells, so that cell–cell signaling of the usual sort has to come into play. A large subset of the segment-polarity genes code for components of two signal transduction pathways, the Wnt pathway and the Hedgehog pathway, including the secreted signal proteins Wingless (a Wnt family member) and Hedgehog. These are expressed in different bands of cells that serve as signaling centers within each parasegment, and they act to maintain and refine the expression of other segment-polarity genes. Moreover, although their initial expression is determined by the pair-rule genes, the two signaling proteins regulate one another's expression in a mutually supportive way, and they proceed to help trigger expression of genes such as *engrailed* in precisely the correct sites.

The *engrailed* expression pattern will persist throughout life, long after the signals that organized its production have disappeared (see Figure 21–40). This example illustrates not only the progressive subdivision of the embryo by means of more and more narrowly localized signals, but also the transition between the transient signaling events of early development and the later stable maintenance of developmental information.

Besides regulating the segment-polarity genes, the products of pair-rule genes collaborate with the products of gap genes to cause the precisely localized activation of a further set of spatial labels—the homeotic selector genes. It is the

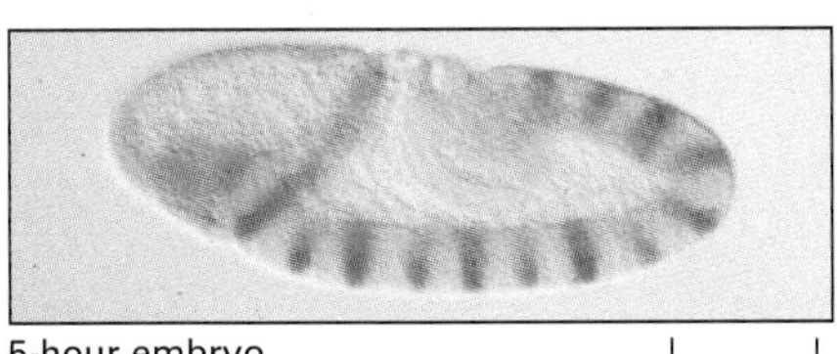

5-hour embryo — 100 μm

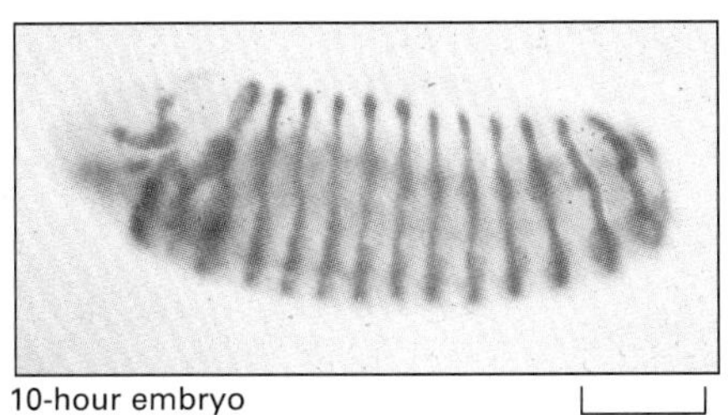

10-hour embryo — 100 μm

adult — 500 μm

Figure 21–40 The pattern of expression of *engrailed*, a segment-polarity gene. The *engrailed* pattern is shown in a 5-hour embryo (at the extended germ-band stage), a 10-hour embryo, and an adult (whose wings have been removed in this preparation). The pattern is revealed by an antibody *(brown)* against the Engrailed protein (for the 5- and 10-hour embryos) or (for the adult) by constructing a strain of *Drosophila* containing the control sequences of the *engrailed* gene coupled to the coding sequence of the reporter *LacZ*, whose product is detected histochemically through the *blue* product of a reaction that it catalyzes. Note that the *engrailed* pattern, once established, is preserved throughout the animal's life. (From C. Hama, Z. Ali, and T.B. Kornberg, *Genes Dev.* 4:1079–1093, 1990. © Cold Spring Harbor Press.)

homeotic selector genes that permanently distinguish one parasegment from another. In the next section we examine these selector genes in detail and consider their role in cell memory.

Summary

The fly Drosophila *has been the foremost model organism for study of the genetics of animal development. Like other insects, it begins its development with a series of nuclear divisions generating a syncytium, and a large amount of early patterning occurs in this single giant multinucleate cell. The pattern originates with asymmetry in the egg, organized both by localized deposits of mRNA inside the egg and by signals from the follicle cells around it. Positional information in the multinucleate embryo is supplied by four intracellular gradients that are set up by the products of four groups of maternal-effect genes called egg-polarity genes. These control four distinctions fundamental to the body plan of animals: dorsal versus ventral, endoderm versus mesoderm and ectoderm, germ cells versus somatic cells, and head versus rear.*

The egg-polarity genes operate by setting up graded distributions of gene regulatory proteins in the egg and early embryo. The gradients along the anteroposterior axis initiate the orderly expression of gap genes, pair-rule genes, segment-polarity genes, and homeotic selector genes. These, through a hierarchy of interactions, become expressed in some regions of the embryo and not others, progressively subdividing the blastoderm into a regular series of repeating modular units called segments. The complex patterns of gene expression reflect the modular organization of the regulatory DNA, with separate enhancers of an individual gene responsible for separate parts of its expression pattern.

The segment-polarity genes come into play toward the end of the segmentation process, soon after the syncytium has become partitioned into separate cells, and they control the internal patterning of each segment through cell–cell signaling via the Wnt (Wingless) and Hedgehog pathways. This leads to persistent localized activation of genes such as engrailed, giving cells a remembered record of their anteroposterior address within the segment. Meanwhile, a new cell–cell signaling gradient is also set up along the dorsoventral axis, with the TGFβ family member Decapentaplegic (Dpp) and its antagonist, Short gastrulation, acting as the morphogens. This gradient helps to refine the assignment of different characters to cells at different dorsoventral levels. Homologous proteins are also known to control the patterning of the ventrodorsal axis in vertebrates.

HOMEOTIC SELECTOR GENES AND THE PATTERNING OF THE ANTEROPOSTERIOR AXIS

As development proceeds, the body becomes more and more complex. In all this growing complexity there is, however, a simplifying feature that puts an understanding of the whole developmental process within our grasp. Again and again, in every species and at every level of organization, we find that complex structures are made by repeating a few basic themes with variations. Thus a limited number of basic differentiated cell types, such as muscle cells or fibroblasts, recur with subtle individual variations in different sites. These cell types are organized into a limited variety of tissue types, such as muscle or tendon, which again are repeated with subtle variations in different regions of the body. From the various tissues, organs such as teeth or digits are built—molars and incisors, fingers and thumbs and toes—a few basic kinds of structure, repeated with variations.

Wherever we find this phenomenon of *modulated repetition*, we can break down the developmental biologist's problem into two kinds of question: what is the basic construction mechanism common to all the objects of the given class, and how is this mechanism modified to give the observed variations? The embryo uses a combinatorial strategy to generate its complexity, and we can use a combinatorial strategy to understand it.

Figure 21–41 A homeotic mutation. The fly shown here is an *Antennapedia* mutant. Its antennae are converted into leg structures by a mutation in the regulatory region of the *Antennapedia* gene that causes it to be expressed in the head. Compare with the normal fly shown in Figure 21–23. (Courtesy of Matthew Scott.)

The segments of the insect body provide a very clear example. We have already sketched the way in which the rudiment of a single typical segment is constructed. We must now consider how one segment is caused to be different from another.

The HOX Code Specifies Anterior-Posterior Differences

The first glimpse of a genetic answer to the question of how each segment acquires its individual identity came over 80 years ago, with the discovery of the first of a set of mutations in *Drosophila* that cause bizarre disturbances of the organization of the adult fly. In the *Antennapedia* mutant, for example, legs sprout from the head in place of antennae (Figure 21–41), while in the *bithorax* mutant, portions of an extra pair of wings appear where normally there should be the much smaller appendages called halteres. These mutations transform parts of the body into structures appropriate to other positions and are called *homeotic*. A whole set of **homeotic selector genes** determines the anteroposterior character of the segments of the fly.

The genes of this set—eight of them in the fly—are all related to one another as members of a multigene family, and they all lie in one or the other of two tight gene clusters known as the **bithorax complex** and the **Antennapedia complex**. The genes in the bithorax complex control the differences among the abdominal and thoracic segments of the body, while those in the Antennapedia complex control the differences among thoracic and head segments. Comparisons with other species show that the same genes are present in essentially all animals, including humans. These comparisons also reveal that the Antennapedia and bithorax complexes are the two halves of a single entity, called the **Hox complex**, that has become split in the course of the fly's evolution, and whose members operate in a coordinated way to exert their control over the head-to-tail pattern of the body.

Homeotic Selector Genes Code for DNA-Binding Proteins That Interact with Other Gene Regulatory Proteins

To a first approximation each homeotic selector gene is normally expressed in just those regions that develop abnormally when the gene is mutated or absent. The products of these genes can thus be viewed as molecular address labels possessed by the cells of each parasegment: they are the physical embodiment of the cells' positional value. If the address labels are changed, the parasegment behaves as though it were located somewhere else, and deletion of the entire complex results in a larva whose body segments are all alike (Figure 21–42).

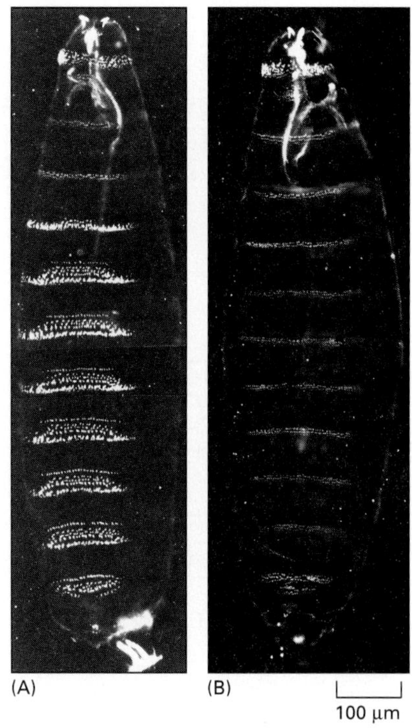

Figure 21–42 The effect of deleting most of the genes of the bithorax complex. (A) A normal *Drosophila* larva shown in dark-field illumination; (B) the mutant larva with the bithorax complex largely deleted. In the mutant the parasegments posterior to P5 all have the appearance of P5. (From G. Struhl, *Nature* 293:36–41, 1981. © Macmillan Magazines Ltd.)

A first problem, therefore, is to understand how the homeotic selector gene products act on the basic segment-patterning machinery to give each parasegment its individuality. The products of the homeotic selector genes are gene regulatory proteins, all related to one another by the possession of a highly conserved DNA-binding *homeodomain* (60 amino acids long), discussed in Chapter 7. The corresponding segment in the DNA sequence is called a *homeobox* from which, by abbreviation, the Hox complex takes its name.

If the products of the homeotic selector genes are similar in their DNA-binding regions, how do they exert different effects so as to make one parasegment different from the next? The answer seems to lie largely in the parts of the proteins that do not bind directly to DNA but interact with other proteins in DNA-bound complexes. The different partners in these complexes act together with the homeotic selector proteins to dictate which DNA-binding sites will be recognized and whether the effect on transcription at those sites will be activation or repression. In this way, the products of the homeotic selector genes combine with other gene regulatory proteins and modulate their actions so as to give each parasegment its characteristic features.

The Homeotic Selector Genes Are Expressed Sequentially According to Their Order in the Hox Complex

To understand how the Hox complex provides cells with positional values, we also need to consider how the expression of the Hox genes themselves is regulated. The coding sequences of the eight homeotic selector genes in the Antennapedia and bithorax complexes are interspersed amid a much larger quantity—a total of about 650,000 nucleotide pairs—of regulatory DNA. This DNA includes binding sites for the products of egg-polarity and segmentation genes. The regulatory DNA in the Hox complex acts as an interpreter of the multiple items of positional information supplied to it by all these gene regulatory proteins. In response, a particular set of homeotic selector genes is transcribed, appropriate to the location.

In the pattern of control there is a remarkable regularity. The sequence in which the genes are ordered along the chromosome, in both the Antennapedia and the bithorax complexes, corresponds almost exactly to the order in which they are expressed along the axis of the body (Figure 21–43). This suggests that

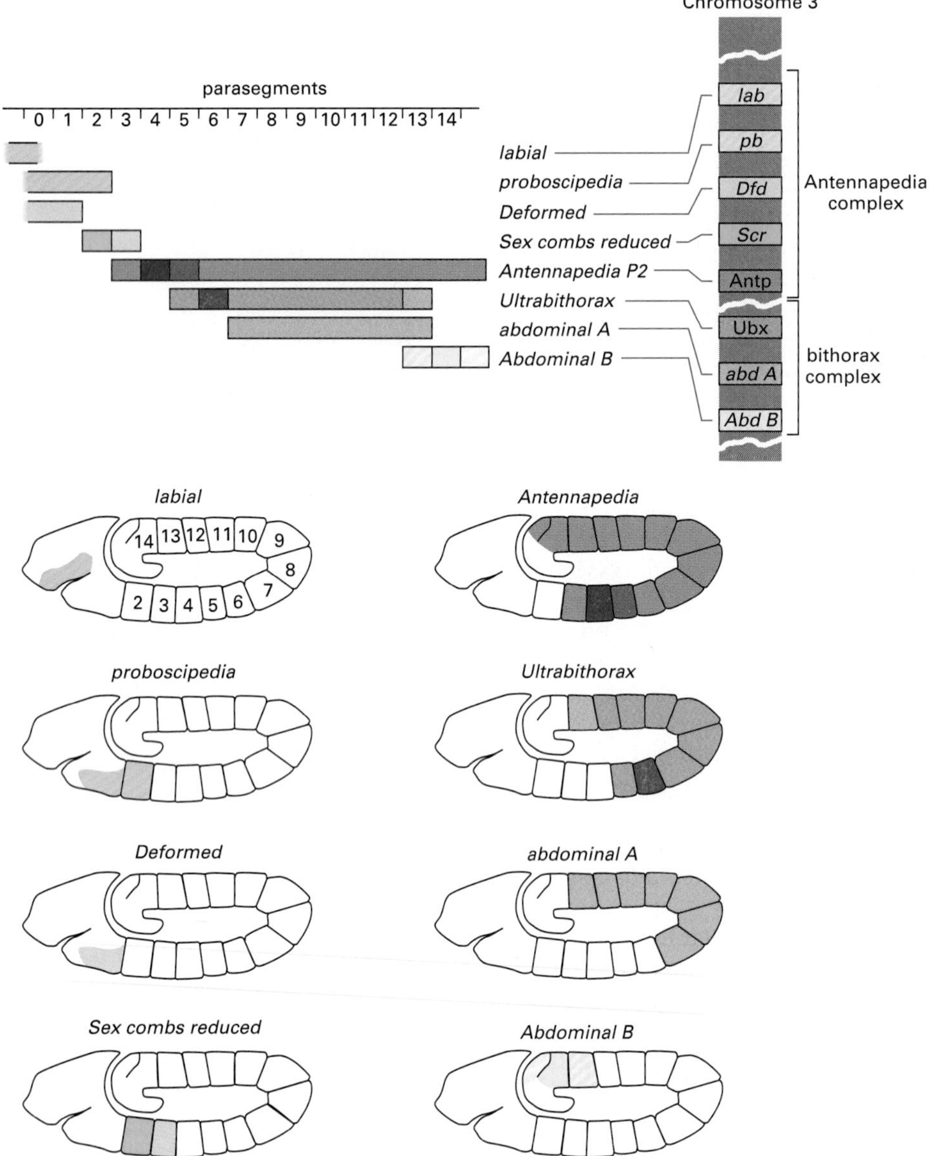

Figure 21–43 The patterns of expression compared to the chromosomal locations of the genes of the Hox complex. The sequence of genes in each of the two subdivisions of the chromosomal complex corresponds to the spatial sequence in which the genes are expressed. Note that most of the genes are expressed at a high level throughout one parasegment *(dark color)* and at a lower level in some adjacent parasegments *(medium color* where the presence of the transcripts is necessary for a normal phenotype, *light color* where it is not). In regions where the expression domains overlap, it is usually the most "posterior" of the locally active genes that determines the local phenotype. The drawings in the lower part of the figure represent the gene expression patterns in embryos at the extended germ band stage, about 5 hours after fertilization.

the genes are activated serially by some process that is graded—in duration or intensity—along the axis of the body, and whose action spreads gradually along the chromosome. The most "posterior" of the genes expressed in a cell generally dominates, driving down expression of the previously activated "anterior" genes and dictating the character of the segment. The gene regulatory mechanisms underlying these phenomena are still not well understood, but their consequences are profound. We shall see that the serial organization of gene expression in the Hox complex is a fundamental feature that has been highly conserved in the course of evolution.

There are hundreds of other homeobox-containing genes in the genome of the fly—and of other animal species—but most of them are scattered and not clustered in complexes such as the Hox complex. They have many different gene regulatory functions, but a substantial proportion of them have roles akin to that of the Hox genes: they control the variations on a basic developmental theme. Different classes of neurons, for example, are often distinguished from one another by expression of specific genes of this large superfamily.

The Hox Complex Carries a Permanent Record of Positional Information

The spatial pattern of expression of the genes in the Hox complex is set up by signals acting early in development, but the consequences are long-lasting. Although the pattern of expression undergoes complex adjustments as development proceeds, the Hox complex behaves in each cell as though stamped with a permanent record of the anteroposterior position that the cell occupied in the early embryo. In this way, the cells of each segment are equipped with a long-term memory of their location along the anteroposterior axis of the body—in other words, with an anteroposterior positional value. As we shall see in the next section, the memory trace imprinted on the Hox complex governs the segment-specific identity not only of the larval segments, but also of the structures of the adult fly, which are generated at a much later stage from the larval imaginal discs and other nests of imaginal precursor cells in the larva.

The molecular mechanism of the cell memory for this positional information relies on two types of regulatory inputs. One is from the homeotic selector genes themselves: many of the Hox proteins autoactivate the transcription of their own genes. Another crucial input is from two large complementary sets of transcriptional regulators called the *Polycomb group* and the *Trithorax group*. If these regulators are defective, the pattern of expression of the homeotic selector genes is set up correctly at first but is not correctly maintained as the embryo grows older.

The two sets of regulators act in opposite ways. Trithorax group proteins are needed to maintain the transcription of Hox genes in cells where transcription has already been switched on. In contrast, Polycomb group proteins form stable complexes that bind to the chromatin of the Hox complex and maintain the repressed state in cells where Hox genes have not been activated at the critical time (Figure 21–44). The developmental memory involves acetylation of histone H4 at specific regulatory sites in the chromatin adjacent to the Hox genes; the proteins of the Trithorax and Polycomb groups act somehow to perpetuate the H4 state—hyperacetylated if the target Hox gene has been transiently exposed at embryonic stages to an activator of gene transcription, not acetylated in this way otherwise (see Figure 4–48).

The Anteroposterior Axis Is Controlled by Hox Selector Genes in Vertebrates Also

Homologs of the *Drosophila* homeotic selector genes have been found in almost every animal species studied, from cnidarians (hydroids) and nematodes to molluscs and mammals. Remarkably, these genes are often grouped in complexes similar to the insect Hox complex. In the mouse there are four such complexes—called the HoxA, HoxB, HoxC, and HoxD complexes—each on a

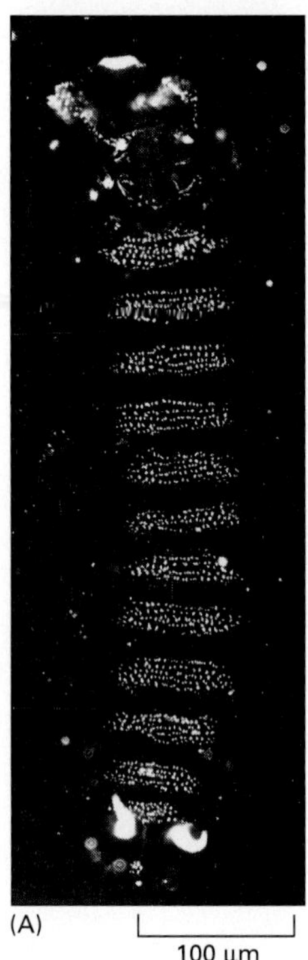

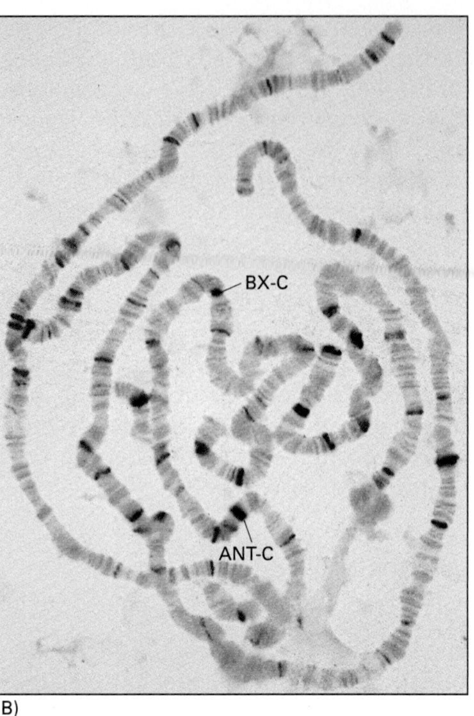

Figure 21–44 Action of genes of the Polycomb group. (A) Photograph of a mutant embryo defective for the gene *extra sex combs (esc)* and derived from a mother also lacking this gene. The gene belongs to the Polycomb group. Essentially all segments have been transformed to resemble the most posterior abdominal segment (compare with Figure 21–42). In the mutant the pattern of expression of the homeotic selector genes, which is roughly normal initially, is unstable in such a way that all these genes soon become switched on all along the body axis. (B) The normal pattern of binding of Polycomb protein to *Drosophila* giant chromosomes, visualized with an antibody against Polycomb. The protein is bound to the Antennapedia complex (ANT-C) and the bithorax complex (BX-C) as well as about 60 other sites. (A, from G. Struhl, *Nature* 293:36–41, 1981. © Macmillan Magazines Ltd; B, courtesy of B. Zink and R. Paro, *Trends Genet.* 6:416–421, 1990. © Elsevier.)

different chromosome. Individual genes in each complex can be recognized by their sequences as counterparts of specific members of the *Drosophila* set. Indeed, mammalian Hox genes can function in *Drosophila* as partial replacements for the corresponding *Drosophila* Hox genes. It appears that each of the four mammalian Hox complexes is, roughly speaking, the equivalent of a complete insect complex (that is, an Antennapedia complex plus a bithorax complex) (Figure 21–45).

The ordering of the genes within each vertebrate Hox complex is essentially the same as in the insect Hox complex, suggesting that all four vertebrate complexes originated by duplications of a single primordial complex and have preserved its basic organization. Most tellingly, when the expression patterns of the Hox genes are examined in the vertebrate embryo by *in situ* hybridization, it turns out that the members of each complex are expressed in a head-to-tail series along the axis of the body, just as they are in *Drosophila* (Figure 21–46). The pattern is most clearly seen in the neural tube, but is also visible in other tissues, especially the mesoderm. With minor exceptions this anatomical ordering matches the chromosomal ordering of the genes in each complex, and corresponding genes in the four different Hox complexes have almost identical anteroposterior domains of expression.

The gene expression domains define a detailed system of correspondences between insect body regions and vertebrate body regions (see Figure 21–45). The parasegments of the fly correspond to a similarly labeled series of segments in the anterior part of the vertebrate embryo. As shown in Figure 21–47, these are most clearly demarcated in the hindbrain, where they are called *rhombomeres.* In the tissues lateral to the hindbrain the segmentation is seen in the series of *branchial arches,* prominent in all vertebrate embryos—the precursors of the system of gills in fish and of the jaws and structures of the neck in mammals; each pair of rhombomeres in the hindbrain corresponds to one branchial arch (Figure 21–47). In the hindbrain, as in *Drosophila,* the boundaries of the expression domains of many of the Hox genes are aligned with the boundaries of the anatomical segments.

The products of the mammalian Hox genes appear to specify positional values that control the anteroposterior pattern of parts in the hindbrain, neck, and trunk. Eliminating the function of a Hox gene in the mouse leads to a defect in a

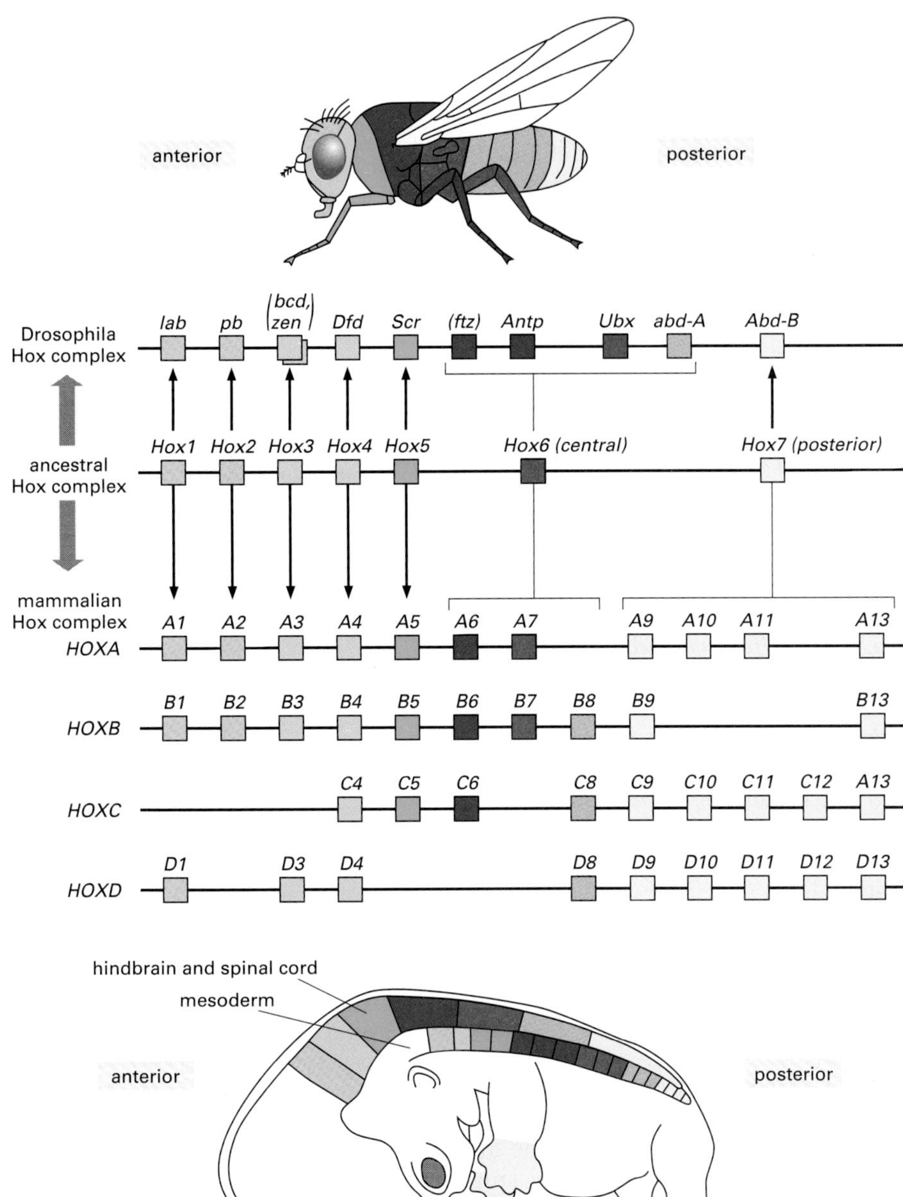

Figure 21–45 The Hox complex of an insect and the Hox complexes of a mammal compared and related to body regions. The genes of the Antennapedia and bithorax complexes of *Drosophila* are shown in their chromosomal order in the *top line*; the corresponding genes of the four mammalian Hox complexes are shown *below*, also in chromosomal order. The gene expression domains in fly and mammal are indicated in a simplified form by color in the cartoons of animals *above* and *below*. However, the details of the patterns depend on developmental stage and vary somewhat from one mammalian Hox complex to another. Also, in many cases, genes shown here as expressed in an anterior domain are also expressed more posteriorly, overlapping the domains of more posterior Hox genes (see, for example, Figure 21–46).

The complexes are thought to have evolved as follows: first, in some common ancestor of worms, flies, and vertebrates, a single primordial homeotic selector gene underwent repeated duplication to form a series of such genes in tandem—the ancestral Hox complex. In the *Drosophila* sublineage this single complex became split into separate Antennapedia and bithorax complexes. Meanwhile, in the lineage leading to the mammals the whole complex was repeatedly duplicated to give four Hox complexes. The parallelism is not perfect because apparently some individual genes have been duplicated, others lost, and still others coopted for different purposes (genes in parentheses in the *top line*) since the complexes diverged. (Based on a diagram kindly supplied by William McGinnis.)

Figure 21–46 Expression domains of Hox genes in a mouse. The photographs show whole embryos displaying the expression domains of two genes of the HoxB complex *(blue* stain). These domains can be revealed by *in situ* hybridization or, as in these examples, by constructing transgenic mice containing the control sequence of a Hox gene coupled to a *LacZ* reporter gene, whose product is detected histochemically. Each gene is expressed in a long expanse of tissue with a sharply defined anterior limit. The earlier the position of the gene in its chromosomal complex, the more anterior the anatomical limit of its expression. Thus, with minor exceptions, the anatomical domains of the successive genes form a nested set, ordered according to the ordering of the genes in the chromosomal complex. (Courtesy of Robb Krumlauf.)

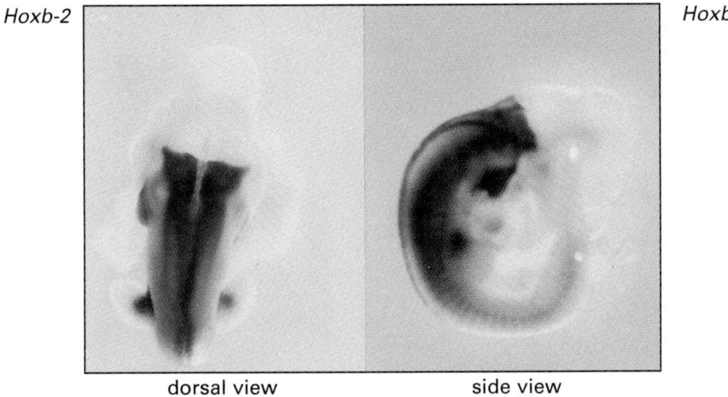

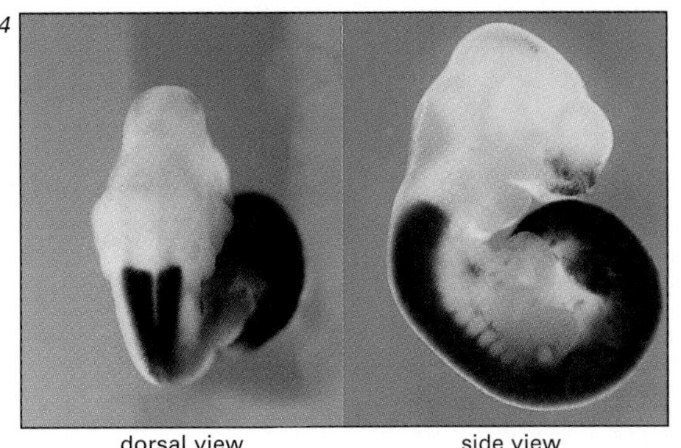

body region that corresponds to the domain of expression of that gene. Sometimes the affected body region is transformed into a more anterior body region, as in *Drosophila* homeotic mutants; sometimes the affected body region dies or fails to grow. The transformations observed in mouse Hox mutants are often incomplete, perhaps because of a redundancy between genes in the four Hox gene clusters. But it seems clear that the fly and the mouse use essentially the same molecular machinery to give individual characters to successive regions along at least a part of their anteroposterior axis.

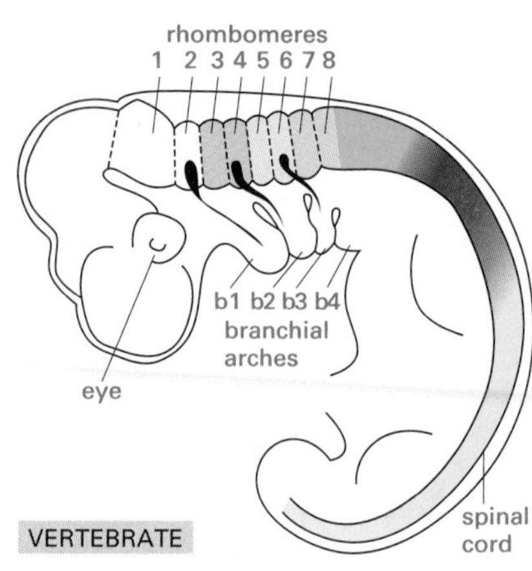

Figure 21–47 Segmentation and Hox gene expression in the hindbrain, as seen in a chick embryo. The pattern of HoxB gene expression is indicated by coloring as in Figure 21–45. For simplicity, the expression in tissues other than the central nervous system is not shown. In regions where the expression domains of two or more Hox genes overlap, the coloring corresponds to the most "posterior" of the genes expressed. Just as the expression domains in the fly are related to parasegments, so the expression domains in the vertebrate are related to the rhombomeres (segments in the hindbrain). Each pair of rhombomeres is associated with a branchial arch (a modified gill rudiment), to which it sends innervation. The pattern of Hox gene expression in the branchial arches (not shown) matches that in the associated rhombomeres. In the spinal cord, there are no rhombomere boundaries, and the expression domains of the different posterior Hox genes overlap.

Summary

The complexity of the adult body of an animal is built up by modulated repetition of a few basic types of structure. Thus, superimposed on the pattern of gene expression that repeats itself in every segment, there is a serial pattern of expression of homeotic selector genes that confer on each segment a different identity. The homeotic selector genes code for DNA-binding proteins of the homeodomain family. They are grouped in the* Drosophila *genome in two clusters, called the Antennapedia and bithorax complexes, believed to be the two parts of a single primordial Hox complex that became split during evolution of the fly. In each complex, the genes are arranged in a sequence that matches their sequence of expression along the axis of the body. Hox gene expression is initiated in the embryo. It is maintained subsequently by the action of DNA-binding proteins of the Polycomb and Trithorax group, which stamp the chromatin of the Hox complex with a heritable record of its embryonic state of activation. Hox complexes homologous to that of* Drosophila *are found in virtually every type of animal that has been examined, from cnidarians to humans, and they appear to have an evolutionarily conserved role in patterning the anteroposterior axis of the body. Mammals have four Hox complexes, each showing a similar relationship between a serial arrangement of the genes in the chromosome and their serial pattern of expression along the body axis.

ORGANOGENESIS AND THE PATTERNING OF APPENDAGES

We have seen that the segments of the insect larva are all variations on the same basic theme, with segmentation genes defining the basic repetitive module and homeotic selector genes giving each segment its individual character. The same applies to the major appendages of the adult insect body—legs, wings, antennae, mouthparts and external genitalia: they too are variations on a common basic theme. At a finer level of detail, we encounter the same wonderful simplification: the appendages—and many other parts of the body—consist of substructures that are themselves variations on a small number of basic evolutionarily conserved themes.

In this section we follow the course of development in *Drosophila* through to its end, narrowing our focus at each step to examine one example of the many related structures that are developing in parallel. As we go along, we shall point out parallels with vertebrate structures that develop similarly, using not only the same general strategies, but many of the same specific molecular mechanims. But to avoid interrupting the narrative later, we must first briefly explain some key experimental methods, required to cope with a special problem that arises when we try to discover how genes control the later stages of development.

Conditional and Induced Somatic Mutations Make it Possible to Analyze Gene Functions Late in Development

As emphasized earlier, the same gene may be used repeatedly in many different situations—in different regions of the body, and at different times. Often, loss-of-function mutations disrupt early development so severely that the embryo or larva dies, depriving us of the opportunity to see how the mutation would affect later processes.

One way around this problem is to study conditional mutations. If we have, for example, a temperature-sensitive mutation in the gene of interest, we can maintain the animal during early development at a low temperature, where the gene product functions normally, and then disable the gene product whenever we please by raising the temperature to discover the late functions.

Other methods involve actually modifying the DNA in subsets of cells at late stages of development—a sort of genetic surgery on individual cells that allows mutant groups of cells of a specified genotype to be generated at a chosen time in development. This remarkable feat can be achieved by *induced somatic recombination*. A current version of this technique uses transgenic flies that have been bred to contain two types of yeast-derived genetic elements: the FLP site-specific recombinase gene, and the FLP Recombinase Target (FRT) sequence. Typically, the animal is homozygous for an insertion of the FRT sequence close to the centromere on a chosen chromosome arm, while a construct consisting of the FLP gene under a heat-shock promoter is inserted elsewhere in the genome. If such a transgenic embryo or larva is given a heat shock (that is, exposed to a high temperature for a few minutes), expression of FLP is induced, and this enzyme catalyzes crossing-over and recombination between the maternal and paternal chromosomes at the FRT site. If the heat shock is adjusted to be sufficiently mild, this event will occur in only one or a few cells, scattered at random. As explained in Figure 21–48, if the animal is also heterozygous for a gene of interest in the crossed-over chromosomal region, the process can result in a pair of daughter cells that are homozygous, the one receiving two copies of the maternal allele of the gene, the other receiving two copies of the paternal allele. Each of these daughter cells will normally grow and divide to give clonal patches of homozygous progeny.

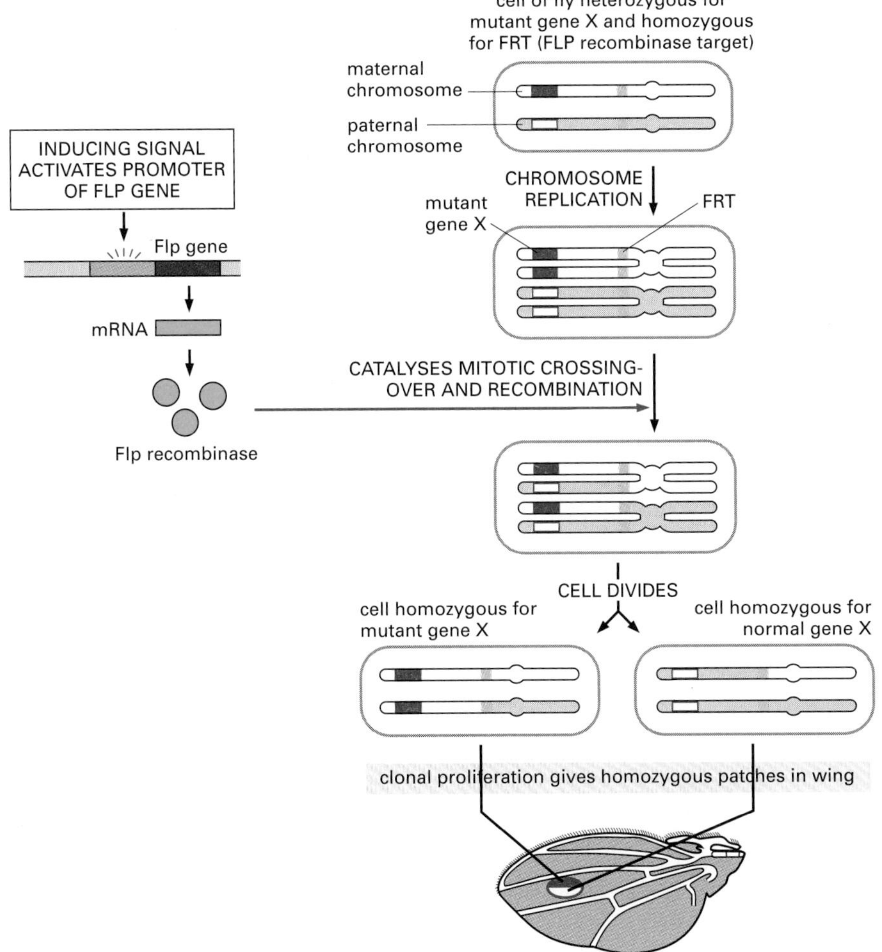

Figure 21–48 Creation of mutant cells by induced somatic recombination. The diagrams follow the fate of a single pair of homologous chromosomes, one from the father *(shaded)*, the other from the mother *(unshaded)*. These chromosomes have an FRT element *(green)* inserted close to their centromere, and contain a locus for a gene of interest—gene X—further out along the same chromosome arm. The paternal chromosome (in this example) carries the wild-type allele of gene X (open red box) while the maternal chromosome carries a recessive mutant allele (filled red box). Recombination by exchange of DNA between the maternal and paternal chromosomes, catalyzed by the FLP recombinase, can give rise to a pair of daughter cells, one containing two wild-type copies of gene X, the other containing two mutant copies. To help identify the cells where recombination has occurred, the maternal and paternal chromosomes can be chosen to carry, respectively, mutant and wild-type versions of an additional gene that provides a visible marker—a pigmentation gene, for example (not shown here). This marker gene is positioned on the chromosome so that recombination involving the marker locus—resulting in a visible alteration in the appearance of the cells—can be taken as a sure sign that gene X has also undergone recombination.

The occurrence of the cross-over can be detected if the animal is chosen to be also heterozygous for a mutation in a marker gene—a pigmentation gene, for example—that lies on the same chromosome arm as the gene of interest and so undergoes crossing over in company with it. In this way clearly marked homozygous mutant clones of cells can be created to order. Either FLP and FRT, or the analogous Cre and Lox pair of recombination elements, can also be used in other configurations to switch expression of a gene on or off (see Figure 5–82). With these techniques, one can discover what happens, for example, when cells are caused to produce a particular signaling molecule at an abnormal site, or are deprived of a particular receptor.

Instead of using a heat-shock promoter to drive expression of the FLP recombinase, one can use a copy of the regulatory sequence of a gene in the fly's normal genome that is expressed at some interesting time and place. The recombination event will then be triggered, and mutant cells created, at just the sites where that gene is normally expressed. A variant of this technique uses transcriptional regulation machinery borrowed from yeast, rather than genetic recombination machinery, to switch expression of a chosen fly gene reversibly on or off according to the normal pattern of expression of some other chosen fly gene (Figure 21–49).

By switching gene functions off or on at specific times and places in these ways, developmental biologists can set about deciphering the system of genetically specified signals and responses that control the patterning of any organ of the body.

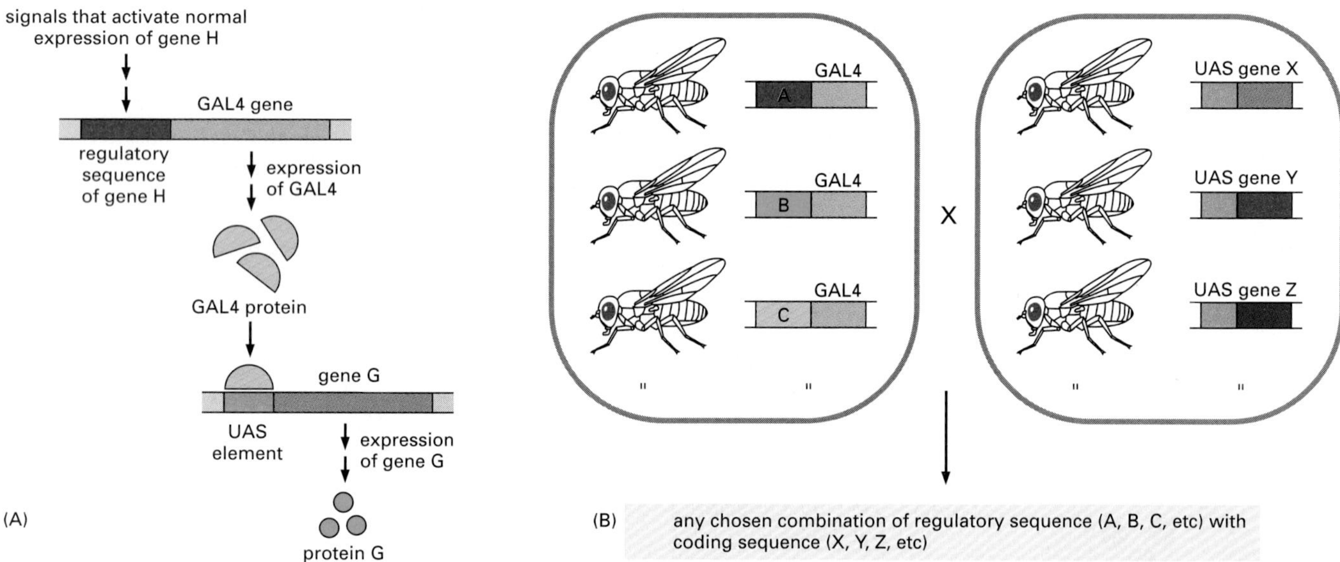

Figure 21–49 The GAL4/UAS technique for controlled gene misexpression in *Drosophila*. The method allows one to drive expression of a chosen gene *G* at the places and times where some other *Drosophila* gene *H* is normally expressed. (A) A transgenic animal is created, with two separate constructs inserted in its genome. One insert consists of a yeast-specific regulatory sequence, called the *UAS* element, coupled to a copy of the coding sequence of gene *G*. The other insert contains the coding sequence of the yeast *GAL4* gene, whose product is a yeast-specific gene regulatory protein that binds to the *UAS* element; this *GAL4* insert is placed next to, and controlled by, the regulatory region of gene H. Wherever gene *H* is normally expressed, GAL4 protein is also made and drives transcription of gene *G*. (B) Although one can achieve the same result by linking a copy of the *H* regulatory sequence directly to the *G* coding sequence, the GAL4/UAS approach allows a strategy that is more efficient in the long run. Two separate "libraries" of transgenic flies are constructed, one containing GAL4 inserts driven by a variety of regulatory sequences of different genes A, B, C, etc., the other containing UAS inserts driving a variety of different coding sequences X, Y, Z, etc. By mating a fly from one library with a fly from the other, any desired coding sequence can be functionally coupled to any desired regulatory sequence.

To generate the library of flies with *GAL4* insertions at useful sites, flies are first produced with *GAL4* insertions at random locations in their genome. These are then mated with flies containing a *UAS* element linked to a reporter gene with an easily detectable product. Expression of the reporter reveals whether *GAL4* has been inserted at a site that brings its expression under the control of an interesting enhancer; flies showing interesting reporter patterns are kept and studied. This is called the *enhancer trap* technique, because it provides a way to hunt out and characterize interesting regulatory sequences in the genome.

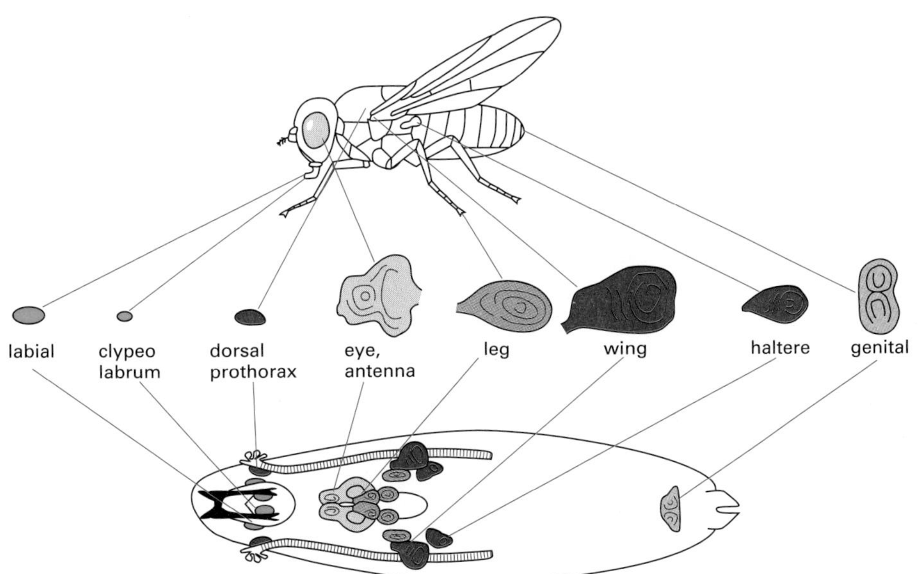

Figure 21–50 The imaginal discs in the *Drosophila* larva and the adult structures they give rise to. (After J.W. Fristrom et al., in Problems in Biology: RNA in Development [E.W. Hanley, ed.], p. 382. Salt Lake City: University of Utah Press, 1969.)

Body Parts of the Adult Fly Develop From Imaginal Discs

The external structures of the adult fly are formed largely from rudiments called **imaginal discs**—groups of cells that are set aside, apparently undifferentiated, in each segment of the larva. The discs are pouches of epithelium, shaped like crumpled and flattened balloons, and continuous with the epidermis (the surface layer) of the larva. There are 19 of them, arranged as 9 pairs on either side of the larva plus 1 disc in the midline (Figure 21–50). They grow and develop their internal pattern as the larva grows, until finally, at metamorphosis, they evert (turn inside out), extend, and differentiate overtly to form the epidermal layer of the adult. The eyes and antennae develop from one pair of discs, the wings and part of the thorax from another, the first pair of legs from another, and so on.

Homeotic Selector Genes Are Essential for the Memory of Positional Information in Imaginal Disc Cells

The cells of one imaginal disc look like those of another, but grafting experiments show that they are in fact already regionally determined and nonequivalent. If one imaginal disc is transplanted into the position of another in the larva and the larva is then left to go through metamorphosis, the grafted disc is found to differentiate autonomously into the structure appropriate to its origin: a wing disc will give wing structures, a haltere disc, haltere structures, regardless of its new site. This shows that the imaginal disc cells are governed by a memory of their original position. By a more complex serial grafting procedure that lets the imaginal disc cells proliferate for an extended period before differentiating, it can be shown that this cell memory is stably heritable (with rare lapses) through an indefinitely large number of cell generations.

The homeotic selector genes are essential components of the memory mechanism. If, at any stage in the long period leading up to differentiation at metamorphosis, both copies of a homeotic selector gene are eliminated by induced somatic recombination from a clone of imaginal disc cells that would normally express that gene, those cells will differentiate into incorrect structures, as though they belonged to a different segment of the body. These and other observations indicate that each cell's memory of positional information depends on the continued activity of the homeotic selector genes. This memory, furthermore, is expressed in a cell-autonomous fashion—each cell appears to maintain its state individually, depending on its own history and genome.

Specific Regulatory Genes Define the Cells That Will Form an Appendage

We must now examine how an appendage develops its internal pattern. We shall take the insect wing as our example.

The process begins with the early patterning mechanisms we have already discussed. The anteroposterior and dorsoventral systems of signals in the early embryo in effect mark out an orthogonal grid in the blastoderm, in the form of dorsoventral, anteroposterior, and periodically spaced segmental gene expression boundaries. At certain points of intersection of these boundaries, the combination of genes expressed is such as to switch a cluster of cells into the imaginal disc pathway.

In molecular terms this corresponds to switching on expression of imaginal-disc-defining regulatory genes. In most of the discs, the gene *Distal-less* is switched on. This codes for a gene regulatory protein that is essential for the sustained growth required to create an elongated appendage such as a leg or an antenna with a proximodistal axis. In its absence, such appendages fail to form, and when it is artificially expressed at abnormal sites, misplaced appendages can be produced. *Distal-less* is expressed in a similar fashion in the developing limbs and other appendages of most species of invertebrates and vertebrates that have been examined (Figure 21–51). For the eye disc, another gene, *eyeless* (together with two closely related genes), performs the corresponding role; it too has homologues with homologous functions—the *Pax-6* genes that drive eye development in other species, as discussed in Chapter 7.

The Insect Wing Disc Is Divided into Compartments

From the outset, the cluster of cells forming the imaginal disc has the rudiments of an internal pattern, inherited from the earlier patterning process. For example, the cells in the posterior half of the wing-disc rudiment (and of most of the other imaginal-disc rudiments) express the segment-polarity gene *engrailed*, while those in the anterior half do not. The initial asymmetries lay the foundations for a subsequent more detailed patterning, just as in the egg and early embryo.

The sectors of the wing disc defined by these early differences of gene expression correspond to specific parts of the future wing. The posterior, *engrailed*-expressing region will form the posterior half of the wing, while the region that does not express *engrailed* will form the anterior half. Meanwhile, the dorsal part of the wing disc expresses a gene called *apterous*, while the ventral half does not. At metamorphosis, the disc folds along the line separating these domains to give a wing whose dorsal sheet of cells is derived from the *apterous*-expressing region and whose ventral sheet is derived from the region that does not express *apterous*. The wing margin, where these two epithelial sheets are joined, corresponds to the boundary of the *apterous* expression domain in the disc (Figure 21–52).

The cells of the disc, having switched on expression of the genes that mark them as anterior or posterior, dorsal or ventral, retain this specification as the

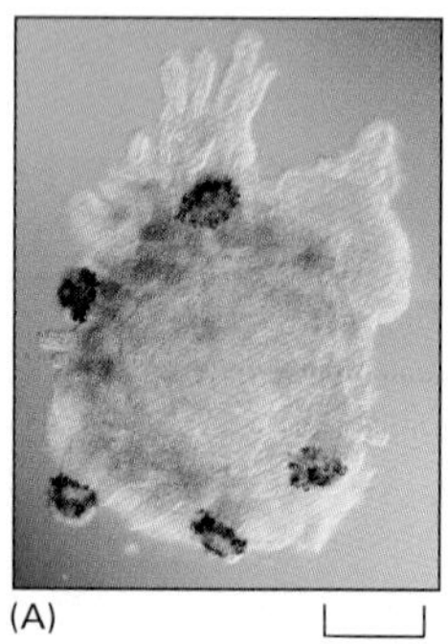

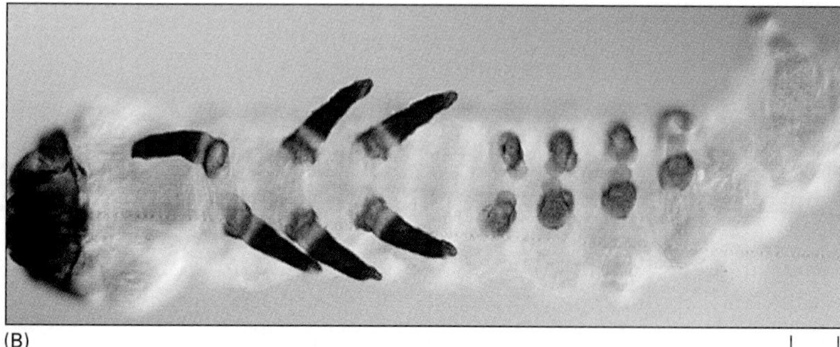

Figure 21–51 Expression of Distal-less in developing legs and related appendages of various species. (A) A sea-urchin larva. (B) A moth larva. (A, from G. Panganiban et al., *Proc. Natl. Acad. Sci. USA* 94:5162–5166, 1997. © National Academy of Sciences; B, from G. Panganiban, L. Nagy, and S.B. Carroll, *Curr. Biol.* 4:671–675, 1994. © Elsevier.)

Figure 21–52 Gene expression domains in the wing imaginal disc, defining quadrants of the future wing. The wing blade itself derives from the oval-shaped domain toward the *right*, and it is divided into four quadrants by the expression of *apterous* and *engrailed*, as shown.

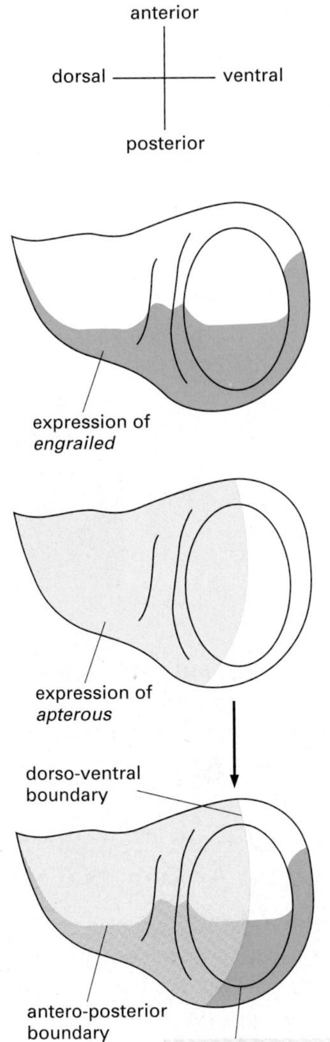

disc grows and develops. Because the cells are sensitive to these differences and selective in their choice of neighbors, sharply defined boundaries are formed between the four resultant sets of cells, with no mixing at the interfaces. The four corresponding quadrants of the disc are called **compartments**, because there is no exchange of cells between them (Figure 21–53).

Four Familiar Signaling Pathways Combine to Pattern the Wing Disc: Wingless, Hedgehog, Dpp, and Notch

Along each of the compartment boundaries—the anteroposterior boundary defined by *engrailed* and the dorsoventral boundary defined by *apterous*—cells in different states confront one another and interact to create narrow bands of specialized cells. These boundary cells produce new signals to organize the subsequent growth and more detailed patterning of the appendage.

Cells in the posterior wing compartment express the Hedgehog signaling protein, but cannot respond to it. Cells in the anterior compartment can respond to Hedgehog. Because Hedgehog acts only over a short distance, the signal reception pathway is activated only in the narrow band of cells just anterior to the compartment boundary, where anterior and posterior cells are juxtaposed. These boundary cells respond by switching on expression of another signaling molecule, Dpp—the same protein that we encountered previously, in the dorsoventral patterning of the early embryo (Figure 21–54). Dpp acts in its new context in much the same way as before: it is thought to diffuse (or somehow spread its effects) outward from the boundary cells, setting up a morphogen gradient to control the subsequent detailed pattern of growth and gene expression.

Analogous events occur at the dorsoventral compartment boundary (see Figure 21–54). Here, at the future wing margin, short-range communication mediated by the Notch pathway creates a band of boundary cells that produce another morphogen, the Wingless protein—the same signaling factor, belonging to the Wnt family, that acted earlier in the anteroposterior patterning of each

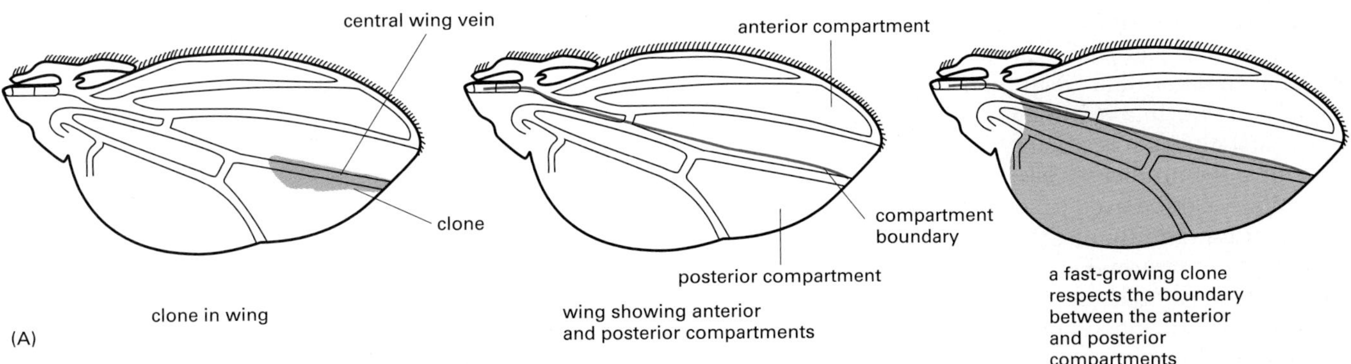

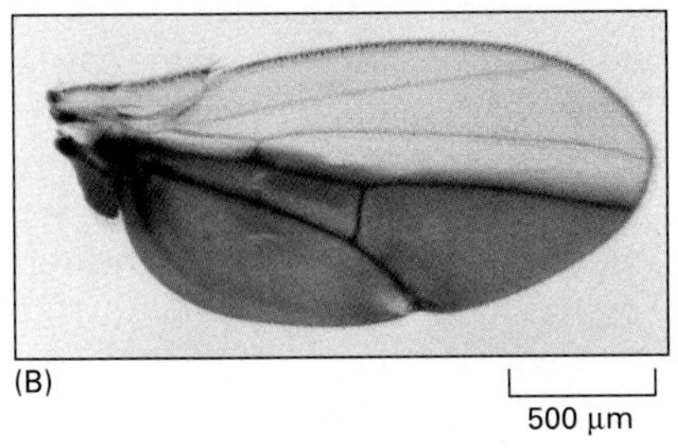

Figure 21–53 Compartments in the adult wing. (A) The shapes of marked clones in the *Drosophila* wing reveal the existence of a compartment boundary. The border of each marked clone is straight where it abuts the boundary. Even when a marked clone has been genetically altered so that it grows more rapidly than the rest of the wing and is therefore very large, it respects the boundary in the same way *(drawing on right)*. Note that the compartment boundary does not coincide with the central wing vein. (B) The pattern of expression of the *engrailed* gene in the wing, revealed by the same technique as for the adult fly shown in Figure 21–40. The compartment boundary coincides with the boundary of *engrailed* gene expression. (A, after F.H.C. Crick and P.A. Lawrence, *Science* 189:340–347, 1975. © AAAS; B, courtesy of Chihiro Hama and Tom Kornberg.)

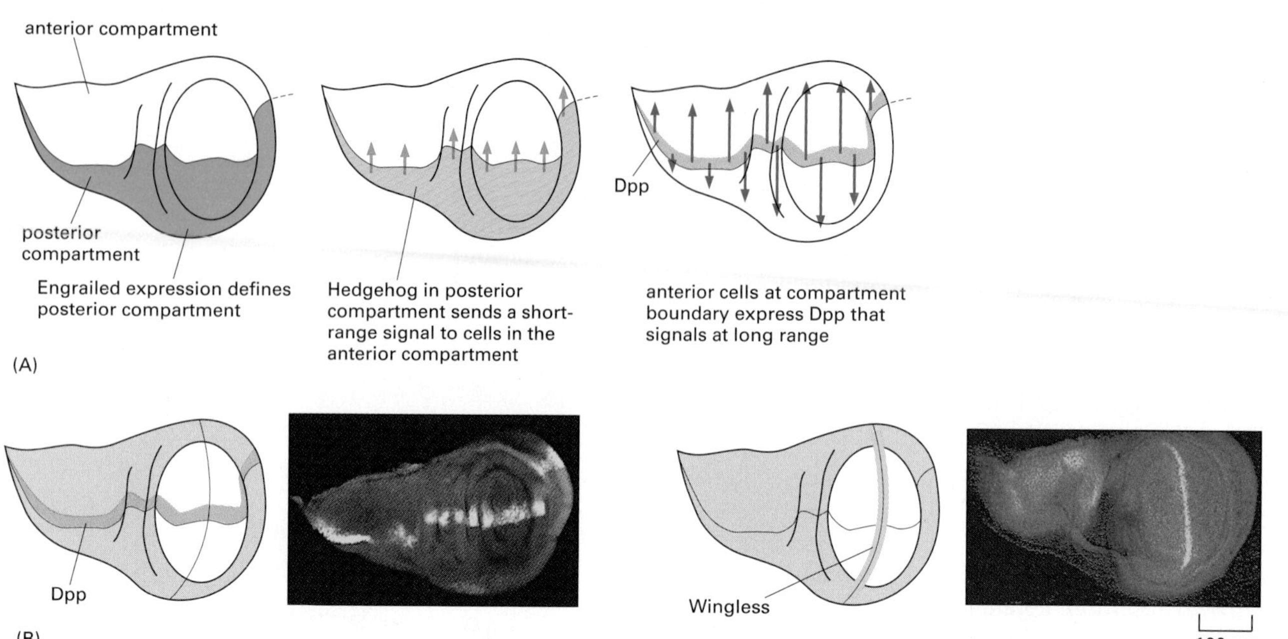
anterior compartment
posterior compartment
Engrailed expression defines posterior compartment
Hedgehog in posterior compartment sends a short-range signal to cells in the anterior compartment
Dpp
anterior cells at compartment boundary express Dpp that signals at long range
(A)
Dpp
Wingless
(B)
100 μm

Figure 21–54 Morphogenetic signals created at compartment boundaries in the wing imaginal disc. (A) Creation of the Dpp signaling region at the anteroposterior compartment boundary through a Hedgehog-mediated interaction between the anterior and posterior cells. In an analogous way, a Notch-mediated interaction between dorsal and ventral cells creates a Wingless (Wnt) signaling region along the dorsoventral boundary. (B) The observed expression patterns of Dpp and Wingless. Although it seems clear that Dpp and Wingless act as morphogens, it is not yet certain whether they spread out from their source by simple diffusion through the extracellular medium or in some other way. Moreover, cells in the imaginal disc have been seen to send out long protrusions, called *cytonemes*, that may allow them to sense signals at a distance. Thus, the receiving cell may send its sensors to the source of the signal, instead of the signal moving to the receiving cell. (B, photographs courtesy of Sean Carroll and Scott Weatherbee, from S.J. Day and P.A. Lawrence, *Development* 127:2977–2987, 2000. © The Company of Biologists.)

embryonic segment. The Dpp and Wingless gradients, together with the other signals and asymmetries of gene expression that we have discussed, combine to drive expression of other genes at precisely defined locations within each compartment.

The Size of Each Compartment Is Regulated by Interactions Among Its Cells

One of the most mysterious and ill-understood aspects of animal development is the control of growth: why does each part of the body grow to a precisely defined size? This problem is exemplified in remarkable way in the imaginal discs of *Drosophila*. By induced somatic recombination, one can, for example, create a clonal patch of cells that proliferate more rapidly than the rest of the cells in the developing organ. The clone may grow to occupy almost the whole of the compartment in which it lies, and yet it does not overstep the boundary of the compartment. Astonishingly, its rapid growth has almost no effect on the compartment's final size, its shape, or even the details of its internal pattern (see Figure 21–53). Somehow, the cells within the compartment interact with one another to determine when their growth should stop, and each compartment behaves as a regulatory unit in this respect.

A first question is whether the size of the compartment is regulated so as to contain a set number of cells. Mutations in components of the cell-cycle control machinery can be used to speed up or slow down the rate of cell division without altering the rate of cell or tissue growth. This results in abnormally large numbers of abnormally small cells, or the converse, but the size—that is, the area—of the compartment is practically unchanged. Thus, the regulatory mechanism seems to depend on signals that indicate the physical distance between one part of the compartment and another, and on cellular responses that somehow read these signals so as to halt growth only when the tissue has attained the correct area.

One clue to how the system works comes from the observation that flies with genetic defects in the signaling pathway mediated by insulin and insulin-like factors are small, with cells that are both small and reduced in numbers, while overactivity of this pathway can produce giant flies, with more and bigger cells. Localized misexpression of the same genes in a single compartment has similar effects on just that compartment. Since insulin is widely used in animals as a regulator of responses to nutrition, the mechanisms controlling the sizes of compartments and organs may have evolved from mechanisms that control cell growth and proliferation according to nutritional conditions.

The deeper problem remains, however: what mechanism ensures that each little piece of the pattern within a compartment grows to its appropriate size, despite local disturbances in growth rate or starting conditions? The morphogen gradients (of Dpp and Wingless, for example) create a pattern by imposing different characters on cells in different positions. Could it be that the cells in each region can somehow sense how close the spacing of the pattern is—how steep the gradient of change in cell character—and continue their growth until the tissue is spread out to the right degree? A striking demonstration of the phenomenon that needs to be understood is seen in the **intercalary regeneration** that occurs when separate parts of a *Drosophila* imaginal disc or of a growing cockroach leg are surgically grafted together. After the graft, the cells in the neighbourhood of the junction proliferate and fill in the parts of the pattern that should normally lie between them, continuing their growth until the normal spacing between landmarks is restored (Figure 21–55). The mechanisms that bring this about are a mystery, but it seems likely that they are similar to the mechanisms that regulate growth during normal development.

Similar Mechanisms Pattern the Limbs of Vertebrates

The limbs of vertebrates seem very different from those of insects. The insect wing, for example, consists mainly of two elaborately patterned sheets of epithelium, with very little tissue in between. In contrast, a limb of a vertebrate consists of an elaborately patterned system of muscles, bones and other connective tissues inside a thin and much more simply structured covering of epidermis. Moreover, the evolutionary evidence suggests that the last common ancestor of insects and vertebrates may have had neither legs, nor arms, nor wings, nor fins and that we have evolved these various appendages independently. And yet, when we examine the molecular mechanisms that control vertebrate limb development, we find a surprising number of similarities with the limbs of insects. We have already mentioned some of these resemblances, but there are many others: almost all the molecules we have already mentioned in the fly wing have their counterparts in the vertebrate limb, although these are expressed in different spatial relationships.

The parallels have been most thoroughly studied in the chick embryo. As we saw earlier, each leg or wing of a chick originates from a tongue-shaped limb bud, consisting of a mass of embryonic connective tissue cells, called mesenchyme cells, encased in a jacket of epithelium. In this structure, one finds expression of homologs of almost all the genes that we have mentioned in our account of *Drosophila* wing patterning, including *Distal-less, wingless, Notch, engrailed, dpp,* and *hedgehog,* mostly performing functions that seem more or less similar to their functions in the *Drosophila* wing disc (Figure 21–56).

The Hox genes likewise make an appearance in the limbs of both insects and vertebrates. In the insect appendage, the anterior and posterior compartments are distinguished by expression of different genes of the Hox complex—a result of the serial expression pattern of these genes along the anteroposterior axis of the body as a whole. In the vertebrate limb, genes of two of the vertebrate Hox complexes (HoxA and HoxD) are expressed in a regular pattern, obedient to the usual rules of serial expression of genes in these complexes. They help, in conjunction with other factors such as the Tbx proteins mentioned earlier (see Figure 21–9), to regulate differences of cell behavior along the proximodistal limb axis.

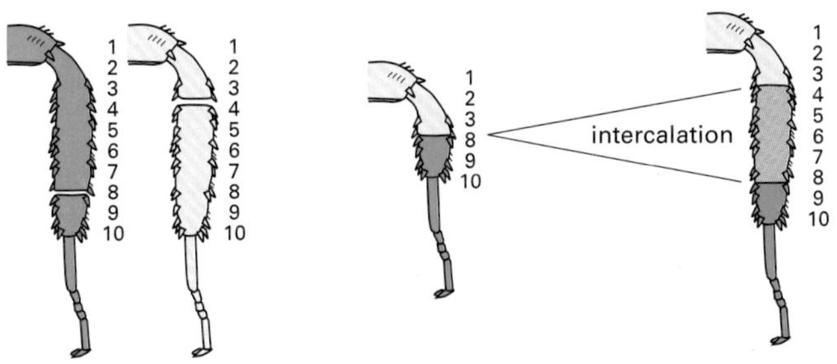

Figure 21–55 Intercalary regeneration. When mismatched portions of the growing cockroach leg are grafted together, new tissue *(green)* is intercalated (by cell proliferation) to fill in the gap in the pattern of leg structures, restoring a leg segment of normal size and pattern.

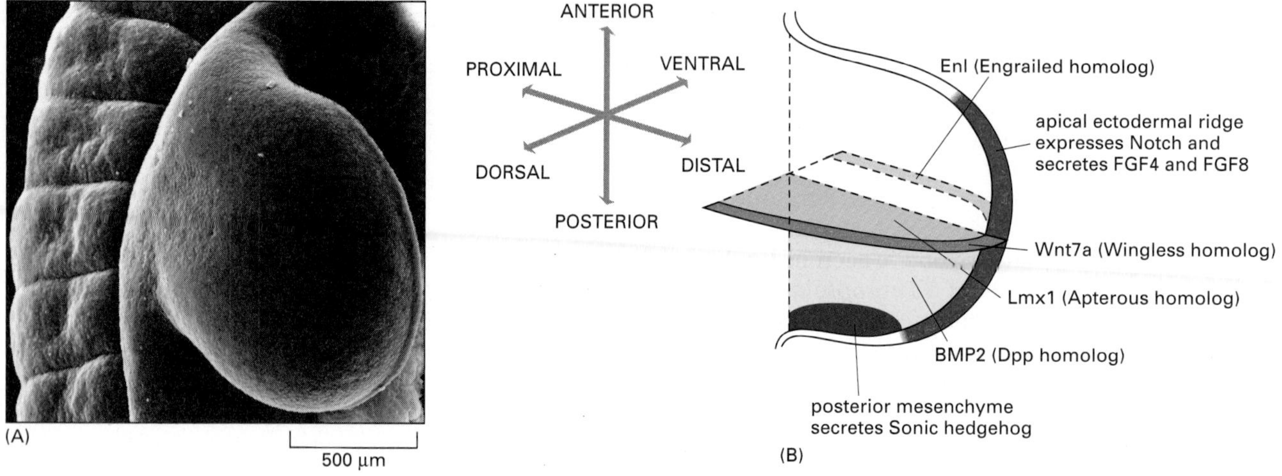

Figure 21–56 Molecules that control patterning in a vertebrate limb bud. (A) A wing bud of a chick embryo at 4 days of incubation. The scanning electron micrograph shows a dorsal view, with somites (the segments of the trunk of the embryo) visible to the *left*. At the distal margin of the limb bud a thickened ridge can just be seen—the apical ectodermal ridge. (B) Expression patterns of key signaling proteins and gene regulatory factors in the chick limb bud. The patterns are depicted schematically in two imaginary planes of section through the limb bud, one (horizontal) to show the dorsoventral system and the other (vertical) to show the anteroposterior and proximodistal systems. Sonic hedgehog, BMP2, and Lmx1 are expressed in the mesodermal core of the limb bud; the other molecules in the diagram are expressed in its epithelial covering. Almost all the molecules shown have homologs that are involved in patterning the *Drosophila* wing disc. (A, courtesy of Paul Martin.)

According to one view, these molecular resemblances between developing limbs in different phyla reflect descent from a common ancestor that, while lacking limbs, had appendages of some sort built on similar principles—antennae, perhaps, or protruding mouthparts for snatching food. Modern limblike appendages, from the wings and legs of the fly to the arms and legs of a human, would then have evolved through activation of the genes for appendage formation at new sites in the body, as a result of changes in gene regulation.

Localized Expression of Specific Classes of Gene Regulatory Proteins Foreshadows Cell Differentiation

We now pick up again the thread of development in the *Drosophila* imaginal disc and follow it through to the final step at which cells become terminally differentiated. Narrowing our focus further, we take as our example the differentiation of just one type of small structure that arises in the imaginal disc epithelium: the **sensory bristle**.

The bristles that cover the body surface of an insect are miniature sense organs. Some respond to chemical stimuli, others to mechanical stimuli, but they are all constructed in a similar way. The structure is seen at its simplest in the mechanosensory bristles. Each of these consists of four cells: a shaft cell, a socket cell, a neural sheath cell, and a neuron (Figure 21–57). Movement of the shaft of the bristle excites the neuron, which sends a signal to the central nervous system.

The cells of the bristle of the adult fly derive from the imaginal disc epithelium, and all four of them are granddaughters or great-granddaughters (see Figure 21–57) of a single *sensory mother cell* that becomes distinct from the neighboring prospective epidermal cells during the last larval instar (Figure 21–58). (A fifth descendant migrates away from the rest to become a glial cell.) To account for the pattern of bristle differentiation, we have to explain first how the genesis of sensory mother cells is controlled and then how the five descendants of each such cell become different from one another.

Two genes, called *achaete* and *scute*, are crucial in initiating the formation of bristles in the imaginal disc epithelium. These genes have similar and overlapping functions and code for closely related gene regulatory proteins of the basic helix–loop–helix class (see Chapter 7). As a result of disc-patterning mechanisms of the type we have already discussed, *achaete* and *scute* are expressed in the imaginal disc in the regions within which bristles will form. Mutations that eliminate the expression of these genes at some of their usual sites block development of bristles at just those sites, and mutations that cause expression in additional, abnormal sites cause bristles to develop there. But expression of *achaete* and *scute* is transient, and only a minority of the cells initially expressing the genes go on to become sensory mother cells; the others become ordinary epidermis. The state that is specified by expression of *achaete* and *scute* is called

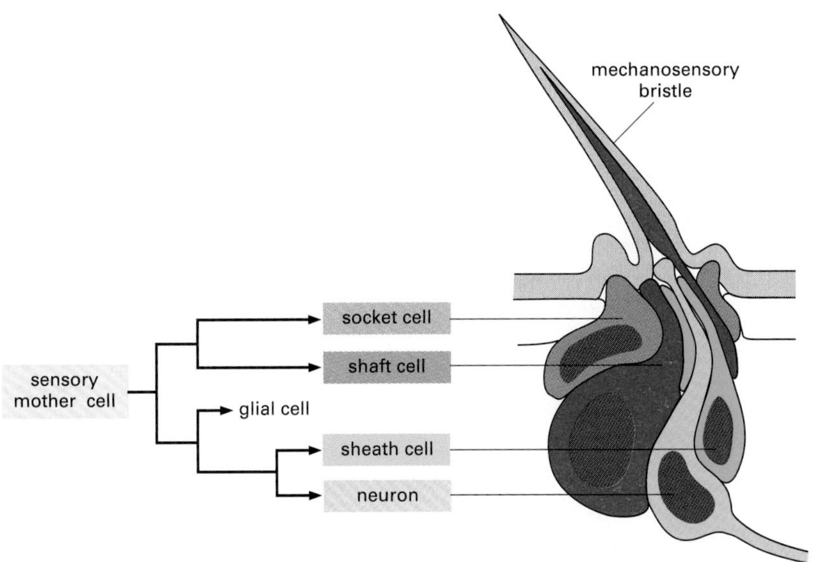

Figure 21–57 The basic structure of a mechanosensory bristle. The lineage of the four cells of the bristle—all descendants of a single sensory mother cell—is shown on the *left*.

proneural, and *achaete* and *scute* are called **proneural genes**. The proneural cells are primed to take the neurosensory pathway of differentiation, but, as we shall see, which of them will actually do so depends on competitive interactions among them.

Lateral Inhibition Singles Out Sensory Mother Cells Within Proneural Clusters

Cells expressing the proneural genes occur in groups in the imaginal disc epithelium—a small, isolated cluster of fewer than 30 cells for a big isolated bristle, a broad, continuous patch of hundreds or thousands of cells for a field of small bristles. In the former case just one member of the cluster becomes a sensory mother cell; in the latter case many cells scattered throughout the proneural region do so. In either case, each sensory mother cell becomes surrounded by cells that switch off expression of the proneural genes and become condemned to differentiate as epidermis instead. Experiments with genetic mosaics show that this is because a cell that becomes committed to the sensory-mother-cell pathway of differentiation sends a signal to its neighbors not to do the same thing: it exerts a *lateral inhibition*. If a cell that would normally become a sensory mother is genetically disabled from doing so, a neighboring proneural cell, freed from lateral inhibition, will become a sensory mother cell instead.

The lateral inhibition is mediated by the Notch signaling pathway. The cells in the cluster initially all express both the transmembrane receptor Notch and its transmembrane ligand Delta. Wherever Delta activates Notch, an inhibitory signal is sent into the Notch-expressing cell; consequently, all the cells in the cluster initially inhibit one another. However, receipt of the signal in a given cell is thought to diminish not only that cell's tendency to specialize as a sensory mother cell, but also its ability to fight back by delivering the inhibitory Delta signal in return. This creates a competitive situation, from which a single cell in each small region—the future sensory mother cell—emerges as winner, sending a strong inhibitory signal to its immediate neighbors but receiving no such signal in return (Figure 21–59). The consequences of a failure of this regulatory mechanism are shown in Figure 21–60.

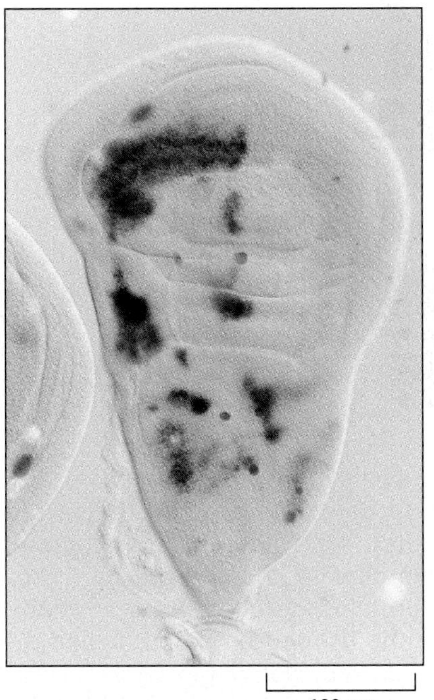

Figure 21–58 Sensory mother cells in the wing imaginal disc. The sensory mother cells (*blue* here) are easily revealed in this special strain of *Drosophila*, which contains an artificial *LacZ* reporter gene that, by chance, has inserted itself in the genome next to a control region that causes it to be expressed selectively in sensory mother cells. The *purple* stain shows the expression pattern of the *scute* gene; this foreshadows the production of sensory mother cells and fades as the sensory mother cells successively develop. (From P. Cubas et al., *Genes Dev.* 5:996–1008, 1991. © Cold Spring Harbor Press.)

Lateral Inhibition Drives the Progeny of the Sensory Mother Cell Toward Different Final Fates

The same lateral inhibition mechanism dependent on Notch operates repeatedly in the formation of bristles—not only to force the neighbors of sensory mother cells to follow a different pathway and become epidermal, and but also later to

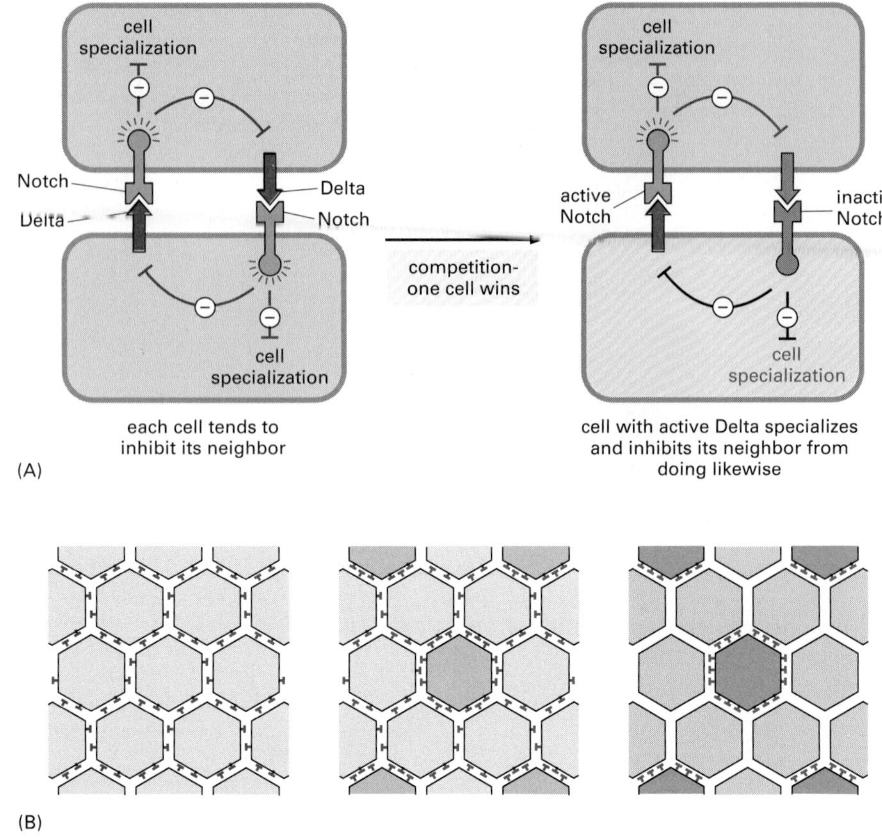

Figure 21–59 Lateral inhibition. (A) The basic mechanism of Notch-mediated competitive lateral inhibition, illustrated for just two interacting cells. In this diagram, the absence of color on proteins or effector lines indicates inactivity. (B) The outcome of the same process operating in a larger patch of cells. At first, all cells in the patch are equivalent, expressing both the transmembrane receptor Notch and its transmembrane ligand Delta. Each cell has a tendency to specialize (as a sensory mother cell), and each sends an inhibitory signal to its neighbors to discourage them from also specializing in that way. This creates a competitive situation. As soon as an individual cell gains any advantage in the competition, that advantage becomes magnified. The winning cell, as it becomes more strongly committed to differentiating as a sensory mother, also inhibits its neighbors more strongly. Conversely, as these neighbors lose their capacity to differentiate as sensory mothers they also lose their capacity to inhibit other cells from doing so. Lateral inhibition thus makes adjacent cells follow different fates. Although the interaction is thought to be normally dependent on cell–cell contacts, the future sensory mother cell may be able to deliver an inhibitory signal to cells that are more than one cell diameter away—for example, by sending out long protrusions to touch them.

make the daughters, the granddaughters, and finally the great-granddaughters of the sensory mother cell express different genes so as to form the different components of the bristle. At each stage, lateral inhibition mediates a competitive interaction that forces adjacent cells to behave in contrasting ways. Using a temperature-sensitive Notch mutation, it is possible to switch off Notch signaling after the sensory mother cell has been singled out, but before it has divided. The progeny then differentiate alike, giving a cluster of neurons in place of the four different cell types of a bristle.

Like many other competitions, those mediated by lateral inhibition are often rigged: one cell starts with an advantage that guarantees it will be the winner. In the development of the different cell types of the sensory bristle, a strong initial bias is provided by an asymmetry in each of the cell divisions of the sensory mother cell and its progeny. A protein called Numb (together with certain other proteins) becomes localized at one end of the dividing cell, so that one daughter inherits the Numb protein and the other does not (Figure 21–61). Numb interacts with Notch, blocking its activity. Thus the Numb-containing cell is deaf to inhibitory signals from its neighbors while its sister remains sensitive. Since both cells initially express the Notch ligand Delta, the cell that has inherited Numb proceeds to become neural, while driving its sister toward a nonneural fate.

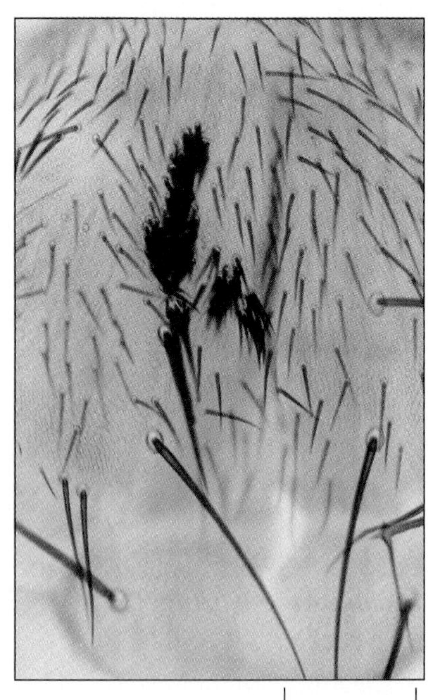

Figure 21–60 The result of switching off lateral inhibition during the singling-out of sensory mother cells. The photograph shows part of the thorax of a fly containing a mutant patch in which the neurogenic gene *Delta* has been partially inactivated. The reduction of lateral inhibition has caused almost all the cells in the mutant patch *(in the center of the picture)* to develop as sensory mother cells, producing a great excess of sensory bristles there. Mutant patches of cells carrying more extreme mutations in the Notch pathway, causing a total loss of lateral inhibition, form no visible bristles because all of the progeny of the sensory mother cells develop as neurons or glial cells instead of diversifying to form both neurons and the external parts of the bristle structure. (Courtesy of P. Heitzler and P. Simpson, *Cell* 64:1083–1093, 1991. © Elsevier.)

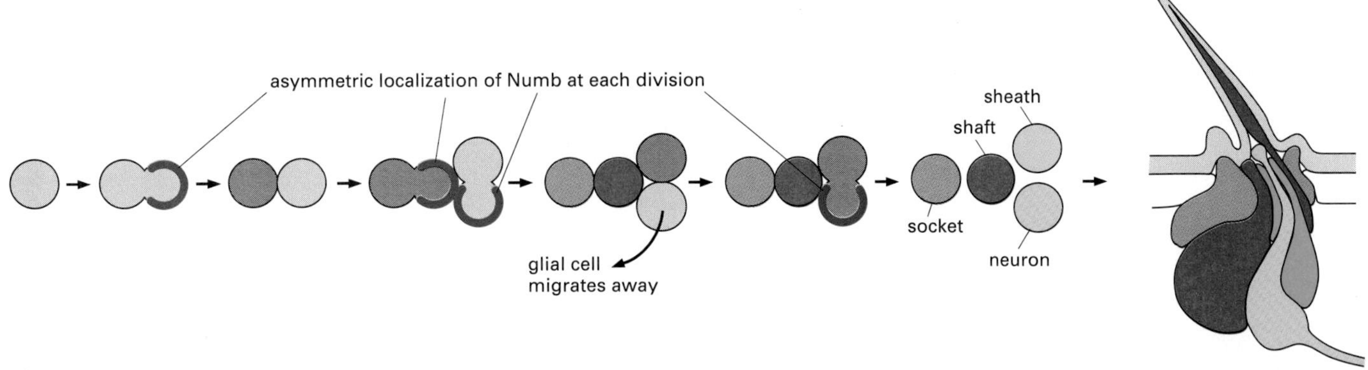

Figure 21–61 Numb biases lateral inhibition during bristle development. At each division of the progeny of the sensory mother cell, Numb protein is asymmetrically localized, producing daughter cells that differ. Note that some of the divisions are oriented with the mitotic spindle in the plane of the epithelium, others at right angles to it; the localization of Numb is controlled in different ways at these different types of division but plays a critical role at each of them in deciding cell fate. (Based on data from M. Gho, Y. Bellaiche, and F. Schweisguth, *Development* 126:3573–3584, 1999.)

Planar Polarity of Asymmetric Divisions is Controlled by Signaling via the Receptor Frizzled

For the Numb mechanism to operate, there must be machinery in the dividing cell to segregate the determinant to one side of the cell before division. In addition, as the cell enters mitosis the mitotic spindle must be aligned with this asymmetry so that the determinant is allocated to just one daughter cell, and not shared out to both daughters at the time of cell division. In the above case, the sensory mother cell, at its first division, regularly divides to give an anterior cell that inherits Numb and a posterior cell that does not. This type of polarity in the plane of the epithelium is called *planar polarity* (in contradistinction to apico-basal polarity, where the cellular asymmetry is perpendicular to the plane of the epithelium). It is manifested in the uniformly backward-pointing orientation of the bristles, giving the fly its wind-swept appearance (Figure 21–62).

The planar polarity in the initial division of the sensory mother cell is controlled by a signaling pathway similar to the one that we encountered controlling asymmetric divisions in the nematode (see Figure 21–20), depending on the receptor Frizzled. Frizzled proteins have been discussed in Chapter 15 as receptors for Wnt proteins, but in the control of planar polarity—in flies and probably in vertebrates too—this pathway functions in a special way: the intracellular relay mechanism exerts its main effects on the actin cytoskeleton, rather than on gene expression. The intracellular protein Dishevelled, downstream from Frizzled, is common to the gene-regulatory and the actin-regulatory branches of the signaling pathway. Separate domains of the Dishevelled molecule can be shown to be responsible for the two functions (Figure 21–63). Frizzled and Dishevelled both take their names from the unkempt look of flies where bristle polarity is disordered.

Figure 21–62 Planar cell polarity manifest in bristle polarity on a fly's back: the bristles all point backwards. (Scanning electron micrograph courtesy of S. Oldham and E. Hafen, from E. Spana and N. Perrimon, *Trends Genet.* 15:301–302, 1999. © Elsevier.)

Lateral Inhibition and Asymmetric Division Combine to Regulate Genesis of Neurons Throughout the Body

The mechanisms we have described for controlling the genesis of neurons of sensory bristles operate also, with minor variations, in the genesis of virtually all other neurons—not only in insects, but also in other phyla. Thus in the embryonic central nervous system, both in flies and in vertebrates, neurons are generated in regions of expression of proneural genes akin to *achaete* and *scute*. The nascent neurons or neuronal precursors express Delta and inhibit their immediate neighbors, which express Notch, from becoming committed to neural differentiation at the same time. When Notch signaling is blocked, inhibition fails,

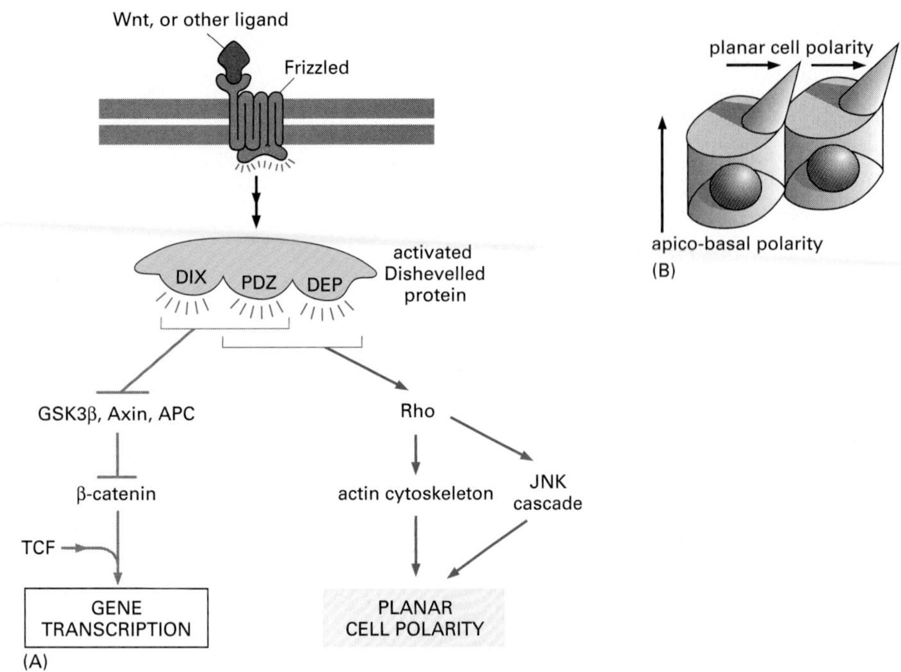

Figure 21–63 The control of planar cell polarity. (A) The two branches of the Wnt/Frizzled signaling pathway. The main branch, discussed in Chapter 15, controls gene expression via β-catenin; the planar-polarity branch controls the actin cytoskeleton via Rho GTPases. Different domains of the Dishevelled protein are responsible for the two effects. It is not yet clear which member of the Wnt signal protein family, if any, is responsible for activating the planar polarity function of Frizzled in *Drosophila*. (B) Cartoon of cells displaying planar polarity. In at least some systems, planar cell polarity is associated with asymmetric localization of the receptor Frizzled itself to one side of each cell.

and in the proneural regions neurons are generated in huge excess at the expense of non-neuronal cells (Figure 21–64).

In many of these processes of neurogenesis, as in the development of the sensory bristle, asymmetric cell divisions play an important part, although the details, and the relationship to Notch signaling, are variable. Thus, in the embryonic central nervous system of *Drosophila,* the nerve-cell precursors, or *neuroblasts,* are singled out from the neurogenic ectoderm by a typical lateral-inhibition mechanism that depends on Notch, but then undergo asymmetric divisions in which the cells are polarized apico-basally. The localization of a set of neuronal cell-fate determinants at one end of the neuroblast is ultimately directed, in this case, by Bazooka, an apically localized protein that has a fundamental role in defining the apico-basal polarity of epithelia in general and is also involved in neurogenesis in the central nervous system of vertebrates. Bazooka is homologous to Par-3, one of the Par proteins that govern the asymmetric cell divisions in the early nematode embryo.

Notch Signaling Regulates the Fine-Grained Pattern of Differentiated Cell Types in Many Different Tissues

The process of lateral inhibition and cell diversification, initiated by expression of proneural genes and mediated by Notch, has turned out to be crucial for the fine-grained patterning of an enormous variety of different tissues. In the fly, it

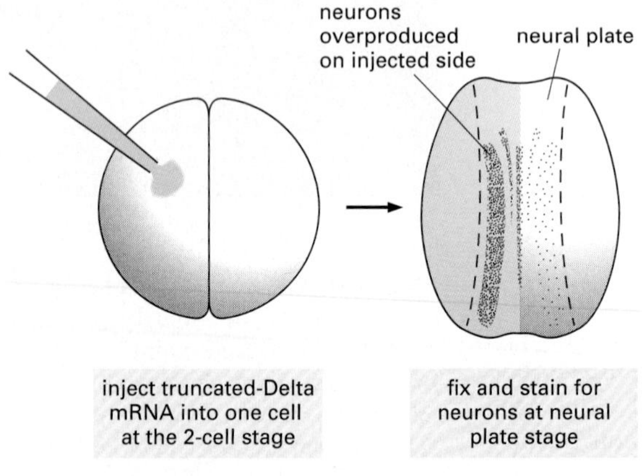

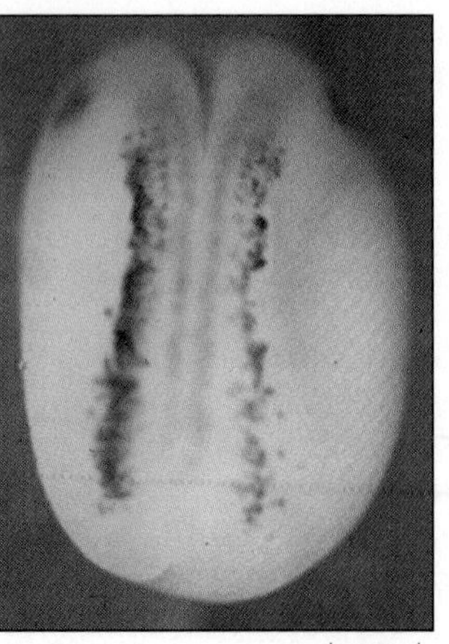

Figure 21–64 Effects of blocking Notch signaling in a *Xenopus* embryo. In the experiment shown, mRNA coding for a truncated form of the Notch ligand Delta is injected, together with *LacZ* mRNA as a marker, into one cell of an embryo at the two-cell stage. The truncated Delta protein produced from the mRNA blocks Notch signaling in the cells descended from the cell that received the injection. These cells lie on the *left* side of the embryo and are identifiable because they contain LacZ protein *(blue* stain) as well as the truncated Delta protein. The *right* side of the embryo is unaffected and serves as a control. The embryo is fixed and stained at a stage when the central nervous system has not yet rolled up to form a neural tube, but is still a more or less flat plate of cells—the neural plate—exposed on the surface of the embryo. The first neurons (stained *purple* in the photograph) have already begun to differentiate in elongated bands (proneural regions) on each side of the midline. On the control *(right)* side, they are a scattered subset of the proneural cell population. On the Notch-blocked *(left)* side, virtually all the cells in the proneural regions have differentiated as neurons, creating a densely stained band of neurons without intervening cells. Injections of mRNA coding for normal, functional Delta have an opposite effect, reducing the number of cells that differentiate as neurons. (Photograph from A. Chitnis et al., *Nature* 375:761–766, 1995. © Macmillan Magazines Ltd.)

controls the production not only of neurons but also of many other differentiated cell types—for example, in muscle, in the lining of the gut, in the excretory system, in the tracheae, and in the eye and other sense organs. In vertebrates, homologs of the proneural genes and of Notch and its ligands are expressed in the corresponding tissues and have similar functions: mutations in the Notch pathway upset the balance not only of neurons and non-neuronal cells in the central nervous system, but also of the different specialized cell types in the lining of the gut, of endocrine and exocrine cells in the pancreas, and of sensory and supporting cells in sense organs such as the ear, to give only a few examples.

In all these tissues, a balanced mixture of different cell types is required. Notch signaling provides the means to generate the mixture, by enabling individual cells expressing one set of genes to direct their immediate neighbors to express another set.

Some Key Regulatory Genes Define a Cell Type; Others Can Activate the Program for Creation of an Entire Organ

The final choice of a particular mode of differentiation is marked by expression of specific cell-differentiation genes in the aftermath of the interactions mediated by Notch. Each cell type must express a whole collection of genes to perform its differentiated function, but the expression of these genes is coordinated by a much smaller set of high-level regulators. These regulators are sometimes called "master regulatory proteins" (though even they can exert their specific effect only in combination with the right partners, in a cell that is adequately primed). An example is the MyoD family of gene regulatory proteins (MyoD, myogenin, Myf5, MRF4 and their homologs in invertebrates). These proteins drive cells to differentiate as muscle, expressing muscle-specific actins and myosins and all the other cytoskeletal, metabolic and membrane proteins that a muscle cell needs (see Figure 7–72).

The gene regulatory proteins that define particular cell types often belong (as do MyoD and its relatives) to the basic helix–loop–helix family, encoded by genes homologous to, and in some cases apparently identical to, the proneural genes that initiate the final phase of development. Their expression is often governed by the Notch pathway via complicated feedback loops.

Terminal cell differentiation has brought us to the end of our sketch of how genes control the making of a fly. Our account necessarily has been simplified. Many more genes than we have mentioned are involved in each of the developmental processes that we have described. Feedback loops, alternative mechanisms operating in parallel, genetic redundancy, and other phenomena complicate the full picture. Despite all this, the overriding message of developmental genetics is one of an unexpected simplicity. A limited number of genes and mechanisms, used repeatedly in different circumstances and combinations, are responsible for controlling the main features of the development of all multicellular animals.

We next turn to an essential aspect of animal development that we have so far neglected: controlled cell movements.

Summary

The external parts of an adult fly develop from epithelial structures called imaginal discs. Each imaginal disc is divided at the outset into a small number of domains expressing different gene regulatory proteins as a result of early embryonic patterning processes. These domains are called compartments, because their cells do not mix. At the compartment boundaries, cells expressing different genes confront one another and interact, inducing localized production of morphogens that govern the further growth and internal patterning of each compartment. Thus, in the wing disc, dorsal and ventral cells interact by the Notch signaling mechanism to create a source of Wingless (Wnt) protein along the dorsoventral compartment boundary, while anterior and posterior cells interact through short-range Hedgehog signaling to create a source of Dpp protein (a TGFβ family member) along the anteroposterior

compartment boundary. All these signaling molecules have homologs that play similar parts in limb patterning in vertebrates.

Within each compartment, the morphogen gradients control the sites of expression of further sets of genes, defining patches of cells that interact with one another yet again to create the finest details of the ultimate pattern of cell differentiation. Thus, proneural gene expression defines the sites where sensory bristles will form, and Notch-mediated interactions among the cells of the proneural cluster, together with asymmetric cell divisions, force the individual cells of the bristle to follow different paths of terminal differentiation.

Each compartment of an imaginal disc, and each substructure within it, grows to a precisely predictable size, even in the face of seemingly drastic disturbances, such as mutations that alter the cell division rate. Although the morphogen gradients in the disc are clearly involved, the critical regulatory mechanisms that control organ size are not understood.

CELL MOVEMENTS AND THE SHAPING OF THE VERTEBRATE BODY

Most cells of the animal body are motile, and in the developing embryo their movements are often extensive, dramatic, and surprising. Controlled changes of gene expression create ordered arrays of cells in different states; cell movements rearrange these cellular building blocks and put them in their proper places. The genes that the cells express determine how they move; in this sense, the control of gene expression is the primary phenomenon. But the cell movements are also crucial, and no less in need of explanation if we want to understand how the architecture of the body is created. In this section, we examine this topic in the context of vertebrate development. We take as our main example the frog *Xenopus laevis* (Figure 21–65), where cell movements have been well studied, though we shall also draw on evidence from chick, zebrafish and mouse.

The Polarity of the Amphibian Embryo Depends on the Polarity of the Egg

The *Xenopus* egg is a large cell, just over a millimeter in diameter (Figure 21–66A). The light-colored lower end of the egg is called the vegetal pole; the dark-colored upper end is called the animal pole. The animal and vegetal hemispheres contain different selections of mRNA molecules and other cell components, which become allocated to separate cells as the egg cell divides after fertilization. Near the vegetal pole, for example, there is an accumulation of mRNAs coding for the gene regulatory protein VegT (a DNA-binding protein of the T-box family) and for signal proteins of the TGFβ superfamily, as well as some ready-made protein components of the Wnt signaling pathway (Figure 21–66B). As a result, the cells that inherit vegetal cytoplasm will produce signals to organize

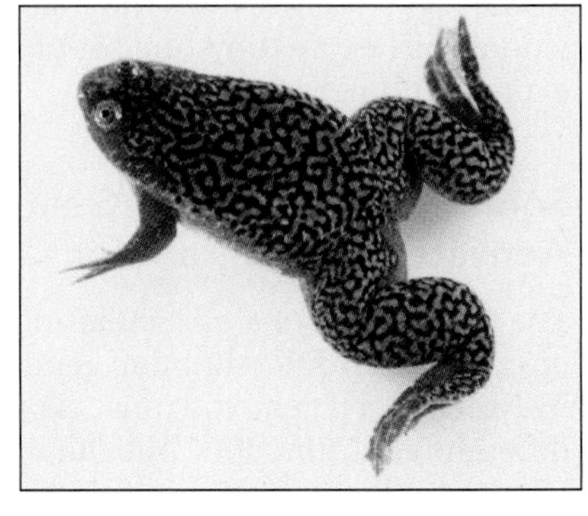

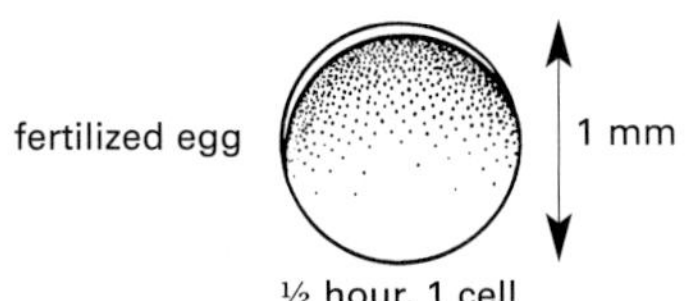

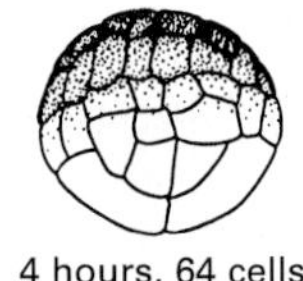

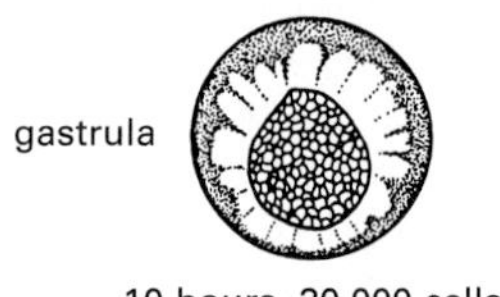

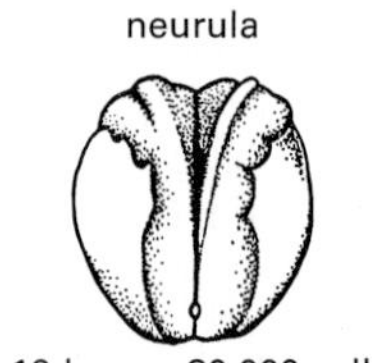

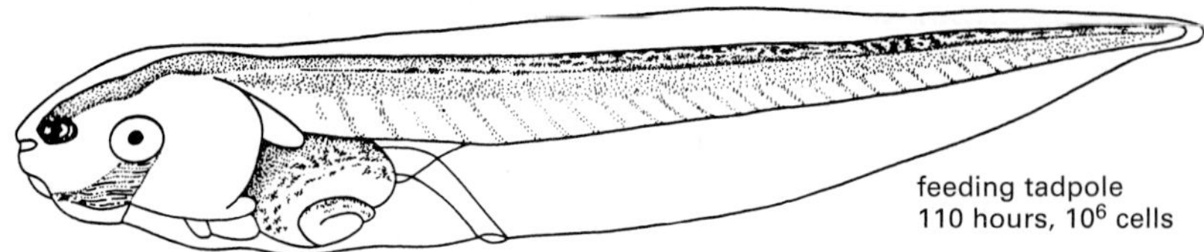

Figure 21–65 Synopsis of the development of *Xenopus laevis* from newly fertilized egg to feeding tadpole. The adult frog is shown in the photograph at the *top*. The developmental stages are viewed from the *side*, except for the 10-hour and 19-hour embryos, which are viewed from *below* and from *above*, respectively. All stages except the adult are shown at the same scale. (Photograph courtesy of Jonathan Slack; drawings after P.D. Nieuwkoop and J. Faber, Normal Table of *Xenopus laevis* [Daudin]. Amsterdam: North-Holland, 1956.)

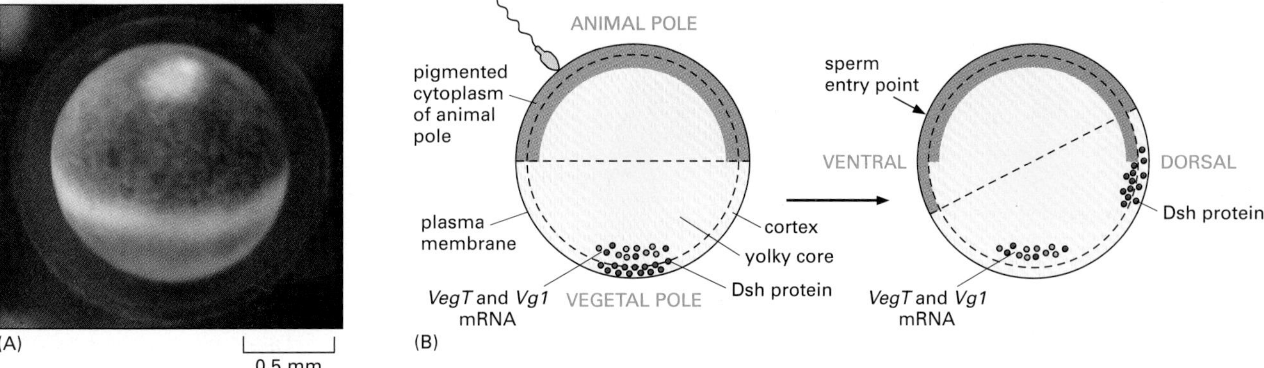

Figure 21–66 The *Xenopus* egg and its asymmetries. (A) Side view of an egg photographed just before fertilization. (B) The asymmetric distribution of molecules inside the egg, and how this changes following fertilization so as to define a dorsoventral as well as an animal-vegetal asymmetry. *Vg1* (not to be confused with *VegT*) codes for a member of the TGFβ protein superfamily. Fertilization triggers two types of intracellular movement, both dependent on microtubules: (1) a rotation of the egg cortex (a layer a few μm deep) through about 30° relative to the core of the egg in a direction determined by the site of sperm entry; and (2) active transport of Dishevelled (Dsh) protein, a component of the Wnt signaling pathway, toward the future dorsal side. The resulting dorsal concentration of Dishevelled protein defines the dorsoventral polarity of the future embryo. (A, courtesy of Tony Mills.)

the behavior of adjacent cells and are committed to form the gut—the innermost tissue of the body; the cells that inherit animal cytoplasm will form the outer tissues.

Fertilization initiates a complex series of movements that will eventually tuck the vegetal cells and cells from the equatorial (middle) region of the animal-vegetal axis into the interior. In the process, the three principal axes of the body are established: *anteroposterior,* from head to tail; *dorsoventral,* from back to belly; and *mediolateral,* from the midline outward to the left or to the right. The animal-vegetal asymmetry of the unfertilized egg is sufficient to define only one of these future body axes—the anteroposterior. Fertilization triggers an intracellular movement that gives the egg an additional asymmetry defining a dorsoventral difference. Following entry of the sperm, the outer, actin-rich cortex of the egg cytoplasm rotates relative to the central core of the egg, so that the animal pole of the cortex is slightly shifted to the future ventral side. Treatments that block the rotation allow cleavage to occur normally but produce an embryo with a central gut and no dorsal structures or dorsoventral asymmetry.

The direction of the cortical rotation is biased according to the point of sperm entry, perhaps through the centrosome that the sperm brings into the egg, and the movement is associated with a reorganization of microtubules in the egg cytoplasm. This leads to a microtubule-based transport of the protein Dishevelled, a downstream component of the Wnt signaling pathway, toward the future dorsal side (see Figure 21–66B). The subcellular region in which Dishevelled thus becomes concentrated gives rise to cells that behave as though they have received a Wnt signal and express a dorsal-specific set of genes as a result. These cells will generate further signals to organize the dorsoventral axis of the body.

Cleavage Produces Many Cells from One

The cortical rotation is completed in about an hour after fertilization and is followed by cleavage, in which the single large egg cell rapidly subdivides by repeated mitosis into many smaller cells, or *blastomeres,* without any change in total mass (Figure 21–67). In this way, the determinants distributed asymmetrically in the egg become partitioned into separate cells, with different fates (Figure 21–68).

These first cell divisions in *Xenopus* have a cycle time of about 30 minutes, with a direct alternation of S and M phases, as discussed in Chapter 17. The very high rate of DNA replication and mitosis seems to preclude gene transcription (although protein synthesis occurs), and the cleaving embryo is almost entirely

Figure 21–67 The stages of cleavage in *Xenopus*. The cleavage divisions rapidly subdivide the egg into many smaller cells. All the cells divide synchronously for the first 12 cleavages, but the divisions are asymmetric, so that the lower, vegetal cells, encumbered with yolk, are fewer and larger.

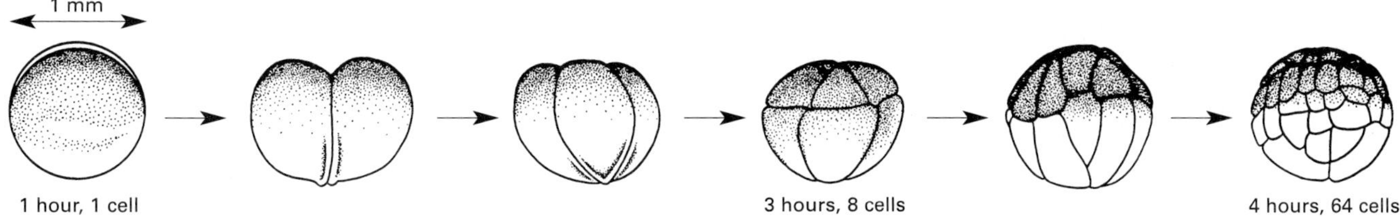

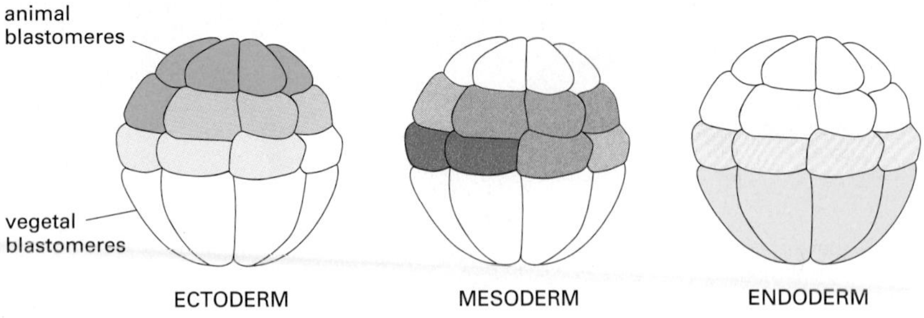

Figure 21–68 The origins of the three germ layers can be traced back to distinct blastomeres of the embryo in its early cleavage stages. The endoderm derives from the most vegetal blastomeres, the ectoderm from the most animal, and the mesoderm from a middle set that contribute also to endoderm and ectoderm. The coloring in each picture is the more intense, the higher the proportion of cell progeny that will contribute to the given germ layer. (After L. Dale, *Curr. Biol.* 9:R812–R815, 1999.)

dependent on reserves of RNA, protein, membrane, and other materials that accumulated in the egg while it developed as an oocyte in the mother. After about 12 cycles of cleavage (7 hours), the cell division rate slows down, the cell cycles begin to follow the standard pattern with G_1 and G_2 phases intervening between the S and M phases, and transcription of the embryo's genome begins. This event is called the *mid-blastula transition*.

Gastrulation Transforms a Hollow Ball of Cells into a Three-Layered Structure with a Primitive Gut

During the period of cleavage, the frog embryo becomes transformed from a solid sphere of cells into something more like a hollow ball, with an internal fluid-filled cavity surrounded by cells that cohere to form an epithelial sheet. The embryo is now termed a **blastula** (Figure 21–69).

Soon after this, the coordinated movements of gastrulation begin. This dramatic process transforms the simple hollow ball of cells into a multilayered structure with a central gut tube and bilateral symmetry: by a more elaborate version of the process outlined earlier for the sea urchin (see Figure 21–3), many of the cells on the outside of the embryo are moved inside it. Subsequent development depends on the interactions of the inner, outer, and middle layers of cells thus formed: the *endoderm* on the inside, consisting of the cells that have moved into the interior to form the primitive gut; the *ectoderm* on the outside, consisting of cells that have remained external; and the *mesoderm* between them, consisting of cells that detach from the epithelium to form a more loosely organized embryonic connective tissue (Figure 21–70). From these three *germ layers,* the tissues of the adult vertebrate body will be generated, preserving the basic body plan established through gastrulation.

The Movements of Gastrulation Are Precisely Predictable

The pattern of gastrulation movements that creates the germ layers and establishes the body axes is described for *Xenopus* in Figure 21–71. The details are complex, but the principles are simple.

Cells of the future endoderm are folded into the interior, or *involuted,* in succession. The cells that were near the vegetal pole of the blastula involute first, turning inward and then moving up toward the animal pole to form the most anterior part of the gut. As they near the animal pole, these leading endoderm

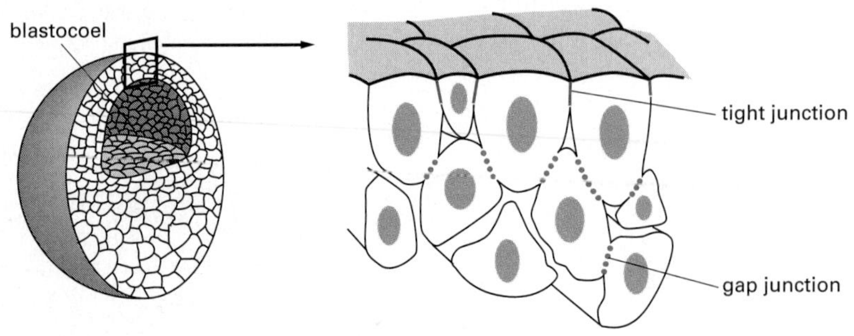

Figure 21–69 The blastula. In the outermost regions of the embryo, tight junctions between the blastomeres begin to create an epithelial sheet that isolates the interior of the embryo from the external medium. Na^+ is pumped across this sheet into the spaces in the interior of the embryo, and water follows into these spaces because of the resulting osmotic pressure gradient. As a result, the intercellular crevices inside the embryo enlarge to form a single cavity, the *blastocoel.* In *Xenopus* the wall of the blastocoel is several cells thick, and only the outermost cells are tightly bound together as an epithelium.

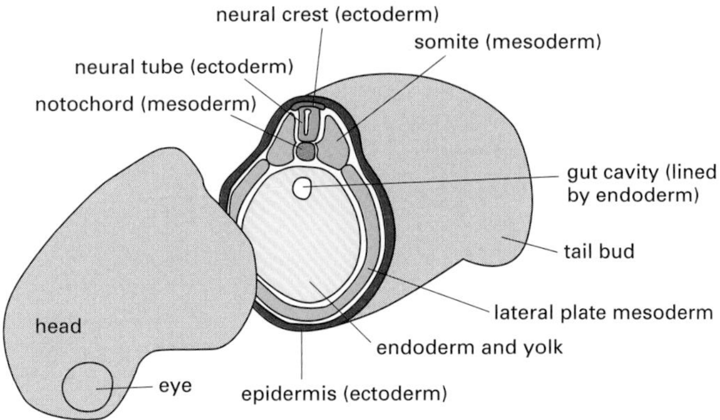

Figure 21–70 A cross section through the trunk of an amphibian embryo after the end of gastrulation, showing the arrangement of endodermal, mesodermal and ectodermal tissues.

The endoderm will form the epithelial lining of the gut, from the mouth to the anus. It gives rise not only to the pharynx, esophagus, stomach, and intestines, but also to many associated glands. The salivary glands, the liver, the pancreas, the trachea, and the lungs, for example, all develop from extensions of the wall of the originally simple digestive tract and grow to become systems of branching tubes that open into the gut or pharynx. The endoderm forms only the epithelial components of these structures—the lining of the gut and the secretory cells of the pancreas, for example. The supporting muscular and fibrous elements arise from the mesoderm.

The mesoderm gives rise to the connective tissues—at first to the loose, space-filling, three-dimensional mesh of cells in the embryo known as mesenchyme and ultimately to cartilage, bone, and fibrous tissue, including the dermis (the inner layer of the skin). The mesoderm also forms the muscles, the entire vascular system—including the heart, the blood vessels, and the blood cells—and the tubules, ducts, and supporting tissues of the kidneys and gonads.

The ectoderm will form the epidermis (the outer, epithelial layer of the skin) and epidermal appendages such as hair, sweat glands, and mammary glands. It will also give rise to the whole of the nervous system, central and peripheral, including not only neurons and glia but also the sensory cells of the nose, the ear, the eye, and other sense organs. (After T. Mohun et al., *Cell* 22:9–15, 1980.)

cells will signal to the overlying ectoderm to define the anterior extremity of the head. The mouth will eventually develop as a hole formed at an anterior site where endoderm and ectoderm come into direct contact. Meanwhile, future mesoderm cells, destined to detach from the epithelial sheet to form the sandwich filling between endoderm and ectoderm, tuck into the interior along with the endoderm cells, and also move up toward the animal pole. The cells that are first to involute go to form parts of the head, and those that are last form parts of the tail. In this way, the anteroposterior axis of the final embryo is laid down sequentially.

The anteroposterior movements go hand in hand with movements that organize the dorsoventral axis of the body. Gastrulation begins on the side of the blastula that has been marked out as dorsal by the cortical rotation. Here, involution of cells into the interior starts with a short indentation that rapidly extends to form the *blastopore*—a line of invagination that curves around to encircle the vegetal pole. The site where the invagination starts defines the *dorsal lip of the blastopore*. As we shall see, this tissue plays a leading part in subsequent events and gives rise to the central dorsal structures of the main body axis.

Chemical Signals Trigger the Mechanical Processes

The movements of gastrulation are triggered by chemical signals from the vegetal blastomeres. Several proteins of the TGFβ superfamily are secreted by these cells and act on the blastomeres above them. If these signals are blocked,

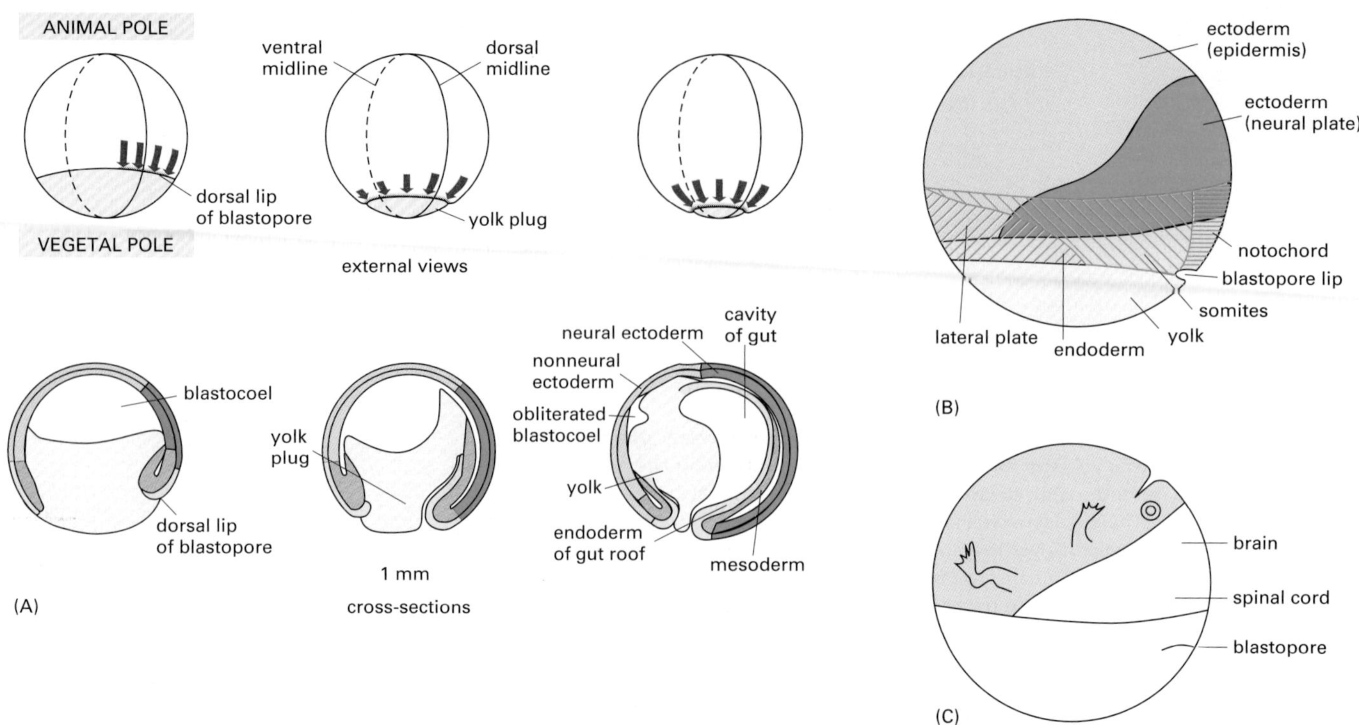

Figure 21–71 Gastrulation in *Xenopus*. (A) The external views *(above)* show the embryo as a semitransparent object, seen from the side; with the directions of cell movement indicated by *red arrows,* cross sections *(below)* are cut in the median plane (the plane of the dorsal and ventral midlines). Gastrulation begins when a short indentation, the beginning of the blastopore, becomes visible in the exterior of the blastula. This indentation gradually extends, curving around to form a complete circle surrounding a plug of very yolky cells (destined to be enclosed in the gut and digested). Sheets of cells meanwhile turn in around the lip of the blastopore and move deep into the interior of the embryo. At the same time the external epithelium in the region of the animal pole actively spreads to take the place of the cell sheets that have turned inward. Eventually, the epithelium of the animal hemisphere spreads in this way to cover the whole external surface of the embryo, and, as gastrulation reaches completion, the blastopore circle shrinks almost to a point. (B) A fate map for the early *Xenopus* embryo (viewed from the side) as it begins gastrulation, showing the origins of the cells that will come to form the three germ layers as a result of the movements of gastrulation. The various parts of the mesoderm (lateral plate, somites, and notochord) derive from deep-lying cells that segregate from the epithelium in the cross-hatched region. The other cells, including the more superficial cells in the cross-hatched region, will give rise to ectoderm *(blue, above)* or endoderm *(yellow, below).* Roughly speaking, the first cells to turn into the interior, or *involute,* will move forward inside the embryo to form the most anterior endodermal and mesodermal structures, while the last to involute will form the most posterior structures. (C) Cartoon to show how the different regions of the ectoderm map into the body surface of the adult animal. (After R.E. Keller, *J. Exp. Zool.* 216:81–101, 1981 and *Dev. Biol.* 42:222–241, 1975.)

gastrulation is disrupted and no mesodermal cell types are generated. The local activation of components of the Wnt signaling pathway on the dorsal side of the embryo (as a result of the earlier cortical rotation, see Figure 21–66) modifies the action of the other signals so as to induce development of the special cells that form the dorsal lip of the blastopore (Figure 21–72).

The dorsal lip of the blastopore plays a central role in gastrulation not just in a geometrical sense, but as a powerful new source of control. If the dorsal lip of the blastopore is excised from an embryo at the beginning of gastrulation and grafted into another embryo but in a different position, the host embryo initiates gastrulation both at the site of its own dorsal lip and at the site of the graft. The movements of gastrulation at the second site entail the formation of a second whole set of body structures, and a double embryo (Siamese twins) results (see Figure 21–6B).

Evidently, the dorsal lip of the blastopore is the source of a signal (or signals) coordinating both the movements of gastrulation and the pattern of specialization of the tissues in its neighborhood. Because of this crucial role in organizing the formation of the main body axis, the dorsal lip of the blastopore is known as

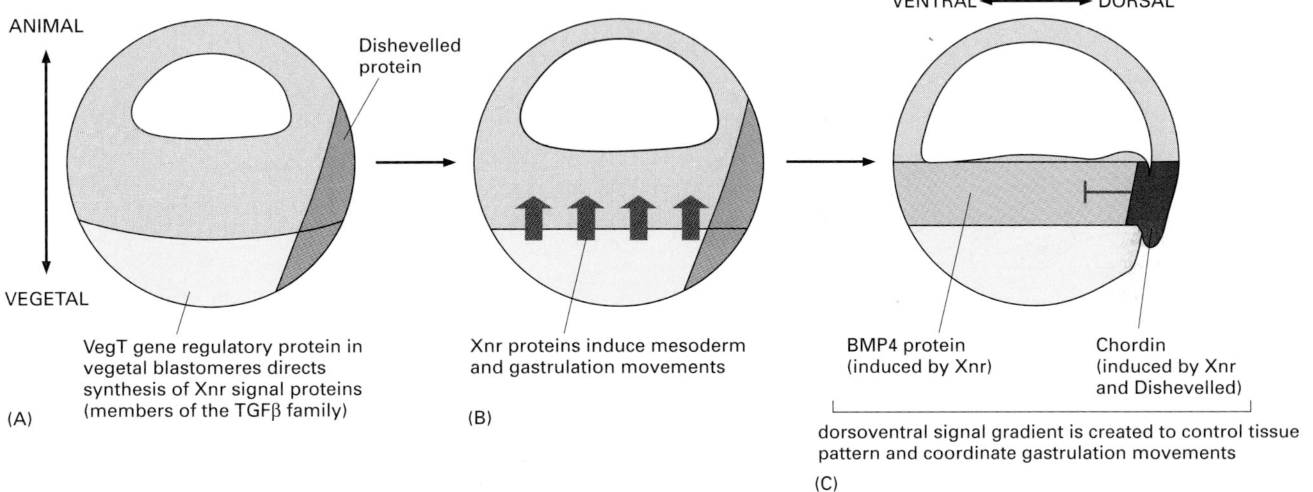
ANIMAL
VEGETAL
Dishevelled protein
VegT gene regulatory protein in vegetal blastomeres directs synthesis of Xnr signal proteins (members of the TGFβ family)
(A)
Xnr proteins induce mesoderm and gastrulation movements
(B)
VENTRAL ↔ DORSAL
BMP4 protein (induced by Xnr)
Chordin (induced by Xnr and Dishevelled)
dorsoventral signal gradient is created to control tissue pattern and coordinate gastrulation movements
(C)

the **Organizer** (or Spemann's Organizer, after its co-discoverer). It is the oldest and most famous example of an *embryonic signaling center*.

Figure 21–72 A current view of the main inductive signals organizing the events of gastrulation. (A) The distribution of axis-determining molecules in the blastula results from inheritance of different parts of the cytoplasm of the fertilized frog egg. The VegT gene regulatory protein in the vegetal blastomeres is translated from VegT mRNA that was localized at the vegetal pole before fertilization. The Dishevelled protein on the future dorsal side is localized there as a result of the cortical rotation that follows fertilization (see Figure 21–66). (B) Signaling from the vegetal cells to the cells just above them, mediated by Xnr *(Xenopus* nodal-related) proteins and other members of the TGFβ superfamily, induces formation of a band of mesoderm in the middle part of the embryo. (C) A morphogen gradient that organizes the dorsoventral axis is set up by a combination of BMP4 (another TGFβ superfamily member) secreted by the mesoderm, and BMP4 antagonists, including chordin, secreted by the cells at the dorsal lip of the blastopore.

Active Changes of Cell Packing Provide a Driving Force for Gastrulation

The Organizer controls the dorsoventral pattern of cell differentiation in its neighborhood by secreting proteins, including chordin that inhibit the action of the TGFβ-like signals (specifically BMP proteins) produced by the more ventral cells (Figure 21–72C). This sets up a gradient of signaling activity—a morphogen gradient, whose local value tells cells how far they lie from the Organizer (see Figure 21–14). But how is the pattern of cell movements organized in mechanical terms, and what are the forces that bring it about?

Gastrulation begins with changes in the shape of the cells at the site of the blastopore. In the amphibian these are called bottle cells: they have broad bodies and narrow necks that anchor them to the surface of the epithelium (Figure 21–73), and they may help to force the epithelium to curve and so to tuck inward, producing the initial indentation seen from outside. Once this first tuck has formed, cells can continue to pass into the interior as a sheet to form the gut and mesoderm. The movement seems to be driven mainly by an active repacking of the cells, especially those in the involuting regions around the Organizer (see Figure 21–73). Here **convergent extension** occurs. Small square fragments of tissue from these regions, isolated in culture, will spontaneously narrow and elongate through a rearrangement of the cells, just as they would in the embryo in the process of converging toward the dorsal midline, turning inward around the blastopore lip, and then elongating to form the main axis of the body.

To bring about this remarkable transformation, the individual cells have to crawl over one another in a coordinated way (Figure 21–74). The alignment of

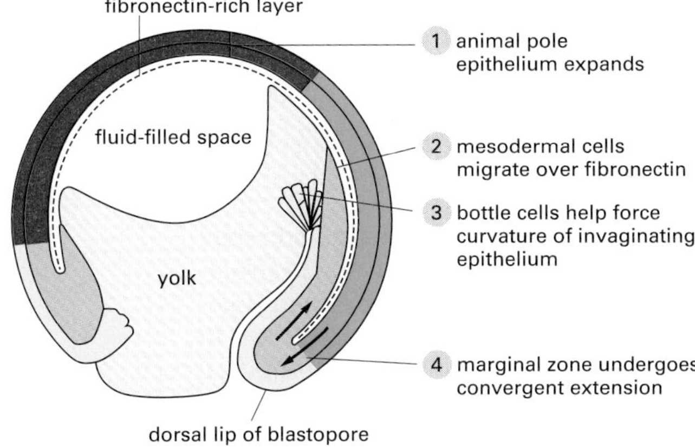
fibronectin-rich layer
fluid-filled space
yolk
dorsal lip of blastopore
1 animal pole epithelium expands
2 mesodermal cells migrate over fibronectin
3 bottle cells help force curvature of invaginating epithelium
4 marginal zone undergoes convergent extension

Figure 21–73 Cell movements in gastrulation. A section through a gastrulating *Xenopus* embryo, cut in the same plane as in Figure 21–71, indicating the four main types of movement that gastrulation involves. The animal pole epithelium expands by cell rearrangement, becoming thinner as it spreads. Migration of mesodermal cells over a fibronectin-rich matrix lining the roof of the blastocoel may help to pull the invaginated tissues forward. But the main driving force for gastrulation in *Xenopus* is convergent extension in the marginal zone. (After R.E. Keller, *J. Exp. Zool.* 216:81–101, 1981.)

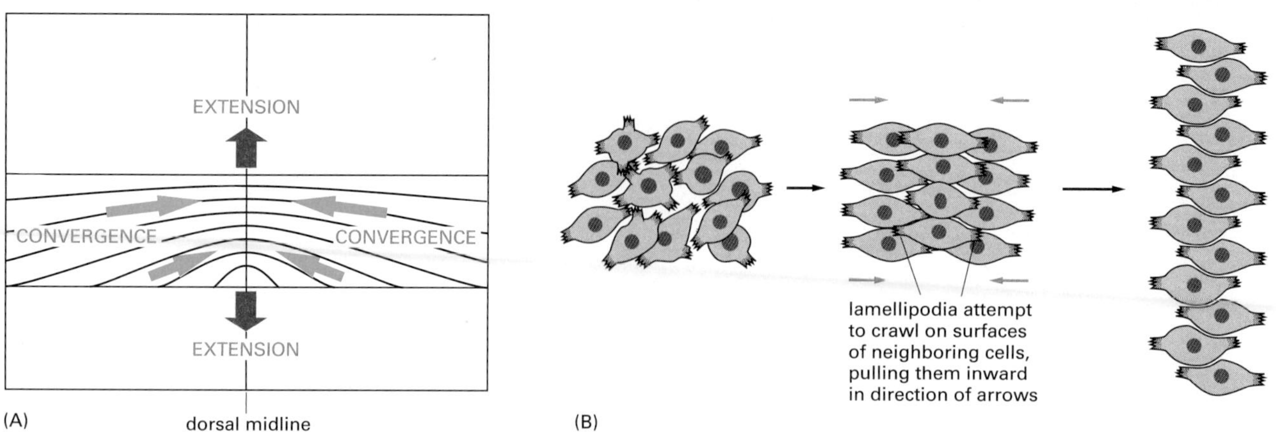

Figure 21–74 Convergent extension and its cellular basis. (A) The pattern of convergent extension in the marginal zone of a gastrula as viewed from the dorsal aspect. *Blue arrows* represent convergence toward the dorsal midline, *red arrows* represent extension of the anteroposterior axis. The simplified diagram does not attempt to show the accompanying movement of involution, whereby the cells are tucking into the interior of the embryo. (B) Schematic diagram of the cell behavior that underlies convergent extension. The cells form lamellipodia, with which they attempt to crawl over one another. Alignment of the lamellipodial movements along a common axis leads to convergent extension. The process depends on the Frizzled/Dishevelled polarity-signaling pathway and is presumably cooperative because cells that are already aligned exert forces that tend to align their neighbors in the same way. (B, after J. Shih and R. Keller, *Development* 116:901–914, 1992.)

their movements appears to depend on the same machinery we encountered in the worm and the fly controlling planar cell polarity: the Frizzled/Dishevelled polarity-signaling pathway. When this pathway is blocked—for example, by a dominant-negative form of Dishevelled—convergent extension fails to occur.

Changing Patterns of Cell Adhesion Molecules Force Cells Into New Arrangements

Patterns of gene expression govern embryonic cell movements in many different ways. They regulate cell motility, cell shape, and the production of signals for guidance. Very importantly, they also determine the sets of adhesion molecules that the cells display on their surfaces. Through changes in its surface molecules, a cell can break old attachments and make new ones. Cells in one region may develop surface properties that make them cohere with one another and become segregated from a neighboring group of cells whose surface chemistry is different.

Experiments done half a century ago on early amphibian embryos showed that the effects of selective cell–cell adhesion can be so powerful that they can bring about an approximate reconstruction of the normal structure of an early postgastrulation embryo even after the cells have been artificially dissociated. When these cells are reaggregated into a random mixture, the cells sort out spontaneously according to their original characters (Figure 21–75). As discussed in Chapter 19, a central role in such phenomena is played by the *cadherins*—a large and varied family of evolutionarily related Ca^{2+}-dependent cell–cell adhesion proteins. These and other cell–cell-adhesion molecules are differentially expressed in the various tissues of the early embryo, and antibodies against them interfere with the normal selective adhesion between cells of a similar type.

Changes in the patterns of expression of the various cadherins correlate closely with the changing patterns of association among cells during gastrulation, neurulation, and somite formation (see Figure 19–25). These rearrangements are likely to be regulated and driven in part by the cadherin pattern. In particular, cadherins appear to have a major role in controlling the formation and dissolution of epithelial sheets and clusters of cells. They not only glue one cell to another, but also provide anchorage for intracellular actin filaments at the

Figure 21–75 Sorting out. Cells from different parts of an early amphibian embryo will sort out according to their origins. In the classical experiment shown here, mesoderm cells *(green)*, neural plate cells *(blue)*, and epidermal cells *(red)* have been disaggregated and then reaggregated in a random mixture. They sort out into an arrangement reminiscent of a normal embryo, with a "neural tube" internally, epidermis externally, and mesoderm in between. (Modified from P.L. Townes and J. Holtfreter, *J. Exp. Zool.* 128:53–120, 1955.)

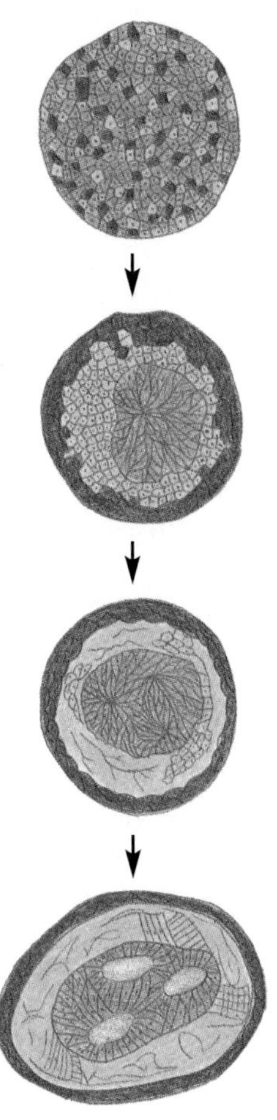

sites of cell–cell adhesion. In this way the pattern of stresses and movements in the developing tissue is regulated according to the pattern of adhesions.

The Notochord Elongates, While the Neural Plate Rolls Up to Form the Neural Tube

Gastrulation is only the first—though perhaps the most dramatic—of a dizzying variety of cell movements that shape the parts of the body. We have space to discuss only a few of these.

In the embryo just after gastrulation, the layer of mesoderm is divided into separate slabs on the left and right sides of the body. Defining the central body axis, and effecting this separation, is the very early specialization of the mesoderm known as the **notochord**. This slender rod of cells, with ectoderm above it, endoderm below it, and mesoderm on either side (see Figure 21–70), derives from the cells of the Organizer itself. The notochordal cells are characterized by expression of a gene regulatory protein called Brachyury (Greek for "short-tail", from the mutant phenotype); this belongs to the same T-box family as the VegT protein in the vegetal blastomeres.

As the notochordal cells pass around the dorsal lip of the blastopore and move into the interior of the embryo, they form a column of tissue that elongates dramatically by convergent extension. The cells of the notochord also become swollen with vacuoles, so that the rod elongates still further and stretches out the embryo. The notochord is the defining peculiarity of the chordates—the phylum to which the vertebrates belong. It is one of the major vertebrate features that do not have any apparent counterpart in *Drosophila*. In the most primitive chordates, which have no vertebrae, the notochord persists as a primitive substitute for a vertebral column. In vertebrates it serves as a core around which other mesodermal cells will eventually gather to form the vertebrae.

In the overlying sheet of ectoderm, meanwhile, other movements are occurring to form the rudiments of the nervous system. In a process known as *neurulation*, a broad central region of ectoderm, called the *neural plate*, thickens, rolls up into a tube, and pinches off from the rest of the cell sheet. The tube thus created from the ectoderm is called the **neural tube**; it will form the brain and the spinal cord (Figure 21–76).

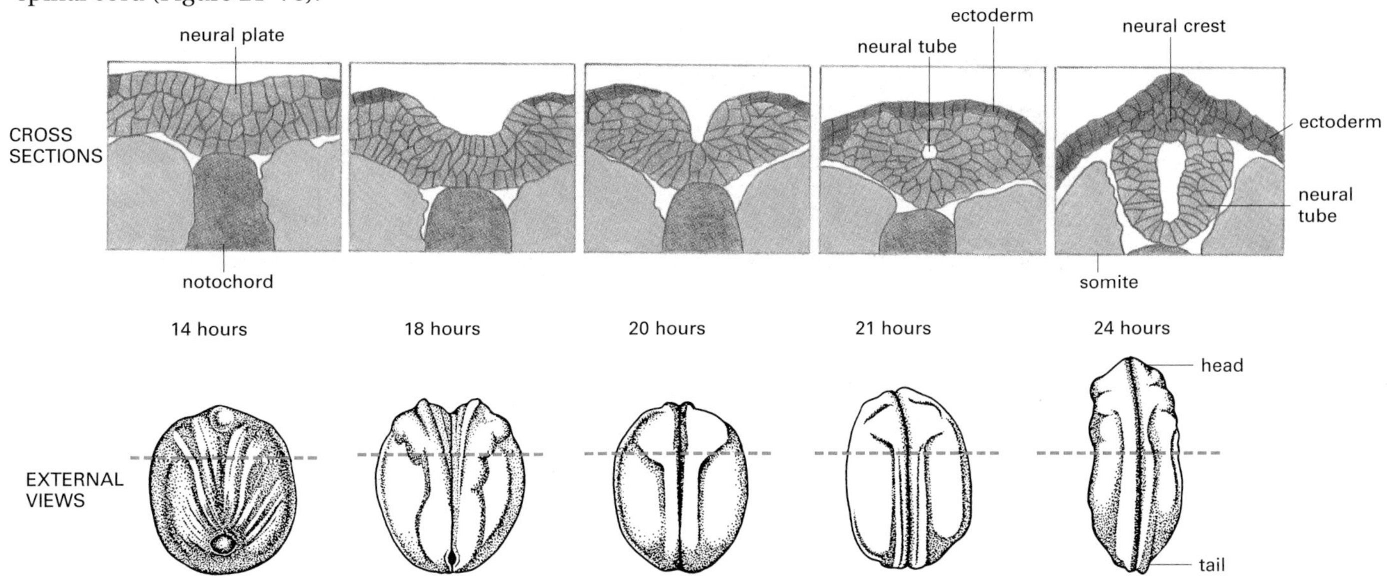

Figure 21–76 Neural tube formation in *Xenopus*. The external views are from the dorsal aspect. The cross sections are cut in a plane indicated by the *broken lines*. (After T.E. Schroeder, *J. Embryol. Exp. Morphol.* 23:427–462, 1970. © The Company of Biologists.)

The mechanics of neurulation depend on changes of cell packing and cell shape that make the epithelium roll up into a tube (Figure 21–77). Signals initially from the Organizer and later from the underlying notochord and mesoderm define the extent of the neural plate, induce the movements that make it roll up, and help to organize the internal pattern of the neural tube. The notochord in particular secretes Sonic hedgehog protein—a homolog of the *Drosophila* signal protein Hedgehog—and this acts as a morphogen to control gene expression in the neighboring tissues (see Figure 21–94).

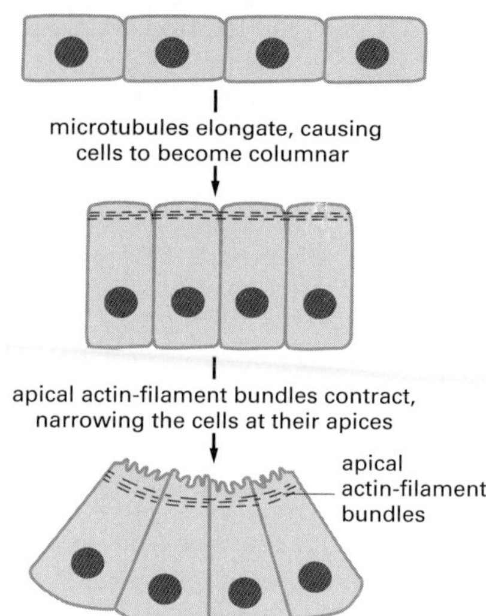

Figure 21–77 The bending of an epithelium through cell shape changes mediated by microtubules and actin filaments. The diagram is based on observations of neurulation in newts and salamanders, where the epithelium is only one cell layer thick. As the apical ends of the cells become narrower, their upper-surface membrane becomes puckered.

A Gene-Expression Oscillator Controls Segmentation of the Mesoderm Into Somites

Genetically regulated changes in cell adhesion underlie one of the most striking and characteristic processes in vertebrate development—the formation of the segments of the body axis.

On either side of the newly formed neural tube lies a slab of mesoderm (see Figure 21–70). To form the repetitive series of vertebrae, ribs, and segmental muscles, this slab breaks up into separate blocks, or **somites**—cohesive groups of cells, separated by clefts. Figure 21–78A shows the process as it occurs in the chick embryo. The somites form one after another, starting in the head and ending in the tail. Depending on the species, the final number of somites ranges from less than 50 (in a frog or a bird) to more than 300 (in a snake). The posterior, most immature part of the mesodermal slab, called the *presomitic mesoderm*, supplies the necessary tissue: as it retreats tailward, extending the embryo, it deposits a trail of somites.

Formation of the cleft between one somite and the next is foreshadowed by an alternating spatial pattern of gene expression in the presomitic mesoderm: cells about to form the posterior part of a new somite switch on expression of one set of genes, while those destined to form the anterior part of the next somite switch on expression of another set. The genes that mark out the somite pattern in this way include at least one member of the cadherin superfamily (called *paraxial protocadherin*), and the intersomitic cleft forms where cells that express it confront cells that do not. Selective cohesion resulting from differential gene expression (either of this gene or some other) thus seems to be the underlying cause of the physical segmentation observed.

The problem then is to understand how the repetitive alternating pattern of gene expression is set up. Studies in the chick embryo have provided an answer.

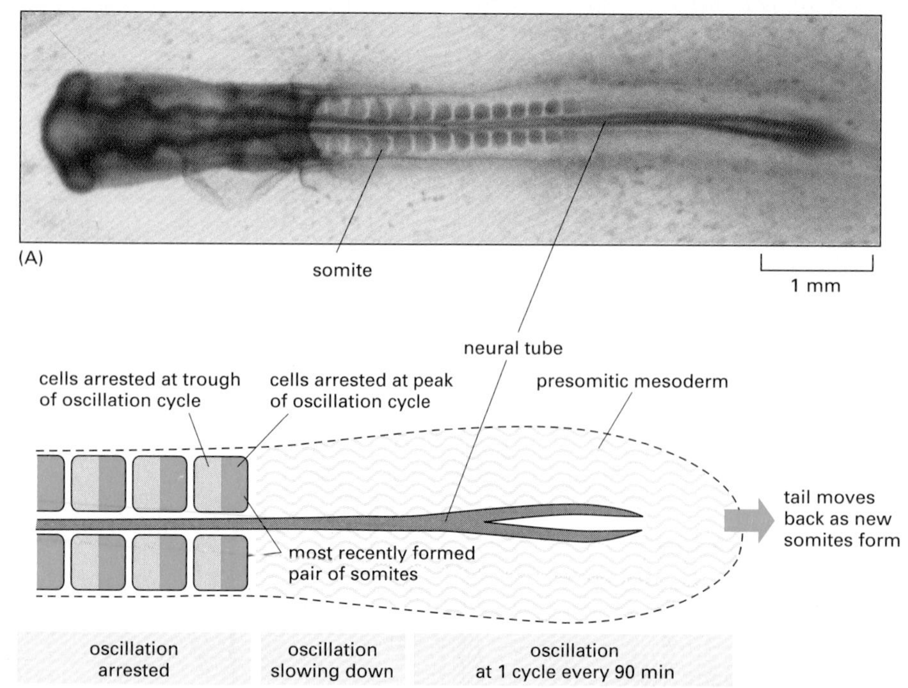

Figure 21–78 Somite formation in the chick embryo. (A) A chick embryo at 40 hours of incubation. (B) How the temporal oscillation of *c-hairy-1* expression in the presomitic mesoderm becomes converted into a spatial pattern of gene expression in the formed somites. In the posterior part of the presomitic mesoderm, each cell oscillates with a cycle time of 90 minutes. As cells mature and emerge from the presomitic region, their oscillation is gradually slowed down and finally brought to a halt in whatever phase of the cycle they happen to be in at the critical moment. In this way, a temporal oscillation of gene expression traces out an alternating spatial pattern. (A, from Y.J. Jiang, L. Smithers and J. Lewis, *Curr. Biol.* 8:868–871, 1998. © Elsevier.)

In the posterior part of the presomitic mesoderm, expression of certain genes—in particular the gene *c-hairy-1*, a homolog of the *Drosophila* pair-rule gene *hairy*—is found to oscillate in time. The length of one complete oscillation cycle of this gene-expression "clock" (90 minutes in the chick) equals the time taken to lay down one further somite. As cells mature and emerge from the presomitic mesoderm to form somites, their oscillation is arrested. *c-hairy-1* encodes a gene regulatory protein, and the cells arrested at the peak of their *c-hairy-1* oscillation cycle express one set of genes, while those arrested at the trough of the cycle express another (Figure 21–78B). In this way, the temporal oscillation of gene expression in the presomitic mesoderm leaves its trace in a spatially periodic pattern of gene expression in the maturing mesoderm, and this in turn dictates how the tissue will break up into physically separate blocks. The mechanism that generates the temporal oscillation itself is unknown.

Most of the cells of each newly formed somite will rapidly differentiate to form a block of muscle, corresponding to one muscle segment of the main body axis. The embryo can (and does) now begin to wriggle. Separate subsets of the somite cells will go to form the vertebrae and other connective tissues such as dermis. A further subset detach from the somite and migrate away into the lateral unsegmented mesoderm, crawling through the spaces between other cells: these emigrants will give rise to almost all the other skeletal muscle cells in the body, including those of the limbs.

Embryonic Tissues Are Invaded in a Strictly Controlled Fashion by Migratory Cells

The muscle-cell precursors, or *myoblasts*, that emigrate from the somites are determined but not overtly differentiated. In the tissues that they colonize they will mingle with other classes of cells from which they appear practically indistinguishable; but they will maintain expression of myoblast-specific gene regulatory proteins (such as members of the MyoD family), and when the time comes for differentiation, they, and they alone, will turn into muscle cells (Figure 21–79).

The eventual pattern of muscles—in the limbs, for example—is determined by the routes that the migrant cells follow and the selection of sites that they colonize. The embryonic connective tissues form the framework through which the myoblasts travel and provide signals that guide their distribution. No matter which somite they come from, myoblasts that migrate into a forelimb bud will form the pattern of muscles appropriate to a forelimb, and those that migrate into a hindlimb bud will form the pattern appropriate to a hindlimb.

Other classes of migrant cells, meanwhile, select different routes for their travels. Along the line where the neural tube pinches off from the future epidermis, a number of ectodermal cells break loose from the epithelium and also migrate as individuals out through the mesoderm (Figure 21–80). These are the cells of the **neural crest**; they will give rise to almost all of the neurons and glial cells of the peripheral nervous system, as well as the pigment cells of the skin and many connective tissues in the head, including bones of the skull and jaws. Other important migrants are the precursors of the blood cells, of the germ cells, and of many groups of neurons within the central nervous system, as well as the *endothelial cells* that form blood vessels. Each of these classes of travelers will colonize a different set of sites. As a result of such invasions, most tissues in the vertebrate body are mixtures of cells of different characters derived from widely separate parts of the embryo.

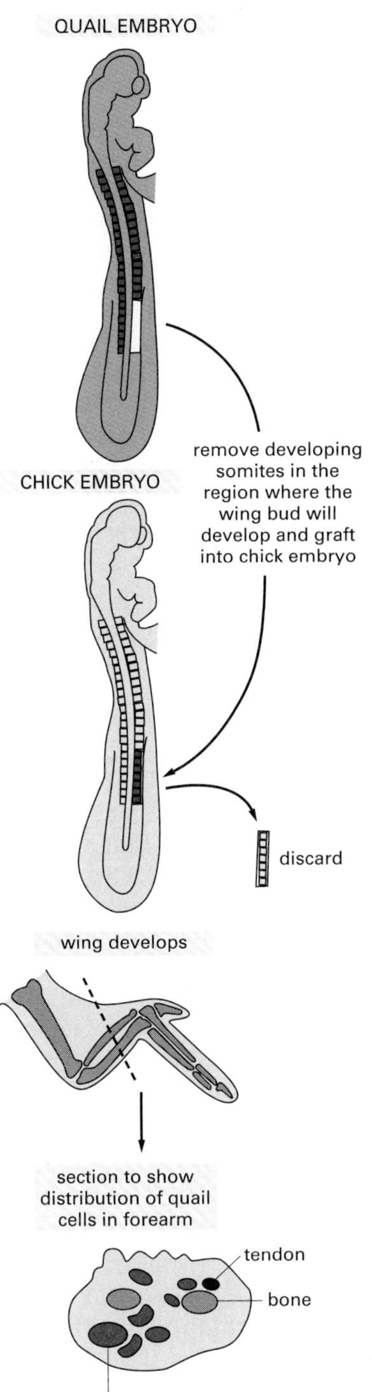
QUAIL EMBRYO
remove developing somites in the region where the wing bud will develop and graft into chick embryo
CHICK EMBRYO
discard
wing develops
section to show distribution of quail cells in forearm
tendon
bone
muscle

Figure 21–79 The migratory origin of limb muscle cells. The migrations can be traced by grafting cells from a quail embryo into a chick embryo; the two species are very similar in their development, but the quail cells are recognizable by the distinctive appearance of their nucleoli. If quail somite cells are substituted for the somite cells of a chick embryo at 2 days of incubation and the wing of the chick is sectioned a week later, it is found that the muscle cells in the chick wing derive from the transplanted quail somites.

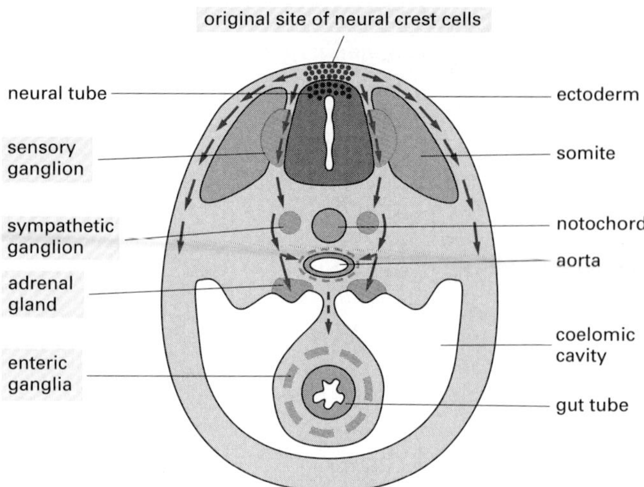

Figure 21–80 The main pathways of neural crest cell migration. A chick embryo is shown in a schematic cross-section through the middle part of the trunk. The cells that take the pathway just beneath the ectoderm will form pigment cells of the skin; those that take the deep pathway via the somites will form the neurons and glial cells of sensory and sympathetic ganglia, and parts of the adrenal gland. The neurons and glial cells of the enteric ganglia, in the wall of the gut, are formed from neural crest cells that migrate along the length of the body, originating from either the neck region or the sacral region. In *Drosophila,* neurons in the wall of the gut originate in a similar way, by migration from the head end of the embryo. (See also Figure 19–23.)

As a migrant cell travels through the embryonic tissues, it repeatedly extends projections that probe its immediate surroundings, testing for subtle cues to which it is particularly sensitive by virtue of its specific assortment of cell-surface receptor proteins. Inside the cell these receptor proteins are connected to the cytoskeleton, which moves the cell along. Some extracellular matrix materials, such as the protein fibronectin, provide adhesive sites that help the cell to advance; others, such as chondroitin sulfate proteoglycan, inhibit locomotion and repel immigration. The nonmigrant cells along the pathway may likewise have inviting or repellent surfaces, or may even extend filopodia that touch the migrant cell and affect its behavior. An incessant tug-of-war between opposing tentative attachments made by the migrant cell leads to a net movement in the most favored direction until the cell finds a site where it can form a lasting attachment. Other factors such as chemotaxis and interactions among the migratory cells may also contribute to their guidance.

The Distribution of Migrant Cells Depends on Survival Factors as Well as Guidance Cues

The final distribution of migrant cells depends not only on the routes they take, but also on whether they survive the journey and thrive in the environment they find at the journey's end. Specific sites provide survival factors needed by specific types of migrant. For example, the neural crest cells that give rise to the pigment cells of the skin and the nerve cells of the gut depend on a peptide factor called *Endothelin-3* that is secreted by tissues on the migration pathways; mutant mice and humans defective in the gene for this factor or its receptor have non-pigmented (albino) patches and potentially lethal gut malformations resulting from the lack of gut innervation (a condition called megacolon, because the colon becomes hugely distended).

Germ cells, blood cell precursors, and neural-crest-derived pigment cells all appear to share at least one common requirement for survival. This involves a transmembrane receptor, called the *Kit protein,* in the membrane of the migrant cells and a ligand, called the *Steel factor,* produced by the cells of the tissue through which the cells migrate and/or in which they come to settle. Individuals with mutations in the genes for either of these proteins are deficient in their pigmentation, their supply of blood cells, and their production of germ cells (Figure 21–81).

Left–Right Asymmetry of the Vertebrate Body Derives From Molecular Asymmetry in the Early Embryo

Vertebrates may look bilaterally symmetrical from the outside, but many of their internal organs—the heart, the stomach, the liver, and so on—are highly

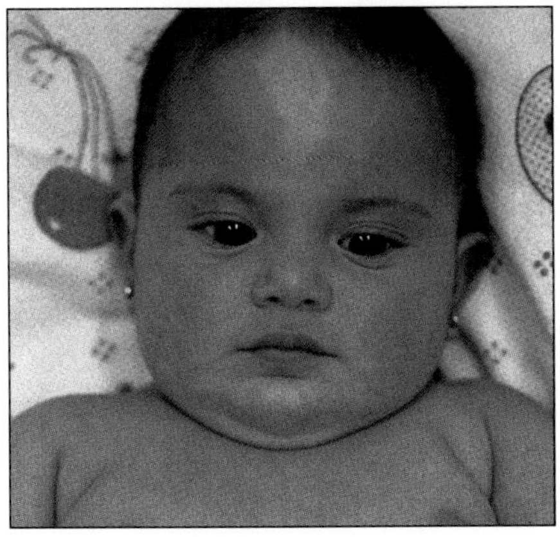

Figure 21–81 Effect of mutations in the *kit* gene. Both the baby and the mouse are heterozygous for a loss-of-function mutation that leaves them with only half the normal quantity of *kit* gene product. In both cases pigmentation is defective because pigment cells depend on the *kit* product as a receptor for a survival factor. (Courtesy of R.A. Fleischman, from *Proc. Natl. Acad. Sci. USA* 88:10885–10889, 1991. © National Academy of Sciences.)

asymmetric. This asymmetry is quite reproducible: 99.98% of people have their heart on the left. We have seen how a vertebrate embryo develops its internal and external tissue layers and its anteroposterior and dorsoventral axes. But how does the left–right asymmetry arise?

Genetic studies in mammals show that this problem can be broken down into two distinct questions—one concerning the creation of asymmetry and the other concerning its orientation. Several mutations are known, in humans and in mice, that cause a randomization of the left–right axis: 50% of the mutant individuals have their internal organs arranged in the normal way, while the other 50% have an inverted anatomy, with the heart on the right. In these individuals, it seems, the mechanism that makes the left and right sides different has functioned correctly, but the mechanism that decides between the two possible orientations of the left–right axis is defective.

A key to the basis of these phenomena comes from the discovery of molecular asymmetries that precede the first gross anatomical asymmetries. The earliest signs are seen in patterns of gene expression in the neighborhood of the *node*—the homolog in mouse and chick of the frog Organizer. Several members of the TGFβ superfamily are expressed asymmetrically in this region (not only in the mouse, but also in chick, frog and zebrafish), as are some other signal proteins including Sonic hedgehog (Figure 21–82). These factors regulate one another's production, producing a cascade of effects through which an asymmetry in the immediate neighborhood of the node apparently becomes self-reinforcing and is relayed outward to create an asymmetric pattern of gene expression over a much larger region.

Mutations in some of these signaling components can have the effect of obliterating the asymmetry, so as to give a mirror-symmetrical anatomy. For example, mice with a knock-out mutation in the *Lefty-1* gene (coding for a Tgfβ-superfamily member) frequently have the right side converted into a mirror image of the left. The mechanism that normally forces the two sides to be different is still not understood. Somehow, cells neighboring the node on opposite sides must become different even if there is very little difference between them at the outset. One possibility is that some sort of lateral inhibition operates: cells

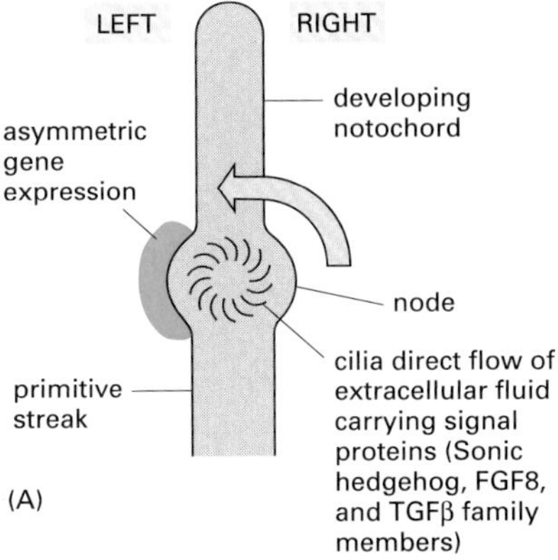

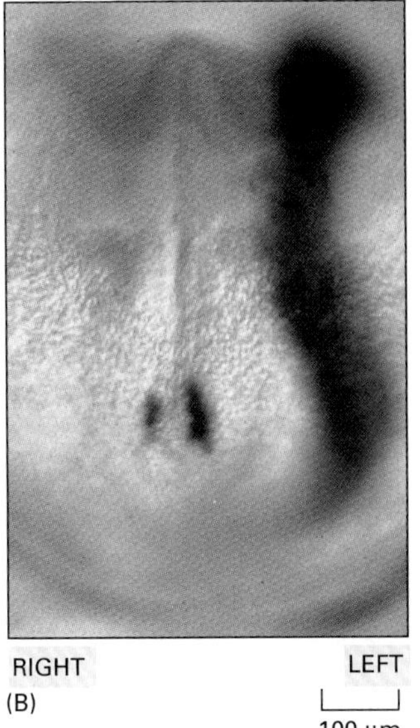

Figure 21–82 Helical beating of cilia at the node, and the origins of left–right asymmetry. (A) The beating of the cilia drives a current towards one side of the node. Various signal proteins are produced in this neighborhood, and the current is thought to sweep them toward one side, provoking asymmetric expression of several genes, including the genes that code for the signal proteins themselves. (B) The asymmetric expression pattern of one such signal protein—a member of the TGFβ superfamily called Nodal—in the neighborhood of the node (lower two *blue* spots) in a mouse embryo at 8 days of gestation, as shown by *in situ* hybridization. At this stage, the asymmetry has already been relayed outward to the lateral plate mesoderm, where Nodal is expressed on the left side (large elongated *blue* patch) but not the right. Note that the embryo is viewed from its ventral side, so that its left appears on the right of the photograph. (B, courtesy of Elizabeth Robertson.)

on opposite sides of the node may exchange mutually inhibitory signals, and thereby compete with one another to adopt a left-side (or a right-side) character. Or perhaps some small but influential group of cells, originating near the midline, is forced to one side by the formation of other midline structures.

Whatever the mechanism, the outcome of events at the node in a normal animal must be biased so that left-specific genes are regularly expressed on the left side: there has to be a link between the mechanism that creates asymmetry and the mechanism that orients it. A clue to the orienting mechanism first came to light in a Swedish infertility clinic. A small subset of infertile men were found to have sperm that were immotile because of a defect in the dynein molecules needed for beating of cilia and flagella. These men also suffered from chronic bronchitis and sinusitis because the cilia in their respiratory tract were defective. And strikingly, 50% of them had their internal organs left–right inverted, with the heart on the right. The findings originally seemed completely mysterious; but similar effects are seen in mammals with other mutations resulting in defective cilia. This suggests that ciliary beating somehow controls which way the left–right axis is oriented.

Time-lapse videomicroscopy in the living mouse embryo reveals that the cells at the node, on its internal face, have cilia that beat in a helical fashion: like a screw-thread, they have a definite handedness, and at the node they are set in a little hollow that is shaped so that their beating drives a current of fluid towards the left side (see Figure 21–82A). Signal proteins carried in this current are thought to provide the bias that orients the left–right axis of the mouse body.

The handedness of the ciliary beating reflects the handedness—the left–right asymmetry—of the organic molecules of which all living things are made. It seems that this, therefore, is the ultimate director of the left–right asymmetry of our anatomy.

Summary

Animal development involves dramatic cell movements. Thus, in gastrulation, cells from the exterior of the early embryo tuck into the interior to form a gut cavity and create the three germ layers—endoderm, mesoderm, and ectoderm—from which higher animals are constructed. In vertebrates, the movements of gastrulation are organized by signals from the Organizer (the dorsal lip of the amphibian blastopore, corresponding to the node in a chick or mouse embryo). These signals specify the dorsoventral axis of the body and govern convergent extension, in which the sheet of cells moving into the interior of the body lengthens along the head-to-tail axis while narrowing at right angles to this axis. The active repacking movements of individual cells that drive convergent extension are coordinated through the Frizzled/Dishevelled planar-polarity signaling pathway—a branch of the Wnt signaling pathway that regulates the actin cytoskeleton.

Subsequent development involves many further cell movements. Part of the ectoderm thickens, rolls up, and pinches off to form the neural tube and neural crest. In the midline, a rod of specialized cells called the notochord elongates to form the central axis of the embryo. The long slabs of mesoderm on either side of the notochord become segmented into somites. Migrant cells, such as those of the neural crest, break loose from their original neighbors and travel through the embryo to colonize new sites. Specific cell-adhesion molecules, such as cadherins and integrins, help to guide the migrations and control the selective cohesion of cells in new arrangements.

Ultimately, the pattern of cell movements is directed by the pattern of gene expression, which determines cell surface properties and motility. Thus, the formation of somites depends on a periodic pattern of gene expression, which is laid down by a biochemical oscillator in the mesoderm and dictates the way the mass of cells will break up into separate blocks. Similarly, the left–right anatomical asymmetry of the vertebrate body is foreshadowed by left–right asymmetry in the pattern of gene expression in the early embryo. This asymmetry, in mammals at least, is thought to be directed ultimately by the handedness of ciliary beating in the neighborhood of the node.

THE MOUSE

The mouse embryo—tiny and inaccessible in its mother's womb—presents a hard challenge to developmental biologists. It has, however, two immediate attractions. First, the mouse is a mammal, and mammals are the animals that we, as humans, care about most. Second, among mammals, it is one of the most convenient for genetic studies, because it is small and breeds rapidly. These two factors have spurred an enormous research effort, resulting in the development of some remarkably powerful experimental tools. In this way, the mouse has become the main model organism for experimentation in mammalian genetics and the most intensively studied surrogate for humans. It is separated from humans by only about 100 million years of evolution. Its genome is the same as ours in size, and there is very nearly a one-to-one correspondence between mouse and human genes. Our proteins are typically 80–90% identical in amino-acid sequence, and large blocks of close nucleotide sequence similarity are also evident when the regulatory DNA sequences are compared.

Through ingenuity and perseverance, developmental biologists have now found ways to gain access to the early mouse embryo without killing it and to generate mice to order with mutations in any chosen gene. Almost any genetic modification that can be made in a worm, a fly, or a zebrafish can now also be made in the mouse, and in some cases made better. The costs of research in the mouse are far greater, but so are the incentives. As a result, the mouse has become a rich source of information about all aspects of the molecular genetics of development—a key model system not only for mammals, but also for other animals. It has provided, for example, much of what we know about Hox genes, left–right asymmetry, cell death controls, the role of Notch signaling, and a host of other topics.

We have already drawn repeatedly upon data from the mouse. We shall make use of it even more in the next chapter, where we discuss adult tissues and the developmental processes that occur in them. In this section, we examine the special features of mouse development that have been exploited to make the genetic manipulations possible. By way of example, we shall also outline how the mouse has been used to illuminate one further important developmental process—the creation of organs such as lungs and glands by interactions between embryonic connective tissue and epithelium.

Mammalian Development Begins With a Specialized Preamble

The mammalian embryo begins its development in an exceptional way. Protected within the uterus, it does not have the same need as the embryos of most other species to complete the early stages of development rapidly. Moreover, the development of a placenta quickly provides nutrition from the mother, so that the egg does not have to contain large stores of raw materials such as yolk. The egg of a mouse has a diameter of only about 80 μm and therefore a volume about 2000 times smaller than that of a typical amphibian egg. Its cleavage divisions occur no more quickly than the divisions of many ordinary somatic cells, and gene transcription has already begun by the two-cell stage. Most importantly, while the later stages of mammalian development are similar to those of other vertebrates such as *Xenopus*, mammals begin by taking a large developmental detour to generate a complicated set of structures—notably the amniotic sac and the placenta—that enclose and protect the embryo proper and provide for the exchange of metabolites with the mother. These structures, like the rest of the body, derive from the fertilized egg but are called *extraembryonic* because they are discarded at birth and form no part of the adult. Similar accessory structures are formed in the development of birds and reptiles.

The early stages of mouse development are summarized in Figure 21–83. The fertilized egg divides to generate 16 cells by three days after fertilization. At first, the cells stick together only loosely, but at the 16-cell stage they become more cohesive and undergo *compaction* to form a solid ball of cells called a *morula* (Latin for "little mulberry") (Figure 21–84). Tight junctions form between the cells, sealing off the interior of the morula from the external medium. Soon

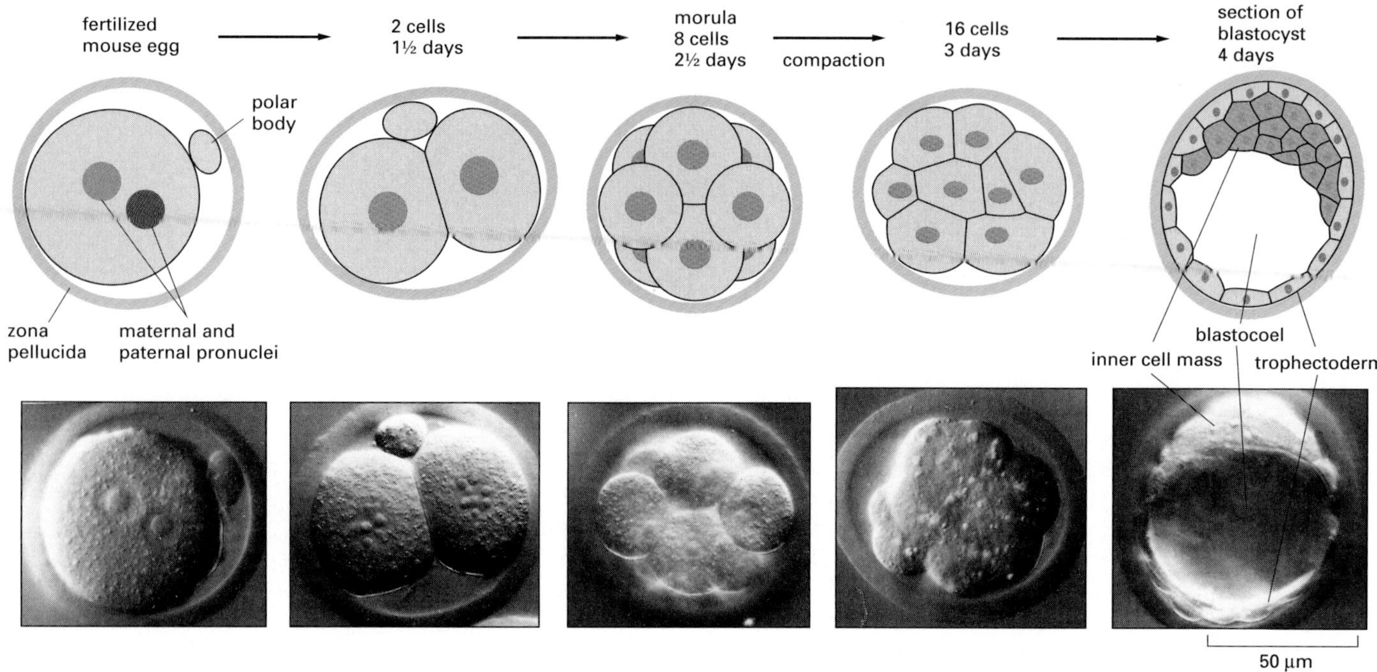

Figure 21–83 The early stages of mouse development. The zona pellucida is a jelly capsule from which the embryo escapes after a few days, allowing it to implant in the wall of the uterus. (Photographs courtesy of Patricia Calarco.)

after this, an internal cavity develops, converting the morula into a *blastocyst*—a hollow sphere. The outer layer of cells, forming the wall of the sphere, is called the *trophectoderm*. It will give rise to extraembryonic tissues. An inner clump of cells, called the *inner cell mass*, is located to one side of the cavity. It will give rise to the whole of the embryo proper.

After the embryo has escaped from its jelly capsule (at about four days), the cells of the trophectoderm make close contact with the wall of the uterus, initiating formation of the placenta. Meanwhile the inner cell mass grows and begins to differentiate. Part of it gives rise to some further extraembryonic structures, such as the yolk sac, while the rest of it goes on to form the embryo proper by processes of gastrulation, neurulation, and so on, that are fundamentally similar to those seen in other vertebrates, although distortions of the geometry make some of the homologies hard to discern at first sight.

The Early Mammalian Embryo Is Highly Regulative

Localized intracellular determinants play only a small part in early mammalian development, and the blastomeres produced by the first few cell divisions are remarkably adaptable. If the early embryo is split in two, a pair of identical twins

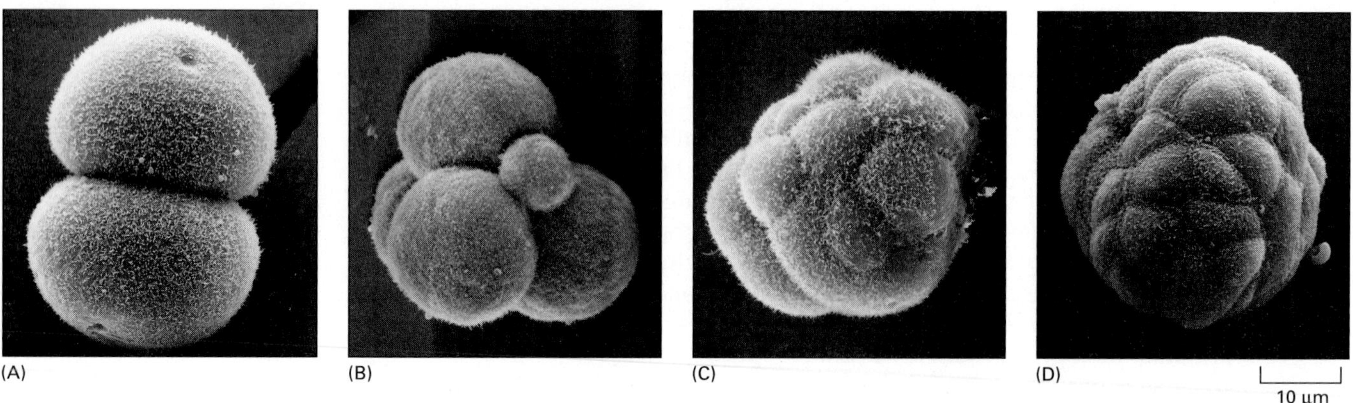

Figure 21–84 Scanning electron micrographs of the early mouse embryo. The zona pellucida has been removed. (A) Two-cell stage. (B) Four-cell stage (a polar body is visible in addition to the four blastomeres—see Figure 20–22). (C) Eight-to-sixteen-cell morula—compaction occurring. (D) Blastocyst. (A–C, courtesy of Patricia Calarco; D, from P. Calarco and C.J. Epstein, *Dev. Biol.* 32:208–213, 1973. © Academic Press.)

can be produced—two complete normal individuals from a single cell. Similarly, if one of the cells in a two-cell mouse embryo is destroyed by pricking it with a needle and the resulting "half-embryo" is placed in the uterus of a foster mother to develop, in many cases a perfectly normal mouse will emerge.

Conversely, two eight-cell mouse embryos can be combined to form a single giant morula, which then develops into a mouse of normal size and structure (Figure 21–85). Such creatures, formed from aggregates of genetically different groups of cells, are called *chimeras*. Chimeras can also be made by injecting cells from an early embryo of one genotype into a blastocyst of another genotype. The injected cells become incorporated into the inner cell mass of the host blastocyst, and a chimeric animal develops. A single injected cell taken from an eight-cell embryo or from the inner cell mass of another early blastocyst can give rise to any combination of cell types in the chimera. Wherever the injected cell may happen to find itself, it responds correctly to cues from its neighbors and follows the appropriate developmental pathway.

These findings have two implications. First, during the early stages, the developmental system is self-adjusting, so that a normal structure emerges even if the starting conditions are perturbed. Embryos or parts of embryos that have this property are said to be **regulative**. Second, the individual cells of the inner cell mass are initially *totipotent:* they can give rise to any part of the adult body, including germ cells.

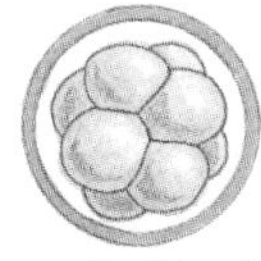
8-cell-stage mouse embryo whose parents are white mice
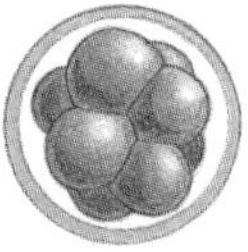
8-cell-stage mouse embryo whose parents are black mice
zona pellucida of each egg is removed by treatment with protease
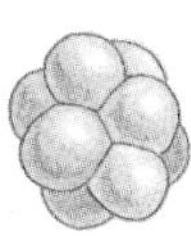
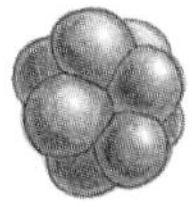
embryos are pushed together and fuse when incubated at 37°
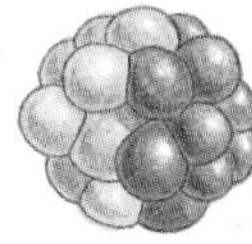
development of fused embryos continues *in vitro* to blastocyst stage
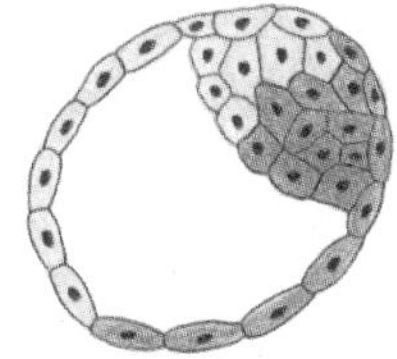
blastocyst transferred to pseudopregnant mouse, which acts as a foster mother

the baby mouse has four parents (but its foster mother is not one of them)

Figure 21–85 A procedure for creating a chimeric mouse. Two morulae of different genotypes are combined.

Totipotent Embryonic Stem Cells Can Be Obtained From a Mammalian Embryo

If a normal early mouse embryo is grafted into the kidney or testis of an adult, its development is disturbed beyond any possibility of proper regulation, but not halted. The result is a bizarre tumorous growth known as a *teratoma*, consisting of a disorganized mass of cells containing many varieties of differentiated tissue—skin, bone, glandular epithelium, and so on—mixed with undifferentiated stem cells that continue to divide and generate yet more of these differentiated tissues.

Investigation of the stem cells in teratomas and related types of tumors led to the discovery that their behavior reflects a remarkable property of the cells of the normal inner cell mass: given a suitable environment, they can be induced to proliferate indefinitely while retaining their totipotent character. Cultured cells with this property are called **embryonic stem cells**, or **ES cells**. They can be derived by placing a normal inner cell mass in culture and dispersing the cells as soon as they proliferate. Separating the cells from their normal neighbors and putting them in the appropriate culture medium evidently arrests the normal program of change of cell character with time and so enables the cells to carry on dividing indefinitely without differentiating. Many tissues of the adult body also contain stem cells that can divide indefinitely without terminally differentiating, as we shall see in the next chapter; but these *adult stem cells,* when allowed to differentiate, normally give rise only to a narrowly restricted range of differentiated cell types.

The state in which the ES cells are arrested seems to be equivalent to that of normal inner-cell-mass cells. This can be shown by taking ES cells from the culture dish and injecting them into a normal blastocyst (Figure 21–86). The injected cells become incorporated in the inner cell mass of the blastocyst and can contribute to the formation of an apparently normal chimeric mouse. Descendants of the injected stem cells can be found in practically any of the tissues of this mouse, where they differentiate in a well-behaved manner appropriate to their location and can even form viable germ cells. The extraordinarily adaptable behavior of ES cells shows that cues from a cell's neighbors not only guide choices between different pathways of differentiation, but can also stop or start the developmental clock—the processes that drive a cell to progress from an embryonic to an adult state.

On a practical level, ES cells have a two-fold importance. First, from a medical point of view, they offer the prospect of a versatile source of cells for repair

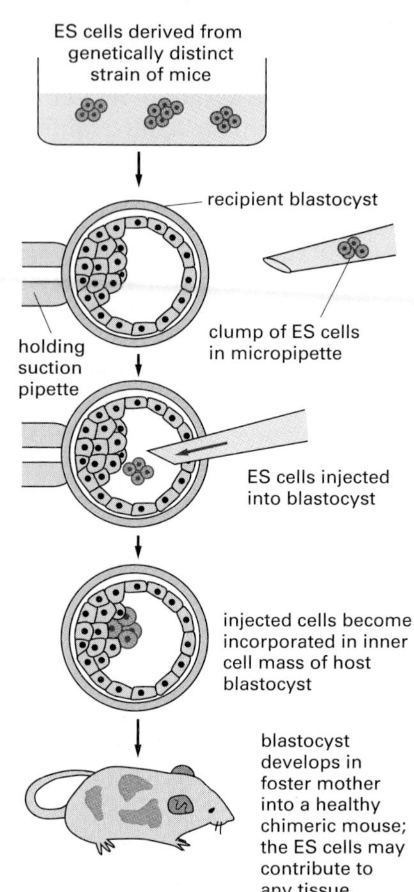
ES cells derived from genetically distinct strain of mice
recipient blastocyst
holding suction pipette
clump of ES cells in micropipette
ES cells injected into blastocyst
injected cells become incorporated in inner cell mass of host blastocyst
blastocyst develops in foster mother into a healthy chimeric mouse; the ES cells may contribute to any tissue

Figure 21–86 Making a chimeric mouse with ES cells. The cultured ES cells can combine with the cells of a normal blastocyst to form a healthy chimeric mouse, and can contribute to any of its tissues, including the germ line. Thus the ES cells are totipotent.

of damaged and defective tissues in the adult body, as we shall discuss at the end of the next chapter. Second, ES cells make possible the most precisely controlled forms of genetic modification, allowing animals to be created with virtually any desired alteration introduced into their genome. The technique uses genetic recombination to substitute an artificially constructed DNA segment for the normal DNA sequence at a chosen site in the genome of an ES cell. Although only a rare cell incorporates the DNA construct correctly, selection procedures have been devised to find this cell among the thousands of cells into which the DNA construct has been transfected. Once selected, the genetically modified ES cells can be injected into a blastocyst to make a chimeric mouse. This mouse will, with luck, have some ES-derived germ cells, capable of acting as founders of a new generation of mice that consist entirely of cells carrying the carefully designed mutation. In this way, an entire mutant mouse can be resurrected from the culture dish (see Figure 8–70).

Interactions Between Epithelium and Mesenchyme Generate Branching Tubular Structures

Vertebrates are comparatively big animals, and they owe much of their bulk to connective tissues. For excretion, absorption of nutrients, and gas exchange, however, they also require large quantities of various specialized types of epithelial surfaces. Many of these take the form of tubular structures created by *branching morphogenesis,* in which an epithelium invades embryonic connective tissue (mesenchyme) to form a composite organ. The lung is a typical example. It originates from the endoderm lining the floor of the foregut. This epithelium buds and grows out into the neighboring mesenchyme to form the *bronchial tree,* a system of tubes that branch repeatedly as they extend (Figure 21–87). The same mesenchyme is also invaded by endothelial cells—the lining cells of blood vessels—to create the system of closely apposed airways and blood vessels required for gas exchange in the lung (discussed in Chapter 22).

The whole process depends on exchanges of signals in both directions between the growing buds of epithelium and the mesenchyme that they are invading. These signals can be analyzed by genetic manipulation in the mouse. A central part is played by signal proteins of the fibroblast growth factor (FGF) family and the receptor tyrosine kinases on which they act. This signaling pathway has various roles in development, but it seems to be especially important in the many interactions that occur between epithelium and mesenchyme.

Mammals have about 20 different FGF genes, as compared with just one in *Drosophila* and in *C. elegans*. The FGF that is most important in the lung is FGF10. This is expressed in clusters of mesenchyme cells near the tips of the growing epithelial tubes, while its receptor is expressed in the epithelial cells themselves. FGF10 or its receptor can be knocked out (by the standard techniques based on recombination in ES cells). In the resulting knock-out mutant mouse, the whole process of branching morphogenesis then fails—a primary bud of lung epithelium is formed but fails to grow out into the mesenchyme to create a bronchial tree. Conversely, a microscopic bead soaked in FGF10 and placed near embryonic lung epithelium in culture will induce a bud to form and grow out toward it. Evidently, the epithelium invades the mesenchyme only by invitation, in response to FGF10.

But what makes the growing epithelial tubes branch repeatedly as they invade? This seems to depend on a Sonic hedgehog signal that is sent in the opposite direction, from the epithelial cells at the tips of the buds back to the mesenchyme. In mice lacking Sonic hedgehog, the lung epithelium grows and differentiates, but forms a sac instead of a branching tree of tubules. Meanwhile, FGF10, instead of being restricted to small clusters of mesenchyme cells, with

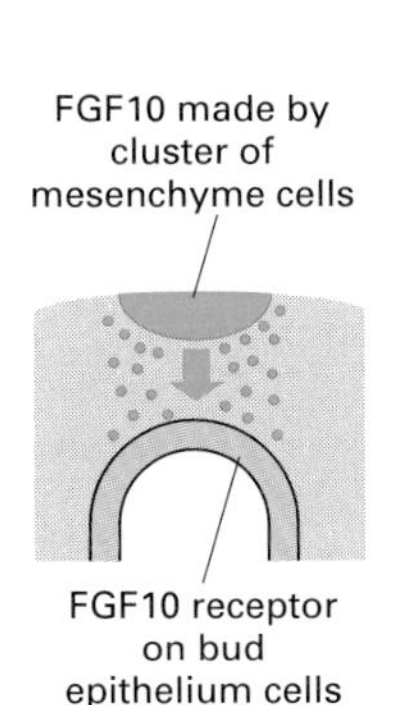
FGF10 made by cluster of mesenchyme cells
FGF10 receptor on bud epithelium cells
(A)

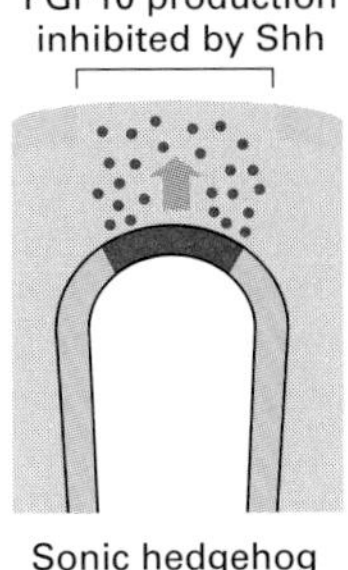
FGF10 production inhibited by Shh
Sonic hedgehog (Shh) produced by epithelial cells at tip of growing bud
two new centers of FGF10 production created

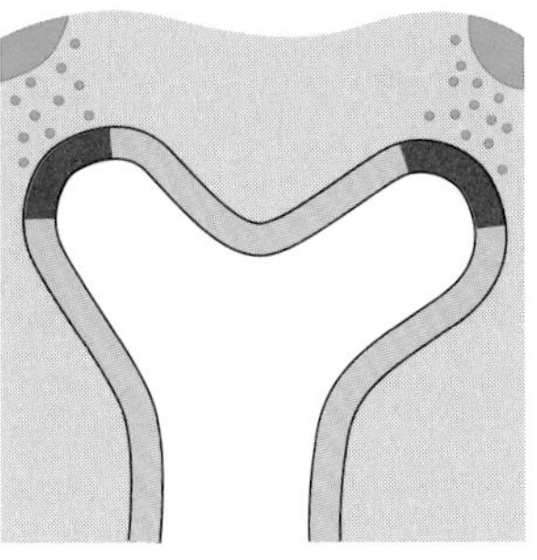
two new buds are formed and the whole process repeats

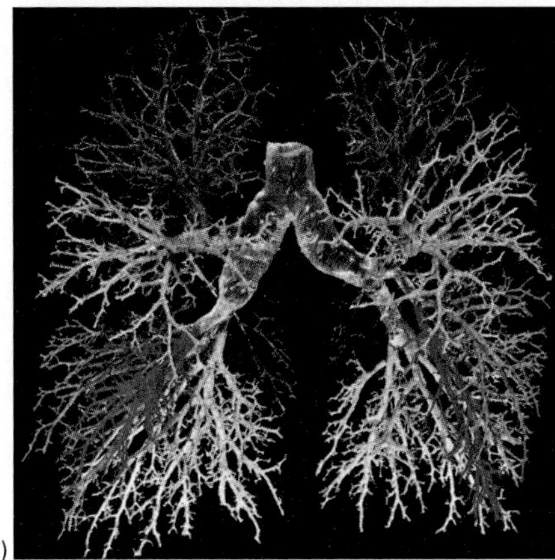
(B)

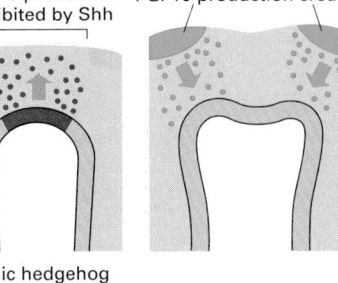

Figure 21–87 Branching morphogenesis of the lung. (A) How FGF10 and Sonic hedgehog are thought to induce the growth and branching of the buds of the bronchial tree. Many other signal molecules, such as BMP4, are also expressed in this system, and the suggested branching mechanism is only one of several possibilities. (B) A cast of the adult human bronchial tree, prepared by injecting resin into the airways; resins of different colors have been injected into different branches of the tree. (B, from R. Warwick and P.L. Williams, Gray's Anatomy, 35th edn. Edinburgh: Longman, 1973.)

each cluster acting as a beacon to direct the outgrowth of a separate epithelial bud, is expressed in broad bands of cells immediately adjacent to the epithelium. This finding suggests that the Sonic hedgehog signal may serve to shut off FGF10 expression in the mesenchyme cells closest to the growing tip of a bud, splitting the FGF10-secreting cluster into two separate clusters, which in turn cause the bud to branch into two (see Figure 21–87A).

Many aspects of this system are still not understood, but it is known that *Drosophila* uses closely related mechanisms to govern the branching morphogenesis of its tracheal system—the tubules that form the airways of an insect. Again, the process depends on the *Drosophila* FGF protein, encoded by the *branchless* gene, and the *Drosophila* FGF receptor, encoded by the *breathless* gene, both operating in much the same way as in the mouse. Indeed, genetic studies of tracheal development in *Drosophila* have also identified other components of the control machinery, and the *Drosophila* genes have led us to their vertebrate homologs. Genetic manipulations in the mouse have given us the means to test whether these genes have similar functions in mammals too; and to a remarkable extent they do.

Summary

The mouse has a central role as model organism for study of the molecular genetics of mammalian development. Mouse development is essentially similar to that of other vertebrates, but begins with a specialized preamble to form extraembryonic structures such as the amnion and placenta. Powerful techniques have been devised for creation of gene knock-outs and other targeted genetic alterations by exploiting the highly regulative properties of the cells of the inner cell mass of the mouse embryo. These cells can be put into culture and maintained as embryonic stem cells (ES cells). Under the right culture conditions, ES cells can proliferate indefinitely without differentiating, while retaining the ability to give rise to any part of the body when injected back into an early mouse embryo.

Many general developmental processes, including most of those discussed elsewhere in the chapter, have been illuminated by studies in the mouse. As just one example, the mouse has been used to investigate the control of branching morphogenesis. This process gives rise to structures such as lungs and glands, and is governed by exchanges of signals, between mesenchyme cells and an invading epithelium. The functions of these signals can be analyzed by gene knock-out experiments.

NEURAL DEVELOPMENT

Nerve cells, or **neurons**, are among the most ancient of all specialized animal cell types. Their structure is like that of no other class of cells, and the development of the nervous system poses problems that have no real parallel in other tissues. A neuron is extraordinary above all for its enormously extended shape, with a long *axon* and branching *dendrites* connecting it through synapses to

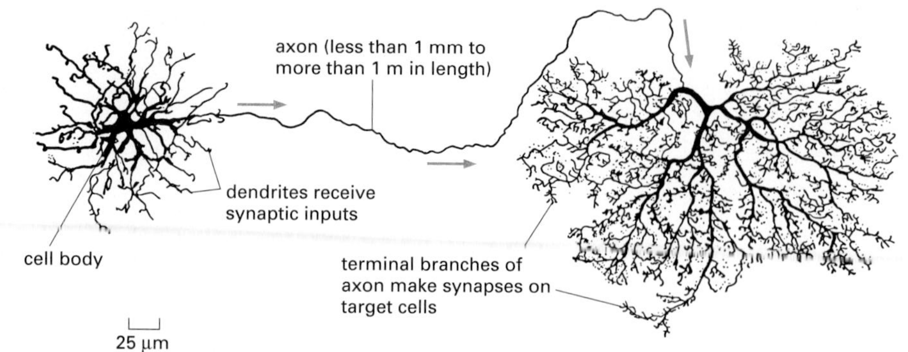

Figure 21–88 A typical neuron of a vertebrate. The *arrows* indicate the direction in which signals are conveyed. The neuron shown is from the retina of a monkey. The longest and largest neurons in a human extend for about 1 million μm and have an axon diameter of 15 μm. (Drawing of neuron from B.B. Boycott, in Essays on the Nervous System [R. Bellairs and E.G. Gray, eds.]. Oxford, UK: Clarendon Press, 1974.)

other cells (Figure 21–88). The central challenge of neural development is to explain how the axons and dendrites grow out, find their right partners, and synapse with them selectively to create a functional network (Figure 21–89). The problem is formidable: the human brain contains more than 10^{11} neurons, each of which, on average, has to make connections with a thousand others, according to a regular and predictable wiring plan. The precision required is not so great as in a man-made computer, for the brain performs its computations in a different way and is more tolerant of vagaries in individual components; but the brain nevertheless outstrips all other biological structures in its organized complexity.

The components of a typical nervous system—the various classes of neurons, *glial cells, sensory cells,* and muscles—originate in a number of widely separate locations in the embryo and are initially unconnected. Thus, in the first phase of neural development (Figure 21–90), the different parts develop according to their own local programs: neurons are born and assigned specific characters according to the place and time of their birth, under the control of inductive signals and gene regulatory mechanisms similar to those we have already discussed for other tissues of the body. The next phase involves a type of morphogenesis unique to the nervous system: axons and dendrites grow out along specific routes, setting up a provisional but orderly network of connections between the separate parts of the system. In the third and final phase, which continues into adult life, the connections are adjusted and refined through interactions among the far-flung components in a way that depends on the electrical signals that pass between them.

Neurons Are Assigned Different Characters According to the Time and Place Where They Are Born

Neurons are almost always produced in association with **glial cells**, which provide a supporting framework and create an enclosed, protected environment in which the neurons can perform their functions. Both cell types, in all animals, develop from the ectoderm, usually as sister cells or cousins derived from a common precursor. Thus, in vertebrates, the neurons and glial cells of the *central nervous system* (including the spinal cord, the brain, and the retina of the eye) derive from the part of the ectoderm that rolls up to form the neural tube, while those of the *peripheral nervous system* derive mainly from the neural crest (Figure 21–91).

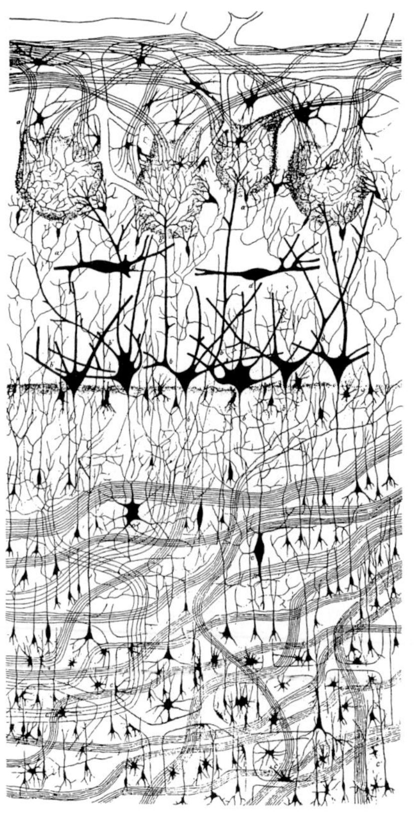

Figure 21–89 The complex organization of nerve cell connections. This drawing depicts a section through a small part of a mammalian brain—the olfactory bulb of a dog, stained by the Golgi technique. The *black* objects are neurons; the *thin lines* are axons and dendrites, through which the various sets of neurons are interconnected according to precise rules. (From C. Golgi, *Riv. sper. freniat. Reggio-Emilia* 1:405–425, 1875; reproduced in M. Jacobson, Developmental Neurobiology, 3rd edn. New York: Plenum, 1992.)

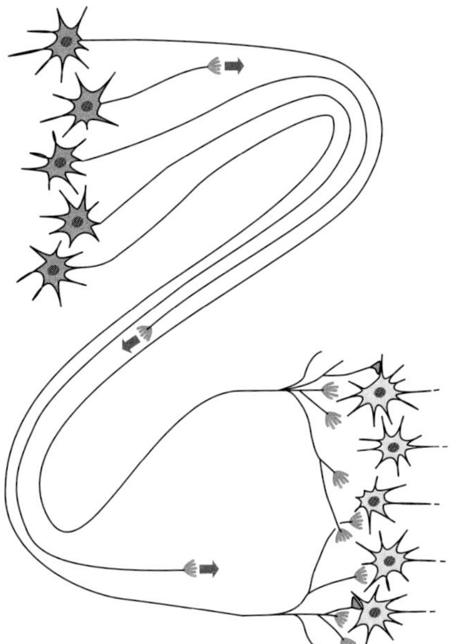

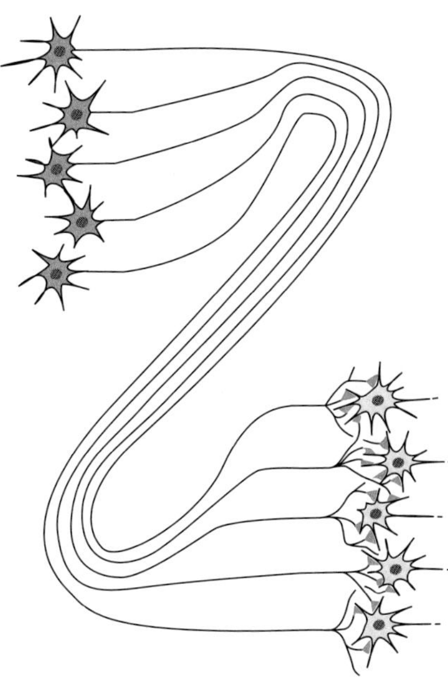

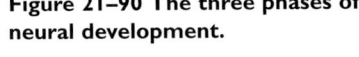

Figure 21–90 The three phases of neural development.

The **neural tube**, with which we shall be mainly concerned, consists initially of a single-layered epithelium (Figure 21–92). The epithelial cells are the progenitors of the neurons and glia. As these cell types are generated, the epithelium becomes thickened and transformed into a more complex structure. Progenitor and, later, glial cells maintain the cohesiveness of the epithelium and form a scaffolding that spans its thickness. Along and between these tall cells, like animals amid the trees of the forest, the new-born neurons migrate, find their resting places, mature, and send out their axons and dendrites (Figure 21–93).

Signal proteins secreted from the ventral and dorsal sides of the neural tube act as opposing morphogens, causing neurons born at different dorsoventral levels to express different gene regulatory proteins (Figure 21–94). There are differences along the head-to-tail axis as well, reflecting the anteroposterior pattern of expression of Hox genes and the actions of yet other morphogens. Moreover, neurons continue to be generated in each region of the central nervous system over many days, weeks, or even months, and this gives rise to still greater diversity, because the cells adopt different characters according to their "birthday"—the time of the terminal mitosis that marks the beginning of neuronal differentiation (Figure 21–95).

The Character Assigned to a Neuron at Its Birth Governs the Connections It Will Form

The differences of gene expression modulate the characters of the neurons and help to cause them to make connections with different partners. In the spinal cord, for example, ventrally located clusters of cells express genes of the *Islet/Lim* homeobox family (coding for gene regulatory proteins) and develop as motor

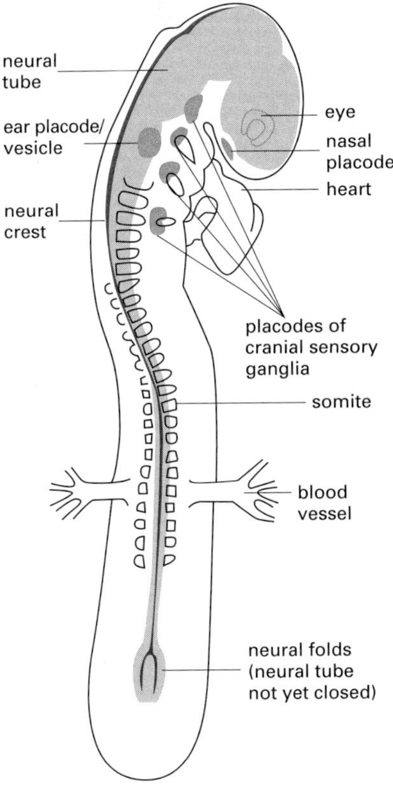

Figure 21–91 Diagram of a 2-day chick embryo, showing the origins of the nervous system. The neural tube *(light green)* has already closed, except at the tail end, and lies internally, beneath the ectoderm, of which it was originally a part (see Figure 21–76). The neural crest *(red)* lies dorsally just beneath the ectoderm, in or above the roof of the neural tube. In addition, thickenings, or *placodes (dark green)*, in the ectoderm of the head give rise to some of the sensory transducer cells and neurons of that region, including those of the ear and the nose. The cells of the retina of the eye, by contrast, originate as part of the neural tube.

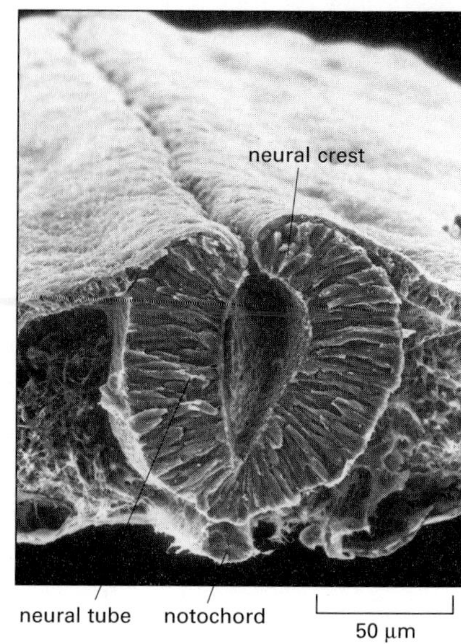

Figure 21–92 Formation of the neural tube. The scanning electron micrograph shows a cross section through the trunk of a 2-day chick embryo. The neural tube is about to close and pinch off from the ectoderm; at this stage it consists (in the chick) of an epithelium that is only one cell thick. (Courtesy of J.P. Revel and S. Brown.)

neurons, sending out axons to connect with specific subsets of muscles—different muscles according to the particular *Islet/Lim* family members expressed. If the pattern of gene expression is artificially altered, the neurons project to different target muscles.

The different destinations reflect different pathway choices that the axons make as they grow out from the nerve cell body, as well as their selective recognition of different target cells at the end of the journey. In the dorsal part of the spinal cord lie neurons that receive and relay sensory information from sensory neurons in the periphery of the body. In intermediate positions, there are various other classes of interneurons, connecting specific sets of nerve cells to one another. Some send their axons dorsally, others ventrally; some up toward the head, others down toward the tail, still others across the floor of the neural tube to the other side of the body (Figure 21–96). In a timelapse film where the developing neurons are stained with a fluorescent dye, one can watch the movements of the growing tips of the axons as they extend: one is reminded of the lights of rush-hour traffic at night, as the cars streak along a network of highways, turning this way or that at busy junctions, each one making its own choice of route.

How are these complex movements guided? Before attempting an answer, we must examine more closely the structure of the growing neuron.

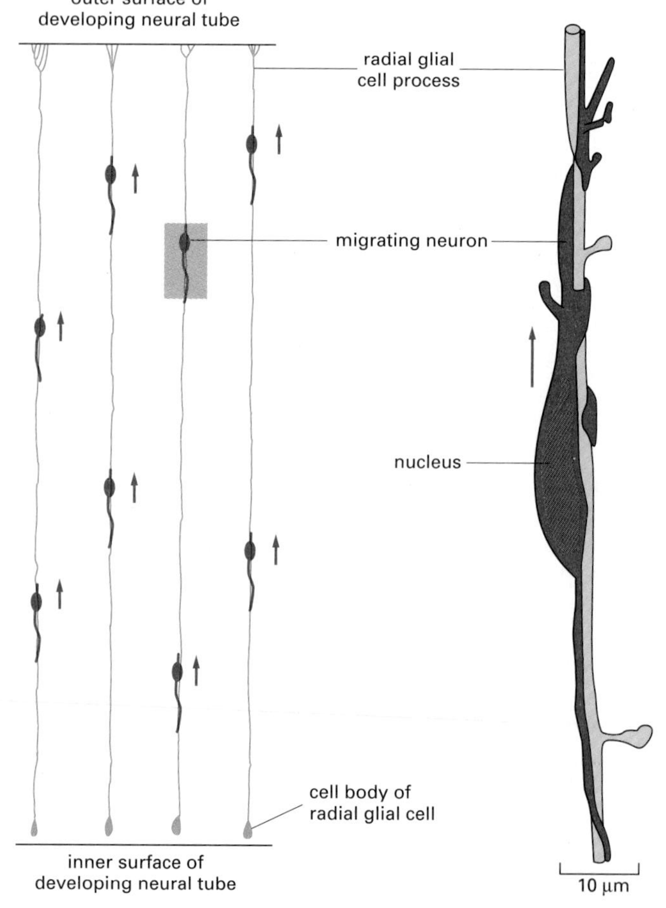

Figure 21–93 Migration of immature neurons. Before sending out axons and dendrites, newborn neurons often migrate from their birthplace and settle in some other location. The diagrams are based on reconstructions from sections of the cerebral cortex of a monkey (part of the neural tube). The neurons go through their final cell division close to the inner, luminal face of the neural tube and then migrate outward by crawling along *radial glial cells.* Each of these cells extends from the inner to the outer surface of the tube, a distance that may be as much as 2 cm in the cerebral cortex of the developing brain of a primate. The radial glial cells can be considered as persisting cells of the original columnar epithelium of the neural tube that become extraordinarily stretched as the wall of the tube thickens. (After P. Rakić, *J. Comp. Neurol.* 145:61–84, 1972.)

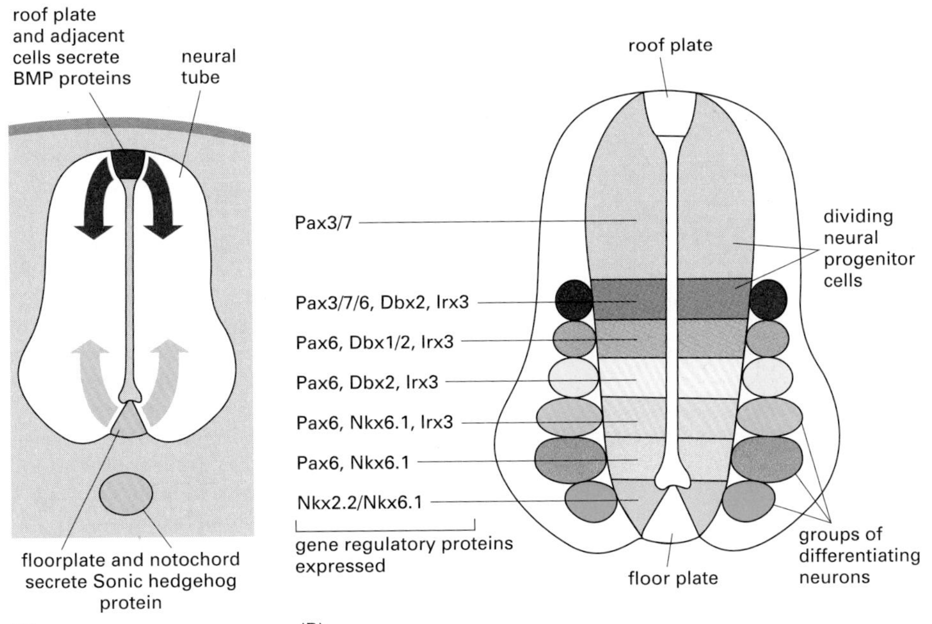

Figure 21–94 A schematic cross section of the spinal cord of a chick embryo, showing how cells at different levels along the dorsoventral axis express different gene regulatory proteins. (A) Signals that direct the dorsoventral pattern: Sonic hedgehog protein from the notochord and the floor plate (the ventral midline of the neural tube) and BMP proteins from the roof plate (the dorsal midline) act as morphogens to control gene expression. (B) The resulting patterns of gene expression in the ventral part of the developing spinal cord. Different groups of proliferating neural progenitor cells (in the *ventricular zone*, close to the lumen of the neural tube) and of differentiating neurons (in the *mantle zone*, further out) express different combinations of gene regulatory proteins. Those indicated in this diagram are almost all members of the homeodomain superfamily; various other genes of the same superfamily (including Islet/Lim proteins) are expressed in the differentiating neurons. Neurons expressing different gene regulatory proteins will form connections with different partners and may make different combinations of neurotransmitters and receptors.

Each Axon or Dendrite Extends by Means of a Growth Cone at Its Tip

A typical neuron sends out one long axon, projecting toward a distant target to which signals are to be delivered, and several shorter dendrites, on which it mainly receives incoming signals from axon terminals of other neurons. Each process extends by growth at its tip, where an irregular, spiky enlargement is seen. This structure, called the **growth cone**, appears to be crawling through the surrounding tissue, trailing a slender axon or dendrite behind it (see Figure 21–96). The growth cone comprises both the engine that produces the movement and the steering apparatus that directs the tip of each process along the proper path (see Figure 16–99).

Much of what we know about the properties of growth cones has come from studies in tissue or cell culture. One can watch as a neuron begins to put out its processes, all at first alike, until one of the growth cones puts on a sudden turn of speed, identifying its process as the axon, with its own axon-specific set of proteins (Figure 21–97). The contrast between axon and dendrite established at this stage involves polarized intracellular transport of different materials into the two types of process. As a result, they will grow out for different distances, follow different paths, and play different parts in synapse formation.

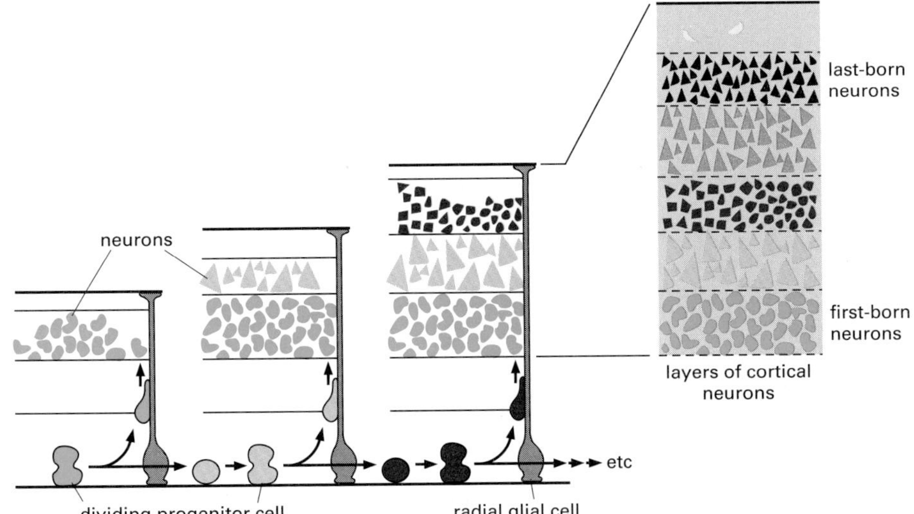

Figure 21–95 Programmed production of different types of neurons at different times from dividing progenitors in the cerebral cortex of the brain of a mammal. Close to one face of the cortical neuroepithelium, progenitor cells divide repeatedly, in stem-cell fashion, to produce neurons. The neurons migrate out toward the opposite face of the epithelium by crawling along the surfaces of radial glial cells, as shown in Figure 21–93. The first-born neurons settle closest to their birthplace, while neurons born later crawl past them to settle farther out. Successive generations of neurons thus occupy different layers in the cortex and have different intrinsic characters according to their birth dates.

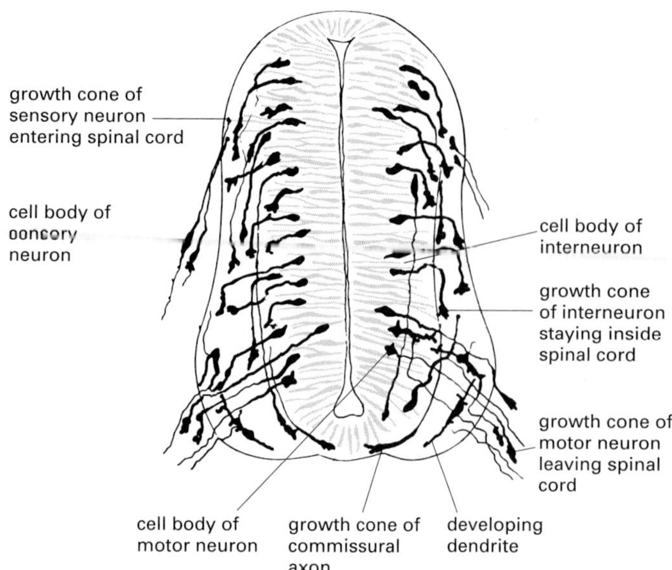

Figure 21–96 Growing axons in the developing spinal cord of a 3-day chick embryo. The drawing shows a cross section stained by the Golgi technique. Most of the neurons, apparently, have as yet only one elongated process—the future axon. An irregularly shaped expansion—a growth cone—is seen at the growing tip of each axon. The growth cones of the motor neurons emerge from the spinal cord (to make their way toward muscles), those of the sensory neurons grow into it from outside (where their cell bodies lie), and those of the interneurons remain inside the spinal cord . Many of the interneurons send their axons down toward the floor plate to cross to the other side of the spinal cord; these axons are called *commissural*. At this early stage, many of the embryonic spinal-cord cells (in the regions shaded *gray*) are still proliferating and have not yet begun to differentiate as neurons or glial cells. (From S. Ramón y Cajal, Histologie du Système Nerveux de l'Homme et des Vertébrés, 1909–1911. Paris: Maloine; reprinted, Madrid: C.S.I.C., 1972.)

The growth cone at the end of a typical growing nerve cell process—either axon or dendrite—moves forward at a speed of about 1 mm per day, continually probing the regions that lie ahead and on either side by putting out filopodia and lamellipodia. When such a protrusion contacts an unfavorable surface, it withdraws; when it contacts a more favorable surface, it persists longer, steering the growth cone as a whole to move in that direction. In this way the growth cone can be guided by subtle variations in the surface properties of the substrata over which it moves. At the same time, it is sensitive to diffusible chemotactic factors in the surrounding medium, which can also encourage or hinder its advance. These behaviors depend on the cytoskeletal machinery inside the growth cone, as discussed in Chapter 16. A multitude of receptors in the growth cone membrane detect the external signals and, through the agency of intracellular regulators such as the monomeric GTPases Rho and Rac, control the assembly and disassembly of actin filaments and other components of the machinery of cell movement.

The Growth Cone Pilots the Developing Neurite Along a Precisely Defined Path *in vivo*

In living animals, growth cones generally travel toward their targets along predictable, stereotyped routes, exploiting a multitude of different cues to find their way, but always requiring a substratum of extracellular matrix or cell surface to crawl over. Often, growth cones take routes that have been pioneered by other neurites, which they follow by contact guidance. As a result, nerve fibers in a mature animal are usually found grouped together in tight parallel bundles (called fascicles or fiber tracts). Such crawling of growth cones along axons is

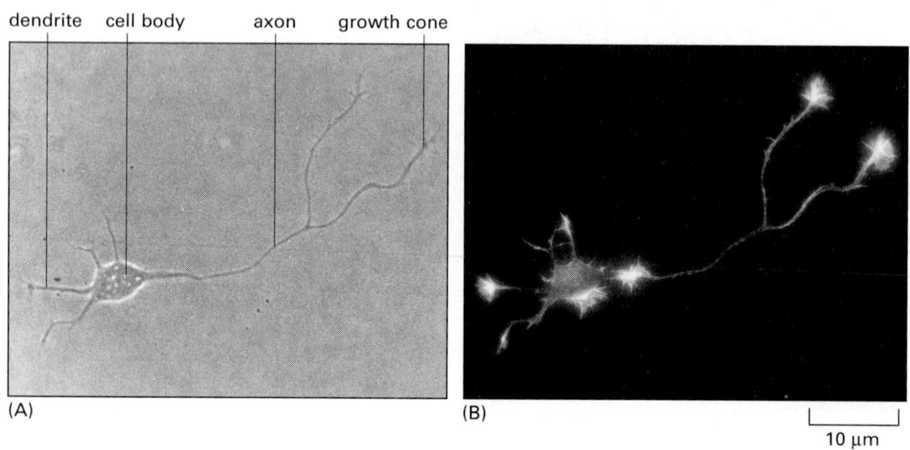

Figure 21–97 Formation of axon and dendrites in culture. A young neuron has been isolated from the brain of a mammal and put to develop in culture, where it sends out processes. One of these processes, the future axon, has begun to grow out faster than the rest (the future dendrites) and has bifurcated. (A) A phase-contrast picture; (B) the pattern of staining with filorescent phalloidin, which binds to filamentous actin. Actin is concentrated in the growth cones at the tips of the processes that are actively extending and at some other sites of lamellipodial activity. (Courtesy of Kimberly Goslin.)

thought to be mediated by homophilic cell–cell-adhesion molecules—membrane glycoproteins that help a cell displaying them to stick to any other cell that also displays them. As discussed in Chapter 19, two of the most important classes of such molecules are those that belong to the immunoglobulin superfamily, such as N-CAM, and those of the Ca^{2+}-dependent cadherin family, such as N-cadherin. Members of both families are generally present on the surfaces of growth cones, of axons, and of various other cell types that growth cones crawl over, including glial cells in the central nervous system and muscle cells in the periphery of the body. The human genome contains more than 100 cadherin genes, for example, and most of them are expressed in the brain (see Figure 19–28). Different sets of cell–cell adhesion molecules, acting in varied combinations, provide a mechanism for selective neuronal guidance and recognition. Growth cones also migrate over components of the extracellular matrix. Some of the matrix molecules, such as *laminin*, favor axon outgrowth, while others, such as chondroitin sulfate proteoglycans, discourage it.

Growth cones are guided by a succession of different cues at different stages of their journey, and the stickiness of the substratum is not the only thing that matters. Another important part is played by chemotactic factors, secreted from cells that act as beacons at strategic points along the path—some attracting, others repelling. The trajectory of *commissural* axons—those that cross from one side of the body to the other—provides a beautiful example of how a combination of guidance signals can specify a complex path. Commissural axons are a general feature of bilaterally symmetrical animals, because the two sides of the body have to be neurally coordinated. Worms, flies and vertebrates use closely related mechanisms to guide their outgrowth.

In the developing spinal cord of a vertebrate, for example, a large number of neurons send their axonal growth cones ventrally toward the floor plate—a specialized band of cells forming the ventral midline of the neural tube (see Figure 21–96). The growth cones cross the floor plate and then turn abruptly through a right angle to follow a longitudinal path up toward the brain, parallel to the floor plate but never again crossing it (Figure 21–98A). The first stage of the journey depends on a concentration gradient of the protein *netrin*, secreted by the cells of the floor plate: the commissural growth cones sniff their way toward its source. Netrin was purified from chick embryos, by assaying extracts of neural tissue for an activity that would attract commissural growth cones in a culture dish. Its sequence revealed that it was the vertebrate homolog of a protein already known from *C. elegans*, through genetic screens for mutant worms with misguided axons—called *unc* mutants because they move in an *unc*oordinated fashion. One of the *unc* genes, *unc-6*, codes for the homolog of netrin. Another, *unc-40*, codes for its transmembrane receptor; and this too has a vertebrate homolog, called DCC that is expressed in the commissural neurons and mediates their response to the netrin gradient.

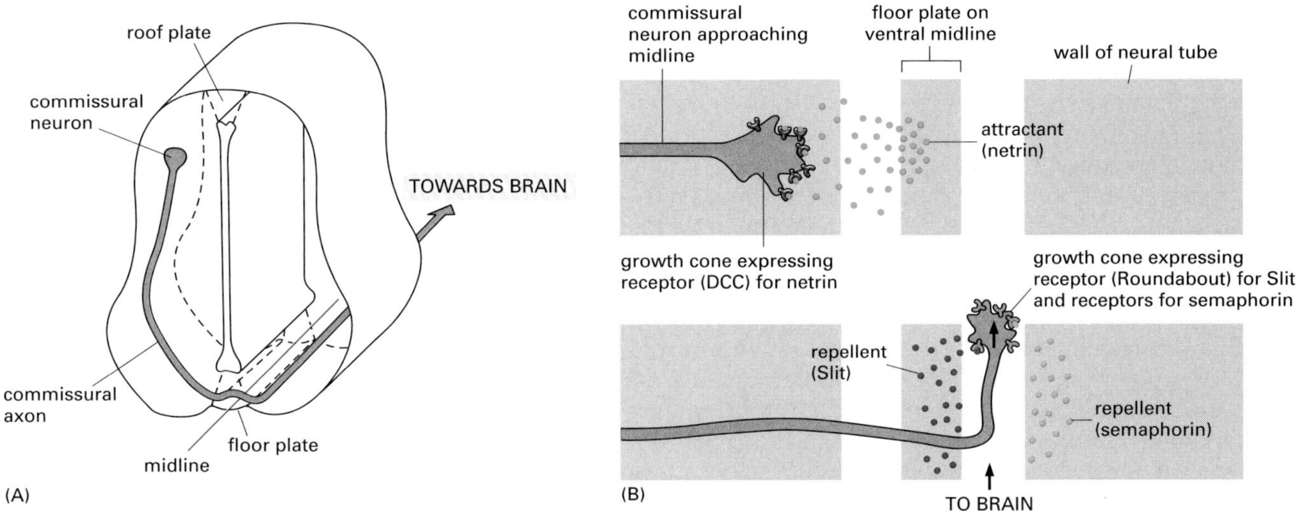

Figure 21–98 The guidance of commissural axons. (A) The pathway taken by commissural axons in the embryonic spinal cord of a vertebrate. (B) The signals that guide them. The growth cones are first attracted to the floorplate by netrin, which is secreted by the floor-plate cells and acts on the receptor DCC in the axonal membrane. As they cross the floor plate, the growth cones upregulate their expression of Roundabout, the receptor for a repellent protein, Slit, that is also secreted by the floor plate. Slit, binding to Roundabout, not only acts as a repellent to keep the cells from re-entering the floor plate, but also blocks responsiveness to the attractant netrin. At the same time, the growth cones switch on expression of receptors for another repellent protein, semaphorin, that is secreted by the cells in the side walls of the neural tube. Trapped between two repellent territories, the growth cones, having crossed the midline, travel in a tight fascicle up toward the brain.

The receptors on each growth cone determine the route it will take: non-commissural neurons in the neural tube, lacking DCC, are not attracted to the floorplate; and neurons expressing a different netrin receptor—called Unc-5H in vertebrates (with a counterpart Unc-5 in the worm)—are actively repelled by the floorplate and send their axons instead toward the roof plate.

Growth Cones Can Change Their Sensibilities as They Travel

If commissural growth cones are attracted to the floor plate, why do they cross it and emerge on the other side, instead of staying in the attractive territory? And having crossed it, why do they never veer back onto it again? The likely answer lies in another set of molecules, several of which are also conserved between vertebrates and invertebrates. Studies of *Drosophila* mutants with misguided commissural axons first identified three of the key proteins: Slit, Roundabout, and Commissureless.

Slit, like netrin, is produced by midline cells of the developing fly, while its receptor, Roundabout, is expressed in the commissural neurons. Slit, acting on Roundabout, has an effect exactly opposite to that of netrin: it repels the growth cones, blocking entry to the midline territory. Commissureless, however, apparently inhibits Roundabout expression initially so as to make the growth cones blind to this "keep-out" signal. Commissural growth cones supplied with Commissureless protein advance to the midline; as they cross it, they seem to lose their blindfold of Commissureless protein and begin to be repelled. Emerging on the far side, they now have functional Roundabout on their surfaces and are thereby prohibited from re-entry.

In vertebrates, a similar mechanism may operate. Commissural growth cones are at first attracted to the midline, and then change their surface receptor proteins as they cross; in this way they may switch their sensibilities, gaining sensitivity to repulsion by Slit—which is expressed in the floor plate—and losing sensitivity to attraction by netrin. Repulsion from the midline now prevents them from straying back across it. At the same time, the growth cones apparently become sensitive to another set of repulsive signals, in the form of proteins called semaphorins, which prevent them from traveling back up into the dorsal regions of the spinal cord. Trapped between the two sets of repulsive signals, the growth cones have no choice but to travel in a narrow track, running parallel to the floor plate but never re-entering it (Figure 21–98B).

Target Tissues Release Neurotrophic Factors That Control Nerve Cell Growth and Survival

Eventually, axonal growth cones reach the target region where they must halt and make synapses. The neurons that sent out the axons can now begin to communicate with their target cells. Although synapses generally transmit signals in one direction, from axon to either dendrite or muscle, the developmental communications are a two-way affair. Signals from the target tissue not only regulate which growth cones shall synapse where (as we discuss below), but also how many of the innervating neurons shall survive.

Most types of neurons in the vertebrate central and peripheral nervous system are produced in excess; up to 50% or more of them then die soon after they reach their target, even though they appear perfectly normal and healthy up to the time of their death. About half of all the motor neurons that send axons to skeletal muscle, for example, die within a few days after making contact with their target muscle cells. A similar proportion of the sensory neurons that innervate the skin die after their growth cones have arrived there.

This large-scale death of neurons is thought to reflect the outcome of a competition. Each type of target cell releases a limited amount of a specific neurotrophic factor that the neurons innervating that target require to survive. The neurons apparently compete for the factor, and those that do not get enough die by programmed cell death. If the amount of target tissue is increased—for example, by grafting an extra limb bud onto the side of the embryo—more

Figure 21–99 NGF effects on neurite outgrowth. Dark-field photomicrographs of a sympathetic ganglion cultured for 48 hours with *(above)* or without *(below)* NGF. Neurites grow out from the sympathetic neurons only if NGF is present in the medium. Each culture also contains Schwann (glial) cells that have migrated out of the ganglion; these are not affected by NGF. Neuronal survival and maintenance of growth cones for neurite extension represent two distinct effects of NGF. The effect on growth cones is local, direct, rapid, and independent of communication with the cell body; when NGF is removed, the deprived growth cones halt their movements within a minute or two. The effect of NGF on cell survival is less immediate and is associated with uptake of NGF by endocytosis and its intracellular transport back to the cell body. (Courtesy of Naomi Kleitman.)

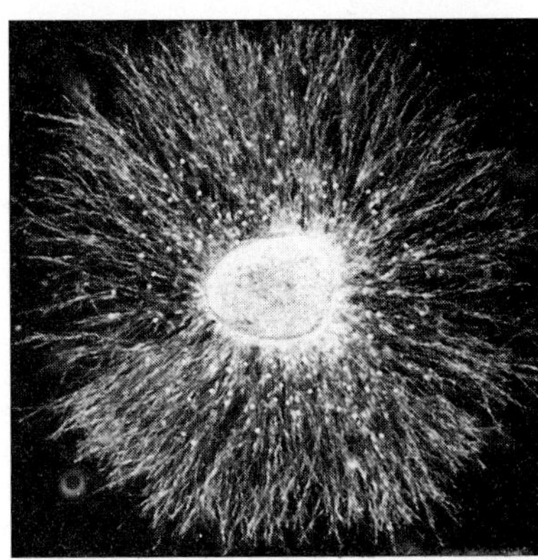

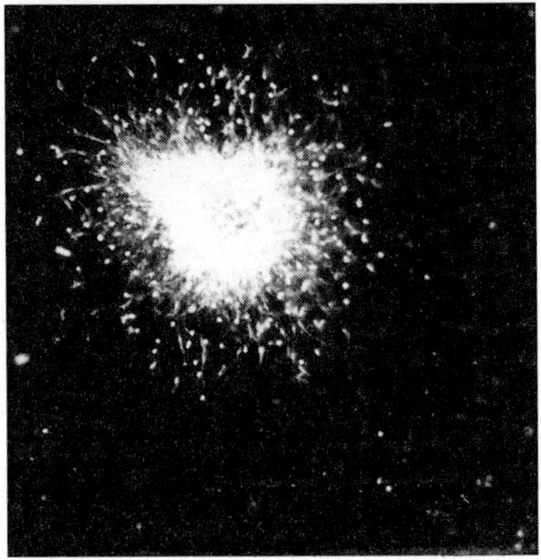

limb-innervating neurons survive; conversely, if the limb bud is cut off, the limb-innervating neurons all die. In this way, although individuals may vary in their bodily proportions, they always retain the right number of motor neurons to innervate all their muscles and the right number of sensory neurons to innervate their whole body surface. The seemingly wasteful strategy of overproduction followed by death of surplus cells operates in almost every region of the nervous system. It provides a simple and effective means to adjust each population of innervating neurons according to the amount of tissue requiring innervation.

The first neurotrophic factor to be identified, and still the best characterized, is known simply as nerve growth factor, or NGF—the founding member of the *neurotrophin* family of signal proteins. It promotes the survival of specific classes of sensory neurons derived from the neural crest and of sympathetic neurons (a subclass of peripheral neurons that control contractions of smooth muscle and secretion from exocrine glands), NGF is produced by the tissues that these neurons innervate. When extra NGF is provided, extra sensory and sympathetic neurons survive, just as if extra target tissue were present. Conversely, in a mouse with a mutation that knocks out the NGF gene or its receptor (a transmembrane tyrosine kinase called TrkA), almost all sympathetic neurons and the NGF-dependent sensory neurons are lost. There are many neurotrophic factors, only a few of which belong to the neurotrophin family, and they act in different combinations to promote survival of different classes of neurons.

NGF and its relatives have an additional role: besides acting on the nerve cell as a whole to control its survival, they regulate the outgrowth of axons and dendrites (Figure 21–99). These can even act locally on just one part of the tree of nerve cell processes, promoting or pruning the growth of individual branches: a growth cone exposed to NGF shows an immediate increase of motility. Conversely, an axon branch that is deprived of NGF, while the rest of the neuron continues to be bathed in the factor, dies back.

The peripheral action of NGF continues to be important after the phase of neuronal death. In the skin, for example, it controls the branching of sensory nerve fibers, ensuring not only that the whole body surface becomes innervated during development but also that it recovers its innervation after damage.

Neuronal Specificity Guides the Formation of Orderly Neural Maps

In many cases, axons originating from neurons of a similar type but located in different positions come together for the journey and arrive at the target in a tight bundle. There they disperse again, to terminate at different sites in the target territory.

The projection from the eye to the brain provides an important example. The neurons in the retina that convey visual information back to the brain are called *retinal ganglion cells*. There are more than a million of them, each one reporting on a different part of the visual field. Their axons converge on the optic nerve head at the back of the eye and travel together along the optic stalk into the brain. Their main site of termination, in most vertebrates other than mammals, is the *optic tectum*—a broad expanse of cells in the midbrain. In connecting with the tectal neurons, the retinal axons distribute themselves in a

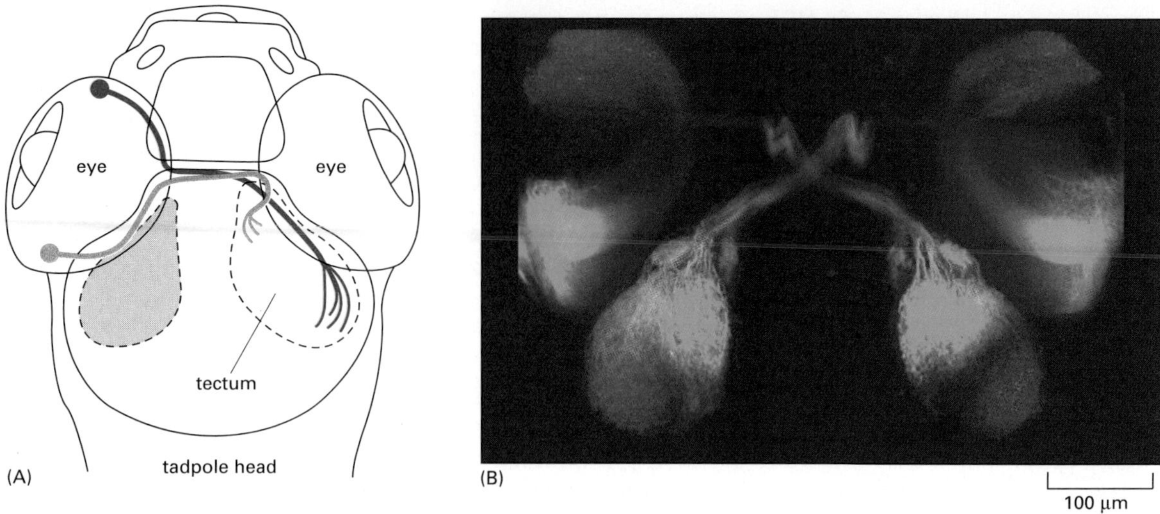

Figure 21–100 The neural map from eye to brain in a young zebrafish. (A) Diagrammatic view, looking down on the top of the head. (B) Fluorescence micrograph. Fluorescent tracer dyes have been injected into each eye—*red* into the anterior part, *green* into the posterior part. The tracer molecules have been taken up by the neurons in the retina and carried along their axons, revealing the paths they take to the optic tectum in the brain and the map that they form there. (Courtesy of Chi-Bin Chien, from D.H. Sanes, T.A. Reh, and W.A. Harris, Development of the Nervous System. San Diego, CA: Academic Press, 2000.)

predictable pattern according to the arrangement of their cell bodies in the retina: ganglion cells that are neighbors in the retina connect with target cells that are neighbors in the tectum. The orderly projection creates a **map** of visual space on the tectum (Figure 21–100).

Orderly maps of this sort are found in many brain regions. In the auditory system, for example, neurons project from the ear to the brain in a tonotopic order, creating a map in which brain cells receiving information about sounds of different pitch are ordered along a line, like the keys of a piano. And in the somatosensory system, neurons conveying information about touch map onto the cerebral cortex so as to mark out a "homunculus"—a small, distorted, two dimensional image of the body surface (Figure 21–101).

The retinotopic map of visual space in the optic tectum is the best characterized of all these maps. How does it arise? In principle, the growth cones could

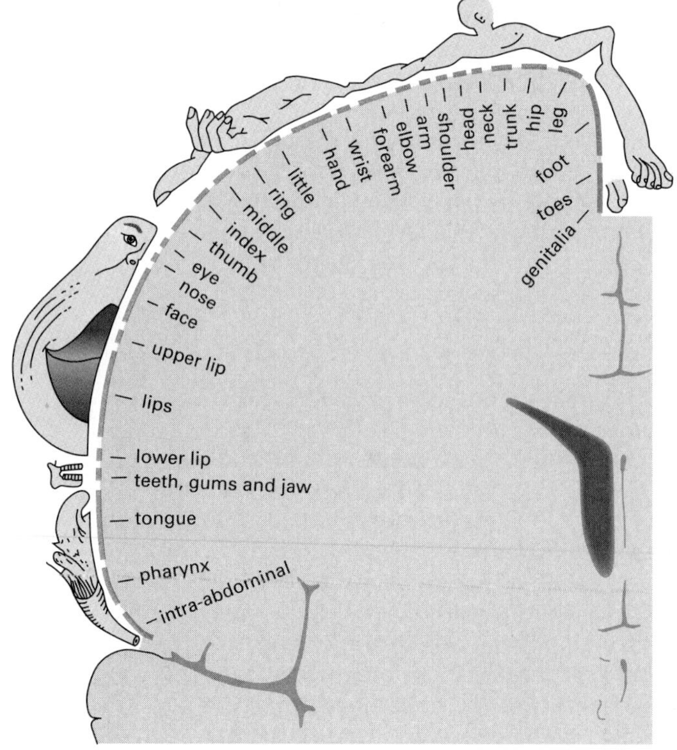

Figure 21–101 A map of the body surface in the human brain. The surface of the body is mapped onto the somatosensory region of the cerebral cortex by an orderly system of nerve cell connections, such that sensory information from neighboring body sites is delivered to neighboring sites in the brain. This means that the map in the brain is largely faithful to the topology of the body surface, even though different body regions are represented at different magnifications according to their density of innervation. The homunculus (the "little man" in the brain) has big lips, for example, because the lips are a particularly large and important source of sensory information. The map was determined by stimulating different points in the cortex of conscious patients during brain surgery and recording what they said they felt. (After W. Penfield and T. Rasmussen, The Cerebral Cortex of Man. New York: Macmillan, 1950.)

be physically channeled to different destinations as a consequence of their different starting positions, like drivers on a multilane highway where it is forbidden to change lanes. This possibility was tested in the visual system by a famous experiment in the 1940s. If the optic nerve of a frog is cut, it will regenerate. The retinal axons grow back to the optic tectum, restoring normal vision. If, in addition, the eye is rotated in its socket at the time of cutting of the nerve, so as to put originally ventral retinal cells in the position of dorsal retinal cells, vision is still restored, but with an awkward flaw: the animal behaves as though it sees the world upside down and left–right inverted. This is because the misplaced retinal cells make the connections appropriate to their original, not their actual, positions. It seems that the cells have positional values—position-specific biochemical properties representing records of their original location. As a result, cells on opposite sides of the retina are intrinsically different, just as the motor neurons in the spinal cord that project to different muscles are intrinsically different.

Such nonequivalence among neurons is referred to as *neuronal specificity*. It is this intrinsic characteristic that guides the retinal axons to their appropriate target sites in the tectum. Those target sites themselves are distinguishable by the retinal axons because the tectal cells also carry positional labels. Thus the neuronal map depends on a correspondence between two systems of positional markers, one in the retina and the other in the tectum.

Axons From Different Regions of the Retina Respond Differently to a Gradient of Repulsive Molecules in the Tectum

Axons from the nasal retina (the side closest to the nose) project to the posterior tectum, and axons from the temporal retina (the side farthest from the nose) project to the anterior tectum, with intermediate regions of retina projecting to intermediate regions of tectum. When nasal and temporal axons are allowed to grow out over a carpet of anterior or posterior tectal membranes in a culture dish, they also show selectivity (Figure 21–102). Temporal axons strongly prefer the anterior tectal membranes, as *in vivo*, whereas nasal axons either prefer posterior tectal membranes, or show no preference (depending on the species of animal). The key difference between anterior and posterior tectum appears to be a repulsive factor on the posterior tectum, to which temporal retinal axons are sensitive but nasal retinal axons are not: if a temporal retinal growth cone touches posterior tectal membrane, it collapses its filopodia and withdraws.

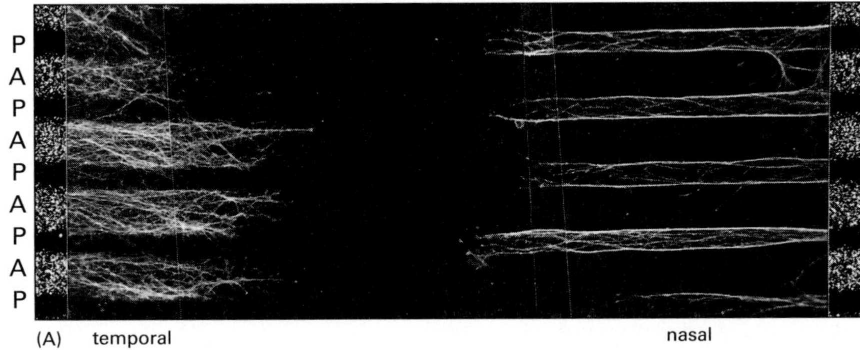

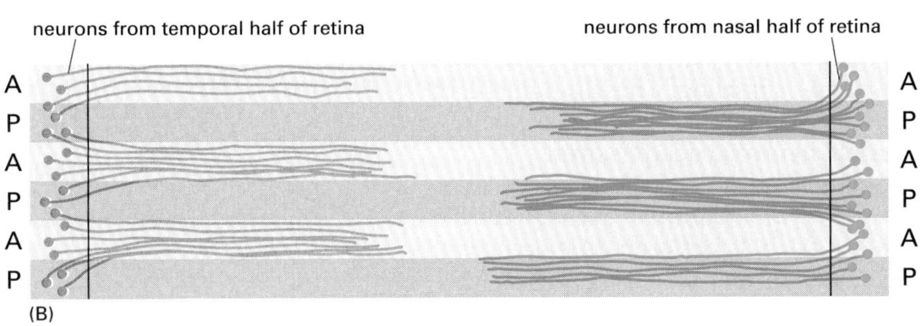

Figure 21–102 Selectivity of retinal axons growing over tectal membranes. (A) A photograph of the experimental observation. (B) A diagram of what is happening. The culture substratum has been coated with alternating stripes of membrane prepared either from posterior tectum (P) or from anterior tectum (A). In the photograph, the anterior tectal stripes are made visible by staining them with a fluorescent marker in the vertical strips at the sides of the picture. Axons of neurons from the temporal half of the retina (growing in from the *left*) follow the stripes of anterior tectal membrane but avoid the posterior tectal membrane, while axons of neurons from the nasal half of the retina (growing in from the *right*) do the converse. Thus anterior tectum differs from posterior tectum and nasal retina from temporal retina, and the differences guide selective axon outgrowth. These experiments were performed with cells from the chick embryo. (From Y. von Boxberg, S. Deiss, and U. Schwarz, *Neuron* 10:345–357, 1993. © Elsevier.)

Assays based on these phenomena *in vitro* have identified some of the molecules responsible. The repulsive factor on posterior tectal membrane seems to be partly or entirely comprised of *ephrin A* proteins, a subset of the family of GPI-linked proteins that act as ligands for the *Eph family* of tyrosine kinase receptors. In the mouse, two different ephrins are expressed to form an anterior-to-posterior gradient on the tectal cells. Anterior cells have little or no ephrin, cells in the center of the tectum express ephrin A2, and cells at the posterior edge of the tectum express ephrin A2 and ephrin A5. Thus there is a gradient of ephrin expression across the tectum. Meanwhile, the incoming axons express Eph receptors, also in a gradient: temporal axons express high Eph levels, making them sensitive to repulsion by ephrin A, whereas nasal axons express low Eph levels.

This system of signals and receptors is enough to produce an orderly map, if we make one further assumption—an assumption supported by experiments *in vivo*: that the retinal axons somehow interact with one another and compete for tectal territory. Thus, temporal axons are restricted to anterior tectum, and drive nasal axons off it; nasal axons, consequently, are restricted to posterior tectum. Between the extremes, a balance is struck, creating a smooth map of the temporo-nasal axis of the retina onto the anteroposterior axis of the tectum. Ephrin gene knockout studies in the mouse are consistent with this picture, although they also indicate that additional cues help to guide the pattern of retino-tectal projections. Moreover, the map also has to be patterned along the dorsoventral axis; it is thought that this depends on similar mechanisms, perhaps even involving some of the same molecules.

Diffuse Patterns of Synaptic Connections Are Sharpened by Activity-Dependent Remodeling

In a normal animal the retinotectal map is initially fuzzy and imprecise: the system of matching markers we have just described is enough to define the broad layout of the map, but not sufficient to specify its fine details. Studies in frogs and fish show that each retinal axon at first branches widely in the tectum and makes a profusion of synapses, distributed over a large area of tectum that overlaps with the territories innervated by other axons. These territories are subsequently trimmed back by selective elimination of synapses and retraction of axon branches. This is accompanied by the formation of new sprouts, through which each axon develops a denser distribution of synapses in the territory that it retains.

A central part in this remodeling and refinement of the map is played by two competition rules that jointly help to create spatial order: (1) axons from separate regions of retina, which tend to be excited at different times, compete to dominate the available tectal territory, but (2) axons from neighboring sites in the retina, which tend to be excited at the same time, innervate neighboring territories in the tectum because they collaborate to retain and strengthen their synapses on shared tectal cells (Figure 21–103). The mechanism underlying both these rules depends on electrical activity and signaling at the synapses that are formed. If all action potentials are blocked by a toxin that binds to voltage-gated Na^+ channels, synapse remodeling is inhibited and the map remains fuzzy.

The phenomenon of activity-dependent synapse elimination is encountered in almost every part of the developing vertebrate nervous system. Synapses are first formed in abundance and distributed over a broad target field; then the system of connections is pruned back and remodeled by competitive processes that depend on electrical activity and synaptic signaling. The elimination of synapses in this way is distinct from the elimination of surplus neurons by cell death, and it occurs after the period of normal neuronal death is over.

Much of what we know about the cellular mechanisms of synapse formation and elimination comes from experiments on the innervation of skeletal muscle in vertebrate embryos. A two-way exchange of signals between the nerve axon terminals and the muscle cells controls the initial formation of synapses. At sites of contact, acetylcholine receptors are clustered in the muscle cell membrane

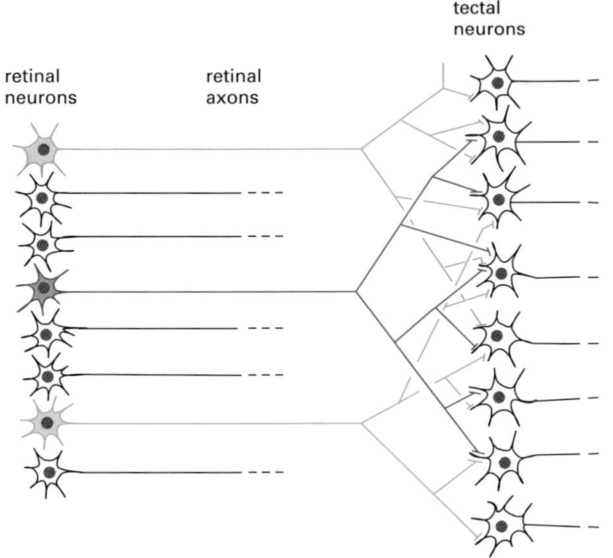
tectal neurons
retinal neurons
retinal axons
FUZZY INITIAL MAP: DIFFUSE CONNECTIONS

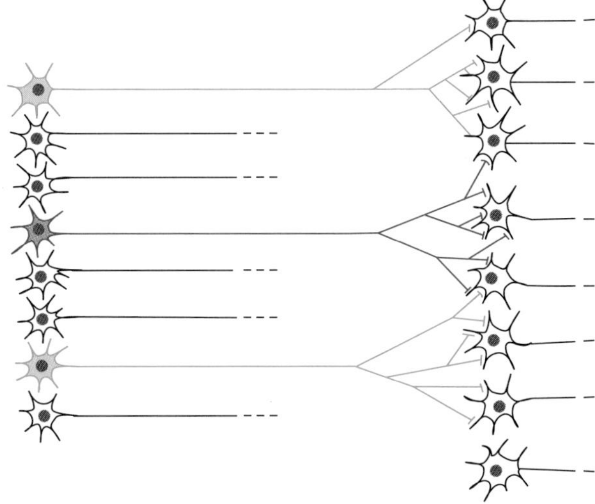
SHARP FINAL MAP: DIFFUSE CONNECTIONS ELIMINATED

Figure 21–103 Sharpening of the retinotectal map by synapse elimination. At first the map is fuzzy because each retinal axon branches widely to innervate a broad region of tectum overlapping the regions innervated by other retinal axons. The map is then refined by synapse elimination. Where axons from separate parts of the retina synapse on the same tectal cell, competition occurs, eliminating the connections made by one of the axons. But axons from cells that are close neighbors in the retina cooperate, maintaining their synapses on shared tectal cells. Thus each retinal axon ends up innervating a small tectal territory, adjacent to and partly overlapping the territory innervated by axons from neighboring sites in the retina.

and the apparatus for secretion of this neurotransmitter becomes organized in the axon terminals (see Chapter 11). Each muscle cell at first receives synapses from several neurons; but in the end, through a process that typically takes a couple of weeks, it is left innervated by only one. The synapse retraction again depends on synaptic communication: if synaptic transmission is blocked by a toxin that binds to the acetylcholine receptors in the muscle cell membrane, the muscle cell retains its multiple innervation beyond the normal time of elimination.

Experiments on the musculoskeletal system, as well as in the retinotectal system, suggest that it is not only the amount of electrical activity at a synapse that is important for its maintenance, but also its temporal coordination. Whether a synapse is strengthened or weakened seems to depend critically on whether or not activity in the presynaptic cell is synchronized with activity of the other presynaptic cells synapsing on the same target (and thus also synchronized with activity of the target cell itself).

These and many other findings have suggested a simple interpretation of the competition rules for synapse elimination in the retinotectal system (Figure 21–104). Axons from different parts of the retina fire at different times and so compete. Each time one of them fires, the synapse(s) made by the other on a shared tectal target cell are weakened, until one of the axons is left in sole command of

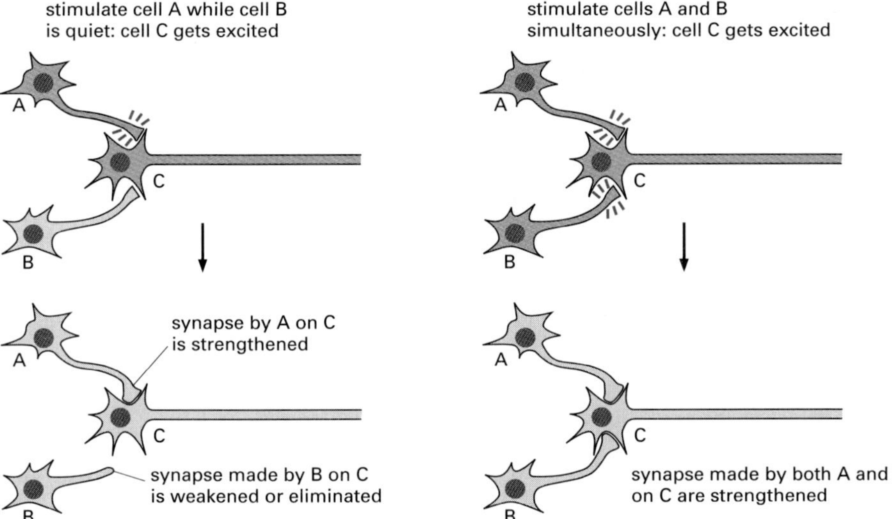
stimulate cell A while cell B is quiet: cell C gets excited
A
B
C
synapse by A on C is strengthened
synapse made by B on C is weakened or eliminated
stimulate cells A and B simultaneously: cell C gets excited
synapse made by both A and B on C are strengthened

Figure 21–104 Synapse modification and its dependence on electrical activity. Experiments in several systems indicate that synapses are strengthened or weakened by electrical activity according to the rule shown in the diagram. The underlying principle appears to be that each excitation of a target cell tends to weaken any synapse where the presynaptic axon terminal has just been quiet but to strengthen any synapse where the presynaptic axon terminal has just been active. As a result, "neurons that fire together, wire together." A synapse that is repeatedly weakened and rarely strengthened is eventually eliminated altogether.

that cell. Axons from neighboring retinal cells, on the other hand, tend to fire in synchrony with one another: they therefore do not compete but instead maintain synapses on shared tectal cells, creating a precisely ordered map in which neighboring cells of the retina project to neighboring sites in the tectum.

Experience Molds the Pattern of Synaptic Connections in the Brain

The phenomenon that we have just described is summed up in the catch-phrase that "neurons that fire together, wire together". The same firing rule relating synapse maintenance to neural activity helps to organize our developing brains in the light of experience.

In the brain of a mammal, axons relaying inputs from the two eyes are brought together in a specific cell layer in the visual region of the cerebral cortex. Here, they form two overlapping maps of the external visual field, one as perceived through the right eye, the other as perceived through the left. Although there is some evidence of a tendency for right- and left-eye inputs to be segregated even before synaptic communication begins, a large proportion of the axons carrying information from the two eyes at early stages synapse together on shared cortical target cells. A period of early signaling activity, however, occurring spontaneously and independently in each retina even before vision begins, leads to a clean segregation of inputs, creating stripes of cells in the cortex that are driven by inputs from the right eye alternating with stripes that are driven by inputs from the left eye (Figure 21–105). The firing rule suggests a simple interpretation: a pair of axons bringing information from neighboring sites in the left eye will frequently fire together, and therefore wire together, as will a pair of axons from neighboring sites in the right eye; but a right-eye axon and a left-eye axon will rarely fire together, and will instead compete. Indeed, if activity from both eyes is silenced using drugs that block action potentials or synaptic transmission, the inputs fail to segregate correctly.

Maintenance of the pattern of connections is extraordinarily sensitive to experience early in life. If, during a certain *critical period* (ending at about the age of five years in humans), one eye is kept covered for a time so as to deprive it of visual stimulation, while the other eye is allowed normal stimulation, the deprived eye loses its synaptic connections to the cortex and becomes almost entirely, and irreversibly, blind. In accordance with what the firing rule would predict, a competition has occurred in which synapses in the visual cortex made by inactive axons are eliminated while synapses made by active axons are consolidated. In this way cortical territory is allocated to axons that carry information and is not wasted on those that are silent.

In establishing the nerve connections that enable us to see, it is not only the quantity of visual stimulation that is important, but also its temporal coordination. For example, the ability to see depth—stereo vision—depends on cells in other layers of the visual cortex that receive inputs relayed from both eyes at once, conveying information about the same part of the visual field as seen from two slightly different angles. These binocularly driven cells allow us to compare

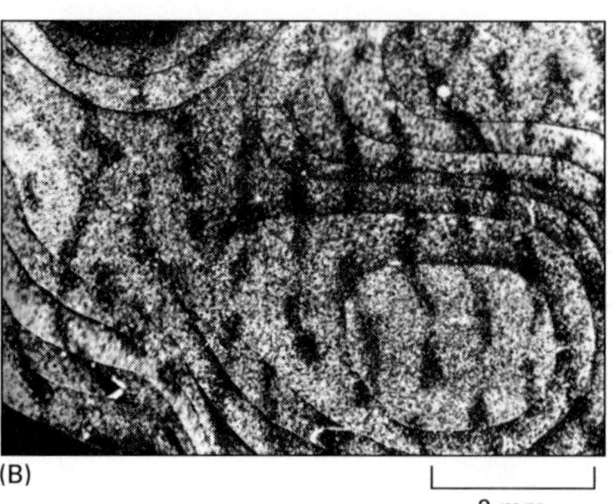

Figure 21–105 Ocular dominance columns in the visual cortex of a monkey's brain, and their sensitivity to visual experience. (A) Normally, stripes of cortical cells driven by the right eye alternate with stripes, of equal width, driven by the left eye. The stripes are revealed here by injecting a radioactive tracer molecule into one eye, allowing time for this tracer to be transported to the visual cortex, and detecting radioactivity there by autoradiography, in sections cut parallel to the cortical surface. (B) If one eye is kept covered during the critical period of development, and thus deprived of visual experience, its stripes shrink and those of the active eye expand. In this way, the deprived eye may lose the power of vision almost entirely. (From D.H. Hubel, T.N. Wiesel, and S. Le Vay, *Philos. Trans. Roy. Soc. (Biol.)* 278:377–409, 1977. © The Royal Society.)

the view through the right eye with that through the left so as to derive information about the relative distances of objects from us. If, however, the two eyes are prevented during the critical period from ever seeing the same scene at the same time—for example, by covering first one eye and then the other on alternate days or simply as a consequence of a childhood squint—almost no binocularly driven cells are retained in the cortex, and the capacity for stereo perception is irretrievably lost. Evidently, in accordance with the firing rule, the inputs from each eye to a binocularly driven neuron are maintained only if the two inputs are frequently triggered to fire in synchrony, as occurs when the two eyes look together at the same scene.

Adult Memory and Developmental Synapse Remodeling May Depend on Similar Mechanisms

We saw in Chapter 11 that synaptic changes underlying memory in at least some parts of the adult brain, notably the hippocampus, hinge on the behavior of a particular type of receptor for the neurotransmitter glutamate—the NMDA receptor. Ca^{2+} flooding into the postsynaptic cell through the channels opened by this receptor triggers lasting changes in the strengths of the synapses on that cell, affecting the presynaptic as well as the postsynaptic structures. The changes that are induced by the NMDA-dependent mechanism in the adult brain obey rules closely akin to the developmental firing rule: events in the external world that cause two neurons to be active at the same time, or in quick succession, favor the making or strengthening of synapses between them. This condition, called the *Hebb rule*, has been suggested to be the fundamental principle underlying associative learning.

Is it possible, then, that adult learning and the more drastic forms of synaptic plasticity seen during development both reflect the operation of the same basic machinery of synapse adjustment? There are many hints that it may be so. For example, inhibitors that specifically block activation of the NMDA receptor have been found to interfere with the refinement and remodeling of synaptic connections in the developing visual system. But the question is still open. The molecular basis of the processes of synapse remodeling through which experience molds our brains remains one of the central challenges that the nervous system presents to cell biology.

Summary

The development of the nervous system proceeds in three phases: first, nerve cells are generated through cell division; then, having ceased dividing, they send out axons and dendrites to form profuse synapses with other, remote cells so that communication can begin; last, the system of synaptic connections is refined and remodeled according to the pattern of electrical activity in the neural network.

The neurons, and the glial cells that always accompany them, are generated from ectodermal precursors, and those born at different times and places express different sets of genes, which help to determine the connections they will form. Axons and dendrites grow out from the neurons by means of growth cones, which follow specific pathways delineated by signals along the way. Structures such as the floor plate of the embryonic spinal cord secrete both chemoattractants and chemorepellents, to which growth cones from different classes of neurons respond differently. On reaching their target area, the axons terminate selectively on a subset of the accessible cells, and in many parts of the nervous system neural maps are set up—orderly projections of one array of neurons onto another. In the retinotectal system, the map is based on the matching of complementary systems of position-specific cell-surface markers—ephrins and Eph receptors—possessed by the two sets of cells.

After the growth cones have reached their targets and initial connections have formed, two major sorts of adjustment occur. First, many of the innervating neurons die as a result of a competition for survival factors such as NGF (nerve growth factor) secreted by the target tissue. This cell death adjusts the quantity of innervation according to the size of the target. Second, individual synapses are pruned away in

some places, reinforced in others, so as to create a more precisely ordered pattern of connections. This latter process depends on electrical activity: synapses that are frequently active are reinforced, and different neurons contacting the same target cell tend to maintain their synapses on the shared target only if they are both frequently active at the same time. In this way the structure of the brain can be adjusted to reflect the connections between events in the external world. The underlying molecular mechanism of this synaptic plasticity may be similar to that responsible for the formation of memories in adult life.

PLANT DEVELOPMENT

Plants and animals are separated by about 1.5 billion years of evolutionary history. They have evolved their multicellular organization independently but using the same initial tool kit—the set of genes inherited from their common unicellular eucaryotic ancestor. Most of the contrasts in their developmental strategies spring from two basic peculiarities of plants. First, they get their energy from sunlight, not by ingesting other organisms. This dictates a body plan different from that of animals. Second, their cells are encased in semirigid cell walls and cemented together, preventing them from moving as animal cells do. This dictates a different set of mechanisms for shaping the body and different developmental processes to cope with a changeable environment.

Animal development is largely buffered against environmental changes, and the embryo generates the same genetically determined body structure unaffected by external conditions. The development of most plants, by contrast, is dramatically influenced by the environment. Because they cannot match themselves to their environment by moving from place to place, plants adapt instead by altering the course of their development. Their strategy is opportunistic. A given type of organ—a leaf, a flower, or a root, say—can be produced from the fertilized egg by many different paths according to environmental cues. A begonia leaf pegged to the ground may sprout a root; the root may throw up a shoot; the shoot, given sunlight, may grow leaves and flowers.

The mature plant is typically made of many copies of a small set of standardized modules, as described in Figure 21–106. The positions and times at which those modules are generated are strongly influenced by the environment, causing the overall structure of the plant to vary. The choices between alternative modules and their organization into a whole plant depend on external cues and long-range hormonal signals that play a much smaller part in the control of animal development.

But although the global structure of a plant—its pattern of roots or branches, its numbers of leaves or flowers—can be highly variable, its detailed organization on a small scale is not. A leaf, a flower, or indeed an early plant embryo, is as precisely specified as any organ of an animal, possessing a *determinate* structure, in contrast with the *indeterminate* pattern of branching and sprouting of the plant as a whole. The internal organization of a plant module raises essentially the same problems in the genetic control of pattern formation as does animal development, and they are solved in analogous ways. In this section we focus on the cellular mechanisms of development in flowering plants. We examine both the contrasts and the similarities with animals.

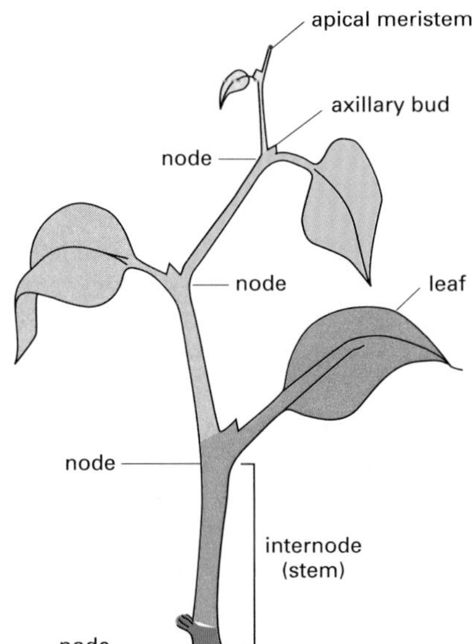

Figure 21–106 A simple example of the modular construction of plants. Each module (shown in different shades of *green*) consists of a stem, a leaf, and a bud containing a potential growth center, or *meristem*. The bud forms at the branch point, or *node*, where the leaf diverges from the stem. Modules arise sequentially from the continuous activity of the apical meristem.

Arabidopsis Serves as a Model Organism for Plant Molecular Genetics

Flowering plants, despite their amazing variety, are of relatively recent origin. The earliest known fossil examples are 125 million years old, as against 350 million years or more for vertebrate animals. Underlying the diversity of form, therefore, there is a high degree of similarity in molecular mechanisms. As we shall see, a small genetic change can transform a plant's large-scale structure; and just as plant physiology allows survival in many different environments, so also it allows survival of many differently structured forms. A mutation that gives

Figure 21–107 *Arabidopsis thaliana.* This small plant is a member of the mustard (or crucifer) family (see also Figure 1–47). It is a weed of no economic use but of great value for genetic studies of plant development. (From M.A. Estelle and C.R. Somerville, *Trends Genet.* 12:89–93, 1986. © Elsevier.)

an animal two heads is generally lethal; one that doubles the number of flowers or branches on a plant is generally not.

To identify the genes that govern plant development and to discover how they function, plant biologists have selected a small weed, the common wall cress *Arabidopsis thaliana* (Figure 21–107), as their primary model organism. Like *Drosophila* or *Caenorhabditis elegans,* it is small, quick to reproduce, and convenient for genetics. It can be grown indoors in Petri dishes or tiny plant pots in large numbers and produces hundreds of seeds per plant after 8 to 10 weeks. It has, in common with *C. elegans,* a significant advantage over *Drosophila* or vertebrate animals for genetics: like many flowering plants, it can reproduce as a hermaphrodite because a single flower produces both eggs and the male gametes that can fertilize them. Therefore, when a flower that is heterozygous for a recessive lethal mutation is self-fertilized, one-fourth of its seeds will display the homozygous embryonic phenotype. This makes it easy to perform genetic screens (Figure 21–108) and so to obtain a catalog of the genes required for specific developmental processes.

The *Arabidopsis* Genome Is Rich in Developmental Control Genes

Arabidopsis has one of the smallest plant genomes—125 million nucleotide pairs, on a par with *C. elegans* and *Drosophila*—and the complete DNA sequence is now known. It contains approximately 26,000 genes. This total includes many recently generated duplicates, however, so that the number of functionally distinct types of protein represented may be considerably less. Cell culture and genetic transformation methods have been established, as well as vast libraries

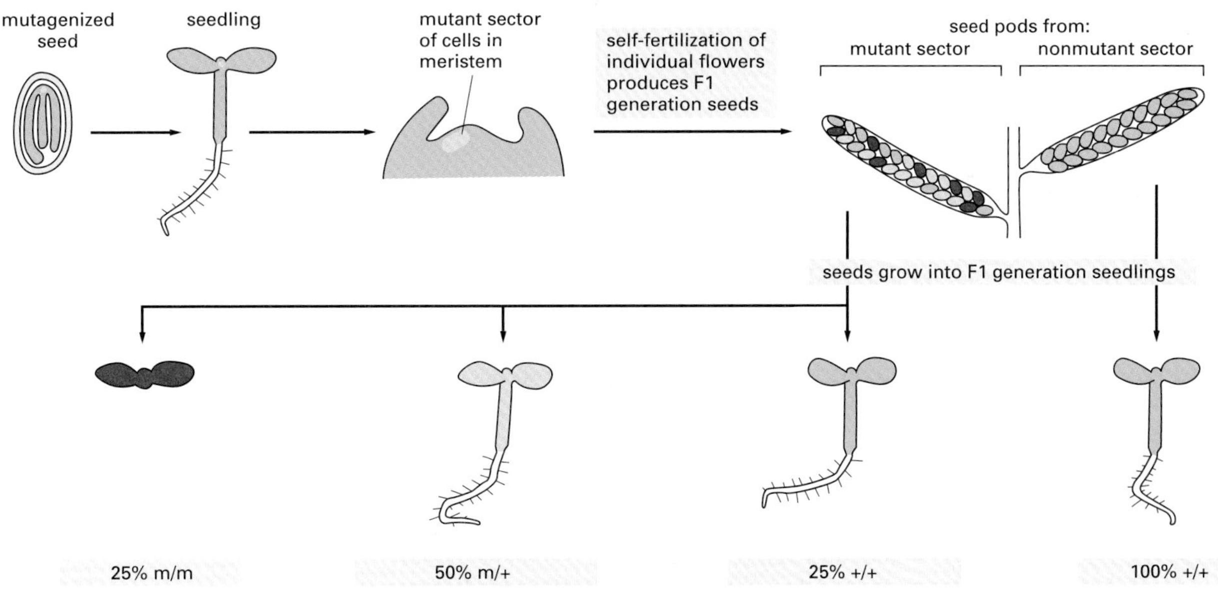

Figure 21–108 Production of mutants in *Arabidopsis.* A seed, containing a multicellular embryo, is treated with a chemical mutagen and left to grow into a plant. In general, this plant will be a mosaic of clones of cells carrying different induced mutations. An individual flower produced by this plant will usually be composed of cells belonging to the same clone, all carrying the same mutation, *m,* in heterozygous form (*m*/+). Self-fertilization of individual flowers by their own pollen results in seed pods, each of which contains a family of embryos of whose members half, on average, will be heterozygous (*m*/+), one quarter will be homozygous mutant (*m*/*m*), and one quarter will be homozygous wild-type (+/+). Often, the mutation will have a recessive lethal effect, as indicated here by the lack of a root in the *m*/*m* seedling. The mutant stock is then maintained by breeding from the heterozygote: it will produce seed pods (F2 generation) that all contain a mixture of +/+, *m*/+, and *m*/*m* seeds.

TABLE 21–2 Some Major Families of Gene Regulatory Proteins in *Arabidopsis*, *Drosophila*, *C. elegans*, and the Yeast *Saccharomyces cerevisiae*

FAMILY	NUMBER OF FAMILY MEMBERS PREDICTED FROM GENOME ANALYSIS			
	ARABIDOPSIS	*DROSOPHILA*	*C. ELEGANS*	YEAST
Myb	190	6	3	10
AP2/EREBP (Apetala2/ethylene-responsive-element binding protein)	144	0	0	0
bHLH (basic helix–loop–helix)	139	46	25	8
NAC	109	0	0	0
C2H2 (Zn finger)	105	291	139	53
Homeobox	89	103	84	9
MADS box	82	2	2	4
bZIP	81	21	25	21
WRKY (Zn finger)	72	0	0	0
GARP	56	0	0	0
C2C2 (Zn finger)/GATA	104	6	9	10
Nuclear hormone receptor	0	21	25	0
C6 (Zn finger)	0	0	0	52
Estimated total (including many not listed above)	1,533	635	669	209
% of genes in genome	5.9	4.5	3.5	3.5

The Table lists only those families that have at least 50 members in at least one organism. (Data from J.L. Riechmann et al., *Science* 290:2105–2110, 2000.)

of seeds carrying mutations produced by random insertions of mobile genetic elements, so that plants with mutations in any chosen gene can be obtained to order. Powerful tools are thus available to analyze gene functions. Although only a small fraction of the total gene set has been characterized experimentally as yet, functions can be tentatively assigned to many genes—about 18,000—on the basis of their sequence similarities to well-characterized genes in *Arabidopsis* and other organisms.

Even more than the genomes of multicellular animals, the *Arabidopsis* genome is rich in genes that code for gene regulatory proteins (Table 21–2). Some major families of animal gene regulatory proteins (such as the Myb family of DNA-binding proteins) are greatly expanded, while others (such as nuclear hormone receptors) seem to be entirely absent, and there are large families of gene regulatory proteins in the plant that have no animal homologs.

Where homologous gene regulatory proteins (such as homeodomain proteins) can be recognized in both plants and animals, they have little in common with regard to the genes they regulate or the types of developmental decisions that they control, and there is very little conservation of protein sequence outside the DNA-binding domains.

Arabidopsis is like multicellular animals in possessing many genes for cell communication and signal transduction (1,900 genes out of 18,000 classified), but the specific details of these gene sets are very different, as discussed in Chapter 15. The Wnt, Hedgehog, Notch, and TGFβ signaling mechanisms are all absent in *Arabidopsis*. In compensation, other signaling pathways peculiar to plants are highly developed. Cell-surface receptors of the tyrosine kinase class seem to be entirely absent, although many of the signaling components downstream of these receptors in animals are present. Conversely, receptors of the serine/threonine kinase class are very plentiful, but they do not act through the same system of intracellular messengers as the receptor serine/threonine kinases in animals. Substantial sets of genes are devoted to developmental processes of special importance in plants: more than 700 for synthesis and remodelling of the plant cell wall, for example, and more than 100 for detecting and responding to light.

We must now examine how the genes of the plant are used to control plant development.

Embryonic Development Starts by Establishing a Root–Shoot Axis and Then Halts Inside the Seed

The basic strategy of sexual reproduction in flowering plants is briefly summarized in Panel 21–1. The fertilized egg, or zygote, of a higher plant begins by dividing asymmetrically to establish the polarity of the future embryo. One product of this division is a small cell with dense cytoplasm, which will become the embryo proper. The other is a large vacuolated cell that divides further and forms a structure called the *suspensor*, which in some ways is comparable to the umbilical cord in mammals. The suspensor attaches the embryo to the adjacent nutritive tissue and provides a pathway for the transport of nutrients.

During the next step in development the diploid embryo cell proliferates to form a ball of cells that quickly acquires a polarized structure. This comprises two key groups of proliferating cells—one at the suspensor end of the embryo that will collaborate with the uppermost suspensor cell to generate a root, and one at the opposite end that will generate a shoot (Figure 21–109). The main root-shoot axis established in this way is analogous to the head-to-tail axis of an animal. At the same time it begins to be possible to distinguish the future *epidermal cells*, forming the outermost layer of the embryo, the future *ground tissue cells*, occupying most of the interior, and the future *vascular tissue cells*, forming the central core (Panel 21–2). These three sets of cells can be compared to the three germ layers of an animal embryo. Slightly later in development, the rudiment of the shoot begins to produce the embryonic seed leaves, or *cotyledons*—one in the case of monocots and two in the case of dicots. Soon after this stage, development usually halts and the embryo becomes packaged in a **seed** (a case formed by tissues of the mother plant), specialized for dispersal and for survival in harsh conditions. The embryo in a seed is stabilized by dehydration, and it can remain dormant for a very long time—even hundreds of years. When rehydrated, the seeds germinate and embryonic development resumes.

Genetic screens can be used in *Arabidopsis*, just as in *Drosophila* or *C. elegans*, to identify the genes that govern the organization of the embryo and to group these into categories according to their homozygous mutant phenotypes. Some are required for formation of the seedling root, some for the seedling stem, and some for the seedling apex with its cotyledons. Another class is required for formation of the three major tissue types—epidermis, ground tissue, and vascular tissue—and yet another class for the organized changes of cell shape that give the embryo and seedling their elongated form (Figure 21–110).

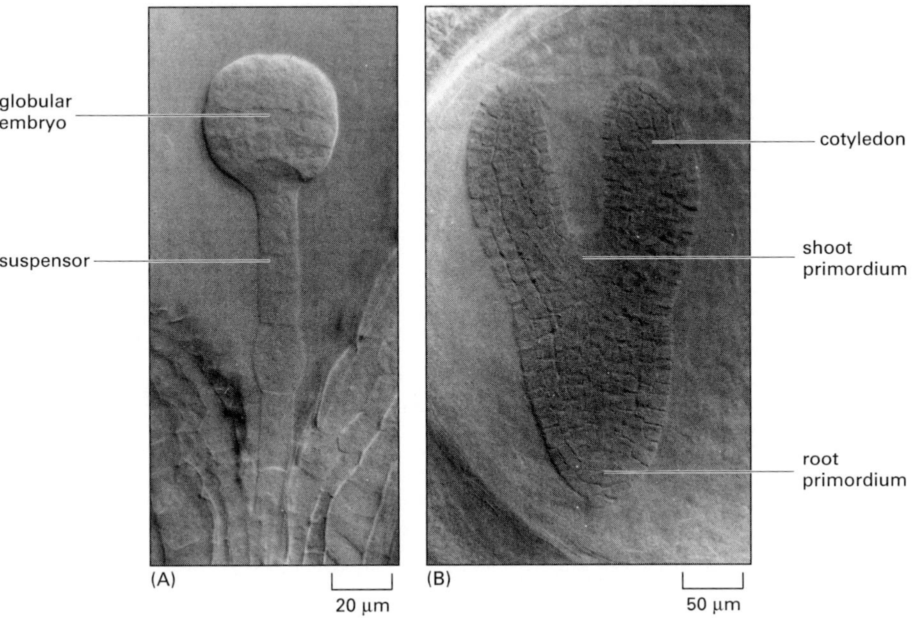

Figure 21–109 Two stages of embryogenesis in *Arabidopsis thaliana*. (From G. Jürgens et al., *Development [Suppl.]* 1:27–38, 1991. © The Company of Biologists.)

PANEL 21–1 Features of Early Development in Flowering Plants

THE FLOWER

Flowers, which contain the reproductive cells of higher plants, arise from vegetative shoot apical meristems (see Figures 21–115 and 21–122). They terminate further vegetative growth from that meristem. Environmental factors, often the rhythms of day length and temperature, trigger the switch from vegetative to floral development. The germ cells thus arise late in plant development from somatic cells rather than from a germ cell line, as in animals.

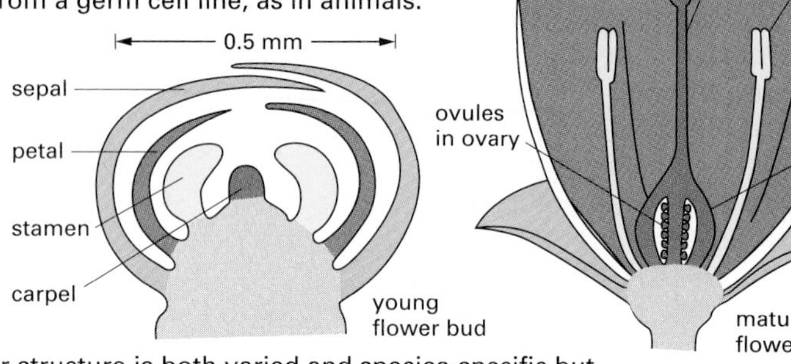

Flower structure is both varied and species-specific but generally comprises four concentrically arranged sets of structures that may each be regarded as modified leaves.

Petal: distinctive leaflike structures, usually brightly colored, facilitate pollination via, for example, attracted insects.

Stamen: an organ containing cells that undergo meiosis and form haploid pollen grains, each of which contains two male sperm cells. Pollen transferred to a stigma germinates, and the pollen tube delivers the two nonmotile sperm to the ovary.

pollen grain
sperm cells
pollen tube nucleus

Carpel: an organ containing one or more ovaries, each of which contains ovules. Each ovule houses cells that undergo meiosis and form an embryo sac containing the female egg cell. At fertilization, one sperm cell fuses with the egg cell and will form the future diploid embryo, while the other fuses with two cells in the embryo sac to form the triploid endosperm tissue.

Sepals: leaflike structures that form a protective covering during early flower development.

THE SEED

A seed contains a dormant embryo, a food store, and a seed coat. By the end of its development a seed's water content can drop from 90% to 5%. The seed is usually protected in a *fruit* whose tissues are of maternal origin.

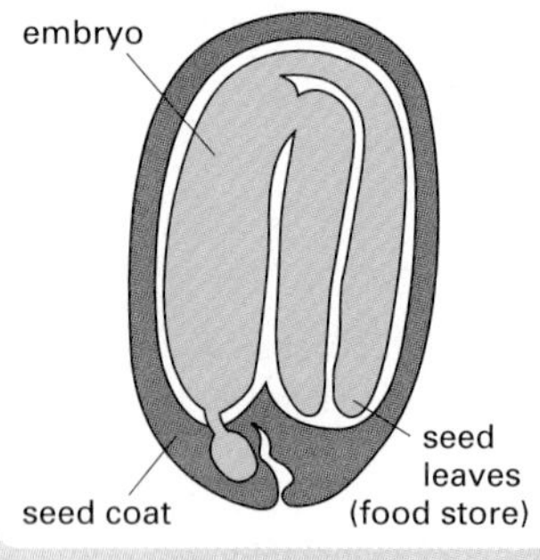

THE EMBRYO

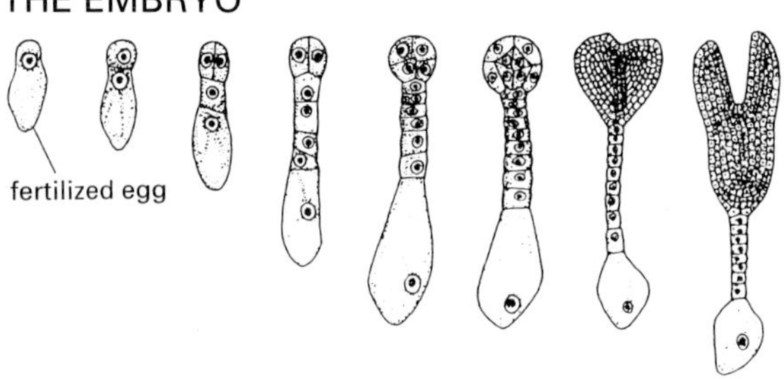

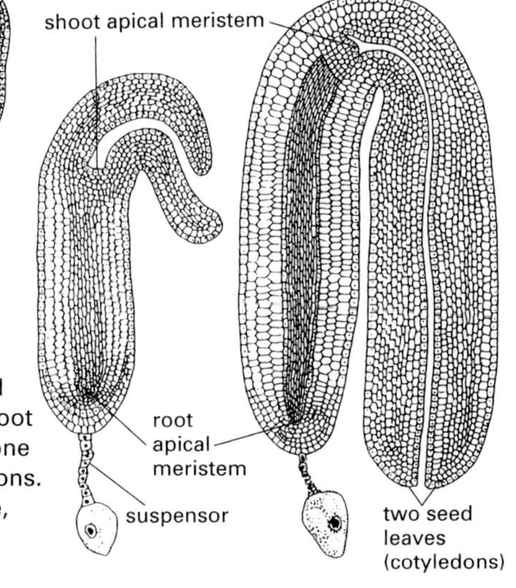

The fertilized egg within the ovule will grow to form an embryo using nutrients transported from the endosperm by the suspensor. A complex series of cell divisions, illustrated here for the common weed called shepherd's purse, produces an embryo with a root apical meristem, a shoot apical meristem, and either one (monocots) or two (dicots) seed leaves, called cotyledons.

Development is arrested at this stage, and the ovule, containing the embryo, now becomes a seed, adapted for dispersal and survival.

GERMINATION

For the embryo to resume its growth the seed must germinate, a process dependent upon both internal factors (dormancy) and environmental factors including water, temperature, and oxygen. The food reserves for the early phase of germination may either be the endosperm (maize) or the cotyledons (pea and bean).

The primary root usually emerges first from the seed to ensure an early water supply for the seedling. The cotyledon(s) may appear above the ground, as in the garden bean shown here, or they may remain in the soil, as in peas. In both cases the cotyledons eventually wither away.

The apical meristem can now show its capacity for continuous growth, producing a typical pattern of nodes, internodes, and buds (see Figure 21–106).

garden bean germination
first foliage leaves
withered cotyledon
seed coat
cotyledons
primary root
lateral roots

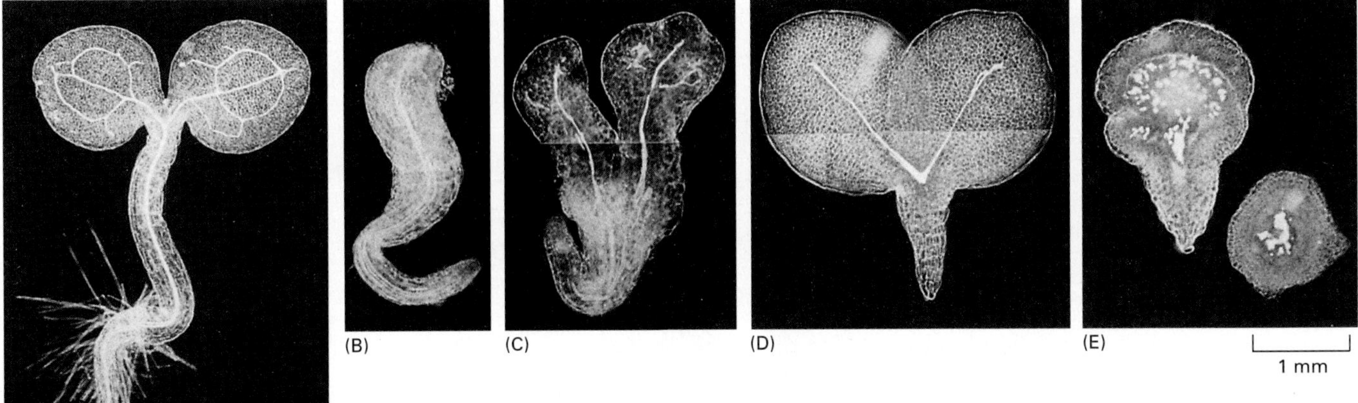

Figure 21–110 Mutant *Arabidopsis* seedlings. A normal seedling (A) compared with four types of mutant (B–E) defective in different parts of their apico-basal pattern: (B) has structures missing at its apex, (C) has an apex and a root but lacks a stem between them, (D) lacks a root, and (E) forms stem tissues but is defective at both ends. The seedlings have been "cleared" so as to show the vascular tissue inside them (pale strands). (From U. Mayer et al., *Nature* 353:402–407, 1991. © Macmillan Magazines Ltd.)

The Parts of a Plant Are Generated Sequentially by Meristems

Roughly speaking, the embryo of an insect or a vertebrate animal is a rudimentary miniature scale model of the later organism, and the details of body structure are filled in progressively as it enlarges. The plant embryo grows into an adult in a quite different way: the parts of the adult plant are created sequentially by groups of cells that proliferate to lay down additional structures at the plant's periphery. These all-important groups of cells are called **apical meristems** (see Figure 21–106). Each meristem consists of a self-renewing population of stem cells. As these divide, they leave behind a trail of progeny that become displaced from the meristem region, enlarge, and finally differentiate. Although the shoot and root apical meristems generate all the basic varieties of cells that are needed to build leaves, roots, and stems, many cells outside the apical meristems also keep a capacity for further proliferation and retain meristem potential. In this way trees and other perennial plants, for example, are able to increase the girth of their stems and roots as the years go by and can sprout new shoots from dormant regions if the plant is damaged.

The rudiments of the apical meristems of root and shoot are already determined in the embryo. As soon as the seed coat ruptures during germination, a dramatic enlargement of nonmeristematic cells occurs, driving the emergence first of a root, to establish an immediate foothold in the soil, and then of a shoot (Figure 21–111). This is accompanied by rapid and continual cell divisions in the apical meristems: in the apical meristem of a maize root, for example, cells divide every 12 hours, producing 5×10^5 cells per day. The rapidly growing root and shoot probe the environment—the root increasing the plant's capacity for taking up water and minerals from the soil, the shoot increasing its capacity for photosynthesis (see Panel 21–1).

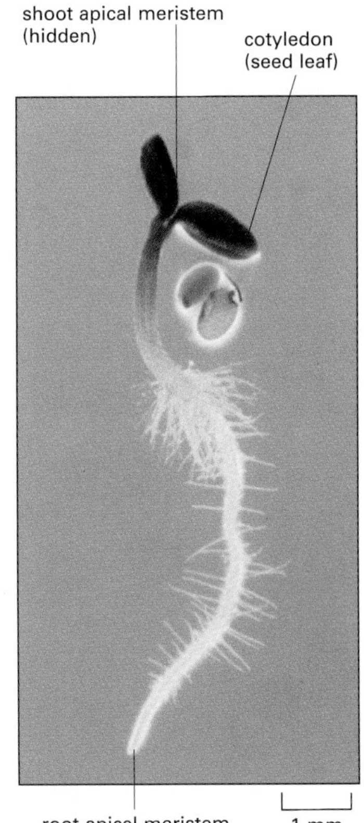

Figure 21–111 A seedling of *Arabidopsis*. The *brown* objects to the *right* of the young seedling are the two halves of the discarded seed coat. (Courtesy of Catherine Duckett.)

Development of the Seedling Depends on Environmental Signals

From germination onward, the course of plant development is powerfully influenced by signals from the environment. The shoot has to push its way rapidly up through the soil, and must open its cotyledons and begin photosynthesis only after it has reached the light. The timing of this transition from rapid subterranean sprouting to illuminated growth cannot be genetically programmed, because the depth at which the seed is buried is unpredictable. The developmental switch is controlled instead by light, which, among other effects, acts on the seedling by inhibiting production of a class of plant hormones called *brassinosteroids,* discussed in Chapter 15. Mutations in genes required for production

THE THREE TISSUE SYSTEMS

Cell division, growth, and differentiation give rise to tissue systems with specialized functions.

DERMAL TISSUE (▬): This is the plant's protective outer covering in contact with the environment. It facilitates water and ion uptake in roots and regulates gas exchange in leaves and stems.

VASCULAR TISSUE: Together the phloem (▭) and the xylem (▬) form a continuous vascular system throughout the plant. This tissue conducts water and solutes between organs and also provides mechanical support.

GROUND TISSUE (▭): This packing and supportive tissue accounts for much of the bulk of the young plant. It also functions in food manufacture and storage.

THE PLANT

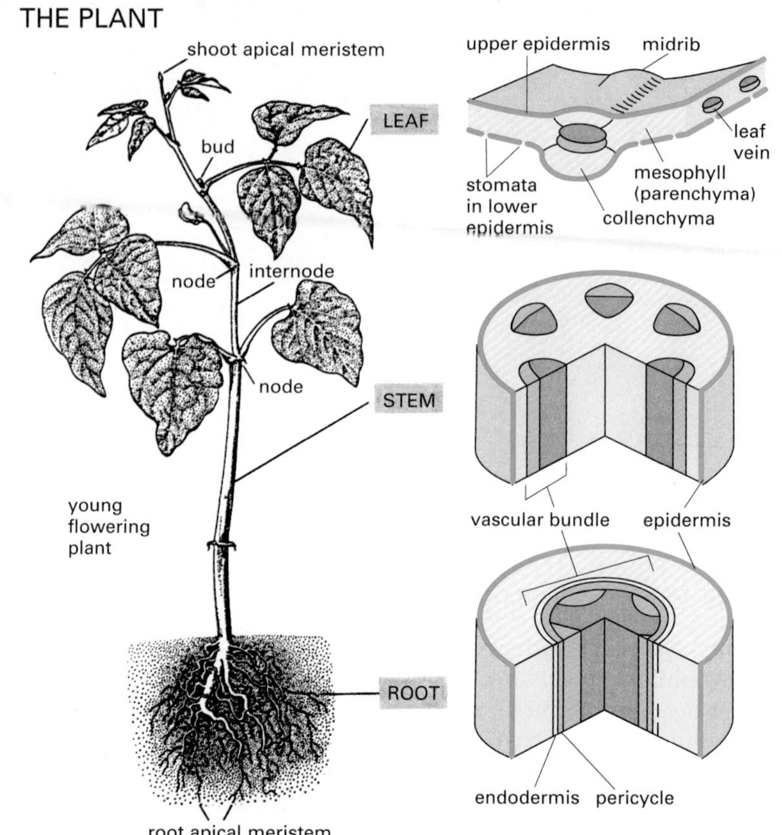

The young flowering plant shown on the *right* is constructed from three main types of organs: leaves, stems, and roots. Each plant organ in turn is made from three tissue systems: ground (▭), dermal (▬), and vascular (▭▬).

All three tissue systems derive ultimately from the cell proliferative activity of the shoot or root apical meristems, and each contains a relatively small number of specialized cell types. These three common tissue systems, and the cells that comprise them, are described in this panel.

GROUND TISSUE

The ground tissue system contains three main cell types called parenchyma, collenchyma, and sclerenchyma.

Parenchyma cells are found in all tissue systems. They are living cells, generally capable of further division, and have a thin primary cell wall. These cells have a variety of functions. The apical and lateral meristematic cells of shoots and roots provide the new cells required for growth. Food production and storage occur in the photosynthetic cells of the leaf and stem (called mesophyll cells); storage parenchyma cells form the bulk of most fruits and vegetables. Because of their proliferative capacity, parenchyma cells also serve as stem cells for wound healing and regeneration.

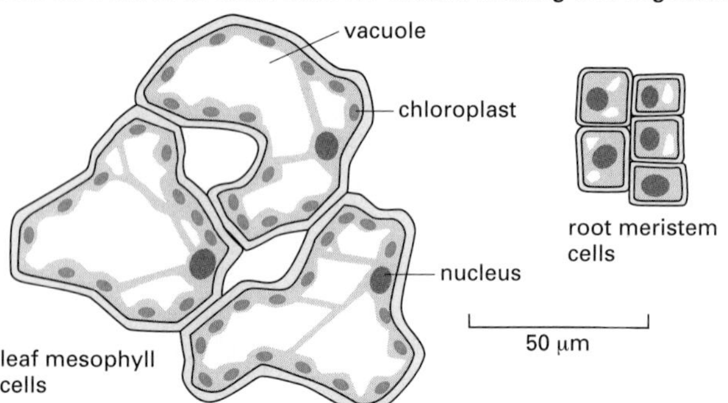

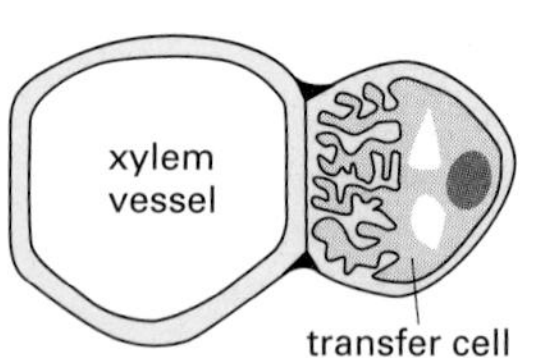

A transfer cell, a specialized form of the parenchyma cell, is readily identified by elaborate ingrowths of the primary cell wall. The increase in the area of the plasma membrane beneath these walls facilitates the rapid transport of solutes to and from cells of the vascular system.

Collenchyma are living cells similar to parenchyma cells except that they have much thicker cell walls and are usually elongated and packed into long ropelike fibers. They are capable of stretching and provide mechanical support in the ground tissue system of the elongating regions of the plant. Collenchyma cells are especially common in subepidermal regions of stems.

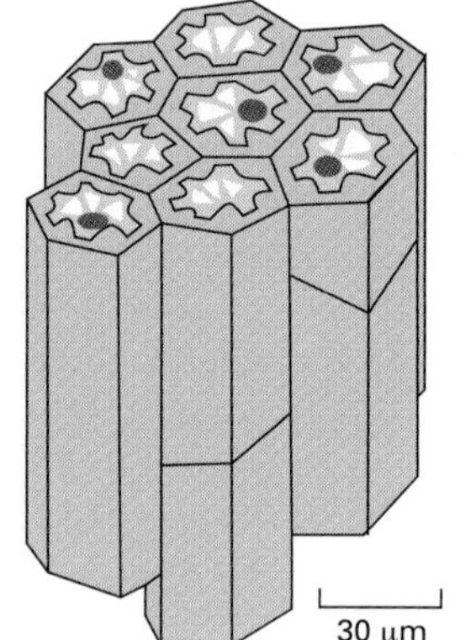

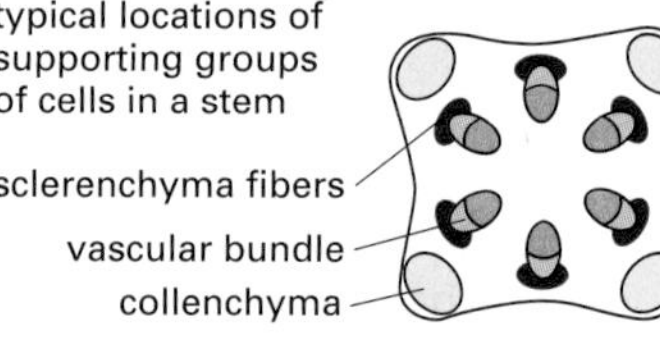

Sclerenchyma, like collenchyma, have strengthening and supporting functions. However, they are usually dead cells with thick, lignified secondary cell walls that prevent them from stretching as the plant grows. Two common types are fibers, which often form long bundles, and sclereids, which are shorter branched cells found in seed coats and fruit.

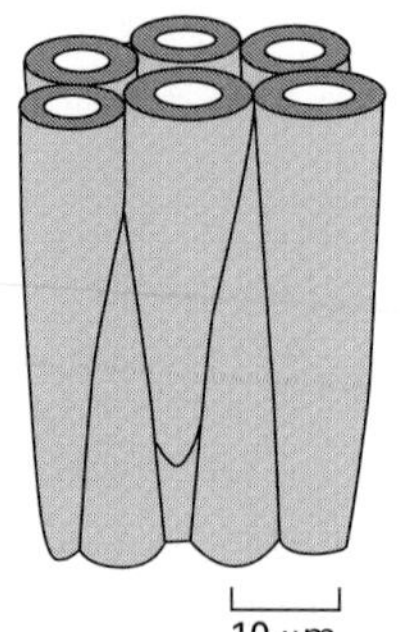

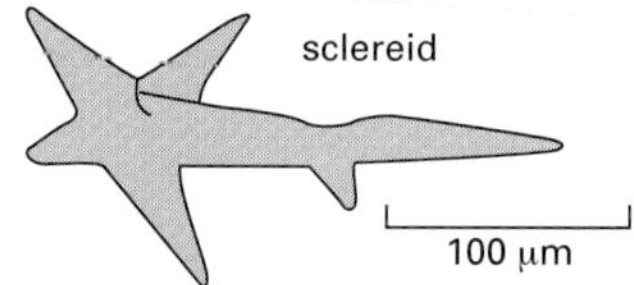

DERMAL TISSUE

The epidermis is the primary outer protective covering of the plant body. Cells of the epidermis are also modified to form stomata and hairs of various kinds.

Epidermis

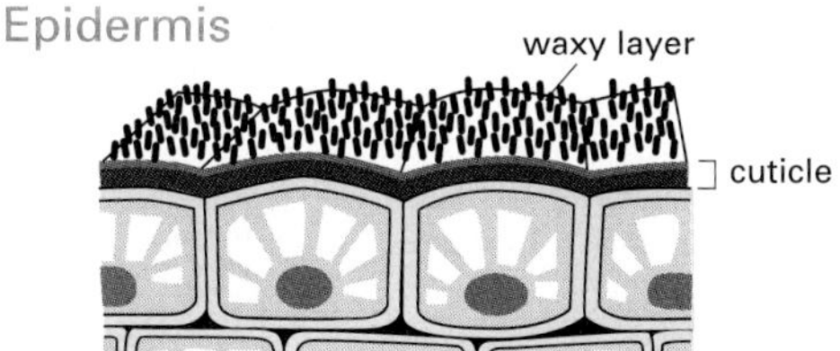

The epidermis (usually one layer of cells deep) covers the entire stem, leaf, and root of the young plant. The cells are living, have thick primary cell walls, and are covered on their outer surface by a special cuticle with an outer waxy layer. The cells are tightly interlocked in different patterns.

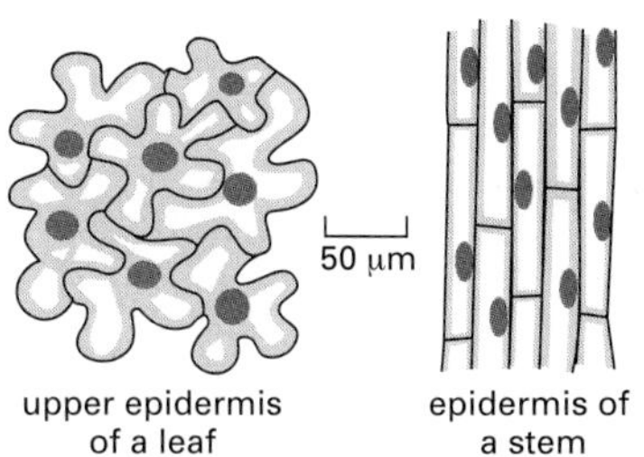

upper epidermis of a leaf

epidermis of a stem

Stomata

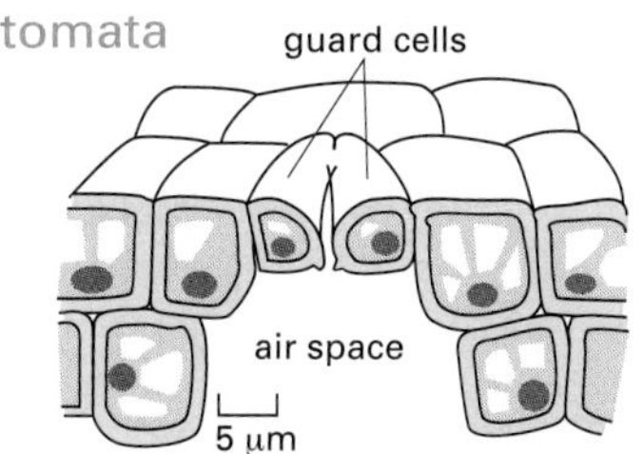

Stomata are openings in the epidermis, mainly on the lower surface of the leaf, that regulate gas exchange in the plant. They are formed by two specialized epidermal cells called *guard cells,* which regulate the diameter of the pore. Stomata are distributed in a distinct species-specific pattern within each epidermis.

Hairs (or trichomes) are appendages derived from epidermal cells. They exist in a variety of forms and are commonly found in all plant parts. Hairs function in protection, absorption, and secretion; for example,

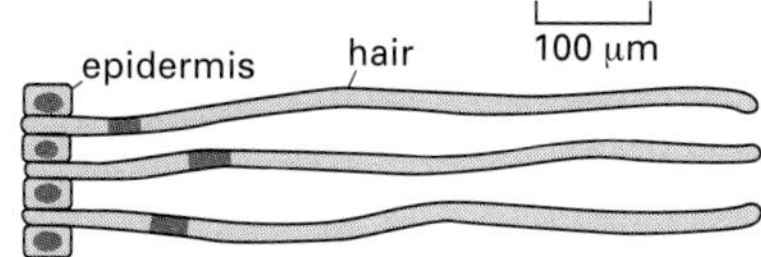

young, single-celled hairs in the epidermis of the cotton seed. When these grow, the walls will be secondarily thickened with cellulose to form cotton fibers.

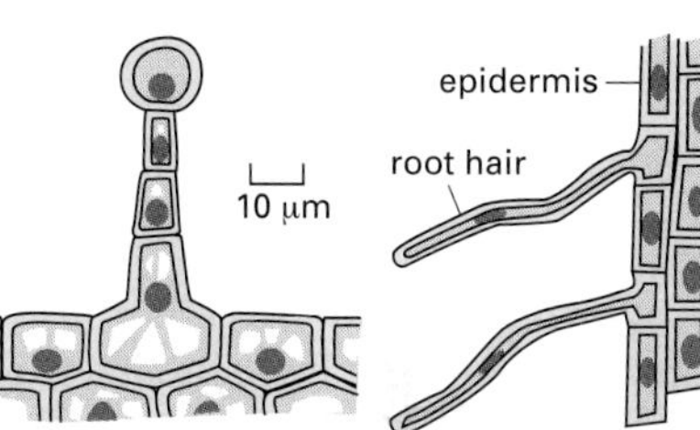

a multicellular secretory hair from a geranium leaf

Single-celled root hairs have an important function in water and ion uptake.

Vascular bundles

Roots usually have a single vascular bundle, but stems have several bundles. These are arranged with strict radial symmetry in dicots, but they are more irregularly dispersed in monocots.

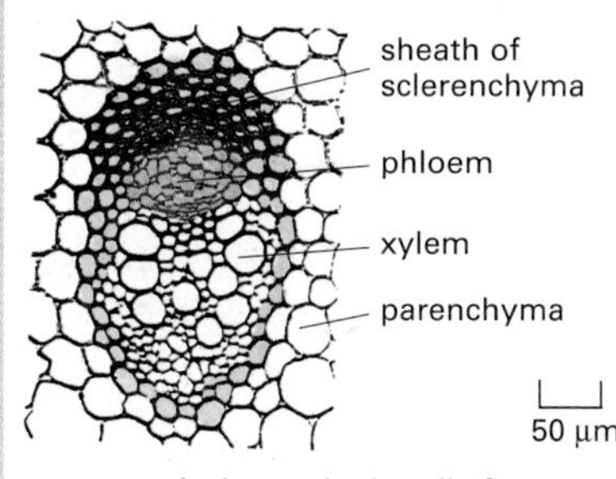

a typical vascular bundle from the young stem of a buttercup

VASCULAR TISSUE

The phloem and the xylem together form a continuous vascular system throughout the plant. In young plants they are usually associated with a variety of other cell types in *vascular bundles*. Both phloem and xylem are complex tissues. Their conducting elements are associated with parenchyma cells that maintain the elements and exchange materials with them. In addition, groups of collenchyma and sclerenchyma cells provide mechanical support.

Phloem

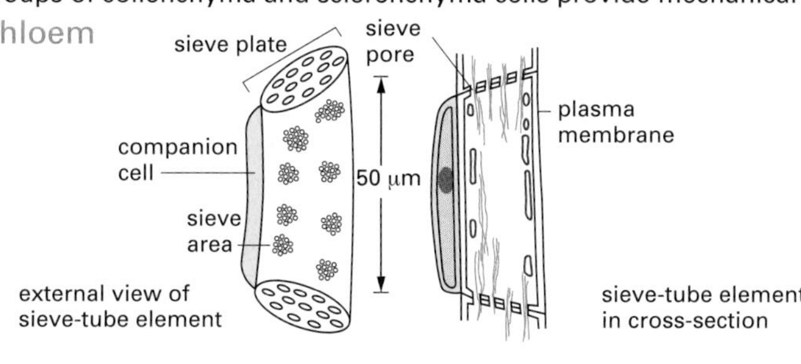

external view of sieve-tube element

sieve-tube element in cross-section

Phloem is involved in the transport of organic solutes in the plant. The main conducting cells (elements) are aligned to form tubes called *sieve tubes*. The sieve-tube elements at maturity are living cells, interconnected by perforations in their end walls formed from enlarged and modified plasmodesmata (sieve plates). These cells retain their plasma membrane, but they have lost their nuclei and much of their cytoplasm; they therefore rely on associated *companion cells* for their maintenance. These companion cells have the additional function of actively transporting soluble food molecules into and out of sieve-tube elements through porous sieve areas in the wall.

Xylem

Xylem carries water and dissolved ions in the plant. The main conducting cells are the vessel elements shown here, which are dead cells at maturity that lack a plasma membrane. The cell wall has been secondarily thickened and heavily lignified. As shown below, its end wall is largely removed, enabling very long, continuous tubes to be formed.

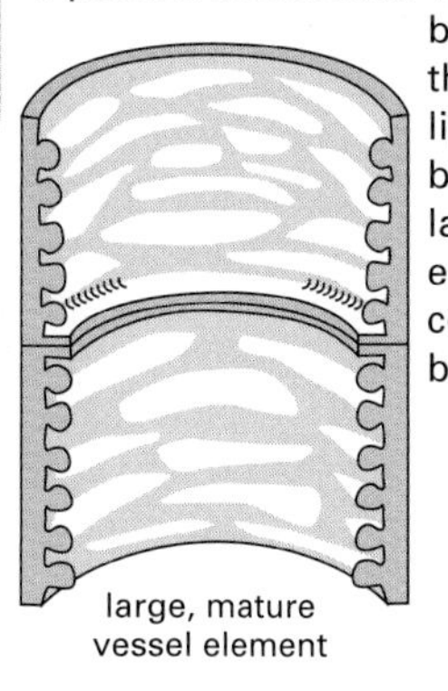

small vessel element in root tip

large, mature vessel element

The vessel elements are closely associated with xylem parenchyma cells, which actively transport selected solutes into and out of the elements across the parenchyma cell plasma membrane.

xylem parenchyma cells

vessel element

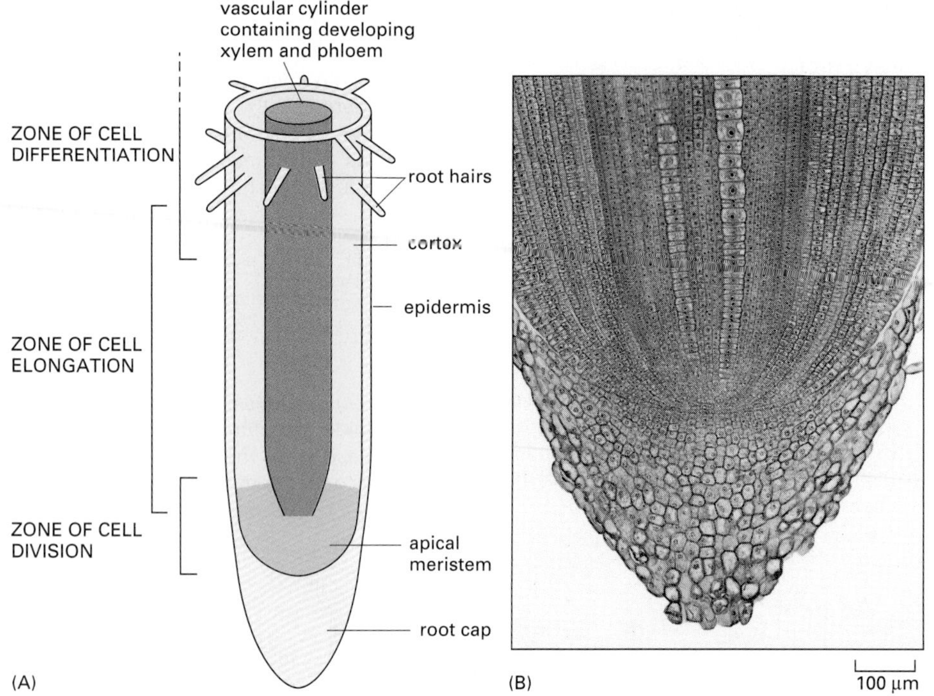

Figure 21–112 A growing root tip. (A) The organization of the final 2 mm of a growing root tip. The approximate zones in which cells can be found dividing, elongating, and differentiating are indicated. (B) The apical meristem and root cap of a corn root tip, showing the orderly files of cells produced. (B, from R.F. Evert, Biology of Plants, 4th edn. New York: Worth, 1986.)

or reception of the brassinosteroid signal cause the stem of the seedling to go green, slow its elongation, and open its cotyledons prematurely, while it is still in the dark.

The Shaping of Each New Structure Depends on Oriented Cell Division and Expansion

Plant cells, imprisoned within their cell walls, cannot crawl about and cannot be shuffled as the plant grows; but they can divide, and they can swell, stretch, and bend. The morphogenesis of a developing plant therefore depends on orderly cell divisions followed by strictly oriented cell expansions. Most cells produced in the root-tip meristem, for example, go through three distinct phases of development—division, growth (elongation), and differentiation. These three steps, which overlap in both space and time, give rise to the characteristic architecture of a root tip. Although the process of cell differentiation often begins while a cell is still enlarging, it is comparatively easy to distinguish in a root tip a zone of cell division, a zone of oriented cell elongation (which accounts for the growth in length of the root), and a zone of cell differentiation (Figure 21–112).

In the phase of controlled expansion that generally follows cell division, the daughter cells may often increase in volume by a factor of 50 or more. This expansion is driven by an osmotically based turgor pressure that presses outward on the plant cell wall, and its direction is determined by the orientation of the cellulose fibrils in the cell wall, which constrain expansion along one axis (see Figure 19–73). The orientation of the cellulose in turn is apparently controlled by the orientation of arrays of microtubules just inside the plasma membrane, which are thought to guide cellulose deposition (discussed in Chapter 19). These orientations can be rapidly changed by plant growth regulators, such as ethylene and gibberellic acid (Figure 21–113), but the molecular mechanisms underlying these dramatic cytoskeletal rearrangements are still unknown.

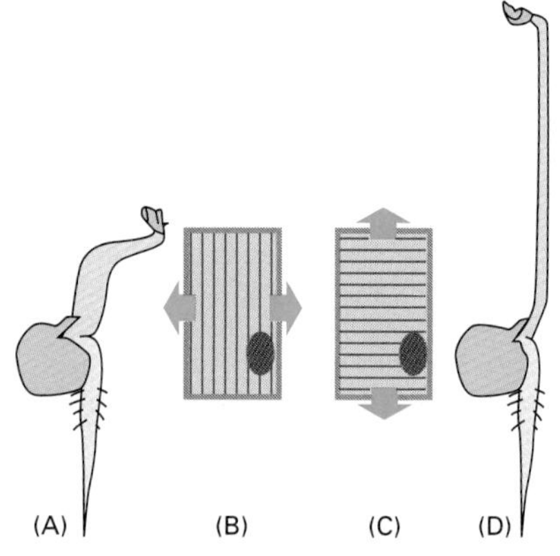

Figure 21–113 The different effects of the plant growth regulators ethylene and gibberellic acid. These regulators exert rapid and opposing effects on the orientation of the cortical microtubule array in cells of young pea shoots. A typical cell in an ethylene-treated plant (B) shows a net longitudinal orientation of microtubules, while a typical cell in a gibberellic-acid-treated plant (C) shows a net transverse orientation. New cellulose microfibrils are deposited parallel to the microtubules. Since this influences the direction of cell expansion, gibberellic acid and ethylene encourage growth in opposing directions: ethylene-treated seedlings will develop short, fat shoots (A), while gibberellic-acid-treated seedlings will develop long, thin shoots (D).

Each Plant Module Grows From a Microscopic Set of Primordia in a Meristem

The apical meristems are self-perpetuating: in a perennial plant, they carry on with their functions indefinitely, as long as the plant survives, and they are responsible for its continuous growth and development. But apical meristems also give rise to a second type of outgrowth, whose development is strictly limited and culminates in the formation of a structure such as a leaf or a flower, with a determinate size and shape and a short lifespan. Thus, as a vegetative (nonflowering) shoot elongates, its apical meristem lays down behind itself an orderly sequence of *nodes*, where leaves have grown out, and *internodes* (segments of stem). In this way the continuous activity of the meristem produces an ever increasing number of similar modules, each consisting of a stem, a leaf, and a bud (see Figure 21–106). The modules are connected to one another by supportive and transport tissue, and successive modules are precisely located relative to each other, giving rise to a repetitively patterned structure. This iterative mode of development is characteristic of plants and is seen in many other structures besides the stem-leaf system (Figure 21–114).

Although the final module is large, its organization, like that of an animal embryo, is mapped out at first on a microscopic scale. At the apex of the shoot, within a space of a millimeter or less, one finds a small, low central dome surrounded by a set of distinctive swellings in various stages of enlargement (Figure 21–115). The central dome is the apical meristem itself; each of the surrounding swellings is the primordium of a leaf. This small region, therefore, contains the already distinct rudiments of several entire modules. Through a well-defined program of cell proliferation and cell enlargement, each leaf primordium and its adjacent cells will grow to form a leaf, a node, and an internode. Meanwhile, the apical meristem itself will give rise to new leaf primordia, so as to generate more and more modules in a potentially unending succession. The serial organization of the modules of the plant is thus controlled by events at the shoot apex. It is the system of local signals within this tiny region that determines the pattern of primordia—the position of one leaf rudiment relative to the next, the spacing between them, and their location relative to the apical meristem itself.

Variations on this basic repetitive theme can give rise to more complex architectures, including structures such as tendrils, thorns, branches, and flowers. Thus, by switching on different sets of genes at the shoot apex, the plant can produce different types of primordia, in different spatial patterns.

Cell Signaling Maintains the Meristem

Central to all these phenomena is the question of how the apical meristem maintains itself. The meristem cells must continue to proliferate for weeks,

(A)

(B)

(C)

Figure 21–114 Repetitive patterning in plants. Accurate placing of successive modules from a single apical meristem produces these elaborate but regular patterns in leaves (A), flowers (B), and fruits (C). (A, from John Sibthorp, Flora Graeca. London: R. Taylor, 1806–1840; B, from Pierre Joseph Redouté, Les Liliacées. Paris: chez l'Auteur, 1807; C, from Christopher Jacob Trew, Uitgezochte planten. Amsterdam: Jan Christiaan Sepp, 1771—all courtesy of the John Innes Foundation.)

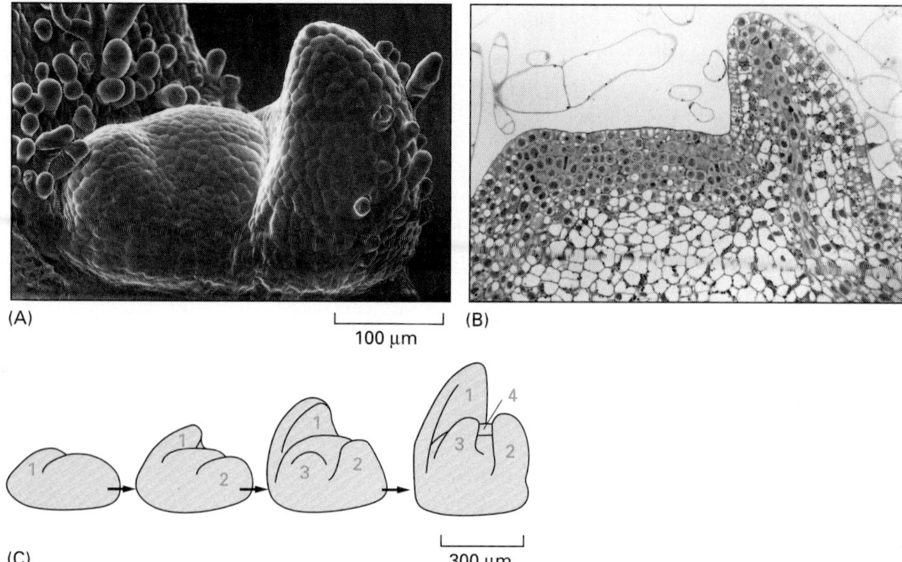

Figure 21–115 A shoot apex from a young tobacco plant. (A) A scanning electron micrograph shows the shoot apex with two sequentially emerging leaf primordia, seen here as lateral swellings on either side of the domed apical meristem. (B) A thin section of a similar apex shows that the youngest leaf primordium arises from a small group of cells (about 100) in the outer four or five layers of cells. (C) A very schematic drawing showing that the sequential appearance of leaf primordia takes place over a small distance and very early in shoot development. Growth of the apex will eventually form internodes that will separate the leaves in order along the stem (see Figure 21–106). (A and B, from R.S. Poethig and I.M. Sussex, *Planta* 165:158–169, 1985. © Springer-Verlag.)

years, or even centuries as a plant grows, replacing themselves while continually generating progeny cells that differentiate. Through all this, the size of the cluster of cells that constitute the meristem remains practically constant (about 100 cells in *Arabidopsis*, for example). New meristems may arise as the plant branches, but they too preserve the same size.

Genetic screens have identified genes required for meristem maintenance. For example, mutations that disrupt the *WUSCHEL* gene, which codes for a homeodomain protein, convert the apical meristem into non-meristematic tissue, so that the seedling fails to sprout. Conversely, mutations in the *CLAVATA* group of genes, coding for components of a cell–cell signaling pathway (see Figure 15–77), make the meristem abnormally big. These genes are expressed in different layers of cells in the meristem region (Figure 21–116A). The two most superficial cell layers, called the L1 and L2 layers, together with the uppermost part of the L3 layer, contain the cells of the meristem proper, capable of dividing indefinitely to give rise to future parts of the plant. The meristematic cells of the L1 and L2 layers express Clavata3, a small secreted signal protein. Just beneath, in the L3 layer, lies a cluster of cells expressing Clavata1 (the receptor for Clavata3). In the center of this Clavata1 patch are cells that express the Wuschel gene regulatory protein.

The pattern of cell divisions implies that the cells expressing Wuschel are not themselves part of the meristem proper; new Wuschel-expressing cells are apparently continually recruited from the meristematic part of the L3 population, just above the Wuschel domain. Nevertheless, the Wuschel-expressing cells are at the heart of the mechanism that maintains the meristem. A signal that they produce maintains meristematic behavior in the cells above, stimulates expression of the *CLAVATA* genes, and, presumably, causes new cells recruited into the Wuschel domain to switch on Wuschel. Negative feedback from the upper meristematic cells, delivered by the Clavata signaling pathway, acts back

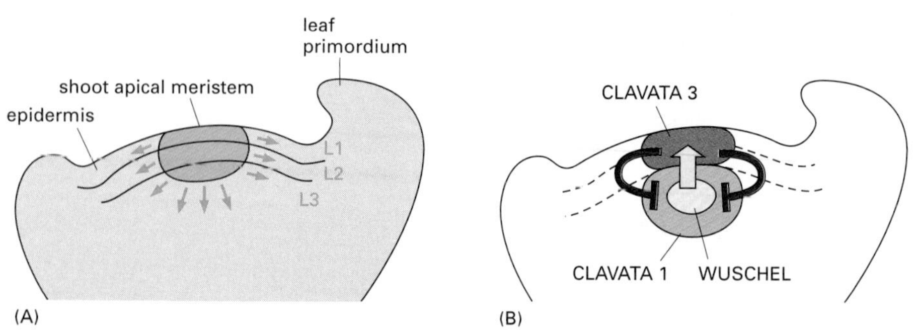

Figure 21–116 The feedback loops that are thought to maintain the shoot apical meristem. (A) The arrangement of cell layers constituting a shoot apical meristem. (B) The pattern of cell–cell communication that maintains the meristem. Artificial overexpression of Wuschel in the L3 region causes an increase in the number of cells in the L1 and L2 layers that behave as meristem cells and express Clavata3; artificial overexpression of Clavata3 in the L1 and L2 layers causes a reduction of Wuschel expression in the L3 region below and a decrease in the number of meristem cells. Clavata3 codes for a small signal protein, while Clavata1 codes for its receptor, a transmembrane protein kinase. Wuschel, which is expressed in the central part of the region that expresses the receptor Clavata1, codes for a gene regulatory protein of the homeodomain class. The size of the meristem is thought to be controlled by a self-regulating balance between a short-range stimulatory signal produced by cells expressing Wuschel *(yellow arrow)*, and a longer-range inhibitory signal delivered by Clavata3 *(red bars)*.

on the regions below to limit the size of the Wuschel domain, thereby preventing the meristem from becoming too big (Figure 21–116B).

This account of the plant meristem is still uncertain in many details and other genes besides those we have mentioned are also involved. Nevertheless, mathematical modeling shows that systems of a similar sort, based on a feedback loop involving a short-range activating signal and a long-range inhibitory signal, can stably maintain a signaling center of a well-defined size even when there is continual proliferation and turnover of the cells that form that center. Analogous systems of signals are thought to operate in animal development to maintain localized signaling centers—such as the Organizer of the amphibian gastrula, or the zone of polarizing activity in a limb bud.

It is still not known how the Wuschel-expressing cells signal to their neighbors. One possibility is that the Wuschel protein itself diffuses directly from cell to cell through plasmodesmata—a signaling pathway peculiar to plants. Some other gene regulatory proteins have in fact been shown to travel in this way in meristems, spreading from cells that contain the corresponding mRNA into neighboring cells that do not.

Regulatory Mutations Can Transform Plant Topology by Altering Cell Behavior in the Meristem

If a stem is to branch, new shoot apical meristems must be created, and this too depends on events in the neighborhood of the shoot apex. At each developing node, in the acute angle (the axil) between the leaf primordium and the stem, a bud is formed (Figure 21–117). This contains a nest of cells, derived from the apical meristem, that keep a meristematic character. They have the capacity to become the apical meristem of a new branch or the primordium of a structure such as a flower; but they also have the alternative option of remaining quiescent as *axillary buds*. The plant's pattern of branching is regulated through this choice of fate, and mutations that affect it can transform the structure of the plant. Maize provides a beautiful example.

Maize represents one of mankind's most remarkable feats of genetic engineering. Native Americans created it by selective breeding, over a period of several centuries or perhaps millennia between 5,000 and 10,000 years ago. They started from a wild grass known as teosinte, with highly branched leafy stems and tiny ears bearing inedible hard kernels. Detailed genetic analysis has identified a handful of genetic loci—about five—as the sites of the mutations that account for most of the difference between this unpromising ancestor and modern corn. One of these loci, with a particularly dramatic effect, corresponds to a gene called *teosinte branched-1* (*tb1*). In maize with loss-of-function mutations in *tb1*, the usual simple unbranched stem, with a few large leaves at intervals along it, is transformed into a dense, branching, leafy mass reminiscent of teosinte (Figure 21–118A). The pattern of branching in the mutant implies that axillary buds, originating in normal positions, have escaped from an inhibition that prevents them, in normal maize, from growing into branches.

In normal maize, the single stem is crowned with a tassel—a male flower—while a few of the axillary buds along the stem develop into female flowers and, upon fertilization, form the ears of corn that we eat. In the mutant maize with a defective *tb1* gene, these fruitful axillary buds are transformed into branches bearing tassels. The wild teosinte plant is like the *tb1*-defective maize in its leafy, highly branched appearance, but unlike this mutant it makes ears on many of its side branches, as though *tb1* were active. DNA analysis reveals the explanation. Both teosinte and normal maize possess a functional *tb1* gene, with an almost identical coding sequence, but in maize the regulatory region has undergone a mutation that boosts the level of gene expression. Thus in normal maize the gene is expressed at a high level in every axillary bud, inhibiting branch formation, while in teosinte the expression in many axillary buds is low, so that branches are permitted to form (Figure 21–118B).

This example shows how simple mutations, by switching the behavior of meristem cells, can transform plant structure—a principle of enormous importance

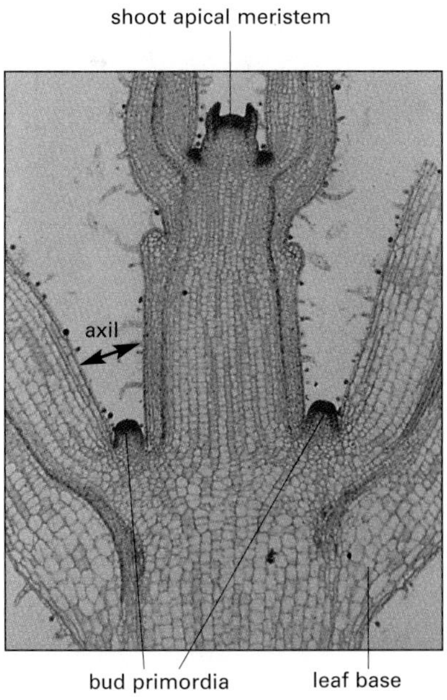

Figure 21–117 Axillary buds in the neighborhood of a shoot apex. The photograph shows a longitudinal section of *Coleus blumei*, a common houseplant. (From P.H. Raven, R.F. Evert, and S.E. Eichhorn, Biology of Plants, 6th edn. New York: Freeman/Worth, 1999, used with permission.)

in the breeding of plants for food. More generally, the case of *tb1* illustrates how new body plans, whether of plant or animal, can evolve through changes in regulatory DNA without change in the characters of the proteins made.

Long-Range Hormonal Signals Coordinate Developmental Events in Separate Parts of the Plant

The fate of an axillary bud is dictated not only by its genes, but also by environmental conditions. Separate parts of a plant experience different environments and react to them individually by changes in their mode of development. The plant, however, must continue to function as a whole. This demands that developmental choices and events in one part of the plant affect developmental choices elsewhere. There must be long-range signals to bring about such coordination.

As gardeners know, for example, by pinching off the tip of a branch one can stimulate side growth: removal of the apical meristem relieves the quiescent axillary meristems of an inhibition and allows them to form new twigs. In this case the long-range signal from the apical meristem, or at least a key component has been identified. It is an auxin, a member of one of six known classes of **plant growth regulators** (sometimes called *plant hormones*), all of which have powerful influences on plant development. The five other known classes are the *gibberellins,* the *cytokinins, abscisic acid,* the gas *ethylene,* and the *brassinosteroids.* As shown in Figure 21–119, all are small molecules that readily penetrate cell walls. They are all synthesized by most plant cells and can either act locally or be transported to influence target cells at a distance. Auxin, for example, is transported from cell to cell at a rate of about 1 cm per hour from the tip of a

Figure 21–118 Transformation of plant architecture by mutation: a comparison of teosinte, normal maize, and *tb1*-defective maize. (A) Photographs of three types of plants. (B) The architecture of teosinte, normal maize and the *tb1*-defective maize compared schematically. The *tb1* gene product is needed for development of ears. It is absent in the *tb1* mutant; it is present in both teosinte and normal maize, but these two plants differ because the gene is differently regulated. (A *(left image),* from J. Doebley and R.-L. Wang, *Cold Spring Harbor Symp.* 62:361–367, 1997. © Cold Spring Harbor Press; A *(middle and right images)* from J. Doebley, A. Stec, and L. Hubbard, *Nature* 386:485–488, 1997. © Macmillan Magazines Ltd.)

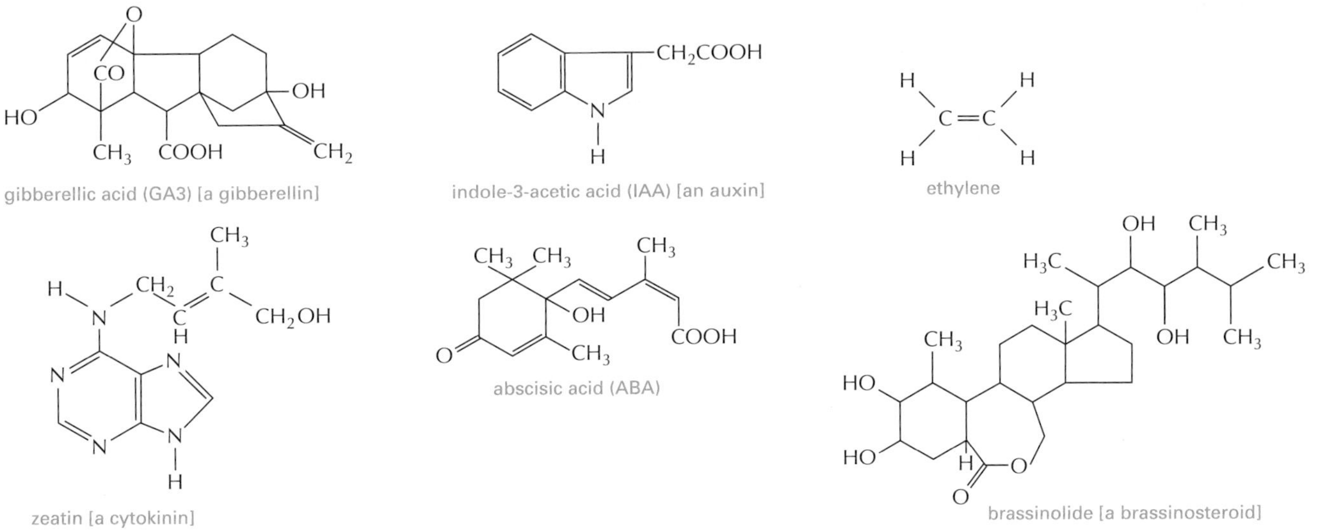

gibberellic acid (GA3) [a gibberellin]

indole-3-acetic acid (IAA) [an auxin]

ethylene

zeatin [a cytokinin]

abscisic acid (ABA)

brassinolide [a brassinosteroid]

Figure 21–119 Plant growth regulators. The formula of one naturally occurring representative molecule from each of the six groups of plant growth regulatory molecules is shown.

shoot toward its base. Each growth regulator has multiple effects, and these are modulated by the other growth regulators, as well as by environmental cues and nutritional status. Thus auxin alone can promote root formation, but in conjunction with gibberellin it can promote stem elongation, with cytokinin, auxin it can suppress lateral shoot outgrowth, and with ethylene it can stimulate lateral root growth. The receptors that recognize some of these growth regulators are discussed in Chapter 15.

Homeotic Selector Genes Specify the Parts of a Flower

Meristems face other developmental choices besides that between quiescence and growth, as we have already seen in our discussion of maize, and these also are frequently regulated by the environment. The most important is the decision to form a flower (Figure 21–120).

The switch from meristematic growth to flower formation is typically triggered by light. By poorly understood mechanisms based on light absorption by phytochrome and cryptochrome proteins (discussed in Chapter 15), the plant can sense very precisely a change in day length. It responds by turning on expression of a set of *floral meristem-identity* genes in the apical meristem. By switching on these genes, the apical meristem abandons its chances of continuing vegetative growth and gambles its future on the production of gametes. Its cells embark on a strictly finite program of growth and differentiation: by a modification of the ordinary mechanisms for generating leaves, a series of whorls of specialized appendages are formed in a precise order—typically sepals first, then petals, then stamens carrying anthers containing pollen, and lastly carpels containing eggs (see Panel 21–1). By the end of this process the meristem has disappeared, but among its progeny it has created germ cells.

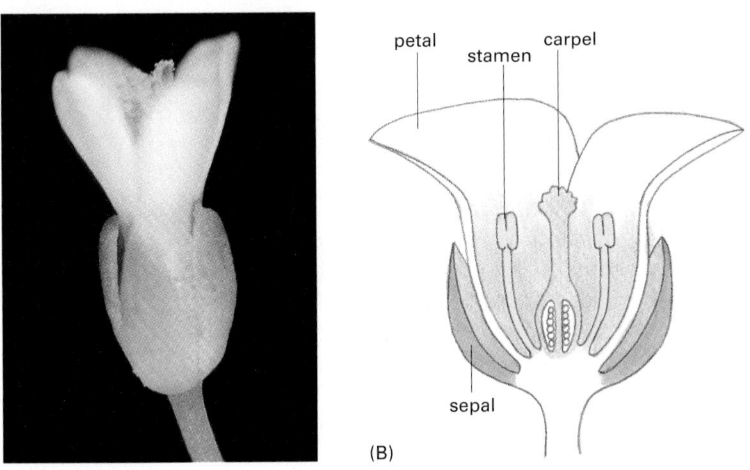

Figure 21–120 The structure of an *Arabidopsis* flower. (A) Photograph. (B) Drawings. (C) Schematic cross-sectional view. The basic plan, as shown in (C), is common to most flowering dicotyledonous plants. (A, courtesy of Leslie Sieburth.)

(A)

(B)

(C)

(D)

Figure 21–121 *Arabidopsis* flowers showing a selection of homeotic mutations. (A) In *apetala2,* sepals are converted into carpels and petals into stamens; (B) In *apetala3,* petals are converted into sepals and stamens into carpels; (C)In *agamous,* stamens are converted into petals and carpels into floral meristem. (D) In a triple mutant where these three functions are defective, all the organs of the flower are converted into leaves. (A–C, courtesy of Leslie Sieburth; D, courtesy of Mark Running.)

The series of modified leaves forming a flower can be compared to the series of body segments forming a fly. In plants, as in flies, one can find homeotic mutations that convert one part of the pattern to the character of another. The mutant phenotypes can be grouped into at least four classes, in which different but overlapping sets of organs are altered (Figure 21–121). The first or 'A' class, exemplified by the *apetala2* mutant of *Arabidopsis,* has its two outermost whorls transformed: the sepals are converted into carpels and the petals into stamens. The second or 'B' class, exemplified by *apetala3,* has its two middle whorls transformed: the petals are converted into sepals and the stamens into carpels. The third or 'C' class, exemplified by *agamous,* has its two innermost whorls transformed, with a more drastic consequence: the stamens are converted into petals, the carpels are missing, and in their place the central cells of the flower behave as a floral meristem, which begins the developmental performance all over again, generating another abnormal set of sepals and petals nested inside the first and, potentially, another nested inside that, and so on, indefinitely. A fourth class, the *sepallata* mutants, has its three inner whorls all transformed into sepals.

These phenotypes identify four classes of homeotic selector genes, which, like the homeotic selector genes of *Drosophila,* all code for gene regulatory proteins. These are expressed in different domains and define the differences of cell state that give the different parts of a normal flower their different characters, as shown in Figure 21–122. The gene products collaborate to form protein complexes that drive expression of the appropriate downstream genes. In a triple mutant where the A, B and C genetic functions are all absent, one obtains in place of a flower an indefinite succession of tightly nested leaves (see Figure 21–121D). Conversely, in a transgenic plant where genes of the A, B and *sepallata* classes are all expressed together outside their normal domains, leaves are transformed into petals. Leaves therefore represent a "ground state" in which none of these homeotic selector genes are expressed, while the other types of organ result from expressing the genes in different combinations.

Similar studies have been carried out in other plant species, and a similar set of phenotypes and genes have been identified: plants, no less than animals, have conserved their homeotic selector gene systems. Gene duplication has played a large part in the evolution of these genes: several of them, required in different organs of the flower, have clearly homologous sequences. These are not of the homeobox class but are members of another family of gene regulatory proteins (the so-called MADS family), also found in yeast and in vertebrates.

Clearly, plants and animals, have independently found very similar solutions to many of the fundamental problems of multicellular development.

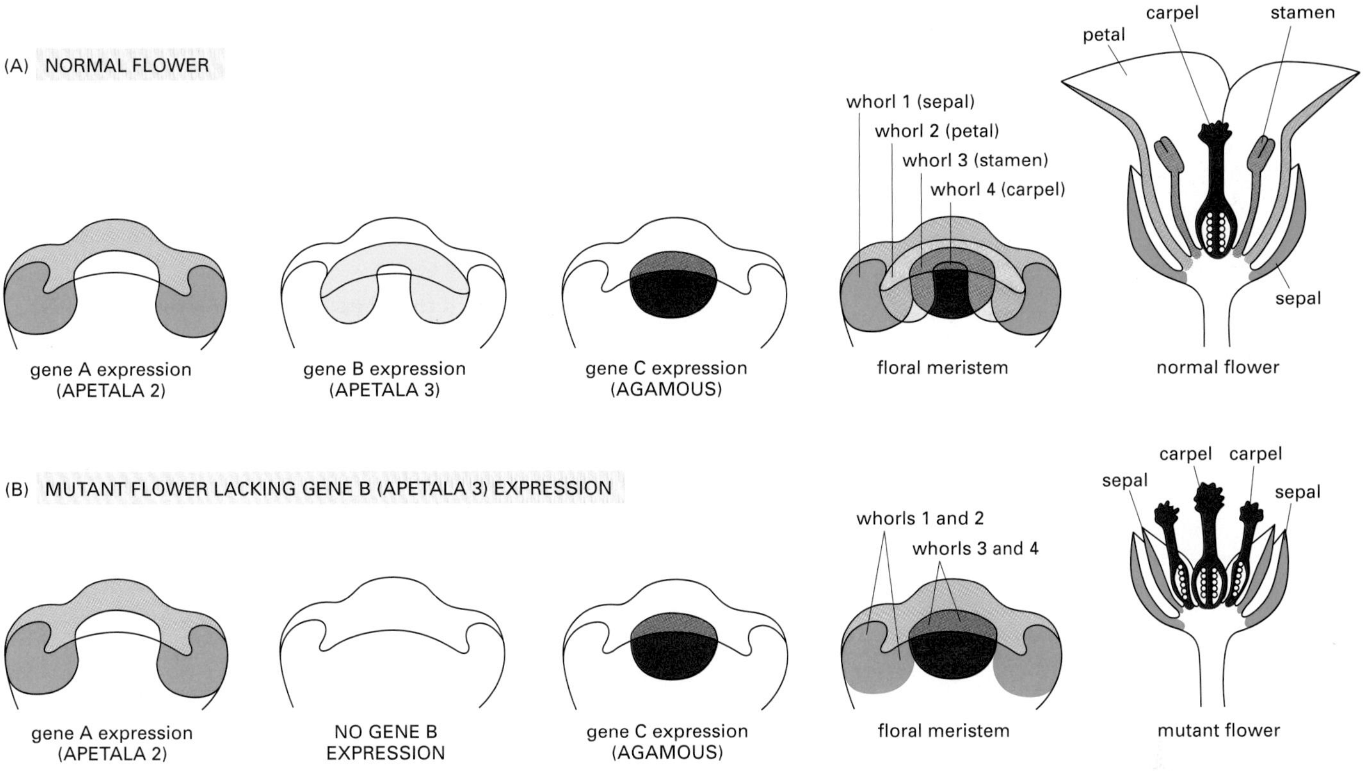

Figure 21–122 Homeotic selector gene expression in an *Arabidopsis* flower. (A) Diagram of the normal expression patterns of the three genes whose mutant phenotypes are illustrated in Figure 21–121A–C. All three genes code for gene regulatory proteins. The colored shading on the flower indicates which organ develops from each whorl of the meristem, but does not imply that the homeotic selector genes are still expressed at this stage. (B) The patterns in a mutant where the *apetala3* gene is defective. Because the character of the organs in each whorl is defined by the set of homeotic selector genes that they express, the stamens and petals are converted into sepals and carpels. The consequence of a deficiency of a gene of class A, such as *apetala2*, is slightly more complex: the absence of this class A gene product allows the class C gene to be expressed in the outer two whorls as well as the inner two, causing these outer whorls to develop as carpels and stamens, respectively. Deficiency of a class C gene prevents the central region from undergoing terminal differentiation as a carpel and causes it instead to continue growth as a meristem, generating more and more sepals and petals.

Summary

The development of a flowering plant, like that of an animal, begins with division of a fertilized egg to form an embryo with a polarized organization: the apical part of the embryo will form the shoot, the basal part, the root, and the middle part, the stem. At first, cell division occurs throughout the body of the embryo. As the embryo grows, however, addition of new cells becomes restricted to small regions known as meristems. Apical meristems, at shoot tips and root tips, will persist throughout the life of the plant, enabling it to grow by sequentially adding new body parts at its periphery. Typically, the shoot generates a repetitive series of modules, each consisting of a segment of stem, a leaf, and an axillary bud. An axillary bud is a potential new meristem, capable of giving rise to a side branch; the environment—and long-range hormonal signals within the plant can control the development of the plant by regulating bud activation. Mutations that alter the rules for activating axillary buds can have a drastic effect on the shape and structure of the plant; a single such mutation—one of about five key genetic alterations—accounts for a large part of the dramatic difference between modern maize and its wild ancestor, teosinte.

The small weed Arabidopsis thaliana *is widely used as a model organism for genetic studies and is the first plant to have had its genome completely sequenced. As in animals, genes governing plant development can be identified through genetic screens and their functions tested by genetic manipulations. Such studies have begun to reveal the molecular mechanisms by which the internal organization of each plant module is sketched out on a microscopic scale through cell–cell interactions in the neighborhood of the apical meristem. The meristem itself appears to be maintained by a local feedback loop, in which cells expressing the gene regulatory protein Wuschel provide a positive stimulus, and a negative feedback dependent on the Clavata cell–cell signaling pathway keeps the meristem from becoming too big.*

Environmental cues—especially light that is appropriately timed—can cause the expression of genes that switch the apical meristem from a leaf-forming to a flower-forming mode. The parts of a flower—its sepals, petals, stamens and carpels—are formed by a modification of the mechanism for development of leaves, and the differences between these parts are controlled by homeotic selector genes that are closely analogous (although not homologous) to those of animals.

References

General

Carroll SB, Grenier JK & Weatherbee SD (2001) From DNA to Diversity: Molecular Genetics and the Evolution of Animal Design. Malden, MA: Blackwell Science.

Gilbert SF (2000) Developmental Biology, 6th edn. Sunderland, MA: Sinauer.

Slack JMW (2001) Essential Developmental Biology. Oxford: Blackwell Science.

Wolpert L, Beddington R, Jessell TM & Lawrence P (2002) Principles of Development, 2nd edn. London/Oxford: Current Biology/Oxford University Press.

Universal Mechanisms of Animal Development

Gilbert SF & Raunio AM (eds) (1997) Embryology: Constructing The Organism. Sunderland, MA: Sinauer.

Scott MP (2000) Development: the natural history of genes. *Cell* 100, 27–40.

Spemann H (1938) Embryonic Development and Induction. New Haven: Yale University Press. Reprinted 1988, New York: Garland Publishing.

The Zebrafish Issue (1996) *Development* 123, 1–481. (A genetic screen)

Caenorhabditis elegans: Development From the Perspective of the Individual Cell

Harris WA & Hartenstein V (1991) Neuronal determination without cell division in *Xenopus* embryos. *Neuron* 6, 499–515.

Lin R, Hill RJ & Priess JR (1998) POP-1 and anterior-posterior fate decisions in *C. elegans* embryos. *Cell* 92, 229–239.

Metzstein MM, Stanfield GM & Horvitz HR (1998) Genetics of programmed cell death in *C. elegans*: past, present and future. *Trends Genet.* 14, 410–416.

Pasquinelli AE, Reinhart BJ, Slack F et al. (2000) Conservation of the sequence and temporal expression of let-7 heterochronic regulatory RNA. *Nature* 408, 86–89.

Drosophila and the Molecular Genetics of Pattern Formation: Genesis of the Body Plan

Bate M & Martinez Arias A (eds) (1993) The Development of *Drosophila melanogaster.* Cold Spring Harbor, NY: Cold Spring Harbor Laboratory Press.

Dearden P & Akam M (1999) Developmental evolution: Axial patterning in insects. *Curr. Biol.* 9, R591–R594.

Lawrence PA (1992) The Making of a Fly: The Genetics of Animal Design. Oxford: Blackwell Scientific.

Nusslein-Volhard C & Wieschaus E (1980) Mutations affecting segment number and polarity in *Drosophila. Nature* 287, 795–801.

Rubin GM, Yandell MD, Wortman JR et al. (2000) Comparative genomics of the eukaryotes. *Science* 287, 2204–2215.

Shulman JM, Benton R & St Johnston D (2000) The *Drosophila* homolog of *C. elegans* PAR-1 organizes the oocyte cytoskeleton and directs oskar mRNA localization to the posterior pole. *Cell* 101, 377–388.

Homeotic Selector Genes and the Patterning of the Anteroposterior Axis

Ferrier DE & Holland PW (2001) Ancient origin of the Hox gene cluster. *Nat. Rev. Genet.* 2, 33–38.

Lewis EB (1978) A gene complex controlling segmentation in *Drosophila. Nature* 276, 565–570.

Maconochie M, Nonchev S, Morrison A & Krumlauf R (1996) Paralogous Hox genes: function and regulation. *Annu. Rev. Genet.* 30, 529–556.

McGinnis W, Garber RL, Wirz J et al. (1984) A homologous protein-coding sequence in *Drosophila* homeotic genes and its conservation in other metazoans. *Cell* 37, 403–408.

Ringrose L & Paro R (2001) Remembering silence. *BioEssays* 23, 566–570.

Organogenesis and the Patterning of Appendages

Capdevila J & Belmonte JCI (2001) Patterning mechanisms controlling vertebrate limb development. *Annu. Rev. Cell Dev. Biol.* 17, 87–132.

Day SJ & Lawrence PA (2000) Measuring dimensions: the regulation of size and shape. *Development* 127, 2977–2987.

Irvine KD & Rauskolb C (2001) Boundaries in development: formation and function. *Annu. Rev. Cell Dev. Biol.* 17, 189–214.

Jan YN & Jan LY (2001) Development: asymmetric cell division in the *Drosophila* nervous system. *Nat. Rev. Neurosci.* 2, 772–779.

Panganiban G, Irvine SM, Lowe C et al. (1997) The origin and evolution of animal appendages. *Proc. Natl. Acad. Sci. USA* 94, 5162–5166.

Teleman AA, Strigini M & Cohen SM (2001) Shaping morphogen gradients. *Cell* 105, 559–562.

Cell Movement and the Shaping of the Vertebrate Body

Harland R & Gerhart J (1997) Formation and function of Spemann's Organizer. *Annu. Rev. Cell Dev. Biol.* 13, 611–667.

Heisenberg CP, Tada M, Rauch GJ et al. (2000) Silberblick/Wnt11 mediates convergent extension movements during zebrafish gastrulation. *Nature* 405, 76–81.

Kalcheim C & Le Douarin NM (1999) The Neural Crest, 2nd edn. Cambridge: Cambridge University Press.

Mercola M & Levin M (2001) Left–right asymmetry determination in vertebrates. *Annu. Rev. Cell Dev. Biol.* 17, 779–132.

Palmeirim I, Henrique D, Ish-Horowicz D & Pourquie O (1997) Avian hairy gene expression identifies a molecular clock linked to vertebrate segmentation and somitogenesis. *Cell* 91, 639–648.

Takeichi M, Nakagawa S, Aono S et al. (2000) Patterning of cell assemblies regulated by adhesion receptors of the cadherin superfamily. *Philos. Trans. R. Soc. Lond. B. Biol. Sci.* 355, 885–890.

The Mouse

Larsen WJ (2001) Human Embryology, 3rd edn: Churchill Livingstone.

Lu CC, Brennan J & Robertson EJ (2001) From fertilization to gastrulation: axis formation in the mouse embryo. *Curr. Opin. Genet. Dev.* 11, 384–392.

Metzger RJ & Krasnow MA (1999) Genetic control of branching morphogenesis. *Science* 284, 1635–1639.

Smith AG (2001) Embryo-derived stem cells: of mice and men. *Annu. Rev. Cell Dev. Biol.* 17, 435–462.

Weaver M, Dunn NR & Hogan BL (2000) Bmp4 and Fgf10 play opposing roles during lung bud morphogenesis. *Development* 127, 2695–2704.

Neural Development

Bi G & Poo M (2001) Synaptic modification by correlated activity: Hebb's postulate revisited. *Annu. Rev. Neurosci.* 24, 139–166.

Hubel DH & Wiesel TN (1965) Binocular interaction in striate cortex of kittens reared with artificial squint. *J. Neurophysiol.* 28, 1041–1059.

Kandel ER, Schwartz JH & Jessell TM (2000) Principles of Neural Science, 4th edn. New York: McGraw-Hill.

Mueller BK (1999) Growth cone guidance: first steps towards a deeper understanding. *Annu. Rev. Neurosci.* 22, 351–388.

Sanes DH, Reh TA & Harris WA (2000) Development of the Nervous System. San Diego: Academic Press.

Sanes JR & Lichtman JW (2001) Development: induction, assembly, maturation and maintenance of a postsynaptic apparatus. *Nat. Rev. Neurosci.* 2, 791–805.

Zou Y, Stoeckli E, Chen H & Tessier-Lavigne M (2000) Squeezing axons out of the gray matter: a role for slit and semaphorin proteins from midline and ventral spinal cord. *Cell* 102, 363–375.

Plant Development

Doebley J, Stec A & Hubbard L (1997) The evolution of apical dominance in maize. *Nature* 386, 485–488.

Howell SH (1998) Molecular Genetics of Plant Development. Cambridge: Cambridge University Press.

Jack T (2001) Relearning our ABCs: new twists on an old model. *Trends Plant Sci.* 6, 310–316.

Schoof H, Lenhard M, Haecker A et al. (2000) The stem cell population of Arabidopsis shoot meristems is maintained by a regulatory loop between the CLAVATA and WUSCHEL genes. *Cell* 100, 635–644.

Westhoff P, Jeske H, Jürgens G et al. (1998) Molecular Plant Development From Gene to Plant. Oxford: Oxford University Press.

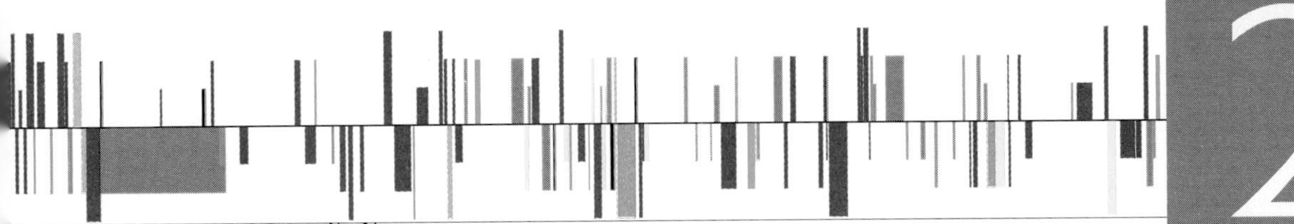

22

HISTOLOGY: THE LIVES AND DEATHS OF CELLS IN TISSUES

Cells evolved originally as free-living individuals, but the cells that matter most to us, as human beings, are specialized members of a multicellular community. They have lost features needed for independent survival and acquired peculiarities that serve the needs of the body as a whole. Although they share the same genome, they are spectacularly diverse: more than 200 different cell types are traditionally recognized in the human body (see our web site for a list). These collaborate with one another to form a multitude of different tissues, arranged into organs performing widely varied functions. To understand them, it is not enough to analyze them in a culture dish: we need also to know how they live, work, and die in their natural habitat.

In Chapters 7 and 21, we saw how the various cell types become different in the embryo and how cell memory and signals from their neighbors enable them to remain different thereafter. In Chapter 19, we discussed the building technology of multicellular tissues—the devices that bind cells together and the extracellular materials that give them support. In this chapter, we consider the functions and lifestyles of the specialized cells in the adult body of a vertebrate. We describe how cells work together to perform their tasks, how new specialized cells are born, how they live and die, and how the architecture of tissues is preserved despite the constant replacement of old cells by new.

We examine these topics through a series of examples—some chosen because they illustrate important general principles, others because they highlight favorite objects of study, still others because they pose intriguing problems that cell biology has yet to solve.

EPIDERMIS AND ITS RENEWAL BY STEM CELLS

To support its specialized functions, the skin has basic requirements that must be satisfied for almost every tissue. It needs mechanical strength, largely provided by a supporting framework of extracellular matrix, mainly secreted by *fibroblasts*. It needs a blood supply to bring nutrients and oxygen and remove waste products and carbon dioxide, and this requires a network of blood vessels, lined with

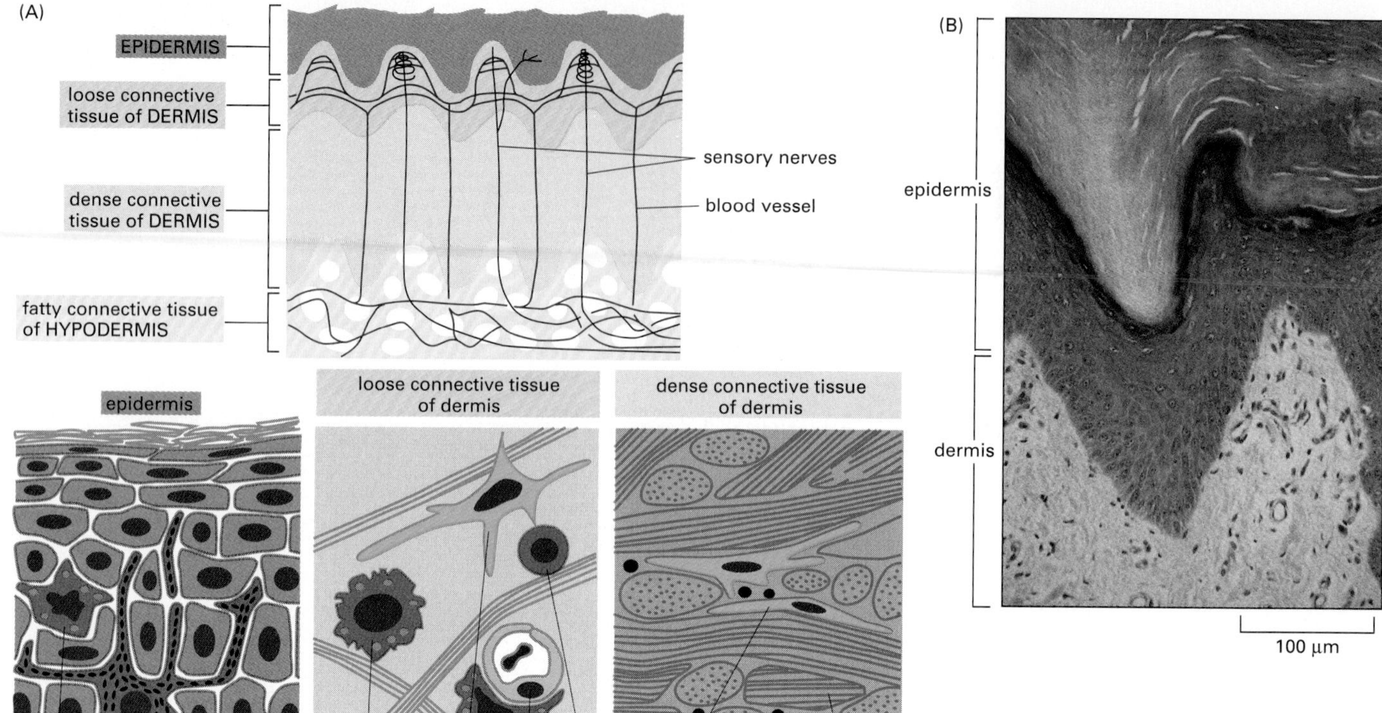

Figure 22–1 Mammalian skin. (A) These diagrams show the cellular architecture of thick skin. (B) Micrograph of a cross section through the sole of a human foot, stained with hematoxylin and eosin. The skin can be viewed as a large organ composed of two main tissues: the epidermis and the underlying connective tissue, which consists of the dermis and the hypodermis. Each tissue is composed of a variety of cell types. The dermis and hypodermis are richly supplied with blood vessels and nerves. Some nerve fibers extend into the epidermis.

endothelial cells. These vessels also provide access routes for cells of the immune system to provide defenses against infection: *macrophages* and *dendritic* cells phagocytose invading pathogens and help activate lymphocytes, which mediate more sophisticated adaptive immune system responses (discussed in Chapter 24). *Nerve fibers* are needed too, to convey sensory information from the tissue to the central nervous system, and to deliver signals in the opposite direction for glandular secretion and smooth muscle contraction.

Figure 22–1 illustrates the architecture of the tissue and shows how it makes provision for all these support services. Skin consists of two main parts: an epithelium, the *epidermis*, lying outermost, and beneath this a layer of connective tissue, which includes the tough collagen-rich *dermis* (from which leather is made) and the underlying fatty *subcutaneous layer* or *hypodermis*. In the skin, as elsewhere, the connective tissue, with vessels and nerves running through it, is responsible for most of the general supportive functions listed above.

The defining component of the skin—the specialized tissue that is peculiar to this organ, even though not the major part of its bulk—is the epidermis. This has a simple organization, and it provides a beautiful introduction to the way in which tissues of the adult body are continually renewed, through processes similar to those that operate in the embryo. We return to connective tissues later.

Epidermal Cells Form a Multilayered Waterproof Barrier

The **epidermis** suffers more direct, frequent, and damaging encounters with the external world than any other tissue in the body. Its need for repair and renewal is central to its organization.

The epidermis is a multilayered *(stratified)* epithelium composed largely of *keratinocytes* (so named because their characteristic differentiated activity is the synthesis of intermediate filament proteins called keratins, which give the epidermis its toughness) (Figure 22–2). These cells change their appearance from one layer to the next. Those in the innermost layer, attached to an underlying

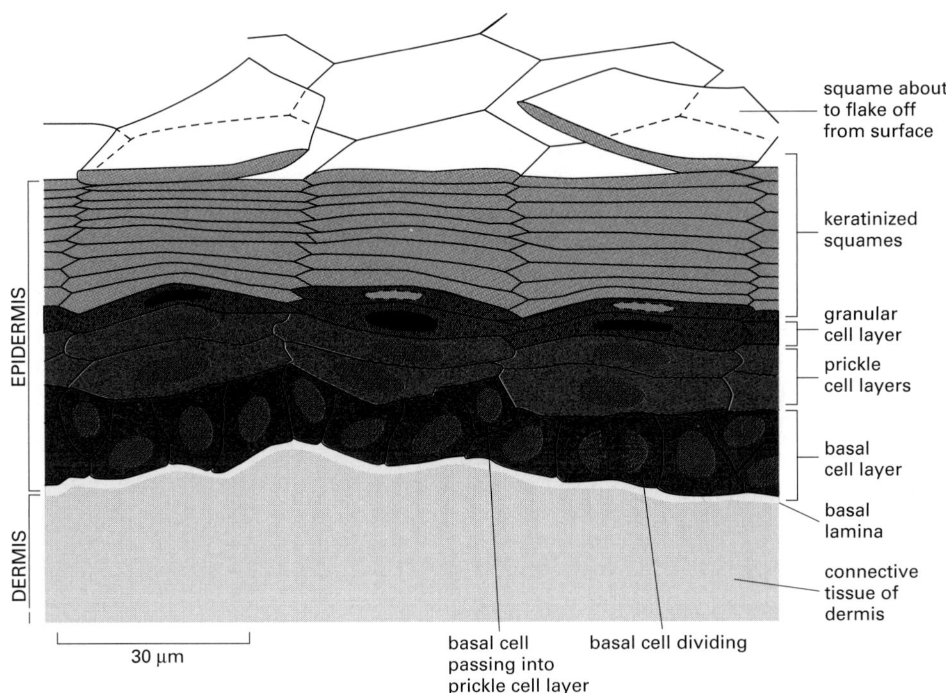

Figure 22–2 The multilayered structure of the epidermis, as seen in a mouse. The outlines of the keratinized squames are revealed by swelling them in a solution containing sodium hydroxide. The highly ordered hexagonal arrangement of interlocking columns of cells shown here occurs only in some sites where the epidermis is thin. In human skin, the stacks of squames are usually many times higher and less regular, and where the skin is very thick mitotic cells are seen not only in the basal layer but also in the first few cell layers above it. In addition to the cells destined for keratinization, the deep layers of the epidermis include small numbers of cells of different characters, as indicated in Figure 21–1—including dendritic cells, called Langerhans cells, derived from bone marrow; melanocytes (pigment cells) derived from the neural crest; and Merkel cells, which are associated with nerve endings in the epidermis.

basal lamina, are termed *basal cells*, and it is usually only these that divide. Above the basal cells are several layers of larger prickle cells (Figure 22–3), whose numerous desmosomes—each a site of anchorage for thick tufts of keratin filaments—are just visible in the light microscope as tiny prickles around the cell surface (hence the name). Beyond the prickle cells lies the thin, darkly staining granular cell layer (see Figure 22–2). It is at this level that the cells are sealed together to form a waterproof barrier, fulfilling the most fundamentally important of all the functions of the epidermis. Mice that fail to form this barrier because of a genetic defect die from rapid fluid loss soon after birth, even though their skin appears normal in other respects.

The granular layer, with its barrier to the movement of water and solutes, marks the boundary between the inner, metabolically active strata and the outermost layer of the epidermis, consisting of dead cells whose intracellular organelles have disappeared. These outermost cells are reduced to flattened scales, or *squames*, filled with densely packed keratin. The plasma membranes of both the squames and the outer granular cells are reinforced on their cytoplasmic surface by a thin (12 nm), tough, cross-linked layer of proteins, including a cytoplasmic protein called involucrin. The squames themselves are normally so compressed and thin that their boundaries are hard to make out in the light microscope, but soaking in sodium hydroxide solution (or a warm bath tub) makes them swell slightly, and their outlines can then be seen (see Figure 22–2).

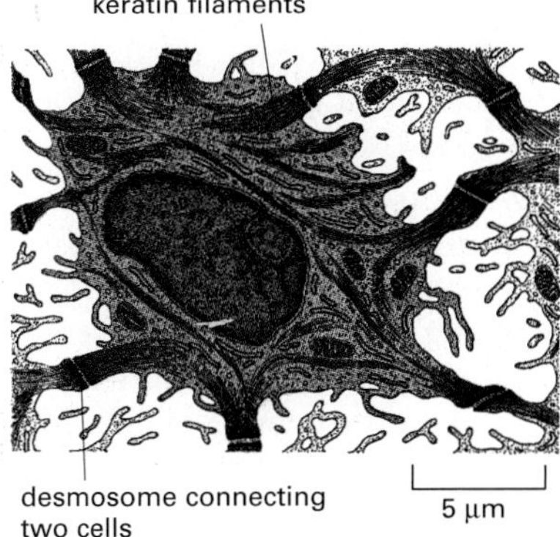

Figure 22–3 A prickle cell. This drawing, from an electron micrograph of a section of the epidermis, shows the bundles of keratin filaments that traverse the cytoplasm and are inserted at the desmosome junctions that bind the prickle cell *(red)* to its neighbors. Nutrients and water diffuse freely through the intercellular spaces in the metabolically active layers of the epidermis occupied by the prickle cells. Farther out, at the level of the granular cells, there is a waterproof barrier that is thought to be created by a sealant material that the granular cells secrete. (From R.V. Krstić, Ultrastructure of the Mammalian Cell: an Atlas. Berlin: Springer-Verlag, 1979.)

Differentiating Epidermal Cells Synthesize a Sequence of Different Keratins as They Mature

Having described the static picture, let us now set it in motion and see how the epidermis is continually renewed by the production of new cells in the basal layer. While some basal cells are dividing, adding to the population in the basal layer, others (their sisters or cousins) are slipping out of the basal cell layer into the prickle cell layer, taking the first step on their outward journey. When they reach the granular layer, the cells start to lose their nucleus and cytoplasmic organelles, through a degradative mechanism that involves partial activation of the machinery of apoptosis; in this way, the cells are transformed into the keratinized squames of the keratinized layer. These finally flake off from the surface of the skin (and become a main constituent of household dust). The period from the time a cell is born in the basal layer of the human skin to the time it is shed from the surface is of the order of a month, depending on the region of the body.

The accompanying molecular transformations can be studied by analyzing either thin slices of epidermis cut parallel to the surface or successive layers of cells stripped off by repeatedly applying and removing strips of adhesive tape. The keratin molecules, for example, which are plentiful in all layers of the epidermis, can be extracted and characterized. They are of many types (discussed in Chapter 16), encoded by a large family of homologous genes, with the variety further increased through alternative RNA splicing. As the new keratinocyte in the basal layer is transformed into the squame in the outermost layers (see Figure 22–3), it switches from one selection of keratins to another. Meanwhile other characteristic proteins, such as involucrin, also begin to be synthesized as part of a coordinated program of **terminal cell differentiation**—the process in which a precursor cell acquires its final specialized characteristics and usually permanently stops dividing. The whole program is initiated in the basal layer. It is here that the fates of the cells are decided.

Epidermis Is Renewed by Stem Cells Lying in Its Basal Layer

The outer layers of the epidermis are replaced a thousand times over in the course of a human lifetime. In the basal layer there have to be cells that can remain undifferentiated and carry on dividing for this whole period, continually throwing off descendants that differentiate, leave the basal layer, and are eventually discarded. The process can be maintained only if the basal cell population is self-renewing. It must therefore contain some cells that generate a mixture of progeny, including daughters that remain undifferentiated like their parent, as well as daughters that differentiate. Cells with this property are called **stem cells**. They have so important a role in such a variety of tissues that it is useful to have a formal definition.

The defining properties of a stem cell are as follows:

1. It is not itself terminally differentiated (that is, it is not at the end of a pathway of differentiation).
2. It can divide without limit (or at least for the lifetime of the animal).
3. When it divides, each daughter has a choice: it can either remain a stem cell, or it can embark on a course that commits it to terminal differentiation (Figure 22–4).

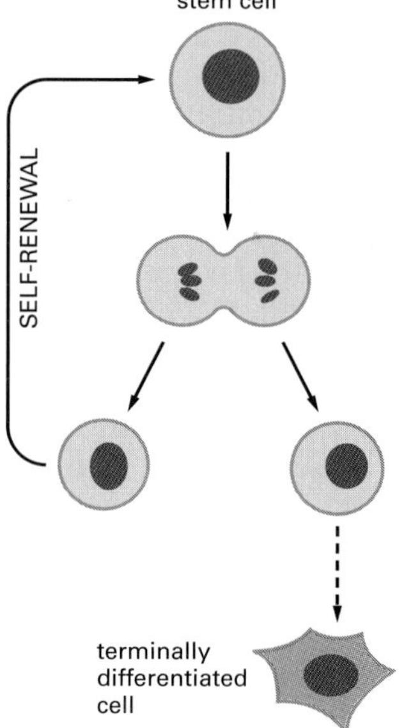

Figure 22–4 The definition of a stem cell. Each daughter produced when a stem cell divides can either remain a stem cell or go on to become terminally differentiated. In many cases, the daughter that opts for terminal differentiation undergoes additional cell divisions before terminal differentiation is completed.

Although it is part of the definition of a stem cell that it should be able to divide, it is not part of the definition that it should divide rapidly; in fact, stem cells usually divide at a relatively low rate. They are required wherever there is a recurring need to replace differentiated cells that cannot themselves divide, and this includes a great variety of tissues. Thus stem cells are of many types, specialized for the genesis of different classes of terminally differentiated cells—epidermal stem cells for epidermis, intestinal stem cells for intestinal epithelium, hemopoietic stem cells for blood, and so on. Each stem-cell system nevertheless raises similar fundamental questions. What factors determine whether the stem cell divides or stays quiescent? What decides whether a given daughter cell differentiates or remains a stem cell? And where the stem cell can give rise to more than one kind of differentiated cell—as is very often the case—what determines which differentiation pathway is followed?

The Two Daughters of a Stem Cell Do Not Always Have to Become Different

At steady state, to maintain a stable stem-cell population, precisely 50% of the daughters of stem cells in each cell generation must remain as stem cells. In principle, this could be achieved in two ways—through *environmental asymmetry* or through *divisional asymmetry* (Figure 22–5). In the one strategy, the division of a stem cell could generate two initially similar daughters whose fates would be governed by their subsequent environment; 50% of the *population* of daughters would remain as stem cells, but the two daughters of an individual

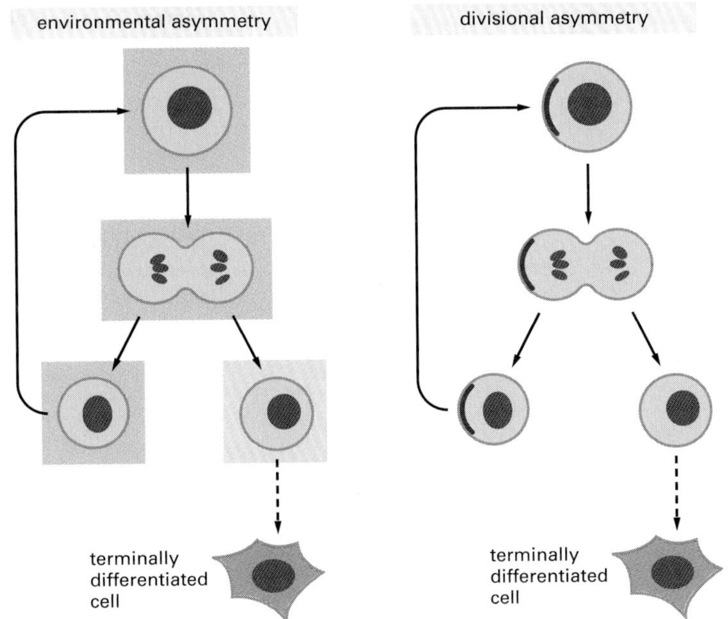

Figure 22–5 Two ways for a stem cell to produce daughters with different fates. In the strategy based on environmental asymmetry, the daughters of the stem cell are initially similar and are directed into different pathways according to the environmental influences that act on them after they are born. The environment is shown as *colored shading* around the cell. With this strategy, the number of stem cells can be increased or reduced to fit the niche available for them. In the strategy based on divisional asymmetry, the stem cell has an internal asymmetry and divides in such a way that its two daughters are already endowed with different determinants at the time of their birth.

stem cell in the population might often have the same fate. At the opposite extreme, the stem cell division could be always strictly asymmetric, producing one daughter that inherits the stem-cell character and another that inherits factors that force it to embark on differentiation. In the latter case, the existing stem cells could never increase their numbers, and any loss of stem cells would be irreparable.

In fact, if a patch of epidermis is destroyed, the damage is repaired by surrounding epidermal cells that migrate in and proliferate to cover the denuded area. In this process, a new self-renewing patch of epidermis is established, implying that additional stem cells have been generated to make up for the loss. These must have been produced by symmetric divisions in which one stem cell gives rise to two. In this way, the stem cell population adjusts its numbers to fit the available niche.

Observations such as these suggest that the maintenance of stem cell character in the epidermis might be controlled by contact with the basal lamina, with a loss of contact triggering the start of terminal differentiation, and maintenance of contact tending to preserve stem cell potential. This idea contains a grain of truth, but it is not the whole truth, as we now explain.

The Basal Layer Contains Both Stem Cells and Transit Amplifying Cells

Basal keratinocytes can be dissociated from intact epidermis and can proliferate in a culture dish, giving rise to new basal cells and to terminally differentiated cells. Even within a population of cultured basal keratinocytes that all seem undifferentiated, there is great variation in the ability to proliferate. When cells are taken singly and tested for their ability to found new colonies, some seem unable to divide at all, others go through only a few division cycles and then halt, and still others divide enough times to form large colonies. This proliferative potential directly correlates with the expression of the β1 subunit of integrin (see Figure 19–64). Clusters of cells with high levels of this molecule can be found in the basal layer of the intact human epidermis also, and they are thought to be the stem cells (Figure 22–6).

Basal cells expressing β1 integrin at a lower level can also divide—indeed, they divide more frequently—but only for a limited number of division cycles, after which they leave the basal layer and differentiate. These latter cells are called **transit amplifying cells**—"transit", because they are in transit from a

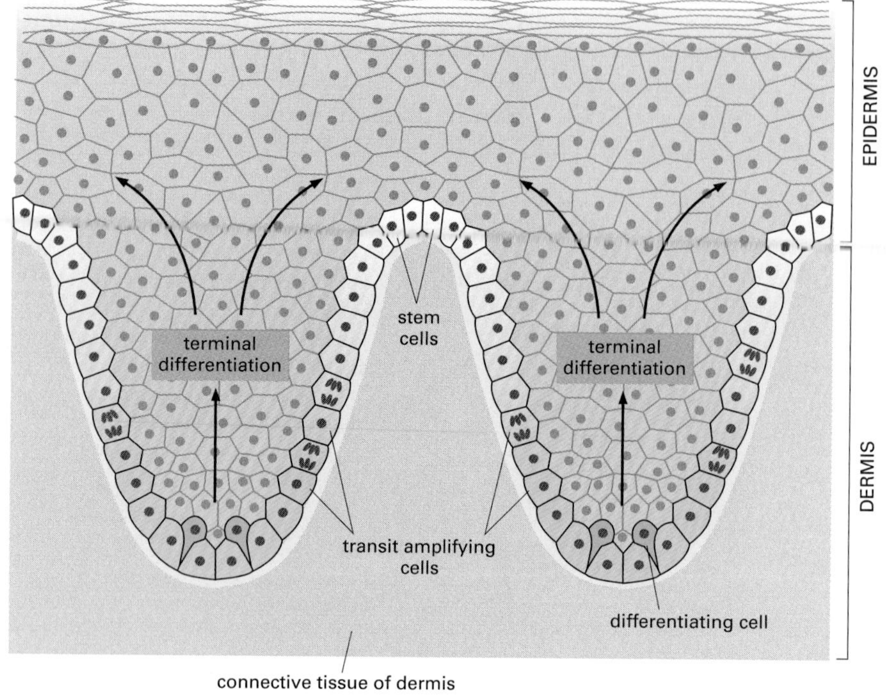

Figure 22–6 The distribution of stem cells in human epidermis, and the pattern of epidermal cell production. The diagram is based on specimens in which the location of the stem cells was identified by staining for β1-integrin, and that of the differentiating cells by staining for keratin-10, a marker of keratinocyte differentiation; dividing cells were identifed by labeling with BrdU, a thymidine analog that is incorporated into cells in S-phase of the cell division cycle. The stem cells seem to be clustered near the tips of the dermal papillae. They divide infrequently, giving rise (through a sideways movement) to transit amplifying cells, which occupy the intervening regions. The transit amplifying cells divide frequently, but for a limited number of division cycles, at the end of which they begin to differentiate and slip out of the basal layer. The precise distribution of stem cells and transit amplifying cells varies from one region of epidermis to another. (Adapted from S. Lowell et al., *Curr. Biol.* 10:491–500, 2000.)

stem-cell character to a differentiated character; "amplifying", because the division cycles they go through have the effect of amplifying the number of differentiated progeny that result from a single stem-cell division (Figure 22–7). Mingled with the population of transit amplifying cells are some cells, still connected with the basal lamina by a thin stalk, that have already stopped dividing and begun to differentiate, as indicated by the types of keratin molecules they express. Contact with the basal lamina, therefore, cannot be the only factor controlling the developmental fate of an epidermal basal cell.

This is not to say that contact with the basal lamina or a similar substratum does not matter. If cultured basal keratinocytes are held in suspension, instead of being allowed to settle and attach to the bottom of the culture dish, they all stop dividing and differentiate. To remain as an epidermal stem cell, it is apparently necessary for it to be attached to the basal lamina or other extracellular

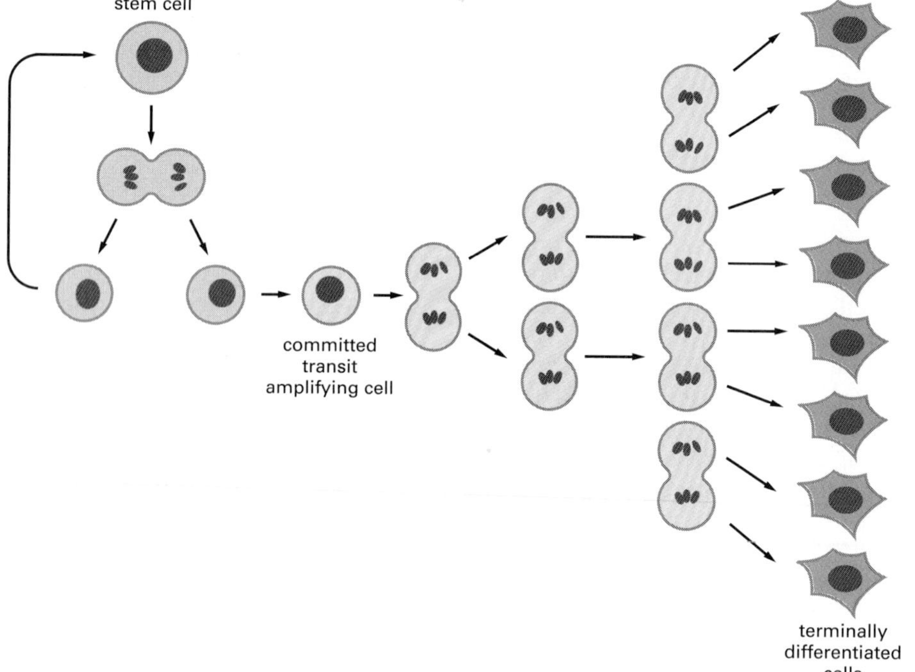

Figure 22–7 Transit amplifying cells. Stem cells in many tissues divide only rarely, but give rise to transit amplifying cells—daughters committed to differentiation that go through a limited series of more rapid divisions before completing the process. In the example shown here, each stem cell division gives rise in this way to eight terminally differentiated progeny.

matrix, even though it is not sufficient. This requirement helps ensure that the size of the stem cell population does not increase without limit. If crowded out of their regular niche on the basal lamina, the cells lose their stem cell character. When this rule is broken, as in some cancers, the result can be an ever-growing tumor.

Epidermal Renewal Is Governed by Many Interacting Signals

Cell turnover in the epidermis seems at first glance a simple matter, but the simplicity, as we have just seen, is deceptive. There are many points in the process that have to be controlled according to circumstances: the rate of stem-cell division; the probability that a stem-cell daughter will remain a stem cell; the number of cell divisions of the transit amplifying cells; the timing of exit from the basal layer, and the time that the cell then takes to complete its differentiation program and be sloughed from the surface. Regulation of these steps must enable the epidermis to respond to rough usage by becoming thick and callused, and to repair itself when wounded. In specialized regions of epidermis, such as those that form hair follicles, with their own specialized subtypes of stem cells, yet more controls are needed to organize the local pattern.

Each control point is important in its own way, and a whole panoply of molecular signals regulate them, so as to keep the body surface always properly covered. Most of the cell communication mechanisms described in Chapter 15 are implicated, either in signaling between cells within the epidermis or in signaling between epidermis and dermis. The EGF, FGF, Wnt, Hedgehog, Notch, BMP/TGFβ, and integrin signaling pathways are all involved (and we shall see that the same is true of most other tissues). Mutations in components of the Hedgehog or Wnt pathways, for example, can lead to the development of epidermal cancers. Components of the Hedgehog, Notch, BMP, and Wnt pathways when misexpressed interfere with the formation of hairs, blocking their development or causing them to develop out of place.

Activation of the Wnt pathway seems to favor maintenance of stem-cell character, inhibiting the switch from stem cell to transit amplifying cell, whereas Notch signaling in the epidermis seems to have a contrary effect, inhibiting neighbors of stem cells from remaining as stem cells. TGFβ has a key role in signaling to the dermis during the repair of skin wounds, promoting the formation of collagen-rich scar tissue. And integrins in the epidermis are not merely markers of cell character, but also regulators of cell fate. Thus, when transgenic mice are engineered to maintain in upper epidermal layers the expression of integrins normally confined to the basal layer, they develop a condition resembling the common human skin disorder *psoriasis*: the rate of basal cell proliferation is greatly increased, the epidermis thickens, and cells are shed from the surface of the skin within as little as a week after emerging from the basal layer, before they have had time to keratinize fully. The precise individual functions of all the various signaling mechanisms in the epidermis are only beginning to be disentangled.

The Mammary Gland Undergoes Cycles of Development and Regression

In specialized regions of the body surface, other types of cells besides the keratinized cells described above develop from the embryonic epidermis. In particular, secretions such as sweat, tears, saliva, and milk are produced by cells segregated in deep-lying glands that originate as ingrowths of the epidermis. These epithelial structures have functions and patterns of renewal quite different from those of keratinizing regions.

The mammary glands are the largest and most remarkable of these secretory organs. They are the defining feature of mammals and an important concern in many ways: not only for nourishment of babies and attraction of the opposite sex, but also as the basis for a large industry—the dairy industry—and as the site of some of the commonest forms of cancer. Mammary tissue illustrates most dramatically that developmental processes continue in the adult body; and it shows how cell death by apoptosis can allow development to go into reverse.

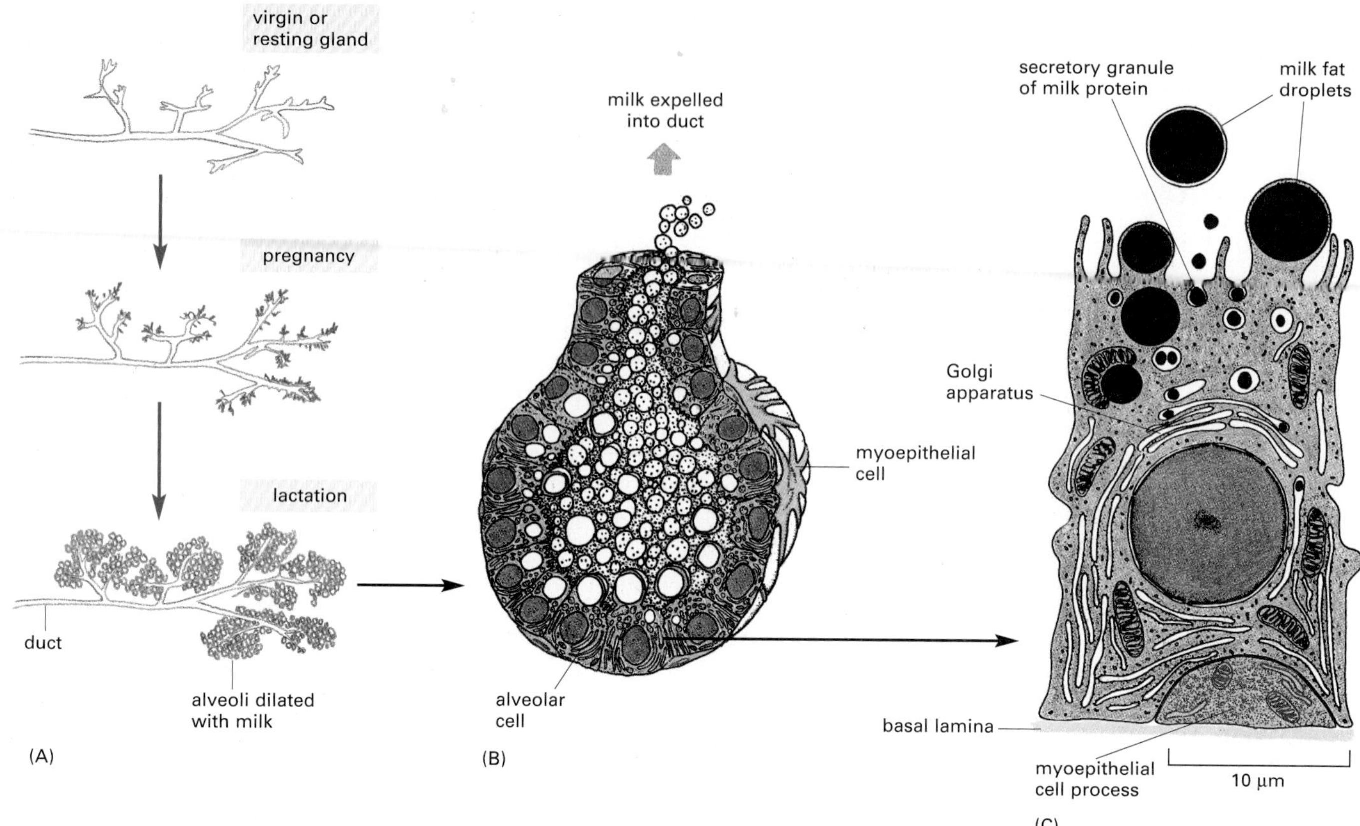

Figure 22–8 The mammary gland. (A) The growth of alveoli from the ducts of the mammary gland during pregnancy and lactation. Only a small part of the gland is shown. The "resting" gland contains a small amount of inactive glandular tissue embedded in a large amount of fatty connective tissue . During pregnancy an enormous proliferation of the glandular tissue takes place at the expense of the fatty connective tissue, with the secretory portions of the gland developing preferentially to create alveoli. (B) One of the milk-secreting alveoli with a basket of contractile myoepithelial cells *(green)* embracing it (see also Figure 22–40E). (C) A single type of secretory alveolar cell produces both the milk proteins and the milk fat. The proteins are secreted in the normal way by exocytosis, while the fat is released as droplets surrounded by plasma membrane detached from the cell. (B, after R. Krstić, Die Gewebe des Menschen und der Säugetiere. Berlin: Springer-Verlag, 1978; C, from D.W. Fawcett, A Textbook of Histology, 12th edn. New York: Chapman and Hall, 1994.)

Milk production must be switched on when a baby is born and switched off when the baby is weaned. A "resting" adult mammary gland consists of branching systems of ducts embedded in fatty connective tissue. The ducts are lined by an epithelium that includes a subpopulation of mammary stem cells (though the precise location of the stem cells is still debated). As a first step toward milk production, the hormones that circulate during pregnancy cause the duct cells to proliferate, increasing their numbers tenfold or twentyfold. The terminal portions of the ducts grow and branch, forming little dilated outpocketings, or alveoli, containing secretory cells (Figure 22–8). Milk secretion begins only when these cells are stimulated by the different combination of hormones that circulate in the mother after the birth of the baby. A further tier of hormonal control governs the actual ejection of milk from the breast: the stimulus of suckling causes cells of the hypothalamus (in the brain) to release the hormone *oxytocin*, which travels via the bloodstream to act on *myoepithelial cells*. These musclelike cells originate from the same epithelial precursor population as the secretory cells of the breast, and they have long spidery processes that embrace the alveoli. In response to oxytocin they contract, thereby squirting milk out of the alveoli into the ducts.

Eventually, when the baby is weaned and suckling stops, the secretory cells die by apoptosis, and most of the alveoli disappear. Macrophages rapidly clear away the dead cells, and the gland reverts to its resting state. This ending of lactation is abrupt and, unlike the events that lead up to it, seems to be induced by the accumulation of milk, rather than by a hormonal mechanism. If one subset of mammary ducts is obstructed so that no milk can be discharged, the secretory cells that supply it commit mass suicide by apoptosis, while other regions of the gland survive and continue to function. The apoptosis is triggered by a combination of factors including TGFβ3, which accumulates where milk secretion is blocked (Figure 22–9).

Cell division in the growing mammary gland is regulated not only by hormones but also by local signals passing between cells within the epithelium and between the epithelial cells and the connective tissue, or *stroma*, in which the epithelial cells are embedded. All the signals listed earlier as important in controlling cell turnover in the epidermis are also implicated in controlling events

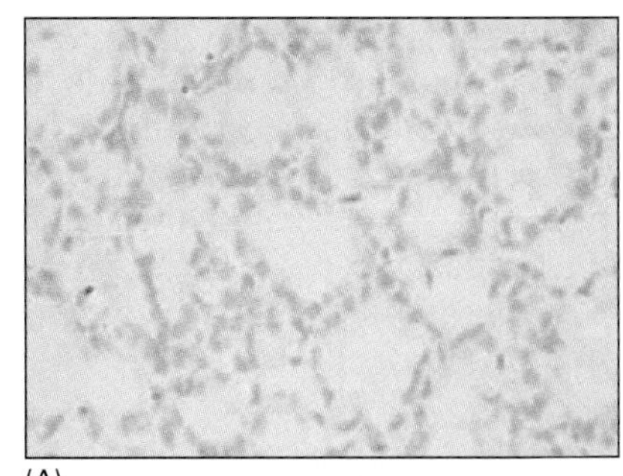
(A)

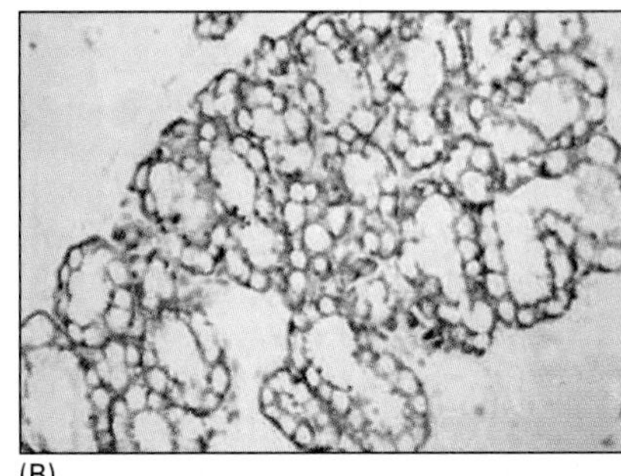
(B)

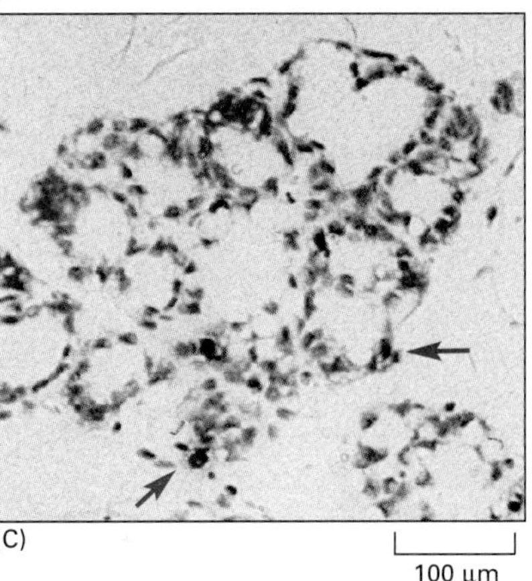

(C)

Figure 22–9 Death of milk-secreting cells when suckling stops. (A) A section of part of a lactating mammary gland of a mouse that had been suckling her newborn babies in the normal way. (B) A corresponding section from a mouse that had not suckled for 9 hours. The sections in (A) and (B) were both stained with an antibody against TGFβ3 *(brown)*. (C) Similar section from a mammary gland whose duct had been sealed for three days, stained to reveal cells dying by apoptosis *(brown)*. Where milk is produced and not drained off by suckling, production of TGFβ3 is induced, and this, together with other signals, triggers apoptosis and leads to regression of the ducts. (From A.V. Nguyen and J.W. Pollard, *Development* 127:3107–3118, 2000. © The Company of Biologists.)

in the mammary gland. Again, signals delivered via integrins play a crucial part: deprived of the basal lamina adhesions that activate integrin signaling, the epithelial cells fail to respond normally to hormonal signals. Faults in these interacting control systems underlie some of the commonest forms of cancer, and we need to understand them better.

Summary

Skin consists of a tough connective tissue, the dermis, overlaid by a multilayered waterproof epithelium, the epidermis. The epidermis is continually renewed from stem cells, with a turnover time, in humans, of the order of a month. Stem cells, by definition, are not terminally differentiated and have the ability to divide throughout the lifetime of the organism, yielding some progeny that differentiate and others that remain stem cells. The epidermal stem cells lie in the basal layer, attached to the basal lamina. Progeny that become committed to differentiation go through several rapid divisions in the basal layer, and then stop dividing and move out toward the surface of the skin. They progressively differentiate, switching from expression of one set of keratins to expression of another until, eventually, their nuclei degenerate, producing an outer layer of dead keratinized cells that are continually shed from the surface.

The fate of the daughters of a stem cell is controlled in part by interactions with the basal lamina and in part by a variety of signals from neighboring cells. These controls allow two stem cells to be generated from one during repair processes, and they regulate the rate of basal cell proliferation according to need. Glands connected to the epidermis, such as the mammary glands, have their own stem cells and their own distinct patterns of cell turnover, governed by other conditions. In the breast, for example, circulating hormones stimulate the cells to proliferate, differentiate, and make milk; the cessation of suckling triggers the milk-secreting cells to die by apoptosis, as a result of a build-up of TGFβ3 where milk fails to be drained away.

SENSORY EPITHELIA

It is through another class of specializations of the epithelium covering our body surface that we sense the smells, sounds and sights of the external world. The sensory tissues of the nose, the ears, and the eyes—and, indeed, if we look back to origins in the early embryo, the whole of the central nervous system—all arise as parts of the same sheet of cells, the *ectoderm*, that gives rise to the epidermis (see Chapter 21). These structures have several basic features in common, and their development is governed by related systems of genes (discussed in Chapter 21). They all retain an epithelial organization, but it is very different from that of the ordinary epidermis or of the glands that derive from it.

The nose, the ear, and the eye are complex organs, with elaborate devices to collect signals from the external world and to deliver them, filtered and concentrated, to the sensory epithelia, where they can trigger effects on the nervous system. The sensory epithelium in each organ is the key component, although it is small relative to all the ancillary apparatus. It is the part that has been most highly conserved in evolution, not only from one vertebrate to another, but also between vertebrates and invertebrates.

Within each sensory epithelium lie sensory cells that act as *transducers*, converting signals from the outside world into an electrical form that can be interpreted by the nervous system. In the nose, the sensory transducers are *olfactory sensory neurons*; in the ear, *auditory hair cells*; and in the eye, *photoreceptors*. All of these cell types are either neurons or neuron-like. Each carries at its apical end a specialized structure that detects the external stimulus and converts it to a change in the membrane potential. At its basal end, each makes synapses with neurons that relay the sensory information to specific sites in the brain.

Olfactory Sensory Neurons Are Continually Replaced

In the olfactory epithelium of the nose (Figure 22–10A), a subset of the epithelial cells differentiate as **olfactory sensory neurons**. These cells have modified, immotile cilia on their free surfaces (see Figure 15–43), containing odorant receptor proteins, and a single axon extending from their basal end towards the brain (Figure 22–10B). The neurons are held in place and separated from one another by *supporting cells* that span the thickened epithelium and have properties similar to those of glial cells in the central nervous system. The sensory surfaces are kept moist and protected by a layer of fluid secreted by cells sequestered in glands that communicate with the exposed surface. Even with this protection, however, each olfactory neuron survives for only a month or two, and so a third class of cells—the *basal cells*—is present in the epithelium to generate replacements for the olfactory neurons that are lost. The basal cells are stem cells for the production of the neurons.

As discussed in Chapter 15, each olfactory neuron expresses most probably only one of the hundreds of odorant receptor genes in the genome. This determines which odorants the cell responds to. But regardless of the odor, the response of every olfactory neuron is the same—a train of action potentials sent back along its axon to the brain. The discriminating sensibility of an individual olfactory neuron is therefore useful only if its axon delivers its messages to the specific relay station in the brain that is dedicated to the particular range of odors that the neuron senses. These relay stations are called *glomeruli*. They are located in structures called the olfactory bulbs (one on each side of the brain), with about 1800 glomeruli in each bulb (in the mouse). Olfactory neurons expressing the same odorant receptor are widely scattered in the olfactory

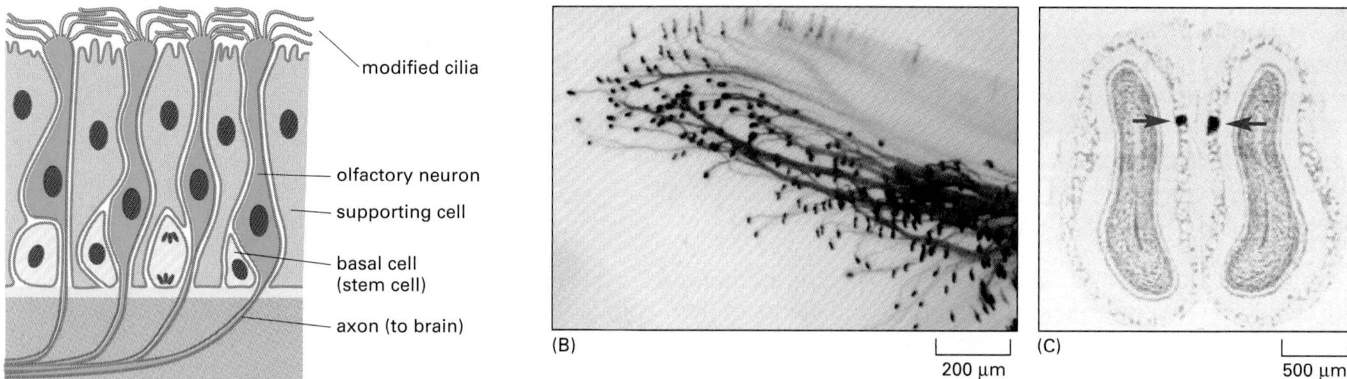

Figure 22–10 Olfactory epithelium and olfactory neurons. (A) Olfactory epithelium consists of supporting cells, basal cells, and olfactory sensory neurons. The basal cells are the stem cells for production of the olfactory neurons. Six to eight modified cilia project from the apex of the olfactory neuron and contain the odorant receptors. (B) This micrograph shows olfactory neurons in the nose of a genetically modified mouse in which the gene encoding LacZ has been inserted into an odorant receptor locus, so that all the cells that would normally express that particular receptor now make the enzyme LacZ in addition. The LacZ is detected through the blue product of the enzymatic reaction that it catalyzes. The cell bodies *(dark blue)* of the marked olfactory neurons, lying scattered in the olfactory epithelium, send their axons *(light blue)* toward the brain (out of the picture to the *right*). (C) A cross section of the *left* and *right* olfactory bulbs, stained for LacZ. Axons of all the olfactory neurons expressing the same odorant receptor converge on the same glomeruli *(red arrows)* symmetrically placed within the bulbs on the *right* and *left* sides of the brain. Other glomeruli (unstained) receive their inputs from olfactory neurons expressing other odorant receptors. (B and C, from P. Mombaerts et al., *Cell* 87:675–686, 1996. © Elsevier.)

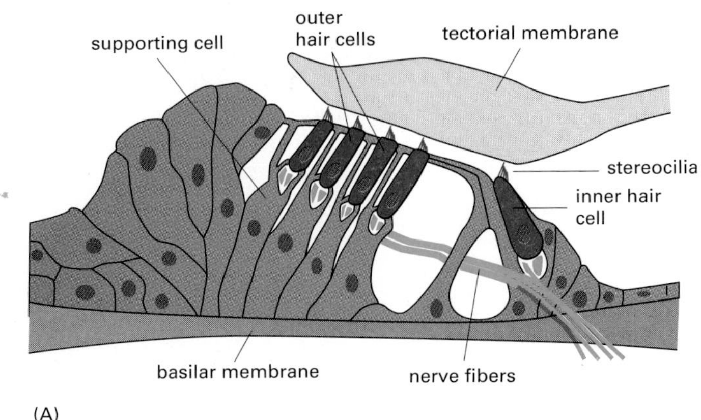

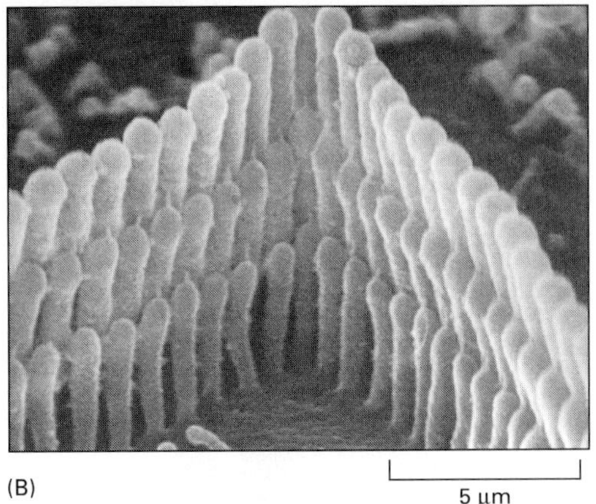

Figure 22–11 Auditory hair cells. (A) A diagrammatic cross section of the auditory apparatus (the organ of Corti) in the inner ear of a mammal shows the auditory hair cells held in an elaborate epithelial structure of supporting cells and overlaid by a mass of extracellular matrix (the tectorial membrane). The epithelium containing the hair cells sits on the basilar membrane—a thin, resilient sheet of tissue that forms a long, narrow dividing partition between two fluid-filled channels. Sound causes pressure changes in these channels and makes the basilar membrane vibrate up and down. (B) This scanning electron micrograph shows the apical surface of an outer auditory hair cell, with the characteristic organ-pipe array of giant microvilli (stereocilia). The inner hair cells, of which there are just 3500 in each human ear, are the principal auditory receptors. The outer hair cells, roughly four times more numerous, are thought to form part of a feedback mechanism that regulates the mechanical stimulus delivered to the inner hair cells. (B, from J.D. Pickles, *Prog. Neurobiol.* 24:1–42, 1985. © Elsevier.)

epithelium, but their axons all converge on the same glomerulus (Figure 22–10C). As new olfactory neurons are generated, replacing those that die, they must in turn send their axons to the correct glomerulus. The odorant receptor proteins thus have been proposed to have a second function: helping to guide the growing tips of the new axons along specific paths to the appropriate target glomeruli in the olfactory bulbs. If it were not for the continual operation of this guidance system, a rose might smell in one month like a lemon, in the next like rotting fish.

Auditory Hair Cells Have to Last a Lifetime

The sensory epithelium responsible for hearing is the most precisely and minutely engineered of all the tissues in the body (Figure 22–11). Its sensory cells, the **auditory hair cells**, are held in a rigid framework of supporting cells and overlaid by a mass of extracellular matrix (the tectorial membrane), in a structure called the *organ of Corti*. The hair cells convert mechanical stimuli into electrical signals. Each has a characteristic organ-pipe array of giant microvilli (called *stereocilia)* protruding from its surface as rigid rods, filled with cross-linked actin filaments, and arranged in ranks of graded height. The dimensions of each such array are specified with extraordinary accuracy according to the location of the hair cell in the ear and the frequency of sound that it has to respond to. Sound vibrations rock the organ of Corti, causing the bundles of stereocilia to tilt (Figure 22–12) and mechanically gated ion channels in the

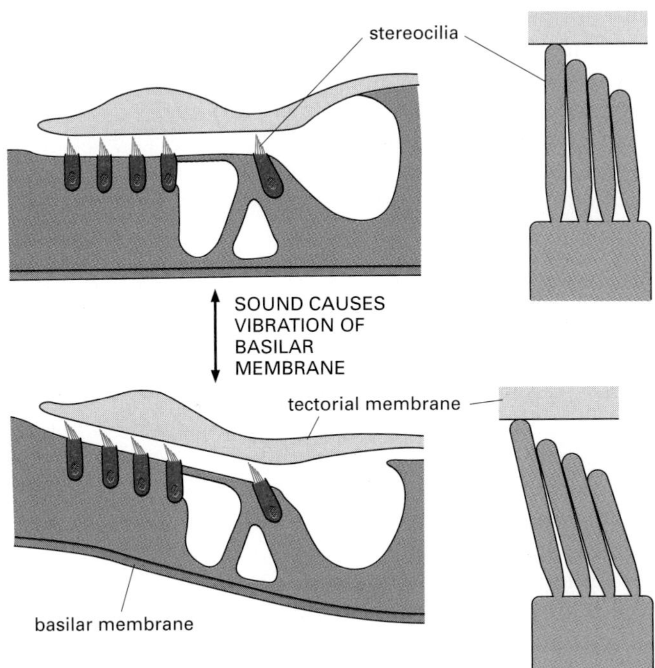

Figure 22–12 How a relative movement of the overlying extracellular matrix (the tectorial membrane) tilts the stereocilia of auditory hair cells in the organ of Corti in the inner ear of a mammal. The stereocilia behave as rigid rods hinged at the base and bundled together at their tips.

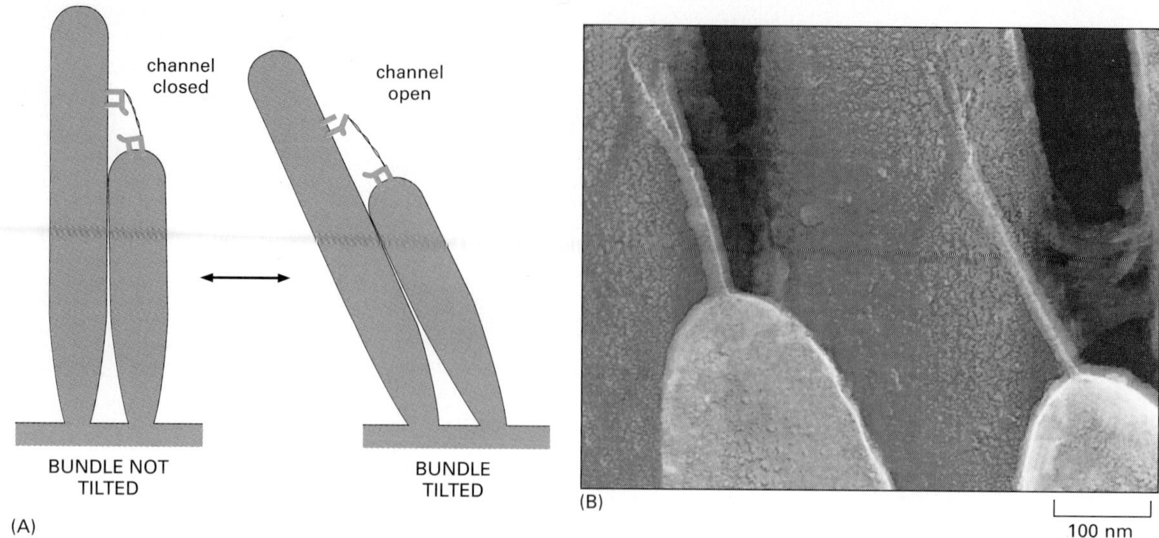

Figure 22–13 How a sensory hair cell works. (A) The cell functions as a transducer, generating an electrical signal in response to sound vibrations that rock the organ of Corti and so cause the stereocilia to tilt. A fine filament runs more or less vertically upward from the tip of each shorter stereocilium to attach at a higher point on its adjacent taller neighbor. Tilting the bundle puts tension on the filaments, which pull on mechanically gated ion channels in the membrane of the stereocilia. Opening of these channels allows an influx of positive charge, depolarizing the hair cell. (B) An electron micrograph of the filaments extending from the tops of two stereocilia. By extraordinarily delicate mechanical measurements, correlated with electrical recordings from a single hair cell as the bundle of stereocilia is deflected by pushing with a flexible glass probe, it is possible to detect an extra "give" of the bundle as the mechanically gated channels yield to the applied force and are pulled open. In this way it can be shown that the force required to open a single one of the hypothesized channels is about 2×10^{-13} newtons and that its gate swings through a distance of about 4 nm as it opens. The mechanism is astonishingly sensitive: the faintest sounds that we can hear have been estimated to stretch the filaments by an average of 0.04 nm, which is just under half the diameter of a hydrogen atom. (B, from B. Kachar et al., *Proc. Natl. Acad. Sci. USA* 97:13336–13341, 2000. © National Academy of Sciences.)

membranes of the stereocilia to open or close (Figure 22–13). The flow of electric charge carried into the cell by the ions alters the membrane potential and thereby controls the release of neurotransmitter at the cell's basal end, where the cell synapses with a nerve ending.

In humans and other mammals, the auditory hair cells, unlike olfactory neurons, have to last a lifetime. If they are destroyed by disease, toxins, or excessively loud noise, they are not regenerated and the resultant hearing loss is permanent. But in other vertebrates, when auditory hair cells are destroyed, the supporting cells are triggered to divide and behave as stem cells, generating progeny that can differentiate as replacements for the hair cells that are lost. With better understanding of how this regeneration process is regulated, we may one day be able to induce the auditory epithelium to repair itself in humans also.

Most Permanent Cells Renew Their Parts: the Photoreceptor Cells of the Retina

The neural retina is the most complex of the sensory epithelia. It consists of several cell layers organized in a way that seems perverse. The neurons that transmit signals from the eye to the brain (called *retinal ganglion cells)* lie closest to the external world, so that the light, focused by the lens, must pass through them to reach the photoreceptor cells. The **photoreceptors**, which are classified as *rods* or *cones*, according to their shape, lie with their photoreceptive ends, or outer segments, partly buried in the *pigment epithelium* (Figure 22–14). Rods and cones contain different *rhodopsins*—photosensitive complexes of *opsin* protein with the visual pigment *retinal*. Rods are especially sensitive at low light levels, while cones (of which there are three types, each with a different opsin, giving a different spectral response) detect color and fine detail.

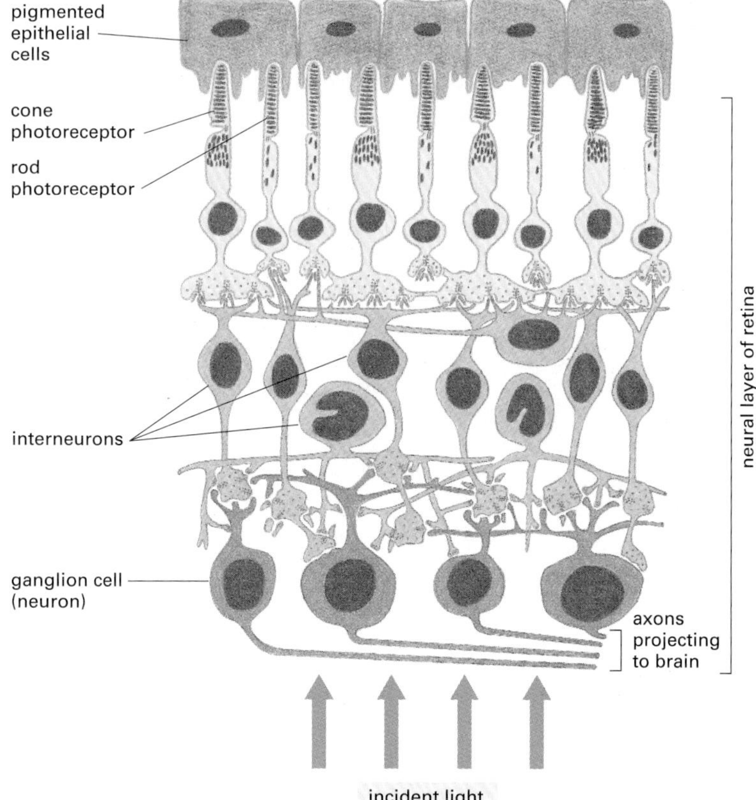

Figure 22–14 The structure of the retina. The stimulation of the photoreceptors by light is relayed via interneurons to the ganglion cells, which convey the signal to the brain. The spaces between neurons and between photoreceptors in the neural retina are occupied by a population of specialized supporting cells (not shown here). (Modified from J.E. Dowling and B.B. Boycott, *Proc. R. Soc. Lond.* B 166:80–111, 1966.)

The outer segment of a photoreceptor appears to be a modified cilium with a characteristic ciliumlike arrangement of microtubules in the region where the outer segment is connected to the rest of the cell (Figure 22–15). The remainder of the outer segment is almost entirely filled with a dense stack of membranes in which the photosensitive complexes are embedded; light absorbed here produces an electrical response, as discussed in Chapter 15. At their opposite ends, the photoreceptors form synapses on interneurons, which relay the signal to the retinal ganglion cells (see Figure 22–14).

Photoreceptors in humans, like human auditory hair cells, are permanent cells that do not divide and are not replaced if destroyed by disease or by a misdirected laser beam. But the photosensitive rhodopsin molecules are not permanent but are continually degraded and replaced. In rods (although not, curiously, in cones), this turnover is organized in an orderly production line, which can be analyzed by following the passage of a cohort of radiolabeled protein molecules through the cell after a short pulse of radioactive amino acid (Figure 22–16). The radiolabeled proteins can be followed from the Golgi apparatus in the inner segment of the cell to the base of the stack of membranes in the outer segment. From here they are gradually displaced toward the tip as new material is fed into the base of the stack. Finally (after about 10 days in the rat), on reaching the tip of the outer segment, the labeled proteins and the layers of membrane in which they are embedded are phagocytosed (chewed off and digested) by the cells of the pigment epithelium.

This example illustrates a general point: even though individual cells of certain types persist, very little of the adult body consists of the same molecules that were laid down in the embryo.

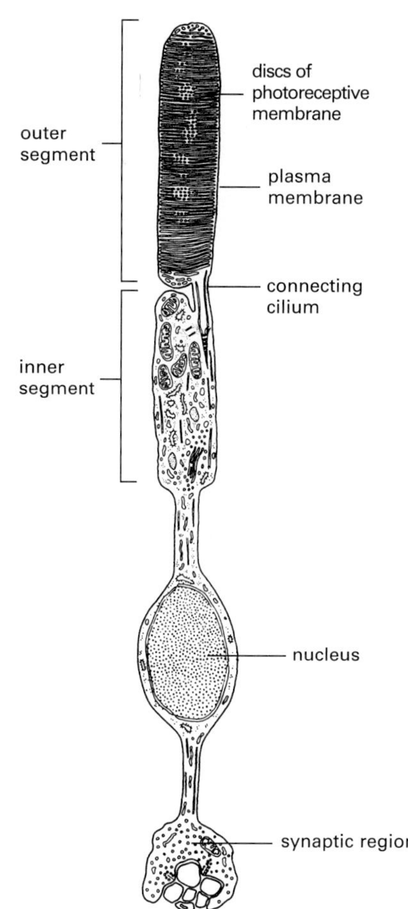

Figure 22–15 A rod photoreceptor.

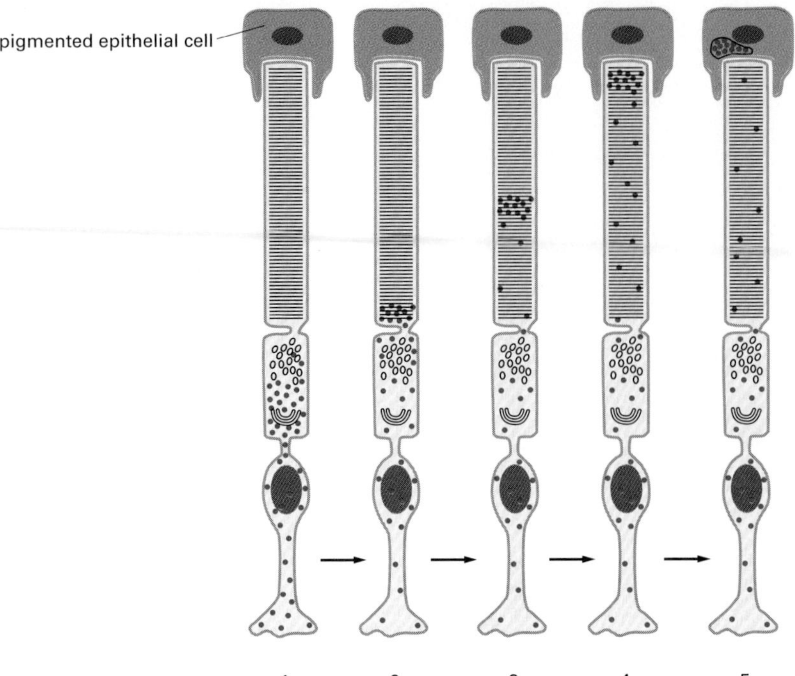

Figure 22–16 Turnover of membrane protein in a rod cell. Following a pulse of ^{3}H-leucine, the passage of radiolabeled proteins through the cell is followed by autoradiography. *Red dots* indicate sites of radioactivity. The method reveals only the ^{3}H-leucine that has been incorporated into proteins; the rest is washed out during the preparation of the tissue. (1) The incorporated leucine is first seen concentrated in the neighborhood of the Golgi apparatus. (2) From there it passes to the base of the outer segment into a newly synthesized disc of photoreceptive membrane. (3–5) New discs are formed at a rate of three or four per hour (in a mammal), displacing the older discs toward the pigment epithelium.

Summary

Most sensory receptor cells, like epidermal cells and nerve cells, derive from the epithelium forming the outer surface of the embryo. They transduce external stimuli into electrical signals, which they relay to neurons via chemical synapses. Olfactory receptor cells in the nose are themselves full-fledged neurons, sending their axons to the brain. They have a lifetime of only a month or two, and are continually replaced by new cells derived from stem cells in the olfactory epithelium. Each olfactory neuron expresses just one of the hundreds of different olfactory receptor proteins for which genes exist in the genome, and the axons from all olfactory neurons expressing the same receptor protein navigate to the same glomeruli in the olfactory bulbs of the brain.

Auditory hair cells—the receptor cells for sound—unlike olfactory receptor cells, have to last a lifetime, in mammals at least. They have no axon but make synaptic contact with nerve terminals in the auditory epithelium. They take their name from the hair-like bundle of stereocilia (giant microvilli) on their outer surface. Sound vibrations tilt the bundle, pulling mechanically gated ion channels on the stereocilia into an open configuration to excite the cell electrically.

Photoreceptor cells in the retina of the eye absorb photons in rhodopsin molecules held in stacks of membrane in the photoreceptor outer segments, triggering an electrical excitation by a more indirect intracellular signaling pathway. Although the photoreceptor cells themselves are permanent and irreplaceable, the stacks of rhodopsin-rich membrane that they contain undergo continual renewal.

THE AIRWAYS AND THE GUT

The examples we have discussed so far represent a small selection of the tissues and cell types that derive from the outer layer of the embryo—the ectoderm. They are enough, however, to give a sense of the amazing variety of ways in which this epithelium becomes specialized for different purposes, and to show how widely adult cells can differ in their lifestyles. The inmost layer of the embryo—the *endoderm*, forming the primitive gut tube—gives rise to another whole zoo of cell types lining the digestive tract and its appendages. We begin with the lungs.

Adjacent Cell Types Collaborate in the Alveoli of the Lungs

The airways of the lungs are formed by repeated branching of a system of tubes that originated in the embryo from an outpocketing of the gut lining, as discussed in Chapter 21. Repeated tiers of branching terminate in several hundred million air-filled sacs—the **alveoli**. Alveoli have thin walls, closely apposed to the walls of blood capillaries so as to allow exchange of O_2 and CO_2 with the blood stream (Figure 22–17).

To survive, the cells lining the alveoli must remain moist. At the same time, they must serve as a gas container that can expand and contract with each breath in and out. This creates a problem. When two wet surfaces touch, they become stuck together by surface tension in the layer of water between them—an effect that operates more powerfully the smaller the scale of the structure. There is a risk, therefore, that the alveoli may collapse and be impossible to reinflate. The problem is solved by the presence of two types of cells in the lining of the alveoli. *Type I alveolar cells* cover most of the wall: they are thin and flat (*squamous*) to allow gas exchange. *Type II alveolar cells* are interspersed among them. These are plump and secrete *surfactant,* a phospholipid-rich material that forms a film on the free water surfaces and reduces surface tension, making the alveoli easy to reinflate even if they collapse. The production of adequate amounts of surfactant in the fetus, starting at about 5 months of pregnancy in humans, marks the beginning of the possibility of independent life. Premature babies born before this stage are unable to inflate their lungs and breathe; those born after it can do so and, with intensive care, can survive.

Goblet Cells, Ciliated Cells, and Macrophages Collaborate to Keep the Airways Clean

Higher up in the airways one finds different combinations of cell types, serving different purposes. The air we breathe is full of dust, dirt, and air-borne microorganisms. To keep the lungs clear and healthy, this debris must be constantly swept out. To perform this task, the larger airways are lined by a relatively thick

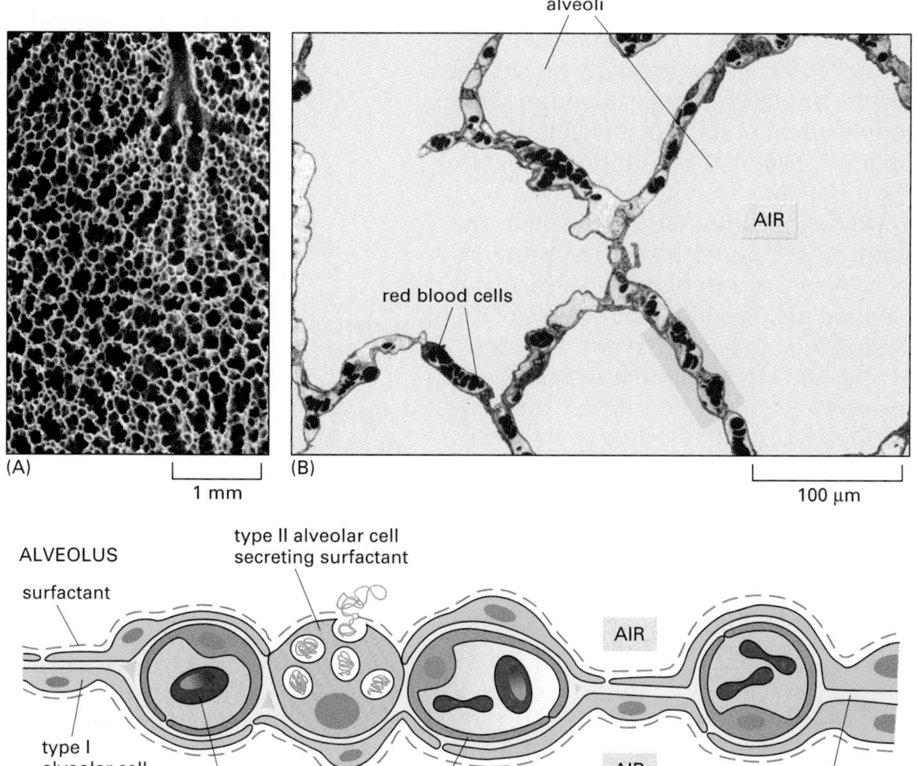

Figure 22–17 Alveolin the lung. (A) Scanning electron micrograph at low magnification, showing the sponge-like texture created by the many air-filled alveoli. A bronchiole (small tubular airway) is seen at the top, communicating with the alveoli. (B) Transmission electron micrograph of a section through a region corresponding to the the *yellow* box in (A) showing the alveolar walls, where gas exchange occurs. (C) Diagram of the cellular architecture of a piece of alveolar wall, corresponding to the *yellow* box in (B). (A, from P. Gehr et al., *Resp. Physiol.* 44:61–69, 1981. © Elsevier; B, courtesy of Peter Gehr, from D.W. Fawcett, A Textbook of Histology, 12th edn. New York: Chapman and Hall, 1994.)

respiratory epithelium (Figure 22–18). This includes three differentiated cell types: *goblet cells* (so named because of their shape), which secrete mucus, *ciliated cells*, with cilia that beat, and a small number of *endocrine cells*, secreting serotonin and peptides that act as local mediators. These signal molecules affect nerve endings and other neighboring cells in the respiratory tract, so as to help regulate the rate of mucus secretion and ciliary beating, the contraction of surrounding smooth muscle cells that can constrict the airways, and other functions. Basal cells are also present, and serve as stem cells for renewal of the epithelium.

The mucus secreted by the goblet cells forms a viscoelastic blanket about 5 μm thick over the tops of the cilia. The cilia, all beating in the same direction, at a rate of about 12 beats per second, sweep the mucus out of the lungs, carrying with it the debris that has become stuck to it. This conveyor belt for the removal of rubbish from the lungs is called the *mucociliary escalator*. Of course, some inhaled particles may reach the alveoli themselves, where there is no escalator. Here, the unwanted matter is removed by yet another class of specialized cells, *macrophages*, which roam the lungs and engulf foreign matter and kill and digest bacteria. Many millions of macrophages, loaded with debris, are swept out of the lungs every hour on the mucociliary escalator.

At the upper end of the respiratory tract, the wet mucus-covered respiratory epithelium gives way abruptly to stratified squamous epithelium. This cell sheet is structured for mechanical strength and protection, and, like epidermis, it consists of many layers of flattened cells densely packed with keratin. It differs from epidermis in that it is kept moist and its cells retain their nucleus even in the outermost layers. Abrupt boundaries of epithelial cell specialization, such as that between the mucous and the stratified squamous epithelium of the respiratory tract, are also found in other parts of the body, but very little is known about how they are created and maintained.

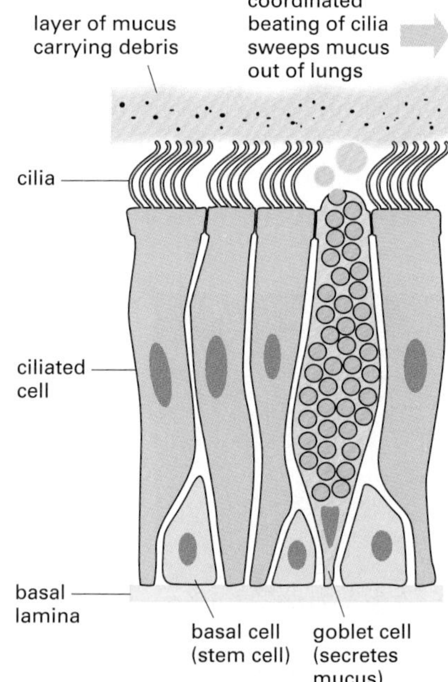

Figure 22–18 Respiratory epithelium. The goblet cells secrete mucus, which forms a blanket over the tops of the ciliated cells. The regular, coordinated beating of the cilia sweeps the mucus up and out of the airways, carrying any debris that is stuck to it. The mechanism that coordinates the ciliary beating is a mystery, but it seems to reflect an intrinsic polarity in the epithelium. If a segment of rabbit trachea is surgically reversed, it carries on sweeping mucus, but in the wrong direction, back down toward the lung, in opposition to adjacent unreversed portions of trachea.

The Lining of the Small Intestine Renews Itself Faster Than Any Other Tissue

Only air-breathing vertebrates have lungs, but all vertebrates, and almost all invertebrate animals, have a gut—that is, a digestive tract lined with cells specialized for the digestion of food and absorption of the nutrient molecules released by the digestion. These two activities are hard to carry on at the same time, as the processes that digest food in the lumen of the gut are liable also to digest the lining of the gut itself, including the cells that absorb the nutrients. The gut uses several strategies to solve the problem.

The fiercest digestive processes, involving acid hydrolysis as well as enzyme action, are conducted in a separate reaction vessel, the stomach. The products are then passed on to the small intestine, where the nutrients are absorbed and enzymatic digestion continues, but at a neutral pH. The different regions of the gut lining consist of correspondingly different mixtures of cell types. The stomach epithelium includes cells that secrete acid, and other cells that secrete digestive enzymes that work at acid pH. Conversely, glands (in particular the pancreas) that discharge into the initial segment of the small intestine contain cells that secrete bicarbonate to neutralize the acidity, along with other cells that secrete digestive enzymes that work at neutral pH. The lining of the intestine, downstream from the stomach, contains both absorptive cells and cells specialized for secretion of mucus, which covers the epithelium with a protective coat. In the stomach, too, the most exposed surfaces are lined with mucous cells. And, in case these measures are not enough, the whole lining of the stomach and intestine is continually renewed and replaced by freshly generated cells, with a turnover time of a week or less.

The renewal process has been studied best in the small intestine (Figure 22–19). The lining of the small intestine (and of most other regions of the gut) is a single-layered epithelium. This epithelium covers the surfaces of the *villi* that project into the lumen, and it lines the *crypts* that descend into the underlying connective tissue. Dividing stem cells lie in a protected position in the depths of

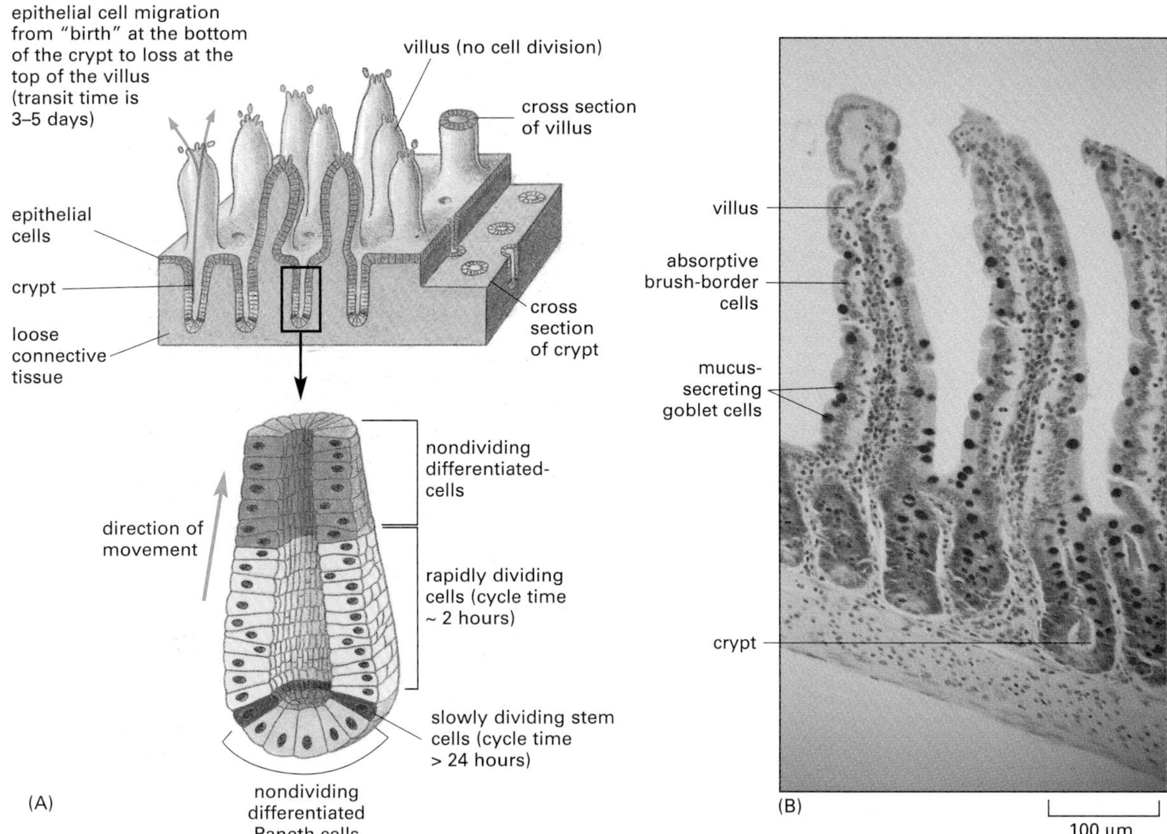

Figure 22–19 Renewal of the gut lining. (A) The pattern of cell turnover and the proliferation of stem cells in the epithelium that forms the lining of the small intestine. The *arrow* shows the general upward direction of cell movement onto the villi, but some cells, including a proportion of the goblet and enteroendocrine cells, stay behind and differentiate while still in the crypts. The nondividing differentiated cells (Paneth cells) at the very bottom of the crypts also have a finite lifetime, and are continually replaced by progeny of the stem cells. (B) Photograph of a section of part of the lining of the small intestine, showing the villi and crypts. Note how mucus-secreting goblet cells (stained *red*) are interspersed among other cell types. Enteroendocrine cells are less numerous and less easy to identify without special stains. See Figure 22–20 for the structure of these cells.

the crypts. These generate four types of differentiated progeny (Figure 22–20): (1) *absorptive cells* (also called *brush-border cells*), with densely packed microvilli on their exposed surfaces to increase their active surface area; (2) *goblet cells* (as in respiratory epithelium), secreting mucus; (3) *Paneth cells*, forming part of the innate immune defense system (discussed in Chapter 25) and secreting (along with some growth factors) *cryptdins*—proteins of the defensin family that kill bacteria (see Figure 25–39); and (4) *enteroendocrine cells*, of more than 15 different subtypes, secreting serotonin and peptide hormones, such as cholecystokinin, that act on neurons and other cell types in the gut wall and regulate the growth, proliferation and digestive activities of cells of the gut and other tissues; many of these gut hormones function also as neuropeptide signal molecules in the nervous system.

The absorptive, goblet, and enteroendocrine cells travel mainly upward from the stem-cell region, by a sliding movement in the plane of the epithelial sheet, to cover the surfaces of the villi. As in the epidermis, there is a transit amplifying stage of cell proliferation: on their way out of the crypt, the precursor cells, already committed to differentiation, go through four to six rapid divisions before they stop dividing and differentiate terminally. Within 2–5 days (in the mouse) after emerging from the crypts, the cells reach the tips of the villi, where they undergo the initial stages of apoptosis and are finally discarded into the gut lumen. The Paneth cells are produced in much smaller numbers and have

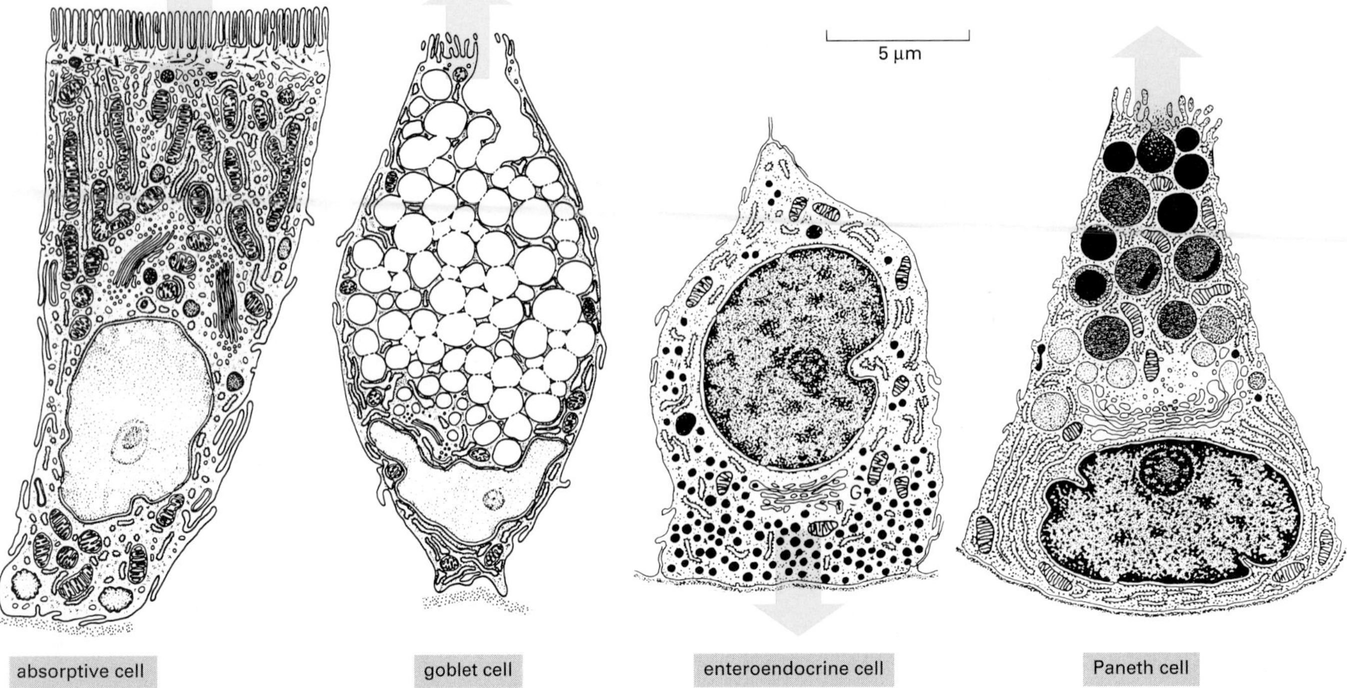

Figure 22–20 The four main differentiated cell types found in the epithelial lining of the small intestine. All of these are generated from undifferentiated multipotent stem cells living near the bottoms of the crypts (see Figure 22–19). The microvilli on the apical surface of the absorptive (brush-border) cell provide a 30-fold increase of surface area, not only for the import of nutrients, but also for the anchorage of enzymes that perform the final stages of extracellular digestion, breaking down small peptides and disaccharides into monomers that can be transported across the cell membrane. (After T.L. Lentz, Cell Fine Structure. Philadelphia: Saunders, 1971; R. Krstić, Illustrated Encyclopedia of Human Histology. Berlin: Springer-Verlag, 1984.)

a different migration pattern. They stay down at the bottom of the crypts, where they too are continually replaced, although not so rapidly, persisting for about 20 days (in the mouse) before undergoing apoptosis and being phagocytosed by their neighbors.

The driving force for the movements of the gut epithelial cells is still a mystery. Their different patterns of migration may be controlled by cell-type-specific responses to molecular cues in the basal lamina on which they sit. Different regions of the lamina are rich in different types of laminin: α1 and α2 laminin subunits are concentrated in the basal lamina of the crypts, while α5 is in the basal lamina of the villi, in a concentration gradient with its maximum at the villus tip.

Components of the Wnt Signaling Pathway Are Required to Maintain the Gut Stem-Cell Population

What controls whether a gut stem cell retains its stem-cell character or embarks on differentiation? What drives the diversification of the stem cell progeny to produce four different cell types? Despite their importance in normal life and in many diseases, including common forms of cancer, the answers are unknown. There are, however, some clues.

The production of enteroendocrine cells, for example, depends on a cell-fate choice governed by the Notch signaling pathway in much the same way as the production of neurons in the embryonic central nervous system (discussed in Chapter 21). Mutations that block Notch signaling cause enteroendocrine cells to be produced in excess at the expense of other cell types, at least in the embryo.

Other experiments indicate that the Wnt signaling pathway has a crucial role in maintaining the stem cell population of the gut. Mice deficient in one of the LEF-1/TCF family of gene regulatory proteins involved in Wnt signaling (discussed in Chapter 15) provide one line of evidence: villi covered with differentiated cells are formed in the mutant fetus, but the epithelium cannot be renewed, because no crypts form and no proliferating stem cells are retained, and the mouse dies soon after birth. Conversely, in adult life, overactivation of the same signaling pathway by mutations in another gene (the *APC* gene—see Figure 15–72) result in an overproliferation of crypt cells and frequently lead to cancer, as we shall see in Chapter 23. All this suggests that in the gut, as in the epidermis, Wnt signaling has a key role in either maintaining stem cells as stem cells or controlling their proliferation.

The Liver Functions as an Interface Between the Digestive Tract and the Blood

As we have just seen, the functions of the gut are divided between a variety of cell types. Some cells are specialized for the secretion of hydrochloric acid, others for the secretion of enzymes, others for the absorption of nutrients, and so on. Some of these cell types are closely intermingled in the wall of the gut, whereas others are segregated in large glands that communicate with the gut and originate in the embryo as outgrowths of the gut epithelium.

The liver is the largest of these glands. It develops at a site where a major vein runs close to the wall of the primitive gut tube, and the adult organ retains a special relationship with the blood. Cells in the liver that derive from the primitive gut epithelium—the **hepatocytes**—are arranged in interconnected sheets and cords, with blood-filled spaces called sinusoids running between them (Figure 22–21). The blood is separated from the surface of the hepatocytes by a single layer of flattened endothelial cells that covers the exposed faces of the hepatocytes. This structure facilitates the chief functions of the liver, which center on the exchange of metabolites between hepatocytes and the blood.

The liver is the main site at which nutrients that have been absorbed from the gut and then transferred to the blood are processed for use by other cells of the body. It receives a major part of its blood supply directly from the intestinal tract (via the portal vein). Hepatocytes are responsible for the synthesis, degradation, and storage of a vast number of substances. They play a central part in the carbohydrate and lipid metabolism of the body as a whole, and they secrete most of the protein found in blood plasma. At the same time, the hepatocytes remain connected with the lumen of the gut via a system of minute channels (or canaliculi) and larger ducts (see Figure 22–21B,C) and secrete into the gut by this route both waste products of their metabolism and an emulsifying agent, *bile*,

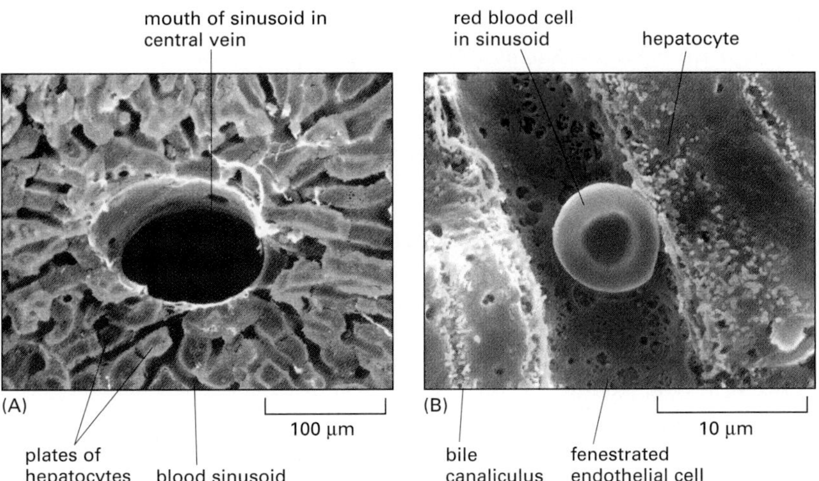

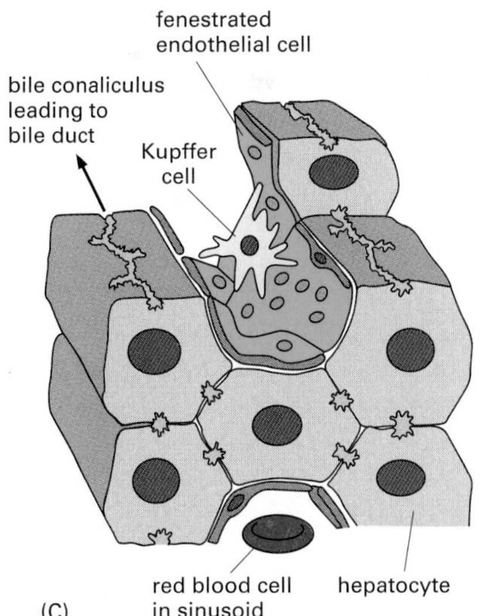

Figure 22–21 The structure of the liver. (A) A scanning electron micrograph of a portion of the liver, showing the irregular sheets and cords of hepatocytes and the many small channels, or sinusoids, for the flow of blood. The larger channels are vessels that distribute and collect the blood that flows through the sinusoids. (B) Detail of a sinusoid (enlargement of region similar to that marked by yellow rectangle at lower right in [A]). (C) Schematized diagram of the fine structure of the liver. The hepatocytes are separated from the bloodstream by a single thin sheet of endothelial cells with interspersed macrophagelike Kupffer cells. Small holes in the endothelial sheet, called fenestrae (Latin for "windows"), allow the exchange of molecules and small particles between the hepatocytes and the bloodstream. Besides exchanging materials with the blood, the hepatocytes form a system of tiny bile canaliculi into which they secrete bile, which is ultimately discharged into the gut via bile ducts. The real structure is less regular than this diagram suggests. (A and B, courtesy of Pietro M. Motta, University of Rome "La Sapienza.")

which helps in the absorption of fats. Hepatocytes are big cells, and about 50% of them (in an adult human) are polyploid, with two, four, eight, or even more times the normal diploid quantity of DNA per cell.

In contrast to the rest of the digestive tract, there seems to be remarkably little division of labor within the population of hepatocytes. Each hepatocyte seems able to perform the same broad range of metabolic and secretory tasks. These fully differentiated cells can also divide repeatedly, when the need arises, as we explain next.

Liver Cell Loss Stimulates Liver Cell Proliferation

The liver illustrates in a striking way one of the great unsolved problems of developmental and tissue biology: what determines the size of an organ of the body, or the quantity of one type of tissue relative to another? For different organs, the answers are almost certainly different, but there is scarcely any case in which the mechanism is well understood.

Hepatocytes normally live for a year or more and are renewed at a slow rate. Even in a slowly renewing tissue, however, a small but persistent imbalance between the rate of cell production and the rate of cell death would lead to disaster. If 2% of the hepatocytes in a human divided each week but only 1% died, the liver would grow to exceed the weight of the rest of the body within 8 years. Homeostatic mechanisms must operate to adjust the rate of cell proliferation and/or the rate of cell death so as to keep the organ at its normal size. This size, moreover, needs to be matched to the size of the rest of the body. Indeed, when the liver of a small dog is grafted into a large dog, it rapidly grows to almost the size appropriate to the host; conversely, when the liver is grafted from a large dog into a small one, it shrinks.

Direct evidence for the homeostatic control of liver cell proliferation comes from experiments in which large numbers of hepatocytes are removed surgically or are intentionally killed by poisoning with carbon tetrachloride. Within a day or so after either sort of damage, a surge of cell division occurs among the surviving hepatocytes, and the lost tissue is quickly replaced. (If the hepatocytes themselves are totally eliminated, another class of cells, located in the bile ducts, can serve as stem cells for the genesis of new hepatocytes, but usually there is no need for this.) If two-thirds of a rat's liver is removed, for example, a liver of nearly normal size can regenerate from the remainder by hepatocyte proliferation within about 2 weeks. Many molecules have been implicated in the triggering of this reaction. One of the most important is a protein called *hepatocyte growth factor*. It stimulates hepatocytes to divide in culture, and its production increases steeply (by poorly understood mechanisms) in response to liver damage.

The balance between cell births and cell deaths in the adult liver (and other organs too) does not depend exclusively on the regulation of cell proliferation: cell survival controls also play a part. If an adult rat is treated with the drug phenobarbital, for example, hepatocytes are stimulated to divide, causing the liver to enlarge. When the phenobarbital treatment is stopped, hepatocyte cell death greatly increases until the liver returns to its original size, usually within a week or so. The mechanism of this type of cell survival control is unknown, but it has been suggested that hepatocytes, like most vertebrate cells, depend on signals from other cells for their survival and that the normal level of these signals can support only a certain standard number of hepatocytes. If the number of hepatocytes rises above this (as a result of phenobarbital treatment, for example), hepatocyte death will automatically increase to bring their number back down. It is not known how the appropriate levels of survival factors are maintained.

Summary

The lung performs a simple function—gas exchange—but its housekeeping systems are complex. Surfactant-secreting cells help to keep the alveoli from collapsing. Macrophages constantly scour the alveoli for dirt and microorganisms. A mucociliary

escalator formed by mucus-secreting goblet cells and beating ciliated cells sweeps debris out of the airways.

In the gut, where more potentially damaging chemical processes occur, the absorptive epithelium is kept in good repair by constant rapid renewal. In the small intestine, stem cells in the crypts generate new absorptive, goblet, enteroendocrine, and Paneth cells, replacing most of the epithelial lining of the intestine every week. The diverse fates of the stem-cell progeny are controlled, in part at least, by the Notch signaling pathway, while the Wnt pathway is required to maintain the stem-cell population.

The liver is a more protected organ, but it too can rapidly adjust its size up or down by cell proliferation or cell death when the need arises. Differentiated hepatocytes remain able to divide throughout life, showing that a specialized class of stem cells is not always needed for tissue renewal.

BLOOD VESSELS AND ENDOTHELIAL CELLS

From the tissues that derive from the embryonic ectoderm and endoderm, we turn now to those derived from *mesoderm*. This middle layer of cells, sandwiched between ectoderm and endoderm, grows and diversifies to provide a wide range of supportive functions. It gives rise to the body's connective tissues, blood cells, and blood vessels, as well as muscle, kidney, and many other structures and cell types. We begin with blood vessels.

Almost all tissues depend on a blood supply, and the blood supply depends on **endothelial cells**, which form the linings of the blood vessels. Endothelial cells have a remarkable capacity to adjust their number and arrangement to suit local requirements. They create an adaptable life-support system, extending by cell migration into almost every region of the body. If it were not for endothelial cells extending and remodeling the network of blood vessels, tissue growth and repair would be impossible. Cancerous tissue is as dependent on a blood supply as is normal tissue, and this has led to a surge of interest in endothelial cell biology. It is hoped that by blocking the formation of new blood vessels through drugs that act on endothelial cells, it may be possible to block the growth of tumors (discussed in Chapter 23).

Endothelial Cells Line All Blood Vessels

The largest blood vessels are arteries and veins, which have a thick, tough wall of connective tissue and and many layers of smooth muscle cells (Figure 22–22). The wall is lined by an exceedingly thin single sheet of endothelial cells, the *endothelium,* separated from the surrounding outer layers by a basal lamina. The amounts of connective tissue and smooth muscle in the vessel wall vary according to the vessel's diameter and function, but the endothelial lining is always present. In the finest branches of the vascular tree—the capillaries and sinusoids—the walls consist of nothing but endothelial cells and a basal lamina (Figure 22–23), together with a few scattered—but functionally important—*pericytes.* These cells of the connective-tissue family, related to vascular smooth muscle cells, that wrap themselves round the small vessels (Figure 22–24).

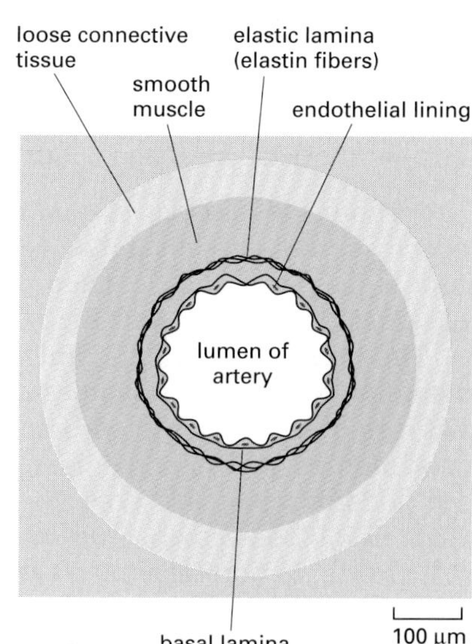

Figure 22–22 Diagram of a small artery in cross section. The endothelial cells, although inconspicuous, are the fundamental component. Compare with the capillary in Figure 22–23.

Thus, endothelial cells line the entire vascular system, from the heart to the smallest capillary, and control the passage of materials—and the transit of white blood cells—into and out of the bloodstream. A study of the embryo reveals, moreover, that arteries and veins develop from small vessels constructed solely of endothelial cells and a basal lamina: pericytes, connective tissue and smooth muscle are added later where required, under the influence of signals from the endothelial cells. The recruitment of pericytes in particular depends on PDGF-B secreted by the endothelial cells, and in mutants lacking this signal protein or its receptor, pericytes in many regions are missing. As a result, the embryonic blood vessels develop microaneurysms—microscopic pathological dilatations—that eventually rupture, as well as other abnormalities, reflecting the importance of signals exchanged in both directions between the pericytes and the endothelial cells.

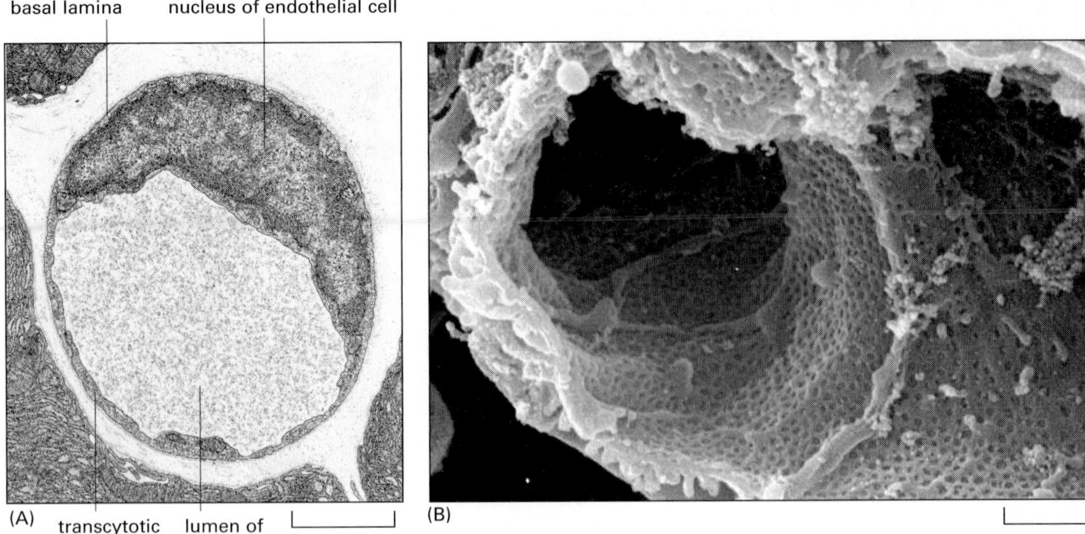

Figure 22–23 Capillaries. (A) Electron micrograph of a cross section of a small capillary in the pancreas. The wall is formed by a single endothelial cell surrounded by a basal lamina. Note the small "transcytotic" vesicles, which according to one theory provide transport of large molecules in and out of this type of capillary: materials are taken up into the vesicles by endocytosis at the luminal surface of the cell and discharged by exocytosis at the external surface, or vice versa. (B) Scanning electron micrograph of the interior of a capillary in a glomerulus of the kidney, where filtration of the blood occurs to produce urine. Here, as in the liver (see Figure 22–20), the endothelial cells are specialized to form a sieve-like structure, with fenestrae ("windows"), constructed rather like the pores in the nuclear envelope of eucaryotic cells, allowing water and most molecules to pass freely out of the bloodstream. (A, from R.P. Bolender, *J. Cell Biol.* 61:269–287, 1974. © The Rockefeller University Press; B, courtesy of Steve Gschmeissner and David Shima.)

Once a vessel has matured, signals from the endothelial cells to the surrounding connective tissue and smooth muscle continue to play a crucial part in regulating the vessel's function and structure. For example, the endothelial cells have mechanoreceptors that allow them to sense the shear stress due to flow of blood over their surface; by signaling this information to the surrounding cells, they enable the blood vessel to adapt its diameter and wall thickness to suit the blood flow. Endothelial cells also mediate rapid responses to neural signals for blood vessel dilation, by releasing the gas NO to make smooth muscle relax in the vessel wall, as discussed in Chapter 15.

New Endothelial Cells Are Generated by Simple Duplication of Existing Endothelial Cells

Throughout the vascular system of the adult body, endothelial cells retain a capacity for cell division and movement. If, for example, a part of the wall of the aorta is damaged and denuded of endothelial cells, neighboring endothelial cells proliferate and migrate in to cover the exposed surface.

The proliferation of endothelial cells can be demonstrated by using ^{3}H-thymidine to label cells synthesizing DNA. In most adult tissues, endothelial cells turn over very slowly, with a cell lifetime ranging, for a mouse, from a couple of months (in liver and lung) to years (in brain and muscle). But endothelial cells not only repair and renew the lining of established blood vessels, they also create new blood vessels. They must do this in embryonic tissues to keep pace with growth, in normal adult tissues to support recurrent cycles of remodeling and reconstruction (as, for example, in the lining of the uterus during the menstrual cycle), and in damaged adult tissues to support repair. In such circumstances, they can be roused to proliferate with a doubling time of just a few days. There is some evidence that, where there is a call for rapid blood-vessel growth, the local population of endothelial cells may also increase by recruitment from the blood stream, which has been reported to contain small numbers of endothelial precursor cells derived from the bone marrow.

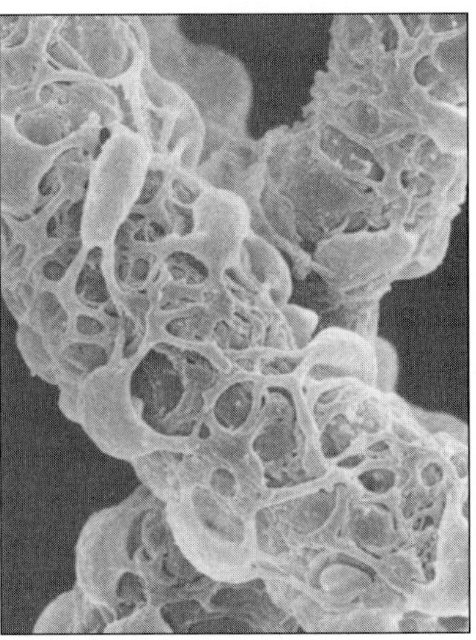

Figure 22–24 Pericytes. The scanning electron micrograph shows pericytes wrapping their processes around a small blood vessel (a post-capillary venule) in the mammary gland of a cat. Pericytes are present also around capillaries, but much more sparsely distributed there. (From T. Fujiwara and Y. Uehara, *Am. J. Anat.* 170:39–54, 1984. © Wiley-Liss.)

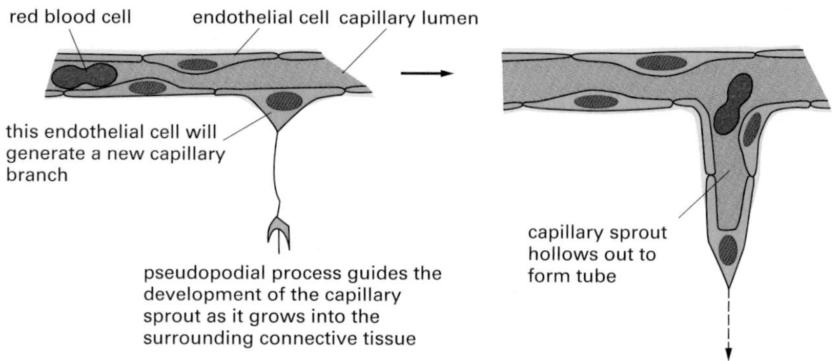
red blood cell
endothelial cell
capillary lumen
this endothelial cell will generate a new capillary branch
pseudopodial process guides the development of the capillary sprout as it grows into the surrounding connective tissue
capillary sprout hollows out to form tube

Figure 22–25 Angiogenesis. A new blood capillary forms by the sprouting of an endothelial cell from the wall of an existing small vessel. This diagram is based on observations of cells in the transparent tail of a living tadpole. (After C.C. Speidel, *Am. J. Anat.* 52:1–79, 1933.)

New Capillaries Form by Sprouting

New vessels in the adult originate as capillaries, which sprout from existing small vessels. This process of **angiogenesis** occurs in response to specific signals and can be conveniently observed in naturally transparent structures, such as the cornea of the eye. Irritants applied to the cornea induce the growth of new blood vessels from the rim of tissue surrounding the cornea, which has a rich blood supply, in toward the center of the cornea, which normally has none. Thus, the cornea becomes vascularized through an invasion of endothelial cells into the tough collagen-packed corneal tissue.

Observations such as these reveal that endothelial cells that are to form a new capillary grow out from the side of an existing capillary or small venule by extending long pseudopodia, pioneering the formation of a capillary sprout that hollows out to form a tube (Figure 22–25). This process continues until the sprout encounters another capillary, with which it connects, allowing blood to circulate. Endothelial cells on the arterial and venous sides of the developing network of vessels differ in their surface properties, in the embryo at least: the plasma membranes of the arterial cells contain the transmembrane protein ephrin-B2 (see Chapter 15), while the membranes of the venous cells contain the corresponding receptor protein, Eph-B4, which is a receptor tyrosine kinase (discussed in Chapter 15). These molecules mediate a signal delivered at sites of cell–cell contact, and they are essential for the development of a properly organized network of vessels. One suggestion is that they somehow define the rules for joining one piece of growing capillary tube to another.

Experiments in culture show that endothelial cells in a medium containing suitable growth factors will spontaneously form capillary tubes, even if they are isolated from all other types of cells (Figure 22–26). The capillary tubes that develop do not contain blood, and nothing travels through them, indicating that blood flow and pressure are not required for the initiation of a new capillary network.

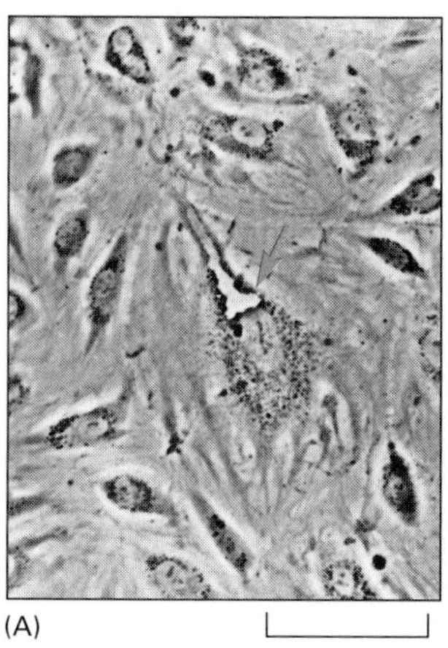
(A)
100 μm

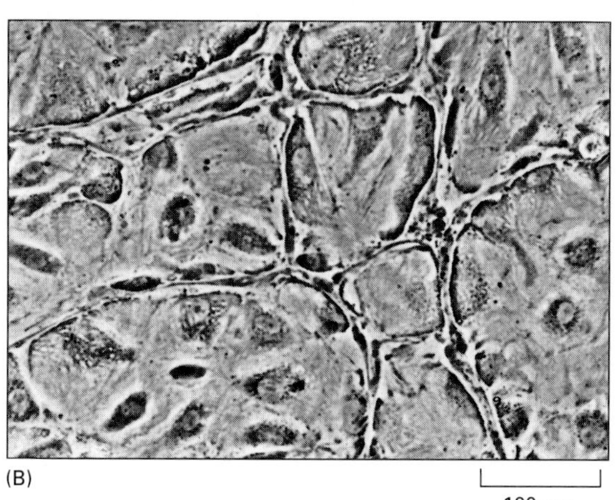
(B)
100 μm

Figure 22–26 Capillary formation *in vitro*. Endothelial cells in culture spontaneously develop internal vacuoles that appear to join up from cell to cell, giving rise to a network of capillary tubes. These photographs show successive stages in the process. The *arrow* in (A) indicates a vacuole forming initially in a single endothelial cell. The cultures are set up from small patches of two to four endothelial cells taken from short segments of capillary. These cells settle on the surface of a collagen-coated culture dish and form a small flattened colony that enlarges gradually as the cells proliferate. The colony spreads across the dish, and eventually, after about 20 days, capillary tubes begin to form in the central regions. Once tube formation has started, branches soon appear, and after 5–10 more days, an extensive network of tubes is visible, as seen in (B). The process is strongly dependent on the nature of the extracellular matrix in the environment of the cells: the formation of capillary tubes is promoted by basal lamina components, such as laminin, which the endothelial cells themselves can secrete. (From J. Folkman and C. Haudenschild, *Nature* 288:551–556, 1980. © Macmillan Magazines Ltd.)

Angiogenesis Is Controlled by Factors Released by the Surrounding Tissues

Almost every cell, in almost every tissue of a vertebrate, is located within 50–100 μm of a capillary. What mechanism ensures that the system of blood vessels ramifies into every nook and cranny? How is it adjusted so perfectly to the local needs of the tissues, not only during normal development but also in all sorts of pathological circumstances? Wounding, for example, induces a burst of capillary growth in the neighborhood of the damage, to satisfy the high metabolic requirements of the repair process (Figure 22–27). Local irritants and infections also cause a proliferation of new capillaries, most of which regress and disappear when the inflammation subsides. Less benignly, a small sample of tumor tissue implanted in the cornea, which normally lacks blood vessels, causes blood vessels to grow quickly toward the implant from the vascular margin of the cornea; the growth rate of the tumor increases abruptly as soon as the vessels reach it.

In all these cases, the invading endothelial cells respond to signals produced by the tissue that they invade. The signals are complex, but a key part is played by a protein known as **vascular endothelial growth factor** (**VEGF**), a distant relative of platelet-derived growth factor (PDGF). The regulation of blood vessel growth to match the needs of the tissue depends on the control of VEGF production, through changes in the stability of its mRNA and in its rate of transcription. The latter control is relatively well understood. A shortage of oxygen, in practically any type of cell, causes an increase in the intracellular concentration of the active form of a gene regulatory protein called **hypoxia-inducible factor 1** (**HIF-1**). HIF stimulates transcription of the VEGF gene (and of other genes whose products are needed when oxygen is in short supply). The VEGF protein is secreted, diffuses through the tissue, and acts on nearby endothelial cells.

The response of the endothelial cells includes at least four components. First, the cells produce proteases to digest their way through the basal lamina of the parent capillary or venule. Second, the endothelial cells migrate toward the source of the signal. Third, the cells proliferate. Fourth, the cells form tubes and differentiate. VEGF acts on endothelial cells selectively to stimulate this entire set of effects. (Other growth factors, including some members of the fibroblast growth factor family, can also stimulate angiogenesis, but they influence other cell types besides endothelial cells.)

As the new vessels form, bringing blood to the tissue, the oxygen concentration rises, HIF-1 activity declines, VEGF production is shut off, and angiogenesis comes to a halt (Figure 22–28). As in all signaling systems, it is as important to switch the signal off correctly as to switch it on. In normal well-oxygenated tissue, the concentration of HIF-1 is kept low by continual degradation of the HIF-1 protein. This degradation depends on ubiquitylation of HIF-1—a process that requires the product of another gene, which is defective in a rare disorder called *von Hippel–Lindau (VHL) syndrome*. People with this condition are born with only one functional copy of the VHL gene; mutations occurring at random

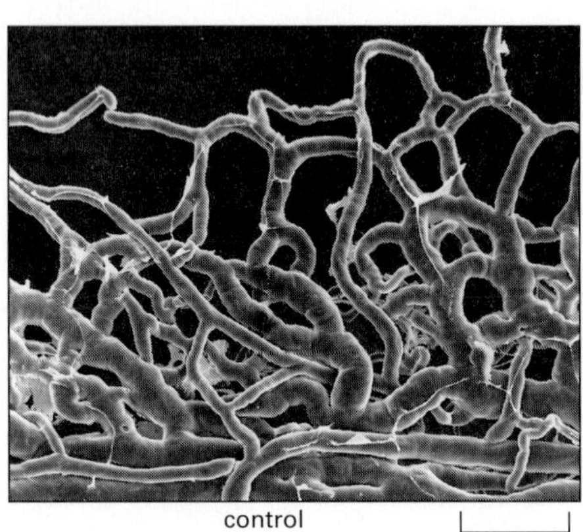

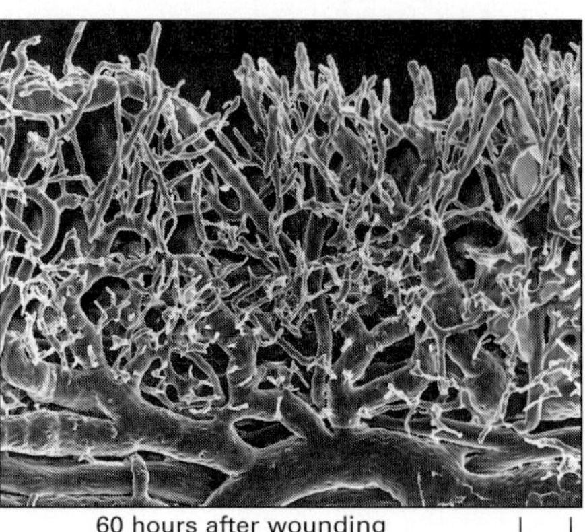

Figure 22–27 New capillary formation in response to wounding. Scanning electron micrographs of casts of the system of blood vessels surrounding the margin of the cornea show the reaction to wounding. The casts are made by injecting a resin into the vessels and letting the resin set; this reveals the shape of the lumen, as opposed to the shape of the cells. Sixty hours after wounding many new capillaries have begun to sprout toward the site of injury, which is just *above* the *top* of the picture. Their oriented outgrowth reflects a chemotactic response of the endothelial cells to an angiogenic factor released at the wound. (Courtesy of Peter C. Burger.)

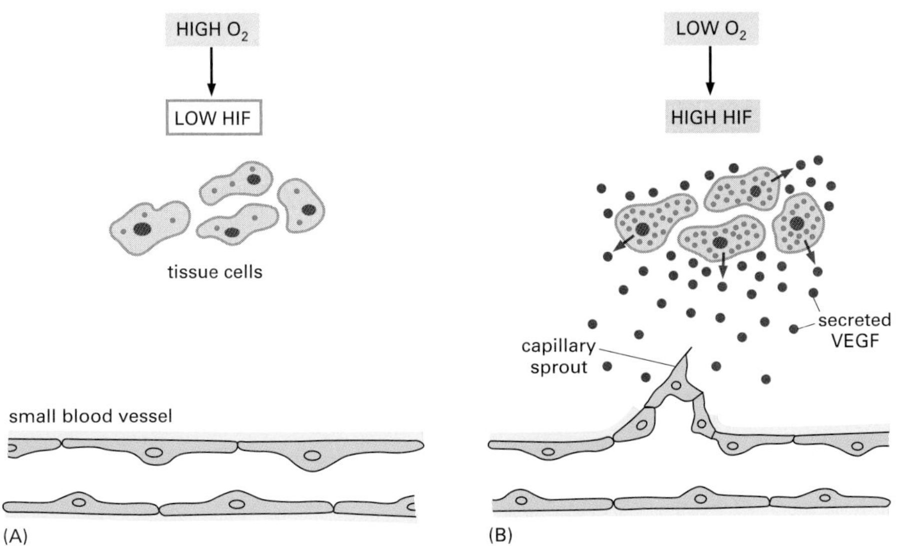

Figure 22–28 The regulatory mechanism controlling blood vessel growth according to a tissue's need for oxygen. Lack of oxygen triggers the secretion of VEGF, which stimulates angiogenesis.

in the body then give rise to cells in which both gene copies are defective. These cells contain large quantities of HIF-1 regardless of oxygen availability, triggering the continual overproduction of VEGF. The result is development of *hemangioblastomas,* tumors that contain dense masses of blood vessels. The mutant cells that produce the VEGF are apparently themselves encouraged to proliferate by the over-rich nourishment provided by the excess blood vessels, creating a vicious cycle that promotes tumor growth. Loss of the VHL gene product also gives rise to other tumors as well as hemangioblastomas, by mechanisms that may be independent of effects on angiogenesis.

Summary

Endothelial cells form a single cell layer that lines all blood vessels and regulates exchanges between the bloodstream and the surrounding tissues. Signals from endothelial cells organize the growth and development of connective tissue cells that form the surrounding layers of the blood-vessel wall. New blood vessels can develop from the walls of existing small vessels by the outgrowth of endothelial cells, which have the capacity to form hollow capillary tubes even when isolated in culture. Endothelial cells of developing arteries and veins express different cell-surface proteins, which may control the way in which they link up to create a capillary bed.

A homeostatic mechanism ensures that blood vessels permeate every region of the body. Cells that are short of oxygen increase their concentration of hypoxia-inducible factor 1 (HIF-1), which stimulates the production of vascular endothelial growth factor (VEGF). VEGF acts on endothelial cells, causing them to proliferate and invade the hypoxic tissue to supply it with new blood vessels.

RENEWAL BY MULTIPOTENT STEM CELLS: BLOOD CELL FORMATION

Blood contains many types of cells with very different functions, ranging from the transport of oxygen to the production of antibodies. Some of these cells function entirely within the vascular system, while others use the vascular system only as a means of transport and perform their function elsewhere. All blood cells, however, have certain similarities in their life history. They all have limited life-spans and are produced throughout the life of the animal. Most remarkably, they are all generated ultimately from a common stem cell in the bone marrow. This *hemopoietic* (blood-forming) *stem cell* is thus multipotent, giving rise to all the types of terminally differentiated blood cells as well as some other types of cells, such as osteoclasts in bone, which we discuss later.

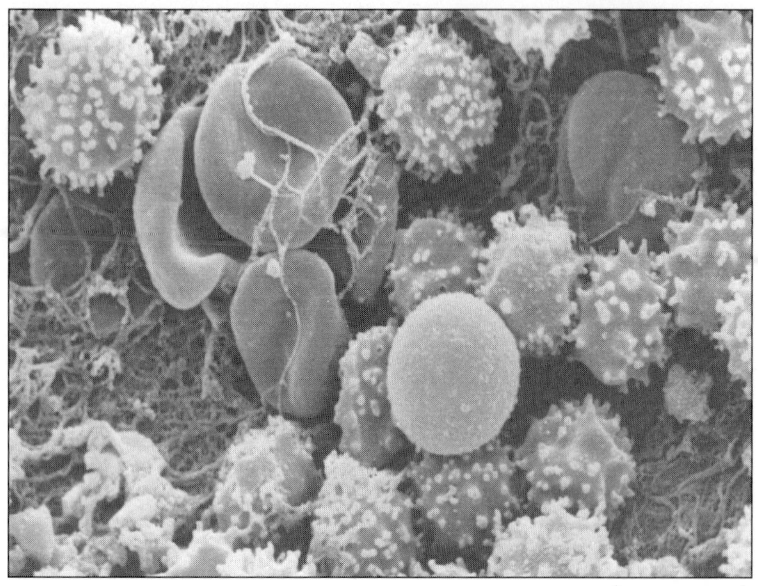

Figure 22–29 Scanning electron micrograph of mammalian blood cells caught in a blood clot. The larger, more spherical cells with a rough surface are white blood cells; the smoother, flattened cells are red blood cells. (Courtesy of Ray Moss.)

Blood cells can be classified as red or white (Figure 22–29). The **red blood cells**, or **erythrocytes**, remain within the blood vessels and transport O_2 and CO_2 bound to hemoglobin. The **white blood cells**, or **leucocytes**, combat infection and in some cases phagocytose and digest debris. Leucocytes, unlike erythrocytes, must make their way across the walls of small blood vessels and migrate into tissues to perform their tasks. In addition, the blood contains large numbers of **platelets**, which are not entire cells but small, detached cell fragments or "minicells" derived from the cortical cytoplasm of large cells called *megakaryocytes*. Platelets adhere specifically to the endothelial cell lining of damaged blood vessels, where they help to repair breaches and aid in the process of blood clotting.

The Three Main Categories of White Blood Cells: Granulocytes, Monocytes, and Lymphocytes

All red blood cells belong in a single class, following the same developmental trajectory as they mature, and the same is true of platelets; but there are many distinct types of white blood cells. White blood cells are traditionally grouped into three major categories—granulocytes, monocytes, and lymphocytes—on the basis of their appearance in the light microscope.

Granulocytes contain numerous lysosomes and secretory vesicles (or granules) and are subdivided into three classes according to the morphology and staining properties of these organelles (Figure 22–30). The differences in staining reflect major differences of chemistry and function. *Neutrophils* (also called *polymorphonuclear leucocytes* because of their multilobed nucleus) are the most common type of granulocyte; they phagocytose and destroy microorganisms, especially bacteria, and thus have a key role in innate immunity to bacterial infection, as discussed in Chapter 25. *Basophils* secrete histamine (and, in some species, serotonin) to help mediate inflammatory reactions; they are closely related in function to *mast cells*, which reside in connective tissues but are also generated from the hemopoietic stem cells. *Eosinophils* help to destroy parasites and modulate allergic inflammatory responses.

Once they leave the bloodstream, **monocytes** (see Figure 22–30D) mature into **macrophages**, which, together with neutrophils, are the main "professional phagocytes" in the body. As discussed in Chapter 13, both types of phagocytic cells contain specialized lysosomes that fuse with newly formed phagocytic vesicles (phagosomes), exposing phagocytosed microorganisms to a barrage of enzymatically produced, highly reactive molecules of superoxide (O_2^-) and hypochlorite (HOCl, the active ingredient in bleach), as well as to a concentrated

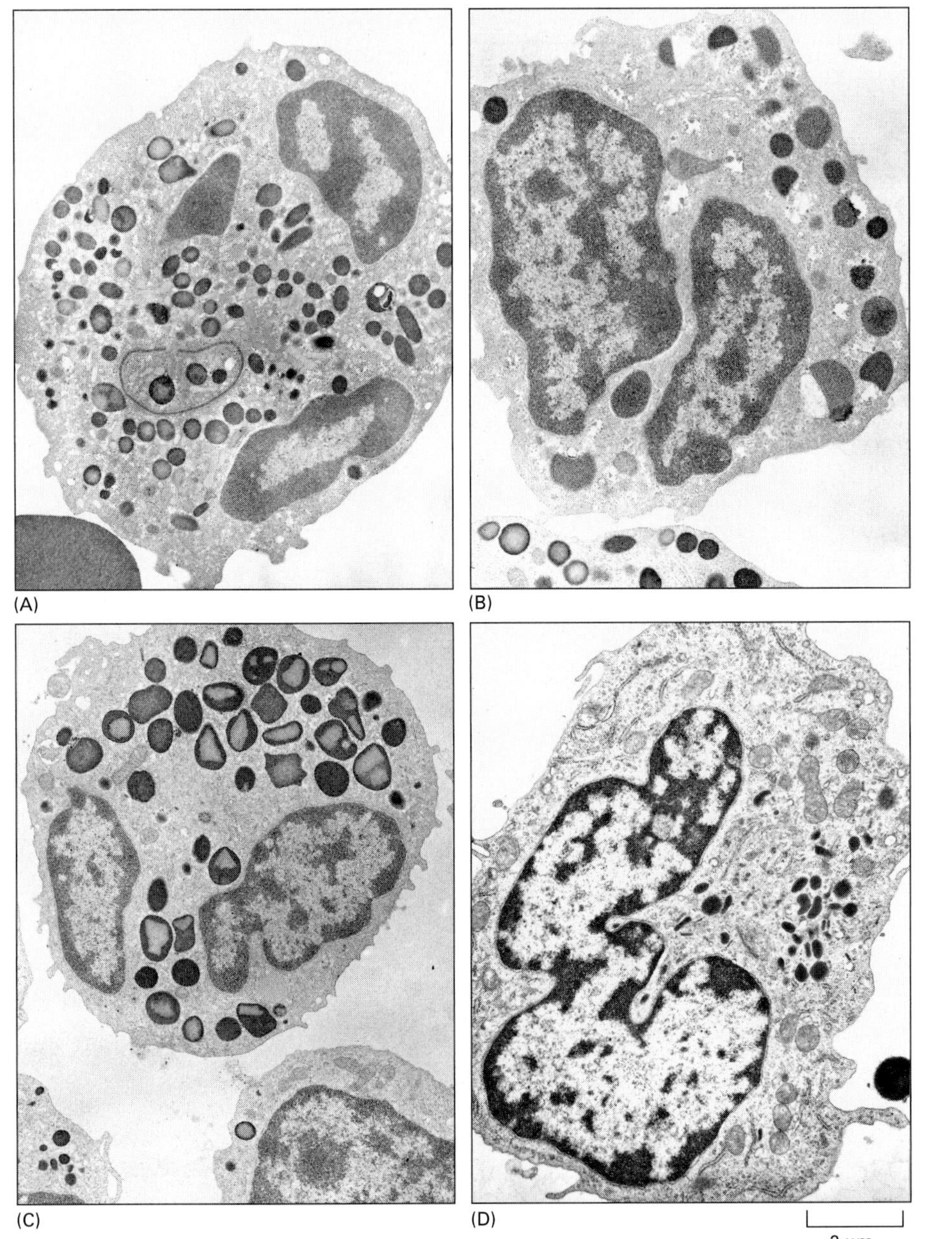

Figure 22–30 White blood cells. (A–D) These electron micrographs show (A) a neutrophil, (B) a basophil, (C) an eosinophil, and (D) a monocyte. Electron micrographs of lymphocytes are shown in Figure 24–7. Each of the cell types shown here has a different function, which is reflected in the distinctive types of secretory granules and lysosomes it contains. There is only one nucleus per cell, but it has an irregular lobed shape, and in (A), (B), and (C) the connections between the lobes are out of the plane of section. (E) A light micrograph of a blood smear stained with the Romanowsky stain, which colors the white blood cells strongly. (A–D, from B.A. Nichols et al., *J. Cell Biol.* 50:498–515, 1971. © The Rockefeller University Press; E, courtesy of David Mason.)

mixture of lysosomal hydrolases. Macrophages, however, are much larger and longer-lived than neutrophils. They are responsible for recognizing and removing senescent, dead, and damaged cells in many tissues, and they are unique in being able to ingest large microorganisms such as protozoa.

Monocytes also give rise to *dendritic cells*, such as the *Langerhans cells* scattered in the epidermis. Like macrophages, dendritic cells are migratory cells that can ingest foreign substances and organisms; but they do not have as active an appetite for phagocytosis and are instead specialized as presenters of foreign antigens to lymphocytes to trigger an immune response. Langerhans cells, for example, ingest foreign antigens in the epidermis and carry these trophies back to present to lymphocytes in lymph nodes.

There are two main classes of **lymphocytes**, both involved in immune responses: *B lymphocytes* make antibodies, while *T lymphocytes* kill virus-infected cells and regulate the activities of other white blood cells. In addition, there are lymphocytelike cells called *natural killer (NK) cells*, which kill some types of tumor cells and virus-infected cells. The production of lymphocytes is a specialized topic discussed in detail in Chapter 24. Here we shall concentrate mainly on the development of the other blood cells, often referred to collectively as **myeloid cells**.

TABLE 22–1 Blood Cells

TYPE OF CELL	MAIN FUNCTIONS	TYPICAL CONCENTRATION IN HUMAN BLOOD (CELLS/LITER)
Red blood cells (erythrocytes)	transport O_2 and CO_2	5×10^{12}
White blood cells (leucocytes)		
Granulocytes		
Neutrophils (polymorphonuclear leucocytes)	phagocytose and destroy invading bacteria	5×10^{9}
Eosinophils	destroy larger parasites and modulate allergic inflammatory responses	2×10^{8}
Basophils	release histamine (and in some species serotonin) in certain immune reactions	4×10^{7}
Monocytes	become tissue macrophages, which phagocytose and digest invading microorganisms and foreign bodies as well as damaged senescent cells	4×10^{8}
Lymphocytes		
B cells	make antibodies	2×10^{9}
T cells	kill virus-infected cells and regulate activities of other leucocytes	1×10^{9}
Natural killer (NK) cells	kill virus-infected cells and some tumor cells	1×10^{8}
Platelets (cell fragments arising from *megakaryocytes* in bone marrow)	initiate blood clotting	3×10^{11}

Humans contain about 5 liters of blood, accounting for 7% of body weight. Red blood cells constitute about 45% of this volume and white blood cells about 1%, the rest being the liquid blood plasma.

The various types of blood cells and their functions are summarized in Table 22–1.

The Production of Each Type of Blood Cell in the Bone Marrow Is Individually Controlled

Most white blood cells function in tissues other than the blood; blood simply transports them to where they are needed. A local infection or injury in any tissue rapidly attracts white blood cells into the affected region as part of the inflammatory response, which helps fight the infection or heal the wound.

The inflammatory response is complex and is mediated by a variety of signal molecules produced locally by mast cells, nerve endings, platelets, and white blood cells, as well as by the activation of complement (discussed in Chapters 24 and 25). Some of these signal molecules act on nearby capillaries, causing the endothelial cells to adhere less tightly to one another but making their surfaces adhesive to passing white blood cells. The white blood cells are thus caught like flies on flypaper and then can escape from the vessel by squeezing between the endothelial cells and crawling across the basal lamina with the aid of digestive enzymes. The initial binding to endothelial cells is mediated by homing receptors called *selectins*, and the stronger binding required for the white blood cells to crawl out of the blood vessel is mediated by *integrins* (see Figure 19–30). Other molecules called *chemokines* are secreted by damaged or inflamed tissue and local endothelial cells; they act as chemoattractants for specific types of white blood cells, causing these cells to become polarized and crawl toward the source

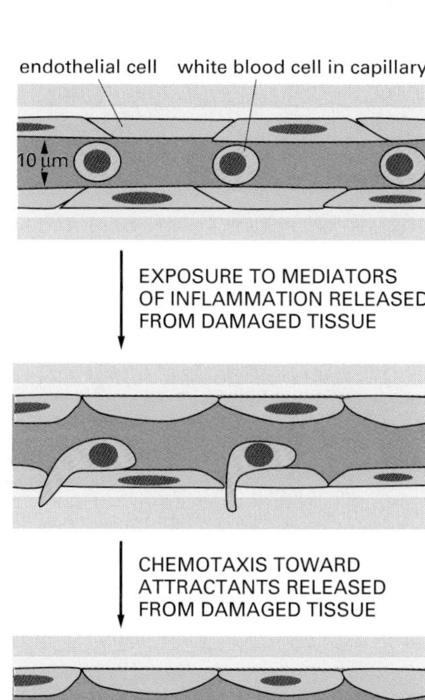

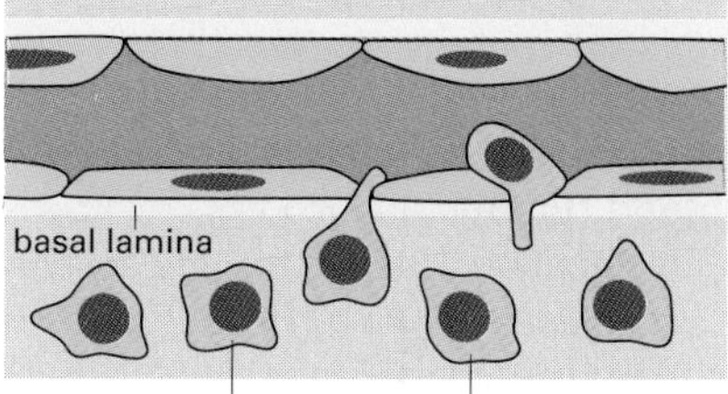

Figure 22–31 The migration of white blood cells out of the bloodstream during an inflammatory response. The response is initiated by a variety of signal molecules produced locally by cells (mainly in the connective tissue) or by complement activation. Some of these mediators act on capillary endothelial cells, causing them to loosen their attachments to their neighbors so that the capillaries become more permeable. Endothelial cells are also stimulated to express selectins, cell-surface molecules that recognize specific carbohydrates that are present on the surface of leucocytes in the blood and cause them to stick to the endothelium. Other mediators called chemokines are secreted by the inflamed tissues and local endothelial cells and act as chemoattractants, causing the bound leucocytes to crawl between the capillary endothelial cells into the tissue.

of the attractant. As a result, large numbers of white blood cells enter the affected tissue (Figure 22–31).

Other signal molecules produced in the course of an inflammatory response escape into the blood and stimulate the bone marrow to produce more leucocytes and release them into the bloodstream. The bone marrow is the key target for such regulation because, with the exception of lymphocytes and some macrophages, most types of blood cells in adult mammals are generated only in the bone marrow. The regulation tends to be cell-type-specific: some bacterial infections, for example, cause a selective increase in neutrophils, while infections with some protozoa and other parasites cause a selective increase in eosinophils. (For this reason, physicians routinely use differential white blood cell counts to aid in the diagnosis of infectious and other inflammatory diseases.)

In other circumstances erythrocyte production is selectively increased—for example, in the process of acclimatization when one goes to live at high altitude, where oxygen is scarce. Thus, blood cell formation, or *hemopoiesis,* necessarily involves complex controls, by which the production of each type of blood cell is regulated individually to meet changing needs. It is a problem of great medical importance to understand how these controls operate, and much progress has been made in this area in recent years.

In intact animals, hemopoiesis is more difficult to analyze than is cell turnover in a tissue such as the epidermal layer of the skin. In the epidermis there is a simple, regular spatial organization that makes it easy to follow the process of renewal and to locate the stem cells. This is not true of the hemopoietic tissues. However, hemopoietic cells have a nomadic life-style that makes them more accessible to experimental study in other ways. Dispersed hemopoietic cells are easily obtained and can be readily transferred, without damage, from one animal to another. Moreover, the proliferation and differentiation of individual cells and their progeny can be observed and analyzed in culture, and numerous molecular markers distinguish the various stages of differentiation. Because of this, more is known about the molecules that control blood cell production than about those that control cell production in other mammalian tissues.

Bone Marrow Contains Hemopoietic Stem Cells

The different types of blood cells and their immediate precursors can be recognized in the bone marrow by routine staining methods (Figure 22–32). They are intermingled with one another, as well as with fat cells and other stromal cells (connective-tissue cells), which produce a delicate supporting meshwork of collagen fibers and other extracellular matrix components. In addition, the whole tissue is richly supplied with thin-walled blood vessels, called *blood sinuses,* into which the new blood cells are discharged. **Megakaryocytes** are also present; these, unlike other blood cells, remain in the bone marrow when mature and are one of its most striking features, being extraordinarily large (diameter up to 60 μm), with a highly polyploid nucleus. They normally lie close beside blood sinuses, and they extend processes through holes in the endothelial lining of

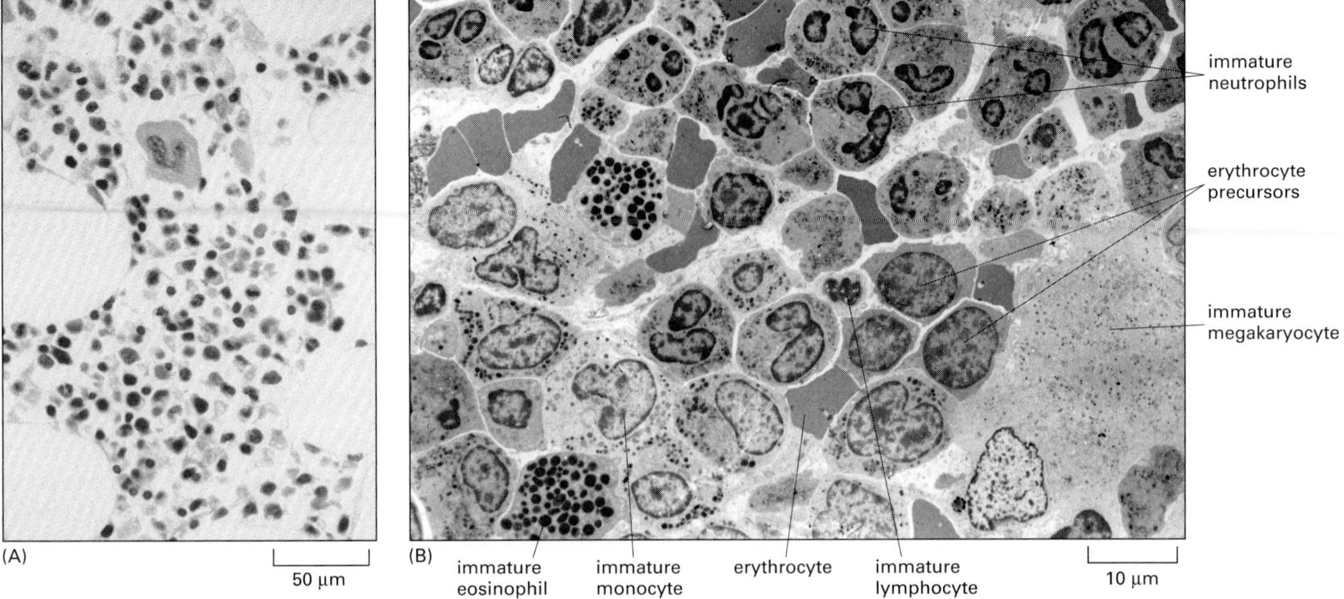

Figure 22–32 Bone marrow. (A) A light micrograph of a stained section. The large empty spaces correspond to fat cells, whose fatty contents have been dissolved away during specimen preparation. The giant cell with a lobed nucleus is a megakaryocyte. (B) A low-magnification electron micrograph. Bone marrow is the main source of new blood cells (except for T lymphocytes, which are produced in the thymus). Note that the immature blood cells of a particular type tend to cluster in "family groups." (A, courtesy of David Mason; B, from J.A.G. Rhodin, Histology: A Text and Atlas. New York: Oxford University Press, 1974.)

these vessels; platelets pinch off from the processes and are swept away into the blood (Figure 22–33).

Because of the complex arrangement of the cells in bone marrow, it is difficult to identify in ordinary tissue sections any but the immediate precursors of the mature blood cells. The corresponding cells at still earlier stages of development, before any overt differentiation has begun, are confusingly similar in appearance, and although the spatial distribution of cell types has some orderly features, there is no obvious visible characteristic by which the ultimate stem cells can be recognized. To identify and characterize the stem cells, one needs a functional assay, which involves tracing the progeny of single cells. As we shall see, this can be done *in vitro* simply by examining the colonies that isolated cells produce in culture. The hemopoietic system, however, can also be manipulated so that such clones of cells can be recognized *in vivo* in the intact animal.

When an animal is exposed to a large dose of x-rays, most of the hemopoietic cells are destroyed and the animal dies within a few days as a result of its inability to manufacture new blood cells. The animal can be saved, however, by a transfusion of cells taken from the bone marrow of a healthy, immunologically compatible donor. Among these cells there are some that can colonize the irradiated host and permanently reequip it with hemopoietic tissue (Figure 22–34). Experiments of this sort prove that the marrow contains hemopoietic stem cells. They also show how one can assay for the presence of hemopoietic stem cells and hence discover the molecular features that distinguish them from other cells.

For this purpose, cells taken from bone marrow are sorted (with the help of a fluorescence-activated cell sorter) according to the surface antigens that they

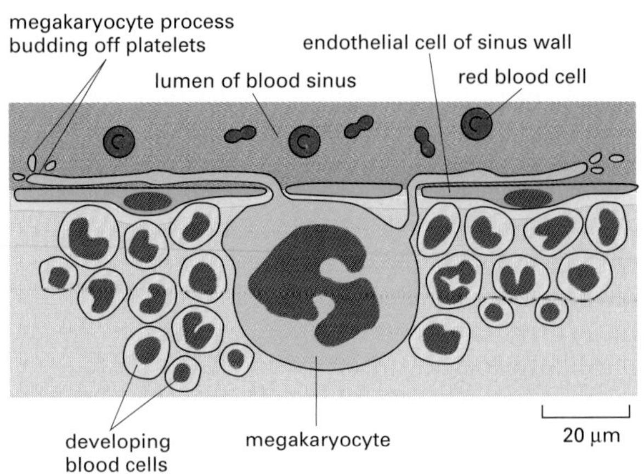

Figure 22–33 A megakaryocyte among other cells in the bone marrow. Its enormous size results from its having a highly polyploid nucleus. One megakaryocyte produces about 10,000 platelets, which split off from long processes that extend through holes in the walls of an adjacent blood sinus.

Figure 22–34 Rescue of an irradiated mouse by a transfusion of bone marrow cells. An essentially similar procedure is used in the treatment of leukemia in human patients by bone marrow transplantation.

x-irradiation halts blood cell production; mouse would die if no further treatment were given

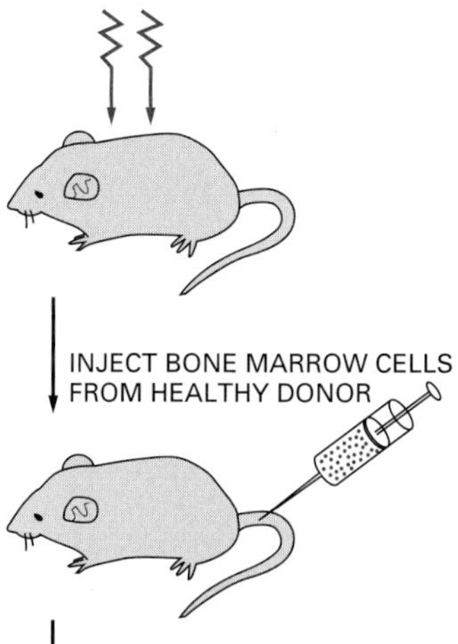

mouse survives; the injected stem cells colonize its hemopoietic tissues and generate a steady supply of new blood cells

display, and the different fractions are transfused back into irradiated mice. If a fraction rescues an irradiated host mouse, it must contain hemopoietic stem cells. In this way, it has been possible to show that the hemopoietic stem cells are characterized by a specific combination of cell-surface proteins, and by appropriate sorting virtually pure stem cell preparations can be obtained. The stem cells turn out to be a tiny fraction of the bone-marrow population—about 1 cell in 10,000; but this is enough. As few as five such cells injected into a host mouse with defective hemopoiesis are sufficient to reconstitute its entire hemopoietic system, generating a complete set of blood-cell types, as well as fresh stem cells.

A Multipotent Stem Cell Gives Rise to All Classes of Blood Cells

To see what range of cell types a single **hemopoietic stem cell** can generate, one needs a way of tracing the fate of its progeny. This can be done by marking the individual stem cell genetically, so that the progeny can be identified even after they have been released into the bloodstream. Although several methods have been used for this, a specially engineered retrovirus (a retroviral vector carrying a marker gene) serves the purpose particularly well. The marker virus, like other retroviruses, can insert its own genome into the chromosomes of the cell it infects, but the genes that would enable it to generate new infectious virus particles have been removed. The marker is therefore confined to the progeny of the cells that were originally infected, and the progeny of one such cell can be distinguished from the progeny of another because the chromosomal sites of insertion of the virus are different. To analyze hemopoietic cell lineages, bone marrow cells are first infected with the retroviral vector *in vitro* and then are transferred into a lethally irradiated recipient; DNA probes can then be used to trace the progeny of individual infected cells in the various hemopoietic and lymphoid tissues of the host. These experiments show that the individual hemopoietic stem cell is *multipotent* and can give rise to the complete range of blood cell types, both myeloid and lymphoid, as well as new stem cells like itself (Figure 22–35).

The same methods that were developed for experimentation in mice can now be used for the treatment of disease in humans. Mice, we have seen, can be irradiated to kill off their hemopoietic cells, and then rescued by a transfusion of new stem cells. In the same way, patients with leukemia, for example, can be irradiated or chemically treated to destroy their cancerous cells along with the rest of their hemopoietic tissue, and then can be rescued by a transfusion of hemopoietic stem cells that are free of the cancer-causing mutations. These healthy stem cells can be obtained either from an immunologically matched donor, or by sorting from a sample of bone marrow previously taken from the leukemic patient himself or herself.

The same technology also opens the way, in principle, to one form of gene therapy: hemopoietic stem cells can be isolated in culture, genetically modified by DNA transfection or some other technique to introduce a desired gene, and then transfused back into a patient in whom the gene was lacking, to provide a self-renewing source of the missing genetic component.

Commitment Is a Stepwise Process

Hemopoietic stem cells do not jump directly from a multipotent state into a commitment to just one pathway of differentiation; instead, they go through a series of progressive restrictions. The first step is commitment to either a myeloid or a lymphoid fate. This is thought to give rise to two kinds of progenitor cells, one capable of generating large numbers of all the different types of myeloid cells, or perhaps of myeloid cells plus B lymphocytes, and the other to

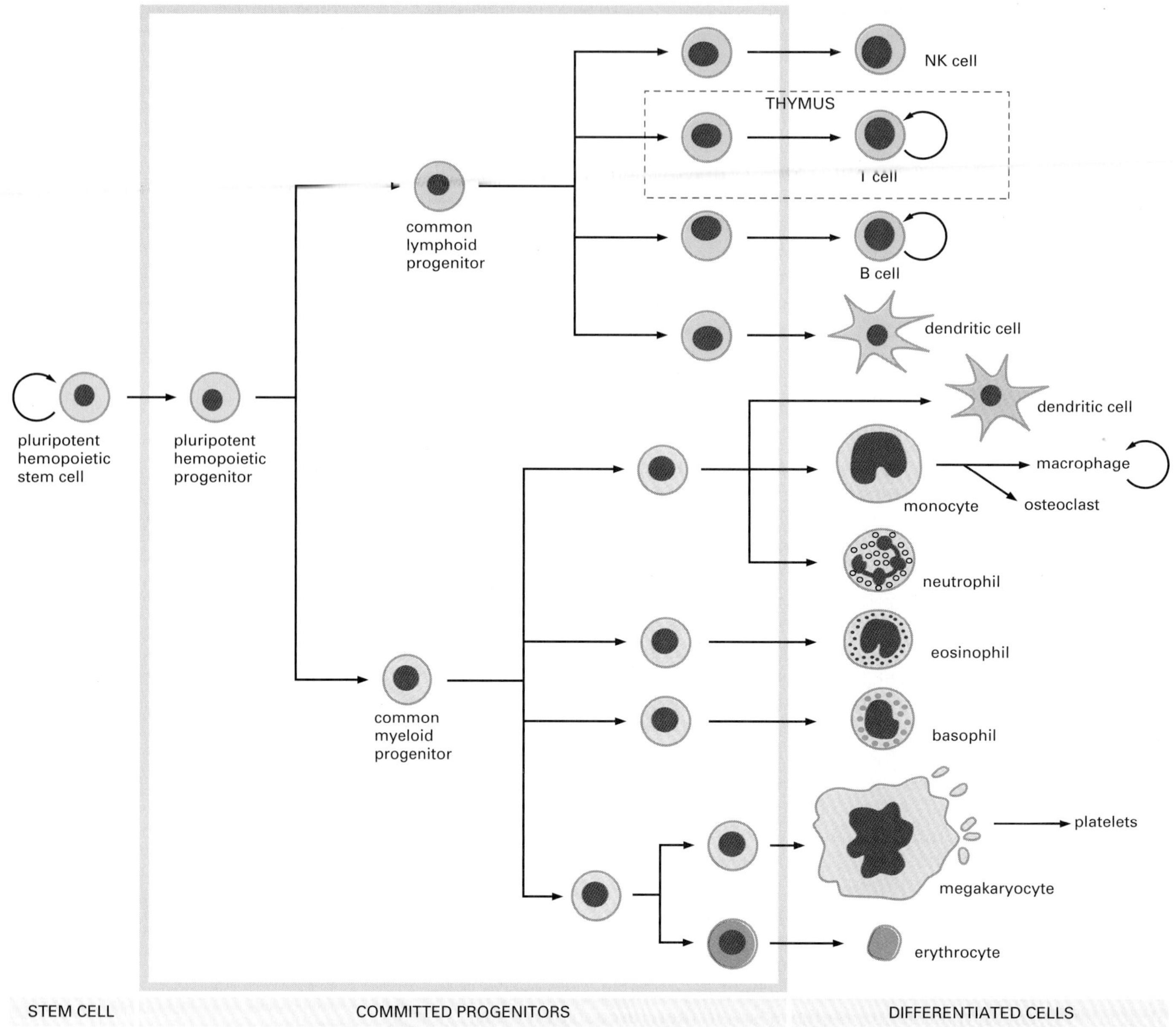

Figure 22–35 A tentative scheme of hemopoiesis. The multipotent stem cell normally divides infrequently to generate either more multipotent stem cells, which are self-renewing, or committed progenitor cells, which are limited in the number of times that they can divide before differentiating to form mature blood cells. As they go through their divisions, the progenitors become progressively more specialized in the range of cell types that they can give rise to, as indicated by the branching of the cell-lineage diagram in the region enclosed in the *gray* box. Many of the details of this part of the lineage diagram are still controversial, however.

In adult mammals, all of the cells shown develop mainly in the bone marrow—except for T lymphocytes, which develop in the thymus, and macrophages and osteoclasts, which develop from blood monocytes. Some dendritic cells may also derive from monocytes. Mast cells (not shown) are thought to develop from circulating basophils.

large numbers of all the different types of lymphoid cells, or at least T lymphocytes. Further steps give rise to progenitors committed to the production of just one cell type. The steps of commitment can be correlated with changes in the expression of specific gene regulatory proteins, needed for the production of different subsets of blood cells. These seem to act in a complicated combinatorial fashion: the *GATA-1 protein*, for example, is needed for the maturation of red blood cells, but is active also at much earlier steps in the hemopoietic pathway.

The meaning of "commitment" in molecular terms is still unclear, but it has at least two aspects: genes for the chosen mode of differentiation begin to be switched on, while access to other developmental pathways is shut off. These processes normally go hand-in-hand, but studies of mutants show that they are in principle distinct. Animals lacking the gene regulatory protein Pax5 have a block in the production of mature B lymphocytes: progenitor cells begin the process leading to B cell restriction but do not complete it. Normal progenitors that have taken the same initial step cannot be induced, no matter what their environment, to differentiate along any other pathway than that of a B lymphocyte. But the Pax5-defective cells show no such restriction: exposure to appropriate conditions can drive them to generate other blood cell types, including T lymphocytes, macrophages, and granulocytes. This is because the Pax5 protein is required not only to activate genes required for B cell development, but also to shut off genes required for development along other blood-cell pathways.

The Number of Specialized Blood Cells Is Amplified by Divisions of Committed Progenitor Cells

Hemopoietic progenitor cells generally become committed to a particular pathway of differentiation long before they cease proliferating and terminally differentiate. The committed progenitors go through many rounds of cell division to amplify the ultimate number of cells of the given specialized type. In this way, a single stem-cell division can lead to the production of thousands of differentiated progeny, which explains why the number of stem cells is so small a fraction of the total population of hemopoietic cells. For the same reason, a high rate of blood cell production can be maintained even though the stem-cell division rate is low. Infrequent division is a common feature of stem cells in several tissues, including epidermis and gut, as well as the hemopoietic system. The amplifying divisions of the committed progenitors reduce the number of division cycles that the stem cells themselves have to undergo in the course of a lifetime, thereby reducing the risk of replicative senescence (discussed in Chapter 17) and damaging mutations.

The stepwise nature of commitment means that the hemopoietic system can be viewed as a hierarchical family tree of cells. Multipotent stem cells give rise to *committed progenitor cells*, which are specified to give rise to only one or a few blood cell types. The committed progenitors divide rapidly, but only a limited number of times, before they terminally differentiate into cells that divide no further and die after several days or weeks. Many cells normally die at the earlier steps in the pathway as well. Studies in culture provide a way to find out how the proliferation, differentiation, and death of the hemopoietic cells are regulated.

Stem Cells Depend on Contact Signals From Stromal Cells

Hemopoietic cells can survive, proliferate, and differentiate in culture if, and only if, they are provided with specific signal proteins or accompanied by cells that produce these proteins. If deprived of such proteins, the cells die. For long-term maintenance, contact with appropriate supporting cells also seems to be necessary: hemopoiesis can be kept going for months or even years *in vitro* by culturing dispersed bone-marrow hemopoietic cells on top of a layer of bone-marrow stromal cells, which mimic the environment in intact bone marrow. Such cultures can generate all the types of myeloid cells, and their long-term continuation implies that stem cells, as well as differentiated progeny, are being continually produced.

There has to be some mechanism, in these cultures as well as *in vivo*, to guarantee that while some stem cell progeny become committed to differentiation, others remain as stem cells. The cell-fate choice seems to be controlled, in part at least, by specific signals generated at contacts with stromal cells. These signals seem to be needed by the hemopoietic stem cells to keep them in their uncommitted state, just as signals from contacts with the basal lamina are needed to maintain stem cells of the epidermis. In both systems, stem cells are normally confined to a particular niche, and when they lose contact with this niche, they tend to lose their stem-cell potential (Figure 22–36).

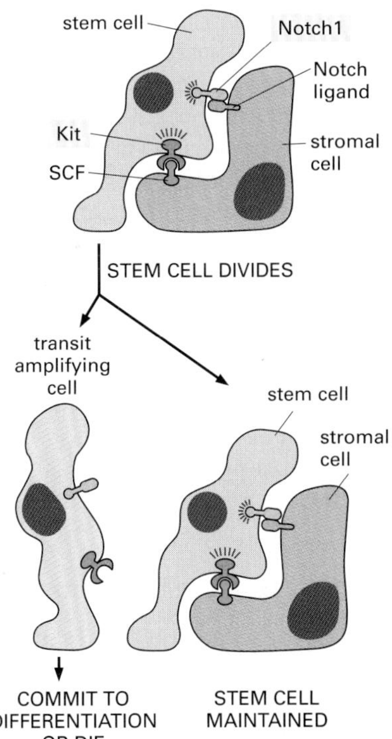

Figure 22–36 Dependence of hemopoietic stem cells on contact with stromal cells. The cartoon shows a dependence on two signaling mechanisms for which there is some evidence. The real system is certainly more complex; the dependence of hemopoietic cells on contact with stromal cells cannot be absolute, since small numbers of the functional stem cells can be found free in the circulation.

The nature of the critical signals from the stromal cells is not yet certain, although several mechanisms have been implicated, involving both secreted and cell-surface-attached factors. For example, hemopoietic precursor cells, including stem cells, in the bone marrow display the transmembrane receptor Notch1 in their plasma membrane, while marrow stromal cells express Notch ligands, and there is evidence that Notch activation helps to keep the stem cells or progenitor cells from embarking on differentiation (see Chapter 15).

Another contact interaction that is important for the maintenance of hemopoiesis came to light through the analysis of mouse mutants with a curious combination of defects: a shortage of red blood cells (anemia), of germ cells (sterility), and of pigment cells (white spotting of the skin; see Figure 21–81). As discussed in Chapter 21, this syndrome results from mutations in either of two genes: one, called *c-kit*, codes for a receptor tyrosine kinase; the other, called *Steel*, codes for its ligand, stem-cell factor (SCF). The cell types affected by the mutations all derive from migratory precursors, and it seems that these precursors in each case must express the receptor (Kit) and be provided with the ligand (SCF) by their environment if they are to survive and produce progeny in normal numbers. Studies in mutant mice suggest that SCF must be membrane-bound to be effective, implying that normal hemopoiesis requires direct cell–cell contact between the hemopoietic cells that express Kit receptor protein and stromal cells that express SCF.

Factors That Regulate Hemopoiesis Can Be Analyzed in Culture

While stem cells depend on contact with stromal cells for long-term maintenance, their committed progeny do not, or not to the same degree. Thus, dispersed bone-marrow hemopoietic cells can be cultured in a semisolid matrix of dilute agar or methylcellulose, and factors derived from other cells can be added artificially to the medium. Because cells in the semisolid matrix cannot migrate, the progeny of each isolated precursor cell remain together as an easily distinguishable colony. A single committed neutrophil progenitor, for example, may give rise to a clone of thousands of neutrophils. Such culture systems have provided a way to assay for the factors that support hemopoiesis and hence to purify them and explore their actions. These substances are found to be glycoproteins and are usually called **colony-stimulating factors (CSFs)**. Of the growing number of CSFs that have been defined and purified, some circulate in the blood and act as hormones, while others act in the bone marrow either as secreted local mediators or, like SCF, as membrane-bound signals that act through cell–cell contact. The best understood of the CSFs that act as hormones is the glycoprotein erythropoietin, which is produced in the kidneys and regulates *erythropoiesis*, the formation of red blood cells.

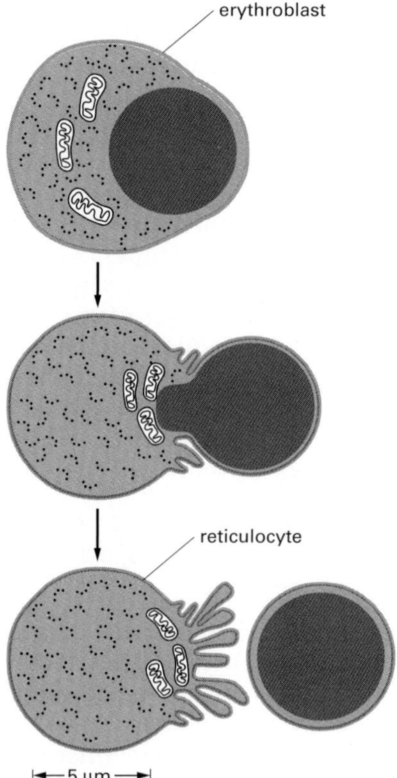

Figure 22–37 A developing red blood cell (erythroblast). The cell is shown extruding its nucleus to become an immature erythrocyte (a reticulocyte), which then leaves the bone marrow and passes into the bloodstream. The reticulocyte will lose its mitochondria and ribosomes within a day or two to become a mature erythrocyte. Erythrocyte clones develop in the bone marrow on the surface of a macrophage, which phagocytoses and digests the nuclei discarded by the erythroblasts.

Erythropoiesis Depends on the Hormone Erythropoietin

The erythrocyte is by far the most common type of cell in the blood (see Table 22–1). When mature, it is packed full of hemoglobin and contains practically none of the usual cell organelles. In an erythrocyte of an adult mammal, even the nucleus, endoplasmic reticulum, mitochondria, and ribosomes are absent, having been extruded from the cell in the course of its development (Figure 22–37). The erythrocyte therefore cannot grow or divide; the only possible way of making more erythrocytes is by means of stem cells. Furthermore, erythrocytes have a limited life-span—about 120 days in humans or 55 days in mice. Worn-out erythrocytes are phagocytosed and digested by macrophages in the liver and spleen, which remove more than 10^{11} senescent erythrocytes in each of us each day. Young erythrocytes actively protect themselves from this fate: they have a protein on their surface that binds to an inhibitory receptor on macrophages and thereby prevents their phagocytosis.

A lack of oxygen or a shortage of erythrocytes stimulates cells in the kidney to synthesize and secrete increased amounts of **erythropoietin** into the bloodstream. The erythropoietin, in turn, stimulates the production of more erythrocytes. Since a change in the rate of release of new erythrocytes into the bloodstream is observed as early as 1–2 days after an increase in erythropoietin levels in the bloodstream, the hormone must act on cells that are very close precursors of the mature erythrocytes.

The cells that respond to erythropoietin can be identified by culturing bone marrow cells in a semisolid matrix in the presence of erythropoietin. In a few days, colonies of about 60 erythrocytes appear, each founded by a single committed erythroid progenitor cell. This cell is known as an *erythrocyte colony-forming cell,* or *CFC-E,* (or colony-forming *unit, CFU-E)* and it gives rise to mature erythrocytes after about six division cycles or less. The CFC-Es do not yet contain hemoglobin, and they are derived from an earlier type of progenitor cell whose proliferation does not depend on erythropoietin. CFC-Es themselves depend on erythropoietin for their survival, as well as for proliferation: if erythropoietin is removed from the cultures, the cells rapidly undergo apoptosis.

A second CSF, called **interleukin-3 (IL-3)**, promotes the survival and proliferation of the earlier erythroid progenitor cells. In its presence, much larger erythroid colonies, each comprising up to 5000 erythrocytes, develop from cultured bone marrow cells in a process requiring a week or 10 days. These colonies derive from erythroid progenitor cells called *erythrocyte burst-forming cells,* or *BFC-Es* (or burst-forming *units, BFU-E*). A BFC-E is distinct from a multipotent stem cell in that it has a limited capacity to proliferate and gives rise to colonies that contain erythrocytes only, even under culture conditions that enable other progenitor cells to give rise to other classes of differentiated blood cells. It is distinct from a CFC-E in that it is insensitive to erythropoietin, and its progeny must go through as many as 12 divisions before they become mature erythrocytes (for which erythropoietin must be present). Thus, the BFC-E is thought to be a progenitor cell committed to erythrocyte differentiation and an early ancestor of the CFC-E.

Multiple CSFs Influence the Production of Neutrophils and Macrophages

The two professional phagocytic cells, neutrophils and macrophages, develop from a common progenitor cell called a **granulocyte/macrophage (GM) progenitor cell.** Like the other granulocytes (eosinophils and basophils), neutrophils circulate in the blood for only a few hours before migrating out of capillaries into the connective tissues or other specific sites, where they survive for only a few days. They then die by apoptosis and are phagocytosed by macrophages. Macrophages, in contrast, can persist for months or perhaps even years outside the bloodstream, where they can be activated by local signals to resume proliferation.

At least seven distinct CSFs that stimulate neutrophil and macrophage colony formation in culture have been defined, and some or all of these are thought to act in different combinations to regulate the selective production of these cells *in vivo.* These CSFs are synthesized by various cell types—including endothelial cells, fibroblasts, macrophages, and lymphocytes—and their concentration in the blood typically increases rapidly in response to bacterial infection in a tissue, thereby increasing the number of phagocytic cells released from the bone marrow into the bloodstream. IL-3 is one of the least specific of the factors, acting on multipotent stem cells as well as on most classes of committed progenitor cells, including GM progenitor cells. Various other factors act more selectively on committed GM progenitor cells and their differentiated progeny (Table 22–2), although in many cases they act on certain other branches of the hemopoietic family tree as well.

All of these CSFs, like erythropoietin, are glycoproteins that act at low concentrations (about 10^{-12} M) by binding to specific cell-surface receptors, as discussed in Chapter 15. A few of these receptors are transmembrane tyrosine kinases but most belong to the large cytokine receptor family, whose members are usually composed of two or more subunits, one of which is frequently shared

TABLE 22–2 Some Colony-stimulating Factors (CSFs) That Influence Blood Cell Formation

FACTOR	TARGET CELLS	PRODUCING CELLS	RECEPTORS
Erythropoietin	CFC-E	kidney cells	cytokine family
Interleukin 3 (IL-3)	multipotent stem cell, most progenitor cells, many terminally differentiated cells	T lymphocytes, epidermal cells	cytokine family
Granulocyte/ macrophage CSF (GM-CSF)	GM progenitor cells	T lymphocytes, endothelial cells, fibroblasts	cytokine family
Granulocyte CSF (G-CSF)	GM progenitor cells and neutrophils	macrophages, fibroblasts	cytokine family
Macrophage CSF (M-CSF)	GM progenitor cells and macrophages	fibroblasts, macrophages, endothelial cells	receptor tyrosine kinase family
Steel factor (stem cell factor)	hemopoietic stem cells	stromal cells in bone marrow and many other cells	receptor tyrosine kinase family

among several receptor types (Figure 22–38). The CSFs not only operate on the precursor cells to promote the production of differentiated progeny, they also activate the specialized functions (such as phagocytosis and target-cell killing) of the terminally differentiated cells. Proteins produced artificially from the cloned genes for these factors (sometimes referred to as *recombinant proteins* because they are made using recombinant DNA technology) are strong stimulators of hemopoiesis in experimental animals. They are now widely used in human patients to stimulate the regeneration of hemopoietic tissue and to boost resistance to infection—an impressive demonstration of how basic cell biological research and animal experiments can lead to better medical treatment.

The Behavior of a Hemopoietic Cell Depends Partly on Chance

Up to this point we have glossed over a central question. CSFs are defined as factors that promote the production of colonies of differentiated blood cells. But precisely what effect does a CSF have on an individual hemopoietic cell? The factor might control the rate of cell division or the number of division cycles that the progenitor cell undergoes before differentiating; it might act late in the hemopoietic lineage to facilitate differentiation; it might act early to influence commitment; or it might simply increase the probability of cell survival (Figure 22–39). By monitoring the fate of isolated individual hemopoietic cells in

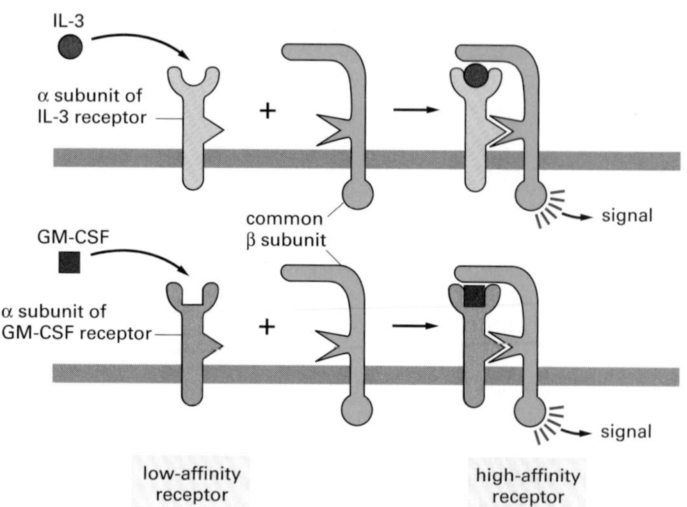

Figure 22–38 Sharing of subunits among CSF receptors. Human IL-3 receptors and GM-CSF receptors have different α subunits and a common β subunit. Their ligands are thought to bind to the free α subunit with low affinity, and this triggers the assembly of the heterodimer that binds the ligand with high affinity.

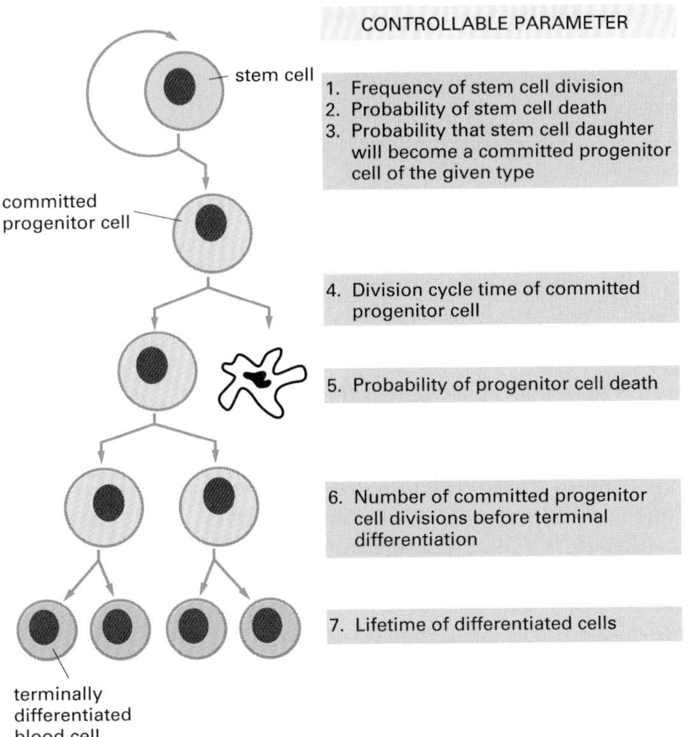

Figure 22–39 Some of the parameters through which the production of blood cells of a specific type might be regulated. Studies in culture suggest that colony-stimulating factors (CSFs) can affect all of these aspects of hemopoiesis.

culture, it has been possible to show that a single CSF, such as GM-CSF, can exert all these effects, although it is still not clear which are most important *in vivo.*

Studies *in vitro* indicate, moreover, that there is a large element of chance in the way a hemopoietic cell behaves. At least some of the CSFs seem to act by regulating probabilities, not by dictating directly what the cell shall do. In hemopoietic cell cultures, even if the cells have been selected to be as homogeneous a population as possible, there is a remarkable variability in the sizes and often in the characters of the colonies that develop. And if two sister cells are taken immediately after a cell division and cultured apart under identical conditions, they frequently give rise to colonies that contain different types of blood cells or the same types of blood cells in different numbers. Thus, both the programming of cell division and the process of commitment to a particular path of differentiation seem to involve random events at the level of the individual cell, even though the behavior of the multicellular system as a whole is regulated in a reliable way.

Regulation of Cell Survival Is as Important as Regulation of Cell Proliferation

While such observations show that CSFs are not strictly required to instruct the hemopoietic cells how to differentiate or how many times to divide, CSFs are required to keep the cells alive: the default behavior of the cells in the absence of CSFs is death by apoptosis (discussed in Chapter 17). In principle, the CSFs could regulate the numbers of the various types of blood cells entirely through selective control of cell survival in this way, and there is increasing evidence that the control of cell survival plays a central part in regulating the numbers of blood cells, just as it does for hepatocytes and many other cell types, as we have already seen. The amount of apoptosis in the vertebrate hemopoietic system is enormous: billions of neutrophils die in this way each day in an adult human, for example. Indeed, the vast majority of neutrophils produced in the bone marrow die there without ever functioning. This futile cycle of production and destruction presumably serves to maintain a reserve supply of cells that can be promptly mobilized to fight infection whenever it flares up, or phagocytosed and digested for recycling when all is quiet. Compared with the life of the organism, the lives of cells are cheap.

Too little cell death can be as dangerous to the health of a multicellular organism as too much proliferation. In the hemopoietic system, mutations that inhibit cell death by causing excessive production of the intracellular apoptosis inhibitor Bcl-2 promote the development of cancer in B lymphocytes. Indeed, the capacity for unlimited self-renewal is a dangerous property for any cell to possess, and many cases of leukemia arise through mutations that confer this capacity on committed hemopoietic precursor cells that would normally be fated to differentiate and die after a limited number of division cycles.

Summary

The many types of blood cells, including erythrocytes, lymphocytes, granulocytes, and macrophages, all derive from a common multipotent stem cell. In the adult, hemopoietic stem cells are found mainly in bone marrow, and they depend on contact-mediated signals from the marrow stromal (connective-tissue) cells to maintain their stem-cell character. The stem cells normally divide infrequently to produce more stem cells (self-renewal) and various committed progenitor cells (transit amplifying cells), each able to give rise to only one or a few types of blood cells. The committed progenitor cells divide extensively under the influence of various protein signal molecules (colony-stimulating factors, or CSFs) and then terminally differentiate into mature blood cells, which usually die after several days or weeks.

Studies of hemopoiesis have been greatly aided by in vitro *assays in which stem cells or committed progenitor cells form clonal colonies when cultured in a semisolid matrix. The progeny of stem cells seem to make their choices between alternative developmental pathways in a partly random manner. Cell death by apoptosis, controlled by the availability of CSFs, also plays a central part in regulating the numbers of mature differentiated blood cells.*

GENESIS, MODULATION, AND REGENERATION OF SKELETAL MUSCLE

The term "muscle" covers a multitude of cell types, all specialized for contraction but in other respects dissimilar. As noted in Chapter 16, a contractile system involving actin and myosin is a basic feature of animal cells in general, but muscle cells have developed this apparatus to a high degree. Mammals possess four main categories of cells specialized for contraction: skeletal muscle cells, heart (cardiac) muscle cells, smooth muscle cells, and myoepithelial cells (Figure 22–40). These differ in function, structure, and development. Although all of them generate contractile forces by means of organized filament systems based on actin and myosin, the actin and myosin molecules employed are somewhat different in amino acid sequence, are differently arranged in the cell, and are associated with different sets of proteins to control contraction.

Skeletal muscle cells are responsible for practically all movements that are under voluntary control. These cells can be very large (2–3 cm long and 100 μm in diameter in an adult human) and are often referred to as *muscle fibers* because of their highly elongated shape. Each one is a syncytium, containing many nuclei within a common cytoplasm. The other types of muscle cells are more conventional, generally having only a single nucleus. **Heart muscle cells** resemble skeletal muscle fibers in that their actin and myosin filaments are aligned in very orderly arrays to form a series of contractile units called *sarcomeres,* so that the cells have a striated (striped) appearance. **Smooth muscle cells** are so named because they do not appear striated. The functions of smooth muscle vary greatly, from propelling food along the digestive tract to erecting hairs in response to cold or fear. **Myoepithelial cells** also have no striations, but unlike all other muscle cells they lie in epithelia and are derived from the ectoderm. They form the dilator muscle of the eye's iris and serve to expel saliva, sweat, and milk from the corresponding glands, as discussed earlier (see Figure 22–8). The four main categories of muscle cells can be further divided into distinctive subtypes, each with its own characteristic features.

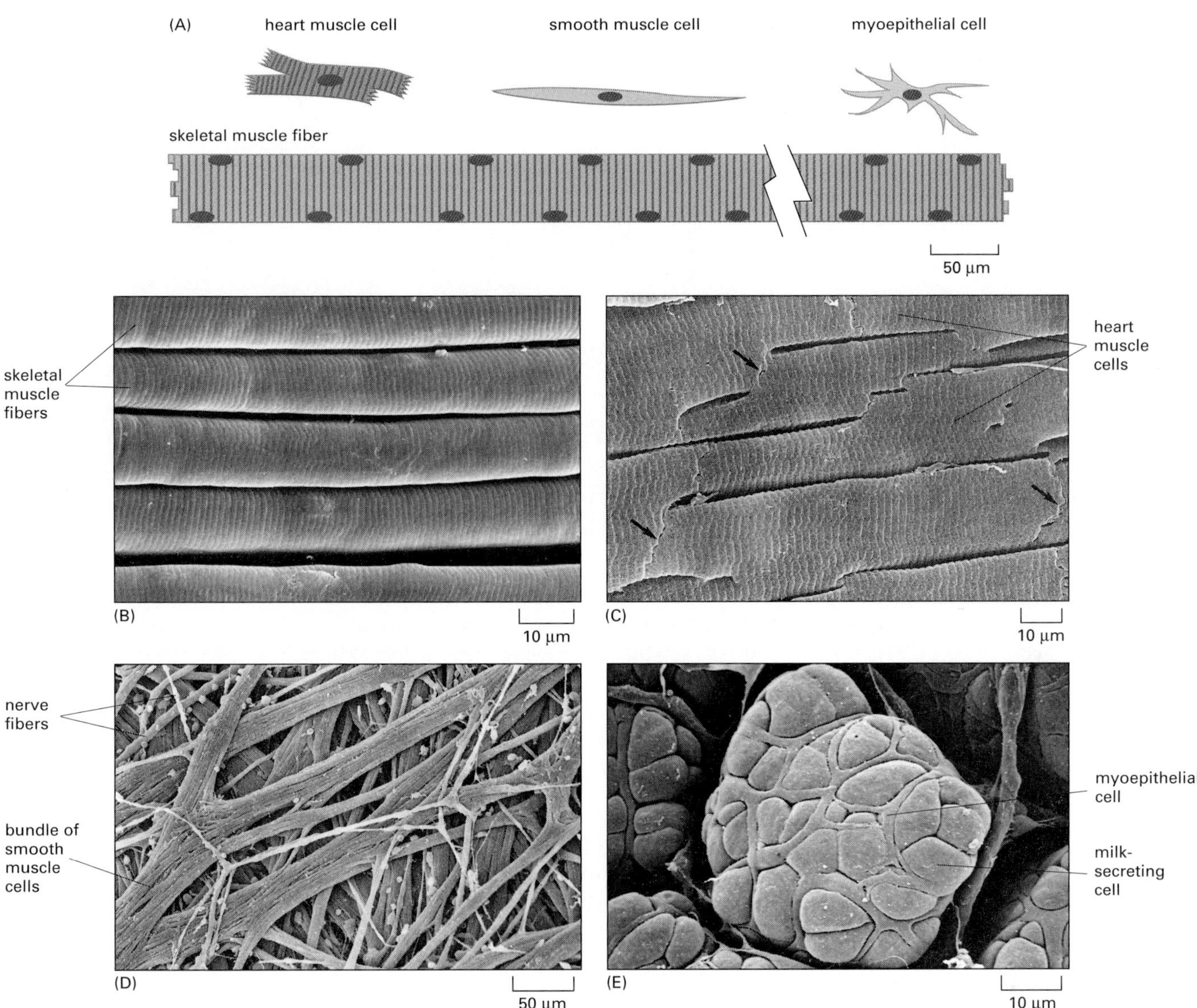

Figure 22–40 The four classes of muscle cells of a mammal. (A) Schematic drawings (to scale). (B–E) Scanning electron micrographs, showing (B) skeletal muscle from the neck of a hamster, (C) heart muscle from a rat, (D) smooth muscle from the urinary bladder of a guinea pig, and (E) myoepithelial cells in a secretory alveolus from a lactating rat mammary gland. The *arrows* in (C) point to intercalated discs—end-to-end junctions between the heart muscle cells; skeletal muscle cells in long muscles are joined end to end in a similar way. Note that the smooth muscle is shown at a lower magnification than the others. (B, courtesy of Junzo Desaki; C, from T. Fujiwara, in Cardiac Muscle in Handbook of Microscopic Anatomy [E.D. Canal, ed.]. Berlin: Springer-Verlag, 1986; D, courtesy of Satoshi Nakasiro; E, from T. Nagato et al., *Cell Tiss. Res.* 209:1–10, 1980. © Springer-Verlag.)

The mechanisms of muscle contraction are discussed in Chapter 16. Here we consider how muscle tissue is generated and maintained. We focus on the skeletal muscle fiber, which has a curious mode of development, a striking ability to modulate its differentiated character, and an unusual strategy for repair.

New Skeletal Muscle Fibers Form by the Fusion of Myoblasts

The previous chapter described how certain cells, originating from the somites of a vertebrate embryo at a very early stage, become determined as **myoblasts**, the precursors of skeletal muscle fibers. As discussed in Chapter 7, the commitment to be a myoblast depends on gene regulatory proteins of at least two families—the *MyoD family* of basic helix–loop–helix proteins, and the *MEF2 family* of MADS box proteins. These act in combination to give the myoblast a memory of its committed state, and, eventually, to regulate the expression of other genes that give the mature muscle cell its specialized character (see Figure 7–72). After a period of proliferation, the myoblasts undergo a dramatic switch of phenotype that depends on the coordinated activation of a whole battery of muscle-specific genes, a process known as myoblast differentiation. As they differentiate, they fuse with one another to form multinucleate skeletal muscle fibers (Figure 22–41). Fusion involves specific cell–cell adhesion molecules that mediate recognition between newly differentiating myoblasts and fibers. Once differentiation has occurred, the cells do not divide and the nuclei never again replicate their DNA.

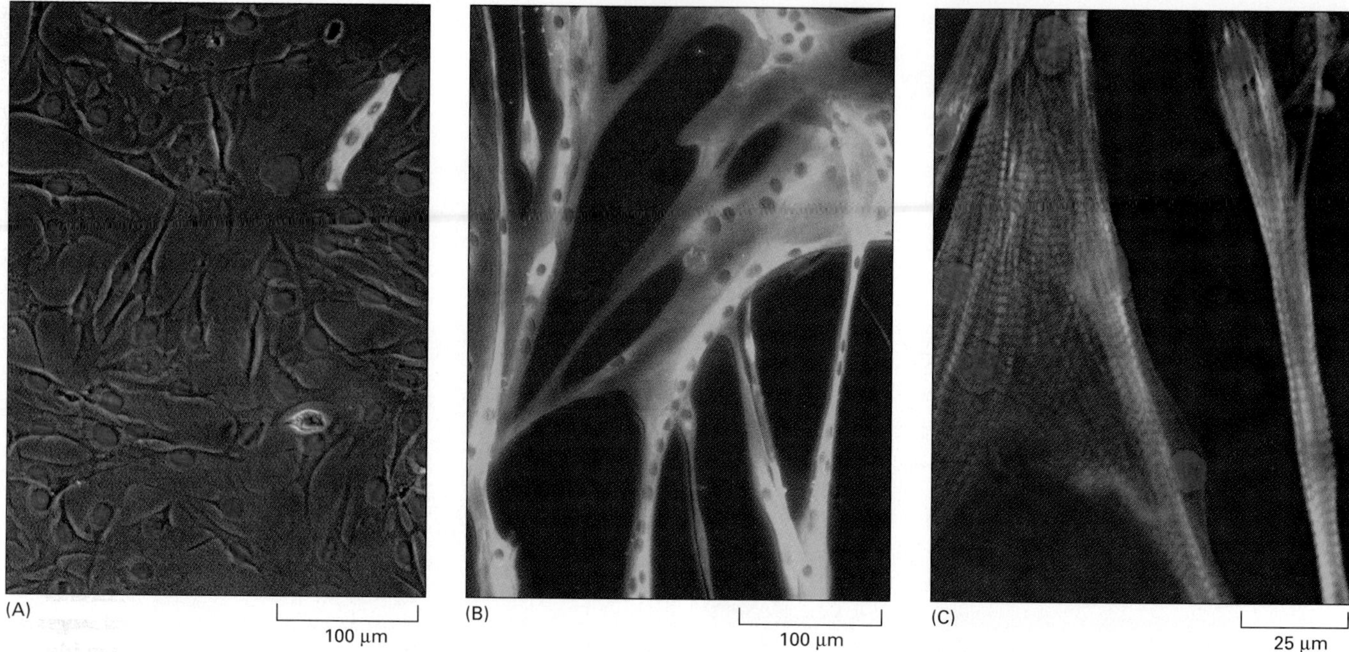

Figure 22–41 Myoblast fusion in culture. The culture is stained with a fluorescent antibody *(green)* against skeletal muscle myosin, which marks differentiated muscle cells, and with a DNA-specific dye *(blue)* to show cell nuclei. (A) A short time after a change to a culture medium that favors differentiation, just two of the many myoblasts in the field of view have switched on myosin production and have fused to form a muscle cell with two nuclei (upper right). (B) Somewhat later, almost all the cells have differentiated and fused. (C) High-magnification view, showing characteristic striations (fine transverse stripes) in two of the multinucleate muscle cells. (Courtesy of Jacqueline Gross and Terence Partridge.)

Myoblasts that have been kept proliferating in culture for as long as two years still retain the ability to differentiate and can fuse to form muscle cells in response to a suitable change in culture conditions. Appropriate signal proteins such as fibroblast or hepatocyte growth factor (FGF or HGF) in the culture medium can maintain myoblasts in the proliferative, undifferentiated state: if these soluble factors are removed, the cells rapidly stop dividing, differentiate, and fuse. The system of controls is complex, however, and attachment to the extracellular matrix is also important for myoblast differentiation. Moreover, the process of differentiation is cooperative: differentiating myoblasts secrete factors that apparently encourage other myoblasts to differentiate. In the intact animal, the myoblasts and muscle fibers are held in the meshes of a connective-tissue framework formed by fibroblasts. This framework guides muscle development and controls the arrangement and orientation of the muscle cells.

Muscle Cells Can Vary Their Properties by Changing the Protein Isoforms They Contain

Once formed, a skeletal muscle fiber grows, matures, and modulates its character according to functional requirements. The genome contains multiple variant copies of the genes encoding many of the characteristic proteins of the skeletal muscle cell, and the RNA transcripts of many of these genes can be spliced in several ways. As a result, a wealth of protein variants (isoforms) can be produced for the components of the contractile apparatus. As the muscle fiber matures, different isoforms are produced, adapted to the changing demands for speed, strength, and endurance in the fetus, the newborn, and the adult. Within a single adult muscle, several distinct types of skeletal muscle fibers, each with different sets of protein isoforms and different functional properties, can be found side by side (Figure 22–42). Slow muscle fibers (for sustained contraction) and fast muscle fibers (for rapid twitch) are innervated by slow and fast motor neurons, respectively, and the innervation can regulate muscle-fiber gene expression and size through the different patterns of electrical stimulation that these neurons deliver.

Skeletal Muscle Fibers Secrete Myostatin to Limit Their own Growth

A muscle can grow in three ways: its fibers can increase in number, in length, or in girth. Because skeletal muscle fibers are unable to divide, more of them can be made only by the fusion of myoblasts, and the adult number of multinucleated skeletal muscle fibers is in fact attained early—before birth, in humans. Once formed, a skeletal muscle fiber generally survives for the entire lifetime of the animal. However, individual muscle nuclei can be added or lost. Thus, the enormous postnatal increase in muscle bulk is achieved by cell enlargement. Growth in length depends on recruitment of more myoblasts into the existing multinucleated fibers, which increases the number of nuclei in each cell. Growth in girth, such as occurs in the muscles of weightlifters, involves both myoblast recruitment and an increase in the size and numbers of the contractile myofibrils that each muscle fiber nucleus supports.

What, then, are the mechanisms that control muscle cell numbers and muscle cell size? One part of the answer lies in an extracellular signal protein called *myostatin*. Mice with a loss-of-function mutation in the myostatin gene have enormous muscles—two to three times larger than normal (Figure 22–43). Both the numbers and the size of the muscle cells seem to be increased. Mutations in the same gene turn out to be present in so-called "double-muscled" breeds of cattle (see Figure 17–51): in selecting for big muscles, cattle breeders have unwittingly selected for myostatin deficiency. Myostatin belongs to the TGFβ superfamily of signal proteins, and it is normally made and secreted by skeletal muscle cells. Its function, evidently, is to provide negative feedback to limit muscle growth. Small amounts of the protein can be detected in the circulation of adult humans, and it has been reported that the amount is raised in AIDS patients who show muscle wasting. Thus, myostatin may act as a negative regulator of muscle growth in adult life as well as during development. The growth of some other organs is similarly controlled by a negative-feedback action of a factor that they themselves produce. We shall encounter another example in a later section.

Figure 22–42 Fast and slow muscle fibers. Two consecutive cross sections of the same piece of adult mouse leg muscle were stained with different antibodies, each specific for a different isoform of myosin heavy chain protein, and images of the two sections were overlaid in false color to show the pattern of muscle fiber types. Fibers stained with antibodies against "fast" myosin *(gray)* are specialized to produce fast-twitch contractions; fibers stained with antibodies against "slow" myosin *(pink)* are specialized to produce slow, sustained contractions. The fast-twitch fibers are known as white muscle fibers because they contain relatively little of the colored oxygen-binding protein myoglobin. The slow muscle fibers are called red muscle fibers because they contain much more of it. (Courtesy of Simon Hughes.)

Some Myoblasts Persist as Quiescent Stem Cells in the Adult

Even though humans do not normally generate new skeletal muscle fibers in adult life, the capacity for doing so is not completely lost. Cells capable of serving as myoblasts are retained as small, flattened, and inactive cells lying in close contact with the mature muscle cell and contained within its sheath of basal lamina (Figure 22–44). If the muscle is damaged, these *satellite cells* are activated to proliferate, and their progeny can fuse to repair the damaged muscle. Satellite cells are thus the stem cells of adult skeletal muscle, normally held in reserve in a quiescent state but available when needed as a self-renewing source of terminally differentiated cells. Athletes who specialize in muscular strength often damage their muscle fibers and are thought to depend on this mechanism for muscle repair, resulting in regenerated fibers that are often highly branched.

The process of muscle repair by means of satellite cells is, nevertheless, limited in what it can achieve. In one form of *muscular dystrophy*, for example, differentiated skeletal muscle cells are damaged because of a genetic defect in the cytoskeletal protein dystrophin. As a result, satellite cells proliferate to repair the damaged muscle fibers. This regenerative response is, however, unable to keep pace with the damage, and the muscle cells are eventually replaced by connective

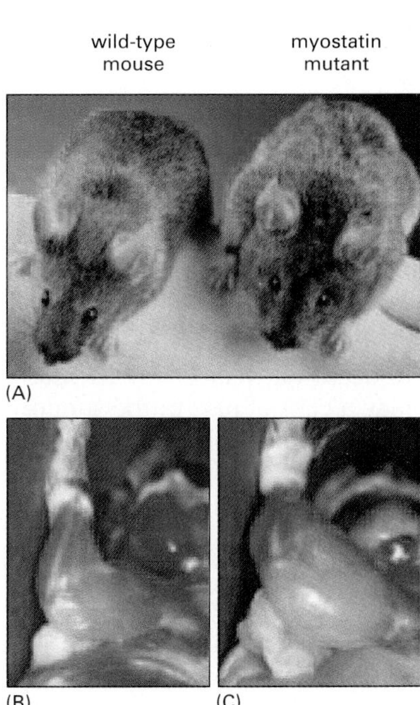

Figure 22–43 Regulation of muscle size by myostatin. (A) A normal mouse compared with a mutant mouse deficient in myostatin. (B) Leg of a normal and (C) of a myostatin-deficient mouse, with skin removed to show the massive enlargement of the musculature in the mutant. (From S.-J. Lee and A.C. McPherron, *Curr. Opin. Genet. Devel.* 9:604–607, 1999. © Elsevier.)

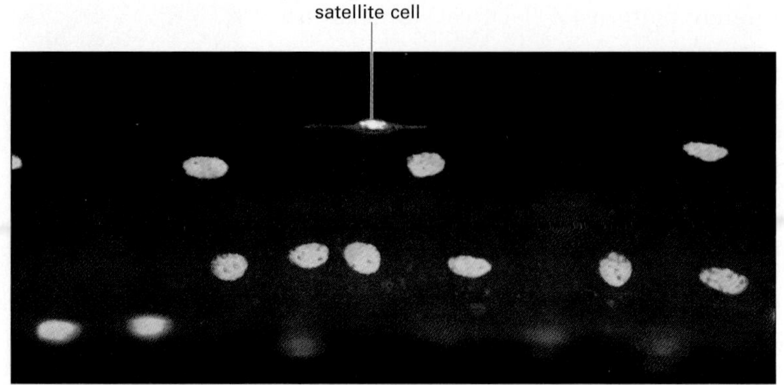

Figure 22–44 A satellite cell on a skeletal muscle fiber. The specimen is stained with an antibody *(red)* against a muscle cadherin, M-cadherin, which is present on both the satellite cell and the muscle fiber and is concentrated at the site where their membranes are in contact. The nuclei of the muscle fiber are stained *green*, and the nucleus of the satellite cell is stained *blue*. (Courtesy of Terence Partridge.)

tissue, blocking any further possibility of regeneration. A similar loss of capacity for repair seems to contribute to the weakening of muscle in the elderly.

In muscular dystrophy, where the satellite cells are constantly called upon to proliferate, their capacity to divide may become exhausted as a result of progressive shortening of their telomeres in the course of each cell cycle (discussed in Chapter 17). Stem cells of other tissues, such as blood, are limited in the same way: they normally divide only at a slow rate, and mutations or exceptional circumstances that cause them to divide more rapidly can lead to premature exhaustion of the stem-cell supply.

Summary

Skeletal muscle fibers are one of the four main categories of vertebrate cells specialized for contraction, and they are responsible for all voluntary movement. Each skeletal muscle fiber is a syncytium and develops by the fusion of many myoblasts. Myoblasts proliferate extensively, but once they have fused, they can no longer divide. Fusion generally follows the onset of myoblast differentiation, in which many genes encoding muscle-specific proteins are switched on coordinately. Some myoblasts persist in a quiescent state as satellite cells in adult muscle; when a muscle is damaged, these cells are reactivated to proliferate and to fuse to replace the muscle cells that have been lost. Muscle bulk is regulated homeostatically by a negative-feedback mechanism, in which existing muscle secretes myostatin, which inhibits further muscle growth.

FIBROBLASTS AND THEIR TRANSFORMATIONS: THE CONNECTIVE-TISSUE CELL FAMILY

Many of the differentiated cells in the adult body can be grouped into families whose members are closely related by origin and by character. An important example is the family of **connective-tissue cells**, whose members are not only related but also unusually interconvertible. The family includes *fibroblasts*, *cartilage cells*, and *bone cells*, all of which are specialized for the secretion of collagenous extracellular matrix and are jointly responsible for the architectural framework of the body. The connective-tissue family also includes *fat cells* and *smooth muscle cells*. These cell types and the interconversions that are thought to occur between them are illustrated in Figure 22–45. Connective-tissue cells play a central part in the support and repair of almost every tissue and organ, and the adaptability of their differentiated character is an important feature of the responses to many types of damage.

Fibroblasts Change Their Character in Response to Chemical Signals

Fibroblasts seem to be the least specialized cells in the connective-tissue family. They are dispersed in connective tissue throughout the body, where they secrete a nonrigid extracellular matrix that is rich in type I and/or type III collagen, as

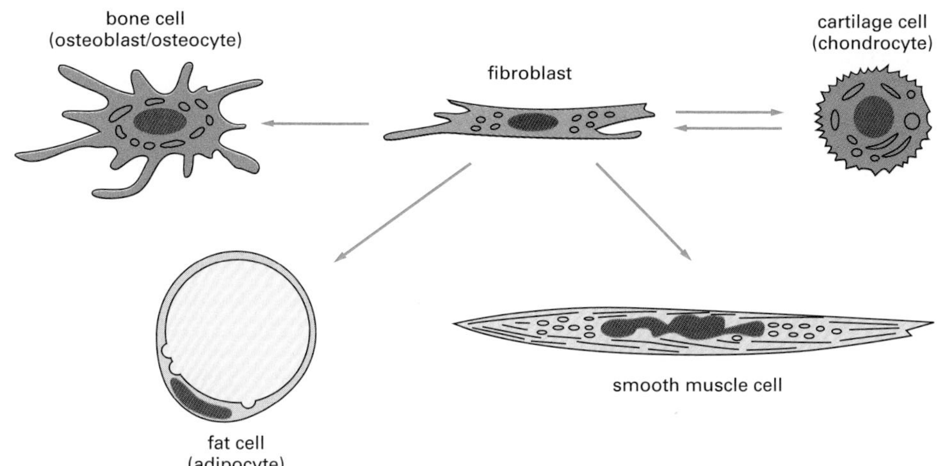

Figure 22–45 The family of connective-tissue cells. *Arrows* show the interconversions that are thought to occur within the family. For simplicity, the fibroblast is shown as a single cell type, but in fact it is uncertain how many types of fibroblasts exist and whether the differentiation potential of different types is restricted in different ways.

discussed in Chapter 19. When a tissue is injured, the fibroblasts nearby proliferate, migrate into the wound, and produce large amounts of collagenous matrix, which helps to isolate and repair the damaged tissue. Their ability to thrive in the face of injury, together with their solitary lifestyle, may explain why fibroblasts are the easiest of cells to grow in culture—a feature that has made them a favorite subject for cell biological studies (Figure 22–46).

As indicated in Figure 22–45, fibroblasts also seem to be the most versatile of connective-tissue cells, displaying a remarkable capacity to differentiate into other members of the family. There are uncertainties about their interconversions, however. Strong evidence indicates that fibroblasts in different parts of the body are intrinsically different, and there may be differences between them even in a single region. "Mature" fibroblasts with a lesser capacity for transformation may, for example, exist side by side with "immature" fibroblasts (often called mesenchymal cells) that can develop into a variety of mature cell types.

The stromal cells of bone marrow, mentioned earlier, provide a good example of connective-tissue versatility. These cells, which can be regarded as a kind of fibroblast, can be isolated from the bone marrow and propagated in culture. Large clones of progeny can be generated in this way from single ancestral stromal cells. According to the signal proteins that are added to the culture medium, the members of such a clone can either continue proliferating to produce more cells of the same type, or can differentiate as fat cells, cartilage cells, or bone cells. Because of their self-renewing, multipotent character, they are referred to as *mesenchymal stem cells*.

Fibroblasts from the skin are different. Placed in the same culture conditions, they do not show the same plasticity. Yet they, too, can be induced to change their character. At a healing wound, for example, they change their actin gene expression and take on some of the contractile properties of smooth muscle cells, thereby helping to pull the wound margins together; such cells are called *myofibroblasts*. More dramatically, if a preparation of bone matrix, made by grinding bone into a fine powder and dissolving away the hard mineral component, is implanted in the dermal layer of the skin, some of the cells there (probably fibroblasts) become transformed into cartilage cells, and a little later, others transform into bone cells, thereby creating a small lump of bone. These experiments suggest that components in the extracellular matrix can dramatically influence the differentiation of connective-tissue cells.

We shall see that similar cell transformations are important in the natural repair of broken bones. In fact, bone matrix contains high concentrations of

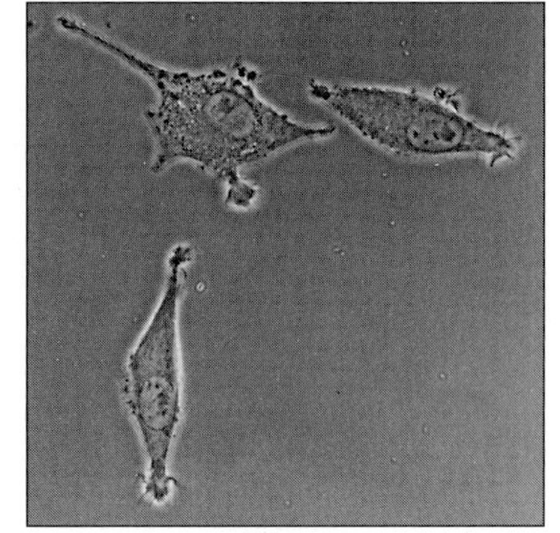

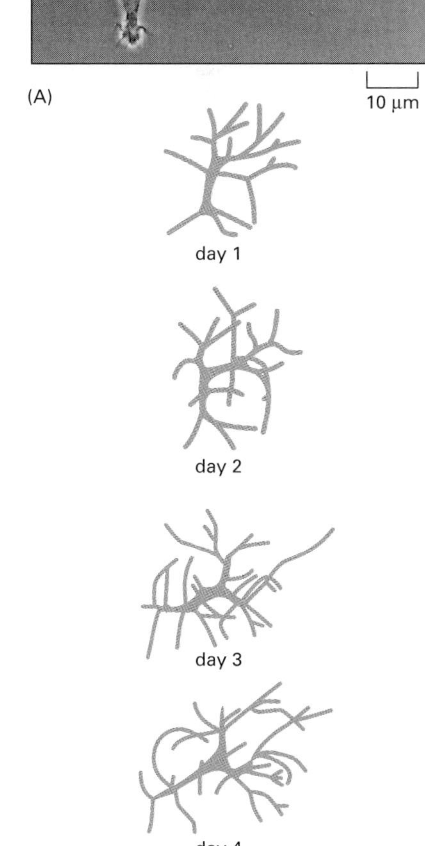

Figure 22–46 The fibroblast. (A) A phase-contrast micrograph of fibroblasts in culture. (B) These drawings of a living fibroblastlike cell in the transparent tail of a tadpole show the changes in its shape and position on successive days. Note that while fibroblasts flatten out in culture, they can have more complex, process-bearing morphologies in tissues. See also Figure 19–35. (A, from E. Pokorna et al., *Cell Motil. Cytoskel.* 28:25–33, 1994. © Wiley-Liss; B, redrawn from E. Clark, *Am. J. Anat.* 13:351–379, 1912.)

several signal proteins that can affect the behavior of connective-tissue cells. These include members of the TGFβ superfamily, including BMPs and TGFβ itself. These factors are powerful regulators of growth, differentiation, and matrix synthesis by connective-tissue cells, exerting a variety of actions depending on the target cell type and the combination of other factors and matrix components that are present. When injected into a living animal, they can induce the formation of cartilage, bone, or fibrous matrix, according to the site and circumstances of injection. TGFβ is especially important in wound healing, where it stimulates conversion of fibroblasts into myofibroblasts and promotes formation of the collagen-rich scar tissue that gives a healed wound its strength.

The Extracellular Matrix May Influence Connective-Tissue Cell Differentiation by Affecting Cell Shape and Attachment

The extracellular matrix may influence the differentiated state of connective-tissue cells through physical as well as chemical effects. This has been shown in studies on cultured cartilage cells, or **chondrocytes**. Under appropriate culture conditions, these cells proliferate and maintain their differentiated character, continuing for many cell generations to synthesize large quantities of highly distinctive cartilage matrix, with which they surround themselves. If, however, the cells are kept at relatively low density and remain as a monolayer on the culture dish, a transformation occurs. They lose their characteristic rounded shape, flatten down on the substratum, and stop making cartilage matrix: they stop producing type II collagen, which is characteristic of cartilage, and start producing type I collagen, which is characteristic of fibroblasts. By the end of a month in culture, almost all the cartilage cells have switched their collagen gene expression and taken on the appearance of fibroblasts. The biochemical change must occur abruptly, since very few cells are observed to make both types of collagen simultaneously.

Several lines of evidence suggest that the biochemical change is induced at least in part by the change in cell shape and attachment. Cartilage cells that have made the transition to a fibroblastlike character, for example, can be gently detached from the culture dish and transferred to a dish of agarose. By forming a gel around them, the agarose holds the cells suspended without any attachment to a substratum, forcing them to adopt a rounded shape. In these circumstances, the cells promptly revert to the character of chondrocytes and start making type II collagen again. Cell shape and anchorage may control gene expression through intracellular signals generated at focal contacts by integrins acting as matrix receptors, as discussed in Chapter 19.

For most types of cells, and especially for a connective-tissue cell, the opportunities for anchorage and attachment depend on the surrounding matrix, which is usually made by the cell itself. Thus, a cell can create an environment that then acts back on the cell to reinforce its differentiated state. Furthermore, the extracellular matrix that a cell secretes forms part of the environment for its neighbors as well as for the cell itself, and thus tends to make neighboring cells differentiate in the same way (see Figure 19–61). A group of chondrocytes forming a nodule of cartilage, for example, either in the developing body or in a culture dish, can be seen to enlarge by the conversion of neighboring fibroblasts into chondrocytes.

Fat Cells Can Develop From Fibroblasts

Fat cells, or **adipocytes**, also derive from fibroblastlike cells, both during normal mammalian development and in various pathological circumstances. In muscular dystrophy, for example, where the muscle cells die, they are gradually replaced by fatty connective tissue, probably by conversion of local fibroblasts. Fat cell differentiation (whether normal or pathological) begins with the expression of two families of gene regulatory proteins: the *C/EBP* (CCAAT/enhancer binding protein) family and the *PPAR* (peroxisome proliferator-activated receptor) family, especially PPARγ. Like the MyoD and MEF2 families in skeletal

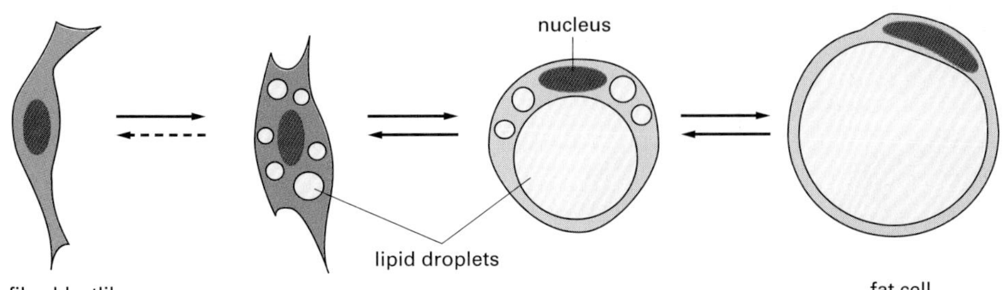

Figure 22–47 The development of a fat cell. A fibroblastlike precursor cell is converted into a mature fat cell by the accumulation and coalescence of lipid droplets. The process is at least partly reversible, as indicated by the *arrows*. The cells in the early and intermediate stages can divide, but the mature fat cell cannot.

muscle development, the C/EBP and PPARγ proteins drive and maintain one another's expression, through various cross-regulatory and autoregulatory control loops. They work together to control the expression of the other genes characteristic of adipocytes.

The production of enzymes for import of fatty acids and glucose and for fat synthesis leads to an accumulation of fat droplets, consisting mainly of triacylglycerol (see Figure 2–77). These then coalesce and enlarge until the cell is hugely distended (up to 120 μm in diameter), with only a thin rim of cytoplasm around the mass of lipid (Figures 22–47 and 22–48). Lipases are also made in the fat cell, giving it the capacity to reverse the process of lipid accumulation, by breaking down the triacylglycerols into fatty acids that can be secreted for consumption by other cells. The fat cell can change its volume by a factor of a thousand as it accumulates and releases lipid.

Leptin Secreted by Fat Cells Provides Negative Feedback to Inhibit Eating

For almost all animals under natural circumstances, food supplies are variable and unpredictable. Fat cells have the vital role of storing reserves of nourishment in times of plenty and releasing them in times of dearth. It is thus essential to the function of adipose tissue that its quantity should be adjustable throughout life, according to the supply of nutrients. For our ancestors, this was a blessing; in the well-fed half of the modern world, it has become also a curse. In the United States, for example, it is estimated that more than 30% of the population suffer from obesity, defined as a body mass index (weight/height^2) more than 30 kg/m^2, equivalent to about 30% above ideal weight.

It is not easy to determine to what extent the changes in the quantity of adipose tissue depend on changes in the numbers of fat cells, as opposed to changes in fat-cell size. Changes in cell size are probably the main factor in normal nonobese adults, but in severe obesity, at least, the number of fat cells also increases. The factors that drive the recruitment of new fat cells are not well understood, although they are thought to include growth hormone and IGF-1 (insulinlike growth factor-1). It is clear, however, that the increase or decrease of fat cell size is regulated directly by levels of circulating nutrients and by hormones, such as insulin, that reflect nutrient levels. The surplus of food intake over energy expenditure thus directly governs the accumulation of adipose tissue.

But how are food intake and energy expenditure themselves regulated? A human adult eats about a million kilocalories per year, equivalent to about 200 kg of pure fat. Clearly, if we are not to get hopelessly fat or hopelessly thin over the course of a lifetime, there must be control mechanisms to adjust our eating and energy expenditure over the long term according to the quantity of our fat reserves. The key signal is a protein hormone called **leptin**, which circulates in the bloodstream. Mutant mice that lack leptin or the appropriate leptin receptor are extremely fat (Figure 22–49). Mutations in the same genes sometimes occur in humans, although very rarely. The consequences are similar: constant hunger, overeating, and crippling obesity.

Leptin is normally made by fat cells; the bigger they are, the more they make. Leptin acts on many tissues, and in particular in the brain, on cells in those regions of the hypothalamus that regulate eating behavior. The effect in the

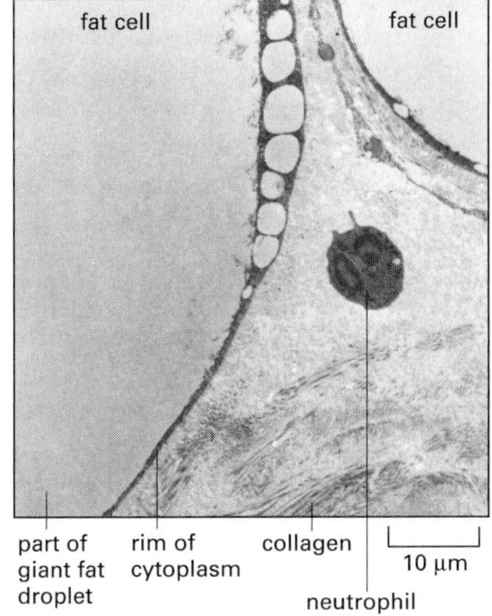

Figure 22–48 Fat cells. This low-magnification electron micrograph shows parts of two fat cells. A neutrophil cell that happens to be present in the adjacent connective tissue provides a sense of scale; each of the fat cells is more than 10 times larger than the neutrophil in diameter and is almost entirely filled with a single large fat droplet. The small fat droplets (pale oval shapes) in the remaining rim of cytoplasm are destined to fuse with the central droplet. The nucleus is not visible in either of the fat cells in the picture. (Courtesy of Don Fawcett, from D.W. Fawcett, A Textbook of Histology, 12th edn. New York: Chapman and Hall, 1994.)

Figure 22–49 Effects of leptin deficiency. Normal mice are here compared with a mouse that has a mutation in the *obese* gene, which codes for leptin. The leptin-deficient mutant fails to limit its eating and becomes grotesquely fat (three times the weight of a normal mouse). (Courtesy of Jeffrey M. Friedman.)

brain is to lessen hunger and discourage eating, which results in a decreased amount of fat tissue. Thus, leptin, like myostatin released from muscle cells, provides a negative-feedback mechanism to regulate the growth of the tissue that secretes it.

In most obese people, leptin levels in the blood stream are persistently high. Although leptin receptors are present and functional, the effect of leptin on food intake is overwhelmed by other influences, which are poorly understood.

Bone Is Continually Remodeled by the Cells Within It

Bone is a very dense, specialized form of connective tissue, as different as could be from adipose tissue, even though closely related in origin. Like reinforced concrete, bone matrix is predominantly a mixture of tough fibers (type I collagen fibrils), which resist pulling forces, and solid particles (calcium phosphate as hydroxyapatite crystals), which resist compression. The volume occupied by the collagen is nearly equal to that occupied by the calcium phosphate. The collagen fibrils in adult bone are arranged in regular plywoodlike layers, with the fibrils in each layer lying parallel to one another but at right angles to the fibrils in the layers on either side.

For all its rigidity, bone is by no means a permanent and immutable tissue. Running through the hard extracellular matrix are channels and cavities occupied by living cells, which account for about 15% of the weight of compact bone. These cells are engaged in an unceasing process of remodeling: one class of cells *(osteoclasts,* related to macrophages) demolishes old bone matrix while another *(osteoblasts,* related to fibroblasts) deposits new bone matrix. This mechanism provides for continuous turnover and replacement of the matrix in the interior of the bone.

Figure 22–50 Achondroplasia. This type of dwarfism has a frequency of one in 10,000–100,000 births; in more than 99% of cases it is caused by a mutation at an identical site in the genome, corresponding to amino-acid 380 in FGF-receptor-3 (a glycine in the transmembrane domain). The mutation is dominant, and almost all cases are due to new, independently occurring mutations, implying an extraordinarily high mutation rate at this particular site in the genome. The defect in FGF signaling causes dwarfism by interfering with the growth of cartilage in developing long bones. (From Velasquez's painting of Sebastian de Morra. © Museo del Prado, Madrid.)

Unlike soft tissues, which can grow by internal expansion, bone can grow only by apposition—that is, by the laying down of additional matrix and cells on the free surfaces of existing bone. During development, this process must occur in coordination with the growth of other tissues, in such a way that the pattern of the body can be scaled up without its proportions being radically disturbed. For most of the skeleton, and in particular for the long bones of the limbs and trunk, the coordinated growth is achieved by a complex strategy. A set of minute "scale models" of these bones are first formed out of cartilage. Each scale model then grows, and as new cartilage is formed, the older cartilage is replaced by bone. Cartilage growth and erosion and bone deposition are so ingeniously coordinated during development that the adult bone, though it may be half a meter long, is almost the same shape as the initial cartilaginous model, which was no more than a few millimeters long.

Defective growth of cartilage during the development of long bones, as a result of a dominant mutation in the gene that codes for an FGF receptor (FGFR3), is responsible for the commonest form of dwarfism, known as *achondroplasia* (Figure 22–50). Conversely, osteoblasts are lacking in individuals with mutations that disrupt production of a gene regulatory protein (called CBFA1)

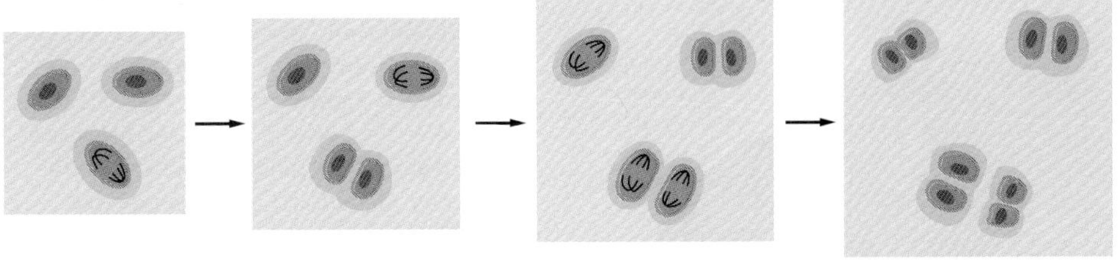

Figure 22–51 The growth of cartilage. The tissue expands as the chondrocytes divide and make more matrix. The freshly synthesized matrix with which each cell surrounds itself is shaded *dark green*. Cartilage may also grow by recruiting fibroblasts from the surrounding tissue and converting them into chondrocytes.

specifically required for osteoblast differentiation: mice homozygous for this genetic defect are born with a skeleton consisting solely of cartilage and die soon after birth as a result.

Osteoblasts Secrete Bone Matrix, While Osteoclasts Erode It

Cartilage is a simple tissue, consisting of cells of a single type—chondrocytes—embedded in a more or less uniform matrix. The cartilage matrix is deformable, and the tissue grows by expanding as the chondrocytes divide and secrete more matrix (Figure 22–51). Bone is more complex. The bone matrix is secreted by **osteoblasts** that lie at the surface of the existing matrix and deposit fresh layers of bone onto it. Some of the osteoblasts remain free at the surface, while others gradually become embedded in their own secretion. This freshly formed material (consisting chiefly of type I collagen) is called *osteoid*. It is rapidly converted into hard bone matrix by the deposition of calcium phosphate crystals in it. Once imprisoned in hard matrix, the original bone-forming cell, now called an **osteocyte**, has no opportunity to divide, although it continues to secrete further matrix in small quantities around itself. The osteocyte, like the chondrocyte, occupies a small cavity, or *lacuna*, in the matrix, but unlike the chondrocyte it is not isolated from its fellows. Tiny channels, or *canaliculi*, radiate from each lacuna and contain cell processes from the resident osteocyte, enabling it to form gap junctions with adjacent osteocytes (Figure 22–52). Although the networks of osteocytes do not themselves secrete or erode substantial quantities of matrix, they probably play a part in controlling the activities of the cells that do.

While bone matrix is deposited by osteoblasts, it is eroded by **osteoclasts** (Figure 22–53). These large multinucleated cells originate, like macrophages, from hemopoietic stem cells in the bone marrow. The precursor cells are released as monocytes into the bloodstream and collect at sites of bone resorption, where they fuse to form the multinucleated osteoclasts, which cling to surfaces of the bone matrix and eat it away. Osteoclasts are capable of tunneling deep into the substance of compact bone, forming cavities that are then invaded by other cells. A blood capillary grows down the center of such a tunnel, and the walls of the tunnel become lined with a layer of osteoblasts (Figure 22–54). To produce the plywoodlike structure of compact bone, these osteoblasts lay

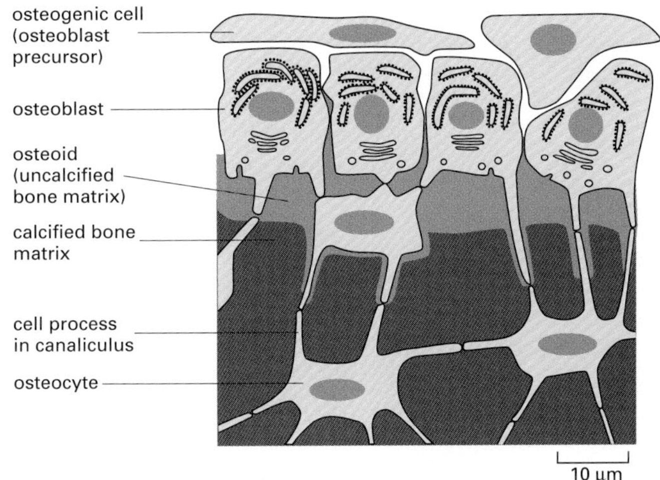

Figure 22–52 Deposition of bone matrix by osteoblasts. Osteoblasts lining the surface of bone secrete the organic matrix of bone (osteoid) and are converted into osteocytes as they become embedded in this matrix. The matrix calcifies soon after it has been deposited. The osteoblasts themselves are thought to derive from osteogenic stem cells that are closely related to fibroblasts.

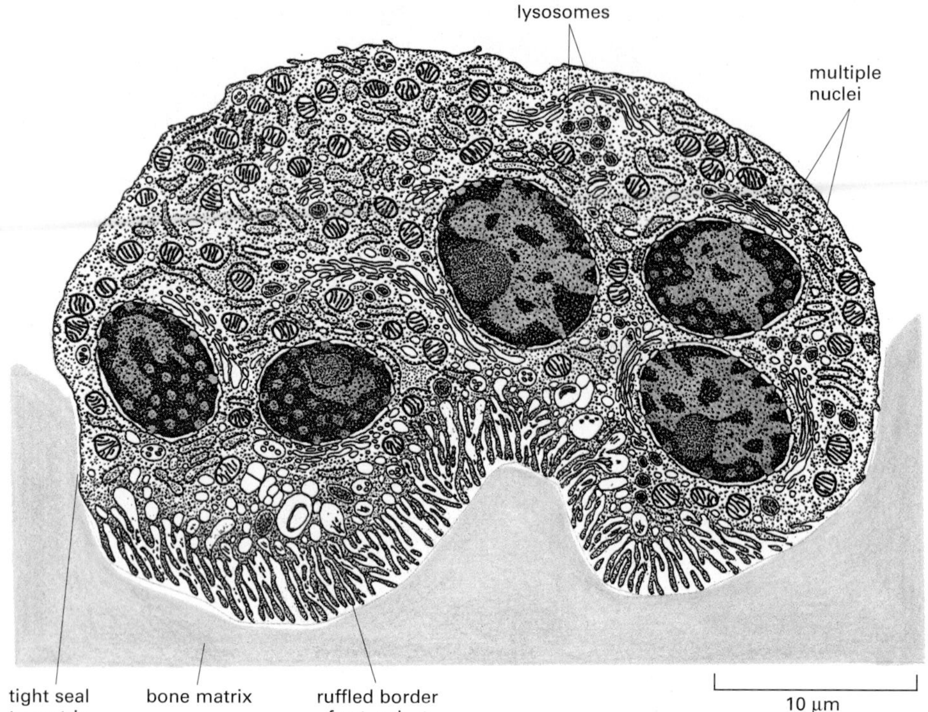

Figure 22–53 An osteoclast shown in cross section. This giant, multinucleated cell erodes bone matrix. The "ruffled border" is a site of secretion of acids (to dissolve the bone minerals) and hydrolases (to digest the organic components of the matrix). Osteoclasts vary in shape, are motile, and often send out processes to resorb bone at multiple sites. They develop from monocytes and can be viewed as specialized macrophages. (From R.V. Krstić, Ultrastructure of the Mammalian Cell: An Atlas. Berlin: Springer-Verlag, 1979.)

down concentric layers of new matrix, which gradually fill the cavity, leaving only a narrow canal surrounding the new blood vessel. Many of the osteoblasts become trapped in the bone matrix and survive as concentric rings of osteocytes. At the same time as some tunnels are filling up with bone, others are being bored by osteoclasts, cutting through older concentric systems. The consequences of this perpetual remodeling are beautifully displayed in the layered patterns of matrix observed in compact bone (Figure 22–55).

Through remodelling, bones are endowed with a remarkable ability to adjust their structure in response to long-term variations in the load imposed on them. This adaptive behavior implies that the deposition and erosion of the matrix are somehow controlled by local mechanical stresses, but the mechanisms involved are not understood. The bone cells secrete signal proteins that become trapped in the matrix, and it is likely that these are released when the matrix is degraded

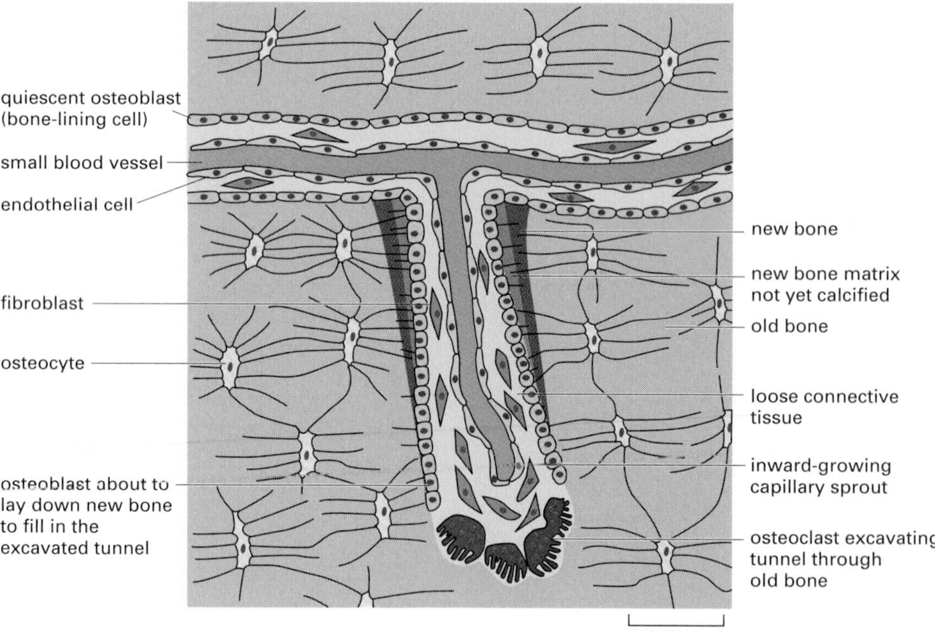

Figure 22–54 The remodeling of compact bone. Osteoclasts acting together in a small group excavate a tunnel through the old bone, advancing at a rate of about 50 μm per day. Osteoblasts enter the tunnel behind them, line its walls, and begin to form new bone, depositing layers of matrix at a rate of 1–2 μm per day. At the same time, a capillary sprouts down the center of the tunnel. The tunnel eventually becomes filled with concentric layers of new bone, with only a narrow central canal remaining. Each such canal, besides providing a route of access for osteoclasts and osteoblasts, contains one or more blood vessels that transport the nutrients the bone cells require for survival. Typically, about 5–10% of the bone in a healthy adult mammal is replaced in this way each year. (After Z.F.G. Jaworski, B. Duck, and G. Sekaly, *J. Anat.* 133:397–405, 1981.)

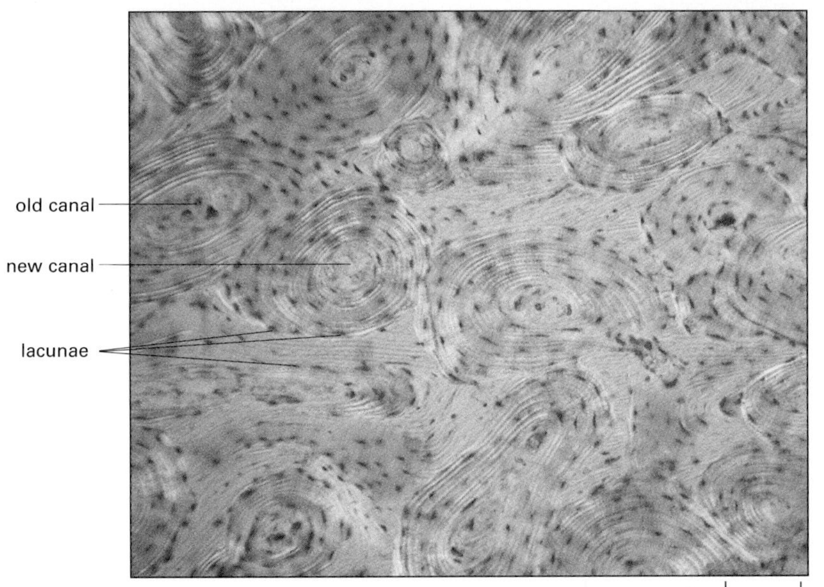

Figure 22–55 A transverse section through a compact outer portion of a long bone. The micrograph shows the outlines of tunnels formed by osteoclasts and then filled in by osteoblasts during successive rounds of bone remodeling. The section has been prepared by grinding. The hard matrix has been preserved, but not the cells. Lacunae and canaliculi that were occupied by osteocytes are clearly visible, however. The alternating bright and dark concentric rings correspond to an alternating orientation of the collagen fibers in the successive layers of bone matrix laid down by the osteoblasts that lined the wall of the canal during life. (This pattern is revealed here by viewing the specimen between partly crossed polarizing filters.) Note how older systems of concentric layers of bone have been partly cut through and replaced by newer systems.

or suitably stressed. The released proteins, especially members of the BMP subfamily of TGFβ proteins, may help to guide the remodelling process.

Remodelling carries a risk: defects in its control can lead to *osteoporosis,* where there is excessive erosion of the bone matrix and weakening of the bone, or to the opposite condition, *osteopetrosis,* where the bone becomes excessively thick and dense.

During Development, Cartilage Is Eroded by Osteoclasts to Make Way for Bone

The replacement of cartilage by bone in the course of development is also thought to depend on the activities of osteoclasts. As the cartilage matures, its cells in certain regions become greatly enlarged at the expense of the surrounding matrix, and the matrix itself becomes mineralized, like bone, by the deposition of calcium phosphate crystals. The swollen chondrocytes die, leaving large empty cavities. Osteoclasts and blood vessels invade the cavities and erode the residual cartilage matrix, while osteoblasts following in their wake begin to deposit bone matrix. The only surviving remnant of cartilage in the adult long bone is a thin layer that forms a smooth covering on the bone surfaces at joints, where one bone articulates with another (Figure 22–56).

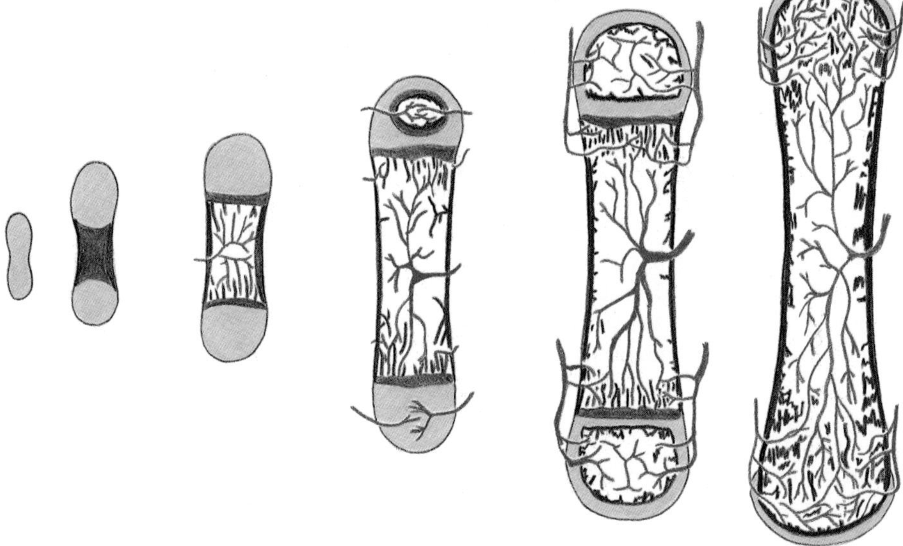

Figure 22–56 The development of a long bone. Long bones, such as the femur or the humerus, develop from a miniature cartilage model. Uncalcified cartilage is shown in *light green*, calcified cartilage in *dark green*, bone in *black*, and blood vessels in *red*. The cartilage is not converted to bone but is gradually replaced by it through the action of osteoclasts and osteoblasts, which invade the cartilage in association with blood vessels. Osteoclasts erode cartilage and bone matrix, while osteoblasts secrete bone matrix. The process of ossification begins in the embryo and is not completed until the end of puberty. The resulting bone consists of a thick-walled hollow cylinder of compact bone enclosing a large central cavity occupied by the bone marrow. Note that not all bones develop in this way. The membrane bones of the skull, for example, are formed directly as bony plates, not from a prior cartilage model. (Adapted from D.W. Fawcett, A Textbook of Histology, 12th edn. New York: Chapman and Hall, 1994.)

Some cells capable of forming new cartilage persist, however, in the connective tissue that surrounds a bone. If the bone is broken, the cells in the neighborhood of the fracture repair it by a sort of recapitulation of the original embryonic process: cartilage is first laid down to bridge the gap and is then replaced by bone.

The capacity for self-repair, so strikingly illustrated by the tissues of the skeleton, is a property of living structures that has no parallel among present-day man-made objects.

Summary

The family of connective-tissue cells includes fibroblasts, cartilage cells, bone cells, fat cells, and smooth muscle cells. Some classes of fibroblasts seem to be able to transform into any of the other members of the family. These transformations of connective-tissue cell type are regulated by the composition of the surrounding extracellular matrix, by cell shape, and by hormones and growth factors. While the chief function of most members of the family is to secrete extracellular matrix, fat cells serve as storage sites for fat. The quantity of fat tissue is regulated in part by negative feedback: fat cells release a hormone, leptin, which acts in the brain to reduce appetite, which leads to a decrease in fat tissue.

Cartilage and bone both consist of cells embedded in a solid matrix. The matrix of cartilage is deformable so that the tissue can grow by swelling, whereas bone is rigid and can grow only by apposition. Bone undergoes perpetual remodeling through which it can adapt to the load it bears; the remodelling depends on the combined action of osteoclasts, which erode matrix, and osteoblasts, which secrete it. Some osteoblasts become trapped in the matrix as osteocytes and play a part in regulating the turnover of bone matrix. Most long bones develop from miniature cartilage "models," which, as they grow, serve as templates for the deposition of bone by the combined action of osteoblasts and osteoclasts. Similarly, in the repair of a bone fracture in the adult, the gap is first bridged by cartilage, which is later replaced by bone.

STEM-CELL ENGINEERING

When cells are removed from the body and maintained in culture, they generally maintain their original character. Keratinocytes continue to behave as keratinocytes, chondrocytes as chondrocytes, liver cells as liver cells, and so on. Each type of specialized cell has a memory of its developmental history and seems fixed in its specialized fate, although some limited transformations can occur, just as in the intact tissues we discussed earlier. Stem cells in culture, as in tissues, may continue to divide, or they may differentiate into one or more cell types, but the cell types they can generate are restricted. Each type of stem cell serves for the renewal of one particular type of tissue. For some tissues, such as the brain, it was long thought that regeneration is impossible in adult life because no stem cells remain. There seemed to be little hope, therefore, of replacing lost nerve cells in the mammalian brain through the genesis of new ones, or of regenerating any other cell type whose normal progenitors are no longer present.

Recent discoveries have overturned this gloomy judgement and have led to a more optimistic perception of what stem cells can do and how we may be able to use them. The change has come from several findings that demonstrate exceptional forms of stem-cell versatility that could scarcely have been suspected from knowledge of the normal life histories of cells in tissues. In this last section of the chapter, we examine these phenomena and consider the new opportunities they create to improve on nature's own mechanisms of damage repair.

ES Cells Can Be Used to Make Any Part of the Body

As described in Chapter 21, it is possible through cell culture to derive from early mouse embryos an extraordinary class of stem cells called **embryonic stem cells**, or **ES cells**. ES cells can be kept proliferating indefinitely in culture and yet

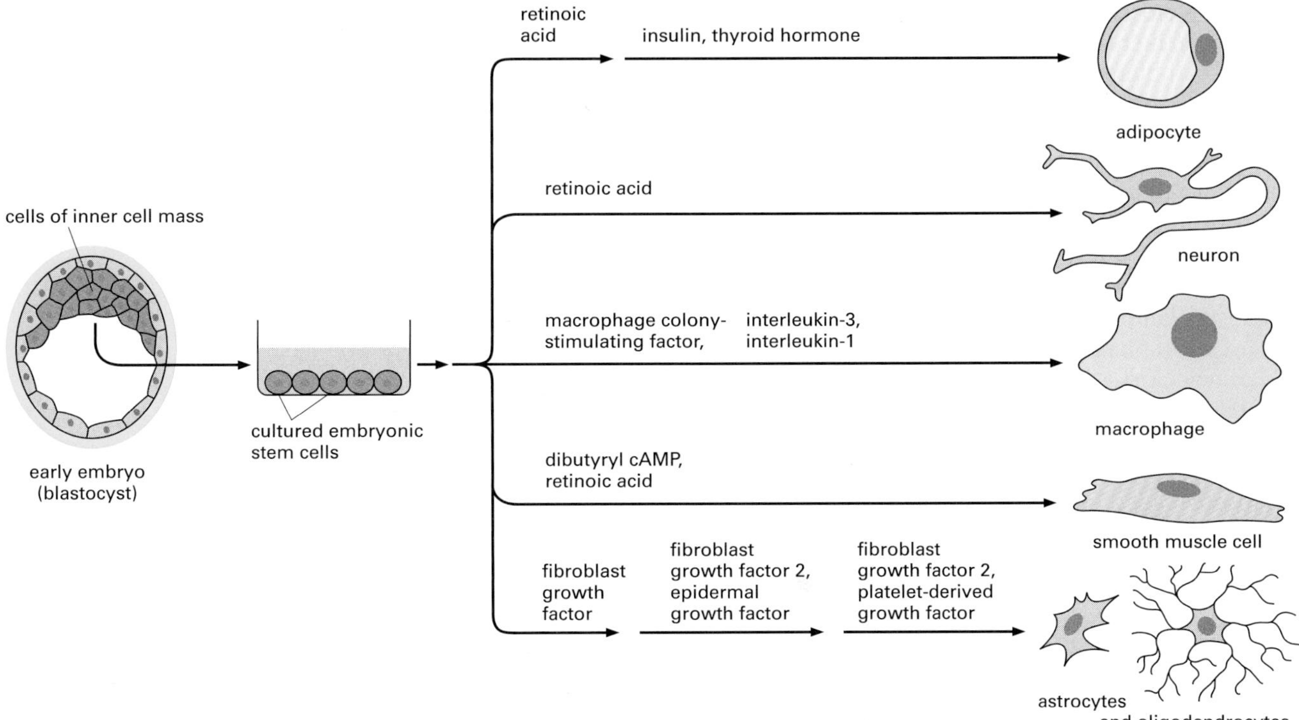

Figure 22–57 Production of differentiated cells from mouse ES cells in culture. ES cells derived from an early mouse embryo can be cultured indefinitely as a monolayer, or allowed to form aggregates called embryoid bodies, in which the cells begin to specialize. Cells from embryoid bodies, cultured in media with different factors added, can then be driven to differentiate in various ways. (Based on E. Fuchs and J.A. Segre, *Cell* 100:143–155, 2000.)

retain an unrestricted developmental potential. If they are put back into an early embryonic environment, they can give rise to all the tissues and cell types in the body, including germ cells. They integrate perfectly into whatever site they may come to occupy, adopting the character and behavior that normal cells would show at that site. One can think of development in terms of a series of choices presented to cells as they follow a road that leads from the fertilized egg to terminal differentiation. After their long sojourn in culture, the ES cell and its progeny can evidently still read the signs at each branch in the highway and respond as normal embryonic cells would.

Cells with properties similar to those of mouse ES cells can now be derived from early human embryos and from human fetal ovaries and testes, creating a potentially inexhaustible supply of cells that might be used for the replacement and repair of mature human tissues that are damaged. Whether or not one has ethical objections to such use of human embryos, it is worth considering the possibilities that are opened up. Setting aside the dream of growing entire organs from ES cells by a recapitulation of embryonic development, experiments in mice suggest that it will be possible in the near future to use ES cells to replace the skeletal muscle fibers that degenerate in victims of muscular dystrophy, the nerve cells that die in patients with Parkinson's disease, the insulin-secreting cells that are lacking in type I diabetics, the heart muscle cells that die in a heart attack, and so on.

Mouse ES cells can be induced to differentiate into a variety of cell types in culture (Figure 22–57). When treated with a carefully chosen combination of signal proteins, for example, the ES cells differentiate into astrocytes and oligodendrocytes, the two main types of glial cells in the central nervous system. If the treated ES cells are injected into a mouse brain, they can serve as progenitors of these cell types. If the host mouse is deficient in myelin-forming oligodendrocytes, for example, the grafted cells can correct the deficiency and form myelin sheaths around axons that lack them.

Epidermal Stem Cell Populations Can Be Expanded in Culture for Tissue Repair

It is a long way still from this sort of success in mice to routine treatments for human diseases. One of the main difficulties lies in immune rejection. If ES-derived cells of one genotype are grafted into an individual of another, the grafted cells are likely to be rejected by the immune system as foreign. Methods of dealing with this problem have been developed for the transplantation of organs such as kidneys and hearts. Immunological problems—and some ethical problems—can, however, be avoided altogether if the right kinds of stem cells can be obtained from the patient's own body.

A simple example is the use of epidermal stem cells for repair of the skin after extensive burns. By culturing cells from undamaged regions of the skin of the burned patient, it is possible to obtain epidermal stem cells quite rapidly in large numbers. These can then be used to repopulate the damaged body surface. For good results after a third-degree burn, however, it is essential to provide first of all an urgent replacement for the lost dermis. For this, dermis taken from a human cadaver can be used, or an artificial dermis substitute. This is still an area of active experimentation. In one technique, an artificial matrix of collagen mixed with a glycosaminoglycan is formed into a sheet, with a thin membrane of silicone rubber covering its external surface as a barrier to water loss, and this skin substitute (called Integra) is laid on the burned body surface after the damaged tissue has been cleaned away. Fibroblasts and blood capillaries from the patient's surviving deep tissues migrate into the artificial matrix and gradually replace it with new connective tissue. Meanwhile, the epidermal cells are cultivated until there are enough to form a thin sheet of adequate extent. Two or more weeks after the original operation, the silicone rubber membrane is carefully removed and replaced with this cultured epidermis, so as to reconstruct a complete skin.

Neural Stem Cells Can Repopulate the Central Nervous System

While the epidermis is one of the simplest and most easily regenerated tissues, the central nervous system (the CNS) is the most complex and seems the most difficult to reconstruct in adult life. The adult mammalian brain and spinal cord have very little capacity for self-repair. Stem cells capable of generating new neurons are hard to find in adult mammals—so hard to find, indeed, that until recently they were thought to be absent.

It is now known, however, that CNS neural stem cells capable of giving rise to both neurons and glial cells do persist in the adult mammalian brain. Moreover, in certain parts of the brain they continually produce new neurons to replace those that die (Figure 22–58). Neuronal turnover occurs on a more dramatic scale in certain songbirds, where large numbers of neurons die each year and are replaced by newborn neurons as part of the process by which a new song is learned in each breeding season.

In experiments with rodents, adult neural stem cells have been harvested from the brain, grown in culture, and then implanted back into the brain of a host animal, where they produce differentiated progeny. Remarkably, it seems that the grafted cells adjust their behavior to match their new location. Stem cells from the hippocampus, for example, implanted in the olfactory-bulb-precursor pathway (see Figure 22–58) give rise to neurons that become correctly incorporated into the olfactory bulb, and vice versa. These findings hold out the hope that, in spite of the extraordinary complexity of nerve cell types and neuronal connections, it may be possible to use neural stem cells to repair at least some types of damage and disease in the central nervous system.

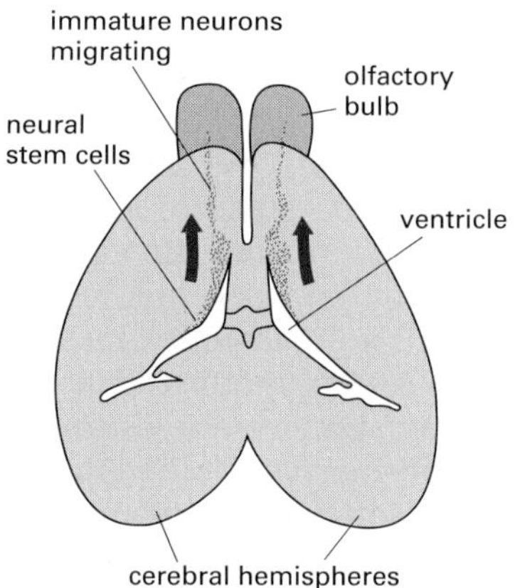

Figure 22–58 The continuing production of neurons in an adult mouse brain. The brain is viewed from above, in a cut-away section, to show the region lining the ventricles of the forebrain where neural stem cells are found. These cells continually produce progeny that migrate to the olfactory bulb, where they differentiate as neurons. The constant turnover of neurons in the olfactory bulb is presumably linked in some way to the turnover of the olfactory receptor neurons that project to it from the olfactory epithelium, as discussed earlier. There is also a continuing turnover of neurons in the adult hippocampus, a region specially concerned with learning and memory, where plasticity of adult function seems to be associated with turnover of a specific subset of neurons (Adapted from B. Barres, *Cell* 97:667–670, 1999.)

The Stem Cells of Adult Tissues May Be More Versatile Than They Seem

Do cells with unexpected stem-cell capabilities lurk in other parts of the body also? If embryonic stem cells can be caused to differentiate along any pathway we please, is it possible to induce other kinds of stem cells, taken from adult tissues, to produce cell types other than their standard range of progeny?

There is now strong evidence that the answer is yes. It has been reported, for example, that neural stem cells, derived from adult brain, can give rise to hemopoietic cells when injected onto a mouse whose own stock of hemopoietic cells has been depleted by x-irradiation. These neural stem cells have also been found to give rise to skeletal muscle cells when injected directly into muscle tissue.

As another important example, cells from adult bone marrow, when injected into x-irradiated recipients, can not only supply the host with fresh hemopoietic cells, but can also give rise to various other cell types, including pneumocytes in the lung and hepatocytes in the liver. Genesis of liver cells (and some other equally surprising cell types) from bone-marrow cells has been demonstrated both in mice and in humans who have received bone-marrow transplants for the treatment of leukemia. Well-differentiated hepatocytes displaying genetic markers proving that they come from the donor are found in the liver of the host. Experiments in mice have shown that the hepatocytes in such cases derive specifically from the hemopoietic stem cells in the bone marrow.

When the host liver is itself defective or damaged, so as to encourage its repopulation by grafted cells, hepatocytes derived from donor hemopoietic cells can be found in plenty. It is not yet clear, however, how readily the conversion of cell fate takes place. It may be that any given hemopoietic stem cell has only a small probability of switching to an alien fate, even when placed in the appropriate alien environment. As yet, we have only a very hazy idea of which transitions are permitted or what factors provoke or inhibit them.

Stem cells taken directly from adult tissues promise to be useful in many ways for tissue repair, but other strategies may have even greater potential. In principle, at least, it should even be possible to use adult tissues to derive "personalized" ES cells—that is, ES cells with the same genome as the adult patient whose body is in need of repair. The cloning of Dolly the sheep and of other mammals has indicated a way to do this. The nucleus of an egg cell can be artificially replaced by a nucleus derived from an adult cell, and the hybrid cell can then go on to develop into an entire individual whose nuclear genome is identical to that of the adult donor (see Figure 7–2). While most of us would not wish to make human beings in this way, it should be possible to obtain ES cells from the immediate descendants of the hybrid cell, by techniques like those used to derive ES cells from early mouse embryos.

Serious ethical issues to need be resolved and enormous technical problems overcome before such an approach can become a reality. Perhaps other, better ways will be found to restore adult cells to an embryonic state of versatility. But by one route or another, it seems that cell biology is beginning to open up new opportunities for improving on Nature's mechanisms of tissue repair, remarkable as those mechanisms are.

Summary

In the normal adult body, different classes of stem cells are responsible for the renewal of different types of tissue. Some tissues, however, seem incapable of repair by the genesis of new cells because no competent stem cells are present. Recent discoveries have opened up new possibilities for manipulating stem-cell behavior artificially so as to repair tissues that previously seemed unrepairable. Epidermal stem cells taken from undamaged skin of a badly burned patient can be rapidly grown in large numbers in culture and grafted back to reconstruct an epidermis to cover the burns. Neural stem cells persist in a few regions of the adult mammalian brain, and when grafted into a developing or damaged brain can generate new neurons and glia appropriate to the site of grafting.

Embryonic stem cells (ES cells) are able to differentiate into any cell type in the body, and they can be induced to differentiate into many cell types in culture. Stem cells of some adult tissues, such as bone marrow, when placed in a suitable environment, seem able to generate a much wider range of differentiated cell types than they produce normally. These findings of stem-cell biology offer the hope of remedy for many serious diseases.

References

General

Fawcett DW (1994) Bloom and Fawcett: A Textbook of Histology, 12th edn. New York/London: Arnold/Chapman & Hall.

Kerr JB (1999) Atlas of Functional Histology. London: Mosby.

Marshak DR, Gardner RL & Gottlieb D (eds) (2001) Stem Cell Biology. Cold Spring Harbor, NY: Cold Spring Harbor Laboratory Press.

Young B & Heath JW (2000) Wheater's Functional Histology; A Text and Colour Atlas, 4th edn. Edinburgh: Churchill Livingstone.

Epidermis and Its Renewal by Stem Cells

Fuchs E (1998) Beauty is skin deep: the fascinating biology of the epidermis and its appendages. *Harvey Lect.* 94, 47–77.

Imagawa W, Yang J, Guzman R & Nandi S (1994) Control of mammary gland development. In The Physiology of Reproduction (Knobil E & Neill JD eds), 2nd edn, pp 1033–1063. New York: Raven Press.

Jensen UB, Lowell S & Watt FM (1999) The spatial relationship between stem cells and their progeny in the basal layer of human epidermis: a new view based on whole-mount labelling and lineage analysis. *Development* 126, 2409–2418.

Nguyen AV & Pollard JW (2000) Transforming growth factor beta3 induces cell death during the first stage of mammary gland involution. *Development* 127, 3107–3118.

Steinert PM (2000) The complexity and redundancy of epithelial barrier function. *J. Cell Biol.* 151, F5–F8.

Watt FM (2001) Stem cell fate and patterning in mammalian epidermis. *Curr. Opin. Genet. Dev.* 11, 410–417.

Sensory Epithelia

Buck LB (2000) The molecular architecture of odor and pheromone sensing in mammals. *Cell* 100, 611–618.

Gillespie PG & Walker RG (2001) Molecular basis of mechanosensory transduction. *Nature* 413, 194–202.

Howard J & Hudspeth AJ (1988) Compliance of the hair bundle associated with gating of mechanoelectrical transduction channels in the bullfrog's saccular hair cell. *Neuron* 1, 189–199.

Masland RH (2001) The fundamental plan of the retina. *Nat. Neurosci.* 4, 877–886.

Mombaerts P, Wang F, Dulac C et al. (1996) Visualizing an olfactory sensory map. *Cell* 87, 675–686.

Morrow EM, Furukawa T & Cepko CL (1998) Vertebrate photoreceptor cell development and disease. *Trends Cell Biol.* 8, 353–358.

Stone JS & Rubel EW (2000) Cellular studies of auditory hair cell regeneration in birds. *Proc. Natl. Acad. Sci. USA* 97, 11714–11721.

The Gut and Its Appendages

Conlon I & Raff M (1999) Size control in animal development. *Cell* 96, 235–244.

Ganz T (2000) Paneth cells—guardians of the gut cell hatchery. *Nat Immunol* 1, 99–100.

Korinek V, Barker N, Moerer P et al. (1998) Depletion of epithelial stem-cell compartments in the small intestine of mice lacking Tcf-4. *Nat. Genet.* 19, 379–383.

Michalopoulos GK & DeFrances MC (1997) Liver regeneration. *Science* 276, 60–66.

Stappenbeck TS, Wong MH, Saam JR et al. (1998) Notes from some crypt watchers: regulation of renewal in the mouse intestinal epithelium. *Curr. Opin. Cell Biol.* 10, 702–709.

Wright NA (2000) Epithelial stem cell repertoire in the gut: clues to the origin of cell lineages, proliferative units and cancer. *Int. J. Exp. Pathol.* 81, 117–143.

Blood Vessels and Endothelial Cells

Folkman J & Haudenschild C (1980) Angiogenesis in vitro. *Nature* 288, 551–556.

Folkman J (1996) Fighting cancer by attacking its blood supply. *Sci. Am.* 275(3), 150–154.

Risau W (1997) Mechanisms of angiogenesis. *Nature* 386, 671–674.

Semenza GL (1998) Hypoxia-inducible factor 1: master regulator of O_2 homeostasis. *Curr. Opin. Genet. Dev.* 8, 588–594.

Shima DT & Mailhos C (2000) Vascular developmental biology: getting nervous. *Curr. Opin. Genet. Dev.* 10, 536–542.

Yancopoulos GD, Davis S, Gale NW et al. (2000) Vascular-specific growth factors and blood vessel formation. *Nature* 407, 242–248.

Renewal by Pluripotent Stem Cells: Blood Cell Formation

Metcalf D (1980) Clonal analysis of proliferation and differentiation of paired daughter cells: action of granulocyte-macrophage colony stimulating factor on granulocyte-macrophage precursors. *Proc. Natl. Acad. Sci. USA* 77, 5327–5330.

Metcalf D (1999) Stem cells, pre-progenitor cells and lineage-committed cells: are our dogmas correct? *Ann. N.Y. Acad. Sci.* 872, 289–303; discussion 303–304.

Nutt SL, Eberhard D, Horcher M et al. (2001) Pax5 determines the identity of B cells from the beginning to the end of B-lymphopoiesis. *Int. Rev. Immunol.* 20, 65–82.

Orkin SH (2000) Diversification of haematopoietic stem cells to specific lineages. *Nat. Rev. Genet.* 1, 57–64.

Weissman I, Anderson DJ & Gage FH (2001) Stem and progenitor cells: origins, phenotypes, lineage commitments, and transdifferentiations. *Annu. Rev. Cell Dev. Biol.* 17, 387–403.

Wintrobe MM (1980) Blood, Pure and Eloquent. New York: McGraw-Hill.

Genesis, Modulation, and Regeneration of Skeletal Muscle

Andersen JL, Schjerling P & Saltin B (2000) Muscle, genes and athletic performance. *Sci. Am.* 283(3), 48–55.

Lee SJ & McPherron AC (1999) Myostatin and the control of skeletal muscle mass. *Curr. Opin. Genet. Dev.* 9, 604–607.

Naya FS & Olson E (1999) MEF2: a transcriptional target for signaling pathways controlling skeletal muscle growth and differentiation. *Curr. Opin. Cell Biol.* 11, 683–688.

Seale P, Sabourin LA, Girgis-Gabardo A et al. (2000) Pax7 is required for the specification of myogenic satellite cells. *Cell* 102, 777–786.

Weintraub H, Davis R, Tapscott S et al. (1991) The myoD gene family: nodal point during specification of the muscle cell lineage. *Science* 251, 761–766.

Fibroblasts and Their Transformations: the Connective-tissue Cell Family

Ahima RS & Flier JS (2000) Leptin. *Annu. Rev. Physiol.* 62, 413–437.

Ducy P, Schinke T & Karsenty G (2000) The osteoblast: a sophisticated fibroblast under central surveillance. *Science* 289, 1501–1504.

Martin P (1997) Wound healing—aiming for perfect skin regeneration. *Science* 276, 75–81.

Rosen ED & Spiegelman BM (2000) Molecular regulation of adipogenesis. *Annu. Rev. Cell Dev. Biol.* 16, 145–171.

Teitelbaum SL (2000) Bone resorption by osteoclasts. *Science* 289, 1504–1508.

Stem-cell Engineering

Brustle O, Jones KN, Learish RD et al. (1999) Embryonic stem cell-derived glial precursors: a source of myelinating transplants. *Science* 285, 754–756.

Clarke D & Frisén J (2001) Differentiation potential of adult stem cells. *Curr. Opin. Genet. Dev.* 11, 575–580.

Donovan PJ & Gearhart J (2001) The end of the beginning for pluripotent stem cells. *Nature* 414, 92–97.

Lagasse E, Connors H, Al-Dhalimy M et al. (2000) Purified hematopoietic stem cells can differentiate into hepatocytes *in vivo*. *Nat. Med.* 6, 1229–1234.

Schulz JT, 3rd, Tompkins RG & Burke JF (2000) Artificial skin. *Annu. Rev. Med.* 51, 231–244.

Suhonen JO, Peterson DA, Ray J & Gage FH (1996) Differentiation of adult hippocampus-derived progenitors into olfactory neurons in vivo. *Nature* 383, 624–627.

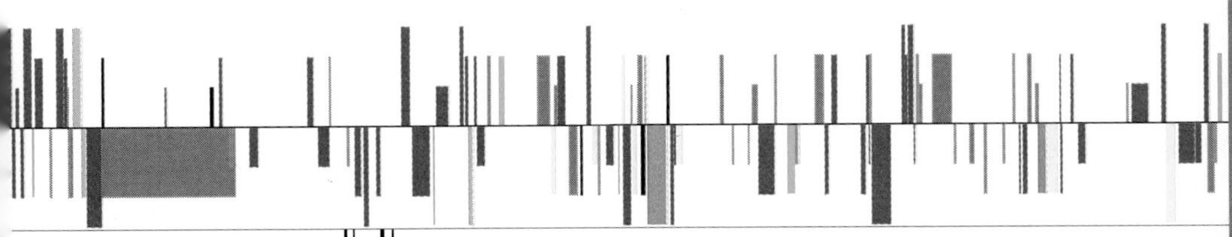

23

CANCER

Cancer cells break the most basic rules of behavior by which multicellular organisms are built and maintained, and they exploit every kind of opportunity to do so. In studying the transgressions, we discover what the normal rules are and how they are enforced. Thus, in the context of cell biology, cancer has a unique importance, and the emphasis given to cancer research has profoundly benefited a much wider area of medical knowledge than that of cancer alone.

We have already discussed many offshoots of cancer research in the preceding chapters. Indeed, the effort to combat cancer has driven many fundamental discoveries in cell biology. Many proteins involved in DNA repair (Chapter 5), cell signaling (Chapter 15), the cell cycle and programmed cell death (Chapter 17), and tissue architecture (Chapter 19) were discovered because abnormalities in their function can lead to the uncontrolled proliferation, genetic mayhem, and other antisocial behavior characteristic of cancer cells.

In this chapter, we first consider more closely what cancer is and examine the natural history of the disease from a cellular standpoint. Next we review the molecular changes that make a cell cancerous. Finally, we discuss how our enhanced understanding of the molecular basis of cancer is leading to improved methods for the prevention and treatment of the disease.

CANCER AS A MICROEVOLUTIONARY PROCESS

The body of an animal operates as a society or ecosystem whose individual members are cells, reproducing by cell division and organized into collaborative assemblies or tissues. In our earlier discussion of the maintenance of tissues (Chapter 22), our interests were similar to those of the ecologist: cell births, deaths, habitats, territorial limitations, and the maintenance of population sizes. The one ecological topic conspicuously absent was that of natural selection: we said nothing of competition or mutation among somatic cells. The reason is that a healthy body is in this respect a very peculiar society, where self-sacrifice—as opposed to survival of the fittest—is the rule. Ultimately, all somatic cell lineages are committed to die: they leave no progeny and instead dedicate their existence to support of the germ cells, which alone have a chance of survival. There is no mystery in this, for the body is a clone, and the genome of the somatic cells is the same as that of the germ cells. By their self-sacrifice for the sake of the germ cells, the somatic cells help to propagate copies of their own genes.

Thus, unlike free-living cells such as bacteria, which compete to survive, the cells of a multicellular organism are committed to collaboration. To coordinate their behavior, the cells send, receive, and interpret an elaborate set of signals that serve as social controls, telling each of them how to act (see Chapter 15). As a result, each cell behaves in socially responsible manner, resting, dividing, differentiating, or dying as needed for the good of the organism. Molecular disturbances that upset this harmony mean trouble for a multicellular society. In a human body with more than 10^{14} cells, billions of cells experience mutations every day, potentially disrupting the social controls. Most dangerously, a mutation may give one cell a selective advantage, allowing it to divide more vigorously than its neighbors and to become a founder of a growing mutant clone. A mutation that gives rise to such selfish behavior by individual members of the cooperative can jeopardize the future of the whole enterprise. Repeated rounds of mutation, competition, and natural selection operating within the population of somatic cells cause matters to go from bad to worse. These are the basic ingredients of cancer: it is a disease in which individual mutant clones of cells begin by prospering at the expense of their neighbors, but in the end destroy the whole cellular society.

In this section we discuss the development of cancer as a microevolutionary process. This process occurs on a time scale of months or years in a population of cells in the body, but it depends on the same principles of mutation and natural selection that govern the long-term evolution of all living organisms.

Cancer Cells Reproduce Without Restraint and Colonize Foreign Tissues

Cancer cells are defined by two heritable properties: they and their progeny (1) reproduce in defiance of the normal restraints on cell division and (2) invade and colonize territories normally reserved for other cells. It is the combination of these actions that makes cancers peculiarly dangerous. An isolated abnormal cell that does not proliferate more than its normal neighbors does no significant damage, no matter what other disagreeable properties it may have; but if its proliferation is out of control, it will give rise to a tumor, or *neoplasm*—a relentlessly growing mass of abnormal cells. As long as the neoplastic cells remain clustered together in a single mass, however, the tumor is said to be **benign**. At this stage, a complete cure can usually be achieved by removing the mass surgically. A tumor is considered a cancer only if it is **malignant**, that is, only if its cells have acquired the ability to invade surrounding tissue. Invasiveness usually implies an ability to break loose, enter the bloodstream or lymphatic vessels, and form secondary tumors, called **metastases**, at other sites in the body (Figure 23–1). The more widely a cancer spreads, the harder it becomes to eradicate.

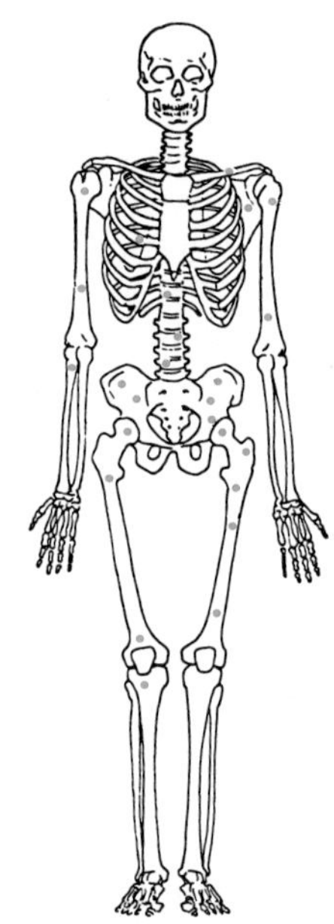

Figure 23–1 Metastasis. Malignant tumors typically give rise to metastases, making the cancer hard to eradicate. The drawing shows common sites in the bone marrow for metastases from carcinoma of the prostate gland. (From Union Internationale Contre le Cancer, TNM Atlas: Illustrated Guide to the Classification of Malignant Tumors, 2nd edn., Berlin: Springer-Verlag, 1986.)

Cancers are classified according to the tissue and cell type from which they arise. Cancers arising from epithelial cells are termed **carcinomas**; those arising from connective tissue or muscle cells are termed **sarcomas**. Cancers that do not fit in either of these two broad categories include the various **leukemias**, derived from hemopoietic cells, and cancers derived from cells of the nervous system. Figure 23–2 shows the types of cancers that are common in the United States, together with their incidence and resulting death rate. Each of the broad categories has many subdivisions according to the specific cell type, the location in the body, and the structure of the tumor; many of the names used are fixed by tradition and have no modern rational basis.

In parallel with the set of names for malignant tumors, there is a related set of names for benign tumors: an *adenoma*, for example, is a benign epithelial tumor with a glandular organization, the corresponding type of malignant tumor being an *adenocarcinoma* (Figure 23–3); similarly a *chondroma* and a *chondrosarcoma* are, respectively, benign and malignant tumors of cartilage. About 90% of human cancers are carcinomas, perhaps because most of the cell

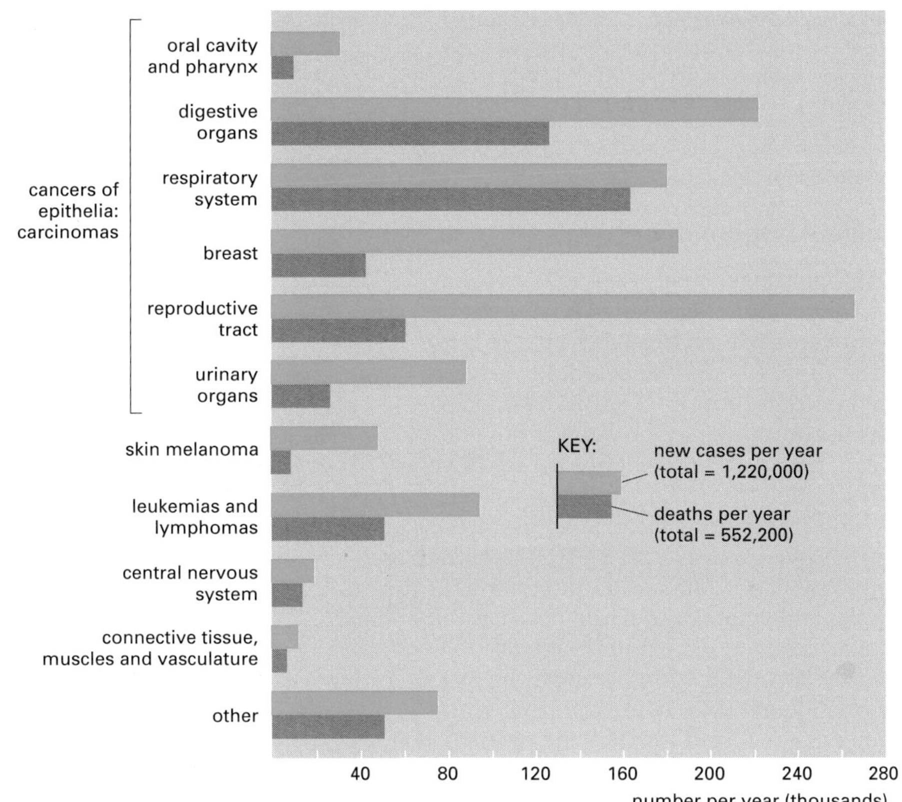

Figure 23–2 Cancer incidence and mortality in the United States. These figures are for the year 2000. Total new cases diagnosed in that year in the United States were 1,220,000, and total cancer deaths were 552,200. Note that only about half the number of people who develop cancer die of it.

In the world as a whole, the five most common cancers are those of the lung, stomach, breast, colon/rectum, and uterine cervix, and the total number of new cancer cases recorded per year is just over 6 million.

Skin cancers other than melanomas are not included in figures, since almost all are cured easily and many go unrecorded. (Data from American Cancer Society, Cancer Facts and Figures, 2000.)

proliferation in the body occurs in epithelia, or because epithelial tissues are most frequently exposed to the various forms of physical and chemical damage that favor the development of cancer.

Each cancer has characteristics that reflect its origin. Thus, for example, the cells of an epidermal *basal-cell carcinoma*, derived from a keratinocyte stem cell in the skin, will generally continue to synthesize cytokeratin intermediate filaments, whereas the cells of a *melanoma*, derived from a pigment cell in the skin, will often (but not always) continue to make pigment granules. Cancers originating from different cell types are, in general, very different diseases. The basal-cell carcinoma, for example, is only locally invasive and rarely forms metastases, whereas the melanoma, if not removed promptly, is much more malignant and rapidly gives rise to many metastases (behavior that recalls the migratory tendencies of the normal pigment-cell precursors during development, discussed in Chapter 21). The basal-cell carcinoma is usually easy to remove by surgery, leading to complete cure; but the malignant melanoma, once it has metastasized widely, is often impossible to eliminate and consequently fatal.

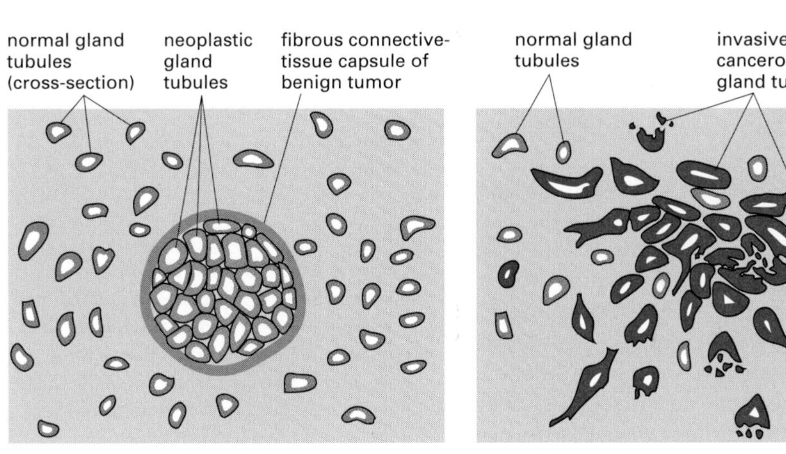

Figure 23–3 Benign versus malignant tumors. A benign glandular tumor (an adenoma) and a malignant glandular tumor (an adenocarcinoma) appear structurally distinct. There are many forms that such tumors may take; the diagram illustrates types that might be found in the breast.

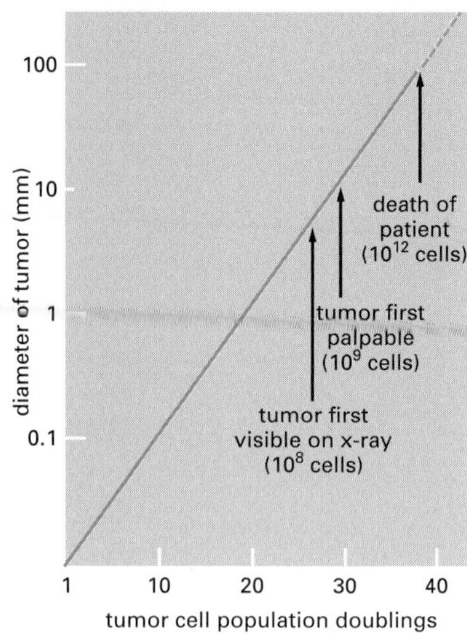

Figure 23–4 The growth of a typical human tumor such as a tumor of the breast. The diameter of the tumor is plotted on a logarithmic scale. Years may elapse before the tumor becomes noticeable. The doubling time of a typical breast tumor, for example, is about 100 days.

Most Cancers Derive From a Single Abnormal Cell

Even when a cancer has metastasized, its origins can usually be traced to a single **primary tumor**, arising in an identified organ and presumed to be derived by cell division from a single cell that has undergone some heritable change that enables it to outgrow its neighbors. By the time it is first detected, however, a typical tumor already contains about a billion cells or more (Figure 23–4), often including many normal cells—fibroblasts, for example, in the supporting connective tissue that is associated with a carcinoma. What evidence do we have that the cancer cells are indeed a clone descended from a single abnormal cell?

One clear demonstration of clonal evolution comes from analysis of the chromosomes in tumor cells. Chromosomal aberrations and rearrangements are present in the cells of most common cancers. In almost all patients with *chronic myelogenous leukemia*, for example, leukemic white blood cells can be distinguished from normal cells by a specific chromosomal abnormality: the so-called Philadelphia chromosome, created by a translocation between the long arms of chromosomes 9 and 22, as shown in Figure 23–5. When the DNA at the site of translocation is cloned and sequenced, it is found that the site of breakage and rejoining of the translocated fragments is identical in all the leukemic cells in any given patient, but differs slightly (by a few hundred or thousand base pairs) from one patient to another, as expected if each case of the leukemia arises from a unique accident occurring in a single cell. We will see later how this Philadelphia translocation leads to leukemia by inappropriately activating a specific gene.

Many other lines of evidence, from a variety of cancers, point to the same conclusion: most cancers originate from a single aberrant cell (Figure 23–6).

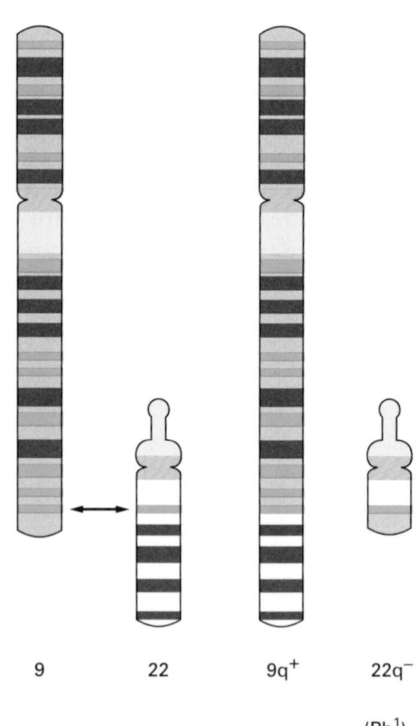

Figure 23–5 The translocation between chromosomes 9 and 22 responsible for chronic myelogenous leukemia. The smaller of the two resulting abnormal chromosomes is called the Philadelphia chromosome, after the city where the abnormality was first recorded.

Cancers Result From Somatic Mutation

If a single abnormal cell is to give rise to a tumor, it must pass on its abnormality to its progeny: the aberration has to be heritable. A first problem in understanding a cancer is to discover whether the heritable aberration is due to a genetic change—that is, an alteration in the cell's DNA sequence—or to an *epigenetic* change—that is, a change in the pattern of gene expression without a change in the DNA sequence. Heritable epigenetic changes, reflecting cell memory, occur during normal development, as manifest in the stability of the differentiated state and in such phenomena as X-chromosome inactivation and imprinting (discussed in Chapter 7). Such epigenetic changes have also been found to play a part in the development of some cancers.

There are, however, good reasons to think that the vast majority of cancers are initiated by genetic changes. First, cells of a variety of cancers can be shown to have a shared abnormality in their DNA sequence that distinguishes them from the normal cells surrounding the tumor, as in the example of chronic myelogenous leukemia that we have just described.

Second, many of the agents known to give rise to cancer also cause genetic changes. Thus **carcinogenesis** (the generation of cancer) appears to be linked with *mutagenesis* (the production of a change in the DNA sequence). This correlation is clear for three classes of agents: **chemical carcinogens** (which typically cause simple local changes in the nucleotide sequence), ionizing radiations such as x-rays (which typically cause chromosome breaks and translocations), and viruses (which introduce foreign DNA into the cell). We will discuss each of these agents in detail later, in the section on the preventable causes of cancer.

Finally, the conclusion that somatic mutations underlie cancer is supported by studies of people who inherit a strong susceptibility to the disease. In a significant proportion of cases, the propensity to cancer can be traced to a genetic defect of some sort in the DNA repair mechanisms of these individuals, which allows them to accumulate mutations at an elevated rate. People with the disease *xeroderma pigmentosum*, for example, have defects in the cellular system that repairs DNA damage induced by UV light, and they experience a hugely increased incidence of skin cancers.

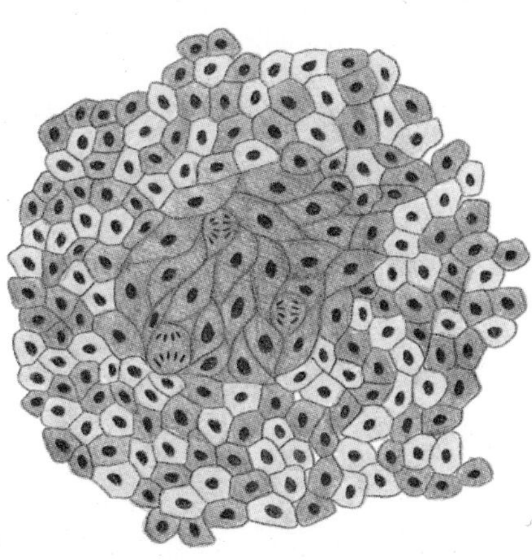

Figure 23–6 Evidence from X-inactivation mosaics demonstrates the monoclonal origin of cancers. As a result of a random process that occurs in the early embryo, practically every normal tissue in a woman's body is a mixture of cells with different X chromosomes heritably inactivated (indicated here by the mixture of *red* cells and *gray* cells in the normal tissue). When the cells of a cancer are tested for their expression of an X-linked marker gene, however, they are usually all found to have the same X chromosome inactivated. This implies that they are all derived from a single cancerous founder cell.

A Single Mutation Is Not Enough to Cause Cancer

An estimated 10^{16} cell divisions take place in a normal human body in the course of a lifetime; in a mouse, with its smaller number of cells and its shorter life span, the number is about 10^{12}. Even in an environment that is free of mutagens, mutations will occur spontaneously at an estimated rate of about 10^{-6} mutations per gene per cell division—a value set by fundamental limitations on the accuracy of DNA replication and repair. Thus, in a lifetime, every single gene is likely to have undergone mutation on about 10^{10} separate occasions in any individual human being, or about 10^6 occasions in a mouse. Among the resulting mutant cells one might expect that there would be many that have disturbances in genes that regulate cell division and that consequently disobey the normal restrictions on cell proliferation. From this point of view, the problem of cancer seems to be not why it occurs but why it occurs so infrequently.

Clearly, if a single mutation were enough to convert a typical healthy cell into a cancer cell that proliferates without restraint, we would not be viable organisms. Many types of evidence indicate that the genesis of a cancer typically requires that several independent, rare accidents occur in the lineage of one cell. One such indication comes from epidemiological studies of the incidence of cancer as a function of age (Figure 23–7). If a single mutation were responsible, occurring with a fixed probability per year, the chance of developing cancer in any given year should be independent of age. In fact, for most types of cancer the incidence rises steeply with age—as would be expected if cancer is caused by a slow accumulation of numerous random mutations in a single line of cells. (An additional reason for the increased incidence of cancer in old age is discussed later, when we come to the topic of replicative cell senescence.)

Now that many of the specific mutations responsible for the development of cancer have been identified, we can test directly for their presence in a particular case of the disease. Such tests have revealed that an individual malignant cell generally harbors multiple mutations. Animal models also confirm that a single one of these genetic alterations is insufficient to cause cancer: when genetically engineered in a mouse, a single such mutation typically produces mild abnormalities in tissue growth, followed occasionally by the formation of randomly scattered benign tumors; but the vast majority of cells in the mutant animal remain non-cancerous.

The concept that the development of a cancer requires mutations in many genes—perhaps ten or more—fits with a large body of information, dating back over many years, concerning the phenomenon of **tumor progression**, whereby an initial mild disorder of cell behavior evolves gradually into a full-blown cancer. As we explain next, these observations of how tumors develop also provide insight into the nature of the changes that must occur for a normal cell to become a cancer cell.

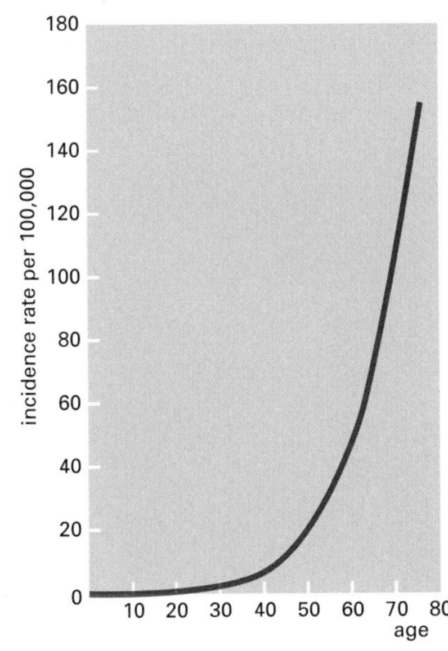

Figure 23–7 Cancer incidence as a function of age. The number of newly diagnosed cases of colon cancer in women in England and Wales in one year is plotted as a function of age at diagnosis and expressed relative to the total number of individuals in each age group. The incidence of cancer rises steeply as a function of age. If only a single mutation were required to trigger the cancer and this mutation had an equal chance of occurring at any time, the incidence would be independent of age. (Data from C. Muir et al., Cancer Incidence in Five Continents, Vol. V. Lyon: International Agency for Research on Cancer, 1987.)

Cancers Develop in Slow Stages From Mildly Aberrant Cells

For those cancers that have a discernible external cause, the disease does not usually become apparent until long after exposure to the causal agent: the incidence of lung cancer does not begin to rise steeply until after 10 or 20 years of heavy smoking; the incidence of leukemias in Hiroshima and Nagasaki did not show a marked rise until about 5 years after the explosion of the atomic bombs; industrial workers exposed for a limited period to chemical carcinogens do not usually develop the cancers characteristic of their occupation until 10, 20, or even more years after the exposure (Figure 23–8). During this long incubation period, the prospective cancer cells undergo a succession of changes. The same applies to cancers where the initial genetic lesion has no such obvious external cause.

Chronic myelogenous leukemia, mentioned earlier, provides a clear and simple example. This disease begins as a disorder characterized by a nonlethal overproduction of white blood cells and continues as such for several years before changing into a much more rapidly progressing illness that usually ends in death within a few months. In the chronic early phase, the leukemic cells in the body are distinguished mainly by their possession of the chromosomal translocation mentioned previously (although there may well be other genetic changes that are not seen so easily). In the subsequent acute phase of the illness, the hemopoietic system is overrun by cells that show not only this chromosomal abnormality but also several others. It appears as though members of the initial mutant clone have undergone further mutations that make them proliferate more rapidly (or divide more times before they die or terminally differentiate), so that they come to outnumber both the normal hemopoietic cells and their relatives that have only the primary chromosomal translocation (the Philadelphia chromosome).

Carcinomas and other solid tumors are thought to evolve in a similar way. Although most such cancers in humans are not diagnosed until a relatively late stage, in a few cases it is possible to observe the early steps in the development of the disease. We shall discuss one example—*colorectal cancer*—toward the end of this chapter. Another example is provided by cancers of the *uterine cervix* (the neck of the womb). These cancers are thought to derive from the epithelium near the opening of the cervix.

This epithelium undergoes physiological changes in structure at different times in a woman's reproductive life. In a cervical region liable to give rise to cancer, under the conditions in which the disease originates, the cells are initially organized as a stratified (multilayered) squamous epithelium (Figure 23–9A,E), similar in structure to the epidermis of the skin or the lining of the inside of the mouth (see p. 1274). In such stratified epithelia, proliferation normally occurs only in the basal layer, generating cells that then move out toward the surface; these cells differentiate as they move, forming flattened, keratin-rich, nondividing cells that are sloughed off as they reach the surface. When specimens of cervical epithelium from different women are examined, however, it is not unusual to find patches in which this organization is disturbed in a way

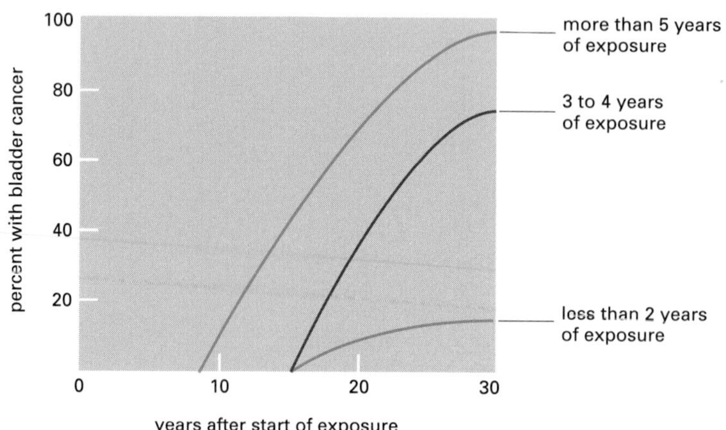

Figure 23–8 Delayed onset of cancer following exposure to a carcinogen. The graph shows the length of the delay before onset of bladder cancer in a set of 78 workers in the chemical industry who had been exposed to the carcinogen 2-naphthylamine, grouped according to the duration of their exposure. (Modified from J. Cairns, Cancer: Science and Society. San Francisco: W.H. Freeman, 1978. After M.H.C. Williams, in Cancer, Vol. III [R.W. Raven, ed.]. London: Butterworth Heinemann, 1958.)

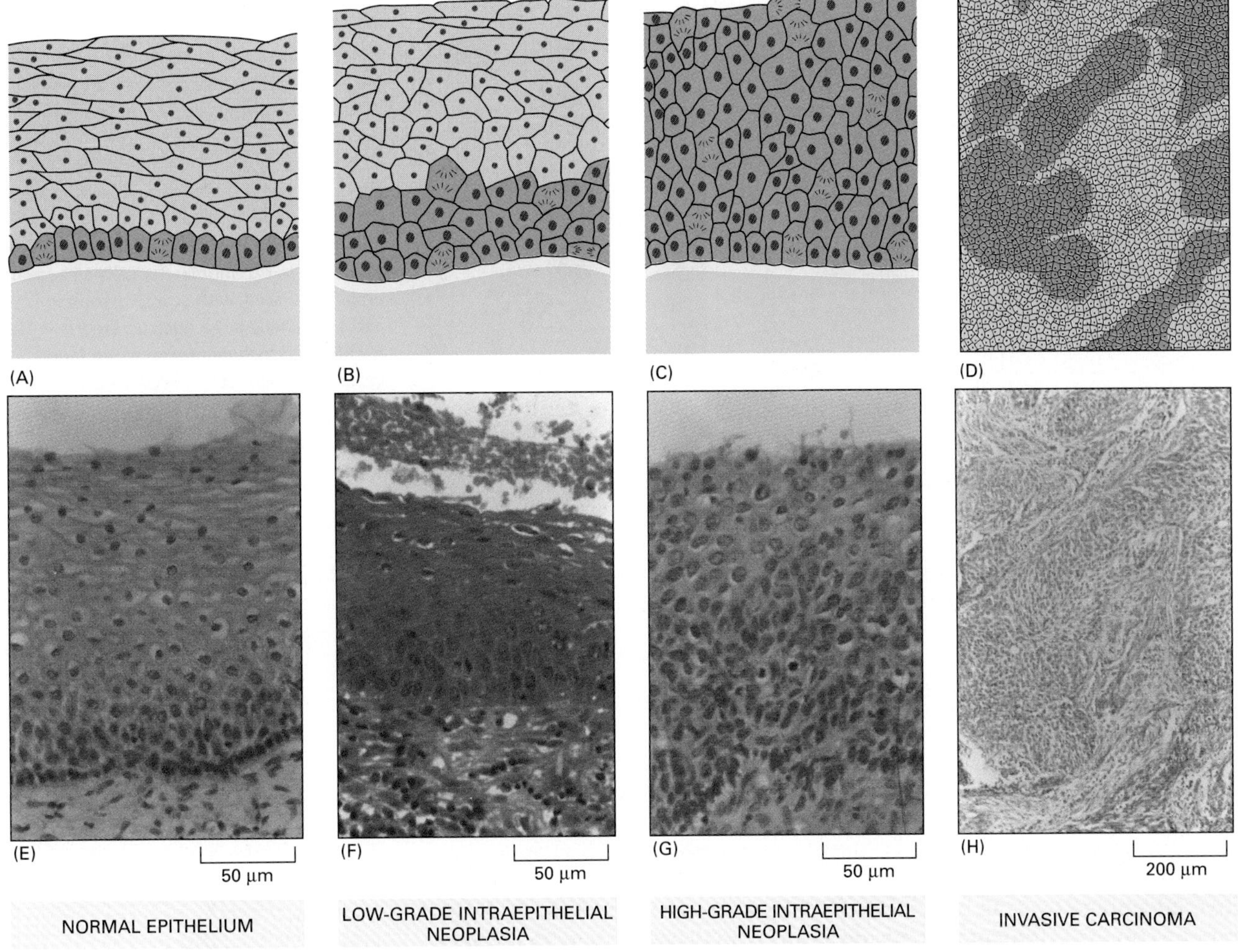

Figure 23–9 The stages of progression in the development of cancer of the epithelium of the uterine cervix. Pathologists use standardized terminology to classify the types of disorder they see, so as to guide the choice of treatment. Schematic diagrams are shown in A–D; sections of cervical epithelium corresponding to these changes are shown in E–H. (A, E) In the normal stratified squamous epithelium, dividing cells are confined to the basal layer. (B, F) In low-grade intraepithelial neoplasia, dividing cells can be found throughout the lower third of the epithelium; the superficial cells are still flattened and show signs of differentiation, but this is incomplete. (C, G) In high-grade intraepithelial neoplasia, cells in all the epithelial layers are proliferating and show no sign of differentiation. (D, H) True malignancy begins when the cells move through or destroy the basal lamina and invade the underlying connective tissue. (Photographs courtesy of Margaret Stanley.)

that suggests the beginnings of a cancerous transformation. Pathologists describe these changes as *intraepithelial neoplasia,* and classify them as low-grade (mild) or high-grade (moderate to severe).

In the low-grade lesions, dividing cells are no longer confined to the basal layer but occupy the lower third of the epithelium; although differentiation proceeds in the upper epithelial layers, it is slightly disordered (Figure 23–9B,F). Left alone, most of these mild lesions will spontaneously regress, but a small number (about 10%) may progress to become high-grade lesions. In these more seriously abnormal patches, most or all of the epithelial layers are occupied by undifferentiated dividing cells, which are usually highly variable in cell and nuclear size and shape. Abnormal mitotic figures are frequently seen and the karyotype is usually abnormal, but the abnormal cells are still confined to the epithelial side of the basal lamina (Figure 23–9C,G). The presence of such lesions can be detected by scraping off a sample of cells from the surface of the cervix and viewing it under the microscope (the "Pap smear" technique—Figure 23–10). At this stage, it is still easy to achieve a complete cure by destroying or removing the abnormal tissue surgically.

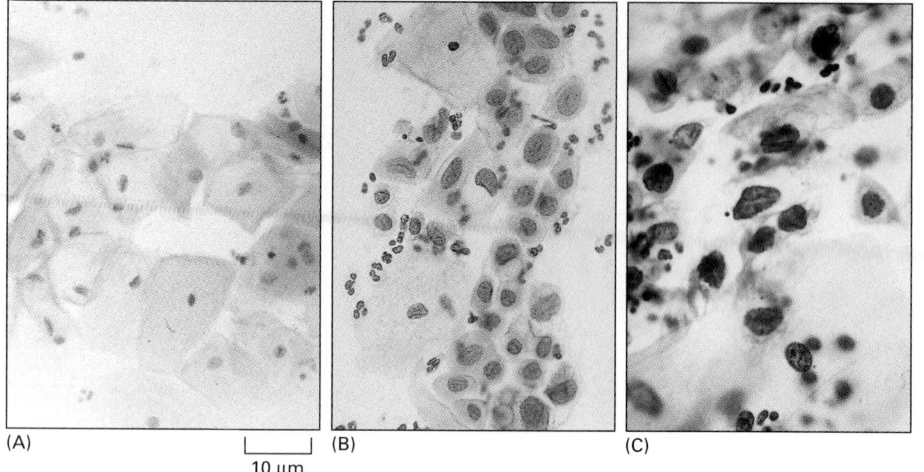

Figure 23–10 Photographs of cells collected by scraping the surface of the uterine cervix (the Papanicolaou or "Pap smear" technique). (A) Normal: the cells are large and well differentiated, with highly condensed nuclei. (B) Precancerous lesion: differentiation and proliferation are abnormal but the lesion is not yet invasive; the cells are in various stages of differentiation, some quite immature. (C) Invasive carcinoma: the cells all appear undifferentiated, with scanty cytoplasm and a relatively large nucleus. Debris in the background includes some white blood cells. (Courtesy of Winifred Gray.)

Without treatment, the abnormal tissue may simply persist and progress no further or may even regress spontaneously; but in at least 30–40% of cases, progression will occur, giving rise, over a period of several years, to a frank invasive carcinoma (Figure 23–9D,H)—a malignant lesion where cells cross or destroy the basal lamina, invade the underlying tissue, and metastasize via the lymphatic vessels. Surgical cure becomes progressively more difficult as the invasive growth spreads.

Tumor Progression Involves Successive Rounds of Mutation and Natural Selection

As we have seen, cancers in general seem to arise by a process in which an initial population of slightly abnormal cells, descendants of a single mutant ancestor, evolves from bad to worse through successive cycles of mutation and natural selection. At each stage, one cell acquires an additional mutation that gives it a selective advantage over its neighbors, making it better able to thrive in its environment—an environment that, inside a tumor, may be harsh, with low levels of oxygen, scarce nutrients, and the natural barriers to growth presented by the surrounding normal tissues. The offspring of this well-adapted cell will continue to divide, eventually taking over the tumor and becoming the dominant clone in the developing lesion (Figure 23–11). Thus, tumors grow in fits and starts, as additional advantageous mutations arise and the cells bearing them flourish. Their evolution involves a large element of chance and usually takes many years; most of us die of other ailments before cancer has had time to develop.

Why are so many mutations needed? One reason is that cellular processes are controlled in complex and interconnected ways; cells employ redundant regulatory mechanisms to help them maintain tight and precise control over their behavior. Thus, many different regulatory systems have to be disrupted before a cell can throw off its normal restraints and behave defiantly as a malignant cancer cell. In addition, tumor cells may meet new barriers to further expansion at each stage of the evolutionary process. For example, oxygen and nutrients do not become limiting until a tumor is one or two millimeters in diameter, at which point the cells in the tumor interior may not have adequate access to such necessary resources. Each new barrier, whether physical or physiological, must be overcome by the acquisition of additional mutations.

In general, the rate of evolution in any population would be expected to depend on four main parameters: (1) the *mutation rate*, that is, the probability per gene per unit time that any given member of the population will undergo genetic change; (2) the *number of individuals* in the population; (3) the *rate of reproduction*, that is, the average number of generations of progeny produced per unit time; and (4) the *selective advantage* enjoyed by successful mutant individuals, that is, the ratio of the number of surviving fertile progeny they produce per unit time to the number of surviving fertile progeny produced by nonmutant individuals. These are the critical factors for the evolution of cancer cells in a

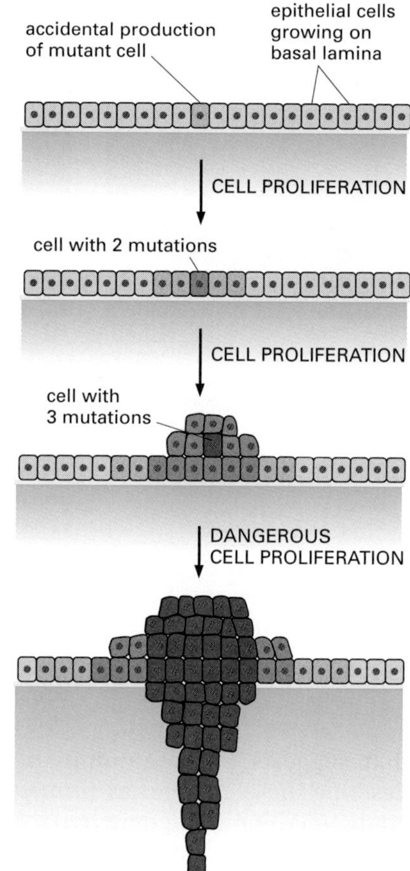

Figure 23–11 Clonal evolution. A tumor develops through repeated rounds of mutation and proliferation, giving rise eventually to a clone of fully malignant cancer cells. At each step, a single cell undergoes a mutation that enhances cell proliferation, so that its progeny become the dominant clone in the tumor. Proliferation of this clone then hastens occurrence of the next step of tumor progression by increasing the size of the cell population at risk of undergoing an additional mutation.

multicellular organism, just as they are for the evolution of organisms on the surface of the Earth.

Clearly the rate of progression toward cancer depends on many things beside the changing genotype of the individual cancer cell. Equally, it is plain that there are a number of quite disparate genetic properties that might help a cancer cell to be evolutionarily successful. In later sections, we shall examine the molecular changes that confer these properties. But first it is helpful to consider in general terms what the key properties actually are: what special capabilities are common to the majority of cancer cells and responsible for their bad behavior?

Most Human Cancer Cells Are Genetically Unstable

The great majority of human cancers show signs of a dramatically enhanced mutation rate: the cells are said to be **genetically unstable**. This instability can take various forms. Some cancer cells are defective in the ability to repair local DNA damage or to correct replication errors that affect individual nucleotides. These cells tend to accumulate more point mutations and small, localized DNA sequence changes than do normal cells. Other cancer cells have trouble maintaining the integrity of their chromosomes and thus display gross abnormalities in their karyotype (Figure 23–12). From an evolutionary perspective this is not a surprise: one might expect that genetic instability would promote the formation of cancer, as it increases the probability that cells will experience a mutation that will lead toward malignancy. In fact, it seems that some degree of genetic instability may be essential for the development of cancer; at the very least it appears to make a powerful contribution to cancer progression.

Different tumors—even from the same tissues—can show different kinds of genetic instability, caused by mutations in one or another of a very specific set of genes whose products are needed to protect the genome from alteration. As mentioned earlier, people who inherit mutations in these genes are found to have a raised incidence of cancer. Although such cancer-prone conditions are relatively rare, they include examples of mutations in practically every type of gene known to be required for genetic stability, confirming that loss of this stability can have a causative role in cancer no matter how it occurs.

Most often, the destabilizing mutations are not inherited, but arise *de novo* as a tumor develops, helping the cancer cell to accumulate mutations much more rapidly than its neighbors do. It is important, however, to note that genetic instability does not, in itself, give a cell a selective advantage. On the contrary, genetic instability actually damages a cell's fitness, as most random mutations are harmful. Thus, a genetically unstable cell will not be favored by selection, unless it has additional properties or manages to accrue additional mutations that confer some competitive advantage. It seems that some "optimum" level of genetic instability exists for the development of cancer, making

(A)

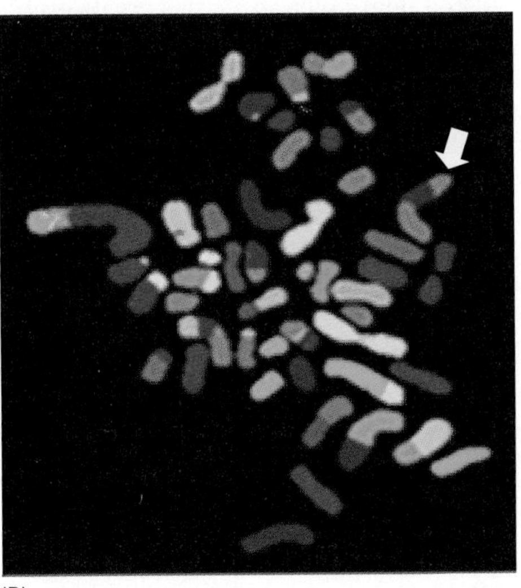

(B)

Figure 23–12 Chromosomes from a breast tumor displaying abnormalities in structure and number. Chromosomes were prepared from a breast tumor cell in metaphase, spread on a glass slide, and stained with (A) a general DNA stain or (B) a combination of fluorescent stains that give a different color for each normal human chromosome. The staining (displayed in false color) shows multiple translocations, including a doubly translocated chromosome *(white arrow)* made up of two pieces of chromosome 8 *(green)* and a piece of chromosome 17 *(purple)*. The karyotype also contains 48 chromosomes, instead of the normal 46. (Courtesy of Joanne Davidson and Paul Edwards.)

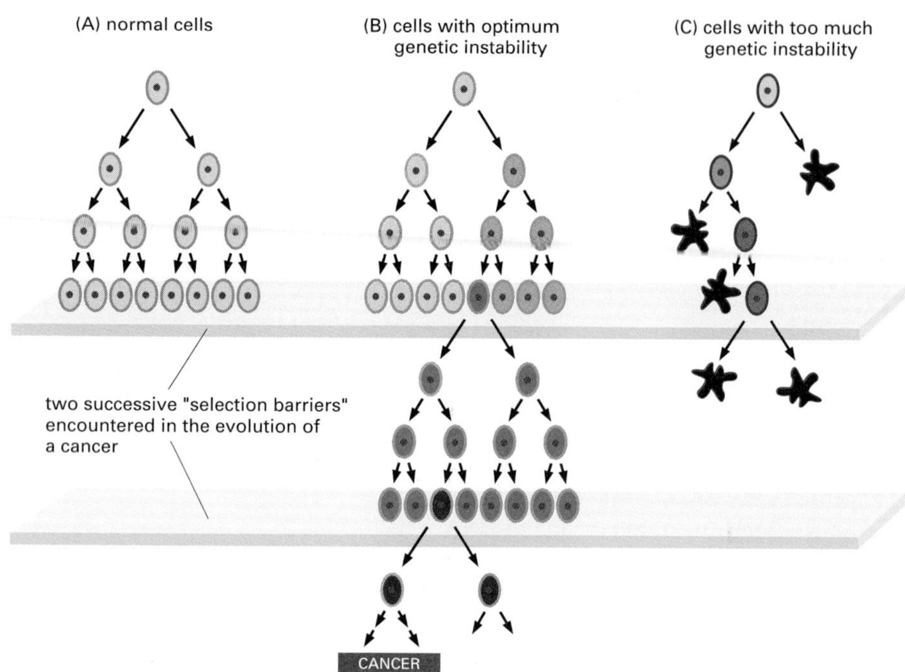

Figure 23–13 Genetic instability and tumor progression. Cells that maintain an optimum level of genetic instability may be the most successful in the race to form a tumor. (A) In normal cells, the intrinsic amount of genetic instability is low. Thus when normal cells hit a selection barrier—low levels of oxygen or a scarcity of proliferation signals, for example—they are very unlikely to be mutable enough to produce a cell that continues proliferating. (B) In tumor cell precursors, an increased level of genetic instability makes it likely that at least one cell will contain the requisite genetic alteration to pass the selection barrier and continue the process of tumor progression. This genetic instability is retained in the lineage, and it can be measured in the resulting tumor. (C) If the level of genetic instability is too high, many of the cells suffer deleterious mutations and either proliferate much more slowly than their neighbors or are eliminated by cell death. This excessive mutability can lead to extinction of the cell lineage. (Adapted from D.P. Cahill et al., *Trends Cell Biol.* 9:M57–M60, 1999.)

a cell mutable enough to evolve dangerously, but not so mutable that it dies (Figure 23–13).

Cancerous Growth Often Depends on Defective Control of Cell Death or Cell Differentiation

Just as an increased mutation rate can raise the probability of cancer progression, so can any circumstance that increases the number of cells available for mutating. The bigger the clone of mutant cells resulting from an early mutation, the greater the chance that an additional mutation will allow the cancer to progress, until its growth is completely out of control and malignant. Thus, at every stage in the development of cancer, mutations that help cells to increase in number are critical.

How can a mutation have this effect? The most obvious way is to increase the rate of cell division. Indeed, mutations that tend to make cells blind to the normal restraints on cell division are a common feature of cancer, as we shall discuss in detail later. It has become increasingly clear, however, that such mutations are not the only—or necessarily the most important—mechanism for increasing cell number. In normal adult tissues, especially those at risk of cancer, cells may proliferate continually; but their numbers remain steady because cell production is balanced by cell loss. Programmed cell death by apoptosis very often has an essential role in this balance, as we saw in Chapters 17 and 22. If too many cells are generated, the rate of apoptosis increases to dispose of the surplus. One of the most important properties of cancer cells, therefore, is that they fail to commit suicide when a normal cell would honorably do so.

Mutations can also increase the size of a clone of mutant cells by altering their ability to differentiate, as illustrated by the situation in the uterine cervix, discussed earlier. Like the epidermis of the skin and many other epithelia, the epithelium of the uterine cervix normally renews itself continually by shedding terminally differentiated cells from its outer surface and generating replacements from stem cells in the basal layer. On average, each normal stem cell division generates one daughter stem cell and one cell that is condemned to terminal differentiation and a cessation of cell division. If the stem cell divides more rapidly, terminally differentiated cells will be produced and shed more rapidly, but a balance of genesis and destruction will still be maintained. Thus, for an abnormal stem cell to generate a steadily growing clone of mutant progeny, the basic rules must be upset: either more than 50% of the daughter cells must remain as stem cells or the process of differentiation must be deranged so that daughter cells that embark on this route somehow retain an ability to carry on dividing indefinitely and avoid dying or being discarded at the end of the production line (Figure 23–14).

Presumably, the development of such properties underlies the progression from low-grade intraepithelial neoplasia of the uterine cervix to high-grade intraepithelial neoplasia and malignant cancer (see Figure 23–9). Similar considerations apply to the development of cancer in other tissues that rely on stem cells, such as the skin, the lining of the gut, and the hemopoietic system. Several forms of leukemia, for example, seem to arise from a disruption of the normal program of differentiation, such that a committed progenitor of a particular type of blood cell continues to divide indefinitely, instead of differentiating terminally in the normal way and dying after a strictly limited number of division cycles (as discussed in Chapter 22).

In conclusion, changes that block the normal maturation of cells toward a nondividing, terminally differentiated state or prevent normal programmed cell death play an essential part in many cancers.

Many Cancer Cells Escape a Built-in Limit to Cell Proliferation

Many normal cells cease to divide when they mature into terminally differentiated, specialized cells. Differentiation is not, however, the only condition that can arrest cell proliferation. Cells also cease to divide when they are stressed or when they detect damage to their DNA. Further, most normal human cells, when removed from a tissue and tested in culture for their ability to proliferate, appear to show a set limit on the number of times they can divide. Once cells progress through a certain number of population doublings—25 to 50 for human fibroblasts, for example—they simply stop proliferating, a process termed **replicative cell senescence** (discussed in Chapter 17). It is possible that, as a person ages, some cells may reach this limit to their proliferation— particularly in tissues requiring constant renewal, such as the epidermis or the lining of the gut. If a cell in such circumstances is to persist in dividing so as to generate a cancer, it must escape the restraints of replicative senescence. In fact, cancer cells, when tested in culture, often behave as "immortalized": they continue dividing indefinitely, unlike their normal counterparts.

While these facts are well established, it is not yet clear precisely how they relate to the cancer disease process. Despite the findings in culture, there is remarkably little solid information about the importance of replicative senescence

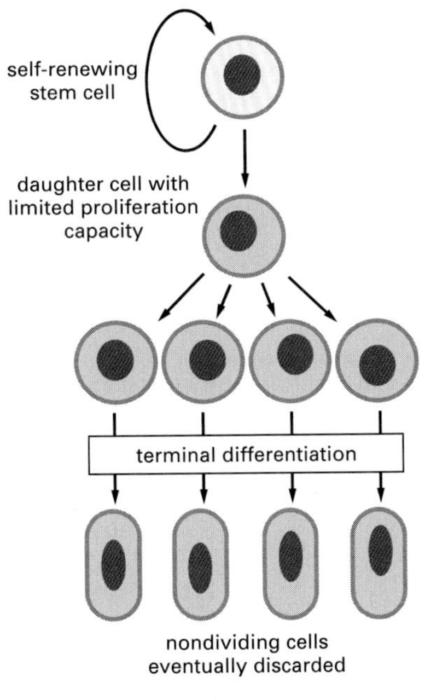

(A) NORMAL PATHWAY

tumor

(B) STEM CELL FAILS TO PRODUCE ONE NON-STEM-CELL DAUGHTER IN EACH DIVISION AND THEREBY PROLIFERATES TO FORM A TUMOR

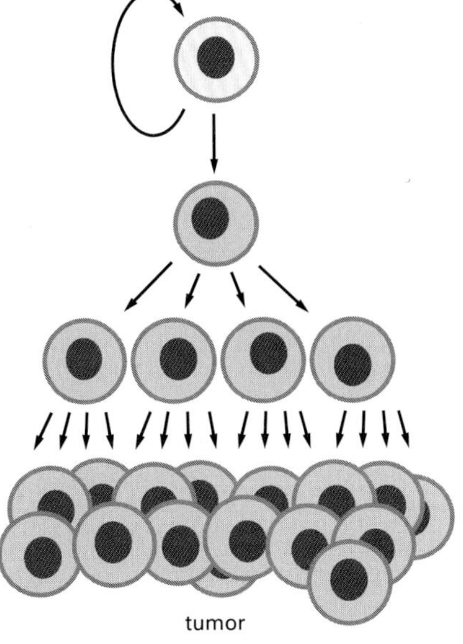

(C) DAUGHTER CELLS FAIL TO DIFFERENTIATE NORMALLY AND INSTEAD PROLIFERATE TO FORM A TUMOR

Figure 23–14 Normal and deranged control of cell production from stem cells. (A) The normal strategy for producing new differentiated cells. (B and C) Two types of derangement that can give rise to the unbridled proliferation characteristic of cancer. Note that an excessive cell-division rate for the stem cells in (A) will not by itself have this effect, as long as each cell division produces only one daughter that is a stem cell.

in normal tissues of intact human beings. According to one view, replicative senescence is a useful mechanism to protect us against cancer—an added barrier that cancer cells have to break through. According to another school of thought, cells in most tissues never actually reach replicative senescence in the course of a human lifetime, and the "immortality" of cancer cells is merely a side-effect of selection for some other property they need to have. Some recent studies, discussed later in the chapter, suggest a third view, radically different from both of the other two. The proposal is that many cells in normal tissues undergo replicative cell senescence and slow down or halt their proliferation as a person ages, and that this creates circumstances in which cancer cells can thrive all the better by continuing to divide at full throttle. On this view, replicative cell senescence, far from protecting us from the growth of tumors, creates a breeding ground for mutant cells that evade the normal controls and overrun the tissue because they face no competition from their senescent normal neighbors. The mutations that allow continued proliferation may confer at the same time other cancerous traits, such as genetic instability or a general disregard for cell-cycle controls, leading on to progressively more disordered behavior. In this way, replicative cell senescence in a self-renewing tissue might be expected to favor the genesis of cancer; it could be a part of the reason why cancer is predominantly a disease of old age.

Studies at a molecular level may soon clarify the picture. Later in the chapter, we see that replicative cell senescence in human cells is related to the shortening of telomeres—the repetitive DNA sequences and associated proteins that cap the ends of each chromosome. And we discuss how some cancers may arise from the dangerous ways in which certain mutations can allow cells to escape the block to cell division that normal cells encounter when their telomeres get too short.

To Metastasize, Malignant Cancer Cells Must Survive and Proliferate in an Alien Environment

Metastasis is the most feared—and least understood—aspect of cancer. By spreading throughout the body a cancer becomes almost impossible to eradicate surgically or by localized irradiation, and thus deadly. Metastasis is itself a multistep process: cells have to break away from the primary tumor, invade local tissues and vessels, and establish new cellular colonies at distant sites. Each of these events is, in itself, fairly complex and the molecular mechanisms involved are not yet clear.

For a cancer cell to metastasize, it must first detach from the parent tumor; key to this escape is an ability to invade neighboring tissues. Such invasiveness is the defining property of malignant tumors, which show a disorganized pattern of growth and ragged borders, with extensions into the surrounding tissue (see, for example, Figure 23–3). Although invasiveness is not thoroughly understood, it almost certainly requires a disruption of the adhesive mechanisms that normally keep cells tethered to their proper neighbors.

The next step in the process—escape from the neighborhood of the primary tumor and establishment of colonies in distant tissues—is a complex, slow, and inefficient operation; few cells in the primary tumor achieve it. To metastasize successfully, a cell must penetrate a blood vessel or a lymphatic vessel, cross the basal lamina and the endothelial lining of the vessel so as to enter the circulation, exit from the circulation elsewhere in the body, and survive and proliferate in the new environment in which it finds itself (Figure 23–15). Many cancers may invade local connective tissues but do not succeed in forming metastases before they are discovered and surgically removed. Only a tiny proportion of malignant cells manage to escape into the circulation; and experiments show that only a tiny proportion of these, less than one in thousands, perhaps one in millions, survive to form metastases. What seems to distinguish cells that are able to establish metastases is their ability to survive and grow after settling in an alien site.

It is thought that normal cells depend on molecular survival signals that are peculiar to their normal environment; when deprived of these signals, they

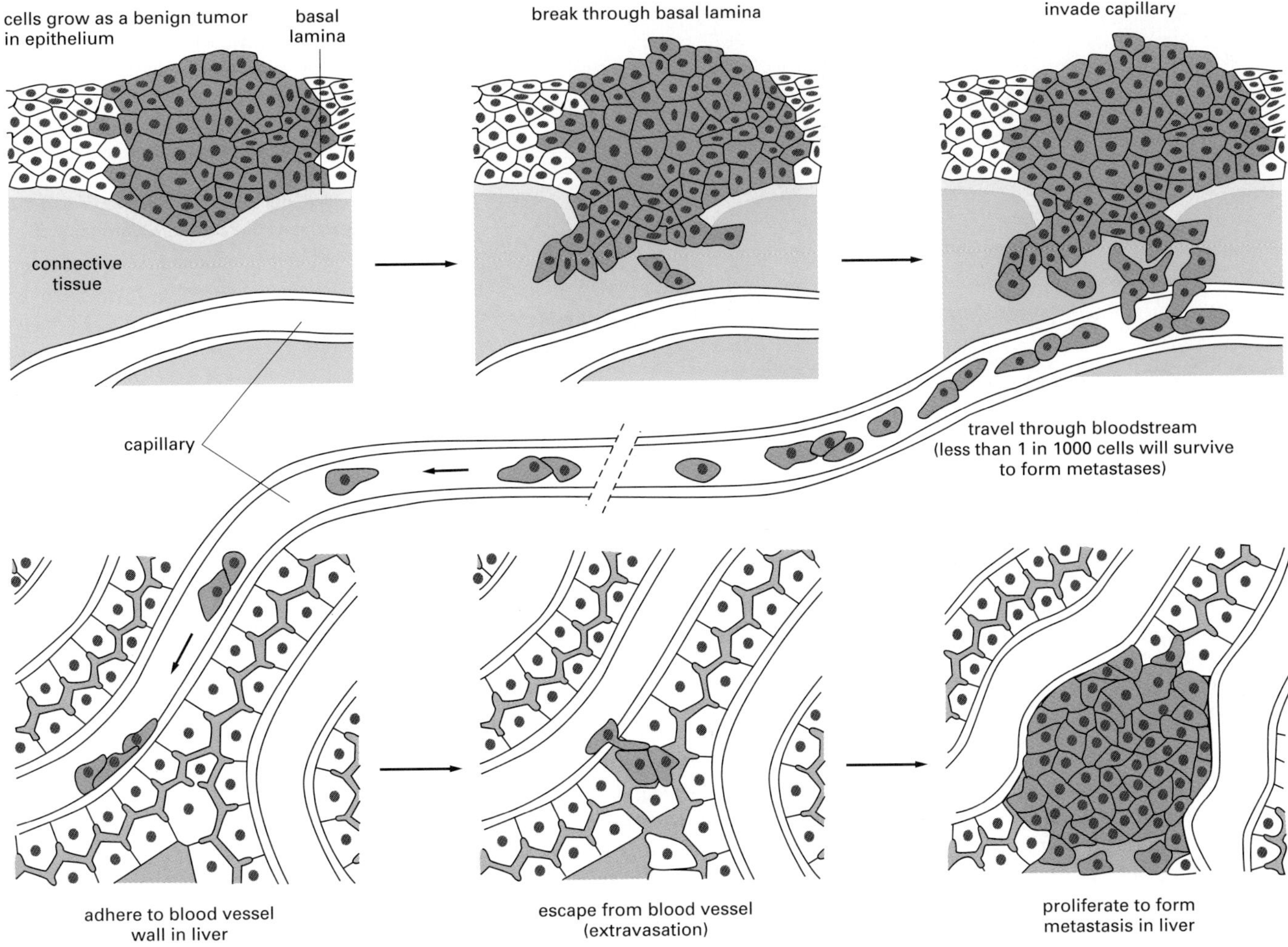
cells grow as a benign tumor in epithelium
basal lamina
connective tissue
break through basal lamina
invade capillary
capillary
travel through bloodstream (less than 1 in 1000 cells will survive to form metastases)
adhere to blood vessel wall in liver
escape from blood vessel (extravasation)
proliferate to form metastasis in liver

Figure 23–15 Steps in the process of metastasis. This example illustrates the spread of a tumor from an organ such as the lung or bladder to the liver. Tumor cells may enter the bloodstream directly by crossing the wall of a blood vessel, as diagrammed here, or, more commonly perhaps, by crossing the wall of a lymphatic vessel that ultimately discharges its contents (lymph) into the bloodstream. Tumor cells that have entered a lymphatic vessel often become trapped in lymph nodes along the way, giving rise to lymph-node metastases. Studies in animals show that typically far less than one in every thousand malignant tumor cells that enter the bloodstream will survive to produce a tumor at a new site.

activate their cell death machinery and undergo apoptosis. Cancer cells capable of metastasis are often relatively resistant to apoptosis compared with normal cells and therefore can survive and continue to grow after escaping from their proper home (Figure 23–16).

In addition to all these requirements, to grow large, a tumor must recruit an adequate blood supply. Thus, angiogenesis, the formation of new blood vessels, is a necessity for growth of both primary tumors and metastases. Although normal tissues have an automatic mechanism to attract an increased blood supply when they require it (as discussed in Chapter 22), many tumors achieve rapid growth by switching on enhanced production of angiogenic signals. The new blood vessels supply the tumor with nutrients and oxygen, and they may also provide an easier escape route for metastatic cells.

Six Key Properties Make Cells Capable of Cancerous Growth

Clearly, to be successful as a cancer a cell must acquire a whole range of aberrant properties—a collection of subversive new skills—as it evolves. Different cancers require different combinations of properties. Nevertheless, we can draw up a short list of the key behaviors of cancer cells in general:

1. They disregard the external and internal signals that regulate cell proliferation.
2. They tend to avoid suicide by apoptosis.
3. They circumvent programmed limitations to proliferation, escaping replicative senescence and avoiding differentiation.
4. They are genetically unstable.
5. They escape from their home tissues (that is, they are invasive).
6. They survive and proliferate in foreign sites (that is, they metastasize).

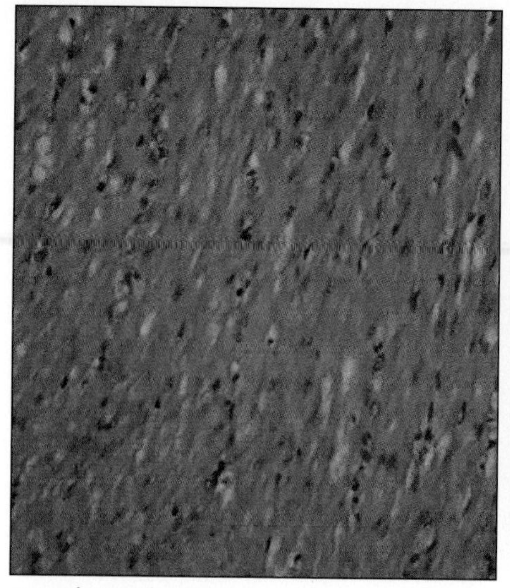

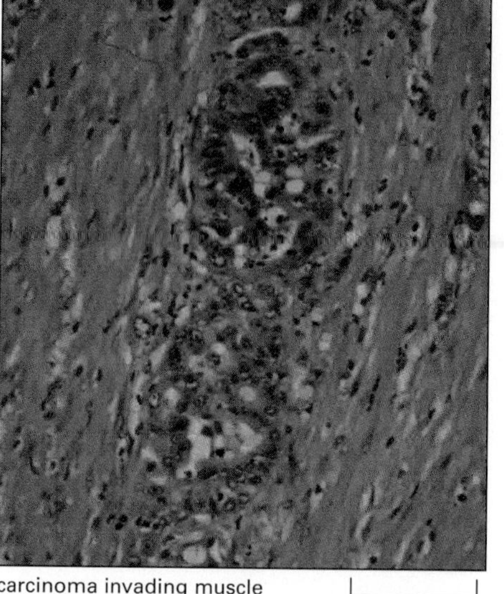

Figure 23–16 Colon cancer invading the smooth muscle layer that lies beneath the colon epithelium. In this tissue slice, colorectal cancer cells can be seen growing between the muscle fibers *(pink)*. Some stages in the development of this type of cancer are shown in Figure 23–38. (Courtesy of Paul Edwards and the Department of Pathology, University of Cambridge.)

In the later sections of the chapter, we shall examine the mutations and molecular mechanisms that underlie these properties. But first, we must consider the external factors that can help to cause the disease.

Summary

Cancer cells, by definition, proliferate in defiance of normal controls (that is, they are neoplastic) and are able to invade and colonize surrounding tissues (that is, they are malignant). By giving rise to secondary tumors, or metastases, they become difficult to eradicate surgically. Most cancers are thought to originate from a single cell that has experienced an initial mutation, but the progeny of this cell must undergo further changes, requiring numerous additional mutations, to become cancerous. This phenomenon of tumor progression, which usually takes many years, reflects the unfortunate operation of evolution by mutation and natural selection among somatic cells.

The rational treatment of cancer requires an understanding of the special properties that cancer cells acquire as they evolve, multiply, and spread. These special properties include alterations in cell signaling pathways, enabling the cells in a tumor to ignore the signals from their environment that normally keep cell proliferation under tight control. In this way, the cells are first able to proliferate abnormally in their original tissue and then to metastasize, surviving and proliferating in foreign tissues. As part of the evolutionary process of tumor progression, cancer cells also acquire an abnormal aversion to suicide, and they avoid or break free of programmed limitations to proliferation—including replicative senescence and the normal pathways of differentiation that would otherwise hamper their ability to grow and divide.

Since many mutations are needed to confer this whole collection of bad properties, it is perhaps not surprising to find that nearly all cancer cells are observed to have the additional property of being abnormally mutable, having acquired one or more defects in various aspects of their DNA metabolism. This genetic instability speeds the cell's acquisition of the complex set of alterations that are required for neoplasia and malignancy.

THE PREVENTABLE CAUSES OF CANCER

The development of a cancer generally requires many steps, each governed by multiple factors—some dependent on the genetic constitution of the individual, others dependent on his or her environment and way of life. A certain irreducible background incidence of cancer is to be expected regardless of circumstances: mutations can never be absolutely avoided, because they are an inescapable consequence of fundamental limitations on the accuracy of DNA replication, as discussed in Chapter 5. If a human could live long enough, it is inevitable that at least one of his or her cells would eventually accumulate a set of mutations sufficient for cancer to develop.

Nevertheless, there is evidence that avoidable environmental factors play some part in the causation of most cases of the disease. This is demonstrated most clearly by a comparison of cancer incidence in different countries: for almost every cancer that is common in one country, there is another country where the incidence is several times lower (Table 23–1). Because migrant populations tend to adopt the pattern of cancer incidence typical of the host country, the differences appear to be due mostly to environmental, not genetic, factors. From such data it is estimated that 80–90% of cancers should be avoidable, or at least postponable. Unfortunately, different cancers have different environmental risk factors, and a population that happens to escape one such danger is usually exposed to another. This is not, however, inevitable. There are some subgroups whose way of life substantially reduces the total cancer death rate among individuals of a given age. Under the current conditions in the United States and Europe, approximately one in five people die of cancer. But the incidence of cancer among strict Mormons in Utah, for example, is only about half that among Americans in general. Cancer incidence is also low in certain relatively affluent populations in Africa.

Although such observations on human populations indicate that cancer can often be avoided, it has been difficult in most cases to identify the specific environmental risk factors or to establish how they act. We will first look at what has been learned about the external agents that are known to cause cancer. We will then consider some of the triumphs, and difficulties, in finding ways in which

TABLE 23–1 Variation Between Countries in the Incidence of Some Common Cancers

SITE OF ORIGIN OF CANCER	HIGH-INCIDENCE POPULATION		LOW-INCIDENCE POPULATION	
	LOCATION	INCIDENCE*	LOCATION	INCIDENCE*
Lung	USA (New Orleans, blacks)	110	India (Madras)	5.8
Breast	Hawaii (Hawaiians)	94	Israel (non-Jews)	14.0
Prostate	USA (Atlanta, blacks)	91	China (Tianjin)	1.3
Uterine cervix	Brazil (Recife)	83	Israel (non-Jews)	3.0
Stomach	Japan (Nagasaki)	82	Kuwait (Kuwaitis)	3.7
Liver	China (Shanghai)	34	Canada (Nova Scotia)	0.7
Colon	USA (Connecticut, whites)	34	India (Madras)	1.8
Melanoma	Australia (Queensland)	31	Japan (Osaka)	0.2
Nasopharynx	Hong Kong	30	UK (Southwestern)	0.3
Esophagus	France (Calvados)	30	Romania (urban Cluj)	1.1
Bladder	Switzerland (Basal)	28	India (Nagpur)	1.7
Uterus	USA (San Francisco Bay Area, whites)	26	India (Nagpur)	1.2
Ovary	New Zealand (Polynesian Islanders)	26	Kuwait (Kuwaitis)	3.3
Rectum	Israel (European and USA born)	23	Kuwait (Kuwaitis)	3.0
Larynx	Brazil (São Paulo)	18	Japan (rural Miyagi)	2.1
Pancreas	USA (Los Angeles, Koreans)	16	India (Poona)	1.5
Lip	Canada (Newfoundland)	15	Japan (Osaka)	0.1
Kidney	Canada (NWT and Yukon)	15	India (Poona)	0.7
Oral cavity	France (Bas-Rhin)	14	India (Poona)	0.4
Leukemia	Canada (Ontario)	12	India (Nagpur)	2.2
Testis	Switzerland (urban Vaud)	10	China (Tianjin)	0.6

*Incidence = number of new cases per year per 100,000 population, adjusted for standardized population age distribution (so as to eliminate effects due merely to differences of population age distribution). Figures for cancers of breast, uterine cervix, uterus, and ovary are for women; other figures are for men. (Adapted from V.T. DeVita, S. Hellman, and S.A. Rosenberg (eds.), Cancer: Principles and Practice of Oncology, 4th edn. Philadelphia: Lippincott, 1993; based on data from C. Muir et al., Cancer Incidence in Five Continents, Vol. 5. Lyon: International Agency for Research on Cancer, 1987.)

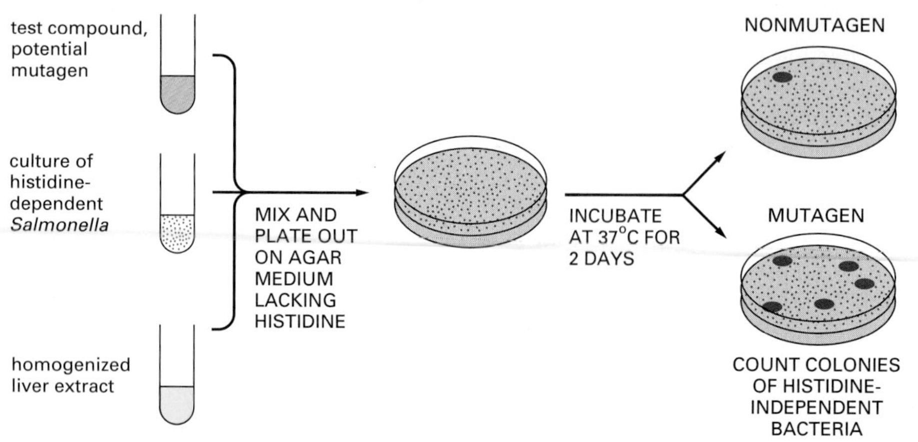

Figure 23–17 The Ames test for mutagenicity. The test uses a strain of *Salmonella* bacteria that require histidine in the medium because of a defect in a gene necessary for histidine synthesis. Mutagens can cause a further change in this gene that reverses the defect, creating revertant bacteria that do not require histidine. To increase the sensitivity of the test, the bacteria also have a defect in their DNA repair machinery that makes them especially susceptible to agents that damage DNA. A majority of compounds that are mutagenic in tests such as this are also carcinogenic and vice versa.

human cancer can be prevented. The problem of treatment will be discussed in the last section, after we have examined the molecular biology of the disease.

Many, But Not All, Cancer-Causing Agents Damage DNA

The agents that can cause cancer are many and varied, but the easiest to understand are those that cause damage to DNA, and so generate mutations. These cancer-causing mutagens include chemical carcinogens, viruses, and various forms of radiation—UV light and ionizing radiation such as gamma rays and alpha particles from radioactive decay.

Many quite disparate chemicals have been shown to be carcinogenic when they are fed to experimental animals or painted repeatedly on their skin. Examples include a range of aromatic hydrocarbons and derivatives of them such as aromatic amines; nitrosamines; and alkylating agents such as mustard gas. Although these chemical carcinogens are diverse in structure, they have at least one property in common—they cause mutations. In one popular test for mutagenicity, the carcinogen is mixed with an activating extract prepared from rat liver cells (to mimic the biochemical processing that occurs in an intact animal) and is added to a culture of specially designed test bacteria; the resulting mutation rate of the bacteria is then measured (Figure 23–17). Most of the compounds scored as mutagenic by this rapid and convenient assay in bacteria also cause mutations or chromosome aberrations when tested on mammalian cells. When mutagenicity data from various sources are analyzed, one finds that the majority of identified carcinogens are mutagens.

A few of these carcinogens act directly on the target DNA, but generally the more potent ones are relatively inert chemically and become damaging only after they have been changed to a more reactive form by metabolic processes—notably by a set of intracellular enzymes known as the cytochrome P-450 oxidases. These enzymes normally help to convert ingested toxins into harmless and easily excreted compounds. Unfortunately, their activity on certain chemicals generates products that are highly mutagenic. Examples of carcinogens activated in this way include the fungal toxin aflatoxin B1 (Figure 23–18) and benzo[a]-pyrene, a cancer-causing chemical present in coal tar and tobacco smoke.

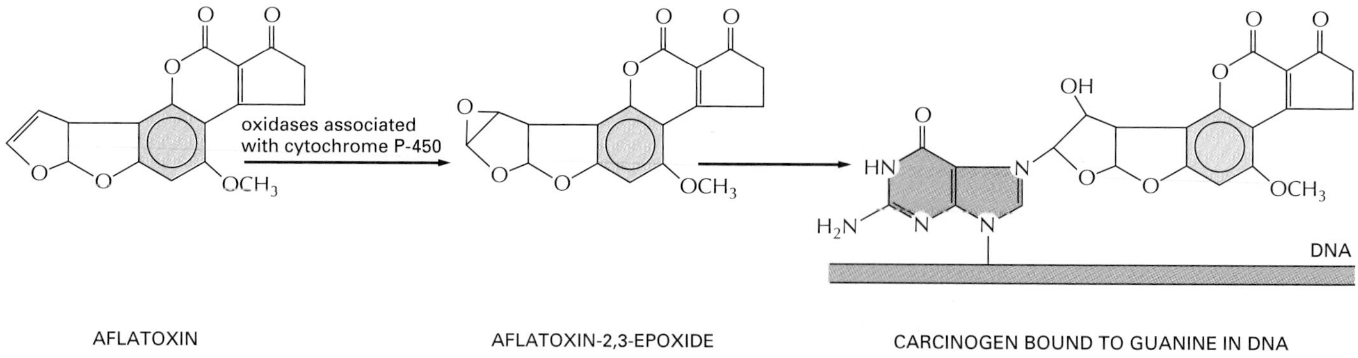

Figure 23–18 Metabolic activation of a carcinogen. Many chemical carcinogens have to be activated by a metabolic transformation before they will cause mutations by reacting with DNA. The compound illustrated here is *aflatoxin B1*, a toxin from a mold *(Aspergillus flavus oryzae)* that grows on grain and peanuts when they are stored under humid tropical conditions. It is thought to be a contributory cause of liver cancer in the tropics and is associated with characteristic mutations of the *p53* gene (discussed later in this chapter).

oxidases associated with cytochrome P-450

AFLATOXIN

AFLATOXIN-2,3-EPOXIDE

CARCINOGEN BOUND TO GUANINE IN DNA

The Development of a Cancer Can Be Promoted by Factors That Do Not Alter the Cell's DNA Sequence

Not all substances that favor the development of cancer are mutagens, however. Some of the clearest evidence comes from studies done long ago on the effects of cancer-causing chemicals on mouse skin, where it is easy to observe the stages of tumor progression. Skin cancers can be elicited in mice by repeatedly painting the skin with a mutagenic chemical carcinogen such as benzo[a]pyrene or the related compound dimethylbenz[a]anthracene (DMBA). A single application of the carcinogen, however, usually does not by itself give rise to a tumor or any other obvious lasting abnormality. Yet it does cause latent genetic damage—mutations that set the stage for a greatly increased incidence of cancer when the cells are exposed either to further treatments with the same substance or to certain other, quite different, insults. A carcinogen that sows the seeds of cancer in this way is said to act as a **tumor initiator**.

Simply wounding skin that has been exposed once to such an initiator can cause cancers to develop from some of the cells at the edge of the wound. Alternatively, repeated exposure over a period of months to certain substances known as **tumor promoters**, which are not themselves mutagenic, can cause cancer selectively in skin previously exposed to a tumor initiator. The most widely studied tumor promoters are *phorbol esters*, such as tetradecanoylphorbol acetate (TPA), which behave as artificial activators of protein kinase C and hence activate part of the phosphatidylinositol intracellular signaling pathway (discussed in Chapter 15). These substances cause cancers at high frequency only if they are applied after a treatment with a mutagenic initiator (Figure 23–19).

As one might expect for genetic damage, the hidden changes caused by a tumor initiator are irreversible: thus, they can be uncovered by treatment with a tumor promoter even after a long delay. The immediate effect of the promoter is apparently to stimulate cell division (or to cause cells that would normally undergo terminal differentiation to continue dividing instead). In the region that had previously been exposed to the initiator, this proliferation results in the growth of many small, benign, wartlike tumors called *papillomas*. The greater the prior dose of initiator, the larger the number of papillomas induced; it is thought that each papilloma (at least for low doses of the initiator) consists of a single clone of cells descended from a mutant cell that the initiator has produced. How tumor promoters work is not certain, and different tumor promoters are likely to work in different ways. One possibility is that they simply induce expression of growth-controlling genes that had been mutated before the promoter was applied: a mutation that makes a gene product hyperactive will not show its effects until the gene is expressed. Another possibility is that the promoter temporarily releases the cell from an inhibitory influence that normally

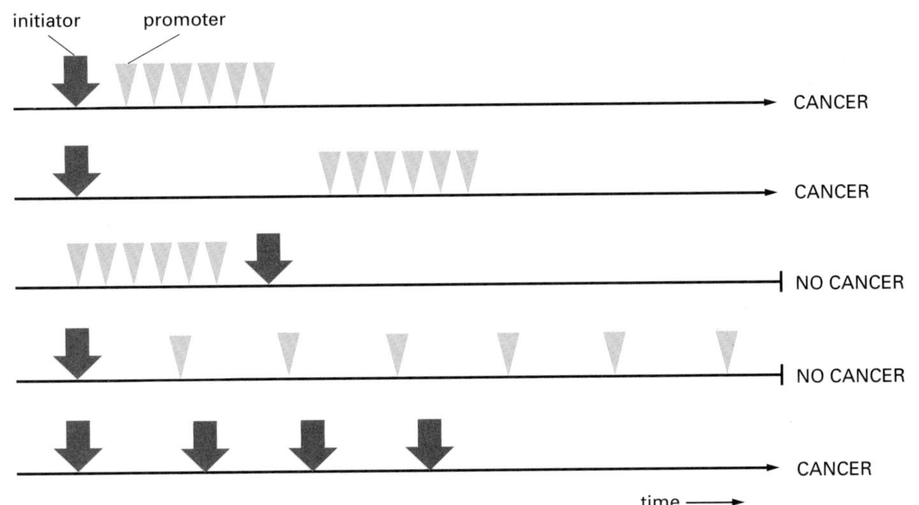

Figure 23–19 Some possible schedules of exposure to a tumor initiator (mutagenic) and a tumor promoter (nonmutagenic) and their outcomes. Cancer ensues only if the exposure to the promoter follows exposure to the initiator and only if the intensity of exposure to the promoter exceeds a certain threshold. Cancer can also occur as a result of repeated exposure to the initiator alone.

overrides the proliferation-inducing effect of the mutation; as a result, the cell is enabled to divide and grow into a large cluster of cells (Figure 23–20).

A typical papilloma might contain about 10^5 cells. If exposure to the tumor promoter is stopped, almost all the papillomas regress, and the skin regains a largely normal appearance. In a few of the papillomas, however, further changes occur that enable growth to continue in an uncontrolled way, even after the promoter has been withdrawn. These changes seem to originate in an occasional single papilloma cell, at about the frequency expected for spontaneous mutations. In this way, a small proportion of the papillomas progress to become cancers. Thus, the tumor promoter apparently favors the development of cancer by expanding the population of cells that carry an initial mutation: the more such cells there are and the more times they divide, the greater the chance that at least one of them will undergo another mutation carrying it one step further toward malignancy.

Although naturally occurring cancers do not necessarily arise through the specific sequence of distinct initiation and promotion steps just described, their evolution must be governed by similar principles. They too will evolve at a rate that depends both on the frequency of mutations and on the influences affecting the survival, proliferation, and spread of selected mutant cells.

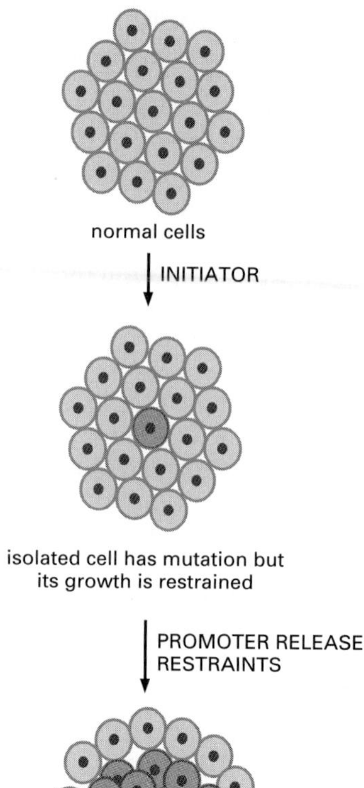
normal cells
INITIATOR
isolated cell has mutation but its growth is restrained
PROMOTER RELEASES RESTRAINTS

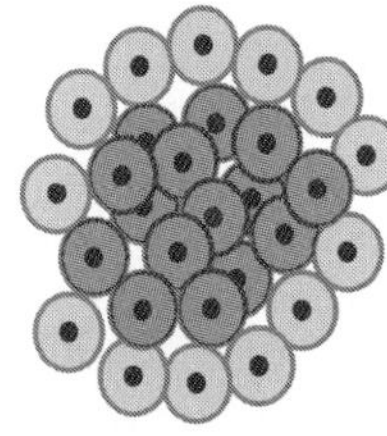
mutant cells grow into a large clone of cells, in which a further mutation may occur

Figure 23–20 The effect of a tumor promoter. The tumor promoter expands the population of mutant cells, thereby increasing the probability of tumor progression by further genetic change.

Viruses and Other Infections Contribute to a Significant Proportion of Human Cancers

As far as we know, viruses and other infections play no part in the majority of human cancers. However, a small but significant proportion of human cancers, perhaps 15% in the world as a whole, are thought to arise by mechanisms that do involve viruses, bacteria or parasites. The main culprits, as shown in Table 23–2, are the DNA viruses. Evidence for their involvement comes partly from the detection of viruses in cancer patients and partly from epidemiology. Liver can-

TABLE 23–2 Viruses Associated with Human Cancers

VIRUS	ASSOCIATED TUMORS	AREAS OF HIGH INCIDENCE
DNA Viruses		
Papovavirus family		
Papillomavirus (many distinct strains)	warts (benign) carcinoma of the uterine cervix	worldwide worldwide
Hepadnavirus family		
Hepatitis-B virus	liver cancer (hepatocellular carcinoma)	Southeast Asia, tropical Africa
Herpesvirus family		
Epstein-Barr virus	Burkitt's lymphoma (cancer of B lymphocytes) nasopharyngeal carcinoma	West Africa, Papua New Guinea southern China, Greenland
RNA viruses		
Retrovirus family		
Human T-cell leukemia virus type I (HTLV-1)	adult T-cell leukemia/ lymphoma	Japan, West Indies
Human immunodeficiency virus (HIV, the AIDS virus)	Kaposi's sarcoma	Central and Southern Africa

For all the above viruses, the number of people infected is much larger than the numbers who develop cancer: the viruses must act in conjunction with other factors. Moreover, some of the viruses contribute to cancer only indirectly; for example, HIV, by upsetting normal cell-mediated immune defenses, allows endothelial cells to be transformed by another virus (a type of herpesvirus) and thrive as a tumor instead of being destroyed by the immune system.

cer, for example, is common in parts of the world (Africa and Southeast Asia) where hepatitis-B viral infections are common, and in those regions the cancer occurs almost exclusively in people who show signs of chronic hepatitis-B infection.

The precise role of a cancer-associated virus is often hard to decipher because there is a delay of many years from the initial viral infection to the development of the cancer. Moreover, the virus is responsible for only one of a series of steps in the progression to cancer, and other environmental factors and genetic accidents are also involved. Like the cancer-causing chemicals we discussed earlier, viruses can either alter a cell's DNA directly or act as tumor promoters. As we shall explain later, DNA viruses frequently carry genes that can subvert the control of cell division in the host cell, causing uncontrolled proliferation. DNA viruses that operate in this manner include the human papillomaviruses (Figure 23–21); some of these viruses cause warts, while others infect the uterine cervix and are implicated in the development of carcinomas of the cervix.

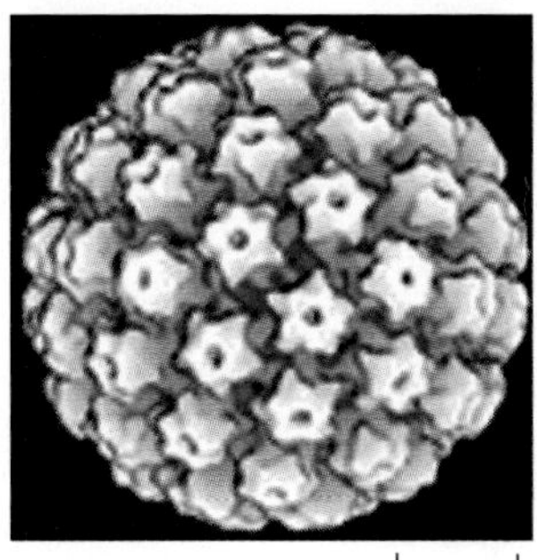

Figure 23–21 A human papillomavirus. (Courtesy of Norman Olson.)

In some other cancers, viruses seem to have additional, indirect tumor-promoting actions; the hepatitis-B virus may, for example, favor the development of liver cancer by doing damage that provokes cell division in the liver, as well as by altering cell growth control directly. In AIDS, the human immunodeficiency virus (HIV) promotes development of an otherwise rare cancer called Kaposi's sarcoma by destroying the immune system, thereby permitting a secondary infection with a human herpes virus (HHV-8) that has a direct carcinogenic action. Chronic infection with parasites and bacteria may also promote the development of some cancers. For example, infection of the stomach with the bacterium *Helicobacter pylori*, which causes ulcers, appears to be a major cause of stomach cancer. And bladder cancer in some parts of the world is associated with infection by the blood fluke, *Schistosoma haematobium*, a parasitic flatworm.

Identification of Carcinogens Reveals Ways to Avoid Cancer

Tobacco smoke is by far the most important environmental cause of cancer in the world today. Other comparably important chemical causes of cancer in humans remain to be identified. It is sometimes thought that the main environmental causes of cancer are the products of a highly industrialized way of life—the rise in pollution, the enhanced use of food additives, and so on—but there is little evidence to support this view. The idea may have come in part from the identification of some highly carcinogenic materials used in industry, such as 2-naphthylamine and asbestos. In fact, except for the increase in cancers caused by smoking, and a remarkable decrease in stomach cancer, the incidence of the most common cancers for individuals of a given age has not changed very much during the course of the twentieth century (Figure 23–22).

Most of the carcinogenic factors that are known to be significant are by no means peculiar to the modern world. The most potent known carcinogen (by certain assays), and an important cause of liver cancer in Africa and Asia, is

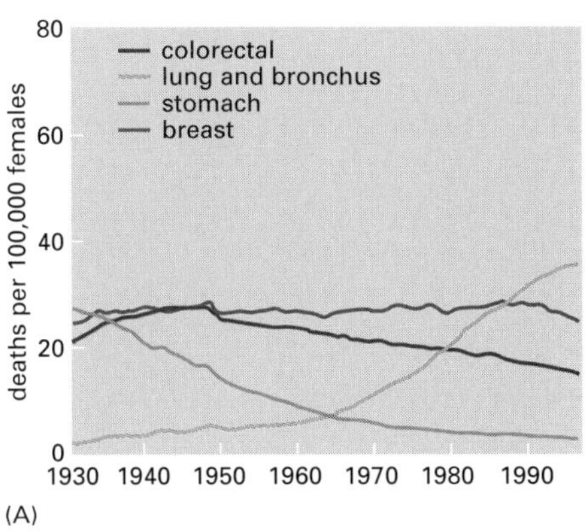

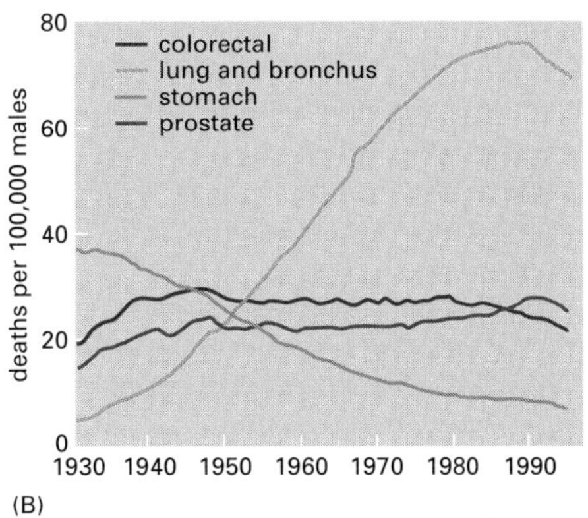

Figure 23–22 Age-adjusted cancer death rates, United States, 1930–1996. Selected death rates adjusted to the age distribution of the US population in 1970, are plotted for (A) females and (B) males. Note the dramatic rise in lung cancer for both sexes, following the pattern of tobacco smoking, and the fall in deaths from stomach cancer, possibly related to changes in diet or in patterns of infection with *Helicobacter*. Recent reductions in other cancer death rates may correspond to improvements in detection and treatment. (Adapted from Cancer Facts and Figures 2000. © American Cancer Society, 2000.)

aflatoxin B1 (see Figure 23–18), a compound produced by fungi that naturally contaminate foods such as tropical peanuts. And for women, the risk of cancer is powerfully influenced by the reproductive hormones that circulate in the body at different stages of life. Thus, a striking correlation exists between reproductive history and the occurrence of breast cancer (Figure 23–23). The reproductive hormones presumably affect breast-cancer incidence through their influence on cell proliferation in the breast. Clearly, when attempting to identify the environmental causes of cancer, we need to keep an open mind.

Epidemiology—the analysis of disease frequency in populations—remains the principal tool for finding environmental causes of human cancer. The approach has enjoyed some notable successes and it promises more to come. Simply by revealing the role of smoking, epidemiology has shown a way to reduce the total cancer death rate in North America and Europe by as much as 30%. The approach works best when applied to a fairly uniform population in which it is easy to distinguish between individuals who were exposed to the agent and those who were not, and when the agent under investigation is responsible for most of the cases of a certain kind of cancer. For example, in the early part of this century, in one British factory, all of the men who had been employed in distilling 2-napthylamine (and were thereby subjected to prolonged exposure) eventually developed bladder cancer (see Figure 23–8); the connection was relatively easy to establish because both the chemical and the form of cancer were uncommon in the general population.

In contrast, it is very hard to identify, by epidemiology alone, everyday environmental factors that favor development of common cancers: most of these factors are probably agents to which we are all exposed to some extent, and many of them probably contribute together to a given cancer's incidence. If, say, eating oranges doubled the risk of colorectal cancer, it is unlikely that we would find it out, unless we had some prior reason to suspect a connection; and, the same goes for the countless other substances that we eat, drink, breathe, and put on our bodies. Even when evidence is obtained that a substance may be carcinogenic, either from epidemiology or from laboratory tests, it may be difficult to decide what level of human exposure is acceptable. Estimating how many cases of human cancer a certain amount of a substance is likely to cause is difficult; balancing this risk against the utility of the substance is more difficult still. For example, certain agricultural fungicides appear to be mildly carcinogenic at high doses in animal tests, but it has been calculated that if they were not used in agriculture the contamination of food by fungal metabolites such as aflatoxin B1 would cause far more cases of cancer than the fungicide residues in food ever could.

Nevertheless, efforts to identify potential carcinogens still have a central place in our struggle with cancer. Prevention of the disease is not only better than a cure; for many types of cancer, it seems also, in our present state of knowledge, to be much more readily attainable.

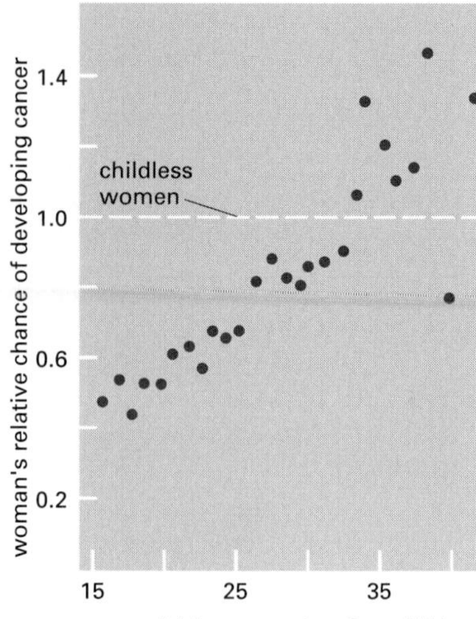

Figure 23–23 Effects of childbearing on the risk of breast cancer. The relative probability of breast cancer developing at some time in a woman's life is plotted as a function of the age at which she gives birth to her first child. The graph shows the value of the probability relative to that for a childless woman. The longer the period of exposure to reproductive hormones up to birth of the first child, the greater the risk. It is thought that the first full-term pregnancy may result in a permanent change in the state of differentiation of the cells of the breast, altering their subsequent responses to hormones. Several other lines of epidemiological evidence also support the view that exposure to certain combinations of reproductive hormones, especially estrogen, can promote development of breast cancer. (From J. Cairns, Cancer: Science and Society. San Francisco: W.H. Freeman, 1978. After B. MacMahon, P. Cole, and J. Brown, *J. Natl. Cancer Inst.* 50:21–42, 1973.)

Summary

The rate of tumor evolution and progression is accelerated both by mutagenic agents (tumor initiators) and by nonmutagenic agents (tumor promoters) that affect gene expression, stimulate cell proliferation, and alter the ecological balance of mutant and nonmutant cells. The majority of known cancer-causing agents are mutagens, including chemical carcinogens, certain viruses, and various forms of radiation—such as UV light and ionizing radiation. Because many factors contribute to the development of a given cancer—some of which are under our control—a large proportion of cancers are in principle preventable.

We remain largely ignorant of the principal environmental factors affecting cancer incidence. Of the environmental risk factors that have been identified, however, many can be avoided. These include smoking tobacco and falling prey to infection with cancer-causing viruses such as papillomaviruses or hepatitis B. Epidemiology can be a powerful tool for identifying such causes of human cancer and revealing ways to prevent the disease. The approach does not require knowing how

cancer-causing agents work, and it can uncover factors that are not simply mutagens, such as viruses and certain patterns of child-bearing.

FINDING THE CANCER-CRITICAL GENES

Cancer is a genetic disease: it results from mutations in somatic cells. To understand it at a molecular level, we need to identify the relevant mutations and to discover how they give rise to cancerous cell behavior. Finding the mutations is easy in one respect: the mutant cells are favored by natural selection and call attention to themselves by giving rise to tumors. The hard task then begins: how are the genes with the carcinogenic mutations to be identified among all the other genes in the cancerous cells? A similar needle-in-haystack problem arises in any search for a gene underlying a given mutant phenotype, but for cancer the task is particularly complex. A typical cancer depends on a whole set of mutations—usually a somewhat different set in each individual patient—and introduction of any single one of these into a normal cell is usually not enough to make it cancerous. This genetic cooperation makes it hard to test the significance of mutations on which suspicion falls. To make matters worse, most cancer cells will contain mutations that are accidental by-products of genetic instability, and it can be difficult to distinguish these from the mutations that have a causative role in the disease.

Despite these difficulties, many genes that are repeatedly altered in human cancer have been identified—more than 100 of them—although it is clear that many more remain to be discovered. We will call such genes, for want of a better term, **cancer-critical genes**, meaning all genes whose mutation contributes to the causation of cancer. Our knowledge of these genes has accumulated piecemeal through many different and sometimes circuitous approaches, ranging from investigations of embryonic development to studies of cancer-causing infections in chickens. Analyses of exceptional but highly revealing forms of the disease have played a large part.

In this section, we discuss both the methods used for identifying cancer-critical genes and the varied kinds of mutations that occur in them in the development of cancer.

Different Methods Are Used to Identify Gain-of-Function and Loss-of-Function Mutations

Cancer-critical genes are grouped into two broad classes, according to whether the cancer risk arises from too much activity of the gene product, or too little. Genes of the first class, for which a gain-of-function mutation drives a cell toward cancer, are called **proto-oncogenes**; their mutant, overactive forms are called **oncogenes**. Genes of the second class, for which a loss-of-function mutation creates the danger, are called **tumor suppressor genes**.

As we will see, both kinds of mutations can have similar effects in enhancing cell proliferation and survival. Thus, from the point of view of a cancer cell, oncogenes and tumor suppressors—and the mutations that affect them—are flip sides of the same coin. The techniques needed to find these genes, however, are different, and are dictated by whether they are made overactive or underactive in cancer.

Mutation of a single copy of a proto-oncogene can have a dominant, growth-promoting effect on a cell (Figure 23–24A). Thus the oncogene can be detected by its effect when it is *added*—by DNA transfection, for example, or through infection with a viral vector—to the genome of a suitable type of tester cell. In the case of the tumor suppressor gene, on the other hand, the cancer-causing mutations are generally recessive: both copies of the normal gene must be removed or inactivated in the diploid somatic cell before an effect is seen (Figure 23–24B). This calls for a different approach, to discover what is missing.

In some cases, a specific gross chromosomal abnormality, visible under the microscope, is repeatedly associated with a particular type of cancer. This can

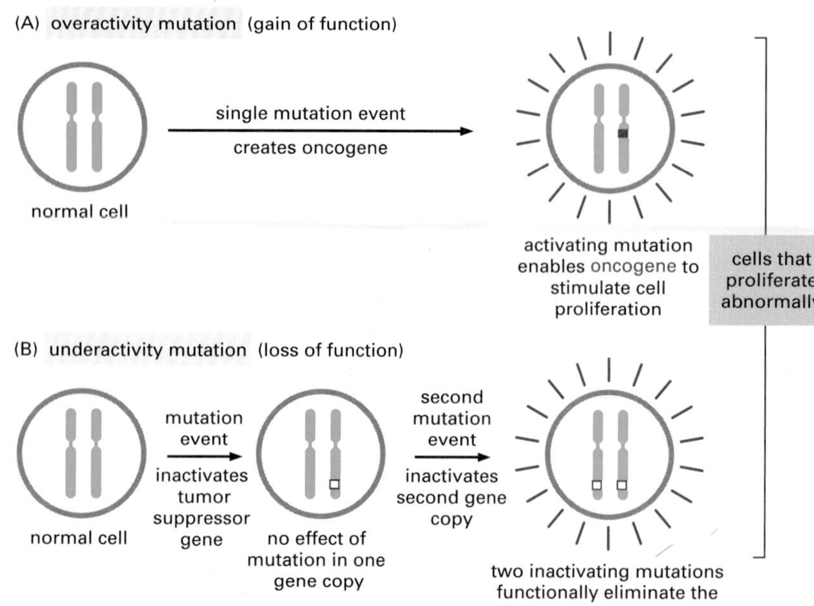
(A) overactivity mutation (gain of function)
single mutation event
creates oncogene
normal cell
activating mutation enables oncogene to stimulate cell proliferation
cells that proliferate abnormally
(B) underactivity mutation (loss of function)
mutation event
inactivates tumor suppressor gene
normal cell
no effect of mutation in one gene copy
second mutation event
inactivates second gene copy
two inactivating mutations functionally eliminate the tumor suppressor gene, stimulating cell proliferation

Figure 23–24 Cancer-critical genes fall into two readily distinguishable categories, dominant and recessive. Oncogenes act in a dominant manner: a gain-of-function mutation in a single copy of the cancer-critical gene can drive a cell toward cancer. Tumor suppressor genes, on the other hand, generally act in a recessive manner: the function of both alleles of the cancer-critical gene must be lost to drive a cell toward cancer. In this diagram, activating mutations are represented by solid red boxes, inactivating mutations by hollow red boxes.

give a clue to the location of either an oncogene that is activated as a result of the chromosomal rearrangement (as in the example of the chromosomal translocation responsible for chronic myelogenous leukemia, mentioned earlier) or a tumor suppressor gene that is deleted; but the types of chromosomal abnormality involved in the two cases, and the circumstances in which they are typically encountered, are again different.

Oncogenes Are Identified Through Their Dominant Transforming Effects

Traditionally, geneticists identify genes by studying patterns of inheritance in families of individuals who show some heritable trait. To find where a mutation responsible for the trait lies in the genome, one looks for genetic linkage between this trait and genetic markers whose chromosomal position is known. This approach exploits an effect of sexual reproduction: genes that lie close to one another tend to be inherited together, whereas genes that are further apart are often separated by recombination during meiosis. In this way, a disease-causing mutation can be located by determining how far it lies from other genetic markers (see Chapter 8 for a detailed discussion).

Cancer cells do not reproduce sexually, so geneticists must use other methods to track down the mutations that make cancer cells different from their normal neighbors in the body. For oncogenes, a direct and conceptually simple (though laborious) approach is to scan the genome of the cancer cell for segments of DNA that will, when introduced into cells of a suitable tester cell line, drive them toward cancerous behavior. A mouse-derived fibroblast cell line is a convenient source of tester cells for the assay; these cells, which were previously selected to thrive in culture, are thought to already contain genetic alterations that take them part of the way toward malignancy. For this reason, addition of a single oncogene can be enough to produce a dramatic effect.

To detect an oncogene in this way, DNA is extracted from tumor cells, broken into fragments, and introduced into these fibroblasts in culture. If any of the fragments contains an oncogene, small colonies of abnormally proliferating—so-called 'transformed'—cells may begin to appear. Each of these colonies will be composed of a clone of cells that originated from a single cell whose growth was promoted by the added gene. Because they have been released from some of the social controls on cell division, the transformed cells outgrow normal ones, piling up in layer upon layer in the culture dish as they proliferate (Figure 23–25).

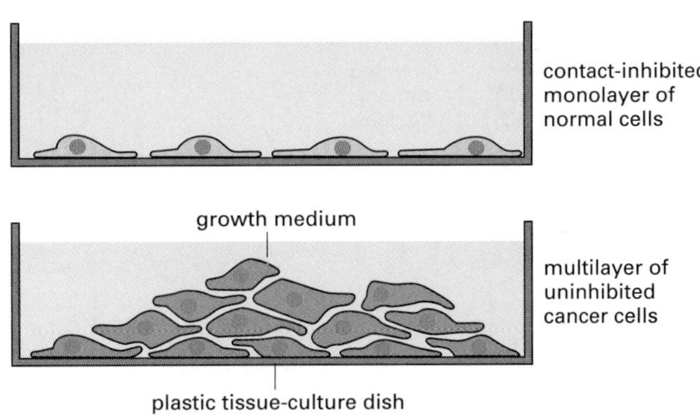

Figure 23–25 Loss of contact inhibition in cell culture. Most normal cells stop proliferating once they have carpeted the dish with a single layer of cells: proliferation seems to depend on contact with the dish, and to be inhibited by contacts with other cells. Cancer cells, in contrast, usually disregard these restraints and continue to grow, so that they pile up on top of one another.

This assay, applied to DNA from human tumors, led to the first direct identification of an oncogenic mutation in a human cancer. Several overactive growth-promoting genes were identified by isolating and sequencing the DNA fragments that had been transferred to these transformed cells. The first of these genes to be sequenced was a mutant version of the ***Ras* gene,** which is now known to be mutated in about one in four human tumors.

This discovery was all the more dramatic because, shortly beforehand, a mutated *Ras* gene had been found to be the tumor-causing gene in a **retrovirus** that causes sarcomas in rodents. Retroviruses, we now know, are capable of picking up, at random, fragments of genetic material from their animal hosts and ferrying them from one infected individual to another. Occasionally, a proto-oncogene is picked up, in a damaged or misregulated form that turns it into an oncogene. Infection with a virus that carries such a cargo can trigger tumors in some animal species, and a large number of oncogenes were first discovered in this way.

Twenty years ago, the discovery of the same mutant gene in human tumor cells and in an animal virus that causes tumors was electrifying. The suggestion that cancers are caused by mutations in a limited number of cancer-critical genes made unraveling the exact nature of cancer seem a soluble scientific problem, and it helped to launch a transformation in our understanding of the molecular biology of cancer.

As we saw in Chapter 15, normal Ras proteins are monomeric GTPases that help to transmit signals from growth factor receptors on the cell surface. The point mutations found in *Ras* genes isolated from human tumors generate a hyperactive Ras protein that persists abnormally in its active state—transmitting an inappropriate signal for cell proliferation. Because this type of mutation makes a gene product hyperactive, the effect is dominant—only one of the cell's two gene copies needs to undergo the change. The *Ras* genes are mutated in a wide range of human cancers, and they remain one of the most important examples of cancer-critical genes.

Tumor Suppressor Genes Can Sometimes Be Identified by Study of Rare Hereditary Cancer Syndromes

Given a cancer cell, identifying a gene that has been inactivated requires a different strategy than finding a gene that has become hyperactive: one cannot, for example, use a cell transformation assay to identify something that simply is not there. Thus, the search for tumor suppressor genes has taken a route quite different from that followed in the hunt for oncogenes. The key insight that led to the discovery of the first tumor suppressor gene came from studies of a rare type of human cancer, **retinoblastoma**, which arises from cells in the body that are converted to a cancerous state by an unusually small number of mutations. As often happens in biology, the discovery arose from examination of a special case, but it turned out to be of universal relevance.

Retinoblastoma occurs in childhood; tumors develop from neural precursor cells in the immature retina. About one child in 20,000 is afflicted. There are two forms of the disease, one hereditary, the other not. In the hereditary form, multiple tumors usually arise independently, affecting both eyes; in the non-hereditary

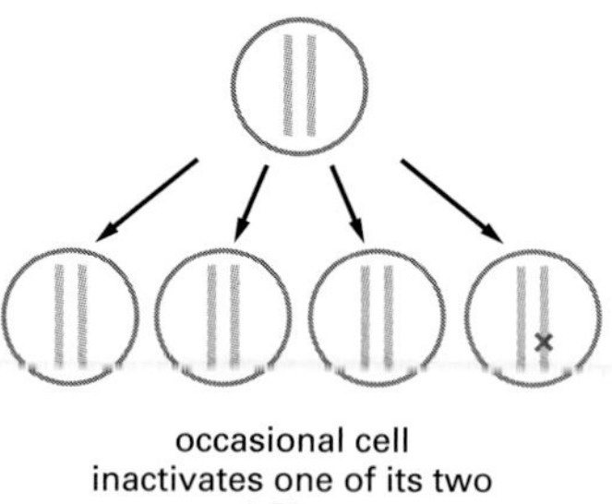
NORMAL, HEALTHY INDIVIDUAL
occasional cell inactivates one of its two good Rb genes
RESULT: NO TUMOR

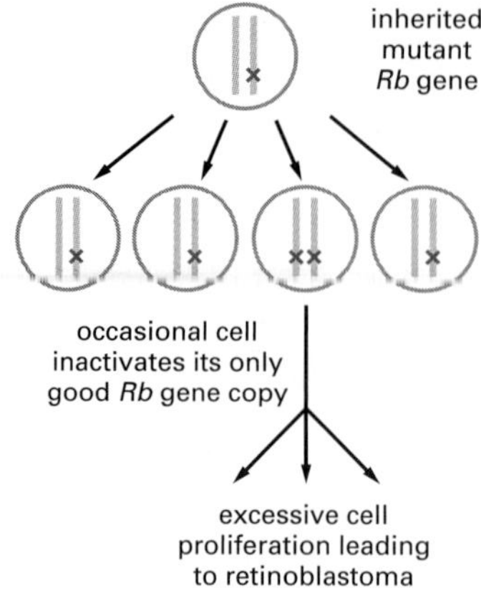
HEREDITARY RETINOBLASTOMA
inherited mutant Rb gene
occasional cell inactivates its only good Rb gene copy
excessive cell proliferation leading to retinoblastoma
RESULT: MOST PEOPLE WITH INHERITED MUTATION DEVELOP TUMOR

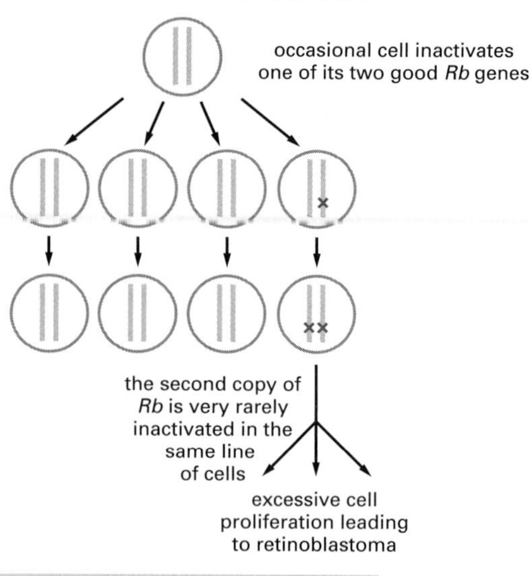
NONHEREDITARY RETINOBLASTOMA
occasional cell inactivates one of its two good Rb genes
the second copy of Rb is very rarely inactivated in the same line of cells
excessive cell proliferation leading to retinoblastoma
RESULT: ONLY ABOUT 1 IN 30,000 NORMAL PEOPLE DEVELOP TUMOR

Figure 23–26 The genetic mechanisms underlying retinoblastoma. In the hereditary form, all cells in the body lack one of the normal two functional copies of the *Rb* tumor suppressor gene, and tumors occur where the remaining copy is lost or inactivated by a somatic mutation. In the nonhereditary form, all cells initially contain two functional copies of the gene, and the tumor arises because both copies are lost or inactivated through the coincidence of two somatic mutations in one line of cells.

form only one eye is affected, and by only one tumor. Some individuals with hereditary retinoblastoma have a visibly abnormal karyotype, with a deletion of a specific band on chromosome 13. Deletions of this same locus are also encountered in tumor cells from some patients with the nonhereditary disease, suggesting that the cancer may be caused by loss of a critical gene in that chromosomal region.

Using the known location of the chromosomal deletion associated with retinoblastoma, it was possible to clone and sequence the gene whose loss appears to be critical for development of the cancer—the *Rb* gene. As would be predicted, in those who suffer from the hereditary form of the disease, a deletion or loss-of-function mutation occurs in one copy of the *Rb* gene in every cell of the body. Thus, these cells are predisposed to becoming cancerous, but are not actually cancerous so long as they retain one good copy of the gene. The retinal cells that do become cancerous are defective in both copies of *Rb* because a somatic mutation has occurred, in addition to the original inherited mutation, and has eliminated the remaining good copy. In patients with the nonhereditary form of the disease, by contrast, the noncancerous cells show no defect in either copy of *Rb*, while the cancerous cells are again defective in both copies. These nonhereditary retinoblastomas are very rare, because they require the coincidence of two somatic mutations in a single retinal cell lineage, so as to destroy both copies of the *Rb* gene (Figure 23–26).

The *Rb* gene subsequently turned out to much more than a gene mutated in a rare childhood tumor: it is also missing in several common types of cancer, including carcinomas of lung, breast, and bladder. These more common cancers arise by a more complex series of genetic changes than does retinoblastoma, and they make their appearance later in life and in other tissues of the body. But in all of them, it seems, loss of *Rb* is frequently a major step in the progression toward malignancy. The *Rb* gene encodes the Rb protein, which is a universal regulator of the cell cycle that is normally expressed in almost all the cells of the body (see Figure 17–30). Because it acts as one of the main brakes on progress through the cell-division cycle, the loss of Rb can allow cells to enter the cell division cycle inappropriately, as we shall discuss in the section on the molecular basis of cancer cell behavior.

Tumor Suppressor Genes Can Be Identified Even Without Clues from Heritable Cancer Syndromes

Hereditary cancer syndromes such as retinoblastoma are very rare, as we have already emphasized, and only a few—though important—tumor suppressor genes have been discovered by studying them. Where there is no such clue from a hereditary syndrome, we face the arduous task of identifying tumor suppressor genes simply by virtue of their absence from tumor cells. This involves comparing the tumor cells with non-cancerous cells from the same patient and discovering what exactly, out of the 3 billion nucleotides of the human genome, is missing or functionally defective. Because of the genetic instability of cancer cells, there is usually a great deal missing. Most of the defects are random and accidental by-products of the genetic instability. The tumor suppressor genes can only be identified by the criterion that they are repeatedly missing or defective in many independent cases of the cancer.

Gene loss often occurs by deletion of a relatively large segment of a chromosome. As a result, one copy of a tumor suppressor gene in a cancer cell may, for example, undergo an inactivating point mutation, while a gross deletion eliminates the other copy along with some neighboring genes (deletion of a large cluster of genes on both chromosomes is likely to kill the cell). The large defect on one chromosome makes detection of the loss much easier.

One deletion detection strategy takes advantage of the normal human genetic variation that makes the maternal and paternal chromosome sets distinguishably different. On average human DNA sequences differ—that is, we are heterozygous—at roughly one in every thousand nucleotides. Where a large segment of one chromosome has been lost, there is consequently a *loss of heterozygosity*: only one version of each variable DNA sequence in that neighborhood remains. A huge number—over a million—common sites of heterozygosity in the human genome have been mapped as part of the Human Genome Project: each of these sites is characterized by a specific DNA sequence that is known to be *polymorphic*—that is, to occur commonly in two or more slightly different versions in the human population. Given a sample of tumor DNA, one can check which of the versions of these polymorphic sequences are present. The same can be done with a sample of DNA from non-cancerous tissue from the same patient, for comparison. Absence of heterozygosity throughout a region of the genome containing many polymorphic sites, or loss of a genetic marker sequence that is seen in the non-cancerous control DNA, points the way toward a chromosomal region that has been deleted.

Techniques for large-scale DNA analysis and for the detection of deletions and other mutations are advancing rapidly, and they can be expected to add many more tumor suppressor genes to the present catalog.

Genes Mutated in Cancer Can Be Made Overactive or Underactive in Many Ways

We now know of at least 100 cancer-critical genes that can be converted into oncogenes by an activating mutation. The collection of genes whose absence or inactivation leads toward cancer, the tumor suppressor genes, is smaller but also growing. The ways in which genes of either class can be mutated to make them more—or less—active are enormously varied: as we might expect, any genetic accident that can increase, decrease, or change the activity of a gene is likely to be found somewhere in the increasingly complex catalogue of gene changes that occur in cancer.

The types of genetic alterations that can make a cancer-critical gene into an oncogene fall into three basic categories, as summarized in Figure 23–27. The gene may be altered by a small change in sequence such as a point mutation, by a larger-scale change such as a partial deletion, or by a chromosomal translocation that involves the breakage and rejoining of the DNA helix. These changes can occur in the protein-coding region so as to yield a hyperactive product, or they can occur in adjacent control regions so that the gene is simply expressed

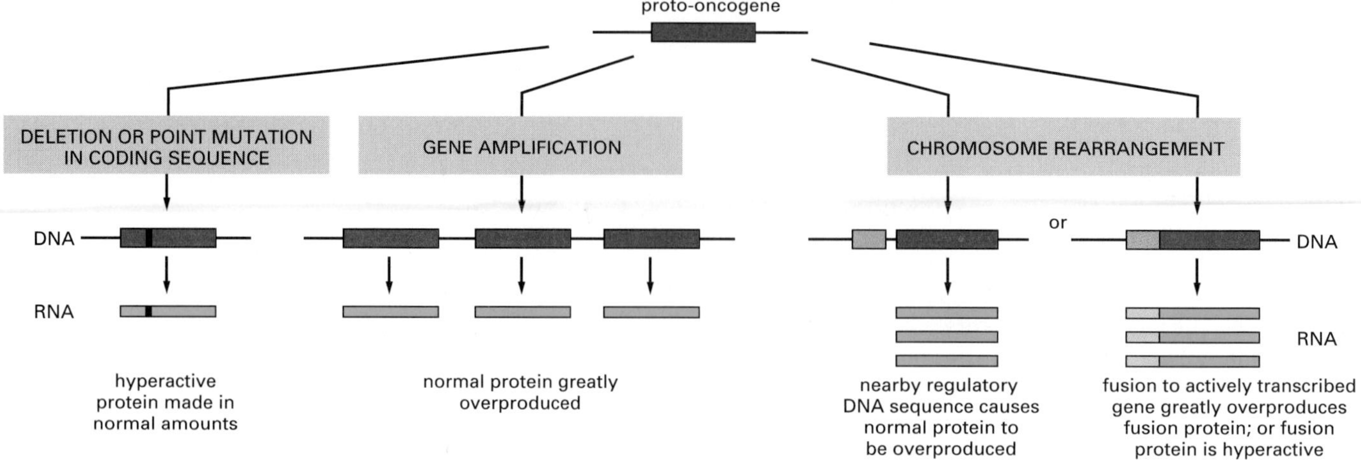
proto-oncogene
DELETION OR POINT MUTATION IN CODING SEQUENCE
GENE AMPLIFICATION
CHROMOSOME REARRANGEMENT
DNA
RNA
or
DNA
RNA
hyperactive protein made in normal amounts
normal protein greatly overproduced
nearby regulatory DNA sequence causes normal protein to be overproduced
fusion to actively transcribed gene greatly overproduces fusion protein; or fusion protein is hyperactive

Figure 23–27 Three ways in which a proto-oncogene can be made overactive to convert it into an oncogene.

at concentrations that are much higher than normal. Alternatively, the cancer-critical gene may be overexpressed because extra copies are present due to *gene amplification* events caused by errors in DNA replication (Figure 23–28).

Specific types of abnormality are characteristic of particular genes and of the responses to particular carcinogens. For example, 90% of the skin tumors evoked in mice by the tumor initiator dimethylbenz[a]anthracene (DMBA) have an A-to-T alteration at exactly the same site in a mutant *Ras* gene; presumably, of the mutations caused by DMBA, only those at this site efficiently activate skin cells to form a tumor.

The receptor for epidermal growth factor (EGF), on the other hand, can be activated by a deletion that removes part of its extracellular domain. These mutant receptors are able to form active dimers even in the absence of EGF, and thus they produce a stimulatory signal inappropriately, like a faulty doorbell that rings even when nobody is pressing the button. Such mutations are found in many human brain tumors of the type called glioblastomas.

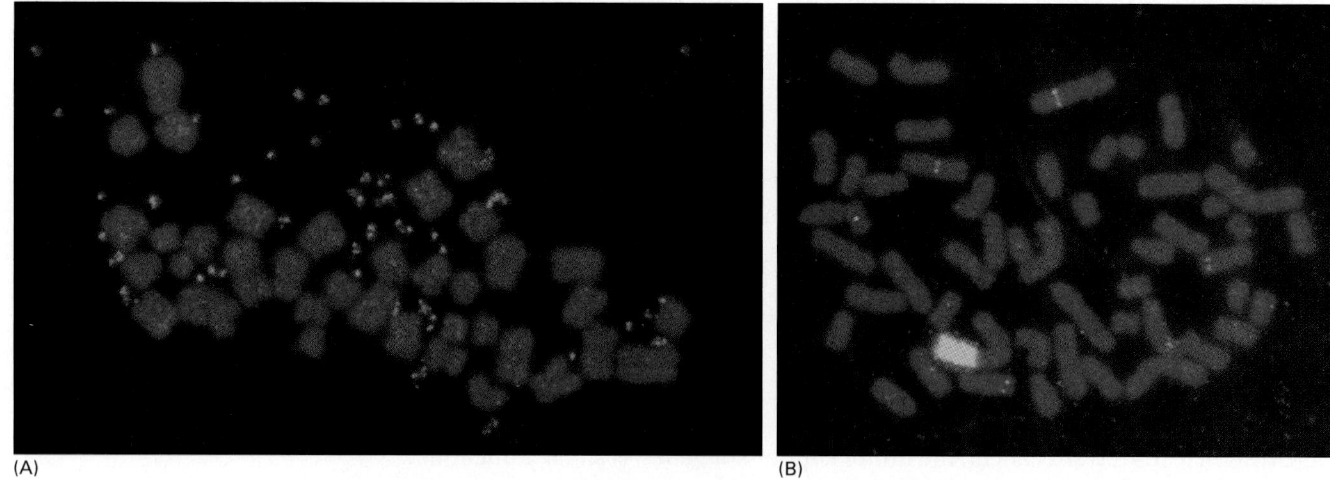
(A)
(B)

Figure 23–28 Chromosomal changes in cancer cells reflecting gene amplification. In these examples, the numbers of copies of a *myc* proto-oncogene have been amplified. Amplification of oncogenes is common in carcinomas and is often visible as a curious change in the karyotype: the cell is seen to contain additional pairs of miniature chromosomes—so-called *double minute chromosomes*—or to have a *homogeneously staining region* interpolated in the normal banding pattern of one of its regular chromosomes. Both these aberrations consist of massively amplified numbers of copies of a small segment of the genome. The chromosomes are stained with a *red* fluorescent dye, while the multiple copies of the *Myc* gene are detected by *in situ* hybridization with a *yellow* fluorescent probe. (A) Karyotype of a cell in which the *Myc* gene copies are present as double minute chromosomes (paired *yellow specks*). (B) Karyotype of a cell in which the multiple *Myc* gene copies appear as a homogeneously staining region *(yellow)* interpolated in one of the regular chromosomes. (Ordinary single-copy *myc* genes can be detected as tiny *yellow dots* elsewhere in the genome.) (Courtesy of Denise Sheer.)

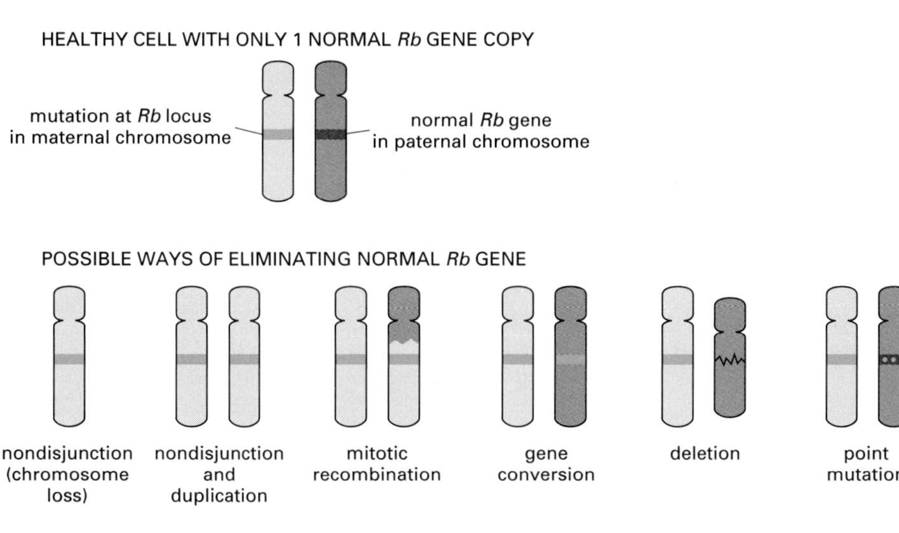

Figure 23–29 Six ways of losing the remaining good copy of a tumor suppressor gene. A cell that is defective in only one of its two copies of a tumor suppressor gene—for example, the *Rb* gene—usually behaves as a normal, healthy cell; the diagrams show how it may come to lose the function of the other gene copy as well and thereby progress toward cancer. Cloned DNA probes can be used in conjunction with tests for polymorphisms (see Chapter 8) to analyze the tumor DNA and discover which type of event has occurred in a given patient. A seventh possibility, encountered with some tumor suppressors, is that the gene may be silenced by an epigenetic change, without alteration of the DNA sequence. (After W.K. Cavenee et al., *Nature* 305:779–784, 1983. © Macmillan Magazines Ltd.)

Members of the *Myc* proto-oncogene family, on the other hand, are not usually activated by mutations in the protein coding region; instead the genes are overexpressed or amplified (see Figure 23–28). In normal cells, the Myc protein acts in the nucleus as a signal for cell proliferation, as discussed in Chapter 17; excessive quantities of Myc can cause the cell to proliferate in circumstances where a normal cell would halt. Although *Myc* is frequently amplified in cancers, it can also be made active by a chromosomal translocation. As a result of this rearrangement, powerful gene regulatory sequences are placed inappropriately next to the Myc protein coding sequence, producing unusually large amounts of *Myc* mRNA. For example, in Burkitt's lymphoma a translocation brings the *Myc* gene under the control of sequences that normally drive the expression of antibodies in B cells. As a result, the mutant B cells proliferate to excess and form a tumor. Similar specific chromosome translocations are common in lymphomas and leukemias.

Tumor suppressor genes can also be inactivated in many different ways, with different combinations of genetic mishaps coming together to eliminate or cripple both gene copies. The first copy may, for example, be lost by a small chromosomal deletion or inactivated by a point mutation. Even epigenetic changes can inactivate a tumor suppressor gene: for example, its promoter can become methylated or the gene packed into heterochromatin, effectively shutting down gene expression. The second copy may be inactivated in a similar way, but more commonly it is eliminated by a less specific mechanism: the chromosome carrying the remaining normal copy may be lost from the cell, or the normal gene can be replaced by a mutant version through mitotic recombination or gene conversion. The range of possibilities for losing the remaining good copy of a tumor suppressor gene is summarized, using the *Rb* gene as an example, in Figure 23–29.

The Hunt for Cancer-Critical Genes Continues

The sequencing of the human genome has opened up new avenues toward the systematic discovery of cancer-critical genes. It is now possible in principle to draw up a practically complete list of the 30,000-odd genes in the human genome and to examine every one of them in a given cancer cell line, looking for potentially significant abnormalities, through automated analysis of either the mRNA that the cells produce or their genomic DNA. By applying this procedure to a reasonable number of different cancers, it should be possible to identify the genes that repeatedly undergo mutation in cancer. To carry out such a project in an exhaustive and systematic way is an enormously costly proposition, but not impossibly so, since less than 2% of the human genome actually codes for proteins. Such an effort is already under way. At the same time, a collaborative enterprise has begun to create an accessible central clearing-house for information about cancer-critical genes—their sequences, mutations, and expression profiles in a variety of cancerous and normal cells. This accumulation of data should lead eventually to an exhaustive catalog of the genes whose mutations contribute to cancer.

To search for cancer-critical genes in the ways we have described, there is in principle no need at the outset to know their normal functions or to understand how mutations in them cause cancer. The strategy, as in a genetic screen for genes involved in any other process, is first to find the culprits, then to work out how they commit their crimes. As our understanding of cell biology improves, however, it becomes easier to guess which genes are likely suspects, and to use these functional clues to track them down. The task of finding the genes is therefore closely entangled with the problem of discovering what they do. This is the topic of the next section.

Summary

Cancer-critical genes in general can be classified into two groups, according to whether the dangerous mutations in them are those that cause loss of function or those that cause gain of function. Loss-of-function mutations of tumor suppressor genes relieve cells of inhibitions that normally help to hold their numbers in check; gain-of-function mutations of proto-oncogenes stimulate cells to increase their numbers when they should not. These latter mutations have a dominant effect, and the mutant genes, known as oncogenes, are sometimes identified by their ability to drive a specialized line of tester cells toward cancerous proliferation. Many were first discovered because they cause cancer in animals when they are introduced by infection with a viral vector that has picked up genetic material from a previous host cell. Oncogenes can also be located by examining human cancer cells for genes targeted by activating mutations or by the chromosomal translocations that can signal the presence of a cancer-critical gene.

Mutations in tumor suppressor genes are generally recessive in their effects on the individual cell: there is no loss of control until both gene copies are put out of action. Current methods for finding these genes depend on scanning the genomes of cancer cells for signs of gene loss, often manifest as a loss of heterozygosity in a specific chromosomal region. Another approach has been to study cancers that run in families. Though these hereditary forms of cancer are rare, they have led to the discovery of tumor suppressor genes whose loss is a common feature of many cancers. Such cancer-prone individuals often inherit one defective and one functional gene copy of a tumor suppressor gene; they have an increased predisposition toward developing cancer because a single mutation in a somatic cell is enough, with the inherited mutation, to create a cell that totally lacks the tumor suppressor gene function. The recent sequencing of the human genome and the availability of increasingly powerful tools for systematically searching DNA for significant mutations should soon lead to a much more complete catalog of cancer-critical genes.

THE MOLECULAR BASIS OF CANCER-CELL BEHAVIOR

Oncogenes and tumor suppressors—and the mutations that affect them—are different beasts from the point of view of the cancer gene hunter. But from a cancer cell's point of view they are two sides of the same target. The same kinds of effects on cell behavior can result from mutations in either class of genes, because most of the control mechanisms in the cell involve both inhibitory (tumor suppressor) and stimulatory (proto-oncogene) components. In terms of function, the important distinction is not the distinction between a tumor suppressor and a proto-oncogene, but between genes lying in different biochemical and regulatory pathways.

Some of the pathways important in cancer carry signals from a cell's environment (as discussed in Chapter 15); others are responsible for the cell's internal programs, such as those that control the cell cycle or cell death (discussed in Chapter 17); still others govern the cell's movements and mechanical interactions with its neighbors (discussed in Chapter 19). The various pathways are linked and interdependent in complex ways. Much of what we know about them has been learned as a byproduct of cancer research; conversely, study of these basic aspects of cell biology has transformed our understanding of cancer.

In the first section of this chapter, we summarized in general terms the properties that make a cancer cell a cancer cell and listed the kinds of misbehavior a cancer cell displays. In this section, we consider how each of the characteristic properties arises from mutations that have been identified in cancer-critical genes affecting specific regulatory pathways. For some parts of this problem the answers are straightforward. For other parts, large mysteries remain.

We begin with a brief general discussion of how we determine the cellular function of cancer-critical genes. We then review what is known about how these cancer-critical genes control the relevant cell behaviors. Finally, we turn to the development of colon cancer as an example of how tumors evolve through the accumulation of mutations that lead from one pattern of bad behavior to another that is worse.

Studies of Developing Embryos and Transgenic Mice Help to Uncover the Function of Cancer-Critical Genes

Given a gene that is mutated in a cancer, we need to understand both how the gene functions in normal cells and how mutations in the gene contribute to the aberrant behaviors characteristic of cancer cells. When *Rb* was originally cloned, for example, all that was known was that it was deleted in a cancer. In the case of *Ras*, the mutant gene was known to direct cells to proliferate excessively and inappropriately in culture, but this observation did not reveal how the Ras protein functions in normal or cancer cells. In both cases, cancer research was the starting point for studies that revealed the key role of these gene products in normal cells—Rb as a cell-cycle regulator, Ras as a central component of cell-signaling pathways.

Today, we know much more about cells, so that when a new gene is identified as critical for cancer, it often turns out to be familiar from studies in another context. For example, many oncogenes and tumor suppressor genes are found to be homologs of genes already known for their role in embryonic development. Examples include components of practically all the major developmental signaling pathways through which cells communicate (discussed in Chapters 15 and 21): the Wnt, Hedgehog, TGFβ, Notch, and receptor tyrosine kinase signaling pathways all include important cancer-critical genes—with *Ras* being part of the last of these pathways.

In hindsight, this is no surprise. As we saw in Chapter 22, the same signaling mechanisms that control embryonic development operate in the normal adult body to control cell turnover and maintain homeostasis. Both the development of a multicellular animal and the maintenance of its adult structure depend on cell-cell communication and on regulated cell proliferation, cell differentiation, cell death, cell movement, and cell adhesion—in other words, on all the aspects of cell behavior whose derangement underlies cancer. Developmental biology, often using model animals such as *Drosophila* and *C. elegans,* thus provides a key to the normal functions of many cancer-critical genes.

Ultimately, however, we want to know what mutations in these genes do to cells in the tissues that give rise to the cancer. A certain amount of information can be obtained by studying cells *in vitro* or by examining human cancer patients. But to investigate how mutations in various cancer-critical genes affect tissues in a whole organism, the transgenic mouse has proved particularly useful.

Transgenic mice that carry an oncogene in all their cells can be generated by methods described in detail in Chapter 8. Oncogenes introduced in this way may be expressed in many tissues or in only a select few, according to the tissue specificity of the associated regulatory DNA. Studies of such transgenic animals reveal that, even in mice, a single oncogene is not usually sufficient to turn a normal cell into a cancer cell. Typically, in mice that are endowed with a *Myc* or *Ras* oncogene, some of the tissues that express the oncogene grow to an exaggerated size, and over time occasional cells undergo further changes to give rise to cancers. The vast majority of the cells in the transgenic mouse that express the *Myc* or *Ras* oncogene, however, do not give rise to cancers, showing that the single oncogene is not enough to cause malignancy. From the point of view of the whole animal, the inherited oncogene, nevertheless, is a serious menace because it increases the risk of developing cancer. Mice that express more than one oncogene can be generated by mating a pair of transgenic mice—one carrying a *Myc* oncogene, the other carrying a *Ras* oncogene, for example. These offspring develop cancers at a much higher rate than either parental strain (Figure 23–30), but again the cancers originate as scattered isolated tumors among noncancerous cells. Thus, even with these two expressed oncogenes, the cells must undergo further, randomly generated changes to become cancerous.

Just as activated oncogenes can be introduced into mouse tissues, so tumor suppressor genes can be inactivated by 'knocking out' the gene in the mouse using reverse genetic techniques (see Figure 8–74). Several tumor suppressor genes have been knocked out in mice, including *Rb*. As anticipated, many of the mutant strains that are missing one copy of a tumor suppressor gene are cancer prone. Deletion of both copies often leads to death at an embryonic stage, reflecting the essential roles these genes play during normal development. To bypass this block and see the effect of homozygous mutations in an adult tissue, one can use the methods described in Chapter 8 to create conditional mutations, such that only one tissue—say the liver—displays the defect. In these ways, transgenic mice have become a key source of information as to the mechanisms of tumor formation. We shall see later that they also provide important models for the development of new cancer therapies.

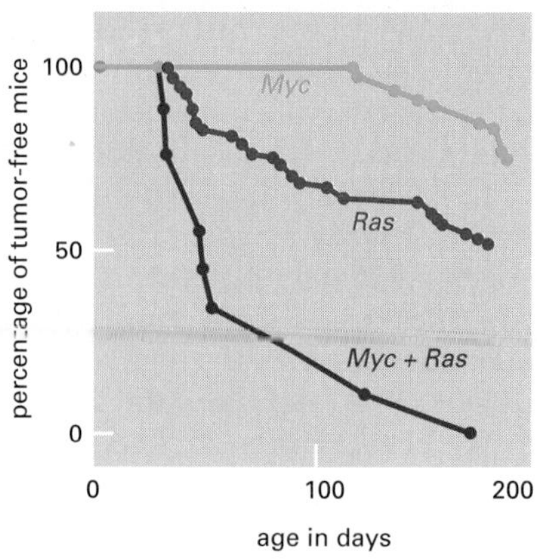

Figure 23–30 Oncogene collaboration in transgenic mice. The graphs show the incidence of tumors in three types of transgenic mice, one carrying a *Myc* oncogene, one carrying a *Ras* oncogene, and one carrying both oncogenes. For these experiments two lines of transgenic mice were first constructed. One carries an inserted copy of an oncogene created by fusing the proto-oncogene *Myc* with the mouse mammary tumor virus regulatory DNA (which then drives *Myc* overexpression in specific tissues such as the mammary gland). The other line carries an inserted copy of the *Ras* oncogene under control of the same regulatory element. Both strains of mice develop tumors much more frequently than normal, most often in the mammary or salivary glands. Mice that carry both oncogenes together are obtained by crossing the two strains. These hybrids develop tumors at a far higher rate still, much greater than the sum of the rates for the two oncogenes separately. Nevertheless, the tumors arise only after a delay and only from a small proportion of the cells in the tissues where the two genes are expressed. Some further accidental change, in addition to the two oncogenes, is apparently required for the development of cancer. (After E. Sinn et al., *Cell* 49:465–475, 1987. © Elsevier.)

Many Cancer-Critical Genes Regulate Cell Division

Most cancer-critical genes code for components of the pathways that regulate the social behavior of cells in the body—in particular, the mechanisms by which signals from a cell's neighbors can impel it to divide, differentiate, or die (Figure 23–31). In fact, many of the components of cell-signaling pathways were first identified through searches for cancer-causing genes, and a full list of proto-oncogene products and tumor suppressors includes examples of practically every type of molecule involved in cell signaling—secreted proteins, transmembrane receptors, GTP-binding proteins, protein kinases, gene regulatory proteins, and so on (discussed in detail in Chapter 15). Many cancer mutations alter signal pathway components in a way that causes them to deliver proliferative signals even when more cells are not needed, switching on cell growth, DNA replication, and cell division inappropriately. Mutations that inappropriately activate a receptor tyrosine kinase, such as the EGF receptor, or proteins in the Ras family, which lie downstream from such growth factor receptors, act in this way.

Other signaling pathways can function to inhibit cell division, the best known example being the antigrowth effect of the TGFβ family of signaling proteins. Loss of growth inhibition through TGFβ–mediated pathways contributes to the genesis of several types of human cancers. The receptor TGFβ-RII is found to be mutated in some cancers of the colon and Smad4—a key intracellular signal transducer in the pathway—is inactivated in cancers of the pancreas and some other tissues.

Ultimately, the cancer-critical genes that regulate cell division exert their effects by acting on the central cell-cycle control machinery (see Figure 17–41). Not surprisingly, mutations in this machinery feature prominently in many cancers. As described in Chapter 17, a key point at which cells make the decision to replicate their DNA and enter the cell division cycle is thought to be controlled by the Rb protein, the product of the tumor suppressor gene *Rb*. Rb serves as a brake that restricts entry into S phase by binding to gene regulatory proteins needed to express genes whose products are required for progress round the cycle. Normally, this inhibition by Rb is relieved at the appropriate time by phosphorylation of Rb, which causes it to release its inhibitory grip.

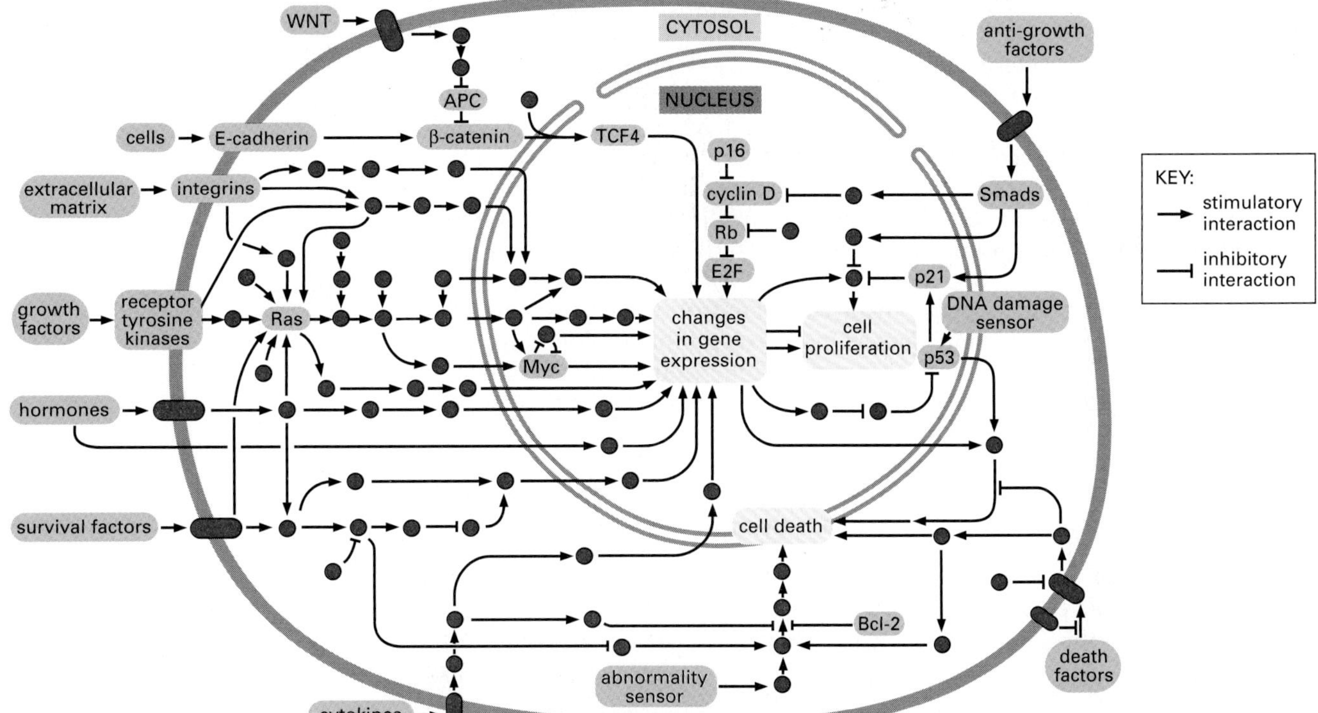

Figure 23–31 Chart of the major signaling pathways relevant to cancer in human cells, indicating the cellular locations of some of the proteins modified by mutation in cancers. Products of both oncogenes and tumor suppressor genes often occur within the same pathways. Individual signaling proteins are indicated by *red circles*, with the cancer-critical components and control mechanisms discussed in this chapter in *green*. Stimulatory and inhibitory interactions between proteins are designated as shown in the key. (Adapted from D. Hanahan and R.A. Weinberg, *Cell* 100:57–70, 2000.)

Many cancer cells proliferate inappropriately by eliminating Rb entirely, as we have already seen. Other tumors achieve the same endpoint by acquiring mutations in other components of the Rb regulatory pathway (Figure 23–32). Thus, in normal cells, a complex of cyclin D1 and the cyclin-dependent kinase Cdk4 (G_1-Cdk) stimulates progression through the cell cycle by phosphorylating Rb (see Figure 17–30). The p16 (INK4) protein—which is produced when cells are stressed—inhibits cell-cycle progression by preventing the formation of an active cyclin D1–Cdk4 complex. Some glioblastomas and breast cancers are found to have amplified the genes encoding Cdk4 or cyclin D1, thus favoring cell proliferation. And deletion or inactivation of the *p16* gene is common in many forms of human cancer. In cancers where it is not inactivated by mutation, this gene is often silenced by methylation of its regulatory DNA.

The variety of ways in which the machinery of cell-cycle control can be altered in cancer illustrates two important points. First, it explains why individual cases of a particular cancer showing the same symptoms may arise from different mutations: in many cases several alternative mutations will have much the same effect on cell proliferation. Second, it reinforces the point that there is no fundamental difference in the processes that are affected by oncogenes—which become activated by mutation—and those affected by tumor suppressor genes—which become inactivated. These two classes of cancer-critical genes merely differ in whether they play a stimulatory or inhibitory role in a pathway (see also Figure 23–31).

Mutations in Genes That Regulate Apoptosis Allow Cancer Cells to Escape Suicide

To achieve net cell proliferation, it is necessary not only to drive cells into division, but also to keep cells from committing suicide by apoptosis. There are many normal situations in which cells proliferate continuously, but the cell division is exactly balanced by cell loss. In the germinal centers of lymph nodes, for example, B cells proliferate rapidly but most of their progeny are eliminated by apoptosis. Apoptosis is thus essential in maintaining the normal balance of cell births and deaths in tissues that undergo cell turnover. It also has a vital role in the cellular reaction to damage and disorder. As described in Chapter 17, cells in a multicellular organism commit suicide when they sense that something has gone wrong—when their DNA is severely damaged or when they are deprived of survival signals that tell them they are in their proper place. Resistance to apoptosis is thus a key characteristic of malignant cells, essential for enabling them to increase in number and survive where they should not.

A number of mutations that inhibit apoptosis have been found in tumors. One protein that blocks apoptosis, called Bcl-2, was discovered because it is the target of a chromosome translocation in a B-cell lymphoma. The effect of the translocation is to place the *Bcl-2* gene under the control of a regulatory element that drives overexpression, which allows survival of B lymphocytes that would normally have died.

Of all the cancer-critical genes involved in control of apoptosis, however, there is one that is implicated in cancers in an exceptionally wide range of tissues. This gene, called *p53*, stands at a crucial intersection in the network of pathways governing a cell's responses to DNA damage and other stressful mishaps. Control of apoptosis is only part of the gene's function—though a very important part. As we shall now explain, when *p53* is defective, genetically damaged cells do not merely fail to die; worse still, they wantonly continue to proliferate, accumulating yet more genetic damage that can lead toward cancer.

Mutations in the *p53* Gene Allow Cancer Cells to Survive and Proliferate Despite DNA Damage

The ***p53*** gene—named for the molecular mass of its protein product—may be the most important gene in human cancer. This tumor suppressor gene is mutated in about half of all human cancers. What makes *p53* so critical? The answer lies in its triple involvement in cell-cycle control, in apoptosis, and in maintenance of genetic stability—all aspects of the fundamental role of the p53 protein in protecting the organism against cellular damage and disorder.

In contrast with Rb, very little p53 protein is found in most of the cells of the body under normal conditions. In fact, p53 is not required for normal development: transgenic mice in which both copies of the gene have been knocked out appear normal in all respects except one—they usually develop cancer by the

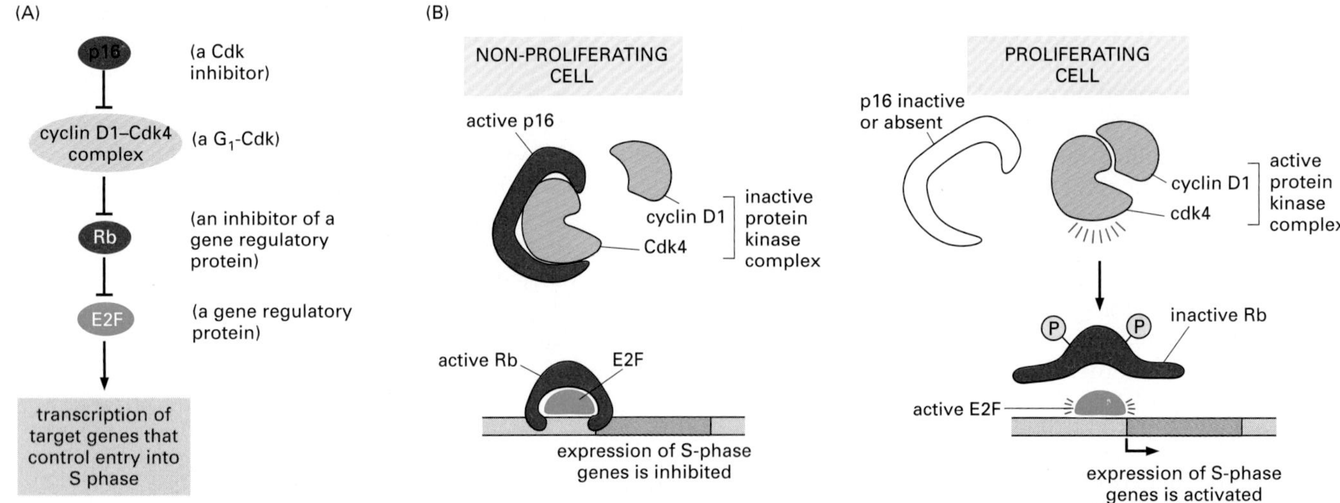

Figure 23–32 The pathway that controls cell cycling via Rb protein. All the components of this pathway have been found to be altered by mutation in human cancers (products of proto-oncogenes, *green;* products of tumor suppressor genes, *red;* E2F shown in *blue* because it has both inhibitory and stimulatory actions, depending on the other proteins that are bound to it). In most cases, only one of the components is altered in any individual tumor. (A) A simplified view of the dependency relationships in this pathway; see Figure 17–30 for further details. (B) The Rb protein inhibits entry into the cell-division cycle when it is unphosphorylated. The complex of Cdk4 and cyclin D1 phosphorylates Rb, thereby encouraging cell proliferation. When a cell is stressed, p16 inhibits the formation of an active Cdk4/cyclin D1 complex, preventing proliferation. Inactivation of Rb or p16 by mutation encourages cell division (thus each can be regarded as a tumor suppressor) while overactivity of Cdk4 or cyclin D1 encourages cell division (thus each can be regarded as a proto-oncogene).

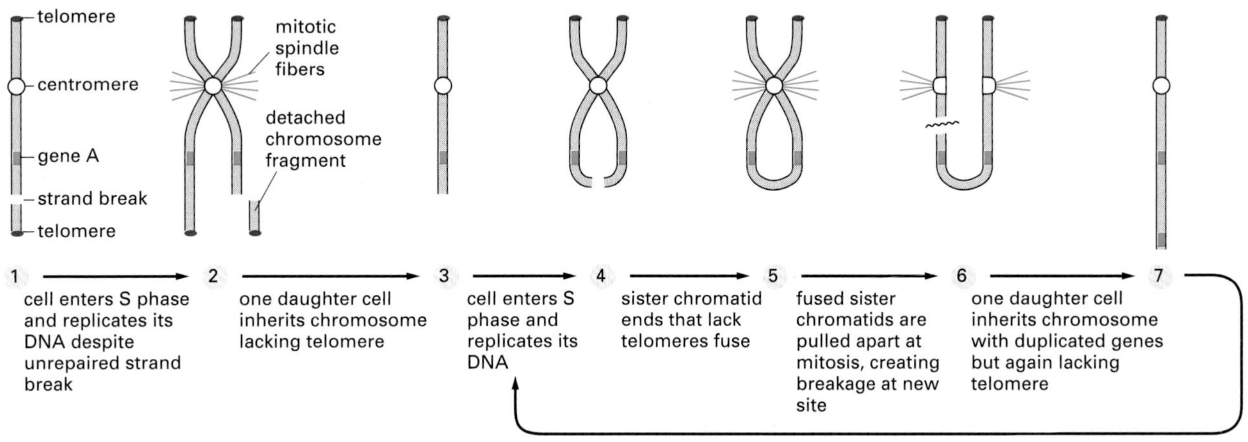
telomere
centromere
gene A
strand break
telomere
mitotic spindle fibers
detached chromosome fragment
1
cell enters S phase and replicates its DNA despite unrepaired strand break
2
one daughter cell inherits chromosome lacking telomere
3
cell enters S phase and replicates its DNA
4
sister chromatid ends that lack telomeres fuse
5
fused sister chromatids are pulled apart at mitosis, creating breakage at new site
6
one daughter cell inherits chromosome with duplicated genes but again lacking telomere
7
BREAKAGE-FUSION-BRIDGE CYCLE

Figure 23–33 How the replication of damaged DNA can lead to chromosome abnormalities, gene amplification and gene loss. The diagram shows one of several possible mechanisms. The process begins with accidental DNA damage in a cell that lacks functional p53 protein. Instead of halting at the p53-dependent checkpoint in the division cycle, where a normal cell with damaged DNA would pause until the damage was repaired, the p53-defective cell enters S phase, with the consequences shown. Once a chromosome carrying a duplication and lacking a telomere has been generated, repeated rounds of replication, chromatid fusion, and unequal breakage (the so-called breakage-fusion-bridge cycle) can increase the number of copies of the duplicated region still further.

Selection in favor of cells with increased numbers of copies of a gene in the affected chromosomal region will thus lead to mutants in which the gene is amplified to a high copy number. The multiple copies may eventually become visible as a homogeneously staining region in the chromosome, or they may—either through a recombination event or through unrepaired DNA strand breakage—become excised from their original locus and so appear as independent double minute chromosomes (see Figure 23–28). The chromosomal disorder can also lead to loss of genes, with selection in favor of cells that have lost tumor suppressors.

age of 3 months. These observations suggest that p53 may have a function that is required only occasionally or in special circumstances. Indeed, when normal cells are deprived of oxygen or exposed to treatments that damage DNA, such as ultraviolet light or gamma rays, they raise their concentration of p53 protein by reducing the normally rapid rate of degradation of the molecule. The p53 response is seen also in cells where oncogenes such as *Ras* and *Myc* are active, generating an abnormal stimulus for cell division.

In all these cases, the high level of p53 protein acts to limit the harm done. Depending on circumstances and the severity of the damage, the p53 may either drive the damaged or mutant cell to commit suicide by apoptosis (see p. 1013)—a relatively harmless event for the multicellular organism—or it may trigger a mechanism that bars the cell from dividing so long as the damage remains unrepaired. The protection provided by p53 is an important part of the reason why mutations that activate oncogenes such as *Ras* and *Myc* are not enough by themselves to create a tumor.

As discussed in Chapter 17, the p53 protein exerts its cell-cycle effects, in part at least, by binding to DNA and inducing the transcription of *p21*—a regulatory gene whose protein product binds to Cdk complexes required for entry into and progress through S-phase. By blocking the kinase activity of these Cdk complexes, the p21 protein prevents the cell from entering S phase and replicating its DNA.

Cells defective in *p53* fail to show these responses. They tend to escape apoptosis, and if their DNA is damaged—by radiation or by some other mishap—they carry on dividing, plunging into DNA replication without pausing to repair the breaks and other DNA lesions that the damage has caused. As a result, they may either die or, far worse, survive and proliferate with a corrupted genome. A common consequence is that chromosomes become fragmented and incorrectly rejoined, creating, through further rounds of cell division, an increasingly disrupted genome as explained in Figure 23–33. Such chromosomal mayhem can lead to both loss of tumor suppressor genes and activation of oncogenes, for example by gene amplification. In addition to being an important mechanism for activating oncogenes, gene amplification can also enable cells to develop resistance to therapeutic drugs, as we see below.

In summary, p53 helps a multicellular organism to cope safely with DNA damage and other stressful cellular events, acting as a check on cell proliferation in circumstances where it would be dangerous. Many cancer cells contain large quantities of mutant p53 protein (of an ineffectual variety), suggesting that the genetic accidents they undergo, or the stresses of growth in an inappropriate environment, have created the signals that normally call the p53 protein into play. The loss of *p53* activity can thus be trebly dangerous in relation to cancer. First, it allows faulty mutant cells to continue through the cell cycle. Second, it allows them to escape apoptosis. Third, it leads to the genetic instability characteristic of cancer cells, allowing further cancer-promoting mutations to accumulate as they divide. Many other mutations can contribute to each of these types of misbehavior, but p53 mutations contribute to them all.

DNA Tumor Viruses Activate the Cell's Replication Machinery by Blocking the Action of Key Tumor Suppressor Genes

DNA tumor viruses cause cancer mainly by interfering with cell-cycle controls, including those that depend on p53. To understand this type of viral carcinogenesis, it is important to understand the life history of the virus. Viruses use the DNA replication machinery of the host cell to replicate their own genomes. To make many infectious virus particles from a single host cell, a DNA virus has to commandeer this machinery and drive it hard, breaking through the normal constraints on DNA replication and usually killing the host cell in the process. Typically, however, the virus also has another option: it can propagate its genome as a quiet, well-behaved passenger in the host cell, replicating in parallel with the host cell's DNA in the course of ordinary cell division cycles. The virus can switch between these two modes of existence, remaining latent and harmless or proliferating to generate infectious particles according to circumstances. No matter which way of life the virus is following, it is not in its interests to cause cancer. But genetic accidents can occur, such that the virus misuses its equipment for commandeering the DNA replication machinery, and instead of switching on rapid replication of its own genome, switches on persistent proliferation of the host cell.

DNA viruses are a diverse group, but something of this sort seems to happen with most of those that are involved in cancer. The *papillomaviruses,* for example, are the cause of human warts and are especially important as a key causative factor in carcinomas of the uterine cervix (about 6% of all human cancers). Papillomaviruses infect the epithelium, and are retained in the basal layer of cells as extrachromosomal plasmids that replicate in step with the chromosomes. Infectious virus particles are generated in the outer epithelial layers, as cells begin to differentiate before being sloughed from the surface. Here, cell division should normally be arrested, but the virus interferes with this arrest so as to allow rapid replication of its own genome. Usually, the effect is restricted to the outer layers of cells and relatively harmless, as in a wart. Occasionally, through a genetic accident causing misregulation of the viral genes whose products prevent cell-cycle arrest, the control of cell division is subverted in the basal layer also, in the stem cells of the epithelium. This can lead to cancer, with the viral genes acting as oncogenes (Figure 23–34).

In papillomaviruses, the viral genes that are mainly to blame are called *E6* and *E7.* The products of these viral oncogenes interact with many host cell proteins, but in particular they bind to the protein products of two key tumor suppressor genes of the host cell, putting them out of action and so permitting the cell to replicate its DNA and divide in an uncontrolled way. One of these host

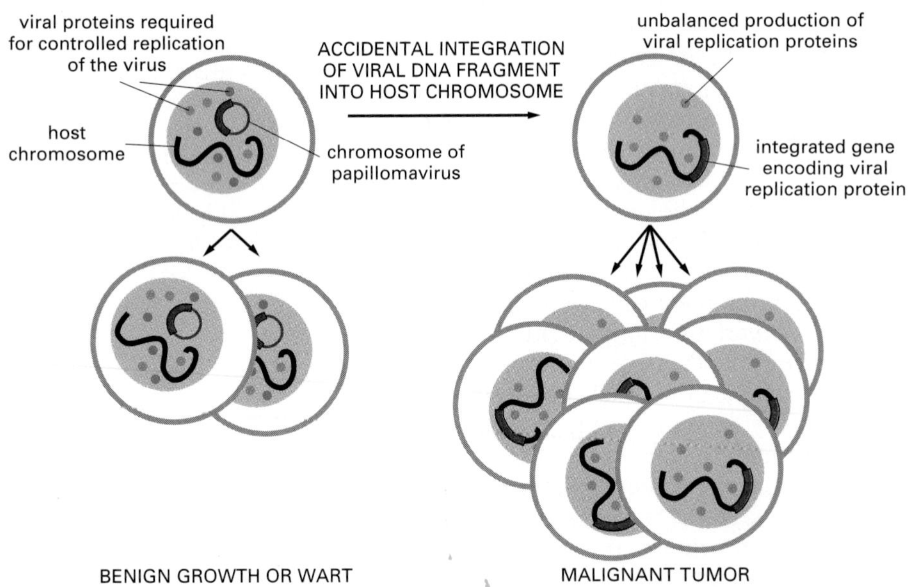

Figure 23–34 How certain papillomaviruses are thought to give rise to cancer of the uterine cervix. Papillomaviruses have double-stranded circular DNA chromosomes of about 8000 nucleotide pairs. In a wart or other benign infection these chromosomes are stably maintained in the basal cells of the epithelium as plasmids whose replication is regulated so as to keep step with the chromosomes of the host *(left).* Rare accidents can cause the integration of a fragment of such a plasmid into a chromosome of the host, altering the environment of the viral genes. This (or possibly some other cause) disrupts the control of viral gene expression. The unregulated production of viral replication proteins interferes with the control of cell division, thereby helping to generate a cancer *(right).*

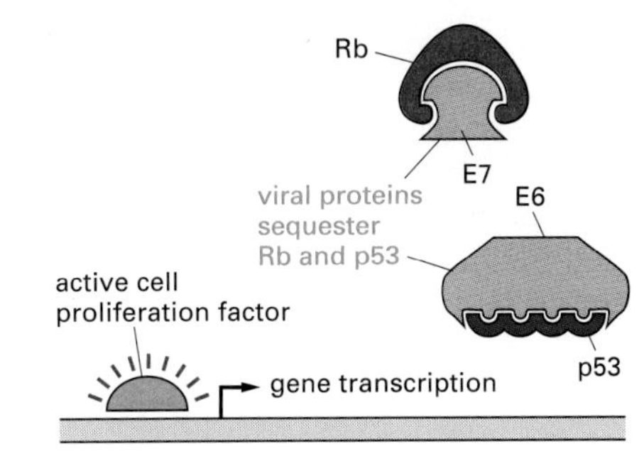

Figure 23–35 Activation of cell proliferation by a DNA tumor virus. Papillomavirus uses two viral proteins, E6 and E7, to sequester the host cell's p53 and Rb respectively. The SV40 virus (a related virus that infects monkeys) uses a single dual-purpose protein called large T antigen, for the same purpose. The E6 protein binding leads to ubiquitylation of its p53 partner, inducing p53 proteolysis (not shown).

proteins is Rb: by binding to Rb, the viral E7 protein prevents it from binding to its normal associates in the cell. The other host protein inactivated by the virus is the tumor suppressor p53, which is bound by the viral E6 protein, triggering p53 destruction (Figure 23–35). Elimination of p53 allows the abnormal cell to survive, divide, and accumulate yet more abnormalities.

Telomere Shortening May Pave the Way to Cancer in Humans

The mouse is the most widely used model organism for the study of cancer, yet the spectrum of cancers seen in mice differs dramatically from that seen in humans. The great majority of mouse cancers are sarcomas and leukemias, whereas more than 80 percent of human cancers are carcinomas—cancers of epithelia where rapid cell turnover occurs (see Figure 23–2). Many therapies have been found to cure cancers in mice; but when the same treatments are tried in humans, they usually fail. What could be the reason for the difference between mouse and human cancer, and what can it tell us about the molecular mechanisms of the disease? An important part of the answer may lie in the behavior of telomeres and the relationship between telomere shortening, replicative cell senescence, and genetic instability.

As we saw earlier, most human cells seem to have a built-in limit to their proliferation: they show replicative senescence, at least when grown in culture. Replicative cell senescence in humans is thought to be caused by changes in the structure of **telomeres**—the repetitive DNA sequences and associated proteins that cap the ends of each chromosome. These telomeric DNA sequences are synthesized and maintained by a special mechanism that requires the enzyme telomerase, as explained in Chapter 5. In most human cells, other than those of the germ line and some stem cells, expression of the gene coding for the catalytic subunit of telomerase is switched off, or at least not fully activated. As a result, the telomeres in these cells tend to become a little shorter with each round of cell division. Eventually, the telomeric cap on the chromosome end can become shortened to the point where a danger signal is generated, arresting the cell cycle. The signal is similar, in function at least, to the one that arrests the cycle when an uncapped DNA end is created by an accidental double-strand chromosome break. The effect in both cases is to prevent cell division so long as the cell contains broken or inadequately capped DNA. In the cell with the chromosome break, this allows time for DNA repair; in the normal senescent cell, it seems that it simply puts a stop to cell proliferation. As we discussed at the beginning of the chapter, it is not clear how often cells in normal human tissues run up against this limit; but if a self-renewing cell population does undergo replicative senescence, any rogue cell that undergoes a mutation that lets it carry on dividing will enjoy a huge competitive advantage—much more than if the same mutation had occurred in a cell in a nonsenescent population. Viewed in this light, replicative senescence might be expected to favor the development of cancer.

Mice have telomeres much longer than those of humans. Moreover, unlike humans, they keep telomerase active in their somatic cells, and mouse telomeres therefore do not tend to shorten with increasing age of the organism. It is possible, however, to use gene knockout technology to make mice that lack functional telomerase. In these mice, the telomeres become shorter with every generation, but no untoward consequences are seen until, in the great-great-grandchildren of the initial mutants, the telomeres become so short that they disappear or cease to function. Beyond this point, the mice begin to show various abnormalities, including an increased incidence of cancer. This raises the possibility that natural telomere shortening helps to engender many human tumors.

In a Population of Telomere-Deficient Cells, Loss of p53 Opens an Easy Gateway to Cancer

In contrast with most normal cells in humans, most human cancer cells express telomerase. This is thought to be the reason why, unlike normal cells, they tend to divide without limit in culture, and it is added evidence that telomere maintenance has a significant role in cancer.

Given that cancer cells contain telomerase, an obvious suggestion is that they arise from mutant precursors that have simply avoided shortening their telomeres, and so have never encountered the telomeric limit to cell division. There is, however, another possibility, highlighted by the observations in the telomerase-deficient mice. A cancer may derive from a cell that has experienced telomere shortening, but has suffered a mutation that lets it disregard the signals that normally arrest cell division when telomeres are too short. A mutation causing loss of p53 activity can have just this effect: occurring in one cell within a population arrested by telomere shortening, it can give that cell and its progeny an immediate competitive advantage over their nondividing neighbors. At the same time, as explained in Figure 23–36, the absence of p53 can bring on the gross chromosomal instability that is characteristic of human carcinomas, allowing the mutant cells to accumulate more mutations and evolve rapidly toward cancer. This is what appears to happen in the telomerase-deficient mice. If they lack not only telomerase but also one of their two copies of the *p53* gene, carcinomas become even more frequent. It is striking that these additional cancers are predominantly carcinomas—cancers of self-renewing epithelia—rather than the sarcomas and lymphomas usually seen in mice. Carcinomas constitute by far the commonest class of human tumors. The differences in telomere behavior could thereby account for the major difference observed between normal mice and humans in the predominant types of cancers that arise. Like human carcinoma cells, the mouse tumor cells are usually found to have inactivated their last remaining *p53* gene; they also show gross chromosomal abnormalities, with many breaks, fusions, and broken chromosome ends.

If this scenario applies to cancers in humans, the suggestion would be that telomerase becomes reactivated after, not before, the genetic catastrophe. The progressive genetic disruption following loss of p53 may be so severe that cells can rarely survive it for more than a few generations. A cell that reactivates telomerase expression will be able to halt the catastrophic cycle and regain enough chromosomal stability to survive (see Figure 23–36). Its progeny will inherit a highly abnormal chromosome set, with many mutations and alterations; these damaged cells can then continue to accumulate further mutations at a more moderate rate, driving tumor progression. This model tallies with the findings that, in breast and colorectal cancers, gross chromosomal abnormalities appear to arise early during tumor development, before telomerase is reactivated. With telomerase turned on, these carcinomas still possess enough genetic instability—due to a loss of p53 or other mutations—to continue to evolve, thereby tending to become metastatic.

It is possible that many of the most common types of human carcinomas originate in the way we have just described. But there are certainly many other ways in which cancers can arise, and the true importance of replicative senescence and telomere behavior in human cancer remains to be determined. The uncertainties highlight how little we still know about the natural history of cancer, despite the dramatic advances in understanding of cancer molecular genetics. At the very least, however, the results suggest that telomere-deficient mice can provide a reasonable model for the study of common human carcinomas. Perhaps, this mouse model will also help us to devise cancer treatments that work as well in humans as they do in mice.

The Mutations That Lead to Metastasis Are Still a Mystery

Perhaps the most serious gap in our understanding of cancer concerns malignancy and metastasis. We have yet to clearly identify mutations that specifically permit cells to invade surrounding tissues, spread through the body, and form metastases. Indeed, it is not even clear exactly what properties a cancer cell must acquire to become metastatic. One extreme view would be that the ability of cancer cells in the body to metastasize requires no further genetic changes beyond those needed for loss of cell division control. An opposite view, more commonly espoused, is that metastasis is a difficult and complex task for a cell, requiring a mass of new mutations—so many, and so varied according to circumstance, that it is hard to discover what they are individually.

Which steps in metastasis are the most difficult is still a matter of debate. But there are experimental findings that throw some light on the issue. It is obvious, first of all, that metastasis presents different problems for different types of cells. For a leukemia cell, already roaming the body via the circulation, metastasis should be easier than for a carcinoma cell that has to escape from an epithelium. As we saw at the beginning of the chapter, it is helpful to distinguish two degrees of malignancy for carcinomas, representing two phases of tumor progression. In the first phase, the tumor cells escape the normal confines of their parent epithelium and begin to invade the tissue immediately beneath—becoming *locally invasive.* In the second phase, they travel to distant sites and settle to form colonies, a process known as *metastasis.*

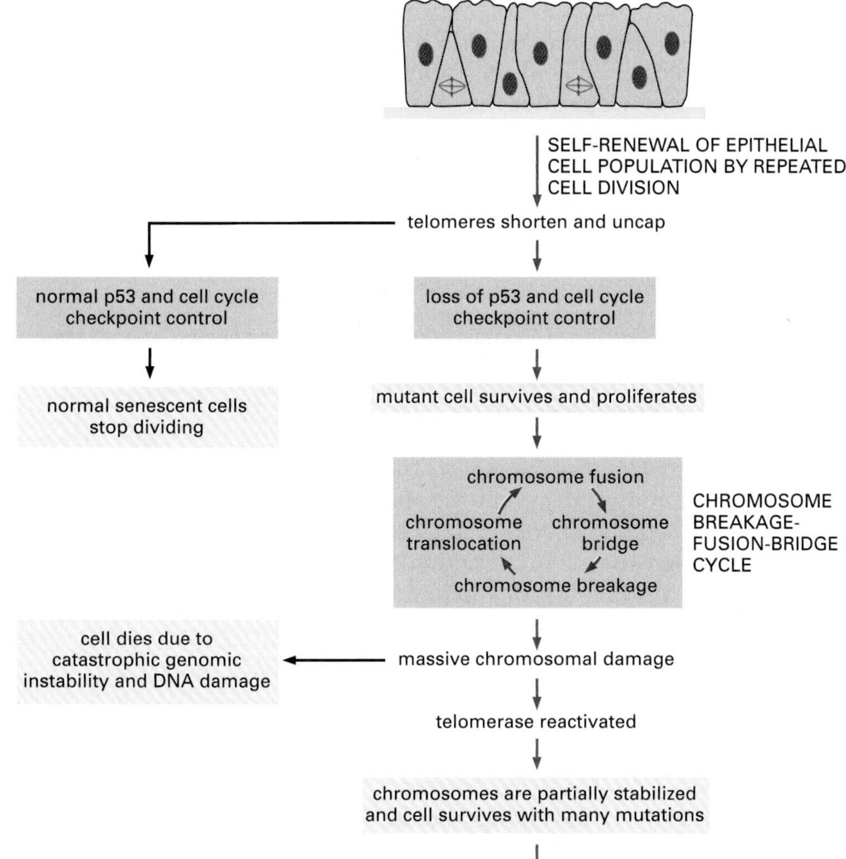

Figure 23–36 A view of how shortened telomeres may lead to chromosomal instability and cancer. Most human cells lack telomerase. As such cells divide, the telomeres that cap the ends of their chromosomes shrink. After many divisions, the telomeres become so short that they cease to function correctly, and the inadequately capped chromosome ends produce an intracellular signal similar to that produced by double-strand breaks. In the normal cells that still produce functional p53 and have their cell-cycle checkpoints intact, this triggers an arrest of cell division (replicative senescence). But a rare cell within the cell population that has acquired a mutation in its p53 gene or in its cell-cycle checkpoints may ignore this signal and continue to divide, entering a breakage–fusion–bridge cycle that causes massive chromosomal damage (see Figure 23–33). Some cells may survive this period of genetic disruption by reactivating telomerase, which halts the catastrophic cycle and restores enough chromosomal stability for cell survival. These damaged cells can then go on to accumulate the additional mutations needed to produce a cancer.

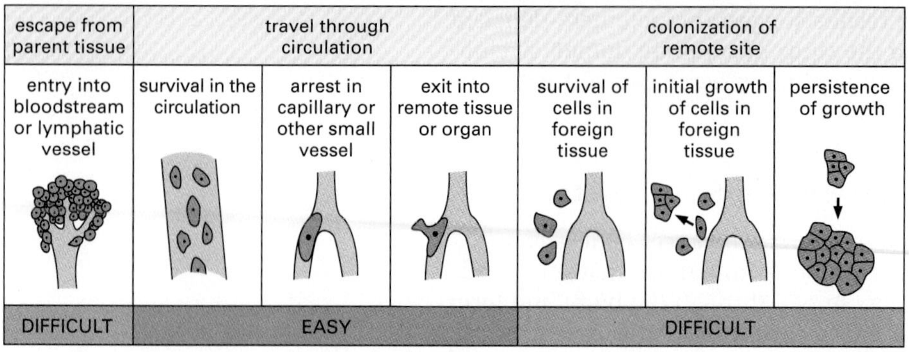

escape from parent tissue	travel through circulation			colonization of remote site		
entry into bloodstream or lymphatic vessel	survival in the circulation	arrest in capillary or other small vessel	exit into remote tissue or organ	survival of cells in foreign tissue	initial growth of cells in foreign tissue	persistence of growth
DIFFICULT	EASY			DIFFICULT		

Figure 23–37 The barriers to metastasis. Studies of labeled tumor cells leaving a tumor site, entering the circulation, and establishing metastases show which steps in the metastatic process, outlined in Figure 23–15, are difficult or 'inefficient,' in the sense that they are steps in which large numbers of cells fail and are lost. It is in these difficult steps that highly metastatic cells are observed to have much greater success than nonmetastatic cells. It seems that the ability to escape from the parent tissue, and an ability to survive and grow in the foreign tissue, are key properties that cells must acquire to become metastatic. (Adapted from A.F. Chambers et al., *Breast Cancer Res.* 2:400–407, 2000.)

Local invasiveness requires a breakdown of the mechanisms that normally hold epithelial cells together. In some carcinomas of the stomach and of the breast, the *E-cadherin* gene has been identified as a tumor suppressor gene. The primary function of the E-cadherin protein is in cell–cell adhesion, where this protein is embedded in two adjacent plasma membranes to bind epithelial cells together (see Figure 19–24). When tumor cells lacking this adhesion molecule are placed in culture, and a functional *E-cadherin* gene is put back into them, they lose some of their invasive characteristics and begin to cohere more like normal cells. Loss of *E-cadherin*, therefore, may favor cancer by specifically contributing to local invasiveness.

The second stage of malignancy, involving entry into the bloodstream or lymphatic vessels, travel via the circulation, and colonization of remote sites, is more enigmatic. The cells must, for example, cross barriers such as the basal lamina of the parent epithelium and of blood vessels. Do they need to acquire additional mutations to become able to do this? The answer is not clear. To discover what steps in metastasis present cells with the greatest difficulties—and thus might be aided by the acquisition of additional mutations—one can label cancer cells with a fluorescent dye, inject them into the circulation of a living animal, and monitor their fate. In these experiments, many cells are found to survive in the circulation, to become lodged in small vessels, and to get out of these into the surrounding tissue, regardless of whether they come from a metastatic or a nonmetastatic tumor. Most of the losses occur after this. Some cells die immediately; others survive entry into the foreign tissue but fail to grow; still others divide a few times and then stop. Here the metastasis-competent cells outperform their nonmetastatic relatives, suggesting that the ability to grow in the foreign tissue is a key property that cells must acquire to become metastatic (Figure 23–37).

To discover the changes that confer metastatic potential, one can use DNA microarrays to look for genes that are selectively switched on in cancer cells that have become highly malignant. These microarrays allow one to monitor the expression of thousands of genes at a time (see Figure 8–64). One such study took human and mouse melanoma cells that had been selected for high metastatic potential and compared them with their poorly metastatic counterparts. Of the dozen or so genes that appeared to be selectively active in the malignant cells, one that showed increased expression repeatedly in the metastatic cells was *RhoC*—a member of a family of genes known to regulate cell motility (discussed in Chapter 16). A greatly expanded use of such methods, made much more powerful with the availability of the human genome sequence, should soon give us a clearer picture of the molecular changes that allow tumor cells to metastasize.

Colorectal Cancers Evolve Slowly Via a Succession of Visible Changes

At the beginning of this chapter, we saw that most cancers develop gradually from a single aberrant cell, progressing from benign to malignant tumors by the accumulation of a number of independent genetic accidents. We have discussed what some of these accidents are in molecular terms and seen how they contribute to cancerous behavior. We now examine one particular class of common human cancers more closely, using it to illustrate and enlarge upon some of the general principles and molecular mechanisms we have introduced, and to see how we can make sense of the natural history of the disease in terms of them. We take **colorectal cancer** as our example, where the steps of tumor progression have been followed *in vivo* and carefully studied at the molecular level.

Colorectal cancers arise from the epithelium lining the colon and rectum (the lower end of the gut). They are common, currently causing over 60,000 deaths a year in the United States, or about 11% of total deaths from cancer. Like most cancers, they are not usually diagnosed until late in life (90% after the age of 55). However, routine examination of normal adults with a colonoscope (a fiber-optic device for viewing the interior of the colon and rectum) often reveals a small benign tumor, or adenoma, of the gut epithelium in the form of a protruding mass of tissue called a polyp (Figure 23–38A). These adenomatous polyps are believed to be the precursors of a large proportion of colorectal cancers. Because progression of the disease is usually very slow, there is typically a period of 10–35 years in which the slowly growing tumor is detectable but has not yet turned malignant. Thus, when people are screened by colonoscopy in their fifties and the polyps are removed—a quick and easy surgical procedure—the subsequent incidence of colorectal cancer is very low—according to some studies, less than a quarter of what it would be otherwise.

Colon cancer provides a clear example of the phenomenon of tumor progression discussed previously. In polyps smaller than 1 cm in diameter, the cells and the local details of their arrangement in the epithelium usually appear almost normal. The larger the polyp, the more likely it is to contain cells that

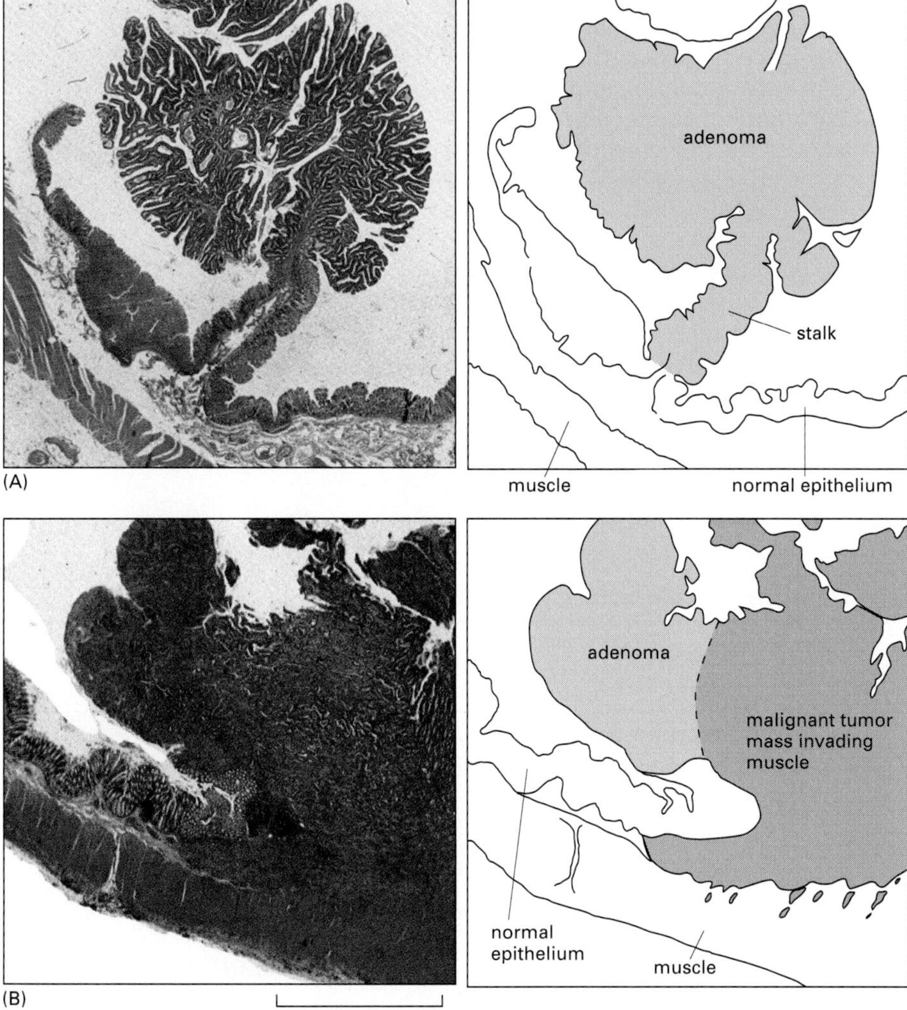

Figure 23–38 Cross-sections showing the stages in development of a typical colon cancer. (A) An adenomatous polyp from the colon. The polyp protrudes into the lumen—the space inside the colon. The rest of the wall of the colon is covered with normal colonic epithelium; the epithelium on the polyp appears mildly abnormal. (B) A carcinoma that is beginning to invade the underlying muscle layer. A high magnification view of this region showing the tumor cells growing between muscle cells is shown in Figure 23–16. (Courtesy of Paul Edwards.)

look abnormally undifferentiated and form abnormally organized structures. Sometimes, two or more sectors can be distinguished within a single polyp, the cells in one sector appearing relatively normal, those in the other appearing frankly cancerous, as though they have arisen as a mutant subclone within the original clone of adenomatous cells. At later stages in the disease, the tumor cells become invasive, first breaking through the epithelial basal lamina, then spreading through the layer of muscle that surrounds the gut (Figure 23–38B), and finally metastasizing to lymph nodes, liver, lung, and other tissues.

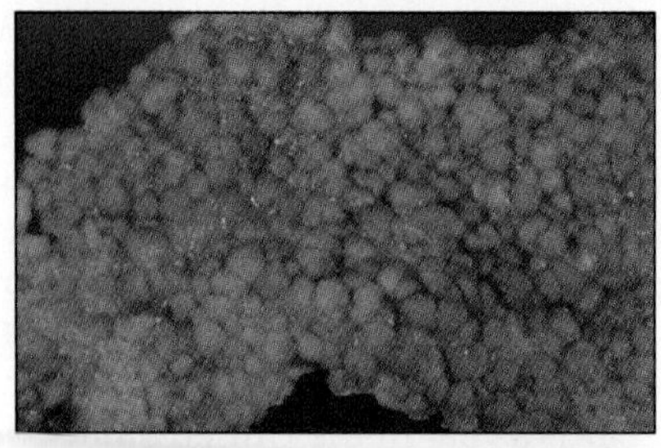

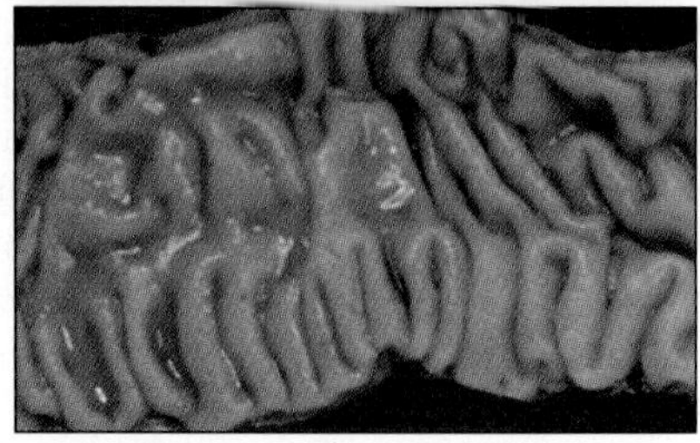

Figure 23–39 Colon of familial adenomatous polyposis coli patient compared to normal colon. (A) The polyposis colon is completely covered by hundreds of projecting polyps (shown in section in Figure 23–38A), each resembling a tiny cauliflower when viewed with the naked eye. (B) The normal colon wall is a gently undulating but smooth surface. (Courtesy of Andrew Wyllie and Mark Arends.)

A Few Key Genetic Lesions Are Common to a Majority of Cases of Colorectal Cancer

What are the mutations that accumulate with time to produce this chain of events? Of those genes so far discovered to be involved in colorectal cancer, three—*K-Ras* (a member of the *Ras* gene family), *p53*, and a third gene, *APC*, to be discussed below (Table 23–3)—stand out as very frequently mutated. Others are involved in smaller numbers of colon cancers. Still other critical genes remain to be identified.

One approach to discovery of the mutations responsible for colorectal cancer is to screen the cells for abnormalities in genes already known or suspected to be involved in cancers elsewhere. This type of genetic screening has revealed that about 40% of colorectal cancers have a specific point mutation in *K-Ras*, activating it as an oncogene, and about 60% have inactivating mutations or deletions of *p53*.

Another approach to finding cancer-critical genes is to track down the genetic defects in those rare families that show a hereditary predisposition to colorectal cancer. The first of these hereditary colorectal cancer syndromes to be elucidated was a condition known as *familial adenomatous polyposis coli (FAP)*, in which hundreds or thousands of polyps develop along the length of the colon (Figure 23–39). These make their appearance in early adult life, and if they are not removed, it is almost inevitable that one or more of them will progress to become malignant; on average, 12 years elapse from the first detection of polyps to the diagnosis of cancer. The disease can be traced to deletion or inactivation of *APC*, a gene on the long arm of chromosome 5. Individuals with FAP have inactivating mutations or deletions of the *APC* gene in all the cells of the body. Of the patients with colorectal cancer who do not have the hereditary condition—the vast majority of cases of the disease—more than 60% have similar mutations in the cells of the cancer but not in their other tissues; in other words, both copies of the *APC* gene have been lost during their lifetime. Thus, by a route similar to that which we have discussed for retinoblastoma, mutation of *APC* is identified as one of the central ingredients of colorectal cancer.

As explained earlier, tumor suppressor genes can also be tracked down, even where there is no known hereditary syndrome, by searching for genetic deletions in the tumor cells. A systematic scan of a large number of colorectal

TABLE 23–3 Some Genetic Abnormalities Detected in Colorectal Cancer Cells

GENE	CLASS	PATHWAY AFFECTED	TUMORS WITH MUTATIONS (%)
K-Ras	oncogene	receptor tyrosine-kinase signaling	40
β-catenin	oncogene	Wnt signaling	5–10
p53	tumor suppressor	stress/genetic-damage response	60
APC	tumor suppressor	Wnt signaling	> 60
Smad4	tumor suppressor	TGFβ signaling	30
TGFβ receptor II	tumor suppressor	TGFβ signaling	10
MLH1 and other DNA mismatch repair genes	tumor suppressor	DNA mismatch repair	15 (often silenced by methylation)

cancers reveals frequent losses of large sections of certain chromosomes, suggesting that those regions may harbor tumor suppressor genes. One of them is the region including *APC*. Another includes the *SMAD4* gene, which is mutated in perhaps 30% of colon cancers. At least one other important tumor suppressor gene is thought to lie in the same neighborhood as *SMAD4* on chromosome 18 but remains to be identified. Specific parts of other chromosomes also show frequent losses—or gains—in colorectal cancers and are now being searched for additional cancer-critical genes.

As our knowledge of genes and their functions has expanded, another fruitful approach has been to look for genes that interact with a known cancer-critical gene in the hope that these, too, may be targets for mutation. The APC protein is now known to be an inhibitory component of the Wnt signaling pathway (discussed in Chapter 15). It acts by binding to β-catenin, another component of the pathway, and thereby preventing activation of TCF4, a gene regulatory protein that stimulates growth of the colonic epithelium when it has β-catenin bound to it. As we saw in Chapter 22, loss of TCF4 causes a depletion of the gut stem-cell population, so that loss of the antagonist APC may cause overgrowth by the opposite effect. When the β-catenin gene was sequenced in a collection of colorectal tumors, it turned out that among the few that did not have *APC* mutations, a high proportion had activating mutations in β-catenin instead. Thus, it is the Wnt signaling pathway, rather than any single oncogene or tumor suppressor gene that it contains, that is critical for the cancer.

Defects in DNA Mismatch Repair Provide an Alternative Route to Colorectal Cancer

In addition to the hereditary disease associated with *APC* mutations, there is a second, and actually commoner, kind of hereditary predisposition to colon carcinoma in which the course of events is quite different from the one we have described for FAP. In patients with this condition, called *hereditary nonpolyposis colorectal cancer*, or *HNPCC*, the probability of colon cancer is increased without any increase in the number of colorectal polyps (adenomas). Moreover, the cancer cells in the tumors that develop are unusual, inasmuch as examination of their chromosomes in a microscope reveals a normal (or almost normal) karyotype and a normal (or almost normal) number of chromosomes. In contrast, the vast majority of colorectal tumors in non-HNPCC patients have gross chromosomal abnormalities, with multiple translocations, deletions, and other aberrations, and a total of 55 to 70 or more chromosomes instead of the normal 46 (Figure 23–40).

The mutations that predispose an individual with HNPCC to colorectal cancer turn out to be in one of several genes that code for central components of the DNA mismatch repair system in humans, homologous in structure and function to the *mutL* and *mutS* genes in bacteria and yeast (see Figure 5–23). Only one of the two copies of the involved gene is defective, so the inevitable DNA replication errors are efficiently removed in most of the patient's cells. However, as discussed previously for other tumor suppressor genes, these individuals are at risk, because the accidental loss or alteration of the remaining good gene copy will immediately elevate the spontaneous mutation rate by a hundred-fold or

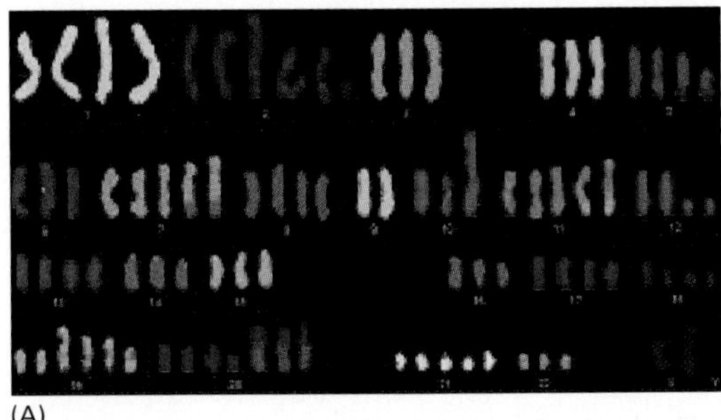
(A)

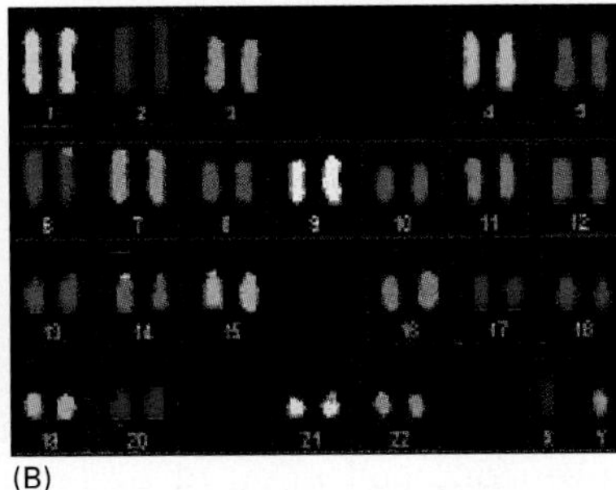
(B)

Figure 23–40 Chromosome complements (karyotypes) of colon cancers showing different kinds of genetic instability. (A) The karyotype of a typical cancer shows many gross abnormalities in chromosome number and structure. Considerable variation can also exist from cell to cell. (B) The karyotype of a tumor that has a stable chromosome complement with few chromosomal anomalies. Its defects are mostly invisible, having been created by defects in DNA mismatch repair. All of the chromosomes in this figure were stained as in Figure 23–12, the DNA of each normal chromosome being stained with a different combination of fluorescent dyes. (Courtesy Wael Abdel-Rahman and Paul Edwards.)

more (discussed in Chapter 5). These genetically unstable cells presumably go speeding through the standard processes of mutation and natural selection that allow clones of cells to progress to a malignancy.

This type of genetic instability produces invisible changes in the chromosomes—most notably changes in individual nucleotides and short expansions and contractions of mono- and dinucleotide repeats such as AAAA… or CACA-CA…. Once the phenomenon was recognized, mutations in mismatch repair genes were found in about 15% of the colorectal cancers occurring in normal people, with no inherited mutation. Again, the chromosomes in these cancers were unusual in having nearly normal karyotypes.

Thus genetic instability is found in practically all colorectal cancers, but it can be acquired in at least two very different ways. The majority of colorectal cancers have become unstable by rearranging their chromosomes—some perhaps as the combined result of a *p53* mutation and telomere shortening, as previously discussed. Others have been able to avoid this type of trauma; their genetic instability occurs on a much smaller scale, being caused by a defect in DNA mismatch repair. The fact that many carcinomas show either chromosomal instability or defective mismatch repair—but rarely both—clearly demonstrates that genetic instability is not an accidental byproduct of malignant behavior, but a contributory cause; it is a property that most cancer cells need in order to become malignant, but one that they can acquire in several alternative ways.

The Steps of Tumor Progression Can Be Correlated with Specific Mutations

In what sequence do *K-Ras, p53, APC,* and the other identified colorectal cancer genes undergo their mutations, and what contribution does each of them make to the eventual unruly behavior of the cancer cell? There cannot be a simple answer to this question, because colorectal cancer can arise by more than one route. Thus, as we have just seen, in some cases the first mutation leading toward the cancer may be in a DNA mismatch repair gene; in other cases, it may be in a gene regulating growth. A general feature, such as genetic instability, can arise in a variety of ways, through mutations in different genes.

Nevertheless, certain patterns of events are particularly common. Thus, mutations inactivating the *APC* gene appear to be the first, or at least a very early, step in most cases. They can be detected already in small benign polyps at the same high frequency as in large malignant tumors. Loss of *APC* seems to increase the rate of cell proliferation in the colonic epithelium relative to the rate of cell loss, without affecting the way the cells differentiate or the details of the histological pattern they form.

Mutations activating the *K-Ras* oncogene—a member of the *Ras* family—appear to take place a little later than those in *APC*; they are rare in small polyps but common in larger ones that show disturbances of cell differentiation and histological pattern. When malignant colorectal carcinoma cells containing such *Ras* mutations are grown in culture, they show typical features of transformed cells, such as the ability to proliferate without anchorage to a substratum. Loss of cancer-critical genes on chromosome 18 and mutations in *p53* may come later still. They are rare in polyps but common in carcinomas, suggesting that they may often occur late in the sequence (Figure 23–41). As discussed earlier, loss of *p53* function is thought to allow the abnormal cells not only to avoid apoptosis and to divide, but also to accumulate additional mutations at a rapid rate by progressing through the cell cycle when they are not fit to do so, creating many abnormal chromosomes.

Each Case of Cancer Is Characterized by Its Own Array of Genetic Lesions

As we have just seen in colorectal cancer, the traditional classification of cancers is simplistic: a single one of the conventional categories of tumor will turn out—on close scrutiny—to be a heterogeneous collection of disorders with

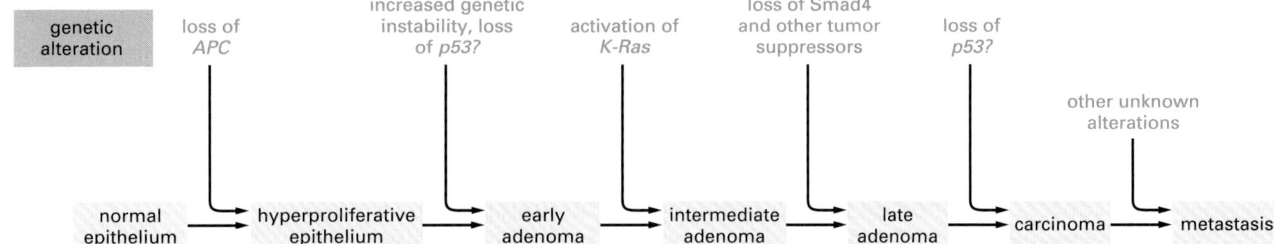

Figure 23–41 Suggested typical sequence of genetic changes underlying the development of a colorectal carcinoma. This over-simplified diagram provides a general idea of the way mutation and tumor development can fit together. But there are certainly other mutant genes of which we are not yet aware, and different colon cancers can progress through different sequences of mutations. There is still debate about the timing of p53 mutations and of the onset of genetic instability in relation to other mutations.

some features in common, but each characterized by its own array of genetic lesions (Figure 23–42). Many types of cancers that have been analyzed genetically show a large variety of genetic lesions and a great deal of variation from one case of the disease to another. In the form of lung cancer known as small-cell lung cancer, for example, one finds mutations not only in *Ras*, *p53*, and *APC*, but also in *Rb*, in members of the *Myc* gene family (in the form of amplification of the number of *Myc* gene copies), and in at least five other known proto-oncogenes and tumor suppressor genes. Different combinations of mutations are encountered in different patients and correspond to cancers that react differently to treatment.

In principle, molecular biology provides the tools to find out precisely which genes are amplified, which are deleted, and which are mutated in the tumor cells of any given patient. As we see in the next section, such information may soon prove to be as important for the diagnosis and treatment of cancer as is the identification of microorganisms in patients with infectious diseases.

Summary

Studies of developing embryos and transgenic mice have helped to reveal the functions of many cancer-critical genes. Most of the genes found to be mutated in cancer, both oncogenes and tumor suppressor genes, code for components of the pathways that regulate the social and proliferative behavior of cells in the body—in particular, the mechanisms by which signals from a cell's neighbors can impel it to divide, differentiate, or die. Other cancer-critical genes are involved in maintaining the integrity of the genome and guarding against damage. The molecular changes that allow cancers to metastasize, however, escaping the parent tumor and growing in foreign tissues, are still largely unknown.

DNA viruses such as papillomaviruses can promote the development of cancer by sequestering the products of tumor suppressor genes—in particular, the Rb protein, which regulates cell division, and the p53 protein, which is thought to act as an emergency brake on cell division in cells that have suffered genetic damage and to call a halt to cell division in senescent cells with shortened telomeres.

The p53 protein has a dual role, regulating both progression through the cell cycle and the initiation of apoptosis. So loss or inactivation of p53, which occurs in about half of all human cancers, is doubly dangerous: it allows genetically damaged and senescent cells to continue to replicate their DNA, increasing the damage, and it allows them to escape apoptosis. The loss of p53 function may contribute to the genetic instability of many full-blown metastasizing cancers.

Generally speaking, the steps of tumor progression can be correlated with mutations that activate specific oncogenes and inactivate specific tumor suppressor genes. But different combinations of mutations are found in different forms of cancer and even in patients that nominally have the same form of the disease, reflecting the random way in which mutations occur. Nevertheless, many of the same types of genetic lesions are encountered repeatedly, suggesting that there is only a limited number of ways in which our defenses against cancer can be breached.

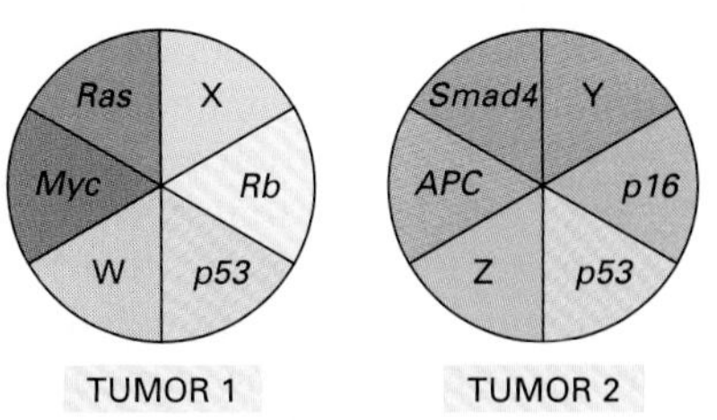

Figure 23–42 Each tumor will generally contain a different set of genetic lesions. In this schematic diagram, W, X, Y, and Z denote alterations in as yet undiscovered tumor suppressor genes or oncogenes. Tumors that arise from different tissues are generally more different in their genetic abnormalities than tumors of similar origin.

CANCER TREATMENT: PRESENT AND FUTURE

How can we apply our growing understanding of the biology of cancer to combat the disease? Prevention is always better than cure, and as we have already discussed in the first part of this chapter, many cancers can indeed be prevented—first and foremost, by avoiding the use of tobacco, a more important hazard by far than any known carcinogen that is a by-product of our industrialized society. Moreover, cancers can often be nipped in the bud by screening: primary tumors can be detected early and removed before they have metastasized, as we saw for cervical cancers, for example. Many opportunities for better prevention and screening remain, some using highly sensitive new molecular assays. Advances in these areas probably offer the most immediate prospects of reducing the cancer death rate substantially. But prevention and screening can never be perfectly effective. It is certain that the full-blown malignant disease will continue to be common—and in need of treatment—for many years to come.

The Search for Cancer Cures Is Difficult but Not Hopeless

The difficulty of curing a cancer is similar to the difficulty of getting rid of weeds. Cancer cells can be removed surgically or destroyed with toxic chemicals or radiation; but it is hard to eradicate every single one of them. Surgery can rarely ferret out every metastasis, and treatments that kill cancer cells are generally toxic to normal cells as well. If even a few cancerous cells remain, they can proliferate to produce a resurgence of the disease; and, unlike the normal cells, they often evolve resistance to the poisons used against them. In spite of the difficulties, effective cures using anticancer drugs (alone or in combination with other treatments) have already been found for some formerly highly lethal cancers—notably Hodgkin's lymphoma, testicular cancer, choriocarcinoma, and some leukemias and other cancers of childhood. Even for types of cancer where a cure at present seems beyond our reach, there are treatments that will prolong life or at least relieve distress. But what prospect is there of doing better, and finding cures for the most common forms of cancer?

Current Therapies Exploit the Loss of Cell-Cycle Control and the Genetic Instability of Cancer Cells

Anticancer therapies need to take advantage of some property of cancer cells that distinguishes them from normal cells. One such property is the genetic instability that results from loss of chromosome maintenance or DNA repair mechanisms. Remarkably, it seems that most existing cancer therapies work because, unknown to the people who developed them, they exploit these molecular defects. Traditional anticancer therapies mostly rely on agents—drugs and ionizing radiation—that damage DNA and the machinery that maintains chromosomal integrity. Such treatments preferentially kill certain kinds of cancer cells because these mutants have a diminished ability to survive the damage. Normal cells, when treated with radiation, for example, will suffer damage to their DNA, but will then arrest their cell cycle until they have repaired it (Figure 23–43). Tumor cells that have defects in various cell-cycle checkpoints, on the other hand, lose the ability to arrest the cell cycle in these circumstances, and so continue to multiply immediately after irradiation. Almost all of these cells will therefore die after a few days as a result of the catastrophic DNA damage they sustain when they attempt to divide with defective chromosomes.

Cancers Can Evolve Resistance to Therapies

Unfortunately, while some of the molecular defects present in cancer cells may enhance their sensitivity to such cytotoxic agents, others may increase their resistance. For example, much of the cell death induced by DNA damage occurs through apoptosis, so cancer cells that harbor defects in their cell death programs can sometimes escape the effects of cytotoxic therapy. Cancer cells vary

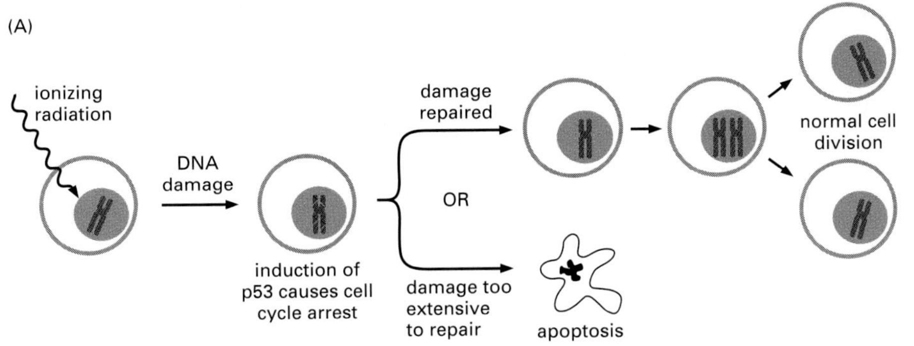

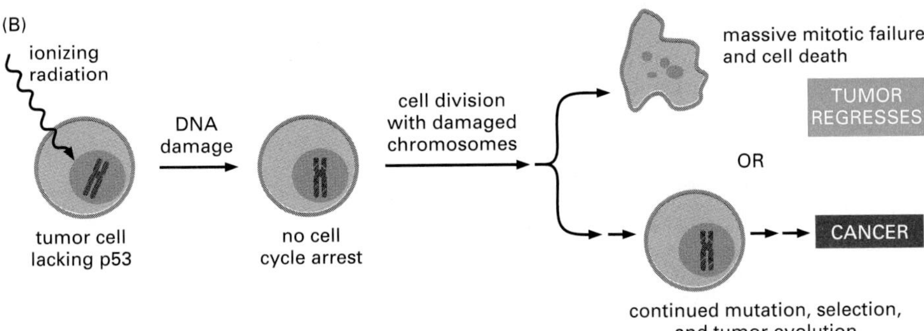

Figure 23–43 Effects of ionizing radiation on normal cells (A) and cancer cells (B). Cancer cells tend to be more susceptible than normal cells to the damaging effects of ionizing radiation because they lack an ability to arrest the cell cycle and make the necessary repairs. Unfortunately, the same genetic defects may render some cancer cells resistant to radiation treatment, as they may also be less adept at activating apoptosis in the face of DNA damage.

widely in their response to radiation and to the different kinds of cytotoxic drugs, and it seems likely that this difference reflects the particular kinds of defects they have in DNA repair, cell-cycle checkpoints, and apoptosis pathways.

Genetic instability itself can be both good and bad for anticancer therapy. Although it seems to provide an Achilles' heel that many conventional therapies exploit, genetic instability can also make eradicating cancer more difficult. Because of the abnormally high mutability of many cancer cells, most malignant tumor cell populations are heterogeneous in many respects, which may make them difficult to target with a single type of treatment. Moreover, this mutability allows many cancers to evolve resistance to therapeutic drugs at an alarming rate.

To make matters worse, cells that are exposed to one anticancer drug often develop a resistance not only to that drug, but also to other drugs to which they have never been exposed. This phenomenon of *multidrug resistance* is frequently correlated with amplification of a part of the genome that contains a gene called *Mdr1*. This gene codes for a plasma-membrane-bound transport ATPase (belonging to the ABC transporter superfamily discussed in Chapter 11). The overproduction of this protein or some other members of the same family can prevent the intracellular accumulation of certain lipophilic drugs by pumping them out of the cell. The amplification of other types of genes can likewise give the cancer cell a selective advantage: thus the gene for the enzyme dihydrofolate reductase (DHFR) often becomes amplified in response to cancer chemotherapy with the folic-acid antagonist methotrexate.

New Therapies May Emerge From Our Knowledge of Cancer Biology

Our growing understanding of cancer cell biology and tumor progression is gradually leading to better methods for treating the disease, and not only by targeting defects in cell cycle arrest and DNA repair processes. As an example, estrogen antagonists (such as tamoxifen) and drugs that block estrogen synthesis are now widely used in patients to prevent or delay recurrence of breast cancer (and they are even being tested as agents to prevent new cancers from arising). Such antiestrogen compounds do not directly kill off the tumor cells.

Nevertheless, they improve the patient's prospect of survival, presumably because estrogens are necessary for the growth of normal mammary epithelium and a proportion of breast cancers retain this hormone dependence.

The greatest hopes lie, however, in finding more powerful and selective ways to directly exterminate cancer cells. Now that we can pinpoint their genetic lesions, can we use our knowledge of cell biology to kill them off? In recent years, a wide variety of adventurous new ways to attack tumor cells have been suggested, many of which have been shown to work in model systems—typically reducing or preventing tumor growth in mice. Many of these protocols will turn out to be of no medical use, because they do not work in humans, have bad side effects, or are simply too difficult to implement. But some seem likely to succeed. For example, some tumor cells are heavily dependent on a particular protein that they overproduce (although it may not be unique to them). Blocking the activity of this protein may be an effective means of treating cancer if it does not unduly damage normal tissues. For example, about 25% of breast tumors express unusually high levels of the Her2 protein, a receptor tyrosine kinase, related to the EGF receptor, that normally plays a part in the development of the mammary epithelium. Thus, shutting off Her2 function might be expected to slow or halt the growth of breast tumors in humans; in fact, this approach is currently being tested with some success in clinical trials, using as the blocking agent a monoclonal antibody that recognizes Her2.

Another approach to destroying tumors targets the delivery of a toxic compound directly to the cancer cells by exploiting proteins like Her2 that are abundant on their surface. Antibodies against such proteins can be armed with a toxin, or made to carry an enzyme that cleaves a harmless 'prodrug' into a toxic molecule. In the latter case, one molecule of enzyme can then generate a large number of toxic molecules. A virtue of this strategy is that the toxic drug generated enzymatically can then diffuse to neighboring tumor cells, increasing the odds that they too will be killed, even if the antibody did not bind to them directly.

Treatments Can Be Designed to Attack Cells That Lack p53

These treatments all target properties or molecules that tumor cells possess—but what about molecules that they lack? One ingenious way to target tumor cells exploits their loss of p53. Certain viruses, including the papillomaviruses and the adenoviruses, encode proteins that bind to and inactivate the host cell's p53 (see Figure 23–35). This enables these viruses to outwit the p53-mediated defenses of the host cell and replicate their own genomes freely inside it. As part of their lytic life style, adenoviruses replicate continuously inside an undefended cell, and then burst out when their numbers are sufficient, killing the cell and infecting its neighbors. An adenovirus has been constructed that lacks the gene that encodes the p53-blocking protein; this defective virus can therefore only replicate in cells in which p53 is already inactivated—including many types of cancer cells. If this modified adenovirus is injected into a tumor, the virus might be expected to replicate in and kill only the cancer cells that lack p53, leaving normal cells unharmed. This strategy is also undergoing clinical trials.

Tumor Growth Can Be Choked by Depriving the Cancer Cells of Their Blood Supply

Another promising approach to destroying tumors does not directly target cancer cells at all. Because tumors require the formation of new blood vessels to grow to more than a millimeter or so in size, treatments that block angiogenesis should block tumor growth in many different types of cancer. As discussed in Chapter 22, growth of new vessels requires local signals—angiogenic growth factors—and the action of these molecules can, in principle, be blocked. Clinical trials with angiogenesis inhibitors are now taking place. Endothelial cells that are in the process of forming new vessels also turn out to express distinctive cell-surface markers, providing a promising way in which they might be attacked without harming the existing blood vessels in non-cancerous tissues.

Small Molecules Can Be Designed to Target Specific Oncogenic Proteins

The new treatments outlined above are still at an experimental stage. For most of them, it remains to be seen how effective they are in curing human cancers or in slowing their progress; past experience has taught us to be cautious. There is at least one recent dramatic success, however, that raises high hopes.

As we saw earlier, chronic myelogenous leukemia is associated, in more than 95% of cases, with a particular chromosomal translocation, visible as a characteristic abnormality in the karyotype—the Philadelphia chromosome (see Figure 23–5). This is the consequence of chromosome breakage and rejoining at the sites of two specific genes, called *Abl* and *Bcr.* The fusion of these genes creates a hybrid gene that codes for a chimeric protein, consisting of the N-terminal fragment of Bcr fused to the C-terminal portion of Abl (Figure 23–44). Abl is a protein tyrosine kinase involved in cell signaling. The substitution of the Bcr fragment for its normal N terminus makes it hyperactive, so that it stimulates inappropriate proliferation of hemopoietic precursor cells that contain it and inhibits these cells from dying by apoptosis, which many of them would normally do. Excessive numbers of white blood cells are consequently made and released into the bloodstream, producing leukemia.

The chimeric Bcr-Abl protein is an obvious target for therapeutic attack. Searches for synthetic drug molecules that can inhibit the activity of protein kinases discovered one, called STI-571, that blocks Bcr-Abl (Figure 23–45). When this molecule, now renamed Gleevec, was given to a series of 54 patients with chronic myeloid leukemia that had resisted other treatments, all but one of them showed an excellent response, with return of their white blood cell counts to normal and, in some cases, an apparent eradication of the cells carrying the Philadelphia chromosome. After a year of treatment, 51 out of the initial 54 patients were still well. Results were not so good for patients who had already progressed through further mutations to the acute phase of myeloid leukemia, where genetic instability has set in and the march of the disease is far more rapid. These patients showed a response at first and then relapsed: the cancer cells were able to evolve a resistance to the Gleevec. Nevertheless, the extraordinary success of Gleevec for the patients in the chronic (early) stage of the disease is enough to prove the principle: once we understand precisely what genetic lesions have occurred in a cancer, we can begin to design effective rational methods to treat it.

Understanding of Cancer Biology Leads Toward Rational, Tailored Medical Treatments

All medical progress depends on accurate diagnosis. If one cannot identify a disease correctly, one cannot discover its causes, predict its outcome, select the

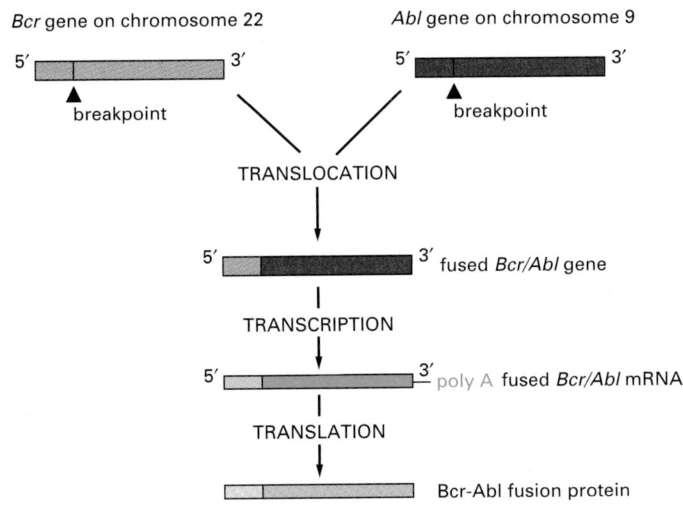

Figure 23-44 The conversion of the *Abl* proto-oncogene into an oncogene in patients with chronic myelogenous leukemia. The chromosome translocation responsible joins the *Bcr* gene on chromosome 22 to the *Abl* gene from chromosome 9, thereby generating a Philadelphia chromosome (see Figure 23–5). The resulting fusion protein has the N-terminus of the Bcr protein joined to the C-terminus of the Abl tyrosine protein kinase; in consequence, the Abl kinase domain becomes inappropriately active, driving excessive proliferation of a clone of hemopoietic cells in the bone marrow.

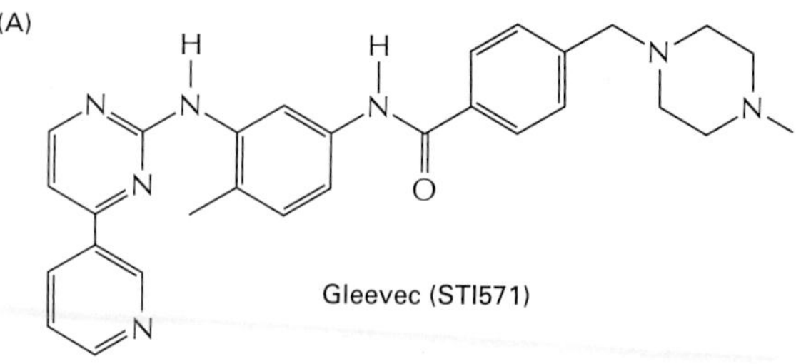

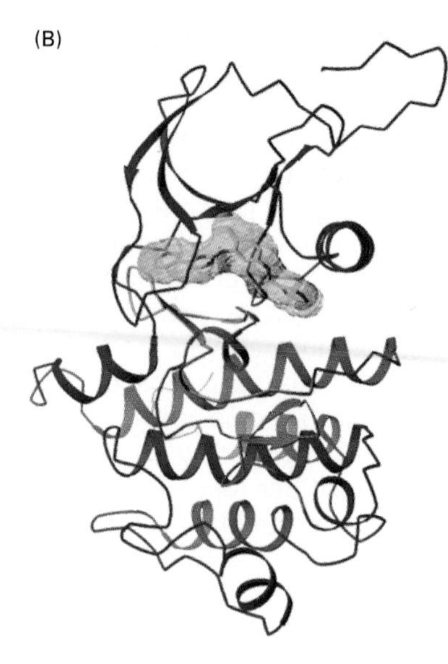

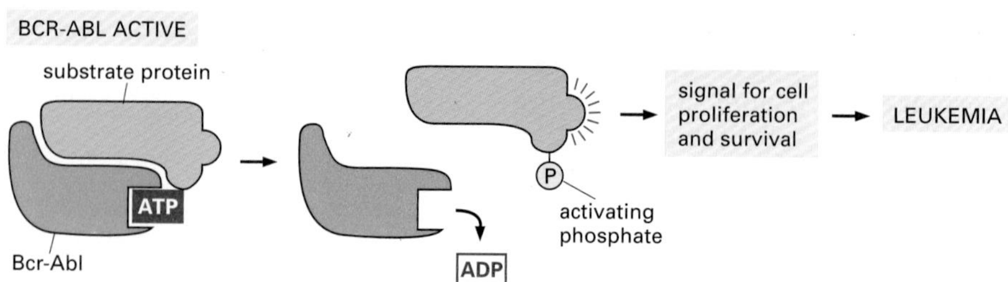

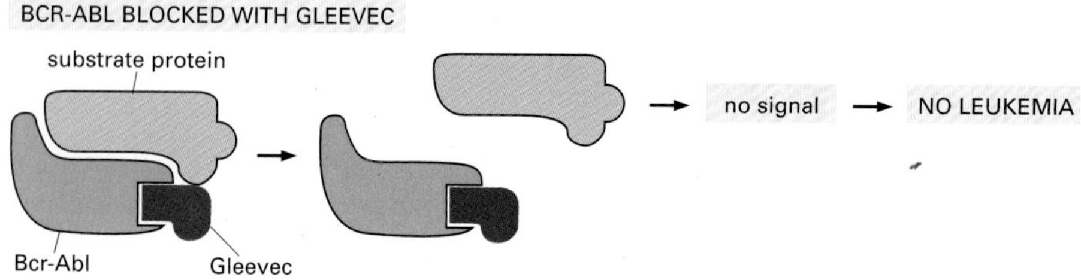

Figure 23–45 How Gleevec (STI-571) blocks the activity of Bcr-Abl protein and halts chronic myeloid leukemia. (A) The chemical structure of Gleevec. The drug can be given by mouth; it has side effects but they are usually quite tolerable. (B) The structure of the complex of Gleevec (solid *green* object) with the tyrosine kinase domain of the Abl protein (ribbon diagram), as determined by X-ray crystallography. (C) Gleevec sits in the ATP-binding pocket of the tyrosine kinase domain of Bcr-Abl and thereby prevents Bcr-Abl from transferring a phosphate group from ATP onto a tyrosine residue in a substrate protein. This blocks onward transmission of a signal for cell proliferation and survival. (B, from T. Schindler et al., *Science* 289:1938–1942, 2000. © AAAS.)

appropriate treatment for a given patient, or conduct trials on a population of patients to judge whether a proposed treatment is effective. Cancers, as we have seen, are an extraordinarily heterogeneous collection of diseases. Nevertheless, new techniques provide tools to make diagnosis precise and specific. We have the means now to characterize each individual tumor at a molecular level in unprecedented detail. For example, by using DNA microarrays (as described in Chapter 8) to analyze the mRNA present in a tissue, the levels of expression of thousands of genes can be determined simultaneously in a single sample and compared with the levels in normal control tissue. Each case of a given form of cancer, such as breast cancer, has its own gene expression profile, but when the profiles of many patients are compared it is found that they can be grouped into a smaller number of distinct classes whose members share common features. The different classes of gene expression profiles, reflecting the consequences of different sets of oncogenic mutations, correlate with different prognoses and different responses to therapy. These correlations are just beginning to be discovered, interpreted, and acted upon.

With our greatly increased understanding of molecular genetic mechanisms, we can for example, aim to determine for each tumor the precise defects in DNA metabolism—the alterations in DNA replication, DNA recombination, DNA repair, chromosome maintenance, and/or checkpoint controls—that have presumably in most cases helped its cells to acquire the multiple mutations required for tumor growth and invasiveness. These defects should make the tumor unusually vulnerable to particular types of attack on its DNA and its DNA-handling machinery. The observation that the cells of certain tumors are killed unusually easily by irradiation or by exposure to drugs that damage DNA supports this view. By designing precisely targeted drugs that exploit a particular weakness more precisely than these traditional treatments, we should be able to attack the cancer cells more effectively. In such ways, the molecular analysis of cancer promises to transform cancer treatment by enabling us to tailor therapy much more accurately to the individual patient.

The discovery of a range of cancer-critical genes has marked the end of an era of groping in the dark for clues to the molecular basis of cancer. It has been encouraging to find that there are, after all, some general principles and that some key genetic abnormalities are shared by many forms of the disease. But we are still far from fully understanding the most common human cancers. We know the DNA sequences of many cancer-critical genes, and the physiological functions of an increasing number of them. It is beginning to be possible to devise precisely targeted, rational treatments. But we still need a better understanding of how the relevant molecules interact to govern the behavior of the individual cell, a better understanding of the sociology of cells in tissues, and a better understanding of the many processes that govern the genesis and spread of cancer cells through mutation and natural selection.

Looking back on the history of cell biology and contemplating the speed of recent progress, we can be hopeful. The desire to understand that drives basic research will surely reveal new ways to use our knowledge of the cell for humanitarian goals, not only in relation to cancer, but also with regard to infectious disease, mental illness, agriculture, and other areas that we can scarcely foresee.

Summary

Our growing understanding of cancer cell biology should lead to better ways of diagnosing and treating this disease. Anticancer therapies can be designed to destroy cancer cells preferentially by exploiting the properties that distinguish them from normal cells, including the defects they harbor in their DNA repair mechanisms, cell-cycle checkpoints, and apoptosis pathways. Tumors can also be attacked through their dependence on their blood supply. By understanding the normal control mechanisms and exactly how they are subverted in specific cancers, it becomes possible to devise drugs to target cancers more precisely. As we become better able to determine which genes are amplified, which are deleted, and which are mutated in the cells of any given tumor, we can begin to tailor treatments more accurately to each individual patient.

References

General

Brugge J, Curran T, Harlow E & McCormick F (eds) (1991) Origins of Human Cancer. Cold Spring Harbor, NY: Cold Spring Harbor Laboratory Press.

Cairns J (1978) Cancer: Science and Society. San Francisco: WH Freeman.

De Vita VT, Hellman S & Rosenberg SA (eds) (2000) Cancer: Principles and Practice of Oncology, 6th edn. Philadelphia: Lippincott, Williams and Wilkins.

Greaves M (2000) Cancer: The Evolutionary Legacy. Oxford: Oxford University Press.

Cancer as a Microevolutionary Process

Cahill DP, Kinzler KW, Vogelstein B & Lengauer C (1999) Genetic instability and darwinian selection in tumours. *Trends Cell Biol.* 9, M57–M60.

Cairns J (1975) Mutation, selection and the natural history of cancer. *Nature* 255, 197–200.

Evan GI & Vousden KH (2001) Proliferation, cell cycle and apoptosis in cancer. *Nature* 411, 342–348.

Fialkow PJ (1976) Clonal origin of human tumors. *Biochim. Biophys. Acta* 458, 283–321.

Hanahan D & Weinberg RA (2000) The hallmarks of cancer. *Cell* 100, 57–70.

Kern SE (2001) Progressive genetic abnormalities in human neoplasia. In Molecular Basis of Cancer (J Mendelsohn, PM Howley, MA Israel, LA Liotta eds), 2nd edn, pp 41–70. Philadelphia: Saunders.

Klein G (1998) Foulds' dangerous idea revisited: the multistep development of tumors 40 years later. *Adv. Cancer Res.* 72, 1–23.

Nowell PC (1976) The clonal evolution of tumor cell populations. *Science* 194, 23–28.

Ruoslahti E (1996) How cancer spreads. *Sci. Am.* 275(3), 72–77.

Solomon E, Borrow J & Goddard AD (1991) Chromosome aberrations and cancers. *Science* 254, 1153–1160.

Wright WE & Shay JW (2001) Cellular senescence as a tumor-protection mechanism: The essential role of counting. *Curr. Opin. Genet. Dev.* 11, 98–103.

The Preventable Causes of Cancer

Ames B, Durston WE, Yamasaki E & Lee FD (1973) Carcinogens are mutagens: a simple test system combining liver homogenates for activation and bacteria for detection. *Proc. Natl. Acad. Sci. USA* 70, 2281–2285.

Berenblum I (1954) A speculative review: the probable nature of promoting action and its significance in the understanding of the mechanism of carcinogenesis. *Cancer Res.* 14, 471–477.

Cairns J (1985) The treatment of diseases and the war against cancer. *Sci. Am.* 253(5), 51–59.

Doll R (1977) Strategy for detection of cancer hazards to man. *Nature* 265, 589–597.

Doll R & Peto R (1981) The causes of cancer: quantitative estimates of avoidable risks of cancer in the United States today. *J. Natl. Cancer Inst.* 66, 1191–1308.

Hsu IC, Metcalf RA, Sun T et al. (1991) Mutational hotspot in the p53 gene in human hepatocellular carcinomas. *Nature* 350, 427–428.

Newton R, Beral V & Weiss RA (eds) (1999) Infections and Human Cancer. Cold Spring Harbor, NY: Cold Spring Harbor Laboratory Press.

Peto J (2001) Cancer epidemiology in the last century and the next decade. *Nature* 411, 390–395.

Finding the Cancer-Critical Genes

Hahn SA, Schutte M, Hoque AT et al. (1996) DPC4, a candidate tumor suppressor gene at human chromosome 18q21.1. *Science* 271, 350–353.

Hilgers W & Kern SE (1999) Molecular genetic basis of pancreatic adenocarcinoma. *Genes Chromosomes Cancer* 26, 1–12.

Varmus H & Weinberg RA (1993) Genes and the Biology of Cancer. New York: Scientific American Library.

Vogelstein B & Kinzler KW (eds) (1998) The Genetic Basis of Human Cancer. New York: McGraw-Hill.

Wallrapp C, Muller-Pillasch F, Micha A et al. (1999) Strategies for the detection of disease genes in pancreatic cancer. *Ann. NY Acad. Sci.* 880, 122–146.

The Molecular Basis of Cancer Cell Behavior

Artandi SE & DePinho RA (2000) Mice without telomerase: what can they teach us about human cancer? *Nat. Med.* 6, 852–855.

Chambers AF, Naumov GN, Vantyghem S & Tuck AB (2000) Molecular biology of breast cancer metastasis: clinical implications of experimental studies on metastatic inefficiency. *Breast Cancer Res.* 2, 400–407.

Edwards PAW (1999) The impact of developmental biology on cancer research: an overview. *Cancer Metastasis Rev.* 18, 175–180.

Hanahan D & Weinberg RA (2000) The hallmarks of cancer. *Cell* 100, 57–70.

Karran P (1996) Microsatellite instability and DNA mismatch repair in human cancer. *Semin. Cancer Biol.* 7, 15–24.

Kinzler KW & Vogelstein B (1996) Lessons from hereditary colorectal cancer. *Cell* 87, 159–170.

Lowe SW & Lin AW (2000) Apoptosis in cancer. *Carcinogenesis* 21, 485–495.

Macleod KF & Jacks T (1999) Insights into cancer from transgenic mouse models. *J. Pathol.* 187, 43–60.

McCormick F (1999) Signaling networks that cause cancer. *Trends Cell Biol.* 9, M53–56.

Mendelsohn J, Howley PM, Israel MA & Liotta LA (eds) (2001) Molecular Basis of Cancer, 2nd edn. Philadelphia: Saunders.

Ridley A (2000) Molecular switches in metastasis. *Nature* 406, 466–467.

Vogelstein B, Lane DP & Levine AJ (2000) Surfing the p53 network. *Nature* 408, 307–310.

Weinberg RA (1995) The retinoblastoma protein and cell-cycle control. *Cell* 81, 323–330.

Cancer Treatment: Present and Future

Baselga J & Albanell J (2001) Mechanism of action of anti-HER2 monoclonal antibodies. *Ann. Oncol.* 12, S35–41.

Druker BJ & Lydon NB (2000) Lessons learned from the development of an abl tyrosine kinase inhibitor for chronic myelogenous leukemia. *J. Clin. Invest.* 105, 3–7.

Folkman J (1996) Fighting cancer by attacking its blood supply. *Sci. Am.* 275(3), 150–154.

Golub TR, Slonim DK, Tamayo P et al. (1999) Molecular classification of cancer: class discovery and class prediction by gene expression monitoring. *Science* 286, 531–537.

Hu Z & Garen A (2001) Targeting tissue factor on tumor vascular endothelial cells and tumor cells for immunotherapy in mouse models of prostatic cancer. *Proc. Natl. Acad. Sci. USA* 98, 12180–12185.

Huang P & Oliff A (2001) Signaling pathways in apoptosis as potential targets for cancer therapy. *Trends Cell Biol.* 11, 343–348.

Kreitman RJ (1999) Immunotoxins in cancer therapy. *Curr. Opin. Immunol.* 11, 570–578.

Malpas J (1997) Chemotherapy. In Introduction to the Cellular and Molecular Biology of Cancer (LM Franks, NM Teich eds), 3rd edn, pp 343–352. Oxford: Oxford University Press.

McCormick F (2000) Interactions between adenovirus proteins and the p53 pathway: the development of ONYX-015. *Semin. Cancer Biol.* 10, 453–459.

Roninson IB, Abelson HT, Housman DE et al. (1984) Amplification of specific DNA sequences correlates with multi-drug resistance in Chinese hamster cells. *Nature* 309, 626–628.

Sidransky D (1997) Nucleic acid-based methods for the detection of cancer. *Science* 278, 1054–1059.

Waldman T, Khang Y, Dillehay L et al. (1997) Cell-cycle arrest versus cell death in cancer therapy. *Nat. Med.* 3, 1034–1036.

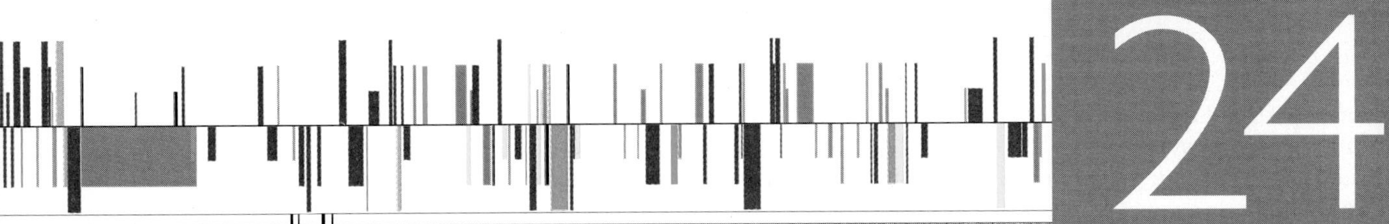

24

THE ADAPTIVE IMMUNE SYSTEM

Our **adaptive immune system** saves us from certain death by infection. An infant born with a severely defective adaptive immune system will soon die unless extraordinary measures are taken to isolate it from a host of infectious agents, including bacteria, viruses, fungi, and parasites. Indeed, all multicellular organisms need to defend themselves against infection by such potentially harmful invaders, collectively called **pathogens**. Invertebrates use relatively simple defense strategies that rely chiefly on protective barriers, toxic molecules, and phagocytic cells that ingest and destroy invading microorganisms *(microbes)* and larger parasites (such as worms). Vertebrates, too, depend on such **innate immune responses** as a first line of defense (discussed in Chapter 25), but they can also mount much more sophisticated defenses, called **adaptive immune responses**. The innate responses call the adaptive immune responses into play, and both work together to eliminate the pathogens (Figure 24–1). Unlike innate immune responses, the adaptive responses are highly specific to the particular pathogen that induced them. They can also provide long-lasting protection. A person who recovers from measles, for example, is protected for life against measles by the adaptive immune system, although not against other common viruses, such as those that cause mumps or chickenpox. In this chapter, we focus mainly on adaptive immune responses, and, unless we indicate otherwise, the term immune responses refers to them. We discuss innate immune responses in detail in Chapter 25.

The function of adaptive immune responses is to destroy invading pathogens and any toxic molecules they produce. Because these responses are destructive, it is crucial that they be made only in response to molecules that are foreign to the host and not to the molecules of the host itself. The ability to distinguish what is *foreign* from what is *self* in this way is a fundamental feature of the adaptive immune system. Occasionally, the system fails to make this distinction and reacts destructively against the host's own molecules. Such *autoimmune diseases* can be fatal.

Of course, many foreign molecules that enter the body are harmless, and it would be pointless and potentially dangerous to mount adaptive immune responses against them. Allergic conditions such as hayfever and asthma are examples of deleterious adaptive immune responses against apparently harmless foreign molecules. Such inappropriate responses are normally avoided because the innate immune system calls adaptive immune responses into play only when it recognizes molecules characteristic of invading pathogens called

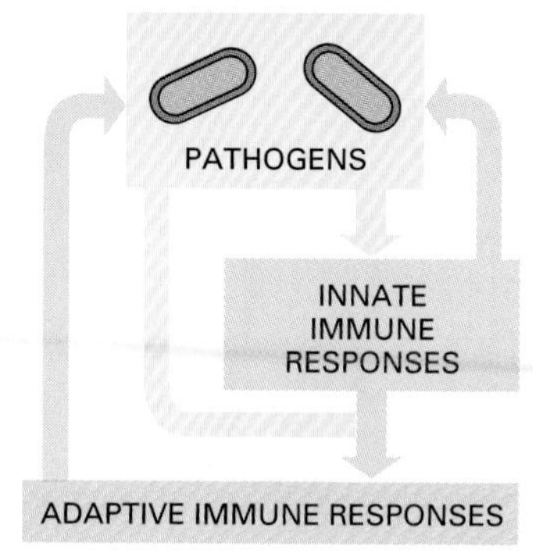

Figure 24–1 Innate and adaptive immune responses. Innate immune responses are activated directly by pathogens and defend all multicellular organisms against infection. In vertebrates, pathogens, together with the innate immune responses they activate, stimulate adaptive immune responses, which then help fight the infection.

pathogen-associated immunostimulants (discussed in Chapter 25). Moreover, the innate immune system can distinguish between different classes of pathogens and recruit the most effective form of adaptive immune response to eliminate them.

Any substance capable of eliciting an adaptive immune response is referred to as an **antigen** *(anti*body *gen*erator). Most of what we know about such responses has come from studies in which an experimenter tricks the adaptive immune system of a laboratory animal (usually a mouse) into responding to a harmless foreign molecule, such as a foreign protein. The trick involves injecting the harmless molecule together with immunostimulants (usually microbial in origin) called *adjuvants,* which activate the innate immune system. This process is called **immunization**. If administered in this way, almost any macromolecule, as long as it is foreign to the recipient, can induce an adaptive immune response that is specific to the administered macromolecule. Remarkably, the adaptive immune system can distinguish between antigens that are very similar—such as between two proteins that differ in only a single amino acid, or between two optical isomers of the same molecule.

Adaptive immune responses are carried out by white blood cells called **lymphocytes**. There are two broad classes of such responses—*antibody responses* and *cell-mediated immune responses,* and they are carried out by different classes of lymphocytes, called **B cells** and **T cells**, respectively. In **antibody responses**, B cells are activated to secrete antibodies, which are proteins called *immunoglobulins.* The antibodies circulate in the bloodstream and permeate the other body fluids, where they bind specifically to the foreign antigen that stimulated their production (Figure 24–2). Binding of antibody inactivates viruses and microbial toxins (such as tetanus toxin or diphtheria toxin) by blocking their ability to bind to receptors on host cells. Antibody binding also marks invading pathogens for destruction, mainly by making it easier for phagocytic cells of the innate immune system to ingest them.

In **cell-mediated immune responses**, the second class of adaptive immune response, activated T cells react directly against a foreign antigen that is presented to them on the surface of a host cell. The T cell, for example, might kill a virus-infected host cell that has viral antigens on its surface, thereby eliminating the infected cell before the virus has had a chance to replicate (see Figure 24–2). In other cases, the T cell produces signal molecules that activate macrophages to destroy the invading microbes that they have phagocytosed.

We begin this chapter by discussing the general properties of lymphocytes. We then consider the functional and structural features of antibodies that enable them to recognize and neutralize extracellular microbes and the toxins they make. Next, we discuss how B cells can produce a virtually unlimited number of different antibody molecules. Finally, we consider the special features of T cells and the cell-mediated immune responses they are responsible for. Remarkably, T cells can detect microbes hiding inside host cells and either kill the infected cells or help other cells to eliminate the microbes.

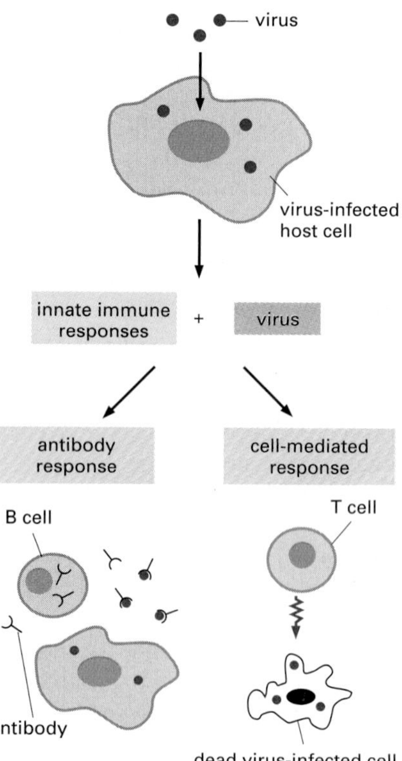

Figure 24–2 The two main classes of adaptive immune responses. Lymphocytes carry out both classes of responses. Here, the lymphocytes are responding to a viral infection. In one class of response, B cells secrete antibodies that neutralize the virus. In the other, a cell-mediated response, T cells kill the virus-infected cells.

LYMPHOCYTES AND THE CELLULAR BASIS OF ADAPTIVE IMMUNITY

Lymphocytes are responsible for the astonishing specificity of adaptive immune responses. They occur in large numbers in the blood and lymph (the colorless fluid in the lymphatic vessels that connect the lymph nodes in the body to each other and to the bloodstream) and in **lymphoid organs**, such as the thymus, lymph nodes, spleen, and appendix (Figure 24–3).

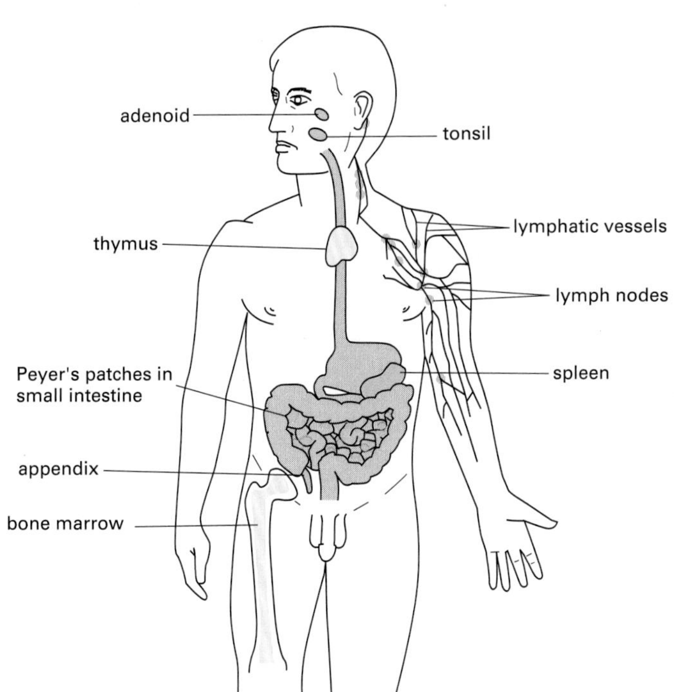

Figure 24–3 Human lymphoid organs. Lymphocytes develop in the thymus and bone marrow *(yellow)*, which are therefore called *central* (or *primary*) *lymphoid organs*. The newly formed lymphocytes migrate from these primary organs to *peripheral* (or *secondary*) *lymphoid organs (blue)*, where they can react with foreign antigen. Only some of the peripheral lymphoid organs and lymphatic vessels are shown; many lymphocytes, for example, are found in the skin and respiratory tract. As we discuss later, the lymphatic vessels ultimately empty into the bloodstream (not shown).

In this section, we discuss the general properties of lymphocytes that apply to both B cells and T cells. We shall see that each lymphocyte is committed to respond to a specific antigen and that its response during its first encounter with an antigen ensures that a more rapid and effective response occurs on subsequent encounters with the same antigen. We consider how lymphocytes avoid responding to self antigens and how they continuously recirculate between the blood and lymphoid organs, ensuring that a lymphocyte will find its specific foreign antigen no matter where the anitgen enters the body.

Lymphocytes Are Required for Adaptive Immunity

There are about 2×10^{12} lymphocytes in the human body, making the immune system comparable in cell mass to the liver or brain. Despite their abundance, their central role in adaptive immunity was not demonstrated until the late 1950s. The crucial experiments were performed in mice and rats that were heavily irradiated to kill most of their white blood cells, including lymphocytes. This treatment makes the animals unable to mount adaptive immune responses. Then, by transferring various types of cells into the animals it was possible to determine which cells reversed the deficiency. Only lymphocytes restored the adaptive immune responses of irradiated animals, indicating that lymphocytes are required for these responses (Figure 24–4).

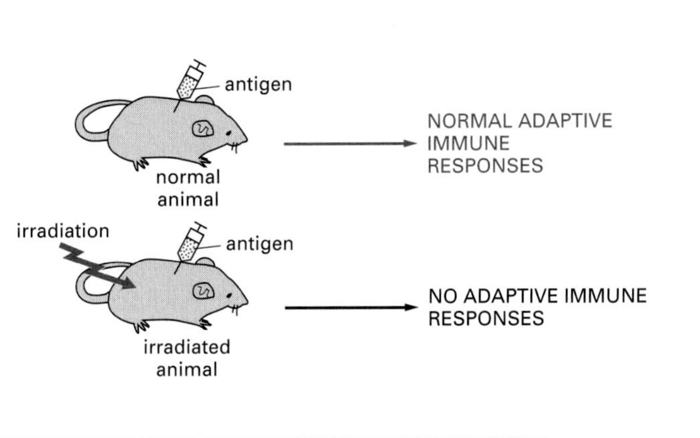

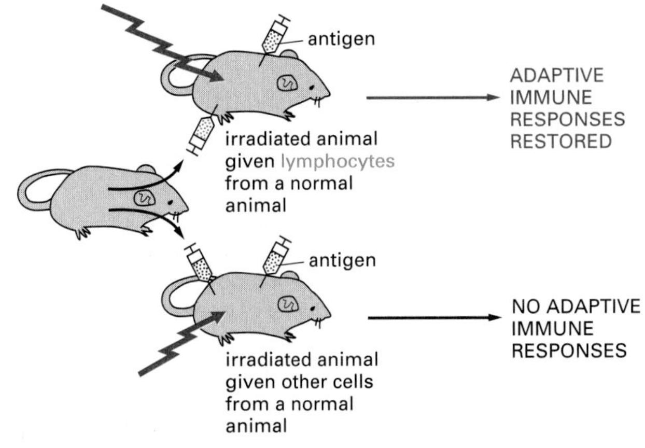

Figure 24–4 A classic experiment showing that lymphocytes are required for adaptive immune responses to foreign antigens. An important requirement of all such cell-transfer experiments is that cells are transferred between animals of the same *inbred strain.* Members of an inbred strain are genetically identical. If lymphocytes are transferred to a genetically different animal that has been irradiated, they react against the "foreign" antigens of the host and can kill the animal. In the experiment shown, the injection of lymphocytes restores both antibody and cell-mediated adaptive immune responses, indicating that lymphocytes are required for both types of responses.

The Innate and Adaptive Immune Systems Work Together

As mentioned earlier, lymphocytes usually respond to foreign antigens only if the **innate immune system** is first activated. As discussed in Chapter 25, the innate immune responses to an infection are rapid. They depend on *pattern recognition receptors* that recognize patterns of pathogen-associated molecules (immunostimulants) that are not present in the host organism, including microbial DNA, lipids, and polysaccharides, and proteins that form bacterial flagella. Some of these receptors are present on the surface of professional phagocytic cells such as macrophages and neutrophils, where they mediate the uptake of pathogens, which are then delivered to lysosomes for destruction. Others are secreted and bind to the surface of pathogens, marking them for destruction by either phagocytes or the complement system. Still others are present on the surface of various types of host cells and activate intracellular signaling pathways in response to the binding of pathogen-associated immunostimulants; this leads to the production of extracellular signal molecules that promote inflammation and help activate adaptive immune responses.

Some cells of the innate immune system directly present microbial antigens to T cells to initiate an adaptive immune response. The cells that do this most efficiently are called *dendritic cells,* which are present in most vertebrate tissues. They recognize and phagocytose invading microbes or their products at a site of infection and then migrate with their prey to a nearby peripheral lymphoid organ. There they act as *antigen-presenting cells,* which directly activate T cells to respond to the microbial antigens. Once activated, some of the T cells then migrate to the site of infection, where they help other phagocytic cells, mainly macrophages, destroy the microbes (Figure 24–5). Other activated T cells remain

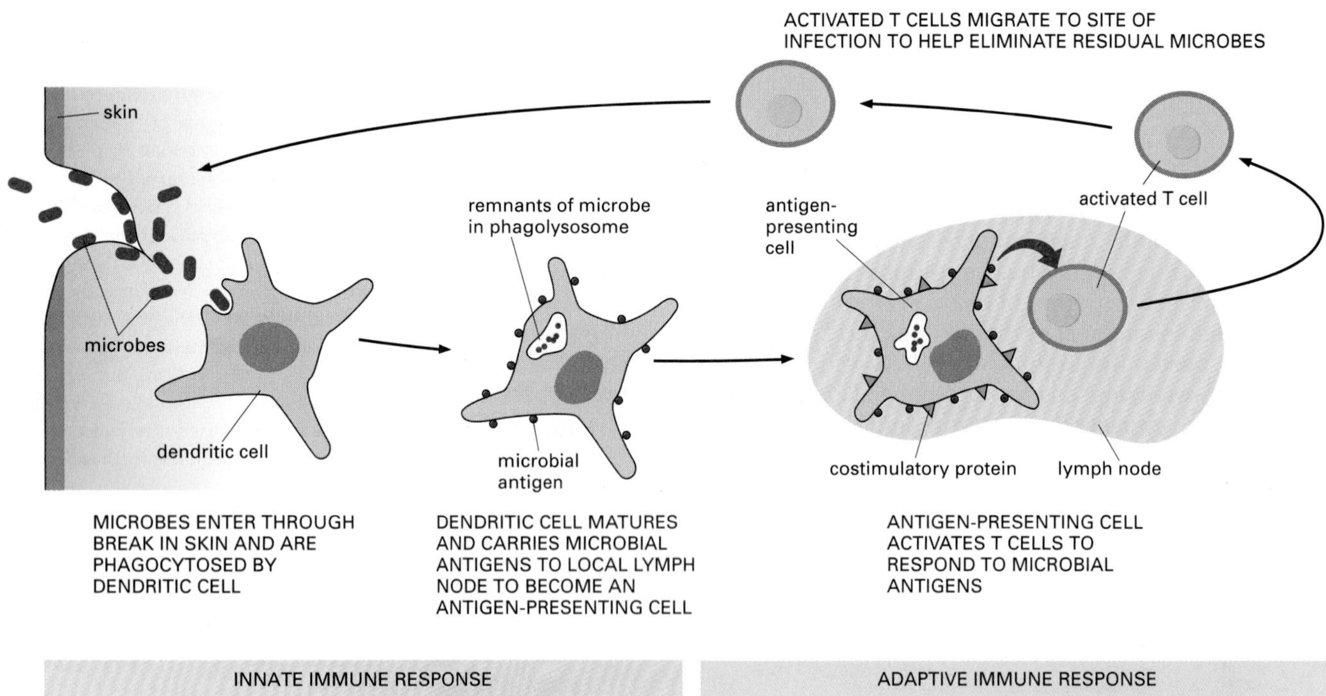

Figure 24–5 One way in which the innate immune system helps activate the adaptive immune system. Specialized phagocytic cells of the innate immune system, including macrophages (not shown) and dendritic cells ingest invading microbes or their products at the site of infection. The dendritic cells then mature and migrate in lymphatic vessels to a nearby lymph node, where they serve as antigen-presenting cells. The antigen-presenting cells activate T cells to respond to the microbial antigens that are displayed on the presenting cells' surface. The antigen-presenting cells also have special proteins on their surface (called *costimulatory molecules*) that help activate the T cells. Some of the activated T cells then migrate to the site of infection where they either help activate macrophages or kill infected cells, thereby helping to eliminate the microbes. As we discuss later, the costimulatory molecules appear on dendritic cells only after these cells mature in response to invading microbes.

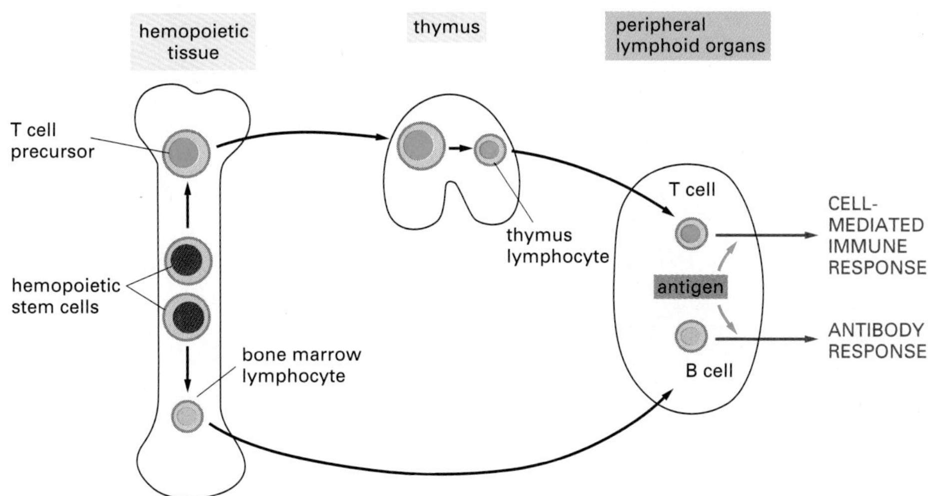

Figure 24–6 The development and activation of T and B cells. The central lymphoid organs, where lymphocytes develop from precursor cells, are labeled in *yellow boxes*. Lymphocytes respond to antigen in peripheral lymphoid organs, such as lymph nodes or spleen.

in the lymphoid organ and help B cells respond to the microbial antigens. The activated B cells secrete antibodies that circulate in the body and coat the microbes, targeting them for efficient phagocytosis.

Thus, innate immune responses are activated mainly at sites of infection, whereas adaptive immune responses are activated in peripheral lymphoid organs. The two types of responses work together to eliminate invading pathogens.

B Lymphocytes Develop in the Bone Marrow; T Lymphocytes Develop in the Thymus

T cells and B cells derive their names from the organs in which they develop. T cells develop in the *thymus,* and B cells, in mammals, develop in the *bone marrow* in adults or the liver in fetuses.

Despite their different origins, both T and B cells develop from the same *pluripotent hemopoietic stem cells,* which give rise to all of the blood cells, including red blood cells, white blood cells, and platelets. These stem cells (discussed in Chapter 22) are located primarily in *hemopoietic* tissues—mainly the liver in fetuses and the bone marrow in adults. T cells develop in the thymus from precursor cells that migrate there from the hemopoietic tissues via the blood. In most mammals, including humans and mice, B cells develop from stem cells in the hemopoietic tissues themselves (Figure 24–6). Because they are sites where lymphocytes develop from precursor cells, the thymus and hemopoietic tissues are referred to as **central (primary) lymphoid organs** (see Figure 24–3).

As we discuss later, most lymphocytes die in the central lymphoid organ soon after they develop, without ever functioning. Others, however, mature and migrate via the blood to the **peripheral (secondary) lymphoid organs**—mainly, the lymph nodes, spleen, and epithelium-associated lymphoid tissues in the gastrointestinal tract, respiratory tract, and skin (see Figure 24–3). As mentioned earlier, it is in the peripheral lymphoid organs that T cells and B cells react with foreign antigens (see Figure 24–6).

T and B cells become morphologically distinguishable from each other only after they have been activated by antigen. Nonactivated T and B cells look very similar, even in an electron microscope. Both are small, only marginally bigger than red blood cells, and contain little cytoplasm (Figure 24–7A). Both are activated by antigen to proliferate and mature into *effector cells.* Effector B cells secrete antibodies. In their most mature form, called *plasma cells,* they are filled with an extensive rough endoplasmic reticulum (Figure 24–7B). In contrast, effector T cells (Figure 24–7C) contain very little endoplasmic reticulum and do not secrete antibodies.

There are two main classes of T cells—*cytotoxic T cells* and *helper T cells.* Cytotoxic T cells kill infected cells, whereas helper T cells help activate

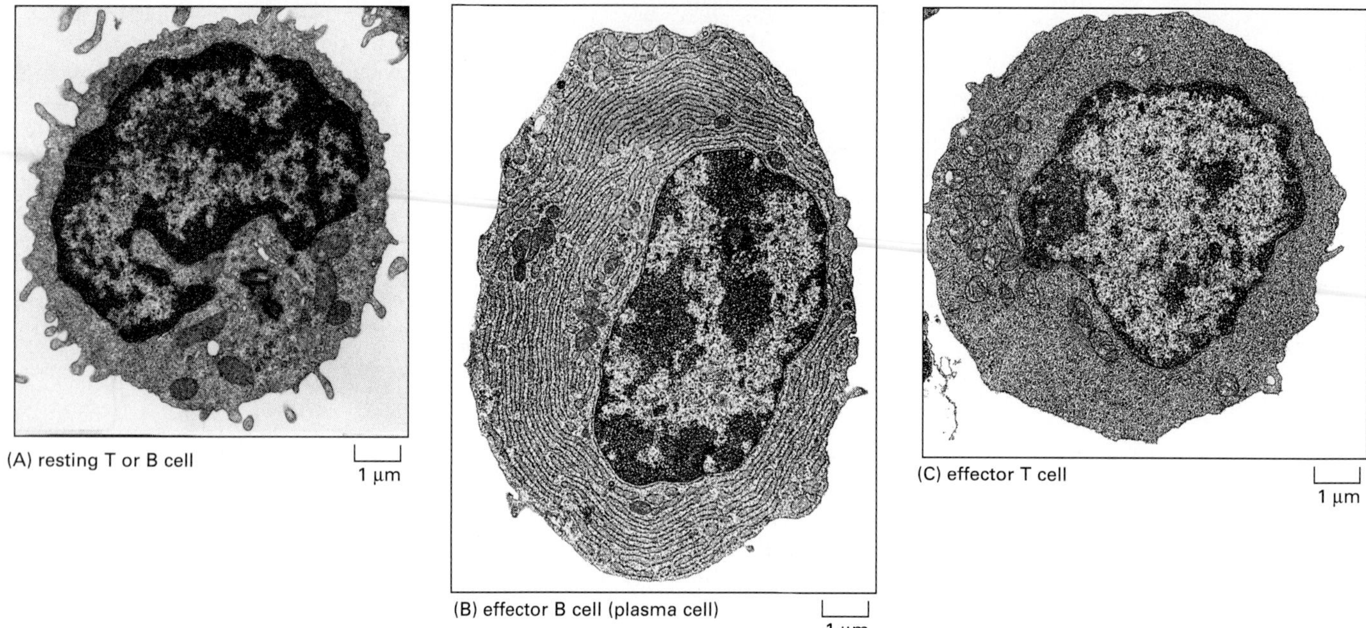

Figure 24–7 Electron micrographs of nonactivated and activated lymphocytes. (A) A resting lymphocyte, which could be a T cell or a B cell, as these cells are difficult to distinguish morphologically until they have been activated to become effector cells. (B) An effector B cell (a plasma cell). It is filled with an extensive rough endoplasmic reticulum (ER), which is distended with antibody molecules. (C) An effector T cell, which has relatively little rough ER but is filled with free ribosomes. Note that the three cells are shown at the same magnification. (A, courtesy of Dorothy Zucker-Franklin; B, courtesy of Carlo Grossi; A and B, from D. Zucker-Franklin et al., Atlas of Blood Cells: Function and Pathology, 2nd edn. Milan, Italy: Edi. Ermes, 1988; C, courtesy of Stefanello de Petris.)

macrophages, B cells, and cytotoxic T cells. Effector helper T cells secrete a variety of signal proteins called **cytokines**, which act as local mediators. They also display a variety of costimulatory proteins on their surface. By means of these cytokines and membrane-bound costimulatory proteins, they can influence the behavior of the various cell types they help. Effector cytotoxic T cells kill infected target cells also by means of proteins that they either secrete or display on their surface. Thus, whereas B cells can act over long distances by secreting antibodies that are distributed by the bloodstream, T cells can migrate to distant sites, but there they act only locally on neighboring cells.

The Adaptive Immune System Works by Clonal Selection

The most remarkable feature of the adaptive immune system is that it can respond to millions of different foreign antigens in a highly specific way. B cells, for example, make antibodies that react specifically with the antigen that induced their production. How do B cells produce such a diversity of specific antibodies? The answer began to emerge in the 1950s with the formulation of the **clonal selection theory**. According to this theory, an animal first randomly generates a vast diversity of lymphocytes, and then those lymphocytes that can react against the foreign antigens that the animal actually encounters are specifically selected for action. As each lymphocyte develops in a central lymphoid organ, it becomes committed to react with a particular antigen before ever being exposed to the antigen. It expresses this commitment in the form of cell-surface receptor proteins that specifically fit the antigen. When a lymphocyte encounters its antigen in a peripheral lymphoid organ, the binding of the antigen to the receptors activates the lymphocyte, causing it both to proliferate and to differentiate into an effector cell. An antigen therefore selectively stimulates those cells that express complementary antigen-specific receptors and are thus already committed to respond to it. This arrangement is what makes adaptive immune responses antigen-specific.

The term "clonal" in clonal selection theory derives from the postulate that the adaptive immune system is composed of millions of different families, or clones, of lymphocytes, each consisting of T or B cells descended from a common ancestor. Each ancestral cell was already committed to make one particular antigen-specific receptor protein, and so all cells in a clone have the same antigen specificity (Figure 24–8). According to the clonal selection theory, then, the immune system functions on the "ready-made" principle rather than the "made-to-order" one.

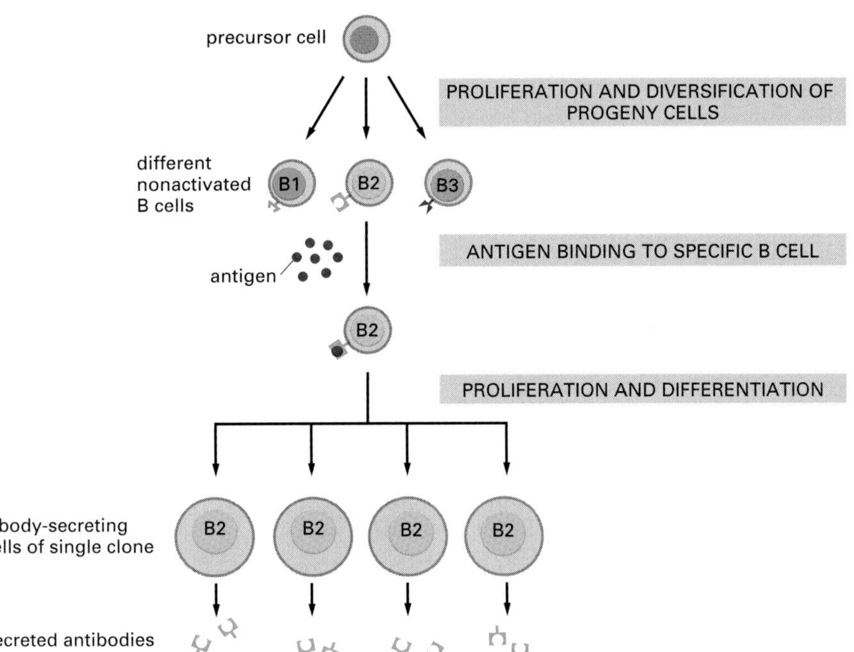

Figure 24–8 The clonal selection theory. An antigen activates only those lymphocyte clones (represented here by single cells) that are already committed to respond to it. A cell committed to respond to a particular antigen displays cell-surface receptors that specifically recognize the antigen, and all cells within a clone display the same receptor. The immune system is thought to consist of millions of different lymphocyte clones. A particular antigen may activate hundreds of different clones. Although only B cells are shown here, T cells operate in a similar way.

There is compelling evidence to support the main tenets of the clonal selection theory. For example, when lymphocytes from an animal that has not been immunized are incubated in a test tube with a number of radioactively labeled antigens, only a very small proportion (less than 0.01%) bind each antigen, suggesting that only a few cells are committed to respond to these antigens. Moreover, when one antigen is made so highly radioactive that it kills any cell that it binds to, the remaining lymphocytes can no longer produce an immune response to that particular antigen, even though they can still respond normally to other antigens. Thus, the committed lymphocytes must have receptors on their surface that specifically bind that antigen. Although most experiments of this kind have involved B cells and antibody responses, other experiments indicate that T cells, like B cells, operate by clonal selection.

How can the adaptive immune system produce lymphocytes that collectively display such an enormous diversity of receptors, including ones that recognize synthetic molecules that never occur in nature? We shall see later that the antigen-specific receptors on both T and B cells are encoded by genes that are assembled from a series of gene segments by a unique form of genetic recombination that occurs early in a lymphocyte's development, before it has encountered antigen. This assembly process generates the enormous diversity of receptors and lymphocytes, thereby enabling the immune system to respond to an almost unlimited diversity of antigens.

Most Antigens Activate Many Different Lymphocyte Clones

Most large molecules, including virtually all proteins and many polysaccharides, can serve as antigens. Those parts of an antigen that combine with the antigen-binding site on either an antibody molecule or a lymphocyte receptor are called **antigenic determinants** (or *epitopes*). Most antigens have a variety of antigenic determinants that can stimulate the production of antibodies, specific T cell responses, or both. Some determinants of an antigen produce a greater response than others, so that the reaction to them may dominate the overall response. Such determinants are said to be *immunodominant.*

The diversity of lymphocytes is such that even a single antigenic determinant is likely to activate many clones, each of which produces an antigen-binding site with its own characteristic affinity for the determinant. Even a relatively simple structure, like the *dinitrophenyl (DNP)* group in Figure 24–9, can be "looked at" in many ways. When it is coupled to a protein, as shown in the figure, it usually stimulates the production of hundreds of species of anti-DNP

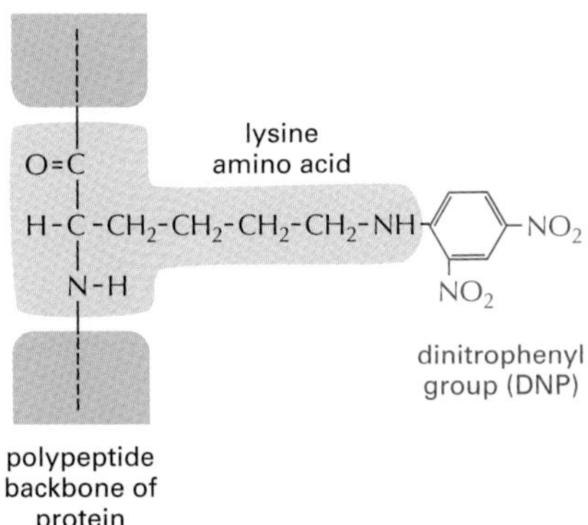

Figure 24–9 The dinitrophenyl (DNP) group. Although it is too small to induce an immune response on its own, when it is coupled covalently to a lysine side chain on a protein, as illustrated, DNP stimulates the production of hundreds of different species of antibodies that all bind specifically to it.

antibodies, each made by a different B cell clone. Such responses are said to be *polyclonal*. When only a few clones are activated, the response is said to be *oligoclonal;* and when the response involves only a single B or T cell clone, it is said to be *monoclonal*. Monoclonal antibodies are widely used as tools in biology and medicine, but they have to be produced in a special way (see Figure 8–6), as the responses to most antigens are polyclonal.

Immunological Memory Is Due to Both Clonal Expansion and Lymphocyte Differentiation

The adaptive immune system, like the nervous system, can remember prior experiences. This is why we develop lifelong immunity to many common infectious diseases after our initial exposure to the pathogen, and it is why vaccination works. The same phenomenon can be demonstrated in experimental animals. If an animal is immunized once with antigen A, an immune response (either antibody or cell-mediated) appears after several days, rises rapidly and exponentially, and then, more gradually, declines. This is the characteristic course of a **primary immune response**, occurring on an animal's first exposure to an antigen. If, after some weeks, months, or even years have elapsed, the animal is reinjected with antigen A, it will usually produce a **secondary immune response** that is very different from the primary response: the lag period is shorter, and the response is greater. These differences indicate that the animal has "remembered" its first exposure to antigen A. If the animal is given a different antigen (for example, antigen B) instead of a second injection of antigen A, the response is typical of a primary, and not a secondary, immune response. The secondary response must therefore reflect antigen-specific **immunological memory** for antigen A (Figure 24–10).

The clonal selection theory provides a useful conceptual framework for understanding the cellular basis of immunological memory. In an adult animal, the peripheral lymphoid organs contain a mixture of cells in at least three stages of maturation: *naïve cells, effector cells* and *memory cells*. When **naïve cells** encounter antigen for the first time, some of them are stimulated to proliferate and differentiate into **effector cells**, which are actively engaged in making a response (effector B cells secrete antibody, while effector T cells kill infected cells or help other cells fight the infection). Instead of becoming effector cells, some naïve cells are stimulated to multiply and differentiate into **memory cells**—cells that are not themselves engaged in a response but are more easily and more quickly induced to become effector cells by a later encounter with the same antigen. Memory cells, like naïve cells, give rise to either effector cells or more memory cells (Figure 24–11).

Thus, immunological memory is generated during the primary response in part because the proliferation of antigen-stimulated naïve cells creates many memory cells—a process known as *clonal expansion*—and in part because memory cells are able to respond more sensitively and rapidly to the same

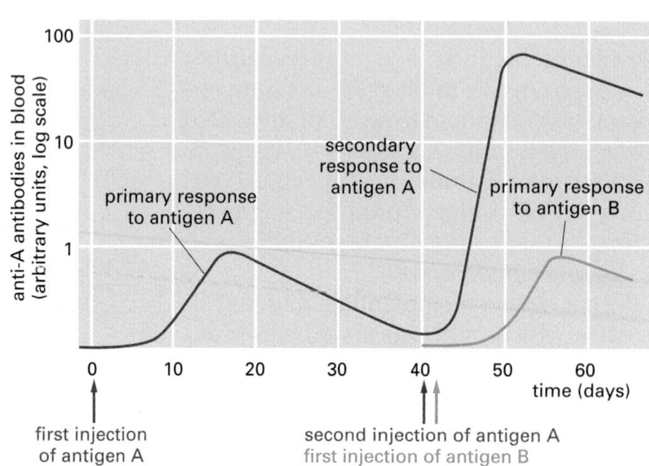

Figure 24–10 Primary and secondary antibody responses. The secondary response induced by a second exposure to antigen A is faster and greater than the primary response and is specific for A, indicating that the adaptive immune system has specifically remembered encountering antigen A before. The same type of immunological memory is observed in T-cell-mediated responses.

Figure 24–11 A model for the cellular basis of immunological memory. When naïve lymphocytes are stimulated by their specific antigen, they proliferate and differentiate. Most become effector cells which function and then die, while others become long-lived memory cells. During a subsequent exposure to the same antigen, the memory cells respond more readily and rapidly than did the naïve cells: they proliferate and give rise to effector cells and to more memory cells. In the case of T cells, memory cells can also develop from effector cells (not shown).

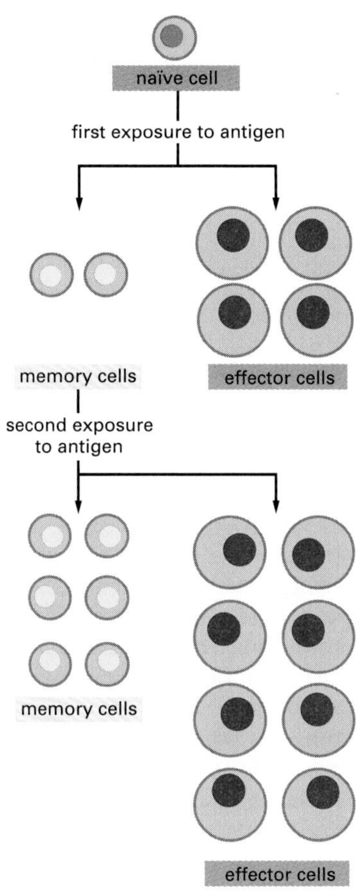

antigen than do naïve cells. And, unlike most effector cells, which die within days or weeks, memory cells can live for the lifetime of the animal, thereby providing lifelong immunological memory.

Acquired Immunological Tolerance Ensures That Self Antigens Are Not Attacked

As discussed in Chapter 25, cells of the innate immune system recognize molecules on the surface of pathogens that are not found in the host. The adaptive immune system has a far more difficult recognition task: it must be able to respond specifically to an almost unlimited number of foreign macromolecules, while avoiding responding to the large number of molecules made by the host organism itself. How does it do it? For one thing, self molecules do not induce the innate immune reactions that are required to activate adaptive immune responses. But even when an infection triggers innate reactions, self molecules still do not normally induce adaptive immune responses. Why not?

One answer is that the adaptive immune system "learns" not to respond to self antigens. Transplantation experiments provide one line of evidence for this learning process. When tissues are transplanted from one individual to another, as long as the two individuals are not identical twins, the immune system of the recipient usually recognizes the donor cells as foreign and destroys them. (For reasons we discuss later, the foreign antigens on the donor cells are so powerful that they can stimulate adaptive immune responses in the absence of infection or an adjuvant.) If, however, cells from one strain of mouse are introduced into a neonatal mouse of another strain, some of these cells survive for most of the recipient animal's life, and the recipient will now accept a graft from the original donor, even though it rejects "third-party" grafts. Apparently, nonself antigens can, in some circumstances, induce the immune system to become specifically unresponsive to them. This antigen-specific unresponsiveness to foreign antigens is known as **acquired immunological tolerance** (Figure 24–12).

The unresponsiveness of an animal's adaptive immune system to its own macromolecules *(natural immunological tolerance)* is acquired in the same way. Normal mice, for example, cannot make an immune response against one of their own protein components of the complement system called C5 (discussed in Chapter 25). Mutant mice, however, that lack the gene encoding C5 (but are otherwise genetically identical to the normal mice) can make a strong immune response to this blood protein when immunized with it. Natural immunological

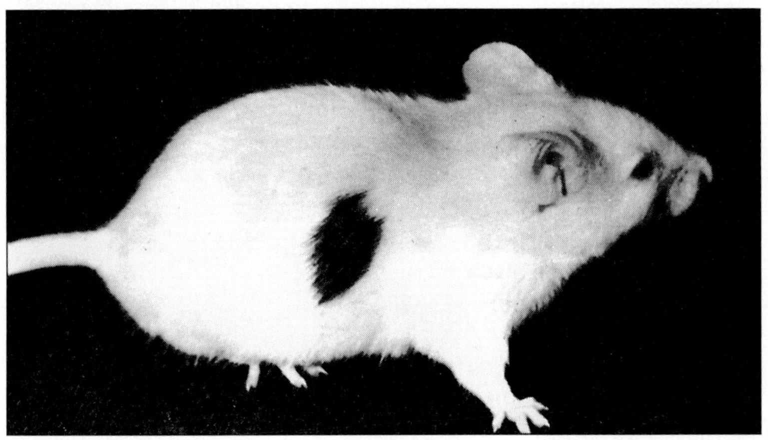

Figure 24–12 Immunological tolerance. The skin graft seen here, transplanted from an adult brown mouse to an adult white mouse, has survived for many weeks only because the white mouse, at the time of its birth, received an injection of cells from the brown mouse and therefore became immunologically tolerant to them. The cells from the brown mouse persist in the adult white mouse and continue to induce tolerance in newly formed lymphocytes that would otherwise react against the brown skin. (Courtesy of Leslie Brent, from I. Roitt, Essential Immunology, 6th edn. Oxford, UK: Blackwell Scientific, 1988.)

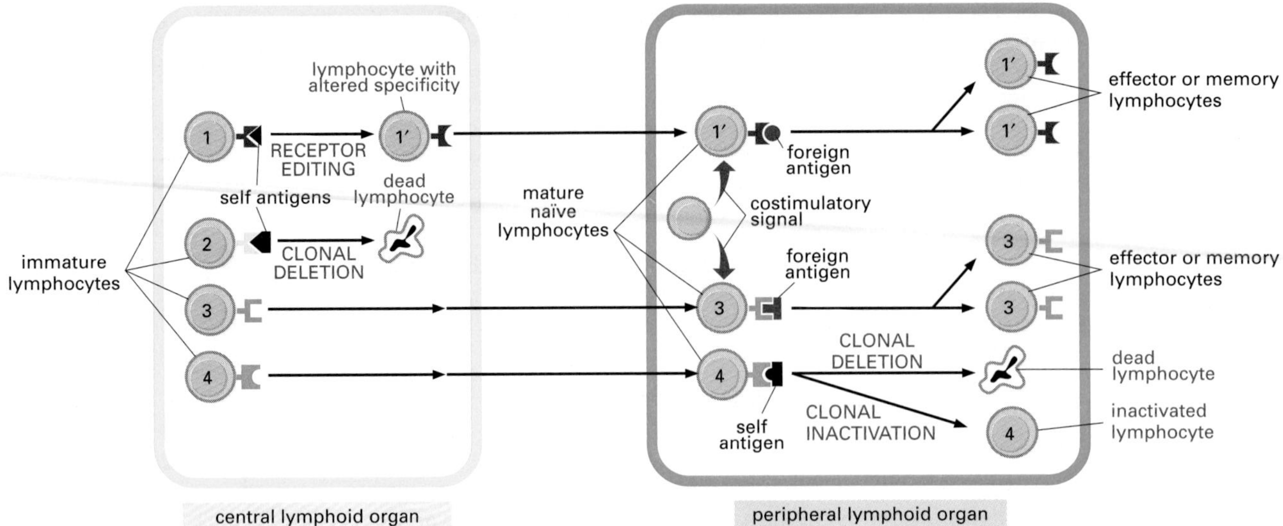

Figure 24–13 Induction of immunological tolerance to self antigens in central and peripheral lymphoid organs. When a self-reactive immature lymphocyte binds its self antigen in the central lymphoid organ where the cell is produced, it may be induced to alter the receptor it makes so that it is no longer self-reactive. This process is called receptor editing and seems to occur only in developing B cells. Alternatively, the cell may die by apoptosis, a process called clonal deletion. When a self-reactive naïve lymphocyte escapes tolerance in the central lymphoid organ and binds its self antigen in a peripheral lymphoid organ, it may either die by apoptosis or be inactivated, as the binding usually occurs in the absence of a costimulatory signal. Although not shown, some self-reactive cells survive and are suppressed by special regulatory T cells.

When a naïve lymphocyte binds a foreign antigen in a peripheral lymphoid organ in the presence of a costimulatory signal, it is stimulated to proliferate and differentiate into an effector or memory cell. As microbes are usually responsible for inducing costimulatory signals, most adaptive immune reactions normally occur only in response to microbes.

tolerance for a particular self molecule persists only for as long as the molecule remains present in the body. If a self molecule such as C5 is removed, an animal gains the ability to respond to it after a few weeks or months. Thus, the immune system is genetically capable of responding to self molecules but learns not to do so.

The learning process that leads to self-tolerance can involve killing the self-reactive lymphocytes *(clonal deletion)*, functionally inactivating them *(clonal anergy* or *inactivation)*, stimulating the cells to produce modified receptors that no longer recognize the self antigen *(receptor editing)*, or the suppression of self-reactive lymphocytes by a special type of regulatory T cell. The process begins in the central lymphoid organs when newly formed self-reactive lymphocytes first encounter their self antigen. Instead of being activated by binding antigen, the immature lymphocytes are induced to either alter their receptors or die by apoptosis. Lymphocytes that could potentially respond to self antigens that are not present in the central lymphoid organs often die or are either inactivated or suppressed after they have matured and migrated to peripheral lymphoid organs.

Why does the binding of self antigen lead to tolerance rather than activation? As we discuss later, for a lymphocyte to be activated in a peripheral lymphoid organ, it must not only bind its antigen but must also receive a *costimulatory signal*. The latter signal is provided by a helper T cell in the case of a B lymphocyte and by an antigen-presenting cell in the case of a T lymphocyte. The production of costimulatory signals usually depends on exposure to pathogens, and so a self-reactive lymphocyte normally encounters its antigen in the absence of such signals. Without a costimulatory signal, an antigen tends to kill or inactivate a lymphocyte rather than activate it (Figure 24–13).

Tolerance to self antigens sometimes breaks down, causing T or B cells (or both) to react against the organism's own tissue antigens. *Myasthenia gravis* is an example of such an **autoimmune disease**. Affected individuals make antibodies against the acetylcholine receptors on their own skeletal muscle cells. These antibodies interfere with the normal functioning of the receptors so that the patients become weak and may die because they cannot breathe. The mechanisms responsible for the breakdown of tolerance to self antigens in autoimmune diseases are unknown. It is thought, however, that activation of the innate immune system by infection may help trigger certain anti-self responses in genetically susceptible individuals.

Lymphocytes Continuously Circulate Through Peripheral Lymphoid Organs

Pathogens generally enter the body through an epithelial surface, usually through the skin, gut, or respiratory tract. How do the microbial antigens travel from these entry points to a peripheral lymphoid organ, such as a lymph node or the spleen, where lymphocytes are activated (see Figure 24–6)? The route and destination depend on the site of entry. Antigens that enter through the skin or respiratory tract are carried via the lymph to local lymph nodes; those that enter through the gut end up in gut-associated peripheral lymphoid organs such as Peyer's patches; and those that enter the blood are filtered out in the spleen. In most cases, dendritic cells carry the antigen from the site of infection to the peripheral lymphoid organ, where they become antigen-presenting cells (see Figure 24–5), specialized for activating T cells (as we discuss later).

But the lymphocytes that can recognize a particular microbial antigen in a peripheral lymph organ are only a tiny fraction of the total lymphocyte population. How do these rare cells find an antigen-presenting cell displaying their antigen? The answer is that they continuously circulate between the lymph and blood until they encounter their antigen. In a lymph node, for example, lymphocytes continually leave the bloodstream by squeezing out between specialized endothelial cells lining small veins called *postcapillary venules.* After percolating through the node, they accumulate in small lymphatic vessels that leave the node and connect with other lymphatic vessels that pass through other lymph nodes downstream (see Figure 24–3). Passing into larger and larger vessels, the lymphocytes eventually enter the main lymphatic vessel (the *thoracic duct),* which carries them back into the blood (Figure 24–14). This continuous recirculation between the blood and lymph ends only if a lymphocyte encounters its specific antigen (and a costimulatory signal) on the surface of an antigen-presenting cell in a peripheral lymphoid organ. Now the lymphocyte is retained in the peripheral lymphoid organ, where it proliferates and differentiates into effector cells. Some of the effector T cells then leave the organ via the lymph and migrate through the blood to the site of infection (see Figure 24–5).

Lymphocyte recirculation depends on specific interactions between the lymphocyte cell surface and the surface of the specialized endothelial cells lining the postcapillary venules in the peripheral lymphoid organs. Many cell types in the blood come into contact with these endothelial cells, but only lymphocytes adhere and then migrate out of the bloodstream. The lymphocytes initially adhere to the endothelial cells via *homing receptors* that bind to specific ligands (often called *counterreceptors)* on the endothelial cell surface. Lymphocyte migration into lymph nodes, for example, depends on a homing receptor protein called *L-selectin,* a member of the selectin family of cell-surface lectins discussed in Chapter 19. This protein binds to specific sugar groups on a counterreceptor that is expressed exclusively on the surface of the specialized endothelial cells in lymph nodes, causing the lymphocytes to adhere weakly to the endothelial cells and to roll slowly along their surface. The rolling continues until another, much stronger adhesion system is called into play by chemoattractant proteins (called *chemokines;* see below) secreted by endothelial cells. This strong adhesion is mediated by members of the *integrin* family of cell adhesion molecules (discussed in Chapter 19), which become activated on the lymphocyte surface. Now the lymphocytes stop rolling and crawl out of the blood vessel into the lymph node (Figure 24–15).

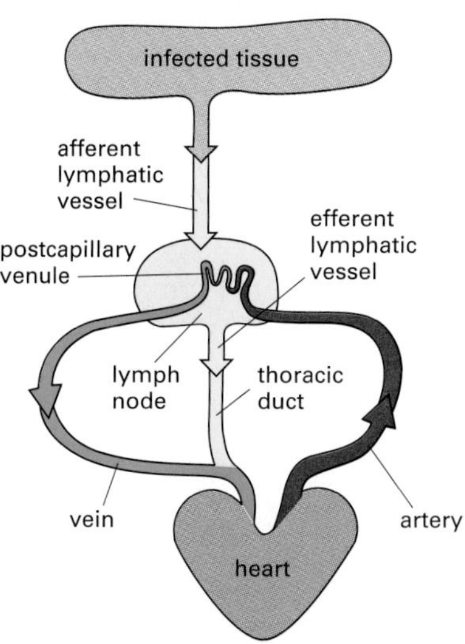

Figure 24–14 The path followed by lymphocytes as they continuously circulate between the lymph and blood. The circulation through a lymph node is shown here. Microbial antigens are carried into the lymph node by dendritic cells, which enter via afferent lymphatic vessels draining an infected tissue. T and B cells, by contrast, enter the lymph node via an artery and migrate out of the bloodstream through postcapillary venules. Unless they encounter their antigen, the T and B cells leave the lymph node via efferent lymphatic vessels, which eventually join the thoracic duct. The thoracic duct empties into a large vein carrying blood to the heart. A typical circulation cycle takes about 12–24 hours.

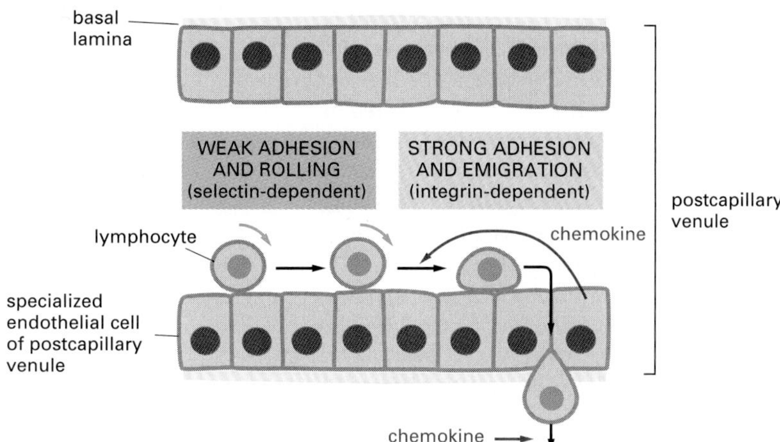

Figure 24–15 Migration of a lymphocyte out of the bloodstream into a lymph node. A circulating lymphocyte adheres weakly to the surface of the specialized endothelial cells lining a postcapillary venule in a lymph node. This initial adhesion is mediated by L-selectin on the lymphocyte surface. The adhesion is sufficiently weak to enable the lymphocyte to roll along the surface of the endothelial cells, pushed along by the flow of blood. Stimulated by chemokines secreted by the endothelial cells, the lymphocyte rapidly activates a stronger adhesion system, mediated by an integrin. This strong adhesion enables the cell to stop rolling and migrate out of the venule between the endothelial cells. The subsequent migration of the lymphocytes in the lymph node also depends on chemokines, which are produced within the node. The migration of other white blood cells out of the bloodstream into sites of infection occurs in a similar way.

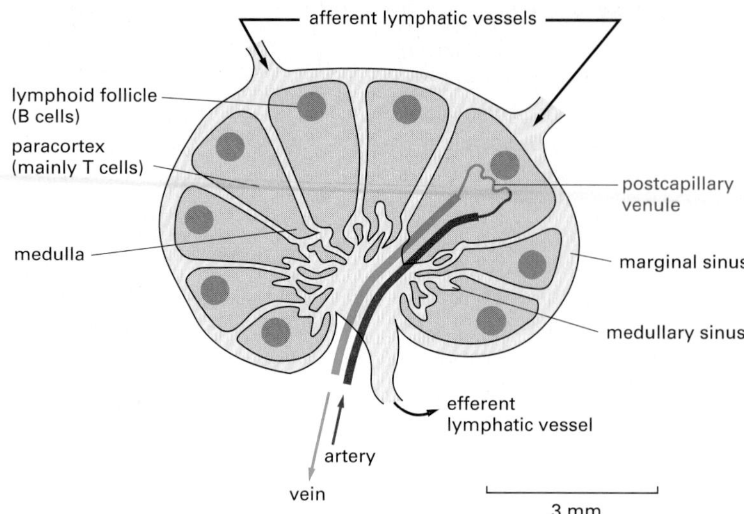

Figure 24–16 A simplified drawing of a human lymph node. B cells are primarily clustered in structures called lymphoid follicles, whereas T cells are found mainly in the paracortex. Both types of lymphocytes are attracted by chemokines to enter the lymph node from the blood via postcapillary venules. They then migrate to their respective areas, attracted by different chemokines. If they do not encounter their specific antigen, both T cells and B cells then enter the medullary sinuses and leave the node via the efferent lymphatic vessel. This vessel ultimately empties into the bloodstream, allowing the lymphocytes to begin another cycle of circulation through a secondary lymphoid organ (see Figure 24–14).

Chemokines are small, secreted, positively charged proteins that have a central role in guiding the migrations of various types of white blood cells. They are all structurally related and bind to the surface of endothelial cells, and to negatively charged proteoglycans of the extracellular matrix in organs. By binding to G-protein-linked receptors (discussed in Chapter 15) on the surface of specific blood cells, chemokines attract these cells from the bloodstream into an organ, guide them to specific locations within the organ, and then help stop migration. (The AIDS virus (HIV) also binds to chemokine receptors, which allows the virus to infect white blood cells.) T and B cells initially enter the same region of a lymph node but are then attracted by different chemokines to separate regions of the node—T cells to the *paracortex* and B cells to *lymphoid follicles* (Figure 24–16). Unless they encounter their antigen, both types of cells soon leave the lymph node via lymphatic vessels. If they encounter their antigen, however, they remain in the node, proliferate, and differentiate into either effector cells or memory cells. Most of the effector cells leave the node, expressing different chemokine receptors that help guide them to their new destinations—T cells to sites of infection and B cells to the bone marrow.

Summary

Innate immune responses are triggered at sites of infection by microbe-specific molecules associated with invading pathogens. In addition to fighting infection directly, these responses help activate adaptive immune responses in peripheral lymphoid organs. Unlike innate immune responses, adaptive responses provide specific and long-lasting protection against the particular pathogen that induced them.

The adaptive immune system is composed of millions of lymphocyte clones, with the cells in each clone sharing a unique cell-surface receptor that enables them to bind a particular antigen. The binding of antigen to these receptors, however, is usually not sufficient to stimulate a lymphocyte to proliferate and differentiate into an effector cell that can help eliminate the pathogen. Costimulatory signals provided by another specialized cell in a peripheral lymphoid organ are also required. Helper T cells provide such signals for B cells, while antigen-presenting dendritic cells usually provide them for T cells. Effector B cells secrete antibodies, which can act over long distances to help eliminate extracellular pathogens and their toxins. Effector T cells, by contrast, act locally at sites of infection to either kill infected host cells or help other cells to eliminate pathogens. As part of the adaptive immune response, some lymphocytes proliferate and differentiate into memory cells, which are able to respond faster and more efficiently the next time the same pathogen invades. Lymphocytes that would react against self molecules are either induced to alter their receptors, induced to kill themselves, inactivated, or suppressed, so that the adaptive immune system normally reacts only against foreign antigens. Both B and T cells

circulate continuously between the blood and lymph. Only if they encounter their specific foreign antigen in a peripheral lymphoid organ do they stop migrating, proliferate, and differentiate into effector cells or memory cells.

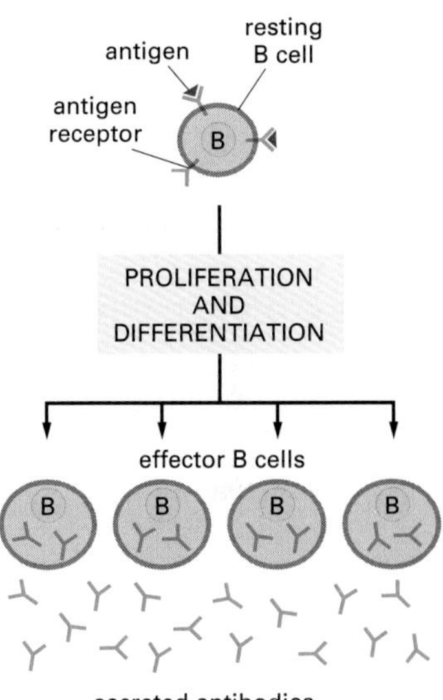

Figure 24–17 B cell activation. When naïve or memory B cells are activated by antigen (and helper T cells—not shown), they proliferate and differentiate into effector cells. The effector cells produce and secrete antibodies with a unique antigen-binding site, which is the same as that of their original membrane-bound antibodies that served as antigen receptors.

B CELLS AND ANTIBODIES

Vertebrates inevitably die of infection if they are unable to make antibodies. Antibodies defend us against infection by binding to viruses and microbial toxins, thereby inactivating them (see Figure 24–2). The binding of antibodies to invading pathogens also recruits various types of white blood cells and a system of blood proteins, collectively called *complement* (discussed in Chapter 25). The white blood cells and activated complement components work together to attack the invaders.

Synthesized exclusively by B cells, antibodies are produced in billions of forms, each with a different amino acid sequence and a different antigen-binding site. Collectively called **immunoglobulins** (abbreviated as **Ig**), they are among the most abundant protein components in the blood, constituting about 20% of the total protein in plasma by weight. Mammals make five classes of antibodies, each of which mediates a characteristic biological response following antigen binding. In this section, we discuss the structure and function of antibodies and how they interact with antigen.

B Cells Make Antibodies as Both Cell-Surface Receptors and Secreted Molecules

As predicted by the clonal selection theory, all antibody molecules made by an individual B cell have the same antigen-binding site. The first antibodies made by a newly formed B cell are not secreted. Instead, they are inserted into the plasma membrane, where they serve as receptors for antigen. Each B cell has approximately 10^5 such receptors in its plasma membrane. As we discuss later, each of these receptors is stably associated with a complex of transmembrane proteins that activate intracellular signaling pathways when antigen binds to the receptor.

Each B cell produces a single species of antibody, each with a unique antigen-binding site. When a naïve or memory B cell is activated by antigen (with the aid of a helper T cell), it proliferates and differentiates into an antibody-secreting effector cell. Such cells make and secrete large amounts of soluble (rather than membrane-bound) antibody, which has the same unique antigen-binding site as the cell-surface antibody that served earlier as the antigen receptor (Figure 24–17). Effector B cells can begin secreting antibody while they are still small lymphocytes, but the end stage of their maturation pathway is a large *plasma cell* (see Figure 24–7B), which continuously secretes antibodies at the astonishing rate of about 2000 molecules per second. Plasma cells seem to have committed so much of their protein-synthesizing machinery to making antibody that they are incapable of further growth and division. Although many die after several days, some survive in the bone marrow for months or years and continue to secrete antibodies into the blood.

A Typical Antibody Has Two Identical Antigen-Binding Sites

The simplest antibodies are Y-shaped molecules with two identical antigen-binding sites, one at the tip of each arm of the Y (Figure 24–18). Because of their two antigen-binding sites, they are described as *bivalent.* As long as an antigen has three or more antigenic determinants, bivalent antibody molecules can cross-link it into a large lattice (Figure 24–19). This lattice can be rapidly phagocytosed and degraded by macrophages. The efficiency of antigen binding and cross-linking is greatly increased by a flexible *hinge region* in most antibodies, which allows the distance between the two antigen-binding sites to vary (Figure 24–20).

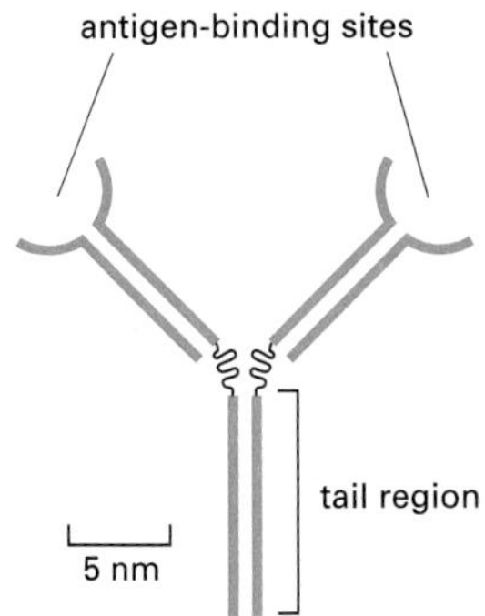

Figure 24–18 A simple representation of an antibody molecule. Note that its two antigen-binding sites are identical.

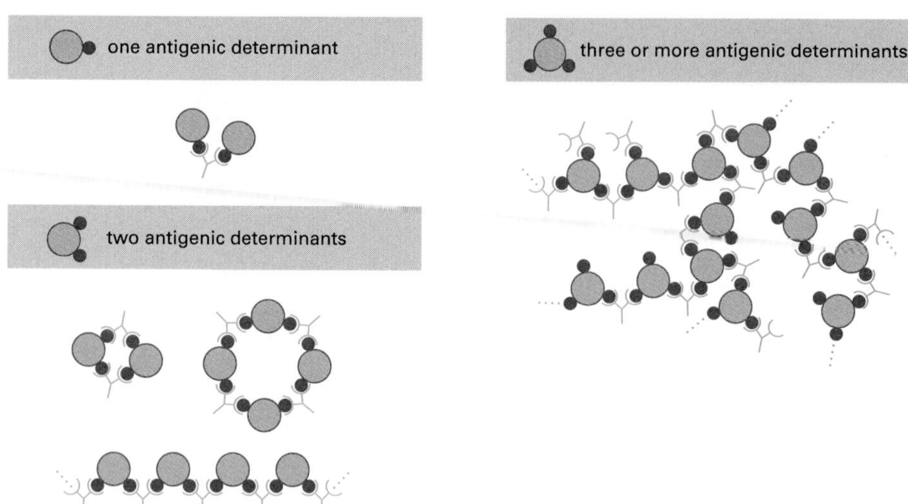

Figure 24–19 Antibody–antigen interactions. Because antibodies have two identical antigen-binding sites, they can cross-link antigens. The types of antibody–antigen complexes that form depend on the number of antigenic determinants on the antigen. Here a single species of antibody (a monoclonal antibody) is shown binding to antigens containing one, two, or three copies of a single type of antigenic determinant. Antigens with two antigenic determinants can form small cyclic complexes or linear chains with antibody, while antigens with three or more antigenic determinants can form large three-dimensional lattices that readily precipitate out of solution. Most antigens have many different antigenic determinants (see Figure 24–29A), and different antibodies that recognize different determinants can cooperate in cross-linking the antigen (not shown).

The protective effect of antibodies is not due simply to their ability to bind antigen. They engage in a variety of activities that are mediated by the tail of the Y-shaped molecule. As we discuss later, antibodies with the same antigen-binding sites can have any one of several different tail regions. Each type of tail region gives the antibody different functional properties, such as the ability to activate the complement system, to bind to phagocytic cells, or to cross the placenta from mother to fetus.

An Antibody Molecule Is Composed of Heavy and Light Chains

The basic structural unit of an antibody molecule consists of four polypeptide chains, two identical **light (L) chains** (each containing about 220 amino acids) and two identical **heavy (H) chains** (each usually containing about 440 amino acids). The four chains are held together by a combination of noncovalent and covalent (disulfide) bonds. The molecule is composed of two identical halves, each with the same antigen-binding site. Both light and heavy chains usually cooperate to form the antigen-binding surface (Figure 24–21).

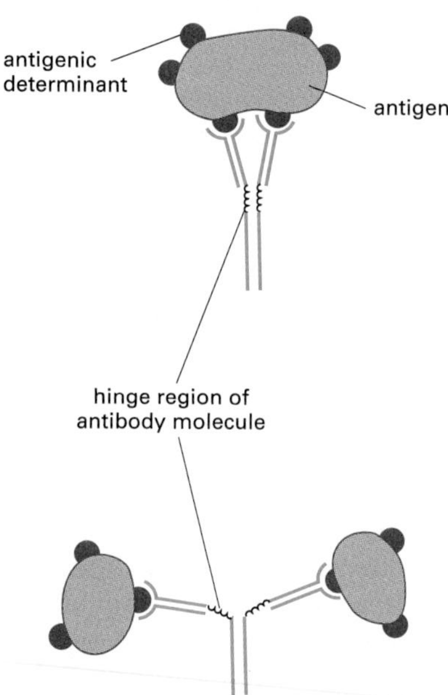

Figure 24–20 The hinge region of an antibody molecule. Because of its flexibility, the hinge region improves the efficiency of antigen binding and cross-linking.

There Are Five Classes of Heavy Chains, Each With Different Biological Properties

In mammals, there are five *classes* of antibodies, IgA, IgD, IgE, IgG, and IgM, each with its own class of heavy chain—α, δ, ε, γ, and μ, respectively. IgA molecules have α chains, IgG molecules have γ chains, and so on. In addition, there are a number of subclasses of IgG and IgA immunoglobulins; for example, there are four human IgG subclasses (IgG1, IgG2, IgG3, and IgG4), having γ_1, γ_2, γ_3, and γ_4 heavy chains, respectively. The various heavy chains give a distinctive conformation to the hinge and tail regions of antibodies, so that each class (and subclass) has characteristic properties of its own.

IgM, which has μ heavy chains, is always the first class of antibody made by a developing B cell, although many B cells eventually switch to making other

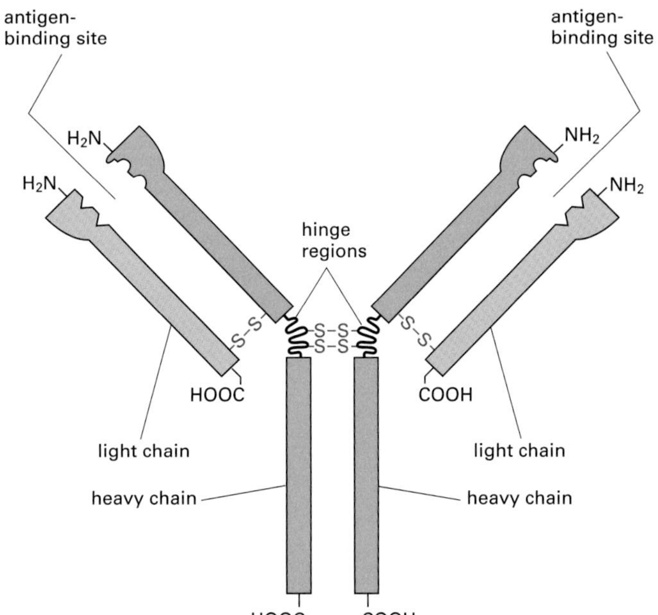

Figure 24–21 A schematic drawing of a typical antibody molecule. It is composed of four polypeptide chains—two identical heavy chains and two identical light chains. The two antigen-binding sites are identical, each formed by the N-terminal region of a light chain and the N-terminal region of a heavy chain. Both the tail (Fc) and hinge region are formed by the two heavy chains.

classes of antibody (discussed below). The immediate precursor of a B cell, called a **pre-B cell**, initially makes μ chains, which associate with so-called *surrogate light chains* (substituting for genuine light chains) and insert into the plasma membrane. The complexes of μ chains and surrogate light chains are required for the cell to progress to the next stage of development, where it makes bona fide light chains. The light chains combine with the μ chains, replacing the surrogate light chains, to form four-chain IgM molecules (each with two μ chains and two light chains). These molecules then insert into the plasma membrane, where they function as receptors for antigen. At this point, the cell is called an *immature naïve B cell.* After leaving the bone marrow, the cell starts to produce cell-surface **IgD** molecules as well, with the same antigen-binding site as the IgM molecules. It is now called a *mature naïve B cell.* It is this cell that can respond to foreign antigen in peripheral lymphoid organs (Figure 24–22).

IgM is not only the first class of antibody to appear on the surface of a developing B cell. It is also the major class secreted into the blood in the early stages of a *primary* antibody response, on first exposure to an antigen. (Unlike IgM, IgD molecules are secreted in only small amounts and seem to function mainly as cell-surface receptors for antigen.) In its secreted form, IgM is a pentamer composed of five four-chain units, giving it a total of 10 antigen-binding sites. Each pentamer contains one copy of another polypeptide chain, called a *J (joining) chain.* The J chain is produced by IgM-secreting cells and is covalently inserted between two adjacent tail regions (Figure 24–23).

The binding of an antigen to a single secreted pentameric IgM molecule can activate the complement system. As discussed in Chapter 25, when the antigen is on the surface of an invading pathogen, this activation of complement can either mark the pathogen for phagocytosis or kill it directly.

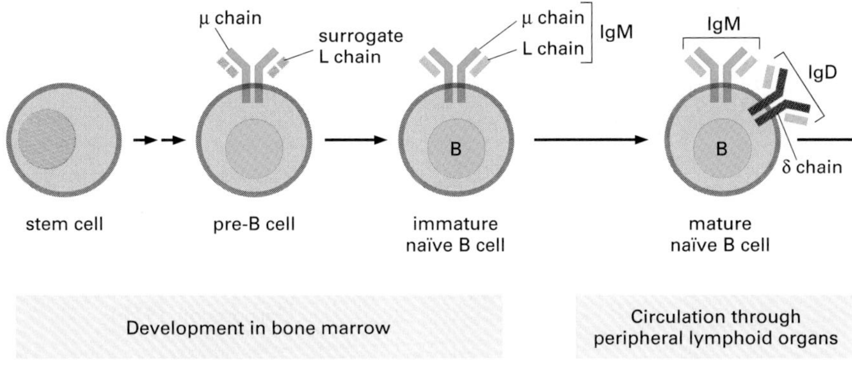

Figure 24–22 The main stages in B cell development. All of the stages shown occur independently of antigen. When they are activated by their specific foreign antigen and helper T cells in peripheral lymphoid organs, mature naïve B cells proliferate and differentiate into effector or memory cells (not shown).

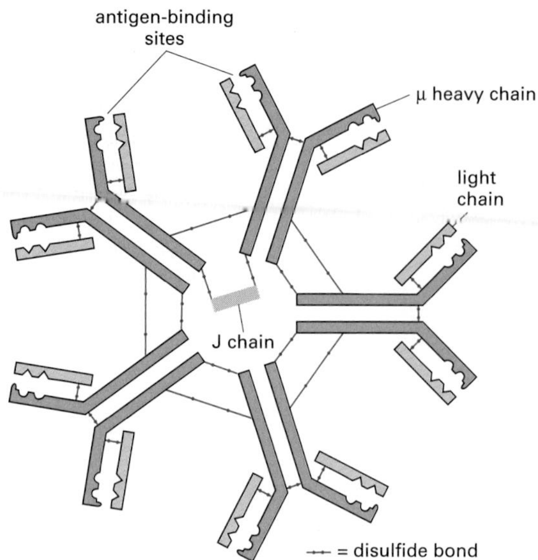

Figure 24–23 A pentameric IgM molecule. The five subunits are held together by disulfide bonds *(red)*. A single J chain, which has a structure similar to that of a single Ig domain (discussed later), is disulfide-bonded between the tails of two μ heavy chains. The J chain is required for pentamer formation. The addition of each successive four chain IgM subunit requires a J chain, which is then discarded, except for the last one, which is retained. Note that IgM molecules do not have hinge regions.

The major class of immunoglobulin in the blood is **IgG**, which is a four-chain monomer produced in large quantities during *secondary* immune responses. Besides activating complement, the tail region of an IgG molecule binds to specific receptors on macrophages and neutrophils. Largely by means of such **Fc receptors** (so-named because antibody tails are called *Fc* regions), these phagocytic cells bind, ingest, and destroy infecting microorganisms that have become coated with the IgG antibodies produced in response to the infection (Figure 24–24).

IgG molecules are the only antibodies that can pass from mother to fetus via the placenta. Cells of the placenta that are in contact with maternal blood have Fc receptors that bind blood-borne IgG molecules and direct their passage to the fetus. The antibody molecules bound to the receptors are first taken into the placental cells by receptor-mediated endocytosis. They are then transported

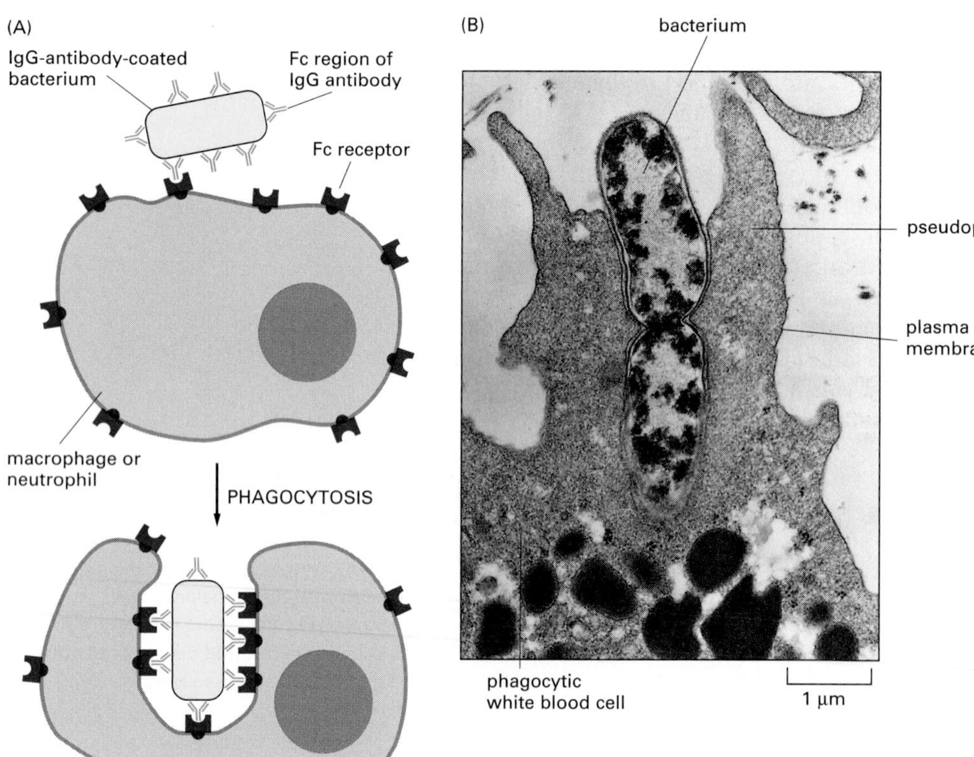

Figure 24–24 Antibody-activated phagocytosis. (A) An IgG-antibody-coated bacterium is efficiently phagocytosed by a macrophage or neutrophil, which has cell-surface receptors that bind the tail (Fc) region of IgG molecules. The binding of the antibody-coated bacterium to these Fc receptors activates the phagocytic process. The tail of an antibody molecule is called an Fc region because, when antibodies are cleaved with the proteolytic enzyme papain, the fragments containing the tail region readily crystallize. (B) Electron micrograph of a neutrophil phagocytosing an IgG-coated bacterium, which is in the process of dividing. (B, courtesy of Dorothy F. Bainton, from R.C. Williams, Jr. and H.H. Fudenberg, Phagocytic Mechanisms in Health and Disease. New York: Intercontinental Book Corporation, 1971.)

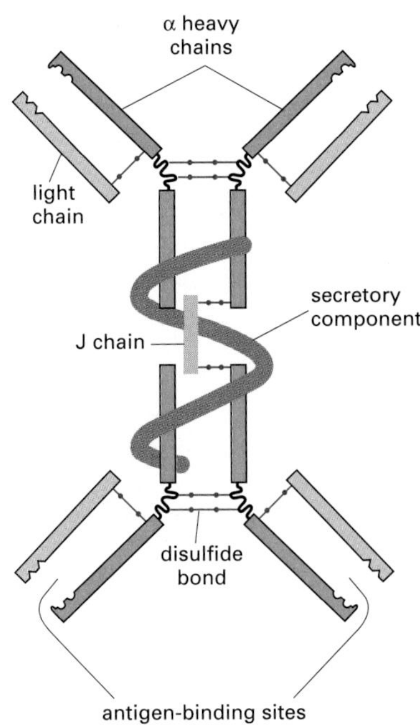

Figure 24–25 A highly schematized diagram of a dimeric IgA molecule found in secretions. In addition to the two IgA monomers, there is a single J chain and an additional polypeptide chain called the *secretory component,* which is thought to protect the IgA molecules from digestion by proteolytic enzymes in secretions.

across the cell in vesicles and released by exocytosis into the fetal blood (a process called *transcytosis,* discussed in Chapter 13). Because other classes of antibodies do not bind to these particular Fc receptors, they cannot pass across the placenta. IgG is also secreted into the mother's milk and is taken up from the gut of the neonate into the blood, providing protection for the baby against infection.

IgA is the principal class of antibody in secretions, including saliva, tears, milk, and respiratory and intestinal secretions. Whereas IgA is a four-chain monomer in the blood, it is an eight-chain dimer in secretions (Figure 24–25). It is transported through secretory epithelial cells from the extracellular fluid into the secreted fluid by another type of Fc receptor that is unique to secretory epithelia (Figure 24–26). This Fc receptor can also transport IgM into secretions (but less efficiently), which is probably why individuals with a selective IgA deficiency, the most common form of antibody deficiency, are only mildly affected by the defect.

The tail region of **IgE** molecules, which are four-chain monomers, binds with unusually high affinity ($K_a \sim 10^{10}$ liters/mole) to yet another class of Fc receptors. These receptors are located on the surface of *mast cells* in tissues and of *basophils* in the blood. The IgE molecules bound to them function as passively acquired receptors for antigen. Antigen binding triggers the mast cell or basophil to secrete a variety of cytokines and biologically active amines, especially *histamine* (Figure 24–27). These molecules cause blood vessels to dilate and become leaky, which in turn helps white blood cells, antibodies, and complement components to enter sites of infection. The same molecules are also largely responsible for the symptoms of such *allergic* reactions as hay fever, asthma, and hives. In addition, mast cells secrete factors that attract and activate white blood cells called *eosinophils.* These cells also have Fc receptors that bind IgE molecules and can kill various types of parasites, especially if the parasites are coated with IgE antibodies.

In addition to the five classes of heavy chains found in antibody molecules, higher vertebrates have two types of light chains, κ and λ, which seem to be functionally indistinguishable. Either type of light chain may be associated with any of the heavy chains. An individual antibody molecule, however, always contains identical light chains and identical heavy chains: an IgG molecule, for

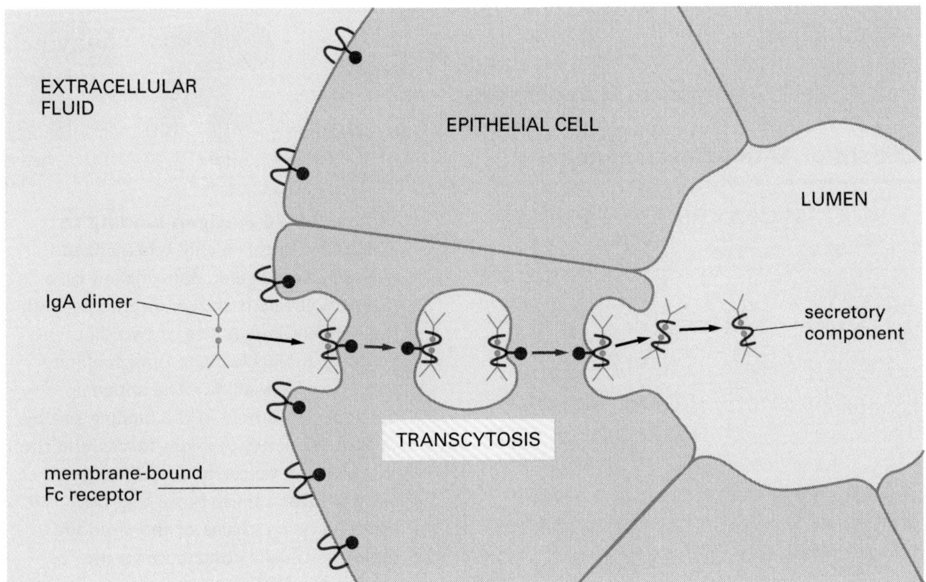

Figure 24–26 The mechanism of transport of a dimeric IgA molecule across an epithelial cell. The IgA molecule, as a J-chain-containing dimer, binds to a transmembrane receptor protein on the nonlumenal surface of a secretory epithelial cell. The receptor–IgA complexes are ingested by receptor-mediated endocytosis, transferred across the epithelial cell cytoplasm in vesicles, and secreted into the lumen on the opposite side of the cell by exocytosis. When exposed to the lumen, the part of the Fc receptor protein that is bound to the IgA dimer (the *secretory component)* is cleaved from its transmembrane tail, thereby releasing the antibody in the form shown in Figure 24–25. The J chain is not shown.

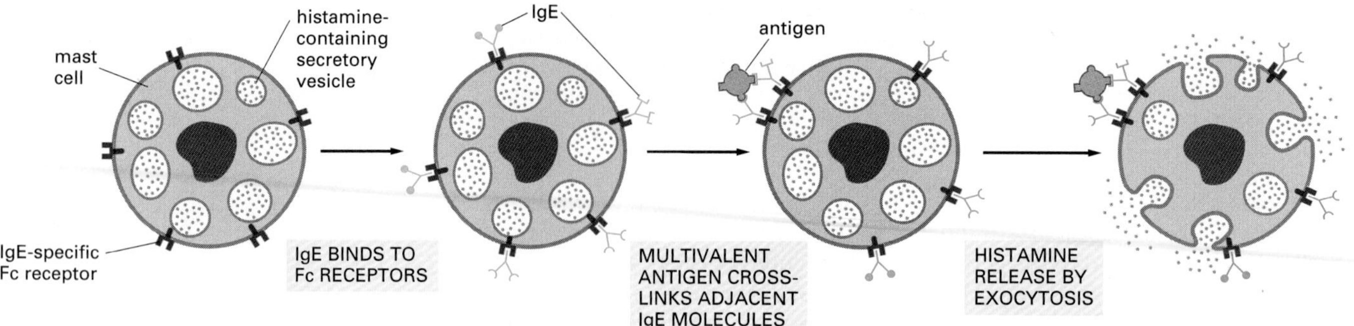

Figure 24–27 The role of IgE in histamine secretion by mast cells. A mast cell (or a basophil) binds IgE molecules after they are secreted by activated B cells. The soluble IgE antibodies bind to Fc receptor proteins on the mast cell surface that specifically recognize the Fc region of these antibodies. The bound IgE molecules serve as cell-surface receptors for antigen. Thus, unlike B cells, each mast cell (and basophil) has a set of cell-surface antibodies with a wide variety of antigen-binding sites. When an antigen molecule binds to these membrane-bound IgE antibodies so as to cross-link them to their neighbors, it signals the mast cell to release its histamine and other local mediators by exocytosis.

instance, may have either κ or λ light chains, but not one of each. As a result of this symmetry, an antibody's antigen-binding sites are always identical. Such symmetry is crucial for the cross-linking function of secreted antibodies (see Figure 24–19).

The properties of the various classes of antibodies in humans are summarized in Table 24–1.

The Strength of an Antibody–Antigen Interaction Depends on Both the Number and the Affinity of the Antigen-Binding Sites

The binding of an antigen to antibody, like the binding of a substrate to an enzyme, is reversible. It is mediated by the sum of many relatively weak noncovalent forces, including hydrogen bonds and hydrophobic van der Waals forces, and ionic interactions. These weak forces are effective only when the antigen molecule is close enough to allow some of its atoms to fit into complementary recesses on the surface of the antibody. The complementary regions of a four-chain antibody unit are its two identical antigen-binding sites; the corresponding region on the antigen is an *antigenic determinant* (Figure 24–28). Most antigenic macromolecules have many different antigenic determinants and are said to be *multivalent;* if two or more of them are identical (as in a polymer with a repeating structure), the antigen is said to be *polyvalent* (Figure 24–29).

The reversible binding reaction between an antigen with a single antigenic determinant (denoted Ag) and a single antigen-binding site (denoted Ab) can be expressed as

$$\text{Ag} + \text{Ab} \leftrightarrow \text{AgAb}$$

The equilibrium point depends both on the concentrations of Ab and Ag and on the strength of their interaction. Clearly, a larger fraction of Ab will become associated with Ag as the concentration of Ag increases. The strength of the interaction is generally expressed as the **affinity constant (K_a)** (see Figure 3–44), where

$$K_a = \frac{[\text{AgAb}]}{[\text{Ag}][\text{Ab}]}$$

(the square brackets indicate the concentration of each component at equilibrium).

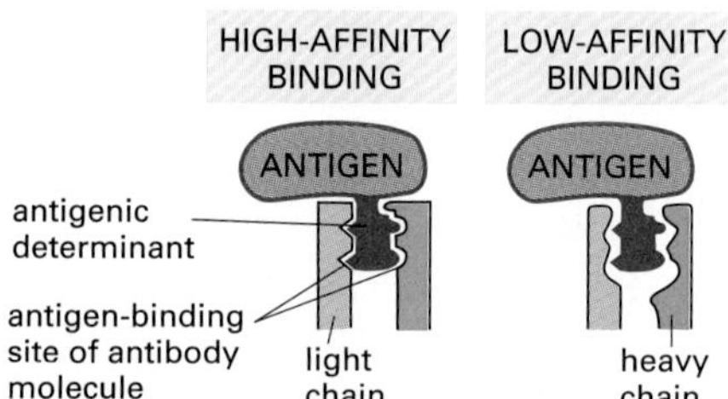

Figure 24–28 Antigen binding to antibody. In this highly schematized diagram, an antigenic determinant on a macromolecule is shown interacting with the antigen-binding site of two different antibody molecules, one of high affinity and one of low affinity. The antigenic determinant is held in the binding site by various weak noncovalent forces, and the site with the better fit to the antigen has a greater affinity. Note that both the light and heavy chains of the antibody molecule usually contribute to the antigen-binding site.

TABLE 24–1 Properties of the Major Classes of Antibodies in Humans

	CLASS OF ANTIBODY				
PROPERTIES	IgM	IgD	IgG	IgA	IgE
Heavy chains	μ	δ	γ	α	ε
Light chains	κ or λ	κ or λ	κ or λ	κ or λ	κ or λ
Number of four-chain units	5	1	1	1 or 2	1
Percentage of total Ig in blood	10	<1	75	15	<1
Activates complement	++++	–	++	–	–
Crosses placenta	–	–	+	–	–
Binds to macrophages and neutrophils	–	–	+	–	–
Binds to mast cells and basophils	–	–	–	–	+

Figure 24–29 Molecules with multiple antigenic determinants. (A) A globular protein is shown with a number of different antigenic determinants. Different regions of a polypeptide chain usually come together in the folded structure to form each antigenic determinant on the surface of the protein. (B) A polymeric structure is shown with many *identical* antigenic determinants.

(A)

multiple different antigenic determinants
(a multivalent antigen)

(B)

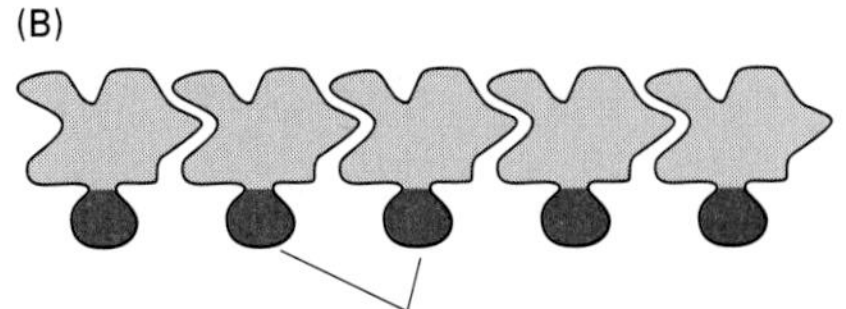

The affinity constant, sometimes called the association constant, can be determined by measuring the concentration of free Ag required to fill half of the antigen-binding sites on the antibody. When half the sites are filled, [AgAb] = [Ab] and $K_a = 1/[Ag]$. Thus, the reciprocal of the antigen concentration that produces half the maximum binding is equal to the affinity constant of the antibody for the antigen. Common values range from as low as 5×10^4 to as high as 10^{11} liters/mole.

The **affinity** of an antibody for an antigenic determinant describes the strength of binding of a single copy of the antigenic determinant to a single antigen-binding site, and it is independent of the number of sites. When, however, a polyvalent antigen, carrying multiple copies of the same antigenic determinant, combines with a polyvalent antibody, the binding strength is greatly increased because all of the antigen-antibody bonds must be broken simultaneously before the antigen and antibody can dissociate. As a result, a typical IgG molecule can bind at least 100 times more strongly to a polyvalent antigen if both antigen-binding sites are engaged than if only one site is engaged. The total binding strength of a polyvalent antibody with a polyvalent antigen is referred to as the **avidity** of the interaction.

If the affinity of the antigen-binding sites in an IgG and an IgM molecule is the same, the IgM molecule (with 10 binding sites) will have a much greater avidity for a multivalent antigen than an IgG molecule (which has two binding sites). This difference in avidity, often 10^4-fold or more, is important because antibodies produced early in an immune response usually have much lower affinities than those produced later. Because of its high total avidity, IgM—the major Ig class produced early in immune responses—can function effectively even when each of its binding sites has only a low affinity.

So far we have considered the general structure and function of antibodies. Next we look at the details of their structure, as revealed by studies of their amino acid sequence and three-dimensional structure.

Light and Heavy Chains Consist of Constant and Variable Regions

Comparison of the amino acid sequences of different antibody molecules reveals a striking feature with important genetic implications. Both light and heavy chains have a variable sequence at their N-terminal ends but a constant sequence at their C-terminal ends. Consequently, when the amino acid sequences of many different κ chains are compared, the C-terminal halves are the same or show only minor differences, whereas the N-terminal halves are all very different. Light chains have a **constant region** about 110 amino acids long and a **variable region** of the same size. The variable region of the heavy chains (at their N-terminus) is also about 110 amino acids long, but the heavy-chain constant region is about three or four times longer (330 or 440 amino acids), depending on the class (Figure 24–30).

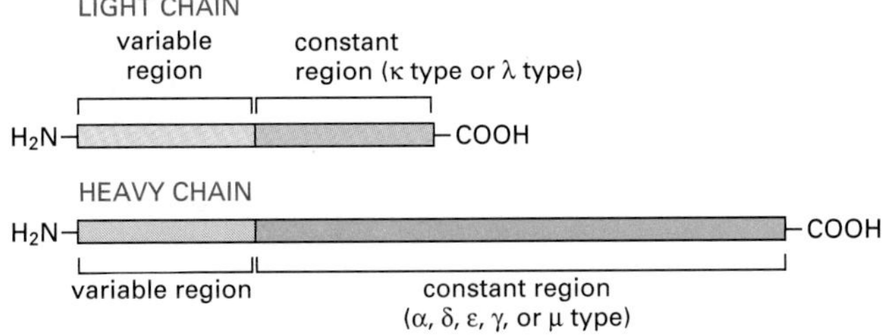

Figure 24–30 Constant and variable regions of immunoglobulin chains. Both light and heavy chains of an antibody molecule have distinct constant and variable regions.

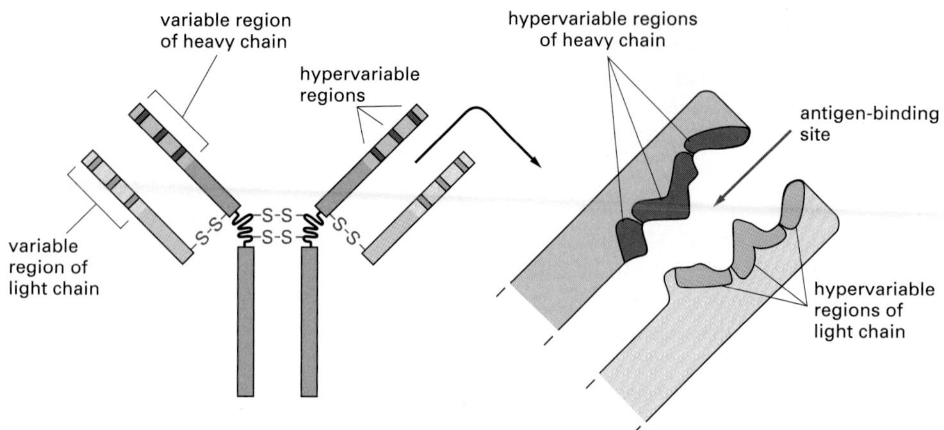

Figure 24–31 Antibody hypervariable regions. Highly schematized drawing of how the three hypervariable regions in each light and heavy chain together form the antigen-binding site of an antibody molecule.

It is the N-terminal ends of the light and heavy chains that come together to form the antigen-binding site (see Figure 24–21), and the variability of their amino acid sequences provides the structural basis for the diversity of antigen-binding sites. The diversity in the variable regions of both light and heavy chains is for the most part restricted to three small **hypervariable regions** in each chain; the remaining parts of the variable region, known as *framework regions,* are relatively constant. Only the 5–10 amino acids in each hypervariable region form the antigen-binding site (Figure 24–31). As a result, the size of the antigenic determinant that an antibody recognizes is generally comparably small. It can consist of fewer than 25 amino acids on the surface of a globular protein, for example.

The Light and Heavy Chains Are Composed of Repeating Ig Domains

Both light and heavy chains are made up of repeating segments—each about 110 amino acids long and each containing one intrachain disulfide bond. These repeating segments fold independently to form compact functional units called **immunoglobulin (Ig) domains**. As shown in Figure 24–32, a light chain consists of one variable (V_L) and one constant (C_L) domain (equivalent to the variable and constant regions shown in the top half of Figure 24–30). These domains pair with the variable (V_H) and first constant (C_H1) domain of the heavy chain to form the antigen-binding region. The remaining constant domains of the heavy chains form the Fc region, which determines the other biological properties of the antibody. Most heavy chains have three constant domains (C_H1, C_H2, and C_H3), but those of IgM and IgE antibodies have four.

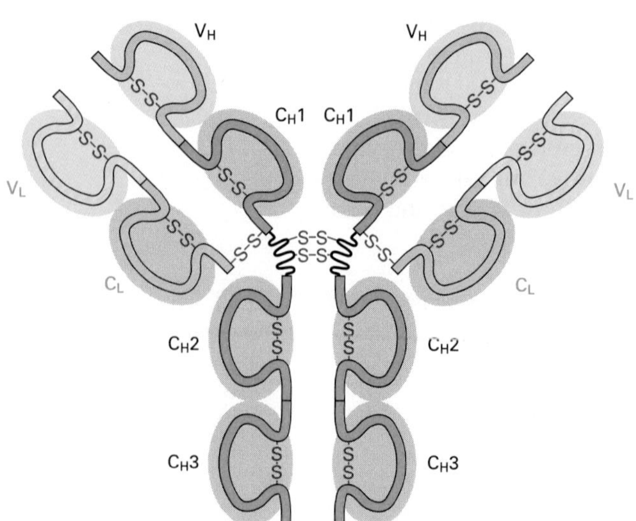

Figure 24–32 Immunoglobulin domains. The light and heavy chains in an antibody molecule are each folded into repeating domains that are similar to one another. The variable domains (shaded in *blue)* of the light and heavy chains (V_L and V_H) make up the antigen-binding sites, while the constant domains of the heavy chains (mainly C_H2 and C_H3) determine the other biological properties of the molecule. The heavy chains of IgM and IgE antibodies do not have a hinge region and have an extra constant domain (C_H4). Hydrophobic interactions between domains on adjacent chains have an important role in holding the chains together in the antibody molecule: V_L binds to V_H, C_L binds to C_H1, and so on (see Figure 24–34).

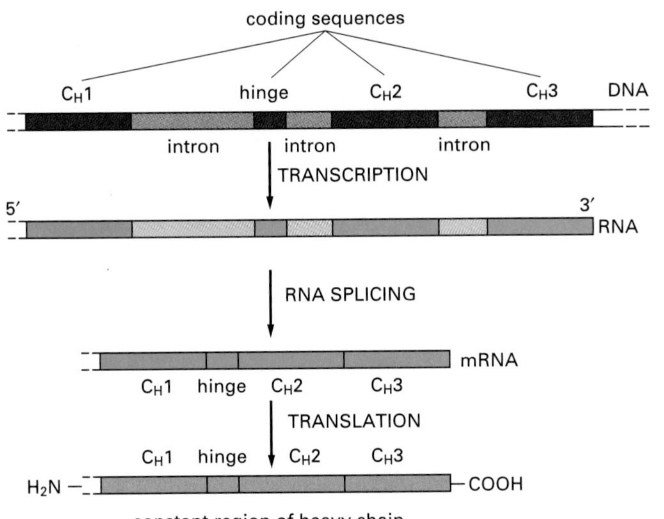

Figure 24–33 The organization of the DNA sequences that encode the constant region of an antibody heavy chain. The coding sequences (exons) for each domain and for the hinge region are separated by noncoding sequences (introns). The intron sequences are removed by splicing the primary RNA transcripts to form mRNA. The presence of introns in the DNA is thought to have facilitated accidental duplications of DNA segments that gave rise to the antibody genes during evolution (discussed in Chapter 7). The DNA and RNA sequences that encode the variable region of the heavy chain are not shown.

The similarity in their domains suggests that antibody chains arose during evolution by a series of gene duplications, beginning with a primordial gene coding for a single 110 amino acid domain of unknown function. This hypothesis is supported by the finding that each domain of the constant region of a heavy chain is encoded by a separate coding sequence (exon) (Figure 24–33).

An Antigen-Binding Site Is Constructed From Hypervariable Loops

A number of fragments of antibodies, as well as intact antibody molecules, have been studied by x-ray crystallography. From these examples, we can understand the way in which billions of different antigen-binding sites are constructed on a common structural theme.

As illustrated in Figure 24–34, each Ig domain has a very similar three-dimensional structure based on what is called the *immunoglobulin fold,* which consists of a sandwich of two β sheets held together by a disulfide bond. We shall see later that many other proteins on the surface of lymphocytes and other cells, many of which function as cell–cell adhesion molecules (discussed in Chapter 19), contain similar domains and hence are members of a very large *immunoglobulin (Ig) superfamily* of proteins.

The variable domains of antibody molecules are unique in that each has its particular set of three hypervariable regions, which are arranged in three *hypervariable loops* (see Figure 24–34). The hypervariable loops of both the light and

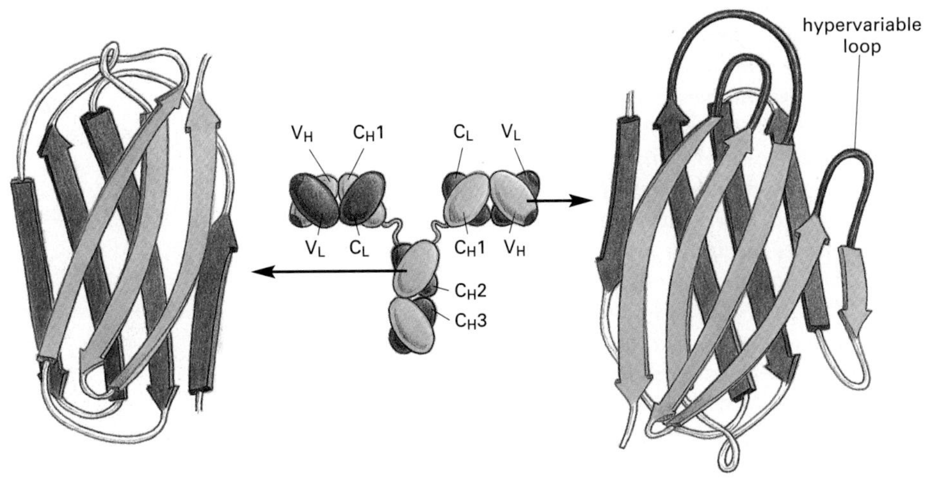

Figure 24–34 The folded structure of an IgG antibody molecule, based on x-ray crystallography studies. The structure of the whole protein is shown in the middle, while the structure of a constant domain is shown on the left and of a variable domain on the right. Both domains consist of two β sheets, which are joined by a disulfide bond (not shown). Note that all the hypervariable regions *(red)* form loops at the far end of the variable domain, where they come together to form part of the antigen-binding site.

Figure 24–35 Antigen-binding sites of antibodies. The hypervariable loops of different V_L and V_H domains can combine to form a large variety of binding surfaces. The antigenic determinants and the antigen-binding site of the antibodies are shown in *red*. Only one antigen-binding site is shown for each antibody.

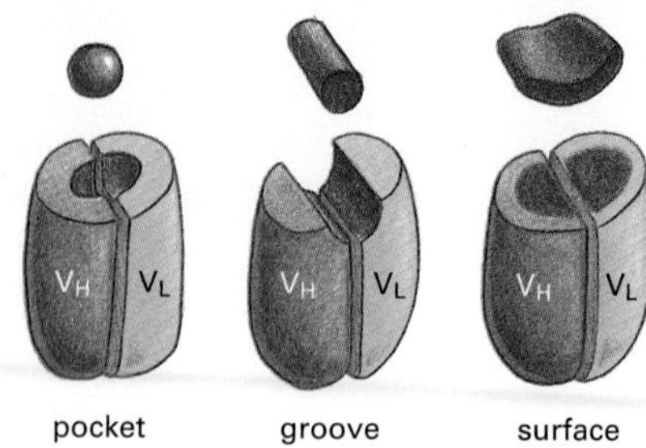

heavy variable domains are clustered together to form the antigen-binding site. Because the variable region of an antibody molecule consists of a highly conserved rigid framework, with hypervariable loops attached at one end, an enormous diversity of antigen-binding sites can be generated by changing only the lengths and amino acid sequences of the hypervariable loops. The overall three-dimensional structure necessary for antibody function remains constant.

X-ray analyses of crystals of antibody fragments bound to an antigenic determinant reveal exactly how the hypervariable loops of the light and heavy variable domains cooperate to form an antigen-binding surface in particular cases. The dimensions and shape of each different site vary depending on the conformations of the polypeptide chain in the hypervariable loops, which in turn are determined by the sequences of the amino acid side chains in the loops. The shapes of binding sites vary greatly—from pockets, to grooves, to undulating flatter surfaces, and even to protrusions—depending on the antibody (Figure 24–35). Smaller ligands tend to bind to deeper pockets, whereas larger ones tend to bind to flatter surfaces. In addition, the binding site can alter its shape after antigen binding to better fit the ligand.

Now that we have discussed the structure and functions of antibodies, we are ready to consider the crucial question that puzzled immunologists for many years—what are the genetic mechanisms that enable each of us to make many billions of different antibody molecules?

Summary

Antibodies defend vertebrates against infection by inactivating viruses and microbial toxins and by recruiting the complement system and various types of white blood cell to kill the invading pathogens. A typical antibody molecule is Y-shaped, with two identical antigen-binding sites at the tips of the Y and binding sites for complement components and/or various cell-surface receptors on the tail of the Y.

Each B cell clone makes antibody molecules with a unique antigen-binding site. Initially, during B cell development in the bone marrow, the antibody molecules are inserted into the plasma membrane, where they serve as receptors for antigen. In peripheral lymphoid organs, antigen binding to these receptors, together with costimulatory signals provided by helper T cells, activates the B cells to proliferate and differentiate into either memory cells or antibody-secreting effector cells. The effector cells secrete antibodies with the same unique antigen-binding site as the membrane-bound antibodies.

A typical antibody molecule is composed of four polypeptide chains, two identical heavy chains and two identical light chains. Parts of both the heavy and light chains usually combine to form the antigen-binding sites. There are five classes of antibodies (IgA, IgD, IgE, IgG, and IgM), each with a distinctive heavy chain (α, δ, ε, γ, and μ, respectively). The heavy chains also form the tail (Fc region) of the antibody, which determines what other proteins will bind to the antibody and therefore what biological properties the antibody class has. Either type of light chain (κ or λ) can be associated with any class of heavy chain, but the type of light chain does not seem to influence the properties of the antibody, other than its specificity for antigen.

Each light and heavy chain is composed of a number of Ig domains—β sheet structures containing about 110 amino acids. A light chain has one variable (V_L) and one constant (C_L) domain, while a heavy chain has one variable (V_H) and three or four constant (C_H) domains. The amino acid sequence variation in the variable domains of both light and heavy chains is mainly confined to several small hypervariable regions, which protrude as loops at one end of the domains to form the antigen-binding site.

THE GENERATION OF ANTIBODY DIVERSITY

Even in the absence of antigen stimulation, a human can probably make more than 10^{12} different antibody molecules—its *preimmune antibody repertoire*. Moreover, the antigen-binding sites of many antibodies can cross-react with a variety of related but different antigenic determinants, making the antibody defense force even more formidable. The preimmune repertoire is apparently large enough to ensure that there will be an antigen-binding site to fit almost any potential antigenic determinant, albeit with low affinity. After repeated stimulation by antigen, B cells can make antibodies that bind their antigen with much higher affinity—a process called *affinity maturation*. Thus, antigen stimulation greatly increases the antibody arsenal.

Antibodies are proteins, and proteins are encoded by genes. Antibody diversity therefore poses a special genetic problem: how can an animal make more antibodies than there are genes in its genome? (The human genome, for example, contains fewer than 50,000 genes.) This problem is not quite as formidable as it might first appear. Recall that the variable regions of both the light and heavy chains of antibodies usually form the antigen-binding site. Thus, an animal with 1000 genes encoding light chains and 1000 genes encoding heavy chains could, in principle, combine their products in 1000×1000 different ways to make 10^6 different antigen-binding sites (although, in reality, not every light chain can combine with every heavy chain to make an antigen-binding site). Nonetheless, the mammalian immune system has evolved unique genetic mechanisms that enable it to generate an almost unlimited number of different light and heavy chains in a remarkably economical way, by joining separate *gene segments* together before they are transcribed. Birds and fish use very different strategies for diversifying antibodies, and even sheep and rabbits use somewhat different strategies from mice and humans. We shall confine our discussion to the mechanisms used by mice and humans.

We begin this section by discussing the mechanisms that B cells use to produce antibodies with an enormous diversity of antigen-binding sites. We then consider how a B cell can alter the tail region of the antibody it makes, while keeping the antigen-binding site unchanged. This ability allows the B cell to switch from making membrane-bound antibody to making secreted antibody, or from making one class of antibody to making another, all without changing the antigen-specificity of the antibody.

Antibody Genes Are Assembled From Separate Gene Segments During B Cell Development

The first direct evidence that DNA is rearranged during B cell development came in the 1970s from experiments in which molecular biologists compared DNA from early mouse embryos, which do not make antibodies, with the DNA of a mouse B cell tumor, which makes a single species of antibody molecule. The specific variable (V)-region and constant (C)-region coding sequences that the tumor cells used were present on the same DNA restriction fragment in the tumor cells but on two different restriction fragments in the embryos. This showed that the DNA sequences encoding an antibody molecule are rearranged at some stage in B cell development (Figure 24–36).

We now know that each type of antibody chain—κ light chains, λ light chains, and heavy chains—has a separate pool of **gene segments** and exons from which a single polypeptide chain is eventually synthesized. Each pool is on a different chromosome and contains a large number of gene segments encoding the V region of an antibody chain and, as we saw in Figure 24–33, a smaller number of exons encoding the C region. During the development of a B cell, a complete coding sequence for each of the two antibody chains to be synthesized is assembled by site-specific genetic recombination (discussed in Chapter 5). In addition to bringing together the separate gene segments and the C-region exons of the antibody gene, these rearrangements also activate transcription from the gene promoter through changes in the relative positions of the

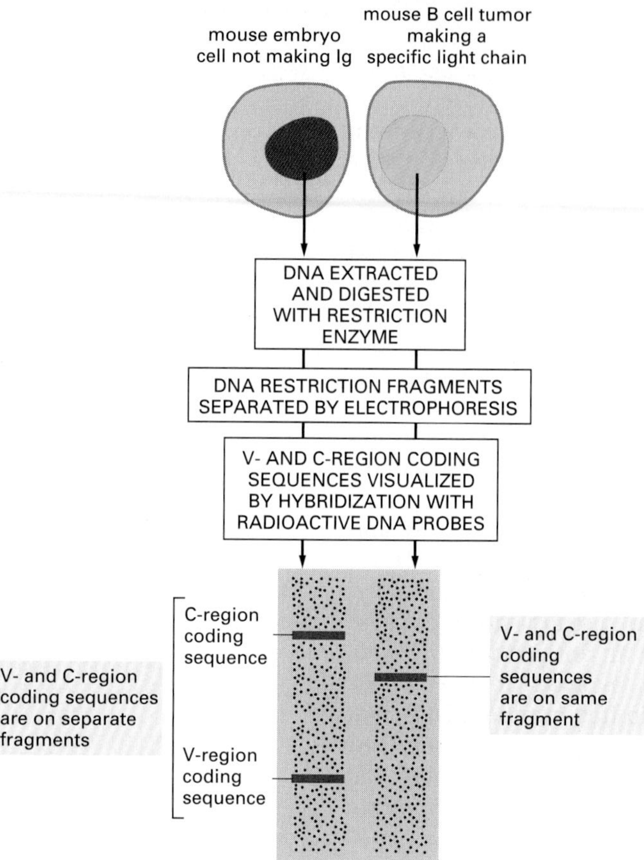

Figure 24–36 Drawing of an experiment that directly demonstrates that DNA is rearranged during B cell development. The B cell tumor arose from a single B cell and therefore makes a single species of antibody molecule. The two radioactive DNA probes used are specific for the DNA sequences encoding the C region and the V region of the light chain that the tumor cells make.

enhancers and silencers acting on the promoter. Thus, a complete antibody chain can be synthesized only after the DNA has been rearranged. As we shall see, the process of joining gene segments contributes to the diversity of antigen-binding sites in several ways.

Each Variable Region Is Encoded by More Than One Gene Segment

When genomic DNA sequences encoding V and C regions were first analyzed, it was found that a single region of DNA encodes the C region of an antibody chain (see Figure 24–33), but two or more regions of DNA have to be assembled to encode each V region. Each light-chain V region is encoded by a DNA sequence assembled from two gene segments—a long ***V* gene segment** and a short *joining*, or ***J* gene segment** (not to be confused with the protein *J chain* (see Figure 24–23), which is encoded elsewhere in the genome). Figure 24–37 illustrates the genetic mechanisms involved in producing a human κ light-chain polypeptide from a C-region exon and separate *V* and *J* gene segments.

Each heavy-chain V region is encoded by a DNA sequence assembled from three gene segments—a *V* segment, a *J* segment, and a *diversity segment*, or ***D* gene segment**. Figure 24–38 shows the number and organization of the gene segments used in making human heavy chains.

The large number of inherited *V, J*, and *D* gene segments available for encoding antibody chains makes a substantial contribution on its own to antibody diversity, but the combinatorial joining of these segments (called *combinatorial diversification)* greatly increases this contribution. Any of the 40 *V* segments in the human κ light-chain gene-segment pool, for example, can be joined to any of the 5 *J* segments (see Figure 24–37), so that at least 200 (40 × 5) different κ-chain V regions can be encoded by this pool. Similarly, any of the 51 *V* segments in the human heavy-chain pool can be joined to any of the 6 *J* segments and any of the 27 *D* segments to encode at least 8262 (51 × 6 × 27) different heavy-chain V regions.

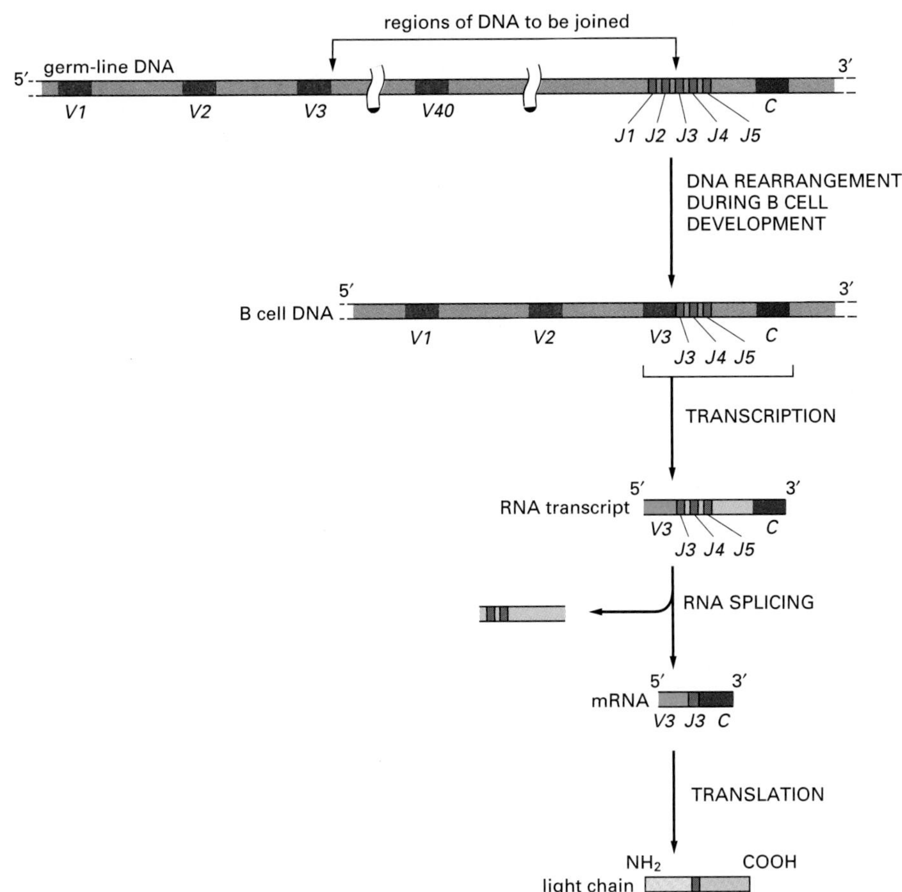

Figure 24–37 The *V–J* joining process involved in making a human κ light chain. In the "germ-line" DNA (where the antibody genes are not being expressed and are therefore not rearranged), the cluster of five *J* gene segments is separated from the *C*-region exon by a short intron and from the 40 *V* gene segments by thousands of nucleotide pairs. During the development of a B cell, the randomly chosen *V* gene segment (*V3* in this case) is moved to lie precisely next to one of the *J* gene segments (*J3* in this case). The "extra" *J* gene segments (*J4* and *J5*) and the intron sequence are transcribed (along with the joined *V3* and *J3* gene segments and the C-region exon) and then removed by RNA splicing to generate mRNA molecules in which the *V3*, *J3*, and *C* sequences are contiguous. These mRNAs are then translated into κ light chains. A *J* gene segment encodes the C-terminal 15 or so amino acids of the V region, and the *V–J* segment junction coincides with the third hypervariable region of the light chain, which is the most variable part of the V region.

The combinatorial diversification resulting from the assembly of different combinations of inherited *V*, *J*, and *D* gene segments just discussed is an important mechanism for diversifying the antigen-binding sites of antibodies. By this mechanism alone, a human can produce 287 different V_L regions (200 κ and 116 λ) and 8262 different V_H regions. In principle, these could then be combined to make about 2.6×10^6 (316×8262) different antigen-binding sites. In addition, as we discuss next, the joining mechanism itself greatly increases this number of possibilities (probably more than 10^8-fold), making it much greater than the total number of B cells (about 10^{12}) in a human.

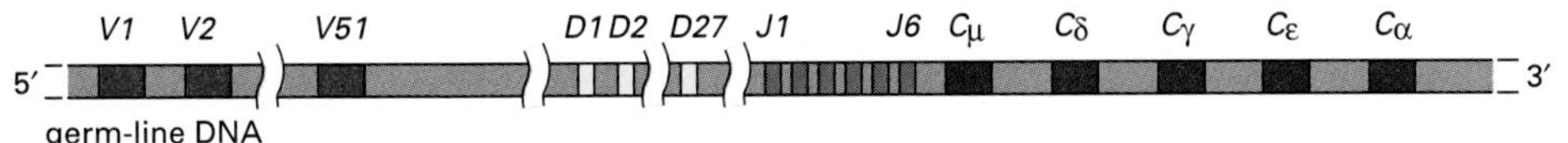

Figure 24–38 The human heavy-chain gene-segment pool. There are 51 *V* segments, 27 *D* segments, 6 *J* segments, and an ordered cluster of *C*-region exons, each cluster encoding a different class of heavy chain. The *D* segment (and part of the *J* segment) encodes amino acids in the third hypervariable region, which is the most variable part of the V region. The figure is not drawn to scale: the total length of the heavy chain locus is over 2 megabases. Moreover, many details are omitted. For instance, each *C* region is encoded by multiple exons (see Figure 24–33); there are four clusters of C_γ-region exons ($C_{\gamma 1}$, $C_{\gamma 2}$, $C_{\gamma 3}$, and $C_{\gamma 4}$), and the V_H gene segments are clustered on the chromosome in groups of homologous families. The genetic mechanisms involved in producing a heavy chain are the same as those shown in Figure 24–37 for light chains except that two DNA rearrangement steps are required instead of one. First a *D* segment joins to a *J* segment, and then a *V* segment joins to the rearranged *DJ* segment.

Imprecise Joining of Gene Segments Greatly Increases the Diversity of V Regions

During B cell development, the *V* and *J* gene segments (for the light chain) and the *V, D,* and *J* gene segments (for the heavy chain) are joined together to form a functional V_L- or V_H-region coding sequence by a process of site-specific recombination called **V(D)J joining**. Conserved DNA sequences flank each gene segment and serve as recognition sites for the joining process, ensuring that only appropriate gene segments recombine. Thus, for example, a *V* segment will always join to a *J* or *D* segment but not to another *V* segment. Joining is mediated by an enzyme complex called the **V(D)J recombinase**. This complex contains two proteins that are specific to developing lymphocytes, as well as enzymes that help repair damaged DNA in all our cells.

The lymphocyte-specific proteins of the V(D)J recombinase are encoded by two closely linked genes called *rag-1* and *rag-2 (rag* = *r*ecombination *a*ctivating *g*enes). The **RAG proteins** introduce double-strand breaks at the flanking DNA sequences, and this is followed by a rejoining process that is mediated by both the RAG proteins and the enzymes involved in general DNA double-strand repair (discussed in Chapter 5). Thus, if both *rag* genes are artificially expressed in a fibroblast, the fibroblast is now able to rearrange experimentally introduced antibody gene segments just as a developing B cell normally does. Moreover, individuals who are deficient in either *rag* gene or in one of the general repair enzymes are highly susceptible to infection because they are unable to carry out *V(D)J* joining and consequently do not have functional B or T cells. (T cells use the same recombinase to assemble the gene segments that encode their antigen-specific receptors.)

In most cases of site-specific recombination, DNA joining is precise. But during the joining of antibody (and T cell receptor) gene segments, a variable number of nucleotides are often lost from the ends of the recombining gene segments, and one or more randomly chosen nucleotides may also be inserted. This random loss and gain of nucleotides at joining sites is called **junctional diversification**, and it enormously increases the diversity of V-region coding sequences created by recombination, specifically in the third hypervariable region. This increased diversification comes at a price, however. In many cases, it will result in a shift in the reading frame that produces a nonfunctional gene. Because roughly two in every three rearrangements are "nonproductive" in this way, many developing B cells never make a functional antibody molecule and consequently die in the bone marrow. B cells making functional antibody molecules that bind strongly to self antigens in the bone marrow are stimulated to re-express the RAG proteins and undergo a second round of *V(D)J* rearrangements, thereby changing the specificity of the cell-surface antibody they make—a process referred to as **receptor editing**. Self-reactive B cells that fail to change their specificity in this way are eliminated through the process of clonal deletion (see Figure 24–13).

Antigen-Driven Somatic Hypermutation Fine-Tunes Antibody Responses

As mentioned earlier, with the passage of time after immunization, there is usually a progressive increase in the affinity of the antibodies produced against the immunizing antigen. This phenomenon, known as **affinity maturation**, is due to the accumulation of point mutations specifically in both heavy-chain and light-chain V-region coding sequences. The mutations occur long after the coding regions have been assembled, when B cells are stimulated by antigen and helper T cells to generate memory cells in a lymphoid follicle in a peripheral lymphoid organ (see Figure 24–16). They occur at the rate of about one per V-region coding sequence per cell generation. Because this is about a million times greater than the spontaneous mutation rate in other genes, the process is called **somatic hypermutation**. The molecular mechanism is still uncertain, but it is believed to involve some form of error-prone DNA repair process targeted to the rearranged V-region coding sequence by specific regions of DNA brought together by V(D)J joining. Surprisingly, an enzyme involved in RNA editing (discussed in Chapter 7) is required, but its function in the hypermutation process is unknown.

Only a small minority of the altered antigen receptors generated by hypermutation have an increased affinity for the antigen. The few B cells expressing these higher-affinity receptors, however, are preferentially stimulated by the antigen to survive and proliferate, whereas most other B cells die by apoptosis. Thus, as a result of repeated cycles of somatic hypermutation, followed by antigen-driven proliferation of selected clones of memory B cells, antibodies of increasingly higher affinity become abundant during an immune response, providing progressively better protection against the pathogen.

The main mechanisms of antibody diversification are summarized in Figure 24–39.

The Control of V(D)J Joining Ensures That B Cells Are Monospecific

As the clonal selection theory predicts, B cells are *monospecific*. That is, all the antibodies that any one B cell produces have identical antigen-binding sites. This property enables antibodies to cross-link antigens into large aggregates, thereby promoting antigen elimination (see Figure 24–19). It also means that an activated B cell secretes antibodies with the same specificity as that of the membrane-bound antibody on the B cell that was originally stimulated.

The requirement of monospecificity means that each B cell can make only one type of V_L region and one type of V_H region. Since B cells, like most other somatic cells, are diploid, each cell has six gene-segment pools encoding antibody chains: two heavy-chain pools (one from each parent) and four light-chain pools (one κ and one λ from each parent). If DNA rearrangements occurred independently in each heavy-chain pool and each light-chain pool, a single cell could make up to eight different antibodies, each with a different antigen-binding site.

In fact, however, each B cell uses only two of the six gene-segment pools: one of the two heavy-chain pools and one of the four light-chain pools. Thus, each B cell must choose not only between its κ and λ light-chain pools, but also between its maternal and paternal light-chain and heavy-chain pools. This second choice is called **allelic exclusion**, and it also occurs in the expression of genes that encode T cell receptors. For most other proteins that are encoded by autosomal genes, both maternal and paternal genes in a cell are expressed about equally.

Allelic exclusion and κ versus λ light-chain choice during B cell development depend on negative feedback regulation of the *V(D)J* joining process. A functional rearrangement in one gene-segment pool suppresses rearrangements in all remaining pools that encode the same type of polypeptide chain (Figure 24–40). In B cell clones isolated from transgenic mice expressing a rearranged μ-chain gene, for example, the rearrangement of endogenous heavy-chain genes is usually suppressed. Comparable results have been obtained for light chains. The suppression does not occur if the product of the rearranged gene fails to assemble into a receptor that inserts into the plasma membrane. It has therefore been proposed that either the receptor assembly process itself or extracellular signals that act on the receptor are involved in the suppression of further gene rearrangements.

Although no biological differences between the constant regions of κ and λ light chains have been discovered, there is an advantage in having two separate pools of gene segments encoding light chain variable regions. Having two

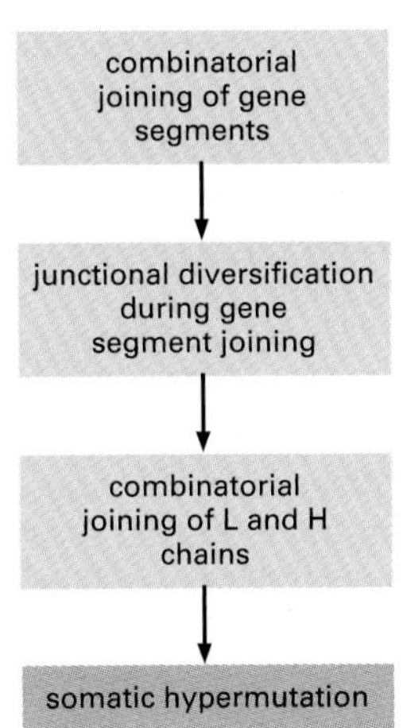

Figure 24–39 The four main mechanisms of antibody diversification. Those shaded in *green* occur during B cell development in the bone marrow (or fetal liver), while the mechanism shaded in *red* occurs when B cells are stimulated by foreign antigen and helper T cells in peripheral lymphoid organs to produce memory B cells.

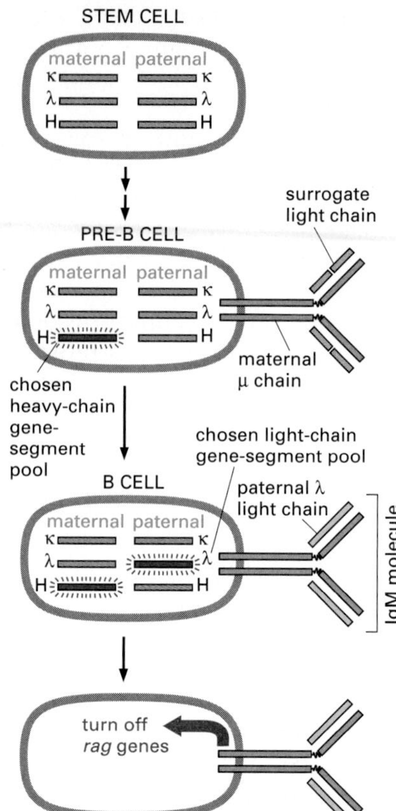

Figure 24–40 Antibody gene-pool selection in B cell development. To produce antibodies with only one type of antigen-binding site, a developing B cell must use only one L-chain gene-segment pool and one H-chain pool. Although the choice between maternal and paternal pools is thought to be random, the assembly of V-region coding sequences in a developing B cell proceeds in an orderly sequence, one segment at a time, usually beginning with the heavy-chain pool. In this pool, *D* segments first join to J_H segments on both parental chromosomes; then V_H to DJ_H joining occurs on one of these chromosomes (not shown). If this rearrangement produces a functional gene, the resulting production of complete μ chains (always the first heavy chains made) leads to their expression on the cell surface in association with surrogate light chains. The cell now shuts down all further rearrangements of V_H-region-encoding gene segments and initiates V_L rearrangement. Although not shown, V_L rearrangement usually occurs first in a κ gene-segment pool, and only if that fails does it occur in the other κ pool or in the λ pools. If, at any point, "in-phase" V_L-to-J_L joining leads to the production of light chains, these combine with preexisting μ chains to form IgM antibody molecules, which insert into the plasma membrane. The IgM cell-surface receptors are thought to enable the newly formed B cell to receive extracellular signals that shut down all further *V(D)J* joining, by turning off the expression of the *rag-1* and *rag-2* genes. If a developing B cell makes a receptor that recognizes a self-antigen, it is stimulated to re-express the *rag* genes and undergo another round of *V(D)J* joining (called receptor editing), thereby changing the specificity of its receptor (not shown). If a cell fails to assemble both a functional V_H-region and a functional V_L-region coding sequence, it is unable to make antibody molecules and dies by apoptosis (not shown).

separate pools increases the chance that a pre-B cell that has successfully assembled a V_H-region coding sequence will go on to assemble successfully a V_L-region coding sequence to become a B cell. This chance is further increased because, before a developing pre-B cell produces ordinary light chains, it makes surrogate light chains (see Figure 24–22), which assemble with μ heavy chains. The resulting receptors are displayed on the cell surface and allow the cell to proliferate, producing large numbers of progeny cells, some of which are likely to succeed in producing bona fide light chains.

When Activated by Antigen, a B Cell Switches From Making a Membrane-Bound Antibody to Making a Secreted Form of the Same Antibody

We now turn from the genetic mechanisms that determine the antigen-binding site of an antibody to those that determine its biological properties—that is, those that determine what form of heavy-chain constant region is synthesized. The choice of the particular gene segments that encode the antigen-binding site is a commitment for the life of a B cell and its progeny, but the type of C_H region that is made changes during B cell development. The changes are of two types: changes from a membrane-bound form to a secreted form of the same C_H region and changes in the class of the C_H region made.

All classes of antibody can be made in a membrane-bound form, as well as in a soluble, secreted form. The membrane-bound form serves as an antigen receptor on the B cell surface, while the soluble form is made only after the cell is activated by antigen to become an antibody-secreting effector cell (see Figure 24–17). The sole difference between the two forms resides in the C-terminus of the heavy chain. The heavy chains of membrane-bound antibody molecules have a hydrophobic C-terminus, which anchors them in the lipid bilayer of the B cell's plasma membrane. The heavy chains of secreted antibody molecules, by contrast, have instead a hydrophilic C-terminus, which allows them to escape from the cell. The switch in the character of the antibody molecules made occurs because the activation of B cells by antigen (and helper T cells) induces a change in the way in which the H-chain RNA transcripts are made and processed in the nucleus (see Figure 7–93).

B Cells Can Switch the Class of Antibody They Make

During B cell development, many B cells switch from making one class of antibody to making another—a process called **class switching**. All B cells begin their antibody-synthesizing lives by making IgM molecules and inserting them into the plasma membrane as receptors for antigen. After the B cells leave the bone marrow, but before they interact with antigen, they switch and make both IgM and IgD molecules as membrane-bound antigen receptors, both with the same antigen-binding sites (see Figure 24–22). On stimulation by antigen and helper T cells, some of these cells are activated to secrete IgM antibodies, which dominate the primary antibody response. Later in the immune response, the combination of antigen and the cytokines that helper T cells secrete induce many B cells to switch to making IgG, IgE, or IgA antibodies. These cells generate both memory cells that express the corresponding classes of antibody molecules on their surface and effector cells that secrete the antibodies. The IgG, IgE, and IgA molecules are collectively referred to as *secondary* classes of antibodies, both because they are produced only after antigen stimulation and because they dominate secondary antibody responses. As we saw earlier, each different class of antibody is specialized to attack microbes in different ways and in different sites.

The constant region of an antibody heavy chain determines the class of the antibody. Thus, the ability of B cells to switch the class of antibody they make without changing the antigen-binding site implies that the same assembled V_H-region coding sequence (which specifies the antigen-binding part of the heavy chain) can sequentially associate with different C_H-coding sequences. This has important functional implications. It means that, in an individual animal, a particular antigen-binding site that has been selected by environmental antigens can be distributed among the various classes of antibodies, thereby acquiring the different biological properties of each class.

When a B cell switches from making IgM and IgD to one of the secondary classes of antibody, an irreversible change at the DNA level occurs—a process called *class switch recombination*. It entails deletion of all the C_H-coding sequences between the assembled VDJ-coding sequence and the particular C_H-coding sequence that the cell is destined to express (Figure 24–41). Switch recombination differs from *V(D)J* joining in several ways: (1) it involves noncoding sequences only and therefore leaves the coding sequence unaffected; (2) it uses different flanking recombination sequences and different enzymes; (3) it happens after antigen stimulation; and (4) it is dependent on helper T cells.

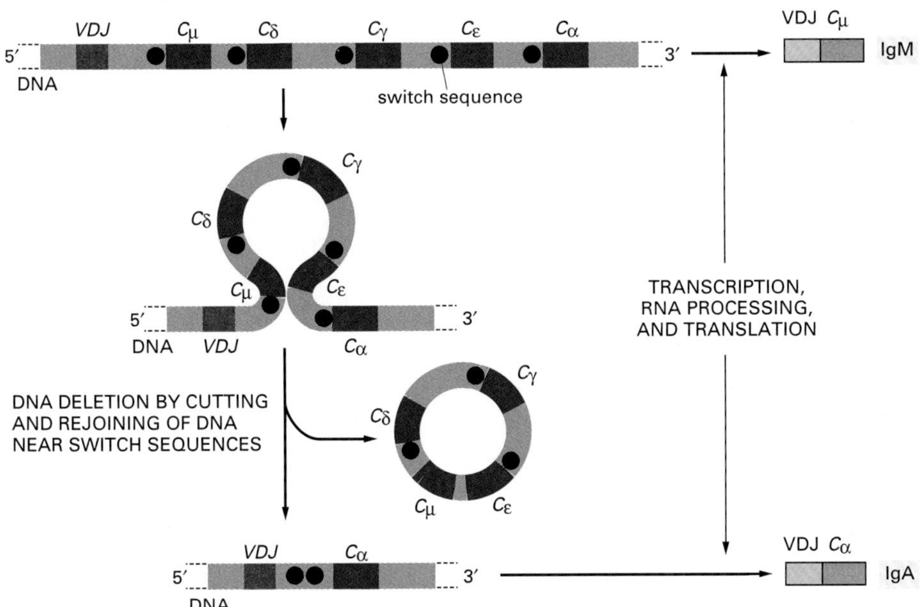

Figure 24–41 An example of the DNA rearrangement that occurs in class switch recombination. A B cell making an IgM antibody from an assembled *VDJ* DNA sequence is stimulated by antigen and the cytokines made by helper T cells to switch to making an IgA antibody. In the process, it deletes the DNA between the *VDJ* sequence and the C_α-coding sequence. Specific DNA sequences *(switch sequences)* located upstream of each C_H-coding sequence recombine with each other to delete the intervening DNA. Class switch recombination is thought to be mediated by a *switch recombinase*, which is directed to the appropriate switch sequences when these become accessible under the influence of cytokines, as we discuss later.

Summary

Antibodies are produced from three pools of gene segments and exons. One pool encodes κ light chains, one encodes λ light chains, and and one encodes heavy chains. In each pool, separate gene segments that code for different parts of the variable region of the light or heavy chains are brought together by site-specific recombination during B cell development. The light-chain pools contain one or more constant- (C-) region exons and sets of variable (V) and joining (J) gene segments. The heavy-chain pool contains sets of C-region exons and sets of V, diversity (D), and J gene segments.

To make an antibody molecule, a V_L gene segment recombines with a J_L gene segment to produce a DNA sequence coding for the V region of a light chain, and a V_H gene segment recombines with a D and a J_H gene segment to produce a DNA sequence coding for the V region of a heavy chain. Each of the assembled V-region coding sequences is then cotranscribed with the appropriate C-region sequence to produce an RNA molecule that codes for the complete polypeptide chain. Cells making functional heavy and light chains turn off the V(D)J joining process to ensure that each B cell makes only one species of antigen-binding site.

By randomly combining inherited gene segments that code for V_L and V_H regions, humans can make hundreds of different light chains and thousands of different heavy chains. Because the antigen-binding site is formed where the hypervariable loops of the V_L and V_H come together in the final antibody, the heavy and light chains can pair to form antibodies with millions of different antigen-binding sites. This number is enormously increased by the loss and gain of nucleotides at the site of gene-segment joining, as well as by somatic mutations that occur with very high frequency in the assembled V-region coding sequences after stimulation by antigen and helper T cells.

All B cells initially make IgM antibodies, and most then make IgD as well. Later many switch and make antibodies of other classes but with the same antigen-binding site as the original IgM and IgD antibodies. Such class switching depends on antigen stimulation and helper T cells, and it allows the same antigen-binding sites to be distributed among antibodies with varied biological properties.

T CELLS AND MHC PROTEINS

The diverse responses of T cells are collectively called *cell-mediated immune reactions*. This is to distinguish them from antibody responses, which, of course, also depend on cells (B cells). Like antibody responses, T cell responses are exquisitely antigen-specific, and they are at least as important as antibodies in defending vertebrates against infection. Indeed, most adaptive immune responses, including antibody responses, require helper T cells for their initiation. Most importantly, unlike B cells, T cells can help eliminate pathogens that reside inside host cells. Much of the rest of this chapter is concerned with how T cells accomplish this feat.

T cell responses differ from B cell responses in at least two crucial ways. First, T cells are activated by foreign antigen to proliferate and differentiate into effector cells only when the antigen is displayed on the surface of antigen-presenting cells in peripheral lymphoid organs. The T cells respond in this manner because the form of antigen they recognize is different from that recognized by B cells. Whereas B cells recognize intact antigen, T cells recognize fragments of protein antigens that have been partly degraded inside the antigen-presenting cell. The peptide fragments are then carried to the surface of the presenting cell on special molecules called *MHC proteins*, which present the fragments to T cells. The second difference is that, once activated, effector T cells act only at short range, either within a secondary lymphoid organ or after they have migrated into a site of infection. They interact directly with another cell in the body, which they either kill or signal in some way (we shall refer to such cells as *target cells)*. Activated B cells, by contrast, secrete antibodies that can act far away.

There are two main classes of T cells—cytotoxic T cells and helper T cells. Effector *cytotoxic T cells* directly kill cells that are infected with a virus or some other intracellular pathogen. Effector *helper T cells,* by contrast, help stimulate the responses of other cells—mainly macrophages, B cells, and cytotoxic T cells.

In this section, we describe these two classes of T cells and their respective functions. We discuss how they recognize foreign antigens on the surface of antigen-presenting cells and target cells and consider the crucial part played by MHC proteins in the recognition process. Finally, we describe how T cells are selected during their development in the thymus to ensure that only cells with potentially useful receptors survive and mature. We begin by considering the nature of the cell-surface receptors that T cells use to recognize antigen.

T Cell Receptors Are Antibodylike Heterodimers

Because T cell responses depend on direct contact with an antigen-presenting cell or a target cell, the antigen receptors made by T cells, unlike antibodies made by B cells, exist only in membrane-bound form and are not secreted. For this reason, T cell receptors were difficult to isolate, and it was not until the 1980s that they were first identified biochemically. On both cytotoxic and helper T cells, the receptors are similar to antibodies. They are composed of two disulfide-linked polypeptide chains (called α and β), each of which contains two Ig-like domains, one variable and one constant (Figure 24–42A). Moreover, the three-dimensional structure of the extracellular part of a T cell receptor has been determined by x-ray diffraction, and it looks very much like one arm of a Y-shaped antibody molecule (Figure 24–42B).

The pools of gene segments that encode the α and β chains are located on different chromosomes. Like antibody heavy-chain pools, the T cell receptor pools contain separate *V, D,* and *J* gene segments, which are brought together by site-specific recombination during T cell development in the thymus. With one exception, all the mechanisms used by B cells to generate antibody diversity are also used by T cells to generate T cell receptor diversity. Indeed, the same *V(D)J* recombinase is used, including the RAG proteins discussed earlier. The mechanism that does not operate in T cell receptor diversification is antigen-driven somatic hypermutation. Thus, the affinity of the receptors remains low (K_a ~ 10^5–10^7 liters/mole), even late in an immune response. We discuss later how various co-receptors and cell–cell adhesion mechanisms greatly strengthen the binding of a T cell to an antigen-presenting cell or a target cell, helping to compensate for the low affinity of the T cell receptors.

A small minority of T cells, instead of making α and β chains, make a different but related type of receptor heterodimer, composed of γ and δ chains. These

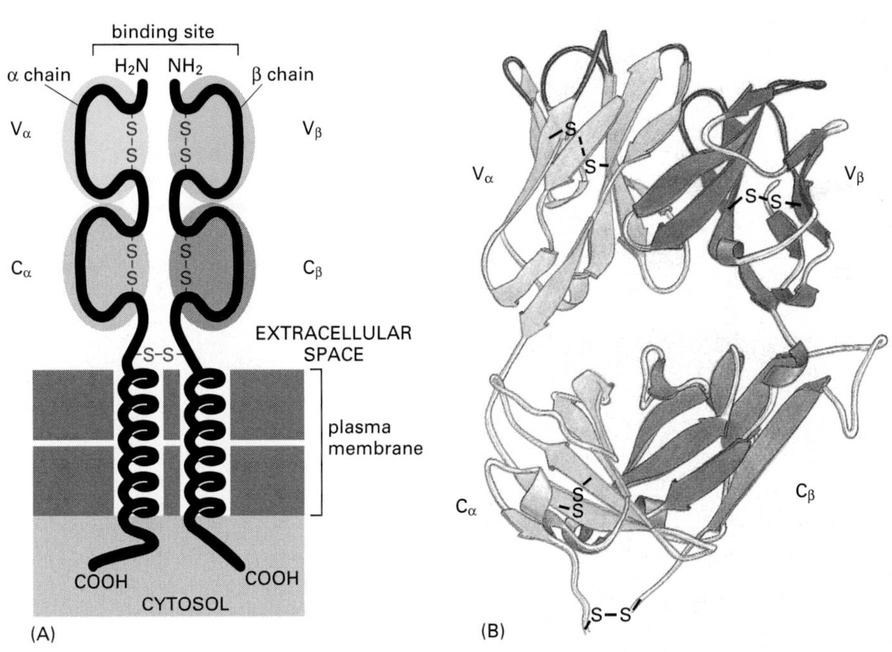

Figure 24–42 A T cell receptor heterodimer. (A) Schematic drawing showing that the receptor is composed of an α and a β polypeptide chain. Each chain is about 280 amino acids long and has a large extracellular part that is folded into two Ig-like domains—one variable (V) and one constant (C). The antigen-binding site is formed by a V_α and a V_β domain (shaded in *blue).* Unlike antibodies, which have two binding sites for antigen, T cell receptors have only one. The $\alpha\beta$ heterodimer is noncovalently associated with a large set of invariant membrane-bound proteins (not shown), which help activate the T cell when the T cell receptors bind to antigen. A typical T cell has about 30,000 such receptor complexes on its surface. (B) The three-dimensional structure of the extracellular part of a T cell receptor. The antigen-binding site is formed by the hypervariable loops of both the V_α and V_β domains *(red),* and it is similar in its overall dimensions and geometry to the antigen-binding site of an antibody molecule. (B, based on K.C. Garcia et al., *Science* 274:209–219, 1996.)

cells arise early in development and are found mainly in epithelia (in the skin and gut, for example). Their functions are uncertain, and we shall not discuss them further.

As with antigen receptors on B cells, the T cell receptors are tightly associated in the plasma membrane with a number of invariant membrane-bound proteins that are involved in passing the signal from an antigen-activated receptor to the cell interior. We discuss these proteins in more detail later. First, however, we need to consider how cytotoxic and helper T cells function and the special ways in which they recognize foreign antigen.

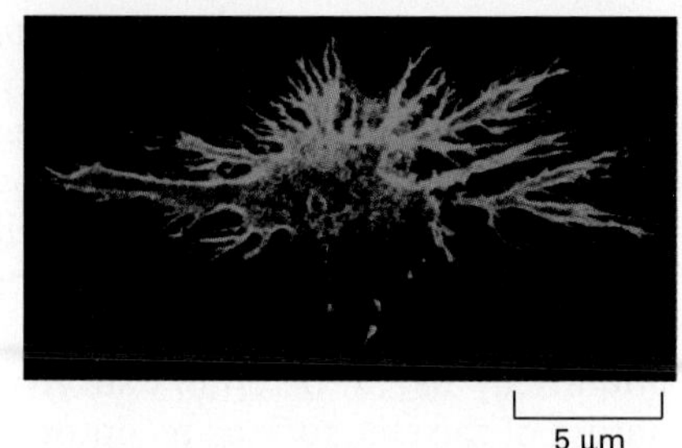

Figure 24–43 Immunofluorescence micrograph of a dendritic cell in culture. These crucial antigen-presenting cells derive their name from their long processes, or "dendrites." The cell has been labelled with a monoclonal antibody that recognizes a surface antigen on these cells. (Courtesy of David Katz.)

Antigen-Presenting Cells Activate T Cells

Before cytotoxic or helper T cells can kill or help their target cells, respectively, they must be activated to proliferate and differentiate into effector cells. This activation occurs in peripheral lymphoid organs on the surface of **antigen-presenting cells** that display foreign antigen complexed with MHC proteins on their surface.

There are three main types of antigen-presenting cells in peripheral lymphoid organs that can activate T cells—dendritic cells, macrophages, and B cells. The most potent of these are **dendritic cells** (Figure 24–43), whose only known function is to present foreign antigens to T cells. Immature dendritic cells are located in tissues throughout the body, including the skin, gut, and respiratory tract. When they encounter invading microbes at these sites, they endocytose the pathogens or their products and carry them via the lymph to local lymph nodes or gut-associated lymphoid organs. The encounter with a pathogen induces the dendritic cell to mature from an antigen-capturing cell to an antigen-presenting cell that can activate T cells (see Figure 24–5).

Antigen-presenting cells display three types of protein molecules on their surface that have a role in activating a T cell to become an effector cell: (1) *MHC proteins,* which present foreign antigen to the T cell receptor, (2) *costimulatory proteins,* which bind to complementary receptors on the T cell surface, and (3) *cell–cell adhesion molecules,* which enable a T cell to bind to the antigen-presenting cell for long enough to become activated (Figure 24–44).

Before discussing the role of MHC proteins in presenting antigen to T cells, we consider the functions of the two major classes of T cells.

Effector Cytotoxic T Cells Induce Infected Target Cells to Kill Themselves

Cytotoxic T cells provide protection against intracellular pathogens such as viruses and some bacteria and parasites that multiply in the host–cell cytoplasm, where they are sheltered from attack by antibodies. They provide this

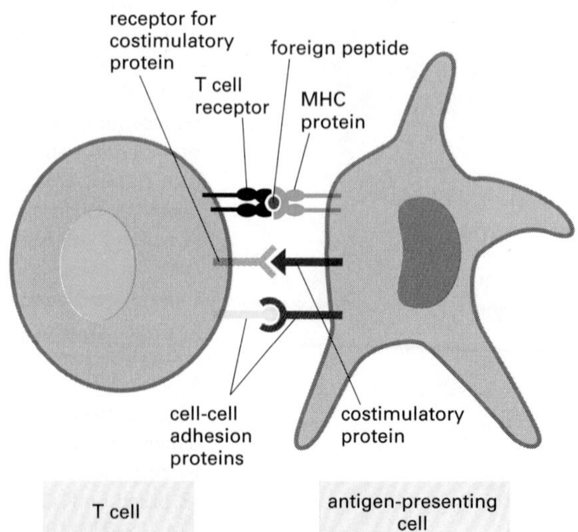

Figure 24–44 Three types of proteins on the surface of an antigen-presenting cell involved in activating a T cell. The invariant polypeptide chains that are stably associated with the T cell receptor are not shown.

protection by killing the infected cell before the microbes can proliferate and escape from the infected cell to infect neighboring cells.

Once a cytotoxic T cell has been activated by an infected antigen-presenting cell to become an effector cell, it can kill any target cell infected with the same pathogen. When the effector T cell recognizes a microbial antigen on the surface of an infected target cell, it focuses its secretory apparatus on the target. We can observe this behavior by studying effector T cells bound to their targets: when labeled with anti-tubulin antibodies, the T cell centrosome is seen to be oriented toward the point of contact with the target cell (Figure 24–45). Moreover, antibody labeling shows that talin and other proteins that help link cell-surface receptors to cortical actin filaments are concentrated in the cortex of the T cell at the contact site. The aggregation of T cell receptors at the contact site apparently leads to a local alteration in the actin filaments in the cell cortex. A microtubule-dependent mechanism then moves the centrosome and its associated Golgi apparatus toward the contact site, focusing the killing machinery on the target cell. A similar cytoskeletal polarization is seen when an effector helper T cell interacts functionally with a target cell.

Once bound to its target cell, a cytotoxic T cell can employ at least two strategies to kill the target, both of which operate by inducing the target cell to kill itself by undergoing apoptosis (discussed in Chapter 17). In killing an infected target cell, the cytotoxic T cell usually releases a pore-forming protein called **perforin**, which is homologous to the complement component C9 (see Figure 25–42) and polymerizes in the target cell plasma membrane to form transmembrane channels. Perforin is stored in secretory vesicles of the cytotoxic T cell and is released by local exocytosis at the point of contact with the target cell. The secretory vesicles also contain serine proteases, which are thought to enter the target cell cytosol through the perforin channels. One of the proteases, called

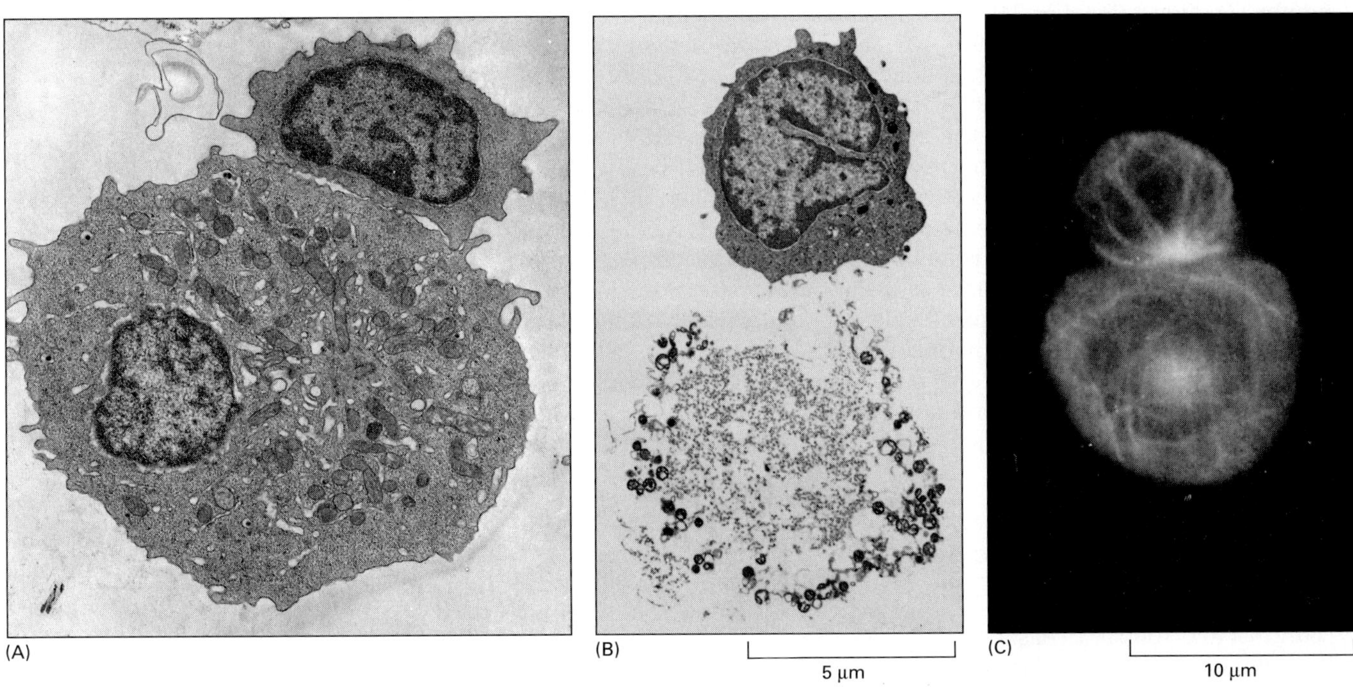

Figure 24–45 Effector cytotoxic T cells killing target cells in culture. (A) Electron micrograph showing an effector cytotoxic T cell binding to the target cell. The cytotoxic T cells were obtained from mice immunized with the target cells, which are foreign tumor cells. (B) Electron micrograph showing a cytotoxic T cell and a tumor cell that the T cell has killed. In an animal, as opposed to in a tissue culture dish, the killed target cell would be phagocytosed by neighboring cells long before it disintegrated in the way that it has here. (C) Immunofluorescence micrograph of a T cell and tumor cell after staining with anti-tubulin antibodies. Note that the centrosome in the T cell and the microtubules radiating from it are oriented toward the point of cell–cell contact with the target cell. See also Figure 16–97A. (A and B, from D. Zagury, J. Bernard, N. Thierness, M. Feldman, and G. Berke, *Eur. J. Immunol.* 5:818–822, 1975; C, reproduced from B. Geiger, D. Rosen, and G. Berke, *J. Cell Biol.* 95:137–143, 1982. © The Rockefeller University Press.)

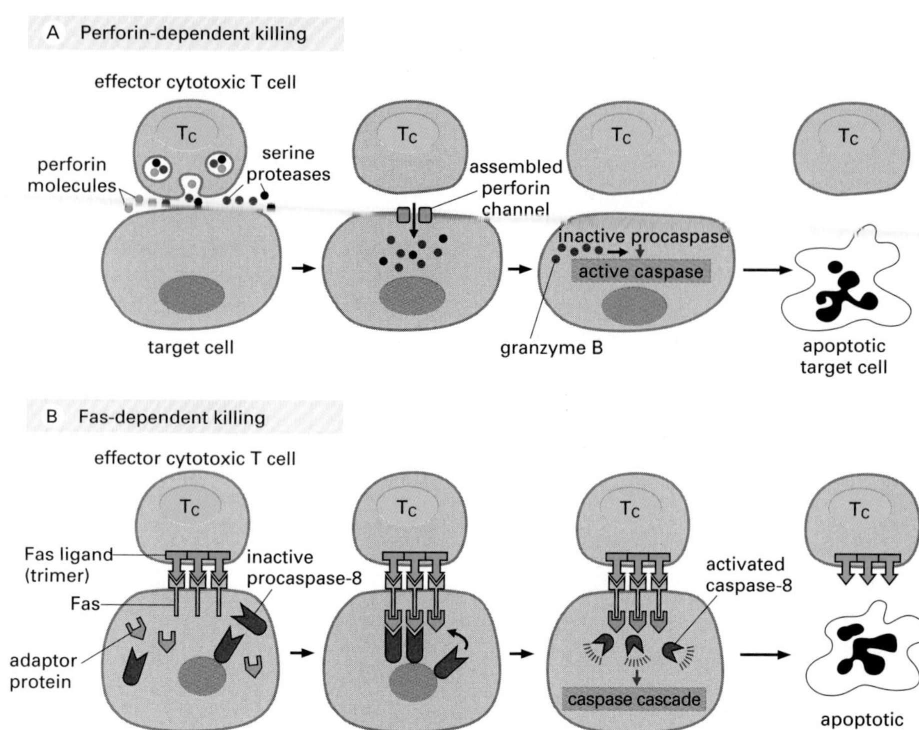

Figure 24–46 Two strategies by which effector cytotoxic T cells kill their target cells. (A) The cytotoxic T cell (T_C) releases perforin and proteolytic enzymes onto the surface of an infected target cell by localized exocytosis. The high concentration of Ca^{2+} in the extracellular fluid causes the perforin to assemble into transmembrane channels, which are thought to allow the proteolytic enzymes to enter the target cell cytosol. One of the enzymes, granzyme B, cleaves and activates specific procaspases, thereby triggering the proteolytic caspase cascade leading to apoptosis. (B) The homotrimeric Fas ligand on the cytotoxic T cell surface binds to and activates Fas receptor protein on the surface of a target cell. The cytosolic tail of Fas contains a *death domain,* which, when activated, binds to an adaptor protein, which in turn recruits a specific procaspase (procaspase-8). Clustered procaspase-8 molecules then cleave one another to produce active caspase-8 molecules that initiate the proteolytic caspase cascade leading to apoptosis.

granzyme B, cleaves, and thereby activates, one or more members of the *caspase family* of proteases that mediate apoptosis. These caspases then activate other caspases, producing a proteolytic cascade that helps kill the cell (discussed in Chapter 17) (Figure 24–46A). Mice in which the perforin gene is inactivated cannot generate microbe-specific cytotoxic T cells and show increased susceptibility to certain viral and intracellular bacterial infections.

In the second killing strategy, the cytotoxic T cell also activates a death-inducing caspase cascade in the target cell but does it less directly. A homotrimeric protein on the cytotoxic T cell surface called **Fas ligand** binds to transmembrane receptor proteins on the target cell called **Fas**. The binding alters the Fas proteins so that their clustered cytosolic tails recruit procaspase-8 into the complex via an adaptor protein. The recruited procaspase-8 molecules cross-cleave and activate each other to begin the caspase cascade that leads to apoptosis (Figure 24–46B). Cytotoxic T cells apparently use this killing strategy to help contain an immune response once it is well underway, by killing excessive effector lymphocytes, especially effector T cells: if the gene encoding either Fas or Fas ligand is inactivated by mutation, effector lymphocytes accumulate in vast numbers in the spleen and lymph nodes, which become enormously enlarged.

Effector Helper T Cells Help Activate Macrophages, B Cells, and Cytotoxic T Cells

In contrast to cytotoxic T cells, **helper T cells** are crucial for defense against both extracellular and intracellular pathogens. They help stimulate B cells to make antibodies that help inactivate or eliminate extracellular pathogens and their toxic products. They activate macrophages to destroy any intracellular pathogen multiplying within the macrophage's phagosomes, and they help activate cytotoxic T cells to kill infected target cells.

Once a helper T cell has been activated by an antigen-presenting cell to become an effector cell, it can then help activate other cells. It does this both by secreting a variety of cytokines and by displaying costimulatory proteins on its surface. When activated by an antigen-presenting cell, a naïve helper T cell can differentiate into either of two distinct types of effector helper cell, called T_H1 and T_H2. *T_H1 cells* mainly help activate macrophages and cytotoxic T cells,

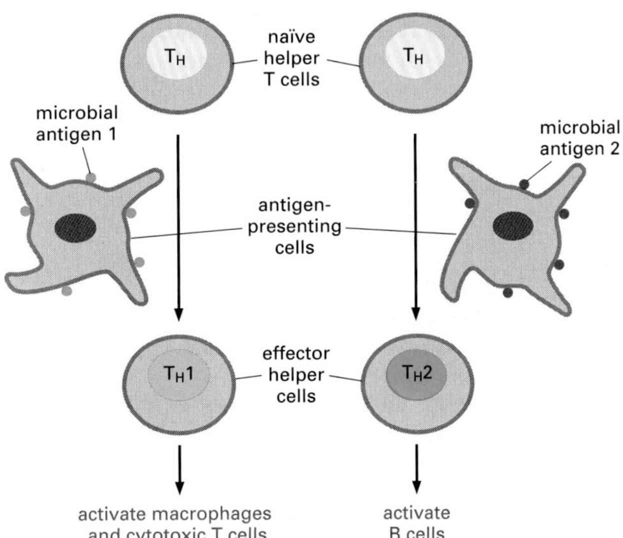

Figure 24–47 Differentiation of naïve helper T cells into either T_H1 or T_H2 effector helper cells in a peripheral lymphoid organ. The antigen-presenting cell and the characteristics of the pathogen that activated it mainly determine which type of effector helper cell develops.

whereas *T_H2 cells* mainly help activate B cells (Figure 24–47). As we discuss later, the nature of the invading pathogen and the types of innate immune responses it elicits largely determine which type of helper T cell develops. This, in turn, determines the nature of the adaptive immune responses mobilized to fight the invaders.

Before discussing how helper T cells function to activate macrophages, cytotoxic T cells, or B cells, we need to consider the crucial role of MHC proteins in T cell responses.

T Cells Recognize Foreign Peptides Bound to MHC Proteins

As discussed earlier, both cytotoxic T cells and helper T cells are initially activated in peripheral lymphoid organs by recognizing foreign antigen on the surface of an antigen-presenting cell, usually a dendritic cell. The antigen is in the form of peptide fragments that are generated by the degradation of foreign protein antigens inside the antigen-presenting cell. The recognition process depends on the presence in the antigen-presenting cell of **MHC proteins**, which bind these fragments, carry them to the cell surface, and present them there, along with a co-stimulatory signal, to the T cells. Once activated, effector T cells then recognize the same peptide–MHC complex on the surface of the target cell they influence, which may be a B cell, a cytotoxic T cell, or an infected macrophage in the case of a helper T cell, or a virus-infected cell in the case of a cytotoxic T cell.

MHC proteins are encoded by a large complex of genes called the **major histocompatibility complex (MHC)**. There are two main structurally and functionally distinct classes of MHC proteins: *class I MHC proteins,* which present foreign peptides to cytotoxic T cells, and *class II MHC proteins,* which present foreign peptides to helper cells (Figure 24–48).

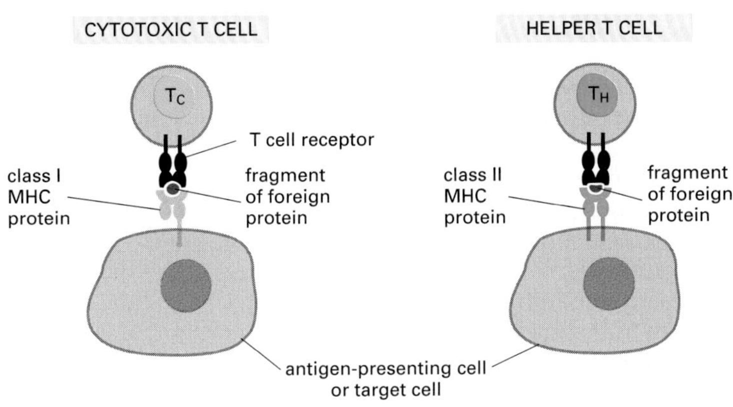

Figure 24–48 Recognition by T cells of foreign peptides bound to MHC proteins. Cytotoxic T cells recognize foreign peptides in association with class I MHC proteins, whereas helper T cells recognize foreign peptides in association with class II MHC proteins. In both cases, the peptide–MHC complexes are recognized on the surface of an antigen-presenting cell or a target cell.

Before examining the different mechanisms by which protein antigens are processed for display to the two main classes of T cells, we must look more closely at the MHC proteins themselves, which have such an important role in T cell function.

MHC Proteins Were Identified in Transplantation Reactions Before Their Functions Were Known

MHC proteins were initially identified as the main antigens recognized in **transplantation reactions.** When organ grafts are exchanged between adult individuals, either of the same species *(allografts)* or of different species *(xenografts),* they are usually rejected. In the 1950s, skin grafting experiments between different strains of mice demonstrated that *graft rejection* is an adaptive immune response to the foreign antigens on the surface of the grafted cells. Rejection is mediated mainly by T cells, which react against genetically "foreign" versions of cell-surface proteins called *histocompatibility molecules* (from the Greek word *histo,* meaning "tissue"). The MHC proteins encoded by the clustered genes of the major histocompatibility complex (MHC) are by far the most important of these. MHC proteins are expressed on the cells of all higher vertebrates. They were first demonstrated in mice, where they are called *H-2 antigens (h*istocompatibility-*2* antigens). In humans they are called *HLA antigens (h*uman-*l*eucocyte-*a*ssociated antigens) because they were first demonstrated on leucocytes (white blood cells).

Three remarkable properties of MHC proteins baffled immunologists for a long time. First, MHC proteins are overwhelmingly the preferred antigens recognized in T-cell-mediated transplantation reactions. Second, an unusually large fraction of T cells are able to recognize foreign MHC proteins: whereas fewer than 0.001% of an individual's T cells respond to a typical viral antigen, more than 0.1% of them respond to a single foreign MHC antigen. Third, some of the genes that code for MHC proteins are the most *polymorphic* known in higher vertebrates. That is, within a species, there is an extraordinarily large number of *alleles* (alternative forms of the same gene) present (in some cases more than 200), without any one allele predominating. As each individual has at least 12 genes encoding MHC proteins (see later), it is very rare for two unrelated individuals to have an identical set of MHC proteins. This makes it very difficult to match donor and recipient for organ transplantation unless they are closely related.

Of course, a vertebrate does not need to protect itself against invasion by foreign vertebrate cells. So the apparent obsession of its T cells with foreign MHC proteins and the extreme polymorphism of these molecules were a great puzzle. The puzzle was solved only after it was discovered that (1) MHC proteins bind fragments of foreign proteins and display them on the surface of host cells for T cells to recognize, and (2) T cells respond to foreign MHC proteins in the same way they respond to self MHC proteins that have foreign antigen bound to them.

Class I and Class II MHC Proteins Are Structurally Similar Heterodimers

Class I and class II MHC proteins have very similar overall structures. They are both transmembrane heterodimers with extracellular N-terminal domains that bind antigen for presentation to T cells.

Class I MHC proteins consist of a transmembrane α chain, which is encoded by a class I MHC gene, and a small extracellular protein called *β_2-microglobulin* (Figure 24–49A). The β_2-microglobulin does not span the membrane and is encoded by a gene that does not lie in the MHC gene cluster. The α chain is folded into three extracellular globular domains (α_1, α_2, α_3), and the α_3 domain and the β_2-microglobulin, which are closest to the membrane, are both similar to an Ig domain. The two N-terminal domains of the α chain, which are farthest from the membrane, contain the polymorphic (variable) amino acids that are

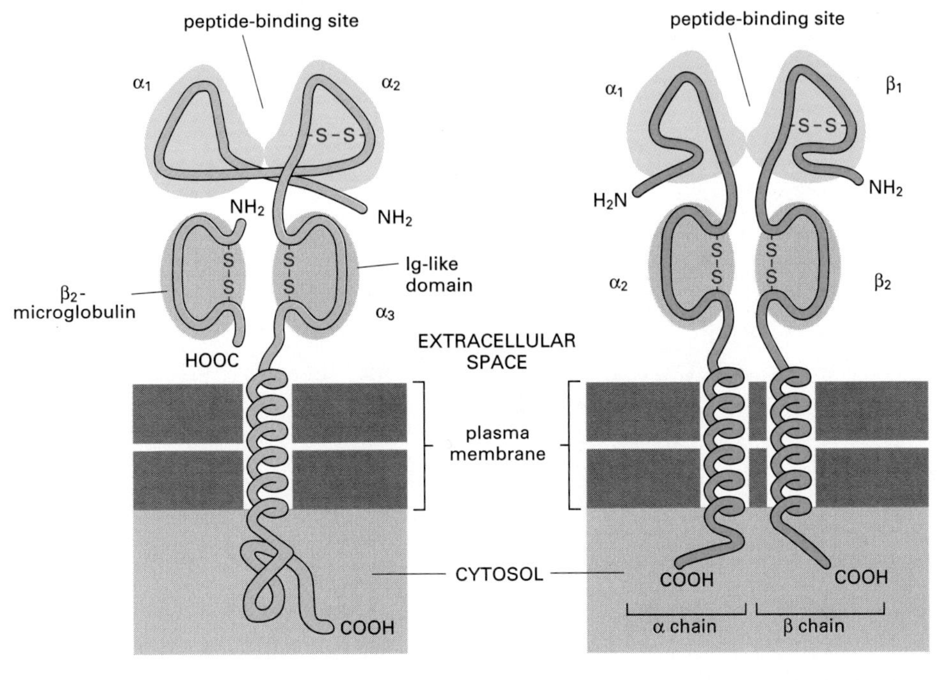

Figure 24–49 Class I and class II MHC proteins. (A) The α chain of the class I molecule has three extracellular domains, α_1, α_2 and α_3, encoded by separate exons. It is noncovalently associated with a smaller polypeptide chain, β_2-microglobulin, which is not encoded within the MHC. The α_3 domain and β_2-microglobulin are Ig-like. While β_2-microglobulin is invariant, the α chain is extremely polymorphic, mainly in the α_1 and α_2 domains. (B) In class II MHC proteins, both chains are polymorphic, mainly in the α_1 and β_1 domains; the α_2 and β_2 domains are Ig-like. Thus, there are striking similarities between class I and class II MHC proteins. In both, the two outermost domains (shaded in *blue*) interact to form a groove that binds peptide fragments of foreign proteins and presents them to T cells

recognized by T cells in transplantation reactions. These domains bind a peptide and present it to cytotoxic T cells.

Like class I MHC proteins, **class II MHC proteins** are heterodimers with two conserved Ig-like domains close to the membrane and two polymorphic (variable) N-terminal domains farthest from the membrane. In these proteins, however, both chains (α and β) are encoded by genes within the MHC, and both span the membrane (Figure 24–49B). The two polymorphic domains bind a peptide and present it to helper T cells.

The presence of Ig-like domains in class I and class II proteins suggests that MHC proteins and antibodies have a common evolutionary history. The locations of the genes that encode class I and class II MHC proteins in humans are shown in Figure 24–50, where we illustrate how an individual can make six types of class I MHC proteins and more than six types of class II proteins.

In addition to the classic class I MHC proteins, there are many *nonclassical class I MHC proteins,* which form dimers with β_2-microglobulin. These proteins are not polymorphic, but some of them present specific microbial antigens, including some lipids and glycolipids, to T cells. The functions of most of them, however, are unknown.

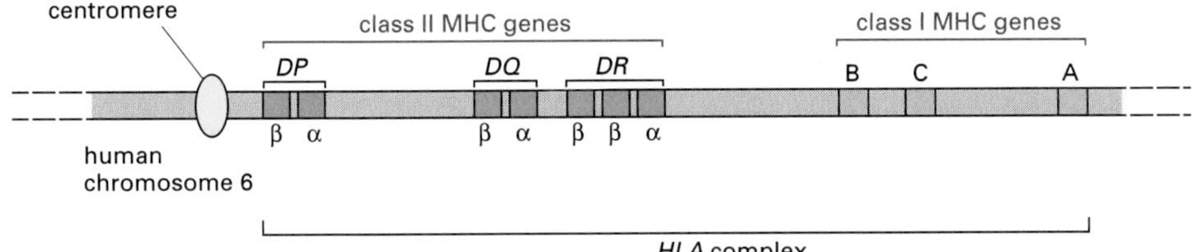

Figure 24–50 Human MHC genes. This simplified schematic drawing shows the location of the genes that encode the transmembrane subunits of class I *(light green)* and class II *(dark green)* MHC proteins. The genes shown encode three types of class I proteins (HLA-A, HLA-B, and HLA-C) and three types of class II MHC proteins (HLA-DP, HLA-DQ, and HLA-DR). An individual can therefore make six types of class I MHC proteins (three encoded by maternal genes and three by paternal genes) and more than six types of class II MHC proteins. The number of class II MHC proteins that can be made is increased because there are two DR β genes and because maternally encoded and paternally encoded polypeptide chains can sometimes pair.

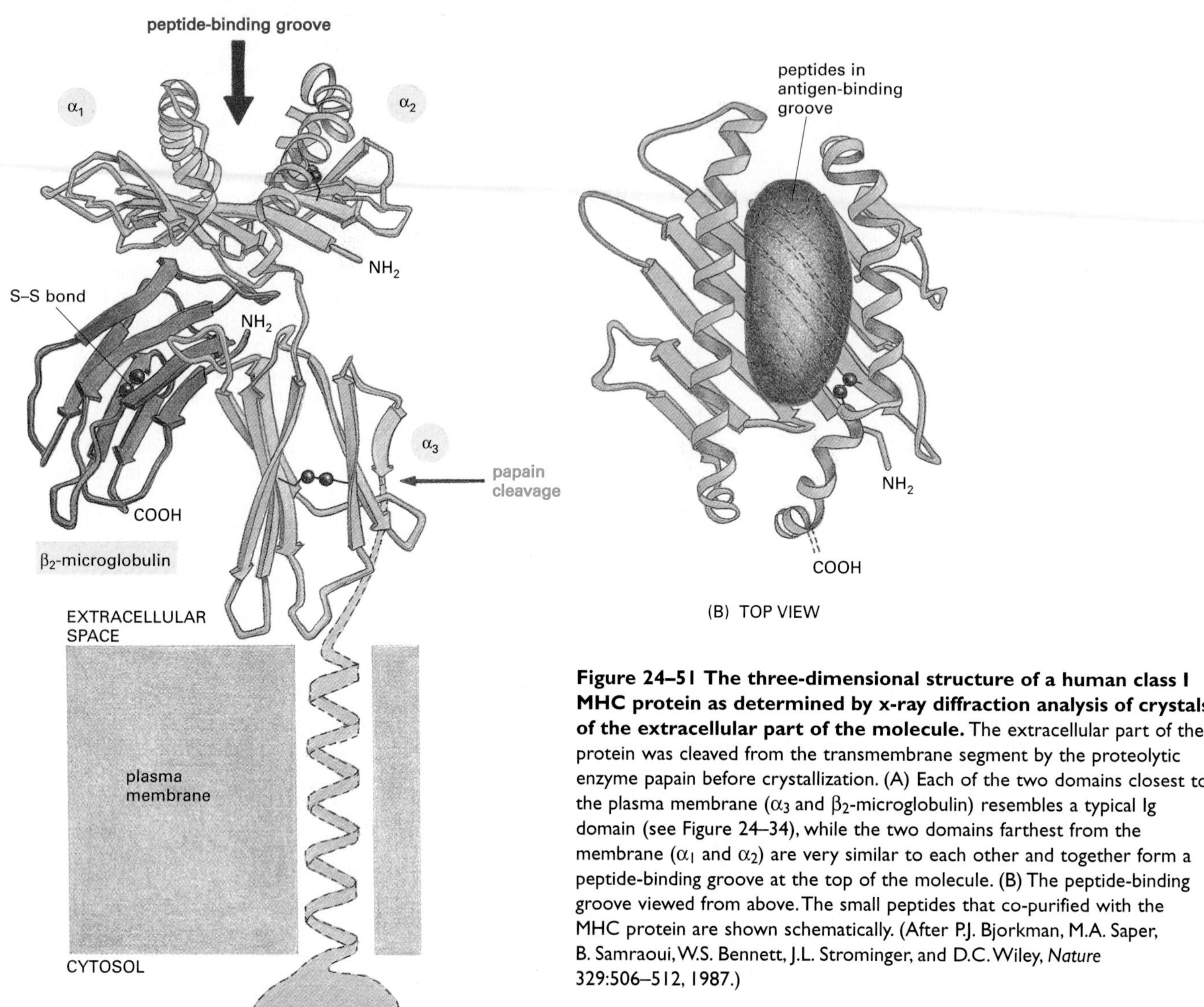

Figure 24–51 The three-dimensional structure of a human class I MHC protein as determined by x-ray diffraction analysis of crystals of the extracellular part of the molecule. The extracellular part of the protein was cleaved from the transmembrane segment by the proteolytic enzyme papain before crystallization. (A) Each of the two domains closest to the plasma membrane (α_3 and β_2-microglobulin) resembles a typical Ig domain (see Figure 24–34), while the two domains farthest from the membrane (α_1 and α_2) are very similar to each other and together form a peptide-binding groove at the top of the molecule. (B) The peptide-binding groove viewed from above. The small peptides that co-purified with the MHC protein are shown schematically. (After P.J. Bjorkman, M.A. Saper, B. Samraoui, W.S. Bennett, J.L. Strominger, and D.C. Wiley, *Nature* 329:506–512, 1987.)

An MHC Protein Binds a Peptide and Interacts with a T Cell Receptor

Any individual can make only a small number of different MHC proteins, which together must be able to present peptide fragments from almost any foreign protein to T cells. Thus, unlike an antibody molecule, each MHC protein has to be able to bind a very large number of different peptides. The structural basis for this versatility has emerged from x-ray crystallographic analyses of MHC proteins.

As shown in Figure 24–51A, a class I MHC protein has a single peptide-binding site located at one end of the molecule, facing away from the plasma membrane. This site consists of a deep groove between two long α helices; the groove narrows at both ends so that it is only large enough to accommodate an extended peptide about 8–10 amino acids long. In fact, when a class I MHC protein was first analyzed by x-ray crystallography in 1987, this groove contained bound peptides that had co-crystallized with the MHC protein (Figure 24–51B), suggesting that once a peptide binds to this site it does not normally dissociate.

A typical peptide binds in the groove of a class I MHC protein in an extended conformation, with its terminal amino group bound to an invariant pocket at one end of the groove and its terminal carboxyl group bound to an invariant pocket at the other end of the groove. Other amino acids (called "anchor amino acids") in the peptide bind to "specificity pockets" in the groove formed by polymorphic portions of the MHC protein (Figure 24–52). The side chains of other amino acids of the peptide point outward, in a position to be recognized by receptors on cytotoxic T cells. Because the conserved pockets at the ends of the binding groove recognize features of the peptide backbone that are common to all peptides, each allelic form of a class I MHC protein can bind a large variety of peptides of diverse sequence. At the same time, the differing specificity pockets along the groove, which bind particular amino acid side chains of the peptide, ensure that each allelic form binds and presents a distinct characteristic set of peptides. Thus, the six types of class I MHC proteins in an individual can present a broad range of foreign peptides to the cytotoxic T cells, but in each individual they do so in slightly different ways.

Class II MHC proteins have a three-dimensional structure that is very similar to that of class I proteins, but their antigen-binding groove does not narrow at the ends, so it can accommodate longer peptides, which are usually 13–17 amino acids long. Moreover, the peptide is not bound at its ends. It is held in the groove by parts of its peptide backbone that bind to invariant pockets formed by conserved amino acids that line all class II MHC peptide-binding grooves, as well as by the side chains of anchor amino acids that bind to variable specificity pockets in the groove (Figure 24–53). A class II MHC binding groove can accommodate a more heterogeneous set of peptides than can a class I MHC groove. Thus, although an individual makes only a small number of types of class II proteins, each with its own unique peptide-binding groove, together these proteins can bind and present an enormous variety of foreign peptides to helper T cells, which have a crucial role in almost all adaptive immune responses.

The way in which the T cell receptor recognizes a peptide fragment bound to an MHC protein is revealed by x-ray crystallographic analyses of complexes formed between a soluble receptor and a soluble MHC protein with peptide in its binding groove. (The soluble proteins for these experiments are produced by recombinant DNA technology.) In each case studied, the T cell receptor fits diagonally across the peptide-binding groove and binds through its V_α and V_β hypervariable loops to both the walls of the groove and the peptide (Figure 24–54). Soluble MHC–peptide complexes are now widely used to detect T cells with a particular specificity; they are usually cross-linked into tetramers to increase their avidity for T cell receptors.

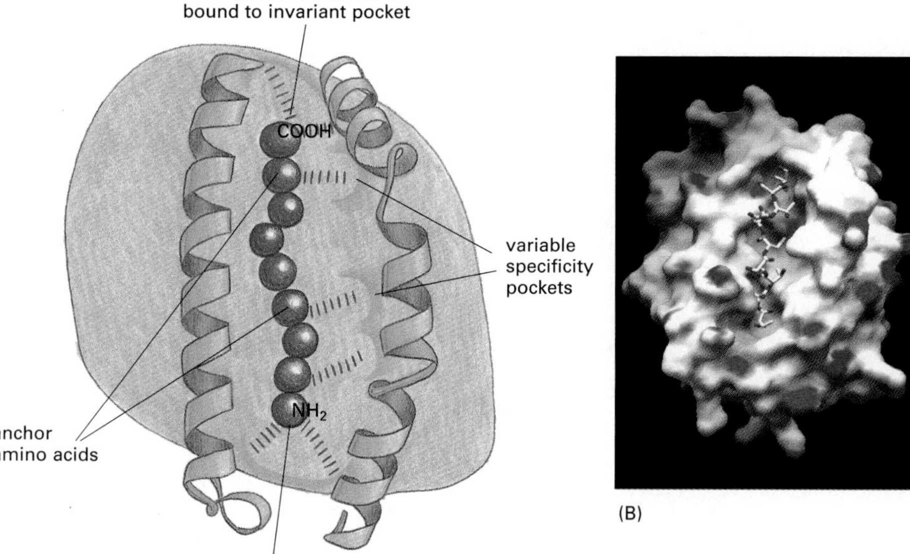

Figure 24–52 A peptide bound in the groove of a class I MHC protein. (A) Schematic drawing of a top view of the groove. The peptide backbone is shown as a string of *red* balls, each of which represents one of the nine amino acids of the peptide. The terminal amino and carboxyl groups of the peptide backbone bind to invariant pockets at the ends of the groove, while the side chains of several anchor amino acids of the peptide bind to variable specificity pockets in the groove. (B) The three-dimensional structure of a peptide bound in the groove of a class I MHC protein, as determined by x-ray diffraction. (B, courtesy of Paul Travers.)

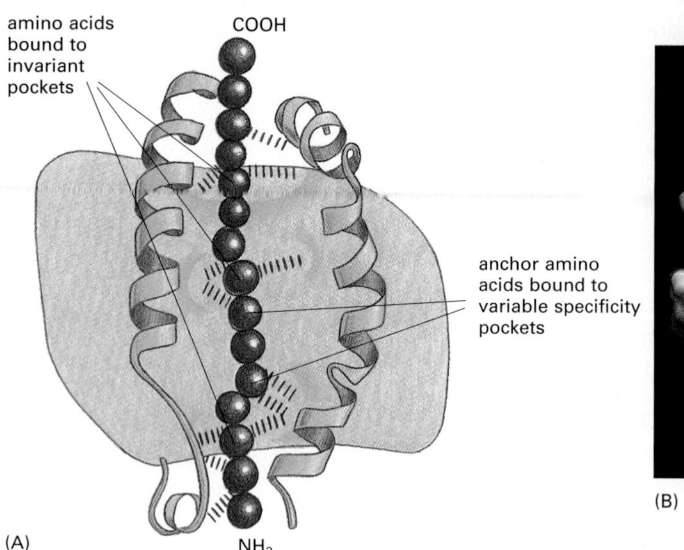

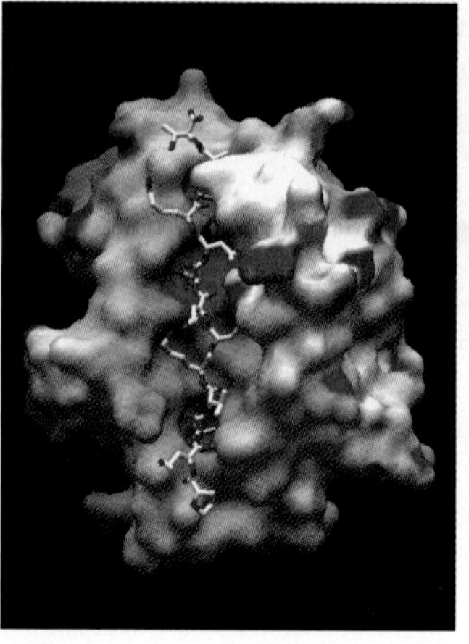

Figure 24–53 A peptide bound in the groove of a class II MHC protein. (A) Schematic drawing similar to that shown in Figure 24–52A. Note that the ends of the peptide are not tightly bound and extend beyond the cleft. The peptide is held in the groove by interactions between parts of its backbone and invariant pockets in the groove and between the side chains of several anchor amino acids of the peptide and variable specificity pockets in the groove. (B) The three-dimensional structure of a peptide bound in the groove of a class II MHC protein, as determined by x-ray diffraction. (B, courtesy of Paul Travers.)

MHC Proteins Help Direct T Cells to Their Appropriate Targets

Class I MHC proteins are expressed on virtually all nucleated cells. This is presumably because effector cytotoxic T cells must be able to focus on and kill any cell in the body that happens to become infected with an intracellular microbe such as a virus. Class II proteins, by contrast, are normally confined largely to cells that take up foreign antigens from the extracellular fluid and interact with helper T cells. These include dendritic cells, which initially activate helper T cells, as well as the targets of effector helper T cells, such as macrophages and B cells. Because dendritic cells express both class I and class II MHC proteins, they can activate both cytotoxic and helper T cells.

It is important that effector cytotoxic T cells focus their attack on cells that *make* the foreign antigens (such as viral proteins), while helper T cells focus their help mainly on cells that have taken up foreign antigens from the extracellular fluid. Since the former type of target cell is always a menace, while the latter type is essential for the body's immune defenses, it is vitally important that T cells never confuse the two target cells and misdirect their cytotoxic and helper functions. Therefore, in addition to the antigen receptor that recognizes a peptide–MHC complex, each of the two major classes of T cells also expresses a *co-receptor* that recognizes a separate, invariant part of the appropriate class of MHC protein. These two co-receptors, called CD4 and CD8, help direct helper T cells and cytotoxic T cells, respectively, to their appropriate targets, as we now discuss. The properies of class I and class II MHC proteins are compared in Table 24–2.

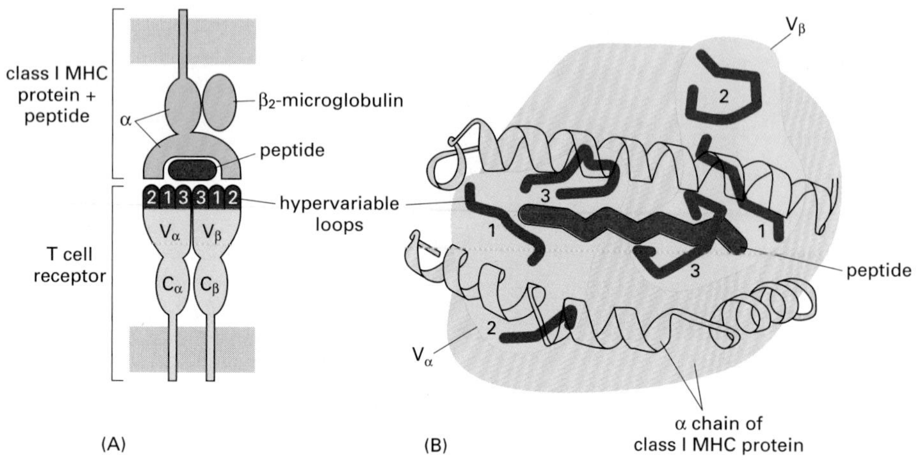

Figure 24–54 The interaction of a T cell receptor with a viral peptide bound to a class I MHC protein. (A) Schematic view of the hypervariable loops of the V_α and V_β domains of the T cell receptor interacting with the peptide and the walls of the peptide-binding groove of the MHC protein. The precise contacts are not illustrated. (B) Drawing of the "footprint" of the V domains *(blue)* and hypervariable loops *(dark blue)* of the receptor over the peptide-binding groove, as determined by X-ray diffraction. The V_α domain covers the amino half of the peptide, while the V_β domain covers the carboxyl half. Note that the receptor is oriented diagonally across the peptide-binding groove. (B, adapted from D.N. Garboczi et al., *Nature* 384:134–141, 1996.)

TABLE 24–2 Properties of Human Class I and Class II MHC Proteins

	CLASS I	CLASS II
Genetic loci	*HLA-A, HLA-B, HLA-C*	*DP, DQ, DR*
Chain structure	α chain + β_2-microglobulin	α chain + β chain
Cell distribution	most nucleated cells	antigen-presenting cells (including B cells), thymus epithelial cells, some others
Involved in presenting antigen to	cytotoxic T cells	helper T cells
Source of peptide fragments	proteins made in cytoplasm	endocytosed plasma membrane and extracellular proteins
Polymorphic domains	$\alpha_1 + \alpha_2$	$\alpha_1 + \beta_1$
Recognition by co-receptor	CD8	CD4

CD4 and CD8 Co-receptors Bind to Nonvariable Parts of MHC Proteins

The affinity of T cell receptors for peptide–MHC complexes on an antigen-presenting cell or target cell is usually too low to mediate a functional interaction between the two cells by itself. T cells normally require *accessory receptors* to help stabilize the interaction by increasing the overall strength of the cell–cell adhesion. Unlike T cell receptors or MHC proteins, the accessory receptors do not bind foreign antigens and are invariant.

When accessory receptors also have a direct role in activating the T cell by generating their own intracellular signals, they are called **co-receptors**. The most important and best understood of the co-receptors on T cells are the CD4 and CD8 proteins, both of which are single-pass transmembrane proteins with extracellular Ig-like domains. Like T cell receptors, they recognize MHC proteins, but, unlike T cell receptors, they bind to nonvariable parts of the protein, far away from the peptide-binding groove. **CD4** is expressed on helper T cells and binds to class II MHC proteins, whereas **CD8** is expressed on cytotoxic T cells and binds to class I MHC proteins (Figure 24–55). Thus, CD4 and CD8 contribute to T cell recognition by helping to focus the cell on particular MHC proteins, and thus on particular types of cells—helper T cells on dendritic cells, macrophages, and B cells, and cytotoxic cells on any nucleated host cell displaying a foreign peptide on a class I MHC protein. The cytoplasmic tail of these transmembrane proteins is associated with a member of the Src family of cytoplasmic tyrosine protein kinases called *Lck*, which phosphorylates various intracellular proteins on tyrosines and thereby participates in the activation of the T cell. Antibodies to CD4 and CD8 are widely used as tools to distinguish between the two main classes of T cells, in both humans and experimental animals.

Ironically, the AIDS virus (HIV) makes use of CD4 molecules (as well as chemokine receptors) to enter helper T cells. It is the eventual depletion of helper T cells that renders AIDS patients susceptible to infection by microbes that are not normally dangerous. As a result, most AIDS patients die of infection within several years of the onset of symptoms, unless they are treated with a combination of powerful anti-HIV drugs. HIV also uses CD4 and chemokine receptors to enter macrophages, which also have both of these receptors on their surface.

Before a cytotoxic or helper T cell can recognize a foreign protein, the protein has to be processed inside an antigen-presenting cell or target cell so that it can be displayed as peptide–MHC complexes on the cell surface. We first consider how a virus-infected antigen-presenting cell or target cell processes viral proteins for presentation to a cytotoxic T cell. We then discuss how ingested foreign proteins are processed for presentation to a helper T cell.

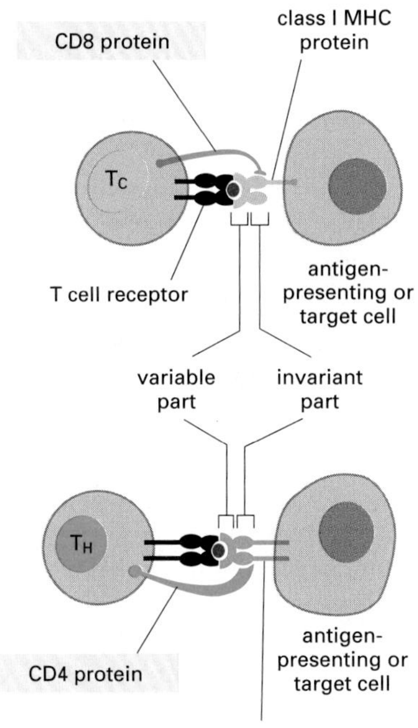

Figure 24–55 CD4 and CD8 co-receptors on the surface of T cells. Cytotoxic T cells (T_C) express CD8, which recognizes class I MHC proteins, whereas helper T cells (T_H) express CD4, which recognizes class II MHC proteins. Note that the co-receptors bind to the same MHC protein that the T cell receptor has engaged, so that they are brought together with T cell receptors during the antigen recognition process. Whereas the T cell receptor binds to the variable (polymorphic) parts of the MHC protein that form the peptide-binding groove, the co-receptor binds to the invariant part, far away from the groove.

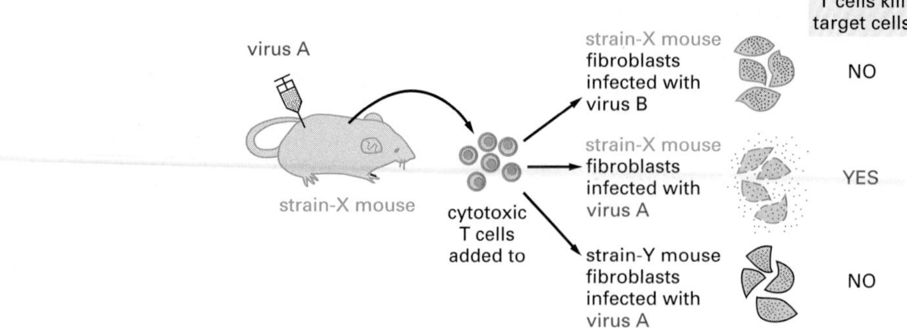

Figure 24–56 The classic experiment showing that an effector cytotoxic T cell recognizes some aspect of the surface of the host target cell in addition to a viral antigen. Mice of strain X are infected with virus A. Seven days later, the spleens of these mice contain effector cytotoxic T cells able to kill virus-infected, strain-X fibroblasts in cell culture. As expected, they kill only fibroblasts infected with virus A and not those infected with virus B. Thus, the cytotoxic T cells are virus-specific. The same T cells, however, are also unable to kill fibroblasts from strain-Y mice infected with the same virus A, indicating that the cytotoxic T cells recognize a genetic difference between the two kinds of fibroblasts and not just the virus. Pinning down the difference required the use of special strains of mice (known as *congenic strains*) that either were genetically identical except for the alleles at their class I MHC loci or were genetically different except for these alleles. In this way, it was found that the killing of infected target cells required that they express at least one of the same class I MHC alleles as expressed by the original infected mouse. This suggested that class I MHC proteins are necessary to present cell-surface-bound viral antigens to effector cytotoxic T cells.

Cytotoxic T Cells Recognize Fragments of Foreign Cytosolic Proteins in Association with Class I MHC Proteins

One of the first, and most dramatic, demonstrations that MHC proteins present foreign antigens to T cells came from an experiment performed in the 1970s. It was found that effector cytotoxic T cells from a virus-infected mouse could kill cultured cells infected with the same virus only if these target cells expressed some of the same class I MHC proteins as the infected mouse (Figure 24–56). This experiment demonstrated that the T cells of any individual that recognize a specific antigen do so only when that antigen is associated with the allelic forms of MHC proteins expressed by that individual, a phenomenon known as *MHC restriction.*

The chemical nature of the viral antigens recognized by cytotoxic T cells was not discovered for another 10 years. In experiments on cells infected with influenza virus, it was unexpectedly found that some of the effector cytotoxic T cells activated by the virus specifically recognize internal proteins of the virus that would not be accessible in the intact virus particle. Subsequent evidence indicated that the T cells were recognizing degraded fragments of the internal viral proteins that were bound to class I MHC proteins on the infected cell surface. Because a T cell can recognize tiny amounts of antigen (as few as one hundred peptide–MHC complexes), only a small fraction of the fragments generated from viral proteins have to bind to class I MHC proteins and get to the cell surface to attract an attack by an effector cytotoxic T cell.

The viral proteins are synthesized in the cytosol of the infected cell. As discussed in Chapter 3, proteolytic degradation in the cytosol is mainly mediated by an ATP- and ubiquitin-dependent mechanism that operates in *proteasomes*—large proteolytic enzyme complexes constructed from many different protein subunits. Although all proteasomes are probably able to generate peptide fragments that can bind to class I MHC proteins, some proteasomes are thought to be specialized for this purpose, as they contain two subunits that are encoded by genes located within the MHC chromosomal region. Even bacterial proteasomes cut proteins into peptides of about the length that fits into the groove of a class I MHC protein, suggesting that the MHC groove evolved to fit this length of peptide.

How do peptides generated in the cytosol make contact with the peptide-binding groove of class I MHC proteins in the lumen of the endoplasmic reticulum (Figure 24–57)? The answer was discovered through observations on mutant cells in which class I MHC proteins are not expressed at the cell surface but are instead degraded within the cell. The mutant genes in these cells proved to encode subunits of a protein belonging to the family of *ABC transporters,* which we discuss in Chapter 11. This transporter protein is located in the ER membrane and uses the energy of ATP hydrolysis to pump peptides from the cytosol into the ER lumen. The genes encoding its two subunits are in the MHC chromosomal region, and, if either gene is inactivated by mutation, cells are unable to supply peptides to class I MHC proteins. The class I MHC proteins in such mutant cells are degraded in the cell because peptide binding is normally required for the proper folding of these proteins. Until it binds a peptide, a class I MHC protein remains in the ER, tethered to an ABC transporter by a chaperone protein (Figure 24–58).

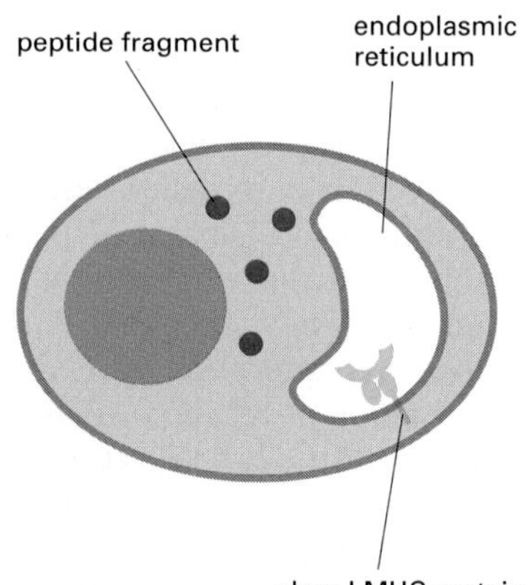

Figure 24–57 The peptide-transport problem. How do peptide fragments get from the cytosol, where they are produced, into the ER lumen, where the peptide-binding grooves of class I MHC proteins are located? A special transport process is required.

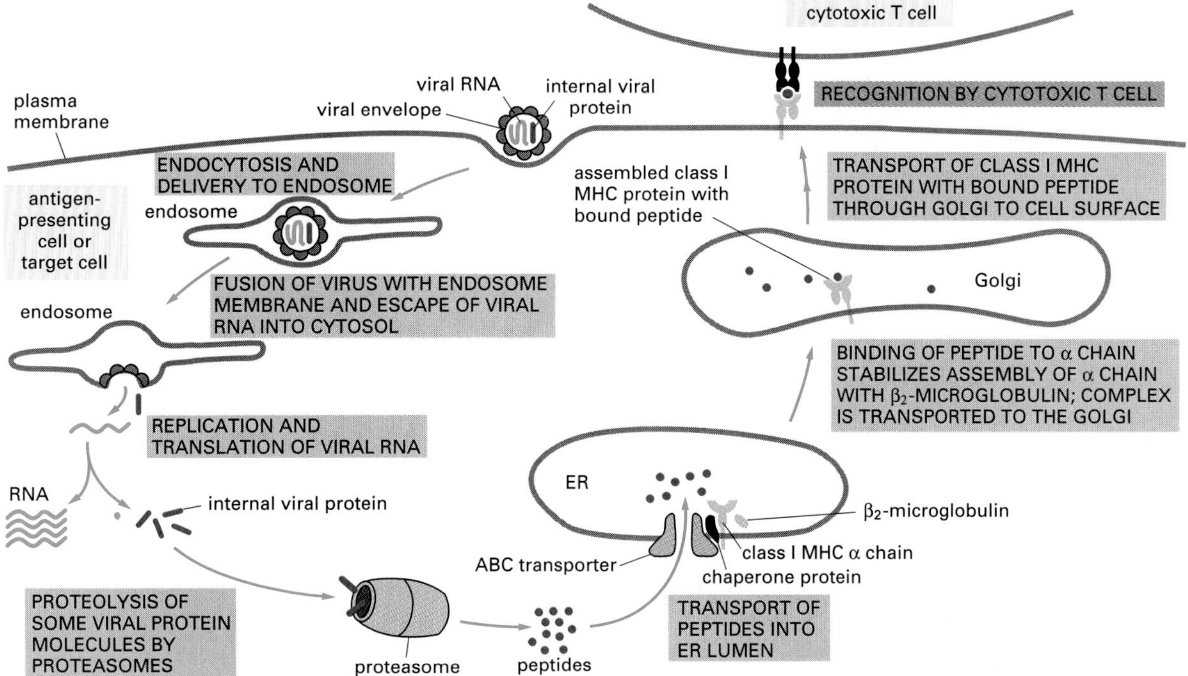

Figure 24–58 The processing of a viral protein for presentation to cytotoxic T cells. An effector cytotoxic T cell kills a virus-infected cell when it recognizes fragments of viral protein bound to class I MHC proteins on the surface of the infected cell. Not all viruses enter the cell in the way that this enveloped RNA virus does, but fragments of internal viral proteins always follow the pathway shown. Some of the viral proteins synthesized in the cytosol are degraded, and this is a sufficient amount to attract an attack by a cytotoxic T cell. The folding and assembly of a class I MHC protein is aided by several chaperone proteins in the ER lumen, only one of which is shown. The chaperones bind to the class I MHCα chain and act sequentially. The last one binds the MHC protein to the ABC transporter, as shown.

In cells that are not infected, peptide fragments come from the cells' own cytosolic and nuclear proteins that are degraded in the processes of normal protein turnover and quality control mechanisms. (Surprisingly, more than 30% of the proteins made by mammalian cells are apparently faulty and are degraded in proteasomes soon after they are synthesized.) These peptides are pumped into the ER and are carried to the cell surface by class I MHC proteins. They are not antigenic because the cytotoxic T cells that could recognize them have been eliminated or inactivated during T cell development, as we discuss later.

When cytotoxic T cells and some helper T cells are activated by antigen to become effector cells, they secrete the cytokine **interferon-γ (IFN-γ)**, which greatly enhances anti-viral responses. The IFN-γ acts on infected cells in two ways. It blocks viral replication, and it increases the expression of many genes within the MHC chromosomal region. These genes include those that encode class I (and class II) MHC proteins, the two specialized proteasome subunits, and the two subunits of the peptide transporter located in the ER (Figure 24–59). Thus, all of the machinery required for presenting viral antigens to cytotoxic T cells is coordinately called into action by IFN-γ, creating a positive feedback that amplifies the immune response and culminates in the death of the infected cells.

Helper T Cells Recognize Fragments of Endocytosed Foreign Protein Associated with Class II MHC Proteins

Unlike cytotoxic T cells, helper T cells do not act directly to kill infected cells so as to eliminate microbes. Instead, they stimulate macrophages to be more effective in destroying intracellular microorganisms, and they help B cells and cytotoxic T cells to respond to microbial antigens.

Like the viral proteins presented to cytotoxic T cells, the proteins presented to helper T cells on antigen-presenting cells or target cells are degraded fragments of foreign proteins. The fragments are bound to class II MHC proteins in much the same way that virus-derived peptides are bound to class I MHC proteins. But both the source of the peptide fragments presented and the route they take to find the MHC proteins are different from those of peptide fragments presented by class I MHC proteins to cytotoxic T cells.

Rather than being derived from foreign protein synthesized in the cytosol of a cell, the foreign peptides presented to helper T cells are derived from endosomes. Some come from extracellular microbes or their products that the

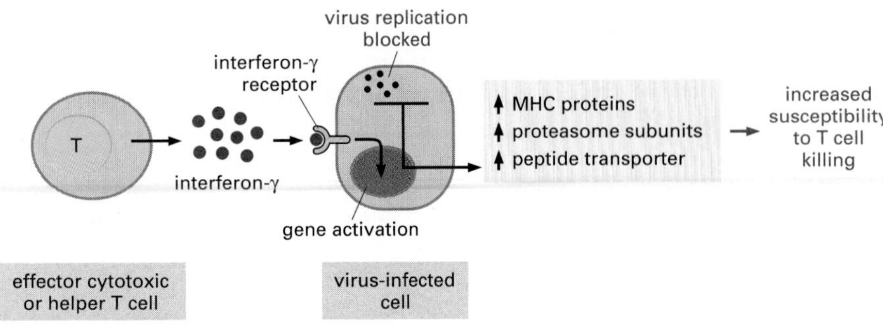

Figure 24–59 Some effects of interferon-γ on infected cells. The activated interferon-γ receptors signal to the nucleus, altering gene transcription, which leads to the effects indicated. The effects shaded in *yellow* tend to make the infected cell a better target for killing by an effector cytotoxic T cell.

antigen-presenting cell has endocytosed and degraded in the acidic environment of its endosomes. Others come from microbes growing within the endocytic compartment of the antigen-presenting cell. These peptides do not have to be pumped across a membrane because they do not originate in the cytosol; they are generated in a compartment that is topologically equivalent to the extracellular space. They never enter the lumen of the ER, where the class II MHC proteins are synthesized and assembled, but instead bind to preassembled class II heterodimers in a special endosomal compartment. Once the peptide has bound, the class II MHC protein alters its conformation, trapping the peptide in the binding groove for presentation at the cell surface to helper T cells.

A newly synthesized class II MHC protein must avoid clogging its binding groove prematurely in the ER lumen with peptides derived from endogenously synthesized proteins. A special polypeptide, called the **invariant chain**, ensures this by associating with newly synthesized class II MHC heterodimers in the ER. Part of its polypeptide chain lies within the peptide-binding groove of the MHC protein, thereby blocking the groove from binding other peptides in the lumen of the ER. The invariant chain also directs class II MHC proteins from the *trans* Golgi network to a late endosomal compartment. Here, the invariant chain is cleaved by proteases, leaving only a short fragment bound in the peptide-binding groove of the MHC protein. This fragment is then released (catalyzed by a class II-MHC-like protein called HLA-DM), freeing the MHC protein to bind peptides derived from endocytosed proteins (Figure 24–60). In this way, the functional differences between class I and class II MHC proteins are ensured—the former presenting molecules that come from the cytosol, the latter presenting molecules that come from the endocytic compartment.

Most of the class I and class II MHC proteins on the surface of a target cell have peptides derived from self proteins in their binding groove. For class I proteins, the fragments derive from degraded cytosolic and nuclear proteins. For class II proteins, they mainly derive from degraded proteins that originate in the plasma membrane or extracellular fluid and are endocytosed. Only a small fraction of the 10^5 or so class II MHC proteins on the surface of an antigen-presenting cell have foreign peptides bound to them. This is sufficient, however, because only a hundred or so of such molecules are required to stimulate a helper T cell, just as in the case of peptide–class-I-MHC complexes stimulating a cytotoxic T cell.

Potentially Useful T Cells Are Positively Selected in the Thymus

We have seen that T cells recognize antigen in association with self MHC proteins but not in association with foreign MHC proteins (see Figure 24–56): that is, T cells show *MHC restriction.* This restriction results from a process of **positive selection** during T cell development in the thymus. In this process, those immature T cells that will be capable of recognizing foreign peptides presented by self MHC proteins are selected to survive, while the remainder, which would be of no use to the animal, undergo apoptosis. Thus, MHC restriction is an acquired property of the immune system that emerges as T cells develop in the thymus.

The most direct way to study the selection process is to follow the fate of a set of developing T cells of known specificity. This can be done by using transgenic mice that express a specific pair of rearranged α and β T cell receptor genes derived from a T cell clone of known antigen and MHC specificity. Such experiments show that the transgenic T cells mature in the thymus and populate the peripheral lymphoid organs only if the transgenic mouse also expresses the same allelic form of MHC protein as is recognized by the transgenic T cell receptor. If the mouse does not express the appropriate MHC protein, the transgenic T cells die in the thymus. Thus, the survival and maturation of a T cell depend on a match between its receptor and the MHC proteins expressed in the thymus. Similar experiments using transgenic mice in which MHC expression is confined to specific cell types in the thymus indicate that it is MHC proteins on epithelial cells in the cortex of the thymus that are responsible for this positive selection process. After positively selected T cells leave the thymus, their continued survival depends on their continual stimulation by self-peptide–MHC complexes; this stimulation is enough to promote cell survival but not enough to activate the T cells to become effector cells.

As part of the positive selection process in the thymus, developing T cells that express receptors recognizing class I MHC proteins are selected to become cytotoxic cells, while T cells that express receptors recognizing class II MHC proteins are selected to become helper cells. Thus, genetically engineered mice that lack cell-surface class I MHC proteins specifically lack cytotoxic T cells, whereas mice that lack class II MHC proteins specifically lack helper T cells. The cells that are undergoing positive selection initially express both CD4 and CD8 co-receptors, and these are required for the selection process: without CD4, helper T cells fail to develop, and without CD8, cytotoxic T cells fail to develop.

Positive selection still leaves a large problem to be solved. If developing T cells with receptors that recognize self peptides associated with self MHC proteins were to mature in the thymus and migrate to peripheral lymphoid tissues, they might wreak havoc. A second, *negative selection* process in the thymus is required to help avoid this potential disaster.

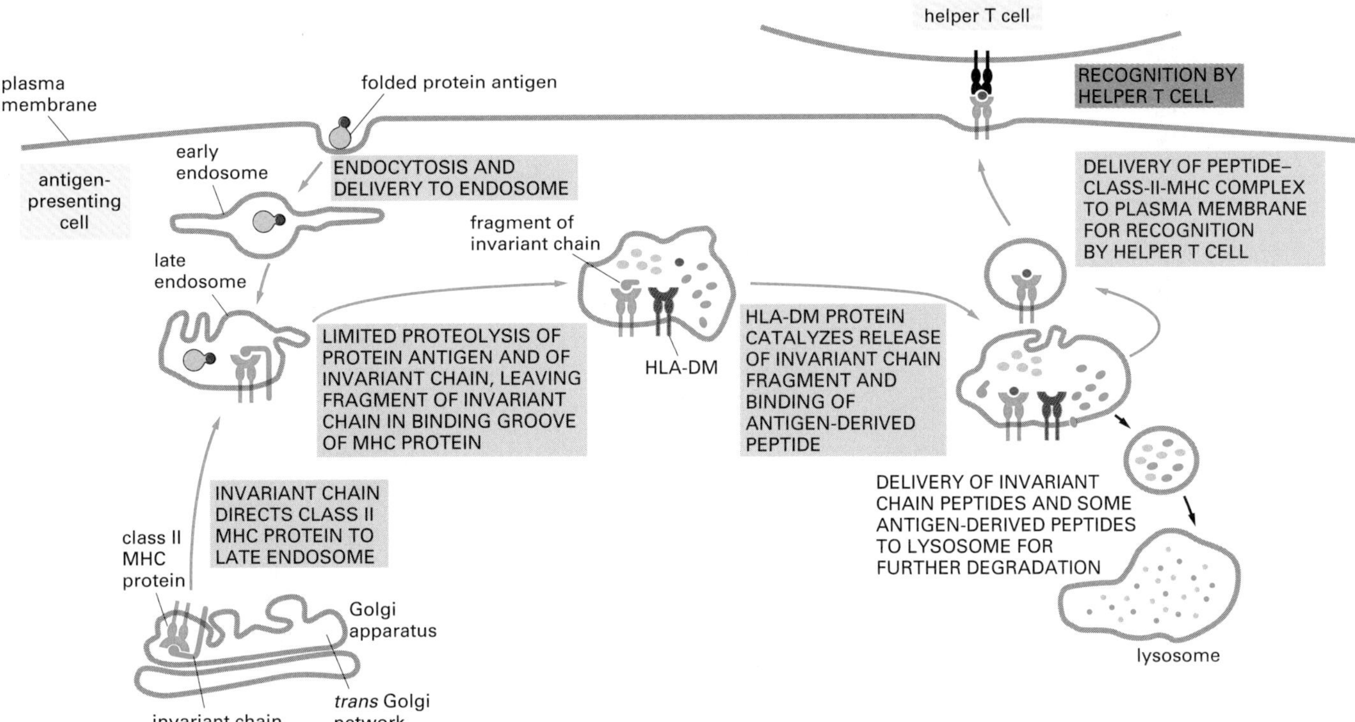

Figure 24–60 The processing of an extracellular protein antigen for presentation to a helper T cell. The drawing shows a simplified view of how peptide–class-II-MHC complexes are formed in endosomes and delivered to the cell surface. Note that the release of the invariant-chain fragment from the binding groove of the class II MHC protein in the endosome is catalyzed by a class-II-MHC-like protein called HLA-DM. Viral glycoproteins can also be processed by this pathway for presentation to helper T cells. They are made in the ER, are transported to the plasma membrane, and can then enter endosomes after endocytosis.

Many Developing T Cells That Could Be Activated by Self Peptides Are Eliminated in the Thymus

As discussed previously, a fundamental feature of the adaptive immune system is that it can distinguish self from nonself and normally does not react against self molecules. An important mechanism in achieving this state of *immunological self tolerance* is the deletion in the thymus of developing self-reactive T cells—that is, T cells whose receptors bind strongly enough to the complex of a self peptide and a self MHC protein to become activated. Because, as we discuss later, most B cells require helper T cells to respond to antigen, the elimination of self-reactive helper T cells also helps ensure that self-reactive B cells that escape B cell tolerance induction are harmless.

It is not enough, therefore, for the thymus to select *for* T cells that recognize self MHC proteins; it must also select *against* T cells that could be activated by self MHC proteins complexed with self peptides. In other words, it must pick out for survival just those T cells that will be capable of responding to self MHC proteins complexed with foreign peptides, even though these peptides are not present in the developing thymus. It is thought that these T cells bind weakly in the thymus to self MHC proteins that are carrying self peptides mismatched to the T cell receptors. Thus, the required goal can be achieved by (1) ensuring the death of T cells that bind *strongly* to the self-peptide–MHC complexes in the thymus while (2) promoting the survival of those that bind weakly and (3) permitting the death of those that do not bind at all. Process 2 is the positive selection we have just discussed. Process 1 is called **negative selection**. In both death processes, the cells that die undergo apoptosis (Figure 24–61).

The most convincing evidence for negative selection derives once again from experiments with transgenic mice. After the introduction of T cell receptor transgenes encoding a receptor that recognizes a male-specific peptide antigen, for example, large numbers of mature T cells expressing the transgenic receptor are found in the thymus and peripheral lymphoid organs of female mice. Very few, however, are found in male mice, where the cells die in the thymus before they have a chance to mature. Like positive selection, negative selection requires the interaction of a T cell receptor and a CD4 or CD8 co-receptor with an appropriate MHC protein. Unlike positive selection, however, which occurs mainly on

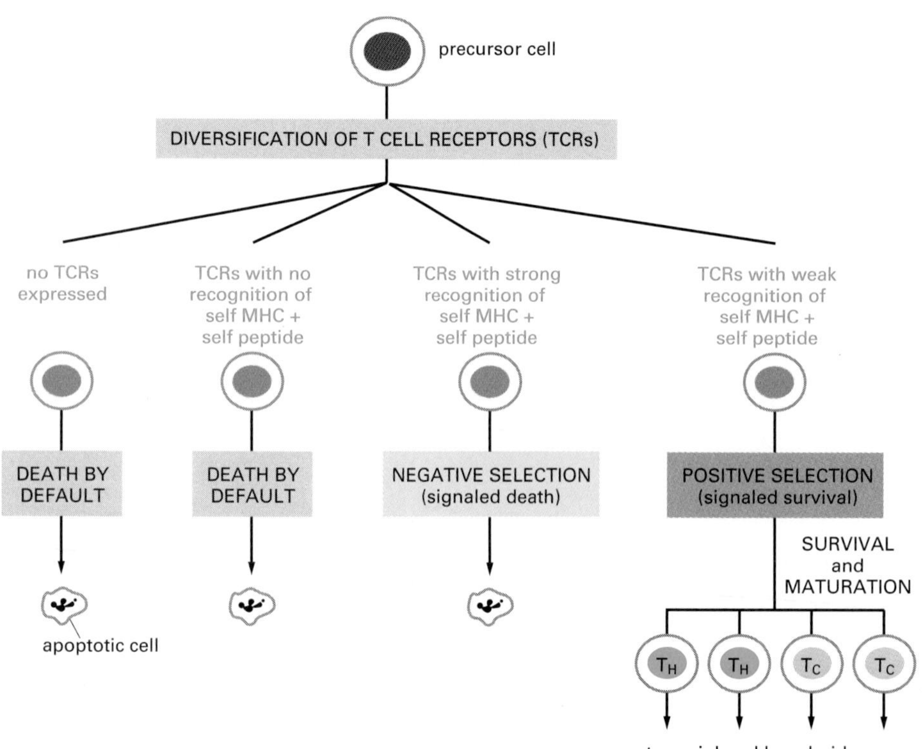

Figure 24–61 Positive and negative selection in the thymus. Cells with receptors that would enable them to respond to foreign peptides in association with self MHC proteins survive, mature, and migrate to peripheral lymphoid organs. All of the other cells undergo apoptosis. The cells undergoing positive selection initially express both CD4 and CD8 co-receptors. During the process of positive selection, helper T cells (T_H) and cytotoxic T cells (T_C) diverge by a poorly understood mechanism. In this process, helper cells develop that express CD4 but not CD8 and recognize foreign peptides in association with class II MHC proteins, while cytotoxic cells develop that express CD8 but not CD4 and recognize foreign peptides in association with class I MHC proteins (not shown).

the surface of thymus epithelial cells, negative selection occurs on the surface of thymus dendritic cells and macrophages, which, as we have seen, function as antigen-presenting cells in peripheral lymphoid organs.

The deletion of self-reactive T cells in the thymus cannot eliminate all potentially self-reactive T cells, as some self molecules are not present in the thymus. Thus, some potentially self-reactive T cells are deleted or functionally inactivated after they leave the thymus, presumably because they recognize self peptides bound to MHC proteins on the surface of dendritic cells that have not been activated by microbes and therefore do not provide a costimulatory signal. As we discuss later, antigen recognition without costimulatory signals can delete or inactivate a T or B cell.

Some potentially self-reactive T cells, however, are not deleted or inactivated. Instead, special *regulatory* (or *suppressor) T cells* are thought to keep them from responding to their self antigens by secreting inhibitory cytokines such as TGF-β (discussed in Chapter 15). These self-reactive T cells may sometimes escape from this suppression and cause autoimmune diseases.

The Function of MHC Proteins Explains Their Polymorphism

The role of MHC proteins in binding foreign peptides and presenting them to T cells provides an explanation for the extensive polymorphism of these proteins. In the evolutionary war between pathogenic microbes and the adaptive immune system, microbes tend to change their antigens to avoid associating with MHC proteins. When a microbe succeeds, it is able to sweep through a population as an epidemic. In such circumstances, the few individuals that produce a new MHC protein that can associate with an antigen of the altered microbe have a large selective advantage. In addition, individuals with two different alleles at any given MHC locus (heterozygotes) have a better chance of resisting infection than those with identical alleles at the locus, as they have a greater capacity to present peptides from a wide range of microbes and parasites. Thus, selection will tend to promote and maintain a large diversity of MHC proteins in the population. Strong support for this hypothesis, that infectious diseases have provided the driving force for MHC polymorphism, has come from studies in West Africa. Here, it is found that individuals with a specific MHC allele have a reduced susceptibility to a severe form of malaria. Although the allele is rare elsewhere, it is found in 25% of the West African population where this form of malaria is common.

If greater MHC diversity means greater resistance to infection, why do we each have so few MHC genes encoding these molecules? Why have we not evolved strategies for increasing the diversity of MHC proteins—by alternative RNA splicing, for example, or by the genetic recombination mechanisms used to diversify antibodies and T cell receptors? Presumably, the limits exist because each time a new MHC protein is added to the repertoire, the T cells that recognize self peptides in association with the new MHC protein must be eliminated to maintain self tolerance. The elimination of these T cells would counteract the advantage of adding the new MHC protein. Thus, the number of MHC proteins we express may represent a balance between the advantages of presenting a wide diversity of foreign peptides to T cells against the disadvantages of severely restricting the T cell repertoire during negative selection in the thymus. This explanation is supported by computer modeling studies.

Summary

There are two main functionally distinct classes of T cells: cytotoxic T cells kill infected cells directly by inducing them to undergo apoptosis, while helper T cells help activate B cells to make antibody responses and macrophages to destroy microorganisms that either invaded the macrophage or were ingested by it. Helper T cells also help activate cytotoxic T cells. Both classes of T cells express cell-surface, antibodylike receptors, which are encoded by genes that are assembled from multiple gene segments during T cell development in the thymus. These receptors recognize fragments

of foreign proteins that are displayed on the surface of host cells in association with MHC proteins. Both cytotoxic and helper T cells are activated in peripheral lymphoid organs by antigen-presenting cells, which express peptide–MHC complexes, costimulatory proteins, and various cell-cell adhesion molecules on their cell surface.

Class I and class II MHC proteins have crucial roles in presenting foreign protein antigens to cytotoxic and helper T cells, respectively. Whereas class I proteins are expressed on almost all vertebrate cells, class II proteins are normally restricted to those cell types that interact with helper T cells, such as dendritic cells, macrophages, and B lymphocytes. Both classes of MHC proteins have a single peptide-binding groove, which binds small peptide fragments derived from proteins. Each MHC protein can bind a large and characteristic set of peptides, which are produced intracellularly by protein degradation: class I MHC proteins generally bind fragments produced in the cytosol, while class II MHC proteins bind fragments produced in the endocytic compartment. After they have formed inside the target cell, the peptide–MHC complexes are transported to the cell surface. Complexes that contain a peptide derived from a foreign protein are recognized by T cell receptors, which interact with both the peptide and the walls of the peptide-binding groove. T cells also express CD4 or CD8 co-receptors, which recognize nonpolymorphic regions of MHC proteins on the target cell: helper cells express CD4, which recognizes class II MHC proteins, while cytotoxic T cells express CD8, which recognizes class I MHC proteins.

The T cell receptor repertoire is shaped mainly by a combination of positive and negative selection processes that operate during T cell development in the thymus. These processes help to ensure that only T cells with potentially useful receptors survive and mature, while the others die by apoptosis. T cells that will be able to respond to foreign peptides complexed with self MHC proteins are positively selected, while many T cells that could react strongly with self peptides complexed with self MHC proteins are eliminated. T cells with receptors that could react strongly with self antigens not present in the thymus are eliminated, functionally inactivated, or actively kept suppressed after they leave the thymus.

HELPER T CELLS AND LYMPHOCYTE ACTIVATION

Helper T cells are arguably the most important cells in adaptive immunity, as they are required for almost all adaptive immune responses. They not only help activate B cells to secrete antibodies and macrophages to destroy ingested microbes, but they also help activate cytotoxic T cells to kill infected target cells. As dramatically demonstrated in AIDS patients, without helper T cells we cannot defend ourselves even against many microbes that are normally harmless.

Helper T cells themselves, however, can only function when activated to become effector cells. They are activated on the surface of antigen-presenting cells, which mature during the innate immune responses triggered by an infection. The innate responses also dictate what kind of effector cell a helper T cell will develop into and thereby determine the nature of the adaptive immune response elicited.

In this final section, we discuss the multiple signals that help activate a T cell and how a helper T cell, once activated to become an effector cell, helps activate other cells. We also consider how innate immune responses determine the nature of adaptive responses by stimulating helper T cells to differentiate into either T_H1 or T_H2 effector cells.

Costimulatory Proteins on Antigen-Presenting Cells Help Activate T Cells

To activate a cytotoxic or helper T cell to proliferate and differentiate into an effector cell, an antigen-presenting cell provides two kinds of signals. *Signal 1* is provided by a foreign peptide bound to an MHC protein on the surface of the presenting cell. This peptide–MHC complex signals through the T cell receptor

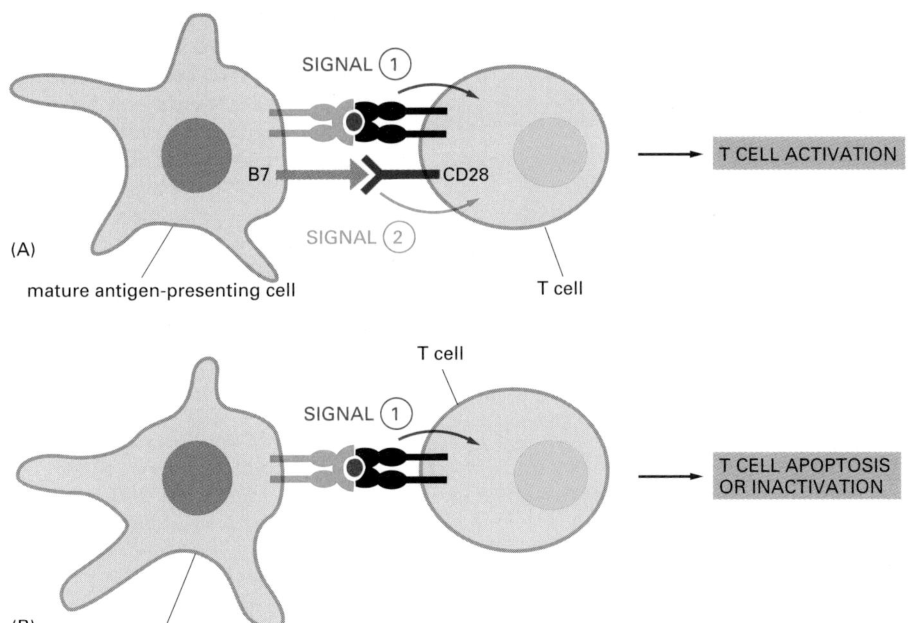

Figure 24–62 The two signals that activate a helper T cell. (A) A mature antigen-presenting cell can deliver both signal 1 and 2 and thereby activate the T cell. (B) An immature antigen-presenting cell delivers signal 1 without signal 2, which can kill or inactivate the T cell; this is one mechanism for immunological tolerance to self antigens. One model for the role of signal 2 is that it induces the active transport of signaling proteins in the T cell plasma membrane to the site of contact between the T cell and the antigen-presenting cell. The accumulation of signaling proteins around the T cell receptor is thought to greatly enhance the intensity and duration of the signaling process activated by signal 1. In this way, "immunological synapses" form in the contact zone, with the T cell receptors (and their associated proteins—see Figure 24–63) and co-receptors in the center and cell–cell adhesion proteins forming a peripheral ring (not shown).

and its associated proteins. *Signal 2* is provided by costimulatory proteins, especially the **B7 proteins** (CD80 and CD86), which are recognized by the co-receptor protein **CD28** on the surface of the T cell. The expression of B7 proteins on an antigen-presenting cell is induced by pathogens during the innate response to an infection. Effector T cells act back to promote the expression of B7 proteins on antigen-presenting cells, creating a positive feedback loop that amplifies the T cell response.

Signal 2 is thought to amplify the intracellular signaling process triggered by signal 1. If a T cell receives signal 1 without signal 2, it may undergo apoptosis or become altered so that it can no longer be activated, even if it later receives both signals (Figure 24–62). This is one mechanism by which a T cell can become *tolerant* to self antigens.

The T cell receptor does not act on its own to transmit signal 1 into the cell. It is associated with a complex of invariant transmembrane proteins called **CD3**, which transduces the binding of the peptide–MHC complex into intracellular signals (Figure 24–63). In addition, the CD4 and CD8 co-receptors play important parts in the signaling process, as illustrated in Figure 24–64.

The combined actions of signal 1 and signal 2 stimulate the T cell to proliferate and begin to differentiate into an effector cell by a curiously indirect mechanism. In culture, they cause the T cells to stimulate their own proliferation and differentiation by inducing the cells to secrete a cytokine called **interleukin-2 (IL-2)** and simultaneously to synthesize high affinity cell-surface receptors that bind it. The binding of IL-2 to the IL-2 receptors activates intracellular signaling pathways that turn on genes that help the T cells to proliferate and differentiate into effector cells (Figure 24–65). As discussed in Chapter 15, there are advantages to such an autocrine mechanism. It helps ensure that T cells differentiate into effector cells only when substantial numbers of them respond to antigen simultaneously in the same location, such as in a lymph node during an infection. Only then do IL-2 levels rise high enough to be effective.

Once bound to the surface of an antigen-presenting cell, a T cell increases the strength of the binding by activating an integrin adhesion protein called *lymphocyte-function-associated protein 1 (LFA-1)*. Activated LFA-1 now binds more strongly to its Ig-like ligand, *intracellular adhesion molecule 1 (ICAM-1)*, on the surface of the presenting cell. This increased adhesion enables the T cell to remain bound to the antigen-presenting cell long enough for the T cell to become activated.

The activation of a T cell is controlled by negative feedback. During the activation process, the cell starts to express another cell-surface protein called

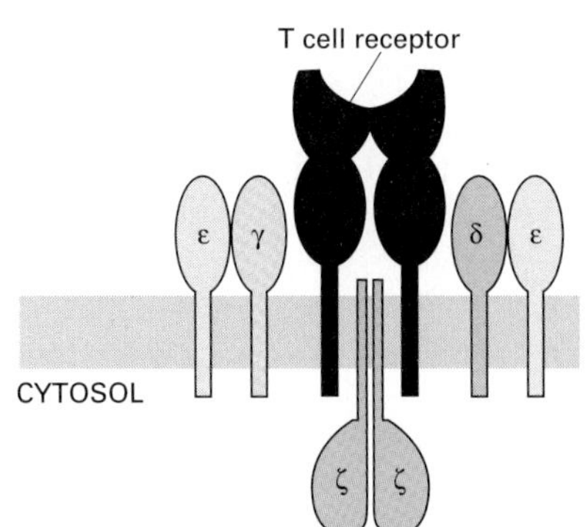

Figure 24–63 The T cell receptor and its associated CD3 complex. All of the CD3 polypeptide chains (shown in *green*), except for the ζ (zeta) chains, have extracellular Ig-like domains and are therefore members of the Ig superfamily.

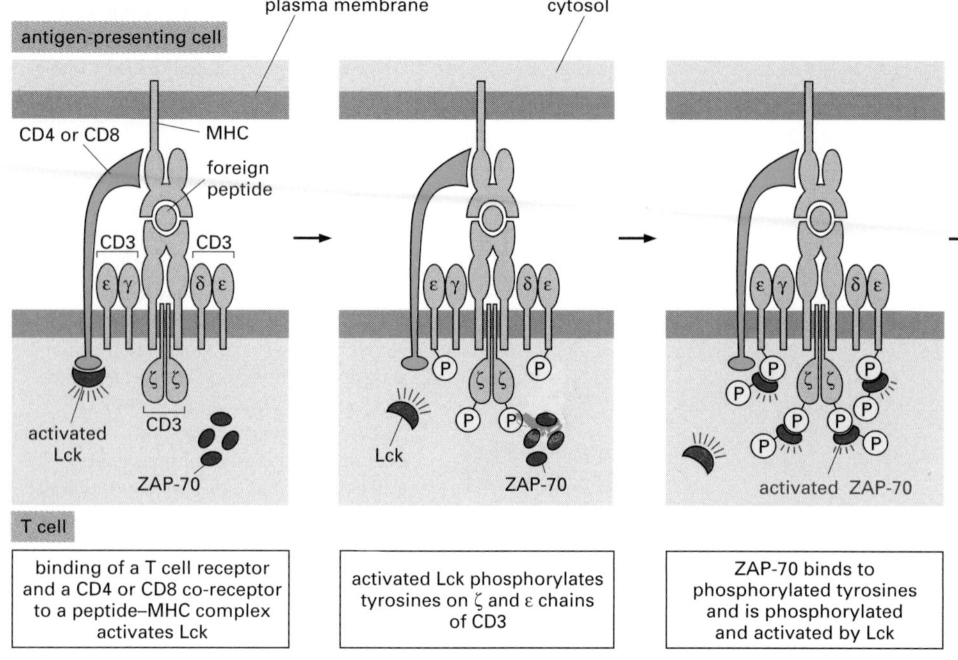

Figure 24–64 The signaling events initiated by the binding of peptide–MHC complexes to T cell receptors (signal 1). When T cell receptors are clustered by binding to peptide–MHC complexes on an antigen-presenting cell, CD4 molecules on helper cells or CD8 molecules on cytotoxic T cells are clustered with them, binding to invariant parts of the same class II or class I MHC proteins, respectively, on the presenting cell. This brings the Src-like cytoplasmic tyrosine kinase Lck into the signaling complex and activates it. Once activated, Lck phosphorylates tyrosines on the ζ and ε chains of the CD3 complex, which now serve as docking sites for yet another cytoplasmic tyrosine kinase called *ZAP-70*. Lck phosphorylates, and thereby activates, ZAP-70. Although not shown, ZAP-70 then phosphorylates tyrosines on the tail of another transmembrane protein, which then serve as docking sites for a variety of adaptor proteins and enzymes. These proteins then help relay the signal to the nucleus and other parts of the cell by activating the inositol phospholipid and MAP kinase signaling pathways (discussed in Chapter 15), as well as a Rho family GTPase that regulates the actin cytoskeleton (discussed in Chapter 16).

CTLA-4, which acts to inhibit intracellular signaling. It resembles CD28, but it binds to B7 proteins on the surface of the antigen-presenting cell with much higher affinity than does CD28, and, when it does, it holds the activation process in check. Mice with a disrupted *CTLA-4* gene die from a massive accumulation of activated T cells.

Most of the T (and B) effector cells produced during an immune response must be eliminated after they have done their job. As antigen levels fall and the response subsides, effector cells are deprived of the antigen and cytokine stimulation that they need to survive, and the majority die by apoptosis. Only memory cells and some long-lived effector cells survive.

Table 24–3 summarizes some of the co-receptors and other accessory proteins found on the surface of T cells.

Before considering how effector helper T cells help activate macrophages and B cells, we need to discuss the two functionally distinct subclasses of effector helper T cells, T_H1 and T_H2 cells, and how they are generated.

The Subclass of Effector Helper T Cell Determines the Nature of the Adaptive Immune Response

When a an antigen-presenting cell activates a naïve helper T cell in a peripheral lymphoid tissue, the T cell can differentiate into either a T_H1 or T_H2 effector helper cell. These two types of functionally distinct subclasses of effector helper T cells can be distinguished by the cytokines they secrete. If the cell differentiates into a **T_H1 cell**, it will secrete *interferon-γ (IFN-γ)* and *tumor necrosis factor-α* (*TNF*-α) and will activate macrophages to kill microbes located within the

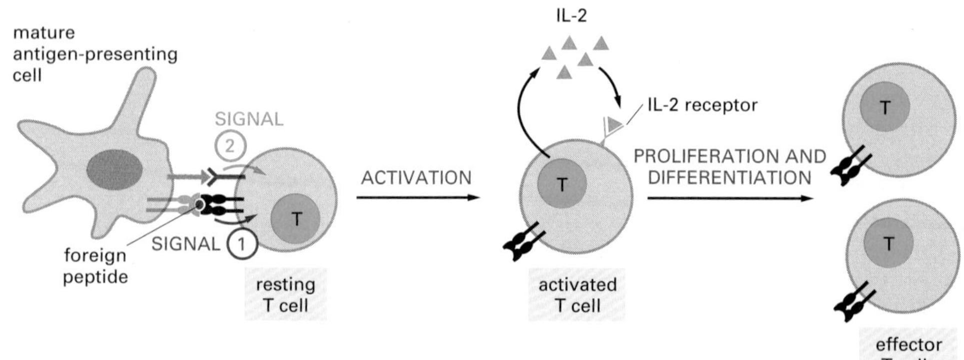

Figure 24–65 The stimulation of T cells by IL-2 in culture. Signals 1 and 2 activate T cells to make high affinity IL-2 receptors and to secrete IL-2. The binding of IL-2 to its receptors helps stimulate the cell to proliferate and differentiate into effector cells. Although some T cells do not make IL-2, as long as they have been activated by their antigen and therefore express IL-2 receptors, they can be helped to proliferate and differentiate by IL-2 made by neighboring T cells (not shown).

macrophages' phagosomes. It will also activate cytotoxic T cells to kill infected cells. Although, in these ways, T_H1 cells mainly defend an animal against intracellular pathogens, they may also stimulate B cells to secrete specific subclasses of IgG antibodies that can coat extracellular microbes and activate complement.

If the naïve T helper cell differentiates into a **T_H2** cell, by contrast, it will secrete *interleukins 4, 5, 10,* and *13 (IL-4, IL-5, IL-10,* and *IL-13)* and will mainly defend the animal against extracellular pathogens. A T_H2 cell can stimulate B cells to make most classes of antibodies, including IgE and some subclasses of IgG antibodies that bind to mast cells, basophils, and eosinophils. These cells release local mediators that cause sneezing, coughing, or diarrhea and help expel extracellular microbes and larger parasites from epithelial surfaces of the body.

Thus, the decision of naïve helper T cells to differentiate into T_H1 or T_H2 effector cells influences the type of adaptive immune response that will be mounted against the pathogen—whether it will be dominated by macrophage activation or by antibody production. The specific cytokines present during the process of helper T cell activation influence the type of effector cell produced. Microbes at a site of infection not only stimulate dendritic cells to make cell-surface B7 costimulatory proteins; they also stimulate them to produce cytokines. The dendritic cells then migrate to a peripheral lymphoid organ and activate naïve helper T cells to differentiate into either T_H1 or T_H2 effector cells, depending on the cytokines the dendritic cells produce. Some intracellular bacteria, for example, stimulate dendritic cells to produce *IL-12,* which encourages T_H1 development, and thereby macrophage activation. As expected, mice that are deficient in either IL-12 or its receptor are much more susceptible to these bacterial infections than are normal mice. Many parasitic protozoa and worms, by contrast, stimulate the production of cytokines that encourage T_H2 development, and thereby antibody production and eosinophil activation, leading to parasite expulsion (Figure 24–66).

Once a T_H1 or T_H2 effector cell develops, it inhibits the differentiation of the other type of helper T cell. IFN-γ produced by T_H1 cells inhibits the development of T_H2 cells, while IL-4 and IL-10 produced by T_H2 cells inhibit the development of T_H1 cells. Thus, the initial choice of response is reinforced as the response proceeds.

TABLE 24–3 Some Accessory Proteins on the Surface of T Cells

PROTEIN*	SUPERFAMILY	EXPRESSED ON	LIGAND ON TARGET CELL	FUNCTIONS
CD3 complex	Ig (except for ζ)	all T cells	—	helps transduce signal when antigen-MHC complexes bind to T cell receptors; helps transport T cell receptors to cell surface
CD4	Ig	helper T cells	class II MHC	promotes adhesion to antigen-presenting cells and to target cells; signals T cell
CD8	Ig	cytotoxic T cells	class I MHC	promotes adhesion to antigen-presenting cells and infected target cells; signals T cell
CD28	Ig	most T cells	B7 proteins (CD80 and CD86)	provides signal 2 to some T cells
CTLA	Ig	activated T cells	B7 proteins (CD80 and CD86)	inhibits T cell activation
CD40 ligand	Fas ligand family	effector helper T cells	CD40	costimulatory protein that helps activate macrophages and B cells
LFA-1	integrin	most white blood cells, including all T cells	ICAM-1	promotes cell–cell adhesion

* CD stands for cluster of differentiation, as each of the CD proteins was originally defined as a blood cell "differentiation antigen" recognized by multiple monoclonal antibodies. Their identification depended on large-scale collaborative studies in which hundreds of such antibodies, generated in many laboratories, were compared and found to consist of relatively few groups (or "clusters"), each recognizing a single cell-surface protein. Since these initial studies, however, more than 150 CD proteins have been identified.

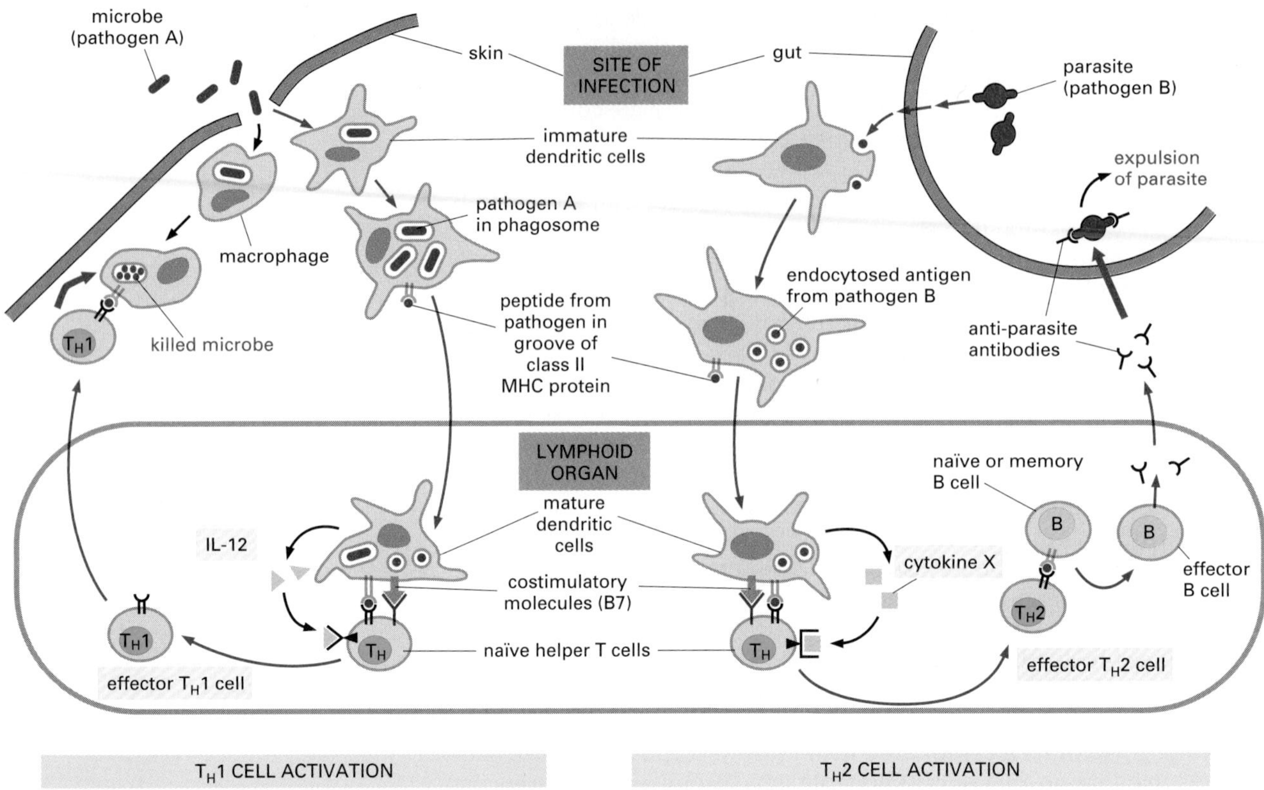

Figure 24–66 The activation of T_H1 and T_H2 cells. The differentiation of helper T cells into either T_H1 or T_H2 effector cells determines the nature of the subsequent adaptive immune responses that the effector cells activate. Whether a naïve helper T cell becomes a T_H1 or T_H2 cell depends mainly on the cytokines present when the helper T cell is activated by a mature dendritic cell in a peripheral lymphoid organ. The types of cytokines produced depend on the local environment and the nature of the microbe or parasite that activated the immature dendritic cell at the site of infection. IL-12 produced by mature dendritic cells promotes T_H1 cell development. The cytokine(s) produced by dendritic cells that promotes T_H2 cell development (cytokine X) is not known, although IL-4 produced by T cells can serve this function. In this figure, the effector T_H1 cell produced in the peripheral lymphoid organ migrates to the site of infection and helps a macrophage kill the microbes it has phagocytosed. The effector T_H2 cell remains in the lymphoid organ and helps activate a B cell to produce antibodies against the parasite. The antibodies arm mast cells, basophils, and eosinophils (not shown), which then can help expel the parasite from the gut.

The importance of the T_H1/T_H2 decision is illustrated by individuals infected with *Mycobacterium leprae,* the bacterium that causes leprosy. The bacterium replicates mainly within macrophages and causes either of two forms of disease, depending mainly on the genetic make-up of the infected individual. In some patients, the *tuberculoid* form of the disease occurs. T_H1 cells develop and stimulate the infected macrophages to kill the bacteria. This produces a local inflammatory response, which damages skin and nerves. The result is a chronic disease that progresses slowly but does not kill the host. In other patients, by contrast, the *lepromatous* form of the disease occurs. T_H2 cells develop and stimulate the production of antibodies. As the antibodies cannot get through the plasma membrane to attack the intracellular bacteria, the bacteria proliferate unchecked and eventually kill the host.

T_H1 Cells Help Activate Macrophages at Sites of Infection

T_H1 cells are preferentially induced by antigen-presenting cells that harbor microbes in intracellular vesicles. The bacteria that cause tuberculosis for example, replicate mainly in phagosomes inside macrophages, where they are protected from antibodies. They are also not readily attacked by cytotoxic T cells, which mainly recognize foreign antigens that are produced in the cytosol (see Figure 24–58). The bacteria can survive in phagosomes because they inhibit both the fusion of the phagosomes with lysosomes and the acidification of the phagosomes that is necessary to activate lysosomal hydrolases. Infected dendritic cells recruit helper T cells to assist in the killing of such microbes. The dendritic cells migrate to peripheral lymphoid organs, where they stimulate the production of T_H1 cells, which then migrate to sites of infection to help activate infected macrophages to kill the microbes harboring in their phagosomes (see Figure 24–66).

T_H1 effector cells use two signals to activate a macrophage. They secrete IFN-γ, which binds to IFN-γ receptors on the macrophage surface, and they display the costimulatory protein **CD40 ligand**, which binds to **CD40** on the macrophage (Figure 24–67). (We see later that CD40 ligand is also used by helper T cells to activate B cells.) Once activated, the macrophage can kill the microbes it contains: lysosomes can now fuse more readily with the phagosomes, unleashing a hydrolytic attack, and the activated macrophage makes oxygen radicals and nitric oxide, both of which are highly toxic to the microbes (discussed in Chapter 25). Because dendritic cells also express CD40, the T_H1 cells at sites of infection can also help activate them. As a result, the dendritic cells increase their production of class II MHC proteins, B7 costimulatory proteins, and various cytokines, especially IL-12. This makes them more effective at stimulating helper T cells to differentiate into T_H1 effector cells in peripheral lymphoid organs, providing a positive feedback loop that increases the production of T_H1 cells and, thereby, the activation of macrophages.

T_H1 effector cells stimulate an inflammatory response by recruiting more phagocytic cells into the infected site. They do so in three ways:

1. They secrete cytokines that act on the bone marrow to increase the production of monocytes (macrophage precursors that circulate in the blood) and neutrophils.
2. They secrete other cytokines that activate endothelial cells lining local blood vessels to express cell adhesion molecules that cause monocytes and neutrophils in the blood to adhere there.
3. They secrete chemokines that direct the migration of the adherent monocytes and neutrophils out of the bloodstream into the site of infection.

T_H1 cells can also help activate cytotoxic T cells in peripheral lymphoid organs by stimulating dendritic cells to produce more costimulatory proteins. In addition, they can help effector cytotoxic T cells kill virus-infected target cells, by secreting IFN-γ, which increases the efficiency with which target cells process viral antigens for presentation to cytotoxic T cells (see Figure 24–59). An effector T_H1 cell can also directly kill some cells itself, including effector lymphocytes: by expressing *Fas ligand* on its surface, it can induce effector T or B cells that express cell-surface *Fas* to undergo apoptosis (see Figure 24–46B).

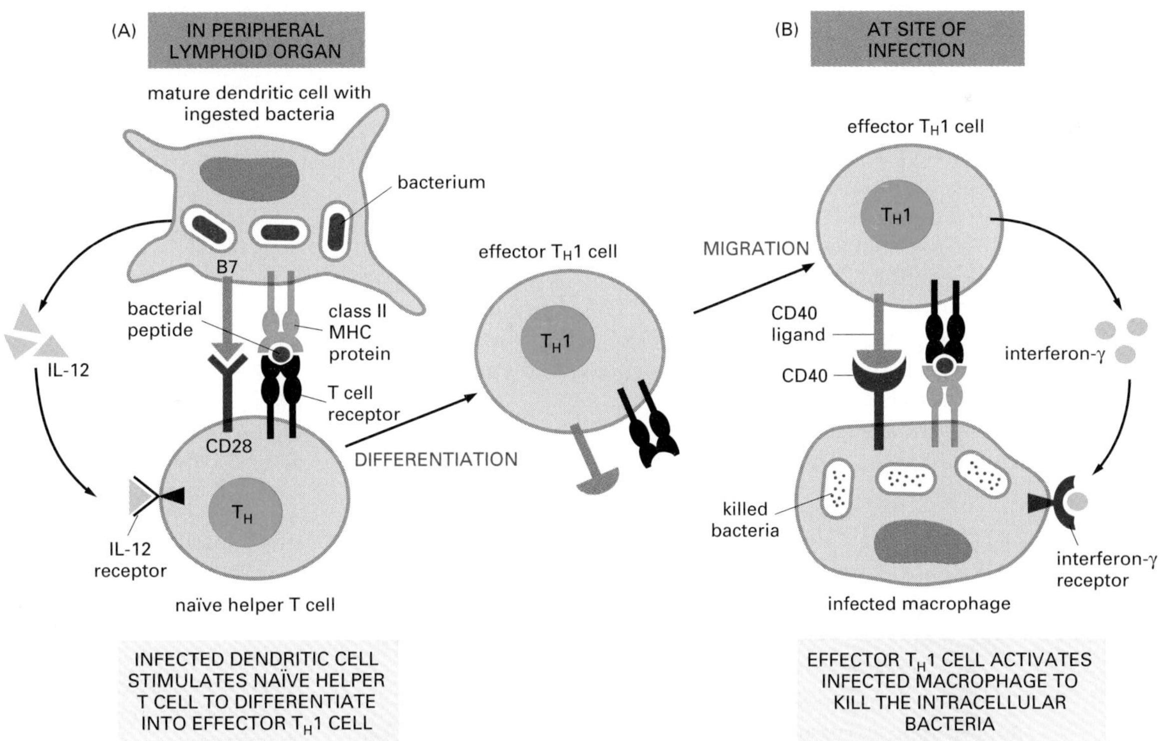

Figure 24–67 The differentiation of T_H1 cells and their activation of macrophages. (A) An infected dendritic cell that has migrated from a site of infection to a peripheral lymphoid organ activates a naïve helper T cell to differentiate into a T_H1 effector cell, using both cell-surface B7 and secreted IL-12. (B) A T_H1 effector cell that has migrated from the peripheral lymphoid organ to an infected site helps activate macrophages to kill the bacteria harboring within the macrophages' phagosomes. The T cell activates the macrophage by means of CD40 ligand on its surface and secreted interferon-γ.

Both T_H1 and T_H2 cells can help stimulate B cells to proliferate and differentiate into either antibody-secreting effector cells or memory cells. They can also stimulate B cells to switch the class of antibody they make, from IgM (and IgD) to one of the secondary classes of antibody. Before considering how helper T cells do this, we need to discuss the role of the B cell antigen receptor in the activation of B cells.

Antigen Binding Provides Signal 1 to B Cells

Like T cells, B cells require two types of extracellular signals to become activated. Signal 1 is provided by antigen binding to the antigen receptor, which is a membrane-bound antibody molecule. Signal 2 is usually provided by a helper T cell. Like a T cell, if a B cell receives the first signal only, it is usually eliminated or functionally inactivated, which is one way in which B cells become tolerant to self antigens.

Signaling through the B cell antigen receptor works in much the same way as signaling through the T cell receptor (see Figure 24–64). The receptor is associated with two invariant protein chains, Igα and Igβ, which help convert antigen binding to the receptor into intracellular signals. When antigen cross-links its receptors on the surface of a B cell, it causes the receptors and its associated invariant chains to cluster into small aggregates. This aggregation leads to the assembly of an intracellular signaling complex at the site of the clustered receptors and to the initiation of a phosphorylation cascade (Figure 24–68).

Just as the CD4 and CD8 co-receptors on T cells enhance the efficiency of signaling through the T cell receptor, so a co-receptor complex that binds complement proteins greatly enhances the efficiency of signaling through the B cell antigen receptor and its associated invariant chains. If a microbe activates the complement system (discussed in Chapter 25), complement proteins are often deposited on the microbe surface, greatly increasing the B cell response to the microbe. Now, when the microbe clusters antigen receptors on a B cell, the *complement-binding co-receptor complexes* are brought into the cluster, increasing the strength of signaling (Figure 24–69A). As expected, antibody responses are greatly reduced in mice lacking either one of the required complement components or complement receptors on B cells.

Later in the immune response, by contrast, when IgG antibodies decorate the surface of the microbe, a different co-receptor comes into play to dampen down the B cell response. These are *Fc receptors,* which bind the tails of the IgG antibodies. They recruit phosphatase enzymes into the signaling complex that decrease the strength of signaling (Figure 24–69B). In this way the Fc receptors on B cells act as inhibitory co-receptors, just as the CTLA-4 proteins do on T cells. Thus, the co-receptors on a T cell or B cell allow the cell to gain additional information about the antigen bound to its receptors and thereby make a more informed decision as to how to respond.

Unlike T cell receptors, the antigen receptors on B cells do more than just bind antigen and transmit signal 1. They deliver the antigen to an endosomal

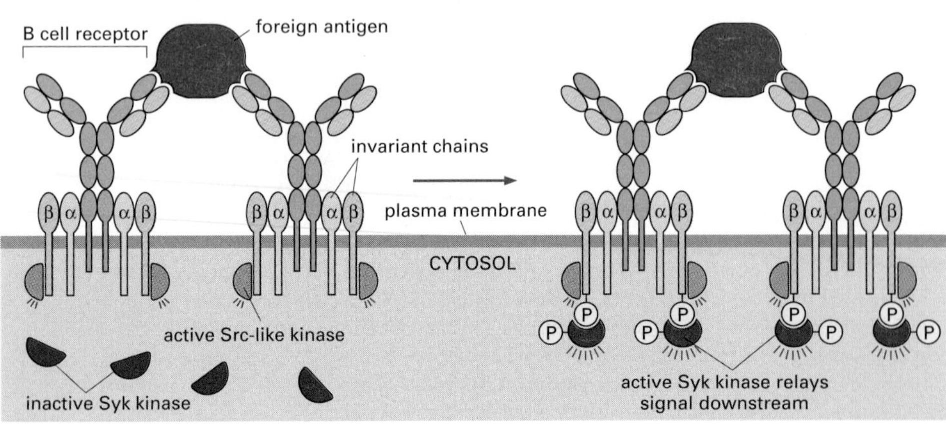

Figure 24–68 Signaling events activated by the binding of antigen to B cell receptors (signal 1). The antigen cross-links adjacent receptor proteins, which are transmembrane antibody molecules, causing the receptors and their associated invariant chains (Igα and Igβ) to cluster. The Src-like tyrosine kinase associated with the cytosolic tail of Igβ joins the cluster and phosphorylates Igα and Igβ (for simplicity, only the phosphorylation on Igβ is shown). The resulting phosphotyrosines on Igα and Igβ serve as docking sites for another Src-like tyrosine kinase called *Syk*, which is homologous to ZAP-70 in T cells (see Figure 24–64). Like ZAP-70, Syk becomes phosphorylated and relays the signal downstream.

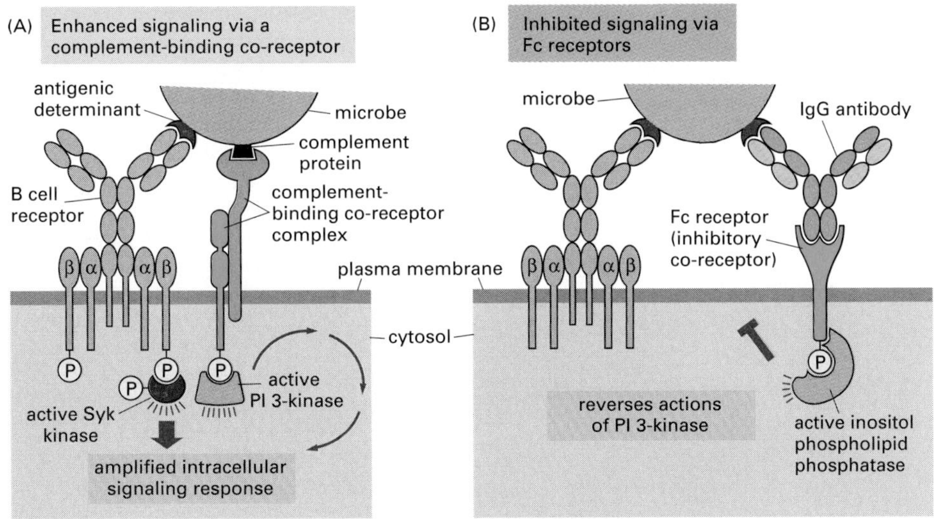

Figure 24–69 The influence of B cell co-receptors on the effectiveness of signal 1. (A) The binding of microbe–complement complexes to a B cell cross-links the antigen receptors to complement-binding, co-receptor complexes. The cytosolic tail of one component of the co-receptor complex becomes phosphorylated on tyrosines, which then serve as docking sites for PI 3-kinase. As discussed in Chapter 15, PI 3-kinase is activated to generate inositol phospholipid docking sites in the plasma membrane, which recruit intracellular signaling proteins (not shown). These signaling proteins act together with the signals generated by the Syk kinase to amplify the response. (B) When IgG antibodies become bound to foreign antigen, usually late in a response, the Fc regions of the antibodies bind to Fc receptors on the B cell surface and are thus recruited into the signaling complex. The Fc receptors become phosphorylated on tyrosines, which then serve as docking sites for an inositol phospholipid phosphatase. The phosphatase dephosphorylates the inositol phopholipid docking sites in the plamsa membrane generated by PI 3-kinase, thereby reversing the activating effects of PI 3-kinase. The Fc receptors also inhibit signaling by recruiting protein tyrosine phosphatases into the signaling complex (not shown).

compartment where the antigen is degraded to peptides, which are returned to the B cell surface bound to class II MHC proteins (see Figure 24–60). The peptide–class-II-MHC complexes are then recognized by effector helper T cells, which can now deliver signal 2. Signal 1 prepares the B cell for its interaction with a helper T cell by increasing the expression of both class II MHC proteins and receptors for signal 2.

Helper T Cells Provide Signal 2 to B Cells

Whereas antigen-presenting cells such as dendritic cells and macrophages are omnivorous and ingest and present antigens nonspecifically, a B cell generally presents only an antigen that it specifically recognizes. In a primary antibody response, naïve helper T cells are activated in a peripheral lymphoid organ by binding to a foreign peptide bound to a class II MHC protein on the surface of a dendritic cell. Once activated, the effector helper T cell can then activate a B cell that specifically displays the same complex of foreign peptide and class II MHC protein on its surface (see Figure 24–66).

The display of antigen on the B cell surface reflects the selectivity with which it takes up foreign proteins from the extracellular fluid. These foreign proteins are selected by the antigen receptors on the surface of the B cell and are ingested by receptor-mediated endocytosis. They are then degraded and recycled to the cell surface in the form of peptides bound to class II MHC proteins. Thus, the helper T cell activates those B cells with receptors that specifically recognize the antigen that initially activated the T cell, although the T and B cells usually recognize distinct antigenic determinants on the antigen (see Figure 24–70). In secondary antibody responses, memory B cells themselves can act as antigen-presenting cells and activate helper T cells, as well as being the subsequent targets of the effector helper T cells. The mutually reinforcing actions of helper T cells and B cells lead to an immune response that is both intense and highly specific.

Once a helper T cell has been activated to become an effector cell and contacts a B cell, the contact initiates an internal rearrangement of the helper cell cytoplasm. The T cell orients its centrosome and Golgi apparatus toward the B cell, as described previously for an effector cytotoxic T cell contacting its target cell (see Figure 24–45). In this case, however, the orientation is thought to enable the effector helper T cell to provide signal 2 by directing both membrane-bound and secreted signal molecules onto the B cell surface. The membrane-bound signal molecule is the transmembrane protein CD40 ligand, which we encountered earlier and is expressed on the surface of effector helper T cell, but not on nonactivated naïve or memory helper T cells. It is recognized by the CD40 protein on the B cell surface. The interaction between CD40 ligand and CD40 is required for helper T cells to activate B cells to proliferate and differentiate into

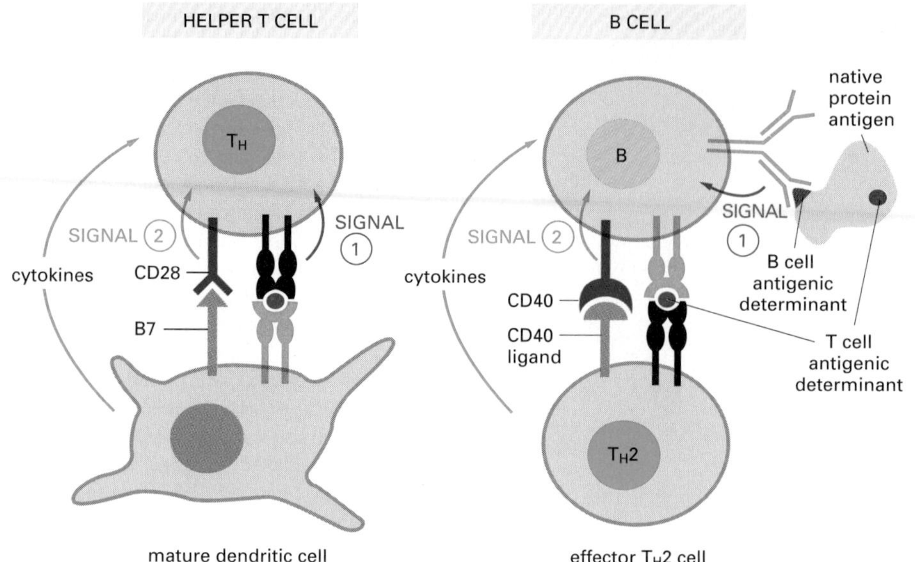

Figure 24–70 Comparison of the signals required to activate a helper T cell and a B cell. Note that in both cases secreted and membrane-bound molecules can cooperate to provide signal 2. Although not shown, CD40 is also expressed on the surface of mature dendritic cells and helps maintain helper T cells in an active state. The native protein antigen is endocytosed by both the dendritic cell and the B cell and is degraded in endosomes (not shown). The T cell antigenic determinant is presented on the surface of both the dendritic cell and the B cell as a peptide fragment bound to a class II MHC protein. By contrast, the B cell recognizes an antigenic determinant on the surface of the folded protein.

memory or antibody-secreting effector cells. Individuals that lack CD40 ligand are severely immunodeficient. They are susceptible to the same infections that affect AIDS patients, whose helper T cells have been destroyed.

Secreted signals from helper T cells also help B cells to proliferate and differentiate and, in some cases, to switch the class of antibody they produce. *Interleukin-4 (IL-4)* is one such signal. Produced by T_H2 cells, it collaborates with CD40 ligand in stimulating B cell proliferation and differentiation, and it promotes switching to IgE antibody production. Mice deficient in IL-4 production are severely impaired in their ability to make IgE.

The signals required for T and B cell activation are compared in Figure 24–70, and some of the cytokines discussed in this chapter are listed in Table 24–4.

Some antigens can stimulate B cells to proliferate and differentiate into antibody-secreting effector cells without help from T cells. Most of these *T-cell-independent antigens* are microbial polysaccharides that do not activate helper T cells. Some activate B cells directly by providing both signal 1 and signal 2. Others are large polymers with repeating, identical antigenic determinants (see Figure 24–29B); their multipoint binding to B cell antigen receptors can generate a strong enough signal 1 to activate the B cell directly, without signal 2. Because T-cell-independent antigens do not activate helper T cells, they fail to induce B

TABLE 24–4 Properties of Some Interleukins

CYTOKINE	SOME SOURCES	SOME TARGETS	SOME ACTIONS
IL-2	all helper T cells; some cytotoxic T cells; activated mast cells	all activated T cells and B cells	stimulates proliferation and differentiation
IL-4	T_H2 cells and mast cells	B cells and T_H cells	stimulates B cell proliferation, maturation, and class switching to IgE and IgG1; inhibits T_H1 cell development
IL-5	T_H2 cells and mast cells	B cells, eosinophils	promotes proliferation and maturation
IL-10	T_H2 cells, macrophages, and dendritic cells	macrophages and T_H1 cells	inhibits macrophages and T_H1 cell development
IL-12	B cells, macrophages, and dendritic cells	naïve T cells	induces T_H2 cell development and inhibits T_H1 cell development
IFN-γ	T_H1 cells	B cells, macrophages, endothelial cells	activates various MHC genes and macrophages; increases MHC expression in many cell types
TNF-α	T_H1 cells and macrophages	endothelial cells	activates

cell memory, affinity maturation, or class switching, all of which require help from T cells. They therefore mainly stimulate the production of low-affinity (but high-avidity) IgM antibodies. Most B cells that make antibodies without T cell help belong to a distinct B cell lineage. They are called *B1 cells* to distinguish them from *B2 cells,* which require T cell help. B1 cells seem to be especially important in defense against intestinal pathogens.

Immune Recognition Molecules Belong to an Ancient Superfamily

Most of the proteins that mediate cell–cell recognition or antigen recognition in the immune system contain Ig or Ig-like domains, suggesting that they have a common evolutionary history. Included in this **Ig superfamily** are antibodies, T cell receptors, MHC proteins, the CD4, CD8, and CD28 co-receptors, and most of the invariant polypeptide chains associated with B and T cell receptors, as well as the various Fc receptors on lymphocytes and other white blood cells. All of these proteins contain one or more Ig or Ig-like domains. In fact, about 40% of the 150 or so polypeptides that have been characterized on the surface of white blood cells belong to this superfamily. Many of these molecules are dimers or higher oligomers in which Ig or Ig-like domains of one chain interact with those in another (Figure 24–71).

The amino acids in each Ig-like domain are usually encoded by a separate exon. It seems likely that the entire gene superfamily evolved from a gene coding for a single Ig-like domain—similar to that encoding β_2-microglobulin (see Figure 24–50A) or the Thy-1 protein (see Figure 24–71)—that may have mediated cell–cell interactions. There is evidence that such a primordial gene arose before vertebrates diverged from their invertebrate ancestors about 400 million years ago. New family members presumably arose by exon and gene duplications.

The multiple gene segments that encode antibodies and T cell receptors may have arisen when a transposable element, or transposon (discussed in

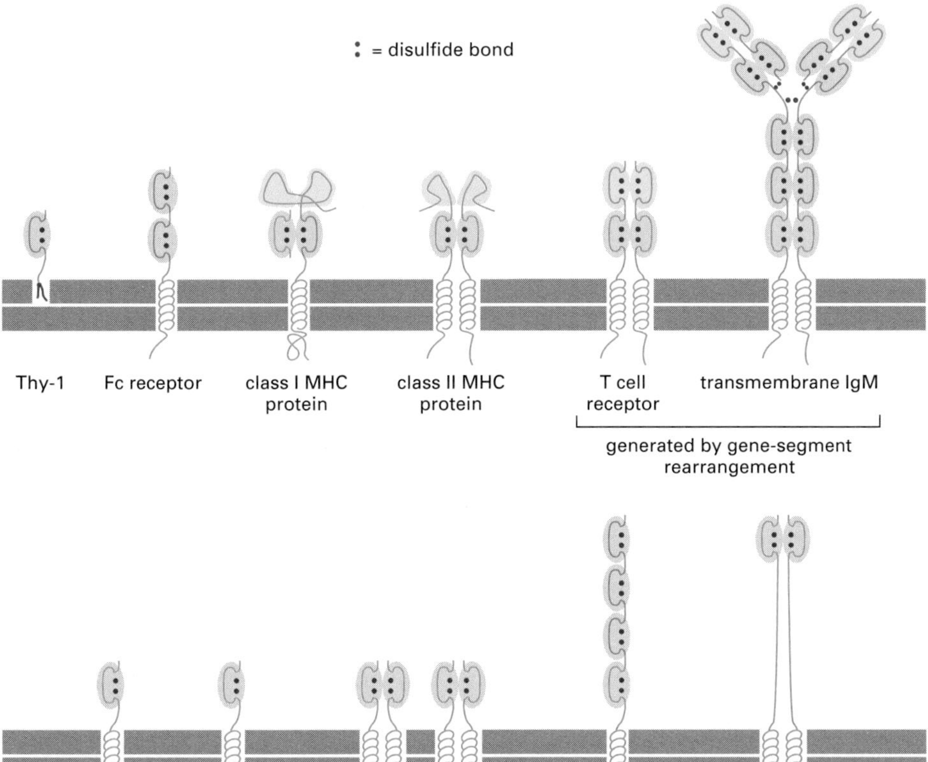

Figure 24–71 Some of the membrane proteins belonging to the Ig superfamily. The Ig and Ig-like domains are shaded in *gray,* except for the antigen-binding domains (not all of which are Ig domains), which are shaded in *blue.* The function of Thy-1 is unknown, but it is widely used to idenitfy T cells in mice. The Ig superfamily also includes many cell-surface proteins involved in cell–cell interactions outside the immune system, such as the neural cell-adhesion molecule (N-CAM) discussed in Chapter 19 and the receptors for various protein growth factors discussed in Chapters 15 and 17 (not shown). There are about 765 members of the Ig superfamily in humans.

Chapter 5), inserted into an exon of a gene encoding an Ig family member in an ancestral lymphocyte-like cell. The transposon may have contained the ancestors of the *rag* genes, which, as discussed earlier, encode the proteins that initiate *V(D)J* joining; the finding that the RAG proteins can act as transposons in a test tube strongly supports this view. Once the transposon had inserted into the exon, the gene could be expressed only if the transposon was excised by the RAG proteins and the two ends of the exon were rejoined, much as occurs when the the *V* and *J* gene segments of an Ig light chain gene are assembled (see Figure 24–37). A second insertion of the transposon into the same exon may then have divided the gene into three segments, equivalent to the present-day *V, D,* and *J* gene segments. Subsequent duplication of either the individual gene segments or the entire split gene may have generated the arrangements of gene segments that characterize the adaptive immune systems of present-day vertebrates.

Adaptive immune systems evolved to defend vertebrates against infection by pathogens. Pathogens, however, evolve more quickly, and they have acquired remarkably sophisticated strategies to counter these defenses, as we discuss in Chapter 25.

Summary

Naïve T cells require at least two signals for activation. Both are provided by an antigen-presenting cell, which is usually a dendritic cell: signal 1 is provided by MHC–peptide complexes binding to T cell receptors, while signal 2 is mainly provided by B7 costimulatory proteins binding to CD28 on the T cell surface. If the T cell receives only signal 1, it is usually deleted or inactivated. When helper T cells are initially activated on a dendritic cell, they can differentiate into either T_H1 or T_H2 effector cells, depending on the cytokines in their environment: T_H1 cells activate macrophages, cytotoxic T cells, and B cells, while T_H2 cells mainly activate B cells. In both cases, the effector helper T cells recognize the same complex of foreign peptide and class II MHC protein on the target cell surface as they initially recognized on the dendritic cell that activated them. They activate their target cells by a combination of membrane-bound and secreted signal proteins. The membrane-bound signal is CD40 ligand. Like T cells, B cells require two simultaneous signals for activation. Antigen binding to the B cell antigen receptors provides signal 1, while effector helper T cells provide signal 2 in the form of CD40 ligand and various cytokines.

Most of the proteins involved in cell–cell recognition and antigen recognition in the immune system, including antibodies, T cell receptors, and MHC proteins, as well as the various co-receptors discussed in this chapter, belong to the ancient Ig superfamily. This superfamily is thought to have evolved from a primordial gene encoding a single Ig-like domain.

References

General

Abbas AK, Lichtman AH & Pober JS (1997) Cellular and Molecular Immunology, 3rd edn. Philadelphia: WB Saunders.

Janeway CA, Jr, Travers P, Walport M & Shlomchik M (2001) Immunobiology: The Immune System in Health and Disease, 5th edn. London: Garland.

Parham P (2001) The Immune System. New York/London: Garland Publishing/Elsevier Science Ltd.

Paul WE (1999) Fundamental Immunology. Philadelphia: Lippincott-Raven.

Lymphocytes and the Cellular Basis of Adaptive Immunity

Billingham RE, Brent L & Medewar PB (1956) Quantitative studies on tissue transplantation immunity. III. Actively acquired tolerance. *Philos. Trans. R. Soc. Lond. B. Biol. Sci.* 239, 357–414.

Butcher EC & Picker LJ (1996) Lymphocyte homing and homeostasis. *Science* 272, 60–66.

Cyster JG (1999) Chemokines and cell migration in secondary lymphoid organs. *Science* 286, 2098–2102.

Fearon DT & Locksley RM (1996) The instructive role of innate immunity in the acquired immune response. *Science* 272, 50–53.

Hoffmann JA, Kafatos FC, Janeway CA, Jr & Ezekowitz RAB (1999) Phylogenetic perspectives in innate immunity. *Science* 284, 1313–1318.

Ikuta K, Uchida N, Friedman J & Weissman IL (1992) Lymphocyte development from stem cells. *Annu. Rev. Immunol.* 10, 759–784.

Janeway CA, Jr, Goodnow CC & Medzhitov R (1996) Danger—pathogen on the premises. *Curr. Biol.* 6, 519–522.

Sprent J (1997) Immunological memory. *Curr. Opin. Immunol.* 9, 371–379.

Zinkernagel RM, Bachmann MF, Kundig TM et al. (1996) On immunological memory. *Annu. Rev. Immunol.* 14, 333–367.

B Cells and Antibodies

Braden BC & Poljak RJ (1995) Structural features of the reactions between antibodies and protein antigens. *FASEB J.* 9, 9–16.

Burton DR & Woof JM (1992) Human antibody effector function. *Adv. Immunol.* 51, 1–84.

Davies DR, Sherrif S & Padlan EA (1988) Antigen–antibody complexes. *J. Biol. Chem.* 263, 10541–10544.

DeFranco AL (1993) Structure and function of the B cell antigen receptor. *Annu. Rev. Cell Biol.* 9, 377–410.

Padlan EA (1994) Anatomy of the antibody molecule. *Mol. Immunol.* 31, 169–217.

Reth M (1994) B cell antigen receptors. *Curr. Opin. Immunol.* 6, 3–8.

Sakano H, Rogers JH, Huppi K et al. (1979) Domains and the hinge region of an immunoglobulin heavy chain are encoded in separate DNA segments. *Nature* 277, 627–633.

Wilson IA & Stanfield RL (1994) Antibody–antigen interactions: new structures and new conformational changes. *Curr. Opin. Struct. Biol.* 4, 857–867.

The Generation of Antibody Diversity

Bergman Y (1999) Allelic exclusion in B and T lymphopoiesis. *Semin. Immunol.* 11, 319–328.

Chen J & Alt FW (1993) Gene rearrangement and B-cell development. *Curr. Opin. Immunol.* 5, 194–200.

Fugmann SD, Lee AI, Shockett PE et al. (2000) The RAG proteins and V(D)J recombination: complexes, ends, and transposition. *Annu. Rev. Immunol.* 18, 495–527.

Green NS, Lin MM & Scharff MD (1998) Somatic hypermutation of antibody genes: a hot spot warms up. *Bioessays* 20, 227–234.

Kinoshita K & Honjo T (2001) Linking class-switch recombination with somatic hypermutation. *Nat. Rev. Mol. Cell Biol.* 2, 493–503.

Rajewsky K (1996) Clonal selection and learning in the antibody system. *Nature* 381, 751–758.

Stavnezer J (1996) Antibody class switching. *Adv. Immunol.* 61, 79–146.

Tonegawa S (1983) Somatic generation of antibody diversity. *Nature* 302, 575–581.

Willerford DM, Swat W & Alt FW (1996) Developmental regulation of V(D)J recombination and lymphocyte differentiation. *Curr. Opin. Genet. Dev.* 6, 603–609.

T Cells and MHC Proteins

Bentley GA & Mariuzza RA (1996) The structure of the T cell antigen receptor. *Annu. Rev. Immunol.* 14, 563–590.

Bjorkman PJ (1997) MHC restriction in three dimensions: a view of T cell receptor/ligand interactions. *Cell* 89, 167–170.

Cresswell P (1998) Proteases, processing, and thymic selection. *Science* 280, 394–395.

Dong C & Flavell RA (2001) Th1 and Th2 cells. *Curr. Opin. Hematol.* 8, 47–51.

Dustin ML & Cooper JA (2000) The immunological synapse and the actin cytoskeleton: molecular hardware for T cell signaling. *Nat. Immunol.* 1, 23–29.

Garcia KC, Teyton L & Wilson IA (1999) Structural basis of T cell recognition. *Annu. Rev. Immunol.* 17, 369–397.

Goldrath AW & Bevan MJ (1999) Selecting and maintaining a diverse T-cell repertoire. *Nature* 402, 255–262.

Hennecke J & Wiley DC (2001) T cell receptor–MHC interactions up close. *Cell* 104, 1–4.

Lanzavecchia A & Sallusto F (2001) The instructive role of dendritic cells on T cell responses: lineages, plasticity and kinetics. *Curr. Opin. Immunol.* 13, 291–298.

McDevitt HO (2000) Discovering the role of the major histocompatibility complex in the immune response. *Annu. Rev. Immunol.* 18, 1–17.

Meyer D & Thomson G (2001) How selection shapes variation of the human major histocompatibility complex: a review. *Ann. Hum. Genet.* 65, 1–26.

Natarajan K, Li H, Mariuzza RA & Margulies DH (1999) MHC class I molecules, structure and function. *Rev. Immunogenet.* 1, 32–46.

Nossal GJ (1994) Negative selection of lymphocytes. *Cell* 76, 229–239.

Pieters J (2000) MHC class II-restricted antigen processing and presentation. *Adv. Immunol.* 75, 159–208.

Rock KL & Goldberg AL (1999) Degradation of cell proteins and the generation of MHC class I-presented peptides. *Annu. Rev. Immunol.* 17, 739–779.

The MHC sequencing consortium (1999) Complete sequence and gene map of human major histocompatibility complex. *Nature* 401, 921–923.

von Boehmer H (1994) Positive selection of lymphocytes. *Cell* 76, 219–228.

Watts C & Powis S (1999) Pathways of antigen processing and presentation. *Rev. Immunogenet.* 1, 60–74.

Wülfing C & Davis MM (1998) A receptor/cytoskeletal movement triggered by costimulation during T cell activation. *Science* 282, 2267–2269.

Zinkernagel RM & Doherty PC (1979) MHC-restricted cytotoxic T cells: studies of the biological role of polymorphic major transplantation antigens determining T-cell restriction-specificity, function and responsiveness. *Adv. Immunol.* 27, 51–177.

Helper T Cells and Lymphocyte Activation

Buck CA (1992) Immunoglobulin superfamily: structure, function and relationship to other receptor molecules. *Semin. Cell Biol.* 3, 179–188.

Carroll MC (2000) The role of complement in B cell activation and tolerance. *Adv. Immunol.* 74, 61–88.

Croft M & Dubey C (1997) Accessory molecule and costimulation requirements for CD4 T cell response. *Crit. Rev. Immunol.* 17, 89–118.

DeFranco AL (1996) The two-headed antigen. *Curr. Biol.* 6, 548–550.

Lichtman AH & Abbas AK (1997) Recruiting the right kind of help. *Curr. Biol.* 7, R242–R244.

Weintraub BC & Goodnow CC (1998) Costimulatory receptors have their say. *Curr. Biol.* 8, R575–R577.

Williams AF, Davis SJ, He Q & Barclay AN (1989) Structural diversity in domains of the immunoglobulin superfamily. *Cold Spring Harb. Symp. Quant. Biol.* 54, 637–647.

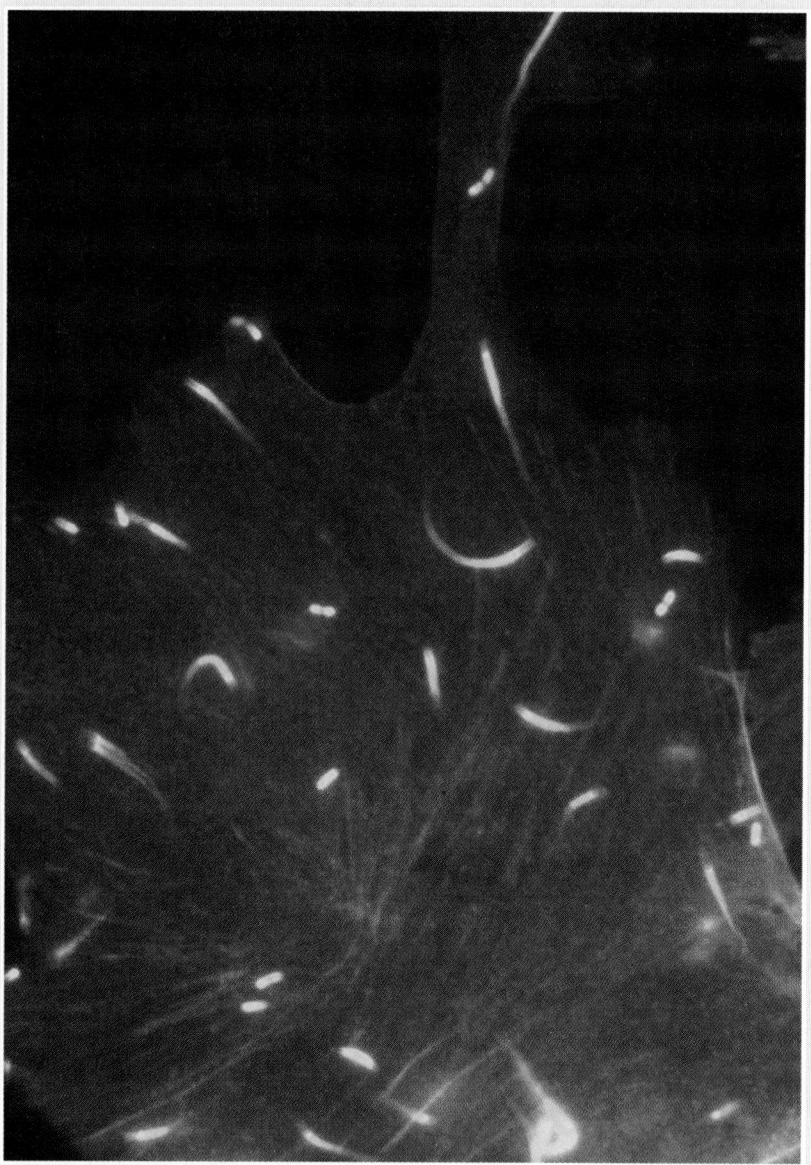

Host–pathogen interactions. The bacterial pathogen *Listeria monocytogenes* grows directly in the cytoplasm of infected mammalian host cells, including the human cell shown here. These bacteria induce the assembly of comet-like tails of host cell actin filaments *(green)* that propel the invaders through the host cytoplasm and enable the infection to spread directly from a cell to its neighbors. (Courtesy of Julie Theriot.)

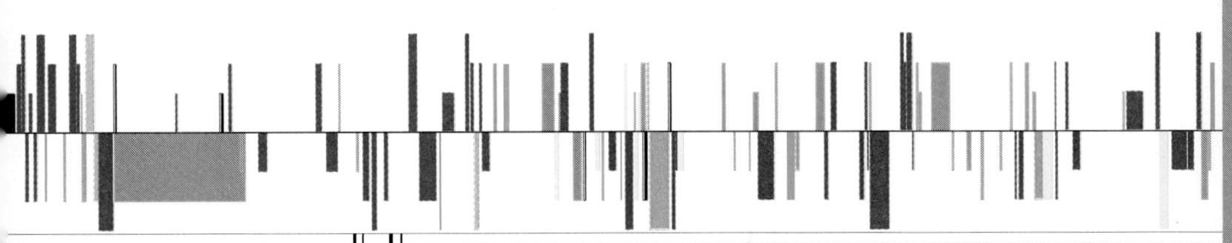

25

PATHOGENS, INFECTION, AND INNATE IMMUNITY

Infectious and parasitic diseases currently cause about one third of all human deaths in the world, more than all forms of cancer combined. In addition to the continuing heavy burden of ancient diseases like tuberculosis and malaria, new infectious diseases are continually emerging, including the current pandemic (worldwide epidemic) of *AIDS (acquired immune deficiency syndrome),* which has already caused more than twenty million deaths worldwide. Moreover, some diseases long thought to result from other causes are now turning out to be associated with infections. Most gastric ulcers, for example, are caused not by stress or spicy food, as was once believed, but by a bacterial infection of the stomach caused by *Helicobacter pylori*.

The burden of infectious and parasitic diseases is not spread equally across the planet. Poorer countries and communities suffer disproportionately. Frequently, there is a correlation between the prevalence of infectious diseases and poor public sanitation and public health systems, which are often further compromised by political upheavals. Some infectious diseases, however, occur primarily or exclusively among industrialized communities: Legionnaire's disease is a recent example.

Humans have long been troubled and fascinated with infectious diseases. The earliest written descriptions of how to limit the spread of rabies date back more than three thousand years. Since the mid-1800s, physicians and scientists have struggled to identify the agents that cause infectious diseases, collectively called **pathogens**. More recently, the advent of microbial genetics and molecular cell biology has greatly enhanced our understanding of the causes and mechanisms of infectious diseases. We now know that pathogens frequently exploit the biological attributes of their host's cells in order to infect them. This understanding can give us new insights into normal cell biology, as well as strategies for treating and preventing infectious diseases.

In a world teeming with hostile, clever, and rapidly evolving pathogens, how does a fragile and slowly evolving human survive? Like all other multicellular organisms, we have developed mechanisms to resist infection by pathogens. These defenses are of two kinds: **innate immune responses**, which spring into action immediately after an infection begins and do not depend on the host's prior exposure to the pathogen, and **adaptive immune responses**, which are more powerful defenses that operate later in an infection and are highly specific for the pathogen that induced them.

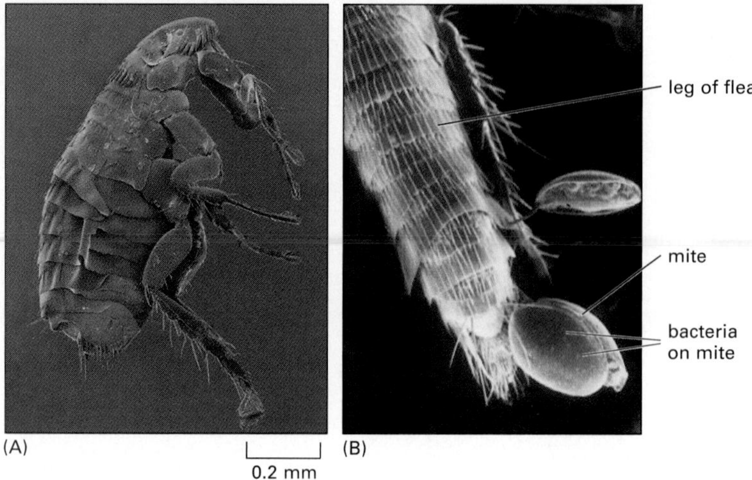

Figure 25–1 Parasitism at many levels. (A) Scanning electron micrograph of a flea. The flea is a common parasite of mammals—including dogs, cats, rats, and humans. It drinks the blood of its host. Flea bites spread bubonic plague by passing the pathogenic bacterium *Yersinia pestis* from the bloodstream of one infected host to that of another.
(B) A close-up view of a flea's leg reveals that this flea also has a parasite, a type of mite. The mite, in turn, is covered with bacteria. It is likely that these bacteria are parasitized by *bacteriophages*, which are bacterial viruses.

A similar observation was reported by Jonathan Swift in 1733:
So, naturalists observe, a flea
Has smaller fleas that on him prey;
And these have smaller still to bite 'em;
And so proceed ad infinitum.
(A, courtesy of Tina Carvalho/ MicroAngela; B, courtesy of Stanley Falkow.)

In this chapter, we begin with an overview of the different kinds of organisms that cause disease. We then discuss the cell biology of infection and, finally, consider innate immunity. Adaptive immunity is the subject of Chapter 24.

INTRODUCTION TO PATHOGENS

We normally think of pathogens in hostile terms—as invaders that attack our bodies. But a pathogen or a parasite, like any other organism, is simply trying to live and procreate. Living at the expense of a host organism is a very attractive strategy, and it is possible that every living organism on earth is subject to some type of infection or parasitism (Figure 25–1). A human host is a nutrient-rich, warm, and moist environment, which remains at a uniform temperature and constantly renews itself. It is not surprising that many microorganisms have evolved the ability to survive and reproduce in this desirable niche. In this section, we discuss some of the common features that microorganisms must have in order to be infectious. We then explore the wide variety of organisms that are known to cause disease in humans.

Pathogens Have Evolved Specific Mechanisms for Interacting with Their Hosts

The human body is a complex and thriving ecosystem. It contains about 10^{13} human cells and also about 10^{14} bacterial, fungal, and protozoan cells, which represent thousands of microbial species. These microbes, called the **normal flora**, are usually limited to certain areas of the body, including the skin, mouth, large intestine, and vagina. In addition, humans are always infected with viruses, most of which rarely, if ever, become symptomatic. If it is normal for us to live in such close intimacy with a wide variety of microbes, how is it that some of them are capable of causing us illness or death?

Pathogens are usually distinct from the normal flora. Our normal microbial inhabitants only cause trouble if our immune systems are weakened or if they gain access to a normally sterile part of the body (for example, when a bowel perforation enables the gut flora to enter the peritoneal cavity of the abdomen, causing *peritonitis*). In contrast, dedicated pathogens do not require that the host be immunocompromised or injured. They have developed highly specialized mechanisms for crossing cellular and biochemical barriers and for eliciting specific responses from the host organism that contribute to the survival and multiplication of the pathogen.

In order to survive and multiply in a host, a successful pathogen must be able to: (1) colonize the host; (2) find a nutritionally compatible niche in the host body; (3) avoid, subvert, or circumvent the host innate and adaptive immune responses; (4) replicate, using host resources; and (5) exit and spread to a new host. Under severe selective pressure to induce only the correct host cell responses to accomplish this complex set of tasks, pathogens have evolved mechanisms that maximally exploit the biology of their host organisms. Many of the pathogens we discuss in this chapter are skillful and practical cell biologists. We can learn a great deal of cell biology by observing them.

The Signs and Symptoms of Infection May Be Caused by the Pathogen or by the Host's Responses

Although we can easily understand why infectious microorganisms would evolve to reproduce in a host, it is less clear why they would evolve to cause disease. One explanation may be that, in some cases, the pathological responses elicited by microorganisms enhance the efficiency of their spread or propagation and hence clearly have a selective advantage for the pathogen. The virus-containing lesions on the genitalia caused by *herpes simplex* infection, for example, facilitate direct spread of the virus from an infected host to an uninfected partner during sexual contact. Similarly, diarrheal infections are efficiently spread from patient to caretaker. In many cases, however, the induction of disease has no apparent advantage for the pathogen.

Many of the symptoms and signs that we associate with infectious disease are direct manifestations of the host's immune responses in action. Some hallmarks of bacterial infection, including the swelling and redness at the site of infection and the production of pus (mainly dead white blood cells), are the direct result of immune system cells attempting to destroy the invading microorganisms. Fever, too, is a defensive response, as the increase in body temperature can inhibit the growth of some microorganisms. Thus, understanding the biology of an infectious disease requires an appreciation of the contributions of both pathogen and host.

Pathogens Are Phylogenetically Diverse

Many types of pathogens cause disease in humans. The most familiar are viruses and bacteria. Viruses cause diseases ranging from AIDS and smallpox to the common cold. They are essentially fragments of nucleic acid (DNA or RNA) instructions, wrapped in a protective shell of proteins and (in some cases) membrane (Figure 25–2A). They use the basic transcription and translation machinery of their host cells for their replication.

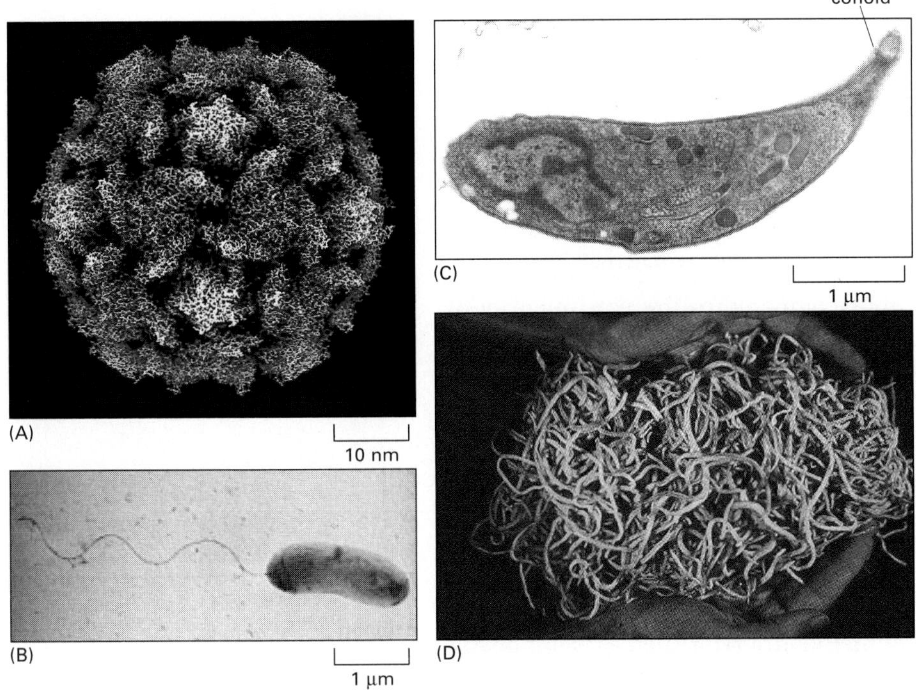

Figure 25–2 Pathogens in many forms. (A) The structure of the protein coat, or *capsid*, of poliovirus. This virus was once a common cause of paralysis, but the disease (poliomyelitis) has been nearly eradicated by widespread vaccination. (B) The bacterium *Vibrio cholerae*, the causative agent of the epidemic, diarrheal disease cholera. (C) The protozoan parasite *Toxoplasma gondii*. This organism is normally a parasite of cats, but it can cause serious infections in the muscles and brains of immunocompromised people with AIDS. (D) This clump of *Ascaris* nematodes was removed from the obstructed intestine of a two-year-old boy. (A, courtesy of Robert Grant, Stephan Crainic, and James M. Hogle; B, all attempts have been made to contact the copyright holder and we would be pleased to hear from them; C, courtesy of John Boothroyd and David Ferguson; D, from J.K. Baird et al., *Amer. J. Trop. Med. Hyg.* 35:314–318, 1986. Photograph by Daniel H. Connor.)

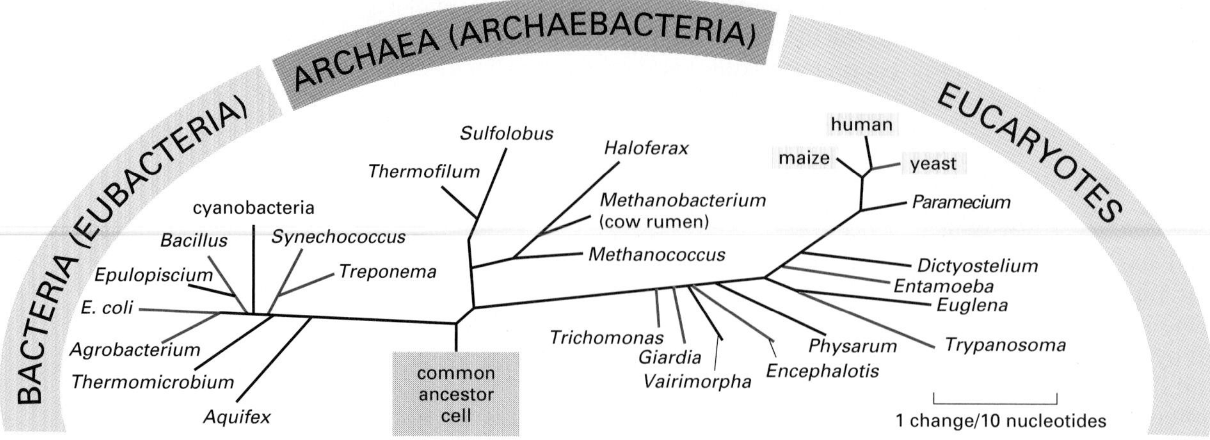

Figure 25–3 Phylogenetic diversity of pathogens. This diagram shows the similarities among 16S ribosomal RNA for cellular life forms (bacteria, archaea, and eucaryotes). Each branch is labeled with the name of a representative member of that group, and the length of the branches corresponds to the degree of difference in the rRNA sequence. Note that all the organisms we can see with the unaided eye (highlighted in *yellow*)—animals, plants, and fungi—represent a small subset of the diversity of life. In the two branches of the tree representing the bacteria and the eucaryotes, the branches that include known pathogens are indicated in *red*. No diseases are currently known to be caused by archaea, though many humans and all cows carry some types of archaea in their normal intestinal flora.

Of all the bacteria we encounter in our lives, only a small minority are dedicated pathogens. Much larger and more complex than viruses, bacteria are usually free-living cells, which perform most of their basic metabolic functions themselves, relying on the host primarily for nutrition (Figure 25–2B).

Some other infectious agents are eucaryotic organisms. These range from single-celled fungi and protozoa (Figure 25–2C), through large complex metazoa such as parasitic worms. One of the most common infectious diseases on the planet, shared by about a billion people at present, is an infestation in the gut by *Ascaris lumbricoides*. This nematode closely resembles its cousin *Caenorhabditis elegans*, which is widely used as a model organism for genetic and developmental biological research (discussed in Chapter 21). *C. elegans*, however, is only about 1 mm in length, whereas *Ascaris* can reach 30 cm (Figure 25–2D).

Some rare neurodegenerative diseases, including mad cow disease, are caused by an unusual type of infectious particle called a *prion*, which is made only of protein. Although the prion contains no genome, it can nevertheless replicate and kill the host.

Even within each class of pathogen, there is striking diversity. Viruses vary tremendously in their size, shape, and content (DNA versus RNA, enveloped or not, and so on), and the same is true for the other pathogens. The ability to cause disease *(pathogenesis)* is a lifestyle choice, not a legacy shared only among close relatives (Figure 25–3).

Each individual pathogen causes disease in a different way, which makes it challenging to understand the basic biology of infection. But, when considering the interactions of infectious agents with their hosts, some common themes of pathogenesis emerge. These common themes are the focus of this chapter. First, we introduce the basic features of each of the major types of pathogens that exploit features of host cell biology. Then, we examine in turn the mechanisms that pathogens use to control their hosts and the innate mechanisms that hosts use to control pathogens.

Bacterial Pathogens Carry Specialized Virulence Genes

Bacteria are small and structurally simple, compared to the vast majority of eucaryotic cells. Most can be classified broadly by their shape as rods, spheres, or spirals and by their cell-surface properties. Although they lack the elaborate morphological variety of eucaryotic cells, they display a surprising array of surface appendages that enable them to swim or to adhere to desirable surfaces (Figure 25–4). Their genomes are correspondingly simple, typically on the order of 1,000,000–5,000,000 nucleotide pairs in size (compared to 12,000,000 for yeast and more than 3,000,000,000 for humans).

As emphasized above, only a minority of bacterial species have developed the ability to cause disease in humans. Some of those that do cause disease can only replicate inside the cells of the human body and are called **obligate pathogens**. Others replicate in an environmental reservoir such as water or soil and only cause disease if they happen to encounter a susceptible host; these are called **facultative pathogens**. Many bacteria are normally benign but have a latent ability to cause disease in an injured or immunocompromised host; these are called **opportunistic pathogens**.

Some bacterial pathogens are fastidious in their choice of host and will only infect a single species or a group of related species, whereas others are generalists. *Shigella flexneri*, for example, which causes epidemic dysentery (bloody diarrhea) in areas of the world lacking a clean water supply, will only infect humans and other primates. By contrast, the closely related bacterium *Salmonella enterica*, which is a common cause of food poisoning in humans, can also infect many other vertebrates, including chickens and turtles. A champion generalist is the opportunistic pathogen *Pseudomonas aeruginosa*, which is capable of causing disease in plants as well as animals.

The significant differences between a virulent pathogenic bacterium and its closest nonpathogenic relative may result from a very small number of genes. Genes that contribute to the ability of an organism to cause disease are called **virulence genes**. The proteins they encode are called **virulence factors**. Virulence genes are frequently clustered together, either in groups on the bacterial chromosome called *pathogenicity islands* or on extrachromosomal *virulence plasmids* (Figure 25–5). These genes may also be carried on mobile *bacteriophages* (bacterial viruses). It seems therefore that a pathogen may arise when groups of virulence genes are transferred together into a previously avirulent bacterium. Consider, for example, *Vibrio cholerae*—the bacterium that causes cholera. Several of the genes encoding the toxins that cause the diarrhea in cholera are carried on a mobile bacteriophage (Figure 25–6). Of the hundreds of strains of *Vibrio cholerae* found in lakes in the wild, the only ones that cause human disease are those that have become infected with this virus.

Many virulence genes encode proteins that interact directly with host cells. Two of the genes carried by the *Vibrio cholerae* phage, for example, encode two subunits of **cholera toxin**. The B subunit of this secreted, toxic protein binds to a glycolipid component of the plasma membrane of the epithelial cells in the gut of a person who has consumed *Vibrio cholerae* in contaminated water. The

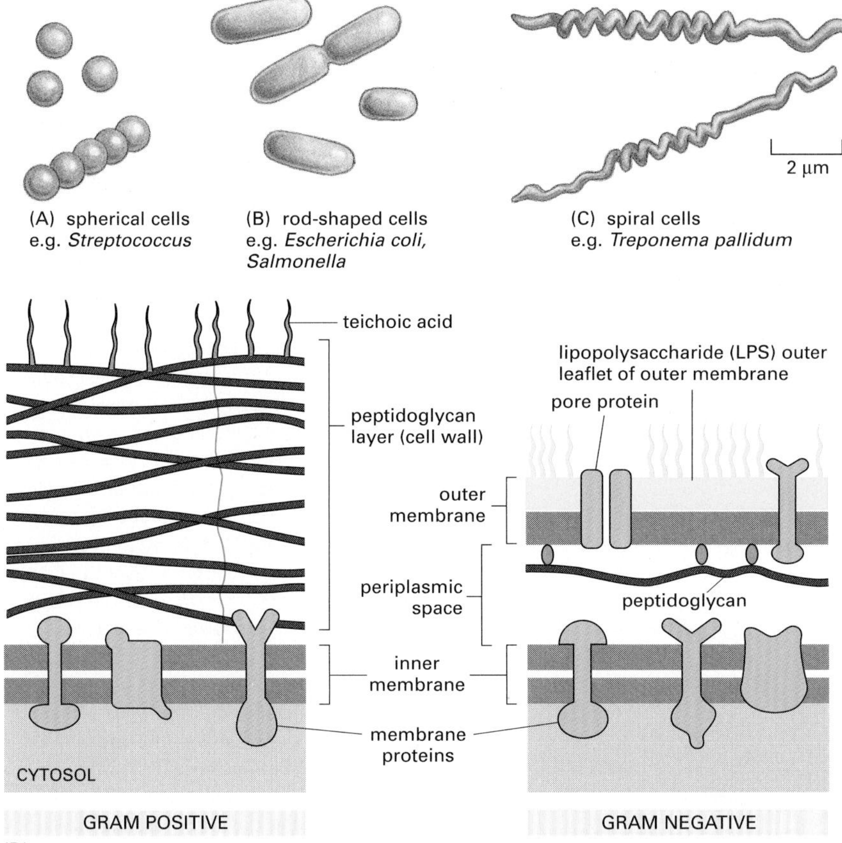

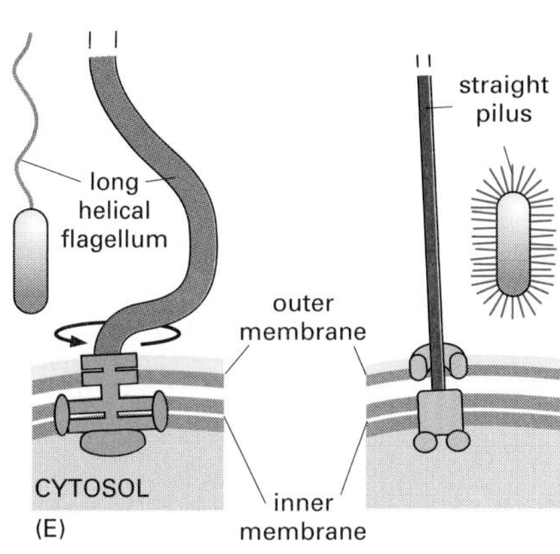

Figure 25–4 Bacterial shapes and cell-surface structures. Bacteria are classified into three different shapes: (A) spheres (cocci), (B) rods (bacilli), and (C) spiral cells (spirochetes). (D) They are also classified as *Gram-positive* or *Gram-negative*. Bacteria such as *Streptococci* and *Staphylococci* have a single membrane and a thick cell wall made of cross-linked *peptidoglycan*. They retain the violet dye used in the Gram staining procedure and are thus called Gram-positive. Gram-negative bacteria such as *E. coli* and *Salmonella* have two membranes, separated by a *periplasmic space* (see Figure 11–17). The peptidoglycan layer in the cell wall of these organisms is located in the periplasmic space and is thinner than in Gram-positives; they therefore fail to retain the dye in the Gram staining procedure. The inner membrane of Gram-negative bacteria is a phospholipid bilayer, and the inner leaflet of the outer membrane is also made primarily of phospholipids; the outer leaflet of the outer membrane, however, is composed of a unique glycosylated lipid called *lipopolysaccharide (LPS)* (see Figure 25–40). (E) Cell-surface projections are important for bacterial behavior. Many bacteria swim using the rotation of helical flagella (see Figure 15–68). The bacterium illustrated has only a single flagellum at one pole; others such as *E. coli* are decorated with multiple flagella all over the surface. Straight *pili* (also called *fimbriae*) are used to adhere to surfaces in the host and to facilitate genetic exchange between bacteria. Both flagella and pili are anchored to the cell surface by large multiprotein complexes.

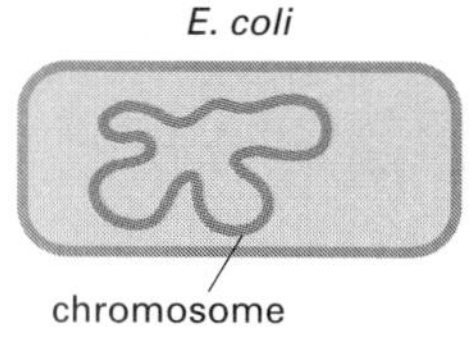

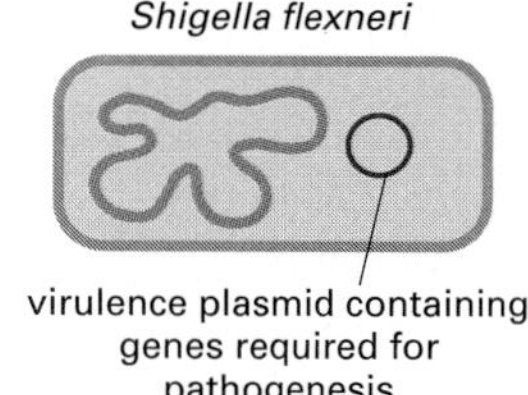

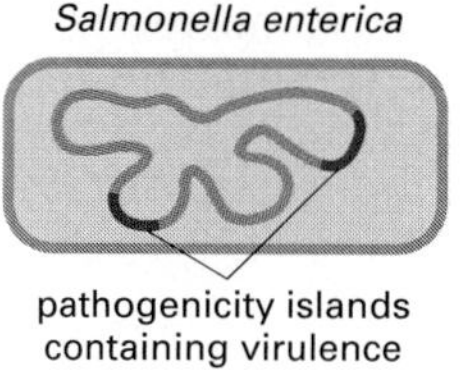

Figure 25–5 Genetic differences between pathogens and nonpathogens. Nonpathogenic *Escherichia coli* has a single circular chromosome. *E. coli* is very closely related to two types of food-borne pathogens—*Shigella flexneri*, which causes dysentery, and *Salmonella enterica*, a common cause of food poisoning. If these three organisms were being named today based on molecular techniques, they would be classified in the same genus, if not the same species. The chromosome of *S. flexneri* differs from that of *E. coli* at only a few loci; most of the genes required for pathogenesis are carried on an extrachromosomal virulence plasmid. The chromosome of *S. enterica* carries two large inserts (pathogenicity islands) not found in the *E. coli* chromosome; these inserts each contain many virulence genes.

B subunit transfers the A subunit through the membrane into the epithelial cell cytoplasm. The A subunit is an enzyme that catalyzes the transfer of an ADP-ribose moiety from NAD to the trimeric G protein G_S, which normally activates adenylyl cyclase to make cyclic AMP (discussed in Chapter 15). ADP-ribosylation of the G protein results in an overaccumulation of cyclic AMP and an ion imbalance, leading to the massive watery diarrhea associated with cholera. The infection is then spread by the fecal-oral route by contaminated food and water.

Some pathogenic bacteria use several independent mechanisms to cause toxicity to the cells of their host. *Anthrax,* for example, is an acute infectious disease of sheep, cattle, other herbivores, and occasionally humans. It is usually caused by contact with spores of the Gram-positive bacterium, *Bacillus anthracis.* Unlike cholera, anthrax has never been observed to spread directly from one infected person to another. Dormant spores can survive in soil for long periods of time and are highly resistant to adverse environmental conditions, including heat, ultraviolet and ionizing radiation, pressure, and chemical agents. After the spores are inhaled, ingested, or rubbed into breaks in the skin, the spores germinate, and the bacteria begin to replicate. Growing bacteria secrete two toxins, called **lethal toxin** and **edema toxin**. Either toxin alone is sufficient to cause signs of infection. Like the A and B subunits of cholera toxin, both toxins are made of two subunits. The B subunit is identical between lethal toxin and edema toxin, and it binds to a host cell-surface receptor to transfer the two different A subunits into host cells. The A subunit of edema toxin is an adenylyl cyclase that directly converts host cell ATP into cyclic AMP. This causes an ion imbalance that can lead to accumulation of extracellular fluid *(edema)* in the infected skin or lung. The A subunit of lethal toxin is a zinc protease that cleaves several members of the MAP kinase kinase family (discussed in Chapter 15). Injection of lethal toxin into the bloodstream of an animal causes shock and death. The molecular mechanisms and the sequence of events leading to death in anthrax remain uncertain.

These examples illustrate a common theme among virulence factors. They are frequently either toxic proteins *(toxins)* that directly interact with important host structural or signaling proteins to elicit a host cell response that is beneficial to pathogen colonization or replication, or they are proteins that are needed to deliver such toxins to their host cell targets. One common and particularly

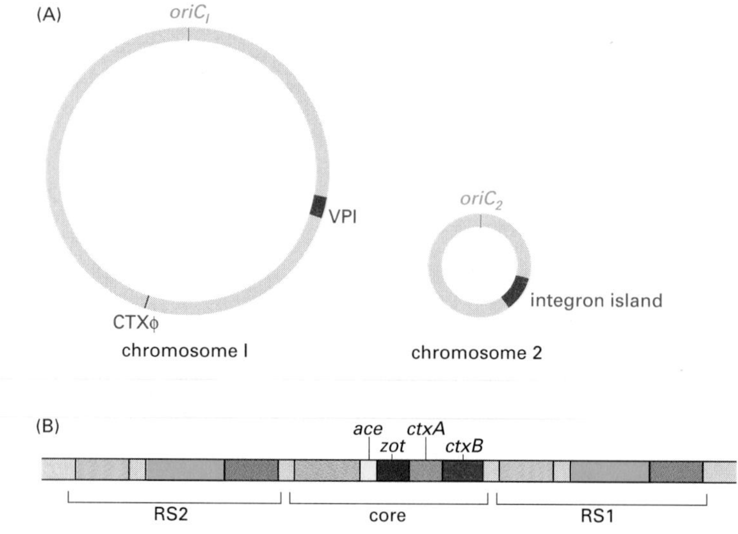

Figure 25–6 Genetic organization of *Vibrio cholerae*. (A) *Vibrio cholerae* is unusual in having two circular chromosomes rather than one. The two chromosomes have distinct origins of replication ($oriC_1$ and $oriC_2$). Three loci in pathogenic strains of *V. cholerae* are absent in nonpathogenic strains and appear to have been acquired relatively recently. CTXϕ on chromosome 1 is an integrated bacteriophage genome that carries the genes for cholera toxin. The pathogenicity island VPI on chromosome 1 includes genes for factors required for intestinal colonization. The *integron island* on chromosome 2 is a structure that enables the sequential acquisition of novel genes by facilitating the insertion of newly acquired DNA fragments downstream of a strong transcriptional promoter. Although this integron island has not yet been shown to be required for virulence in *V. cholerae*, similar integron islands in many other pathogens contain virulence genes, as well as genes involved in antibiotic resistance. (B) Map of the CTXϕ locus. The genes encoding the two subunits of cholera toxin are ctxA and ctxB. Other genes in the core region *(ace* and *zot)* are also involved in virulence. The two repeated flanking sequences RS2 and RS1 are involved in the chromosomal insertion of the bacteriophage genome.

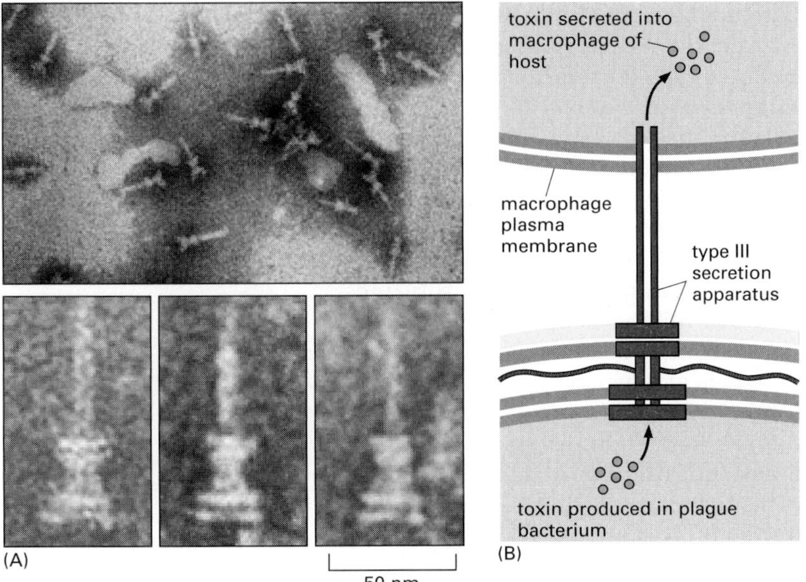

Figure 25–7 Type III secretion systems that can deliver virulence factors into the cytoplasm of host cells. (A) Electron micrographs of purified type III apparatuses. About two dozen proteins are necessary to make the complete structure, which is seen in the three enlarged micrographs below. The large lower ring is embedded in the inner membrane, and the smaller upper ring is embedded in the outer membrane. The long projection at the top is a hollow tube, through which secreted proteins can travel. (B) During infection, contact of the tube tip with the plasma membrane of a host cell triggers secretion. Here, the plague bacillus *Yersinia pestis* delivers toxins to a macrophage. (A, from K. Tamano et al., *EMBO J.* 19:3876–3887, 2000, by permission of Oxford University Press.)

efficient delivery mechanism, called the **type III secretion system**, acts like a tiny syringe that injects toxic proteins from the cytoplasm of an extracellular bacterium directly into the cytoplasm of an adjacent host cell (Figure 25–7). There is a remarkable degree of structural similarity between the type III syringe and the base of a bacterial flagellum (see Figure 15–67), and many of the proteins in the two structures are clearly homologous.

Because bacteria form a kingdom distinct from the eucaryotes they infect (see Figure 25–3), much of their basic machinery for DNA replication, transcription, translation, and fundamental metabolism is quite different from that of their host. These differences enable us to find antibacterial drugs that specifically inhibit these processes in bacteria, without disrupting them in the host. Most of the **antibiotics** that we use to treat bacterial infections are small molecules that inhibit macromolecular synthesis in bacteria by targeting bacterial enzymes that are either distinct from their eucaryotic counterparts or that are involved in pathways, such as cell wall biosynthesis, that are absent in humans (Figure 25–8 and Table 6–3).

Fungal and Protozoan Parasites Have Complex Life Cycles with Multiple Forms

Pathogenic fungi and protozoan parasites are eucaryotes. It is therefore more difficult to find drugs that will kill them without killing the host. Consequently, antifungal and antiparasitic drugs are often less effective and more toxic than antibiotics. A second characteristic of fungal and parasitic infections that makes them difficult to treat is the tendency of the infecting organisms to switch among several different forms during their life cycles. A drug that is effective at killing one form is often ineffective at killing another form, which therefore survives the treatment.

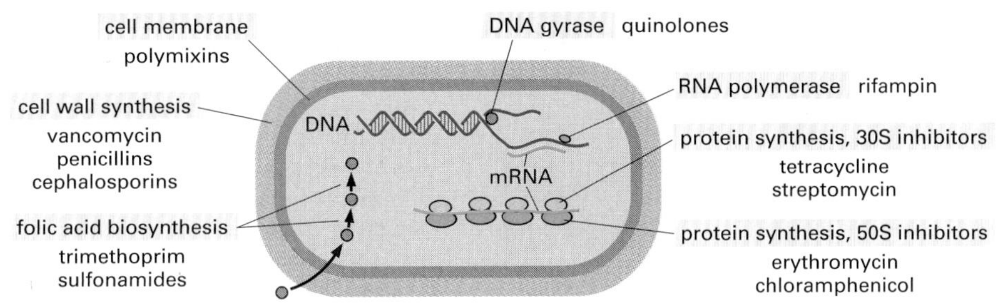

Figure 25–8 Antibiotic targets. Despite the large number of antibiotics available, they have a narrow range of targets, which are highlighted in *yellow*. A few representative antibiotics in each class are listed. All antibiotics used to treat human infections fall into one of these categories. The vast majority inhibit either bacterial protein synthesis or bacterial cell wall synthesis.

The **fungal** branch of the eucaryotic kingdom includes both unicellular *yeasts* (such as *Saccharomyces cerevisiae* and *Schizosaccharomyces pombe)* and filamentous, multicellular *molds* (like those found on moldy fruit or bread). Most of the important pathogenic fungi exhibit *dimorphism*—the ability to grow in either yeast or mold form. The yeast-to-mold or mold-to-yeast transition is frequently associated with infection. *Histoplasma capsulatum,* for example, grows as a mold at low temperature in the soil, but it switches to a yeast form when inhaled into the lung, where it can cause the disease histoplasmosis (Figure 25–9).

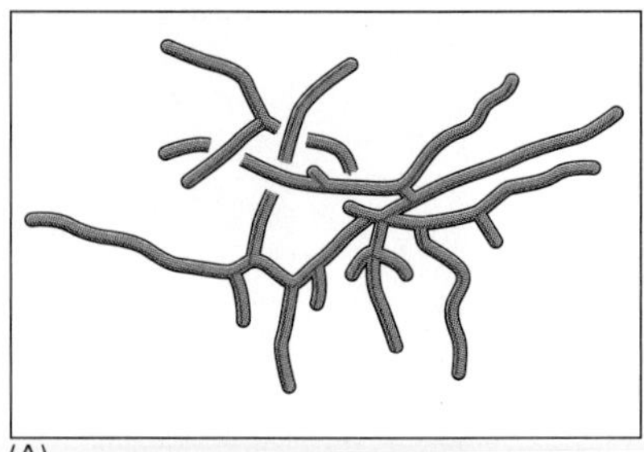

(A)

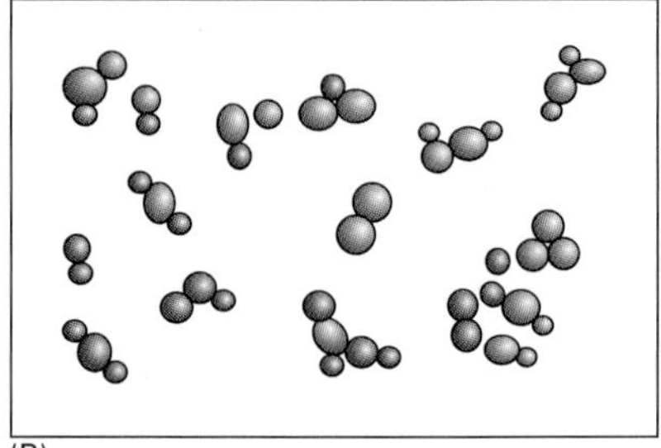

(B)

Figure 25–9 Dimorphism in the pathogenic fungus *Histoplasma capsulatum*. (A) At low temperature in the soil, *Histoplasma* grows as a filamentous fungus. (B) After being inhaled into the lung of a mammal, *Histoplasma* undergoes a morphological switch triggered by the change in temperature. In this yeast-like form, it closely resembles *Saccharomyces cerevisiae*.

Protozoan parasites have more elaborate life cycles than do fungi. These cycles frequently require the services of more than one host. **Malaria** is the most common protozoal disease, infecting 200–300 million people every year and killing 1–3 million of them. It is caused by four species of *Plasmodium,* which are transmitted to humans by the bite of the female of any of 60 species of *Anopheles* mosquito. *Plasmodium falciparum*—the most intensively studied of the malaria-causing parasites—exists in no fewer than eight distinct forms, and it requires both the human and mosquito hosts to complete its sexual cycle (Figure 25–10). Gametes are formed in the bloodstream of infected humans, but they can only fuse to form a zygote in the gut of the mosquito. Three of the *Plasmodium* forms are highly specialized to invade and replicate in specific tissues—the insect gut lining, the human liver, and the human red blood cell.

Because malaria is so widespread and devastating, it has acted as a strong selective pressure on human populations in areas of the world that harbor the *Anopheles* mosquito. *Sickle cell anemia,* for example, is a recessive genetic disorder caused by a point mutation in the gene that encodes the hemoglobin β

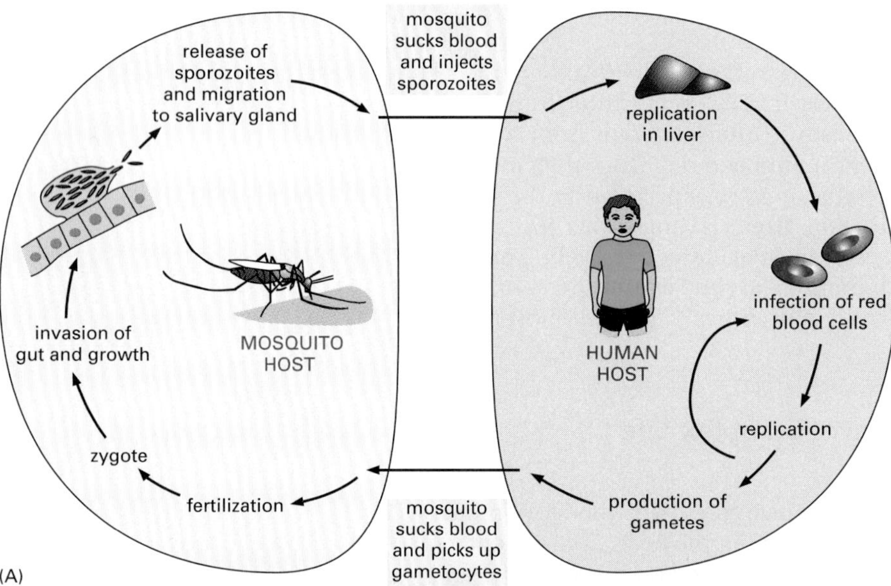

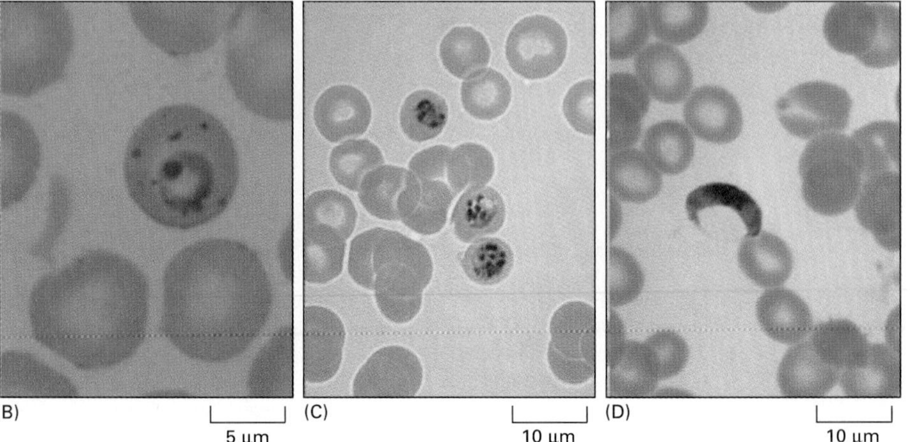

Figure 25–10 The complex life cycle of malaria. (A) The sexual cycle of *Plasmodium falciparum* requires passage between a human host and an insect host. (B)–(D) Blood smears from people infected with malaria, showing three different forms of the parasite that appear in red blood cells: (B) trophozoite (ring form); (C) schizont; (D) gametocyte. (Micrographs courtesy of the Centers for Disease Control, Division of Parasitic Diseases, DPDx.)

chain, and it is common in areas of Africa with a high incidence of the most serious form of malaria (caused by *Plasmodium falciparum*). The malarial parasites grow poorly in red blood cells from either homozygous sickle cell patients or healthy heterozygous carriers, and, as a result, malaria is seldom found among carriers of this mutation. For this reason, malaria has maintained the sickle cell mutation at high frequency in these regions of Africa.

Viruses Exploit Host Cell Machinery for All Aspects of Their Multiplication

Bacteria, fungi, and eucaryotic parasites are cells themselves. Even when they are obligate parasites, they use their own machinery for DNA replication, transcription, and translation, and they provide their own sources of metabolic energy. **Viruses**, by contrast, are the ultimate hitchhikers, carrying little more than information in the form of nucleic acid. The information is largely replicated, packaged, and preserved by the host cells (Figure 25–11). Viruses have a small genome, made up of a single nucleic acid type—either DNA or RNA—which, in either case, may be single-stranded or double-stranded. The genome is packaged in a protein coat, which in some viruses is further enclosed by a lipid envelope.

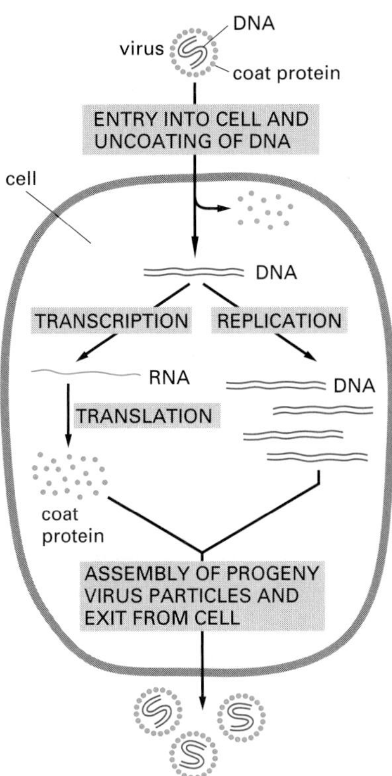

Figure 25–11 A simple viral life cycle. The hypothetical virus shown consists of a small double-stranded DNA molecule that codes for only a single viral capsid protein. No known virus is this simple.

Viruses replicate in various ways. In general, replication involves (1) disassembly of the infectious virus particle, (2) replication of the viral genome, (3) synthesis of the viral proteins by the host cell translation machinery, and (4) reassembly of these components into progeny virus particles. A single virus particle (a *virion*) that infects a single host cell can produce thousands of progeny in the infected cell. Such prodigious viral multiplication is often enough to kill the host cell: the infected cell breaks open (lyses) and thereby allows the progeny viruses access to nearby cells. Many of the clinical manifestations of viral infection reflect this *cytolytic effect* of the virus. Both the cold sores formed by *herpes simplex* virus and the lesions caused by the *smallpox* virus, for example, reflect the killing of the epidermal cells in a local area of infected skin.

Viruses come in a wide variety of shapes and sizes, and, unlike cellular life forms, they cannot be systematically classified by their relatedness into a single phylogenetic tree. Because of their tiny sizes, complete genome sequences have been obtained for nearly all clinically important viruses. *Poxviruses* are among the largest, up to 450 nm long, which is about the size of some small bacteria. Their genome of double-stranded DNA consists of about 270,000 nucleotide pairs. At the other end of the size scale are parvoviruses, which are less than 20 nm long and have a single-stranded DNA genome of under 5000 nucleotides (Figure 25–12). The genetic information in a virus can be carried in a variety of unusual nucleic acid forms (Figure 25–13).

The **capsid** that encloses the viral genome is made of one or several proteins, arranged in regularly repeating layers and patterns. In *enveloped viruses*, the capsid itself is enclosed by a lipid bilayer membrane that is acquired in the process of budding from the host cell plasma membrane (Figure 25–14). Whereas *nonenveloped viruses* usually leave an infected cell by lysing it, an enveloped virus can leave the cell by budding, without disrupting the plasma membrane and, therefore, without killing the cell. These viruses can cause chronic infections, and some can help transform an infected cell into a cancer cell.

Despite this variety, all viral genomes encode three types of proteins: proteins for replicating the genome, proteins for packaging the genome and delivering it to more host cells, and proteins that modify the structure or function of the host cell to suit the needs of the virus (Figure 25–15). In the second section of this chapter, we focus primarily on this third class of viral proteins.

Since most of the critical steps in viral replication are performed by host cell machinery, the identification of effective antiviral drugs is particularly problematic. Whereas the antibiotic tetracycline specifically poisons bacterial ribosomes, for example, it will not be possible to find a drug that specifically poisons viral ribosomes, as viruses use the ribosomes of the host cell to make their proteins. The best strategy for containing viral diseases is to prevent them by

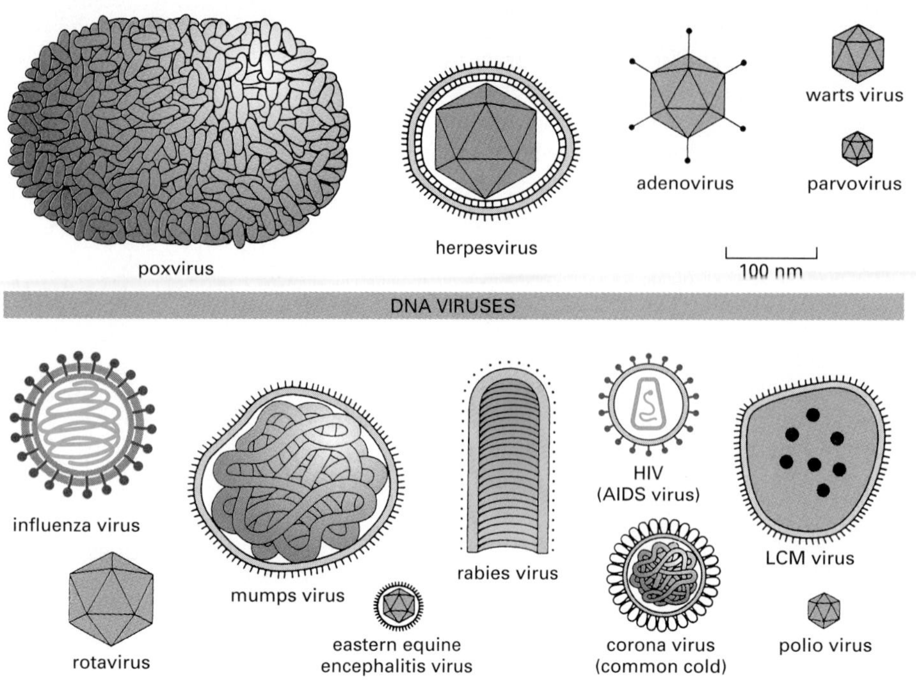

Figure 25–12 Examples of viral morphology. As shown, viruses vary greatly in both size and shape.

vaccination of the potential hosts. Highly successful vaccination programs have effectively eliminated smallpox from the planet, and the eradication of poliomyelitis is imminent (Figure 25–16).

Prions Are Infectious Proteins

All information in biological systems is encoded by structure. We are used to thinking of biological information in the form of nucleic acid sequences (as in our description of viral genomes), but the sequence itself is a shorthand code for describing nucleic acid structure. The replication and expression of the information encoded in DNA and RNA are strictly dependent on the structure of these nucleic acids and their interactions with other macromolecules. The propagation of genetic information primarily requires that the information be stored in a structure that can be duplicated from unstructured precursors. Nucleic acid sequences are the simplest and most robust solution that organisms have found to the problem of faithful structural replication.

Nucleic acids are not the only solution, however. **Prions** are infectious agents that are replicated in the host by copying an aberrant protein structure.

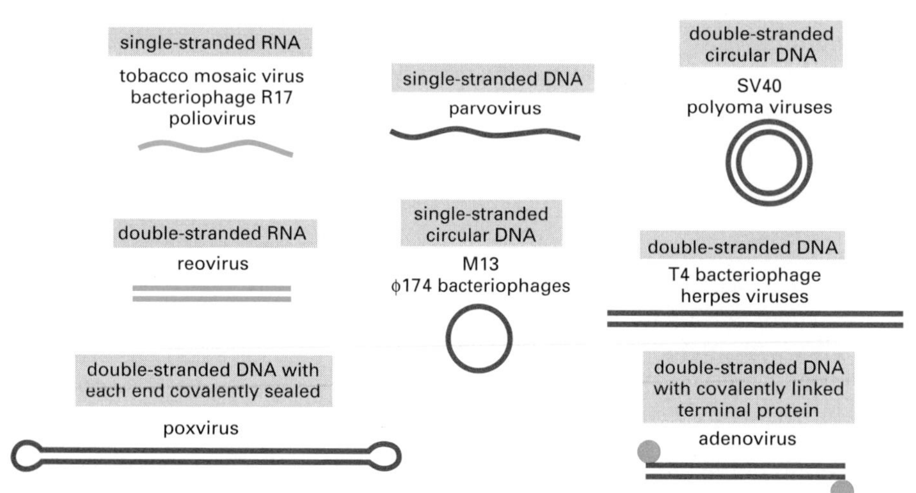

Figure 25–13 Schematic drawings of several types of viral genomes. The smallest viruses contain only a few genes and can have an RNA or a DNA genome. The largest viruses contain hundreds of genes and have a double-stranded DNA genome. The peculiar ends (as well as the circular forms) overcome the difficulty of replicating the last few nucleotides at the end of a DNA strand (discussed in Chapter 5).

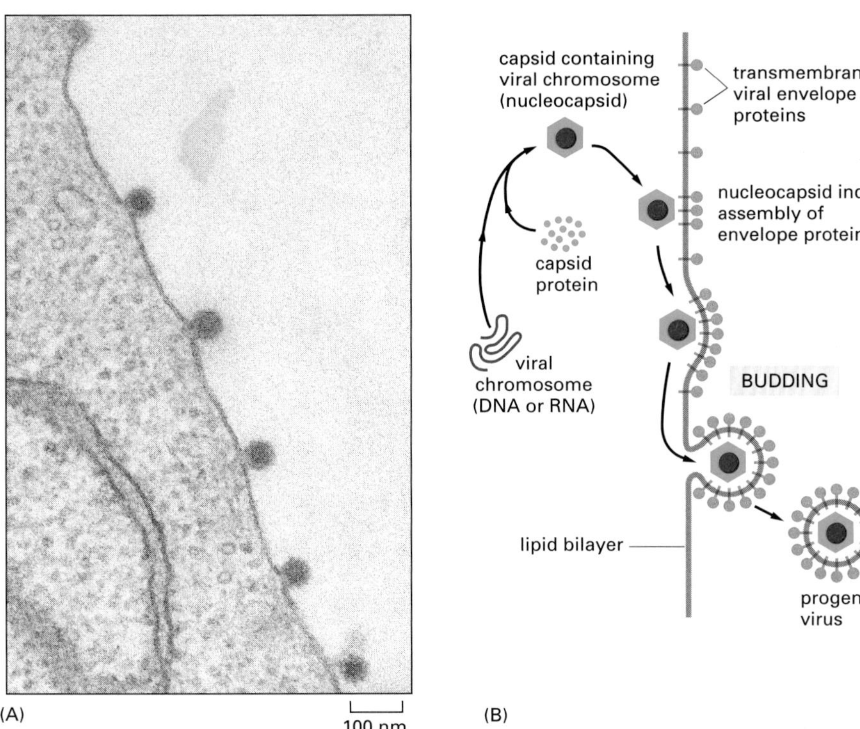

Figure 25–14 Acquisition of a viral envelope. (A) Electron micrograph of an animal cell from which six copies of an enveloped virus (*Semliki forest virus*) are budding. (B) Schematic view of the envelope assembly and budding processes. The lipid bilayer that surrounds the viral capsid is derived directly from the plasma membrane of the host cell. In contrast, the proteins in this lipid bilayer (shown in *green*) are encoded by the viral genome. (A, courtesy of M. Olsen and G. Griffith.)

They can occur in yeasts, and they cause various neurodegenerative diseases in mammals. The most well-known infection caused by prions is *bovine spongiform encephalopathy* (BSE, or mad cow disease), which occasionally spreads to humans who eat infected parts of the cow (Figure 25–17). Isolation of the infectious prions that cause the disease *scrapie* in sheep, followed by years of painstaking laboratory characterization of scrapie-infected mice, eventually established that the protein itself is infectious.

Intriguingly, the infectious prion protein is made by the host, and its amino acid sequence is identical to a normal host protein. Moreover, the prion and normal forms of the protein are indistinguishable in their posttranslational modifications. The only difference between them appears to be in their folded three-dimensional structure. The misfolded prion protein tends to aggregate, and it has the remarkable capacity to cause the normal protein to adopt its misfolded prion conformation and thereby to become infectious (see Figure 6–89). This ability of the prion to convert the normal host protein to misfolded prion protein is equivalent to the prion's having replicated itself in the host. If eaten by another susceptible host, these newly-misfolded prions can transmit the infection.

It is not known how normal proteins are usually able to find the single, correct, folded conformation, among the billions of other possibilities, without becoming stuck in dead-end intermediates (discussed in Chapters 3 and 6). Prions are a good example of how protein folding can go dangerously wrong. But, why are the prion diseases so uncommon? What are the constraints that determine whether a misfolded protein will behave like a prion, or simply get refolded

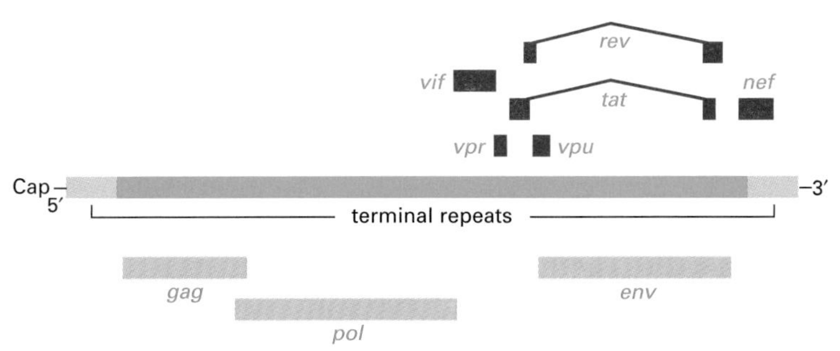

Figure 25–15 A map of the HIV genome. This retroviral genome consists of about 9000 nucleotides and contains nine genes, the locations of which are shown in *green* and *red*. Three of the genes *(green)* are common to all retroviruses: *gag* encodes capsid proteins, *env* encodes envelope proteins, and *pol* encodes both the reverse transcriptase and the integrase proteins (discussed in Chapter 5). The HIV genome is unusually complex, because it contains six small genes *(red)*, in addition to the three large genes *(green)* normally required for the retrovirus life cycle. At least some of these small genes encode proteins that regulate viral gene expression *(tat* and *rev*—see Figure 7–97); others encode proteins that modify host cell processes, including protein trafficking *(vpu* and *nef)* and progression through the cell cycle *(vpr)*. As indicated by the *red* lines, RNA splicing (using the host cell spliceosome) is required to produce the Rev and Tat proteins.

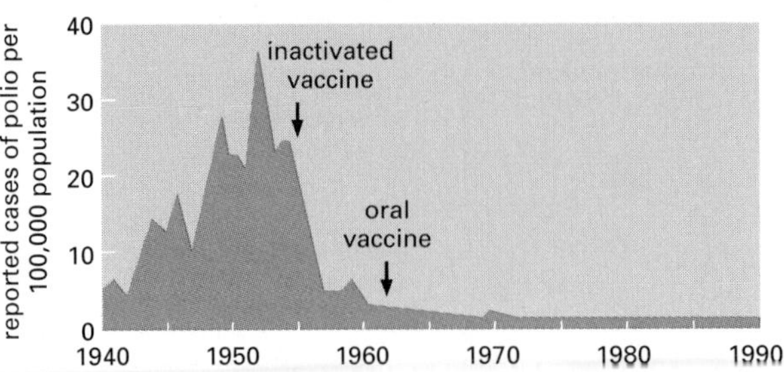

Figure 25–16 Eradication of a viral disease through vaccination. The graph shows number of cases of poliomyelitis reported per year in the United States. The *arrows* indicate the timing of the introduction of the Salk vaccine (inactivated virus given by injection) and the Sabin vaccine (live attenuated virus given orally).

or degraded by the cell that made it? We do not yet have answers to these questions, and the study of prions remains an area of intense research.

Summary

Infectious diseases are caused by pathogens, which include bacteria, fungi, protozoa, worms, viruses, and even infectious proteins called prions. Pathogens of all classes must have mechanisms for entering their host and for evading immediate destruction by the host immune system. Most bacteria are not pathogenic. Those that are contain specific virulence genes that mediate interactions with the host, eliciting particular responses from the host cells that promote the replication and spread of the pathogen. Pathogenic fungi, protozoa, and other eucaryotic parasites typically pass through several different forms during the course of infection; the ability to switch among these forms is usually required for the parasites to be able to survive in a host and cause disease. In some cases, such as malaria, parasites must pass sequentially through several host species to complete their life cycles. Unlike bacteria and eucaryotic parasites, viruses have no metabolism of their own and no intrinsic ability to produce the proteins encoded by their DNA or RNA genomes. They rely entirely on subverting the machinery of the host cell to produce their proteins and to replicate their genomes. Prions, the smallest and simplest infectious agents, contain no nucleic acid; instead, they are rare, aberrantly folded proteins that happen to catalyze the misfolding of proteins in the host that share their primary amino acid sequence.

CELL BIOLOGY OF INFECTION

We have just seen that pathogens constitute a diverse set of agents. There are correspondingly diverse ranges of mechanisms by which pathogens cause disease. But the survival and success of all pathogens require that they colonize the host, reach an appropriate niche, avoid host defenses, replicate, and exit the infected host to spread to an uninfected one. In this section, we examine the common strategies that are used by many pathogens to accomplish these tasks.

Pathogens Cross Protective Barriers to Colonize the Host

The first step in infection is for the pathogen to colonize the host. Most parts of the human body are well-protected from the environment by a thick and fairly tough covering of skin. The protective boundaries in some other human tissues (eyes, nasal passages and respiratory tract, mouth and digestive tract, urinary tract, and female genital tract) are less robust. For example, in the lungs and small intestine where oxygen and nutrients, respectively, are absorbed from the environment, the barrier is just a single monolayer of epithelial cells.

Skin and many other barrier epithelial surfaces are usually densely populated by normal flora. Some bacterial and fungal pathogens also colonize these surfaces and attempt to outcompete the normal flora, but most of them (as well as all viruses) avoid such competition by crossing these barriers to gain access to unoccupied niches within the host.

Wounds in barrier epithelia, including the skin, allow pathogens direct access to the interior of the host. This avenue of entry requires little in the way of specialization on the part of the pathogen. Indeed, many members of the

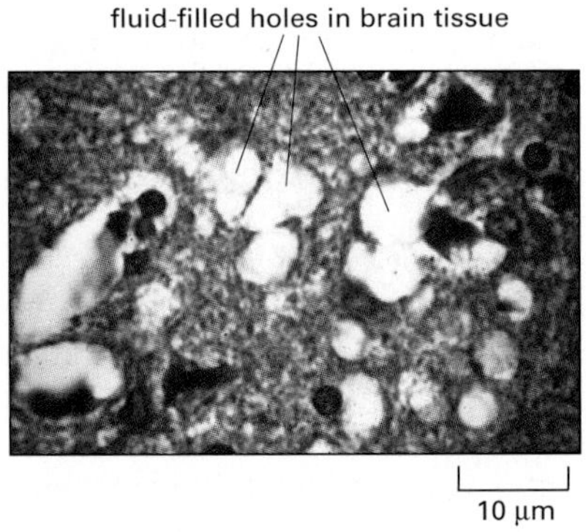

fluid-filled holes in brain tissue

10 μm

Figure 25–17 Neural degeneration in a prion infection. This micrograph shows a slice from the brain of a person who died of kuru. Kuru is a human prion disease, very similar to BSE, that was spread from one person to another by ritual mortuary practices in New Guinea. The large fluid-filled holes are places where neurons have died. These characteristic holes give the syndrome the name of spongiform encephalopathy. (Courtesy of Gary Baumbach.)

normal flora can cause serious illness if they enter through such wounds. Anaerobic bacteria of the genus *Bacteroides*, for example, are carried as harmless flora at very high density in the large intestine, but they can cause life-threatening peritonitis if they enter the peritoneal cavity through a perforation in the intestine caused by trauma, surgery, or infection in the intestinal wall. *Staphylococcus* from the skin and nose, or *Streptococcus* from the throat and mouth, are also responsible for many serious infections resulting from breaches in epithelial barriers.

Dedicated pathogens, however, need not wait for a well-timed wound to allow them access to their host. A particularly efficient way for a pathogen to cross the skin is to catch a ride in the saliva of a biting insect. Many arthropods nourish themselves by sucking blood, and a diverse group of bacteria, viruses, and protozoa have developed the ability to survive in the arthropod so that they can use these biting animals as *vectors* to spread from one mammalian host to another. As discussed earlier, the *Plasmodium* protozoan that causes malaria develops through several forms in its life cycle, including some that are specialized for survival in a human and some that are specialized for survival in a mosquito (see Figure 25–10). Viruses that are spread by insect bites include the causative agents for several types of hemorrhagic fever, including yellow fever and Dengue fever, as well as the causative agents for many kinds of viral encephalitis (inflammation of the brain). All these viruses have acquired the ability to replicate in both insect cells and mammalian cells, as is required for a virus to be transmitted by an insect vector. Bloodborne viruses such as HIV that are not capable of replicating in insect cells are rarely, if ever, spread from insect to human.

The efficient spread of a pathogen via an insect vector requires that individual insects consume blood meals from numerous mammalian hosts. In a few striking cases, the pathogen appears to alter the behavior of the insect so that its transmission is more likely. Like most animals, the tsetse fly (whose bite spreads the protozoan parasite *Trypanosoma brucei* which causes sleeping sickness in Africa) stops eating when it is satiated. But tsetse flies carrying trypanosomes bite much more frequently and ingest more blood than do uninfected flies. The presence of trypanosomes impairs the function of the insect mechanoreceptors that measure blood flow through the gullet to assess the fullness of the stomach, effectively fooling the tsetse fly into thinking that it is still hungry. The bacterium *Yersinia pestis*, which causes bubonic plague, uses a different mechanism to ensure that a flea carrying it bites repeatedly: it multiplies in the flea's foregut to form aggregated masses that eventually enlarge and physically block the digestive tract. The insect is then unable to feed normally and begins to starve. During repeated attempts to satisfy its appetite, some of the bacteria in the foregut are flushed into the bite site, thus transmitting plague to a new host (Figure 25–18).

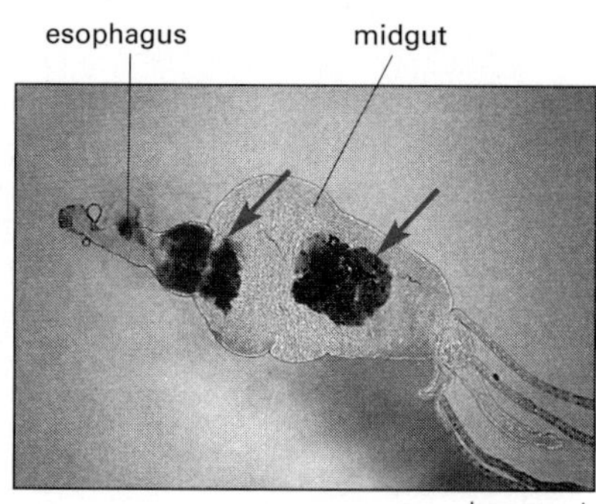

Figure 25–18 The spread of plague. This micrograph shows the digestive tract dissected from a flea that had dined about two weeks previously on the blood of an animal infected with the plague bacterium, *Yersinia pestis*. The bacteria multiplied in the flea gut to produce large cohesive aggregates, indicated by the *red* arrows; the bacterial mass on the left is occluding the passage between the esophagus and the midgut. This type of blockage prevents a flea from digesting its blood meals, thereby causing it to bite repeatedly, disseminating the infection. (From B.J. Hinnebusch, E.R. Fischer and T.G. Schwann, *J. Infect. Dis.* 178:1406–1415, 1998. © The University of Chicago Press.)

Pathogens That Colonize Epithelia Must Avoid Clearance by the Host

Hitching a ride through the skin on an insect proboscis is just one strategy that pathogens use to pass through the initial barriers of host defense. Whereas many barrier zones such as the skin, mouth, and large intestine, are densely populated by normal flora, others including the lower lung, the small intestine, and the bladder, are normally kept nearly sterile, despite relatively direct access to the environment. The epithelium in these zones actively resists bacterial colonization. As discussed in Chapter 22, the respiratory epithelium is covered with a layer of protective mucus, and the coordinated beating of cilia sweeps the mucus and trapped bacteria and debris up and out of the lung. The epithelia lining the bladder and the upper gastrointestinal tract also have a thick layer of mucus, and these organs are periodically flushed by micturition and peristalsis, respectively, to wash away undesirable microbes. The pathogenic bacteria and parasites that infect these epithelial surfaces have specific mechanisms for overcoming these host cleaning mechanisms. Those that infect the urinary tract, for

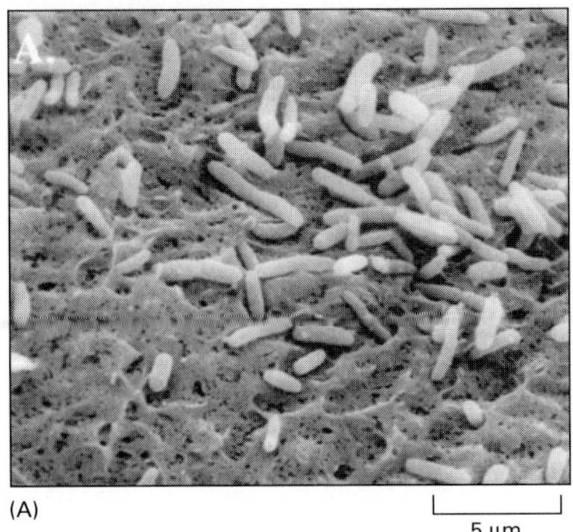

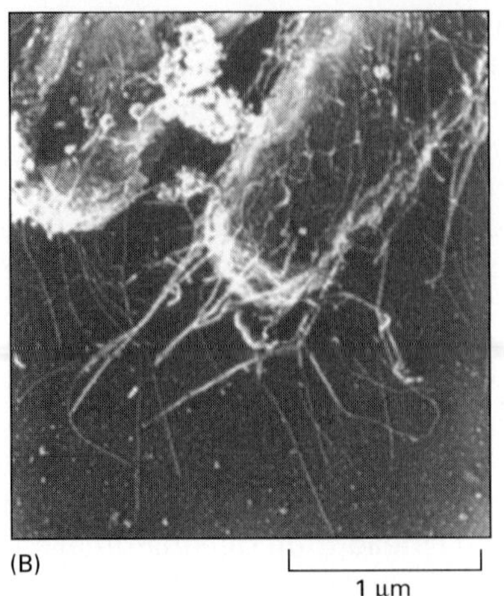

Figure 25–19 Uropathogenic *E. coli* and P pili. (A) Scanning electron micrograph of uropathogenic *E. coli*, a common cause of bladder and kidney infections, attached to the surface of epithelial cells in the bladder of an infected mouse. (B) A close-up view of the bacteria shows the P pili on the surface. (A, from G.E. Soto and S.J. Hultgren, *J. Bact.* 181:1059–1071, 1999; B, courtesy of D.G. Thanassi and S.J. Hultgren, *Meth. Comp. Meth. Enzym.* 20:111–126, 2000. © Academic Press.)

example, resist the washing action of urine by adhering tightly to the bladder epithelium via specific **adhesins**, proteins or protein complexes that recognize and bind to host cell-surface molecules. An important group of adhesins in uropathogenic *E. coli* strains are components of the *P pili* (see Figure 25–4E) that help the bacteria infect the kidney. These surface projections can be several micrometers long and are thus able to span the thickness of the protective mucus layer. At the tip of each pilus is a protein that binds tightly to a particular disaccharide linked to a glycolipid that is found on the surface of cells in the bladder and kidney (Figure 25–19)

One of the hardest organs for a microbe to colonize is the stomach. Besides peristaltic washing and protection by a thick layer of mucus, the stomach is filled with acid (average pH ~2). This extreme environment is lethal to almost all bacteria ingested in food. Nonetheless, it is colonized by the hardy and enterprising bacterium *Helicobacter pylori*, which was recognized only recently as a major causative agent of stomach ulcers and possibly stomach cancer. Although the older treatments for ulcers (acid-reducing drugs and bland diets) are still used to reduce inflammation, a short and relatively cheap course of antibiotics can now effectively cure a patient of recurrent stomach ulcers. The hypothesis that stomach ulcers could be caused by a persistent bacterial infection of the stomach lining initially met with considerable skepticism. The point was finally proven by the young Australian doctor who made the initial discovery: he drank a flask of a pure culture of *H. pylori* and developed a typical ulcer. One way that *H. pylori* survives in the stomach is by producing the enzyme *urease*, which converts urea to ammonia and carbon dioxide; in this way, the bacterium surrounds itself with a layer of ammonia, which neutralizes stomach acid in its immediate vicinity. The bacteria also express at least five types of adhesins, which enable them to adhere to the stomach epithelium, and they produce several cytotoxins that destroy the stomach epithelial cells, creating painful ulcers. The resulting chronic inflammation promotes cell proliferation and thus predisposes the infected individual to stomach cancer.

A more extreme example of active colonization is provided by *Bordetella pertussis*, the bacterium that causes whooping cough. The first step in a *B. pertussis* infection is colonization of the respiratory epithelium. The bacteria circumvent the normal clearance mechanism (the *mucociliary escalator* described in Chapter 22) by binding tightly to the surface of ciliated cells and multiplying on them. *B. pertussis* expresses at least four types of adhesins that bind tightly to particular glycolipids on the ciliated cells. The adherent bacteria produce various toxins which eventually kill the ciliated cells, compromising the host's ability to clear the infection. The most familiar of these is *pertussis toxin*, which—like cholera toxin—is an ADP-ribosylating enzyme. It ADP-ribosylates the α subunit of the G protein G_i, causing deregulation of the host cell adenylyl cyclase and overproduction of cyclic AMP (discussed in Chapter 15). Not content with this, *B. pertussis* also produces an adenylyl cyclase of its own, which is inactive unless bound to the eucaryotic Ca^{2+}-binding protein calmodulin. The bacterially-produced enzyme is therefore active only in the cytoplasm of a eucaryotic cell. Although both *B. pertussis* and *V. cholerae* have the similar effect of drastically raising cAMP levels in the host cells to which they adhere, the symptoms of the diseases are very different because the two bacteria colonize different sites in the host: *B. pertussis* colonizes the respiratory tract and causes paroxysmal coughing, whereas *V. cholerae* colonizes the gut and causes watery diarrhea.

Not all examples of specific colonization require that the bacterium express adhesins that bind to host cell glycolipids or proteins. Enteropathogenic *E. coli*, which causes diarrhea in young children, instead uses a type III secretion system (see Figure 25–7) to deliver its own bacterially-expressed receptor protein (called Tir) into its host cell (Figure 25–20A). After Tir is inserted into the host cell membrane, a bacterial surface protein binds to the extracellular domain of Tir, triggering a remarkable series of events inside the host cell. First, the Tir receptor protein is phosphorylated on tyrosine residues by a host protein tyrosine kinase. Unlike eucaryotic cells, bacteria generally do not phosphorylate tyrosine residues, yet Tir contains a peptide domain that is a specific recognition motif for a eucaryotic tyrosine kinase. The phosphorylated Tir then is thought to recruit a member of the Rho family of small GTPases, which promotes actin polymerization through a series of intermediate steps (discussed in Chapter 16). The polymerized actin forms a unique cell surface protrusion, called a *pedestal*, that pushes the tightly adherent bacteria up about 10 μm from the host cell surface (Figure 25–20B, C).

These examples of host colonization illustrate the importance of host–pathogen communication in the infection process. Pathogenic organisms have acquired genes that encode proteins that interact specifically with particular molecules of the host cells. In some cases, such as the *B. pertussis* adenylyl cyclase, an ancestor of the pathogen may have acquired the gene from its host, whereas in others, such as Tir, random mutation may have given rise to protein motifs that are recognized by a eucaryotic protein partner.

Intracellular Pathogens Have Mechanisms for Both Entering and Leaving Host Cells

Many pathogens, including *V. cholerae* and *B. pertussis*, infect their host without entering host cells. Others, however, including all viruses and many bacteria and protozoa, are **intracellular pathogens**. Their preferred niche for replication and survival is within the cytoplasm or intracellular compartments of particular host cells. This strategy has several advantages. The pathogens are not accessible to *antibodies* (discussed in Chapter 24), and they are not easy targets for phagocytic cells (discussed below). This lifestyle, however, does require that the pathogen develop mechanisms for entering host cells, for finding a suitable subcellular niche where it can replicate, and for exiting the infected cell to spread the infection. In the remainder of this section, we consider some of the myriad ways that individual intracellular pathogens exploit and modify host cell biology to satisfy these requirements.

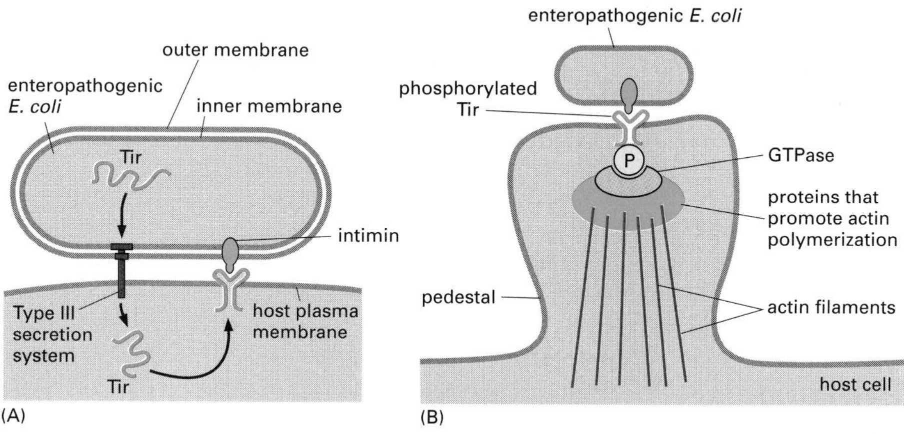

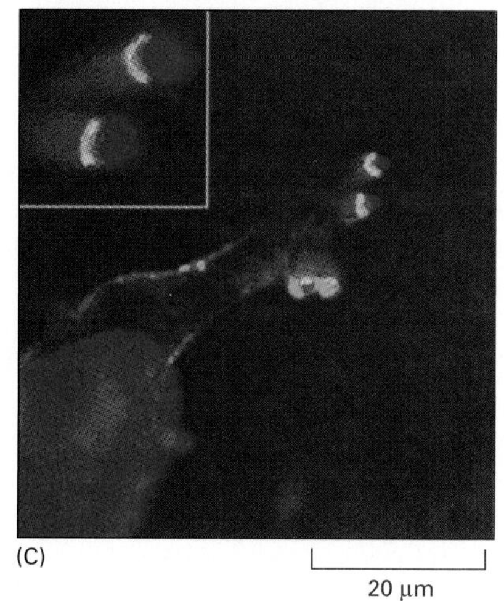

Figure 25–20 Interaction of enteropathogenic *E. coli* (EPEC) with host cells. (A) When EPEC contacts an epithelial cell in the lining of the human gut, it delivers a bacterial protein, Tir, into the host cell through a type III secretion system. Tir then inserts into the plasma membrane of the host cell, where it functions as a receptor for the bacterial adhesin intimin. (B) The intracellular domain of Tir is phosphorylated on a tyrosine residue by a host cell protein tyrosine kinase. Phosphorylated Tir then probably recruits a Rho family GTPase, which triggers actin polymerization; it also recruits other host cell cytoskeletal factors that interact with actin. Consequently, a bundle of actin filaments is assembled underneath the bacterium to form an actin pedestal. (C) EPEC on a pedestal. In this fluorescence micrograph, the DNA of the EPEC and host cell are labeled in *blue*, Tir protein is labeled in *green*, and host cell actin filaments are labeled in *red*. The inset shows a close up view of two of the bacteria on pedestals. (C, from D. Goosney et al., *Annu. Rev. Cell Dev. Biol.* 16:173–189, 2000. © Annual Reviews.)

Viruses Bind to Molecules Displayed on the Host Cell Surface

The first step for any intracellular pathogen is to bind to the surface of the host target cell. For viruses, binding is accomplished through the association of a viral surface protein with a specific receptor on the host cell surface. Of course, no host cell receptor evolved for the sole purpose of allowing a pathogen to bind to it; these receptors all have other functions. The first such "virus receptor" identified was the *E. coli* surface protein that allows the bacteriophage lambda to bind to the bacterium. Its normal function is as a transport protein responsible for the uptake of maltose.

Viruses that infect animal cells generally use cell-surface receptor molecules that are either very abundant (such as sialic-acid-containing oligosaccharides, which are used by the influenza virus) or uniquely found on those cell types in which the virus can replicate (such as the nerve growth factor receptor, the nicotinic acetylcholine receptor, or the cell–cell adhesion protein N-CAM, all of which are used by the rabies virus to specifically infect neurons). Often, a single type of receptor is used by many types of virus, and some viruses can use several different receptors. Moreover, different viruses that infect the same cell type may each use a different receptor. Hepatitis, for example, is caused by at least six viruses, all of which preferentially replicate in liver cells. Receptors for four of the hepatitis viruses have been identified, and they are all different. Receptors need not be proteins; herpes simplex virus, for example, binds to heparan sulfate proteoglycans through specific viral membrane proteins.

Frequently, viruses require both a primary receptor and a secondary co-receptor for efficient attachment and entry into host cells. An important example is HIV. Its primary receptor is CD4, a protein involved in immune recognition which is found on the surface of many T cells and macrophages (discussed in Chapter 24). Viral entry also requires the presence of a co-receptor, either CCR5 (a receptor for β-chemokines) or CXCR4 (a receptor for α-chemokines), depending on the particular variant of the virus (Figure 25–21). Macrophages are susceptible only to HIV variants that use CCR5 for entry, whereas T cells are most efficiently infected by variants that use CXCR4. The viruses that are found within the first few months after HIV infection almost invariably require CCR5, which presumably explains why individuals who carry a defective *ccr5* gene are not susceptible to HIV infection. In the later stages of infection, viruses may either switch to use the CXCR4 co-receptor or adapt to use both co-receptors; in this way, the virus can change the cell types it infects as the disease progresses.

Viruses Enter Host Cells by Membrane Fusion, Pore Formation, or Membrane Disruption

After recognition and attachment to the host cell surface, the virus must next enter the host cell and release its nucleic acid genome from its protective protein coat or lipid envelope. In most cases, the liberated nucleic acid remains complexed with some viral proteins. **Enveloped viruses** enter the host cell by

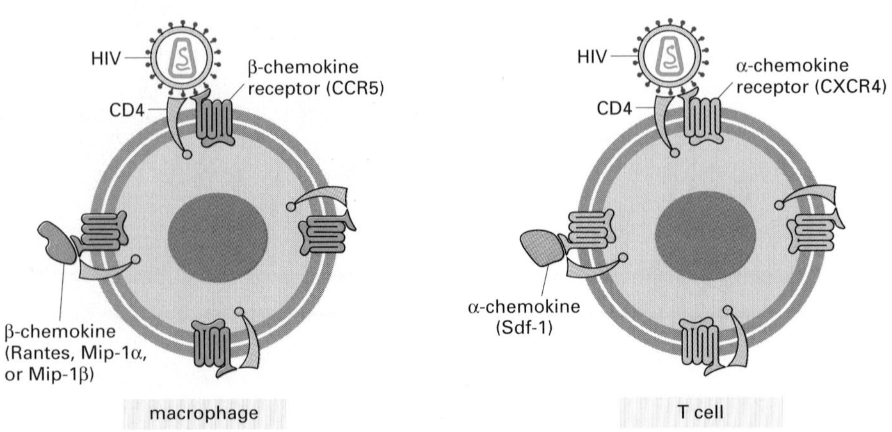

Figure 25–21 Receptor and co-receptors for HIV. All strains of HIV require CD4 as a primary receptor. Early in an infection, most of the viruses use CCR5 as a co-receptor, allowing them to infect macrophages and their precursors, monocytes. As the infection progresses, mutant variants arise that now use CXCR4 as a co-receptor, enabling them to efficiently infect T cells. Invasion of either virus type can be blocked by the natural ligand for the chemokine receptors (Sdf-1 for CXCR4, and Rantes, Mip-1α, or Mip-1β for CCR5).

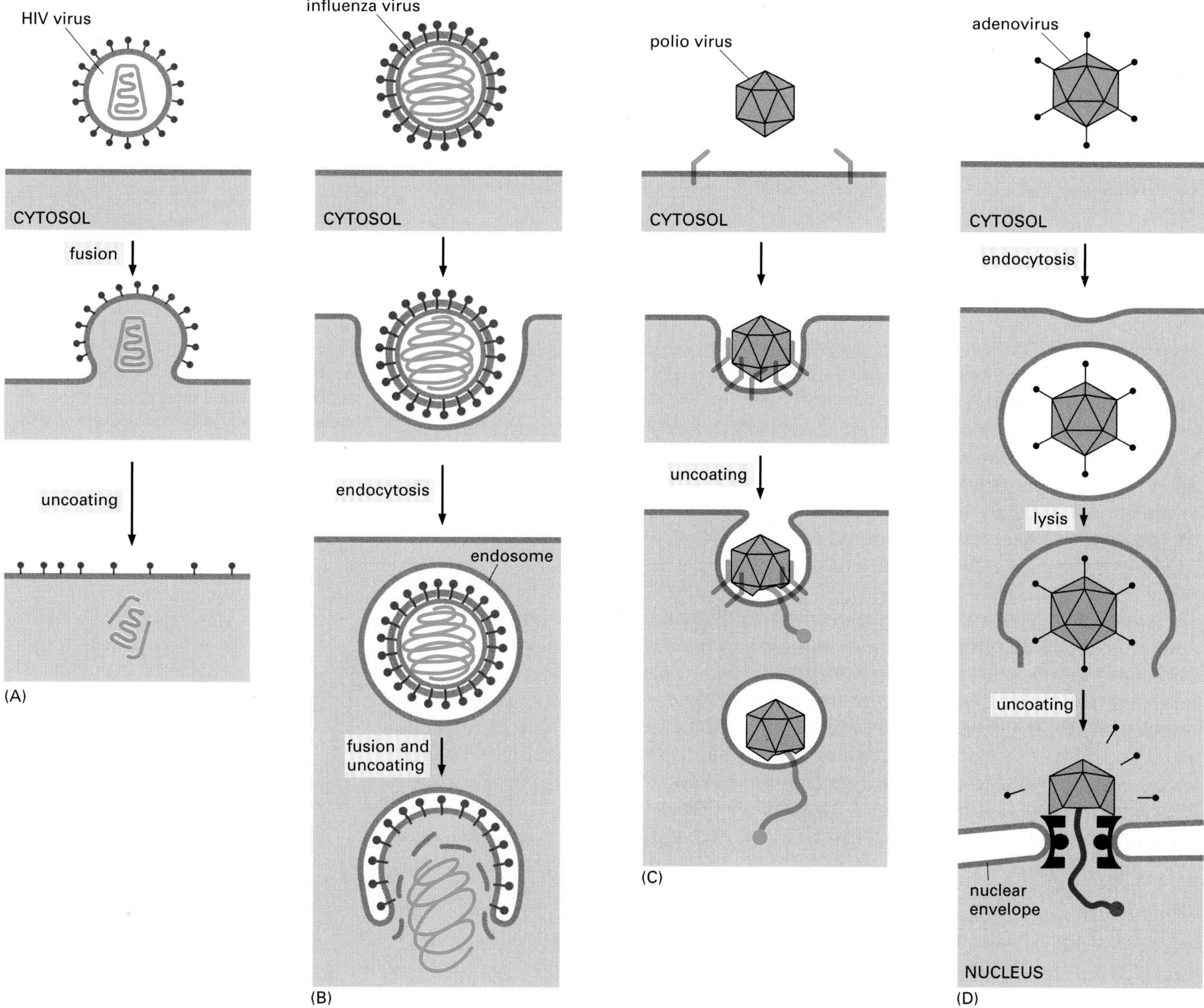

Figure 25–22 Four virus uncoating strategies. (A) Some enveloped viruses, such as HIV, fuse directly with the host cell plasma membrane to release their capsid *(green)* into the cytosol. (B) Other enveloped viruses, such as influenza virus, first bind to cell-surface receptors, triggering receptor-mediated endocytosis. When the endosome acidifies, the virus envelope fuses with the endosomal membrane, releasing the nucleocapsid *(blue)* into the cytosol. (C) Poliovirus, a nonenveloped virus, binds to a receptor *(green)* on the host cell surface and then forms a pore in the host cell membrane to extrude its RNA genome *(blue)*.
(D) Adenovirus, another nonenveloped virus, uses a more complicated strategy. It induces receptor-mediated endocytosis and then disrupts the endosomal membrane, releasing part of the capsid into the cytosol. The capsid eventually docks onto a nuclear pore and releases its DNA genome *(red)* directly into the nucleus.

fusing either with the plasma membrane or with the endosomal membrane following endocytosis (Figure 25–22A,B). Fusion is thought to proceed via a mechanism similar to SNARE-mediated fusion of vesicles during normal intracellular vesicular traffic (discussed in Chapter 13).

Fusion is regulated both to ensure that virus particles fuse only with the appropriate host cell membrane and to prevent virus particles from fusing with one another. For viruses such as HIV that fuse at neutral pH at the plasma membrane, binding to receptors or co-receptors usually triggers a conformational change in the viral envelope protein to expose a normally buried fusion peptide (see Figure 13–16). Other enveloped viruses, such as influenza, postpone fusion until after endocytosis; in this case, it is frequently the acid environment in the early endosome that triggers the conformational change in a viral surface protein that exposes the fusion peptide (Figure 25–23). The H^+ pumped into the early endosome enters the influenza particle through an ion channel and triggers the *uncoating* of the viral nucleic acid, which is directly released into the cytosol as the virus fuses with the endosomal membrane. For some viruses, uncoating occurs after release into the cytosol. In the case of Semliki forest virus, for example, the binding of host ribosomes to the capsid causes the capsid proteins to separate from the viral genome.

It is more difficult to envision how **nonenveloped viruses** enter host cells, as it is not obvious how large assemblies of protein and nucleic acid can cross the plasma or endosomal membrane. Where the entry mechanism is understood,

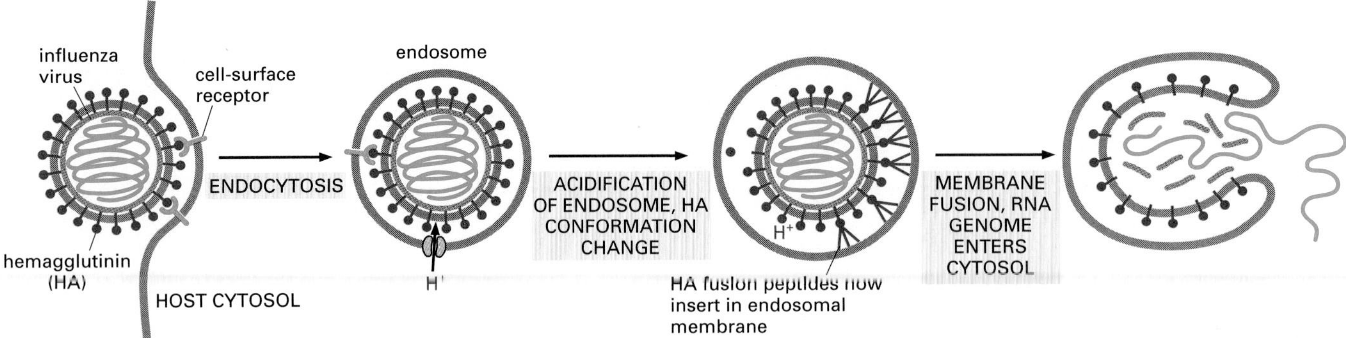

Figure 25–23 The entry strategy used by the influenza virus. The globular heads of the viral hemagglutinin (HA) mediate binding of the virus to sialic-acid-containing cell-surface receptors. The virus-receptor complexes are endocytosed and, in the acidic environment of the endosome, the loop region of the HA becomes a coiled-coil, moving the fusion peptide to the top of the molecule, near the endosomal membrane. To allow release of the viral genome into the cytosol, H^+ ions in the endosome enter the virus through an ion channel in the viral membrane, releasing the RNA genome *(blue)* from the capsid coat. The fusion of the viral and endocytic membranes allows the viral genome to enter the cytosol.

nonenveloped viruses generally either form a pore in the cell membrane to deliver the viral genome into the cytoplasm or they disrupt the endosomal membrane after endocytosis.

Poliovirus uses the first strategy. Binding of poliovirus to its receptor triggers both receptor-mediated endocytosis and a conformational change in the viral particle. The conformational change exposes a hydrophobic projection on one of the capsid proteins, which apparently inserts into the endosomal membrane to form a pore. The viral genome then enters the cytoplasm through the pore, leaving the capsid either in the endosome or on the cell surface, or in both places (see Figure 25–22C).

Adenovirus uses the second strategy. It is initially taken up by receptor-mediated endocytosis. As the endosome matures and becomes more acidic, the virus undergoes multiple uncoating steps in which structural proteins are sequentially removed from the capsid. Some of these steps require the action of a viral protease, which is inactive in the extracellular virus particle (probably because of intrachain disulfide bonds) but which is activated in the reducing environment of the endosome. One of the proteins released from the capsid lyses the endosomal membrane, releasing the remainder of the virus into the cytosol. This trimmed-down virus then docks onto the nuclear pore complex, and the viral DNA genome is released through the pore into the nucleus, where it is transcribed (see Figure 25–22D).

In these various entry strategies, viruses exploit a variety of host cell molecules and processes, including cell-surface components, receptor-mediated endocytosis, and endosomal maturation steps. These strategies again illustrate the sophisticated ways that pathogens have evolved to utilize the basic cell biology of their hosts.

Bacteria Enter Host Cells by Phagocytosis

Bacteria are much larger than viruses, and they are too large to be taken up by receptor-mediated endocytosis. Instead, they enter host cells through phagocytosis. Phagocytosis of bacteria is a normal function of macrophages. They patrol the tissues of the body and ingest and destroy unwanted microbes. Some pathogens, however, have acquired the ability to survive and replicate within macrophages after they have been phagocytosed.

Tuberculosis, a serious lung infection that is widespread in some urban populations, is caused by one such pathogen, *Mycobacterium tuberculosis*. This bacterium is usually acquired by inhalation into the lungs, where it is phagocytosed by alveolar macrophages. Although the microbe can survive and replicate within macrophages, the macrophages of most healthy individuals, with the help of the adaptive immune system, contain the infection within a lesion called a *tubercle*. In most cases, the lesion becomes walled off within a fibrous capsule that undergoes calcification, after which it can easily be seen on an X-ray of the lungs of an infected person. An unusual feature of *M. tuberculosis* is its ability to survive for decades within macrophages contained in such lesions. Later in life, especially if the immune system becomes weakened by disease or drugs, the infection may be reactivated, spreading in the lung and even to other organs.

Tuberculosis has been an obvious presence in human populations for thousands of years, but another bacterium that lives within alveolar macrophages was first recognized as a human pathogen only in 1976. *Legionella pneumophila* is normally a parasite of freshwater amoebae, which take it up by phagocytosis. When droplets of water containing *L. pneumophila* or infected amoebae are inhaled into the lung, the bacteria can invade and live inside alveolar macrophages (Figure 25–24), which, to the bacteria, must seem just like large amoebae. This infection leads to the type of pneumonia known as **Legionnaire's disease**. The pathogen can be efficiently spread by central air conditioning systems, as the amoebae that are the bacterium's normal host are particularly adept at growing in air-conditioning cooling towers; moreover, these cooling systems produce microdroplets of water that are easily inhaled. The incidence of Legionnaire's disease has increased dramatically in recent decades, and outbreaks are frequently traced to the air conditioning systems in office buildings, hospitals, and hotels. Other forms of modern aerosolization, including decorative fountains and produce sprayers in supermarkets, have also been implicated in outbreaks of this disease.

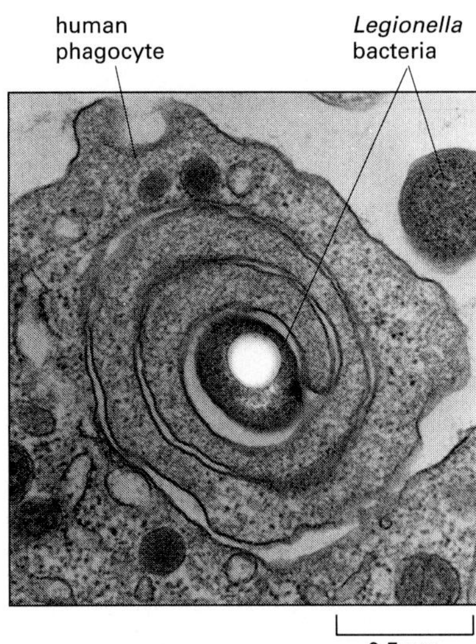

Figure 25–24 Uptake of *Legionella pneumophila* by a human mononuclear phagocyte. This electron micrograph shows the unusual coil structure induced on the surface of the phagocyte by the bacterium. (From M.A. Horwitz, *Cell* 36:27–33, 1984. © Elsevier.)

Some bacteria invade cells that are normally nonphagocytic. One way that bacteria can induce such a cell to phagocytose them is by expressing an adhesin that binds with high affinity to a cell adhesion protein that the cell normally uses to adhere to another cell or to the extracellular matrix (discussed in Chapter 19). For example, a bacterium that causes diarrhea, *Yersinia pseudotuberculosis* (a close relative of the plague bacterium *Yersinia pestis*), expresses a protein called *invasin* that binds to β1 integrins, and *Listeria monocytogenes*, which causes a rare but serious form of food poisoning, expresses a protein that binds to E-cadherin. Binding to these transmembrane adhesion proteins fools the host cell into attempting to form a cell junction, and it begins moving actin and other cytoskeletal components to the site of bacterial attachment. Since the bacterium is small relative to the host cell, the host cell's attempt to spread over the adhesive surface of the bacterium results in the phagocytic uptake of the bacterium—a process known as the zipper mechanism of invasion (Figure 25–25A). The similarity of this form of invasion to the natural process of cell adhesion was underscored by the determination of the three-dimensional structure of invasin. This bacterial protein has an RGD motif, the structure of which is almost identical to that of the normal β1-integrin-binding site in laminin (discussed in Chapter 19).

A second pathway by which bacteria can invade nonphagocytic cells is known as the trigger mechanism (Figure 25–25B). It is used by various pathogens, including *Salmonella enterica*, which causes food poisoning. This dramatic form of invasion is initiated when the bacterium injects a set of effector molecules into the host cell cytoplasm through a type III secretion system. Some of these

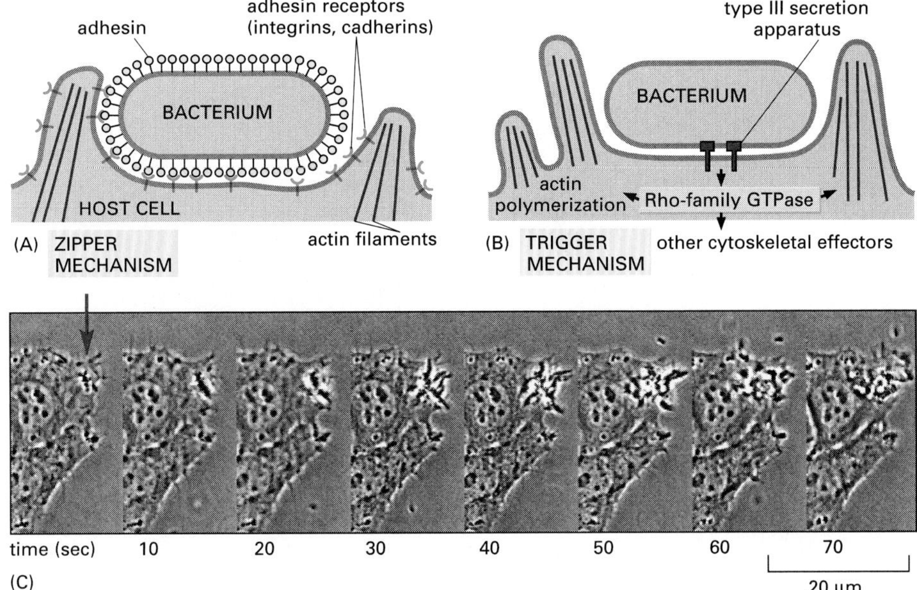

Figure 25–25 Mechanisms used by bacteria to induce phagocytosis by nonphagocytic host cells. (A) The zipper and (B) trigger mechanisms for pathogen-induced phagocytosis both require polymerization of actin at the site of bacterial entry. (C) Frames from a time-lapse movie, recorded ten seconds apart, showing the formation of a giant membrane ruffle *(arrow)* when *Salmonella enterica* contacts a host cell in culture. This leads to the phagocytosis of the bacterium by the trigger mechanism. (C, courtesy of Julie Theriot and Jorge Galàn.)

effector molecules activate Rho-family GTPases, which stimulate actin polymerization (discussed in Chapter 16). Others interact with cytoskeletal elements more directly, severing actin filaments and causing the rearrangement of cross-linking proteins. The net effect is to cause dramatic localized ruffling on the surface of the host cell (Figure 25–25C), which throws up large actin-rich protrusions that fold over and trap the bacterium within large endocytic vesicles called *macropinosomes*. The overall appearance of cells being invaded by the trigger mechanism is similar to the dramatic ruffling induced by some growth factors, suggesting that similar intracellular signaling cascades are activated in both cases.

Intracellular Parasites Actively Invade Host Cells

The uptake of viruses by receptor-mediated endocytosis and bacteria by phagocytosis requires energy provided by the host cell. The pathogen is a relatively passive participant, usually providing a trigger to initiate the invasion process, but not contributing any metabolic energy. In contrast, the invasion of intracellular eucaryotic parasites, which are typically much larger than bacteria, proceeds through a variety of complex pathways that usually require significant energy expenditure on the part of the parasite.

Toxoplasma gondii, the cat parasite that also causes occasional serious human infections, is an instructive example. When this protozoan contacts a host cell, it protrudes a microtubule-ribbed structure called a *conoid* and then reorients so that the conoid is touching the host cell surface. The parasite then slowly pushes its way into the host cell. The energy for invasion seems to come entirely from the parasite, and the process can be disrupted by depolymerizing the parasite's—but not the host's—actin cytoskeleton. As the parasite moves into the host cell, it secretes lipids from a pair of specialized organelles, generating a vacuolar membrane that is made primarily of parasite lipids and proteins. In this way, the organism protects itself in a membrane-enclosed compartment within the host cell that does not participate in host cell membrane trafficking pathways and does not fuse with lysosomes (Figure 25–26). The specialized vacuolar membrane allows the parasite to take up metabolic intermediates and nutrients from the host cell cytosol but excludes larger molecules.

An entirely different, but no less peculiar, invasion strategy is used by the protozoan *Trypanosoma cruzi*, which causes a multiorgan disease (Chagas disease) mainly in Mexico and Central and South America. After attachment to β1 integrins on the host cell surface, this parasite induces a local elevation of Ca^{2+} in the host cell cytoplasm. The Ca^{2+} signal recruits lysosomes to the site of parasite attachment, which then fuse with the plasma membrane, allowing the

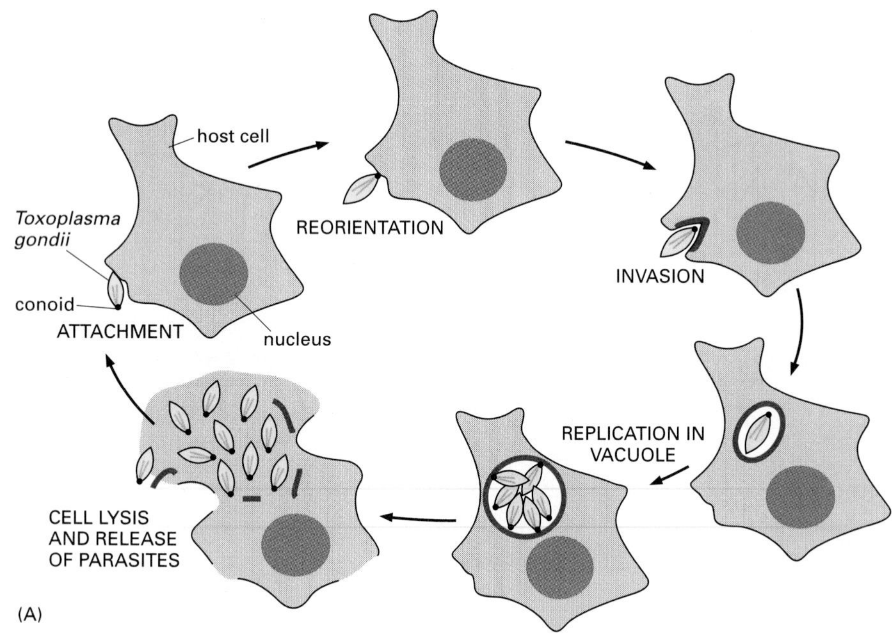

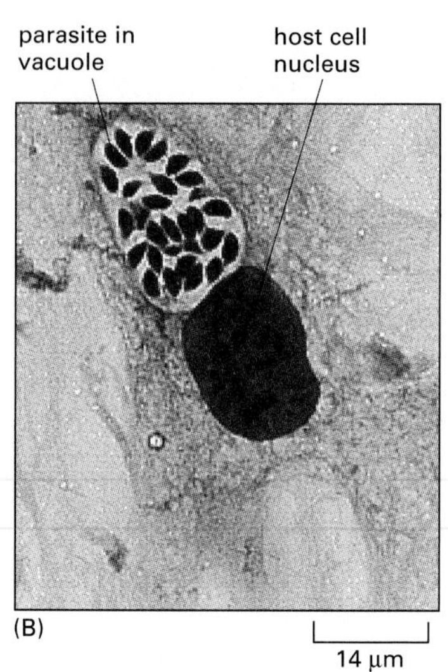

Figure 25–26 The life cycle of the intracellular parasite *Toxoplasma gondii*. (A) After attachment to a host cell, *T. gondii* secretes lipids *(red)* from internal organelles. After invasion, the parasite replicates in the special vacuole formed from parasite lipids. This vacuole lacks the host cell proteins associated with normal endosomes, and the parasitic vacuole does not fuse with lysosomes. After several rounds of replication, the parasite causes the host cell to lyse, and progeny are released to infect other host cells. (B) Light micrograph of *T. gondii* replicating in a vacuole in a cultured cell. (B, courtesy of Manuel Camps and John Boothroyd.)

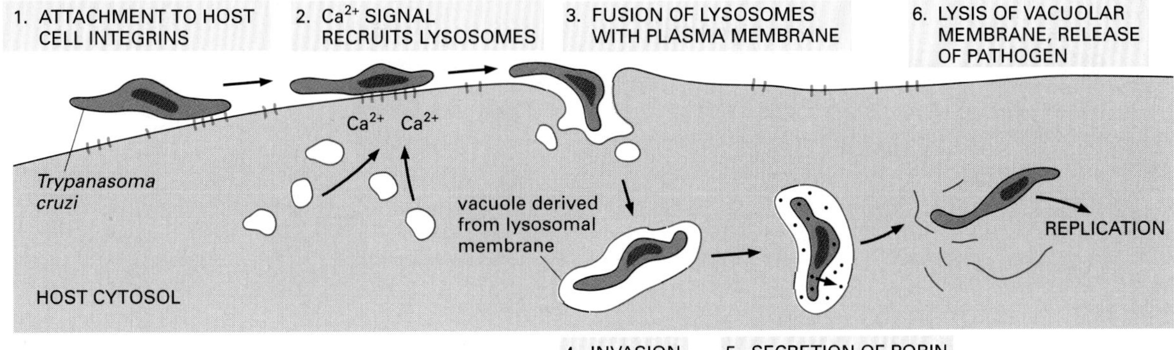

Figure 25–27 Invasion of *Trypansoma cruzi*. This parasite recruits host cell lysosomes to its site of attachment. The lysosomes fuse with the plasma membrane to create a vacuole constructed almost entirely of lysosomal membrane. After a brief stay in the vacuole, the parasite secretes a protein (porin) that disrupts the vacuolar membrane, allowing the parasite to escape into the host cell cytosol and replicate.

parasite to enter a vacuole derived almost entirely from lysosomal membrane (Figure 25–27). As we discuss below, most intracellar pathogens go to great lengths to avoid exposure to the hostile, proteolytic environment of the lysosome, but this trypanosome uses the lysosome as its means of entry. In the vacuole, the parasite secretes a transsialidase enzyme that removes sialic acid from lysosomal glycoproteins and transfers it to its own surface molecules, thereby coating itself with host cell sugars. Next, the parasite secretes a pore-forming toxin that lyses the vacuolar membrane, releasing the parasite into the host cell cytosol, where it replicates.

Many Pathogens Alter Membrane Traffic in the Host Cell

The two examples of intracellular parasites discussed in the preceding section raise a general problem that faces all pathogens with an intracellular lifestyle, including viruses, bacteria, and parasites. They must deal in some way with membrane traffic in the host cell. After endocytosis by a host cell, they usually find themselves in an endosomal compartment that normally would fuse with lysosomes. They therefore must either modify the compartment to prevent its fusion, escape from the compartment before getting digested, or find ways to survive in the hostile environment of the phagolysosome (Figure 25–28).

Most pathogens use either the first or second strategy. It is much less common to find pathogens that can survive in the phagolysosome. As we have seen, *Trypanosoma cruzi* uses the escape route, as do essentially all viruses (see Figure 25–22). The bacterium *Listeria monocytogenes* also uses this strategy. It is taken up into cells via the zipper mechanism discussed earlier and secretes a protein called *hemolysin* that forms large pores in the phagosomal membrane, eventually disrupting the membrane and releasing the bacteria into the cytosol. Once in the cytosol, the bacteria continue to secrete hemolysin, but it does not destroy the plasma membrane. The hemolysin selectively destroys only the phagosomal membrane for two reasons: first, it is ten times more active at the acidic pH found in the phagosome than at neutral pH found in the cytosol, and second, the hemolysin contains a PEST sequence (a peptide motif rich in proline, glutamate, serine, and threonine), which is recognized by the protein degradation

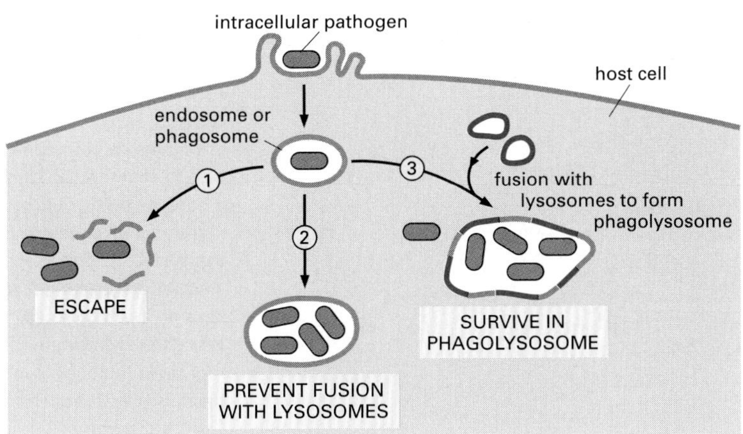

Figure 25–28 The choices faced by an intracellular pathogen. After entry, generally through endocytosis or phagocytosis into a membrane-enclosed compartment, intracellular pathogens may use one of three strategies to survive and replicate. Pathogens that follow strategy (1) include all viruses, *Trypansoma cruzi*, *Listeria monocytogenes*, and *Shigella flexneri*. Those that follow strategy (2) include *Plasmodium falciparum*, *Mycobacterium tuberculosis*, *Salmonella enterica*, *Legionella pneumophila*, and *Chlamydia trachomatis*. Those that follow strategy (3) include *Coxiella burnetii* and *Leishmania*.

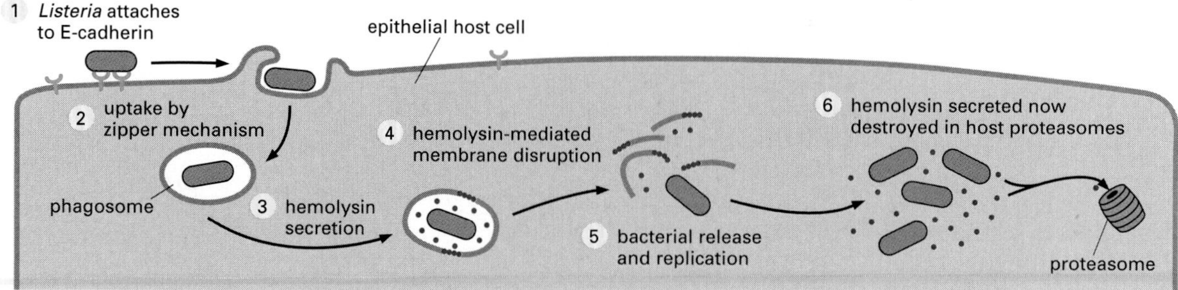

Figure 25–29 Selective destruction of the phagosomal membrane by *Listeria monocytogenes*. *L. monocytogenes* attaches to E-cadherin on the surface of epithelial cells and induces its own uptake by the zipper mechanism (see Figure 25–25A). Within the phagosome, the bacterium secretes the hydrophobic protein hemolysin, which forms oligomers in the host cell membrane, thereby creating large pores and eventually disrupting the membrane. Once in the host cell cytosol, the bacteria begin to replicate and continue to secrete hemolysin. Because the hemolysin is rapidly degraded by proteasomes, the host cell plasma membrane remains intact.

machinery of the host cell, leading to the rapid degradation of the bacterial hemolysin in the proteasome (see Figure 6–86). The hemolysin is stable in the phagosome, as the degradation machinery does not have access to it there (Figure 25–29). The hemolysin secreted by *L. monocytogenes* is closely related to hemolysins secreted by other bacteria that are not intracellular pathogens. These related hemolysins, however, all lack PEST sequences. It seems that the *L. monocytogenes* hemolysin has acquired an essentially eucaryotic protein domain expressly to allow its activity to be regulated in the host cell.

By far the most common strategy used by intracellular bacteria and parasites is to modify the endosomal compartment so that they can remain there and replicate. They must modify the compartment in at least two ways: first, they must prevent lysosomal fusion, and second, they must provide a pathway for importing nutrients from the host cytosol. Different pathogens have distinct strategies for doing this. As we have seen, *Toxoplasma gondii* creates its own membrane-enclosed compartment that does not participate in normal host cell membrane traffic and allows nutrient transport. This option is not open to bacteria, which are too small to carry in their own supply of membrane. *Mycobacterium tuberculosis* somehow prevents the very early endosome that contains it from maturing, so that it never acidifies or acquires the characteristics of a late endosome. Endosomes containing *Salmonella enterica,* in contrast, do acidify and acquire markers of late endosomes, but they arrest their maturation at a stage prior to lysosomal fusion. Other bacteria seem to find shelter in intracellular compartments that are completely distinct from the usual endocytic pathway. *Legionella pneumophila,* for example, replicates in compartments that are wrapped in layers of rough endoplasmic reticulum. *Chlamydia trachomatis,* a sexually transmitted bacterial pathogen that can cause sterility and blindness, replicates in a compartment that seems to be most similar to part of the exocytic pathway (Figure 25–30). The mechanisms

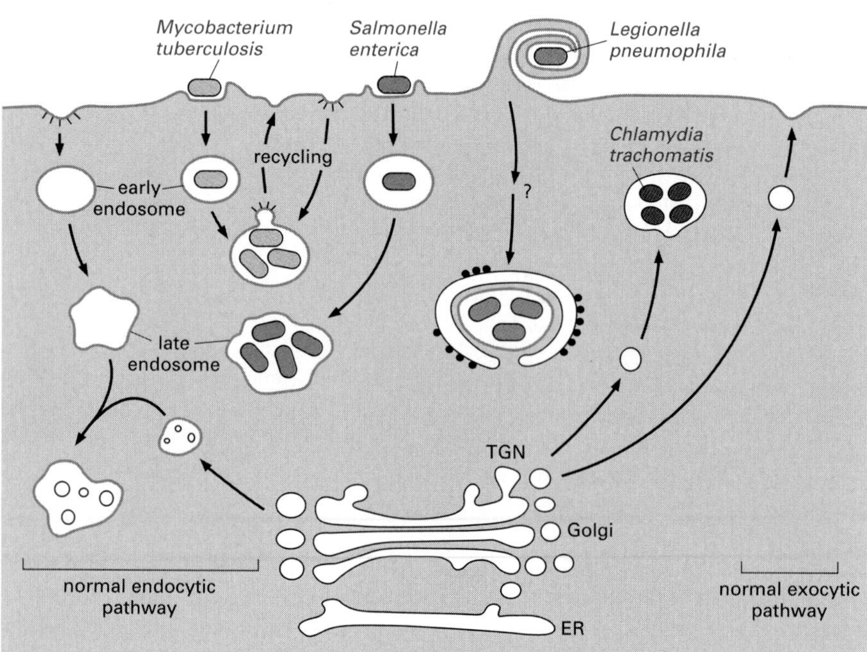

Figure 25–30 Modifications of intracellular membrane trafficking by bacterial pathogens. Four intracellular bacterial pathogens, *Mycobacterium tuberculosis*, *Salmonella enterica*, *Legionella pneumophila*, and *Chlamydia trachomatis*, all replicate in membrane-enclosed compartments, but the four compartments differ. *M. tuberculosis* remains in a compartment that has early endosomal markers and continues to communicate with the plasma membrane. *S. enterica* replicates in a compartment that has late endosomal markers and does not communicate with the surface. *L. pneumophila* replicates in an unusual compartment that is wrapped in several layers of rough endoplasmic reticulum membrane. *C. trachomatis* replicates in an exocytic compartment that fuses with vesicles coming from the Golgi apparatus.

used by these organisms to alter their membrane compartments are not yet understood.

Viruses also often alter membrane traffic in the host cell. Enveloped viruses must acquire their membrane from host cell phospholipids. In the simplest cases, virally encoded proteins are inserted into the ER membrane and follow the usual path through the Golgi apparatus to the plasma membrane, undergoing various posttranslational modifications *en route*. The viral capsid then assembles at the plasma membrane and buds off from the cell surface. This is the mechanism used by HIV. Other enveloped viruses interact in complex ways with membrane trafficking pathways in the host cell (Figure 25–31). Even some nonenveloped viruses alter membrane traffic in the host cell to suit their own purposes. The replication of poliovirus, for example, is carried out by a membrane-associated, virus-encoded RNA polymerase. Replication can proceed more quickly if the surface area of the host cell membranes is increased. To accomplish this, the virus induces increased lipid synthesis in the host cell and blocks secretion from the ER. Intracellular membranes thereby accumulate, expanding the surface area on which RNA replication can occur (Figure 25–32).

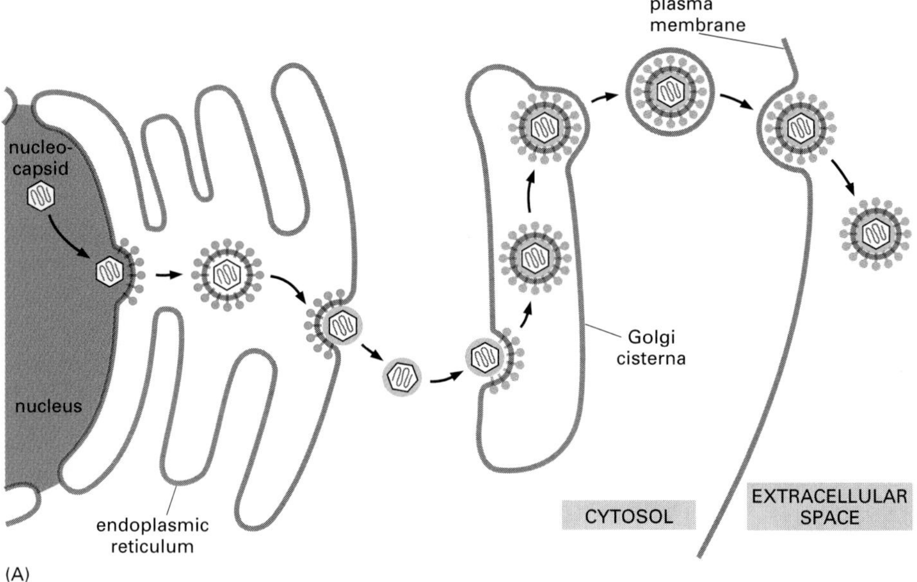

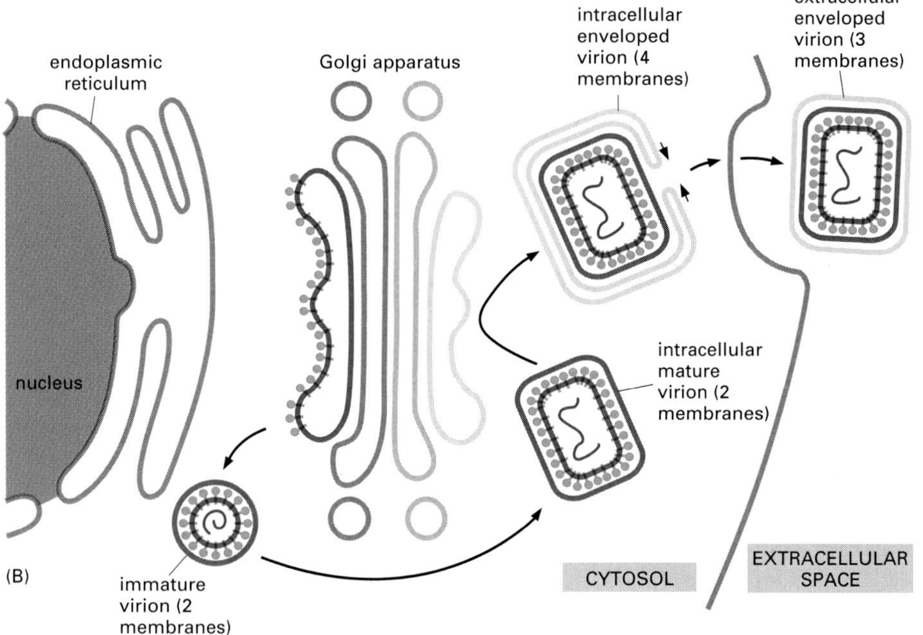

Figure 25–31 Complicated strategies for viral envelope acquisition. (A) Herpes virus capsids are assembled in the nucleus and then bud through the inner nuclear membrane, acquiring a membrane coat. They then apparently lose this coat when they fuse with the outer nuclear membrane to escape into the cytosol. Subsequently, they bud into the Golgi apparatus, and bud out again on the other side, acquiring two new membrane coats. The virus then buds from the cell with a single membrane when its outer membrane fuses with the plasma membrane. (B) Vaccinia virus, which is closely related to the virus that causes smallpox, is assembled in "replication factories" deep within the cytosol. The first structure assembled contains two membranes acquired from the Golgi apparatus by a poorly defined wrapping mechanism. A variable proportion of these viral particles are then engulfed by the membranes of a second intracellular membrane-enclosed compartment. These viral particles have a total of four layers of membrane envelope. After fusion at the plasma membrane, the virus escapes with three membrane layers.

Viruses and Bacteria Exploit the Host Cell Cytoskeleton for Intracellular Movement

The cytoplasm of mammalian cells is extremely viscous. It is crowded with organelles and supported by networks of cytoskeletal filaments, all of which inhibit the diffusion of particles the size of a bacterium or a viral capsid. If a pathogen must reach a particular part of the cell to carry out part of its replication cycle, it must move there actively. As with transport of intracellular organelles, pathogens generally use the cytoskeleton for active movement.

Several bacteria that replicate in the host cell cytosol (rather than in membrane-enclosed compartments) have adopted a remarkable mechanism for moving, which depends on actin polymerization. These bacteria, including *Listeria monocytogenes, Shigella flexneri,* and *Rickettsia rickettsii* (which causes Rocky Mountain spotted fever) induce the nucleation and assembly of host cell actin filaments at one pole of the bacterium. The growing filaments generate substantial force and push the bacteria through the cytoplasm at rates up to 1 μm/sec. New filaments form at the rear of each bacterium and are left behind like a rocket trail as the bacterium advances, depolymerizing again within a minute or so as they encounter depolymerizing factors in the cytosol. When a moving bacterium reaches the plasma membrane, it continues to move outward, inducing the formation of a long, thin protrusion with the bacterium at its tip. This projection is often engulfed by a neighboring cell, allowing the bacterium to enter the neighbor's cytoplasm without exposure to the extracellular environment, thereby avoiding recognition by antibodies produced by the host's adaptive immune system (Figure 25–33).

The molecular mechanism of pathogen-induced actin assembly has been determined for two of these bacteria. The mechanisms are different, suggesting that they evolved independently. Although both make use of the same host cell regulatory pathway that normally controls the nucleation of actin filaments, they exploit different points in the pathway. As discussed in Chapter 16, activation of the small GTPase Cdc42 by growth factors or other external signals causes the activation of a protein called N-WASp, which in turn activates the

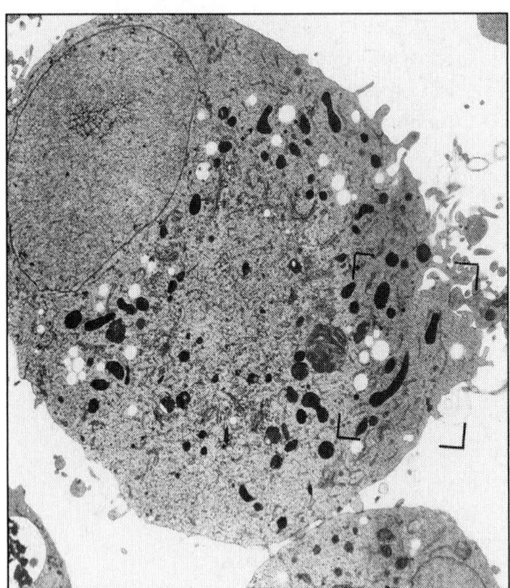

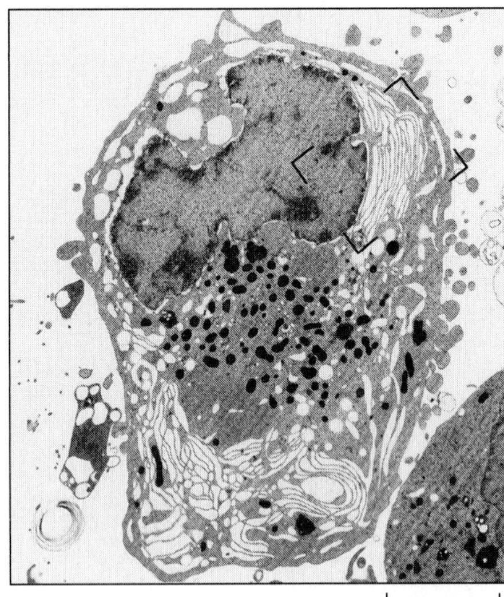

Figure 25–32 Intracellular membrane alterations induced by a poliovirus protein. Poliovirus, like other positive-stranded RNA viruses, replicates its RNA genome using a polymerase that associates with intracellular membranes. Several of the proteins encoded in its genome alter the structure or dynamic behavior of the membrane-enclosed organelles in the host cell. These electron micrographs show a normal Cos-7 cell *(left)* and a cell expressing the 3A protein from poliovirus *(right)*. In the transfected cell, the ER is swollen and traffic from the ER to the Golgi is inhibited. (From J.J.R. Doedens, T.H. Giddings Jr., and K. Kirkegaard, *J. Virol.* 71:9054–9064, 1997.)

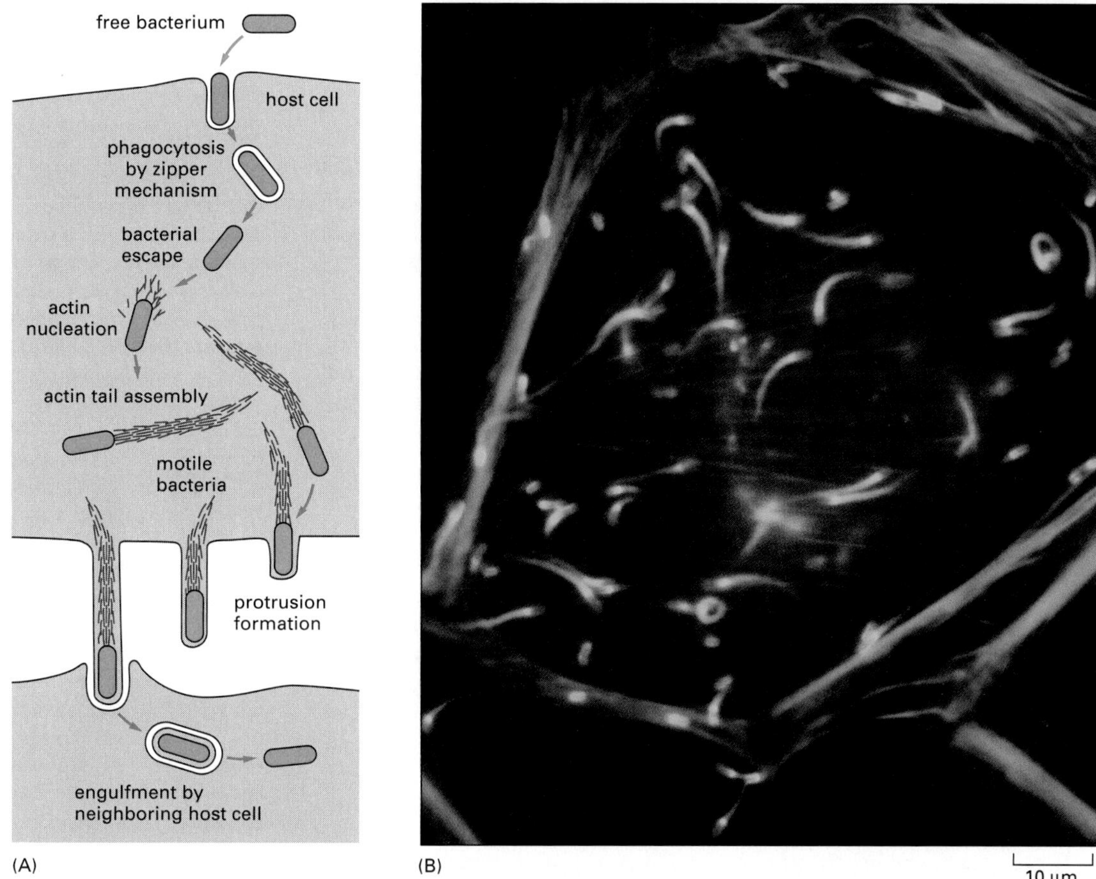

Figure 25–33 The actin-based movement of *Listeria monocytogenes* within and between host cells. (A) These bacteria induce the assembly of actin-rich tails in the host cell cytoplasm, which enable them to move rapidly. Motile bacteria spread from cell to cell by forming membrane-enclosed protrusions that are engulfed by neighboring cells. (B) Fluorescence micrograph of the bacteria moving in a cell that has been stained to reveal both bacteria and actin filaments. Note the comet-like tail of actin filaments *(green)* behind each moving bacterium *(red)*. Regions of overlap of red and green fluorescence appear *yellow*. (B, courtesy of Julie Theriot and Tim Mitchison.)

ARP complex that can nucleate the growth of a new actin filament. An *L. monocytogenes* surface protein directly binds to and activates the ARP complex to initiate the formation of an actin tail, while an unrelated surface protein on *S. flexneri* binds to and activates N-WASp, which then activates the ARP complex. Remarkably, vaccinia virus uses yet another mechanism to move intracellularly by inducing actin polymerization, again exploiting this same regulatory pathway (Figure 25–34).

Other pathogens rely primarily on microtubule-based transport to move within the host cell. This movement is particularly well-illustrated by viruses that infect neurons. An important example is provided by the neurotropic alpha herpes viruses, a group that includes the virus that causes chicken pox. Upon infection of sensory neurons, the virus particles are transported to the nucleus by microtubule-based transport, probably mediated by attachment of the capsids to the motor protein dynein. After replication and assembly in the nucleus, the enveloped virus is transported along microtubules away from the neuronal cell body down the axon, presumably by attachment to a kinesin motor protein (Figure 25–35).

One bacterium that is known to associate with microtubules is *Wolbachia*. This fascinating genus includes many species that are parasites or symbionts of insects and other invertebrates, living in the cytosol of each cell in the animal. The infection is spread vertically from mother to offspring, as *Wolbachia* are also present in eggs. The bacteria ensure their transmission into every cell by binding to microtubules, and they are therefore segregated by the mitotic

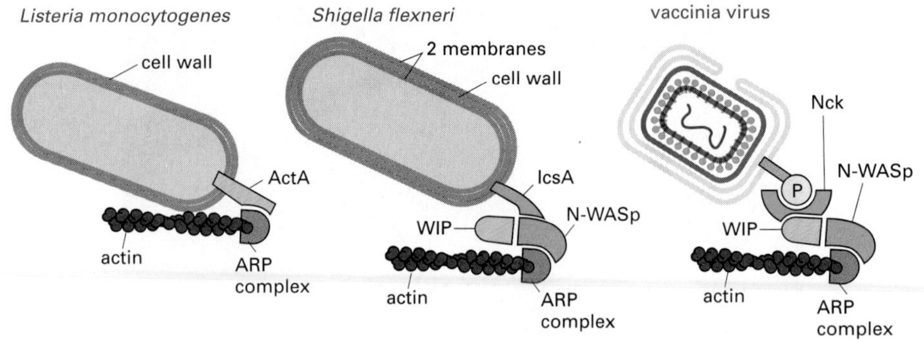

Figure 25–34 Molecular mechanisms for actin nucleation by various pathogens. The bacteria *Listeria monocytogenes* and *Shigella flexneri* and the virus vaccinia all move intracellularly using actin polymerization. To induce actin nucleation, all of these pathogens recruit and activate the ARP complex (see Figure 16–28), although each pathogen uses a different recruitment strategy. *L. monocytogenes* expresses a surface protein, ActA, that directly binds to and activates the ARP complex. *S. flexneri* expresses a surface protein, IcsA (unrelated to ActA), that recruits the host cell signaling protein N-WASp. N-WASp in turn recruits the ARP complex, along with other host proteins, including WIP (WASp-interacting protein). Vaccinia virus expresses an envelope protein that is phosphorylated on tyrosine by a host cell protein tyrosine kinase. The phosphorylated protein then recruits Nck, which binds WIP. WIP binds N-WASp, which recruits and activates the ARP complex. The more complicated cascade used by vaccinia is thought to closely resemble how the ARP complex is activated by chemotactic factors in motile eucaryotic cells. Despite these distinct molecular strategies, the actin comet tails formed by all three pathogens look very similar, and the pathogens move at similar speeds inside infected cells.

spindle simultaneously with chromosome segregation when an infected cell divides (Figure 25–36). As we discuss later, *Wolbachia* infection can significantly alter the reproductive behavior of its insect hosts.

Viruses Take Over the Metabolism of the Host Cell

Most intracellular bacteria and parasites carry the basic genetic information required for their own metabolism and replication, and they rely on their host cells only for nutrients. Viruses, in contrast, use the basic host cell machinery for most aspects of their reproduction: they all depend on host cell ribosomes to produce their proteins, and some also use host cell DNA and RNA polymerases for replication and transcription, respectively.

Many viruses encode proteins that modify the host transcription or translation apparatus to favor the synthesis of viral proteins over those of the host cell. As a result, the synthetic capability of the host cell is directed principally to the production of new virus particles. Poliovirus, for example, encodes a protease that specifically cleaves the TATA-binding factor component of TFIID (see Figure 6–16), effectively shutting off all host cell transcription via RNA polymerase II. Influenza virus produces a protein that blocks both the splicing and the polyadenylation of mRNA transcripts, which therefore fail to be exported from the nucleus (see Figure 6–40).

Translation initiation of most host cell mRNAs depends on recognition of their 5′ cap by a group of translation initiation factors (see Figure 6–71). Translation initiation of host mRNAs is often inhibited during viral infection, so that the host cell ribosomes can be used more efficiently for synthesis of viral proteins.

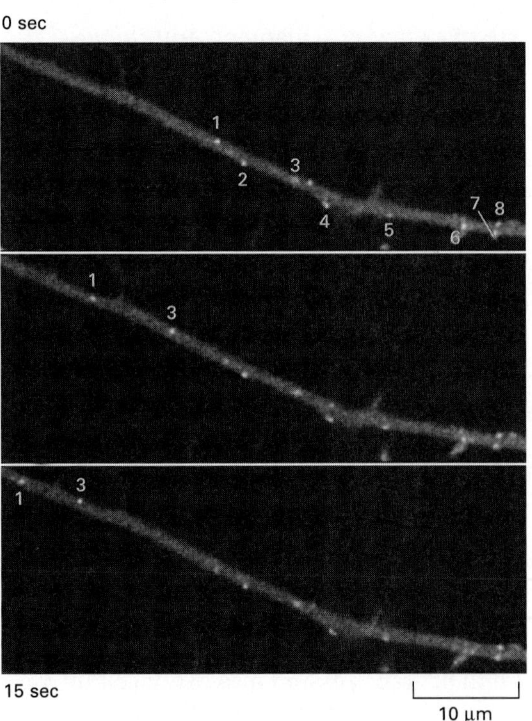

Figure 25–35 Trafficking of herpes virus in an axon. This nerve cell has been infected with an alpha herpes virus that has been genetically engineered to express green fluorescent protein (GFP) fused to one of its capsid proteins. In this segment of the axon, several viral particles are visible, and two of them (numbered one and three) are moving toward the cell body. (From G.A. Smith, S.P. Gross, and L.W. Enquist, *Proc. Natl. Acad. Sci. USA* 98:3466–3470, 2001. © National Academy of Sciences.)

Some viruses encode endonucleases that cleave the 5′ cap from host cell mRNAs. Some even then use the liberated 5′ caps as primers to synthesize viral mRNAs, a process called *cap snatching*. Several other viral RNA genomes encode proteases that cleave certain translation initiation factors. These viruses rely on 5′ cap-independent translation of the viral RNA, using internal ribosome entry sites (IRESs) (see Figure 7–102).

A few DNA viruses use host cell DNA polymerase to replicate their genome. These viruses need to solve a problem: DNA polymerase is expressed at high levels only during S phase of the cell cycle, whereas most cells that the viruses infect spend most of their time in G_1 phase. Adenovirus solves this problem by inducing the host cell to enter S phase. Its genome encodes proteins that inactivate both Rb and p53, two key suppressors of cell-cycle progression (discussed in Chapter 17). As might be expected for any mechanism that induces unregulated DNA replication, these viruses are frequently oncogenic.

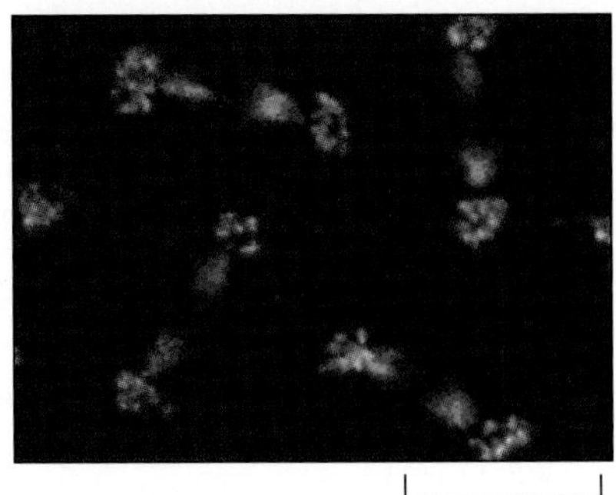

Figure 25–36 *Wolbachia* associates with microtubules. Fluorescence micrograph of *Wolbachia (red)* associated with the microtubules *(green)* of mitotic spindles in a *Drosophila* embryo. The clumps of bacteria at the spindle poles will be segregated into the two daughter cells when each infected cell divides. (From H. Kose and T.L. Karr, *Mech. Cell Dev.* 51:275–288, 1995. © Elsevier.)

Pathogens Can Alter the Behavior of the Host Organism to Facilitate the Spread of the Pathogen

As we have seen, pathogens often alter the behavior of the host cell in ways that benefit the survival and replication of the pathogen. Similarly, pathogens often alter the behavior of the whole host organism to facilitate pathogen spread, as we saw earlier for *Trypanosoma brucei* and *Yersinia pestis*. In some cases, it is difficult to tell whether a particular host response is more for the benefit of the host or for the pathogen. Pathogens like *Salmonella enterica* that cause diarrhea, for example, usually produce self-limiting infections because the diarrhea efficiently washes out the pathogen. The bacteria-laden diarrhea, however, can spread the infection to a new host. Similarly, coughing and sneezing help to clear pathogens from the respiratory tract, but they also spread the infection to new individuals. A person with a common cold may produce 20,000 droplets in a single sneeze, all carrying rhinovirus or corona virus.

A frightening example of a pathogen modifying host behavior is seen in rabies, as first described in Egyptian writings over 3000 years ago. This virus replicates in neurons and causes infected people or animals to become "rabid": they are unusually aggressive and develop a strong desire to bite. The virus is shed in the saliva and transmitted through the bite wound into the bloodstream of the victim, spreading the infection to a new host.

Toxoplasma gondii, a eucaryotic parasite that forms lesions in muscle and brain tissue, provides an equally remarkable example. It can complete its life cycle only in its normal host—cats. If it infects an intermediate host, such as a rodent or human, the infection is a dead end for the parasite, unless the intermediate host is eaten by a cat. Behavioral studies show that rats infected with *T. gondii* lose their innate fear of cats and instead preferentially seek out locations perfumed with cat urine over locations perfumed with rabbit urine, exactly the opposite of normal rat behavior.

But the most dramatic example of pathogens modifying host behavior belongs to *Wolbachia*. These bacteria manipulate the sexual behavior of their host to maximize their dissemination. As described earlier, *Wolbachia* are passed vertically into offspring through eggs. If they live in a male, however, they hit a dead end, as they are excluded from sperm. In some species of *Drosophila*, *Wolbachia* modify the sperm of their host so that they can fertilize the eggs only of infected females. This modification creates a reproductive advantage for infected females over uninfected females, so that the overall proportion of *Wolbachia* carriers increases. In other host species, a *Wolbachia* infection kills males but spares females, increasing the number of females in the population and thus the number of individuals that can produce eggs to pass on the infection. In a few types of wasp, *Wolbachia* infections enable the females to produce eggs that develop parthenogenetically without the need for fertilization by sperm; in this species, males have been completely eliminated. For some of its hosts, *Wolbachia* has become an indispensable symbiont, and curing the infection causes death of the host. In one case, humans are making use of this dependence: the filarial nematode that causes African river blindness is difficult to kill with antiparasite medications, but when people with river blindness are treated with antibiotics that cure the nematode's *Wolbachia* infection, the nematode infection is also arrested.

Pathogens Evolve Rapidly

The complexity and specificity of the molecular interactions between pathogens and their host cells might suggest that virulence would be difficult to acquire by random mutation. Yet, new pathogens are constantly emerging, and old pathogens are constantly changing in ways that make familiar infections difficult to treat. Pathogens have two great advantages that enable them to evolve rapidly. First, they replicate very quickly, providing a great deal of material for the engine of natural selection. Whereas humans and chimpanzees have acquired a 2% difference in genome sequences over about 8 million years of divergent evolution, poliovirus manages a 2% change in its genome in 5 days, about the time it takes the virus to pass from the human mouth to the gut. Second, this rapid genetic variation is acted on by powerful selective pressures provided by the host's adaptive immune system and by modern medicine, which destroy pathogens that fail to change.

A small-scale example of the constant battle between infection and immunity is the phenomenon of **antigenic variation**. An important adaptive immune response against many pathogens is the production of antibodies that recognize specific antigens on the pathogen surface (discussed in Chapter 24). Many pathogens evade complete elimination by antibodies by changing these antigens during the course of an infection. Some parasites, for example, undergo programmed rearrangements of the genes encoding their surface antigens. The most striking example occurs in African trypanosomes such as *Trypanosoma brucei*, a parasite that causes sleeping sickness and is spread by an insect vector (*T. brucei* is a close relative of *T. cruzi* (see Figure 25–27), but it replicates extracellularly rather than inside cells). *T. brucei* is covered with a single type of glycoprotein, called *variant-specific glycoprotein* (*VSG*), which elicits a protective antibody response that rapidly clears most of the parasites. The trypanosome genome, however, contains about 1000 *VSG* genes, each encoding a VSG with distinct antigenic properties. Only one of these genes is expressed at any one time, by being copied into an active expression site in the genome. The *VSG* gene expressed can be changed repeatedly by gene rearrangements that copy new alleles into the expression site. In this way, a few trypanosomes with an altered VSG escape the antibody-mediated clearance, replicate, and cause a recurrence of disease, leading to a chronic cyclic infection (Figure 25–37).

A different type of antigenic variation occurs during the course of infection by viruses that have error-prone replication mechanisms. Retroviruses, for example, acquire on average one point mutation every replication cycle, because the viral reverse transcriptase that produces DNA from the viral RNA genome cannot correct nucleotide misincorporation errors. A typical, untreated HIV infection may eventually involve HIV genomes with every possible point mutation. In some ways, the high mutation rate is beneficial for the pathogen. By a microevolutionary process of mutation and selection within each host, the virus can switch from infecting macrophages to infecting T-cells (as described earlier), and, once treatment is begun, it can quickly acquire resistance to drugs. If the reverse transcriptase error rate were too high, however, deleterious mutations might accumulate too rapidly to allow the virus to survive. Furthermore, rapid viral diversification in one host does not necessarily lead to rapid evolution of the virus in the population, as a mutated virus may not be able to infect a new host. For HIV-1, the extent of this constraint can be estimated by examining the sequence diversity among different individuals infected with the virus. Remarkably, only about one-third of the nucleotide positions in the coding sequence of the virus are invariant, and nucleotide sequences in some parts of the genome, such as the *env* gene, can differ by as much as 30%. This extraordinary genomic plasticity greatly complicates attempts to develop vaccines

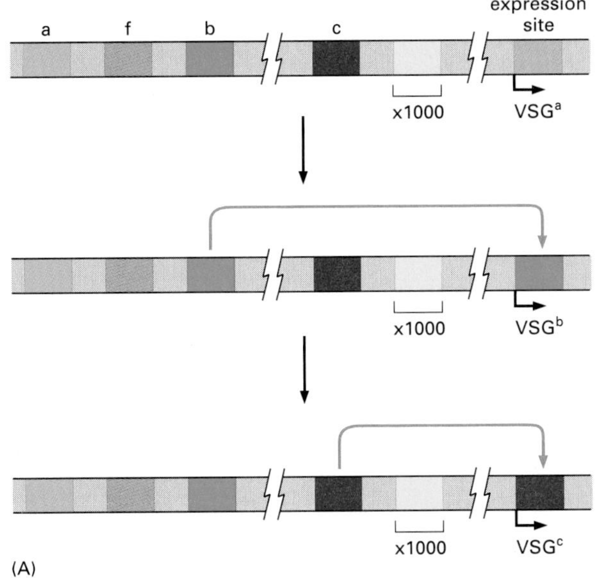

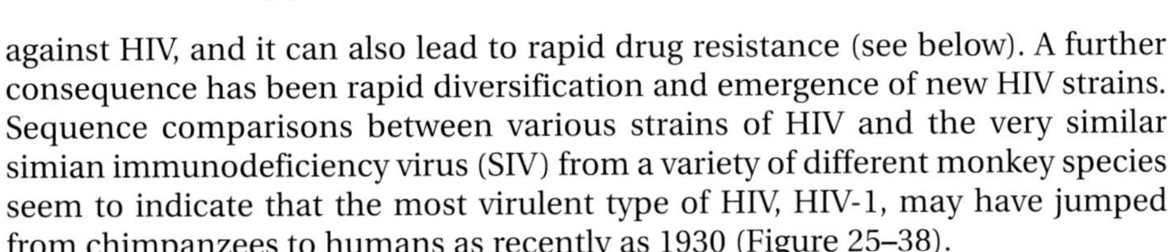

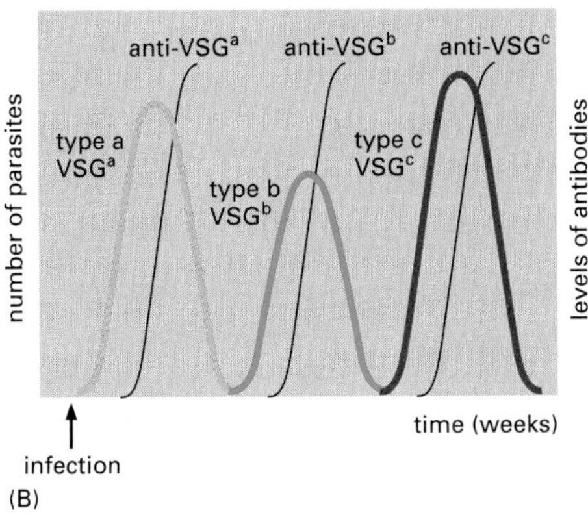

Figure 25–37 Antigenic variation in trypanosomes. (A) There are about 1000 distinct *VSG* genes in *Trypanosoma brucei* but only one site for *VSG* gene expression. An inactive gene is copied into the expression site by gene conversion, where it is now expressed. Rare switching events allow the trypanosome to change repeatedly the surface antigen it expresses. (B) A person infected with trypanosomes expressing VSG^a quickly mounts a protective antibody response, which results in clearance of most of the parasites expressing this antigen. However, a few of the trypanosomes may have switched to expression of VSG^b, which can now proliferate until they are cleared by anti-VSG^b antibodies. By that time, however, some parasites will have switched to VSG^c, and so the cycle repeats seemingly indefinitely.

against HIV, and it can also lead to rapid drug resistance (see below). A further consequence has been rapid diversification and emergence of new HIV strains. Sequence comparisons between various strains of HIV and the very similar simian immunodeficiency virus (SIV) from a variety of different monkey species seem to indicate that the most virulent type of HIV, HIV-1, may have jumped from chimpanzees to humans as recently as 1930 (Figure 25–38).

Rapid evolution in bacteria frequently takes place by horizontal gene transfer rather than by point mutation. Most of this horizontal gene transfer is mediated by the acquisition of plasmids and bacteriophages. Bacteria readily pick up pathogenicity islands and virulence plasmids (see Figure 25–5) from other bacteria. Once a bacterium has acquired a new set of virulence-related genes, it may quickly establish itself as a new cause of human epidemics. *Yersinia pestis*, for example, is an infection endemic to rats and other rodents that first appeared in human history in 542 A.D., when the city of Constantinople was devastated by plague. Sequence comparisons of *Y. pestis* to its close relative *Y. pseudotuberculosis*, which causes a severe diarrheal disease, suggest that *Y. pestis* may have emerged as a distinct strain only a few thousand years ago, not long before its debut in the field of urban devastation.

The emergence and evolution of new infectious disease is in many cases exacerbated by changes in human behavior. For example, crowded and filthy living conditions in medieval cities contributed to the spread of plague to humans from the natural rodent host. The tendency of modern humans to live at high population densities in large cities has also created the opportunity for infectious organisms to initiate epidemics, such as influenza, tuberculosis, and AIDS, which could not have spread so rapidly or so far among sparser human populations. Air travel can, in principle, allow an asymptomatic, newly infected host to carry an epidemic to any previously unexposed population within a few hours or days. Modern agricultural practices also foster the emergence of certain types of infectious agents that could not have easily developed in the hunter-gatherer societies of early humans. Influenza viruses, for example, are unusual in that their genome consists of several (usually eight) strands of RNA. When two strains of influenza infect the same host, the strands of the two strains can reshuffle to form a new and distinct type of influenza virus. Prior to 1900, the influenza strain that infected humans caused a very mild disease; a separate influenza strain infected fowl such as ducks and chickens but could not infect humans. Both the avian and the human strains, however, are capable of infecting pigs, and can recombine in a pig host to form new strains that can cause serious human disease. In communities where pigs are farmed together with chickens or turkeys, newly recombined influenza strains periodically emerge and cause worldwide epidemics. The first and most serious of these epidemics was the "Spanish flu" epidemic of 1918, which killed more people than did World War I.

Figure 25–38 Diversification of HIV-1, HIV-2, and related strains of SIV. The genetic distance between any two viral isolates is found by following the shortest path joining them in the tree. HIV-1 is divided into two groups, major (M) and outlier (O). The HIV-1 M group is responsible for the global AIDS epidemic. HIV-1 M is further subdivided into several subtypes, A through G. Subtype B is dominant in America and Europe, B, C, and E predominate in Asia, and all subtypes are found in Africa. At least two monkey viruses, chimpanzee and mandrill, are more closely related to HIV-1 than is HIV-2, suggesting that HIV-1 and HIV-2 arose independently of one another. HIV-1 and SIV-CPZ (chimpanzee) are estimated to have diverged about 1930. This tree was constructed from the nucleotide sequences of the *gag* gene, using a database containing about 16,000 sequences from virus samples isolated around the world.

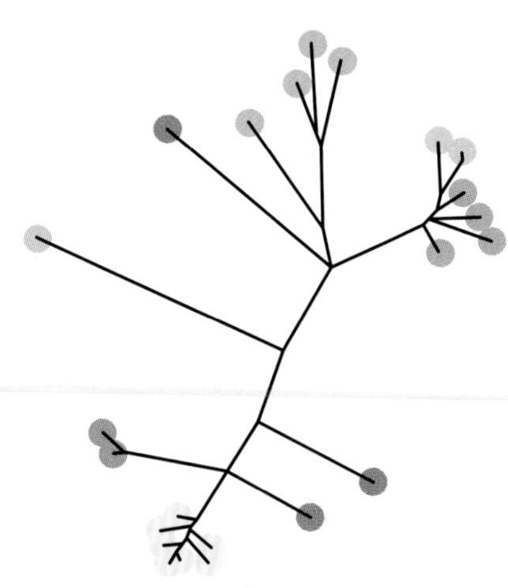

HIV-1 M group
HIV-1 O group
SIV chimpanzee
SIV mandrill
SIV Sykes' monkey
SIV African green monkey
SIV sooty mangabey
HIV-2

Drug Resistant Pathogens Are a Growing Problem

While some human activities have promoted the spread of certain infectious diseases, advances in public sanitation and in medicine have prevented or alleviated the suffering caused by many others. Effective vaccines and worldwide vaccination programs have eliminated smallpox and severely reduced poliomyelitis, and many deadly childhood infections such as mumps and measles are now rarities in wealthy industrialized nations. However, there are still many widespread and devastating infectious diseases, such as malaria, for which no effective vaccines are available. Of equal importance is the development of drugs that cure rather than prevent infections. The most successful class are the *antibiotics,* which kill bacteria. Penicillin was one of the first antibiotics used to treat infections in humans, introduced into clinical use just in time to prevent tens of thousands of deaths from infected battlefield wounds in World War II. The rapid evolution of pathogens, however, has enabled targeted bacteria to develop resistance to antibiotics very quickly; the typical lag between introduction of an antibiotic into clinical use and the appearance of resistant strains is only a few years. Similar drug resistance arises rapidly among viruses when infections are treated with antiviral agents. The virus population in an HIV-infected person treated with the reverse transcriptase inhibitor AZT, for example, will acquire complete resistance to the drug within a few months. The current protocol for treatment of HIV infections involves the simultaneous use of three drugs, which helps to minimize the acquisition of resistance.

There are three general strategies by which pathogens develop drug resistance. Pathogens can (1) produce an enzyme that destroys the drug, (2) alter the molecular target of the drug so that it is no longer sensitive to the drug, or (3) prevent access to the target by, for example, actively pumping the drug out of the pathogen. Once a pathogen has chanced upon an effective strategy, the newly acquired or mutated genes that confer that resistance are frequently spread throughout the pathogen population and may even be transferred to pathogens of different species that are treated with the same drug. The highly effective and very expensive antibiotic *vancomycin,* for example, has been used as a treatment of last resort for many severe hospital-acquired bacterial infections that are already resistant to most other known antibiotics. Vancomycin prevents one step in bacterial cell wall biosynthesis, by binding to part of the growing peptidoglycan chain and preventing it from being cross-linked to other chains. Resistance can arise if the bacterium synthesizes a different type of cell wall, using different subunits that do not bind vancomycin. The most effective type of vancomycin resistance depends on a transposon that encodes seven genes. The products of these genes work together to sense the presence of vancomycin, shut down the normal pathway for bacterial cell wall synthesis, and generate a different type of cell wall. Although the joining of these genes into a single transposon must have been a difficult evolutionary task (it took 15 years for vancomycin resistance to develop, rather than the typical lag of a year or two), the transposon can now be readily transmitted to many other pathogenic bacterial species.

Like most other aspects of infectious disease, the problem of drug resistance has been exacerbated by human behavior. Many patients take antibiotics for viral diseases that are not helped by the drugs, including influenza, colds, and earaches. Persistent and chronic misuse of antibiotics in this way can eventually result in the antibiotic resistance of the normal flora, which can then transfer the resistance to pathogens. Several antibiotic-resistant outbreaks of infectious diarrhea caused by *Shigella flexneri*, for example, originated in this way. The problem is particularly severe in countries where antibiotics are available without a physician's prescription, as in Brazil, where more than 80% of the strains of *S. flexneri* found in infected patients are resistant to four or more antibiotics.

Antibiotics are also misused in agriculture, where they are commonly employed as food additives to promote the growth of healthy animals. An antibiotic closely related to vancomycin was commonly added to cattle feed in Europe. Resistance arising in the normal flora of these animals is widely believed to be one of the original sources for the vancomycin resistance that is now seen in life-threatening human infections.

Summary

All pathogens share the ability to interact with host cells in ways that promote replication and spread of the pathogen, but these host-pathogen interactions are diverse. Pathogens often colonize the host by adhering to or invading through the epithelial surfaces that line the lungs, gut, bladder, and other surfaces in direct contact with the environment. Intracellular pathogens, including all viruses and many bacteria and protozoa, replicate inside a host cell, which they invade by one of a variety of mechanisms. Viruses rely largely on receptor-mediated endocytosis for host cell entry, while bacteria exploit cell adhesion and phagocytic pathways. Protozoa employ unique invasion strategies that usually require significant metabolic expense. Once inside, intracellular pathogens seek out a niche that is favorable for their replication, frequently altering host cell membrane traffic and exploiting the cytoskeleton for intracellular movement. Besides altering the behavior of individual host cells, pathogens frequently alter the behavior of the host organism in ways that favor spread to a new host. Pathogens evolve rapidly, so new infectious diseases frequently emerge, and old diseases acquire new ways to evade human attempts at treatment, prevention, and eradication.

INNATE IMMUNITY

Humans are exposed to millions of potential pathogens daily, through contact, ingestion, and inhalation. Our ability to avoid infection depends in part on the adaptive immune system (discussed in Chapter 24), which remembers previous encounters with specific pathogens and destroys them when they attack again. Adaptive immune responses, however, are slow to develop on first exposure to a new pathogen, as specific clones of B and T cells have to become activated and expand; it can therefore take a week or so before the responses are effective. By contrast, a single bacterium with a doubling time of one hour can produce almost 20 million progeny, a full-blown infection, in a single day. Therefore, during the first critical hours and days of exposure to a new pathogen, we rely on our **innate immune system** to protect us from infection.

Innate immune responses are not specific to a particular pathogen in the way that the adaptive immune responses are. They depend on a group of proteins and phagocytic cells that recognize conserved features of pathogens and become quickly activated to help destroy invaders. Whereas the adaptive immune system arose in evolution less than 500 million years ago and is confined to vertebrates, innate immune responses have been found among both vertebrates and invertebrates, as well as in plants, and the basic mechanisms that regulate them are conserved. As discussed in Chapter 24, the innate immune responses in vertebrates are also required to activate adaptive immune responses.

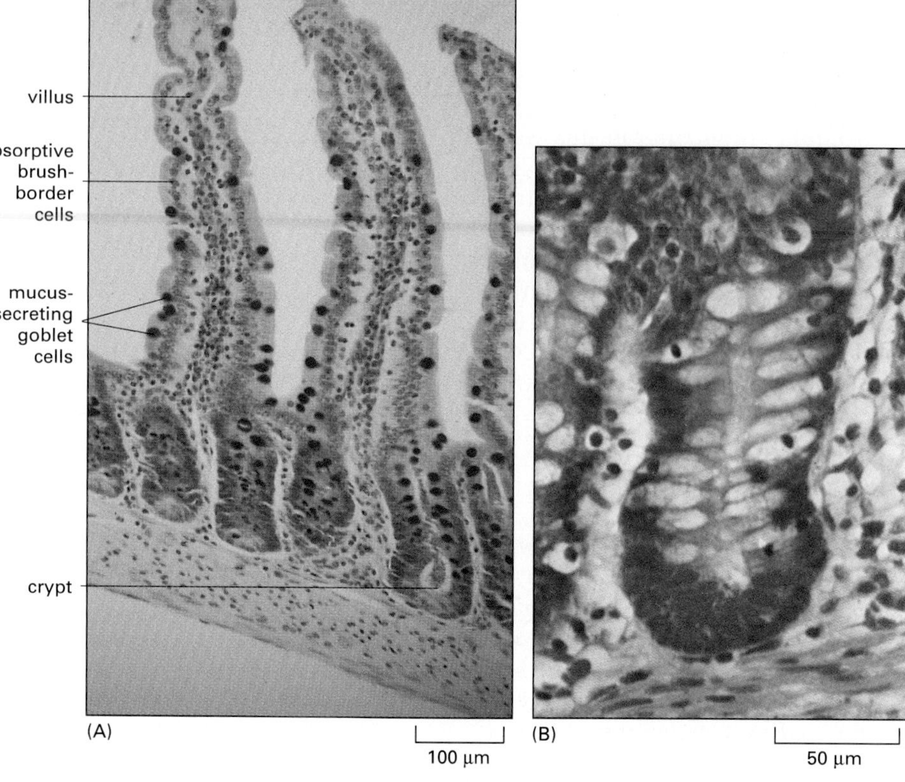

Figure 25–39 Epithelial defenses against microbial invasion. (A) Cross section through the wall of the human small intestine, showing three villi. Goblet cells secreting mucus are stained *magenta*. The protective mucus layer covers the exposed surfaces of the villi. At the base of the villi lie the *crypts* where the epithelial cells proliferate. (B) Close-up view of a crypt, stained using a method that renders the granules in the Paneth cells *scarlet*. These cells secrete large quantities of antimicrobial peptides and defensins into the intestinal lumen. (B, courtesy of H.G. Burkitt, from P.R. Wheater, Functional Histology, 2nd edn. London: Churchill–Livingstone, 1987.)

Epithelial Surfaces Help Prevent Infection

In vertebrates, the skin and other epithelial surfaces, including those lining the lung and gut (Figure 25–39), provide a physical barrier between the inside of the body and the outside world. Tight junctions (discussed in Chapter 19) between neighboring cells prevent easy entry by potential pathogens. The interior epithelial surfaces are also covered with a mucus layer that protects these surfaces against microbial, mechanical, and chemical insults; many amphibians and fish also have a mucus layer covering their skin. The slimy mucus coating is made primarily of secreted mucin and other glycoproteins, and it physically helps prevent pathogens from adhering to the epithelium. It also facilitates their clearance by beating cilia on the epithelial cells (discussed in Chapter 22).

The mucus layer also contains substances that kill pathogens or inhibit their growth. Among the most abundant of these are antimicrobial peptides, called **defensins**, which are found in all animals and plants. They are generally short (12–50 amino acids), positively charged, and have hydrophobic or amphipathic domains in their folded structure. They constitute a diverse family with a broad spectrum of antimicrobial activity, including the ability to kill or inactivate Gram-negative and Gram-positive bacteria, fungi (including yeasts), parasites (including protozoa and nematodes), and even enveloped viruses like HIV. Defensins are also the most abundant protein type in neutrophils (see below), which use them to kill phagocytosed pathogens.

It is still uncertain how defensins kill pathogens. One possibility is that they use their hydrophobic or amphipathic domains to insert into the membrane of their victims, thereby disrupting membrane integrity. Some of their selectivity for pathogens over host cells may come from their preference for membranes that do not contain cholesterol. After disrupting the membrane of the pathogen, the positively-charged peptides may also interact with various negatively-charged targets within the microbe, including DNA. Because of the relatively nonspecific nature of the interaction between defensins and the microbes they kill, it is difficult for the microbes to acquire resistance to the defensins. Thus, in principle, defensins might be useful therapeutic agents to combat infection, either alone or in combination with more traditional drugs.

Human Cells Recognize Conserved Features of Pathogens

Microorganisms do occasionally breach the epithelial barricades. It is then up to the innate and adaptive immune systems to recognize and destroy them, without harming the host. Consequently, the immune systems must be able to distinguish self from nonself. We discuss in Chapter 24 how the adaptive immune system does this. The innate immune system relies on the recognition of particular types of molecules that are common to many pathogens but are absent in the host. These pathogen-associated molecules (called *pathogen-associated immunostimulants*) stimulate two types of innate immune responses—*inflammatory responses* (discussed below) and phagocytosis by cells such as neutrophils and macrophages. Both of these responses can occur quickly, even if the host has never been previously exposed to a particular pathogen.

The **pathogen-associated immunostimulants** are of various types. Procaryotic translation initiation differs from eucaryotic translation initiation in that *formylated methionine*, rather than regular methionine, is generally used as the first amino acid. Therefore, any peptide containing formylmethionine at the N-terminus must be of bacterial origin. Formylmethionine-containing peptides act as very potent chemoattractants for neutrophils, which migrate quickly to the source of such peptides and engulf the bacteria that are producing them (see Figure 16–96).

In addition, the outer surface of many microorganisms is composed of molecules that do not occur in their multicellular hosts, and these molecules also act as immunostimulants. They include the peptidoglycan cell wall and flagella of bacteria, as well as lipopolysaccharide (LPS) on Gram-negative bacteria (Figure 25–40) and teichoic acids on Gram-positive bacteria (see Figure 25–4D). They also include molecules in the cell walls of fungi such as zymosan, glucan, and chitin. Many parasites also contain unique membrane components that act as immunostimulants, including glycosylphosphatidylinositol in *Plasmodium*.

Short sequences in bacterial DNA can also act as immunostimulants. The culprit is a "CpG motif", which consists of the unmethylated dinucleotide CpG flanked by two 5′ purine residues and two 3′ pyrimidines. This short sequence is at least twenty times less common in vertebrate DNA than in bacterial DNA, and it can activate macrophages, stimulate an inflammatory response, and increase antibody production by B cells.

Figure 25–40 Structure of lipopolysaccharide (LPS). On the *left* is the 3-dimensional structure of a molecule of LPS with the fatty acids shown in *yellow* and the sugars in *blue*. The molecular structure of the base of LPS is shown on the *right*. The hydrophobic membrane anchor is made up of two linked glucosamine sugars attached to three phosphates and six fatty acid tails. This basic structure is elaborated by attachment of a long, usually highly branched, chain of sugars. This drawing shows the simplest type of LPS that will allow *E. coli* to live; it has just two sugar molecules in the chain, both 3-deoxy-D-manno-octulosonic acid. At the position marked by the *arrow*, wild-type Gram-negative bacteria also attach a *core saccharide* made up of eight to twelve linked sugars and a long *O antigen*, which is made up of an oligosaccharide unit that is repeated many (up to 40) times. The sugars making up the core saccharide and O antigen vary from one bacterial species to another and even among different strains of the same species. All forms of LPS are highly immunogenic.

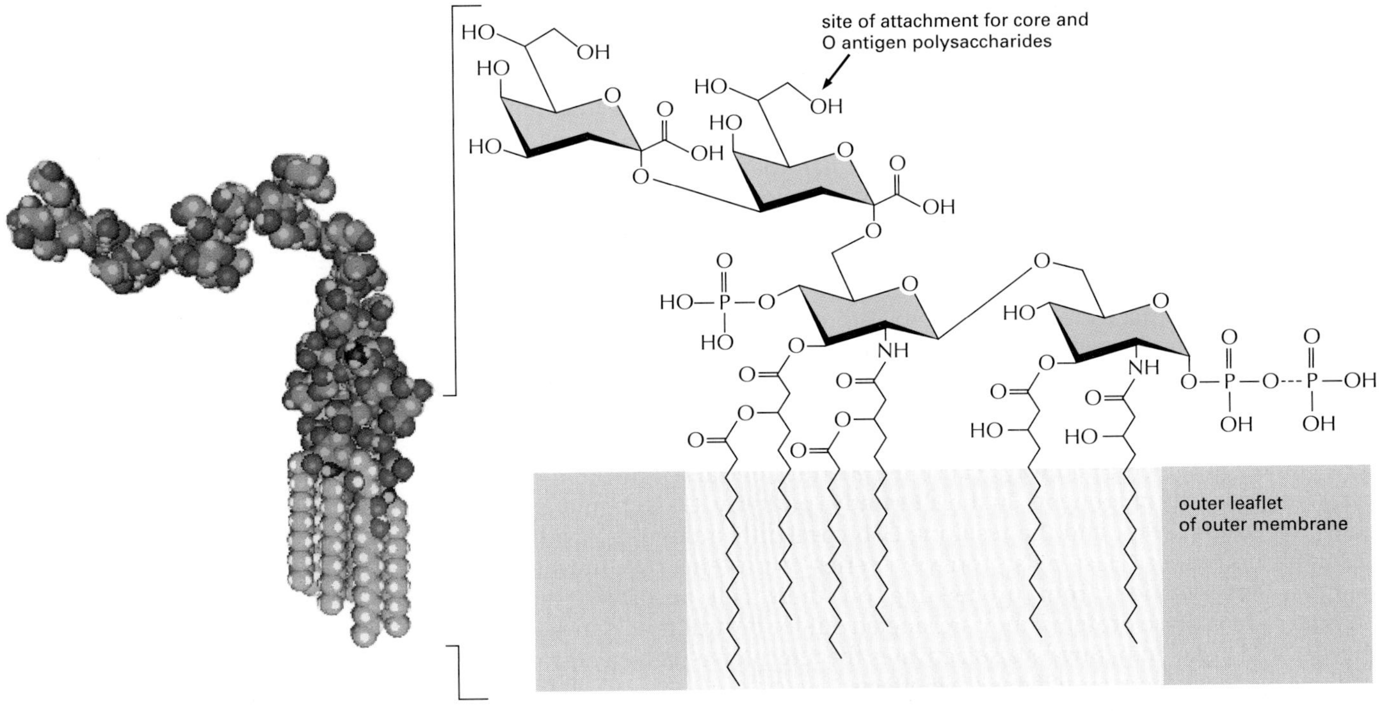

site of attachment for core and O antigen polysaccharides

outer leaflet of outer membrane

The various classes of pathogen-associated immunostimulants often occur on the pathogen surface in repeating patterns. They are recognized by several types of dedicated receptors in the host, that are collectively called *pattern recognition receptors*. These receptors include soluble receptors in the blood (components of the complement system) and membrane-bound receptors on the surface of host cells (members of the Toll-like receptor family). The cell-surface receptors have two functions: they initiate the phagocytosis of the pathogen, and they stimulate a program of gene expression in the host cell for stimulating innate immune responses. The soluble receptors also aid in the phagocytosis and, in some cases, the direct killing of the pathogen.

Complement Activation Targets Pathogens for Phagocytosis or Lysis

The **complement system** consists of about 20 interacting soluble proteins that are made mainly by the liver and circulate in the blood and extracellular fluid. Most are inactive until they are triggered by an infection. They were originally identified by their ability to amplify and "complement" the action of antibodies, but some components of complement are also pattern recognition receptors that can be activated directly by pathogen-associated immunostimulants.

The *early complement components* are activated first. There are three sets of these, belonging to three distinct pathways of complement activation—the *classical pathway*, the *lectin pathway*, and the *alternative pathway*. The early components of all three pathways act locally to activate C3, which is the pivotal component of complement (Figure 25–41). Individuals with a deficiency in C3 are subject to repeated bacterial infections. The early components and C3 are all proenzymes, which are activated sequentially by proteolytic cleavage. The cleavage of each proenzyme in the series activates the next component to generate a serine protease, which cleaves the next proenzyme in the series, and so on. Since each activated enzyme cleaves many molecules of the next proenzyme in the chain, the activation of the early components consists of an amplifying, proteolytic cascade.

Many of these cleavages liberate a biologically active small peptide fragment and a membrane-binding larger fragment. The binding of the large fragment to a cell membrane, usually the surface of a pathogen, helps to carry out the next reaction in the sequence. In this way, complement activation is confined largely to the particular cell surface where it began. The larger fragment of C3, called C3b, binds covalently to the surface of the pathogen. Once in place, it not only acts as a protease to catalyze the subsequent steps in the complement cascade, but it also is recognized by specific receptors on phagocytic cells that enhance the ability of these cells to phagocytose the pathogen. The smaller fragment of C3 (called C3a), as well as fragments of C4 and C5 (see Figure 25–41), act independently as diffusible signals to promote an inflammatory response by recruiting phagocytes and lymphocytes to the site of infection.

The **classical pathway** is activated by IgG or IgM antibody molecules (discussed in Chapter 24) bound to the surface of a microbe. **Mannan-binding lectin**, the protein that initiates the second pathway of complement activation, is a serum protein that forms clusters of six carbohydrate-binding heads around a central collagen-like stalk. This assembly binds specifically to mannose and fucose residues in bacterial cell walls that have the correct spacing and orientation to match up perfectly with the six carbohydrate-binding sites, providing a good example of a pattern recognition receptor. These initial binding events in the classical and lectin pathways cause the recruitment and activation of the early complement components. In the **alternative pathway**, C3 is spontaneously activated at low levels, and the resulting C3b covalently attaches to both host cells and pathogens. Host cells produce a series of proteins that prevent the complement reaction from proceeding on their cell surfaces. Because pathogens lack these proteins, they are singled out for destruction. Activation of the classical or lectin pathways also activates the alternative pathway through a positive feedback loop, amplifying their effects.

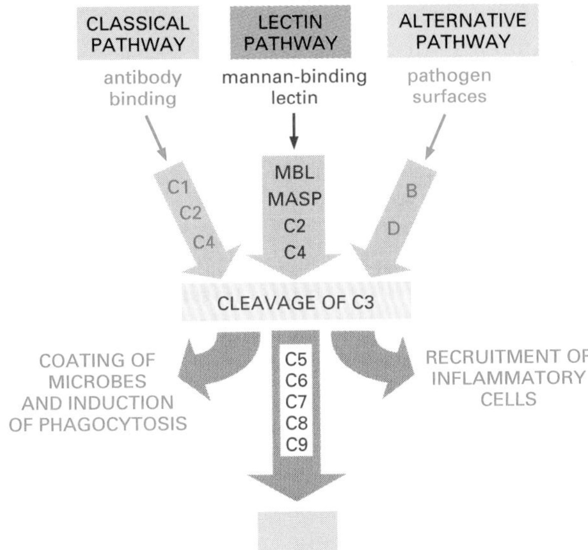

Figure 25–41 The principal stages in complement activation by the classical, lectin, and alternative pathways. In all three pathways, the reactions of complement activation usually take place on the surface of an invading microbe, such as a bacterium. C1–C9 and factors B and D are the reacting components of the complement system; various other components regulate the system. The early components are shown within *gray arrows*, while the late components are shown within a *brown arrow*.

Membrane-immobilized C3b, produced by any of the three pathways, triggers a further cascade of reactions that leads to the assembly of the late components to form *membrane attack complexes* (Figure 25–42). These complexes assemble in the pathogen membrane near the site of C3 activation and have a characteristic appearance in negatively stained electron micrographs, where they are seen to form aqueous pores through the membrane (Figure 25–43). For this reason, and because they perturb the structure of the bilayer in their vicinity, they make the membrane leaky and can, in some cases, cause the microbial cell to lyse, much like the defensins mentioned earlier.

The self-amplifying, inflammatory, and destructive properties of the complement cascade make it essential that key activated components be rapidly inactivated after they are generated to ensure that the attack does not spread to nearby host cells. Deactivation is achieved in at least two ways. First, specific inhibitor proteins in the blood or on the surface of host cells terminate the cascade, by either binding or cleaving certain components once they have been activated by proteolytic cleavage. Second, many of the activated components in the cascade are unstable; unless they bind immediately to either an appropriate component in the cascade or to a nearby membrane, they rapidly become inactive.

Toll-like Proteins Are an Ancient Family of Pattern Recognition Receptors

Many of the mammalian cell-surface pattern recognition receptors responsible for triggering host cell gene expression in response to pathogens are members of the **Toll-like receptor** (**TLR**) family. *Drosophila* Toll is a transmembrane protein with a large extracellular domain consisting of a series of leucine-rich repeats

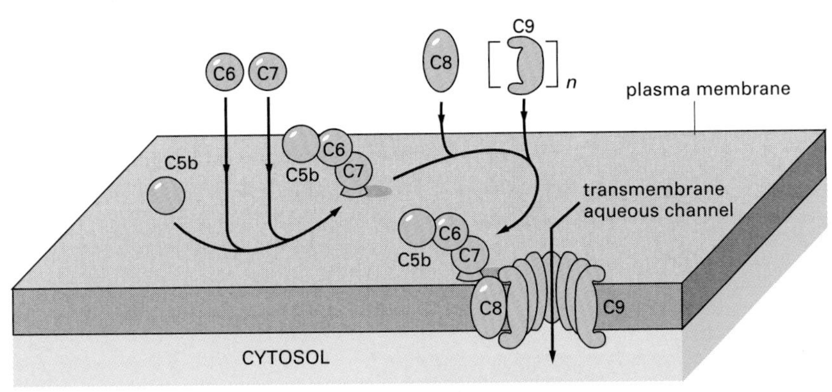

Figure 25–42 Assembly of the late complement components to form a membrane attack complex. When C3b is produced by any of the three activation pathways, it is immobilized on a membrane, where it causes the cleavage of the first of the late components, C5, to produce C5a (not shown) and C5b. C5b remains loosely bound to C3b (not shown) and rapidly assembles with C6 and C7 to form C567, which then binds firmly via C7 to the membrane, as illustrated. To this complex is added one molecule of C8 to form C5678. The binding of a molecule of C9 to C5678 induces a conformational change in C9 that exposes a hydrophobic region and causes C9 to insert into the lipid bilayer of the target cell. This starts a chain reaction in which the altered C9 binds a second molecule of C9, where it can bind another molecule of C9, and so on. In this way, a large transmembrane channel is formed by a chain of C9 molecules.

(see Figure 15–76). It was originally identified as a protein involved in the establishment of dorso-ventral polarity in developing fly embryos (discussed in Chapter 21). It is also involved, however, in the adult fly's resistance to fungal infections. The intracellular signal transduction pathway activated downstream of Toll when a fly is exposed to a pathogenic fungus leads to the translocation of the NF-κB protein (discussed in Chapter 15) into the nucleus, where it activates the transcription of various genes, including those encoding antifungal defensins. Another member of the Toll family in *Drosophila* is activated by exposure to pathogenic bacteria, leading to the production of an antibacterial defensin.

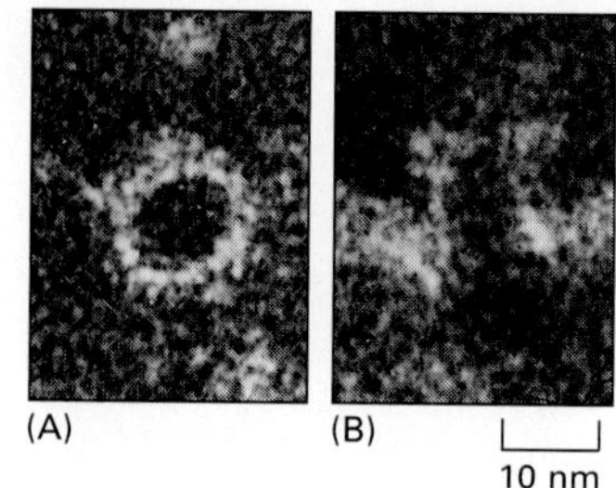

Figure 25–43 Electron micrographs of negatively stained complement lesions in the plasma membrane of a red blood cell. The lesion in (A) is seen *en face*, while that in (B) is seen from the side as an apparent transmembrane channel. The negative stain fills the channels, which therefore look black. (From R. Dourmashkin, *Immunology* 35:205–212, 1978. © Blackwell Scientific.)

Humans have at least ten TLRs, several of which have been shown to play important parts in innate immune recognition of pathogen-associated immunostimulants, including lipopolysaccharide, peptidoglycan, zymosan, bacterial flagella, and CpG DNA. As with *Drosophila* Toll family members, the different human TLRs are activated in response to different ligands, although many of them use the NF-κB signaling pathway (Figure 25–44). In mammals, TLR activation stimulates the expression of molecules that both initiate an inflammatory response (discussed below) and help induce adaptive immune responses. TLRs are abundant on the surface of macrophages and neutrophils, as well as on the epithelial cells lining the lung and gut. They act as an alarm system to alert both the innate and adaptive immune systems that an infection is brewing.

Molecules related to Toll and TLRs are apparently involved in innate immunity in all multicellular organisms. In plants, proteins with leucine-rich repeats and with domains homologous to the cytosolic portion of the TLRs are required for resistance to fungal, bacterial, and viral pathogens (Figure 25–45). Thus, at least two parts of the innate immune system—the defensins and the TLRs—seem to be evolutionarily very ancient, perhaps predating the split between

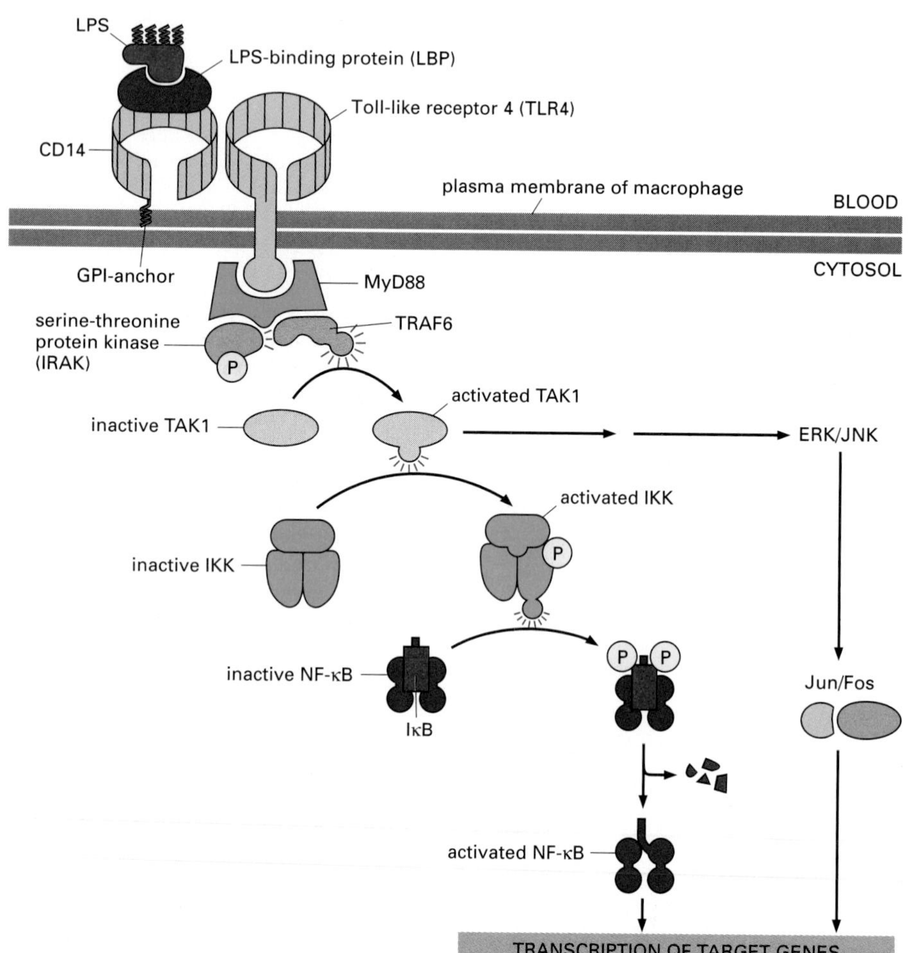

Figure 25–44 The activation of a macrophage by lipopolysaccharide (LPS). LPS is bound by LPS-binding protein (LBP) in the blood, and the complex binds to the GPI-anchored receptor CD14 on the macrophage surface. The ternary complex then activates Toll-like receptor 4 (TLR4). Activated TLR4 recruits the adaptor protein MyD88, which interacts with the serine-threonine protein kinase IRAK. Recruitment of IRAK to the activated receptor complex results in its autophosphorylation and association with another adaptor protein, TRAF6. TRAF6, in turn, associates with and activates a MAP kinase kinase kinase, TAK1. Via several intermediate steps, TAK1 activation leads to the phosphorylation and activation of the IκB kinase (IKK). IKK phosphorylates the NF-κB inhibitor, IκB, inducing its degradation and releasing NF-κB. By way of additional MAP kinases (ERK and JNK), TAK1 also activates the AP-1 transcription family members Jun and Fos, which, together with NF-κB, activate the transcription of genes that promote immune and inflammatory responses (see also Figure 15–74).

Figure 25–45 Microbial disease in a plant. These tomato leaves are infected with the leaf mold fungus *Cladosporium fulvum*. Resistance to this type of infection depends on recognition of a fungal protein by a host receptor that is structurally related to the TLRs. (Courtesy of Jonathan Jones.)

animals and plants over a billion years ago. Their conservation during evolution indicates the importance of these innate responses in the defense against microbial pathogens.

Phagocytic Cells Seek, Engulf, and Destroy Pathogens

In all animals, invertebrate as well as vertebrate, the recognition of a microbial invader is usually quickly followed by its engulfment by a phagocytic cell. Plants, however, lack this type of innate immune response. In vertebrates, *macrophages* reside in tissues throughout the body and are especially abundant in areas where infections are likely to arise, including the lungs and gut. They are also present in large numbers in connective tissues, the liver, and the spleen. These long-lived cells patrol the tissues of the body and are among the first cells to encounter invading microbes. The second major family of phagocytic cells in vertebrates, the *neutrophils*, are short-lived cells, which are abundant in blood but are not present in normal, healthy tissues. They are rapidly recruited to sites of infection both by activated macrophages and by molecules such as formylmethionine-containing peptides released by the microbes themselves.

Macrophages and neutrophils display a variety of cell-surface receptors that enable them to recognize and engulf pathogens. These include pattern recognition receptors such as TLRs. In addition, they have cell-surface receptors for the Fc portion of antibodies produced by the adaptive immune system, as well as for the C3b component of complement. Ligand binding to any of these receptors induces actin polymerization at the site of pathogen attachment, causing the phagocyte's plasma membrane to surround the pathogen and engulf it in a large membrane-enclosed phagosome (Figure 25–46).

Once the pathogen has been phagocytosed, the macrophage or neutrophil unleashes an impressive armory of weapons to kill it. The phagosome is acidified and fuses with lysosomes, which contain lysozyme and acid hydrolases that can degrade bacterial cell walls and proteins. The lysosomes also contain defensins, which make up about 15% of the total protein in neutrophils. In addition, the phagocytes assemble an *NADPH oxidase complex* on the phagosomal membrane that catalyzes the production of a series of highly toxic oxygen-derived compounds, including superoxide (O_2^-), hypochlorite (HOCl, the active ingredient in bleach), hydrogen peroxide, hydroxyl radicals, and nitric oxide (NO). The production of these toxic compounds is accompanied by a transient increase in oxygen consumption by the cells, called the *respiratory burst*. Whereas macrophages will generally survive this killing frenzy and continue to patrol tissues for other pathogens, neutrophils usually die. Dead and dying neutrophils are a major component of the pus that forms in acutely infected wounds. The distinctive greenish tint of pus is due to the abundance in neutrophils of the copper-containing enzyme myeloperoxidase, which is one of the components active in the respiratory burst.

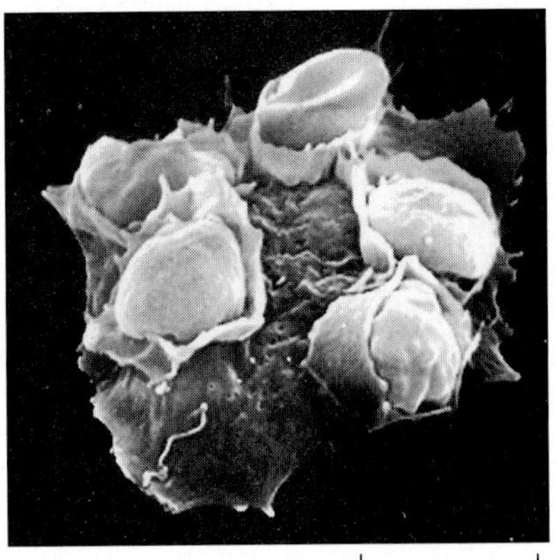

Figure 25–46 Phagocytosis. This scanning electron micrograph shows a macrophage in the midst of consuming five red blood cells that have been coated with an antibody against a surface glycoprotein. (From E.S. Gold et al., *J Exp. Med.* 190:1849–1856, 1999. © The Rockefeller University Press.)

If a pathogen is too large to be successfully phagocytosed (if it is a large parasite such as a nematode, for example), a group of macrophages, neutrophils, or eosinophils (discussed in Chapter 22) will gather around the invader. They will secrete their defensins and other lysosomal products by exocytosis and will also release the toxic products of the respiratory burst (Figure 25–47). This barrage is generally sufficient to destroy the pathogen.

Many pathogens have developed strategies that allow them to avoid being ingested by phagocytes. Some Gram-positive bacteria coat themselves with a very thick, slimy polysaccharide coat, or *capsule*, that is not recognized by complement or any phagocyte receptor. Other pathogens are phagocytosed but avoid being killed; as we saw earlier, *Mycobacterium tuberculosis* prevents the maturation of the phagosome and thereby survives. Some pathogens escape the phagosome entirely, and yet others secrete enzymes that detoxify the products of the respiratory burst. For such wily pathogens, these first lines of defense are insufficient to clear the infection, and adaptive immune responses are required to contain them.

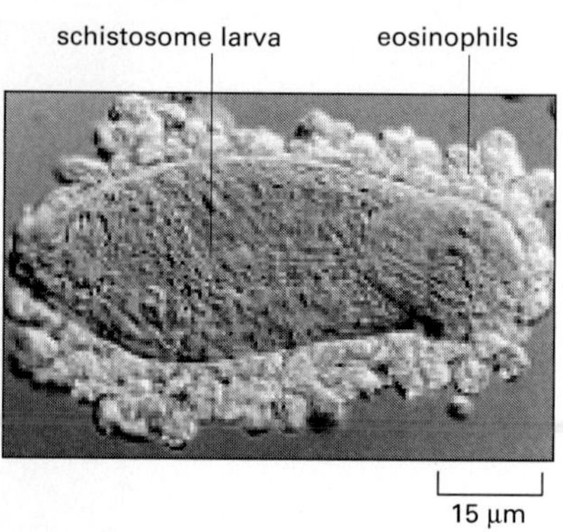

Figure 25–47 Eosinophils attacking a schistosome larva. Large parasites, such as worms, cannot be ingested by phagocytes. When the worm is coated with antibody or complement, however, eosinophils and other white blood cells can recognize and attack it. (Courtesy of Anthony Butterworth.)

Activated Macrophages Recruit Additional Phagocytic Cells to Sites of Infection

When a pathogen invades a tissue, it almost always elicits an **inflammatory response**. This response is characterized by pain, redness, heat, and swelling at the site of infection, all caused by changes in local blood vessels. The blood vessels dilate and become permeable to fluid and proteins, leading to local swelling and an accumulation of blood proteins that aid in defense, including the components of the complement cascade. At the same time, the endothelial cells lining the local blood vessels are stimulated to express cell adhesion proteins (discussed in Chapter 19) that facilitate the attachment and extravasion of white blood cells, including neutrophils, lymphocytes, and monocytes (the precursors of macrophages).

The inflammatory response is mediated by a variety of signaling molecules. Activation of TLRs results in the production of both lipid signaling molecules such as prostaglandins and protein (or peptide) signaling molecules such as cytokines (discussed in Chapter 15), all of which contribute to the inflammatory response. The proteolytic release of complement fragments also contribute. Some of the cytokines produced by activated macrophages are chemoattractants (known as *chemokines*). Some of these attract neutrophils, which are the first cells recruited in large numbers to the site of the new infection. Others later attract monocytes and dendritic cells. The dendritic cells pick up antigens from the invading pathogens and carry them to nearby lymph nodes, where they present the antigens to lymphocytes to marshal the forces of the adaptive immune system (discussed in Chapter 24). Other cytokines trigger *fever*, a rise in body temperature. On balance, fever helps the immune system in the fight against infection, since most bacterial and viral pathogens grow better at lower temperatures, whereas adaptive immune responses are more potent at higher temperatures.

Some proinflammatory signaling molecules stimulate endothelial cells to express proteins that trigger blood clotting in local small vessels. By occluding the vessels and cutting off blood flow, this response can help prevent the pathogen from entering the bloodstream and spreading the infection to other parts of the body.

The same inflammatory responses, however, which are so effective at controlling local infections, can have disastrous consequences when they occur in a disseminated infection in the bloodstream, a condition called *sepsis*. The systemic release of proinflammatory signaling molecules into the blood causes dilation of blood vessels, loss of plasma volume, and widespread blood clotting, which is an often fatal condition known as *septic shock*. Inappropriate or overzealous inflammatory responses are also associated with some chronic conditions, such as *asthma* (Figure 25–48).

Just as with phagocytosis, some pathogens have developed mechanisms to either prevent the inflammatory response or, in some cases, take advantage of it to spread the infection. Many viruses, for example, encode potent cytokine antagonists that block aspects of the inflammatory response. Some of these are simply modified forms of cytokine receptors, encoded by genes acquired by the viral genome from the host. They bind the cytokines with high affinity and block their activity. Some bacteria, such as *Salmonella*, induce an inflammatory response in the gut at the initial site of infection, thereby recruiting macrophages and neutrophils that they then invade. In this way, the bacteria hitch a ride to other tissues in the body.

Virus-Infected Cells Take Drastic Measures to Prevent Viral Replication

The pathogen-associated immunostimulants on the surface of bacteria and parasites that are so important in eliciting innate immune responses are generally not present on the surface of viruses. Viral proteins are constructed by the host cell ribosomes, and the membranes of enveloped viruses are composed of host cell lipids. The only unusual molecule associated with viruses is the double-stranded RNA (dsRNA) that is an intermediate in the life cycle of many viruses. Host cells can detect the presence of dsRNA and initiate a program of drastic responses in attempt to eliminate it.

The program occurs in two steps. First, the cells degrade the dsRNA into small fragments (about 21–25 nucleotide pairs in length). These fragments bind to any single-stranded RNA (ssRNA) in the host cell with the same sequence as either strand of the dsRNA fragment, leading to the destruction of the ssRNA. This dsRNA-directed ssRNA destruction is the basis of the technique of *RNA interference* (RNAi) that is used by researchers to block specific gene expression (discussed in Chapter 8). Second, the dsRNA induces the host cell to produce and secrete two cytokines—*interferon* α *(IFN-*α*)* and *interferon* β *(IFN-*β*)*, which act in both an autocrine fashion on the infected cell and a paracrine fashion on uninfected neighbors. The binding of the interferons to their cell-surface receptors stimulates specific gene transcription by the Jak/STAT intracellular signaling pathway (see Figure 15–63), leading to the activation of a latent ribonuclease, which nonspecifically degrades ssRNA. It also leads to the activation of a protein kinase that phosphorylates and inactivates the protein synthesis initiation factor eIF-2, shutting down most protein synthesis in the embattled host cell. Apparently, by destroying most of the RNA it contains and transiently halting most protein synthesis, the cell inhibits viral replication without killing itself. In some cases, however, a cell infected with a virus is persuaded by white blood cells to destroy itself to prevent the virus from replicating.

Natural Killer Cells Induce Virus-Infected Cells to Kill Themselves

Another way that the interferons help vertebrates defend themselves against viruses is by stimulating both innate and adaptive cellular immune responses. In Chapter 24, we discuss how interferons enhance the expression of class I MHC proteins, which present viral antigens to cytotoxic T lymphocytes (see Figure 24–48). Here, we consider how interferons enhance the activity of **natural killer cells** (**NK cells**), which are part of the innate immune system. Like cytotoxic T cells, NK cells destroy virus-infected cells by inducing the infected cell to kill itself by undergoing apoptosis. Unlike T cells, however, NK cells do not express antigen-specific receptors. How, then, do they distinguish virus-infected cells from uninfected cells?

NK cells monitor the level of class I MHC proteins, which are expressed on the surface of most vertebrate cells. The presence of high levels of these proteins inhibits the killing activity of NK cells, so that the NK cells selectively kill cells expressing low levels, including both virally-infected cells and some cancer cells (Figure 25–49). Many viruses have developed mechanisms to inhibit the expression of class I MHC molecules on the surface of the cells they infect, in order to avoid detection by cytotoxic T lymphocytes. Adenovirus and HIV, for example, encode proteins that block class I MHC gene transcription. Herpes simplex virus and cytomegalovirus block the peptide translocators in the ER membrane that

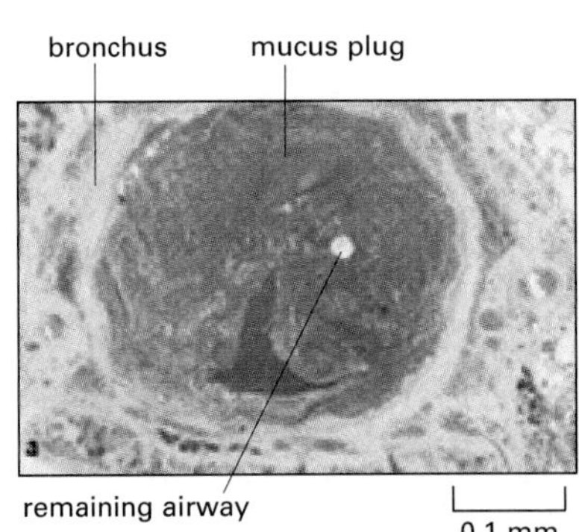

Figure 25–48 Inflammation of the airways in chronic asthma restricts breathing. Light micrograph of a section through the bronchus of a patient who died of asthma. There is almost total occlusion of the airway by a mucus plug. The mucus plug is a dense inflammatory infiltrate that includes eosinophils, neutrophils, and lymphocytes. (Courtesy of Thomas Krausz.)

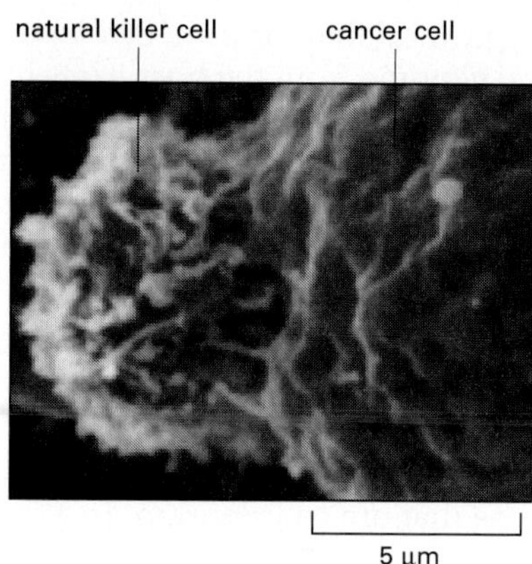

Figure 25–49 A natural killer (NK) cell attacking a cancer cell. The NK cell is the smaller cell on the *left*. This scanning electron micrograph was taken shortly after the NK cell attached, but before it induced the cancer cell to kill itself. (Courtesy of J.C. Hiserodt, in Mechanisms of Cytotoxicity by Natural Killer Cells [R.B. Herberman and D. Callewaert, eds.]. New York: Academic Press, 1995.)

transport proteasome-derived peptides from the cytosol into the lumen of the ER; such peptides are required for newly-made class I MHC proteins to assemble in the ER membrane and be transported through the Golgi apparatus to the cell surface (see Figure 24–58). Cytomegalovirus causes the retrotranslocation of class I MHC proteins from the ER membrane into the cytosol, where they are rapidly degraded by proteasomes. Proteins encoded by still other viruses prevent the delivery of assembled class I MHC proteins from the ER to the Golgi apparatus, or from the Golgi apparatus to the plasma membrane. By evading recognition by cytotoxic T cells in these ways, however, a virus incurs the wrath of NK cells. The local production of IFN-α and IFN-β activates the killing activity of NK cells and also increases the expression of class I MHC proteins in uninfected cells. The cells infected with a virus that blocks class I MHC expression are thereby exposed and become the victims of the activated NK cells. Thus, it is difficult or impossible for viruses to hide from both the innate and adaptive immune systems simultaneously.

Both NK cells and cytotoxic T lymphocytes kill infected target cells by inducing them to undergo apoptosis before the virus has had a chance to replicate. It is not surprising, then, that many viruses have acquired mechanisms to inhibit apoptosis, particularly early in infection. As discussed in Chapter 17, apoptosis depends on an intracellular proteolytic cascade, which the cytotoxic cell can trigger either through the activation of cell-surface death receptors or by injecting a proteolytic enzyme into the target cell (see Figure 24–46). Viral proteins can interfere with nearly every step in these pathways. In some cases, however, viruses encode proteins that act late in their replication cycle to induce apoptosis in the host cell, thereby releasing progeny virus that can infect neighboring cells.

The battle between pathogens and host defenses is remarkably balanced. At present, humans seem to be gaining a slight advantage, using public sanitation measures, vaccines, and drugs to aid the efforts of our innate and adaptive immune systems. However, infectious and parasitic diseases are still the leading cause of death worldwide, and new epidemics such as AIDS continue to emerge. The rapid evolution of pathogens and the almost infinite variety of ways that they can invade the human body and elude immune responses will prevent us from ever winning the battle completely.

Summary

The innate immune responses are the first line of defense against invading pathogens. They are also required to initiate specific adaptive immune responses. Innate immune responses rely on the body's ability to recognize conserved features of pathogens that are not present in the uninfected host. These include many types of molecules on microbial surfaces and the double-stranded RNA of some viruses. Many of these pathogen-specific molecules are recognized by Toll-like receptor proteins, which are found in plants and in invertebrate and vertebrate animals. In vertebrates, microbial surface molecules also activate complement, a group of blood proteins that act together to disrupt the membrane of the microorganism, to target microorganisms for phagocytosis by macrophages and neutrophils, and to produce an inflammatory response. The phagocytic cells use a combination of degradative enzymes, antimicrobial peptides, and reactive oxygen species to kill the invading microorganisms. In addition, they release signaling molecules that trigger an inflammatory response and begin to marshal the forces of the adaptive immune system. Cells infected with viruses produce interferons, which induce a series of cell responses to inhibit viral replication and activate the killing activities of natural killer cells and cytotoxic T lymphocytes.

References

General

Cossart P, Boquet P, Normark S & Rappuoli R (eds) (2000) Cellular Microbiology. Washington: ASM Press.

Flint SJ, Enquist LW, Krug RM et al. (2000) Principles of Virology: Molecular Biology, Pathogenesis, and Control. Washington: ASM Press.

Janeway CA, Travers P, Walport M & Shlomchik M (2001) Immunobiology: The Immune System in Health and Disease, 5th edn. New York: Garland Science.

Salyers A & Whitt DD (1994) Bacterial Pathogenesis: A Molecular Approach. Washington: ASM Press.

Schaechter M, Engleberg NC, Isenstein BI & Medoff G (eds) (1988) Mechanisms of Microbial Disease. Philadelphia: Lippincott, Williams & Wilkins.

Introduction to Pathogens

Baltimore D (1971) Expression of animal virus genomes. *Bacteriol. Rev.* 35, 235–241.

Crick FHC & Watson JD (1956) Structure of small viruses. *Nature* 177, 374–475.

Galan JE & Collmer A (1999) Type III secretion machines: bacterial devices for protein delivery into host cells. *Science* 284, 1322–1328.

Hacker J & Kaper JB (2000) Pathogenicity islands and the evolution of microbes. *Annu. Rev. Microbiol.* 54, 641–679.

Heidelberg JF, Eisen JA, Nelson WC et al. (2000) DNA sequence of both chromosomes of the cholera pathogen *Vibrio cholerae*. *Nature* 406, 477–483.

Lang-Unnasch N & Murphy AD (1988) Metabolic changes of the malaria parasite during the transition from the human to the mosquito host. *Annu. Rev. Microbiol.* 52, 561–590.

Lorber B (1996) Are all diseases infectious? *Ann. Intern. Med.* 125, 844–51.

Madhani HD & Fink GR (1998) The control of filamentous differentiation and virulence in fungi. *Trends Cell Biol.* 1998, 348–353.

Poulin R & Morand S (2000) The diversity of parasites. *Q. Rev. Biol.* 75, 277–293.

Prusiner SB (1996) Molecular biology and genetics of prion diseases. *Cold Spring Harb. Symp. Quant. Biol.* 61, 473–493.

Rixon FJ (1990) Structure and assembly of herpesviruses. *Semin. Virol.* 1, 477–487.

Cell Biology of Infection

Baranowski E, Ruiz-Jarabo CM & Domingo E (2001) Evolution of cell recognition by viruses. *Science* 292, 1102–1105.

Berdoy M, Webster JP & Macdonald DW (2000) Fatal attraction in rats infected with *Toxoplasma gondii*. *Proc. R. Soc. Lond. B. Biol. Sci.* 267, 1591–1594.

Berger EA, Murphy PM & Farber JM (1999) Chemokine receptors as HIV-1 coreceptors: roles in viral entry, tropism, and disease. *Annu. Rev. Immunol.* 17, 657–700.

Bliska JB, Galan JE & Falkow S (1993) Signal transduction in the mammalian cell during bacterial attachment and entry. *Cell* 73, 903–920.

Deitsch KW, Moxon ER & Wellems TE (1997) Shared themes of antigenic variation and virulence in bacterial, protozoal, and fungal infections. *Microbiol. Mol. Biol. Rev.* 61, 281–293.

Dramsi S & Cossart P (1988) Intracellular pathogens and the actin cytoskeleton. *Annu. Rev. Cell Dev. Biol.* 14, 137–166.

Finlay BB & Cossart P (1997) Exploitation of mammalian host cell functions by bacterial pathogens. *Science* 276, 718–725.

Galan JE (1996) Molecular genetic bases of Salmonella entry into host cells. *Mol. Microbiol.* 20, 263–271.

Garoff H, Hewson R & Opstelten DJ (1998) Virus maturation by budding. *Microbiol. Mol. Biol. Rev.* 62, 1171–1190.

Hacker J & Carniel E (2001) Ecological fitness, genomic islands and bacterial pathogenicity. A Darwinian view of the evolution of microbes. *EMBO Rep.* 2, 376–381.

Hackstadt T (2000) Redirection of host vesicle trafficking pathways by intracellular parasites. *Traffic* 1, 93–99.

Jones NC (1990) Transformation by the human adenoviruses. *Semin. Cancer Biol.* 1, 425–435.

Kaariainen L & Ranki M (1984) Inhibition of cell functions by RNA-virus infections. *Annu. Rev. Microbiol.* 38, 91–109.

Kenny B, DeVinney R, Stein M et al. (1997) Enteropathogenic *E. coli* (EPEC) transfers its receptor for intimate adherence into mammalian cells. *Cell* 91, 511–520.

Koch AL (1981) Evolution of antibiotic resistance gene function. *Microbiol. Rev.* 45, 355–378.

Lyles DS (2000) Cytopathogenesis and inhibition of host gene expression by RNA viruses. *Microbiol. Mol. Biol. Rev.* 64, 709–724.

Neu HC (1992) The crisis in antibiotic resistance. *Science* 257, 1064–1073.

Overbaugh J & Bangham CR (2001) Selection forces and constraints on retroviral sequence variation. *Science* 292, 1106–1109.

Rosenshine I & Finlay BB (1993) Exploitation of host signal transduction pathways and cytoskeletal functions by invasive bacteria. *Bioessays* 15, 17–24.

Sibley LD & Andrews NW (2000) Cell invasion by un-palatable parasites. *Traffic* 1, 100–106.

Skehel JJ & Wiley DC (2000) Receptor binding and membrane fusion in virus entry: the influenza hemagglutinin. *Annu. Rev. Biochem.* 69, 531–569.

Sodeik B (2000) Mechanisms of viral transport in the cytoplasm. *Trends Microbiol.* 8, 465–472.

Stephens EB & Compans RW (1988) Assembly of animal viruses at cellular membranes. *Annu. Rev. Microbiol.* 42, 489–516.

Tilney LG & Portnoy DA (1989) Actin filaments and the growth, movement, and spread of the intracellular bacterial parasite, *Listeria monocytogenes*. *J. Cell Biol.* 109, 1597–1608.

Innate Immunity

Aderem A & Ulevitch RJ (2000) Toll-like receptors in the induction of the innate immune response. *Nature* 406, 782–787.

Aderem A & Underhill DM (1999) Mechanisms of phagocytosis in macrophages. *Annu. Rev. Immunol.* 17, 593–623.

Ganz T & Lehrer RI (1998) Antimicrobial peptides of vertebrates. *Curr. Opin. Immunol.* 10, 41–44.

Guidotti LG & Chisari FV (2001) Noncytolytic control of viral infections by the innate and adaptive immune response. *Annu. Rev. Immunol.* 19, 65–91.

Hampton MB, Kettle AJ & Winterbourn CC (1998) Inside the neutrophil phagosome: oxidants, myeloperoxidase, and bacterial killing. *Blood* 92, 3007–3017.

Hancock RE & Scott MG (2000) The role of antimicrobial peptides in animal defenses. *Proc. Natl. Acad. Sci. USA* 97, 8856–8861.

Imler JL & Hoffmann JA (2000) Toll and Toll-like proteins: an ancient family of receptors signaling infection. *Rev. Immunogenet.* 2, 294–304.

Kimbrell DA & Beutler B (2001) The evolution and genetics of innate immunity. *Nat. Rev. Genet.* 2, 256–267.

Medzhitov R & Janeway C Jr (2000) Innate immune recognition: mechanisms and pathways. *Immunol. Rev.* 173, 89–97.

Medzhitov R & Janeway C Jr (2000) Innate immunity. *N. Engl. J. Med.* 343, 338–344.

Muller-Eberhard HJ (1988) Molecular organization and function of the complement system. *Annu. Rev. Biochem.* 57, 321–347.

Murphy PM (2001) Viral exploitation and subversion of the immune system through chemokine mimicry. *Nat. Immunol.* 2, 116–22.

Ploegh HL (1998) Viral strategies of immune evasion. *Science* 280, 248–253.

Super M & Ezekowitz RA (1992) The role of mannose-binding proteins in host defense. *Infect. Agents Dis.* 1, 194–199.

Timonen T & Helander TS (1997) Natural killer cell-target cell interactions. *Curr. Opin. Cell Biol.* 9, 667–673.

Tomlinson S (1993) Complement defense mechanisms. *Curr. Opin. Immunol.* 5, 83–89.

Watanabe S & Arai Ki (1996) Roles of the JAK-STAT system in signal transduction via cytokine receptors. *Curr. Opin. Genet. Dev.* 6, 587–596.

Yang RB, Mark MR, Gray A et al. (1998) Toll-like receptor-2 mediates lipopolysaccharide-induced cellular signalling. *Nature* 395, 284–288.

Glossary

α helix—*see* **alpha helix**

ABC transporter proteins
Large superfamily of membrane transport proteins that use the energy of hydrolysis of ATP to transfer peptides and a variety of small molecules across membranes.

acetyl
Chemical group derived from acetic acid. Acetyl groups are important in metabolism and are added covalently to some proteins as a posttranslational modification.

acetyl CoA
Small water-soluble molecule that carries acetyl groups in cells. It consists of an acetyl group linked to coenzyme A (CoA) by an easily hydrolyzable thioester bond. (*See* Figure 2–62.)

acetylcholine receptor
Ion channel that opens in response to binding of acetylcholine, thereby converting a chemical signal into an electrical one. Best understood example of a transmitter-gated channel. Sometimes called the nicotinic acetylcholine receptor to distinguish it from a muscarinic acetylcholine receptor, which is a G-protein-linked cell-surface receptor.

acetylcholine
Neurotransmitter that functions at a class of chemical synapses known as cholinergic synapses. Found both in the brain and in the peripheral nervous system. It is the neurotransmitter at vertebrate neuromuscular junctions. (*See* Figure 15–9.)

acid
Substance that releases protons when dissolved in water, forming a hydronium ion (H_3O^+).

acid hydrolase
Any of a group of diverse hydrolytic enzymes (including proteases, nucleases, glycosidases, etc.) that have their optimal activity at acid pH (around 5.0) and are found in lysosomes.

acquired immunological tolerance
Unresponsiveness of the immune system to a given foreign antigen that can develop in some circumstances.

acrosomal vesicle
Region at the head end of a sperm cell that contains a sac of hydrolytic enzymes used to digest the protective coating of the egg.

acrosome reaction
Reaction that occurs when a sperm starts to enter an egg, in which the contents of the acrosomal vesicle are released, helping the sperm to penetrate the zona pellucida.

actin
Abundant protein that forms actin filaments in all eucaryotic cells. The monomeric form is sometimes called globular or G-actin; the polymeric form is filamentous or F-actin.

actin-binding protein
Protein that associates with either actin monomers or actin filaments in cells and modifies their properties. Examples include myosin, α-actinin, and profilin.

actin filament (microfilament)
Helical protein filament formed by the polymerization of globular actin molecules. A major constituent of the cytoskeleton of all eucaryotic cells and part of the contractile apparatus of skeletal muscle. (*See* Panel 16–1, p. 909.)

action potential
Rapid, transient, self-propagating electrical excitation in the plasma membrane of a cell such as a neuron or muscle cell. Action potentials, or nerve impulses, make possible long-distance signaling in the nervous system.

activated carrier
Small diffusible molecule in cells that stores easily-exchangeable energy in the form of one or more energy-rich covalent bonds. Examples are ATP and NADPH. Also called a coenzyme.

activation energy
Extra energy that must be possessed by atoms or molecules in addition to their ground-state energy in order to undergo a particular chemical reaction. (*See* Figure 2–44.)

active site
Region of an enzyme surface to which a substrate molecule binds in order to undergo a catalyzed reaction.

active transport
Movement of a molecule across a membrane or other barrier driven by energy other than that stored in the electrochemical gradient of the transported molecule.

acyl group
Functional group derived from a carboxylic acid ($R{-}C{\lessgtr}^{O}_{OH}$). (R represents an alkyl group, such as methyl.)

adaptation
Adjustment of sensitivity following repeated stimulation. This is the mechanism that allows a neuron, a photodetector, or a bacterium to react to small changes in stimuli even against a high background level of stimulation.

adaptin
Protein that binds clathrin to the membrane surface in clathrin-coated vesicles.

adaptive immune response
Response of the vertebrate immune system to a specific antigen that typically generates immunological memory.

adaptor protein
General term for proteins in intracellular signaling pathways that link different proteins in the pathway directly together.

adenomatous polyposis coli (APC)
Tumor suppressor protein that forms part of a protein complex that recruits free cytoplasmic β-catenin and degrades it.

adenosine triphosphate—*see* **ATP**

adenylyl cyclase (adenylate cyclase)
Membrane-bound enzyme that catalyzes the formation of cyclic AMP from ATP. An important component of some intracellular signaling pathways.

adherens junction
Cell junction in which the cytoplasmic face of the plasma membrane is attached to actin filaments. Examples include the adhesion belts linking adjacent epithelial cells and the focal contacts on the lower surface of cultured fibroblasts.

adhesion belt
Beltlike adherens junction that encircles the apical end of an epithelial cell and attaches it to the adjoining cell. Also known as the **zonula adherens**.

adhesion plaque—*see* **focal adhesion.**

adipocyte
A fat cell.

ADP (adenosine 5′-diphosphate)
Nucleotide that is produced by hydrolysis of the terminal phosphate of ATP. It regenerates ATP when phosphorylated by an energy-generating process such as oxidative phosphorylation. (*See* Figure 2–57.)

adrenaline (epinephrine)
Hormone released by chromaffin cells (in the adrenal gland) and by some neurons in response to stress. Produces "fight or flight" responses, including increased heart rate and blood sugar levels.

aerobic
Describes a process that requires, or occurs in the presence of, gaseous oxygen (O_2).

affinity chromatography
Type of chromatography in which the protein mixture to be purified is passed over a matrix to which specific ligands for the required protein are attached, so that the protein is retained on the matrix.

affinity constant (association constant) (K_a)
Measure of the strength of binding of the components in a complex. For components A and B and a binding equilibrium $A + B \rightleftharpoons AB$, the association constant is given by $[AB]/[A][B]$, and is larger the tighter the binding between A and B. (*See also* **dissociation constant**.)

affinity maturation
Progressive increase in the affinity of antibodies for the immunizing antigen with the passage of time after immunization.

Akt—*see* **protein kinase B**

alcohol
Polar organic molecule that contains a functional hydroxyl group (–OH) bound to a carbon atom that is not in an aromatic ring. An example is ethyl alcohol (CH_3CH_2OH).

aldehyde
Organic compound that contains the $-C\begin{smallmatrix}{}^{\nearrow O}\\{}_{\searrow H}\end{smallmatrix}$ group. An example is glyceraldehyde. Can be oxidized to an acid or reduced to an alcohol.

alga (plural **algae**)
Informal term used to describe a wide range of simple unicellular and multicellular eucaryotic photosynthetic organisms. Examples include *Nitella, Volvox*, and *Fucus*.

alkaloid
Small but chemically complex nitrogen-containing metabolite produced by plants as a defense against herbivores. Examples include caffeine, morphine, and colchicine.

alkane (adjective **aliphatic**)
Compound of carbon and hydrogen that has only single covalent bonds. An example is ethane (CH_3CH_3).

alkene
Hydrocarbon with one or more carbon-carbon double bonds. An example is ethylene (CH_2CH_2).

alkyl group
General term for a group of covalently linked carbon and hydrogen atoms such as methyl ($-CH_3$) or ethyl ($-CH_2CH_3$) groups. These groups can be formed by removing a hydrogen atom from an alkane.

allele
One of a set of alternative forms of a gene. In a diploid cell each gene will have two alleles, each occupying the same position (locus) on homologous chromosomes.

allelic exclusion
The expression of an immunoglobulin chain (or T cell receptor chain) gene from only one of the two homologous loci present for that gene in the lymphocyte.

allosteric protein
Protein that changes from one conformation to another when it binds another molecule or when it is covalently modified. The change in conformation alters the activity of the protein and can form the basis of directed movement.

alpha helix (α helix)
Common folding pattern in proteins in which a linear sequence of amino acids folds into a right-handed helix stabilized by internal hydrogen bonding between backbone atoms.

alternative RNA splicing
The production of different proteins from the same RNA transcript by splicing it in different ways.

amide
Molecule containing a carbonyl group linked to an amine.

amine
Chemical group containing nitrogen and hydrogen. It becomes positively charged in water.

amino acid
Organic molecule containing both an amino group and a carboxyl group. Those that serve as the building blocks of proteins are alpha amino acids, having both the amino and carboxyl groups linked to the same carbon atom. (*See* Panel 3–1, pp. 132–133.)

aminoacyl-tRNA synthetase
Enzyme that attaches the correct amino acid to a tRNA molecule to form an aminoacyl-tRNA. (*See* Figure 6–57.)

amino group
Weakly basic functional group derived from ammonia (NH_3) in which one or more hydrogen atoms are replaced by another atom. In aqueous solution it can accept a proton and carry a positive charge.

amino terminus (N terminus)
The end of a polypeptide chain that carries a free α-amino group.

aminoacyl tRNA
Activated form of amino acid used in protein synthesis. Consists of an amino acid linked through a labile ester bond from its carboxyl group to a hydroxyl group on tRNA. (*See* Figure 6–57.)

AMP (adenosine 5′-monophosphate)
One of the four nucleotides in an RNA molecule. Two phosphates are added to AMP to form ATP. (*See* Panel 2–6, pp. 120–121.)

amphipathic
Having both hydrophobic and hydrophilic regions, as in a phospholipid or a detergent molecule.

anabolism
System of biosynthetic reactions in a cell by which large molecules are made from smaller ones.

anaerobic
Describes a cell, organism, or metabolic process that functions in the absence of air or, more precisely, in the absence of molecular oxygen (O_2).

anaphase
Stage of mitosis during which the two sets of chromosomes separate and move away from each other. Composed of **anaphase A** (chromosomes move toward the two spindle poles) and **anaphase B** (spindle poles move apart).

anaphase-promoting complex (APC)
Ubiquitin ligase that promotes the destruction of a set of proteins, some of which initiate the separation of sister chromatids during the metaphase-to-anaphase transition during mitosis.

anchorage dependence
Dependence of cell growth on attachment to a substratum.

anchoring junction
Type of cell junction that attaches cells to neighboring cells or to the extracellular matrix.

angiogenesis
Growth of new blood vessels by sprouting from existing ones.

Ångstrom (Å)
Unit of length used to measure atoms and molecules. Equal to 10^{-10} meter or 0.1 nanometer (nm).

animal pole
In yolky eggs, that end free of yolk that cleaves more rapidly than the vegetal pole.

ankyrin
Protein mainly responsible for attaching the spectrin cytoskeleton to the red blood cell plasma membrane.

antenna complex
Part of a photosystem that captures light energy and channels it into the photochemical reaction center. It consists of protein complexes that bind large numbers of chlorophyll molecules and other pigments.

anterior
Situated toward the head end of the body.

anteroposterior
Describes the axis running from the head to the tail of the animal body.

antibiotic
Substance such as penicillin or streptomycin that is toxic to microorganisms. Usually a product of a particular microorganism or plant.

antibody (immunoglobulin)
Protein produced by B cells in response to a foreign molecule or invading microorganism. Often binds to the foreign molecule or cell extremely tightly, thereby inactivating it or marking it for destruction by phagocytosis or complement-induced lysis.

anticodon
Sequence of three nucleotides in a transfer RNA molecule that is complementary to a three-nucleotide codon in a messenger RNA molecule.

antigen
Molecule that is able to provoke an immune response.

antigenic determinant (epitope)
Specific region of an antigenic molecule that binds to an antibody or a T cell receptor.

antigenic variation
The ability to change the antigens displayed on the cell surface; a property of some pathogenic microorganisms that enables them to evade attack by the immune system.

antigen-presenting cell
Cell that displays foreign antigen complexed with MHC molecules on its surface.

antiparallel
Describes the relative orientation of the two strands in a DNA double helix; the polarity of one strand is oriented in the opposite direction to that of the other.

antiporter
Carrier protein that transports two different ions or small molecules across a membrane in opposite directions, either simultaneously or in sequence.

antisense RNA
RNA complementary to a specific RNA transcript of a gene that can hybridize to the specific RNA and block its function.

APC—*see* **adenomatous polyposis coli; anaphase-promoting complex**

apical
Describes the tip of a cell, a structure, or an organ. The apical surface of an epithelial cell is the exposed free surface, opposite to the basal surface. The basal surface rests on the basal lamina that separates the epithelium from other tissue.

apoptosis
Form of cell death, also known as programmed cell death, in which a 'suicide' program is activated within the cell, leading to fragmentation of the DNA, shrinkage of the cytoplasm, membrane changes and cell death without lysis or damage to neighboring cells. It is a normal phenomenon, occurring frequently in a multicellular organism.

aqueous
Pertaining to water, as for example, in an aqueous solution.

archea (singular **archeon**)
Members of one of the two major divisions of procaryotes (the Archea), the other being the Bacteria.

ARF protein
Monomeric GTPase responsible for regulating both COPI coat assembly and clathrin coat assembly at Golgi membranes.

aromatic
Describes a molecule that contains carbon atoms in a ring, commonly drawn as linked through alternating single and double bonds. Often a molecule related to benzene.

ARP complex (ARP2/3 complex)
Complex of proteins that nucleates actin filament growth from the minus end.

asexual reproduction
Any type of reproduction (such as budding in *Hydra*, binary fission in bacteria, or mitotic division in eucaryotic microorganisms) that does not involve gamete formation and fusion. It produces an individual genetically identical to the parent.

association constant—*see* **affinity constant**

aster
Star-shaped system of microtubules emanating from a centrosome or from a pole of a mitotic spindle.

astral microtubule
In the mitotic spindle, any of the microtubules radiating from the aster which are not attached to a kinetochore of a chromosome.

asymmetric cell division
Cell division that produces two daughter cells that differ, for example in size or in the presence or absence of some cytoplasmic constituent.

atomic weight
Mass of an atom relative to the mass of a hydrogen atom. Essentially equal to the number of protons plus neutrons.

ATP (adenosine 5′-triphosphate)
Nucleoside triphosphate composed of adenine, ribose, and three phosphate groups that is the principal carrier of chemical energy in cells. The terminal phosphate groups are highly reactive in the sense that their hydrolysis, or transfer to another molecule, takes place with release of a large amount of free energy. (*See* Figure 2–26.)

ATP synthase
Enzyme complex in the inner membrane of a mitochondrion and the thylakoid membrane of a chloroplast that catalyzes the formation of ATP from ADP and inorganic phosphate during oxidative phosphorylation and photosynthesis, respectively. Also present in the plasma membrane of bacteria.

ATPase
Enzyme that catalyzes a process involving the hydrolysis of ATP. A large number of different proteins have ATPase activity.

autoantibody
Antibody produced by an individual against a protein, or other potential antigen, of its own cells and tissues. Autoantibodies can cause autoimmune disease.

autocatalysis
Reaction that is catalyzed by one of its products, creating a positive feedback (self-amplifying) effect on the reaction rate.

autocrine signaling
Type of cell signaling in which a cell secretes signal molecules that act on itself or on other adjacent cells of the same type.

autoimmune disease
A pathological state in which the body mounts an immune response against one or more of its own potential antigens.

autophagy
Digestion of worn-out organelles by the cell's own lysosomes.

autoradiography
Technique in which a radioactive object produces an image of itself on a photographic film. The image is called an autoradiograph or autoradiogram.

autosome
Any chromosome other than a sex chromosome.

avidity
Total binding strength of a polyvalent antibody with a polyvalent antigen.

Avogadro's number
6×10^{23}. This is the number of atoms in 1 gram of hydrogen, and thus in the atomic or molecular weight equivalent in grams of any element or molecule.

axon
Long nerve cell process that is capable of rapidly conducting nerve impulses over long distances so as to deliver signals to other cells.

axonal transport
Directed transport of organelles and molecules along a nerve cell axon. It can be anterograde (outward from the cell body) or retrograde (back toward the cell body).

axoneme
Bundle of microtubules and associated proteins that forms the core of a cilium or flagellum in a eucaryotic cell and is responsible for their movements.

β sheet—*see* **beta sheet**

B cell (B lymphocyte)
Type of lymphocyte that makes antibodies.

bacteria (singular **bacterium)**
Members of the Bacteria, one of the two major divisions of procaryotes, the other being the Archea. Most exist as single cells and some cause disease.

bacterial artificial chromosome (BAC)
Cloning vector that can accommodate large pieces of DNA up to 1 million base pairs.

bacteriophage (phage)
Any virus that infects bacteria. Bacteriophages were the first entities used for the study of molecular genetics and are now widely used as cloning vectors.

bacteriorhodopsin
Pigmented protein found in the plasma membrane of a salt-loving bacterium, *Halobacterium halobium*. It pumps protons out of the cell in response to light.

basal
Situated near the base. The basal surface of a cell is opposite the apical surface.

basal body
Short cylindrical array of microtubules plus their associated proteins found at the base of a eucaryotic cell cilium or flagellum. Serves as a nucleation site for the growth of the axoneme. Closely similar in structure to a centriole.

basal lamina (plural **basal laminae)**
Thin mat of extracellular matrix that separates epithelial sheets, and many other types of cells such as muscle or fat cells, from connective tissue.

base
A substance that can accept a proton in solution. The purines and pyrimidines in DNA and RNA are organic nitrogenous bases and are often referred to simply as bases.

base pair
Two nucleotides in an RNA or DNA molecule that are held together by hydrogen bonds—for example, G pairs with C, and A with T or U.

basic
Having the properties of a base.

benign
Describes tumors that are self-limiting in their growth and noninvasive.

benzene
Molecule composed of a six-membered ring of carbon atoms, commonly drawn containing three alternating double bonds. The benzene ring occurs as part of many biological molecules.

beta-catenin (β-catenin)
Multifunctional cytoplasmic protein that is involved in cadherin-mediated cell–cell adhesion, linking cadherins to the actin cytoskeleton. Can also act independently as a gene regulatory protein. Has an important role in animal development as part of a Wnt signaling pathway.

beta sheet (β sheet)
Common structural motif in proteins in which different sections of the polypeptide chain run alongside each other, joined together by hydrogen bonding between atoms of the polypeptide backbone. Also known as a β-pleated sheet.

binding site
A region on the surface of one molecule (usually a protein or nucleic acid) that can interact with another molecule through noncovalent bonding.

biosphere
The world of living organisms.

biotin
Low-molecular-weight compound used as a coenzyme. Useful technically as a covalent label for proteins, allowing them to be detected by the egg protein avidin, which binds extremely tightly to biotin. (*See* Figure 2–63.)

bivalent
A duplicated chromosome paired with its homologous duplicated chromosome at the beginning of meiosis.

black membrane
Artificial planar lipid bilayer membrane.

blastomere
One of the cells formed by the cleavage of a fertilized egg.

blastula
Early stage of an animal embryo, usually consisting of a hollow ball of cells, before gastrulation begins.

blotting
Biochemical technique in which macromolecules separated on an agarose or polyacrylamide gel are transferred to a nylon membrane or sheet of paper, thereby immobilizing them for further analysis. (*See* **Northern blotting, Southern blotting, Western blotting**.)

bond energy
Strength of the chemical linkage between two atoms, measured by the energy in kilocalories or kilojoules needed to break it.

bright-field microscope
The normal light microscope in which the image is obtained by simple transmission of light through the object being viewed.

brush border
Dense covering of microvilli on the apical surface of epithelial cells in the intestine and kidney. The microvilli aid absorption by increasing the surface area of the cell.

budding yeast
Common name often given to the baker's yeast *Saccharomyces cerevisiae*, a common experimental organism, which divides by budding off a smaller cell.

C terminus—*see* **carboxyl terminus**

Ca^{2+}/calmodulin-dependent protein kinase (CaM kinase)
Protein kinase whose activity is regulated by the binding of Ca^{2+}-activated calmodulin (Ca^{2+}/calmodulin), and which indirectly mediates the effects of Ca^{2+} by phosphorylation of other proteins.

cadherin
A member of a family of proteins that mediates Ca^{2+}-dependent cell–cell adhesion in animal tissues.

caged molecule
Organic molecule designed to change into an active form when irradiated with light of a specific wavelength. An example is caged ATP.

calcium pump (Ca^{2+} ATPase)
Transport protein in the membrane of the sarcoplasmic reticulum of muscle cells (and elsewhere) that pumps Ca^{2+} out of the cytoplasm into the sarcoplasmic reticulum using the energy of ATP hydrolysis.

calmodulin
Ubiquitous calcium-binding protein whose binding to other proteins is governed by changes in intracellular Ca^{2+} concentration. Its binding modifies the activity of many target enzymes and membrane transport proteins.

calorie
Unit of heat. One calorie (small "c") is the amount of heat needed to raise the temperature of 1 gram of water by 1°C. A kilocalorie (1000 calories) is the unit used to describe the energy content of foods.

Calvin cycle (Calvin-Benson cycle)
Major metabolic pathway by which CO_2 is incorporated into carbohydrate during the second stage of photosynthesis (carbon fixation) in plants. Also called the carbon-fixation cycle.

CAM—*see* **cell adhesion molecule**

CaM kinase—*see* **Ca^{2+}/calmodulin-dependent protein kinase**

CaM-kinase II
Multifunctional Ca^{2+}/calmodulin-dependent protein kinase found in all animal cells that undergoes autophosphorylation when activated. It is especially abundant in brain and is thought to have a role in learning and memory in vertebrates.

cAMP—*see* **cyclic AMP**

capacitation
Poorly understood process that sperm must go through in the female reproductive tract before they are competent for fertilization.

capsid
Protein coat of a virus, formed by the self-assembly of one or more protein subunits into a geometrically regular structure.

carbohydrate
General term for sugars and related compounds containing carbon, hydrogen, and oxygen, usually with the empirical formula $(CH_2O)n$.

carbon fixation
Process by which green plants incorporate carbon atoms from atmospheric carbon dioxide into sugars. The second stage of photosynthesis.

carbon-fixation cycle—*see* **Calvin cycle**

carbonyl group
Pair of atoms consisting of a carbon atom linked to an oxygen atom by a double bond (C=O).

carboxyl group
Carbon atom linked both to an oxygen atom by a double bond and to a hydroxyl group. Molecules containing a carboxyl group are weak acids—**carboxylic acids** ($-C\genfrac{}{}{0pt}{}{\diagup O}{\diagdown OH}$).

carboxyl terminus (C terminus)
The end of a polypeptide chain that carries a free α-carbonyl group.

carcinogen
Any agent, such as a chemical or a form of radiation, that causes cancer.

carcinogenesis
Generation of a cancer.

carcinoma
Cancer of epithelial cells. The most common form of human cancer.

cardiac muscle
Specialized form of striated muscle found in the heart, consisting of individual heart muscle cells linked together by cell junctions.

carrier protein
Membrane transport protein that binds to a solute and transports it across the membrane by undergoing a series of conformational changes.

cartilage
Form of connective tissue composed of cells (chondrocytes) embedded in a matrix rich in type II collagen and chondroitin sulfate.

caspase
Any of a family of intracellular proteases that are involved in initiating the cellular events of apoptosis.

catabolism
General term for the enzyme-catalyzed reactions in a cell by which complex molecules are degraded to simpler ones with release of energy. Intermediates in these reactions are sometimes called catabolites.

catalyst
Substance that can lower the activation energy of a reaction, thus increasing its rate.

caveola (plural **caveolae**)
Invaginations at the cell surface that bud off internally to form pinocytic vesicles. Thought to form from lipid rafts, regions of membrane rich in certain lipids.

CD28
Cell-surface protein on T cells that binds the co-stimulatory B7 protein on "professional" antigen-presenting cells, providing an additional signal required for the activation of a naïve T cell by antigen.

CD4
Co-receptor protein found on helper T cells that binds to class II MHC molecules outside the antigen-binding site.

CD8
Co-receptor protein found on cytotoxic T cells that binds to class I MHC molecules outside the antigen-binding site.

***cdc* gene**—*see* **cell-division-cycle gene**

Cdk inhibitor protein (CKI)
Protein that binds to and inhibits cyclin-Cdk complexes, primarily involved in the control of G_1 and S phases.

Cdk-activating kinase (CAK)
Protein kinase that phosphorylates Cdks in cyclin-Cdk complexes, activating the Cdk.

Cdk—*see* **cyclin-dependent kinase**

cDNA
DNA molecule made as a copy of messenger RNA and therefore lacking the introns that are present in genomic DNA. **cDNA clones** represent DNA cloned from cDNA and a collection of such clones, usually representing the genes expressed in a particular cell type or tissue, is a **cDNA library**.

cell adhesion molecule (CAM)
Protein on the surface of an animal cell that mediates cell–cell binding or cell–matrix binding.

cell body
Main part of a nerve cell that contains the nucleus. The other parts are the axons and dendrites.

cell coat—*see* **glycocalyx**

cell cortex
Specialized layer of cytoplasm on the inner face of the plasma membrane. In animal cells it is an actin-rich layer responsible for movements of the cell surface.

cell cycle (cell-division cycle)
Reproductive cycle of a cell: the orderly sequence of events by which a cell duplicates its contents and divides into two.

cell division
Separation of a cell into two daughter cells. In eucaryotic cells it entails division of the nucleus (mitosis) closely followed by division of the cytoplasm (cytokinesis).

cell fate
In developmental biology, describes what a particular cell at a given stage of development will normally give rise to.

cell fusion
Process in which the plasma membranes of two cells fuse down at the point of contact between them, allowing the two cytoplasms to mingle.

cell junction
Specialized region of connection between two cells or between a cell and the extracellular matrix.

cell line
Population of cells of plant or animal origin capable of dividing indefinitely in culture.

cell plate
Flattened membrane-bounded structure that forms by fusing vesicles in the cytoplasm of a dividing plant cell and is the precursor of the new cell wall.

cell wall
Mechanically strong extracellular matrix deposited by a cell outside its plasma membrane. It is prominent in most plants, bacteria, algae, and fungi. Not present in most animal cells.

cell-cycle control system
Network of regulatory proteins that governs progression of a eucaryotic cell through the cell cycle.

cell-division-cycle gene (*cdc* gene)
Gene that controls a specific step or set of steps in the eucaryotic cell cycle. Originally identified in yeasts.

cell-free system
Fractionated cell homogenate that retains a particular biological function of the intact cell, and in which biochemical reactions and cell processes can be more easily studied.

cell-mediated immune response
That part of an adaptive immune response in which antigen-specific T cells are activated to perform various functions such as killing infected cells and activating macrophages.

cellularization
The formation of cells around each nucleus in a multinucleate cytoplasm, transforming it into a multicellular structure.

cellulose
Structural polysaccharide consisting of long chains of covalently linked glucose units. It provides tensile strength in plant cell walls.

centimorgan—*see* **genetic map distance**

central lymphoid organ (primary lymphoid organ)
Lymphoid organ in which lymphocytes develop. In adult mammals these are the thymus and bone marrow.

central nervous system (CNS)
Main information-processing organ of the nervous system. In vertebrates it consists of the brain and spinal cord.

centriole
Short cylindrical array of microtubules, closely similar in structure to a basal body. A pair of centrioles is usually found at the center of a centrosome in animal cells.

centromere
Constricted region of a mitotic chromosome that holds sister chromatids together. It is also the site on the DNA where the kinetochore forms that captures microtubules from the mitotic spindle.

centrosome cycle
Duplication of the centrosome (during interphase) and separation of the two new centrosomes (at the beginning of mitosis), which provides two centrosomes to form the poles of the mitotic spindle.

centrosome
Centrally located organelle of animal cells that is the primary microtubule-organizing center and acts as the spindle pole during mitosis. In most animal cells it contains a pair of centrioles.

CG island
Region of DNA with a greater than average density of CG sequences; these regions generally remain unmethylated.

CGN—*see* ***cis*** **Golgi network**

channel protein
Membrane transport protein that forms an aqueous pore in the membrane through which a specific solute, usually an ion, can pass.

chaperone (molecular chaperone)
Protein that helps other proteins avoid misfolding pathways that produce inactive or aggregated polypeptides.

checkpoint
Point in the eucaryotic cell-division cycle where progress through the cycle can be halted until conditions are suitable for the cell to proceed to the next stage.

chelate
Combine reversibly, usually with high affinity, with a metal ion such as iron, calcium, or magnesium.

chemical group
Set of covalently linked atoms, such as a hydroxyl group (–OH) or an amino group ($-NH_2$), the chemical behavior of which is well characterized.

chemiosmotic coupling
Mechanism in which a gradient of hydrogen ions (a pH gradient) across a membrane is used to drive an energy-requiring process, such as ATP production or the rotation of bacterial flagella.

chemokine
Small secreted protein that attracts cells, such as white blood cells, to move towards its source. Important in the functioning of the immune system.

chemotaxis
Directed movement of a cell or organism towards or away from a diffusible chemical.

chiasma (plural **chiasmata)**
X-shaped connection visible between paired homologous chromosomes in division I of meiosis, and which represents a site of crossing-over.

chlorophyll
Light-absorbing green pigment that plays a central part in photosynthesis in bacteria, plants, and algae.

chloroplast
Organelle in green algae and plants that contains chlorophyll and carries out photosynthesis. It is a specialized form of plastid.

cholesterol
Lipid molecule with a characteristic four-ring steroid structure that is an important component of the plasma membranes of animal cells. (*See* Figure 10–10.)

chondrocyte (cartilage cell)
Connective-tissue cell that secretes the matrix of cartilage.

chromaffin cell
Cell that stores adrenaline in secretory vesicles and secretes it in times of stress when stimulated by the nervous system.

chromatid
One copy of a chromosome formed by DNA replication that is still joined at the centromere to the other copy. The two identical chromatids are called sister chromatids.

chromatin
Complex of DNA, histones, and nonhistone proteins found in the nucleus of a eucaryotic cell. The material of which chromosomes are made.

chromatography
Biochemical technique in which a mixture of substances is separated by charge, size, or some other property by allowing it to partition between a moving phase and a stationary phase. (*See* **affinity chromatography, DNA affinity chromatography, high-performance liquid chromatography**.)

chromosomal crossing-over
The exchange of DNA between paired homologous chromosomes in division I of meiosis. It is a sign of genetic recombination and the crossovers (chiasmata) are visible in the light microscope. (*See* Figure 20–10.)

chromosome
Structure composed of a very long DNA molecule and associated proteins that carries part (or all) of the hereditary information of an organism. Especially evident in plant and animal cells undergoing mitosis or meiosis, where each chromosome becomes condensed into a compact rodlike structure visible under the light microscope.

chromosome condensation
Process by which a chromosome becomes packed up into a more compact structure prior to M phase of the cell cycle.

cilium (plural **cilia)**
Hairlike extension of a eucaryotic cell containing a core bundle of microtubules and capable of performing repeated beating movements. Cilia are found in large numbers on the surface of many cells, and are responsible for the swimming of many single-celled organisms.

circadian clock
Internal cyclical process that produces a particular change in a cell or organism with a period of around 24 hours, for example the sleep-wakefulness cycle in humans.

***cis* face**
Face of a Golgi stack at which material enters the organelle. It is adjacent to the *cis* Golgi network.

***cis* Golgi network (CGN)**
Network of interconnected cisternae and tubules which receives vesicles from the endoplasmic reticulum and transfers material to the *cis* face of the Golgi apparatus.

cisterna (plural **cisternae)**
Flattened membrane-bounded compartment, as found in the endoplasmic reticulum or Golgi apparatus.

citric acid cycle (tricarboxylic acid (TCA) cycle, Krebs cycle)
Central metabolic pathway found in aerobic organisms. Oxidizes acetyl groups derived from food molecules to CO_2 and H_2O. In eucaryotic cells it occurs in the mitochondria.

CKI—*see* **Cdk inhibitor protein**

class I MHC molecule
One of the two classes of MHC molecule. It is present on almost all cell types and presents viral peptides on the surface of virus-infected cells, where they are recognized by cytotoxic T cells. (*See* Figure 24–49.)

class II MHC molecule
One of the two classes of MHC molecule. It is present on "professional" antigen-presenting cells and presents foreign peptides to helper T cells. (*See* Figure 24–49.)

class switching
The change from making one class of immunoglobulin (for example IgM) to making another class (for example IgG) that many B cells undergo during the course of an immune response.

classical pathway
A pathway for activating the complement system that is initiated by IgG or IgM antibodies bound to the surface of a microbe.

clathrin
Protein that assembles into a polyhedral cage on the cytosolic side of a membrane so as to form a clathrin-coated pit, which buds off by endocytosis to form an intracellular clathrin-coated vesicle.

clathrin-coated pit
Region of plasma membrane of animal cells that is coated with the protein clathrin on its cytosolic face. Such regions are continually forming and budding off by endocytosis to form intracellular clathrin-coated vesicles containing extracellular fluid and the materials dissolved in it.

cleavage
(1) Physical splitting of a cell into two. (2) Specialized type of cell division seen in many early embryos whereby a large cell becomes subdivided into many smaller cells without growth.

clonal selection theory
Theory that explains how the adaptive immune system can respond to millions of different antigens in a highly specific way. From a population of lymphocytes with a vast repertoire of randomly generated antigen specificities, a given foreign antigen activates (selects) only those cells with the corresponding antigen specificity.

clone
Population of cells or organisms formed by repeated (asexual) division from a common cell or organism. Also used as a verb: "to clone a gene" means to produce many copies of a gene by repeated cycles of replication.

cloning vector
A small DNA molecule, usually derived from a bacteriophage or plasmid, which is used to carry the fragment of DNA to be cloned into the recipient cell, and which enables the DNA fragment to be replicated.

coated vesicle
Small membrane-bounded organelle with a cage of proteins (the coat) on its cytosolic surface. It is formed by the pinching off of a coated region of membrane (**coated pit**). Some coats are made of clathrin, whereas others are made from other proteins.

codon
Sequence of three nucleotides in a DNA or messenger RNA molecule that represents the instruction for incorporation of a specific amino acid into a growing polypeptide chain.

coenzyme
Small molecule tightly associated with an enzyme that participates in the reaction that the enzyme catalyzes, often by forming a covalent bond to the substrate. Examples include biotin, NAD^+, and coenzyme A.

coenzyme A
Small molecule used in the enzymatic transfer of acyl groups in the cell. (*See also* **acetyl CoA**.)

cofactor
Inorganic ion or coenzyme that is required for an enzyme's activity.

cohesin, cohesin complex
Complex of proteins that holds sister chomatids together along their length before their separation.

coiled-coil
Especially stable rodlike structure in proteins which is formed by two of these α helices coiled around each other.

collagen fibril
Extracellular structure formed by self-assembly of secreted fibrillar collagen subunits. An abundant constituent of the extracellular matrix in many animal tissues.

collagen
Fibrous protein rich in glycine and proline that is a major component of the extracellular matrix and connective tissues. Exists in many forms: type I, the most common, is found in skin, tendon, and bone; type II is found in cartilage; type IV is present in basal laminae.

colony-stimulating factor (CSF)
General name for the numerous signal molecules that control the differentiation of blood cells.

colorectal tumor
Common carcinoma of the epithelium lining the colon and rectum.

combinatorial control
Describes the control of a step in a cellular process, such as the initiation of DNA transcription, by a combination of proteins rather than by any individual one.

communicationg junction
Type of cell junction that allows the passage of chemical or electrical signals from one cell to another.

compartment
Regions in the embryo that are formed exclusively from the descendants of a few founder cells; there is no cell movement beween compartments once delimited.

complement system
System of serum proteins activated by antibody–antigen complexes or by microorganisms. Helps eliminate pathogenic microorganisms by directly causing their lysis or by promoting their phagocytosis.

complementary DNA—*see* **cDNA**

complementary
Two nucleic acid sequences are said to be complementary if they can form a perfect base-paired double helix with each other.

complex oligosaccharide
Chain of sugars attached to a glycoprotein that is generated by trimming of the original oligosaccharide attached in the endoplasmic reticulum and subsequent addition of further sugars. (*See* Figure 13–25.)

complex
Assembly of molecules that are held together by noncovalent bonds. Protein complexes perform most cell functions.

condensation reaction
Chemical reaction in which two molecules are covalently linked through –OH groups with the removal of a molecule of water.

condensin, condensin complex
Complex of proteins involved in chromosome condensation prior to mitosis. Target for the M-Cdk.

conditional mutation
A mutation that changes a protein or RNA molecule so that its function is altered only under some conditions, such as at an unusually high or an unusually low temperature.

confocal microscope
Type of light microscope that produces a clear image of a given plane within a solid object. It uses a laser beam as a pinpoint source of illumination and scans across the plane to produce a two-dimensional 'optical section.'

conformation
The spatial arrangement of atoms in three dimensions in a macromolecule such as a protein or nucleic acid.

connective tissue
Any supporting tissue that lies between other tissues and consists of cells embedded in a relatively large amount of extracellular matrix. Includes bone, cartilage, and loose connective tissue.

connective-tissue cell
Any of the various cell types found in connective tissue, e.g. fibroblasts, cartilage cells (chondrocytes), bone cells (osteoblasts and osteocytes), fat cells (adipocytes) and smooth muscle cells.

connexon
Water-filled pore in the plasma membrane formed by a ring of six protein subunits. Part of a gap junction: connexons from two adjoining cells join to form a continuous channel between the two cells.

consensus sequence
Average or most typical form of a sequence that is reproduced with minor variations in a group of related DNA, RNA, or protein sequences. The consensus sequence shows the nucleotide or amino acid most often found at each position. The preservation of a consensus implies that the sequence is functionally important. (*See* Figure 6–12.)

constitutive secretory pathway
Pathway present in all cells by which molecules such as plasma membrane proteins are continually delivered to the plasma membrane from the Golgi apparatus in vesicles that fuse with the plasma membrane. (*See also* **default pathway**.)

constitutive
Produced in constant amount; opposite of regulated.

contact-dependent signaling
Cell–cell communication in which the signal molecule remains bound to the signaling cell and only influences cells that physically contact it.

contractile ring
Ring containing actin and myosin that forms under the surface of animal cells undergoing cell division and contracts to pinch the two daughter cells apart.

convergent extension
Cellular rearrangement within a tissue that causes it to extend in one dimension (e.g. length) and shrink in another (e.g. width).

cooperativity
Phenomenon in which the binding of one ligand molecule to a target molecule promotes the binding of successive ligand molecules. Seen in the assembly of large complexes, as well as in enzymes and receptors composed of multiple allosteric subunits, where it sharpens the response to a ligand. (*See* Figure 15–22.)

cortical granule
Specialized secretory vesicle present under the plasma membrane of unfertilized eggs, including those of mammals; after fertilization it is involved in preventing the entry of further sperm.

co-translational
Describes import of a protein into the endoplasmic reticulum before the polypeptide chain is completely synthesized.

co-transport (coupled transport)
Membrane transport process in which the transfer of one molecule depends on the simultaneous or sequential transfer of a second molecule.

coupled reaction
Linked pair of chemical reactions in which the free energy released by one of the reactions serves to drive the other.

covalent bond
Stable chemical link between two atoms produced by sharing one or more pairs of electrons.

crista (plural **cristae**)
(1) One of the folds of the inner mitochondrial membrane. (2) A sensory structure in the inner ear.

critical concentration
Concentration of a protein monomer, such as actin or tubulin, that is in equilibrium with the assembled form of the protein (i.e. assembled into actin filaments or microtubules respectively). (*See* Panel 16–2, pp. 912–913.)

crossing-over—*see* **chromosomal crossing-over**

cryoelectron microscopy
Electron microscopy technique in which the objects to be viewed, such as macromolecules and viruses, are rapidly frozen.

cryptochrome
Flavoprotein responsive to blue light, found in both plants and animals. In animals it is involved in circadian rhythms.

cut-and-paste transposition
Type of movement of a transposable element in which it is cut out of the DNA and inserted into a new site by a special transposase enzyme.

cyclic AMP (cAMP)
Nucleotide that is generated from ATP by adenylyl cyclase in response to stimulation of many types of cell-surface receptors. cAMP acts as an intracellular signaling molecule by activating cyclic-AMP-dependent kinase (protein kinase A, PKA). It is hydrolyzed to AMP by a phosphodiesterase.

cyclic AMP-dependent protein kinase (protein kinase A, PKA)
Enzyme that phosphorylates target proteins in response to a rise in intracellular cyclic AMP.

cyclic GMP
Small soluble intracellular signaling molecule formed from GTP by the enzyme guanylyl cyclase in response to photoreceptor stimulation in the retina.

cyclin
Protein that periodically rises and falls in concentration in step with the eucaryotic cell cycle. Cyclins activate crucial protein kinases (called a cyclin-dependent protein kinase, or Cdk) and thereby help control progression from one stage of the cell cycle to the next.

cyclin-Cdk complex
Protein complexes that are formed periodically during the eucaryotic cell cycle as the level of cyclin increases, and in which the cyclin-dependent kinase (Cdk) becomes partially activated.

cyclin-dependent kinase (Cdk)
Protein kinase that has to be complexed with a cyclin protein in order to act. Different Cdk-cyclin complexes trigger different steps in the cell-division cycle by phosphorylating specific target proteins.

cytochrome *b*-*c*$_1$ complex
Second of the three electron-driven proton pumps in the respiratory chain. It accepts electrons from ubiquinone.

cytochrome oxidase complex
Third of the three electron-driven proton pumps in the respiratory chain. It accepts electrons from cytochrome *c* and generates water using molecular oxygen as an electron acceptor.

cytochrome
Colored, heme-containing protein that transfers electrons during cellular respiration and photosynthesis.

cytokine
Extracellular signal protein or peptide that acts as a local mediator in cell–cell communication.

cytokine receptor
Type of cell-surface receptor whose ligands are cytokines such as interferons, growth hormone and prolactin, and which acts through the Jak-STAT pathway.

cytokinesis
Division of the cytoplasm of a plant or animal cell into two, as distinct from the division of its nucleus (which is mitosis).

cytoplasm
Contents of a cell that are contained within its plasma membrane but, in the case of eucaryotic cells, outside the nucleus.

cytoskeleton
System of protein filaments in the cytoplasm of a eucaryotic cell that gives the cell shape and the capacity for directed movement. Its most abundant components are actin filaments, microtubules, and intermediate filaments.

cytosol
Contents of the main compartment of the cytoplasm, excluding membrane-bounded organelles such as endoplasmic reticulum and mitochondria. Originally defined operationally as the cell fraction remaining after membranes, cytoskeletal components, and other organelles have been removed by low-speed centrifugation.

cytotoxic T cell
Type of T cell responsible for killing infected cells.

$\Delta G°$—*see* **standard free-energy change**

ΔG—*see* **free-energy change**

dalton
Unit of molecular mass. Approximately equal to the mass of a hydrogen atom (1.66×10^{-24} g).

default pathway
Constitutive secretory pathway that automatically delivers material from the Golgi apparatus to the plasma membrane if no other sorting signals are present.

degenerate
Not a moral judgment but an adjective that describes multiple states that amount to the same thing: different triplet combinations of nucleotide bases (codons) that code for the same amino acid, for example.

deletion
Type of mutation in which a single nucleotide or sequence of nucleotides has been removed from the DNA.

denaturation
Dramatic change in conformation of a protein or nucleic acid caused by heating or by exposure to chemicals and usually resulting in the loss of biological function.

dendrite
Extension of a nerve cell, typically branched and relatively short, that receives stimuli from other nerve cells.

dendritic cell
Cell derived from bone marrow and present in lymphoid and other tissues that is specialized for the uptake of particulate material by phagocytosis and which acts as a "professional" antigen-presenting cell in immune responses.

deoxyribonucleic acid—*see* **DNA**

desensitization—*see* **adaptation**

desmosome
Type of anchoring cell–cell junction, usually formed between two epithelial cells, characterized by dense plaques of protein into which intermediate filaments in the two adjoining cells insert.

detergent
Type of small amphipathic molecule that tends to coalesce in water, with its hydrophobic tails buried and its hydrophilic heads exposed. It is widely used to solubilize membrane proteins.

determined
In developmental biology, an embryonic cell is said to be determined if it has become committed to a particular specialized path of development. This **determination** reflects a change in the internal character of the cell, and it precedes the much more readily detected process of cell differentiation.

development
Succession of changes that take place in an organism as a fertilized egg gives rise to an adult plant or animal.

diacylglycerol
Lipid produced by the cleavage of inositol phospholipids in response to extracellular signals. Composed of two fatty acid chains linked to glycerol, it serves as a signaling molecule to help activate protein kinase C.

dideoxy method
The standard method of DNA sequencing.

differentiation
Process by which a cell undergoes a change to an overtly specialized cell type.

diffraction pattern
Pattern set up by wave interference between radiation transmitted or scattered by different parts of an object.

diffusion
Net drift of molecules in the direction of lower concentration due to random thermal movement.

diploid
Containing two sets of homologous chromosomes and hence two copies of each gene or genetic locus.

diplotene
Fourth stage of division I of meiosis, in which chiasmata are first seen.

disaccharide
Carbohydrate molecule consisting of two covalently joined monosaccharide units. (*See* Panel 2–4, p. 116–117.)

dissociation constant (K_d)
Measure of the tendency of a complex to dissociate. For components A and B and the binding equilibrium $A + B \rightleftharpoons AB$, the dissociation constant is given by [A][B]/[AB], and it is smaller the tighter the binding between A and B. (*See also* **association constant**.)

disulfide bond (–S–S–)
Covalent linkage formed between two sulfhydryl groups on cysteines. For extracellular proteins, a common way of joining two proteins together or linking different parts of the same protein. Formed in the endoplasmic reticulum of eucaryotic cells.

division I of meiosis
The first cell division of meiosis, in which the members of each pair of (duplicated) homologous chromosomes are segregated to opposite poles of the dividing cell.

division II of meiosis
The second cell division of meiosis, in which the chromatids of each duplicated chromosome are segregated to opposite poles of the dividing cell.

DNA (deoxyribonucleic acid)
Polynucleotide formed from covalently linked deoxyribonucleotide units. It serves as the store of hereditary information within a cell and the carrier of this information from generation to generation.

DNA affinity chromatography
Technique for purifying sequence-specific DNA-binding proteins by their binding to a matrix to which the appropriate DNA fragments are attached.

DNA footprinting
Technique for determining the DNA sequence to which a DNA-binding protein binds.

DNA helicase
Enzyme that is involved in opening the DNA helix into its single strands for DNA replication.

DNA library
Collection of cloned DNA molecules, representing either an entire genome (genomic library) or DNA copies of the messenger RNA produced by a cell (cDNA library).

DNA ligase
Enzyme that joins the ends of two strands of DNA together with a covalent bond to make a continuous DNA strand.

DNA methylation
Addition of a methyl group to DNA. Extensive methylation of the cytosine base in CG sequences is used in vertebrates to keep genes in an inactive state.

DNA microarray
Technique for analyzing the simultaneous expression of large numbers of genes in cells, in which isolated cellular RNA is hybridized to a large array of short DNA probes immobilized on glass slides.

DNA polymerase
Enzyme that synthesizes DNA by joining nucleotides together using a DNA template as a guide.

DNA primase
Enzyme that synthesizes a short strand of RNA on a DNA template, producing a primer for DNA synthesis.

DNA repair
Collective name for those biochemical processes that correct accidental changes in the DNA.

DNA sequencing
Determination of the order of nucleotides in a DNA molecule. (*See* Figure 8–36.)

DNA supercoiling
Additional twisting of the DNA helix that occurs in response to the superhelical tension created when, for example, a circular DNA is partly unwound (*See* Figure 6–20.)

DNA topoisomerase
Enzyme that binds to DNA and reversibly breaks a phosphodiester bond in one or both strands, allowing the DNA to rotate at that point. It prevents DNA tangling during replication.

DNA transcription—*see* transcription

DNA tumor virus
A general term for a variety of different DNA viruses that can cause tumors.

DNA-only transposon
Type of transposable element that exists as DNA throughout its life cycle. Many types move by cut-and-paste transposition.

domain—*see* protein domain

dominant negative mutation
Mutation that dominantly affects the phenotype by means of a defective protein or RNA molecule that interferes with the function of the normal gene product in the same cell.

dominant
In genetics, refers to the member of a pair of alleles that is expressed in the phenotype of the organism while the other allele is not, even though both alleles are present. Opposite of recessive.

dorsal
Relating to the back of an animal. Also the upper surface of a leaf, wing, etc.

dorsoventral
Describes the axis running from the back to the belly of an animal or from the upper side to the underside of a structure.

double helix
The three-dimensional structure of DNA, in which two DNA chains held together by hydrogen bonding between the bases are wound into a helix.

Drosophila melanogaster
Species of small fly, commonly called a fruit fly, much used in genetic studies of development.

dynamic instability
The property of sudden conversion from growth to shrinkage, and vice versa, in a protein filament such as a microtubule or actin filament. (*See* Panel 16–2, pp. 912–913.)

dynamin
Cytosolic GTPase that binds to the neck of a clathrin-coated vesicle in the process of budding from the membrane, and which is involved in completing vesicle formation.

dynein
Member of a family of large motor proteins that undergo ATP-dependent movement along microtubules. In cilia, dynein forms the side arms in the axoneme that cause adjacent microtubule doublets to slide past one another.

dysplasia
A change in cell growth and behavior in a tissue in which the structure becomes disordered.

ectoderm
Embryonic tissue that is the precursor of the epidermis and nervous system.

effector cell
A cell that carries out the final response or function of a particular process. The main effector cells of the immune system, for example, are activated lymphocytes and phagocytes—the cells involved in destroying pathogens and removing them from the body.

egg
The mature female gamete in sexually reproducing organisms. It is usually a large and immobile cell.

elastin
Hydrophobic protein that forms extracellular extensible fibres (elastic fibres) that give tissues their stretchability and resilience.

electrochemical gradient
The combined influence of a difference in the concentration of an ion on the two sides of the membrane and the electrical charge difference across the membrane (membrane potential). It produces a driving force that causes the ion to move across the membrane.

electrochemical proton gradient
The result of a combined pH gradient (proton gradient) and the membrane potential.

electron
Negatively charged subatomic particle that generally occupies orbitals surrounding the nucleus in an atom.

electron acceptor
Atom or molecule that takes up electrons readily, thereby gaining an electron and becoming reduced.

electron carrier
Molecule such as cytochrome *c*, which transfers an electron from a donor molecule to an acceptor molecule.

electron donor
Molecule that easily gives up an electron, becoming oxidized in the process.

electron microscope
Type of microscope that uses a beam of electrons to create the image.

electron-transport chain
Series of electron carrier molecules along which electrons move from a higher to a lower energy level to a final acceptor molecule. The energy released during electron movement can be used to power various processes. Electron-transport chains present in the inner mitochondrial membrane and in the thylakoid membrane of chloroplasts generate a proton gradient across the membrane that is used to drive ATP synthesis.

elongation factor
Protein required for the addition of amino acids to growing polypeptide chains on ribosomes.

embryogenesis
Development of an embryo from a fertilized egg, or zygote.

embryonic stem cell (ES cell)
Cell derived from the inner cell mass of the early mammalian embryo that can give rise to all the cells in the body. It can be grown in culture, genetically modified and inserted into a blastocyst to develop a transgenic animal.

endocrine cell
Specialized animal cell that secretes a hormone into the blood. Usually part of a gland, such as the thyroid or pituitary gland.

endocytic-exocytic cycle
The processes of endocytosis and exocytosis that, respectively, add and remove plasma membrane from the cell, resulting in no overall change in the cell's surface area and volume.

endocytosis
Uptake of material into a cell by an invagination of the plasma membrane and its internalization in a membrane-bounded vesicle. (*See also* **pinocytosis** and **phagocytosis**.)

endoderm
Embryonic tissue that is the precursor of the gut and associated organs.

endoplasmic reticulum (ER)
Labyrinthine membrane-bounded compartment in the cytoplasm of eucaryotic cells, where lipids are synthesized and membrane-bound proteins and secretory proteins are made.

endosome
Membrane-bounded organelle in animal cells that carries materials newly ingested by endocytosis and passes many of them on to lysosomes for degradation.

endothelial cell
Flattened cell type that forms a sheet (the endothelium) lining all blood vessels.

enhancer
Regulatory DNA sequence to which gene regulatory proteins bind, influencing the rate of transcription of a structural gene that can be many thousands of base pairs away.

entropy
Thermodynamic quantity that measures the degree of disorder in a system; the higher the entropy, the greater the disorder.

enveloped virus
Virus with a capsid surrounded by a lipid membrane (the envelope), which is derived from the host cell plasma membrane when the virus buds from the cell.

enzyme
Protein that catalyzes a specific chemical reaction.

enzyme-linked receptor
Major type of cell-surface receptor in which the cytoplasmic domain either has enzymatic activity itself or is associated with an intracellular enzyme. In both cases enzymatic activity is stimulated by ligand binding to the receptor.

epidermis
Epithelial layer covering the outer surface of the body. It has different structures in different animal groups. The outer layer of plant tissue is also called the epidermis.

epimerization
Reaction that alters the steric arrangement around one atom, as in a sugar molecule.

epinephrine—*see* **adrenaline**

epithelial tissue—*see* **epithelium**

epithelium (plural **epithelia**)
Coherent cell sheet formed from one or more layers of cells covering an external surface or lining a cavity.

epitope—*see* **antigenic determinant**

equilibrium constant (*K*)
Ratio of forward and reverse rate constants for a reaction and equal to the association constant. (*See* Figure 3–44.)

equilibrium
State where there is no net change in a system. For example, equilibrium is reached in a chemical reaction when the forward and reverse rates are equal.

ER lumen
The space enclosed by the membrane of the endoplasmic reticulum (ER).

ER resident protein
Protein that remains in the endoplasmic reticulum (ER) or its membranes and carries out its function there, as opposed to proteins that are present in the ER only in transit.

ER retention signal
Short amino acid sequence on a protein that prevents it moving out of the endoplasmic reticulum (ER). Found on proteins that are resident in the ER and function there.

ER signal sequence
N-terminal signal sequence that directs proteins to enter the endoplasmic reticulum (ER). It is cleaved off by signal peptidase after entry.

ER—*see* **endoplasmic reticulum**

erythrocyte (red blood cell)
Small, hemoglobin-containing blood cell of vertebrates that transports oxygen and carbon dioxide to and from tissues.

erythropoietin
Growth factor that stimulates the production of red blood cells. It is produced by the kidney and acts on precursor cells in bone marrow.

ES cell—*see* **embryonic stem cell**

Escherichia coli (E. coli)
Rodlike bacterium normally found in the colon of humans and other mammals and widely used in biomedical research.

ester
Molecule formed by the condensation reaction of an alcohol group with an acidic group. Phosphate groups usually form esters when linked to a second molecule. (*See* Panel 2–1, 110–111.)

ethyl ($-CH_2CH_3$)
Hydrophobic chemical group derived from ethane (CH_3CH_3).

eucaryote (eukaryote)
Organism composed of one or more cells with a distinct nucleus and cytoplasm. Includes all forms of life except viruses and procaryotes (bacteria and archea).

euchromatin
Region of an interphase chromosome that stains diffusely; "normal" chromatin, as opposed to the more condensed heterochromatin.

exocytosis
Process by which most molecules are secreted from a eucaryotic cell. These molecules are packaged in membrane-bounded vesicles that fuse with the plasma membrane, releasing their contents to the outside.

exon
Segment of a eucaryotic gene that consists of a sequence of nucleotides that will be represented in messenger RNA or the final transfer RNA or ribosomal RNA. In protein-coding genes, exons encode amino acids in the protein. An exon is usually adjacent to a noncoding DNA segment called an intron.

expression vector
A virus or plasmid that carries a DNA sequence into a suitable host cell and there directs the synthesis of the protein encoded by the sequence.

expression
Production of an observable phenotype by a gene—usually by directing the synthesis of a protein.

extracellular matrix
Complex network of polysaccharides (such as glycosaminoglycans or cellulose) and proteins (such as collagen) secreted by cells. Serves as a structural element in tissues and also influences their development and physiology.

facilitated diffusion—*see* **passive transport**

$FADH_2$ (reduced flavin adenine dinucleotide)
Activated carrier molecule that is produced by the citric acid cycle.

FAK—*see* **focal adhesion kinase**

Fas protein (Fas)
Membrane-bound receptor that initiates apoptosis in the receptor-bearing cell after binding to its ligand (Fas ligand).

fat
Energy-storage lipid in cells. It is composed of triglycerides—fatty acids esterified with glycerol.

fat cell
Connective-tissue cell that produces and stores fat in animals.

fatty acid
Compound such as palmitic acid that has a carboxylic acid attached to a long hydrocarbon chain. Used as a major source of energy during metabolism and as a starting point for the synthesis of phospholipids. (*See* Panel 2–5, pp. 118–119.)

Fc receptor
One of a family of receptors specific for the invariant constant region (Fc region) of immunoglobulins (other than IgM and IgD); different Fc receptors are specific for IgG, IgA, IgE and their subclasses.

feedback inhibition
Type of regulation of metabolism in which an enzyme acting early in a reaction pathway is inhibited by a late product of that pathway.

fermentation
Anaerobic energy-yielding metabolic pathway in which pyruvate produced by glycolysis is converted, for example, into lactate or ethanol, with the conversion of NADH to NAD^+.

fertilization
Fusion of a male and a female gamete (both haploid) to form a diploid zygote, which develops into a new individual.

fibrillar collagen
Type of collagen molecule which assembles into rope-like structures. Collagens type I (common in skin), II, III, V and XI are of this type.

fibroblast
Common cell type found in connective tissue. Secretes an extracellular matrix rich in collagen and other extracellular matrix macromolecules. Migrates and proliferates readily in wounded tissue and in tissue culture.

fibronectin
Extracellular matrix protein that is involved in adhesion of cells to the matrix and the guidance of migrating cells during embryogenesis. Integrins on the cell surface are receptors for fibronectin.

filopodium (plural **filopodia**)
Thin, spike-like protrusion with an actin filament core, generated on the leading edge of a crawling animal cell.

fission yeast
Common name often given to the yeast *Schizosaccharomyces pombe*, a common experimental organism. It divides to give two equal-sized cells.

fixative
Chemical reagent such as formaldehyde or osmium tetroxide used to preserve cells for microscropy. Samples treated with these reagents are said to be "fixed," and the process is called fixation.

flagellum (plural **flagella**)
Long, whiplike protrusion whose undulations drive a cell through a fluid medium. Eucaryotic flagella are longer versions of cilia. Bacterial flagella are smaller and completely different in construction and mechanism of action.

fluid-phase endocytosis
Type of endocytosis in which small vesicles bud off internally from the plasma membrane, carrying extracellular fluid and dissolved material into the cell. (*See also* **pinocytosis**.)

fluorescein
Fluorescent dye that fluoresces green when illuminated with blue light or ultraviolet light.

fluorescence microscope
Microscope designed to view material stained with fluorescent dyes. Similar to a light microscope but the illuminating light is passed through one set of filters before the specimen, to select those wavelengths that excite the dye, and through another set of filters before it reaches the eye, to select only those wavelengths emitted when the dye fluoresces.

fluorescent dye
Molecule that absorbs light at one wavelength and responds by emitting light at another wavelength. The emitted light is of longer wavelength (and hence of lower energy) than the light absorbed.

fluorescent resonance energy transfer (FRET)
Technique for monitoring the closeness of two fluorescently labeled molecules (and thus their interaction) in cells.

focal adhesion kinase (FAK)
Cytoplasmic tyrosine kinase present at cell-matrix junctions (focal adhesions) in association with the cytoplasmic tails of integrins.

focal adhesion, focal contact (adhesion plaque)
A type of anchoring cell junction, forming a small region on the surface of a fibroblast or other cell that is anchored to the extracellular matrix. Attachment is mediated by transmembrane proteins such as integrins, which are linked, through other proteins, to actin filaments in the cytoplasm.

follicle cell
One of the cell types that surround a developing oocyte or egg.

free energy (*G*)
The energy that can be extracted from a system to drive reactions. Takes into account changes in both energy and entropy.

free ribosome
Ribosome that is free in the cytosol, unattached to any membrane. It is the site of synthesis of all proteins encoded by the nuclear genome other than those destined to enter the endoplasmic reticulum.

free-energy change (ΔG)
Change in the free energy during a reaction: the free energy of the product molecules minus the free energy of the starting molecules. A large negative value of ΔG indicates that the reaction has a strong tendency to occur. (*See* Panel 2–7, pp. 122–123.)

freeze-fracture electron microscopy
Technique for studying membrane structure, in which the membrane of a frozen cell is fractured along the interior of the bilayer, separating it into the two monolayers with the interior faces exposed.

FRET—*see* **fluorescent resonance energy transfer**

fungus (plural **fungi**)
Kingdom of eucaryotic organisms that includes the yeasts, molds, and mushrooms. Many plant diseases and a relatively small number of animal diseases are caused by fungi.

γ-tubulin ring complex (γTuRC)
Protein complex containing γ-tubulin and other proteins that is an efficient nucleator of microtubules.

G—*see* **free energy**

G_0
G-"zero" phase. State of withdrawal from the eucaroytic cell-division cycle by entry into a quiescent G_1 phase. A common state for differentiated cells.

G_1 phase
Gap 1 phase of the eucaryotic cell-division cycle, between the end of cytokinesis and the start of DNA synthesis.

G_1/S-Cdk
Complex formed in vertebrate cells by a G_1/S-cyclin and the corresponding cyclin-dependent kinase (Cdk).

G_1-Cdk
Complex formed in vertebrate cells by a G_1-cyclin and the corresponding cyclin-dependent kinase (Cdk).

G_2 phase
Gap 2 phase of the eucaryotic cell-division cycle, between the end of DNA synthesis and the beginning of mitosis.

GAG—*see* **glycosaminoglycan**

gamete
Specialized haploid cell, either a sperm or an egg, serving for sexual reproduction.

ganglion (plural **ganglia**)
Cluster of nerve cells and associated glial cells located outside the central nervous system.

ganglioside
Any glycolipid having one or more sialic acid residues in its structure. Found in the plasma membrane of eucaryotic cells and especially abundant in nerve cells.

gap junction
Communicating cell–cell junction that allows ions and small molecules to pass from the cytoplasm of one cell to the cytoplasm of the next.

gastrulation
The stage in animal embryogenesis during which the embryo is transformed from a ball of cells to a structure with a gut (a gastrula).

gene activator protein
A gene regulatory protein that when bound to its regulatory sequence in DNA activates transcription.

gene control region
DNA sequences required to initiate transcription of a given gene and control the rate of initiation.

gene conversion
Process by which DNA sequence information can be transferred from one DNA helix (which remains unchanged) to another DNA helix whose sequence is altered. It occurs occasionally during general recombination.

gene regulatory protein
General name for any protein that binds to a specific DNA sequence to alter the expression of a gene.

gene repressor protein
A gene regulatory protein that prevents the initiation of transcription.

gene
Region of DNA that controls a discrete hereditary characteristic, usually corresponding to a single protein or RNA. This definition includes the entire functional unit, encompassing coding DNA sequences, noncoding regulatory DNA sequences, and introns.

general recombination, general genetic recombination
Recombination that takes place between two homologous chromosomes (as in meiosis).

general transcription factor
Any of the proteins whose assembly around the TATA box is required for the initiation of transcription of most eucaryotic genes.

genetic code
Set of rules specifying the correspondence between nucleotide triplets (codons) in DNA or RNA and amino acids in proteins.

genetic map
Map of the chromosomes in which the distance of genes relative to each other is determined by the amount of genetic recombination that occurs between them.

genetic recombination—*see* **recombination**

genetic screen
A search through a large collection of mutants for a mutant with a particular phenotype.

genome
The totality of genetic information belonging to a cell or an organism; in particular, the DNA that carries this information.

genomic DNA
DNA constituting the genome of a cell or an organism. Often used in contrast to cDNA (DNA prepared by reverse transcription from messenger RNA). **Genomic DNA clones** represent DNA cloned directly from chromosomal DNA, and a collection of such clones from a given genome is a **genomic DNA library**.

genomic imprinting
Situation where a gene is either expressed or not expressed in the embryo depending on which parent it is inherited from.

genomics
The science of studying the DNA sequences and properties of entire genomes.

genotype
Genetic constitution of an individual cell or organism.

germ cell
Precursor cell that will give rise to gametes.

germ line
The lineage of germ cells (which contribute to the formation of a new generation of organisms), as distinct from somatic cells (which form the body and leave no descendants).

GFP—*see* **green fluorescent protein**

giga-
Prefix denoting 10^9. (From Greek *gigas,* giant.)

G_i—*see* **inhibitory G protein**

glial cell
Supporting cell of the nervous system, including oligodendrocytes and astrocytes in the vertebrate central nervous system and Schwann cells in the peripheral nervous system.

globular protein
Any protein with an approximately rounded shape. Such proteins are contrasted with highly elongated, fibrous proteins such as collagen.

glucose
Six-carbon sugar that plays a major role in the metabolism of living cells. Stored in polymeric form as glycogen in animal cells and as starch in plant cells. (*See* Panel 2–4, pp. 116–117.)

glutaraldehyde
Small reactive molecule with two aldehyde groups that is often used as a cross-linking fixative.

glycerol
Small organic molecule that is the parent compound of many small molecules in the cell, including phospholipids.

glycocalyx (cell coat)
Carbohydrate-rich layer that forms the outer coat of a eucaryotic cell. Composed of the oligosaccharides linked to intrinsic plasma membrane glycoproteins and glycolipids, as well as glycoproteins and proteoglycans that have been secreted and reabsorbed onto the cell surface.

glycogen
Polysaccharide composed exclusively of glucose units used to store energy in animal cells. Large granules of glycogen are especially abundant in liver and muscle cells.

glycolipid
Membrane lipid molecule with a sugar residue or oligosaccharide attached to the polar headgroup. (*See* Panel 2–5, pp. 118–119.)

glycolysis
Ubiquitous metabolic pathway in the cytosol in which sugars are incompletely degraded with production of ATP. (Literally, "sugar splitting.")

glycoprotein
Any protein with one or more oligosaccharide chains covalently linked to amino-acid side chains. Most secreted proteins and most proteins exposed on the outer surface of the plasma membrane are glycoproteins.

glycosaminoglycan (GAG)
Long, linear, highly charged polysaccharide composed of a repeating pair of sugars, one of which is always an amino sugar. Mainly found covalently linked to a protein core in extracellular matrix proteoglycans. Examples include chondroitin sulfate, hyaluronic acid, and heparin.

glycosylation
The process of adding one or more sugars to a protein or lipid molecule. (*See also* ***O*-linked glycosylation**, ***N*-linked glycosylation**.)

glycosylphosphatidylinositol anchor (GPI anchor)
Type of lipid linkage by which some membrane proteins are bound to the membrane. It is formed as the proteins travel through the endoplasmic reticulum.

Golgi apparatus (Golgi complex)
Membrane-bounded organelle in eucaryotic cells in which proteins and lipids transferred from the endoplasmic reticulum are modified and sorted. It is the site of synthesis of many cell wall polysaccharides in plants and extracellular matrix glycosaminoglycans in animal cells.

GPI anchor—*see* **glycosylphosphatidylinositol anchor**

G-protein—*see* **GTP-binding protein**

G-protein-linked receptor
Cell-surface receptor that associates with an intracellular trimeric GTP-binding protein (G protein) after receptor activation by an extracellular ligand. These receptors are seven-pass transmembrane proteins.

G_q
Class of receptor-coupled G protein that activates phospholipase C-β and originates the inositol phospholipid signaling pathway.

grana (singular **granum**)
Stacked membrane discs (thylakoids) in chloroplasts that contain chlorophyll and are the site of the light-trapping reactions of photosynthesis.

granulocyte
Category of white blood cell distinguished by conspicuous cytoplasmic granules. Includes neutrophils, basophils, and eosinophils.

gray crescent
Band of pale pigmentation that appears in the egg of some species of amphibian opposite the site of sperm entry following fertilization. Caused by rotation of the egg cortex and associated pigment granules. Marks the future dorsal side.

green fluorescent protein (GFP)
Fluorescent protein isolated from a jellyfish. Widely used as a marker in cell biology.

growth cone
Migrating motile tip of a growing nerve cell axon or dendrite.

growth factor
Extracellular polypeptide signal molecule that can stimulate a cell to grow or proliferate. Examples are epidermal growth factor (EGF) and platelet-derived growth factor (PDGF). Most growth factors also have other actions.

growth regulator—*see* **plant growth regulator**

G_s—*see* **stimulatory G protein**

GTP (guanosine 5′-triphosphate)
Nucleoside triphosphate produced by phosphorylating GDP (guanosine diphosphate). Like ATP it releases a large amount of free energy on hydrolysis of its terminal phosphate group. It has a special role in microtubule assembly, protein synthesis, and cell signaling.

GTPase
Enzyme activity that converts GTP to GDP. Also the common name used for monomeric GTP-binding proteins. (*See* GTP-binding protein.)

GTPase-activating protein (GAP)
Protein that binds to a GTP-binding protein and inactivates it by stimulating its GTPase activity so that it hydrolyzes its bound GTP to GDP.

GTP-binding protein, G protein
Protein with GTPase activity that binds GTP, which activates the protein. The intrinsic GTPase activity eventually converts the GTP to GDP which inactivates the protein. These GTPases act as molecular switches in, for example, intracellular signaling pathways. One family is composed of three different subunits (heterotrimeric GTP-binding proteins). The members of the other, very large family are monomeric GTP-binding proteins; these are commonly referred to as monomeric GTPases.

guanine nucleotide exchange factor (GEF)
Protein that binds to a GTP-binding protein and activates it by stimulating it to release its tightly bound GDP, thereby allowing it to bind GTP in its place.

H^+—*see* **proton**

haploid
Having only one set of chromosomes, as in a sperm cell or a bacterium, as distinct from diploid (having two sets of chromosomes).

heat shock protein (stress-response protein)
Protein synthesized in increased amounts in response to an elevated temperature or other stressful treatment, and which usually helps the cell to survive the stress. Prominent examples are hsp60 and hsp70.

heavy chain (H chain)
The larger of the two types of polypeptide in an immunoglobulin molecule.

HeLa cell
Line of human epithelial cells that grows vigorously in culture. Derived from a human cervical carcinoma.

helix-loop-helix (HLH)
DNA-binding structural motif present in many gene regulatory proteins. Should not be confused with the helix-turn-helix.

helper T cell
Type of T cell that helps stimulate B cells to make antibodies and activates macrophages to kill ingested microorganisms.

heme
Cyclic organic molecule containing an iron atom that carries oxygen in hemoglobin and carries an electron in cytochromes. (*See* Figure 14–22.)

hemidesmosome
Specialized anchoring cell junction between an epithelial cell and the underlying basal lamina.

hemoglobin
The major protein in red blood cells that associates with O_2 in the lungs by means of a bound heme group.

hemopoiesis
Generation of blood cells, mainly in the bone marrow.

hepatocyte
Liver cell.

heterocaryon
Cell with two or more genetically different nuclei; produced by the fusion of two or more different cells.

heterochromatin
Region of a chromosome that remains unusually condensed chromatin; transcriptionally inactive during interphase.

heterodimer
Protein complex composed of two different polypeptide chains.

heterozygote
Diploid cell or individual having two different alleles of one or more specified genes.

high-energy bond
Covalent bond whose hydrolysis releases an unusually large amount of free energy under the conditions existing in a cell. A group linked to a molecule by such a bond is readily transferred from one molecule to another. Examples include the phosphodiester bonds in ATP and the thioester linkage in acetyl CoA.

high-mannose oligosaccharide
Chain of sugars attached to a glycoprotein which contains many mannose residues. It is generated by a trimming of the original mannose-rich oligosaccharide that leaves most of the mannose residues with no subsequent addition of further sugars. (*See* Figure 13–26.)

high-performance liquid chromatography (HPLC)
Type of chromatography that uses columns packed with tiny beads of matrix; the solution to be separated is pushed through under high pressure.

histidine-kinase-associated receptor
Type of transmembrane receptor found in the plasma membrane of bacteria, yeast and plant cells, and involved, for example, in sensing stimuli that cause bacterial chemotaxis. Associated with a histidine protein kinase on its cytoplasmic side.

histone
One of a group of small abundant proteins, rich in arginine and lysine, four of which form the nucleosome on the DNA in eucaryotic chromosomes.

HIV
Human immunodeficiency virus, the retrovirus that is the cause of AIDS.

HLH—*see* **helix-loop-helix**

hnRNP protein (heterogeneous nuclear ribonuclear protein)
Any of a group of proteins that assemble on newly synthesized RNA, organizing it into a more compact form.

Holliday junction
X-shaped structure observed in DNA undergoing recombination, in which the two DNA molecules are held together at the site of crossing-over, also called a cross-strand exchange.

homeobox
Short (180 base pairs long) conserved DNA sequence that encodes a DNA-binding protein motif (homeodomain) famous for its presence in genes that are involved in orchestrating development in a wide range of organisms.

homeodomain
DNA-binding domain that defines a class of gene regulatory proteins important in animal development.

homeotic mutation
Mutation that causes cells in one region of the body to behave as though they were located in another, causing a bizarre disturbance of the body plan.

homolog
(1) One of two or more genes that are similar in sequence as a result of derivation from the same ancestral gene. The term covers both orthologs and paralogs. (2) *See* **homologous chromosome**.

homologous
Describes organs or molecules that are similar because of their common evolutionary origin. Specifically it describes similarities in protein or nucleic acid sequence.

homologous chromosome (homolog)
One of the two copies of a particular chromosome in a diploid cell, each copy being derived from a different parent.

homozygote
Diploid cell or organism having two identical alleles of a specified gene or set of genes.

hormone
Signal molecule secreted by an endocrine cell into the bloodstream, which can then carry it to distant target cells.

housekeeping gene
Gene serving a function required in all the cell types of an organism, regardless of their specialized role.

Hox complex
Two tightly linked clusters of genes in *Drosophila* (the bithorax and Antennapedia complexes) that control the differences between the different segments of the body. Homologous Hox complexes are found in other animals, where they also determine pattern along the anteroposterior axis.

HPLC—*see* **high-performance liquid chromatography**

hybridization
In molecular biology, the process whereby two complementary nucleic acid strands form a double helix. Forms the basis of a powerful technique for detecting specific nucleotide sequences.

hybridoma
Cell line used in the production of monoclonal antibodies. Obtained by fusing antibody-secreting B cells with cells of a lymphocyte tumor.

hydrocarbon
Compound that has only carbon and hydrogen atoms. (*See* Panel 2–1, p 110–111.)

hydrogen bond
Noncovalent bond in which an electropositive hydrogen atom is partially shared by two electronegative atoms.

hydrolysis (adjective **hydrolytic**)
Cleavage of a covalent bond with accompanying addition of water, –H being added to one product of the cleavage and –OH to the other.

hydronium ion (H_3O)
Water molecule associated with an additional proton.

hydrophilic
Describes a polar molecule or part of a molecule that forms enough energetically favorable interactions with water molecules to dissolve readily in water. (Literally, "water loving.")

hydrophobic (lipophilic)
Describes a nonpolar molecule or part of a molecule that cannot form energetically favorable interactions with water molecules and therefore does not dissolve in water. (Literally, "water hating.")

hydrophobic force
Force exerted by the hydrogen-bonded network of water molecules that brings two nonpolar surfaces together by excluding water between them.

hydroxyl (–OH)
Chemical group consisting of a hydrogen atom linked to an oxygen, as in an alcohol.

hypertonic
Describes any medium with a sufficiently high concentration of solutes to cause water to move out of a cell due to osmosis.

hypervariable region
Any of three small regions within the variable region of an immunoglobulin light or heavy chain that show the highest variability from molecule to molecule. These regions determine the specificity of the antigen-binding site.

hypotonic
Describes any medium with a sufficiently low concentration of solutes to cause water to move into a cell due to osmosis.

IAP family
Intracellular protein inhibitors of apoptosis.

Ig—*see* **immunoglobulin**

Ig superfamily
Large family of proteins that contain immunoglobulin domains or immunoglobulin-like domains. Most are involved in cell-cell interactions or antigen recognition.

image processing
Computer treatment of images gained from microscopy that reveal information not immediately visible to the eye.

imaginal disc
Group of cells that are set aside in the *Drosophila* embryo and which will develop into an adult structure, e.g. eye, leg, wing.

immature secretory vesicle
Secretory vesicle that appears to have just pinched off the Golgi stack. Its structure resembles that of a cisterna of the *trans* Golgi network.

immortalization
Production of a cell line capable of an unlimited number of cell divisions. Can be the result of a chemical or viral transformation or of fusion of the original cells with cells of a tumor line.

immune response
Response made by the immune system when a foreign substance or microorganism enters its body. (*See also* **innate immune response, adaptive immune response, primary immune response, secondary immune response**.)

immune system
Population of lymphocytes and other white blood cells in the vertebrate body that defends it against infection.

immunoglobulin (Ig)
An antibody molecule. Higher vertebrates have five classes of immunoglobulin—IgA, IgD, IgE, IgG, and IgM—each with a different role in the immune response.

immunoglobulin domain (Ig domain)
Characteristic protein domain of about 100 amino acids that is found in immunoglobulin light and heavy chains. Similar domains, known as immunoglobulin-like (Ig-like) domains, are present in many other proteins involved in cell–cell interactions and antigen recognition and define the Ig superfamily.

immunogold electron microscopy
Electron microscopy technique in which cellular structures or molecules of interest are labeled with antibodies tagged with electron-dense gold particles. These show up as black spots on the image.

immunological memory
Long-lived state that follows a primary immune response to many antigens, in which subsequent encounter with that antigen will provoke a rapid secondary immune response.

immunoprecipitation
Use of a specific antibody to draw the corresponding protein antigen out of solution. The technique can identify complexes of interacting proteins in cell extracts by using an antibody specific for one of the proteins to precipitate the complex.

***in situ* hybridization**
Technique in which a single-stranded RNA or DNA probe is used to locate a gene or a messenger RNA molecule in a cell or tissue by hybridization.

in vitro
Term used by biochemists to describe a process taking place in an isolated cell-free extract. Also used by cell biologists to refer to cells growing in culture (*in vitro*), as opposed to in an organism (*in vivo*). (Latin for "in glass.")

in vivo
In an intact cell or organism. (Latin for "in life.")

induction
In developmental biology, a change in the developmental fate of one tissue caused by an interaction with another tissue. Such an interaction is called an **inductive interaction**.

inflammatory response
Local response of a tissue to injury or infection—characterized by tissue redness, swelling, heat, and pain. Caused by invasion of white blood cells, which release various local mediators such as histamine.

inhibitor of apoptosis family—*see* **IAP family**

inhibitory G protein (G_i)
G protein that can regulate ion channels and inhibit the enzyme adenylyl cyclase.

inhibitory neurotransmitter
Neurotransmitter that opens transmitter-gated Cl^- or K^+ channels in the postsynaptic membrane of a nerve or muscle cell and thus tends to inhibit the generation of an action potential.

initiation factor
Protein that promotes the proper association of ribosomes with messenger RNA and is required for the initiation of protein synthesis.

initiator tRNA
Special tRNA that intiates translation. It always carries the amino acid methionine.

innate immune response
Immune response (of both vertebrates and invertebrates) to a pathogen that involves the pre-existing defenses of the body—the **innate immune system**—such as barriers formed by skin and mucosa, antimicrobial molecules and phagocytes. Such a response is not specific for the pathogen.

inner membrane
The innermost of two membranes surrounding an organelle. In the mitochondrion, it encloses the matrix and contains the respiratory electron transport chains.

inner nuclear membrane
The innermost of the two nuclear membranes. It contains binding sites for chromatin and the nuclear lamina on its internal face.

inositol phospholipids (phosphoinositides)
One of a family of lipids containing phosphorylated inositol derivatives. Although minor components of the plasma membrane, they are important in signal transduction in eucaryotic cells. (*See* Figure 15–34.)

insulator element
DNA sequence that prevents a gene regulatory protein bound to DNA in the control region of one gene from influencing the transcription of adjacent genes.

insulin
Polypeptide hormone that is secreted by β cells in the pancreas and helps regulate glucose metabolism in animals.

integral membrane protein
Protein that is held tightly in a membrane and can only be removed by treatments that disrupt the lipid bilayer.

integrin
Member of a large family of transmembrane proteins involved in the adhesion of cells to the extracellular matrix and to each other.

intercalary regeneration
Type of regeneration that fills in the missing tissues when two mismatched parts of a structure are grafted together.

interferon-γ (IFN-γ)
Cytokine secreted by certain types of T cells after activation, and which enhances the anti-viral response and macrophage activation.

interleukin
Secreted peptide or protein that mainly mediates local interactions between white blood cells (leucocytes) during inflammation and immune responses.

intermediate filament
Fibrous protein filament (about 10 nm in diameter) that forms ropelike networks in animal cells. One of the three most prominent types of cytoskeletal filaments. (*See* Panel 16–1, p. 909.)

intermembrane space
(1) The subcompartment formed between the inner and outer mitochondrial membranes. (2) The corresponding compartment in a chloroplast.

internal membrane
Eucaryotic cell membrane other than the plasma membrane. The membranes of the endoplasmic reticulum and the Golgi apparatus are examples.

interphase
Long period of the cell cycle between one mitosis and the next. Includes G_1 phase, S phase, and G_2 phase.

intracellular signaling protein
Protein that relays a signal as part of an intracellular signaling pathway. It may either activate the next protein in the pathway or generate a small intracellular mediator.

intron
Noncoding region of a eucaryotic gene that is transcribed into an RNA molecule but is then excised by RNA splicing during production of the messenger RNA or other functional structural RNA.

inversion
Type of mutation in which a segment of chromosome is inverted.

ion channel
Transmembrane protein complex that forms a water-filled channel across the lipid bilayer through which specific inorganic ions can diffuse down their electrochemical gradients.

ion
An atom that has either gained or lost electrons to acquire a charge; for example Na^+ and Cl^-.

ionic bond
Cohesion between two atoms, one with a positive charge, the other with a negative charge. One type of noncovalent bond.

ionophore
Small hydrophobic molecule that dissolves in lipid bilayers and increases their permeability to specific inorganic ions.

iron-sulfur center
Electron-transporting group consisting of either two or four iron atoms bound to an equal number of sulfur atoms, found in a class of electron-transport proteins.

isoelectric point
The pH at which a charged molecule in solution has no net electric charge and therefore does not move in an electric field.

isomers
Molecules that are formed from the same atoms in the same chemical linkages but have different three-dimensional conformations. (*See* Panel 2–4, pp. 116–117.)

isoprenoid (polyisoprenoid)
Member of a large family of lipid molecules with a carbon skeleton based on multiple five-carbon isoprene units. Examples include retinoic acid and dolichol.

isotope
One of a number of forms of an atom that differ in atomic weight but have the same number of protons and electrons, and therefore the same chemistry. May be either stable or radioactive.

Jak-STAT signaling pathway
Rapid signaling pathway by which some extracellular signals (for example interferon) activate gene expression. Involves cell-surface receptors and cytoplasmic Janus kinases (Jaks) plus signal transducers and activators of transcription (STATs).

joule
Standard unit of energy in the meter-kilogram system. One joule is the energy delivered in one second by a one-watt power source. Approximately equal to 0.24 calories.

K—*see* **equilibrium constant**

K^+ leak channel
A K^+-transporting ion channel in the plasma membrane of animals cells that remains open even in a "resting" cell.

karyotype
Full set of chromosomes of a cell arranged with respect to size, shape, and number.

K_a—*see* **affinity constant**

keratin
Member of the family of proteins that form keratin intermediate filaments, mainly in epithelial cells. Specialized keratins are found in hair, nails, and feathers.

ketone
Organic molecule containing a carbonyl group linked to two alkyl groups.

kilo-
Prefix denoting 10^3.

kilocalorie (kcal)
Unit of heat energy equal to 1000 calories. Often used to express the energy content of food or molecules: bond strengths, for example, are measured in kcal/mole. An alternative unit in wide use is the kilojoule, equal to 0.24 kcal.

kilojoule
Standard unit of energy equal to 1000 joules, or 0.24 kilocalories.

kinesin
One type of motor protein that uses the energy of ATP hydrolysis to move along a microtubule.

kinetochore
Complex structure formed from proteins on a mitotic chromosome to which microtubules attach and which plays an active part in the movement of chromosomes to the poles. The kinetochore forms on the part of the chromosome known as the centromere.

kinetochore microtubule
In a mitotic or meiotic spindle, a microtubule with one end attached to the kinetochore on a chromosome.

Krebs cycle—*see* **citric acid cycle**

label
Chemical group, radioactive atom or fluorescent dye added to a molecule in order to follow its progress through a biochemical reaction or to locate it spatially. Also, as a verb, to add such a group or atom to a cell or molecule.

lagging strand
One of the two newly synthesized strands of DNA found at a replication fork. The lagging strand is made in discontinuous lengths that are later joined covalently.

lambda bacteriophage (λ bacteriophage)
Virus that infects *E. coli*. Widely used as a DNA cloning vector

llamellipodium (plural **lamellipodia)**
Flattened, sheetlike protrusion supported by a meshwork of actin filaments, which is extended at the leading edge of a crawling animal cell.

laminin
Extracellular matrix protein found in basal laminae, where it forms a sheetlike network.

lamin—*see* **nuclear lamin**

lampbrush chromosome
Paired chromosome in meiosis in immature amphibian eggs, in which the chromatin forms large stiff loops extending out from the linear axis of the chromosome.

leading strand
One of the two newly synthesized strands of DNA found at a replication fork. The leading strand is made by continuous synthesis in the 5′-to-3′ direction.

lectin
Protein that binds tightly to a specific sugar. Abundant lectins from plant seeds are often used as affinity reagents to purify glycoproteins or to detect them on the surface of cells.

leptotene
The first phase of division I of meiosis, in which the paired duplicated homologous chromosomes condense and become visible in the light microscope.

lethal mutation
A mutation that causes the death of the cell or the organism that contains it.

leucine zipper
Structural motif seen in many DNA-binding proteins in which two α helices from separate proteins are joined together in a coiled-coil (rather like a zipper), forming a protein dimer.

leucine-rich repeat protein (LRR protein)
Common type of receptor serine/threonine kinase in plants. Characterized by a tandem array of leucine-rich repeat sequences in the extracellular portion.

leucocyte—*see* **white blood cell**

leukemia
Cancer of white blood cells.

ligand
Any molecule that binds to a specific site on a protein or other molecule. (From Latin *ligare,* to bind.)

ligase
Enzyme that joins together (ligates) two molecules in an energy-dependent process. DNA ligase, for example, joins two DNA molecules together end to end through phosphodiester bonds.

light chain
One of the smaller polypeptides of a multisubunit protein such as myosin or immunoglobulin. Abbreviated as **L chain** in immunoglobulins.

lineage analysis
Tracing the ancestry of individual cells in a developing embryo.

linkage
(1) Mutual effect of the binding of one ligand on the binding of another that is a central feature of the behavior of all allosteric proteins. (2) Co-inheritance of two genetic loci that lie near each other on the same chromosome. The closer together the two loci, that is, the greater the linkage, the lower the frequency of recombination between them.

lipase
Enzyme that catalyzes the cleavage of fatty acids from the glycerol moiety of a triglyceride.

lipid
Organic molecule that is insoluble in water but tends to dissolve in nonpolar organic solvents. A special class, the phospholipids, forms the structural basis of biological membranes.

lipid bilayer
Thin bimolecular sheet of mainly phospholipid molecules that forms the core structure of all cell membranes. The two layers of lipid molecules are packed with their hydrophobic tails pointing inward and their hydrophilic heads outward, exposed to water.

lipid raft
Small region of the plasma membrane enriched in sphingolipids and cholesterol.

lipophilic—*see* **hydrophobic**

liposome
Artificial phospholipid bilayer vesicle formed from an aqueous suspension of phospholipid molecules.

local mediator
Secreted signal molecule that acts at short range on adjacent cells.

locus
In genetics, the position of a gene on a chromosome. Different alleles of the same gene all occupy the same locus.

long-term potentiation
Long-lasting increase (days to weeks) in the sensitivity of certain synapses in the hippocampus. Induced by a short burst of repetitive firing in the presynaptic neurons.

low-density lipoprotein (LDL)
Large complex composed of a single protein molecule and many esterified cholesterol molecules, together with other lipids. The form in which cholesterol is transported in the blood and taken up into cells.

LTP—*see* long-term potentiation

lumen
Cavity enclosed by an epithelial sheet (in a tissue) or by a membrane (in a cell).

lymph
Colorless fluid derived from blood by filtration through capillary walls. Carries lymphocytes in a special system of ducts and vessels—the lymphatic vessels.

lymphocyte
Type of white blood cell responsible for the specificity of adaptive immune responses. There are two main types: B cells, which produce antibody, and T cells, which interact directly with other effector cells of the immune system and with infected cells. T cells develop in the thymus and are responsible for cell-mediated immunity. B cells develop in the bone marrow in mammals and are responsible for the production of circulating antibodies.

lymphoid organ
Organs involved in the production or function of lymphocytes, such as thymus, spleen, lymph nodes, and tonsils.

lysis
Rupture of a cell's plasma membrane, leading to the release of cytoplasm and the death of the cell.

lysogeny
State of a bacterium in which it carries the DNA of an inactive virus integrated into its genome. The virus can subsequently be activated to replicate and lyse the cell.

lysosome
Membrane-bounded organelle in eucaryotic cells containing digestive enzymes, which are typically most active at the acid pH found in the lumen of lysosomes.

lysozyme
Enzyme that catalyzes the cutting of polysaccharide chains in the cell walls of bacteria.

M phase
Period of the eucaryotic cell cycle during which the nucleus and cytoplasm divide.

M6P—*see* mannose 6-phosphate

macromolecule
Molecule such as a protein, nucleic acid, or polysaccharide with a molecular mass greater than a few thousand daltons.

macrophage
Phagocytic cell derived from blood monocytes, typically resident in most tissues. It has both scavenger and antigen-presenting functions in immune responses.

major histocompatibility complex (MHC)
Complex of highly polymorphic genes in vertebrates. They code for a large family of cell-surface glycoproteins (MHC molecules) that bind peptide fragments of foreign proteins and present them to T cells to induce an immune response. (*See* Figure 24–50.)

malaria
Potentially fatal human disease caused by the protozoan parasite *Plasmodium*, which is transmitted by the bite of an infected mosquito.

malignant
Describes tumors and tumor cells that are invasive and/or able to undergo metasis. A malignant tumor is a cancer.

mannose 6-phosphate (M6P)
Unique marker attached to the oligosaccharides on some glycoproteins destined for lysosomes.

map unit—*see* genetic map distance

MAP—*see* microtubule-associated protein

MAP-kinase (mitogen-activated protein kinase)
Protein kinase that performs a crucial step in relaying signals from the plasma membrane to the nucleus. Turned on by a wide range of proliferation- or differentiation-inducing signals.

mating-type locus (MAT locus)
In budding yeast, the locus that determines the mating type (α or a) of the haploid yeast cell.

matrix space
(1) Central subcompartment of a mitochondrion, bounded by the inner mitochondrial membrane. (2) The corresponding compartment in a chloroplast, which is more commonly known as the stroma.

M-Cdk—*see* M-phase Cdk

Mcm proteins
Proteins in the eucaryotic cell that bind to origin recognition complexes in DNA in early G_1 and are involved in forming the pre-replicative complex.

M-cyclin
Type of cyclin found in all eucaryotic cells that promotes the events of mitosis.

MDR protein—*see* multidrug resistance protein

mega-
Prefix denoting 10^6. (From Greek *megas,* huge, powerful.)

megakaryocyte
Large myeloid cell with a multilobed nucleus that remains in the bone marrow when mature. It buds off platelets from long cytoplasmic processes.

meiosis
Special type of cell division by which eggs and sperm cells are produced. It comprises two successive nuclear divisions with only one round of DNA replication, which produces four haploid daughter cells from an initial diploid cell.

melanocyte
Cell that produces the dark pigment melanin. Responsible for the pigmentation of skin and hair.

membrane
The lipid bilayer plus associated proteins that encloses all cells and, in eucaryotic cells, many organelles as well.

membrane-bound ribosome
Ribosome attached to the cytosolic face of the endoplasmic reticulum. The site of synthesis of proteins that enter the endoplasmic reticulum.

membrane channel
Transmembrane protein complex that allows inorganic ions or other small molecules to diffuse passively across the lipid bilayer.

membrane potential
Voltage difference across a membrane due to a slight excess of positive ions on one side and of negative ions on the other. A typical membrane potential for an animal cell plasma membrane is –60 mV (inside negative relative to the surrounding fluid).

membrane protein
Protein that is normally closely associated with a cell membrane. (*See* Figure 10–17.)

membrane transport
Movement of molecules across a membrane mediated by a membrane transport protein.

membrane transport protein
Membrane protein that mediates the passage of ions or molecules across a membrane. Examples are ion channels and carrier proteins.

meristem
An organized group of dividing cells whose derivatives give rise to the tissues and organs of a flowering plant. Key examples are the apical meristems at the tips of shoots and roots.

mesenchyme
Immature, unspecialized form of connective tissue in animals, consisting of cells embedded in a thin extracellular matrix.

mesoderm
Embryonic tissue that is the precursor to muscle, connective tissue, skeleton and many of the internal organs.

messenger RNA (mRNA)
RNA molecule that specifies the amino acid sequence of a protein. Produced by RNA splicing (in eucaryotes) from a larger RNA molecule made by RNA polymerase as a complementary copy of DNA. It is translated into protein in a process catalyzed by ribosomes.

metabolism
The sum total of the chemical processes that take place in living cells.

metaphase
Stage of mitosis at which chromosomes are firmly attached to the mitotic spindle at its equator but have not yet segregated toward opposite poles.

metaphase plate
Imaginary plane at right angles to the mitotic spindle and midway between the spindle poles; the plane in which chromosomes are positioned at metaphase.

metaplasia
A change in the pattern of cell differentiation in a tissue.

metastasis
Spread of cancer cells from their site of origin to other sites in the body.

methyl ($-CH_3$)
Hydrophobic chemical group derived from methane (CH_4).

MHC molecule
One of a large family of ubiquitous cell-surface glycoproteins encoded by genes of the major histocompatibility complex (MHC). They bind peptide fragments of foreign antigens and present them to T cells to induce an immune response. (*See also* **class I MHC molecule, class II MHC molecule**.)

MHC—*see* **major histocompatibility complex**

micro-
Prefix denoting 10^{-6}.

microelectrode, micropipette
Piece of fine glass tubing pulled to an even finer tip. Used to penetrate a cell to study its physiology or to inject electric current or molecules.

microfilament—*see* **actin filament**

micrograph
Photograph of an image seen through a microscope. May be either a light micrograph or an electron micrograph depending on the type of microscope employed.

microinjection
Injection of molecules into a cell using a micropipette.

micron (μm or micrometer)
Unit of measurement often applied to cells and organelles. Equal to 10^{-6} meter or 10^{-4} centimeter.

micropipette—*see* **microelectrode**

microsome
Small vesicle that is derived from fragmented endoplasmic reticulum produced when cells are homogenized.

microtubule
Long hollow cylindrical structure composed of the protein tubulin. It is one of the three major classes of filaments of the cytoskeleton. (*See* Panel 16–1, p. 909.)

microtubule-associated protein (MAP)
Any protein that binds to microtubules and modifies their properties. Many different kinds have been found, including structural proteins, such as MAP-2, and motor proteins, such as dynein.

microtubule-organizing center (MTOC)
Region in a cell, such as a centrosome or a basal body, from which microtubules grow.

microvillus (plural microvilli)
Thin cylindrical membrane-covered projection on the surface of an animal cell containing a core bundle of actin filaments. Present in especially large numbers on the absorptive surface of intestinal epithelial cells.

midbody
Structure formed at the end of cleavage that can persist for some time as a tether between the two daughter cells in animals.

milli-
Prefix denoting 10^{-3}.

minus end
The end of a microtubule or actin filament at which the addition of monomers occurs least readily; the "slow-growing" end of the microtubule or actin filament. The minus end of an actin filament is also known as the pointed end. (*See* Panel 16–2, pp. 912–913.)

mismatch repair
DNA repair process that corrects mismatched nucleotides inserted during DNA replication. A short stretch of newly synthesized DNA including the mismatched nucleotide is removed and replaced with the correct sequence with reference to the template strand.

mitochondrial precursor protein
Mitochondrial protein encoded by a nuclear gene, synthesized in the cytosol, and subsequently transported into mitochondria.

mitochondrion (plural **mitochondria)**
Membrane-bounded organelle, about the size of a bacterium, that carries out oxidative phosphorylation and produces most of the ATP in eucaryotic cells.

mitogen
An extracellular substance, such as a growth factor, that stimulates cell proliferation.

mitogen-activated protein kinase—*see* **MAP-kinase**

mitosis
Division of the nucleus of a eucaryotic cell, involving condensation of the DNA into visible chromosomes, and separation of the duplicated chromosomes to form two identical sets. (From Greek *mitos*, a thread, referring to the threadlike appearance of the condensed chromosomes.)

mitotic chromosome
Highly condensed duplicated chromosome with the two new chromosomes still held together at the centromere as sister chromatids.

mitotic spindle
Array of microtubules and associated molecules that forms between the opposite poles of a eucaryotic cell during mitosis and serves to move the duplicated chromosomes apart.

model organism
A species, such as *Drosophila melanogaster* or *Escherichia coli*, that has been studied intensively over a long period and thus serves as a "model" of the biology of a particular type of organism.

module
In proteins or nucleic acids, a unit of structure or function that is found in a variety of different contexts in different molecules.

molar
Describes a solution with a concentration of 1 mole of a substance dissolved in 1 liter of solution (abbreviated as 1 M).

mole
X grams of a substance, where X is its relative molecular mass (molecular weight). A mole consists of 6×10^{23} molecules of the substance.

molecular chaperone—*see* **chaperone**

molecular weight
Numerically, the same as the relative molecular mass of a molecule expressed in daltons. For example, a protein of relative molecular mass 20,000 has a molecular weight of 20,000.

molecule
Group of atoms joined together by covalent bonds.

monoclonal antibody
Antibody secreted by a hybridoma clone. Because each such clone is derived from a single B cell, all of the antibody molecules produced are identical.

monocyte
Type of white blood cell that leaves the bloodstream and matures into a macrophage in tissues.

monomer
Small molecular building block that can serve as a subunit, being linked to others of the same type to form a larger molecule (a polymer).

monosaccharide
Simple sugar with the general formula $(CH_2O)_n$, where $n = 3$ to 8.

morphogen
Signal molecule that can impose a pattern on a field of cells by causing cells in different places to adopt different fates.

mosaic
In developmental biology, an organism made of a mixture of cells with different genotypes.

motif
Element of structure or pattern that recurs in many contexts. Specifically, a small structural domain that can be recognized in a variety of proteins.

motor protein
Protein that uses energy derived from nucleoside triphosphate hydrolysis to propel itself along a protein filament or another polymeric molecule.

M-phase Cdk (M-Cdk)
Complex formed in vertebrate cells by an M-cyclin and the corresponding cyclin-dependent kinase (Cdk).

mRNA—*see* **messenger RNA**

MTOC—*see* **microtubule-organizing center**

multidrug resistance protein (MDR protein)
Type of ABC transporter protein that can pump hydrophobic drugs (such as some anti-cancer drugs) out of the cytoplasm of eucaryotic cells.

multipass transmembrane protein
Membrane protein in which the polypeptide chain crosses the lipid bilayer more than once.

mutant
Organism in which a mutation has occurred that makes it different from wild-type or from the 'normal' extent of variation in the population.

mutation rate
The rate at which observable changes occur in a DNA sequence.

mutation
Heritable change in the nucleotide sequence of a chromosome.

myelin sheath
Insulating layer of specialized cell membrane wrapped around vertebrate axons. Produced by oligodendrocytes in the central nervous system and by Schwann cells in the peripheral nervous system.

myeloid cell
Any white blood cell other than lymphocytes.

myoblast
Mononucleated, undifferentiated muscle precursor cell. A skeletal muscle cell is formed by the fusion of multiple myoblasts.

myoepithelial cell
Type of unstriated muscle cell found in epithelia, e.g. in the iris of the eye and in glandular tissue.

myofibril
Long, highly organized bundle of actin, myosin, and other proteins in the cytoplasm of muscle cells that contracts by a sliding filament mechanism.

N terminus—*see* **amino terminus**

Na^+-K^+ pump (Na^+-K^+ ATPase)
Transmembrane carrier protein found in the plasma membrane of most animal cells that pumps Na^+ out of and K^+ into the cell, using energy derived from ATP hydrolysis.

NAD^+ (nicotine adenine dinucleotide)
Activated carrier that participates in an oxidation reaction by accepting a hydride ion (H^-) from a donor molecule. The **NADH** formed is an important carrier of electrons for oxidative phosphorylation.

NADH dehydrogenase complex
First of the three electron-driven proton pumps in the mitochondrial respiratory chain. It accepts electrons from NADH.

$NADP^+$ (nicotine adenine dinucleotide phosphate)
Activated carrier closely related to NAD^+ that is used extensively in biosynthetic, rather than catabolic, pathways. The reduced form is **NADPH**.

nano-
Prefix denoting 10^{-9}.

nanometer (nm)
Unit of length commonly used to measure molecules and cell organelles. 1 nm = 10^{-3} micrometer (μm) = 10^{-9} meter.

natural killer cell (NK cell)
Cytotoxic cell of the innate immune system that can kill virus-infected cells.

N-CAM—*see* **neural cell adhesion molecule**

negative control
Type of control of gene expression in which the active DNA-binding form of the regulatory protein turns the gene off.

negative staining
Staining technique for use in the electron electron microscope in which a reverse, or negative, image of the object is created.

Nernst equation
Quantitative expression that relates the equilibrium ratio of concentrations of an ion on either side of a permeable membrane to the voltage difference across the membrane. (*See* Panel 11–2, p. 634.)

nerve cell—*see* **neuron**

neural cell adhesion molecule (N-CAM)
Cell adhesion molecule of the immunoglobulin superfamily, expressed by many cell types including most nerve cells. It mediates Ca^{2+}-independent cell-cell attachment in vertebrates.

neural tube
Tube of ectoderm that will form the brain and spinal cord in a vertebrate embryo.

neurite
Long process growing from a nerve cell in culture. A generic term that does not specify whether the process is an axon or a dendrite.

neurofilament
Type of intermediate filament found in nerve cells.

neuromuscular junction
Specialized chemical synapse between an axon terminal of a motor neuron and a skeletal muscle cell.

neuron (nerve cell)
Cell with long processes specialized to receive, conduct, and transmit signals in the nervous system.

neuropeptide
Peptide secreted by neurons as a signaling molecule either at synapses or elsewhere.

neurotransmitter
Small signal molecule secreted by the presynaptic nerve cell at a chemical synapse to relay the signal to the postsynaptic cell. Examples include acetylcholine, glutamate, GABA, glycine, and many neuropeptides.

neutron
Uncharged subatomic particle that forms part of an atomic nucleus.

neutrophil
White blood cell that is specialized for the uptake of particulate material by phagocytosis and which enters tissues that become infected or inflamed.

nicotine adenine dinucleotide phosphate—*see* **$NADP^+$**

nicotine adenine dinucleotide—*see* **NAD^+**

nitric oxide (NO)
Gaseous signal molecule in both animals and plants. In animals it regulates smooth muscle contraction, for example; in plants it is involved in responses to injury or infection.

nitrogen cycle
The natural circulation of nitrogen between molecular nitrogen in the atmosphere, inorganic molecules in the soil, and organic molecules in living organisms.

nitrogen fixation
Biochemical process carried out by certain bacteria that reduces atmospheric nitrogen (N_2) to ammonia, leading eventually to various nitrogen-containing metabolites.

nitrogenase complex
Complex of enzymes in nitrogen-fixing bacteria that catalyzes the reduction of atmospheric N_2 to ammonia.

NK cell—*see* **natural killer cell**

***N*-linked oligosaccharide**
Chain of sugars attached to a protein through the NH_2 group of the side chain of an asparagine residue.

NMR (nuclear magnetic resonance)
Resonant absorption of electromagnetic radiation at a specific frequency by atomic nuclei in a magnetic field, due to flipping of the orientation of their magnetic dipole moments. The NMR spectrum provides information about the chemical environment of the nuclei. Two-dimensional NMR is used widely to determine the three-dimensional structure of small proteins.

nm—*see* **nanometer**

noncovalent attraction
Chemical bond in which, in contrast to a covalent bond, no electrons are shared. Noncovalent bonds are relatively weak, but they can sum together to produce strong, highly specific interactions between molecules.

noncyclic photophosphorylation
Photosynthetic process that produces both ATP and NADPH in plants and cyanobacteria.

nonenveloped virus
Virus consisting of a nucleic acid core and protein capsid only.

nonpolar (apolar)
Lacking any asymmetric accumulation of positive and negative charge. Nonpolar molecules are generally insoluble in water.

nonsense-mediated mRNA decay
Mechanism for removing aberrant mRNAs containing in-frame internal stop codons before they can be translated.

Northern blotting
Technique in which RNA fragments separated by electrophoresis are immobilized on a paper sheet. A specific RNA is then detected by hybridization with a labeled nucleic acid probe.

NO—*see* **nitric oxide**

Notch
Receptor protein involved in many instances of choice of cell fate in animal development, for example in the specification of nerve cells from ectodermal epithelium. Its ligands are cell-surface proteins such as Delta and Serrate.

notochord
Stiff rod of mesoderm that runs along the back of all chordate embryos. In vertebrates it does not persist and becomes incorporated into the vertebral column.

NSF
Protein with ATPase activity that disassembles a complex of a v-SNARE and a t-SNARE.

nuclear envelope
Double membrane surrounding the nucleus. Consists of an outer and inner membrane and is perforated by nuclear pores.

nuclear export signal
Sorting signal contained in the structure of molecules and complexes, such as RNA and new ribosomal subunits, that are transported from the nucleus to the cytosol through nuclear pore complexes.

nuclear lamin
Protein subunit of the intermediate filaments of the nuclear lamina.

nuclear lamina
Fibrous meshwork of proteins on the inner surface of the inner nuclear membrane. It is made up of a network of intermediate filaments formed from nuclear lamins.

nuclear localization signal (NLS)
Signal sequences or signal patches found in proteins destined for the nucleus and which enable their selective transport into the nucleus from the cytosol through the nuclear pore complexes.

nuclear magnetic resonance—*see* NMR

nuclear pore complex
Large multiprotein structure forming a channel (the **nuclear pore**) through the nuclear envelope that allows selected molecules to move between nucleus and cytoplasm.

nuclear receptor superfamily
Intracellular receptors for hydrophobic signal molecules such as steroids and retinoic acid. The receptor-ligand complex acts as a transcription factor in the nucleus.

nuclear transport
Movement of macromolecules into or out of the nucleus mediated by nuclear transport receptors.

nucleation
Critical stage in the assembly of a polymeric structure, such as a microtubule, at which a small cluster of monomers aggregates in the correct arrangement to initiate rapid polymerization. (*See* Panel 16–2, pp. 912–913.) More generally, the rate-limiting step in an assembly process.

nucleic acid
RNA or DNA, a macromolecule consisting of a chain of nucleotides joined together by phosphodiester bonds.

nucleolar organizer
Region of a chromosome containing a cluster of ribosomal RNA genes that gives rise to a nucleolus.

nucleolus
Structure in the nucleus where ribosomal RNA is transcribed and ribosomal subunits are assembled.

nucleoporin
Any of a number of different proteins that make up nuclear pore complexes.

nucleoside
Molecule composed of a purine or pyrimidine base covalently linked to a ribose or deoxyribose sugar. (*See* Panel 2–6, pp. 120–121.)

nucleosome
Beadlike structure in eucaryotic chromatin. It is composed of a short length of DNA wrapped around a core of histone proteins, and is the fundamental structural unit of chromatin.

nucleotide
Nucleoside with one or more phosphate groups joined in ester linkages to the sugar moiety. DNA and RNA are polymers of nucleotides. (*See* Panel 2–6, pp. 120–121.)

nucleus
Prominent membrane-bounded organelle in a eucaryotic cell, containing DNA organized into chromosomes.

nurse cell
Cell connected by cytoplasmic bridges to a developing oocyte and which thereby supplies it with ribosomes, mRNAs, and proteins needed for the development of the early embryo.

occluding junction
Type of cell junction that seals cells together in an epithelium, forming abarrier through which even small molecules cannot pass.

Okazaki fragments
Short lengths of DNA produced on the lagging strand during DNA replication. They are rapidly joined by DNA ligase to form a continuous DNA strand.

oligodendrocyte
Type of glial cell in the vertebrate central nervous system that forms a myelin sheath around axons.

oligomer
Short polymer, usually consisting (in a cell) of amino acids (oligopeptides), sugars (oligosaccharides), or nucleotides (oligonucleotides). (From Greek *oligos,* few, little.)

oligosaccharide
Short linear or branched chain of covalently linked sugars (*see* Panel 2–4, pp. 116–117.) (*See also* **complex oligosaccharide, high-mannose oligosaccharide, *N*-linked oligosacharide, *O*-linked glycosylation.**)

***O*-linked glycosylation**
Addition of an oligosaccharide chain to a protein through the OH group of a serine or threonine side chain.

oncogene
An altered gene whose product can act in a dominant fashion to help make a cell cancerous. Typically, an oncogene is a mutant form of a normal gene (proto-oncogene) involved in the control of cell growth or division.

oocyte
The developing egg. It is usually a large and immobile cell.

oogenesis
Formation and maturation of oocytes in the ovary.

operator
Short region of DNA in a bacterial chromosome that controls the transcription of an adjacent gene.

operon
In a bacterial chromosome, a group of contiguous genes that are transcribed into a single mRNA molecule.

ORC—*see* origin recognition complex

organelle
Membrane-enclosed compartment in a eucaryotic cell that has a distinct structure, macromolecular composition, and function. Examples are nucleus, mitochondrion, chloroplast, Golgi apparatus.

Organizer—*see* Spemann's Organizer

origin recognition complex (ORC)
Large protein complex that is bound to the DNA at origins of replication in eucaryotic chromosomes throughout the cell cycle.

osmolarity
A term used to describe the concentration of a solute in terms of the osmotic pressure it can exert.

osmosis
Net movement of water molecules across a semipermeable membrane driven by a difference in concentration of solute on either side. The membrane must be permeable to water but not to the solute molecules.

osteoblast
Cell that secretes matrix of bone.

osteoclast
Macrophage-like cell that erodes bone, enabling it to be remodeled during growth and in response to stresses throughout life.

outer membrane
Outermost of the two membranes surrounding an organelle; the membrane adjacent to the cytosol.

outer nuclear membrane
The outermost of the two nuclear membranes. It is continuous with the endoplasmic reticulum and is studded with ribosomes on its cytosolic face.

overlap microtubule
In the mitotic or meiotic spindle, a microtubule interdigitating at the equator with the microtubules emanating from the other pole.

ovulation
Release of an egg from the ovary.

ovum—*see* **egg**

oxidation (verb **oxidize**)
Loss of electrons from an atom, as occurs during the addition of oxygen to a molecule or when a hydrogen is removed. Opposite of reduction. (*See* Figure 2–43.)

oxidative phosphorylation
Process in bacteria and mitochondria in which ATP formation is driven by the transfer of electrons from food molecules to molecular oxygen. Involves the intermediate generation of a proton gradient (pH gradient) across a membrane and chemiosmotic coupling.

p53
Tumor suppressor gene found mutated in about half of human cancers. It encodes a gene regulatory protein that is activated by damage to DNA and is involved in blocking further progression through the cell cycle.

pachytene
Third stage of division I of meiosis, in which synapsis is complete.

palindromic sequence
Nucleotide sequence that is identical to its complementary strand when each is read in the same chemical direction—for example, GATC.

paracrine signaling
Short-range cell-cell communication via secreted signal molecules that act on adjacent cells.

parthenogenesis
Production of a new individual from an egg cell in the absence of fertilization by sperm.

passive transport
Transport of a solute across a membrane down its concentration gradient or its electrochemical gradient, using only the energy stored in the gradient.

patch-clamp recording
Electrophysiological technique in which a tiny electrode tip is sealed onto a patch of cell membrane, thereby making it possible to record the flow of current through individual ion channels in the patch.

pathogen (*adjective* **pathogenic)**
An organism or other agent that causes disease.

PCR (polymerase chain reaction)
Technique for amplifying specific regions of DNA by the use of sequence-specific primers and multiple cycles of DNA synthesis, each cycle being followed by a brief heat treatment to separate complementary strands.

peptide bond
Chemical bond between the carbonyl group of one amino acid and the amino group of a second amino acid—a special form of amide linkage. Peptide bonds link amino acids together in proteins. (*See* Panel 3–1, pp. 132–133.)

peptide map
Characteristic two-dimensional pattern (on paper or gel) formed by the separation of the mixture of peptides produced by the partial digestion of a protein.

peripheral lymphoid organ (secondary lymphoid organ)
Lymphoid organ in which T cells and B cells interact with foreign antigens. Examples are spleen, lymph nodes, and mucosal-associated lymphoid tissue.

peripheral membrane protein
Protein that is attached to one face of a membrane by noncovalent interactions with other membrane proteins, and which can be removed by relatively gentle treatments that leave the lipid bilayer intact.

peroxisome
Small membrane-bounded organelle that uses molecular oxygen to oxidize organic molecules. Contains some enzymes that produce and others that degrade hydrogen peroxide (H_2O_2).

pH
Common measure of the acidity of a solution: "p" refers to power of 10, "H" to hydrogen. Defined as the negative logarithm of the hydrogen ion concentration in moles per liter (M). Thus on the **pH scale**, pH 3 (10^{-3} M H^+) is acidic and pH 9 (10^{-9} M H^+) is alkaline.

PH domain—*see* **pleckstrin homology domain**

phage display
Technique for detecting proteins that interact with each other by screening a protein against a library of genetically modified phage, each displaying a potential binding protein on their surface.

phage—*see* **bacteriophage**

phagocyte
General term for a professional phagocytic cell—that is, a cell such as a macrophage or neutrophil that is specialized to take up particles and microorganisms by phagocytosis.

phagocytosis
Process by which particulate material is endocytosed ("eaten") by a cell. Prominent in carnivorous cells, such as *Amoeba proteus*, and in vertebrate macrophages and neutrophils. (From Greek *phagein*, to eat.)

phagosome
Large intracellular membrane-bounded vesicle that is formed as a result of phagocytosis. Contains ingested extracellular material.

phase-contrast microscope
Type of light microscope that exploits the interference effects that occur when light passes through material of different refractive indexes. Used to view living cells.

phenotype
The observable character of a cell or an organism.

phosphatase
Enzyme that removes phosphate groups from a molecule.

phosphatidylinositol 3-kinase (PI 3-kinase)
A kinase involved in intracellular signaling pathways activated by a variety of cell-surface receptors. It phosphorylates inositol phospholipids at the 3 position of the inositol ring. (*See* Figure 15–58.)

phosphatidylinositol
An inositol phospholipid. (*See* Figure 15–34.)

phosphodiester linkage
Set of covalent chemical bonds formed when two hydroxyl groups are linked in ester linkage to the same phosphate group. This linkage joins adjacent nucleotides in RNA or DNA. (*See* Figure 2–28.)

phosphoinositide—*see* **inositol phospholipid**

phospholipase C-β (PLC-β)
Enzyme bound to the cytoplasmic face of the plasma membrane that converts membrane phosphatidylinositol 4,5-bisphosphate to diacylglycerol (which remains in the plasma membrane) and inositol 1,4,5-trisphosphate (IP_3). It is activated by certain G proteins to trigger the inositol phospholipid signaling pathway.

phospholipase C-γ (PLC-γ)
Like phospholipase C-β, an enzyme that cleaves inositol phospholipids to diacylglycerol and IP_3 to trigger the inositol phospholipid signaling pathway. Activated by certain receptor tyrosine kinases.

phospholipid exchange protein
Water-soluble carrier protein that transfers a phospholipid molecule from one membrane to another.

phospholipid
The main category of lipid molecules used to construct biological membranes. Generally composed of two fatty acids linked through glycerol phosphate to one of a variety of polar groups.

phosphoprotein phosphatase
Enzyme that removes a phosphate group from a protein by hydrolysis.

phosphorylation
Reaction in which a phosphate group becomes covalently coupled to another molecule.

photochemical reaction center
The part of a photosystem that converts light energy into chemical energy.

photon
Elementary particle of light and other electromagnetic radiation.

photoreceptor
Cell or molecule that is sensitive to light.

photosynthesis
Process by which plants, algae and some bacteria use the energy of sunlight to drive the synthesis of organic molecules from carbon dioxide and water.

photosynthetic electron-transfer
Light-driven reactions in photosynthesis in which electrons move along the electron-transport chain in the thylakoid membrane, generating ATP and NADPH.

photosystem
Multiprotein complex involved in photosynthesis that captures the energy of sunlight and converts it to useful forms of energy.

phragmoplast
Structure made of microtubules and actin filaments that forms in the prospective plane of division of a plant cell and guides formation of the cell plate.

phylogeny
Evolutionary history of an organism or group of organisms, often presented in chart form as a phylogenetic tree.

pinocytosis
Type of endocytosis in which soluble materials are taken up from the environment and incorporated into vesicles for digestion. Literally, "cell drinking." (*See also* **fluid-phase endocytosis**.)

PKA—*see* **cyclic AMP-dependent protein kinase**

PKC—*see* **protein kinase C**

plant growth regulator
Signal molecule (also known as a plant hormone) that helps coordinate growth and development. Examples are ethylene, auxins, gibberellins, cytokines, abscisic acid, and the brassinosteroids.

plasma membrane
Membrane that surrounds a living cell.

plasmid
Small circular DNA molecule that replicates independently of the genome. Modified plasmids are used extensively as **plasmid vectors** for DNA cloning.

plasmodesma (plural **plasmodesmata)**
Communicating cell–cell junction in plants in which a channel of cytoplasm lined by plasma membrane connects two adjacent cells through a small pore in their cell walls.

plastid
Cytoplasmic organelle in plants, bounded by a double membrane, that carries its own DNA and is often pigmented. Chloroplasts are plastids.

platelet
Cell fragment, lacking a nucleus, that breaks off from a megakaryocyte in the bone marrow and is found in large numbers in the bloodstream. It helps initiate blood clotting when blood vessels are injured.

PLC-β—*see* **phospholipase C-β**

PLC-γ—*see* **phospholipase C-γ**

pleckstrin homology domain (PH domain)
Protein domain found in intracellular signaling proteins by which they bind to inositol phospholipids phosphorylated by PI 3-kinase.

ploidy
The number of complete sets of chromosomes in a genome. Diploid organisms have two sets in their somatic cells, polyploid organisms more than two. Natural polyploidy is the result of previous duplications of the whole genome or the introduction of complete genomes from another species during evolution.

plus end
The end of a microtubule or actin filament at which addition of monomers occurs most readily; the "fast-growing" end of a microtubule or actin filament. The plus end of an actin filament is also known as the barbed end. (*See* Panel 16–2, pp. 912–913.)

point mutation
Change of a single nucleotide in DNA, especially in a region of DNA coding for protein.

polar
In the electrical sense, describes a structure (for example, a chemical bond, chemical group, or molecule) with positive charge concentrated toward one end and negative charge toward the other as a result of an uneven distribution of electrons. Polar molecules are likely to be soluble in water.

polyisoprenoid–*see* **isoprenoid**

polymer
Large molecule made by covalently linking multiple identical or similar units (monomers) together.

polymerase chain reaction—*see* **PCR**

polymorphic
Describes a gene with many different alleles, none of which is predominant in the population.

polypeptide
Linear polymer composed of multiple amino acids. Proteins are large polypeptides, and the two terms can be used interchangeably.

polypeptide backbone
The chain of repeating carbon and nitrogen atoms, linked by peptide bonds, in a polypeptide or protein. The side chains of the amino acids project from this backbone.

polyploid
Describes a cell or an organism that contains more than two sets of homologous chromosomes.

polyribosome (polysome)
Messenger RNA molecule to which are attached a number of ribosomes engaged in protein synthesis.

polysaccharide
Linear or branched polymer of monosaccharides. They include glycogen, starch, hyaluronic acid, and cellulose.

polytene chromosome
Giant chromosome in which the DNA has undergone repeated replication without separation into new chromosomes.

position effect
Differences in gene expression that depend on the position of the gene on the chromosome and probably reflect differences in the state of the chromatin along the chromosome.

positional information
Information supplied to or possessed by cells according to their position in a multicellular organism. A cell's internal record of its positional information is called its **positional value**.

positive control
Type of control of gene expression in which the active DNA-binding form of the regulatory protein turns the gene on.

posterior
Situated toward the tail end of the body.

posttranscriptional control
Any control on gene expression that is exerted at a stage after transcription has begun.

posttranslational
Describes any process involving a protein that occurs after protein synthesis is completed.

posttranslational modification
The enzyme-catalyzed change to a protein made after it is synthesized. Examples are acetylation, cleavage, glycosylation, methylation, phosphorylation, and prenylation.

pre-B cell
Immediate precursor of a B cell.

prenylation
Covalent attachment of an isoprenoid lipid group to a protein.

preprophase band
Circumferential band of microtubules and actin filaments that forms around a plant cell under the plasma membrane prior to mitosis and cell division.

primary immune response
Adaptive immune response to an antigen that is made on first encounter with that antigen.

primary structure
Sequence of monomer units in a linear polymer, such as the amino acid sequence of a protein.

primordial germ cell
Cell set aside early in embryonic development that is a precursor to germ cells that give rise to gametes.

primosome
A complex of DNA primase and DNA helicase that is formed on the lagging strand during DNA replication, improving the efficiency of replication.

prion
An infectious abnormal form of a normal protein that is replicated in the host by forcing the normal proteins of the same type to adopt the aberrant structure.

prion disease
Transmissible spongiform encephalopathies such as Kreutzfeld–Jacob disease in humans, scrapie in sheep and bovine spongiform encephalopathy (BSE) in cattle, that are apparently caused and transmitted by abnormal forms of a protein (prions).

probe
Defined fragment of RNA or DNA, radioactively or chemically labeled, used to locate specific nucleic acid sequences by hybridization.

procaryote (prokaryote)
Single-celled microorganism whose cells lack a well-defined, membrane-enclosed nucleus. The procaryotes comprise two of the major domains of living organisms—the Bacteria and the Archaea.

programmed cell death—*see* **apoptosis**

prometaphase
Phase of mitosis preceding metaphase in which the nuclear envelope breaks down and chromosomes first attach to the spindle.

promoter
Nucleotide sequence in DNA to which RNA polymerase binds to begin transcription.

prophase
First stage of mitosis, during which the chromosomes are condensed but not yet attached to a mitotic spindle.

protease (proteinase, proteolytic enzyme)
Enzyme such as trypsin that degrades proteins by hydrolyzing some of their peptide bonds.

proteasome
Large protein complex in the cytosol with proteolytic activity that is responsible for degrading proteins that have been marked for destruction by ubiquitylation or by some other means.

protein
The major macromolecular constituent of cells. A linear polymer of amino acids linked together by peptide bonds in a specific sequence.

protein domain
Portion of a protein that has a tertiary structure of its own. Larger proteins are generally composed of several domains, each connected to the next by short flexible regions of polypeptide chain.

protein glycosylation
Posttranslational addition of oligosaccharide side chains to a protein.

protein kinase
Enzyme that transfers the terminal phosphate group of ATP to a specific amino acid of a target protein.

protein kinase C (PKC)
Ca^{2+}-dependent protein kinase that, when activated by diacylglycerol and an increase in the concentration of Ca^{2+}, phosphorylates target proteins on specific serine and threonine residues.

protein module—*see* module

protein phosphatase—*see* phosphoprotein phosphatase

protein phosphorylation
The covalent addition of a phosphate group to a side chain of a protein catalyzed by a protein kinase.

protein translocator
Membrane-bound protein that mediates the transport of another protein across an organelle membrane.

proteoglycan
Molecule consisting of one or more glycosaminoglycan (GAG) chains attached to a core protein.

proteolysis
Degradation of a protein by hydrolysis at one or more of its peptide bonds.

proteolytic enzyme—*see* protease

protofilament
A linear chain of protein subunits joined end to end, which associates laterally with other protofilaments to form cytoskeletal components such as microtubules and intermediate filaments.

proton
Positively charged subatomic particle that forms part of an atomic nucleus. Hydrogen has a nucleus composed of a single proton (H^+).

proton-motive force
Driving force that moves protons across a membrane as a result of an electrochemical proton gradient.

proto-oncogene
Normal gene, usually concerned with the regulation of cell proliferation, that can be converted into a cancer-promoting oncogene by mutation.

protozoa
Free-living or parasitic, nonphotosynthetic, single-celled, motile eucaryotic organisms, such as *Paramecium* and *Amoeba*. Free-living protozoa feed on bacteria or other microorganisms.

pseudogene
Gene that has accumulated multiple mutations that has rendered it inactive and nonfunctional.

pseudopodium (plural **pseudopodia)**
Large cell-surface protrusion formed by amoeboid cells as they crawl. More generally, any dynamic actin-rich extension of the surface of an animal cell.

pulse-chase
Technique for following the movement of a substance through a biochemical or cellular pathway, by briefly adding the radioactively labeled substance (the pulse) followed by the unlabeled substance (the chase).

pump
Transmembrane protein that drives the active transport of ions or small molecules across the lipid bilayer.

purine
One of the two categories of nitrogen-containing ring compounds found in DNA and RNA. Examples are adenine and guanine. (*See* Panel 2–6, pp. 120–121.)

pyrimidine
One of the two categories of nitrogen-containing ring compounds found in DNA and RNA. Cytosine, thymine and uracil are pyrimidines. (*See* Panel 2–6, pp. 120–121.)

quaternary structure
Three-dimensional relationship of the different polypeptide chains in a multisubunit protein or protein complex.

quinone (Q)
Small, lipid soluble, mobile electron carrier molecule found in the respiratory and photosynthetic electron-transport chains. (*See* Figure 14–24.)

Rab protein
Any of a large family of monomeric GTPases present in the plasma membrane and organelle membranes that are involved in conferring specificity on vesicle docking.

radioactive isotope
Form of an atom with an unstable nucleus that emits radiation as it decays.

Ran
Monomeric GTPase present in both cytosol and nucleus that is required for the active transport of macromolecules into and out of the nucleus through nculear pore complexes. Hydrolysis of GTP to GDP is thought to provide the energy required for this transport.

Ras protein
The most famous member of a large family of GTP-binding proteins (called monomeric GTPases) that help relay signals from cell-surface receptors to the nucleus. Named for the *ras* gene, first identified in viruses that cause rat sarcomas.

reaction
In chemistry, any process in which one molecule is converted into another by the removal or addition of atoms, or in which the arrangement of atoms in a molecule or molecules is altered by a change in chemical bonds.

reading frame
The phase in which nucleotides are read in sets of three to encode a protein. A messenger RNA molecule can be read in any one of three reading frames, only one of which will give the required protein. (*See* Figure 6–51.)

RecA protein
The prototype for a class of DNA-binding proteins that catalyze synapsis of DNA strands during genetic recombination.

receptor
Protein that binds a specific extracellular signal molecule (ligand) and initiates a response in the cell. Cell-surface receptors, such as the acetylcholine receptor and the insulin receptor, are located in the plasma membrane, with their ligand-binding site exposed to the external medium. Intracellular receptors, such as steroid hormone receptors, bind ligands that diffuse into the cell across the plasma membrane.

receptor-mediated endocytosis
Internalization of receptor-ligand complexes from the plasma membrane by endocytosis, It is used to take up some macromolecules, such as cholesterol-containing lipoproteins, from the extracellular fluid, and is also a means of recycling receptor proteins once they have bound their ligands.

recessive
In genetics, refers to the member of a pair of alleles that fails to be expressed in the phenotype of the organism when the dominant allele is present. Also refers to the phenotype of an individual that has only the recessive allele.

recombinant DNA
Any DNA molecule formed by joining DNA segments from different sources. Recombinant DNAs are widely used in the cloning of genes, in the genetic modification of organisms, and in molecular biology generally.

recombination
Process in which DNA molecules are broken and the fragments are rejoined in new combinations. Can occur in the living cell—for example, through crossing-over during meiosis—or *in vitro* using purified DNA and enzymes that break and ligate DNA strands.

recycling endosomes
Large intracellular membrane-bounded vesicle formed from a fragment of an endosome that is an intermediate stage on the passage of recycled receptors back to the cell membrane.

red blood cell—*see* **erythrocyte**

redox pair
Pair of molecules in which one acts as an electron donor and one as an electron acceptor in an oxidation-reduction reaction; for example, NADH (electron donor) and NAD^+ (electron acceptor).

redox potential
The affinity of a redox pair for electrons, generally measured as the voltage difference between an equimolar mixture of the pair and a standard reference. $NADH/NAD^+$ has a low redox potential and O_2/H_2 has a high redox potential (high affinity for electrons).

redox reaction
A reaction in which one component becomes oxidized and the other reduced; an oxidation-reduction reaction.

reduction (verb **reduce**)
Addition of electrons to an atom, as occurs during the addition of hydrogen to a molecule or the removal of oxygen from it. Opposite of oxidation. (*See* Figure 2–43.)

regulatory sequence
DNA sequence to which a gene regulatory protein binds to control the rate of assembly of the transcirptional complex at the promoter.

regulatory site
Site on an enzyme, other than the active site, that binds a molecule that affects enzyme activity.

replication fork
Y-shaped region of a replicating DNA molecule at which the two daughter strands are formed and separate.

replication origin
Location on a DNA molecule at which duplication of the DNA begins.

replicative cell senescence
Phenomenon observed in primary cell cultures as they age, in which cell proliferation slows down and finally halts.

repressor
Protein that binds to a specific region of DNA to prevent transcription of an adjacent gene.

residue
General term for the unit of a polymer. That portion of a sugar, amino acid, or nucleotide that is retained as part of the polymer chain during the process of polymerization.

respiration
General term for a process in a cell involving the oxidative breakdown of sugars or other organic molecules, and requiring the uptake of O_2 while producing CO_2 and H_2O as waste products.

respiratory chain
Electron-transport chain in the inner mitochondrial membrane that receives high-energy electrons derived from the citric acid cycle and generates the proton gradient across the membrane that is used to power ATP synthesis.

respiratory control
Regulatory mechanism that controls the rate of electron transport in the respiratory chain according to need via a direct influence of the electrochemical proton gradient.

respiratory enzyme complex
Any of the major protein complexes of the mitochondrial respiratory chain that act as electron-driven proton pumps to generate the proton gradient across the inner membrane.

resting membrane potential
The membrane potential in equilibrium conditions in which there is no net flow of ions across the plasma membrane.

restriction map
Diagrammatic representation of a DNA molecule indicating the sites of cleavage by various restriction enzymes.

restriction nuclease (restriction enzyme)
One of a large number of nucleases that can cleave a DNA molecule at any site where a specific short sequence of nucleotides occurs. Extensively used in recombinant DNA technology.

restriction point
Important checkpoint in the mammalian cell cycle. Passage through the restriction point commits the cell to enter S phase. It corresponds to Start in the yeast cell cycle.

retrotransposon
Type of transposable element that moves by being first transcribed into an RNA copy that is then reconverted to DNA by reverse transcriptase and inserted elsewhere in the chromosomes.

retrovirus
RNA-containing virus that replicates in a cell by first making a double-stranded DNA intermediate.

reverse genetics
Approach to discovering gene function that starts from the DNA (gene) and protein and then creates mutants to analyze the gene's function.

reverse transcriptase
Enzyme first discovered in retroviruses that makes a double-stranded DNA copy from a single-stranded RNA template molecule.

rhodopsin
G-protein-linked light-sensitive receptor protein in the rod photoreceptor cells of the retina.

ribonuclease
Enzyme that cuts an RNA molecule by hydrolyzing one or more of its phosphodiester bonds.

ribonucleic acid—*see* RNA

ribosomal RNA (rRNA)
Any one of a number of specific RNA molecules that form part of the structure of a ribosome and participate in the synthesis of proteins. Often distinguished by their sedimentation coefficient, such as 28S rRNA or 5S rRNA.

ribosome
Particle composed of ribosomal RNAs and ribosomal proteins that associates with messenger RNA and catalyzes the synthesis of protein.

ribozyme
RNA with catalytic activity.

RNA (ribonucleic acid)
Polymer formed from covalently linked ribonucleotide monomers. (*See also* **messenger RNA, ribosomal RNA, transfer RNA.**)

RNA editing
Production of a functional mRNA by insertion or alteration of individual nucleotides in an RNA molecule after it is synthesized.

RNA interference (RNAi)
Selective intracellular degradation of RNA that is intended to remove foreign RNAs, such as those of viruses. Fragments cleaved from free double-stranded RNA direct the degradative mechanism to other similar RNA sequences. Widely exploited in a technique used to silence the expression of selected genes.

RNA polymerase II holoenzyme
Large pre-assumbled complex of RNA polymerase II, most of the general transcription factors required for its function, and the mediator protein complex.

RNA polymerase
Enzyme that catalyzes the synthesis of an RNA molecule on a DNA template from nucleoside triphosphate precursors. (*See* Figure 6–8.)

RNA primer
Short stretch of RNA synthesized on a DNA template. It is required by DNA polymerases to start their DNA synthesis.

RNA processing control
Control of gene expression by controlling how the RNA transcript is spliced or otherwise processed.

RNA splicing
Process in which intron sequences are excised from RNA transcripts in the nucleus during formation of messenger and other RNAs.

RNAi—*see* RNA interference

rod photoreceptor (rod)
Photoreceptor cell type in the retina that is responsible for noncolor vision in dim light.

rough endoplasmic reticulum (rough ER)
Endoplasmic reticulum with ribosomes on its cytosolic surface. Involved in the synthesis of secreted and membrane-bound proteins.

rRNA—*see* ribosomal RNA

rRNA gene
Gene that specifies a ribosomal RNA (rRNA).

S phase
Period of a eucaryotic cell cycle in which DNA is synthesized.

Saccharomyces
Genus of yeasts that reproduce asexually by budding or sexually by conjugation. Economically important in brewing and baking, they are also widely used in genetic engineering and as simple model organisms in the study of eucaryotic cell biology.

sarcoma
Cancer of connective tissue.

sarcomere
Repeating unit of a myofibril in a muscle cell, composed of an array of overlapping thick (myosin) and thin (actin) filaments between two adjacent Z discs.

sarcoplasmic reticulum
Network of internal membranes in the cytoplasm of a muscle cell that contains high concentrations of sequestered Ca^{2+} which is released into the cytosol during muscle excitation.

satellite DNA
Regions of highly repetitive DNA from a eucaryotic chromosome, usually identifiable by its unusual nucleotide composition. Satellite DNA is not transcribed and has no known function.

saturated
Describes a molecule containing carbon–carbon bonds that has only single covalent bonds.

scaffold protein
Protein that organizes groups of interacting intracellular signaling proteins into signaling complexes.

scanning electron microscope
Type of electron microscope that produces an image of the surface of an object.

S-Cdk
Complex formed in vertebrate cells by an S-cyclin and the corresponding cyclin-dependent kinase (Cdk).

Schwann cell
Glial cell responsible for forming myelin sheaths in the peripheral nervous system.

SDS polyacrylamide gel electrophoresis (SDS-PAGE)
Type of electrophoresis in which the protein mixture to be separated is run through a gel containing the detergent sodium dodecyl sulfate (SDS) which unfolds the proteins and frees them from association with other molecules.

second messenger
Small molecule that is formed in or released into the cytosol in response to an extracellular signal and helps to relay the signal to the interior of the cell. Examples include cAMP, IP_3, and Ca^{2+}.

secondary immune response
Adaptive immune response to an antigen that is made on a second or subsequent encounter with a given antigen. It is more rapid in onset, stronger, and more specific than the primary immune response.

secondary structure
Regular local folding pattern of a polymeric molecule. In proteins, α helices and β sheets.

secretory vesicle
Membrane-bounded organelle in which molecules destined for secretion are stored prior to release. Sometimes called secretory granule because darkly staining contents make the organelle visible as a small solid object.

section
A very thin slice of tissue, suitable for viewing under the microscope.

selectin
Member of a family of cell-surface carbohydrate-binding proteins that mediate transient, Ca^{2+}-dependent cell-cell adhesion in the bloodstream, for example between white blood cells and the endothelium of the blood vessel wall.

selectivity filter
That part of an ion channel structure that determines which ions it can transport.

septate junction
Main type of occluding cell junction in invertebrates; their structure is distinct from that of vertebrate tight junctions.

serine protease
Type of protease that has a reactive serine in the active site.

sex chromosome
Chromosome that may be present or absent, or present in a variable number of copies, according to the sex of the individual. In mammals, the X and Y chromosomes.

sexual reproduction
Type of reproduction in which the genomes of two individuals are mixed in the formation of a new organism. Individuals produced by sexual reproduction differ from either of their parents and from each other.

SH2 domain
Src homology region 2, a protein domain present in many signaling proteins; it binds a short amino acid sequence containing a phosphotyrosine.

side chain
The part of an amino acid that differs between different amino acids, giving the amino acid its unique physical and chemical properties.

signal molecule
Extracellular or intracellular molecule that cues the response of a cell to the behavior of other cells or objects in the environment.

signal patch
Protein sorting signal that consists of a specific three-dimensional arrangement of atoms on the folded protein's surface.

signal peptidase
Enzyme that removes a terminal signal sequence from a protein once the sorting process is complete.

signal-recognition particle (SRP)
Ribonucleoprotein particle that binds an ER signal sequence on a partially synthesized polypeptide chain and directs the polypeptide and its attached ribosome to the endoplasmic reticulum.

signal sequence
Short continuous sequence of amino acids that determines the eventual location of a protein in the cell. An example is the N-terminal sequence of 20 or so amino acids that directs nascent secretory and transmembrane proteins to the endoplasmic reticulum.

signal transduction
Relaying of a signal by conversion from one physical or chemical form to another. In cell biology, the process by which a cell converts an extracellular signal into a response.

single-nucleotide polymorphism (SNP)
Variation between individuals at certain nucleotide positions in the genome.

single-pass transmembrane protein
Membrane protein in which the polypeptide chain crosses the lipid bilayer only once.

single-strand DNA-binding protein
Protein that binds to the single strands of the opened-up DNA double helix, preventing helical structures from reforming while the DNA is being replicated.

sister chromatid—*see* **chromatid**

site-directed mutagenesis
Technique by which a mutation can be made at a particular site in DNA.

site-specific recombination
Type of recombination that does not require extensive similarity in the two DNA sequences undergoing recombination. Can occur between two different DNA molecules or within a single DNA molecule.

small intracellular mediator—*see* **second messenger**

small nuclear RNA (snRNA)
Small RNA molecules that are complexed with proteins to form the ribonucleoprotein particles involved in RNA splicing.

smooth endoplasmic reticulum (smooth ER)
Region of the endoplasmic reticulum not associated with ribosomes. It is involved in lipid synthesis.

smooth muscle cell
Type of long, spindle-shaped mononucleate muscle cell making up the muscular tissue found in the walls of arteries and of the intestine and other viscera, and in some other locations of the vertebrate body. Called "smooth" because it lacks the striated myofibrils of skeletal and cardiac muscle cells.

SNAREs
Large family of transmembrane proteins present in organelle membranes and the vesicles derived from them. They are involved in guiding vesicles to their correct destinations. They exist in pairs—a v-SNARE in the vesicle membrane that binds specifically to a complementary t-SNARE in the target membrane.

SNP—*see* **single-nucleotide polymorphism**

snRNA—*see* **small nuclear RNA**

solute
Any molecule that is dissolved in a liquid. The liquid is called a solvent.

somatic cell
Any cell of a plant or animal other than a germ cell or germ-cell precursor. (From Greek *soma*, body.)

somite
One of a series of paired blocks of mesoderm that form during early development and lie on either side of the notochord in a vertebrate embryo. They give rise to the vertebral column, muscles and associated connective tissue. Each somite produces the musculature of one vertebral segment, plus associated connective tissue.

sorting signal
Amino acid sequence that directs the delivery of a protein to a specific location outside the cytosol.

Southern blotting
Technique in which DNA fragments separated by electrophoresis are immobilized on a paper sheet. Specific fragments are then detected with a labeled nucleic acid probe. (Named after E.M. Southern, inventor of the technique.)

spectrin
Abundant protein associated with the cytosolic side of the plasma membrane in red blood cells, forming a rigid network that supports the membrane.

Spemann's Organizer
Specialized tissue at the dorsal lip of the blastopore in an amphibian embryo; a source of signals that help to orchestrate formation of the embryonic body axis. (After H. Spemann and H. Mangold, co-discoverers.)

sperm (spermatozoon, plural **spermatozoa)**
The mature male gamete in animals. It is motile and usually small compared with the egg.

spermatogenesis
Development of sperm.

spindle-attachment checkpoint
Checkpoint that operates during mitosis to ensure that all chromosomes are properly attached to the spindle before sister-chromatid separation starts.

spliceosome
Large assembly of RNA and protein molecules that performs pre-mRNA splicing in eucaryotic cells.

Src family
Family of cytoplasmic tyrosine kinases (pronounced "sark") that associate with the cytoplasmic domains of some enzyme-linked receptors (for example, the T cell antigen receptor) that lack intrinsic tyrosine kinase activity. They transmit a signal onwards by phosphorylating the receptor itself and other signaling proteins.

SRP—*see* **signal-recognition particle**

standard free-energy change ($\Delta G°$)
Free-energy change of two reacting molecules at standard temperature and pressure when all components are present at a concentration of 1 mole per liter.

starch
Polysaccharide composed exclusively of glucose units, used as an energy storage material in plant cells.

start-transfer signal
Short amino-acid sequence that enables a polypeptide chain to start being translocated across the endoplasmic reticulum membrane through a protein translocator. Multipass membrane proteins have both N-terminal (signal sequence) and internal start-transfer signals.

stem cell
Relatively undifferentiated cell that can continue dividing indefinitely, throwing off daughter cells that can undergo terminal differentiation into particular cell types.

stereocilium
A large, rigid microvillus found in "organ pipe" arrays on the apical surface of hair cells in the ear. A stereocilium contains a bundle of actin filaments, rather than microtubules, and is thus not a true cilium.

steroid
Hydrophobic lipid molecule with a characteristic four-ringed structure. Many important hormones such as estrogen and testosterone are steroids. (*See* Panel 2–5, pp. 118–119.)

stimulatory G protein (G_s)
G protein that, when activated, activates the enzyme adenylyl cyclase and thus stimulates the production of cyclic AMP.

stop-transfer signal
Hydrophobic amino acid sequence that halts translocation of a polypeptide chain through the endoplasmic reticulum membrane, thus anchoring the protein chain in the membrane (*See* Figure 12–49).

strand-directed mismatch repair—*see* **mismatch repair**

striated muscle
Muscle composed of transversely striped (striated) myofibrils. Skeletal and heart muscle of vertebrates are the best-known examples.

stroma
(1) The connective tissue in which a glandular or other epithelium is embedded. (2) The large interior space of a chloroplast, containing enzymes that incorporate CO_2 into sugars.

structural gene
Region of DNA that codes for a protein or for an RNA molecule that forms part of a structure or has an enzymatic function. Distinguished from regions of DNA that regulate gene expression.

substrate
Molecule on which an enzyme acts.

substratum
Solid surface to which a cell adheres.

subunit
Component of a multicomponent complex—for example, one protein component of a protein complex or one polypeptide chain of a multichain protein.

sucrose
Disaccharide composed of one glucose unit and one fructose unit. The major form in which glucose is transported between plant cells.

sugar
Small carbohydrates with a monomer unit of general formula $(CH_2O)n$. Examples are the monosaccharides glucose, fructose and mannose, and the disacharide sucrose (composed of a molecule of glucose and one of fructose linked together).

sulfhydryl (thiol, –SH)
Chemical group containing sulfur and hydrogen found in the amino acid cysteine and other molecules. Two sulfhydryls can join to produce a disulfide bond.

supercoiled DNA
Region of DNA in which the double helix is further twisted on itself. (*See* Figure 6–20.)

survival factor
Extracellular signal required for a cell to survive; in its absence the cell will undergo apoptosis and die.

symbiosis
Intimate association between two organisms of different species from which both derive a long-term selective advantage.

symporter
Carrier protein that transports two types of solute across the membrane in the same direction.

synapse
Communicating cell–cell junction that allows signals to pass from a nerve cell to another cell. In a chemical synapse the signal is carried by a diffusible neurotransmitter; in an electrical synapse a direct connection is made between the cytoplasms of the two cells via gap junctions.

synapsis
(1) In genetic recombination, the initial formation of base pairs between complementary DNA strands in different DNA molecules that occurs at sites of crossing-over between chromosomes. (2) In meiosis, the pairing of maternal and paternal copies of a chromosome as they become attached to each other along their length.

synaptic signaling
Type of cell–cell communication that occurs across chemical synapses in the nervous system.

synaptic vesicle
Small neurotransmitter-filled secretory vesicle formed at the axon terminals of nerve cells and whose contents are released into the synaptic cleft by exocytosis when an action potential reaches the axon terminal.

synaptonemal complex
Structure that holds paired chromosomes together during prophase I of meiosis and promotes genetic recombination.

syncytium
Mass of cytoplasm containing many nuclei enclosed by a single plasma membrane. Typically the result either of cell fusion or of a series of incomplete division cycles in which the nuclei divide but the cell does not.

synteny
The presence in different species of of regions of chromosomes with the same genes in the same order.

T cell (T lymphocyte)
Type of lymphocyte responsible for cell-mediated immunity; includes both cytotoxic T cells and helper T cells.

TATA box
Consensus sequence in the promoter region of many eucaryotic genes that binds a general transcription factor and hence specifies the position at which transcription is initiated.

TCA cycle—*see* citric acid cycle

telomerase
Enzyme that elongates telomere sequences in DNA.

telomere
End of a chromosome, associated with a characteristic DNA sequence that is replicated in a special way. Counteracts the tendency of the chromosome otherwise to shorten with each round of replication. (From Greek *telos*, end.)

telophase
Final stage of mitosis in which the two sets of separated chromosomes decondense and become enclosed by nuclear envelopes.

temperature-sensitive (ts) mutant
Organism or cell carrying a genetically altered protein (or RNA molecule) that performs normally at one temperature but is abnormal at another (usually higher) temperature.

template
A single strand of DNA or RNA whose nucleotide sequence acts as a guide for the synthesis of a complementary strand.

terminator
Signal in bacterial DNA that halts transcription.

tertiary structure
Complex three-dimensional form of a folded polymer chain, especially a protein or RNA molecule.

TGF-β superfamily—*see* transforming growth factor-β superfamily

TGN—*see* trans Golgi network (TGN)

thioester bond
High-energy bond formed by a condensation reaction between an acid (acyl) group and a thiol group (–SH); seen, for example, in acetyl CoA and in many enzyme-substrate complexes.

thiol—*see* sulfhydryl

thylakoid
Flattened sac of membrane in a chloroplast that contains chlorophyll and other pigments and carries out the light-trapping reactions of photosynthesis. Stacks of thylakoids form the grana of chloroplasts.

tight junction
Cell–cell junction that seals adjacent epithelial cells together, preventing the passage of most dissolved molecules from one side of the epithelial sheet to the other.

TIM complexes
Protein translocators in the mitochondrial inner membrane. The TIM23 complex mediates the transport of proteins into the matrix and the insertion of some proteins into the inner membrane; the TIM22 complex mediates the insertion of a subgroup of proteins into the inner membrane.

Toll-like receptor family (TLR)
Important family of mammalian pattern recognition receptors abundant on macrophages, neutrophils and the epithelial cells of the gut. They recognize pathogen-associated immunostimulants such as lipopolysacharide and peptidoglycan.

TOM complex
Multisubunit protein complex that transports proteins across the mitochondrial outer membrane.

topoisomerase (DNA topoisomerase)
Enzyme that makes reversible cuts in a double-helical DNA molecule for the purpose of removing knots or unwinding excessive twists.

tracer
Molecule or atom that has been labeled either chemically or radioactively so that it can be followed in a biochemical process or readily located in a cell or tissue.

***trans* face**
Face of a Golgi stack at which material leaves the organelle for the cell surface or another cell compartment. It is adjacent to the *trans* Golgi network.

***trans* Golgi network (TGN)**
Network of interconnected cisternae and tubules at the *trans* face of the Golgi apparatus, through which material is transferred out of the Golgi.

transcellular transport
Transport of solutes, such as nutrients, across an epithelium, by means of membrane transport proteins in the apical and basal faces of the epithelial cells.

transcript
RNA product of DNA transcription.

transcription (DNA transcription)
Copying of one strand of DNA into a complementary RNA sequence by the enzyme RNA polymerase.

transcription attenuation
Inhibition of gene expression in bacteria by the premature termination of transcription.

transcription factor
Term loosely applied to any protein required to initiate or regulate transcription in eucaryotes. Includes both gene regulatory proteins as well as the general transcription factors.

transcriptional control
Control of of gene expression by controlling when and how often the gene is transcribed.

transcytosis
The uptake of material at one face of a cell by endocytosis, its transfer across a cell in vesicles, and its discharge from another face by exocytosis.

transfection
Introduction of a foreign DNA molecule into a eucaryotic cell. It is usually followed by expression of one or more genes in the newly introduced DNA.

transfer RNA (tRNA)
Set of small RNA molecules used in protein synthesis as an interface (adaptor) between messenger RNA and amino acids. Each type of tRNA molecule is covalently linked to a particular amino acid.

transforming growth factor-β superfamily (TGF-β superfamily)
Large family of structurally related, secreted proteins that act as hormones and local mediators to control a wide range of functions in animals, including during development. It includes TGF-βs, activins, and bone morphogenetic proteins (BMPs).

transgenic organism
Plant or animal that has stably incorporated one or more genes from another cell or organism and can pass them on to successive generations.

transition state
Structure that forms transiently in the course of a chemical reaction and has the highest free energy of any reaction intermediate. Its formation is a rate-limiting step in the reaction.

translation (RNA translation)
Process by which the sequence of nucleotides in a messenger RNA molecule directs the incorporation of amino acids into protein. It occurs on a ribosome.

translational control
Control of gene expression by selection of which mRNAs in the cytoplasm are translated by ribosomes.

translocation
Type of mutation in which a portion of one chromosome is broken off and attached to another.

transmembrane protein
Membrane protein that extends through the lipid bilayer, with part of its mass on either side of the membrane.

transmitter-gated ion channel
Ion channel in the postsynaptic plasma membranes of nerve and muscle cells that opens only in response to the binding of a specific extracellular neurotransmitter. The resulting inflow of ions leads to the generation of a local electrical signal in the postsynaptic cell.

transposable element
Segment of DNA that can move from one position in a genome to another. Also called a transposon.

transposition
The movement of a DNA sequence from one site to another within the genome. *See also* **cut-and-paste transposition**.

trans-splicing
Type of RNA splicing present in a few eucaryotic organisms in which exons from two separate RNA molecules are joined together to form an mRNA.

treadmilling
The process by which a polymeric protein filament is maintained at constant length by addition of protein subunits at one end and loss of subunits at the other. (*See* Panel 16–2, pp. 912–913.)

triacylglycerol
Molecule composed of three fatty acids esterified to glycerol. The main constituent of fat droplets in animal tissues (where the fatty acids are saturated) and of vegetable oils (where the fatty acids are mainly unsaturated). Also known as triglyceride. (*See* Panel 2–5, pp. 118–119.)

tricarboxylic acid (TCA) cycle—*see* **citric acid cycle**

trimeric GTP-binding protein—*see* **GTP-binding protein**

tRNA—*see* **transfer RNA**

t-SNARE—*see* **SNAREs**

tubulin
The protein subunit of microtubules.

tumor progression
The process by which an initial mildly disordered cell behavior gradually evolves into a full-blown cancer.

tumor suppressor gene
Gene that appears to prevent formation of a cancer. Loss-of-function mutations in such genes enhance susceptibility to cancer.

two-dimensional gel electrophoresis
Type of electrophoresis in which the protein mixture is run first in one direction and then in a direction at right angles to the first. It enables better separation of individual proteins.

two-hybrid system
Technique for identifying interacting proteins using genetically engineered yeast cells.

type III secretion system
A bacterial system for delivering toxic proteins into the cells of their host.

ubiquitin
Small, highly conserved protein present in all eucaryotic cells that becomes covalently attached to lysines of other proteins. Attachment of a short chain of ubiquitins to such a lysine tags a protein for intracellular proteolytic destruction by a proteasome.

ubiquitin ligase
Any one of a large number of enzymes that attach ubiquitin to a protein, thus marking it for destruction in a proteasome. The process catalyzed by a ubiquitin ligase is called ubiquitylation.

unfolded protein response
Cellular response triggered by an accumulation of misfolded proteins in the endoplasmic reticulum. It involves increased transcription of ER chaperones and degradative enzymes.

uniporter
Carrier protein that transports a single solute from one side of the membrane to the other.

unsaturated
Describes a molecule that contains one or more double or triple carbon-carbon bonds, such as isoprene or benzene.

V gene segment
Gene segment encoding most of the variable region of the polypeptide chains of immunoglobulins and T cell receptors.

V(D)J joining
Recombination process by which gene segments are brought together to form a functional gene for a polypeptide chain of an immunoglobulin or T cell receptor.

vacuole
Very large fluid-filled vesicle found in most plant and fungal cells, typically occupying more than a third of the cell volume.

van der Waals attraction
Type of (individually weak) noncovalent bond that is formed at close range between nonpolar atoms.

variable region
Region of an immunoglobulin light or heavy chain that differs from molecule to molecule; it comprises the antigen-binding site.

vector
In cell biology, the DNA of an agent (virus or plasmid) used to transmit genetic material to a cell or organism. (*See also* **cloning vector**, **expression vector**.)

vegetal pole
The end at which most of the yolk is located in an animal egg. The end opposite the animal pole.

ventral
Situated toward the belly surface of an animal, or towards the underside of a wing or leaf.

vesicle
Small, membrane-bounded, spherical organelle in the cytoplasm of a eucaryotic cell.

vesicular transport
Transport of proteins from one cellular compartment to another by means of membrane-bounded intermediaries such as vesicles or organelle fragments.

virulence gene
Gene that contributes to an organism's ability to cause disease.

virus
Particle consisting of nucleic acid (RNA or DNA) enclosed in a protein coat and capable of replicating within a host cell and spreading from cell to cell. Many viruses cause disease.

voltage-gated cation channel
Type of ion channel found in the membranes of excitable cells (such as nerve cells and muscle) which opens in response to a shift in membrane potential past a threshold value.

v-SNARE—*see* **SNAREs**

Western blotting
Technique by which proteins are separated by electrophoresis and immobilized on a paper sheet and then analyzed, usually by means of a labeled antibody.

white blood cell (leucocyte)
General name for all the nucleated blood cells lacking hemoglobin. Includes lymphocytes, neutrophils, eosinophils, basophils, and monocytes.

wild-type
Normal, nonmutant form of an organism; the form found in nature (in the wild).

Xenopus laevis **(South African clawed toad)**
Species of frog (not toad) frequently used in studies of early vertebrate development.

XIC—*see* **X-inactivation center**

X-inactivation
Inactivation of one copy of the X chromosome in the somatic cells of female mammals.

X-inactivation center (XIC)
Site in an X chromosome at which inactivation is initiated and spreads outwards.

X-ray crystallography
Technique for determining the three-dimensional arrangement of atoms in a molecule based on the diffraction pattern of X-rays passing through a crystal of the molecule.

yeast
Common term for several families of unicellular fungi. Includes species used for brewing beer and making bread, as well as pathogenic species (that is, species that cause disease).

yolk
Nutritional reserves rich in lipids, proteins and polysaccharides, present in the eggs of many animals.

Z disc (Z line)
Platelike region of a muscle sarcomere to which the plus ends of actin filaments are attached. Seen as a dark transverse line in micrographs.

zinc finger
DNA-binding structural motif present in many gene regulatory proteins. Composed of a loop of polypeptide chain held in a hairpin bend bound to a zinc atom.

zona pellucida
Glycoprotein layer on the surface of the unfertilized egg. It is often a barrier to fertilization across species.

zygote
Diploid cell produced by fusion of a male and female gamete. A fertilized egg.

zygotene
Second stage of division I of meiosis, in which the synaptonemal complex begins to form between the two sets of sister chromatids in each bivalent chromosome.

Index

Page numbers in **boldface** refer to a major text discussion of the entry; page numbers with an F refer to a figure, with an FF to figures that follow consecutively; page numbers with a T refer to a table; cf. means compare.

Page numbers in **boldface** refer to a major text discussion of the entry; page numbers with an F refer to a figure, with an FF to figures that follow consecutively; page numbers with a T refer to a table; cf. means compare.

C

Page numbers in **boldface** refer to a major text discussion of the entry; page numbers with an F refer to a figure, with an FF to figures that follow consecutively; page numbers with a T refer to a table; cf. means compare.

Page numbers in **boldface** refer to a major text discussion of the entry; page numbers with an F refer to a figure, with an FF to figures that follow consecutively; page numbers with a T refer to a table; cf. means compare.

Page numbers in **boldface** refer to a major text discussion of the entry; page numbers with an F refer to a figure, with an FF to figures that follow consecutively; page numbers with a T refer to a table; cf. means compare.

E

Page numbers in **boldface** refer to a major text discussion of the entry; page numbers with an F refer to a figure, with an FF to figures that follow consecutively; page numbers with a T refer to a table; cf. means compare.

F

G

I

J

K

L

Page numbers in **boldface** refer to a major text discussion of the entry; page numbers with an F refer to a figure, with an FF to figures that follow consecutively; page numbers with a T refer to a table; cf. means compare.

M

Page numbers in **boldface** refer to a major text discussion of the entry; page numbers with an F refer to a figure, with an FF to figures that follow consecutively; page numbers with a T refer to a table; cf. means compare.

Page numbers in **boldface** refer to a major text discussion of the entry; page numbers with an F refer to a figure, with an FF to figures that follow consecutively; page numbers with a T refer to a table; cf. means compare.

Page numbers in **boldface** refer to a major text discussion of the entry; page numbers with an F refer to a figure, with an FF to figures that follow consecutively; page numbers with a T refer to a table; cf. means compare.

Q

R

Page numbers in **boldface** refer to a major text discussion of the entry; page numbers with an F refer to a figure, with an FF to figures that follow consecutively; page numbers with a T refer to a table; cf. means compare.

S

T

Tables

The Genetic Code

1st position (5′ end)	2nd Position: U	C	A	G	3rd Position (3′ end)
U	Phe	Ser	Tyr	Cys	U
	Phe	Ser	Tyr	Cys	C
	Leu	Ser	**STOP**	**STOP**	A
	Leu	Ser	**STOP**	Trp	G
C	Leu	Pro	His	Arg	U
	Leu	Pro	His	Arg	C
	Leu	Pro	Gln	Arg	A
	Leu	Pro	Gln	Arg	G
A	Ile	Thr	Asn	Ser	U
	Ile	Thr	Asn	Ser	C
	Ile	Thr	Lys	Arg	A
	Met	Thr	Lys	Arg	G
G	Val	Ala	Asp	Gly	U
	Val	Ala	Asp	Gly	C
	Val	Ala	Glu	Gly	A
	Val	Ala	Glu	Gly	G

Amino Acids Codons

		Amino Acids	Codons
A	Ala	Alanine	GCA GCC GCG GCU
C	Cys	Cysteine	UGC UGU
D	Asp	Aspartic acid	GAC GAU
E	Glu	Glutamic acid	GAA GAG
F	Phe	Phenylalanine	UUC UUU
G	Gly	Glycine	GGA GGC GGG GGU
H	His	Histidine	CAC CAU
I	Ile	Isoleucine	AUA AUC AUU
K	Lys	Lysine	AAA AAG
L	Leu	Leucine	UUA UUG CUA CUC CUG CUU
M	Met	Methionine	AUG
N	Asn	Asparagine	AAC AAU
P	Pro	Proline	CCA CCC CCG CCU
Q	Gln	Glutamine	CAA CAG
R	Arg	Arginine	AGA AGG CGA CGC CGG CGU
S	Ser	Serine	AGC AGU UCA UCC UCG UCU
T	Thr	Threonine	ACA ACC ACG ACU
V	Val	Valine	GUA GUC GUG GUU
W	Trp	Tryptophan	UGG
Y	Tyr	Tyrosine	UAC UAU